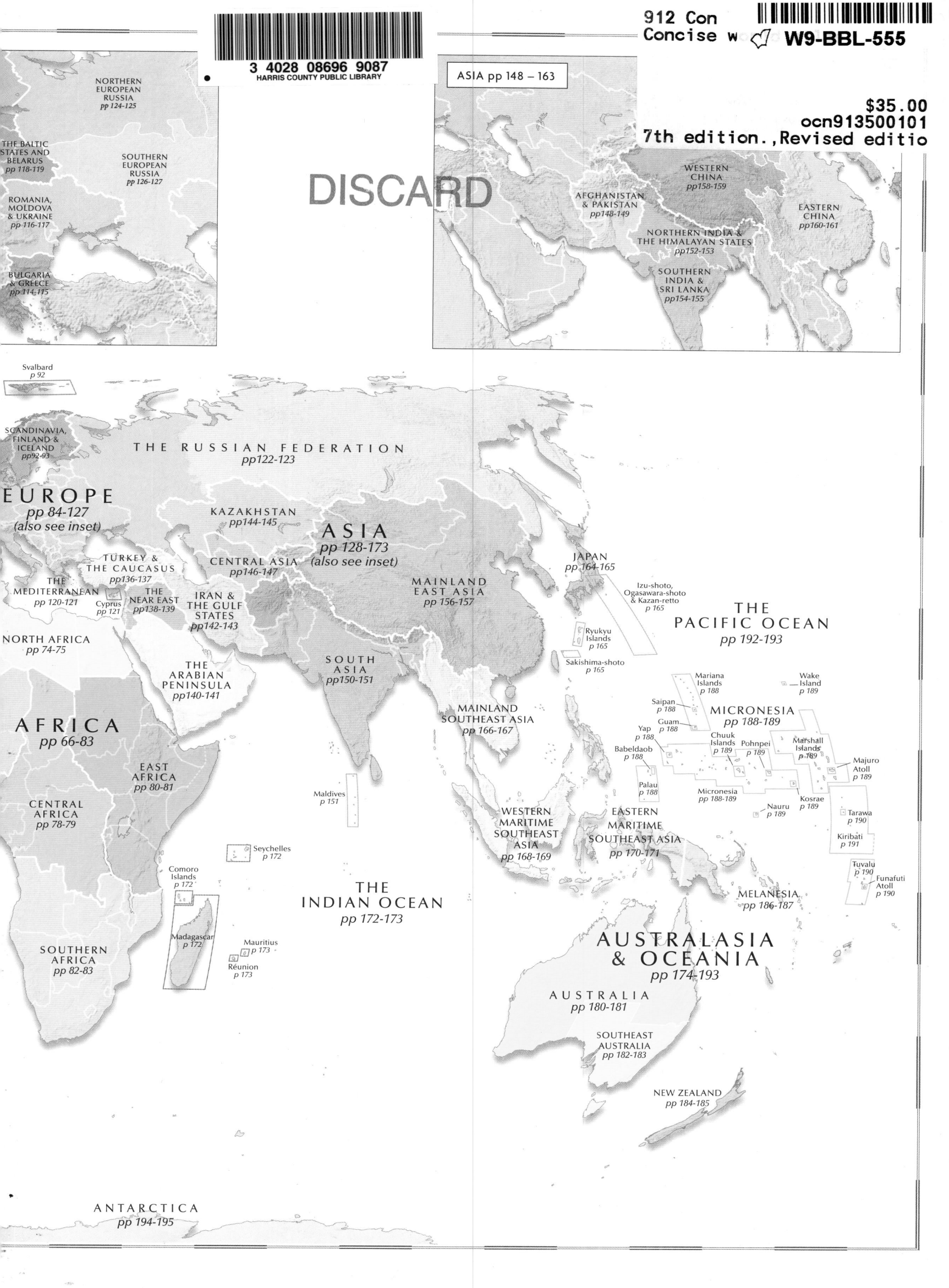

CONCISE
WORLD
ATLAS

CONCISE
WORLD
ATLAS

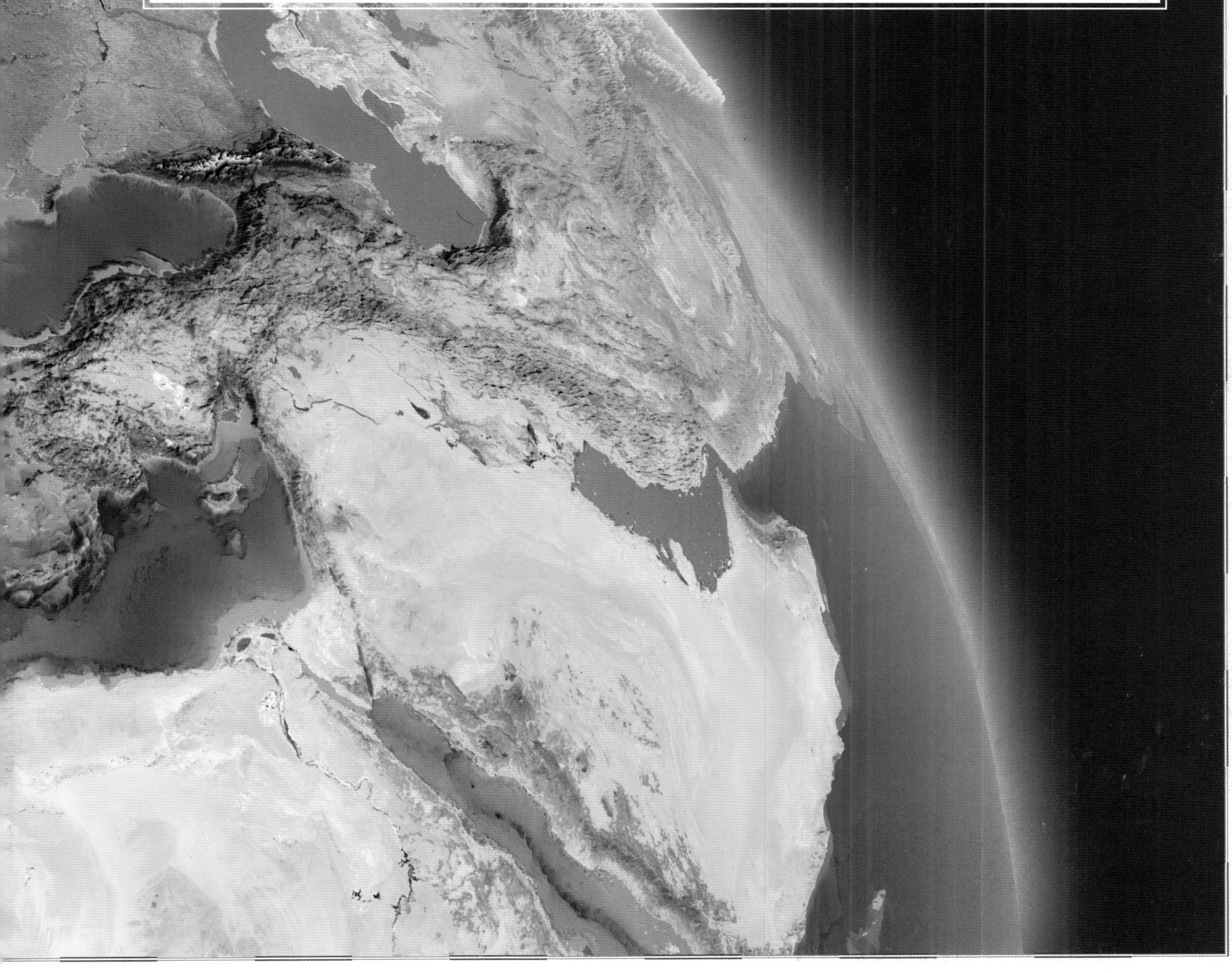

FOR THE SEVENTH EDITION

Senior Cartographic Editor Simon Mumford
Producer, Pre-Production Luca Frassinetti **Producer** Vivienne Yong
Jacket Design Development Manager Sophia MTT
Publishing Director Jonathan Metcalf **Associate Publishing Director** Liz Wheeler **Art Director** Karen Self

General Geographical Consultants

Physical Geography Denys Brunsden, Emeritus Professor, Department of Geography, King's College, London
Human Geography Professor J Malcolm Wagstaff, Department of Geography, University of Southampton
Place Names Caroline Burgess, CartoConsulting Ltd, Reading
Boundaries International Boundaries Research Unit, Mountjoy Research Centre, University of Durham

Digital Mapping Consultants

DK Cartopia developed by George Galfalvi and XMap Ltd, London
Professor Jan-Peter Muller, Department of Photogrammetry and Surveying, University College, London
Planets and information on the Solar System provided by Philip Eales and Kevin Tildsley, Planetary Visions Ltd, London

Regional Consultants

North America Dr David Green, Department of Geography, King's College, London • Jim Walsh, Head of Reference, Wessell Library, Tufts University, Medford, Massachussetts
South America Dr David Preston, School of Geography, University of Leeds **Europe** Dr Edward M Yates, formerly of the Department of Geography, King's College, London
Africa Dr Philip Amis, Development Administration Group, University of Birmingham • Dr Ieuan Ll Griffiths, Department of Geography, University of Sussex
Dr Tony Binns, Department of Geography, University of Sussex
Central Asia Dr David Turnock, Department of Geography, University of Leicester **South and East Asia** Dr Jonathan Rigg, Department of Geography, University of Durham
Australasia and Oceania Dr Robert Allison, Department of Geography, University of Durham

Acknowledgments

Digital terrain data created by Eros Data Center, Sioux Falls, South Dakota, USA. Processed by GVS Images Inc, California, USA and Planetary Visions Ltd, London, UK
Cambridge International Reference on Current Affairs (CIRCA), Cambridge, UK • Digitization by Robertson Research International, Swanley, UK • Peter Clark
British Isles maps generated from a dataset supplied by Map Marketing Ltd/European Map Graphics Ltd in combination with DK Cartopia copyright data

DORLING KINDERSLEY CARTOGRAPHY

Editor-in-Chief Andrew Heritage **Managing Cartographer** David Roberts **Senior Cartographic Editor** Roger Bullen
Editorial Direction Louise Cavanagh **Database Manager** Simon Lewis **Art Direction** Chez Picthall

Cartographers

Pamela Alford • James Anderson • Caroline Bowie • Dale Buckton • Tony Chambers • Jan Clark • Bob Croser • Martin Darlison • Damien Demaj • Claire Ellam • Sally Gable
Jeremy Hepworth • Geraldine Horner • Chris Jackson • Christine Johnston • Julia Lunn • Michael Martin • Ed Merritt • James Mills-Hicks • Simon Mumford • John Plumer
John Scott • Ann Stephenson • Gail Townsley • Julie Turner • Sarah Vaughan • Jane Voss • Scott Wallace • Iorwerth Watkins • Bryony Webb • Alan Whitaker • Peter Winfield

Digital Maps Created in DK Cartopia by
Tom Coulson • Thomas Robertshaw
Philip Rowles • Rob Stokes
Managing Editor
Lisa Thomas
Editors
Thomas Heath • Wim Jenkins • Jane Oliver
Siobhan Ryan • Elizabeth Wyse
Editorial Research
Helen Dangerfield • Andrew Rebeiro-Hargrave
Additional Editorial Assistance
Debra Clapson • Robert Damon • Ailsa Heritage
Constance Novis • Jayne Parsons • Chris Whitwell

Placenames Database Team
Natalie Clarkson • Ruth Duxbury • Caroline Falce • John Featherstone • Dan Gardiner
Ciárán Hynes • Margaret Hynes • Helen Rudkin • Margaret Stevenson • Annie Wilson
Senior Managing Art Editor
Philip Lord
Designers
Scott David • Carol Ann Davis • David Douglas • Rhonda Fisher
Karen Gregory • Nicola Liddiard • Paul Williams
Illustrations
Ciárán Hughes • Advanced Illustration, Congleton, UK
Picture Research
Melissa Albany • James Clarke • Anna Lord
Christine Rista • Sarah Moule • Louise Thomas

First American edition, 2001. This revised edition, 2016.

Published in the United States by DK Publishing, 345 Hudson Street, New York, New York 10014

16 17 18 19 20 10 9 8 7 6 5 4 3 2 1

265171 – March 2016

Published in Great Britain by Dorling Kindersley Ltd. A Penguin Random House company.

DK Publishing books are available at special discounts when purchased in
bulk for sales promotion, premiums, fundraising, or educational use.
For details, contact:
DK Publishing Special Markets, 345 Hudson Street,
New York, New York 10014 or specialsales@dk.com

A catalog record for this book is avaiable from the Library of Congress.

ISBN 978-1-4654-4499-8

Printed and bound in Hong Kong.

A WORLD OF IDEAS:
SEE ALL THERE IS TO KNOW
www.dk.com

Introduction

EVERYTHING YOU NEED TO KNOW ABOUT OUR PLANET TODAY

For many, the outstanding legacy of the twentieth century was the way in which the Earth shrank. In the third millennium, it is increasingly important for us to have a clear vision of the world in which we live. The human population has increased fourfold since 1900. The last scraps of *terra incognita*—the polar regions and ocean depths—have been penetrated and mapped. New regions have been colonized and previously hostile realms claimed for habitation. The growth of air transportation and mass tourism allows many of us to travel further, faster, and more frequently than ever before. In doing so we are given a bird's-eye view of the Earth's surface denied to our forebears.

At the same time, the amount of information about our world has grown enormously. Our multi-media environment hurls uninterrupted streams of data at us, on the printed page, through the airwaves, and across our television, computer, and phone screens; events from all corners of the globe reach us instantaneously and are witnessed as they unfold. Our sense of stability and certainty has been eroded; instead, we are aware that the world is in a constant state of flux and change. Natural disasters, man-made cataclysms, and conflicts between nations remind us daily of the enormity and fragility of our domain. The ongoing threat of international terrorism throws into very stark relief the difficulties that arise when trying to "know" or "understand" our planet and its many cultures.

The current crisis in our "global" culture has made the need greater than ever before for everyone to possess an atlas. DK's **CONCISE WORLD ATLAS** has been conceived to meet this need. At its core, like all atlases, it seeks to define where places are located, to describe their main characteristics, and to map them in relation to other places. Every attempt has been made to produce information and maps that are as clear, accurate, and accessible as possible using the latest digital cartographic techniques. In addition, each page of the atlas provides a wealth of further information, bringing the maps to life. Using photographs, diagrams, at-a-glance maps, introductory texts, and captions, the atlas builds up a detailed portrait of those features—cultural, political, economic, and geomorphological—that make each region unique, and which are also the main agents of change.

This seventh edition of the **CONCISE WORLD ATLAS** incorporates hundreds of revisions and updates affecting every map and every page, distilling the burgeoning mass of information available through modern technology into an extraordinarily detailed and reliable view of our world.

CONTENTS

THE WORLD

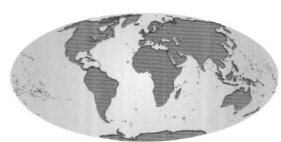

ATLAS OF THE WORLD

North America

South America

Africa

Europe

Asia

Australasia & Oceania

INDEX–GAZETTEER

Key to maps

Regional

Physical features

elevation

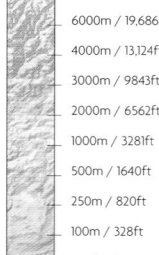

	6000m / 19,686ft
	4000m / 13,124ft
	3000m / 9843ft
	2000m / 6562ft
	1000m / 3281ft
	500m / 1640ft
	250m / 820ft
	100m / 328ft
	sea level
	below sea level

▲ elevation above sea level (mountain height)

▲ volcano

✕ pass

▼ elevation below sea level (depression depth)

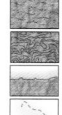

	sand desert
	lava flow
	coastline
	reef
	atoll

sea depth

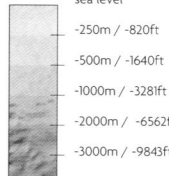

	sea level
	-250m / -820ft
	-500m / -1640ft
	-1000m / -3281ft
	-2000m / -6562ft
	-3000m / -9843ft

▲ seamount / guyot symbol

▼ undersea spot depth

Drainage features

main river

secondary river

tertiary river

minor river

main seasonal river

secondary seasonal river

canal

waterfall

rapids

dam

perennial lake

seasonal lake

perennial salt lake

seasonal salt lake

reservoir

 salt flat / salt pan

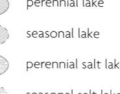 marsh / salt marsh

 mangrove

 wadi

○ spring / well / waterhole / oasis

Ice features

	ice cap / sheet
	ice shelf
	glacier / snowfield

• • • • summer pack ice limit

○ ○ ○ ○ winter pack ice limit

Communications

motorway / highway

motorway / highway (under construction)

major road

minor road

tunnel (road)

main railroad

minor railroad

tunnel (railroad)

✈ international airport

Borders

full international border

undefined international border

disputed de facto border

disputed territorial claim border

indication of country extent (Pacific only)

indication of dependent territory extent (Pacific only)

demarcation / cease fire line

autonomous / federal region border

other 1st order internal administrative border

2nd order internal administrative border

Settlements

 built up area

settlement population symbols

▣ more than 5 million

◪ 1 million to 5 million

◉ 500,000 to 1 million

◎ 100,000 to 500,000

⊙ 50,000 to 100,000

○ 10,000 to 50,000

○ fewer than 10,000

■ ● ⊕ country/dependent territory capital city

■ ● ⊕ autonomous / federal region / other 1st order internal administrative center

■ ● ⊕ 2nd order internal administrative center

Miscellaneous features

═══════ ancient wall

◇ site of interest

○ scientific station

Graticule features

lines of latitude and longitude / Equator

Tropics / Polar circles

45° degrees of longitude / latitude

Typographic key

Physical features

landscape features ... *Namib Desert*
Massif Central
ANDES

headland .... *Nordkapp*

elevation / volcano / pass .... Mount Meru
4556 m

drainage features .... *Lake Geneva*

rivers / canals
spring / well /
waterhole / oasis /
waterfall /
rapids / dam .... *Mekong*

ice features .... *Vatnajökull*

sea features .... *Golfe de Lion*
Andaman Sea
INDIAN OCEAN

undersea features ...... *Barracuda Fracture Zone*

Regions

country .... **ARMENIA**

dependent territory with parent state ..... NIUE (to NZ)

region outside feature area .... ANGOLA

autonomous / federal region .... MINAS GERAIS

other 1st order internal administrative region .... MINSKAYA VOBLASTS'

2nd order internal administrative region .... Vaucluse

cultural region .... New England

Settlements

capital city .... **BEIJING**

dependent territory capital city .... FORT-DE-FRANCE

other settlements ... Chicago
Adana
Tizi Ozou
Yonezawa
Farnham

Miscellaneous

sites of interest / miscellaneous .... *Valley of the Kings*

Tropics / Polar circles .... *Antarctic Circle*

How to use this Atlas

The atlas is organized by continent, moving eastward from the International Date Line. The opening section describes the world's structure, systems, and its main features. The Atlas of the World which follows, is a continent-by-continent guide to today's world, starting with a comprehensive insight into the physical, political, and economic structure of each continent, followed by integrated mapping and descriptions of each region or country.

The world

The introductory section of the Atlas deals with every aspect of the planet, from physical structure to human geography, providing an overall picture of the world we live in. Complex topics such as the landscape of the Earth, climate, oceans, population, and economic patterns are clearly explained with the aid of maps and diagrams drawn from the latest information.

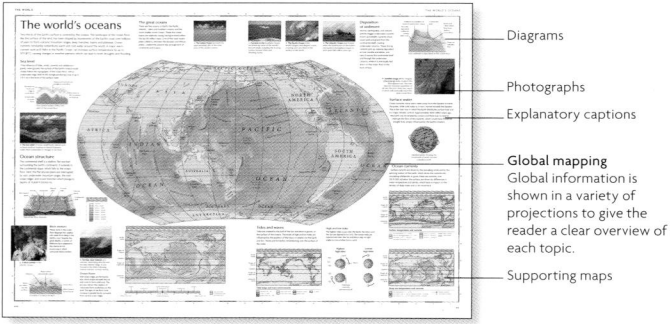

Diagrams

Photographs

Explanatory captions

Global mapping
Global information is shown in a variety of projections to give the reader a clear overview of each topic.

Supporting maps

The political continent

The political portrait of the continent is a vital reference point for every continental section, showing the position of countries relative to one another, and the relationship between human settlement and geographic location. The complex mosaic of languages spoken in each continent is mapped, as is the effect of communications networks on the pattern of settlement.

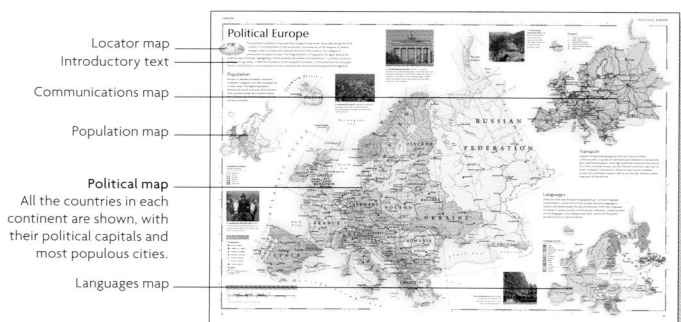

Locator map
Introductory text

Communications map

Population map

Political map
All the countries in each continent are shown, with their political capitals and most populous cities.

Languages map

Continental resources

The Earth's rich natural resources, including oil, gas, minerals, and fertile land, have played a key role in the development of society. These pages show the location of minerals and agricultural resources on each continent, and how they have been instrumental in dictating industrial growth and the varieties of economic activity across the continent.

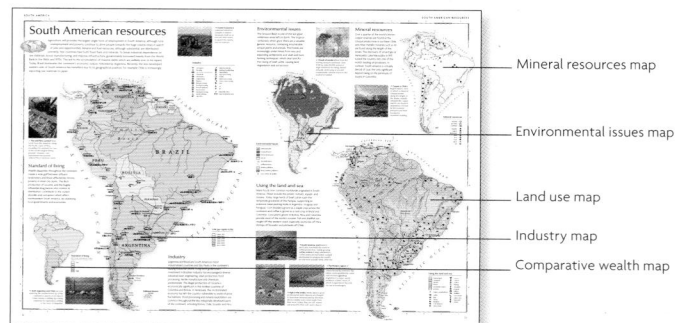

Mineral resources map

Environmental issues map

Land use map

Industry map

Comparative wealth map

The physical continent

The astonishing variety of landforms, and the dramatic forces that created and continue to shape the landscape, are explained in the continental physical spread. Cross-sections, illustrations, and terrain maps highlight the different parts of the continent, showing how nature's forces have produced the landscapes we see today.

Climate charts
Rainfall and temperature charts clearly show the continental patterns of rainfall and temperature.

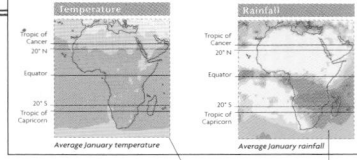

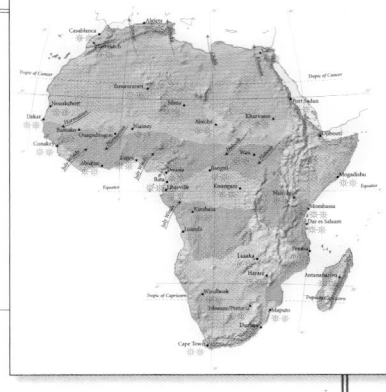

Climate map
Climatic regions vary across each continent. The map displays the differing climatic regions, as well as daily hours of sunshine at selected weather stations.

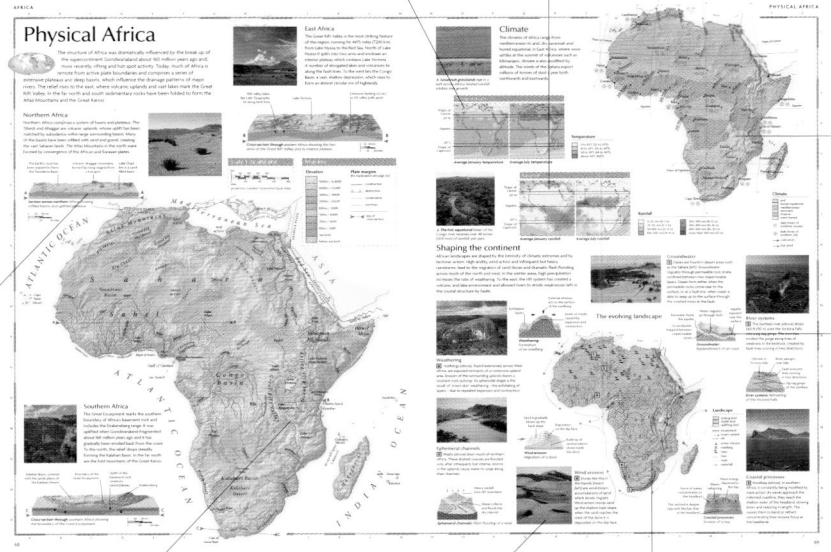

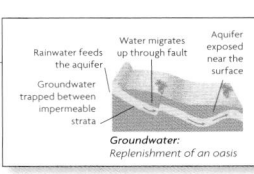

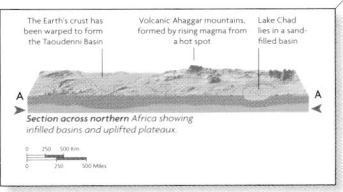

Section across northern Africa showing infilled basins and uplifted plateaus.

Cross-sections
Detailed cross-sections through selected parts of the continent show the underlying geomorphic structure.

Landform diagrams
The complex formation of many typical landforms is summarized in these easy-to-understand illustrations.

Groundwater:
Replenishment of an oasis

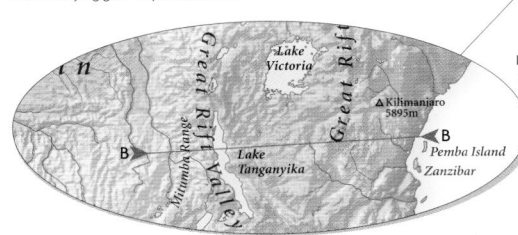

Main physical map
Detailed satellite data has been used to create an accurate and visually striking picture of the surface of the continent.

Photographs
A wide range of beautiful photographs bring the world's regions to life.

Landscape evolution map
The physical shape of each continent is affected by a variety of forces which continually sculpt and modify the landscape. This map shows the major processes which affect different parts of the continent.

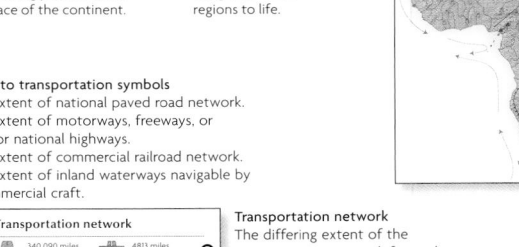

Key to transportation symbols
❶ Extent of national paved road network.
❷ Extent of motorways, freeways, or major national highways.
❸ Extent of commercial railroad network.
❹ Extent of inland waterways navigable by commercial craft.

Regional mapping

The main body of the Atlas is a unique regional map set, with detailed information on the terrain, the human geography of the region, and its infrastructure. Around the edge of the map, additional "at-a-glance" maps, give an instant picture of regional industry, land use, and agriculture. The detailed terrain map (shown in perspective), focuses on the main physical features of the region, and is enhanced by annotated illustrations, and photographs of the physical structure.

Transportation network

❶	340,090 miles (544,344 km)	4813 miles (7700 km)	❷
❸	12,872 miles (20,592 km)	2108 miles (3389 km)	❹

New York's commercial success is tied historically to its transportation connections. The Erie Canal, completed in 1825, opened up the Great Lakes and the interior to New York's markets and carried a stream of immigrants into the Midwest.

Transportation network
The differing extent of the transportation network for each region is shown here, along with key facts about the transportation system.

Regional Locator
This small map shows the location of each country in relation to its continent.

Key to main map
A key to the population symbols and land heights accompanies the main map.

World locator
This locates the continent in which the region is found on a small world map.

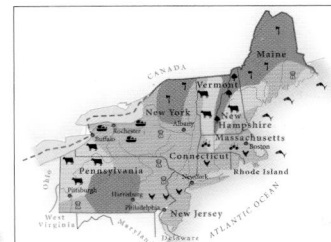

Land use map
This shows the different types of land use which characterize the region, as well as indicating the principal agricultural activities.

Map keys
Each supporting map has its own key.

Grid reference
The framing grid provides a location reference for each place listed in the Index.

USA: NORTHEASTERN STATES

Connecticut, Maine, Massachusetts, New Hampshire, New Jersey, New York, Pennsylvania, Rhode Island, Vermont

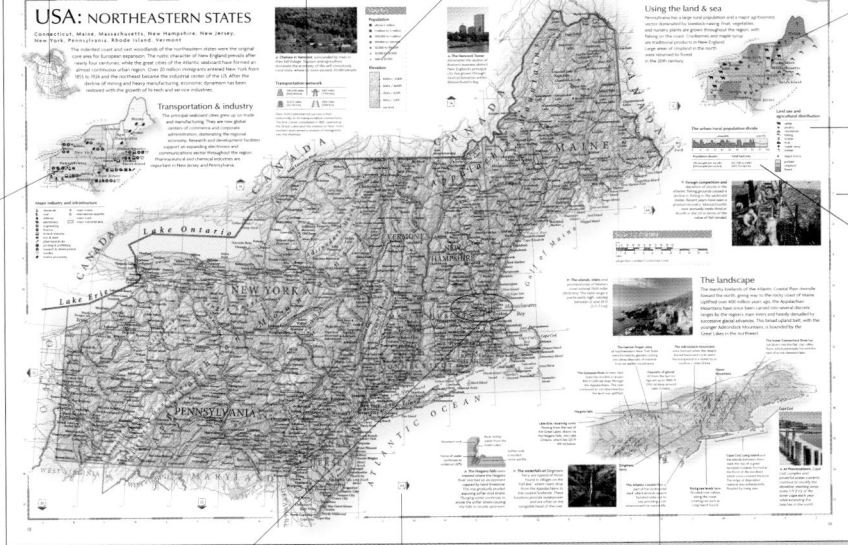

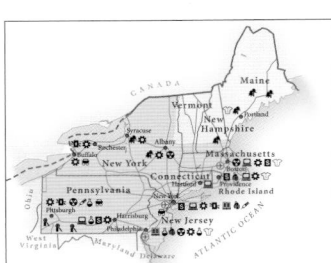

Transportation and industry map
The main industrial areas are mapped, and the most important industrial and economic activities of the region are shown.

The urban/rural population divide

urban 83%	rural 17%

Population density	Total land area
335 people per sq mile (120 people per sq km)	162,258 sq miles (420,232 sq km)

Urban/rural population divide
The proportion of people in the region who live in urban and rural areas, as well as the overall population density and land area are clearly shown in these simple graphics.

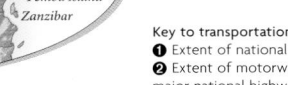

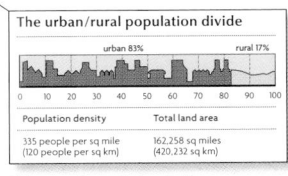

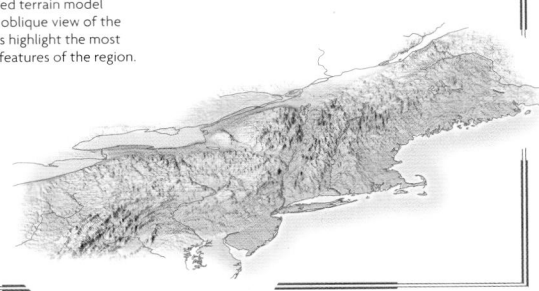

Continuation symbols
These symbols indicate where adjacent maps can be found.

Main regional map
A wealth of information is displayed on the main map, building up a rich portrait of the interaction between the physical landscape and the human and political geography of each region. The key to the regional maps can be found on page viii.

Landscape map
The computer-generated terrain model accurately portrays an oblique view of the landscape. Annotations highlight the most important geographic features of the region.

The Solar System

Nine major planets, their satellites, and countless minor planets (asteroids) orbit the Sun to form the Solar System. The Sun, our nearest star, creates energy from nuclear reactions deep within its interior, providing all the light and heat which make life on Earth possible. The Earth is unique in the Solar System in that it supports life: its size, gravitational pull and distance from the Sun have all created the optimum conditions for the evolution of life. The planetary images seen here are composites derived from actual spacecraft images (not shown to scale).

Orbits

All the Solar System's planets and dwarf planets orbit the Sun in the same direction and (apart from Pluto) roughly in the same plane. All the orbits have the shapes of ellipses (stretched circles). However, in most cases, these ellipses are close to being circular: only Pluto and Eris have very elliptical orbits. Orbital period (the time it takes an object to orbit the Sun) increases with distance from the Sun. The more remote objects not only have further to travel with each orbit, they also move more slowly.

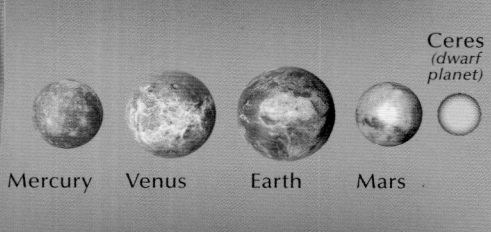

Mercury   Venus   Earth   Mars

Ceres
(dwarf planet)

Jupiter

The Sun

⊖ *Diameter: 864,948 miles (1,392,000 km)*
⊙ *Mass: 1990 million million million million tons*

The Sun was formed when a swirling cloud of dust and gas contracted, pulling matter into its center. When the temperature at the center rose to 1,000,000°C, nuclear fusion – the fusing of hydrogen into helium, creating energy – occurred, releasing a constant stream of heat and light.

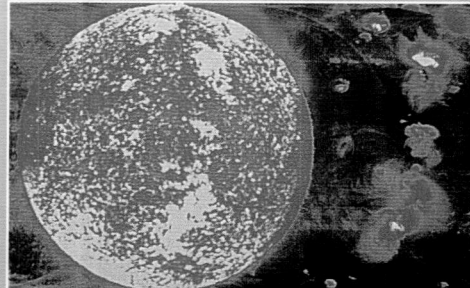

▲ *Solar flares are sudden bursts of energy from the Sun's surface. They can be 125,000 miles (200,000 km) long.*

The formation of the Solar System

The cloud of dust and gas thrown out by the Sun during its formation cooled to form the Solar System. The smaller planets nearest the Sun are formed of minerals and metals. The outer planets were formed at lower temperatures, and consist of swirling clouds of gases.

Solar eclipse

A solar eclipse occurs when the Moon passes between Earth and the Sun, casting its shadow on Earth's surface. During a total eclipse *(below)*, viewers along a strip of Earth's surface, called the area of totality, see the Sun totally blotted out for a short time, as the umbra (Moon's full shadow) sweeps over them. Outside this area is a larger one, where the Sun appears only partly obscured, as the penumbra (partial shadow) passes over.

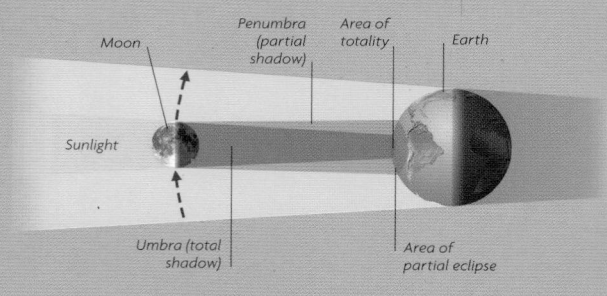

Moon

Penumbra *(partial shadow)*

Area of totality

Earth

Sunlight

Umbra *(total shadow)*

Area of partial eclipse

PLANETS

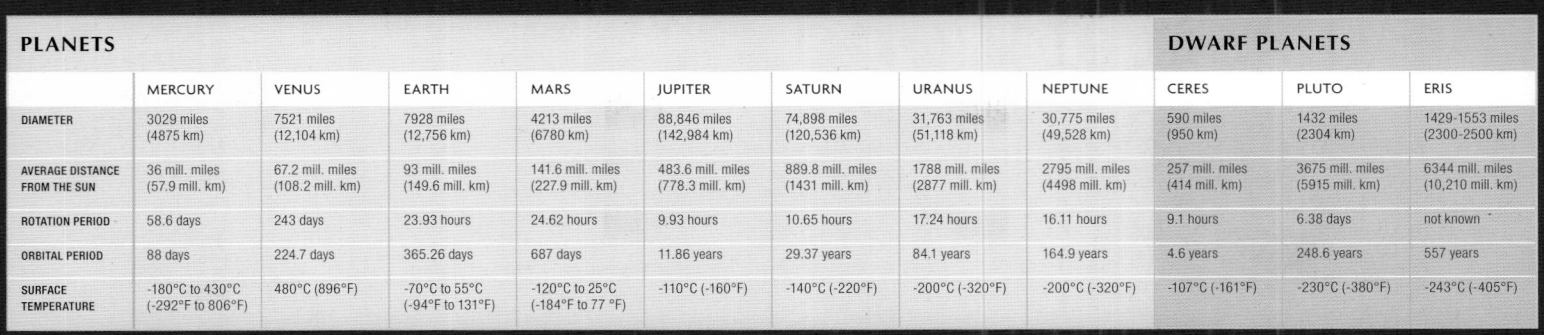

	MERCURY	VENUS	EARTH	MARS	JUPITER	SATURN	URANUS	NEPTUNE
DIAMETER	3029 miles (4875 km)	7521 miles (12,104 km)	7928 miles (12,756 km)	4213 miles (6780 km)	88,846 miles (142,984 km)	74,898 miles (120,536 km)	31,763 miles (51,118 km)	30,775 miles (49,528 km)
AVERAGE DISTANCE FROM THE SUN	36 mill. miles (57.9 mill. km)	67.2 mill. miles (108.2 mill. km)	93 mill. miles (149.6 mill. km)	141.6 mill. miles (227.9 mill. km)	483.6 mill. miles (778.3 mill. km)	889.8 mill. miles (1431 mill. km)	1788 mill. miles (2877 mill. km)	2795 mill. miles (4498 mill. km)
ROTATION PERIOD	58.6 days	243 days	23.93 hours	24.62 hours	9.93 hours	10.65 hours	17.24 hours	16.11 hours
ORBITAL PERIOD	88 days	224.7 days	365.26 days	687 days	11.86 years	29.37 years	84.1 years	164.9 years
SURFACE TEMPERATURE	-180°C to 430°C (-292°F to 806°F)	480°C (896°F)	-70°C to 55°C (-94°F to 131°F)	-120°C to 25°C (-184°F to 77 °F)	-110°C (-160°F)	-140°C (-220°F)	-200°C (-320°F)	-200°C (-320°F)

DWARF PLANETS

	CERES	PLUTO	ERIS
DIAMETER	590 miles (950 km)	1432 miles (2304 km)	1429-1553 miles (2300-2500 km)
AVERAGE DISTANCE FROM THE SUN	257 mill. miles (414 mill. km)	3675 mill. miles (5915 mill. km)	6344 mill. miles (10,210 mill. km)
ROTATION PERIOD	9.1 hours	6.38 days	not known
ORBITAL PERIOD	4.6 years	248.6 years	557 years
SURFACE TEMPERATURE	-107°C (-161°F)	-230°C (-380°F)	-243°C (-405°F)

AVERAGE DISTANCE FROM THE SUN

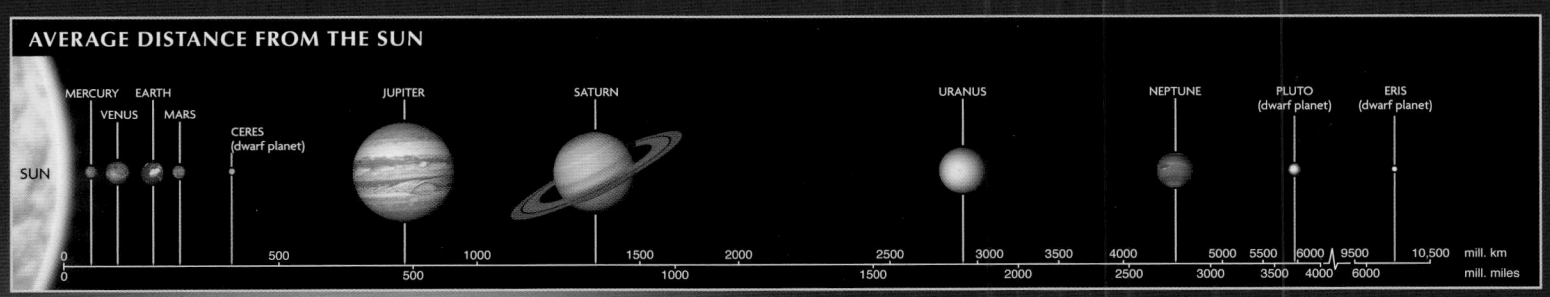

SUN — MERCURY — VENUS — EARTH — MARS — CERES (dwarf planet) — JUPITER — SATURN — URANUS — NEPTUNE — PLUTO (dwarf planet) — ERIS (dwarf planet)

mill. km: 0, 500, 1000, 1500, 2000, 2500, 3000, 3500, 4000, 5000 5500, 6000 9500, 10,500
mill. miles: 0, 500, 1000, 1500, 2000, 2500, 3000, 3500, 4000, 6000

Saturn

Uranus

Neptune

Pluto (dwarf planet)

Eris (dwarf planet)

Space Debris

Millions of objects, remnants of planetary formation, circle the Sun in a zone lying between Mars and Jupiter: the asteroid belt. Fragments of asteroids break off to form meteoroids, which can reach the Earth's surface. Comets, composed of ice and dust, originated outside our Solar System. Their elliptical orbit brings them close to the Sun and into the inner Solar System.

▲ *Meteor Crater in Arizona is 4200 ft (1300 m) wide and 660 ft (200 m) deep. It was formed over 10,000 years ago.*

Possible and actual meteorite craters

Map key
- Possible impact craters
- Meteorite impact craters

The Earth's Atmosphere

During the early stages of the Earth's formation, ash, lava, carbon dioxide, and water vapor were discharged onto the surface of the planet by constant volcanic eruptions. The water formed the oceans, while carbon dioxide entered the atmosphere or was dissolved in the oceans. Clouds, formed of water droplets, reflected some of the Sun's radiation back into space. The Earth's temperature stabilized and early life forms began to emerge, converting carbon dioxide into life-giving oxygen.

▲ *It is thought that the gases that make up the Earth's atmosphere originated deep within the interior, and were released many millions of years ago during intense volcanic actvity, similar to this eruption at Mount St. Helens.*

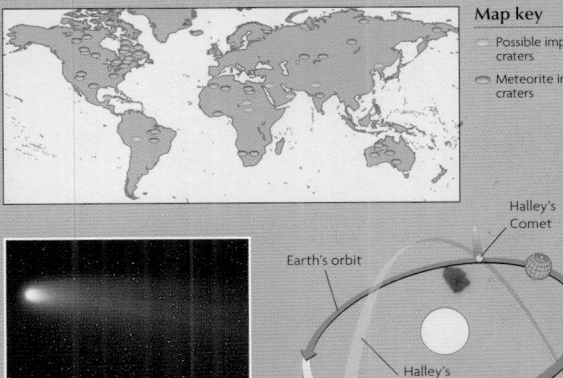

▲ *The orbit of Halley's Comet brings it close to the Earth every 76 years. It last visited in 1986.*

Halley's Comet

Earth's orbit

Halley's orbit

Orbit of Halley's Comet around the Sun

The physical world

The Earth's surface is constantly being transformed: it is uplifted, folded, and faulted by tectonic forces; weathered and eroded by wind, water, and ice. Sometimes change is dramatic, the spectacular results of earthquakes or floods. More often it is a slow process lasting millions of years. A physical map of the world represents a snapshot of the ever-evolving architecture of the Earth. This terrain map shows the whole surface of the Earth, both above and below the sea.

The world in section

These cross-sections around the Earth, one in the northern hemisphere; one straddling the Equator, reveal the limited areas of land above sea level in comparison with the extent of the sea floor. The greater erosive effects of weathering by wind and water limit the upward elevation of land above sea level, while the deep oceans retain their dramatic mountain and trench profiles.

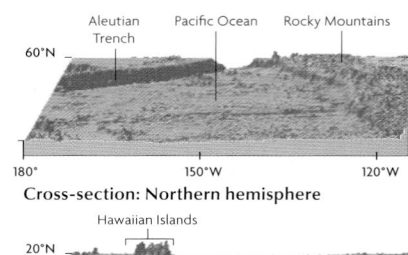

Cross-section: Northern hemisphere

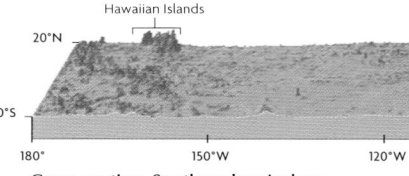

Cross-section: Southern hemisphere

Map key

Elevation

- 6000m / 19,686ft
- 4000m / 13,124ft
- 3000m / 9843ft
- 2000m / 6562ft
- 1000m / 3281ft
- 500m / 1640ft
- 250m / 820ft
- 100m / 328ft
- sea level
- below sea level

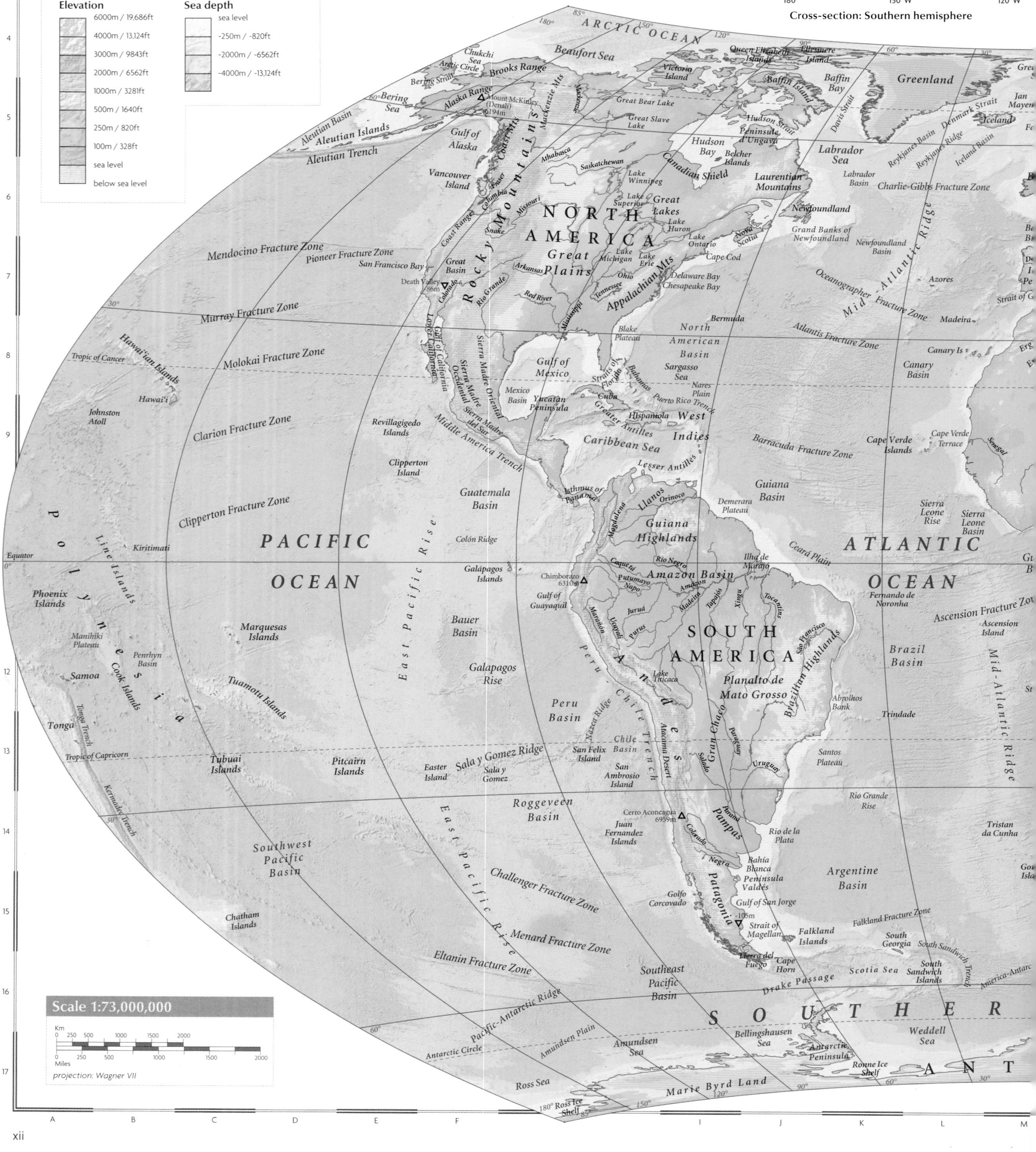

Sea depth

- sea level
- -250m / -820ft
- -2000m / -6562ft
- -4000m / -13,124ft

Scale 1:73,000,000

Km
0 250 500 1000 1500 2000

Miles
0 250 500 1000 1500 2000

projection: Wagner VII

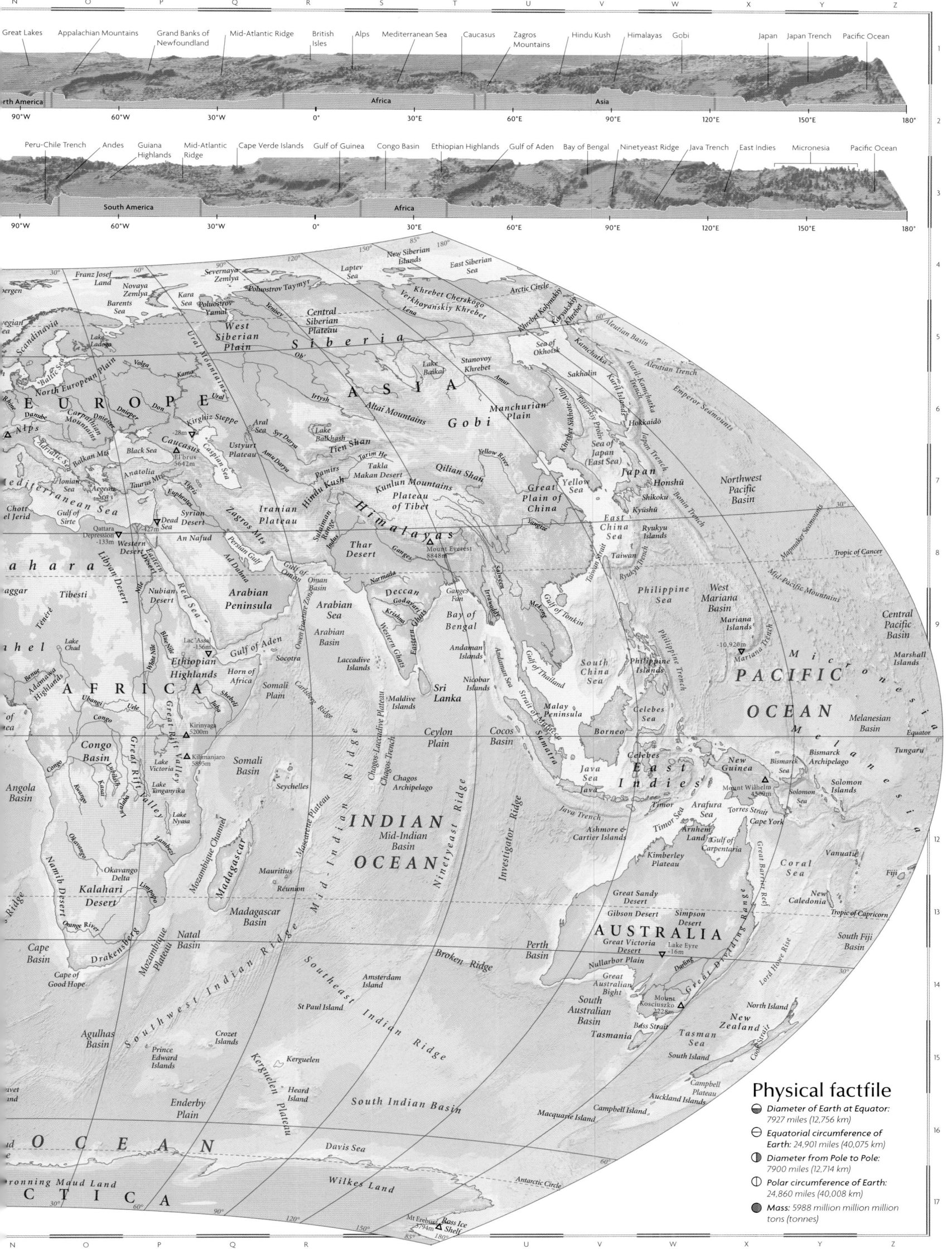

Physical factfile

- **Diameter of Earth at Equator:** 7927 miles (12,756 km)
- **Equatorial circumference of Earth:** 24,901 miles (40,075 km)
- **Diameter from Pole to Pole:** 7900 miles (12,714 km)
- **Polar circumference of Earth:** 24,860 miles (40,008 km)
- **Mass:** 5988 million million million tons (tonnes)

Structure of the Earth

The Earth as it is today is just the latest phase in a constant process of evolution which has occurred over the past 4.5 billion years. The Earth's continents are neither fixed nor stable; over the course of the Earth's history, propelled by currents rising from the intense heat at its center, the great plates on which they lie have moved, collided, joined together, and separated. These processes continue to mold and transform the surface of the Earth, causing earthquakes and volcanic eruptions and creating oceans, mountain ranges, deep ocean trenches, and island chains.

Inside the Earth

The Earth's hot inner core is made up of solid iron, while the outer core is composed of liquid iron and nickel. The mantle nearest the core is viscous, whereas the rocky upper mantle is fairly rigid. The crust is the rocky outer shell of the Earth. Together, the upper mantle and the crust form the lithosphere.

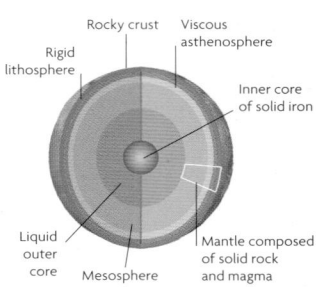

Rocky crust
Viscous asthenosphere
Rigid lithosphere
Inner core of solid iron
Liquid outer core
Mesosphere
Mantle composed of solid rock and magma

The dynamic Earth

The Earth's crust is made up of eight major (and several minor) rigid continental and oceanic tectonic plates, which fit closely together. The positions of the plates are not static. They are constantly moving relative to one another. The type of movement between plates affects the way in which they alter the structure of the Earth. The oldest parts of the plates, known as shields, are the most stable parts of the Earth and little tectonic activity occurs here.

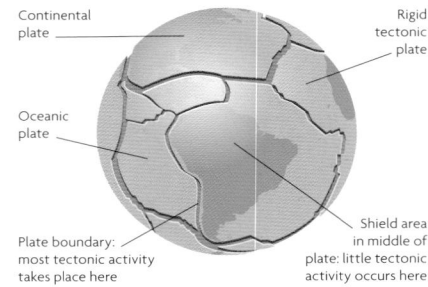
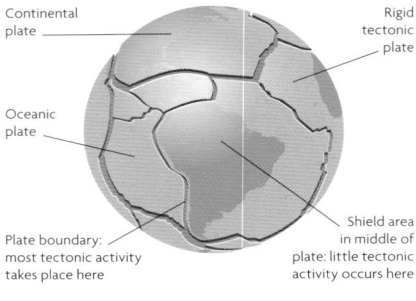

Continental plate
Oceanic plate
Plate boundary: most tectonic activity takes place here
Rigid tectonic plate
Shield area in middle of plate: little tectonic activity occurs here

Convection currents

Deep within the Earth, at its inner core, temperatures may exceed 8,100°F (4,500°C). This heat warms rocks in the mesosphere which rise through the partially molten mantle, displacing cooler rocks just below the solid crust, which sink, and are warmed again by the heat of the mantle. This process is continuous, creating convection currents which form the moving force beneath the Earth's crust.

Outer core
Inner core
Subduction zone
Ocean crust
Movement of plate
Mid-ocean ridge
Lithosphere
Asthenosphere
Mesosphere
Continental crust

Plate boundaries

The boundaries between the plates are the areas where most tectonic activity takes place. Three types of movement occur at plate boundaries: the plates can either move toward each other, move apart, or slide past each other. The effect this has on the Earth's structure depends on whether the margin is between two continental plates, two oceanic plates, or an oceanic and continental plate.

▲ *The Mid-Atlantic Ridge* rises above sea level in Iceland, producing geysers and volcanoes.

Mid-ocean ridges

Mid-ocean ridges are formed when two adjacent oceanic plates pull apart, allowing magma to force its way up to the surface, which then cools to form solid rock. Vast amounts of volcanic material are discharged at these mid-ocean ridges which can reach heights of 10,000 ft (3000 m).

Ocean floor
Earthquake zone
Magma pushed upwards along centre of ridge
Solid mantle

Formation of a mid-ocean ridge

▲ *Mount Pinatubo is an active volcano, lying on the Pacific "Ring of Fire."*

Ocean plates meeting

Oceanic crust is denser and thinner than continental crust; on average it is 3 miles (5 km) thick, while continental crust averages 18–24 miles (30–40 km). When oceanic plates of similar density meet, the crust is contorted as one plate overrides the other, forming deep sea trenches and volcanic island arcs above sea level.

Overriding plate
Chain of islands
Ocean trench
Diving plate
Volcanic activity

Ocean plates meeting to form an island arc

Tectonic activity

- - - - - uncertain plate boundary
▲ volcanic zone
● earthquake zone
● hot spot
ᐯᐯᐯᐯ rift valley

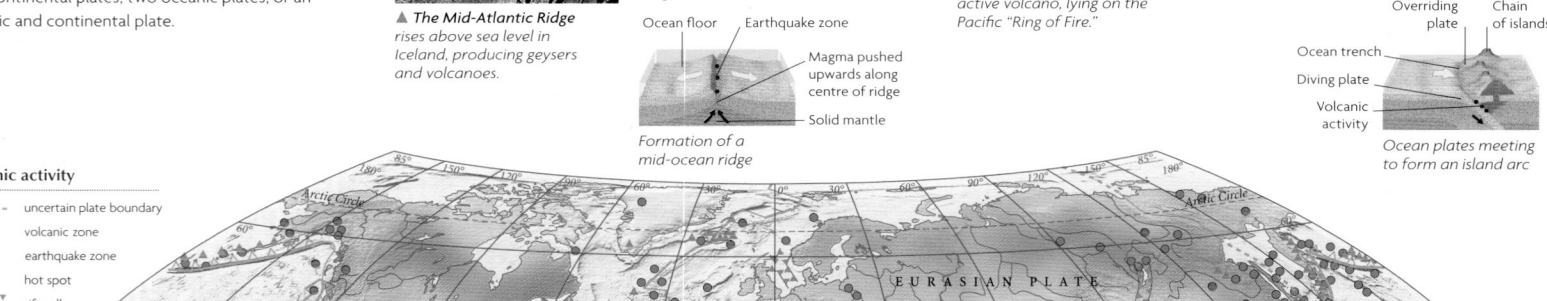

JUAN DE FUCA PLATE
NORTH AMERICAN PLATE
EURASIAN PLATE
ANATOLIAN PLATE
IRANIAN PLATE
PACIFIC PLATE
ARABIAN PLATE
PHILIPPINE PLATE
CARIBBEAN PLATE
COCOS PLATE
CAROLINE PLATE
BISMARCK PLATE
PACIFIC PLATE
AFRICAN PLATE
SOUTH AMERICAN PLATE
NAZCA PLATE
SOLOMON PLATE
FIJI PLATE
INDO AUSTRALIAN PLATE
SCOTIA PLATE
ANTARCTIC PLATE

Diving plates

ᐃ ᐃ When an oceanic and a continental plate meet, the denser oceanic plate is driven underneath the continental plate, which is crumpled by the collision to form mountain ranges. As the ocean plate plunges downward, it heats up, and molten rock (magma) is forced up to the surface.

◀ *The Andean mountain chain is the typical result of the impact of a diving plate.*

Oceanic plate dives under continental plate
Mountains thrust up by collision
Earthquake zone
Continental plate

Diving plate

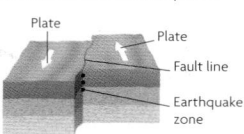

▲ *The deep fracture caused by the sliding plates of the San Andreas Fault can be clearly seen in parts of California.*

Sliding plates

When two plates slide past each other, friction is caused along the fault line which divides them. The plates do not move smoothly, and the uneven movement causes earthquakes.

Plate
Plate
Fault line
Earthquake zone

Sliding plates

▶ *The Alps were formed when the African Plate collided with the Eurasian Plate, about 65 million years ago.*

Plate buckles as it collides
Mountains thrust upwards
Earthquake zone
Crust thickens in response to the impact

Continental plates colliding to form a mountain range

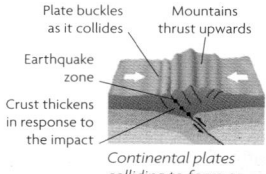

Colliding plates

ᐃ ᐃ ᐃ When two continental plates collide, great mountain chains are thrust upward as the crust buckles and folds under the force of the impact.

Continental drift

Although the plates which make up the Earth's crust move only a few inches in a year, over the millions of years of the Earth's history, its continents have moved many thousands of miles, to create new continents, oceans, and mountain chains

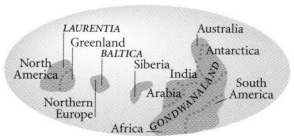

1: Cambrian period

570–510 million years ago. Most continents are in tropical latitudes. The supercontinent of Gondwanaland reaches the South Pole.

2: Devonian period

408–362 million years ago. The continents of Gondwanaland and Laurentia are drifting northward.

3: Carboniferous period

362–290 million years ago. The Earth is dominated by three continents; Laurentia, Angaraland, and Gondwanaland.

4: Triassic period

245–208 million years ago. All three major continents have joined to form the super-continent of Pangea.

5: Jurassic period

208–145 million years ago. The super-continent of Pangea begins to break up, causing an overall rise in sea levels.

6: Cretaceous period

145–65 million years ago. Warm, shallow seas cover much of the land: sea levels are about 80 ft (25 m) above present levels.

7: Tertiary period

65–2 million years ago. Although the world's geography is becoming more recognizable, major events such as the creation of the Himalayan mountain chain, are still to occur during this period.

Continental shields

The centers of the Earth's continents, known as shields, were established between 2500 and 500 million years ago; some contain rocks over three billion years old. They were formed by a series of turbulent events: plate movements, earthquakes, and volcanic eruptions. Since the Pre-Cambrian period, over 570 million years ago, they have experienced little tectonic activity, and today, these flat, low-lying slabs of solidified molten rock form the stable centers of the continents. They are bounded or covered by successive belts of younger sedimentary rock.

The Hawai'ian island chain

A hot spot lying deep beneath the Pacific Ocean pushes a plume of magma from the Earth's mantle up through the Pacific Plate to form volcanic islands. While the hot spot remains stationary, the plate on which the islands sit is moving slowly. A long chain of islands has been created as the plate passes over the hot spot.

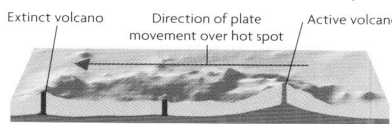

Cross-section through the Hawai'ian Islands

Evolution of the Hawai'ian Islands

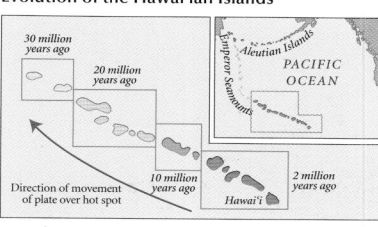

Creation of the Himalayas

Between 10 and 20 million years ago, the Indian subcontinent, part of the ancient continent of Gondwanaland, collided with the continent of Asia. The Indo-Australian Plate continued to move northward, displacing continental crust and uplifting the Himalayas, the world's highest mountain chain.

Movements of India

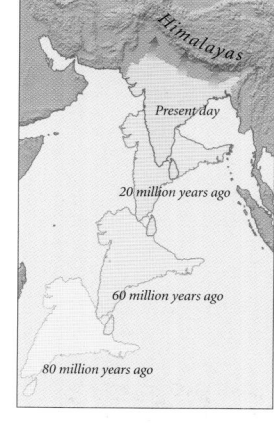

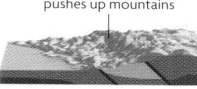

Cross-section through the Himalayas

▲ **The Himalayas were** uplifted when the Indian subcontinent collided with Asia.

The Earth's geology

The Earth's rocks are created in a continual cycle. Exposed rocks are weathered and eroded by wind, water, and chemicals and deposited as sediments. If they pass into the Earth's crust they will be transformed by high temperatures and pressures into metamorphic rocks or they will melt and solidify as igneous rocks.

Sandstone

8 Sandstones are sedimentary rocks formed mainly in deserts, beaches, and deltas. Desert sandstones are formed of grains of quartz which have been well rounded by wind erosion.

▲ **Rock stacks** of desert sandstone, at Bryce Canyon National Park, Utah, US.

◄ **Extrusive igneous rocks** are formed during volcanic eruptions, as here in Hawai'i.

Andesite

7 Andesite is an extrusive igneous rock formed from magma which has solidified on the Earth's crust after a volcanic eruption.

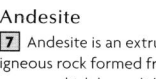
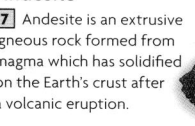
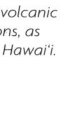

Geological regions

- continental shield
- sedimentary cover
- coral formation
- igneous rock types

Mountain ranges

- Alpine (new)
- Hercynian (old)
- Caledonian (ancient)

Schist

1 Schist is a metamorphic rock formed during mountain building, when temperature and pressure are comparatively high. Both mudstones and shales reform into schist under these conditions.

▶ **Schist formations** in the Atlas Mountains, northwestern Africa.

Gneiss

1 Gneiss is a metamorphic rock made at great depth during the formation of mountain chains, when intense heat and pressure transform sedimentary or igneous rocks.

▲ **Gneiss formations** in Norway's Jotunheimen Mountains.

▲ **Basalt columns at** Giant's Causeway, Northern Ireland, UK.

Basalt

2 Basalt is an igneous rock, formed when small quantities of magma lying close to the Earth's surface cool rapidly.

Granite

5 Granite is an intrusive igneous rock formed from magma which has solidified deep within the Earth's crust. The magma cools slowly, producing a coarse-grained rock.

▶ **Namibia's Namaqualand Plateau** is formed of granite.

Limestone

3 Limestone is a sedimentary rock, which is formed mainly from the calcite skeletons of marine animals which have been compressed into rock.

▲ **Limestone hills, Guilin, China.**

Coral

4 Coral reefs are formed from the skeletons of millions of individual corals.

▲ **Great Barrier Reef, Australia.**

Shaping the landscape

The basic material of the Earth's surface is solid rock: valleys, deserts, soil, and sand are all evidence of the powerful agents of weathering, erosion, and deposition which constantly shape and transform the Earth's landscapes. Water, either flowing continually in rivers or seas, or frozen and compacted into solid sheets of ice, has the most clearly visible impact on the Earth's surface. But wind can transport fragments of rock over huge distances and strip away protective layers of vegetation, exposing rock surfaces to the impact of extreme heat and cold.

Coastal water

The world's coastlines are constantly changing: every day, tides deposit, sift and sort sand, and gravel on the shoreline. Over longer periods, powerful wave action erodes cliffs and headlands and carves out bays.

▶ *A low, wide* sandy beach on *South Africa's Cape Peninsula is continually re-shaped by the action of the Atlantic waves.*

▲ *The sheer chalk* cliffs at Seven Sisters in *southern England are constantly under attack from waves.*

Water

Less than 2% of the world's water is on the land, but it is the most powerful agent of landscape change. Water, as rainfall, groundwater, and rivers, can transform landscapes through both erosion and deposition. Eroded material carried by rivers forms the world's most fertile soils.

▲ *Waterfalls such as* the Iguaçu Falls on *the border between Argentina and southern Brazil, erode the underlying rock, causing the falls to retreat.*

Groundwater

In regions where there are porous rocks such as chalk, water is stored underground in large quantities; these reservoirs of water are known as aquifers. Rain percolates through topsoil into the underlying bedrock, creating an underground store of water. The limit of the saturated zone is called the water table.

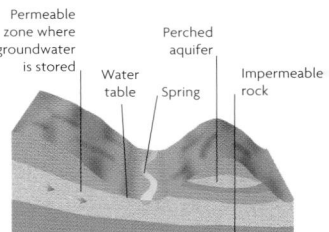

Storage of groundwater in an aquifer

World river systems

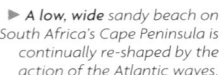

drainage basin

World river systems: Sediment deposited annually per drainage basin

tons per sq mile per year

[map of the world showing river systems]

Rivers

Rivers erode the land by grinding and dissolving rocks and stones. Most erosion occurs in the river's upper course as it flows through highland areas. Rock fragments are moved along the river bed by fast-flowing water and deposited in areas where the river slows down, such as flat plains, or where the river enters seas or lakes.

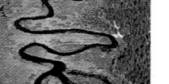

▲ *The Mississippi River forms meanders as it flows across the southern US.*

Meanders

In their lower courses, rivers flow slowly. As they flow across the lowlands, they form looping bends called meanders.

▲ *The meanders of Utah's San Juan River have become deeply incised.*

River valleys

Over long periods of time rivers erode uplands to form characteristic V-shaped valleys with smooth sides.

Resistant rock
River
Chemical erosion cuts valley in softer rock

River valley erosion

Deposition

When rivers have deposited large quantities of fertile alluvium, they are forced to find new channels through the alluvium deposits, creating braided river systems.

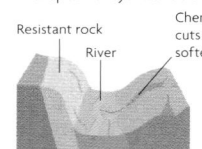

Deltas

When a river deposits its load of silt and sediment (alluvium) on entering the sea, it may form a delta. As this material accumulates, it chokes the mouth of the river, forcing it to create new channels to reach the sea.

▶ *The Nile forms a broad delta as it flows into the Mediterranean.*

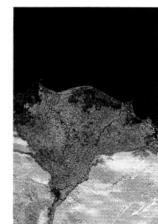

◀ *Mud is deposited by China's Yellow River in its lower course.*

Landslides

Heavy rain and associated flooding on slopes can loosen underlying rocks, which crumble, causing the top layers of rock and soil to slip.

▶ *A huge landslide in the Swiss Alps has left massive piles of rocks and pebbles called scree.*

Drainage basins

The drainage basin is the area of land drained by a major trunk river and its smaller branch rivers or tributaries. Drainage basins are separated from one another by natural boundaries known as watersheds.

Watershed
Major trunk river
Alps
Dolomites
Apennines
Tributary river
Delta
River mouth
Po Valley

The drainage basin of the Po river, northern Italy.

Gullies

In areas where soil is thin, rainwater is not effectively absorbed, and may flow overland. The water courses downhill in channels, or gullies, and may lead to rapid erosion of soil.

▲ *A deep gully in the French Alps caused by the scouring of upper layers of turf.*

Ice

During its long history, the Earth has experienced a number of glacial episodes when temperatures were considerably lower than today. During the last Ice Age, 18,000 years ago, ice covered an area three times larger than it does today. Over these periods, the ice has left a remarkable legacy of transformed landscapes.

Glaciers

Glaciers are formed by the compaction of snow into "rivers" of ice. As they move over the landscape, glaciers pick up and carry a load of rocks and boulders which erode the landscape they pass over, and are eventually deposited at the end of the glacier.

▲ *A massive glacier* advancing down a valley in southern Argentina.

Post-glacial features

When a glacial episode ends, the retreating ice leaves many features. These include depositional ridges called moraines, which may be eroded into low hills known as drumlins; sinuous ridges called eskers; kames, which are rounded hummocks; depressions known as kettle holes; and windblown loess deposits.

Glacial valleys

Glaciers can erode much more powerfully than rivers. They form steep-sided, flat-bottomed valleys with a typical U-shaped profile. Valleys created by tributary glaciers, whose floors have not been eroded to the same depth as the main glacial valley floor, are called hanging valleys

▲ *The U-shaped profile* and piles of morainic debris are characteristic of a valley once filled by a glacier.

▲ *A series of* hanging valleys high up in the Chilean Andes.

Past and present world ice-cover and glacial features

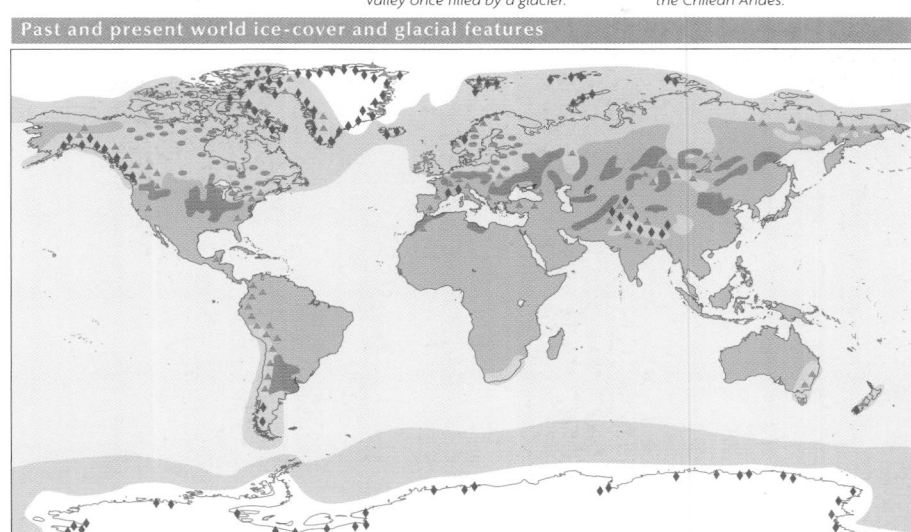

Past and present world ice cover and glacial features

- extent of last Ice Age
- loess deposits
- post-glacial feature
- glacial feature
- present day ice cover
- glacial field

Post-glacial landscape features

Labels: Kame terrace, Retreating glacier, Kettle hole, Esker, Drumlin, Braided river, Terminal moraine, Windblown loess, Glacial till, Bedrock

Ice shattering

Water drips into fissures in rocks and freezes, expanding as it does so. The pressure weakens the rock, causing it to crack, and eventually to shatter into polygonal patterns.

▲ *Irregular polygons show* through the sedge-grass tundra in the Yukon, Canada.

▲ *The profile of* the Matterhorn has been formed by three cirques lying "back-to-back."

Cirques

Cirques are basin-shaped hollows which mark the head of a glaciated valley. Where neighboring cirques meet, they are divided by sharp rock ridges called arêtes. It is these arêtes which give the Matterhorn its characteristic profile.

Fjords

Fjords are ancient glacial valleys flooded by the sea following the end of a period of glaciation. Beneath the water, the valley floor can be 4000 ft (1300 m) deep.

▲ *A fjord fills* a former glacial valley in southern New Zealand.

Periglaciation

Periglacial areas occur near to the edge of ice sheets. A layer of frozen ground lying just beneath the surface of the land is known as permafrost. When the surface melts in the summer, the water is unable to drain into the frozen ground, and so "creeps" downhill, a process known as solifluction.

Wind

Strong winds can transport rock fragments great distances, especially where there is little vegetation to protect the rock. In desert areas, wind picks up loose, unprotected sand particles, carrying them over great distances. This powerfully abrasive debris is blasted at the surface by the wind, eroding the landscape into dramatic shapes.

Deposition

The rocky, stony floors of the world's deserts are swept and scoured by strong winds. The smaller, finer particles of sand are shaped into surface ripples, dunes, or sand mountains, which rise to a height of 650 ft (200 m). Dunes usually form single lines, running perpendicular to the direction of the prevailing wind. These long, straight ridges can extend for over 100 miles (160 km).

Dunes

Dunes are shaped by wind direction and sand supply. Where sand supply is limited, crescent-shaped barchan dunes are formed.

Prevailing winds and dust trajectories

Prevailing winds
- northeast trade
- southeast trade
- westerly
- westerly
- polar easterly
- polar easterly

Dust trajectories
- trajectory of aeolian dust

Hot and cold deserts

Main desert types
- hot arid
- semi-arid
- cold polar

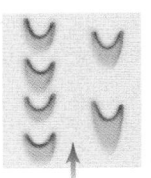

▲ *Barchan dunes in the* Arabian Desert.

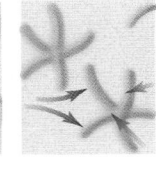

▲ *Complex dune system in* the Sahara.

Types of dune

Wind direction

Transverse dune

Barchan dune

Linear dune

Star dune

Heat

Fierce sun can heat the surface of rock, causing it to expand more rapidly than the cooler, underlying layers. This creates tensions which force the rock to crack or break up. In arid regions, the evaporation of water from rock surfaces dissolves certain minerals within the water, causing salt crystals to form in small openings in the rock. The hard crystals force the openings to widen into cracks and fissures.

▲ *The cracked and* parched floor of Death Valley, California. This is one of the hottest deserts on Earth.

Temperature

Most of the world's deserts are in the tropics. The cold deserts which occur elsewhere are arid because they are a long way from the rain-giving sea. Rock in deserts is exposed because of lack of vegetation and is susceptible to changes in temperature; extremes of heat and cold can cause both cracks and fissures to appear in the rock.

Desert abrasion

Abrasion creates a wide range of desert landforms from faceted pebbles and wind ripples in the sand, to large-scale features such as yardangs (low, streamlined ridges), and scoured desert pavements.

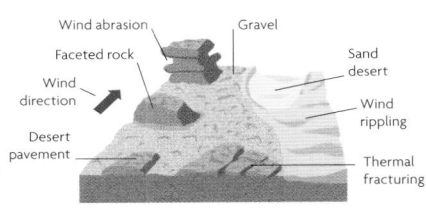

Features of a desert surface

Labels: Wind abrasion, Gravel, Faceted rock, Sand desert, Wind direction, Wind rippling, Desert pavement, Thermal fracturing

◄ *This dry valley* at Ellesmere Island in the Canadian Arctic is an example of a cold desert. The cracked floor and scoured slopes are features also found in hot deserts.

The world's oceans

Two-thirds of the Earth's surface is covered by the oceans. The landscape of the ocean floor, like the surface of the land, has been shaped by movements of the Earth's crust over millions of years to form volcanic mountain ranges, deep trenches, basins, and plateaus. Ocean currents constantly redistribute warm and cold water around the world. A major warm current, such as El Niño in the Pacific Ocean, can increase surface temperature by up to 10°F (8°C), causing changes in weather patterns which can lead to both droughts and flooding.

The great oceans

There are five oceans on Earth: the Pacific, Atlantic, Indian, and Southern oceans, and the much smaller Arctic Ocean. These five ocean basins are relatively young, having evolved within the last 80 million years. One of the most recent plate collisions, between the Eurasian and African plates, created the present-day arrangement of continents and oceans.

▲ *The Indian Ocean* accounts for approximately 20% of the total area of the world's oceans.

Sea level

If the influence of tides, winds, currents, and variations in gravity were ignored, the surface of the Earth's oceans would closely follow the topography of the ocean floor, with an underwater ridge 3000 ft (915 m) high producing a rise of up to 3 ft (1 m) in the level of the surface water.

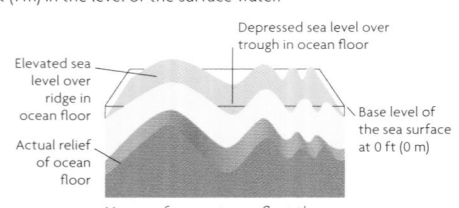

How surface waters reflect the relief of the ocean floor

▲ *The low relief* of many small Pacific islands such as these atolls at Huahine in French Polynesia makes them vulnerable to changes in sea level.

Ocean structure

The continental shelf is a shallow, flat seabed surrounding the Earth's continents. It extends to the continental slope, which falls to the ocean floor. Here, the flat abyssal plains are interrupted by vast, underwater mountain ranges, the mid-ocean ridges, and ocean trenches which plunge to depths of 35,828 ft (10,920 m).

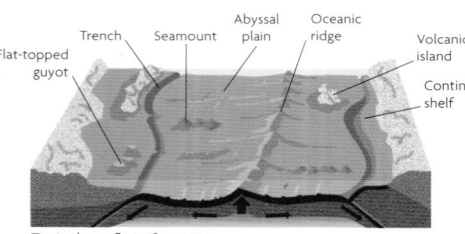

Typical sea-floor features

Ocean depth

	Sea level
	200m / 656ft
	1000m / 3281ft
	2000m / 6562ft
	3000m / 9843ft
	4000m / 13,124ft
	5000m / 16,400ft
	6000m / 19,686ft

Black smokers

These vents in the ocean floor disgorge hot, sulfur-rich water from deep in the Earth's crust. Despite the great depths, a variety of lifeforms have adapted to the chemical-rich environment which surrounds black smokers.

▲ *A black smoker* in the Atlantic Ocean.

▲ *Surtsey, near Iceland,* is a volcanic island lying directly over the Mid-Atlantic Ridge. It was formed in the 1960s following intense volcanic activity nearby.

Ocean floors

Mid-ocean ridges are formed by lava which erupts beneath the sea and cools to form solid rock. This process mirrors the creation of volcanoes from cooled lava on the land. The ages of sea floor rocks increase in parallel bands outward from central ocean ridges.

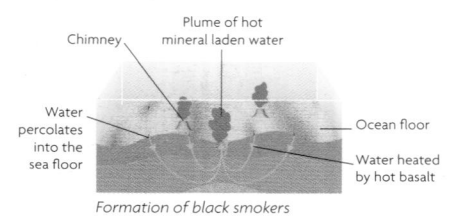

Formation of black smokers

Ages of the ocean floor

Jurassic	Cretaceous	Tertiary (Paleogene)		Cretaceous	Jurassic
		Quaternary			
208	145	65	23 0 23	65 145	208
million years old					*million years old*

Tertiary (Neogene)

Age uncertain
Continental shelf and island arcs

N O P Q R S T U V W X Y Z

▲ *Currents in the Southern Ocean* are driven by some of the world's fiercest winds, including the Roaring Forties, Furious Fifties, and Shrieking Sixties.

▲ *The Pacific Ocean is the* world's largest and deepest ocean, covering over one-third of the surface of the Earth.

▲ *The Atlantic Ocean* was formed when the landmasses of the eastern and western hemispheres began to drift apart 180 million years ago.

Deposition of sediment

Storms, earthquakes, and volcanic activity trigger underwater currents known as turbidity currents which scour sand and gravel from the continental shelf, creating underwater canyons. These strong currents pick up material deposited at river mouths and deltas, and carry it across the continental shelf and through the underwater canyons, where it is eventually laid down on the ocean floor in the form of fans.

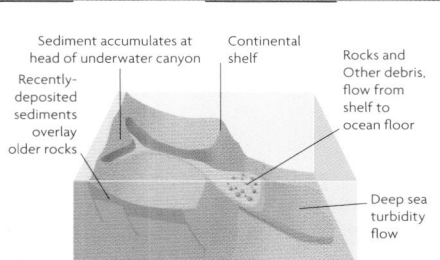

How sediment is deposited on the ocean floor

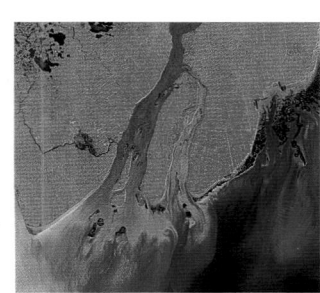

▶ *Satellite image of the Yangtze (Chang Jiang) Delta, in which the land appears red. The river deposits immense quantities of silt into the East China Sea, much of which will eventually reach the deep ocean floor.*

Surface water

Ocean currents move warm water away from the Equator toward the poles, while cold water is, in turn, moved towards the Equator. This is the main way in which the Earth distributes surface heat and is a major climatic control. Approximately 4000 million years ago, the Earth was dominated by oceans and there was no land to interrupt the flow of the currents, which would have flowed as straight lines, simply influenced by the Earth's rotation.

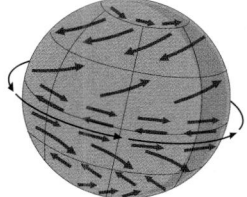

Idealized globe showing the movement of water around a landless Earth.

Ocean currents

Surface currents are driven by the prevailing winds and by the spinning motion of the Earth, which drives the currents into circulating whirlpools, or gyres. Deep sea currents, over 330 ft (100 m) below the surface, are driven by differences in water temperature and salinity, which have an impact on the density of deep water and on its movement.

Surface temperature and currents

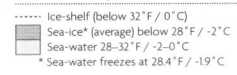

Surface temperature and currents

Map

NORTH AMERICA

Beaufort Sea
Gulf of Alaska
Aleutian Trench
Hawaiian Ridge
Mendocino Fracture Zone
Murray Fracture Zone
Molokai Fracture Zone
Clarion Fracture Zone
Clipperton Fracture Zone

PACIFIC

Central Pacific Basin

OCEAN

Southwest Pacific Basin

East Pacific Rise

Pacific-Antarctic Ridge

OCEAN

Amundsen Sea
Bellingshausen Sea
Southeast Pacific Basin
Weddell Sea
Antarctic Circle

Greenland Sea
Baffin Bay
Davis Strait
Hudson Strait
Hudson Bay
Labrador Sea
Arctic Circle
Newfoundland Basin
Mid-Atlantic Ridge
North American Basin

ATLANTIC

Gulf of Mexico
Yucatan Basin
Sargasso Sea
Caribbean Sea
Middle America Trench
Guatemala Basin

SOUTH AMERICA

Peru Basin
Peru-Chile Trench
Nazca Ridge
Chile Basin
Sala y Gomez Ridge

Canary Basin
Tropic of Cancer
Barracuda Fracture Zone
Brazil Basin
Equator
Tropic of Capricorn
Rio Grande Rise
Argentine Basin
Mid-Atlantic Ridge

OCEAN

Scotia Sea
South Sandwich Trench

Tides and waves

Tides are created by the pull of the Sun and Moon's gravity on the surface of the oceans. The levels of high and low tides are influenced by the position of the Moon in relation to the Earth and Sun. Waves are formed by wind blowing over the surface of the water.

High and low tides

The highest tides occur when the Earth, the Moon and the Sun are aligned *(below left)*. The lowest tides are experienced when the Sun and Moon align at right angles to one another *(below right)*.

Tidal range and wave environments

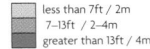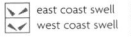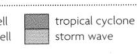

Tidal range and wave environments

Highest high tides
Lowest high tides

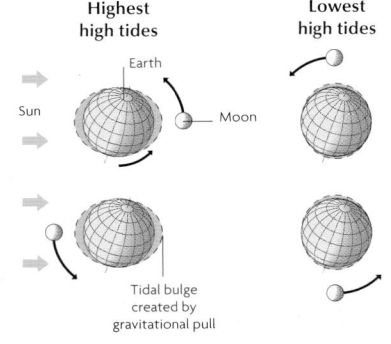

Sun
Earth
Moon
Tidal bulge created by gravitational pull

Deep sea temperature and currents

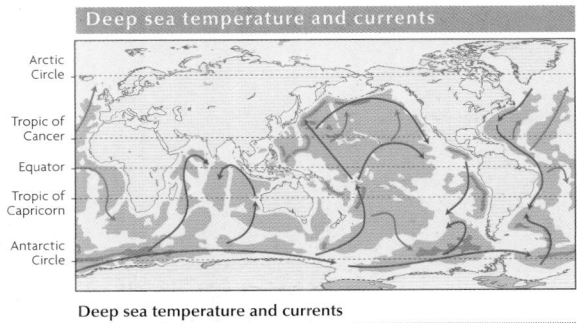

Deep sea temperature and currents

N O P Q R S T U V W X Y Z

The global climate

The Earth's climatic types consist of stable patterns of weather conditions averaged out over a long period of time. Different climates are categorized according to particular combinations of temperature and humidity. By contrast, weather consists of short-term fluctuations in wind, temperature, and humidity conditions. Different climates are determined by latitude, altitude, the prevailing wind, and circulation of ocean currents. Longer-term changes in climate, such as global warming or the onset of ice ages, are punctuated by shorter-term events which comprise the day-to-day weather of a region, such as frontal depressions, hurricanes, and blizzards.

The atmosphere, wind and weather

The Earth's atmosphere has been compared to a giant ocean of air which surrounds the planet. Its circulation patterns are similar to the currents in the oceans and are influenced by three factors; the Earth's orbit around the Sun and rotation about its axis, and variations in the amount of heat radiation received from the Sun. If both heat and moisture were not redistributed between the Equator and the poles, large areas of the Earth would be uninhabitable.

◄ *Heavy fogs, as here in southern England, form as moisture-laden air passes over cold ground.*

Temperature

The world can be divided into three major climatic zones, stretching like large belts across the latitudes: the tropics which are warm; the cold polar regions and the temperate zones which lie between them. Temperatures across the Earth range from above 86°F (30°C) in the deserts to as low as -70°F (-55°C) at the poles. Temperature is also controlled by altitude; because air becomes cooler and less dense the higher it gets, mountainous regions are typically colder than those areas which are at, or close to, sea level.

Global air circulation

Air does not simply flow from the Equator to the poles, it circulates in giant cells known as Hadley and Ferrel cells. As air warms it expands, becoming less dense and rising; this creates areas of low pressure. As the air rises it cools and condenses, causing heavy rainfall over the tropics and slight snowfall over the poles. This cool air then sinks, forming high pressure belts. At surface level in the tropics these sinking currents are deflected poleward as the westerlies and toward the equator as the trade winds. At the poles they become the polar easterlies.

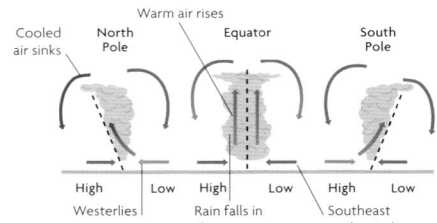

▲ *The Antarctic pack ice expands its area by almost seven times during the winter as temperatures drop and surrounding seas freeze.*

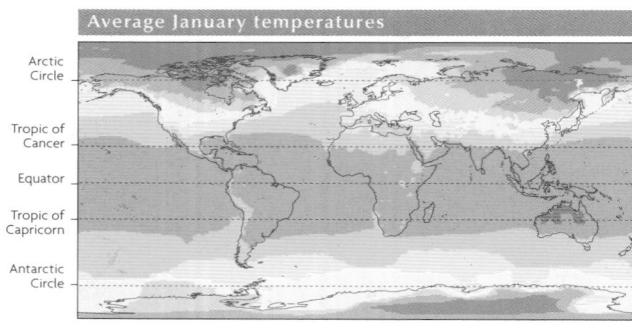

Average January temperatures

Arctic Circle
Tropic of Cancer
Equator
Tropic of Capricorn
Antarctic Circle

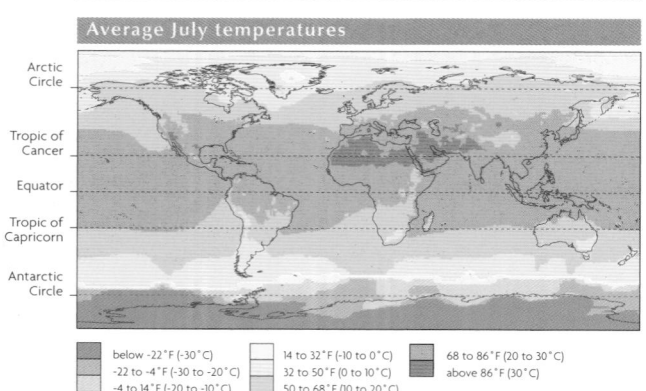

Average July temperatures

Arctic Circle
Tropic of Cancer
Equator
Tropic of Capricorn
Antarctic Circle

below -22°F (-30°C)	14 to 32°F (-10 to 0°C)	68 to 86°F (20 to 30°C)
-22 to -4°F (-30 to -20°C)	32 to 50°F (0 to 10°C)	above 86°F (30°C)
-4 to 14°F (-20 to -10°C)	50 to 68°F (10 to 20°C)	

Climatic change

The Earth is currently in a warm phase between ice ages. Warmer temperatures result in higher sea levels as more of the polar ice caps melt. Most of the world's population lives near coasts, so any changes which might cause sea levels to rise, could have a potentially disastrous impact.

▲ *This ice fair, painted by Pieter Brueghel the Younger in the 17th century, shows the Little Ice Age which peaked around 300 years ago.*

The greenhouse effect

Gases such as carbon dioxide are known as "greenhouse gases" because they allow shortwave solar radiation to enter the Earth's atmosphere, but help to stop longwave radiation from escaping. This traps heat, raising the Earth's temperature. An excess of these gases, such as that which results from the burning of fossil fuels, helps trap more heat and can lead to global warming.

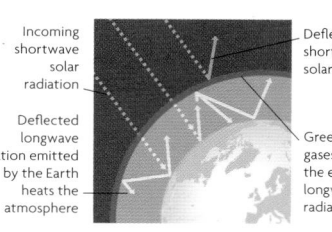

Incoming shortwave solar radiation

Deflected shortwave solar radiation

Deflected longwave radiation emitted by the Earth heats the atmosphere

Greenhouse gases prevent the escape of longwave radiation

Map labels: POLAR · WESTERLIES · NORTH EAST TRADES · SOUTH EAST TRADES · SOUTH EAST TRADE · WESTERLIES · POLAR · Arctic Circle · Tropic of Cancer · Equator · Tropic of Capricorn · Antarctic Circle · Doldrums · El Niño · Equatorial Counter Current · North Equatorial Current · Northern Equatorial Current · South Equatorial Current · Alaska Current · North Pacific Current · California Current · Gulf Stream · North Atlantic Drift · Labrador Current · Canary Current · Peru (Humboldt) Current · Brazil Current · Falkland Current · West Wind Drift · Chinook · Blizzards · Tornadoes · Pampero

Diagram labels: Cooled air sinks · North Pole · Warm air rises · Equator · South Pole · High · Low · High · Low · High · Low · Westerlies · Rain falls in the tropics · Southeast trade winds

◀ *The islands of the Caribbean, Mexico's Gulf coast and the southeastern US are often hit by hurricanes formed far out in the Atlantic.*

Oceanic water circulation

In general, ocean currents parallel the movement of winds across the Earth's surface. Incoming solar energy is greatest at the Equator and least at the poles. So, water in the oceans heats up most at the Equator and flows poleward, cooling as it moves north or south toward the Arctic or Antarctic. The flow is eventually reversed and cold water currents move back toward the Equator. These ocean currents act as a vast system for moving heat from the Equator toward the poles and are a major influence on the distribution of the Earth's climates.

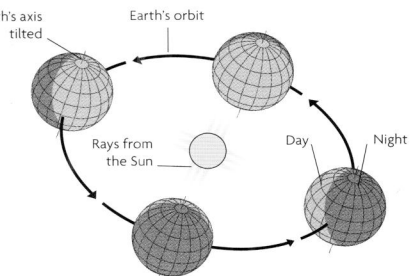

▲ *In marginal climatic zones years of drought can completely dry out the land and transform grassland to desert.*

Tilt and rotation

The tilt and rotation of the Earth during its annual orbit largely control the distribution of heat and moisture across its surface, which correspondingly controls its large-scale weather patterns. As the Earth annually rotates around the Sun, half its surface is receiving maximum radiation, creating summer and winter seasons. The angle of the Earth means that on average the tropics receive two and a half times as much heat from the Sun each day as the poles.

Earth's axis tilted   Earth's orbit
Rays from the Sun   Day   Night

Map key

Climate zones
- ice cap
- subarctic
- tundra
- continental
- temperate
- warm temperate
- mediterranean
- semi-arid
- arid
- hot humid
- humid equatorial
- tropical

Ocean currents
- warm
- cold

Prevailing winds
- → warm
- → cold

Local winds
- → warm
- → cold
- June ···· seasonal*
- * (seasonal winds which can either be warm or cold)

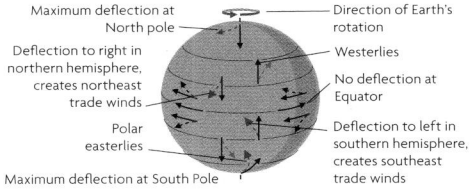
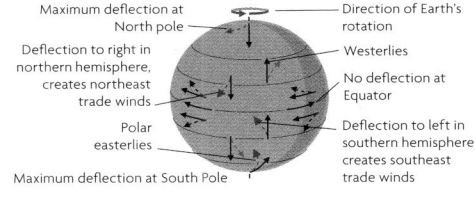

▲ *The wide range of environments found in the Andes is strongly related to their altitude, which modifies climatic influences. While the peaks are snow-capped, many protected interior valleys are semi-tropical.*

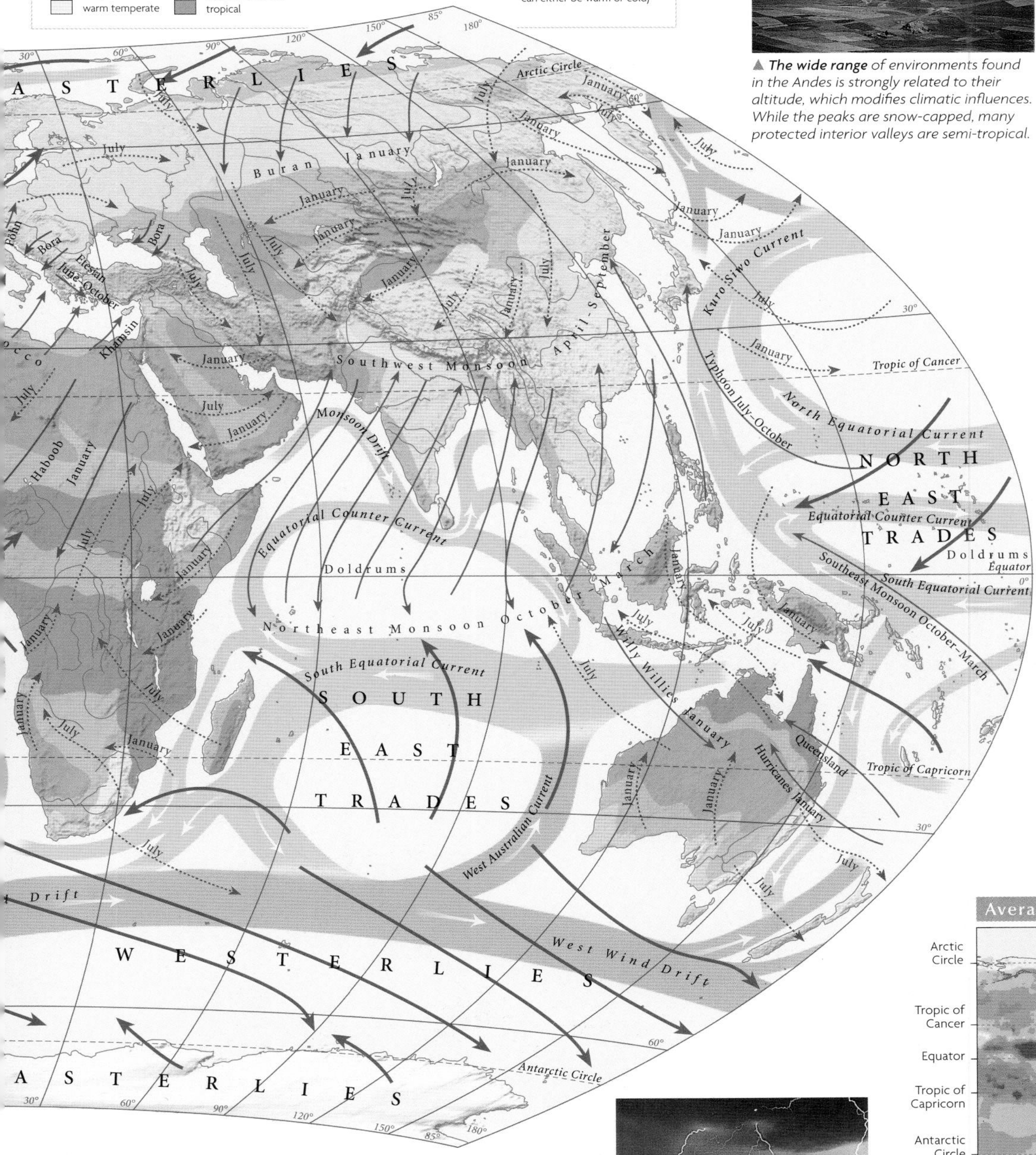

The Coriolis effect

The rotation of the Earth influences atmospheric circulation by deflecting winds and ocean currents. Winds blowing in the northern hemisphere are deflected to the right and those in the southern hemisphere are deflected to the left, creating large-scale patterns of wind circulation, such as the northeast and southeast trade winds and the westerlies. This effect is greatest at the poles and least at the Equator.

Maximum deflection at North pole
Deflection to right in northern hemisphere, creates northeast trade winds
Polar easterlies
Maximum deflection at South Pole
Direction of Earth's rotation
Westerlies
No deflection at Equator
Deflection to left in southern hemisphere, creates southeast trade winds

Precipitation

When warm air expands, it rises and cools, and the water vapor it carries condenses to form clouds. Heavy, regular rainfall is characteristic of the equatorial region, while the poles are cold and receive only slight snowfall. Tropical regions have marked dry and rainy seasons, while in the temperate regions rainfall is relatively unpredictable.

▲ *Monsoon rains, which affect southern Asia from May to September, are caused by sea winds blowing across the warm land.*

▲ *Heavy tropical rainstorms occur frequently in Papua New Guinea, often causing soil erosion and landslides in cultivated areas.*

Average January rainfall

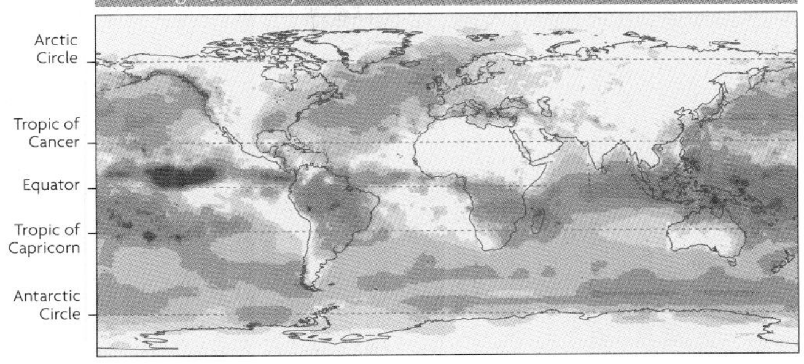

Arctic Circle
Tropic of Cancer
Equator
Tropic of Capricorn
Antarctic Circle

Average July rainfall

Arctic Circle
Tropic of Cancer
Equator
Tropic of Capricorn
Antarctic Circle

- 0–1 in (0–25 mm)
- 1–2 in (25–50 mm)
- 2–4 in (50–100 mm)
- 4–8 in (100–200 mm)
- 8–12 in (200–300 mm)
- 12–16 in (300–400 mm)
- 16–20 in (400–500 mm)
- above 20 in (500 mm)

▲ *The intensity of some blizzards in Canada and the northern US can give rise to snowdrifts as high as 10 ft (3 m).*

▲ *The Atacama Desert in Chile is one of the driest places on Earth, with an average rainfall of less than 2 inches (50 mm) per year.*

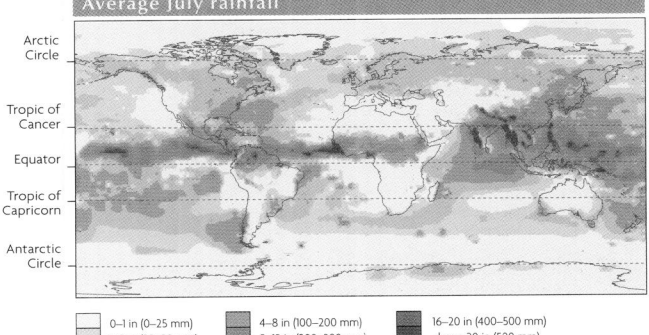

▲ *Violent thunderstorms occur along advancing cold fronts, when cold, dry air masses meet warm, moist air, which rises rapidly, its moisture condensing into thunderclouds. Rain and hail become electrically charged, causing lightning.*

The rainshadow effect

When moist air is forced to rise by mountains, it cools and the water vapor falls as precipitation, either as rain or snow. Only the dry, cold air continues over the mountains, leaving inland areas with little or no rain. This is called the rainshadow effect and is one reason for the existence of the Mojave Desert in California, which lies east of the Coast Ranges.

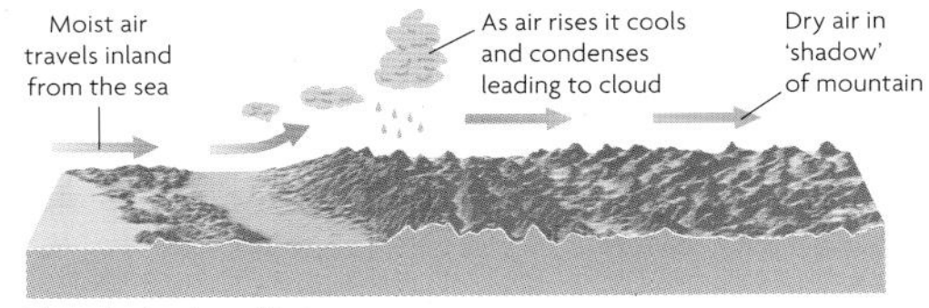

Moist air travels inland from the sea
As air rises it cools and condenses leading to cloud
Dry air in 'shadow' of mountain

The rainshadow effect

Map labels: WESTERLIES, Arctic Circle, Buran, Kuro-Siwo Current, North Equatorial Current, NORTH EAST TRADES, Equatorial Counter Current, Doldrums, Equator, Southeast Monsoon October–March, South Equatorial Current, Tropic of Cancer, Tropic of Capricorn, SOUTH EAST TRADES, Northeast Monsoon, Southwest Monsoon, Monsoon Drift, Equatorial Counter Current, Typhoon July–October, Willy Willies, Hurricanes January, Queensland, West Australian Current, West Wind Drift, Antarctic Circle, Haboob, Khamsin, Föhn, Bora, Mistral, Drift, Doldrums

Life on Earth

A unique combination of an oxygen-rich atmosphere and plentiful water is the key to life on Earth. Apart from the polar ice caps, there are few areas which have not been colonized by animals or plants over the course of the Earth's history. Plants process sunlight to provide them with their energy, and ultimately all the Earth's animals rely on plants for survival. Because of this reliance, plants are known as primary producers, and the availability of nutrients and temperature of an area is defined as its primary productivity, which affects the quantity and type of animals which are able to live there. This index is affected by climatic factors – cold and aridity restrict the quantity of life, whereas warmth and regular rainfall allow a greater diversity of species.

Biogeographical regions

The Earth can be divided into a series of biogeographical regions, or biomes, ecological communities where certain species of plant and animal coexist within particular climatic conditions. Within these broad classifications, other factors including soil richness, altitude, and human activities such as urbanization, intensive agriculture, and deforestation, affect the local distribution of living species within each biome.

Polar regions
A layer of permanent ice at the Earth's poles covers both seas and land. Very little plant and animal life can exist in these harsh regions.

Tundra
A desolate region, with long, dark freezing winters and short, cold summers. With virtually no soil and large areas of permanently frozen ground known as permafrost, the tundra is largely treeless, though it is briefly clothed by small flowering plants in the summer months.

Needleleaf forests
With milder summers than the tundra and less wind, these areas are able to support large forests of coniferous trees.

Broadleaf forests
Much of the northern hemisphere was once covered by deciduous forests, which occurred in areas with marked seasonal variations. Most deciduous forests have been cleared for human settlement.

Temperate rain forests
In warmer wetter areas, such as southern China, temperate deciduous forests are replaced by evergreen forest.

Deserts
Deserts are areas with negligible rainfall. Most hot deserts lie within the tropics; cold deserts are dry because of their distance from the moisture-providing sea.

Mediterranean
Hot, dry summers and short winters typify these areas, which were once covered by evergreen shrubs and woodland, but have now been cleared by humans for agriculture.

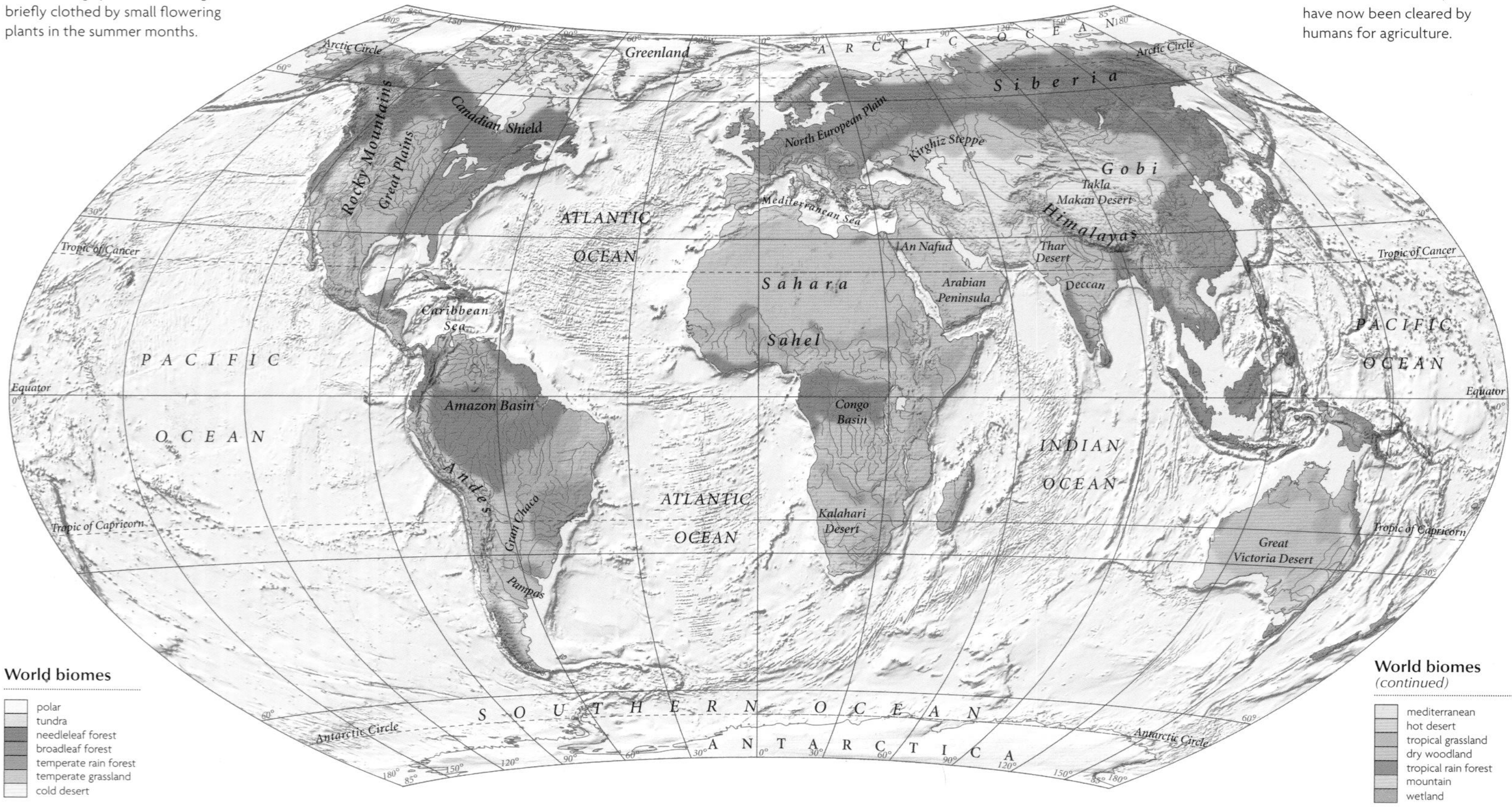

World biomes
- polar
- tundra
- needleleaf forest
- broadleaf forest
- temperate rain forest
- temperate grassland
- cold desert

World biomes (continued)
- mediterranean
- hot desert
- tropical grassland
- dry woodland
- tropical rain forest
- mountain
- wetland

Tropical and temperate grasslands
The major grassland areas are found in the centers of the larger continental landmasses. In Africa's tropical savannah regions, seasonal rainfall alternates with drought. Temperate grasslands, also known as steppes and prairies are found in the northern hemisphere, and in South America, where they are known as the pampas.

Dry woodlands
Trees and shrubs, adapted to dry conditions, grow widely spaced from one another, interspersed by savannah grasslands.

Tropical rain forests
Characterized by year-round warmth and high rainfall, tropical rain forests contain the highest diversity of plant and animal species on Earth.

Mountains
Though the lower slopes of mountains may be thickly forested, only ground-hugging shrubs and other vegetation will grow above the tree line which varies according to both altitude and latitude.

Wetlands
Rarely lying above sea level, wetlands are marshes, swamps, and tidal flats. Some, with their moist, fertile soils, are rich feeding grounds for fish and breeding grounds for birds. Others have little soil structure and are too acidic to support much plant and animal life.

Biodiversity

The number of plant and animal species, and the range of genetic diversity within the populations of each species, make up the Earth's biodiversity. The plants and animals which are endemic to a region – that is, those which are found nowhere else in the world – are also important in determining levels of biodiversity. Human settlement and intervention have encroached on many areas of the world once rich in endemic plant and animal species. Increasing international efforts are being made to monitor and conserve the biodiversity of the Earth's remaining wild places.

Animal adaptation

The degree of an animal's adaptability to different climates and conditions is extremely important in ensuring its success as a species. Many animals, particularly the largest mammals, are becoming restricted to ever-smaller regions as human development and modern agricultural practices reduce their natural habitats. In contrast, humans have been responsible – both deliberately and accidentally – for the spread of some of the world's most successful species. Many of these introduced species are now more numerous than the indigenous animal populations.

Polar animals

The frozen wastes of the polar regions are able to support only a small range of species which derive their nutritional requirements from the sea. Animals such as the walrus *(left)* have developed insulating fat, stocky limbs, and double-layered coats to enable them to survive in the freezing conditions.

Desert animals

Many animals which live in the extreme heat and aridity of the deserts are able to survive for days and even months with very little food or water. Their bodies are adapted to lose heat quickly and to store fat and water. The Gila monster *(above)* stores fat in its tail.

Amazon rain forest

The vast Amazon Basin is home to the world's greatest variety of animal species. Animals are adapted to live at many different levels from the treetops to the tangled undergrowth which lies beneath the canopy. The sloth *(below)* hangs upside down in the branches. Its fur grows from its stomach to its back to enable water to run off quickly.

Diversity of animal species

Number of animal species per country

- more than 2000
- 1000–1999
- 700–999
- 400–699
- 200–399
- 100–199
- 0–99
- data not available

High altitudes

Few animals exist in the rarefied atmosphere of the highest mountains. However, birds of prey such as eagles and vultures *(above)*, with their superb eyesight can soar as high as 23,000 ft (7000 m) to scan for prey below.

Urban animals

The growth of cities has reduced the amount of habitat available to many species. A number of animals are now moving closer into urban areas to scavenge from the detritus of the modern city *(left)*. Rodents, particularly rats and mice, have existed in cities for thousands of years, and many insects, especially moths, quickly develop new coloring to provide them with camouflage.

Endemic species

Isolated areas such as Australia and the island of Madagascar, have the greatest range of endemic species. In Australia, these include marsupials such as the kangaroo *(below)*, which carry their young in pouches on their bodies. Destruction of habitat, pollution, hunting, and predators introduced by humans, are threatening this unique biodiversity.

Marine biodiversity

The oceans support a huge variety of different species, from the world's largest mammals like whales and dolphins down to the tiniest plankton. The greatest diversities occur in the warmer seas of continental shelves, where plants are easily able to photosynthesize, and around coral reefs, where complex ecosystems are found. On the ocean floor, nematodes can exist at a depth of more than 10,000 ft (3000 m) below sea level.

Plant adaptation

Environmental conditions, particularly climate, soil type, and the extent of competition with other organisms, influence the development of plants into a number of distinctive forms. Similar conditions in quite different parts of the world create similar adaptations in the plants, which may then be modified by other, local, factors specific to the region.

Ancient plants

Some of the world's most primitive plants still exist today, including algae, cycads, and many ferns *(above)*, reflecting the success with which they have adapted to changing conditions.

Resisting predators

A great variety of plants have developed devices including spines *(above)*, poisons, stinging hairs, and an unpleasant taste or smell to deter animal predators.

Cold conditions

In areas where temperatures rarely rise above freezing, plants such as lichens *(left)* and mosses grow densely, close to the ground.

Rain forests

Most of the world's largest and oldest plants are found in rain forests; warmth and heavy rainfall provide ideal conditions for vast plants like the world's largest flower, the rafflesia *(left)*.

Hot, dry conditions

Arid conditions lead to the development of plants whose surface area has been reduced to a minimum to reduce water loss. In cacti *(above)*, which can survive without water for months, leaves are minimal or not present at all.

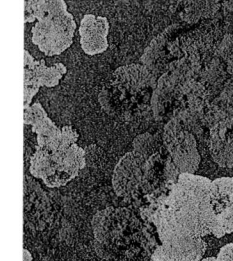

Weeds

Weeds such as bindweed *(above)* are fast-growing, easily dispersed, and tolerant of a number of different environments, enabling them to quickly colonize suitable habitats. They are among the most adaptable of all plants.

Diversity of plant species

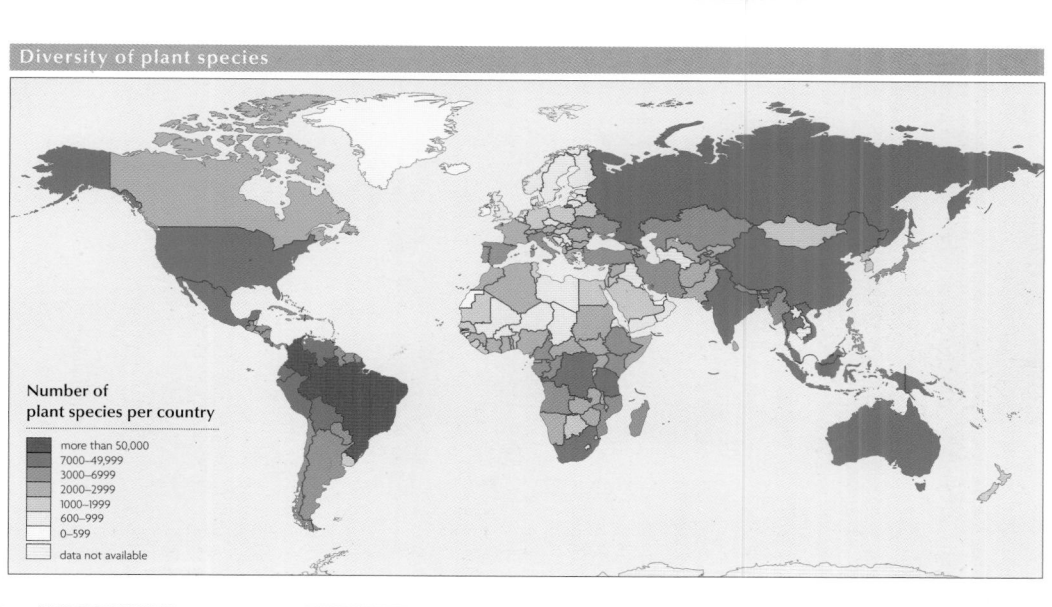

Number of plant species per country

- more than 50,000
- 7000–49,999
- 3000–6999
- 2000–2999
- 1000–1999
- 600–999
- 0–599
- data not available

A B C D E F G H I J K L M

Population and settlement

The Earth's population is projected to rise from its current level of about 7.2 billion to reach some 10.5 billion by 2050. The global distribution of this rapidly growing population is very uneven, and is dictated by climate, terrain, and natural and economic resources. The great majority of the Earth's people live in coastal zones, and along river valleys. Deserts cover over 20% of the Earth's surface, but support less than 5% of the world's population. It is estimated that over half of the world's population live in cities – most of them in Asia – as a result of mass migration from rural areas in search of jobs. Many of these people live in the so-called "megacities," some with populations as great as 40 million.

Patterns of settlement

The past 200 years have seen the most radical shift in world population patterns in recorded history.

Nomadic life

All the world's peoples were hunter-gatherers 10,000 years ago. Today nomads, who live by following available food resources, account for less than 0.0001% of the world's population. They are mainly pastoral herders, moving their livestock from place to place in search of grazing land.

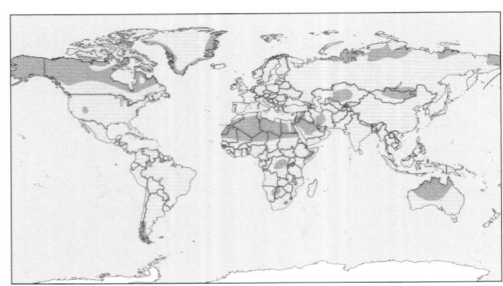

Nomadic population

▨ Nomadic population area

The growth of cities

In 1900 there were only 14 cities in the world with populations of more than a million, mostly in the northern hemisphere. Today, as more and more people in the developing world migrate to towns and cities, there are over 70 cities whose population exceeds 5 million, and around 490 "million-cities."

Million-cities in 1900

Million-cities in 1900

• Cities over 1 million population

Million-cities in 2005

Million-cities in 2005

• Cities over 1 million population

North America

The eastern and western seaboards of the US, with huge expanses of interconnected cities, towns, and suburbs, are vast, densely-populated megalopolises. Central America and the Caribbean also have high population densities. Yet, away from the coasts and in the wildernesses of northern Canada the land is very sparsely settled.

▲ **Vancouver on Canada's** west coast, grew up as a port city. In recent years it has attracted many Asian immigrants, particularly from the Pacific Rim.

▲ **North America's central** plains, the continent's agricultural heartland, are thinly populated and highly productive.

Europe

With its temperate climate, and rich mineral and natural resources, Europe is generally very densely settled. The continent acts as a magnet for economic migrants from the developing world, and immigration is now widely restricted. Birthrates in Europe are generally low, and in some countries, such as Germany, the populations have stabilized at zero growth, with a fast-growing elderly population.

▲ **Many European cities,** like Siena, once reflected the "ideal" size for human settlements. Modern technological advances have enabled them to grow far beyond the original walls.

▲ **Within the densely-populated** Netherlands the reclamation of coastal wetlands is vital to provide much-needed land for agriculture and settlement.

Population density
(inhabitants per sq mile)

520–2600
260–520
130–260
52–130
26–52
13–26
3–13
Fewer than 3

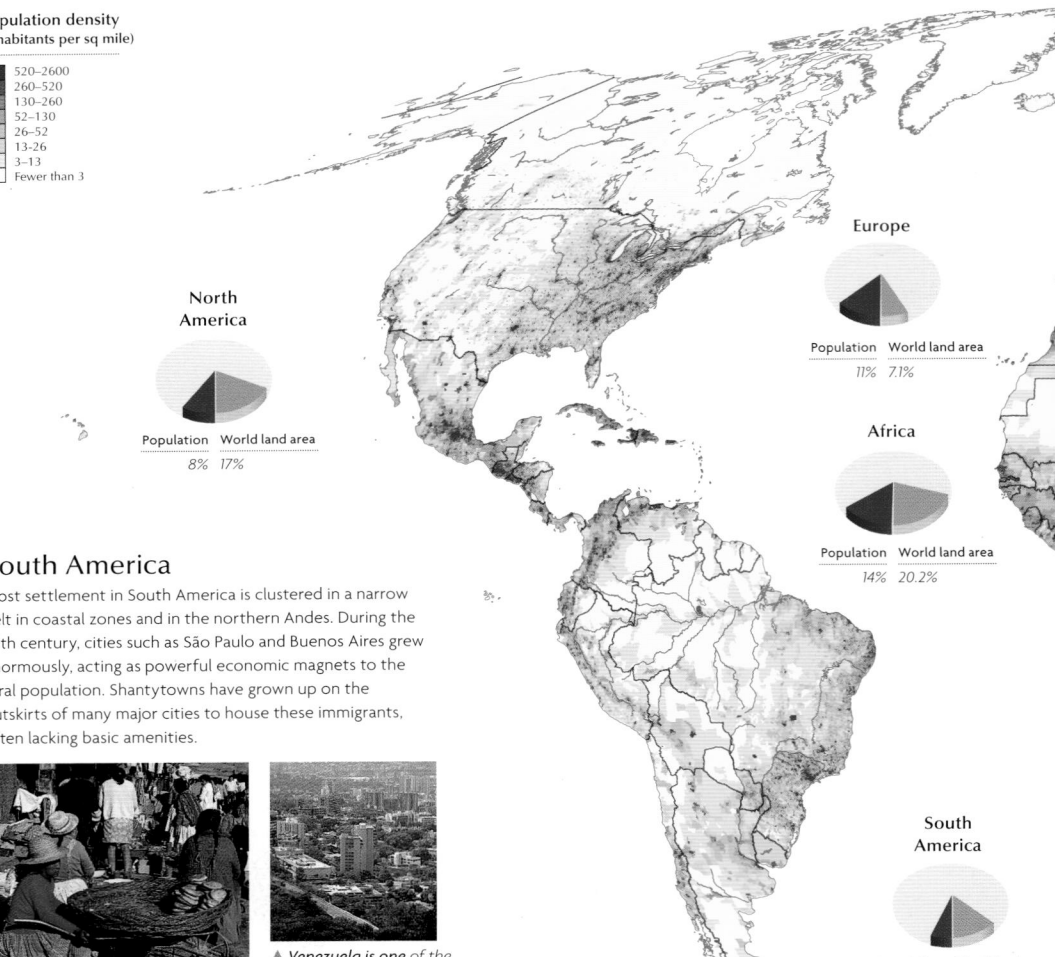

North America

Population 8% World land area 17%

Europe

Population 11% World land area 7.1%

Africa

Population 14% World land area 20.2%

South America

Population 6% World land area 11.8%

South America

Most settlement in South America is clustered in a narrow belt in coastal zones and in the northern Andes. During the 20th century, cities such as São Paulo and Buenos Aires grew enormously, acting as powerful economic magnets to the rural population. Shantytowns have grown up on the outskirts of many major cities to house these immigrants, often lacking basic amenities.

▲ **Many people in** western South America live at high altitudes in the Andes, both in cities and in villages such as this one in Bolivia.

▲ **Venezuela is one** of the most highly urbanized countries in South America, with nearly 90% of the population living in cities such as Caracas.

Africa

The arid climate of much of Africa means that settlement of the continent is sparse, focusing in coastal areas and fertile regions such as the Nile Valley. Africa still has a high proportion of nomadic agriculturalists, although many are now becoming settled, and the population is predominantly rural.

▲ **Cities such as** Nairobi (above), Cairo, and Johannesburg have grown rapidly in recent years, although only Cairo has a significant population on a global scale.

▲ **Traditional lifestyles and** homes persist across much of Africa, which has a higher proportion of rural or village-based population than any other continent.

Asia

Most Asian settlement originally centered around the great river valleys such as the Indus, the Ganges, and the Yangtze. Today, almost 60% of the world's population lives in Asia, many in burgeoning cities – particularly in the economically-buoyant Pacific Rim countries. Even rural population densities are high in many countries; practices such as terracing in Southeast Asia making the most of the available land.

▲ **Many of China's** cities are now vast urban areas with populations of more than 5 million people.

▲ **This stilt village** in Bangladesh is built to resist the regular flooding. Pressure on land, even in rural areas, forces many people to live in marginal areas.

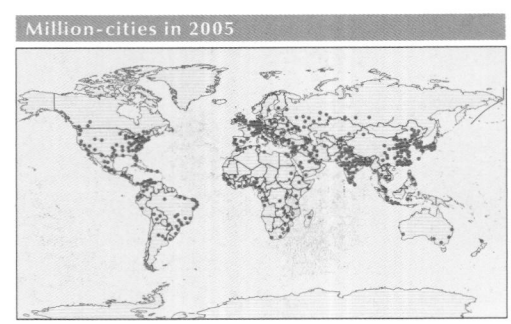

Population structures

Population pyramids are an effective means of showing the age structures of different countries, and highlighting changing trends in population growth and decline. The typical pyramid for a country with a growing, youthful population, is broad-based *(left)*, reflecting a high birthrate and a far larger number of young rather than elderly people. In contrast, countries with populations whose numbers are stabilizing have a more balanced distribution of people in each age band, and may even have lower numbers of people in the youngest age ranges, indicating both a high life expectancy, and that the population is now barely replacing itself *(right)*. The Russian Federation *(center)* is suffering from a declining population, forcing the government to consider a number of measures, including tax incentives and immigration, in an effort to stabilize the population.

Youthful population
(India)

Declining population
(Russian Federation)

Ageing population
(United States of America)

Population growth

Improvements in food supply and advances in medicine have both played a major role in the remarkable growth in global population, which has increased five-fold over the last 150 years. Food supplies have risen with the mechanization of agriculture and improvements in crop yields. Better nutrition, together with higher standards of public health and sanitation, have led to increased longevity and higher birthrates.

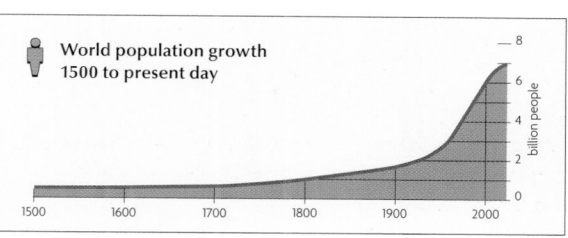

World population growth 1500 to present day

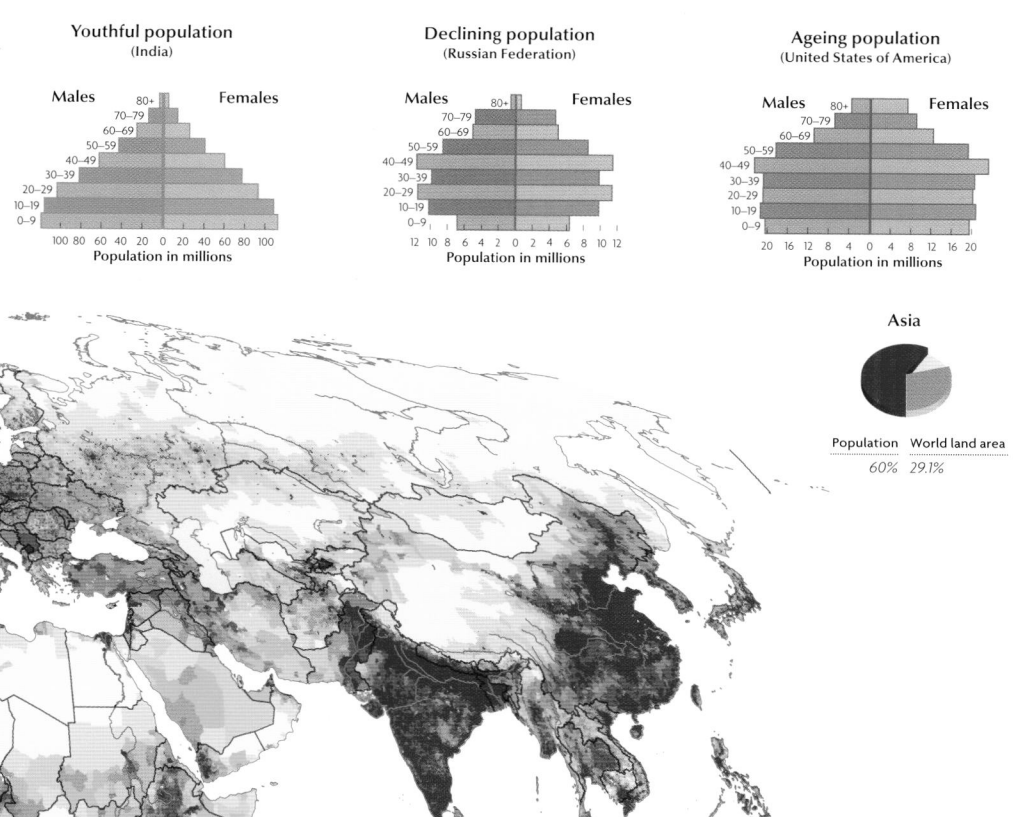

Asia
Population 60% World land area 29.1%

Australasia & Oceania
Population 1% World land area 5.9%

Antarctica
Population 0% World land area 8.9%

World nutrition

Two-thirds of the world's food supply is consumed by the industrialized nations, many of which have a daily calorific intake far higher than is necessary for their populations to maintain a healthy body weight. In contrast, in the developing world, about 800 million people do not have enough food to meet their basic nutritional needs.

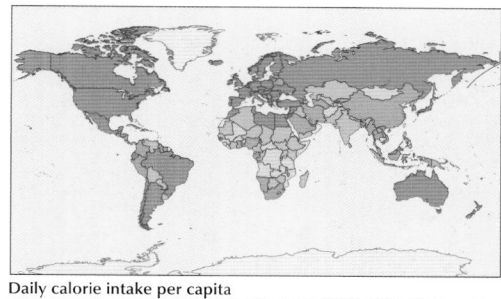

Daily calorie intake per capita
above 3000 · 2500–2999 · 2000–2499 · below 2000 · data not available

World life expectancy

Improved public health and living standards have greatly increased life expectancy in the developed world, where people can now expect to live twice as long as they did 100 years ago. In many of the world's poorest nations, inadequate nutrition and disease, means that the average life expectancy still does not exceed 45 years.

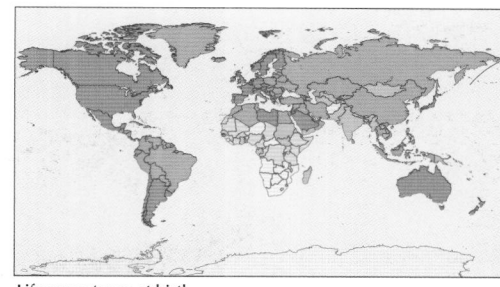

Life expectancy at birth
above 75 years · 65–74 years · 55–64 years · 45–54 years · below 44 years · data not available

Australasia and Oceania

This is the world's most sparsely settled region. The peoples of Australia and New Zealand live mainly in the coastal cities, with only scattered settlements in the arid interior. The Pacific islands can only support limited populations because of their remoteness and lack of resources.

▶ *Brisbane, on Australia's Gold Coast is the most rapidly expanding city in the country. The great majority of Australia's population lives in cities near the coasts.*

◀ *The remote highlands of Papua New Guinea are home to a wide variety of peoples, many of whom still subsist by traditional hunting and gathering.*

Average world birth rates

Birthrates are much higher in Africa, Asia, and South America than in Europe and North America. Increased affluence and easy access to contraception are both factors which can lead to a significant decline in a country's birthrate.

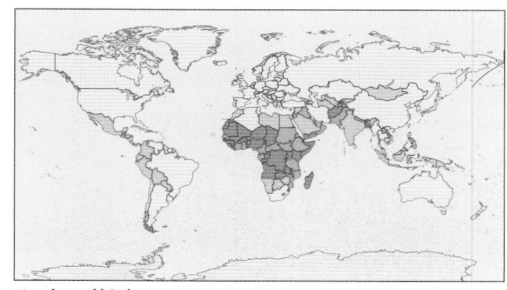

Number of births (per 1000 people)
above 40 · 30–39 · 20–29 · below 20 · data not available

World infant mortality

In parts of the developing world infant mortality rates are still high; access to medical services such as immunization, adequate nutrition, and the promotion of breast-feeding have been important in combating infant mortality.

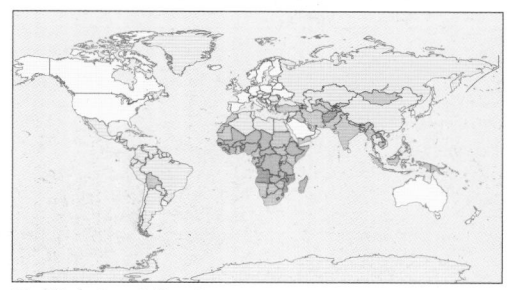

World infant mortality rates (deaths per 1000 live births)
above 125 · 75–124 · 35–74 · 15–34 · below 15 · data not available

The economic system

The wealthy countries of the developed world, with their aggressive, market-led economies and their access to productive new technologies and international markets, dominate the world economic system. At the other extreme, many of the countries of the developing world are locked in a cycle of national debt, rising populations, and unemployment. In 2008 a major financial crisis swept the world's banking sector leading to a huge downturn in the global economy. Despite this, China overtook Japan in 2010 to become the world's second largest economy.

Trade blocs

Trade blocs

EU
CACM
NAFTA
SADC
ASEAN
ECOWAS
LAIA
CEEAC

International trade blocs are formed when groups of countries, often already enjoying close military and political ties, join together to offer mutually preferential terms of trade for both imports and exports. Increasingly, global trade is dominated by three main blocs: the EU, NAFTA, and ASEAN. They are supplanting older trade blocs such as the Commonwealth, a legacy of colonialism.

International trade flows

World trade acts as a stimulus to national economies, encouraging growth. Over the last three decades, as heavy industries have declined, services – banking, insurance, tourism, airlines, and shipping – have taken an increasingly large share of world trade. Manufactured articles now account for nearly two-thirds of world trade; raw materials and food make up less than a quarter of the total.

Shipping
Ships carry 80% of international cargo, and extensive container ports, where cargo is stored, are vital links in the international transportation network.

Multinationals
Multinational companies are increasingly penetrating inaccessible markets. The reach of many American commodities is now global.

Primary products
Many countries, particularly in the Caribbean and Africa, are still reliant on primary products such as rubber and coffee, which makes them vulnerable to fluctuating prices.

Service industries
Service industries such as banking, tourism and insurance were the fastest-growing industrial sector in the last half of the 20th century. Lloyds of London is the center of the world insurance market.

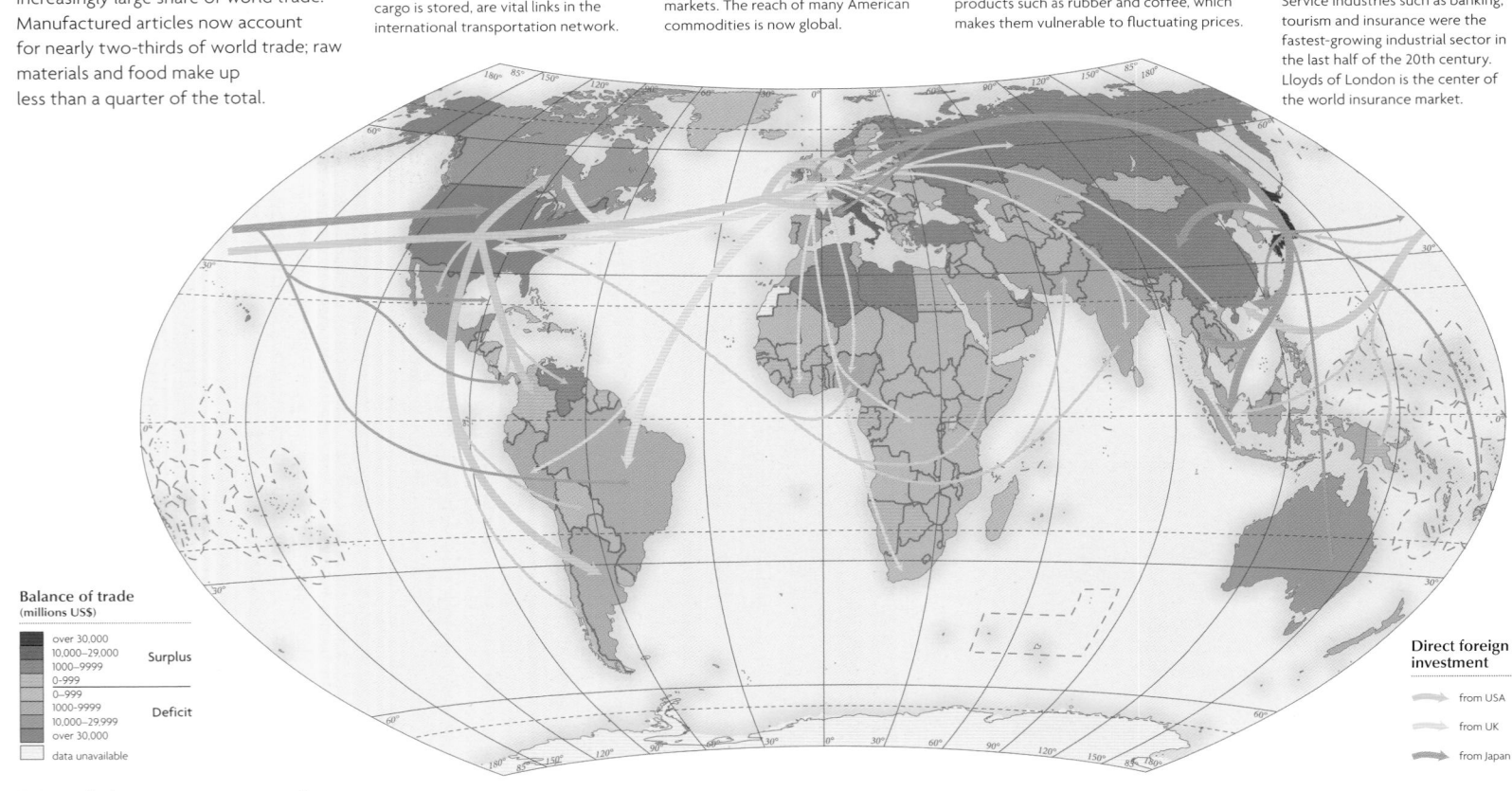

Balance of trade
(millions US$)

over 30,000
10,000–29,000
1000–9999
0-999
Surplus

0-999
1000–9999
10,000–29,999
over 30,000
Deficit

data unavailable

Direct foreign investment

from USA
from UK
from Japan

World money markets

The financial world has traditionally been dominated by three major centers – Tokyo, New York, and London, which house the headquarters of stock exchanges, multinational corporations and international banks. Their geographic location means that, at any one time in a 24-hour day, one major market is open for trading in shares, currencies, and commodities. Since the late 1980s, technological advances have enabled transactions between financial centers to occur at ever-greater speed, and new markets have sprung up throughout the world.

New stock markets
New stock markets are now opening in many parts of the world, where economies have recently emerged from state controls. In Moscow and Beijing, and several countries in eastern Europe, newly-opened stock exchanges reflect the transition to market-driven economies.

The developing world
International trade in capital and currency is dominated by the rich nations of the northern hemisphere. In parts of Africa and Asia, where exports of any sort are extremely limited, home-produced commodities are simply sold in local markets.

Major money markets

London
New York
Kolkata
Tokyo

Location of major stock markets

● Major stock markets

▲ **The Tokyo Stock Market** crashed in 1990, leading to a slow-down in the growth of the world's most powerful economy, and a refocusing on economic policy away from export-led growth and toward the domestic market.

▲ **Dealers at the** Kolkata Stock Market. The Indian economy has been opened up to foreign investment and many multinationals now have bases there.

▲ **Markets have thrived** in communist Vietnam since the introduction of a liberal economic policy.

World wealth disparity

A global assessment of Gross Domestic Product (GDP) by nation reveals great disparities. The developed world, with only a quarter of the world's population, has 80% of the world's manufacturing income. Civil war, conflict, and political instability further undermine the economic self-sufficiency of many of the world's poorest nations.

Urban sprawl

Cities are expanding all over the developing world, attracting economic migrants in search of work and opportunities. In cities such as Rio de Janeiro, housing has not kept pace with the population explosion, and squalid shanty towns (favelas) rub shoulders with middle-class housing.

▲ *The favelas of Rio de Janeiro sprawl over the hills surrounding the city.*

Agricultural economies

In parts of the developing world, people survive by subsistence farming – only growing enough food for themselves and their families. With no surplus product, they are unable to exchange goods for currency, the only means of escaping the poverty trap. In other countries, farmers have been encouraged to concentrate on growing a single crop for the export market. This reliance on cash crops leaves farmers vulnerable to crop failure and to changes in the market price of the crop.

▲ *Cities such as Detroit have been badly hit by the decline in heavy industry.*

Urban decay

Although the US still dominates the global economy, it faces deficits in both the federal budget and the balance of trade. Vast discrepancies in personal wealth, high levels of unemployment, and the dismantling of welfare provisions throughout the 1980s have led to severe deprivation in several of the inner cities of North America's industrial heartland.

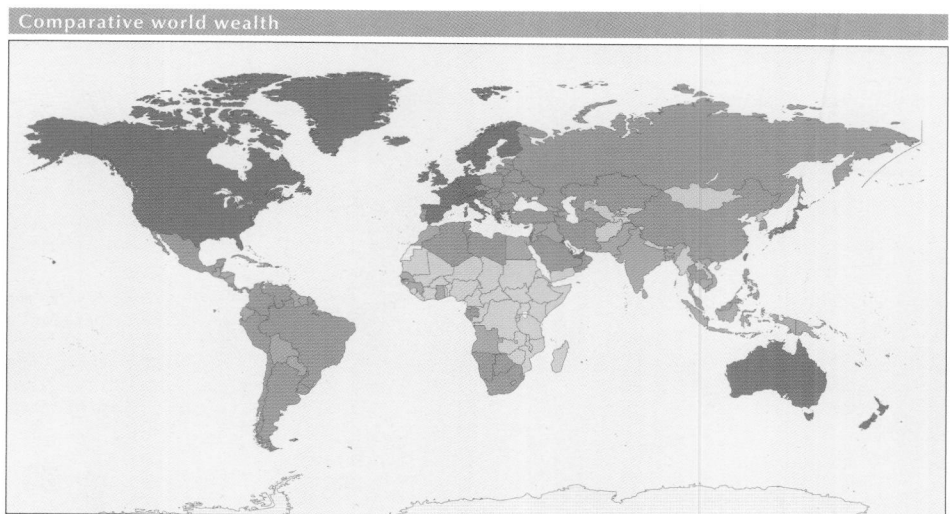

Comparative world wealth

▲ *The Ugandan uplands are fertile, but poor infrastructure hampers the export of cash crops.*

World economies – average GDP per capita (US$)

- above 20,000
- 5000–20,000
- 2000–5000
- below 2000
- data unavailable

Booming cities

Since the 1980s the Chinese government has set up special industrial zones, such as Shanghai, where foreign investment is encouraged through tax incentives. Migrants from rural China pour into these regions in search of work, creating "boomtown" economies.

◀ *Foreign investment has encouraged new infrastructure development in cities like Shanghai.*

Economic "tigers"

The economic "tigers" of the Pacific Rim – China, Singapore, and South Korea – have grown faster than Europe and the US over the last decade. Their export- and service-led economies have benefited from stable government, low labor costs, and foreign investment.

▲ *Hong Kong, with its fine natural harbor, is one of the most important ports in Asia.*

The affluent West

The capital cities of many countries in the developed world are showcases for consumer goods, reflecting the increasing importance of the service sector, and particularly the retail sector, in the world economy. The idea of shopping as a leisure activity is unique to the western world. Luxury goods and services attract visitors, who in turn generate tourist revenue.

▲ *A shopping arcade in Paris displays a great profusion of luxury goods.*

Tourism

In 2004, there were over 940 million tourists worldwide. Tourism is now the world's biggest single industry, employing over 130 million people, though frequently in low-paid unskilled jobs. While tourists are increasingly exploring inaccessible and less-developed regions of the world, the benefits of the industry are not always felt at a local level. There are also worries about the environmental impact of tourism, as the world's last wildernesses increasingly become tourist attractions.

▲ *Botswana's Okavango Delta is an area rich in wildlife. Tourists go on safaris to the region, but the impact of tourism is controlled.*

Money flows

In 2008 a global financial crisis swept through the world's economic system. The crisis triggered the failure of several major financial institutions and lead to increased borrowing costs known as the "credit crunch". A consequent reduction in economic activity together with rising inflation forced many governments to introduce austerity measures to reduce borrowing and debt, particulary in Europe where massive "bailouts" were needed to keep some European single currency (Euro) countries solvent.

◀ *In rural Southeast Asia, babies are given medical checks by UNICEF as part of a global aid program sponsored by the UN.*

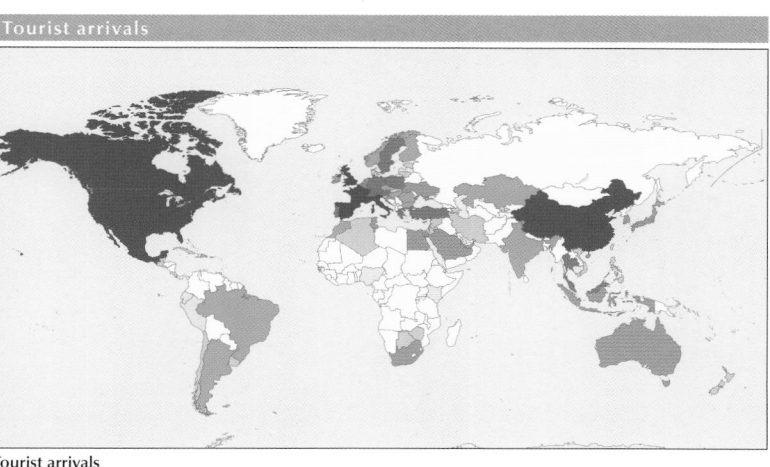

Tourist arrivals

Tourist arrivals

- over 20 million
- 10–20 million
- 5–10 million
- 2.5–5 million
- 1–2.5 million
- 700,000–999,000
- under 700,000
- data unavailable

International debt

International debt (as percentage of GNI)

- over 100%
- 70–99%
- 50–69%
- 30–49%
- 10–29%
- below 10%
- data unavailable

The political world

There are 196 independent countries in the world today. With the exception of Antarctica, where territorial claims have been deferred by international treaty, every land area of the Earth's surface either belongs to, or is claimed by, one country or another. The largest country in the world is the Russian Federation, the smallest is Vatican City. Some 60 overseas dependent territories remain, administered variously by France, Australia, Denmark, New Zealand, Norway, Portugal, the UK, the US, and the Netherlands.

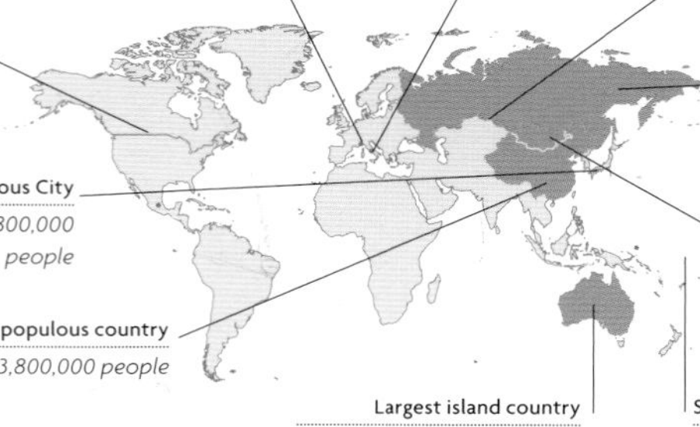

International borders

The map shows three main types of boundary between states. Full borders represent internationally agreed and recognized territorial boundaries. Undefined borders exist where no fixed boundary between states has been demarcated; the boundaries indicated in this way show approximate areas of sovereignty. A disputed border is indicated where a *de facto* territorial boundary exists, which is not agreed or is subject to arbitration.

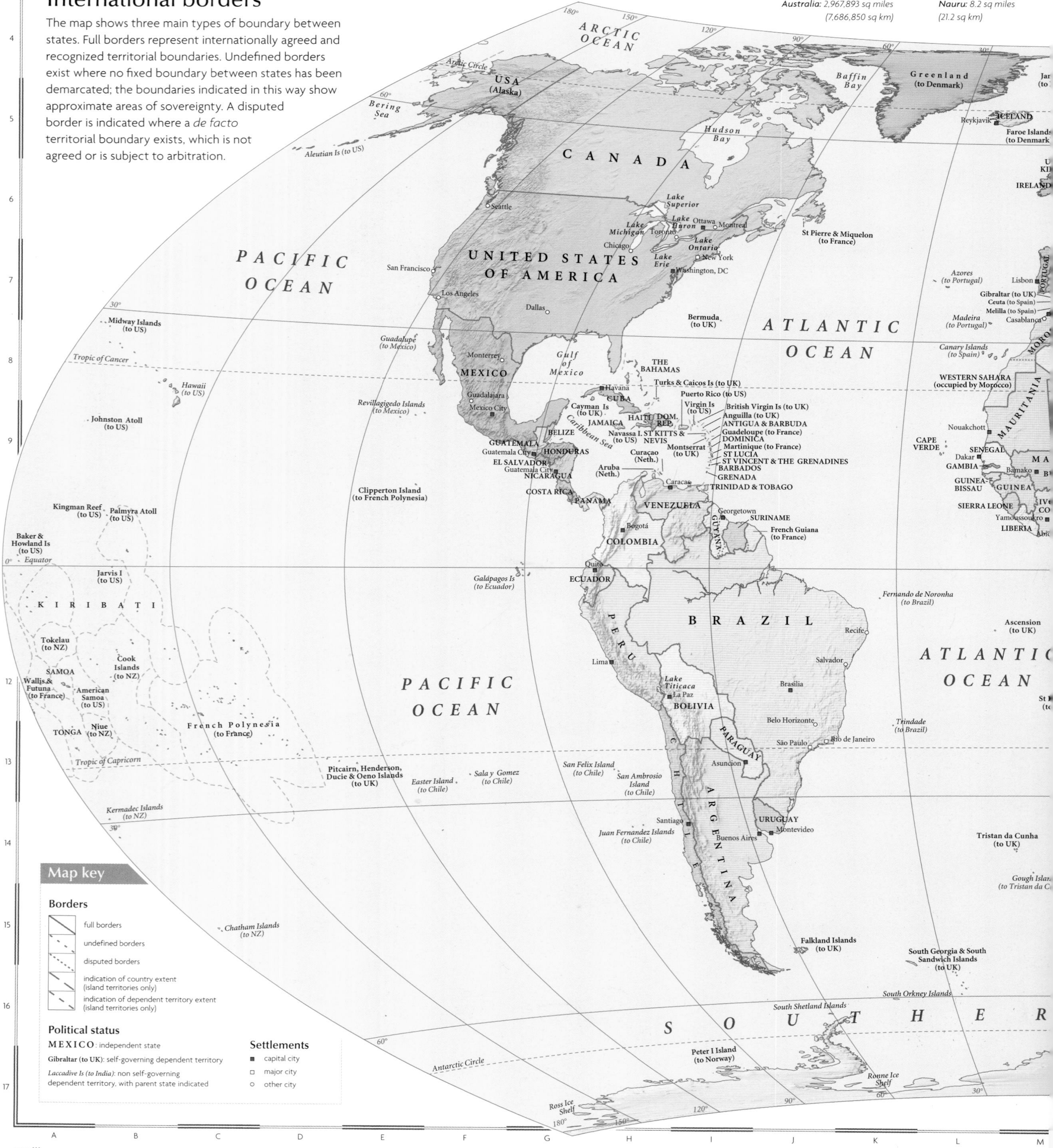

Most densely populated country
Monaco: 49,267 people per sq mile
(18,949 people per sq km)

Smallest country
Vatican City: 0.17 sq miles (0.44 sq km)

Longest land borders
Russian Federation:
12,427 miles (20,000 km)

Longest single land border
Canada/USA: 5526 miles
(8893 km)

Largest country
Russian Federation:
6,592,735 sq miles
(17,075,200 sq km)

Most populous City
Tokyo: 37,800,000
people

Most sparsely
populated country
Mongolia:
5 people per sq mile
(2 people per sq km)

Most populous country
China: 1,393,800,000 people

Largest island country
Australia: 2,967,893 sq miles
(7,686,850 sq km)

Smallest island country
Nauru: 8.2 sq miles
(21.2 sq km)

Map key

Borders

full borders

undefined borders

disputed borders

indication of country extent
(island territories only)

indication of dependent territory extent
(island territories only)

Political status

MEXICO: independent state

Gibraltar (to UK): self-governing dependent territory

Laccadive Is (to India): non self-governing
dependent territory, with parent state indicated

Settlements

■ capital city

□ major city

○ other city

The world in 1914

The early years of the 20th century saw the mainly European colonial empires reaching their greatest extents by 1914. Two world wars inaugurated their disintegration, but even in 1950 there were only 82 independent countries. Since then, over 100 have gained their independence, culminating in the breakup of the Soviet Union and former Yugoslavia in the early 1990s.

Percentage of Earth's land surface controlled by colonial empires in 1914

Independent: 29.8%
Chinese: 6%
Ottoman: 1.5%
Russian: 15%
Portuguese: 1%
Spanish: 1%
British: 21.5%
Dutch: 1.4%
Danish: 1.5%
German: 1.6%
Japanese: 0.4%
Italian: 1.8%
Belgian: 1.6%
French: 7.7%
United States: 7.6%

Colonial empires in 1914

Colonial Empires in 1914

- Belgian
- British
- Chinese
- Danish
- Dutch
- French
- German
- Italian
- Japanese
- Ottoman
- Portuguese
- Russian
- Spanish
- United States
- Independent
- Disputed

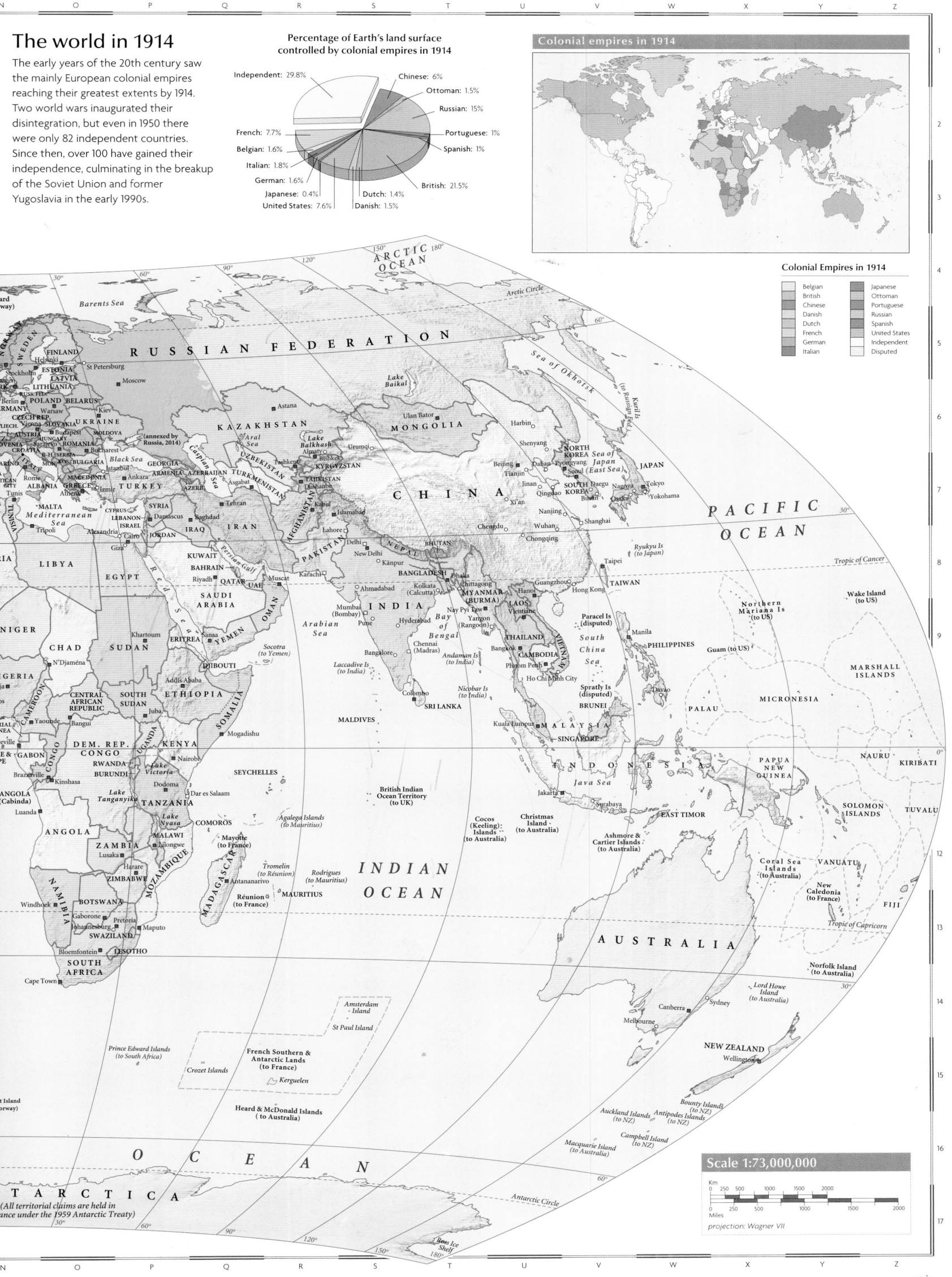

Scale 1:73,000,000

projection: Wagner VII

States and boundaries

There are almost 200 sovereign states in the world today; in 1950 there were only 82. Over the last 65 years national self-determination has been a driving force for many states with a history of colonialism and oppression. As more borders have been added to the world map, the number of international border disputes has increased.

In many cases, where the impetus toward independence has been religious or ethnic, disputes with minority groups have also caused violent internal conflict. While many newly-formed states have moved peacefully toward independence, successfully establishing government by multiparty democracy, dictatorship by military regime or individual despot is often the result of the internal power-struggles which characterize the early stages in the lives of new nations.

The nature of politics

Democracy is a broad term: it can range from the ideal of multiparty elections and fair representation to, in countries such as Singapore, a thin disguise for single-party rule. In despotic regimes, on the other hand, a single, often personal authority has total power; institutions such as parliament and the military are mere instruments of the dictator.

◀ The stars and stripes of the US flag are a potent symbol of the country's status as a federal democracy.

Types of government

- Multiparty democracy for more than 10 yrs
- Multiparty democracy within last 10 yrs
- Single-party government
- Military regime
- Theocracy
- Monarchy
- Non-party system
- Transitional regime

⚑ Current civil unrest

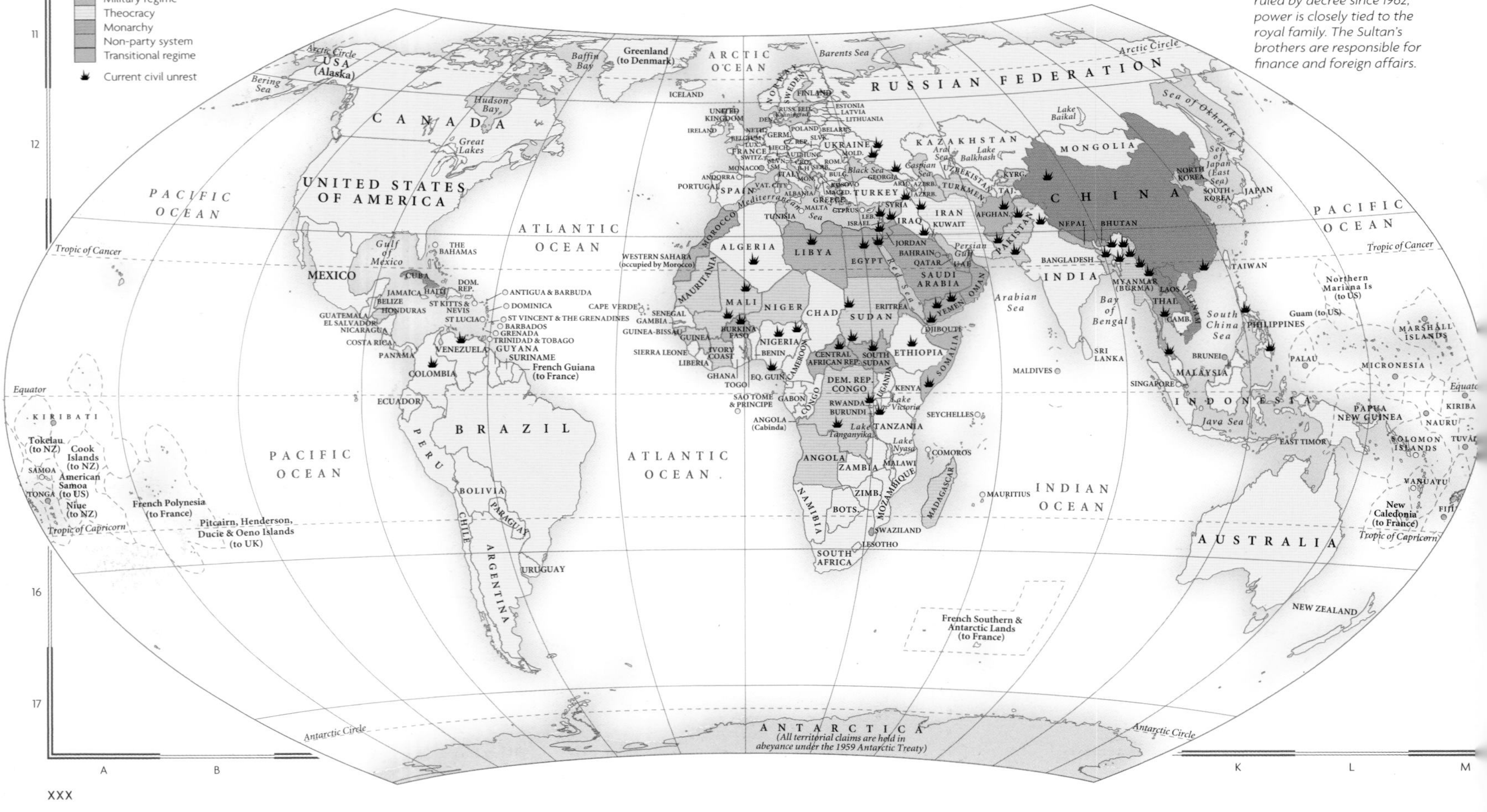

The changing world map

Decolonization

In 1950, large areas of the world remained under the control of a handful of European countries (page xxix). The process of decolonization had begun in Asia, where, following the Second World War, much of southern and southeastern Asia sought and achieved self-determination. In the 1960s, a host of African states achieved independence, so that by 1965, most of the larger tracts of the European overseas empires had been substantially eroded. The final major stage in decolonization came with the breakup of the Soviet Union and the Eastern bloc after 1990. The process continues today as the last toeholds of European colonialism, often tiny island nations, press increasingly for independence.

▲ Icons of communism, including statues of former leaders such as Lenin and Stalin, were destroyed when the Soviet bloc was dismantled in 1989, creating several new nations.

▲ Iran has been one of the modern world's few true theocracies; Islam has an impact on every aspect of political life.

◀ Afghanistan has suffered decades of war and occupation resulting in widespread destruction. The hardline Taliban government were ousted by a US-led coalition in 2001 but efforts to stabilize the country are still continuing.

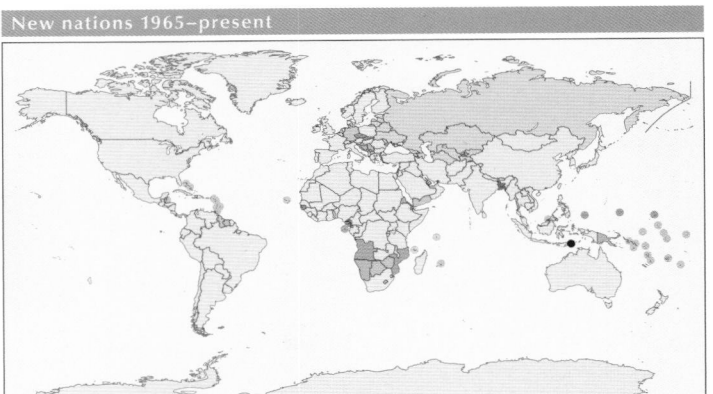

New nations 1945–1965

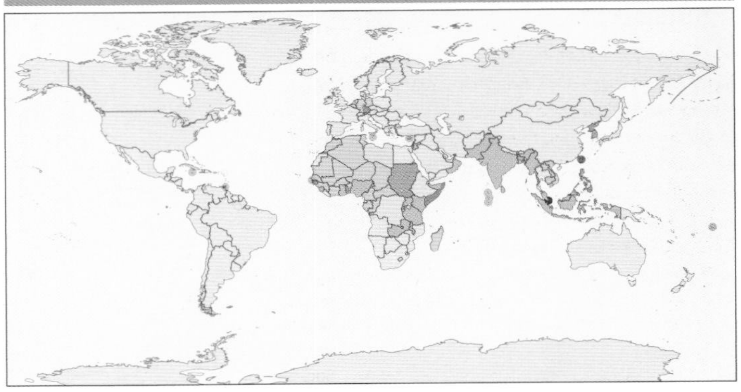

New nations 1965–present

▲ North Korea is an independent communist republic. Power was transferred directly to Kim Jong-un in 2012 following the death of his father Kim Jong-il.

◀ In early 2011, Egypt underwent a revolution, part of the so called "Arab Spring," which resulted in the ousting of President Hosni Mubarak after nearly 30 years in power.

Administration at the time of independence

Australia	Netherlands
Aust/NZ/UK	New Zealand
Belgium	Pakistan
China	Portugal
Czechoslovakia	South Africa
Egypt/UK	Spain
Ethiopia	Sudan
France	UK
France/UK	Unified country
Indonesia	USA
Italy	USSR
Japan	Yugoslavia
Malaysia	

▲ In Brunei the Sultan has ruled by decree since 1962; power is closely tied to the royal family. The Sultan's brothers are responsible for finance and foreign affairs.

Lines on the map

The determination of international boundaries can use a variety of criteria. Many of the borders between older states follow physical boundaries; some mirror religious and ethnic differences; others are the legacy of complex histories of conflict and colonialism, while others have been imposed by international agreements or arbitration.

Post-colonial borders

When the European colonial empires in Africa were dismantled during the second half of the 20th century, the outlines of the new African states mirrored colonial boundaries. These boundaries had been drawn up by colonial administrators, often based on inadequate geographical knowledge. Such arbitrary boundaries were imposed on people of different languages, racial groups, religions, and customs. This confused legacy often led to civil and international war.

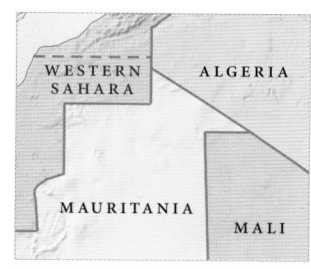

▲ The conflict that has plagued many African countries since independence has caused millions of people to become refugees.

Physical borders

Many of the world's countries are divided by physical borders: lakes, rivers, mountains. The demarcation of such boundaries can, however, lead to disputes. Control of waterways, water supplies, and fisheries are frequent causes of international friction.

Enclaves

The shifting political map over the course of history has frequently led to anomalous situations. Parts of national territories may become isolated by territorial agreement, forming an enclave. The West German part of the city of Berlin, which until 1989 lay a hundred miles (160km) within East German territory, was a famous example

Antarctica

When Antarctic exploration began a century ago, seven nations, Australia, Argentina, Britain, Chile, France, New Zealand, and Norway, laid claim to the new territory. In 1961 the Antarctic Treaty, now signed by 45 nations, agreed to hold all territorial claims in abeyance.

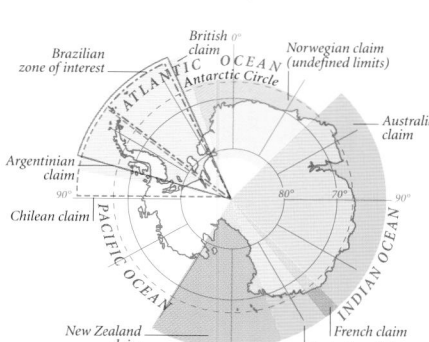

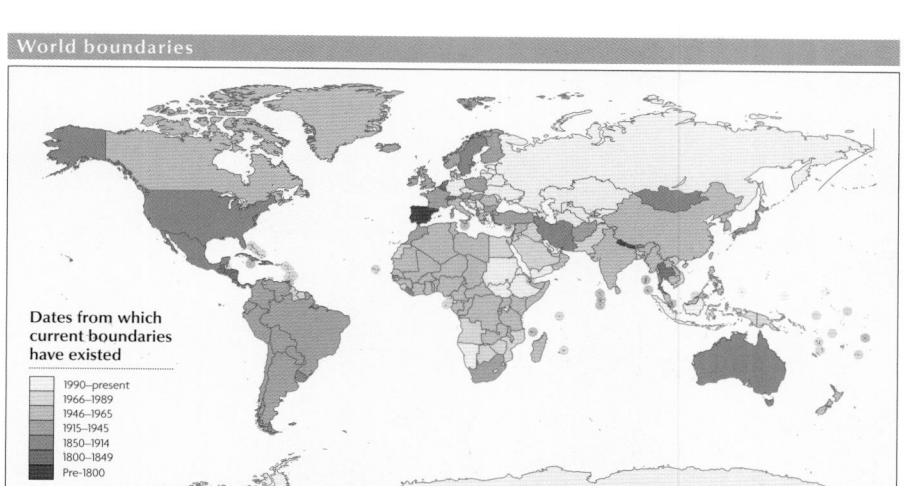

World boundaries

Dates from which current boundaries have existed
- 1990–present
- 1966–1989
- 1946–1965
- 1915–1945
- 1850–1914
- 1800–1849
- Pre-1800

▲ Since the independence of Lithuania and Belarus, the peoples of the Russian enclave of Kaliningrad have become physically isolated.

Geometric borders

Straight lines and lines of longitude and latitude have occasionally been used to determine international boundaries; and indeed the world's second longest continuous international boundary, between Canada and the USA follows the 49th Parallel for over one-third of its course. Many Canadian, American, and Australian internal administrative boundaries are similarly determined using a geometric solution.

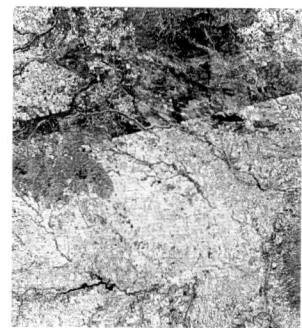

▲ Different farming techniques in Canada and the US clearly mark the course of the international boundary in this satellite map.

Lake borders

Countries which lie next to lakes usually fix their borders in the middle of the lake. Unusually the Lake Nyasa border between Malawi and Tanzania runs along Tanzania's shore.

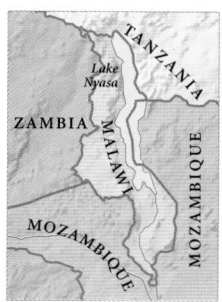

▲ Complicated agreements between colonial powers led to the awkward division of Lake Nyasa.

River borders

Rivers alone account for one-sixth of the world's borders. Many great rivers form boundaries between a number of countries. Changes in a river's course and interruptions of its natural flow can lead to disputes, particularly in areas where water is scarce. The center of the river's course is the nominal boundary line.

▲ The Danube forms all or part of the border between nine European nations.

Mountain borders

Mountain ranges form natural barriers and are the basis for many major borders, particularly in Europe and Asia. The watershed is the conventional boundary demarcation line, but its accurate determination is often problematic.

▲ The Pyrenees form a natural mountain border between France and Spain.

Shifting boundaries – Poland

Borders between countries can change dramatically over time. The nations of eastern Europe have been particularly affected by changing boundaries. Poland is an example of a country whose boundaries have changed so significantly that it has literally moved around Europe. At the start of the 16th century, Poland was the largest nation in Europe. Between 1772 and 1795, it was absorbed into Prussia, Austria, and Russia, and it effectively ceased to exist. After the First World War, Poland became an independent country once more, but its borders changed again after the Second World War following invasions by both Soviet Russia and Nazi Germany.

▲ In 1634, Poland was the largest nation in Europe, its eastern boundary reaching toward Moscow.

▲ From 1772–1795, Poland was gradually partitioned between Austria, Russia, and Prussia. Its eastern boundary receded by over 100 miles (160 km).

▲ Following the First World War, Poland was reinstated as an independent state, but it was less than half the size it had been in 1634.

▲ After the Second World War, the Baltic Sea border was extended westward, but much of the eastern territory was annexed by Russia.

International disputes

There are more than 60 disputed borders or territories in the world today. Although many of these disputes can be settled by peaceful negotiation, some areas have become a focus for international conflict. Ethnic tensions have been a major source of territorial disagreement throughout history, as has the ownership of, and access to, valuable natural resources. The turmoil of the postcolonial era in many parts of Africa is partly a result of the 19th century "carve-up" of the continent, which created potential for conflict by drawing often arbitrary lines through linguistic and cultural areas.

Jammu and Kashmir

Disputes over Jammu and Kashmir have caused three serious wars between India and Pakistan since 1947. Pakistan wishes to annex the largely Muslim territory, while India refuses to cede any territory or to hold a referendum, and also lays claim to the entire territory. Most international maps show the "line of control" agreed in 1972 as the *de facto* border. In addition, India has territorial disputes with neighboring China. The situation is further complicated by a Kashmiri independence movement, active since the late 1980s.

▲ *Indian army troops* maintain their positions in the mountainous terrain of northern Kashmir.

North and South Korea

Since 1953, the *de facto* border between North and South Korea has been a cease-fire line which straddles the 38th Parallel and is designated as a demilitarized zone. Both countries have heavy fortifications and troop concentrations behind this zone.

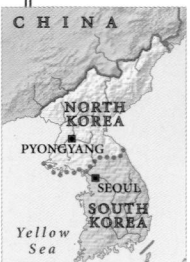

▲ *Heavy fortifications on* the border between North and South Korea.

Cyprus

Cyprus was partitioned in 1974, following an invasion by Turkish troops. The south is now the Greek Cypriot Republic of Cyprus, while the self-proclaimed Turkish Republic of Northern Cyprus is recognized only by Turkey.

▲ *The so-called "green line"* divides Cyprus into Greek and Turkish sectors.

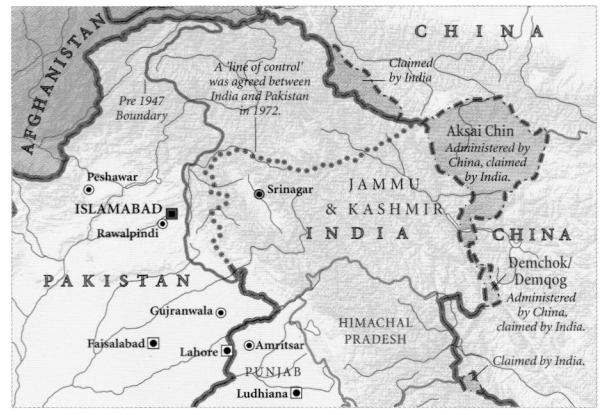

Conflicts and international disputes

- UN peacekeeping missions 2005–2015
- Major active land based territorial or border disputes
- Countries involved in internal conflict
- Active land based territorial or border disputes and internal conflict

The Falkland Islands

The British dependent territory of the Falkland Islands was invaded by Argentina in 1982, sparking a full-scale war with the UK. Tensions ran high during 2012 in the build up to the thirtieth anniversary of the conflict.

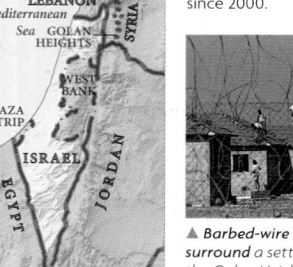
◄ *British warships in Falkland Sound during the 1982 war with Argentina.*

Israel

Israel was created in 1948 following the 1947 UN Resolution (147) on Palestine. Until 1979 Israel had no borders, only cease-fire lines from a series of wars in 1948, 1967, and 1973. Treaties with Egypt in 1979 and Jordan in 1994 led to these borders being defined and agreed. Negotiations over Israeli settlements and Palestinian self-government seen little effective progress since 2000.

- Palestinian control
- Mixed control
- Israeli settlement block
- □ Israeli settlement
- ○ Palestinian settlement
- — West Bank fence

Former Yugoslavia

Following the disintegration in 1991 of the communist state of Yugoslavia, the breakaway states of Croatia and Bosnia and Herzegovina came into conflict with the "parent" state (consisting of Serbia and Montenegro). Warfare focused on ethnic and territorial ambitions in Bosnia. The tenuous Dayton Accord of 1995 sought to recognize the post-1990 borders, whilst providing for ethnic partition and required international peace-keeping troops to maintain the terms of the peace.

▲ *Barbed-wire fences surround* a settlement in the Golan Heights.

- □ Republika Srpska
- □ Federacija Bosne i Hercegovine
- □ Brčko Distrikt

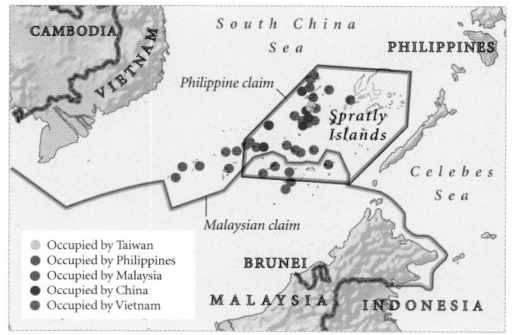
▲ *Most claimant states* have small military garrisons on the Spratly Islands.

The Spratly Islands

The site of potential oil and natural gas reserves, the Spratly Islands in the South China Sea have been claimed by China, Vietnam, Taiwan, Malaysia, and the Philippines since the Japanese gave up a wartime claim in 1951.

- Occupied by Taiwan
- Occupied by Philippines
- Occupied by Malaysia
- Occupied by China
- Occupied by Vietnam

ATLAS
OF THE WORLD

THE MAPS IN THIS ATLAS ARE ARRANGED CONTINENT BY CONTINENT, STARTING

FROM THE INTERNATIONAL DATE LINE, AND MOVING EASTWARD. THE MAPS PROVIDE

A UNIQUE VIEW OF TODAY'S WORLD, COMBINING TRADITIONAL CARTOGRAPHIC

TECHNIQUES WITH THE LATEST REMOTE-SENSED AND DIGITAL TECHNOLOGY.

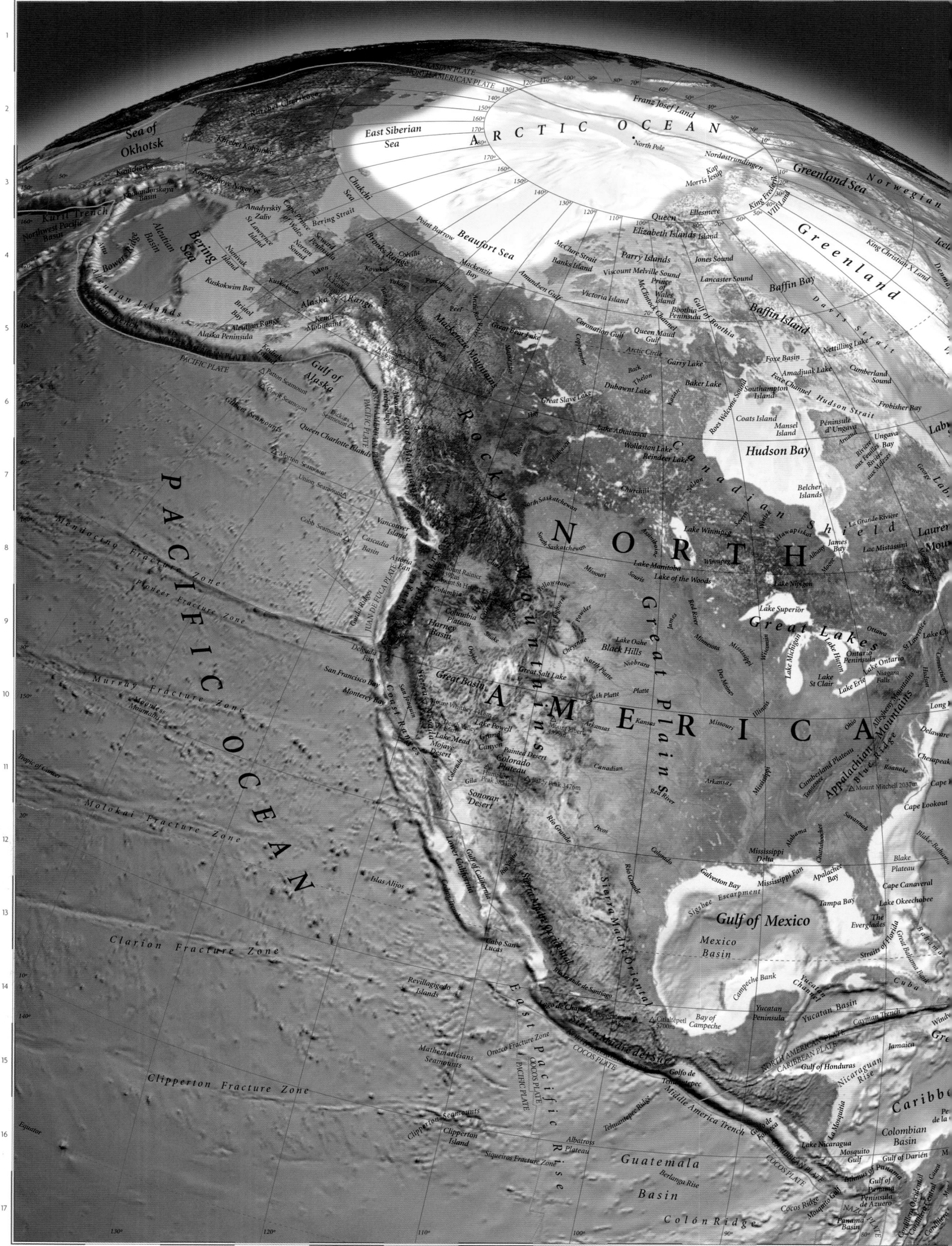

North America

North America is the world's third largest continent with a total area of 9,358,340 sq miles

(24,238,000 sq km) including Greenland and the Caribbean islands.

It lies wholly within the Northern Hemisphere.

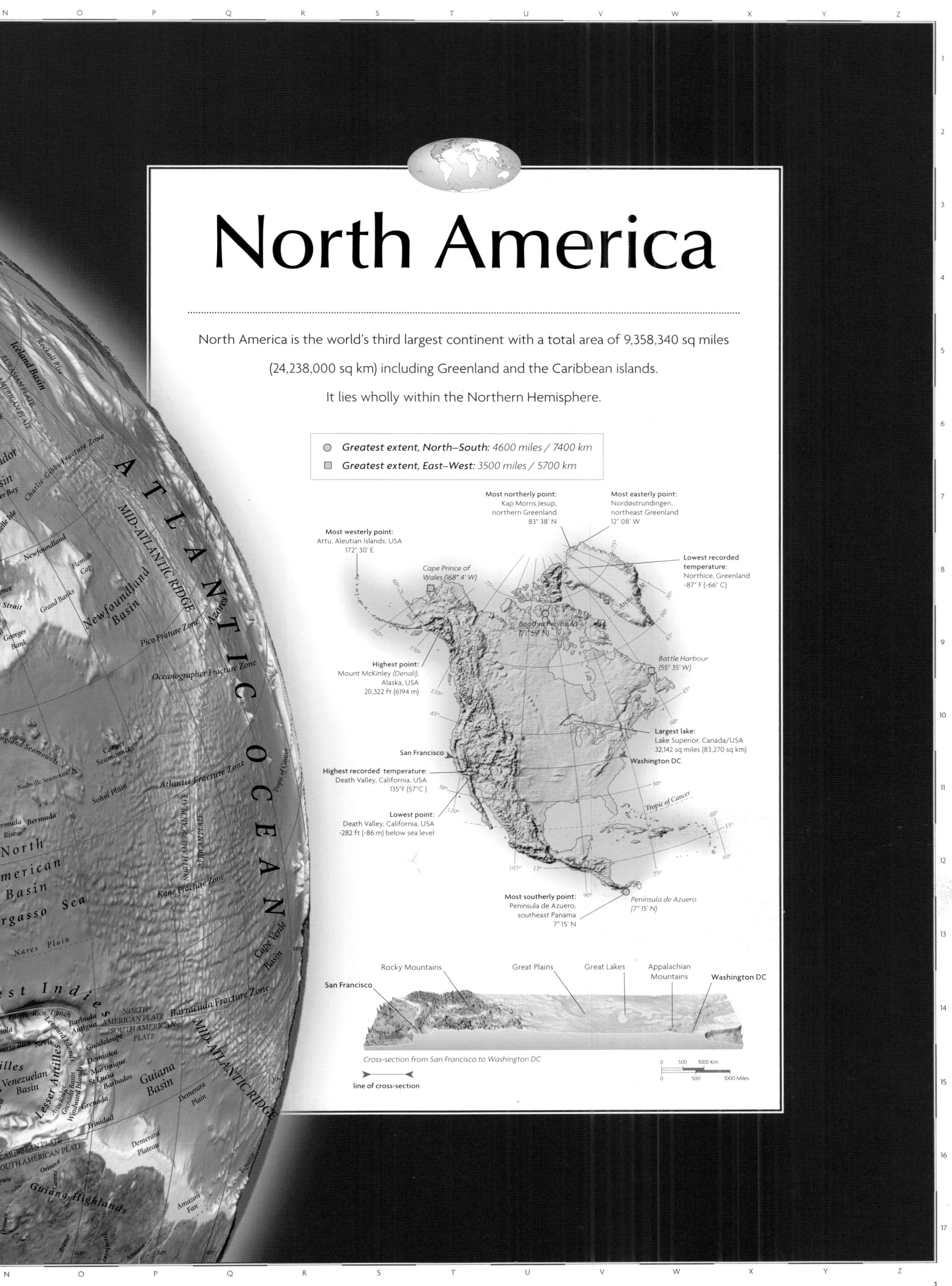

- **Greatest extent, North–South:** *4600 miles / 7400 km*
- **Greatest extent, East–West:** *3500 miles / 5700 km*

Most northerly point:
Kap Morris Jesup,
northern Greenland
83° 38' N

Most easterly point:
Nordøstrundingen,
northeast Greenland
12° 08' W

Most westerly point:
Attu, Aleutian Islands, USA
172° 30' E

**Lowest recorded
temperature:**
Northice, Greenland
-87° F (-66° C)

*Cape Prince of
Wales (168° 4' W)*

*Boothia Peninsula
(71° 59' N)*

*Battle Harbour
(55° 35' W)*

Highest point:
Mount McKinley *(Denali)*,
Alaska, USA
20,322 ft (6194 m)

San Francisco

Highest recorded temperature:
Death Valley, California, USA
135°F (57°C)

Lowest point:
Death Valley, California, USA
-282 ft (-86 m) below sea level

Largest lake:
Lake Superior, Canada/USA
32,142 sq miles (83,270 sq km)

Washington DC

Tropic of Cancer

Most southerly point:
Peninsula de Azuero,
southeast Panama
7° 15' N

*Peninsula de Azuero
(7° 15' N)*

Rocky Mountains

Great Plains

Great Lakes

Appalachian
Mountains

Washington DC

San Francisco

Cross-section from San Francisco to Washington DC

line of cross-section

0 500 1000 Km
0 500 1000 Miles

Physical North America

The North American continent can be divided into a number of major structural areas: the Western Cordillera, the Canadian Shield, the Great Plains, and Central Lowlands, and the Appalachians. Other smaller regions include the Gulf Atlantic Coastal Plain which borders the southern coast of North America from the southern Appalachians to the Great Plains. This area includes the expanding Mississippi Delta. A chain of volcanic islands, running in an arc around the margin of the Caribbean Plate, lie to the east of the Gulf of Mexico.

The Canadian Shield

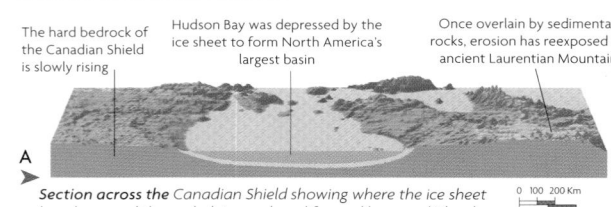

Spanning northern Canada and Greenland, this geologically stable plain forms the heart of the continent, containing rocks more than two billion years old. A long history of weathering and repeated glaciation has scoured the region, leaving flat plains, gentle hummocks, numerous small basins and lakes, and the bays and islands of the Arctic.

The Western Cordillera

About 80 million years ago the Pacific and North American plates collided, uplifting the Western Cordillera. This consists of the Aleutian, Coast, Cascade, and Sierra Nevada mountains, and the inland Rocky Mountains. These run parallel from the Arctic to Mexico.

The weight of the ice sheet, 1.8 miles (3 km) thick, has depressed the land to 0.6 miles (1 km) below sea level

▲ This computer-generated view shows the ice-covered island of Greenland without its ice cap.

The hard bedrock of the Canadian Shield is slowly rising

Hudson Bay was depressed by the ice sheet to form North America's largest basin

Once overlain by sedimentary rocks, erosion has reexposed the ancient Laurentian Mountains

Section across the Canadian Shield showing where the ice sheet has depressed the underlying rock and formed bays and islands.

Volcanic rock

Strata have been thrust eastward along fault lines

The Rocky Mountain Trench is the longest linear fault on the continent

Cross-section through the Western Cordillera showing direction of mountain building.

Map key

Elevation

3500m / 11,484ft
3000m / 9843ft
2500m / 8203ft
2000m / 6562ft
1500m / 4922ft
1000m / 3281ft
500m / 1640ft
250m / 820ft
100m / 328ft
sea level

Plate margins
(for explanation see page xiv)

constructive
destructive
conservative
uncertain

physiographic regions

line of cross-section

Scale 1:42,000,000

projection: Lambert Azimuthal Equal Area

The Great Plains & Central Lowlands

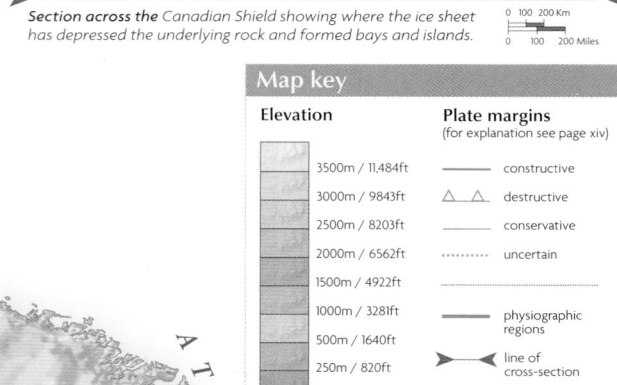

Deposits left by retreating glaciers and rivers have made this vast flat area very fertile. In the north this is the result of glaciation, with deposits up to one mile (1.7 km) thick, covering the basement rock. To the south and west, the massive Missouri/Mississippi river system has for centuries deposited silt across the plains, creating broad, flat floodplains and deltas.

The Appalachians

The Appalachian Mountains, uplifted about 400 million years ago, are some of the oldest in the world. They have been lowered and rounded by erosion and now slope gently toward the Atlantic across a broad coastal plain.

Horizontal strata

Sedimentary strata folded and faulted into ridges and valleys

Softer strata has been crumpled against the harder basement rock

Hard basement rock

Cross-section through the Appalachians showing the numerous folds, which have subsequently been weathered to create a rounded relief.

Sedimentary layers overlay domed basement rock

Upland rivers drain south toward the Mississippi Basin

Confluence of the Missouri and Mississippi Rivers

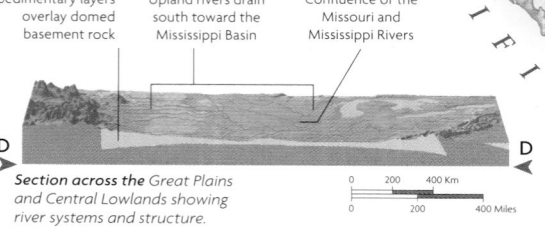

Section across the Great Plains and Central Lowlands showing river systems and structure.

ASIA
Bering Strait
Beaufort Sea
Greenland
ATLANTIC OCEAN
Bering Sea
Brooks Range
Mackenzie Delta
Mount McKinley 6194m
Aleutian Range Alaska Range
Aleutian Islands
Gulf of Alaska
NORTH AMERICAN PLATE
PACIFIC PLATE
Coast Mountains
Mackenzie Mountains
Mackenzie
Great Bear Lake
Great Slave Lake
Lake Athabasca
Reindeer Lake
CANADIAN
Baffin Bay
Baffin Island
Foxe Basin
Hudson Strait
Hudson Bay
SHIELD
Davis Strait
Labrador Sea
Labrador
Laurentian Mountains
Newfoundland
WESTERN
CORDILLERA
ROCKY MOUNTAINS
JUAN DE FUCA PLATE
Mount Rainier 4392m
Mount St Helens 2549m
Cascade Range
Sierra Nevada
San Joaquin Valley
Great Basin
San Andreas Fault
Death Valley -86m
Mojave Desert
Great Salt Lake
Grand Canyon
Colorado Plateau
Colorado
Sonoran Desert
GREAT PLAINS
CENTRAL LOWLANDS
Missouri
Lake Winnipeg
Lake Manitoba
Lake Superior
Lake Huron
Lake Michigan
Great Lakes
Lake Ontario
Lake Erie
St Lawrence
Nova Scotia
Cape Cod
APPALACHIAN MOUNTAINS
APPALACHIANS
Arkansas
Ohio
Mississippi
GULF ATLANTIC COASTAL PLAIN
Mississippi Delta
Gulf of Mexico
Rio Grande
Sierra Madre Occidental
Sierra Madre Oriental
Lower California
Gulf of California
PACIFIC OCEAN
Sierra Madre del Sur
Volcán Pico de Orizaba 5700m
Yucatan Peninsula
NORTH AMERICAN PLATE
CARIBBEAN PLATE
Greater Antilles
West Indies
Lesser Antilles
Caribbean Sea
Lake Nicaragua
Isthmus of Panama
CARIBBEAN PLATE
SOUTH AMERICAN PLATE
SOUTH AMERICA

Climate

North America's climate includes extremes ranging from freezing Arctic conditions in Alaska and Greenland, to desert in the southwest, and tropical conditions in southeastern Florida, the Caribbean, and Central America. Central and southern regions are prone to severe storms including tornadoes and hurricanes.

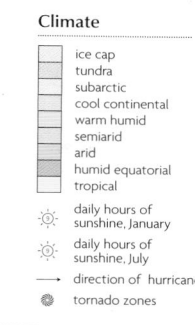

▲ *"Tornado alley" in the Mississippi Valley suffers frequent tornadoes.*

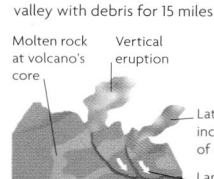

▲ *Much of the southwest is semi-desert; receiving less than 12 inches (300 mm) of rainfall a year.*

Climate

- ice cap
- tundra
- subarctic
- cool continental
- warm humid
- semiarid
- arid
- humid equatorial
- tropical
- ☼ daily hours of sunshine, January
- ☼ daily hours of sunshine, July
- → direction of hurricanes
- ⊚ tornado zones

Temperature

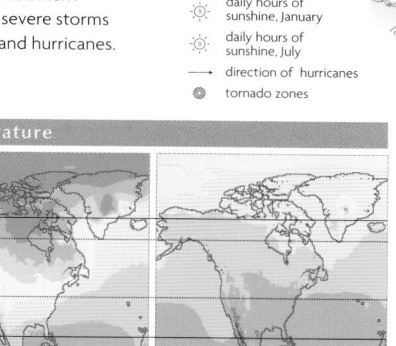

Average January temperature *Average July temperature*

Temperature

- -22°F (below -30°C)
- -22 to -4°F (-30 to -20°C)
- -4 to 14°F (-20 to -10°C)
- 14 to 32°F (-10 to 0°C)
- 32 to 50°F (0 to 10°C)
- 50 to 68°F (10 to 20°C)
- 68 to 86°F (20 to 30°C)
- 86°F (above 30°C)

Rainfall

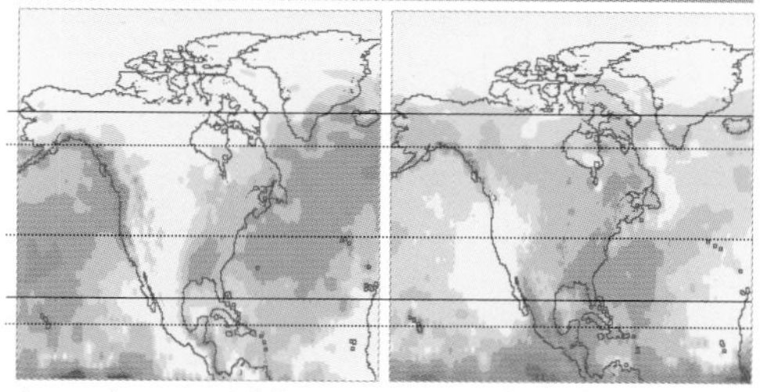

Average January rainfall *Average July rainfall*

Rainfall

- 0–1 in (0–25 mm)
- 1–2 in (25–50 mm)
- 2–4 in (50–100 mm)
- 4–8 in (100–200 mm)
- 8–12 in (200–300 mm)
- 12–16 in (300–400 mm)
- 16–20 in (400–500 mm)
- more than 20 in (500 mm)

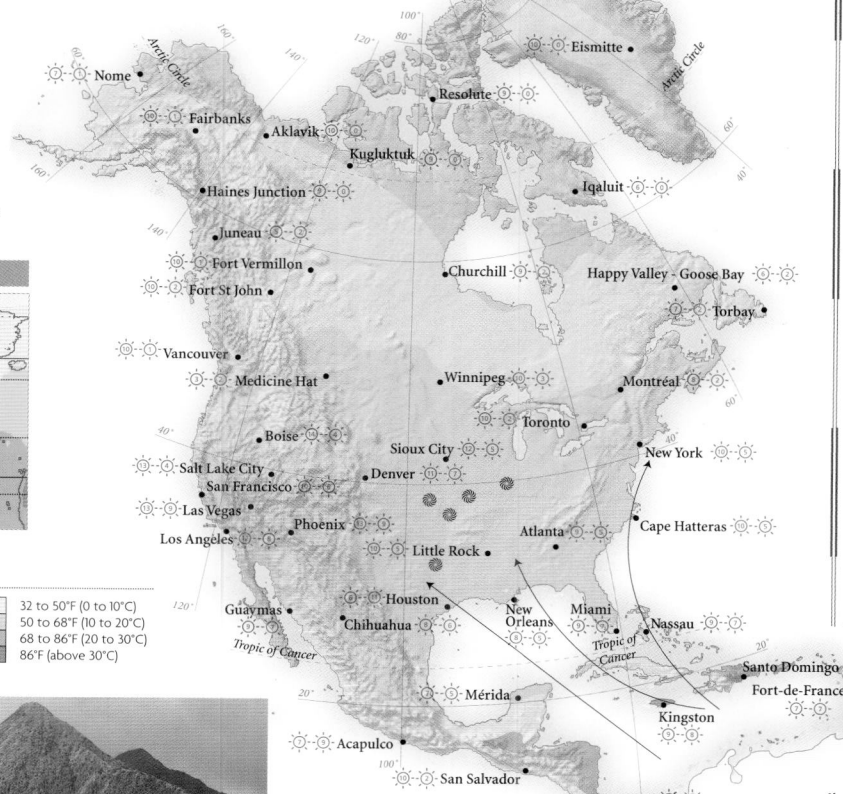

◄ *The lush, green mountains of the Lesser Antilles receive annual rainfalls of up to 360 inches (9000 mm).*

Shaping the continent

Glacial processes affect much of northern Canada, Greenland, and the Western Cordillera. Along the western coast of North America, Central America, and the Caribbean, underlying plates moving together lead to earthquakes and volcanic eruptions. The vast river systems, fed by mountain streams, constantly erode and deposit material along their paths.

Volcanic activity

1 Mount St. Helens volcano *(right)* in the Cascade Range erupted violently in May 1980, killing 57 people and leveling large areas of forest. The lateral blast filled a valley with debris for 15 miles (25 km).

Molten rock at volcano's core | Vertical eruption | Lateral explosion increases extent of damage | Landslide fills valley

Volcanic activity: Eruption of Mount St Helens

Seismic activity

5 The San Andreas Fault *(above)* places much of the North America's west coast under constant threat from earthquakes. It is caused by the Pacific Plate grinding past the North American Plate at a faster rate, though in the same direction.

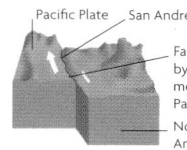

Pacific Plate | San Andreas Fault | Fault is caused by faster movement of Pacific Plate | North American Plate

Seismic activity: Action of the San Andreas Fault

River erosion

6 The Grand Canyon *(above)* in the Colorado Plateau was created by the downward erosion of the Colorado River, combined with the gradual uplift of the plateau, over the past 30 million years. The contours of the canyon formed as the softer rock layers eroded into gentle slopes, and the hard rock layers into cliffs. The depth varies from 3855–6560 ft (1175–2000 m).

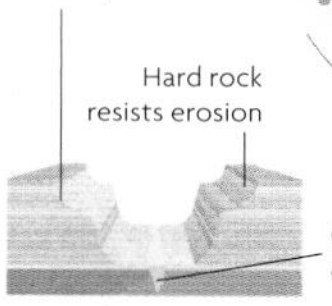

Soft rock is easily eroded into gentle slopes | Hard rock resists erosion | Colorado River cuts down through rock

River Erosion: Formation of the Grand Canyon

Periglaciation

2 The ground in the far north is nearly always frozen: the surface thaws only in summer. This freeze-thaw process produces features such as pingos *(left)*, formed by the freezing of groundwater. With each successive winter ice accumulates producing a mound with a core of ice.

Ice core pushes up ground to form pingo | Unfrozen lake | Groundwater attracted to ice core

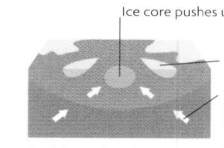
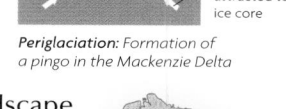

Periglaciation: Formation of a pingo in the Mackenzie Delta

The evolving landscape

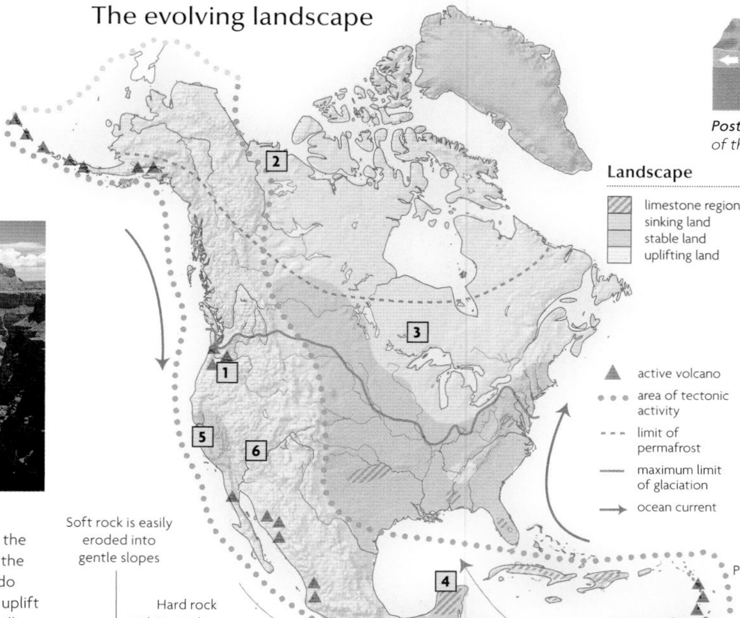

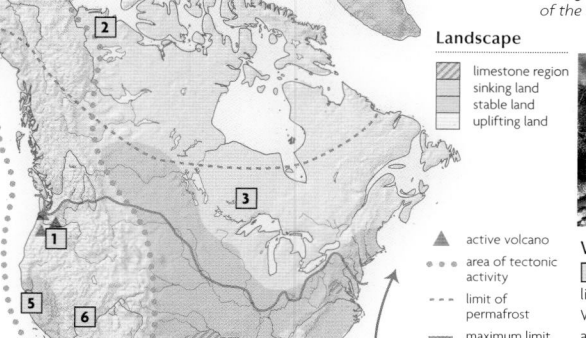

Landscape

- limestone region
- sinking land
- stable land
- uplifting land

- ▲ active volcano
- ⁙ area of tectonic activity
- --- limit of permafrost
- — maximum limit of glaciation
- → ocean current

Post-glacial lakes

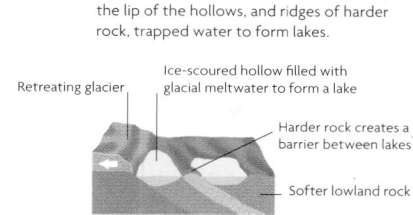

3 A chain of lakes from Great Bear Lake to the Great Lakes *(above)* was created as the ice retreated northward. Glaciers scoured hollows in the softer lowland rock. Glacial deposits at the lip of the hollows, and ridges of harder rock, trapped water to form lakes.

Retreating glacier | Ice-scoured hollow filled with glacial meltwater to form a lake | Harder rock creates a barrier between lakes | Softer lowland rock

Post-glacial lakes: Formation of the Great Lakes

Weathering

4 The Yucatan Peninsula is a vast, flat limestone plateau in southern Mexico. Weathering action from both rainwater and underground streams has enlarged fractures in the rock to form caves and hollows, called sinkholes *(above)*.

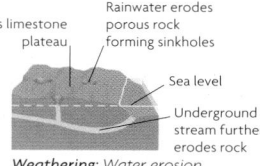

Porous limestone plateau | Rainwater erodes porous rock forming sinkholes | Sea level | Underground stream further erodes rock

Weathering: Water erosion on the Yucatan Peninsula

Map city labels

Nome, Fairbanks, Aklavik, Kugluktuk, Resolute, Eismitte, Iqaluit, Haines Junction, Juneau, Fort Vermillon, Churchill, Happy Valley - Goose Bay, Torbay, Fort St John, Vancouver, Medicine Hat, Winnipeg, Montréal, Boise, Toronto, New York, Salt Lake City, Sioux City, Denver, San Francisco, Las Vegas, Phoenix, Atlanta, Cape Hatteras, Los Angeles, Little Rock, Houston, Miami, Nassau, Guaymas, New Orleans, Chihuahua, Santo Domingo, Fort-de-France, Mérida, Kingston, Acapulco, San Salvador, San José

Political North America

Democracy is well established in some parts of the continent but is a recent phenomenon in others. The economically dominant nations of Canada and the US have a long democratic tradition but elsewhere, notably in the countries of Central America, political turmoil has been more common. In Nicaragua and Haiti, harsh dictatorships have only recently been superseded by democratically elected governments. North America's largest countries, Canada, Mexico, and the US have federal state systems, sharing political power between national and state governments. The US has intervened militarily on several occasions in Central America and the Caribbean to protect its strategic interests.

Transportation

In the 19th century, railroads opened up the North American continent. Air transportation is now more common for long distance passenger travel, although railroads are still extensively used for bulk freight transportation. Waterways like the Mississippi River are important for the transportation of bulk materials, and the Panama Canal is a vital link between the Pacific and Atlantic Oceans. In the 20th century, road transportation increased massively, with the introduction of cheap, mass-produced motor cars and extensive highway construction.

◄ *This busy suburban* interchange in Los Angeles is part of the US's Interstate freeway system. Construction of the 55,000 mile (88,500 km) freeway network began in the 1950s, and it now connects most major cities, and carries one-fifth of the US's road traffic.

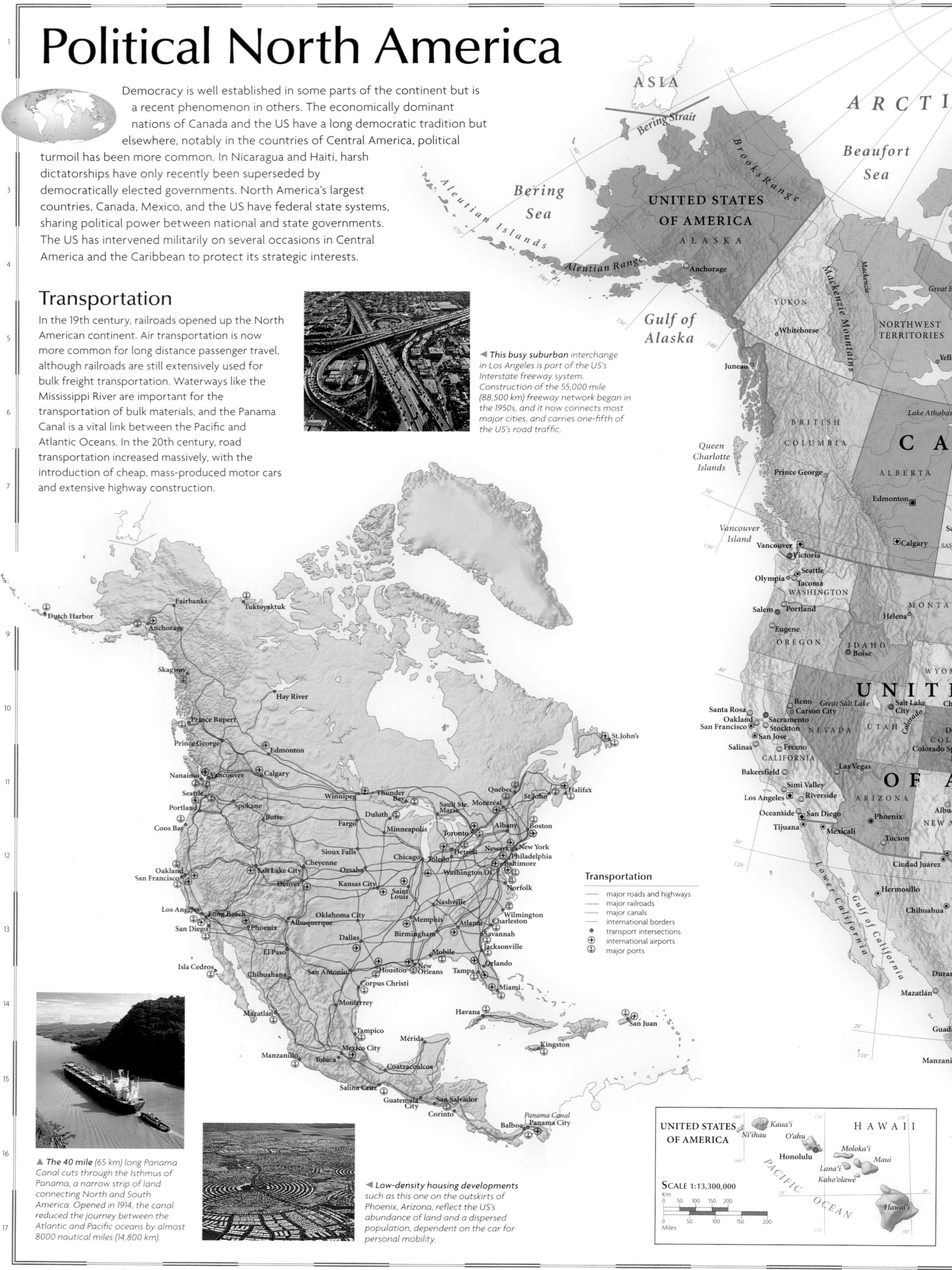

Transportation

— major roads and highways
— major railroads
— major canals
— international borders
• transport intersections
⊕ international airports
⊕ major ports

▲ *The 40 mile* (65 km) long Panama Canal cuts through the Isthmus of Panama, a narrow strip of land connecting North and South America. Opened in 1914, the canal reduced the journey between the Atlantic and Pacific oceans by almost 8000 nautical miles (14,800 km).

◄ *Low-density housing developments* such as this one on the outskirts of Phoenix, Arizona, reflect the US's abundance of land and a dispersed population, dependent on the car for personal mobility.

UNITED STATES OF AMERICA

HAWAI'I

SCALE 1:13,300,000

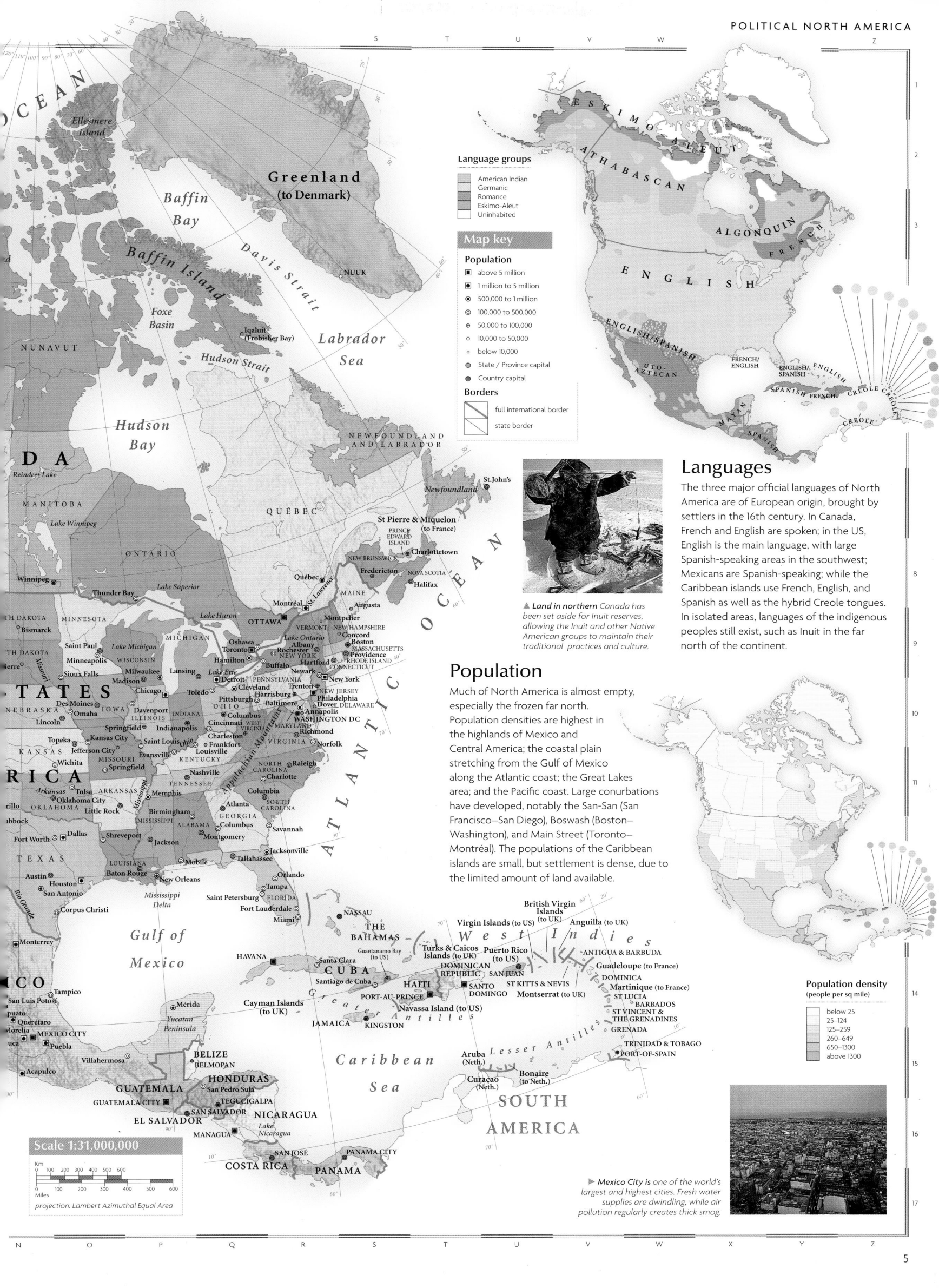

Language groups

- American Indian
- Germanic
- Romance
- Eskimo-Aleut
- Uninhabited

Map key

Population
- ■ above 5 million
- ⊡ 1 million to 5 million
- ◉ 500,000 to 1 million
- ◎ 100,000 to 500,000
- ⊕ 50,000 to 100,000
- ⊙ 10,000 to 50,000
- ∘ below 10,000
- ● State / Province capital
- ● Country capital

Borders
- full international border
- state border

Languages

The three major official languages of North America are of European origin, brought by settlers in the 16th century. In Canada, French and English are spoken; in the US, English is the main language, with large Spanish-speaking areas in the southwest; Mexicans are Spanish-speaking; while the Caribbean islands use French, English, and Spanish as well as the hybrid Creole tongues. In isolated areas, languages of the indigenous peoples still exist, such as Inuit in the far north of the continent.

▲ *Land in northern* Canada has been set aside for Inuit reserves, allowing the Inuit and other Native American groups to maintain their traditional practices and culture.

Population

Much of North America is almost empty, especially the frozen far north. Population densities are highest in the highlands of Mexico and Central America; the coastal plain stretching from the Gulf of Mexico along the Atlantic coast; the Great Lakes area; and the Pacific coast. Large conurbations have developed, notably the San-San (San Francisco–San Diego), Boswash (Boston–Washington), and Main Street (Toronto–Montréal). The populations of the Caribbean islands are small, but settlement is dense, due to the limited amount of land available.

Population density
(people per sq mile)
- below 25
- 25–124
- 125–259
- 260–649
- 650–1300
- above 1300

▶ *Mexico City is* one of the world's largest and highest cities. Fresh water supplies are dwindling, while air pollution regularly creates thick smog.

Scale 1:31,000,000

projection: Lambert Azimuthal Equal Area

5

A B C D E F G H I J K L M

North American resources

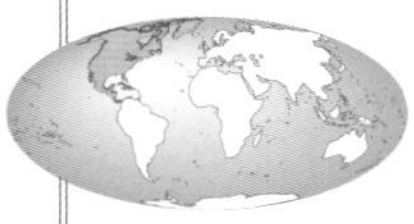

The two northern countries of Canada and the US are richly endowed with natural resources that have helped to fuel economic development. The US is the world's largest economy, although today it is facing stiff competition from the Far East. Mexico has relied on oil revenues but there are hopes that the North American Free Trade Agreement (NAFTA), will encourage trade growth with Canada and the US. The poorer countries of Central America and the Caribbean depend largely on cash crops and tourism.

Industry

The modern, industrialized economies of the US and Canada contrast sharply with those of Mexico, Central America, and the Caribbean. Manufacturing is especially important in the US; vehicle production is concentrated around the Great Lakes, while electronic and hi-tech industries are increasingly found in the western and southern states. Mexico depends on oil exports and assembly work, taking advantage of cheap labor. Many Central American and Caribbean countries rely heavily on agricultural exports.

◀ After its purchase from Russia in 1867, Alaska's frozen lands were largely ignored by the US. Oil reserves similar in magnitude to those in eastern Texas were discovered in Prudhoe Bay, Alaska in 1968. Freezing temperatures and a fragile environment hamper oil extraction.

Standard of living

The US and Canada have one of the highest overall standards of living in the world. However, many people still live in poverty, especially in urban ghettos and some rural areas. Central America and the Caribbean are markedly poorer than their wealthier northern neighbors. Haiti is the poorest country in the western hemisphere.

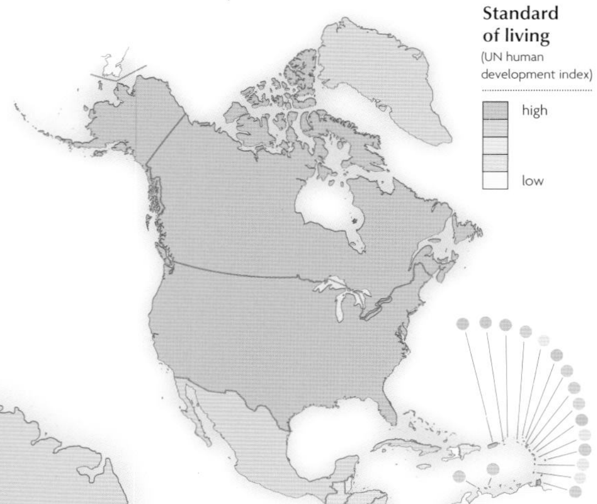

Standard of living
(UN human development index)

high

low

▲ Fish such as cod, flounder, and plaice are caught in the Grand Banks, off the Newfoundland coast, and processed in many North Atlantic coastal settlements.

▲ South of San Francisco, "Silicon Valley" is both a national and international center for hi-tech industries, electronic industries, and research institutions.

▲ Multinational companies rely on cheap labor and tax benefits to facilitate the assembly of vehicle parts in Mexican factories.

▲ The health of the Wall Street stock market in New York is the standard measure of the state of the world's economy.

Industry

- ✈ aerospace
- brewing
- car/vehicle manufacture
- chemicals
- defense
- electronics
- engineering
- film industry
- finance
- food processing
- hi-tech industry
- iron & steel
- pharmaceuticals
- printing & publishing
- research & development
- shipbuilding
- sugar processing
- textiles
- timber processing
- tobacco processing
- coal
- oil
- gas
- industrial cities
- major industrial areas

GNI per capita (US$)

- below 1999
- 2000–4999
- 5000–9999
- 10,000–19,999
- 20,000–24,999
- above 25,000

Map labels

ARCTIC OCEAN
RUSS. FED.
Bering Strait
Bering Sea
Beaufort Sea
Prudhoe Bay
Baffin Bay
Greenland (to Denmark)
USA
Gulf of Alaska
Hudson Strait
Labrador Sea
Hudson Bay
CANADA
Vancouver
Calgary
Seattle
Winnipeg
Portland
Montréal
Minneapolis
Toronto
Boston
Milwaukee
Buffalo
Albany
Detroit
New York
UNITED STATES
Chicago
Cleveland
OF AMERICA
Pittsburgh
Baltimore
Philadelphia
San Francisco
Dayton
Cincinnati
Denver
Saint Louis
Kansas City
Greensboro
Wichita
Nashville
Charlotte
Los Angeles
Tulsa
San Diego
Phoenix
Birmingham
Atlanta
Tijuana
Dallas
Ciudad Juárez
El Paso
Houston
Jacksonville
New Orleans
Orlando
Tampa
Miami
PACIFIC OCEAN
ATLANTIC OCEAN
Monterrey
Guadalajara
Mexico City
MEXICO
Gulf of Mexico
Havana
CUBA
BAHAMAS
West Indies
Virgin Islands (to US)
Turks & Caicos Islands (to UK)
British Virgin Islands (to UK)
Anguilla (to UK)
ST KITTS & NEVIS
ANTIGUA & BARBUDA
Montserrat (to UK)
Puerto Rico (to US)
Guadeloupe (to France)
San Juan
DOMINICA
DOMINICAN REPUBLIC
Martinique (to France)
HAITI
Port-au-Prince
Santo Domingo
ST LUCIA
Cayman Islands (to UK)
JAMAICA
Greater Antilles
BARBADOS
ST VINCENT & THE GRENADINES
Navassa Island (to US)
Lesser Antilles
GRENADA
Aruba (to Neth.)
TRINIDAD & TOBAGO
Port-of-Spain
BELIZE
Caribbean Sea
Curaçao (to Neth.)
Bonaire (to Neth.)
VENEZUELA
GUATEMALA
Guatemala City
HONDURAS
Tegucigalpa
EL SALVADOR
San Salvador
NICARAGUA
Managua
San José
Panama City
COSTA RICA
PANAMA
COLOMBIA

Environmental issues

Many fragile environments are under threat throughout the region. In Haiti, all the primary rain forest has been destroyed, while air pollution from factories and cars in Mexico City is among the worst in the world. Elsewhere, industry and mining pose threats, particularly in the delicate arctic environment of Alaska where oil spills have polluted coastlines and decimated fish stocks.

Mineral resources

Fossil fuels are exploited in considerable quantities throughout the continent. Coal mining in the Appalachians is declining but vast open pits exist further west in Wyoming. Oil and natural gas are found in Alaska, Texas, the Gulf of Mexico, and the Canadian West. Canada has large quantities of nickel, while Jamaica has considerable deposits of bauxite, and Mexico has large reserves of silver.

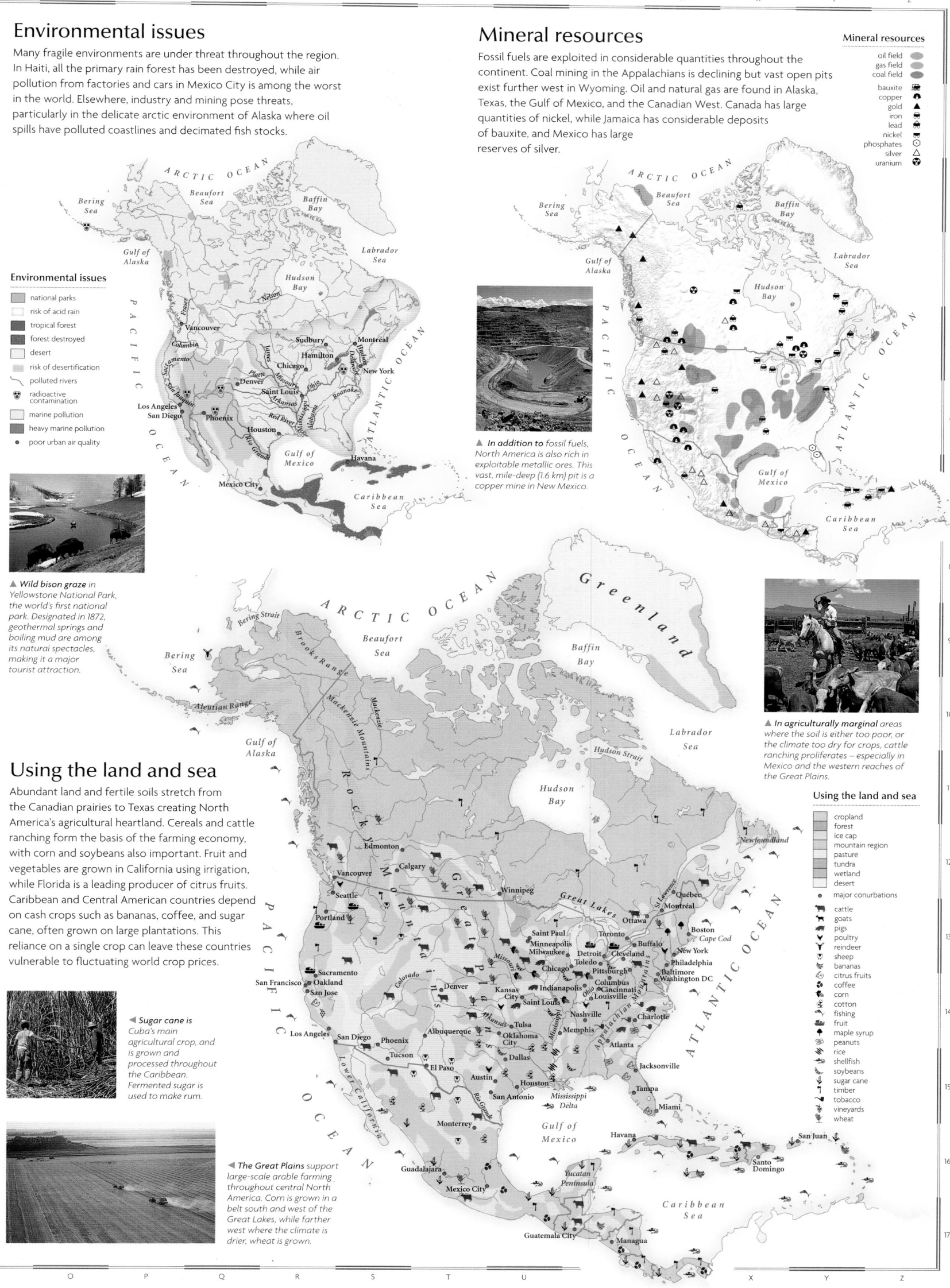

Mineral resources
- oil field
- gas field
- coal field
- bauxite
- copper
- gold
- iron
- lead
- nickel
- phosphates
- silver
- uranium

▲ *In addition to fossil fuels, North America is also rich in exploitable metallic ores. This vast, mile-deep (1.6 km) pit is a copper mine in New Mexico.*

Environmental issues
- national parks
- risk of acid rain
- tropical forest
- forest destroyed
- desert
- risk of desertification
- polluted rivers
- radioactive contamination
- marine pollution
- heavy marine pollution
- poor urban air quality

▲ *Wild bison graze in Yellowstone National Park, the world's first national park. Designated in 1872, geothermal springs and boiling mud are among its natural spectacles, making it a major tourist attraction.*

▲ *In agriculturally marginal areas where the soil is either too poor, or the climate too dry for crops, cattle ranching proliferates – especially in Mexico and the western reaches of the Great Plains.*

Using the land and sea

Abundant land and fertile soils stretch from the Canadian prairies to Texas creating North America's agricultural heartland. Cereals and cattle ranching form the basis of the farming economy, with corn and soybeans also important. Fruit and vegetables are grown in California using irrigation, while Florida is a leading producer of citrus fruits. Caribbean and Central American countries depend on cash crops such as bananas, coffee, and sugar cane, often grown on large plantations. This reliance on a single crop can leave these countries vulnerable to fluctuating world crop prices.

◀ *Sugar cane is Cuba's main agricultural crop, and is grown and processed throughout the Caribbean. Fermented sugar is used to make rum.*

◀ *The Great Plains support large-scale arable farming throughout central North America. Corn is grown in a belt south and west of the Great Lakes, while farther west where the climate is drier, wheat is grown.*

Using the land and sea
- cropland
- forest
- ice cap
- mountain region
- pasture
- tundra
- wetland
- desert
- major conurbations
- cattle
- goats
- pigs
- poultry
- reindeer
- sheep
- bananas
- citrus fruits
- coffee
- corn
- cotton
- fishing
- fruit
- maple syrup
- peanuts
- rice
- shellfish
- soybeans
- sugar cane
- timber
- tobacco
- vineyards
- wheat

Canada

Canada is the second largest country in the world, and with only about one-tenth of its land area inhabited, it is one of the most sparsely populated. Canada became a confederation in 1867, though Newfoundland did not join until 1949. As a founding member of the UN and of the Commonwealth, Canada has played an important role in international affairs. A constitutional crisis, focusing on the French-speaking Québécois, and Inuit, and Native American land rights, dominated politics in the 1990s. In 1999, part of the Northwest Territories, Nunavut, became a self-governing homeland for the Inuit.

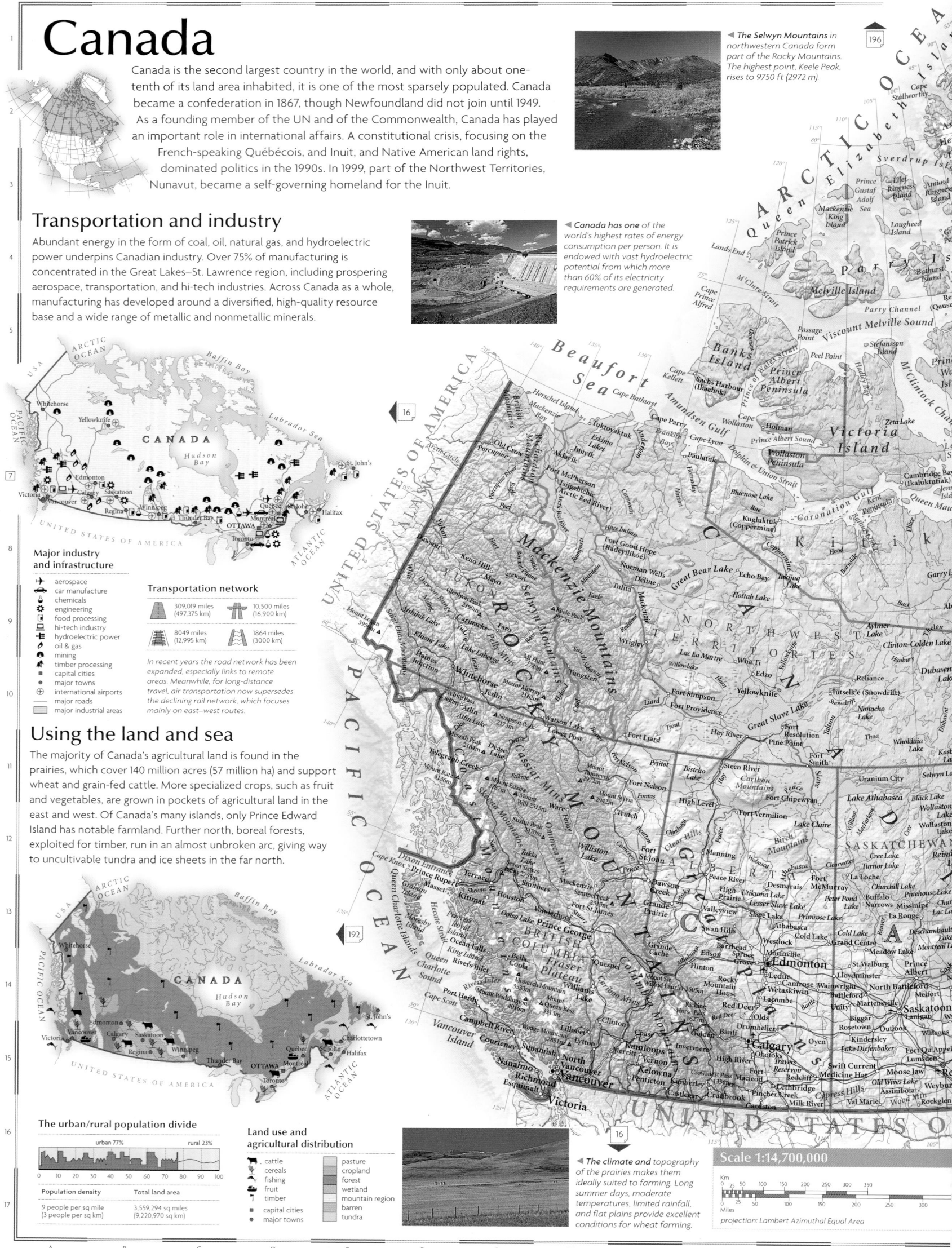

The Selwyn Mountains in northwestern Canada form part of the Rocky Mountains. The highest point, Keele Peak, rises to 9750 ft (2972 m).

Transportation and industry

Abundant energy in the form of coal, oil, natural gas, and hydroelectric power underpins Canadian industry. Over 75% of manufacturing is concentrated in the Great Lakes–St. Lawrence region, including prospering aerospace, transportation, and hi-tech industries. Across Canada as a whole, manufacturing has developed around a diversified, high-quality resource base and a wide range of metallic and nonmetallic minerals.

Canada has one of the world's highest rates of energy consumption per person. It is endowed with vast hydroelectric potential from which more than 60% of its electricity requirements are generated.

Major industry and infrastructure

- aerospace
- car manufacture
- chemicals
- engineering
- food processing
- hi-tech industry
- hydroelectric power
- oil & gas
- mining
- timber processing
- capital cities
- major towns
- international airports
- major roads
- major industrial areas

Transportation network

309,019 miles (497,375 km)	10,500 miles (16,900 km)
8049 miles (12,995 km)	1864 miles (3000 km)

In recent years the road network has been expanded, especially links to remote areas. Meanwhile, for long-distance travel, air transportation now supersedes the declining rail network, which focuses mainly on east–west routes.

Using the land and sea

The majority of Canada's agricultural land is found in the prairies, which cover 140 million acres (57 million ha) and support wheat and grain-fed cattle. More specialized crops, such as fruit and vegetables, are grown in pockets of agricultural land in the east and west. Of Canada's many islands, only Prince Edward Island has notable farmland. Further north, boreal forests, exploited for timber, run in an almost unbroken arc, giving way to uncultivable tundra and ice sheets in the far north.

The urban/rural population divide

urban 77%   rural 23%

Population density	Total land area
9 people per sq mile (3 people per sq km)	3,559,294 sq miles (9,220,970 sq km)

Land use and agricultural distribution

- cattle
- cereals
- fishing
- fruit
- timber
- capital cities
- major towns

- pasture
- cropland
- forest
- wetland
- mountain region
- barren
- tundra

The climate and topography of the prairies makes them ideally suited to farming. Long summer days, moderate temperatures, limited rainfall, and flat plains provide excellent conditions for wheat farming.

Scale 1:14,700,000

projection: Lambert Azimuthal Equal Area

The landscape

Glaciers on islands in the Arctic Ocean are the last remnants of the ice sheet that once covered and shaped Canada. Hudson Bay is the center of the Canadian Shield, a huge, eroded plateau marked at its southern extremity by a string of lakes running southeastward from Great Bear Lake to the Great Lakes. In contrast to the rolling relief of the Shield and the central lowland region, the Rocky Mountains rise to peaks of over 13,000 ft (4000 m), stretching 500 miles (800 km) along the west coast.

▶ **Permanently frozen ground** known as permafrost is common in Canada's northern tundra. It thickens farther north, becoming hundreds of yards deep in parts of the Arctic.

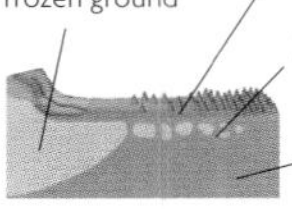

Permanently frozen ground

Top layer thaws in the summer

Marginal areas of permafrost thaw in summer

Unfrozen ground where temperature is more moderate

The Mackenzie river, flowing north over the permafrost, forms a wide river channel with many tributaries. Together with the Peel river it has created a long, narrow delta at its mouth. The entire river freezes during the winter.

Fertile prairies stretch from the southern rim of the Canadian Shield, south into the US.

Exposure to three phases of mountain-building and subsequent erosion over millions of years has molded the ancient Canadian Shield into a series of basins and ridges.

▲ **Along the northeastern** coast of Baffin Island the mountains rise to 8000 ft (2440 m). Glaciers move down through the valleys to the sea, eroding wide U-shaped valleys.

The Rocky Mountains were formed some 80 million years ago, when the Pacific plate was driven under the North American plate, forcing up the land.

The Great Lakes lie on the Canada–US border. The basins they now occupy were fashioned by repeated ice advance. At one time, Lakes Superior, Huron, and Michigan formed a single large lake, Lake Nipissing.

The St. Lawrence River is 2350 miles (3782 km) long. It flows from the western shore of Lake Superior through the Great Lakes and on to the Atlantic Ocean. From December to April, the St. Lawrence Seaway freezes between Lake Ontario and Montréal.

▶ **The Great Lakes** are drained by the St. Lawrence River which flows down through a wide tectonic depression. It forms a broad estuary for much of its course, the width varying from 1.2 miles (1.9 km) in the upper reaches to 90 miles (145 km) at its mouth.

◀ **Isolated pillars, known** as hoodoos near Red Deer river in the badlands of Alberta are a product of wind and water erosion, especially flash floods. The badlands lie in the rain shadow of the Rocky Mountains, which creates a semiarid climate.

Map key

Population
- 1 million to 5 million
- 500,000 to 1 million
- 100,000 to 500,000
- 50,000 to 100,000
- 10,000 to 50,000
- below 10,000

Elevation
- 6000m / 19,686ft
- 4000m / 13,124ft
- 3000m / 9843ft
- 2000m / 6562ft
- 1000m / 3281ft
- 500m / 1640ft
- 250m / 820ft
- 100m / 328ft
- sea level

Canada:
WESTERN PROVINCES

Alberta, British Columbia, Manitoba,
Saskatchewan, Yukon

The mountains of the west coast, incorporating British Columbia and the Yukon, descend into the vast, flat prairies of Alberta, Saskatchewan, and Manitoba. The empty lands and fertile soils of the prairie provinces attracted migrants, and the descendants of early European immigrants still make up a large proportion of the population. The mechanization of agriculture has reduced the need for labor, and rural population densities remain low. The majority of the people live within 100 miles (160 km) of the southern Canada–US border, and in British Columbia, one of the leading Canadian provinces in terms of economic wealth. The Yukon, in the far north, remains a relatively unspoiled wilderness, containing large, untapped mineral reserves. This province has a significant population of Native American people, many of whom maintain a traditional lifestyle.

Using the land and sea

Wheat farming is the economic mainstay of Alberta, Manitoba, and Saskatchewan, which contain 82% of farmland in Canada. Cattle are also raised on the prairies. Forestry and fishing are the most prominent resource-based industries in British Columbia. Despite the mountainous terrain, fruit and specialized grains can be grown in the Okanagan and Fraser valleys.

Land use and agricultural distribution

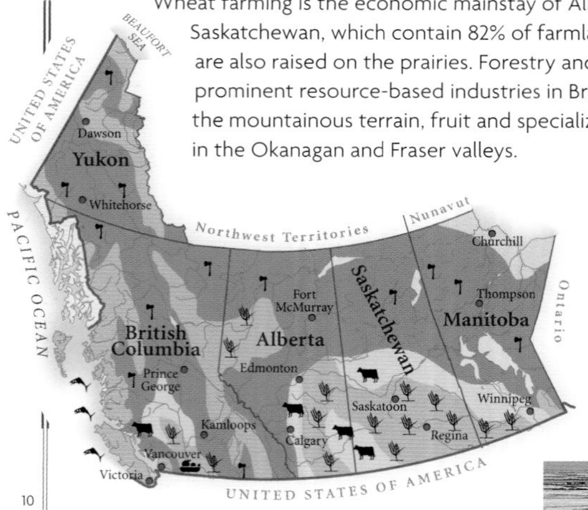

- cattle
- cereals
- fishing
- fruit
- timber
- major towns
- pasture
- cropland
- forest
- wetland
- barren
- tundra

The urban/rural population divide

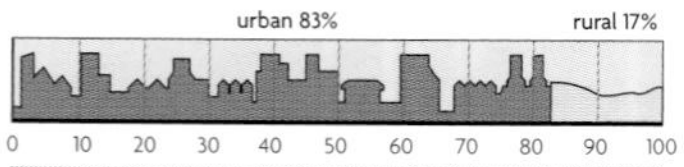

urban 83%   rural 17%

0 10 20 30 40 50 60 70 80 90 100

Population density	Total land area
8 people per sq mile (3 people per sq km)	1,230,547 sq miles (3,187,120 sq km)

▲ Large, highly-mechanized and often very specialized farms, requiring huge investment but little labor, characterize modern farming in the prairies.

Transportation & industry

The western provinces contain a wealth of mineral resources. Alberta holds the bulk of Canada's fossil fuels; the other provinces contain reserves of metallic ores, such as zinc, lead, and silver. Isolation from markets has slowed the development of manufacturing, restricting it to the large cities like Vancouver, Winnipeg, and Calgary. Hydroelectric power is widely exploited, although there is increasing concern about potential ecological damage.

Major industry and infrastructure

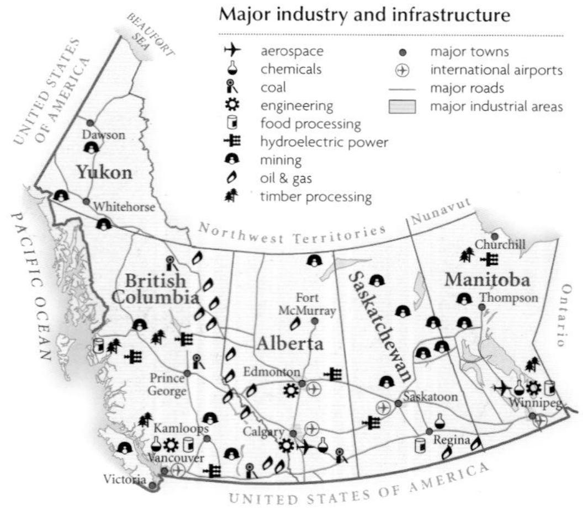

- aerospace
- chemicals
- coal
- engineering
- food processing
- hydroelectric power
- mining
- oil & gas
- timber processing
- major towns
- international airports
- major roads
- major industrial areas

Transportation network

82,438 miles (135,145 km)	
6459 miles (10,401 km)	
24,041 miles (38,694 km)	
None	

The transportation network of the western provinces is dominated by east–west routes that weave through mountain passes and spread across the plains. Access to some northern areas is restricted to air travel.

◄ Much of the Yukon is uninhabited tundra. Industry is based on the extraction of mineral resources, and to a lesser extent, on the scattered forests of the south.

▲ The Fraser River valley is a major area of settlement in British Columbia. Railroads cross the Rocky Mountains via this valley.

▲ Established in 1907, Jasper National Park lies in the heart of the Rocky Mountains. It is noted for its spectacular alpine scenery and contains part of the large Columbia Icefield.

The landscape

The massive Rocky Mountains form a continental divide between rivers flowing eastward and westward. The interior plains lie east of the mountains, stretching from the Arctic Circle south into the US. Covered with glacial deposits from the last Ice Age, these are interspersed with hilly regions and long, steep escarpments.

Vegetated island
Bar
River flow is diverted by deposited sediments
Sand flat

▲ ***Braided rivers are** shallow and fast-flowing. The interlaced branches are formed when excess sediments, which can no longer be transported, are deposited. The sediments collect in the river channel forming bars and sand flats. Islands form when the bars are colonized by vegetation.*

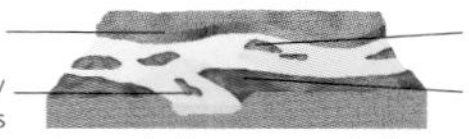

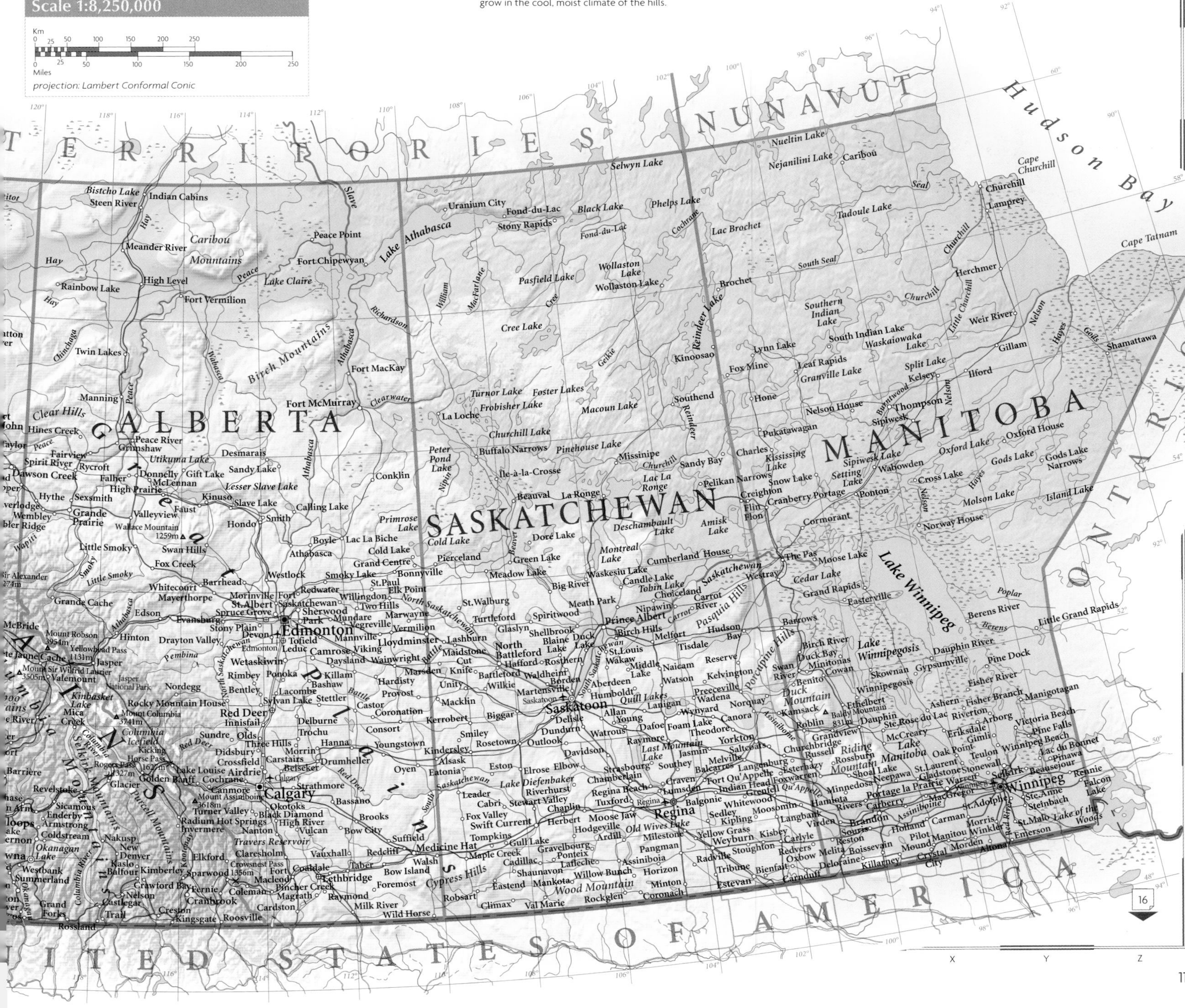

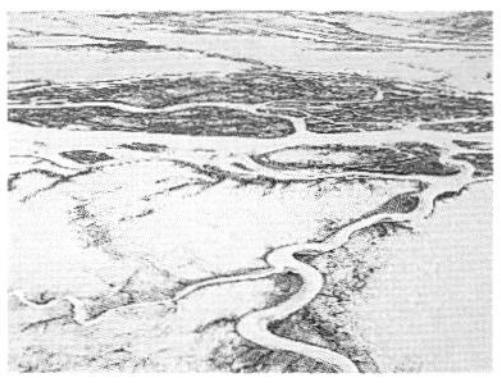

▲ ***Across the tundra** of northern Manitoba, widespread permafrost inhibits water from permeating the soil. This causes rivers like the Churchill to flow in many channels, which can be frozen for up to six months during the winter.*

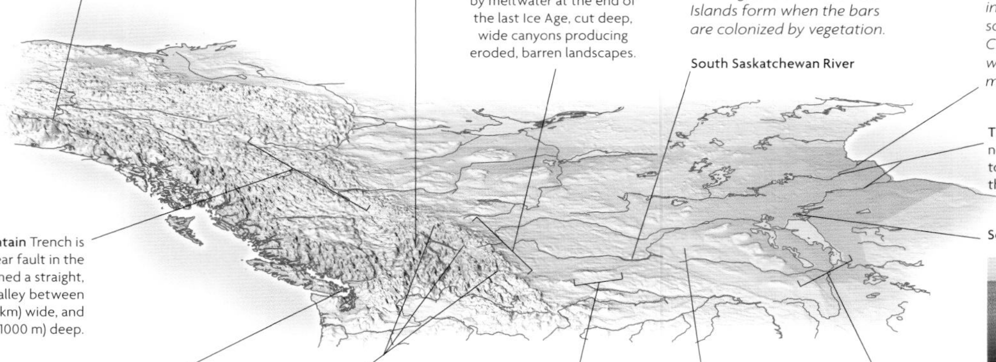

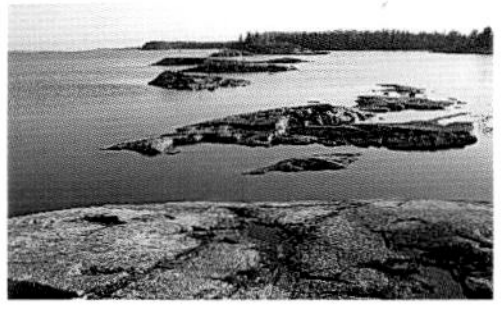

Mount Logan rises 19,551 ft (5959 m). It is the highest peak in Canada.
The Columbia Icefield in the Rocky Mountains is the source of two major rivers, the Athabasca and the North Saskatchewan.
The badlands of Alberta were created when east-flowing rivers, swollen by meltwater at the end of the last Ice Age, cut deep, wide canyons producing eroded, barren landscapes.
South Saskatchewan River
The Nelson and Churchill rivers drain northward across the Canadian Shield to Hudson Bay. The shield covers three-fifths of Saskatchewan.
Setting Lake
The Rocky Mountain Trench is the longest linear fault in the world. It has formed a straight, flat-bottomed valley between 2–9 miles (4–15 km) wide, and up to 3280 ft (1000 m) deep.
Hundreds of islands dot the fjord-indented coast of British Columbia; the largest is Vancouver Island.
Three major passes cut through the Rocky Mountains: Yellowhead, Kicking Horse, and Crowsnest. They are all used as transportation routes through the mountains.
The Cypress Hills rise to 4806 ft (1465 m) above the surrounding plain. Having escaped the last glaciation they contain unique plant and animal life. The silvery lupine, bunchberry, and lodgepole pine all grow in the cool, moist climate of the hills.
The Alberta and Saskatchewan plains bear strong testament to past glaciations. The Assiniboine, Saskatchewan and Qu'Appelle rivers occupy flat-bottomed, steep-sided valleys eroded during the last Ice Age by glacial meltwater.
The lowlands of Manitoba are a basin that once held the vast post-glacial Lake Agassiz, remnants of which include Lake Winnipeg, Lake Winnipegosis, and Lake Manitoba.

***Ancient granite outcrops**, part of the Canadian Shield, rise above the surface of Setting Lake, which was initially formed by meltwater from the last Ice Age.*

Map key

Population

- ◉ 500,000 to 1 million
- ◎ 100,000 to 500,000
- ⊕ 50,000 to 100,000
- ○ 10,000 to 50,000
- ∘ below 10,000

Elevation

- 6000m / 19,686ft
- 4000m / 13,124ft
- 3000m / 9843ft
- 2000m / 6562ft
- 1000m / 3281ft
- 500m / 1640ft
- 250m / 820ft
- 100m / 328ft
- sea level

Scale 1:8,250,000

Km 0 25 50 100 150 200 250

Miles 0 25 50 100 150 200 250

projection: Lambert Conformal Conic

N O P Q R S T U V W X Y
TERRITORIES
NUNAVUT
Hudson Bay
ALBERTA
SASKATCHEWAN
MANITOBA
ONTARIO
UNITED STATES OF AMERICA
Lake Athabasca
Lake Winnipeg
Lake Winnipegosis
Lake Claire
Reindeer Lake
Wollaston Lake
Caribou Mountains
Birch Mountains
Cypress Hills
Edmonton
Calgary
Regina
Saskatoon
Winnipeg
Red Deer
Lethbridge
Medicine Hat
Moose Jaw
Prince Albert
Brandon
Fort McMurray
Fort Chipewyan
Uranium City
Churchill
Cape Churchill
Cape Tatnam
Thompson
The Pas
Flin Flon
Lloydminster
Grande Prairie
Peace River
Fort Vermilion
High Level
Jasper
Banff
Nelson
Kelowna
Dauphin
Portage la Prairie
Selkirk
Steinbach
Estevan
Weyburn
Yorkton
Swift Current
North Battleford
Mount Robson 3954m
Mount Columbia 3741m
Mount Assiniboine 3618m
Swan Hills
Porcupine Hills
Riding Mountain
Duck Mountain
Pasquia Hills
16

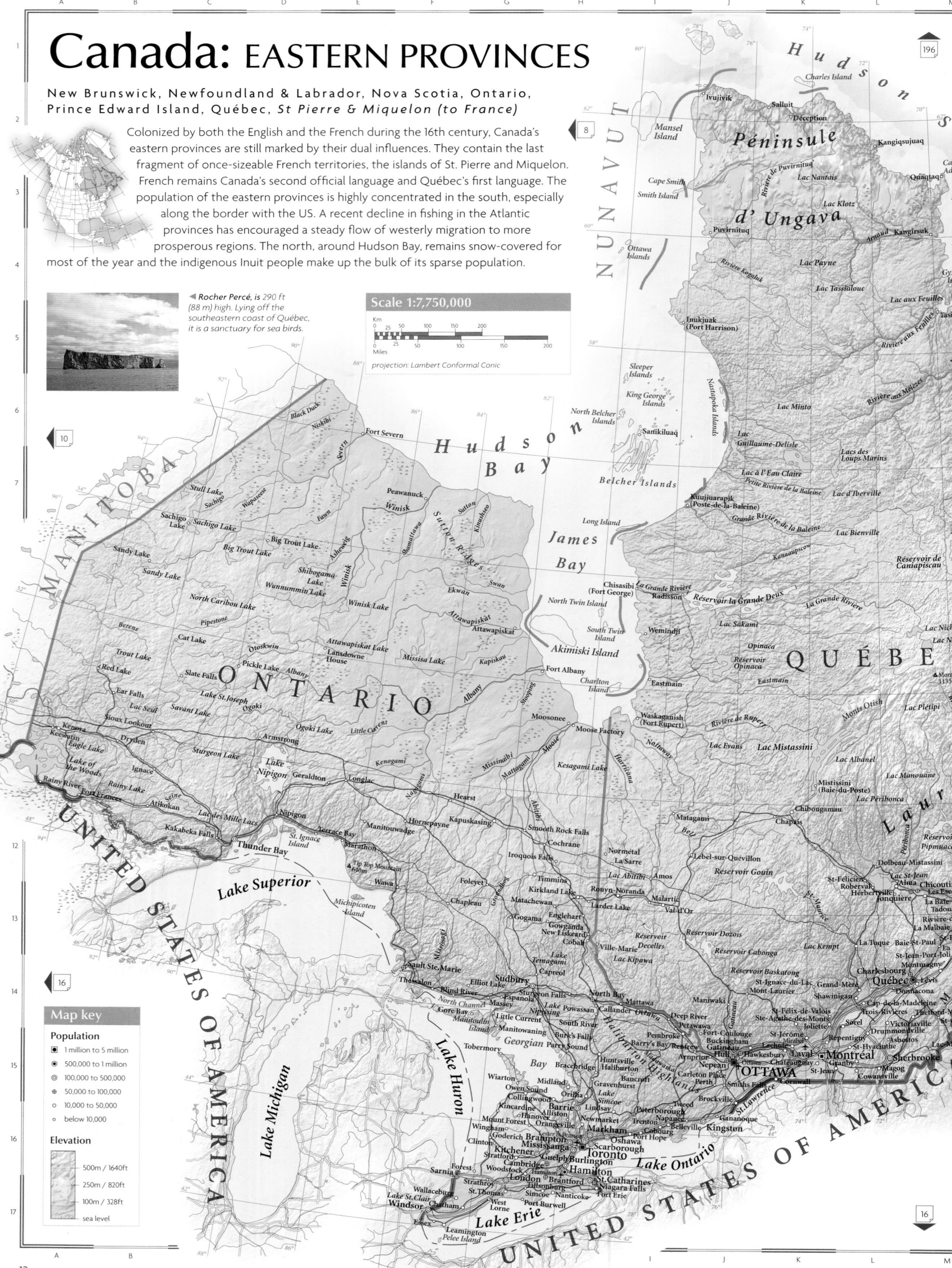

Canada: EASTERN PROVINCES

New Brunswick, Newfoundland & Labrador, Nova Scotia, Ontario, Prince Edward Island, Québec, *St Pierre & Miquelon (to France)*

Colonized by both the English and the French during the 16th century, Canada's eastern provinces are still marked by their dual influences. They contain the last fragment of once-sizeable French territories, the islands of St. Pierre and Miquelon. French remains Canada's second official language and Québec's first language. The population of the eastern provinces is highly concentrated in the south, especially along the border with the US. A recent decline in fishing in the Atlantic provinces has encouraged a steady flow of westerly migration to more prosperous regions. The north, around Hudson Bay, remains snow-covered for most of the year and the indigenous Inuit people make up the bulk of its sparse population.

◀ *Rocher Percé, is 290 ft (88 m) high. Lying off the southeastern coast of Québec, it is a sanctuary for sea birds.*

Scale 1:7,750,000

Km
0 25 50   100   150   200

Miles
0 25   50   100   150   200

projection: Lambert Conformal Conic

Map key

Population
- ▣ 1 million to 5 million
- ◉ 500,000 to 1 million
- ◎ 100,000 to 500,000
- ⊕ 50,000 to 100,000
- ○ 10,000 to 50,000
- ∘ below 10,000

Elevation
- 500m / 1640ft
- 250m / 820ft
- 100m / 328ft
- sea level

12

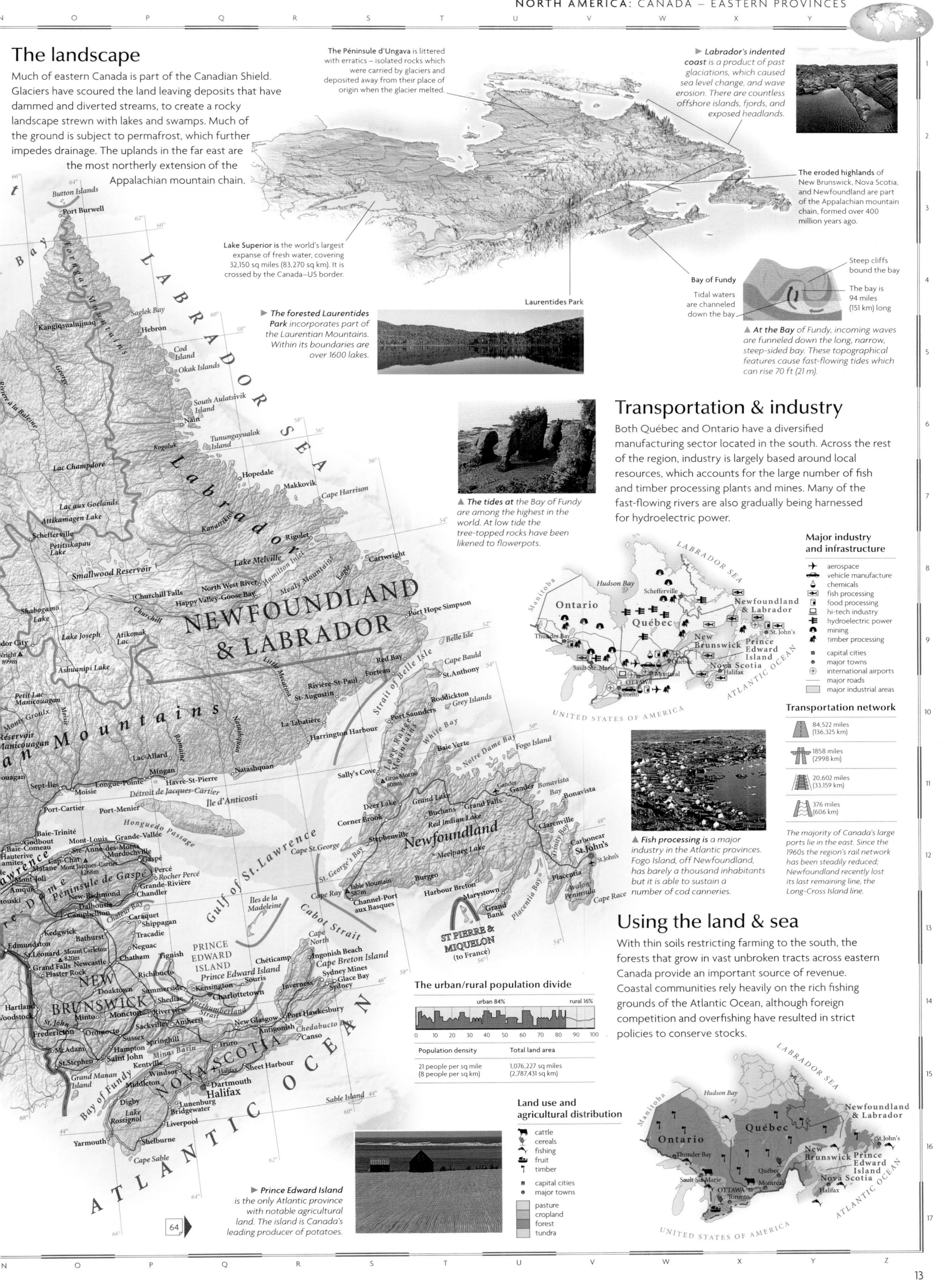

The landscape

Much of eastern Canada is part of the Canadian Shield. Glaciers have scoured the land leaving deposits that have dammed and diverted streams, to create a rocky landscape strewn with lakes and swamps. Much of the ground is subject to permafrost, which further impedes drainage. The uplands in the far east are the most northerly extension of the Appalachian mountain chain.

The **Péninsule d'Ungava** is littered with erratics – isolated rocks which were carried by glaciers and deposited away from their place of origin when the glacier melted.

► **Labrador's indented coast** is a product of past glaciations, which caused sea level change, and wave erosion. There are countless offshore islands, fjords, and exposed headlands.

The **eroded highlands** of New Brunswick, Nova Scotia, and Newfoundland are part of the Appalachian mountain chain, formed over 400 million years ago.

Lake Superior is the world's largest expanse of fresh water, covering 32,150 sq miles (83,270 sq km). It is crossed by the Canada–US border.

Laurentides Park

► **The forested Laurentides Park** incorporates part of the Laurentian Mountains. Within its boundaries are over 1600 lakes.

Bay of Fundy
Tidal waters are channeled down the bay

Steep cliffs bound the bay

The bay is 94 miles (151 km) long

▲ **At the Bay** of Fundy, incoming waves are funneled down the long, narrow, steep-sided bay. These topographical features cause fast-flowing tides which can rise 70 ft (21 m).

▲ **The tides at** the Bay of Fundy are among the highest in the world. At low tide the tree-topped rocks have been likened to flowerpots.

Transportation & industry

Both Québec and Ontario have a diversified manufacturing sector located in the south. Across the rest of the region, industry is largely based around local resources, which accounts for the large number of fish and timber processing plants and mines. Many of the fast-flowing rivers are also gradually being harnessed for hydroelectric power.

Major industry and infrastructure
- ✈ aerospace
- 🚗 vehicle manufacture
- chemicals
- 🐟 fish processing
- food processing
- hi-tech industry
- hydroelectric power
- mining
- timber processing

- ● capital cities
- ● major towns
- ✈ international airports
- — major roads
- ▢ major industrial areas

Transportation network
- 84,522 miles (136,325 km)
- 1858 miles (2998 km)
- 20,602 miles (33,159 km)
- 376 miles (606 km)

The majority of Canada's large ports lie in the east. Since the 1960s the region's rail network has been steadily reduced; Newfoundland recently lost its last remaining line, the Long-Cross Island line.

▲ **Fish processing is** a major industry in the Atlantic provinces. Fogo Island, off Newfoundland, has barely a thousand inhabitants but it is able to sustain a number of cod canneries.

Using the land & sea

With thin soils restricting farming to the south, the forests that grow in vast unbroken tracts across eastern Canada provide an important source of revenue. Coastal communities rely heavily on the rich fishing grounds of the Atlantic Ocean, although foreign competition and overfishing have resulted in strict policies to conserve stocks.

The urban/rural population divide
urban 84%   rural 16%

0 10 20 30 40 50 60 70 80 90 100

Population density	Total land area
21 people per sq mile (8 people per sq km)	1,076,227 sq miles (2,787,431 sq km)

Land use and agricultural distribution
- cattle
- cereals
- fishing
- fruit
- timber
- ■ capital cities
- ● major towns
- pasture
- cropland
- forest
- tundra

► **Prince Edward Island** is the only Atlantic province with notable agricultural land. The island is Canada's leading producer of potatoes.

Southeastern Canada

Southern Ontario, Southern Québec

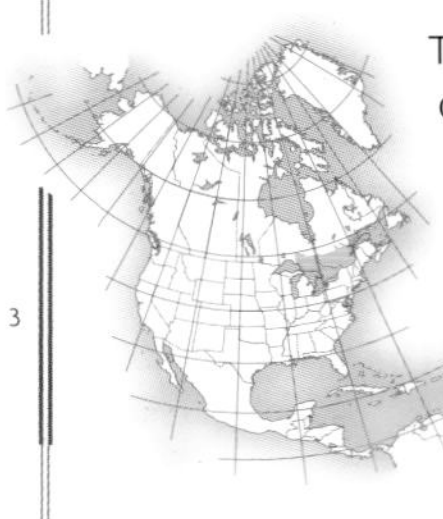

The southern parts of Québec and Ontario form the economic heart of Canada. The two provinces are divided by their language and culture; in Québec, French is the main language, whereas English is spoken in Ontario. Separatist sentiment in Québec has led to a provincial referendum on the question of a sovereignty association with Canada. The region contains Canada's capital, Ottawa, and its two largest cities: Toronto, the center of commerce, and Montréal, the cultural and administrative heart of French Canada.

▲ The port at Montréal is situated on the St. Lawrence Seaway. A network of 16 locks allows oceangoing vessels access to routes once plied by fur-trappers and early settlers.

▶ Niagara Falls lies on the border between Canada and the US. It comprises a system of two falls: American Falls, in New York, is separated from Horseshoe Falls, in Ontario, by Goat Island. Horseshoe Falls, seen here, plunges 184 ft (56 m) and is 2500 ft (762 m) wide.

Transportation & industry

The cities of southern Québec and Ontario, and their hinterlands, form the heart of Canadian manufacturing industry. Toronto is Canada's leading financial center, and Ontario's motor and aerospace industries have developed around the city. A major center for nickel mining lies to the north of Toronto. Most of Québec's industry is located in Montréal, the oldest port in North America. Chemicals, paper manufacture, and the construction of transportation equipment are leading industrial activities.

Major industry and infrastructure

- car manufacture
- chemicals
- engineering
- finance
- food processing
- hi-tech industry
- mining
- iron & steel
- textiles
- paper industry
- timber processing
- capital cities
- major towns
- international airports
- major roads
- major industrial areas

Transportation network

The opening of the St. Lawrence Seaway in 1959 finally allowed oceangoing ships (up to 24,000 tons [tonnes]) access to the interior of Canada, creating a vital trading route.

Map key

Population

- 1 million to 5 million
- 500,000 to 1 million
- 100,000 to 500,000
- 50,000 to 100,000
- 10,000 to 50,000
- below 10,000

Elevation

- 500m / 1640ft
- 250m / 820ft
- 100m / 328ft
- sea level

▶ Montréal, on the banks of the St. Lawrence River, is Québec's leading metropolitan center and one of Canada's two largest cities – Toronto is the other. Montréal clearly reflects French culture and traditions.

14

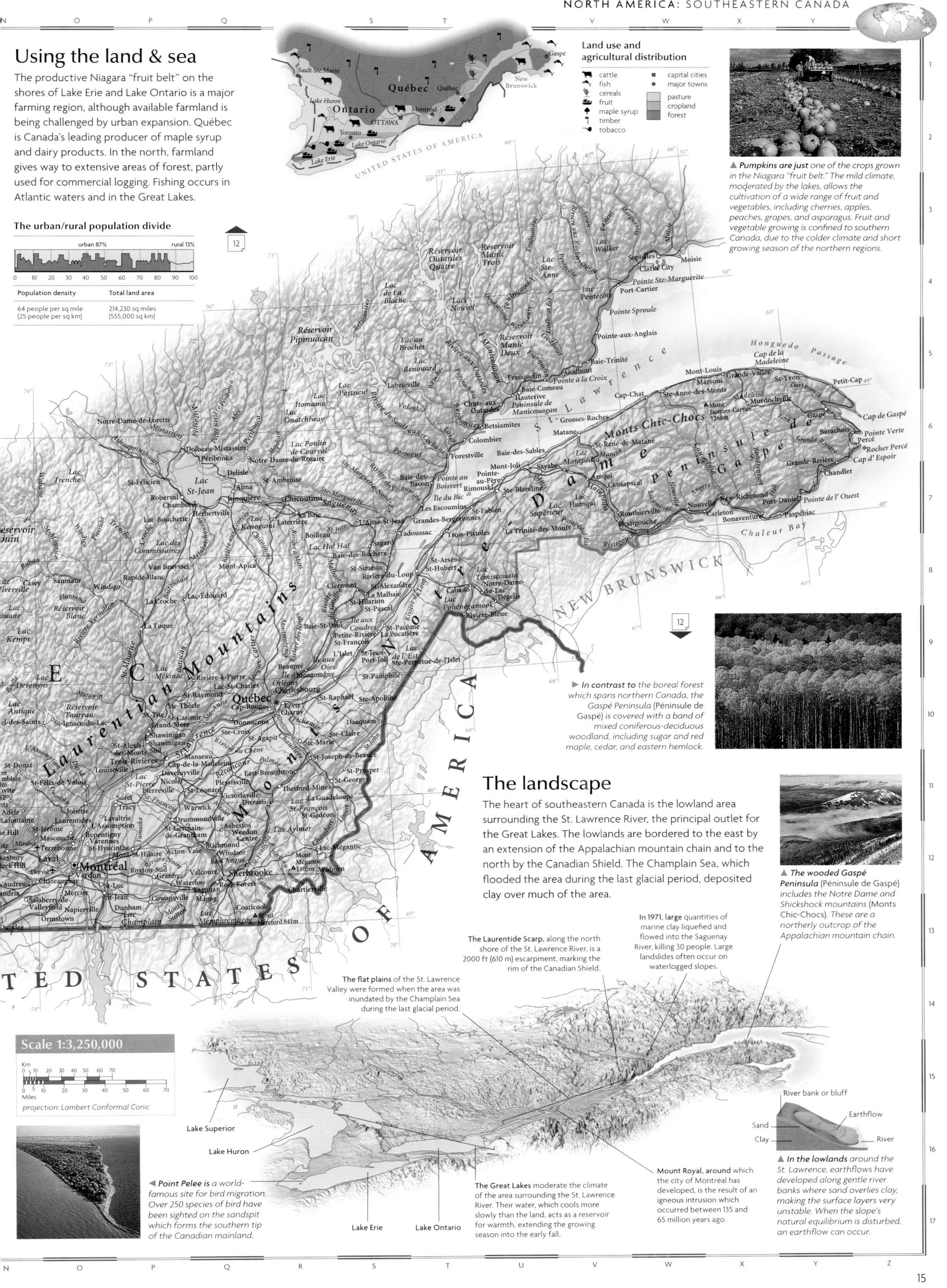

Using the land & sea

The productive Niagara "fruit belt" on the shores of Lake Erie and Lake Ontario is a major farming region, although available farmland is being challenged by urban expansion. Québec is Canada's leading producer of maple syrup and dairy products. In the north, farmland gives way to extensive areas of forest, partly used for commercial logging. Fishing occurs in Atlantic waters and in the Great Lakes.

The urban/rural population divide

urban 87%   rural 13%

0 10 20 30 40 50 60 70 80 90 100

Population density	Total land area
64 people per sq mile (25 people per sq km)	214,230 sq miles (555,000 sq km)

Land use and agricultural distribution

- cattle
- fish
- cereals
- fruit
- maple syrup
- timber
- tobacco
- capital cities
- major towns
- pasture
- cropland
- forest

▲ **Pumpkins are just** one of the crops grown in the Niagara "fruit belt." The mild climate, moderated by the lakes, allows the cultivation of a wide range of fruit and vegetables, including cherries, apples, peaches, grapes, and asparagus. Fruit and vegetable growing is confined to southern Canada, due to the colder climate and short growing season of the northern regions.

▶ **In contrast to** the boreal forest which spans northern Canada, the Gaspé Peninsula (Péninsule de Gaspé) is covered with a band of mixed coniferous-deciduous woodland, including sugar and red maple, cedar, and eastern hemlock.

The landscape

The heart of southeastern Canada is the lowland area surrounding the St. Lawrence River, the principal outlet for the Great Lakes. The lowlands are bordered to the east by an extension of the Appalachian mountain chain and to the north by the Canadian Shield. The Champlain Sea, which flooded the area during the last glacial period, deposited clay over much of the area.

▲ **The wooded Gaspé Peninsula** (Péninsule de Gaspé) includes the Notre Dame and Shickshock mountains (Monts Chic-Chocs). These are a northerly outcrop of the Appalachian mountain chain.

In 1971, large quantities of marine clay liquefied and flowed into the Saguenay River, killing 30 people. Large landslides often occur on waterlogged slopes.

The Laurentide Scarp, along the north shore of the St. Lawrence River, is a 2000 ft (610 m) escarpment, marking the rim of the Canadian Shield.

The flat plains of the St. Lawrence Valley were formed when the area was inundated by the Champlain Sea during the last glacial period.

Scale 1:3,250,000

projection: Lambert Conformal Conic

◀ **Point Pelee is** a world-famous site for bird migration. Over 250 species of bird have been sighted on the sandspit which forms the southern tip of the Canadian mainland.

The Great Lakes moderate the climate of the area surrounding the St. Lawrence River. Their water, which cools more slowly than the land, acts as a reservoir for warmth, extending the growing season into the early fall.

Mount Royal, around which the city of Montreal has developed, is the result of an igneous intrusion which occurred between 135 and 65 million years ago.

▲ **In the lowlands** around the St. Lawrence, earthflows have developed along gentle river banks where sand overlies clay, making the surface layers very unstable. When the slope's natural equilibrium is disturbed, an earthflow can occur.

15

The United States of America

COTERMINOUS US (FOR ALASKA AND HAWAII SEE PAGES 38-39)

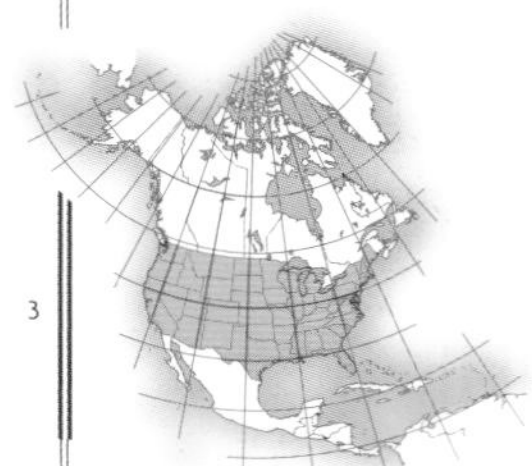

The US's progression from frontier territory to economic and political superpower has taken less than 200 years. The 48 coterminous states, along with the outlying states of Alaska and Hawaii, are part of a federal union, held together by the guiding principles of the US Constitution, which embodies the ideals of democracy and liberty for all. Abundant fertile land and a rich resource base fueled and sustained US economic development. With the spread of agriculture and the growth of trade and industry came the need for a larger workforce, which was supplied by millions of immigrants, many seeking an escape from poverty and political or religious persecution. Immigration continues today, particularly from Central America and Asia.

▲ *Washington DC was* established as the site for the nation's capital in 1790. It is home to the seat of national government, on Capitol Hill, as well as the President's official residence, the White House.

▲ *Mount Rainier is a* dormant volcano in the Cascade Range, Washington. This 14,090 ft (4392 m) peak is flanked by the most extensive glacier outside Alaska.

▶ *The clear waters* of Niagara Falls cascade 190 ft (58 m) into the gorge below. It is one of America's most famous spectacles and a leading tourist attraction. The falls are slowly receding and the gorge may one day stretch from Lake Ontario to Lake Erie.

Scale 1:12,700,000

projection: Lambert Azimuthal Equal Area

Transportation & industry

The US has been the industrial powerhouse of the world since the Second World War, pioneering mass-production and the consumer lifestyle. Initially, heavy engineering and manufacturing in the northeast led the economy. Today, heavy industry has declined and the US economy is driven by service and financial industries, with the most important being defense, hi-tech, and electronics.

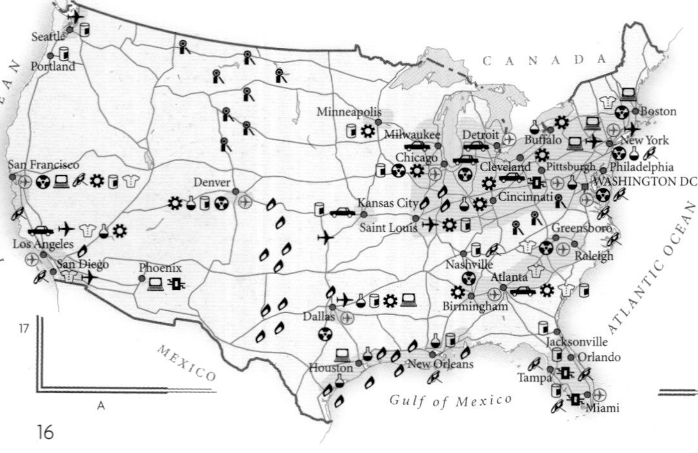

Transportation network

🛣	3,875,040 miles (6,240,000 km)	🛣	52,388 miles (84,361 km)
🚂	148,308 miles (235,238 km)	✈	25,467 miles (41,009 km)

Transportation in the US is dominated by the car which, with the extensive Interstate Highway system, allows great personal mobility. Today, internal air flights between major cities provide the most rapid cross-country travel.

Major industry and infrastructure

- ✈ aerospace
- 🚗 car manufacture
- 🧪 chemicals
- ⛏ coal
- 💻 electronics
- ⚙ engineering
- 🍴 food processing
- 🖥 hi-tech industry
- ⛽ oil & gas
- ⊗ research & development
- textiles
- tourism
- ■ capital cities
- ⊕ major towns
- ✈ international airports
- major roads
- major industrial areas

The landscape

The high, rugged mountain ranges of the west are about 80 million years old, geologically young compared to the old, eroded, Appalachian mountain chain, which dates from when North America and Europe were joined together as part of the supercontinent Pangaea, 400 million years ago. In contrast, the Great Plains and Mississippi Basin have a low relief and fertile soils.

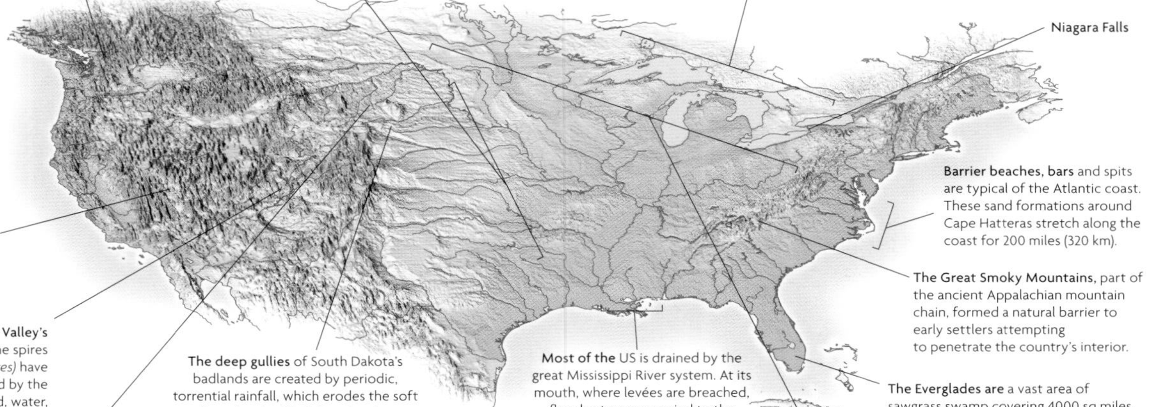

Mount Rainier

Great Plains

The Great Lakes

Niagara Falls

Death Valley, California, 282 ft (86 m) below sea level, is the lowest point in the western hemisphere, and one of the hottest places on Earth. Temperatures of 135° F (57° C) have been recorded here.

Monument Valley's striking sandstone spires and pillars *(buttes)* have been formed by the action of wind, water, heat, and cold.

The deep gullies of South Dakota's badlands are created by periodic, torrential rainfall, which erodes the soft soils and rocks. Their form has been greatly affected by changes in land use.

Most of the US is drained by the great Mississippi River system. At its mouth, where levées are breached, floodwaters are carried to the swamps through a series of channels. This region is known as the bayou.

Barrier beaches, bars and spits are typical of the Atlantic coast. These sand formations around Cape Hatteras stretch along the coast for 200 miles (320 km).

The Great Smoky Mountains, part of the ancient Appalachian mountain chain, formed a natural barrier to early settlers attempting to penetrate the country's interior.

The Everglades are a vast area of sawgrass swamp covering 4000 sq miles (10,300 sq km) of southern Florida.

Missouri River
Ohio River
Mississippi River
Mississippi Delta

▶ *Devils Tower, in* Wyoming is a 1280 ft (390 m) intrusion of basalt rock, which cooled to form octagonal pillars. In 1906 it became the first US National Monument.

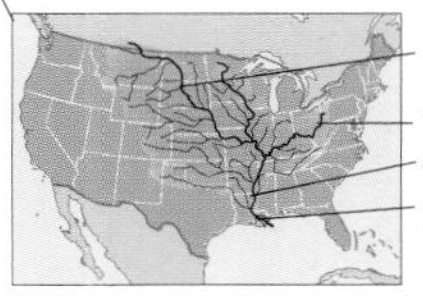

▲ *The massive drainage* basin of the Mississippi covers 1,250,000 sq miles (3,200,000 sq km). It includes all areas drained by the Mississippi and its chief tributaries, the Missouri and Ohio Rivers, and drains the entire region from the Appalachians to the Rockies.

Map key

Population

■ above 5 million
■ 1 million to 5 million
◉ 500,000 to 1 million
◎ 100,000 to 500,000
⊕ 50,000 to 100,000
○ 10,000 to 50,000
○ below 10,000

Elevation

4000m / 13,124ft
3000m / 9843ft
2000m / 6562ft
1000m / 3281ft
500m / 1640ft
250m / 820ft
100m / 328ft
sea level

Using the land and sea

Over half of the US is used for agriculture, typified by the large cereal grain farms and cattle ranches of the Great Plains and Midwest prairie regions. Although wheat and corn are still primary crops, a diverse range of fruits and vegetables are grown in the fertile areas, particularly near the east and west coasts. Despite the abundance of cultivable land, inadequate soil management has resulted in a third of the topsoil being lost through wind and water erosion.

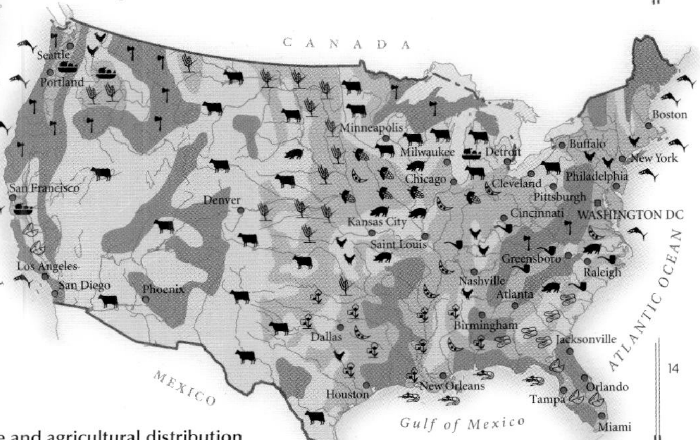

Land use and agricultural distribution

- 🐄 cattle
- 🐖 pigs
- 🦃 poultry
- 🍊 citrus fruits
- cotton
- 🐟 fishing
- 🍎 fruit
- 🌽 corn
- peanuts
- 🦪 shellfish
- soybeans
- 🌲 timber
- tobacco
- 🌾 wheat

- ● capital cities
- ● major towns

- pasture
- cropland
- forest
- wetland
- desert
- mountain region

The urban/rural population divide

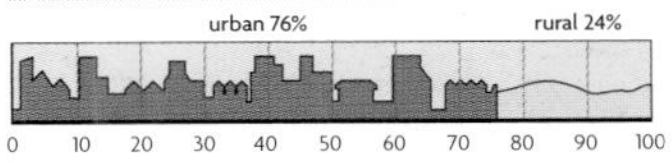

urban 76%   rural 24%

0 10 20 30 40 50 60 70 80 90 100

Population density	Total land area
98 people per sq mile (38 people per sq km)	2,959,045 sq miles (7,663,631 sq km)

◀ *Farming on the* Great Plains and in the Midwest is characterized by large-scale, mechanized wheat farms.

▶ *Fakahatchee Strand is* part of the extensive subtropical swamps in the Florida Everglades. The swamps support a wide variety of animal life, including many rare birds, fish, alligators, and crocodiles.

17

USA: NORTHEASTERN STATES

Connecticut, Maine, Massachusetts, New Hampshire, New Jersey, New York, Pennsylvania, Rhode Island, Vermont

The indented coast and vast woodlands of the northeastern states were the original core area for European expansion. The rustic character of New England prevails after nearly four centuries, while the great cities of the Atlantic seaboard have formed an almost continuous urban region. Over 20 million immigrants entered New York from 1855 to 1924 and the northeast became the industrial center of the US. After the decline of mining and heavy manufacturing, economic dynamism has been restored with the growth of hi-tech and service industries.

▲ *Chelsea in Vermont,* surrounded by trees in their fall foliage. Tourism and agriculture dominate the economy of this self-consciously rural state, where no town exceeds 30,000 people.

Transportation & industry

The principal seaboard cities grew up on trade and manufacturing. They are now global centers of commerce and corporate administration, dominating the regional economy. Research and development facilities support an expanding electronics and communications sector throughout the region. Pharmaceutical and chemical industries are important in New Jersey and Pennsylvania.

Map key

Population
- ◼ above 5 million
- ◼ 1 million to 5 million
- ⊙ 500,000 to 1 million
- ⊚ 100,000 to 500,000
- ⊙ 50,000 to 100,000
- ○ 10,000 to 50,000
- ◦ below 10,000

Elevation
- 1000m / 3281ft
- 500m / 1640ft
- 250m / 820ft
- 100m / 328ft
- sea level

Transportation network

- 340,090 miles (544,144 km)
- 4813 miles (7700 km)
- 12,872 miles (20,592 km)
- 2108 miles (3389 km)

New York's commercial success is tied historically to its transportation connections. The Erie Canal, completed in 1825, opened up the Great Lakes and the interior to New York's markets and carried a stream of immigrants into the Midwest.

Major industry and infrastructure

- ⚒ chemicals
- coal
- defense
- electronics
- engineering
- finance
- hi-tech industry
- iron & steel
- pharmaceuticals
- printing & publishing
- research & development
- textiles
- timber processing
- ● major towns
- ⊕ international airports
- — major roads
- ▢ major industrial area

18

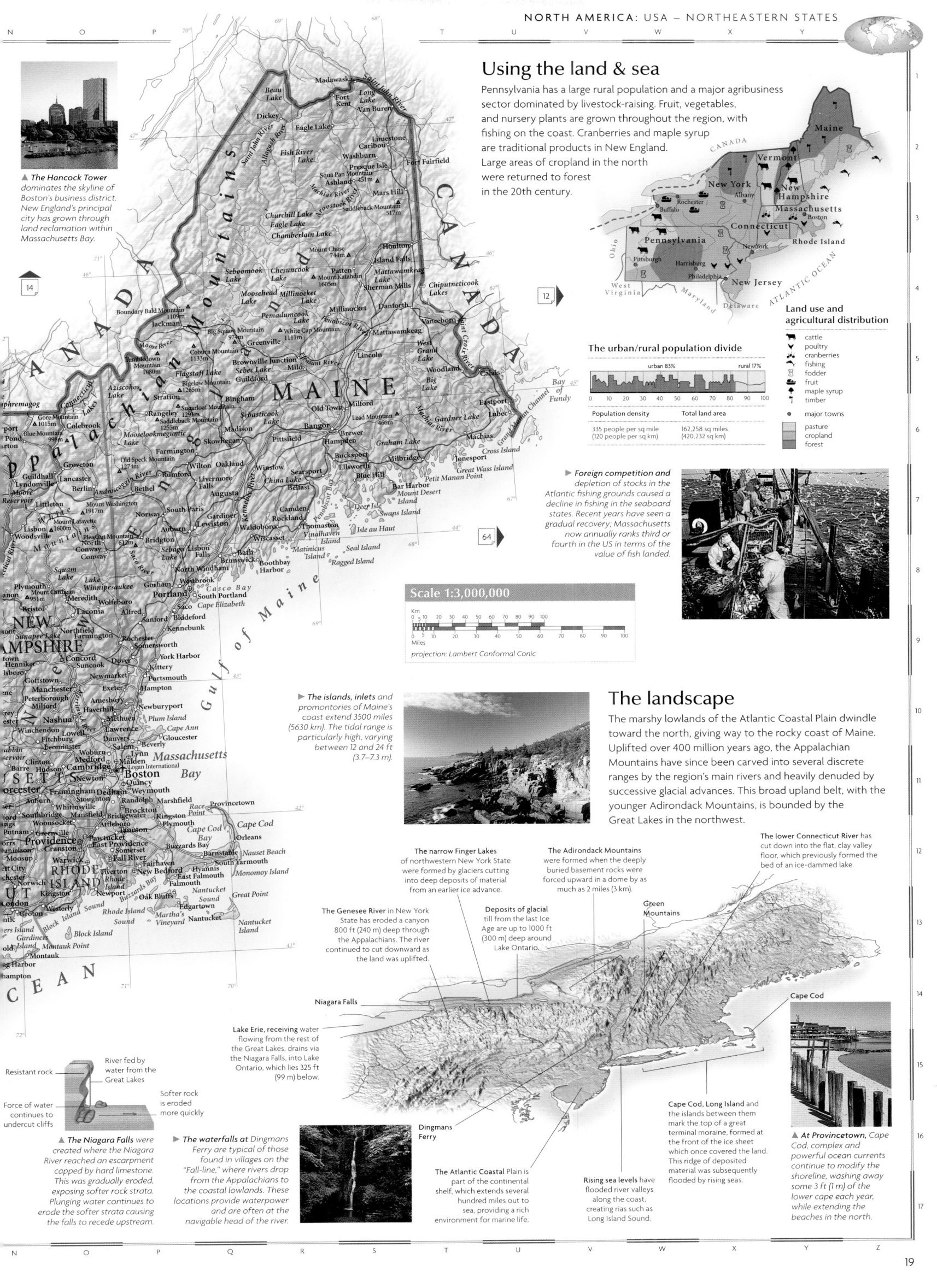

▲ The Hancock Tower dominates the skyline of Boston's business district. New England's principal city has grown through land reclamation within Massachusetts Bay.

Using the land & sea

Pennsylvania has a large rural population and a major agribusiness sector dominated by livestock-raising. Fruit, vegetables, and nursery plants are grown throughout the region, with fishing on the coast. Cranberries and maple syrup are traditional products in New England. Large areas of cropland in the north were returned to forest in the 20th century.

Land use and agricultural distribution

- 🐄 cattle
- 🐓 poultry
- cranberries
- 🐟 fishing
- fodder
- 🍎 fruit
- maple syrup
- timber
- • major towns
- pasture
- cropland
- forest

The urban/rural population divide

urban 83%   rural 17%

0 10 20 30 40 50 60 70 80 90 100

Population density	Total land area
335 people per sq mile (120 people per sq km)	162,258 sq miles (420,232 sq km)

▶ Foreign competition and depletion of stocks in the Atlantic fishing grounds caused a decline in fishing in the seaboard states. Recent years have seen a gradual recovery; Massachusetts now annually ranks third or fourth in the US in terms of the value of fish landed.

Scale 1:3,000,000

Km
0 5 10 20 30 40 50 60 70 80 90 100

Miles
0 5 10 20 30 40 50 60 70 80 90 100

projection: Lambert Conformal Conic

▶ The islands, inlets and promontories of Maine's coast extend 3500 miles (5630 km). The tidal range is particularly high, varying between 12 and 24 ft (3.7–7.3 m).

The landscape

The marshy lowlands of the Atlantic Coastal Plain dwindle toward the north, giving way to the rocky coast of Maine. Uplifted over 400 million years ago, the Appalachian Mountains have since been carved into several discrete ranges by the region's main rivers and heavily denuded by successive glacial advances. This broad upland belt, with the younger Adirondack Mountains, is bounded by the Great Lakes in the northwest.

The narrow Finger Lakes of northwestern New York State were formed by glaciers cutting into deep deposits of material from an earlier ice advance.

The Adirondack Mountains were formed when the deeply buried basement rocks were forced upward in a dome by as much as 2 miles (3 km).

The lower Connecticut River has cut down into the flat, clay valley floor, which previously formed the bed of an ice-dammed lake.

The Genesee River in New York State has eroded a canyon 800 ft (240 m) deep through the Appalachians. The river continued to cut downward as the land was uplifted.

Deposits of glacial till from the last Ice Age are up to 1000 ft (300 m) deep around Lake Ontario.

Green Mountains

Niagara Falls

Cape Cod

Lake Erie, receiving water flowing from the rest of the Great Lakes, drains via the Niagara Falls, into Lake Ontario, which lies 325 ft (99 m) below.

Cape Cod, Long Island and the islands between them mark the top of a great terminal moraine, formed at the front of the ice sheet which once covered the land. This ridge of deposited material was subsequently flooded by rising seas.

Dingmans Ferry

The Atlantic Coastal Plain is part of the continental shelf, which extends several hundred miles out to sea, providing a rich environment for marine life.

Rising sea levels have flooded river valleys along the coast, creating rias such as Long Island Sound.

Resistant rock

River fed by water from the Great Lakes

Force of water continues to undercut cliffs

Softer rock is eroded more quickly

▲ The Niagara Falls were created where the Niagara River reached an escarpment capped by hard limestone. This was gradually eroded, exposing softer rock strata. Plunging water continues to erode the softer strata causing the falls to recede upstream.

▶ The waterfalls at Dingmans Ferry are typical of those found in villages on the "Fall-line," where rivers drop from the Appalachians to the coastal lowlands. These locations provide waterpower and are often at the navigable head of the river.

▲ At Provincetown, Cape Cod, complex and powerful ocean currents continue to modify the shoreline, washing away some 3 ft (1 m) of the lower cape each year, while extending the beaches in the north.

USA: MID-EASTERN STATES

Delaware, District of Columbia, Kentucky, Maryland, North Carolina, South Carolina, Tennessee, Virginia, West Virginia

Key events in American history took place in this diverse region, which became the front line between the North and the South during the Civil War of the 1860s. Strong regional contrasts exist between the fertile coastal plains, the isolated upcountry of the Appalachian Mountains, and the cotton-growing areas of the Mississippi lowlands to the west. While coal mining, a traditional industry in the Appalachians, has declined in recent years leaving much rural poverty, service industries elsewhere have increased, especially in Washington DC, the nation's capital.

Map key

Population

- ⊙ 500,000 to 1 million
- ⊚ 100,000 to 500,000
- ⊕ 50,000 to 100,000
- ○ 10,000 to 50,000
- ∘ below 10,000

Elevation

- 6000m / 19,686ft
- 4000m / 13,124ft
- 3000m / 9843ft
- 2000m / 6562ft
- 1000m / 3281ft
- 500m / 1640ft
- 250m / 820ft
- 100m / 328ft
- sea level

Scale 1:3,250,000

Km 0 10 20 30 40 50 60 70 80
Miles 0 5 10 20 30 40 50 60 70 80

projection: Lambert Conformal Conic

▲ The Bluegrass region of Kentucky centers on the town of Lexington. This exceptionally fertile rolling plain is well known for its thoroughbred horse-breeding ranches.

Transportation & industry

In the urbanized northeast, manufacturing remains important, alongside a burgeoning service sector. North Carolina is a major center for industrial research and development. Traditional industries include Tennessee whiskey and textiles in South Carolina. The decline of open-pit coal mining in the Appalachians has been hastened by environmental controls, although adventure-tourism is a flourishing new industry.

Major industry and infrastructure

- adventure-tourism
- car manufacture
- coal
- electronics
- engineering
- finance
- food processing
- hi-tech industry
- mining
- research & development
- textiles
- capital cities
- major towns
- international airports
- major roads
- major industrial areas

Transportation network

452,218 miles (723,548 km)		5737 miles (8267 km)	
18,336 miles (29,503 km)		4404 miles (7081 km)	

Tennessee's rivers are part of an important inland bulk transportation network. Memphis connects with New Orleans in the south, and with cities as distant as Minneapolis, Sioux City, Chicago, and Pittsburgh, via the Mississippi and its tributaries.

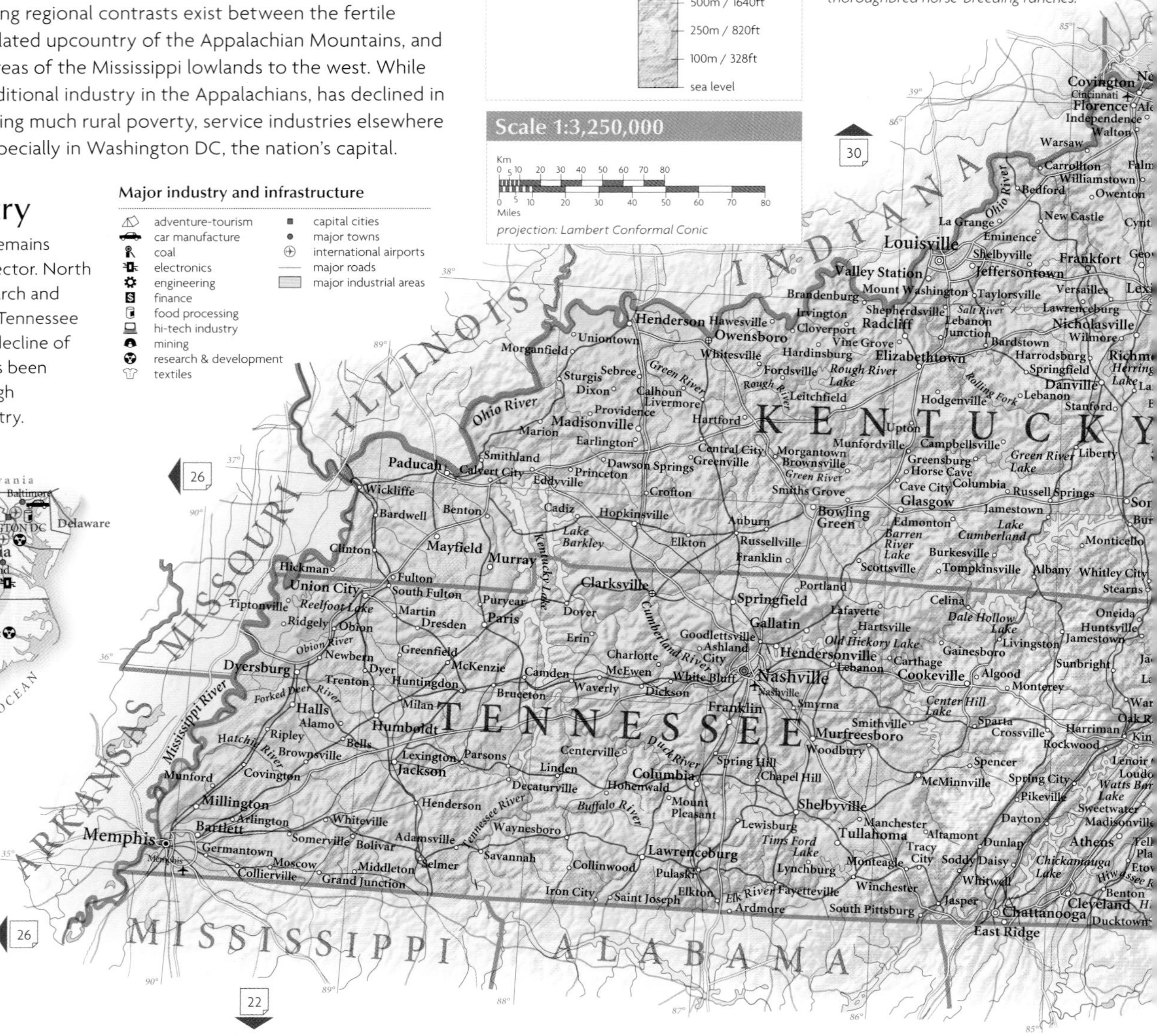

The landscape

The eastern tributaries of the Mississippi drain the interior lowlands. The Cumberland Plateau and the parallel ranges of the Appalachians have been successively uplifted and eroded over time, with the eastern side reduced to a series of foothills known as the Piedmont. The broad coastal plain gradually falls away into salt marshes, lagoons, and offshore bars, broken by flooded estuaries along the shores of the Atlantic.

Natural Bridge in eastern Kentucky is an arch 78 ft (26 m) long and 65 ft (20 m) high. It has been shaped from resistant sandstone by gradual weathering processes, which removed the softer rock lying underneath.

The Allegheny Mountains form the northwestern edge of the Appalachian mountain chain. Continuous folding has formed rich seams of bituminous coal.

Appalachian Mountains

◀ Farmland on the eastern shores of Chesapeake Bay is sustained by artificial drainage. The area also provides refuge for a variety of waterfowl.

The many inlets of Chesapeake Bay are the flooded tributaries of the main river valley, which have been inundated by rising sea levels.

Salt marshes such as Great Dismal Swamp, develop where the coast is sheltered. Vast areas of such marshland have been reclaimed for farmland and settlement.

Cape Hatteras is the easternmost point of an offshore barrier island, a wave-deposited sand-bar which has become permanent, establishing its own vegetation.

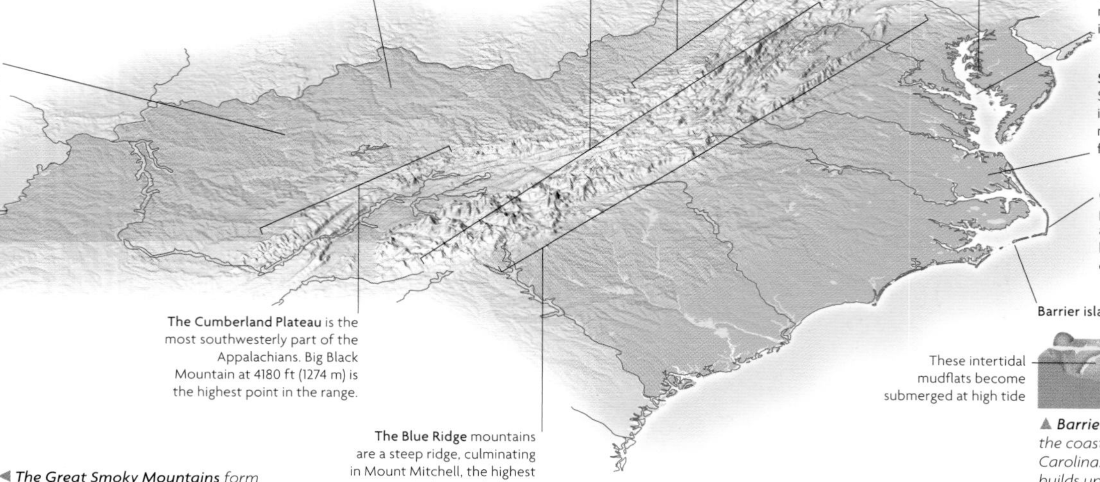

The Mammoth Cave is part of an extensive cave system in the limestone region of southwestern Kentucky. It stretches for over 300 miles (485 km) on five different levels and contains three rivers and three lakes.

The Mississippi River and its tributary the Ohio River form the western border of the region.

The Cumberland Plateau is the most southwesterly part of the Appalachians. Big Black Mountain at 4180 ft (1274 m) is the highest point in the range.

The Blue Ridge mountains are a steep ridge, culminating in Mount Mitchell, the highest point in the Appalachians, at 6684 ft (2037 m).

Barrier islands

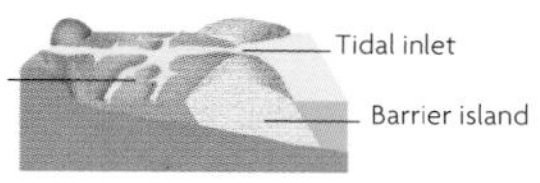

These intertidal mudflats become submerged at high tide

Tidal inlet

Barrier island

▲ Barrier islands are common along the coasts of North and South Carolina. As sea levels rise, wave action builds up ridges of sand and pebbles parallel to the coast, separated by lagoons or intertidal mud flats, which are flooded at high tide.

◀ The Great Smoky Mountains form the western escarpment of the Appalachians. The region is heavily forested, with over 130 species of tree.

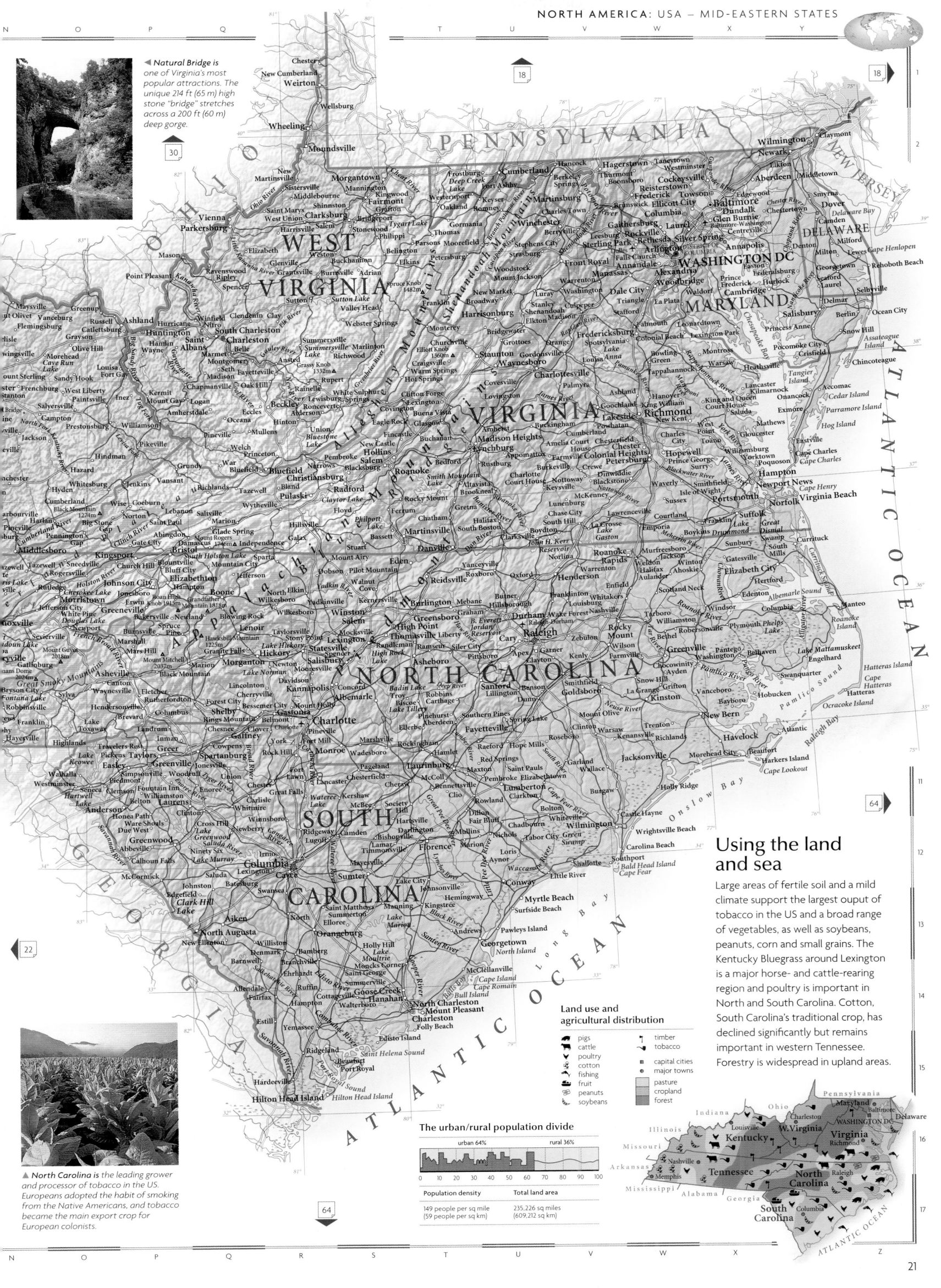

◄ *Natural Bridge* is one of Virginia's most popular attractions. The unique 214 ft (65 m) high stone "bridge" stretches across a 200 ft (60 m) deep gorge.

▲ *North Carolina is* the leading grower and processor of tobacco in the US. Europeans adopted the habit of smoking from the Native Americans, and tobacco became the main export crop for European colonists.

Using the land and sea

Large areas of fertile soil and a mild climate support the largest ouput of tobacco in the US and a broad range of vegetables, as well as soybeans, peanuts, corn and small grains. The Kentucky Bluegrass around Lexington is a major horse- and cattle-rearing region and poultry is important in North and South Carolina. Cotton, South Carolina's traditional crop, has declined significantly but remains important in western Tennessee. Forestry is widespread in upland areas.

Land use and agricultural distribution

- pigs
- cattle
- poultry
- cotton
- fishing
- fruit
- peanuts
- soybeans
- timber
- tobacco
- capital cities
- major towns
- pasture
- cropland
- forest

The urban/rural population divide

urban 64%   rural 36%

Population density	Total land area
149 people per sq mile (59 people per sq km)	235,226 sq miles (609,212 sq km)

USA: SOUTHERN STATES

Alabama, Florida, Georgia, Louisiana, Mississippi

The South has maintained a separate identity and outlook throughout the history of the US. Defeat in the Civil War (1861–65) brought chronic poverty to the former confederate states, while the subsequent liberation of four million slaves began a struggle not resolved until the 1960s, when the Civil Rights movement achieved an end to legal racial segregation. Many parts of the South have experienced rapid change. Tourism and retirement communities, together with agriculture, have fueled growth in Florida, while defense-related industries have boosted the growth of cities such as Miami and Atlanta. Many people retain a strong attachment to their history and culture, evidenced by Creole-speaking Cajuns in Louisiania and Hispanic communities in South Florida.

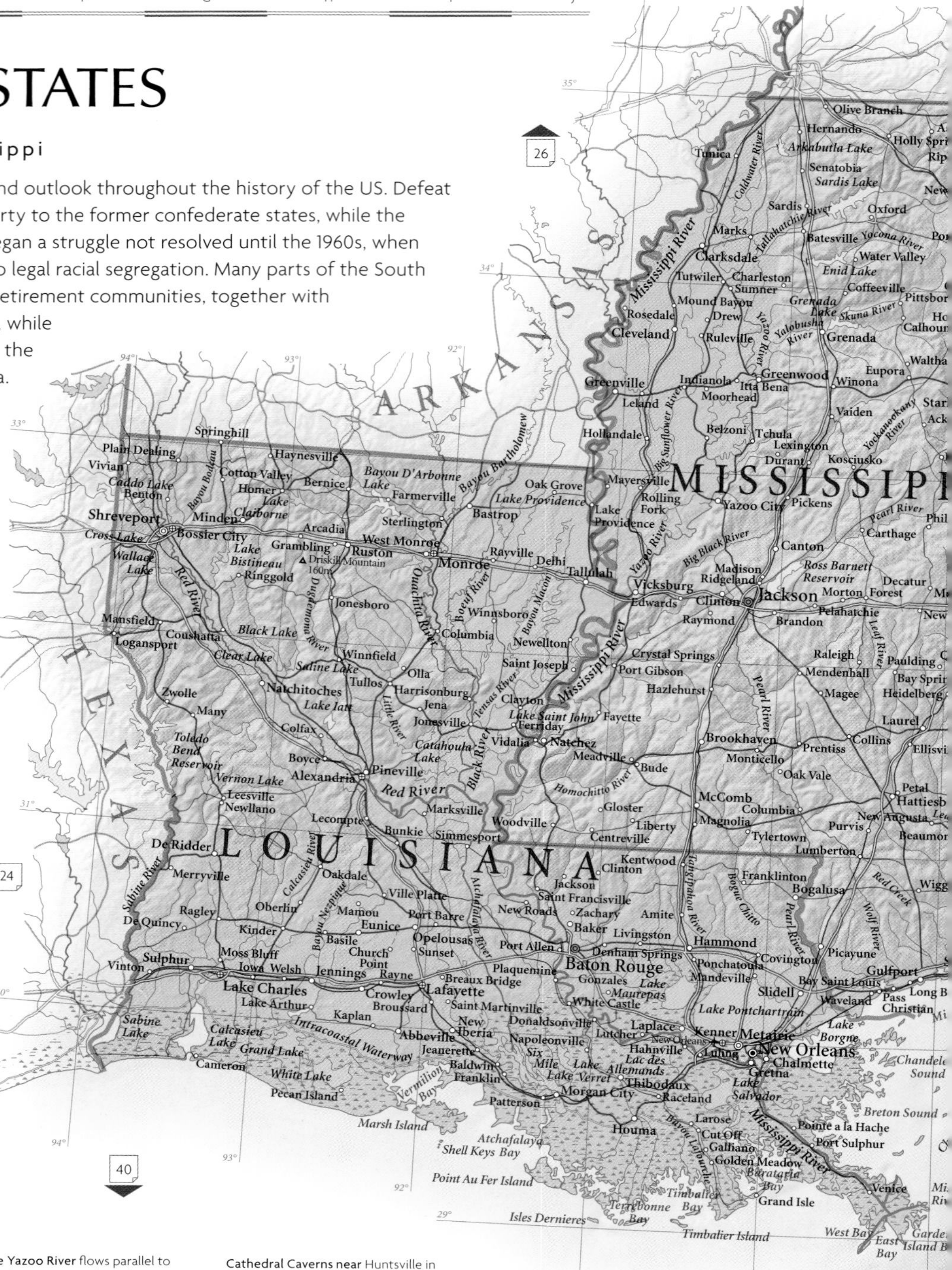

Transportation & industry

Florida's tourist trade is only part of a flourishing service sector, which has swelled the principal cities of the south. Petroleum and mineral extraction has made the Gulf Coast a major industrial region. Traditional textile production remains important in Georgia, while advanced new industries have grown from the NASA Space Program.

Transportation network

441,625 miles (706,600 km)	
5116 miles (8186 km)	
16,597 miles (26,555 km)	
6179 miles (9942 km)	

Atlanta's Hartsfield International airport is one of the busiest in the world. A dramatic rise in the use of regional air transportation has helped to integrate the major cities of the southern states.

◄ *The French Quarter is the traditional cultural center of New Orleans. The city, extensively damaged by Hurricane Katrina in 2005, once thrived on the cotton trade but now relies mainly on tourism and on oil from the Gulf of Mexico.*

Major industry and infrastructure

- ✈ aerospace
- 🚗 car manufacture
- chemicals
- coal
- defense
- electronics
- engineering
- food processing
- ♦ oil
- textiles
- tourism
- • major towns
- ⊕ international airports
- major roads
- major industrial areas

▲ *The cypress swamps of the Mississippi Delta form in the backswamps behind the levées of the river and in the multitude of subsiding delta basins.*

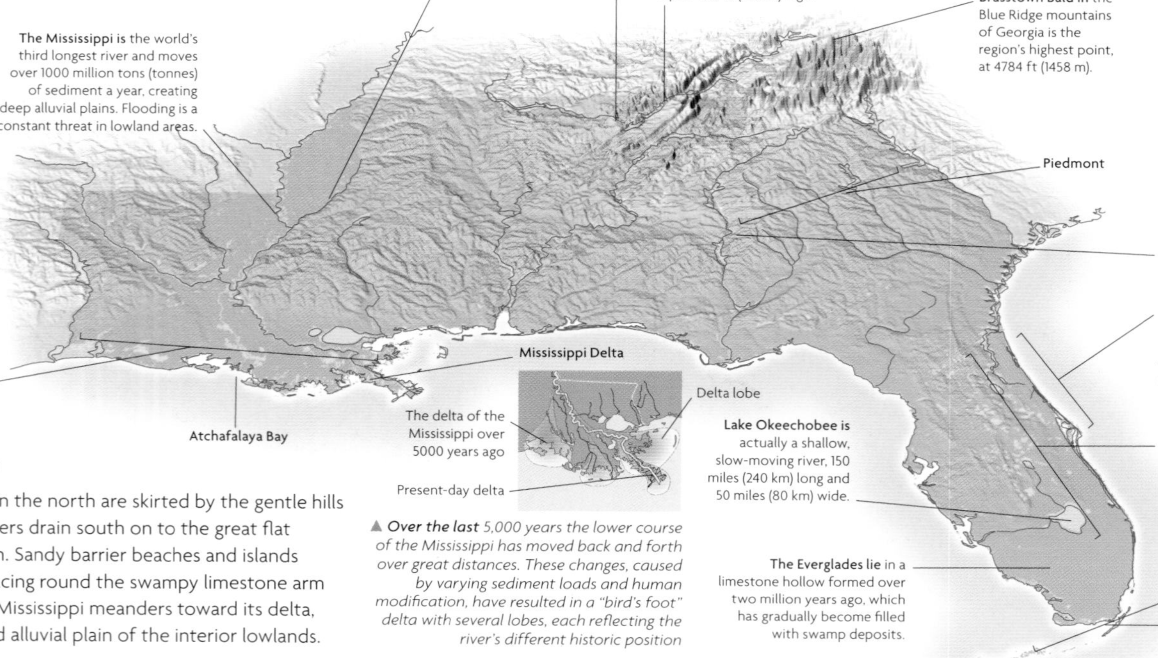

The landscape

The Blue Ridge mountains in the north are skirted by the gentle hills of the Piedmont, whose rivers drain south on to the great flat expanse of the coastal plain. Sandy barrier beaches and islands dominate the sea shore, tracing round the swampy limestone arm of Florida. In the west, the Mississippi meanders toward its delta, crossing the thickly mantled alluvial plain of the interior lowlands.

The Yazoo River flows parallel to the Mississippi through a common floodplain. The confluence of the rivers is deferred downstream because flood deposition has built the Mississippi channel up above the level of the Yazoo.

Cathedral Caverns near Huntsville in Alabama is a system of vast limestone caves, with a main opening 1000 ft (300 m) high and 150 ft (50 m) wide.

At De Soto Falls, Alabama, the Little River descends into the deepest canyon east of the Mississippi, with sheer cliff walls up to 700 ft (230 m) high.

Brasstown Bald in the Blue Ridge mountains of Georgia is the region's highest point, at 4784 ft (1458 m).

The Mississippi is the world's third longest river and moves over 1000 million tons (tonnes) of sediment a year, creating deep alluvial plains. Flooding is a constant threat in lowland areas.

Piedmont

▲ *In Providence Canyon, Georgia, the Chattahoochee River has cut straight down through the sandy bedrock, to leave sheer rock faces and pinnacles, which have been smoothed by subsequent weathering.*

Sandbars, deposited by waves breaking offshore, form barrier beaches along much of the coastline, creating sheltered lagoons and salt marshes behind them.

Across Florida the coastal plain is mostly less than 75 ft (25 m) above sea level. The land is underlain by limestone, pitted with hollows which have been filled by over 10,000 lakes.

Mississippi Delta

Atchafalaya Bay

The delta of the Mississippi over 5000 years ago

Delta lobe

Present-day delta

Lake Okeechobee is actually a shallow, slow-moving river, 150 miles (240 km) long and 50 miles (80 km) wide.

▲ *Over the last 5,000 years the lower course of the Mississippi has moved back and forth over great distances. These changes, caused by varying sediment loads and human modification, have resulted in a "bird's foot" delta with several lobes, each reflecting the river's different historic position*

The Everglades lie in a limestone hollow formed over two million years ago, which has gradually become filled with swamp deposits.

Florida Keys

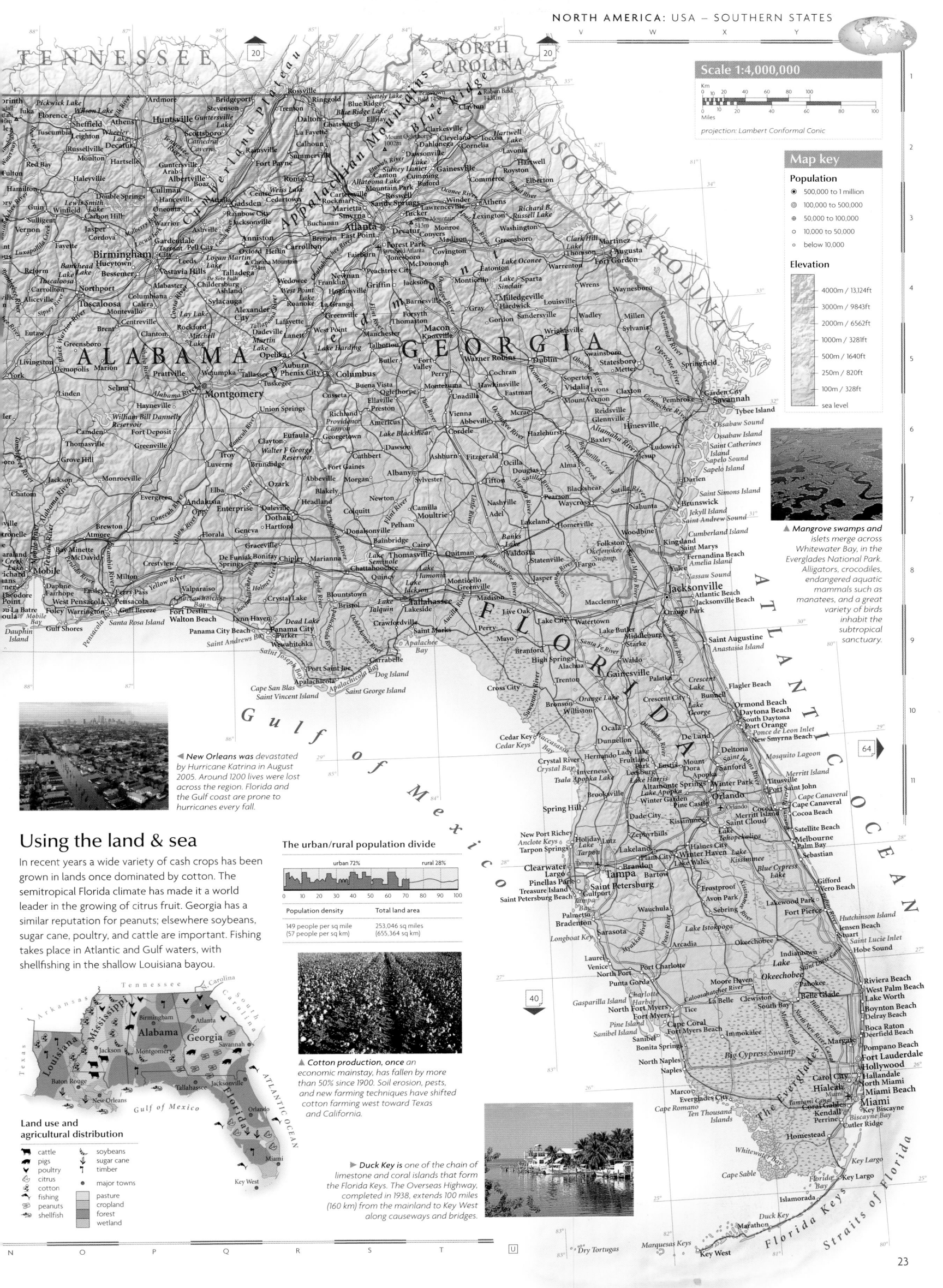

Scale 1:4,000,000

projection: Lambert Conformal Conic

Map key

Population

- 500,000 to 1 million
- 100,000 to 500,000
- 50,000 to 100,000
- 10,000 to 50,000
- below 10,000

Elevation

- 4000m / 13,124ft
- 3000m / 9843ft
- 2000m / 6562ft
- 1000m / 3281ft
- 500m / 1640ft
- 250m / 820ft
- 100m / 328ft
- sea level

▲ *Mangrove swamps and islets merge across Whitewater Bay, in the Everglades National Park. Alligators, crocodiles, endangered aquatic mammals such as manatees, and a great variety of birds inhabit the subtropical sanctuary.*

◀ *New Orleans was devastated by Hurricane Katrina in August 2005. Around 1200 lives were lost across the region. Florida and the Gulf coast are prone to hurricanes every fall.*

Using the land & sea

In recent years a wide variety of cash crops has been grown in lands once dominated by cotton. The semitropical Florida climate has made it a world leader in the growing of citrus fruit. Georgia has a similar reputation for peanuts; elsewhere soybeans, sugar cane, poultry, and cattle are important. Fishing takes place in Atlantic and Gulf waters, with shellfishing in the shallow Louisiana bayou.

The urban/rural population divide

urban 72%   rural 28%

0 10 20 30 40 50 60 70 80 90 100

Population density	Total land area
149 people per sq mile (57 people per sq km)	253,046 sq miles (655,364 sq km)

▲ *Cotton production, once an economic mainstay, has fallen by more than 50% since 1900. Soil erosion, pests, and new farming techniques have shifted cotton farming west toward Texas and California.*

▶ *Duck Key is one of the chain of limestone and coral islands that form the Florida Keys. The Overseas Highway, completed in 1938, extends 100 miles (160 km) from the mainland to Key West along causeways and bridges.*

Land use and agricultural distribution

- cattle
- pigs
- poultry
- citrus
- cotton
- fishing
- peanuts
- shellfish
- soybeans
- sugar cane
- timber
- major towns

- pasture
- cropland
- forest
- wetland

USA: TEXAS

First explored by Spaniards moving north from Mexico in search of gold, Texas was controlled by Spain and then by Mexico, before becoming an independent republic in 1836, and joining the Union of States in 1845. During the 19th century, many migrants who came to Texas raised cattle on the abundant land; in the 20th century, they were joined by prospectors attracted by the promise of oil riches. Today, although natural resources, especially oil, still form the basis of its wealth, the diversified Texan economy includes thriving hi-tech and financial industries. The major urban centers, home to 80% of the population, lie in the south and east, and include Houston, the "oil-city," and Dallas–Fort Worth. Hispanic influences remain strong, especially in southern and western Texas.

▲ *Dallas was founded* in 1841 as a prairie trading post and its development was stimulated by the arrival of railroads. Cotton and then oil funded the town's early growth. Today, the modern, high rise skyline of Dallas reflects the city's position as a leading center of banking, insurance, and the petroleum industry in the southwest.

Using the land

Cotton production and livestock-raising, particularly cattle, dominate farming, although crop failures and the demands of local markets have led to some diversification. Following the introduction of modern farming techniques, cotton production spread out from the east to the plains of western Texas. Cattle ranches are widespread, while sheep and goats are raised on the dry Edwards Plateau.

Land use and agricultural distribution

- cattle
- goats
- sheep
- cereals
- cotton
- • major towns
- pasture
- cropland
- forest
- barren

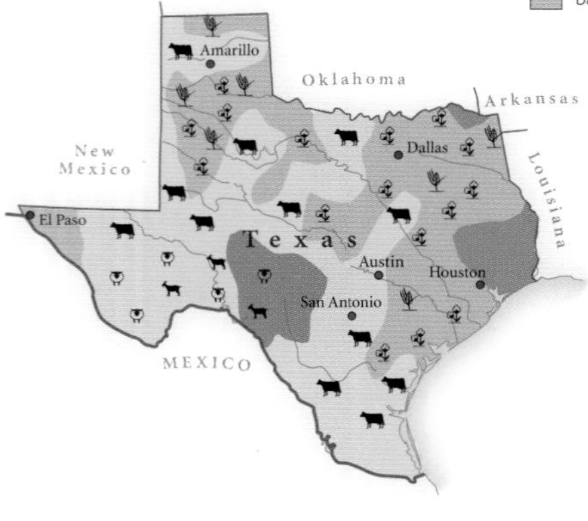

The urban/rural population divide

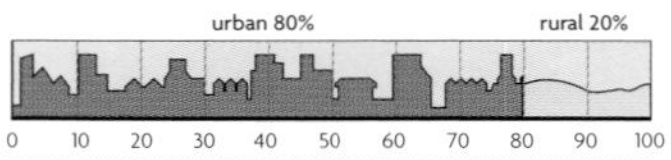

urban 80%   rural 20%

0 10 20 30 40 50 60 70 80 90 100

Population density	Total land area
84 people per sq mile (33 people per sq km)	261,797 sq miles (678,028 sq km)

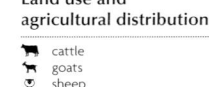

▲ *The huge cattle* ranches of Texas developed during the 19th century when land was plentiful and could be acquired cheaply. Today, more cattle and sheep are raised in Texas than in any other state.

The landscape

Texas is made up of a series of massive steps descending from the mountains and high plains of the west and northwest to the coastal lowlands in the southeast. Many of the state's borders are delineated by water. The Rio Grande flows from the Rocky Mountains to the Gulf of Mexico, marking the border with Mexico.

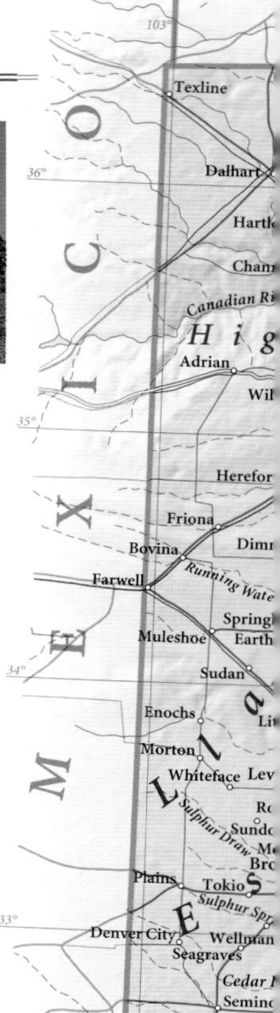

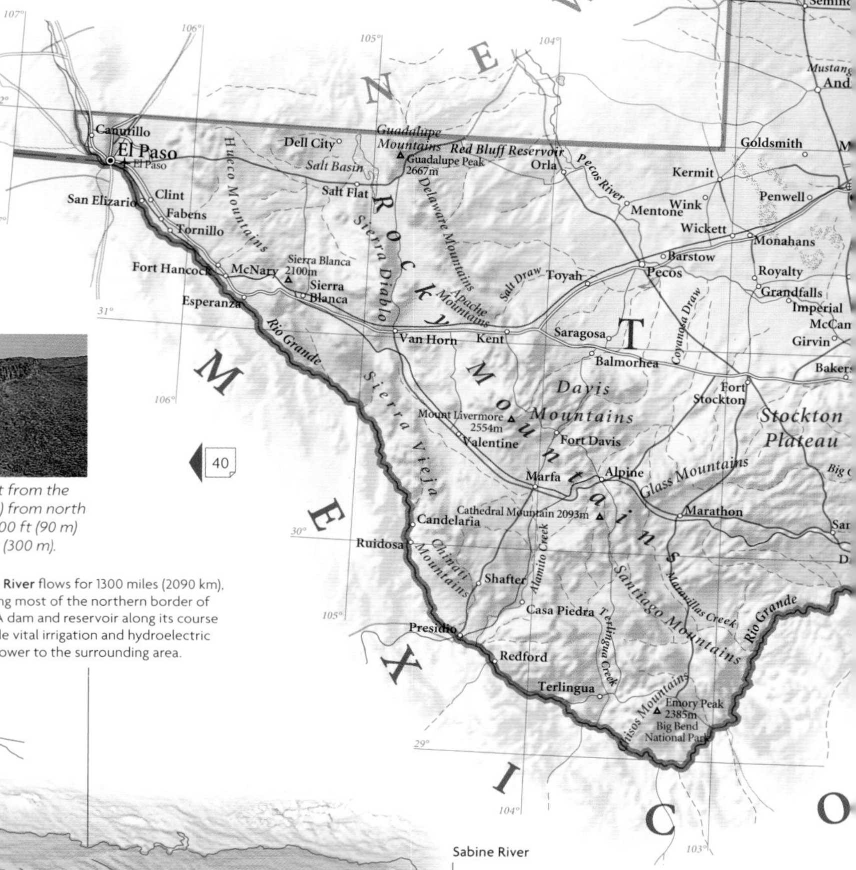

▲ *Cap Rock Escarpment* juts out from the plains, running 200 miles (320 km) from north to south. Its height varies from 300 ft (90 m) rising to sheer cliffs up to 1000 ft (300 m).

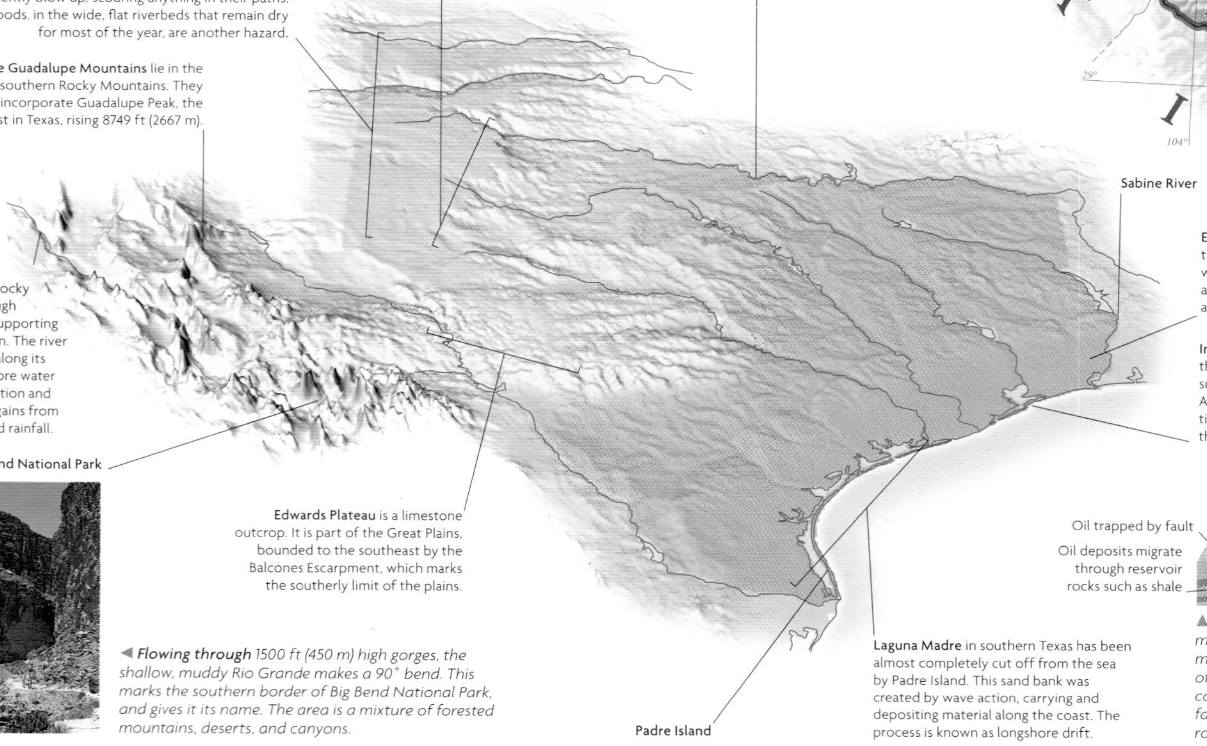

The Llano Estacado or Staked Plain in northern Texas is known for its harsh environment. In the north, freezing winds carrying ice and snow sweep down from the Rocky Mountains. To the south, sandstorms frequently blow up, scouring anything in their paths. Flash floods, in the wide, flat riverbeds that remain dry for most of the year, are another hazard.

The Guadalupe Mountains lie in the southern Rocky Mountains. They incorporate Guadalupe Peak, the highest in Texas, rising 8749 ft (2667 m).

The Red River flows for 1300 miles (2090 km), marking most of the northern border of Texas. A dam and reservoir along its course provide vital irrigation and hydroelectric power to the surrounding area.

The Rio Grande flows from the Rocky Mountains through semi-arid land, supporting sparse vegetation. The river actually shrinks along its course, losing more water through evaporation and seepage than it gains from its tributaries and rainfall.

Big Bend National Park

Extensive forests of pine and cypress grow in the eastern corner of the coastal lowlands where the average rainfall is 45 inches (1145 mm) a year. This is higher than the rest of the state and over twice the average in the west.

In the coastal lowlands of southeastern Texas the Earth's crust is warping, causing the land to subside and allowing the sea to invade. Around Galveston, the rate of downward tilting is 6 inches (15 cm) per year. Erosion of the coast is also exacerbated by hurricanes.

◀ *Flowing through* 1500 ft (450 m) high gorges, the shallow, muddy Rio Grande makes a 90° bend. This marks the southern border of Big Bend National Park, and gives it its name. The area is a mixture of forested mountains, deserts, and canyons.

Edwards Plateau is a limestone outcrop. It is part of the Great Plains, bounded to the southeast by the Balcones Escarpment, which marks the southerly limit of the plains.

Laguna Madre in southern Texas has been almost completely cut off from the sea by Padre Island. This sand bank was created by wave action, carrying and depositing material along the coast. The process is known as longshore drift.

Padre Island

Oil deposits

Oil trapped by fault
Oil deposits migrate through reservoir rocks such as shale
Oil accumulates beneath impermeable cap rock
Impermeable rock strata
Salt dome

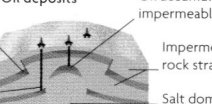

▲ *Oil deposits are* found beneath much of Texas. They collect as oil migrates upward through porous layers of rock until it is trapped, either by a cap of rock above a salt dome, or by a fault line which exposes impermeable rock through which the oil cannot rise.

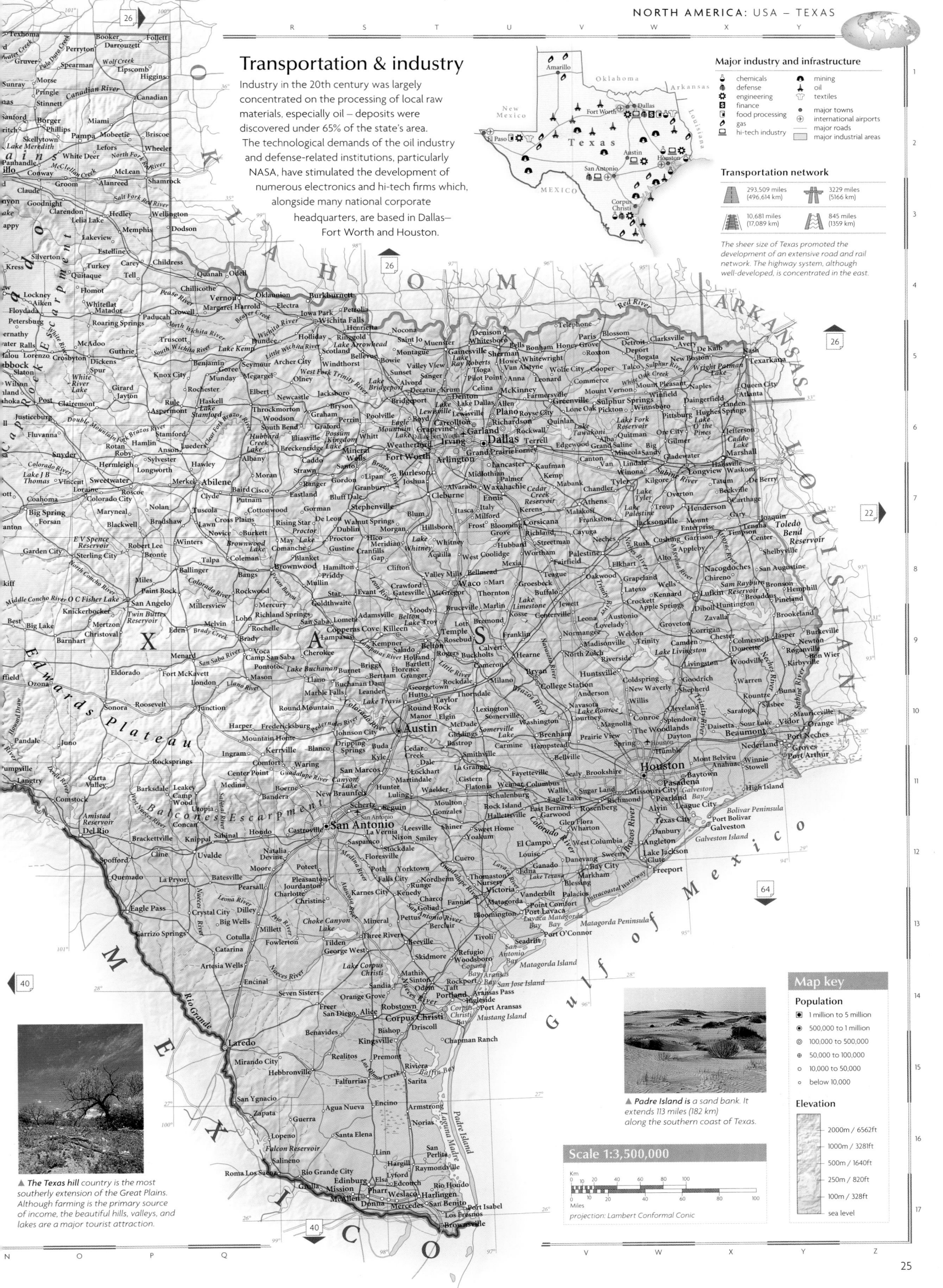

Transportation & industry

Industry in the 20th century was largely concentrated on the processing of local raw materials, especially oil – deposits were discovered under 65% of the state's area. The technological demands of the oil industry and defense-related institutions, particularly NASA, have stimulated the development of numerous electronics and hi-tech firms which, alongside many national corporate headquarters, are based in Dallas– Fort Worth and Houston.

Major industry and infrastructure

- chemicals
- defense
- engineering
- finance
- food processing
- gas
- hi-tech industry
- mining
- oil
- textiles
- major towns
- international airports
- major roads
- major industrial areas

Transportation network

293,509 miles (496,614 km)	3229 miles (5166 km)
10,681 miles (17,089 km)	845 miles (1359 km)

The sheer size of Texas promoted the development of an extensive road and rail network. The highway system, although well-developed, is concentrated in the east.

▲ *The Texas hill country is the most southerly extension of the Great Plains. Although farming is the primary source of income, the beautiful hills, valleys, and lakes are a major tourist attraction.*

▲ *Padre Island is a sand bank. It extends 113 miles (182 km) along the southern coast of Texas.*

Map key

Population

- 1 million to 5 million
- 500,000 to 1 million
- 100,000 to 500,000
- 50,000 to 100,000
- 10,000 to 50,000
- below 10,000

Elevation

- 2000m / 6562ft
- 1000m / 3281ft
- 500m / 1640ft
- 250m / 820ft
- 100m / 328ft
- sea level

Scale 1:3,500,000

projection: Lambert Conformal Conic

USA: SOUTH MIDWESTERN STATES

Arkansas, Kansas, Missouri, Oklahoma

The expansion of the US focused on this region in the mid-19th century. Settlers spread from the confluence of the Missouri and Mississippi rivers up onto the Great Plains. This treeless expanse, which early explorers had called the Great American Desert was turned into one of the world's richest agricultural regions. But periodic droughts, coupled with overintensive farming, led to the "dustbowl" soil erosion crisis of the 1930s, the abandonment of many farms, and a mass exodus to the west coast. The land has since recovered, although the mechanization of agriculture has led to a decline in the rural population. In recent years, suburban residential development has spread rapidly across the wooded Ozark Plateau in the east of the region.

Transportation & industry

The processing of agricultural products, such as brewing and meatpacking, has been traditionally important in these states. In Kansas and Oklahoma, diversified manufacturing now supplements income from fossil fuels; Wichita has become a world center for aeronautical engineering, an industry which also employs many people in neighboring Missouri.

Major industry and infrastructure

- ✈ aerospace
- ✿ engineering
- $ finance
- food processing
- ∂ gas
- mining
- oil
- vehicle manufacture
- ⊕ major towns
- ⊕ international airports
- — major roads
- major industrial areas

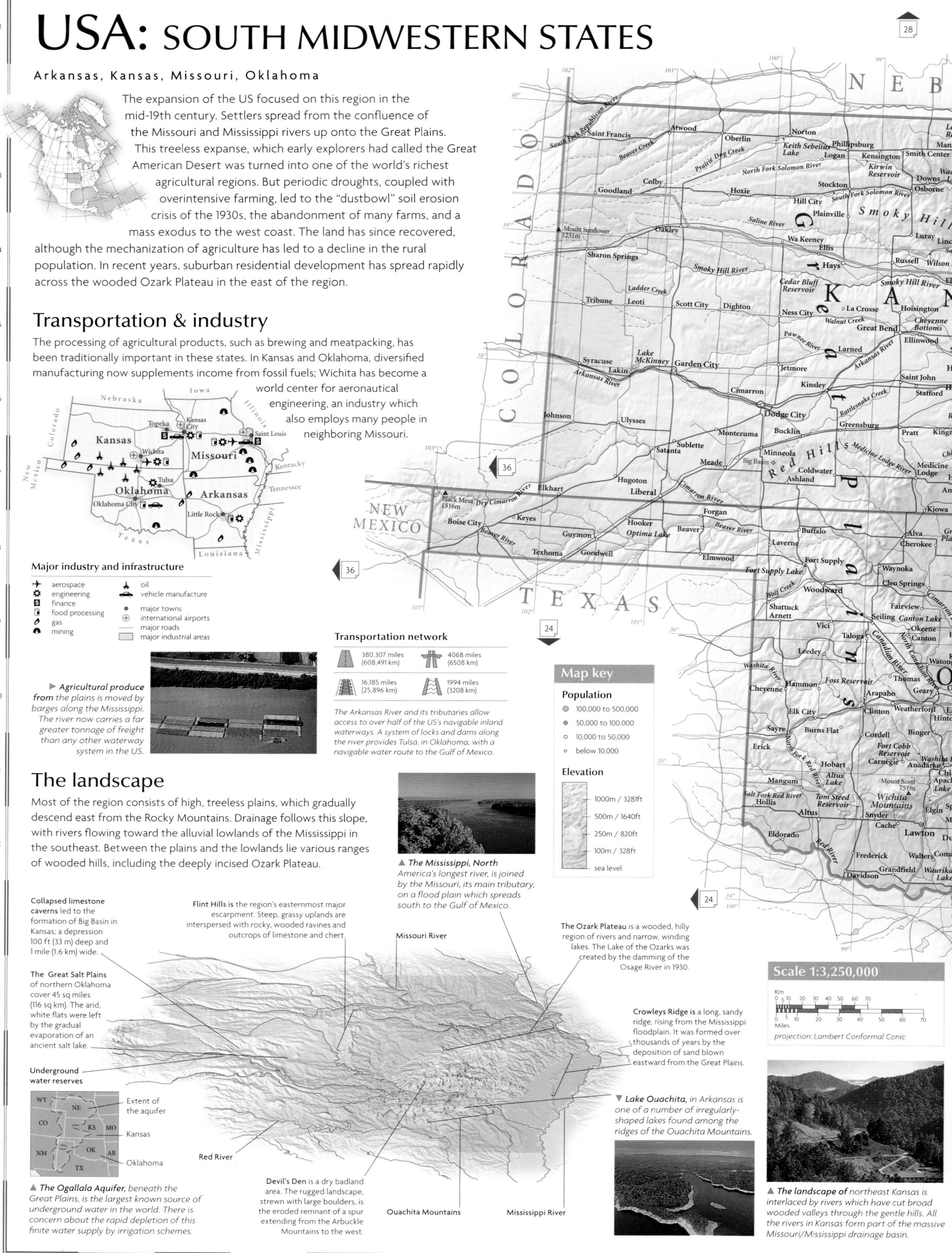

▶ Agricultural produce from the plains is moved by barges along the Mississippi. The river now carries a far greater tonnage of freight than any other waterway system in the US.

Transportation network

380,307 miles (608,491 km)		4068 miles (6508 km)	
16,185 miles (25,896 km)		1994 miles (3208 km)	

The Arkansas River and its tributaries allow access to over half of the US's navigable inland waterways. A system of locks and dams along the river provides Tulsa, in Oklahoma, with a navigable water route to the Gulf of Mexico.

The landscape

Most of the region consists of high, treeless plains, which gradually descend east from the Rocky Mountains. Drainage follows this slope, with rivers flowing toward the alluvial lowlands of the Mississippi in the southeast. Between the plains and the lowlands lie various ranges of wooded hills, including the deeply incised Ozark Plateau.

▲ The Mississippi, North America's longest river, is joined by the Missouri, its main tributary, on a flood plain which spreads south to the Gulf of Mexico.

Map key

Population
- ◉ 100,000 to 500,000
- ⊕ 50,000 to 100,000
- ○ 10,000 to 50,000
- ○ below 10,000

Elevation
- 1000m / 3281ft
- 500m / 1640ft
- 250m / 820ft
- 100m / 328ft
- sea level

Collapsed limestone caverns led to the formation of Big Basin in Kansas; a depression 100 ft (33 m) deep and 1 mile (1.6 km) wide.

Flint Hills is the region's easternmost major escarpment. Steep, grassy uplands are interspersed with rocky, wooded ravines and outcrops of limestone and chert.

Missouri River

The Ozark Plateau is a wooded, hilly region of rivers and narrow, winding lakes. The Lake of the Ozarks was created by the damming of the Osage River in 1930.

The Great Salt Plains of northern Oklahoma cover 45 sq miles (116 sq km). The arid, white flats were left by the gradual evaporation of an ancient salt lake.

Underground water reserves

Crowleys Ridge is a long, sandy ridge, rising from the Mississippi floodplain. It was formed over thousands of years by the deposition of sand blown eastward from the Great Plains.

Scale 1:3,250,000

Km
0 5 10 20 30 40 50 60 70

Miles
0 5 10 20 30 40 50 60 70

projection: Lambert Conformal Conic

▼ Lake Ouachita, in Arkansas, is one of a number of irregularly-shaped lakes found among the ridges of the Ouachita Mountains.

Red River

▲ The Ogallala Aquifer, beneath the Great Plains, is the largest known source of underground water in the world. There is concern about the rapid depletion of this finite water supply by irrigation schemes.

Devil's Den is a dry badland area. The rugged landscape, strewn with large boulders, is the eroded remnant of a spur extending from the Arbuckle Mountains to the west.

Ouachita Mountains

Mississippi River

▲ The landscape of northeast Kansas is interlaced by rivers which have cut broad wooded valleys through the gentle hills. All the rivers in Kansas form part of the massive Missouri/Mississippi drainage basin.

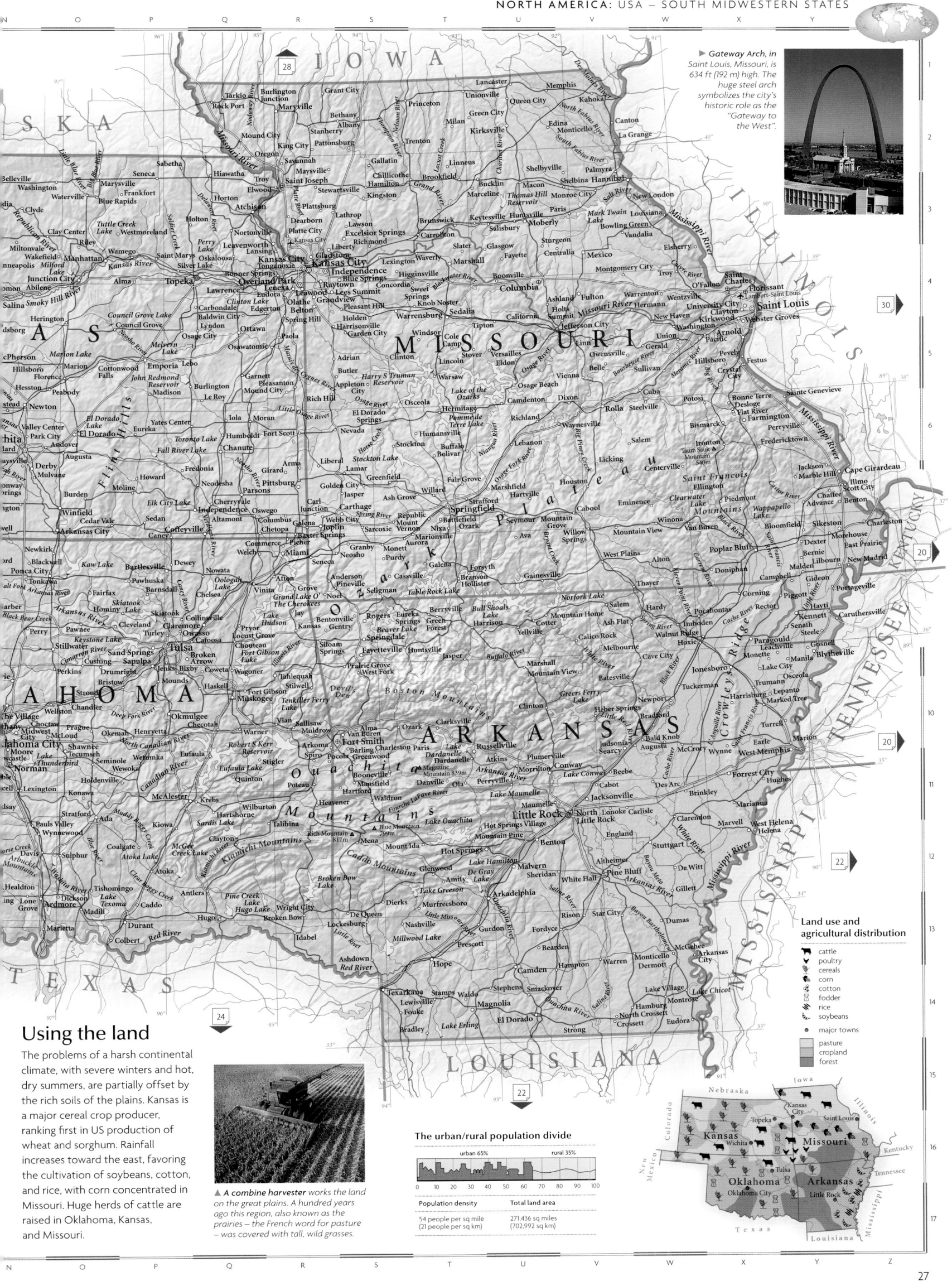

▶ *Gateway Arch, in Saint Louis, Missouri, is 634 ft (192 m) high. The huge steel arch symbolizes the city's historic role as the "Gateway to the West".*

Using the land

The problems of a harsh continental climate, with severe winters and hot, dry summers, are partially offset by the rich soils of the plains. Kansas is a major cereal crop producer, ranking first in US production of wheat and sorghum. Rainfall increases toward the east, favoring the cultivation of soybeans, cotton, and rice, with corn concentrated in Missouri. Huge herds of cattle are raised in Oklahoma, Kansas, and Missouri.

▲ *A combine harvester works the land on the great plains. A hundred years ago this region, also known as the prairies – the French word for pasture – was covered with tall, wild grasses.*

The urban/rural population divide

urban 65% | rural 35%

0 10 20 30 40 50 60 70 80 90 100

Population density	Total land area
54 people per sq mile (21 people per sq km)	271,436 sq miles (702,992 sq km)

Land use and agricultural distribution

- cattle
- poultry
- cereals
- corn
- cotton
- fodder
- rice
- soybeans
- major towns

pasture
cropland
forest

27

USA: UPPER PLAINS STATES

Iowa, Minnesota, Nebraska, North Dakota, South Dakota

Lying at the very heart of the North American continent, much of this region was acquired from France as part of the Louisiana Purchase in 1803. The area was largely bypassed by the early waves of westward migrants. When Europeans did settle, during the 19th century, they displaced the Native Americans who lived on the plains. The settlers planted arable crops and raised cattle on the immensely fertile prairie land, founding an agrarian tradition which flourishes today. Most of this region remains rural; of the five states, only in Minnesota has there been significant diversification away from agriculture and resource-based industries into the hi-tech and service sectors.

Using the land

The popular image of these states as agricultural is entirely justified; prairies stretch uninterrupted across most of the area. Croplands fall into two regions: the wheat belt of the plains, and the corn belt of the central US. Cash crops, such as soybeans, are grown to supplement incomes. Livestock, particularly pigs and cattle, are raised throughout this region.

▶ Dark, fertile prairie soils in the southeast provide Minnesota's most productive farmland. Hot, humid summers create a long growing season for corn cultivation.

The urban/rural population divide

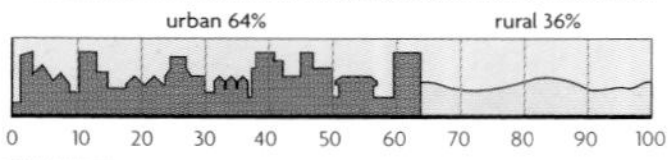

urban 64%   rural 36%

0 10 20 30 40 50 60 70 80 90 100

Population density	Total land area
31 people per sq mile (12 people per sq km)	357,212 sq miles (925,143 sq km)

Land use and agricultural distribution

- cattle
- pigs
- corn
- soybeans
- wheat
- major towns
- pasture
- cropland
- forest
- wetland

Transportation & industry

Food processing and the production of farm machinery are supported by the large agricultural sector. Mineral exploitation is also an important activity: gold is mined in the ore-rich Black Hills of South Dakota, and both North Dakota and Nebraska are emerging as major petroleum producers.

▶ Water erosion along the Little Missouri River has carried away sedimentary deposits, creating rugged landscapes known as badlands.

Transportation network

504,522 miles (807,235 km)		3422 miles (5475 km)	
16,940 miles (27,104 km)		683 miles (1098 km)	

Nebraska's central location has made it an important transportation artery for east–west traffic. Minnesota's road network radiates out from the hub of the twin cities, Minneapolis–Saint Paul.

Major industry and infrastructure

- coal
- engineering
- electronics
- finance
- food processing
- oil & gas
- mining
- major towns
- international airports
- major roads
- major industrial areas

The landscape

These states straddle the Great Plains and the lowlands of the central US, with Minnesota lying in a transition zone between the eastern forests and the prairies. The region was shaped by repeated ice advances and retreats, leaving a flat relief, broken only by the numerous lakes and broad river networks that drain the prairies.

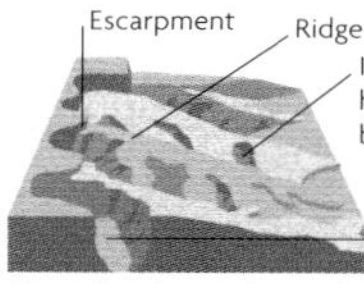

Escarpment   Ridge

In permeable strata hollows are formed by small mudslides

Water flowing into gullies erodes back the escarpment

▲ Badlands are formed by stormwater run-off. This flows down the impermeable strata of the escarpment and saturates the permeable strata, leading to mudslides and the formation of gullies.

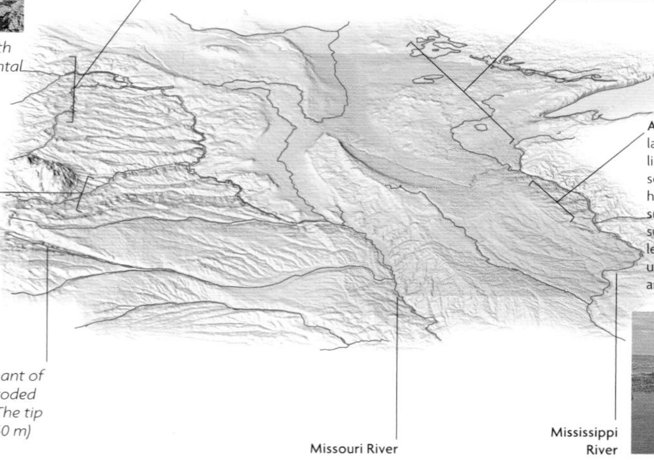

The Minnesota landscape contains many post-glacial features, including its numerous lakes, boulder-strewn hills, and mineral-rich deposits.

North Dakota Badlands

▲ In the badlands of North and South Dakota, horizontal layers of sandstone have been eroded by rivers, leaving a landscape of narrow gullies, sharp crests and pinnacles.

South Dakota Badlands

Although it escaped the last glaciation, the limestone bedrock of southeastern Minnesota has been eroded by surface and subterranean streams, leaving a network of underground caverns and steepsided valleys.

▲ Chimney Rock is a remnant of an ancient land surface, eroded by the North Platte River. The tip of its spire stands 500 ft (150 m) above the plain.

Missouri River

Mississippi River

◀ In northeastern Iowa, the Mississippi and its tributaries have deeply incised the underlying bedrock creating a hilly terrain, with bluffs standing 300 ft (90 m) above the valley.

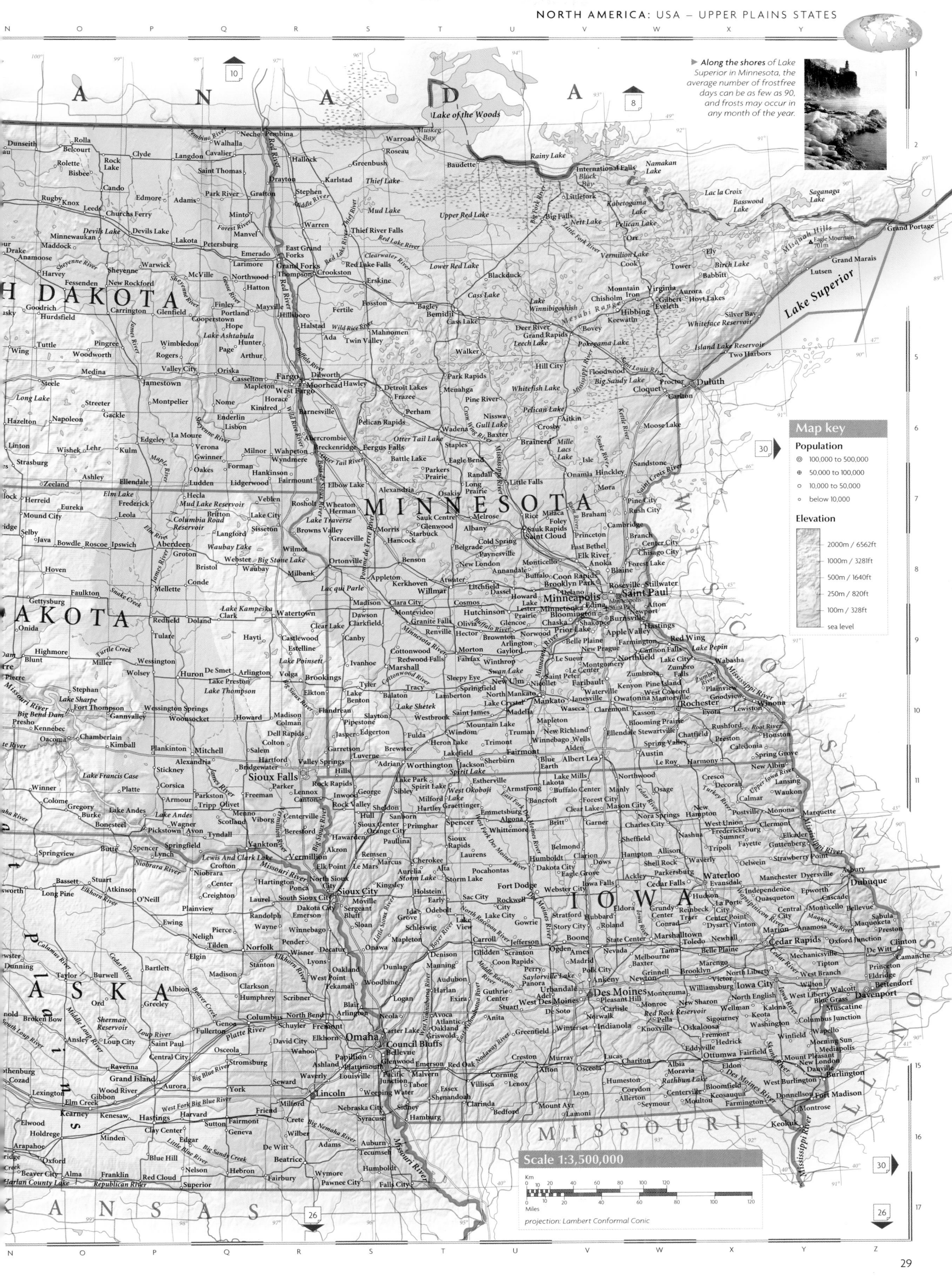

▶ **Along the shores** of Lake Superior in Minnesota, the average number of frostfree days can be as few as 90, and frosts may occur in any month of the year.

USA: GREAT LAKES STATES

Illinois, Indiana, Michigan, Ohio, Wisconsin

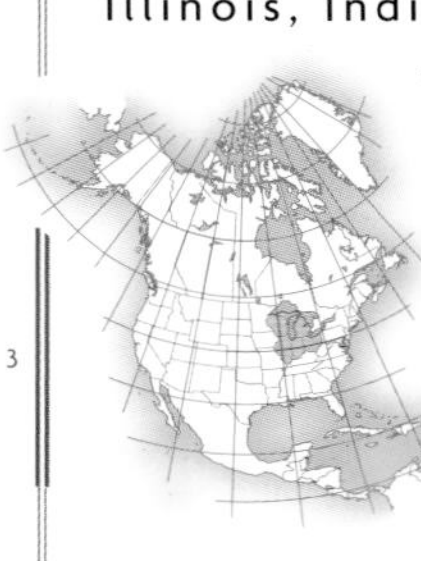

The states bordering the Great Lakes developed rapidly in the second half of the 19th century as a result of improvements in communications: railroads to the west and waterways to the south and east. Fertile land and good links with growing eastern seaboard cities encouraged the development of agriculture and food processing. Migrants from Europe and other parts of the US flooded into the region and for much of the 20th century the region's economy boomed. However, in recent years heavy industry has declined, earning the region the unwanted label the "Rustbelt."

Transportation & industry

The Great Lakes region is the center of the US car industry. Since the early part of the 20th century, its prosperity has been closely linked to the fortunes of automobile manufacturing. Iron and steel production has expanded to meet demand from this industry. In the 1970s, nationwide recession, cheaper foreign competition in the automobile sector, pollution in and around the Great Lakes, and the collapse of the meatpacking industry, centered on Chicago, forced these states to diversify their industrial base. New industries have emerged, notably electronics, service, and finance industries.

Transportation network

540,682 miles (865,091 km)	6550 miles (10,480 km)
24,928 miles (39,884 km)	2330 miles (3748 km)

Few areas of the US have a comparable system. Chicago is a principal transportation terminus with a dense network of roads, railroads, and Interstate freeways that radiates out from the city.

▶ *Ever since Ransom Olds and Henry Ford started mass-producing automobiles in Detroit early in the 20th century, the city's name has become synonymous with the American automotive industry.*

Major industry and infrastructure

- car manufacture
- coal
- electronics
- engineering
- finance
- food processing
- iron & steel
- oil
- research & development
- textiles
- major towns
- international airports
- major roads
- major industrial areas

The landscape

Much of this region shows the impact of glaciation which lasted until about 10,000 years ago, and extended as far south as Illinois and Ohio. Although the relief of the region slopes toward the Great Lakes, because the ice sheets blocked northerly drainage, most of the rivers today flow southward, forming part of the massive Mississippi/Missouri drainage basin.

The many lakes and marshes of Wisconsin and Michigan are the result of glacial erosion and deposition which occurred during the last Ice Age.

Southwestern Wisconsin is known as a "driftless" area. Unlike most of the region, low hills protected it from erosion by the advancing ice sheet.

Most of the water used in northern Illinois is pumped from underground reservoirs. Due to increased demand, many areas now face a water shortage. Around Joliet, the water table was lowered by more than 700 ft (210 m) over the last century.

◀ *The dunes near Sleeping Bear Point rise 400 ft (120 m) from the banks of Lake Michigan. They are constantly being resculpted by wind action.*

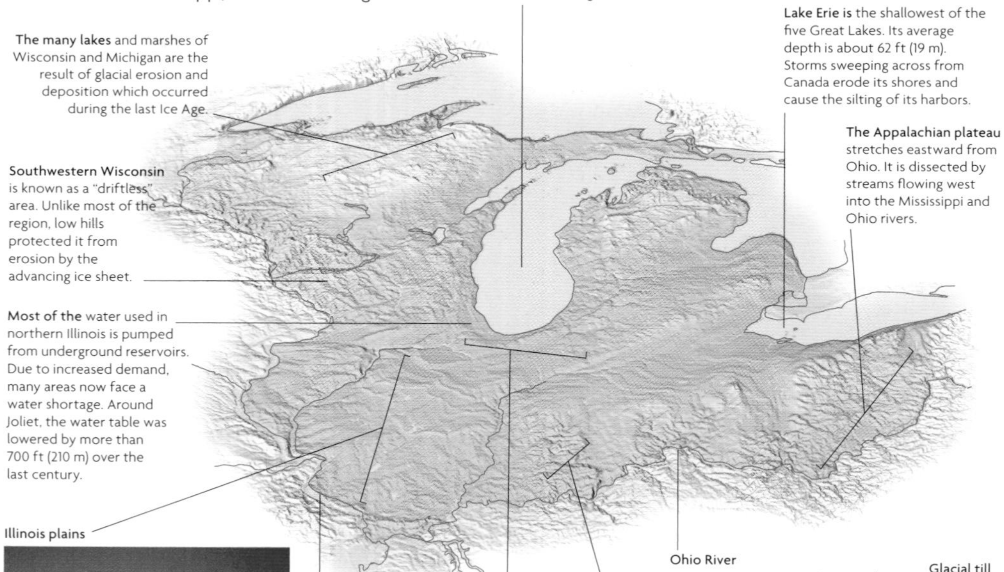

Lake Michigan

Lake Erie is the shallowest of the five Great Lakes. Its average depth is about 62 ft (19 m). Storms sweeping across from Canada erode its shores and cause the silting of its harbors.

The Appalachian plateau stretches eastward from Ohio. It is dissected by streams flowing west into the Mississippi and Ohio rivers.

Illinois plains

▲ *The plains of Illinois are characteristic of drift landscapes, scoured and flattened by glacial erosion and covered with fertile glacial deposits.*

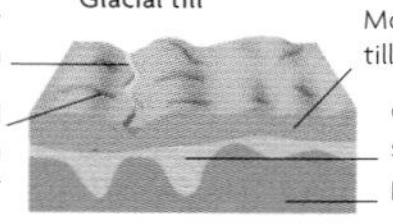

Mississippi River

Relic landforms from the last glaciation, such as shallow basins and ridges, cover all but the south of this region. Ridges, known as moraines, up to 300 ft (100 m) high, lie to the south of Lake Michigan.

Unlike the level prairie to the north, southern Indiana is relatively rugged. Limestone in the hills has been dissolved by water, producing features such as sinkholes and underground caves.

Ohio River

Glacial till

Present-day river or stream / Channels caused by outwash from melting glacier / Most recent till deposits / Older till sheet / Bedrock

▲ *As a result of successive glacial depositions, the total depth of till along the former southern margin of the Laurentide ice sheet can exceed 1300 ft (400 m).*

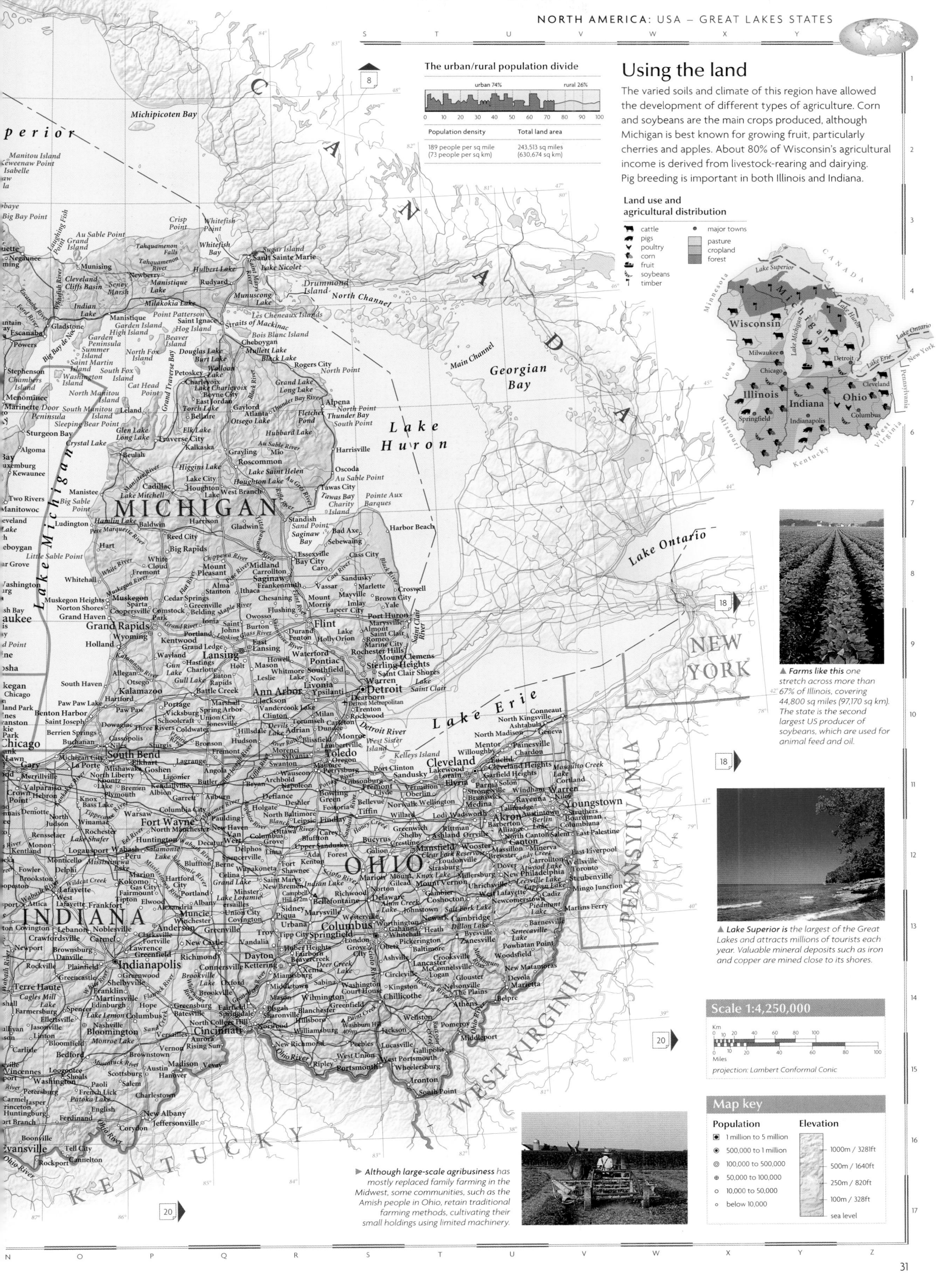

The urban/rural population divide

urban 74%   rural 26%

0 10 20 30 40 50 60 70 80 90 100

Population density	Total land area
189 people per sq mile (73 people per sq km)	243,513 sq miles (630,674 sq km)

Using the land

The varied soils and climate of this region have allowed the development of different types of agriculture. Corn and soybeans are the main crops produced, although Michigan is best known for growing fruit, particularly cherries and apples. About 80% of Wisconsin's agricultural income is derived from livestock-rearing and dairying. Pig breeding is important in both Illinois and Indiana.

Land use and agricultural distribution

- cattle
- pigs
- poultry
- corn
- fruit
- soybeans
- timber
- major towns
- pasture
- cropland
- forest

▲ *Farms like this* one stretch across more than 67% of Illinois, covering 44,800 sq miles (97,170 sq km). The state is the second largest US producer of soybeans, which are used for animal feed and oil.

▲ *Lake Superior is* the largest of the Great Lakes and attracts millions of tourists each year. Valuable mineral deposits such as iron and copper are mined close to its shores.

Scale 1:4,250,000

Km
0 10 20 40 60 80 100

Miles
0 20 40 60 80 100

projection: Lambert Conformal Conic

Map key

Population

- 1 million to 5 million
- 500,000 to 1 million
- 100,000 to 500,000
- 50,000 to 100,000
- 10,000 to 50,000
- below 10,000

Elevation

- 1000m / 3281ft
- 500m / 1640ft
- 250m / 820ft
- 100m / 328ft
- sea level

▶ *Although large-scale agribusiness* has mostly replaced family farming in the Midwest, some communities, such as the Amish people in Ohio, retain traditional farming methods, cultivating their small holdings using limited machinery.

USA: NORTH MOUNTAIN STATES

Idaho, Montana, Oregon, Washington, Wyoming

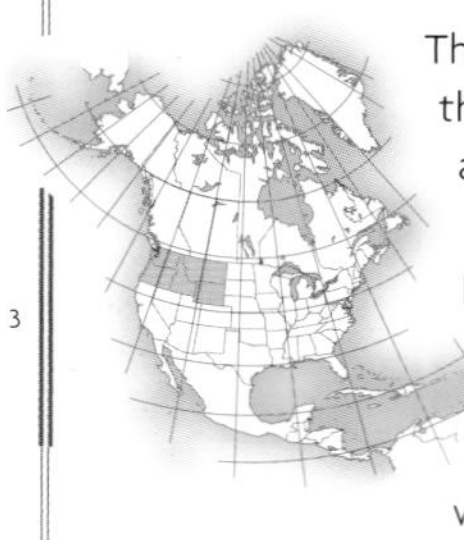

The remoteness of the northwestern states, coupled with the rugged landscape, ensured that this was one of the last areas settled by Europeans in the 19th century. Fur-trappers and gold-prospectors followed the Snake River westward as it wound its way through the Rocky Mountains. The states of the northwest have pioneered many conservationist policies, with the first US National Park opened at Yellowstone in 1872. More recently, the Cascades and Rocky Mountains have become havens for adventure tourism. The mountains still serve to isolate the western seaboard from the rest of the continent. This isolation has encouraged West Coast cities to expand their trade links with countries of the Pacific Rim.

▲ *The Snake River* has cut down into the basalt of the Columbia Basin to form Hells Canyon, the deepest in the US, with cliffs up to 7900 ft (2408 m) high.

Map key

Population
- ◉ 500,000 to 1 million
- ◎ 100,000 to 500,000
- ⊕ 50,000 to 100,000
- ⊙ 10,000 to 50,000
- ○ below 10,000

Elevation
- 4000m / 13,124ft
- 3000m / 9843ft
- 2000m / 6562ft
- 1000m / 3281ft
- 500m / 1640ft
- 250m / 820ft
- 100m / 328ft
- sea level

Using the land

Wheat farming in the east gives way to cattle ranching as rainfall decreases. Irrigated farming in the Snake River valley produces large yields of potatoes and other vegetables. Dairying and fruit-growing take place in the wet western lowlands between the mountain ranges.

The urban/rural population divide

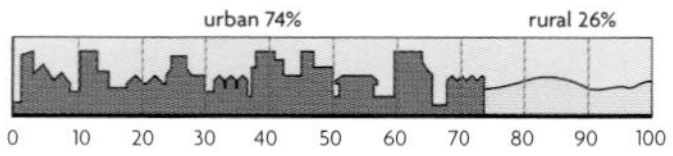

urban 74%   rural 26%

0 10 20 30 40 50 60 70 80 90 100

Population density	Total land area
26 people per sq mile (10 people per sq km)	487,970 sq miles (1,263,716 sq km)

Scale 1:4,250,000

Km
0 10 20 40 60 80 100
Miles
0 20 40 60 80 100

projection: Lambert Conformal Conic

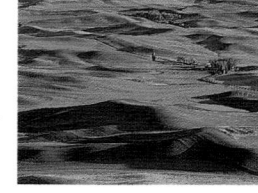

▶ *Fine-textured, volcanic* soils in the hilly Palouse region of eastern Washington are susceptible to erosion.

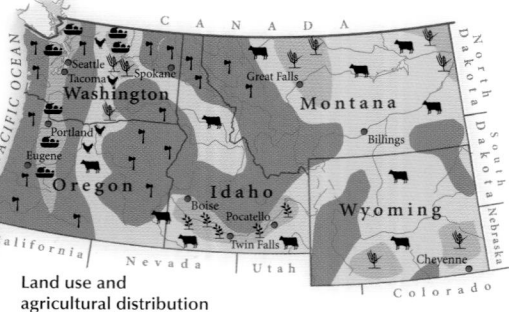

Land use and agricultural distribution

- 🐄 cattle
- 🦃 poultry
- 🌾 cereals
- 🍎 fruit
- 🥔 potatoes
- 🌲 timber
- ● major towns

pasture
cropland
forest

Transportation & industry

Minerals and timber are extremely important in this region. Uranium, precious metals, copper, and coal are all mined, the latter in vast open-cast pits in Wyoming; oil and natural gas are extracted further north. Manufacturing, notably related to the aerospace and electronics industries, is important in western cities.

Transportation network

- 347,857 miles (556,571 km)
- 4200 miles (6720 km)
- 12,354 miles (19,766 km)
- 1108 miles (1782 km)

Major industry and infrastructure

- △ adventure tourism
- ✈ aerospace
- ⚒ coal
- ⚗ chemicals
- ⚡ electronics
- 🍴 food processing
- ⛏ mining
- ⬮ oil & gas
- 🪵 timber processing
- ● major towns
- ⊕ international airports
- — major roads
- ▦ major industrial areas

The Union Pacific Railroad has been in service across Wyoming since 1867. The route through the Rocky Mountains is now shared with the Interstate 80, a major east–west highway.

◀ *Seattle lies in* one of Puget Sound's many inlets. The city receives oil and other resources from Alaska, and benefits from expanding trade across the Pacific.

◀ *Crater Lake, Oregon,* is 6 miles (10 km) wide and 1800 ft (600 m) deep. It marks the site of a volcanic cone, which collapsed after an eruption within the last 7000 years.

The landscape

The Rocky Mountains are flanked by lower parallel ranges, which spread onto the Great Plains in the east and surmount the broad lava plateau which extends westward. The Cascade Range divides the Columbia Basin from the coastlands, where the low areas around Puget Sound are broken by the steep, volcanic Olympic Mountains and the wooded hills of the Coast Ranges.

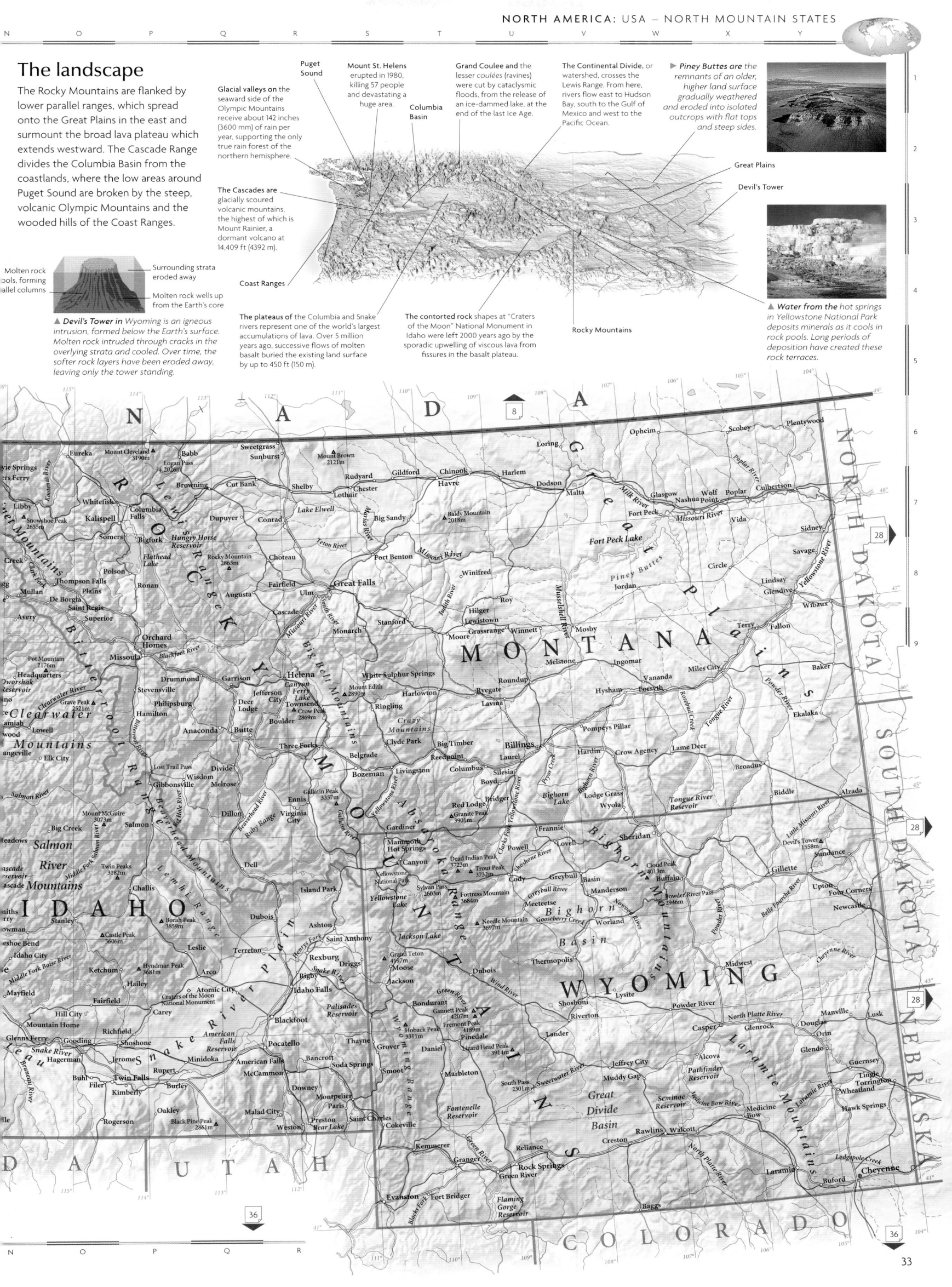

Glacial valleys on the seaward side of the Olympic Mountains receive about 142 inches (3600 mm) of rain per year, supporting the only true rain forest of the northern hemisphere.

Puget Sound

Mount St. Helens erupted in 1980, killing 57 people and devastating a huge area.

Columbia Basin

Grand Coulee and the lesser coulées (ravines) were cut by cataclysmic floods, from the release of an ice-dammed lake, at the end of the last Ice Age.

The Continental Divide, or watershed, crosses the Lewis Range. From here, rivers flow east to Hudson Bay, south to the Gulf of Mexico and west to the Pacific Ocean.

▶ *Piney Buttes are* the remnants of an older, higher land surface gradually weathered and eroded into isolated outcrops with flat tops and steep sides.

Great Plains

Devil's Tower

The Cascades are glacially scoured volcanic mountains, the highest of which is Mount Rainier, a dormant volcano at 14,409 ft (4392 m).

Coast Ranges

Molten rock pools, forming parallel columns

Surrounding strata eroded away

Molten rock wells up from the Earth's core

▲ *Devil's Tower in* Wyoming is an igneous intrusion, formed below the Earth's surface. Molten rock intruded through cracks in the overlying strata and cooled. Over time, the softer rock layers have been eroded away, leaving only the tower standing.

The plateaus of the Columbia and Snake rivers represent one of the world's largest accumulations of lava. Over 5 million years ago, successive flows of molten basalt buried the existing land surface by up to 450 ft (150 m).

The contorted rock shapes at "Craters of the Moon" National Monument in Idaho were left 2000 years ago by the sporadic upwelling of viscous lava from fissures in the basalt plateau.

Rocky Mountains

▲ *Water from the* hot springs in Yellowstone National Park deposits minerals as it cools in rock pools. Long periods of deposition have created these rock terraces.

USA: CALIFORNIA & NEVADA

The Gold Rush of 1849 attracted the first major wave of European settlers to the West Coast. The pleasant climate, beautiful scenery, and dynamic economy continue to attract immigrants – despite the ever-present danger of earthquakes – and California has become the US's most populous state. The overwhelmingly urban population is concentrated in the vast conurbations of Los Angeles, San Francisco, and San Diego; new immigrants include people from South Korea, the Philippines, Vietnam, and Mexico. Nevada's arid lands were initially exploited for minerals; in recent years, revenue from mining has been superseded by income from the tourist and gambling centers of Las Vegas and Reno.

Map key

Population
- ◼ 1 million to 5 million
- ◉ 500,000 to 1 million
- ◎ 100,000 to 500,000
- ◍ 50,000 to 100,000
- ○ 10,000 to 50,000
- ○ below 10,000

Elevation
- 4000m / 13,124ft
- 3000m / 9843ft
- 2000m / 6562ft
- 1000m / 3281ft
- 500m / 1640ft
- 250m / 820ft
- 100m / 328ft
- sea level

Scale 1:3,250,000

Km
0 5 10 20 30 40 50 60 70 80

Miles
0 5 10 20 30 40 50 60 70 80

projection: Lambert Conformal Conic

Transportation & industry

Nevada's rich mineral reserves ushered in a period of mining wealth which has now been replaced by revenue generated from gambling. California supports a broad set of activities including defense-related industries and research and development facilities. "Silicon Valley," near San Francisco, is a world leading center for micro-electronics, while tourism and the Los Angeles film industry also generate large incomes.

◀ *Gambling was legalized in Nevada in 1931. Las Vegas has since become the center of this multimillion dollar industry.*

Major industry and infrastructure
- ✈ aerospace
- 🚗 car manufacture
- defense
- film industry
- 💲 finance
- food processing
- gambling
- 🖥 hi-tech industry
- ⛏ mining
- pharmaceuticals
- 🔬 research & development
- textiles
- tourism
- • major towns
- ⊕ international airports
- — major roads
- ▭ major industrial areas

Transportation network

211,459 miles (338,334 km)	2944 miles (4710 km)
7822 miles (12,595 km)	190 miles (360 km)

In California, the motor vehicle is a vital part of daily life, and an extensive freeway system runs throughout the state, cementing its position as the most important mode of transport.

The landscape

The broad Central Valley divides California's coastal mountains from the Sierra Nevada. The San Andreas Fault, running beneath much of the state, is the site of frequent earth tremors and sometimes more serious earthquakes. East of the Sierra Nevada, the landscape is characterized by the basin and range topography with stony deserts and many salt lakes.

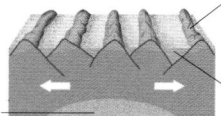

Rising molten rock causes stretching of the Earth's crust

Extensive cracking (faulting) uplifted a series of ridges

As ridges are eroded they fill intervening valleys with sediments

▲ *Molten rock (magma) welling up to form a dome in the Earth's interior, causes the brittle surface rocks to stretch and crack. Some areas were uplifted to form mountains (ranges), while others sunk to form flat valleys (basins).*

◀ *The General Sherman sequoia tree in Sequoia National Park is around 2500 years old and at 275 ft (84 m) is one of the largest living things on earth.*

Most of California's agriculture is confined to the fertile and extensively irrigated Central Valley, running between the Coast Ranges and the Sierra Nevada. It incorporates the San Joaquin and Sacramento valleys.

The dramatic granitic rock formations of Half Dome and El Capitan, and the verdant coniferous forests, attract millions of visitors annually to Yosemite National Park in the Sierra Nevada.

Sierra Nevada

The Great Basin dominates most of Nevada's topography containing large open basins, punctuated by eroded features such as *buttes* and *mesas*. River flow tends to be seasonal, dependent upon spring showers and winter snow melt.

Wheeler Peak is home to some of the world's oldest trees, bristlecone pines, which live for up to 5000 years.

When the Hoover Dam across the Colorado River was completed in 1936, it created Lake Mead, one of the largest artificial lakes in the world, extending for 115 miles (285 km) upstream.

The San Andreas Fault is a transverse fault which extends for 650 miles (1050 km) through California. Major earthquakes occur when the land either side of the fault moves at different rates. San Francisco was devastated by an earthquake in 1906.

Death Valley

Amargosa Desert

▲ *The Sierra Nevada create a "rainshadow," preventing rain from reaching much of Nevada. Pacific air masses, passing over the mountains, are stripped of their moisture.*

▶ *Named by migrating settlers in 1849, Death Valley is the driest, hottest place in North America, as well as being the lowest point on land in the western hemisphere, at 282 ft (86 m) below sea level.*

The sparsely populated Mojave Desert receives less than 8 inches (200 mm) of rainfall a year. It is used extensively for weapons-testing and military purposes.

The Salton Sea was created accidentally between 1905 and 1907 when an irrigation channel from the Colorado River broke out of its banks and formed this salty 300 sq mile (777 sq km), landlocked lake.

Using the land

California is the leading agricultural producer in the US, although low rainfall makes irrigation essential. The long growing season and abundant sunshine allow many crops to be grown in the fertile Central Valley including grapes, citrus fruits, vegetables, and cotton. Almost 17 million acres (6.8 million hectares) of California's forests are used commercially. Nevada's arid climate and poor soil are largely unsuitable for agriculture; 85% of its land is state owned and large areas are used for underground testing of nuclear weapons.

Land use and agricultural distribution
- 🐄 cattle
- citrus fruits
- fruit
- irrigation
- timber
- vineyards
- • major towns
- pasture
- cropland
- forest
- desert

▲ *Without considerable irrigation, this fertile valley at Palm Springs would still be part of the Sonoran Desert. California's farmers account for about 80% of the state's total water usage.*

The urban/rural population divide

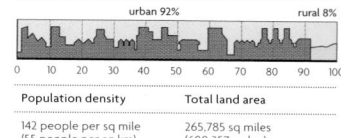

urban 92%   rural 8%

0   10   20   30   40   50   60   70   80   90   100

Population density	Total land area
142 people per sq mile (55 people per sq km)	265,785 sq miles (688,357 sq km)

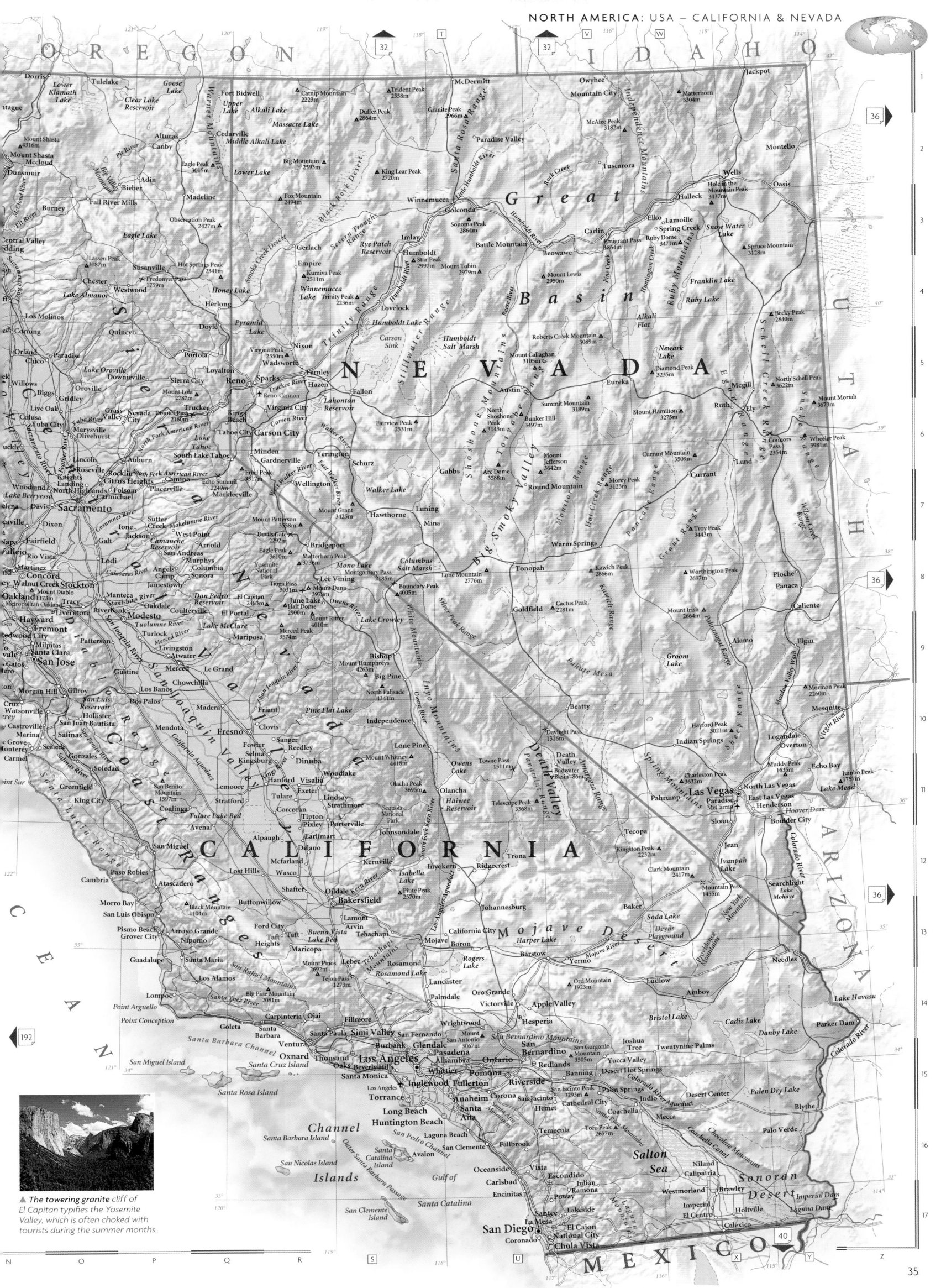

The towering granite cliff of El Capitan typifies the Yosemite Valley, which is often choked with tourists during the summer months.

USA: SOUTH MOUNTAIN STATES

Arizona, Colorado, New Mexico, Utah

This arid region, characterized by expansive plateaus and spectacular canyons is home to several distinct peoples. The ruins of cliff dwellings built a thousand years ago by the Anasazi people still exist today, and native Americans own one-third of the land in Arizona. Spanish and Mexican conquest and settlement left a hispanic presence which is strongest in New Mexico. The Mormons, who came to the Great Salt Lake seeking religious freedom in 1847, were among the earliest Anglo-American settlers and now make up over 70% of Utah's population. The region's mineral wealth drove rapid development in the 20th century, yet the constraints of a fragile environment, including widespread water shortages, may limit prospects for growth.

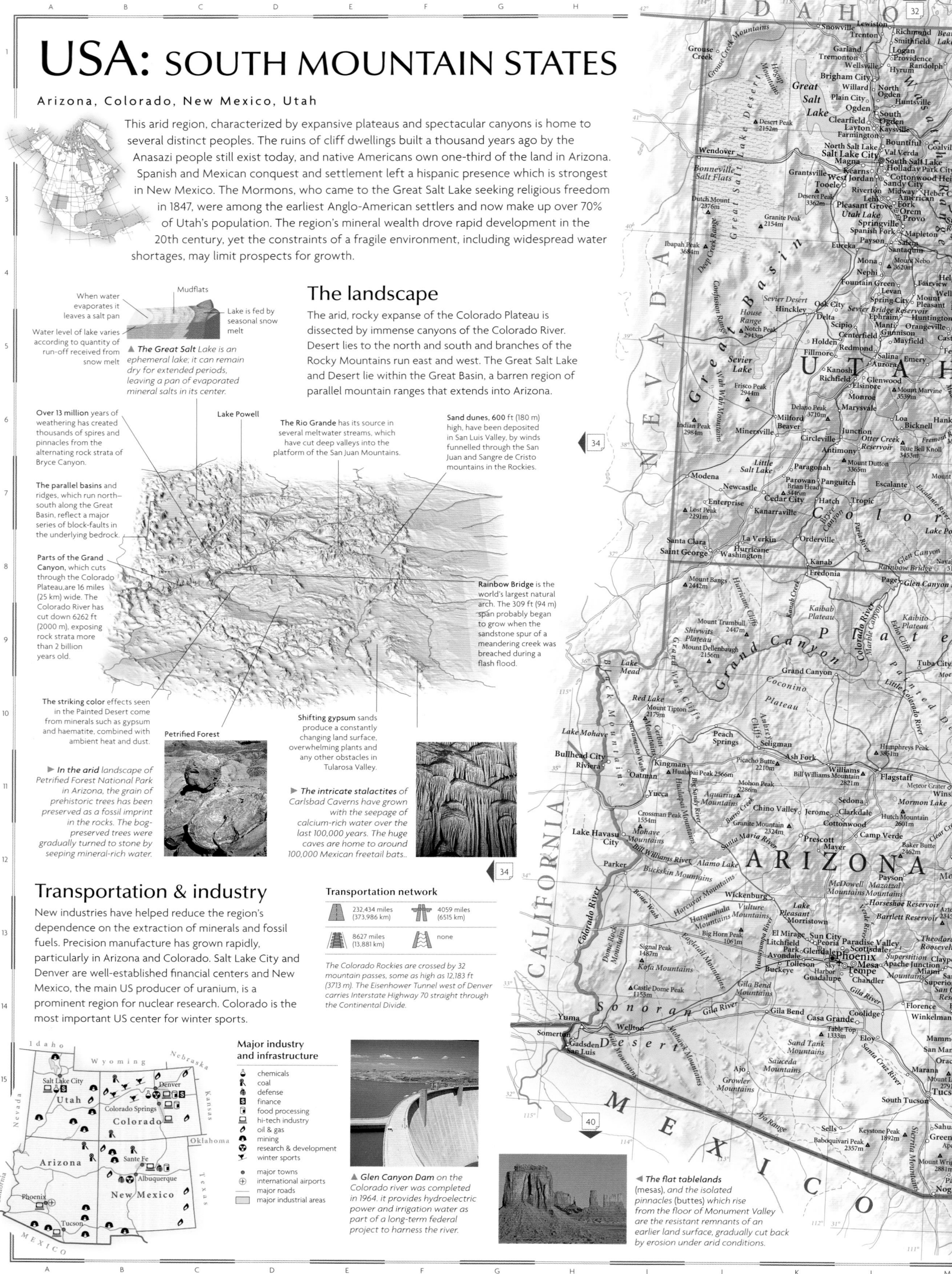

The landscape

The arid, rocky expanse of the Colorado Plateau is dissected by immense canyons of the Colorado River. Desert lies to the north and south and branches of the Rocky Mountains run east and west. The Great Salt Lake and Desert lie within the Great Basin, a barren region of parallel mountain ranges that extends into Arizona.

When water evaporates it leaves a salt pan

Water level of lake varies according to quantity of run-off received from snow melt

Mudflats

Lake is fed by seasonal snow melt

▲ **The Great Salt** Lake is an ephemeral lake; it can remain dry for extended periods, leaving a pan of evaporated mineral salts in its center.

Over 13 million years of weathering has created thousands of spires and pinnacles from the alternating rock strata of Bryce Canyon.

The parallel basins and ridges, which run north–south along the Great Basin, reflect a major series of block-faults in the underlying bedrock.

Parts of the Grand Canyon, which cuts through the Colorado Plateau,are 16 miles (25 km) wide. The Colorado River has cut down 6262 ft (2000 m), exposing rock strata more than 2 billion years old.

Lake Powell

The Rio Grande has its source in several meltwater streams, which have cut deep valleys into the platform of the San Juan Mountains.

Sand dunes, 600 ft (180 m) high, have been deposited in San Luis Valley, by winds funnelled through the San Juan and Sangre de Cristo mountains in the Rockies.

Rainbow Bridge is the world's largest natural arch. The 309 ft (94 m) span probably began to grow when the sandstone spur of a meandering creek was breached during a flash flood.

The striking color effects seen in the Painted Desert come from minerals such as gypsum and haematite, combined with ambient heat and dust.

Petrified Forest

▶ **In the arid** landscape of Petrified Forest National Park in Arizona, the grain of prehistoric trees has been preserved as a fossil imprint in the rocks. The bog-preserved trees were gradually turned to stone by seeping mineral-rich water.

Shifting gypsum sands produce a constantly changing land surface, overwhelming plants and any other obstacles in Tularosa Valley.

▶ **The intricate stalactites** of Carlsbad Caverns have grown with the seepage of calcium-rich water over the last 100,000 years. The huge caves are home to around 100,000 Mexican freetail bats..

Transportation & industry

New industries have helped reduce the region's dependence on the extraction of minerals and fossil fuels. Precision manufacture has grown rapidly, particularly in Arizona and Colorado. Salt Lake City and Denver are well-established financial centers and New Mexico, the main US producer of uranium, is a prominent region for nuclear research. Colorado is the most important US center for winter sports.

Transportation network

232,434 miles (373,986 km)	4059 miles (6515 km)
8627 miles (13,881 km)	none

The Colorado Rockies are crossed by 32 mountain passes, some as high as 12,183 ft (3713 m). The Eisenhower Tunnel west of Denver carries Interstate Highway 70 straight through the Continental Divide.

Major industry and infrastructure

- chemicals
- coal
- defense
- finance
- food processing
- hi-tech industry
- oil & gas
- mining
- research & development
- winter sports

- major towns
- international airports
- major roads
- major industrial areas

▲ **Glen Canyon Dam** on the Colorado river was completed in 1964. it provides hydroelectric power and irrigation water as part of a long-term federal project to harness the river.

◀ **The flat tablelands** (mesas), and the isolated pinnacles (buttes) which rise from the floor of Monument Valley are the resistant remnants of an earlier land surface, gradually cut back by erosion under arid conditions.

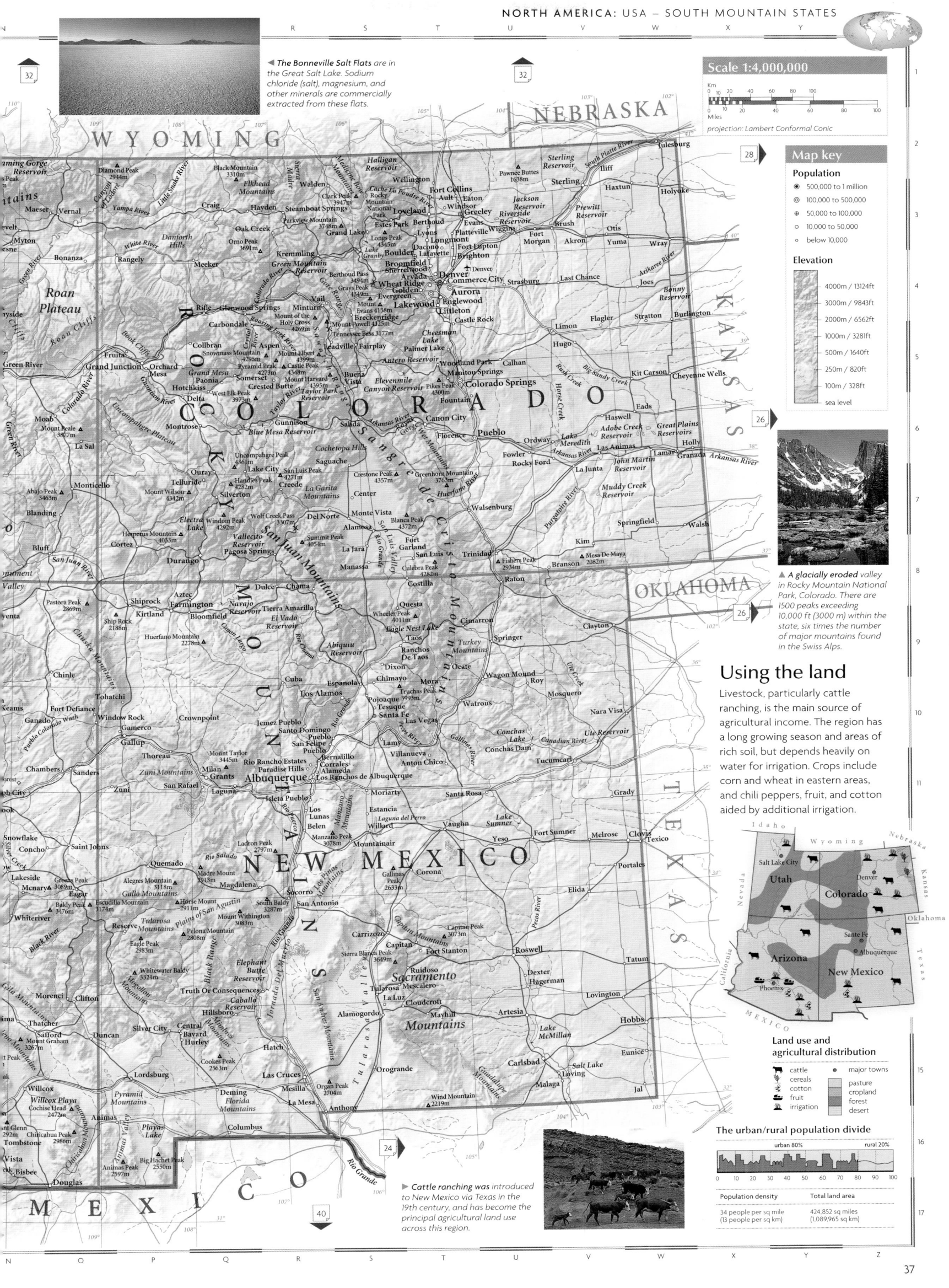

◀ **The Bonneville Salt Flats** are in the Great Salt Lake. Sodium chloride (salt), magnesium, and other minerals are commercially extracted from these flats.

Scale 1:4,000,000

projection: Lambert Conformal Conic

Map key

Population
- 500,000 to 1 million
- 100,000 to 500,000
- 50,000 to 100,000
- 10,000 to 50,000
- below 10,000

Elevation
- 4000m / 13124ft
- 3000m / 9843ft
- 2000m / 6562ft
- 1000m / 3281ft
- 500m / 1640ft
- 250m / 820ft
- 100m / 328ft
- sea level

▲ A glacially eroded valley in Rocky Mountain National Park, Colorado. There are 1500 peaks exceeding 10,000 ft (3000 m) within the state, six times the number of major mountains found in the Swiss Alps.

Using the land

Livestock, particularly cattle ranching, is the main source of agricultural income. The region has a long growing season and areas of rich soil, but depends heavily on water for irrigation. Crops include corn and wheat in eastern areas, and chili peppers, fruit, and cotton aided by additional irrigation.

Land use and agricultural distribution
- cattle
- cereals
- cotton
- fruit
- irrigation
- major towns
- pasture
- cropland
- forest
- desert

The urban/rural population divide

urban 80%   rural 20%

Population density	Total land area
34 people per sq mile (13 people per sq km)	424,852 sq miles (1,089,965 sq km)

▶ **Cattle ranching was** introduced to New Mexico via Texas in the 19th century, and has become the principal agricultural land use across this region.

USA: HAWAII

The 122 islands of the Hawai'ian archipelago – which are part of Polynesia – are the peaks of the world's largest volcanoes. They rise approximately 6 miles (9.7 km) from the floor of the Pacific Ocean. The largest, the island of Hawai'i, remains highly active. Hawaii became the US's 50th state in 1959. A tradition of receiving immigrant workers is reflected in the islands' ethnic diversity, with peoples drawn from around the rim of the Pacific. Only 2% of the current population are native Polynesians.

▲ The island of Moloka'i is formed from volcanic rock. Mature sand dunes cover the rocks in coastal areas.

Transportation & industry

Tourism dominates the economy, with over 90% of the population employed in services. The naval base at Pearl Harbor is also a major source of employment. Industry is concentrated on the island of O'ahu and relies mostly on imported materials, while agricultural produce is processed locally.

Transportation network

🛣 4102 miles (6600 km)		🛣 43 miles (69 km)	
🚊 none		🚂 none	

Hawaii relies on ocean-surface transportation. Honolulu is the main focus of this network, bringing foreign trade and the markets of mainland US to Hawaii's outer islands.

Major industry and infrastructure

🏭 food processing	● major towns		
⚓ military base	⊕ international airports		
textiles	— major roads		
tourism	major industrial areas		

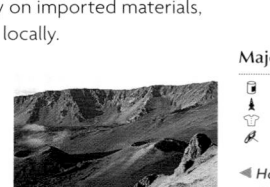

◄ Haleakala's extinct volcanic crater is the world's largest. The giant caldera, containing many secondary cones, is 2000 ft (600 m) deep and 20 miles (32 km) in circumference.

Using the land & sea

The volcanic soils are extremely fertile and the climate hot and humid on the lower slopes, supporting large commercial plantations growing sugar cane, bananas, pineapples, and other tropical fruit, as well as nursery plants and flowers. Some land is given to pasture, particularly for beef and dairy cattle.

Land use and agricultural distribution

🐄	cattle
🐟	fishing
🍍	fruit
⚓	sugar cane
●	major towns

pasture
cropland
forest
mountain region

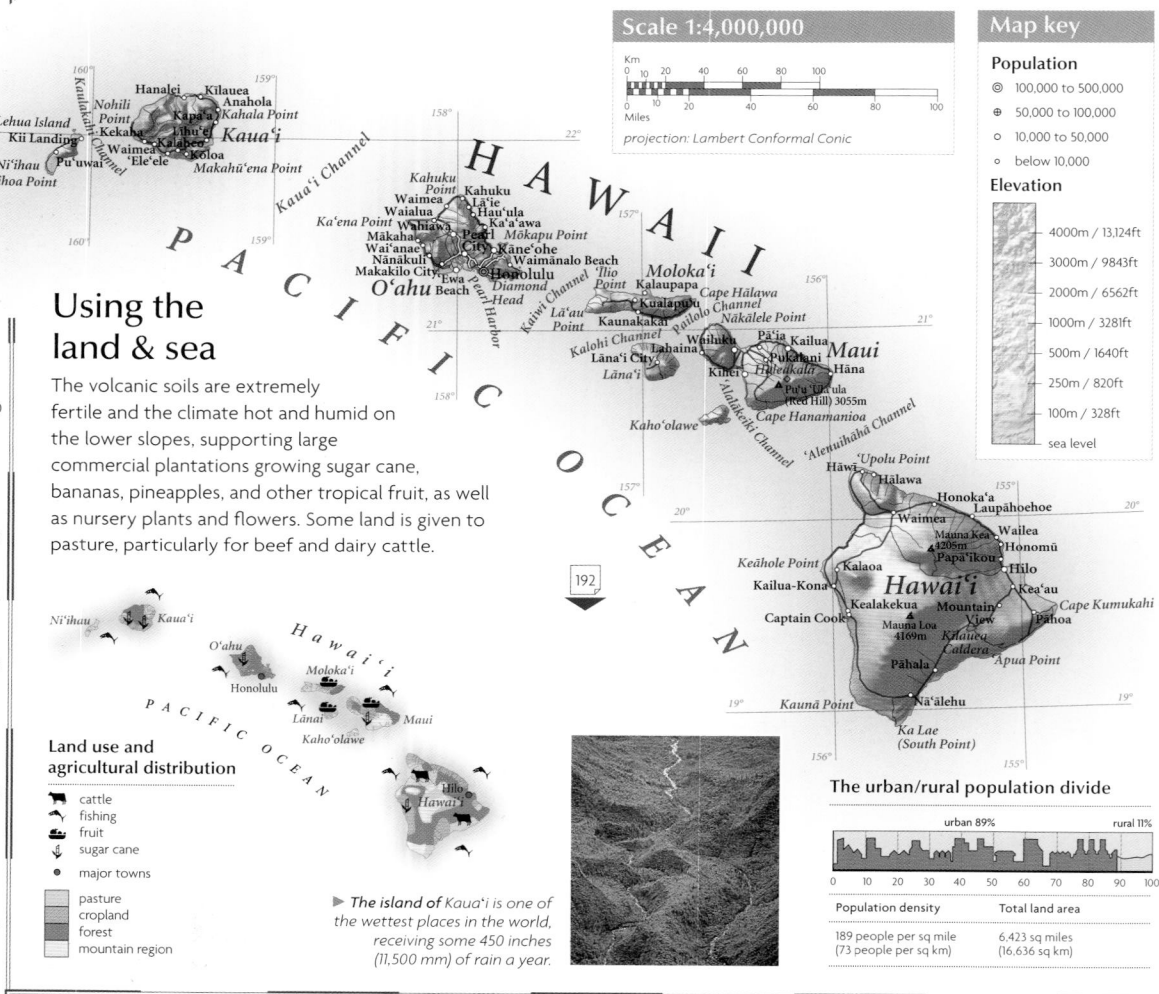

▶ The island of Kaua'i is one of the wettest places in the world, receiving some 450 inches (11,500 mm) of rain a year.

Scale 1:4,000,000

Km
0 10 20 40 60 80 100
Miles
0 10 20 40 60 80 100

projection: Lambert Conformal Conic

Map key

Population
- ◎ 100,000 to 500,000
- ⊕ 50,000 to 100,000
- ○ 10,000 to 50,000
- ○ below 10,000

Elevation
- 4000m / 13,124ft
- 3000m / 9843ft
- 2000m / 6562ft
- 1000m / 3281ft
- 500m / 1640ft
- 250m / 820ft
- 100m / 328ft
- sea level

The urban/rural population divide

urban 89% | rural 11%
0 10 20 30 40 50 60 70 80 90 100

Population density	Total land area
189 people per sq mile (73 people per sq km)	6,423 sq miles (16,636 sq km)

Using the land & sea

The ice-free coastline of Alaska provides access to salmon fisheries and more than 129 million acres (52.2 million ha) of forest. Most of Alaska is uncultivable, and around 90% of food is imported. Barley, hay, and hothouse products are grown around Anchorage, where dairy farming is also concentrated.

The urban/rural population divide

urban 68% | rural 32%
0 10 20 30 40 50 60 70 80 90 100

Population density	Total land area
1 person per sq mile (0.4 people per sq km)	571,951 sq miles (1,481,296 sq km)

◄ A raft of timber from the Tongass forest is hauled by a tug, bound for the pulp mills of the Alaskan coast between Juneau and Ketchikan.

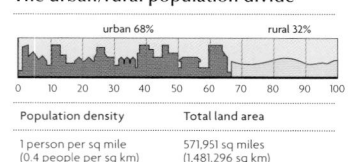

CHUKCHI SEA

RUSSIAN FEDERATION

BERING SEA

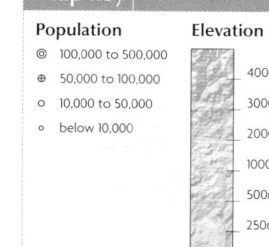
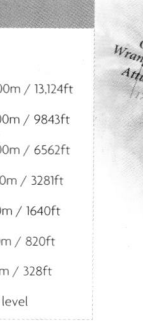

Map key

Population
- ◎ 100,000 to 500,000
- ⊕ 50,000 to 100,000
- ○ 10,000 to 50,000
- ○ below 10,000

Elevation
- 4000m / 13,124ft
- 3000m / 9843ft
- 2000m / 6562ft
- 1000m / 3281ft
- 500m / 1640ft
- 250m / 820ft
- 100m / 328ft
- sea level

Scale 1:9,000,000

Km
0 25 50 100 150 200 250
Miles
0 25 50 100 150 200 250

projection: Lambert Conformal Conic

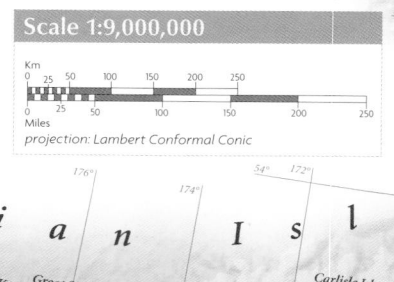
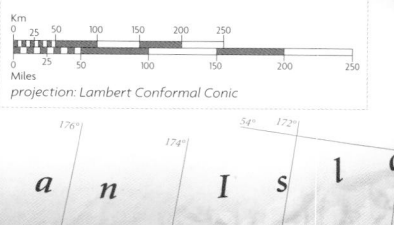

Aleutian Islands

PACIFIC OCEAN

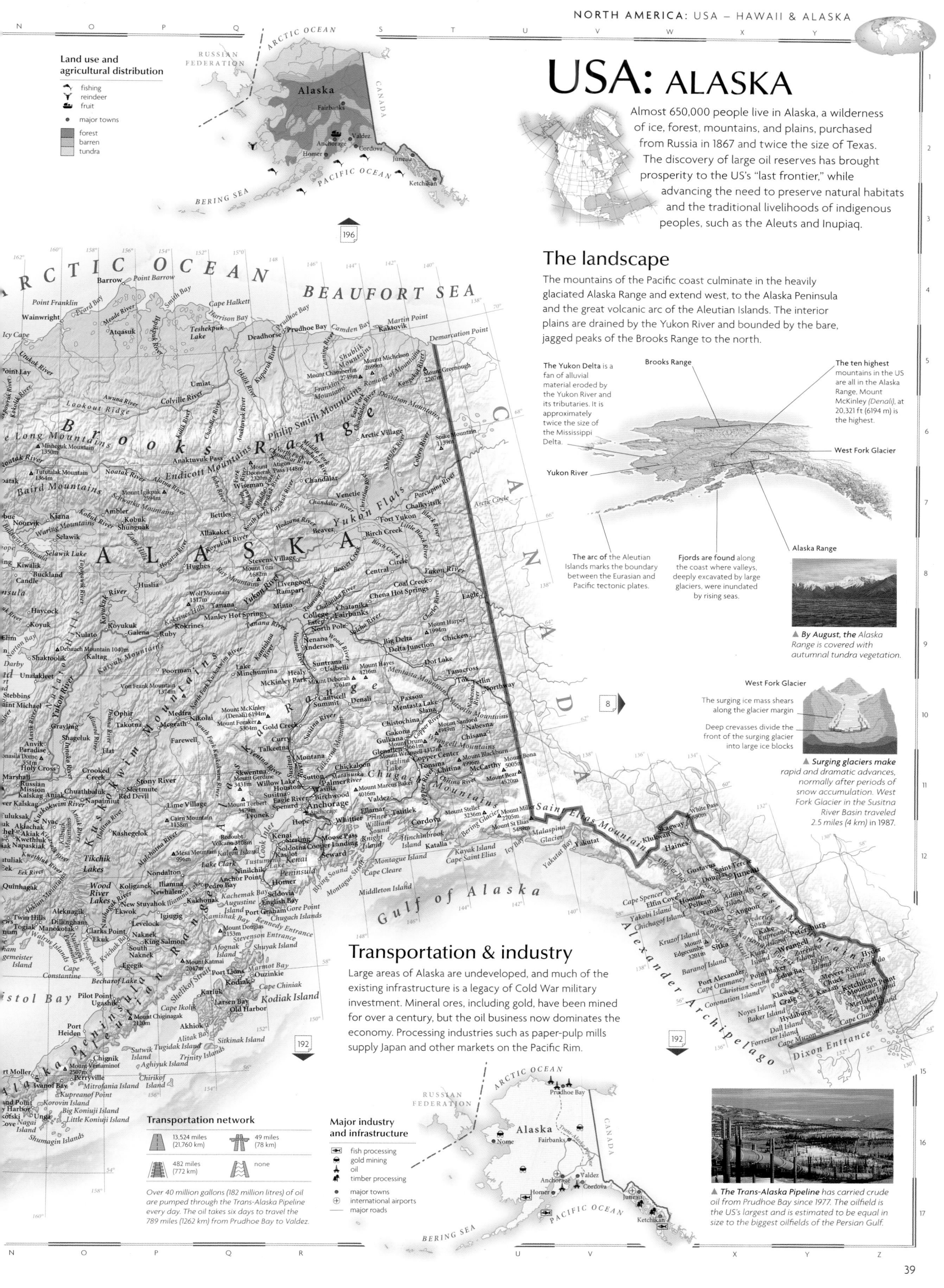

USA: ALASKA

Almost 650,000 people live in Alaska, a wilderness of ice, forest, mountains, and plains, purchased from Russia in 1867 and twice the size of Texas. The discovery of large oil reserves has brought prosperity to the US's "last frontier," while advancing the need to preserve natural habitats and the traditional livelihoods of indigenous peoples, such as the Aleuts and Inupiaq.

The landscape

The mountains of the Pacific coast culminate in the heavily glaciated Alaska Range and extend west, to the Alaska Peninsula and the great volcanic arc of the Aleutian Islands. The interior plains are drained by the Yukon River and bounded by the bare, jagged peaks of the Brooks Range to the north.

Brooks Range

The Yukon Delta is a fan of alluvial material eroded by the Yukon River and its tributaries. It is approximately twice the size of the Mississippi Delta.

The ten highest mountains in the US are all in the Alaska Range, Mount McKinley (Denali), at 20,321 ft (6194 m) is the highest.

West Fork Glacier

Yukon River

The arc of the Aleutian Islands marks the boundary between the Eurasian and Pacific tectonic plates.

Fjords are found along the coast where valleys, deeply excavated by large glaciers, were inundated by rising seas.

Alaska Range

▲ By August, the Alaska Range is covered with autumnal tundra vegetation.

West Fork Glacier

The surging ice mass shears along the glacier margin

Deep crevasses divide the front of the surging glacier into large ice blocks

▲ Surging glaciers make rapid and dramatic advances, normally after periods of snow accumulation. West Fork Glacier in the Susitna River Basin traveled 2.5 miles (4 km) in 1987.

Transportation & industry

Large areas of Alaska are undeveloped, and much of the existing infrastructure is a legacy of Cold War military investment. Mineral ores, including gold, have been mined for over a century, but the oil business now dominates the economy. Processing industries such as paper-pulp mills supply Japan and other markets on the Pacific Rim.

Land use and agricultural distribution

- fishing
- reindeer
- fruit
- major towns
- forest
- barren
- tundra

Transportation network

13,524 miles (21,760 km)	49 miles (78 km)
482 miles (772 km)	none

Over 40 million gallons (182 million litres) of oil are pumped through the Trans-Alaska Pipeline every day. The oil takes six days to travel the 789 miles (1262 km) from Prudhoe Bay to Valdez.

Major industry and infrastructure

- fish processing
- gold mining
- oil
- timber processing
- major towns
- ⊕ international airports
- major roads

▲ The Trans-Alaska Pipeline has carried crude oil from Prudhoe Bay since 1977. The oilfield is the US's largest and is estimated to be equal in size to the biggest oilfields of the Persian Gulf.

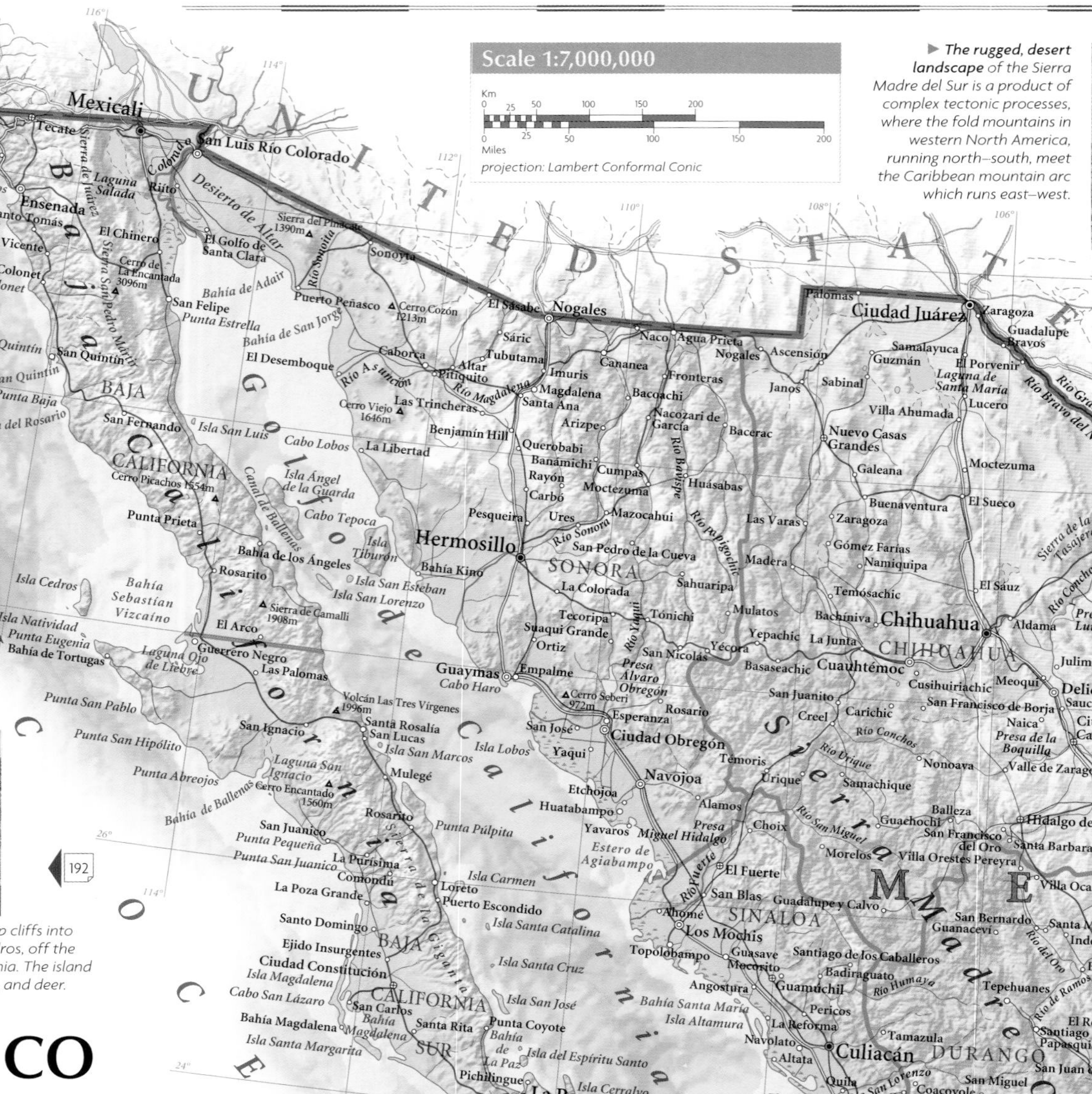

Mexico

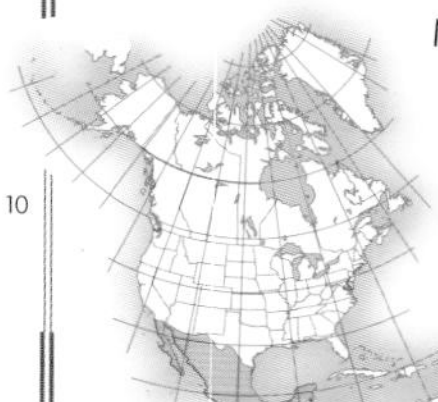

Mexico possesses rich mineral resources, limited agricultural land and the world's largest Spanish-speaking population. Most Mexicans are *mestizo*, although Amerindian communities still exist in the south, almost 500 years after Spain destroyed the Aztec empire at its height. Much of the arid north is sparsely inhabited, while Mexico City is one of the world's most populous cities. Conflict with the US has long overshadowed Mexico's development, but the North American Free Trade Agreement offers the chance for a more benign relationship, which may help to offset Mexico's problems of hyperinflation, foreign debt, unequal wealth distribution, and political instability.

Using the land & sea

Corn occupies much of the cultivated area. Commercial plantations of coffee, sugar, vanilla, and cotton are found along the Gulf coastal plain and in irrigated parts of the arid north, which is otherwise used for extensive ranching. Fishing is important, particularly shellfish for export. A soaring population has created the need for grain imports since 1980.

Scale 1:7,000,000

Km
0 25 50   100   150   200

Miles
0   25   50   100   150   200

projection: Lambert Conformal Conic

▶ The rugged, desert landscape of the Sierra Madre del Sur is a product of complex tectonic processes, where the fold mountains in western North America, running north–south, meet the Caribbean mountain arc which runs east–west.

▲ Wave action has cut steep cliffs into the igneous rocks of Isla Cedros, off the Pacific coast of Baja California. The island is home to sea lions, reptiles, and deer.

The urban/rural population divide

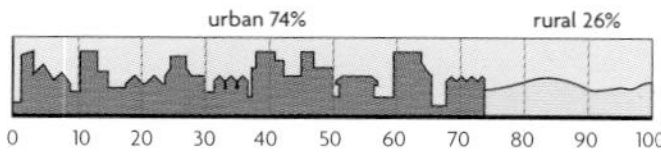

urban 74%   rural 26%

0   10   20   30   40   50   60   70   80   90   100

Population density	Total land area
140 people per sq mile (54 people per sq km)	755,865 sq miles (1,958,200 sq km)

Land use and agricultural distribution

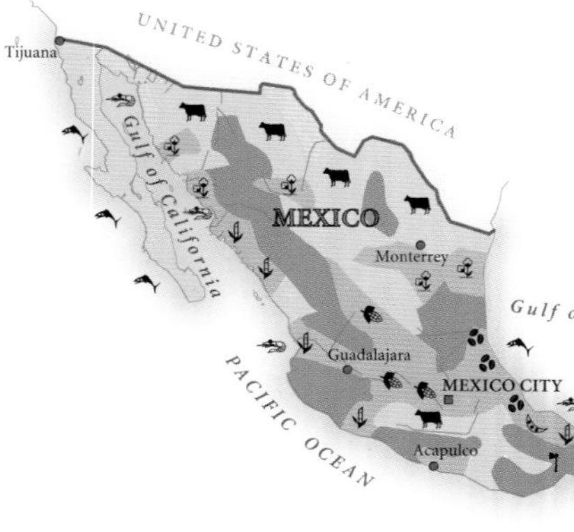

- cattle
- coffee
- corn
- cotton
- fishing
- shellfish
- sugar cane
- timber
- vanilla

- ■ capital cities
- ● major towns

- pasture
- cropland
- forest
- desert

▶ Coffee beans spread out to dry in the sun. Coffee, grown mainly on the Gulf coastal plain, is Mexico's most valuable export crop.

Map key

Mexico: Administrative regions

⊕ Distrito Federal

Population

- ■ above 5 million
- ▣ 1 million to 5 million
- ◉ 500,000 to 1 million
- ◎ 100,000 to 500,000
- ⊕ 50,000 to 100,000
- ⊙ 10,000 to 50,000
- ○ below 10,000

Elevation

- 4000m / 13,124ft
- 3000m / 9843ft
- 2000m / 6562ft
- 1000m / 3281ft
- 500m / 1640ft
- 250m / 820ft
- 100m / 328ft
- sea level

The landscape

The great central plateau rises gently southward from the Rio Grande, isolated from the coastal plains by the Sierra Madre Oriental and Occidental. The two ranges converge from east and west respectively, culminating in high volcanic peaks around Mexico City. Further ranges of the Sierra Madre rise to the south of the Balsas basin, skirted by the low-lying Isthmus of Tehuantepec (*Istmo de Tehuantepec*) and Yucatan Peninsula.

The long, narrow, extremely arid peninsula of Baja (lower) California is an elongated granite block, separated from the mainland by the flooded rift valley of the Gulf of California (*Golfo de California*).

Wave action has constructed sand bars which shelter lagoons along the shore of the Gulf coastal plain.

The dormant cone of Volcán Pico de Orizaba is, at 18,700 ft (5700 m), the highest peak in Mexico. In North America, only Mount McKinley and Mount Logan are taller.

Sierra Madre Oriental

Rio Grande

The heavily-forested Isthmus of Tehuantepec (*Istmo de Tehuantepec*) is a graben; a low-lying trough created by downward movement of the bedrock between two fault lines.

▲ *Tropical rainforest abounds* in the Yucatan Peninsula, a broad, low limestone shelf. Rivers are rare due to the porous nature of limestone, so the forest is mostly fed by streams and underground water.

Formation of the Gulf of California

— Direction of plate movement
Baja California
Transform fault
Gulf of California
Edge of continental crust
Spreading oceanic ridge

Sierra Madre Occidental

Río Balsas

Popocatépetl

▲ *The Gulf of* California (Golfo de California) began to open out about 4 million years ago as a result of rifting and plate displacement along transform faults.

▲ *Popocatépetl is a* dormant volcano, part of the Pacific "Ring of Fire." The crater is over half a mile (1 km) wide.

The unstable, earthquake-prone, upland basin around Mexico City was once a region of shallow lakes. Flood control measures and domestic consumption over the last four centuries have caused the virtual disappearance of this surface water.

The highlands of Chiapas are a series of *horsts*, blocks of land thrust upward between two fault lines. Volcanic cones have developed where lava has flowed out from the faults.

Transportation & industry

Oil and gas on the Gulf coast are Mexico's main sources of export income. Metal mining has declined but the country remains a leading global producer of silver. Manufacturing is heavily concentrated around the metropolitan area of Mexico City, while the duty-free movement of goods in the US border region, under the *Maquiladora* (twin plant) scheme, has created new hi-tech and service growth centers.

Major industry and infrastructure

- brewing
- car manufacture
- chemicals
- electronics
- fish processing
- maquiladoras
- mining
- oil & gas
- textiles
- capital cities
- major towns
- international airports
- major roads
- major industrial areas

Transportation network

67,564 miles (108,746 km)

3994 miles (6429 km)

16,561 miles (26,656 km)

1801 miles (2900 km)

Fast, modern highways or autopistas now link Mexico City with Toluca, Puebla and other satellite cities, yet distant centers like Chihuahua are still served by narrow roads and an outdated railroad network.

▲ *A stone figure* reclines by the Temple of Warriors, within the Mayan city of Chichén-Itzá. The Maya civilization flourished across the Yucatan Peninsula between 200 and 900 AD.

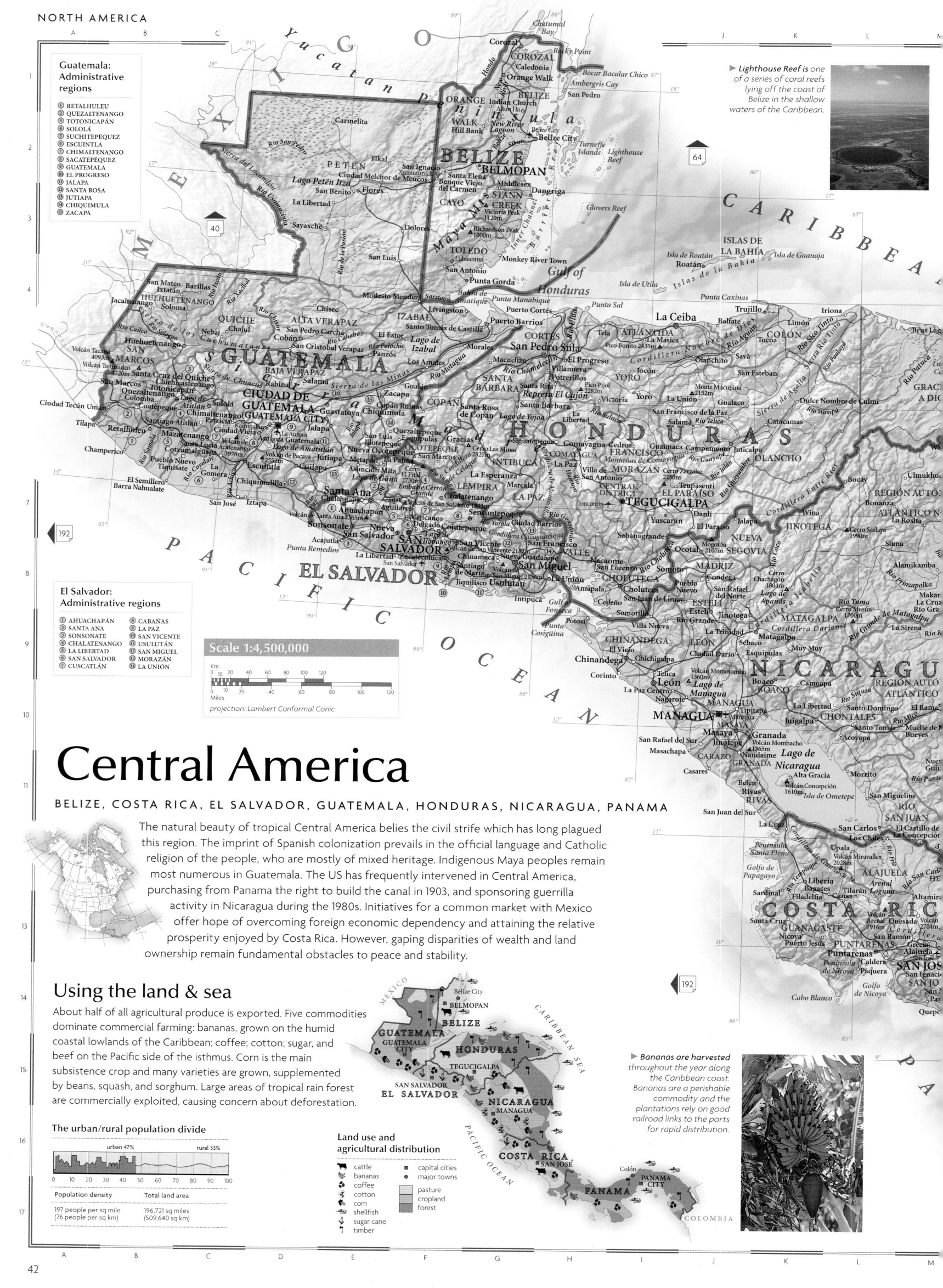

Guatemala: Administrative regions
1. RETALHULEU
2. QUEZALTENANGO
3. TOTONICAPÁN
4. SOLOLÁ
5. SUCHITEPÉQUEZ
6. ESCUINTLA
7. CHIMALTENANGO
8. SACATEPÉQUEZ
9. GUATEMALA
10. EL PROGRESO
11. JALAPA
12. SANTA ROSA
13. JUTIAPA
14. CHIQUIMULA
15. ZACAPA

El Salvador: Administrative regions
1. AHUACHAPÁN
2. SANTA ANA
3. SONSONATE
4. CHALATENANGO
5. LA LIBERTAD
6. SAN SALVADOR
7. CUSCATLÁN
8. CABAÑAS
9. LA PAZ
10. SAN VICENTE
11. USULUTÁN
12. SAN MIGUEL
13. MORAZÁN
14. LA UNIÓN

▶ Lighthouse Reef is one of a series of coral reefs lying off the coast of Belize in the shallow waters of the Caribbean.

Scale 1:4,500,000
Km 0 20 40 60 80 100 120
Miles 0 10 20 40 60 80 100 120
projection: Lambert Conformal Conic

Central America

BELIZE, COSTA RICA, EL SALVADOR, GUATEMALA, HONDURAS, NICARAGUA, PANAMA

The natural beauty of tropical Central America belies the civil strife which has long plagued this region. The imprint of Spanish colonization prevails in the official language and Catholic religion of the people, who are mostly of mixed heritage. Indigenous Maya peoples remain most numerous in Guatemala. The US has frequently intervened in Central America, purchasing from Panama the right to build the canal in 1903, and sponsoring guerrilla activity in Nicaragua during the 1980s. Initiatives for a common market with Mexico offer hope of overcoming foreign economic dependency and attaining the relative prosperity enjoyed by Costa Rica. However, gaping disparities of wealth and land ownership remain fundamental obstacles to peace and stability.

Using the land & sea

About half of all agricultural produce is exported. Five commodities dominate commercial farming: bananas, grown on the humid coastal lowlands of the Caribbean; coffee; cotton; sugar, and beef on the Pacific side of the isthmus. Corn is the main subsistence crop and many varieties are grown, supplemented by beans, squash, and sorghum. Large areas of tropical rain forest are commercially exploited, causing concern about deforestation.

The urban/rural population divide
urban 47% | rural 53%
0 10 20 30 40 50 60 70 80 90 100

Population density
197 people per sq mile
(76 people per sq km)

Total land area
196,721 sq miles
(509,640 sq km)

Land use and agricultural distribution
- cattle
- bananas
- coffee
- cotton
- corn
- shellfish
- sugar cane
- timber
- capital cities
- major towns
- pasture
- cropland
- forest

▶ Bananas are harvested throughout the year along the Caribbean coast. Bananas are a perishable commodity and the plantations rely on good railroad links to the ports for rapid distribution.

The landscape

The Sierra Madre range spreads west from Mexico, between the narrow Pacific coastal plain and the limestone lowland of Petén. Parallel hill ranges sweep across Honduras and extend south, past the Caribbean Mosquito Coast, to lakes Managua and Nicaragua. The Cordillera Central rises to the south, gradually descending to Lake Gatún (Lago Gatún). A highly active volcanic belt runs along the Pacific seaboard from Mexico to Costa Rica.

Over 40 active volcanoes line the Pacific coast north of Panama, including Volcán Tajumulco which, at 13,846 ft (4220 m), is the highest point in Central America.

The high plateau of the Sierra de los Cuchumatanes is a *horst*, an upthrusted block of land. The limestone rock is deeply incised with canyons along the plateau edge.

Lake Petén Itzá is typical of the swampy depressions or *bajos* of the Petén region, formed by intense weathering of limestone in the hot and humid climate.

Low, white limestone cliffs, mangrove swamps and coral reefs characterize the coast of Belize, which is part of the Yucatan Peninsula.

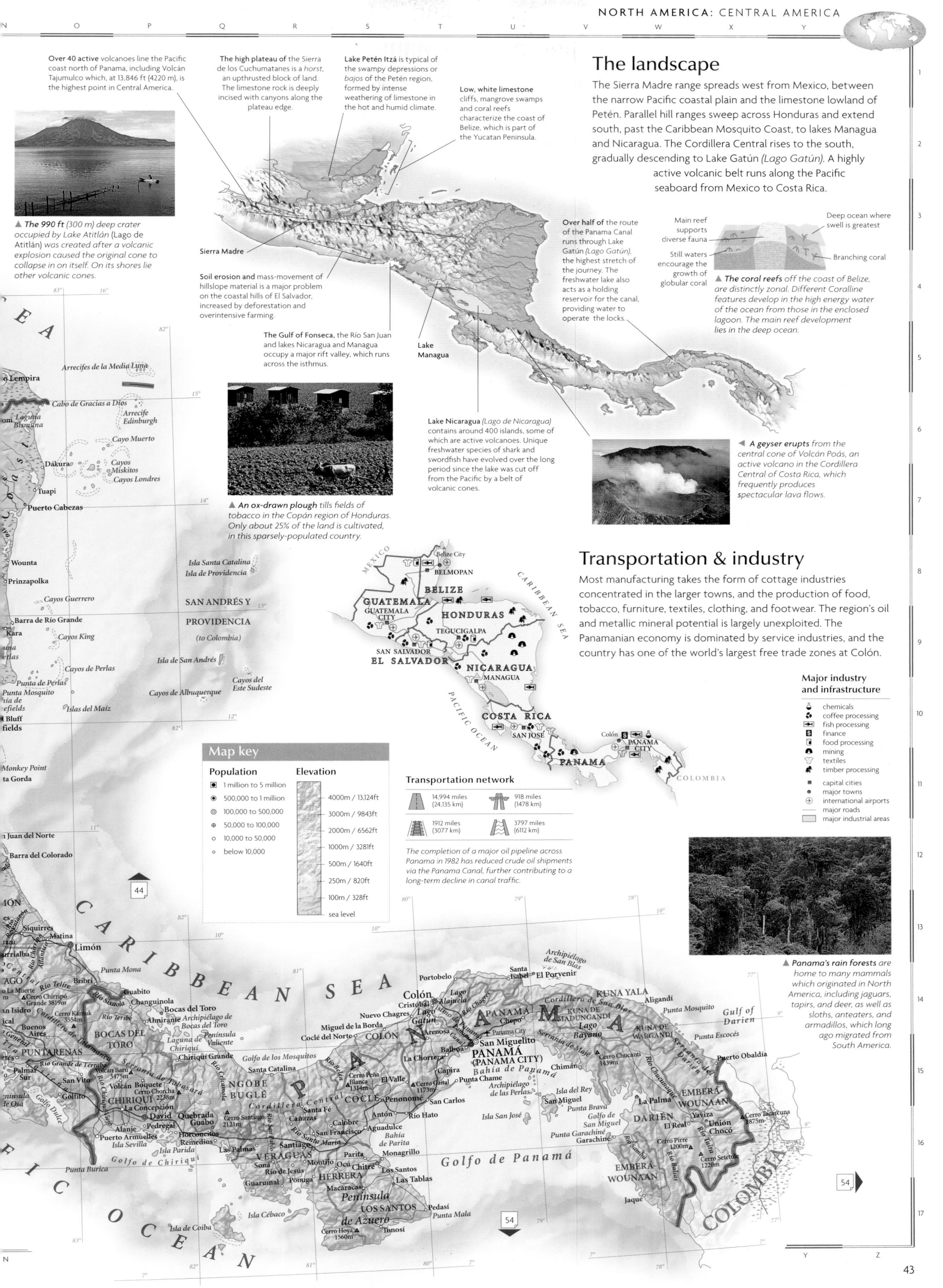

▲ *The 990 ft (300 m) deep crater occupied by Lake Atitlán (Lago de Atitlán) was created after a volcanic explosion caused the original cone to collapse in on itself. On its shores lie other volcanic cones.*

Sierra Madre

Soil erosion and mass-movement of hillslope material is a major problem on the coastal hills of El Salvador, increased by deforestation and overintensive farming.

The Gulf of Fonseca, the Río San Juan and lakes Nicaragua and Managua occupy a major rift valley, which runs across the isthmus.

Lake Managua

Over half of the route of the Panama Canal runs through Lake Gatún (Lago Gatún), the highest stretch of the journey. The freshwater lake also acts as a holding reservoir for the canal, providing water to operate the locks.

Main reef supports diverse fauna

Deep ocean where swell is greatest

Still waters encourage the growth of globular coral

Branching coral

▲ *The coral reefs off the coast of Belize, are distinctly zonal. Different Coralline features develop in the high energy water of the ocean from those in the enclosed lagoon. The main reef development lies in the deep ocean.*

Lake Nicaragua (Lago de Nicaragua) contains around 400 islands, some of which are active volcanoes. Unique freshwater species of shark and swordfish have evolved over the long period since the lake was cut off from the Pacific by a belt of volcanic cones.

◄ *A geyser erupts from the central cone of Volcán Poás, an active volcano in the Cordillera Central of Costa Rica, which frequently produces spectacular lava flows.*

▲ *An ox-drawn plough tills fields of tobacco in the Copán region of Honduras. Only about 25% of the land is cultivated, in this sparsely-populated country.*

Transportation & industry

Most manufacturing takes the form of cottage industries concentrated in the larger towns, and the production of food, tobacco, furniture, textiles, clothing, and footwear. The region's oil and metallic mineral potential is largely unexploited. The Panamanian economy is dominated by service industries, and the country has one of the world's largest free trade zones at Colón.

Major industry and infrastructure

- ⚗ chemicals
- ☕ coffee processing
- 🐟 fish processing
- $ finance
- 🍴 food processing
- ⛏ mining
- 👕 textiles
- 🌲 timber processing

- ■ capital cities
- ● major towns
- ⊕ international airports
- — major roads
- ▨ major industrial areas

Map key

Population

- ◉ 1 million to 5 million
- ◉ 500,000 to 1 million
- ◎ 100,000 to 500,000
- ⊕ 50,000 to 100,000
- ○ 10,000 to 50,000
- ○ below 10,000

Elevation

	4000m / 13,124ft
	3000m / 9843ft
	2000m / 6562ft
	1000m / 3281ft
	500m / 1640ft
	250m / 820ft
	100m / 328ft
	sea level

Transportation network

🛣 14,994 miles (24,135 km)		🛣 918 miles (1478 km)	
🚂 1912 miles (3077 km)		🚂 3797 miles (6112 km)	

The completion of a major oil pipeline across Panama in 1982 has reduced crude oil shipments via the Panama Canal, further contributing to a long-term decline in canal traffic.

▲ *Panama's rain forests are home to many mammals which originated in North America, including jaguars, tapirs, and deer, as well as sloths, anteaters, and armadillos, which long ago migrated from South America.*

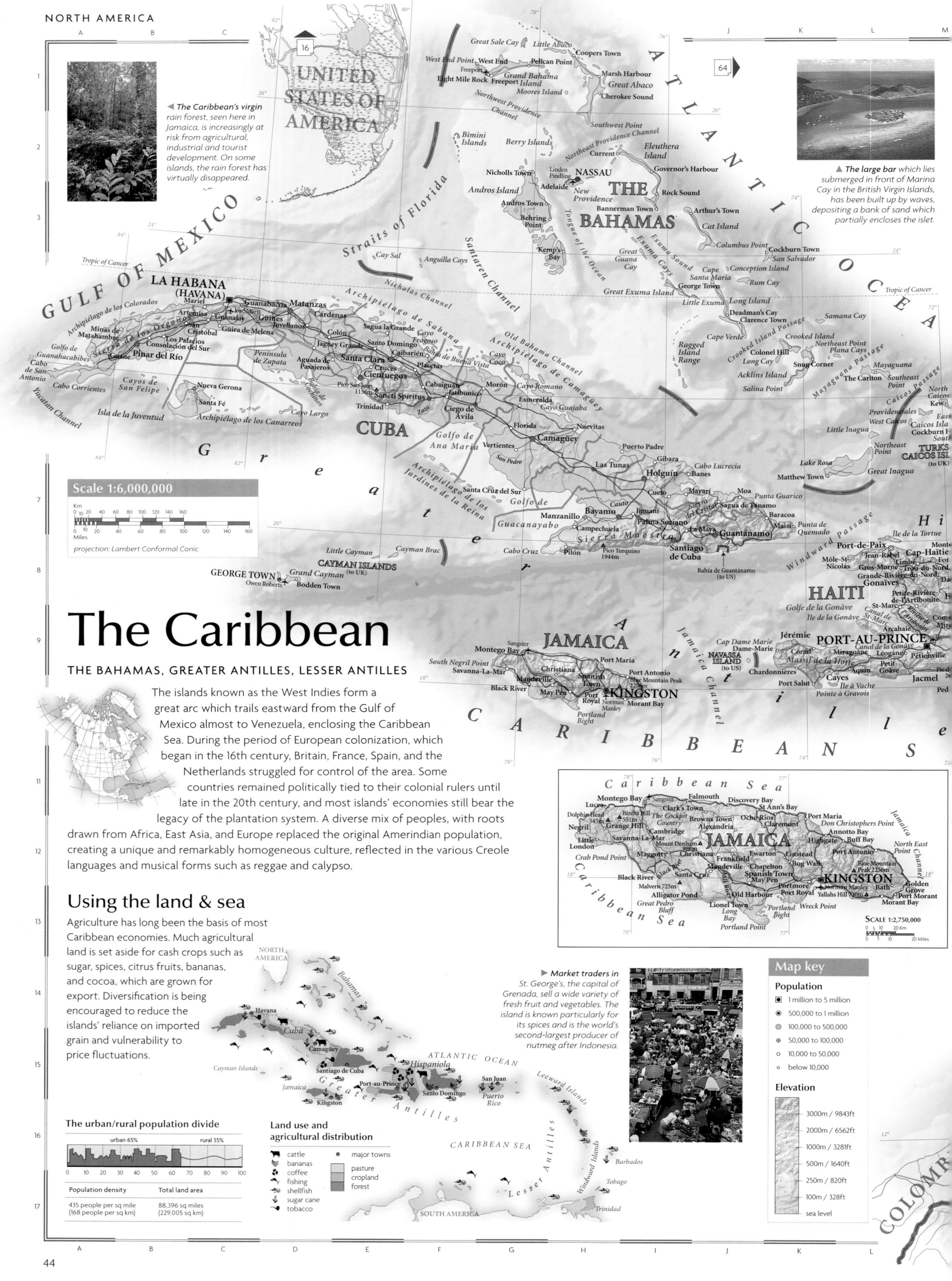

The Caribbean

THE BAHAMAS, GREATER ANTILLES, LESSER ANTILLES

The islands known as the West Indies form a great arc which trails eastward from the Gulf of Mexico almost to Venezuela, enclosing the Caribbean Sea. During the period of European colonization, which began in the 16th century, Britain, France, Spain, and the Netherlands struggled for control of the area. Some countries remained politically tied to their colonial rulers until late in the 20th century, and most islands' economies still bear the legacy of the plantation system. A diverse mix of peoples, with roots drawn from Africa, East Asia, and Europe replaced the original Amerindian population, creating a unique and remarkably homogeneous culture, reflected in the various Creole languages and musical forms such as reggae and calypso.

Using the land & sea

Agriculture has long been the basis of most Caribbean economies. Much agricultural land is set aside for cash crops such as sugar, spices, citrus fruits, bananas, and cocoa, which are grown for export. Diversification is being encouraged to reduce the islands' reliance on imported grain and vulnerability to price fluctuations.

◄ *The Caribbean's virgin rain forest, seen here in Jamaica, is increasingly at risk from agricultural, industrial and tourist development. On some islands, the rain forest has virtually disappeared.*

▲ *The large bar which lies submerged in front of Marina Cay in the British Virgin Islands, has been built up by waves, depositing a bank of sand which partially encloses the islet.*

▶ *Market traders in St. George's, the capital of Grenada, sell a wide variety of fresh fruit and vegetables. The island is known particularly for its spices and is the world's second-largest producer of nutmeg after Indonesia.*

Scale 1:6,000,000
projection: Lambert Conformal Conic

SCALE 1:2,750,000

The urban/rural population divide

urban 65%   rural 35%

Population density	Total land area
435 people per sq mile (168 people per sq km)	88,396 sq miles (229,005 sq km)

Land use and agricultural distribution

- cattle
- bananas
- coffee
- fishing
- shellfish
- sugar cane
- tobacco
- major towns
- pasture
- cropland
- forest

Map key

Population
- 1 million to 5 million
- 500,000 to 1 million
- 100,000 to 500,000
- 50,000 to 100,000
- 10,000 to 50,000
- below 10,000

Elevation
- 3000m / 9843ft
- 2000m / 6562ft
- 1000m / 3281ft
- 500m / 1640ft
- 250m / 820ft
- 100m / 328ft
- sea level

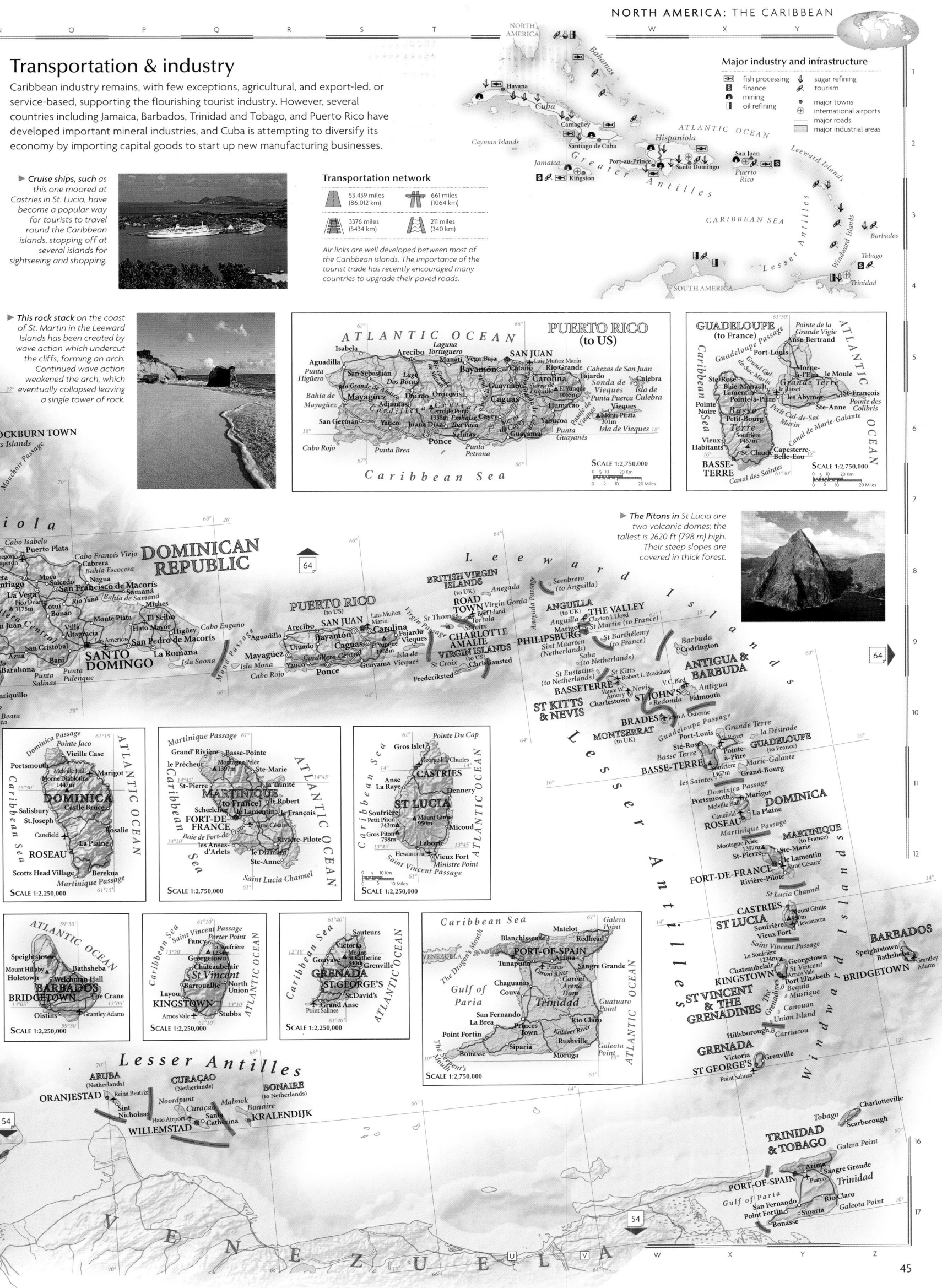

Transportation & industry

Caribbean industry remains, with few exceptions, agricultural, and export-led, or service-based, supporting the flourishing tourist industry. However, several countries including Jamaica, Barbados, Trinidad and Tobago, and Puerto Rico have developed important mineral industries, and Cuba is attempting to diversify its economy by importing capital goods to start up new manufacturing businesses.

▶ **Cruise ships,** such as this one moored at Castries in St. Lucia, have become a popular way for tourists to travel round the Caribbean islands, stopping off at several islands for sightseeing and shopping.

▶ **This rock stack** on the coast of St. Martin in the Leeward Islands has been created by wave action which undercut the cliffs, forming an arch. Continued wave action weakened the arch, which eventually collapsed leaving a single tower of rock.

Major industry and infrastructure

- fish processing
- finance
- mining
- oil refining
- sugar refining
- tourism
- major towns
- international airports
- major roads
- major industrial areas

Transportation network

53,439 miles (86,012 km)		661 miles (1064 km)	
3376 miles (5434 km)		211 miles (340 km)	

Air links are well developed between most of the Caribbean islands. The importance of the tourist trade has recently encouraged many countries to upgrade their paved roads.

▶ **The Pitons in** St Lucia are two volcanic domes; the tallest is 2620 ft (798 m) high. Their steep slopes are covered in thick forest.

PUERTO RICO (to US) SCALE 1:2,750,000

GUADELOUPE (to France) SCALE 1:2,750,000

DOMINICA SCALE 1:2,250,000

MARTINIQUE (to France) SCALE 1:2,750,000

ST LUCIA SCALE 1:2,250,000

BARBADOS SCALE 1:2,250,000

ST VINCENT SCALE 1:2,250,000

GRENADA SCALE 1:2,250,000

Trinidad / PORT-OF-SPAIN SCALE 1:2,750,000

South America

Reaching from the humid tropics down into the cold south Atlantic, South America has an area of 6,886,000 sq miles (17,835,000 sq km). There are 12 separate countries, with the largest, Brazil, covering almost half the continent.

- ⬤ *Greatest extent, North–South:* 4750 miles / 7640 km
- ⬛ *Greatest extent, East–West:* 3100 miles / 4990 km

Punta Gallinas, Colombia 12° 28' N

Most northerly point:
Punta Gallinas, Colombia
12° 28' N

Most westerly point:
Galapagos Islands,
Ecuador 92° W

Equator

Punta Pariñas, Peru
81° 20' W

Cabo Branco /
Ponta do Seixas,
Brazil 34° 50' W

Largest lake: Lake Titicaca,
Bolivia/Peru 3220 sq miles
(8340 sq km)

Highest recorded temperature:
Rivadavia, Argentina
120°F (49°C)

Tropic of Capricorn

Antofagasta

São Paulo

Most easterly point:
Ilhas Martin Vaz, Brazil
28° 51' W

Highest point:
Cerro Aconcagua, Argentina
22,833 ft (6959 m)

Lowest recorded temperature:
Sarmiento, Argentina
-27°F (-33°C)

Lowest point:
Laguna del Carbón, Argentina
-344 ft (-105 m)

Cape Froward, Chile
53° 54' S

Most southerly point:
Cape Horn, Chile
55° 59' S

Antofagasta, Chile | Atacama Desert | Andes | Paraguay river | Planalto de Mato Grosso | São Paulo, Brazil

Cross-section from Antofagasta, Chile to São Paulo, Brazil

line of cross-section

0   250   500   750   1000 Km
0   250   500   750   1000 Miles

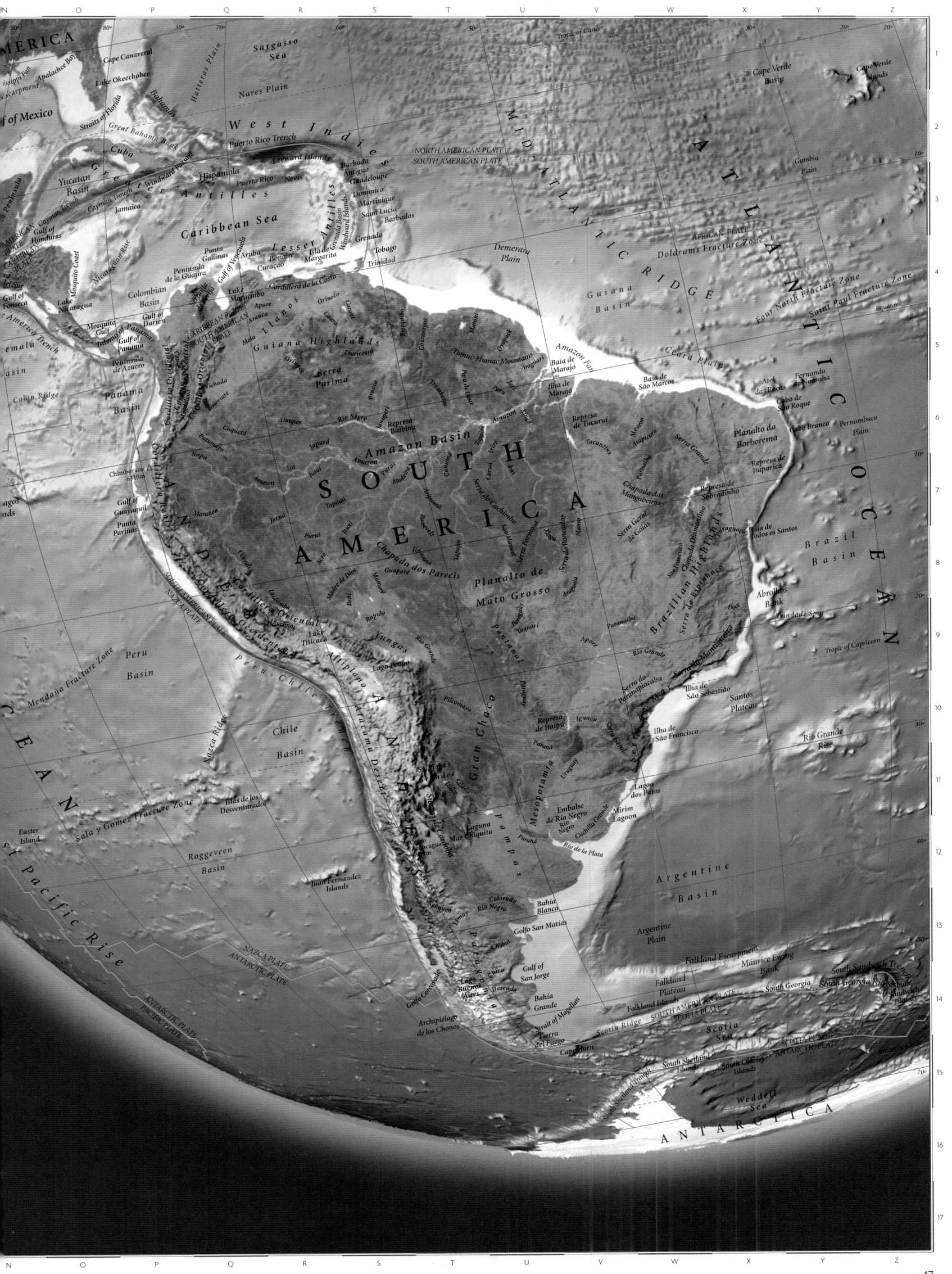

Cape Canaveral
Apalachee Bay
Lake Okeechobee
Hatteras Plain
Sargasso
Sea
Cape Verde
Basin
Cape Verde
Islands

Mississippi Fan
Apalachee Bay
Straits of Florida
Nares Plain

Gulf of Mexico
Great Bahama Bank
Bahamas
West Indies
Gumbia
Plain

Cuba
Puerto Rico Trench
Leeward Islands
Barbuda
NORTH AMERICAN PLATE
SOUTH AMERICAN PLATE

Yucatan
Basin
Hispaniola
Puerto Rico
Nevis
Antigua
Guadeloupe

Cayman Trough
Jamaica
Lesser Antilles
Dominica
Martinique
Saint Lucia
Barbados

Gulf of
Honduras
Caribbean Sea
Punta
Gallinas
Isla de
Margarita
Grenada
AFRICAN PLATE
Doldrums Fracture Zone

Lake
Nicaragua
Colombian
Basin
Peninsula
de la Guajira
Gulf of Venezuela
Aruba
Bonaire
Curaçao
Windward Islands
Tobago
Trinidad
Demerara
Plain
Four North Fracture Zone
Saint Paul Fracture Zone

Gulf of
Fonseca
Mosquito
Gulf
Cordillera de la Costa
Orinoco
Apure
Guiana
Basin
Equator

America Trench
Isthmus of
Panama
Gulf of
Darien
Llanos
Caura
Ceara Plain

Panama
Basin
Peninsula
de Azuero
Guiana Highlands
Casiquiare
Branco
Tumuc-Humac Mountains
Baia de
Marajó
Ceara Plain
Atol
das Rocas
Fernando
de Noronha

Colón Ridge
Serra
Parima
Ilha de
Marajó
Amazon Fan
Cabo de
São Roque

Chimborazo
6310m
Rio Negro
Uaupes
Represa
Balbina
Amazon Basin
Represa
de Tucuruí
Planalto da
Borborema
Cabo Branco
Pernambuco
Plain

Gulf of
Guayaquil
Napo
Içá
SOUTH
Amazon
Tocantins
Serra Grande
Represa de
Itaparica

Punta
Parinas
Marañón
Jurua
Purus
AMERICA
Chapada das
Mangabeiras
Represa de
Sobradinho

Peru
Basin
Madre de Dios
Chapada dos Parecis
Planalto de
Mato Grosso
Serra Geral
de Goiás
Brazilian Highlands
Brazil
Basin

Mendaña Fracture Zone
Cordillera Oriental
Lake
Titicaca
Altiplano
Yungas
Pantanal
Serra do Espinhaco
Abrolhos
Bank
Tropic of Capricorn

Chile
Basin
Atacama Desert
Rio Grande
Gran Chaco
Serra da Mantiqueira
Rio Grande
Rise

Nazca Ridge
Sala y Gomez Fracture Zone
Islas de los
Desventurados
Represa
de Itaipu
Iguaçu
Ilha de
São Sebastião
Santos
Plateau

Easter
Island
Roggeveen
Basin
Juan Fernandez
Islands
Laguna
Mar Chiquita
Pampas
Uruguay
Ilha de
São Francisco
Lagoa
dos Patos
Mirim
Lagoon

East Pacific Rise
Colorada
Rio Negro
Bahia
Blanca
Embalse
de Río Negro
Rio de la Plata
Argentine
Basin

NAZCA PLATE
ANTARCTIC PLATE
Limay
Chubut
Golfo San Matías
Argentine
Plain
Falkland Escarpment
Maurice Ewing
Bank
South Sandwich Trench

ANTARCTIC PLATE
PACIFIC PLATE
Gulf of
San Jorge
Falkland
Plateau
South Georgia
South Georgia Ridge
Sandwich
Islands

Lago
Buenos
Aires
Bahia
Grande
SOUTH AMERICA PLATE
Scotia Ridge
SCOTIA PLATE
Scotia
Sea

Archipiélago
de los Chonos
Strait of Magellan
Tierra
del Fuego
Cape Horn
South Shetland
Islands
South Orkney
Islands
ANTARCTIC PLATE

Weddell
Sea
ANTARCTICA

Physical South America

Three major physiographic regions characterize South America. The oldest, the ancient Brazilian Shield and the smaller Guiana and Patagonian shields, form the stable core of the continent. Stretching along the entire west coast are the younger Andean fold mountains with many summits rising to 20,000 ft (6100 m). These two diverse regions are separated by a number of sedimentary basins carrying South America's large river systems to the sea. These include the massive Amazon Basin and the basin of the Gran Chaco.

The Amazon Basin and Guiana Shield

The Amazon river occupies a large depression in the Earth's crust, formed by the uplift of the Andes. It is covered by thick volcanic deposits and layers of alluvium – these have been laid down by the Amazon's many tributaries. To the north is the smaller Guiana Shield.

Headwaters of the Amazon rise in the Andes — Thick alluvium deposits — Mouths of the Amazon

A ——— A

Section across northern South America showing Amazon Basin and its drainage pattern.

0 500 1000 Km
0 500 1000 Miles

Scale 1:30,500,000

Km 0 200 400 600 800
Miles 0 200 400 600 800

projection: Lambert Azimuthal Equal Area

The Andean Uplands

The Andean Uplands run along the west coast of South America. They are being uplifted as the Nazca Plate is subducted beneath the South American Plate. They contain some of the world's largest volcanoes, such as Cotopaxi, and Lake Titicaca which occupies a dormant site. The far south has many large ice-sheets and a fragmented coastline.

Nazca Plate — South American Plate — Volcanic intrusions

B ——— B

Cross-section through the Andes showing the subduction of the Nazca Plate beneath the South American Plate.

0 200 400 Km
0 200 400 Miles

The Brazilian Shield and Gran Chaco

The immense Brazilian Shield underlies more than one-third of South America. It is pitted with numerous volcanic intrusions, and a large basaltic plateau exists between the Paraná river and the Atlantic Ocean. The flat Gran Chaco lies to the west of the shield, covered by sedimentary deposits eroded from the Andes, and transported by South America's mighty rivers.

Young, folded Andes mountains — Volcanic intrusions — Major rivers drain to the south through the Gran Chaco — Ancient resistant shield

C ——— C

Section across central South America showing the flat basin of the Gran Chaco and the ancient Brazilian Shield.

0 200 400 Km
0 200 400 Miles

Map key

Elevation

- 6000m / 19,686ft
- 4000m / 13,124ft
- 3000m / 9843ft
- 2000m / 6562ft
- 1000m / 3281ft
- 500m / 1640ft
- 250m / 820ft
- 100m / 328ft
- sea level

Plate margins
(for explanation see page xiv)

- ——— constructive
- △ △ destructive
- ——— conservative
- .... uncertain
- ——— physiographic regions
- ▶—◀ line of cross-section

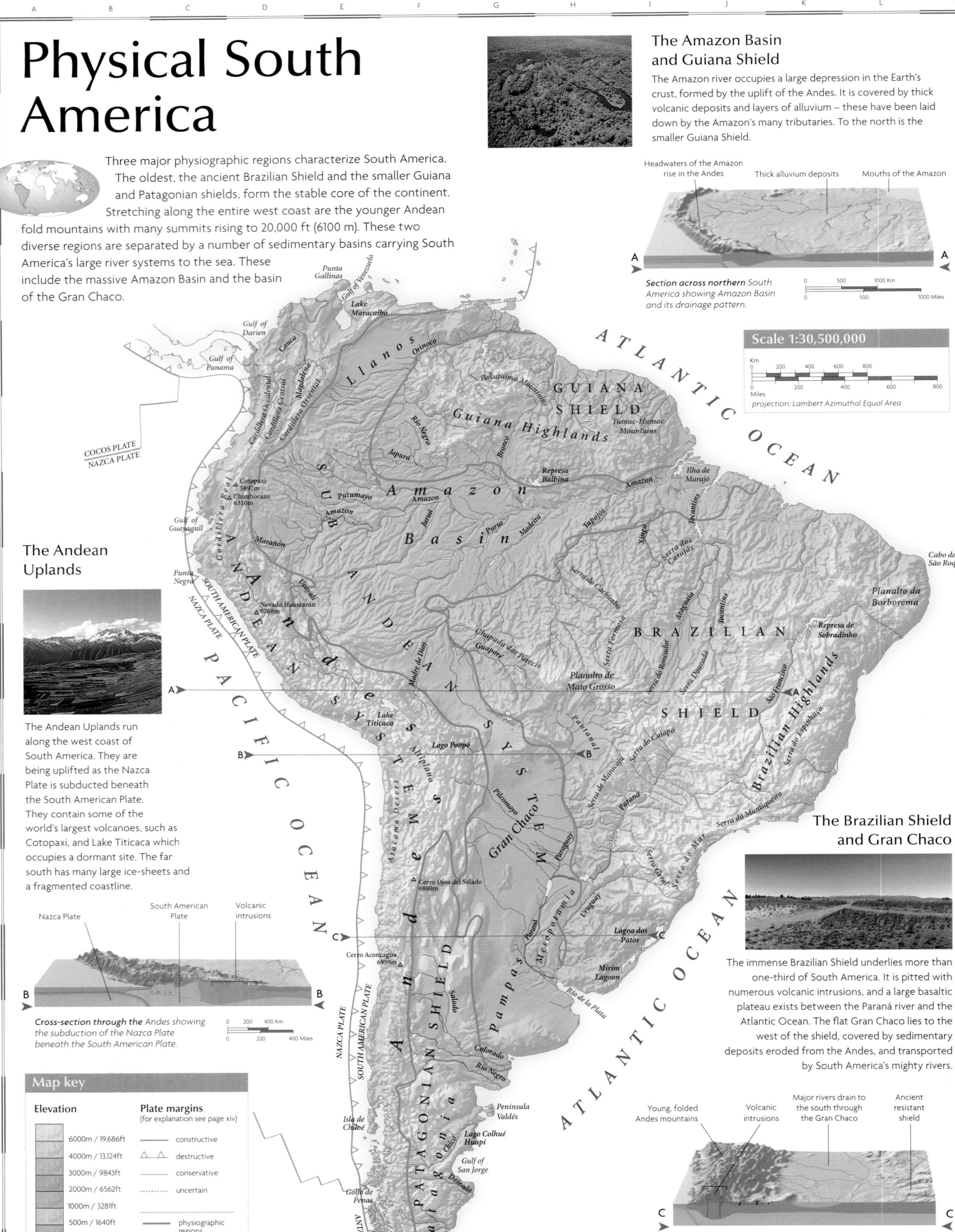

Map labels:
Punta Gallinas, Gulf of Venezuela, Lake Maracaibo, Gulf of Darién, Cauca, Magdalena, Gulf of Panama, Gulf of Guayaquil, Cordillera Occidental, Cordillera Central, Cordillera Oriental, Llanos, Orinoco, Río Negro, Japurá, Pakaraima Mountains, GUIANA SHIELD, Guiana Highlands, Tumuc-Humac Mountains, Cordillera Real, Cotopaxi 5897m, Chimborazo 6310m, Putumayo, Amazon, Marañón, Ucayali, Amazon, Jurua, Purus, Madeira, Tapajós, Xingu, Represa Balbina, Amazon, Ilha de Marajó, Tocantins, Amazon Basin, Punta Negra, Nevado Huascarán 6768m, Madre de Dios, Guaporé, Chapada dos Parecis, Serra do Cachimbo, Serra dos Carajás, Araguaia, Tocantins, Cabo de São Roque, Planalto da Borborema, Serra Formosa, Planalto de Mato Grosso, Serra do Roncador, Serra Dourada, São Francisco, BRAZILIAN SHIELD, Represa de Sobradinho, Lake Titicaca, Pantanal, Serra do Espinhaço, Lago Poopó, Altiplano, Brazilian Highlands, Serra do Caiapó, Serra de Maracaju, Atacama Desert, Pilcomayo, Gran Chaco, Paraná, Paraguay, Serra Geral, Serra do Mar, Serra da Mantiqueira, Cerro Ojos del Salado 6880m, Uruguay, Mesopotamia, Lagoa dos Patos, Cerro Aconcagua 6959m, Mirim Lagoon, Rio de la Plata, Pampas, Colorado, Río Negro, Salado, Isla de Chiloé, Península Valdés, Lago Colhué Huapí, Chico, Gulf of San Jorge, Deseado, Golfo de Peñas, Patagonian Shield, Bahía Grande, Strait of Magellan, Falkland Islands, Tierra del Fuego, Cape Horn

PACIFIC OCEAN, ATLANTIC OCEAN, Andean System, Sub-Andean System, Patagonia

COCOS PLATE, NAZCA PLATE, SOUTH AMERICAN PLATE, ANTARCTIC PLATE, SCOTIA PLATE

Climate

The climate of South America is influenced by three principal factors: the seasonal shift of high pressure air masses over the tropics, cold ocean currents along the western coast, affecting temperature and precipitation, and the mountain barrier produced by by the Andes, which creates a rain shadow over much of the south.

▲ *Mild winters and cool summers typify the extensive Pampas grasslands of Argentina.*

▲ *Chile's hyperarid Atacama Desert is renowned as one of the driest places on Earth.*

Climate

- tundra
- cool continental
- warm humid
- semiarid
- arid
- humid equatorial
- tropical
- ☀ daily hours of sunshine, January
- ☀ daily hours of sunshine, July
- → cold wind

Temperature

Average January temperature

Average July temperature

Temperature

- below -22°F (-30°C)
- -22 to -4°F (-30 to -20°C)
- -4 to 14°F (-20 to -10°C)
- 14 to 32°F (-10 to 0°C)
- 32 to 50°F (0 to 10°C)
- 50°F to 68°F (10 to 20°C)
- 68 to 86°F (20 to 30°C)
- above 86°F (30°C)

Rainfall

Average January rainfall

Average July rainfall

Rainfall

- 0–1 in (0–25 mm)
- 1–2 in (25–50 mm)
- 2–4 in (50–100 mm)
- 4–8 in (100–200 mm)
- 8–12 in (200–300 mm)
- 12–16 in (300–400 mm)
- 16–20 in (400–500 mm)
- more than 20 in (500 mm)

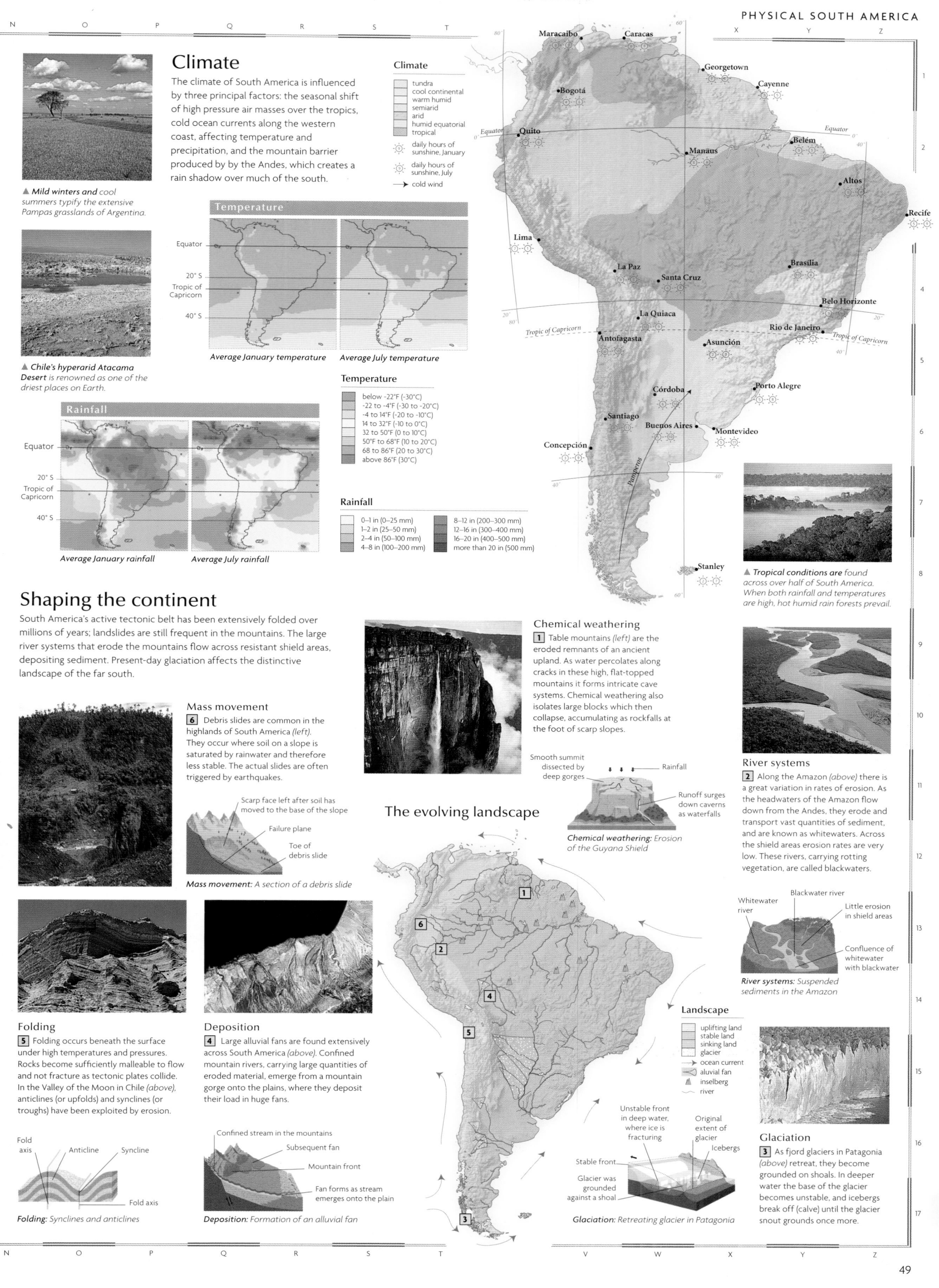

▲ *Tropical conditions are found across over half of South America. When both rainfall and temperatures are high, hot humid rain forests prevail.*

Shaping the continent

South America's active tectonic belt has been extensively folded over millions of years; landslides are still frequent in the mountains. The large river systems that erode the mountains flow across resistant shield areas, depositing sediment. Present-day glaciation affects the distinctive landscape of the far south.

Mass movement

6 Debris slides are common in the highlands of South America (*left*). They occur where soil on a slope is saturated by rainwater and therefore less stable. The actual slides are often triggered by earthquakes.

- Scarp face left after soil has moved to the base of the slope
- Failure plane
- Toe of debris slide

Mass movement: *A section of a debris slide*

Chemical weathering

1 Table mountains (*left*) are the eroded remnants of an ancient upland. As water percolates along cracks in these high, flat-topped mountains it forms intricate cave systems. Chemical weathering also isolates large blocks which then collapse, accumulating as rockfalls at the foot of scarp slopes.

- Smooth summit dissected by deep gorges
- Rainfall
- Runoff surges down caverns as waterfalls

Chemical weathering: *Erosion of the Guyana Shield*

The evolving landscape

River systems

2 Along the Amazon (*above*) there is a great variation in rates of erosion. As the headwaters of the Amazon flow down from the Andes, they erode and transport vast quantities of sediment, and are known as whitewaters. Across the shield areas erosion rates are very low. These rivers, carrying rotting vegetation, are called blackwaters.

- Whitewater river
- Blackwater river
- Little erosion in shield areas
- Confluence of whitewater with blackwater

River systems: *Suspended sediments in the Amazon*

Folding

5 Folding occurs beneath the surface under high temperatures and pressures. Rocks become sufficiently malleable to flow and not fracture as tectonic plates collide. In the Valley of the Moon in Chile (*above*), anticlines (or upfolds) and synclines (or troughs) have been exploited by erosion.

- Fold axis
- Anticline
- Syncline
- Fold axis

Folding: *Synclines and anticlines*

Deposition

4 Large alluvial fans are found extensively across South America (*above*). Confined mountain rivers, carrying large quantities of eroded material, emerge from a mountain gorge onto the plains, where they deposit their load in huge fans.

- Confined stream in the mountains
- Subsequent fan
- Mountain front
- Fan forms as stream emerges onto the plain

Deposition: *Formation of an alluvial fan*

Landscape

- uplifting land
- stable land
- sinking land
- glacier
- ocean current
- alluvial fan
- inselberg
- river

Glaciation

3 As fjord glaciers in Patagonia (*above*) retreat, they become grounded on shoals. In deeper water the base of the glacier becomes unstable, and icebergs break off (calve) until the glacier snout grounds once more.

- Unstable front in deep water, where ice is fracturing
- Original extent of glacier
- Icebergs
- Stable front
- Glacier was grounded against a shoal

Glaciation: *Retreating glacier in Patagonia*

Political South America

Modern South America's political boundaries have their origins in the territorial endeavors of explorers during the 16th century, who claimed almost the entire continent for Portugal and Spain. The Portuguese land in the east later evolved into the federal state of Brazil, while the Spanish vice-royalties eventually emerged as separate independent nation-states in the early 19th century. South America's growing population has become increasingly urbanized, with the growth of coastal cities into large conurbations like Rio de Janeiro and Buenos Aires. In Brazil, Argentina, Chile, and Uruguay, a succession of military dictatorships has given way to fragile, but strengthening, democracies.

◀ *Europe retains a* small foothold in South America. Kourou in French Guiana was the site chosen by the European Space Agency to launch the Ariane rocket. As a result of its status as a French overseas department, French Guiana is actually part of the European Union.

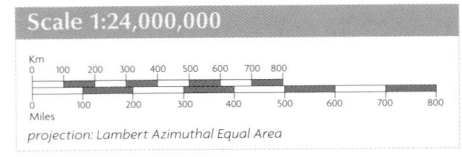

Scale 1:24,000,000

projection: Lambert Azimuthal Equal Area

Transportation

Most major road and rail routes are confined to the coastal regions by the forbidding natural barriers of the Andes mountains and the Amazon Basin. Few major cross-continental routes exist, although Buenos Aires serves as a transportation center for the main rail links to La Paz and Valparaíso, while the construction of the Trans-Amazon and Pan-American Highways have made direct road travel possible from Recife to Lima and from Puerto Montt up the coast into central America. A new waterway project is proposed to transform the River Paraguay into a major shipping route, although it involves considerable wetland destruction.

▶ *South America's most* extensive rail network is centered on the Argentinian capital, Buenos Aires. The construction of new rail lines ouward from this important port, allowed the colonization of the Pampas lands for agriculture.

Languages

Prior to European exploration in the 16th century, a diverse range of indigenous languages were spoken across the continent. With the arrival of Iberian settlers, Spanish became the dominant language, with Portuguese spoken in Brazil, and Native American languages such as Quechua and Guaraní, becoming concentrated in the continental interior. Today this pattern persists, although successive European colonization has led to Dutch being spoken in Suriname, English in Guyana, and French in French Guiana, while in large urban areas, Japanese and Chinese are increasingly common.

Transportation
— major roads and highways
— major railroads
— international borders
• transport intersections
⊕ international airports
⊕ major ports

Language groups
☐ American Indian
☐ Germanic
☐ Romance

▶ *Chile's main port,* Valparaíso, is a vital national shipping center, in addition to playing a key role in the growing trade with Pacific nations. The country's awkward, elongated shape means that sea transportation is frequently used for internal travel and communications in Chile.

▲ *Indigenous South American* lifestyles have not been totally submerged by European cultures and languages. The continental interior, and particularly the Amazon Basin, is still home to many different ethnic peoples.

▶ *Lima's magnificent* cathedral reflects South America's colonial past with its unmistakably Spanish style. In July 1821, Peru became the last Spanish colony on the mainland to declare independence.

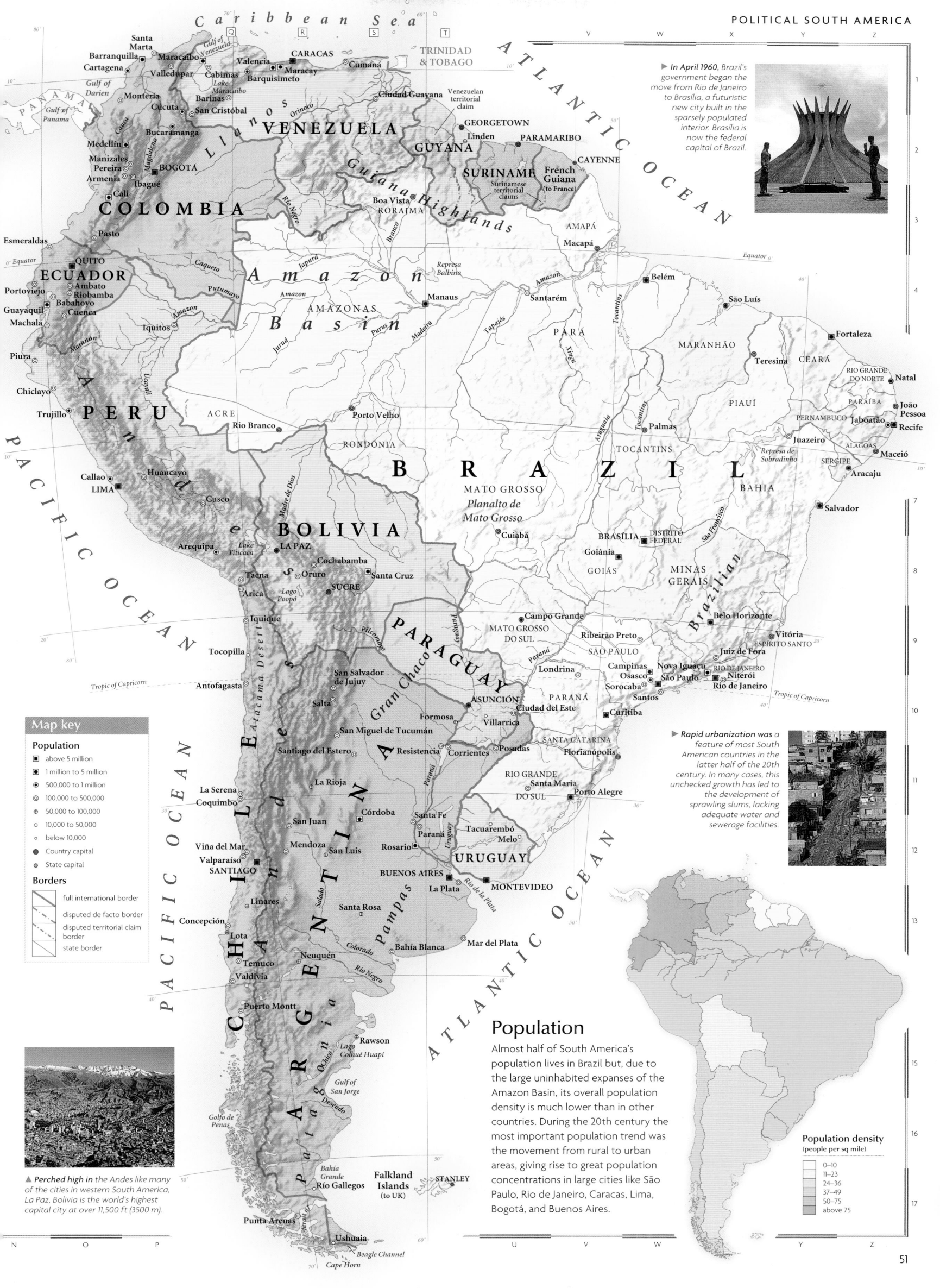

In April 1960, Brazil's government began the move from Rio de Janeiro to Brasília, a futuristic new city built in the sparsely populated interior. Brasília is now the federal capital of Brazil.

Rapid urbanization was a feature of most South American countries in the latter half of the 20th century. In many cases, this unchecked growth has led to the development of sprawling slums, lacking adequate water and sewerage facilities.

▲ *Perched high in the Andes like many of the cities in western South America, La Paz, Bolivia is the world's highest capital city at over 11,500 ft (3500 m).*

Map key

Population
- ▣ above 5 million
- ◼ 1 million to 5 million
- ◉ 500,000 to 1 million
- ◎ 100,000 to 500,000
- ⊕ 50,000 to 100,000
- ○ 10,000 to 50,000
- ・ below 10,000
- ● Country capital
- ◻ State capital

Borders
- full international border
- disputed de facto border
- disputed territorial claim border
- state border

Population

Almost half of South America's population lives in Brazil but, due to the large uninhabited expanses of the Amazon Basin, its overall population density is much lower than in other countries. During the 20th century the most important population trend was the movement from rural to urban areas, giving rise to great population concentrations in large cities like São Paulo, Rio de Janeiro, Caracas, Lima, Bogotá, and Buenos Aires.

Population density
(people per sq mile)
- 0–10
- 11–23
- 24–36
- 37–49
- 50–75
- above 75

South American resources

Agriculture still provides the largest single form of employment in South America, although rural unemployment and poverty continue to drive people towards the huge coastal cities in search of jobs and opportunities. Mineral and fuel resources, although substantial, are distributed unevenly; few countries have both fossil fuels and minerals. To break industrial dependence on raw materials, boost manufacturing, and improve infrastructure, governments borrowed heavily from the World Bank in the 1960s and 1970s. This led to the accumulation of massive debts which are unlikely ever to be repaid. Today, Brazil dominates the continent's economic output, followed by Argentina. Recently, the less-developed western side of South America has benefited due to its geographical position; for example Chile is increasingly exporting raw materials to Japan.

◀ *Ciudad Guayana is a planned industrial complex in eastern Venezuela, built as an iron and steel center to exploit the nearby iron ore reserves.*

Industry

✈ aerospace	pharmaceuticals
⚗ brewing	printing & publishing
🚗 car/vehicle manufacture	shipbuilding
⚗ chemicals	sugar processing
⬚ electronics	textiles
⚙ engineering	timber processing
💲 finance	tobacco processing
🖭 fish processing	wine
food processing	oil
hi-tech industry	gas
iron & steel	
meat processing	● industrial cities
△ metal refining	▨ major industrial areas
narcotics	

Caribbean Sea

PANAMA

Gulf of Panama

Barranquilla
Cartagena
Maracaibo
Barquisimeto
Caracas
Valencia

VENEZUELA
Ciudad Guayana

Georgetown
Paramaribo
GUYANA

SURINAME
French Guiana
(to France)

Medellín

Bogotá
Cali
COLOMBIA

ATLANTIC OCEAN

Quito
ECUADOR

Guayaquil

Iquitos

Belém

Amazon Basin

Manaus

Fortaleza

Natal

Chiclayo

Chimbote

PERU
Lima

Cusco

BRAZIL

Recife

Maceió

Arequipa

BOLIVIA
La Paz

Santa Cruz
Sucre

Salvador

Brasília

Arica
Iquique

Belo Horizonte

Chuquicamata

Antofagasta

PARAGUAY

Asunción
Ciudad del Este

São Paulo

Rio de Janeiro

Curitiba

San Miguel de Tucumán
Corrientes

Porto Alegre

Standard of living

Wealth disparities throughout the continent create a wide gulf between affluent landowners and those afflicted by chronic poverty in inner city slums. The illicit production of cocaine, and the hugely influential drug barons who control its distribution, contribute to the violent disorder and corruption which affect northwestern South America, destabilizing local governments and economies.

Córdoba

Valparaíso
Mendoza

Rosario
Santa Fe
URUGUAY

Rio Grande

GNI per capita (US$)

	below 999
	1000–1999
	2000–2999
	3000–3999
	4000–4999
	above 5000

Santiago

Talca
Concepción

Buenos Aires
Montevideo

ARGENTINA

Neuquén
Bahía Blanca

Valdivia

Standard of living
(UN human development index)

	low
	high

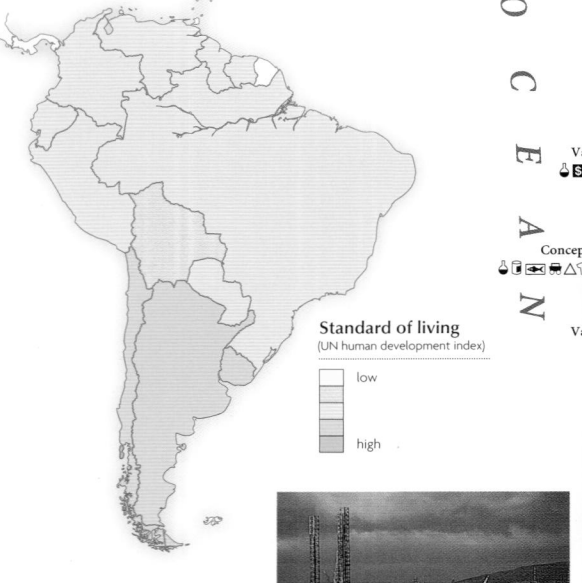

▶ *Both Argentina and Chile are now exploring the southernmost tip of the continent in search of oil. Here in Punta Arenas, a drilling rig is being prepared for exploratory drilling in the Strait of Magellan.*

Comodoro Rivadavia
Gulf of San Jorge

Falkland Islands
(to UK)

Bahía Grande

Punta Arenas

Cape Horn

Industry

Argentina and Brazil are South America's most industrialized countries and São Paulo is the continent's leading industrial center. Long-term government investment in Brazilian industry has encouraged a diverse industrial base; engineering, steel production, food processing, textile manufacture, and chemicals predominate. The illegal production of cocaine is economically significant in the Andean countries of Colombia and Bolivia. In Venezuela, the oil-dominated economy has left the country vulnerable to world oil price fluctuations. Food processing and mineral exploitation are common throughout the less industrially developed parts of the continent, including Bolivia, Chile, Ecuador, and Peru.

PACIFIC OCEAN

▲ *The cold Peru Current flows north from the Antarctic along the Pacific coast of Peru, providing rich nutrients for one of the world's largest fishing grounds. However, over exploitation has severely reduced Peru's anchovy catch.*

Environmental issues

The Amazon Basin is one of the last great wilderness areas left on Earth. The tropical rain forests which grow there are a valuable genetic resource, containing innumerable unique plants and animals. The forests are increasingly under threat from new and expanding settlements and "slash-and-burn" farming techniques, which clear land for the raising of beef cattle, causing land degradation and soil erosion.

▲ *Clouds of smoke* billow from the burning Amazon rainforest. Over 11,500 sq miles (30,000 sq km) of virgin rainforest are being cleared annually, destroying an ancient, irreplaceable, natural resource and biodiverse habitat.

Environmental issues

- national parks
- tropical forest
- forest destroyed
- desert
- risk of desertification
- polluted rivers
- marine pollution
- heavy marine pollution
- • poor urban air quality

Mineral resources

Over a quarter of the world's known copper reserves are found at the Chuquicamata mine in northern Chile, and other metallic minerals such as tin are found along the length of the Andes. The discovery of oil and gas at Venezuela's Lake Maracaibo in 1917 turned the country into one of the world's leading oil producers. In contrast, South America is virtually devoid of coal, the only significant deposit being on the peninsula of Guajira in Colombia.

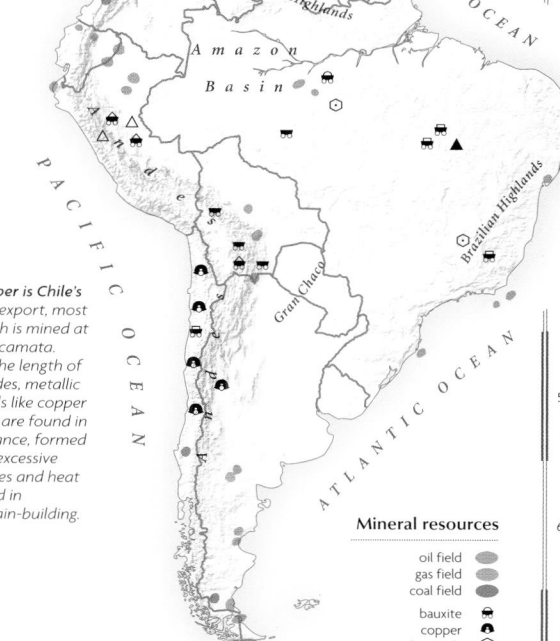

▲ *Copper is Chile's* largest export, most of which is mined at Chuquicamata. Along the length of the Andes, metallic minerals like copper and tin are found in abundance, formed by the excessive pressures and heat involved in mountain-building.

Mineral resources

- oil field
- gas field
- coal field
- bauxite
- copper
- diamonds
- gold
- iron
- lead
- silver
- tin

Using the land and sea

Many foods now common worldwide originated in South America. These include the potato, tomato, squash, and cassava. Today, large herds of beef cattle roam the temperate grasslands of the Pampas, supporting an extensive meatpacking trade in Argentina, Uruguay and Paraguay. Corn is grown as a staple crop across the continent and coffee is grown as a cash crop in Brazil and Colombia. Coca plants grown in Bolivia, Peru, and Colombia provide most of the world's cocaine. Fish and shellfish are caught off the western coast, especially anchovies off Peru, shrimps off Ecuador and pilchards off Chile.

◀ *South America, and* Brazil in particular, now leads the world in coffee production, mainly growing *Coffea arabica* in large plantations. Coffee beans are harvested, roasted and brewed to produce the world's second most popular drink, after tea.

▶ *The Pampas region* of southeast South America is characterized by extensive, flat plains, and populated by cattle and ranchers (gauchos). Argentina is a major world producer of beef, much of which is exported to the US for use in hamburgers.

◀ *High in the Andes,* hardy alpacas graze on the barren land. Alpacas are thought to have been domesticated by the Incas, whose nobility wore robes made from their wool. Today, they are still reared and prized for their soft, warm fleeces.

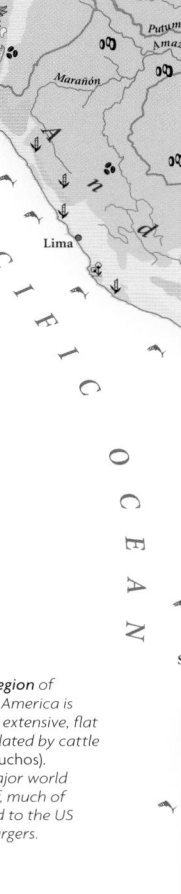

Using the land and sea

- barren land
- cropland
- desert
- forest
- mountain region
- pasture
- • major conurbations
- cattle
- pigs
- sheep
- bananas
- corn
- citrus fruits
- cocoa
- cotton
- coffee
- fishing
- oil palms
- peanuts
- rubber
- shellfish
- soybeans
- sugar cane
- vineyards
- wheat

Northern South America

COLOMBIA, GUYANA, SURINAME, VENEZUELA, French Guiana (to France)

Fringed by the Pacific and Atlantic oceans and the Caribbean Sea, South America's northern region has a rich range of natural resources, some exploited for centuries by colonial powers including the Spanish, French, Dutch, and British, others still to be fully explored. The prospects for further economic development in Colombia, Guyana, and Suriname are blighted by drug-related violence and political instability. Venezuela, despite huge incomes from its oil reserves, remains less developed in other industrial sectors. French Guiana is an overseas *département* of France, now seeking greater autonomy. Most of the major population centers, such as Bogotá, have grown up in the temperate conditions of the high Andes or, like Caracas, at strategic points along the Caribbean coast.

▶ *Flowers grown in* Colombia are exported all over the world, and include fine carnations and roses. Here, workers are cutting roses which have been grown in plastic greenhouses.

Map key

Population
- ▪ 1 million to 5 million
- ◉ 500,000 to 1 million
- ◎ 100,000 to 500,000
- ⊕ 50,000 to 100,000
- ○ 10,000 to 50,000
- ○ below 10,000

Elevation
- 4000m / 13,124ft
- 3000m / 9843ft
- 2000m / 6562ft
- 1000m / 3281ft
- 500m / 1640ft
- 250m / 820ft
- 100m / 328ft
- sea level

▲ *Large open squares* like the Plaza de Bolívar in Bogotá are characteristic of many cities founded by the Spanish.

◀ *Scattered farms and* villages have grown up on the gentle slopes of this Colombian river valley, utilizing the fertile soils for farming.

Scale 1:7,250,000

Km
0 25 50 100 150 200

Miles
0 25 50 100 150 200

projection: Lambert Azimuthal Equal Area

▲ *The Orinoco river* flows from its source in the southern Guiana Highlands to form a broad delta on Venezuela's Atlantic coast. One of its distributary channels opens into a wide bay called the Serpent's Mouth.

N O P Q R S T U V W X Y

Transportation & industry

Many mineral resources are mined in Colombia, including fuels, gold, and precious and semiprecious stones. Revenues from coffee and exports of illegal narcotics are crucial to the economy. Venezuela's major economic activity is the oil industry around Lake Maracaibo *(Lago de Maracaibo)*. Sugar and bauxite are exported from Guyana and Suriname.

Transportation network

31,720 miles (51,054 km)

3411 miles (5490 km)

2448 miles (3940 km)

22,429 miles (36,100 km)

Rivers are an important means of transportation in Colombia: many are extensively navigable. The Pan-American Highway runs through Colombia. In Venezuela, much infrastructure investment is linked to the oil industry.

Major industry and infrastructure

- chemicals
- finance
- food processing
- iron & steel
- narcotics
- mining
- oil
- oil refining
- pharmaceuticals
- textiles
- timber processing
- capital cities
- major towns
- international airports
- major roads
- major industrial areas

▲ *Vast oil reserves around Lake Maracaibo* (Lago de Maracaibo) *form the focus of Venezuelan industry. Incomes from oil are used to invest in other industries and in the development of infrastructure.*

Using the land

The Andean basins support cereals and potatoes. Livestock graze at higher altitudes and on the drier tropical grasslands known as the *llanos;* hardy goats are reared in scrubland areas. Grown at higher elevations, coffee is an important cash crop, as is cotton, sugar cane, bananas, citrus fruits, cocoa, and rice, farmed on the Caribbean lowlands. Coca is the most widely grown narcotic plant, with heroin poppies grown in Colombia and marijuana in lowland areas throughout the region.

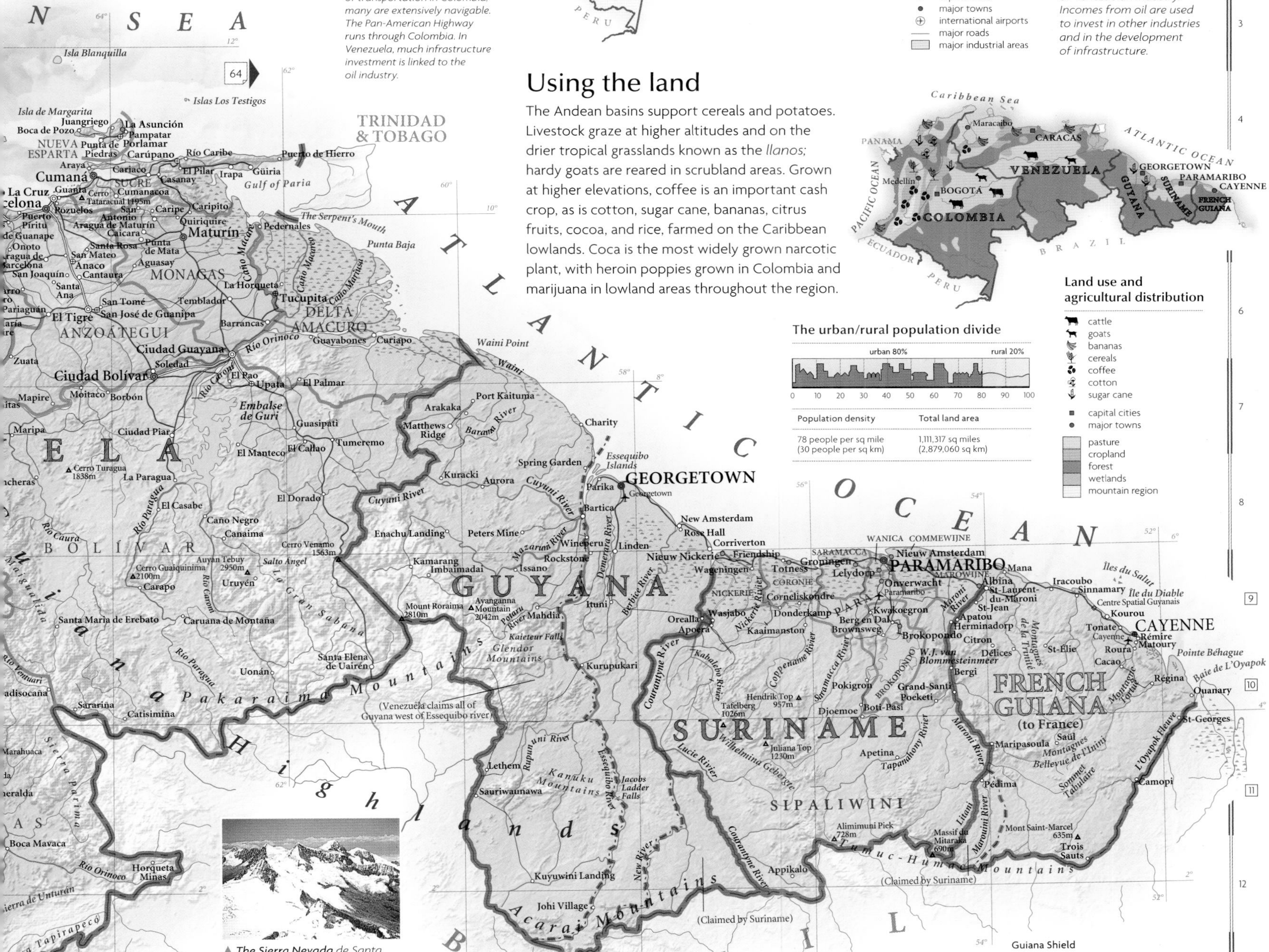

Land use and agricultural distribution

- cattle
- goats
- bananas
- cereals
- coffee
- cotton
- sugar cane
- capital cities
- major towns
- pasture
- cropland
- forest
- wetlands
- mountain region

The urban/rural population divide

urban 80% rural 20%

0 10 20 30 40 50 60 70 80 90 100

Population density	Total land area
78 people per sq mile (30 people per sq km)	1,111,317 sq miles (2,879,060 sq km)

64

58

▲ *The Sierra Nevada de Santa Marta is a granite massif which rises sharply from the Caribbean lowlands to snow-covered peaks, the tallest of which is 18,947 ft (5775 m) high.*

The landscape

At its northernmost reaches, in western Colombia and Venezuela, the great Andean mountain chain splits into three distinct ranges: the Cordillera Oriental, Cordillera Central, and Cordillera Occidental, intercut by a complex series of lesser ranges and basins. The relief becomes lower toward the coast and the interior plains of the northern Amazon Basin, rising again into the tropical hills of the Guiana Highlands.

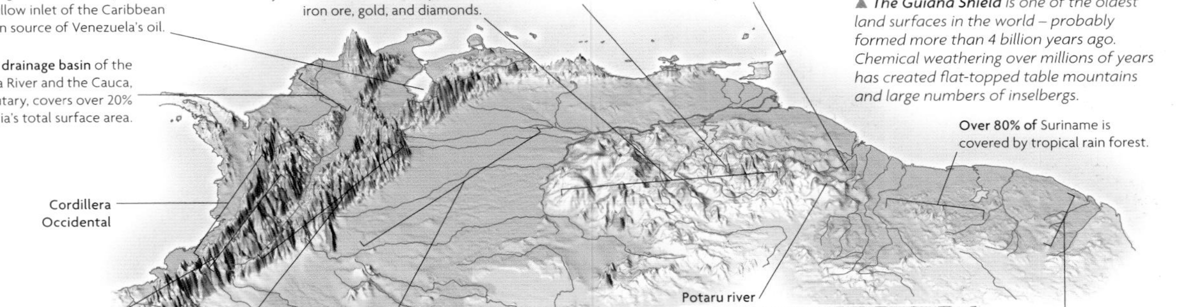

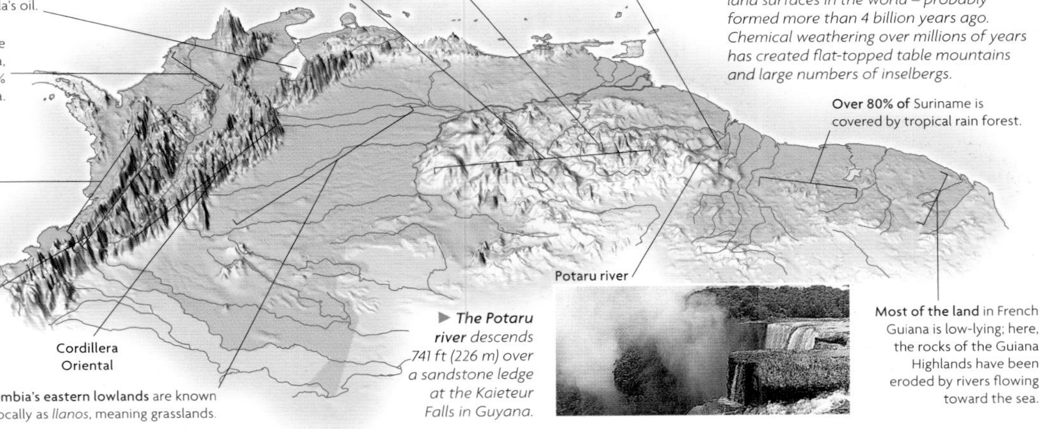

Lake Maracaibo *(Lago de Maracaibo)* is not a true lake but a shallow inlet of the Caribbean Sea. It is the main source of Venezuela's oil.

The drainage basin of the Magdalena River and the Cauca, its main tributary, covers over 20% of Colombia's total surface area.

In the Guiana Highlands, Venezuela's most remote region, the ancient crystalline rocks contain deposits of iron ore, gold, and diamonds.

Angel Falls *(Salto Ángel)*, at 3212 ft (979 m), is the world's highest waterfall.

Igneous intrusions into the crystalline plateau which forms most of central Guyana have led to the formation of the many rapids that characterize Guyana's rivers.

Colombia's eastern lowlands are known locally as *llanos*, meaning grasslands.

Over 80% of Suriname is covered by tropical rain forest.

Most of the land in French Guiana is low-lying; here, the rocks of the Guiana Highlands have been eroded by rivers flowing toward the sea.

Guiana Shield

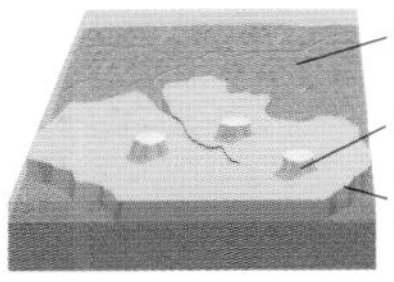

▲ **The Guiana Shield** *is one of the oldest land surfaces in the world – probably formed more than 4 billion years ago. Chemical weathering over millions of years has created flat-topped table mountains and large numbers of inselbergs.*

▶ *The Potaru river descends 741 ft (226 m) over a sandstone ledge at the Kaieteur Falls in Guyana.*

N O P Q R S T U V W X Y Z

Western South America

BOLIVIA, ECUADOR, PERU

The three states of Western South America share a similar geography and recent history. Dominated by the Inca empire until Spanish conquest in the 16th century, they achieved independence from Spain in the early 19th century. The precipitous terrain of the Andes presents severe difficulties for overland transportation and continues to be a barrier to national unity and stability. Although Ecuador is now a relatively stable democracy, the military is highly influential in Peru and Bolivia, while the drug trade and associated corruption discourages external aid and economic progress. Wealth and power are still largely concentrated in the hands of a small elite of families, who attained their position during the Spanish colonial period. Energy resources and political recognition for the indigenous peoples are becoming increasingly important issues, particularly in Bolivia.

The landscape

Bolivia, Peru, and Ecuador each possess a high Andean mountain region and an eastern mountain region consisting of tropical lowlands and the Andean slope leading down to them. Toward the south of the region, the mountains widen to form the high plateau of the Altiplano. Peru and Ecuador also have fertile, lowland coastal plains. A wide variety of environments include *selva* (tropical rain forest), *montaña* (mountain forest), and grassland.

Eruption
column
Subduction zone
Zone of magma
generation
Falling ash
Lava flows
Magma
chamber

▲ **There are many** *large and active volcanoes in the Andes. Magma generated in the heart of the volcano erupts in a huge cloud of ash. Ashfall deposits are common throughout the Andes and the rock produced is known as andesite. This is rapidly soaked by heavy rain, causing massive debris flows.*

The Bolivian *oriente* covers more than two-thirds of the country; it includes *llanos* – low alluvial plains, massive swamps, flooded bottomlands, savannah grassland, and tropical forests.

Fast-flowing tributaries of the Amazon, which rise in the Andes, run eastward through the front ranges to reach the tropical lowlands. They cut valleys so deep that tropical environments can be found extending well into mountainous areas.

Much of eastern Ecuador is covered by the tropical rain forest of the Amazon Basin.

Rolling hills and level plains typify the *montaña* and *selva* region, which makes up more than 65% of Peru.

Cotopaxi is the world's highest active volcano, with a peak 19,347 ft (5897 m) high. A massive eruption in 1877 caused a mudflow which destroyed everything in its path for 150 miles (240 km).

The coastal floodplains are the source of Ecuador's richest soils, enabling the cultivation of a wide range of crops.

The steepness of the Andean slopes means that avalanches and debris flows are an ever-present danger. A landslide starting from Nevado Huascarán in Peru in 1970 killed 20,000 people in 2.5 minutes when it engulfed an inhabited valley.

The Peruvian Andes are relatively young mountains which are continually being uplifted, making the area very unstable, with frequent earthquakes. The transportation difficulties that they present continue to form a barrier to national unity.

The Altiplano is a flat, high plateau lying between the Cordillera Oriental and the Cordillera Occidental at a height of up to 12,500 ft (3800 m). At its margins lie many spurs and alluvial fans.

Bolivian Andes

▲ **Lake Titicaca**, *which forms part of the border between Peru and Bolivia, is the largest lake in South America and the most significant body of water in the world at an altitude of 12,507 ft (3812 m).*

Lake Titicaca

▲ **Nevado de Illampu**, *at 21,275 ft (6485 m) and 21,490 ft (6550 m) respectively, form Illampu, the highest mountain in the Bolivian Andes.*

▲ **Ecuador's capital city,** *Quito, lies high in the Andes, nestling between snowcapped peaks. At 9350 ft (2850 m), Quito is the second highest capital in the world – La Paz in Bolivia is the highest.*

Scale 1:8,500,000

projection: Lambert Azimuthal Equal Area

Map key

Population
- ■ above 5 million
- ◉ 1 million to 5 million
- ◎ 500,000 to 1 million
- ⊕ 100,000 to 500,000
- ⊙ 50,000 to 100,000
- ○ 10,000 to 50,000
- ○ below 10,000

Elevation
- 6000m / 19,686ft
- 4000m / 13,124ft
- 3000m / 9843ft
- 2000m / 6562ft
- 1000m / 3281ft
- 500m / 1640ft
- 250m / 820ft
- 100m / 328ft
- sea level

Ecuador: Administrative regions
- ① CARCHI
- ② TUNGURAHUA
- ③ BOLIVAR
- ④ CHIMBORAZO
- ⑤ ZAMORA CHINCHIPE

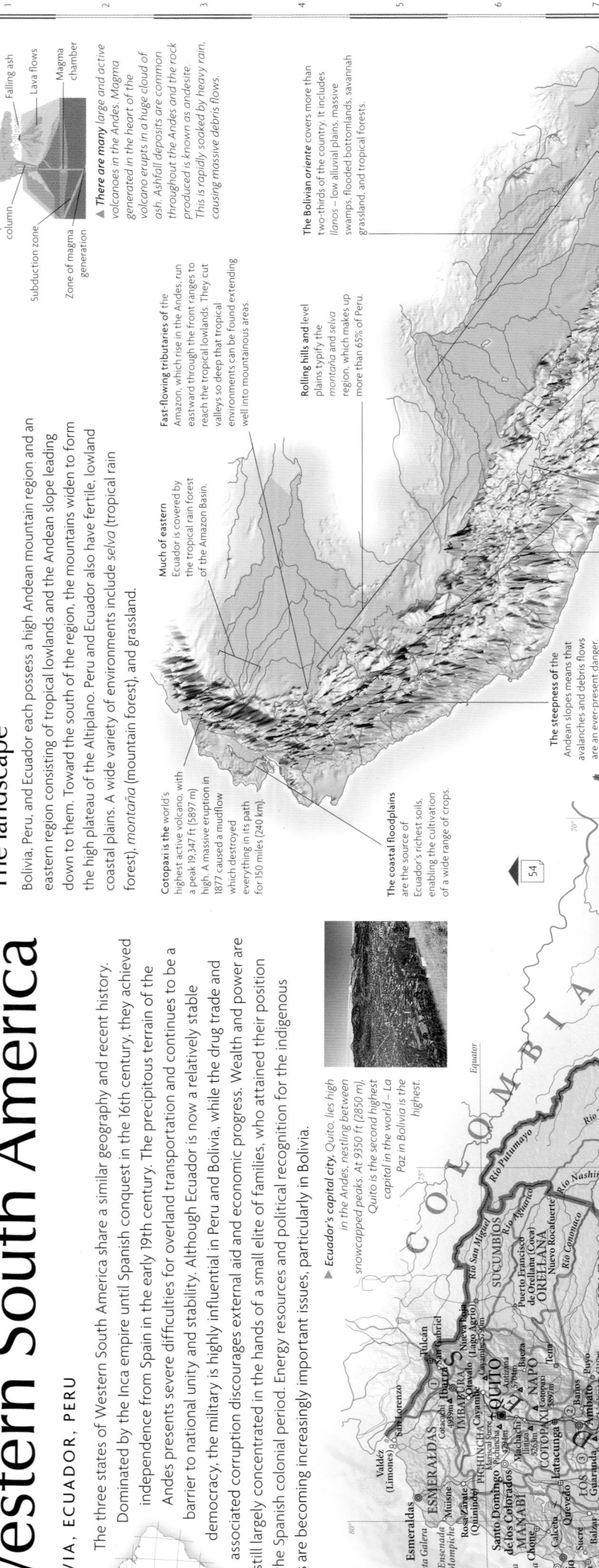

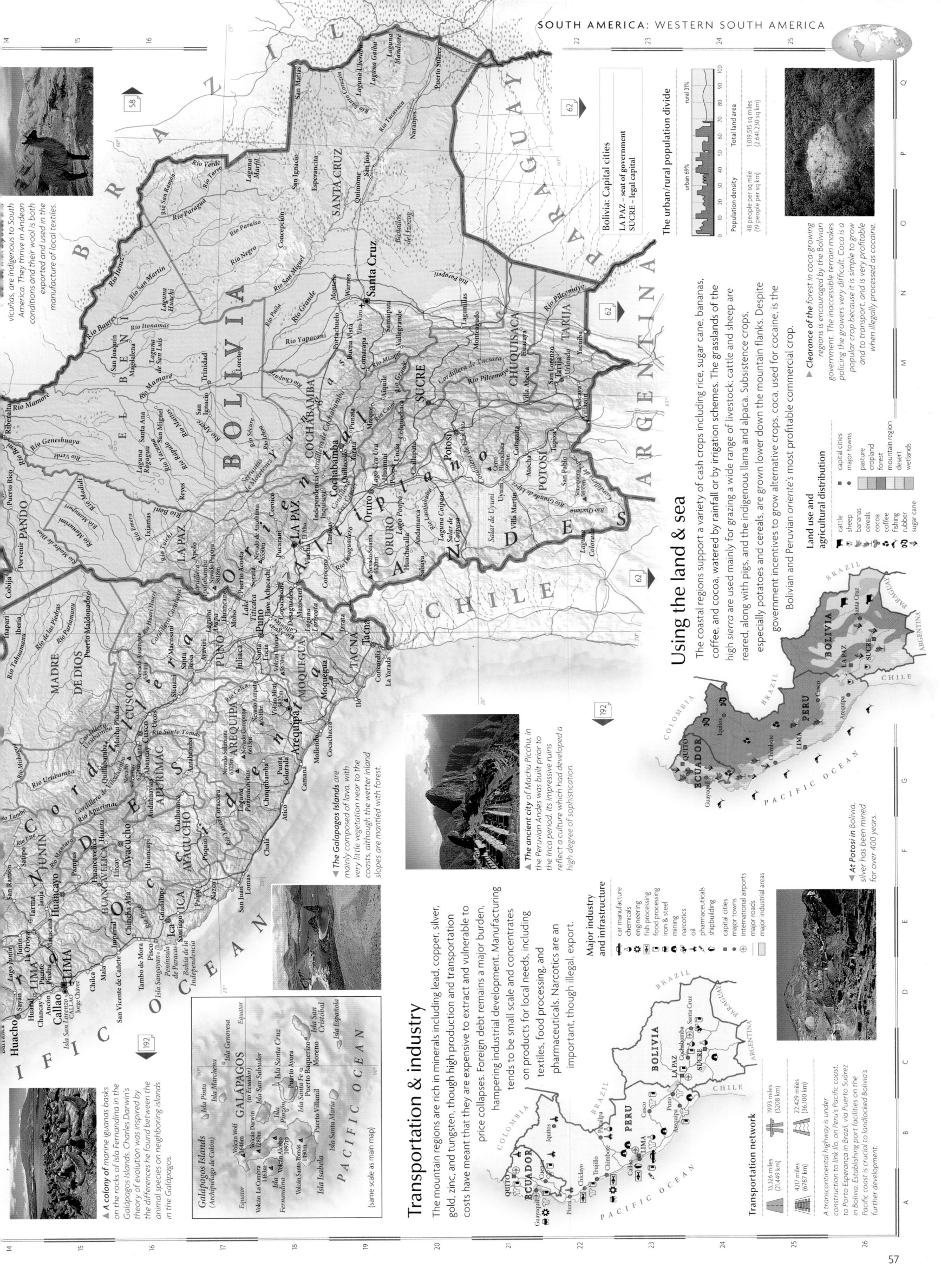

...vicuñas, are indigenous to South America. They thrive in Andean conditions and their wool is both exported and used in the manufacture of local textiles.

Bolivia: Capital cities
LA PAZ – seat of government
SUCRE – legal capital

The urban/rural population divide

rural 31%
urban 69%

Population density	Total land area
48 people per sq mile	1,019,515 sq miles
(19 people per sq km)	(2,641,230 sq km)

▲ *Clearance of the forest in coca-growing regions is encouraged by the Bolivian government. The inaccessible terrain makes policing the growers very difficult. Coca is a popular crop because it is simple to grow and to transport, and is very profitable when illegally processed as cocaine.*

Using the land & sea

The coastal regions support a variety of cash crops including rice, sugar cane, bananas, coffee, and cocoa, watered by rainfall or by irrigation schemes. The grasslands of the high *sierra* are used mainly for grazing a wide range of livestock; cattle and sheep are reared, along with pigs, and the indigenous llama and alpaca. Subsistence crops, especially potatoes and cereals, are grown lower down the mountain flanks. Despite government incentives to grow alternative crops, coca, used for cocaine, is the Bolivian and Peruvian *oriente's* most profitable commercial crop.

Land use and agricultural distribution

- cattle
- sheep
- bananas
- cereals
- cocoa
- coffee
- fishing
- rubber
- sugar cane

- capital cities
- major towns
- pasture
- cropland
- forest
- mountain region
- desert
- wetlands

▲ *At Potosí in Bolivia, silver has been mined for over 400 years.*

▲ *The ancient city of Machu Picchu, in the Peruvian Andes was built prior to the Inca period. Its impressive ruins reflect a culture which had developed a high degree of sophistication.*

▼ *The Galápagos Islands are mainly composed of lava, with very little vegetation near to the coasts, although the wetter inland slopes are mantled with forest.*

Transportation & industry

The mountain regions are rich in minerals including lead, copper, silver, gold, zinc, and tungsten, though high production and transportation costs have meant that they are expensive to extract and vulnerable to price collapses. Foreign debt remains a major burden, hampering industrial development. Manufacturing tends to be small scale and concentrates on products for local needs, including textiles, food processing, and pharmaceuticals. Narcotics are an important, though illegal, export.

Major industry and infrastructure

- car manufacture
- chemicals
- engineering
- fish processing
- food processing
- iron & steel
- mining
- narcotics
- oil
- pharmaceuticals
- shipbuilding
- capital cities
- major towns
- international airports
- major roads
- major industrial areas

Transportation network

13,326 miles (21,449 km)	1993 miles (3208 km)
4217 miles (6787 km)	22,429 miles (36,100 km)

A transcontinental highway is under construction to link Ilo, on Peru's Pacific coast, to Porto Esperança in Brazil, via Puerto Suárez in Bolivia. Establishing port facilities on the Pacific coast is crucial to landlocked Bolivia's further development.

▲ *A colony of marine iguanas basks on the rocks of Isla Fernandina in the Galápagos Islands. Charles Darwin's theory of evolution was inspired by the differences he found between the animal species on neighboring islands in the Galápagos.*

Galápagos Islands
(Archipiélago de Colón)
(same scale as main map)

Brazil

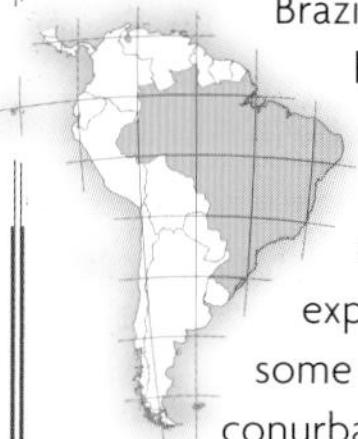

Brazil is the largest country in South America, with a population of 191 million – almost half the combined total of the continent. The 26 states which make up the federal republic of Brazil are administered from the purpose-built capital, Brasília. Tropical rain forest, covering more than one-third of the country, contains rich natural resources, but great tracts are sacrificed to agriculture, industry and urban expansion on a daily basis. Most of Brazil's multiethnic population now live in cities, some of which are vast areas of urban sprawl; São Paulo is one of the world's biggest conurbations, with more than 20 million inhabitants. Although prosperity is a reality for some, many people still live in great poverty, and mounting foreign debts continue to damage Brazil's prospects of economic advancement.

Using the land

Brazil has immense natural resources, including minerals and hardwoods, many of which are found in the fragile rain forest. Brazil is the world's leading coffee grower and a major producer of livestock, sugar, and orange juice concentrate. Soybeans for animal feed, particularly for poultry feed, have become the country's most significant crop.

Land use and agricultural distribution

- cattle
- pigs
- sheep
- citrus fruits
- coffee
- cotton
- soybeans
- sugar cane
- timber

- capital cities
- ● ▪ major towns
- pasture
- cropland
- forest

The landscape

The Amazon Basin, containing the largest area of tropical rain forest on Earth, covers nearly half of Brazil. It is bordered by two shield areas: in the south by the Brazilian Highlands, and in the north by the Guiana Highlands. The east coast is dominated by a great escarpment which runs for 1600 miles (2565 km).

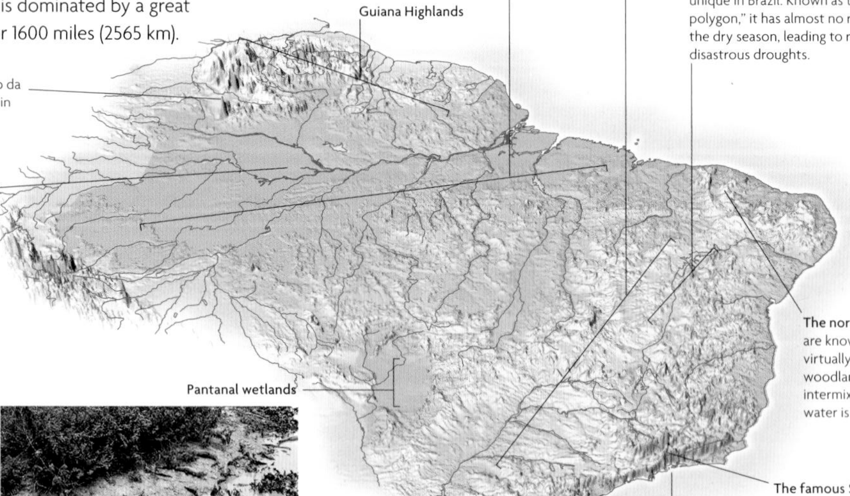

The ancient Brazilian Highlands have a varied topography. Their plateaus, hills, and deep valleys are bordered by highly-eroded mountains containing important mineral deposits. They are drained by three great river systems, the Amazon, the Paraguay–Paraná, and the São Francisco.

The Amazon Basin is the largest river basin in the world. The Amazon river and over a thousand tributaries drain an area of 2,375,000 sq miles (6,150,000 sq km) and carry one-fifth of the world's fresh water out to sea

The São Francisco Basin has a climate unique in Brazil. Known as the "drought polygon," it has almost no rain during the dry season, leading to regular disastrous droughts.

The northeastern scrublands are known as the *caatinga*, a virtually impenetrable thorny woodland, sometimes intermixed with cacti where water is scarce.

The famous Sugar Loaf Mountain (*Pão de Açúcar*) which overlooks Rio de Janeiro is a fine example of a volcanic plug a domed core of solidified lava left after the slopes of the original volcano have eroded away.

Deep natural harbors such as Baía de Guanabara were created where the steep slopes of the Serra da Mantiqueira plunge directly into the ocean.

Guiana Highlands

Brazil's highest mountain is the Pico da Neblina which was only discovered in 1962. It is 9888 ft (3014 m) high.

The floodplains which border the Amazon river are made up of a variety of different features including shallow lakes and swamps, mangrove forests in the tidal delta area, and fertile levees on river banks and point bars.

Pantanal wetlands

▲ **The Pantanal region** in the south of Brazil is an extension of the Gran Chaco plain. The swamps and marshes of this area are renowned for their beauty, and abundant and unique wildlife, including wildfowl and these caimans, a type of crocodile.

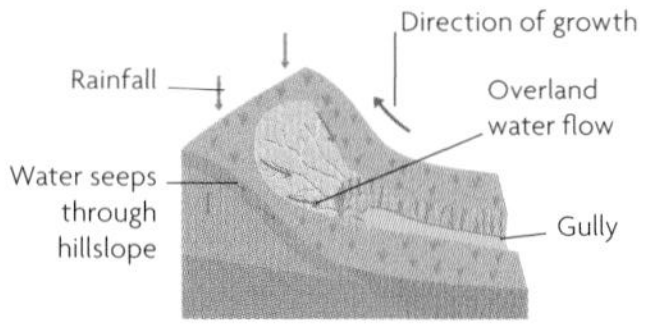

▼ **Large-scale gullies** are common in Brazil, particularly on hillslopes from which vegetation has been removed. Gullies grow headwards (up the slope), aided by a combination of erosion through water seepage and rainwater runoff.

Hillslope gullying

Direction of growth

Overland water flow

Gully

Rainfall

Water seeps through hillslope

▲ **The Iguaçu river** surges over the spectacular Iguaçu Falls (Saltos do Iguaçu) toward the Paraná river. Falls like these are increasingly under pressure from large-scale hydroelectric projects such as that at Itaipú.

▲ **The fecundity of** parts of Brazil's rain forest results from exceptionally high levels of rainfall and the quantities of silt deposited by the Amazon river system.

The urban/rural population divide

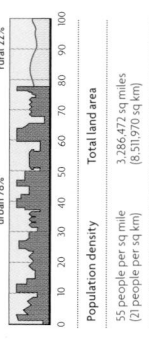

urban 78%

rural 22%

Population density	Total land area
55 people per sq mile (21 people per sq km)	3,286,472 sq miles (8,511,970 sq km)

Map key

Population

- ■ above 5 million
- ◉ 1 million to 5 million
- ◎ 500,000 to 1 million
- ⊕ 100,000 to 500,000
- ⊙ 50,000 to 100,000
- ○ 10,000 to 50,000
- ○ below 10,000

Elevation

- 3000m / 9843ft
- 2000m / 6562ft
- 1000m / 3281ft
- 500m / 1640ft
- 250m / 820ft
- 100m / 328ft
- sea level

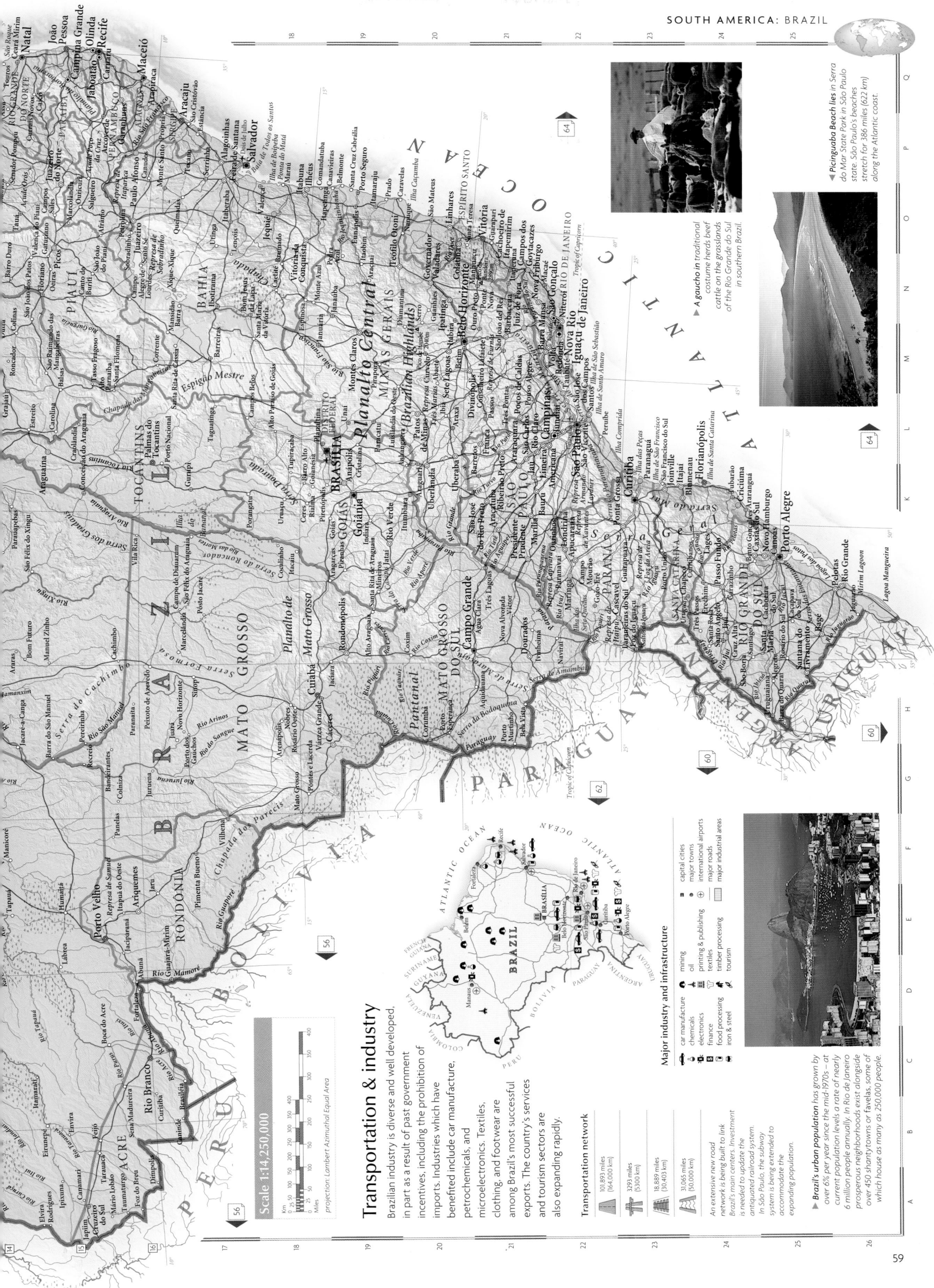

▼ *Picinguaba Beach* lies in Serra do Mar State Park in São Paulo state. São Paulo's beaches stretch for 386 miles (622 km) along the Atlantic coast.

▲ *A gaucho* in traditional costume herds beef cattle on the grasslands of the Rio Grande do Sul in southern Brazil.

Transportation & industry

Brazilian industry is diverse and well developed, in part as a result of past government incentives, including the prohibition of imports. Industries which have benefited include car manufacture, petrochemicals, and microelectronics. Textiles, clothing, and footwear are among Brazil's most successful exports. The country's services and tourism sectors are also expanding rapidly.

Scale 1:14,250,000

projection: Lambert Azimuthal Equal Area

Transportation network

101,893 miles (164,000 km)

3293 miles (5300 km)

18,889 miles (30,403 km)

31,065 miles (50,000 km)

An extensive new road network is being built to link Brazil's main centers. Investment is needed to update the antiquated railroad system. In São Paulo, the subway system is being extended to accommodate the expanding population.

Major industry and infrastructure

- car manufacture
- chemicals
- electronics
- finance
- food processing
- iron & steel
- mining
- oil
- printing & publishing
- textiles
- timber processing
- tourism

- capital cities
- major towns
- international airports
- major roads
- major industrial areas

▲ *Brazil's urban population* has grown by over 6% per year since the mid-1970s – at current population levels a rate of nearly 6 million people annually. In Rio de Janeiro prosperous neighborhoods exist alongside over 450 shantytowns or favelas, some of which house as many as 250,000 people.

Eastern South America

URUGUAY, NORTHEAST ARGENTINA, SOUTHEAST BRAZIL

The vast conurbations of Rio de Janeiro, São Paulo, and Buenos Aires form the core of South America's highly-urbanized eastern region. São Paulo state, with over 40 million inhabitants, is among the world's 20 most powerful economies, and São Paulo is the fastest growing city on the continent. Rio de Janeiro and Buenos Aires, transformed in the last hundred years from port cities to great metropolitan areas each with more than 10 million inhabitants, typify the unstructured growth and wealth disparities of South America's great cities. In Uruguay, over two fifths of the population lives in the capital, Montevideo, which faces Buenos Aires across the Plate River (Río de la Plata).

Immigration from the countryside has created severe pressure on the urban infrastructure, particularly on available housing, leading to a profusion of crowded shanty settlements (favelas or barrios).

Using the land

Most of Uruguay and the Pampas of northern Argentina are devoted to the rearing of livestock, especially cattle and sheep, which are central to both countries' economies. Soybeans, first produced in Brazil's Rio Grande do Sul, are now more widely grown for large-scale export, as are cereals, sugar cane, and grapes. Subsistence crops, including potatoes, corn and sugar beets, are grown on the remaining arable land.

Land use and agricultural distribution

- cattle
- sheep
- cereals
- coffee
- fruit
- soybeans
- sugar cane
- capital cities
- major towns
- pasture
- cropland
- forest
- wetlands
- barren land

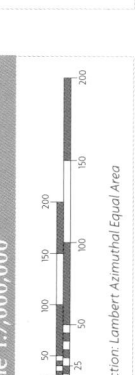
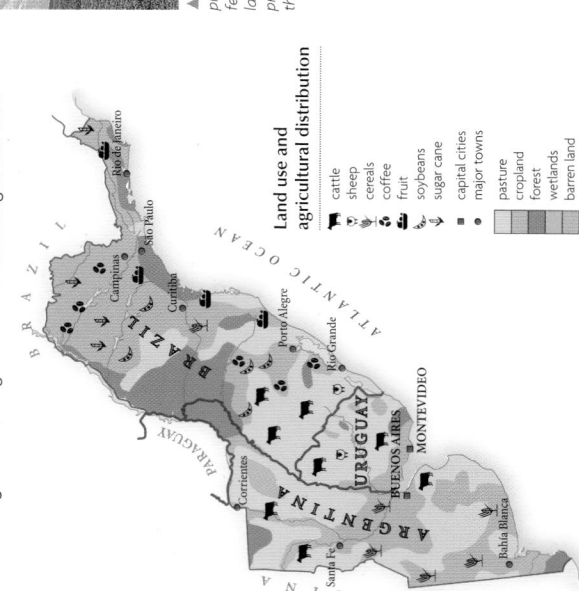

▼ The rolling grasslands of Uruguay are ideally suited to the rearing of cattle. Beef is the country's main export commodity, valued at over one billion US dollars in 2006.

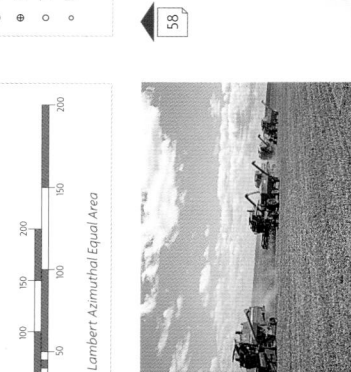

▼ Soybeans are harvested, pressed, and processed into soycake, which is used as animal feed. The cake is fed mainly to chickens on large-scale factory farms, and the growth in soy production has been an important factor in the expansion of the Brazilian poultry trade.

Transportation & industry

Southeast Brazil is home to much of the important motor and capital goods industry, largely based around São Paulo; iron and steel production is also concentrated in this region. Uruguay's economy continues to be based mainly on the export of livestock products including meat and leather goods. Buenos Aires is Argentina's chief port, and the region has a varied and sophisticated economic base including service-based industries such as finance and publishing, as well as primary processing.

Major industry and infrastructure

- car manufacture
- chemicals
- engineering
- finance
- food processing
- iron & steel
- meat processing
- printing & publishing
- shipbuilding
- textiles
- timber processing
- capital cities
- major towns
- international airports
- major roads
- major industrial areas

Transportation network

Throughout the region, road networks need to be expanded to cope with urban development. Plans are underway to build a bridge over the Plate River (Río de la Plata) to link Colonia and Buenos Aires.

Map key

Population
- ■ above 5 million
- ■ 1 million to 5 million
- ◉ 500,000 to 1 million
- ⊚ 100,000 to 500,000
- ⊕ 50,000 to 100,000
- ○ 10,000 to 50,000
- ○ below 10,000

Elevation
- 2000m / 6562ft
- 1000m / 3281ft
- 500m / 1640ft
- 250m / 820ft
- 100m / 328ft
- sea level

Scale 1:7,000,000

Km 0 25 50 100 150 200
Miles 0 25 50 100 150 200

projection: Lambert Azimuthal Equal Area

▲ The Itaipú dam on the Paraná river is one of the largest hydroelectric projects in the world, jointly financed by Brazil and Paraguay.

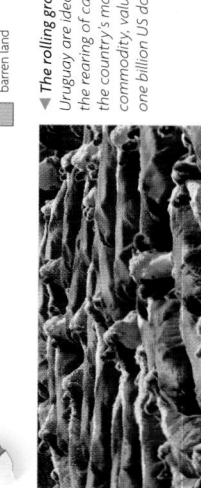

▲ Rio de Janeiro's annual carnival, Mardi Gras, which ushers in the start of Lent, is an extravagant five-day parade through the city, characterized by fantastically decorated floats, exuberant

The landscape

The southern reaches of the Brazilian Highlands follow the Atlantic coast to form low, rolling hills in the northeast of Uruguay. Much of South America's mid-eastern region and all of Uruguay has a gentle relief with land rarely rising above 300 ft (100 m). Argentina's northeast comprises two main regions: a long, narrow lowland known as Mesopotamia; and part of the Pampas grasslands.

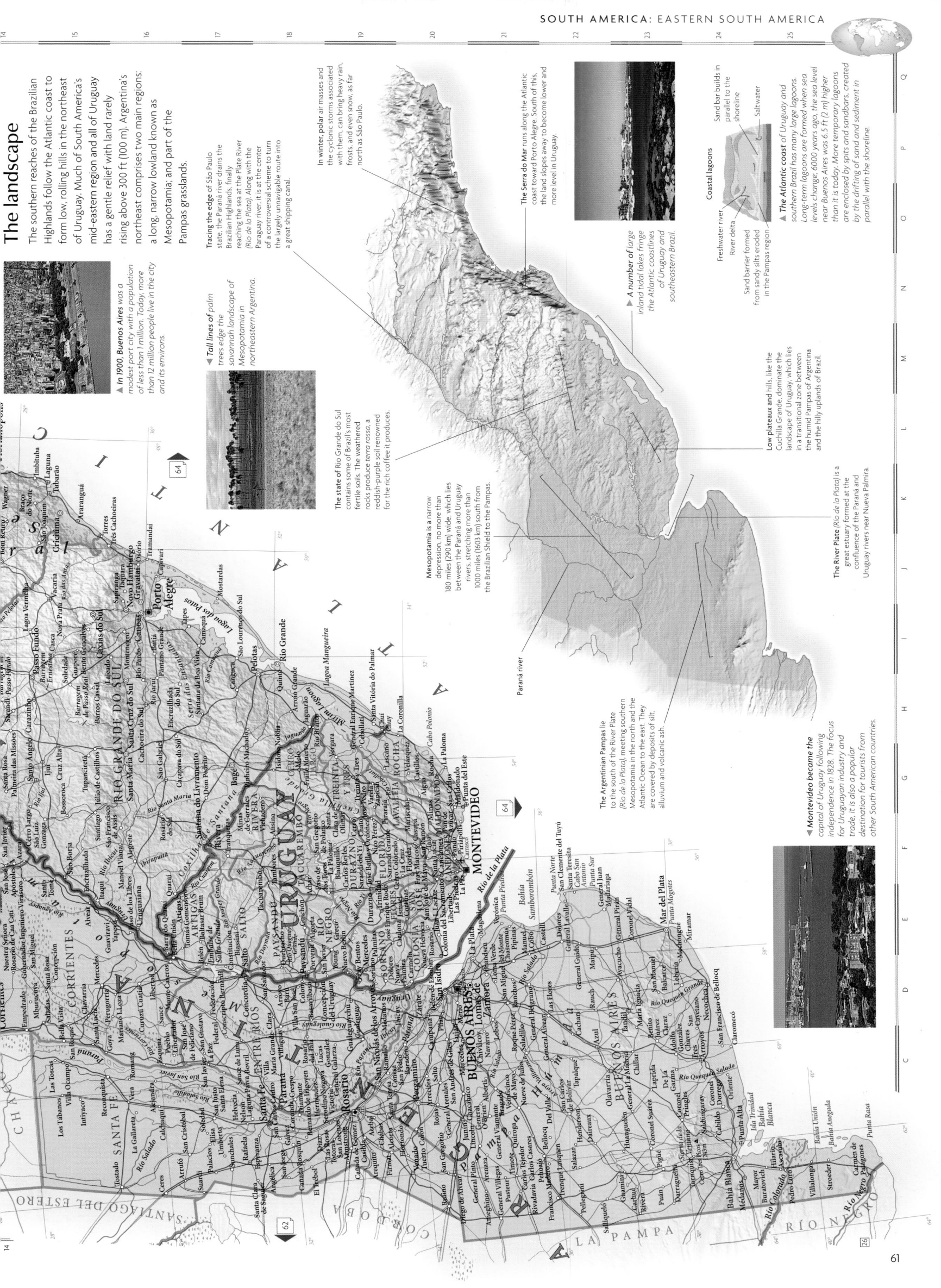

▲ *In 1900, Buenos Aires was a modest port city with a population of less than 1 million. Today, more than 12 million people live in the city and its environs.*

Tracing the edge of São Paulo state, the Paraná river drains the Brazilian Highlands, finally reaching the sea at the Plate River (*Río de la Plata*). Along with the Paraguay river, it is at the center of a controversial scheme to turn the largely unnavigable route into a great shipping canal.

▼ *Tall lines of palm trees edge the savannah landscape of Mesopotamia in northeastern Argentina.*

In winter, polar air masses and the cyclonic storms associated with them, can bring heavy rain, frosts, and even snow, as far north as São Paulo.

The Serra do Mar runs along the Atlantic coast toward Porto Alegre. South of this, the land slopes away to become lower and more level in Uruguay.

▲ *A number of large inland tidal lakes fringe the Atlantic coastlines of Uruguay and southeastern Brazil.*

Coastal lagoons

Sand builds in parallel to the shoreline
Saltwater
Freshwater river
River delta
Sand barrier formed from sandy silts eroded in the Pampas region

▲ *The Atlantic coast of Uruguay and southern Brazil has many large lagoons. Long-term lagoons are formed when sea levels change; 6000 years ago, the sea level near Buenos Aires was 6.5 ft (2 m) higher than it is today. More temporary lagoons are enclosed by spits and sandbars, created by the drifting of sand and sediment in parallel with the shoreline.*

The state of Rio Grande do Sul contains some of Brazil's most fertile soils. The weathered rocks produce *terra rossa*, a reddish-purple soil renowned for the rich coffee it produces.

Low plateaux and hills, like the Cuchilla Grande, dominate the landscape of Uruguay, which lies in a transitional zone between the humid Pampas of Argentina and the hilly uplands of Brazil.

Mesopotamia is a narrow depression, no more than 180 miles (290 km) wide, which lies between the Paraná and Uruguay rivers, stretching more than 1000 miles (1603 km) south from the Brazilian Shield to the Pampas.

The River Plate (*Río de la Plata*) is a great estuary formed at the confluence of the Paraná and Uruguay rivers near Nueva Palmira.

Paraná river

The Argentinian Pampas lie to the south of the River Plate (*Río de la Plata*), meeting southern Mesopotamia in the north and the Atlantic Ocean to the east. They are covered by deposits of silt, alluvium and volcanic ash.

▼ *Montevideo became the capital of Uruguay following independence in 1828. The focus for Uruguayan industry and trade, it is also a popular destination for tourists from other South American countries.*

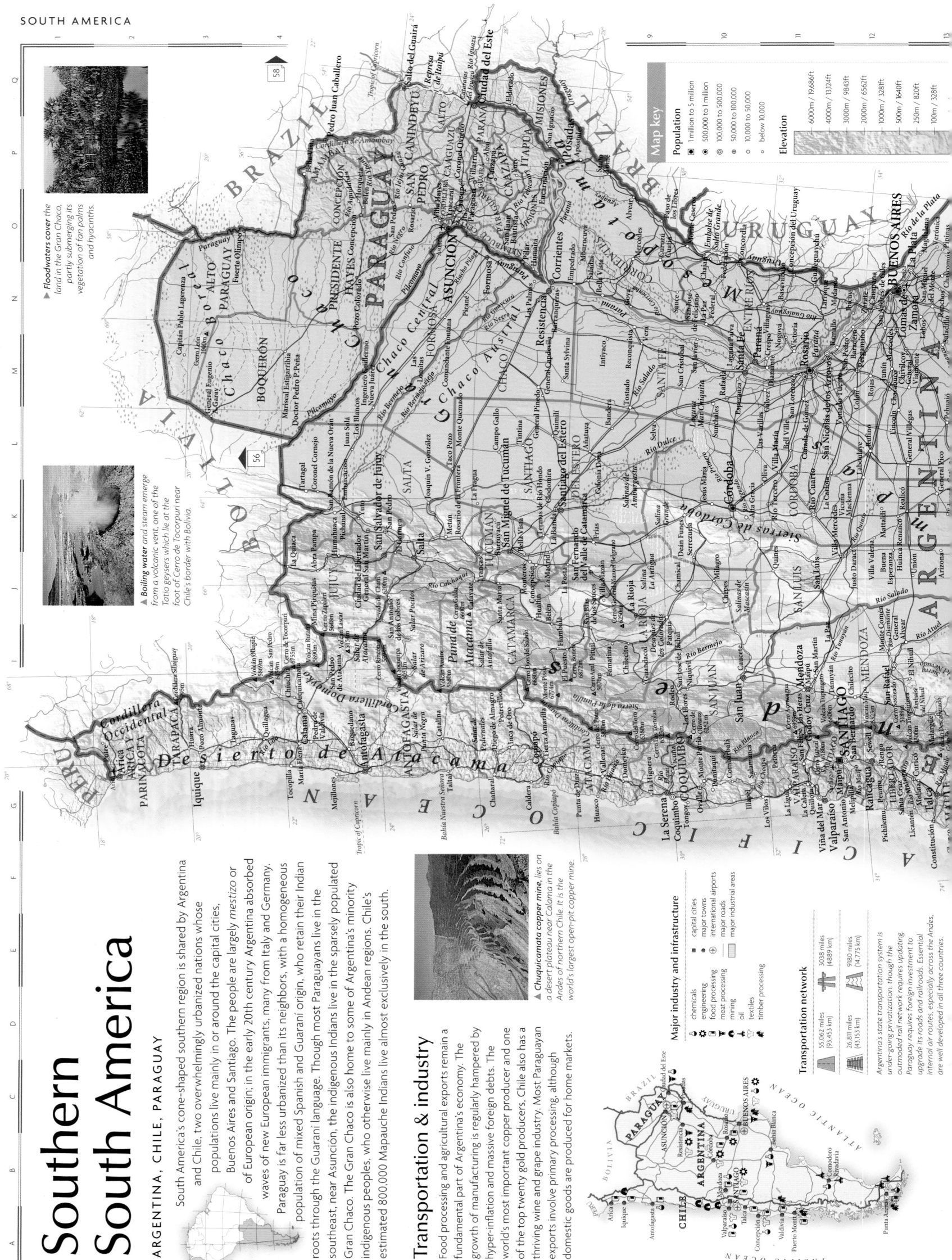

Southern South America

ARGENTINA, CHILE, PARAGUAY

South America's cone-shaped southern region is shared by Argentina and Chile, two overwhelmingly urbanized nations whose populations live mainly in or around the capital cities, Buenos Aires and Santiago. The people are largely *mestizo* or of European origin; in the early 20th century Argentina absorbed waves of new European immigrants, many from Italy and Germany. Paraguay is far less urbanized than its neighbors, with a homogeneous population of mixed Spanish and Guarani origin, who retain their Indian roots through the Guarani language. Though most Paraguayans live in the southeast, near Asunción, the indigenous Indians live in the sparsely populated Gran Chaco. The Gran Chaco is also home to some of Argentina's minority indigenous peoples, who otherwise live mainly in Andean regions. Chile's estimated 800,000 Mapauche Indians live almost exclusively in the south.

Transportation & industry

Food processing and agricultural exports remain a fundamental part of Argentina's economy. The growth of manufacturing is regularly hampered by hyper-inflation and massive foreign debts. The world's most important copper producer and one of the top twenty gold producers, Chile also has a thriving wine and grape industry. Most Paraguayan exports involve primary processing, although domestic goods are produced for home markets.

▲ Floodwaters cover the land in the Gran Chaco, partly submerging its vegetation of fan palms and hyacinths.

▲ Boiling water and steam emerge from a volcanic vent, one of the Tatio geysers which lie at the foot of Cerro de Tocorpuri near Chile's border with Bolivia.

▲ Chuquicamata copper mine, lies on a desert plateau near Calama in the Andes of northern Chile. It is the world's largest open-pit copper mine.

Map key

Population
- 1 million to 5 million
- 500,000 to 1 million
- 100,000 to 500,000
- 50,000 to 100,000
- 10,000 to 50,000
- below 10,000

Elevation
- 6000m / 19,686ft
- 4000m / 13,124ft
- 3000m / 9843ft
- 2000m / 6562ft
- 1000m / 328ft
- 500m / 1640ft
- 250m / 820ft
- 100m / 328ft

Major industry and infrastructure
- chemicals
- engineering
- food processing
- meat processing
- mining
- oil
- textiles
- timber processing

- capital cities
- major towns
- international airports
- major roads
- major industrial areas

Transportation network

| 55,062 miles (93,453 km) | 3038 miles (4889 km) |
| 26,811 miles (43,153 km) | 9180 miles (14,775 km) |

Argentina's state transportation system is under-going privatization, though the outmoded rail network requires updating. Paraguay requires foreign investment to upgrade its roads and railroads. Essential internal routes, especially across the Andes, are well serviced in all three countries.

The landscape

The Andes run from north to south, forming a precipitous natural border between Chile and Argentina. East of the Andes are the scrublands of the Gran Chaco and the plains of the Pampas, which extend northward toward Paraguay. In the far southwest, Chile's indented Pacific coastline has many features typical of areas which have been affected by glaciation.

▲ *The Atacama Desert* (Desierto de Atacama) *in Chile is one of the driest places on Earth where some areas have never recorded any rain. It contains a number of salt lakes.*

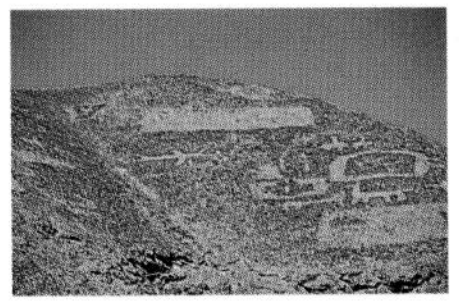

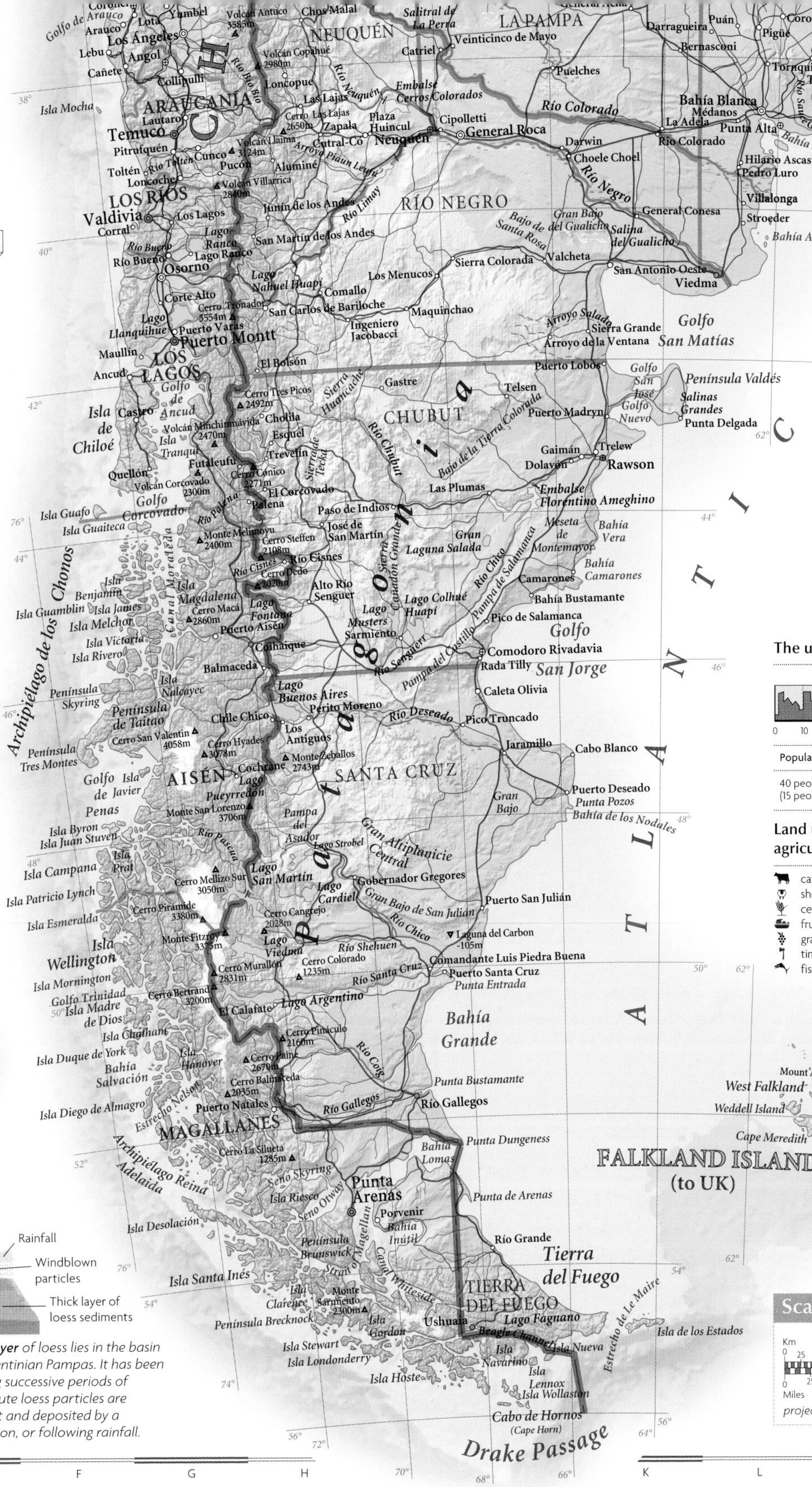

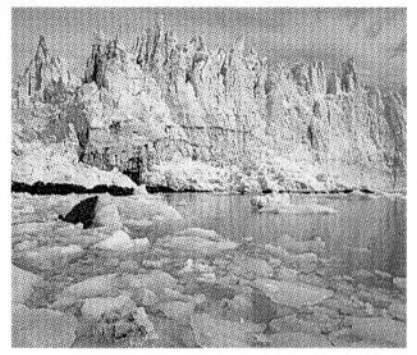

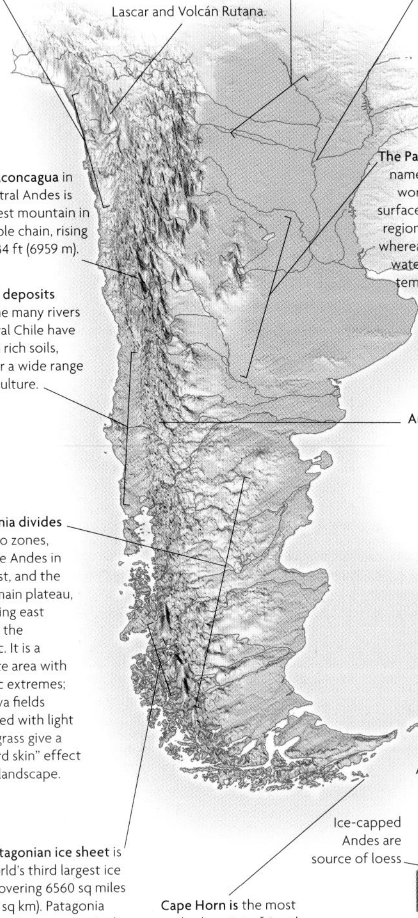

The Gran Chaco combines poor drainage, extremely hot temperatures and thorn-infested scrub to make it one of South America's most inhospitable regions.
Landlocked Paraguay relies on its river system for access to the sea and to produce hydroelectric power. The most important river system is the Paraguay–Paraná which provides links into neighboring countries including Brazil, Uruguay, and Argentina.
Most of the highest mountains in Chile's northern Andes are volcanoes like Volcán Lascar and Volcán Rutana.
The Pampas derive their name from an Indian word meaning flat surface. The dry western region is largely desert, whereas the east is well-watered, supporting temperate grasses.
Cerro Aconcagua in the central Andes is the tallest mountain in the whole chain, rising to 22,834 ft (6959 m).
Alluvial deposits from the many rivers in central Chile have created rich soils, ideal for a wide range of agriculture.
Andes
Patagonia divides into two zones, with the Andes in the west, and the lower main plateau, extending east toward the Atlantic. It is a desolate area with climatic extremes; dark lava fields scattered with light bunchgrass give a "leopard skin" effect to the landscape.
The Patagonian ice sheet is the world's third largest ice field, covering 6560 sq miles (17,000 sq km). Patagonia also contains many typical features from past glaciations. These include glacial lakes, U-shaped valleys, fjords, and deep-cut channels.
Cape Horn is the most southerly point of South America. The severity of the "Roaring Forties" winds makes the Horn one of the world's most treacherous shipping regions.
Ice-capped Andes are source of loess

192

Argentinian Pampas
Jet stream
Rainfall
Windblown particles
Thick layer of loess sediments

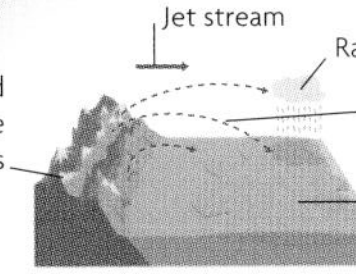

▲ *A thick, fertile layer of loess lies in the basin underlying the Argentinian Pampas. It has been laid down following successive periods of glaciation. The minute loess particles are transported as dust and deposited by a downward air motion, or following rainfall.*

ATLANTIC OCEAN
Patagonia
NEUQUÉN
LA PAMPA
RÍO NEGRO
CHUBUT
SANTA CRUZ
ARAUCANÍA
LOS RÍOS
LOS LAGOS
AISÉN
MAGALLANES
TIERRA DEL FUEGO
Mar del Plata
Bahía Blanca
Neuquén
General Roca
Viedma
Puerto Montt
Valdivia
Temuco
Osorno
Rawson
Trelew
Comodoro Rivadavia
Puerto Deseado
Puerto San Julián
Río Gallegos
Punta Arenas
Ushuaia
Isla de Chiloé
Archipiélago de los Chonos
Golfo San Matías
Golfo San Jorge
Bahía Grande
Estrecho de Magallanes
Isla de los Estados
Cabo de Hornos (Cape Horn)
Drake Passage
FALKLAND ISLANDS (to UK)
West Falkland
East Falkland
STANLEY

▲ *Great blocks of ice break away from the jagged blue peaks of these ice mountains to form icebergs off the coast of Patagonia, Argentina's most southerly region.*

Using the land & sea

The rich plains of the Pampas support massive herds of cattle, producing meat, milk, and hides essential to the domestic and export markets of both Argentina and Paraguay. Wheat and fruit are Argentina's other major agricultural products. A wide range of soft fruits, citrus fruits, and more specialized crops such as walnuts, and grapes for wine and the table, are grown in Chile's fertile Central Valley, while the landscape to the south is dominated by forestry, mainly growing commercial radiata pine. Paraguay is self-sufficient in wheat and other staples. Cotton, coffee, tobacco, and oil sources such as soybeans, are the major export crops.

The urban/rural population divide

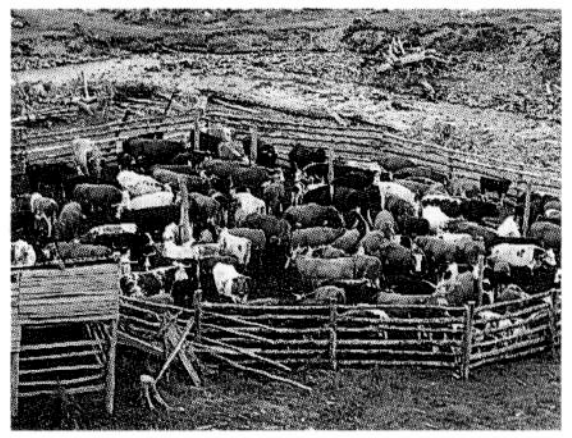

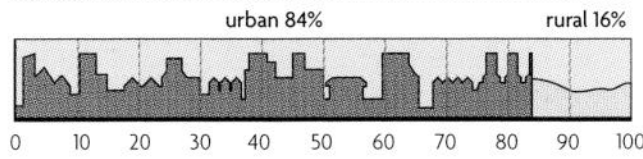

urban 84%
rural 16%
0
10
20
30
40
50
60
70
80
90
100

Population density	Total land area
40 people per sq mile (15 people per sq km)	1,498,757 sq miles (3,882,790 sq km)

Land use and agricultural distribution

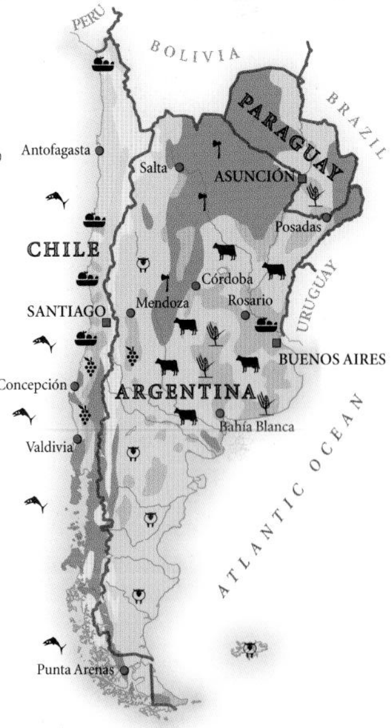

cattle
sheep
cereals
fruit
grapes
timber
fishing
capital cities
major towns
pasture
cropland
forest
barren land
mountain region
desert
PACIFIC OCEAN
ATLANTIC OCEAN
PERU
BOLIVIA
BRAZIL
PARAGUAY
ASUNCIÓN
URUGUAY
CHILE
ARGENTINA
SANTIAGO
BUENOS AIRES
Antofagasta
Salta
Posadas
Córdoba
Rosario
Mendoza
Concepción
Valdivia
Bahía Blanca
Punta Arenas

64

Scale 1:9,750,000

Km 0 25 50 100 150 200
Miles 0 25 50 100 150 200

projection: Lambert Azimuthal Equal Area

▲ *Charred tree stumps surround a cattle enclosure on the island of Tierra del Fuego in southern Argentina. Forest clearance to provide grazing land for cattle is of major environmental concern.*

The Atlantic Ocean

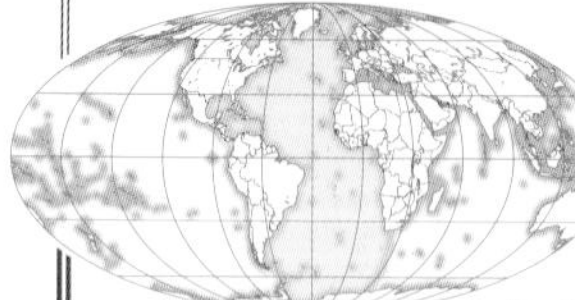

The Atlantic is the youngest of the world's oceans, formed about 180 million years ago when the landmasses of the eastern and western hemispheres separated. Its underwater topography is dominated by the Mid-Atlantic Ridge, a huge mountain system running north to south along the center of the ocean. Although most of the ridge's peaks lie below the sea, some emerge as volcanic islands, like Iceland and the Azores. The Atlantic contains a wealth of resources, including substantial oil and gas reserves and rich fishing grounds. Until the 1950s, the north Atlantic was the world's busiest shipping route; cheaper air transportation and alternative routes have shifted patterns of world trade.

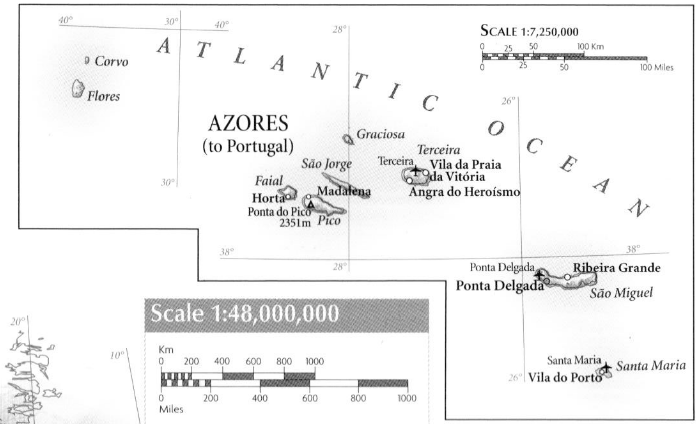

Resources

Development of the oil and gas reserves in the Atlantic began in the 1940s around the Gulf of Mexico. Since then other areas have been exploited, including the North Sea, the west coast of Africa and the area east of Newfoundland and Nova Scotia. There is also extensive mining of sand, gravel, and shell deposits by the US and UK. For centuries, the north Atlantic's fishing grounds have been utilized more heavily than other oceans, leading to a serious decline in many fish stocks.

Resources (including wildlife)
- fish
- whales
- aggregates
- oil & gas
- major towns
- major ports

▲ *Fishing in the seas around northwestern Europe dates back over 1500 years. The high nutrient content of the seas makes them ideal breeding grounds for many species of fish.*

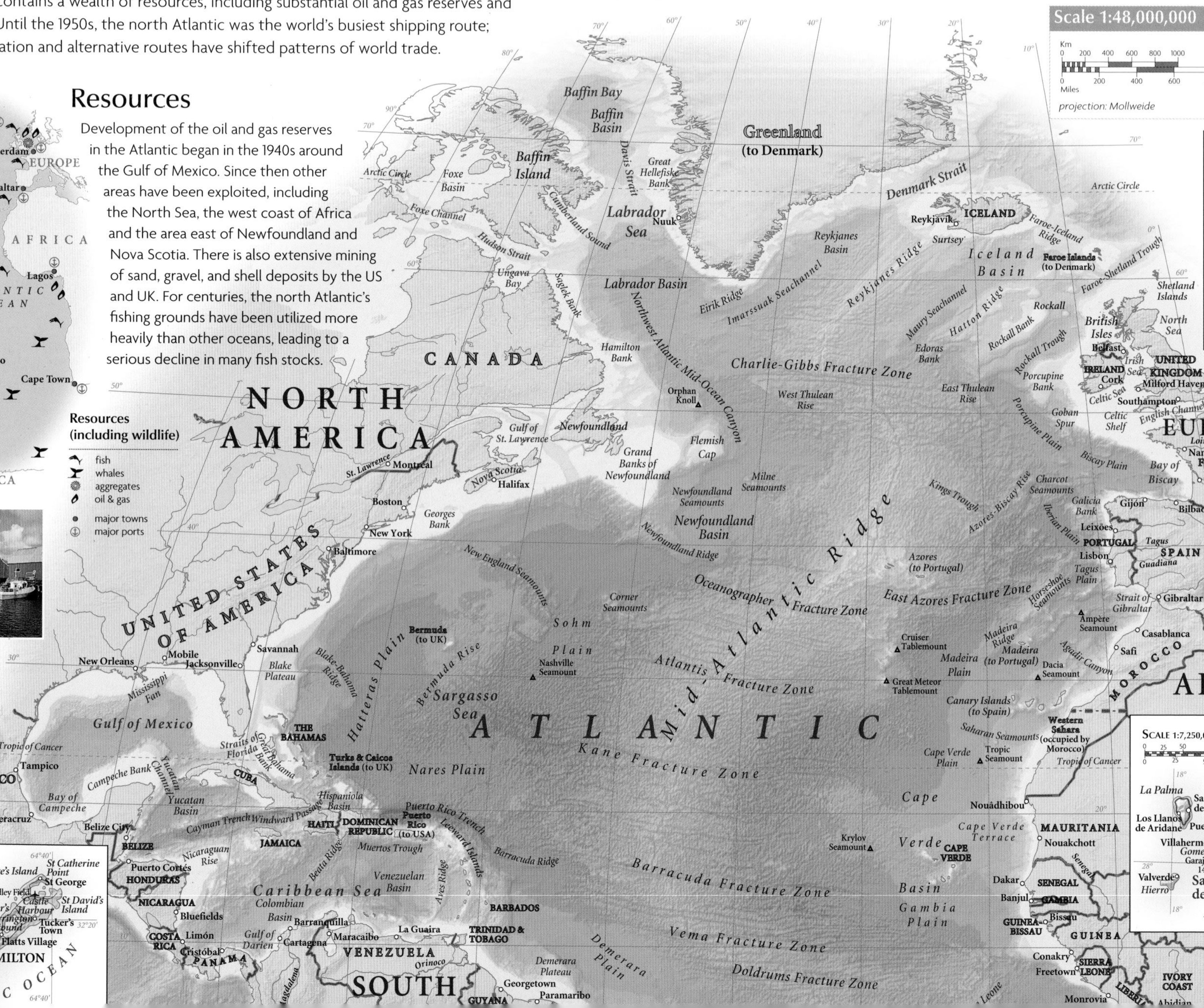

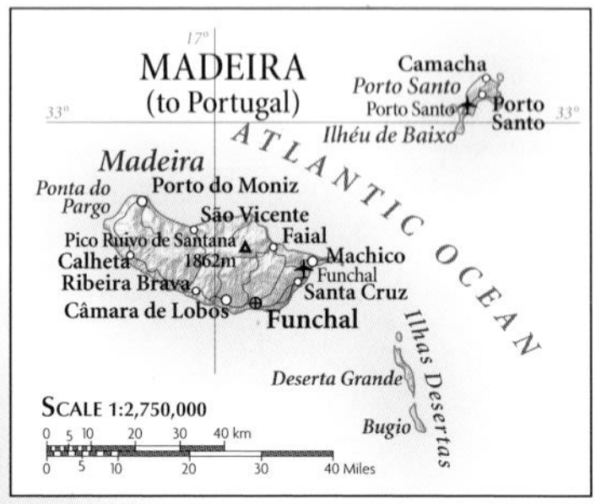

▲ *Surtsey near Iceland, lies on the Mid-Atlantic Ridge. The island was formed in 1963 following a volcanic eruption caused by sea-floor spreading.*

▲ *On January 5 1993, the oil tanker Braer ran aground in the Shetland Islands, spilling 83,660 tons (85,000 tonnes) of light crude oil into the ocean, devastating the local marine ecosystem.*

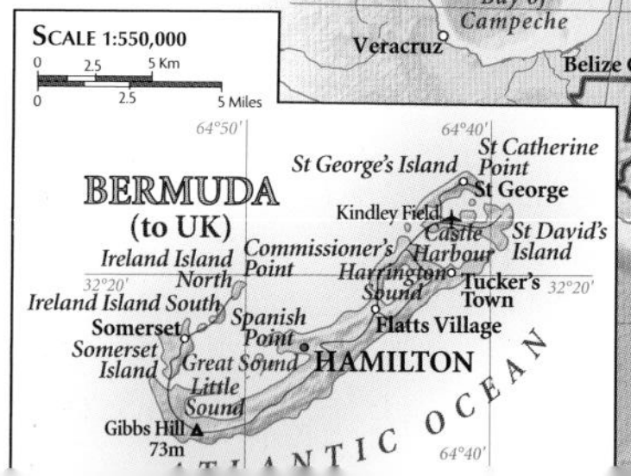

AZORES (to Portugal)

SCALE 1:7,250,000

Corvo, Flores, Graciosa, Terceira, Vila da Praia da Vitória, Angra do Heroismo, São Jorge, Faial, Madalena, Horta, Ponta do Pico, Pico, 2351m, Ribeira Grande, São Miguel, Ponta Delgada, Santa Maria, Vila do Porto

ATLANTIC OCEAN

MADEIRA (to Portugal)

Camacha, Porto Santo, Porto Santo, Ilhéu de Baixo, São Vicente, Porto do Moniso, Pico Ruivo de Santana, 1862m, Galhetas, Ribeira Brava, Câmara de Lobos, Machico, Funchal, Santa Cruz, Ilhas Desertas, Bugio, Deserta Grande, Ponta do Pargo

ATLANTIC OCEAN

SCALE 1:2,750,000

ISLAS CANARIAS (CANARY ISLANDS) (to Spain)

Alegranza, Graciosa, Arrecife, Teguise, Tinajo, Lanzarote, La Oliva, Puerto del Rosario, Fuerteventura, Antigua, Punta de Jandia, Las Palmas, Isla de Gran Canaria, Gáldar, Santa Cruz de Tenerife, La Laguna, La Orotava, Pico de Teide, 3718m, Santa Cruz de la Palma, Los Llanos de Aridane, La Palma, Puerto de la Cruz, Gomera, Valverde, Hierro, Santa Cruz de Tenerife

ATLANTIC OCEAN

SCALE 1:7,250,000

BERMUDA (to UK)

SCALE 1:550,000

St Catherine Point, St George's Island, St George, St David's Island, Commissioner's Harbour, Tucker's Town, Kindley Field, Harrington Sound, Hamilton, Ireland Island North, Ireland Island South, Somerset, Spanish Point, Great Sound, Little Sound, Hatts Village

ATLANTIC OCEAN

Scale 1:48,000,000

projection: Mollweide

Km / Miles

Map labels (main map)

EUROPE, AFRICA, NORTH AMERICA, SOUTH AMERICA, UNITED STATES OF AMERICA, CANADA, GREENLAND (to Denmark), ICELAND

ATLANTIC OCEAN, Mid-Atlantic Ridge, Sargasso Sea, Caribbean Sea, North Sea, Celtic Sea, Irish Sea, Baffin Bay, Baffin Basin, Labrador Sea, Labrador Basin, Davis Strait, Denmark Strait, Hudson Strait, Foxe Basin, Foxe Channel, Hudson Bay, Ungava Bay, Cumberland Sound, Gulf of St. Lawrence, Gulf of Mexico

UNITED KINGDOM, IRELAND, FRANCE, SPAIN, PORTUGAL, MOROCCO, ALGERIA, WESTERN SAHARA (occupied by Morocco), MAURITANIA, SENEGAL, GAMBIA, GUINEA-BISSAU, GUINEA, SIERRA LEONE, LIBERIA, IVORY COAST, GHANA, TOGO, BENIN, NIGERIA

MEXICO, BELIZE, GUATEMALA, HONDURAS, NICARAGUA, COSTA RICA, PANAMA, CUBA, JAMAICA, HAITI, DOMINICAN REPUBLIC, THE BAHAMAS, Turks and Caicos Islands (to UK), Puerto Rico (to USA), TRINIDAD & TOBAGO, BARBADOS, VENEZUELA, GUYANA, Leeward Islands, Windward Islands, Bermuda (to UK)

Shetland Islands, Faroe Islands (to Denmark), Rockall, Azores (to Portugal), Madeira (to Portugal), Canary Islands (to Spain), Cape Verde, Reykjavík, Surtsey, Nuuk

Rotterdam, Amsterdam, Hamburg, Southampton, Milford Haven, Belfast, Cork, Nantes, Bordeaux, Bilbao, Gijón, Leixões, Lisbon, Casablanca, Safi, Nouâdhibou, Nouakchott, Dakar, Banjul, Bissau, Conakry, Freetown, Monrovia, Abidjan, Lagos, Porto-Novo

Boston, New York, Baltimore, Savannah, Jacksonville, Mobile, New Orleans, Tampico, Veracruz, Montreal, Halifax, St. Lawrence, St. John's

Maracaibo, Barranquilla, Cartagena, Cristóbal, Limón, Puerto Cortés, Belize City, Bluefields, Santiago, Barahona, Kingston

Physiographic features

Iceland Basin, Iceland-Faroe Ridge, Faroe-Iceland Ridge, Reykjanes Ridge, Reykjanes Basin, Eirik Ridge, Irminger Basin, Labrador Basin, Hatton Ridge, Rockall Bank, Rockall Trough, Porcupine Bank, Porcupine Plain, Porcupine Seabight, Biscay Plain, Celtic Shelf, Goban Spur, Galicia Bank, Iberian Plain, Charcot Seamounts, Horseshoe Seamounts, Ampère Seamount, Gorringe Ridge, Strait of Gibraltar, Tagus Plain, Tagus, Guadiana, Iberian Abyssal Plain, Azores-Biscay Rise, Kings Trough, Azores Fracture Zone, East Azores Fracture Zone, Cruiser Tablemount, Great Meteor Tablemount, Madeira Ridge, Madeira Plain, Canary Islands, Sahara Seamounts, Tropic Seamount, Cape Verde Terrace, Cape Verde Plain, Cape Verde Basin, Gambia Plain, Gambia Basin, Sierra Leone Rise

Maury Channel, Maury Seachannel, Charlie-Gibbs Fracture Zone, West Thulean Rise, East Thulean Rise, Milne Seamounts, Newfoundland Ridge, Newfoundland Seamounts, Newfoundland Basin, Orphan Knoll, Flemish Cap, Grand Banks of Newfoundland, Hamilton Bank, Saglek Bank, Great Hellefiske Bank, Northwest Atlantic Mid-Ocean Canyon, Sohm Plain, Nashville Seamount, New England Seamounts, Corner Seamounts, Oceanographer Fracture Zone, Atlantis Fracture Zone, Kane Fracture Zone, Barracuda Fracture Zone, Vema Fracture Zone, Doldrums Fracture Zone, Demerara Plain, Demerara Plateau, Georges Bank, Nares Plain, Hatteras Plain, Blake Plateau, Blake-Bahama Ridge, Bermuda Rise, Puerto Rico Trench, Mineros Trough, Venezuelan Basin, Aves Ridge, Colombian Basin, Cayman Trough, Nicaragua Rise, Yucatan Basin, Yucatan Channel, Campeche Bank, Bay of Campeche, Sigsbee Plain, Florida Plain, Straits of Florida, Great Bahama Bank, Little Bahama Bank

Tropic of Cancer, Tropic of Cancer, Arctic Circle, Arctic Circle

64

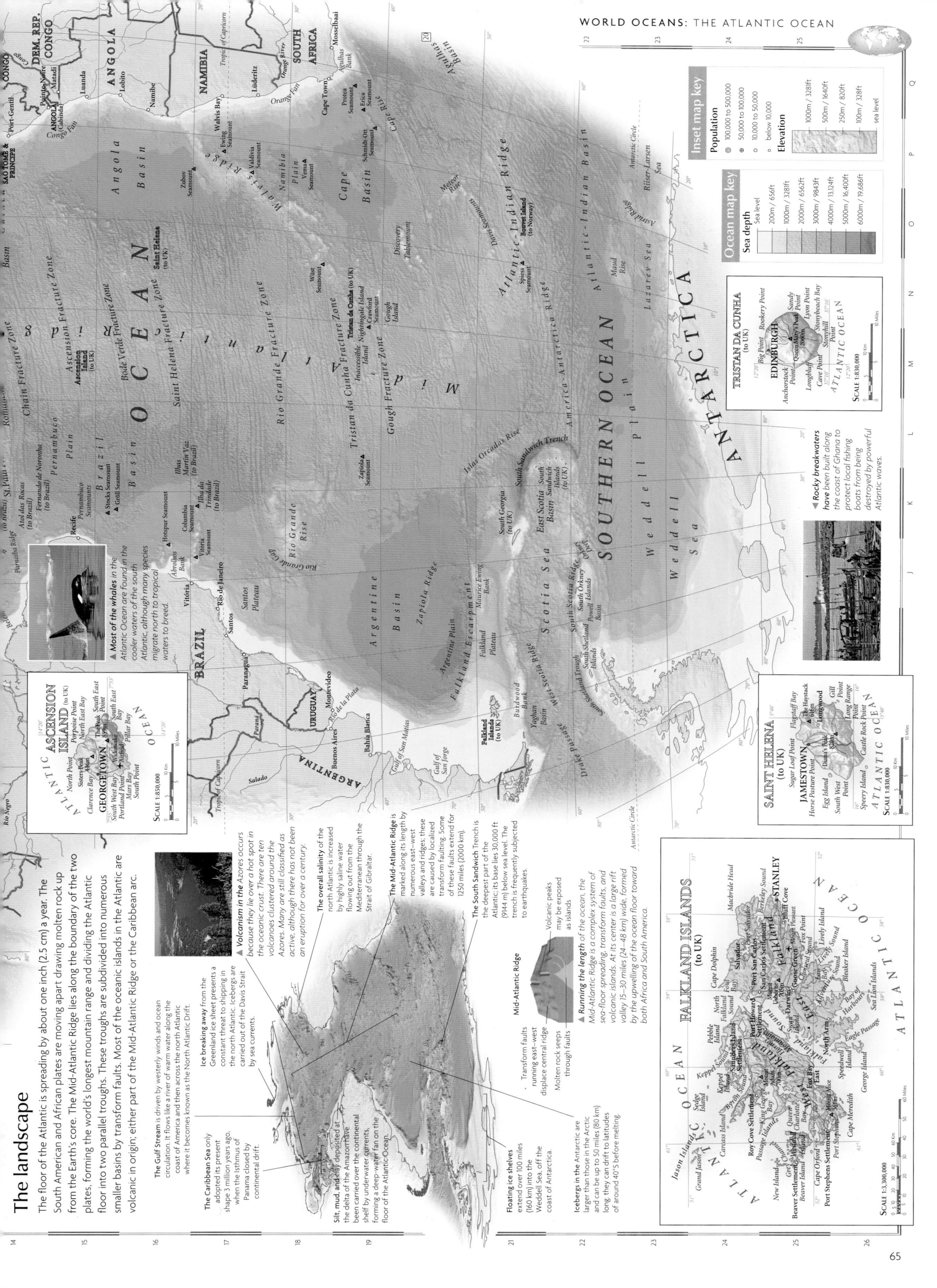

The landscape

The floor of the Atlantic is spreading by about one inch (2.5 cm) a year. The South American and African plates are moving apart drawing molten rock up from the Earth's core. The Mid-Atlantic Ridge lies along the boundary of the two plates, forming the world's longest mountain range and dividing the Atlantic floor into two parallel troughs. These troughs are subdivided into numerous smaller basins by transform faults. Most of the oceanic islands in the Atlantic are volcanic in origin; either part of the Mid-Atlantic Ridge or the Caribbean arc.

▲ Most of the whales in the Atlantic Ocean are found in the cooler waters of the south Atlantic, although many species migrate north to tropical waters to breed.

The Gulf Stream is driven by westerly winds and ocean circulation. It flows like a river of warm water along the coast of America and then across the north Atlantic where it becomes known as the North Atlantic Drift.

The Caribbean Sea only adopted its present shape 3 million years ago, when the Isthmus of Panama closed by continental drift.

Ice breaking away from the Greenland ice sheet presents a constant threat to shipping in the north Atlantic. Icebergs are carried out of the Davis Strait by sea currents.

Silt, mud, and clay deposited at the delta of the Amazon have been carried over the continental shelf by underwater currents, forming a deep-water fan on the floor of the Atlantic Ocean.

▲ Volcanism in the Azores occurs because they lie over a hot spot in the oceanic crust. There are ten volcanoes clustered around the Azores. Many are still classified as active, although there has not been an eruption for over a century.

The overall salinity of the north Atlantic is increased by highly saline water flowing out from the Mediterranean through the Strait of Gibraltar.

The Mid-Atlantic Ridge is marked along its length by numerous east–west valleys and ridges; these are caused by localized transform faulting. Some of these faults extend for 1250 miles (2000 km).

The South Sandwich Trench is the deepest part of the Atlantic. Its base lies 30,000 ft (9144 m) below sea level. The trench is frequently subjected to earthquakes.

Volcanic peaks may be exposed as islands.

▲ Running the length of the ocean, the Mid-Atlantic Ridge is a complex system of sea-floor spreading, transform faults, and volcanic islands. At its center is a large rift valley 15–30 miles (24–48 km) wide, formed by the upwelling of the ocean floor toward both Africa and South America.

Mid-Atlantic Ridge

- Transform faults running east–west displace central ridge
- Molten rock seeps through faults

Floating ice shelves extend over 100 miles (160 km) into the Weddell Sea, off the coast of Antarctica.

Icebergs in the Antarctic are larger than those in the Arctic and can be up to 50 miles (80 km) long; they can drift to latitudes of around 40°S before melting.

▲ Rocky breakwaters have been built along the coast of Ghana to protect local fishing boats from being destroyed by powerful Atlantic waves.

Inset map key

Population
- ⊙ 100,000 to 500,000
- ⊕ 50,000 to 100,000
- ○ 10,000 to 50,000
- below 10,000

Elevation
1000m / 328ft
500m / 1640ft
250m / 820ft
100m / 328ft
sea level

Ocean map key

Sea depth
- Sea level
- 200m / 656ft
- 1000m / 328ft
- 2000m / 6562ft
- 3000m / 9843ft
- 4000m / 13,124ft
- 5000m / 16,400ft
- 6000m / 19,686ft

ASCENSION ISLAND (to UK)
GEORGETOWN
North Point, Sisters Peak, Porpoise Point, The Peak, South East Point, North East Bay, Clarence Bay, Weather Post, South West Bay, Portland Point, Mars Bay, Pillar Bay, South Point
ATLANTIC OCEAN
SCALE 1:850,000

TRISTAN DA CUNHA (to UK)
EDINBURGH
Big Point, Rookery Point, Sandy Point, Queen Mary's Peak 2060m, Stonyhill Point, Anchorstock Point, Longbluff, Lyon Point, Cave Point, Stonybeach Bay
ATLANTIC OCEAN
SCALE 1:830,000

SAINT HELENA (to UK)
JAMESTOWN
Sugar Loaf Point, Flagstaff Bay, The Haystack 610m, Horse Pasture Point, Longwood, Gill Point, Egg Island, Diana's Peak 820m, South West Point, Long Range Point, Sperry Island, Castle Rock Point
ATLANTIC OCEAN
SCALE 1:830,000

FALKLAND ISLANDS (to UK)
STANLEY, Macbride Head, Cape Dolphin, Berkeley Sound, Port Salvador, Bluff Cove, North Falkland Sound, San Carlos, Mount Pleasant, Fox Point, Pebble Island, Keppel Island, Port Howard Settlement, Goose Green, Lively Island, Saunders Island, Bleaker Island, Port San Carlos Settlement, Darwin, Bayol Harbour, Sea Lion Islands, New Island, Fox Bay East, Speedwell Island, George Island, Beaver Island Settlement, Mount Alice, Cape Orford, Port Stephens Settlement, Cape Meredith
ATLANTIC OCEAN
SCALE 1:3,300,000

Africa

The world's second largest continent, Africa covers an area of 11,712,434 sq miles (30,355,000 sq km). It has 54 separate countries, including Madagascar, Comoros, Mauritius, and the Seychelles in the Indian Ocean – the highest number of any continent.

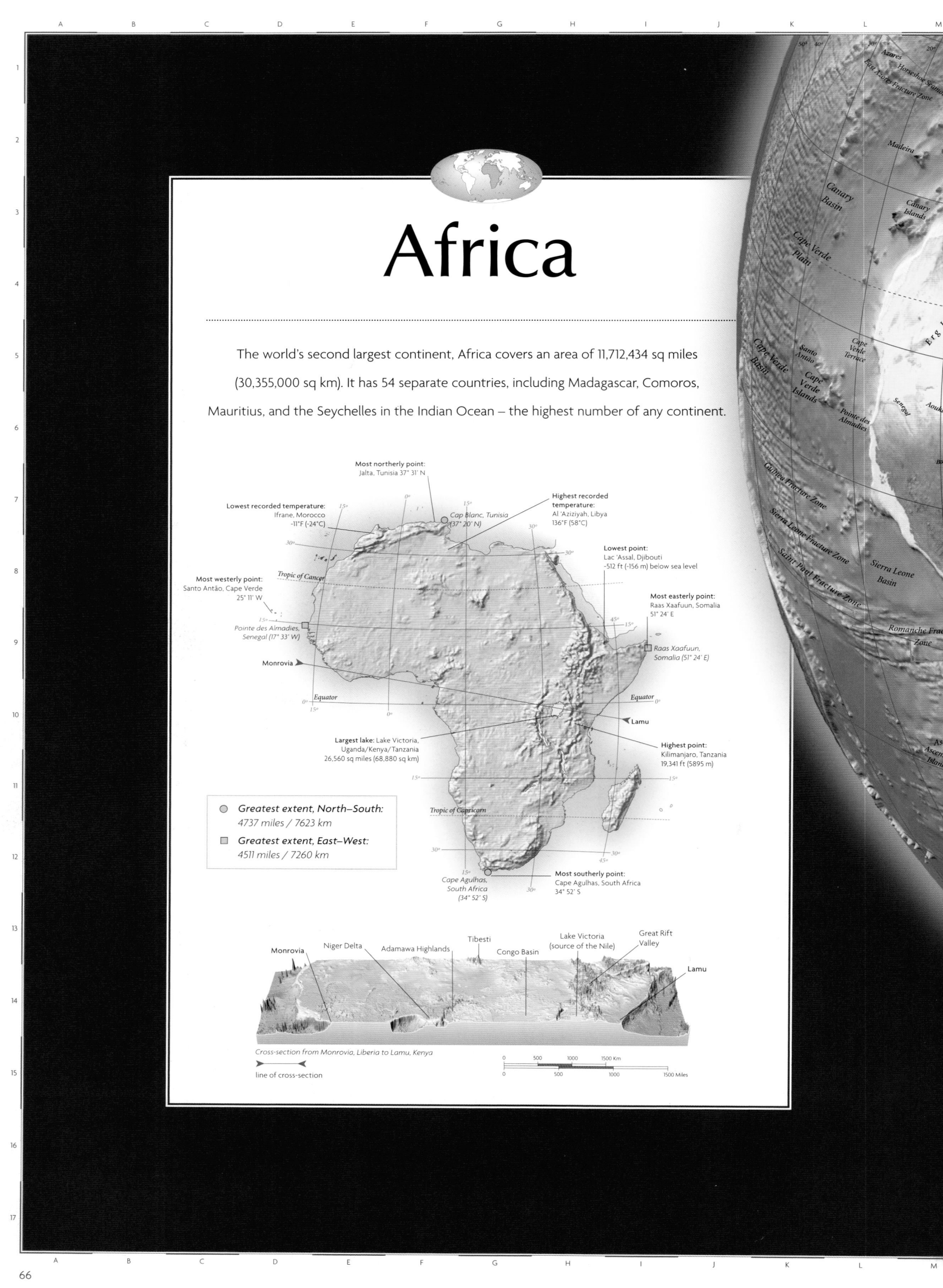

Most northerly point:
Jalta, Tunisia 37° 31' N

Lowest recorded temperature:
Ifrane, Morocco
-11°F (-24°C)

Cap Blanc, Tunisia
(37° 20' N)

Highest recorded temperature:
Al 'Aziziyah, Libya
136°F (58°C)

Lowest point:
Lac 'Assal, Djibouti
-512 ft (-156 m) below sea level

Most westerly point:
Santo Antão, Cape Verde
25° 11' W

Tropic of Cancer

Most easterly point:
Raas Xaafuun, Somalia
51° 24' E

*Pointe des Almadies,
Senegal (17° 33' W)*

*Raas Xaafuun,
Somalia (51° 24' E)*

Monrovia

Equator

Equator

Lamu

Largest lake: Lake Victoria,
Uganda/Kenya/Tanzania
26,560 sq miles (68,880 sq km)

Highest point:
Kilimanjaro, Tanzania
19,341 ft (5895 m)

Tropic of Capricorn

○ *Greatest extent, North–South:*
4737 miles / 7623 km

□ *Greatest extent, East–West:*
4511 miles / 7260 km

*Cape Agulhas,
South Africa
(34° 52' S)*

Most southerly point:
Cape Agulhas, South Africa
34° 52' S

Monrovia | Niger Delta | Adamawa Highlands | Tibesti | Congo Basin | Lake Victoria (source of the Nile) | Great Rift Valley | Lamu

Cross-section from Monrovia, Liberia to Lamu, Kenya

line of cross-section

| 0 | 500 | 1000 | 1500 Km |
| 0 | 500 | 1000 | 1500 Miles |

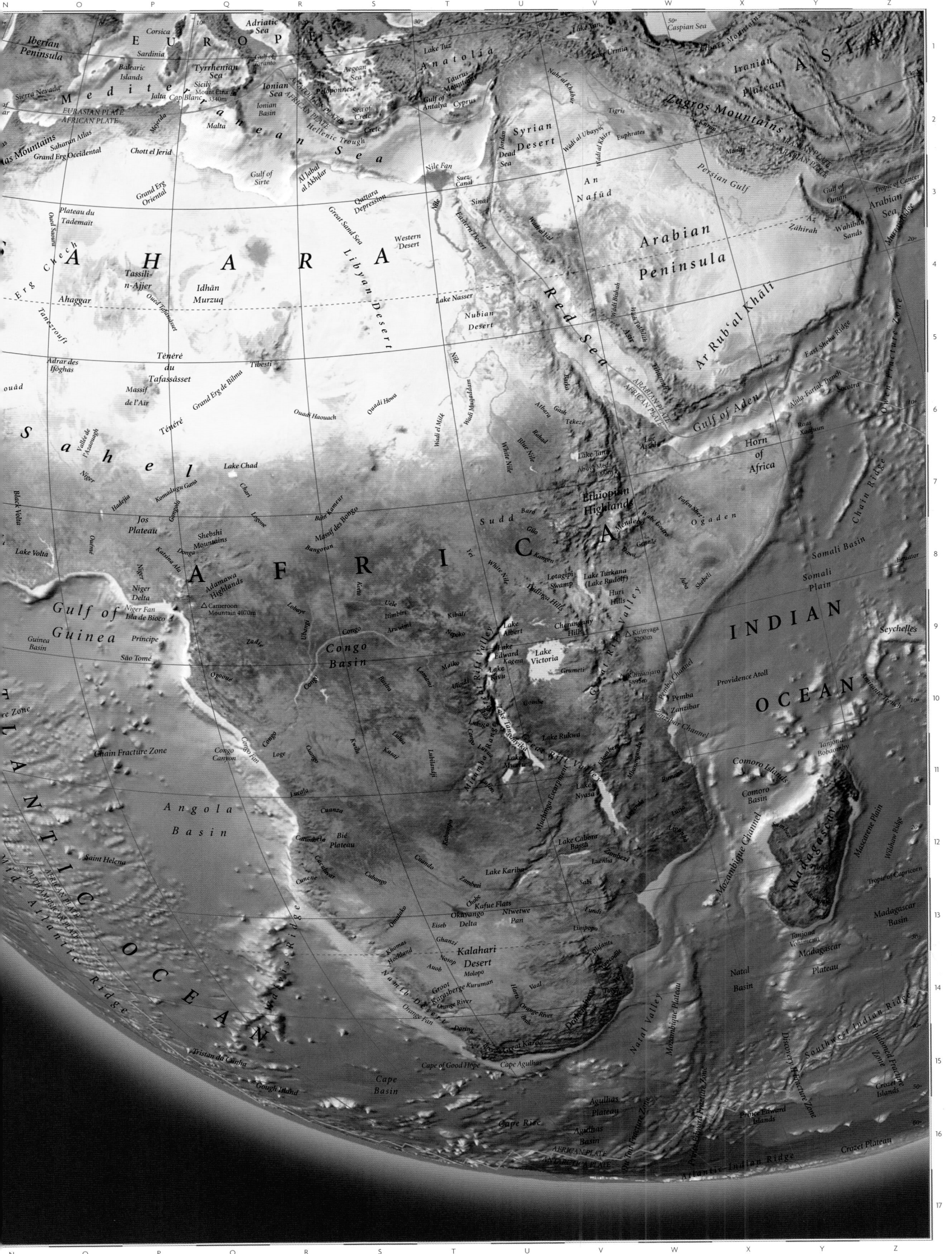

EUROPE

Iberian
Peninsula

Corsica

Adriatic
Sea

Sardinia

Sierra Nevada

Balearic
Islands

Tyrrhenian
Sea

Sicily
Mount Etna
3340m

Gulf of
Taranto

Lake Tuz

Anatolia

Caspian Sea

Elburz Mountains

ASIA

Iranian
Plateau

Aegean
Sea

Peloponnese

Ionian
Sea

Malta

Ionian
Basin

Sea of
Crete

Taurus
Mountains

Gulf of
Antalya

Cyprus

Crete

Hellenic Trough

Lake Van

Lake Urmia

Tigris

Nahr al Khabur

Euphrates

Zagros Mountains

Cap Blanc

Al Jabal
al Akhdar

Gulf of
Sirte

Maydar

Atlas Mountains

Saharan Atlas

Grand Erg Occidental

Chott el Jerid

EURASIAN PLATE
AFRICAN PLATE

M e d i t e r r a n e a n

Qattara
Depression

Nile Fan

Great Sand Sea

Suez
Canal

Sinai

Jordan

Dead
Sea

Syrian
Desert

Wadi al Ubayyd

Wadi al Khirr

Karun

Persian Gulf

Tropic of Cancer

Gulf of
Oman

Arabian
Sea

Grand Erg
Oriental

Plateau du
Tademaït

Oued Saoura

Western
Desert

Eastern Desert

Nile

An
Nafūd

Wadi Birah

Arabian
Peninsula

Az
Zāhirah

Wahibah
Sands

Murray Ridge

Erg Chech

SAHARA

Ahaggar

Tassili
n-Ajjer

Idhān
Murzuq

Tanezrouft

Oued Tiféssasset

Lake Nasser

Nubian
Desert

Red Sea

Asir

Wadi Hadhlil

Najrān

Ar Rub' al Khālī

East Sheba Ridge

Owen Fracture Zone

Sahel

Adrar des
Ifôghas

Ténéré
du
Tafassâsset

Tibesti

ARABIAN PLATE
AFRICAN PLATE

Alula-Fartak Trench

Socotra

Massif
de l'Aïr

Grand Erg de Bilma

Ouadi Howa

Wadi al Milk

Tarba

Gash

Gulf of Aden

Raas
Xaafuun

Niger

Valle de
l'Azaouagh

Ténéré

Ouadi Haouach

Wadi Magadam

Atbara

Tekeze

Reball

Lac
Assal

Horn
of
Africa

Chain Ridge

Black Volta

Hadejia

Komadugu Gana

Lake Chad

Chari

Logone

Bahr Kameur

White Nile

Blue Nile

Lake Tana
Abaya Meda
4000m

Mendebo

Wabe Shebele

Ogaden

Somali Basin

Lake Volta

Oueme

Jos
Plateau

Shebshi
Mountains

Katsina Ala

Donga

Massif des Bongo

Bangoran

Sudd

Baro

Gilo

Ethiopian
Highlands

Kangen

Genale

AFRICA

Somali
Plain

Equator

Niger

Niger
Delta

Adamawa
Highlands

Lobaye

Uele

Itimbiri

Aruwimi

Kibali

Lotagipi
Swamp

Ilemi
Triangle

Lake Turkana
(Lake Rudolf)

Huri
Hills

Juba

Shebeli

Gulf of
Guinea

Isla de Bioco

Cameroon
Mountain 4070m

Zadie

Congo

Ubangi

Lomami

Negoko

Maiko

Ulindi

Charangeny
Hills

Lake
Albert

Kirinyaga
5200m

INDIAN

Seychelles

Guinea
Basin

Príncipe

São Tomé

Ogooué

Congo
Basin

Lake
Edward

Lake
Kivu

Lake
Victoria

Lake Kagera

Grumeti

Kilimanjaro
5895m

OCEAN

Chain Fracture Zone

Congo

Congo

Loge

Bussira

Kasai

Lukuga

Lubilandji

Lake Tanganyika

Gombe

Lake Rukwa

Pemba Channel

Pemba

Zanzibar

Zanzibar Channel

Providence Atoll

Amirante Trench

Congo
Canyon

Congo Fan

Luala

Kwango

Kwilu

Luapula

Lake Mweru

Ruvuma

Comoro Islands

Tanjona
Bobaomby

Angola
Basin

Cuanza

Carimbela

Bié
Plateau

Cuito

Cassai

Cubango

Kasongo

Kabompo

Lake
Nyasa

Lunga

Comoro
Basin

Saint Helena

Cuando

Zambezi

Lake Cabora
Bassa

Luenha

Lake Karibu

Sabi

Mozambique Channel

Madagascar

Mascarene Plain

Mascarene Ridge

Wilshaw Ridge

Canene

Chobe

Okavango
Delta

Kafue Flats

Eiseb

Ntwetwe
Pan

Limpopo

Lundi

Tanjona
Volimena

Madagascar
Basin

Omuramba

Ghanzi

Khomas
Hochland

Nosop

Molopo

Kalahari
Desert

Tsodilo

Njoko

Tanjona
Volimena

Madagascar
Plateau

Namib Desert

Groot

Auob

Kuruman

Koesberge

Vaal

Natal
Basin

Orange River

Doring

Orange Fan

Harts

Orange River

Great Karoo

Tugela

Drakensberg

Natal Valley

Mozambique Plateau

Tristan da Cunha

Mid-Atlantic Ridge

ATLANTIC OCEAN

Walvis Ridge

Cape
Basin

Cape of Good Hope

Cape Agulhas

Southwest Indian Ridge

Discovery Fracture Zone

Indomed Fracture Zone

Gough Island

Cape Rise

Agulhas
Plateau

Agulhas
Basin

AFRICAN PLATE
ANTARCTIC PLATE

Prince Edward
Islands

Crozet
Islands

Atlantic-Indian Ridge

Crozet Plateau

Physical Africa

The structure of Africa was dramatically influenced by the break up of the supercontinent Gondwanaland about 160 million years ago and, more recently, rifting and hot spot activity. Today, much of Africa is remote from active plate boundaries and comprises a series of extensive plateaus and deep basins, which influence the drainage patterns of major rivers. The relief rises to the east, where volcanic uplands and vast lakes mark the Great Rift Valley. In the far north and south sedimentary rocks have been folded to form the Atlas Mountains and the Great Karoo.

East Africa

The Great Rift Valley is the most striking feature of this region, running for 4475 miles (7200 km) from Lake Nyasa to the Red Sea. North of Lake Nyasa it splits into two arms and encloses an interior plateau which contains Lake Victoria. A number of elongated lakes and volcanoes lie along the fault lines. To the west lies the Congo Basin, a vast, shallow depression, which rises to form an almost circular rim of highlands.

Northern Africa

Northern Africa comprises a system of basins and plateaus. The Tibesti and Ahaggar are volcanic uplands, whose uplift has been matched by subsidence within large surrounding basins. Many of the basins have been infilled with sand and gravel, creating the vast Saharan lands. The Atlas Mountains in the north were formed by convergence of the African and Eurasian plates.

Rift valley lakes, like Lake Tanganyika, lie along fault lines

Lake Victoria

Extensive faulting occurs as rift valley pulls apart

B ——— B

Cross-section through eastern Africa showing the two arms of the Great Rift Valley and its interior plateau.

0   50   100 Km
0   50   100 Miles

The Earth's crust has been warped to form the Taoudenni Basin

Volcanic Ahaggar mountains, formed by rising magma from a hot spot

Lake Chad lies in a sand-filled basin

A ——— A

Section across northern Africa showing infilled basins and uplifted plateaus.

0   250   500 Km
0   250   500 Miles

Scale 1:40,000,000

Km
0   200   400   600   800

Miles
0   200   400   600   800

projection: Lambert Azimuthal Equal Area

Map key

Elevation

5000m / 16,405ft
4000m / 13,124ft
3000m / 9843ft
2000m / 6562ft
1000m / 3281ft
500m / 1640ft
250m / 820ft
100m / 328ft
sea level
below sea level

Plate margins
(for explanation see page xiv)

——— constructive
△   △ destructive
——— conservative
······· uncertain
▷——◁ line of cross-section

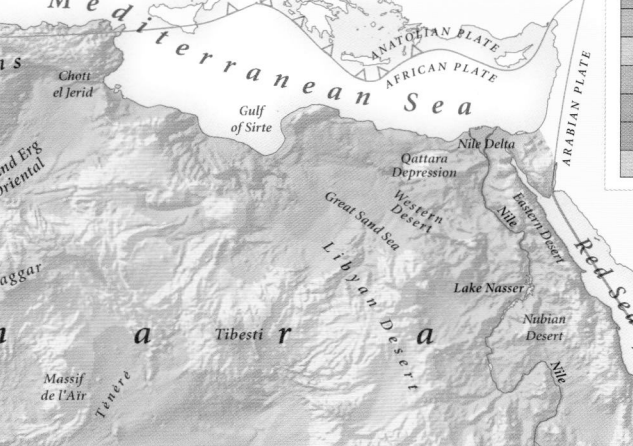

Southern Africa

The Great Escarpment marks the southern boundary of Africa's basement rock and includes the Drakensberg range. It was uplifted when Gondwanaland fragmented about 160 million years ago and it has gradually been eroded back from the coast. To the north, the relief drops steadily, forming the Kalahari Basin. In the far south are the fold mountains of the Great Karoo.

Kalahari Basin, covered with the sandy plains of the Kalahari Desert

Boundary of the Great Escarpment

Uplift of the basement rock created a raised plateau

Drakensberg

C ——— C

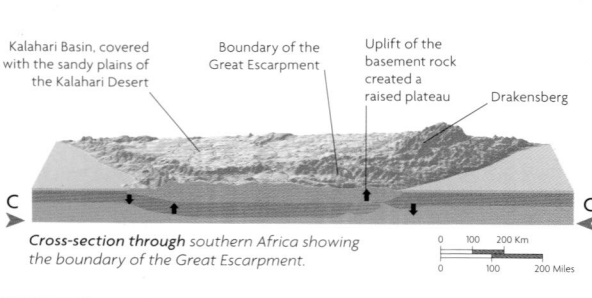

Cross-section through southern Africa showing the boundary of the Great Escarpment.

0   100   200 Km
0   100   200 Miles

Mediterranean Sea

EURASIAN PLATE
AFRICAN PLATE

ANATOLIAN PLATE
AFRICAN PLATE

ARABIAN PLATE

ATLANTIC OCEAN

Atlas Mountains

Chott el Jerid

Gulf of Sirte

Grand Erg Occidental

Grand Erg Oriental

Erg Iguidi

Erg Chech

Ahaggar

S a h a r a

Taoudenni Basin

Niger

Sahel

Niger

White Volta

Cape Verde Islands

Senegal

Massif de l'Aïr

Ténéré

Tibesti

Qattara Depression

Western Desert

Great Sand Sea

Libyan Desert

Nile Delta

Nile

Lake Nasser

Nubian Desert

Red Sea

AFRICAN PLATE

ARABIAN PLATE

ASIA

Lake Chad

Blue Nile

Lake Tana

Gulf of Aden

Horn of Africa

Ethiopian Highlands

A ——— A

Lake Volta

Niger

Benue

Grain Coast

Ivory Coast   Gold Coast

Slave Coast

Bight of Benin

Niger Delta

△ Cameroon Mountain 4070m

Adamawa Highlands

Gulf of Guinea

São Tomé

ATLANTIC OCEAN

Ubangi

Massif des Bongo

Congo

Congo Basin

Congo

Lake Albert

Sudd

White Nile

Shebeli

Juba

Lake Turkana (Lake Rudolf)

Great Rift Valley

Lake Victoria

△ Kilimanjaro 5895m

Mitumba Range

Lake Tanganyika

B ——— B

Pemba Island
Zanzibar

Seychelles

Bié Plateau

Lake Nyasa

Zambezi

Comoro Islands

Mozambique Channel

Madagascar

Mauritius

Réunion

INDIAN OCEAN

Namib Desert

Okavango Delta

Kalahari Basin

Kalahari Desert

Zambezi

Limpopo

Orange River

Drakensberg

Great Karoo

C ——— C

Cape of Good Hope

A   B   C   D   E   F   G   H   I   J   K   L   M

Climate

The climates of Africa range from mediterranean to arid, dry savannah, and humid equatorial. In East Africa, where snow settles at the summit of volcanoes such as Kilimanjaro, climate is also modified by altitude. The winds of the Sahara export millions of tonnes of dust a year both northward and eastward.

▲ *Savannah grasslands run* in a belt across Africa; limited rainfall inhibits tree growth.

Temperature

Tropic of Cancer
20° N
Equator
20° S
Tropic of Capricorn

Average January temperature   *Average July temperature*

Temperature

☐ 32 to 50°F (0 to 10°C)
☐ 50 to 68°F (10 to 20°C)
☐ 68 to 86°F (20 to 30°C)
☐ above 86°F (30°C)

Rainfall

Tropic of Cancer
20° N
Equator
20° S
Tropic of Capricorn

Average January rainfall   *Average July rainfall*

Rainfall

☐ 0–1 in (0–25 mm)   ☐ 8–12 in (200–300 mm)
☐ 1–2 in (25–50 mm)   ☐ 12–16 in (300–400 mm)
☐ 2–4 in (50–100 mm)   ☐ 16–20 in (400–500 mm)
☐ 4–8 in (100–200 mm)   ☐ more than 20 in (500 mm)

▲ *The hot, equatorial basin of the Congo river receives over 48 inches (1200 mm) of rainfall per year.*

Climate

☐ arid
☐ humid equatorial
☐ mediterranean
☐ semi-arid
☐ tropical
☐ warm humid
☼ daily hours of sunshine, January
☼ daily hours of sunshine, July
→ cold wind
→ hot wind

Shaping the continent

African landscapes are shaped by the intensity of climatic extremes and by tectonic action. High aridity, wind action, and infrequent but heavy rainstorms, lead to the migration of sand dunes and dramatic flash flooding across much of the north and west. In the wetter areas, high precipitation increases the rate of weathering. To the east, the rift system has created a volcanic and lake environment and allowed rivers to erode weaknesses left in the crustal structure by faults.

Groundwater

1 Oases are found in desert areas such as the Sahara (*left*). Groundwater migrates through permeable rock strata, confined between two impermeable layers. Oases form either when the permeable rocks come near to the surface, or at a fault line, when water is able to seep up to the surface through the crushed rocks at the fault.

Groundwater: Replenishment of an oasis

Rainwater feeds the aquifer
Water migrates up through fault
Aquifer exposed near the surface
Groundwater trapped between impermeable strata

River systems

2 The Zambezi river (*above*) drops 360 ft (110 m) over the Victoria Falls into a zigzag gorge. The river has eroded the gorge along lines of weakness in the bedrock, created by fault lines running in two directions.

River systems: Retreating of the Victoria Falls

Old site of Victoria Falls
River plunges over falls
Fault and joint lines running in two directions
Zigzag gorge of the Zambezi

The evolving landscape

External stresses act on the surface of the inselberg
Exfoliated layers
Joints or cracks caused by expansion and contraction

Weathering: Formation of an inselberg

Weathering

6 Inselbergs (*above*), found extensively across West Africa, are exposed remnants of an extensive upland area. Erosion of the surrounding uplands leaves a resistant rock outcrop. Its spheroidal shape is the result of "onion-skin" weathering – the exfoliating of layers – due to repeated expansion and contraction.

Ephemeral channels

5 Wadis (*above*) drain much of northern Africa. These drybed courses are flooded only after infrequent, but intense, storms in the uplands cause water to surge along their channels.

Heavy rainfall runs off mountains
Water collects and floods the dry channel

Ephemeral channels: Flash flooding of a wadi

Sand is gradually blown up the back slope
Deposition on the slip face
Build up of sand produces strata inside the dune

Wind erosion: Migration of a dune

Wind erosion

4 Dunes like this in the Namib Desert (*left*) are wind-blown accumulations of sand, which slowly migrate. Wind action moves sand up the shallow back slope; when the sand reaches the crest of the dune it is deposited on the slip face.

Landscape

☐ sinking land
☐ stable land
☐ uplifting land
▽▽ escarpment
→ ocean current
— rift
▲ active volcano
⛰ inselberg
oasis
～ river
～ wadi
～ waterfall

Coastal processes

3 Houtbaai (*above*), in southern Africa, is constantly being modified by wave action. As waves approach the indented coastline, they reach the shallow water of the headland, slowing down and reducing in length. This causes them to bend or refract, concentrating their erosive force at the headlands.

Waves refracting
Wave energy dispersed in the bay
Force of waves concentrates on the headland
The sea bed is deeper opposite the bay than at the headland

Coastal processes: Erosion of a bay

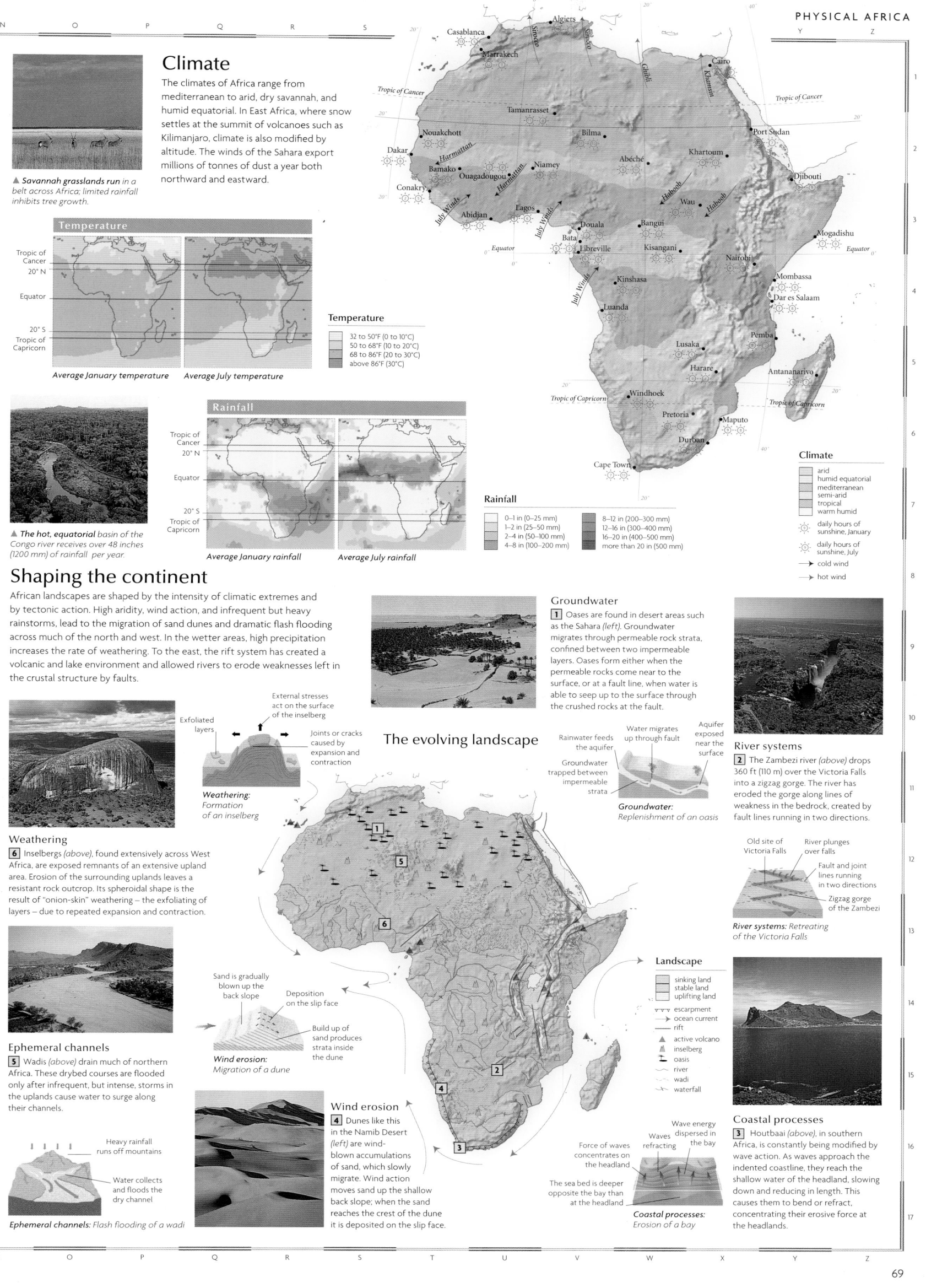

Political Africa

The political map of modern Africa only emerged following the end of the Second World War. Over the next half-century, all of the countries formerly controlled by European powers gained independence from their colonial rulers – only Liberia and Ethiopia were never colonized. The postcolonial era has not been an easy period for many countries, but there have been moves toward multiparty democracy across much of the continent. In South Africa, democratic elections replaced the internationally-condemned apartheid system only in 1994. Other countries have still to find political stability; corruption in government, and ethnic tensions are serious problems. National infrastructures, based on the colonial transportation systems built to exploit Africa's resources, are often inappropriate for independent economic development.

Languages

Three major world languages act as *lingua francas* across the African continent: Arabic in North Africa; English in southern and eastern Africa and Nigeria; and French in Central and West Africa, and in Madagascar. A huge number of African languages are spoken as well – over 2000 have been recorded, with more than 400 in Nigeria alone – reflecting the continuing importance of traditional cultures and values. In the north of the continent, the extensive use of Arabic reflects Middle Eastern influences while Bantu languages are widely-spoken across much of southern Africa.

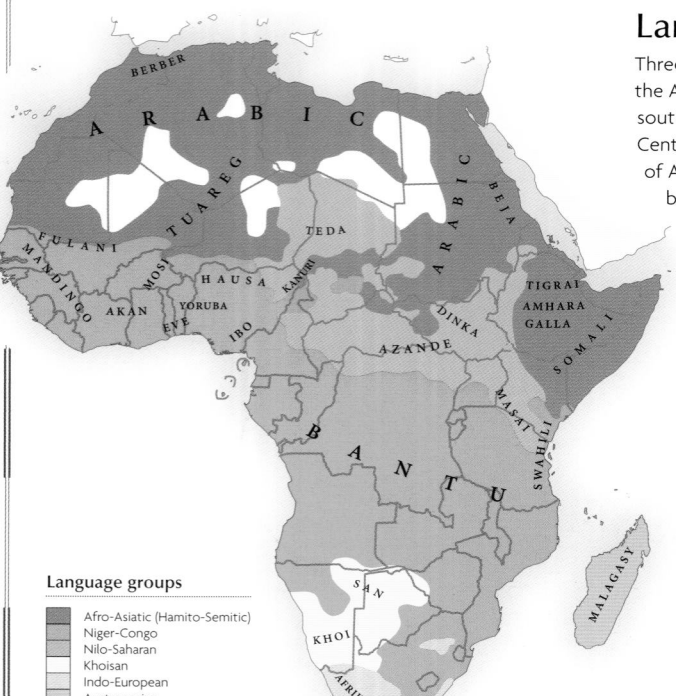

Language groups
- Afro-Asiatic (Hamito-Semitic)
- Niger-Congo
- Nilo-Saharan
- Khoisan
- Indo-European
- Austronesian

Official African languages

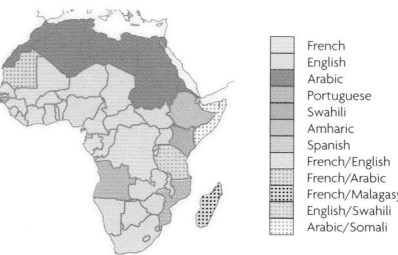

- French
- English
- Arabic
- Portuguese
- Swahili
- Amharic
- Spanish
- French/English
- French/Arabic
- French/Malagasy
- English/Swahili
- Arabic/Somali

▲ *Islamic influences are* evident throughout North Africa. The Great Mosque at Kairouan, Tunisia, is Africa's holiest Islamic place.

▲ *In northeastern Nigeria,* people speak Kanuri – a dialect of the Nilo-Saharan language group.

Transportation

African railroads were built to aid the exploitation of natural resources, and most offer passage only from the interior to the coastal cities, leaving large parts of the continent untouched – five landlocked countries have no railroads at all. The Congo, Nile, and Niger river networks offer limited access to land within the continental interior, but have a number of waterfalls and cataracts which prevent navigation from the sea. Many roads were developed in the 1960s and 1970s, but economic difficulties are making the maintenance and expansion of the networks difficult.

▶ *South Africa has the* largest concentration of railroads in Africa. Over 20,000 miles (32,000 km) of routes have been built since 1870.

▲ *Traditional means of* transportation, such as the camel, are still widely used across the less accessible parts of Africa.

◀ *The Congo river,* though not suitable for river transportation along its entire length, forms a vital link for people and goods in its navigable inland reaches.

Transportation
- major roads and highways
- major railroads
- major canal
- international borders
- ● transport intersections
- ⊕ international airports
- ⊕ major ports

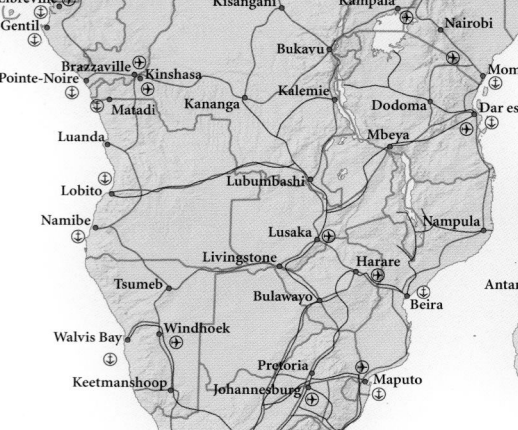

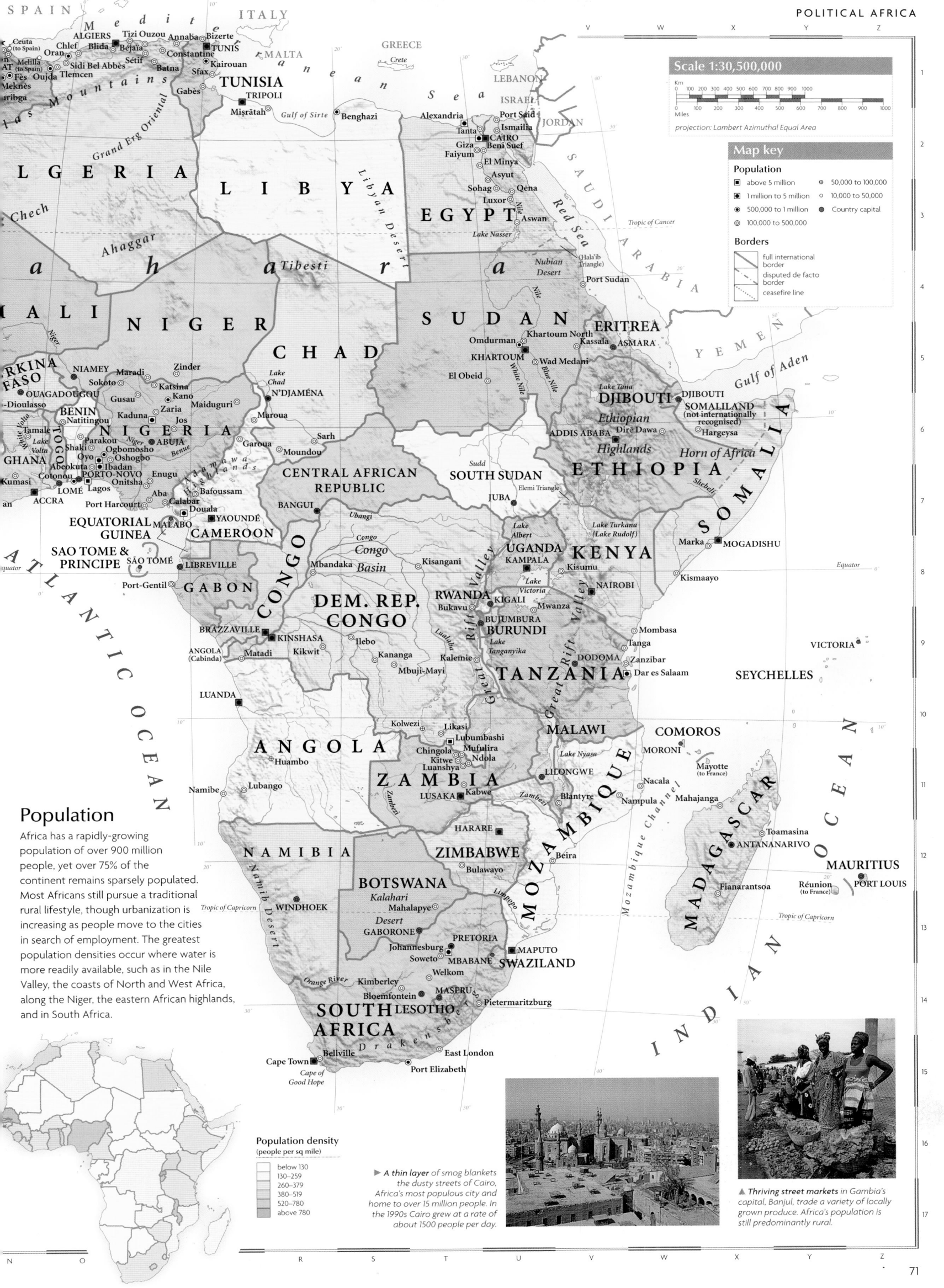

ITALY
GREECE
Crete
MALTA
LEBANON
ISRAEL
JORDAN

Mediterranean Sea

Ceuta (to Spain)
ALGIERS
Tizi Ouzou
Annaba
Bizerte
TUNIS
Chlef
Blida
Bejaïa
Constantine
Oran
Sidi Bel Abbès
Sétif
Batna
Kairouan
Melilla (to Spain)
Fès
Oujda
Tlemcen
Sfax
Meknès
rribga
Gabès
TUNISIA
TRIPOLI
Mişrātah
Gulf of Sirte
Benghazi
Alexandria
Port Said
Ismailia
Tanta
CAIRO
Giza
Beni Suef
Faiyum
El Minya
Asyut
Sohag
Qena
Luxor
Aswan
Lake Nasser

Grand Erg Oriental

A L G E R I A
L I B Y A
E G Y P T

Chech
Ahaggar
Tibesti
Libyan Desert
Nubian Desert

Tropic of Cancer

SAUDI ARABIA

Port Sudan
(Hala'ib Triangle)

Red Sea

M A L I
N I G E R
C H A D
S U D A N
ERITREA
ASMARA
Kassala
Khartoum North
Omdurman
KHARTOUM
Wad Medani
El Obeid
White Nile
Blue Nile

YEMEN

Gulf of Aden

RKINA FASO
NIAMEY
Maradi
Zinder
Sokoto
Gusau
Katsina
Kano
Zaria
Maiduguri
OUAGADOUGOU
Dioulasso
BENIN
Natitingou
Tamale
Parakou
GHANA
Oyo
Ogbomosho
Oshogbo
Ibadan
Abeokuta
Kaduna
Jos
Maroua
Garoua
Sarh
Moundou
N'DJAMÉNA
Lake Chad

DJIBOUTI
DJIBOUTI
SOMALILAND (not internationally recognised)
Hargeysa
ADDIS ABABA
Dire Dawa
Ethiopian Highlands
Horn of Africa

NIGERIA
ABUJA
Benue
Adamawa Highlands
CENTRAL AFRICAN REPUBLIC
SOUTH SUDAN
JUBA
Elemi Triangle
ETHIOPIA
SOMALIA
Lake Tana

Kumasi
Cotonou
PORTO-NOVO
LOMÉ
Lagos
ACCRA
Aba
Enugu
Onitsha
Calabar
Bafoussam
Douala
Port Harcourt
YAOUNDÉ
BANGUI
Ubangi
CAMEROON
EQUATORIAL GUINEA
MALABO
SAO TOME & PRINCIPE
SÃO TOMÉ
LIBREVILLE
Mbandaka
Congo Basin
Kisangani
UGANDA
KAMPALA
Lake Albert
Lake Turkana (Lake Rudolf)
KENYA
Marka
MOGADISHU

Equator

Port-Gentil
GABON
CONGO
Congo
DEM. REP. CONGO
RWANDA
Bukavu
KIGALI
Lake Victoria
Kisumu
NAIROBI
Kismaayo
Equator

BRAZZAVILLE
KINSHASA
ANGOLA (Cabinda)
Matadi
Kikwit
Ilebo
Kananga
Mbuji-Mayi
Kalemie
BUJUMBURA
Lake Tanganyika
BURUNDI
Mwanza
DODOMA
Dar es Salaam
Zanzibar
Tanga
Mombasa
VICTORIA
SEYCHELLES

Great Rift Valley
Lualaba
TANZANIA

LUANDA
Huambo
A N G O L A
Kolwezi
Likasi
Lubumbashi
Mufulira
Chingola
Kitwe
Ndola
Luanshya
Kabwe
MALAWI
LILONGWE
Lake Nyasa
COMOROS
MORONI
Mayotte (to France)

Namibe
Lubango
Z A M B I A
LUSAKA
Kabwe
Kwa
Zambezi
Nacala
Nampula
Mahajanga

HARARE
Beira
MOZAMBIQUE
Blantyre
Mozambique Channel

Namib Desert
Zambezi
ZIMBABWE
Bulawayo
Limpopo
MADAGASCAR
Toamasina
ANTANANARIVO
MAURITIUS

N A M I B I A
WINDHOEK
Tropic of Capricorn
Kalahari Desert
BOTSWANA
Mahalapye
GABORONE
PRETORIA
Johannesburg
Soweto
MAPUTO
MBABANE
SWAZILAND
Welkom
Fianarantsoa
Réunion (to France)
PORT LOUIS

Orange River
Kimberley
Bloemfontein
MASERU
LESOTHO
Pietermaritzburg
Tropic of Capricorn

SOUTH AFRICA
Drakensberg
Bellville
East London
Cape Town
Cape of Good Hope
Port Elizabeth

INDIAN OCEAN
ATLANTIC OCEAN

Population

Africa has a rapidly-growing population of over 900 million people, yet over 75% of the continent remains sparsely populated. Most Africans still pursue a traditional rural lifestyle, though urbanization is increasing as people move to the cities in search of employment. The greatest population densities occur where water is more readily available, such as in the Nile Valley, the coasts of North and West Africa, along the Niger, the eastern African highlands, and in South Africa.

Population density
(people per sq mile)

- below 130
- 130–259
- 260–379
- 380–519
- 520–780
- above 780

► A thin layer of smog blankets the dusty streets of Cairo, Africa's most populous city and home to over 15 million people. In the 1990s Cairo grew at a rate of about 1500 people per day.

▲ Thriving street markets in Gambia's capital, Banjul, trade a variety of locally grown produce. Africa's population is still predominantly rural.

African resources

The economies of most African countries are dominated by subsistence and cash crop agriculture, with limited industrialization. Manufacturing is largely confined to South Africa. Many countries depend on a single resource, such as copper or gold, or a cash crop, such as coffee, for export income, which can leave them vulnerable to fluctuations in world commodity prices. In order to diversify their economies and develop a wider industrial base, investment from overseas is being actively sought by many African governments.

Industry

Many African industries concentrate on the extraction and processing of raw materials. These include the oil industry, food processing, mining, and textile production. South Africa accounts for over half of the continent's industrial output with much of the remainder coming from the countries along the northern coast. Over 60% of Africa's workforce is employed in agriculture.

◄ The unspoiled natural splendor of wildlife reserves, like the Serengeti National Park in Tanzania, attract tourists to Africa from around the globe. The tourist industry in Kenya and Tanzania is particularly well developed, where it accounts for almost 10% of GNI.

Standard of living

Since the 1960s most countries in Africa have seen significant improvements in life expectancy, healthcare, and education. However, 28 of the 30 most deprived countries in the world are African, and the continent as a whole lies well behind the rest of the world in terms of meeting many basic human needs.

Standard of living
(UN human development index)

- high
- low

GNI per capita (US $)

- below 499
- 500–999
- 1000–1999
- 2000–2999
- 3000–3999
- above 4000

Industry

- brewing
- car/vehicle manufacture
- cement
- chemicals
- coffee processing
- electronics
- engineering
- finance
- fish processing
- food processing
- iron & steel
- mining
- palm oil processing
- peanut processing
- pharmaceuticals
- rice milling
- shipbuilding
- sugar processing
- tea processing
- textiles
- timber processing
- tobacco processing

- coal
- oil
- gas

- industrial cities
- major industrial areas

◄ The discovery of oil in the swampy Niger Delta during the 1960s made Nigeria one of Africa's richer nations. As world oil prices fell in the 1980s, the Nigerian economy faltered.

► Exotic rugs and brightly colored textiles are sold in a street market along the banks of the river Nile in Luxor, Egypt.

◄ The Rössing uranium mines in Namibia are one of the largest in the world. Canada and Australia produce over half the world's uranium ore, used to fuel nuclear power plants. Elsewhere, South Africa and Niger also mine uranium on a large scale.

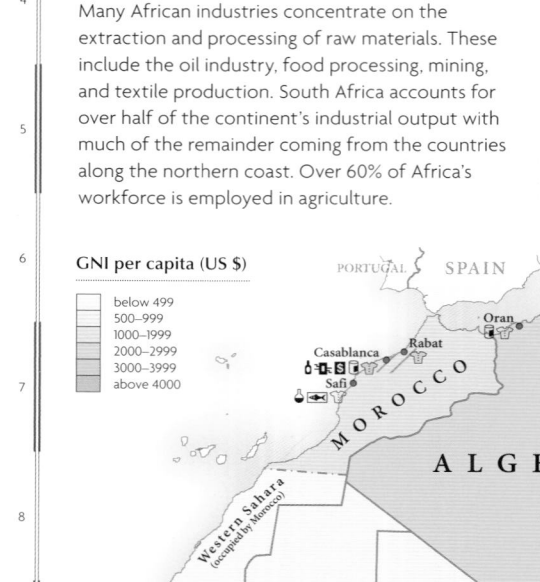

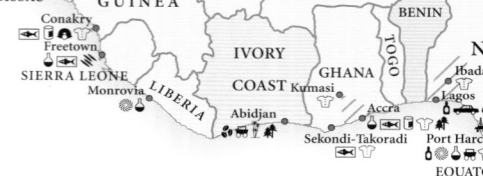

Environmental issues

One of Africa's most serious environmental problems occurs in marginal areas such as the Sahel where scrub and forest clearance, often for cooking fuel, combined with overgrazing, are causing desertification. Game reserves in southern and eastern Africa have helped to preserve many endangered animals, although the needs of growing populations have led to conflict over land use, and poaching is a serious problem.

Environmental issues

- national parks
- tropical forest
- forest destroyed
- desert
- desertification
- polluted rivers
- radioactive contamination
- marine pollution
- heavy marine pollution
- • poor urban air quality

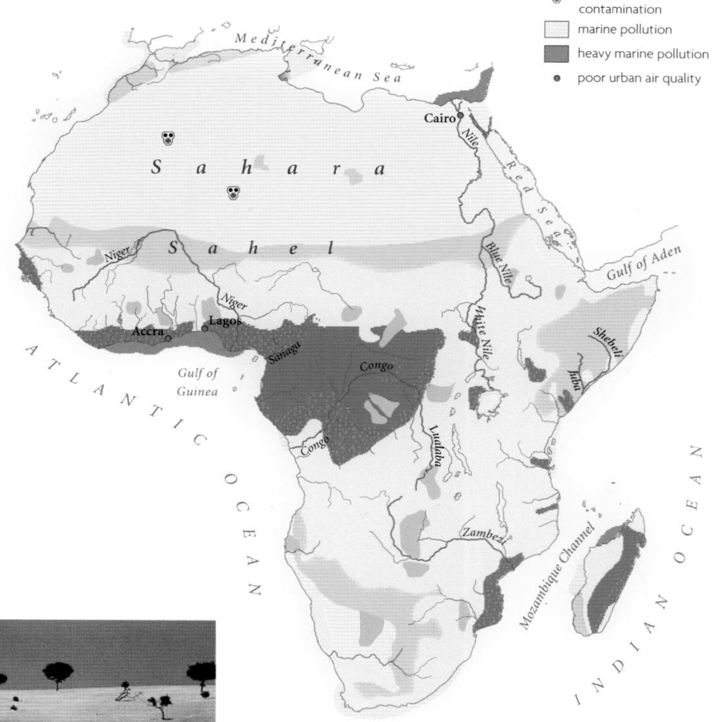

▲ *The Sahel's delicate* natural equilibrium is easily destroyed by the clearing of vegetation, drought, and overgrazing. This causes the Sahara to advance south, engulfing the savannah grasslands.

Mineral resources

Africa's ancient plateaus contain some of the world's most substantial reserves of precious stones and metals. About 15% of the world's gold is mined in South Africa; Zambia has great copper deposits; and diamonds are mined in Botswana, Dem. Rep. Congo, and South Africa. Oil has brought great economic benefits to Algeria, Libya, and Nigeria.

Mineral resources

- oil field
- gas field
- coal field
- bauxite
- copper
- diamonds
- gold
- iron
- phosphates
- tin
- uranium

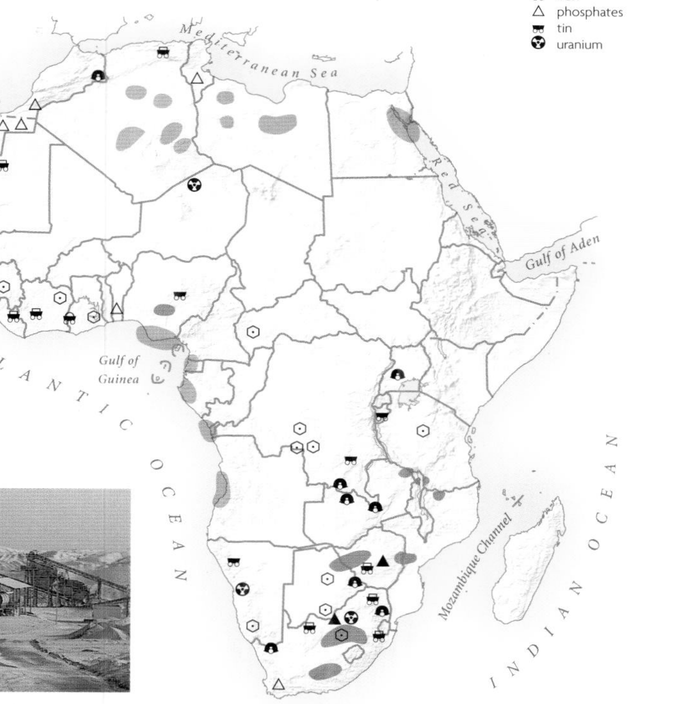

▲ *North and West* Africa have large deposits of white phosphate minerals, which are used in making fertilizers. Morocco, Senegal, and Tunisia are among the continent's leading producers.

▲ *Workers on a* tea plantation gather one of Africa's most important cash crops, providing a valuable source of income. Coffee, rubber, bananas, cotton, and cocoa are also widely grown as cash crops.

◄ *Surrounded by desert*, the fertile floodplains of the Nile Valley and Delta have been extensively irrigated, farmed, and settled since 3000 BC.

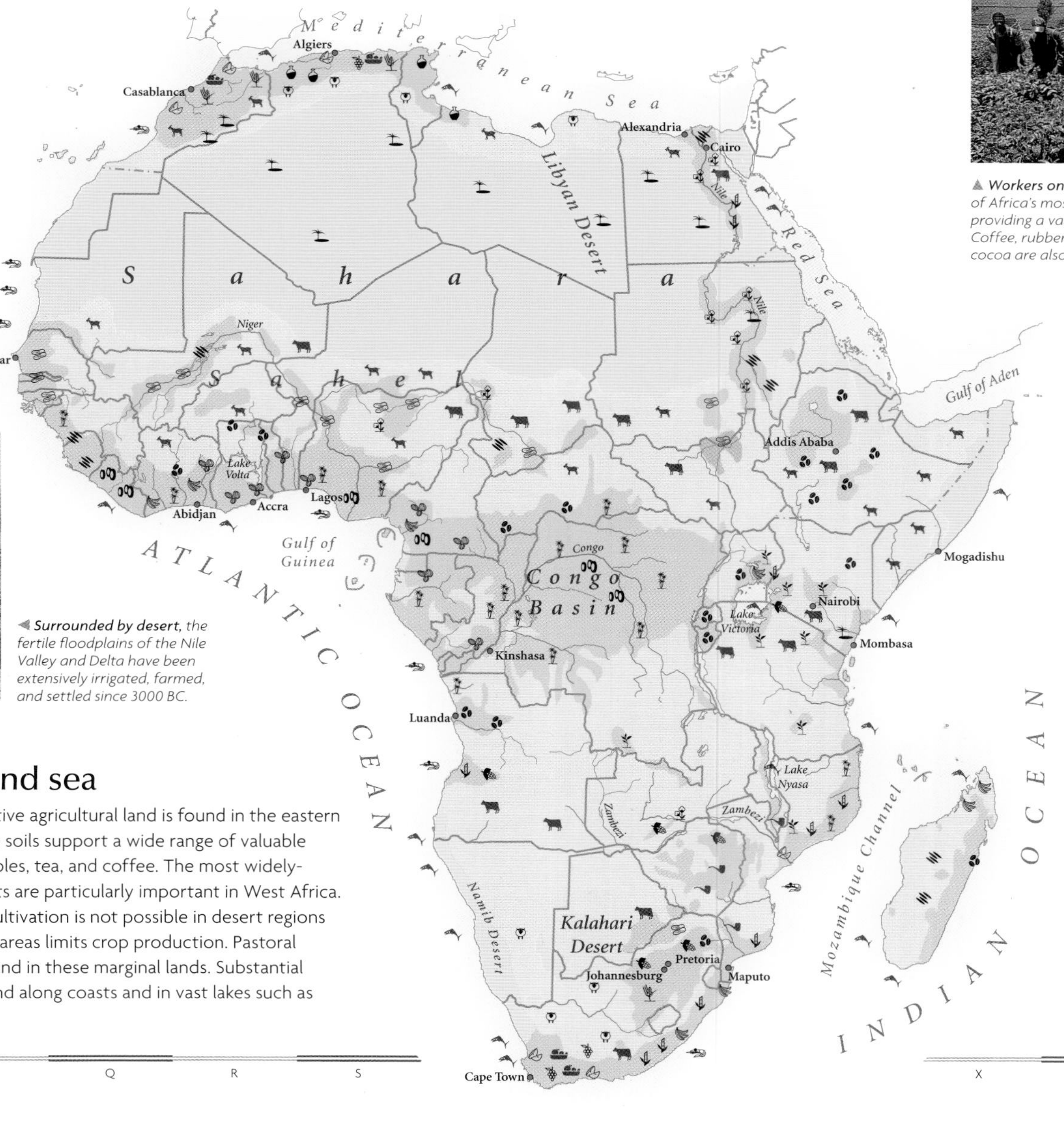

Using the land and sea

Some of Africa's most productive agricultural land is found in the eastern volcanic uplands, where fertile soils support a wide range of valuable export crops including vegetables, tea, and coffee. The most widely-grown grain is corn and peanuts are particularly important in West Africa. Without intensive irrigation, cultivation is not possible in desert regions and unreliable rainfall in other areas limits crop production. Pastoral herding is most commonly found in these marginal lands. Substantial local fishing industries are found along coasts and in vast lakes such as Lake Nyasa and Lake Victoria.

Using the land and sea

- cropland
- desert
- forest
- pasture
- wetland
- • major conurbations
- cattle
- goats
- cereals
- sheep
- bananas
- corn
- citrus fruits
- cocoa
- cotton
- coffee
- dates
- fishing
- fruit
- oil palms
- olives
- peanuts
- rice
- rubber
- shellfish
- sugar cane
- tea
- tobacco
- vineyards
- wheat

North Africa

ALGERIA, EGYPT, LIBYA, MOROCCO, TUNISIA, WESTERN SAHARA

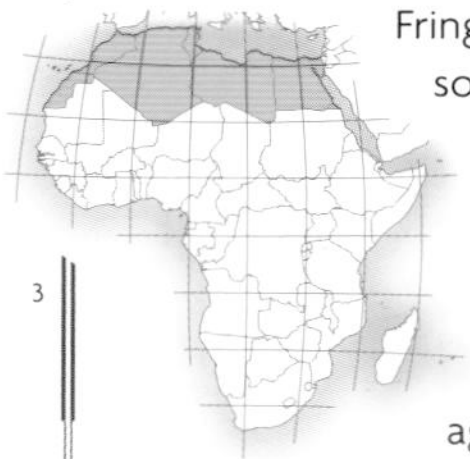

Fringed by the Mediterranean along the northern coast and by the arid Sahara in the south, North Africa reflects the influence of many invaders, both European and, most importantly, Arab, giving the region an almost universal Islamic flavor and a common Arabic language. The countries lying to the west of Egypt are often referred to as the Maghreb, an Arabic term for "west." Today, Morocco and Tunisia exploit their culture and landscape for tourism, while rich oil and gas deposits aid development in Libya and Algeria, despite political turmoil. Egypt, with its fertile, Nile-watered agricultural land and varied industrial base, is the most populous nation.

▲ These rock piles in Algeria's Ahaggar mountains are the result of weathering caused by extremes of temperature. Great cracks or joints appear in the rocks, which are then worn and smoothed by the wind.

The landscape

The Atlas Mountains, which extend across much of Morocco, northern Algeria, and Tunisia, are part of the fold mountain system which also runs through much of southern Europe. They recede to the south and east, becoming a steppe landscape before meeting the Sahara desert which covers more than 90% of the region. The sediments of the Sahara overlie an ancient plateau of crystalline rock, some of which is more than four billion years old.

Map key

Population
- above 5 million
- 1 million to 5 million
- 500,000 to 1 million
- 100,000 to 500,000
- 50,000 to 100,000
- 10,000 to 50,000
- below 10,000

Elevation
- 4000m / 13,124ft
- 3000m / 9843ft
- 2000m / 6562ft
- 1000m / 3281ft
- 500m / 1640ft
- 250m / 820ft
- 100m / 328ft
- sea level

Scale 1:12,250,000

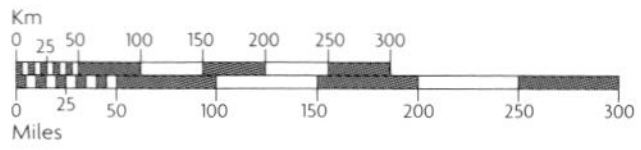

Km
0 25 50 100 150 200 250 300
Miles
0 50 100 150 200 250 300

projection: Lambert Azimuthal Equal Area

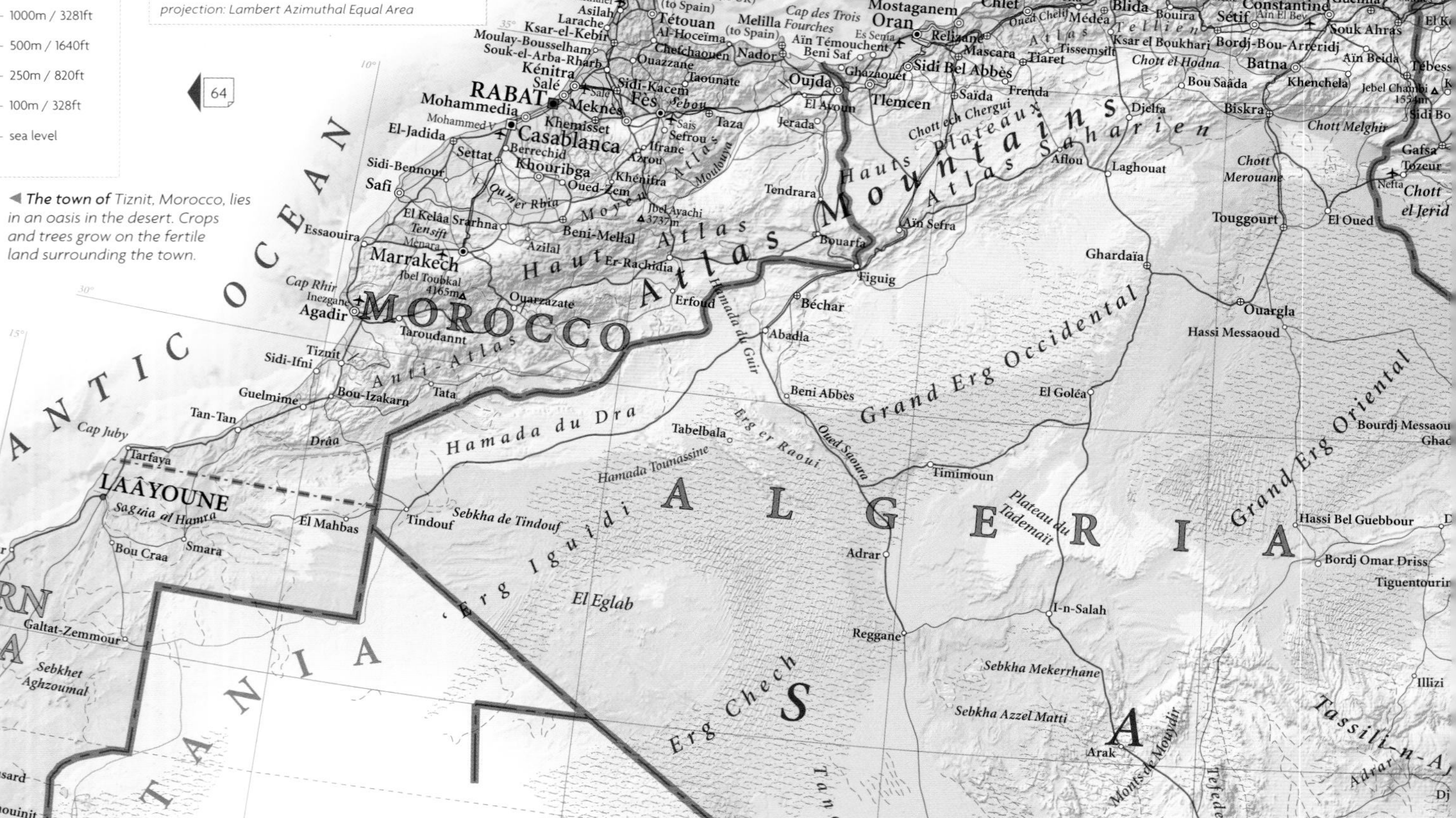

◄ The town of Tiznit, Morocco, lies in an oasis in the desert. Crops and trees grow on the fertile land surrounding the town.

▶ The Grand Erg Occidental is one of Algeria's great Saharan sand seas. Wind force and direction determines the nature of landforms such as the linear or seif dunes in the foreground.

Using the land & sea

Sheltered valleys in the Atlas Mountains, the Nile Valley and Delta, and the Mediterranean coast are the main sources of good farming land. A wide variety of valuable crops including cereals, rice, and cotton, and woods such as cedar and cork, are grown. Typical Mediterranean crops such as olives, figs, dates, and citrus fruits also thrive in these areas. The Nile Valley is particularly fertile, and most of Egypt's population lives close to the river. Elsewhere, irrigation is essential to improve crop yields on the desert margins.

The urban/rural population divide

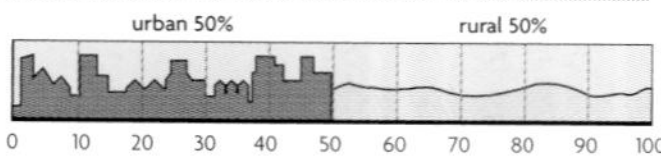

urban 50%   rural 50%

0 10 20 30 40 50 60 70 80 90 100

Population density
65 people per sq mile
(25 people per sq km)

Total land area
2,215,020 sq miles
(5,738,394 sq km)

Land use and agricultural distribution
- goats
- sheep
- cereals
- citrus fruits
- cork
- cotton
- dates
- fishing
- olives
- vineyards
- capital cities
- major towns
- pasture
- cropland
- forest
- desert

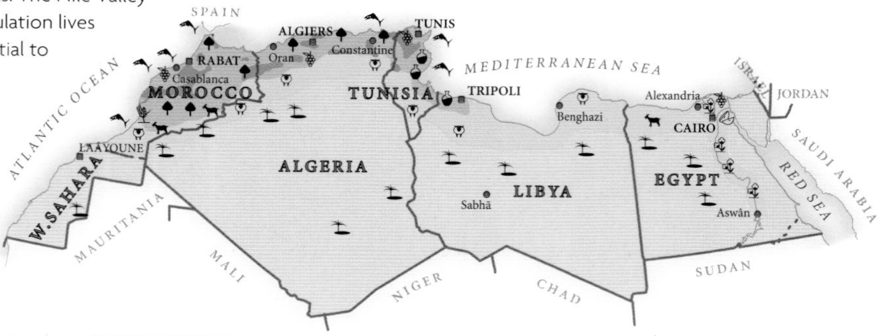

▲ Many North African nomads, such as the Bedouin, maintain a traditional pastoral lifestyle on the desert fringes, moving their herds of sheep, goats, and camels from place to place – crossing country borders in order to find sufficient grazing land.

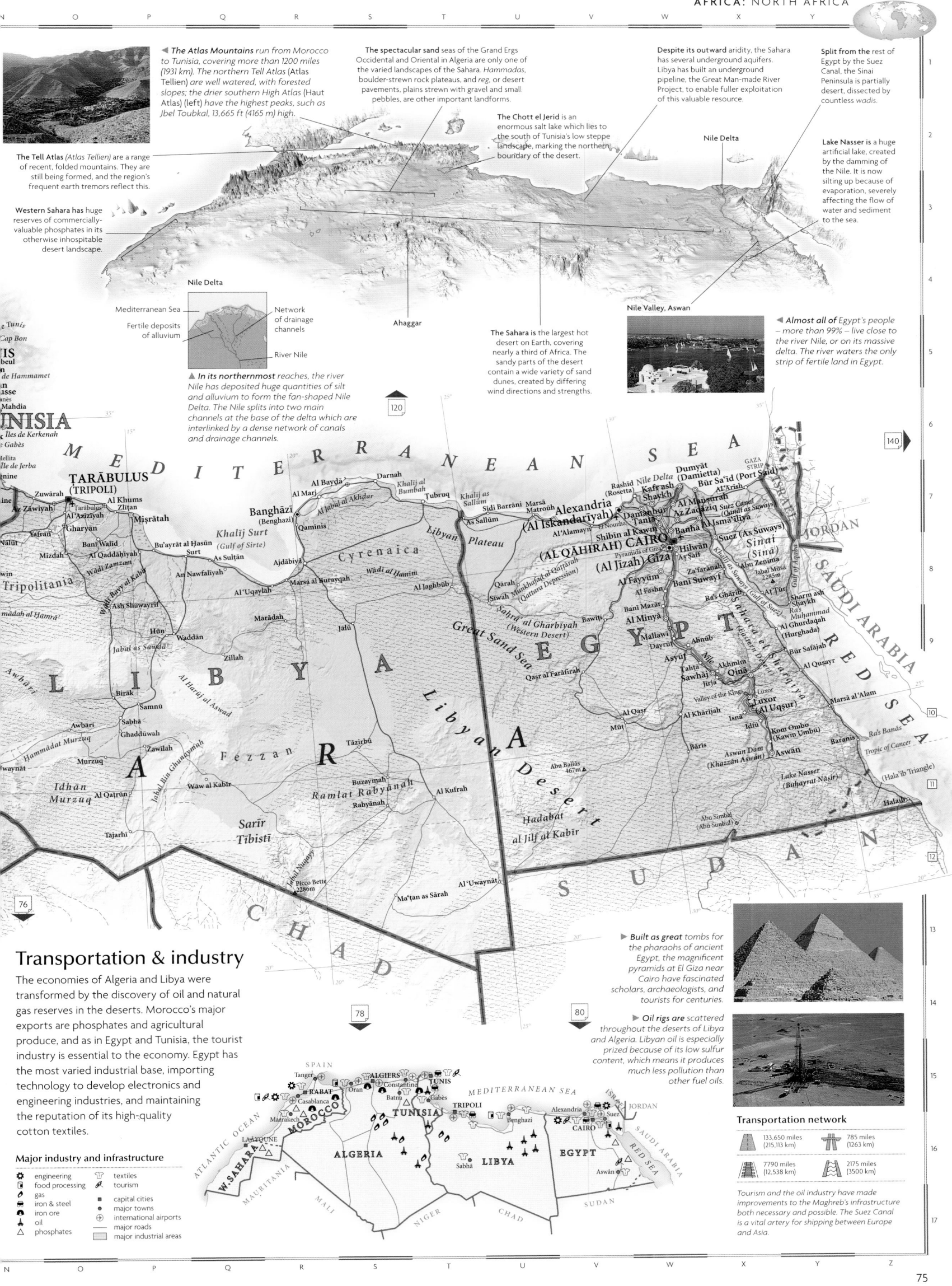

◀ *The Atlas Mountains* run from Morocco to Tunisia, covering more than 1200 miles (1931 km). The northern Tell Atlas (Atlas Tellien) are well watered, with forested slopes; the drier southern High Atlas (Haut Atlas) (left) have the highest peaks, such as Jbel Toubkal, 13,665 ft (4165 m) high.

The spectacular sand seas of the Grand Ergs Occidental and Oriental in Algeria are only one of the varied landscapes of the Sahara. *Hammadas*, boulder-strewn rock plateaus, and *reg*, or desert pavements, plains strewn with gravel and small pebbles, are other important landforms.

Despite its outward aridity, the Sahara has several underground aquifers. Libya has built an underground pipeline, the Great Man-made River Project, to enable fuller exploitation of this valuable resource.

Split from the rest of Egypt by the Suez Canal, the Sinai Peninsula is partially desert, dissected by countless *wadis*.

The Tell Atlas (Atlas Tellien) are a range of recent, folded mountains. They are still being formed, and the region's frequent earth tremors reflect this.

The Chott el Jerid is an enormous salt lake which lies to the south of Tunisia's low steppe landscape, marking the northern boundary of the desert.

Lake Nasser is a huge artificial lake, created by the damming of the Nile. It is now silting up because of evaporation, severely affecting the flow of water and sediment to the sea.

Western Sahara has huge reserves of commercially-valuable phosphates in its otherwise inhospitable desert landscape.

Nile Delta

Mediterranean Sea
Fertile deposits of alluvium
Network of drainage channels
River Nile

▲ *In its northernmost* reaches, the river Nile has deposited huge quantities of silt and alluvium to form the fan-shaped Nile Delta. The Nile splits into two main channels at the base of the delta which are interlinked by a dense network of canals and drainage channels.

Ahaggar

The Sahara is the largest hot desert on Earth, covering nearly a third of Africa. The sandy parts of the desert contain a wide variety of sand dunes, created by differing wind directions and strengths.

Nile Valley, Aswan

◀ *Almost all of* Egypt's people – more than 99% – live close to the river Nile, or on its massive delta. The river waters the only strip of fertile land in Egypt.

▶ *Built as great* tombs for the pharaohs of ancient Egypt, the magnificent pyramids at El Giza near Cairo have fascinated scholars, archaeologists, and tourists for centuries.

▶ *Oil rigs are* scattered throughout the deserts of Libya and Algeria. Libyan oil is especially prized because of its low sulfur content, which means it produces much less pollution than other fuel oils.

Transportation & industry

The economies of Algeria and Libya were transformed by the discovery of oil and natural gas reserves in the deserts. Morocco's major exports are phosphates and agricultural produce, and as in Egypt and Tunisia, the tourist industry is essential to the economy. Egypt has the most varied industrial base, importing technology to develop electronics and engineering industries, and maintaining the reputation of its high-quality cotton textiles.

Major industry and infrastructure

- engineering
- food processing
- gas
- iron & steel
- iron ore
- oil
- phosphates
- textiles
- tourism
- capital cities
- major towns
- international airports
- major roads
- major industrial areas

Transportation network

133,650 miles (215,113 km)		785 miles (1263 km)	
7790 miles (12,538 km)		2175 miles (3500 km)	

Tourism and the oil industry have made improvements to the Maghreb's infrastructure both necessary and possible. The Suez Canal is a vital artery for shipping between Europe and Asia.

West Africa

BENIN, BURKINA FASO, CAPE VERDE, GAMBIA, GHANA, GUINEA, GUINEA-BISSAU, IVORY COAST, LIBERIA, MALI, MAURITANIA, NIGER, NIGERIA, SENEGAL, SIERRA LEONE, TOGO

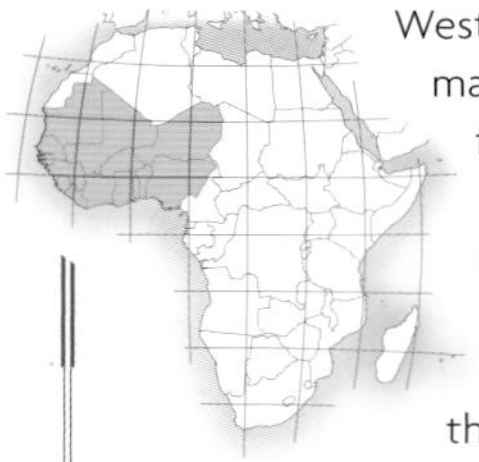

West Africa is an immensely diverse region, encompassing the desert landscapes and mainly Muslim populations of the southern Saharan countries, and the tropical rain forests of the more humid south, with a great variety of local languages and cultures. The rich natural resources and accessibility of the area were quickly exploited by Europeans; most of the Africans taken by slave traders came from this region, causing serious depopulation. The very different influences of West Africa's leading colonial powers, Britain and France, remain today, reflected in the languages and institutions of the countries they once governed.

▶ The dry scrub of the Sahel is only suitable for grazing herd animals like these cattle in Mali.

Transportation & industry

Abundant natural resources including oil and metallic minerals are found in much of West Africa, although investment is required for their further exploitation. Nigeria experienced an oil boom during the 1970s but subsequent growth has been sporadic. Most industry in other countries has a primary basis, including mining, logging, and food processing.

Transportation network

🛣	62,154 miles (100,038 km)	🛤	1037 miles (1669 km)
🚆	6752 miles (10,867 km)	✈	10,192 miles (16,405 km)

The road and rail systems are most developed near the coasts. Some of the landlocked countries remain disadvantaged by the difficulty of access to ports, and their poor road networks.

Major industry and infrastructure

- ⚗ chemicals
- cotton spinning
- 🏭 food processing
- mining
- oil
- palm oil processing
- peanut processing
- textiles
- vehicle manufacture
- ■ capital cities
- ⊕ major towns
- ✈ international airports
- major roads
- major industrial areas

Map key

Population
- ▣ Above 5 million
- ▣ 1 million to 5 million
- ◉ 500,000 to 1 million
- ◎ 100,000 to 500,000
- ⊕ 50,000 to 100,000
- ○ 10,000 to 50,000
- ∘ below 10,000

Elevation
- 2000m / 6562ft
- 1000m / 3281ft
- 500m / 1640ft
- 250m / 820ft
- 100m / 328ft
- sea level

CAPE VERDE

Santo Antão, Pombas, Mindelo, São Vicente, Ribeira Brava, São Nicolau, Boa Vista, Ilhas de Barlavento, Pedra Lume, Amílcar Cabral, Sal, João Barrosa

ATLANTIC OCEAN

Fogo, Tarrafal, Maio, São Filipe, Santiago, PRAIA, Ilhas de Sotavento

(same scale as main map)

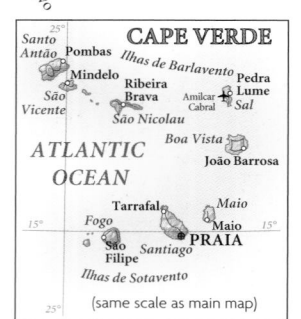

◀ *The southern regions of West Africa still contain great swathes of tropical rainforest, including some of the world's most prized hardwood trees, such as mahogany and iroko.*

Using the land & sea

The humid southern regions are most suitable for cultivation; in these areas, cash crops such as coffee, cotton, cocoa, and rubber are grown in large quantities. Peanuts are grown throughout West Africa. In the north, advancing desertification has made the Sahel increasingly uncultivable, and pastoral farming is more common. Great herds of sheep, cattle, and goats are grazed on the savannah grasses. Fishing is important in coastal and delta areas.

▲ *The Gambia, mainland Africa's smallest country, produces great quantities of peanuts. Winnowing is used to separate the nuts from their stalks.*

Land use and agricultural distribution

- goats
- sheep
- cocoa
- coffee
- cotton
- oil palms
- peanuts
- rubber
- shellfish
- ■ capital cities
- ● major towns
- pasture
- cropland
- forest
- desert

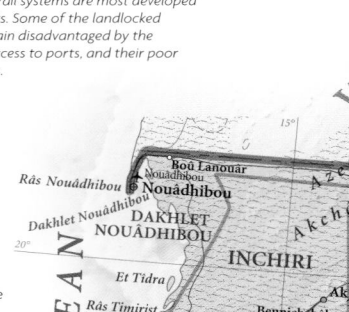

The urban/rural population divide

urban 36%   rural 64%

0 10 20 30 40 50 60 70 80 90 100

Population density	Total land area
104 people per sq mile (40 people per sq km)	2,337,137 sq miles (6,054,760 sq km)

Scale 1:10,000,000

Km 0 25 50 100 150 200 250
Miles 0 25 50 100 150 200 250

projection: Lambert Azimuthal Equal Area

[Map area with labels including:]

WESTERN SAHARA (occupied by Morocco), TIRIS ZEMMOUR, Yetti, 'Aïn Ben Tili, Bir Mogrein, 'Ayoûn 'Abd el Mâlek, Kâghet, El H, El Mreïti, Zouérat, El Hammâmi, Fdérik, Tourine, Touâjil, Char, Maqteïr, El Mrâyer, Er, Tropic of Cancer, Ouâdâne, Ouarâne, Choûm, Atâr, Chinguetti, Oujeft, ADRAR, INCHIRI, Bou Lanouâr, Nouâdhibou, Râs Nouâdhibou, Dakhlet Nouâdhibou, DAKHLET NOUÂDHIBOU, Akjoujt, El Mreyyé, Rachid, TAGANT, S, Râs Timirist, Nonâmghâr, Bennichâb, Bou Rjeimât, Tijikja, Tichît, HODH, ECH CHARG, MAURITANIA, Sebkhet Te-n-Dghâmcha, Nouakchott, TRARZA, Moudjéria, Aoukâr, Tâmchekket, Ouâl, NOUAKCHOTT, Idini, Boutilimit, Boumdeid, Bassik, Tiguent, Magta' Lahjar, BRAKNA, HODH EL, Néma, Mederdra, Rkiz, Aleg, Guérou, Kiffa, Fïntâne, 'Ayoûn el 'Atroûs, GHARBI, Adel, Rosso, Bogué, Bababé, Mônguel, ASSABA, Kobenni, Timbedgha, Amourj, Saint Louis, Dagana, Podor, Kaédi, Mbout, Kankossa, Richard Toll, Lac de Guier, Kankossa, Ould Yenjé, Selibabi, Yélimané, Nioro, Ballé, Louga, Kébémer, Matam, GORGOL, Maghama, GUIDIMAKA, KAYES, Nara, Tivaouane, Dara, Linguère, Ranérou, KOULIKORO, DAKAR, Thiès, Bambey, Mbaké, Vélingara, Bakel, Kidira, Diéma, Mourdiah, Mbour, Diourbel, Fatick, Saloum, Goudiri, Diamou, Sadiola, Bafoulabé, Kolokani, Banamba, Joal-Fadiout, Kaolack, Diré, Koungheul, Maka, Tambacounda, Dialakoto, Kita, Sebekoro, Kati, BAMAKO, SENEGAL, Nioro du Rip, Kaffrine, Georgetown, Koundara, Kédougou, Saraya, Kokofata, Niagassola, Kangaba, BANJUL, Mansa Konko, Basse Santa Su, Médina Gounas, Mali, Maléa, Doko, Ouéléssébougou, GAMBIA, Dioulouloulou, Bignona, Kolda, Vélingara, Gambia, Kangaré, Bougouni, Ziguinchor, Sédhiou, Saraya, SIKASSO, Cacheu, Farim, Bissorā, Gabú, Fouta, Dinguiraye, Tikinso, Siguiri, Yanfolila, Garalo, GUINEA-BISSAU, Mansôa, Bafatá, Fulacunda, Koundâra, Fouta Djallon, Siguiri, Mandiana, Quinhámel, BISSAU, Bolama, Gaoual, Labé, Tougué, Kouroussa, Manankoro, Arquipélago dos Bijagós, Catió, Rio Geba, Pita, Kamsar, Télimélé, Dalaba, Dabola, Kérouané, Odienné, Boké, Boffa, Fria, Kindia, Mamou, Faranah, GUINEA, Kankan, Samatiguila, Kouto, Cap Verga, Dubréka, Coyah, Kissidougou, Kabala, Mandiana, Madinani, Conakry, Forécariah, Moggo, Falaba, Kérouané, Beyla, Bako, CONAKRY, Kambia, Pendembu, Binimani, Tokounou, Pic de Tibé, Sifié, Port Loko, Makeni, Kissidougou, Koidu, Guéckédou, Macenta, Biankouma, Man, Lungi, Pepel, Lunsar, Magburaka, Moyamba, Koidu, Nzérékoré, Lola, FREETOWN, Shenge, Kenema, Mano, Zorzor, Yomou, Yekepa, Toba, SIERRA LEONE, Bo, Matru, Pujehun, Gbanga, Voinjama, Boola, Sherbro Island, Bonthe, Sulima, Robertsport, Tubmanburg, Harbel, Saint John, Toulépleu, Guiglo, IVORY, MONROVIA, Marshall, Kakata, Zwedru, Tapeta, Duékoué, LIBERIA, Buchanan, River Cess, Lac de Buyo, Taï, Greenville, Cestos, Soubré, Grand Cess, Plibo, Harper, Cape Palmas, Grabo, Sassandra, Tabou, ATLANTIC OCEAN

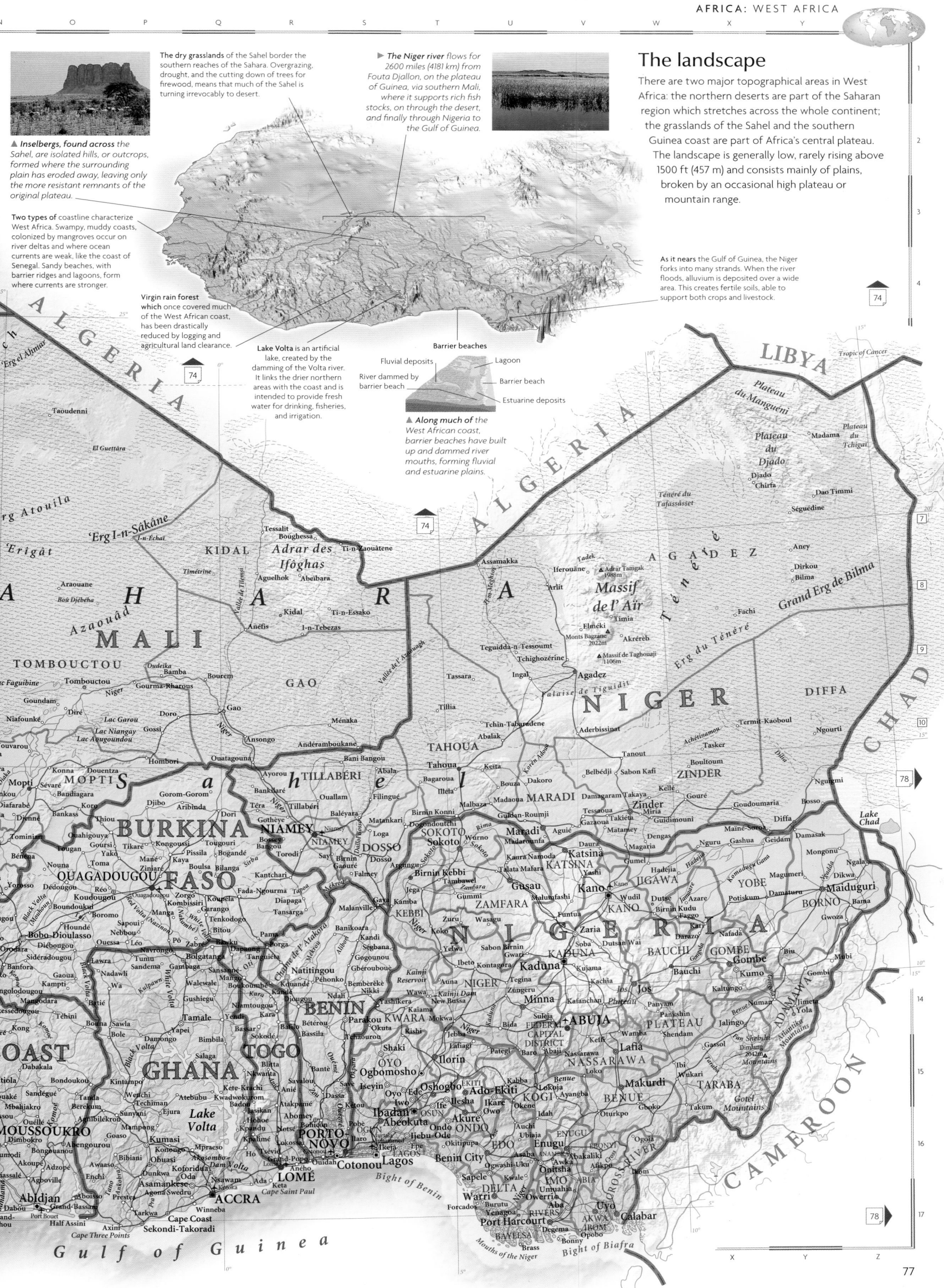

Inselbergs, found across the Sahel, are isolated hills, or outcrops, formed where the surrounding plain has eroded away, leaving only the more resistant remnants of the original plateau.

The dry grasslands of the Sahel border the southern reaches of the Sahara. Overgrazing, drought, and the cutting down of trees for firewood, means that much of the Sahel is turning irrevocably to desert.

▶ **The Niger river** flows for 2600 miles (4181 km) from Fouta Djallon, on the plateau of Guinea, via southern Mali, where it supports rich fish stocks, on through the desert, and finally through Nigeria to the Gulf of Guinea.

The landscape

There are two major topographical areas in West Africa: the northern deserts are part of the Saharan region which stretches across the whole continent; the grasslands of the Sahel and the southern Guinea coast are part of Africa's central plateau. The landscape is generally low, rarely rising above 1500 ft (457 m) and consists mainly of plains, broken by an occasional high plateau or mountain range.

Two types of coastline characterize West Africa. Swampy, muddy coasts, colonized by mangroves occur on river deltas and where ocean currents are weak, like the coast of Senegal. Sandy beaches, with barrier ridges and lagoons, form where currents are stronger.

Virgin rain forest which once covered much of the West African coast, has been drastically reduced by logging and agricultural land clearance.

Lake Volta is an artificial lake, created by the damming of the Volta river. It links the drier northern areas with the coast and is intended to provide fresh water for drinking, fisheries, and irrigation.

As it nears the Gulf of Guinea, the Niger forks into many strands. When the river floods, alluvium is deposited over a wide area. This creates fertile soils, able to support both crops and livestock.

Barrier beaches

Fluvial deposits — Lagoon
River dammed by barrier beach — Barrier beach
Estuarine deposits

▲ **Along much of** the West African coast, barrier beaches have built up and dammed river mouths, forming fluvial and estuarine plains.

Central Africa

CAMEROON, CENTRAL AFRICAN REPUBLIC, CHAD, CONGO, DEM. REP. CONGO, EQUATORIAL GUINEA, GABON, SAO TOME & PRINCIPE

The great rain forest basin of the Congo river embraces most of remote Central Africa. The interior was largely unknown to Europeans until late in the 19th century, when its tribal kingdoms were split – principally between France and Belgium – with Sao Tome and Principe the lone Portuguese territory, and Equatorial Guinea controlled by Spain. Open democracy and regional economic integration are important goals for these nations – several of which have only recently emerged from restrictive regimes – and investment is needed to improve transportation infrastructures. Many of the small, but fast-growing and increasingly urban population, speak French, the regional *lingua franca*, along with several hundred Pygmy, Bantu, and Sudanic dialects.

The landscape

Lake Chad lies in a desert basin bounded by the volcanic Tibesti mountains in the north, plateaus in the east and, in the south, the broad watershed of the Congo basin. The vast circular depression of the Congo is isolated from the coastal plain by the granite Massif du Chaillu. To the northwest, the volcanoes and fold mountains of the Cameroon Ridge (*Dorsale Camerounaise*) extend as islands into the Gulf of Guinea. The high fold mountains fringing the east of the Congo Basin fall steeply into the lakes of the Great Rift Valley.

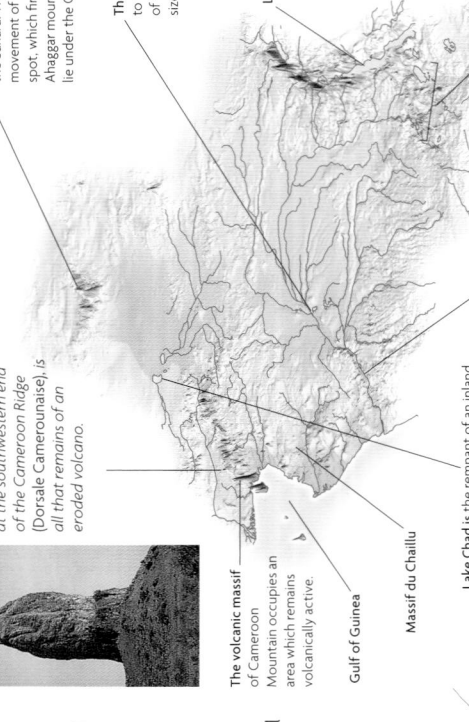

Transportation & industry

Large reserves of valuable minerals are found in Central Africa: copper, cobalt, zinc, and diamonds are mined in Dem. Rep. Congo and manganese in Gabon. Congo, Cameroon, Gabon, and Equatorial Guinea have oil deposits and oil has also been recently discovered in Chad. Goods such as palm oil and rubber are processed for export.

The Tibesti mountains are the highest in the Sahara. They were pushed up by the movement of the African Plate over a hot spot, which first formed the northern Ahaggar mountains and is now thought to lie under the Great Rift Valley.

The Congo river is second only to the Amazon in the volume of water it carries, and in the size of its drainage basin.

Lake Tanganyika, the world's second deepest lake, is the largest of a series of linear "ribbon" lakes occupying a trench within the Great Rift Valley.

Rich mineral deposits in the "Copper Belt" of Dem. Rep. Congo were formed under intense heat and pressure when the ancient African Shield was uplifted to form the region's mountains.

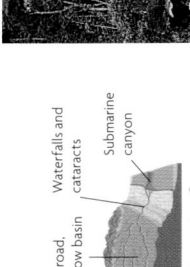

▼ **Virgin tropical rain forest** covers the Ruwenzori range on the borders of Dem. Rep. Congo and Uganda.

The lakelike expansion of the Congo river at Stanley Pool is the lowest point of the interior basin, although the river still descends more than 1000 ft (300 m) to reach the sea.

Lake Chad is the remnant of an inland sea, which once occupied much of the surrounding basin. A series of droughts since the 1970s has reduced the area of this shallow freshwater lake to about 1000 sq miles (2599 sq km).

▲ **The Congo river flows** sluggishly through the rain forest of the interior basin. Toward the coast, the river drops steeply in a series of waterfalls and cataracts. At this point, the erosional power of the river becomes so great that it has formed a deep submarine canyon offshore.

Broad, shallow basin | Waterfalls and cataracts | Submarine canyon

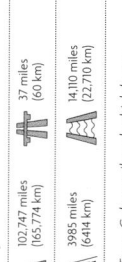

▲ **The vast sandflats** surrounding Lake Chad were once covered by water. Changing climatic patterns caused the lake to shrink, and desert now covers much of its previous area.

▼ **A plug of resistant lava,** at the southwestern end of the Cameroon Ridge (*Dorsale Camerounaise*), is all that remains of an eroded volcano.

The volcanic massif of Cameroon Mountain occupies an area which remains volcanically active.

Gulf of Guinea

Massif du Chaillu

Map key

Population

⊙	1 million to 5 million
◉	500,000 to 1 million
⊕	100,000 to 500,000
⊕	50,000 to 100,000
⊙	10,000 to 50,000
○	below 10,000

Elevation

	4000m / 13124ft
	3000m / 9843ft
	2000m / 6562ft
	1000m / 3281ft
	500m / 1640ft
	250m / 820ft
	100m / 328ft
	sea level

Scale 1:10,500,000

Km 0 25 50 100 150 200 250

miles 0 25 50 100 150 200 250

projection: Lambert Azimuthal Equal Area

Major industry and infrastructure

- ◆ brewing
- ⬡ chemicals
- ⬢ cobalt
- ⬢ copper
- ✦ diamonds
- ⬢ food processing
- ⬢ manganese
- ⬢ oil
- ⬢ palm oil processing
- ⬢ textiles
- ⬢ tin
- ⊙ capital cities
- ⊙ major towns
- ✈ international airports
- — major roads
- ⬢ major industrial areas

▲ **The ancient rocks** of Dem. Rep. Congo hold immense and varied mineral reserves. This open pit copper mine is at Kolwezi in the far south.

Transportation network

102,747 miles (165,774 km)	37 miles (60 km)
3985 miles (6414 km)	14,110 miles (22,710 km)

The Trans-Gabon railroad, which began operating in 1987, has opened up new sources of timber and manganese. Elsewhere, much investment is needed to update and improve road, rail, and water transportation.

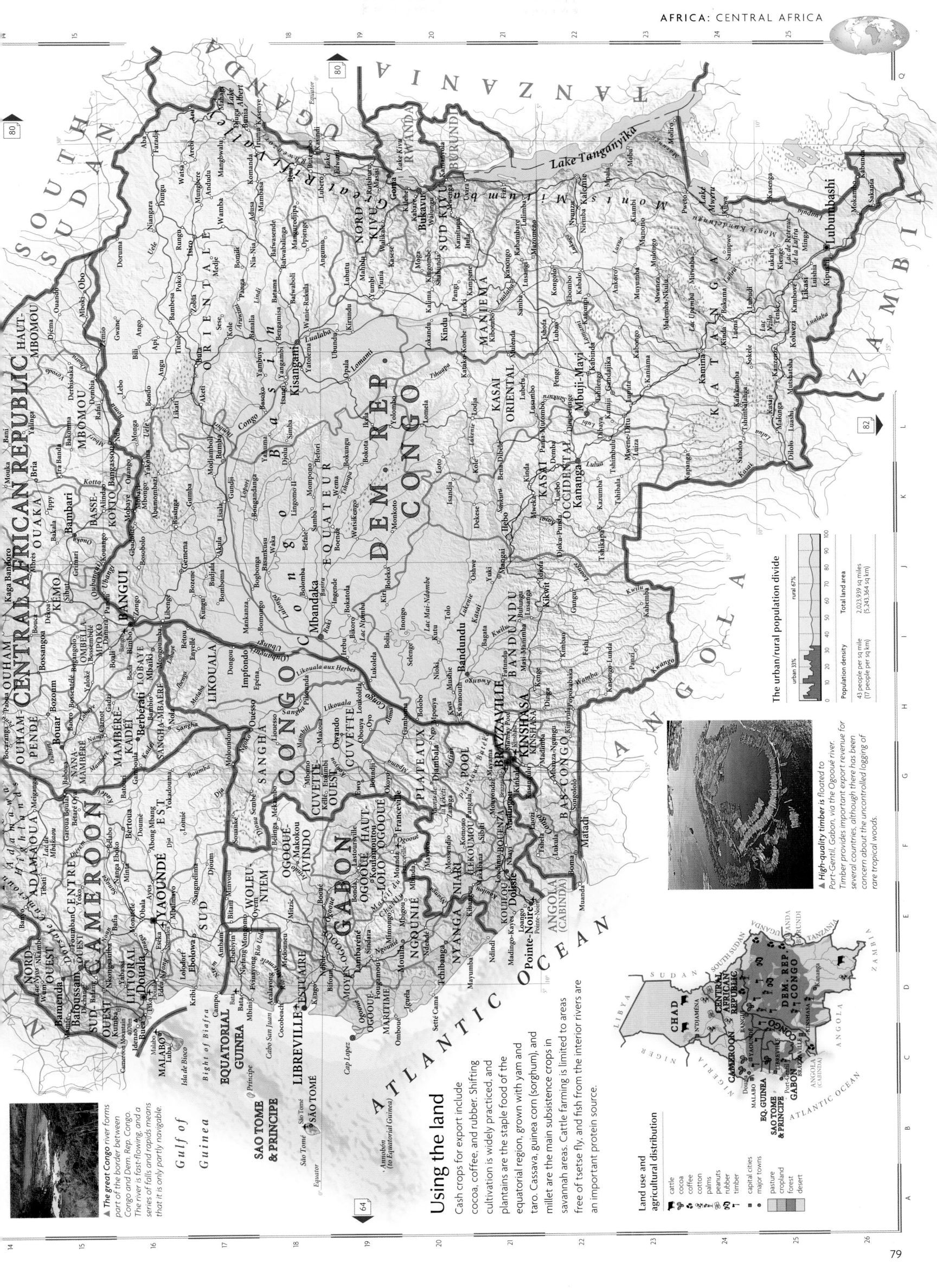

Using the land

Cash crops for export include cocoa, coffee, and rubber. Shifting cultivation is widely practiced, and plantains are the staple food of the equatorial region, grown with yam and taro. Cassava, guinea corn (sorghum), and millet are the main subsistence crops in savannah areas. Cattle farming is limited to areas free of tsetse fly, and fish from the interior rivers are an important protein source.

▲ The great Congo river forms part of the border between Congo and Dem. Rep. Congo. The river is fast-flowing, and a series of falls and rapids means that it is only partly navigable.

▲ High-quality timber is floated to Port-Gentil, Gabon, via the Ogooué river. Timber provides important export revenue for several countries, although there has been concern about the uncontrolled logging of rare tropical woods.

The urban/rural population divide

urban 33%   rural 67%

Population density
43 people per sq mile
(17 people per sq km)

Total land area
2,023,939 sq miles
(5,243,364 sq km)

Land use and agricultural distribution
cattle, cocoa, coffee, cotton, palms, peanuts, rubber, timber
capital cities, major towns
pasture, cropland, forest, desert

East Africa

BURUNDI, DJIBOUTI, ERITREA, ETHIOPIA, KENYA, RWANDA, SOMALIA, SOUTH SUDAN, SUDAN, TANZANIA, UGANDA

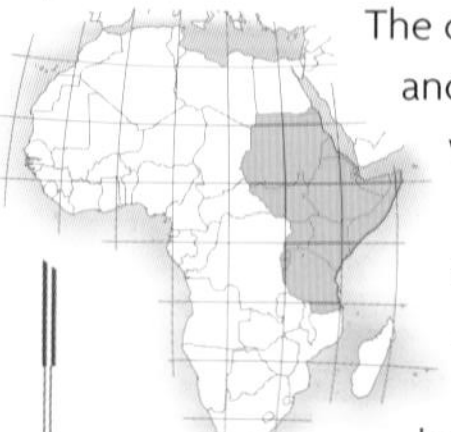

The countries of East Africa divide into two distinct cultural regions. Sudan and the "Horn" nations have been influenced by the Middle East; Ethiopia was the home of one of the earliest Christian civilizations, and Sudan reflects both Muslim and Christian influences. The southern countries share a closer cultural affinity with other sub-Saharan nations. Some of Africa's most densely populated countries lie in this region, and the needs of a growing number of people have put pressure on marginal lands and fragile environments. Although most East African economies remain strongly agricultural, Kenya has developed a varied industrial base.

The landscape

East Africa's most significant landscape feature is the Great Rift Valley, which formed during the most recent phase of continental movement when the rigid basement rocks cracked and buckled. Great blocks of land were raised and lowered, creating huge flat-bottomed valleys and steep escarpments, sometimes covered by volcanic extrusions in highland areas.

Ephemeral lake forms at far edge of slope

Central block slopes towards main fault

Boundary fault

▲ This dome at Gonder, in Ethiopia, is a volcanic intrusion, formed when molten rock pushed up the surface of the Earth and then solidified, leaving an outcrop of igneous rock.

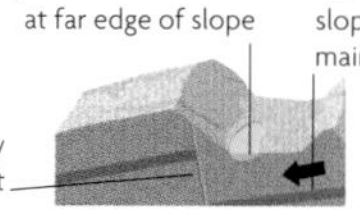

▲ The eastern arm of the Great Rift Valley is gradually being pulled apart; however the forces on one side are greater than the other causing the land to slope. This affects regional drainage which migrates down the slope.

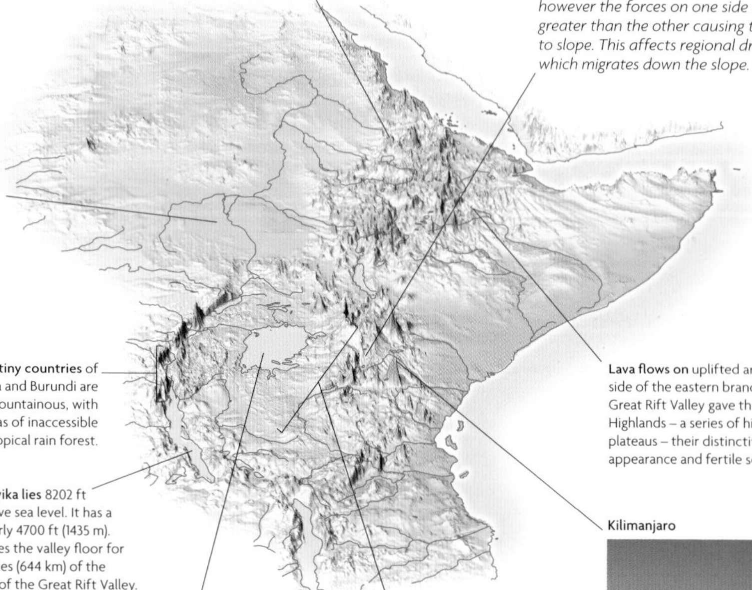

Lava flows on uplifted areas either side of the eastern branch of the Great Rift Valley gave the Ethiopian Highlands – a series of high, wide plateaus – their distinctive rounded appearance and fertile soils.

Kilimanjaro

▲ An extinct volcano, Kilimanjaro is Africa's highest mountain, rising 19,340 ft (5895 m). Once famed for its snow-capped peak, this has almost completely melted due to changing climatic conditions.

A vast plateau lies between the eastern and western rift valleys in Kenya, Uganda, and western Tanzania. It has been leveled by long periods of erosion to form a peneplain, but is dotted with inselbergs – outcrops of more resistant rocks.

Lake Victoria occupies a vast basin between the two arms of the Great Rift Valley. It is the world's second largest lake in terms of surface area, extending 26,560 sq miles (68,880 sq km). The lake contains numerous islands and coral reefs.

Lake Tanganyika lies 820? ft (2500 m) above sea level. It has a depth of nearly 4700 ft (1435 m). The lake traces the valley floor for some 400 miles (644 km) of the western arm of the Great Rift Valley.

The tiny countries of Rwanda and Burundi are mainly mountainous, with large areas of inaccessible tropical rain forest.

In contrast to the desert conditions that prevail in much of Sudan to the north, annual rainfall in the tropical wetlands of the southern Sudd region in South Sudan, can sometimes exceed 40 inches (1000 mm).

▲ The Kassala region in eastern Sudan is watered by the Atbara River, an important tributary of the Nile. Most of the population is engaged in agriculture, growing cotton and cereals.

Scale 1:10,500,000

projection: Lambert Azimuthal Equal Area

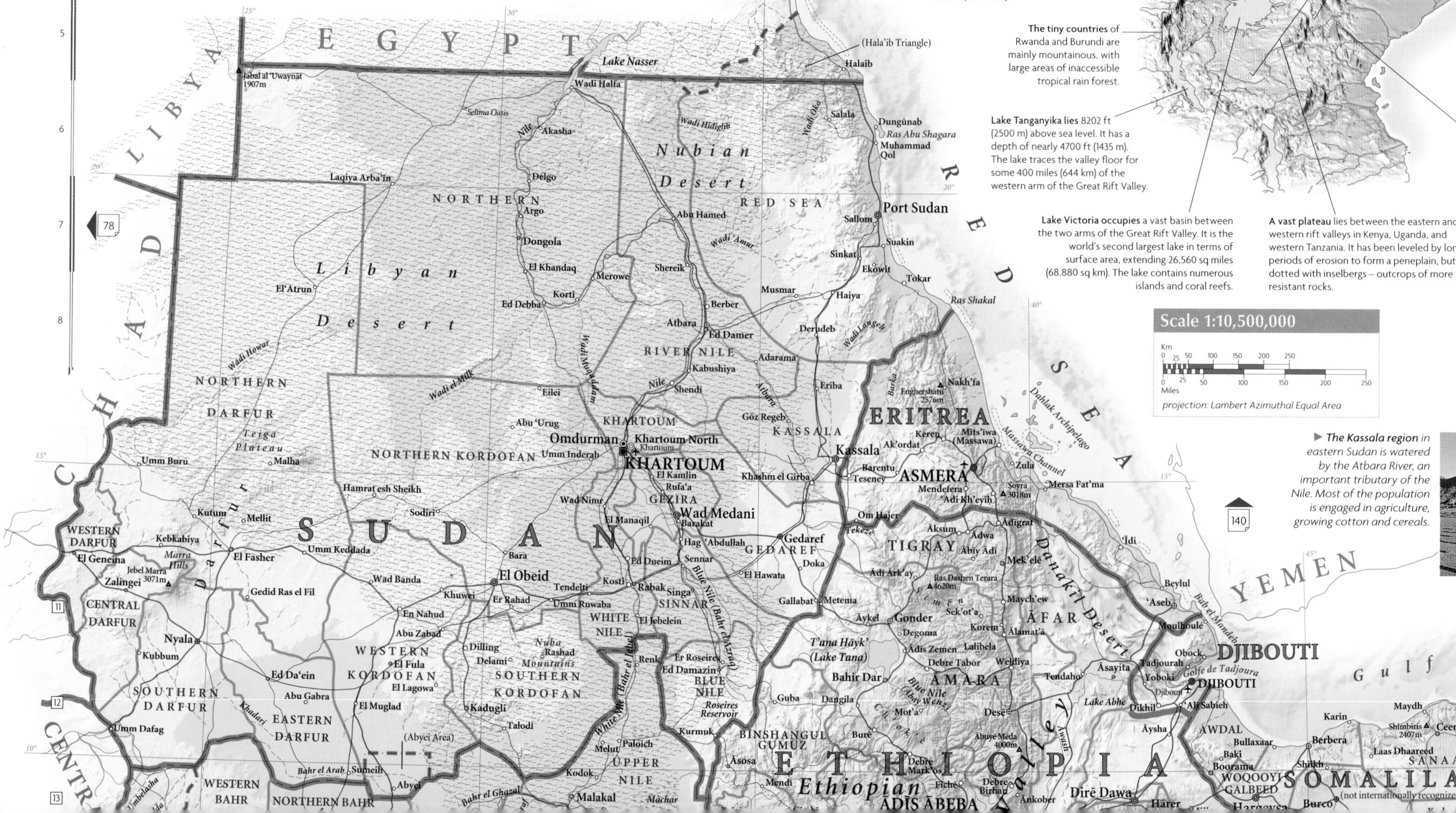

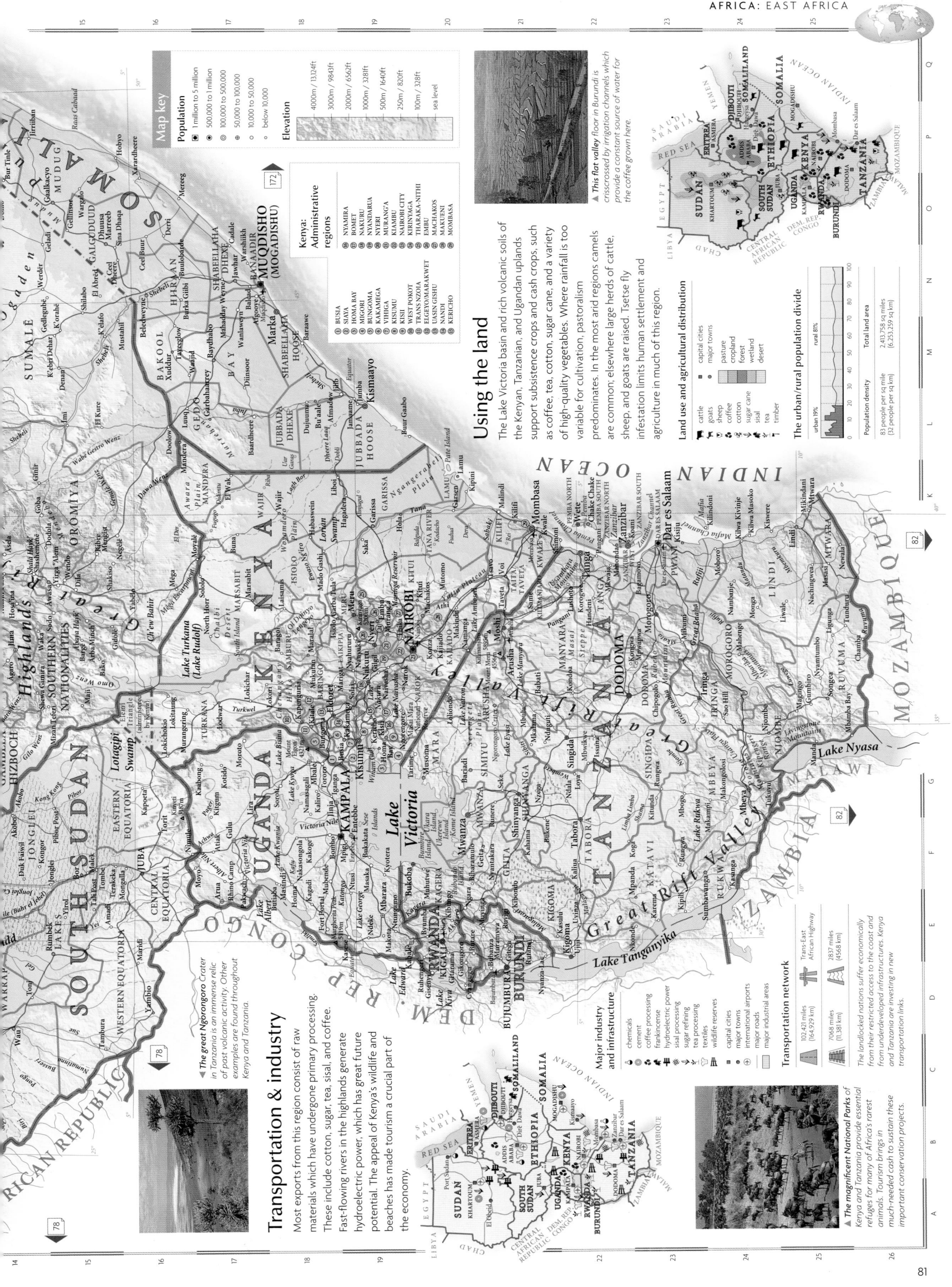

Using the land

The Lake Victoria basin and rich volcanic soils of the Kenyan, Tanzanian, and Ugandan uplands support subsistence crops and cash crops, such as coffee, tea, cotton, sugar cane, and a variety of high-quality vegetables. Where rainfall is too variable for cultivation, pastoralism predominates. In the most arid regions camels are common; elsewhere large herds of cattle, sheep, and goats are raised. Tsetse fly infestation limits human settlement and agriculture in much of this region.

▲ This flat valley floor in Burundi is crisscrossed by irrigation channels which provide a constant source of water for the coffee grown here.

Land use and agricultural distribution

- capital cities
- major towns
- pasture
- cropland
- forest
- wetland
- desert

cattle · goats · sheep · coffee · cotton · sugar cane · sisal · tea · timber

The urban/rural population divide

urban 19%   rural 81%

Population density	Total land area
83 people per sq mile (32 people per sq km)	2,413,758 sq miles (6,253,259 sq km)

Transportation & industry

Most exports from this region consist of raw materials which have undergone primary processing. These include cotton, sugar, tea, sisal, and coffee. Fast-flowing rivers in the highlands generate hydroelectric power, which has great future potential. The appeal of Kenya's wildlife and beaches has made tourism a crucial part of the economy.

▲ The great Ngorongoro Crater in Tanzania is an immense relic of past volcanic activity. Other examples are found throughout Kenya and Tanzania.

Major industry and infrastructure

- chemicals
- cement
- coffee processing
- frankincense
- hydroelectric power
- sisal processing
- sugar refining
- tea processing
- textiles
- wildlife reserves
- capital cities
- major towns
- international airports
- major roads
- major industrial areas

Transportation network

Trans-East African Highway	
102,421 miles (164,929 km)	2837 miles (4568 km)
7068 miles (11,381 km)	

The landlocked nations suffer economically from their restricted access to the coast and from underdeveloped infrastructures. Kenya and Tanzania are investing in new transportation links.

▲ The magnificent National Parks of Kenya and Tanzania provide essential refuges for many of Africa's rarest animals. Tourism brings in much-needed cash to sustain these important conservation projects.

Map key

Population
- 1 million to 5 million
- 500,000 to 1 million
- 100,000 to 500,000
- 50,000 to 100,000
- 10,000 to 50,000
- below 10,000

Elevation
- 4000m / 13124ft
- 3000m / 9843ft
- 2000m / 6562ft
- 1000m / 3281ft
- 500m / 1640ft
- 250m / 820ft
- 100m / 328ft
- sea level

Kenya: Administrative regions

1. BUSIA
2. SIAYA
3. HOMA BAY
4. MIGORI
5. BUNGOMA
6. KAKAMEGA
7. VIHIGA
8. KISUMU
9. KISII
10. WEST POKOT
11. TRANS NZOIA
12. ELGEYO/MARAKWET
13. UASIN GISHU
14. NANDI
15. KERICHO
16. NYAMIRA
17. BOMET
18. NAKURU
19. NYANDARUA
20. NYERI
21. MURANGA
22. KIAMBU
23. NAIROBI CITY
24. KIRINYAGA
25. THARAKA-NITHI
26. EMBU
27. MACHAKOS
28. MAKUENI
29. MOMBASA

Southern Africa

ANGOLA, BOTSWANA, LESOTHO, MALAWI, MOZAMBIQUE, NAMIBIA, SOUTH AFRICA, SWAZILAND, ZAMBIA, ZIMBABWE

Africa's vast southern plateau has been a contested homeland for disparate peoples for many centuries. The European incursion began with the slave trade and quickened in the 19th century, when the discovery of enormous mineral wealth secured South Africa's regional economic dominance. The struggle against white minority rule led to strife in Namibia, Zimbabwe, and the former Portuguese territories of Angola and Mozambique. South Africa's notorious apartheid laws, which denied basic human rights to more than 75% of the people, led to the state being internationally ostracized until 1994, when the first fully democratic elections inaugurated a new era of racial justice.

The landscape

Most of southern Africa rests on a concave plateau comprising the Kalahari basin and a mountainous fringe, skirted by a coastal plain which widens out in Mozambique. The plateau extends north, toward the Planalto de Bié in Angola, the Congo Basin and the lake-filled troughs of the Great Rift Valley. The eastern region is drained by the Zambezi and Limpopo rivers, and the Orange is the major western river.

Transportation & industry

South Africa, the world's largest exporter of gold, has a varied economy which generates about 75% of the region's income and draws migrant labor from neighboring states. Angola exports petroleum; Botswana and Namibia rely on diamond mining; and Zambia is seeking to diversify its economy to compensate for declining copper reserves.

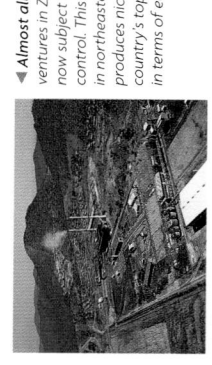

▼ *Almost all new mining ventures in Zimbabwe are now subject to government control. This mine at Bindura in northeastern Zimbabwe produces nickel, one of the country's top three minerals in terms of economic value*

Major industry and infrastructure

- 🚗 car manufacture
- coal
- copper
- ◆ diamonds
- food processing
- gold
- oil
- textiles
- uranium
- ▲ wildlife reserves

- ■ capital cities
- ■ major towns
- ✈ international airports
- major roads
- major industrial areas

At Victoria Falls, the Zambezi river has cut a spectacular gorge taking advantage of large joints in the basalt, which were first formed as the lava cooled and contracted

Lake Nyasa occupies one of the deep troughs of the Great Rift Valley, where the land has been displaced downward by as much as 3000 ft (920 m).

Great Rift Valley

Limpopo river

▲ *The fast-flowing Zambezi river cuts a deep, wide channel as it flows along the Zimbabwe/Zambia border.*

Bushveld intrusion

Volcanic lava, over 250 million years old, caps the peaks of the Drakensberg range, which lie on the mountainous rim of southern Africa's interior plateau.

The Okavango/Cubango River flows from the Planalto de Bié to the swamplands of the Okavango Delta, one of the world's largest inland deltas, where it divides into countless distributary channels, feeding out into the desert.

Broad, flat-topped mountains characterize the Great Karoo, which have been cut from level rock strata under extremely arid conditions.

Planalto de Bié

The mountains of the Little Karoo are composed of sedimentary rocks which have been substantially folded and faulted.

Khorixas, Namibia

Namib Desert

The Orange River, one of the longest in Africa, rises in Lesotho and is the only major river in the south which flows westward, rather than to the east coast.

The Kalahari desert is the largest continuous sand surface in the world. Iron oxide gives a distinctive red color to the windblown sand, which, in eastern areas covers the bedrock by over 200 ft (60 m).

Thousands of years of evaporating water have produced the Etosha Pan, one of the largest salt flats in the world. Lake and river sediments in the area indicate that the region was once less arid and

▲ *Finger Rock, near Khorixas, Namibia is a remnant of a former land surface, which has been denuded by erosion over the last 5 million years. These occasional stacks of partially weathered rocks interrupt the plains of the dry southern interior.*

Transportation network

🚗 84,213 miles (135,609 km)		✈ 746 miles (1202 km)	
🛣 23,208 miles (37,372 km)		🚂 3385 miles (6144 km)	

Southern Africa's Cape-gauge rail network is by far the largest in the continent. About two-thirds of the 20,000 mile (32,000 km) system lies within South Africa. Lines such as the Harare–Bulawayo route have become corridors for industrial growth.

▲ *Following a series of droughts, this baobab tree in Zimbabwe now stands alone in a field once filled by sugar cane. The thick trunk and small leaves of the baobab help it to conserve water, enabling it to survive even in drought conditions.*

Map key

Population
- ◉ 1 million to 5 million
- ◉ 500,000 to 1 million
- ⊚ 100,000 to 500,000
- ⊕ 50,000 to 100,000
- ⊙ 10,000 to 50,000
- ○ below 10,000

Elevation
- 3000m / 9843ft
- 2000m / 6562ft
- 1000m / 3281ft
- 500m / 1640ft
- 250m / 820ft
- 100m / 328ft
- sea level

Granite
Chromite
Gabbro and peridotite
Magnetite
Platinum minerals

Bushveld intrusion

▲ *The Bushveld intrusion lies on South Africa's high "veld." Molten magma intruded into the Earth's crust creating a saucer-shaped feature, more than 180 miles (300 km) across, containing regular layers of precious minerals, overlain by a dome of granite.*

South Africa: Capital cities
- PRETORIA – administrative capital
- CAPE TOWN – legislative capital
- BLOEMFONTEIN – judicial capital

Scale 1:10,500,000

projection: Lambert Azimuthal Equal Area

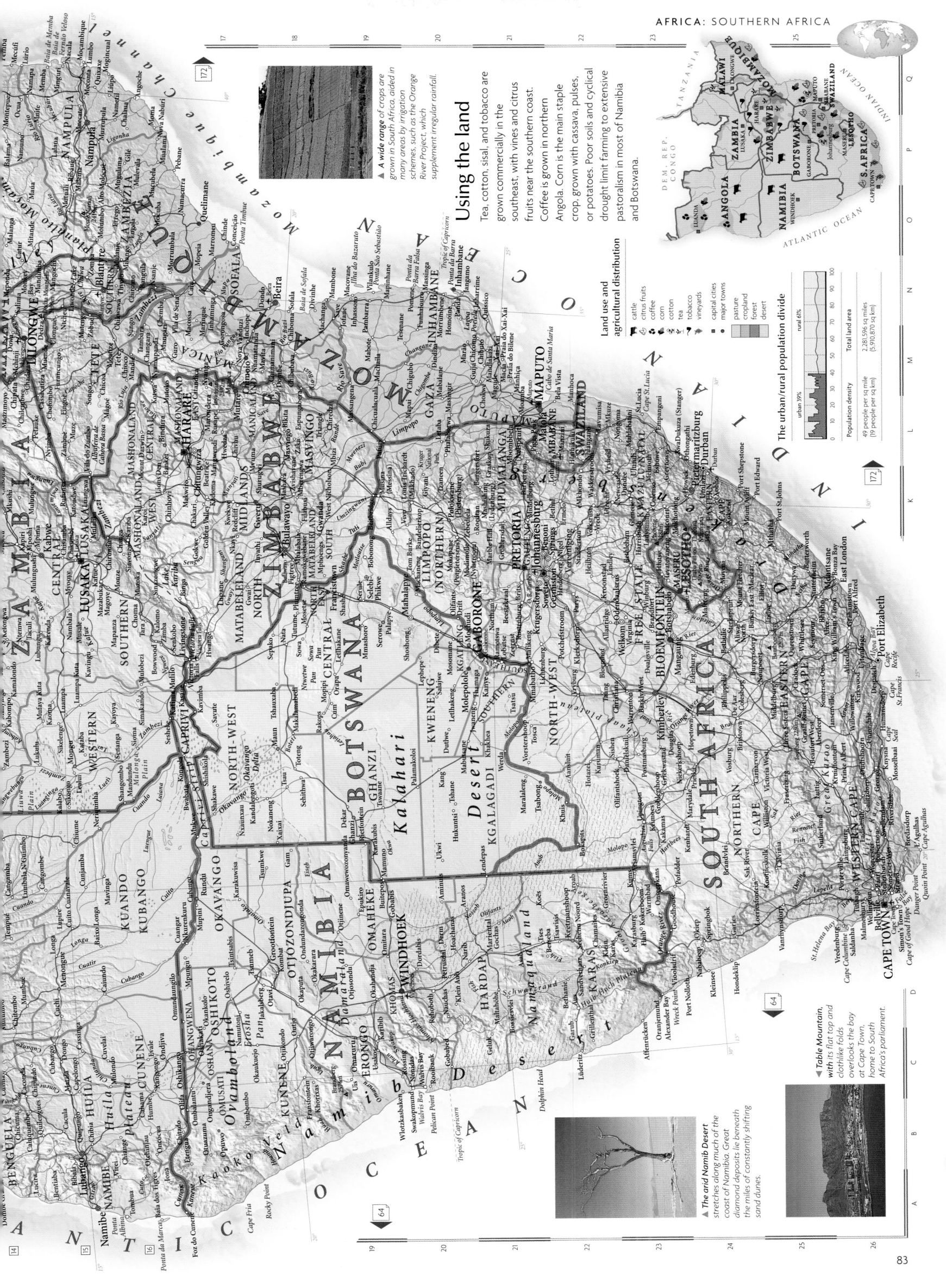

Using the land

Tea, cotton, sisal, and tobacco are grown commercially in the southeast, with vines and citrus fruits near the southern coast. Coffee is grown in northern Angola. Corn is the main staple crop, grown with cassava, pulses, or potatoes. Poor soils and cyclical drought limit farming to extensive pastoralism in most of Namibia and Botswana.

▲ *A wide range* of crops are grown in South Africa, aided in many areas by irrigation schemes, such as the Orange River Project, which supplement irregular rainfall.

Land use and agricultural distribution

- cattle
- citrus fruits
- coffee
- corn
- cotton
- tea
- tobacco
- vineyards
- capital cities
- major towns

- pasture
- cropland
- forest
- desert

The urban/rural population divide

urban 39%   rural 61%

Population density
49 people per sq mile
(19 people per sq km)

Total land area
2,281,596 sq miles
(5,910,870 sq km)

▲ *The arid Namib Desert* stretches along much of the coast of Namibia. Great diamond deposits lie beneath the miles of constantly shifting sand dunes.

▲ *Table Mountain,* with its flat top and clothlike folds overlooks the bay at Cape Town, home to South Africa's parliament.

83

ARCTIC OCEAN

North Pole

Ellesmere Island

Greenland

King Frederik VIII Land

King Christian X Land

Greenland Sea

Spitsbergen

Laptev Sea

Poluostrov Taymyr

Ostrov Rudolfa

Franz Josef Land

Severnaya Zemlya

Mys Flissingskiy

Kara Sea

Novaya Zemlya

Barents Sea

Bjørnøya

Barents Trough

Baydaratskaya Guba

Gulf of Ob

Yenisey

U R A L S

Jan Mayen Fracture Zone

Jan Mayen

Kolbeinsey Ridge

Iceland Plateau

Norwegian Sea

Tromsøflaket North Cape Nordkinn

Fugløya Bank

Murmansk Rise

Ostrov Kolguyev

Poluostrov Kanin

Pechora

West Siberian Plain

Ob

Arctic Circle

Denmark Strait

Bjargtangar

Reykjanes Ridge

Iceland

Vatnajökull

Faroe-Iceland Ridge

Jan Mayen Ridge

Vøring Plateau

Vesterålen

Lofoten

Kebnekaise 2117m

Inarijärvi

Kola Peninsula

White Sea

Ozero Imandra

Murmanskiy Rise

Timanskiy Kryazh

Mezen

Northern Dvina

Iceland Basin

Faroe Islands

Bill Baileys Bank

Norwegian Basin

Traena Bank

Scandinavia

Tornealven

Ljungan

Umealven

Oulujoki

Ozero Vygozero

Onega Bay

Lake Ladoga

Vyga

Hatton Ridge

Faroe-Shetland Trough

Shetland Islands

Viking Bank

Galdhøpiggen 2469m

Ljungan

Ljusdan

Gulf of Bothnia

Åland

Åland

Lake Onega

Lake Ladoga

Svir

Ozero Beloye

Sukhona

Rybinsk Reservoir

Rockall Rise

Feni Ridge

Rockall Trough

Outer Hebrides

Orkney Islands

Ben Nevis 1343m

Grampian Mountains

North Channel

North Sea

Jutland Bank

Skagerrak

Kattegat

Vänern

Vättern

Gotland

Gulf of Finland

Lake Peipus

Lake Pskov

Gulf of Riga

Baltic Sea

Lake Ilmen

Msta

Western Dvina

Gor'ky Reservoir

Moskva

Kuybyshev Reservoir

ATLANTIC OCEAN

Porcupine Plain

British Isles

Ireland

Irish Sea

Shannon

Celtic Sea

Celtic Shelf

St. George's Channel

Bristol Channel

Pennines

Snowdon 1085m

Severn

Trent

Britain

Thames

Land's End

English Channel

Channel Islands

Dogger Bank

Great Fisher Bank

Jutland

Sjaelland

Frisian Islands

Elbe

Oder

Warta

Vistula

Bug

North European Plain

Neman

Pripet Marshes

Dnieper Lowlands

Desna

Bug

Western Dvina

Byerezina

Dnieper

Seym

Sozh

Don

Central Russian Upland

Khopyor

Volga Upland

Sura

EUROPE

Strait of Dover

Ardennes

Rhine

Seine

Marne

Moselle

Meuse

Danube

Harz

Lake Constance

Kiev Reservoir

Kremenchuk Reservoir

Dniester

Podil's'ka Vysochina

Piydennyy Buh

Dnieper

Carpathian Mountains

Tsimlyansk Reservoir

Don

Black Sea Lowland

Kirghiz

Volga

Manych

Yergeni

Loire

Vienne

Cher

Massif Central

Vosges

Jura

Saône

Lake Geneva

Black Forest

Alps

Lake Garda

Danube

Tisza

Tisza

Balaton

Lake Balaton

Great Hungarian Plain

Prut

Transylvanian Alps

Siret

Danube

Balkan Mountains

Sea of Azov

Crimea

Kerch Strait

Kuban

Dordogne

Lot

Garonne

Bay of Biscay

Biscay Plain

Galicia Bank

Theta Gap

Charcot Seamounts

Azores-Biscay Rise

Iberian Plain

Cordillera Cantabrica

Aragón

Ebro

Miño

Douro

Gulf of Lion

Ligurian Sea

Corsica

Po

Apennines

Corno Grande 2912m

Adriatic Sea

Dinaric Alps

Adriatic Basin

Rhodope Mountains

Maritsa

Black Sea

Bosporus

Sea of Marmara

EURASIAN PLATE
ANATOLIAN PLATE

Iberian Peninsula

Tagus Plain

Sierra Morena

Guadalquivir

Douro

Guadiana

Júcar

Segura

Sistema Central

Sistema Ibérica

Gulf of Valencia

Balearic Islands

Sardinia

Strait of Bonifacio

Tyrrhenian Sea

Tyrrhenian Basin

Gulf of Taranto

Strait of Otranto

Lake Scutari

Lake Ohrid

Lake Prespa

Ionian Sea

Pindus Mountains

Aegean Sea

Peloponnese

Lake Tuz

Anatolia

Cabo da Roca

Cape Saint Vincent

Punta de Tarifa

Sistemas Béticos

Sierra Nevada

Alborán Sea

Strait of Gibraltar

Rif

Sebou

Oued Chelif

Mediterranean Sea

EURASIAN PLATE
AFRICAN PLATE

Mount Etna 3340m

Sicily

Malta

Strait of Sicily

Ionian Basin

Mirtoan Sea

Sea of Crete

Mediterranean Ridge

Taurus Mountains

Gulf of Antalya

Cyprus

Cyprus Basin

AFRICAN PLATE

Madeira

Horseshoe Seamounts

Ampère Seamount

Seine Plain

Seine Seamount

Tell Atlas

Middle Atlas

Oum er Rbia

Atlas Mountains

High Atlas

Saharan Atlas

Gávdos

Levantine Basin

Dacia Seamount

Canary Islands

Agadir Canyon

Chott el Jerid

Gulf of Sirte

Nile Fan

Suez Canal

'Erg Iguidi

Grand Erg Occidental

Grand Erg Oriental

Qattara Depression -133m

Western Desert

Libyan Desert

Erg Chech

S A H A R A

A F R I C A

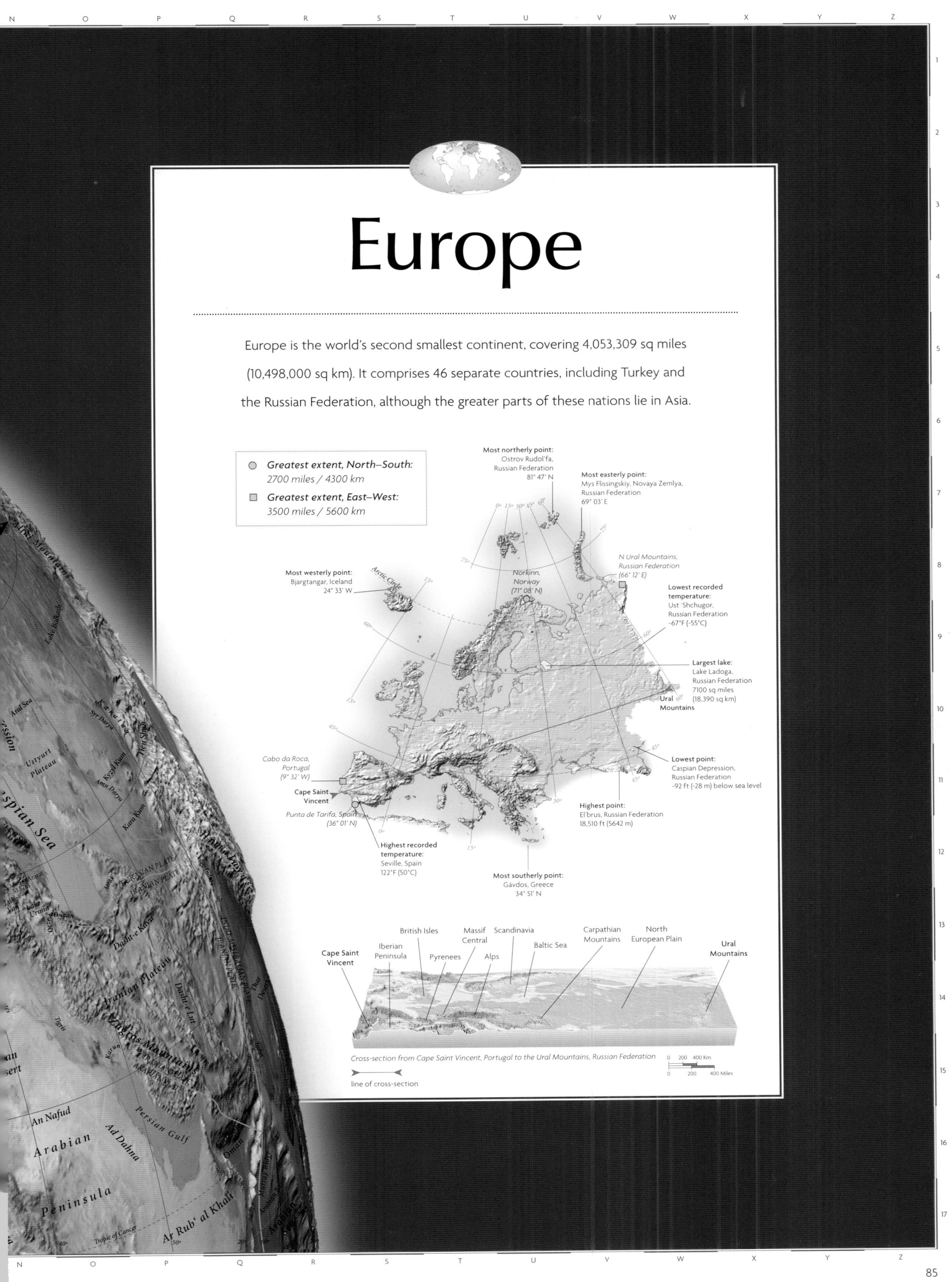

Europe

Europe is the world's second smallest continent, covering 4,053,309 sq miles
(10,498,000 sq km). It comprises 46 separate countries, including Turkey and
the Russian Federation, although the greater parts of these nations lie in Asia.

● **Greatest extent, North–South:**
2700 miles / 4300 km

■ **Greatest extent, East–West:**
3500 miles / 5600 km

Most northerly point:
Ostrov Rudol'fa,
Russian Federation
81° 47' N

Most easterly point:
Mys Flissingskiy, Novaya Zemlya,
Russian Federation
69° 03' E

*N Ural Mountains,
Russian Federation
(66° 12' E)*

Most westerly point:
Bjargtangar, Iceland
24° 33' W

*Norkinn,
Norway
(71° 08' N)*

Lowest recorded
temperature:
Ust 'Shchugor,
Russian Federation
-67°F (-55°C)

Largest lake:
Lake Ladoga,
Russian Federation
7100 sq miles
(18,390 sq km)

Ural
Mountains

*Cabo da Roca,
Portugal
(9° 32' W)*

Cape Saint
Vincent

*Punta de Tarifa, Spain
(36° 01' N)*

Lowest point:
Caspian Depression,
Russian Federation
-92 ft (-28 m) below sea level

Highest point:
El'brus, Russian Federation
18,510 ft (5642 m)

Highest recorded
temperature:
Seville, Spain
122°F (50°C)

Most southerly point:
Gávdos, Greece
34° 51' N

Cape Saint
Vincent

Iberian
Peninsula

British Isles

Pyrenees

Massif
Central

Scandinavia

Alps

Baltic Sea

Carpathian
Mountains

North
European Plain

Ural
Mountains

Cross-section from Cape Saint Vincent, Portugal to the Ural Mountains, Russian Federation

0   200   400 Km

0   200   400 Miles

line of cross-section

Physical Europe

The physical diversity of Europe belies its relatively small size. To the northwest and south it is enclosed by mountains. The older, rounded Atlantic Highlands of Scandinavia and the British Isles lie to the north and the younger, rugged peaks of the Alpine Uplands to the south. In between lies the North European Plain, stretching 2485 miles (4000 km) from The Fens in England to the Ural Mountains in Russia. South of the plain lies a series of gently folded sedimentary rocks separated by ancient plateaus, known as massifs.

The North European Plain

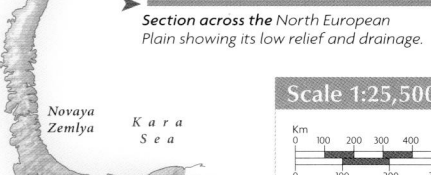

Rising less than 1000 ft (300 m) above sea level, the North European Plain strongly reflects past glaciation. Ridges of both coarse moraine and finer, windblown deposits have accumulated over much of the region. The ice sheet also diverted a number of river channels from their original courses.

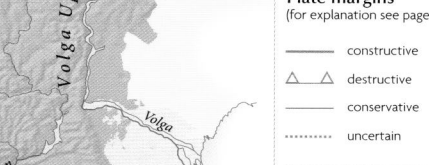

Glacial lakes | Rivers were diverted from their original course by the ice sheet | A layer of glacial sediments covers the North European Plain

Section across the North European Plain showing its low relief and drainage.

0 100 200 Km
0 100 200 Miles

The Atlantic Highlands

The Atlantic Highlands were formed by compression against the Scandinavian Shield during the Caledonian mountain-building period over 500 million years ago. The highlands were once part of a continuous mountain chain, now divided by the North Sea and a submerged rift valley.

The Atlantic Highlands continue in the British Isles | Rift valley buried by sediments | North Sea | Atlantic Highlands in Norway | Rocks affected by ancient mountain-building | Scandinavian Shield

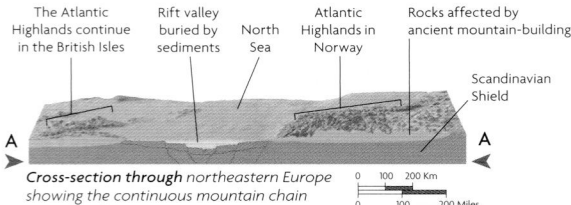
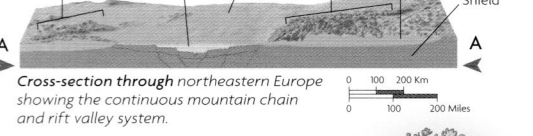
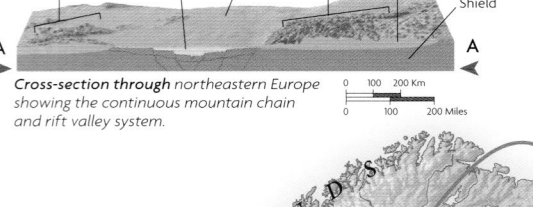
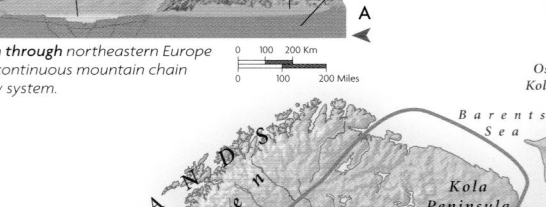

Cross-section through northeastern Europe showing the continuous mountain chain and rift valley system.

0 100 200 Km
0 100 200 Miles

Scale 1:25,500,000

Km
0 100 200 300 400 500 600

Miles
0 100 200 300 400 500 600

projection: Lambert Azimuthal Equal Area

Map key

Elevation

4000m / 13,124ft
3000m / 9843ft
2000m / 6562ft
1000m / 3281ft
500m / 1640ft
250m / 820ft
100m / 328ft
sea level

Plate margins
(for explanation see page xiv)

— constructive
△ △ destructive
— conservative
···· uncertain

— physiographic regions
◄—► line of cross-section

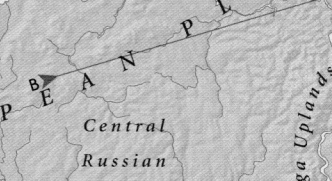

The plateaus and lowlands

The uplifted plateaus or massifs of southern central Europe are the result of long-term erosion, later followed by uplift. They are the source areas of many of the rivers which drain Europe's lowlands. In some of the higher reaches, fractures have enabled igneous rocks from deep in the Earth to reach the surface.

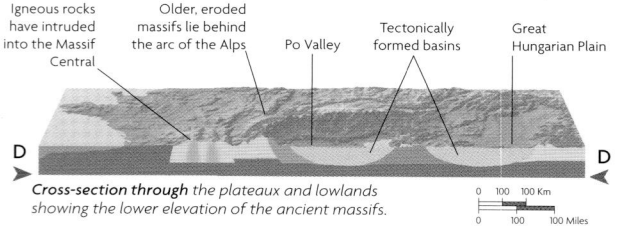

Igneous rocks have intruded into the Massif Central | Older, eroded massifs lie behind the arc of the Alps | Po Valley | Tectonically formed basins | Great Hungarian Plain

Cross-section through the plateaux and lowlands showing the lower elevation of the ancient massifs.

0 100 100 Km
0 100 100 Miles

The Alpine Uplands

The collision of the African and European continents, which began about 65 million years ago, folded and then uplifted a series of mountain ranges running across southern Europe and into Asia. Two major lines of folding can be traced: one includes the Pyrenees, the Alps, and the Carpathian Mountains; the other incorporates the Apennines and the Dinaric Alps.

European basement rock | Alps | Weak sedimentary strata have been folded | African Plate moved northwards | The Apennines

Cross-section through the Alps showing folding and faulting caused by plate tectonics.

0 50 100 Km
0 50 100 Miles

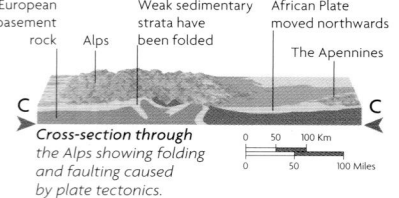

Climate

Europe experiences few extremes in either rainfall or temperature, with the exception of the far north and south. Along the west coast, the warm currents of the North Atlantic Drift moderate temperatures. Although east–west air movement is relatively unimpeded by relief, the Alpine Uplands halt the progress of north–south air masses, protecting most of the Mediterranean from cold, north winds.

▲ *Frost grips northern and eastern Europe during the long cold winters. Lakes and rivers frequently freeze.*

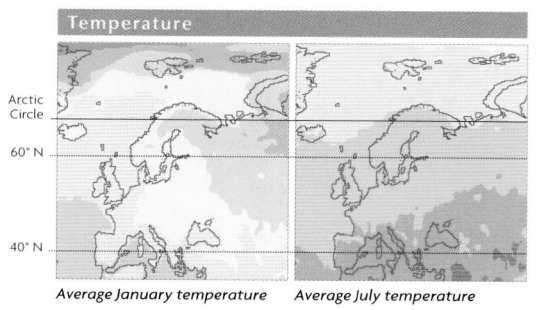

Temperature

Temperature
- below -22°F (-30°C)
- -22 to -4°F (-30 to -20°C)
- -4 to 14°F (-20 to -10°C)
- 14 to 32°F (-10 to 0°C)
- 32 to 50°F (0 to 10°C)
- 50 to 68°F (10 to 20°C)
- 68 to 86°F (20 to 30°C)
- above 86°F (30°C)

Average January temperature *Average July temperature*

▲ *Mild temperatures and frequent rainfall contribute to the fertile farming land found over much of northwestern Europe.*

Rainfall

Rainfall
- 0–1 in (0–25 mm)
- 1–2 in (25–50 mm)
- 2–4 in (50–100 mm)
- 4–8 in (100–200 mm)
- 8–12 in (200–300 mm)
- 12–16 in (300–400 mm)
- 16–20 in (400–500 mm)
- more than 20 in (500 mm)

Average January rainfall *Average July rainfall*

Climate
- tundra
- subarctic
- cool continental
- warm humid
- mediterranean
- semi-arid
- daily hours of sunshine, January
- daily hours of sunshine, July
- cold wind
- hot wind

▶ *Dusty Sirocco winds from Africa help create the semiarid scrubland common across the Mediterranean coastlands of southern Europe.*

Shaping the continent

Successive Ice Ages have left many relict landforms across Europe. Present glaciers continue to carve peaks and valleys in the northern Atlantic Highlands and Alpine Uplands. Tectonic activity, both past and present, has shaped southern Europe and Iceland. Active volcanoes and earthquakes still occur in Italy and Greece. Europe's extensive coastline, particularly in the northwest, is constantly modified by wave action and fluvial deposits.

Glaciation

1 Valley glaciers, such as this one *(left)* in Iceland, form in hollows at the top of valleys and flow downward, drawn by gravity. Their growth is dynamic; new snowfall constantly accumulates at the head of the glacier, while the snout melts, depositing material eroded and carried by the glacier.

Snow accumulates at the head of glacier
Glacier movement erodes valley
Glacier snout melts depositing eroded debris

Glaciation: Development of a glacier

Landscape
- uplifting land
- stable land
- sinking land
- limestone region
- glacier
- active volcano
- ocean current
- area of tectonic activity
- maximum limit of glaciation

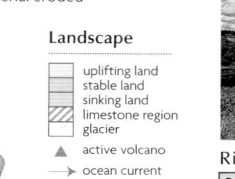

River systems

2 Rivers are continuously transporting eroded material toward the sea. Slow-moving, low-gradient rivers, like this one in western Russia *(above)*, deposit their alluvium load, infilling valleys creating a floodplain. Subsequent climatic and tectonic fluctuations may erode the floodplain to form terraces.

Terrace created by erosion
Flood plain
Deposited alluvium
River channel

River systems: Formation of a flood plain and terraces

Coastal processes

5 Spits are narrow bands of sand or shingle, formed by longshore drift; a process whereby waves carry material along the beach. They usually form where the coastline changes direction, and their growth is then halted by an opposing river current, as at Spurn Head, in the British Isles *(left)*. Coastal features such as these are constantly being created and destroyed.

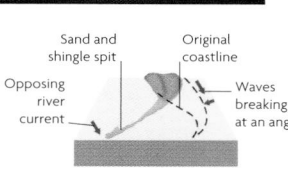

Sand and shingle spit
Original coastline
Opposing river current
Waves breaking at an angle

Coastal processes: Formation of a spit

The evolving landscape

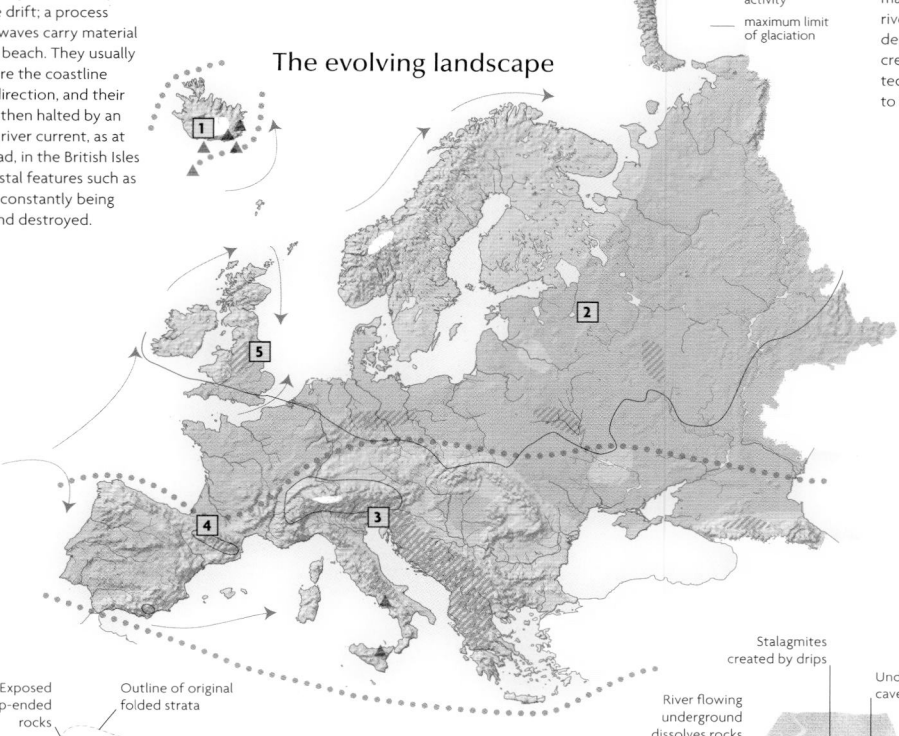

Erosion and weathering

4 Much of Europe was once subjected to folding and faulting, exposing hard and soft rock layers. Subsequent erosion and weathering has worn away the softer strata, leaving up-ended layers of hard rock as in the French Pyrenees *(above)*.

Exposed up-ended rocks
Outline of original folded strata
Soft rock
Hard rock
Fault line
Folded rock strata

Erosion and weathering: Modification of a fold

Weathering

3 As surface water filters through permeable limestone, the rock dissolves to form underground caves, like Postojna in the Karst region of Slovenia *(above)*. Stalactites grow downward as lime-enriched water seeps from roof fractures; stalagmites grow upward where drips splash down.

Stalagmites created by drips
Underground cavern
River flowing underground dissolves rocks and creates caves
Stalactites formed by seeping water

Weathering: Formation of a cave

Political Europe

The political boundaries of Europe have changed many times, especially during the 20th century in the aftermath of two world wars, the breakup of the empires of Austria-Hungary, Nazi Germany and, toward the end of the century, the collapse of communism in eastern Europe. The fragmentation of Yugoslavia has again altered the political map of Europe, highlighting a trend toward nationalism and devolution. In contrast, economic federalism is growing. In 1958, the formation of the European Economic Community (now the European Union or EU) started a move toward economic and political union and increasing internal migration.

▲ The Brandenburg Gate in Berlin is a potent symbol of German reunification. From 1961, the road beneath it ended in a wall, built to stop the flow of refugees to the West. It was opened again in 1989 when the wall was destroyed and East and West Germany were reunited.

Population

Europe is a densely populated, urbanized continent; in Belgium over 90% of people live in urban areas. The highest population densities are found in an area stretching east from southern Britain and northern France, into Germany. The northern fringes are only sparsely populated.

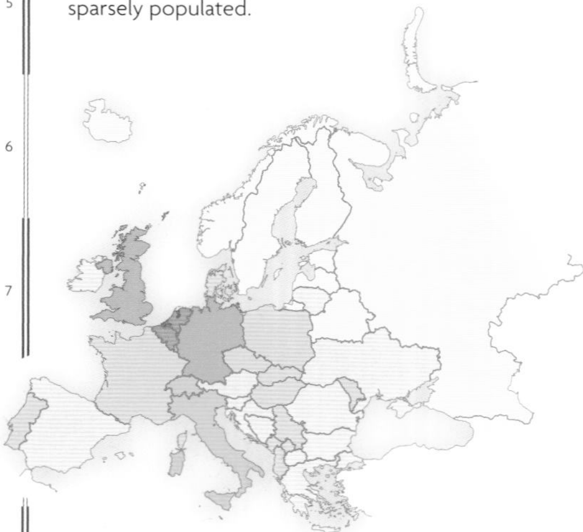

▲ Demand for space in densely populated European cities like London has led to the development of high-rise offices and urban sprawl.

Population density
(people per sq mile)

	below 130
	130–259
	260–379
	380–519
	520–780
	above 780

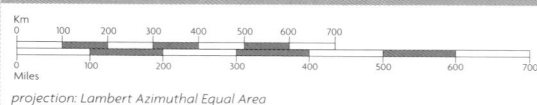

▲ Traditional lifestyles still persist in many remote and rural parts of Europe, especially in the south, east, and in the far north.

Map key

Population

- ■ above 5 million
- ■ 1 million to 5 million
- ◉ 500,000 to 1 million
- ◎ 100,000 to 500,000
- ⊕ 50,000 to 100,000
- ○ 10,000 to 50,000
- ● Country capital

Borders

full international border

Scale 1:17,250,000

Km
0 100 200 300 400 500 600 700

Miles
0 100 200 300 400 500 600 700

projection: Lambert Azimuthal Equal Area

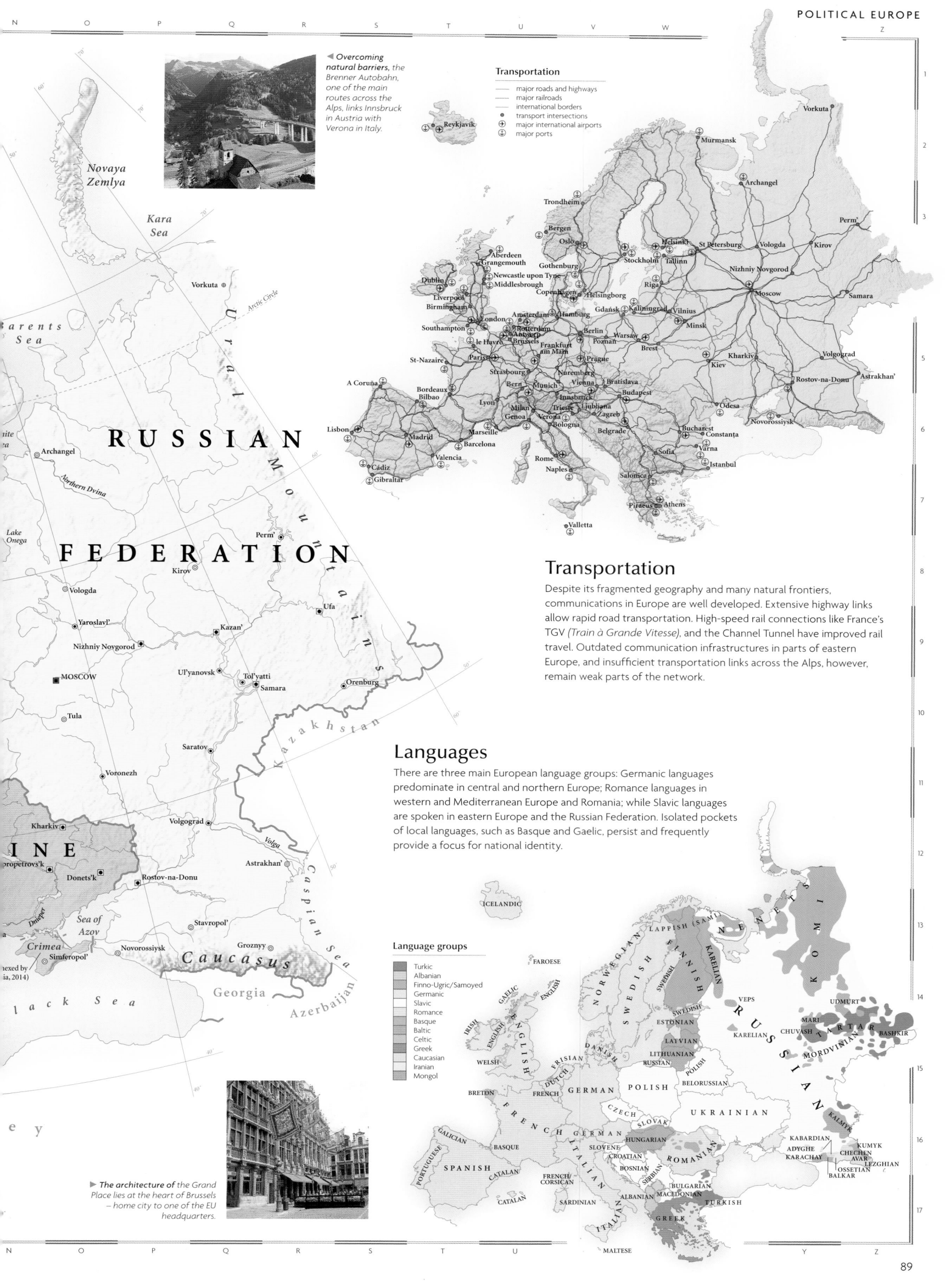

Overcoming natural barriers, the Brenner Autobahn, one of the main routes across the Alps, links Innsbruck in Austria with Verona in Italy.

Transportation

- major roads and highways
- major railroads
- international borders
- transport intersections
- major international airports
- major ports

Novaya Zemlya

Kara Sea

Reykjavik

Vorkuta

Murmansk

Archangel

Trondheim

Perm'

Barents Sea

Bergen   Oslo   Helsinki   St Petersburg   Vologda   Kirov

Aberdeen   Stockholm   Tallinn   Nizhniy Novgorod   Samara

Grangemouth   Gothenburg

Dublin   Newcastle upon Tyne   Copenhagen   Helsingborg   Riga   Moscow

Liverpool   Middlesbrough   Gdańsk   Kaliningrad   Vilnius

Birmingham   Amsterdam   Hamburg   Berlin   Warsaw   Minsk   Astrakhan'

Southampton   London   Rotterdam   Poznań   Brest   Kharkiv   Volgograd

le Havre   Antwerp   Brussels   Frankfurt am Main   Prague   Kiev   Rostov-na-Donu

St-Nazaire   Paris   Strasbourg   Nuremberg   Vienna   Bratislava   Odesa

A Coruña   Bern   Munich   Budapest   Novorossiysk

Bordeaux   Lyon   Milan   Innsbruck   Ljubljana   Zagreb   Bucharest   Constanţa

Bilbao   Genoa   Verona   Trieste   Bologna   Belgrade   Varna

Lisbon   Marseille   Rome   Sofia   Istanbul

Madrid   Barcelona   Naples   Salonica

Cádiz   Valencia   Piraeus   Athens

Gibraltar   Valletta

Transportation

Despite its fragmented geography and many natural frontiers, communications in Europe are well developed. Extensive highway links allow rapid road transportation. High-speed rail connections like France's TGV *(Train à Grande Vitesse)*, and the Channel Tunnel have improved rail travel. Outdated communication infrastructures in parts of eastern Europe, and insufficient transportation links across the Alps, however, remain weak parts of the network.

Languages

There are three main European language groups: Germanic languages predominate in central and northern Europe; Romance languages in western and Mediterranean Europe and Romania; while Slavic languages are spoken in eastern Europe and the Russian Federation. Isolated pockets of local languages, such as Basque and Gaelic, persist and frequently provide a focus for national identity.

RUSSIAN FEDERATION

Northern Dvina

Lake Onega

Archangel

Vologda

Yaroslavl'   Kirov

Nizhniy Novgorod   Perm'

Kazan'   Ufa

U r a l   M o u n t a i n s

MOSCOW   Ul'yanovsk   Tol'yatti   Orenburg

Tula   Samara

Kazakhstan

Saratov

Voronezh

Kharkiv   Volgograd

...propetrovs'k   Donets'k   *Volga*   Astrakhan'

...INE   Rostov-na-Donu

Dnieper   Stavropol'   *Caspian Sea*

Sea of Azov

Crimea   Novorossiysk   Groznyy

...nexed by   Simferopol'   *Caucasus*

...ia, 2014)   Georgia   Azerbaijan

...lack Sea

...ey

Language groups

- Turkic
- Albanian
- Finno-Ugric/Samoyed
- Germanic
- Slavic
- Romance
- Basque
- Baltic
- Celtic
- Greek
- Caucasian
- Iranian
- Mongol

ICELANDIC

FAROESE

NORWEGIAN   SWEDISH   SWEDISH   FINNISH   KARELIAN   NENETS   KOMI

LAPPISH (SAMI)   SWEDISH   VEPS   UDMURT

GAELIC   ENGLISH   ESTONIAN   KARELIAN   MARI   BASHKIR

IRISH   ENGLISH   LATVIAN   CHUVASH   TARTAR

ENGLISH   DANISH   LITHUANIAN   RUSSIAN   MORDVINIAN

WELSH   FRISIAN   RUSSIAN   POLISH

BRETON   DUTCH   GERMAN   POLISH   BELORUSSIAN   KALMYK

FRENCH   GERMAN   CZECH   UKRAINIAN

GALICIAN   SLOVAK   KABARDIAN   KUMYK

BASQUE   ITALIAN   SLOVENE   HUNGARIAN   ROMANIAN   ADYGHE   CHECHEN

FRENCH   CROATIAN   KARACHAY   AVAR   LEZGHIAN

PORTUGUESE   SPANISH   CATALAN   FRENCH CORSICAN   BOSNIAN   SERBIAN   OSSETIAN   BALKAR

CATALAN   SARDINIAN   ALBANIAN   MACEDONIAN   BULGARIAN   TURKISH

ITALIAN   GREEK

MALTESE

The architecture of the Grand Place lies at the heart of Brussels — home city to one of the EU headquarters.

European resources

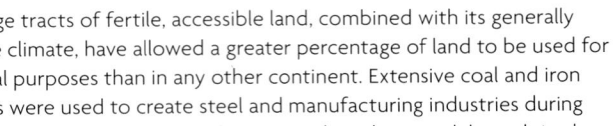

Europe's large tracts of fertile, accessible land, combined with its generally temperate climate, have allowed a greater percentage of land to be used for agricultural purposes than in any other continent. Extensive coal and iron ore deposits were used to create steel and manufacturing industries during the 19th and 20th centuries. Today, although natural resources have been widely exploited, and heavy industry is of declining importance, the growth of hi-tech and service industries has enabled Europe to maintain its wealth.

Standard of living

Living standards in western Europe are among the highest in the world, although there is a growing sector of homeless, jobless people. Eastern Europeans have lower overall standards of living – a legacy of stagnated economies.

Standard of living
(UN human development index)
- low
- high
- data not available

Industry

Europe's wealth was generated by the rise of industry and colonial exploitation during the 19th century. The mining of abundant natural resources made Europe the industrial center of the world. Adaptation has been essential in the changing world economy, and a move to service-based industries has been widespread except in eastern Europe, where heavy industry still dominates.

▲ *Countries like Hungary* are still struggling to modernize inefficient factories left over from extensive, centrally-planned industrialization during the communist era.

◀ *Frankfurt am Main* is an example of a modern service-based city. The skyline is dominated by headquarters from the worlds of banking and commerce.

▲ *Other power sources* are becoming more attractive as fossil fuels run out; 16% of Europe's electricity is now provided by hydroelectric power.

▶ *Skiing brings millions* of tourists to the slopes each year, which means that even unproductive, marginal land is used to create wealth in the French, Swiss, Italian, and Austrian Alps.

GNI per capita (US $)
- below 1999
- 2000–4999
- 5000–9999
- 10,000–19,999
- 20,000–24,999
- above 25,000

Industry
✈ aerospace	🗇 food processing	🍷 wine
🍺 brewing	🗇 hi-tech industry	coal
car/vehicle manufacture	iron & steel	oil
chemicals	pharmaceuticals	gas
defense	printing & publishing	
electronics	shipbuilding	• industrial cities
engineering	textiles	major industrial areas
finance	timber processing	

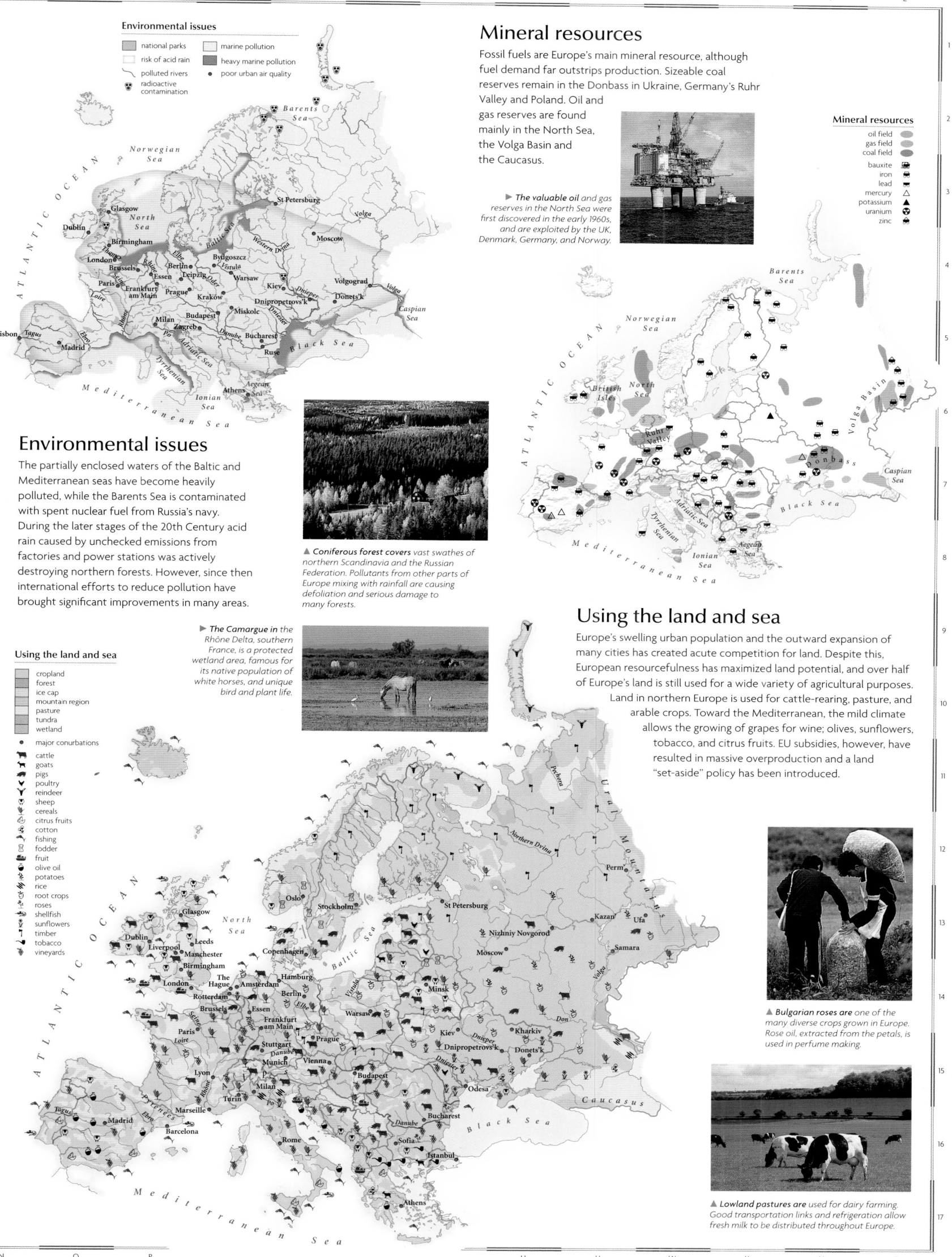

Environmental issues

national parks
risk of acid rain
polluted rivers
radioactive contamination
marine pollution
heavy marine pollution
poor urban air quality

Mineral resources

Fossil fuels are Europe's main mineral resource, although fuel demand far outstrips production. Sizeable coal reserves remain in the Donbass in Ukraine, Germany's Ruhr Valley and Poland. Oil and gas reserves are found mainly in the North Sea, the Volga Basin and the Caucasus.

▶ *The valuable oil* and gas reserves in the North Sea were first discovered in the early 1960s, and are exploited by the UK, Denmark, Germany, and Norway.

Mineral resources

oil field
gas field
coal field
bauxite
iron
lead
mercury
potassium
uranium
zinc

Environmental issues

The partially enclosed waters of the Baltic and Mediterranean seas have become heavily polluted, while the Barents Sea is contaminated with spent nuclear fuel from Russia's navy. During the later stages of the 20th Century acid rain caused by unchecked emissions from factories and power stations was actively destroying northern forests. However, since then international efforts to reduce pollution have brought significant improvements in many areas.

▲ *Coniferous forest covers* vast swathes of northern Scandinavia and the Russian Federation. Pollutants from other parts of Europe mixing with rainfall are causing defoliation and serious damage to many forests.

▶ *The Camargue in* the Rhône Delta, southern France, is a protected wetland area, famous for its native population of white horses, and unique bird and plant life.

Using the land and sea

Europe's swelling urban population and the outward expansion of many cities has created acute competition for land. Despite this, European resourcefulness has maximized land potential, and over half of Europe's land is still used for a wide variety of agricultural purposes. Land in northern Europe is used for cattle-rearing, pasture, and arable crops. Toward the Mediterranean, the mild climate allows the growing of grapes for wine; olives, sunflowers, tobacco, and citrus fruits. EU subsidies, however, have resulted in massive overproduction and a land "set-aside" policy has been introduced.

Using the land and sea

cropland
forest
ice cap
mountain region
pasture
tundra
wetland

major conurbations
cattle
goats
pigs
poultry
reindeer
sheep
cereals
citrus fruits
cotton
fishing
fodder
fruit
olive oil
potatoes
rice
root crops
roses
shellfish
sunflowers
timber
tobacco
vineyards

▲ *Bulgarian roses are* one of the many diverse crops grown in Europe. Rose oil, extracted from the petals, is used in perfume making.

▲ *Lowland pastures are* used for dairy farming. Good transportation links and refrigeration allow fresh milk to be distributed throughout Europe.

Scandinavia, Finland & Iceland

DENMARK, NORWAY, SWEDEN, FINLAND, ICELAND

Jutting into the Arctic Circle, this northern swath of Europe has some of the continent's harshest environments, but benefits from great reserves of oil, gas, and natural evergreen forests. While most early settlers came from the south, migrants to Finland came from the east, giving it a distinct language and culture. Since the late 19th century, the Scandinavian states have developed strong egalitarian traditions. Today, their welfare benefits systems are among the most extensive in the world, and standards of living are high. The Lapps, or Sami, maintain their traditional lifestyle in the northern regions of Norway, Sweden, and Finland.

The landscape

Glaciers up to 10,000 ft (3000 m) deep covered most of Scandinavia and Finland during the last Ice Age. The effects of glaciation mark the entire landscape, from the mountains to the lowlands, across the tundra landscape of Lapland, and the lake districts of Sweden and Finland.

Geysers are a by-product of Iceland's volcanic activity Geysir, Iceland's largest spring, gives them their name.

The Lofoten Islands were one of the first areas exposed as the ice sheet melted.

Halti Mountain is Finland's highest point, at 4356 ft (1328 m).

Fjords

▲ The fjords on the western coast of Norway were once gentle river valleys. Their deep floors and steep sides were carved out by glaciers during the last Ice Age, and they were later flooded by the sea.

▲ On the coast of Sjælland, these cliffs have been eroded by the sea, exposing layers of chalk and limestone.

Sjælland coast

Area of maximum yearly uplift 0.3 in/yr (9 mm/yr)

Slower rates of uplift 0.1 in/yr (3 mm/yr)

▲ Scandinavia is still recovering from the last Ice Age, when ice depressed the land by 2000 ft (600 m). This gradual uplift is known as isostatic rebound.

Using the land & sea

The cold climate, short growing season, poorly developed soil, steep slopes, and exposure to high winds across northern regions means that most agriculture is concentrated, with the population, in the south. Most of Finland and much of Norway and Sweden are covered by dense forests of pine, spruce, and birch, which supply the timber industries.

Land use and agricultural distribution

- fishing
- pigs
- reindeer
- sheep
- timber

- capital cities
- major towns
- pasture
- cropland
- forest
- mountain region
- tundra

The urban/rural population divide

urban 77%   rural 23%

Population density
Total land area   473,970 sq miles

Lapland, north of the Arctic Circle, is an area of undulating fells and plains known as tundra. The subsoil is permanently frozen and therefore impermeable. There are many peat bogs. Pools reappear in the summer when the surface thaws.

▲ Finland's landscape was fashioned by ice action. Glaciers gouged out its distinctive shallow lake basins, such as Oulujärvi, and left debris called moraines in their wake.

Oulujärvi

SCALE 1:9,000,000

(same scale as main map)

Scale 1:5,500,000

projection: Lambert Conformal Conic

Map key

Population

- ◉ 1 million to 5 million
- ◉ 500,000 to 1 million
- ⊚ 100,000 to 500,000
- ⊕ 50,000 to 100,000
- ⊕ 10,000 to 50,000
- ○ below 10,000

Elevation

- 2000m / 6562ft
- 1000m / 3281ft
- 500m / 1640ft
- 250m / 820ft
- 100m / 328ft
- sea level

Transportation & industry

Norway derives its premier industry, the production of oil and gas, from the North Sea, while Denmark exploits its own oil and gas reserves. Hydroelectric power is a major industry, particularly in Sweden and Iceland. Timber processing remains significant in Finland and Sweden, but metal and engineering industries are increasingly important. In Iceland, fish products are the main source of export earnings.

Transportation network

226,735 miles (364,936 km)	
2042 miles (3286 km)	
13,704 miles (22,057 km)	
6,661 miles (10,721 km)	

Although roads now reach most areas, the railroads are markedly less developed. Much of the north is not served by rail and must rely on air and sea services for long distance travel and freight transportation.

Major industry and infrastructure

- car manufacture
- engineering
- fish processing
- hydroelectric power
- nuclear power
- oil & gas
- timber processing
- capital cities
- major towns
- international airports
- major roads
- major industrial areas

▲ *The use of geothermal power in Iceland began half a century ago. Today geothermal power stations supply 89% of the country's domestic heating requirements.*

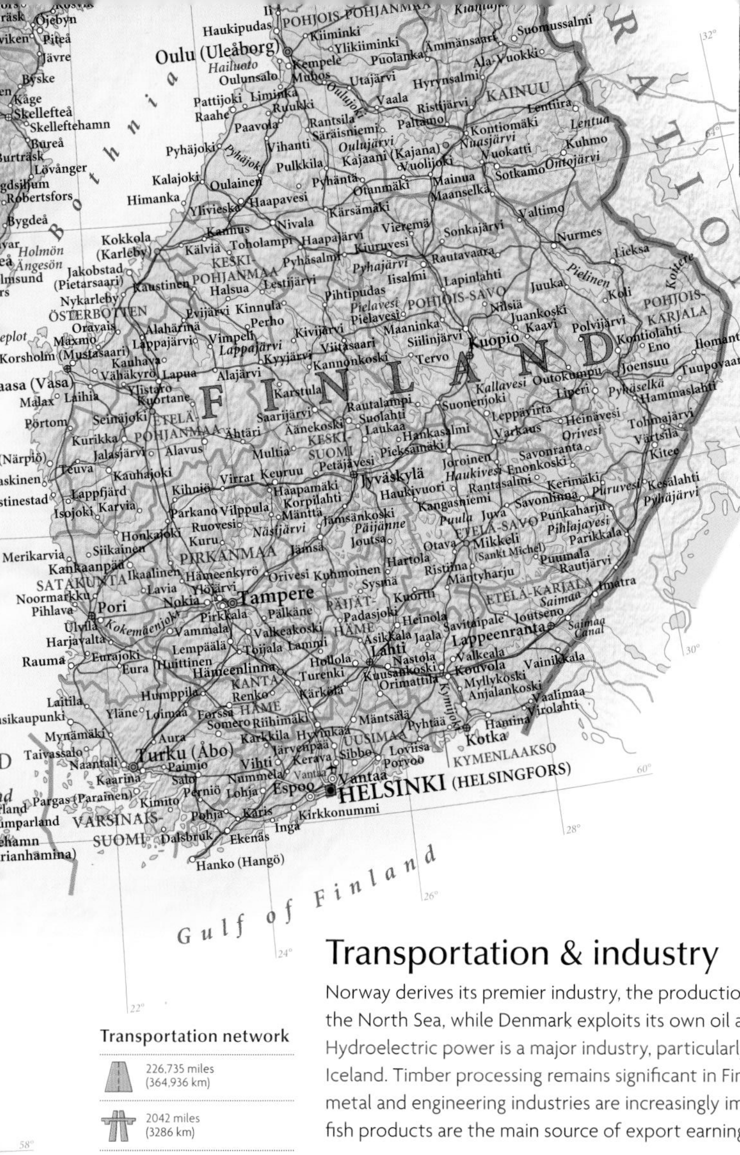

▲ *Sweden is one of the world's largest producers of wood and wood-based products. The traditional movement of logs by floating them down rivers has now been largely replaced by the use of trucks.*

▲ *Many Lappish people, in addition to traditional reindeer herding, now also make their living from fishing and farming, or working in cities. Tourism provides some with an extra source of income.*

Southern Scandinavia

SOUTHERN NORWAY, SOUTHERN SWEDEN, DENMARK

Scandinavia's economic and political hub is the more habitable and accessible southern region. Many of the area's major cities are on the southern coasts, including Oslo and Stockholm, the capitals of Norway and Sweden. In Denmark, most of the population and the capital, Copenhagen, are located on its many islands. A cultural unity links the three Scandinavian countries. Their main languages, Danish, Swedish, and Norwegian, are mutually intelligible, and they all retain their monarchies, although the parliaments have legislative control.

Using the land

Agriculture in southern Scandinavia is highly mechanized although farms are small. Denmark is the most intensively farmed country and its western pastureland is used mainly for pig farming. Cereal crops including wheat, barley, and oats, predominate in eastern Denmark and in the far south of Sweden. Southern Norway, and Sweden have large tracts of forest which are exploited for logging.

Land use and agricultural distribution

- cattle
- pigs
- sheep
- cereals
- fodder
- root crops
- timber
- capital cities
- major towns
- pasture
- cropland
- forest
- mountain region

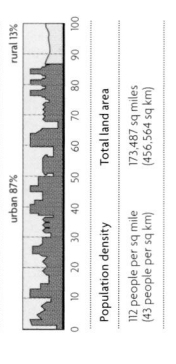

The urban/rural population divide

urban 87%   rural 13%

Population density	112 people per sq mile (43 people per sq km)
Total land area	173,487 sq miles (456,564 sq km)

The landscape

Southern Scandinavia, with the exception of Norway, has a flatter terrain than the rest of the region. Denmark and southern Sweden are both extensions of the North European Plain. In this area, because of glacial deposition rather than erosion, the soils are deeper and more fertile.

Acid rain, caused by industrial pollution carried north from elsewhere in Europe, harms plant and animal life in Scandinavian forests and lakes. The region's surface rocks lack lime to neutralize the acid, so making the problem more serious.

▲ *In the past*, glaciers such as this one in Olden, Norway, were much larger. Today, many are retreating to yield the spectacular glacial scenery.

Distinctive low ridges, called eskers, are found across southern Sweden. They are formed from sand and gravel deposits left by retreating glaciers.

The peak of Glittertind in the Jotunheimen mountains is 8110 ft (2472 m) high.

The lakes of southern Sweden remain from a period when the land was completely flooded. As the ice which covered the area melted, the land rose, leaving lakes in shallow, ice-scoured depressions. Sweden has over 90,000 lakes.

Limestone pillars eroded by the sea dot the coast of Gotland and surrounding islands.

Vänern in Sweden is the largest lake in Scandinavia. It covers an area of 2080 sq miles (5390 sq km).

Denmark's flat and fertile soils are formed on glacial deposits between 100–160 ft (30–50 m) deep.

When the ice retreated the valley was flooded by the sea.

Old valley floor

Sea level

Erosion by glaciers deepened existing river valleys.

Sognefjorden

▲ **Sognefjorden is the deepest** of Norway's many fjords. It drops to 4291 ft (1308 m) below sea level.

▲ **In Norway winters** are longer and colder inland than in coastal areas, where the warm current of the North Atlantic Drift moderates the climate.

Map key

Population
- ◉ 1 million to 5 million
- ◎ 500,000 to 1 million
- ⊕ 100,000 to 500,000
- ⊙ 50,000 to 100,000
- ○ 10,000 to 50,000
- ○ below 10,000

Elevation
- 2000m / 6562ft
- 1000m / 3281ft
- 500m / 1640ft
- 250m / 820ft
- 100m / 328ft

Scale 1:3,250,000

projection: Lambert Conformal Conic

Gulf of Bothnia

NORWEGIAN SEA

NORTH SEA

BALTIC SEA

GERMANY

NORWAY

SWEDEN

DENMARK

COPENHAGEN

STOCKHOLM

OSLO

Uppsala

Örebro

Gothenburg

Malmö

Ålborg

Bergen

Trondheim

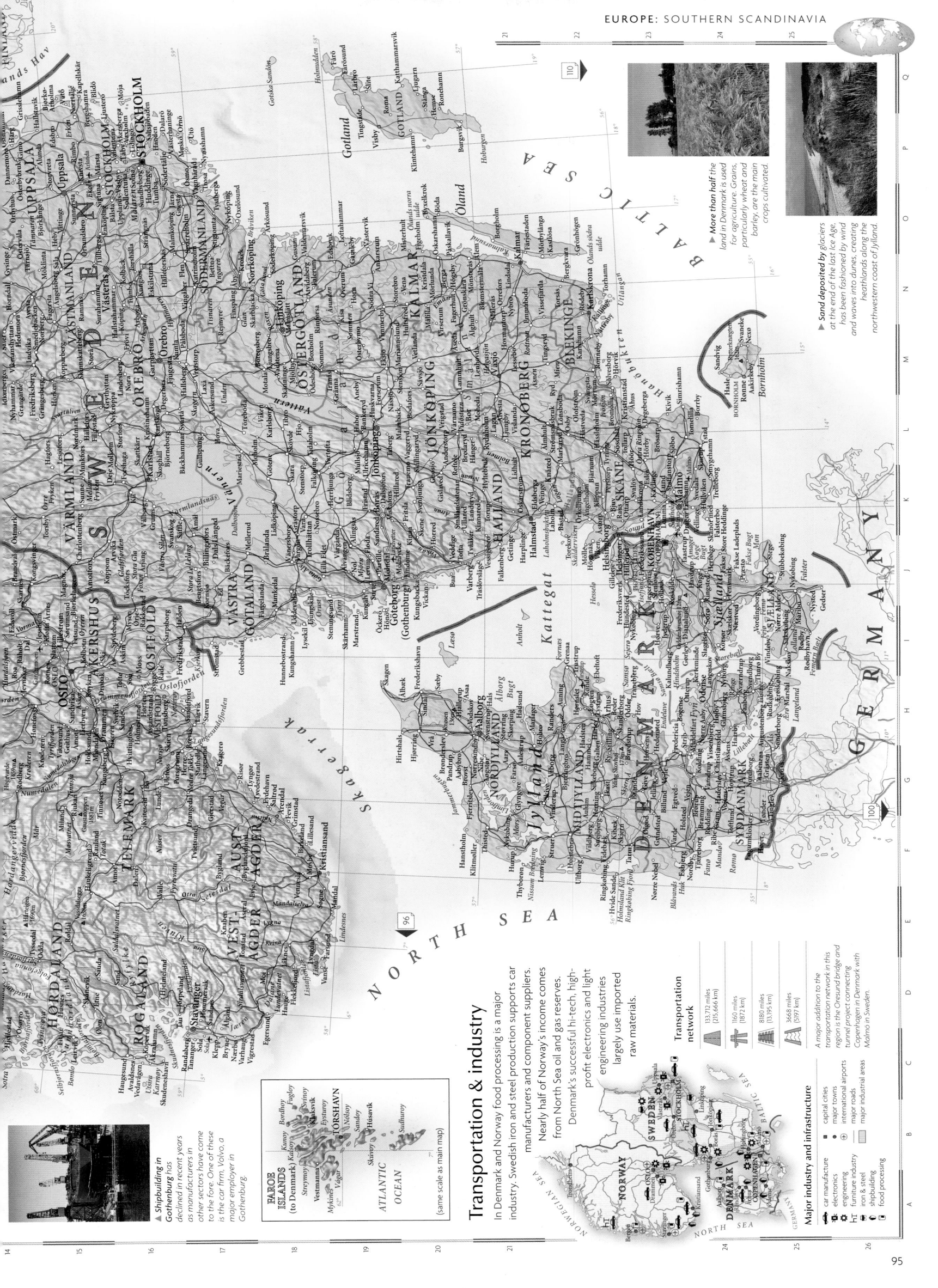

▲ *More than half the land in Denmark is used for agriculture. Grains, particularly wheat and barley, are the main crops cultivated.*

▲ *Sand deposited by glaciers at the end of the last Ice Age, has been fashioned by wind and waves into dunes, creating heathlands along the northwestern coast of Jylland.*

Transportation & industry

In Denmark and Norway food processing is a major industry. Swedish iron and steel production supports car manufacturers and component suppliers. Nearly half of Norway's income comes from North Sea oil and gas reserves. Denmark's successful hi-tech, high-profit electronics and light engineering industries largely use imported raw materials.

Transportation network

133,712 miles (215,666 km)	
1160 miles (1872 km)	
8180 miles (13,195 km)	
3668 miles (5197 km)	

A major addition to the transportation network in this region is the Øresund bridge and tunnel project connecting Copenhagen in Denmark with Malmö in Sweden.

Major industry and infrastructure

- capital cities
- major towns
- ✈ international airports
- major industrial areas

- car manufacture
- electronics
- engineering
- furniture industry
- iron & steel
- shipbuilding
- food processing

▲ *Shipbuilding in Gothenburg has declined in recent years as manufacturers in other sectors have come to the fore. One of these is the car firm, Volvo, a major employer in Gothenburg.*

FAROE ISLANDS (to Denmark)

TÓRSHAVN

Streymoy · Eysturoy · Bordhoy · Kunoy · Kalsoy · Fugloy · Svínoy · Vidhoy

Vágar · Mykines · Koltur · Sandoy · Nólsoy · Skúvoy · Stóra Dímun · Húsavík · Suduroy

ATLANTIC OCEAN

(same scale as main map)

The British Isles

UNITED KINGDOM, IRELAND

The British Isles have for centuries played a central role in European and world history. England, Wales, Scotland, and Northern Ireland together form the United Kingdom (UK), while the southern portion of Ireland is an independent country, self-governing since 1921. Although England has tended to be the politically and economically dominant partner in the UK, the Scots, Welsh, and Irish maintain independent cultures, distinct national identities and languages. Southeastern England is the most densely populated part of this crowded region, with over eight million people living in and around the London area.

▲ The valley of Glen Coe in the Scottish Highlands is a U-shaped valley, typical of the north and west of the British Isles, where glaciers shaped much of the landscape.

Transportation & industry

The British Isles' industrial base was founded primarily on coal, iron, and textiles, based largely in the north. Today, the most productive sectors include hi-tech industries clustered mainly in southeastern England, chemicals, finance, and the service sector, particularly tourism.

Major industry and infrastructure

- car manufacture
- chemicals
- engineering
- hi-tech industry
- iron & steel
- tourism
- capital cities
- major towns
- international airports
- major roads
- major industrial areas

Transportation network

285,947 miles (460,240 km)	2023 miles (3578 km)
11,825 miles (19,032km)	3976 miles (6400 km)

The UK's congested roads have become a major focus of environmental concern in recent years. No longer an island, the UK was finally linked to continental Europe by the Channel Tunnel in 1994.

▼ Clew Bay in western Ireland, is characteristic of the heavily indented west coast, where deep wide-mouthed bays separate the mountains of Mayo, Donegal, and Kerry as they thrust out into the Atlantic Ocean.

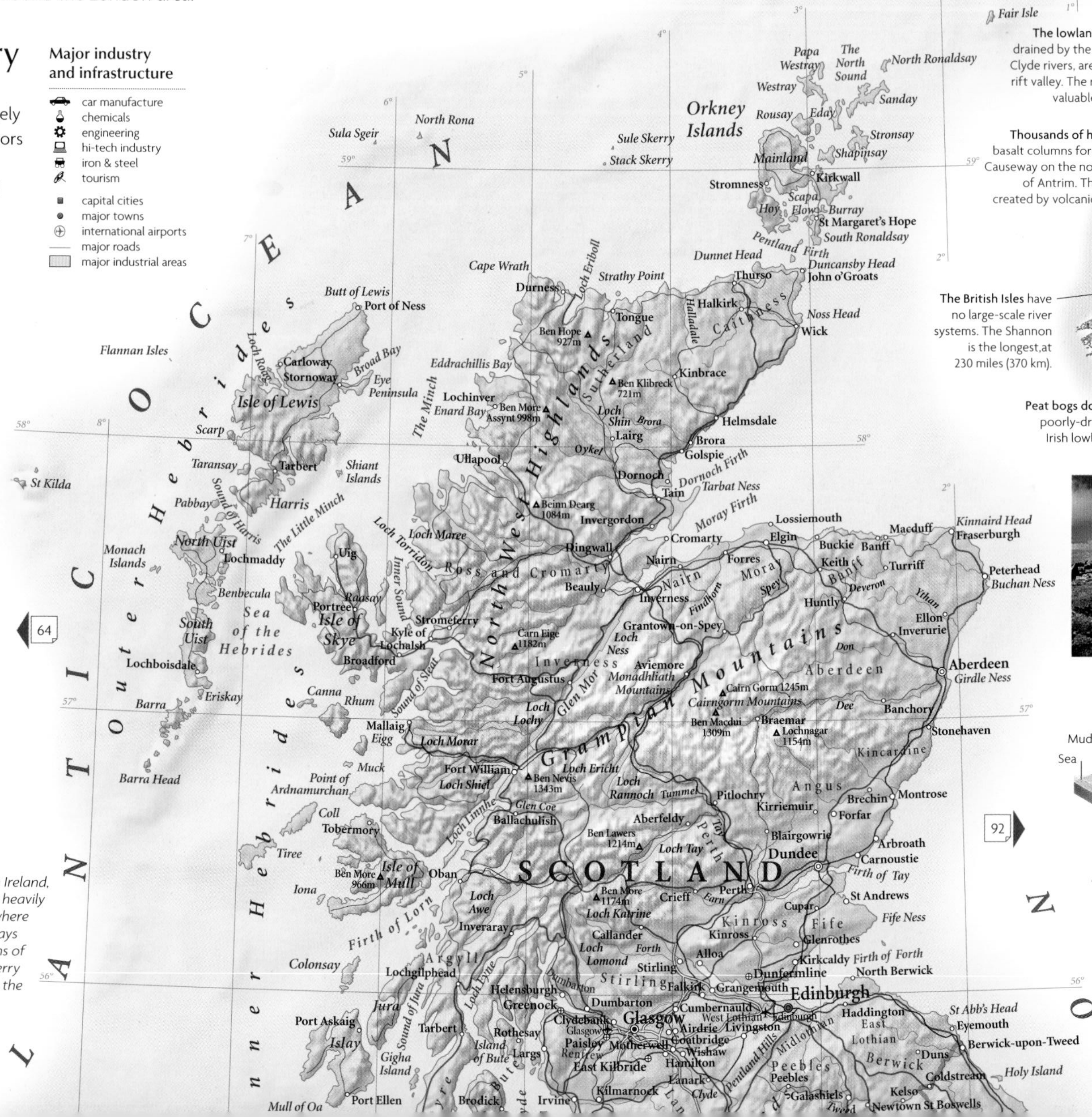

The landscape

Rugged uplands dominate the landscape of Scotland, Wales, and northern England. All the peaks in the British Isles over 4000 ft (1219 m) lie in highland Scotland. Lowland England rises into several ranges of rolling hills, including the older Mendips, and the Cotswolds and the Chilterns, which were formed at the same time as the Alps in southern Europe.

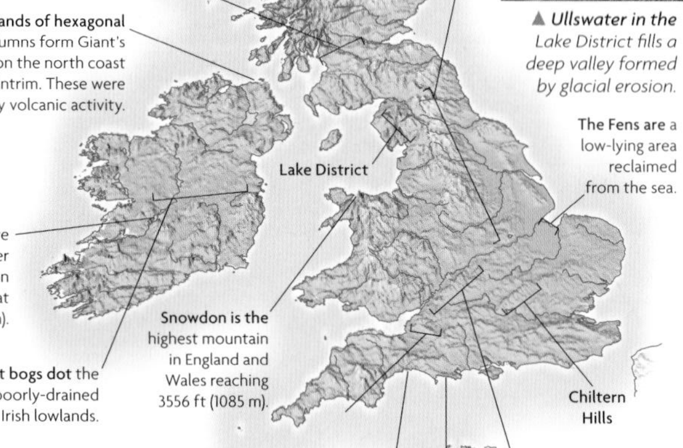

The Pennines, sometimes called "the backbone of England," are formed of limestones and grits.

▲ Ullswater in the Lake District fills a deep valley formed by glacial erosion.

The Fens are a low-lying area reclaimed from the sea.

The Cotswold Hills are characterized by a series of limestone ridges overlooking clay vales.

Chiltern Hills

Ben Nevis at 4409 ft (1343 m) is the highest peak in the UK.

Over 600 islands, mostly uninhabited, lie west and north of the Scottish mainland.

Lake District

Snowdon is the highest mountain in England and Wales reaching 3556 ft (1085 m).

The lowlands of Scotland, drained by the Tay, Forth, and Clyde rivers, are centered on a rift valley. The region contains valuable coal reserves.

Thousands of hexagonal basalt columns form Giant's Causeway on the north coast of Antrim. These were created by volcanic activity.

▲ Coastal erosion around the British Isles forms striking features such as this limestone arch, Durdle Door in Dorset.

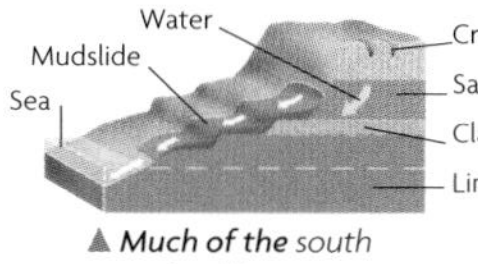

Durdle Door

▼ Dartmoor, studded with tors, is an exposed part of a vast granite dome, formed when molten rock intruded into the Earth's crust.

Peat bogs dot the poorly-drained Irish lowlands.

The British Isles have no large-scale river systems. The Shannon is the longest at 230 miles (370 km).

Black Ven, Lyme Regis

▲ Much of the south coast is subject to landslides. Following rain, porous sandstones feed water into the underlying, less permeable clays which then crumble and slide into the sea.

Cracks
Sandstone
Clay
Limestone
Water
Mudslide
Sea

Map key

Population
- above 5 million
- 1 million to 5 million
- 500,000 to 1 million
- 100,000 to 500,000
- 50,000 to 100,000
- 10,000 to 50,000
- below 10,000

Elevation
- 1000m / 3281ft
- 500m / 1640ft
- 250m / 820ft
- 100m / 328ft
- sea level

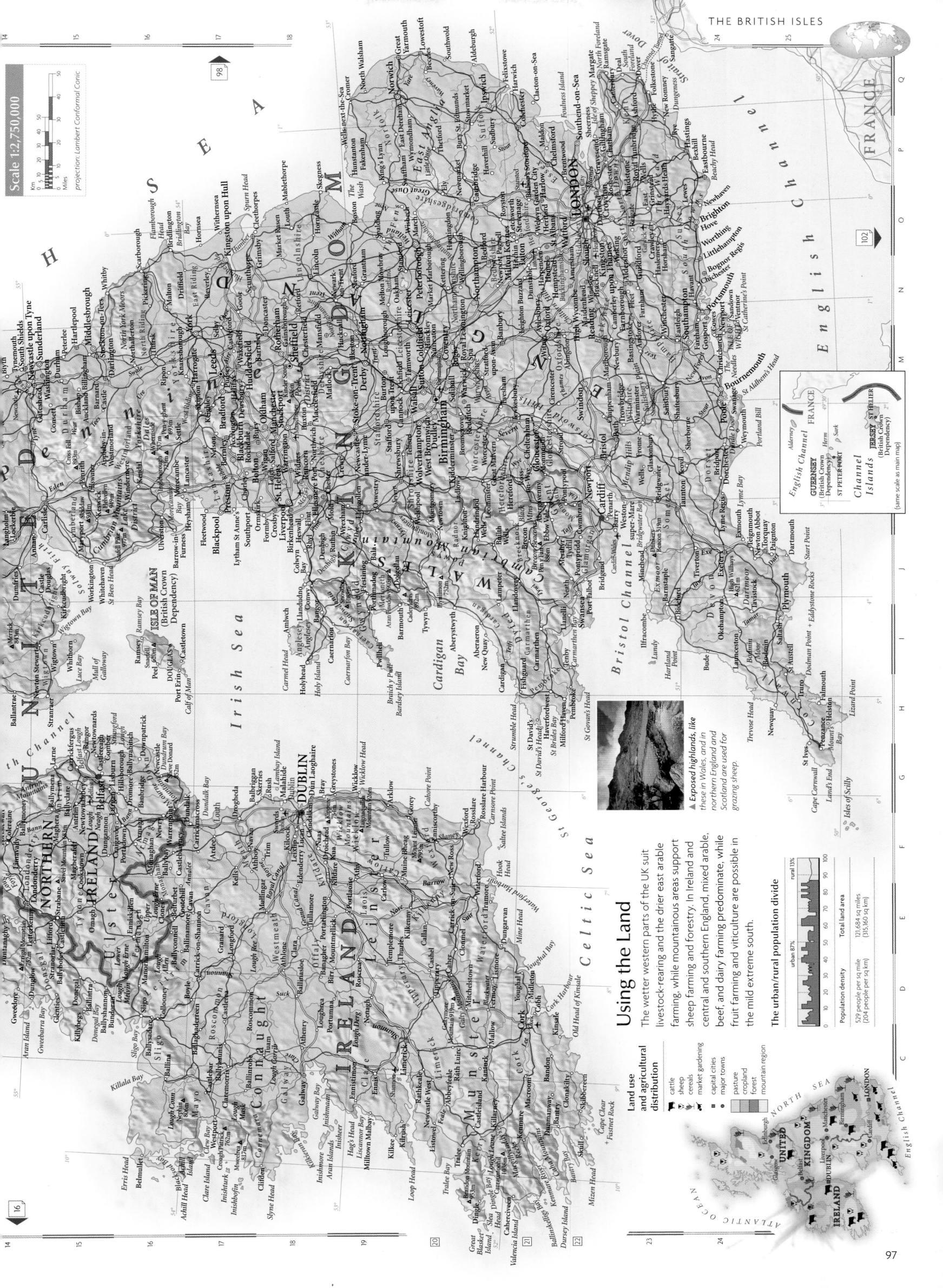

Scale 1:2,750,000

projection: Lambert Conformal Conic

Using the Land

The wetter western parts of the UK suit livestock-rearing and the drier east arable farming, while mountainous areas support sheep farming and forestry. In Ireland and central and southern England, mixed arable, beef, and dairy farming predominate, while fruit farming and viticulture are possible in the mild extreme south.

▲ Exposed highlands, like these in Wales, and in northern England and Scotland are used for grazing sheep.

Land use and agricultural distribution

- cattle
- sheep
- cereals
- market gardening
- capital cities
- major towns
- pasture
- cropland
- forest
- mountain region

The urban/rural population divide

urban 87%

rural 13%

Population density
529 people per sq mile
(204 people per sq km)

Total land area
121,684 sq miles
(315,160 sq km)

GUERNSEY
(British Crown Dependency)
ST PETER PORT

Channel Islands
(same scale as main map)

JERSEY ST HELIER
(British Crown Dependency)

English Channel   FRANCE

Alderney
Herm
Sark

97

The Low Countries

BELGIUM, LUXEMBOURG, NETHERLANDS

One of northwestern Europe's strategic crossroads, the Low Countries are united by a common history in which they have often been a battleground in European wars. For over a thousand years they were ruled by foreign powers. Even after they achieved independence, the three countries maintained close links, later forming the world's first totally free labor and goods market, the Benelux Economic Union, which became the core of the European Community (now the European Union or EU). These states have remained at the forefront of wider European cooperation; Brussels, The Hague, and Luxembourg are hosts to major institutions of the EU.

The landscape

The main geographical regions of the Netherlands are the northern glacial heathlands, the low-lying lands of the Rhine and Maas/Meuse, the reclaimed polders, and the dune coast and islands. Belgium includes part of the Ardennes, together with the coalfields on its northern flanks, and the fertile Flanders plain.

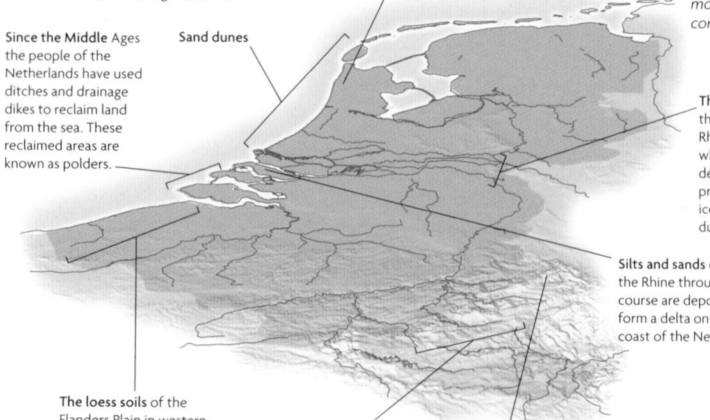

Since the Middle Ages the people of the Netherlands have used ditches and drainage dikes to reclaim land from the sea. These reclaimed areas are known as polders.

The loess soils of the Flanders Plain in western Belgium provide excellent conditions for arable farming.

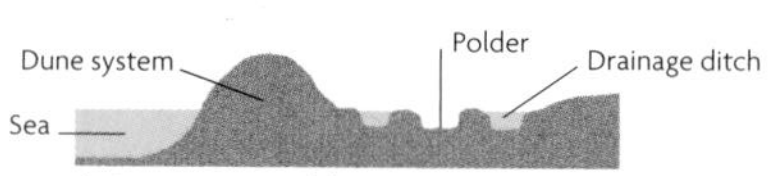

▲ **Extensive sand dune** systems along the coast have prevented flooding of the land. Behind the dunes, marshy land is drained to form polders, usable land suitable for agriculture.

Dune system
Sea
Polder
Drainage ditch

▲ **Uplifted and folded** 220 million years ago, the Ardennes have since been reduced to relatively level plateaus, then sharply incised by rivers such as the Maas/Meuse.

Hautes Fagnes is the highest part of Belgium. The bogs and streams in this upland region result from high rainfall and low temperatures.

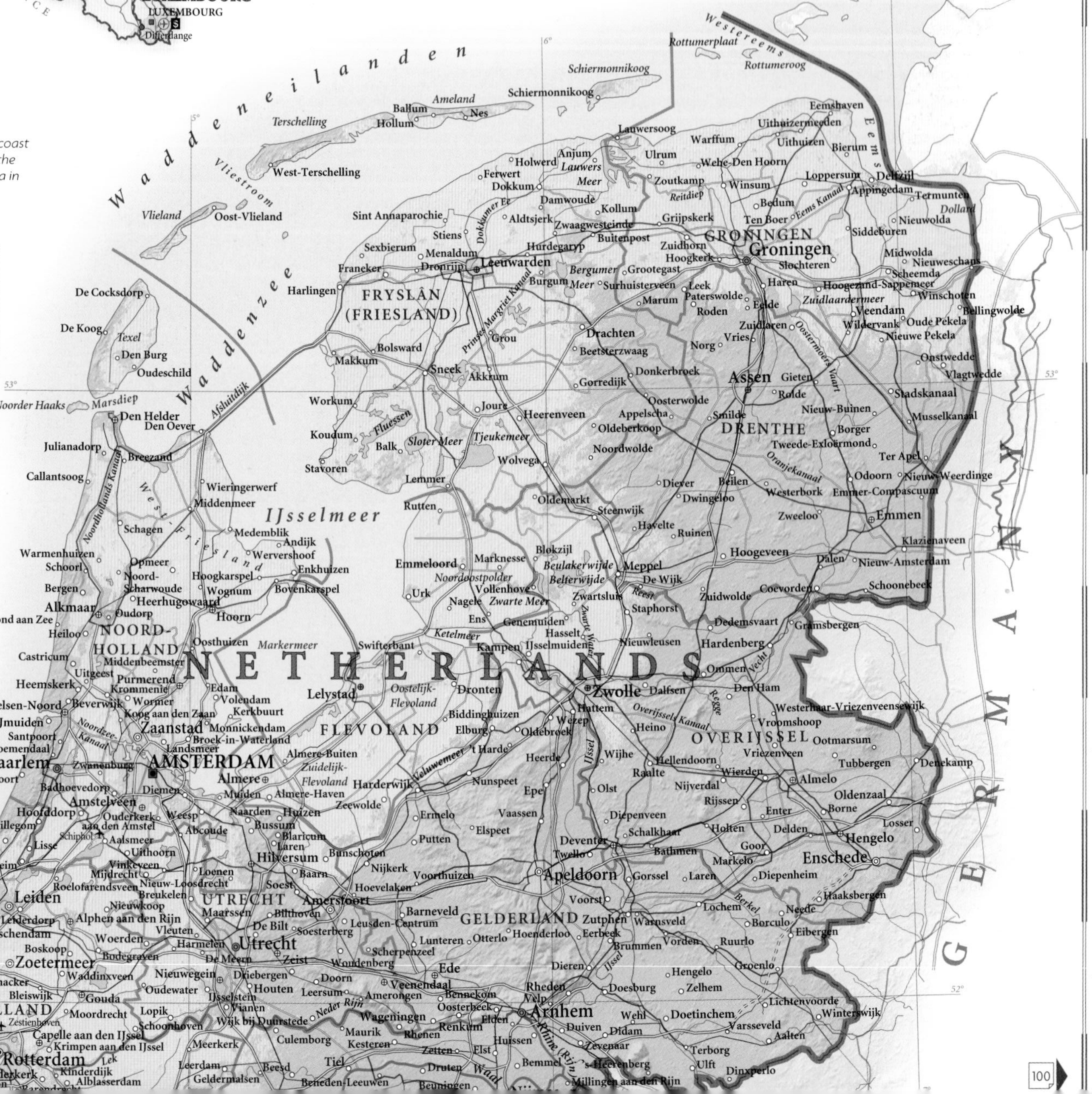

Ardennes

Silts and sands eroded by the Rhine throughout its course are deposited to form a delta on the west coast of the Netherlands.

The parallel valleys of the Maas/Meuse and Rhine rivers were created when the Rhine was deflected from its previous course by the ice sheet which formed during the last Ice Age.

▲ **One-third of the** Netherlands lies below sea level and flooding is a constant threat. Barrages have been built across the mouths of many rivers to contain floodwaters.

▼ **Heathlands,** like these at Schoorl, are found along the coast of the Netherlands. Much of the coast was breached by the sea in the 5th century, creating its distinctive inlets and islands.

Schoorl

Transportation & industry

In the western Netherlands, a massive, sprawling industrialized zone encompasses many new hi-tech and service industries. Belgium's central region has emerged as the country's light manufacturing and services center. Luxembourg city is home to more than 160 banks and the European headquarters of many international companies.

The Low Countries hold a key position on the North Sea, containing Europe's two largest ports, Rotterdam and Antwerp, which are connected to a comprehensive system of inland waterways.

Transportation network

✈	140,588 miles (226.281 km)	✈	2565 miles (4129 km)
▓	4099 miles (6598 km)	⛴	4134 miles (6653 km)

Major industry and infrastructure

- ✈ aerospace
- ⚙ finance
- ⚙ engineering
- ▢ hi-tech industry
- ▽ pharmaceuticals
- ▪ textiles
- ■ capital cities
- ▪ major cities
- ○ major towns
- ⊕ international airports
- — major roads
- ▨ major industrial areas

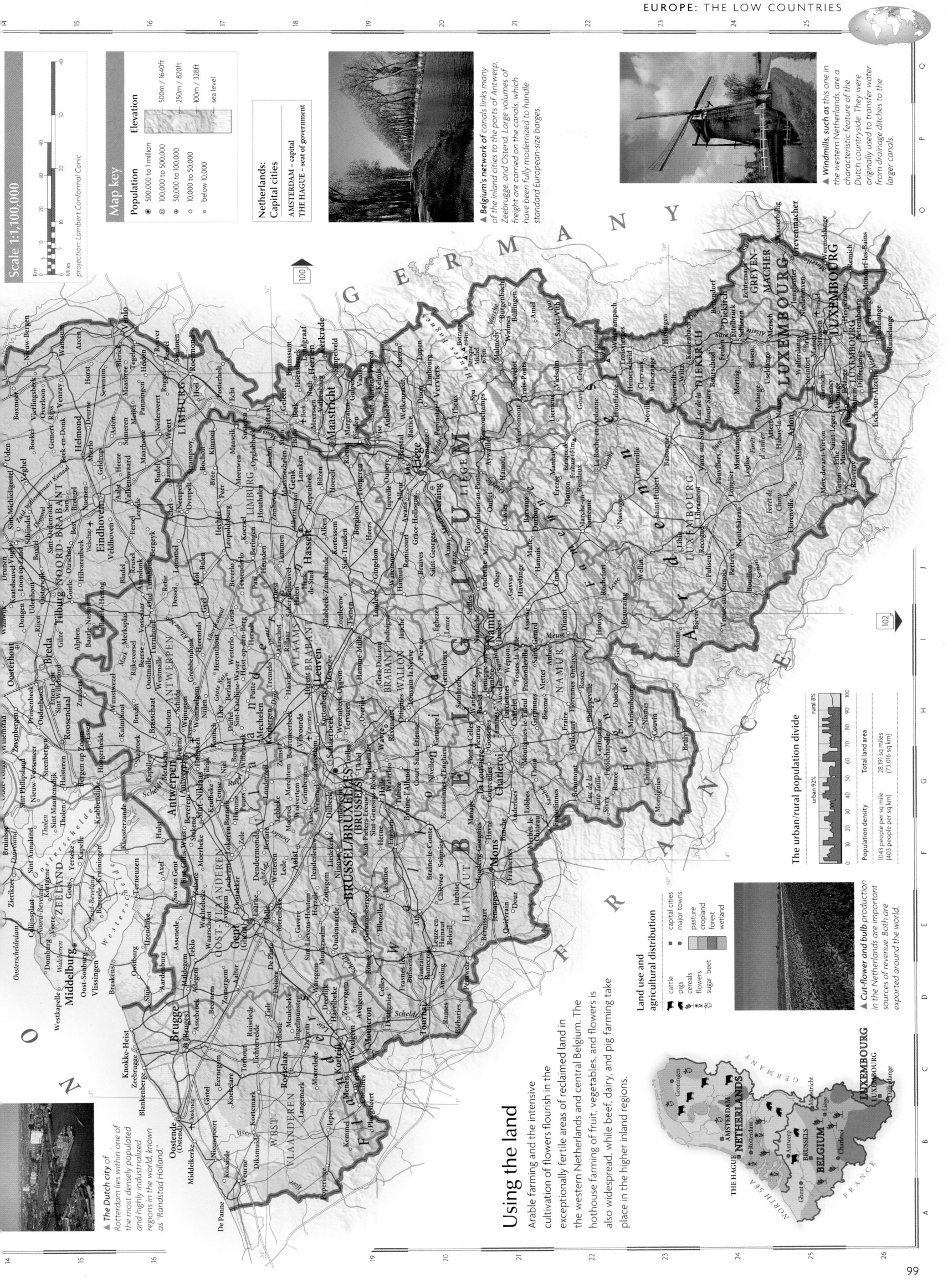

Scale 1:1,100,000

projection: Lambert Conformal Conic

Map key

Population

- ◉ 500,000 to 1 million
- ◎ 100,000 to 500,000
- ⊕ 50,000 to 100,000
- ○ 10,000 to 50,000
- ○ below 10,000

Elevation

- 500m / 1640ft
- 250m / 820ft
- 100m / 328ft
- sea level

Netherlands:
Capital cities

AMSTERDAM – capital
THE HAGUE – seat of government

▲ Belgium's network of canals links many of the inland ports to the ports of Antwerp, Zeebrugge, and Ostend. Large volumes of freight are carried on the canals, which have been fully modernized to handle standard European-size barges.

▲ Windmills, such as this one in the western Netherlands, are a characteristic feature of the Dutch countryside. They were originally used to transfer water from drainage ditches to the larger canals.

▲ The Dutch city of Rotterdam lies within one of the most densely populated and highly industrialized regions in the world, known as "Randstad Holland."

Using the land

Arable farming and the intensive cultivation of flowers flourish in the exceptionally fertile areas of reclaimed land in the western Netherlands and central Belgium. The hothouse farming of fruit, vegetables, and flowers is also widespread, while beef, dairy, and pig farming take place in the higher inland regions.

▲ Cut-flower and bulb production in the Netherlands are important sources of revenue. Both are exported around the world.

Land use and agricultural distribution

- ● capital cities
- ● major towns
- cattle
- pigs
- cereals
- flowers
- sugar beet
- pasture
- cropland
- forest
- wetland

The urban/rural population divide

	urban 92%	rural 8%

Population density	Total land area
1043 people per sq mile (403 people per sq km)	28,191 sq miles (73,016 sq km)

99

Germany

Despite the devastation of its industry and infrastructure during the Second World War and its separation from eastern Germany during the Cold War, West Germany made a rapid recovery in the following generation to become Europe's most formidable economic power. When the Berlin Wall was dismantled in 1989, the two halves of Germany were politically united for the first time in 40 years. Complete social and economic unity remain a longer term goal, as East German industry and society adapt to a free market. Germany has been a key player in the creation of the European Union (EU) and in moves toward a single European currency.

Using the land

Germany has a large, efficient agricultural sector, and produces more than three-quarters of its own food. The major crops grown are cereals and sugar beet on the more fertile soils, and root crops, rye, oats, and fodder on the poorer soils of the northern plains and central uplands. Southern Germany is also a principal producer of high quality wines. Vineyards cover the slopes surrounding the Rhine and its tributaries.

Land use and agricultural distribution

- cattle
- pigs
- cereals
- sugar beet
- vineyards
- capital cities
- major towns
- pasture
- cropland
- forest

The urban/rural population divide

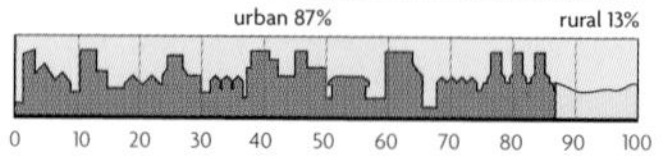

urban 87%   rural 13%

Population density	612 people per sq mile (236 people per sq km)
Total land area	137,804 sq miles (356,910 sq km)

▲ *The Moselle river flows through the Rhine State Uplands (Rheinisches Schiefergebirge). During a period of uplift, preexisting river meanders were deeply incised, to form its present dramatic contours.*

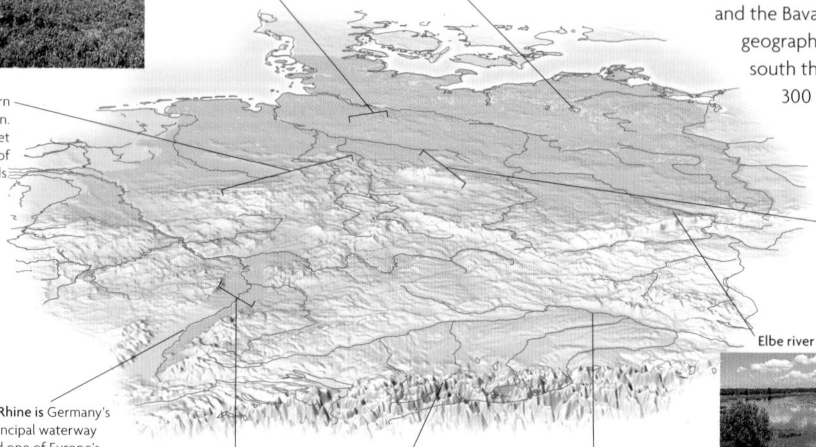

The landscape

The plains of northern Germany, the volcanic plateaus and mountains of the central uplands, and the Bavarian Alps are the three principal geographic regions in Germany. North to south the land rises steadily from barely 300 ft (90 m) in the plains to 6500 ft (2000 m) in the Bavarian Alps, which are a small but distinct region in the far south.

Müritz lake covers 45 sq miles (117 sq km), but is only 108 ft (33 m) deep. It lies in a shallow valley formed by meltwater flowing out from a retreating ice sheet. These valleys are known as *Urstromtäler*.

The Harz Mountains were formed 300 million years ago. They are block-faulted mountains, formed when a section of the Earth's crust was thrust up between two faults.

▼ *The Elbe flows in wide meanders across the north German plain to the North Sea. At its mouth it is 10 miles (16 km) wide.*

Elbe river

Scale 1:2,500,000

projection: Lambert Conformal Conic

The Danube rises in the Black Forest (Schwarzwald) and flows east, across a wide valley, on its course to the Black Sea.

Zugspitze, the highest peak in Germany at 9719 ft (2962 m), was formed during the Alpine mountain-building period, 30 million years ago.

The Rhine is Germany's principal waterway and one of Europe's longest rivers, flowing 820 miles (1320 km).

Rhine Rift Valley

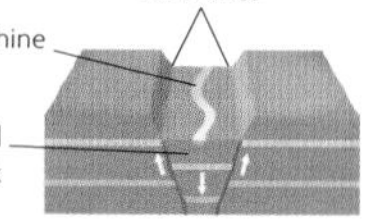

▲ *Part of the floor of the Rhine Rift Valley was let down between two parallel faults in the Earth's crust.*

Fault lines
Rhine
Downfaulted block

Much of the landscape of northern Germany has been shaped by glaciation. During the last Ice Age, the ice sheet advanced as far the northern slopes of the central uplands.

Lüneburg Heath (*Lüneburger Heide*)

▲ *The heathlands of northern Germany are covered by glacial deposits of sandy outwash soil which makes them largely infertile. They support only sheep and solitary trees.*

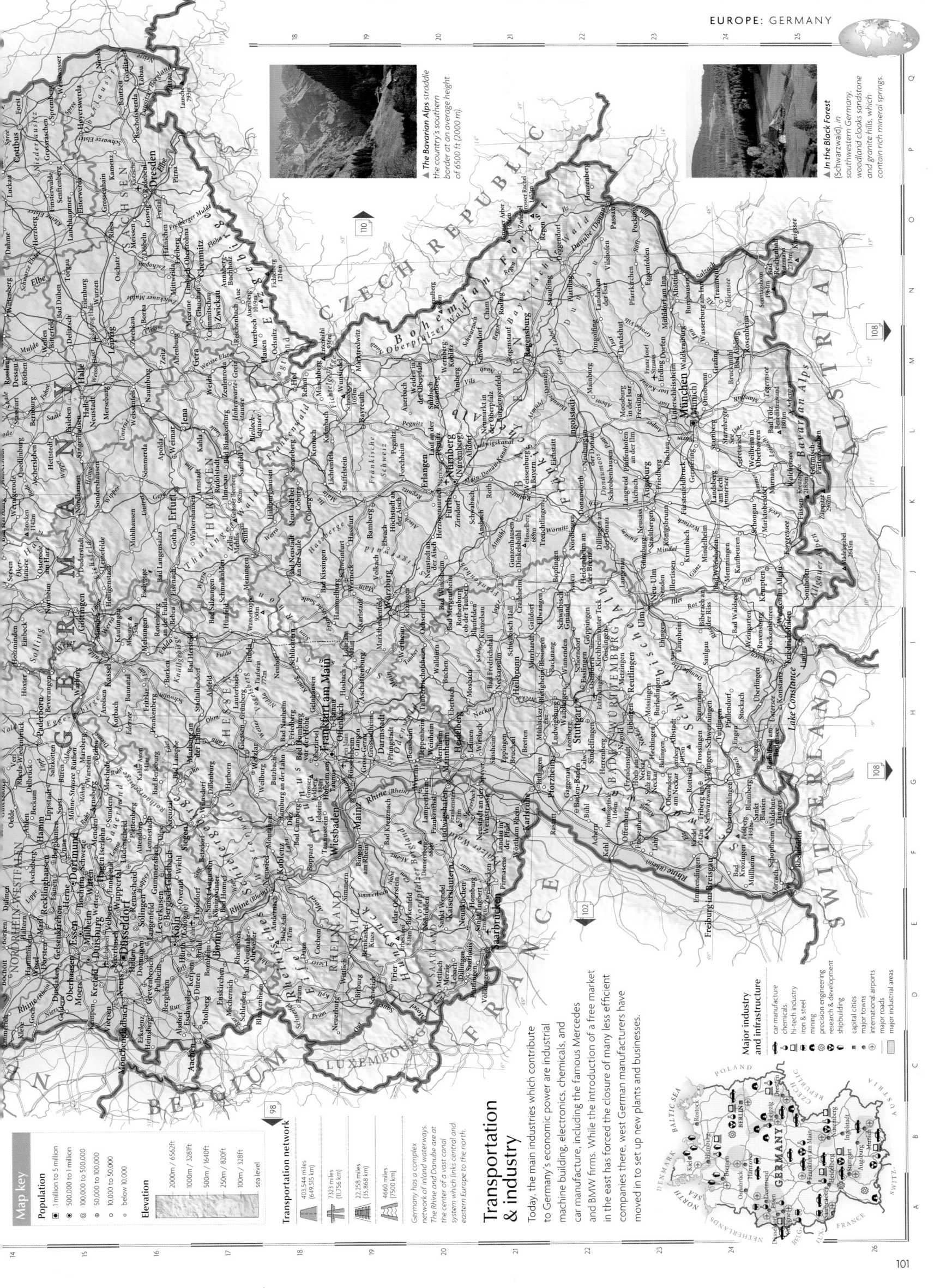

▲ *The Bavarian Alps* straddle the country's southern border at an average height of 6500 ft (2000 m).

▲ *In the Black Forest* (Schwarzwald), in southwestern Germany, woodland clocks sandstone and granite hills, which contain rich mineral springs.

Transportation & industry

Today, the main industries which contribute to Germany's economic power are industrial machine building, electronics, chemicals, and car manufacture, including the famous Mercedes and BMW firms. While the introduction of a free market in the east has forced the closure of many less efficient companies there, west German manufacturers have moved in to set up new plants and businesses.

Major industry and infrastructure

- car manufacture
- chemicals
- hi-tech industry
- iron & steel
- mining
- precision engineering
- research & development
- shipbuilding
- capital cities
- major cities
- major towns
- international airports
- major roads
- major industrial areas

Germany has a complex network of inland waterways. The Rhine and Danube are at the center of a vast canal system which links central and eastern Europe to the north.

Transportation network

403,544 miles (649,515 km)	
7323 miles (11,756 km)	
22,258 miles (35,868 km)	
4660 miles (7500 km)	

Map key

Population
- 1 million to 5 million
- 500,000 to 1 million
- 100,000 to 500,000
- 50,000 to 100,000
- 10,000 to 50,000
- below 10,000

Elevation
- 2000m / 6562ft
- 1000m / 3281ft
- 500m / 1640ft
- 250m / 820ft
- 100m / 328ft
- sea level
- below 10,000

France

FRANCE, MONACO

Europe's second largest nation and the founder of modern Republican government, France is a major center of culture and fashion, and a leading producer of both agricultural and industrial goods. It has played a leading role in European events for centuries, and remains a key player in the push toward European unity. The Paris Basin is the most highly populated area; Île de France is home to over 11 million people. Large parts of France remain thinly populated, particularly the mountainous Massif Central, Pyrenees, and southern Alps.

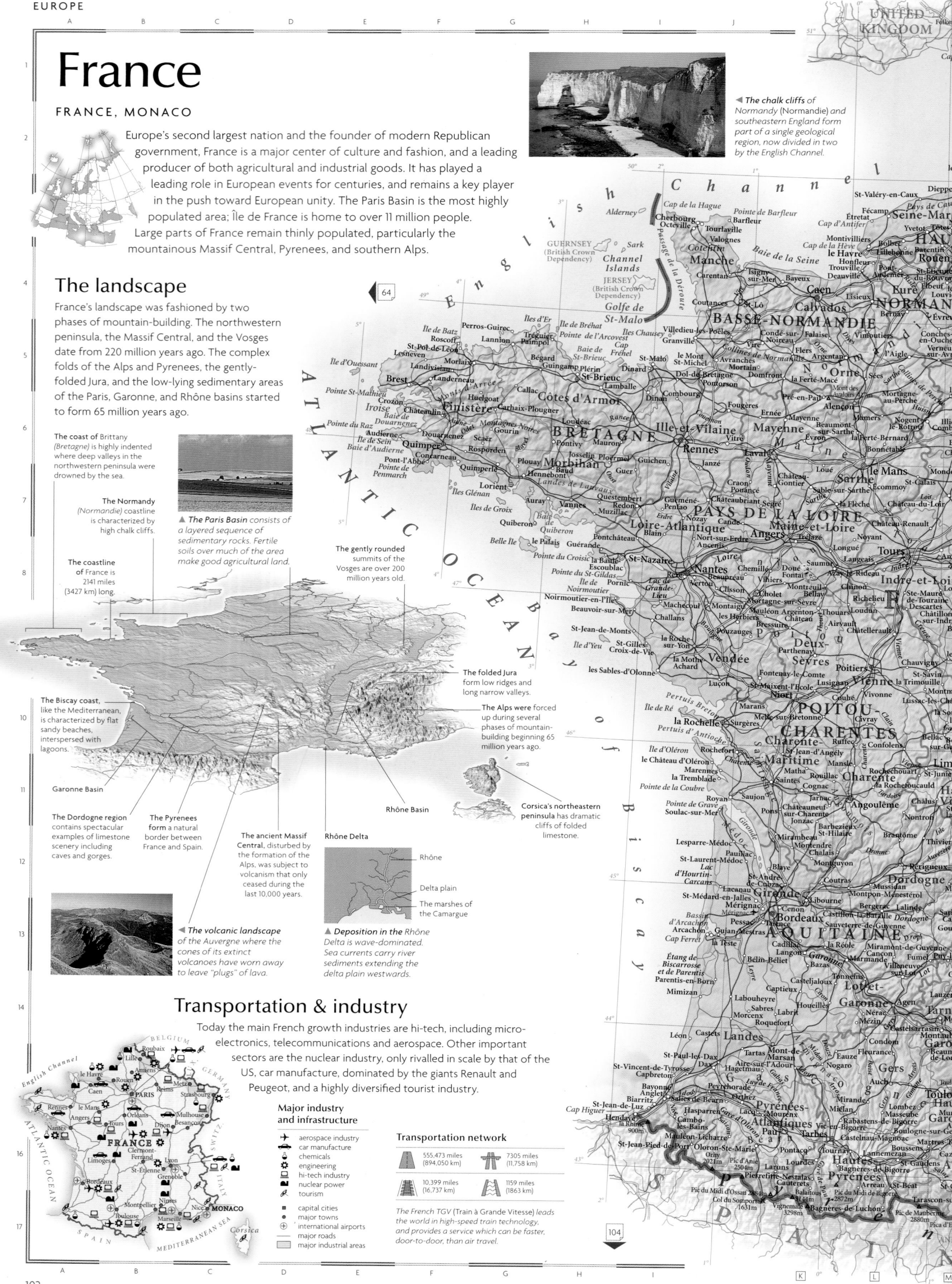

◀ *The chalk cliffs* of Normandy (Normandie) and southeastern England form part of a single geological region, now divided in two by the English Channel.

The landscape

France's landscape was fashioned by two phases of mountain-building. The northwestern peninsula, the Massif Central, and the Vosges date from 220 million years ago. The complex folds of the Alps and Pyrenees, the gently-folded Jura, and the low-lying sedimentary areas of the Paris, Garonne, and Rhône basins started to form 65 million years ago.

The coast of Brittany (Bretagne) is highly indented where deep valleys in the northwestern peninsula were drowned by the sea.

The Normandy (Normandie) coastline is characterized by high chalk cliffs.

The coastline of France is 2141 miles (3427 km) long.

▲ *The Paris Basin* consists of a layered sequence of sedimentary rocks. Fertile soils over much of the area make good agricultural land.

The gently rounded summits of the Vosges are over 200 million years old.

The folded Jura form low ridges and long narrow valleys.

The Alps were forced up during several phases of mountain-building beginning 65 million years ago.

The Biscay coast, like the Mediterranean, is characterized by flat sandy beaches, interspersed with lagoons.

Garonne Basin

The Dordogne region contains spectacular examples of limestone scenery including caves and gorges.

The Pyrenees form a natural border between France and Spain.

The ancient Massif Central, disturbed by the formation of the Alps, was subject to volcanism that only ceased during the last 10,000 years.

Rhône Basin

Rhône Delta

Rhône

Delta plain

The marshes of the Camargue

Corsica's northeastern peninsula has dramatic cliffs of folded limestone.

▲ *The volcanic landscape* of the Auvergne where the cones of its extinct volcanoes have worn away to leave "plugs" of lava.

▲ *Deposition in the Rhône* Delta is wave-dominated. Sea currents carry river sediments extending the delta plain westwards.

Transportation & industry

Today the main French growth industries are hi-tech, including micro-electronics, telecommunications and aerospace. Other important sectors are the nuclear industry, only rivalled in scale by that of the US, car manufacture, dominated by the giants Renault and Peugeot, and a highly diversified tourist industry.

Major industry and infrastructure

✈ aerospace industry
🚗 car manufacture
⚗ chemicals
⚙ engineering
🖥 hi-tech industry
☢ nuclear power
♨ tourism

■ capital cities
■ major towns
✈ international airports
— major roads
▨ major industrial areas

Transportation network

555,473 miles (894,050 km)	7305 miles (11,758 km)
10,399 miles (16,737 km)	1159 miles (1863 km)

The French TGV (Train à Grande Vitesse) leads the world in high-speed train technology, and provides a service which can be faster, door-to-door, than air travel.

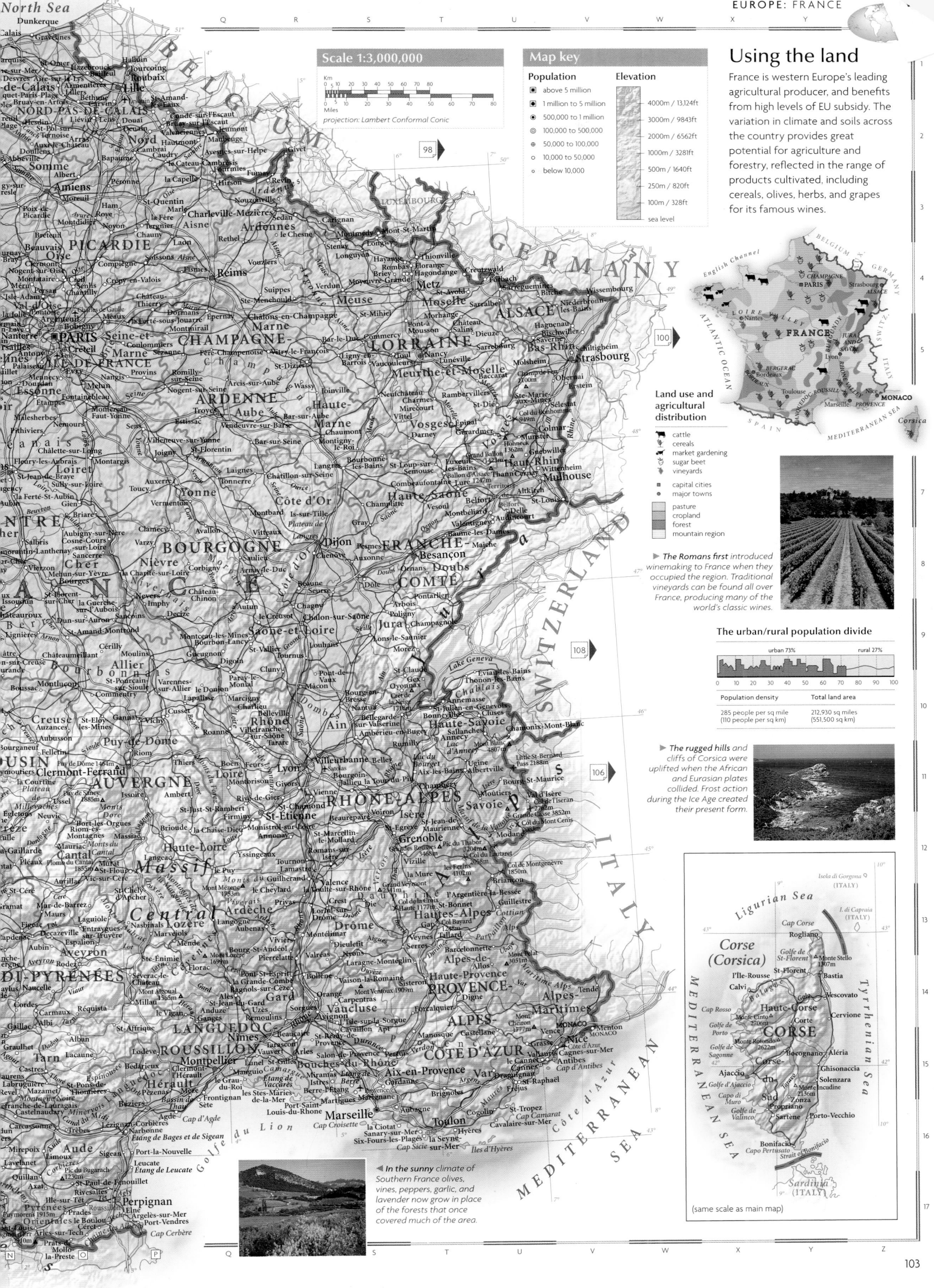

EUROPE: FRANCE

Using the land

France is western Europe's leading agricultural producer, and benefits from high levels of EU subsidy. The variation in climate and soils across the country provides great potential for agriculture and forestry, reflected in the range of products cultivated, including cereals, olives, herbs, and grapes for its famous wines.

Land use and agricultural distribution

- cattle
- cereals
- market gardening
- sugar beet
- vineyards
- capital cities
- major towns
- pasture
- cropland
- forest
- mountain region

▶ **The Romans first** introduced winemaking to France when they occupied the region. Traditional vineyards can be found all over France, producing many of the world's classic wines.

The urban/rural population divide

urban 73%   rural 27%

Population density	Total land area
285 people per sq mile (110 people per sq km)	212,930 sq miles (551,500 sq km)

▶ **The rugged hills** and cliffs of Corsica were uplifted when the African and Eurasian plates collided. Frost action during the Ice Age created their present form.

◀ **In the sunny** climate of Southern France olives, vines, peppers, garlic, and lavender now grow in place of the forests that once covered much of the area.

Scale 1:3,000,000
projection: Lambert Conformal Conic

Map key

Population
- ▪ above 5 million
- ▣ 1 million to 5 million
- ◉ 500,000 to 1 million
- ◎ 100,000 to 500,000
- ⊕ 50,000 to 100,000
- ○ 10,000 to 50,000
- ∘ below 10,000

Elevation
- 4000m / 13,124ft
- 3000m / 9843ft
- 2000m / 6562ft
- 1000m / 3281ft
- 500m / 1640ft
- 250m / 820ft
- 100m / 328ft
- sea level

Corse (Corsica)

(same scale as main map)

103

The Iberian peninsula

ANDORRA, GIBRALTAR, PORTUGAL,
SPAIN (Azores, Canary Islands, Madeira on p.64)

The Iberian peninsula is separated from the rest of
Europe by the Pyrenees, and at its most southerly
point is only 5 miles (8 km) from North Africa.
The location of Iberia has been central to its
diverse history. The Greeks, Carthaginians, Romans,
Visigoths, and most recently the Moors, invaded
Iberia at various times. For much of the 20th century,
both Spain and Portugal were governed by right-wing
dictators. Since the establishment of democratic governments in the
mid-1970s, modernization has been rapid and both countries are now
among the most popular of European holiday destinations.

Using the land

The principal crops grown in Iberia are
cereals, especially wheat and barley. Both
countries are major wine producers, most
notably of Rioja, sherry, and port. Sheep
are kept throughout the region, and citrus
fruits thrive on the Mediterranean coast.
The successful forest industry in Iberia
produces 84% of the world's cork.

▲ The steep, terraced slopes of the
Douro Valley in northern Portugal,
are used to cultivate vines. The
grapes harvested produce
Portugal's famous port wine.

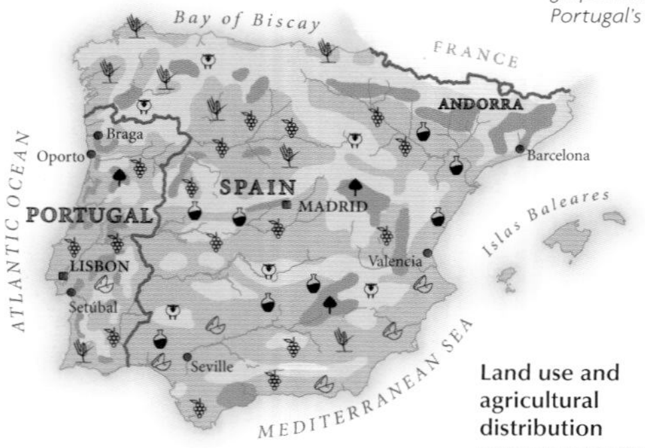

Land use and agricultural distribution

- sheep
- cereals
- citrus fruit
- olives
- vineyards
- cork

- capital cities
- major towns

- pasture
- cropland
- forest
- mountain region

The urban/rural population divide

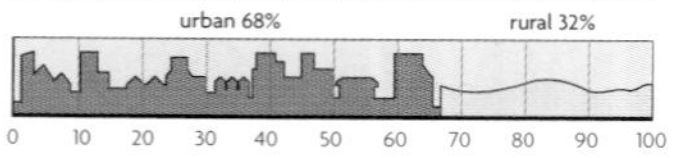

urban 68%   rural 32%

0 10 20 30 40 50 60 70 80 90 100

Population density	Total land area
215 people per sq mile (83 people per sq km)	230,569 sq miles (597,170 sq km)

Transportation & industry

Since the 1970s, the economies of Spain and Portugal
have expanded and diversified. In both countries,
tourism has outstripped agriculture in economic
importance. Spain's resource base is varied, including
coal, iron, and the world's largest reserves of mercury.
Portugal is a leading producer of tungsten ore.

Major industry and infrastructure

- car manufacture
- chemicals
- engineering
- fish processing
- mining
- textiles
- tourism

- capital cities
- major towns
- international airports
- major roads
- major industrial areas

Transportation network

241,720 miles (388,990 km)	1552 miles (2529 km)
11,793 miles (18,979 km)	1159 miles (1865 km)

Radiating from Madrid, the road network in
Spain dates from the 18th century, but now
includes many highways. Portugal's road
system has been completely modernized in
recent years.

◄ The eroded cliffs of the
Algarve in southern Portugal
were carved by Atlantic waves.
The numerous rocky bays and
beaches, and the region's
pleasant climate, have made it
a popular tourist destination.

▶ **The climate in** northwestern Spain is milder in both summer and winter than in the rest of the country, creating a verdant environment, more commonly associated with northwestern Europe.

Map key

Population
- ▣ 1 million to 5 million
- ◉ 500,000 to 1 million
- ◎ 100,000 to 500,000
- ⊕ 50,000 to 100,000
- ⊙ 10,000 to 50,000
- ○ below 10,000

Elevation
- 3000m / 9843ft
- 2000m / 6562ft
- 1000m / 3281ft
- 500m / 1640ft
- 250m / 820ft
- 100m / 328ft
- sea level

Scale 1:3,000,000

Km
Miles

projection: Lambert Conformal Conic

The landscape

A vast plateau, the Meseta dominates the centre of the peninsula, enclosed by the Cordillera Cantábrica to the north and the Sierra Morena to the south. It is drained by three major rivers, the Douro/Duero, the Tagus, and the Guadalquivir. The peninsula experiences great variations in climate and rainfall, both regionally and locally.

▲ **The Pyrenees form** Iberia's northeastern boundary, running for 270 miles (440 km), dividing the peninsula from the rest of Europe.

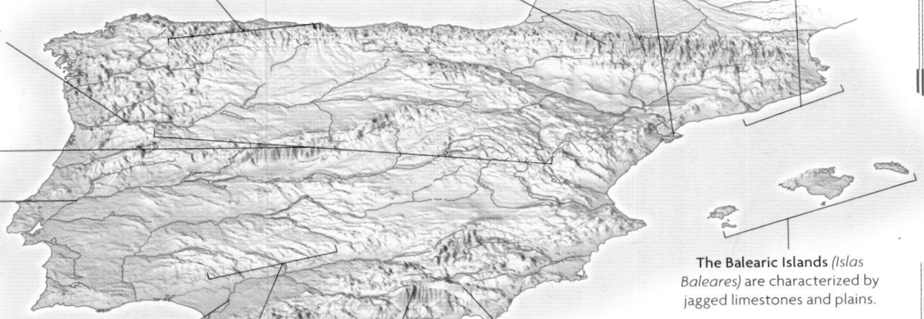

The Ebro river has formed the peninsula's largest delta. Recently, sediment flows have been seriously disturbed by nearby reservoirs.

On the northeastern coast sea level changes are evident from wave-cut beaches which rise up to 200 ft (60 m) above the present sea level.

Cordillera Cantábrica

Douro/Duero river

The Meseta plateau averages 1970 ft (600 m) in height and is now largely dry and treeless.

Tagus River

The Balearic Islands (Islas Baleares) are characterized by jagged limestones and plains.

Mountain front
Weathered material
Pediment

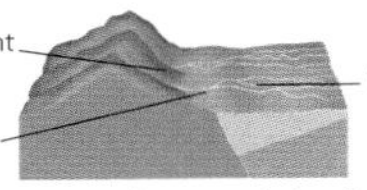

▲ **Pediments are characteristic of** semiarid lands across Iberia. A pediment is a flat, low-lying, eroded platform, cut into the bedrock. Weathered material is transported by streams and deposited in broad fan shapes on the pediment.

The Guadalquivir river brings vital irrigation water to the plains, and like many of Iberia's rivers, is prone to flooding.

Sierra Morena

The Sierra Nevada in southern Spain contain Iberia's highest peak, Mulhacén, which rises 11,418 ft (3481 m).

▶ **In the Sierra de los Filabres** deforestation and overgrazing, which cause soil erosion, have created semidesert badlands.

The Italian peninsula

ITALY, SAN MARINO, VATICAN CITY

The landscape

The Italian peninsula is a land of great contrasts. Until unification in 1861, Italy was a collection of independent states, whose competitiveness during the Renaissance resulted in the architectural and artistic magnificence of cities such as Rome, Florence, and Venice. The majority of Italy's population and economic activity is concentrated in the north, centered on the sophisticated industrial city of Milan. Southern Italy, the *Mezzogiorno*, has a harsh terrain, and remains far less developed than the north. Attempts to attract industry and investment in the south are frequently deterred by the entrenched network of organized crime and corruption.

The mainly mountainous and hilly Italian peninsula took its present form following a collision between the African and Eurasian tectonic plates. The Alps in the northwest rise to a high point of 15,772 ft (4807 m) at Mont Blanc (*Monte Bianco*) on the French border, while the Apennines (*Appennino*) form a rugged backbone, running along the entire length of the country.

▲ *The island of Sardinia is an ancient land mass, an uplifted section of very old igneous rocks. Its rugged mountainous regions provide pasture for sheep and goats, while its valleys support some agriculture.*

Mont Blanc
(*Monte Bianco*)

Costa Smeralda

▲ *The Dolomites* (Alpi Dolomitiche) *are formed of thick limestones, overlying weaker marine strata. They have distinctive serrated peaks and many massive landslides occur.*

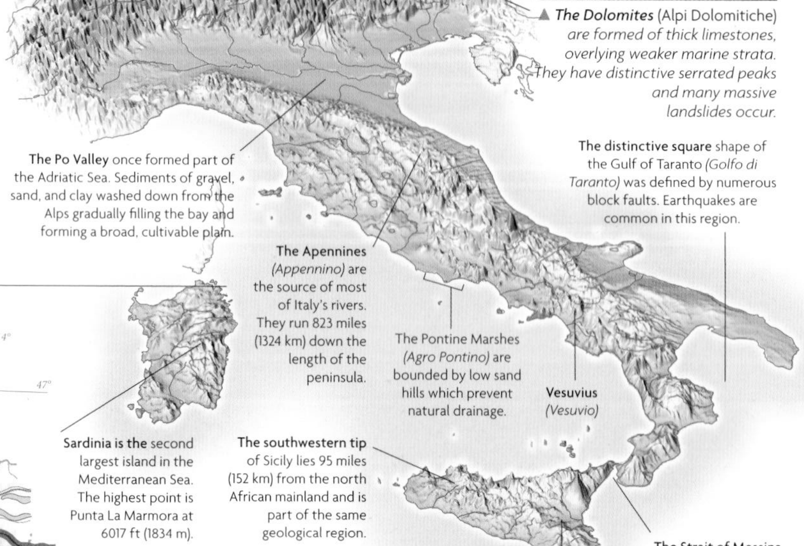

The **distinctive square shape** of the Gulf of Taranto (*Golfo di Taranto*) was defined by numerous block faults. Earthquakes are common in this region.

Vesuvius (*Vesuvio*)

The **Pontine Marshes** (*Agro Pontino*) are bounded by low sand hills which prevent natural drainage.

The **Apennines** (*Appennino*) are the source of most of Italy's rivers. They run 823 miles (1324 km) down the length of the peninsula.

The Strait of Messina (*Stretto di Messina*) is between 2 and 12 miles (3–19 km) wide, and is a rich fishing ground.

Sicily is the largest island in the Mediterranean at 9926 sq miles (25,708 sq km).

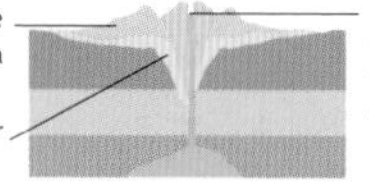

The **southwestern tip** of Sicily lies 95 miles (152 km) from the north African mainland and is part of the same geological region.

Sardinia is the second largest island in the Mediterranean Sea. The highest point is Punta La Marmora at 6017 ft (1834 m).

The **Po Valley** once formed part of the Adriatic Sea. Sediments of gravel, sand, and clay washed down from the Alps gradually filling the bay and forming a broad, cultivable plain.

Present-day crater has developed within the old crater of Monte Somma

▲ *There have been four volcanoes on the site of Vesuvius since volcanic activity began here more than 10,000 years ago.*

Vesuvius (*Vesuvio*)

Monte Somma

Old crater

Using the land

Italy produces 95% of its own food. The best farming land is in the Po Valley in northern Italy, where soft wheat and rice are grown. Irrigation is essential to agriculture in much of the south. Italy is a major producer and exporter of citrus fruits, olives, tomatoes, and wine.

The urban/rural population divide

urban 67%   rural 33%

Population density
506 people per sq mile
(195 people per sq km)

Total land area
116,320 sq miles
(301,270 sq km)

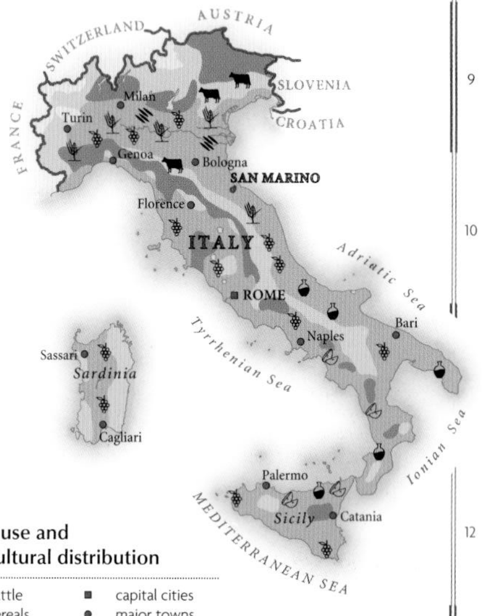

Land use and agricultural distribution

- ● capital cities
- ● major towns
- pasture
- cropland
- forest

- 🐄 cattle
- 🌾 cereals
- citrus fruits
- olive oil
- rice
- vineyards

Scale 1:2,750,000

projection: Lambert Conformal Conic

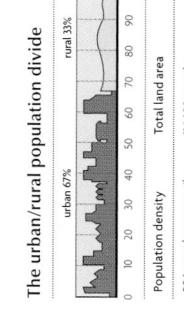
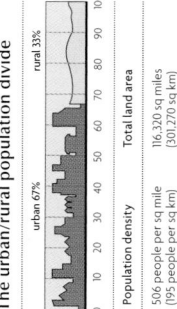
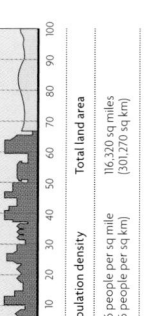

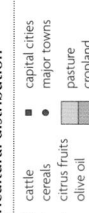
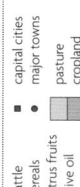
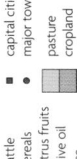

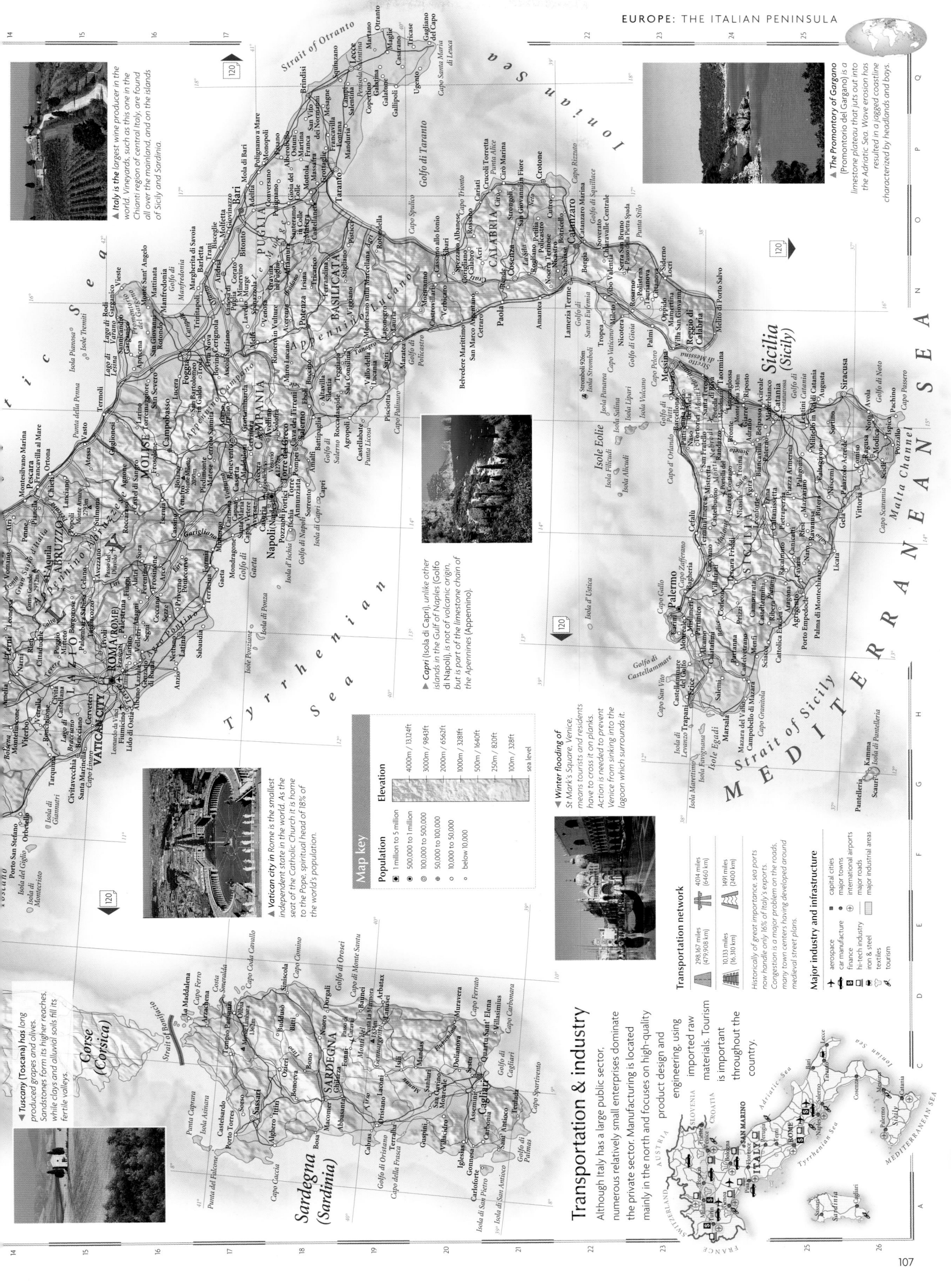

▲ *Italy is the largest wine producer in the world. Vineyards, such as this one in the Chianti region of central Italy, are found all over the mainland, and on the islands of Sicily and Sardinia.*

▲ *The Promontorio del Gargano (Promontorio del Gargano) is a limestone plateau that juts out into the Adriatic Sea. Wave erosion has resulted in a jagged coastline characterized by headlands and bays.*

▲ *Capri (Isola di Capri), unlike other islands in the Gulf of Naples (Golfo di Napoli), is not of volcanic origin, but is part of the limestone chain of the Apennines (Appennino).*

▼ *Winter flooding in St Mark's Square, Venice, means tourists and residents have to cross it on planks. Action is needed to prevent Venice from sinking into the lagoon which surrounds it.*

▲ *Vatican city in Rome is the smallest independent state in the world. As the seat of the Catholic Church it is home to the Pope, spiritual head of 18% of the world's population.*

▲ *Tuscany (Toscana) has long produced grapes and olives. Sandstones form its higher reaches, while clays and alluvial soils fill its fertile valleys.*

Map key

Population

- ◉ 1 million to 5 million
- ◎ 500,000 to 1 million
- ⊕ 100,000 to 500,000
- ⊕ 50,000 to 100,000
- ○ 10,000 to 50,000
- ○ below 10,000

Elevation

4000m / 13,124ft
3000m / 9843ft
2000m / 6562ft
1000m / 3281ft
500m / 1640ft
250m / 820ft
100m / 328ft
sea level

Transportation network

- 298,167 miles (479,908 km)
- 4014 miles (6460 km)
- 10,133 miles (16,310 km)
- 1491 miles (2400 km)

Historically of great importance, sea ports now handle only 16% of Italy's exports. Congestion is a major problem on the roads, many town centers having developed around medieval street plans.

Major industry and infrastructure

- aerospace
- car manufacture
- finance
- hi-tech industry
- iron & steel
- textiles
- tourism
- capital cities
- major cities
- major towns
- international airports
- major roads
- major industrial areas

Transportation & industry

Although Italy has a large public sector, numerous relatively small enterprises dominate the private sector. Manufacturing is located mainly in the north and focuses on high-quality product design and engineering, using imported raw materials. Tourism is important throughout the country.

The Alpine states

AUSTRIA, LIECHTENSTEIN, SLOVENIA, SWITZERLAND

The Alpine countries of Austria, Switzerland, Liechtenstein, and Slovenia form a narrow strip across western Europe's geographical core, lying on the main north–south trading routes across the Alps. Switzerland, politically neutral since 1815, is an important international meeting place and houses one of the headquarters of the United Nations, it only became a member in 2002. Austria, once at the heart of the great Habsburg Empire has been a fully independent nation since 1955, and maintains a deserved reputation as an international center of culture. Slovenia declared independence from the former Yugoslavia in 1991 and despite initial economic hardship, is now starting to achieve the prosperity enjoyed by its Alpine neighbors.

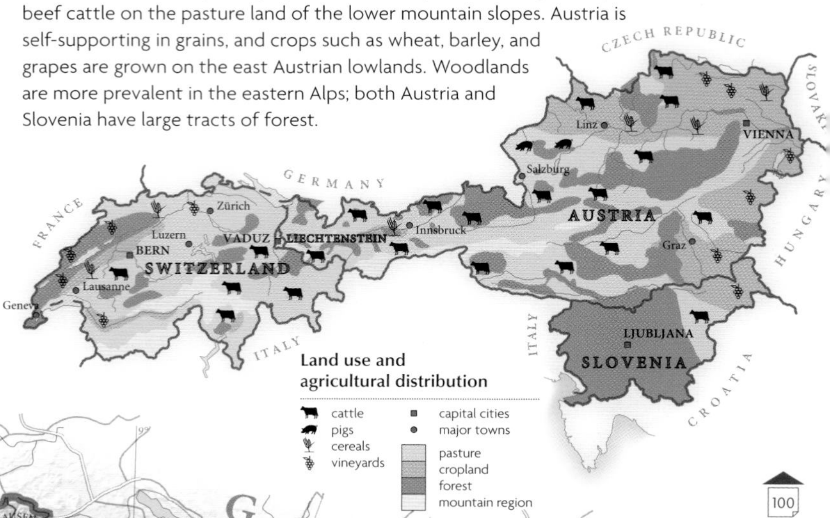

◀ **The Matterhorn**, on the Swiss-Italian border, is one of the highest mountains in the Alps, at 14,692 ft (4478 m). The term "horn" refers to its distinctive peak, formed by three glaciers eroding hollows, known as cirques, in each of its sides.

Using the land

The Alpine region's mountainous terrain discourages cultivation over much of the land area. The primary agricultural activity is the raising of dairy and beef cattle on the pasture land of the lower mountain slopes. Austria is self-supporting in grains, and crops such as wheat, barley, and grapes are grown on the east Austrian lowlands. Woodlands are more prevalent in the eastern Alps; both Austria and Slovenia have large tracts of forest.

Land use and agricultural distribution

- cattle
- pigs
- cereals
- vineyards
- capital cities
- major towns
- pasture
- cropland
- forest
- mountain region

The landscape

The Alps occupy three-fifths of Switzerland, most of southern Austria and the northwest of Slovenia. They were formed by the collision of the African and Eurasian tectonic plates, which began 65 million years ago. Their complex geology is reflected in the differing heights and rock types of the various ranges. The Rhine flows along Liechtenstein's border with Switzerland, creating a broad floodplain in the north and west of Liechtenstein. In the far northeast and east are a number of lowland regions, including the Vienna Basin, Burgenland, and the plain of the Danube. Slovenia's major rivers largely flow across the lower eastern regions; in the west, the rivers flow underground through the limestone Karst region.

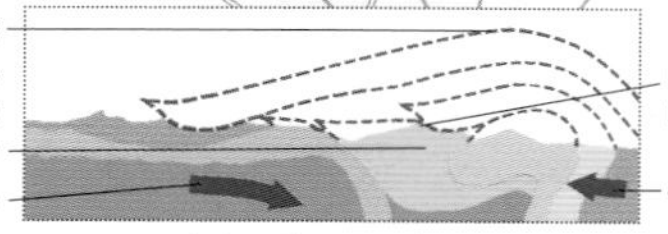

Original height after uplift and folding
Folded strata are overturned creating a *nappe*
Eurasian Plate
Present-day height of Alps
African Plate

▲ **The convergence of** the African and Eurasian plates compressed and folded huge masses of rock strata. As the plates continued to move together, the folded strata were overturned, creating complex nappes. Much of the rock strata has since been eroded, resulting in the current topography of the Alps.

▲ **Constricted as it** cuts through ridges in the Alps, the Danube meanders across the lowlands, where uplift combined with river erosion has deepened meanders.

The Vienna Basin lies mainly below 390 ft (120 m). It gradually subsided and filled with sediment as the Alps were uplifted.

Neusiedler See straddles the border of Austria and Hungary; the area around it provides some of the best wine-growing land in Austria.

The Austrian Alps comprise three distinct mountain ranges, separated by deep trenches. The northern and southern ranges are rugged limestones, while the Tauern range is formed of crystalline rocks.

The mountains of the Jura form a natural border between Switzerland and France. Their marine limestones date from over 200 million years ago. When the Alps were formed the Jura were folded into a series of parallel ridges and troughs.

Tectonic activity has resulted in dramatic changes in land height over very short distances. Lake Geneva, lying at 1221 ft (372 m) is only 43 miles (70 km) away from the 15,772 ft (4807 m) peak of Mont Blanc, on the France–Italy border.

The Bernese Alps (*Berner Alpen*) contain the Aletsch, which at 15 miles (24 km) is the longest Alpine glacier.

The Rhine, like other major Alpine rivers, follows a broad, flat trough between the mountains. Along part of its course, the Rhine forms the boundary between Switzerland and Liechtenstein.

The first road through the Brenner Pass was built in 1772, although it has been used as a mountain route since Roman times. It is the lowest of the main Alpine passes at 4298 ft (1374 m).

▶ **The deep, blue** lakes of the Karst region are part of a drainage network which runs largely underground through this limestone area.

Karst region

The limestone cave system at Postojna extends for more than 10 miles (16 km) and includes caverns reaching 125 ft (40 m) in height and width.

The Tauern range in the central Austrian Alps contains the highest mountain in Austria, the towering Grossglockner, rising 12,461 ft (3798 m).

The urban/rural population divide

urban 66%   rural 34%

0 10 20 30 40 50 60 70 80 90 100

Population density	Total land area
314 people per sq mile (121 people per sq km)	56,135 sq miles (145,390 sq km)

◄ *In this mountainous region, the flatter, more accessible areas are often used for both cattle grazing and recreation.*

◄ *These converging glaciers are marked by dark lines of moraine. This eroded material is carried by glaciers, and deposited as the ice melts.*

Scale 1:2,000,000

Km
0 5 10   20   30   40   50   60

Miles
0 5 10   20   30   40   50   60

projection: Lambert Conformal Conic

Transportation & industry

All four nations concentrate on high-quality manufacturing and services. Austrian iron and steel production is complemented by construction industries; and Slovenia, traditionally the industrial powerhouse of the western Balkans has increasingly diversified industries. Liechtenstein and Switzerland, lacking raw materials, produce pharmaceuticals and precision instruments, such as watches, and act as international banking centers. The spectacular scenery of the region encourages tourism all year round.

Transportation network

181,107 miles (291,497 km)	2116 miles (3405 km)
6368 miles (10,249 km)	993 miles (1598 km)

Tunnels and passes through the Alps are an important feature of this region. The NEAT project, providing two new high-speed rail links between Basel and Milan, was given approval in 1992.

► *The Austrian Tirol contains some of the most spectacular Alpine scenery. Snow cover is a permanent feature in the highest reaches.*

Map key

Population
◉ 1 million to 5 million
◎ 500,000 to 1 million
⊚ 100,000 to 500,000
⊕ 50,000 to 100,000
⊙ 10,000 to 50,000
∘ below 10,000

Elevation
4000m / 13,124ft
3000m / 9843ft
2000m / 6562ft
1000m / 3281ft
500m / 1640ft
250m / 820ft
100m / 328ft
sea level

Major industry and infrastructure
- car manufacture
- chemicals
- engineering
- finance
- food processing
- iron & steel
- pharmaceuticals
- textiles
- tourism
- watch making
- winter sports

◉ capital cities
• major towns
✈ international airports
— major roads
▨ major industrial areas

▲ *The Schönbrunn Palace in Vienna was the summer residence of the Habsburg monarchy. Today, it is a major tourist attraction.*

Central Europe

CZECH REPUBLIC, HUNGARY, POLAND, SLOVAKIA

When Slovakia and the Czech Republic became separate countries in 1993, they joined Hungary and Poland in a new role as independent nation states, following centuries of shifting boundaries and imperial strife. This turbulent history bequeathed the region a rich cultural heritage, shared through the works of its many great writers and composers, and celebrated in the vibrant historic capitals of Prague, Budapest, and Warsaw. Having shaken off years of Soviet domination in 1989, these states are confronting the challenge of winning commercial investment to modernize outmoded industries as they integrate their economies with those of the European Union.

The landscape

The forested Carpathian Mountains, uplifted with the Alps, lie southeast of the older Bohemian Massif, which contains the Sudeten and Krušné Hory (Erzgebirge) ranges. They divide the fertile plains of the Danube to the south and the Vistula (Wisła), which flows north across vast expanses of glacial deposits into the Baltic Sea.

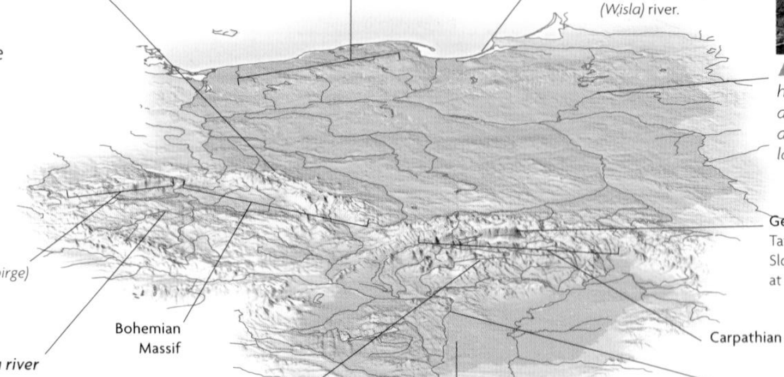

Transportation & industry

Heavy industry has dominated postwar life in Central Europe. Poland has large coal reserves, having inherited the Silesian coalfield from Germany after the Second World War, allowing the export of large quantities of coal, along with other minerals. Hungary specializes in consumer goods and services, while Slovakia's industrial base is still relatively small. The Czech Republic's traditional glassworks and breweries bring some stability to its precarious Soviet-built manufacturing sector.

Longshore currents moving east along the Baltic coast have built a 40 mile (65 km) spit composed of material from the Vistula (Wisła) river.

Pomerania is a sandy coastal region of glacially-formed lakes stretching west from the Vistula (Wisła).

Hot mineral springs occur where geothermally heated water wells up through faults and fractures in the rocks of the Sudeten Mountains.

▲ The Biebrza river has left meanders and oxbow lakes as it flows across low-lying ground

Gerlachovský Štít, in the Tatra Mountains, is Slovakia's highest mountain, at 8711ft (2655 m).

Carpathian Mountains

Danube river

The Great Hungarian Plain formed by the floodplain of the Danube is a mixture of steppe and cultivated land, covering nearly half of Hungary's total area.

The Slovak Ore Mountains (Slovenské Rudohorie) are noted for their mineral resources, including high-grade iron ore.

Bohemian Massif

Krušné Hory (Erzgebirge)

▼ The Berounka river cuts through the precipitous wooded landscape of the Bohemian Massif, banked by a broad floodplain.

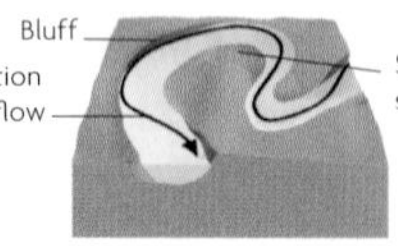

▲ Meanders form as rivers flow across plains at a low gradient. A steep cliff or bluff, forms on the outside curve, and a gentler slip-off slope on the inside bend.

Slip-off slope

Bluff

Direction of flow

Major industry and infrastructure

- car manufacture
- chemicals
- engineering
- food processing
- mining
- shipbuilding
- tourism
- capital cities
- major towns
- international airports
- major industrial areas

Transportation network

213,992 miles (344,600 km)
817 miles (315 km)
27,479 miles (44,249 km)
3784 miles (6094 km)

The huge growth of tourism and business has prompted major investment in the transportation infrastructure, with new roadbuilding schemes within and between the main cities of the region.

▲ Budapest, the capital of Hungary, straddles the Danube. It comprises the historic towns of Buda, on the west bank, and Pest, which contains the Parliament Building, seen here on the far bank.

BELARUS
LITHUANIA
RUSSIAN FEDERATION (Kaliningrad)
BALTIC SEA
Gulf of Danzig
Pomeranian Bay
POLAND
WARSZAWA / WARSAW

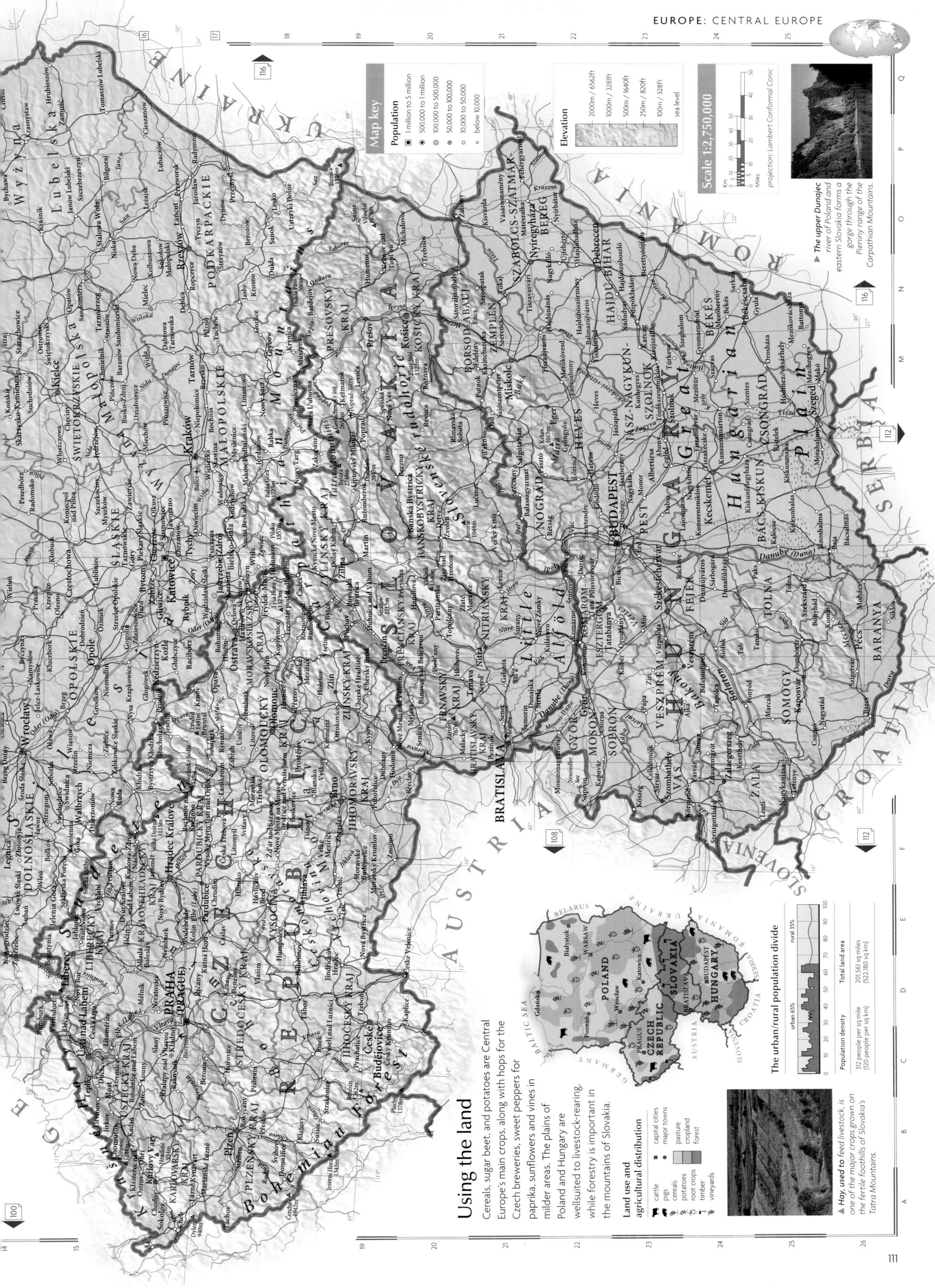

Using the land

Cereals, sugar beet, and potatoes are Central Europe's main crops, along with hops for the Czech breweries, sweet peppers for paprika, sunflowers and vines in milder areas. The plains of Poland and Hungary are well suited to livestock-rearing, while forestry is important in the mountains of Slovakia.

Land use and agricultural distribution

- cattle
- pigs
- cereals
- potatoes
- root crops
- timber
- vineyards
- capital cities
- major towns
- pasture
- cropland
- forest

▲ Hay, used to feed livestock, is one of the major crops grown on the fertile foothills of Slovakia's Tatra Mountains.

▲ The upper Dunajec river of Poland and eastern Slovakia forms a gorge through the Pieniny range of the Carpathian Mountains.

Map key

Population
- 1 million to 5 million
- 500,000 to 1 million
- 100,000 to 500,000
- 50,000 to 100,000
- 10,000 to 50,000
- below 10,000

Elevation
- 2000m / 6562ft
- 1000m / 3281ft
- 500m / 1640ft
- 250m / 820ft
- 100m / 328ft
- sea level

Scale 1:2,750,000
projection Lambert Conformal Conic

The urban/rural population divide

- urban 65%
- rural 35%

Population density
312 people per sq mile
(120 people per sq km)

Total land area
201,561 sq miles
(522,180 sq km)

111

Southeast Europe

ALBANIA, BOSNIA & HERZEGOVINA, CROATIA, KOSOVO, MACEDONIA, MONTENEGRO, SERBIA

For 46 years the federation of Yugoslavia held together the most diverse ethnic region in Europe, along the picturesque mountain hinterland of the Dalmatian coast. Economic collapse resulted in internal tensions. In the early 1990s, civil war broke out in both Croatia and Bosnia as the ethnic populations struggled to establish their own exclusive territories. Peace was only restored by the UN after NATO launched air strikes in 1995. Montenegro voted to split from Serbia in 2006. More recently, Kosovo controversially declared independence from Serbia in 2008, although this may take some time to be fully recognized. Neighboring Albania is slowly improving its fragile economy but remains one of Europe's poorest nations.

The landscape

The Tisza, Sava, and Drava Rivers drain the broad northern lowland, meeting the Danube after it crosses the Hungarian border. In the west, the Dinaric Alps divide the Adriatic Sea from the interior. Mainland valleys and elongated islands run parallel to the steep Dalmatian (Dalmacija) coastline, following alternating bands of resistant limestone.

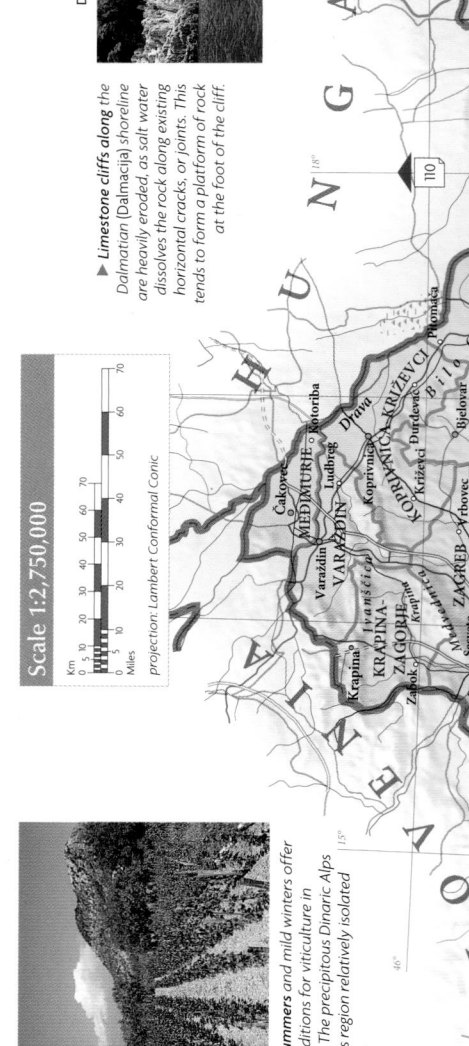

▲ Hot, dry summers and mild winters offer excellent conditions for viticulture in Montenegro. The precipitous Dinaric Alps have kept this region relatively isolated for centuries.

Scale 1:2,750,000

Km
0 10 20 30 40 50 60 70
Miles
0 10 20 30 40 50 60 70

projection: Lambert Conformal Conic

Poljes in the Kosovo region

Sheer limestone walls enclose all sides

Flat polje floor

▲ *Rain and underground water dissolve limestone along massive vertical joints (cracks). This creates poljes: depressions several miles across with steep walls and broad, flat floors.*

Underground drainage along joints in the rock

Spring at foot of cliff

At least 70% of the fresh water in the western Balkans drains eastward into the Black Sea, mostly via the Danube (Dunav).

The river floodplains of the Pannonian Basin are flanked by terraces of gravel and wind-blown glacial deposits known as loess.

Tisza river

Drava river

Sava river

The elongated islands, promontories and straits of the Dalmatian (Dalmacija) coast were formed as the Adriatic Sea rose to flood valleys running parallel to the shore.

Dalmatian (Dalmacija) coast

▶ *Limestone cliffs along the Dalmatian (Dalmacija) shoreline are heavily eroded, as salt water dissolves the rock along existing horizontal cracks, or joints. This tends to form a platform of rock at the foot of the cliff.*

A series of river valleys breaking through the Dinaric Alps from the lowlands of western Albania, give access to the interior.

At Iron Gate (Đerdap), on the border with Romania, the Danube narrows and cuts through foothills of the Balkan and Carpathian mountains, forming the deepest gorge in Europe.

A major earthquake at Skopje, Macedonia, in 1963 killed 1000 people. The whole region lies on an active crustal plate margin.

Lake Ohrid

▲ *Lake Ohrid borders Albania and Macedonia. Ohrid is the deepest lake in the western Balkans, reaching depths of 938 ft (286 m).*

Transportation & industry

Processing industries based on the region's wealth of mineral reserves predominate in Albania and Macedonia. In other regions, industrial plants have been commandeered, if not destroyed in the war and mineral extraction has severely declined. The fast-flowing rivers found throughout the Dinaric Alps are exploited to generate hydroelectric power.

Major industry and infrastructure

- aluminum refining
- car manufacture
- chemicals
- engineering
- food processing
- hydroelectric power
- mining
- shipbuilding
- textiles
- timber processing
- capital cities
- major towns
- international airports
- major roads

Transportation network

46,996 miles (75,642 km)	685 miles (1103 km)
5413 miles (8713 km)	879 miles (1415 km)

The war has resulted in the destruction or disintegration of infrastructure for transportation, communications, and power supply, though this is now in the process of recovery.

▲ *Industrial processing plants were established throughout Albania by the Hoxha regime, which collapsed in 1992. They remain incongruous among the villages of one of Europe's most conservative rural societies.*

Using the land

Crops of wheat, maize, sugar beet, vegetables, and fruit are widely grown. The hilly terrain is suited to forestry and livestock farming. The mild, Mediterranean climate of the coastal regions provides ideal conditions for growing vines and olives. Albania's largely agricultural economy has been adversely affected by the recent dismantling of state farms.

▼ *Sweet red peppers are dried in the sun, ready to make paprika. Macedonia's economy is mainly agricultural and its fertile soils support a broad range of crops.*

Land use and agricultural distribution

- pigs
- sheep
- cereals
- fruit
- olives
- sugar beet
- timber
- tobacco
- vineyards
- capital cities
- major towns
- pasture
- cropland
- forest
- mountain region

The urban/rural population divide

urban 51% / rural 49%

Population density	Total land area
240 people per sq mile (93 people per sq km)	95,038 sq miles (246,278 sq km)

▲ *The historic center of Mostar in southern Bosnia, with its famous 16th-century Turkish bridge, was destroyed by shelling during 1993. The bridge was rebuilt and opened again in 2004.*

▲ *The ancient Croatian port of Dubrovnik was one of the former Yugoslavia's most popular tourist resorts and an important point of access to the sea along the Dalmatian (Dalmacija) coast. Shelling of the old city by Serb forces in 1991 provoked international condemnation.*

▲ *The Tara river is one of Montenegro's major rivers. It flows into the Danube via the Drina and Sava rivers. Along its course the Tara has eroded spectacular gorges up to 3280 ft (1000 m) deep.*

In February 2008, Kosovo (a UN Protectorate within Serbia since 1999) declared independence. Although now recognized by numerous countries, this decision has proved controversial with other states wary of setting a precedent for separatist groups within their own borders. It is therefore likely to be some time before Kosovo becomes universally recognized.

Map key

Population
- 1 million to 5 million
- 500,000 to 1 million
- 100,000 to 500,000
- 50,000 to 100,000
- 10,000 to 50,000
- below 10,000

Elevation
- 2000m / 6562ft
- 1000m / 3281ft
- 500m / 1640ft
- 250m / 820ft
- 100m / 328ft
- sea level

Bulgaria & Greece

Including EUROPEAN TURKEY

Greece is renowned as the original hearth of western civilization. The rugged terrain and numerous islands have profoundly affected its development, creating a strong agricultural and maritime tradition.

In the past 50 years, this formerly rural society has rapidly urbanized, with one third of the population now living in the capital, Athens, and in the northern city of Salonica. Bulgaria, dominated for centuries by the Ottoman Turks, became part of the eastern bloc after the Second World War, only slowly emerging from Soviet influence in 1989. Moves toward democracy led to some instability in Bulgaria and Greece, now outweighed by the challenge of integration with the European Union.

The landscape

Bulgaria's Balkan mountains divide the Danubian Plain (*Dunavska Ravnina*) and Maritsa Basin, meeting the Black Sea in the east along sandy beaches. The steep Rhodope Mountains form a natural barrier with Greece, while the younger Pindus form a rugged central spine which descends into the Aegean Sea to give a vast archipelago of over 2000 islands, the largest of which is Crete.

▲ *The Arda river cuts through the Rhodope Mountains in rugged, rocky gorges.*

The Danube, Europe's second longest river, forms most of Bulgaria's northern border. The Danubian plain (*Dunavska Ravnina*), extending from the southern bank, is extremely fertile.

The islands of Crete, Kythira, Karpathos, and Rhodes are part of an arc which bends southeastward from the Peloponnese, forming the southern boundary of the Aegean.

▲ *Layers of black volcanic ash still cover the island of Santorini. This volcano last erupted 3500 years ago, but still shows signs of volcanic activity.*

Mount Olympus is the mythical home of the Greek Gods and, at 9570 ft (2917 m), is the highest mountain in Greece.

The Peloponnese consist of several mountainous peninsulas, linked to the mainland by the Isthmus of Corinth. The Corinth Canal (*Dioryga Korinthou*), built in 1893, cuts through the isthmus, linking the Aegean and Ionian Seas.

Ancient metamorphic rock, formed miles below the surface

Limestone rocks exposed by erosion of metamorphic rocks

Younger limestones created in shallow seas

▲ *Mount Olympus is a composite of rocks formed by two major tectonic events. First the older metamorphic rocks were thrust over the limestones, then two million years ago regional warping and subsequent erosion, reexposed the limestone.*

Transportation & industry

Soviet investment introduced heavy industry into Bulgaria, and the processing of agricultural produce, such as tobacco, is important throughout the country. Both countries have substantial shipyards and Greece has one of the world's largest merchant fleets. Many small craft workshops, producing textiles and processed foods, are clustered around Greek cities. The service and construction sectors have profited from the successful tourist industry.

Major industry and infrastructure
- chemicals
- engineering
- food processing
- shipbuilding
- textiles
- tourism
- capital cities
- major towns
- international airports
- major roads
- major industrial areas

Transportation network
- 103,930 miles (167,630 km)
- 345 miles (557 km)
- 4346 miles (6995 km)
- 294 miles (474 km)

Bulgaria's railroads require investment to revive an outdated infrastructure. In Greece, despite a developing road network, ferry-boats remain the most effective form of transportation in many areas.

Scale 1:2,750,000

projection Lambert Conformal Conic

▲ *A towering pinnacle at Meteora in central Greece is home to the monastery of Roussanou. The 24 rock towers which dominate the plain of Thessaly are remnants of an old plateau. Long-term weathering along fissures in the rock have worn away the rest of the plateau.*

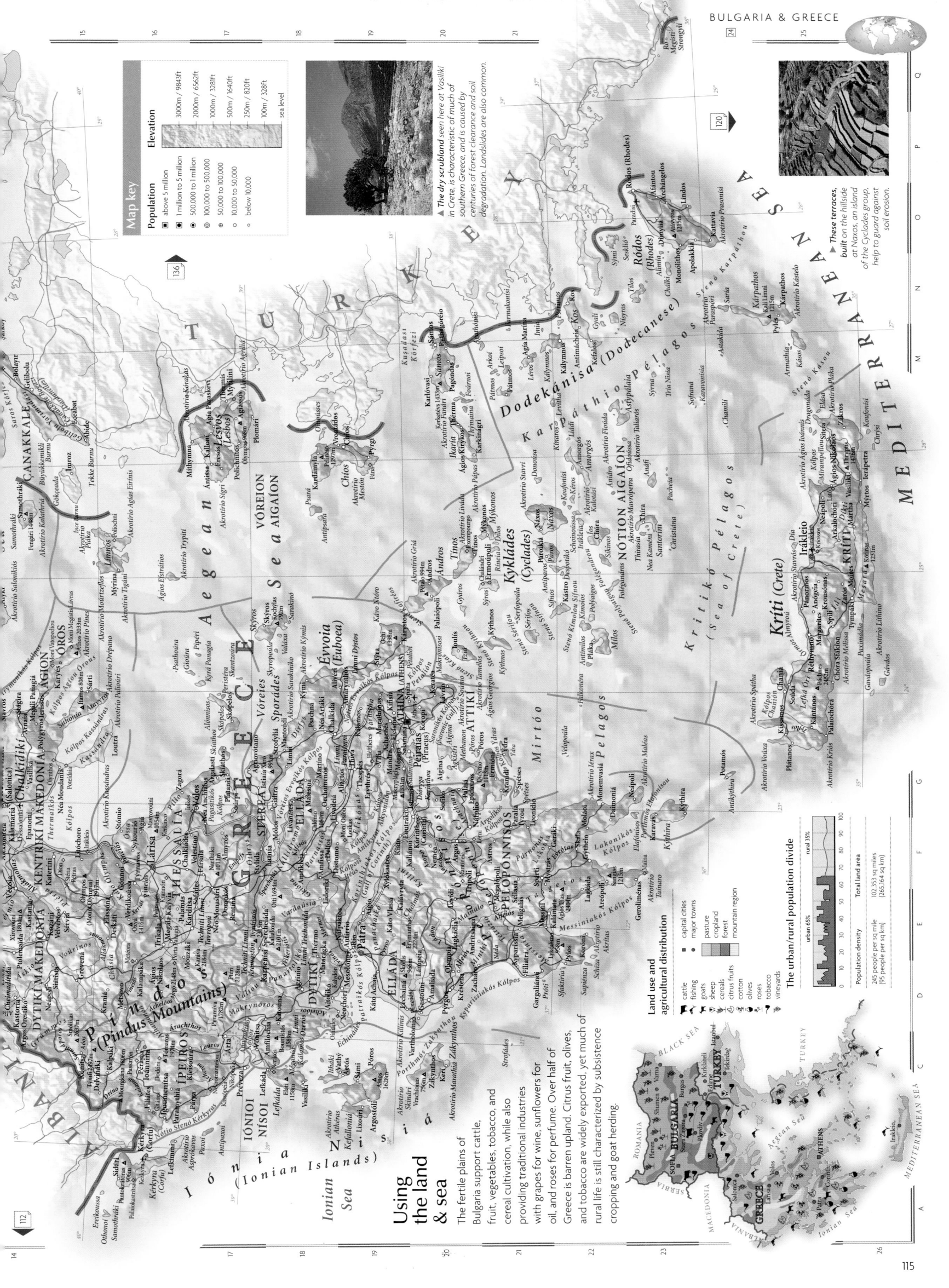

Map key

Population
- ■ above 5 million
- ■ 1 million to 5 million
- ◉ 500,000 to 1 million
- ◎ 100,000 to 500,000
- ⊕ 50,000 to 100,000
- ○ 10,000 to 50,000
- ○ below 10,000

Elevation
- 3000m / 9843ft
- 2000m / 6562ft
- 1000m / 3281ft
- 500m / 1640ft
- 250m / 820ft
- 100m / 328ft
- sea level

▲ The dry scrubland seen here at Vasilikí in Crete, is characteristic of much of southern Greece, and is caused by centuries of forest clearance and soil degradation. Landslides are also common.

120

▲ These terraces, built on the hillside at Naxos, an island of the Cyclades group, help to guard against soil erosion.

MEDITERRANEAN SEA

Using the land & sea

The fertile plains of Bulgaria support cattle, fruit, vegetables, tobacco, and cereal cultivation, while also providing traditional industries with grapes for wine, sunflowers for oil, and roses for perfume. Citrus fruit, olives, and tobacco are widely exported, yet much of Greece is barren upland. Citrus fruit, olives, and roses for perfume. Over half of Greece is barren upland. Citrus fruit, olives, and tobacco are widely exported, yet much of rural life is still characterized by subsistence cropping and goat herding.

Land use and agricultural distribution
- cattle
- fishing
- goats
- sheep
- cereals
- citrus fruits
- cotton
- olives
- roses
- tobacco
- vineyards

- ■ capital cities
- ● major towns
- pasture
- cropland
- forest
- mountain region

The urban/rural population divide

urban 65%   rural 35%

Population density	Total land area
245 people per sq mile (95 people per sq km)	102,353 sq miles (265,164 sq km)

GREECE

PÍNDOS (Pindus Mountains)

Aegean Sea

Kykládes (Cyclades)

Dodekánisa (Dodecanese)

Kárpáthio Pélagos

Kríti (Crete)

Kritikó Pélagos (Sea of Crete)

NÓTION AIGAÍON

VÓREION AIGAÍON

TURKEY

ALBANÍA

Ionian Sea (Ionian Islands)

MEDITERRANEAN SEA

BLACK SEA

ROMANIA

BULGARIA

SERBIA

MACEDONIA

TURKEY

GREECE

Aegean Sea

ATHENS

SOFIA

115

Romania, Moldova & Ukraine

The industrial, social, and cultural make-up of Romania and the former Soviet states of Moldova and Ukraine still bear the imprint of their communist past. As part of the USSR, Ukraine was a leading agricultural, industrial, and energy producer. These industries, like those in Moldova and Romania, are now being reoriented more firmly toward western markets. As a result of shifting borders, and Soviet policy actively encouraging Russian immigration into other Soviet states like Ukraine and Moldova, all three countries now contain large numbers of foreign nationals. In 2014, the Russian Federation drew international condemnation by annexing the Ukrainian territory of Crimea.

Using the land

The fertile black soils of Ukraine, often called "the breadbasket of Europe," have enabled the cultivation of a variety of cereals and vegetables, which are widely exported. Romania and Moldova also grow cereals, sunflowers, and vegetables, and are noted for the quality of their wines.

◄ *The fertile lands and tolerant climate of Moldova are ideally suited to growing grapes for wine.*

Land use and agricultural distribution

- cattle
- pigs
- poultry
- sheep
- cereals
- cotton
- sugar beet
- sunflowers
- vineyards
- capital cities
- major towns

- pasture
- cropland
- forest
- wetland

The urban/rural population divide

urban 65%   rural 35%

0 10 20 30 40 50 60 70 80 90 100

Population density	Total land area
222 people per sq mile (86 people per sq km)	334,947 sq miles (867,740 sq km)

◄ *Glacial lakes are found throughout the Transylvanian Alps (Carpatii Meridionali), although the mountains no longer have any permanent snow cover.*

Transportation & industry

Heavy industry using local raw materials characterizes much of this region. The industrial heartland of Ukraine, specializing in metal and machine-building industries, is based around its vast mineral reserves in the Donbass region. In Moldova, food processing draws on produce from its agricultural sector. Romanian industry relies both on local raw materials and imported iron, steel, and oil.

Major industry and infrastructure

- car manufacture
- chemicals
- coal
- engineering
- food processing
- mining
- oil & gas
- textiles
- tourism

- capital cities
- major towns
- international airports
- major roads
- major industrial areas

Transportation network

170,707 miles (274,757 km)	1170 miles (1883 km)
21,474 miles (34,563 km)	4130 miles (6647 km)

Increased industrialization has necessitated the upgrading of road and rail networks in all three countries. Modernization has tended to focus only on major cities and industrial areas.

▶ *During the 1960s and 1970s, many industries, like this carbon factory, developed using the mineral resources on the flanks of the Transylvanian Alps (Carpatii Meridionali).*

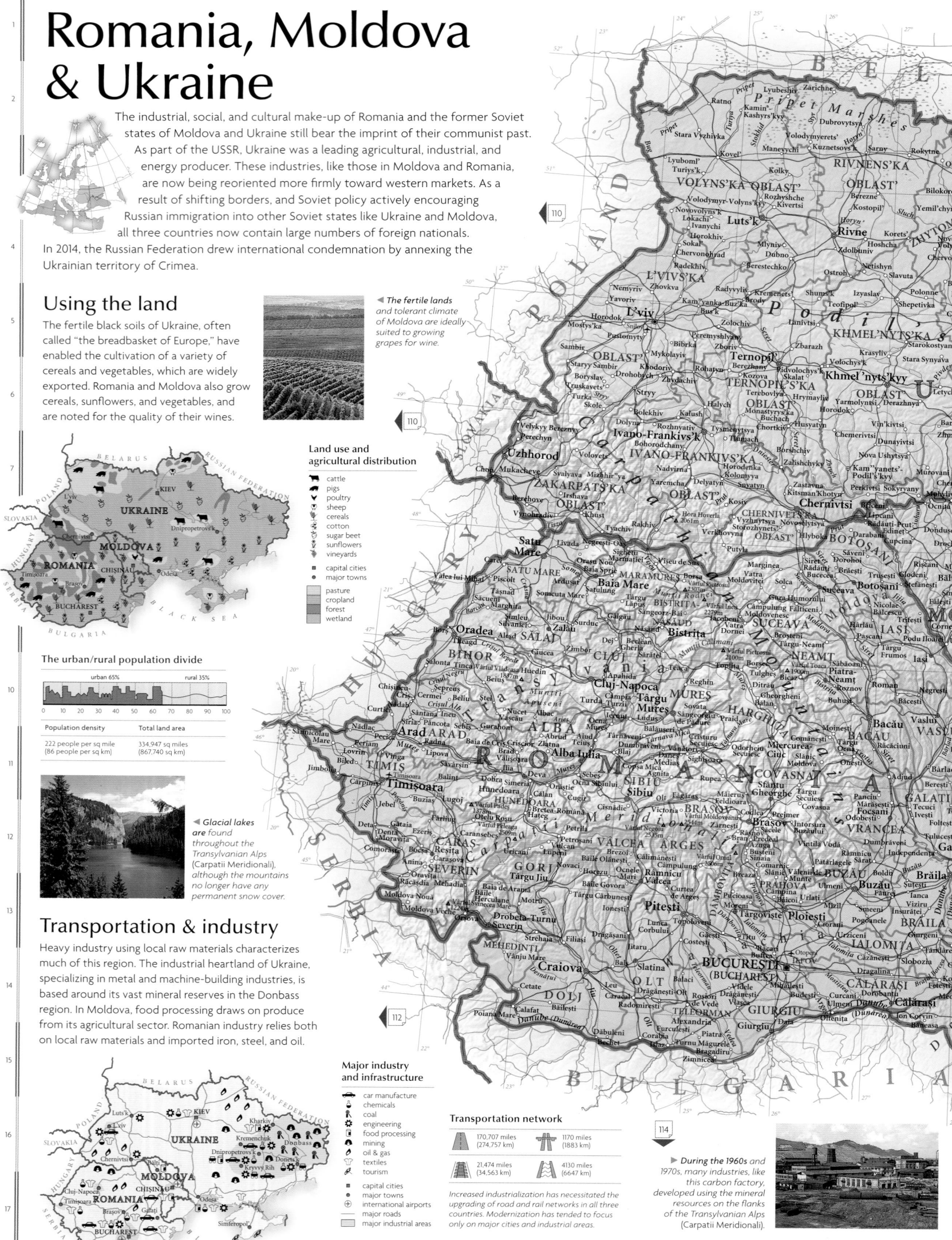

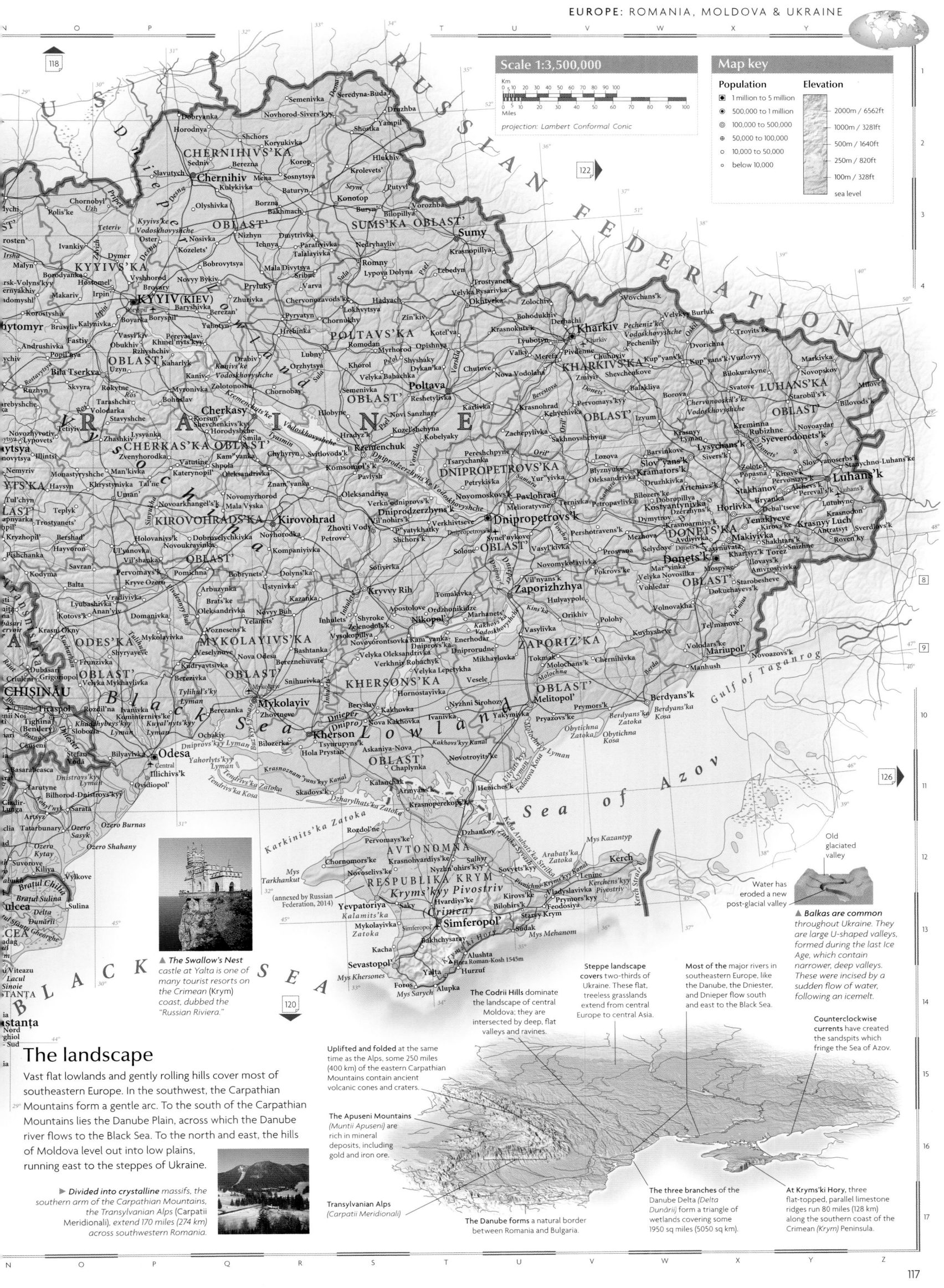

Scale 1:3,500,000

projection: Lambert Conformal Conic

Map key

Population
- ◼ 1 million to 5 million
- ◉ 500,000 to 1 million
- ⊚ 100,000 to 500,000
- ⊕ 50,000 to 100,000
- ○ 10,000 to 50,000
- ○ below 10,000

Elevation
- 2000m / 6562ft
- 1000m / 3281ft
- 500m / 1640ft
- 250m / 820ft
- 100m / 328ft
- sea level

▲ The Swallow's Nest castle at Yalta is one of many tourist resorts on the Crimean (Krym) coast, dubbed the "Russian Riviera."

Old glaciated valley

Water has eroded a new post-glacial valley

▲ Balkas are common throughout Ukraine. They are large U-shaped valleys, formed during the last Ice Age, which contain narrower, deep valleys. These were incised by a sudden flow of water, following an icemelt.

Counterclockwise currents have created the sandspits which fringe the Sea of Azov.

The landscape

Vast flat lowlands and gently rolling hills cover most of southeastern Europe. In the southwest, the Carpathian Mountains form a gentle arc. To the south of the Carpathian Mountains lies the Danube Plain, across which the Danube river flows to the Black Sea. To the north and east, the hills of Moldova level out into low plains, running east to the steppes of Ukraine.

▶ Divided into crystalline massifs, the southern arm of the Carpathian Mountains, the Transylvanian Alps (Carpatii Meridionali), extend 170 miles (274 km) across southwestern Romania.

The Codrii Hills dominate the landscape of central Moldova; they are intersected by deep, flat valleys and ravines.

Steppe landscape covers two-thirds of Ukraine. These flat, treeless grasslands extend from central Europe to central Asia.

Most of the major rivers in southeastern Europe, like the Danube, the Dniester, and Dnieper flow south and east to the Black Sea.

Uplifted and folded at the same time as the Alps, some 250 miles (400 km) of the eastern Carpathian Mountains contain ancient volcanic cones and craters.

The Apuseni Mountains (Muntii Apuseni) are rich in mineral deposits, including gold and iron ore.

Transylvanian Alps (Carpatii Meridionali)

The Danube forms a natural border between Romania and Bulgaria.

The three branches of the Danube Delta (Delta Dunării) form a triangle of wetlands covering some 1950 sq miles (5050 sq km).

At Kryms'ki Hory, three flat-topped, parallel limestone ridges run 80 miles (128 km) along the southern coast of the Crimean (Krym) Peninsula.

The Baltic states & Belarus

BELARUS, ESTONIA, LATVIA, LITHUANIA, Kaliningrad

Occupying Europe's main corridor to Russia, the four distinct cultures of Estonia, Latvia, Lithuania, and Belarus share a history of struggle for nationhood against the interests of more powerful neighbors. As the first republics to declare their independence from the Soviet Union in 1990–91, the Baltic states of Estonia, Latvia, and Lithuania sought an economic role in the EU, while reaffirming their European cultural roots through the church and a strong musical tradition. Meanwhile, Belarus has shown economic and political allegiance to Russia by joining the Commonwealth of Independent States.

▲ The seaport of Riga is Latvia's capital and the center of economic and cultural life. With a 32% Russian minority in Latvia, language and the right to national citizenship are key issues.

Using the land

Across the four nations cattle and pig farming are widespread, together with diverse arable crops, including flax for making linen, potatoes used to produce vodka, cereals, and other vegetables. Almost a third of the land is forested; demand for timber has increased the importance of forest management.

Land use and agricultural distribution

- cattle
- pigs
- cereals
- flax
- potatoes
- timber
- capital cities
- major towns
- pasture
- cropland
- forest
- wetland

The urban/rural population divide

urban 69%   rural 31%

Population density
122 people per sq mile
(47 people per sq km)

Total land area
145,006 sq miles
(375,656 sq km)

▲ A pine forest in northern Belarus. Conifers in the north give way to hardwood forest farther south. Timber milling is supplied with logs floated along the country's many navigable waterways.

▲ The Western Dvina river provides hydroelectric power and, during the summer months, access to the Baltic Sea. The lower course of the river freezes from December to April.

Map key

Population
- 1 million to 5 million
- 500,000 to 1 million
- 100,000 to 500,000
- 50,000 to 100,000
- 10,000 to 50,000
- below 10,000

Elevation
- 250m / 820ft
- 100m / 328ft
- sea level

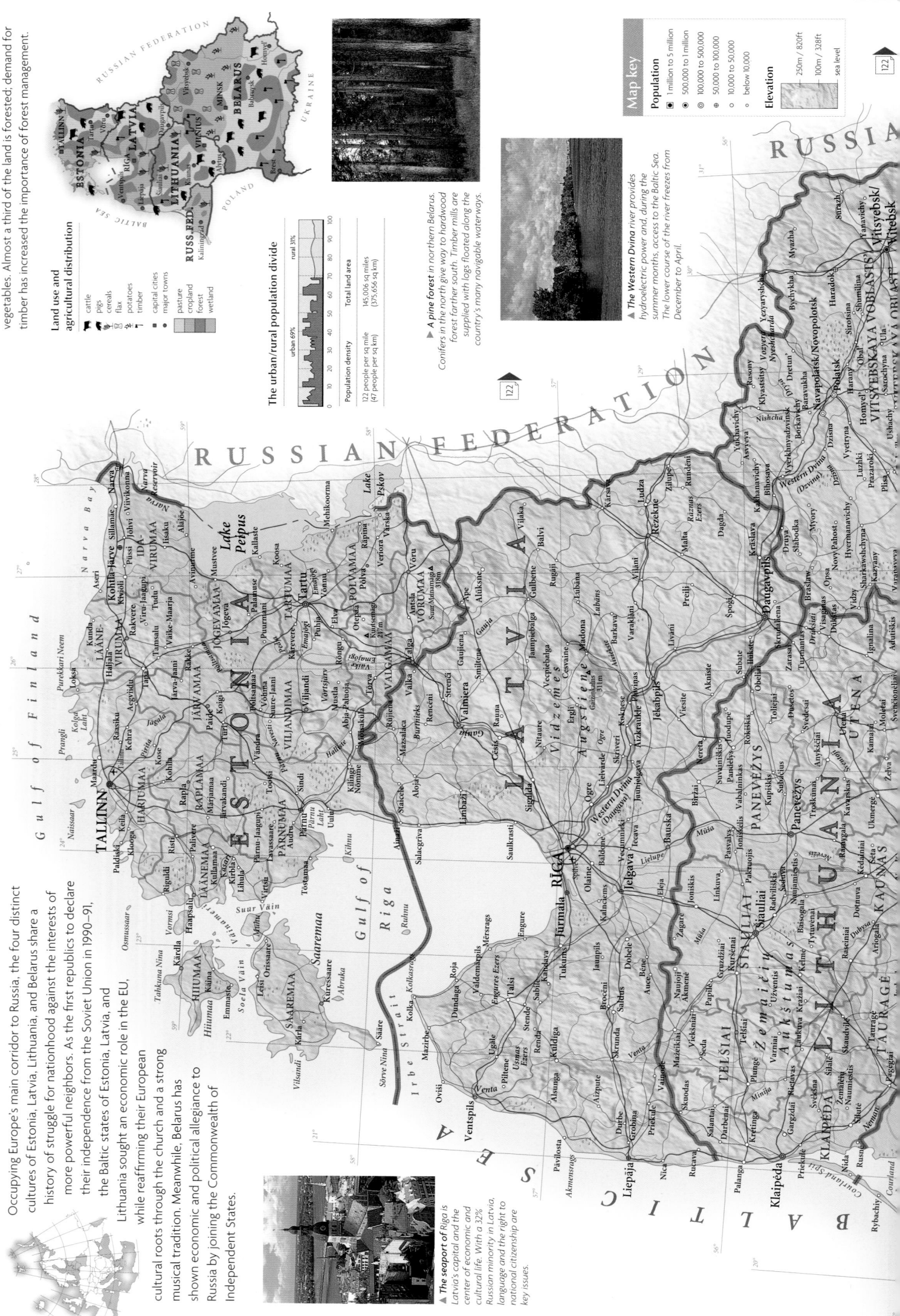

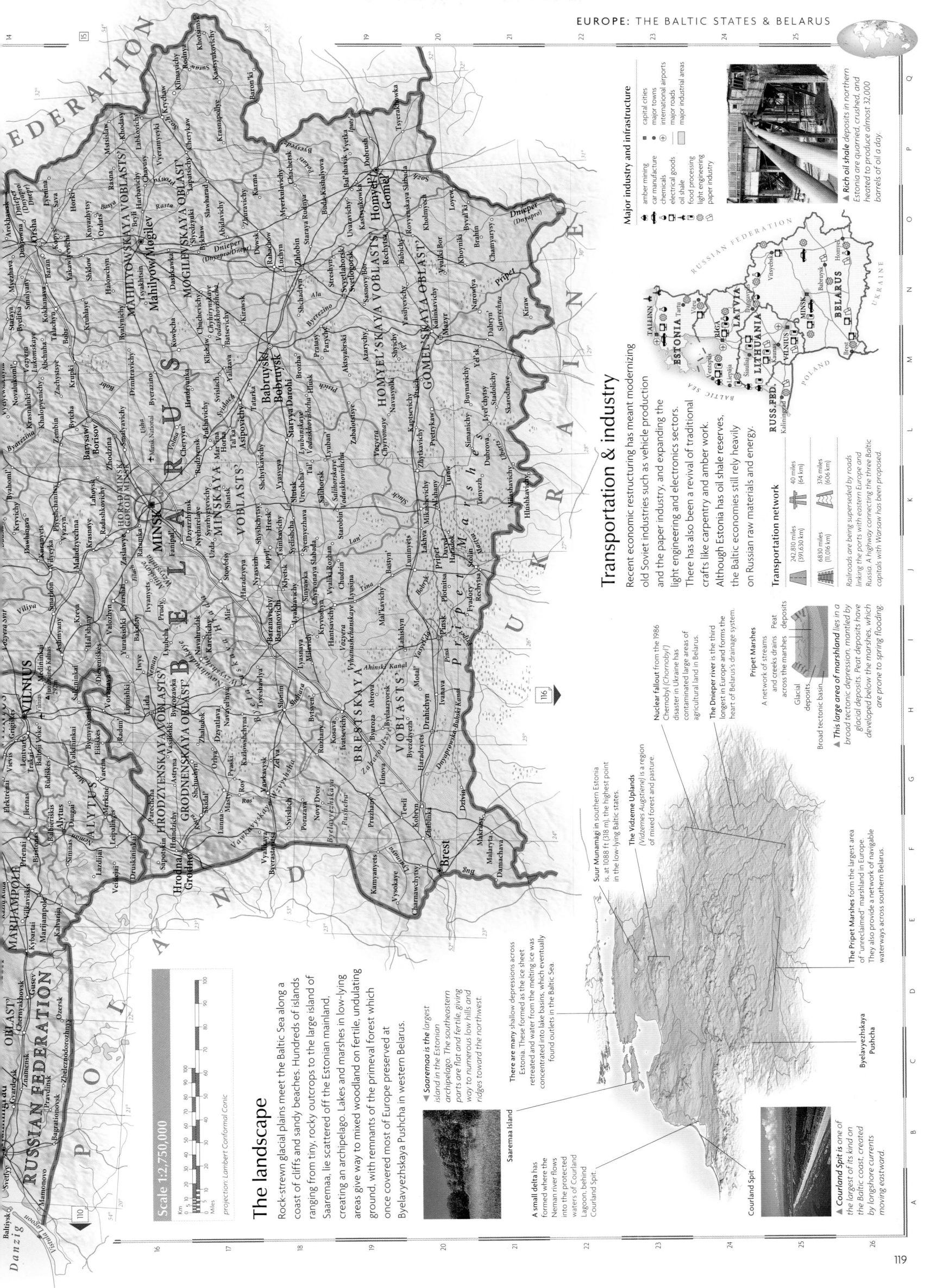

The landscape

Rock-strewn glacial plains meet the Baltic Sea along a coast of cliffs and sandy beaches. Hundreds of islands ranging from tiny, rocky outcrops to the large island of Saaremaa, lie scattered off the Estonian mainland, creating an archipelago. Lakes and marshes in low-lying areas give way to mixed woodland on fertile, undulating ground, with remnants of the primeval forest which once covered most of Europe preserved at Byelavyezhskaya Pushcha in western Belarus.

Saaremaa is the largest island in the Estonian archipelago. The southeastern parts are flat and fertile, giving way to numerous low hills and ridges toward the northwest.

Saaremaa Island

A small delta has formed where the Neman river flows into the protected waters of Courland Lagoon, behind Courland Spit.

Courland Spit is one of the largest of its kind on the Baltic coast, created by longshore currents moving eastward.

Courland Spit

There are many shallow depressions across Estonia. These formed as the ice sheet retreated and water from the melting ice was concentrated into lake basins, which eventually found outlets in the Baltic Sea.

Suur Munamagi in southern Estonia is, at 1088 ft (318 m), the highest point in the low-lying Baltic states.

The Vidzeme Uplands (Vidzemes Augstiene) is a region of mixed forest and pasture.

Nuclear fallout from the 1986 Chernobyl (Chornobyl) disaster in Ukraine has contaminated large areas of agricultural land in Belarus.

The Dnieper river is the third longest in Europe and forms the heart of Belarus's drainage system.

Pripet Marshes
A network of streams and creeks drains across the marshes

Peat deposits
Glacial deposits
Broad tectonic basin.

This large area of marshland lies in a broad tectonic depression, mantled by glacial deposits. Peat deposits have developed below the marshes, which are prone to spring flooding.

The Pripet Marshes form the largest area of "unreclaimed" marshland in Europe. They also provide a network of navigable waterways across southern Belarus.

Byelavyezhskaya Pushcha

Transportation & industry

Recent economic restructuring has meant modernizing old Soviet industries such as vehicle production and the paper industry, and expanding the light engineering and electronics sectors. There has also been a revival of traditional crafts like carpentry and amber work. Although Estonia has oil shale reserves, the Baltic economies still rely heavily on Russian raw materials and energy.

Railroads are being superseded by roads linking the ports with eastern Europe and Russia. A highway connecting the three Baltic capitals with Warsaw has been proposed.

Transportation network

Major industry and infrastructure

Rich oil shale deposits in northern Estonia are quarried, crushed, and heated to produce almost 32,000 barrels of oil a day.

Scale 1:2,750,000
projection: Lambert Conformal Conic

119

A   B   C   D   E   F   G   H   I   J   K   L   M

The Mediterranean

The Mediterranean Sea stretches over 2500 miles (4000 km) east to west, separating Europe from Africa. At its westernmost point it is connected to the Atlantic Ocean through the Strait of Gibraltar. In the east, the Suez canal, opened in 1869, gives passage to the Indian Ocean. In the northeast, linked by the Sea of Marmara, lies the Black Sea. The Mediterranean is bordered by almost 30 states and territories, and more than 100 million people live on its shores and islands. Throughout history, the Mediterranean has been a focal area for many great empires and civilizations, reflected in the variety of cultures found on its shores. Since the 1960s, development along the southern coast of Europe has expanded rapidly to accommodate increasing numbers of tourists and to enable the exploitation of oil and gas reserves. This has resulted in rising levels of pollution, threatening the future of the sea.

▲ *Monte Carlo is* just one of the luxurious resorts scattered along the Riviera, which stretches along the coast from Cannes in France to La Spezia in Italy. The region's mild winters and hot summers have attracted wealthy tourists since the early 19th century.

The landscape

The Mediterranean Sea is almost totally landlocked, joined to the Atlantic Ocean through the Strait of Gibraltar, which is only 8 miles (13 km) wide. Lying on an active plate margin, sea floor movements have formed a variety of basins, troughs, and ridges. A submarine ridge running from Tunisia to the island of Sicily divides the Mediterranean into two distinct basins. The western basin is characterized by broad, smooth abyssal (or ocean) plains. In contrast, the eastern basin is dominated by a large ridge system, running east to west.

The narrow Strait of Gibraltar inhibits water exchange between the Mediterranean Sea and the Atlantic Ocean, producing a high degree of salinity and a low tidal range within the Mediterranean. The lack of tides has encouraged the build-up of pollutants in many semienclosed bays.

Main surface current

Dense currents sink below surface

Denser, more saline currents flow back to Atlantic

▲ *Because the Mediterranean* is almost enclosed by land, its circulation is quite different to the oceans. There is one major current which flows in from the Atlantic and moves east. Currents flowing back to the Atlantic are denser and flow below the main current.

Industrial pollution flowing from the Dnieper and Danube rivers has destroyed a large proportion of the fish population that used to inhabit the upper layers of the Black Sea.

The Ionian Basin is the deepest in the Mediterranean, reaching depths of 16,800 ft (5121 m).

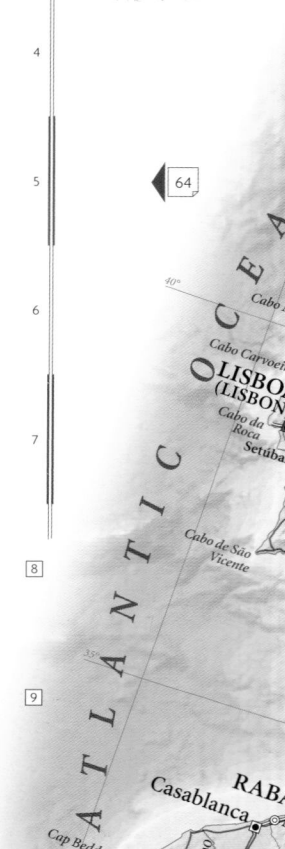

The edge of the Eurasian Plate is edged by a continental shelf. In the Mediterranean Sea this is widest at the Ebro Fan where it extends 60 miles (96 km).

◄ **The Atlas Mountains** are a range of fold mountains that lie in Morocco and Algeria. They run parallel to the Mediterranean, forming a topographical and climatic divide between the Mediterranean coast and the western Sahara.

An arc of active submarine, island and mainland volcanoes, including Etna and Vesuvius, lie in and around southern Italy. The area is also susceptible to earthquakes and landslides.

Nutrient flows into the eastern Mediterranean, and sediment flows to the Nile Delta have been severely lowered by the building of the Aswan Dam across the Nile in Eygpt. This is causing the delta to shrink.

Oxygen in the Black Sea is dissolved only in its upper layers; at depths below 230–300 ft (70–100 m) the sea is "dead" and can support no lifeforms other than specially adapted bacteria.

The Suez Canal, opened in 186_ extends 100 miles (160 km) fro_ Port Said to the Gulf of Suez.

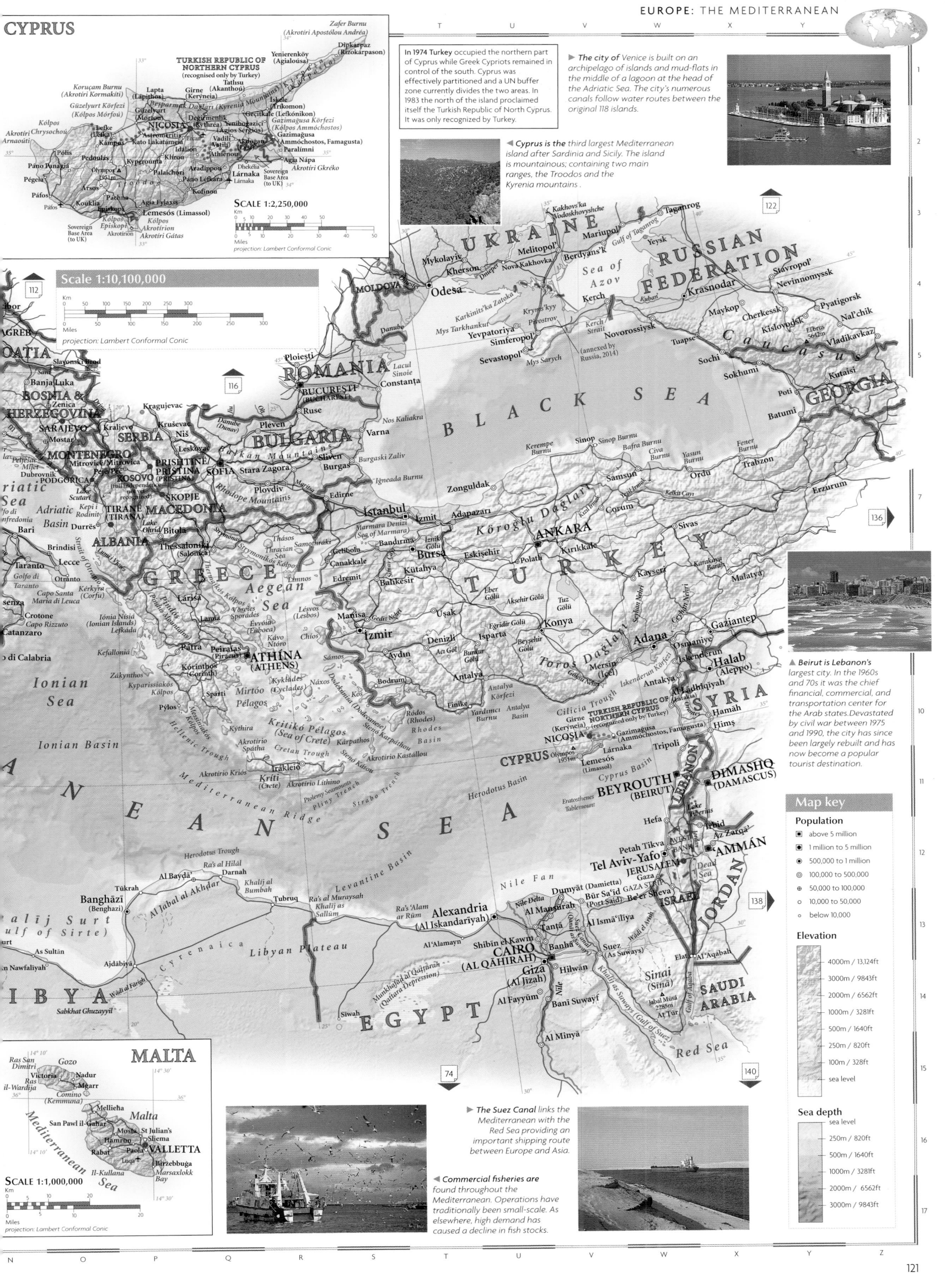

CYPRUS

TURKISH REPUBLIC OF NORTHERN CYPRUS
(recognised only by Turkey)

Zafer Burnu
(Akrotiri Apostólou Andréa)
Dipkarpaz
(Rizokarpason)
Yenierenköy
(Agialoúsa)
Koruçam Burnu
(Akrotiri Kormakíti)
Lapta
(Lápithos)
Girne
(Kerýneia)
Tatlisu
(Akanthoú)
Güzelyurt Körfezi
(Kólpos Mórfou)
Beşparmak Daĝlari (Kyrenia Mountains)
Iskele
(Trikomon)
Geçitkale (Lefkónikon)
Güzelyurt
(Mórfou)
Lefke
(Léfka)
Değirmenlik
Kythréa
Yeniboğaziçi
Agios Sergios
Kólpos Chrysochoú
Astromeritis
Kato Lakatámeia
Gazimağusa Körfezi
(Kólpos Ammóchostos)
NICOSIA
Akrotiri Arnaoúti
Pólis
Kámpos
Kyperounta
Vatili
Aredíppou
Gazimağusa
(Ammóchostos, Famagusta)
Pedoulás
Klírou
Athíenou
Agia Nápa
Pégeia
Páno Panagiá
Olýmpos ▲
1951m
Palaichóri
Pano Léfkara
Lárnaka
(Lárnaca)
Akrotiri Gkréko
Arsos
Pachna
Agia Fylaxis
Kofinou
Sovereign Base Area (to UK)
Páfos
Koúklia
Episkopí
Lemesós (Limassol)
Kólpos Akrotírion
Sovereign Base Area (to UK)
Dhekélia Sovereign Base Area (to UK)
Akrotírion
Akrotíri Gátas

SCALE 1:2,250,000

projection: Lambert Conformal Conic

In 1974 Turkey occupied the northern part of Cyprus while Greek Cypriots remained in control of the south. Cyprus was effectively partitioned and a UN buffer zone currently divides the two areas. In 1983 the north of the island proclaimed itself the Turkish Republic of North Cyprus. It was only recognized by Turkey.

► **The city of** Venice is built on an archipelago of islands and mud-flats in the middle of a lagoon at the head of the Adriatic Sea. The city's numerous canals follow water routes between the original 118 islands.

◄ **Cyprus is the** third largest Mediterranean island after Sardinia and Sicily. The island is mountainous; containing two main ranges, the Troodos and the Kyrenia mountains.

Scale 1:10,100,000

projection: Lambert Conformal Conic

▲ Beirut is Lebanon's largest city. In the 1960s and 70s it was the chief financial, commercial, and transportation center for the Arab states. Devastated by civil war between 1975 and 1990, the city has since been largely rebuilt and has now become a popular tourist destination.

Map key

Population
- above 5 million
- 1 million to 5 million
- 500,000 to 1 million
- 100,000 to 500,000
- 50,000 to 100,000
- 10,000 to 50,000
- below 10,000

Elevation
- 4000m / 13,124ft
- 3000m / 9843ft
- 2000m / 6562ft
- 1000m / 3281ft
- 500m / 1640ft
- 250m / 820ft
- 100m / 328ft
- sea level

Sea depth
- sea level
- 250m / 820ft
- 500m / 1640ft
- 1000m / 3281ft
- 2000m / 6562ft
- 3000m / 9843ft

MALTA

Ras San Dimitri
Gozo
Victoria
Nadur
Ras il-Wardija
Mgarr
Comino
(Kemmuna)
Malta
Mellieħa
San Pawl il-Baħar
Mosta
St Julian's
Sliema
Ħamrun
Paola
Rabat
Luqa
VALLETTA
Birżebbuġa
Marsaxlokk Bay
Il-Kullana
Mediterranean Sea

SCALE 1:1,000,000

projection: Lambert Conformal Conic

► **The Suez Canal** links the Mediterranean with the Red Sea providing an important shipping route between Europe and Asia.

◄ **Commercial fisheries are** found throughout the Mediterranean. Operations have traditionally been small-scale. As elsewhere, high demand has caused a decline in fish stocks.

The Russian Federation

The Cold War era of global relations was concluded in 1991 with the formal dissolution of the Soviet Union. The Russian Federation declared its separate sovereignty from the foundering communist empire following independence declarations from a number of former Soviet republics. As the leading member of the Commonwealth of Independent States, the Russian Federation has a central role in the development of post-Soviet Eurasia. Crossing 11 time zones, the Russian Federation is almost twice the size of the US, and with more than 150 ethnic minorities and 21 autonomous republics, regionalist dissent within its own territory remains a danger.

THE RUSSIAN FEDERATION: ADMINISTRATIVE REGIONS

The administrative area names in European Russia have been omitted west of the Ural Mountains. Please refer to pages 124–125 and 126–127 where these areas are shown at a larger scale.

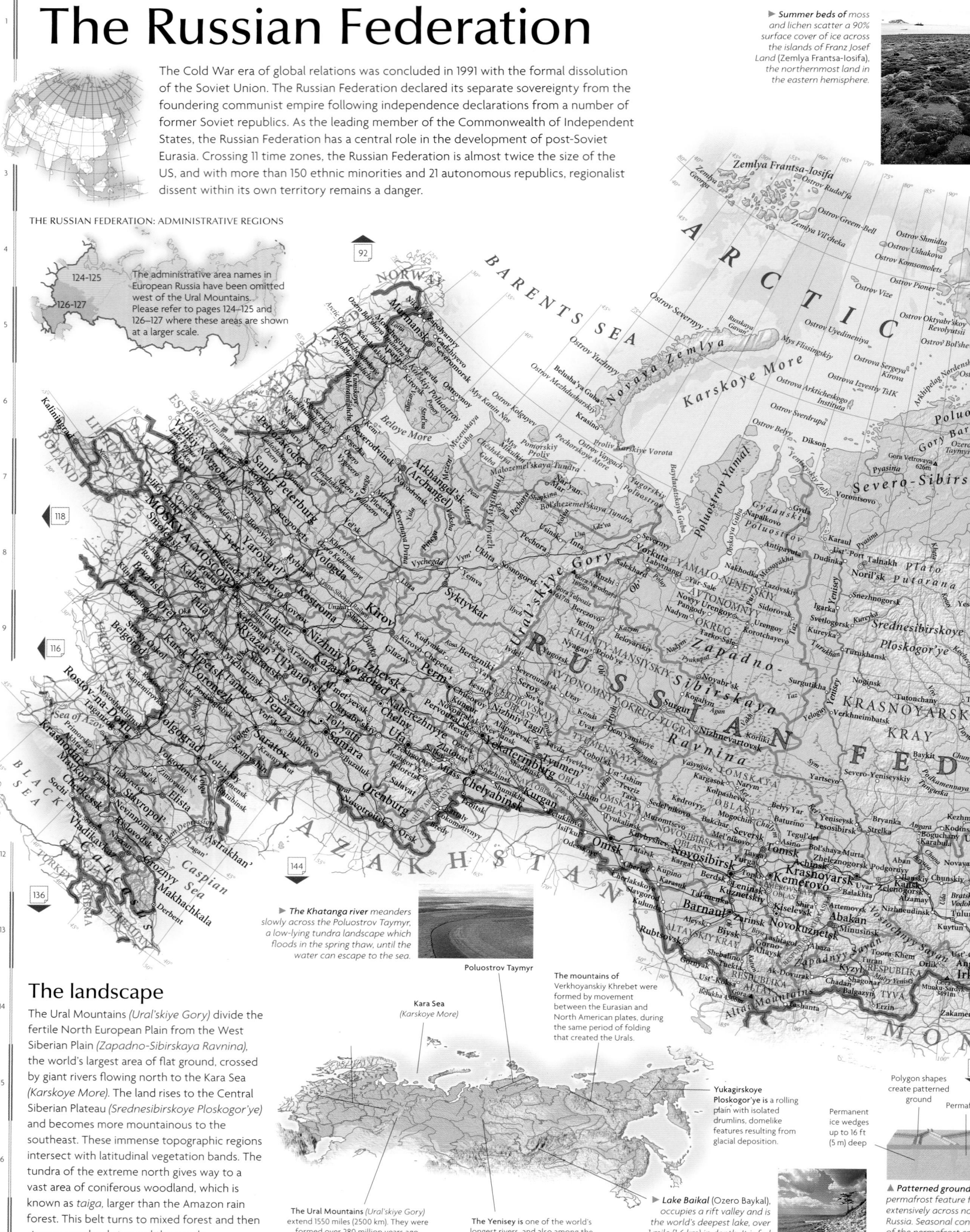

▶ Summer beds of moss and lichen scatter a 90% surface cover of ice across the islands of Franz Josef Land (Zemlya Frantsa-Iosifa), the northernmost land in the eastern hemisphere.

▶ The Khatanga river meanders slowly across the Poluostrov Taymyr, a low-lying tundra landscape which floods in the spring thaw, until the water can escape to the sea.

Poluostrov Taymyr

Kara Sea (Karskoye More)

The mountains of Verkhoyanskiy Khrebet were formed by movement between the Eurasian and North American plates, during the same period of folding that created the Urals.

Yukagirskoye Ploskogor'ye is a rolling plain with isolated drumlins, domelike features resulting from glacial deposition.

Polygon shapes create patterned ground

Permanent ice wedges up to 16 ft (5 m) deep

▲ Patterned ground is [a] permafrost feature fou[nd] extensively across nort[hern] Russia. Seasonal contr[action] of the permafrost cre[ates] polygonal cracks, whi[ch are] filled by ice wedges.

The landscape

The Ural Mountains (Ural'skiye Gory) divide the fertile North European Plain from the West Siberian Plain (Zapadno-Sibirskaya Ravnina), the world's largest area of flat ground, crossed by giant rivers flowing north to the Kara Sea (Karskoye More). The land rises to the Central Siberian Plateau (Srednesibirskoye Ploskogor'ye) and becomes more mountainous to the southeast. These immense topographic regions intersect with latitudinal vegetation bands. The tundra of the extreme north gives way to a vast area of coniferous woodland, which is known as taiga, larger than the Amazon rain forest. This belt turns to mixed forest and then steppe grasslands toward the south.

The Ural Mountains (Ural'skiye Gory) extend 1550 miles (2500 km). They were formed over 280 million years ago, folded as the East European and Siberian plates moved closer together.

The Yenisey is one of the world's longest rivers, and also among the most languid, dropping only 500 ft (152 m) over 1200 miles (2000 km).

▶ Lake Baikal (Ozero Baykal), occupies a rift valley and is the world's deepest lake, over 1 mile (1.6 km) in depth. It is fed by over 300 rivers and drained by just one, the Angara.

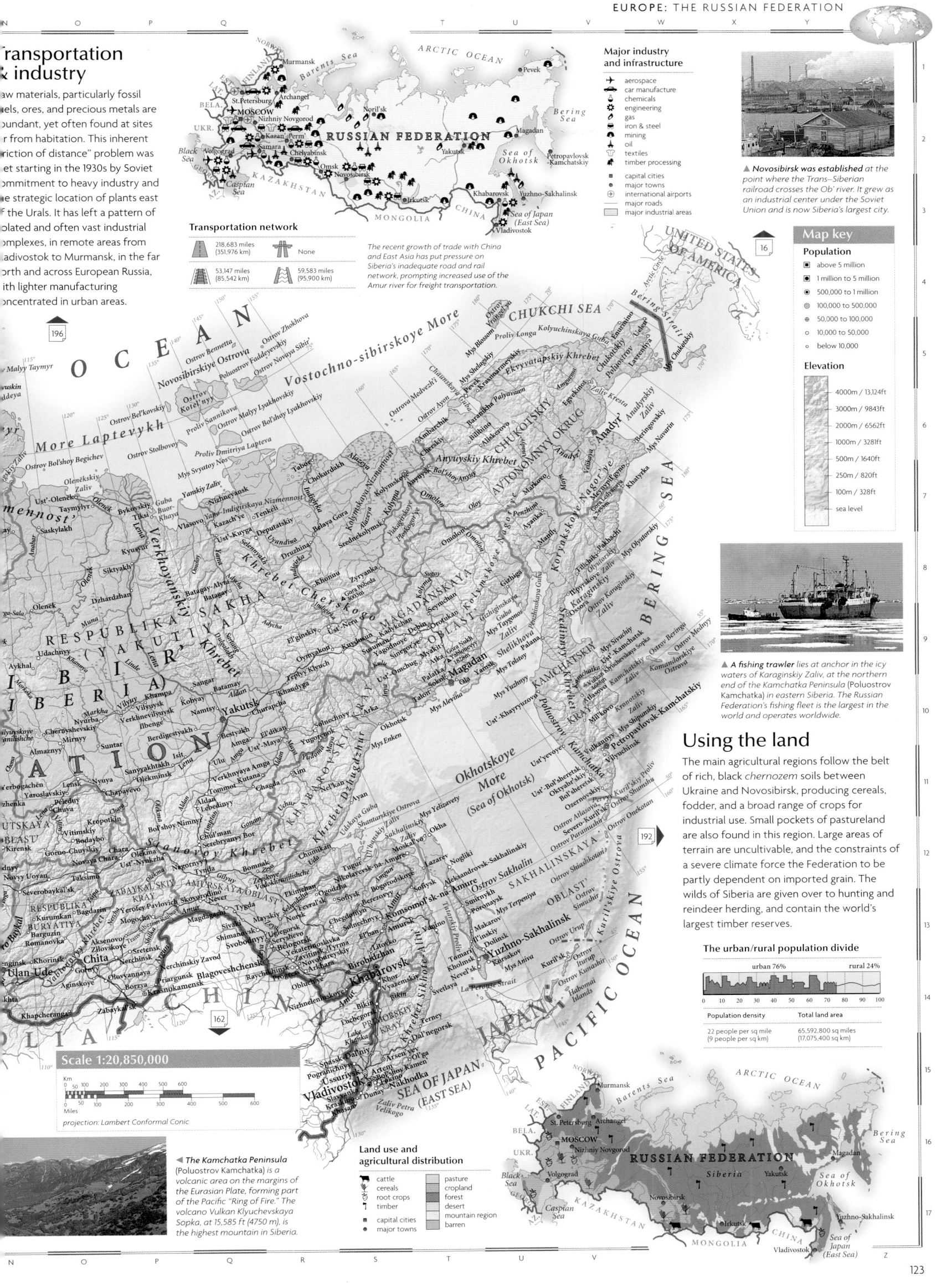

Transportation & industry

Raw materials, particularly fossil fuels, ores, and precious metals are abundant, yet often found at sites far from habitation. This inherent "friction of distance" problem was met starting in the 1930s by Soviet commitment to heavy industry and the strategic location of plants east of the Urals. It has left a pattern of isolated and often vast industrial complexes, in remote areas from Vladivostok to Murmansk, in the far north and across European Russia, with lighter manufacturing concentrated in urban areas.

Major industry and infrastructure

- aerospace
- car manufacture
- chemicals
- engineering
- gas
- iron & steel
- mining
- oil
- textiles
- timber processing
- ● capital cities
- ○ major towns
- ✈ international airports
- major roads
- major industrial areas

Transportation network

218,683 miles (351,976 km)		None	
53,147 miles (85,542 km)		59,583 miles (95,900 km)	

The recent growth of trade with China and East Asia has put pressure on Siberia's inadequate road and rail network, prompting increased use of the Amur river for freight transportation.

▲ Novosibirsk was established at the point where the Trans–Siberian railroad crosses the Ob' river. It grew as an industrial center under the Soviet Union and is now Siberia's largest city.

Map key

Population

- ■ above 5 million
- ■ 1 million to 5 million
- ◉ 500,000 to 1 million
- ◎ 100,000 to 500,000
- ⊕ 50,000 to 100,000
- ○ 10,000 to 50,000
- ○ below 10,000

Elevation

- 4000m / 13,124ft
- 3000m / 9843ft
- 2000m / 6562ft
- 1000m / 3281ft
- 500m / 1640ft
- 250m / 820ft
- 100m / 328ft
- sea level

▲ A fishing trawler lies at anchor in the icy waters of Karaginskiy Zaliv, at the northern end of the Kamchatka Peninsula (Poluostrov Kamchatka) in eastern Siberia. The Russian Federation's fishing fleet is the largest in the world and operates worldwide.

Using the land

The main agricultural regions follow the belt of rich, black *chernozem* soils between Ukraine and Novosibirsk, producing cereals, fodder, and a broad range of crops for industrial use. Small pockets of pastureland are also found in this region. Large areas of terrain are uncultivable, and the constraints of a severe climate force the Federation to be partly dependent on imported grain. The wilds of Siberia are given over to hunting and reindeer herding, and contain the world's largest timber reserves.

The urban/rural population divide

urban 76%   rural 24%

0 10 20 30 40 50 60 70 80 90 100

Population density	Total land area
22 people per sq mile (9 people per sq km)	65,592,800 sq miles (17,075,400 sq km)

Scale 1:20,850,000

Km
0 50 100 200 300 400 500 600

Miles
0 50 100 200 300 400 500 600

projection: Lambert Conformal Conic

◄ The Kamchatka Peninsula (Poluostrov Kamchatka) is a volcanic area on the margins of the Eurasian Plate, forming part of the Pacific "Ring of Fire." The volcano Vulkan Klyuchevskaya Sopka, at 15,585 ft (4750 m), is the highest mountain in Siberia.

Land use and agricultural distribution

- cattle
- cereals
- root crops
- timber
- ■ capital cities
- ● major towns
- pasture
- cropland
- forest
- desert
- mountain region
- barren

Northern European Russia

Reaching into the Arctic Circle, this region of lakeland, forest and tundra is historically bound to Europe by St Petersburg, the old imperial capital of Tsarist Russia and home to a third of the region's population. Communist rule from Moscow left the north politically marginalized, contributing to the present problems of outmoded industry, poor infrastructure and serious environmental neglect. However, with borders embracing Finland, Norway, the Baltic and the northern sea route to the Atlantic, the region's success in foreign trade is now of prime importance to the Russian economy.

▶ *St. Peter and Paul Fortress* is the oldest building in St Petersburg, founded by Peter the Great in 1703 as a modern, European capital for Russia.

The landscape

The ancient bedrock of the Scandinavian Shield lies exposed across the glacially scoured Khibiny Mountains of the Kola Peninsula (*Kol'skiy Poluostrov*), becoming mantled with till toward the North European Plain. The Valdai Hills (*Valdayskaya Vozvyshennost'*) form an important watershed for the plain's rivers, while thick forest veils a complicated topography of moraines, lakes, and ground disturbed by frost action. The Ural Mountains (*Ural'skiye Gory*) form a border with Asia in the east.

◀ *The Kola Peninsula* (Kol'skiy Poluostrov) is part of the Scandinavian Shield, an area of ancient bedrock underlying Scandinavia. Rocks in excess of 2500 million years old are exposed across the peninsula.

▲ *The Khibiny mountains* were formed by volcanic intrusions into the Scandinavian Shield, over 570 million years ago.

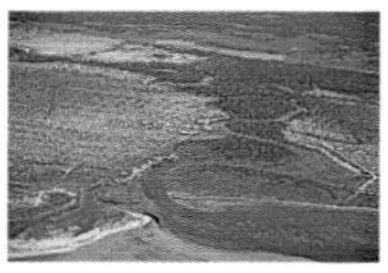

Kola Peninsula (*Kol'skiy Poluostrov*)

Karst features, including sinkholes, lakes, and caverns, are found in limestone outcrops across the plain of the Severnaya Dvina and Mezen' rivers.

The low-lying plains of the Pechora, Mezen', and Severnaya Dvina rivers were flooded by the sea while the land was still isostatically depressed following the last Ice Age, a process which has hidden the landforms created by glacial deposition.

Retreating glacier / Meltwater channels / Terminal moraine

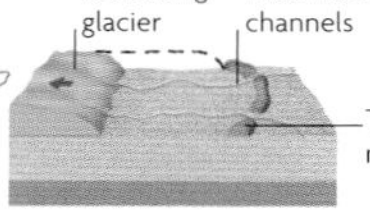

▲ *Terminal moraines are* crescent-shaped ridges of glacial deposits, widely found in central Russia. Detritus is carried by the glacier and deposited at its terminus (snout) as it melts, marking the limit of the ice advance.

Ural Mountains (*Ural'skiye Gory*)

Two of Europe's biggest rivers, the Volga and Western Dvina, rise in the swampy uplands of the Valdai Hills (*Valdayskaya Vozvyshennost'*.)

▶ *Lake Onega* (Onezhskoye Ozero) is the remnant of a body of water which, 12,000 years ago, connected the White Sea (Beloye More) with the Gulf of Finland and the Baltic Sea.

Using the land & sea

The cold climate confines agriculture mainly to southern and western provinces, where dairy farming predominates and arable land is given over to fodder crops as well as flax, potatoes, oats, and rye. Areas beyond the northern margins of cultivation are used for forestry, hunting, herding, and fishing, with some vegetables grown in hothouses around urban areas.

Land use and agricultural distribution
- cattle
- fishing
- reindeer
- timber
- fodder
- major towns
- pasture
- cropland
- forest
- mountain region
- wetland
- tundra
- barren
- ice

RUSSIAN FEDERATION

The urban/rural population divide

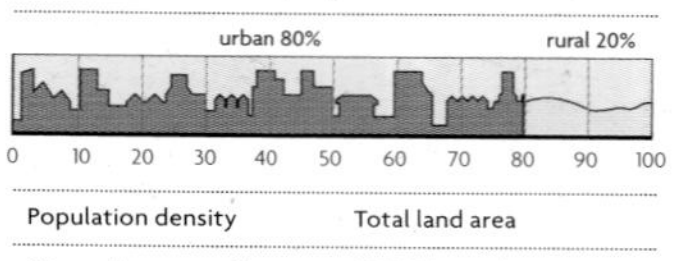

urban 80% / rural 20%

Population density	Total land area
26 people per sq mile (10 people per sq km)	829,398 sq miles (2,148,700 sq km)

◀ *Many rapids are* found along the 175 mile (280 km) course of the Suna river.

The Ural Mountains (Ural'skiye Gory) form the traditional boundary between Europe and Asia. Elevations rarely exceed 6000 ft (1830 m). The region is extremely barren in the far northern latitudes.

Scale 1:6,000,000

Km
0 20 40 60 80 100 140

Miles
0 20 40 60 80 100 120 140

projection: Lambert Conformal Conic

Map key

Population
- 1 million to 5 million
- 500,000 to 1 million
- 100,000 to 500,000
- 50,000 to 100,000
- 10,000 to 50,000
- below 10,000

Elevation
- 1000m / 3281ft
- 500m / 1640ft
- 250m / 820ft
- 100m / 328ft
- sea level

Transportation & industry

The ports of St. Petersburg, Murmansk, and Archangel serve a regional economy led by large-scale resource extraction. Nickel, iron ore, and apatite are mined in the Kola Peninsula (Kol'skiy Poluostrov), and fossil fuels in the Pechora Basin. Paper production is central to Archangel's vast timber industry, while St. Petersburg, drawing on ample labor, has become a major manufacturing center.

Major industry and infrastructure
- chemicals
- coal
- defense
- engineering
- food processing
- hydroelectric power
- mining
- oil & gas
- textiles
- timber processing
- major towns
- international airports
- major roads
- major industrial areas

Transportation network
- 53,700 miles (85,920 km)
- None
- 10,300 miles (16,572 km)
- 12,500 miles (20,000 km)

Railroads linking remote industrial centers with the region's ports are the principal means of supply, although the impressive system of canals, linking natural waterways, is used for freight haulage during the summer.

► *Ice forces the port at St. Petersburg to close in winter, yet Murmansk, on the Barents Sea, remains open, its waters prevented from freezing by warmer ocean currents extending from the North Atlantic Drift.*

► *Kaliningrad has been a Russian enclave since 1945. The port is an important center for the Russian Federation's Baltic fishing fleet.*

◄ *St Basil's Cathedral, completed in 1561, stands in Moscow's Red Square next to the Kremlin; the original fortified stronghold of the city.*

Southern European Russia

This region, divided from Asia by desert, seas, and mountains, has exerted a powerful influence both east and west since the 13th century. Over 70 years of Communist rule produced a highly urbanized, industrial society dominated by Moscow, which was the capital of the Soviet Union until 1991. Almost two-thirds of the Russian Federation's population live in this core area, with a relatively high per capita share of its wealth. However, the rapid growth of a market economy has caused great social upheaval, with rising crime and political instability.

The landscape

Ancient folds in the deep sedimentary strata of the North European Plain have created a sequence of high and low regions. The Central Russian Upland (*Srednerusskaya Vozvyshennost'*) in the west is deeply incised by rivers draining into the lowland of the Oka and Don rivers. In the east the Volga, Europe's longest river, flows south to the Caspian Sea, dividing the Volga Uplands (*Privolzhskaya Vozvyshennost'*) from the foothills of the Ural Mountains (*Ural'skiye Gory*). The Caucasus mountains and the Black Sea form a natural border to the southwest.

▲ *A plantation of Scots pine helps consolidate the loose sandy soils of the Meshchera Lowland (Meshcherskaya Nizmennost), which lies on the bed of an old glacial lake.*

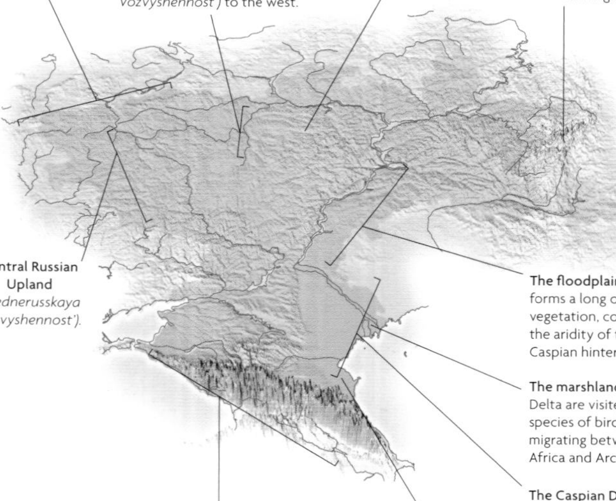

The Smolensk-Moscow Upland (*Smolensko-Moskovskaya Vozvyshennost'*) is a series of terminal moraine ridges marking the southern extent of the last glaciation.

Glacial till covers the bedrock to the north of the North European Plain, giving a gentle surface relief.

The lowland of the Oka and Don rivers lies over a broad trough, between the upfolds of the Volga Uplands (*Privolzhskaya Vozvyshennost'*) to the east, and the Central Russian Upland (*Srednerusskaya Vozvyshennost'*) to the west.

The southern Ural mountains (*Ural'skiye Gory*) consist of several parallel ranges of ancient fold mountains running from north to south.

Central Russian Upland (*Srednerusskaya Vozvyshennaya*).

The floodplain of the Volga forms a long oasis of verdant vegetation, contrasting with the aridity of the surrounding Caspian hinterland.

The marshlands of the Volga Delta are visited by over 260 species of bird each year, migrating between South Africa and Arctic Siberia.

The Caspian Depression is a large downfold (or syncline) which became flooded, forming the Caspian Sea. The shoreline is 98 ft (30 m) below sea level.

◄ *The Caucasus mountains run from the Black Sea to the Caspian Sea. They include El' brus which, at 18,511 ft (5642 m), is the highest point in Europe. It is still uplifting at a rate of 0.4 inches (10 mm) per year.*

Drifting sand occupies large areas of the south, forming dunes up to 50 ft (15 m) high.

Salt dome

Salt dome is forced up and through the rock strata

Sedimentary strata

Salts are forced upwards by denser overlying strata

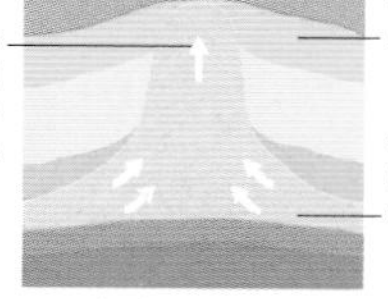

▲ *Salt domes, rounded hills up to 500 ft (150 m) high, are produced as less dense rock salts are displaced under the extreme pressure of denser, overlying strata and forced up toward the surface creating domes. They are widespread in the Caspian Depression.*

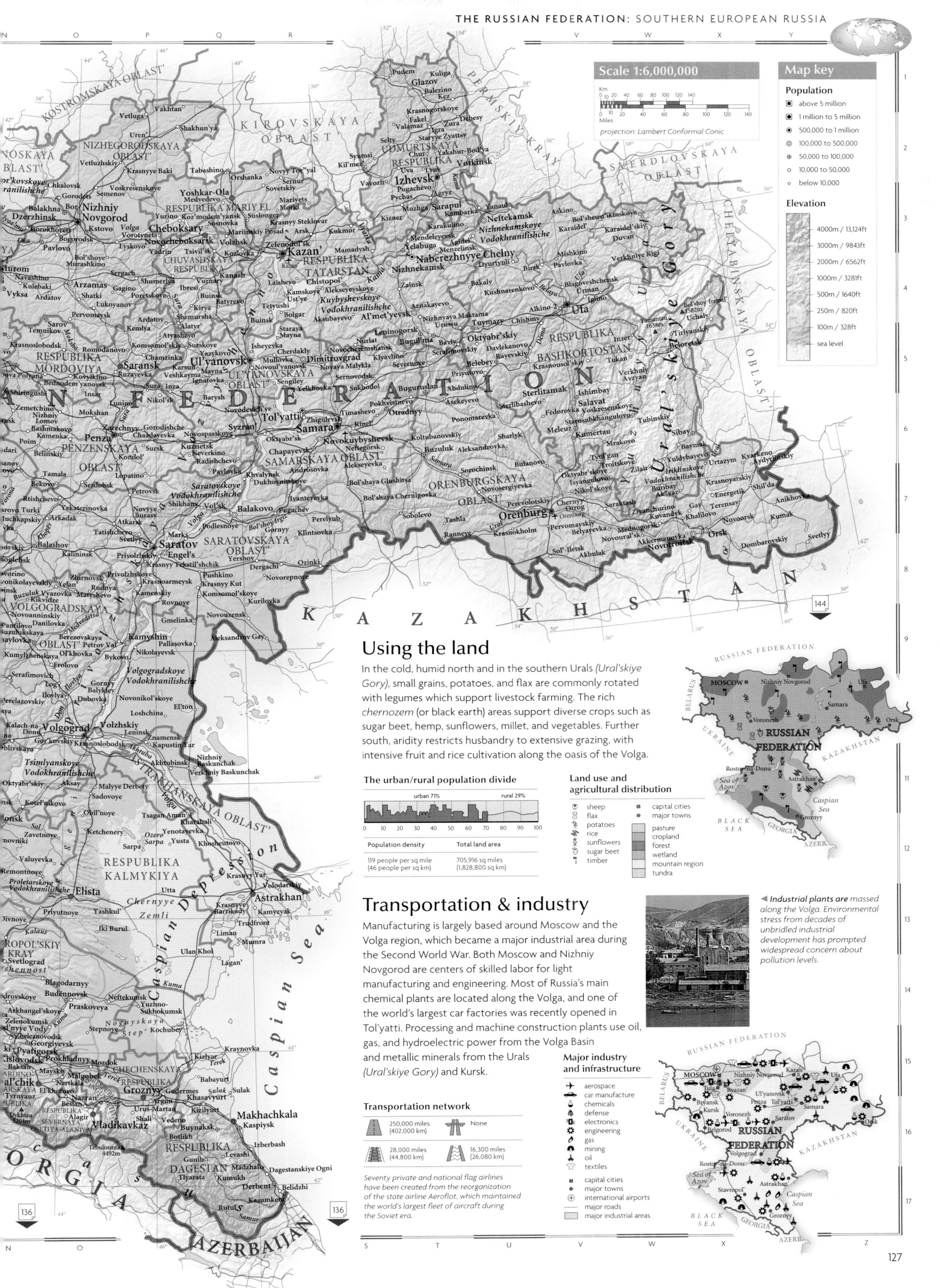

Scale 1:6,000,000

Km 0 20 40 60 80 100 120 140
Miles 0 10 20 40 60 80 100 120 140

projection: Lambert Conformal Conic

Map key

Population

- ■ above 5 million
- ■ 1 million to 5 million
- ◉ 500,000 to 1 million
- ⊚ 100,000 to 500,000
- ⊕ 50,000 to 100,000
- ⊙ 10,000 to 50,000
- ○ below 10,000

Elevation

- 4000m / 13,124ft
- 3000m / 9843ft
- 2000m / 6562ft
- 1000m / 3281ft
- 500m / 1640ft
- 250m / 820ft
- 100m / 328ft
- sea level

Using the land

In the cold, humid north and in the southern Urals (Ural'skiye Gory), small grains, potatoes, and flax are commonly rotated with legumes which support livestock farming. The rich chernozem (or black earth) areas support diverse crops such as sugar beet, hemp, sunflowers, millet, and vegetables. Further south, aridity restricts husbandry to extensive grazing, with intensive fruit and rice cultivation along the oasis of the Volga.

The urban/rural population divide

urban 71%   rural 29%

0 10 20 30 40 50 60 70 80 90 100

Population density
119 people per sq mile
(46 people per sq km)

Total land area
705,916 sq miles
(1,828,800 sq km)

Land use and agricultural distribution

- sheep
- flax
- potatoes
- rice
- sunflowers
- sugar beet
- timber
- capital cities
- major towns
- pasture
- cropland
- forest
- wetland
- mountain region
- tundra

Transportation & industry

Manufacturing is largely based around Moscow and the Volga region, which became a major industrial area during the Second World War. Both Moscow and Nizhniy Novgorod are centers of skilled labor for light manufacturing and engineering. Most of Russia's main chemical plants are located along the Volga, and one of the world's largest car factories was recently opened in Tol'yatti. Processing and machine construction plants use oil, gas, and hydroelectric power from the Volga Basin and metallic minerals from the Urals (Ural'skiye Gory) and Kursk.

◄ Industrial plants are massed along the Volga. Environmental stress from decades of unbridled industrial development has prompted widespread concern about pollution levels.

Transportation network

- 250,000 miles (402,000 km)
- None
- 28,000 miles (44,800 km)
- 16,300 miles (26,080 km)

Seventy private and national flag airlines have been created from the reorganization of the state airline Aeroflot, which maintained the world's largest fleet of aircraft during the Soviet era.

Major industry and infrastructure

- aerospace
- car manufacture
- chemicals
- defense
- electronics
- engineering
- gas
- mining
- oil
- textiles
- capital cities
- major towns
- international airports
- major roads
- major industrial areas

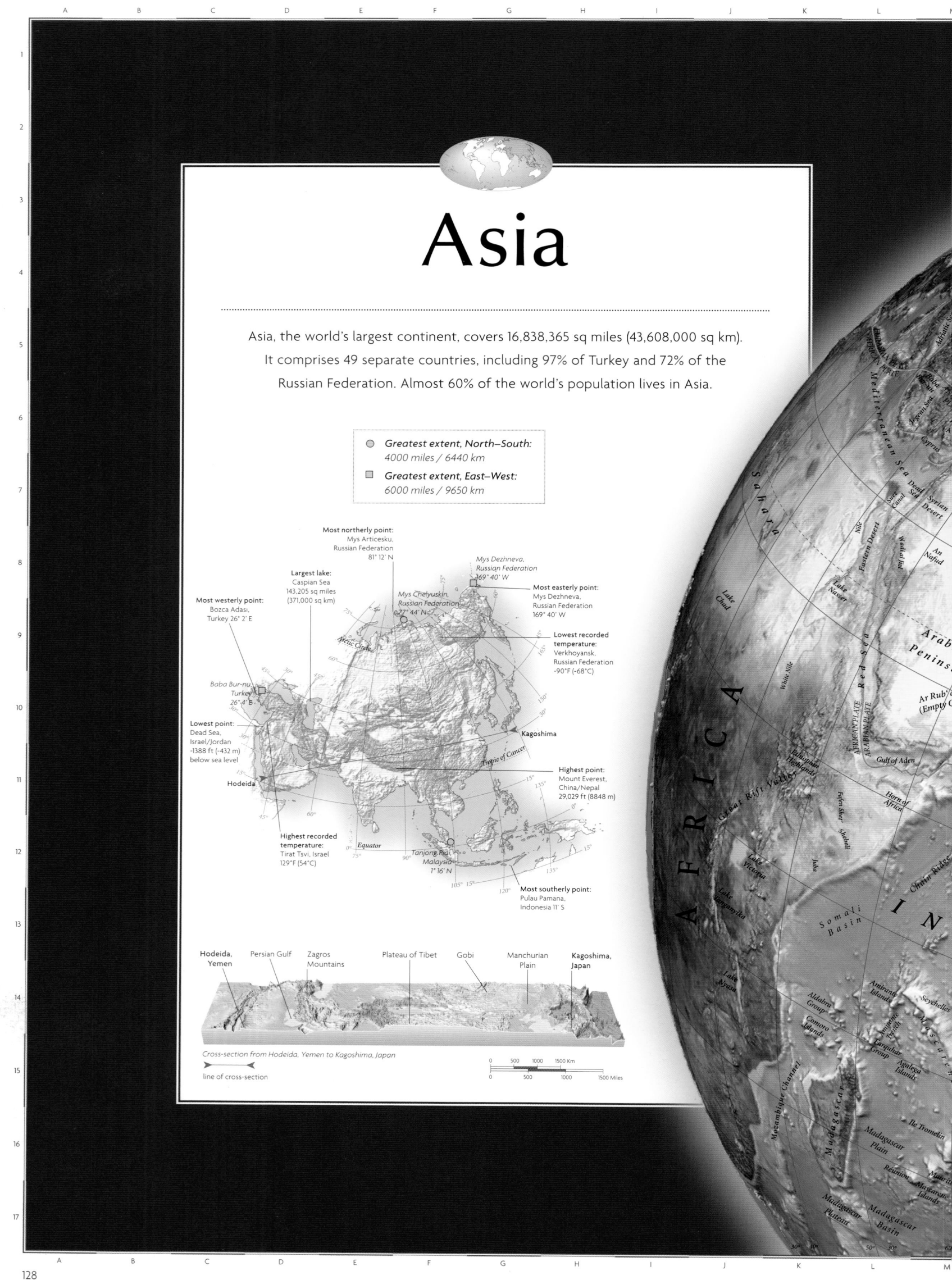

Asia

Asia, the world's largest continent, covers 16,838,365 sq miles (43,608,000 sq km).
It comprises 49 separate countries, including 97% of Turkey and 72% of the
Russian Federation. Almost 60% of the world's population lives in Asia.

- ○ *Greatest extent, North–South:*
  4000 miles / 6440 km
- ◻ *Greatest extent, East–West:*
  6000 miles / 9650 km

Most northerly point:
Mys Articesku,
Russian Federation
81° 12' N

Mys Dezhneva,
Russian Federation
169° 40' W

Largest lake:
Caspian Sea
143,205 sq miles
(371,000 sq km)

Mys Chelyuskin,
Russian Federation
77° 44' N

Most easterly point:
Mys Dezhneva,
Russian Federation
169° 40' W

Most westerly point:
Bozca Adasi,
Turkey 26° 2' E

**Lowest recorded
temperature:**
Verkhoyansk,
Russian Federation
-90°F (-68°C)

Baba Bur-nu
Turkey
26° 4' E

Lowest point:
Dead Sea,
Israel/Jordan
-1388 ft (-432 m)
below sea level

Kagoshima

Hodeida

Highest point:
Mount Everest,
China/Nepal
29,029 ft (8848 m)

**Highest recorded
temperature:**
Tirat Tsvi, Israel
129°F (54°C)

Equator

Tanjong Piai,
Malaysia
1° 16' N

Most southerly point:
Pulau Pamana,
Indonesia 11° S

Hodeida, Yemen | Persian Gulf | Zagros Mountains | Plateau of Tibet | Gobi | Manchurian Plain | **Kagoshima, Japan**

Cross-section from Hodeida, Yemen to Kagoshima, Japan

► ◄ line of cross-section

0 500 1000 1500 Km
0 500 1000 1500 Miles

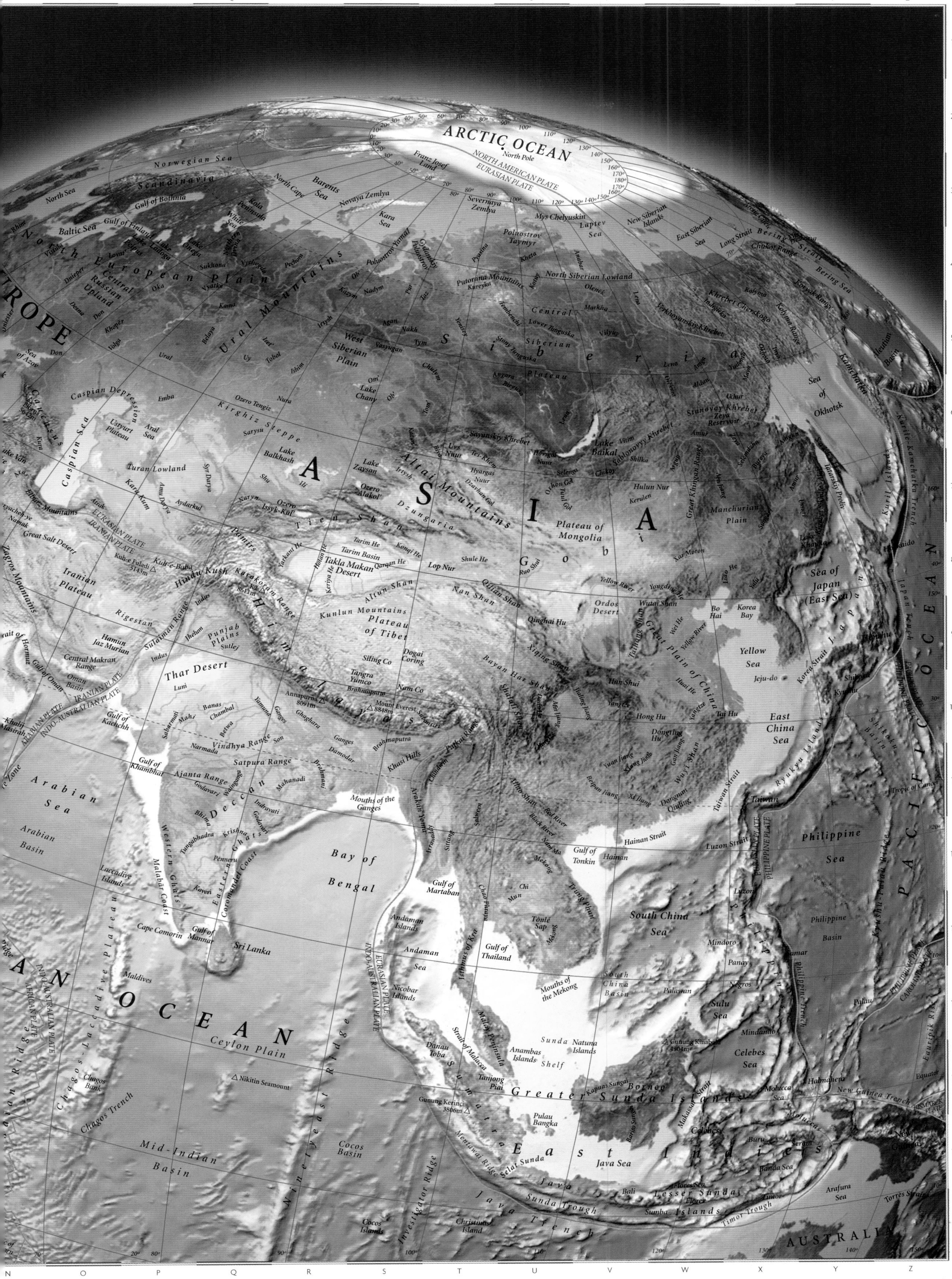

ARCTIC OCEAN

North Pole

NORTH AMERICAN PLATE
EURASIAN PLATE

Norwegian Sea

North Sea

Scandinavia

Gulf of Bothnia

Baltic Sea

North Cape

Barents Sea

Novaya Zemlya

Kola Peninsula

White Sea

Severnaya Zemlya

Kara Sea

Franz Josef Land

Mys Chelyuskin

Laptev Sea

New Siberian Islands

East Siberian Sea

Long Strait

Chukar Range

Bering Strait

Bering Sea

EUROPE

North European Plain

Russian Upland

Central

Ural Mountains

West Siberian Plain

S i b e r i a

North Siberian Lowland

Central Siberian Plateau

Putorana Mountains

Verkhoyansky Khrebet

Khrebet Cherskogo

Kolyma Range

Koryak Range

Kamchatka Basin

Caspian Depression

Kirghiz Steppe

Altai Mountains

Sayunskiy Khrebet

Lake Baikal

Stanovoy Khrebet

Zeya Reservoir

Sea of Okhotsk

Caspian Sea

Turan Lowland

Lake Balkhash

A S I A

Ozero Zaysan

Dzungaria

Tien Shan

Plateau of Mongolia

G o b i

Hulun Nur

Manchurian Plain

Lake Khanka

Sea of Japan (East Sea)

Hokkaido

Iranian Plateau

Hindu Kush

Karakoram Range

Takla Makan Desert

Altun Shan

Nan Shan

Qilian Shan

Ordos Desert

Wutai Shan

Great Plain of China

Bo Hai

Korea Bay

Yellow Sea

Korea Strait

Jeju-do

Kyushu

Zagros Mountains

Iranian Plateau

Punjab Plains

Kunlun Mountains

Plateau of Tibet

Qinghai Hu

Yellow River

Han Shui

Yangtze

Hong Hu

Dongting Hu

Tai Hu

East China Sea

Ryukyu Islands

Thar Desert

H i m a l a y a s

Mount Everest 8848m

Siling Co

Dogai Coring

Tangra Yumco

Nam Co

Brahmaputra

Bayan Har Shan

Hainan Strait

Luzon Strait

Taiwan Strait

Taiwan

Arabian Sea

Gulf of Khambhat

Deccan

Vindhya Range

Satpura Range

Ganges

Damodar

Khasi Hills

Hainan

Gulf of Tonkin

Philippine Sea

Arabian Basin

Western Ghats

Eastern Ghats

Bay of Bengal

Gulf of Martaban

South China Sea

Luzon

Philippine Basin

Laccadive Islands

Coromandel Coast

Mindoro

Panay

Cape Comorin

Gulf of Mannar

Sri Lanka

Andaman Islands

Gulf of Thailand

Mouths of the Mekong

South China Basin

Palawan

Negros

Mindanao

Maldives

O C E A N

Andaman Sea

Nicobar Islands

Tonlé Sap

Sulu Sea

Celebes Sea

Ceylon Plain

Sunda Shelf

Natuna Islands

Anambas Islands

Borneo

Molucca Sea

Halmahera

New Guinea Trench

Nikitin Seamount

Pulau Bangka

Greater Sunda Islands

Celebes

Buru

Seram

Banda Sea

Mid-Indian Basin

Cocos Basin

Java Sea

Bali

Lesser Sunda Islands

Arafura Sea

Cocos Islands

Christmas Island

Java

Sunda Trough

Java Trench

Sumba Islands

Timor

Torres Strait

PACIFIC OCEAN

AUSTRALIA

Physical Asia

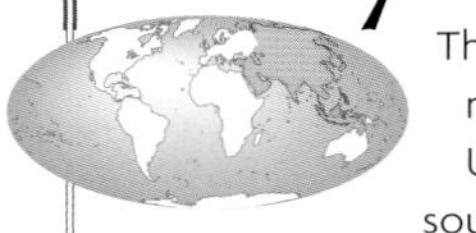

The structure of Asia can be divided into two distinct regions. The landscape of northern Asia consists of old mountain chains, shields, plateaus, and basins, like the Ural Mountains in the west and the Central Siberian Plateau to the east. To the south of this region, are a series of plateaus and basins, including the vast Plateau of Tibet and the Tarim Basin. In contrast, the landscapes of southern Asia are much younger, formed by tectonic activity beginning about 65 million years ago, leading to an almost continuous mountain chain running from Europe, across much of Asia, and culminating in the mighty Himalayan mountain belt, formed when the Indo-Australian Plate collided with the Eurasian Plate. They are still being uplifted today. North of the mountains lies a belt of deserts, including the Gobi and the Takla Makan. In the far south, tectonic activity has formed narrow island arcs, extending over 4000 miles (7000 km). To the west lies the Arabian Shield, once part of the African Plate. As it was rifted apart from Africa, the Arabian Plate collided with the Eurasian Plate, uplifting the Zagros Mountains.

Coastal Lowlands and Island Arcs

The coastal plains that fringe Southeast Asia contain many large delta systems, caused by high levels of rainfall and erosion of the Himalayas, the Plateau of Tibet, and relict loess deposits. To the south is an extensive island archipelago, lying on the drowned Sunda Shelf. Most of these islands are volcanic in origin, caused by the subduction of the Indo-Australian Plate beneath the Eurasian Plate.

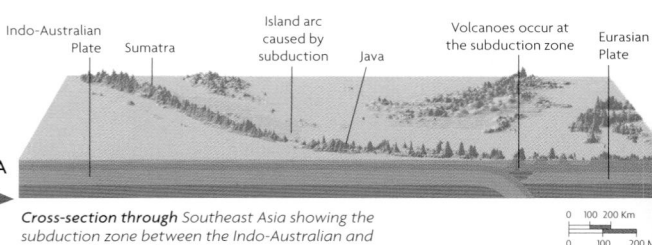

Cross-section through *Southeast Asia showing the subduction zone between the Indo-Australian and Eurasian plates and the island arc.*

The Indian Shield and Himalayan System

The large shield area beneath the Indian subcontinent is between 2.5 and 3.5 billion years old. As the floor of the southern Indian Ocean spread, it pushed the Indian Shield north. This was eventually driven beneath the Plateau of Tibet. This process closed up the ancient Tethys Sea and uplifted the world's highest mountain chain, the Himalayas. Much of the uplifted rock strata was from the seabed of the Tethys Sea, partly accounting for the weakness of the rocks and the high levels of erosion found in the Himalayas.

Cross-section through *the Himalayas showing thrust faulting of the rock strata.*

East Asian Plains and Uplands

Several, small, isolated shield areas, such as the Shandong Peninsula, are found in east Asia. Between these stable shield areas, large river systems like the Yangtze and the Yellow River have deposited thick layers of sediment, forming extensive alluvial plains. The largest of these is the Great Plain of China, the relief of which does not rise above 300 ft (100 m).

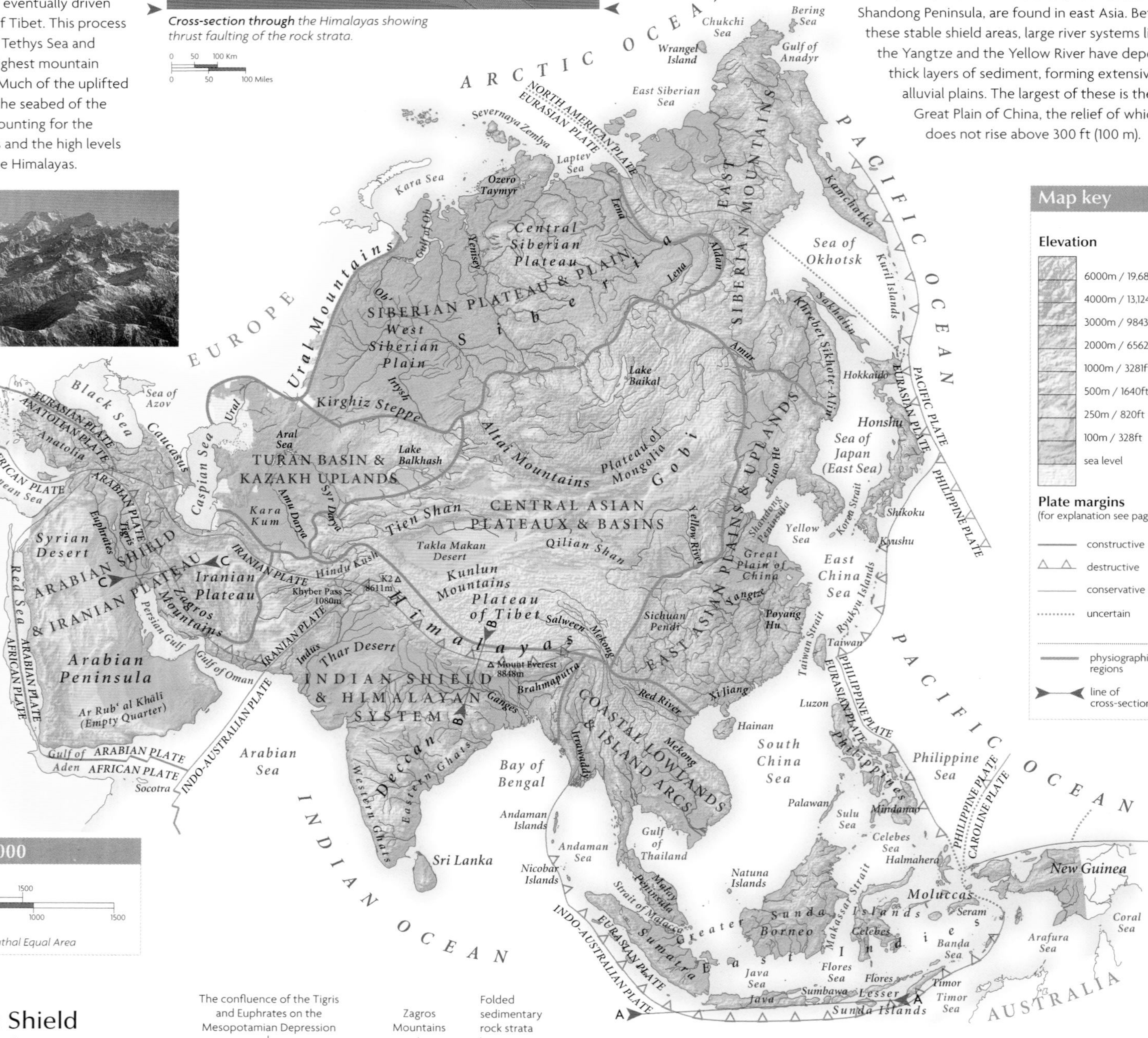

Map key

Elevation

- 6000m / 19,686ft
- 4000m / 13,124ft
- 3000m / 9843ft
- 2000m / 6562ft
- 1000m / 3281ft
- 500m / 1640ft
- 250m / 820ft
- 100m / 328ft
- sea level

Plate margins
(for explanation see page xiv)

- —— constructive
- △ △ destructive
- —— conservative
- ·········· uncertain
- —— physiographic regions
- ►◄ line of cross-section

Scale 1:63,000,000

projection: Lambert Azimuthal Equal Area

The Arabian Shield and Iranian Plateau

Approximately five million years ago, rifting of the continental crust split the Arabian Plate from the African Plate and flooded the Red Sea. As this rift spread, the Arabian Plate collided with the Eurasian Plate, transforming part of the Tethys seabed into the Zagros Mountains which run northwest-southeast across western Iran.

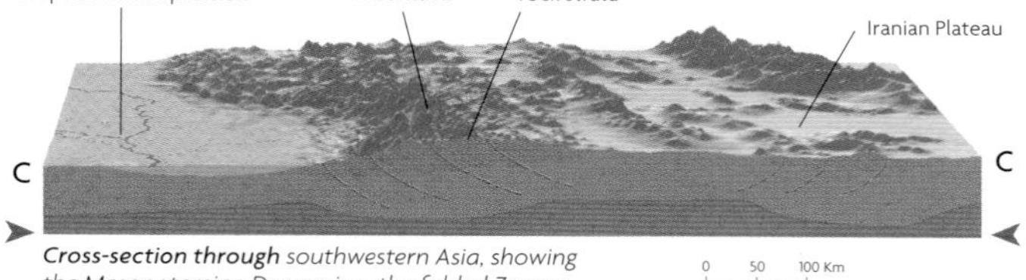

Cross-section through *southwestern Asia, showing the Mesopotamian Depression, the folded Zagros Mountains, and the Iranian Plateau.*

Climate

The climate of Asia exhibits marked differences from region to region, with freezing polar conditions in the north, hot and cold deserts in central regions and subtropical conditions throughout the south. Much of this variation can be attributed to enormous mountain barriers and internal depressions found across the continent. Monsoon winds, which reverse semiannually, cause alternate wet and dry seasons across southern Asia. These air masses moving north from the ocean are stripped of their moisture over the Himalayas causing arid conditions across the Plateau of Tibet. Both the south and east are susceptible to tropical cyclones or typhoons.

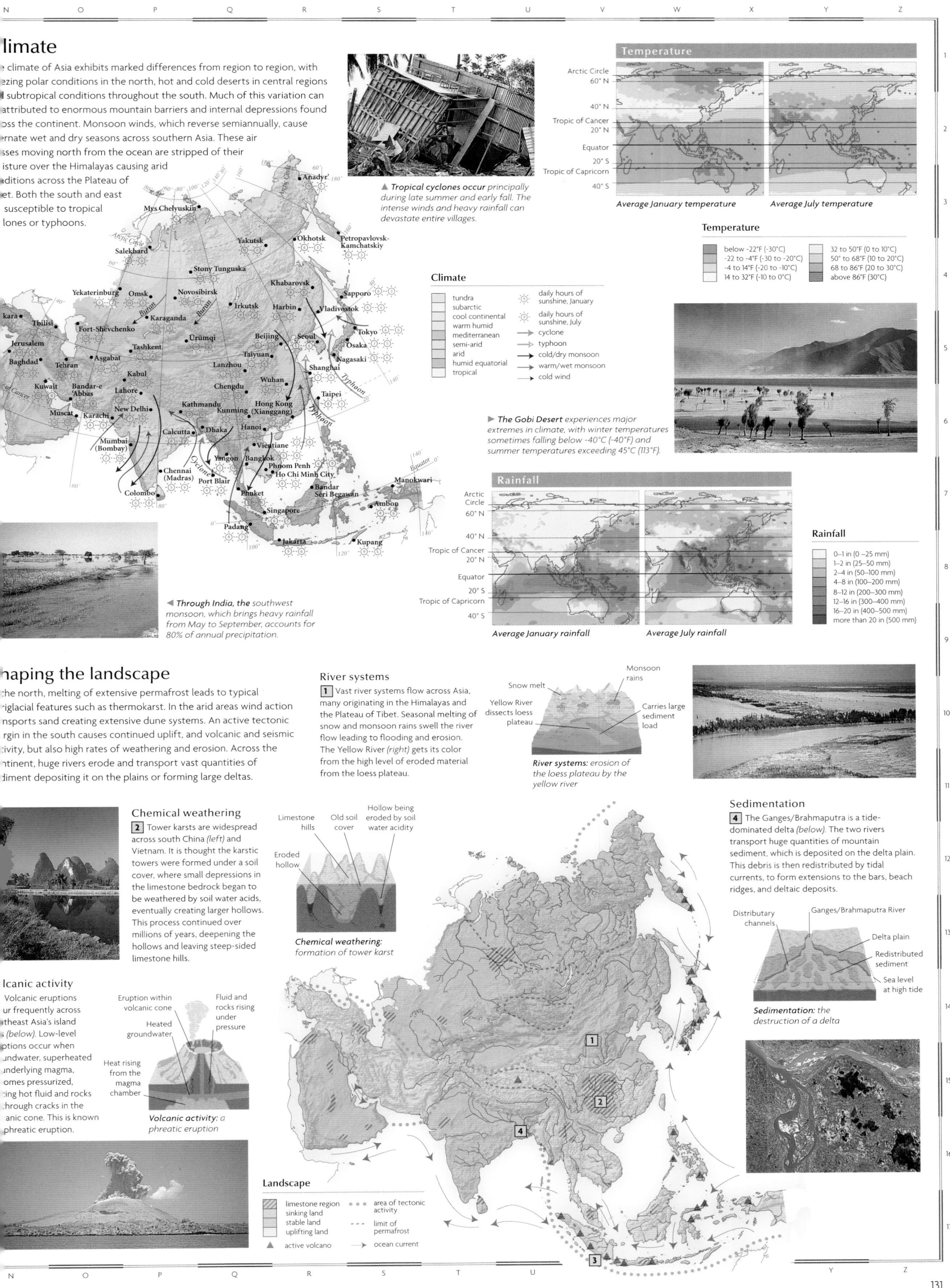

▲ *Tropical cyclones occur* principally during late summer and early fall. The intense winds and heavy rainfall can devastate entire villages.

Temperature

Average January temperature

Average July temperature

Temperature

below -22°F (-30°C)	32 to 50°F (0 to 10°C)
-22 to -4°F (-30 to -20°C)	50 to 68°F (10 to 20°C)
-4 to 14°F (-20 to -10°C)	68 to 86°F (20 to 30°C)
14 to 32°F (-10 to 0°C)	above 86°F (30°C)

Climate

tundra	daily hours of sunshine, January
subarctic	daily hours of sunshine, July
cool continental	cyclone
warm humid	typhoon
mediterranean	
semi-arid	cold/dry monsoon
arid	warm/wet monsoon
humid equatorial	cold wind
tropical	

▶ *The Gobi Desert* experiences major extremes in climate, with winter temperatures sometimes falling below -40°C (-40°F) and summer temperatures exceeding 45°C (113°F).

Rainfall

Average January rainfall

Average July rainfall

Rainfall

0–1 in (0 –25 mm)
1–2 in (25–50 mm)
2–4 in (50–100 mm)
4–8 in (100–200 mm)
8–12 in (200–300 mm)
12–16 in (300–400 mm)
16–20 in (400–500 mm)
more than 20 in (500 mm)

◀ *Through India, the* southwest monsoon, which brings heavy rainfall from May to September, accounts for 80% of annual precipitation.

Shaping the landscape

In the north, melting of extensive permafrost leads to typical periglacial features such as thermokarst. In the arid areas wind action transports sand creating extensive dune systems. An active tectonic margin in the south causes continued uplift, and volcanic and seismic activity, but also high rates of weathering and erosion. Across the continent, huge rivers erode and transport vast quantities of sediment depositing it on the plains or forming large deltas.

River systems

1 Vast river systems flow across Asia, many originating in the Himalayas and the Plateau of Tibet. Seasonal melting of snow and monsoon rains swell the river flow leading to flooding and erosion. The Yellow River *(right)* gets its color from the high level of eroded material from the loess plateau.

River systems: erosion of the loess plateau by the yellow river

Chemical weathering

2 Tower karsts are widespread across south China *(left)* and Vietnam. It is thought the karstic towers were formed under a soil cover, where small depressions in the limestone bedrock began to be weathered by soil water acids, eventually creating larger hollows. This process continued over millions of years, deepening the hollows and leaving steep-sided limestone hills.

Chemical weathering: formation of tower karst

Volcanic activity

Volcanic eruptions occur frequently across southeast Asia's island arcs *(below)*. Low-level eruptions occur when groundwater, superheated by underlying magma, becomes pressurized, forcing hot fluid and rocks up through cracks in the volcanic cone. This is known as a phreatic eruption.

Volcanic activity: a phreatic eruption

Sedimentation

4 The Ganges/Brahmaputra is a tide-dominated delta *(below)*. The two rivers transport huge quantities of mountain sediment, which is deposited on the delta plain. This debris is then redistributed by tidal currents, to form extensions to the bars, beach ridges, and deltaic deposits.

Sedimentation: the destruction of a delta

Landscape

limestone region	area of tectonic activity
sinking land	limit of permafrost
stable land	
uplifting land	ocean current
active volcano	

A B C D E F G H I J

Political Asia

Asia is the world's largest continent, encompassing many different and discrete realms, from the desert Arab lands of the southwest to the subtropical archipelago of Indonesia; from the vast barren wastes of Siberia to the fertile river valleys of China and South Asia, seats of some of the world's most ancient civilizations. The collapse of the Soviet Union has fragmented the north of the continent into the Siberian portion of the Russian Federation, and the new republics of Central Asia. Strong religious traditions heavily influence the politics of South and Southwest Asia. Hindu and Muslim rivalries threaten to upset the political equilibrium in South Asia where India – in terms of population – remains the world's largest democracy. Communist China another population giant, is reasserting its position as a world and political power, while on its doorstep, the economically progressive and dynamic Pacific Rim countries, led by Japan, continue to assert their worldwide economic force.

Population density
(people per sq mile)

- below 25
- 25–124
- 125–259
- 260–649
- 650–10,400
- above 10,400

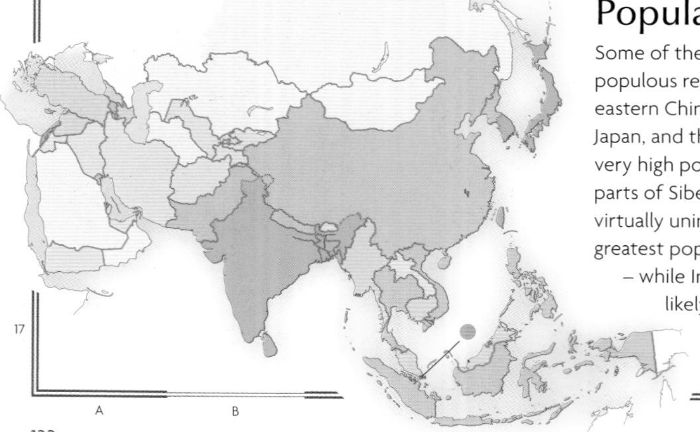

Population

Some of the world's most populous and least populous regions are in Asia. The plains of eastern China, the Ganges river plains in India, Japan, and the Indonesian island of Java, all have very high population densities; by contrast parts of Siberia and the Plateau of Tibet are virtually uninhabited. China has the world's greatest population – 20% of the globe's total – while India, with the second largest, is likely to overtake China within 30 years.

◄ *Over 13 million people bustle through Kolkata's maze of crowded, narrow streets. Population densities in India's largest city reach almost 85,000 per sq mile (33,000 per sq km).*

Map labels

ARCTIC OCEAN

East Siberian Sea
Laptev Sea
Kara Sea
Arctic Circle

Noril'sk
Central Siberian Plateau
Lower Tunguska
Yakutsk

RUSSIAN FEDERATION

Yekaterinburg
Chelyabinsk
West Siberian Plain
Omsk
Tomsk
Krasnoyarsk
Novosibirsk
Novokuznetsk
Irkutsk
Lake Baikal

Istanbul
Black Sea
ANKARA
Sokhumi
GEORGIA
K'ut'aisi
Bat'umi
TBILISI
TURKEY
Anatolia
ASTANA
Karaganda
Semipalatinsk
Choybalsan
Erdenet
ULAN BATOR
Sühbaatar

KAZAKHSTAN
Zhezkazgan
Balkhash
Lake Balkhash
MONGOLIA
Gobi

CYPRUS
NICOSIA
LEBANON
BEIRUT
Tripoli
Adana
Gaziantep
Aleppo
ARMENIA
YEREVAN
AZERB.
Gäncä
BAKU
Aktau
Aral Sea
Kyzylorda
Taraz
BISHKEK
Almaty
Karakol
Ürümqi
Datong
Baotou
Shijiazhuang
Taiyuan

Haifa
Tel Aviv-Yafo
Gaza
JERUSALEM
ISRAEL
AMMAN
JORDAN
DAMASCUS
SYRIA
Mosul
Tabriz
TURKMENISTAN
Dasoguz
Amu Darya
UZBEKISTAN
TASHKENT
Syr Darya
KYRGYZSTAN
Osh
Tien Shan
Tarim He

Kirkuk
BAGHDAD
An Najaf
IRAQ
Basra
Esfahan
TEHRAN
Gorgan
Qom
Mashhad
AŞGABAT
DUSHANBE
TAJIKISTAN
Balkh
Qal'eh-ye Now
Takla Makan Desert
(claimed by India)
Kunlun Mountains
CHINA
Lanzhou
Xi'an
Luoya

Ahvaz
IRAN
Iranian Plateau
Herat
AFGHANISTAN
KABUL
(line of control)
(administered by China, claimed by India)

KUWAIT
SAUDI ARABIA
Shiraz
Kerman
Kandahar
Peshawar
Srinagar
ISLAMABAD
Jammu
Gujranwala
Faisalabad
Lahore
Plateau of Tibet
(Much of Arunachal Pradesh is claimed by China)
Himalayas
Brahmaputra
Mianyang
Chengdu
Leshan
Chong

Jedda
At Ta'if
RIYADH
MANAMA
BAHRAIN
QATAR
DOHA
ABU DHABI
UAE
Zahedan
Bandar-e 'Abbas
Quetta
Multan
Ludhiana
PAKISTAN
Larkana
Shikarpur
Delhi
NEW DELHI
Bareilly
Agra
NEPAL
KATHMANDU
THIMPHU
BHUTAN
Guwahati
Kunming
Liuzho
Nanning

Red Sea
AFRICA
Tropic of Cancer
Persian Gulf
Ar Rustaq
MUSCAT
Sur
Gulf of Oman
Karachi
Hyderabad
Jaipur
Lucknow
Kanpur
Varanasi
INDIA
Patna
Rajshahi
BANGLADESH
DHAKA
Brahmanbaria
Rangpur
Chittagong
Pakokku
MYANMAR (BURMA)
Mandalay
Taunggyi
HANOI
Hai
Guiy

Arabian Peninsula
Ar Rub' al Khali (Empty Quarter)
OMAN
Ganges
Ahmadabad
Vadodara
Indore
Bhopal
Jamshedpur
Khulna
Kolkata (Calcutta)

SANA
YEMEN
Ta'izz
Aden
Gulf of Aden
Socotra (to Yemen)
Arabian Sea
Mumbai (Bombay)
Surat
Nagpur
Pune
Solapur
Hyderabad
Vijayawada
Bhubaneshwar
Bay of Bengal
NAY PYI TAW
Prome
Pegu
Yangon (Rangoon)
Bassein
Bogale
Chiang Mai
VIENTIANE
LAOS
Pakxe
THAILAND
BANGKOK
CAMBOD
Batdambang

Godavari
Krishna
Narmada
Hubli
Bangalore
Mysore
Chennai (Madras)
Coimbatore
Andaman Islands (to India)
Andaman Sea
PHNOM PENH
Gulf of Thailand
Ho
Min

INDIAN OCEAN
Kochi
Thiruvananthapuram
Jaffna
SRI LANKA
COLOMBO
SRI JAYEWARDENAPURA KOTTE
Nicobar Islands (to India)
Kota Bharu
MAL
Taiping
KUALA LUMPUR
PUTRAJAYA
SINGAPO
Medan
Equator
Sumatra
Padang
Palembang
JAK

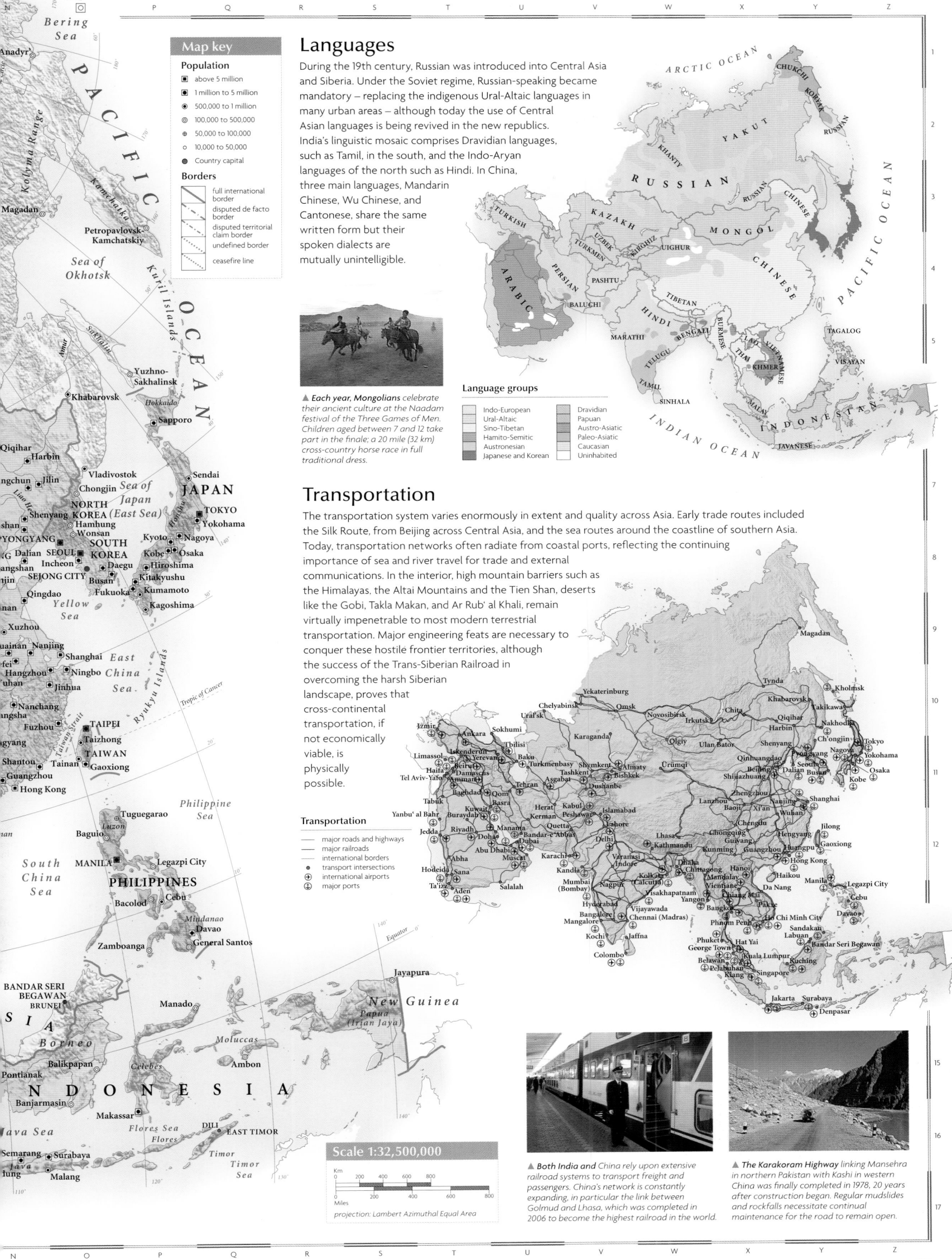

Languages

During the 19th century, Russian was introduced into Central Asia and Siberia. Under the Soviet regime, Russian-speaking became mandatory – replacing the indigenous Ural-Altaic languages in many urban areas – although today the use of Central Asian languages is being revived in the new republics. India's linguistic mosaic comprises Dravidian languages, such as Tamil, in the south, and the Indo-Aryan languages of the north such as Hindi. In China, three main languages, Mandarin Chinese, Wu Chinese, and Cantonese, share the same written form but their spoken dialects are mutually unintelligible.

▲ *Each year, Mongolians celebrate their ancient culture at the Naadam festival of the Three Games of Men. Children aged between 7 and 12 take part in the finale; a 20 mile (32 km) cross-country horse race in full traditional dress.*

Language groups

- Indo-European
- Ural-Altaic
- Sino-Tibetan
- Hamito-Semitic
- Austronesian
- Japanese and Korean
- Dravidian
- Papuan
- Austro-Asiatic
- Paleo-Asiatic
- Caucasian
- Uninhabited

Transportation

The transportation system varies enormously in extent and quality across Asia. Early trade routes included the Silk Route, from Beijing across Central Asia, and the sea routes around the coastline of southern Asia. Today, transportation networks often radiate from coastal ports, reflecting the continuing importance of sea and river travel for trade and external communications. In the interior, high mountain barriers such as the Himalayas, the Altai Mountains and the Tien Shan, deserts like the Gobi, Takla Makan, and Ar Rub' al Khali, remain virtually impenetrable to most modern terrestrial transportation. Major engineering feats are necessary to conquer these hostile frontier territories, although the success of the Trans-Siberian Railroad in overcoming the harsh Siberian landscape, proves that cross-continental transportation, if not economically viable, is physically possible.

Transportation

- major roads and highways
- major railroads
- international borders
- ● transport intersections
- ⊕ international airports
- ⊕ major ports

Map key

Population
- ■ above 5 million
- ▣ 1 million to 5 million
- ◉ 500,000 to 1 million
- ◎ 100,000 to 500,000
- ⊕ 50,000 to 100,000
- ○ 10,000 to 50,000
- ● Country capital

Borders
- full international border
- disputed de facto border
- disputed territorial claim border
- undefined border
- ceasefire line

Scale 1:32,500,000

Km 0 200 400 600 800
Miles 0 200 400 600 800

projection: Lambert Azimuthal Equal Area

▲ *Both India and China rely upon extensive railroad systems to transport freight and passengers. China's network is constantly expanding, in particular the link between Golmud and Lhasa, which was completed in 2006 to become the highest railroad in the world.*

▲ *The Karakoram Highway linking Mansehra in northern Pakistan with Kashi in western China was finally completed in 1978, 20 years after construction began. Regular mudslides and rockfalls necessitate continual maintenance for the road to remain open.*

Asian resources

Although agriculture remains the economic mainstay of most Asian countries, the number of people employed in agriculture has steadily declined, as new industries have been developed during the past 30 years. China, Indonesia, Malaysia, Thailand, and Turkey have all experienced far-reaching structural change in their economies, while the breakup of the Soviet Union has created a new economic challenge in the Central Asian republics. The countries of The Persian Gulf illustrate the rapid transformation from rural nomadism to modern, urban society which oil wealth has brought to parts of the continent. Asia's most economically dynamic countries, Japan, Singapore, South Korea, and Taiwan, fringe the Pacific Ocean and are known as the Pacific Rim. In contrast, other Southeast Asian countries like Laos and Cambodia remain both economically and industrially underdeveloped.

Industry

East Asian industry leads the continent in both productivity and efficiency; electronics, hi-tech industries, car manufacture, and shipbuilding are important. The so-called economic "tigers" of the Pacific Rim are Japan, South Korea, and Taiwan and in recent years China has rediscovered its potential as an economic superpower. Heavy industries such as engineering, chemicals, and steel typify the industrial complexes along the corridor created by the Trans-Siberian Railroad, the Fergana Valley in Central Asia, and also much of the huge industrial plain of east China. The discovery of oil in the Persian Gulf has brought immense wealth to countries that previously relied on subsistence agriculture on marginal desert land.

Industry

- aerospace
- brewing
- car/vehicle manufacture
- cement
- chemicals
- electronics
- engineering
- finance
- fish processing
- food processing
- hi-tech industry
- iron & steel
- pharmaceuticals
- printing & publishing
- shipbuilding
- sugar processing
- tea processing
- textiles
- timber processing
- tobacco processing
- coal
- oil
- gas
- industrial cities
- major industrial areas

Standard of living

Despite Japan's high standards of living, and Southwest Asia's oil-derived wealth, immense disparities exist across the continent. Afghanistan remains one of the world's most underdeveloped nations, as do the mountain states of Nepal and Bhutan. Further rapid population growth is exacerbating poverty and overcrowding in many parts of India and Bangladesh.

Standard of living
(UN human development index)

low

high

▲ On a small island at the southern tip of the Malay Peninsula lies Singapore, one of the Pacific Rim's most vibrant economic centers. Multinational banking and finance form the core of the city's wealth.

GNI per capita (US$)

- below 1999
- 2000–4999
- 5000–9999
- 10,000–19,999
- 20,000–24,999
- above 25,000

▲ Iron and steel, engineering, and shipbuilding typify the heavy industry found in eastern China's industrial cities, especially the nation's leading manufacturing center, Shanghai.

◄ Traditional industries are still crucial to many rural economies across Asia. Here, on the Vietnamese coast, salt has been extracted from seawater by evaporation and is being loaded into a van to take to market.

ARCTIC OCEAN

PACIFIC OCEAN

RUSSIAN FEDERATION

Sea of Okhotsk

JAPAN

Yakutsk

Bratsk

Khabarovsk

Yekaterinburg
Chelyabinsk
Magnitogorsk
Omsk
Novosibirsk
Kemerovo
Krasnoyarsk
Novokuznetsk
Irkutsk

Trans-Siberian Railway

Vladivostok

Harbin

Istanbul
Izmir
Ankara
GEORGIA
Tbilisi
TURKEY
ARMENIA
Yerevan
AZERB.
Baku
KAZAKHSTAN
Karaganda
MONGOLIA
Ulan Bator
Shenyang
NORTH KOREA
Pyongyang
Tokyo
Nagoya
Kobe

CYPRUS
LEBANON
Beirut
SYRIA
Damascus
Tel Aviv-Yafo
ISRAEL
JORDAN
Amman
Kirkuk
Baghdad
IRAQ
Basra
Tehran
Isfahan
IRAN
Caspian Sea
Aral Sea
UZBEKISTAN
Tashkent
Almaty
Ürümqi
KYRGYZSTAN
Farghona
TURKMENISTAN
Asgabat
Dushanbe
TAJIKISTAN
Beijing
Tianjin
Dalian
Seoul
SOUTH KOREA
Busan
Taiyuan
Jinan
Qingdao

SAUDI ARABIA
Kuwait
KUWAIT
Ad Damman
BAHRAIN
QATAR
Abu Dhabi
Dubai
UAE
Jedda
Riyadh
Persian Gulf
AFGHANISTAN
Rawalpindi
Lahore
PAKISTAN
Lanzhou
Zhengzhou
Xi'an
CHINA
Nanjing
Shanghai
Wuhan
Chengdu
Chongqing

Red Sea
Gulf of Oman
Delhi
Kanpur
NEPAL
BHUTAN
Kunming
Guangzhou
Taipei
TAIWAN
Hong Kong

YEMEN
OMAN
Karachi
Ahmadabad
INDIA
Indore
Jamshedpur
BANGLADESH
Dhaka
Chittagong
Mandalay
Hanoi
Gulf of Aden
Arabian Sea
Mumbai (Bombay)
Nagpur
Kolkata (Calcutta)
MYANMAR (BURMA)
Yangon (Rangoon)
LAOS
VIETNAM
Da Nang
South China Sea
Manila
PHILIPPINES

Bangalore
Chennai (Madras)
THAILAND
Bangkok
CAMBODIA

SRI LANKA
INDIAN OCEAN
Ho Chi Minh City

BRUNEI
MALAYSIA

Kuala Lumpur
SINGAPORE
Singapore
INDONESIA

Jakarta
Surabaya
EAST TIMOR

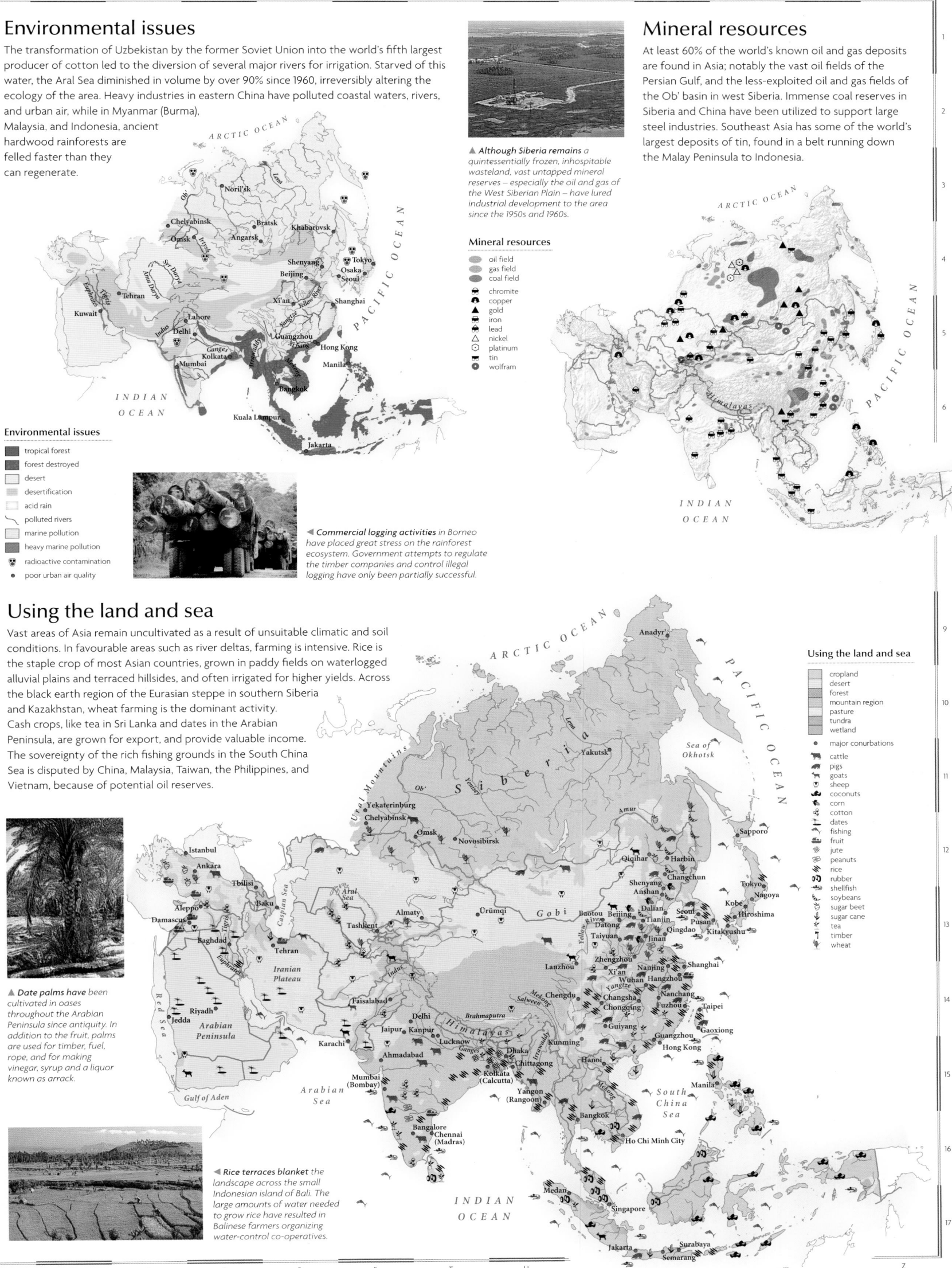

Environmental issues

The transformation of Uzbekistan by the former Soviet Union into the world's fifth largest producer of cotton led to the diversion of several major rivers for irrigation. Starved of this water, the Aral Sea diminished in volume by over 90% since 1960, irreversibly altering the ecology of the area. Heavy industries in eastern China have polluted coastal waters, rivers, and urban air, while in Myanmar (Burma), Malaysia, and Indonesia, ancient hardwood rainforests are felled faster than they can regenerate.

▲ **Although Siberia remains** a quintessentially frozen, inhospitable wasteland, vast untapped mineral reserves – especially the oil and gas of the West Siberian Plain – have lured industrial development to the area since the 1950s and 1960s.

Environmental issues

- tropical forest
- forest destroyed
- desert
- desertification
- acid rain
- polluted rivers
- marine pollution
- heavy marine pollution
- radioactive contamination
- poor urban air quality

◀ **Commercial logging activities** in Borneo have placed great stress on the rainforest ecosystem. Government attempts to regulate the timber companies and control illegal logging have only been partially successful.

Mineral resources

At least 60% of the world's known oil and gas deposits are found in Asia; notably the vast oil fields of the Persian Gulf, and the less-exploited oil and gas fields of the Ob' basin in west Siberia. Immense coal reserves in Siberia and China have been utilized to support large steel industries. Southeast Asia has some of the world's largest deposits of tin, found in a belt running down the Malay Peninsula to Indonesia.

Mineral resources

- oil field
- gas field
- coal field
- chromite
- copper
- gold
- iron
- lead
- nickel
- platinum
- tin
- wolfram

Using the land and sea

Vast areas of Asia remain uncultivated as a result of unsuitable climatic and soil conditions. In favourable areas such as river deltas, farming is intensive. Rice is the staple crop of most Asian countries, grown in paddy fields on waterlogged alluvial plains and terraced hillsides, and often irrigated for higher yields. Across the black earth region of the Eurasian steppe in southern Siberia and Kazakhstan, wheat farming is the dominant activity. Cash crops, like tea in Sri Lanka and dates in the Arabian Peninsula, are grown for export, and provide valuable income. The sovereignty of the rich fishing grounds in the South China Sea is disputed by China, Malaysia, Taiwan, the Philippines, and Vietnam, because of potential oil reserves.

Using the land and sea

- cropland
- desert
- forest
- mountain region
- pasture
- tundra
- wetland
- major conurbations
- cattle
- pigs
- goats
- sheep
- coconuts
- corn
- cotton
- dates
- fishing
- fruit
- jute
- peanuts
- rice
- rubber
- shellfish
- soybeans
- sugar beet
- sugar cane
- tea
- timber
- wheat

▲ **Date palms have** been cultivated in oases throughout the Arabian Peninsula since antiquity. In addition to the fruit, palms are used for timber, fuel, rope, and for making vinegar, syrup and a liquor known as arrack.

◀ **Rice terraces blanket** the landscape across the small Indonesian island of Bali. The large amounts of water needed to grow rice have resulted in Balinese farmers organizing water-control co-operatives.

Turkey & the Caucasus

ARMENIA, AZERBAIJAN, GEORGIA, TURKEY

This region occupies the fragmented junction between Europe, Asia, and the Russian Federation. Sunni Islam provides a common identity for the secular state of Turkey, which the revered leader Kemal Atatürk established from the remnants of the Ottoman Empire after the First World War. Turkey has a broad resource base and expanding trade links with Europe, but the east is relatively undeveloped and strife between the state and a large Kurdish minority has yet to be resolved. Georgia is similarly challenged by ethnic separatism, while the Christian state of Armenia and the mainly Muslim and oil-rich Azerbaijan are locked in conflict over the territory of Nagorno-Karabakh.

Using the land & sea

Turkey is largely self-sufficient in food. The irrigated Black Sea coastlands have the world's highest yields of hazelnuts. Tobacco, cotton, sultanas, tea, and figs are the region's main cash crops and a great range of fruit and vegetables are grown. Wine grapes are among the labor-intensive crops which allow full use of limited agricultural land in the Caucasus. Sturgeon fishing is particularly important in Azerbaijan.

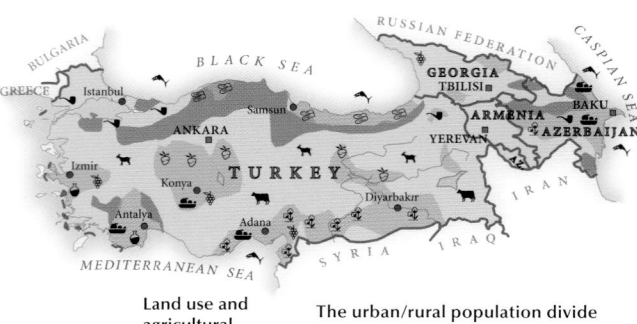

Transportation & industry

Turkey leads the region's well diversified economy. Petrochemicals, textiles, engineering, and food processing are the main industries. Azerbaijan is able to export oil, while the other states rely heavily on hydroelectric power and imported fuel. Georgia produces precision machinery. War and earthquake damage have devastated Armenia's infrastructure.

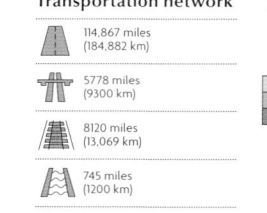

▲ Azerbaijan has substantial oil reserves, located in and around the Caspian Sea. They were some of the earliest oilfields in the world to be exploited.

Land use and agricultural distribution

- cattle
- goats
- cotton
- fishing
- fruit
- hazelnuts
- olives
- sugar beet
- tobacco
- vineyards

- ◼ capital cities
- • major towns

- pasture
- cropland
- forest

The urban/rural population divide

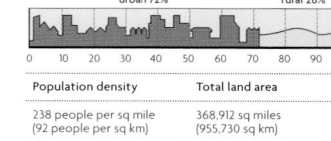

urban 72%   rural 28%

0 10 20 30 40 50 60 70 80 90 100

Population density	Total land area
238 people per sq mile (92 people per sq km)	368,912 sq miles (955,730 sq km)

Major industry and infrastructure

- carpet weaving
- cement
- chemicals
- coal
- engineering
- food processing
- oil
- textiles
- tourism
- vehicle manufacture

- ◼ capital cities
- • major towns
- ⊕ international airports
- — major roads
- major industrial areas

Transportation network

114,867 miles (184,882 km)	
5778 miles (9300 km)	
8120 miles (13,069 km)	
745 miles (1200 km)	

Physical and political barriers have severely limited communications between Armenia, Georgia and Azerbaijan. Turkey has a relatively well-developed transportation network.

▲ For many centuries, Istanbul has held tremendous strategic importance as a crucial gateway between Europe and Asia. Founded by the Greeks as Byzantium, the city became the center of the East Roman Empire and was known as Constantinople to the Romans. From the 15th century onward the city became the center of the great Ottoman Empire.

The landscape

The deeply eroded hills and salty basins of the Anatolian Plateau are bordered by several mountain ranges along the Black Sea coast, and the limestone Taurus Mountains (Toros Daglari) in the south. A lowland trough divides the Caucasus and the Lesser Caucasus, which form a formidable barrier of peaks in the north.

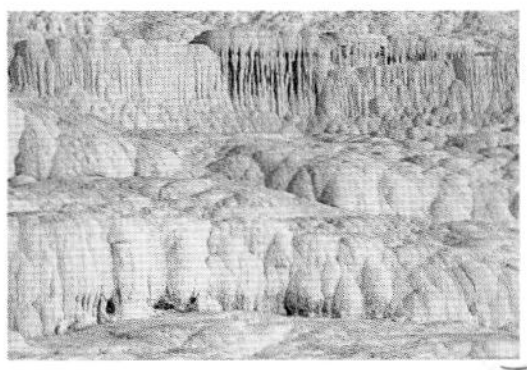

Limestone weathering in the Anatolian Plateau
- Eroded gully
- High plateau
- Layers of tephra
- Remnant landforms

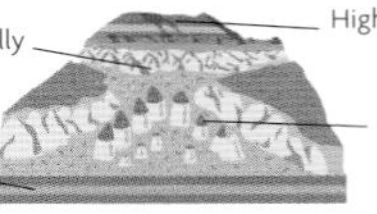

▲ In central Turkey, rainwater has chemically weathered away numerous layers of limestone, leaving isolated outcrops and pinnacles and deep eroded gullies.

▶ The Caucasus are fold mountains, which formed around the same time as the Taurus Mountains (Toros Daglari) around 65 million years ago and have since been modified by volcanic erruptions.

▲ The white rock terraces at Pamukkale in western Turkey were formed when underground water, heated by volcanic activity, dissolved minerals in the rocks. When the water reached the surface and evaporated the minerals were left behind in these extraordinary formations.

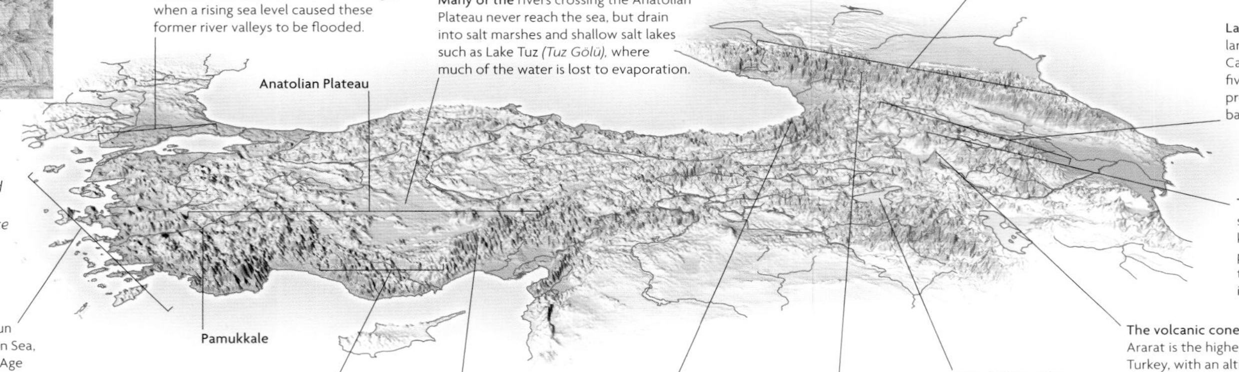

The straits of the Bosporus and the Dardanelles, respectively linking the Black and Mediterranean seas with the Sea of Marmara, formed after the last Ice Age, when a rising sea level caused these former river valleys to be flooded.

Many of the rivers crossing the Anatolian Plateau never reach the sea, but drain into salt marshes and shallow salt lakes such as Lake Tuz (Tuz Gölü), where much of the water is lost to evaporation.

Anatolian Plateau

Lava has flowed over large areas of the Lesser Caucasus within the last five million years, producing extensive basalt plateaus.

The earthquake that struck Armenia in 1988 killed over 55,000 people and devastated the country's infrastructure.

Long, parallel mountain ranges run from east to west into the Aegean Sea, which has risen since the last Ice Age to form a drowned coastline of numerous islands and extended inlets.

Pamukkale

The folded peaks of the Taurus Mountains (Toros Daglari) were formed 60–65 million years ago, at the same time as the Alps. The rock is mainly limestone, with deep caves, gorges, and underground rivers.

The Cilician Gates (Gülek Bogazi), a major pass through the Taurus Mountains (Toros Daglari), is the point where streams flow from the interior plateau onto the lowland of Adana.

Thick, temperate forest veils the seaward slopes of the Kaçkar Daglari. The southern slopes, which lie in a rainshadow, are dry and barren.

The granite massif near Surami divides the lowlands of Georgia from the oil-rich basin of Azerbaijan's Kura river, which has built a large delta into the Caspian Sea.

The shallow, saline Lake Van (Van Gölü) is the largest lake in Turkey. Dry terraces mark a previous shoreline 181 ft (55 m) above the present water level.

The volcanic cone of Mount Ararat is the highest peak in Turkey, with an altitude of 16,853 ft (5137 m).

▶ Since the 6th century BC, the pinnacles and caves of east-central Anatolia have been utilized as dwellings. Many are still inhabited today.

Map key

Population
- ▣ above 5 million
- ▣ 1 million to 5 million
- ◎ 500,000 to 1 million
- ◉ 100,000 to 500,000
- ⊕ 50,000 to 100,000
- ○ 10,000 to 50,000
- ○ below 10,000

Elevation
- 4000m / 13,124ft
- 3000m / 9843ft
- 2000m / 6562ft
- 1000m / 3281ft
- 500m / 1640ft
- 250m / 820ft
- 100m / 328ft
- sea level

Scale 1:4,500,000
Km 0 10 20 40 60 80 100 120
Miles
projection: Lambert Conformal Conic

▲ The fisheries of Azerbaijan are noted for their hauls of sturgeon, and the Caspian Sea accounts for 80% of the world's total catch. However, stocks are now under serious threat due to overfishing.

▲ Traditional steam baths are found throughout the region, and are used for socializing as well as for bathing.

The Near East

IRAQ, ISRAEL, JORDAN, LEBANON, SYRIA

Some of the world's oldest civilizations developed in this region – the Fertile Crescent – which is venerated by Jews, Muslims, and Christians, but torn by competing religious, ethnic, and national claims to the land. Turkish Ottoman rule ended with the First World War and the region was divided into areas administered by Britain and France. The UN endorsed calls for a Jewish homeland in what was then Palestine and in 1948 the state of Israel was declared. Hostility towards the Jewish state led to a series of wars with its Arab neighbors. After 2000, attempts to broker peaceful resolutions with both the Palestinian population and with adjacent Arab states were hampered by a revival of Islamic militarism and conflicting international interests in the oil-rich region. This led to an Israeli retrenchment and culminated in a US-led invasion of Iraq in 2003, which toppled the Ba'athist regime of Saddam Hussein in the name of a "war on terror".

Using the land & sea

Water scarcity limits cropland to the north and to areas watered principally by the Tigris, Euphrates, and Jordan rivers. In Israel, new irrigation techniques are allowing cultivation in the arid Negev. Wheat is the chief grain and large areas of scrub support livestock herding. Commercial produce includes dates, tobacco, citrus fruits, olives, grapes, and cotton, which is Syria's main export crop. Fishing is still important in the Mediterranean.

The urban/rural population divide

urban 70%   rural 30%

0 10 20 30 40 50 60 70 80 90 100

Population density	Total land area
217 people per sq mile (84 people per sq km)	325,460 sq miles (843,160 sq km)

Land use and agricultural distribution

- sheep
- cereals
- citrus fruits
- cotton
- dates
- fishing
- rice
- tobacco

- capital cities
- major towns
- pasture
- cropland
- wetland
- desert

Transportation & industry

The petrochemical industry is well established, and central to the economies of Syria and Iraq, which was the world's second largest oil exporter before the war with Iran which began in 1980. Lebanon has traditionally been a center for commerce, while Israel has a well-diversified economy with an expanding tourist industry, despite few natural resources.

Transportation network

- 49,859 miles (80,249 km)
- 1365 miles (2197 km)
- 3826 miles (6158 km)
- 1171 miles (1885 km)

Jordan's seaport of Al 'Aqaba is connected to Damascus in Syria by road and rail. This route to the Red Sea provides for large exports of phosphate and trade with states in the Persian Gulf.

Major industry and infrastructure

- car manufacture
- cement
- chemicals
- electronics
- finance
- food processing
- iron & steel
- oil
- oil refining
- textiles
- capital cities
- major towns
- international airports
- major roads
- major industrial areas

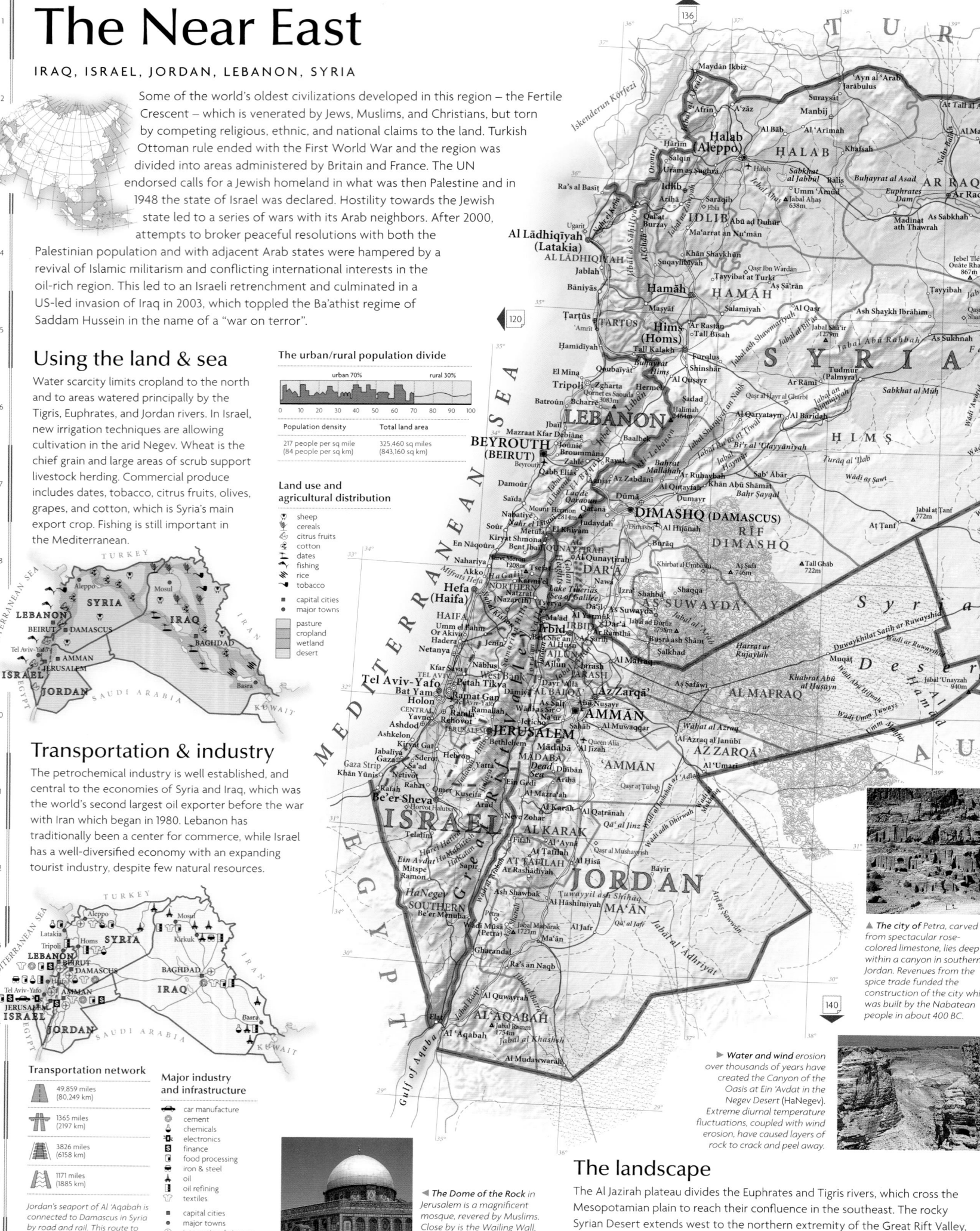

▲ The Dome of the Rock in Jerusalem is a magnificent mosque, revered by Muslims. Close by is the Wailing Wall, the city's most sacred Jewish landmark and the Church of the Holy Sepulchre, a famous Christian place of worship.

▲ The city of Petra, carved from spectacular rose-colored limestone, lies deep within a canyon in southern Jordan. Revenues from the spice trade funded the construction of the city which was built by the Nabatean people in about 400 BC.

▶ Water and wind erosion over thousands of years have created the Canyon of the Oasis at Ein 'Avdat in the Negev Desert (HaNegev). Extreme diurnal temperature fluctuations, coupled with wind erosion, have caused layers of rock to crack and peel away.

The landscape

The Al Jazirah plateau divides the Euphrates and Tigris rivers, which cross the Mesopotamian plain to reach their confluence in the southeast. The rocky Syrian Desert extends west to the northern extremity of the Great Rift Valley, which runs from the mountains of Lebanon to the Gulf of Aqaba. The Jordan river flows south along this trough into the Dead Sea, divided from the Mediterranean coastal plain by a steep-sided plateau.

► The island of El Hlayaye near Saida in southern Lebanon is linked to the mainland by a bridge built as part of the fort in the 12th century.

Map key

Population
- 1 million to 5 million
- 500,000 to 1 million
- 100,000 to 500,000
- 50,000 to 100,000
- 10,000 to 50,000
- below 10,000

Elevation
- 4000m / 13,124ft
- 3000m / 9843ft
- 2000m / 6562ft
- 1000m / 3281ft
- 500m / 1640ft
- 250m / 820ft
- 100m / 328ft
- sea level

Scale 1: 3,500,000

Km 0 10 20 40 60 80 100
Miles 0 10 20 40 60 80 100

projection: Lambert Conformal Conic

▲ The marshlands of the Tigris/Euphrates Delta were for centuries home to the Marsh Arabs, who for centuries maintained a traditional and unique lifestyle. Attempts to destroy this by Saddam Hussein's regime through drainage and genocide have now been halted.

◄ The shores of the Dead Sea are the lowest land on the Earth's surface – 1401 ft (427 m) below sea level. This highly saline lake is fed by the Jordan river but has no outlet to the sea. The water level has continued to fall in recent years, due to increased use of the Jordan river for irrigation.

Ancient eruptions of lava formed the plateau of Jabal ad Duruz which is deeply weathered and eroded along the edge of the Great Rift Valley. The lava impounded the waters of the Jordan river to form the Sea of Galilee (Lake Tiberias).

The Nahr el Litani, Lebanon's only permanent river, flows along the fertile El Beqaa Valley, which runs for 110 miles (175 km), between the Jebel Liban and Anti-Lebanon mountains.

Dead Sea

The gravel-strewn terrain of the Syrian Desert is interrupted by wadis – river valleys which remain dry for most of the year.

Iraq Marshlands

Great quantities of sediment, deposited by the Tigris and Euphrates rivers, have infilled the head of the Persian Gulf, shifting the coastline south by more than 150 miles (250 km) in the last 5000 years.

Extensive marshlands surround the lake of Hawr al Hammar, which is 70 miles (110 km) long.

Lake
Tigris
Dried salt marsh
Euphrates
Salt-covered alluvial plain

▲ The floodplains of southern Iraq are crossed by the Tigris and Euphrates rivers. Salt marshes and alluvial plains crusted with salt cover much of the area. The many small lakes are filled with brackish water and the marshes are colonized by reeds.

The Arabian Peninsula

BAHRAIN, KUWAIT, OMAN, QATAR, SAUDI ARABIA,
UNITED ARAB EMIRATES (UAE), YEMEN

Huge expanses of desert cover much of the Arabian Peninsula, limiting settlement to oases, the mountains along the Red Sea, and coastal belts. The most populous area is the fertile highlands of Yemen. The Islamic faith and Arabic language give the region a cultural and religious unity, and the Saudi city of Mecca *(Makkah)* is Islam's most holy place, visited by over two million pilgrims each year. More than half the world's oil reserves are contained in this region, and the exploitation of oil and gas has brought great wealth, particularly to Saudi Arabia. Yemen and Oman are the least developed of the Arabian states, with large rural populations. Within Saudi Arabia over 86% of the people live in urban areas.

Using the land

Most of the Arabian Peninsula is unsuited to settled agriculture, making irrigation and land reclamation projects essential. The narrow coastal plain and isolated oases, commonly amounting to less than 1% of the land area, are used to cultivate grains, coffee, and exotic fruits. Goats, sheep, and camels are widespread throughout the region.

The urban/rural population divide

urban 64%   rural 36%

0 10 20 30 40 50 60 70 80 90 100

Population density | Total land area
50 people per sq mile (19 people per sq km) | 1,147,856 sq miles (2,973,720 sq km)

Land use and agricultural distribution

- goats
- sheep
- cereals
- coffee
- dates
- fruit
- capital cities
- major towns
- pasture
- cropland
- desert

◀ *The fertile soils* of Yemen have encouraged settlement of almost all of the land from sea level up to the mountains at 10,000 ft (3050 m). In the higher reaches elaborate terraces have been constructed to facilitate crop cultivation.

The landscape

A plateau more than 2500 ft (760 m) high extends across much of the Arabian Peninsula. The plateau slopes eastward from the massive, rifted escarpment along the coast of the Red Sea, to the shallow waters of the Persian Gulf. The interior is characterized by *cuestas* and valleys, drained by a system of *wadis*. A crescent of sand and gravel deserts lies to the east.

The An Nafud Desert is covered with *barchan* dunes varying between 30–100 ft (10–30 m) high. The "horns" of the crescent-shaped dunes reflect the direction in which they are being moved by the wind.

Inselbergs are dotted over a wide area of the Najd Plateau. These resistant remnants of the ancient basement rock are left standing when the softer weathered rock has been worn away.

Evaporation | Crusted layer left behind
Storm surge flooding
Normal level of tidal range
Salt wedge penetrates inland water

▲ *A sabkha is a* flat, salt-encrusted plain which occurs near the coast just above the high water mark. Flooding by sea water leads to saturation of the land with saline-rich groundwater. As this evaporates, a cracked layer of sand, cemented together with salt, gypsum, and calcium carbonate is left behind.

Across the Najd Plateau the flat relief is broken by *mesas*; steep-sided rock plateaus and *cuestas*; ridges with one steep and one gentle slope.

Few areas in the Arabian Peninsula have rivers flowing through them. Most are drained by ephemeral watercourses called *wadis*.

The Hejaz *(Al Hijaz)* and Asir mountains form part of the same geological region as the highlands of Sudan and Eritrea, to which they were once joined. They were separated when faulting opened the Red Sea, over 50 million years ago.

▲ *Ar Rub' al Khali*, also known as the Empty Quarter, is the most arid part of the Arabian Peninsula. It is the largest uninterrupted sand desert in the world. Ridges of sand up to 25 miles (40 km) long, run northeast–southwest, giving characteristic linear dunes.

The Jabal an Nabi Shu'ayb in Yemen is the highest point on the peninsula, rising to 12,336 ft (3760 m).

The Arabian Shield underpins the west of the peninsula. It is a fragment of the ancient continent, Gondwanaland, which was separated by rifting millions of years ago.

◀ *Every Muslim must* make at least one pilgrimage or hajj to Mecca *(Makkah)*, in Saudi Arabia, during their lifetime. The cloth-covered shrine is called the Ka'bah, and is regarded by Muslims as the most sacred place on Earth.

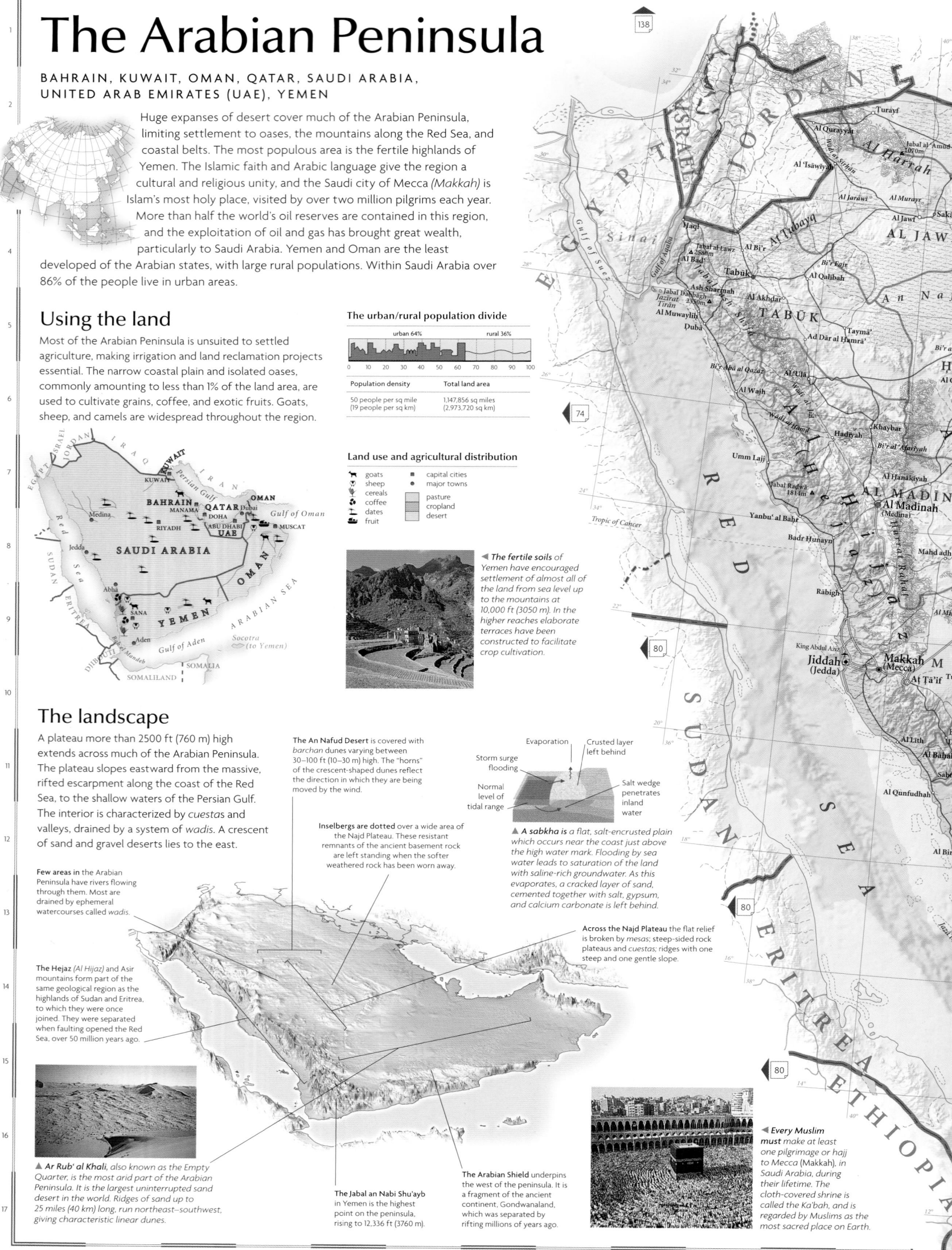

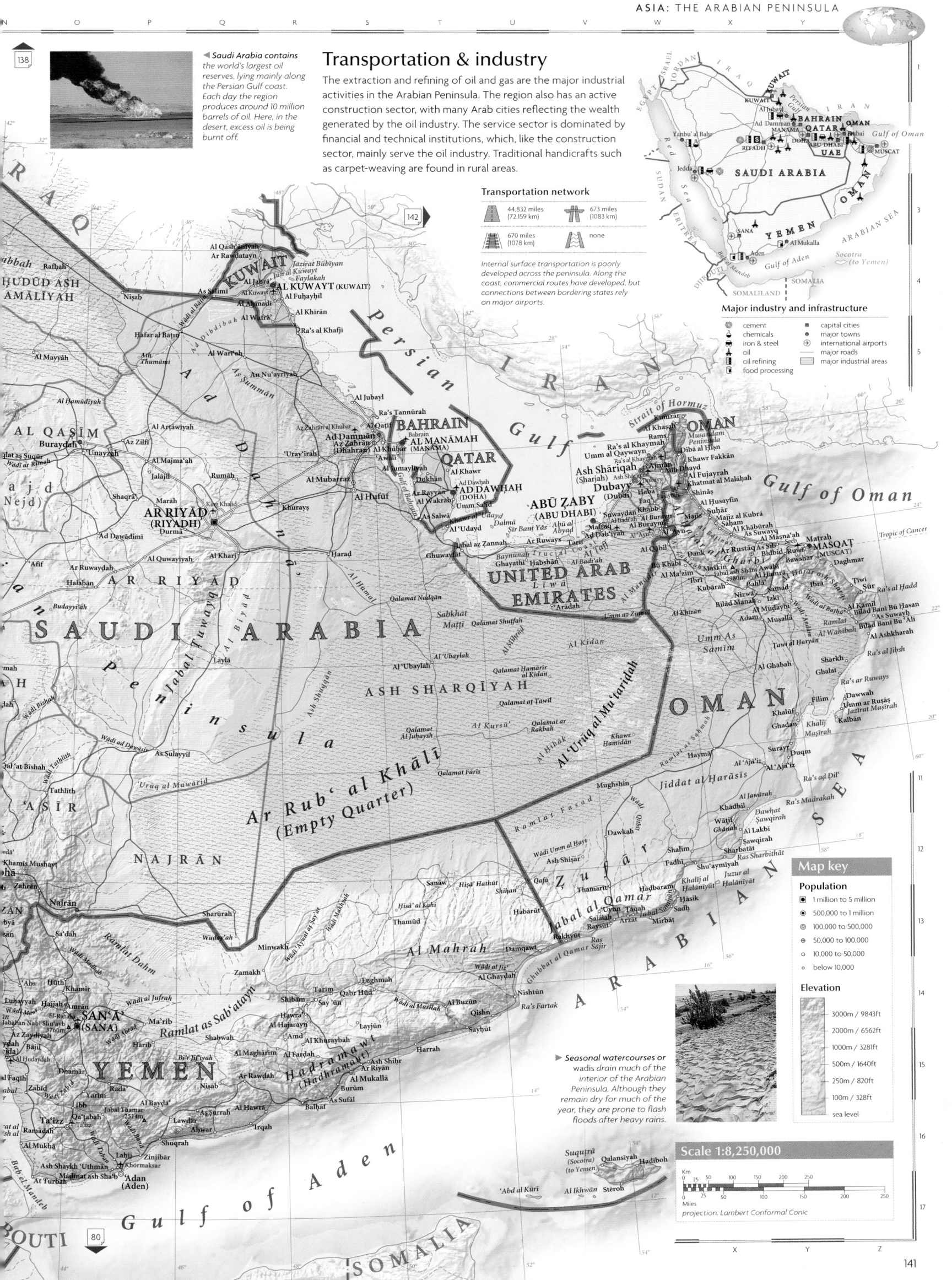

Saudi Arabia contains the world's largest oil reserves, lying mainly along the Persian Gulf coast. Each day the region produces around 10 million barrels of oil. Here, in the desert, excess oil is being burnt off.

Transportation & industry

The extraction and refining of oil and gas are the major industrial activities in the Arabian Peninsula. The region also has an active construction sector, with many Arab cities reflecting the wealth generated by the oil industry. The service sector is dominated by financial and technical institutions, which, like the construction sector, mainly serve the oil industry. Traditional handicrafts such as carpet-weaving are found in rural areas.

Transportation network

44,832 miles (72,159 km)	673 miles (1083 km)
670 miles (1078 km)	none

Internal surface transportation is poorly developed across the peninsula. Along the coast, commercial routes have developed, but connections between bordering states rely on major airports.

Major industry and infrastructure

- cement
- chemicals
- iron & steel
- oil
- oil refining
- food processing
- capital cities
- major towns
- international airports
- major roads
- major industrial areas

Seasonal watercourses or wadis drain much of the interior of the Arabian Peninsula. Although they remain dry for much of the year, they are prone to flash floods after heavy rains.

Map key

Population
- 1 million to 5 million
- 500,000 to 1 million
- 100,000 to 500,000
- 50,000 to 100,000
- 10,000 to 50,000
- below 10,000

Elevation
- 3000m / 9843ft
- 2000m / 6562ft
- 1000m / 3281ft
- 500m / 1640ft
- 250m / 820ft
- 100m / 328ft
- sea level

Scale 1:8,250,000

projection: Lambert Conformal Conic

Iran & the Gulf states

BAHRAIN, IRAN, KUWAIT, QATAR, UNITED ARAB EMIRATES (UAE)

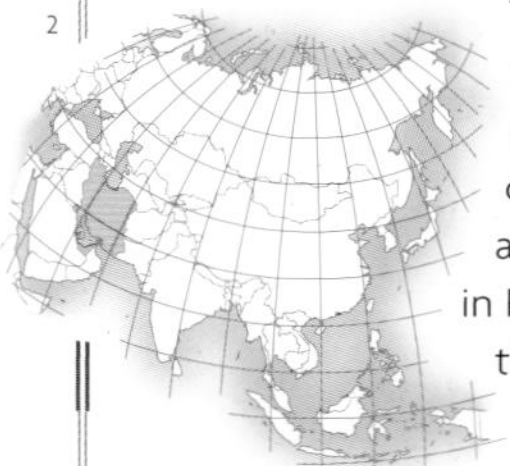

The discovery of oil in the Persian Gulf in the 1930s brought great wealth to the surrounding states. The revenue was largely used to modernize industry and infrastructure, initiating great social change in these formerly agrarian countries. Today, over 90% of the people in the Gulf states live in urban areas, and foreign nationals make up a sizeable proportion of the population in Kuwait, Qatar, and the United Arab Emirates. The importance of control of the oil reserves has led to a number of territorial disputes, including most recently the Iran–Iraq War (1980-88) and the First Gulf War (1991). Islam is practiced almost exclusively throughout the region and two distinct strands are found; Sunni Muslims in Qatar, Kuwait, and UAE, and Shi'a Muslims in Iran and Bahrain. In 1979 Iran became the world's largest theocracy.

The landscape

The land rises steeply from the fragmented coastal lowlands bordering the Persian Gulf, to reach Iran's interior plateau, bounded by heavily eroded mountain chains. An unstable plate boundary runs northwest to southeast across Iran causing frequent earthquakes. On the sandy west coast of the Persian Gulf, the relief is generally flat, with patches of salt marsh. Bahrain consists of two groups of islands, which are mostly small and rocky.

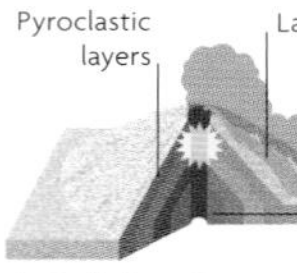

Pyroclastic layers

Lava flow

Lava flow layers

▲ *Qolleh-ye Damavand* in the Elburz Mountains is a composite volcano. It comprises layers of lava and pyroclasts fragmentary rocks which accumulate on the slopes of the volcano after being ejected into the air.

▲ *Marine sediments from* deep beneath the ancient Tethys Sea have been uplifted to form the Elburz Mountains, which stretch along the shores of the Caspian Sea, northern Iran.

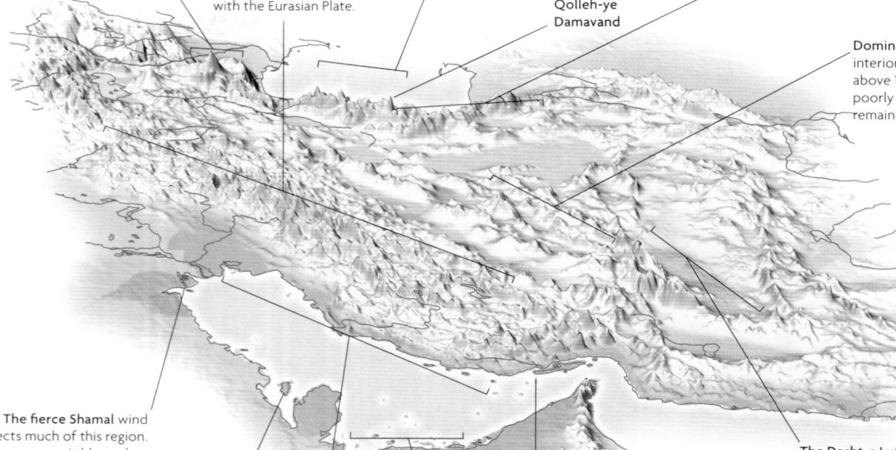

Lava and ash from previous volcanic activity covers a 200 mile (320 km) stretch from the border with Azerbaijan to the Caspian Sea.

Iran's two mountain chains, the Zagros and Elburz, were uplifted at the same time as the Alps in Europe, when the African Plate collided with the Eurasian Plate.

Caspian Sea

Qolleh-ye Damavand

Dominated by a vast, semi-arid interior plateau, most of Iran lies above 1640 ft (500 m). The region is poorly drained with many of its basins remaining dry for months at a time.

The fierce Shamal wind affects much of this region. Every summer it blows dust south from the flood plains of the Tigris and Euphrates, reducing visibility to such an extent that Kuwait International Airport is frequently forced to close.

The Dasht-e Lut

Autumn winds blowing across the Persian Gulf can reach speeds of up to 95 mph (150 kmph) causing severe storms, squalls, and waterspouts.

Prolific springs tapping artesian water make cultivation possible across the north of Bahrain's main island. This provides a sharp contrast to the sandy plains in the south and west.

The oilfields of the Persian Gulf are formed from marine shale deposits lying in sedimentary basins at the margins of the Zagros Mountains.

Numerous islands lie along the southern coast of the Persian Gulf. Some of these are salt domes, created when less dense salts were displaced and forced up to the surface by denser, overlying strata.

◄ *The Dasht-e Lut* covers a large portion of eastern Iran with its dry, wind-eroded plain of scattered sandstone pillars and salty depressions. During the summer, temperatures soar, making it one of the world's hottest, driest places.

Using the land & sea

Along the coast of the Caspian Sea, desalinated water allows fruits and vegetables to be produced, although water shortages and desert soils still limit farming. Sheep are the most important livestock raised in Iran and commercial forests cover the northwest of the country. Shrimp stocks were decimated by pollution during the Gulf War, but fishing remains important for domestic and export markets.

◄ *All of the* Gulf states have commercial fishing fleets. Before the discovery of oil, fishing was the region's leading industry.

Land use and agricultural distribution

- goats
- sheep
- cereals
- citrus fruits
- cotton
- dates
- fishing
- timber

- ■ capital cities
- ● major towns

pasture
cropland
forest
desert
wetland

The urban/rural population divide

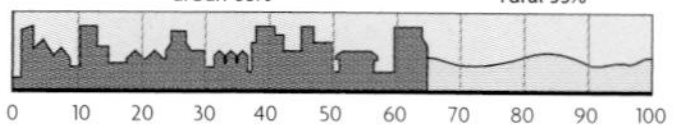

urban 65%   rural 35%

0 10 20 30 40 50 60 70 80 90 100

Population density	Total land area
112 people per sq mile (43 people per sq km)	642,883 sq miles (1,665,500 sq km)

◄ *The Kuwait Towers* in the center of Kuwait are symbols of the vast wealth oil has brought to the country. Before 1960, the city had only one main street and was surrounded by a mud wall.

◀ *Many volcanoes lie* in Iran's 1200 mile (1930 km) volcanic belt, including the country's highest peak, the now-extinct Qolleh-ye Damavand at 18,600 ft (5671 m).

▶ *Extensive oil and* gas exploitation in the Gulf region has allowed the economic transformation of the Gulf states. Consequently, many of these states have a hugely improved per capita income compared to the 1960's.

Transportation & industry

Both onshore and offshore oil reserves are exploited throughout the region. Kuwait not only extracts but also refines 80% of its oil. Bahrain has diversified its economy to become the main commercial and financial center in the Persian Gulf. Iran produces a wide range of products: textile mills are widespread and carpet weaving is an important export industry.

Major industry and infrastructure

- carpet manufacture
- chemicals
- finance
- food processing
- oil
- oil refining
- textiles
- capital city
- major towns
- international airports
- major roads
- major industrial areas

Transportation network

63,543 miles (102,274 km)	884 miles (1423 km)
3822 miles (6151 km)	562 miles (904 km)

Major towns and neighboring countries are linked by adequate road networks, although rural areas are less well served. Bahrain is linked to the mainland by a 15 mile (25 km) long causeway.

Map key

Population

- ■ above 5 million
- ◪ 1 million to 5 million
- ◉ 500,000 to 1 million
- ◎ 100,000 to 500,000
- ⊕ 50,000 to 100,000
- ⊙ 10,000 to 50,000
- ○ below 10,000

Elevation

- 4000m / 13,124ft
- 3000m / 9843ft
- 2000m / 6562ft
- 1000m / 3281ft
- 500m / 1640ft
- 250m / 820ft
- 100m / 328ft
- sea level

Scale 1:6,000,000

projection: Lambert Conformal Conic

Map labels

TURKMENISTAN
AFGHANISTAN
PAKISTAN
IRAN
OMAN
QATAR
BAHRAIN
UNITED ARAB EMIRATES
Gulf of Oman
Makran Coast
Strait of Hormuz
Dasht-e Kavir
Dasht-e Lut
Iranian Plateau
Zagros Mountains
Alborz (Elburz Mountains)
Kavir-e Namak

GOLESTAN
KHORĀSĀN-E SHOMĀLĪ
KHORĀSĀN-E RAZAVĪ
KHORĀSĀN-E JONŪBĪ
MAZANDARĀN
SEMNĀN
ESFAHĀN
YAZD
KERMĀN
FĀRS
HORMOZGĀN
SĪSTĀN VA BALŪCHESTĀN
ČAHĀR MAHALL VA BAKHTĪĀRĪ
QOM

Tehrān, Mashhad, Eşfahān, Shīrāz, Kermān, Yazd, Zāhedān, Bandar-e ʿAbbās, Gorgan, Bojnūrd, Qom, Shiraz, Abu Zaby (Abu Dhabi), Ad Dawhah (Doha), Dubayy (Dubai), Ash Shāriqah (Sharjah), Ra's al Khaymah

Kazakhstan

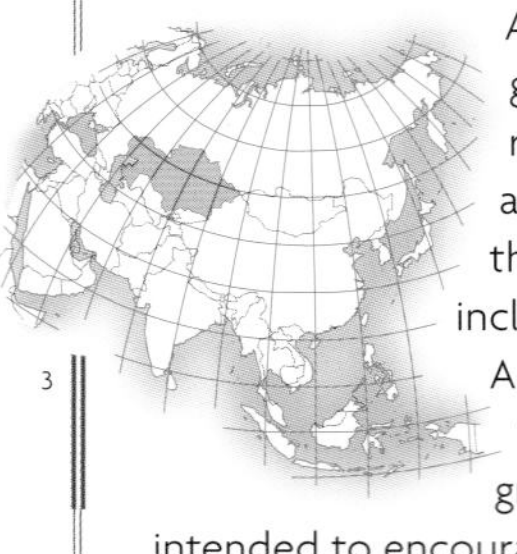

Abundant natural resources lie in the immense steppe grasslands, deserts, and central plateau of the former Soviet republic of Kazakhstan. An intensive program of industrial and agricultural development to exploit these resources during the Soviet era resulted in catastrophic industrial pollution, including fallout from nuclear testing and the shrinkage of the Aral Sea. Since independence, the government has encouraged foreign investment and liberalized the economy to promote growth. The adoption of Kazakh as the national language is intended to encourage a new sense of national identity in a state where living conditions for the majority remain harsh, both in cramped urban centers and impoverished rural areas.

Transportation & industry

The single most important industry in Kazakhstan is mining, based around extensive oil deposits near the Caspian Sea, the world's largest chromium mine, and vast reserves of iron ore. Recent foreign investment has helped to develop industries including food processing and steel manufacture, and to expand the exploitation of mineral resources. The Russian space program is still based at Baykonyr, near Kyzylorda in central Kazakhstan.

Major industry and infrastructure

- chemicals
- engineering
- fish processing
- food processing
- iron & steel
- metallurgy
- mining
- oil
- capital cities
- major towns
- international airports
- major roads
- major industrial areas

Transportation network

- 48,263 miles (77,680 km)
- none
- 8483 miles (13,660 km)
- 3900 miles (2423 km)

Industrial areas in the north and east are well-connected to Russia. Air and rail links with Germany and China have been established through foreign investment. Better access to Baltic ports is being sought.

◄ *An open-pit coal mine in Kazakhstan. Foreign investment is being actively sought by the Kazakh government in order to fully exploit the potential of the country's rich mineral reserves.*

Map key

Population
- ▣ 1 million to 5 million
- ◉ 500,000 to 1 million
- ⊙ 100,000 to 500,000
- ⊕ 50,000 to 100,000
- ○ 10,000 to 50,000
- ∘ below 10,000

Elevation
- 4000m / 13,124ft
- 3000m / 9843ft
- 2000m / 6562ft
- 1000m / 3281ft
- 500m / 1640ft
- 250m / 820ft
- 100m / 328ft
- sea level

Using the land & sea

The rearing of large herds of sheep and goats on the steppe grasslands forms the core of Kazakh agriculture. Arable cultivation and cotton-growing in pasture and desert areas was encouraged during the Soviet era, but relative yields are low. The heavy use of fertilizers and the diversion of natural water sources for irrigation has degraded much of the land.

Land use and agricultural distribution

- cattle
- goats
- sheep
- cotton
- fishing
- wheat
- capital cities
- major towns
- pasture
- cropland
- forest
- mountain region
- desert

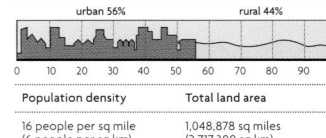

The urban/rural population divide

urban 56%	rural 44%

Population density	Total land area
16 people per sq mile (6 people per sq km)	1,048,878 sq miles (2,717,300 sq km)

◄ *The nomadic peoples who moved their herds around the steppe grasslands are now largely settled, although echoes of their traditional lifestyle, in particular their superb riding skills, remain.*

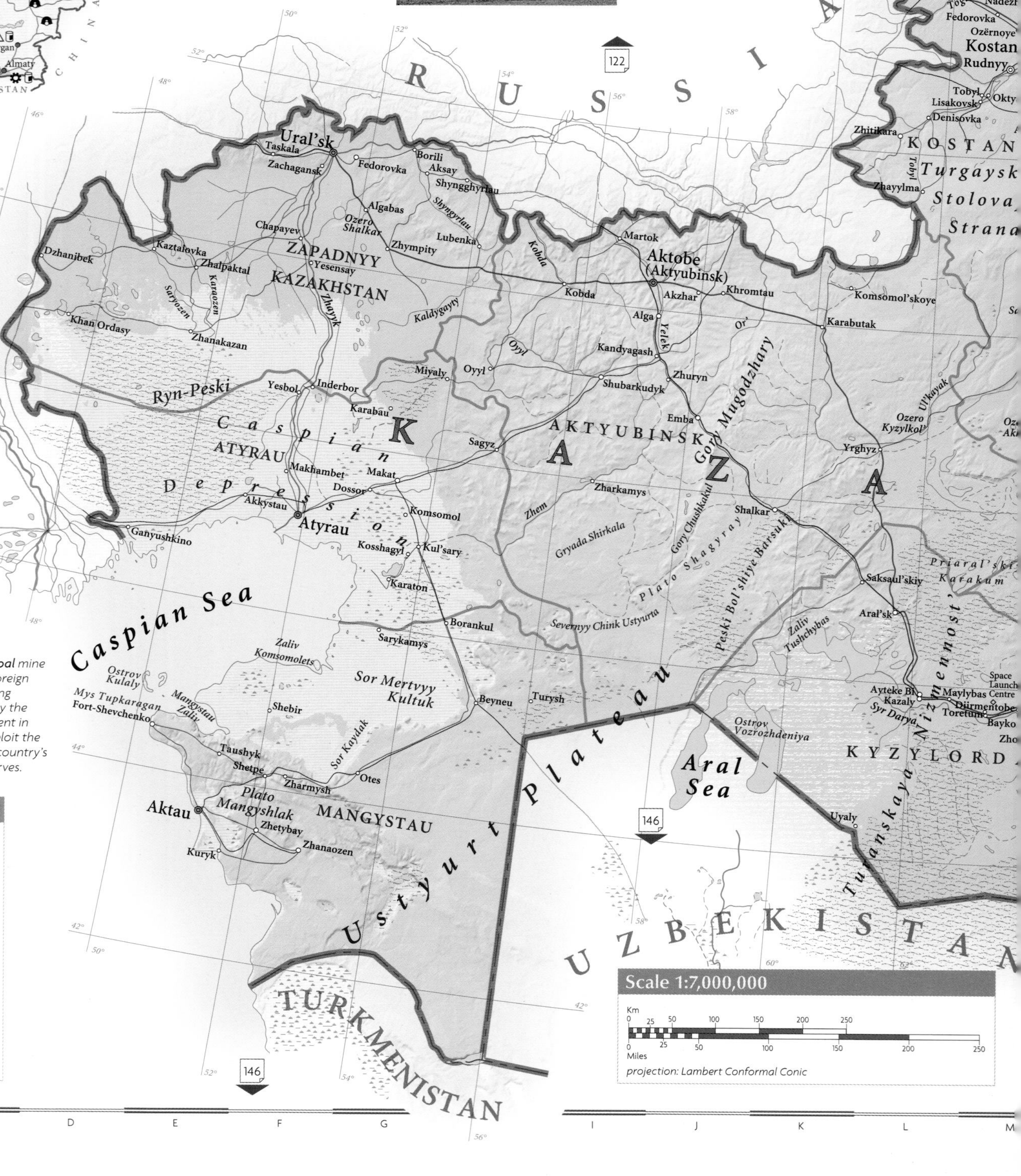

Scale 1:7,000,000

projection: Lambert Conformal Conic

The landscape

Stretching more than 1250 miles (2000 km) from the Caspian Sea in the west to China in the east, more than 40% of Kazakhstan is covered by steppe grasslands which give way to barren desert in the south. The land rises eastward towards the mineral-rich central plateau, to form the Altai Mountains.

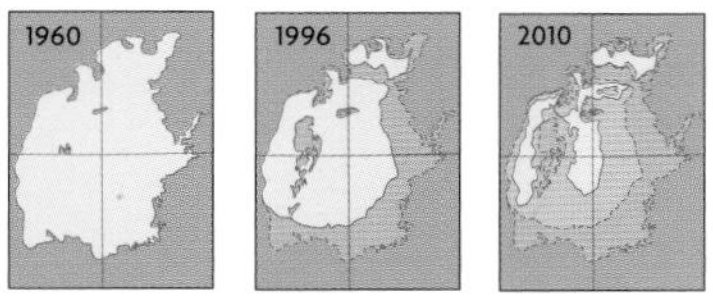

1960   1996   2010

▲ Since 1960, the Aral Sea has shrunk by 75%, become extremely saline, and lost all but five of its once-abundant fish species. Factors in this ecological disaster include the excessive use of fertilizers, defoliants and the diversion of its main source rivers for the irrigation of desert lands.

The Caspian Sea is the largest body of inland water in the world.

The desert of Peski Bol'shiye Barsuki is mainly sandy, displaying a number of classic dune formations. Groundwater supports a small amount of vegetation.

A large number of salt lakes fill depressions in the rolling uplands of central Kazakhstan.

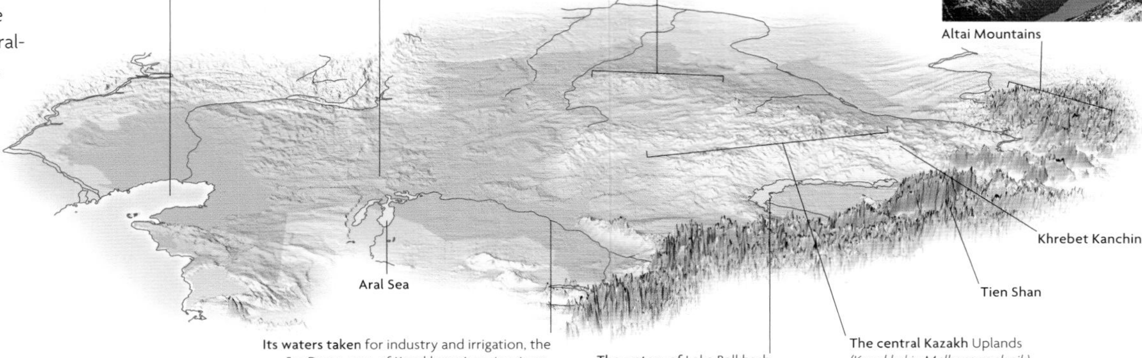

▶ The Altai Mountains lie on Kazakhstan's eastern borders with China and the Russian Federation. Cold and largely barren, they are the source of many of the rivers which flow across the steppe.

Altai Mountains

Khrebet Kanchingiz

Tien Shan

Aral Sea

Its waters taken for industry and irrigation, the Syr Darya, one of Kazakhstan's major rivers, now barely reaches the Aral Sea which it used to fill. Like many Kazakh rivers it has been heavily polluted with chemicals and its flow has been restricted by up to 60%.

The waters of Lake Balkhash (Ozero Balkhash), unlike those of the Aral Sea, are still able to support a fishing industry.

The central Kazakh Uplands (Kazakhskiy Melkosopochnik) contain much of the country's mineral riches. The landscape is largely flat with occasional rocky outcrops and hillocks.

▶ Immense stretches of steppe grasslands characterize much of the Kazakh landscape. These lowland areas have been used for arable cultivation in recent years, although problems with irrigation have meant that much of the land is being allowed to revert to its natural vegetation and pastoral usage.

▲ Rows of pine trees edge this valley near Almaty. The snow-covered slopes in the background are used for skiing.

145

Central Asia

KYRGYZSTAN, TAJIKISTAN, TURKMENISTAN, UZBEKISTAN

The four republics that declared independence in 1991 were created in the early years of the Soviet Union, promoting ethnic divisions in a region whose common focus, since the 8th century, has been Islam. Traditional rural, nomadic ways of life have survived the Soviet era, while the benefits of modern industry and grand irrigation schemes have resulted in severe pollution in the delicate, arid environment of the steppe, particularly in Uzbekistan. Many ethnic minority groups are scattered among the four republics, with isolated communities in the mountains of Kyrgyzstan.

The current Islamic revival has brought hope of greater regional unity, in spite of religious factionalism which, in 1992, plunged Tajikistan into civil war.

◀ **The desert** of the Kara Kum (Garagum) occupies over 70% of Turkmenistan; its wind-scoured surface of dune ridges and depressions severely limits human settlement.

▲ **The southern shoreline** of the Aral Sea has retreated over 30 miles (48 km) since 1960. A major cause is the diversion of water from the Amu Darya river for irrigation via the Kara Kum Canal (Garagum Kanaly).

Map key

Population
- ◙ 1 million to 5 million
- ◉ 500,000 to 1 million
- ◎ 100,000 to 500,000
- ⊕ 50,000 to 100,000
- ○ 10,000 to 50,000
- ∘ below 10,000

Elevation
- 6000m / 19,686ft
- 4000m / 13,124ft
- 3000m / 9843ft
- 2000m / 6562ft
- 1000m / 3281ft
- 500m / 1640ft
- 250m / 820ft
- 100m / 328ft
- sea level

Transportation & industry

Fossil fuels are extracted and processed in all four states, with scope for further exploitation. Agriculture provides raw materials for many industries, including food and textiles processing, and the manufacture of leather goods, clothing, and carpets. Farm machinery is also produced.

Transportation network

🛣 73,658 miles (118,555 km)		✈ 87 miles (140 km)	
🚆 4773 miles (7683 km)		⚓ 1180 miles (1900 km)	

The Kara Kum Canal (Garagum Kanaly) runs for 870 miles (1400 km) from the Amu Darya river to the Caspian Sea. The canal is principally used for irrigation but is navigable for 280 miles (450 km).

Major industry and infrastructure

- 🐫 carpet weaving
- 🜂 chemicals
- ⚙ engineering
- 🍴 food processing
- 🛢 oil & gas
- ⊤ textiles

- ■ capital cities
- ■ major towns
- ⊕ international airports
- — major roads
- ▨ major industrial areas

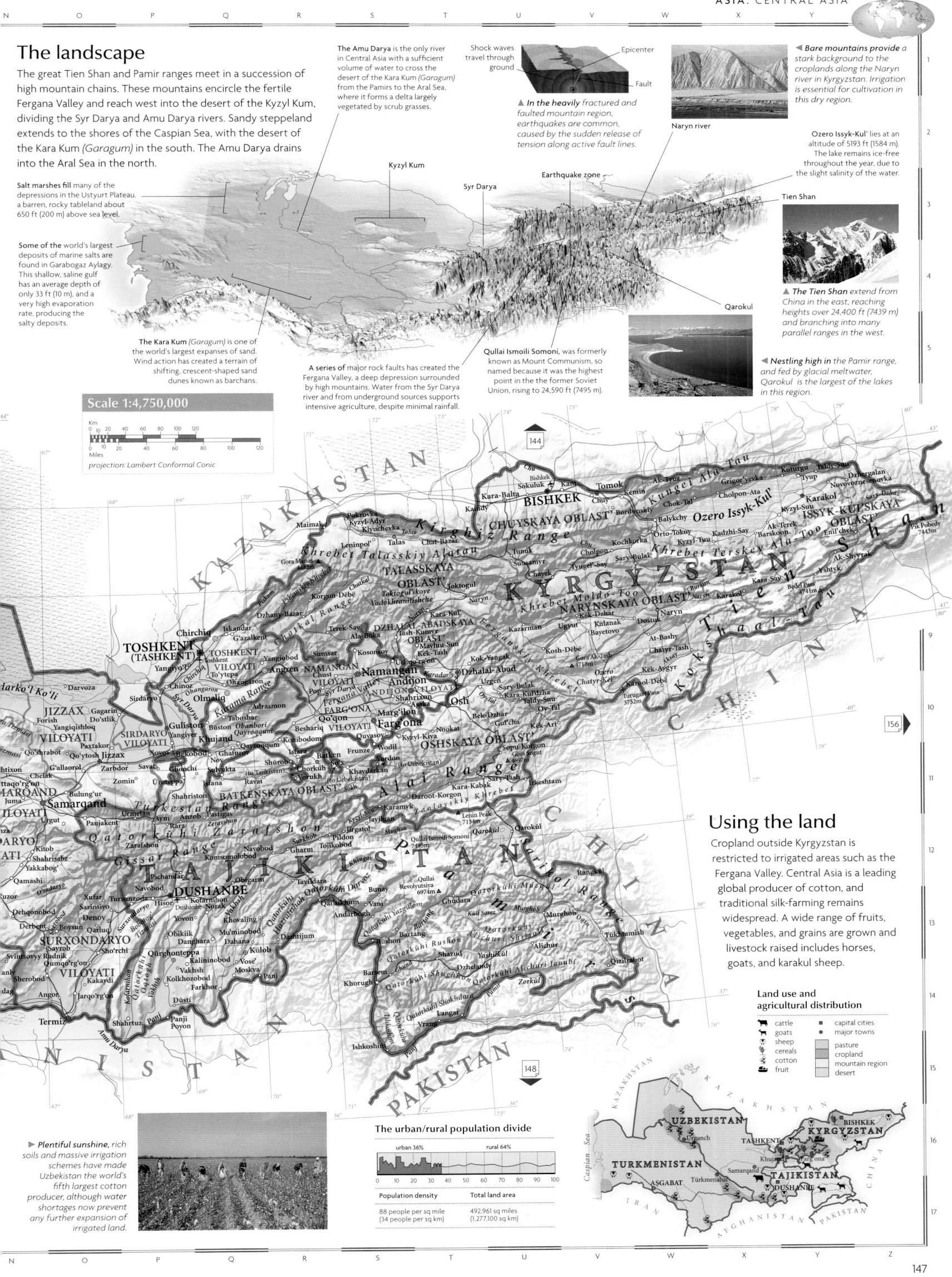

The landscape

The great Tien Shan and Pamir ranges meet in a succession of high mountain chains. These mountains encircle the fertile Fergana Valley and reach west into the desert of the Kyzyl Kum, dividing the Syr Darya and Amu Darya rivers. Sandy steppeland extends to the shores of the Caspian Sea, with the desert of the Kara Kum (Garagum) in the south. The Amu Darya drains into the Aral Sea in the north.

Salt marshes fill many of the depressions in the Ustyurt Plateau, a barren, rocky tableland about 650 ft (200 m) above sea level.

Some of the world's largest deposits of marine salts are found in Garabogaz Aylagy. This shallow, saline gulf has an average depth of only 33 ft (10 m), and a very high evaporation rate, producing the salty deposits.

The Kara Kum (Garagum) is one of the world's largest expanses of sand. Wind action has created a terrain of shifting, crescent-shaped sand dunes known as barchans.

A series of major rock faults has created the Fergana Valley, a deep depression surrounded by high mountains. Water from the Syr Darya river and from underground sources supports intensive agriculture, despite minimal rainfall.

The Amu Darya is the only river in Central Asia with a sufficient volume of water to cross the desert of the Kara Kum (Garagum) from the Pamirs to the Aral Sea, where it forms a delta largely vegetated by scrub grasses.

Qullai Ismoili Somoni, was formerly known as Mount Communism, so named because it was the highest point in the the former Soviet Union, rising to 24,590 ft (7495 m).

▲ **In the heavily** fractured and faulted mountain region, earthquakes are common, caused by the sudden release of tension along active fault lines.

◀ **Bare mountains provide** a stark background to the croplands along the Naryn river in Kyrgyzstan. Irrigation is essential for cultivation in this dry region.

Ozero Issyk-Kul' lies at an altitude of 5193 ft (1584 m). The lake remains ice-free throughout the year, due to the slight salinity of the water.

▲ **The Tien Shan** extend from China in the east, reaching heights over 24,400 ft (7439 m) and branching into many parallel ranges in the west.

◀ **Nestling high in** the Pamir range, and fed by glacial meltwater, Qarokul is the largest of the lakes in this region.

Scale 1:4,750,000
projection: Lambert Conformal Conic

Using the land

Cropland outside Kyrgyzstan is restricted to irrigated areas such as the Fergana Valley. Central Asia is a leading global producer of cotton, and traditional silk-farming remains widespread. A wide range of fruits, vegetables, and grains are grown and livestock raised includes horses, goats, and karakul sheep.

Land use and agricultural distribution

- cattle
- goats
- sheep
- cereals
- cotton
- fruit
- capital cities
- major towns
- pasture
- cropland
- mountain region
- desert

▶ **Plentiful sunshine,** rich soils and massive irrigation schemes have made Uzbekistan the world's fifth largest cotton producer, although water shortages now prevent any further expansion of irrigated land.

The urban/rural population divide

urban 36%   rural 64%

Population density
88 people per sq mile
(34 people per sq km)

Total land area
492,961 sq miles
(1,277,100 sq km)

147

A B C D E F G H I J K L M

Afghanistan & Pakistan

Pakistan was created by the partition of British India in 1947, becoming the western arm of a new Islamic state for Indian Muslims; the eastern sector, in Bengal, seceded to become the separate country of Bangladesh in 1971. Over half of Pakistan's 158 million people live in the Punjab, at the fertile head of the great Indus Basin. The river sustains a national economy based on irrigated agriculture, including cotton for the vital textiles industry. Afghanistan, a mountainous, landlocked country, with an ancient and independent culture, has been wracked by war since 1979. Factional strife escalated into an international conflict in late 2001, as US-led troops ousted the militant and fundamentally Islamist *taliban* regime as part of their "war on terror."

◄ *The town of* Bamian lies high inthe Hindu Kush west of Kabul. Between the 2nd and 5th centuries two huge statues of Buddha were carved into the nearby rock,the largest of which stood 125 ft (38 m) high. The statues were destroyed by the taliban regime in March 2001.

Transportation & industry

Pakistan is highly dependent on the cotton textiles industry, although diversified manufacture is expanding around cities such as Karachi and Lahore. Afghanistan's limited industry is based mainly on the processing of agricultural raw materials and includes traditional crafts such as carpet weaving.

Major industry and infrastructure

- 🧵 carpet weaving
- ⚗ chemicals
- ⚙ engineering
- Ⓢ finance
- food processing
- iron & steel
- ⬧ oil & gas
- textiles
- ■ capital cities
- ● major towns
- ⊕ international airports
- — major roads
- ▢ major industrial areas

Transportation network

🛣	96,154 miles (154,763 km)
🛣	211 miles (340 km)
🚆	4852 miles (7814 km)
✈	745 miles (1200 km)

The Karakoram Highway was completed after 20 years of construction in 1978. It breaches the Himalayan mountain barrier providing a commercial motor route linking lowland Pakistan and China.

► *The Karakoram Highway* is one of the highest major roads in the world. It took over 24,000 workers almost 20 years to complete.

The landscape

Afghanistan's topography is dominated by the mountains of the Hindu Kush, which spread south and west into numerous mountain spurs. The dry plateau of southwestern Afghanistan extends into Pakistan and the hills which overlook the great Indus Basin. In northern Pakistan the Hindu Kush, Himalayan, and Karakoram ranges meet to form one of the world's highest mountain regions.

◄ *The Hunza river* rises in the northern Karakoram Range, running for 120 miles (193 km) before joining the Gilgit river.

Hunza river

The plains and foothills which extend from the northern slopes of the Hindu Kush are part of the great grassy steppe lands of Central Asia.

Hindu Kush

K2 (Mount Godwin Austen), in the Karakoram Range, is the second highest mountain in the world, at an altitude of 28,251 ft (8611 m).

► *The arid Hindu Kush* makes much of Afghanistan uninhabitable, with over 50% of the land lying above 6500 ft (2000 m).

Some of the largest glaciers outside the polar regions are found in the Karakoram Range, including Siachen Glacier (Siachen Muztagh), which is 40 miles (72 km) long.

Himalayas

Frequent earthquakes mean that mountain-building processes are continuing in this region, as the Indo-Australian Plate drifts northward, colliding with the Eurasian Plate.

The soils of the Punjab plain are nourished by enormous quantities of sediment, carried from the Himalayas by the five tributaries of the Indus river.

Mountain chains running southwest from the Hindu Kush into Pakistan form a barrier to the humid winds which blow from the Indian Ocean, creating arid conditions across southern Afghanistan.

Glacis covered by coarse-grained sediment

Sediments washed down from mountains accumulate on glacis slopes

Fine sediments deposited on salt flats are removed by wind erosion.

The Indus Basin is part of the Indus-Ganges lowland, a vast depression which has been filled with layers of sediment over the last 50 million years. These deposits are estimated to be over 16,400 ft (5000 m) deep.

The Indus Delta is prone to heavy flooding and high levels of salinity. It remains a largely uncultivated wilderness area.

Bedrock

▲ *Glacis are gentle,* debris-covered slopes which lead into saltflats or deserts. They typically occur at the base of mountains in arid regions such as Afghanistan.

Map labels

TURKMENISTAN · UZBEKISTAN · TAJIKISTAN · CHINA · IRAN · AFGHANISTAN · PAKISTAN · INDIA · ARABIAN SEA

Mazar-e Sharif · Herat · KABUL · Peshawar · ISLAMABAD · Rawalpindi · Kandahar · Lahore · Quetta · Faisalabad · Multan · Bahawalpur · Sukkur · Karachi · Hyderabad

Bālā Murghāb · Selseleh-... · BADGHIS · Kāriz-e Elyās · Towraghoudi · Qarah Bāgh · Kushk · Qal'ah-ye Now · Qādis · Eslām Qal'eh · Kūhestān · Dasht-e Hamdam Āb · Zindah Jān · Ghōriān · Herāt · HERĀT · GHOR · Namakzar · Darya-ye M... · AFGH · Dak · Shindand · Kūh-e Chehel Abd... · Farāh Rūd · Dasht-e Bābūs · FARĀH · Anār Darah · Farāh · Dilārām · Now Zāc · Hāmūn-e Şāberī · Dasht-e Khāsh · Gereshk · Hāmūn-e Pūzak · NĪMRŌZ · Lashkar Gāh · Chakhānsūr · Shelleh-ye Pūdeh Tal · Darwēshān · Kūchna · Zaranj · Dasht-e Mārgow · Darwēsh... · HELMAND · Daryā-ye Helmand · Dīshū · Dasht-e Gowd-e Zereh · Chāgai Hills · Hāmūn Lo... · BAĪ · Nok Kundi · Yakmach · Dasht-i Tāhlāb · Dālbandin · Tāhlāb · Hāmūn-i Māshkel · Kamarod · Sīāhān Ra... · Tag... · Panjgūr · Central Mak... · Ispikān · Nihing · Māla... · Mand · Nasirābād · Kech · Hoshab · Dasht · Turbat · Suntsar · Khor Kalamat · Jiwani · Gwādar · Gwādar West Bay · Gwādar East Bay · Pasni · Astola Island · Ormār...

146 · 142 · 142

36° · 62° · 60° · 34° · 32° · 30° · 28° · 26° · 62° · 64°

A B C D E F G H I J K L M

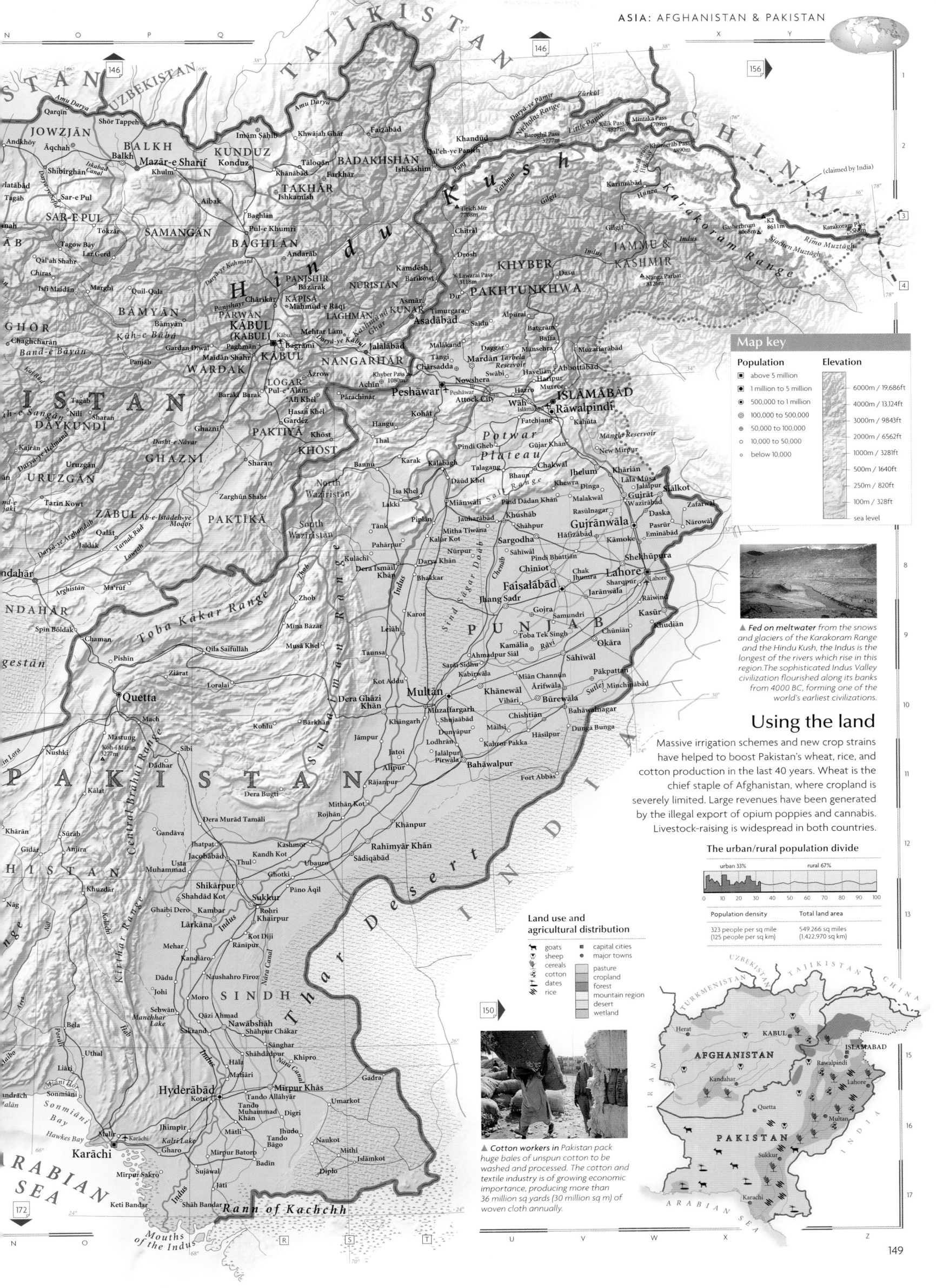

Map key

Population

- ■ above 5 million
- ■ 1 million to 5 million
- ◉ 500,000 to 1 million
- ◎ 100,000 to 500,000
- ⊙ 50,000 to 100,000
- ○ 10,000 to 50,000
- ○ below 10,000

Elevation

- 6000m / 19,686ft
- 4000m / 13,124ft
- 3000m / 9843ft
- 2000m / 6562ft
- 1000m / 3281ft
- 500m / 1640ft
- 250m / 820ft
- 100m / 328ft
- sea level

▲ *Fed on meltwater* from the snows and glaciers of the Karakoram Range and the Hindu Kush, the Indus is the longest of the rivers which rise in this region. The sophisticated Indus Valley civilization flourished along its banks from 4000 BC, forming one of the world's earliest civilizations.

Using the land

Massive irrigation schemes and new crop strains have helped to boost Pakistan's wheat, rice, and cotton production in the last 40 years. Wheat is the chief staple of Afghanistan, where cropland is severely limited. Large revenues have been generated by the illegal export of opium poppies and cannabis. Livestock-raising is widespread in both countries.

The urban/rural population divide

urban 33%   rural 67%

0 10 20 30 40 50 60 70 80 90 100

Population density	Total land area
323 people per sq mile (125 people per sq km)	549,266 sq miles (1,422,970 sq km)

Land use and agricultural distribution

- 🐐 goats
- 🐑 sheep
- 🌾 cereals
- cotton
- dates
- rice
- ● capital cities
- • major towns
- pasture
- cropland
- forest
- mountain region
- desert
- wetland

▲ *Cotton workers in* Pakistan pack huge bales of unspun cotton to be washed and processed. The cotton and textile industry is of growing economic importance, producing more than 36 million sq yards (30 million sq m) of woven cloth annually.

149

South Asia

BANGLADESH, BHUTAN, INDIA, MALDIVES,
NEPAL, PAKISTAN, SRI LANKA

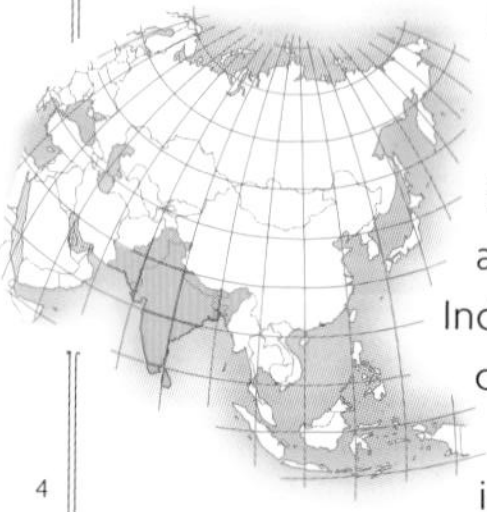

More than one-fifth of the world's population lives in the south Asian subcontinent. Great cultural diversity has come from a long succession of foreign invaders, including Hindu Aryans, Islamic Moguls, and the British, whose empire incorporated the princely states of the Maharajas and extended to the borders of Nepal and Bhutan in the Himalayas. Independent since 1947, India is the world's largest democracy, and at the current rate of growth, may overtake China as the world's most populous country during the 21st century. There are points of tension in the region over claims for independence by the Sikhs in the Indian Punjab and the long-standing dispute with Pakistan over Jammu and Kashmir in the north.

▼ The towering Karakoram and Hindu Kush ranges, formed at the same time as the Himalayas, dominate Pakistan's northern borders. K2 on the border of northern Pakistan is the second highest mountain on Earth, at 28,251 ft (8611 m).

The landscape

South Asia is effectively isolated from the rest of Asia by desert along the western flank of Pakistan, and a continuous wall of mountains, dominated by the Himalayas, to the north and east. The great basins of the Indus and Ganges separate this mountain fringe from the rolling plateau of the Indian peninsula, which is bordered by a line of coastal hills, the Eastern and Western Ghats.

The Indus river flows more than 1970 miles (3180 km) from southwestern Tibet to its mouth on the Arabian Sea. It has an estimated catchment area of 450,000 sq miles (1,165,500 sq km).

The coast of western Pakistan is a staircase of folded rock strata caused by successive periods of rapid uplift.

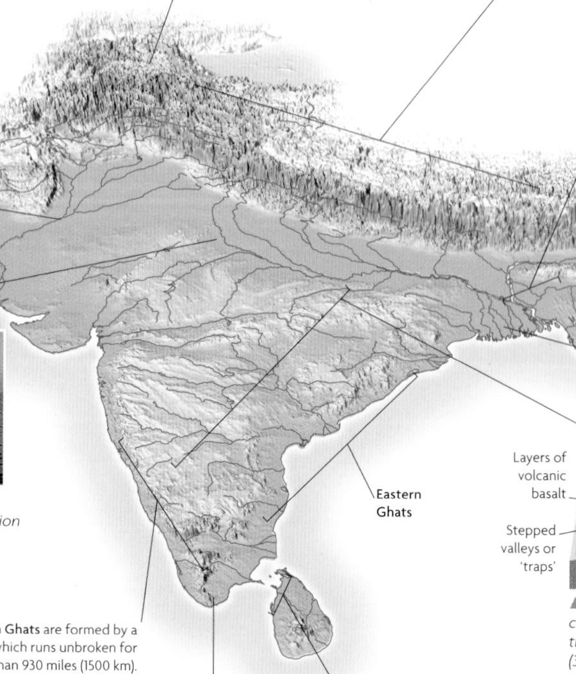

▼ The Indus valley near Skardu in northern Pakistan has been partially infilled by great quantities of eroded sediment. Most of this is carried from the region's bare slopes by swollen rivers during the spring thaw and mass movement activity.

The Himalayas are the highest and most extensive mountain system in the world. They were formed when the Indo-Australian Plate collided with the Eurasian Plate about 40 million years ago, thrusting up huge masses of land and creating a "ripple" effect, which formed lesser mountain ranges in Tibet and Southeast Asia. Mount Everest is the world's tallest mountain at 29,029 ft (8848 m).

Almost all of Bangladesh lies in the immense delta formed by the Ganges and the Brahmaputra which merge and flow out into the Bay of Bengal.

Ganges delta

Deccan plateau

Layers of volcanic basalt

Stepped valleys or 'traps'

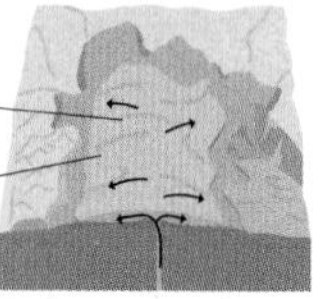

▲ The Deccan plateau covers an area of more than 123,553 sq miles (320,000 sq km). It is formed of deep layers of volcanic basalt, reaching thicknesses of more than 9800 ft (3000 m) toward the coast. Distinctive stepped valleys cut in the basalt plateau by rivers are known as "traps."

Eastern Ghats

Coastal deposition has formed many typical features along the western coast of Sri Lanka. These include spits and bars, sometimes enclosing lagoons.

Trivandrum in southern India normally receives the first of the monsoon rains, which are essential to south Asian agriculture and moderate the extreme summer heat. The monsoon then moves northward over a period of about two months.

The Western Ghats are formed by a fault scarp which runs unbroken for more than 930 miles (1500 km). They reach their highest point at the southern Cardamom Hills.

Bharatpur

▲ Rivers flowing from the Himalayas into a broad depression in northern India have formed marshes around Bharatpur. They are now a sanctuary for numerous bird species.

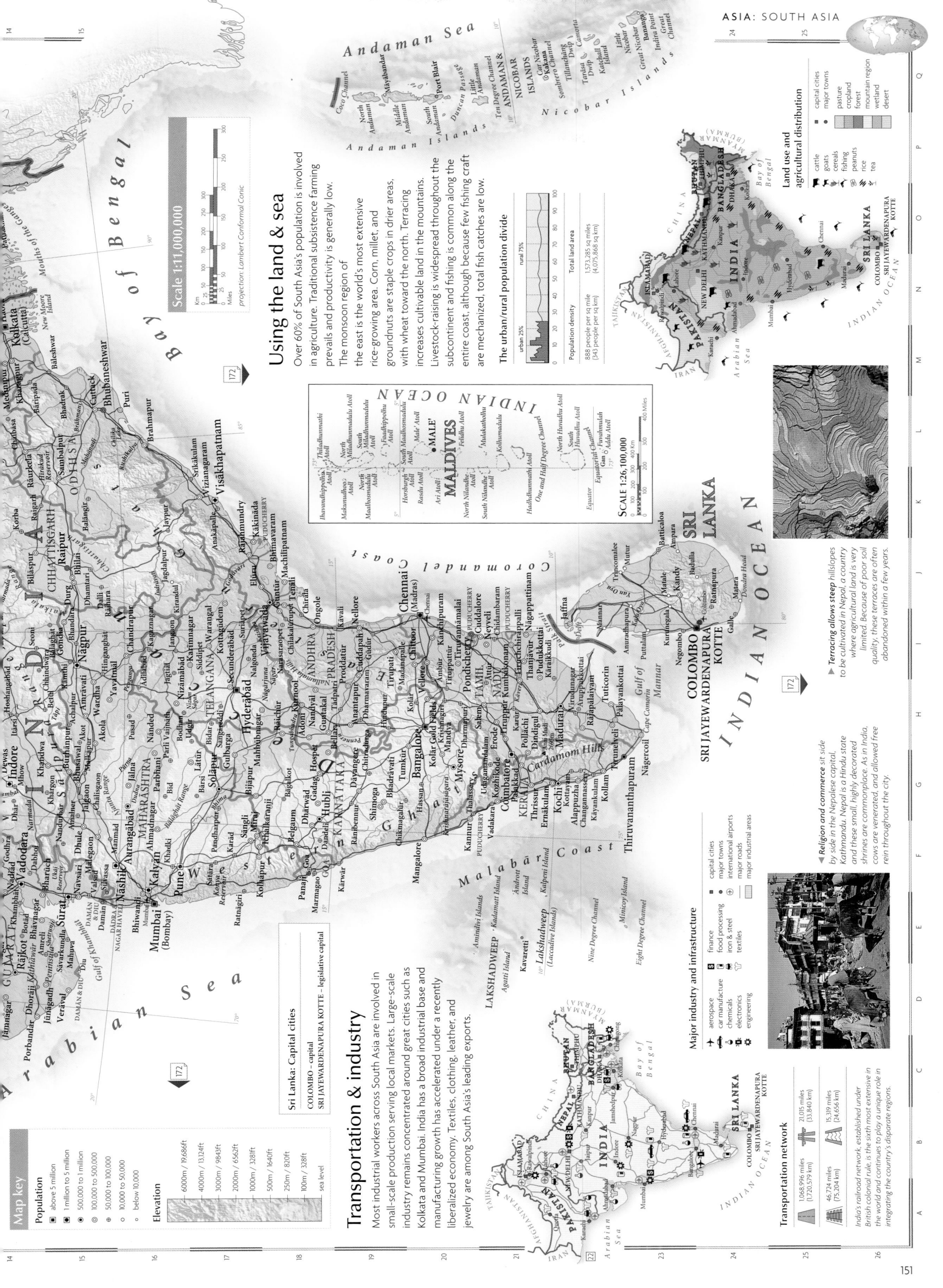

Using the land & sea

Over 60% of South Asia's population is involved in agriculture. Traditional subsistence farming prevails and productivity is generally low. The monsoon region of the east is the world's most extensive rice-growing area. Corn, millet, and groundnuts are staple crops in drier areas, with wheat toward the north. Terracing increases cultivable land in the mountains. Livestock-raising is widespread throughout the subcontinent and fishing is common along the entire coast, although because few fishing craft are mechanized, total fish catches are low.

The urban/rural population divide

urban 25%
rural 75%

Population density   Total land area

888 people per sq mile   1,573,285 sq miles
(343 people per sq km)   (4,075,868 sq km)

Land use and agricultural distribution

- cattle
- goats
- cereals
- fishing
- peanuts
- rice
- tea

- capital cities
- major towns
- pasture
- cropland
- forest
- mountain region
- wetland
- desert

▲ Terracing allows steep hillslopes to be cultivated in Nepal, a country where agricultural land is very limited. Because of poor soil quality, these terraces are often abandoned within a few years.

▼ Religion and commerce sit side by side in the Nepalese capital, Kathmandu. Nepal is a Hindu state and these small, highly decorated shrines are commonplace. As in India, cows are venerated, and allowed free rein throughout the city.

Transportation & industry

Most industrial workers across South Asia are involved in small-scale production serving local markets. Large-scale industry remains concentrated around great cities such as Kolkata and Mumbai. India has a broad industrial base and manufacturing growth has accelerated under a recently liberalized economy. Textiles, clothing, leather, and jewelry are among South Asia's leading exports.

Sri Lanka: Capital cities

COLOMBO – capital
SRI JAYEWARDENAPURA KOTTE – legislative capital

Major industry and infrastructure

- aerospace
- car manufacture
- chemicals
- electronics
- engineering
- finance
- food processing
- iron & steel
- textiles

- capital cities
- major towns
- international airports
- major roads
- major industrial areas

Transportation network

1,068,996 miles (1,720,579 km)	21,015 miles (33,840 km)
46,724 miles (75,204 km)	15,319 miles (24,656 km)

India's railroad network, established under British colonial rule, is the sixth most extensive in the world and continues to play a unique role in integrating the country's disparate regions.

Map key

Population
- above 5 million
- 1 million to 5 million
- 500,000 to 1 million
- 100,000 to 500,000
- 50,000 to 100,000
- 10,000 to 50,000
- below 10,000

Elevation
- 6000m / 19,686ft
- 4000m / 13,124ft
- 3000m / 9843ft
- 2000m / 6562ft
- 1000m / 3281ft
- 500m / 1640ft
- 250m / 820ft
- 100m / 328ft
- sea level

Scale 1:11,000,000
projection: Lambert Conformal Conic

SCALE 1:26,100,000

MALDIVES
MALE'

INDIAN OCEAN

Andaman Sea

Bay of Bengal

Arabian Sea

ANDAMAN & NICOBAR ISLANDS

Nicobar Islands

Andaman Islands

Northern India & the Himalayan states

BANGLADESH, BHUTAN, NEPAL, Arunachal Pradesh, Assam, Bihar, Chandigarh, Delhi, Haryana, Himachal Pradesh, Jammu & Kashmir, Jharkhand, Manipur, Meghalaya, Mizoram, Nagaland, Punjab, Rajasthan, Sikkim, Tripura, Uttarakhand, Uttar Pradesh, West Bengal

The Ganges and Brahmaputra river basins and the massive mountain barrier of the Himalayas define this region's landscape and have served to reinforce potent cultural and religious differences among its people. Hinduism pervades most aspects of national life and is a growing political force within India, a secular country which also encompasses the center of Sikhism at Amritsar and the world's largest Muslim minority. Nepal is a crowded mountain state, which faces severe ecological problems from deforestation, while the tiny Himalayan Buddhist kingdom of Bhutan is emerging from long-term isolation, to welcome selected visitors. The Muslim state of Bangladesh, formerly East Pakistan, is one of the world's most densely populated countries and one of the poorest, with more than 145 million people living largely on the massive Ganges/Brahmaputra delta. Many Bangladeshis live under threat of repeated, catastrophic floods.

◀ *The Golden Temple in Amritsar, the most sacred shrine of the Sikh religion, was the scene of violent clashes between Sikh separatists and government forces in 1984.*

Map key

Population

- ▣ 1 million to 5 million
- ◉ 500,000 to 1 million
- ◉ 100,000 to 500,000
- ⊕ 50,000 to 100,000
- ○ 10,000 to 50,000
- ∘ below 10,000

Elevation

- 6000m / 19,686ft
- 4000m / 13,124ft
- 3000m / 9843ft
- 2000m / 6562ft
- 1000m / 3281ft
- 500m / 1640ft
- 250m / 820ft
- 100m / 328ft
- sea level

Transportation & industry

Textiles, engineering, chemicals, and electronics are leading industries in north India. The plateau of Chota Nagpur provides ore for iron and steel production in the major industrial region northeast of Kolkata. Bangladesh processes jute and Nepal has a small manufacturing sector based on agricultural produce, while Bhutan's limited industry is concentrated in the southern lowland area.

Scale 1:6,500,000

projection: Lambert Conformal Conic

Major industry and infrastructure

- ⚓ adventure tourism
- 🚗 car manufacture
- ⚗ chemicals
- ⛏ coal
- ⚡ electronics
- ⚙ engineering
- 💲 finance
- 🍴 food processing
- ⛓ iron & steel
- jute processing
- ⛽ oil
- 🍵 tea processing
- textiles

- ● capital cities
- ∘ major towns
- ⊕ international airports
- major roads
- major industrial areas

Transportation network

Over 60% of Bangladesh's internal trade is carried by boat. The country has a very disjointed land transportation network, with no bridges over the Brahmaputra and few road crossings on the Ganges river.

The landscape

Most of the region is drained by the Ganges river, which meets the Brahmaputra in Bangladesh to form an immense delta before flowing into the Bay of Bengal. The Himalayas extend eastward over 1500 miles (2400 km), from the parallel ranges running through Jammu and Kashmir. The Thar Desert occupies the southwest.

The Indian Punjab lies mainly to the west of the Ganges watershed and its rivers flow into the Indus. Control of this water resource has been a source of great friction with neighboring Pakistan.

The border between India and Pakistan runs through the Thar Desert, an area of sandy seif dunes 50–100 ft (15–30 m) in height. Fossils found in the desert indicate that the dunes, stabilized by vegetation, have been in their current position for about 3000 years.

Sambhar Salt Lake in Rajasthan is India's largest lake. Unlike most of the Himalayan lakes which are glacial in origin – formed in ice-scoured basins or as the result of depositional damming – it is an ephemeral salt lake filled periodically by flash flooding.

▶ *The Pir Panjal Range* in southwestern Kashmir rises to elevations of 12,500 ft (3810 m). Despite the freezing conditions, settlements and extensive pastures are found above the tree line.

The northern ranges of the Himalayas contain the highest mountains in the world, with average heights of more than 23,000 ft (7000 m) and many peaks higher than 26,000 ft (8000 m).

In the last 40 million years, the course of the Brahmaputra has been diverted hundreds of miles to the east by the rising landmass of the Himalayas.

The Khasi Hills are an example of a *horst*, a fractured block of bedrock which has been thrust upward.

▲ *The summit of* Machhapuchhre rises to 22,942 ft (6993 m). It is also known as the "Fish's Tail" because of its distinctive peak.

Debris slides in the middle Himalayas

Debris fans at base of slope

Soil blocks

Slide plain

▲ *Soil loss in* the middle Himalayas has largely been attributed to debris slides, where large blocks of soil are mobilized by saturation along a slide plane. Once mobile, the soil slides down the slope, gaining speed and thinning to form a fan at the base of the slope.

The Ganges river, sacred to the Hindu people, drains a vast lowland area at the base of the Himalayas. The northern plains are covered by sandy deposits, broken by mud banks formed when the river floods.

The rapid deforestation of Himalayan valleys has led to acute soil erosion and increased rates of rainwater runoff, both cited as possible causes of the worsening floods downstream in the Ganges/Brahmaputra delta, although natural rates are high and may be the real cause.

Over half of the great Ganges/Brahmaputra delta floods each year during the monsoon as rivers, swollen by meltwater from the Himalayas and by excess rainwater, break their banks and fertilize the land with nutrient-rich sediment.

Using the land

Grain production dominates land use. Rice is most widely grown in the east. Irrigation and new crop strains have dramatically increased yields in the Punjab, a major wheat-producing area. River floodplains are intensively farmed and livestock herding is widespread, particularly in Bhutan. Regional crops include jute in Bangladesh, tea in Assam, cardamom in Sikkim, and saffron in Kashmir.

The urban/rural population divide

urban 23%   rural 77%

0 10 20 30 40 50 60 70 80 90 100

Population density	Total land area
993 people per sq mile (384 people per sq km)	665,104 sq miles (1,723,068 sq km)

▲ *An adverse climate,* steep slopes, and poor soils limit crop cultivation in Bhutan, which is a largely agrarian economy. Rice, corn, and wheat are the main staples, although orchards are being established as the soil and climate suit this type of farming.

Land use and agricultural distribution

- cattle
- goats
- sheep
- cereals
- jute
- rice
- tea
- capital cities
- major towns
- pasture
- cropland
- forest
- mountain region
- wetland
- desert

▲ *Flooded streets in* Dhaka, Bangladesh are a testament to the region's vulnerability to flooding. In 1988 alone, 75% of the country was flooded, leaving thousands of people dead and over 25 million homeless.

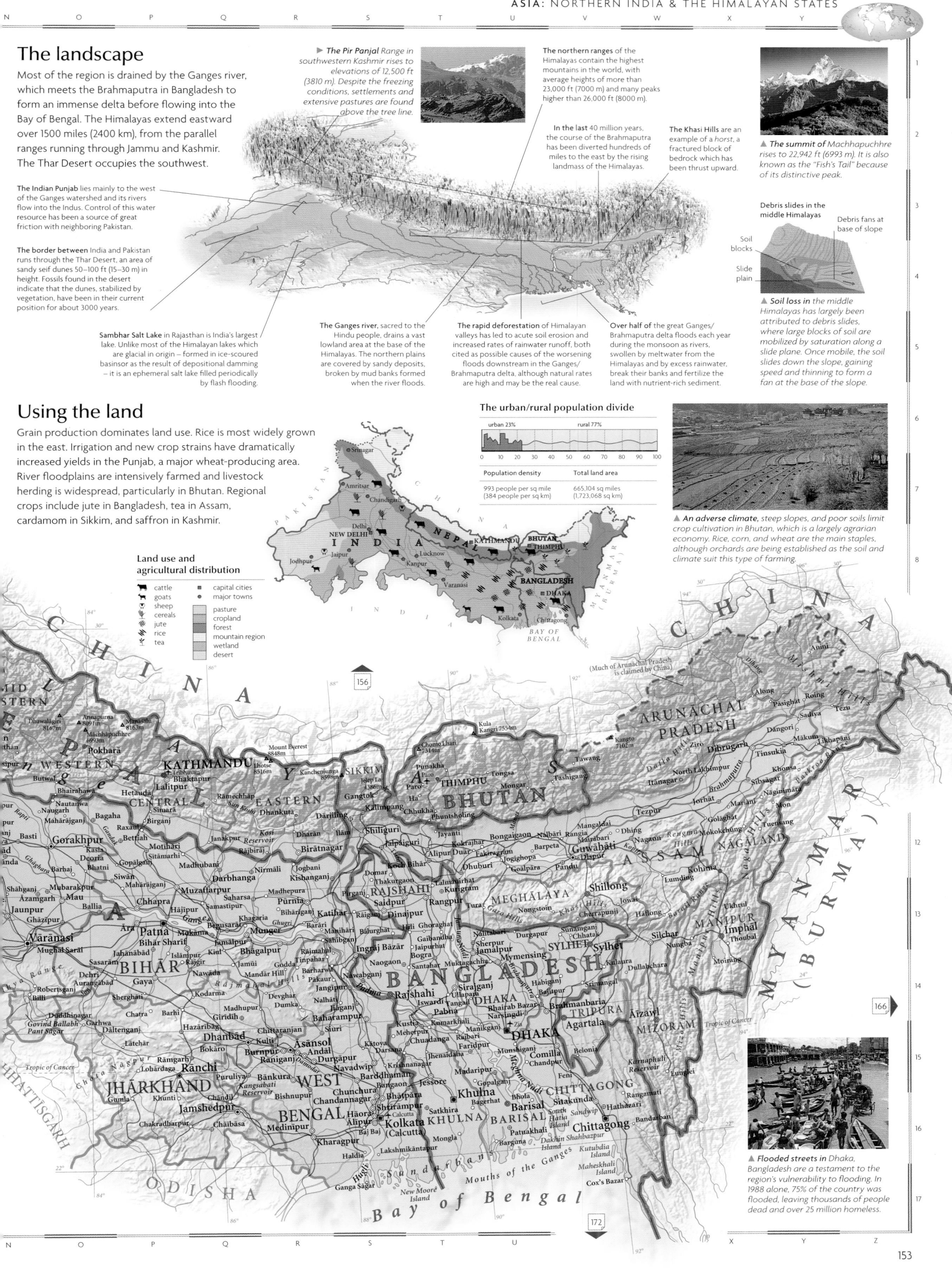

Southern India & Sri Lanka

SRI LANKA, Andhra Pradesh, Chhattisgarh, Dadra & Nagar Haveli, Daman & Diu, Goa, Gujarat, Karnataka, Kerala, Lakshadweep, Madhya Pradesh, Maharashtra, Odisha, Puducherry, Tamil Nadu, Telangana

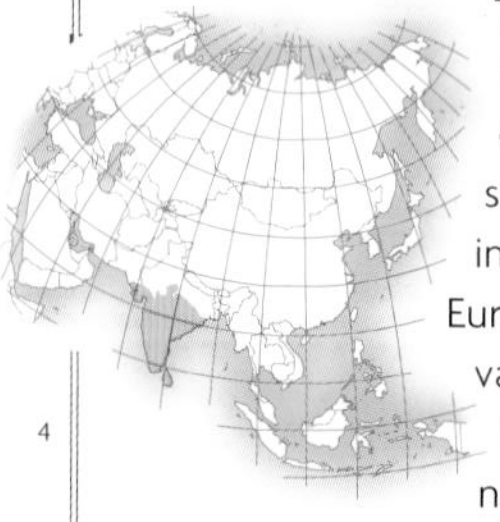

The unique and highly independent southern states reflect the diverse and decentralized nature of India, which has fourteen official languages. The southern half of the peninsula lay beyond the reach of early invaders from the north and retained the distinct and ancient culture of Dravidian peoples such as the Tamils, whose language is spoken in preference to Hindi throughout southern India. The interior plateau of southern India is less densely populated than the coastal lowlands, where the European colonial imprint is strongest. Urban and industrial growth is accelerating, but southern India's vast population remains predominantly rural. The island of Sri Lanka has two distinct cultural groups; the mainly Buddhist Sinhalese majority, and the Tamil minority whose struggle for a homeland in the northeast led to prolonged civil war.

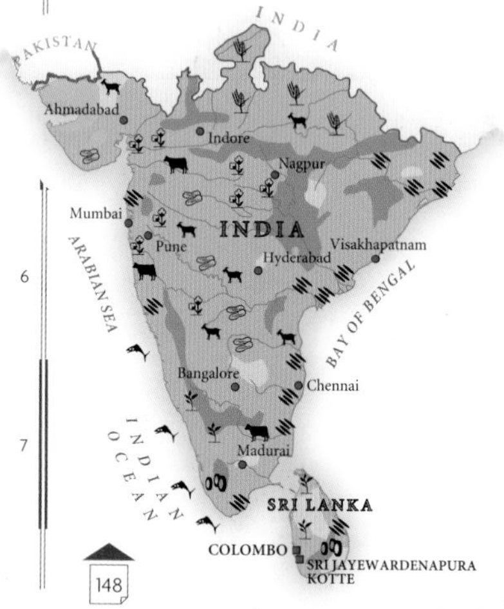

The landscape

The undulating Deccan plateau underlies most of southern India; it slopes gently down toward the east and is largely enclosed by the Ghats coastal hill ranges. The Western Ghats run continuously along the Arabian Sea coast, while the Eastern Ghats are interrupted by rivers which follow the slope of the plateau and flow across broad lowlands into the Bay of Bengal. The plateaus and basins of Sri Lanka's central highlands are surrounded by a broad plain.

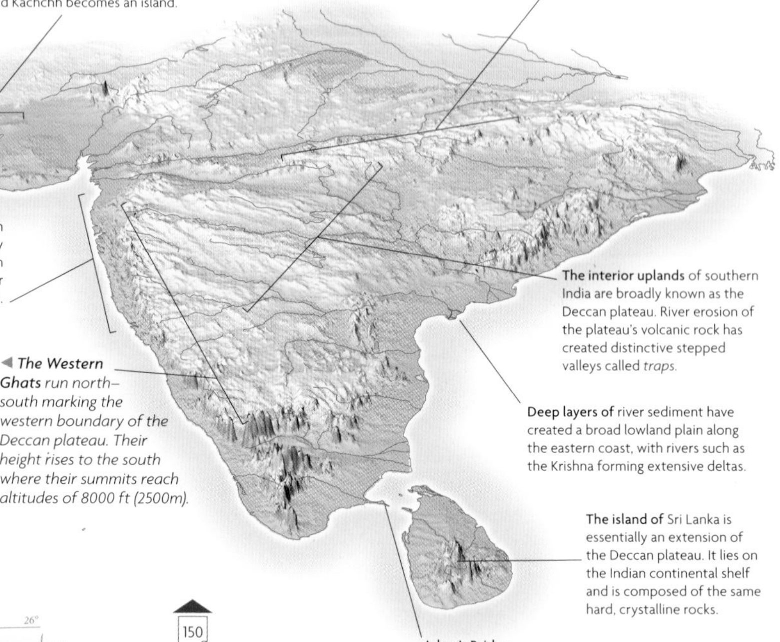

Along the northern boundary of the Deccan plateau, old basement rocks are interspersed with younger sedimentary strata. This creates spectacular scarplands, cut by numerous waterfalls along the softer sedimentary strata.

The interior uplands of southern India are broadly known as the Deccan plateau. River erosion of the plateau's volcanic rock has created distinctive stepped valleys called traps.

Deep layers of river sediment have created a broad lowland plain along the eastern coast, with rivers such as the Krishna forming extensive deltas.

The island of Sri Lanka is essentially an extension of the Deccan plateau. It lies on the Indian continental shelf and is composed of the same hard, crystalline rocks.

The Rann of Kachchh tidal marshes encircle the low-lying Kachchh peninsula. For several months during the rainy season the water level of the marshes rises and Kachchh becomes an island.

The Konkan coast, which runs between Daman and Goa, is characterized by rocky headlands, and bays with crescent-shaped beaches. Flooded river valleys known as *rias* extend inland.

▼ **The Western Ghats** run north–south marking the western boundary of the Deccan plateau. Their height rises to the south where their summits reach altitudes of 8000 ft (2500m).

Adam's Bridge

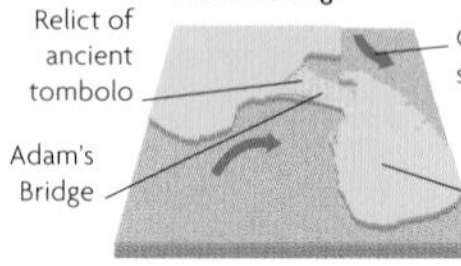

Relict of ancient tombolo

Ocean currents cause sediment build up

Sri Lanka

Adam's Bridge

▲ **Adam's Bridge (Rama's Bridge)** is a chain of sandy shoals lying about 4 ft (1.2 m) under the sea between India and Sri Lanka. They once formed the world's longest tombolo, or land bridge, before the sea level began to rise several thousand years ago.

Using the land and sea

Rice is the main staple in the east, in Sri Lanka and along the humid Malabar Coast. Peanuts are grown on the Deccan plateau, with wheat, corn, and chickpeas, toward the north. Sri Lanka is a leading exporter of tea, coconuts and rubber. Cotton plantations supply local mills around Nagpur and Mumbai. Fishing supports many communities in Kerala and the Laccadive Islands.

Land use and agricultural distribution

- cattle
- goats
- cereals
- cotton
- fishing
- peanuts
- rice
- rubber
- tea
- pasture
- cropland
- forest
- wetland
- capital cities
- major towns

The urban/rural population divide

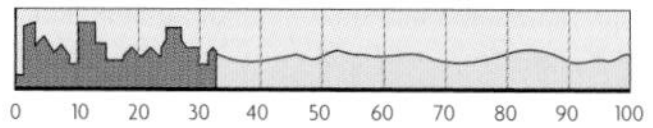

urban 33%   rural 67%

Population density
730 people per sq mile
(282 people per sq km)

Total land area
698,295 sq miles
(1,809,054 sq km)

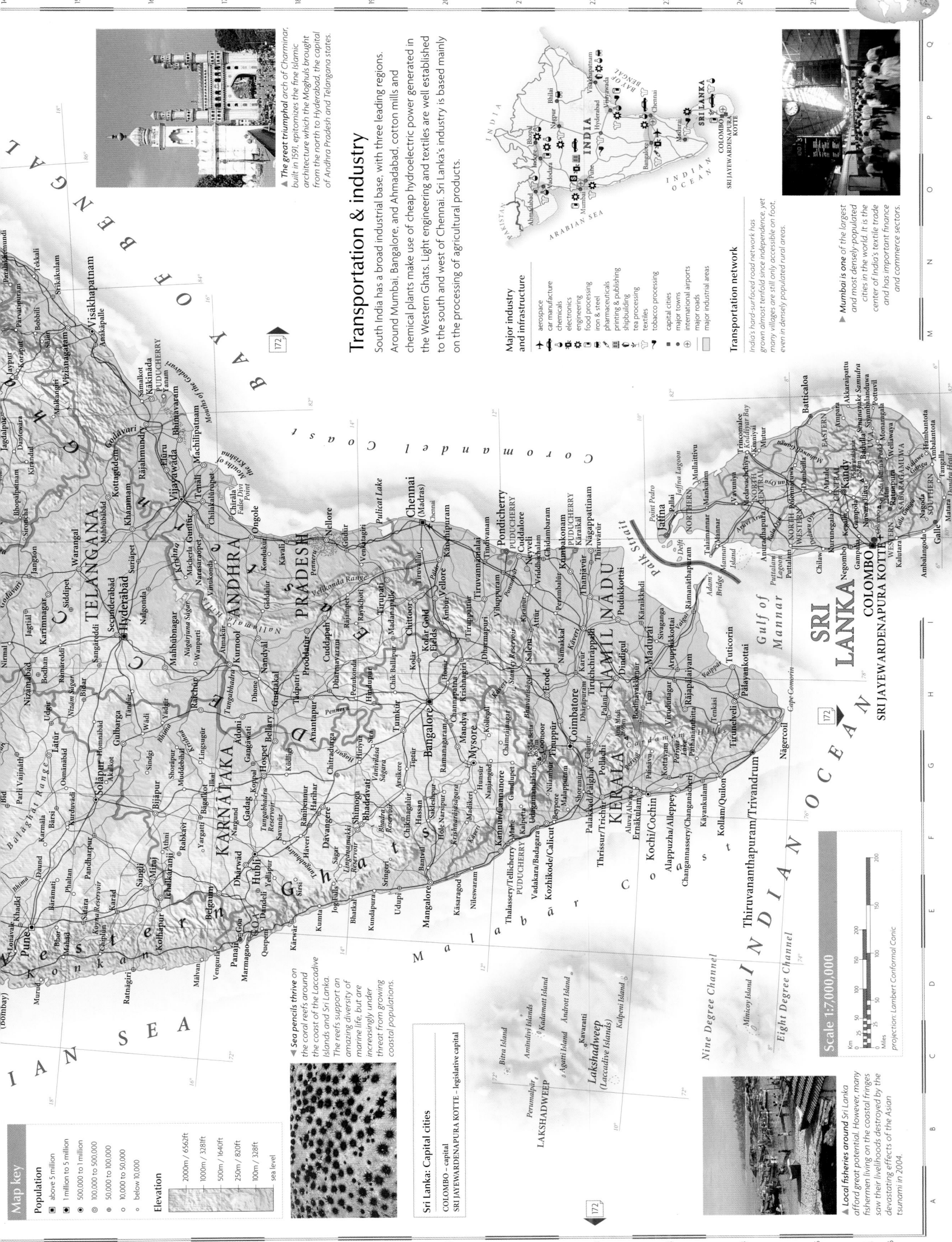

▲ The great triumphal arch of Charminar, built in 1591, epitomize the fine Islamic architecture which the Moghuls brought from the north to Hyderabad, the capital of Andhra Pradesh and Telangana states.

Transportation & industry

South India has a broad industrial base, with three leading regions. Around Mumbai, Bangalore, and Ahmadabad. Light engineering and textiles are well established to the south and west of Chennai. Sri Lanka's industry is based mainly on the processing of agricultural products. Use of cheap hydroelectric power generated in the Western Ghats. Light engineering and textiles are well established to the south and west of Chennai. Sri Lanka's industry is based mainly on the processing of agricultural products.

Major industry and infrastructure

- aerospace
- car manufacture
- chemicals
- electronics
- engineering
- food processing
- iron & steel
- pharmaceuticals
- printing & publishing
- shipbuilding
- tea processing
- textiles
- tobacco processing
- capital cities
- major towns
- international airports
- major roads
- major industrial areas

Transportation network

India's hard-surfaced road network has grown almost tenfold since independence, yet many villages are still only accessible on foot, even in densely populated rural areas.

▲ Mumbai is one of the largest and most densely-populated cities in the world. It is the center of India's textile trade and has important finance and commerce sectors.

▼ Sea pencils thrive on the coral reefs around the coast of the Laccadive Islands and Sri Lanka. The reefs support an amazing diversity of marine life, but are increasingly under threat from growing coastal populations.

Sri Lanka: Capital cities

COLOMBO – capital
SRI JAYEWARDENAPURA KOTTE – legislative capital

Map key

Population
- ■ above 5 million
- ■ 1 million to 5 million
- ◉ 500,000 to 1 million
- ◎ 100,000 to 500,000
- ◉ 50,000 to 100,000
- ○ 10,000 to 50,000
- ○ below 10,000

Elevation
- 2000m / 6562ft
- 1000m / 3281ft
- 500m / 1640ft
- 250m / 820ft
- 100m / 328ft
- sea level

Scale 1:7,000,000

Km 0 25 50 100 150 200
Miles 0 25 50 100 150 200

projection: Lambert Conformal Conic

▲ Local fisheries around Sri Lanka afford great potential. However, many fishermen living on the coastal fringes saw their livelihoods destroyed by the devastating effects of the Asian tsunami in 2004.

Mainland East Asia

CHINA, MONGOLIA, NORTH KOREA, SOUTH KOREA, TAIWAN

China, the world's most populous nation, has an unbroken cultural history, longer than that of any other country, and is rapidly emerging as a leading world power. When Mao Zedong established Communist rule in 1949, China had become a backward feudal empire, stricken by civil war and over a century of European and Japanese incursions. The closed regime withstood the traumas of rapid industrialization, communal farming, and the brutal purges of the Cultural Revolution but, since the 1980s has introduced economic reforms, led by expanded foreign trade. China's population is heavily concentrated in the east and, despite accelerating urban growth, remains predominantly rural. One cultural group, the Han, make up over 90% of the people, while five "Autonomous Regions" have been established in the south and west for the main ethnic minorities.

Transportation & industry

Large-scale industrial growth has always been a priority of the Communist government. Metals and machine production, chemicals, and engineering are among the leading industries, concentrated in the major cities of the east coast. Textiles and clothing manufacture, the main consumer goods sector, is relatively well dispersed, with a few significant centers such as Shanghai, Beijing, and Hong Kong.

Major industry and infrastructure

- 🚗 car manufacture
- ⚗ chemicals
- ⚙ electronics
- ✿ engineering
- $ finance
- 🍴 food processing
- ⚒ iron & steel
- ⚓ shipbuilding
- ⊤ textiles
- ■ capital cities
- ● major towns
- ⊕ international airports
- — major roads
- ▢ major industrial areas

Transportation network

829,790 miles (1,335,571 km)	12,740 miles (20,506 km)
43,976 miles (70,780 km)	70,991 miles (114,262 km)

Ever-increasing demand for rail transportation has led to major improvement and expansion of the network, notably the 690 mile (1100 km) link between Golmud and Lhasa opened in 2006.

◄ *Coal is China's* most abundant mineral resource. This mine at Fuxin in Liaoning province is used to provide coal for a nearby power station.

The landscape

The East Asian landmass is arranged in three distinct levels, the highest of which is the Plateau of Tibet in the southwest. The arid uplands of northwestern China form a barren middle step. The main rivers flow eastward from these two platforms to the East China and South China sea coasts, across a broad region of alluvial lowlands and low hills.

◄ *Gansu province, through* which the ancient Silk Route passes on its way to the west, is characterized by extensive loess deposits which are terraced and used for crop cultivation.

◄ *Paektu-san, at 9023* ft (2750 m), is North Korea's highest peak; an extinct volcanic cone now filled by a crater lake.

The Gobi Desert extends across the Nei Mongol Gaoyuan; a vast saucer-shaped upland surrounded by a rim of higher mountains.

The loess plateau of northern China is the world's greatest expanse of loess, a loose soil made up of wind-blown material. The plateau has been heavily eroded by tributaries of the Yellow River.

Shifting sand dunes are found in the arid west of the northeast China Plain, while the eastern part of this great expanse is wet and swampy.

River-eroded fine soils
Thick blanket of loess

▲ *Because of its* very small grain-size, loess has been easily transported and deposited by winds which scour the plains, and in northern China, deposits of loess can be up to 3000 ft (1000 m) thick. Loess-based soils are very fertile, but clearing land for agriculture quickly destabilizes the soil and allows it to be eroded.

▲ *The Plateau of Tibet* occupies about a quarter of China's total area. The Yangtze, Mekong, Indus, and Brahmaputra rivers all originate in the south and east of the plateau.

The Himalayas extend along the southwestern edge of the Plateau of Tibet, forming a continuous mountain barrier over 1500 miles (2500 km) long.

Warm, humid conditions have caused intensive erosion of south China's karst areas, producing spectacular jagged peaks and vast caves in the limestone.

Tarim Basin (Tarim Pendi)
Plateau of Tibet
Paektu-san
North China Plain
The Yangtze is China's longest river and the principal navigable waterway.
Sichuan Pendi

◄ *Although it is* over 30 years since his death, the legacy of Chairman Mao Zedong, architect of the Great Proletariat Cultural Revolution, is still very much in evidence across China's landscape. In 1959 Mao launched a 20-year period of industrialization and socioeconomic realignment, rejecting western ideals and social codes.

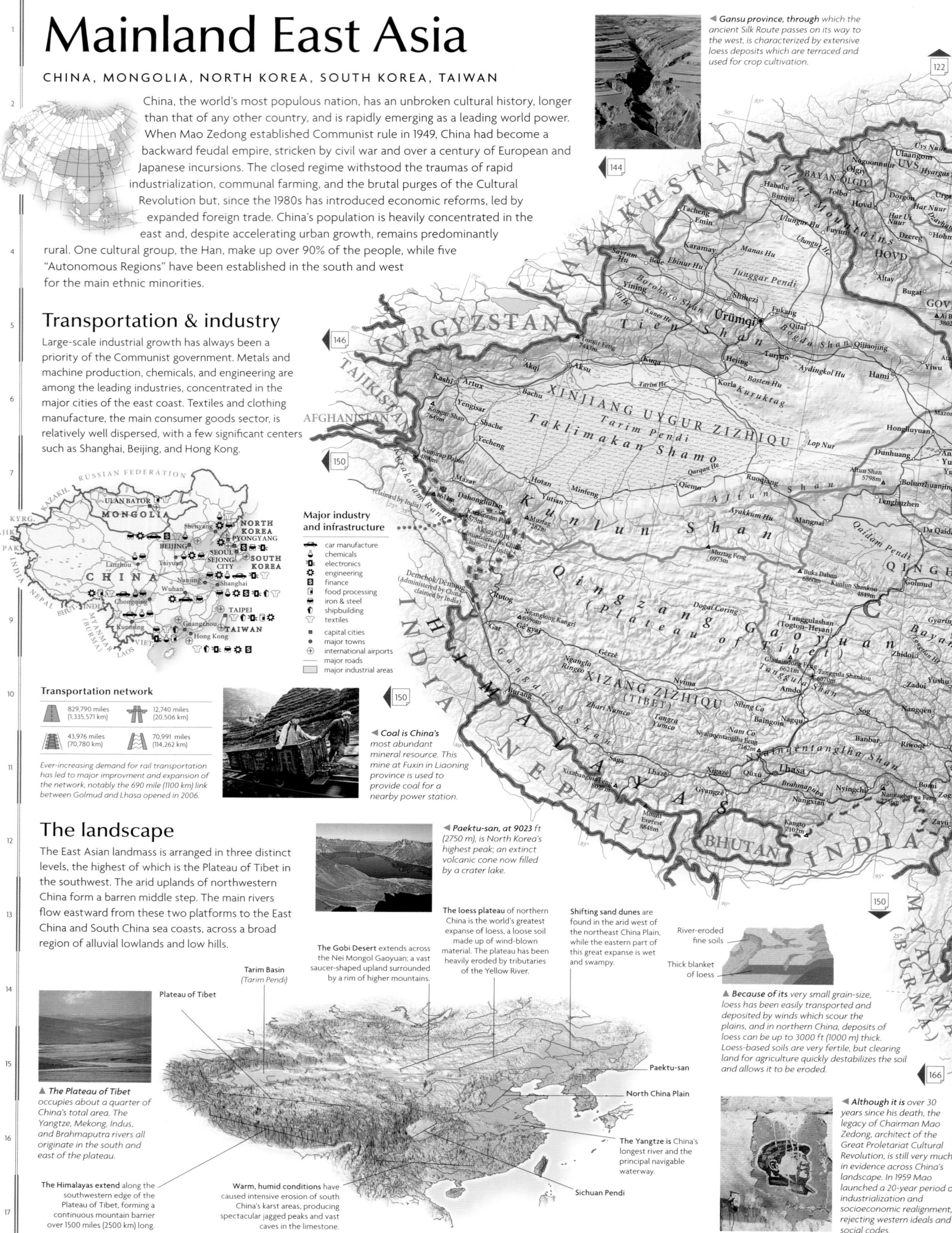

Scale 1:14,000,000

Km
0 25 50 100 150 200 250 300 350 400

Miles
0 25 50 100 150 200 250 300 400

projection: Lambert Conformal Conic

Map key

Population

- ■ above 5 million
- ▪ 1 million to 5 million
- ◙ 500,000 to 1 million
- ◉ 100,000 to 500,000
- ⊕ 50,000 to 100,000
- ○ 10,000 to 50,000
- ∘ below 10,000

Elevation

	6000m / 19,686ft
	4000m / 13,124ft
	3000m / 9843ft
	2000m / 6562ft
	1000m / 3281ft
	500m / 1640ft
	250m / 820ft
	100m / 328ft
	sea level

Using the land & sea

Around 90% of China is unsuitable for cultivation, being either climatically or topographically adverse, or lacking sufficiently fertile soils. Most of the west is used for nomadic herding, while farmland is concentrated in the eastern monsoon region, with rice grown in the tropical and subtropical south. Cereals and soybeans predominate as rainfall and temperatures decline further north.

Land use and agricultural distribution

- 🐖 pigs
- 🐑 sheep
- 🌽 corn
- cotton
- 🎣 fishing
- 🍓 fruit
- rice
- sugar cane
- soybeans

- ■ capital cities
- ● major towns

- pasture
- cropland
- forest
- mountain region

▲ **The Great Wall** of China remains one of the world's largest-ever construction projects, and is so vast that it is visible from space. Sections were added as late as 1640 and it runs for over 4000 miles (6400 km) from the Yellow Sea to Central Asia.

The urban/rural population divide

urban 32%   rural 68%

0 10 20 30 40 50 60 70 80 90 100

Population density	Total land area
325 people per sq mile (125 people per sq km)	4,288,672 sq miles (11,110,550 sq km)

Western China

Gansu, Ningxia, Qinghai, Tibet, Xinjiang

The plateaus and basins of China's dry, desolate western domain are sparsely populated and largely undeveloped, although they have rich mineral reserves; they also form a critical buffer zone for China, in a geographically important and culturally sensitive part of the Asian continent. Across most of the west, the Han Chinese are outnumbered by a range of cultural groups, including the Uygur, the largest group of the various seminomadic Muslim peoples from Central Asia. The remote, inhospitable Plateau of Tibet is the world's coldest and highest plateau. It has been occupied by the Chinese since 1950. Tibet is one of western China's five "Autonomous Regions," but its reclusive Buddhist culture has been systematically undermined by the Chinese government.

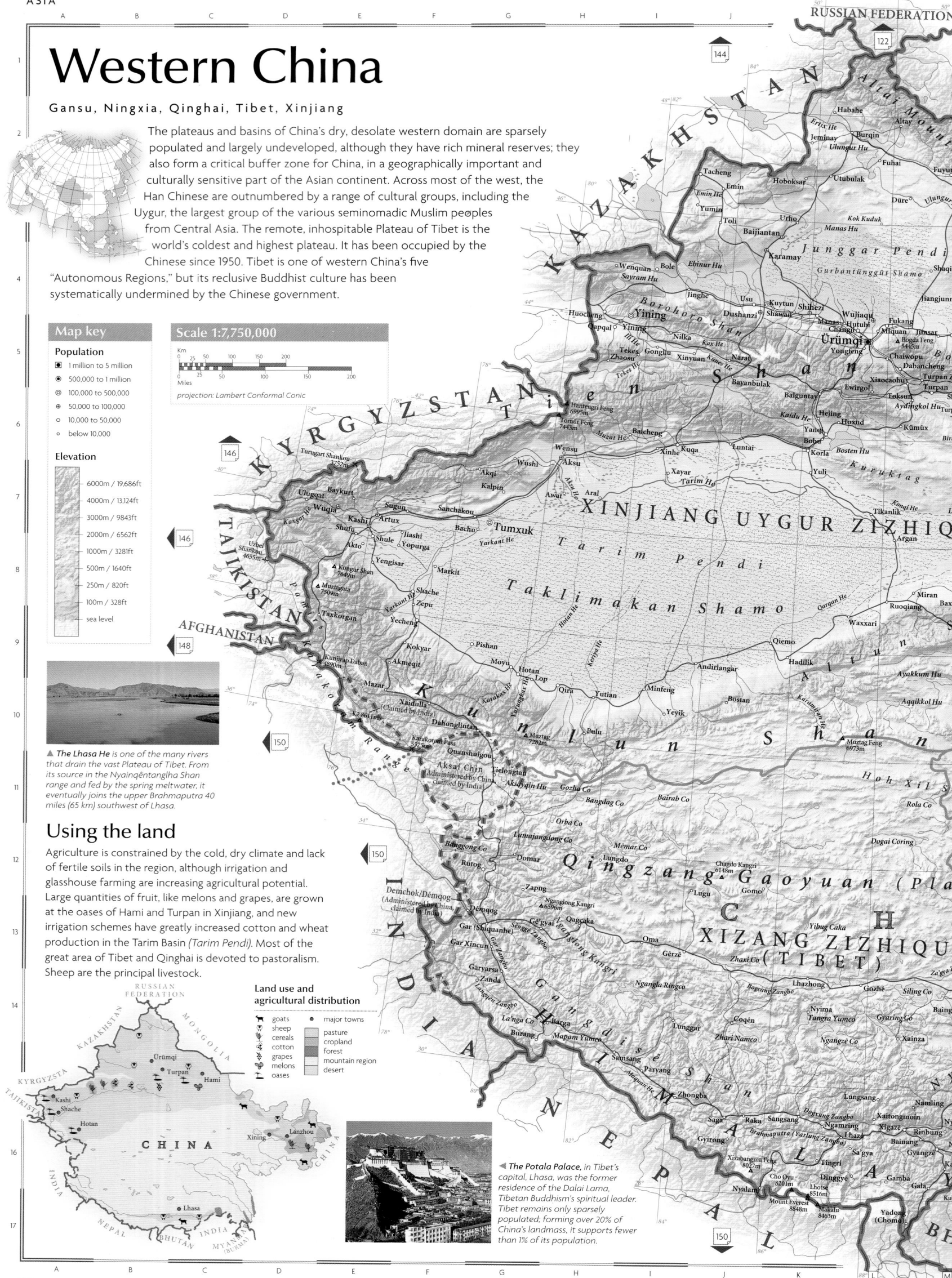

Map key

Population

- ▣ 1 million to 5 million
- ◉ 500,000 to 1 million
- ◎ 100,000 to 500,000
- ⊙ 50,000 to 100,000
- ○ 10,000 to 50,000
- ∘ below 10,000

Elevation

- 6000m / 19,686ft
- 4000m / 13,124ft
- 3000m / 9843ft
- 2000m / 6562ft
- 1000m / 3281ft
- 500m / 1640ft
- 250m / 820ft
- 100m / 328ft
- sea level

Scale 1:7,750,000

projection: Lambert Conformal Conic

▲ *The Lhasa He is one of the many rivers that drain the vast Plateau of Tibet. From its source in the Nyainqêntanglha Shan range and fed by the spring meltwater, it eventually joins the upper Brahmaputra 40 miles (65 km) southwest of Lhasa.*

Using the land

Agriculture is constrained by the cold, dry climate and lack of fertile soils in the region, although irrigation and glasshouse farming are increasing agricultural potential. Large quantities of fruit, like melons and grapes, are grown at the oases of Hami and Turpan in Xinjiang, and new irrigation schemes have greatly increased cotton and wheat production in the Tarim Basin *(Tarim Pendi)*. Most of the great area of Tibet and Qinghai is devoted to pastoralism. Sheep are the principal livestock.

Land use and agricultural distribution

- goats
- sheep
- cereals
- cotton
- grapes
- melons
- oases
- • major towns
- pasture
- cropland
- forest
- mountain region
- desert

◀ *The Potala Palace, in Tibet's capital, Lhasa, was the former residence of the Dalai Lama, Tibetan Buddhism's spiritual leader. Tibet remains only sparsely populated; forming over 20% of China's landmass, it supports fewer than 1% of its population.*

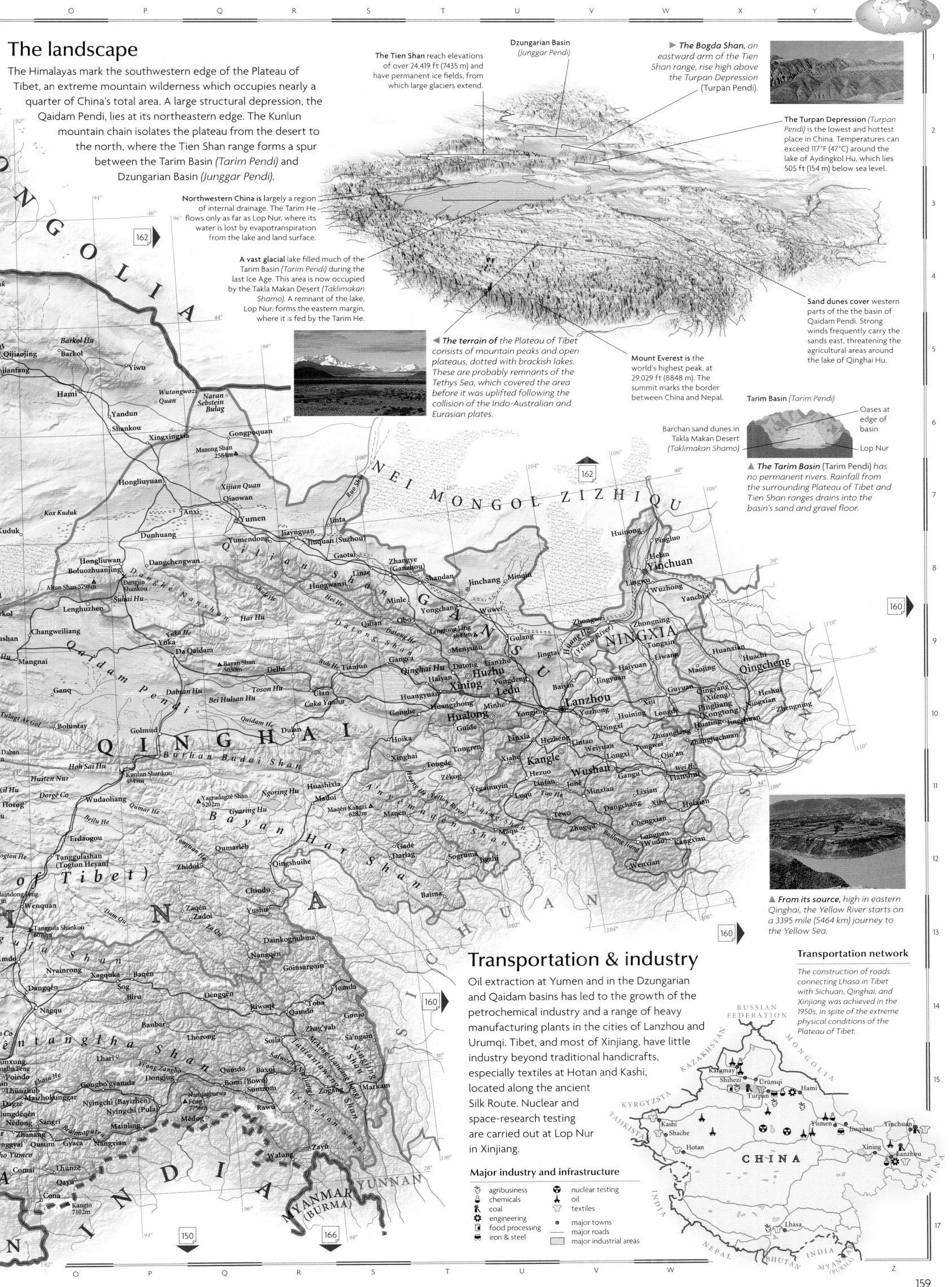

The landscape

The Himalayas mark the southwestern edge of the Plateau of Tibet, an extreme mountain wilderness which occupies nearly a quarter of China's total area. A large structural depression, the Qaidam Pendi, lies at its northeastern edge. The Kunlun mountain chain isolates the plateau from the desert to the north, where the Tien Shan range forms a spur between the Tarim Basin (Tarim Pendi) and Dzungarian Basin (Junggar Pendi).

Northwestern China is largely a region of internal drainage. The Tarim He flows only as far as Lop Nur, where its water is lost by evapotranspiration from the lake and land surface.

A vast glacial lake filled much of the Tarim Basin (Tarim Pendi) during the last Ice Age. This area is now occupied by the Takla Makan Desert (Taklimakan Shamo). A remnant of the lake, Lop Nur, forms the eastern margin, where it is fed by the Tarim He.

◀ **The terrain of** the Plateau of Tibet consists of mountain peaks and open plateaus, dotted with brackish lakes. These are probably remnants of the Tethys Sea, which covered the area before it was uplifted following the collision of the Indo-Australian and Eurasian plates.

The Tien Shan reach elevations of over 24,419 ft (7435 m) and have permanent ice fields, from which large glaciers extend.

Dzungarian Basin (Junggar Pendi)

▶ **The Bogda Shan,** an eastward arm of the Tien Shan range, rise high above the Turpan Depression (Turpan Pendi).

The Turpan Depression (Turpan Pendi) is the lowest and hottest place in China. Temperatures can exceed 117°F (47°C) around the lake of Aydingkol Hu, which lies 505 ft (154 m) below sea level.

Mount Everest is the world's highest peak, at 29,029 ft (8848 m). The summit marks the border between China and Nepal.

Sand dunes cover western parts of the the basin of Qaidam Pendi. Strong winds frequently carry the sands east, threatening the agricultural areas around the lake of Qinghai Hu.

Tarim Basin (Tarim Pendi)

Barchan sand dunes in Takla Makan Desert (Taklimakan Shamo)

Oases at edge of basin

Lop Nur

▲ **The Tarim Basin** (Tarim Pendi) has no permanent rivers. Rainfall from the surrounding Plateau of Tibet and Tien Shan ranges drains into the basin's sand and gravel floor.

▲ **From its source,** high in eastern Qinghai, the Yellow River starts on a 3395 mile (5464 km) journey to the Yellow Sea.

Transportation & industry

Oil extraction at Yumen and in the Dzungarian and Qaidam basins has led to the growth of the petrochemical industry and a range of heavy manufacturing plants in the cities of Lanzhou and Urumqi. Tibet, and most of Xinjiang, have little industry beyond traditional handicrafts, especially textiles at Hotan and Kashi, located along the ancient Silk Route. Nuclear and space-research testing are carried out at Lop Nur in Xinjiang.

Transportation network

The construction of roads connecting Lhasa in Tibet with Sichuan, Qinghai, and Xinjiang was achieved in the 1950s, in spite of the extreme physical conditions of the Plateau of Tibet.

Major industry and infrastructure

- agribusiness
- chemicals
- coal
- engineering
- food processing
- iron & steel
- nuclear testing
- oil
- textiles
- • major towns
- — major roads
- major industrial areas

Eastern China

TAIWAN, Anhui, Beijing, Chongqing, Fujian, Guangdong, Guangxi,
Guizhou, Hainan, Hebei, Henan, Hubei, Hunan, Jiangsu, Jiangxi, Shaanxi,
Shandong, Shanghai, Shanxi, Sichuan, Tianjin, Yunnan, Zhejiang

The east is China's heartland. Massive industrial development since 1949 has
transformed much of the densely populated rural landscape, in a region still prone
to flooding and drought. Over 30 cities have populations of over a million, including
the giant metropolis of Shanghai and the capital Beijing, which has been China's
cultural and political center since the 13th century. The ethnically diverse southwest
and the oil-rich interior provinces of Sichuan and Shaanxi have largely missed out on
the remarkable economic growth occurring in designated free-trade areas along the
coasts of the South and East China seas. The republic of Taiwan was established in
1949 by Chinese nationalists ousted from the mainland by the victorious Communist
forces. Taiwan now has one of the strongest economies in the world but its sovereignty is not
recognized by China. Hong Kong provides a major international trade link for China; a 99-year "lease"
period of British control was concluded in 1997.

▲ North of the Qin Ling range in Shaanxi
province, is an agriculturally fertile region
covered with fine, wind-blown deposits and
known as the loess plateau. The loose
sediments are vulnerable to water erosion.

Using the land & sea

This is a region of intensive cultivation. Wheat, millet,
sorghum, and cotton are the main crops of the
Yellow River basin. South from Sichuan, rice
becomes the principal crop, grown with wheat,
corn, and cotton along the Yangtze river. Tea is
produced in the hills and sugar cane along the coast
of the southeast, where flat land is limited. Pigs and
poultry are raised in great numbers.

**Land use and
agricultural distribution**

- cattle
- pigs
- cereals
- corn
- cotton
- fishing
- peanuts
- rice
- sugar cane
- tea
- capital cities
- major towns
- pasture
- cropland
- forest
- mountain region

▲ On the hills above the North China
Plain, slopes are terraced to utilize the rich
loess soils of the Taihang Shan range.

Map key

Population
- above 5 million
- 1 million to 5 million
- 500,000 to 1 million
- 100,000 to 500,000
- 50,000 to 100,000
- 10,000 to 50,000
- below 10,000

Elevation
- 6000m / 19,686ft
- 4000m / 13,124ft
- 3000m / 9843ft
- 2000m / 6562ft
- 1000m / 3281ft
- 500m / 1640ft
- 250m / 820ft
- 100m / 328ft
- sea level

Scale 1:8,500,000

Km
0 25 50 100 150 200 250 300
Miles
0 25 50 100 150 200 250 300

projection: Lambert Conformal Conic

◀ The former
Portuguese territory of
Macao, with its colonial
architecture, bars and
casinos, reverted to
Chinese rule in 1999.

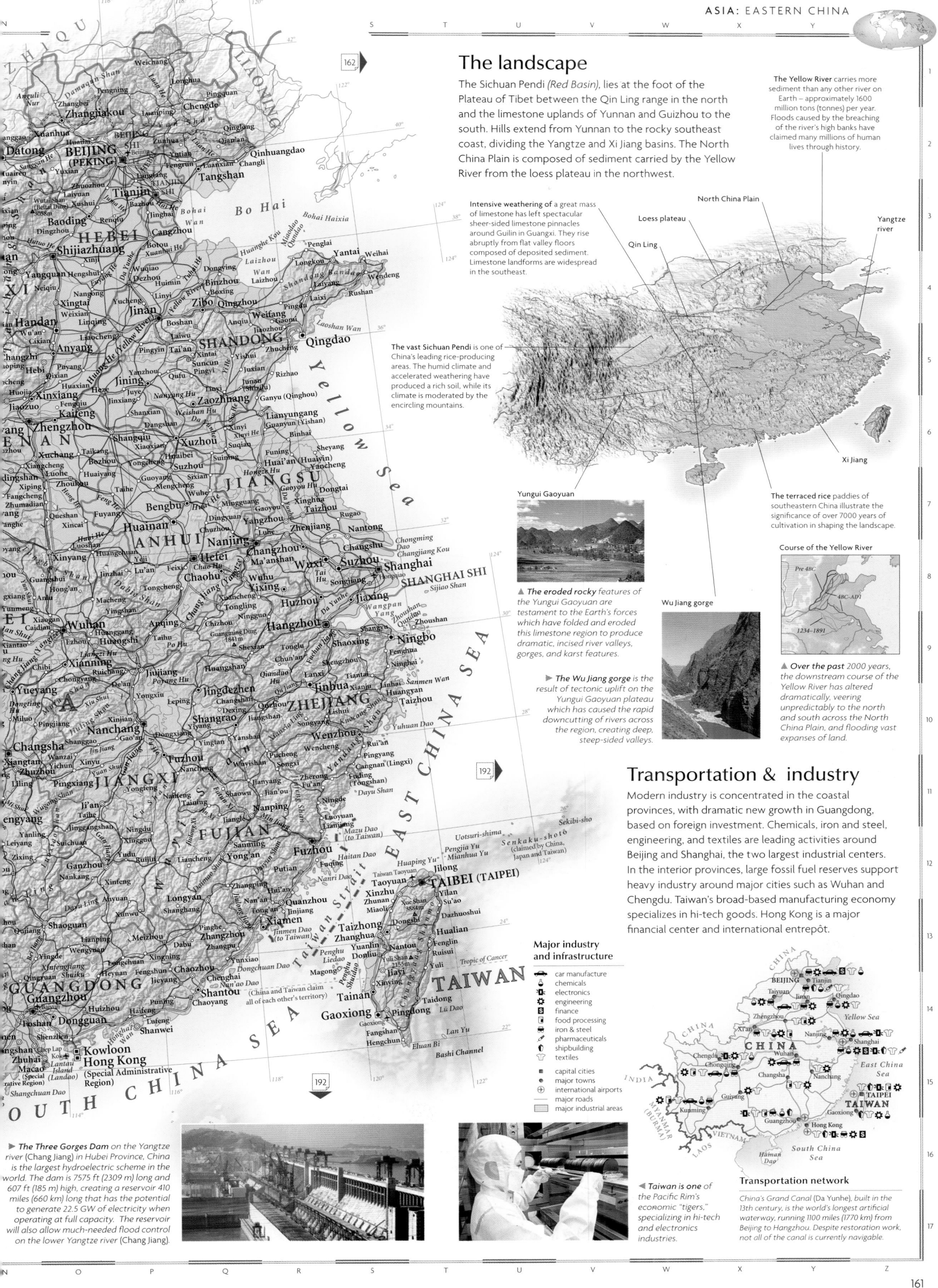

The landscape

The Sichuan Pendi *(Red Basin)*, lies at the foot of the Plateau of Tibet between the Qin Ling range in the north and the limestone uplands of Yunnan and Guizhou to the south. Hills extend from Yunnan to the rocky southeast coast, dividing the Yangtze and Xi Jiang basins. The North China Plain is composed of sediment carried by the Yellow River from the loess plateau in the northwest.

The Yellow River carries more sediment than any other river on Earth – approximately 1600 million tons (tonnes) per year. Floods caused by the breaching of the river's high banks have claimed many millions of human lives through history.

Intensive weathering of a great mass of limestone has left spectacular sheer-sided limestone pinnacles around Guilin in Guangxi. They rise abruptly from flat valley floors composed of deposited sediment. Limestone landforms are widespread in the southeast.

The vast Sichuan Pendi is one of China's leading rice-producing areas. The humid climate and accelerated weathering have produced a rich soil, while its climate is moderated by the encircling mountains.

The terraced rice paddies of southeastern China illustrate the significance of over 7000 years of cultivation in shaping the landscape.

Yungui Gaoyuan

▲ *The eroded rocky* features of the Yungui Gaoyuan are testament to the Earth's forces which have folded and eroded this limestone region to produce dramatic, incised river valleys, gorges, and karst features.

Wu Jiang gorge

▶ *The Wu Jiang gorge* is the result of tectonic uplift on the Yungui Gaoyuan plateau which has caused the rapid downcutting of rivers across the region, creating deep, steep-sided valleys.

Course of the Yellow River

Pre 48C
48C–AD1
1234–1891

▲ *Over the past* 2000 years, the downstream course of the Yellow River has altered dramatically, veering unpredictably to the north and south across the North China Plain, and flooding vast expanses of land.

Transportation & industry

Modern industry is concentrated in the coastal provinces, with dramatic new growth in Guangdong, based on foreign investment. Chemicals, iron and steel, engineering, and textiles are leading activities around Beijing and Shanghai, the two largest industrial centers. In the interior provinces, large fossil fuel reserves support heavy industry around major cities such as Wuhan and Chengdu. Taiwan's broad-based manufacturing economy specializes in hi-tech goods. Hong Kong is a major financial center and international entrepôt.

Major industry and infrastructure

- car manufacture
- chemicals
- electronics
- engineering
- finance
- food processing
- iron & steel
- pharmaceuticals
- shipbuilding
- textiles
- capital cities
- major towns
- international airports
- major roads
- major industrial areas

▼ *The Three Gorges Dam* on the Yangtze river (Chang Jiang) in Hubei Province, China is the largest hydroelectric scheme in the world. The dam is 7575 ft (2309 m) long and 607 ft (185 m) high, creating a reservoir 410 miles (660 km) long that has the potential to generate 22.5 GW of electricity when operating at full capacity. The reservoir will also allow much-needed flood control on the lower Yangtze river (Chang Jiang).

◀ *Taiwan is one* of the Pacific Rim's economic "tigers," specializing in hi-tech and electronics industries.

Transportation network

China's Grand Canal (Da Yunhe), built in the 13th century, is the world's longest artificial waterway, running 1100 miles (1770 km) from Beijing to Hangzhou. Despite restoration work, not all of the canal is currently navigable.

Northeastern China, Mongolia & Korea

MONGOLIA, NORTH KOREA, SOUTH KOREA, Heilongjiang, Inner Mongolia, Jilin, Liaoning

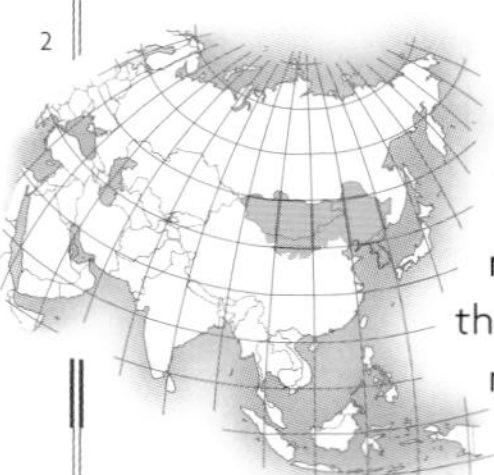

This northerly region has been a domain of shifting borders and competing colonial powers for centuries. Mongolia was the heartland of Chinghiz Khan's vast Mongol empire in the 13th century, while northeastern China was home to the Manchus, China's last ruling dynasty (1644–1911). The mineral and forest wealth of the northeast helped make this China's principal region of heavy industry, although the outdated state factories now face decline. South Korea's state-led market economy has grown dramatically and Seoul is now one of the world's largest cities. The austere communist regime of North Korea has isolated itself from the expanding markets of the Pacific Rim and faces continuing economic stagnation.

▲ **The Eurasian steppe** stretches from the mouth of the Danube in Europe, to Mongolia. In Mongolia, nomadic people have lived in felt huts called yurts or gers, for thousands of years.

Map key

Population
- ■ above 5 million
- ◉ 1 million to 5 million
- ◉ 500,000 to 1 million
- ◉ 100,000 to 500,000
- ⊕ 50,000 to 100,000
- ⊙ 10,000 to 50,000
- ○ below 10,000

Elevation
- 4000m / 13,124ft
- 3000m / 9843ft
- 2000m / 6562ft
- 1000m / 3281ft
- 500m / 1640ft
- 250m / 820ft
- 100m / 328ft
- sea level

Scale 1:7,750,000

Km
0 50 100 150 200

Miles
0 25 50 100 150 200

projection: Lambert Conformal Conic

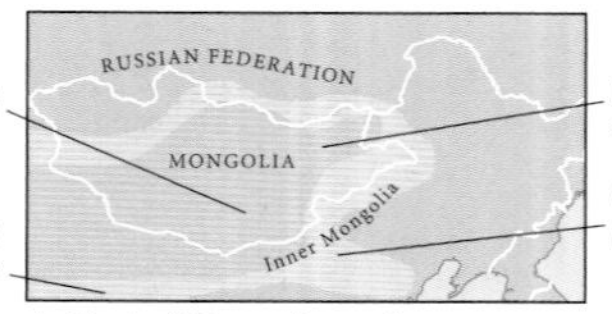

The landscape

The great North China Plain is largely enclosed by mountain ranges including the Great and Lesser Khingan Ranges (Da Hinggan Ling and Xiao Hinggan Ling) in the north, and the Changbai Shan, which extend south into the rugged peninsula of Korea. The broad steppeland plateau of Nei Mongol Gaoyuan borders the southeastern edge of the great cold desert of the Gobi which extends west across the southern reaches of Mongolia. In northwest Mongolia the Altai Mountains and various lesser ranges are interspersed with lakeland basins.

Gobi

Semiarid zone

Desert zone

Ordos Desert (Mu Us Shadi)

RUSSIAN FEDERATION

MONGOLIA

Inner Mongolia

▲ **Much of Mongolia** and Inner Mongolia is a vast desert area. To the south and east, a semiarid region extends into China proper.

▲ **The Gobi desert** stretches from Central Asia, through Mongolia and into China. Bare rock surfaces, rather than sand dunes, typify the cold desert landscape of the Gobi.

Tributaries of the Amur river follow U-shaped valleys through the Great Khingan Range (Da Hinggan Ling). These were cut by ice-age glaciers between 3 and 10 million years ago.

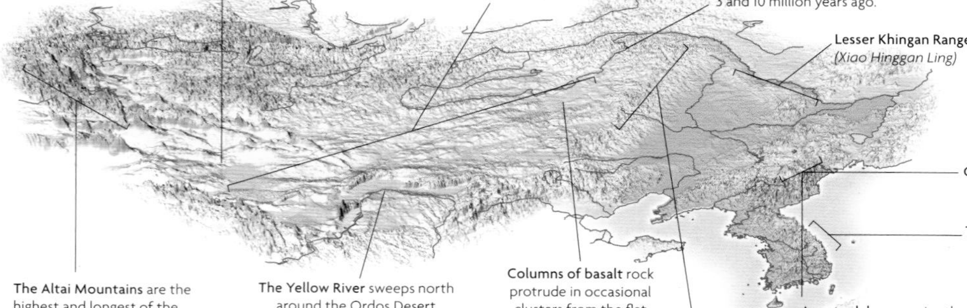

Lesser Khingan Range (Xiao Hinggan Ling)

Changbai Shan

T'aebaek-sanmaek

The Altai Mountains are the highest and longest of the mountain ranges that extend into Mongolia from the northwest. These mountains provide one of the last refuges for the endangered snow leopard.

The Yellow River sweeps north around the Ordos Desert (Mu Us Shadi), bringing water to an otherwise barren region.

Columns of basalt rock protrude in occasional clusters from the flat surface of the eastern Gobi. Their regular, six-sided form was produced when the rock cooled and contracted from its molten state.

Great Khingan Range (Da Hinggan Ling)

A crater lake occupies the 9023 ft (2750 m) snowy summit of the extinct volcano Paektu-san, the highest peak in the mountains of the Changbai Shan.

◄ **The wooded mountain** range of T'aebaek-sanmaek forms the backbone of the Korean peninsula, running north–south along the eastern coastline.

Transportation & industry

North Korea's centrally-planned economy is strongly oriented toward heavy industry, while South Korea has a broad manufacturing base which includes textiles, steel, electronics, and one of the world's largest shipbuilding industries. Mongolia and Inner Mongolia's great mineral resource potential is largely undeveloped. The heavy industrial region around Shenyang produces iron, steel, chemicals, and cement on a massive scale.

Major industry and infrastructure

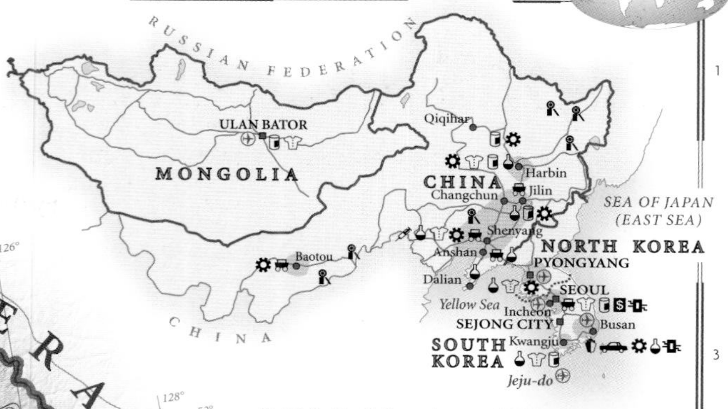

- car manufacture
- chemicals
- coal
- electronics
- engineering
- finance
- food processing
- iron & steel
- pharmaceuticals
- shipbuilding
- textiles
- ◾ capital cities
- • major towns
- ✈ international airports
- major roads
- major industrial areas

Transportation network

Liaoning has China's most comprehensive railroad network, the legacy of the Japanese occupation of Manchuria in the 20th century. The railroads are used primarily for freight transportation.

▲ *Ulan Bator, the Mongolian capital bears many of the hallmarks of Soviet-style central planning, the result of economic and industrial assistance from the Soviet Union following Mongolian independence in 1921.*

▶ *While North Korea has remained politically and economically isolated from the rest of the world, South Korea has enjoyed immense economic growth. It has benefited considerably from US economic aid in the aftermath of the Korean war of 1950–1953.*

South Korea: Capital cities

SEOUL – capital
SEJONG CITY – administrative capital

Using the land & sea

Mongolia and Inner Mongolia rely heavily on livestock farming, with only about 1% of the land area cultivated. Northeastern China produces wheat, corn, soybeans, and sugar beet. The cool climate limits the range of crops and large upland areas of the northeast remain forested. Rice is the staple food of North and South Korea. The latter has become a leading ocean-fishing nation.

Land use and agricultural distribution

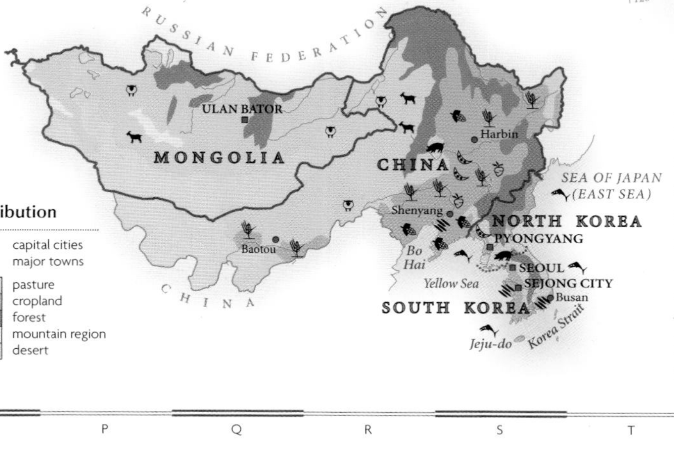

- goats
- pigs
- sheep
- corn
- fishing
- rice
- soybeans
- sugar beet
- wheat
- ◾ capital cities
- • major towns
- pasture
- cropland
- forest
- mountain region
- desert

Japan

In the years since the end of the Second World War, Japan has become the world's most dynamic industrial nation. The country comprises a string of over 4000 islands which lie in a great northeast to southwest arc in the northwest Pacific. Four major islands: Hokkaido, Honshu, Shikoku, and Kyushu are home to the great majority of Japan's population of 128 million people, although the mountainous terrain of the central region means that most cities are situated on the coast. A densely populated industrial belt stretches along much of Honshu's southern coast, including Japan's crowded capital, Tokyo. Alongside its spectacular economic growth and the increasing westernization of its cities, Japan still maintains a highly individual culture, reflected in its traditional food, formal behavioral codes, unique Shinto religion, and a deep reverence for the emperor.

Using the land & sea

Although only about 11% of Japan is suitable for cultivation, substantial government support, a favorable climate and intensive farming methods enable the country to be virtually self-sufficient in rice produc[...]. Northern Hokkaido, the largest and most productive farming region, has an open terrain and climate similar to that of the American Midwest, and produces over half of Japan's cereal requirements. Farme[...] are being encouraged to diversify by growing fruit, vegetables, and wheat, as well as raising livestock.

Land use and agricultural distribution

- cattle
- pigs
- fishing
- cereals
- citrus fruits
- fruit
- herbs
- rice
- root crops
- tobacco

■ capital cities
● major towns

pasture
cropland
forest

The urban/rural population divide

urban 78%　rural 22%

0 10 20 30 40 50 60 70 80 90 100

Population density	Total land area
885 people per sq mile (342 people per sq km)	145,869 sq miles (377,800 sq km)

The landscape

The islands of Japan lie on the Pacific "Ring of Fire," and form a series of clearly defined arcs. The largely mountainous landscape was formed very recently in geological terms. Volcanic eruptions and earthquakes continue to reshape the terrain and shake the country's complex infrastructure. There is no single continuous mountain range; the mountains divide into many small land blocks separated by lowlands and dissected by numerous river valleys.

Sea of Japan (East Sea)
Active volcanic island
Japan Trench (subduction zone)

▲ Japan is part of an arc of volcanic islands, formed by the Pacific Plate diving under the Eurasian Plate. This process generates intense stress which is periodically released as earthquakes.

◀ Mount Fuji is Japan's highest mountain, rising 12,388 ft (3776 m) above the Kanto Plain in the central region of Honshu. The flat land below is suitable for growing crops such as tea. Like many Japanese mountains, it is revered as a sacred site.

Mount Fuji

A number of rivers which emerge from the volcanic parts of northwestern Honshu are so highly acidic that their water is unsuitable for irrigation and consumption.

▶ Cutting terraces maximizes the limited agricultural land, enabling Japan to produce large quantities of rice.

▶ Trees cling to the sheer slopes of the waterfalls on the northern island of Hokkaido. The island's climate is similar to that in northern Europe, with long, cold winters and short, warm summers.

In much of Kyushu the coast is subsiding, giving a highly indented coastline. In some places, former hilltops are barely visible above the current sea level.

There are over 60 active volcanoes – like Asahi-dake, Hokkaido's highest peak – throughout Japan. This accounts for more than 10% of the world's total.

The Inland Sea (Seto-naikai) has resulted from the depression of faulted blocks which has allowed sea water to invade the region between northern Shikoku and western Honshu.

Strong southeasterly winds blowing onshore during the winter create sand dunes which extend for miles along the eastern coasts.

Biwa-ko is the largest lake in Japan, covering 260 sq miles (673 sq km) in central Honshu. The depression in which it lies was created by recent faulting of the underlying rocks.

Rising land on the Pacific coast of Honshu leads to typical features such as raised beaches, some lying over 1000 ft (300 m) above sea level.

▼ Autumnal trees near Gifu, on central Honshu, create a spectacular display. Native trees on this island include camphor, pasania, Japanese evergreen oak, camellia, and holly.

▶ The Kobe earthquake in January 1995 highlighted Japan's vulnerability to earthquakes, despite technological advances. It shattered much of the infrastructure of this important port. More than 5000 people died as buildings and overhead highways collapsed and fires broke out.

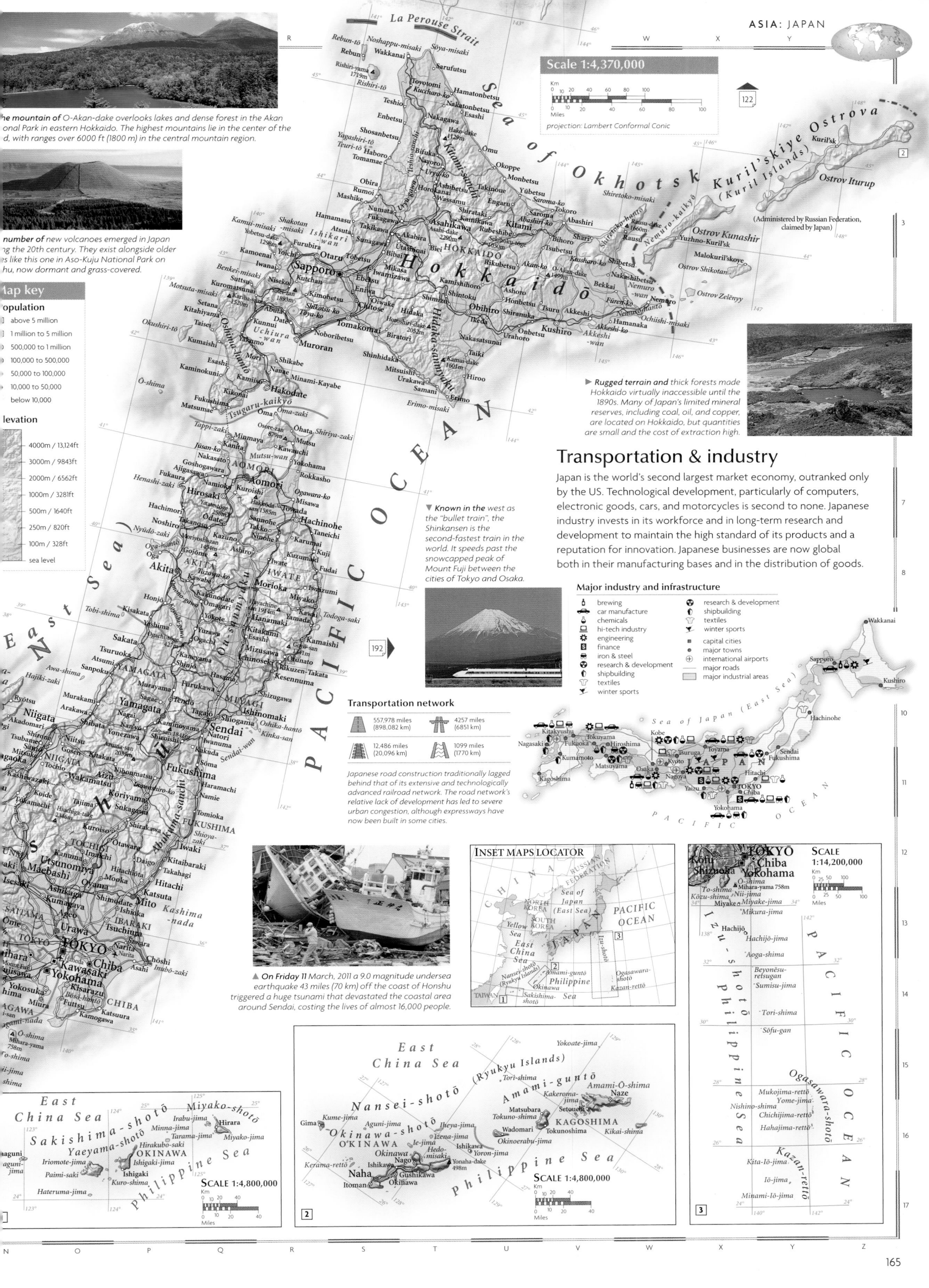

Scale 1:4,370,000

projection: Lambert Conformal Conic

The mountain of O-Akan-dake overlooks lakes and dense forest in the Akan [Nati]onal Park in eastern Hokkaido. The highest mountains lie in the center of the [islan]d, with ranges over 6000 ft (1800 m) in the central mountain region.

A number of new volcanoes emerged in Japan [duri]ng the 20th century. They exist alongside older [on]es like this one in Aso-Kuju National Park on [Kyus]hu, now dormant and grass-covered.

Map key

Population

- above 5 million
- 1 million to 5 million
- 500,000 to 1 million
- 100,000 to 500,000
- 50,000 to 100,000
- 10,000 to 50,000
- below 10,000

Elevation

- 4000m / 13,124ft
- 3000m / 9843ft
- 2000m / 6562ft
- 1000m / 3281ft
- 500m / 1640ft
- 250m / 820ft
- 100m / 328ft
- sea level

▶ Rugged terrain and thick forests made Hokkaido virtually inaccessible until the 1890s. Many of Japan's limited mineral reserves, including coal, oil, and copper, are located on Hokkaido, but quantities are small and the cost of extraction high.

Transportation & industry

Japan is the world's second largest market economy, outranked only by the US. Technological development, particularly of computers, electronic goods, cars, and motorcycles is second to none. Japanese industry invests in its workforce and in long-term research and development to maintain the high standard of its products and a reputation for innovation. Japanese businesses are now global both in their manufacturing bases and in the distribution of goods.

▼ Known in the west as the "bullet train", the Shinkansen is the second-fastest train in the world. It speeds past the snowcapped peak of Mount Fuji between the cities of Tokyo and Osaka.

Major industry and infrastructure

- brewing
- car manufacture
- chemicals
- hi-tech industry
- engineering
- finance
- iron & steel
- research & development
- shipbuilding
- textiles
- winter sports
- research & development
- shipbuilding
- textiles
- winter sports
- capital cities
- major towns
- international airports
- major roads
- major industrial areas

Transportation network

557,978 miles (898,082 km)	4257 miles (6851 km)
12,486 miles (20,096 km)	1099 miles (1770 km)

Japanese road construction traditionally lagged behind that of its extensive and technologically advanced railroad network. The road network's relative lack of development has led to severe urban congestion, although expressways have now been built in some cities.

▲ On Friday 11 March, 2011 a 9.0 magnitude undersea earthquake 43 miles (70 km) off the coast of Honshu triggered a huge tsunami that devastated the coastal area around Sendai, costing the lives of almost 16,000 people.

INSET MAPS LOCATOR

TOKYO SCALE 1:14,200,000

SCALE 1:4,800,000

SCALE 1:4,800,000

(Administered by Russian Federation, claimed by Japan)

Mainland Southeast Asia

CAMBODIA, LAOS, MYANMAR (BURMA), THAILAND, VIETNAM

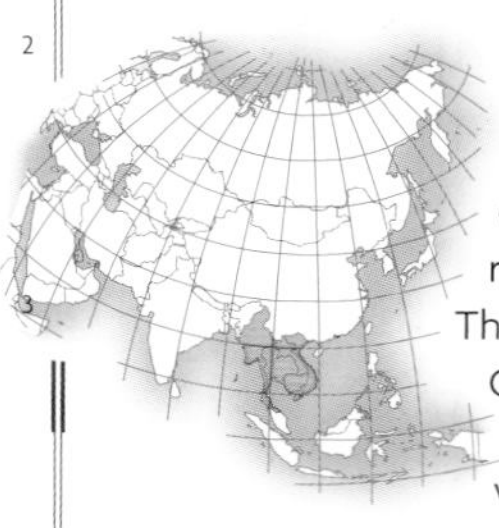

Thickly forested mountains, intercut by the broad valleys of five great rivers characterize the landscape of Southeast Asia's mainland countries. Agriculture remains the main activity for much of the population, which is concentrated in the river flood plains and deltas. Linked ethnic and cultural roots give the region a distinct identity. Most people on the mainland are Theravada Buddhists, and the Philippines is the only predominantly Christian country in Southeast Asia. Foreign intervention began in the 16th century with the opening of the spice trade; Cambodia, Laos and Vietnam were French colonies until the end of the Second World War, Myanmar (Burma) was under British control. Only Thailand was never colonized. Today, Thailand is poised to play a leading role in the economic development of the Pacific Rim, and Laos and Vietnam continues to mend the devastation of the Vietnam War, and to develop their economies. With ongoing political instability and a shattered infrastructure, Cambodia faces an uncertain future, while Myanmar (Burma) is seeking investment and the ending of its long isolation from the world community.

▲ *The Irrawaddy river* is Myanmar's (Burma) vital central artery, watering the ricefields and providing a rich source of fish, as well as an important transport link, particularly for local traffic.

The landscape

A series of mountain ranges runs north–south through the mainland, formed as the result of the collision between the Eurasian Plate and the Indian subcontinent, which created the Himalayas. They are interspersed by the valleys of a number of great rivers. On their passage to the sea these rivers have deposited sediment, forming huge, fertile flood plains and deltas.

The coastline of the Isthmus of Kra

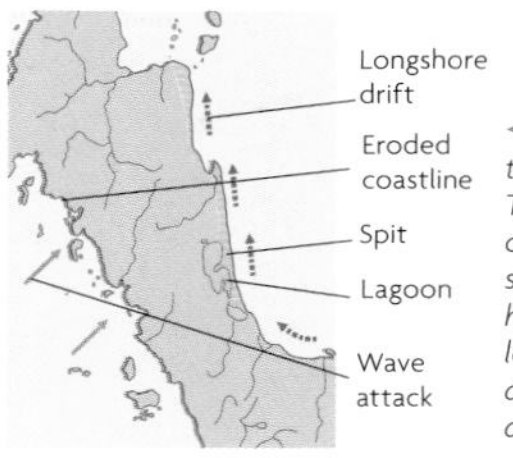

- Longshore drift
- Eroded coastline
- Spit
- Lagoon
- Wave attack

◀ *The east and* west coasts of the Isthmus of Kra differ greatly. The tectonically uplifting west coast is exposed to the harsh south-westerly monsoon and is heavily eroded. On the east coast, longshore currents produce depositional features such as spits and lagoons.

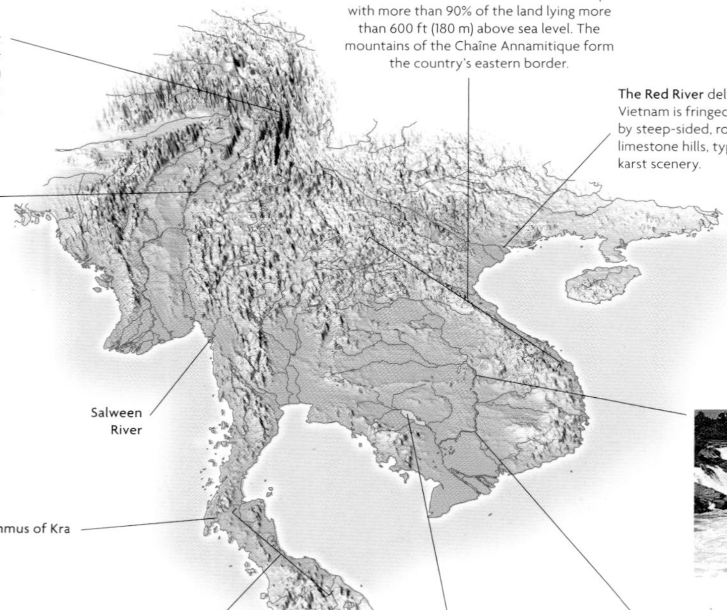

Hkakabo Razi is the highest point in mainland Southeast Asia. It rises 19,300 ft (5885 m) at the border between China and Myanmar (Burma).

Mountains dominate the Laotian landscape with more than 90% of the land lying more than 600 ft (180 m) above sea level. The mountains of the Chaîne Annamitique form the country's eastern border.

The Red River delta in northern Vietnam is fringed to the north by steep-sided, round-topped limestone hills, typical of karst scenery.

The Irrawaddy river runs virtually north–south, draining the plains of northern Myanmar (Burma). The Irrawaddy delta is the country's main rice-growing area.

Salween River

◀ *The fast-flowing waters* of the Mekong river cascade over this waterfall in Champasak province in Laos. The force of the water erodes rocks at the base of the fall.

Isthmus of Kra

▲ *The coast of* the Isthmus of Kra, in southeast Thailand has many small, precipitous islands like these, formed by chemical erosion on limestone, which is weathered along vertical cracks. The humidity of the climate in Southeast Asia increases the rate of weathering.

Malay Peninsula

Tonle Sap, a freshwater lake, drains into the Mekong delta via the Mekong river. It is the largest lake in Southeast Asia.

The Mekong river flows through southern China and Myanmar (Burma), then for much of its length forms the border between Laos and Thailand, flowing through Cambodia before terminating in a vast delta on the southern Vietnamese coast.

Using the land and sea

The fertile flood plains of rivers such as the Mekong and Salween, and the humid climate, enable the production of rice throughout the region. Cambodia, Laos, and Myanmar (Burma) still have substantial forests, producing hardwoods such as teak and rosewood. Cash crops include tropical fruits such as coconuts, bananas and pineapples, rubber, oil palm, sugar cane and the jute substitute, kenaf. Pigs and cattle are the main livestock raised. Large quantities of marine and freshwater fish are caught throughout the region.

▲ *Commercial logging* – still widespread in Myanmar (Burma) – has now been stopped in Thailand because of over-exploitation of the tropical rainforest.

The urban/rural population divide

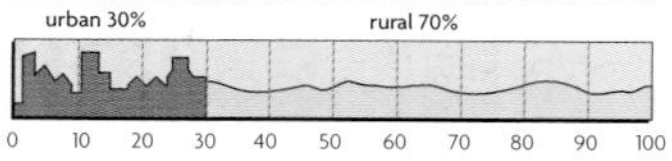

urban 30%   rural 70%

0	10	20	30	40	50	60	70	80	90	100

Population density	Total land area
345 people per sq mile (133 people per sq km)	733,828 sq miles (1,901,110 sq km)

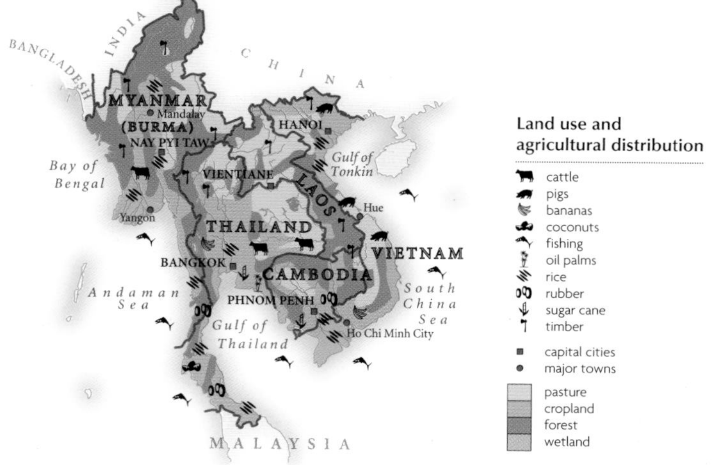

Land use and agricultural distribution

- cattle
- pigs
- bananas
- coconuts
- fishing
- oil palms
- rice
- rubber
- sugar cane
- timber
- capital cities
- major towns
- pasture
- cropland
- forest
- wetland

Transportation & industry

Industrial manufacturing has become increasingly important in Thailand and Vietnam in recent years. The assembling of component-based electrical and electronic goods is becoming more common throughout this region, with foreign companies benefiting from low labour costs and the upgrading of technology. The economies of Myanmar (Burma) and Cambodia are still based on agricultural produce and the processing of raw materials. Tin is the region's most important metal, and nickel, copper and chromite are also mined, although the quantities produced are not significant on a global scale. Thailand's successful tourist industry is the country's highest earner of foreign exchange.

Transportation network

🛣	82,958 miles (133,524 km)	🛣	267 miles (430 km)
🚂	7500 miles (12,071 km)	✈	28,585 miles (46,008 km)

Transportation development has concentrated on the building of road networks. Water and sea transport remain important, although air links have improved, particularly in Thailand and the Philippines.

Major industry and infrastructure

⚗	chemicals	⬙	oil & gas	■	capital cities
⚡	electronics	⛏	mining	■	major towns
⚙	engineering	⚓	shipbuilding	⊕	international airports
$	finance	⟁	textiles	—	major roads
⚒	food processing	♣	timber processing	▨	major industrial areas
⬢	iron & steel				

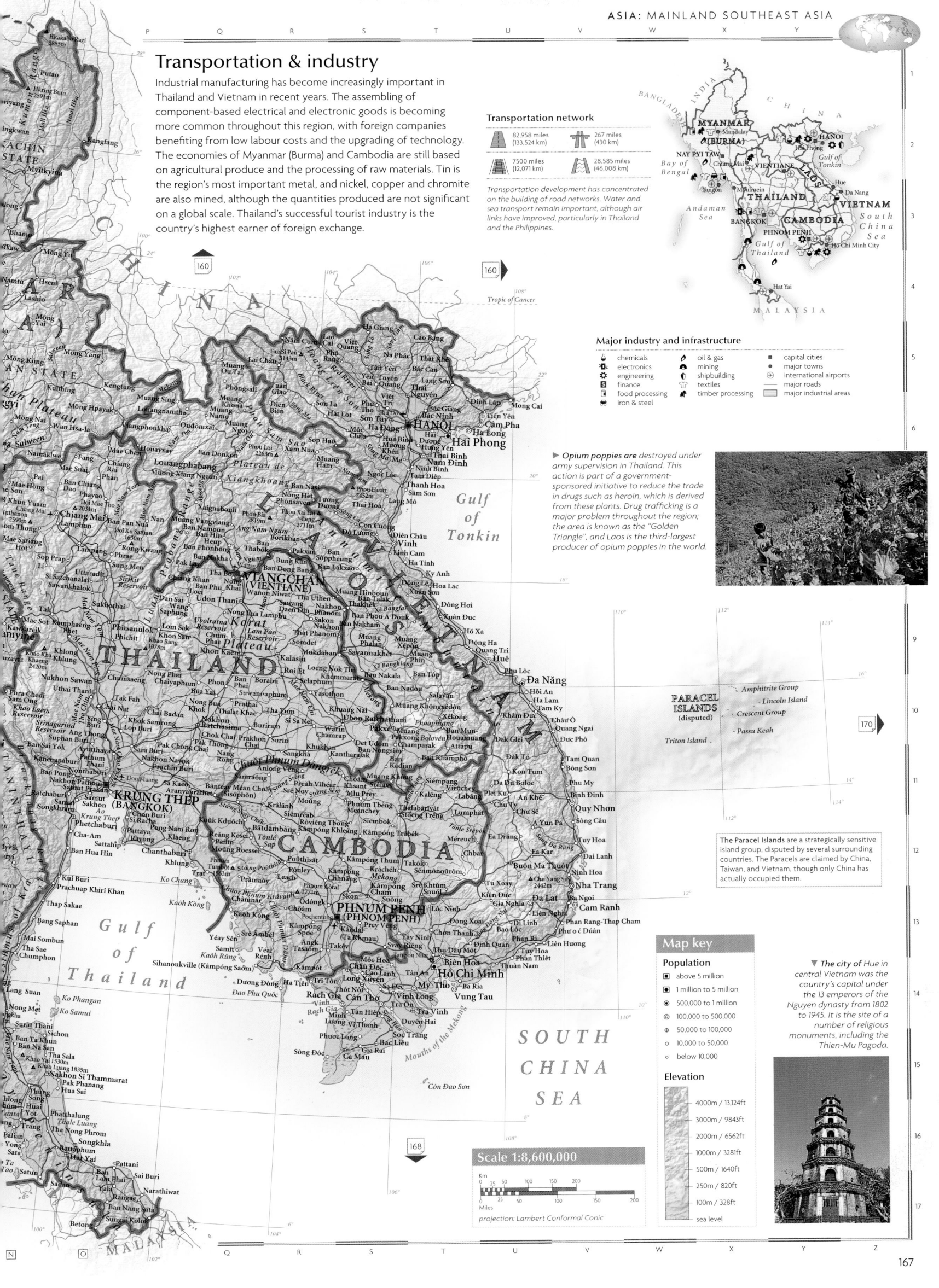

▶ **Opium poppies are** destroyed under army supervision in Thailand. This action is part of a government-sponsored initiative to reduce the trade in drugs such as heroin, which is derived from these plants. Drug trafficking is a major problem throughout the region; the area is known as the "Golden Triangle", and Laos is the third-largest producer of opium poppies in the world.

The Paracel Islands are a strategically sensitive island group, disputed by several surrounding countries. The Paracels are claimed by China, Taiwan, and Vietnam, though only China has actually occupied them.

Map key

Population

▣	above 5 million
◪	1 million to 5 million
◉	500,000 to 1 million
◎	100,000 to 500,000
⊕	50,000 to 100,000
⊙	10,000 to 50,000
∘	below 10,000

Elevation

	4000m / 13,124ft
	3000m / 9843ft
	2000m / 6562ft
	1000m / 3281ft
	500m / 1640ft
	250m / 820ft
	100m / 328ft
	sea level

▼ **The city of** Hue in central Vietnam was the country's capital under the 13 emperors of the Nguyen dynasty from 1802 to 1945. It is the site of a number of religious monuments, including the Thien-Mu Pagoda.

Scale 1:8,600,000

projection: Lambert Conformal Conic

Western Maritime Southeast Asia

BRUNEI, INDONESIA, MALAYSIA, SINGAPORE

The world's largest archipelago, Indonesia's myriad islands stretch 3100 miles (5000 km) eastward across the Pacific, from the Malay Peninsula to western New Guinea. Only about 1500 of the 13,677 islands are inhabited and the huge, predominently Muslim population is unevenly distributed, with some two-thirds crowded onto the western islands of Java, Madura, and Bali. The national government is trying to resettle large numbers of people from these islands to other parts of the country to reduce population pressure there. Malaysia, split between the mainland and the east Malaysian states of Sabah and Sarawak on Borneo, has a diverse population, as well as a fast-growing economy, although the pace of its development is still far outstripped by that of Singapore. This small island nation is the financial and commercial capital of Southeast Asia. The Sultanate of Brunei in northern Borneo, one of the world's last princely states, has an extremely high standard of living, based on its oil revenues.

The landscape

Indonesia's western islands are characterized by rugged volcanic mountains cloaked with dense tropical forest, which slope down to coastal plains covered by thick alluvial swamps. The Sunda Shelf, an extension of the Eurasian Plate, lies between Java, Bali, Sumatra, and Borneo. These islands' mountains rise from a base below the sea, and they were once joined together by dry land, which has since been submerged by rising sea levels.

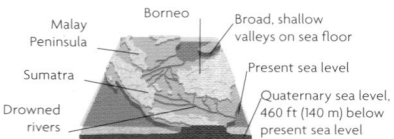

▲ The Sunda Shelf underlies this whole region. It is one of the largest submarine shelves in the world, covering an area of 714,285 sq miles (1,850,000 sq km). During the early Quaternary period, when sea levels were lower, the shelf was exposed.

◄ On January 24, 2005 a 9.2 magnitude earthquake off the coast of Sumatra triggered a devastating tsunami that was up to 90 ft (30 m) high in places. The death toll was estimated to be around 230,000 people from fourteen different countries around the Indian Ocean.

Malay Peninsula has a rugged east coast, but the west coast, fronting the Strait of Malacca, has many sheltered beaches and bays. The two coasts are divided by the Banjaran Titiwangsa, which run the length of the peninsula.

Gunung Kinabalu is the highest peak in Malaysia, rising 13,455 ft (4101 m).

◄ The river of Sungai Mahakam cuts through the central highlands of Borneo, the third largest island in the world, with a total area of 290,000 sq miles (757,050 sq km). Although mountainous, Borneo is one of the most stable of the Indonesian islands, with little volcanic activity.

The island of Krakatau (Pulau Rakata), lying between Sumatra and Java, was all but destroyed in 1883, when the volcano erupted. The release of gas and dust into the atmosphere disrupted cloud cover and global weather patterns for several years.

Indonesia has more than 220 volcanoes, most of which are still active. They are strung out along the island arc from Sumatra through the Lesser Sunda Islands, into the Moluccas and Celebes.

Transportation & industry

Singapore has a thriving economy based on international trade and finance. Annual trade through the port is among the highest of any in the world. Indonesia's western islands still depend on natural resources, particularly petroleum, gas, and wood, although the economy is rapidly diversifying with manufactured exports including garments, consumer electronics, and footwear. A high-profile aircraft industry has developed in Bandung on Java. Malaysia has a fast-growing and varied manufacturing sector, although oil, gas, and timber remain important resource-based industries.

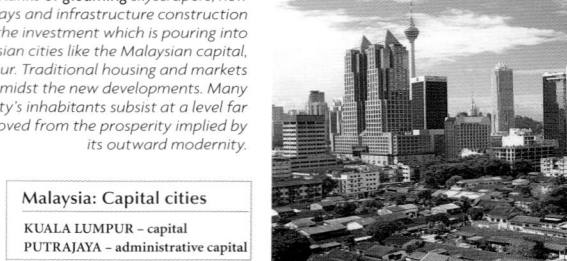

► Ranks of gleaming skyscrapers, new motorways and infrastructure construction reflect the investment which is pouring into Southeast Asian cities like the Malaysian capital, Kuala Lumpur. Traditional housing and markets still exist amidst the new developments. Many of the city's inhabitants subsist at a level far removed from the prosperity implied by its outward modernity.

Malaysia: Capital cities
KUALA LUMPUR – capital
PUTRAJAYA – administrative capital

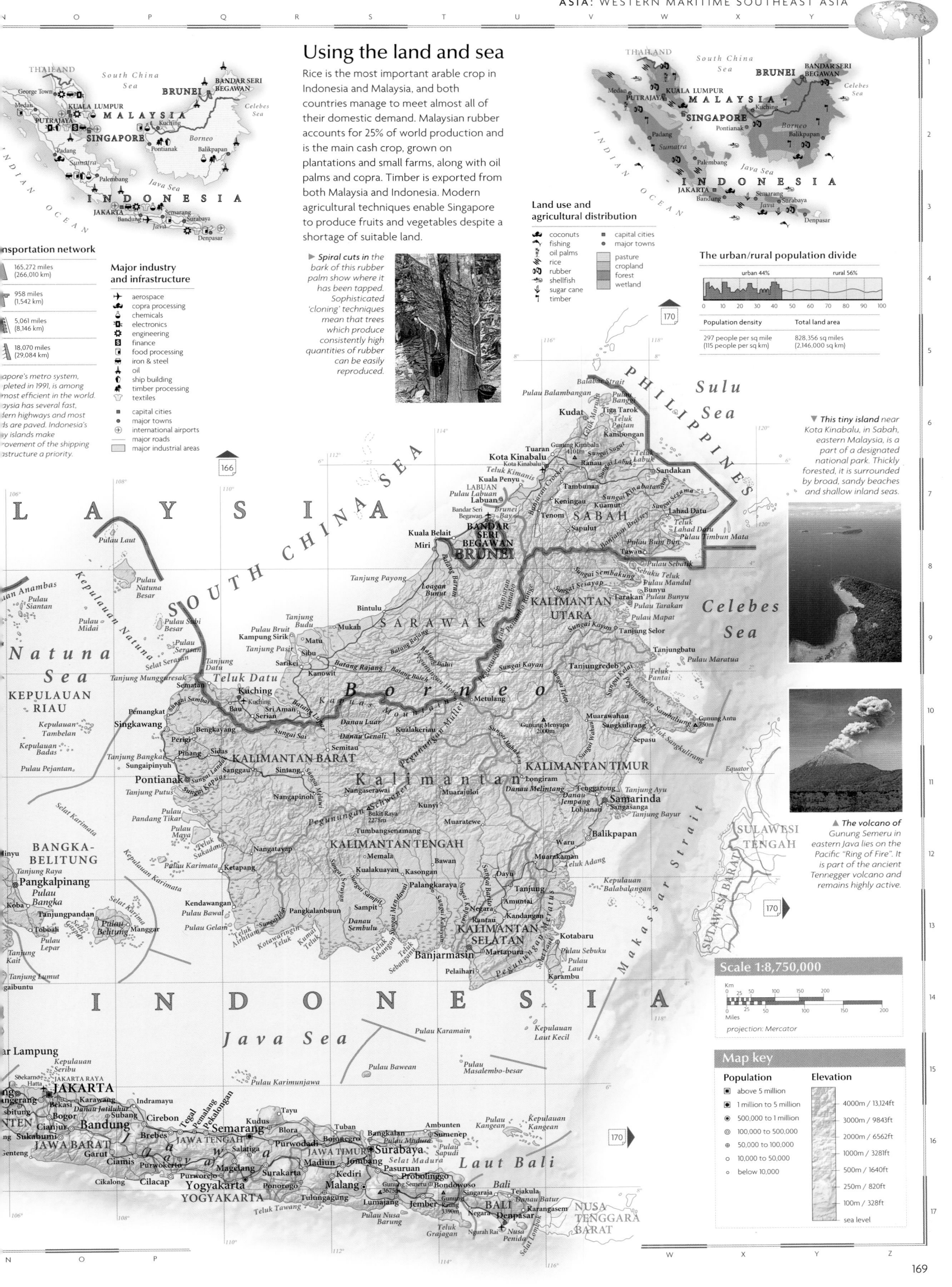

Using the land and sea

Rice is the most important arable crop in Indonesia and Malaysia, and both countries manage to meet almost all of their domestic demand. Malaysian rubber accounts for 25% of world production and is the main cash crop, grown on plantations and small farms, along with oil palms and copra. Timber is exported from both Malaysia and Indonesia. Modern agricultural techniques enable Singapore to produce fruits and vegetables despite a shortage of suitable land.

► Spiral cuts in the bark of this rubber palm show where it has been tapped. Sophisticated 'cloning' techniques mean that trees which produce consistently high quantities of rubber can be easily reproduced.

Transportation network

165,272 miles (266,010 km)	
958 miles (1,542 km)	
5,061 miles (8,146 km)	
18,070 miles (29,084 km)	

Singapore's metro system, completed in 1991, is among the most efficient in the world. Malaysia has several fast, modern highways and most roads are paved. Indonesia's many islands make improvement of the shipping infrastructure a priority.

Major industry and infrastructure

- aerospace
- copra processing
- chemicals
- electronics
- engineering
- finance
- food processing
- iron & steel
- oil
- ship building
- timber processing
- textiles

- ■ capital cities
- ● major towns
- ✈ international airports
- major roads
- major industrial areas

Land use and agricultural distribution

- coconuts
- fishing
- oil palms
- rice
- rubber
- shellfish
- sugar cane
- timber

- ■ capital cities
- ● major towns
- pasture
- cropland
- forest
- wetland

The urban/rural population divide

urban 44%   rural 56%

0 10 20 30 40 50 60 70 80 90 100

Population density	Total land area
297 people per sq mile (115 people per sq km)	828,356 sq miles (2,146,000 sq km)

▼ This tiny island near Kota Kinabalu, in Sabah, eastern Malaysia, is a part of a designated national park. Thickly forested, it is surrounded by broad, sandy beaches and shallow inland seas.

▲ The volcano of Gunung Semeru in eastern Java lies on the Pacific "Ring of Fire". It is part of the ancient Tennegger volcano and remains highly active.

Scale 1:8,750,000

Km
0 25 50   100   150   200

Miles
0 25 50   100   150   200

projection: Mercator

Map key

Population
- ■ above 5 million
- ■ 1 million to 5 million
- ◉ 500,000 to 1 million
- ◉ 100,000 to 500,000
- ⊙ 50,000 to 100,000
- ○ 10,000 to 50,000
- ○ below 10,000

Elevation
- 4000m / 13,124ft
- 3000m / 9843ft
- 2000m / 6562ft
- 1000m / 3281ft
- 500m / 1640ft
- 250m / 820ft
- 100m / 328ft
- sea level

A   B   C   D   E   F   G   H   I   J   K   L   M

Eastern Maritime Southeast Asia

EAST TIMOR, INDONESIA, PHILIPPINES

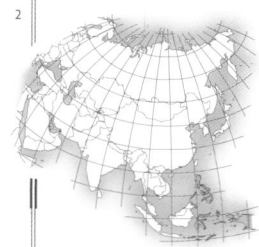

The Philippines takes its name from Philip II of Spain who was king when the islands were colonized during the 16th century. Almost 400 years of Spanish, and later US, rule have left their mark on the country's culture; English is widely spoken and over 90% of the population is Christian. The Philippines' economy is agriculturally based – inadequate infrastructure and electrical power shortages have so far hampered faster industrial growth. Indonesia's eastern islands are less economically developed than the rest of the country. Papua, which constitutes the western portion of New Guinea, is one of the world's last great wildernesses. After a long struggle, East Timor gained full autonomy from Indonesia in 2002.

▲ *The traditional boat-shaped* houses of the Toraja people in Sulawesi. Although now Christian, the Toraja still practice the animist traditions and rituals of their ancestors. They are famous for their elaborate funeral ceremonies and burial sites in cliffside caves.

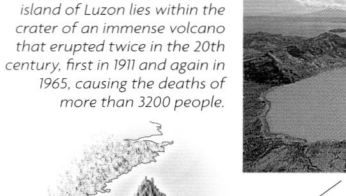

▶ *Lake Taal on* the Philippines island of Luzon lies within the crater of an immense volcano that erupted twice in the 20th century, first in 1911 and again in 1965, causing the deaths of more than 3200 people.

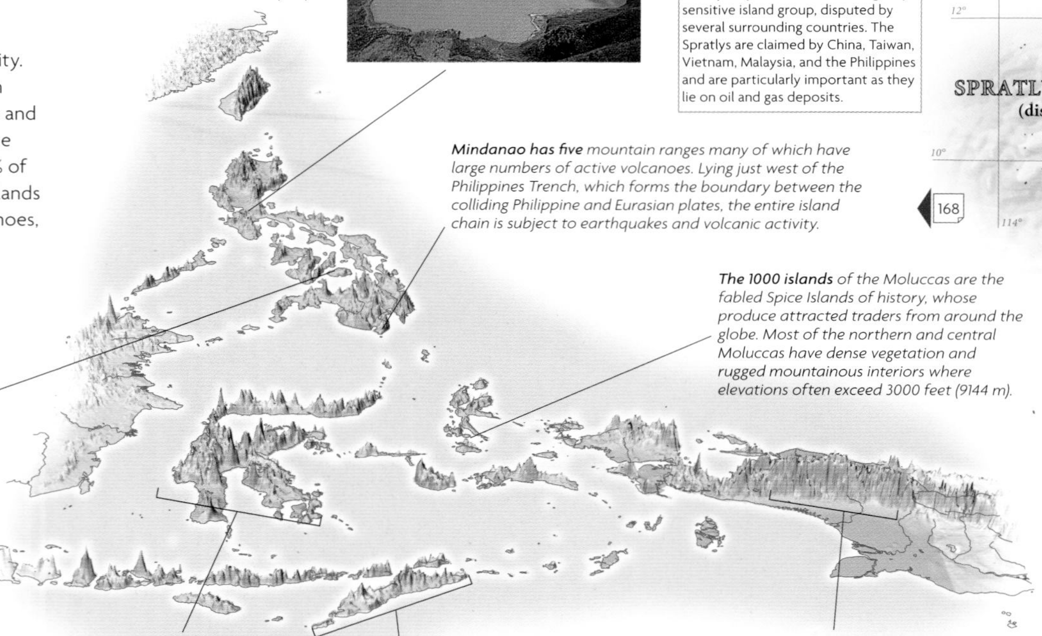

The landscape

Located on the Pacific "Ring of Fire" the Philippines' 7100 islands are subject to frequent earthquakes and volcanic activity. Their terrain is largely mountainous, with narrow coastal plains and interior valleys and plains. Luzon and Mindanao are by far the largest islands and comprise roughly 66% of the country's area. Indonesia's eastern islands are mountainous and dotted with volcanoes, both active and dormant.

The Spratly Islands are a strategically sensitive island group, disputed by several surrounding countries. The Spratlys are claimed by China, Taiwan, Vietnam, Malaysia, and the Philippines and are particularly important as they lie on oil and gas deposits.

Mindanao has five mountain ranges many of which have large numbers of active volcanoes. Lying just west of the Philippines Trench, which forms the boundary between the colliding Philippine and Eurasian plates, the entire island chain is subject to earthquakes and volcanic activity.

The 1000 islands of the Moluccas are the fabled Spice Islands of history, whose produce attracted traders from around the globe. Most of the northern and central Moluccas have dense vegetation and rugged mountainous interiors where elevations often exceed 3000 feet (9144 m).

▲ *Bohol in the* southern Philippines is famous for its so-called "chocolate hills". There are more than 1000 of these regular mounds on the island. The hills are limestone in origin, the smoothed remains of an earlier cycle of erosion. Their brown appearance in the dry season gives them their name.

The four-pronged island of Celebes is the product of complex tectonic activity which ruptured and then reattached small fragments of the Earth's crust to form the island's many peninsulas.

Coral islands such as Timor in eastern Indonesia show evidence of very recent and dramatic movements of the Earth's plates. Reefs in Timor have risen by as much as 4000 ft (1300 m) in the last million years.

The Pegunungan Jayawijaya range in central Papua contains the world's highest range of limestone mountains, some with peaks more than 16,400 ft (5000 m) high. Heavy rainfall and high temperatures, which promote rapid weathering, have led to the creation of large underground caves and river systems such as the river of Sungai Baliem.

Using the land and sea

Indonesia's eastern islands are less intensively cultivated than those in the west. Coconuts, coffee and spices such as cloves and nutmeg are the major commercial crops while rice, corn and soybeans are grown for local consumption. The Philippines' rich, fertile soils support year-round production of a wide range of crops. The country is one of the world's largest producers of coconuts and a major exporter of coconut products, including one-third of the world's copra. Although much of the arable land is given over to rice and corn, the main staple food crops, tropical fruits such as bananas, pineapples and mangos, and sugar cane are also grown for export.

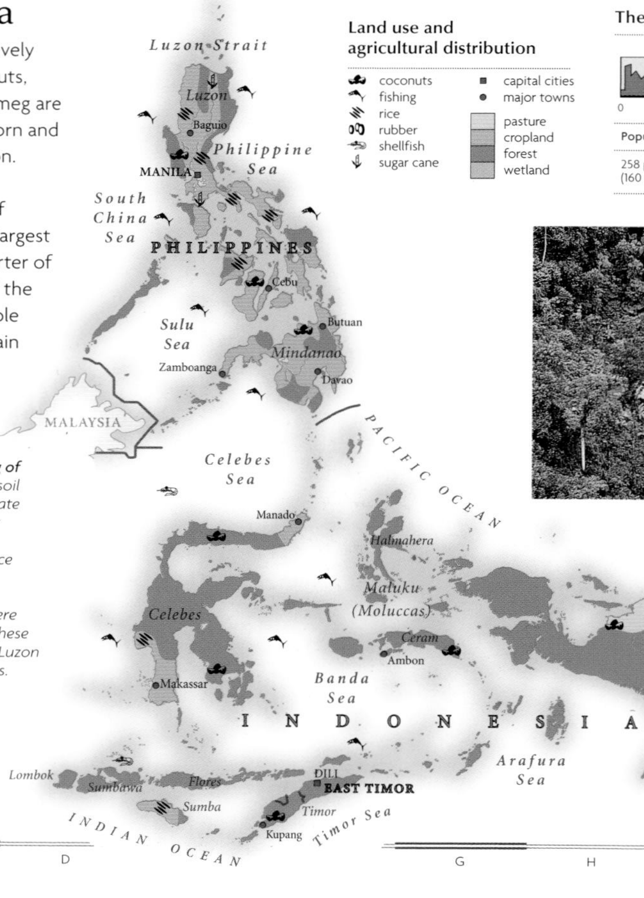

Land use and agricultural distribution

- 🥥 coconuts
- 🐟 fishing
- 🌾 rice
- rubber
- shellfish
- 🌾 sugar cane

- ■ capital cities
- ● major towns

- pasture
- cropland
- forest
- wetland

The urban/rural population divide

urban 45%	rural 55%

0   10   20   30   40   50   60   70   80   90   100

Population density	Total land area
258 people per sq mile (160 people per sq km)	654,771 sq miles (1,053,755 sq km)

◀ *The terracing of* land to restrict soil erosion and create flat surfaces for agriculture is a common practice throughout Southeast Asia, particularly where land is scarce. These terraces are on Luzon in the Philippines.

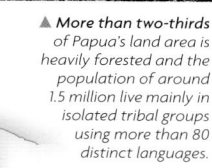

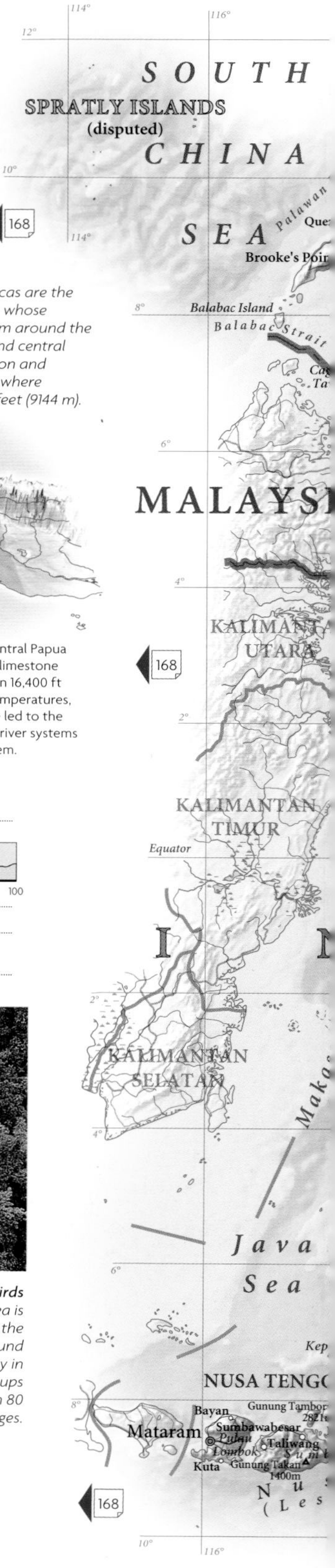

▲ *More than two-thirds* of Papua's land area is heavily forested and the population of around 1.5 million live mainly in isolated tribal groups using more than 80 distinct languages.

Map labels

SOUTH CHINA SEA

SPRATLY ISLANDS (disputed)

168

Palawan
Brooke's Point
Balabac Island
Balabac Strait

MALAYSIA

KALIMANTAN UTARA

168

KALIMANTAN TIMUR

Equator

NUSA TENGGARA

Mataram
Bayan
Gunung Tambor
Sumbawabesar
Lombok
Kuta

Luzon Strait
Luzon
Baguio
Philippine Sea
MANILA
South China Sea
PHILIPPINES
Cebu
Butuan
Sulu Sea
Mindanao
Zamboanga
Davao
MALAYSIA
Celebes Sea
Manado
Halmahera
Maluku (Moluccas)
Celebes
Ceram
Ambon
Makassar
Banda Sea
Jayapura
New Guinea
PAPUA NEW GUINEA
INDONESIA
Arafura Sea
Lombok
Sumbawa
Flores
Sumba
DILI
EAST TIMOR
Timor
Kupang
Timor Sea
INDIAN OCEAN
PACIFIC OCEAN
Java Sea

A   B   C   D   E   F   G   H   I   J   K   L   M

Transportation & industry

The Philippines' economy is primarily a mixture of agriculture and light industry. The manufacturing sector is still developing; many factories are licensees of foreign companies producing finished goods for export. Mining is also important – the country's chromite, nickel, and copper deposits are among the largest in the world. Agriculture is the main activity in eastern Indonesia. Most industry has a primary basis, including logging, food-processing, and mining. Nickel, the most important metal, is produced on Sulawesi, in Papua, and in the Moluccas.

Major industry and infrastructure

- copra processing
- chemicals
- finance
- food processing
- mining
- oil
- timber processing
- textiles
- capital cities
- major towns
- international airports
- major roads
- major industrial areas

Transportation network

	16,652 miles (26,800 km)
	None
	500 miles (805 km)
	8704 miles (14,008 km)

Sulawesi has some good roads, but on Papua and the Moluccas there are few road interconnections between major settled areas. Water and sea transportation remain important although air links have improved in the Philippines.

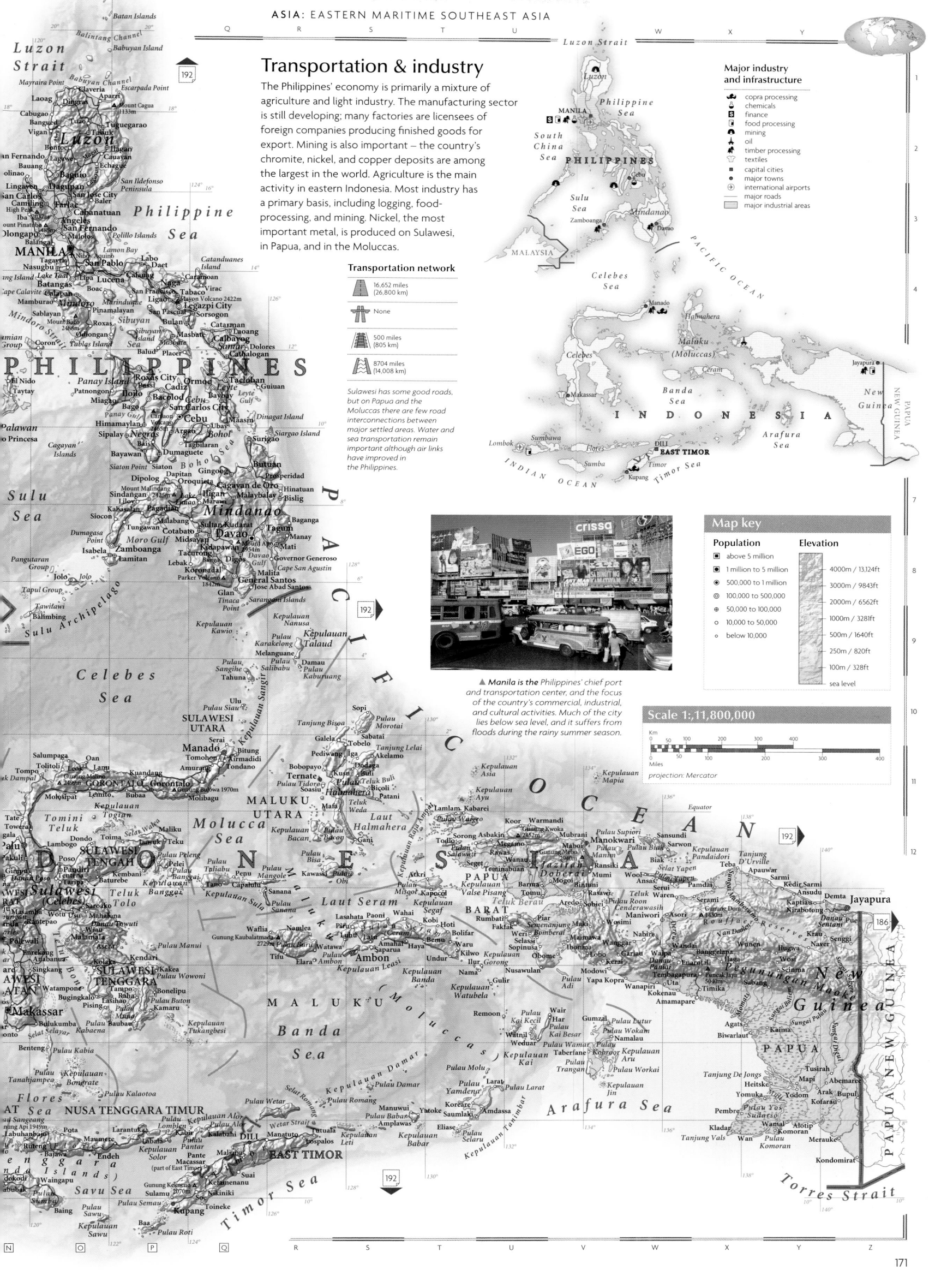

▲ **Manila is the** Philippines' chief port and transportation center, and the focus of the country's commercial, industrial, and cultural activities. Much of the city lies below sea level, and it suffers from floods during the rainy summer season.

Map key

Population
- ■ above 5 million
- ■ 1 million to 5 million
- ◉ 500,000 to 1 million
- ◉ 100,000 to 500,000
- ⊕ 50,000 to 100,000
- ⊕ 10,000 to 50,000
- ○ below 10,000

Elevation
- 4000m / 13,124ft
- 3000m / 9843ft
- 2000m / 6562ft
- 1000m / 3281ft
- 500m / 1640ft
- 250m / 820ft
- 100m / 328ft
- sea level

Scale 1:11,800,000

projection: Mercator

The Indian Ocean

Despite being the smallest of the three major oceans, the evolution of the Indian Ocean was the most complex. The ocean basin was formed during the breakup of the supercontinent Gondwanaland, when the Indian subcontinent moved northeast, Africa moved west, and Australia separated from Antarctica. Like the Pacific Ocean, the warm waters of the Indian Ocean are punctuated by coral atolls and islands. About one-fifth of the world's population – over a billion people – live on its shores. In 2004, over 290,000 died and millions more were left homeless after a tsunami devastated large stretches of the ocean's coastline.

The landscape

The Indian Ocean began forming about 150 million years ago, but in its present form it is relatively young, only about 36 million years old. Along the three subterranean mountain chains of its mid-ocean ridge the seafloor is still spreading. The Indian Ocean has fewer trenches than other oceans and only a narrow continental shelf around most of its surrounding land.

Sediments come from Ganges/Brahmaputra river system

Submarine canyons transport sediment to fan – some of these are more than 1500 miles (2500 km) long

Sri Lanka

▲ *The Ganges Fan* is one of the world's largest submarine accumulations of sediment, extending far beyond Sri Lanka. It is fed by the Ganges/Brahmaputra river system, whose sediment is carried through a network of underwater canyons at the edge of the continental shelf.

The Ninetyeast Ridge takes its name from the line of longitude it follows. It is the world's longest and straightest under-sea ridge.

Two of the world's largest rivers flow into the Indian Ocean; the Indus and the Ganges/Brahmaputra. Both have deposited enormous fans of sediment.

The mid-oceanic ridge runs from the Arabian Sea. It diverges east of Madagascar. One arm runs southwest to join the Mid-Atlantic Ridge, the other branches southeast, joining the Pacific-Antarctic Ridge, southeast of Tasmania.

Indus River

▶ *A large proportion* of the coast of Thailand, on the Isthmus of Kra, is stabilized by mangrove thickets. They act as an important breeding ground for wildlife.

The relief of Madagascar rises from a low-lying coastal strip in the east, to the central plateau. The plateau is also a major watershed separating Madagascar's three main river basins.

▶ *The central group* of the Seychelles are mountainous, granite islands. They have a narrow coastal belt and lush, tropical vegetation cloaks the highlands.

The Kerguelen Islands in the Southern Ocean were created by a hot spot in the Earth's crust. The islands were formed in succession as the Antarctic Plate moved slowly over the hot spot.

The Java Trench is the world's longest, it runs 1600 miles (2570 km) from the southwest of Java, but is only 50 miles (80 km) wide.

The circulation in the northern Indian Ocean is controlled by the monsoon winds. Biannually these winds reverse their pattern, causing a reversal in the surface currents and alternative high and low pressure conditions over Asia and Australia.

Resources

Many of the small islands in the Indian Ocean rely exclusively on tuna-fishing and tourism to maintain their economies. Most fisheries are artisanal, although large-scale tuna-fishing does take place in the Seychelles, Mauritius and the western Indian Ocean. Other resources include oil in the Persian Gulf, pearls in the Red Sea, and tin from deposits off the shores of Myanmar, Thailand, and Indonesia.

▶ *The recent use* of large dragnets for tuna-fishing has not only threatened the livelihoods of many small-scale fisheries, but also caused widespread environmental concern about the potential impact on other marine species.

Resources (including wildlife)

- ⌇ fish
- 🐧 penguins
- 🐚 shellfish
- 🐋 whales
- ⬗ oil & gas
- △ tin deposits
- ◔ tourism
- ● major towns
- ⊕ major ports

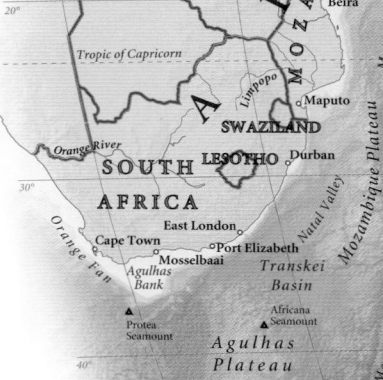

SCALE 1:12,250,000

MADAGASCAR

Îles Glorieuses (to France)
Tanjona Bobaomby
Antsiraùana
Nosy Be
Ambilobe
Nosy Be
Ambanja
ANTSIRAÙANA
Iharaùa
Maromokotro 2876m
Analalava
Bealanana
Sambava
Antsohihy
Andapa
Antalaha
Mahajanga
Befandriana
Ambohitralanana
Maevatanana
Mampikony
Maroantsetra
Marovoay
Soalala
Mitsinjo
Manarana Avaratra
Soanierana-Ivongo
Tsaratanana
Andilamena
Tanjona Masoala
MAHAJANGA
TOAMASINA
Tanjona Vilanandro
Besalampy
Maevatanana
Farihy Alaotra
Nosy Boraha
Kandreho
Ambatomainty
Amparafaravola
Ambodifotatra
Ambatondrazaka
Maintirano
Soavinandriana
Ankazobe
Ambatolampy
Toamasina
Anjozorobe
Toamasina
Antsalova
Tsiroanomandidy
Arivonimamo
ANTANANARIVO
Tsaratanana
Ambohidratrimo
Manjakandriana
Ampasimanolotra
Belo Tsiribihina
Zsiribihiny
Ambohimahasoa
Ambatolampy
Anosibe
Morondava
Antsirabe
An'ala
Vatomandry
Miandrivazo
Faratsiho
Antanambao
Manampotsy
Mahanoro
Mahabo
Fandriana
Marolambo
Mandabe
Ambatofinandrahana
Ambositra
Nosy Varika
Belo sur Tsiribihina
Ikalamavony
Mananjary
Beroroha
Fianarantsoa
Ifanadiana
Tanjona Ankaboa
Morombe
Mangoky
Ambalavao
Ikongo
Manakara
Ankazoabo
FIANARANTSOA
Ihosy
Ivohibe
Vohipeno
Sakaraha
Vondrozo
Farafangana
Betroka
Jakora
Toliara
Onilahy
Midongy Atsimo
Vangaindrano
Betioky
TOLIARA
Befotaka
Bekily
Mandrare
Ampanihy
Amboasary
Tôlaùaro
Beloha
Ambovombe
Tanjona Vohimena
Tsiombe

SCALE 1:5,000,000

Ngazidja (Grande Comore)
Mitsamiouli
Saondzou 1087m
Hahaya
Mbéni
Koimbani
MORONI
Itsandra
Le Kartala 2361m
Mitsoudjé
Foumbouni
Dembéni
COMOROS
INDIAN OCEAN
Nzwani (Anjouan)
Mwali (Mohéli)
Moutsamoudou
Ouani
Miringoni
Fomboni Sima
Domoni
Nioumachoua
Ouanani Moya
Mramani
MAYOTTE (to France)
Mozambique Channel
Comoro Islands
Dzaoudzi
Pamandzi
MAMOUDZOU
Bandrélé

Inner Islands
Île Aride
Praslin
Les Sœurs
Curieuse
Grand Sœur
Cousin
Félicité
Cousine
Marianne
Île du Nord
La Digue
Mount Dauban 740m
SEYCHELLES
INDIAN OCEAN
Silhouette
Mahé
North Point
Île aux Récifs
VICTORIA
Sainte Anne
Frégate
Morne Seychellois 905m
Île au Cerf
Cascade
Île Thérèse
Anse Boileau
Pointe Lazare
Baie Lazare
Quatre Bornes
Pointe Police

SCALE 1:2,250,000

EGYPT
Suez
Yanbu' al Bahr
Red Sea
Tropic of Cancer
Port Sudan
Massawa
SUDAN
ERITREA
DJIBOUTI
Djibouti
ETHIOPIA
KENYA
Lake Victoria
Equator
Tanga
Mombasa
Pemba
Zanzibar
TANZANIA
Dar es Salaam
Mafia
Lake Nyasa
Ruvuma
COMOR
Mayo (to Fra
Nacala
Quelimane
MOZAMBIQUE
Beira
Tropic of Capricorn
Bassas da India
Île Europa
Davie Fracture Zone
Limpopo
SWAZILAND
Maputo
Natal Basin
Orange River
SOUTH
LESOTHO
Durban
AFRICA
East London
Natal Valley
Cape Town
Port Elizabeth
Transkei Basin
Agulhas Bank
Mosselbaai
Mozambique Escarpment
Protea Seamount
Africana Seamount
Agulhas Plateau
Agulhas Basin
Prince E
Atlantic-Indian Ridge
Prince Edward Fracture Zone
Atlantic-Ind Basin

ASIA
Suez
Kuwait
Mumbai
Rangoon
Arabian Sea
Bay of Bengal
South China Sea
Singapore
Mombasa
Java Sea
INDIAN OCEAN
Timor Sea
AFRICA
Toamasina
AUSTRALIA
Fremantle
SOUTHERN OCEAN
ANTARCTICA

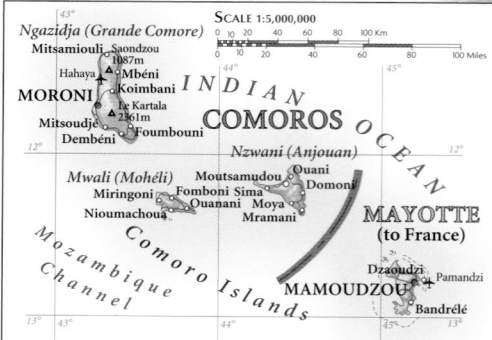

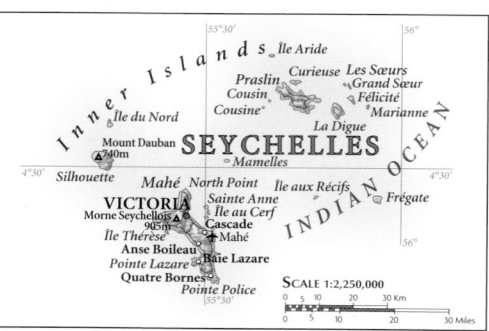

▲ *Coral reefs support* an enormous diversity of animal and plant life. Many species of tropical fish, like these squirrel fish, live and feed around the profusion of reefs and atolls in the Indian Ocean.

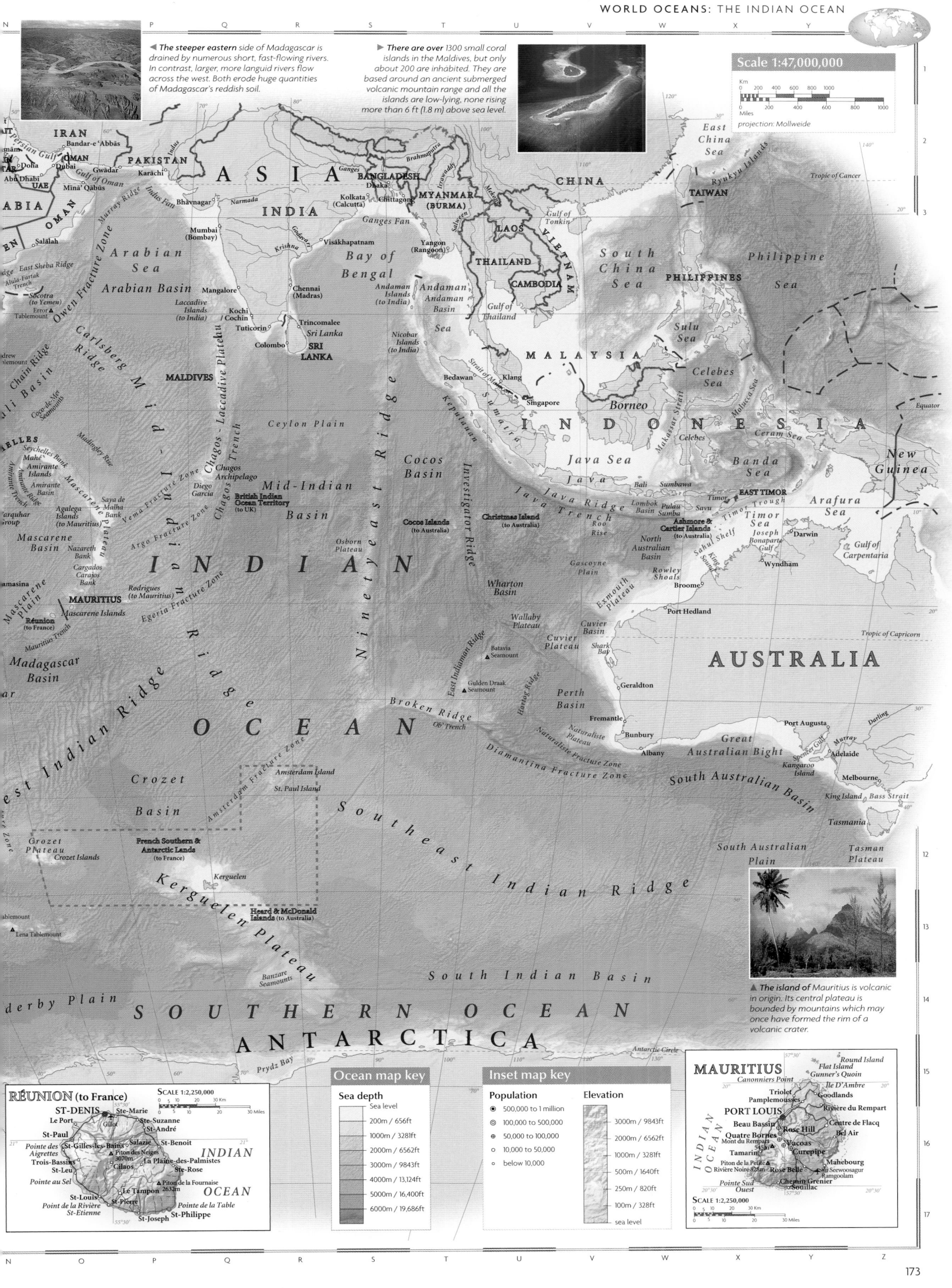

The steeper eastern side of Madagascar is drained by numerous short, fast-flowing rivers. In contrast, larger, more languid rivers flow across the west. Both erode huge quantities of Madagascar's reddish soil.

There are over 1300 small coral islands in the Maldives, but only about 200 are inhabited. They are based around an ancient submerged volcanic mountain range and all the islands are low-lying, none rising more than 6 ft (1.8 m) above sea level.

The island of Mauritius is volcanic in origin. Its central plateau is bounded by mountains which may once have formed the rim of a volcanic crater.

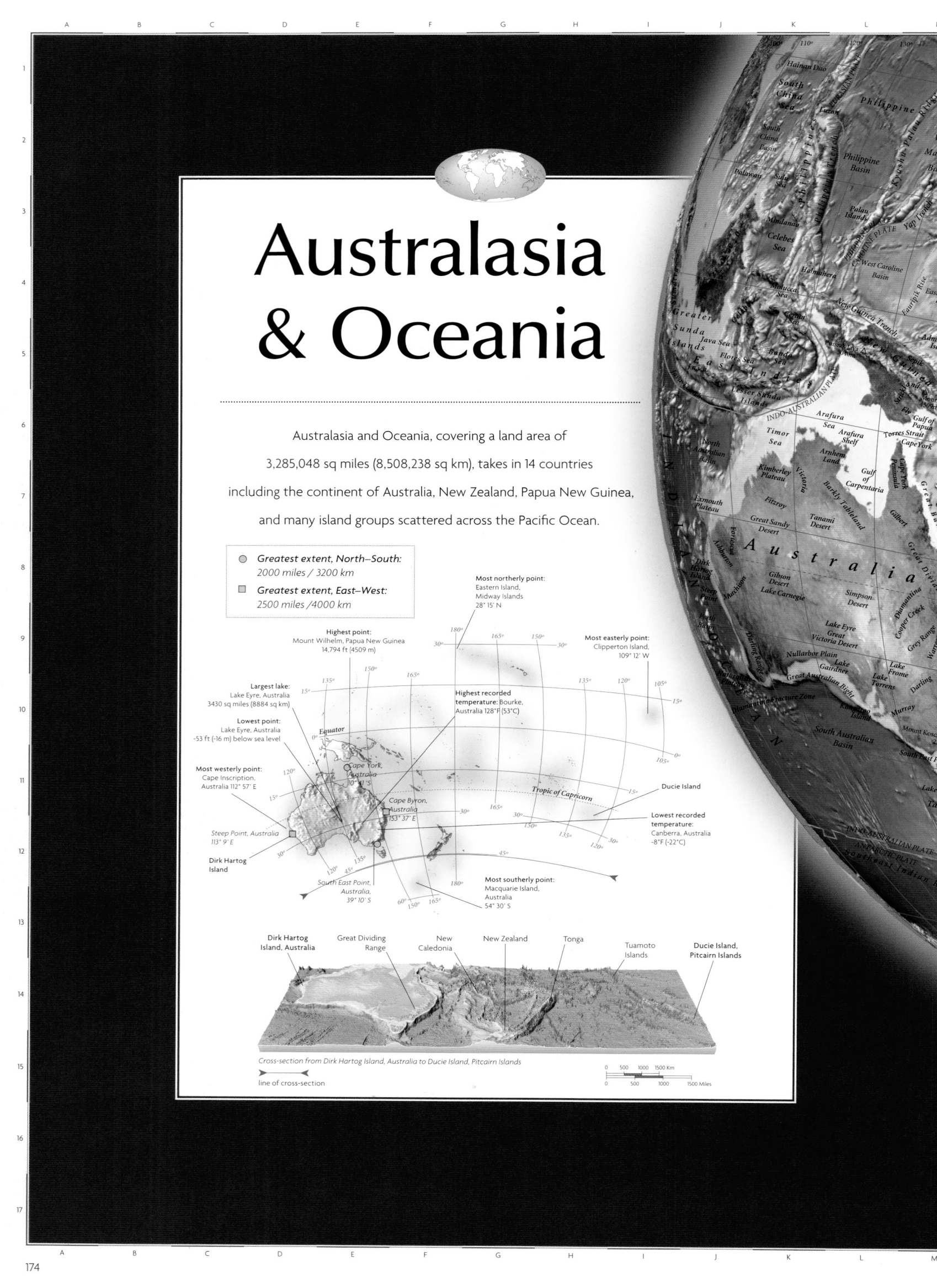

Australasia & Oceania

Australasia and Oceania, covering a land area of
3,285,048 sq miles (8,508,238 sq km), takes in 14 countries
including the continent of Australia, New Zealand, Papua New Guinea,
and many island groups scattered across the Pacific Ocean.

● **Greatest extent, North–South:**
2000 miles / 3200 km

■ **Greatest extent, East–West:**
2500 miles /4000 km

Most northerly point:
Eastern Island,
Midway Islands
28° 15' N

Highest point:
Mount Wilhelm, Papua New Guinea
14,794 ft (4509 m)

Most easterly point:
Clipperton Island,
109° 12' W

Largest lake:
Lake Eyre, Australia
3430 sq miles (8884 sq km)

Highest recorded
temperature: Bourke,
Australia 128°F (53°C)

Lowest point:
Lake Eyre, Australia
-53 ft (-16 m) below sea level

Equator

Most westerly point:
Cape Inscription,
Australia 112° 57' E

Cape York,
Australia
10° 41' S

Ducie Island

Tropic of Capricorn

Cape Byron,
Australia
153° 37' E

Lowest recorded
temperature:
Canberra, Australia
-8°F (-22°C)

Steep Point, Australia
113° 9' E

Dirk Hartog
Island

South East Point,
Australia,
39° 10' S

Most southerly point:
Macquarie Island,
Australia
54° 30' S

Dirk Hartog
Island, Australia

Great Dividing
Range

New
Caledonia

New Zealand

Tonga

Tuamoto
Islands

Ducie Island,
Pitcairn Islands

Cross-section from Dirk Hartog Island, Australia to Ducie Island, Pitcairn Islands

line of cross-section

0   500   1000   1500 Km

0   500   1000   1500 Miles

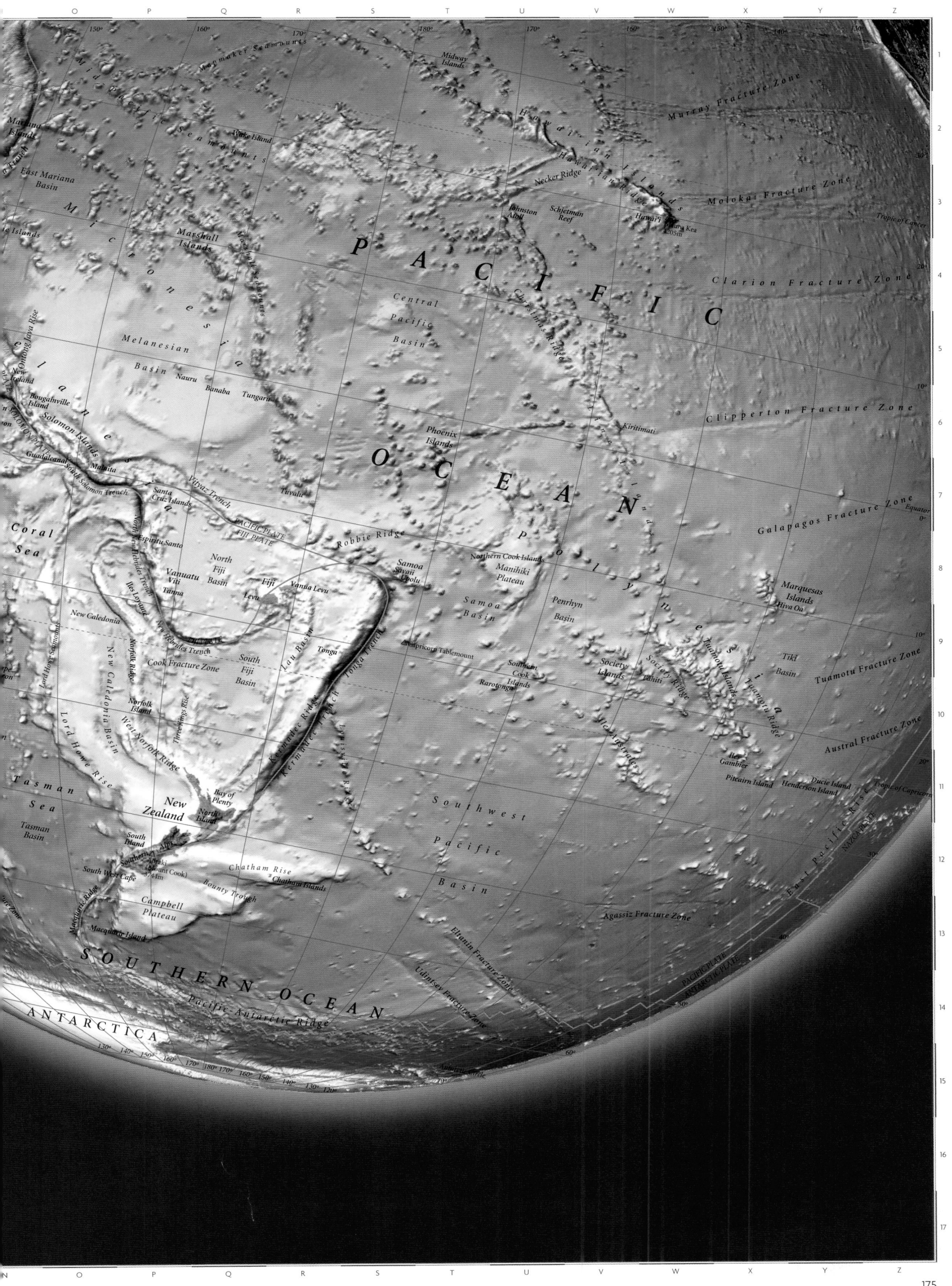

Marcus Seamounts
Midway Islands
Hawaiian Islands
Murray Fracture Zone
Mariana Islands
Necker Ridge
East Mariana Basin
Molokai Fracture Zone
Johnston Atoll
Schjetman Reef
Hawai'i
Mauna Kea 4205m
Tropic of Cancer
Micronesia
P A C I F I C
Islands
Marshall Islands
Clarion Fracture Zone
Marshall Seamounts
Wake Island
Central Pacific Basin
Melanesian Basin
Nauru
Banaba
Tungaru
Clipperton Fracture Zone
Ontong Java Rise
New Ireland
Bougainville Island
Solomon Islands
Kiritimati
O C E A N
Phoenix Islands
Guadalcanal
North Solomon Trench
Malaita
Vityaz Trench
Tuvalu
Santa Cruz Islands
Coral Sea
New Hebrides Trench
Espiritu Santo
PACIFIC PLATE
FIJI PLATE
Robbie Ridge
Galapagos Fracture Zone
Equator
Vanuatu
North Fiji Basin
Fiji
Samoa
Savai'i
Upolu
Northern Cook Islands
Manihiki Plateau
Viti Levu
Vanua Levu
Marquesas Islands
Hiva Oa
Tanna
Ile Loyauté
New Hebrides Trench
Samoa Basin
Penrhyn Basin
New Caledonia
Tonga
Cook Fracture Zone
South Fiji Basin
Lau Basin
Capricorn Tablemount
P o l y n e s i a
Society Islands
Society Ridge
Tahiti
Tuamotu Islands
Tuamotu Ridge
Tiki Basin
Tuamotu Fracture Zone
Norfolk Ridge
Three Kings Rise
Southern Cook Islands
Rarotonga
Lord Howe Seamounts
Norfolk Island
New Caledonia Basin
Kermadec Ridge
Kermadec Trench
Louisville Ridge
Tasman Sea
Lord Howe Rise
West Norfolk Ridge
Ile Gambier
Austral Fracture Zone
Bay of Plenty
Pitcairn Island
Henderson Island
Ducie Island
Tropic of Capricorn
New Zealand
North Island
Southwest
Tasman Basin
South Island
Southern Alps
Aoraki (Mount Cook) 3744m
Chatham Rise
Chatham Islands
Pacific
East Pacific Rise
NAZCA PLATE
South West Cape
Bounty Trough
Basin
Agassiz Fracture Zone
Macquarie Ridge
Campbell Plateau
Eltanin Fracture Zone
Macquarie Island
Udintsev Fracture Zone
PACIFIC PLATE
ANTARCTIC PLATE
S O U T H E R N   O C E A N
A N T A R C T I C A
Pacific-Antarctic Ridge
Antarctic Circle

Political Australasia & Oceania

Vast expanses of ocean separate this geographically fragmented realm, characterized more by each country's isolation than by any political unity. Australia's and New Zealand's traditional ties with the United Kingdom, as members of the Commonwealth, are now being called into question as Australasian and Oceanian nations are increasingly looking to forge new relationships with neighboring Asian countries like Japan. External influences have featured strongly in the politics of the Pacific Islands; the various territories of Micronesia were largely under US control until the late 1980s, and France, New Zealand, the US, and the UK still have territories under colonial rule in Polynesia. Nuclear weapons-testing by Western superpowers was widespread during the Cold War period, but has now been discontinued.

◄ *Western Australia's mineral* wealth has transformed its state capital, Perth, into one of Australia's major cities. Perth is one of the world's most isolated cities – over 2500 miles (4000 km) from the population centers of the eastern seaboard.

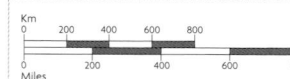
Scale 1:35,500,000

Km 0 200 400 600 800
Miles 0 200 400 600 800
projection: Lambert Azimuthal Equal Area

Population

Density of settlement in the region is generally low. Australia is one of the least densely populated countries on Earth with over 80% of its population living within 25 miles (40 km) of the coast – mostly in the southeast of the country. New Zealand, and the island groups of Melanesia, Micronesia, and Polynesia, are much more densely populated, although many of the smaller islands remain uninhabited.

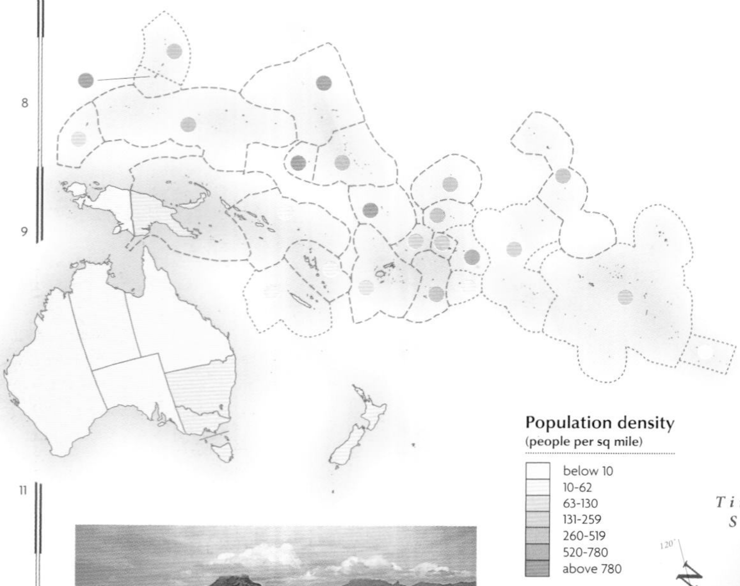

Population density
(people per sq mile)

- below 10
- 10-62
- 63-130
- 131-259
- 260-519
- 520-780
- above 780

▲ *The myriad of* small coral islands that are scattered across the Pacific Ocean are often uninhabited, as they offer little shelter from the weather, often no fresh water, and only limited food supplies.

◄ *The planes of* the Australian Royal Flying Doctor Service are able to cover large expanses of barren land quickly, bringing medical treatment to the most inaccessible and far-flung places.

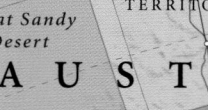

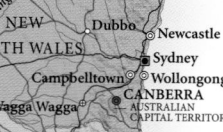

Philippine Sea

Northern Mariana Islands (to US)

Mariana Islands

Saipan

Guam (to US)

HAGÅTÑA

Bikini Atoll

Yap

Caroline Islands

Chuuk

Pohnpei ● PALIKIR

Kosrae

MELEKEOK
Babeldaob

PALAU

M i c r o n

MICRONESIA

M e l a n e s i

NAUR
YAREN

PAPUA NEW GUINEA

Bismarck Sea

New Ireland

Wewak

New Britain

Rabaul

New Guinea

Madang

Ubai

Bougainville Island

Arawa

Solomon Islands

SOLOM
ISLANI

Mount Hagen

Lae

Solomon Sea

New Georgia Islands

HONIARA

Tapini

Guadalcanal

PORT MORESBY

Arafura Sea

Torres Strait

Coral Sea

Santa Cruz Islands

VANU
Espiritu Santo
Malekul

Equator

Darwin

Arnhem Land

Katherine

Cape York Peninsula

Gulf of Carpentaria

Cairns

Coral Sea Islands (to Australia)

New Caledonia (to France)

NOUMÉA

P

Timor Sea

Joseph Bonaparte Gulf

Wyndham

Normanton

Townsville

Great Barrier Reef

Mackay

Kimberley Plateau

Derby

Broome

NORTHERN

Tennant Creek

Tanami Desert

Mount Isa

Hughenden

QUEENSLAND

Barcaldine

Rockhampton

Great Dividing Range

Port Hedland

Great Sandy Desert

TERRITORY

Alice Springs

Simpson Desert

Charleville

Cunnamulla

Miles

Toowoomba

Brisbane

Norfol
(to A

Hamersley Range

Gibson Desert

AUSTRALIA

Lake Eyre North

Bourke

Barwon

Grey Range

Darling

Grafton

Lord Howe Island (to Australia)

Carnarvon

WESTERN AUSTRALIA

Great Victoria Desert

SOUTH AUSTRALIA

Lake Torrens

Lake Everard

Wilcannia

Flinders Ranges

NEW SOUTH WALES

Dubbo

Newcastle

Mount Magnet

Lake Gairdner

Ceduna

Whyalla

Port Augusta

Murray

Campbelltown

Sydney

Wollongong

Geraldton

Kalgoorlie

Nullarbor Plain

Great Australian Bight

Adelaide

Wagga Wagga

CANBERRA

AUSTRALIAN CAPITAL TERRITORY

INDIAN OCEAN

Tropic of Capricorn

Kangaroo Island

Bendigo

Horsham

VICTORIA

Ballarat

Melbourne

Geelong

Tasman Sea

Perth

Esperance

Mount Gambier

Bass Strait

Launceston

TASMANIA

Albany

Tasmania

Hobart

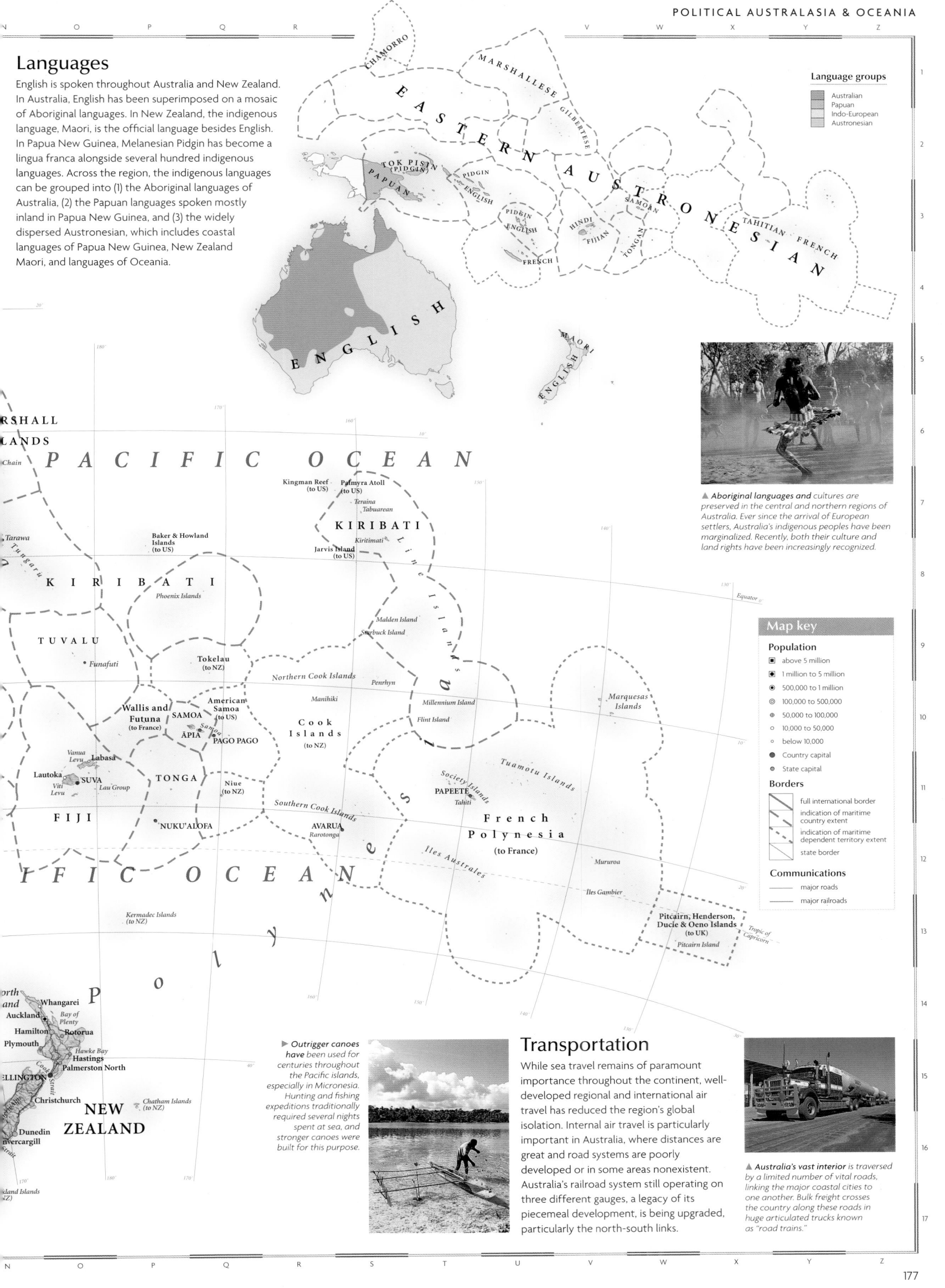

Languages

English is spoken throughout Australia and New Zealand. In Australia, English has been superimposed on a mosaic of Aboriginal languages. In New Zealand, the indigenous language, Maori, is the official language besides English. In Papua New Guinea, Melanesian Pidgin has become a lingua franca alongside several hundred indigenous languages. Across the region, the indigenous languages can be grouped into (1) the Aboriginal languages of Australia, (2) the Papuan languages spoken mostly inland in Papua New Guinea, and (3) the widely dispersed Austronesian, which includes coastal languages of Papua New Guinea, New Zealand Maori, and languages of Oceania.

Language groups
- Australian
- Papuan
- Indo-European
- Austronesian

CHAMORRO

MARSHALLESE GILBERTESE

EASTERN AUSTRONESIAN

TOK PISIN (PIDGIN)
PAPUAN
PIDGIN ENGLISH
PIDGIN ENGLISH
SAMOAN
HINDI
FIJIAN
TONGAN
TAHITIAN FRENCH
FRENCH

ENGLISH

MAORI
ENGLISH

▲ *Aboriginal languages and cultures are preserved in the central and northern regions of Australia. Ever since the arrival of European settlers, Australia's indigenous peoples have been marginalized. Recently, both their culture and land rights have been increasingly recognized.*

PACIFIC OCEAN

MARSHALL ISLANDS
Ratak Chain
Tungaru

Kingman Reef (to US)
Palmyra Atoll (to US)
Teraina
Tabuaerean

KIRIBATI
Kiritimati

Tarawa

Baker & Howland Islands (to US)
Jarvis Island (to US)

KIRIBATI
Phoenix Islands

TUVALU
Funafuti

Malden Island
Starbuck Island

Tokelau (to NZ)

Line Islands

Equator

Northern Cook Islands
Penrhyn

Manihiki

Marquesas Islands

Wallis and Futuna (to France)
SAMOA
APIA
American Samoa (to US)
PAGO PAGO
Samoa
Cook Islands (to NZ)

Millennium Island
Flint Island

Vanua Levu
Labasa
Lautoka
Viti Levu
SUVA
Lau Group
TONGA
Niue (to NZ)

Tuamotu Islands

Society Islands
PAPEETE
Tahiti

FIJI
NUKU'ALOFA
Southern Cook Islands
AVARUA
Rarotonga

French Polynesia (to France)

Iles Australes
Mururoa

PACIFIC OCEAN

Iles Gambier

Kermadec Islands (to NZ)

Pitcairn, Henderson, Ducie & Oeno Islands (to UK)
Pitcairn Island

Tropic of Capricorn

Polynesia

Map key

Population
- ▣ above 5 million
- ▣ 1 million to 5 million
- ◉ 500,000 to 1 million
- ◎ 100,000 to 500,000
- ⊕ 50,000 to 100,000
- ○ 10,000 to 50,000
- ○ below 10,000
- ● Country capital
- ● State capital

Borders
- full international border
- indication of maritime country extent
- indication of maritime dependent territory extent
- state border

Communications
- major roads
- major railroads

North Island
Whangarei
Auckland
Bay of Plenty
Hamilton
Rotorua
Plymouth
Hawke Bay
Hastings
Palmerston North
WELLINGTON
Chatham Islands (to NZ)

NEW ZEALAND

Christchurch
Dunedin
Invercargill

▶ *Outrigger canoes have been used for centuries throughout the Pacific islands, especially in Micronesia. Hunting and fishing expeditions traditionally required several nights spent at sea, and stronger canoes were built for this purpose.*

Transportation

While sea travel remains of paramount importance throughout the continent, well-developed regional and international air travel has reduced the region's global isolation. Internal air travel is particularly important in Australia, where distances are great and road systems are poorly developed or in some areas nonexistent. Australia's railroad system still operating on three different gauges, a legacy of its piecemeal development, is being upgraded, particularly the north-south links.

▲ *Australia's vast interior is traversed by a limited number of vital roads, linking the major coastal cities to one another. Bulk freight crosses the country along these roads in huge articulated trucks known as "road trains."*

Australasian & Oceanian resources

Natural resources are of major economic importance throughout Australasia and Oceania. Australia in particular is a major world exporter of raw materials such as coal, iron ore, and bauxite, while New Zealand's agricultural economy is dominated by sheep-raising. Trade with western Europe has declined significantly in the last 20 years, and the Pacific Rim countries of Southeast Asia are now the main trading partners, as well as a source of new settlers to the region. Australasia and Oceania's greatest resources are its climate and environment; tourism increasingly provides a vital source of income for the whole continent.

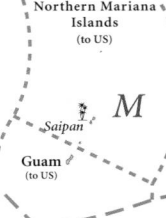

▲ *The largely unpolluted* waters of the Pacific Ocean support rich and varied marine life, much of which is farmed commercially. Here, oysters are gathered for market off the coast of New Zealand's South Island.

▶ *Huge flocks of* sheep are a common sight in New Zealand, where they outnumber people by 12 to 1. New Zealand is one of the world's largest exporters of wool and frozen lamb.

Standard of living

In marked contrast to its neighbor, Australia, with one of the world's highest life expectancies and standards of living, Papua New Guinea is one of the world's least developed countries. In addition, high population growth and urbanization rates throughout the Pacific islands contribute to overcrowding. In Australia and New Zealand, the Aboriginal and Maori people have been isolated, although recently their traditional land ownership rights have begun to be legally recognized in an effort to ease their social and economic isolation, and to improve living standards.

Standard of living
(UN human development index)

- low
- high
- figures unavailable

Environmental issues

The prospect of rising sea levels poses a threat to many low-lying islands in the Pacific. The testing of nuclear weapons, once common throughout the region, was finally discontinued in 1996. Australia's ecological balance has been irreversibly altered by the introduction of alien species. Although it has the world's largest underground water reserve, the Great Artesian Basin, the availability of fresh water in Australia remains critical. Periodic droughts combined with overgrazing lead to desertification and increase the risk of devastating bush fires, and occasional flash floods.

Environmental issues

- national parks
- tropical forest
- forest destroyed
- desert
- desertification
- polluted rivers
- radioactive contamination
- marine pollution
- heavy marine pollution
- • poor urban air quality

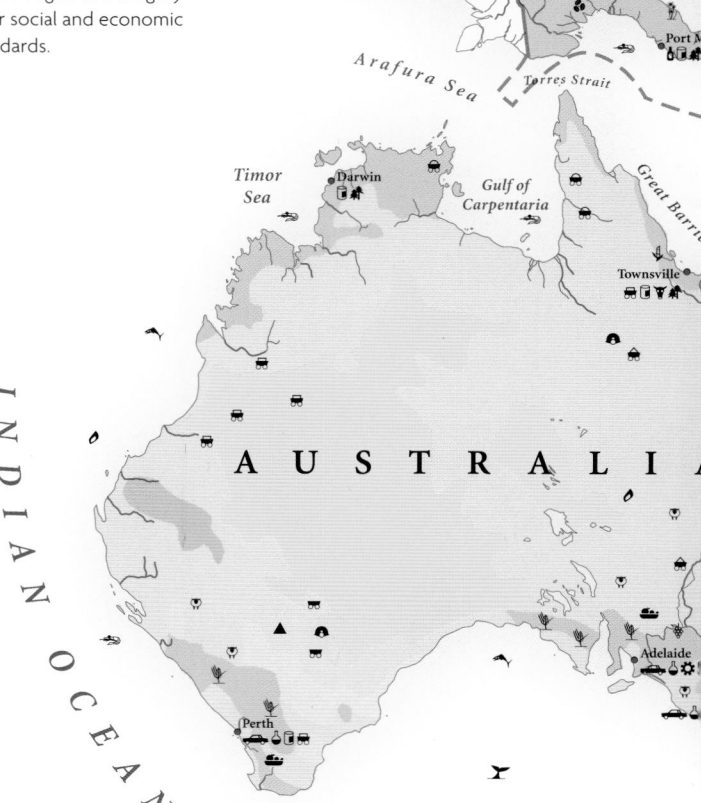

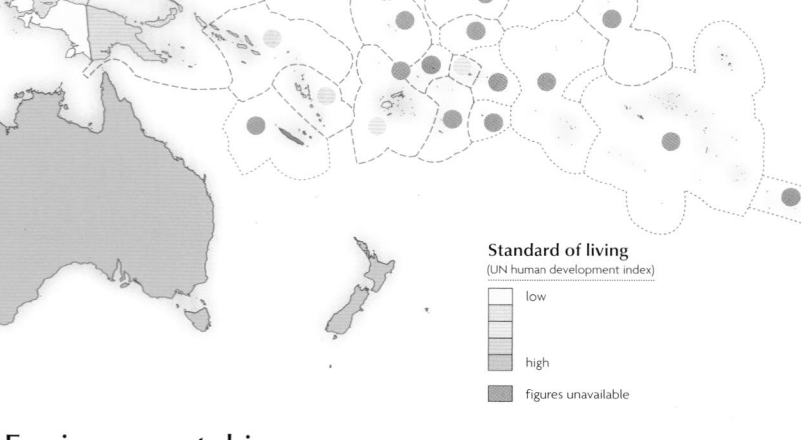

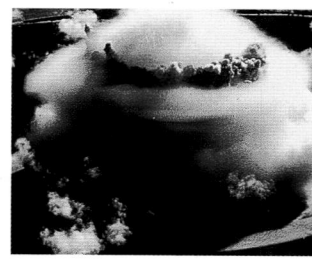

▲ *In 1946 Bikini Atoll,* in the Marshall Islands, was chosen as the site for Operation Crossroads – investigating the effects of atomic bombs upon naval vessels. Further nuclear tests continued until the early 1990s. The long-term environmental effects are unknown.

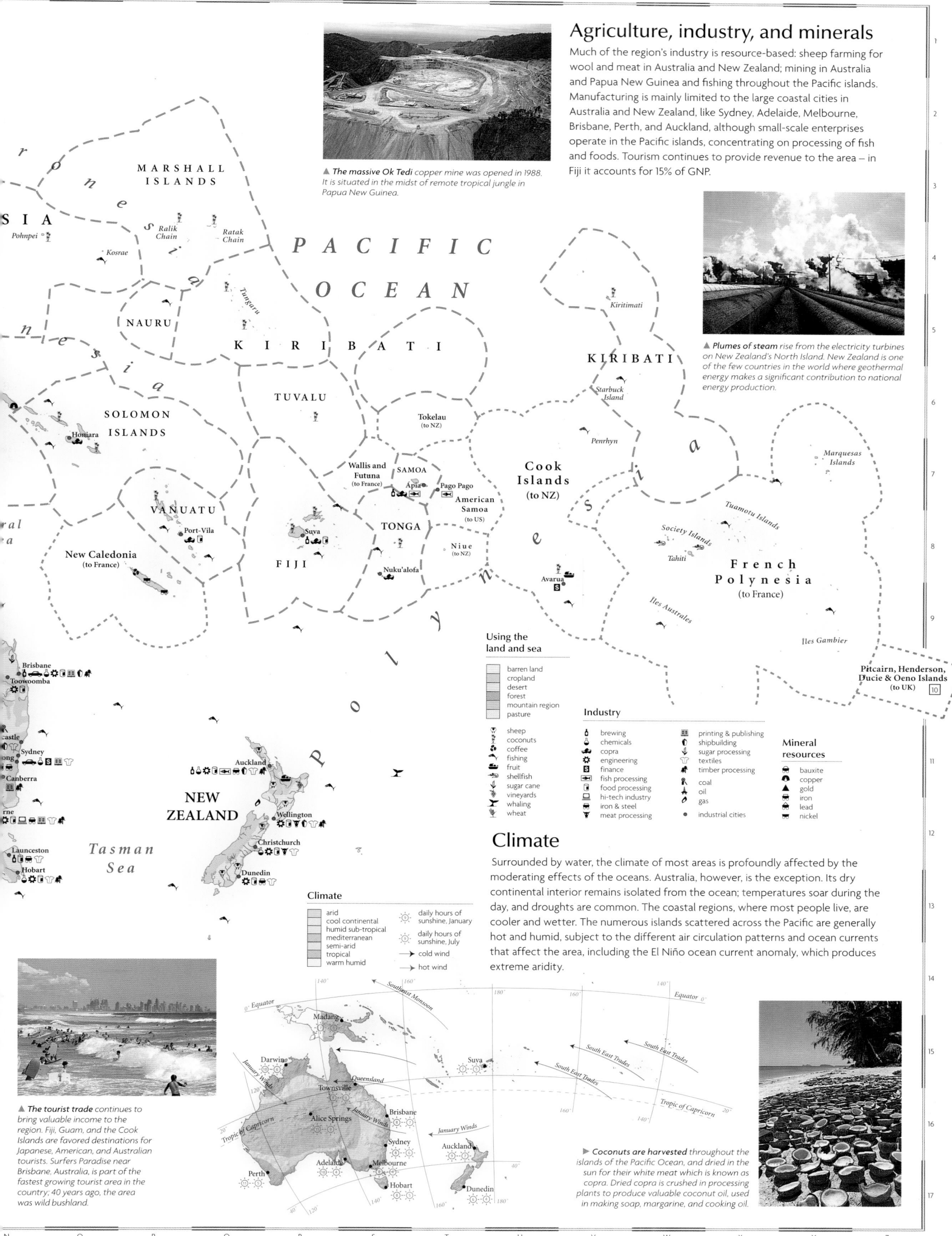

Agriculture, industry, and minerals

Much of the region's industry is resource-based: sheep farming for wool and meat in Australia and New Zealand; mining in Australia and Papua New Guinea and fishing throughout the Pacific islands. Manufacturing is mainly limited to the large coastal cities in Australia and New Zealand, like Sydney, Adelaide, Melbourne, Brisbane, Perth, and Auckland, although small-scale enterprises operate in the Pacific islands, concentrating on processing of fish and foods. Tourism continues to provide revenue to the area – in Fiji it accounts for 15% of GNP.

▲ *The massive Ok Tedi* copper mine was opened in 1988. It is situated in the midst of remote tropical jungle in Papua New Guinea.

▲ *Plumes of steam* rise from the electricity turbines on New Zealand's North Island. New Zealand is one of the few countries in the world where geothermal energy makes a significant contribution to national energy production.

Using the land and sea

barren land
cropland
desert
forest
mountain region
pasture

sheep
coconuts
coffee
fishing
fruit
shellfish
sugar cane
vineyards
whaling
wheat

Industry

brewing
chemicals
copra
engineering
finance
fish processing
food processing
hi-tech industry
iron & steel
meat processing

printing & publishing
shipbuilding
sugar processing
textiles
timber processing
coal
oil
gas
industrial cities

Mineral resources

bauxite
copper
gold
iron
lead
nickel

Climate

Surrounded by water, the climate of most areas is profoundly affected by the moderating effects of the oceans. Australia, however, is the exception. Its dry continental interior remains isolated from the ocean; temperatures soar during the day, and droughts are common. The coastal regions, where most people live, are cooler and wetter. The numerous islands scattered across the Pacific are generally hot and humid, subject to the different air circulation patterns and ocean currents that affect the area, including the El Niño ocean current anomaly, which produces extreme aridity.

Climate

arid
cool continental
humid sub-tropical
mediterranean
semi-arid
tropical
warm humid

daily hours of sunshine, January
daily hours of sunshine, July
cold wind
hot wind

▲ *The tourist trade* continues to bring valuable income to the region. Fiji, Guam, and the Cook Islands are favored destinations for Japanese, American, and Australian tourists. Surfers Paradise near Brisbane, Australia, is part of the fastest growing tourist area in the country; 40 years ago, the area was wild bushland.

▶ *Coconuts are harvested* throughout the islands of the Pacific Ocean, and dried in the sun for their white meat which is known as copra. Dried copra is crushed in processing plants to produce valuable coconut oil, used in making soap, margarine, and cooking oil.

Australia

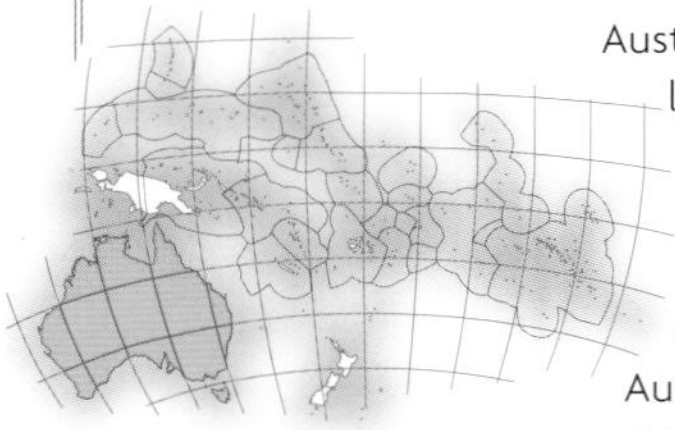

Australia is the world's smallest continent, a stable landmass lying between the Indian and Pacific oceans. Previously home to its aboriginal peoples only, since the end of the 18th century immigration has transformed the face of the country. Initially settlers came mainly from western Europe, particularly the UK, and for years Australia remained wedded to its British colonial past. More recent immigrants have come from eastern Europe, and from Asian countries such as Japan, South Korea, and Indonesia. Australia is now forging strong trading links with these "Pacific Rim" countries and its economic future seems to lie with Asia and the Americas, rather than Europe, its traditional partner.

Using the land

Over 104 million sheep are dispersed in vast herds around the country, contributing to a major export industry. Cattle-ranching is important, particularly in the west. Wheat, and grapes for Australia's wine industry, are grown mainly in the south. Much of the country is desert, unsuitable for agriculture unless irrigation is used.

The urban/rural population divide

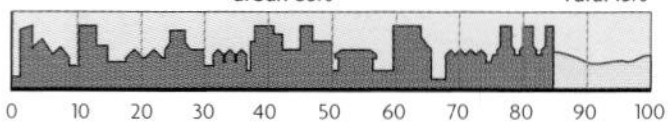

urban 85%   rural 15%

0 10 20 30 40 50 60 70 80 90 100

Population density	Total land area
6 people per sq mile (2 people per sq km)	2,967,893 sq miles (7,686,850 sq km)

Land use and agricultural distribution

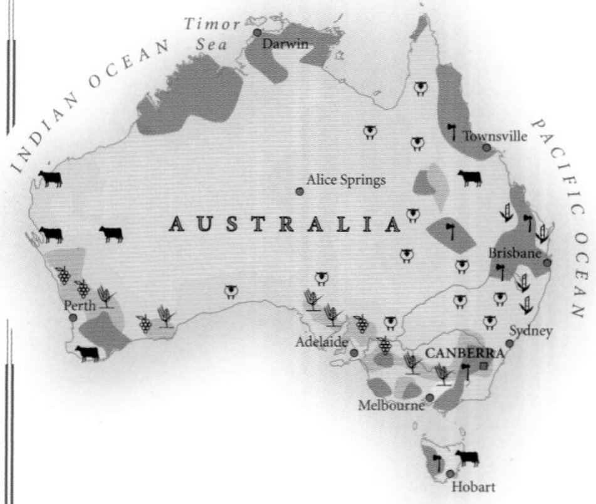

- cattle
- sheep
- cereals
- sugar cane
- timber
- vineyards
- capital cities
- major towns
- pasture
- cropland
- forest
- desert
- mountain region

▲ *Lines of ripening* vines stretch for miles in Barossa Valley, a major wine-growing region near Adelaide.

The landscape

Australia consists of many eroded plateaus, lying firmly in the middle of the Indo-Australian Plate. It is the world's flattest continent, and the driest, after Antarctica. The coasts tend to be more hilly and fertile, especially in the east. The mountains of the Great Dividing Range form a natural barrier between the eastern coastal areas and the flat, dry plains and desert regions of the Australian "outback."

▲ *The Great Barrier Reef* is the world's largest area of coral islands and reefs. It runs for about 1240 miles (2000 km) along the Queensland coast.

▲ *The Pinnacles are* a series of rugged sandstone pillars. Their strange shapes have been formed by water and wind erosion.

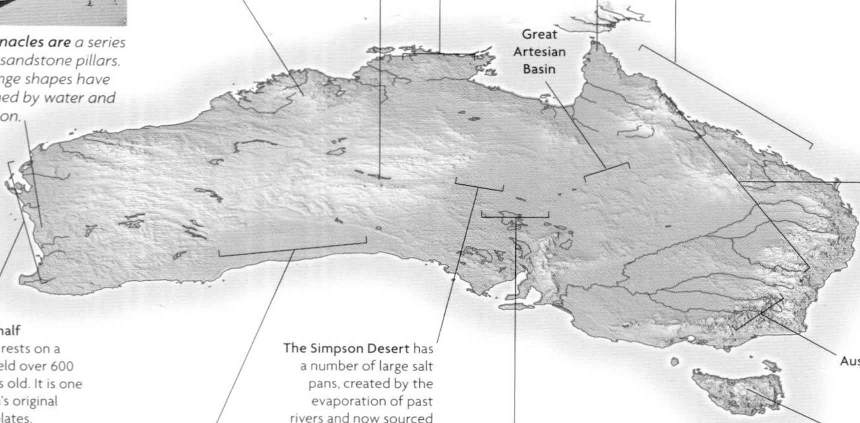

The ancient Kimberley Plateau is the source of some of Australia's richest mineral deposits, including diamonds.

Uluru (Ayers Rock)

Arnhem Land

The tropical rain forest of the Cape York Peninsula contains more than 600 different varieties of tree.

Great Artesian Basin

More than half of Australia rests on a uniform shield over 600 million years old. It is one of the Earth's original geological plates.

The Simpson Desert has a number of large salt pans, created by the evaporation of past rivers and now sourced by seasonal rains. Some are crusted with gypsum, but most are covered by common salt crystals.

The Nullarbor Plain is a low-lying limestone plateau which is so flat that the Trans-Australian Railway runs through it in a straight line for more than 300 miles (483 km).

The Lake Eyre basin, lying 51 ft (16 m) below sea level, is one of the largest inland drainage systems in the world, covering an area of more than 500,000 sq miles (1,300,000 sq km).

The Great Dividing Range forms a watershed between east- and west-flowing rivers. Erosion has created deep valleys, gorges, and waterfalls where rivers tumble over escarpments on their way to the sea.

Australian Alps

Tasmania has the same geological structure as the Australian Alps. During the last period of glaciation, 18,000 years ago, sea levels were some 300 ft (100 m) lower and it was joined to the mainland.

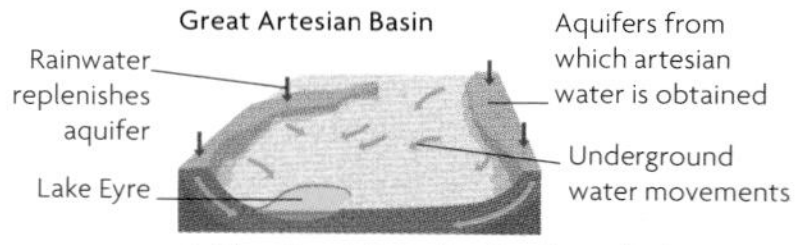

Great Artesian Basin

Rainwater replenishes aquifer

Lake Eyre

Aquifers from which artesian water is obtained

Underground water movements

▲ *The Great Artesian Basin* underlies nearly 20% of the total area of Australia, providing a valuable store of underground water, essential to Australian agriculture. The ephemeral rivers which drain the northern part of the basin have highly braided courses and, in consequence, the area is known as "channel country."

◄ *Uluru (Ayers Rock)*, the world's largest free-standing rock, is a massive outcrop of red sandstone in Australia's desert center. Wind and sandstorms have ground the rock into the smooth curves seen here. Uluru is revered as a sacred site by many aboriginal peoples.

Scale 1:11,500,000

Km
0 25 50 100 150 200 250 300 350
0 25 50 100 150 200 250 300 350
Miles

projection: Lambert Conformal Conic

Map key

Population	Elevation
■ 1 million to 5 million	2000m / 6562ft
◉ 500,000 to 1 million	1000m / 3281ft
◎ 100,000 to 500,000	500m / 1640ft
⊕ 50,000 to 100,000	250m / 820ft
⊙ 10,000 to 50,000	100m / 328ft
○ below 10,000	sea level

INDIAN OCEAN

PACIFIC OCEAN

Timor Sea
Darwin

AUSTRALIA

Alice Springs

Townsville

Brisbane

Perth

Adelaide

CANBERRA

Sydney

Melbourne

Hobart

154

Cape London
Cape Bougainville
Bigge Island
Bonaparte Archipelago
Heywood Islands
Adele Island
Mount Hann 779m
Collier Bay
Lombadina
Derby
Broome
Fitzroy River

Great Sandy Desert

De Grey River
Percival Lakes
Port Hedland
Wickham
Marble Bar
Lake Dora
Lake Auld
Dampier Archipelago
Dampier
Karratha
Roebourne
Whim Creek
Fortescue River
Hamersley Range
Wittenoom
Barrow Island
Onslow
Tom Price
Paraburdoo
Mount Meharry 1251m
Newman
Little Sandy Desert
Lake Disappointment
North West Cape
Exmouth
Learmonth
Kenneth Range
Kumarina Roadhouse
Gibson Desert
Coral Bay
WESTERN
Minilya
Mount Augustus 1105m
Carnarvon Range
Lake Carnegie
Barlee Range
Waldburg Range
Tropic of Capricorn
Lake Macleod
Gascoyne River
Robinson Range
Lake Gregory
Bernier Island
Gascoyne Junction
Wiluna
Lake Way
Lake Wells
Carnarvon
Meekatharra
Dorre Island
Denham
AUSTRALIA
Dirk Hartog Island
Lake Annean
Lake Austin
Leonora
Lake Carey
Kalbarri
Mount Magnet
Yalgoo
Lake Ballard
Menzies
Geraldton
Mongers Lake
Lake Moore
Lake Rebecca
Wubin
Pithara
Kalgoorlie
Coolgardie
Kitchener
Moora
Kambalda
Lake Lefroy
The Pinnacles
Southern Cross
Gingin
Northam
Merredin
Lake Johnston
Lake Cowan
Norseman
Balladonia
Wanneroo
York
Lake Dundas
Perth
Fremantle
Brookton
Kondinin
Lake Hope
Rockingham
Narrogin
Lake King
Ravensthorpe
Tower Peak 594m
Mandurah
Wagin
Katanning
Esperance
Bunbury
Collie
Busselton
Bridgetown
Manjimup
Margaret River
Cape Leeuwin
Augusta
Pemberton
Mount Barker
Albany

180

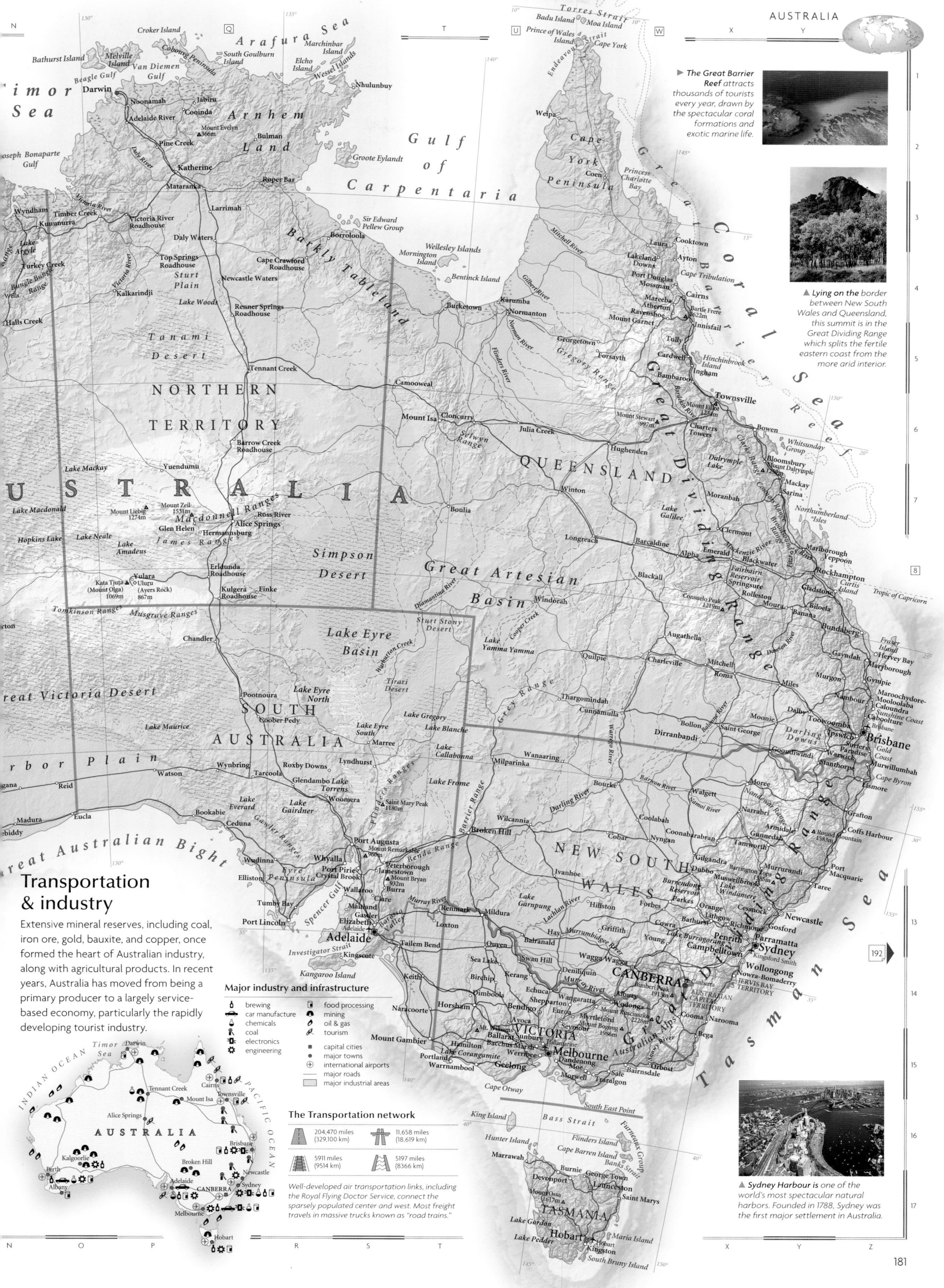

The Great Barrier Reef attracts thousands of tourists every year, drawn by the spectacular coral formations and exotic marine life.

▲ *Lying on the* border between New South Wales and Queensland, this summit is in the Great Dividing Range which splits the fertile eastern coast from the more arid interior.

Transportation & industry

Extensive mineral reserves, including coal, iron ore, gold, bauxite, and copper, once formed the heart of Australian industry, along with agricultural products. In recent years, Australia has moved from being a primary producer to a largely service-based economy, particularly the rapidly developing tourist industry.

Major industry and infrastructure

- brewing
- car manufacture
- chemicals
- coal
- electronics
- engineering
- food processing
- mining
- oil & gas
- tourism
- ■ capital cities
- ⊕ major towns
- ✈ international airports
- major roads
- major industrial areas

The Transportation network

204,470 miles (329,100 km)	11,658 miles (18,619 km)
5911 miles (9514 km)	5197 miles (8366 km)

Well-developed air transportation links, including the Royal Flying Doctor Service, connect the sparsely populated center and west. Most freight travels in massive trucks known as "road trains."

▲ *Sydney Harbour is* one of the world's most spectacular natural harbors. Founded in 1788, Sydney was the first major settlement in Australia.

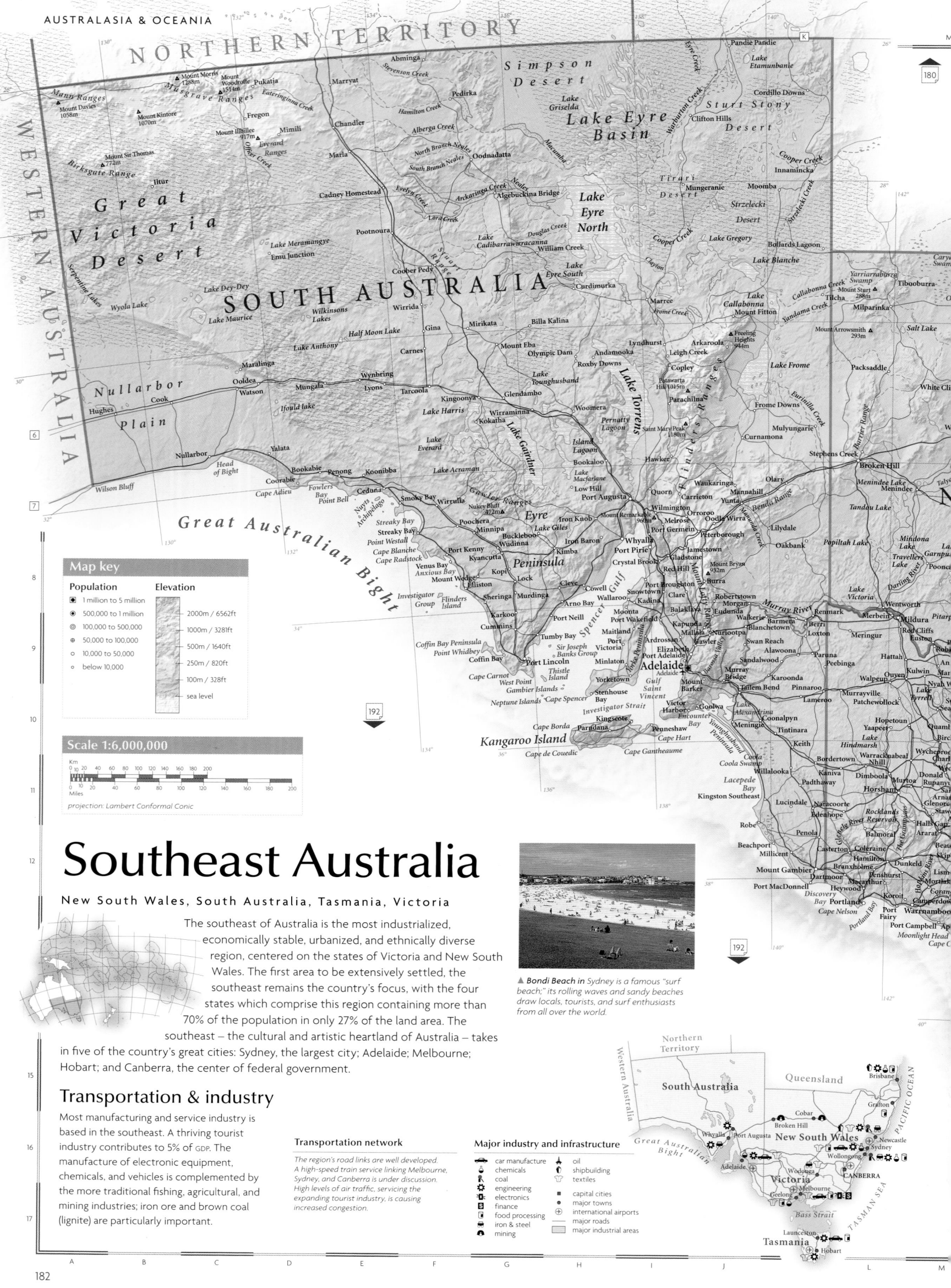

Map key

Population
- ◙ 1 million to 5 million
- ◉ 500,000 to 1 million
- ⊙ 100,000 to 500,000
- ⊕ 50,000 to 100,000
- ○ 10,000 to 50,000
- ○ below 10,000

Elevation
- 2000m / 6562ft
- 1000m / 3281ft
- 500m / 1640ft
- 250m / 820ft
- 100m / 328ft
- sea level

Scale 1:6,000,000

Km
0 10 20 40 60 80 100 120 140 160 180 200

Miles
0 10 20 40 60 80 100 120 140 160 180 200

projection: Lambert Conformal Conic

Southeast Australia

New South Wales, South Australia, Tasmania, Victoria

The southeast of Australia is the most industrialized, economically stable, urbanized, and ethnically diverse region, centered on the states of Victoria and New South Wales. The first area to be extensively settled, the southeast remains the country's focus, with the four states which comprise this region containing more than 70% of the population in only 27% of the land area. The southeast – the cultural and artistic heartland of Australia – takes in five of the country's great cities: Sydney, the largest city; Adelaide; Melbourne; Hobart; and Canberra, the center of federal government.

▲ **Bondi Beach in** Sydney is a famous "surf beach;" its rolling waves and sandy beaches draw locals, tourists, and surf enthusiasts from all over the world.

Transportation & industry

Most manufacturing and service industry is based in the southeast. A thriving tourist industry contributes to 5% of GDP. The manufacture of electronic equipment, chemicals, and vehicles is complemented by the more traditional fishing, agricultural, and mining industries; iron ore and brown coal (lignite) are particularly important.

Transportation network

The region's road links are well developed. A high-speed train service linking Melbourne, Sydney, and Canberra is under discussion. High levels of air traffic, servicing the expanding tourist industry, is causing increased congestion.

Major industry and infrastructure

- 🚗 car manufacture
- ⚗ chemicals
- coal
- ⚙ engineering
- electronics
- $ finance
- food processing
- iron & steel
- ⚓ mining
- ♦ oil
- ⛴ shipbuilding
- ▽ textiles
- ■ capital cities
- ⊕ major towns
- ✈ international airports
- — major roads
- ▨ major industrial areas

Using the land & sea

The western flanks of the Great Dividing Range and the northern deserts of South Australia support massive herds of sheep and cattle, while more intensive stockrearing occurs near the cities. Sugar cane is the most important industrial crop, and cereal grains including wheat, corn, barley, and sorghum are also grown. Grapes, citrus, and orchard fruits are among the wide range of fruit and vegetables cultivated in this region. Tasmania's forestry and fishing contributes to over one-third of the state's exports.

▲ *The fertile Darling Downs,* known as the "breadbasket of Australia," support a wide range of crops including cereals, sugar cane, and fruit.

▶ *The Murray River* has its source in the eastern uplands of the Great Dividing Range. Fed by melting snow, it runs for 1609 miles (2589 km), and has sufficient volume to reach the ocean southeast of Adelaide despite a minimal gradient for most of its lower reaches.

The urban/rural population divide

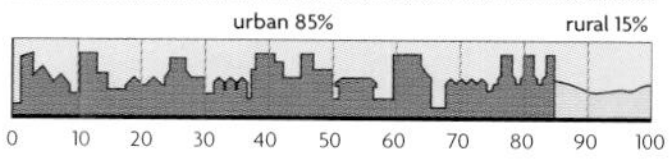

urban 85%   rural 15%

0 10 20 30 40 50 60 70 80 90 100

Population density	Total land area
18 people per sq mile (7 people per sq km)	778,022 sq miles (2,015,600 sq km)

Land use and agricultural distribution

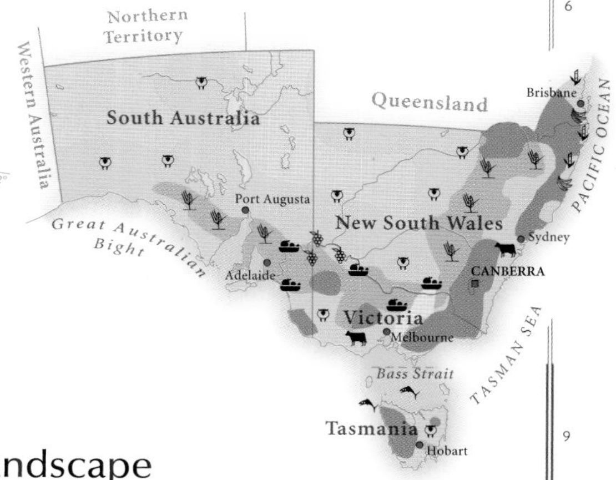

- cattle
- sheep
- bananas
- fishing
- fruit
- sugar cane
- vineyards
- wheat

- capital cities
- major towns
- pasture
- cropland
- forest
- desert
- mountain region

The landscape

The southern half of the Great Dividing Range runs parallel to the eastern coast of Victoria and New South Wales as far as Tasmania, which, though divided from the mainland is part of the same mountain chain. South Australia comprises the Australian shield and half of the dry, flat Nullarbor Plain. The Murray/Darling river basin is the only major river system.

◀ *The heavily folded* Flinders Ranges is part of an arc of sedimentary rocks reaching northward from Kangaroo Island.

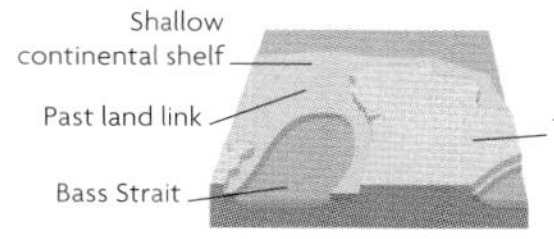

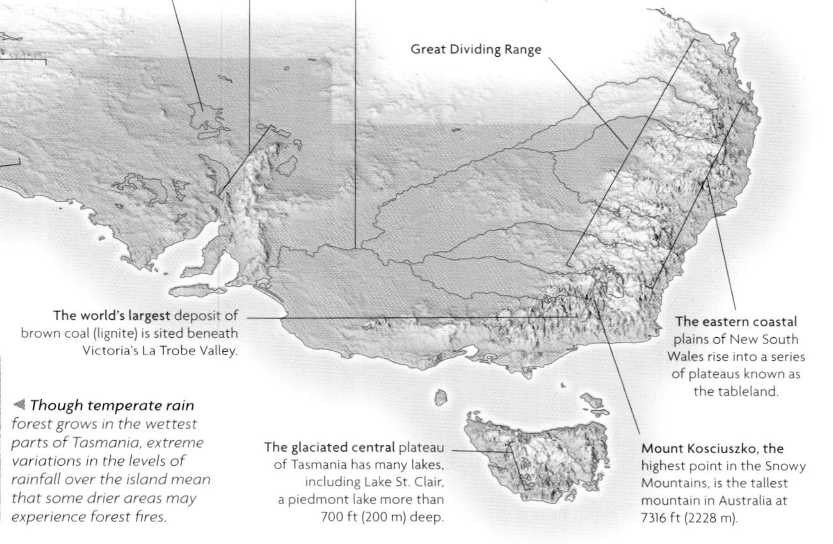

Lake Eyre is the largest of southern Australia's dry lakes. Lying -51 ft (-16 m) below sea level, it has flooded only three times in the last century.

The Musgrave and Everard ranges form bare, rounded hills made up of ancient granite and gneiss.

The Murray/Darling is Australia's longest river at 1703 miles (2739 km).

▲ *Tasmania is part* of Australia's eastern highlands, separated from the mainland by 155 miles (250 km) of the Bass Strait. In the recent geological past, dry land links between Tasmania and Victoria would have been possible during periods of world-wide glaciation, when the sea level was more than 180 ft (55 m) below that of present sea levels.

Shallow continental shelf
Past land link
Tasmania
Bass Strait

Great Dividing Range

The eastern part of the Nullarbor Plain has many sinkholes, eroded by rainwater, which run underground to form a system of long caves in the limestone rocks.

The world's largest deposit of brown coal (lignite) is sited beneath Victoria's La Trobe Valley.

The eastern coastal plains of New South Wales rise into a series of plateaus known as the tableland.

◀ *Though temperate rain* forest grows in the wettest parts of Tasmania, extreme variations in the levels of rainfall over the island mean that some drier areas may experience forest fires.

The glaciated central plateau of Tasmania has many lakes, including Lake St. Clair, a piedmont lake more than 700 ft (200 m) deep.

Mount Kosciuszko, the highest point in the Snowy Mountains, is the tallest mountain in Australia at 7316 ft (2228 m).

183

New Zealand

Lying 1500 miles east-southeast of Australia, New Zealand was originally settled by the Maori people of Polynesia. It was visited by Europeans for the first time only as recently as the 1770s. The islands' rugged topography means that most settlement has concentrated in coastal areas. People of European origin make up about 70% of the population of 4 million, following immigration which began in the 1920s. Many recent settlers have come from Asia, including India and China, and a number of the Pacific islands. The Maori now make up a minority of less than half a million. Their ancient claims to at least half of national territory, however, are gaining increasing legal credence.

The landscape

New Zealand comprises two large islands and many scattered smaller islands. On South Island the Alpine Fault marks the boundary between the Pacific and Indo-Australian plates.

Tectonic activity has strongly influenced the formation of the Southern Alps, snowcapped mountains with several peaks over 9800 ft (3000 m). North Island has a lower and less extensive mountain region, containing forested hills, a central volcanic plateau, and downlands.

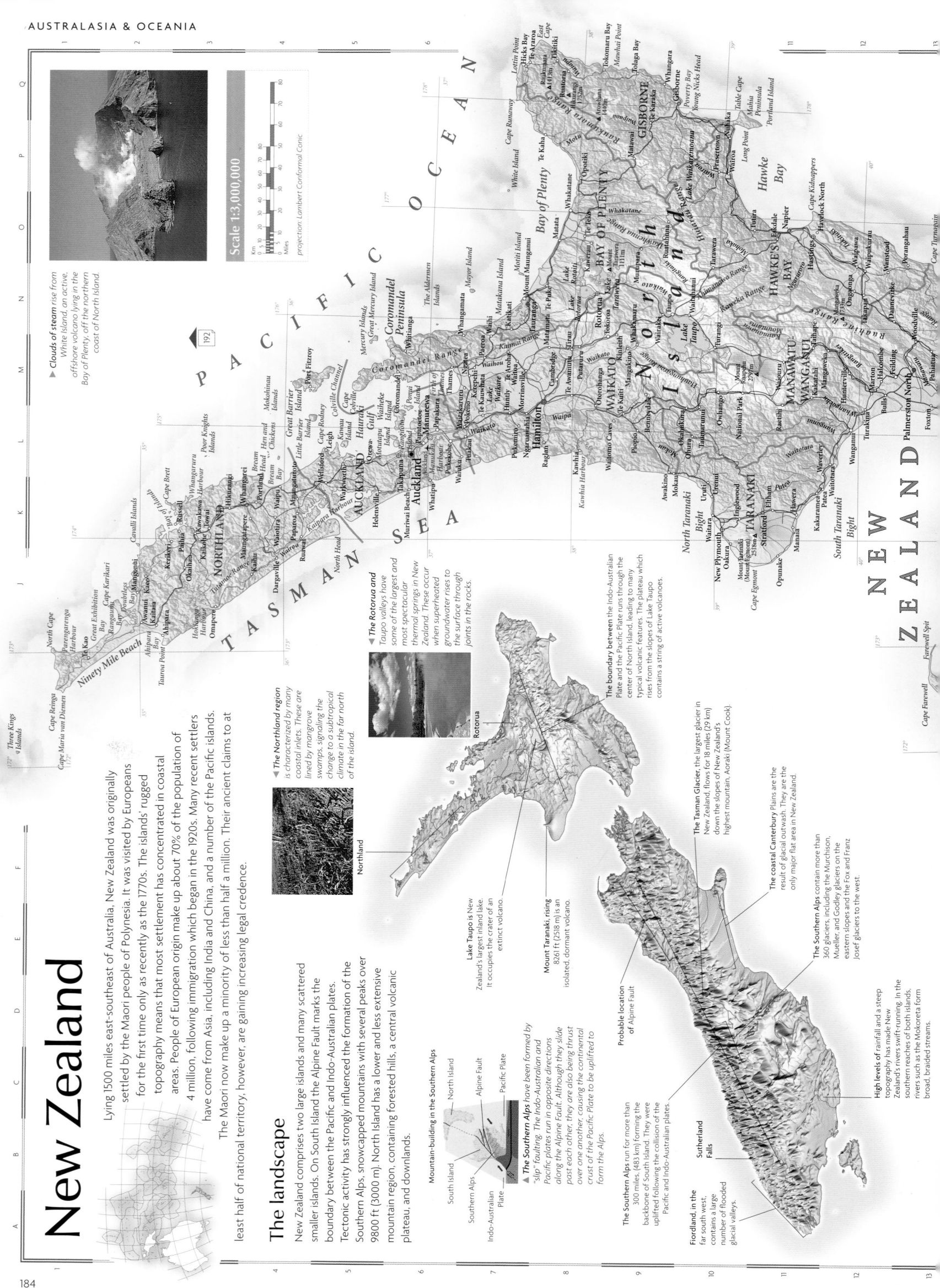

▲ *Clouds of steam* rise from White Island, an active, offshore volcano lying in the Bay of Plenty, off the northern coast of North Island.

Scale 1:3,000,000

projection: Lambert Conformal Conic

▼ *The Northland region* is characterized by many coastal inlets. These are lined by mangrove swamps, signaling the change to a subtropical climate in the far north of the island.

Northland

▼ *The Rotorua and Taupo valleys* have some of the largest and most spectacular thermal springs in New Zealand. These occur when superheated groundwater rises to the surface through joints in the rocks.

Rotorua

The boundary between the Indo-Australian Plate and the Pacific Plate runs through the center of North Island, leading to many typical volcanic features. The plateau which rises from the slopes of Lake Taupo contains a string of active volcanoes.

Lake Taupo is New Zealand's largest inland lake. It occupies the crater of an extinct volcano.

Mount Taranaki, rising 8261 ft (2518 m) is an isolated, dormant volcano.

The Tasman Glacier, the largest glacier in New Zealand, flows for 18 miles (29 km) down the slopes of New Zealand's highest mountain, Aoraki (Mount Cook).

The coastal Canterbury Plains are the result of glacial outwash. They are the only major flat area in New Zealand.

The Southern Alps contain more than 360 glaciers, including the Murchison, Mueller, and Godley glaciers on the eastern slopes and the Fox and Franz Josef glaciers to the west.

Probable location of Alpine Fault

High levels of rainfall and a steep topography has made New Zealand's rivers swift-running. In the southern reaches of both islands, rivers such as the Mokoreta form broad, braided streams.

Fiordland, in the far south west, contains a large number of flooded glacial valleys.

Sutherland Falls

The Southern Alps run for more than 300 miles, (483 km) forming the backbone of South Island. They were uplifted following the collision of the Pacific and Indo-Australian plates.

Mountain-building in the Southern Alps

▲ *The Southern Alps* have been formed by "slip" faulting. The Indo-Australian and Pacific plates run in opposite directions along the Alpine Fault. Although they slide post each other, they are also being thrust over one another, causing the continental crust of the Pacific Plate to be uplifted to form the Alps.

North Island

Alpine Fault

Pacific Plate

South Island

Southern Alps

Indo-Australian Plate

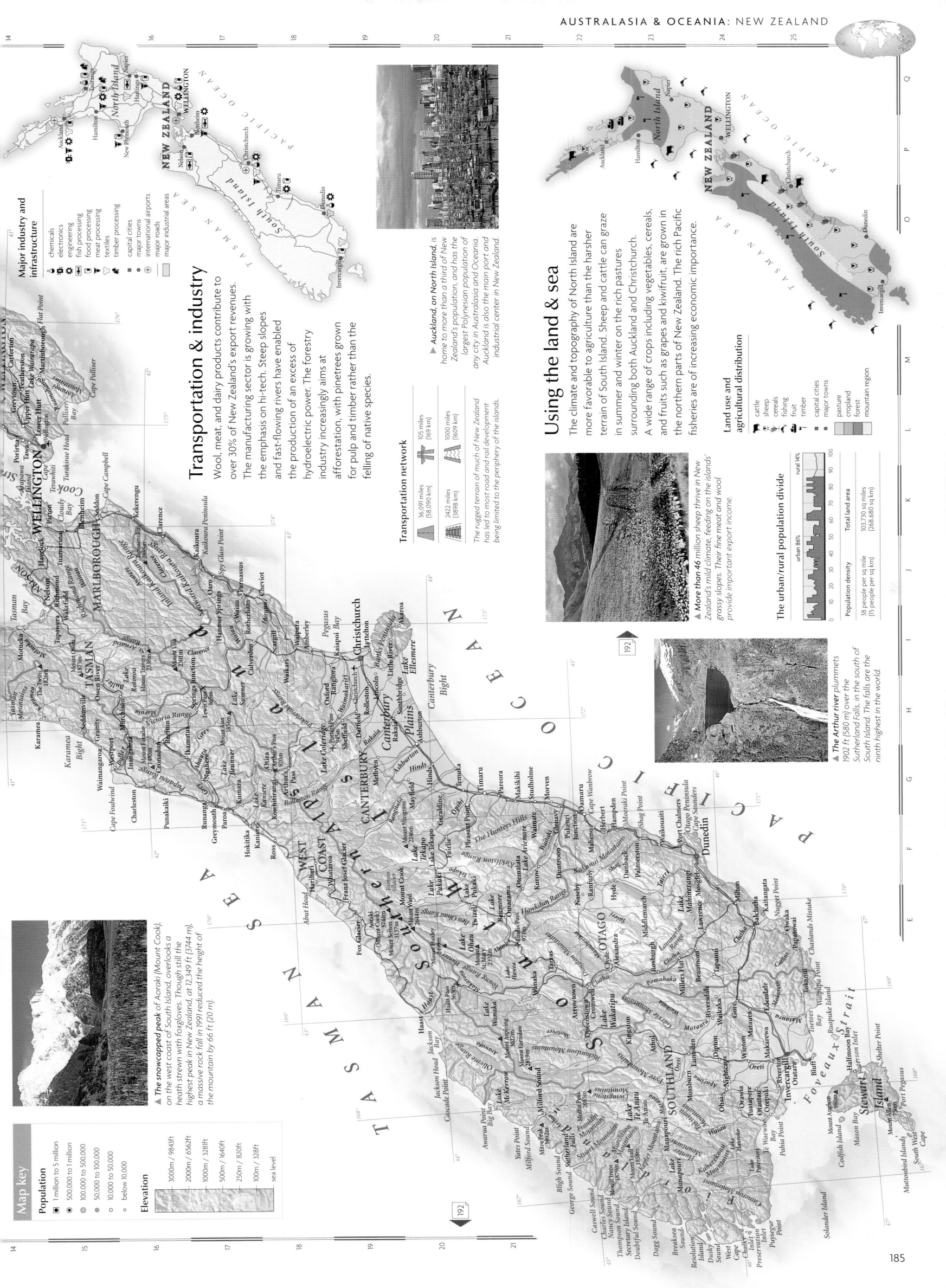

Transportation & industry

Wool, meat, and dairy products contribute to over 30% of New Zealand's export revenues. The manufacturing sector is growing with the emphasis on hi-tech. Steep slopes and fast-flowing rivers have enabled the production of an excess of hydroelectric power. The forestry industry increasingly aims at afforestation, with pinetrees grown for pulp and timber rather than the felling of native species.

▲ *Auckland, on North Island, is home to more than a third of New Zealand's population, and has the largest Polynesian population of any city in Australasia and Oceania. Auckland is also the main port and industrial center in New Zealand.*

Transportation network

36,091 miles (58,090 km)	105 miles (169 km)
2422 miles (3898 km)	1000 miles (1609 km)

The rugged terrain of much of New Zealand has led to most road and rail development being limited to the periphery of the islands.

Major industry and infrastructure

- chemicals
- electronics
- engineering
- fish processing
- food processing
- meat processing
- textiles
- timber processing
- capital cities
- major towns
- international airports
- major roads
- major industrial areas

Using the land & sea

The climate and topography of North Island are more favorable to agriculture than the harsher terrain of South Island. Sheep and cattle can graze in summer and winter on the rich pastures surrounding both Auckland and Christchurch. A wide range of crops including vegetables, cereals, and fruits such as grapes and kiwifruit, are grown in the northern parts of New Zealand. The rich Pacific fisheries are of increasing economic importance.

Land use and agricultural distribution

- cattle
- sheep
- cereals
- fruit
- timber
- capital cities
- major towns
- pasture
- cropland
- forest
- mountain region

▲ *More than 46 million sheep thrive in New Zealand's mild climate, feeding on the islands' grassy slopes. Their fine meat and wool provide important export income.*

▲ *The Arthur river plummets 1902 ft (580 m) over the Sutherland Falls, in the south of South Island. The falls are the ninth highest in the world.*

The urban/rural population divide

urban 86% rural 14%

Population density	Total land area
38 people per sq mile (15 people per sq km)	103,730 sq miles (268,680 sq km)

▲ *The snowcapped peak of Aoraki (Mount Cook), on the west coast of South Island, overlooks a heath strewn with foxgloves. Though still the highest peak in New Zealand, at 12,349 ft (3744 m), a massive rock fall in 1991 reduced the height of the mountain by 66 ft (20 m).*

Map key

Population

- 1 million to 5 million
- 500,000 to 1 million
- 100,000 to 500,000
- 50,000 to 100,000
- 10,000 to 50,000
- below 10,000

Elevation

- 3000m / 9843ft
- 2000m / 6562ft
- 1000m / 3281ft
- 500m / 1640ft
- 250m / 820ft
- 100m / 328ft
- sea level

Melanesia

FIJI, New Caledonia *(to France)*, PAPUA NEW GUINEA, SOLOMON ISLANDS, VANUATU

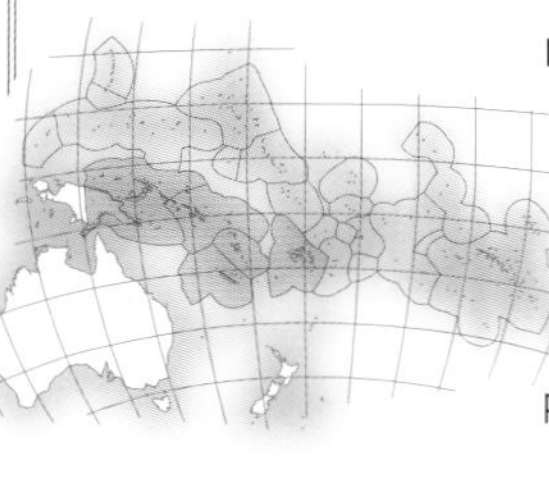

Lying in the southwest Pacific Ocean, northeast of Australia and south of the Equator, the islands of Melanesia form one of the three geographic divisions (along with Polynesia and Micronesia) of Oceania. Melanesia's name derives from the Greek *melas*, "black," and *nesoi*, "islands." Most of the larger islands are volcanic in origin. The smaller islands tend to be coral atolls and are mainly uninhabited. Rugged mountains, covered by dense rain forest, take up most of the land area. Melanesian's cultivate yams, taro, and sweet potatoes for local consumption and live in small, usually dispersed, homesteads.

▲ *Huli tribesmen from* Southern Highlands Province in Papua New Guinea parade in ceremonial dress, their powdered wigs decorated with exotic plumage and their faces and bodies painted with colored pigments.

Map key

Population
- ◉ 100,000 to 500,000
- ⊕ 50,000 to 100,000
- ○ 10,000 to 50,000
- ○ below 10,000

Elevation
- 4000m / 13,124ft
- 3000m / 9843ft
- 2000m / 6562ft
- 1000m / 3281ft
- 500m / 1640ft
- 250m / 820ft
- 100m / 328ft
- sea level

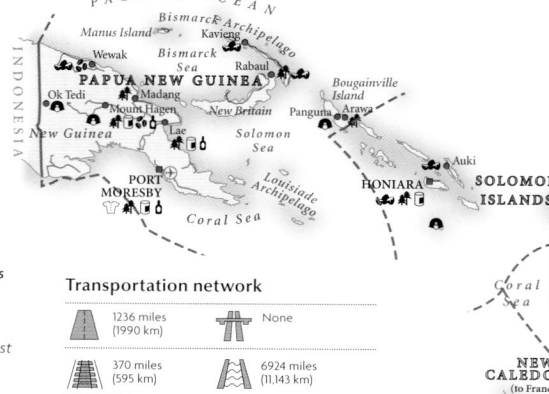

◀ *Lying close to* the banks of the Sepik river in northern Papua New Guinea, this building is known as the Spirit House. It is constructed from leaves and twigs, ornately woven and trimmed into geometric patterns. The house is decorated with a mask and topped by a carved statue.

▲ *On one of* Vanuatu's many islands, beach houses stand at the water's edge, surrounded by coconut palms and other tropical vegetation. The unspoilt beaches and tranquillity of its islands are drawing ever-larger numbers of tourists to Vanuatu.

Transportation & Industry

The processing of natural resources generates significant export revenue for the countries of Melanesia. The region relies mainly on copra, tuna, and timber exports, with some production of cocoa and palm oil. The islands have substantial mineral resources including the world's largest copper reserves on Bougainville Island; gold, and potential oil and natural gas. Tourism has become the fastest growing sector in most of the countries' economies.

◀ *On New Caledonia's* main island, relatively high interior plateaus descend to coastal plains. Nickel is the most important mineral resource, but the hills also harbor metallic deposits including chrome, cobalt, iron, gold, silver, and copper.

Transportation network

1236 miles (1990 km)		None
370 miles (595 km)		6924 miles (11,143 km)

As most of the islands of Melanesia lie off the major sea and air routes, services to and from the rest of the world are infrequent. Transportation by road on rugged terrain is difficult and expensive.

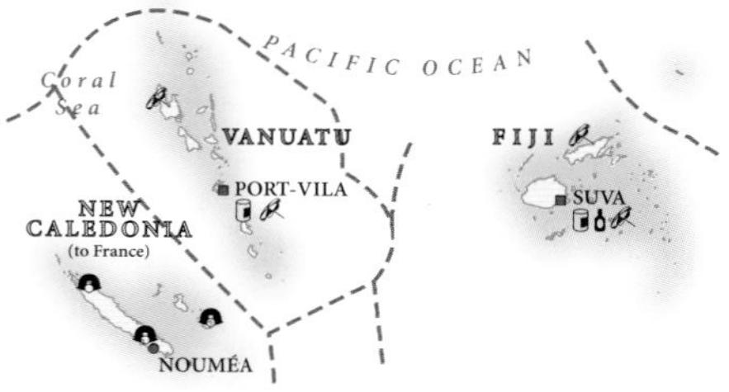

Major industry and infrastructure
- beverages
- coffee processing
- copra processing
- food processing
- mining
- textiles
- timber processing
- tourism
- capital cities
- major towns
- international airports
- major roads

The Landscape

Melanesia comprises high, volcanic islands, low coral islands and continental islands. New Guinea is part of the Australian continental platform, and is separated from it only by the shallow flooding of the Torres Strait. The plate margin of the Pacific and Indo-Australian plates cuts through mainland Papua New Guinea. Volcanic activity, resulting from the collision of these plates, has sculpted much of Melanesia's landscape.

◀ *The slopes of this extinct volcano near Talasea on the island of New Britain have been almost entirely colonized by rain forest vegetation.*

▲ *A series of coral reefs can be seen in the clear waters off Cape Esperance on the island of Guadalcanal in the Solomons.*

The Star Mountains include some of the most remote terrain on Earth. The area is rich in gold and copper.

The Sepik river drains the lowlands north of the Central Range, flowing eastward into the Bismarck Sea.

The Bismarck Range is precipitous, rugged and covered in dense vegetation, rising to 14,793 ft (4509 m) at Mount Wilhelm in central Papua New Guinea.

Most of Papua New Guinea's outlying islands, including New Britain, Bougainville Island and New Ireland, are precipitous and of volcanic origin.

Kavachi is an active submarine volcano near New Georgia, which erupts every few years.

The physical landscapes of the islands of Vanuatu range from rugged mountains and high plateaus, to rolling hills and low plateaus and offshore coral reefs.

The lowland plains in the south and north of Papua New Guinea's main island are swampy, and contain some fertile alluvial soils. This contrasts with the mountainous lands in the rest of the country where soils are generally thin and nutrients are retained in the existing vegetation.

Huon Peninsula

Kikori river

Southern Papua New Guinea is part of the Indo-Australian Plate. New Guinea only became separated physically from Australia about 8000 years ago following the flooding of the Torres Strait.

The Owen Stanley Range contains several of Papua New Guinea's highest peaks, the greatest of which is Mount Victoria at 13,200 ft (4035 m).

The Louisiade Archipelago contains 10 volcanic islands and numerous coral islets. Tagula Island is the largest of the islands, containing the archipelago's highest peak at 2645 ft (806 m).

The Solomon Islands are mountainous continental-type islands with largely andesitic volcanoes.

New Caledonia's main island is surrounded by coral reef that extends from the Huon island group in the north, to Île des Pins in the south.

Viti Levu, the largest of Fiji's islands, contains the country's highest mountain, Mount Victoria at 4339 ft (1323 m).

▶ *Papua New Guinea's rivers, though fairly short, carry extremely high sediment loads, largely due to soil erosion. This is caused by a combination of very steep slopes and heavy rainfall, and is made worse by forest clearance, particularly "slash and burn" techniques and road or mine operations.*

Huon Peninsula

Caves and undercut cliffs mark former shoreline

Former level of beach

Current beach

Uplift of the land in tectonically active regions can lead to former coastlines being lifted beyond the reach of the sea. New cliffs and caves are formed at a lower level, and rivers cut down through the lower land to reach sea level once more.

Stream cuts down through recently exposed land

Using the land and sea

Almost 60% of the population of Melanesia is engaged in agriculture and animal husbandry at a subsistence level. Coconuts and cocoa are grown for export revenue. Over 80% of the land area is cloaked by tropical forest and woodlands, which have proved to be a rich timber source. In coastal areas, fishing, mainly for tuna, is a staple industry.

The urban/rural population divide

urban 32%   rural 68%

0 10 20 30 40 50 60 70 80 90 100

Population density	Total land area
32 people per sq mile (12 people per sq km)	205,354 sq miles (332,008 sq km)

◀ *Abaca Eco-tourist Park near Lautoka on the island of Viti Levu in western Fiji is one of a number of projects aimed at combining tourism with awareness about the environment. The government and people of Fiji are keen to protect the unique ecology of the islands and prevent further damage to the coral reefs. Until the recent ending of nuclear testing in the Pacific by Western nations, Fiji lay downwind of some of the main testing sites.*

Land use and agricultural distribution

- 🍌 bananas
- 🌰 cocoa
- 🌴 coconuts
- 🐟 fishing
- 🌴 oil palms
- 🌿 rubber
- 🌲 timber
- ■ capital cities
- • major towns
- cropland
- forest
- wetland

Scale 1:9,800,000

Km
0 25 50   100   150   200   250   300

Miles
0 25 50   100   150   200   250   300

projection: Mercator

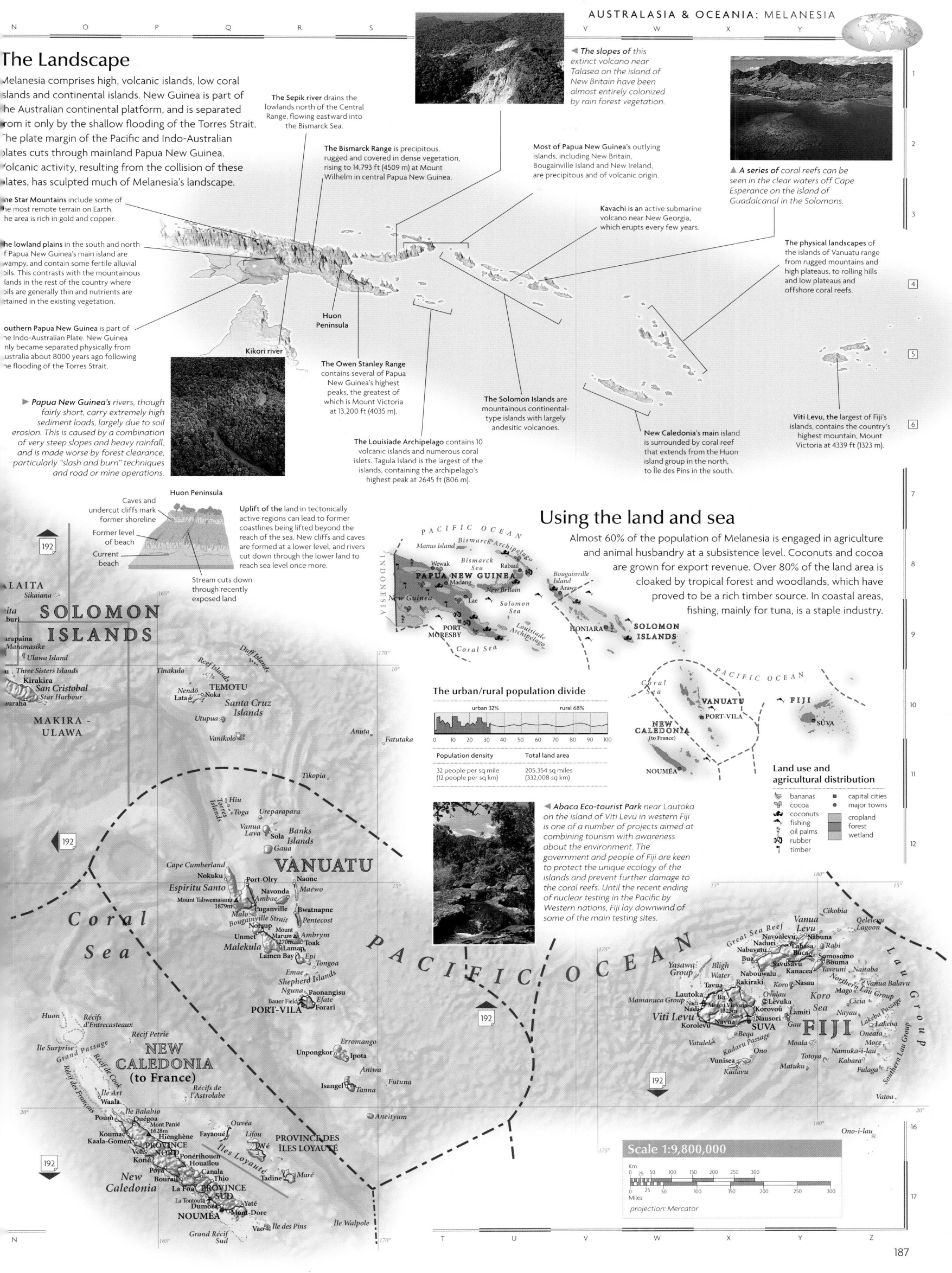

Micronesia

**MARSHALL ISLANDS, MICRONESIA, NAURU, PALAU,
Guam, Northern Mariana Islands, Wake Island**

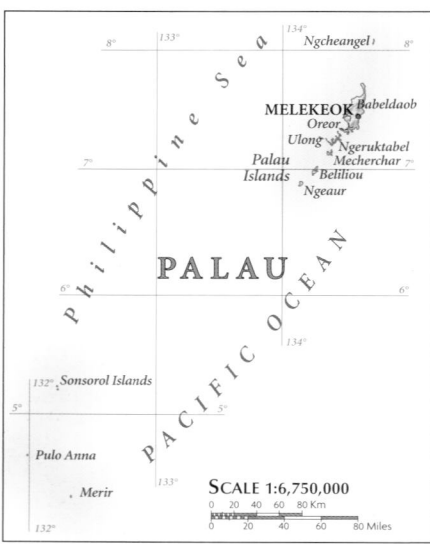

The Micronesian islands lie in the western reaches of the Pacific Ocean and are all part of the same volcanic zone. The Federated States of Micronesia is the largest group, with more than 600 atolls and forested volcanic islands in an area of more than 1120 sq miles (2900 sq km). Micronesia is a mixture of former colonies, overseas territories, and dependencies. Most of the region still relies on aid and subsidies to sustain economies limited by resources, isolation, and an emigrating population, drawn to New Zealand and Australia by the attractions of a western lifestyle.

Palau

Palau is an archipelago of over 200 islands, only eight of which are inhabited. It was the last remaining UN trust territory in the Pacific, controlled by the US until 1994, when it became independent. The economy operates on a subsistence level, with coconuts and cassava the principal crops. Fishing licenses and tourism provide foreign currency.

SCALE 1:825,000

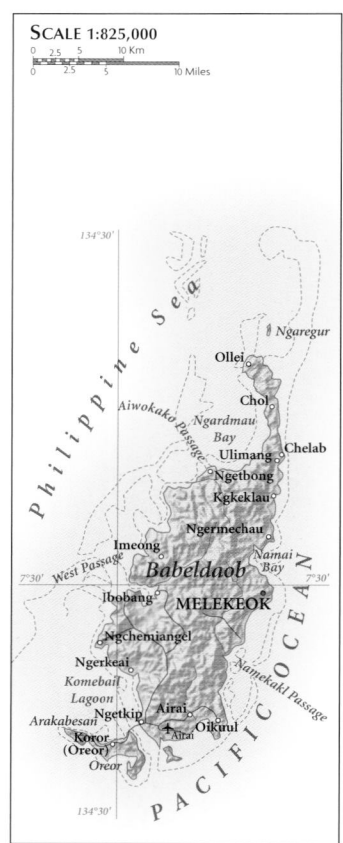

SCALE 1:6,750,000

Guam (to US)

Lying at the southern end of the Mariana Islands, Guam is an important US military base and tourist destination. Social and political life is dominated by the indigenous Chamorro, who make up just under half the population, although the increasing prevalence of western culture threatens Guam's traditional social stability.

◄ The tranquility of these coastal lagoons, at Inarajan in southern Guam, belies the fact that the island lies in a region where typhoons are common.

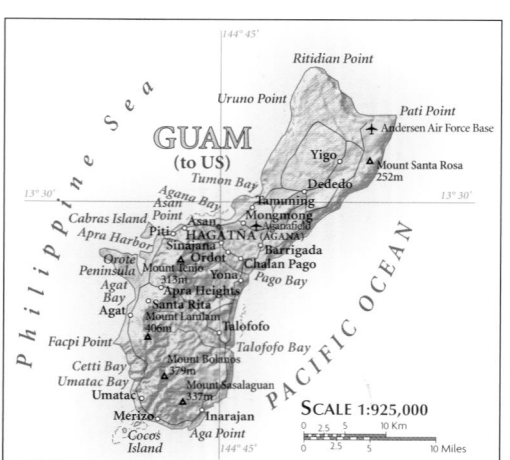

SCALE 1:925,000

Northern Mariana Islands (to US)

A US Commonwealth territory, the Northern Marianas comprise the whole of the Mariana archipelago except for Guam. The islands retain their close links with the US and continue to receive American aid. Tourism, though bringing in much-needed revenue, has speeded the decline of the traditional subsistence economy. Most of the population lives on Saipan.

SCALE 1:550,000

Northern Mariana Islands: capital cities

CAPITOL HILL – executive & legislative capital
SUSUPE – judicial capital

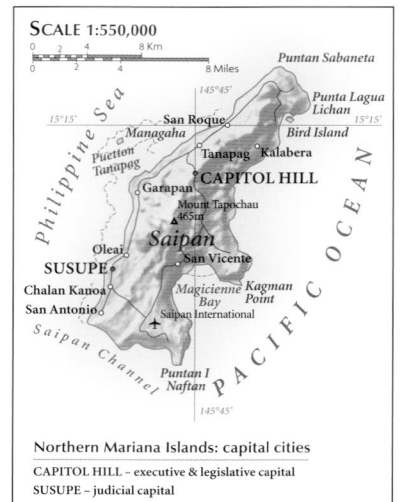

▲ The Palau Islands have numerous hidden lakes and lagoons. These sustain their own ecosystems which have developed in isolation. This has produced adaptations in the animals and plants that are often unique to each lake.

SCALE 1:5,500,000

Micronesia

A mixture of high volcanic islands and low-lying coral atolls, the Federated States of Micronesia include all the Caroline Islands except Palau. Pohnpei, Kosrae, Chuuk, and Yap are the four main island cluster states, each of which has its own language, with English remaining the official language. Nearly half the population is concentrated on Pohnpei, the largest island. Independent since 1986, the islands continue to receive considerable aid from the US which supplements an economy based primarily on fishing and copra processing.

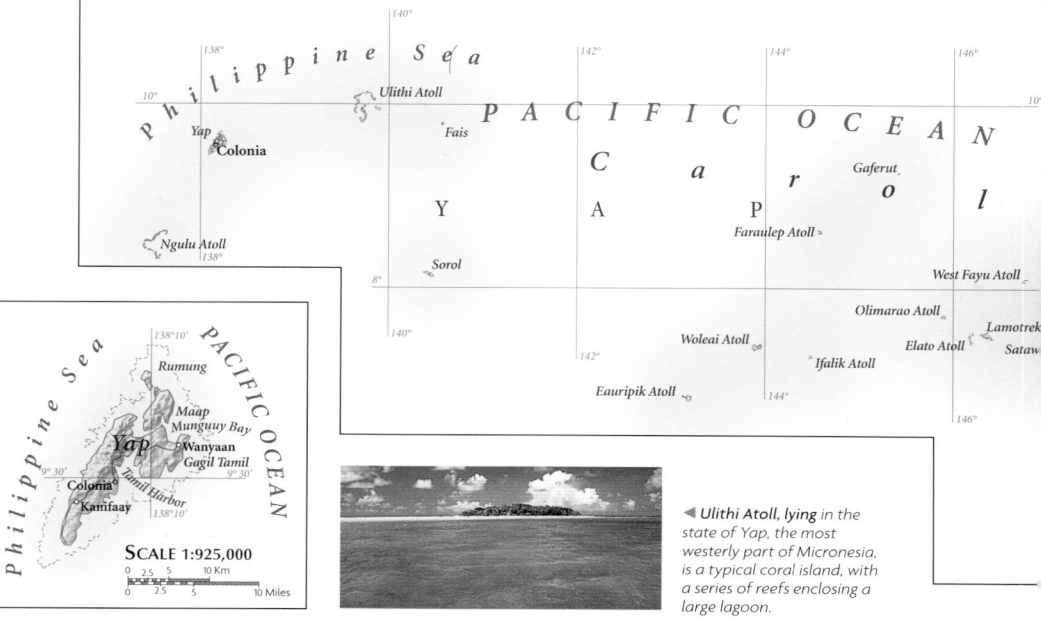

SCALE 1:925,000

◄ Ulithi Atoll, lying in the state of Yap, the most westerly part of Micronesia, is a typical coral island, with a series of reefs enclosing a large lagoon.

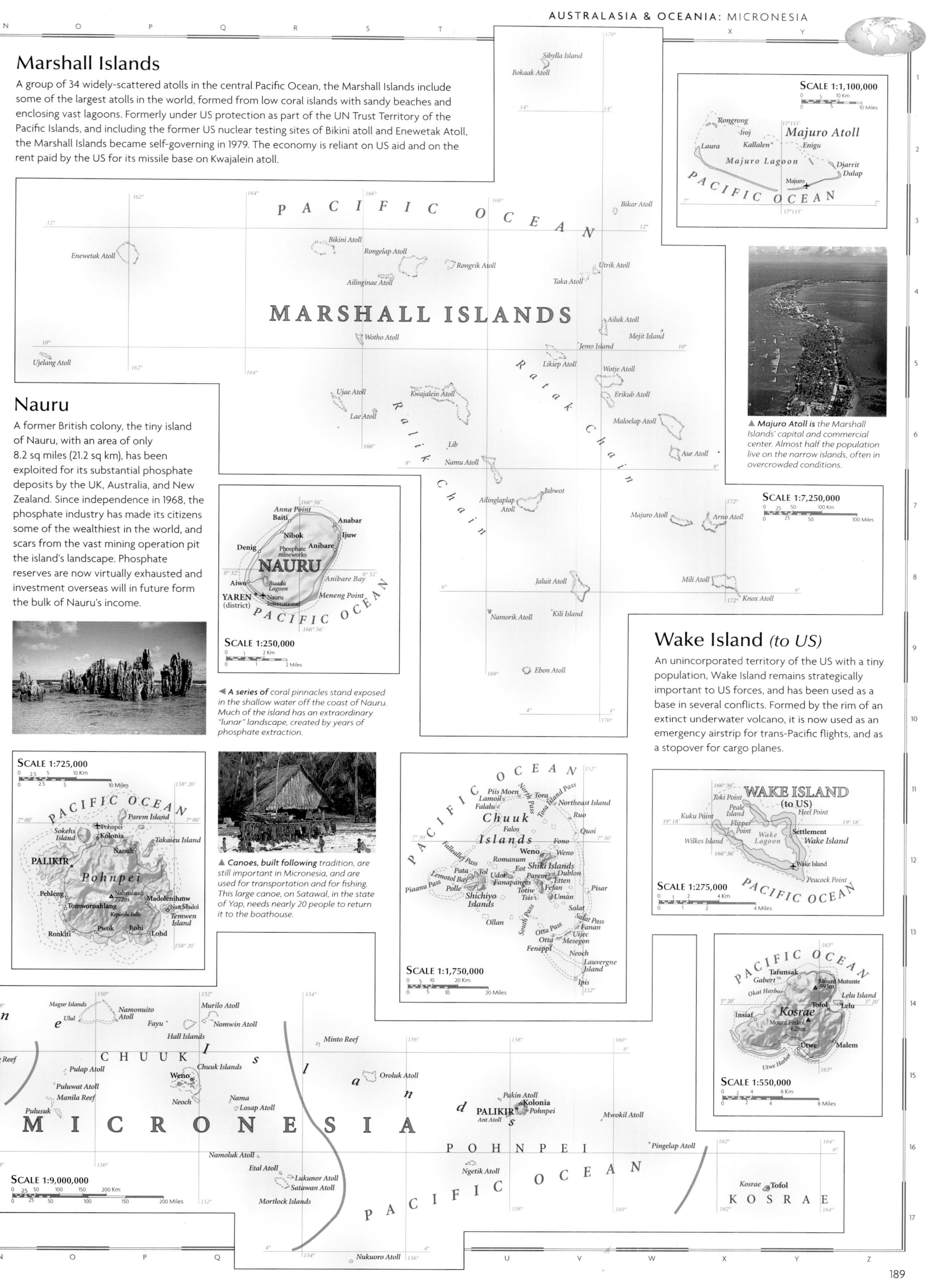

Marshall Islands

A group of 34 widely-scattered atolls in the central Pacific Ocean, the Marshall Islands include some of the largest atolls in the world, formed from low coral islands with sandy beaches and enclosing vast lagoons. Formerly under US protection as part of the UN Trust Territory of the Pacific Islands, and including the former US nuclear testing sites of Bikini atoll and Enewetak Atoll, the Marshall Islands became self-governing in 1979. The economy is reliant on US aid and on the rent paid by the US for its missile base on Kwajalein atoll.

Nauru

A former British colony, the tiny island of Nauru, with an area of only 8.2 sq miles (21.2 sq km), has been exploited for its substantial phosphate deposits by the UK, Australia, and New Zealand. Since independence in 1968, the phosphate industry has made its citizens some of the wealthiest in the world, and scars from the vast mining operation pit the island's landscape. Phosphate reserves are now virtually exhausted and investment overseas will in future form the bulk of Nauru's income.

▲ *Majuro Atoll is* the Marshall Islands' capital and commercial center. Almost half the population live on the narrow islands, often in overcrowded conditions.

◄ *A series of* coral pinnacles stand exposed in the shallow water off the coast of Nauru. Much of the island has an extraordinary "lunar" landscape, created by years of phosphate extraction.

▲ *Canoes, built following* tradition, are still important in Micronesia, and are used for transportation and for fishing. This large canoe, on Satawal, in the state of Yap, needs nearly 20 people to return it to the boathouse.

Wake Island *(to US)*

An unincorporated territory of the US with a tiny population, Wake Island remains strategically important to US forces, and has been used as a base in several conflicts. Formed by the rim of an extinct underwater volcano, it is now used as an emergency airstrip for trans-Pacific flights, and as a stopover for cargo planes.

SCALE 1:1,100,000
SCALE 1:7,250,000
SCALE 1:250,000
SCALE 1:725,000
SCALE 1:1,750,000
SCALE 1:275,000
SCALE 1:550,000
SCALE 1:9,000,000

Polynesia

KIRIBATI, TUVALU, Cook Islands, Easter Island, French Polynesia, Niue, Pitcairn Islands, Tokelau, Wallis & Futuna

The numerous island groups of Polynesia lie to the east of Australia, scattered over a vast area in the south Pacific. The islands are a mixture of low-lying coral atolls, some of which enclose lagoons, and the tips of great underwater volcanoes. The populations on the islands are small, and most people are of Polynesian origin, as are the Maori of New Zealand. Local economies remain simple, relying mainly on subsistence crops, mineral deposits, many now exhausted, fishing, and tourism.

SCALE 1:1,100,000

Kiribati

A former British colony, Kiribati became independent in 1979. Banaba's phosphate deposits ran out in 1980, following decades of exploitation by the British. Economic development remains slow and most agriculture is at a subsistence level, though coconuts provide export income, and underwater agriculture is being developed.

▶ *With the exception of Banaba all the islands in Kiribati's three groups are low-lying, coral atolls. This aerial view shows the sparsely vegetated islands, intercut by many small lagoons.*

Tuvalu

A chain of nine coral atolls, 360 miles (579 km) long with a land area of just over 9 sq miles (23 sq km), Tuvalu is one of the world's smallest and most isolated states. As the Ellice Islands, Tuvalu was linked to the Gilbert Islands (now part of Kiribati) as a British colony until independence in 1978. Politically and socially conservative, Tuvaluans live by fishing and subsistence farming.

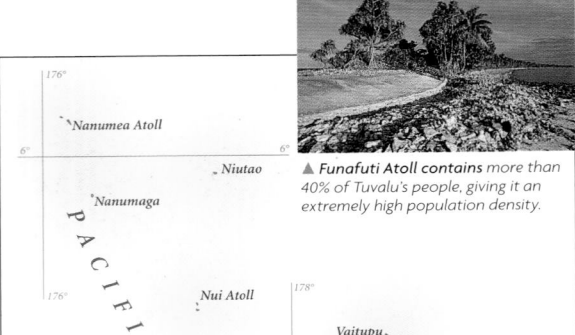

▲ *Funafuti Atoll contains more than 40% of Tuvalu's people, giving it an extremely high population density.*

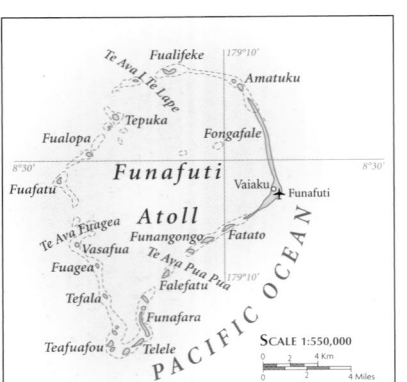

SCALE 1:550,000

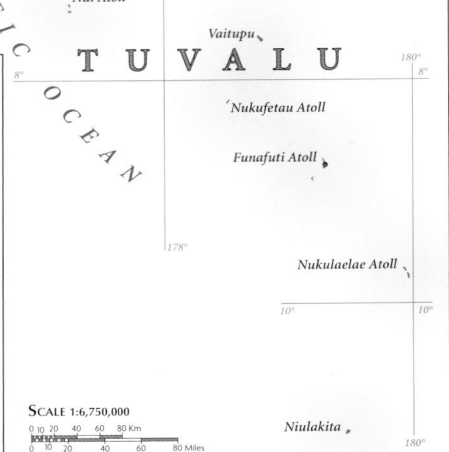

SCALE 1:6,750,000

Tokelau (to New Zealand)

A low-lying coral atoll, Tokelau is a dependent territory of New Zealand with few natural resources. Although a 1990 cyclone destroyed crops and infrastructure, a tuna cannery and the sale of fishing licenses have raised revenue and a catamaran link between the islands has increased their tourism potential. Tokelau's small size and economic weakness makes independence from New Zealand unlikely.

▲ *Fishermen cast their nets to catch small fish in the shallow waters off Atafu Atoll, the most westerly island in Tokelau.*

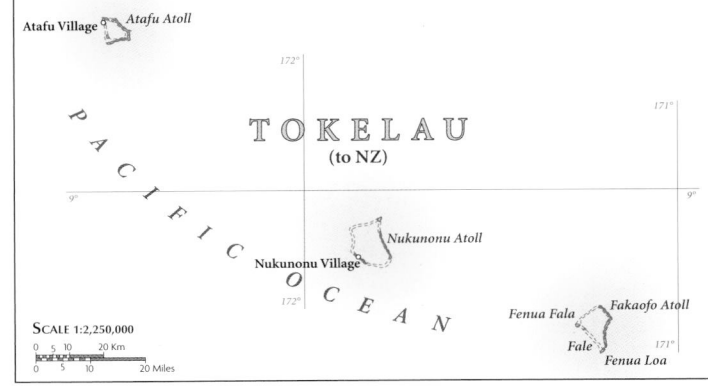

SCALE 1:2,250,000

Wallis & Futuna
(to France)

In contrast to other French overseas territories in the south Pacific, the inhabitants of Wallis and Futuna have shown little desire for greater autonomy. A subsistence economy produces a variety of tropical crops, while foreign currency remittances come from expatriates and from the sale of licenses to Japanese and Korean fishing fleets.

SCALE 1:1,100,000

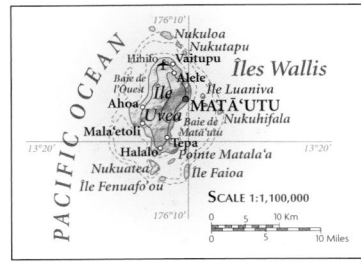

SCALE 1:1,100,000

Cook Islands (to New Zealand)

A mixture of coral atolls and volcanic peaks, the Cook Islands achieved self-government in 1965 but exist in free association with New Zealand. A diverse economy includes pearl and giant clam farming, and an ostrich farm, plus tourism and banking. A 1991 friendship treaty with France provides for French surveillance of territorial waters.

Niue (to New Zealand)

Niue, the world's largest coral island, is self-governing but exists in free association with New Zealand. Tropical fruits are grown for local consumption; tourism and the sale of postage stamps provide foreign currency. The lack of local job prospects has led more than 10,000 Niueans to emigrate to New Zealand, which has now invested heavily in Niue's economy in the hope of reversing this trend.

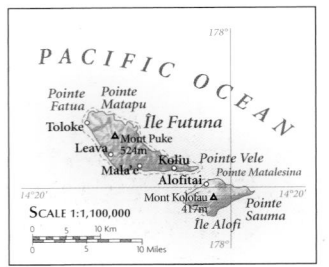

▲ *Palm trees fringe the white sands of a beach at Aitutaki in the Southern Cook Islands, where tourism is of increasing economic importance.*

SCALE 1:1,100,000

▲ *Waves have cut back the original coastline, exposing a sandy beach, near Mutalau in the northeast corner of Niue.*

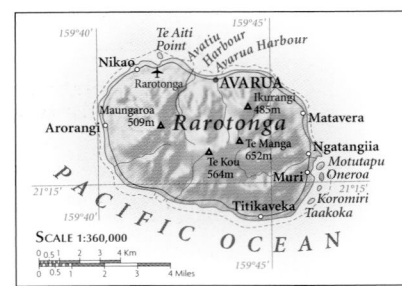

SCALE 1:360,000

COOK ISLANDS
(to NZ)

SCALE 1:22,250,000

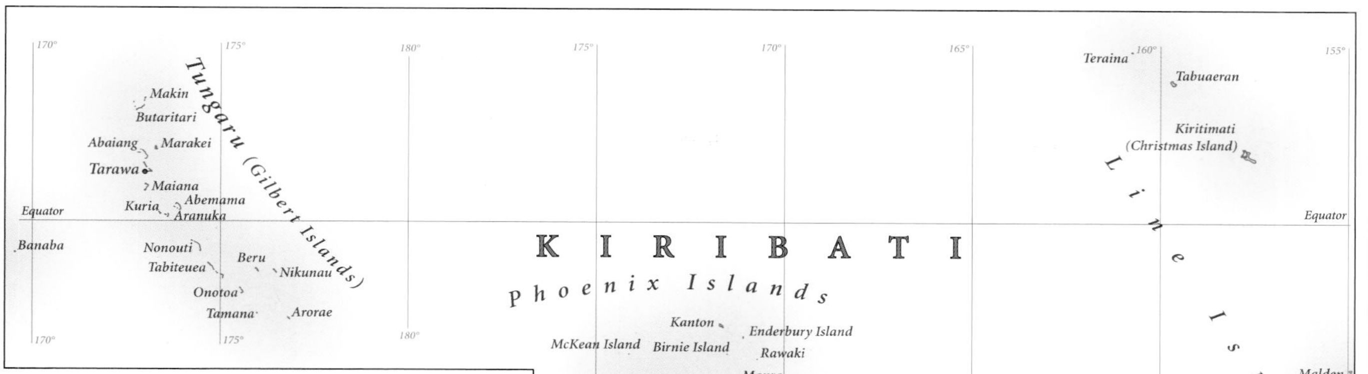

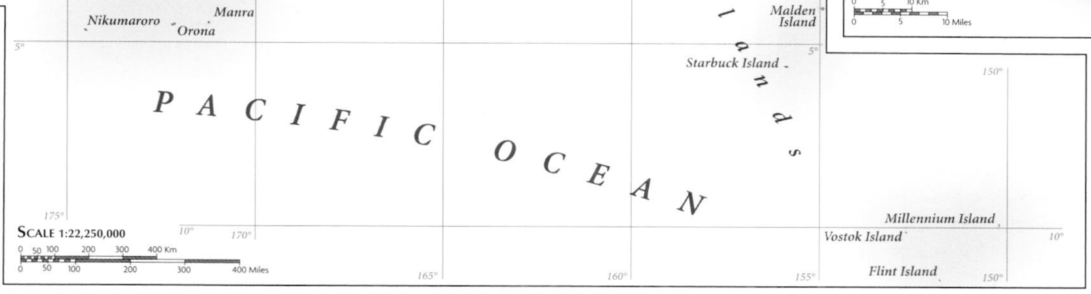

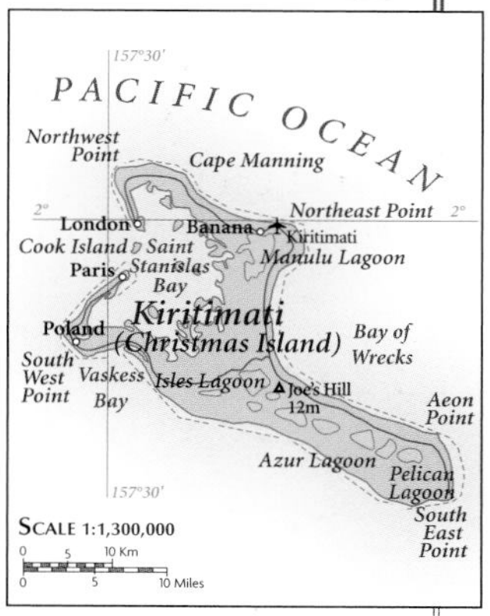

French Polynesia *(to France)*

The 130 islands of French Polynesia cover 4 million sq miles (10.5 million sq km). Nearly 75% of the people live on Tahiti. The use of Mururoa as a nuclear testing site by the French military transformed the economy, creating many jobs. The end of testing led to calls from the Polynesian majority for greater autonomy from France, the rebuilding of indigenous trade, and a reduction in tourism to stop the erosion of the islands' traditional culture.

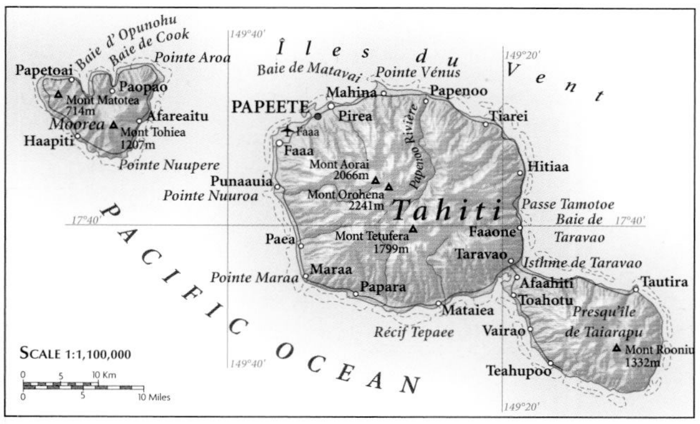

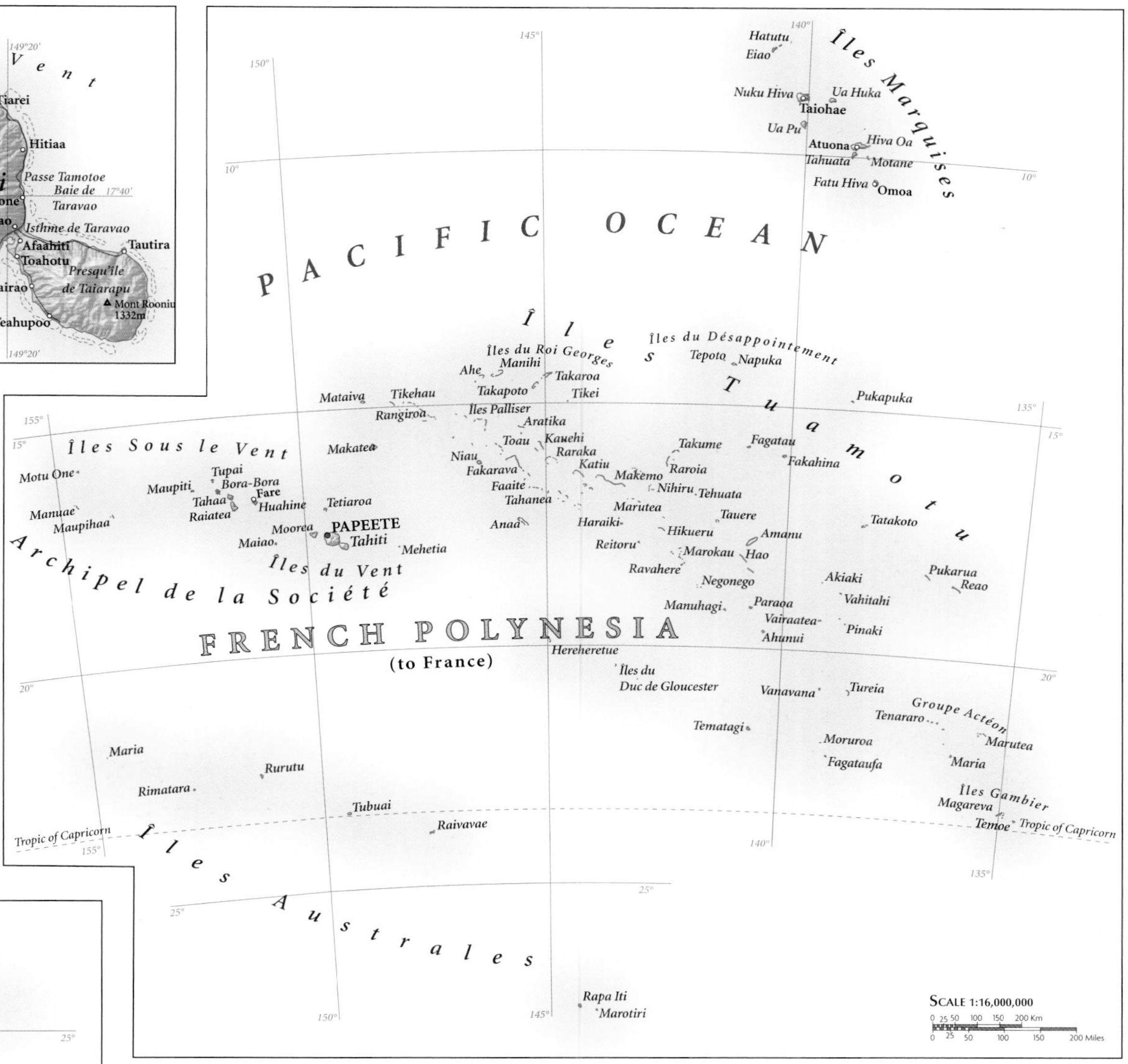

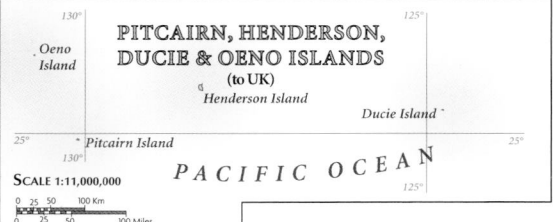

◄ **The traditional Tahitian** welcome for visitors, who are greeted by parties of canoes, has become a major tourist attraction.

Pitcairn Group of Islands *(to UK)*

Britain's most isolated dependency, Pitcairn Island was first populated by mutineers from the HMS *Bounty* in 1790. Emigration is further depleting the already limited gene pool of the island's inhabitants, with associated social and health problems. Barter, fishing and subsistence farming form the basis of the economy whilst offshore mineral exploitation may boost the economy in future.

◄ **The Pitcairn Islanders** rely on regular airdrops from New Zealand and periodic visits by supply vessels to provide them with basic commodities.

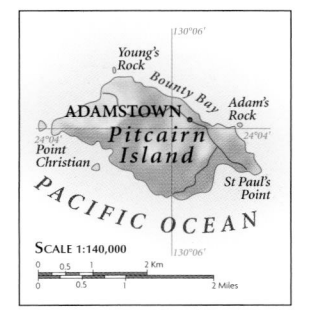

Easter Island *(to Chile)*

One of the most easterly islands in Polynesia, Easter Island *(Isla de Pascua)* – also known as Rapa Nui, is part of Chile. The mainly Polynesian inhabitants support themselves by farming, which is mainly of a subsistence nature, and includes cattle rearing and crops such as sugar cane, bananas, corn, gourds, and potatoes. In recent years, tourism has become the most important source of income and the island sustains a small commercial airport.

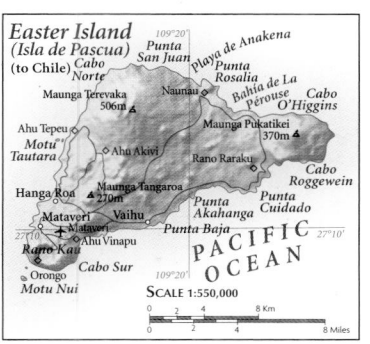

▲ **The Naunau,** a series of huge stone statues overlook Playa de Anakena, on Easter Island. Carved from a soft volcanic rock, they were erected between 400 and 900 years ago.

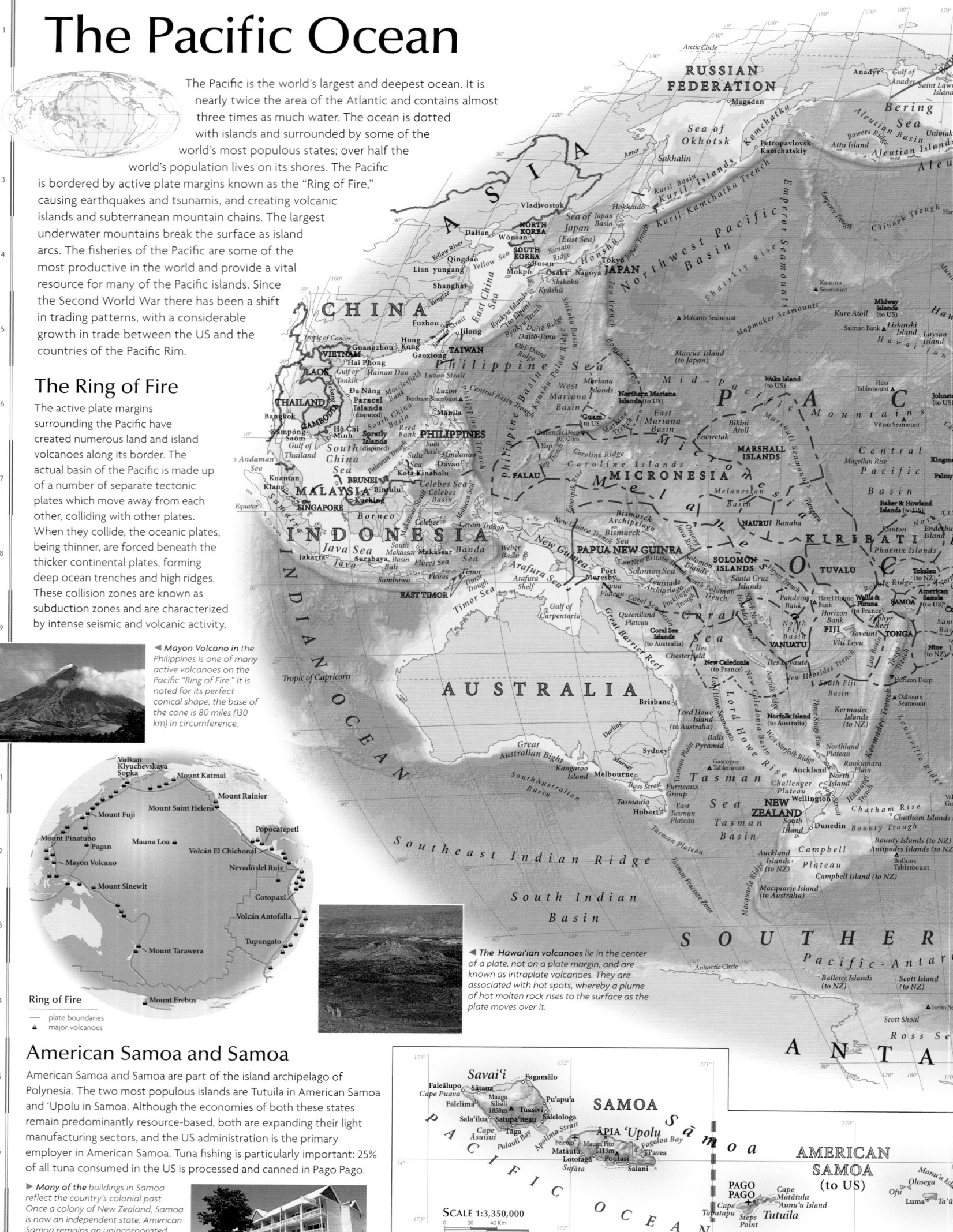

The Pacific Ocean

The Pacific is the world's largest and deepest ocean. It is nearly twice the area of the Atlantic and contains almost three times as much water. The ocean is dotted with islands and surrounded by some of the world's most populous states; over half the world's population lives on its shores. The Pacific is bordered by active plate margins known as the "Ring of Fire," causing earthquakes and tsunamis, and creating volcanic islands and subterranean mountain chains. The largest underwater mountains break the surface as island arcs. The fisheries of the Pacific are some of the most productive in the world and provide a vital resource for many of the Pacific islands. Since the Second World War there has been a shift in trading patterns, with a considerable growth in trade between the US and the countries of the Pacific Rim.

The Ring of Fire

The active plate margins surrounding the Pacific have created numerous land and island volcanoes along its border. The actual basin of the Pacific is made up of a number of separate tectonic plates which move away from each other, colliding with other plates. When they collide, the oceanic plates, being thinner, are forced beneath the thicker continental plates, forming deep ocean trenches and high ridges. These collision zones are known as subduction zones and are characterized by intense seismic and volcanic activity.

◀ *Mayon Volcano in the Philippines is one of many active volcanoes on the Pacific "Ring of Fire." It is noted for its perfect conical shape; the base of the cone is 80 miles (130 km) in circumference.*

Ring of Fire
— plate boundaries
• major volcanoes

◀ *The Hawai'ian volcanoes lie in the center of a plate, not on a plate margin, and are known as intraplate volcanoes. They are associated with hot spots, whereby a plume of hot molten rock rises to the surface as the plate moves over it.*

American Samoa and Samoa

American Samoa and Samoa are part of the island archipelago of Polynesia. The two most populous islands are Tutuila in American Samoa and 'Upolu in Samoa. Although the economies of both these states remain predominantly resource-based, both are expanding their light manufacturing sectors, and the US administration is the primary employer in American Samoa. Tuna fishing is particularly important: 25% of all tuna consumed in the US is processed and canned in Pago Pago.

▶ *Many of the buildings in Samoa reflect the country's colonial past. Once a colony of New Zealand, Samoa is now an independent state; American Samoa remains an unincorporated territory of the United States.*

SCALE 1:3,350,000

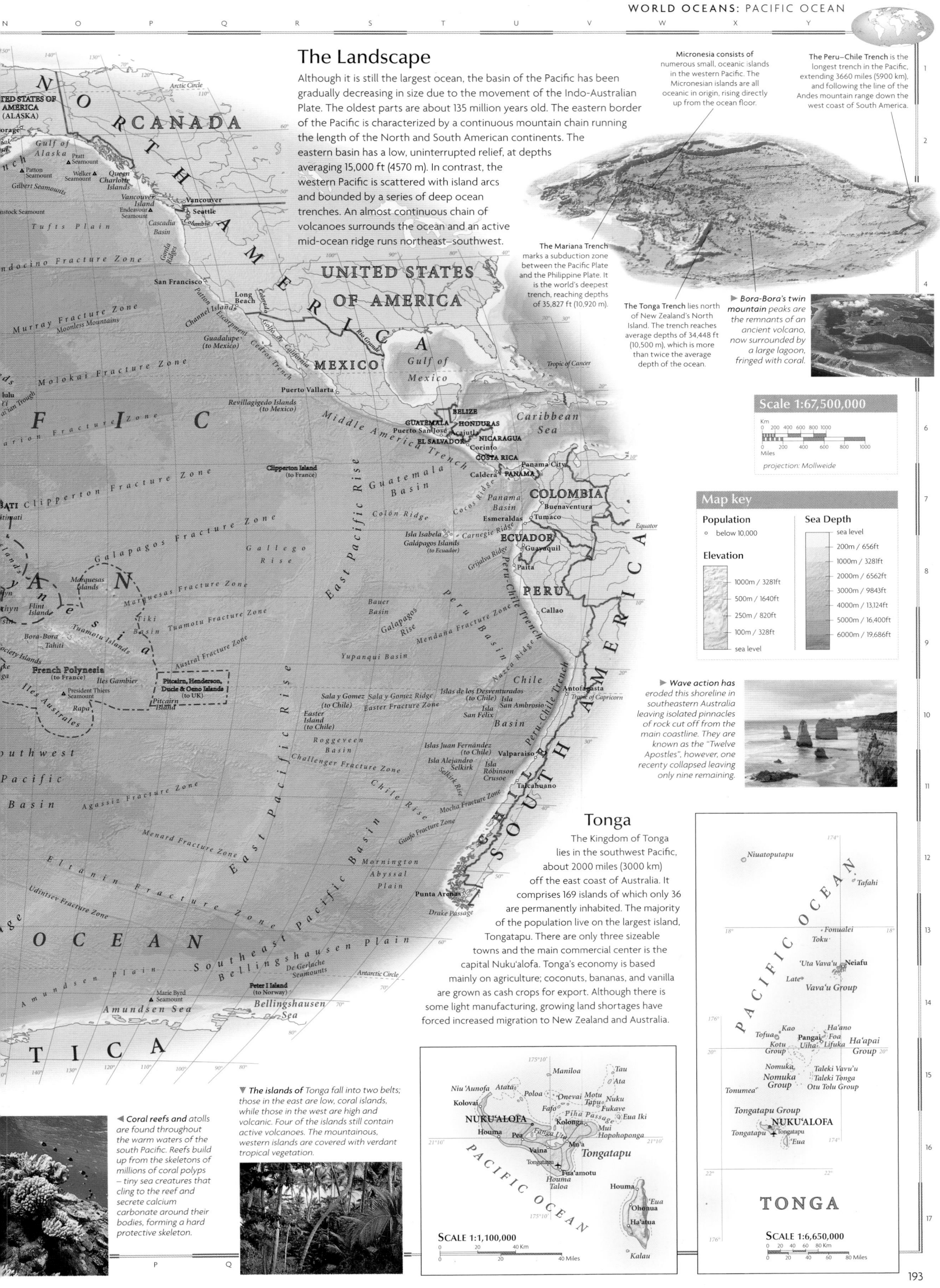

The Landscape

Although it is still the largest ocean, the basin of the Pacific has been gradually decreasing in size due to the movement of the Indo-Australian Plate. The oldest parts are about 135 million years old. The eastern border of the Pacific is characterized by a continuous mountain chain running the length of the North and South American continents. The eastern basin has a low, uninterrupted relief, at depths averaging 15,000 ft (4570 m). In contrast, the western Pacific is scattered with island arcs and bounded by a series of deep ocean trenches. An almost continuous chain of volcanoes surrounds the ocean and an active mid-ocean ridge runs northeast–southwest.

Micronesia consists of numerous small, oceanic islands in the western Pacific. The Micronesian islands are all oceanic in origin, rising directly up from the ocean floor.

The Peru–Chile Trench is the longest trench in the Pacific, extending 3660 miles (5900 km), and following the line of the Andes mountain range down the west coast of South America.

The Mariana Trench marks a subduction zone between the Pacific Plate and the Philippine Plate. It is the world's deepest trench, reaching depths of 35,827 ft (10,920 m).

The Tonga Trench lies north of New Zealand's North Island. The trench reaches average depths of 34,448 ft (10,500 m), which is more than twice the average depth of the ocean.

▶ **Bora-Bora's twin mountain** peaks are the remnants of an ancient volcano, now surrounded by a large lagoon, fringed with coral.

Scale 1:67,500,000

Km
0 200 400 600 800 1000

Miles
0 200 400 600 800 1000

projection: Mollweide

Map key

Population
○ below 10,000

Elevation
1000m / 3281ft
500m / 1640ft
250m / 820ft
100m / 328ft
sea level

Sea Depth
sea level
200m / 656ft
1000m / 3281ft
2000m / 6562ft
3000m / 9843ft
4000m / 13,124ft
5000m / 16,400ft
6000m / 19,686ft

▶ **Wave action has** eroded this shoreline in southeastern Australia leaving isolated pinnacles of rock cut off from the main coastline. They are known as the "Twelve Apostles", however, one recently collapsed leaving only nine remaining.

Tonga

The Kingdom of Tonga lies in the southwest Pacific, about 2000 miles (3000 km) off the east coast of Australia. It comprises 169 islands of which only 36 are permanently inhabited. The majority of the population live on the largest island, Tongatapu. There are only three sizeable towns and the main commercial center is the capital Nuku'alofa. Tonga's economy is based mainly on agriculture; coconuts, bananas, and vanilla are grown as cash crops for export. Although there is some light manufacturing, growing land shortages have forced increased migration to New Zealand and Australia.

◀ **Coral reefs and atolls** are found throughout the warm waters of the south Pacific. Reefs build up from the skeletons of millions of coral polyps – tiny sea creatures that cling to the reef and secrete calcium carbonate around their bodies, forming a hard protective skeleton.

▼ **The islands of** Tonga fall into two belts; those in the east are low, coral islands, while those in the west are high and volcanic. Four of the islands still contain active volcanoes. The mountainous, western islands are covered with verdant tropical vegetation.

TONGA

SCALE 1:1,100,000
0 20 40 Km
0 20 40 Miles

SCALE 1:6,650,000
0 20 40 60 80 Km
0 20 40 60 80 Miles

Antarctica

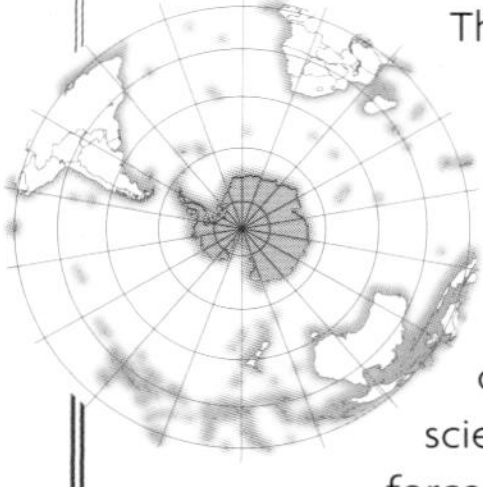

The ice-covered continent of Antarctica, which is the Earth's most southerly region, has drawn explorers and entrepreneurs seeking challenge and riches in its wintry lands for over 200 years. The extreme climate has deterred any large-scale settlement of the continent, and though commercial hunters built outposts in the past, habitation is now limited to scientific bases. The Antarctic Treaty, which came into force in 1961, provides for international governance and scientific cooperation in place of potential territorial conflict.

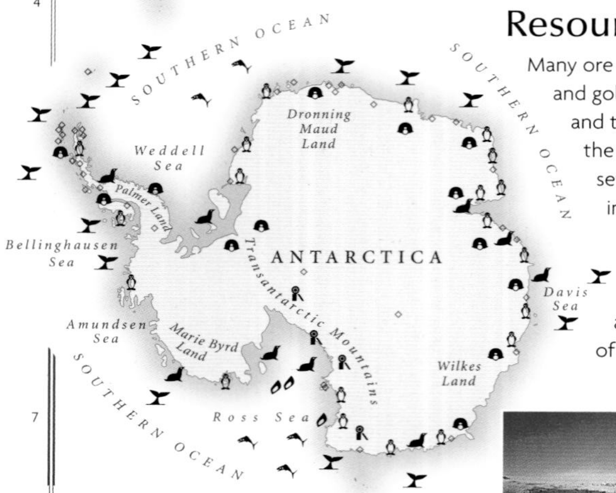

Resources

Many ore minerals, including iron and gold, are found in the Antarctic, and there are also coal reserves in the Transantarctic Mountains. The severe conditions and environmental importance of the region mean that exploitation of potential mineral resources is both uneconomic and undesirable. The unique wildlife and landscape draw a small number of tourists annually.

Resources (including wildlife)

- coal
- fish
- minerals
- oil & gas
- penguins
- seals
- whales
- polar research base

◀ Most settlements in Antarctica are research bases such as this one at Rothera on Adelaide Island, although there is a small Chilean settlement on King George Island.

The landscape

There are two distinct parts to Antarctica: West Antarctica, a series of ice-covered, mountainous islands, joined together by the ice; and the high plateau of East Antarctica. The Ross Sea and the Weddell Sea are outliers of the Southern Ocean – deep bays partially covered by thick ice shelves.

Grease ice   Pancake ice   Sea-ice sheet   Ice floe

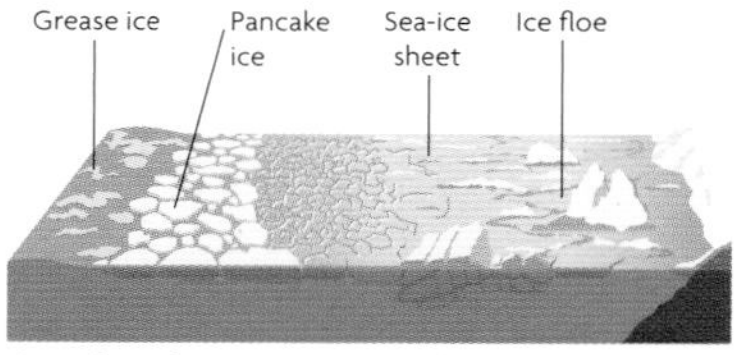

▲ Pack ice forms out at sea in freezing temperatures. At the outer limits, grease ice congeals on the surface of the ocean. This is then spun around by wind and waves into irregular "pancakes," freezing and breaking up several times before bonding together again to form sea-ice sheets, which finally cement into enormous ice floes.

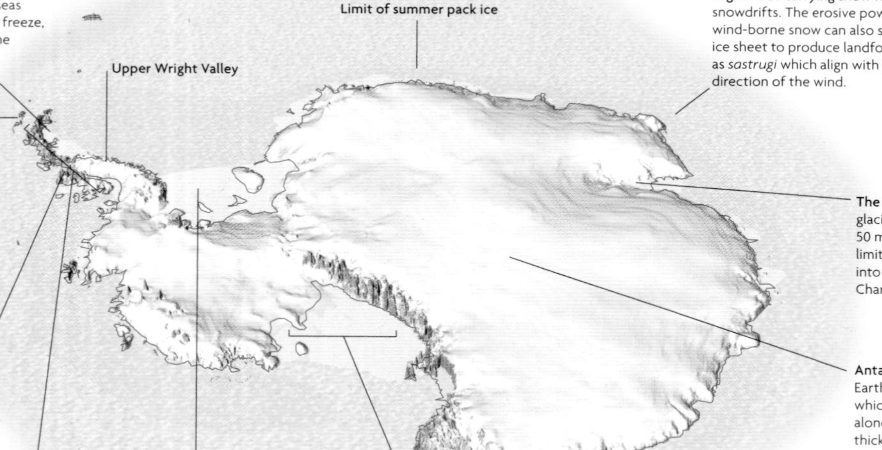

◀ On Elephant Island, the coast is edged by glaciers, although the land is not permanently covered by ice.

During the winter the seas surrounding Antarctica freeze, increasing the size of the continent by 100%.

Limit of winter pack ice

Elephant Island

Upper Wright Valley

Limit of summer pack ice

High winds carrying snow form huge snowdrifts. The erosive power of the wind-borne snow can also sculpt the ice sheet to produce landforms known as *sastrugi* which align with the direction of the wind.

Many volcanoes, some of them still active, can be found in the mountains of the Antarctic Peninsula.

The Lambert Glacier is the largest glacier system in the world, up to 50 miles (80 km) wide at its seaward limit, and reaching 180 miles (300 km) into the interior by way of the Prince Charles Mountains.

Antarctica is the highest continent on Earth, because of the great thickness of ice which overlays the land. In places the ice alone can each up to 15,700 ft (4800 m) thick. Much of the basement rock of west Antarctica lies below sea level, pushed down by the weight of the ice.

The mountainous Antarctic Peninsula is formed of rocks 65–225 million years old, overlain by more recent rocks and glacial deposits. It is connected to the Andes in South America by a submarine ridge.

Nearly half – 44% – of the Antarctic coastline is bounded by ice shelves, like the Ronne Ice Shelf, which float on the Ocean. These are joined to the inland ice sheet by dome-shaped ice "rises."

More than 30% of Antarctic ice is contained in the Ross Ice Shelf.

◀ The barren, flat-bottomed Upper Wright Valley was once filled by a glacier, but is now dry, strewn with boulders and pebbles. In some dry valleys, there has been no rain for over 2 million years.

▲ Large colonies of seabirds live in the extremely harsh Antarctic climate. The Emperor penguins seen here, the smaller Adélie penguin, the Antarctic petrel, and the South Polar skua are the only birds that breed exclusively on the continent.

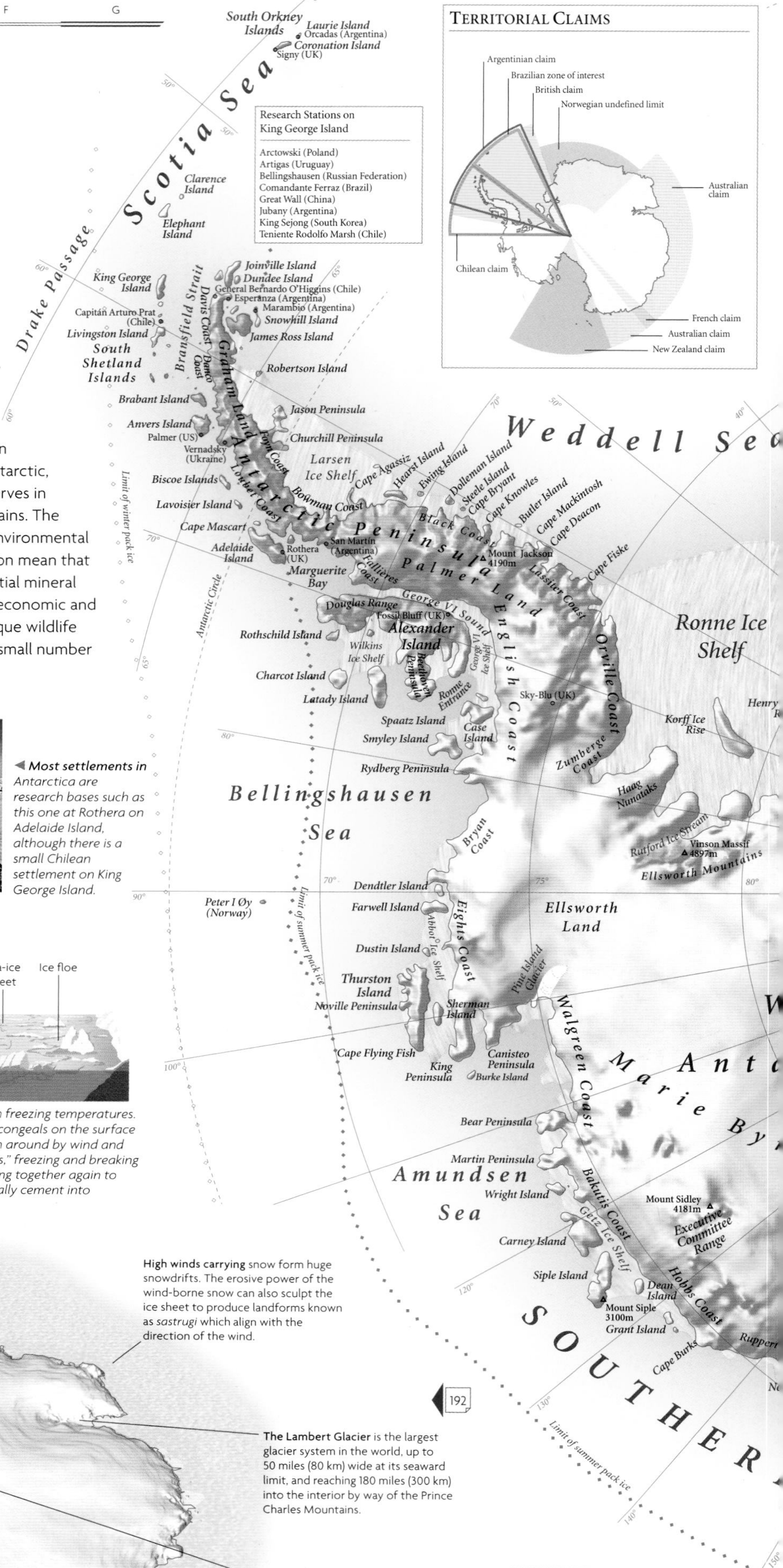

TERRITORIAL CLAIMS

- Argentinian claim
- Brazilian zone of interest
- British claim
- Norwegian undefined limit
- Australian claim
- Chilean claim
- French claim
- Australian claim
- New Zealand claim

Research Stations on King George Island

Arctowski (Poland)
Artigas (Uruguay)
Bellingshausen (Russian Federation)
Comandante Ferraz (Brazil)
Great Wall (China)
Jubany (Argentina)
King Sejong (South Korea)
Teniente Rodolfo Marsh (Chile)

◄ **The sun sets** over the Antarctic Peninsula for more than six months during the winter. However, there are more hours of sunshine during the brief Antarctic summer than most equatorial countries experience in a whole year.

▲ **Immense, flat-topped icebergs** are formed when blocks of ice break away from the main ice sheet. Though the exposed area is enormous, the volume of ice concealed beneath the water may be many times greater.

Map key

Elevation

ice cap

ice shelf

exposed land

Scale 1:16,500,000

projection: Lambert Azimuthal Equal Area

The Arctic

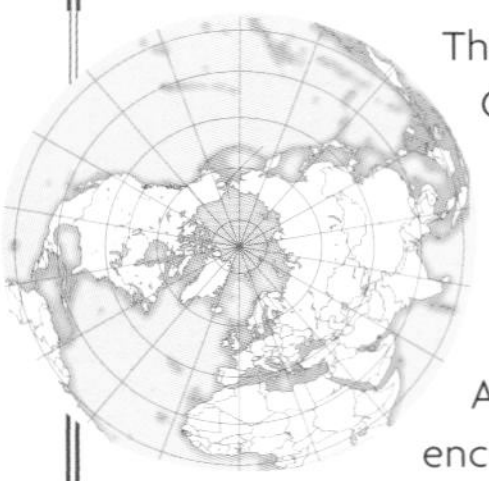

Three continents, Asia, North America, and Europe, reach into the Arctic Circle at their northernmost limits, almost entirely encircling the Arctic Ocean. Despite the region's extraordinarily harsh climate, it has been inhabited for thousands of years by peoples such as the European Lapps, the Russian Nenet, and the North American Inuit, who draw a living from fishing, herding, and hunting. More recently, particularly in the Russian Arctic, opportunities to exploit oil and other mineral reserves have encouraged immigration. Pollution of the Arctic's unique ecology and damage to the traditional lifestyles of many native peoples have been the unfortunate results of this activity, and international cooperation is needed to safeguard the future of the region.

Map key

Population
■ above 5 million
▣ 1 million to 5 million
◉ 500,000 to 1 million
◎ 100,000 to 500,000
⊕ 50,000 to 100,000
○ 10,000 to 50,000
○ below 10,000

Sea depth
Sea level
200m / 656ft
1000m / 3281ft
2000m / 6562ft
3000m / 9843ft
4000m / 13,124ft
5000m / 16,400ft
6000m / 19,686ft

Scale 1:23,500,000

Km 0 100 200 300 400 500 600
Miles 0 100 200 300 400 500 600

projection: Lambert Azimuthal Equal Area

192

▲ **Windblown snow etches** deep patterns in the ice sheet known as sastrugi. They align with the direction of the wind

Resources

Large quantities of coal, oil, and natural gas are to be found in the basins of the Arctic Ocean, and in northern Canada, Alaska, and the Russian Federation. The cost and difficulty of extraction and, more recently, awareness of damage to the environment, have limited exploitation to coastal regions. The unfrozen waters have stocks of fish including cod, flounder, and haddock. Quotas have now been put in place to restrict the number of fish caught annually. Reindeer are herded in large numbers by many of the native Arctic peoples. Most grain and vegetables are imported from elsewhere.

Bering Sea

NORTH AMERICA
Inuvik

ASIA
Tiksi

ARCTIC OCEAN

Noril'sk

Qaanaaq

Murmansk

Reykjavik

ATLANTIC OCEAN

EUROPE

▲ **Icebreakers are ships** with specially strengthened hulls, designed to break a path through the ice. They are used to keep important routes open during the winter, when falling temperatures cause much of the Arctic Ocean to freeze over.

Resources
🏴 coal
🐟 fish
⛏ mining
🛢 oil & gas
☢ radioactive contamination
● major towns
⊕ major ports

8

The landscape

The Arctic Ocean comprises two large ocean basins divided by three submarine ridges, the greatest of which, the Lomonosov Ridge, is a huge underwater mountain range which has an average height of more than 10,000 ft (3000 m). The lands which encircle the Arctic Ocean are underlain by great shield areas of ancient rocks, which were heavily glaciated during the last Ice Age.

◀ **Icebergs are constantly** broken up and reshaped by wind and the oceans. This flat-topped iceberg has been undercut, leaving a craggy ice cliff.

The Canadian Shield underlies almost all of the Canadian Arctic. It is a very stable plateau of ancient rock, now covered by glacial lakes and sediment, which supports tundra vegetation.

The Arctic Ocean is the world's smallest ocean with a total area of 5,440,000 sq miles (15,100,000 sq km).

At a latitude of more than 75° N, the Arctic Ocean is almost permanently covered by pack ice, though high winds and the movement of the seas may cause the ice to crack and break up.

In the more southerly reaches of the Arctic, like Siberia, much of the land is covered by permafrost. In the summer, higher temperatures warm the frozen ground, causing a number of typical phenomena. These include solifluction, the fast downhill movement of top soil layers; freeze/thaw activity, which patterns the ground into regular polygonal shapes, and the formation of large domes with a frozen ice core, known as pingos.

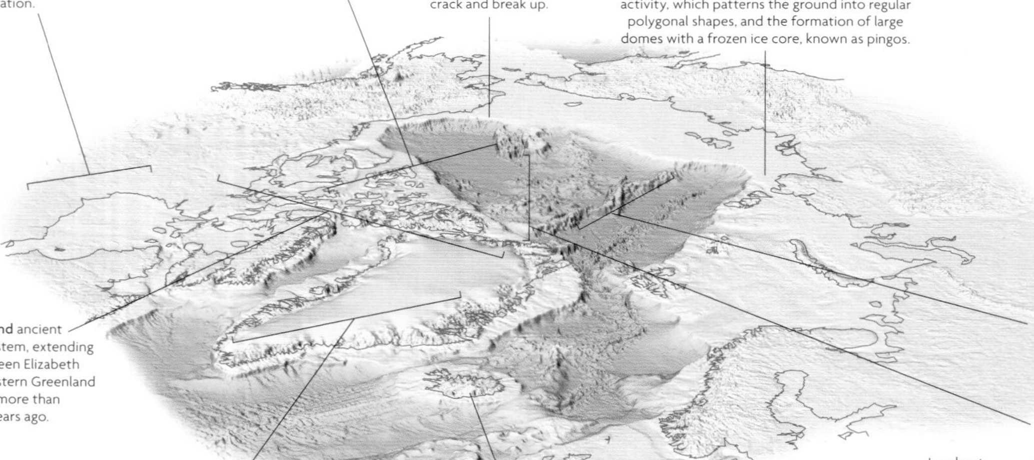

A complex and ancient mountain system, extending from the Queen Elizabeth Islands to eastern Greenland was formed more than 245 million years ago.

Lomonosov Ridge

Arctic ice shelf

◀ **Much of Greenland is** covered by a massive ice sheet more than 650,000 sq miles (1,683,400 sq km) in extent. The weight of the ice has depressed the central land area to form a basin lying more than 1000 ft (300 m) below sea level. Only at the edges of the island is bare rock visible.

Iceland has five major glaciers, sustained by heavy snowfall. Parts of the ice cap cover active volcanoes, such as Bárðharbunga, which periodically erupt causing the melted ice to form a great lake at the glacier margins.

Ice sheet
Iceberg
Crevasses occur at the edge of the ice sheet
Sea water melts the edge of the ice sheet

▲ **At the boundary** of the Arctic ice shelves, sea water flows under the ice causing melting and forming crevasses on the surface. This eventually weakens blocks of ice which break away as icebergs. This process is known as calving.

64

Map (right side)

NORTH
CANADA
AMERICA

Great Bear Lake

Great Slave Lake

Kugluktuk (Coppermine)

Bathurst Inlet

Cambridge Bay (Ikaluktutiak)

Queen Maud

Nelson

Back

Churchill

Southampton Island

Repulse Bay

Melville Peninsula

Hudson Bay

Coats Island

Mansel Island

Foxe Basin

Prince Charles Island

Foxe Peninsula

Baffin Isl

Ivujivik

Inukjuak (Port Harrison)

Hudson Strait

Kimmirut (Lake Harbour)

Iqaluit (Frobisher Bay)

Frobisher Bay

Cumberland Sound

Ungava Bay

Davis Str

Cape Chidley

Maniit

Labrador Sea

NUUK

Nain

Paamiut

Ivittuut

Labrador Basin

Qaqortoq

Narsarsua

Nanortalik

Nunap Isua (Kap Farvel)

Eirik Ridge

ATLANTIC

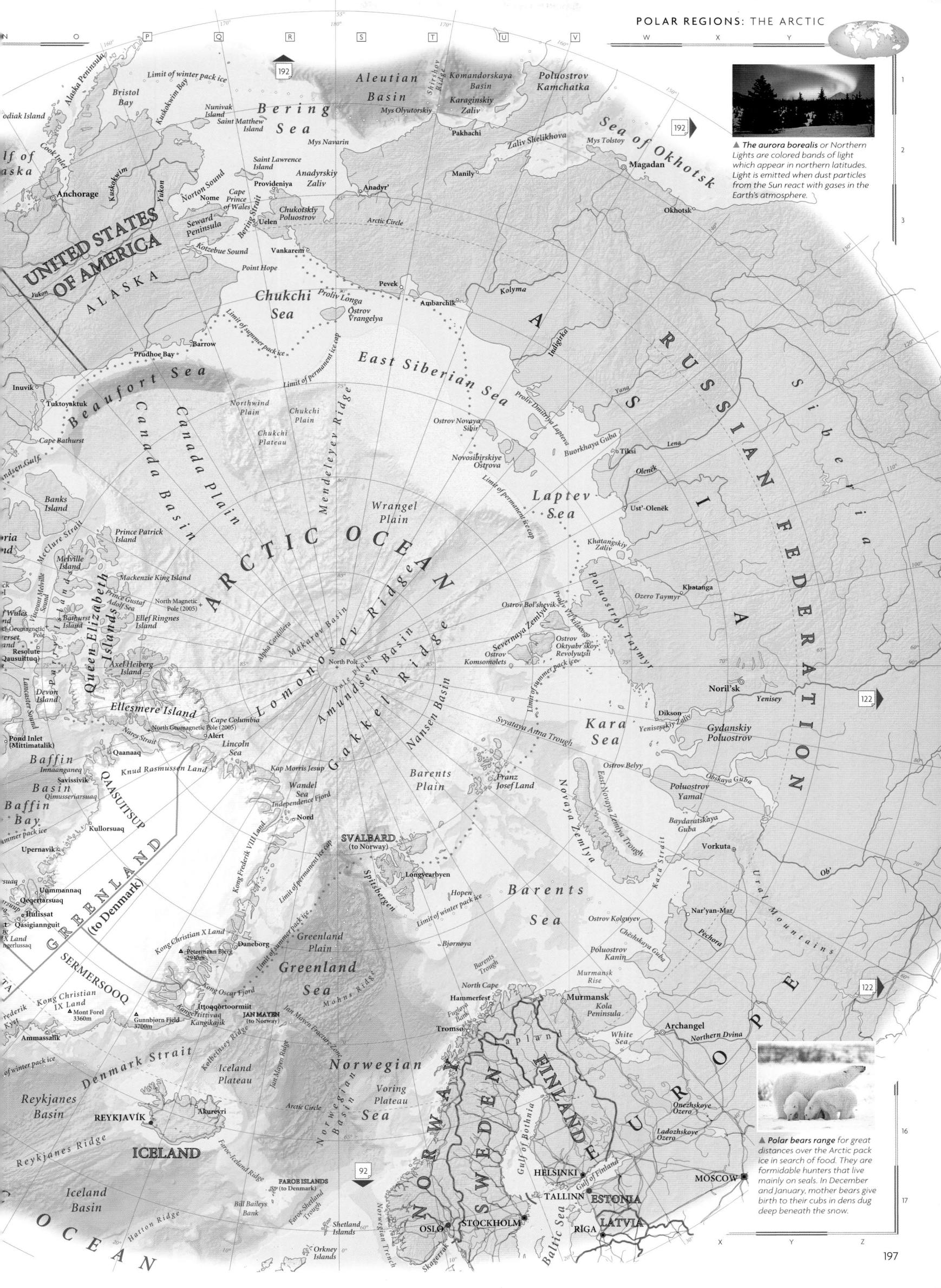

▲ The aurora borealis or Northern Lights are colored bands of light which appear in northern latitudes. Light is emitted when dust particles from the Sun react with gases in the Earth's atmosphere.

▲ Polar bears range for great distances over the Arctic pack ice in search of food. They are formidable hunters that live mainly on seals. In December and January, mother bears give birth to their cubs in dens dug deep beneath the snow.

Geographical comparisons

Largest countries

Russian Federation	6,592,735 sq miles	(17,075,200 sq km)
Canada	3,855,171 sq miles	(9,984,670 sq km)
USA	3,794,100 sq miles	(9,826,675 sq km)
China	3,705,386 sq miles	(9,596,960 sq km)
Brazil	3,286,470 sq miles	(8,511,965 sq km)
Australia	2,967,893 sq miles	(7,686,850 sq km)
India	1,269,339 sq miles	(3,287,590 sq km)
Argentina	1,068,296 sq miles	(2,766,890 sq km)
Kazakhstan	1,049,150 sq miles	(2,717,300 sq km)
Algeria	919,590 sq miles	(2,381,740 sq km)

Smallest countries

Vatican City	0.17 sq miles	(0.44 sq km)
Monaco	0.75 sq miles	(1.95 sq km)
Nauru	8.2 sq miles	(21.2 sq km)
Tuvalu	10 sq miles	(26 sq km)
San Marino	24 sq miles	(61 sq km)
Liechtenstein	62 sq miles	(160 sq km)
Marshall Islands	70 sq miles	(181 sq km)
St. Kitts & Nevis	101 sq miles	(261 sq km)
Maldives	116 sq miles	(300 sq km)
Malta	124 sq miles	(320 sq km)

Largest islands

	To the nearest 1000 – or 100,000 for the largest	
Greenland	849,400 sq miles	(2,200,000 sq km)
New Guinea	312,000 sq miles	(808,000 sq km)
Borneo	292,222 sq miles	(757,050 sq km)
Madagascar	229,300 sq miles	(594,000 sq km)
Sumatra	202,300 sq miles	(524,000 sq km)
Baffin Island	183,800 sq miles	(476,000 sq km)
Honshu	88,800 sq miles	(230,000 sq km)
Britain	88,700 sq miles	(229,800 sq km)
Victoria Island	81,900 sq miles	(212,000 sq km)
Ellesmere Island	75,700 sq miles	(196,000 sq km)

Richest countries

	GNI per capita, in US$
Monaco	186,950
Liechtenstein	136,770
Norway	102,610
Switzerland	90,760
Qatar	86,790
Luxembourg	69,900
Australia	65,390
Sweden	61,760
Denmark	61,680
Singapore	54,040

Poorest countries

	GNI per capita, in US$
Burundi	260
Malawi	270
Somalia	288
Central African Republic	320
Niger	400
Liberia	410
Dem. Rep. Congo	430
Madagascar	440
Guinea	460
Ethiopia	470
Eritrea	490
Gambia	500

Most populous countries

China	1,393,800,000
India	1,267,400,000
USA	322,600,000
Indonesia	252,800,000
Brazil	202,120,000
Pakistan	185,100,000
Nigeria	178,500,000
Bangladesh	159,000,000
Russian Federation	142,500,000
Japan	127,000,000

Least populous countries

Vatican City	842
Nauru	9488
Tuvalu	10,782
Palau	21,186
San Marino	32,742
Monaco	36,950
Liechtenstein	37,313
St Kitts & Nevis	51,538
Marshall Islands	70,983
Dominica	73,449
Andorra	85,458
Antigua & Barbuda	91,295

Most densely populated countries

Monaco	49,267 people per sq mile	(18,949 per sq km)
Singapore	23,305 people per sq mile	(9016 per sq km)
Vatican City	4953 people per sq mile	(1914 per sq km)
Bahrain	4762 people per sq mile	(1841 per sq km)
Maldives	3448 people per sq mile	(1333 per sq km)
Malta	3226 people per sq mile	(1250 per sq km)
Bangladesh	3066 people per sq mile	(1184 per sq km)
Taiwan	1879 people per sq mile	(725 per sq km)
Barbados	1807 people per sq mile	(698 per sq km)
Mauritius	1671 people per sq mile	(645 per sq km)

Most sparsely populated countries

Mongolia	5 people per sq mile	(2 per sq km)
Namibia	7 people per sq mile	(3 per sq km)
Australia	8 people per sq mile	(3 per sq km)
Suriname	8 people per sq mile	(3 per sq km)
Iceland	8 people per sq mile	(3 per sq km)
Botswana	9 people per sq mile	(4 per sq km)
Libya	9 people per sq mile	(4 per sq km)
Mauriania	10 people per sq mile	(4 per sq km)
Canada	10 people per sq mile	(4 per sq km)
Guyana	11 people per sq mile	(4 per sq km)

Most widely spoken languages

1. Chinese (Mandarin)	6. Arabic
2. English	7. Bengali
3. Hindi	8. Portuguese
4. Spanish	9. Malay-Indonesian
5. Russian	10. French

Largest conurbations

	Urban area population
Tokyo	37,800,000
Jakarta	30,500,000
Manila	24,100,000
Delhi	24,000,000
Karachi	23,500,000
Seoul	23,500,000
Shanghai	23,400,000
Beijing	21,000,000
New York City	20,600,000
Guangzhou	20,600,000
São Paulo	20,300,000
Mexico City	20,000,000
Mumbai	17,700,000
Osaka	17,400,000
Lagos	17,000,000
Moscow	16,100,000
Dhaka	15,700,000
Lahore	15,600,000
Los Angeles	15,000,000
Bangkok	15,000,000
Kolkatta	14,700,000
Buenos Aires	14,100,000
Tehran	13,500,000
Istanbul	13,300,000
Shenzhen	12,000,000

Countries with the most land borders

14: China	(Afghanistan, Bhutan, India, Kazakhstan, Kyrgyzstan, Laos, Mongolia, Myanmar (Burma), Nepal, North Korea, Pakistan, Russian Federation, Tajikistan, Vietnam)	
14: Russian Federation	(Azerbaijan, Belarus, China, Estonia, Finland, Georgia, Kazakhstan, Latvia, Lithuania, Mongolia, North Korea, Norway, Poland, Ukraine)	
10: Brazil	(Argentina, Bolivia, Colombia, French Guiana, Guyana, Paraguay, Peru, Suriname, Uruguay, Venezuela)	
9: Congo, Dem. Rep.	(Angola, Burundi, Central African Republic, Congo, Rwanda, South Sudan, Tanzania, Uganda, Zambia)	
9: Germany	(Austria, Belgium, Czech Republic, Denmark, France, Luxembourg, Netherlands, Poland, Switzerland)	
8: Austria	(Czech Republic, Germany, Hungary, Italy, Liechtenstein, Slovakia, Slovenia, Switzerland)	
8: France	(Andorra, Belgium, Germany, Italy, Luxembourg, Monaco, Spain, Switzerland)	
8: Tanzania	(Burundi, Dem. Rep. Congo, Kenya, Malawi, Mozambique, Rwanda, Uganda, Zambia)	
8: Turkey	(Armenia, Azerbaijan, Bulgaria, Georgia, Greece, Iran, Iraq, Syria)	
8: Zambia	(Angola, Botswana, Dem. Rep.Congo, Malawi, Mozambique, Namibia, Tanzania, Zimbabwe)	

Longest rivers

Nile (NE Africa)	4160 miles	(6695 km)
Amazon (South America)	4049 miles	(6516 km)
Yangtze (China)	3915 miles	(6299 km)
Mississippi/Missouri (USA)	3710 miles	(5969 km)
Ob'-Irtysh (Russian Federation)	3461 miles	(5570 km)
Yellow River (China)	3395 miles	(5464 km)
Congo (Central Africa)	2900 miles	(4667 km)
Mekong (Southeast Asia)	2749 miles	(4425 km)
Lena (Russian Federation)	2734 miles	(4400 km)
Mackenzie (Canada)	2640 miles	(4250 km)
Yenisey (Russian Federation)	2541 miles	(4090km)

Highest mountains

	Height above sea level	
Everest	29,029 ft	(8848 m)
K2	28,253 ft	(8611 m)
Kangchenjunga I	28,210 ft	(8598 m)
Makalu I	27,767 ft	(8463 m)
Cho Oyu	26,907 ft	(8201 m)
Dhaulagiri I	26,796 ft	(8167 m)
Manaslu I	26,783 ft	(8163 m)
Nanga Parbat I	26,661 ft	(8126 m)
Annapurna I	26,547 ft	(8091 m)
Gasherbrum I	26,471 ft	(8068 m)

Largest bodies of inland water

	With area and depth	
Caspian Sea	143,243 sq miles (371,000 sq km)	3215 ft (980 m)
Lake Superior	31,151 sq miles (83,270 sq km)	1289 ft (393 m)
Lake Victoria	26,828 sq miles (69,484 sq km)	328 ft (100 m)
Lake Huron	23,436 sq miles (60,700 sq km)	751 ft (229 m)
Lake Michigan	22,402 sq miles (58,020 sq km)	922 ft (281 m)
Lake Tanganyika	12,703 sq miles (32,900 sq km)	4700 ft (1435 m)
Great Bear Lake	12,274 sq miles (31,790 sq km)	1047 ft (319 m)
Lake Baikal	11,776 sq miles (30,500 sq km)	5712 ft (1741 m)
Great Slave Lake	10,981 sq miles (28,440 sq km)	459 ft (140 m)
Lake Erie	9,915 sq miles (25,680 sq km)	197 ft (60 m)

Deepest ocean features

Challenger Deep, Mariana Trench (Pacific)	35,827 ft	(10,920 m)
Vityaz III Depth, Tonga Trench (Pacific)	35,704 ft	(10,882 m)
Vityaz Depth, Kuril-Kamchatka Trench (Pacific)	34,588 ft	(10,542 m)
Cape Johnson Deep, Philippine Trench (Pacific)	34,441 ft	(10,497 m)
Kermadec Trench (Pacific)	32,964 ft	(10,047 m)
Ramapo Deep, Japan Trench (Pacific)	32,758 ft	(9984 m)
Milwaukee Deep, Puerto Rico Trench (Atlantic)	30,185 ft	(9200 m)
Argo Deep, Torres Trench (Pacific)	30,070 ft	(9165 m)
Meteor Depth, South Sandwich Trench (Atlantic)	30,000 ft	(9144 m)
Planet Deep, New Britain Trench (Pacific)	29,988 ft	(9140 m)

Greatest waterfalls

	Mean flow of water	
Boyoma (Dem. Rep. Congo)	600,400 cu. ft/sec	(17,000 cu.m/sec)
Khône (Laos/Cambodia)	410,000 cu. ft/sec	(11,600 cu.m/sec)
Niagara (USA/Canada)	195,000 cu. ft/sec	(5500 cu.m/sec)
Grande, Salto (Uruguay)	160,000 cu. ft/sec	(4500 cu.m/sec)
Paulo Afonso (Brazil)	100,000 cu. ft/sec	(2800 cu.m/sec)
Urubupungá, Salto do (Brazil)	97,000 cu. ft/sec	(2750 cu.m/sec)
Iguaçu (Argentina/Brazil)	62,000 cu. ft/sec	(1700 cu.m/sec)
Maribondo, Cachoeira do (Brazil)	53,000 cu. ft/sec	(1500 cu.m/sec)
Victoria (Zimbabwe)	39,000 cu. ft/sec	(1100 cu.m/sec)
Murchison Falls (Uganda)	42,000 cu. ft/sec	(1200 cu.m/sec)
Churchill (Canada)	35,000 cu. ft/sec	(1000 cu.m/sec)
Kaveri Falls (India)	33,000 cu. ft/sec	(900 cu.m/sec)

Highest waterfalls

	* Indicates that the total height is a single leap	
Angel (Venezuela)	3212 ft	(979 m)
Tugela (South Africa)	3110 ft	(948 m)
Utigard (Norway)	2625 ft	(800 m)
Mongefossen (Norway)	2539 ft	(774 m)
Mtarazi (Zimbabwe)	2500 ft	(762 m)
Yosemite (USA)	2425 ft	(739 m)
Ostre Mardola Foss (Norway)	2156 ft	(657 m)
Tyssestrengane (Norway)	2119 ft	(646 m)
*Cuquenan (Venezuela)	2001 ft	(610 m)
Sutherland (New Zealand)	1903 ft	(580 m)
*Kjellfossen (Norway)	1841 ft	(561 m)

Largest deserts

	NB – Most of Antarctica is a polar desert, with only 50mm of precipitation annually	
Sahara	3,450,000 sq miles	(9,065,000 sq km)
Gobi	500,000 sq miles	(1,295,000 sq km)
Ar Rub al Khali	289,600 sq miles	(750,000 sq km)
Great Victorian	249,800 sq miles	(647,000 sq km)
Sonoran	120,000 sq miles	(311,000 sq km)
Kalahari	120,000 sq miles	(310,800 sq km)
Kara Kum	115,800 sq miles	(300,000 sq km)
Takla Makan	100,400 sq miles	(260,000 sq km)
Namib	52,100 sq miles	(135,000 sq km)
Thar	33,670 sq miles	(130,000 sq km)

Hottest inhabited places

Djibouti (Djibouti)	86° F	(30 °C)
Tombouctou (Mali)	84.7° F	(29.3 °C)
Tirunelveli (India)		
Tuticorin (India)		
Nellore (India)	84.5° F	(29.2 °C)
Santa Marta (Colombia)		
Aden (Yemen)	84° F	(28.9 °C)
Madurai (India)		
Niamey (Niger)		
Hodeida (Yemen)	83.8° F	(28.8 °C)
Ouagadougou (Burkina Faso)		
Thanjavur (India)		
Tiruchchirappalli (India)		

Driest inhabited places

Aswân (Egypt)	0.02 in	(0.5 mm)
Luxor (Egypt)	0.03 in	(0.7 mm)
Arica (Chile)	0.04 in	(1.1 mm)
Ica (Peru)	0.1 in	(2.3 mm)
Antofagasta (Chile)	0.2 in	(4.9 mm)
Al Minya (Egypt)	0.2 in	(5.1 mm)
Asyut (Egypt)	0.2 in	(5.2 mm)
Callao (Peru)	0.5 in	(12.0 mm)
Trujillo (Peru)	0.55 in	(14.0 mm)
Al Fayyum (Egypt)	0.8 in	(19.0 mm)

Wettest inhabited places

Mawsynram (India)	467 in	(11,862 mm)
Mount Waialeale (Hawaii, USA)	460 in	(11,684 mm)
Cherrapunji (India)	450 in	(11,430 mm)
Cape Debundsha (Cameroon)	405 in	(10,290 mm)
Quibdo (Colombia)	354 in	(8892 mm)
Buenaventura (Colombia)	265 in	(6743 mm)
Monrovia (Liberia)	202 in	(5131 mm)
Pago Pago (American Samoa)	196 in	(4990 mm)
Mawlamyine (Myanmar [Burma])	191 in	(4852 mm)
Lae (Papua New Guinea)	183 in	(4645 mm)

Standard time zones

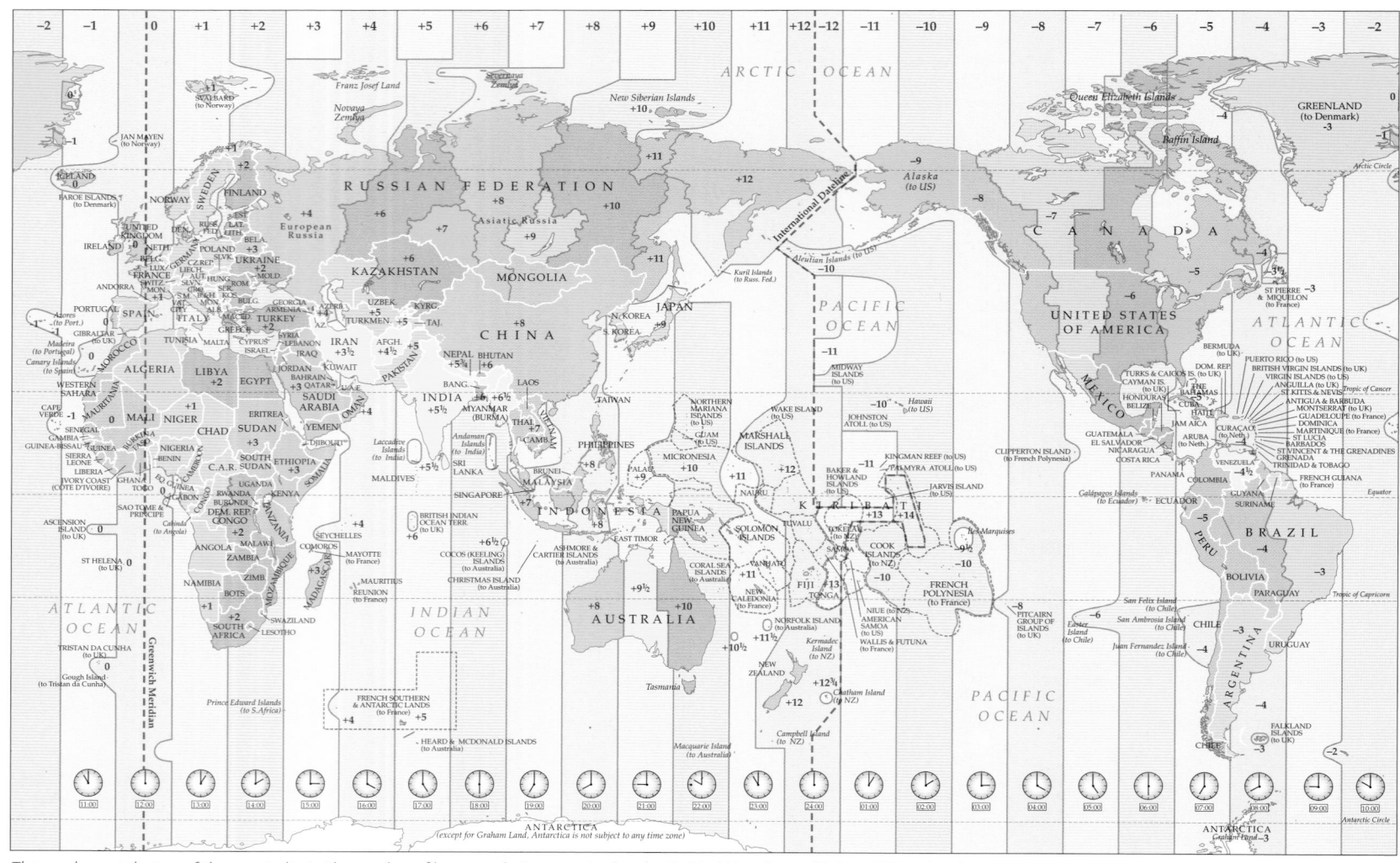

The numbers at the top of the map indicate the number of hours each time zone is ahead or behind Coordinated Universal Time (UTC).
The clocks and 24-hour times given at the bottom of the map show the time in each time zone when it is 12:00 hours noon (UTC)

Time Zones

Because Earth is a rotating sphere, the Sun shines on only half of its surface at any one time. Thus, it is simultaneously morning, evening and night time in different parts of the world (see diagram below). Because of these disparities, each country or part of a country adheres to a local time.

A region of Earth's surface within which a single local time is used is called a time zone. There are 24 one hour time zones around the world, arranged roughly in longitudinal bands.

Standard Time

Standard time is the official local time in a particular country or part of a country. It is defined by the

Day and night around the world

time zone or zones associated with that country or region. Although time zones are arranged roughly in longitudinal bands, in many places the borders of a zone do not fall exactly on longitudinal meridians, as can be seen on the map (above), but are determined by geographical factors or by borders between countries or parts of countries. Most countries have just one time zone and one standard time, but some large countries (such as the US, Canada, and Russia) are split between several time zones, so standard time varies across those countries. For example, the coterminous United States straddles four time zones and so has four standard times, called the Eastern, Central, Mountain, and Pacific standard times. China is unusual in that just one standard time is used for the whole country, even though it extends across 60° of longitude from west to east.

Coordinated Universal Time (UTC)

Coordinated Universal Time (UTC) is a reference by which the local time in each time zone is set. For example, Australian Western Standard Time (the local time in Western Australia) is set 8 hours ahead of UTC (it is

UTC+8) whereas Eastern Standard Time in the United States is set 5 hours behind UTC (it is UTC-5). UTC is a successor to, and closely approximates, Greenwich Mean Time (GMT). However, UTC is based on an atomic clock, whereas GMT is determined by the Sun's position in the sky relative to the 0° longitudinal meridian, which runs through Greenwich, UK.

The International Dateline

The International Dateline is an imaginary line from pole to pole that roughly corresponds to the 180° longitudinal meridian. It is an arbitrary marker between calendar days. The dateline is needed because of the use of local times around the world rather than a single universal time. When moving from west to east across the dateline, travelers have to set their watches back one day. Those traveling in the opposite direction, from east to west, must add a day.

Daylight Saving Time

Daylight saving is a summertime adjustment to the local time in a country or region, designed to cause a higher proportion of its citizens' waking hours to pass during daylight. To follow the system, timepieces are advanced by an hour on a pre-decided date in spring and reverted back in the fall. About half of the world's nations use daylight saving.

Countries of the World

There are currently 196 independent countries in the world and almost 60 dependencies. Antarctica is the only land area on Earth that is not officially part of, and does not belong to, any single country.

In 1950, the world comprised 82 countries. In the decades following, many more states came into being as they achieved independence from their former colonial rulers. Most recent additions were caused by the breakup of the former Soviet Union in 1991, and the former Yugoslavia in 1992, which swelled the ranks of independent states. In July 2011, South Sudan became the latest country to be formed after declaring independence from Sudan.

Country factfile key

Formation Date of political origin or independence/ date current borders were established

Population Total population / population density – based on total *land* area

Languages An asterisk (*) denotes the official language(s)

Calorie consumption Average number of kilocalories consumed daily per person

AFGHANISTAN
Central Asia

Official name Islamic Republic of Afghanistan
Formation 1919 / 1919
Capital Kabul
Population 31.3 million / 124 people per sq mile (48 people per sq km)
Total area 250,000 sq. miles (647,500 sq. km)
Languages Pashtu*, Tajik, Dari*, Farsi, Uzbek, Turkmen
Religions Sunni Muslim 80%, Shi'a Muslim 19%, Other 1%
Ethnic mix Pashtun 38%, Tajik 25%, Hazara 19%, Uzbek and Turkmen 15%, Other 3%
Government Nonparty system
Currency Afghani = 100 puls
Literacy rate rate 32%
Calorie consumption 2090 kilocalories

ALBANIA
Southeast Europe

Official name Republic of Albania
Formation 1912 / 1921
Capital Tirana
Population 3.2 million / 302 people per sq mile (117 people per sq km)
Total area 11,100 sq. miles (28,748 sq. km)
Languages Albanian*, Greek
Religions Sunni Muslim 70%, Albanian Orthodox 20%, Roman Catholic 10%
Ethnic mix Albanian 98%, Greek 1%, Other 1%
Government Parliamentary system
Currency Lek = 100 qindarka (qintars)
Literacy rate 97%
Calorie consumption 3023 kilocalories

ALGERIA
North Africa

Official name People's Democratic Republic of Algeria
Formation 1962 / 1962
Capital Algiers
Population 39.9 million / 43 people per sq mile (17 people per sq km)
Total area 919,590 sq. miles (2,381,740 sq. km)
Languages Arabic*, Tamazight (Kabyle, Shawia, Tamashek), French
Religions Sunni Muslim 99%, Christian and Jewish 1%
Ethnic mix Arab 75%, Berber 24%, European and Jewish 1%
Government Presidential system
Currency Algerian dinar = 100 centimes
Literacy rate 73%
Calorie consumption 3296 kilocalories

ANDORRA
Southwest Europe

Official name Principality of Andorra
Formation 1278 / 1278
Capital Andorra la Vella
Population 85,485 / 475 people per sq mile (184 people per sq km)
Total area 181 sq. miles (468 sq. km)
Languages Spanish, Catalan*, French, Portuguese
Religions Roman Catholic 94%, Other 6%
Ethnic mix Spanish 46%, Andorran 28%, Other 18%, French 8%
Government Parliamentary system
Currency Euro = 100 cents
Literacy rate 99%
Calorie consumption Not available

ANGOLA
Southern Africa

Official name Republic of Angola
Formation 1975 / 1975
Capital Luanda
Population 22.1 million / 46 people per sq mile (18 people per sq km)
Total area 481,351 sq. miles (1,246,700 sq. km)
Languages Portuguese*, Umbundu, Kimbundu, Kikongo
Religions Roman Catholic 68%, Protestant 20%, Indigenous beliefs 12%
Ethnic mix Ovimbundu 37%, Kimbundu 25%, Other 25%, Bakongo 13%
Government Presidential system
Currency Readjusted kwanza = 100 lwei
Literacy rate 71%
Calorie consumption 2473 kilocalories

ANTIGUA & BARBUDA
West Indies

Official name Antigua and Barbuda
Formation 1981 / 1981
Capital St. John's
Population 91,295 / 537 people per sq mile (207 people per sq km)
Total area 170 sq. miles (442 sq. km)
Languages English*, English patois
Religions Anglican 45%, Other Protestant 42%, Roman Catholic 10%, Other 2%, Rastafarian 1%
Ethnic mix Black African 95%, Other 5%
Government Parliamentary system
Currency East Caribbean dollar = 100 cents
Literacy rate 99%
Calorie consumption 2396 kilocalories

ARGENTINA
South America

Official name Argentine Republic
Formation 1816 / 1816
Capital Buenos Aires
Population 41.8 million / 40 people per sq mile (15 people per sq km)
Total area 1,068,296 sq. miles (2,766,890 sq. km)
Languages Spanish*, Italian, Amerindian languages
Religions Roman Catholic 70%, Other 18%, Protestant 9%, Muslim 2%, Jewish 1%
Ethnic mix Indo-European 97%, Mestizo 2%, Amerindian 1%
Government Presidential system
Currency Argentine peso = 100 centavos
Literacy rate 98%
Calorie consumption 3155 kilocalories

ARMENIA
Southwest Asia

Official name Republic of Armenia
Formation 1991 / 1991
Capital Yerevan
Population 3 million / 261 people per sq mile (101 people per sq km)
Total area 11,506 sq. miles (29,800 sq. km)
Languages Armenian*, Azeri, Russian
Religions Armenian Apostolic Church (Orthodox) 88%, Armenian Catholic Church 6%, Other 6%
Ethnic mix Armenian 98%, Other 1%, Yezidi 1%
Government Parliamentary system
Currency Dram = 100 luma
Literacy rate 99%
Calorie consumption 2809 kilocalories

AUSTRALIA
Australasia & Oceania

Official name Commonwealth of Australia
Formation 1901 / 1901
Capital Canberra
Population 23.6 million / 8 people per sq mile (3 people per sq km)
Total area 2,967,893 sq. miles (7,686,850 sq. km)
Languages English*, Italian, Cantonese, Greek, Arabic, Vietnamese, Aboriginal languages
Religions Roman Catholic 26%, Nonreligious 19%, Anglican 19%, Other 17%, Other Christian 13%, United Church 6%
Ethnic mix European origin 50%, Australian 25.5%, other 19%, Asian 5%, Aboriginal 0.5%
Government Parliamentary system
Currency Australian dollar = 100 cents
Literacy rate 99%
Calorie consumption 3265 kilocalories

AUSTRIA
Central Europe

Official name Republic of Austria
Formation 1918 / 1919
Capital Vienna
Population 8.5 million / 266 people per sq mile (103 people per sq km)
Total area 32,378 sq. miles (83,858 sq. km)
Languages German*, Croatian, Slovenian, Hungarian (Magyar)
Religions Roman Catholic 78%, Nonreligious 9%, Other (including Jewish and Muslim) 8%, Protestant 5%
Ethnic mix Austrian 93%, Croat, Slovene, and Hungarian 6%, Other 1%
Government Parliamentary system
Currency Euro = 100 cents
Literacy rate 99%
Calorie consumption 3784 kilocalories

AZERBAIJAN
Southwest Asia

Official name Republic of Azerbaijan
Formation 1991 / 1991
Capital Baku
Population 9.5 million / 284 people per sq mile (110 people per sq km)
Total area 33,436 sq. miles (86,600 sq. km)
Languages Azeri*, Russian
Religions Shi'a Muslim 68%, Sunni Muslim 26%, Russian Orthodox 3%, Armenian Apostolic Church (Orthodox) 2%, Other 1%
Ethnic mix Azeri 91%, Other 3%, Lazs 2%, Armenian 2%, Russian 2%
Government Presidential system
Currency New manat = 100 gopik
Literacy rate 99%
Calorie consumption 2952 kilocalories

THE BAHAMAS
West Indies

Official name Commonwealth of The Bahamas
Formation 1973 / 1973
Capital Nassau
Population 400,000 / 103 people per sq mile (40 people per sq km)
Total area 5382 sq. miles (13,940 sq. km)
Languages English*, English Creole, French Creole
Religions Baptist 32%, Anglican 20%, Roman Catholic 19%, Other 17%, Methodist 6%, Church of God 6%
Ethnic mix Black African 85%, European 12%, Asian and Hispanic 3%
Government Parliamentary system
Currency Bahamian dollar = 100 cents
Literacy rate 96%
Calorie consumption 2575 kilocalories

BAHRAIN
Southwest Asia

Official name Kingdom of Bahrain
Formation 1971 / 1971
Capital Manama
Population 1.3 million / 4762 people per sq mile (1841 people per sq km)
Total area 239 sq. miles (620 sq. km)
Languages Arabic*
Religions Muslim (mainly Shi'a) 99%, Other 1%
Ethnic mix Bahraini 63%, Asian 19%, Other Arab 10%, Iranian 8%
Government Mixed monarchical–parliamentary system
Currency Bahraini dinar = 1000 fils
Literacy rate 95%
Calorie consumption Not available

BANGLADESH
South Asia

Official name People's Republic of Bangladesh
Formation 1971 / 1971
Capital Dhaka
Population 159 million / 3066 people per sq mile (1184 people per sq km)
Total area 55,598 sq. miles (144,000 sq. km)
Languages Bengali*, Urdu, Chakma, Marma (Magh), Garo, Khasi, Santhali, Tripuri, Mro
Religions Muslim (mainly Sunni) 88%, Hindu 11%, Other 1%
Ethnic mix Bengali 98%, Other 2%
Government Parliamentary system
Currency Taka = 100 poisha
Literacy rate 59%
Calorie consumption 2450 kilocalories

BARBADOS
West Indies

Official name Barbados
Formation 1966 / 1966
Capital Bridgetown
Population 300,000 / 1807 people per sq mile (698 people per sq km)
Total area 166 sq. miles (430 sq. km)
Languages Bajan (Barbadian English), English*
Religions Anglican 40%, Other 24%, Nonreligious 17%, Pentecostal 8%, Methodist 7%, Roman Catholic 4%
Ethnic mix Black African 92%, White 3%, Other 3%, Mixed race 2%
Government Parliamentary system
Currency Barbados dollar = 100 cents
Literacy rate 99%
Calorie consumption 3047 kilocalories

BELARUS
Eastern Europe

Official name Republic of Belarus
Formation 1991 / 1991
Capital Minsk
Population 9.3 million / 116 people per sq mile (45 people per sq km)
Total area 80,154 sq. miles (207,600 sq. km)
Languages Belarussian*, Russian*
Religions Orthodox Christian 80%, Roman Catholic 14%, Other 4%, Protestant 2%
Ethnic mix Belarussian 81%, Russian 11%, Polish 4%, Ukrainian 2%, Other 2%
Government Presidential system
Currency Belarussian rouble = 100 kopeks
Literacy rate 99%
Calorie consumption 3253 kilocalories

BELGIUM
Northwest Europe

Official name Kingdom of Belgium
Formation 1830 / 1919
Capital Brussels
Population 11.1 million / 876 people per sq mile (338 people per sq km)
Total area 11,780 sq. miles (30,510 sq. km)
Languages Dutch*, French*, German*
Religions Roman Catholic 88%, Other 10%, Muslim 2%
Ethnic mix Fleming 58%, Walloon 33%, Other 6%, Italian 2%, Moroccan 1%
Government Parliamentary system
Currency Euro = 100 cents
Literacy rate 99%
Calorie consumption 3793 kilocalories

BELIZE
Central America

Official name Belize
Formation 1981 / 1981
Capital Belmopan
Population 300,000 / 34 people per sq mile (13 people per sq km)
Total area 8867 sq. miles (22,966 sq. km)
Languages English Creole, Spanish, English*, Mayan, Garifuna (Carib)
Religions Roman Catholic 62%, Other 13%, Anglican 12%, Methodist 6%, Mennonite 4%, Seventh-day Adventist 3%
Ethnic mix Mestizo 49%, Creole 25%, Maya 11%, Garifuna 6%, Other 6%, Asian Indian 3%
Government Parliamentary system
Currency Belizean dollar = 100 cents
Literacy rate 75%
Calorie consumption 2751 kilocalories

BENIN
West Africa

Official name Republic of Benin
Formation 1960 / 1960
Capital Porto-Novo
Population 10.6 million / 248 people per sq mile (96 people per sq km)
Total area 43,483 sq. miles (112,620 sq. km)
Languages Fon, Bariba, Yoruba, Adja, Houeda, Somba, French*
Religions Indigenous beliefs and Voodoo 50%, Christian 30%, Muslim 20%
Ethnic mix Fon 41%, Other 21%, Adja 16%, Yoruba 12%, Bariba 10%
Government Presidential system
Currency CFA franc = 100 centimes
Literacy rate 29%
Calorie consumption 2594 kilocalories

BHUTAN
South Asia

Official name Kingdom of Bhutan
Formation 1656 / 1865
Capital Thimphu
Population 800,000 / 44 people per sq mile (17 people per sq km)
Total area 18,147 sq. miles (47,000 sq. km)
Languages Dzongkha*, Nepali, Assamese
Religions Mahayana Buddhist 75%, Hindu 25%
Ethnic mix Drukpa 50%, Nepalese 35%, Other 15%
Government Mixed monarchical–parliamentary system
Currency Ngultrum = 100 chetrum
Literacy rate 53%
Calorie consumption Not available

BOLIVIA
South America

Official name Plurinational State of Bolivia
Formation 1825 / 1938
Capital La Paz (administrative); Sucre (judicial)
Population 10.8 million / 26 people per sq mile (10 people per sq km)
Total area 424,162 sq. miles (1,098,580 sq. km)
Languages Aymara*, Quechua*, Spanish*
Religions Roman Catholic 93%, Other 7%
Ethnic mix Quechua 37%, Aymara 32%, Mixed race 13%, European 10%, Other 8%
Government Presidential system
Currency Boliviano = 100 centavos
Literacy rate 94%
Calorie consumption 2254 kilocalories

BOSNIA & HERZEGOVINA
Southeast Europe

Official name Bosnia and Herzegovina
Formation 1992 / 1992
Capital Sarajevo
Population 3.8 million / 192 people per sq mile (74 people per sq km)
Total area 19,741 sq. miles (51,129 sq. km)
Languages Bosnian*, Serbian*, Croatian*
Religions Muslim (mainly Sunni) 40%, Orthodox Christian 31%, Roman Catholic 15%, Other 10%, Protestant 4%
Ethnic mix Bosniak 48%, Serb 34%, Croat 16%, Other 2%
Government Parliamentary system
Currency Marka = 100 pfeninga
Literacy rate 98%
Calorie consumption 3130 kilocalories

BOTSWANA
Southern Africa

Official name Republic of Botswana
Formation 1966 / 1966
Capital Gaborone
Population 2 million / 9 people per sq mile (4 people per sq km)
Total area 231,803 sq. miles (600,370 sq. km)
Languages Setswana, English*, Shona, San, Khoikhoi, isiNdebele
Religions Christian (mainly Protestant) 70%, Nonreligious 20%, Traditional beliefs 6%, Other (including Muslim) 4%
Ethnic mix Tswana 79%, Kalanga 11%, Other 10%
Government Presidential system
Currency Pula = 100 thebe
Literacy rate 87%
Calorie consumption 2285 kilocalories

BRAZIL
South America

Official name Federative Republic of Brazil
Formation 1822 / 1828
Capital Brasília
Population 202 million / 62 people per sq mile (24 people per sq km)
Total area 3,286,470 sq. miles (8,511,965 sq. km)
Languages Portuguese*, German, Italian, Spanish, Polish, Japanese, Amerindian languages
Religions Roman Catholic 74%, Protestant 15%, Atheist 7%, Other 3%, Afro-American Spiritist 1%
Ethnic mix White 54%, Mixed race 38%, Black 6%, Other 2%
Government Presidential system
Currency Real = 100 centavos
Literacy rate 91%
Calorie consumption 3263 kilocalories

BRUNEI
Southeast Asia

Official name Brunei Darussalam
Formation 1984 / 1984
Capital Bandar Seri Begawan
Population 400,000 / 197 people per sq mile (76 people per sq km)
Total area 2228 sq. miles (5770 sq. km)
Languages Malay*, English, Chinese
Religions Muslim (mainly Sunni) 66%, Buddhist 14%, Other 10%, Christian 10%
Ethnic mix Malay 67%, Chinese 16%, Other 11%, Indigenous 6%
Government Monarchy
Currency Brunei dollar = 100 cents
Literacy rate 95%
Calorie consumption 2949 kilocalories

BULGARIA
Southeast Europe

Official name Republic of Bulgaria
Formation 1908 / 1947
Capital Sofia
Population 7.2 million / 169 people per sq mile (65 people per sq km)
Total area 42,822 sq. miles (110,910 sq. km)
Languages Bulgarian*, Turkish, Romani
Religions Bulgarian Orthodox 83%, Muslim 12%, Other 4%, Roman Catholic 1%
Ethnic mix Bulgarian 84%, Turkish 9%, Roma 5%, Other 2%
Government Parliamentary system
Currency Lev = 100 stotinki
Literacy rate 98%
Calorie consumption 2877 kilocalories

BURKINA FASO
West Africa

Official name Burkina Faso
Formation 1960 / 1960
Capital Ouagadougou
Population 17.4 million / 165 people per sq mile (64 people per sq km)
Total area 105,869 sq. miles (274,200 sq. km)
Languages Mossi, Fulani, French*, Tuare g, Dyula, Songhai
Religions Muslim 55%, Christian 25%, Traditional beliefs 20%
Ethnic mix Mossi 48%, Other 21%, Peul 10%, Lobi 7%, Bobo 7%, Mandé 7%
Government Transitional regime
Currency CFA franc = 100 centimes
Literacy rate 29%
Calorie consumption 2720 kilocalories

BURUNDI
Central Africa

Official name Republic of Burundi
Formation 1962 / 1962
Capital Bujumbura
Population 10.5 million / 1060 people per sq mile (409 people per sq km)
Total area 10,745 sq. miles (27,830 sq. km)
Languages Kirundi*, French*, Kiswahili
Religions Roman Catholic 62%, Traditional beliefs 23%, Muslim 10%, Protestant 5%
Ethnic mix Hutu 85%, Tutsi 14%, Twa 1%
Government Presidential system
Currency Burundian franc = 100 centimes
Literacy rate 87%
Calorie consumption 1604 kilocalories

CAMBODIA
Southeast Asia

Official name Kingdom of Cambodia
Formation 1953 / 1953
Capital Phnom Penh
Population 15.4 million / 226 people per sq mile (87 people per sq km)
Total area 69,900 sq. miles (181,040 sq. km)
Languages Khmer*, French, Chinese, Vietnamese, Cham
Religions Buddhist 93%, Muslim 6%, Christian 1%
Ethnic mix Khmer 90%, Vietnamese 5%, Other 4%, Chinese 1%
Government Parliamentary system
Currency Riel = 100 sen
Literacy rate 74%
Calorie consumption 2411 kilocalories

CAMEROON
Central Africa

Official name Republic of Cameroon
Formation 1960 / 1961
Capital Yaoundé
Population 22.8 million / 127 people per sq mile (49 people per sq km)
Total area 183,567 sq. miles (475,400 sq. km)
Languages Bamileke, Fang, Fulani, French*, English*
Religions Roman Catholic 35%, Traditional beliefs 25%, Muslim 22%, Protestant 18%
Ethnic mix Cameroon highlanders 31%, Other 21%, Equatorial Bantu 19%, Kirdi 11%, Fulani 10%, Northwestern Bantu 8%
Government Presidential system
Currency CFA franc = 100 centimes
Literacy rate 71%
Calorie consumption 2586 kilocalories

CANADA
North America

Official name Canada
Formation 1867 / 1949
Capital Ottawa
Population 35.5 million / 10 people per sq mile (4 people per sq km)
Total area 3,855,171 sq. miles (9,984,670 sq. km)
Languages English*, French*, Chinese, Italian, German, Ukrainian, Portuguese, Inuktitut, Cree
Religions Roman Catholic 44%, Protestant 29%, Other and nonreligious 27%
Ethnic mix European origin 66%, other 27%, Asian 5%, Amerindian 2%
Government Parliamentary system
Currency Canadian dollar = 100 cents
Literacy rate 99%
Calorie consumption 3419 kilocalories

CAPE VERDE
Atlantic Ocean

Official name Republic of Cape Verde
Formation 1975 / 1975
Capital Praia
Population 500,000 / 321 people per sq mile (124 people per sq km)
Total area 1557 sq. miles (4033 sq. km)
Languages Portuguese Creole, Portuguese*
Religions Roman Catholic 97%, Other 2%, Protestant (Church of the Nazarene) 1%
Ethnic mix Mestiço 71%, African 28%, European 1%
Government Mixed presidential–parliamentary system
Currency Escudo = 100 centavos
Literacy rate 85%
Calorie consumption 2716 kilocalories

CENTRAL AFRICAN REPUBLIC
Central Africa

Official name Central African Republic
Formation 1960 / 1960
Capital Bangui
Population 4.7 million / 20 people per sq mile (8 people per sq km)
Total area 240,534 sq. miles (622,984 sq. km)
Languages Sango, Banda, Gbaya, French*
Religions Traditional beliefs 35%, Roman Catholic 25%, Protestant 25%, Muslim 15%
Ethnic mix Baya 33%, Banda 27%, Other 17%, Mandjia 13%, Sara 10%
Government Transitional regime
Currency CFA franc = 100 centimes
Literacy rate 37%
Calorie consumption 2154 kilocalories

CHAD
Central Africa

Official name Republic of Chad
Formation 1960 / 1960
Capital N'Djaména
Population 13.2 million / 27 people per sq mile (10 people per sq km)
Total area 495,752 sq. miles (1,284,000 sq. km)
Languages French*, Sara, Arabic*, Maba
Religions Muslim 51%, Christian 35%, Animist 7%, Traditional beliefs 7%
Ethnic mix Other 30%, Sara 28%, Mayo-Kebbi 12%, Arab 12%, Ouaddaï 9%, Kanem-Bornou 9%
Government Presidential system
Currency CFA franc = 100 centimes
Literacy rate 37%
Calorie consumption 2110 kilocalories

CHILE
South America

Official name Republic of Chile
Formation 1818 / 1883
Capital Santiago
Population 17.8 million / 62 people per sq mile (24 people per sq km)
Total area 292,258 sq. miles (756,950 sq. km)
Languages Spanish*, Amerindian languages
Religions Roman Catholic 89%, Other and nonreligious 11%
Ethnic mix Mestizo and European 90%, Other Amerindian 9%, Mapuche 1%
Government Presidential system
Currency Chilean peso = 100 centavos
Literacy rate 99%
Calorie consumption 2989 kilocalories

CHINA
East Asia

Official name People's Republic of China
Formation 960 / 1999
Capital Beijing
Population 1.39 billion / 387 people per sq mile (149 people per sq km)
Total area 3,705,386 sq. miles (9,596,960 sq. km)
Languages Mandarin*, Wu, Cantonese, Hsiang, Min, Hakka, Kan
Religions Nonreligious 59%, Traditional beliefs 20%, Other 13%, Buddhist 6%, Muslim 2%
Ethnic mix Han 92%, Other 4%, Hui 1%, Miao 1%, Manchu 1%, Zhuang 1%
Government One-party state
Currency Renminbi (known as yuan) = 10 jiao = 100 fen
Literacy rate 95%
Calorie consumption 3108 kilocalories

COLOMBIA
South America

Official name Republic of Colombia
Formation 1819 / 1903
Capital Bogotá
Population 48.9 million / 122 people per sq mile (47 people per sq km)
Total area 439,733 sq. miles (1,138,910 sq. km)
Languages Spanish*, Wayuu, Páez, and other Amerindian languages
Religions Roman Catholic 95%, Other 5%
Ethnic mix Mestizo 58%, White 20%, European–African 14%, African 4%, African–Amerindian 3%, Amerindian 1%
Government Presidential system
Currency Colombian peso = 100 centavos
Literacy rate 94%
Calorie consumption 2804 kilocalories

COMOROS
Indian Ocean

Official name Union of the Comoros
Formation 1975 / 1975
Capital Moroni
Population 800,000 / 929 people per sq mile (359 people per sq km)
Total area 838 sq. miles (2170 sq. km)
Languages Arabic*, Comoran*, French*
Religions Muslim (mainly Sunni) 98%, Other 1%, Roman Catholic 1%
Ethnic mix Comoran 97%, Other 3%
Government Presidential system
Currency Comoros franc = 100 centimes
Literacy rate 76%
Calorie consumption 2139 kilocalories

CONGO
Central Africa

Official name Republic of the Congo
Formation 1960 / 1960
Capital Brazzaville
Population 4.6 million / 35 people per sq mile (13 people per sq km)
Total area 132,046 sq. miles (342,000 sq. km)
Languages Kongo, Teke, Lingala, French*
Religions Traditional beliefs 50%, Roman Catholic 35%, Protestant 13%, Muslim 2%
Ethnic mix Bakongo 51%, Teke 17%, Other 16%, Mbochi 11%, Mbédé 5%
Government Presidential system
Currency CFA franc = 100 centimes
Literacy rate 79%
Calorie consumption 2195 kilocalories

CONGO, DEM. REP.
Central Africa

Official name Democratic Republic of the Congo
Formation 1960 / 1960
Capital Kinshasa
Population 69.4 million / 79 people per sq mile (31 people per sq km)
Total area 905,563 sq. miles (2,345,410 sq. km)
Languages Kiswahili, Tshiluba, Kikongo, Lingala, French*
Religions Roman Catholic 50%, Protestant 20%, Traditional beliefs and other 10%, Muslim 10%, Kimbanguist 10%
Ethnic mix Other 55%, Mongo, Luba, Kongo, and Mangbetu-Azande 45%
Government Presidential system
Currency Congolese franc = 100 centimes
Literacy rate 61%
Calorie consumption 1585 kilocalories

COSTA RICA
Central America

Official name Republic of Costa Rica
Formation 1838 / 1838
Capital San José
Population 4.9 million / 249 people per sq mile (96 people per sq km)
Total area 19,730 sq. miles (51,100 sq. km)
Languages Spanish*, English Creole, Bribri, Cabecar
Religions Roman Catholic 71%, Evangelical 14%, Nonreligious 11%, Other 4%
Ethnic mix Mestizo and European 94%, Black 3%, Other 1%, Chinese 1%, Amerindian 1%
Government Presidential system
Currency Costa Rican colón = 100 céntimos
Literacy rate 97%
Calorie consumption 2898 kilocalories

CROATIA
Southeast Europe

Official name Republic of Croatia
Formation 1991 / 1991
Capital Zagreb
Population 4.3 million / 197 people per sq mile (76 people per sq km)
Total area 21,831 sq. miles (56,542 sq. km)
Languages Croatian*
Religions Roman Catholic 88%, Other 7%, Orthodox Christian 4%, Muslim 1%
Ethnic mix Croat 90%, Other 5%, Serb 5%
Government Parliamentary system
Currency Kuna = 100 lipa
Literacy rate 99%
Calorie consumption 3052 kilocalories

CUBA
West Indies

Official name Republic of Cuba
Formation 1902 / 1902
Capital Havana
Population 11.3 million / 264 people per sq mile (102 people per sq km)
Total area 42,803 sq. miles (110,860 sq. km)
Languages Spanish*
Religions Nonreligious 49%, Roman Catholic 40%, Atheist 6%, Other 4%, Protestant 1%
Ethnic mix Mulatto (mixed race) 51%, White 37%, Black 11%, Chinese 1%
Government One-party state
Currency Cuban peso = 100 centavos
Literacy rate 99%
Calorie consumption 3277 kilocalories

CYPRUS
Southeast Europe

Official name Republic of Cyprus
Formation 1960 / 1960
Capital Nicosia
Population 1.2 million / 336 people per sq mile (130 people per sq km)
Total area 3571 sq. miles (9250 sq. km)
Languages Greek*, Turkish*
Religions Orthodox Christian 78%, Muslim 18%, Other 4%
Ethnic mix Greek 81%, Turkish 11%, Other 8%
Government Presidential system
Currency Euro = 100 cents; (TRNC: new Turkish lira = 100 kurus)
Literacy rate 99%
Calorie consumption 2661 kilocalories

CZECH REPUBLIC
Central Europe

Official name Czech Republic
Formation 1993 / 1993
Capital Prague
Population 10.7 million / 351 people per sq mile (136 people per sq km)
Total area 30,450 sq. miles (78,866 sq. km)
Languages Czech*, Slovak, Hungarian (Magyar)
Religions Roman Catholic 39%, Atheist 38%, Other 18%, Protestant 3%, Hussite 2%
Ethnic mix Czech 90%, Moravian 4%, Other 4%, Slovak 2%
Government Parliamentary system
Currency Czech koruna = 100 haleru
Literacy rate 99%
Calorie consumption 3292 kilocalories

DENMARK
Northern Europe

Official name Kingdom of Denmark
Formation 950 / 1944
Capital Copenhagen
Population 5.6 million / 342 people per sq mile (132 people per sq km)
Total area 16,639 sq. miles (43,094 sq. km)
Languages Danish*
Religions Evangelical Lutheran 95%, Roman Catholic 3%, Muslim 2%
Ethnic mix Danish 96%, Other (including Scandinavian and Turkish) 3%, Faeroese and Inuit 1%
Government Parliamentary system
Currency Danish krone = 100 øre
Literacy rate 99%
Calorie consumption 3363 kilocalories

DJIBOUTI
East Africa

Official name Republic of Djibouti
Formation 1977 / 1977
Capital Djibouti
Population 900,000 / 101 people per sq mile (39 people per sq km)
Total area 8494 sq. miles (22,000 sq. km)
Languages Somali, Afar, French*, Arabic*
Religions Muslim (mainly Sunni) 94%, Christian 6%
Ethnic mix Issa 60%, Afar 35%, Other 5%
Government Presidential system
Currency Djibouti franc = 100 centimes
Literacy rate 70%
Calorie consumption 2526 kilocalories

DOMINICA
West Indies

Official name Commonwealth of Dominica
Formation 1978 / 1978
Capital Roseau
Population 73,449 / 253 people per sq mile (98 people per sq km)
Total area 291 sq. miles (754 sq. km)
Languages French Creole, English*
Religions Roman Catholic 77%, Protestant 15%, Other 8%
Ethnic mix Black 87%, Mixed race 9%, Carib 3%, Other 1%
Government Parliamentary system
Currency East Caribbean dollar = 100 cents
Literacy rate 88%
Calorie consumption 3047 kilocalories

DOMINICAN REPUBLIC
West Indies

Official name Dominican Republic
Formation 1865 / 1865
Capital Santo Domingo
Population 10.5 million / 562people per sq mile (217 people per sq km)
Total area 18,679 sq. miles (48,380 sq. km)
Languages Spanish*, French Creole
Religions Roman Catholic 95%, Other and nonreligious 5%
Ethnic mix Mixed race 73%, European 16%, Black African 11%
Government Presidential system
Currency Dominican Republic peso = 100 centavos
Literacy rate 91%
Calorie consumption 2614 kilocalories

EAST TIMOR
Southeast Asia

Official name Democratic Republic of Timor-Leste
Formation 2002 / 2002
Capital Dili
Population 1.2 million / 213 people per sq mile (82 people per sq km)
Total area 5756 sq. miles (14,874 sq. km)
Languages Tetum (Portuguese/Austronesian)*, Bahasa Indonesia, Portuguese*
Religions Roman Catholic 95%, Other (including Muslim and Protestant) 5%
Ethnic mix Papuan groups approx 85%, Indonesian approx 13%, Chinese 2%
Government Parliamentary system
Currency US dollar = 100 cents
Literacy rate 58%
Calorie consumption 2083 kilocalories

ECUADOR
South America

Official name Republic of Ecuador
Formation 1830 / 1942
Capital Quito
Population 16 million / 150 people per sq mile (58 people per sq km)
Total area 109,483 sq. miles (283,560 sq. km)
Languages Spanish*, Quechua, other Amerindian languages
Religions Roman Catholic 95%, Protestant, Jewish, and other 5%
Ethnic mix Mestizo 77%, White 11%, Amerindian 7%, Black African 5%
Government Presidential system
Currency US dollar = 100 cents
Literacy rate 93%
Calorie consumption 2477 kilocalories

EGYPT
North Africa

Official name Arab Republic of Egypt
Formation 1936 / 1982
Capital Cairo
Population 83.4 million / 217 people per sq mile (84 people per sq km)
Total area 386,660 sq. miles (1,001,450 sq. km)
Languages Arabic*, French, English, Berber
Religions Muslim (mainly Sunni) 90%, Coptic Christian and other 9%, Other Christian 1%
Ethnic mix Egyptian 99%, Nubian, Armenian, Greek, and Berber 1%
Government Transitional regime
Currency Egyptian pound = 100 piastres
Literacy rate 74%
Calorie consumption 3557 kilocalories

EL SALVADOR
Central America

Official name Republic of El Salvador
Formation 1841 / 1841
Capital San Salvador
Population 6.4 million / 800 people per sq mile (309 people per sq km)
Total area 8124 sq. miles (21,040 sq. km)
Languages Spanish*
Religions Roman Catholic 80%, Evangelical 18%, Other 2%
Ethnic mix Mestizo 90%, White 9%, Amerindian 1%
Government Presidential system
Currency Salvadorean colón = 100 centavos; and US dollar = 100 cents
Literacy rate 86%
Calorie consumption 2513 kilocalories

EQUATORIAL GUINEA
Central Africa

Official name Republic of Equatorial Guinea
Formation 1968 / 1968
Capital Malabo
Population 800,000 / 74 people per sq mile (29 people per sq km)
Total area 10,830 sq. miles (28,051 sq. km)
Languages Spanish*, Fang, Bubi, French*
Religions Roman Catholic 90%, Other 10%
Ethnic mix Fang 85%, Other 11%, Bubi 4%
Government Presidential system
Currency CFA franc = 100 centimes
Literacy rate 94%
Calorie consumption Not available

ERITREA
East Africa

Official name State of Eritrea
Formation 1993 / 2002
Capital Asmara
Population 6.5 million / 143 people per sq mile (55 people per sq km)
Total area 46,842 sq. miles (121,320 sq. km)
Languages Tigrinya*, English*, Tigre, Afar, Arabic*, Saho, Bilen, Kunama, Nara, Hadareb
Religions Christian 50%, Muslim 48%, Other 2%
Ethnic mix Tigray 50%, Tigre 31%, Other 9%, Afar 5%, Saho 5%
Government Mixed presidential–parliamentary system
Currency Nakfa = 100 cents
Literacy rate 70%
Calorie consumption 1640 kilocalories

ESTONIA
Northeast Europe

Official name Republic of Estonia
Formation 1991 / 1991
Capital Tallinn
Population 1.3 million / 75 people per sq mile (29 people per sq km)
Total area 17,462 sq. miles (45,226 sq. km)
Languages Estonian*, Russian
Religions Evangelical Lutheran 56%, Orthodox Christian 25%, Other 19%
Ethnic mix Estonian 69%, Russian 25%, Other 4%, Ukrainian 2%
Government Parliamentary system
Currency Euro = 100 cents
Literacy rate 99%
Calorie consumption 3214 kilocalories

ETHIOPIA
East Africa

Official name Federal Democratic Republic of Ethiopia
Formation 1896 / 2002
Capital Addis Ababa
Population 96.5 million / 225 people per sq mile (87 people per sq km)
Total area 435,184 sq. miles (1,127,127 sq. km)
Languages Amharic*, Tigrinya, Galla, Sidamo, Somali, English, Arabic
Religions Orthodox Christian 40%, Muslim 40%, Traditional beliefs 15%, Other 5%
Ethnic mix Oromo 40%, Amhara 25%, Other 13%, Sidama 9%, Tigray 7%, Somali 6%
Government Parliamentary system
Currency Birr = 100 cents
Literacy rate 39%
Calorie consumption 2131 kilocalories

FIJI
Australasia & Oceania

Official name Republic of Fiji
Formation 1970 / 1970
Capital Suva
Population 900,000 / 128 people per sq mile (49 people per sq km)
Total area 7054 sq. miles (18,270 sq. km)
Languages Fijian, English*, Hindi, Urdu, Tamil, Telugu
Religions Hindu 38%, Methodist 37%, Roman Catholic 9%, Muslim 8%, Other 8%
Ethnic mix Melanesian 51%, Indian 44%, Other 5%
Government Parliamentary system
Currency Fiji dollar = 100 cents
Literacy rate 94%
Calorie consumption 2930 kilocalories

FINLAND
Northern Europe

Official name Republic of Finland
Formation 1917 / 1947
Capital Helsinki
Population 5.4 million / 46 people per sq mile (18 people per sq km)
Total area 130,127 sq. miles (337,030 sq. km)
Languages Finnish*, Swedish*, Sámi
Religions Evangelical Lutheran 83%, Other 15%, Orthodox Christian 1%, Roman Catholic 1%
Ethnic mix Finnish 93%, Other (including Sámi) 7%
Government Parliamentary system
Currency Euro = 100 cents
Literacy rate 99%
Calorie consumption 3285 kilocalories

FRANCE
Western Europe

Official name French Republic
Formation 987 / 1919
Capital Paris
Population 64.6 million / 304 people per sq mile (117 people per sq km)
Total area 211,208 sq. miles (547,030 sq. km)
Languages French*, Provençal, German, Breton, Catalan, Basque
Religions Roman Catholic 88%, Muslim 8%, Protestant 2%, Buddhist 1%, Jewish 1%
Ethnic mix French 90%, North African (mainly Algerian) 6%, German (Alsace) 2%, Breton 1%, Other (including Corsicans) 1%
Government Mixed presidential–parliamentary system
Currency Euro = 100 cents
Literacy rate 99%
Calorie consumption 3524 kilocalories

GABON
Central Africa

Official name Gabonese Republic
Formation 1960 / 1960
Capital Libreville
Population 1.7 million / 17 people per sq mile (7 people per sq km)
Total area 103,346 sq. miles (267,667 sq. km)
Languages Fang, French*, Punu, Sira, Nzebi, Mpongwe
Religions Christian (mainly Roman Catholic) 55%, Traditional beliefs 40%, Other 4%, Muslim 1%
Ethnic mix Fang 26%, Shira-punu 24%, Other 16%, Foreign residents 15%, Nzabi-duma 11%, Mbédé-Teke 8%
Government Presidential system
Currency CFA franc = 100 centimes
Literacy rate 82%
Calorie consumption 2781 kilocalories

GAMBIA
West Africa

Official name Republic of the Gambia
Formation 1965 / 1965
Capital Banjul
Population 1.9 million / 492 people per sq mile (190 people per sq km)
Total area 4363 sq. miles (11,300 sq. km)
Languages Mandinka, Fulani, Wolof, Jola, Soninke, English*
Religions Sunni Muslim 90%, Christian 8%, Traditional beliefs 2%
Ethnic mix Mandinka 42%, Fulani 18%, Wolof 16%, Jola 10%, Serahuli 9%, Other 5%
Government Presidential system
Currency Dalasi = 100 butut
Literacy rate 52%
Calorie consumption 2849 kilocalories

GEORGIA
Southwest Asia

Official name Georgia
Formation 1991 / 1991
Capital Tbilisi
Population 4.3 million / 160 people per sq mile (62 people per sq km)
Total area 26,911 sq. miles (69,700 sq. km)
Languages Georgian*, Russian, Azeri, Armenian, Mingrelian, Ossetian, Abkhazian* (in Abkhazia)
Religions Georgian Orthodox 74%, Muslim 10%, Russian Orthodox 10%, Armenian Apostolic Church (Orthodox) 4%, Other 2%
Ethnic mix Georgian 84%, Azeri 6%, Armenian 6%, Russian 2%, Ossetian 1%, Other 1%
Government Presidential system
Currency Lari = 100 tetri
Literacy rate 99%
Calorie consumption 2731 kilocalories

GERMANY
Northern Europe

Official name Federal Republic of Germany
Formation 1871 / 1990
Capital Berlin
Population 82.7 million / 613 people per sq mile (237 people per sq km)
Total area 137,846 sq. miles (357,021 sq. km)
Languages German*, Turkish
Religions Protestant 34%, Roman Catholic 33%, Other 30%, Muslim 3%
Ethnic mix German 92%, Other European 3%, Other 3%, Turkish 2%
Government Parliamentary system
Currency Euro = 100 cents
Literacy rate 99%
Calorie consumption 3539 kilocalories

GHANA
West Africa

Official name Republic of Ghana
Formation 1957 / 1957
Capital Accra
Population 26.4 million / 297 people per sq mile (115 people per sq km)
Total area 92,100 sq. miles (238,540 sq. km)
Languages Twi, Fanti, Ewe, Ga, Adangbe, Gurma, Dagomba (Dagbani), English*
Religions Christian 69%, Muslim 16%, Traditional beliefs 9%, Other 6%
Ethnic mix Akan 49%, Mole-Dagbani 17%, Ewe 13%, Other 9%, Ga and Ga-Adangbe 8%, Guan 4%
Government Presidential system
Currency Cedi = 100 pesewas
Literacy rate 72%
Calorie consumption 3003 kilocalories

GREECE
Southeast Europe

Official name Hellenic Republic
Formation 1829 / 1947
Capital Athens
Population 11.1 million / 220 people per sq mile (85 people per sq km)
Total area 50,942 sq. miles (131,940 sq. km)
Languages Greek*, Turkish, Macedonian, Albanian
Religions Orthodox Christian 98%, Muslim 1%, Other 1%
Ethnic mix Greek 98%, Other 2%
Government Parliamentary system
Currency Euro = 100 cents
Literacy rate 97%
Calorie consumption 3433 kilocalories

GRENADA
West Indies

Official name Grenada
Formation 1974 / 1974
Capital St. George's
Population 110,152 / 841 people per sq mile (324 people per sq km)
Total area 131 sq. miles (340 sq. km)
Languages English*, English Creole
Religions Roman Catholic 68%, Anglican 17%, Other 15%
Ethnic mix Black African 82%, Mulatto (mixed race) 13%, East Indian 3%, Other 2%
Government Parliamentary system
Currency East Caribbean dollar = 100 cents
Literacy rate 96%
Calorie consumption 2453 kilocalories

GUATEMALA
Central America

Official name Republic of Guatemala
Formation 1838 / 1838
Capital Guatemala City
Population 15.9 million / 380 people per sq mile (147 people per sq km)
Total area 42,042 sq. miles (108,890 sq. km)
Languages Quiché, Mam, Cakchiquel, Kekchí, Spanish*
Religions Roman Catholic 65%, Protestant 33%, Other and nonreligious 2%
Ethnic mix Amerindian 60%, Mestizo 30%, Other 10%
Government Presidential system
Currency Quetzal = 100 centavos
Literacy rate 78%
Calorie consumption 2419 kilocalories

GUINEA
West Africa

Official name Republic of Guinea
Formation 1958 / 1958
Capital Conakry
Population 12 million / 126 people per sq mile (49 people per sq km)
Total area 94,925 sq. miles (245,857 sq. km)
Languages Pulaar, Malinké, Soussou, French*
Religions Muslim 85%, Christian 8%, Traditional beliefs 7%
Ethnic mix Peul 40%, Malinké 30%, Soussou 20%, Other 10%
Government Presidential system
Currency Guinea franc = 100 centimes
Literacy rate 25%
Calorie consumption 2553 kilocalories

GUINEA-BISSAU
West Africa

Official name Republic of Guinea-Bissau
Formation 1974 / 1974
Capital Bissau
Population 1.7 million / 157 people per sq mile (60 people per sq km)
Total area 13,946 sq. miles (36,120 sq. km)
Languages Portuguese Creole, Balante, Fulani, Malinké, Portuguese*
Religions Traditional beliefs 50%, Muslim 40%, Christian 10%
Ethnic mix Balante 30%, Fulani 20%, Other 16%, Mandyako 14%, Mandinka 13%, Papel 7%
Government Presidential system
Currency CFA franc = 100 centimes
Literacy rate 57%
Calorie consumption 2304 kilocalories

GUYANA
South America

Official name Cooperative Republic of Guyana
Formation 1966 / 1966
Capital Georgetown
Population 800,000 / 11 people per sq mile (4 people per sq km)
Total area 83,000 sq. miles (214,970 sq. km)
Languages English Creole, Hindi, Tamil, Amerindian languages, English*
Religions Christian 57%, Hindu 28%, Muslim 10%, Other 5%
Ethnic mix East Indian 43%, Black African 30%, Mixed race 17%, Amerindian 9%, Other 1%
Government Presidential system
Currency Guyanese dollar = 100 cents
Literacy rate 85%
Calorie consumption 2648 kilocalories

HAITI
West Indies

Official name Republic of Haiti
Formation 1804 / 1844
Capital Port-au-Prince
Population 10.5 million / 987 people per sq mile (381 people per sq km)
Total area 10,714 sq. miles (27,750 sq. km)
Languages French Creole*, French*
Religions Roman Catholic 55%, Protestant 28%, Other (including Voodoo) 16%, Nonreligious 1%
Ethnic mix Black African 95%, Mulatto (mixed race) and European 5%
Government Presidential system
Currency Gourde = 100 centimes
Literacy rate 49%
Calorie consumption 2091 kilocalories

HONDURAS
Central America

Official name Republic of Honduras
Formation 1838 / 1838
Capital Tegucigalpa
Population 8.3 million / 192 people per sq mile (74 people per sq km)
Total area 43,278 sq. miles (112,090 sq. km)
Languages Spanish*, Garifuna (Carib), English Creole
Religions Roman Catholic 97%, Protestant 3%
Ethnic mix Mestizo 90%, Black African 5%, Amerindian 4%, White 1%
Government Presidential system
Currency Lempira = 100 centavos
Literacy rate 85%
Calorie consumption 2651 kilocalories

HUNGARY
Central Europe

Official name Hungary
Formation 1918 / 1947
Capital Budapest
Population 9.9 million / 278 people per sq mile (107 people per sq km)
Total area 35,919 sq. miles (93,030 sq. km)
Languages Hungarian (Magyar)*
Religions Roman Catholic 52%, Calvinist 16%, Other 15%, Nonreligious 14%, Lutheran 3%
Ethnic mix Magyar 90%, Roma 4%, German 3%, Serb 2%, Other 1%
Government Parliamentary system
Currency Forint = 100 fillér
Literacy rate 99%
Calorie consumption 2968 kilocalories

ICELAND
Northwest Europe

Official name Republic of Iceland
Formation 1944 / 1944
Capital Reykjavík
Population 300,000 / 8 people per sq mile (3 people per sq km)
Total area 39,768 sq. miles (103,000 sq. km)
Languages Icelandic*
Religions Evangelical Lutheran 84%, Other (mostly Christian) 10%, Roman Catholic 3%, Nonreligious 3%
Ethnic mix Icelandic 94%, Other 5%, Danish 1%
Government Parliamentary system
Currency Icelandic króna = 100 aurar
Literacy rate 99%
Calorie consumption 3339 kilocalories

INDIA
South Asia

Official name Republic of India
Formation 1947 / 1947
Capital New Delhi
Population 1.27 billion / 1104 people per sq mile (426 people per sq km)
Total area 1,269,339 sq. miles (3,287,590 sq. km)
Languages Hindi*, English*, Urdu, Bengali, Marathi, Telugu, Tamil, Bihari, Gujarati, Kanarese
Religions Hindu 81%, Muslim 13%, Christian 2%, Sikh 2%, Buddhist 1%, Other 1%
Ethnic mix Indo-Aryan 72%, Dravidian 25%, Mongoloid and other 3%
Government Parliamentary system
Currency Indian rupee = 100 paise
Literacy rate 63%
Calorie consumption 2459 kilocalories

INDONESIA
Southeast Asia

Official name Republic of Indonesia
Formation 1949 / 1999
Capital Jakarta
Population 253 million / 364 people per sq mile (141 people per sq km)
Total area 741,096 sq. miles (1,919,440 sq. km)
Languages Javanese, Sundanese, Madurese, Bahasa Indonesia*, Dutch
Religions Sunni Muslim 86%, Protestant 6%, Roman Catholic 3%, Hindu 2%, Other 2%, Buddhist 1%
Ethnic mix Javanese 41%, Other 29%, Sundanese 15%, Coastal Malays 12%, Madurese 3%
Government Presidential system
Currency Rupiah = 100 sen
Literacy rate 93%
Calorie consumption 2777 kilocalories

IRAN
Southwest Asia

Official name Islamic Republic of Iran
Formation 1502 / 1990
Capital Tehran
Population 78.5 million / 124 people per sq mile (48 people per sq km)
Total area 636,293 sq. miles (1,648,000 sq. km)
Languages Farsi*, Azeri, Luri, Gilaki, Mazanderani, Kurdish, Turkmen, Arabic, Baluchi
Religions Shi'a Muslim 89%, Sunni Muslim 9%, Other 2%
Ethnic mix Persian 51%, Azari 24%, Other 10%, Lur and Bakhtiari 8%, Kurdish 7%
Government Islamic theocracy
Currency Iranian rial = 100 dinars
Literacy rate 84%
Calorie consumption 3058 kilocalories

IRAQ
Southwest Asia

Official name Republic of Iraq
Formation 1932 / 1990
Capital Baghdad
Population 34.8 million / 206 people per sq mile (80 people per sq km)
Total area 168,753 sq. miles (437,072 sq. km)
Languages Arabic*, Kurdish*, Turkic languages, Armenian, Assyrian
Religions Shi'a Muslim 60%, Sunni Muslim 35%, Other (including Christian) 5%
Ethnic mix Arab 80%, Kurdish 15%, Turkmen 3%, Other 2%
Government Parliamentary system
Currency New Iraqi dinar = 1000 fils
Literacy rate 79%
Calorie consumption 2489 kilocalories

IRELAND
Northwest Europe

Official name Ireland
Formation 1922 / 1922
Capital Dublin
Population 4.7 million / 177 people per sq mile (68 people per sq km)
Total area 27,135 sq. miles (70,280 sq. km)
Languages English*, Irish*
Religions Roman Catholic 87%, Other and nonreligious 10%, Anglican 3%
Ethnic mix Irish 99%, Other 1%
Government Parliamentary system
Currency Euro = 100 cents
Literacy rate 99%
Calorie consumption 3591 kilocalories

ISRAEL
Southwest Asia

Official name State of Israel
Formation 1948 / 1994
Capital Jerusalem (not internationally recognized)
Population 7.8 million / 994 people per sq mile (384 people per sq km)
Total area 8019 sq. miles (20,770 sq. km)
Languages Hebrew*, Arabic*, Yiddish, German, Russian, Polish, Romanian, Persian
Religions Jewish 76%, Muslim (mainly Sunni) 16%, Other 4%, Druze 2%, Christian 2%
Ethnic mix Jewish 76%, Arab 20%, Other 4%
Government Parliamentary system
Currency Shekel = 100 agorot
Literacy rate 98%
Calorie consumption 3619 kilocalories

ITALY
Southern Europe

Official name Italian Republic
Formation 1861 / 1947
Capital Rome
Population 61.1 million / 538 people per sq mile (208 people per sq km)
Total area 116,305 sq. miles (301,230 sq. km)
Languages Italian*, German, French, Rhaeto-Romanic, Sardinian
Religions Roman Catholic 85%, Other and nonreligious 13%, Muslim 2%
Ethnic mix Italian 94%, Other 4%, Sardinian 2%
Government Parliamentary system
Currency Euro = 100 cents
Literacy rate 99%
Calorie consumption 3539 kilocalories

IVORY COAST
West Africa

Official name Republic of Côte d'Ivoire
Formation 1960 / 1960
Capital Yamoussoukro
Population 20.8 million / 169 people per sq mile (65 people per sq km)
Total area 124,502 sq. miles (322,460 sq. km)
Languages Akan, French*, Krou, Voltaique
Religions Muslim 38%, Traditional beliefs 25%, Roman Catholic 25%, Other 6%, Protestant 6%
Ethnic mix Akan 42%, Voltaique 18%, Mandé du Nord 17%, Krou 11%, Mandé du Sud 10%, Other 2%
Government Presidential system
Currency CFA franc = 100 centimes
Literacy rate 41%
Calorie consumption 2799 kilocalories

JAMAICA
West Indies

Official name Jamaica
Formation 1962 / 1962
Capital Kingston
Population 2.8 million / 670 people per sq mile (259 people per sq km)
Total area 4243 sq. miles (10,990 sq. km)
Languages English Creole, English*
Religions Other and nonreligious 45%, Other Protestant 20%, Church of God 18%, Baptist 10%, Anglican 7%
Ethnic mix Black 91%, Mulatto (mixed race) 7%, European and Chinese 1%, East Indian 1%
Government Parliamentary system
Currency Jamaican dollar = 100 cents
Literacy rate 88%
Calorie consumption 2746 kilocalories

JAPAN
East Asia

Official name Japan
Formation 1590 / 1972
Capital Tokyo
Population 127 million / 874 people per sq mile (337 people per sq km)
Total area 145,882 sq. miles (377,835 sq. km)
Languages Japanese*, Korean, Chinese
Religions Shinto and Buddhist 76%, Buddhist 16%, Other (including Christian) 8%
Ethnic mix Japanese 99%, Other (mainly Korean) 1%
Government Parliamentary system
Currency Yen = 100 sen
Literacy rate 99%
Calorie consumption 2719 kilocalories

JORDAN
Southwest Asia

Official name Hashemite Kingdom of Jordan
Formation 1946 / 1967
Capital Amman
Population 7.5 million / 218 people per sq mile (84 people per sq km)
Total area 35,637 sq. miles (92,300 sq. km)
Languages Arabic*
Religions Sunni Muslim 92%, Christian 6%, Other 2%
Ethnic mix Arab 98%, Circassian 1%, Armenian 1%
Government Monarchy
Currency Jordanian dinar = 1000 fils
Literacy rate 98%
Calorie consumption 3149 kilocalories

KAZAKHSTAN
Central Asia

Official name Republic of Kazakhstan
Formation 1991 / 1991
Capital Astana
Population 16.6 million / 16 people per sq mile (6 people per sq km)
Total area 1,049,150 sq. miles (2,717,300 sq. km)
Languages Kazakh*, Russian, Ukrainian, German, Uzbek, Tatar, Uighur
Religions Muslim (mainly Sunni) 47%, Orthodox Christian 44%, Other 7%, Protestant 2%
Ethnic mix Kazakh 57%, Russian 27%, Other 8%, Uzbek 3%, Ukrainian 3%, German 2%
Government Presidential system
Currency Tenge = 100 tiyn
Literacy rate 99%
Calorie consumption 3107 kilocalories

KENYA
East Africa

Official name Republic of Kenya
Formation 1963 / 1963
Capital Nairobi
Population 45.5 million / 208 people per sq mile (80 people per sq km)
Total area 224,961 sq. miles (582,650 sq. km)
Languages Kiswahili*, English*, Kikuyu, Luo, Kalenjin, Kamba
Religions Christian 80%, Muslim 10%, Traditional beliefs 9%, Other 1%
Ethnic mix Other 28%, Kikuyu 22%, Luo 14%, Luhya 14%, Kalenjin 11%, Kamba 11%
Government Presidential system
Currency Kenya shilling = 100 cents
Literacy rate 72%
Calorie consumption 2206 kilocalories

KIRIBATI
Australasia & Oceania

Official name Republic of Kiribati
Formation 1979 / 1979
Capital Tarawa Atoll
Population 104,488 / 381 people per sq mile (147 people per sq km)
Total area 277 sq. miles (717 sq. km)
Languages English*, Kiribati
Religions Roman Catholic 55%, Kiribati Protestant Church 36%, Other 9%
Ethnic mix Micronesian 99%, Other 1%
Government Presidential system
Currency Australian dollar = 100 cents
Literacy rate 99%
Calorie consumption 3022 kilocalories

KOSOVO (not yet recognised)
Southeast Europe

Official name Republic of Kosovo
Formation 2008 / 2008
Capital Pristina
Population 1.9 million / 451 people per sq mile (174 people per sq km)
Total area 4212 sq. miles (10,908 sq. km)
Languages Albanian*, Serbian*, Bosniak, Gorani, Roma, Turkish
Religions Muslim 92%, Roman Catholic 4%, Orthodox Christian 4%
Ethnic mix Albanian 92%, Serb 4%, Bosniak and Gorani 2%, Turkish 1%, Roma 1%
Government Parliamentary system
Currency Euro = 100 cents
Literacy rate 92%
Calorie consumption Not available

KUWAIT
Southwest Asia

Official name State of Kuwait
Formation 1961 / 1961
Capital Kuwait City
Population 3.5 million / 509 people per sq mile (196 people per sq km)
Total area 6880 sq. miles (17,820 sq. km)
Languages Arabic*, English
Religions Sunni Muslim 45%, Shi'a Muslim 40%, Christian, Hindu, and other 15%
Ethnic mix Kuwaiti 45%, Other Arab 35%, South Asian 9%, Other 7%, Iranian 4%
Government Monarchy
Currency Kuwaiti dinar = 1000 fils
Literacy rate 96%
Calorie consumption 3471 kilocalories

KYRGYZSTAN
Central Asia

Official name Kyrgyz Republic
Formation 1991 / 1991
Capital Bishkek
Population 5.6 million / 73 people per sq mile (28 people per sq km)
Total area 76,641 sq. miles (198,500 sq. km)
Languages Kyrgyz*, Russian*, Uzbek, Tatar, Ukrainian
Religions Muslim (mainly Sunni) 70%, Orthodox Christian 30%
Ethnic mix Kyrgyz 69%, Uzbek 14%, Russian 9%, Other 6%, Dungan 1%, Uighur 1%
Government Presidential system
Currency Som = 100 tyiyn
Literacy rate 99%
Calorie consumption 2828 kilocalories

LAOS
Southeast Asia

Official name Lao People's Democratic Republic
Formation 1953 / 1953
Capital Vientiane
Population 6.9 million / 77 people per sq mile (30 people per sq km)
Total area 91,428 sq. miles (236,800 sq. km)
Languages Lao*, Mon-Khmer, Yao, Vietnamese, Chinese, French
Religions Buddhist 65%, Other (including animist) 34%, Christian 1%
Ethnic mix Lao Loum 66%, Lao Theung 30%, Lao Soung 2%, Other 2%
Government One-party state
Currency Kip = 100 at
Literacy rate 73%
Calorie consumption 2356 kilocalories

LATVIA
Northeast Europe

Official name Republic of Latvia
Formation 1991 / 1991
Capital Riga
Population 2 million / 80 people per sq mile (31 people per sq km)
Total area 24,938 sq. miles (64,589 sq. km)
Languages Latvian*, Russian
Religions Other 43%, Lutheran 24%, Roman Catholic 18%, O rthodox Christian 15%
Ethnic mix Latvian 62%, Russian 27%, Other 4%, Belarussian 3%, Ukrainian 2%, Polish 2%
Government Parliamentary system
Currency Euro = 100 cents
Literacy rate 99%
Calorie consumption 3293 kilocalories

LEBANON
Southwest Asia

Official name Lebanese Republic
Formation 1941 / 1941
Capital Beirut
Population 5 million / 1266 people per sq mile (489 people per sq km)
Total area 4015 sq. miles (10,400 sq. km)
Languages Arabic*, French, Armenian, Assyrian
Religions Muslim 60%, Christian 39%, Other 1%
Ethnic mix Arab 95%, Armenian 4%, Other 1%
Government Parliamentary system
Currency Lebanese pound = 100 piastres
Literacy rate 90%
Calorie consumption 3181 kilocalories

LESOTHO
Southern Africa

Official name Kingdom of Lesotho
Formation 1966 / 1966
Capital Maseru
Population 2.1 million / 179 people per sq mile (69 people per sq km)
Total area 11,720 sq. miles (30,355 sq. km)
Languages English*, Sesotho*, isiZulu
Religions Christian 90%, Traditional beliefs 10%
Ethnic mix Sotho 99%, European and Asian 1%
Government Parliamentary system
Currency Loti = 100 lisente; and South African rand = 100 cents
Literacy rate 76%
Calorie consumption 2595 kilocalories

LIBERIA
West Africa

Official name Republic of Liberia
Formation 1847 / 1847
Capital Monrovia
Population 4.4 million / 118 people per sq mile (46 people per sq km)
Total area 43,000 sq. miles (111,370 sq. km)
Languages Kpelle, Vai, Bassa, Kru, Grebo, Kissi, Gola, Loma, English*
Religions Christian 40%, Traditional beliefs 40%, Muslim 20%
Ethnic mix Indigenous tribes (12 groups) 49%, Kpellé 20%, Bassa 16%, Gio 8%, Krou 7%
Government Presidential system
Currency Liberian dollar = 100 cents
Literacy rate 43%
Calorie consumption 2251 kilocalories

LIBYA
North Africa

Official name State of Libya
Formation 1951 / 1951
Capital Tripoli
Population 6.3 million / 9 people per sq mile (4 people per sq km)
Total area 679,358 sq. miles (1,759,540 sq. km)
Languages Arabic*, Tuareg
Religions Muslim (mainly Sunni) 97%, Other 3%
Ethnic mix Arab and Berber 97%, Other 3%
Government Transitional regime
Currency Libyan dinar = 1000 dirhams
Literacy rate 90%
Calorie consumption 3211 kilocalories

LIECHTENSTEIN
Central Europe

Official name Principality of Liechtenstein
Formation 1719 / 1719
Capital Vaduz
Population 37,313 / 602 people per sq mile (233 people per sq km)
Total area 62 sq. miles (160 sq. km)
Languages German*, Alemannish dialect, Italian
Religions Roman Catholic 79%, Other 13%, Protestant 8%
Ethnic mix Liechtensteiner 66%, Other 12%, Swiss 10%, Austrian 6%, German 3%, Italian 3%
Government Parliamentary system
Currency Swiss franc = 100 rappen/centimes
Literacy rate 99%
Calorie consumption Not available

LITHUANIA
Northeast Europe

Official name Republic of Lithuania
Formation 1991 / 1991
Capital Vilnius
Population 3 million / 119 people per sq mile (46 people per sq km)
Total area 25,174 sq. miles (65,200 sq. km)
Languages Lithuanian*, Russian
Religions Roman Catholic 77%, Other 17%, Russian Orthodox 4%, Protestant 1%, Old believers 1%
Ethnic mix Lithuanian 85%, Polish 7%, Russian 6%, Belarussian 1%, Other 1%
Government Parliamentary system
Currency Euro = 100 cents
Literacy rate 99%
Calorie consumption 3463 kilocalories

LUXEMBOURG
Northwest Europe

Official name Grand Duchy of Luxembourg
Formation 1867 / 1867
Capital Luxembourg-Ville
Population 500,000 / 501 people per sq mile (193 people per sq km)
Total area 998 sq. miles (2586 sq. km)
Languages Luxembourgish*, German*, French*
Religions Roman Catholic 97%, Protestant, Orthodox Christian, and Jewish 3%
Ethnic mix Luxembourger 62%, Foreign residents 38%
Government Parliamentary system
Currency Euro = 100 cents
Literacy rate 99%
Calorie consumption 3568 kilocalories

MACEDONIA
Southeast Europe

Official name Republic of Macedonia
Formation 1991 / 1991
Capital Skopje
Population 2.1 million / 212 people per sq mile (82 people per sq km)
Total area 9781 sq. miles (25,333 sq. km)
Languages Macedonian*, Albanian*, Turkish, Romani, Serbian
Religions Orthodox Christian 65%, Muslim 29%, Roman Catholic 4%, Other 2%
Ethnic mix Macedonian 64%, Albanian 25%, Turkish 4%, Roma 3%, Serb 2%, Other 2%
Government Mixed presidential–parliamentary system
Currency Macedonian denar = 100 deni
Literacy rate 98%
Calorie consumption 2923 kilocalories

MADAGASCAR
Indian Ocean

Official name Republic of Madagascar
Formation 1960 / 1960
Capital Antananarivo
Population 23.6 million / 105 people per sq mile (41 people per sq km)
Total area 226,656 sq. miles (587,040 sq. km)
Languages Malagasy*, French*, English*
Religions Traditional beliefs 52%, Christian (mainly Roman Catholic) 41%, Muslim 7%
Ethnic mix Other Malay 46%, Merina 26%, Betsimisaraka 15%, Betsileo 12%, Other 1%
Government Mixed presidential–parliamentary system
Currency Ariary = 5 iraimbilanja
Literacy rate 64%
Calorie consumption 2052 kilocalories

MALAWI
Southern Africa

Official name Republic of Malawi
Formation 1964 / 1964
Capital Lilongwe
Population 16.8 million / 463 people per sq mile (179 people per sq km)
Total area 45,745 sq. miles (118,480 sq. km)
Languages Chewa, Lomwe, Yao, Ngoni, English*
Religions Protestant 55%, Roman Catholic 20%, Muslim 20%, Traditional beliefs 5%
Ethnic mix Bantu 99%, Other 1%
Government Presidential system
Currency Malawi kwacha = 100 tambala
Literacy rate 61%
Calorie consumption 2334 kilocalories

MALAYSIA
Southeast Asia

Official name Malaysia
Formation 1963 / 1965
Capital Kuala Lumpur; Putrajaya (administrative)
Population 30.2 million / 238 people per sq mile (92 people per sq km)
Total area 127,316 sq. miles (329,750 sq. km)
Languages Bahasa Malaysia*, Malay, Chinese, Tamil, English
Religions Muslim (mainly Sunni) 61%, Buddhist 19%, Christian 9%, Hindu 6%, Other 5%
Ethnic mix Malay 53%, Chinese 26%, Indigenous tribes 12%, Indian 8%, Other 1%
Government Parliamentary system
Currency Ringgit = 100 sen
Literacy rate 93%
Calorie consumption 2855 kilocalories

MALDIVES
Indian Ocean

Official name Republic of Maldives
Formation 1965 / 1965
Capital Male'
Population 400,000 / 3448 people per sq mile (1333 people per sq km)
Total area 116 sq. miles (300 sq. km)
Languages Dhivehi (Maldivian), Sinhala, Tamil, Arabic
Religions Sunni Muslim 100%
Ethnic mix Arab–Sinhalese–Malay 100%
Government Presidential system
Currency Rufiyaa = 100 laari
Literacy rate 98%
Calorie consumption 2722 kilocalories

MALI
West Africa

Official name Republic of Mali
Formation 1960 / 1960
Capital Bamako
Population 15.8 million / 34 people per sq mile (13 people per sq km)
Total area 478,764 sq. miles (1,240,000 sq. km)
Languages Bambara, Fulani, Senufo, Soninke, French*
Religions Muslim (mainly Sunni) 90%, Traditional beliefs 6%, Christian 4%
Ethnic mix Bambara 52%, Other 14%, Fulani 11%, Saracolé 7%, Soninka 7%, Tuareg 5%, Mianka 4%
Government Presidential system
Currency CFA franc = 100 centimes
Literacy rate 34%
Calorie consumption 2833 kilocalories

MALTA
Southern Europe

Official name Republic of Malta
Formation 1964 / 1964
Capital Valletta
Population 400,000 / 3226 people per sq mile (1250 people per sq km)
Total area 122 sq. miles (316 sq. km)
Languages Maltese*, English*
Religions Roman Catholic 98%, Other and nonreligious 2%
Ethnic mix Maltese 96%, Other 4%
Government Parliamentary system
Currency Euro = 100 cents
Literacy rate 92%
Calorie consumption 3389 kilocalories

MARSHALL ISLANDS
Australasia & Oceania

Official name Republic of the Marshall Islands
Formation 1986 / 1986
Capital Majuro
Population 70,983 / 1014 people per sq mile (392 people per sq km)
Total area 70 sq. miles (181 sq. km)
Languages Marshallese*, English*, Japanese, German
Religions Protestant 90%, Roman Catholic 8%, Other 2%
Ethnic mix Micronesian 90%, Other 10%
Government Presidential system
Currency US dollar = 100 cents
Literacy rate 91%
Calorie consumption Not available

MAURITANIA
West Africa

Official name Islamic Republic of Mauritania
Formation 1960 / 1960
Capital Nouakchott
Population 4 million / 10 people per sq mile (4 people per sq km)
Total area 397,953 sq. miles (1,030,700 sq. km)
Languages Arabic*, Hassaniyah Arabic, Wolof, French
Religions Sunni Muslim 100%
Ethnic mix Maure 81%, Wolof 7%, Tukolor 5%, Other 4%, Soninka 3%
Government Presidential system
Currency Ouguiya = 5 khoums
Literacy rate 46%
Calorie consumption 2791 kilocalories

MAURITIUS
Indian Ocean

Official name Republic of Mauritius
Formation 1968 / 1968
Capital Port Louis
Population 1.2 million / 1671 people per sq mile (645 people per sq km)
Total area 718 sq. miles (1860 sq. km)
Languages French Creole, Hindi, Urdu, Tamil, Chinese, English*, French
Religions Hindu 48%, Roman Catholic 24%, Muslim 17%, Protestant 9%, Other 2%
Ethnic mix Indo-Mauritian 68%, Creole 27%, Sino-Mauritian 3%, Franco-Mauritian 2%
Government Parliamentary system
Currency Mauritian rupee = 100 cents
Literacy rate 89%
Calorie consumption 3055 kilocalories

MEXICO
North America

Official name United Mexican States
Formation 1836 / 1848
Capital Mexico City
Population 124 million / 168 people per sq mile (65 people per sq km)
Total area 761,602 sq. miles (1,972,550 sq. km)
Languages Spanish*, Nahuatl, Mayan, Zapotec, Mixtec, Otomi, Totonac, Tzotzil, Tzeltal
Religions Roman Catholic 77%, Other 14%, Protestant 6%, Nonreligious 3%
Ethnic mix Mestizo 60%, Amerindian 30%, European 9%, Other 1%
Government Presidential system
Currency Mexican peso = 100 centavos
Literacy rate 94%
Calorie consumption 3072 kilocalories

MICRONESIA
Australasia & Oceania

Official name Federated States of Micronesia
Formation 1986 / 1986
Capital Palikir (Pohnpei Island)
Population 105,681 / 390 people per sq mile (151 people per sq km)
Total area 271 sq. miles (702 sq. km)
Languages Trukese, Pohnpeian, Kosraean, Yapese, English*
Religions Roman Catholic 50%, Protestant 47%, Other 3%
Ethnic mix Chuukese 49%, Pohnpeian 24%, Other 14%, Kosraean 6%, Yapese 5%, Asian 2%
Government Nonparty system
Currency US dollar = 100 cents
Literacy rate 81%
Calorie consumption Not available

MOLDOVA
Southeast Europe

Official name Republic of Moldova
Formation 1991 / 1991
Capital Chisinau
Population 3.5 million / 269 people per sq mile (104 people per sq km)
Total area 13,067 sq. miles (33,843 sq. km)
Languages Moldovan*, Ukrainian, Russian
Religions Orthodox Christian 93%, Other 6%, Baptist 1%
Ethnic mix Moldovan 84%, Ukrainian 7%, Gagauz 5%, Russian 2%, Bulgarian 1%, Other 1%
Government Parliamentary system
Currency Moldovan leu = 100 bani
Literacy rate 99%
Calorie consumption 2837 kilocalories

MONACO
Southern Europe

Official name Principality of Monaco
Formation 1861 / 1861
Capital Monaco-Ville
Population 36,950 / 49,267 people per sq mile (18,949 people per sq km)
Total area 0.75 sq. miles (1.95 sq. km)
Languages French*, Italian, Monégasque, English
Religions Roman Catholic 89%, Protestant 6%, Other 5%
Ethnic mix French 47%, Other 21%, Italian 16%, Monégasque 16%
Government Mixed monarchical–parliamentary system
Currency Euro = 100 cents
Literacy rate 99%
Calorie consumption Not available

MONGOLIA
East Asia

Official name Mongolia
Formation 1924 / 1924
Capital Ulan Bator
Population 2.9 million / 5 people per sq mile (2 people per sq km)
Total area 604,247 sq. miles (1,565,000 sq. km)
Languages Khalkha Mongolian, Kazakh, Chinese, Russian
Religions Tibetan Buddhist 50%, Nonreligious 40%, Shamanist and Christian 6%, Muslim 4%
Ethnic mix Khalkh 95%, Kazakh 4%, Other 1%
Government Mixed presidential–parliamentary system
Currency Tugrik (tögrög) = 100 möngö
Literacy rate 98%
Calorie consumption 2463 kilocalories

MONTENEGRO
Southeast Europe

Official name Montenegro
Formation 2006 / 2006
Capital Podgorica
Population 600,000 / 113 people per sq mile (43 people per sq km)
Total area 5332 sq. miles (13,812 sq. km)
Languages Montenegrin*, Serbian, Albanian, Bosniak, Croatian
Religions Orthodox Christian 74%, Muslim 18%, Roman Catholic 4%, Other 4%
Ethnic mix Montenegrin 43%, Serb 32%, Other 12%, Bosniak 8%, Albanian 5%
Government Parliamentary system
Currency Euro = 100 cents
Literacy rate 98%
Calorie consumption 3568 kilocalories

MOROCCO
North Africa

Official name Kingdom of Morocco
Formation 1956 / 1969
Capital Rabat
Population 35.5 million / 194 people per sq mile (75 people per sq km)
Total area 172,316 sq. miles (446,300 sq. km)
Languages Arabic*, Tamazight (Berber), French, Spanish
Religions Muslim (mainly Sunni) 99%, Other (mostly Christian) 1%
Ethnic mix Arab 70%, Berber 29%, European 1%
Government Mixed monarchical–parliamentary system
Currency Moroccan dirham = 100 centimes
Literacy rate 67%
Calorie consumption 3334 kilocalories

MOZAMBIQUE
Southern Africa

Official name Republic of Mozambique
Formation 1975 / 1975
Capital Maputo
Population 26.5 million / 88 people per sq mile (34 people per sq km)
Total area 309,494 sq. miles (801,590 sq. km)
Languages Makua, Xitsonga, Sena, Lomwe, Portuguese*
Religions Traditional beliefs 56%, Christian 30%, Muslim 14%
Ethnic mix Makua Lomwe 47%, Tsonga 23%, Malawi 12%, Shona 11%, Yao 4%, Other 3%
Government Presidential system
Currency New metical = 100 centavos
Literacy rate 51%
Calorie consumption 2283 kilocalories

MYANMAR (BURMA)
Southeast Asia

Official name Republic of the Union of Myanmar
Formation 1948 / 1948
Capital Nay Pyi Taw
Population 53.7 million / 212 people per sq mile (82 people per sq km)
Total area 261,969 sq. miles (678,500 sq. km)
Languages Myanmar (Burmese)*, Shan, Karen, Rakhine, Chin, Yangbye, Kachin, Mon
Religions Buddhist 89%, Christian 4%, Muslim 4%, Other 2%, Animist 1%
Ethnic mix Burman (Bamah) 68%, Other 12%, Shan 9%, Karen 7%, Rakhine 4%
Government Presidential system
Currency Kyat = 100 pyas
Literacy rate 93%
Calorie consumption 2571 kilocalories

NAMIBIA
Southern Africa

Official name Republic of Namibia
Formation 1990 / 1994
Capital Windhoek
Population 2.3 million / 7 people per sq mile (3 people per sq km)
Total area 318,694 sq. miles (825,418 sq. km)
Languages Ovambo, Kavango, English*, Bergdama, German, Afrikaans
Religions Christian 90%, Traditional beliefs 10%
Ethnic mix Ovambo 50%, Other tribes 22%, Kavango 9%, Damara 7%, Herero 7%, Other 5%
Government Presidential system
Currency Namibian dollar = 100 cents; and South African rand = 100 cents
Literacy rate 76%
Calorie consumption 2086 kilocalories

NAURU
Australasia & Oceania

Official name Republic of Nauru
Formation 1968 / 1968
Capital None
Population 9488 / 1171 people per sq mile (452 people per sq km)
Total area 8.1 sq. miles (21 sq. km)
Languages Nauruan*, Kiribati, Chinese, Tuvaluan, English
Religions Nauruan Congregational Church 60%, Roman Catholic 35%, Other 5%
Ethnic mix Nauruan 93%, Chinese 5%, European 1%, Other Pacific islanders 1%
Government Nonparty system
Currency Australian dollar = 100 cents
Literacy rate 95%
Calorie consumption Not available

NEPAL
South Asia

Official name Federal Democratic Republic of Nepal
Formation 1769 / 1769
Capital Kathmandu
Population 28.1 million / 532 people per sq mile (205 people per sq km)
Total area 54,363 sq. miles (140,800 sq. km)
Languages Nepali*, Maithili, Bhojpuri
Religions Hindu 81%, Buddhist 11%, Muslim 4%, Other (including Christian) 4%
Ethnic mix Other 52%, Chhetri 16%, Hill Brahman 13%, Tharu 7%, Magar 7%, Tamang 5%
Government Transitional regime
Currency Nepalese rupee = 100 paisa
Literacy rate 57%
Calorie consumption 2673 kilocalories

NETHERLANDS
Northwest Europe

Official name Kingdom of the Netherlands
Formation 1648 / 1839
Capital Amsterdam; The Hague (administrative)
Population 16.8 million / 1283 people per sq mile (495 people per sq km)
Total area 16,033 sq. miles (41,526 sq. km)
Languages Dutch*, Frisian
Religions Roman Catholic 36%, Other 34%, Protestant 27%, Muslim 3%
Ethnic mix Dutch 82%, Other 12%, Surinamese 2%, Turkish 2%, Moroccan 2%
Government Parliamentary system
Currency Euro = 100 cents
Literacy rate 99%
Calorie consumption 3147 kilocalories

NEW ZEALAND
Australasia & Oceania

Official name New Zealand
Formation 1947 / 1947
Capital Wellington
Population 4.6 million / 44 people per sq mile (17 people per sq km)
Total area 103,737 sq. miles (268,680 sq. km)
Languages English*, Maori*
Religions Anglican 24%, Other 22%, Presbyterian 18%, Nonreligious 16%, Roman Catholic 15%, Methodist 5%
Ethnic mix European 75%, Maori 15%, Other 7%, Samoan 3%
Government Parliamentary system
Currency New Zealand dollar = 100 cents
Literacy rate 99%
Calorie consumption 3170 kilocalories

NICARAGUA
Central America

Official name Republic of Nicaragua
Formation 1838 / 1838
Capital Managua
Population 6.2 million / 135 people per sq mile (52 people per sq km)
Total area 49,998 sq. miles (129,494 sq. km)
Languages Spanish*, English Creole, Miskito
Religions Roman Catholic 80%, Protestant Evangelical 17%, Other 3%
Ethnic mix Mestizo 69%, White 17%, Black 9%, Amerindian 5%
Government Presidential system
Currency Córdoba oro = 100 centavos
Literacy rate 78%
Calorie consumption 2564 kilocalories

NIGER
West Africa

Official name Republic of Niger
Formation 1960 / 1960
Capital Niamey
Population 18.5 million / 38 people per sq mile (15 people per sq km)
Total area 489,188 sq. miles (1,267,000 sq. km)
Languages Hausa, Djerma, Fulani, Tuareg, Teda, French*
Religions Muslim 99%, Other (including Christian) 1%
Ethnic mix Hausa 53%, Djerma and Songhai 21%, Tuareg 11%, Fulani 7%, Kanuri 6%, Other 2%
Government Presidential system
Currency CFA franc = 100 centimes
Literacy rate 16%
Calorie consumption 2546 kilocalories

NIGERIA
West Africa

Official name Federal Republic of Nigeria
Formation 1960 / 1961
Capital Abuja
Population 179 million / 508 people per sq mile (196 people per sq km)
Total area 356,667 sq. miles (923,768 sq. km)
Languages Hausa, English*, Yoruba, Ibo
Religions Muslim 50%, Christian 40%, Traditional beliefs 10%
Ethnic mix Other 29%, Hausa 21%, Yoruba 21%, Ibo 18%, Fulani 11%
Government Presidential system
Currency Naira = 100 kobo
Literacy rate 51%
Calorie consumption 2700 kilocalories

NORTH KOREA
East Asia

Official name Democratic People's Republic of Korea
Formation 1948 / 1953
Capital Pyongyang
Population 25 million / 538 people per sq mile (208 people per sq km)
Total area 46,540 sq. miles (120,540 sq. km)
Languages Korean*
Religions Atheist 100%
Ethnic mix Korean 100%
Government One-party state
Currency North Korean won = 100 chon
Literacy rate 99%
Calorie consumption 2094 kilocalories

NORWAY
Northern Europe

Official name Kingdom of Norway
Formation 1905 / 1905
Capital Oslo
Population 5.1 million / 43 people per sq mile (17 people per sq km)
Total area 125,181 sq. miles (324,220 sq. km)
Languages Norwegian* (Bokmål "book language" and Nynorsk "new Norsk"), Sámi
Religions Evangelical Lutheran 88%, Other and nonreligious 8%, Muslim 2%, Pentecostal 1%, Roman Catholic 1%
Ethnic mix Norwegian 93%, Other 6%, Sámi 1%
Government Parliamentary system
Currency Norwegian krone = 100 øre
Literacy rate 99%
Calorie consumption 3484 kilocalories

OMAN
Southwest Asia

Official name Sultanate of Oman
Formation 1951 / 1951
Capital Muscat
Population 3.9 million / 48 people per sq mile (18 people per sq km)
Total area 82,031 sq. miles (212,460 sq. km)
Languages Arabic*, Baluchi, Farsi, Hindi, Punjabi
Religions Ibadi Muslim 75%, Other Muslim and Hindu 25%
Ethnic mix Arab 88%, Baluchi 4%, Persian 3%, Indian and Pakistani 3%, African 2%
Government Monarchy
Currency Omani rial = 1000 baisa
Literacy rate 87%
Calorie consumption 3143 kilocalories

PAKISTAN
South Asia

Official name Islamic Republic of Pakistan
Formation 1947 / 1971
Capital Islamabad
Population 185 million / 622 people per sq mile (240 people per sq km)
Total area 310,401 sq. miles (803,940 sq. km)
Languages Punjabi, Sindhi, Pashtu, Urdu*, Baluchi, Brahui
Religions Sunni Muslim 77%, Shi'a Muslim 20%, Hindu 2%, Christian 1%
Ethnic mix Punjabi 56%, Pathan (Pashtun) 15%, Sindhi 14%, Mohajir 7%, Baluchi 4%, Other 4%
Government Parliamentary system
Currency Pakistani rupee = 100 paisa
Literacy rate 55%
Calorie consumption 2440 kilocalories

PALAU
Australasia & Oceania

Official name Republic of Palau
Formation 1994 / 1994
Capital Ngerulmud
Population 21,186 / 108 people per sq mile (42 people per sq km)
Total area 177 sq. miles (458 sq. km)
Languages Palauan*, English*, Japanese, Angaur, Tobi, Sonsorolese
Religions Christian 66%, Modekngei 34%
Ethnic mix Palauan 74%, Filipino 16%, Other 6%, Chinese and other Asian 4%
Government Nonparty system
Currency US dollar = 100 cents
Literacy rate 99%
Calorie consumption Not available

PANAMA
Central America

Official name Republic of Panama
Formation 1903 / 1903
Capital Panama City
Population 3.9 million / 133 people per sq mile (51 people per sq km)
Total area 30,193 sq. miles (78,200 sq. km)
Languages English Creole, Spanish*, Amerindian languages, Chibchan languages
Religions Roman Catholic 84%, Protestant 15%, Other 1%
Ethnic mix Mestizo 70%, Black 14%, White 10%, Amerindian 6%
Government Presidential system
Currency Balboa = 100 centésimos; and US dollar = 100 cents
Literacy rate 94%
Calorie consumption 2733 kilocalories

PAPUA NEW GUINEA
Australasia & Oceania

Official name Independent State of Papua New Guinea
Formation 1975 / 1975
Capital Port Moresby
Population 7.5 million / 43 people per sq mile (17 people per sq km)
Total area 178,703 sq. miles (462,840 sq. km)
Languages Pidgin English, Papuan, English*, Motu, 800 (est.) native languages
Religions Protestant 60%, Roman Catholic 37%, Other 3%
Ethnic mix Melanesian and mixed race 100%
Government Parliamentary system
Currency Kina = 100 toea
Literacy rate 63%
Calorie consumption 2193 kilocalories

PARAGUAY
South America

Official name Republic of Paraguay
Formation 1811 / 1938
Capital Asunción
Population 6.9 million / 45 people per sq mile (17 people per sq km)
Total area 157,046 sq. miles (406,750 sq. km)
Languages Guaraní*, Spanish*, German
Religions Roman Catholic 90%, Protestant (including Mennonite) 10%
Ethnic mix Mestizo 91%, Other 7%, Amerindian 2%
Government Presidential system
Currency Guaraní = 100 céntimos
Literacy rate 94%
Calorie consumption 2589 kilocalories

PERU
South America

Official name Republic of Peru
Formation 1824 / 1941
Capital Lima
Population 30.8 million / 62 people per sq mile (24 people per sq km)
Total area 496,223 sq. miles (1,285,200 sq. km)
Languages Spanish*, Quechua*, Aymara
Religions Roman Catholic 81%, Other 19%
Ethnic mix Amerindian 45%, Mestizo 37%, White 15%, Other 3%
Government Presidential system
Currency New sol = 100 céntimos
Literacy rate 94%
Calorie consumption 2700 kilocalories

PHILIPPINES
Southeast Asia

Official name Republic of the Philippines
Formation 1946 / 1946
Capital Manila
Population 100 million / 870 people per sq mile (336 people per sq km)
Total area 115,830 sq. miles (300,000 sq. km)
Languages Filipino*, English*, Tagalog, Cebuano, Ilocano, Hiligaynon, many other local languages
Religions Roman Catholic 81%, Protestant 9%, Muslim 5%, Other (including Buddhist) 5%
Ethnic mix Other 34%, Tagalog 28%, Cebuano 13%, Ilocano 9%, Hiligaynon 8%, Bisaya 8%
Government Presidential system
Currency Philippine peso = 100 centavos
Literacy rate 95%
Calorie consumption 2570 kilocalories

POLAND
Northern Europe

Official name Republic of Poland
Formation 1918 / 1945
Capital Warsaw
Population 38.2 million / 325 people per sq mile (125 people per sq km)
Total area 120,728 sq. miles (312,685 sq. km)
Languages Polish*
Religions Roman Catholic 93%, Other and nonreligious 5%, Orthodox Christian 2%
Ethnic mix Polish 98%, Other 2%
Government Parliamentary system
Currency Zloty = 100 groszy
Literacy rate 99%
Calorie consumption 3485 kilocalories

PORTUGAL
Southwest Europe

Official name Portuguese Republic
Formation 1139 / 1640
Capital Lisbon
Population 10.6 million / 299 people per sq mile (115 people per sq km)
Total area 35,672 sq. miles (92,391 sq. km)
Languages Portuguese*
Religions Roman Catholic 92%, Protestant 4%, Nonreligious 3%, Other 1%
Ethnic mix Portuguese 98%, African and other 2%
Government Parliamentary system
Currency Euro = 100 cents
Literacy rate 94%
Calorie consumption 3456 kilocalories

QATAR
Southwest Asia

Official name State of Qatar
Formation 1971 / 1971
Capital Doha
Population 2.3 million / 542 people per sq mile (209 people per sq km)
Total area 4416 sq. miles (11,437 sq. km)
Languages Arabic*
Religions Muslim (mainly Sunni) 95%, Other 5%
Ethnic mix Qatari 20%, Indian 20%, Other Arab 20%, Nepalese 13%, Filipino 10%, Other 10%, Pakistani 7%
Government Monarchy
Currency Qatar riyal = 100 dirhams
Literacy rate 97%
Calorie consumption Not available

ROMANIA
Southeast Europe

Official name Romania
Formation 1878 / 1947
Capital Bucharest
Population 21.6 million / 243 people per sq mile (94 people per sq km)
Total area 91,699 sq. miles (237,500 sq. km)
Languages Romanian*, Hungarian (Magyar), Romani, German
Religions Romanian Orthodox 87%, Protestant 5%, Roman Catholic 5%, Greek Orthodox 1%, Greek Catholic (Uniate) 1%, Other 1%
Ethnic mix Romanian 89%, Magyar 7%, Roma 3%, Other 1%
Government Presidential system
Currency New Romanian leu = 100 bani
Literacy rate 99%
Calorie consumption 3363 kilocalories

RUSSIAN FEDERATION
Europe / Asia

Official name Russian Federation
Formation 1480 / 1991
Capital Moscow
Population 143 million / 22 people per sq mile (8 people per sq km)
Total area 6,592,735 sq. miles (17,075,200 sq. km)
Languages Russian*, Tatar, Ukrainian, Chavash, various other national languages
Religions Orthodox Christian 75%, Muslim 14%, Other 11%
Ethnic mix Russian 80%, Other 12%, Tatar 4%, Ukrainian 2%, Bashkir 1%, Chavash 1%
Government Mixed Presidential–Parliamentary system
Currency Russian rouble = 100 kopeks
Literacy rate 99%
Calorie consumption 3358 kilocalories

RWANDA
Central Africa

Official name Republic of Rwanda
Formation 1962 / 1962
Capital Kigali
Population 12.1 million / 1256 people per sq mile (485 people per sq km)
Total area 10,169 sq. miles (26,338 sq. km)
Languages Kinyarwanda*, French*, Kiswahili, English*
Religions Christian 94%, Muslim 5%, Traditional beliefs 1%
Ethnic mix Hutu 85%, Tutsi 14%, Other (including Twa) 1%
Government Presidential system
Currency Rwanda franc = 100 centimes
Literacy rate 66%
Calorie consumption 2148 kilocalories

ST KITTS & NEVIS
West Indies

Official name Federation of Saint Christopher and Nevis
Formation 1983 / 1983
Capital Basseterre
Population 51,538 / 371 people per sq mile (143 people per sq km)
Total area 101 sq. miles (261 sq. km)
Languages English*, English Creole
Religions Anglican 33%, Methodist 29%, Other 22%, Moravian 9%, Roman Catholic 7%
Ethnic mix Black 95%, Mixed race 3%, White 1%, Other and Amerindian 1%
Government Parliamentary system
Currency East Caribbean dollar = 100 cents
Literacy rate 98%
Calorie consumption 2507 kilocalories

ST LUCIA
West Indies

Official name Saint Lucia
Formation 1979 / 1979
Capital Castries
Population 200,000 / 847 people per sq mile
(328 people per sq km)
Total area 239 sq. miles (620 sq. km)
Languages English*, French Creole
Religions Roman Catholic 90%, Other 10%
Ethnic mix Black 83%, Mulatto (mixed race) 13%,
Asian 3%, Other 1%
Government Parliamentary system
Currency East Caribbean dollar = 100 cents
Literacy rate 95%
Calorie consumption 2629 kilocalories

ST VINCENT & THE GRENADINES
West Indies

Official name Saint Vincent and the Grenadines
Formation 1979 / 1979
Capital Kingstown
Population 102,918 / 786 people per sq mile
(303 people per sq km)
Total area 150 sq. miles (389 sq. km)
Languages English*, English Creole
Religions Anglican 47%, Methodist 28%,
Roman Catholic 13%, Other 12%
Ethnic mix Black 66%, Mulatto (mixed race) 19%,
Other 12%, Carib 2%, Asian 1%
Government Parliamentary system
Currency East Caribbean dollar = 100 cents
Literacy rate 88%
Calorie consumption 2960 kilocalories

SAMOA
Australasia & Oceania

Official name Independent State of Samoa
Formation 1962 / 1962
Capital Apia
Population 200,000 / 183 people per sq mile
(71 people per sq km)
Total area 1104 sq. miles (2860 sq. km)
Languages Samoan*, English*
Religions Christian 99%, Other 1%
Ethnic mix Polynesian 91%, Euronesian 7%,
Other 2%
Government Parliamentary system
Currency Tala = 100 sene
Literacy rate 99%
Calorie consumption 2872 kilocalories

SAN MARINO
Southern Europe

Official name Republic of San Marino
Formation 1631 / 1631
Capital San Marino
Population 32,742 / 1364 people per sq mile
(537 people per sq km)
Total area 23.6 sq. miles (61 sq. km)
Languages Italian*
Religions Roman Catholic 93%, Other and
nonreligious 7%
Ethnic mix Sammarinese 88%, Italian 10%,
Other 2%
Government Parliamentary system
Currency Euro = 100 cents
Literacy rate 99%
Calorie consumption Not available

SAO TOME & PRINCIPE
West Africa

Official name Democratic Republic of
Sao Tome and Principe
Formation 1975 / 1975
Capital São Tomé
Population 200,000 / 539 people per sq mile
(208 people per sq km)
Total area 386 sq. miles (1001 sq. km)
Languages Portuguese Creole, Portuguese*
Religions Roman Catholic 84%, Other 16%
Ethnic mix Black 90%, Portuguese and Creole 10%
Government Presidential system
Currency Dobra = 100 céntimos
Literacy rate 70%
Calorie consumption 2676 kilocalories

SAUDI ARABIA
Southwest Asia

Official name Kingdom of Saudi Arabia
Formation 1932 / 1932
Capital Riyadh
Population 29.4 million / 36 people per sq mile
(14 people per sq km)
Total area 756,981 sq. miles (1,960,582 sq. km)
Languages Arabic*
Religions Sunni Muslim 85%, Shi'a Muslim 15%
Ethnic mix Arab 72%, Foreign residents (mostly
south and southeast Asian) 20%, Afro-Asian 8%
Government Monarchy
Currency Saudi riyal = 100 halalat
Literacy rate 94%
Calorie consumption 3122 kilocalories

SENEGAL
West Africa

Official name Republic of Senegal
Formation 1960 / 1960
Capital Dakar
Population 14.5 million / 195 people per sq mile
(75 people per sq km)
Total area 75,749 sq. miles (196,190 sq. km)
Languages Wolof, Pulaar, Serer, Diola, Mandinka,
Malinké, Soninké, French*
Religions Sunni Muslim 95%, Christian (mainly
Roman Catholic) 4%, Traditional beliefs 1%
Ethnic mix Wolof 43%, Serer 15%, Peul 14%,
Other 14%, Toucouleur 9%, Diola 5%
Government Presidential system
Currency CFA franc = 100 centimes
Literacy rate 52%
Calorie consumption 2426 kilocalories

SERBIA
Southeast Europe

Official name Republic of Serbia
Formation 2006 / 2008
Capital Belgrade
Population 9.5 million / 318 people per sq mile
(123 people per sq km)
Total area 29,905 sq. miles (77,453 sq. km)
Languages Serbian*, Hungarian (Magyar)
Religions Orthodox Christian 85%,
Roman Catholic 6%, Other 6%, Muslim 3%
Ethnic mix Serb 83%, Other 10%, Magyar 4%,
Bosniak 2%, Roma 1%
Government Parliamentary system
Currency Serbian dinar = 100 para
Literacy rate 98%
Calorie consumption 2724 kilocalories

SEYCHELLES
Indian Ocean

Official name Republic of Seychelles
Formation 1976 / 1976
Capital Victoria
Population 91,650 / 881 people per sq mile
(339 people per sq km)
Total area 176 sq. miles (455 sq. km)
Languages French Creole*, English*, French*
Religions Roman Catholic 82%, Anglican 6%, Other
(including Muslim) 6%, Other Christian 3%,
Hindu 2%, Seventh-day Adventist 1%
Ethnic mix Creole 89%, Indian 5%, Other 4%,
Chinese 2%
Government Presidential system
Currency Seychelles rupee = 100 cents
Literacy rate 92%
Calorie consumption 2426 kilocalories

SIERRA LEONE
West Africa

Official name Republic of Sierra Leone
Formation 1961 / 1961
Capital Freetown
Population 6.2 million / 224 people per sq mile
(87 people per sq km)
Total area 27,698 sq. miles (71,740 sq. km)
Languages Mende, Temne, Krio, English*
Religions Muslim 60%, Christian 30%,
Traditional beliefs 10%
Ethnic mix Mende 35%, Temne 32%, Other 21%,
Limba 8%, Kuranko 4%
Government Presidential system
Currency Leone = 100 cents
Literacy rate 44%
Calorie consumption 2333 kilocalories

SINGAPORE
Southeast Asia

Official name Republic of Singapore
Formation 1965 / 1965
Capital Singapore
Population 5.5 million / 23,305 people per sq mile
(9016 people per sq km)
Total area 250 sq. miles (648 sq. km)
Languages Mandarin*, Malay*, Tamil*, English*
Religions Buddhist 55%, Taoist 22%, Muslim 16%,
Hindu, Christian, and Sikh 7%
Ethnic mix Chinese 74%, Malay 14%, Indian 9%,
Other 3%
Government Parliamentary system
Currency Singapore dollar = 100 cents
Literacy rate 96%
Calorie consumption Not available

SLOVAKIA
Central Europe

Official name Slovak Republic
Formation 1993 / 1993
Capital Bratislava
Population 5.5 million / 290 people per sq mile
(112 people per sq km)
Total area 18,859 sq. miles (48,845 sq. km)
Languages Slovak*, Hungarian (Magyar), Czech
Religions Roman Catholic 69%, Nonreligious
13%, Other 13%, Greek Catholic (Uniate) 4%,
Orthodox Christian 1%
Ethnic mix Slovak 86%, Magyar 10%, Roma 2%,
Czech 1%, Other 1%
Government Parliamentary system
Currency Euro = 100 cents
Literacy rate 99%
Calorie consumption 2902 kilocalories

SLOVENIA
Central Europe

Official name Republic of Slovenia
Formation 1991 / 1991
Capital Ljubljana
Population 2.1 million / 269 people per sq mile
(104 people per sq km)
Total area 7820 sq. miles (20,253 sq. km)
Languages Slovenian*
Religions Roman Catholic 58%, Other 28%, Atheist
10%, Orthodox Christian 2%, Muslim 2%
Ethnic mix Slovene 83%, Other 12%, Serb 2%,
Croat 2%, Bosniak 1%
Government Parliamentary system
Currency Euro = 100 cents
Literacy rate 99%
Calorie consumption 3173 kilocalories

SOLOMON ISLANDS
Australasia & Oceania

Official name Solomon Islands
Formation 1978 / 1978
Capital Honiara
Population 600,000 / 56 people per sq mile
(21 people per sq km)
Total area 10,985 sq. miles (28,450 sq. km)
Languages English*, Pidgin English, Melanesian
Pidgin, 120 (est.) native languages
Religions Church of Melanesia (Anglican) 34%,
Roman Catholic 19%, South Seas Evangelical
Church 17%, Methodist 11%, Seventh-day
Adventist 10%, Other 9%
Ethnic mix Melanesian 93%, Polynesian 4%,
Micronesian 2%, Other 1%
Government Parliamentary system
Currency Solomon Islands dollar = 100 cents
Literacy rate 77%
Calorie consumption 2473 kilocalories

SOMALIA
East Africa

Official name Federal Republic of Somalia
Formation 1960 / 1960
Capital Mogadishu
Population 10.8 million / 45 people per sq mile
(17 people per sq km)
Total area 246,199 sq. miles (637,657 sq. km)
Languages Somali*, Arabic*, English, Italian
Religions Sunni Muslim 99%, Christian 1%
Ethnic mix Somali 85%, Other 15%
Government Non-party system
Currency Somali shilin = 100 senti
Literacy rate 24%
Calorie consumption 1696 kilocalories

SOUTH AFRICA
Southern Africa

Official name Republic of South Africa
Formation 1934 / 1994
Capital Pretoria; Cape Town; Bloemfontein
Population 53.1 million / 113 people per sq mile
(43 people per sq km)
Total area 471,008 sq. miles (1,219,912 sq. km)
Languages English, isiZulu, isiXhosa, Afrikaans,
Sepedi, Setswana, Sesotho, Xitsonga, siSwati,
Tshivenda, isiNdebele
Religions Christian 68%, Traditional beliefs and
animist 29%, Muslim 2%, Hindu 1%
Ethnic mix Black 80%, Mixed race 9%,
White 9%, Asian 2%
Government Presidential system
Currency Rand = 100 cents
Literacy rate 94%
Calorie consumption 3007 kilocalories

SOUTH KOREA
East Asia

Official name Republic of Korea
Formation 1948 / 1953
Capital Seoul; Sejong City (administrative)
Population 49.5 million / 1299 people per sq mile
(501 people per sq km)
Total area 38,023 sq. miles (98,480 sq. km)
Languages Korean*
Religions Mahayana Buddhist 47%, Protestant 38%,
Roman Catholic 11%, Confucianist 3%, Other 1%
Ethnic mix Korean 100%
Government Presidential system
Currency South Korean won = 100 chon
Literacy rate 99%
Calorie consumption 3329 kilocalories

SOUTH SUDAN
East Africa

Official name Republic of South Sudan
Formation 2011 / 2011
Capital Juba
Population 11.7 million / 47 people per sq mile
(18 people per sq km)
Total area 248,777 sq. miles (644,329 sq. km)
Languages Arabic, Dinka, Nuer, Zande, Bari,
Shilluk, Lotuko, English*
Religions Over half of the population follow
Christian or traditional beliefs.
Ethnic mix Dinka 40%, Nuer 15%, Bari 10%, Shilluk/
Anwak 10%, Azande 10%, Arab 10%, Other 5%
Government Transitional regime
Currency South Sudan pound = 100 piastres
Literacy rate 37%
Calorie consumption Not available

SPAIN
Southwest Europe

Official name Kingdom of Spain
Formation 1492 / 1713
Capital Madrid
Population 47.1 million / 244 people per sq mile
(94 people per sq km)
Total area 194,896 sq. miles (504,782 sq. km)
Languages Spanish*, Catalan*, Galician*, Basque*
Religions Roman Catholic 96%, Other 4%
Ethnic mix Castilian Spanish 72%, Catalan 17%,
Galician 6%, Basque 2%, Other 2%, Roma 1%
Government Parliamentary system
Currency Euro = 100 cents
Literacy rate 98%
Calorie consumption 3183 kilocalories

SRI LANKA
South Asia

Official name Democratic Socialist Republic of
Sri Lanka
Formation 1948 / 1948
Capital Colombo; Sri Jayewardenapura Kotte
Population 21.4 million / 856 people per sq mile
(331 people per sq km)
Total area 25,332 sq. miles (65,610 sq. km)
Languages Sinhala*, Tamil*, Sinhala-Tamil, English
Religions Buddhist 69%, Hindu 15%, Muslim 8%,
Christian 8%
Ethnic mix Sinhalese 74%, Tamil 18%, Moor 7%,
Other 1%
Government Mixed presidential–
parliamentary system
Currency Sri Lanka rupee = 100 cents
Literacy rate 91%
Calorie consumption 2539 kilocalories

SUDAN
East Africa

Official name Republic of the Sudan
Formation 1956 / 2011
Capital Khartoum
Population 38.8 million / 54 people per sq mile
(21 people per sq km)
Total area 718,722 sq. miles (1,861,481 sq. km)
Languages Arabic, Nubian, Beja, Fur
Religions Nearly the whole population is Muslim
(mainly Sunni)
Ethnic mix Arab 60%, Other 18%, Nubian 10%,
Beja 8%, Fur 3%, Zaghawa 1%
Government Presidential system
Currency New Sudanese pound = 100 piastres
Literacy rate 73%
Calorie consumption 2346 kilocalories

SURINAME
South America

Official name Republic of Suriname
Formation 1975 / 1975
Capital Paramaribo
Population 500,000 / 8 people per sq mile
(3 people per sq km)
Total area 63,039 sq. miles (163,270 sq. km)
Languages Sranan (creole), Dutch*, Javanese,
Sarnami Hindi, Saramaccan, Chinese, Carib
Religions Hindu 27%, Protestant 25%, Roman
Catholic 23%, Muslim 20%, Traditional beliefs 5%
Ethnic mix East Indian 27%, Creole 18%, Black
15%, Javanese 15%, Mixed race 13%, Other 6%,
Amerindian 4%, Chinese 2%
Government Mixed presidential–
parliamentary system
Currency Surinamese dollar = 100 cents
Literacy rate 95%
Calorie consumption 2727 kilocalories

SWAZILAND
Southern Africa

Official name Kingdom of Swaziland
Formation 1968 / 1968
Capital Mbabane
Population 1.3 million / 196 people per sq mile
(76 people per sq km)
Total area 6704 sq. miles (17,363 sq. km)
Languages English*, siSwati*, isiZulu, Xitsonga
Religions Traditional beliefs 40%, Other 30%,
Roman Catholic 20%, Muslim 10%
Ethnic mix Swazi 97%, Other 3%
Government Monarchy
Currency Lilangeni = 100 cents
Literacy rate 83%
Calorie consumption 2275 kilocalories

SWEDEN
Northern Europe

Official name Kingdom of Sweden
Formation 1523 / 1921
Capital Stockholm
Population 9.6 million / 60 people per sq mile
(23 people per sq km)
Total area 173,731 sq. miles (449,964 sq. km)
Languages Swedish*, Finnish, Sámi
Religions Evangelical Lutheran 75%, Other 13%,
Muslim 5%, Other Protestant 5%,
Roman Catholic 2%
Ethnic mix Swedish 86%, Foreign-born or
first-generation immigrant 12%, Finnish and
Sámi 2%
Government Parliamentary system
Currency Swedish krona = 100 öre
Literacy rate 99%
Calorie consumption 3160 kilocalories

SWITZERLAND
Central Europe

Official name Swiss Confederation
Formation 1291 / 1857
Capital Bern
Population 8.2 million / 534 people per sq mile
(206 people per sq km)
Total area 15,942 sq. miles (41,290 sq. km)
Languages German*, Swiss-German, French*,
Italian*, Romansch
Religions Roman Catholic 42%, Protestant 35%,
Other and nonreligious 19%, Muslim 4%
Ethnic mix German 64%, French 20%, Other 9.5%,
Italian 6%, Romansch 0.5%
Government Parliamentary system
Currency Swiss franc = 100 rappen/centimes
Literacy rate 99%
Calorie consumption 3487 kilocalories

SYRIA
Southwest Asia

Official name Syrian Arab Republic
Formation 1941 / 1967
Capital Damascus
Population 22 million / 310 people per sq mile
(120 people per sq km)
Total area 71,498 sq. miles (184,180 sq. km)
Languages Arabic*, French, Kurdish, Armenian,
Circassian, Turkic languages, Assyrian, Aramaic
Religions Sunni Muslim 74%, Alawi 12%, Christian
10%, Druze 3%, Other 1%
Ethnic mix Arab 90%, Kurdish 9%, Armenian,
Turkmen, and Circassian 1%
Government Presidential system
Currency Syrian pound = 100 piastres
Literacy rate 85%
Calorie consumption 3106 kilocalories

TAIWAN
East Asia

Official name Republic of China (ROC)
Formation 1949 / 1949
Capital Taibei (Taipei)
Population 23.4 million / 1879 people per sq mile
(725 people per sq km)
Total area 13,892 sq. miles (35,980 sq. km)
Languages Amoy Chinese, Mandarin Chinese*,
Hakka Chinese
Religions Buddhist, Confucianist, and Taoist 93%,
Christian 5%, Other 2%
Ethnic mix Han Chinese (pre-20th-century
migration) 84%, Han Chinese (20th-century
migration) 14%, Aboriginal 2%
Government Presidential system
Currency Taiwan dollar = 100 cents
Literacy rate 98%
Calorie consumption 2997 kilocalories

TAJIKISTAN
Central Asia

Official name Republic of Tajikistan
Formation 1991 / 1991
Capital Dushanbe
Population 8.4 million / 152 people per sq mile
(59 people per sq km)
Total area 55,251 sq. miles (143,100 sq. km)
Languages Tajik*, Uzbek, Russian
Religions Sunni Muslim 95%, Shi'a Muslim 3%,
Other 2%
Ethnic mix Tajik 80%, Uzbek 15%, Other 3%,
Russian 1%, Kyrgyz 1%
Government Presidential system
Currency Somoni = 100 diram
Literacy rate 99%
Calorie consumption 2101 kilocalories

TANZANIA
East Africa

Official name United Republic of Tanzania
Formation 1964 / 1964
Capital Dodoma
Population 50.8 million / 148 people per sq mile
(57 people per sq km)
Total area 364,898 sq. miles (945,087 sq. km)
Languages Kiswahili*, Sukuma, Chagga, Nyamwezi,
Hehe, Makonde, Yao, Sandawe, English*
Religions Christian 63%, Muslim 35%, Other 2%
Ethnic mix Native African (over 120 tribes) 99%,
European, Asian, and Arab 1%
Government Presidential system
Currency Tanzanian shilling = 100 cents
Literacy rate 68%
Calorie consumption 2208 kilocalories

THAILAND
Southeast Asia

Official name Kingdom of Thailand
Formation 1238 / 1907
Capital Bangkok
Population 67.2 million / 341 people per sq mile
(132 people per sq km)
Total area 198,455 sq. miles (514,000 sq. km)
Languages Thai*, Chinese, Malay, Khmer, Mon,
Karen, Miao
Religions Buddhist 95%, Muslim 4%, Other
(including Christian) 1%
Ethnic mix Thai 83%, Chinese 12%, Malay 3%,
Khmer and Other 2%
Government Transitional regime
Currency Baht = 100 satang
Literacy rate 96%
Calorie consumption 2784 kilocalories

TOGO
West Africa

Official name Togolese Republic
Formation 1960 / 1960
Capital Lomé
Population 7 million / 333 people per sq mile (129 people per sq km)
Total area 21,924 sq. miles (56,785 sq. km)
Languages Ewe, Kabye, Gurma, French*
Religions Christian 47%, Traditional beliefs 33%, Muslim 14%, Other 6%
Ethnic mix Ewe 46%, Other African 41%, Kabye 12%, European 1%
Government Presidential system
Currency CFA franc = 100 centimes
Literacy rate 60%
Calorie consumption 2366 kilocalories

TONGA
Australasia & Oceania

Official name Kingdom of Tonga
Formation 1970 / 1970
Capital Nuku'alofa
Population 106,440 / 383 people per sq mile (148 people per sq km)
Total area 289 sq. miles (748 sq. km)
Languages English*, Tongan*
Religions Free Wesleyan 41%, Other 17%, Roman Catholic 16%, Church of Jesus Christ of Latter-day Saints 14%, Free Church of Tonga 12%
Ethnic mix Tongan 98%, Other 2%
Government Monarchy
Currency Pa'anga (Tongan dollar) = 100 seniti
Literacy rate 99%
Calorie consumption Not available

TRINIDAD & TOBAGO
West Indies

Official name Republic of Trinidad and Tobago
Formation 1962 / 1962
Capital Port-of-Spain
Population 1.3 million / 656 people per sq mile (253 people per sq km)
Total area 1980 sq. miles (5128 sq. km)
Languages English Creole, English*, Hindi, French, Spanish
Religions Roman Catholic 26%, Hindu 23%, Other and nonreligious 23%, Anglican 8%, Baptist 7%, Pentecostal 7%, Muslim 6%
Ethnic mix East Indian 40%, Black 38%, Mixed race 20%, White and Chinese 1%, other 1%
Government Parliamentary system
Currency Trinidad and Tobago dollar = 100 cents
Literacy rate 99%
Calorie consumption 2889 kilocalories

TUNISIA
North Africa

Official name Tunisian Republic
Formation 1956 / 1956
Capital Tunis
Population 11.1 million / 185 people per sq mile (71 people per sq km)
Total area 63,169 sq. miles (163,610 sq. km)
Languages Arabic*, French
Religions Muslim (mainly Sunni) 98%, Christian 1%, Jewish 1%
Ethnic mix Arab and Berber 98%, Jewish 1%, European 1%
Government Mixed presidential–parliamentary system
Currency Tunisian dinar = 1000 millimes
Literacy rate 80%
Calorie consumption 3362 kilocalories

TURKEY
Asia / Europe

Official name Republic of Turkey
Formation 1923 / 1939
Capital Ankara
Population 75.8 million / 255 people per sq mile (98 people per sq km)
Total area 301,382 sq. miles (780,580 sq. km)
Languages Turkish*, Kurdish, Arabic, Circassian, Armenian, Greek, Georgian, Ladino
Religions Muslim (mainly Sunni) 99%, Other 1%
Ethnic mix Turkish 70%, Kurdish 20%, Other 8%, Arab 2%
Government Parliamentary system
Currency Turkish lira = 100 kurus
Literacy rate 95%
Calorie consumption 3680 kilocalories

TURKMENISTAN
Central Asia

Official name Turkmenistan
Formation 1991 / 1991
Capital Ashgabat
Population 5.3 million / 28 people per sq mile (11 people per sq km)
Total area 188,455 sq. miles (488,100 sq. km)
Languages Turkmen*, Uzbek, Russian, Kazakh, Tatar
Religions Sunni Muslim 89%, Orthodox Christian 9%, Other 2%
Ethnic mix Turkmen 85%, Other 6%, Uzbek 5%, Russian 4%
Government Presidential system
Currency New manat = 100 tenge
Literacy rate 99%
Calorie consumption 2883 kilocalories

TUVALU
Australasia & Oceania

Official name Tuvalu
Formation 1978 / 1978
Capital Funafuti Atoll
Population 10,782 / 1078 people per sq mile (415 people per sq km)
Total area 10 sq. miles (26 sq. km)
Languages Tuvaluan, Kiribati, English*
Religions Church of Tuvalu 97%, Baha'i 1%, Seventh-day Adventist 1%, Other 1%
Ethnic mix Polynesian 96%, Micronesian 4%
Government Nonparty system
Currency Australian dollar = 100 cents; and Tuvaluan dollar = 100 cents
Literacy rate 95%
Calorie consumption Not available

UGANDA
East Africa

Official name Republic of Uganda
Formation 1962 / 1962
Capital Kampala
Population 38.8 million / 504 people per sq mile (194 people per sq km)
Total area 91,135 sq. miles (236,040 sq. km)
Languages Luganda, Nkole, Chiga, Lango, Acholi, Teso, Lugbara, English*
Religions Christian 85%, Muslim (mainly Sunni) 12%, Other 3%
Ethnic mix Other 50%, Baganda 17%, Banyakole 10%, Basoga 9%, Iteso 7%, Bakiga 7%
Government Presidential system
Currency Uganda shilling = 100 cents
Literacy rate 74%
Calorie consumption 2279 kilocalories

UKRAINE
Eastern Europe

Official name Ukraine
Formation 1991 / 1991
Capital Kiev
Population 44.9 million / 193 people per sq mile (74 people per sq km)
Total area 223,089 sq. miles (603,700 sq. km)
Languages Ukrainian*, Russian, Tatar
Religions Christian (mainly Orthodox) 95%, Other 5%
Ethnic mix Ukrainian 78%, Russian 17%, Other 5%
Government Presidential system
Currency Hryvna = 100 kopiykas
Literacy rate 99%
Calorie consumption 3142 kilocalories

UNITED ARAB EMIRATES
Southwest Asia

Official name United Arab Emirates
Formation 1971 / 1972
Capital Abu Dhabi
Population 9.4 million / 291 people per sq mile (112 people per sq km)
Total area 32,000 sq. miles (82,880 sq. km)
Languages Arabic*, Farsi, Indian and Pakistani languages, English
Religions Muslim (mainly Sunni) 96%, Christian, Hindu, and other 4%
Ethnic mix Asian 60%, Emirian 25%, Other Arab 12%, European 3%
Government Monarchy
Currency UAE dirham = 100 fils
Literacy rate 90%
Calorie consumption 3215 kilocalories

UNITED KINGDOM
Northwest Europe

Official name United Kingdom of Great Britain and Northern Ireland
Formation 1707 / 1922
Capital London
Population 63.5 million / 681 people per sq mile (263 people per sq km)
Total area 94,525 sq. miles (244,820 sq. km)
Languages English*, Welsh*, Scottish Gaelic, Irish
Religions Anglican 45%, Other and nonreligious 36%, Roman Catholic 9%, Presbyterian 4%, Muslim 3%, Methodist 2%, Hindu 1%
Ethnic mix English 80%, Scottish 9%, West Indian, Asian, and other 5%, Northern Irish 3%, Welsh 3%
Government Parliamentary system
Currency Pound sterling = 100 pence
Literacy rate 99%
Calorie consumption 3414 kilocalories

UNITED STATES
North America

Official name United States of America
Formation 1776 / 1959
Capital Washington D.C.
Population 323 million / 91 people per sq mile (35 people per sq km)
Total area 3,794,100 sq. miles (9,826,675 sq. km)
Languages English*, Spanish, Chinese, French, German, Tagalog, Vietnamese, Italian, Korean, Russian, Polish
Religions Protestant 52%, Roman Catholic 25%, Other and nonreligious 20%, Jewish 2%, Muslim 1%
Ethnic mix White 60%, Hispanic 17%, Black American/African 14%, Asian 6%, American Indians & Alaksa Natives 2%, Pacific Islanders 1%
Government Presidential system
Currency US dollar = 100 cents
Literacy rate 99%
Calorie consumption 3639 kilocalories

URUGUAY
South America

Official name Oriental Republic of Uruguay
Formation 1828 / 1828
Capital Montevideo
Population 3.4 million / 50 people per sq mile (19 people per sq km)
Total area 68,039 sq. miles (176,220 sq. km)
Languages Spanish*
Religions Roman Catholic 66%, Other and nonreligious 30%, Jewish 2%, Protestant 2%
Ethnic mix White 90%, Mestizo 6%, Black 4%
Government Presidential system
Currency Uruguayan peso = 100 centésimos
Literacy rate 98%
Calorie consumption 2939 kilocalories

UZBEKISTAN
Central Asia

Official name Republic of Uzbekistan
Formation 1991 / 1991
Capital Tashkent
Population 29.3 million / 170 people per sq mile (65 people per sq km)
Total area 172,741 sq. miles (447,400 sq. km)
Languages Uzbek*, Russian, Tajik, Kazakh
Religions Sunni Muslim 88%, Orthodox Christian 9%, Other 3%
Ethnic mix Uzbek 80%, Russian 6%, Other 6%, Tajik 5%, Kazakh 3%
Government Presidential system
Currency Som = 100 tiyin
Literacy rate 99%
Calorie consumption 2675 kilocalories

VANUATU
Australasia & Oceania

Official name Republic of Vanuatu
Formation 1980 / 1980
Capital Port Vila
Population 300,000 / 64 people per sq mile (25 people per sq km)
Total area 4710 sq. miles (12,200 sq. km)
Languages Bislama (Melanesian pidgin)*, English*, French*, other indigenous languages
Religions Presbyterian 37%, Other 19%, Anglican 15%, Roman Catholic 15%, Traditional beliefs 8%, Seventh-day Adventist 6%
Ethnic mix ni-Vanuatu 94%, European 4%, Other 2%
Government Parliamentary system
Currency Vatu = 100 centimes
Literacy rate 83%
Calorie consumption 2820 kilocalories

VATICAN CITY
Southern Europe

Official name State of the Vatican City
Formation 1929 / 1929
Capital Vatican City
Population 842 / 4953 people per sq mile (1914 people per sq km)
Total area 0.17 sq. miles (0.44 sq. km)
Languages Italian*, Latin*
Religions Roman Catholic 100%
Ethnic mix The current pope is Argentinian, though most popes for the last 500 years have been Italian. Cardinals are from many nationalities, but Italians form the largest group. Most of the resident lay persons are Italian.
Government Papal state
Currency Euro = 100 cents
Literacy rate 99%
Calorie consumption Not available

VENEZUELA
South America

Official name Bolivarian Republic of Venezuela
Formation 1830 / 1830
Capital Caracas
Population 30.9 million / 91 people per sq mile (35 people per sq km)
Total area 352,143 sq. miles (912,050 sq. km)
Languages Spanish*, Amerindian languages
Religions Roman Catholic 96%, Protestant 2%, Other 2%
Ethnic mix Mestizo 69%, White 20%, Black 9%, Amerindian 2%
Government Presidential system
Currency Bolívar fuerte = 100 céntimos
Literacy rate 96%
Calorie consumption 2880 kilocalories

VIETNAM
Southeast Asia

Official name Socialist Republic of Vietnam
Formation 1976 / 1976
Capital Hanoi
Population 92.5 million / 736 people per sq mile (284 people per sq km)
Total area 127,243 sq. miles (329,560 sq. km)
Languages Vietnamese*, Chinese, Thai, Khmer, Muong, Nung, Miao, Yao, Jarai
Religions Other 74%, Buddhist 14%, Roman Catholic 7%, Cao Dai 3%, Protestant 2%
Ethnic mix Vietnamese 86%, Other 8%, Muong 2%, Tay 2%, Thai 2%
Government One-party state
Currency Dông = 10 hao = 100 xu
Literacy rate 94%
Calorie consumption 2745 kilocalories

YEMEN
Southwest Asia

Official name Republic of Yemen
Formation 1990 / 1990
Capital Sana
Population 25 million / 115 people per sq mile (44 people per sq km)
Total area 203,849 sq. miles (527,970 sq. km)
Languages Arabic*
Religions Sunni Muslim 55%, Shi'a Muslim 42%, Christian, Hindu, and Jewish 3%
Ethnic mix Arab 99%, Afro-Arab, Indian, Somali, and European 1%
Government Transitional regime
Currency Yemeni rial = 100 fils
Literacy rate 66%
Calorie consumption 2223 kilocalories

ZAMBIA
Southern Africa

Official name Republic of Zambia
Formation 1964 / 1964
Capital Lusaka
Population 15 million / 52 people per sq mile (20 people per sq km)
Total area 290,584 sq. miles (752,614 sq. km)
Languages Bemba, Tonga, Nyanja, Lozi, Lala-Bisa, Nsenga, English*
Religions Christian 63%, Traditional beliefs 36%, Muslim and Hindu 1%
Ethnic mix Bemba 34%, Other African 26%, Tonga 16%, Nyanja 14%, Lozi 9%, European 1%
Government Presidential system
Currency New Zambian kwacha = 100 ngwee
Literacy rate 61%
Calorie consumption 1930 kilocalories

ZIMBABWE
Southern Africa

Official name Republic of Zimbabwe
Formation 1980 / 1980
Capital Harare
Population 14.6 million / 98 people per sq mile (38 people per sq km)
Total area 150,803 sq. miles (390,580 sq. km)
Languages Shona, isiNdebele, English*
Religions Syncretic (Christian/traditional beliefs) 50%, Christian 25%, Traditional beliefs 24%, Other (including Muslim) 1%
Ethnic mix Shona 71%, Ndebele 16%, Other African 11%, White 1%, Asian 1%
Government Presidential system
Currency US $, South African rand, Euro, UK £, Botswana pula, Australian $, Chinese yuan, Indian rupee, and Japanese yen are legal tender
Literacy rate 84%
Calorie consumption 2110 kilocalories

GLOSSARY

This glossary lists all geographical, technical, and foreign language terms which appear in the text, followed by a brief definition of the term. Any acronyms used in the text are also listed in full. Terms in italics are for cross-reference and indicate that the word is separately defined in the glossary.

A

Aboriginal The original (*indigenous*) inhabitants of a country or continent. Especially used with reference to Australia.

Abyssal plain A broad *plain* found in the depths of the ocean, more than 10,000 ft (3,000 m) below sea level.

Acid rain Rain, sleet, snow, or mist which has absorbed waste gases from fossil-fueled power stations and vehicle exhausts, becoming more acid. It causes severe environmental damage.

Adaptation The gradual evolution of plants and animals so that they become better suited to survive and reproduce in their *environment*.

Afforestation The planting of new forest in areas that were once forested but have been cleared.

Agribusiness A term applied to activities such as the growing of crops, rearing of animals, or the manufacture of farm machinery, which eventually leads to the supply of agricultural produce at market.

Air mass A huge, homogeneous mass of air, within which horizontal patterns of temperature and *humidity* are consistent. Air masses are separated by *fronts*.

Alliance An agreement between two or more states, to work together to achieve common purposes.

Alluvial fan A large fan-shaped deposit of fine sediments deposited by a river as it emerges from a narrow, mountain valley onto a broad, open *plain*.

Alluvium Material deposited by rivers. Nowadays usually only applied to finer particles of silt and clay.

Alpine Mountain *environment*, between the *treeline* and the level of permanent snow cover.

Alpine mountains Ranges of mountains formed between 30 and 65 million years ago, by *folding*, in western and central Europe.

Amerindian A term applied to people *indigenous* to North, Central, and South America.

Animal husbandry The business of rearing animals.

Antarctic circle The parallel which lies at *latitude* of 66° 32' S.

Anticline A geological *fold* that forms an arch shape, curving upward in the rock *strata*.

Anticyclone An area of relatively high atmospheric pressure.

Aquaculture Collective term for the farming of produce derived from the sea, including fish-farming, the cultivation of shellfish, and plants such as seaweed.

Aquifer A body of rock that can absorb water. Also applied to any rock strata that have sufficient porosity to yield *groundwater* through wells or springs.

Arable Land which has been plowed and is being used, or is suitable, for growing crops.

Archipelago A group or chain of islands.

Arctic Circle The parallel that lies at a *latitude* of 66° 32' N.

Arête A thin, jagged mountain ridge that divides two adjacent *cirques*, found in regions where *glaciation* has occurred.

Arid Dry. An area of low rainfall, where the rate of *evaporation* may be greater than that of *precipitation*. Often defined as those areas that receive less than one inch (25 mm) of rain a year. In these areas only drought-resistant plants can survive.

Artesian well A naturally occurring source of underground water, stored in an *aquifer*.

Artisanal Small-scale, manual operation, such as fishing, using little or no machinery.

ASEAN Association of Southeast Asian Nations. Established in 1967 to promote economic, social, and cultural cooperation. Its members include Brunei, Indonesia, Malaysia, Philippines, Singapore, and Thailand.

Aseismic A region where *earthquake* activity has ceased.

Asteroid A minor planet circling the Sun, mainly between the orbits of Mars and Jupiter.

Asthenosphere A zone of hot, partially melted rock, which underlies the *lithosphere*, within the Earth's *crust*.

Atmosphere The envelope of odorless, colorless and tasteless gases surrounding the Earth, consisting of *oxygen* (23%), *nitrogen* (75%), argon (1%), *carbon dioxide* (0.03%), as well as tiny proportions of other gases.

Atmospheric pressure The pressure created by the action of gravity on the gases surrounding the Earth.

Atoll A ring-shaped island or *coral reef* often enclosing a *lagoon* of sea water.

Avalanche The rapid movement of a mass of snow and ice down a steep slope. Similar movements of other materials are described as *rock avalanches* or *landslides* and *sand avalanches*.

B

Badlands A landscape that has been heavily eroded and dissected by rainwater, and which has little or no vegetation.

Back slope The gentler windward slope of a sand *dune* or gentler slope of a *cuesta*.

Bajos An *alluvial fan* deposited by a river at the base of mountains and hills that encircle *desert* areas.

Bar, coastal An offshore strip of sand or shingle, either above or below the water. Usually parallel to the shore but sometimes crescent-shaped or at an oblique angle.

Barchan A crescent-shaped sand *dune*, formed where wind direction is very consistent. The horns of the crescent point downwind and where there is enough sand the barchan is mobile.

Barrio A Spanish term for the shantytowns – settlements of shacks – that are clustered around many South and Central American cities (*see also Favela*).

Basalt Dark, fine-grained *igneous rock* that is formed near the Earth's surface from fast-cooling *lava*.

Base level The level below which flowing water cannot erode the land.

Basement rock A mass of ancient rock often of *PreCambrian age*, covered by a layer of more recent *sedimentary rocks*. Commonly associated with *shield* areas.

Beach Lake or sea shore where waves break and there is an accumulation of loose sand, mud, gravel, or pebbles.

Bedrock Solid, consolidated and relatively unweathered rock, found on the surface of the land or just below a layer of soil or *weathered* rock.

Biodiversity The quantity of animal or plant species in a given area.

Biomass The total mass of organic matter – plants and animals – in a given area. It is usually measured in kilogrammes per square meter. Plant biomass is proportionally greater than that of animals, except in cities.

Biosphere The zone just above and below the Earth's surface, where all plants and animals live.

Blizzard A severe windstorm with snow and sleet. Visibility is often severely restricted.

Bluff The steep bank of a *meander*, formed by the erosive action of a river.

Boreal forest Tracts of mainly coniferous forest found in northern *latitudes*.

Breccia A type of rock composed of sharp fragments, cemented by a fine-grained material such as clay.

Butte An isolated, flat-topped hill with steep or vertical sides, buttes are the eroded remnants of a former land surface

C

Caatinga Portuguese (Brazilian) term for thorny woodland growing in areas of pale granitic soils.

CACM Central American Common Market. Established in 1960 to further economic ties between its members, which are Costa Rica, El Salvador, Guatemala, Honduras, and Nicaragua.

Calcite Hexagonal crystals of calcium carbonate.

Caldera A huge volcanic vent, often containing a number of smaller vents, and sometimes a crater lake.

Carbon cycle The transfer of carbon to and from the *atmosphere*. This occurs on land through *photosynthesis*. In the sea, *carbon dioxide* is absorbed, some returning to the air and some taken up into the bodies of sea creatures.

Carbon dioxide A colorless, odorless gas (CO_2) that makes up 0.03% of the *atmosphere*.

Carbonation The process whereby rocks are broken down by carbonic acid. Carbon dioxide in the air dissolves in rainwater, forming carbonic acid. *Limestone* terrain can be rapidly eaten away.

Cash crop A single crop grown specifically for export sale, rather than for local use. Typical examples include coffee, tea, and citrus fruits.

Cassava A type of grain meal, used to produce tapioca. A staple crop in many parts of Africa.

Castle kopje Hill or rock outcrop, especially in southern Africa, where steep sides, and a summit composed of blocks, give a castle-like appearance.

Cataracts A series of stepped waterfalls created as a river flows over a band of hard, resistant rock.

Causeway A raised route through marshland or a body of water.

CEEAC Economic Community of Central African States. Established in 1983 to promote regional cooperation and if possible, establish a common market between 16 Central African nations.

Chemical weathering The chemical reactions leading to the decomposition of rocks. Types of chemical weathering include *carbonation*, *hydrolysis*, and *oxidation*.

Chernozem A fertile soil, also known as "black earth" consisting of a layer of dark topsoil, rich in decaying vegetation, overlying a lighter chalky layer.

Cirque Armchair-shaped basin, found in mountain regions, with a steep back, or rear, wall and a raised rock lip, often containing a lake (or *tarn*). The cirque floor has been eroded by a *glacier*, while the back wall is eroded both by the *glacier* and by *weathering*.

Climate The average weather conditions in a given area over a period of years, sometimes defined as 30 years or more.

Cold War A period of hostile relations between the US and the Soviet Union and their allies after the Second World War.

Composite volcano Also known as a strato-volcano, the volcanic cone is composed of alternating deposits of *lava* and *pyroclastic* material.

Compound A substance made up of *elements* chemically combined in a consistent way.

Condensation The process whereby a gas changes into a liquid. For example, water vapor in the *atmosphere* condenses around tiny airborne particles to form droplets of water.

Confluence The point at which two rivers meet.

Conglomerate Rock composed of large, water-worn or rounded pebbles, held together by a natural cement.

Coniferous forest A forest type containing trees which are generally, but not necessarily, *evergreen* and have slender, needlelike leaves. Coniferous trees reproduce by means of seeds contained in a cone.

Continental drift The theory that the continents of today are fragments of one or more prehistoric *supercontinents* which have moved across the Earth's surface, creating ocean basins. The theory has been superseded by a more sophisticated one – *plate tectonics*.

Continental shelf An area of the continental crust, below sea level, which slopes gently. It is separated from the deep ocean by a much more steeply inclined *continental slope*.

Continental slope A steep slope running from the edge of the *continental shelf* to the ocean floor.

Conurbation A vast metropolitan area created by the expansion of towns and cities into a virtually continuous urban area.

Cool continental A rainy *climate* with warm summers [warmest month below 76°F (22°C)] and often severe winters [coldest month below 32°F (0°C)].

Copra The dried, white kernel of a coconut, from which coconut oil is extracted.

Coral reef An underwater barrier created by colonies of the coral polyp. Polyps secrete a protective skeleton of calcium carbonate, and reefs develop as live polyps build on the skeletons of dead generations.

Core The center of the Earth, consisting of a dense mass of iron and nickel. It is thought that the outer core is molten or liquid, and that the hot inner core is solid due to extremely high pressures.

Coriolis effect A deflecting force caused by the rotation of the Earth. In the northern hemisphere a body, such as an *air mass* or ocean current, is deflected to the right, and in the southern hemisphere to the left. This prevents winds from blowing straight from areas of high to low pressure.

Coulées A US / Canadian term for a ravine formed by river *erosion*.

Craton A large block of the Earth's *crust* which has remained stable for a long period of *geological time*. It is made up of ancient *shield* rocks.

Cretaceous A period of *geological time* beginning about 145 million years ago and lasting until about 65 million years ago.

Crevasse A deep crack in a *glacier*.

Crust The hard, thin outer shell of the Earth. The crust floats on the *mantle*, which is softer and more dense. Under the oceans (oceanic crust) the crust is 3.7–6.8 miles (6–11 km) thick. Continental crust averages 18–24 miles (30–40 km).

Crystalline rock Rocks formed when molten *magma* crystallizes (*igneous rocks*) or when heat or pressure cause re-crystallization (*metamorphic rocks*). Crystalline rocks are distinct from *sedimentary rocks*.

Cuesta A hill which rises into a steep slope on one side but has a gentler gradient on its other slope.

Cyclone An area of low *atmospheric pressure*, occurring where the air is warm and relatively low in density, causing low level winds to spiral. *Hurricanes* and *typhoons* are tropical cyclones.

D

De facto
1 Government or other activity that takes place, or exists in actuality if not by right.
2 A border, which exists in practice, but which is not officially recognized by all the countries it adjoins.

Deciduous forest A forest of trees that shed their leaves annually at a particular time or season. In *temperate* climates the fall of leaves occurs in the autumn. Some *coniferous* trees, such as the larch, are deciduous. Deciduous vegetation contrasts with *evergreen*, which keeps its leaves for more than a year.

Defoliant Chemical spray used to remove foliage (leaves) from trees.

Deforestation The act of cutting down and clearing large areas of forest for human activities, such as agricultural land or urban development.

Delta Low-lying, fan-shaped area at a river mouth, formed by the *deposition* of successive layers of *sediment*. Slowing as it enters the sea, a river deposits sediment and may, as a result, split into numerous smaller channels, known as *distributaries*.

Denudation The combined effect of *weathering*, *erosion*, and *mass movement*, which, over long periods, exposes underlying rocks.

Deposition The laying down of material that has accumulated:
(1) after being *eroded* and then transported by physical forces such as wind, ice, or water;
(2) as organic remains, such as coal and coral;
(3) as the result of *evaporation* and chemical *precipitation*.

Depression
1 In climatic terms it is a large low pressure system.
2 A complex *fold*, producing a large valley, which incorporates both a *syncline* and an *anticline*.

Desert An *arid* region of low rainfall, with little vegetation or animal life, which is adapted to the dry conditions. The term is now applied not only to hot tropical and subtropical regions, but to arid areas of the continental interiors and to the ice deserts of the *Arctic* and *Antarctic*.

Desertification The gradual extension of *desert* conditions in *arid* or *semiarid* regions, as a result of climatic change or human activity, such as over-grazing and *deforestation*.

Despot A ruler with absolute power. Despots are often associated with oppressive regimes.

Detritus Piles of rock deposited by an erosive agent such as a river or *glacier*.

Distributary A minor branch of a river, which does not rejoin the main stream, common at *deltas*.

Diurnal Daily, something that occurs each day. Diurnal temperature refers to the variation in temperature over the course of a full day and night.

Divide A US term describing the area of high ground separating two *drainage basins*.

Donga A steep-sided *gully*, resulting from *erosion* by a river or by floods.

Dormant A term used to describe a *volcano* which is not currently erupting. They differ from extinct volcanoes as dormant volcanoes are still considered likely to erupt in the future.

Drainage basin The area drained by a single river system, its boundary is marked by a *watershed* or *divide*.

Drought A long period of continuously low rainfall.

Drumlin A long, streamlined hillock composed of material deposited by a *glacier*. They often occur in groups known as swarms.

Dune A mound or ridge of sand, shaped, and often moved, by the wind. They are found in hot *deserts* and on low-lying coasts where onshore winds blow across sandy beaches.

Dyke A wall constructed in low-lying areas to contain floodwaters or protect from high tides.

E

Earthflow The rapid movement of soil and other loose surface material down a slope, when saturated by water. Similar to a mudflow but not as fast-flowing, due to a lower percentage of water.

Earthquake Sudden movements of the Earth's *crust*, causing the ground to shake. Frequently occurring at *tectonic plate* margins. The shock, or series of shocks, spreads out from an *epicenter*.

EC The European Community (*see* EU).

Ecosystem A system of living organisms – plants and animals – interacting with their *environment*.

ECOWAS Economic Community of West African States. Established in 1975, it incorporates 16 West African states and aims to promote closer regional and economic cooperation.

Element
1 A constituent of the *climate* – *precipitation*, *humidity*, temperature, *atmospheric pressure*, or wind.
2 A substance that cannot be separated into simpler substances by chemical means.

El Niño A climatic phenomenon, the El Niño effect occurs about 14 times each century and leads to major shifts in global air circulation. It is associated with unusually warm currents off the coasts of Peru, Ecuador and Chile. The anomaly can last for up to two years.

Environment The conditions created by the surroundings (both natural and artificial) within which an organism lives. In human geography the word includes the surrounding economic, cultural, and social conditions.

Eon (aeon) Traditionally a long, but indefinite, period of *geological time*.

Ephemeral A nonpermanent feature, often used in connection with seasonal rivers or lakes in dry areas.

Epicenter The point on the Earth's surface directly above the underground origin – or focus – of an *earthquake*.

Equator The line of *latitude* which lies equidistant between the North and South Poles.

Erg An extensive area of sand *dunes*, particularly in the Sahara Desert.

Erosion The processes which wear away the surface of the land. *Glaciers*, wind, rivers, waves, and currents all carry debris which causes *erosion*. Some definitions also include *mass movement* due to gravity as an agent of erosion.

Escarpment A steep slope at the margin of a level, upland surface. In a landscape created by *folding*, escarpments (or scarps) frequently lie behind a more gentle backward slope.

Esker A narrow, winding ridge of sand and gravel deposited by streams of water flowing beneath or at the edge of a *glacier*.

Erratic A rock transported by a *glacier* and deposited some distance from its place of origin.

Eustacy A world-wide fall or rise in ocean levels.

EU The European Union. Established in 1965, it was formerly known as the EEC (European Economic Community) and then the EC (European Community). Its members are Austria, Belgium, Denmark, Finland, France, Germany, Greece, Ireland, Italy, Luxembourg, Netherlands, Portugal, Spain, Sweden, and UK. It seeks to establish an integrated European common market and eventual federation.

Evaporation The process whereby a liquid or solid is turned into a gas or vapor. Also refers to the diffusion of water vapor into the *atmosphere* from exposed water surfaces such as lakes and seas.

Evapotranspiration The loss of moisture from the Earth's surface through a combination of *evaporation*, and *transpiration* from the leaves of plants.

Evergreen Plants with long-lasting leaves, which are not shed annually or seasonally.

Exfoliation A kind of *weathering* whereby scalelike flakes of rock are peeled or broken off by the development of salt crystals in water within the rocks. *Groundwater*, which contains dissolved salts, seeps to the surface and evaporates, precipitating a film of salt crystals, which expands causing fine cracks. As these grow, flakes of rock break off.

Extrusive rock *Igneous* rock formed when molten material (*magma*) pours forth at the Earth's surface and cools rapidly. It usually has a glassy texture.

F

Factionalism The actions of one or more minority political group acting against the interests of the majority government.

Fault A fracture or crack in rock, where strains (*tectonic* movement) have caused blocks to move, vertically or laterally, relative to each other.

Fauna Collective name for the animals of a particular period of time, or region.

Favela Brazilian term for the shantytowns or temporary huts that have grown up around the edge of many South and Central American cities.

Ferrel cell A component in the global pattern of air circulation, which rises in the colder *latitudes* (60° N and S) and descends in warmer *latitudes* (30° N and S). The Ferrel cell forms part of the world's three-cell air circulation pattern, with the *Hadley* and Polar cells.

Fissure A deep crack in a rock or a *glacier*.

Fjord A deep, narrow inlet, created when the sea inundates the *U-shaped valley* created by a *glacier*.

Flash flood A sudden, short-lived rise in the water level of a river or stream, or surge of water down a dry river channel, or *wadi*, caused by heavy rainfall.

Flax A plant used to make linen.

Floodplain The broad, flat part of a river valley, adjacent to the river itself, formed by *sediment* deposited during flooding.

Flora The collective name for the plants of a particular period of time or region.

Flow The movement of a river within its banks, particularly in terms of the speed and volume of water.

Fold A bend in the rock *strata* of the Earth's *crust*, resulting from compression.

Fossil The remains, or traces,of a dead organism preserved in the Earth's *crust*.

Fossil dune A *dune* formed in a once-*arid* region which is now wetter. *Dunes* normally move with the wind, but in these cases vegetation makes them stable.

Fossil fuel Fuel – coal, natural gas or oil – composed of the fossilized remains of plants and animals.

Front The boundary between two *air masses*, which contrast sharply in temperature and *humidity*.

Frontal depression An area of low pressure caused by rising warm air. They are generally 600–1,200 miles (1,000–2,000 km) in diameter. Within *depressions* there are both warm and cold fronts.

Frost shattering A form of *weathering* where water freezes in cracks, causing expansion. As temperatures fluctuate and the ice melts and refreezes, it eventually causes the rocks to shatter and fragments of rock to break off.

G

Gaucho South American term for a stock herder or cowboy who works on the grassy *plains* of Paraguay, Uruguay, and Argentina.

Geological timescale The chronology of the Earth's history as revealed in its rocks. Geological time is divided into a number of periods: *eon*, era, period, epoch, age, and chron (the shortest). These units are not of uniform length.

Geosyncline A concave fold (*syncline*) or large depression in the Earth's *crust*, extending hundreds of miles. This basin contains a deep layer of sediment, especially at its center, from the land masses around it.

Geothermal energy Heat derived from hot rocks within the Earth's *crust* and resulting in hot springs, steam, or hot rocks at the surface. The energy is generated by rock movements, and from the breakdown of radioactive elements occurring under intense pressure.

GDP Gross Domestic Product. The total value of goods and services produced by a country excluding income from foreign countries.

Geyser A jet of steam and hot water that intermittently erupts from vents in the ground in areas that are, or were, *volcanic*. Some geysers occasionally reach heights of 196 ft (60 m).

Ghetto An area of a city or region occupied by an overwhelming majority of people from one racial or religious group, who may be subject to persecution or containment.

Glaciation The growth of *glaciers* and *ice sheets*, and their impact on the landscape.

Glacier A body of ice moving downslope under the influence of gravity and consisting of compacted and frozen snow. A glacier is distinct from an *ice sheet*, which is wider and less confined by features of the landscape.

Glacio-eustacy A world-wide change in the level of the oceans, caused when the formation of *ice sheets* takes up water or when their melting returns water to the ocean. The formation of ice sheets in the *Pleistocene* epoch, for example, caused sea level to drop by about 320 ft (100-m).

Glaciofluvial To do with glacial *meltwater*, the landforms it creates and its processes; *erosion*, transportation, and *deposition*. Glaciofluvial effects are more powerful and rapid where they occur within or beneath the *glacier*, rather than beyond its edge.

Glacis A gentle slope or *pediment*.

Global warming An increase in the average temperature of the Earth. At present the *greenhouse effect* is thought to contribute to this.

GNP Gross National Product. The total value of goods and services produced by a country.

Gondwanaland The *supercontinent* thought to have existed over 200 million years ago in the southern hemisphere. Gondwanaland is believed to have comprised today's Africa, Madagascar, Australia, parts of South America, *Antarctica*, and the Indian subcontinent.

Graben A block of rock let down between two parallel *faults*. Where the graben occurs within a valley, the structure is known as a *rift valley*.

Grease ice Slicks of ice which form in *Antarctic* seas, when ice crystals are bonded together by wind and wave action.

Greenhouse effect A change in the temperature of the *atmosphere*. Short-wave solar radiation travels through the *atmosphere* unimpeded to the Earth's surface, whereas outgoing, long-wave terrestrial radiation is absorbed by materials that reradiate it back to the Earth. Radiation trapped in this way, by water vapor, carbon dioxide, and other "greenhouse gases," keeps the Earth warm. As more *carbon dioxide* is released into the atmosphere by the burning of *fossil fuels*, the greenhouse effect may cause a global increase in temperature.

Groundwater Water that has seeped into the pores, cavities, and cracks of rocks or into soil and water held in an *aquifer*.

Gully A deep, narrow channel eroded in the landscape by *ephemeral* streams.

Guyot A small, flat-topped submarine mountain, formed as a result of subsidence which occurs during *sea-floor spreading*.

Gypsum A soft mineral *compound* (hydrated calcium sulphate), used as the basis of many forms of plaster, including plaster of Paris.

H

Hadley cell A large-scale component in the global pattern of air circulation. Warm air rises over the *Equator* and blows at high altitude toward the poles, sinking in subtropical regions (30° N and 30° S) and creating high pressure. The air then flows at the surface toward the *Equator* in the form of trade winds. There is one cell in each hemisphere. Named after G. Hadley, who published his theory in 1735.

Hamada An Arabic word for a plateau of bare rock in a *desert*.

Hanging valley A tributary valley that ends suddenly, high above the bed of the main valley. The effect is found where the main valley has been more deeply eroded by a *glacier*, than has the tributary valley. A stream in a hanging valley will descend to the floor of the main valley as a waterfall or *cataract*.

Headwards The action of a river eroding back upstream, as opposed to the normal process of downstream *erosion*. Headwards erosion is often associated with *gullying*.

Hoodos Pinnacles of rock that have been worn away by *weathering* in *semiarid* regions.

Horst A block of the Earth's *crust* which has been left upstanding by the sinking of adjoining blocks along fault lines.

Hot spot A region of the Earth's *crust* where high thermal activity occurs, often leading to volcanic eruptions. Hot spots often occur far from plate boundaries, but their movement is associated with *plate tectonics*.

Humid equatorial Rainy *climate* with no winter, where the coolest month is generally above 64°F (18°C).

Humidity The relative amount of moisture held in the Earth's *atmosphere*.

Hurricane
1 A tropical *cyclone* occurring in the Caribbean and western North Atlantic.
2 A wind of more than 65 knots (75 kmph).

Hydroelectric power Energy produced by harnessing the rapid movement of water down steep mountain slopes to drive turbines to generate electricity.

Hydrolysis The chemical breakdown of rocks in reaction with water, forming new compounds.

I

Ice Age A period in the Earth's history when surface temperatures in the temperate *latitudes* were much lower and *ice sheets* expanded considerably. There have been *ice ages* from *Pre-Cambrian* times onward. The most recent began two million years ago and ended 10,000 years ago.

Ice cap A permanent dome of ice in highland areas. The term ice cap is often seen as distinct from *ice sheet*, which denotes a much wider covering of ice; and is also used refer to the very extensive polar and Greenland ice caps.

Ice floe A large, flat mass of ice floating free on the ocean surface. It is usually formed after the break-up of winter ice by heavy storms.

Ice sheet A continuous, very thick layer of ice and snow. The term is usually used of ice masses which are continental in extent.

Ice shelf A floating mass of ice attached to the edge of a coast. The seaward edge is usually a sheer cliff up to 100 ft (30-m) high.

Ice wedge Massive blocks of ice up to 6.5-ft (2-m) wide at the top and extending 32-ft (10-m) deep. They are found in cracks in *polygonally-patterned* ground in *periglacial* regions.

Iceberg A large mass of ice in a lake or a sea, which has broken off from a floating *ice sheet* (an *ice shelf*) or from a *glacier*.

Igneous rock Rock formed when molten material, *magma*, from the hot, lower layers of the Earth's *crust*, cools, solidifies, and crystallizes, either within the Earth's *crust* (*intrusive*) or on the surface (*extrusive*).

IMF International Monetary Fund. Established in 1944 as a UN agency, it contains 182 members around the world and is concerned with world monetary stability and economic development.

Incised meander A *meander* where the river, following its original course, cuts deeply into *bedrock*. This may occur when a mature, meandering river begins to erode its bed much more vigorously after the surrounding land has been uplifted.

Indigenous People, plants, or animals native to a particular region.

Infrastructure The communications and services – roads, railroads, and telecommunications – necessary for the functioning of a country or region.

Inselberg An isolated, steep-sided hill, rising from a low *plain* in *semiarid* and *savannah* landscapes. Inselbergs are usually composed of a rock, such as granite, which resists *erosion*.

Interglacial A period of global *climate*, between two *ice ages*, when temperatures rise and *ice sheets* and *glaciers* retreat.

Intraplate volcano A *volcano* which lies in the centre of one of the Earth's *tectonic plates*, rather than, as is more common, at its edge. They are thought to have been formed by a *hot spot*.

Intrusion (intrusive igneous rock) Rock formed when molten material, *magma*, penetrates existing rocks below the Earth's surface before cooling and solidifying. These rocks cool more slowly than extrusive rock and therefore tend to have coarser grains.

Irrigation The artificial supply of agricultural water to dry areas, often involving the creation of canals and the diversion of natural watercourses.

Island arc A curved chain of islands. Typically, such an arc fringes an ocean trench, formed at the margin between two *tectonic plates*. As one plate overrides another, *earthquakes* and volcanic activity are common and the islands themselves are often volcanic cones.

Isostasy The state of equilibrium that the Earth's *crust* maintains as its lighter and heavier parts float on the denser underlying mantle.

Isthmus A narrow strip of land connecting two larger landmasses or islands.

J

Jet stream A narrow belt of westerly winds in the *troposphere*, at altitudes above 39,000 ft (12,000 m). Jet streams tend to blow more strongly in winter and include: the subtropical jet stream; the *polar* front jet stream in mid-*latitudes*; the *Arctic* jet stream; and the polar-night jet stream.

Joint A crack in a rock, formed where blocks of rock have not shifted relative to each other, as is the case with a *fault*. Joints are created by *folding*; by shrinkage in *igneous rock* as it cools or *sedimentary rock* as it dries out; and by the release of pressure in a rock mass when overlying materials are removed by *erosion*.

Jute A plant fiber used to make coarse ropes, sacks, and matting.

K

Kame A mound of stratified sand and gravel with steep sides, deposited in a *crevasse* by *meltwater* running over a *glacier*. When the ice retreats, this forms an undulating terrain of hummocks.

Karst A barren *limestone* landscape created by carbonic acid in streams and rainwater, in areas where *limestone* is close to the surface. Typical features include caverns, towerlike hills, *sinkholes*, and flat limestone pavements.

Kettle hole A round hollow formed in a glacial deposit by a detached block of glacial ice, which later melted. They can fill with water to form kettle-lakes.

L

Lagoon A shallow stretch of coastal salt-water behind a partial barrier such as a sandbank or *coral reef*. Lagoon is also used to describe the water encircled by an *atoll*.

LAIA Latin American Integration Association. Established in 1980, its members are Argentina, Bolivia, Brazil, Chile, Colombia, Ecuador, Mexico, Paraguay, Peru, Uruguay, and Venezuala. It aims to promote economic cooperation between member states.

Landslide The sudden downslope movement of a mass of rock or earth on a slope, caused either by heavy rain; the impact of waves; an *earthquake* or human activity.

Laterite A hard red deposit left by *chemical weathering* in tropical conditions, and consisting mainly of oxides of iron and aluminium.

Latitude The angular distance from the *Equator*, to a given point on the Earth's surface. Imaginary lines of *latitude* running parallel to the Equator encircle the Earth, and are measured in degrees north or south of the Equator. The Equator is 0°, the poles 90° South and North respectively. Also called parallels.

Laurasia In the theory of *continental drift*, the northern part of the great *supercontinent* of *Pangaea*. Laurasia is said to consist of N America, Greenland and all of Eurasia north of the Indian subcontinent.

Lava The molten rock, *magma*, which erupts onto the Earth's surface through a *volcano*, or through a *fault* or crack in the Earth's *crust*. Lava refers to the rock both in its molten and in its later, solidified form.

Leaching The process whereby water dissolves minerals and moves them down through layers of soil or rock.

Levée A raised bank alongside the channel of a river. Levées are either human-made or formed in times of flood when the river overflows its channel, slows and deposits much of its *sediment* load.

Lichen An organism which is the symbiotic product of an algae and a fungus. Lichens form in tight crusts on stones and trees, and are resistant to extreme cold. They are often found in tundra regions.

Lignite Low-grade coal, also known as brown coal. Found in large deposits in eastern Europe.

Limestone A porous *sedimentary* rock formed from carbonate materials.

Lingua franca The language adopted as the common language between speakers whose native languages are different. This is common in former colonial states.

Lithosphere The rigid upper layer of the Earth, comprising the *crust* and the upper part of the *mantle*.

Llanos Vast grassland *plains* of northern South America.

Loess Fine-grained, yellow deposits of unstratified silts and sands. Loess is believed to be wind-carried *sediment* created in the last *Ice Age*. Some deposits may later have been redistributed by rivers. Loess-derived soils are of high quality, fertile, and easy to work.

Longitude A division of the Earth which pinpoints how far east or west a given place is from the Prime Meridian (0°) which runs through the Royal Observatory at Greenwich, England (UK). Imaginary lines of longitude are drawn around the world from pole to pole. The world is divided into 360 degrees.

Longshore drift The movement of sand and silt along the coast, carried by waves hitting the beach at an angle.

M

Magma Underground, molten rock, which is very hot and highly charged with gas. It is generated at great pressure, at depths 10 miles (16 km) or more below the Earth's surface. It can issue as *lava* at the Earth's surface or, more often, solidify below the surface as *intrusive igneous rock*.

Mantle The layer of the Earth between the *crust* and the *core*. It is about 1,800 miles (2,900-km) thick. The uppermost layer of the mantle is the soft, 125-mile (200 km) thick *asthenosphere* on which the more rigid *lithosphere* floats.

Maquiladoras Factories on the Mexico side of the Mexico/US border, that are allowed to import raw materials and components duty-free and use low-cost labor to assemble the goods, finally exporting them for sale in the US.

Market gardening The intensive growing of fruit and vegetables close to large local markets.

Mass movement Downslope movement of weathered materials such as rock, often helped by rainfall or glacial *meltwater*. Mass movement may be a gradual process or rapid, as in a *landslide* or rockfall.

Massif A single very large mountain or an area of mountains with uniform characteristics and clearly-defined boundaries.

Meander A looplike bend in a river, which is found typically in the lower, mature reaches of a river but can form wherever the valley is wide and the slope gentle.

Mediterranean climate A temperate *climate* of hot, dry summers and warm, damp winters. This is typical of the western fringes of the world's continents in the warm temperate regions between *latitudes* of 30° and 40° (north and south).

Meltwater Water resulting from the melting of a *glacier* or *ice sheet*.

Mesa A broad, flat-topped hill, characteristic of *arid* regions.

Mesosphere A layer of the Earth's *atmosphere*, between the *stratosphere* and the *thermosphere*. Extending from about 25–50 miles (40–80 km) above the surface of the Earth.

Mestizo A person of mixed *Amerindian* and European origin.

Metallurgy The refining and working of metals.

Metamorphic rocks Rocks that have been altered from their original form, in terms of texture, composition, and structure by intense heat, pressure, or by the introduction of new chemical substances – or a combination of more than one of these.

Meteor A body of rock, metal or other material, that travels through space at great speeds. Meteors are visible as they enter the Earth's *atmosphere* as shooting stars and fireballs.

Meteorite The remains of a *meteor* that has fallen to Earth.

Meteoroid A *meteor* that is still traveling in space, outside the Earth's *atmosphere*.

Mezzogiorno A term applied to the southern portion of Italy.

Milankovitch hypothesis A theory suggesting that there are a series of cycles that slightly alter the Earth's position when rotating about the Sun. The cycles identified all affect the amount of *radiation* the Earth receives at different *latitudes*. The theory is seen as a key factor in the cause of *ice ages*.

Millet A grain-crop, forming part of the staple diet in much of Africa.

Mistral A strong, dry, cold northerly or north-westerly wind, which blows from the Massif Central of France to the Mediterranean Sea. It is common in winter and its cold blasts can cause crop damage in the Rhône Delta, in France.

Mohorovicic discontinuity (Moho) The structural divide at the margin between the Earth's *crust* and the *mantle*. On average it is 20 miles (35-km) below the continents and 6-miles (10 km) below the oceans. The different densities of the *crust* and the mantle cause *earthquake* waves to accelerate at this point.

Monarchy A form of government in which the head of state is a single hereditary monarch. The monarch may be a mere figurehead, or may retain significant authority.

Monsoon A wind that changes direction biannually. The change is caused by the reversal of pressure over landmasses and the adjacent oceans. Because the inflowing moist winds bring rain, the term monsoon is also used to refer to the rains themselves. The term is derived from and most commonly refers to the seasonal winds of south and east Asia.

Montaña Mountain areas along the west coast of South America.

Moraine Debris, transported and deposited by a *glacier* or *ice sheet* in unstratified, mixed, piles of rock, boulders, pebbles, and clay.

Mountain-building The formation of *fold* mountains by tectonic activity. Also known as orogeny, mountain-building often occurs on the margin where two *tectonic plates* collide. The periods when most mountain-building occurred are known as orogenic phases and lasted many millions of years.

Mudflow An *avalanche* of mud that occurs when a mass of soil is drenched by rain or melting snow. It is a type of *mass movement*, faster than an *earthflow* because it is lubricated by water.

N

Nappe A mass of rocks which has been overfolded by repeated thrust *faulting*.

NAFTA The North American Free Trade Association. Established in 1994 between Canada, Mexico, and the US to set up a free-trade zone.

NASA The National Aeronautical and Space Administration. It is a US government agency, established in 1958 to develop manned and unmanned space programs.

NATO The North Atlantic Treaty Organization. Established in 1949 to promote mutual defense and cooperation between its members, which are Belgium, Canada, Czech Republic, Denmark, France, Germany, Greece, Iceland, Italy, Luxembourg, the Netherlands, Norway, Portugal, Poland, Spain, Turkey, UK, and US.

Nitrogen The odorless, colorless gas that makes up 78% of the atmosphere. Within the soil, it is a vital nutrient for plants.

Nomads (nomadic) Wandering communities that move around in search of suitable pasture for their herds of animals.

Nuclear fusion A technique used to create a new nucleus by the merging of two lighter ones, resulting in the release of large quantities of energy.

O

Oasis A fertile area in the midst of a *desert*, usually watered by an underground *aquifer*.

Oceanic ridge A mid-ocean ridge formed, according to the theory of *plate tectonics*, when plates drift apart and hot *magma* pours through to form new oceanic *crust*.

Oligarchy The government of a state by a small, exclusive group of people – such as an elite class or a family group.

Onion-skin weathering The *weathering* away or *exfoliation* of a rock or outcrop by the peeling off of surface layers.

Oriente A flatter region lying to the east of the Andes in South America.

Outwash plain *Glaciofluvial* material (typically clay, sand, and gravel) carried beyond an ice sheet by *meltwater* streams, forming a broad, flat deposit.

Oxbow lake A crescent-shaped lake formed on a river *floodplain* when a river erodes the outside bend of a *meander*, making the neck of the *meander* narrower until the river cuts across the neck. The meander is cut off and is dammed off with sediment, creating an oxbow lake. Also known as a cut-off or mortlake.

Oxidation A form of *chemical weathering* where *oxygen* dissolved in water reacts with minerals in rocks – particularly iron – to form oxides. Oxidation causes brown or yellow staining on rocks, and eventually leads to the break down of the rock.

Oxygen A colorless, odorless gas which is one of the main constituents of the Earth's *atmosphere* and is essential to life on Earth.

GLOSSARY

Ozone layer A layer of enriched *oxygen* (O_3) within the stratosphere, mostly between 18–50 miles (30–80 km) above the Earth's surface. It is vital to the existence of life on Earth because it absorbs harmful shortwave ultraviolet radiation, while allowing beneficial longer wave ultraviolet radiation to penetrate to the Earth's surface.

P

Pacific Rim The name given to the economically-dynamic countries bordering the Pacific Ocean.

Pack ice Ice masses more than 10 ft (3-m) thick that form on the sea surface and are not attached to a landmass.

Pancake ice Thin discs of ice, up to 8 ft (2.4 m) wide which form when slicks of *grease ice* are tossed together by winds and stormy seas.

Pangaea In the theory of continental drift, Pangaea is the original great land mass which, about 190 million years ago, began to split into Gondwanaland in the south and Laurasia in the north, separated by the Tethys Sea.

Pastoralism Grazing of livestock– usually sheep, goats, or cattle. Pastoralists in many drier areas have traditionally been *nomadic*.

Parallel *see Latitude.*

Peat Ancient, partially-decomposed vegetation found in wet, boggy conditions where there is little *oxygen*. It is the first stage in the development of coal and is often dried for use as fuel. It is also used to improve soil quality.

Pediment A gently-sloping ramp of *bedrock* below a steeper slope, often found at mountain edges in *desert* areas, but also in other climatic zones. Pediments may include depositional elements such as *alluvial fans*.

Peninsula A thin strip of land surrounded on three of its sides by water. Large examples include Florida and Korea.

Per capita Latin term meaning "for each person."

Periglacial Regions on the edges of *ice sheets* or *glaciers* or, more commonly, cold regions experiencing intense frost action, *permafrost* or both. Periglacial climates bring long, freezing winters and short, mild summers.

Permafrost Permanently frozen ground, typical of *Arctic* regions. Although a layer of soil above the permafrost melts in summer, the melted water does not drain through the permafrost.

Permeable rocks Rocks through which water can seep, because they are either porous or cracked.

Pharmaceuticals The manufacture of medicinal drugs.

Phreatic eruption A volcanic eruption which occurs when *lava* combines with *groundwater*, superheating the water and causing a sudden emission of steam at the surface.

Physical weathering (mechanical weathering) The breakdown of rocks by physical, as opposed to chemical, processes. Examples include: changes in pressure or temperature; the effect of windblown sand; the pressure of growing salt crystals in cracks within rock; and the expansion and contraction of water within rock as it freezes and thaws.

Pingo A dome of earth with a core of ice, found in *tundra* regions. Pingos are formed either when *groundwater* freezes and expands, pushing up the land surface, or when trapped, freezing water in a lake expands and pushes up lake *sediments* to form the pingo dome.

Placer A belt of mineral-bearing rock *strata* lying at or close to the Earth's surface, from which minerals can be easily extracted.

Plain A flat, level region of land, often relatively low-lying.

Plateau A highland tract of flat land.

Plate *see Tectonic plates.*

Plate tectonics The study of *tectonic plates*, that helps to explain *continental drift*, mountain formation and volcanic activity. The movement of tectonic plates may be explained by the currents of rock rising and falling from within the Earth's *mantle*, as it heats up and then cools. The boundaries of the plates are known as plate margins and most mountains, *earthquakes*, and *volcanoes* occur at these margins. Constructive margins are moving apart; destructive margins are crunching together and conservative margins are sliding past one another.

Pleistocene A period of *geological time* spanning from about 5.2 million years ago to 1.6 million years ago.

Plutonic rock *Igneous* rocks found deep below the surface. They are coarse-grained because they cooled and solidified slowly.

Polar The zones within the *Arctic* and *Antarctic* circles.

Polje A long, broad *depression* found in *karst* (*limestone*) regions.

Polygonal patterning Typical ground patterning, found in areas where the soil is subject to severe frost action, often in *periglacial* regions.

Porosity A measure of how much water can be held within a rock or a soil. Porosity is measured as the percentage of holes or pores in a material, compared to its total volume. For example, the porosity of slate is less than 1%, whereas that of gravel is 25–35%.

Prairies Originally a French word for grassy *plains* with few or no trees.

Pre-Cambrian The earliest period of *geological time* dating from over 570-million years ago.

Precipitation The fall of moisture from the *atmosphere* onto the surface of the Earth, whether as dew, hail, rain, sleet, or snow.

Pyramidal peak A steep, isolated mountain summit, formed when the back walls of three or more *cirques* are cut back and move toward each other. The cliffs around such a horned peak, or horn, are divided by sharp *arêtes*. The Matterhorn in the Swiss Alps is an example.

Pyroclasts Fragments of rock ejected during volcanic eruptions.

Q

Quaternary The current period of *geological time*, which started about 1.6-million years ago.

R

Radiation The emission of energy in the form of particles or waves. Radiation from the sun includes heat, light, ultraviolet rays, gamma rays, and X-rays. Only some of the solar energy radiated into space reaches the Earth.

Rainforest Dense forests in tropical zones with high rainfall, temperature and *humidity*. Strictly, the term applies to the equatorial rain forest in tropical lowlands with constant rainfall and no seasonal change. The Congo and Amazon basins are examples. The term is applied more loosely to lush forest in other climates. Within rain forests organic life is dense and varied: at least 40% of all plant and animal species are found here and there may be as many as 100 tree species per hectare.

Rainshadow An area which experiences low rainfall, because of its position on the leeward side of a mountain range.

Reg A large area of stony *desert*, where tightly-packed gravel lies on top of clayey sand. A reg is formed where the wind blows away the finer sand.

Remote-sensing Method of obtaining information about the *environment* using unmanned equipment, such as a satellite, that relays the information to a point where it is collected and used.

Resistance The capacity of a rock to resist *denudation*, by processes such as *weathering* and *erosion*.

Ria A flooded *V-shaped river valley* or estuary, flooded by a rise in sea level (*eustacy*) or sinking land. It is shorter than a *fjord* and gets deeper as it meets the sea.

Rift valley A long, narrow depression in the Earth's *crust*, formed by the sinking of rocks between two *faults*.

River channel The trough which contains a river and is molded by the flow of water within it.

Roche moutonée A rock found in a glaciated valley. The side facing the flow of the *glacier* has been smoothed and rounded, while the other side has been left more rugged because the *glacier*, as it flows over it, has plucked out frozen fragments and carried them away.

Runoff Water draining from a land surface by flowing across it.

S

Sabkha The floor of an isolated *depression* that occurs in an *arid environment* – usually covered by salt deposits and devoid of vegetation.

SADC Southern African Development Community. Established in 1992 to promote economic integration between its member states, which are Angola, Botswana, Lesotho, Malawi, Mauritius, Mozambique, Namibia, South Africa, Swaziland, Tanzania, Zambia, and Zimbabwe.

Salt plug A rounded hill produced by the upward doming of rock *strata* caused by the movement of salt or other evaporite deposits under intense pressure.

Sastrugi Ice ridges formed by wind action. They lie parallel to the direction of the wind.

Savannah Open grassland found between the zone of *deserts*, and that of tropical *rain forests* in the tropics and subtropics. Scattered trees and shrubs are found in some kinds of savannah. A savannah *climate* usually has wet and dry seasons.

Scarp *see Escarpment.*

Scree Piles of rock fragments beneath a cliff or rock face, caused by mechanical *weathering*, especially *frost shattering*, where the expansion and contraction of freezing and thawing water within the rock, gradually breaks it up.

Sea-floor spreading The process whereby *tectonic plates* move apart, allowing hot *magma* to erupt and solidify. This forms a new sea floor and, ultimately, widens the ocean.

Seamount An isolated, submarine mountain or hill, probably of volcanic origin.

Season A period of time linked to regular changes in the weather, especially the intensity of solar *radiation*.

Sediment Grains of rock transported and deposited by rivers, sea, ice, or wind.

Sedimentary rocks Rocks formed from the debris of preexisting rocks or of organic material. They are found in many *environments* – on the ocean floor, on beaches, rivers, and *deserts*. Organically-formed sedimentary rocks include coal and chalk. Other sedimentary rocks, such as flint, are formed by chemical processes. Most of these rocks contain *fossils*, which can be used to date them.

Seif A sand *dune* which lies parallel to the direction of the prevailing wind. Seifs form steep-sided ridges, sometimes extending for miles.

Seismic activity Movement within the Earth, such as an *earthquake* or *tremor*.

Selva A region of wet forest found in the Amazon Basin.

Semiarid, semidesert The *climate* and landscape which lies between *savannah* and *desert* or between savannah and a *mediterranean* climate. In semiarid conditions there is a little more moisture than in a true *desert*; and more patches of drought-resistant vegetation can survive.

Shale (marine shale) A compacted *sedimentary rock*, with fine-grained particles. Marine shale is formed on the seabed. Fuel such as oil may be extracted from it.

Sheetwash Water that runs downhill in thin sheets without forming channels. It can cause *sheet erosion*.

Sheet erosion The washing away of soil by a thin film or sheet of water, known as *sheetwash*.

Shield A vast stable block of the Earth's *crust*, which has experienced little or no *mountain-building*.

Sierra The Spanish word for mountains.

Sinkhole A circular *depression* in a *limestone* region. They are formed by the collapse of an underground cave system or the *chemical weathering* of the *limestone*.

Sisal A plant-fiber used to make matting.

Slash and burn A farming technique involving the cutting down and burning of scrub forest, to create agricultural land. After a number of seasons this land is abandoned and the process is repeated. This practice is common in Africa and South America.

Slip face The steep leeward side of a sand *dune* or slope. Opposite side to a *back slope*.

Soil A thin layer of rock particles mixed with the remains of dead plants and animals. This occurs naturally on the surface of the Earth and provides a medium for plants to grow.

Soil creep The very gradual downslope movement of rock debris and soil, under the influence of gravity. This is a type of *mass movement*.

Soil erosion The wearing away of soil more quickly than it is replaced by natural processes. Soil can be carried away by wind as well as by water. Human activities, such as overgrazing and the clearing of land for farming, accelerate the process in many areas.

Solar energy Energy derived from the Sun. Solar energy is converted into other forms of energy. For example, the wind and waves, as well as the creation of plant material in photosynthesis, depend on solar energy.

Solifluction A kind of *soil creep*, where water in the surface layer has saturated the soil and rock debris which slips slowly downhill. It often happens where frozen top-layer deposits thaw, leaving frozen layers below them.

Sorghum A type of grass found in South America, similar to sugar cane. When refined it is used to make molasses.

Spit A thin linear deposit of sand or shingle extending from the sea shore. Spits are formed as angled waves shift sand along the beach, eventually extending a ridge of sand beyond a change in the angle of the coast. Spits are common where the coastline bends, especially at estuaries.

Squash A type of edible gourd.

Stack A tall, isolated pillar of rock near a coastline, created as wave action erodes away the adjacent rock.

Stalactite A tapering cylinder of mineral deposit, hanging from the roof of a cave in a *karst* area. It is formed by calcium carbonate, dissolved in water, which drips through the roof of a *limestone* cavern.

Stalagmite A cone of calcium carbonate, similar to a *stalactite*, rising from the floor of a *limestone* cavern and formed when drops of water fall from the roof of a *limestone* cave. If the water has dripped from a *stalactite* above the stalagmite, the two may join to form a continuous pillar.

Staple crop The main crop on which a country is economically and or physically reliant. For example, the major crop grown for large-scale local consumption in South Asia is rice.

Steppe Large areas of dry grassland in the northern hemisphere – particularly found in southeast Europe and central Asia.

Strata The plural of stratum, a distinct, virtually horizontal layer of deposited material, lying parallel to other layers.

Stratosphere A layer of the *atmosphere*, above the *troposphere*, extending from about 7–30 miles (11–50 km) above the Earth's surface. In the lower part of the stratosphere, the temperature is relatively stable and there is little moisture.

Strike-slip fault Occurs where plates move sideways past each other and blocks of rocks move horizontally in relation to each other, not up or down as in normal *faults*.

Subduction zone A region where two *tectonic plates* collide, forcing one beneath the other. Typically, a dense oceanic plate dips below a lighter continental plate, melting in the heat of the *asthenosphere*. This is why the zone is also called a destructive margins (*see Plate tectonics*). These zones are characterized by *earthquakes*, volcanoes, *mountain–building*, and the development of oceanic trenches and *island arcs*.

Submarine canyon A steep-sided valley, that extends along the *continental shelf* to the ocean floor. Often formed by *turbidity currents*.

Submarine fan Deposits of silt and *alluvium*, carried by large rivers forming great fan-shaped deposits on the ocean floor.

Subsistence agriculture An agricultural practice in which enough food is produced to support the farmer and his dependents, but not providing any surplus to generate an income.

Subtropical A term applied loosely to *climates* which are nearly tropical or tropical for a part of the year – areas north or south of the *tropics* but outside the *temperate zone*.

Supercontinent A large continent that breaks up to form smaller continents or that forms when smaller continents merge. In the theory of *continental drift*, the supercontinents are *Pangaea*, *Gondwanaland*, and *Laurasia*.

Sustainable development An approach to development, especially applied to economies across the world which exploit natural resources without destroying them or the *environment*.

Syncline A basin-shaped downfold in rock *strata*, created when the *strata* are compressed, for example where *tectonic plates* collide.

T

Tableland A highland area with a flat or gently undulating surface.

Taiga The belt of *coniferous* forest found in the north of Asia and North America. The conifers are adapted to survive low temperatures and long periods of snowfall.

Tarn A Scottish term for a small mountain lake, usually found at the head of a *glacier*.

Tectonic plates Plates, or tectonic plates, are the rigid slabs which form the Earth's outer shell, the *lithosphere*. Eight big plates and several smaller ones have been identified.

Temperate A moderate *climate* without extremes of temperature, typical of the mid-*latitudes* between the *tropics* and the *polar* circles.

Theocracy A state governed by religious laws – today Iran is the world's largest theocracy.

Thermokarst Subsidence created by the thawing of ground ice in *periglacial* areas, creating depressions.

Thermosphere A layer of the Earth's *atmosphere* which lies above the *mesosphere*, about 60–300 miles (100–500 km) above the Earth

Terraces Steps cut into steep slopes to create flat surfaces for cultivating crops. They also help reduce soil *erosion* on unconsolidated slopes. They are most common in heavily-populated parts of Southeast Asia.

Till Unstratified glacial deposits or drift left by a *glacier* or *ice sheet*. Till includes mixtures of clay, sand, gravel, and boulders.

Topography The typical shape and features of a given area such as land height and terrain.

Tombolo A large sand *spit* which attaches part of the mainland to an island.

Tornado A violent, spiraling windstorm, with a center of very low pressure. Wind speeds reach 200 mph (320 kmph) and there is often thunder and heavy rain.

Transform fault In *plate tectonics*, a *fault* of continental scale, occurring where two plates slide past each other, staying close together for example, the San Andreas Fault, USA. The jerky, uneven movement creates *earthquakes* but does not destroy or add to the Earth's *crust*

Transpiration The loss of water vapor through the pores (or stomata) of plants. The process helps to return moisture to the *atmosphere*.

Trap An area of fine-grained *igneous rock* that has been extruded and cooled on the Earth's surface in stages, forming a series of steps or terraces.

Treeline The line beyond which trees cannot grow, dependent on *latitude* and altitude, as well as local factors such as soil.

Tremor A slight *earthquake*.

Trench (oceanic trench) A long, deep trough in the ocean floor, formed, according to the theory of *plate tectonics*, when two plates collide and one dives under the other, creating a *subduction zone*.

Tropics The zone between the *Tropic of Cancer* and the *Tropic of Capricorn* where the *climate* is hot. Tropical climate is also applied to areas rather further north and south of the *Equator* where the climate is similar to that of the true tropics.

Tropic of Cancer A line of *latitude* or imaginary circle round the Earth, lying at 23° 28' N.

Tropic of Capricorn A line of *latitude* or imaginary circle round the Earth, lying at 23° 28' S.

Troposphere The lowest layer of the Earth's *atmosphere*. From the surface, it reaches a height of between 4–10 miles (7–16 km). It is the most turbulent zone of the atmosphere and accounts for the generation of most of the world's weather. The layer above it is called the *stratosphere*.

Tsunami A huge wave created by shock waves from an *earthquake* under the sea. Reaching speeds of up to 600 mph (960-kmph), the wave may increase to heights of 50 ft (15 m) on entering coastal waters; and it can cause great damage.

Tundra The treeless *plains* of the *Arctic Circle*, found south of the *polar* region of permanent ice and snow, and north of the belt of *coniferous* forests known as *taiga*. In this region of long, very cold winters, vegetation is usually limited to mosses, *lichens*, sedges, and rushes, although flowers and dwarf shrubs blossom in the brief summer.

Turbidity current An oceanic feature. A turbidity current is a mass of *sediment*-laden water thathas substantial erosive power. Turbidity currents are thought to contribute to the formation of *submarine canyons*.

Typhoon A kind of *hurricane* (or tropical cyclone) bringing violent winds and heavy rain, a typhoon can do great damage. They occur in the South China Sea, especially around the Philippines.

U

U-shaped valley A river valley that has been deepened and widened by a *glacier*. They are characteristically flat-bottomed and steep-sided and generally much deeper than river valleys.

UN United Nations. Established in 1945, it contains 188 nations and aims to maintain international peace and security, and promote cooperation over economic, social, cultural, and humanitarian problems.

UNICEF United Nations Children's Fund. A UN organization set up to promote family and child related programs.

Urstromtäler A German word used to describe *meltwater* channels that flowed along the front edge of the advancing *ice sheet* during the last Ice Age, 18,000–20,000 years ago.

V

V-shaped valley A typical valley eroded by a river in its upper course.

Virgin rain forest Tropical *rain-forest* in its original state, untouched by human activity such as logging, clearance for agriculture, settlement, or roadbuilding.

Viticulture The cultivation of grapes for wine.

Volcano An opening or vent in the Earth's *crust* where molten rock, *magma*, erupts. Volcanoes tend to be conical but may also be a crack in the Earth's surface or a hole blasted through a mountain. The magma is accompanied by other materials such as gas, steam, and fragments of rock, or *pyroclasts*. They tend to occur on destructive or constructive tectonic*plate* margins.

W-Z

Wadi The dry bed left by a torrent of water. Also classified as a *ephemeral* stream, found in *arid* and *semiarid* regions, which are subject to sudden and often severe flash flooding.

Warm humid climate A rainy climate with warm summers and mild winters.

Water cycle The continuous circulation of water between the Earth's surface and the *atmosphere*. The processes include *evaporation* and *transpiration* of moisture into the atmosphere, and its return as *precipitation*, some of which flows into lakes and oceans.

Water table The upper level of *groundwater* saturation in permeable rock *strata*.

Watershed The dividing line between one *drainage basin* – an area where all streams flow into a single river system – and another. In the US, watershed also means the whole drainage basin of a single river system – its catchment area.

Waterspout A rotating column of water in the form of cloud, mist, and spray which form on open water. Often has the appearance of a small *tornado*.

Weathering The decay and breakup of rocks at or near the Earth's surface, caused by water, wind, heat or ice, organic material, or the *atmosphere*. *Physical weathering* includes the effects of frost and temperature changes. Biological weathering includes the effects of plant roots, burrowing animals and the acids produced by animals, especially as they decay after death. *Carbonation* and *hydrolysis* are among many kinds of *chemical weathering*.

Geographical names

The following glossary lists all geographical terms occurring on the maps and in main-entry names in the Index-Gazetteer. These terms may precede, follow, or be run together with the proper element of the name; where they precede it the term is reversed for indexing purposes - thus Poluostrov Yamal is indexed as Yamal, Poluostrov.

Key

Geographical term
Language, Term

A

Å *Danish, Norwegian*, River
Āb *Persian*, River
Adrar *Berber*, Mountains
Agía, Ágios *Greek*, Saint
Air *Indonesian*, River
Akrotírio *Greek*, Cape, point
Alpen *German*, Alps
Alt- *German*, Old
Altiplanicie *Spanish*, Plateau
Älv, -älven *Swedish*, River
-ån *Swedish*, River
Anse *French*, Bay
'Aqabat *Arabic*, Pass
Archipiélago *Spanish*, Archipelago
Arcipelago *Italian*, Archipelago
Arquipélago *Portuguese*, Archipelago
Arrecife(s) *Spanish*, Reef(s)
Aru *Tamil*, River
Augstiene *Latvian*, Upland
Aukštuma *Lithuanian*, Upland
Aust- *Norwegian*, Eastern
Avtonomnyy Okrug *Russian*, Autonomous district
Āw *Kurdish*, River
'Ayn *Arabic*, Spring, well
'Ayoûn *Arabic*, Wells

B

Baelt *Danish*, Strait
Bahía *Spanish*, Bay
Baḥr *Arabic*, River
Baía *Portuguese*, Bay
Baie *French*, Bay
Bañado *Spanish*, Marshy land
Bandao *Chinese*, Peninsula
Banjaran *Malay*, Mountain range
Barajı *Turkish*, Dam
Barragem *Portuguese*, Reservoir
Bassin *French*, Basin
Batang *Malay*, Stream
Beinn, Ben *Gaelic*, Mountain
-berg *Afrikaans, Norwegian*, Mountain
Besar *Indonesian, Malay*, Big
Birkat, Birket *Arabic*, Lake, well, cistern
Boğazı *Turkish*, Strait, defile
Boka *Serbo-Croatian*, Bay
Bol'sh-aya, -iye, -oy, -oye *Russian*, Big
Botigh(i) *Uzbek*, Depression basin
-bre(en) *Norwegian*, Glacier
Bredning *Danish*, Bay
Bucht *German*, Bay
Bugt(en) *Danish*, Bay
Buḥayrat *Arabic*, Lake, reservoir
Buḥeiret *Arabic*, Lake
Bukit *Malay*, Mountain
-bukta *Norwegian*, Bay
bukten *Swedish*, Bay
Bulag *Mongolian*, Spring
Bulak *Uighur*, Spring
Burnu *Turkish*, Cape, point
Buuraha *Somali*, Mountains

C

Cabo *Portuguese*, Cape
Caka *Tibetan*, Salt lake
Canal *Spanish*, Channel
Cap *French*, Cape
Capo *Italian*, Cape, headland
Cascada *Portuguese*, Waterfall
Cayo(s) *Spanish*, Islet(s), rock(s)
Cerro *Spanish*, Hill
Chaîne *French*, Mountain range
Chapada *Portuguese*, Hills, upland
Chau *Cantonese*, Island
Chay *Turkish*, River
Chhâk *Cambodian*, Bay
Chhu *Tibetan*, River
-chōsuji *Korean*, Reservoir
Chott *Arabic*, Depression, salt lake
Chŭli *Uzbek*, Grassland, steppe
Ch'ün-tao *Chinese*, Island group
Chuŏr Phnum *Cambodian*, Mountains
Ciudad *Spanish*, City, town
Co *Tibetan*, Lake
Colline(s) *French*, Hill(s)
Cordillera *Spanish*, Mountain range
Costa *Spanish*, Coast
Côte *French*, Coast
Coxilha *Portuguese*, Mountains
Cuchilla *Spanish*, Mountains

D

Daban *Mongolian, Uighur*, Pass
Dağı *Azerbaijani, Turkish*, Mountain
Dağları *Azerbaijani, Turkish*, Mountains
-dake *Japanese*, Peak
-dal(en) *Norwegian*, Valley
Danau *Indonesian*, Lake
Dao *Chinese*, Island
Đao *Vietnamese*, Island
Daryā *Persian*, River
Daryācheh *Persian*, Lake
Dasht *Persian*, Desert, plain
Dawḥat *Arabic*, Bay
Denizi *Turkish*, Sea
Dere *Turkish*, Stream
Desierto *Spanish*, Desert
Dili *Azerbaijani*, Spit
-do *Korean*, Island
Dooxo *Somali*, Valley
Düzü *Azerbaijani*, Steppe
-dwīp *Bengali*, Island

E

-eilanden *Dutch*, Islands
Embalse *Spanish*, Reservoir
Ensenada *Spanish*, Bay
Erg *Arabic*, Dunes
Estany *Catalan*, Lake
Estero *Spanish*, Inlet
Estrecho *Spanish*, Strait
Étang *French*, Lagoon, lake
-ey *Icelandic*, Island
Ezero *Bulgarian, Macedonian*, Lake
Ezers *Latvian*, Lake

F

Feng *Chinese*, Peak
-fjella *Norwegian*, Mountain
Fjord *Danish*, Fjord
-fjord(en) *Danish, Norwegian, Swedish*, fjord
-fjördhur *Icelandic*, Fjord
Fleuve *French*, River
Fliegu *Maltese*, Channel
-fljór *Icelandic*, River
-flói *Icelandic*, Bay
Forêt *French*, Forest

G

-gan *Japanese*, Rock
-gang *Korean*, River
Ganga *Hindi, Nepali, Sinhala*, River
Gaoyuan *Chinese*, Plateau
Garagumy *Turkmen*, Sands
-gawa *Japanese*, River
Gebel *Arabic*, Mountain
-gebirge *German*, Mountain range
Ghadīr *Arabic*, Well
Ghubbat *Arabic*, Bay
Gjiri *Albanian*, Bay
Gol *Mongolian*, River
Golfe *French*, Gulf
Golfo *Italian, Spanish*, Gulf
Göl(ü) *Turkish*, Lake
Golyam, -a *Bulgarian*, Big
Gora *Russian, Serbo-Croatian*, Mountain
Góra *Polish*, mountain
Gory *Russian*, Mountain
Gryada *Russian*, ridge
Guba *Russian*, Bay
-gundo *Korean*, island group
Gunung *Malay*, Mountain

H

Ḥadd *Arabic*, Spit
-haehyŏp *Korean*, Strait
Haff *German*, Lagoon
Hai *Chinese*, Bay, lake, sea
Haixia *Chinese*, Strait
Ḥammādah *Arabic*, Desert
Ḥammādat *Arabic*, Rocky plateau
Hāmūn *Persian*, Lake
-hantō *Japanese*, Peninsula
Har, Haré *Hebrew*, Mountain
Ḥarrat *Arabic*, Lava-field
Hav(et) *Danish, Swedish*, Sea
Hawr *Arabic*, Lake
Hāyk' *Amharic*, Lake
He *Chinese*, River
-hegység *Hungarian*, Mountain range
Heide *German*, Heath, moorland
Helodrano *Malagasy*, Bay
Higashi- *Japanese*, East(ern)
Ḥiṣā' *Arabic*, Well
Hka *Burmese*, River
-ho *Korean*, Lake
Ḥolot *Hebrew*, Dunes
Hora *Belarussian, Czech*, Mountain
Hrada *Belarussian*, Mountain, ridge
Hsi *Chinese*, River
Hu *Chinese*, Lake
Huk *Danish*, Point

I

Île(s) *French*, Island(s)
Ilha(s) *Portuguese*, Island(s)
Ilhéu(s) *Portuguese*, Islet(s)
-isen *Norwegian*, Ice shelf
Imeni *Russian*, In the name of
Inish- *Gaelic*, Island
Insel(n) *German*, Island(s)
Irmağı, Irmak *Turkish*, River
Isla(s) *Spanish*, Island(s)
Isola (Isole) *Italian*, Island(s)

J

Jabal *Arabic*, Mountain
Jāl *Arabic*, Ridge
-järv *Estonian*, Lake
-järvi *Finnish*, Lake
Jazā'ir *Arabic*, Islands
Jazīrat *Arabic*, Island
Jazīreh *Persian*, Island
Jebel *Arabic*, Mountain
Jezero *Serbo-Croatian*, Lake
Jezioro *Polish*, Lake
Jiang *Chinese*, River
-jima *Japanese*, Island
Jižní *Czech*, Southern
-jõgi *Estonian*, River
-joki *Finnish*, River
-jökull *Icelandic*, Glacier
Jūn *Arabic*, Bay
Juzur *Arabic*, Islands

K

Kaikyō *Japanese*, Strait
-kaise *Lappish*, Mountain
Kali *Nepali*, River
Kalnas *Lithuanian*, Mountain
Kalns *Latvian*, Mountain
Kang *Chinese*, Harbor
Kangri *Tibetan*, Mountain(s)
Kaôh *Cambodian*, Island
Kapp *Norwegian*, Cape
Káto *Greek*, Lower
Kavīr *Persian*, Desert
K'edi *Georgian*, Mountain range
Kediet *Arabic*, Mountain
Kepi *Albanian*, Cape, point
Kepulauan *Indonesian, Malay*, Island group
Khalig, Khalīj *Arabic*, Gulf
Khawr *Arabic*, Inlet
Khola *Nepali*, River
Khrebet *Russian*, Mountain range
Ko *Thai*, Island
-ko *Japanese*, Inlet, lake
Kólpos *Greek*, Bay
-kopf *German*, Peak
Körfäzi *Azerbaijani*, Bay
Körfezi *Turkish*, Bay
Kõrgustik *Estonian*, Upland
Kosa *Russian, Ukrainian*, Spit
Koshi *Nepali*, River
Kou *Chinese*, River-mouth
Kowtal *Persian*, Pass
Kray *Russian*, Region, territory
Kryazh *Russian*, Ridge
Kuduk *Uighur*, Well
Kūh(hā) *Persian*, Mountain(s)
-kul' *Russian*, Lake
Kŭl(i) *Tajik, Uzbek*, Lake
-kundo *Korean*, Island group
-kysten *Norwegian*, Coast
Kyun *Burmese*, Island

L

Laaq *Somali*, Watercourse
Lac *French*, Lake
Lacul *Romanian*, Lake
Lagh *Somali*, Stream
Lago *Italian, Portuguese, Spanish*, Lake
Lagoa *Portuguese*, Lagoon
Laguna *Italian, Spanish*, Lagoon, lake
Laht *Estonian*, Bay
Laut *Indonesian*, Bay
Lembalemba *Malagasy*, Plateau
Lerr *Armenian*, Mountain
Lernashght'a *Armenian*, Mountain range
Les *Czech*, Forest
Lich *Armenian*, Lake
Liehtao *Chinese*, Island group
Liqeni *Albanian*, Lake
Límni *Greek*, Lake
Ling *Chinese*, Mountain range
Llano *Spanish*, Plain, prairie
Lumi *Albanian*, River
Lyman *Ukrainian*, Estuary

M

Madīnat *Arabic*, City, town
Mae Nam *Thai*, River
-mägi *Estonian*, Hill
Maja *Albanian*, Mountain
Mal *Albanian*, Mountains
Mal-aya, -oye, -yy, *Russian*, Small
-man *Korean*, Bay
Mar *Spanish*, Sea
Marios *Lithuanian*, Lake
Massif *French*, Mountains
Meer *German*, Lake
-meer *Dutch*, Lake
Melkosopochnik *Russian*, Plain
-meri *Estonian*, Sea
Mifraz *Hebrew*, Bay
Minami- *Japanese*, South(ern)
-misaki *Japanese*, Cape, point
Monkhafad *Arabic*, Depression
Montagne(s) *French*, Mountain(s)
Montañas *Spanish*, Mountains
Mont(s) *French*, Mountain(s)
Monte *Italian, Portuguese*, Mountain
More *Russian*, Sea
Mörön *Mongolian*, River
Mys *Russian*, Cape, point

N

-nada *Japanese*, Open stretch of water
Nadi *Bengali*, River
Nagor'ye *Russian*, Upland
Naḥal *Hebrew*, River
Nahr *Arabic*, River
Nam *Laotian*, River
Namakzār *Persian*, Salt desert
Né-a, -on, -os *Greek*, New
Nedre- *Norwegian*, Lower
-neem *Estonian*, Cape, point
Nehri *Turkish*, River
-nes *Norwegian*, Cape, point
Nevado *Spanish*, Mountain (snow-capped)
Nieder- *German*, Lower
Nishi- *Japanese*, West(ern)
-nísi *Greek*, Island
Nisoi *Greek*, Islands
Nizhn-eye, -iy, -iye, -yaya *Russian*, Lower
Nizmennost' *Russian*, Lowland, plain
Nord *Danish, French, German*, North
Norte *Portuguese, Spanish*, North
Nos *Bulgarian*, Point, spit
Nosy *Malagasy*, Island
Nov-a, -i, *Bulgarian, Serbo-Croatian*, New
Nov-aya, -o, -oye, -yy, -yye *Russian*, New
Now-a, -e, -y *Polish*, New
Nur *Mongolian*, Lake
Nuruu *Mongolian*, Mountains
Nuur *Mongolian*, Lake
Nyzovyna *Ukrainian*, Lowland, plain

O

-ø *Danish*, Island
Ober- *German*, Upper
Oblast' *Russian*, Province
Órmos *Greek*, Bay
Orol(i) *Uzbek*, Island
Øster- *Norwegian*, Eastern
Ostrov(a) *Russian*, Island(s)
Otok *Serbo-Croatian*, Island
Oued *Arabic*, Watercourse
-oy *Faeroese*, Island
-øy(a) *Norwegian*, Island
Oya *Sinhala*, River
Ozero *Russian, Ukrainian*, Lake

P

Passo *Italian*, Pass
Pegunungan *Indonesian, Malay*, Mountain range
Pélagos *Greek*, Sea
Pendi *Chinese*, Basin
Penisola *Italian*, Peninsula
Pertuis *French*, Strait
Peski *Russian*, Sands
Phanom *Thai*, Mountain
Phou *Laotian*, Mountain
Pi *Chinese*, Point
Pic *Catalan, French*, Peak
Pico *Portuguese, Spanish*, Peak
-piggen *Danish*, Peak
Pik *Russian*, Peak
Pivostriv *Ukrainian*, Peninsula
Planalto *Portuguese*, Plateau
Planina, Planini *Bulgarian, Macedonian, Serbo-Croatian*, Mountain range
Plato *Russian*, Plateau
Ploskogor'ye *Russian*, Upland
Poluostrov *Russian*, Peninsula
Ponta *Portuguese*, Point
Porthmós *Greek*, Strait
Pótamos *Greek*, River
Presa *Spanish*, Dam
Prokhod *Bulgarian*, Pass
Proliv *Russian*, Strait
Pulau *Indonesian Malay*, Island
Pulu *Malay*, Island
Punta *Spanish*, Point
Pushcha *Belorussian*, Forest
Puszcza *Polish*, Forest

Q

Qā' *Arabic*, Depression
Qalamat *Arabic*, Well
Qatorkŭh(i) *Tajik*, Mountain
Qiuling *Chinese*, Hills
Qolleh *Persian*, Mountain
Qu *Tibetan*, Stream
Quan *Chinese*, Well
Qulla(i) *Tajik*, Peak
Qundao *Chinese*, Island group

R

Raas *Somali*, Cape
-rags *Latvian*, Cape
Ramlat *Arabic*, Sands
Ra's *Arabic*, Cape, headland, point
Ravnina *Bulgarian, Russian*, Plain
Récif *French*, Reef
Recife *Portuguese*, Reef
Reka *Bulgarian*, River
Represa (Rep.) *Portuguese, Spanish*, Reservoir
Reshteh *Persian*, Mountain range
Respublika *Russian*, Republic, first-order administrative division
Respublika(si) *Uzbek*, Republic, first-order administrative division
-retsugan *Japanese*, Chain of rocks
-rettō *Japanese*, Island chain
Riacho *Spanish*, Stream
Riban' *Malagasy*, Mountains
Rio *Portuguese*, River
Río *Spanish*, River
Riu *Catalan*, River
Rivier *Dutch*, River
Rivière *French*, River
Rowd *Pashtu*, River
Rt *Serbo-Croatian*, Point
Rūd *Persian*, River
Rūdkhāneh *Persian*, River
Rudohorie *Slovak*, Mountains
Ruisseau *French*, Stream

S

-saar *Estonian*, Island
-saari *Finnish*, Island
Sabkhat *Arabic*, Salt marsh
Sāgar(a) *Hindi*, Lake, reservoir
Ṣaḥrā' *Arabic*, Desert
Saint, Sainte *French*, Saint
Salar *Spanish*, Salt-pan
Salto *Portuguese, Spanish*, Waterfall
Samudra *Sinhala*, Reservoir
-san *Japanese, Korean*, Mountain
-sanchi *Japanese*, Mountains
-sandur *Icelandic*, Beach
Sankt *German, Swedish*, Saint
-sanmaek *Korean*, Mountain range
-sanmyaku *Japanese*, Mountain range
San, Santa, Santo *Italian, Portuguese, Spanish*, Saint
São *Portuguese*, Saint
Sarīr *Arabic*, Desert
Sebkha, Sebkhet *Arabic*, Depression, salt marsh
Sedlo *Czech*, Pass
See *German*, Lake
Selat *Indonesian*, Strait
Selatan *Indonesian*, Southern
-selkä *Finnish*, Lake, ridge
Selseleh *Persian*, Mountain range
Serra *Portuguese*, Mountain
Serranía *Spanish*, Mountain
-seto *Japanese*, Channel, strait
Sever-naya, -noye, -nyy, -o *Russian*, Northern
Sha'ib *Arabic*, Watercourse
Shākh *Kurdish*, Mountain
Shamo *Chinese*, Desert
Shan *Chinese*, Mountain(s)
Shankou *Chinese*, Pass
Shanmo *Chinese*, Mountain range
Shaṭṭ *Arabic*, Distributary
Shet' *Amharic*, River
Shi *Chinese*, Municipality
-shima *Japanese*, Island
Shiqqat *Arabic*, Depression
-shotō *Japanese*, Group of islands
Shuiku *Chinese*, Reservoir
Shūrkhog(i) *Uzbek*, Salt marsh
Sierra *Spanish*, Mountains
Sint *Dutch*, Saint
-sjø(en) *Norwegian*, Lake
-sjön *Swedish*, Lake
Solonchak *Russian*, Salt lake
Solonchakovyye Vpadiny *Russian*, Salt basin, wetlands
Son *Vietnamese*, Mountain
Sông *Vietnamese*, River
Sør- *Norwegian*, Southern
-spitze *German*, Peak
Star-á, -é *Czech*, Old
Star-aya, -oye, -yy, -yye *Russian*, Old
Stenó *Greek*, Strait
Step' *Russian*, Steppe
Štít *Slovak*, Peak
Stœng *Cambodian*, River
Stolovaya Strana *Russian*, Plateau
Stredné *Slovak*, Middle
Střední *Czech*, Middle
Stretto *Italian*, Strait
Su Anbarı *Azerbaijani*, Reservoir
-suidō *Japanese*, Channel, strait
Sund *Swedish*, Sound, strait
Sungai *Indonesian, Malay*, River
Suu *Turkish*, River

T

Tal *Mongolian*, Plain
Tandavan' *Malagasy*, Mountain range
Tangorombohitr' *Malagasy*, Mountain massif
Tanjung *Indonesian, Malay*, Cape, point
Tao *Chinese*, Island
Ṭaraq *Arabic*, Hills
Tassili *Berber*, Mountain, plateau
Tau *Russian*, Mountain(s)
Taungdan *Burmese*, Mountain range
Techníti Límni *Greek*, Reservoir
Tekojärvi *Finnish*, Reservoir
Teluk *Indonesian, Malay*, Bay
Tengah *Indonesian*, Middle
Terara *Amharic*, Mountain
Timur *Indonesian*, Eastern
-tind(an) *Norwegian*, Peak
Tizma(si) *Uzbek*, Mountain range, ridge
-tō *Japanese*, island
Tog *Somali*, Valley
-tōge *Japanese*, pass
Togh(i) *Uzbek*, mountain
Tônlé *Cambodian*, Lake
Top *Dutch*, Peak
-tunturi *Finnish*, Mountain
Ṭurāq *Arabic*, hills
Tur'at *Arabic*, Channel

U

Udde(n) *Swedish*, Cape, point
'Uqlat *Arabic*, Well
Utara *Indonesian*, Northern
Uul *Mongolian*, Mountains

V

Väin *Estonian*, Strait
Vallée *French*, Valley
Varful *Romanian*, Peak
-vatn *Icelandic*, Lake
-vatnet *Norwegian*, Lake
Velayat *Turkmen*, Province
-vesi *Finnish*, Lake
Vestre- *Norwegian*, Western
-vidda *Norwegian*, Plateau
-vík *Icelandic*, Bay
-viken *Swedish*, Bay, inlet
Vinh *Vietnamese*, Bay
Víztárloló *Hungarian*, Reservoir
Vodaskhovishcha *Belarussian*, Reservoir
Vodokhranilishche (Vdkhr.) *Russian*, Reservoir
Vodoskhovyshche (Vdskh.) *Ukrainian*, Reservoir
Volcán *Spanish*, Volcano
Vostochn-o, yy *Russian*, Eastern
Vozvyshennost' *Russian*, Upland, plateau
Vozyera *Belarussian*, Lake
Vpadina *Russian*, Depression
Vrchovina *Czech*, Mountains
Vrh *Croat, Slovene*, Peak
Vychodné *Slovak*, Eastern
Vysochyna *Ukrainian*, Upland
Vysočina *Czech*, Upland

W

Waadi *Somali*, Watercourse
Wādī *Arabic* Watercourse
Wāḥat, Wâhat *Arabic*, Oasis
Wald *German*, Forest
Wan *Chinese*, Bay
Way *Indonesian*, River
Webi *Somali*, River
Wenz *Amharic*, River
Wiloyat(i) *Uzbek*, Province
Wyżyna *Polish*, Upland
Wzgórza *Polish*, Upland
Wzvyshsha *Belarussian*, Upland

X

Xé *Laotian*, River
Xi *Chinese*, Stream

Y

-yama *Japanese*, Mountain
Yanchi *Chinese*, Salt lake
Yanhu *Chinese*, Salt lake
Yarımadası *Azerbaijani, Turkish*, Peninsula
Yaylası *Turkish*, Plateau
Yazovir *Bulgarian*, Reservoir
Yoma *Burmese*, Mountains
Ytre- *Norwegian*, Outer
Yu *Chinese*, Islet
Yunhe *Chinese*, Canal
Yuzhn-o, -yy *Russian*, Southern

Z

-zaki *Japanese*, Cape, point
Zaliv *Bulgarian, Russian*, Bay
-zan *Japanese*, Mountain
Zangbo *Tibetan*, River
Zapadn-aya, -o, -yy *Russian*, Western
Západné *Slovak*, Western
Západní *Czech*, Western
Zatoka *Polish, Ukrainian*, Bay
-zee *Dutch*, Sea
Zemlya *Russian*, Earth, land
Zizhiqu *Chinese*, Autonomous region

100 Mile House – Ābuyē Mēda

Index

Glossary of Abbreviations

This glossary provides a comprehensive guide to the abbreviations used in this Atlas, and in the Index.

A
abbrev. abbreviated
AD Anno Domini
Afr. Afrikaans
Alb. Albanian
Amh. Amharic
anc. ancient
approx. approximately
Ar. Arabic
Arm. Armenian
ASEAN Association of South East Asian Nations
ASSR Autonomous Soviet Socialist Republic
Aust. Australian
Az. Azerbaijani
Azerb. Azerbaijan

B
Basq. Basque
BC before Christ
Bel. Belorussian
Ben. Bengali
Ber. Berber
B-H Bosnia-Herzegovina
bn billion (one thousand million)
BP British Petroleum
Bret. Breton
Brit. British
Bul. Bulgarian
Bur. Burmese

C
C central
C. Cape
°C degrees Centigrade
CACM Central America Common Market
Cam. Cambodian
Cant. Cantonese
CAR Central African Republic
Cast. Castilian
Cat. Catalan
CEEAC Central America Common Market
Chin. Chinese
CIS Commonwealth of Independent States
cm centimetre(s)
Cro. Croat
Cz. Czech
Czech Rep. Czech Republic

D
Dan. Danish
Div. Divehi
Dom. Rep. Dominican Republic
Dut. Dutch

E
E east
EC see EU
EEC see EU
ECOWAS Economic Community of West African States
ECU European Currency Unit
EMS European Monetary System
Eng. English
est estimated
Est. Estonian
EU European Union (previously European Community [EC], European Economic Community [EEC])

F
°F degrees Fahrenheit
Faer. Faeroese
Fij. Fijian
Fin. Finnish
Fr. French
Fris. Frisian
ft foot/feet
FYROM Former Yugoslav Republic of Macedonia

G
g gram(s)
Gael. Gaelic
Gal. Galician
GDP Gross Domestic Product (the total value of goods and services produced by a country excluding income from foreign countries)
Geor. Georgian
Ger. German
Gk Greek
GNP Gross National Product (the total value of goods and services produced by a country)

H
Heb. Hebrew
HEP hydro-electric power
Hind. Hindi
hist. historical
Hung. Hungarian

I
I. Island
Icel. Icelandic
in inch(es)
In. Inuit (Eskimo)
Ind. Indonesian
Intl International
Ir. Irish
Is Islands
It. Italian

J
Jap. Japanese

K
Kaz. Kazakh
kg kilogram(s)
Kir. Kirghiz
km kilometre(s)
km² square kilometre (singular)
Kor. Korean
Kurd. Kurdish

L
L. Lake
LAIA Latin American Integration Association
Lao. Laotian
Lapp. Lappish
Lat. Latin
Latv. Latvian
Liech. Liechtenstein
Lith. Lithuanian
Lus. Lusatian
Lux. Luxembourg

M
m million/metre(s)
Mac. Macedonian
Maced. Macedonia
Mal. Malay
Malg. Malagasy
Malt. Maltese
mi. mile(s)
Mong. Mongolian
Mt. Mountain
Mts Mountains

N
N north
NAFTA North American Free Trade Agreement
Nep. Nepali
Neth. Netherlands
Nic. Nicaraguan
Nor. Norwegian
NZ New Zealand

P
Pash. Pashtu
PNG Papua New Guinea
Pol. Polish
Poly. Polynesian
Port. Portuguese
prev. previously

R
Rep. Republic
Res. Reservoir
Rmsch Romansch
Rom. Romanian
Rus. Russian
Russ. Fed. Russian Federation

S
S south
SADC Southern Africa Development Community
SCr. Serbian, Croatian
Sinh. Sinhala
Slvk Slovak
Slvn. Slovene
Som. Somali
Sp. Spanish
St., St Saint
Strs Straits
Swa. Swahili
Swe. Swedish
Switz. Switzerland

T
Taj. Tajik
Th. Thai
Thai. Thailand
Tib. Tibetan
Turk. Turkish
Turkm. Turkmenistan

U
UAE United Arab Emirates
Uigh. Uighur
UK United Kingdom
Ukr. Ukrainian
UN United Nations
Urd. Urdu
US/USA United States of America
USSR Union of Soviet Socialist Republics
Uzb. Uzbek

V
var. variant
Vdkhr. Vodokhranilishche (Russian for reservoir)
Vdskh. Vodoskhovyshche (Ukrainian for reservoir)
Vtn. Vietnamese

W
W west
Wel. Welsh

This index lists all the placenames and features shown on the regional and continental maps in this Atlas. Placenames are referenced to the largest scale map on which they appear. The policy followed throughout the Atlas is to use the local spelling or local name at regional level; commonly-used English language names may occasionally be added (in parentheses) where this is an aid to identification e.g. Firenze (Florence). English names, where they exist, have been used for all international features e.g. oceans and country names; they are also used on the continental maps and in the introductory World Today section; these are then fully cross-referenced to the local names found on the regional maps. The index also contains commonly-found alternative names and variant spellings, which are also fully cross-referenced.

All main entry names are those of settlements unless otherwise indicated by the use of italicized definitions or representative symbols, which are keyed at the foot of each page.

1

10 *M16* **100 Mile House** *var.* Hundred Mile House. British Columbia, SW Canada *51°39´N 121°19´W*
25 de Mayo *see* Veinticinco de Mayo
26 Bakinskikh Komissarov *see* Häsänabad
26 Baku Komissarlary Adyndaky *see* Uzboý

A

Aa *see* Gauja
95 *G24* **Aabenraa** *var.* Åbenrå, *Ger.* Apenrade. Syddanmark, SW Denmark *55°03´N 09°26´E*
95 *G20* **Aabybro** *var.* Åbybro. Nordjylland, N Denmark *57°09´N 09°32´E*
101 *C16* **Aachen** *Dut.* Aken, *Fr.* Aix-la-Chapelle; *anc.* Aquae Grani, Aquisgranum. Nordrhein-Westfalen, W Germany *50°47´N 06°06´E*
Aaiún *see* Laâyoune
95 *M24* **Aakirkeby** *var.* Åkirkeby. Bornholm, E Denmark *55°04´N 14°56´E*
95 *G20* **Aalborg** *var.* Ålborg, Ålborg-Nørresundby; *anc.* Alburgum. Nordjylland, N Denmark *57°03´N 09°56´E*
Aalborg Bugt *see* Ålborg Bugt
101 *J21* **Aalen** Baden-Württemberg, S Germany *48°50´N 10°06´E*
95 *G21* **Aalestrup** *var.* Ålestrup. Midtjylland, NW Denmark *56°42´N 09°31´E*
98 *I11* **Aalsmeer** Noord-Holland, C Netherlands *52°17´N 04°43´E*
99 *F18* **Aalst** Oost-Vlaanderen, C Belgium *50°57´N 04°03´E*
99 *K18* **Aalst** *Fr.* Alost. Noord-Brabant, S Netherlands *51°23´N 05°29´E*
98 *O12* **Aalten** Gelderland, E Netherlands *51°56´N 06°35´E*
99 *D17* **Aalter** Oost-Vlaanderen, NW Belgium *51°05´N 03°28´E*
Aanaar *see* Inari
Aanaarjävri *see* Inarijärvi
93 *M17* **Äänekoski** Keski-Suomi, W Finland *62°34´N 25°45´E*
138 *H7* **Aanjar** *var.* ʻAnjar. C Lebanon *33°45´N 35°56´E*
83 *G21* **Aansluit** Northern Cape, N South Africa *26°41´S 22°24´E*
Aar *see* Aare
108 *F7* **Aarau** Aargau, N Switzerland *47°22´N 08°00´E*
108 *D8* **Aarberg** Bern, W Switzerland *47°19´N 07°54´E*
99 *D16* **Aardenburg** Zeeland, SW Netherlands *51°16´N 03°27´E*
108 *D8* **Aare** *var.* Aar. ≋ W Switzerland
108 *F7* **Aargau** *Fr.* Argovie. ◐ *canton* N Switzerland
Aarhus *see* Århus
Aarlen *see* Arlon
95 *G21* **Aars** *var.* Års. Nordjylland, N Denmark *56°49´N 09°32´E*
99 *I17* **Aarschot** Vlaams Brabant, C Belgium *50°59´N 04°50´E*
Aassi, Nahr el *see* Orontes
Aat *see* Ath
160 *G7* **Aba** *prev.* Ngawa. Sichuan, C China *32°51´N 101°46´E*
79 *P16* **Aba** Orientale, NE Dem. Rep. Congo *03°52´N 30°14´E*
77 *V17* **Aba** Abia, S Nigeria *05°06´N 07°22´E*
140 *J6* **Abā al Qazāz, Bi'r** *well* NW Saudi Arabia
Abā as Suʻūd *see* Najrān
59 *G14* **Abacaxis, Rio** ≋ NW Brazil
Abaco Island *see* Great Abaco/Little Abaco
Abaco Island *see* Great Abaco, N Bahamas
142 *K10* **Ābādān** Khūzestān, SW Iran *30°24´N 48°18´E*
146 *F13* **Abadan** *prev.* Bezmein, Büzmeýin, *Rus.* Byuzmeyin. Ahal Welaýaty, C Turkmenistan *38°08´N 57°53´E*
143 *O10* **Ābādeh** Fārs, C Iran *31°06´N 52°40´E*
74 *H8* **Abadla** W Algeria *31°04´N 02°39´W*
59 *M20* **Abaeté** Minas Gerais, SE Brazil *19°10´S 45°24´W*
62 *P7* **Abaí** Caazapá, S Paraguay *25°58´S 55°54´W*
Abai *see* Blue Nile
191 *O2* **Abaiang** *var.* Apia; *prev.* Charlotte Island. *atoll* Tungaru, W Kiribati
Abaj *see* Abay
77 *U15* **Abaji** Federal Capital District, C Nigeria *08°35´N 06°54´E*
37 *O7* **Abajo Peak** ▲ Utah, W USA *37°51´N 109°28´W*
77 *V16* **Abakaliki** Ebonyi, SE Nigeria *06°18´N 08°07´E*
122 *K13* **Abakan** Respublika Khakasiya, S Russian Federation *53°43´N 91°25´E*
77 *S11* **Abala** Tillabéri, SW Niger *14°55´N 03°27´E*
77 *U11* **Abalak** Tahoua, C Niger *15°28´N 06°18´E*
119 *N14* **Abalyanka** *Rus.* Obolyanka. ≋ N Belarus
122 *L12* **Aban** Krasnoyarskiy Kray, S Russian Federation *56°41´N 96°04´E*
143 *P9* **Āb Anbār-e Kān Sorkh** Yazd, C Iran *31°22´N 53°38´E*
57 *G16* **Abancay** Apurímac, SE Peru *13°37´S 72°52´W*
190 *H2* **Abaokoro** *atoll* Tungaru, W Kiribati
Abariringa *see* Kanton
143 *P10* **Abarkūh** Yazd, C Iran *31°07´N 53°17´E*
165 *V3* **Abashiri** *var.* Abasiri. Hokkaidō, NE Japan *44°N 144°15´E*
165 *U3* **Abashiri-ko** ⊚ Hokkaidō, NE Japan
Abasiri *see* Abashiri
41 *P10* **Abasolo** Tamaulipas, C Mexico *24°02´N 98°18´W*
186 *F9* **Abau** Central, S Papua New Guinea *10°04´S 148°34´E*
145 *R10* **Abay** *var.* Abaj. Karaganda, C Kazakhstan *49°38´N 72°50´E*
81 *I15* **Ābaya Hāyk'** *Eng.* Lake Margherita, *It.* Abbaia. ⊚ SW Ethiopia
Ābay Wenz *see* Blue Nile
122 *K13* **Abaza** Respublika Khakasiya, S Russian Federation *52°40´N 89°58´E*
Abbaia *see* Ābaya Hāyk'
143 *Q13* **Āb Bārīk** Fārs, S Iran
107 *C18* **Abbasanta** Sardegna, Italy, C Mediterranean Sea *40°08´N 08°49´E*
Abbatis Villa *see* Abbeville
30 *M3* **Abbaye, Point** *headland* Michigan, N USA *46°58´N 88°08´W*
Abbazia *see* Opatija
Abbé, Lake *see* Abhe, Lake
103 *N2* **Abbeville** *anc.* Abbatis Villa. Somme, N France *50°06´N 01°50´E*
23 *R7* **Abbeville** Alabama, S USA *31°35´N 85°16´W*
23 *U6* **Abbeville** Georgia, SE USA *31°58´N 83°18´W*
22 *I9* **Abbeville** Louisiana, S USA *29°58´N 92°08´W*
21 *P12* **Abbeville** South Carolina, SE USA *34°10´N 82°23´W*
97 *B20* **Abbeyfeale** *Ir.* Mainistir na Féile. SW Ireland *52°24´N 09°21´W*
106 *D8* **Abbiategrasso** Lombardia, NW Italy *45°24´N 08°55´E*
93 *I14* **Abborrträsk** Norrbotten, N Sweden *65°24´N 19°33´E*
194 *J9* **Abbot Ice Shelf** *ice shelf* Antarctica
10 *M17* **Abbotsford** British Columbia, SW Canada *49°02´N 122°18´W*
30 *K6* **Abbotsford** Wisconsin, N USA *44°57´N 90°19´W*
149 *U5* **Abbottābād** Khyber Pakhtunkhwa, NW Pakistan *34°12´N 73°15´E*
119 *M14* **Abchuha** *Rus.* Obchuga. Minskaya Voblasts', NW Belarus *54°30´N 29°22´E*
98 *I10* **Abcoude** Utrecht, C Netherlands *52°17´N 04°59´E*
139 *N2* **ʻAbd al ʻAzīz, Jabal** ▲▲ NE Syria
141 *U17* **ʻAbd al Kūrī** *island* SE Yemen
127 *U6* **Abdulino** Orenburgskaya Oblast', W Russian Federation *53°37´N 53°39´E*
78 *J10* **Abéché** *var.* Abécher, Abeshr. Ouaddaï, SE Chad *13°49´N 20°49´E*
Abécher *see* Abéché
143 *S8* **Āb-e Garm va Sard** Yazd, E Iran
77 *R8* **Abeïbara** Kidal, NE Mali *19°07´N 01°52´E*
105 *P5* **Abejar** Castilla y León, N Spain *41°48´N 02°47´W*
54 *E9* **Abejorral** Antioquia, W Colombia *05°48´N 75°28´W*
Abela *see* Ávila
Abellinum *see* Avellino
92 *Q2* **Abeløya** *island* Kong Karls Land, E Svalbard
80 *I13* **Ābeltī** Oromīya, C Ethiopia *08°09´N 37°31´E*
191 *O2* **Abemama** *var.* Apamama; *prev.* Roger Simpson Island. *atoll* Tungaru, W Kiribati
171 *Y15* **Abemaree** *var.* Abermarre. Papua, E Indonesia *07°03´S 140°10´E*
77 *O16* **Abengourou** E Ivory Coast *06°42´N 03°27´W*
Åbenrå *see* Aabenraa
101 *L22* **Abens** ≋ SE Germany
77 *S16* **Abeokuta** Ogun, SW Nigeria *07°07´N 03°21´E*
97 *I20* **Aberaeron** SW Wales, United Kingdom *52°15´N 04°15´W*
Aberbrothock *see* Arbroath
Abercorn *see* Mbala
29 *R6* **Abercrombie** North Dakota, N USA *46°25´N 96°42´W*
183 *T7* **Aberdeen** New South Wales, SE Australia *32°09´S 150°55´E*
11 *T15* **Aberdeen** Saskatchewan, S Canada *52°15´N 106°19´W*
83 *H25* **Aberdeen** Eastern Cape, S South Africa *32°30´S 24°00´E*
96 *L9* **Aberdeen** *anc.* Devana. NE Scotland, United Kingdom *57°10´N 02°04´W*
21 *X2* **Aberdeen** Maryland, NE USA *39°28´N 76°09´W*
23 *N3* **Aberdeen** Mississippi, S USA *33°49´N 88°32´W*
21 *T10* **Aberdeen** North Carolina, SE USA *35°07´N 79°25´W*
29 *P8* **Aberdeen** South Dakota, N USA *45°27´N 98°29´W*
32 *F8* **Aberdeen** Washington, NW USA *46°57´N 123°48´W*
96 *K9* **Aberdeen** *cultural region* NE Scotland, United Kingdom
8 *L8* **Aberdeen Lake** ⊚ Nunavut, NE Canada
96 *J10* **Aberfeldy** C Scotland, United Kingdom *56°38´N 03°49´W*
97 *K21* **Abergavenny** *anc.* Gobannium. SE Wales, United Kingdom *51°50´N 03°W*
Abergwaun *see* Fishguard
Abermarre *see* Abemaree
25 *N5* **Abernathy** Texas, SW USA *33°49´N 101°50´W*
Abersee *see* Wolfgangsee
Abertawe *see* Swansea
Aberteifi *see* Cardigan
32 *I15* **Abert, Lake** ⊚ Oregon, NW USA
97 *I20* **Aberystwyth** W Wales, United Kingdom *52°25´N 04°05´W*
Abeshr *see* Abéché
Ábeskovvu *see* Abisko
106 *F10* **Abetone** Toscana, C Italy *44°09´N 10°42´E*
125 *V5* **Abez'** Respublika Komi, NW Russian Federation *66°32´N 61°41´E*
142 *M5* **Āb Garm** Qazvīn, N Iran
141 *N12* **Abhā** ʻAsīr, SW Saudi Arabia *18°16´N 42°32´E*
142 *M5* **Abhar** Zanjān, NW Iran *36°05´N 49°18´E*
Abhé Bad/Ābhē Bid Hāyk' *see* Abhe, Lake
80 *K12* **Abhe, Lake** *var.* Lake Abbé, *Amh.* Ābhē Bid Hāyk', *Som.* Abhé Bad. ⊚ Djibouti/Ethiopia
77 *V17* **Abia** ◐ *state* SE Nigeria
139 *V9* **ʻAbīd ʻAlī** Wāsiṭ, E Iraq *32°20´N 45°58´E*
119 *O17* **Abidavichy** *Rus.* Obidovichi. Mahilyowskaya Voblasts', E Belarus *53°20´N 30°25´E*
115 *L15* **Abide** Çanakkale, NW Turkey *40°04´N 26°13´E*
77 *N17* **Abidjan** S Ivory Coast *05°19´N 04°01´W*
Āb-i-Istāda *see* Īstādeh-ye Moqor, Āb-e-
27 *N4* **Abilene** Kansas, C USA *38°55´N 97°14´W*
25 *Q7* **Abilene** Texas, SW USA *32°27´N 99°44´W*
Abindonia *see* Abingdon
97 *M21* **Abingdon** *anc.* Abindonia. S England, United Kingdom *51°41´N 01°17´W*
30 *K12* **Abingdon** Illinois, N USA *40°48´N 90°24´W*
21 *P8* **Abingdon** Virginia, NE USA *36°42´N 81°59´W*
Abingdon *see* Pinta, Isla
18 *J15* **Abington** Pennsylvania, NE USA *40°06´N 75°05´W*
126 *K14* **Abinsk** Krasnodarskiy Kray, SW Russian Federation *44°51´N 38°12´E*
37 *R9* **Abiquiu Reservoir** ⊠ New Mexico, SW USA
Āb-i-safed *see* Sefīd, Darya-ye
92 *I10* **Abisko** *Lapp.* Ábeskovvu. Norrbotten, N Sweden *68°21´N 18°50´E*
12 *G12* **Abitibi** ≋ Ontario, S Canada
12 *H12* **Abitibi, Lac** ⊚ Ontario/Québec, S Canada
80 *J10* **Ābīy Ādī** Tigray, N Ethiopia *13°40´N 38°57´E*
118 *H6* **Abja-Paluoja** Viljandimaa, S Estonia *58°08´N 25°20´E*
Abkhazia *see* Apkhazeti
182 *F1* **Abminga** South Australia *26°07´S 134°49´E*
75 *W9* **Abnūb** *var.* Abnûb. C Egypt *27°18´N 31°09´E*
Abnûb *see* Abnūb
Åbo *see* Turku
152 *G9* **Abohar** Punjab, N India *30°11´N 74°14´E*
77 *O17* **Aboisso** SE Ivory Coast *05°26´N 03°13´W*
78 *H5* **Abo, Massif d'** ▲▲ NW Chad
78 *H5* **Abo, Massif d'** ▲▲ E Chad
77 *R16* **Abomey** S Benin *07°14´N 02°00´E*
79 *F16* **Abong Mbang** Est, SE Cameroon *03°58´N 13°10´E*
111 *L23* **Abony** Pest, C Hungary *47°10´N 20°00´E*
78 *J11* **Abou-Déïa** Salamat, SE Chad *11°30´N 19°18´E*
Aboudouhour *see* Abū aḍ Ḍuhūr
Abou Kémal *see* Abū Kamāl
Abou Simbel *see* Abū Sunbul
137 *T12* **Abovyan** C Armenia *40°16´N 44°33´E*
141 *P15* **Abrād, Wādī** *seasonal river* W Yemen
Abraham Bay *see* The Carlton
104 *G10* **Abrantes** *var.* Abrántes. Santarém, C Portugal *39°28´N 08°12´W*
62 *J4* **Abra Pampa** Jujuy, N Argentina *22°47´S 65°41´W*
Abrashlare *see* Brezovo
54 *G7* **Abrego** Norte de Santander, N Colombia *08°08´N 73°14´W*
Abrene *see* Pytalovo
40 *C7* **Abreojos, Punta** *headland* NW Mexico *26°43´N 113°36´W*
65 *J16* **Abrolhos Bank** *undersea feature* W Atlantic Ocean *18°30´S 38°45´W*
119 *H19* **Abrova** *Rus.* Obrovo. Brestskaya Voblasts', SW Belarus *52°30´N 25°34´E*
116 *G11* **Abrud** *Ger.* Gross-Schlatten, *Hung.* Abrudbánya. Alba, SW Romania *46°16´N 23°05´E*
Abrudbánya *see* Abrud
118 *E6* **Abruka** *island* SW Estonia
107 *J15* **Abruzzese, Appennino** ▲▲ C Italy
107 *J14* **Abruzzo** ◐ *region* C Italy
141 *N14* **ʻAbs** *var.* Sūq ʻAbs. W Yemen *16°42´N 42°55´E*
33 *T12* **Absaroka Range** ▲▲ Montana/Wyoming, NW USA
137 *Z11* **Abşeron Yarımadası** *Rus.* Apsheronskiy Poluostrov. *peninsula* E Azerbaijan
143 *N6* **Āb Shīrīn** Eşfahān, C Iran *34°17´N 51°17´E*
139 *X10* **Abtān** Maysān, SE Iraq *31°37´N 47°06´E*
109 *R6* **Abtenau** Salzburg, NW Austria *47°33´N 13°21´E*
152 *E14* **Ābu** Rājasthān, N India *24°41´N 72°50´E*
164 *E12* **Abu** Yamaguchi, Honshū, SW Japan *34°30´N 131°26´E*
138 *I4* **Abū aḍ Ḍuhūr** *Fr.* Aboudouhour. Idlib, NW Syria *35°30´N 37°00´E*
143 *P17* **Abū al Abyaḍ** *island* C United Arab Emirates
138 *K10* **Abū al Ḥuṣayn, Khabrat** ⊚ N Jordan
139 *R8* **Abū al Jīr** Al Anbār, C Iraq *33°16´N 42°55´E*
139 *Y12* **Abū al Khaṣīb** *var.* Abul Khasib. Al Baṣrah, SE Iraq *30°26´N 48°00´E*
139 *U12* **Abū at Tubrah, Thaqb** *well* S Iraq
Abu Balâs *see* Abu Ballās
75 *V11* **Abu Ballās** *var.* Abu Balâs. ▲ SW Egypt *24°28´N 27°36´E*
Abu Dhabi *see* Abū Ẓabī
139 *R8* **Abū Farūkh** Al Anbār, C Iraq *33°06´N 43°18´E*
80 *C12* **Abu Gabra** Eastern Darfur, W Sudan *11°02´N 26°50´E*
139 *P10* **Abū Ghār, Shaʻīb** *dry watercourse* S Iraq
80 *G7* **Abu Hamed** River Nile, N Sudan *19°32´N 33°20´E*
139 *O5* **Abū Ḥardān** *var.* Hajîne. Dayr az Zawr, E Syria *34°45´N 40°49´E*
139 *T7* **Abū Ḥasāwīyah** Diyālá, E Iraq *33°52´N 44°47´E*
138 *K10* **Abū Ḥifnah, Wādī** *dry watercourse* N Jordan
77 *V15* **Abuja** ● (Nigeria) Federal Capital District, C Nigeria *09°04´N 07°28´E*
139 *R9* **Abū Jahaf, Wādī** *dry watercourse* C Iraq
56 *F12* **Abujao, Río** ≋ E Peru
139 *U12* **Abū Jasrah** Al Muthanná, S Iraq *30°43´N 44°50´E*
139 *O6* **Abū Kamāl** *Fr.* Abou Kémal. Dayr az Zawr, E Syria *34°29´N 40°56´E*
165 *P12* **Abukuma-sanchi** ▲▲ Honshū, C Japan
Abula *see* Ávila
Abul Khasib *see* Abū al Khaṣīb
79 *K16* **Abumombazi** *var.* Abumonbazi. Equateur, N Dem. Rep. Congo *03°43´N 22°06´E*
Abumonbazi *see* Abumombazi
59 *D15* **Abunã** Rondônia, W Brazil *09°41´S 65°20´W*
56 *K13* **Abunã, Rio** *var.* Río Abuná. ≋ Bolivia/Brazil
138 *G10* **Abū Nuṣayr** *var.* Abu Nuseir. ʻAmmān, W Jordan *32°03´N 35°58´E*
Abu Nuseir *see* Abū Nuṣayr
139 *T12* **Abū Qabr** Al Muthanná, S Iraq *31°03´N 44°34´E*
138 *K5* **Abū Raḥbah, Jabal** ▲▲ C Syria
139 *S5* **Abū Rajāsh** Şalāḩ ad Dīn, N Iraq *34°47´N 43°36´E*
139 *W13* **Abū Raqrāq, Ghadīr** *well* S Iraq
152 *E14* **Ābu Road** Rājasthān, N India *24°29´N 72°47´E*
80 *I6* **Abu Shagara, Ras** *headland* NE Sudan *18°04´N 38°31´E*
Abu Simbel *see* Abū Sunbul
139 *U12* **Abū Sudayrah** Al Muthanná, S Iraq *30°55´N 44°58´E*
139 *T10* **Abū Şukhayr** Al Qādisīyah, S Iraq *31°54´N 44°27´E*
Abū Sunbul *see* Abū Sunbul
185 *E18* **Abut Head** *headland* South Island, New Zealand *43°06´S 170°16´E*
80 *E9* **Abu ʻUrug** Northern Kordofan, C Sudan *15°52´N 30°25´E*
80 *K12* **Ābuyē Mēda** ▲ C Ethiopia *10°28´N 39°44´E*

◆ Country
● Country Capital
◇ Dependent Territory
○ Dependent Territory Capital
◐ Administrative Regions
✈ International Airport
▲ Mountain
▲▲ Mountain Range
Volcano
≋ River
⊚ Lake
⊠ Reservoir

80 *D11* **Abu Zabad** Western Kordofan, C Sudan *12°21′N 29°16′E*
143 *P16* **Abū Ẓabī** *var.* Abū Ẓabī, *Eng.* Abu Dhabi. ● (United Arab Emirates) Abū Ẓaby, C United Arab Emirates *24°30′N 54°20′E*
Abū Ẓabī *see* Abū Ẓabī
75 *X8* **Abu Zenīma** E Egypt *29°01′N 33°08′E*
95 *N17* **Åby** Östergötland, S Sweden *58°40′N 16°10′E*
Abyaḍ, Al Baḥr al *see* White Nile
Åbybro *see* Aabybro
80 *D13* **Abyei** Southern Kordofan, S Sudan *09°35′N 28°28′E*
Abyla *see* Ávila
Abymes *see* les Abymes
Abyssinia *see* Ethiopia
Açâba *see* Assaba
54 *F11* **Acacias** Meta, C Colombia *03°59′N 73°46′W*
58 *L13* **Açailândia** Maranhão, E Brazil *04°51′S 47°26′W*
Acaill *see* Achill Island
42 *E8* **Acajutla** Sonsonate, W El Salvador *13°34′N 89°50′W*
79 *D17* **Acalayong** SW Equatorial Guinea *01°05′N 09°34′E*
41 *N13* **Acámbaro** Guanajuato, C Mexico *20°01′N 100°42′W*
54 *C6* **Acandí** Chocó, NW Colombia *08°32′N 77°20′W*
104 *H4* **A Cañiza** *var.* La Cañiza. Galicia, NW Spain *42°13′N 08°16′W*
40 *J11* **Acaponeta** Nayarit, C Mexico *22°30′N 105°21′W*
40 *J11* **Acaponeta, Río de** ≈ C Mexico
41 *O16* **Acapulco** *var.* Acapulco de Juárez. Guerrero, S Mexico *16°51′N 99°53′W*
Acapulco de Juárez *see* Acapulco
55 *T13* **Acarai Mountains** *Sp.* Serra Acaraí. ▲▲ Brazil/Guyana
Acaraí, Serra *see* Acarai Mountains
58 *O13* **Acaraú** Ceará, NE Brazil *04°35′S 37°37′W*
54 *J6* **Acarigua** Portuguesa, N Venezuela *09°35′N 69°12′W*
104 *H2* **A Carreira** Galicia, NW Spain *43°21′N 08°12′W*
42 *C6* **Acatenango, Volcán de** ⛰ S Guatemala *14°30′N 90°52′W*
41 *Q15* **Acatlán** *var.* Acatlán de Osorio. Puebla, S Mexico *18°12′N 98°02′W*
Acatlán de Osorio *see* Acatlán
41 *S15* **Acayucan** *var.* Acayucán. Veracruz-Llave, E Mexico *17°59′N 94°58′W*
Accho *see* Akko
21 *Y5* **Accomac** Virginia, NE USA *37°43′N 75°41′N*
77 *Q17* **Accra** ● (Ghana)SE Ghana *05°33′N 00°15′W*
97 *L17* **Accrington** NW England, United Kingdom *53°46′N 02°21′W*
61 *B19* **Acebal** Santa Fe, C Argentina *33°14′S 60°50′W*
168 *H8* **Aceh** *off.* Daerah Istimewa Aceh, *var.* Acheen, Achin, Atchin, Atjeh. ◈ *autonomous district* NW Indonesia
107 *M18* **Acerenza** Basilicata, S Italy *40°46′N 15°51′E*
107 *K17* **Acerra** *anc.* Acerrae. Campania, S Italy *40°56′N 14°22′E*
Acerrae *see* Acerra
57 *J17* **Achacachi** La Paz, W Bolivia *16°01′S 68°44′W*
54 *K7* **Achaguas** Apure, C Venezuela *07°46′N 68°14′W*
154 *H12* **Achalpur** *prev.* Elichpur, Ellichpur. Mahārāshtra, C India *21°19′N 77°30′E*
61 *F18* **Achar** Tacuarembó, C Uruguay *32°20′S 56°15′W*
137 *R10* **Ach'ara** *prev.* Achara, *var.* Ajaria. ◈ *autonomous republic* SW Georgia
Achara *see* Ach'ara
115 *H19* **Acharnés** *var.* Aharnes; *prev.* Akharnaí. Attikí, C Greece *38°09′N 23°58′E*
Ach'asar Lerr *see* Achk'asari, Mta
Acheen *see* Aceh
99 *K16* **Achel** Limburg, NE Belgium *51°15′N 05°31′E*
115 *D16* **Achelóos** *var.* Akhelóös, Aspropótamos; *anc.* Achelous. ≈ W Greece
Achelous *see* Achelóos
163 *W8* **Acheng** Heilongjiang, NE China *45°32′N 126°56′E*
109 *N6* **Achenkirch** Tirol, W Austria *47°31′N 11°42′E*
101 *L24* **Achenpass** *pass* Austria/Germany
109 *N7* **Achensee** ◎ W Austria
101 *F22* **Achern** Baden-Württemberg, SW Germany *48°37′N 08°04′E*
115 *C16* **Acherón** ≈ W Greece
77 *W11* **Achétinamou** ≈ S Niger
152 *J12* **Achhnera** Uttar Pradesh, N India *27°10′N 77°45′E*
42 *C7* **Achiguate, Río** ≈ S Guatemala
97 *A16* **Achill Head** *Ir.* Ceann Acla. *headland* W Ireland *53°58′N 10°14′W*
97 *A16* **Achill Island** *Ir.* Acaill. *island* W Ireland
100 *H11* **Achim** Niedersachsen, NW Germany *53°01′N 09°01′E*
149 *S5* **Achīn** Nangarhār, E Afghanistan *34°04′N 70°41′E*
Achin *see* Aceh
122 *K12* **Achinsk** Krasnoyarskiy Kray, S Russian Federation *56°21′N 90°25′E*
162 *E5* **Achit Nuur** ◎ NW Mongolia
137 *T11* **Achk'asari, Mta** *Arm.* Ach'asar Lerr. ▲ Armenia/Georgia *41°09′N 43°55′E*
126 *K13* **Achuyevo** Krasnodarskiy Kray, SW Russian Federation *46°00′N 38°01′E*
81 *F16* **Achwa** *var.* Aswa. ≈ N Uganda
136 *E15* **Acıgöl** *salt lake* SW Turkey
107 *L24* **Acireale** Sicilia, Italy, C Mediterranean Sea *37°36′N 15°10′E*
Aciris *see* Agri
25 *N7* **Ackerly** Texas, SW USA *32°31′N 101°43′W*
22 *M4* **Ackerman** Mississippi, S USA *33°18′N 89°10′W*
29 *W13* **Ackley** Iowa, C USA *42°33′N 93°03′W*
44 *J5* **Acklins Island** *island* SE The Bahamas
Acla, Ceann *see* Achill Head
62 *H11* **Aconcagua, Cerro** ▲ W Argentina *32°36′S 69°53′W*
Açores/Açores, Arquipélago dos/Açores, Ilhas dos *see* Azores
104 *H2* **A Coruña** *Cast.* La Coruña, *Eng.* Corunna; *anc.* Caronium. Galicia, NW Spain *43°22′N 08°24′W*
104 *G2* **A Coruña** *Cast.* La Coruña. ◈ *province* Galicia, NW Spain
42 *L10* **Acoyapa** Chontales, S Nicaragua *11°58′N 85°10′W*
106 *H13* **Acquapendente** Lazio, C Italy *42°44′N 11°52′E*
106 *J13* **Acquasanta Terme** Marche, C Italy *42°46′N 13°24′E*
106 *I13* **Acquasparta** Lazio, C Italy *42°41′N 12°31′E*
106 *C9* **Acqui Terme** Piemonte, NW Italy *44°41′N 08°28′E*
Acrae *see* Palazzola Acreide
182 *F7* **Acraman, Lake** *salt lake* South Australia
59 *A15* **Acre** *off.* Estado do Acre. ◈ *state* W Brazil
Acre *see* Akko
59 *C16* **Acre, Rio** ≈ W Brazil
107 *N20* **Acri** Calabria, SW Italy *39°30′N 16°22′E*
Acte *see* Ágion Óros
191 *Y12* **Actéon, Groupe** *island group* Îles Tuamotu, SE French Polynesia
15 *P12* **Acton-Vale** Québec, SE Canada *45°39′N 72°31′W*
41 *P13* **Actopan** *var.* Actopán. Hidalgo, C Mexico *20°19′N 98°59′W*
Açu *see* Assu
Acunum Acusio *see* Montélimar
77 *Q17* **Ada** SE Ghana *05°47′N 00°42′E*
112 *L8* **Ada** Vojvodina, N Serbia *45°48′N 20°08′E*
29 *R5* **Ada** Minnesota, N USA *47°18′N 96°31′W*
31 *R12* **Ada** Ohio, N USA *40°46′N 83°49′W*
27 *O12* **Ada** Oklahoma, C USA *34°47′N 96°41′W*
162 *L8* **Adaatsag** *var.* Tavin. Dundgovĭ, C Mongolia *46°27′N 105°43′E*
Ada Bazar *see* Adapazarı
40 *D3* **Adair, Bahía de** *bay* NW Mexico
104 *M7* **Adaja** ≈ N Spain
38 *H17* **Adak Island** *island* Aleutian Islands, Alaska, USA
Adalia *see* Antalya
Adalia, Gulf of *see* Antalya Körfezi
141 *X9* **Adam** N Oman *22°22′N 57°30′E*
Adama *see* Nazrēt
60 *I8* **Adamantina** São Paulo, S Brazil *21°41′S 51°04′W*
79 *E14* **Adamaoua** *Eng.* Adamawa. ◈ *province* N Cameroon
68 *F11* **Adamaoua, Massif d'** *Eng.* Adamawa Highlands. *plateau* NW Cameroon
77 *Y14* **Adamawa** ◈ *state* E Nigeria
Adamawa *see* Adamaoua
Adamawa Highlands *see* Adamaoua, Massif d'
106 *F6* **Adamello** ▲ N Italy *46°09′N 10°33′E*
81 *J14* **Ādamī Tulu** Oromīya, C Ethiopia *07°52′N 38°39′E*
63 *M23* **Adam, Mount** *var.* Monte Independencia. ▲ West Falkland, Falkland Islands *51°36′S 60°00′W*
29 *R16* **Adams** Nebraska, C USA *40°25′N 96°30′W*
18 *H8* **Adams** New York, NE USA *43°48′N 75°57′W*
29 *Q3* **Adams** North Dakota, N USA *48°23′N 98°01′W*
155 *I23* **Adam's Bridge** *chain of shoals* NW Sri Lanka
32 *H10* **Adams, Mount** ▲ Washington, NW USA *46°12′N 121°29′W*
Adam's Peak *see* Sri Pada
191 *R16* **Adam's Rock** *island* Pitcairn Island, Pitcairn Islands
191 *P16* **Adamstown** ○ (Pitcairn Islands)Pitcairn Island, Pitcairn Islands *25°04′S 130°05′W*
20 *G10* **Adamsville** Tennessee, S USA *35°14′N 88°23′W*
25 *S9* **Adamsville** Texas, SW USA *31°15′N 98°09′W*
141 *O17* **'Adan** *Eng.* Aden. SW Yemen *12°51′N 45°05′E*
136 *K16* **Adana** *var.* Seyhan. Adana, S Turkey *37°N 35°19′E*
136 *K16* **Adana** *var.* Seyhan. ◈ *province* S Turkey
Adâncata *see* Horlivka
169 *V12* **Adang, Teluk** *bay* Borneo, C Indonesia
136 *F11* **Adapazarı** *prev.* Ada Bazar. Sakarya, NW Turkey *40°49′N 30°24′E*
80 *H8* **Adarama** River Nile, NE Sudan *17°04′N 34°57′E*
195 *Q16* **Adare, Cape** *cape* Antarctica
Ādavāni *see* Ādoni
106 *E6* **Adda** *anc.* Addua. ≈ N Italy
80 *A13* **Adda** ≈ W South Sudan
143 *Q17* **Aḍ Ḍab'īyah** Abū Ẓaby, C United Arab Emirates *24°17′N 54°08′E*
143 *O18* **Aḍ Ḍafrah** *desert* S United Arab Emirates
141 *Q6* **Ad Dahnā'** *desert* E Saudi Arabia
74 *A11* **Ad Dakhla** *var.* Dakhla. SW Western Sahara *23°46′N 15°56′W*
Ad Dalanj *see* Dilling
Ad Damar *see* Ed Damer
Ad Damazīn *see* Ed Damazin
Ad Dāmir *see* Ed Damer
173 *N2* **Ad Dammām** *desert* NE Saudi Arabia
141 *R6* **Ad Dammān** *var.* Dammām. Ash Sharqīyah, NE Saudi Arabia *26°23′N 50°05′E*
Ad Dāmūr *see* Damoûr
140 *K5* **Ad Dār al Ḥamrā'** Tabūk, NW Saudi Arabia *27°22′N 37°46′E*
140 *M13* **Ad Darb** Jīzān, SW Saudi Arabia *17°45′N 42°15′E*
141 *O8* **Ad Dawādimī** Ar Riyāḍ, C Saudi Arabia *24°32′N 44°21′E*
143 *N16* **Ad Dawḥah** *Eng.* Doha. ● (Qatar) C Qatar *25°15′N 51°36′E*
143 *N16* **Ad Dawḥah** *Eng.* Doha. ✕ C Qatar *25°11′N 51°37′E*
139 *S6* **Ad Dawr** Ṣalāḥ ad Dīn, N Iraq *34°30′N 43°49′E*
139 *Y12* **Ad Dayr** *var.* Dayr, Shahbān. Al Baṣrah, E Iraq *30°45′N 47°36′E*
Addi Arkay *see* Ādī Ārk'ay
Ad Dībakah *see* Dîbege
139 *X15* **Ad Dibdibah** *physical region* Iraq/Kuwait
Aḍ Ḍiffah *see* Libyan Plateau
Addis Ababa *see* Ādīs Ābeba
Addison *see* Webster Springs
139 *U10* **Ad Dīwānīyah** *var.* Diwaniyah. C Iraq *32°00′N 44°57′E*
Ad Dīwānīyah *see* Al Qādisīyah
Addoo Atoll *see* Addu Atoll
Addua *see* Adda
151 *K22* **Addu Atoll** *var.* Addoo Atoll, Seenu Atoll. *atoll* S Maldives
Ad Dujail *see* Ad Dujayl
139 *T7* **Ad Dujayl** *var.* Ad Dujail. Ṣalāḥ ad Dīn, N Iraq *33°49′N 44°16′E*
Ad Dulaym *see* Al Anbār
Ad Duwaym/Ad Duwēm *see* Ed Dueim
99 *D16* **Adegem** Oost-Vlaanderen, NW Belgium *51°12′N 03°31′E*
23 *U7* **Adel** Georgia, SE USA *31°08′N 83°25′W*
29 *U14* **Adel** Iowa, C USA *41°36′N 94°01′W*
182 *I9* **Adelaide** *state capital* South Australia *34°56′S 138°36′E*
44 *H2* **Adelaide** New Providence, N The Bahamas *24°59′N 77°30′W*
182 *I9* **Adelaide** ✕ South Australia *34°55′S 138°31′E*
194 *H6* **Adelaide Island** *island* Antarctica
181 *P2* **Adelaide River** Northern Territory, N Australia *13°12′S 131°06′E*
76 *M10* **'Adel Bagrou** Hodh ech Chargui, SE Mauritania *15°33′N 07°04′W*
186 *D6* **Adelbert Range** ▲▲ N Papua New Guinea
180 *K3* **Adele Island** *island* Western Australia
107 *O17* **Adelfia** Puglia, SE Italy *41°01′N 16°52′E*
195 *V16* **Adélie Coast** *physical region* Antarctica
195 *V14* **Adélie, Terre** *physical region* Antarctica
Adelnau *see* Odolanów
Adelsberg *see* Postojna
Aden *see* 'Adan
Aden *see* Khormaksar
141 *Q17* **Aden, Gulf of** *gulf* SW Arabian Sea
77 *V10* **Aderbissinat** Agadez, C Niger *15°30′N 07°57′E*
Adhaim *see* Al 'Uẓaym
143 *R16* **Adh Dhayd** *var.* Al Dhaid. Ash Shāriqah, NE United Arab Emirates *25°19′N 55°51′E*
140 *M4* **'Adhfā'** *spring/well* NW Saudi Arabia *29°15′N 41°24′E*
138 *I13* **'Ādhriyāt, Jabāl al** ▲▲ S Jordan
80 *I10* **Ādī Ārk'ay** *var.* Addi Arkay. Āmara, N Ethiopia *13°18′N 37°56′E*
182 *C7* **Adieu, Cape** *headland* South Australia *32°01′S 132°12′E*
106 *H8* **Adige** *Ger.* Etsch. ≈ N Italy
80 *J10* **Ādīgrat** Tigray, N Ethiopia *14°17′N 39°27′E*
154 *I13* **Ādilabād** *var.* Ādilābād. Telangana, C India *19°40′N 78°31′E*
35 *P2* **Adin** California, W USA *41°10′N 120°57′W*
171 *V14* **Adi, Pulau** *island* E Indonesia
18 *K8* **Adirondack Mountains** ▲▲ New York, NE USA
80 *J13* **Ādīs Ābeba** *Eng.* Addis Ababa. ● (Ethiopia) Ādīs Ābeba, C Ethiopia *8°59′N 38°43′E*
80 *J13* **Ādīs Ābeba** ✕ Ādīs Ābeba, C Ethiopia *08°58′N 38°53′E*
80 *I11* **Ādīs Zemen** Āmara, N Ethiopia *12°00′N 37°43′E*
Adi Ugri *see* Mendefera
137 *N15* **Adıyaman** Adıyaman, SE Turkey *37°46′N 38°15′E*
137 *N15* **Adıyaman** ◈ *province* S Turkey
116 *L11* **Adjud** Vrancea, E Romania *46°07′N 27°10′E*
45 *T6* **Adjuntas** C Puerto Rico *18°10′N 66°42′W*
Adjuntas, Presa de las *see* Vicente Guerrero, Presa
Ādkup *see* Erikub Atoll
126 *L15* **Adler** Krasnodarskiy Kray, SW Russian Federation *43°25′N 39°58′E*
Adler *see* Orlice
108 *G7* **Adliswil** Zürich, NW Switzerland *47°19′N 08°32′E*
32 *G7* **Admiralty Inlet** *inlet* Washington, NW USA
39 *X13* **Admiralty Island** *island* Alexander Archipelago, Alaska, USA
186 *E5* **Admiralty Islands** *island group* N Papua New Guinea
136 *B14* **Adnan Menderes** ✕ (İzmir) İzmir, W Turkey *38°16′N 27°09′E*
37 *V6* **Adobe Creek Reservoir** ▣ Colorado, C USA
77 *T16* **Ado-Ekiti** Ekiti, SW Nigeria *07°42′N 05°13′E*
Adola *see* Kibre Mengist
61 *C23* **Adolfo González Chaues** Buenos Aires, E Argentina
155 *H17* **Ādoni** *var.* Ādavāni. Andhra Pradesh, C India *15°38′N 77°16′E*
102 *K15* **Adour** *anc.* Aturus. ≈ SW France
Adowa *see* Ādwa
105 *O15* **Adra** Andalucía, S Spain *36°45′N 03°01′W*
107 *L24* **Adrano** Sicilia, Italy, C Mediterranean Sea *37°39′N 14°49′E*
74 *I9* **Adrar** C Algeria *27°56′N 00°12′W*
76 *K7* **Adrar** ◈ *region* C Mauritania
74 *L11* **Adrar** ▲▲ SE Algeria
74 *A12* **Adrar Souttouf** ▲▲ SW Western Sahara
Adrasman *see* Adrasmon
147 *Q10* **Adrasmon** *Rus.* Adrasman. NW Tajikistan *40°38′N 69°56′E*
78 *K10* **Adré** Ouaddaï, E Chad *13°26′N 22°14′E*
106 *H9* **Adria** *anc.* Atria, Hadria, Hatria. Veneto, NE Italy *45°03′N 12°04′E*
31 *R10* **Adrian** Michigan, N USA *41°54′N 84°02′W*
29 *S11* **Adrian** Minnesota, N USA *43°38′N 95°55′W*
27 *R5* **Adrian** Missouri, C USA *38°24′N 94°21′W*
24 *M2* **Adrian** Texas, SW USA *35°16′N 102°39′W*
21 *S4* **Adrian** West Virginia, NE USA *38°53′N 80°14′W*
Adrianople/Adrianopolis *see* Edirne
121 *P7* **Adriatic Basin** *undersea feature* Adriatic Sea, N Mediterranean Sea *42°00′N 17°30′E*
Adriatico, Mare *see* Adriatic Sea
106 *L13* **Adriatic Sea** *Alb.* Deti Adriatik, *It.* Mare Adriatico, *SCr.* Jadransko More, *Slvn.* Jadransko Morje. *sea* N Mediterranean Sea
Adriatik, Deti *see* Adriatic Sea
Adua *see* Ādwa
Aduana del Sásabe *see* El Sásabe
79 *O17* **Adusa** Orientale, NE Dem. Rep. Congo *01°25′N 28°05′E*
118 *J13* **Adutiškis** Vilnius, E Lithuania *55°09′N 26°34′E*
27 *Y7* **Advance** Missouri, C USA *37°06′N 89°54′W*
65 *D25* **Adventure Sound** *bay* East Falkland, Falkland Islands
80 *J10* **Ādwa** *var.* Adowa, *It.* Adua. Tigray, N Ethiopia *14°08′N 38°51′E*
123 *Q8* **Adycha** ≈ NE Russian Federation
126 *L14* **Adygeya, Respublika** ◈ *autonomous republic* SW Russian Federation
Adzhikui *see* Ajyguýy
77 *N17* **Adzopé** SE Ivory Coast *06°07′N 03°49′W*
125 *U4* **Adz'va** ≈ NW Russian Federation
125 *U5* **Adz'vavom** Respublika Komi, NW Russian Federation *66°35′N 59°13′E*
Ædua *see* Autun
115 *K19* **Aegean Islands** *island group* Greece/Turkey
Aegean North *see* Vóreion Aigaíon
115 *I17* **Aegean Sea** *Gk.* Aigaíon Pelagos, Aigaío Pélagos, *Turk.* Ege Denizi. *sea* NE Mediterranean Sea
Aegean South *see* Nótion Aigaíon
118 *H3* **Aegviidu** *Ger.* Charlottenhof. Harjumaa, NW Estonia *59°17′N 25°37′E*
Aegyptus *see* Egypt
Aelana *see* Al 'Aqabah
Aelok *see* Ailuk Atoll
Aelōninae *see* Ailinginae Atoll
Aelōnlaplap *see* Ailinglaplap Atoll
Æmilia *see* Emilia-Romagna
Æmilianum *see* Millau
Aemona *see* Ljubljana
Aenaria *see* Ischia
Aeolian Islands *see* Eolie, Isole
191 *Z3* **Aeon Point** *headland* Kiritimati, NE Kiribati *01°46′N 157°11′W*
Æsernia *see* Isernia
104 *G3* **A Estrada** Galicia, NW Spain *42°41′N 08°29′W*
115 *C18* **Aetós** Itháki, Iónia Nísoi, Greece, C Mediterranean Sea *38°21′N 20°40′E*
191 *Q8* **Afaahiti** Tahiti, W French Polynesia *17°43′S 149°18′W*
139 *U10* **'Afak** Al Qādisīyah, C Iraq *32°04′N 45°17′E*
Afanasjevo *see* Afanas'yevo
125 *T14* **Afanas'yevo** *var.* Afanasjevo. Kirovskaya Oblast', NW Russian Federation *58°55′N 53°13′E*
Afándou *see* Afántou
115 *O23* **Afántou** *var.* Afándou. Ródos, Dodekánisa, Greece, Aegean Sea *36°17′N 28°10′E*
80 *K11* **Āfar** ◈ *region* NE Ethiopia
Afar Depression *see* Danakil Desert
191 *O7* **Afareaitu** Moorea, W French Polynesia *17°33′S 149°47′W*
140 *L7* **'Afarīyah, Bi'r al** *well* NW Saudi Arabia
Afars et des Issas, Territoire Français des *see* Djibouti
83 *D22* **Affenrücken** Karas, SW Namibia *28°05′S 15°49′E*
Afghānestān, Dowlat-e Eslāmī-ye *see* Afghanistan
148 *M6* **Afghanistan** *off.* Islamic Republic of Afghanistan, *Per.* Dowlat-e Eslāmī-ye Afghānestān; *prev.* Republic of Afghanistan. ◆ *islamic state* C Asia
Afgoi *see* Afgooye
81 *N17* **Afgooye** *It.* Afgoi. Shabeellaha Hoose, S Somalia *02°09′N 45°07′E*
141 *N8* **'Afif** Ar Riyāḍ, C Saudi Arabia *23°57′N 42°57′E*
77 *V17* **Afikpo** Ebonyi, S Nigeria *05°52′N 07°58′E*
Afiun Karahissar *see* Afyon
109 *V6* **Aflenz Kurort** Steiermark, E Austria *47°33′N 15°14′E*
74 *J6* **Aflou** N Algeria *34°09′N 02°06′E*
81 *L18* **Afmadow** Jubbada Hoose, S Somalia *0°24′N 42°04′E*
39 *Q14* **Afognak Island** *island* Alaska, USA
104 *J2* **A Fonsagrada** Galicia, NW Spain *43°09′N 07°03′W*
186 *E9* **Afore** Northern, S Papua New Guinea *09°10′S 148°30′E*
59 *O15* **Afrânio** Pernambuco, E Brazil *08°32′S 40°54′W*
66-67 **Africa** *continent*
68 *L11* **Africa, Horn of** *physical region* Ethiopia/Somalia
172 *K11* **Africana Seamount** *undersea feature* SW Indian Ocean *37°10′S 29°10′E*
86 *A14* **African Plate** *tectonic plate*
138 *I2* **'Afrīn** Ḥalab, N Syria *36°31′N 36°51′E*
136 *M15* **Afşin** Kahramanmaraş, C Turkey *38°14′N 36°54′E*
98 *J7* **Afsluitdijk** *dam* N Netherlands
29 *U15* **Afton** Iowa, C USA *41°01′N 94°12′W*
29 *W9* **Afton** Minnesota, N USA *44°54′N 92°46′W*
27 *R8* **Afton** Oklahoma, C USA *36°41′N 94°57′W*
136 *F14* **Afyon** *prev.* Afyonkarahisar. Afyon, W Turkey *38°46′N 30°32′E*
136 *F14* **Afyon** *var.* Afiun Karahissar, Afyonkarahisar. ◈ *province* W Turkey
Afyonkarahisar *see* Afyon
Agadès *see* Agadez
77 *V10* **Agadez** *prev.* Agadès. Agadez, C Niger *16°57′N 07°56′E*
77 *W8* **Agadez** ◈ *department* N Niger
74 *E8* **Agadir** SW Morocco *30°30′N 09°37′W*
64 *M9* **Agadir Canyon** *undersea feature* SE Atlantic Ocean *32°30′N 12°50′W*
145 *R12* **Agadyr'** Karaganda, C Kazakhstan *48°15′N 72°55′E*
173 *O7* **Agalega Islands** *island group* N Mauritius
42 *K6* **Agalta, Sierra de** ▲▲ E Honduras
122 *I10* **Agan** ≈ C Russian Federation
Agana/Agaña *see* Hagåtña
188 *B15* **Agana Bay** *bay* NW Guam
171 *Kk13* **Agano-gawa** ≈ Honshū, C Japan
188 *B17* **Aga Point** *headland* S Guam
154 *G9* **Agar** Madhya Pradesh, C India *23°44′N 76°01′E*
81 *I14* **Āgaro** Oromīya, C Ethiopia *07°52′N 36°36′E*
153 *V15* **Agartala** *state capital* Tripura, NE India *23°49′N 91°15′E*
194 *I5* **Agassiz, Cape** *headland* Antarctica *68°29′S 62°59′W*
175 *V13* **Agassiz Fracture Zone** *tectonic feature* S Pacific Ocean
9 *N2* **Agassiz Ice Cap** *Ice feature* Nunavut, N Canada
188 *B16* **Agat** W Guam *13°20′N 144°38′E*
188 *B16* **Agat Bay** *bay* W Guam
Agatha *see* Agde
115 *M20* **Agathónisi** *island* Dodekánisa, Greece, Aegean Sea
171 *X14* **Agats** Papua, E Indonesia *05°33′S 138°07′E*
155 *C21* **Agatti Island** *island* Lakshadweep, India, N Indian Ocean
38 *D16* **Agattu Island** *island* Aleutian Islands, Alaska, USA
38 *D16* **Agattu Strait** *strait* Aleutian Islands, Alaska, USA
14 *B8* **Agawa** ≈ Ontario, S Canada
14 *B8* **Agawa Bay** *lake bay* Ontario, S Canada
77 *N17* **Agboville** SE Ivory Coast *05°55′N 04°15′W*
137 *V12* **Ağdam** *Rus.* Agdam. SW Azerbaijan *40°04′N 46°00′E*
Agdam *see* Ağdam
103 *P16* **Agde** *anc.* Agatha. Hérault, S France *43°19′N 03°29′E*
103 *P16* **Agde, Cap d'** *headland* S France *43°17′N 03°30′E*
Agedabia *see* Ajdābiyā
102 *L14* **Agen** *anc.* Aginnum. Lot-et-Garonne, SW France *44°12′N 00°37′E*
Agendicum *see* Sens
165 *O13* **Ageo** Saitama, Honshū, S Japan *35°58′N 139°36′E*
109 *R5* **Ager** ≈ N Austria
Agere Hiywet *see* Hāgere Hiywet
108 *G8* **Ägerisee** ◎ W Switzerland
142 *M10* **Āghā Jārī** Khūzestān, SW Iran *30°48′N 49°45′E*
39 *P15* **Aghiyuk Island** *island* Alaska, USA
74 *B12* **Aghouinit** SE Western Sahara *22°14′N 13°10′W*
Aghri Dagh *see* Büyükağrı Dağı
74 *B10* **Aghzoumal, Sebkhet** *var.* Sebjet Agsumal. *salt lake* E Western Sahara
115 *F15* **Agiá** *var.* Ayiá. Thessalía, C Greece *39°43′N 22°45′E*
40 *G7* **Agiabampo, Estero de** *estuary* NW Mexico
121 *P3* **Agía Fýlaxis** *var.* Ayia Phyla. S Cyprus *34°43′N 33°02′E*
Agialoúsa *see* Yenierenköy
115 *M21* **Agía Marína** Léros, Dodekánisa, Greece, Aegean Sea *37°09′N 26°51′E*
121 *Q2* **Agía Nápa** *var.* Ayia Napa. E Cyprus *34°59′N 34°00′E*
115 *L16* **Agía Paraskeví** Lésvos, E Greece *39°13′N 26°19′E*
115 *J15* **Agías Eirínis, Akrotírio** *headland* Límnos, E Greece *39°47′N 25°21′E*
Agiasós *see* Agiassós
115 *L17* **Agiassós** *var.* Agiasós, Ayiásos, Ayiássos. Lésvos, E Greece *39°05′N 26°23′E*
Aginnum *see* Agen
123 *O14* **Aginskoye** Zabaykal'skiy Kray, S Russian Federation *51°10′N 114°32′E*
115 *I14* **Ágion Óros** *Eng.* Mount Athos. ◈ *monastic republic* NE Greece
115 *H14* **Ágion Óros** *var.* Akte, Aktí; *anc.* Acte. *peninsula* NE Greece
114 *D13* **Ágios Achílleios** *religious building* Dytikí Makedonía, N Greece
115 *J16* **Ágios Efstrátios** *var.* Áyios Evstrátios, Hagios Evstrátios. *island* E Greece *34°00′N 25°00′E*
115 *H20* **Ágios Geórgios** *island* Kykládes, Greece, Aegean Sea
Ágios Geórgios *see* Ro
115 *E21* **Ágios Ilías** ▲ S Greece *36°57′N 22°19′E*
115 *K25* **Ágios Ioánnis, Akrotírio** *headland* Kríti, Greece, E Mediterranean Sea *35°19′N 25°46′E*
115 *L20* **Ágios Kírykos** *var.* Áyios Kírikos. Ikaría, Dodekánisa, Greece, Aegean Sea *37°34′N 26°15′E*
115 *K25* **Ágios Nikólaos** *var.* Áyios Nikólaos. Kríti, Greece, E Mediterranean Sea *35°12′N 25°43′E*
115 *D16* **Ágios Nikolaos** Thessalía, C Greece *39°33′N 21°21′E*
115 *H14* **Agíou Órous, Kólpos** *gulf* N Greece
107 *K24* **Agira** *anc.* Agyrium. Sicilia, Italy, C Mediterranean Sea *37°39′N 14°32′E*
115 *G20* **Agkístri** *island* S Greece
114 *G12* **Ágkistro** *var.* Angistro. ▲ NE Greece *41°21′N 23°29′E*
103 *O17* **Agly** ≈ S France
Agnetheln *see* Agnita
14 *E10* **Agnew Lake** ◎ Ontario, S Canada
77 *O16* **Agnibilékrou** E Ivory Coast *07°10′N 03°11′W*
116 *I11* **Agnita** *Ger.* Agnetheln, *Hung.* Szentágota. Sibiu, SW Romania *45°59′N 24°40′E*
107 *K15* **Agnone** Molise, C Italy *41°49′N 14°21′E*
164 *K14* **Ago** Mie, Honshū, SW Japan *34°18′N 136°50′E*
106 *C8* **Agogna** ≈ N Italy
Agoitz *see* Aoiz
Agoiz *see* Aoiz
77 *P17* **Agona Swedru** *var.* Swedru. SE Ghana *05°31′N 00°42′W*
Agordat *see* Ak'ordat
Agosta *see* Augusta
103 *N15* **Agout** ≈ S France
152 *J12* **Āgra** Uttar Pradesh, N India *27°09′N 78°E*
Agra and Oudh, United Provinces of *see* Uttar Pradesh
Agram *see* Zagreb
105 *U5* **Agramunt** Cataluña, NE Spain *41°48′N 01°07′E*
105 *Q5* **Agreda** Castilla y León, N Spain *41°51′N 01°55′W*
137 *S13* **Ağrı** *var.* Karaköse; *prev.* Karakılısse. Ağrı, NE Turkey *39°44′N 43°04′E*
137 *S13* **Ağrı** ◈ *province* NE Turkey
107 *N19* **Agri** *anc.* Aciris. ≈ S Italy
Agri Dagi *see* Büyükağrı Dağı
107 *J24* **Agrigento** *Gk.* Akragas; *prev.* Girgenti. Sicilia, Italy, C Mediterranean Sea *37°19′N 13°33′E*
188 *K4* **Agrihan** *island* N Northern Mariana Islands
115 *L17* **Agriliá, Akrotírio** *prev.* Ákra Maléas. *cape* Lésvos, E Greece
115 *D18* **Agrínio** *prev.* Agrinion. Dytikí Elláda, W Greece *38°38′N 21°25′E*
Agrinion *see* Agrínio
115 *G17* **Agriovótano** Évvoia, C Greece *39°00′N 23°18′E*
107 *L18* **Agropoli** Campania, S Italy *40°22′N 14°59′E*
127 *T3* **Agryz** Udmurtskaya Respublika, NW Russian Federation *56°27′N 52°58′E*
137 *U11* **Ağstafa** *Rus.* Akstafa. NW Azerbaijan *41°06′N 45°28′E*
137 *X11* **Ağsu** *Rus.* Akhsu. C Azerbaijan
Agsumal, Sebjet *see* Aghzoumal, Sebkhet
40 *J11* **Agua Brava, Laguna** *lagoon* W Mexico
54 *F7* **Aguachica** Cesar, N Colombia *08°16′N 73°35′W*
59 *J20* **Água Clara** Mato Grosso do Sul, SW Brazil *20°21′S 52°58′W*
44 *D5* **Aguada de Pasajeros** Cienfuegos, C Cuba *22°23′N 80°51′W*
54 *J5* **Aguada Grande** Lara, N Venezuela *10°38′N 69°29′W*
45 *S5* **Aguadilla** W Puerto Rico *18°27′N 67°08′W*
43 *S16* **Aguadulce** Coclé, S Panama *08°16′N 80°31′W*
104 *L14* **Aguadulce** Andalucía, S Spain *37°15′N 04°59′W*
41 *O8* **Agualeguas** Nuevo León, NE Mexico *26°17′N 99°30′W*
40 *L9* **Aguanaval, Río** ≈ C Mexico
42 *J5* **Aguán, Río** ≈ N Honduras
25 *R16* **Agua Nueva** Texas, SW USA *26°57′N 98°34′W*
60 *J8* **Aguapeí, Rio** ≈ S Brazil
61 *E14* **Aguapey, Río** ≈ NE Argentina
40 *G3* **Agua Prieta** Sonora, NW Mexico *31°16′N 109°33′W*
104 *G5* **A Guarda** *var.* A Guardia, Laguardia, La Guardia. Galicia, NW Spain *41°54′N 08°53′W*
A Guardia *see* A Guarda
56 *E6* **Aguarico, Río** ≈ Ecuador/Peru
55 *O6* **Aguasay** Monagas, NE Venezuela *09°25′N 63°44′W*
40 *M12* **Aguascalientes** Aguascalientes, C Mexico *21°54′N 102°17′W*
40 *L12* **Aguascalientes** ◈ *state* C Mexico
57 *I18* **Aguas Calientes, Río** ≈ S Peru
105 *R7* **Aguasvivas** ≈ NE Spain
60 *J7* **Água Vermelha, Represa de** ▣ S Brazil
56 *E12* **Aguaytía** Ucayali, C Peru *09°02′S 75°30′W*
104 *I5* **A Gudiña** *var.* La Gudiña. Galicia, NW Spain *42°04′N 07°08′W*
104 *G7* **Águeda** Aveiro, N Portugal *40°34′N 08°28′W*
104 *J8* **Águeda** ≈ Portugal/Spain
77 *Q8* **Aguelhok** Kidal, NE Mali *19°18′N 00°50′E*
77 *V12* **Aguié** Maradi, S Niger *13°28′N 07°43′E*
188 *K8* **Aguijan** *island* S Northern Mariana Islands
104 *M14* **Aguilar** *var.* Aguilar de la Frontera. Andalucía, S Spain *37°31′N 04°40′W*
104 *M3* **Aguilar de Campóo** Castilla y León, N Spain *42°47′N 04°15′W*
Aguilar de la Frontera *see* Aguilar
42 *F7* **Aguilares** San Salvador, C El Salvador *13°56′N 89°09′W*
105 *Q14* **Aguilas** Murcia, SE Spain *37°24′N 01°36′W*
40 *L15* **Aguililla** Michoacán, SW Mexico *18°43′N 102°45′W*
Agulhas *see* L'Agulhas
172 *J11* **Agulhas Bank** *undersea feature* SW Indian Ocean *35°30′S 21°00′E*
172 *K11* **Agulhas Basin** *undersea feature* SW Indian Ocean *47°00′S 27°00′E*
83 *F26* **Agulhas, Cape** *Afr.* Kaap Agulhas. *headland* SW South Africa *34°51′S 19°59′E*
Agulhas, Kaap *see* Agulhas, Cape
60 *O9* **Agulhas Negras, Pico das** ▲ SE Brazil *22°21′S 44°39′W*
172 *K11* **Agulhas Plateau** *undersea feature* SW Indian Ocean *40°00′S 26°00′E*
165 *S16* **Aguni-jima** *island* Nansei-shotō, SW Japan
Agurain *see* Salvatierra
54 *G5* **Agustín Codazzi** *var.* Codazzi. Cesar, N Colombia *10°02′N 73°15′W*
Agyrium *see* Agira
74 *L12* **Ahaggar** *high plateau region* SE Algeria
146 *E12* **Ahal Welaýaty** *Rus.* Akhalskiy Velayat. ◈ *province* C Turkmenistan
142 *K2* **Ahar** Āẕarbāyjān-e Sharqī, NW Iran *38°25′N 47°07′E*
Aharnes *see* Acharnés
138 *J3* **Aḥaṣ, Jabal** ▲ NW Syria
138 *J3* **Aḥaṣ, Jebal** ▲▲ W Syria
185 *G16* **Ahaura** ≈ South Island, New Zealand
100 *E13* **Ahaus** Nordrhein-Westfalen, NW Germany *52°04′N 07°01′E*
191 *U9* **Ahe** *atoll* Îles Tuamotu, C French Polynesia
184 *N10* **Ahimanawa Range** ▲▲ North Island, New Zealand
119 *I19* **Ahinski Kanal** *Rus.* Oginskiy Kanal. *canal* SW Belarus
186 *G10* **Ahioma** SE Papua New Guinea *10°20′S 150°35′E*
184 *I2* **Ahipara** Northland, North Island, New Zealand *35°11′S 173°07′E*
184 *I2* **Ahipara Bay** *bay* SE Tasman Sea
Áhkájávrre *see* Akkajaure
Áhkká *see* Akka
39 *N13* **Ahklun Mountains** ▲▲ Alaska, USA
137 *R14* **Ahlat** Bitlis, E Turkey *38°45′N 42°28′E*
101 *F14* **Ahlen** Nordrhein-Westfalen, W Germany *51°46′N 07°53′E*
154 *D10* **Ahmadābād** *var.* Ahmedabad. Gujarāt, W India *23°03′N 72°40′E*
143 *R10* **Ahmadābād** Kermān, C Iran *35°51′N 59°36′E*
Ahmadi *see* Al Aḥmadī
Ahmad Khel *see* Ḥasan Khēl
155 *F14* **Ahmadnagar** *var.* Ahmednagar. Mahārāshtra, W India *19°08′N 74°48′E*
149 *T9* **Ahmadpur Siāl** Punjab, E Pakistan *30°40′N 71°52′E*
77 *N5* **Aḥmar, 'Erg el** *desert* N Mali
80 *K13* **Ahmar Mountains** ▲▲ C Ethiopia
Ahmedabad *see* Ahmadābād
Ahmednagar *see* Ahmadnagar
114 *N12* **Ahmetbey** Kırklareli, NW Turkey *41°26′N 27°35′E*
14 *H12* **Ahmic Lake** ◎ Ontario, S Canada
190 *G12* **Ahoa** Île Uvea, E Wallis and Futuna *13°17′S 176°12′W*
40 *G8* **Ahomé** Sinaloa, C Mexico *25°55′N 109°10′W*
21 *X8* **Ahoskie** North Carolina, SE USA *36°17′N 76°59′W*
101 *D17* **Ahr** ≈ W Germany
143 *N12* **Ahram** *var.* Ahrom. Būshehr, S Iran *28°52′N 51°18′E*
100 *J9* **Ahrensburg** Schleswig-Holstein, N Germany *53°41′N 10°14′E*
Ahrom *see* Ahram
93 *L17* **Ähtäri** Etelä-Pohjanmaa, W Finland *62°34′N 24°08′E*
114 *N10* **Ahtopol** *var.* Akhtopol. Burgas, E Bulgaria *42°06′N 27°57′E*
40 *K12* **Ahuacatlán** Nayarit, C Mexico *21°02′N 104°30′W*
42 *E7* **Ahuachapán** Ahuachapán, W El Salvador *13°55′N 89°51′W*
42 *A9* **Ahuachapán** ◈ *department* W El Salvador
191 *V16* **Ahu Akivi** *var.* Siete Moai. *ancient monument* Easter Island, Chile, E Pacific Ocean
191 *W11* **Ahunui** *atoll* Îles Tuamotu, C French Polynesia
185 *E20* **Ahuriri** ≈ South Island, New Zealand
95 *L22* **Åhus** Skåne, S Sweden *55°55′N 14°18′E*
Ahu Tahira *see* Ahu Vinapu
191 *V16* **Ahu Tepeu** *ancient monument* Easter Island, Chile, E Pacific Ocean
191 *V17* **Ahu Vinapu** *var.* Ahu Tahira. *ancient monument* Easter Island, Chile, E Pacific Ocean
142 *L9* **Ahvāz** *var.* Ahwāz; *prev.* Nāsiri. Khūzestān, SW Iran *31°20′N 48°38′E*
Ahvenanmaa *see* Åland
141 *Q16* **Aḥwar** SW Yemen *13°34′N 46°41′E*
Ahwāz *see* Ahvāz
94 *H7* **Åi Åfjord** *var.* Åi Åfjord, Årnes. Sør-Trøndelag, C Norway *63°57′N 10°12′E* *see also* Åi Åfjord
Åi Åfjord *see* Åi Åfjord
149 *P3* **Aibak** *var.* Haibak; *prev.* Āybak, Samangān. Samangān, NE Afghanistan *36°16′N 68°04′E*
101 *K22* **Aichach** Bayern, SE Germany *48°26′N 11°06′E*
164 *L14* **Aichi** *off.* Aichi-ken, *var.* Aiti. ◈ *prefecture* Honshū, SW Japan
Aïdin *see* Aydın
Aidussina *see* Ajdovščina
Aifir, Clochán an *see* Giant's Causeway
Aigaíon Pelagos/Aigaío Pélagos *see* Aegean Sea
109 *S3* **Aigen im Mülkreis** Oberösterreich, N Austria *48°39′N 13°57′E*
115 *G20* **Aígina** *var.* Aíyina, Egina. Aígina, C Greece *37°45′N 23°26′E*
115 *G20* **Aígina** *island* S Greece
115 *E18* **Aígio** *var.* Egio; *prev.* Aíyion. Dytikí Elláda, S Greece *38°15′N 22°05′E*
108 *C10* **Aigle** Vaud, SW Switzerland *46°20′N 06°58′E*
103 *P14* **Aigoual, Mont** ▲ S France *44°09′N 03°34′E*
173 *O16* **Aigrettes, Pointe des** *headland* W Réunion *21°02′S 55°14′E*
61 *G19* **Aiguá** *var.* Aigua. Maldonado, S Uruguay *34°13′S 54°46′W*
103 *S13* **Aigues** ≈ SE France
103 *N10* **Aigurande** Indre, C France *46°26′N 01°49′E*
Ai-hun *see* Heihe
163 *N10* **Aikawa** Niigata, Sado, C Japan *38°04′N 138°15′E*
21 *Q13* **Aiken** South Carolina, SE USA *33°34′N 81°44′W*
25 *N4* **Aiken** Texas, SW USA *34°06′N 101°31′W*

◆ Country
● Country Capital
◇ Dependent Territory
○ Dependent Territory Capital
◈ Administrative Regions
✕ International Airport
▲ Mountain
▲▲ Mountain Range
⛰ Volcano
≈ River
◎ Lake
▣ Reservoir

◆ Country | ◇ Dependent Territory | ▲ Administrative Regions | ▲ Mountain | ☒ Volcano | ⊚ Lake
● Country Capital | ○ Dependent Territory Capital | ✕ International Airport | ▲ Mountain Range | ♒ River | ☒ Reservoir

ALCAÑICES – ALMONTE

104 J5 **Alcañices** Castilla y León, N Spain 41°41′N 06°21′W
105 T7 **Alcañiz** Aragón, NE Spain 41°03′N 00°09′W
104 I9 **Alcántara** Extremadura, W Spain 39°42′N 06°54′W
104 J9 **Alcántara, Embalse de** ⊠ W Spain
105 R13 **Alcantarilla** Murcia, SE Spain 37°59′N 01°12′W
105 P11 **Alcaraz** Castilla-La Mancha, C Spain 38°40′N 02°29′W
105 P12 **Alcaraz, Sierra de** ▲▲ C Spain
104 I12 **Alcarrache** ⇝ SW Spain
105 T6 **Alcarràs** Cataluña, NE Spain 41°34′N 00°31′E
105 N14 **Alcaudete** Andalucía, S Spain 37°35′N 04°05′W
Alcázar *see* Ksar-el-Kebir
105 O10 **Alcázar de San Juan** *anc.* Alce. Castilla-La Mancha, C Spain 39°24′N 03°12′W
Alcazarquivir *see* Ksar-el-Kebir
Alce *see* Alcázar de San Juan
57 B17 **Alcebo, Volcán** ⛰ Galapagos Islands, Ecuador, E Pacific Ocean 0°25′S 91°06′W
139 X12 **Al Chabā'ish** *var.* Al Kaba'ish. Dhī Qār, SE Iraq 30°58′N 47°02′E
117 Y7 **Alchevs'k** *prev.* Kommunarsk, Voroshilovsk. Luhans'ka Oblast', E Ukraine 48°29′N 38°52′E
Alcira *see* Alzira
21 N9 **Alcoa** Tennessee, S USA 35°47′N 83°58′W
104 F9 **Alcobaça** Leiria, C Portugal 39°32′N 08°59′W
105 N8 **Alcobendas** Madrid, C Spain 40°32′N 03°38′W
Alcoi *see* Alcoy
105 P7 **Alcolea del Pinar** Castilla-La Mancha, C Spain 41°02′N 02°28′W
104 I11 **Alconchel** Extremadura, W Spain 38°31′N 07°04′W
Alcora *see* L'Alcora
105 N8 **Alcorcón** Madrid, C Spain 40°20′N 03°52′W
105 S7 **Alcorisa** Aragón, NE Spain 40°53′N 00°23′W
61 B19 **Alcorta** Santa Fe, C Argentina 33°32′S 61°07′W
104 H14 **Alcoutim** Faro, S Portugal 37°28′N 07°29′W
33 W15 **Alcova** Wyoming, C USA 42°33′N 106°40′W
105 S11 **Alcoy** *Cat.* Alcoi. Valenciana, E Spain 38°42′N 00°29′W
105 Y9 **Alcúdia** Mallorca, Spain, W Mediterranean Sea 39°51′N 03°06′E
105 Y9 **Alcúdia, Badia d'** *bay* Mallorca, Spain, W Mediterranean Sea
172 M7 **Aldabra Group** *island group* SW Seychelles
139 U10 **Al Daghghārah** Bābil, C Iraq 32°10′N 44°57′E
40 J5 **Aldama** Chihuahua, N Mexico 28°50′N 105°52′W
41 P11 **Aldama** Tamaulipas, C Mexico 22°54′N 98°05′W
123 Q11 **Aldan** Respublika Sakha (Yakutiya), NE Russian Federation 58°31′N 125°15′E
123 Q10 **Aldan** ⇝ NE Russian Federation
Aldar *see* Aldarhaan
al Dar al Baida *see* Rabat
162 G7 **Aldarhaan** *var.* Aldar. Dzavhan, W Mongolia 47°43′N 96°36′E
97 Q20 **Aldeburgh** E England, United Kingdom 52°12′N 01°36′E
105 P5 **Aldehuela de Calatañazor** Castilla y León, N Spain 41°42′N 02°46′W
Aldeia Nova *see* Aldeia Nova de São Bento
104 H13 **Aldeia Nova de São Bento** *var.* Aldeia Nova. Beja, S Portugal 37°55′N 07°24′W
29 V11 **Alden** Minnesota, N USA 43°40′N 93°34′W
184 N6 **Aldermen Islands, The** *island group* N New Zealand
97 L25 **Alderney** *island* Channel Islands
97 N22 **Aldershot** S England, United Kingdom 51°15′N 00°47′W
21 R6 **Alderson** West Virginia, NE USA 37°43′N 80°38′W
Al Dhaid *see* Adh Dhayd
98 L5 **Aldtsjerk** *Dutch.* Oudkerk. Fryslân, N Netherlands 53°16′N 05°52′E
30 J11 **Aledo** Illinois, N USA 41°12′N 90°45′W
76 H9 **Aleg** Brakna, SW Mauritania 17°03′N 13°53′W
64 Q10 **Alegranza** *island* Islas Canarias, Spain, NE Atlantic Ocean
37 P12 **Alegres Mountain** ▲ New Mexico, SW USA 34°09′N 108°11′W
61 F15 **Alegrete** Rio Grande do Sul, S Brazil 29°46′S 55°46′W
61 C16 **Alejandra** Santa Fe, C Argentina 29°54′S 59°50′W
193 T11 **Alejandro Selkirk, Isla** *island* Islas Juan Fernández, Chile, E Pacific Ocean
124 I12 **Alëkhovshchina** Leningradskaya Oblast', NW Russian Federation 60°22′N 33°57′E
39 O13 **Aleknagik** Alaska, USA 59°16′N 158°37′W
Aleksandriya *see* Oleksandriya
Aleksandropol' *see* Gyumri
126 L3 **Aleksandrov** Vladimirskaya Oblast', W Russian Federation 56°24′N 38°42′E
113 N14 **Aleksandrovac** Serbia, C Serbia 43°28′N 21°05′E
127 R9 **Aleksandrov Gay** Saratovskaya Oblast', W Russian Federation 50°08′N 48°34′E
127 U6 **Aleksandrovka** Orenburgskaya Oblast', W Russian Federation 52°47′N 54°14′E
Aleksandrovka *see* Oleksandrivka
125 V13 **Aleksandrovsk** Permskiy Kray, NW Russian Federation 59°12′N 57°27′E
Aleksandrovsk *see* Zaporizhzhya
127 N14 **Aleksandrovskoye** Stavropol'skiy Kray, SW Russian Federation 44°43′N 42°56′E
123 T12 **Aleksandrovsk-Sakhalinskiy** Ostrov Sakhalin, Sakhalinskaya Oblast', SE Russian Federation 50°55′N 142°12′E
110 J10 **Aleksandrów Kujawski** Kujawsko-pormorskie, C Poland 52°52′N 18°40′E
110 K12 **Aleksandrów Łódzki** Łódzkie, C Poland 51°49′N 19°19′E
114 J8 **Aleksandŭr Stamboliyski, Yazovir** ⊠ N Bulgaria
Alekseevka *see* Akkol', Akmola, Kazakhstan
Alekseevka *see* Terekty
145 P7 **Alekseyevka** *Kaz.* Alekseevka. Akmola, N Kazakhstan 53°32′N 69°30′E
126 L9 **Alekseyevka** Belgorodskaya, W Russian Federation 50°35′N 38°41′E
127 S7 **Alekseyevka** Samarskaya Oblast', W Russian Federation 52°37′N 51°20′E
Alekseyevka *see* Akkol', Akmola, Kazakhstan
Alekseyevka *see* Terekty, Vostochnyy Kazakhstan, Kazakhstan
127 R4 **Alekseyevskoye** Respublika Tatarstan, W Russian Federation 55°18′N 50°11′E
126 K5 **Aleksin** Tul'skaya Oblast', W Russian Federation 54°30′N 37°08′E
113 O14 **Aleksinac** Serbia, SE Serbia 43°33′N 21°43′E
190 G11 **Alele** Île Uvea, E Wallis and Futuna 13°14′S 176°09′W
95 N20 **Ålem** Kalmar, S Sweden 56°57′N 16°25′E
102 L6 **Alençon** Orne, N France 48°26′N 00°04′E
58 I12 **Alenquer** Pará, NE Brazil 01°58′S 54°45′W
38 G10 **'Alenuihaha Channel** *var.* Alenuihaha Channel. *channel* Hawai'i, USA, C Pacific Ocean
Alep/Aleppo *see* Ḥalab
103 Y15 **Aléria** Corse, France, C Mediterranean Sea 42°06′N 09°29′E
197 Q11 **Alert** Ellesmere Island, Nunavut, N Canada 82°28′N 62°13′W
103 Q14 **Alès** *prev.* Alais. Gard, S France 44°08′N 04°05′E
116 G9 **Aleşd** *Hung.* Élesd. Bihor, SW Romania 47°03′N 22°22′E
106 C9 **Alessandria** *Fr.* Alexandrie. Piemonte, N Italy 44°54′N 08°37′E
Ålestrup *see* Aalestrup
94 D9 **Ålesund** Møre og Romsdal, S Norway 62°28′N 06°11′E
108 E10 **Aletschhorn** ▲ SW Switzerland 46°33′N 08°01′E
197 S1 **Aleutian Basin** *undersea feature* Bering Sea 57°00′N 177°00′E
38 H17 **Aleutian Islands** *island group* Alaska, USA
39 P14 **Aleutian Range** ▲▲ Alaska, USA
0 B5 **Aleutian Trench** *undersea feature* S Bering Sea 57°00′N 177°00′W
123 T10 **Alevina, Mys** *cape* E Russian Federation
15 Q6 **Alex** ⇝ Québec, SE Canada
28 J3 **Alexander** North Dakota, N USA 47°48′N 103°38′W
39 W14 **Alexander Archipelago** *island group* Alaska, USA
Alexanderbaai *see* Alexander Bay
83 D23 **Alexander Bay** *Afr.* Alexanderbaai. Northern Cape, W South Africa 28°40′S 16°30′E
23 Q5 **Alexander City** Alabama, S USA 32°56′N 85°57′W
194 J6 **Alexander Island** *island* Antarctica
Alexander Range *see* Kirghiz Range
183 O12 **Alexandra** Victoria, SE Australia 37°12′S 145°43′E
185 D22 **Alexandra** Otago, South Island, New Zealand 45°15′S 169°25′E
115 F14 **Alexándreia** *var.* Alexándria. Kentrikí Makedonía, N Greece 40°38′N 22°27′E
Alexandretta *see* İskenderun
Alexandretta, Gulf of *see* İskenderun Körfezi
15 N13 **Alexandria** Ontario, SE Canada 45°19′N 74°37′W
121 U13 **Alexandria** *Ar.* Al Iskandarīyah. N Egypt 31°07′N 29°51′E
44 J12 **Alexandria** C Jamaica 18°18′N 77°21′W
116 J15 **Alexandria** Teleorman, S Romania 43°58′N 25°18′E
31 P13 **Alexandria** Indiana, N USA 40°15′N 85°40′W
20 M4 **Alexandria** Kentucky, S USA 38°59′N 84°22′W
22 H7 **Alexandria** Louisiana, S USA 31°19′N 92°27′W
29 T7 **Alexandria** Minnesota, N USA 45°54′N 95°22′W
29 Q11 **Alexandria** South Dakota, N USA 43°39′N 97°46′W
21 W4 **Alexandria** Virginia, NE USA 38°49′N 77°06′W
Alexándria *see* Alexándreia
18 I7 **Alexandria Bay** New York, NE USA 44°20′N 75°54′W
Alexandrie *see* Alessandria
182 J10 **Alexandrina, Lake** ◎ South Australia
114 K13 **Alexandroúpoli** *var.* Alexandroúpolis, *Turk.* Dedeagaç, Dedeagach. Anatolikí Makedonía kai Thráki, NE Greece 40°52′N 25°53′E
Alexandroúpolis *see* Alexandroúpoli
10 L15 **Alexis Creek** British Columbia, SW Canada 52°06′N 123°25′W
122 I13 **Aleysk** Altayskiy Kray, S Russian Federation 52°32′N 82°46′E
139 S8 **Al Fallūjah** *var.* Falluja. Al Anbār, C Iraq 33°21′N 43°46′E
105 R8 **Alfambra** ⇝ E Spain
Al Faqa *see* Faq'
141 R15 **Al Fardah** C Yemen 14°51′N 48°33′E
105 Q4 **Alfaro** La Rioja, N Spain 42°13′N 01°45′W
105 U5 **Alfarràs** Cataluña, NE Spain 41°50′N 00°34′E
Al Fāshir *see* El Fasher
75 W8 **Al Fashn** *var.* El Fashn. C Egypt 28°49′N 30°54′E
114 M7 **Alfatar** Silistra, NE Bulgaria 43°56′N 27°17′E
139 S5 **Al Fatḥah** Ṣalāḥ ad Dīn, C Iraq 35°06′N 43°34′E
139 Q3 **Al Fatsī** Nīnawá, N Iraq 36°04′N 42°39′E
139 Z13 **Al Fāw** *var.* Fao. Al Baṣrah, SE Iraq 29°55′N 48°26′E
75 W8 **Al Fayyūm** *var.* El Faiyûm. N Egypt 29°19′N 30°50′E
115 D20 **Alfeiós** *prev.* Alfiós; *anc.* Alpheius, Alpheus. ⇝ S Greece
100 I13 **Alfeld** Niedersachsen, C Germany 51°58′N 09°49′E
Alfiós *see* Alfeiós
Alföld *see* Great Hungarian Plain
94 C11 **Ålfotbreen** *glacier* S Norway
19 P9 **Alfred** Maine, NE USA 43°29′N 70°44′W
18 F11 **Alfred** New York, NE USA 42°15′N 77°47′W
61 K14 **Alfredo Wagner** Santa Catarina, S Brazil 27°40′S 49°22′W
94 M12 **Alfta** Gävleborg, C Sweden 61°20′N 16°05′E
140 K12 **Al Fuḥayḥīl** *var.* Fahaheel. SE Kuwait 29°01′N 48°05′E
139 Q6 **Al Fuḥaymī** Al Anbār, C Iraq 34°18′N 42°09′E
143 S16 **Al Fujayrah** *Eng.* Fujairah. Al Fujayrah, NE United Arab Emirates 25°09′N 56°18′E
143 S16 **Al Fujayrah** *Eng.* Fujairah. ✈ Al Fujayrah, NE United Arab Emirates 25°04′N 56°12′E
Al-Furāt *see* Euphrates
144 I10 **Alga** *Kaz.* Algha. Aktyubinsk, NW Kazakhstan 49°56′N 57°19′E
144 G9 **Algabas** *Kaz.* Alghabas. Zapadnyy Kazakhstan, NW Kazakhstan 50°43′N 52°09′E
95 C17 **Ålgård** Rogaland, S Norway 58°45′N 05°52′E
104 G14 **Algarve** *cultural region* S Portugal
182 G3 **Algebuckina Bridge** South Australia 28°03′S 135°48′E
104 K16 **Algeciras** Andalucía, SW Spain 36°08′N 05°27′W
105 S10 **Algemesí** Valenciana, E Spain 39°11′N 00°27′W
Al-Genain *see* El Geneina
120 F9 **Alger** *var.* Algiers, El Djazaïr, Al Jazair. ● (Algeria) N Algeria 36°47′N 02°58′E
74 H9 **Algeria** *off.* Democratic and Popular Republic of Algeria. ◆ *republic* N Africa
Algeria, Democratic and Popular Republic of *see* Algeria
120 J8 **Algerian Basin** *var.* Balearic Plain. *undersea feature* W Mediterranean Sea
Algha *see* Alga
138 I4 **Al Ghāb** *Valley* NW Syria
141 X10 **Al Ghābah** *var.* Ghaba. C Oman 21°22′N 57°14′E
Alghabas *see* Algabas
141 U13 **Al Ghayḍah** E Yemen 16°15′N 52°13′E
140 M6 **Al Ghazālah** Ḥā'il, NW Saudi Arabia 26°55′N 41°23′E
107 B17 **Alghero** Sardegna, Italy, C Mediterranean Sea 40°34′N 08°19′E
95 M20 **Älghult** Kronoberg, S Sweden 57°00′N 15°34′E
75 X9 **Al Ghurdaqah** *var.* Ghurdaqah, Hurghada. E Egypt 27°17′N 33°47′E
Algiers *see* Alger
105 S10 **Alginet** Valenciana, E Spain 39°16′N 00°28′W
83 I26 **Algoa Bay** *bay* S South Africa
104 L15 **Algodonales** Andalucía, S Spain 36°54′N 05°24′W
105 N9 **Algodor** ⇝ C Spain
Al Golea *see* El Goléa
31 N6 **Algoma** Wisconsin, N USA 44°41′N 87°24′W
29 U12 **Algona** Iowa, C USA 43°04′N 94°13′W
20 L8 **Algood** Tennessee, S USA 36°12′N 85°27′W
105 O2 **Algorta** País Vasco, N Spain 43°20′N 03°00′W
61 E18 **Algorta** Río Negro, W Uruguay 32°26′S 57°18′W
Al Haba *see* Haba
139 Q10 **Al Habbārīyah** Al Anbār, S Iraq 32°16′N 42°12′E
Al Hadhar *see* Al Ḥaḍr
139 Q4 **Al Ḥaḍr** *var.* Al Hadhar; *anc.* Hatra. Nīnawýa, NW Iraq 35°34′N 42°44′E
139 T13 **Al Ḥajarah** *desert* S Iraq
141 W8 **Al Ḥajar al Gharbī** ▲▲ N Oman
141 Y8 **Al Ḥajar ash Sharqī** ▲▲ NE Oman
141 R15 **Al Ḥajarayn** C Yemen 15°29′N 48°24′E
138 L10 **Al Ḥamād** *desert* Jordan/Saudi Arabia
Al Hamad *see* Syrian Desert
75 N9 **Al Ḥamādah al Ḥamrā'** *var.* Al Ḥamrā'. *desert* NW Libya
105 N15 **Alhama de Granada** Andalucía, S Spain 37°00′N 03°59′W
105 R13 **Alhama de Murcia** Murcia, SE Spain 37°51′N 01°25′W
35 T15 **Alhambra** California, W USA 34°08′N 118°06′W
139 T12 **Al Ḥammām** An Najaf, S Iraq 31°09′N 44°04′E
141 X8 **Al Ḥamrā'** NE Oman 23°07′N 57°23′E
Al Ḥamrā' *see* Al Ḥamādah al Ḥamrā'
141 O6 **Al Ḥamūdīyah** *spring/well* N Saudi Arabia 27°05′N 44°24′E
140 M7 **Al Ḥanākīyah** Al Madīnah, W Saudi Arabia 24°55′N 40°31′E
139 W14 **Al Ḥanīyah** *escarpment* Iraq/Saudi Arabia
139 Y12 **Al Ḥārithah** Al Baṣrah, SE Iraq 30°43′N 47°44′E
140 L3 **Al Ḥarrah** *desert* NW Saudi Arabia
75 Q10 **Al Harūj al Aswad** *desert* C Libya
Al Hasaifin *see* Al Ḥusayfin
139 N2 **Al Ḥasakah** *var.* Al Hasijah, El Haseke, *Fr.* Hassetché. Al Ḥasakah, NE Syria 36°22′N 40°44′E
139 O2 **Al Ḥasakah** *off.* Muḥāfaẓat al Ḥasakah, *var.* Al Hasakah, Āl Hasakah, Hasakah, Hassakeh. ◆ *governorate* NE Syria
Al Hasakah *see* Al Ḥasakah
Al Hasakah *see* 'Āmūdah
139 T9 **Al Hāshimīyah** Bābil, C Iraq 32°24′N 44°39′E
138 G13 **Al Hāshimīyah** Ma'ān, S Jordan 30°31′N 35°46′E
Al Hasijah *see* Al Ḥasakah
104 M15 **Alhaurín el Grande** Andalucía, S Spain 36°39′N 04°41′W
141 Q16 **Al Ḥawrā** S Yemen 13°54′N 47°36′E
139 V10 **Al Ḥayy** *var.* Kut al Hai, Kūt al Ḥayy. Wāsiṭ, E Iraq 32°11′N 46°03′E
141 U11 **Al Ḥibāk** *desert* E Saudi Arabia
138 H8 **Al Hījānah** *var.* Hejanah, Hijanah. Rīf Dimashq, W Syria 33°23′N 36°34′E
140 K7 **Al Ḥijāz** *Eng.* Hejaz. *physical region* NW Saudi Arabia
Al Hilbeh *see* 'Ulayyānīyah, Bi'r al
139 T9 **Al Ḥillah** *var.* Hilla. Bābil, C Iraq 32°28′N 44°29′E
139 T9 **Al Hindīyah** *var.* Hindiya. Bābil, C Iraq 32°32′N 44°14′E
138 G12 **Al Ḥisā** Aṭ Ṭafīlah, W Jordan 30°49′N 35°58′E
74 G5 **Al-Hoceïma** *var.* al Hoceima, Al-Hoceima, Alhucemas; *prev.* Villa Sanjurjo. N Morocco 35°14′N 03°56′W
Alhucemas *see* Al-Hoceïma
105 N17 **Alhucemas, Peñon de** *island group* S Spain
141 N15 **Al Ḥudaydah** *Eng.* Hodeida. W Yemen 15°N 42°50′E
141 N15 **Al Ḥudaydah** *var.* Hodeida. ✈ W Yemen 14°45′N 43°01′E
140 M4 **Al Ḥudūd ash Shamālīyah** *var.* Minṭaqat al Ḥudūd ash Shamālīyah, *Eng.* Northern Border Region. ◆ *province* N Saudi Arabia
141 S7 **Al Hufūf** *var.* Hofuf. Ash Sharqīyah, NE Saudi Arabia 25°21′N 49°34′E
al-Hurma *see* Al Khurmah
141 X7 **Al Ḥusayfin** *var.* Al Hasaifin. N Oman 24°33′N 56°33′E
138 G9 **Al Ḥuṣn** *var.* Husn. Irbid, N Jordan 32°29′N 35°53′E
139 U9 **'Alī** Wāsiṭ, E Iraq 32°43′N 45°21′E
104 L10 **Alia** Extremadura, W Spain 39°25′N 05°12′W
143 P9 **'Alīābād** Yazd, C Iran 36°55′N 54°33′E
'Alīābad *see* Qā'emshahr
105 S7 **Aliaga** Aragón, NE Spain 40°40′N 00°42′W
136 B13 **Aliağa** İzmir, W Turkey 38°49′N 26°59′E
Aliákmon *see* Aliákmonas
115 F14 **Aliákmonas** *prev.* Aliákmon; *anc.* Haliacmon. ⇝ N Greece
139 W9 **'Alī al Gharbī** Maysān, E Iraq 32°28′N 46°42′E
139 U11 **'Alī al Ḥassūnī** Al Qādisīyah, S Iraq 31°25′N 44°50′E
115 G18 **Alíartos** Stereá Elláda, C Greece 38°23′N 23°06′E
Äli-Bayramli *see* Şirvan
114 P12 **Alibey Barajı** ⊠ NW Turkey
77 S13 **Alibori** ⇝ N Benin
112 M10 **Alibunar** Vojvodina, NE Serbia 45°06′N 20°59′E
105 S12 **Alicante** *Cat.* Alacant, *Lat.* Lucentum. Valenciana, SE Spain 38°21′N 00°29′W
105 S12 **Alicante** ◆ *province* Valenciana, SE Spain
105 S12 **Alicante** ✈ Murcia, E Spain 38°21′N 00°29′W
83 I25 **Alice** Eastern Cape, S South Africa 32°47′S 26°50′E
25 S14 **Alice** Texas, SW USA 27°45′N 98°06′W
83 I25 **Alicedale** Eastern Cape, S South Africa 33°19′S 26°05′E
65 B25 **Alice, Mount** *hill* West Falkland, Falkland Islands
107 P20 **Alice, Punta** *headland* S Italy 39°24′N 17°09′E
181 Q7 **Alice Springs** Northern Territory, C Australia 23°42′S 133°52′E
23 N4 **Aliceville** Alabama, S USA 33°07′N 88°09′W
147 U13 **Alichur** SE Tajikistan 37°49′N 73°45′E
147 U14 **Alichuri Janubí, Qatorkŭhi** *Rus.* Yuzhno-Alichurskiy Khrebet. ▲▲ SE Tajikistan
147 U13 **Alichuri Shimolí, Qatorkŭhi** *Rus.* Severo-Alichurskiy Khrebet. ▲▲ SE Tajikistan
107 K22 **Alicudi, Isola** *island* Isole Eolie, S Italy
Alifu Atoll *see* Ari Atoll
43 W14 **Aligandí** Kuna Yala, NE Panama 09°15′N 78°05′W
152 J11 **Aligarh** Uttar Pradesh, N India 27°54′N 78°04′E
142 M7 **Alīgūdarz** Lorestān, W Iran 33°24′N 49°19′E
163 U5 **Alihe** *var.* Oroqen Zizhiqi. Nei Mongol Zizhiqu, N China 50°34′N 123°40′E
0 F12 **Alijos, Islas** *islets* California, SW USA
149 R6 **'Alī Kbel** *Pash.* 'Alī Khēl. Paktīkā, E Afghanistan 33°55′N 69°49′E
149 R6 **'Alī Khēl** *var.* Ali Khel, Jaji; *prev.* 'Alī Kheyl. Paktīyā, SE Afghanistan 33°55′N 69°46′E
'Alī Khēl *see* 'Alī Kbel, Paktīkā, Afghanistan
Ali Khel *see* 'Alī Kheyl, Paktīā, Afghanistan
'Alī Kheyl *see* 'Alī Khēl
147 V17 **Al Ikhwān** *island group* SE Yemen
Aliki *see* Alykí
79 H19 **Alima** ⇝ C Congo
Al Imārāt al 'Arabīyah al Muttaḥidah *see* United Arab Emirates
115 N23 **Alimía** *island* Dodekánisa, Greece, Aegean Sea
55 V12 **Alimimuni Piek** ▲ S Suriname 02°26′N 55°46′W
79 K15 **Alindao** Basse-Kotto, S Central African Republic 04°58′N 21°16′E
95 J18 **Alingsås** Västra Götaland, S Sweden 57°55′N 12°30′E
81 K18 **Alinjugul** *spring/well* E Kenya 0°03′S 40°31′E
149 S11 **Alīpur** Punjab, E Pakistan 29°22′N 70°59′E
153 T12 **Alīpur Duār** West Bengal, NE India 26°28′N 89°25′E
18 B14 **Aliquippa** Pennsylvania, NE USA 40°36′N 80°15′W
80 L12 **'Ali Sabieh** *var.* 'Alī Sabīḥ. S Djibouti 11°07′N 42°44′E
'Alī Sabīḥ *see* 'Ali Sabieh
140 K3 **Al 'Īsāwīyah** Al Jawf, NW Saudi Arabia 30°41′N 37°58′E
104 J10 **Aliseda** Extremadura, W Spain 39°25′N 06°42′W
139 T8 **Al Iskandarīyah** Bābil, C Iraq 32°53′N 44°22′E
Al Iskandarīyah *see* Alexandria
123 T6 **Aliskerovo** Chukotskiy Avtonomnyy Okrug, NE Russian Federation 67°40′N 167°37′E
75 W7 **Al Ismā'īlīya** *var.* Ismailia, Ismâ'ilîya. N Egypt 30°32′N 32°13′E
114 H13 **Alistráti** Kentrikí Makedonía, NE Greece 41°03′N 23°58′E
39 P15 **Alitak Bay** *bay* Kodiak Island, Alaska, USA
Al Ittiḥād *see* Madīnat ash Sha'b
115 H18 **Alivéri** *var.* Alivérion. Évvoia, C Greece 38°24′N 24°02′E
Alivérion *see* Alivéri
Aliwal-Noord *see* Aliwal North
83 I24 **Aliwal North** *Afr.* Aliwal-Noord. Eastern Cape, SE South Africa 30°42′S 26°43′E
121 Q13 **Al Jabal al Akhḍar** ▲▲ NE Libya
138 H13 **Al Jafr** Ma'ān, S Jordan 30°18′N 36°13′E
75 T8 **Al Jaghbūb** NE Libya 29°45′N 24°31′E
142 K11 **Al Jahrā'** *var.* Al Jahrah, Jahra. C Kuwait 29°18′N 47°36′E
Al Jahrah *see* Al Jahrā'
Al Jamāhīrīyah al 'Arabīyah al Lībīyah ash Sha'bīyah al Ishtirākīy *see* Libya
140 K3 **Al Jarāwī** *spring/well* NW Saudi Arabia 30°12′N 38°48′E
141 X11 **Al Jawārah** *oasis* SE Oman
140 L3 **Al Jawf** *off.* Jauf. Al Jawf, NW Saudi Arabia 29°51′N 39°49′E
140 L4 **Al Jawf** *var.* Minṭaqat al Jawf. ◆ *province* N Saudi Arabia
Al Jawlān *see* Golan Heights
Al Jazair *see* Alger
139 N4 **Al Jazīrah** *physical region* Iraq/Syria
104 F14 **Aljezur** Faro, S Portugal 37°18′N 08°49′W
139 S13 **Al Jil** An Najaf, S Iraq 30°28′N 43°57′E
138 G11 **Al Jīzah** *var.* Jiza. 'Ammān, N Jordan 31°42′N 35°57′E
Al Jīzah *see* Giza
Al Jubail *see* Al Jubayl
141 S6 **Al Jubayl** *var.* Al Jubail. Ash Sharqīyah, NE Saudi Arabia 27°N 49°36′E
141 T10 **Al Juḥaysh, Qalamat** *well* SE Saudi Arabia
143 N15 **Al Jumaylīyah** N Qatar 25°37′N 51°05′E
Al Junaynah *see* El Geneina
104 G13 **Aljustrel** Beja, S Portugal 37°52′N 08°10′W
Al Kaba'ish *see* Al Chabā'ish
Al-Kadhimain *see* Al Kāẓimīyah
Al Kāf *see* El Kef
Alkal'a *see* Alcalá de Henares
35 W4 **Alkali Flat** *salt flat* Nevada, W USA
35 Q1 **Alkali Lake** ◎ Nevada, W USA
141 Z9 **Al Kāmil** NE Oman 22°14′N 58°15′E
138 G11 **Al Karak** *var.* El Kerak, Karak, Kerak; *anc.* Kir Moab, Kir of Moab. Al Karak, W Jordan 31°11′N 35°42′E
138 G12 **Al Karak** *off.* Muḥāfaẓat al Karak. ◆ *governorate* W Jordan
139 W8 **Al Karmashīyah** Wāsiṭ, E Iraq 32°57′N 46°10′E
Al-Kashaniya *see* Al Qash'ānīyah
Al-Kasr al-Kebir *see* Ksar-el-Kebir
139 T8 **Al Kāẓimīyah** *var.* Al-Kadhimain, Kadhimain. Baghdād, C Iraq 33°22′N 44°20′E
99 J18 **Alken** Limburg, NE Belgium 50°52′N 05°19′E
141 X8 **Al Khābūrah** *var.* Khabura. N Oman 23°57′N 57°06′E
Al Khalīl *see* Hebron
139 T7 **Al Khāliṣ** Diyālá, C Iraq 33°51′N 44°33′E
Al Khaluf *see* Khalūf
75 W10 **Al Khārijah** *var.* El Khârga. C Egypt 25°31′N 30°36′E
141 Q8 **Al Kharj** Ar Riyāḍ, C Saudi Arabia 24°12′N 47°12′E
141 W6 **Al Khaṣab** *var.* Khasab. N Oman 26°11′N 56°18′E
Al Khaur *see* Al Khawr
143 N15 **Al Khawr** *var.* Al Khaur, Khor. N Qatar 25°40′N 51°33′E
142 K12 **Al Khīrān** *var.* Al Khiran. SE Kuwait 28°34′N 48°21′E
141 W9 **Al Khīrān** *spring/well* NW Oman 22°31′N 55°42′E
Al Khiyām *see* El Khiyam
Al-Khobar *see* Al Khubar
Al Khor *see* Al Khawr
141 S6 **Al Khubar** *var.* Al-Khobar. Ash Sharqīyah, NE Saudi Arabia 26°00′N 50°00′E
120 M12 **Al Khums** *var.* Homs, Khoms, Khums. NW Libya 32°39′N 14°16′E
141 R15 **Al Khuraybah** C Yemen 15°05′N 48°17′E
140 M9 **Al Khurmah** *var.* al-Hurma. Makkah, W Saudi Arabia 21°59′N 42°00′E
141 V9 **Al Kidan** *desert* NE Saudi Arabia
127 V4 **Alkino-2** Respublika Bashkortostan, W Russian Federation 54°30′N 55°40′E
98 H9 **Alkmaar** Noord-Holland, NW Netherlands 52°37′N 04°45′E
139 T10 **Al Kūfah** *var.* Kufa. An Najaf, S Iraq 32°02′N 44°25′E
75 T11 **Al Kufrah** SE Libya 24°11′N 23°19′E
141 T10 **Al Kursū'** *desert* E Saudi Arabia
139 V9 **Al Kūt** *var.* Kūt al 'Amārah, Kut al Imara. Wāsiṭ, E Iraq 32°30′N 45°51′E
Al-Kuwait *see* Al Kuwayt
Al Kuwayr *see* Guwēr
142 K11 **Al Kuwayt** *var.* Al-Kuwait, *Eng.* Kuwait, Kuwait City; *prev.* Qurein. ● (Kuwait) E Kuwait 29°23′N 48°00′E
142 K11 **Al Kuwayt** ✈ C Kuwait 29°13′N 47°57′E
115 G19 **Alkyonidón, Kólpos** *gulf* C Greece
141 N4 **Al Labbah** *physical region* N Saudi Arabia
138 G4 **Al Lādhiqīyah** *Eng.* Latakia, *Fr.* Lattaquié; *anc.* Laodicea, Laodicea ad Mare. Al Lādhiqīyah, W Syria 35°31′N 35°47′E
138 H4 **Al Lādhiqīyah** *off.* Muḥāfaẓat al Lādhiqīyah, *var.* Al Lathqiyah, Latakia, Lattakia. ◆ *governorate* W Syria
19 R2 **Allagash River** ⇝ Maine, NE USA
152 M13 **Allahābād** Uttar Pradesh, N India 25°27′N 81°50′E
143 S3 **Allāh Dāgh, Reshteh-ye** ▲▲ NE Iran
39 Q8 **Allakaket** Alaska, USA 66°34′N 152°39′W
141 X12 **Al Lakbī** S Oman 18°27′N 56°37′E
11 T15 **Allan** Saskatchewan, S Canada 51°50′N 105°59′W
Allanmyo *see* Aunglan
83 I22 **Allanridge** Free State, C South Africa 27°45′S 26°40′E
104 H4 **Allariz** Galicia, NW Spain 42°11′N 07°48′W
139 R11 **Al Laṣaf** *var.* Al Lussuf. An Najaf, S Iraq 31°38′N 43°16′E
Al Lathqiyah *see* Al Lādhiqīyah
23 S2 **Allatoona Lake** ⊠ Georgia, SE USA
83 J19 **Alldays** Limpopo, NE South Africa 22°39′S 29°04′E
Alle *see* Łyna
31 P10 **Allegan** Michigan, N USA 42°31′N 85°51′W
18 E14 **Allegheny Mountains** ▲▲ NE USA
18 E12 **Allegheny Plateau** ▲▲ New York/Pennsylvania, NE USA
18 D11 **Allegheny Reservoir** ⊠ New York/Pennsylvania, NE USA
18 E12 **Allegheny River** ⇝ New York/Pennsylvania, NE USA
22 K9 **Allemands, Lac des** ◎ Louisiana, S USA
25 U6 **Allen** Texas, SW USA 33°06′N 96°40′W
21 R14 **Allendale** South Carolina, SE USA 33°01′N 81°19′W
41 N6 **Allende** Coahuila, NE Mexico 28°22′N 100°50′W
41 O9 **Allende** Nuevo León, NE Mexico 25°20′N 100°01′W
97 D16 **Allen, Lough** *Ir.* Loch Aillionn. ◎ NW Ireland
185 B26 **Allen, Mount** ▲ Stewart Island, Southland, SW New Zealand 47°05′S 167°49′E
109 V2 **Allensteig** Niederösterreich, N Austria 48°40′N 15°24′E
Allenstein *see* Olsztyn
18 I14 **Allentown** Pennsylvania, NE USA 40°37′N 75°30′W
155 G23 **Alleppey** *var.* Alappuzha. Kerala, SW India 09°30′N 76°22′E *see also* Alappuzha
100 J12 **Aller** ⇝ NW Germany
29 V16 **Allerton** Iowa, C USA 40°42′N 93°22′W
99 K19 **Alleur** Liège, E Belgium 50°40′N 05°33′E
101 J25 **Allgäuer Alpen** ▲▲ Austria/Germany
28 J13 **Alliance** Nebraska, C USA 42°08′N 102°54′W
31 U12 **Alliance** Ohio, N USA 40°55′N 81°06′W
103 O10 **Allier** ◆ *department* N France
139 R13 **Al Līfīyah** *well* An Najaf, S Iraq
44 J13 **Alligator Pond** C Jamaica 17°52′N 77°34′W
21 Y9 **Alligator River** ⇝ North Carolina, SE USA
29 W12 **Allison** Iowa, C USA 42°45′N 92°48′W
14 G14 **Alliston** Ontario, S Canada 44°09′N 79°51′W
140 L11 **Al Līth** Makkah, SW Saudi Arabia 21°N 41°E
Al Liwā' *see* Līwā
96 J12 **Alloa** C Scotland, United Kingdom 56°07′N 03°49′W
103 U14 **Allos** Alpes-de-Haute-Provence, SE France 44°16′N 06°37′E
108 D6 **Allschwil** Basel Landschaft, NW Switzerland 47°34′N 07°32′E
Al Lubnān *see* Lebanon
141 N14 **Al Luḥayyah** W Yemen 15°44′N 42°45′E
15 Q12 **Allumettes, Île des** *island* Québec, SE Canada
Al Lussuf *see* Al Laṣaf
109 S5 **Alm** ⇝ N Austria
15 Q7 **Alma** Québec, SE Canada 48°32′N 71°41′W
27 S10 **Alma** Arkansas, C USA 35°28′N 94°13′W
23 V7 **Alma** Georgia, SE USA 31°32′N 82°27′W
27 P4 **Alma** Kansas, C USA 39°01′N 96°17′W
31 Q8 **Alma** Michigan, N USA 43°22′N 84°39′W
29 O17 **Alma** Nebraska, C USA 40°06′N 99°21′W
30 I7 **Alma** Wisconsin, N USA 44°21′N 91°54′W
139 R12 **Al Ma'ānīyah** *well* An Najaf, S Iraq
Alma-Ata *see* Almaty
Alma-Atinskaya Oblast' *see* Almaty
105 T5 **Almacelles** Cataluña, NE Spain 41°44′N 00°26′E
104 F11 **Almada** Setúbal, W Portugal 38°40′N 09°09′W
104 L11 **Almadén** Castilla-La Mancha, C Spain 38°47′N 04°50′W
66 L6 **Almadies, Pointe des** *headland* W Senegal 14°44′N 17°31′W
140 L7 **Al Madīnah** *Eng.* Medina. Al Madīnah, W Saudi Arabia 24°25′N 39°29′E
140 L7 **Al Madīnah** *off.* Minṭaqat al Madīnah. ◆ *province* W Saudi Arabia
138 H9 **Al Mafraq** *var.* Mafraq. Al Mafraq, N Jordan 32°20′N 36°12′E
138 J10 **Al Mafraq** *off.* Muḥāfaẓat al Mafraq. ◆ *governorate* NW Jordan
141 R15 **Al Maghārīm** C Yemen 15°00′N 47°49′E
105 N11 **Almagro** Castilla-La Mancha, C Spain 38°54′N 03°43′W
Al Maḥallah al Kubrá *see* El Mahalla el Kubra
139 T9 **Al Maḥāwīl** *var.* Khān al Maḥāwīl. Bābil, C Iraq 32°39′N 44°28′E
Al Mahdīyah *see* Mahdia
139 T8 **Al Maḥmūdīyah** *var.* Mahmudiya. Baghdād, C Iraq 33°04′N 44°22′E
141 T14 **Al Mahrah** ▲▲ E Yemen
141 P7 **Al Majma'ah** Ar Riyāḍ, C Saudi Arabia 25°55′N 45°19′E
139 Q11 **Al Makmin** *well* S Iraq
139 Q1 **Al Mālikīyah** *var.* Malkiye. Al Ḥasakah, N Syria 37°12′N 42°13′E
Almalyk *see* Olmaliq
Al Mamlakah *see* Morocco
Al Mamlaka al Urdunīya al Hashemīyah *see* Jordan
143 Q18 **Al Manādir** *var.* Al Manadir. *desert* Oman/United Arab Emirates
142 L15 **Al Manāmah** *Eng.* Manama. ● (Bahrain) N Bahrain 26°13′N 50°33′E
139 O5 **Al Manāṣif** ▲▲ E Syria
35 O4 **Almanor, Lake** ⊠ California, W USA
105 R11 **Almansa** Castilla-La Mancha, C Spain 38°52′N 01°06′W
75 W7 **Al Manṣūrah** *var.* Manṣûra, El Manṣûra. N Egypt 31°03′N 31°23′E
104 L3 **Almanza** Castilla y León, N Spain 42°40′N 05°01′W
104 L8 **Almanzor** ▲ W Spain 40°13′N 05°18′W
105 P14 **Almanzora** ⇝ SE Spain
139 S9 **Al Mardah** Karbalā', C Iraq 32°35′N 43°30′E
Al-Mariyya *see* Almería
75 R7 **Al Marj** *var.* Barka, *It.* Barce. NE Libya 32°30′N 20°54′E
138 L2 **Al Mashrafah** Ar Raqqah, N Syria 36°25′N 39°07′E
141 X8 **Al Maṣna'ah** *var.* Al Muṣana'a. NE Oman 23°45′N 57°38′E
Almassora *see* Almazora
Almatinskaya Oblast' *see* Almaty
145 U15 **Almaty** *var.* Alma-Ata. Almaty, SE Kazakhstan 43°19′N 76°55′E
145 S14 **Almaty** *off.* Almatinskaya Oblast', *Kaz.* Almaty Oblysy; *prev.* Alma-Atinskaya Oblast'. ◆ *province* SE Kazakhstan
145 U15 **Almaty** ✈ Almaty, SE Kazakhstan 43°15′N 76°57′E
Almaty Oblysy *see* Almaty
al-Mawailih *see* Al Muwayliḥ
139 R3 **Al Mawṣil** *Eng.* Mosul. Nīnawá, N Iraq 36°21′N 43°08′E
139 N5 **Al Mayādīn** *var.* Mayadin, *Fr.* Meyadine. Dayr az Zawr, E Syria 35°00′N 40°31′E
139 X10 **Al Maymūnah** *var.* Maimuna. Maysān, SE Iraq 31°43′N 46°55′E
141 N5 **Al Mayyāh** Ḥā'il, N Saudi Arabia 27°57′N 42°53′E
Al Ma'zam *see* Al Ma'zim
105 P6 **Almazán** Castilla y León, N Spain 41°29′N 02°31′W
141 W8 **Al Ma'zim** *var.* Al Ma'zam. NW Oman 23°22′N 56°16′E
123 N11 **Almaznyy** Respublika Sakha (Yakutiya), NE Russian Federation 62°19′N 114°14′E
105 T9 **Almazora** *Cat.* Almassora. Valenciana, E Spain 39°55′N 00°02′W
Al Mazra' *see* Al Mazra'ah
138 G11 **Al Mazra'ah** *var.* Al Mazra', Mazra'a. Al Karak, W Jordan 31°18′N 35°32′E
101 G15 **Alme** ⇝ W Germany
104 I7 **Almeida** Guarda, N Portugal 40°43′N 06°53′W
104 G10 **Almeirim** Santarém, C Portugal 39°12′N 08°37′W
98 O10 **Almelo** Overijssel, E Netherlands 52°22′N 06°42′E
105 S9 **Almenara** Valenciana, E Spain 39°46′N 00°14′W
105 P12 **Almenaras** ▲ S Spain 38°31′N 02°27′W
105 P5 **Almenar de Soria** Castilla y León, N Spain 41°41′N 02°12′W
104 J6 **Almendra, Embalse de** ⊠ Castilla y León, NW Spain
104 J11 **Almendralejo** Extremadura, W Spain 38°41′N 06°25′W
98 J10 **Almere** *var.* Almere-stad. Flevoland, C Netherlands 52°22′N 05°12′E
98 J10 **Almere-Buiten** Flevoland, C Netherlands 52°24′N 05°15′E
98 J10 **Almere-Haven** Flevoland, C Netherlands 52°20′N 05°13′E
Almere-stad *see* Almere
105 P15 **Almería** *Ar.* Al-Mariyya; *anc.* Unci, *Lat.* Portus Magnus. Andalucía, S Spain 36°50′N 02°26′W
105 P14 **Almería** ◆ *province* Andalucía, S Spain
105 P15 **Almería, Golfo de** *gulf* S Spain
127 S5 **Al'met'yevsk** Respublika Tatarstan, W Russian Federation 54°53′N 52°20′E
95 L21 **Älmhult** Kronoberg, S Sweden 56°32′N 14°10′E
141 U9 **Al Miḥrāḍ** *desert* NE Saudi Arabia
Al Mīnā' *see* El Mina
104 L17 **Almina, Punta** *headland* Ceuta, Spain, N Africa 35°54′N 05°16′W
75 W9 **Al Minyā** *var.* El Minya, Minya. C Egypt 28°06′N 30°40′E
Al Miqdādīyah *see* Al Muqdādīyah
43 P14 **Almirante** Bocas del Toro, NW Panama 09°20′N 82°22′W
Almirós *see* Almyrós
140 M9 **Al Mislaḥ** *spring/well* W Saudi Arabia 22°46′N 40°47′E
Almissa *see* Omiš
104 G13 **Almodôvar** *var.* Almodóvar. Beja, S Portugal 37°31′N 08°03′W
104 M11 **Almodóvar del Campo** Castilla-La Mancha, C Spain 38°43′N 04°10′W
105 Q9 **Almodóvar del Pinar** Castilla-La Mancha, C Spain 39°44′N 01°55′W
31 S9 **Almont** Michigan, N USA 42°53′N 83°02′W
14 L13 **Almonte** Ontario, SE Canada 45°13′N 76°12′W

◆ Country
● Country Capital
◇ Dependent Territory
○ Dependent Territory Capital
◆ Administrative Regions
✈ International Airport
▲ Mountain
▲▲ Mountain Range
⛰ Volcano
⇝ River
◎ Lake
⊠ Reservoir

Almonte – Amraoti

104 J14 **Almonte** Andalucía, S Spain 37°16´N 06°31´W
104 K9 **Almonte** ≈ W Spain
152 K9 **Almora** Uttarakhand, N India 29°36´N 79°40´E
104 M8 **Almorox** Castilla-La Mancha, C Spain 40°13´N 04°22´W
141 S7 **Al Mubarraz** Ash Sharqīyah, E Saudi Arabia 25°28´N 49°34´E
Al Muḍaibī see Al Muḍaybī
138 G15 **Al Mudawwarah** Ma'ān, SW Jordan 29°20´N 36°E
141 Y9 **Al Muḍaybī** var. Al Muḍaibī. NE Oman 22°35´N 58°08´E
Almudébar see Almudévar
105 S5 **Almudévar** var. Almudébar. Aragón, NE Spain 42°03´N 00°34´W
141 S15 **Al Mukallā** var. Mukalla. SE Yemen 14°36´N 49°07´E
141 N16 **Al Mukhā** Eng. Mocha. SW Yemen 13°18´N 43°17´E
105 N15 **Almuñécar** Andalucía, S Spain 36°44´N 03°41´W
139 U7 **Al Muqdādīyah** var. Al Miqdādīyah. Diyālá, C Iraq 33°58´N 44°58´E
140 L3 **Al Murayr** spring/well NW Saudi Arabia 30°06´N 39°54´E
136 M12 **Almus** Tokat, N Turkey 40°22´N 36°54´E
Al Muṣana'a see Al Maṣna'ah
139 T9 **Al Musayyib** var. Musaiyib. Bābil, C Iraq 32°47´N 44°20´E
139 V13 **Al Muthannā** off. Muḥāfa at al Muthannā, var. As Samāwah. ◈ governorate S Iraq
139 V9 **Al Muwaffaqīyah** Wāsiṭ, S Iraq 32°19´N 45°22´E
138 H10 **Al Muwaqqar** var. El Muwaqqar. 'Ammān, W Jordan 31°49´N 36°06´E
140 J5 **Al Muwayliḥ** var. al-Mawailih. Tabūk, NW Saudi Arabia 27°39´N 35°33´E
115 F17 **Almyrós** var. Almirós. Thessalía, C Greece 39°11´N 22°45´E
115 I24 **Almyroú, Órmos** bay Kríti, Greece, E Mediterranean Sea
Al Nūwfalīyah see An Nawfalīyah
96 L13 **Alnwick** N England, United Kingdom 55°27´N 01°44´W
Al Obayyid see El Obeid
Al Odaid see Al 'Udayd
190 B16 **Alofi** ○ (Niue) W Niue 19°01´S 169°55´E
190 A16 **Alofi Bay** bay W Niue, C Pacific Ocean
190 E13 **Alofi, Île** island S Wallis and Futuna
190 E13 **Alofitai** Île Alofi, W Wallis and Futuna 14°21´S 178°03´W
Aloha State see Hawai'i
118 G7 **Aloja** N Latvia 57°47´N 24°53´E
153 X10 **Along** Arunāchel Pradesh, NE India 28°15´N 94°56´E
115 H16 **Alónnisos** island Vóreies Sporádes, Greece, Aegean Sea
104 M15 **Álora** Andalucía, S Spain 36°50´N 04°43´W
171 Q16 **Alor, Kepulauan** island group E Indonesia
171 Q16 **Alor, Pulau** prev. Ombai. island Kepulauan Alor, E Indonesia
171 O16 **Alor, Selat** strait Flores Sea/ Savu Sea
168 I7 **Alor Setar** var. Alor Star, Alur Setar. Kedah, Peninsular Malaysia 06°06´N 100°23´E
Alor Star see Alor Setar
Alost see Aalst
154 F9 **Ālot** Madhya Pradesh, C India 23°56´N 75°40´E
186 G10 **Alotau** Milne Bay, SE Papua New Guinea 10°20´S 150°23´E
171 Y16 **Alotip** Papua, E Indonesia 08°07´S 140°06´E
Al Oued see El Oued
35 R12 **Alpaugh** California, W USA 35°52´N 119°29´W
Alpen see Alps
31 R6 **Alpena** Michigan, N USA 45°04´N 83°27´W
Alpes see Alps
103 S14 **Alpes-de-Haute-Provence** ◈ department SE France
103 U14 **Alpes-Maritimes** ◈ department SE France
181 W8 **Alpha** Queensland, E Australia 23°40´S 146°38´E
197 R9 **Alpha Cordillera** var. Alpha Ridge. undersea feature Arctic Ocean 85°30´N 125°00´W
Alpha Ridge see Alpha Cordillera
Alpheius see Alfeiós
99 I15 **Alphen** Noord-Brabant, S Netherlands 51°29´N 04°57´E
Alphen see Alphen aan den Rijn
98 H11 **Alphen aan den Rijn** var. Alphen. Zuid-Holland, C Netherlands 52°08´N 04°40´E
Alpheus see Alfeiós
Alpi see Alps
104 G10 **Alpiarça** Santarém, C Portugal 39°15´N 08°35´W
24 K10 **Alpine** Texas, SW USA 30°22´N 103°40´W
108 F8 **Alpnach** Unterwalden, W Switzerland 46°56´N 08°17´E
108 D11 **Alps** Fr. Alpes, Ger. Alpen, It. Alpi. ▲▲ C Europe
141 W8 **Al Qābil** var. Qabil. N Oman 23°55´N 55°50´E
Al Qaḍārif see Gedaref
75 P8 **Al Qaddāḥīyah** N Libya 31°21´N 15°16´E
139 V10 **Al Qādisīyah** off. Muḥāfa at al Qādisīyah, var. Ad Dīwānīyah. ◈ governorate S Iraq
Al Qāhirah see Cairo
140 K4 **Al Qalībah** Tabūk, NW Saudi Arabia 28°29´N 37°40´E
139 O1 **Al Qāmishlī** var. Kamishli, Qamishly. Al Ḥasakah, NE Syria 37°N 41°E
138 I6 **Al Qaryatayn** var. Qaryatayn, Fr. Qariateïne. Ḥimṣ, C Syria 34°13´N 37°13´E
142 K11 **Al Qash'ānīyah** var. Al-Kashaniya. NE Kuwait 29°59´N 47°42´E
141 N7 **Al Qaṣīm** var. Minṭaqat Qaṣīm, Qassim. ◈ province C Saudi Arabia
75 V10 **Al Qaṣr** var. Al Qaṣr var. El Qaṣr. C Egypt 25°43´N 28°54´E
138 J5 **Al Qaṣr** Ḥimṣ, C Syria 35°06´N 37°39´E
Al Qaṣr see Al Qaṣr
Al Qaṣrayn see Kasserine
141 S6 **Al Qaṭīf** Ash Sharqīyah, NE Saudi Arabia 26°27´N 50°01´E
138 G11 **Al Qaṭrānah** var. El Qatrani, Qatrana. Al Karak, W Jordan 31°14´N 36°03´E
75 P11 **Al Qaṭrūn** SW Libya 24°57´N 14°40´E
Al Qayrawān see Kairouan
Al-Qsar al-Kbir see Ksar-el-Kebir
Al Qubayyāt see Qoubaïyât
Al Quds/Al Quds ash Sharīf see Jerusalem
104 H12 **Alqueva, Barragem do** ⊠ Portugal/Spain
138 G8 **Al Qunayṭirah** var. El Kuneitra, El Quneitra, Kuneitra, Qunaytra. Al Qunayṭirah, SW Syria 33°08´N 35°49´E
138 G8 **Al Qunayṭirah** off. Muḥāfaẓat al Qunayṭirah, var. El Q'unayṭirah, Qunaytirah, Fr. Kuneitra. ◈ governorate SW Syria
140 M11 **Al Qunfudhah** Makkah, SW Saudi Arabia 19°19´N 41°03´E
140 K2 **Al Qurayyāt** Al Jawf, NW Saudi Arabia 31°25´N 37°26´E
139 Y11 **Al Qurnah** var. Kurna. Al Baṣrah, SE Iraq 31°01´N 47°27´E
75 Y10 **Al Quṣayr** var. Al Quṣayr var. Qusair, Quseir. E Egypt 26°05´N 34°16´E
139 V12 **Al Quṣayr** Al Muthanná, S Iraq 30°36´N 45°52´E
138 I6 **Al Quṣayr** var. El Quseir, Quṣayr, Fr. Kousseir. Ḥimṣ, W Syria 34°36´N 36°36´E
Al Quṣayr see Al Quṣayr
138 H7 **Al Quṭayfah** var. Quṭayfah, Qutayfe, Quteife, Fr. Kouteifé. Rīf Dimashq, W Syria 33°44´N 36°33´E
141 P8 **Al Quwayīyah** Ar Riyāḍ, C Saudi Arabia 24°06´N 45°18´E
Al Quwayr see Guwēr
138 F14 **Al Quwayrah** var. El Quweira. Al 'Aqabah, SW Jordan 29°47´E 35°18´E
Al Rayyan see Ar Rayyān
Al Ruweis see Ar Ruways
95 G24 **Als** Ger. Alsen. island SW Denmark
103 U5 **Alsace** Ger. Elsass; anc. Alsatia. ◈ region NE France
11 R16 **Alsask** Saskatchewan, S Canada 51°24´N 109°55´W
Alsasua see Altsasu
Alsatia see Alsace
101 C16 **Alsdorf** Nordrhein-Westfalen, W Germany 50°52´N 06°09´E
10 G8 **Alsek** ≈ Canada/USA
Alsen see Als
101 F19 **Alsenz** ≈ W Germany
101 H17 **Alsfeld** Hessen, C Germany 50°45´N 09°14´E
119 K20 **Al'shany** Rus. Ol'shany. Brestskaya Voblasts', SW Belarus 52°05´N 27°21´E
Alsókubin see Dolný Kubín
118 C9 **Alsunga** W Latvia 56°59´N 21°31´E
Alt see Olt
92 K9 **Alta** Fin. Alattio. Finnmark, N Norway 69°58´N 23°17´E
29 T12 **Alta** Iowa, C USA 42°40´N 95°17´W
108 I7 **Altach** Vorarlberg, W Austria 47°22´N 09°39´E
92 K9 **Altaelva** Lapp. Álaheaieatnu. ≈ N Norway
92 J8 **Altafjorden** fjord NE Norwegian Sea
62 K10 **Alta Gracia** Córdoba, C Argentina 31°42´S 64°25´W
42 K11 **Alta Gracia** Rivas, SW Nicaragua 11°35´N 85°38´W
54 H4 **Altagracia** Zulia, NW Venezuela 10°44´N 71°30´W
54 M5 **Altagracia de Orituco** Guárico, N Venezuela 09°54´N 66°24´W
Altai see Altai Mountains
129 T7 **Altai Mountains** var. Altai, Chin. Altay Shan, Rus. Altay. ▲▲ Asia/Europe
23 V6 **Altamaha River** ≈ Georgia, SE USA
58 J13 **Altamira** Pará, NE Brazil 03°13´S 52°15´W
54 D12 **Altamira** Huila, S Colombia 02°04´N 75°47´W
42 M13 **Altamira** Alajuela, N Costa Rica 10°25´N 84°21´W
41 Q11 **Altamira** Tamaulipas, C Mexico 22°25´N 97°55´W
30 L15 **Altamont** Illinois, N USA 39°03´N 88°45´W
27 Q7 **Altamont** Kansas, C USA 37°11´N 95°18´W
32 H16 **Altamont** Oregon, NW USA 42°12´N 121°44´W
20 K10 **Altamont** Tennessee, S USA 35°28´N 85°42´W
23 X11 **Altamonte Springs** Florida, SE USA 28°39´N 81°22´W
107 O17 **Altamura** anc. Lupatia. Puglia, SE Italy 40°50´N 16°33´E
40 H9 **Altamura, Isla** island C Mexico
Altan see Erdenehayrhan
Altanbulag see Bayanhayrhan
163 Q7 **Altan Emel** var. Xin Barag Youqi. Nei Mongol Zizhiqu, N China 48°37´N 116°40´E
Altan-Ovoo see Tsenher
163 N9 **Altanshiree** var. Chamdmanī. Dornigovī, SE Mongolia 45°36´N 110°30´E
Altanteel see Dzereg
162 D5 **Altantsögts** var. Tsagaantüngi. Bayan-Ölgiy, NW Mongolia 49°06´N 90°26´E
40 F3 **Altar** Sonora, NW Mexico 30°41´N 111°53´W
40 D2 **Altar, Desierto de** var. Sonoran Desert. desert Mexico/USA see also Sonoran Desert
Altar, Desierto de see Sonoran Desert
105 Q8 **Alta, Sierra** ▲ N Spain 40°29´N 01°36´W
40 H9 **Altata** Sinaloa, C Mexico 24°40´N 107°54´W
42 D4 **Alta Verapaz** off. Departamento de Alta Verapaz. ◈ department C Guatemala
Alta Verapaz, Departamento de see Alta Verapaz
107 L18 **Altavilla Silentia** Campania, S Italy 40°32´N 15°06´E
21 T7 **Altavista** Virginia, NE USA 37°06´N 79°17´W
158 L2 **Altay** Xinjiang Uygur Zizhiqu, NW China 47°51´N 88°06´E
162 D6 **Altay** var. Chihertey. Bayan-Ölgiy, W Mongolia 48°10´N 89°35´E
162 G8 **Altay** prev. Yösönbulag. Govĭ-Altay, W Mongolia 46°23´N 96°17´E
162 E8 **Altay** var. Bor-Üdzüür. Hovd, W Mongolia 45°46´N 92°13´E
Altay see Altai Mountains, Asia/Europe
Altay see Bayantes, Mongolia
122 J14 **Altay, Respublika** var. Gornyy Altay; prev. Gorno-Altayskaya Respublika. ◈ autonomous republic S Russian Federation
Altay Shan see Altai Mountains
123 I13 **Altayskiy Kray** ◈ territory S Russian Federation
Altbetsche see Bečej
101 L20 **Altdorf** Bayern, SE Germany 49°23´N 11°22´E
108 G8 **Altdorf** var. Altorf. Uri, C Switzerland 46°53´N 08°38´E
105 T11 **Altea** Valenciana, E Spain 38°37´N 00°03´W
100 L10 **Alte Elde** ≈ N Germany
101 M16 **Altenburg** Thüringen, E Germany 50°59´N 12°27´E
Altenburg see București, Romania
Altenburg see Baia de Criş, Romania
100 P12 **Alte Oder** ≈ NE Germany
104 H10 **Alter do Chão** Portalegre, C Portugal 39°12´N 07°40´W
92 I10 **Altevatnet** Lapp. Álttesjávri. ◎ N Norway
27 V12 **Altheimer** Arkansas, C USA 34°19´N 91°51´W
109 T9 **Althofen** Kärnten, S Austria 46°52´N 14°27´E
114 H7 **Altimir** Vratsa, NW Bulgaria 43°33´N 23°48´E
136 K11 **Altınkaya Barajī** ⊠ N Turkey
139 S3 **Altin Köprü** var. Altun Kupri. Kirkūk, N Iraq 35°50´N 44°10´E
Altin Köprü see Āltūn Kūbrî
136 E13 **Altıntaş** Kütahya, W Turkey 39°05´N 30°07´E
57 K18 **Altiplano** physical region W South America
Altkanischa see Kanjiža
103 U7 **Altkirch** Haut-Rhin, NE France 47°37´N 07°14´E
Altlublau see Stará L'ubovňa
100 L12 **Altmark** cultural region N Germany
Altmoldowa see Moldova Veche
25 W8 **Alto** Texas, SW USA 31°39´N 95°04´W
104 H11 **Alto Alentejo** physical region S Portugal
59 I19 **Alto Araguaia** Mato Grosso, C Brazil 17°19´S 53°10´W
58 L12 **Alto Bonito** Pará, NE Brazil 01°48´S 46°18´W
83 O15 **Alto Molócuè** Zambézia, NE Mozambique 15°38´S 37°42´E
30 K15 **Alton** Illinois, N USA 38°53´N 90°10´W
27 W8 **Alton** Missouri, C USA 36°41´N 91°25´W
11 X17 **Altona** Manitoba, S Canada 49°06´N 94°35´W
18 E14 **Altoona** Pennsylvania, NE USA 40°32´N 78°23´W
30 J6 **Altoona** Wisconsin, N USA 44°49´N 91°22´W
62 N3 **Alto Paraguay** off. Departamento del Alto Paraguay. ◈ department N Paraguay
Alto Paraguay, Departamento del see Alto Paraguay
59 L17 **Alto Paraíso de Goiás** Goiás, S Brazil 14°04´S 47°15´W
62 P6 **Alto Paraná** off. Departamento del Alto Paraná. ◈ department E Paraguay
Alto Paraná see Paraná
Alto Paraná, Departamento del see Alto Paraná
59 L15 **Alto Parnaíba** Maranhão, E Brazil 09°08´S 45°56´W
56 H13 **Alto Purús, Río** ≈ E Peru
Altorf see Altdorf
63 H19 **Alto Río Senguer** var. Alto Río Senguerr. Chubut, S Argentina 45°01´S 70°55´W
Alto Río Senguerr see Alto Río Senguer
41 Q13 **Altotonga** Veracruz-Llave, E Mexico 19°46´N 97°14´W
101 N23 **Altötting** Bayern, SE Germany 48°12´N 12°37´E
Altpasua see Stara Pazova
Altraga see Bayandzürh
105 P3 **Altsasu** Cast. Alsasua. Navarra, N Spain 42°54´N 02°10´W
Alt-Schwanenburg see Gulbene
Altsohl see Zvolen
108 I7 **Altstätten** Sankt Gallen, NE Switzerland 47°22´N 09°33´E
Álttesjávri see Altevatnet
42 G1 **Altun Ha** ruins Belize, N Belize
139 T4 **Āltūn Kūbrî** prev. Kirkūk, var. Altin Köprü, Karkūk, Kerkuk. Kirkūk, N Iraq 35°28´N 44°26´E
Altun Kupri see Altin Köprü
158 D8 **Altun Shan** ▲ C China 39°19´N 93°37´E
158 L9 **Altun Shan** var. Altyn Tagh. ▲▲ NW China
35 P2 **Alturas** California, W USA 41°28´N 120°32´W
26 K12 **Altus** Oklahoma, C USA 34°39´N 99°21´W
26 K11 **Altus Lake** ⊠ Oklahoma, C USA
Altvater see Praděd
Altyn Tagh see Altun Shan
Alu see Shortland Island
al-'Ubaila see Al 'Ubaylah
139 O6 **Al 'Ubaydī** Al Anbār, W Iraq 34°22´N 41°15´E
141 T9 **Al 'Ubaylah** var. al-'Ubaila. Ash Sharqīyah, E Saudi Arabia 22°02´N 50°57´E
141 T9 **Al 'Ubaylah** spring/well E Saudi Arabia 22°02´N 50°56´E
Al Ubayyiḍ see El Obeid
141 T7 **Al 'Udayd** var. Al Odaid. Abū Ẓaby, W United Arab Emirates 24°34´N 51°27´E
118 J8 **Alūksne** Ger. Marienburg. NE Latvia 57°26´N 27°02´E
140 K6 **Al 'Ulā** Al Madīnah, NW Saudi Arabia 26°39´N 37°55´E
173 N4 **Alula-Fartak Trench** var. Illaue Fartak Trench. undersea feature W Indian Ocean 14°04´N 51°47´E
138 I11 **Al 'Umarī** 'Ammān, E Jordan 37°30´N 31°30´E
31 S13 **Alum Creek Lake** ⊠ Ohio, N USA
63 H15 **Aluminé** Neuquén, C Argentina 39°15´S 71°00´W
95 O14 **Alunda** Uppsala, C Sweden 60°04´N 18°04´E
117 T14 **Alupka** Avtonomna Respublika Krym, S Ukraine 44°24´N 34°01´E
75 P8 **Al 'Uqaylah** N Libya 30°13´N 19°12´E
Al Uqṣur see Luxor
Al Urdunn see Jordan
168 J9 **Alur Panal** bay Sumatera, W Indonesia
Alur Setar see Alor Setar
141 V10 **Al 'Urūq al Mu'tariḍah** salt lake SE Saudi Arabia
139 Q7 **Ālūs** Al Anbār, C Iraq 34°05´N 42°27´E
117 T13 **Alushta** Avtonomna Respublika Krym, S Ukraine 44°41´N 34°24´E
151 G22 **Aluva** var. Alwaye. Kerala, SW India 10°06´N 76°23´E see also Alwaye
75 N11 **Al 'Uwaynāt** var. Al Awaynāt. SW Libya 25°47´N 10°34´E
139 T6 **Al 'Uẓaym** var. Adhaim. Diyālá, E Iraq 34°12´N 44°31´E
26 L8 **Alva** Oklahoma, C USA 36°48´N 98°40´W
104 H8 **Alva** ≈ N Portugal
95 J18 **Älvängen** Västra Götaland, S Sweden 57°56´N 12°09´E
14 F14 **Alvanley** Ontario, S Canada 44°33´N 81°05´W
41 S14 **Alvarado** Veracruz-Llave, E Mexico 18°47´N 95°45´W
25 T7 **Alvarado** Texas, SW USA 32°24´N 97°12´W
58 D13 **Alvarães** Amazonas, NW Brazil 03°13´S 64°53´W
40 G6 **Álvaro Obregón, Presa** ⊠ W Mexico
94 H10 **Alvdal** Hedmark, S Norway 62°07´N 10°39´E
94 K12 **Älvdalen** Dalarna, C Sweden 61°13´N 14°04´E
61 E15 **Alvear** Corrientes, NE Argentina 29°05´S 56°35´W
104 F10 **Alverca do Ribatejo** Lisboa, C Portugal 38°56´N 09°01´W
95 L20 **Alvesta** Kronoberg, S Sweden 56°52´N 14°34´E
25 W12 **Alvin** Texas, SW USA 29°25´N 95°14´W
94 O13 **Älvkarleby** Uppsala, C Sweden 60°34´N 17°30´E
25 S5 **Alvord** Texas, SW USA 33°22´N 97°39´W
93 G18 **Älvros** Jämtland, C Sweden 62°04´N 14°30´E
92 J13 **Älvsbyn** Norrbotten, N Sweden 65°41´N 21°00´E
142 K12 **Al Wafrā'** SE Kuwait 28°38´N 47°57´E
140 J6 **Al Wajh** Tabūk, NW Saudi Arabia 26°16´N 36°30´E
143 N16 **Al Wakrah** var. Wakra. C Qatar 25°09´N 51°36´E
138 M8 **al Walaj, Sha'īb** dry watercourse W Iraq
152 I11 **Alwar** Rājasthān, N India 27°32´N 76°35´E
141 Q5 **Al Warī'ah** Ash Sharqīyah, N Saudi Arabia 27°54´N 47°23´E
155 G22 **Alwaye** var. Aluva. Kerala, SW India 10°06´N 76°23´E see also Aluva
Alxa Zuoqi see Bayan Hot
Alx Youqi see Ehen Hudag
Al Yaman see Yemen
138 G9 **Al Yarmūk** Irbid, N Jordan 32°41´N 35°55´E
Alyat/Alyaty-Pristan' see Älät
115 I14 **Alykí** var. Aliki. Thásos, N Greece 40°36´N 24°45´E
119 F14 **Alytus** Pol. Olita. Alytus, S Lithuania 54°24´N 24°02´E
119 F15 **Alytus** ◈ province S Lithuania
101 N23 **Alz** ≈ SE Germany
33 Y11 **Alzada** Montana, NW USA 45°00´N 104°24´W
122 L12 **Alzamay** Irkutskaya Oblast', S Russian Federation 55°33´N 98°36´E
99 M25 **Alzette** ≈ S Luxembourg
105 S10 **Alzira** var. Alcira ; anc. Saetabicula, Suero. Valenciana, E Spain 39°10´N 00°27´W
Al Zubair see Az Zubayr
181 O8 **Amadeus, Lake** seasonal lake Northern Territory, C Australia
81 E15 **Amadi** Western Equatoria, SW South Sudan 05°32´N 30°20´E
9 R7 **Amadjuak Lake** ◎ Baffin Island, Nunavut, N Canada
95 J23 **Amager** island E Denmark
165 N14 **Amagi-san** ▲ Honshū, S Japan 34°51´N 138°57´E
171 S13 **Amahai** var. Masohi. Palau Seram, E Indonesia 03°19´S 128°56´E
38 M16 **Amak Island** island Alaska, USA
164 C15 **Amakusa** prev. Hondo. Kumamoto, Shimo-jima, SW Japan 32°28´N 130°12´E
164 B14 **Amakusa-nada** gulf SW Japan
95 J16 **Åmål** Västra Götaland, S Sweden 59°04´N 12°41´E
54 E8 **Amalfi** Antioquia, N Colombia 06°54´N 75°04´W
107 L18 **Amalfi** Campania, S Italy 40°37´N 14°35´E
115 D19 **Amaliáda** var. Amaliás. Dytikí Elláda, S Greece 37°48´N 21°21´E
Amaliás see Amaliáda
154 F12 **Amalner** Mahārāshtra, C India 21°03´N 75°04´E
171 W14 **Amamapare** Papua, E Indonesia 04°51´S 136°44´E
59 H21 **Amambaí, Serra de** var. Cordillera de Amambay, Serra de Amambay. ▲ Brazil/ Paraguay see also Amambay, Cordillera de
62 P4 **Amambay** off. Departamento del Amambay. ◈ department E Paraguay
62 P5 **Amambay, Cordillera de** var. Serra de Amambaí, Serra de Amambay. ▲ Brazil/ Paraguay see also Amambaí, Serra de
Amambay, Departamento del see Amambay
Amambay, Serra de see Amambaí, Serra de/ Amambay, Cordillera de
165 U16 **Amami-guntō** island group SW Japan
165 V15 **Amami-Ō-shima** island S Japan
186 A5 **Amanab** West Sepik, NW Papua New Guinea 03°38´S 141°16´E
Amānat al 'Āṣimah see Baghdād
106 J13 **Amandola** Marche, C Italy 42°58´N 13°22´E
107 N21 **Amantea** Calabria, SW Italy 39°06´N 16°05´E
191 W10 **Amanu** island Îles Tuamotu, C French Polynesia
58 J10 **Amapá** Amapá, NE Brazil 02°00´N 50°50´W
58 J11 **Amapá** off. Estado de Amapá; prev. Território de Amapá. ◈ state NE Brazil
Amapá, Estado de see Amapá
42 H8 **Amapala** Valle, S Honduras 13°16´N 87°39´W
Amapá, Território de see Amapá
80 J12 **Āmara** var. Amhara. ◈ N Ethiopia
Amara see Al 'Amārah
'Amārah, Al see Maysān
104 H6 **Amarante** Porto, N Portugal 41°16´N 08°05´W
166 M5 **Amarapura** Mandalay, C Myanmar (Burma) 21°54´N 96°01´E
Amardalay see Delgertsogt
104 I12 **Amareleja** Beja, S Portugal 38°12´N 07°13´W
35 V11 **Amargosa Range** ▲▲ California, W USA
25 N2 **Amarillo** Texas, SW USA 35°13´N 101°50´W
Amarinthos see Amárynthos
107 K15 **Amaro, Monte** ▲ C Italy 42°03´N 14°06´E
115 H18 **Amárynthos** var. Amarinthos. Évvoia, C Greece 38°24´N 23°53´E
Amasia see Amasya
136 K12 **Amasya** anc. Amasia. Amasya, N Turkey 40°37´N 35°50´E
136 K11 **Amasya** ◈ province N Turkey
42 F4 **Amatique, Bahía de** bay Gulf of Honduras, W Caribbean Sea
42 D6 **Amatitlán, Lago de** ◎ S Guatemala
107 J14 **Amatrice** Lazio, C Italy 42°38´N 13°19´E
190 C8 **Amatuku** atoll C Tuvalu
99 J20 **Amay** Liège, E Belgium 50°33´N 05°19´E
48 F7 **Amazon** Sp. Amazonas. ≈ Brazil/Peru
59 C14 **Amazonas** off. Estado do Amazonas. ◈ state N Brazil
54 G15 **Amazonas** off. Comisaria del Amazonas. ◈ province SE Colombia
56 C10 **Amazonas** off. Departamento de Amazonas. ◈ department N Peru
54 M12 **Amazonas** off. Territorio Amazonas. ◈ federal territory S Venezuela
Amazonas see Amazon
Amazonas, Comisaria del see Amazonas
Amazonas, Departamento de see Amazonas
Amazonas, Estado do see Amazonas
Amazonas, Territorio see Amazonas
48 F7 **Amazon Basin** basin N South America
47 V5 **Amazon Fan** undersea feature W Atlantic Ocean 05°00´N 47°30´W
58 K11 **Amazon, Mouths of the** delta NE Brazil
187 R13 **Ambae** var. Aoba, Omba. island C Vanuatu
152 I9 **Ambāla** Haryāna, NW India 30°19´N 76°49´E
155 J26 **Ambalangoda** Southern Province, SW Sri Lanka 06°14´N 80°03´E
155 K26 **Ambalantota** Southern Province, S Sri Lanka 06°07´N 81°01´E
172 I6 **Ambalavao** Fianarantsoa, C Madagascar 21°50´S 46°56´E
54 E10 **Ambalema** Tolima, C Colombia 04°49´N 74°48´W
79 E17 **Ambam** Sud, S Cameroon 02°23´N 11°17´E
172 J2 **Ambanja** Antsirañana, N Madagascar 13°40´S 48°27´E
123 T6 **Ambarchik** Respublika Sakha (Yakutiya), NE Russian Federation 69°33´N 162°08´E
62 K9 **Ambargasta, Salinas de** salt lake C Argentina
124 J6 **Ambarnyy** Respublika Kareliya, NW Russian Federation 65°53´N 33°44´E
56 C7 **Ambato** Tungurahua, C Ecuador 01°18´S 78°39´W
172 I6 **Ambato Finandrahana** Fianarantsoa, SE Madagascar
172 I5 **Ambatolampy** Antananarivo, C Madagascar 19°21´S 47°27´E
172 H4 **Ambatomainty** Mahajanga, W Madagascar 17°40´S 45°39´E
172 J4 **Ambatondrazaka** Toamasina, C Madagascar 17°49´S 48°28´E
101 L20 **Amberg** var. Amberg in der Oberpfalz. Bayern, SE Germany 49°26´N 11°52´E
Amberg in der Oberpfalz see Amberg
42 H1 **Ambergris Cay** island NE Belize
103 S11 **Ambérieu-en-Bugey** Ain, E France 45°57´N 05°21´E
185 I18 **Amberley** Canterbury, South Island, New Zealand 43°09´S 172°43´E
103 P11 **Ambert** Puy-de-Dôme, C France 45°33´N 03°45´E
Ambianum see Amiens
76 J11 **Ambidédi** Kayes, SW Mali 14°37´N 11°49´W
154 M10 **Ambikāpur** Chhattīsgarh, C India 23°09´N 83°12´E
172 J2 **Ambilobe** Antsirañana, N Madagascar 13°10´S 49°03´E
39 O7 **Ambler** Alaska, USA 67°05´N 157°51´W
Amblève see Amel
Ambo see Hāgere Hiywet
172 I8 **Amboasary** Toliara, S Madagascar 25°01´S 46°23´E
172 J4 **Ambodifotatra** var. Ambodifototra. Toamasina, E Madagascar 16°59´S 49°51´E
Amboenten see Ambunten
172 I5 **Ambohidratrimo** Antananarivo, C Madagascar 18°48´S 47°26´E
172 I6 **Ambohimahasoa** Fianarantsoa, SE Madagascar 21°07´S 47°13´E
172 K3 **Ambohitralanana** Antsirañana, NE Madagascar 15°13´S 50°28´E
Amboina see Ambon
102 M8 **Amboise** Indre-et-Loire, C France 47°25´N 01°00´E
171 S13 **Ambon** prev. Amboina, Amboyna. Pulau Ambon, E Indonesia 03°41´S 128°10´E
171 S13 **Ambon, Pulau** island E Indonesia
81 I20 **Amboseli, Lake** ◎ Kenya/ Tanzania
172 I6 **Ambositra** Fianarantsoa, SE Madagascar 20°31´S 47°15´E
172 I8 **Ambovombe** Toliara, S Madagascar 25°10´S 46°06´E
35 W14 **Amboy** California, W USA 34°33´N 115°44´W
30 L11 **Amboy** Illinois, N USA 41°42´N 89°19´W
Amboyna see Ambon
Ambracia see Árta
Ambre, Cap d' see Bobaomby, Tanjona
18 B14 **Ambridge** Pennsylvania, NE USA 40°33´N 80°11´W
Ambrim see Ambrym
82 A11 **Ambriz** Bengo, NW Angola 07°55´S 13°11´E
Ambrizete see N'Zeto
187 R13 **Ambrym** var. Ambrim. island C Vanuatu
169 T16 **Ambunten** prev. Amboenten. Pulau Madura, E Indonesia 06°55´S 113°45´E
186 B6 **Ambunti** East Sepik, NW Papua New Guinea 04°12´S 142°49´E
155 I20 **Āmbūr** Tamil Nādu, SE India 12°48´N 78°44´E
38 E17 **Amchitka Island** island Aleutian Islands, Alaska, USA
38 F17 **Amchitka Pass** strait Aleutian Islands, Alaska, USA
141 R15 **'Amd** C Yemen 15°10´N 47°58´E
78 J10 **Am Dam** Sila, E Chad 12°46´N 20°29´E
171 U16 **Amdassa** Pulau Yamdena, E Indonesia 07°40´S 131°24´E
125 U1 **Amderma** Nenetskiy Avtonomnyy Okrug, NW Russian Federation 69°45´N 61°36´E
78 L9 **Am Djarass** Ennedi-Est, NE Chad 17°14´N 21°32´E
159 N14 **Amdo** Xizang Zizhiqu, W China 32°15´N 91°43´E
40 K13 **Ameca** Jalisco, SW Mexico 20°34´N 104°03´W
41 P14 **Amecameca** var. Amecameca de Juárez. México, C Mexico 19°08´N 98°48´W
Amecameca de Juárez see Amecameca
139 R1 **Amêdî** Ar. Al 'Amādiyah. Dahūk, N Iraq 37°09´N 43°27´E
61 A20 **Ameghino** Buenos Aires, E Argentina 34°51´S 62°28´W
99 M21 **Amel** Fr. Amblève. Liège, E Belgium 50°20´N 06°13´E
98 K4 **Ameland** Fris. It Amelân. island Waddeneilanden, N Netherlands
Amelân, It see Ameland
107 H14 **Amelia** Umbria, C Italy 42°33´N 12°26´E
21 V6 **Amelia Court House** Virginia, NE USA 37°20´N 77°59´W
23 W8 **Amelia Island** island Florida, SE USA
18 L12 **Amenia** New York, NE USA 41°51´N 73°31´W
America see United States of America
65 M21 **America-Antarctica Ridge** undersea feature S Atlantic Ocean
America in Miniature see Maryland
60 L9 **Americana** São Paulo, S Brazil 22°44´S 47°19´W
33 Q15 **American Falls** Idaho, NW USA 42°47´N 112°51´W
33 Q15 **American Falls Reservoir** ⊠ Idaho, NW USA
36 L3 **American Fork** Utah, W USA 40°24´N 111°47´W
192 K16 **American Samoa** ◇ US unincorporated territory W Polynesia
23 S6 **Americus** Georgia, SE USA 32°04´N 84°13´W
98 K12 **Amerongen** Utrecht, C Netherlands 52°00´N 05°30´E
98 K11 **Amersfoort** Utrecht, C Netherlands 52°09´N 05°23´E
97 N21 **Amersham** SE England, United Kingdom 51°40´N 00°37´W
30 I5 **Amery** Wisconsin, N USA 45°18´N 92°20´W
195 W6 **Amery Ice Shelf** ice shelf Antarctica
29 V13 **Ames** Iowa, C USA 42°01´N 93°37´W
19 P10 **Amesbury** Massachusetts, NE USA 42°51´N 70°55´W
Amestratus see Mistretta
115 F18 **Amfíkleia** var. Amfíklia. Stereá Elláda, C Greece 38°38´N 22°35´E
Amfíklia see Amfíkleia
115 D17 **Amfilochía** var. Amfilokhía. Dytikí Elláda, C Greece 38°52´N 21°09´E
Amfilokhía see Amfilochía
114 H13 **Amfípoli** anc. Amphipolis. site of ancient city Kentrikí Makedonía, NE Greece
115 F18 **Ámfissa** Stereá Elláda, C Greece 38°32´N 22°22´E
123 Q10 **Amga** Respublika Sakha (Yakutiya), NE Russian Federation 60°55´N 131°45´E
123 Q11 **Amga** ≈ NE Russian Federation
163 R7 **Amgalang** var. Xin Barag Zuoqi. Nei Mongol Zizhiqu, N China 48°12´N 118°15´E
123 V5 **Amguema** ≈ NE Russian Federation
123 S12 **Amgun'** ≈ SE Russian Federation
Amhara see Āmara
13 P15 **Amherst** Nova Scotia, SE Canada 45°50´N 64°14´W
18 M11 **Amherst** Massachusetts, NE USA 42°22´N 72°31´W
18 D10 **Amherst** New York, NE USA 42°57´N 78°47´W
24 M4 **Amherst** Texas, SW USA 33°59´N 102°24´W
21 U6 **Amherst** Virginia, NE USA 37°35´N 79°04´W
Amherst see Kyaikkami
14 C18 **Amherstburg** Ontario, S Canada 42°05´N 83°06´W
21 Q6 **Amherstdale** West Virginia, NE USA 37°46´N 81°46´W
14 K15 **Amherst Island** island Ontario, SE Canada
Amida see Diyarbakır
28 J6 **Amidon** North Dakota, N USA 46°29´N 103°19´W
103 O3 **Amiens** anc. Ambianum, Samarobriva. Somme, N France 49°54´N 02°18´E
139 P8 **'Āmij, Wādī** var. Wadi 'Amīq. dry watercourse W Iraq
136 L17 **Amik Ovasī** ◎ S Turkey
76 E9 **Amilcar Cabral** ✈ Sal, NE Cape Verde
Amilḥayt, Wādī see Umm al Ḥayt, Wādī
Amíndaion/Amindeo see Amýntaio
155 C21 **Amīndīvi Islands** island group Lakshadweep, India, N Indian Ocean
139 U6 **Amīn Ḥabīb** Diyālá, E Iraq 34°17´N 45°10´E
83 E20 **Aminuis** Omaheke, E Namibia 23°43´S 19°21´E
'Amīq, Wadi see 'Āmij, Wādī
142 J7 **Amīrābād** Īlām, NW Iran 33°20´N 46°16´E
Amirante Bank see Amirante Ridge
173 N6 **Amirante Basin** undersea feature W Indian Ocean 07°00´S 54°00´E
173 N6 **Amirante Islands** var. Amirantes Group. island group C Seychelles
173 N7 **Amirante Ridge** var. Amirante Bank. undersea feature W Indian Ocean 06°00´S 53°10´E
Amirantes Group see Amirante Islands
173 N7 **Amirante Trench** undersea feature W Indian Ocean 08°00´S 52°30´E
11 U13 **Amisk Lake** ◎ Saskatchewan, C Canada
Amistad, Presa de la see Amistad Reservoir
25 O12 **Amistad Reservoir** var. Presa de la Amistad. ⊠ Mexico/USA
Amisus see Samsun
22 K8 **Amite** var. Amite City. Louisiana, S USA 30°40´N 90°30´W
Amite City see Amite
27 T12 **Amity** Arkansas, C USA 34°15´N 93°27´W
154 H11 **Amla** prev. Amulla. Madhya Pradesh, C India 21°53´N 78°10´E
38 I17 **Amlia Island** island Aleutian Islands, Alaska, USA
97 I18 **Amlwch** NW Wales, United Kingdom 53°25´N 04°23´W
Ammaia see Portalegre
138 H10 **'Amman** var. Amman; anc. Philadelphia, Bibl. Rabbah Ammon, Rabbath Ammon. ● (Jordan) 'Ammān, NW Jordan 31°57´N 35°56´E
138 H10 **'Ammān** off. Muḥāfazat 'Ammān; prev. Al 'Aṣimah. ◈ governorate NW Jordan
'Ammān, Muḥāfazat see 'Ammān
93 N14 **Ämmänsaari** Kainuu, E Finland 64°51´N 28°58´E
92 H13 **Ammarnäs** Västerbotten, N Sweden 65°58´N 16°10´E
197 O15 **Ammassalik** var. Angmagssalik. Sermersooq, S Greenland 65°51´N 37°30´W
101 K24 **Ammer** ≈ SE Germany
101 K24 **Ammersee** ◎ SE Germany
98 J13 **Ammerzoden** Gelderland, C Netherlands 51°46´N 05°07´E
Ammóchostos see Gazimağusa
Ammóchostos, Kólpos see Gazimağusa Körfezi
Amnok-kang see Yalu
Amoea see Portalegre
Amoentai see Amuntai
Amoerang see Amurang
143 O4 **Āmol** var. Amul. Māzandarān, N Iran 36°31´N 52°24´E
115 K21 **Amorgós** Amorgós, Kykládes, Greece, Aegean Sea 36°49´N 25°54´E
115 K22 **Amorgós** island Kykládes, Greece, Aegean Sea
23 N3 **Amory** Mississippi, S USA 33°58´N 88°29´W
12 I13 **Amos** Québec, SE Canada 48°34´N 78°08´W
95 G15 **Åmot** Buskerud, S Norway 59°52´N 09°55´E
95 E15 **Åmot** Telemark, S Norway 59°34´N 07°59´E
95 J15 **Åmotfors** Värmland, C Sweden 59°46´N 12°24´E
76 L10 **Amourj** Hodh ech Chargui, SE Mauritania 16°04´N 07°12´W
Amoy see Xiamen
172 H7 **Ampanihy** Toliara, SW Madagascar 24°40´S 44°45´E
155 L25 **Ampara** var. Amparai. Eastern Province, E Sri Lanka 07°17´N 81°41´E
172 J4 **Amparafaravola** Toamasina, E Madagascar 17°33´S 48°13´E
Amparai see Ampara
60 M9 **Amparo** São Paulo, S Brazil 22°40´S 46°49´W
172 J5 **Ampasimanolotra** Toamasina, E Madagascar 18°49´S 49°04´E
57 H17 **Ampato, Nevado** ▲ S Peru 15°52´S 71°51´W
101 L23 **Amper** ≈ SE Germany
64 M9 **Ampère Seamount** undersea feature E Atlantic Ocean 35°05´N 13°00´W
Amphipolis see Amfípoli
167 X10 **Amphitrite Group** Chin. Xuande Qundao, Viet. Nhom An Vinh. island group N Paracel Islands
171 T16 **Amplawas** var. Emplawas. Pulau Babar, E Indonesia 08°01´S 129°42´E
105 U7 **Amposta** Cataluña, NE Spain 40°43´N 00°34´E
15 V7 **Amqui** Québec, SE Canada 48°28´N 67°27´W
141 O14 **'Amrān** W Yemen 15°39´N 43°59´E
Amraoti see Amrāvati

◆ Country
● Country Capital
◇ Dependent Territory
○ Dependent Territory Capital
◈ Administrative Regions
✈ International Airport
▲ Mountain
▲▲ Mountain Range
Volcano
≈ River
◎ Lake
⊠ Reservoir

AMRĀVATI – AN TIÚR

154 H12 **Amrāvati** *prev.* Amraoti. Mahārāshtra, C India *20°56′N 77°45′E*
154 C11 **Amreli** Gujarāt, W India *21°36′N 71°20′E*
108 H6 **Amriswil** Thurgau, NE Switzerland *47°33′N 09°18′E*
138 H5 **'Amrīt** *ruins* Ṭarṭūs, W Syria
152 H7 **Amritsar** Punjab, N India *31°38′N 74°55′E*
152 J10 **Amroha** Uttar Pradesh, N India *28°54′N 78°29′E*
100 G7 **Amrum** *island* NW Germany
93 I15 **Åmsele** Västerbotten, N Sweden *64°31′N 19°24′E*
98 I10 **Amstelveen** Noord-Holland, C Netherlands *52°18′N 04°50′E*
98 I10 **Amsterdam** ● (Netherlands) Noord-Holland, C Netherlands *52°22′N 04°54′E*
18 K10 **Amsterdam** New York, NE USA *42°56′N 74°11′W*
173 Q11 **Amsterdam Fracture Zone** *tectonic feature* S Indian Ocean
173 R11 **Amsterdam Island** *island* NE French Southern and Antarctic Territories
109 U4 **Amstetten** Niederösterreich, N Austria *48°08′N 14°52′E*
78 J11 **Am Timan** Salamat, SE Chad *11°02′N 20°17′E*
146 L12 **Amu-Buxoro Kanali** *var.* Aral-Bukhorskiy Kanal. *canal* C Uzbekistan
139 O1 **'Āmūdah** *var.* Amude. Al Ḥasakah, N Syria *37°06′N 40°56′E*
147 O15 **Amu Darya** *Rus.* Amudar'ya, *Taj.* Dar"yoi Amu, *Turkm.* Amyderya, *Uzb.* Amudaryo; *anc.* Oxus. ≈ C Asia
Amu-Dar'ya *see* Amyderýa
Amudar'ya/Amudaryo/Amu, Dar"yoi *see* Amu Darya
Amude *see* 'Āmūdah
140 L3 **'Amūd, Jabal al** ▲ NW Saudi Arabia *30°59′N 39°17′E*
38 J17 **Amukta Island** *island* Aleutian Islands, Alaska, USA
38 I17 **Amukta Pass** *strait* Aleutian Islands, Alaska, USA
Amul *see* Āmol
Amulla *see* Amla
197 S10 **Amundsen Basin** *var.* Fram Basin. *undersea basin* Arctic Ocean
195 X3 **Amundsen Bay** *bay* Antarctica
195 P10 **Amundsen Coast** *physical region* Antarctica
193 O14 **Amundsen Plain** *undersea feature* S Pacific Ocean
195 Q9 **Amundsen-Scott** *US research station* Antarctica *89°59′S 10°00′E*
194 J11 **Amundsen Sea** *sea* S Pacific Ocean
94 M12 **Amungen** ◎ C Sweden
169 U13 **Amuntai** *prev.* Amoentai. Borneo, C Indonesia *02°24′S 115°14′E*
129 W6 **Amur** *Chin.* Heilong Jiang. ≈ China/Russian Federation
171 Q11 **Amurang** *prev.* Amoerang. Sulawesi, C Indonesia *01°12′N 124°37′E*
105 O3 **Amurrio** País Vasco, N Spain *43°03′N 03°00′W*
123 S13 **Amurskiy** Khabarovskiy Kray, SE Russian Federation *50°13′N 136°34′E*
123 Q12 **Amurskaya Oblast'** ◈ *province* SE Russian Federation
80 G7 **'Amur, Wadi** ≈ NE Sudan
115 C17 **Amvrakikós Kólpos** *gulf* W Greece
Amvrosiyevka *see* Amvrosiyivka
117 X8 **Amvrosiyivka** *Rus.* Amvrosiyevka. Donets'ka Oblast', SE Ukraine *47°46′N 38°30′E*
146 M14 **Amyderýa** *Rus.* Amu-Dar'ya. Lebap Welaýaty, NE Turkmenistan *37°58′N 65°14′E*
Amyderya *see* Amu Darya
114 E13 **Amýntaio** *var.* Amindeo; *prev.* Amíndaion. Dytikí Makedonía, N Greece *40°42′N 21°42′E*
14 B6 **Amyot** Ontario, S Canada *48°28′N 84°58′W*
191 U10 **Anaa** *atoll* Îles Tuamotu, C French Polynesia
Anabanoea *see* Anabanua
171 N14 **Anabanua** *prev.* Anabanoea. Sulawesi, C Indonesia *03°58′S 120°07′E*
189 R8 **Anabar** NE Nauru *0°30′S 166°56′E*
123 N8 **Anabar** ≈ NE Russian Federation
An Abhainn Mhór *see* Blackwater
55 O6 **Anaco** Anzoátegui, NE Venezuela *09°30′N 64°28′W*
33 Q10 **Anaconda** Montana, NW USA *46°09′N 112°56′W*
32 H7 **Anacortes** Washington, NW USA *48°30′N 122°36′W*
26 M11 **Anadarko** Oklahoma, C USA *35°04′N 98°16′W*
114 N12 **Ana Dere** ≈ NW Turkey
104 G8 **Anadia** Aveiro, N Portugal *40°26′N 08°27′W*
Anadolu Dağları *see* Doğu Karadeniz Dağları
123 V6 **Anadyr'** Chukotskiy Avtonomnyy Okrug, NE Russian Federation *64°41′N 177°22′E*
123 V6 **Anadyr'** ≈ NE Russian Federation
Anadyr, Gulf of *see* Anadyrskiy Zaliv
129 X4 **Anadyrskiy Khrebet** *var.* Chukot Range. ▲▲ NE Russian Federation
123 W6 **Anadyrskiy Zaliv** *Eng.* Gulf of Anadyr. *gulf* NE Russian Federation
115 K22 **Anáfi** *anc.* Anaphe. *island* Kykládes, Greece, Aegean Sea
107 J15 **Anagni** Lazio, C Italy *41°43′N 13°12′E*
'Ānah *see* 'Annah
35 T15 **Anaheim** California, W USA *33°50′N 117°54′W*
10 L15 **Anahim Lake** British Columbia, SW Canada *52°26′N 125°13′W*
38 B8 **Anahola** Kaua'i, Hawai'i, USA, C Pacific Ocean *22°09′N 159°19′W*
41 O7 **Anáhuac** Nuevo León, NE Mexico *27°13′N 100°09′W*
25 X11 **Anahuac** Texas, SW USA *29°44′N 94°41′W*
155 G22 **Anai Mudi** ▲ S India *10°16′N 77°08′E*
Anaiza *see* 'Unayzah
155 M15 **Anakāpalle** Andhra Pradesh, E India *17°42′N 83°06′E*
191 W15 **Anakena, Playa de** *beach* Easter Island, Chile, E Pacific Ocean
39 Q7 **Anaktuvuk Pass** Alaska, USA *68°08′N 151°44′W*
39 Q6 **Anaktuvuk River** ≈ Alaska, USA
172 J3 **Analalava** Mahajanga, NW Madagascar *14°38′S 47°46′E*
44 F6 **Ana María, Golfo de** *gulf* N Caribbean Sea
Anambas Islands *see* Anambas, Kepulauan
169 N8 **Anambas, Kepulauan** *var.* Anambas Islands. *island group* W Indonesia
77 U17 **Anambra** ◈ *state* SE Nigeria
29 N4 **Anamoose** North Dakota, N USA *47°50′N 100°14′W*
29 Y13 **Anamosa** Iowa, C USA *42°06′N 91°17′W*
136 H17 **Anamur** İçel, S Turkey *36°06′N 32°49′E*
136 H17 **Anamur Burnu** *headland* S Turkey *36°03′N 32°49′E*
154 O12 **Ānandapur** *var.* Anandpur. Odisha, E India *21°14′N 86°10′E*
Anandpur *see* Ānandapur
155 H18 **Anantapur** Andhra Pradesh, S India *14°41′N 77°36′E*
152 H5 **Anantnāg** *var.* Islamabad. Jammu and Kashmir, NW India *33°44′N 75°11′E*
Ananyev *see* Anan'yiv
117 O9 **Anan'yiv** *Rus.* Ananyev. Odes'ka Oblast', SW Ukraine *47°43′N 29°51′E*
126 J14 **Anapa** Krasnodarskiy Kray, SW Russian Federation *44°55′N 37°20′E*
Anaphe *see* Anáfi
59 K18 **Anápolis** Goiás, C Brazil *16°19′S 48°58′W*
143 R10 **Anār** Kermān, C Iran *30°49′N 55°18′E*
Anár *see* Inari
143 P7 **Anārak** Eṣfahān, C Iran *33°21′N 53°43′E*
Anar Dara *see* Anār Darreh
148 J7 **Anār Darreh** *var.* Anar Dara. Farāh, W Afghanistan *32°45′N 61°38′E*
Anárjohka *see* Inarijoki
23 X9 **Anastasia Island** *island* Florida, SE USA
188 K7 **Anatahan** *island* C Northern Mariana Islands
128 M6 **Anatolia** *plateau* C Turkey
86 F14 **Anatolian Plate** *tectonic feature* Asia/Europe
114 H13 **Anatolikí Makedonía kai Thráki** *Eng.* Macedonia East and Thrace. ◈ *region* NE Greece
Anatom *see* Aneityum
62 L8 **Añatuya** Santiago del Estero, N Argentina *28°28′S 62°52′W*
An Baile Meánach *see* Ballymena
Anbār, Muḥāfa at al *see* Al Anbār
An Bheanú *see* Barrow
An Bhóinn *see* Boyne
An Blascaod Mór *see* Great Blasket Island
An Cabhán *see* Cavan
An Caisleán Nua *see* Newcastle
An Caisleán Riabhach *see* Castlerea, Ireland
An Caisleán Riabhach *see* Castlereagh
56 C13 **Ancash** *off.* Departamento de Ancash. ◈ *department* W Peru
Ancash, Departamento de *see* Ancash
An Cathair *see* Caher
102 J8 **Ancenis** Loire-Atlantique, NW France *47°23′N 01°10′W*
An Chanáil Ríoga *see* Royal Canal
An Cheacha *see* Caha Mountains
39 R11 **Anchorage** Alaska, USA *61°13′N 149°52′W*
39 R12 **Anchorage** ✈ Alaska, USA *61°08′N 150°00′W*
39 Q13 **Anchor Point** Alaska, USA *59°46′N 151°49′W*
An Chorr Chríochach *see* Cookstown
65 M24 **Anchorstock Point** *headland* W Tristan da Cunha *37°07′S 12°21′W*
An Clár *see* Clare
An Clochán *see* Clifden
An Clochán Liath *see* Dunglow
23 U12 **Anclote Keys** *island group* Florida, SE USA
An Cóbh *see* Cobh
57 J17 **Ancohuma, Nevado de** ▲ W Bolivia *15°51′S 68°33′W*
An Comar *see* Comber
57 D14 **Ancón** Lima, W Peru *11°45′S 77°08′W*
106 J12 **Ancona** Marche, C Italy *43°38′N 13°30′E*
Ancuabe *see* Ancuabi
82 Q13 **Ancuabi** *var.* Ancuabe. Cabo Delgado, NE Mozambique *13°00′S 39°50′E*
63 F17 **Ancud** *prev.* San Carlos de Ancud. Los Lagos, S Chile *41°53′S 73°50′W*
63 G17 **Ancud, Golfo de** *gulf* S Chile
Ancyra *see* Ankara
163 V8 **Anda** Heilongjiang, NE China *46°25′N 125°20′E*
57 G16 **Andahuaylas** Apurímac, S Peru *13°39′S 73°24′W*
An Daingean *see* Dingle
153 R15 **Andāl** West Bengal, NE India *23°35′N 87°14′E*
94 E9 **Åndalsnes** Møre og Romsdal, S Norway *62°33′N 07°42′E*
104 K13 **Andalucía** *Eng.* Andalusia. ◈ *autonomous community* S Spain
23 P7 **Andalusia** Alabama, S USA *31°18′N 86°29′W*
Andalusia *see* Andalucía
151 Q21 **Andaman and Nicobar Islands** *var.* Andamans and Nicobars. ◈ *union territory* India, NE Indian Ocean
173 T4 **Andaman Basin** *undersea feature* NE Indian Ocean *10°00′N 94°00′E*
151 P19 **Andaman Islands** *island group* India, NE Indian Ocean
Andamans and Nicobars *see* Andaman and Nicobar Islands
173 T4 **Andaman Sea** *sea* NE Indian Ocean
57 K19 **Andamarca** Oruro, C Bolivia *18°46′S 67°31′W*
182 H5 **Andamooka** South Australia *30°26′S 137°12′E*
141 Y9 **'Andām, Wādī** *seasonal river* NE Oman
172 J3 **Andapa** Antsirañana, NE Madagascar *14°39′S 49°40′E*
149 R4 **Andarāb** *var.* Banow. Baghlān, NE Afghanistan *35°36′N 69°18′E*
Andarbag *see* Andarbogh
147 S13 **Andarbogh** *Rus.* Andarbag, Anderbak. S Tajikistan *37°51′N 71°45′E*
109 Z5 **Andau** Burgenland, E Austria *47°47′N 17°02′E*
108 I10 **Andeer** Graubünden, S Switzerland *46°36′N 09°24′E*
92 H9 **Andenes** Nordland, C Norway *69°18′N 16°10′E*
99 J20 **Andenne** Namur, SE Belgium *50°29′N 05°06′E*
77 S11 **Andéramboukane** Gao, E Mali *15°24′N 03°03′E*
Anderbak *see* Andarbogh
99 G18 **Anderlecht** Brussels, C Belgium *50°50′N 04°18′E*
99 G21 **Anderlues** Hainaut, S Belgium *50°24′N 04°16′E*
108 G9 **Andermatt** Uri, C Switzerland *46°39′N 08°36′E*
101 E17 **Andernach** *anc.* Antunnacum. Rheinland-Pfalz, SW Germany *50°26′N 07°24′E*
188 D15 **Andersen Air Force Base** *air base* NE Guam *13°34′N 144°55′E*
39 R9 **Anderson** Alaska, USA *64°20′N 149°11′W*
35 N4 **Anderson** California, W USA *40°26′N 122°21′W*
31 P13 **Anderson** Indiana, N USA *40°06′N 85°40′W*
27 R8 **Anderson** Missouri, C USA *36°39′N 94°26′W*
21 P11 **Anderson** South Carolina, SE USA *34°30′N 82°39′W*
25 V10 **Anderson** Texas, SW USA *30°29′N 96°00′W*
95 K20 **Anderstorp** Jönköping, S Sweden *57°17′N 13°38′E*
54 D9 **Andes** Antioquia, W Colombia *05°40′N 75°56′W*
47 P7 **Andes** ▲▲ W South America
29 P12 **Andes, Lake** ◎ South Dakota, N USA
92 H10 **Andfjorden** *fjord* E Norwegian Sea
155 H16 **Andhra Pradesh** ◈ *state* E India
98 J8 **Andijk** Noord-Holland, NW Netherlands *52°38′N 05°00′E*
147 S10 **Andijon** *Rus.* Andizhan. Andijon Viloyati, E Uzbekistan *40°46′N 72°19′E*
147 S10 **Andijon Viloyati** *Rus.* Andizhanskaya Oblast'. ◈ *province* E Uzbekistan
Andikíthira *see* Antikýthira
172 J4 **Andilamena** Toamasina, C Madagascar *17°00′S 48°35′E*
142 L8 **Andīmeshk** *var.* Andimishk; *prev.* Salehābād. Khūzestān, SW Iran *32°30′N 48°26′E*
Andimishk *see* Andīmeshk
Andipáros *see* Antíparos
Andipaxi *see* Antipaxoí
Andípsara *see* Antípsara
136 L16 **Andırın** Kahramanmaraş, S Turkey *37°33′N 36°18′E*
158 J8 **Andirlangar** Xinjiang Uygur Zizhiqu, NW China *37°38′N 83°40′E*
Andírrion *see* Antírrio
Ándissa *see* Ántissa
Andizhan *see* Andijon
Andizhanskaya Oblast' *see* Andijon Viloyati
149 N2 **Andkhvoy** *prev.* Andkhvoy. Fāryāb, N Afghanistan *36°56′N 65°08′E*
105 Q2 **Andoain** País Vasco, N Spain *43°13′N 02°02′W*
163 Y15 **Andong** *Jap.* Antō. E South Korea *36°34′N 128°44′E*
109 R4 **Andorf** Oberösterreich, N Austria *48°22′N 13°33′E*
105 S7 **Andorra** Aragón, NE Spain *40°59′N 00°27′W*
105 V4 **Andorra** *off.* Principality of Andorra, *Cat.* Valls d'Andorra, *Fr.* Vallée d'Andorre. ◆ *monarchy* SW Europe
Andorra *see* Andorra la Vella
105 V4 **Andorra la Vella** *var.* Andorra, *Fr.* Andorre la Vielle, *Sp.* Andorra la Vieja. ● (Andorra) C Andorra *42°30′N 01°30′E*
Andorra la Vieja *see* Andorra la Vella
Andorra, Principality of *see* Andorra
Andorra, Valls d'/Andorra, Vallée d' *see* Andorra
Andorre la Vielle *see* Andorra la Vella
97 M22 **Andover** S England, United Kingdom *51°13′N 01°28′W*
27 N6 **Andover** Kansas, C USA *37°42′N 97°08′W*
92 G10 **Andøya** *island* C Norway
60 I8 **Andradina** São Paulo, S Brazil *20°54′S 51°19′W*
105 X9 **Andratx** Mallorca, Spain, W Mediterranean Sea *39°35′N 02°25′E*
39 N10 **Andreafsky River** ≈ Alaska, USA
38 H17 **Andreanof Islands** *island group* Aleutian Islands, Alaska, USA
124 H16 **Andreapol'** Tverskaya Oblast', W Russian Federation *56°38′N 32°17′E*
Andreas, Cape *see* Zafer Burnu
Andreevka *see* Kabanbay
21 N10 **Andrews** North Carolina, SE USA *35°19′N 84°01′W*
21 T13 **Andrews** South Carolina, SE USA *33°27′N 79°33′W*
24 M7 **Andrews** Texas, SW USA *32°19′N 102°34′W*
173 N5 **Andrew Tablemount** *var.* Gora Andryu. *undersea feature* W Indian Ocean *06°45′N 50°30′E*
Andreyevka *see* Kabanbay
107 N17 **Andria** Puglia, SE Italy *41°13′N 16°17′E*
113 K16 **Andrijevica** E Montenegro *42°45′N 19°45′E*
115 E20 **Andrítsaina** Pelopónnisos, S Greece *37°29′N 21°52′E*
An Droichead Nua *see* Newbridge
Andropov *see* Rybinsk
115 J19 **Ándros** Ándros, Kykládes, Greece, Aegean Sea *37°49′N 24°54′E*
115 J20 **Ándros** *island* Kykládes, Greece, Aegean Sea
19 O7 **Androscoggin River** ≈ Maine/New Hampshire, NE USA
44 F3 **Andros Island** *island* NW The Bahamas
127 R7 **Androsovka** Samarskaya Oblast', W Russian Federation *52°41′N 49°34′E*
44 G3 **Andros Town** Andros Island, NW The Bahamas *24°40′N 77°47′W*
155 D21 **Āndrott Island** *island* Lakshadweep, India, N Indian Ocean
117 N5 **Andrushivka** Zhytomyrs'ka Oblast', N Ukraine *50°01′N 29°02′E*
111 K17 **Andrychów** Małopolskie, S Poland *49°51′N 19°18′E*
Andryu, Gora *see* Andrew Tablemount
92 I10 **Andselv** Troms, N Norway *69°05′N 18°30′E*
79 O17 **Andudu** Orientale, NE Dem. Rep. Congo *02°26′N 28°39′E*
105 N13 **Andújar** *anc.* Illiturgis. Andalucía, SW Spain *38°02′N 04°03′W*
82 C12 **Andulo** Bié, W Angola *11°29′S 16°43′E*
103 Q14 **Anduze** Gard, S France *44°03′N 03°59′E*
An Earagail *see* Errigal Mountain
95 L19 **Aneby** Jönköping, S Sweden *57°50′N 14°45′E*
Anécho *see* Aného
77 Q9 **Anéfis** Kidal, NE Mali *18°05′N 00°38′E*
45 U8 **Anegada** *island* NE British Virgin Islands
61 B25 **Anegada, Bahía** *bay* E Argentina
45 U9 **Anegada Passage** *passage* Anguilla/British Virgin Islands
77 R17 **Aného** *var.* Anécho; *prev.* Petit-Popo. S Togo *06°14′N 01°36′E*
197 D17 **Aneityum** *var.* Anatom; *prev.* Kéamu. *island* S Vanuatu
117 N10 **Anenii Noi** *Rus.* Novyye Aneny. C Moldova *46°52′N 29°10′E*
186 F7 **Anepmete** New Britain, E Papua New Guinea *05°47′S 148°37′E*
105 U4 **Aneto** ▲ NE Spain *42°36′N 00°37′E*
146 F13 **Änew** *Rus.* Annau. Ahal Welaýaty, C Turkmenistan *37°51′N 58°22′E*
Änewetak *see* Enewetak Atoll
77 Y8 **Aney** Agadez, NE Niger *19°22′N 13°00′E*
An Fheoir *see* Nore
122 L12 **Angara** ≈ C Russian Federation
122 M13 **Angarsk** Irkutskaya Oblast', S Russian Federation *52°31′N 103°55′E*
Angaur *see* Ngeaur
93 G17 **Ånge** Västernorrland, C Sweden *62°31′N 15°40′E*
Angel *see* Úhlava
40 D4 **Ángel de la Guarda, Isla** *island* NW Mexico
171 O3 **Angeles** *off.* Angeles City. Luzon, N Philippines *15°16′N 120°37′E*
Angeles City *see* Angeles
Angel Falls *see* Ángel, Salto
95 J22 **Ängelholm** Skåne, S Sweden *56°14′N 12°52′E*
61 A17 **Angélica** Santa Fe, C Argentina *31°33′S 61°33′W*
25 W8 **Angelina River** ≈ Texas, SW USA
55 Q9 **Ángel, Salto** *Eng.* Angel Falls. *waterfall* E Venezuela
95 M15 **Ängelsberg** Västmanland, C Sweden *59°57′N 16°01′E*
35 P8 **Angels Camp** California, W USA *38°03′N 120°31′W*
109 W7 **Anger** Steiermark, SE Austria *47°16′N 15°41′E*
Angerapp *see* Ozersk
Angerburg *see* Węgorzewo
93 H15 **Ångermanälven** ≈ N Sweden
100 P11 **Angermünde** Brandenburg, NE Germany *53°02′N 13°59′E*
102 K7 **Angers** *anc.* Juliomagus. Maine-et-Loire, NW France *47°30′N 00°33′W*
15 W7 **Angers** ≈ Québec, SE Canada
93 J16 **Ångesön** *island* N Sweden
Angistro *see* Ágkistro
114 H13 **Angítis** ≈ NE Greece
167 R13 **Ângk Tasaôm** *prev.* Angtassom. Takêv, S Cambodia
185 C25 **Anglem, Mount** ▲ Stewart Island, Southland, SW New Zealand *46°44′S 167°56′E*
97 I18 **Anglesey** *cultural region* NW Wales, United Kingdom
97 I18 **Anglesey** *island* NW Wales, United Kingdom
102 I15 **Anglet** Pyrénées-Atlantiques, SW France *43°29′N 01°30′W*
25 W12 **Angleton** Texas, SW USA *29°10′N 95°27′W*
Anglia *see* England
14 H9 **Angliers** Québec, SE Canada *47°33′N 79°17′W*
Anglo-Egyptian Sudan *see* Sudan
Angmagssalik *see* Ammassalik
167 Q7 **Ang Nam Ngum** ◎ C Laos
79 N16 **Ango** Orientale, N Dem. Rep. Congo *04°01′N 25°52′E*
83 Q15 **Angoche** Nampula, E Mozambique *16°10′S 39°58′E*
63 G14 **Angol** Araucanía, C Chile *37°47′S 72°45′W*
31 Q11 **Angola** Indiana, N USA *41°37′N 85°00′W*
82 A9 **Angola** *off.* Republic of Angola; *prev.* People's Republic of Angola, Portuguese West Africa. ◆ *republic* SW Africa
65 P15 **Angola Basin** *undersea feature* E Atlantic Ocean *15°00′S 03°00′E*
Angola, People's Republic of *see* Angola
Angola, Republic of *see* Angola
39 X13 **Angoon** Admiralty Island, Alaska, USA *57°33′N 134°30′W*
147 O14 **Angor** Surkhondaryo Viloyati, S Uzbekistan *37°30′N 67°06′E*
Angora *see* Ankara
186 C6 **Angoram** East Sepik, NW Papua New Guinea *04°04′S 144°04′E*
40 H8 **Angostura** Sinaloa, C Mexico *25°18′N 108°10′W*
Angostura *see* Ciudad Bolívar
41 U17 **Angostura, Presa de la** ⊠ SE Mexico
28 J11 **Angostura Reservoir** ⊠ South Dakota, N USA
102 L11 **Angoulême** *anc.* Iculisma. Charente, W France *45°39′N 00°10′E*
102 K11 **Angoumois** *cultural region* W France
64 O2 **Angra do Heroísmo** Terceira, Azores, Portugal, NE Atlantic Ocean *38°40′N 27°14′W*
60 O10 **Angra dos Reis** Rio de Janeiro, SE Brazil *22°59′S 44°17′W*
Angra Pequena *see* Lüderitz
147 Q10 **Angren** Toshkent Viloyati, E Uzbekistan *41°05′N 70°18′E*
Angtassom *see* Ângk Tasaôm
167 O10 **Ang Thong** *var.* Angthong. Ang Thong, C Thailand *14°35′N 100°25′E*
Angthong *see* Ang Thong
79 M16 **Angu** Orientale, N Dem. Rep. Congo *03°38′N 24°14′E*
105 S5 **Angüés** Aragón, NE Spain *42°07′N 00°10′W*
45 U9 **Anguilla** ◇ *UK dependent territory* E West Indies
45 V9 **Anguilla** *island* E West Indies
44 F4 **Anguilla Cays** *islets* SW The Bahamas
Angul *see* Anugul
161 N1 **Anguli Nur** ◎ E China
79 O18 **Angumu** Orientale, E Dem. Rep. Congo *00°10′S 27°42′E*
14 G14 **Angus** Ontario, S Canada *44°19′N 79°52′W*
96 J10 **Angus** *cultural region* E Scotland, United Kingdom
59 K19 **Anhangüera** Goiás, S Brazil *18°12′S 48°19′W*
99 I21 **Anhée** Namur, S Belgium *50°18′N 04°52′E*
95 I21 **Anholt** *island* C Denmark
160 M11 **Anhua** *var.* Dongping. Hunan, S China *28°25′N 111°10′E*
161 P8 **Anhui** *var.* Anhui Sheng, Anhwei, Wan. ◈ *province* E China
AnhuiSheng/Anhwei Wan *see* Anhui
39 O11 **Aniak** Alaska, USA *61°34′N 159°31′W*
39 O12 **Aniak River** ≈ Alaska, USA
An Iarmhí *see* Westmeath
189 R8 **Anibare** E Nauru *0°31′S 166°56′E*
189 R8 **Anibare Bay** *bay* E Nauru, W Pacific Ocean
Anicium *see* le Puy
115 K22 **Ánidro** *island* Kykládes, Greece, Aegean Sea
77 R15 **Anié** C Togo *07°48′N 01°12′E*
77 Q15 **Anié** ≈ C Togo
102 J16 **Anie, Pic d'** ▲ SW France *42°56′N 00°44′W*
127 Y7 **Anikhovka** Orenburgskaya Oblast', W Russian Federation *51°27′N 60°17′E*
14 G9 **Anima Nipissing Lake** ◎ Ontario, S Canada
37 O16 **Animas** New Mexico, SW USA *31°55′N 108°49′W*
37 P16 **Animas Peak** ▲ New Mexico, SW USA *31°34′N 108°46′W*
37 P16 **Animas Valley** *valley* New Mexico, SW USA
116 F13 **Anina** *Ger.* Steierdorf, *Hung.* Stájerlakanina; *prev.* Ştaierdorf-Anina, Steierdorf-Anina, Steyerlak-Anina. Caraş-Severin, SW Romania *45°05′N 21°51′E*
29 U14 **Anita** Iowa, C USA *41°27′N 94°45′W*
123 U14 **Aniva, Mys** *headland* Ostrov Sakhalin, SE Russian Federation *46°02′N 143°25′E*
187 S15 **Aniwa** *island* S Vanuatu
93 M19 **Anjalankoski** Kymenlaakso, S Finland *60°39′N 26°54′E*
'Anjar *see* Aanjar
Anjiangying *see* Luanping
14 B8 **Anjigami Lake** ◎ Ontario, S Canada
149 N13 **Anjīra** Baluchistān, SW Pakistan *28°19′N 66°19′E*
164 K14 **Anjō** *var.* Anzyô. Aichi, Honshū, SW Japan *34°56′N 137°05′E*
102 J8 **Anjou** *cultural region* NW France
Anjouan *see* Nzwani
172 J4 **Anjozorobe** Antananarivo, C Madagascar *18°22′S 47°52′E*
163 W13 **Anju** W North Korea *39°36′N 125°44′E*
98 M5 **Anjum** *Fris.* Eanjum. Fryslân, N Netherlands *53°22′N 06°09′E*
172 G6 **Ankaboa, Tanjona** *prev./Fr.* Cap Saint-Vincent. *headland* W Madagascar *21°57′S 43°16′E*
160 L7 **Ankang** *prev.* Xing'an. Shaanxi, C China *32°45′N 109°00′E*
136 I12 **Ankara** *prev.* Angora; *anc.* Ancyra. ● (Turkey) Ankara, C Turkey *39°55′N 32°50′E*
136 H12 **Ankara** ◈ *province* C Turkey
95 N19 **Ankarsrum** Kalmar, S Sweden *57°40′N 16°19′E*
172 H6 **Ankazoabo** Toliara, SW Madagascar *22°18′S 44°30′E*
172 I4 **Ankazobe** Antananarivo, C Madagascar *18°20′S 47°07′E*
29 V14 **Ankeny** Iowa, C USA *41°43′N 93°37′W*
167 V11 **An Khê** Gia Lai, C Vietnam *13°57′N 108°39′E*
100 O9 **Anklam** Mecklenburg-Vorpommern, NE Germany *53°51′N 13°42′E*
80 K13 **Ānkober** Āmara, N Ethiopia *09°36′N 39°44′E*
77 O17 **Ankobra** ≈ S Ghana
79 N22 **Ankoro** Katanga, SE Dem. Rep. Congo *06°45′S 26°58′E*
99 L24 **Anlier, Forêt d'** *forest* SE Belgium
160 I13 **Anlong** *var.* Xin'an. Guizhou, S China *25°05′N 105°29′E*
An Longfort *see* Longford
167 R11 **Ânlong Vêng** Siĕmréab, NW Cambodia *14°16′N 104°08′E*
An Lorgain *see* Lurgan
161 N8 **Anlu** Hubei, C China *31°15′N 113°41′E*
An Mhí *see* Meath
An Mhuir Cheilteach *see* Celtic Sea
An Muileann gCearr *see* Mullingar
93 F16 **Ånn** Jämtland, C Sweden *63°19′N 12°34′E*
126 M8 **Anna** Voronezhskaya Oblast', W Russian Federation *51°31′N 40°23′E*
30 L17 **Anna** Illinois, N USA *37°27′N 89°15′W*
25 U5 **Anna** Texas, SW USA *33°21′N 96°33′W*
74 L5 **Annaba** *prev.* Bône. NE Algeria *36°55′N 07°47′E*
An Nabatīyah at Taḥtā *see* Nabatîyé
101 N17 **Annaberg-Buchholz** Sachsen, E Germany *50°35′N 13°01′E*
109 T9 **Annabichl** ✈ (Klagenfurt) Kärnten, S Austria *46°39′N 14°21′E*
140 M5 **An Nafūd** *desert* NW Saudi Arabia
139 P6 **'Annah** *var.* 'Ānah. Al Anbār, NW Iraq *34°50′N 42°00′E*
139 P6 **An Nāḥiyah** Al Anbār, W Iraq *34°24′N 41°33′E*
139 T10 **An Najaf** *off.* Muḥāfa at an Najaf, *var.* Najaf. An Najaf, S Iraq *31°59′N 44°19′E*
21 V5 **Anna, Lake** ⊠ Virginia, NE USA
97 F16 **Annalee** ≈ N Ireland
167 S9 **Annamite Mountains** *var.* annamsescordillera, *Fr.* Chaîne Annamitique, *Lao.* Phou Louang. ▲▲ C Laos
Annamitique, Chaîne *see* Annamite Mountains
Annamses Cordillera *see* Annamite Mountains
97 J14 **Annan** S Scotland, United Kingdom *55°N 03°20′W*
29 U8 **Annandale** Minnesota, N USA *45°15′N 94°07′W*
21 W4 **Annandale** Virginia, NE USA *38°48′N 77°10′W*
189 Q7 **Anna Point** *headland* N Nauru *0°30′S 166°56′E*
21 X3 **Annapolis** *state capital* Maryland, NE USA *38°59′N 76°30′W*
188 A10 **Anna, Pulo** *island* S Palau
153 O10 **Annapurṇa** ▲ C Nepal *28°30′N 83°50′E*
An Nāqūrah *see* En Nâqoûra
31 R10 **Ann Arbor** Michigan, N USA *42°17′N 83°45′W*
An Nás *see* Naas
139 W12 **An Nāṣirīyah** *var.* Nasiriya. Dhī Qār, SE Iraq *31°04′N 46°17′E*
139 W11 **An Naṣr** Dhī Qār, E Iraq *31°34′N 46°08′E*
Annau *see* Änew
121 O13 **An Nawfalīyah** *var.* Al Nūwfalīyah. N Libya *30°46′N 17°48′E*
19 P10 **Ann, Cape** *headland* Massachusetts, NE USA *42°39′N 70°35′W*
180 I10 **Annean, Lake** ◎ Western Australia
Anneciacum *see* Annecy
103 T11 **Annecy** *anc.* Anneciacum. Haute-Savoie, E France *45°53′N 06°09′E*
103 T11 **Annecy, Lac d'** ◎ E France
103 T10 **Annemasse** Haute-Savoie, E France *46°10′N 06°13′E*
39 Z14 **Annette Island** *island* Alexander Archipelago, Alaska, USA
An Nhon *see* Binh Đinh
An Nil al Abyaḍ *see* White Nile
An Nil al Azraq *see* Blue Nile
23 Q3 **Anniston** Alabama, S USA *33°39′N 85°49′W*
79 A19 **Annobón** *island* W Equatorial Guinea
103 R12 **Annonay** Ardèche, E France *45°15′N 04°40′E*
44 K12 **Annotto Bay** C Jamaica *18°16′N 76°47′W*
141 R5 **An Nu'ayrīyah** *var.* Nariya. Ash Sharqīyah, NE Saudi Arabia *27°30′N 48°30′E*
182 M9 **Annuello** Victoria, SE Australia *34°54′S 142°50′E*
139 Q10 **An Nukhayb** Al Anbār, S Iraq *32°02′N 42°15′E*
139 U9 **An Nu'mānīyah** Wāsiṭ, E Iraq *32°34′N 45°23′E*
Áno Arkhánai *see* Archánes
115 J25 **Anógeia** *var.* Anogia, Anóyia. Kríti, Greece, E Mediterranean Sea *35°17′N 24°53′E*
Anogia *see* Anógeia
29 V8 **Anoka** Minnesota, N USA *45°15′N 93°26′W*
An Ómaigh *see* Omagh
172 I1 **Anorontany, Tanjona** *Fr.* Cap Saint-Sébastien. *headland* N Madagascar
172 J5 **Anosibe An'Ala** Toamasina, E Madagascar *19°24′S 48°11′E*
Anóyia *see* Anógeia
An Pointe *see* Warrenpoint
161 P9 **Anqing** Anhui, E China *30°32′N 116°59′E*
161 Q5 **Anqiu** Shandong, E China *36°25′N 119°10′E*
An Ráth *see* Ráth Luirc
An Ribhéar *see* Kenmare River
An Ros *see* Rush
99 K19 **Ans** Liège, E Belgium *50°39′N 05°32′E*
Ansāb *see* Niṣāb
171 W12 **Ansas** Papua, E Indonesia *01°44′S 135°52′E*
101 J20 **Ansbach** Bayern, SE Germany *49°18′N 10°36′E*
An Sciobairín *see* Skibbereen
An Scoil *see* Skull
An Seancheann *see* Old Head of Kinsale
45 Y5 **Anse-Bertrand** Grande Terre, N Guadeloupe *16°28′N 61°31′W*
172 H17 **Anse Boileau** Mahé, NE Seychelles *04°43′S 55°29′E*
45 S11 **Anse La Raye** NW Saint Lucia *13°57′N 61°01′W*
54 D9 **Anserma** Caldas, W Colombia *05°15′N 75°47′W*
109 T4 **Ansfelden** Oberösterreich, N Austria *48°12′N 14°17′E*
163 U12 **Anshan** Liaoning, NE China *41°06′N 122°55′E*
160 J12 **Anshun** Guizhou, S China *26°15′N 105°58′E*
61 F17 **Ansina** Tacuarembó, C Uruguay *31°58′S 55°28′W*
29 O15 **Ansley** Nebraska, C USA *41°16′N 99°22′W*
25 P6 **Anson** Texas, SW USA *32°45′N 99°55′W*
77 Q10 **Ansongo** Gao, E Mali *15°39′N 00°33′E*
An Srath Bán *see* Strabane
21 R5 **Ansted** West Virginia, NE USA *38°08′N 81°06′W*
171 Y13 **Ansudu** Papua, E Indonesia *02°09′S 139°19′E*
57 G15 **Anta** Cusco, S Peru *13°30′S 72°08′W*
57 G16 **Antabamba** Apurímac, C Peru *14°23′S 72°54′W*
Antafalva *see* Kovačica
136 L17 **Antakya** *anc.* Antioch, Antiochia. Hatay, S Turkey *36°12′N 36°10′E*
172 K3 **Antalaha** Antsirañana, NE Madagascar *14°53′S 50°16′E*
136 F17 **Antalya** *prev.* Adalia; *anc.* Attaleia, *Bibl.* Attalia. Antalya, SW Turkey *36°53′N 30°42′E*
136 F17 **Antalya** ◈ *province* SW Turkey
136 F16 **Antalya** ✈ Antalya, SW Turkey *36°53′N 30°45′E*
121 U10 **Antalya Basin** *undersea feature* E Mediterranean Sea
Antalya, Gulf of *see* Antalya Körfezi
136 F16 **Antalya Körfezi** *var.* Gulf of Adalia, *Eng.* Gulf of Antalya. *gulf* SW Turkey
172 J5 **Antanambao Manampotsy** Toamasina, E Madagascar *19°30′S 48°36′E*
172 I5 **Antananarivo** *prev.* Tananarive. ● (Madagascar) Antananarivo, C Madagascar *18°52′S 47°30′E*
172 I4 **Antananarivo** ◈ *province* C Madagascar
172 I5 **Antananarivo** ✈ Antananarivo, C Madagascar *18°52′S 47°30′E*
An tAonach *see* Nenagh
194-195 **Antarctica** *continent*
194 I5 **Antarctic Peninsula** *peninsula* Antarctica
61 J15 **Antas, Rio das** ≈ S Brazil
189 U16 **Ant Atoll** *atoll* Caroline Islands, E Micronesia
An Teampall Mór *see* Templemore
Antep *see* Gaziantep
104 M15 **Antequera** *anc.* Anticaria, Antiquaria. Andalucía, S Spain *37°01′N 04°34′W*
Antequera *see* Oaxaca
37 S5 **Antero Reservoir** ⊠ Colorado, C USA
26 M7 **Anthony** Kansas, C USA *37°10′N 98°02′W*
37 R16 **Anthony** New Mexico, SW USA *32°00′N 106°36′W*
182 D5 **Anthony, Lake** *salt lake* South Australia
74 E8 **Anti-Atlas** ▲▲ SW Morocco
103 U15 **Antibes** *anc.* Antipolis. Alpes-Maritimes, SE France *43°35′N 07°07′E*
103 U15 **Antibes, Cap d'** *headland* SE France *43°33′N 07°08′E*
Anticaria *see* Antequera
13 Q11 **Anticosti, Île d'** *Eng.* Anticosti Island. *island* Québec, E Canada
Anticosti Island *see* Anticosti, Île d'
102 K3 **Antifer, Cap d'** *headland* N France *49°43′N 00°10′E*
30 L6 **Antigo** Wisconsin, N USA *45°10′N 89°10′W*
13 Q15 **Antigonish** Nova Scotia, SE Canada *45°39′N 62°00′W*
64 P11 **Antigua** Fuerteventura, Islas Canarias, NE Atlantic Ocean
45 X10 **Antigua** *island* S Antigua and Barbuda, Leeward Islands
Antigua *see* Antigua Guatemala
45 W9 **Antigua and Barbuda** ◆ *commonwealth republic* E West Indies
42 C6 **Antigua Guatemala** *var.* Antigua. Sacatepéquez, SW Guatemala *14°33′N 90°42′W*
41 P11 **Antiguo Morelos** *var.* Antiguo-Morelos. Tamaulipas, C Mexico *22°35′N 99°08′W*
115 F19 **Antíkyras, Kólpos** *gulf* C Greece
115 G24 **Antikýthira** *var.* Andikíthira. *island* S Greece
138 I7 **Anti-Lebanon** *var.* Jebel esh Sharqi, *Ar.* Al Jabal ash Sharqī, *Fr.* Anti-Liban. ▲▲ Lebanon/Syria
Anti-Liban *see* Anti-Lebanon
115 M22 **Antimácheia** Kos, Dodekánisa, Greece *47°19′N 07°54′E*
115 I22 **Antímilos** *island* Kykládes, Greece, Aegean Sea
36 L6 **Antimony** Utah, W USA *38°07′N 112°00′W*
An tInbhear Mór *see* Arklow
30 M10 **Antioch** Illinois, N USA *42°28′N 88°06′W*
Antioch *see* Antakya
102 I10 **Antioche, Pertuis d'** *inlet* W France
Antiochia *see* Antakya
54 D8 **Antioquia** Antioquia, C Colombia *06°36′N 75°53′W*
54 E8 **Antioquia** *off.* Departamento de Antioquia. ◈ *province* C Colombia
Antioquia, Departamento de *see* Antioquia
115 J21 **Antíparos** *var.* Andipáros. *island* Kykládes, Greece, Aegean Sea
115 B17 **Antipaxoí** *var.* Andipaxi. *island* Iónia Nisiá, Greece, C Mediterranean Sea
122 J8 **Antipayuta** Yamalo-Nenetskiy Avtonomnyy Okrug, N Russian Federation *69°08′N 76°43′E*
192 L12 **Antipodes Islands** *island group* S New Zealand
Antipolis *see* Antibes
115 J18 **Antípsara** *var.* Andípsara. *island* E Greece
Antiquaria *see* Antequera
15 N10 **Antique, Lac** ◎ Québec, SE Canada
115 E18 **Antírrio** *var.* Andírrion. Dytikí Elláda, C Greece *38°20′N 21°46′E*
115 K16 **Ántissa** *var.* Ándissa. Lésvos, E Greece *39°15′N 26°00′E*
An tIúr *see* Newry

◆ Country
● Country Capital
◇ Dependent Territory
○ Dependent Territory Capital
◈ Administrative Regions
✈ International Airport
▲ Mountain
▲▲ Mountain Range
Volcano
≈ River
◎ Lake
⊠ Reservoir

ANTIVARI – ARGUUT

Antivari see Bar
56 C6 **Antizana** ▲ N Ecuador 0°29′S 78°08′W
27 Q13 **Antlers** Oklahoma, C USA 34°15′N 95°38′W
93 J14 **Antnäs** Norrbotten, N Sweden 65°32′N 21°53′E
Antō see Andong
62 G5 **Antofagasta** Antofagasta, N Chile 23°40′S 70°23′W
62 G6 **Antofagasta** off. Región de Antofagasta. ◈ region N Chile
Antofagasta, Región de see Antofagasta
62 I7 **Antofalla, Salar de** salt lake NW Argentina
99 D20 **Antoing** Hainaut, SW Belgium 50°34′N 03°26′E
43 S16 **Antón** Coclé, C Panama 08°23′N 80°15′W
24 M5 **Anton** Texas, SW USA 33°48′N 102°09′W
37 T11 **Anton Chico** New Mexico, SW USA 35°12′N 105°09′W
60 K12 **Antonina** Paraná, S Brazil 25°28′S 48°43′W
188 C16 **Antonio B. Won Pat International** ✈ (Agana) C Guam 13°28′N 144°48′E
103 O5 **Antony** Hauts-de-Seine, N France 48°45′N 02°17′E
Antratsit see Antratsyt
117 Y8 **Antratsyt** Rus. Antratsit. Luhans'ka Oblast', E Ukraine 48°07′N 39°05′E
97 G15 **Antrim** Ir. Aontroim. NE Northern Ireland, United Kingdom 54°43′N 06°13′W
97 G14 **Antrim** Ir. Aontroim. cultural region NE Northern Ireland, United Kingdom
97 G14 **Antrim Mountains** ▲▲ NE Northern Ireland, United Kingdom
172 H5 **Antsalova** Mahajanga, W Madagascar 18°40′S 44°37′E
Antserana see Antsirañana
An tSionainn see Shannon
172 J2 **Antsirañana** var. Antserana; prev. Antsirane, Diégo-Suarez. Antsirañana, N Madagascar 12°19′S 49°17′E
172 J2 **Antsirañana** ◈ province N Madagascar
Antsirane see Antsirañana
An tSiúir see Suir
118 I7 **Antsla** Ger. Anzen. Võrumaa, SE Estonia 57°52′N 26°33′E
An tSláine see Slaney
172 J3 **Antsohihy** Mahajanga, NW Madagascar 14°50′S 47°58′E
63 G14 **Antuco, Volcán** ⛰ C Chile 37°29′S 71°25′W
169 W10 **Antu, Gunung** ▲ Borneo, N Indonesia 0°57′N 118°51′E
An Tullach see Tullow
An-tung see Dandong
Antunnacum see Andernach
Antwerp see Antwerpen
99 G16 **Antwerpen** Eng. Antwerp, Fr. Anvers. Antwerpen, N Belgium 51°13′N 04°25′E
99 H16 **Antwerpen** Eng. Antwerp. ◈ province N Belgium
An Uaimh see Navan
154 N12 **Anugul** var. Angul. Odisha, E India 20°51′N 84°59′E
152 F9 **Anūpgarh** Rājasthān, NW India 29°10′N 73°14′E
154 K10 **Anūppur** Madhya Pradesh, C India 23°05′N 81°45′E
155 K24 **Anuradhapura** North Central Province, C Sri Lanka 08°20′N 80°25′E
Anvers see Antwerpen
194 G4 **Anvers Island** island Antarctica
39 N11 **Anvik** Alaska, USA 62°39′N 160°12′W
39 N10 **Anvik River** ≈ Alaska, USA
38 F17 **Anvil Peak** ▲ Semisopochnoi Island, Alaska, USA 51°59′N 179°36′E
An Vinh, Nhom see Amphitrite Group
159 P7 **Anxi** var. Yuanquan. Gansu, N China 40°32′N 95°50′E
182 F8 **Anxious Bay** bay South Australia
161 O5 **Anyang** Henan, C China 36°11′N 114°18′E
159 S11 **A'nyêmaqên Shan** ▲▲ C China
118 H12 **Anykščiai** Utena, E Lithuania 55°30′N 25°34′E
161 P13 **Anyuan** var. Xinshan. Jiangxi, S China 25°10′N 115°25′E
123 T7 **Anyuysk** Chukotskiy Avtonomnyy Okrug, NE Russian Federation 68°22′N 161°33′E
123 T7 **Anyuyskiy Khrebet** ▲▲ NE Russian Federation
54 D8 **Anzá** Antioquia, C Colombia 06°18′N 75°54′W
Anzen see Antsla
107 I16 **Anzio** Lazio, C Italy 41°28′N 12°38′E
55 O6 **Anzoátegui** off. Estado Anzoátegui. ◈ state NE Venezuela
Anzoátegui, Estado see Anzoátegui
147 P12 **Anzob** W Tajikistan 39°24′N 68°55′E
Anzyô see Anjō
Aoba see Ambae
165 X13 **Aoga-shima** island Izu-shotō, SE Japan
Aohan Qi see Xinhui
105 R3 **Aoiz** Bas. Agoitz, var. Agoiz. Navarra, N Spain 42°47′N 01°23′W
167 O11 **Ao Krung Thep** var. Krung Thep Mahanakhon, Eng. Bangkok. ● (Thailand) Bangkok, C Thailand 13°44′N 100°30′E
186 M9 **Aola** var. Tenaghau. Guadalcanal, C Solomon Islands 09°32′S 160°28′E
166 M15 **Ao Luk Nua** Krabi, SW Thailand 08°21′N 98°43′E
Aomen Tebie Xingzhengqu see Macao
172 N8 **Aomori** Aomori, Honshū, C Japan 40°50′N 140°43′E
172 N8 **Aomori** off. Aomori-ken. ◈ prefecture Honshū, C Japan
Aomori-ken see Aomori
Aontroim see Antrim
115 C15 **Aóos** var. Vijosa, Vijosë, Alb. Lumi i Vjosës. ≈ Albania/Greece see also Vjosës, Lumi i
Aóos see Vjosës, Lumi i
191 Q7 **Aorai, Mont** ⛰ Tahiti, W French Polynesia 17°36′S 149°29′W
185 E19 **Aoraki** prev. Aorangi, Mount Cook. ▲ South Island, New Zealand 43°38′S 170°05′E
167 R13 **Aôral, Phnum** prev. Phnom Aural. ▲ W Cambodia 12°01′N 104°10′E
Aorangi see Aoraki
185 L15 **Aorangi Mountains** ▲▲ North Island, New Zealand
184 H13 **Aorere** ≈ South Island, New Zealand
106 A7 **Aosta** anc. Augusta Praetoria. Valle d'Aosta, NW Italy 45°43′N 07°20′E
77 O11 **Aougoundou, Lac** ◎ S Mali
76 K9 **Aoukâr** var. Aouker. plateau C Mauritania
78 J13 **Aouk, Bahr** ≈ Central African Republic/Chad
Aouker see Aoukâr
74 B11 **Aousard** SE Western Sahara 22°42′N 14°22′W
164 H12 **Aoya** Tottori, Honshū, SW Japan 35°31′N 134°01′E
Aoyang see Shanggao
78 H5 **Aozou** Tibesti, N Chad 22°01′N 17°11′E
26 M11 **Apache** Oklahoma, C USA 34°57′N 98°21′W
36 L14 **Apache Junction** Arizona, SW USA 33°25′N 111°33′W
24 J9 **Apache Mountains** ▲▲ Texas, SW USA
36 M16 **Apache Peak** ▲ Arizona, SW USA 31°50′N 110°25′W
116 H10 **Apahida** Cluj, NW Romania 46°49′N 23°45′E
23 T9 **Apalachee Bay** bay Florida, SE USA
23 T3 **Apalachee River** ≈ Georgia, SE USA
23 S10 **Apalachicola** Florida, SE USA 29°43′N 84°58′W
23 S10 **Apalachicola Bay** bay Florida, SE USA
23 R9 **Apalachicola River** ≈ Florida, SE USA
Apam see Apan
Apamama see Abemama
41 P14 **Apan** var. Apam. Hidalgo, C Mexico 19°48′N 98°25′W
42 J8 **Apanás, Lago de** ◎ NW Nicaragua
54 H14 **Apaporis, Río** ≈ Brazil/Colombia
185 C23 **Aparima** ≈ South Island, New Zealand
171 O1 **Aparri** Luzon, N Philippines 18°16′N 121°42′E
112 J9 **Apatin** Vojvodina, NW Serbia 45°40′N 19°01′E
124 J4 **Apatity** Murmanskaya Oblast', NW Russian Federation 67°34′N 33°26′E
55 X9 **Apatou** NW French Guiana 05°10′N 54°22′W
40 M14 **Apatzingán** var. Apatzingán de la Constitución. Michoacán, SW Mexico 19°05′N 102°20′W
Apatzingán de la Constitución see Apatzingán
171 X12 **Apauwar** Papua, E Indonesia 01°36′S 138°10′E
Apaxtla see Apaxtla de Castrejón
41 O15 **Apaxtla de Castrejón** var. Apaxtla. Guerrero, S Mexico 18°06′N 99°55′W
118 J7 **Ape** NE Latvia 57°32′N 26°42′E
98 L11 **Apeldoorn** Gelderland, E Netherlands 52°13′N 05°57′E
Apennines see Appennino
Apenrade see Aabenraa
57 L17 **Apere, Río** ≈ C Bolivia
55 W11 **Apetina** Sipaliwini, SE Suriname 03°30′N 55°03′W
21 U9 **Apex** North Carolina, SE USA 35°43′N 78°51′W
79 M16 **Api** Orientale, N Dem. Rep. Congo 03°40′N 25°26′E
152 M9 **Api** ▲ NW Nepal 30°07′N 80°57′E
192 H16 **Āpia** ● (Samoa) Upolu, SE Samoa 13°50′S 171°47′W
Apia see Abaiang
60 K11 **Apiaí** São Paulo, S Brazil 24°29′S 48°51′W
170 M16 **Api, Gunung** ▲ Pulau Sangeang, S Indonesia 08°09′S 119°03′E
187 N9 **Apio** Maramasike Island, N Solomon Islands 09°36′S 161°25′E
41 O15 **Apipilulco** Guerrero, S Mexico 18°11′N 99°40′W
41 P14 **Apizaco** Tlaxcala, S Mexico 19°26′N 98°09′W
137 Q8 **Apkhazeti** var. Abkhazia; prev. Ap'khazet'i. ◈ autonomous republic NW Georgia
Ap'khazet'i see Apkhazeti
104 I4 **A Pobla de Trives** Cast. Puebla de Trives. Galicia, NW Spain 42°21′N 07°16′W
55 U9 **Apoera** Sipaliwini, NW Suriname 05°12′N 57°13′W
115 O23 **Apolakkiá** Ródos, Dodekánisa, Greece, Aegean Sea 36°02′N 27°48′E
101 L16 **Apolda** Thüringen, C Germany 51°02′N 11°31′E
192 H16 **Apolima Strait** strait C Pacific Ocean
182 M13 **Apollo Bay** Victoria, SE Australia 38°40′S 143°44′E
Apollonia see Sozopol
57 J16 **Apolo** La Paz, W Bolivia 14°48′S 68°31′W
57 J16 **Apolobamba, Cordillera** ▲▲ Bolivia/Peru
171 Q8 **Apo, Mount** ⛰ Mindanao, S Philippines 06°54′N 125°16′E
23 W11 **Apopka** Florida, SE USA 28°40′N 81°30′W
23 W11 **Apopka, Lake** ◎ Florida, SE USA
59 J19 **Aporé, Rio** ≈ SW Brazil
30 K2 **Apostle Islands** island group Wisconsin, N USA
Apostolas Andreas, Cape see Zafer Burnu
61 F14 **Apóstoles** Misiones, NE Argentina 27°54′S 55°45′W
Apostólou Andréa, Akrotíri see Zafer Burnu
117 S9 **Apostolove** Rus. Apostolovo. Dnipropetrovs'ka Oblast', E Ukraine 47°40′N 33°45′E
Apostolovo see Apostolove
17 S10 **Appalachian Mountains** ▲▲ E USA
95 K14 **Äppelbo** Dalarna, C Sweden 60°30′N 14°00′E
98 N7 **Appelscha** Fris. Appelskea. Fryslân, N Netherlands 52°57′N 06°19′E
Appelskea see Appelscha
106 G11 **Appennino** Eng. Apennines. ▲▲ Italy/San Marino
107 L17 **Appennino Campano** ▲▲ C Italy
108 I7 **Appenzell** Inner-Rhoden, NW Switzerland 47°20′N 09°25′E
55 V12 **Appikalo** Sipaliwini, S Suriname 02°07′N 56°16′W
98 O5 **Appingedam** Groningen, NE Netherlands 53°18′N 06°52′E
25 X8 **Appleby** Texas, SW USA 31°43′N 94°36′W
97 L15 **Appleby-in-Westmorland** Cumbria, NW England, United Kingdom 54°35′N 02°26′W
30 K10 **Apple River** ≈ Illinois, N USA
30 I5 **Apple River** ≈ Wisconsin, N USA
25 W9 **Apple Springs** Texas, SW USA 31°13′N 94°57′W
29 S8 **Appleton** Minnesota, N USA 45°12′N 96°01′W
30 M7 **Appleton** Wisconsin, N USA 44°17′N 88°24′W
27 S5 **Appleton City** Missouri, C USA 38°11′N 94°01′W
35 U14 **Apple Valley** California, W USA 34°30′N 117°11′W
29 V9 **Apple Valley** Minnesota, N USA 44°43′N 93°13′W
21 U6 **Appomattox** Virginia, NE USA 37°21′N 78°51′W
188 B16 **Apra Harbor** harbor W Guam
188 B16 **Apra Heights** W Guam
106 F6 **Aprica, Passo dell'** pass N Italy
107 M15 **Apricena** anc. Hadria Picena. Puglia, SE Italy 41°47′N 15°27′E
114 I9 **Apriltsi** Lovech, N Bulgaria 42°50′N 24°54′E
126 L14 **Apsheronsk** Krasnodarskiy Kray, SW Russian Federation 44°27′N 39°45′E
Apsheronskiy Poluostrov see Abşeron Yarımadası
103 S15 **Apt** anc. Apta Julia. Vaucluse, SE France 43°54′N 05°24′E
Apta Julia see Apt
38 H12 **'Āpua Point** var. Apua Point. headland Hawai'i, USA, C Pacific Ocean 19°15′N 155°13′W
60 I10 **Apucarana** Paraná, S Brazil 23°34′S 51°28′W
Apulia see Puglia
54 K8 **Apure** off. Estado Apure. ◈ state C Venezuela
Apure, Estado see Apure
54 J7 **Apure, Río** ≈ W Venezuela
57 F16 **Apurímac** off. Departamento de Apurímac. ◈ department C Peru
Apurímac, Departamento de see Apurímac
57 F15 **Apurímac, Río** ≈ S Peru
116 G10 **Apuseni, Munții** ▲▲ W Romania
Aqaba/'Aqaba see Al 'Aqabah
138 F15 **Aqaba, Gulf of** var. Gulf of Elat, Ar. Khalīj al 'Aqabah; anc. Sinus Aelaniticus. gulf NE Red Sea
139 R7 **'Aqabah** Al Anbār, C Iraq 33°33′N 42°55′E
'Aqabah, Khalīj al see Aqaba, Gulf of
'Aqabah, Muḥāfa at al see Al 'Aqabah
149 O2 **Āqchah** var. Āqcheh. Jowzjān, N Afghanistan 37°N 66°07′E
Āqcheh see Āqchah
Aqkengse see Akkense
Aqköl see Akkol'
Aqmola see Astana
Aqmola Oblysy see Akmola
158 L10 **Aqqikkol Hu** ◎ NW China
Aqqū see Akku
Aqqystaū see Akkystau
'Aqrah see Ākrē
Aqsay see Aksay
Aqshataū see Akshatau
Aqsū see Aksu
Aqsūat see Aksuat
Aqtas see Aktas
Aqtaū see Aktau
Aqtöbe see Aktobe
Aqtöbe Oblysy see Aktyubinsk
Aqtoghay see Aktogay
Aquae Augustae see Dax
Aquae Calidae see Bath
Aquae Flaviae see Chaves
Aquae Grani see Aachen
Aquae Panoniae see Baden
Aquae Sextiae see Aix-en-Provence
Aquae Solis see Bath
Aquae Tarbelicae see Dax
36 J11 **Aquarius Mountains** ▲▲ Arizona, SW USA
62 O5 **Aquidabán, Río** ≈ E Paraguay
59 H20 **Aquidauana** Mato Grosso do Sul, S Brazil 20°27′S 55°45′W
40 L15 **Aquila** Michoacán, S Mexico 18°36′N 103°32′W
Aquila/Aquila degli Abruzzi see L'Aquila
25 T8 **Aquilla** Texas, SW USA 31°51′N 97°13′W
44 L9 **Aquin** S Haiti 18°16′N 73°24′W
Aquisgranum see Aachen
102 J13 **Aquitaine** ◈ region SW France
Aqzhar see Akzhar
153 P13 **Āra** prev. Arrah. Bihār, N India 25°34′N 84°40′E
105 S4 **Ara** ≈ NE Spain
23 P2 **Arab** Alabama, S USA 34°19′N 86°30′W
Araba see Álava
138 G12 **'Arabah, Wādī al** Heb. Ha'Arava. dry watercourse Israel/Jordan
117 U12 **Arabats'ka Strilka, Kosa** spit S Ukraine
117 U12 **Arabats'ka Zatoka** gulf S Ukraine
'Arab, Baḥr al see Arab, Bahr el
80 C12 **Arab, Bahr el** var. Baḥr al 'Arab. ≈ S Sudan
56 E7 **Arabela, Río** ≈ N Peru
173 O4 **Arabian Basin** undersea feature N Arabian Sea 11°30′N 65°00′E
Arabian Desert see Sahara el Sharqiya
141 N9 **Arabian Peninsula** peninsula SW Asia
85 P15 **Arabian Plate** tectonic feature Africa/Asia/Europe
141 W14 **Arabian Sea** sea NW Indian Ocean
Arabicus, Sinus see Red Sea
'Arabī, Khalīj al see Persian Gulf
Arabistan see Khūzestān
'Arabīyah as Su'ūdīyah, Al Mamlakah al see Saudi Arabia
'Arabīyah Jumhūrīyah, Mişr al see Egypt
138 I9 **'Arab, Jabal al** ▲▲ S Syria
Arab Republic of Egypt see Egypt
139 Y12 **'Arab, Shaṭṭ al** Eng. Shatt al Arab, Per. Arvand Rūd. ≈ Iran/Iraq
136 I11 **Araç** Kastamonu, N Turkey 41°14′N 33°20′E
59 P16 **Aracaju** state capital Sergipe, E Brazil 10°45′S 37°07′W
54 F5 **Aracataca** Magdalena, N Colombia 10°38′N 74°09′W
58 P13 **Aracati** Ceará, E Brazil 04°32′S 37°45′W
60 J8 **Araçatuba** São Paulo, S Brazil 21°12′S 50°24′W
136 I11 **Araç Çayı** ≈ N Turkey
104 J13 **Aracena** Andalucía, S Spain 37°54′N 06°33′W
115 F20 **Arachnaío** ▲▲ S Greece
115 D16 **Árachthos** var. Arta, prev. Árakhthos; anc. Arachthus. ≈ W Greece
Arachthus see Árachthos
59 N19 **Araçuai** Minas Gerais, SE Brazil 16°52′S 42°03′W
138 F11 **Arad** Southern, S Israel 31°16′N 35°09′E
116 F11 **Arad** Arad, W Romania 46°12′N 21°20′E
116 F11 **Arad** ◈ county W Romania
78 J9 **Arada** Wadi Fira, NE Chad 15°00′N 20°38′E
143 P18 **'Arādah** Abū Ẓaby, S United Arab Emirates 22°57′N 53°24′E
Aradhippou see Aradíppou
121 Q3 **Aradíppou** var. Aradhippou. SE Cyprus 34°57′N 33°37′E
174 K6 **Arafura Sea** Ind. Laut Arafuru. sea W Pacific Ocean
174 L6 **Arafura Shelf** undersea feature C Arafura Sea
Arafuru, Laut see Arafura Sea
59 J18 **Aragarças** Goiás, C Brazil 15°55′S 52°12′W
Aragats, Gora see Aragats Lerr
137 T12 **Aragats Lerr** Rus. Gora Aragats. ▲ W Armenia 40°31′N 44°06′E
32 E14 **Arago, Cape** headland Oregon, NW USA 43°17′N 124°25′W
105 R6 **Aragón** autonomous community E Spain
105 Q4 **Aragón** ≈ NE Spain
107 I24 **Aragona** Sicilia, Italy, C Mediterranean Sea 37°25′N 13°37′E
105 Q7 **Aragoncillo** ▲ C Spain 40°59′N 02°01′W
54 L5 **Aragua** off. Estado Aragua. ◈ state N Venezuela
55 N6 **Aragua de Barcelona** Anzoátegui, NE Venezuela 09°30′N 64°51′W
55 O5 **Aragua de Maturín** Monagas, NE Venezuela 09°58′N 63°30′W
Aragua, Estado see Aragua
59 K15 **Araguaia, Río** var. Araguaya. ≈ C Brazil
59 K19 **Araguari** Minas Gerais, SE Brazil 18°38′S 48°13′W
58 J11 **Araguari, Rio** ≈ SW Brazil
Araguaya see Araguaia, Río
104 K14 **Arahal** Andalucía, S Spain 37°15′N 05°33′W
165 N11 **Arai** Niigata, Honshū, C Japan 37°02′N 138°17′E
Árainn see Inishmore
Árainn Mhór see Aran Island
Ara Jovis see Aranjuez
74 J11 **Arak** C Algeria 25°17′N 03°45′E
171 Y15 **Arak** Papua, E Indonesia 07°14′S 139°40′E
142 M7 **Arāk** prev. Sultānābād. Markazī, W Iran 34°07′N 49°39′E
188 D10 **Arakabesan** island Palau Islands, N Palau
55 S7 **Arakaka** NW Guyana 07°37′N 59°58′W
Arakan State see Rakhine State
166 K5 **Arakan Yoma** ▲▲ W Myanmar (Burma)
165 O10 **Arakawa** Niigata, Honshū, C Japan 38°06′N 139°25′E
Árakhthos see Árachthos
Araks/Arak's see Aras
158 H7 **Aral** Xinjiang Uygur Zizhiqu, NW China 40°40′N 81°19′E
Aral see Aralsk, Kazakhstan
Aral see Vose', Tajikistan
Aral-Bukhorskiy Kanal see Amu-Buxoro Kanali
137 T12 **Aralik** İgdır, E Turkey 39°54′N 44°28′E
146 H5 **Aral Sea** Kaz. Aral Tengizi, Rus. Aral'skoye More, Uzb. Orol Dengizi. inland sea Kazakhstan/Uzbekistan
144 L13 **Aral'sk** Kaz. Aral. Kzylorda, SW Kazakhstan 46°48′N 61°40′E
Aral'skoye More/Aral Tengizi see Aral Sea
41 O10 **Aramberri** Nuevo León, NE Mexico 24°05′N 99°52′W
186 B8 **Aramia** ≈ SW Papua New Guinea
105 N5 **Aranda de Duero** Castilla y León, N Spain 41°40′N 03°41′W
112 M12 **Aranđelovac** prev. Arandjelovac. Serbia, C Serbia 44°18′N 20°32′E
Arandjelovac see Aranđelovac
97 J19 **Aran Fawddwy** ▲ NW Wales, United Kingdom 52°48′N 03°42′W
97 C14 **Aran Island** Ir. Árainn Mhór. island NW Ireland
97 A18 **Aran Islands** island group W Ireland
105 N9 **Aranjuez** anc. Ara Jovis. Madrid, C Spain 40°02′N 03°37′W
83 E20 **Aranos** Hardap, SE Namibia 24°10′S 19°08′E
25 U14 **Aransas Bay** inlet Texas, SW USA
25 T14 **Aransas Pass** Texas, SW USA 27°54′N 97°09′W
191 O3 **Aranuka** prev. Nanouki. atoll Tungaru, W Kiribati
143 N6 **Ārān-va-Bidgol** var. Golārā. Eşfahān, C Iran 34°03′N 51°30′E
167 Q11 **Aranyaprathet** Prachin Buri, S Thailand 13°42′N 102°32′E
Aranyosasztal see Zlatý Stôl
Aranyosgyéres see Câmpia Turzii
Aranyosmarót see Zlaté Moravce
164 C14 **Arao** Kumamoto, Kyūshū, SW Japan 32°58′N 130°26′E
77 O8 **Araouane** Tombouctou, N Mali 18°53′N 03°31′W
26 L10 **Arapaho** Oklahoma, C USA 35°34′N 98°57′W
29 N16 **Arapahoe** Nebraska, C USA 40°18′N 99°54′W
57 I16 **Arapa, Laguna** ◎ SE Peru
185 K14 **Arapawa Island** island C New Zealand
61 E17 **Arapey Grande, Río** ≈ N Uruguay
59 P16 **Arapiraca** Alagoas, E Brazil 09°45′S 36°40′W
140 M3 **'Ar'ar** Al Ḥudūd ash Shamālīyah, NW Saudi Arabia 31°N 41°E
54 G15 **Araracuara** Caquetá, S Colombia 0°36′S 72°24′W
61 K15 **Araranguá** Santa Catarina, S Brazil 28°56′S 49°30′W
60 L8 **Araraquara** São Paulo, S Brazil 21°46′S 48°08′W
59 O13 **Araras** Ceará, E Brazil 04°08′S 40°30′W
58 I14 **Araras** Pará, N Brazil 06°04′S 54°34′W
60 L9 **Araras** São Paulo, S Brazil 22°21′S 47°21′W
60 H11 **Araras, Serra das** ▲▲ S Brazil
137 U12 **Ararat** S Armenia 39°49′N 44°45′E
182 M12 **Ararat** Victoria, SE Australia 37°20′S 143°00′E
Ararat, Mount see Büyükağrı Dağı
140 M3 **'Ar'ar, Wādī** dry watercourse Iraq/Saudi Arabia
129 N7 **Aras** Arm. Arak's, Az. Araz Nehri, Per. Rūd-e Aras, Rus. Araks; prev. Araxes. ≈ SW Asia
Aras de Alpuente see Aras de los Olmos
105 R9 **Aras de los Olmos** prev. Aras de Alpuente. Valenciana, E Spain 39°55′N 01°09′W
137 S13 **Aras Güneyi Dağları** ▲▲ NE Turkey
Aras, Rūd-e see Aras
191 U9 **Aratika** atoll Îles Tuamotu, C French Polynesia
Aratürük see Yiwu
54 I8 **Arauca** Arauca, NE Colombia 07°03′N 70°47′W
54 I8 **Arauca** off. Intendencia de Arauca. ◈ province NE Colombia
Arauca, Intendencia de see Arauca
63 G15 **Araucanía** off. Región de la Araucanía. ◈ region C Chile
Araucanía, Región de la see Araucanía
54 L7 **Arauca, Río** ≈ Colombia/Venezuela
63 F14 **Arauco** Bío Bío, C Chile 37°15′S 73°22′W
63 F14 **Arauco, Golfo de** gulf S Chile
54 H8 **Arauquita** Arauca, C Colombia 06°57′N 71°19′W
Arausio see Orange
152 F13 **Arāvali Range** ▲▲ N India
186 J7 **Arawa** Bougainville, NE Papua New Guinea 06°15′S 155°35′E
185 C20 **Arawata** ≈ South Island, New Zealand
186 F7 **Arawe Islands** island group E Papua New Guinea
59 L20 **Araxá** Minas Gerais, SE Brazil 19°37′S 46°50′W
Araxes see Aras
55 O5 **Araya** Sucre, N Venezuela 10°34′N 64°15′W
Araz Nehri see Aras
105 R4 **Arba** ≈ N Spain
81 I15 **Ārba Minch'** Southern Nationalities, S Ethiopia 06°02′N 37°34′E
Arbat see Erbet
'Arbat see Erbet
107 D19 **Arbatax** Sardegna, Italy, C Mediterranean Sea 39°57′N 09°42′E
Arbe see Rab
Arbela see Arbīl
139 S3 **Arbīl** Kurd. Hewlêr, var. Erbil, Irbīl, Hawlēr; anc. Arbela. Arbīl, N Iraq 36°12′N 44°01′E
139 S3 **Arbīl** off. Muḥāfa at Arbīl, var. Irbil, off. Kurd. Parêzga-î Hewlêr, Kurd. Hewlêr. ◈ governorate N Iraq
Arbīl, Muḥāfa at see Arbīl
95 M16 **Arboga** Västmanland, C Sweden 59°24′N 15°50′E
95 M16 **Arbogaån** ≈ C Sweden
103 S9 **Arbois** Jura, E France 46°54′N 05°45′E
54 D6 **Arboletes** Antioquia, NW Colombia 08°52′N 76°25′W
11 X15 **Arborg** Manitoba, S Canada 50°52′N 97°20′W
94 N12 **Arbrå** Gävleborg, C Sweden 61°27′N 16°21′E
96 K10 **Arbroath** anc. Aberbrothock. E Scotland, United Kingdom 56°34′N 02°35′W
35 N6 **Arbuckle** California, W USA 39°00′N 122°05′W
27 N12 **Arbuckle Mountains** ▲▲ Oklahoma, C USA
162 I5 **Arbulag** var. Mandal. Hövsgöl, N Mongolia 49°55′N 99°21′E
Arbuzinka see Arbuzynka
117 Q8 **Arbuzynka** Rus. Arbuzinka. Mykolayivs'ka Oblast', S Ukraine 47°54′N 31°19′E
103 U12 **Arc** ≈ E France
102 J13 **Arcachon** Gironde, SW France 44°40′N 01°11′W
102 J13 **Arcachon, Bassin d'** inlet SW France
18 E10 **Arcade** New York, NE USA 42°32′N 78°19′W
23 W14 **Arcadia** Florida, SE USA 27°13′N 81°51′W
22 H5 **Arcadia** Louisiana, S USA 32°33′N 92°55′W
30 J7 **Arcadia** Wisconsin, N USA 44°15′N 91°30′W
Arcae Remorum see Châlons-en-Champagne
44 L9 **Arcahaie** C Haiti 18°46′N 72°32′W
34 K3 **Arcata** California, W USA 40°51′N 124°06′W
35 U6 **Arc Dome** ▲ Nevada, W USA 38°52′N 117°20′W
107 J16 **Arce** Lazio, C Italy 41°35′N 13°34′E
41 O15 **Arcelia** Guerrero, S Mexico 18°20′N 100°16′W
99 M15 **Arcen** Limburg, SE Netherlands 51°28′N 06°10′E
115 J25 **Archánes** var. Áno Arkhánai, Epáno Archánes; prev. Epáno Arkhánai. Kríti, Greece, E Mediterranean Sea 35°12′N 25°10′E
Archangel see Arkhangel'sk
Archangel Bay see Chëshskaya Guba
115 O23 **Archángelos** var. Arhangelos, Arkhángelos. Ródos, Dodekánisa, Greece, Aegean Sea 36°13′N 28°07′E
114 F7 **Archar** ≈ NW Bulgaria
31 R11 **Archbold** Ohio, N USA 41°31′N 84°18′W
105 R12 **Archena** Murcia, SE Spain 38°07′N 01°17′W
25 R5 **Archer City** Texas, SW USA 33°36′N 98°37′W
104 M14 **Archidona** Andalucía, S Spain 37°06′N 04°23′W
65 B25 **Arch Islands** island group SW Falkland Islands
106 G13 **Arcidosso** Toscana, C Italy 42°52′N 11°30′E
103 Q5 **Arcis-sur-Aube** Aube, N France 48°32′N 04°09′E
182 F3 **Arckaringa Creek** seasonal river South Australia
106 G7 **Arco** Trentino-Alto Adige, N Italy 45°53′N 10°52′E
33 Q14 **Arco** Idaho, NW USA 43°38′N 113°18′W
30 M14 **Arcola** Illinois, N USA 39°39′N 88°19′W
105 P6 **Arcos de Jalón** Castilla y León, N Spain 41°12′N 02°13′W
104 K15 **Arcos de la Frontera** Andalucía, S Spain 36°45′N 05°49′W
104 G5 **Arcos de Valdevez** Viana do Castelo, N Portugal 41°51′N 08°25′W
59 P15 **Arcoverde** Pernambuco, E Brazil 08°23′S 37°W
102 H5 **Arcovest, Pointe de l'** headland NW France 48°49′N 02°58′W
Arctic Mid Oceanic Ridge see Gakkel Ridge
197 R8 **Arctic Ocean** ocean
8 G7 **Arctic Red River** ≈ Northwest Territories/Yukon, NW Canada
Arctic Red River see Tsiigehtchic
39 S6 **Arctic Village** Alaska, USA 68°07′N 145°32′W
194 H1 **Arctowski** Polish research station South Shetland Islands, Antarctica 61°57′S 58°23′W
114 I12 **Arda** var. Ardhas, Gk. Ardas. ≈ Bulgaria/Greece see also Ardas
142 L2 **Ardabīl** var. Ardebil. Ardabīl, NW Iran 38°15′N 48°18′E
142 L2 **Ardabīl** off. Ostān-e Ardabīl. ◈ province NW Iran
Ardabīl, Ostān-e see Ardabīl
137 R11 **Ardahan** Ardahan, NE Turkey 41°08′N 42°41′E
137 S11 **Ardahan** ◈ province Ardahan, Turkey
143 P8 **Ardakān** Yazd, C Iran 32°20′N 53°59′E
94 E12 **Årdalstangen** Sogn Og Fjordane, S Norway 61°14′N 07°45′E
137 R11 **Ardanuç** Artvin, NE Turkey 41°07′N 42°04′E
114 L12 **Ardas** var. Ardhas, Bul. Arda. ≈ Bulgaria/Greece see also Arda
138 I13 **Arḑ aş Şawwān** var. Ardh es Suwwān. plain S Jordan
127 P5 **Ardatov** Respublika Mordoviya, W Russian Federation 54°49′N 46°13′E
14 G12 **Ardbeg** Ontario, S Canada 45°38′N 80°05′W
Ardeal see Transylvania
Ardebil see Ardabīl
103 Q13 **Ardèche** ◈ department E France
103 Q13 **Ardèche** ≈ C France
97 F17 **Ardee** Ir. Baile Átha Fhirdhia. Louth, NE Ireland 53°52′N 06°33′W
103 Q3 **Ardennes** ◈ department NE France
99 J23 **Ardennes** physical region Belgium/France
137 Q11 **Ardeşen** Rize, NE Turkey 41°14′N 41°00′E
143 O7 **Ardestān** var. Ardistan. Eşfahān, C Iran 33°29′N 52°17′E
108 J9 **Ardez** Graubünden, SE Switzerland 46°47′N 10°09′E
Ardhas see Arda/Ardas
Ardh es Suwwān see Arḑ aş Şawwān
Ardila see Ardila, Ribeira de
104 I12 **Ardila, Ribeira de** Sp. Ardilla. ≈ Portugal/Spain see also Ardilla
Ardila, Ribeira de see Ardilla
11 T17 **Ardill** Saskatchewan, S Canada 49°56′N 105°49′W
104 I12 **Ardilla** Port. Ribeira de Ardila. ≈ Portugal/Spain see also Ardila, Ribeira de
40 M11 **Ardilla, Cerro la** ▲ C Mexico 22°15′N 102°33′W
114 J12 **Ardino** Kardzhali, S Bulgaria 41°38′N 25°22′E
Ardistan see Ardestān
183 P9 **Ardlethan** New South Wales, SE Australia 34°24′S 146°52′E
Ard Mhacha see Armagh
23 P1 **Ardmore** Alabama, S USA 34°59′N 86°51′W
27 N13 **Ardmore** Oklahoma, C USA 34°11′N 97°08′W
20 J10 **Ardmore** Tennessee, S USA 35°00′N 86°48′W
96 G10 **Ardnamurchan, Point of** headland N Scotland, United Kingdom 56°42′N 06°15′W
Árdni see Arnøya
99 C17 **Ardooie** West-Vlaanderen, W Belgium 50°59′N 03°10′E
182 I9 **Ardrossan** South Australia 34°27′S 137°54′E
116 H9 **Ardusat** Hung. Erdőszáda. Maramureş, N Romania 47°36′N 23°25′E
93 F16 **Åre** Jämtland, C Sweden 63°25′N 13°04′E
79 P16 **Arebi** Orientale, NE Dem. Rep. Congo 02°47′N 29°35′E
45 T5 **Arecibo** C Puerto Rico 18°29′N 66°44′W′N
171 V13 **Aredo** Papua Barat, E Indonesia 02°27′S 133°59′E
59 P14 **Areia Branca** Rio Grande do Norte, E Brazil 04°53′S 37°03′W
119 O14 **Arekhawsk** Rus. Orekhovsk. Vitsyebskaya Voblasts', N Belarus 54°42′N 30°30′E
Arel see Arlon
Arelas/Arelate see Arles
Arenal, Embalse de see Arenal Laguna
42 L12 **Arenal Laguna** var. Embalse de Arenal. ◎ NW Costa Rica
42 L13 **Arenal, Volcán** ⛰ NW Costa Rica 10°21′N 84°42′W
34 K6 **Arena, Point** headland California, W USA 38°57′N 123°44′W
59 H17 **Arenápolis** Mato Grosso, W Brazil 14°25′S 56°52′W
40 G10 **Arena, Punta** headland NW Mexico 23°28′N 109°24′W
104 L8 **Arenas de San Pedro** Castilla y León, N Spain 40°12′N 05°05′W
63 I24 **Arenas, Punta de** headland S Argentina 53°10′S 68°15′W
61 B20 **Arenaza** Buenos Aires, E Argentina 34°55′S 61°45′W
95 F17 **Arendal** Aust-Agder, S Norway 58°27′N 08°45′E
99 J16 **Arendonk** Antwerpen, N Belgium 51°18′N 05°06′E
43 T15 **Arenosa** Panamá, N Panama 09°02′N 79°57′W
Arensburg see Kuressaare
105 W5 **Arenys de Mar** Cataluña, NE Spain 41°35′N 02°33′E
106 C9 **Arenzano** Liguria, NW Italy 44°25′N 08°43′E
115 F22 **Areópoli** prev. Areópolis. Pelopónnisos, S Greece 36°40′N 22°24′E
Areópolis see Areópoli
57 H18 **Arequipa** Arequipa, SE Peru 16°24′S 71°33′W
57 G17 **Arequipa** off. Departamento de Arequipa. ◈ department Arequia SW Peru
Arequipa, Departamento de see Arequipa
61 B19 **Arequito** Santa Fe, C Argentina 33°09′S 61°28′W
104 M7 **Arévalo** Castilla y León, N Spain 41°04′N 04°44′W
106 H12 **Arezzo** anc. Arretium. Toscana, C Italy 43°28′N 11°50′E
105 Q4 **Arga** ≈ N Spain
Argaeus see Erciyes Dağı
115 G17 **Argalastí** Thessalía, C Greece 39°13′N 23°13′E
105 O10 **Argamasilla de Alba** Castilla-La Mancha, C Spain 39°08′N 03°05′W
158 L8 **Argan** Xinjiang Uygur Zizhiqu, NW China 40°09′N 88°16′E
105 O8 **Arganda** Madrid, C Spain 40°19′N 03°26′W
104 H8 **Arganil** Coimbra, N Portugal 40°13′N 08°03′W
171 P6 **Argao** Cebu, C Philippines 09°52′N 123°33′E
153 V15 **Argartala** Tripura, NE India
123 N9 **Arga-Sala** ≈ Respublika Sakha (Yakutiya),NE Russian Federation
103 P17 **Argelès-sur-Mer** Pyrénées-Orientales, S France 42°33′N 03°01′E
103 T15 **Argens** ≈ SE France
106 H9 **Argenta** Emilia-Romagna, N Italy 44°37′N 11°49′E
102 K5 **Argentan** Orne, N France 48°45′N 00°01′W
103 N12 **Argentat** Corrèze, C France 45°06′N 01°57′E
106 A9 **Argentera** Piemonte, NE Italy 44°25′N 06°57′E
103 N5 **Argenteuil** Val-d'Oise, N France 48°57′N 02°13′E
62 K13 **Argentina** off. Argentine Republic. ◆ republic S South America
Argentina Basin see Argentine Basin
Argentine Abyssal Plain see Argentine Plain
65 I19 **Argentine Basin** var. Argentina Basin. undersea feature SW Atlantic Ocean 45°00′S 45°00′W
65 I20 **Argentine Plain** var. Argentine Abyssal Plain. undersea feature SW Atlantic Ocean 47°31′S 50°00′W
Argentine Republic see Argentina
Argentine Rise see Falkland Plateau
63 H22 **Argentino, Lago** ◎ S Argentina
102 K8 **Argenton-Château** Deux-Sèvres, W France 46°59′N 00°22′W
102 M9 **Argenton-sur-Creuse** Indre, C France 46°34′N 01°32′E
Argentoratum see Strasbourg
116 I12 **Argeş** ◈ county S Romania
116 K14 **Argeş** ≈ S Romania
149 O8 **Arghandāb, Daryā-ye** ≈ SE Afghanistan
Arghastān see Arghistān
Arghestān see Arghistān
149 O8 **Arghistān** Pash. Arghastān; prev. Arghestān. ≈ SE Afghanistan
Argirocastro see Gjirokastër
80 E7 **Argo** Northern, N Sudan 19°31′N 30°25′E
173 P7 **Argo Fracture Zone** tectonic feature C Indian Ocean
115 F20 **Argolikós Kólpos** gulf S Greece
103 R4 **Argonne** physical region NE France
115 F20 **Árgos** Pelopónnisos, S Greece 37°38′N 22°43′E
Argōsh see Argush
115 D14 **Árgos Orestikó** Dytikí Makedonía, N Greece 40°27′N 21°15′E
115 B19 **Argostóli** var. Argostólion. Kefallonía, Iónia Nisiá, Greece, C Mediterranean Sea 38°13′N 20°29′E
Argostólion see Argostóli
Argovie see Aargau
35 O14 **Arguello, Point** headland California, W USA 34°34′N 120°39′W
127 P16 **Argun** Chechenskaya Respublika, SW Russian Federation 43°16′N 45°53′E
157 T2 **Argun** Chin. Ergun He, Rus. Argun'. ≈ China/Russian Federation
77 T12 **Argungu** Kebbi, NW Nigeria 12°45′N 04°24′E
139 S1 **Argush** Ar. Argūsh, var. Argōsh. Dahūk, N Iraq 37°07′N 44°13′E
Argūsh see Argush
Arguut see Guchin-Us

◆ Country
● Country Capital
◇ Dependent Territory
○ Dependent Territory Capital
◈ Administrative Regions
✈ International Airport
▲ Mountain
▲▲ Mountain Range
⛰ Volcano
≈ River
◎ Lake
⊠ Reservoir

ARGYLE, LAKE – AS SAMĀWAH

181 N3 **Argyle, Lake** *salt lake* Western Australia
96 G12 **Argyll** *cultural region* W Scotland, United Kingdom
Argyrokastron *see* Gjirokastër
162 I7 **Arhangay** ◈ *province* C Mongolia
Arhangelos *see* Archángelos
95 G22 **Århus** *var.* Aarhus. Midtjylland, C Denmark *56°09′N 10°11′E*
139 T1 **Arî** *Ar.* Ārī. Arbīl, E Iraq *37°07′N 44°34′E*
Ārī *see* Arî
Aria *see* Herāt
83 F22 **Ariamsvlei** Karas, SE Namibia *28°08′S 19°50′E*
107 L17 **Ariano Irpino** Campania, S Italy *41°08′N 15°00′E*
54 F11 **Ariari, Río** ≈ C Colombia
151 K19 **Ari Atoll** *var.* Alifu Atoll. *atoll* C Maldives
77 P11 **Aribinda** N Burkina Faso *14°12′N 00°50′W*
62 G2 **Arica** *hist.* San Marcos de Arica. Arica y Parinacota, N Chile *18°31′S 70°18′W*
54 H16 **Arica** Amazonas, S Colombia *02°09′S 71°48′W*
62 G2 **Arica** ✈ Arica y Parinacota, N Chile *18°30′S 70°20′W*
62 H2 **Arica y Parinacota** ◈ *region* N Chile
114 E13 **Aridaía** *var.* Aridea, Aridhaía. Dytikí Makedonía, N Greece *40°59′N 22°04′E*
Aridea *see* Aridaía
172 I15 **Aride, Île** *island* Inner Islands, NE Seychelles
Aridhaía *see* Aridaía
103 N17 **Ariège** ◈ *department* S France
102 M16 **Ariège** *var.* la Riege. ≈ Andorra/France
116 H11 **Arieș** ≈ W Romania
149 U10 **Ārifwāla** Punjab, E Pakistan *30°15′N 73°08′E*
Ariguaní *see* El Difícil
138 G11 **Arīhā** Al Karak, W Jordan *31°25′N 35°47′E*
138 I3 **Arīḩā** *var.* Arīhā. Idlib, W Syria *35°50′N 36°36′E*
Arīhā *see* Arīḩā
Arīḩā *see* Jericho
37 W4 **Arikaree River** ≈ Colorado/Nebraska, C USA
112 L13 **Arilje** Serbia, W Serbia *43°45′N 20°06′E*
45 U14 **Arima** Trinidad, Trinidad and Tobago *10°38′N 61°17′W*
Arime *see* Al ʿArīmah
Ariminum *see* Rimini
59 H16 **Arinos, Rio** ≈ W Brazil
40 M14 **Ario de Rosales** *var.* Ario de Rosáles. Michoacán, SW Mexico *19°12′N 101°42′W*
Ario de Rosáles *see* Ario de Rosales
118 F12 **Ariogala** Kaunas, C Lithuania *55°16′N 23°30′E*
47 T7 **Aripuanã** ≈ W Brazil
59 E15 **Ariquemes** Rondônia, W Brazil *09°55′S 63°06′W*
121 W13 **ʿArīsh, Wādi el** ≈ NE Egypt
54 K6 **Arismendi** Barinas, C Venezuela *08°29′N 68°22′W*
10 J14 **Aristazabal Island** *island* SW Canada
60 F13 **Aristóbulo del Valle** Misiones, NE Argentina *27°09′S 54°54′W*
172 I5 **Arivonimamo** ✈ (Antananarivo) Antananarivo, C Madagascar *19°00′S 47°11′E*
Arixang *see* Wenquan
105 Q6 **Ariza** Aragón, NE Spain *41°19′N 02°03′W*
62 I6 **Arizaro, Salar de** *salt lake* NW Argentina
105 O2 **Arizgoiti** *var.* Basauri. País Vasco, N Spain *43°13′N 02°54′W*
62 K13 **Arizona** San Luis, C Argentina *35°44′S 65°16′W*
36 J12 **Arizona** *off.* State of Arizona, *also known as* Copper State, Grand Canyon State. ◈ *state* SW USA
40 G4 **Arizpe** Sonora, NW Mexico *30°20′N 110°11′W*
95 J16 **Årjäng** Värmland, C Sweden *59°24′N 12°09′E*
143 P8 **Arjenān** Yazd, C Iran *32°19′N 53°48′E*
92 I13 **Arjeplog** *Lapp.* Árjepluovve. Norrbotten, N Sweden *66°04′N 18°E*
Árjepluovve *see* Arjeplog
54 E5 **Arjona** Bolívar, N Colombia *10°14′N 75°22′W*
105 N13 **Arjona** Andalucía, S Spain *37°56′N 04°04′W*
123 S10 **Arka** Khabarovskiy Kray, E Russian Federation *60°04′N 142°17′E*
22 L2 **Arkabutla Lake** ⊠ Mississippi, S USA
127 O7 **Arkadak** Saratovskaya Oblast', W Russian Federation *51°55′N 43°29′E*
27 T13 **Arkadelphia** Arkansas, C USA *34°07′N 93°06′W*
115 J25 **Arkalochóri**; *prev.* Arkalokhórion. Kríti, Greece, E Mediterranean Sea *35°09′N 25°15′E*
Arkalohori/Arkalokhórion *see* Arkalochóri
145 O10 **Arkalyk** *Kaz.* Arqalyq. Kostanay, N Kazakhstan *50°17′N 66°51′E*
27 U10 **Arkansas** *off.* State of Arkansas, *also known as* The Land of Opportunity. ◈ *state* S USA
27 W14 **Arkansas City** Arkansas, C USA *33°36′N 91°12′W*
27 O7 **Arkansas City** Kansas, C USA *37°03′N 97°02′W*
16 K11 **Arkansas River** ≈ C USA
182 J5 **Arkaroola** South Australia *30°21′S 139°20′E*
Arkhángelos *see* Archángelos
124 L8 **Arkhangel'sk** *Eng.* Archangel. Arkhangel'skaya Oblast', NW Russian Federation *64°32′N 40°40′E*
124 L9 **Arkhangel'skaya Oblast'** ◈ *province* NW Russian Federation
127 O14 **Arkhangel'skoye** Stavropol'skiy Kray, SW Russian Federation *44°37′N 44°03′E*
123 R14 **Arkhara** Amurskaya Oblast', S Russian Federation *49°20′N 130°04′E*
97 G19 **Arklow** *Ir.* An tInbhear Mór. SE Ireland *52°48′N 06°09′W*
115 M20 **Arkoí** *island* Dodekánisa, Greece, Aegean Sea
27 R11 **Arkoma** Oklahoma, C USA *35°19′N 94°27′W*
100 O7 **Arkona, Kap** *headland* NE Germany *54°40′N 13°24′E*
95 N17 **Arkösund** Östergötland, S Sweden *58°28′N 16°55′E*
122 J6 **Arkticheskogo Instituta, Ostrova** *island* N Russian Federation
95 O15 **Arlanda** ✈ (Stockholm) Stockholm, C Sweden *59°40′N 17°58′E*
146 C11 **Arlandag** *Rus.* Gora Arlan. ▲ W Turkmenistan *39°39′N 54°28′E*
Arlan, Gora *see* Arlandag
105 O5 **Arlanza** ≈ N Spain
105 N5 **Arlanzón** ≈ N Spain
103 R15 **Arles** *var.* Arles-sur-Rhône; *anc.* Arelas, Arelate. Bouches-du-Rhône, SE France *43°41′N 04°38′E*
Arles-sur-Rhône *see* Arles
103 O17 **Arles-sur-Tech** Pyrénées-Orientales, S France *42°27′N 02°37′E*
29 U9 **Arlington** Minnesota, N USA *44°36′N 94°04′W*
29 R15 **Arlington** Nebraska, C USA *41°27′N 96°21′W*
32 J11 **Arlington** Oregon, NW USA *45°43′N 120°10′W*
29 R10 **Arlington** South Dakota, N USA *44°21′N 97°07′W*
20 E10 **Arlington** Tennessee, S USA *35°17′N 89°40′W*
25 T6 **Arlington** Texas, SW USA *32°44′N 97°05′W*
21 W4 **Arlington** Virginia, NE USA *38°54′N 77°09′W*
32 H7 **Arlington** Washington, NW USA *48°12′N 122°07′W*
30 M10 **Arlington Heights** Illinois, N USA *42°08′N 88°03′W*
77 U8 **Arlit** Agadez, C Niger *18°54′N 07°25′E*
99 L24 **Arlon** *Dut.* Aarlen, *Ger.* Arel, *Lat.* Orolaunum. Luxembourg, SE Belgium *49°41′N 05°49′E*
27 R7 **Arma** Kansas, C USA *37°32′N 94°42′W*
97 F16 **Armagh** *Ir.* Ard Mhacha. S Northern Ireland, United Kingdom *54°15′N 06°33′W*
97 F16 **Armagh** *cultural region* S Northern Ireland, United Kingdom
102 K15 **Armagnac** *cultural region* S France
103 Q7 **Armançon** ≈ C France
60 K10 **Armando Laydner, Represa** ⊠ S Brazil
115 M24 **Armathía** *island* SE Greece
137 T12 **Armavir** *prev.* Hoktemberyan, *Rus.* Oktemberyan. SW Armenia *40°09′N 43°58′E*
126 M14 **Armavir** Krasnodarskiy Kray, SW Russian Federation *44°59′N 41°07′E*
54 E10 **Armenia** Quindío, W Colombia *04°32′N 75°40′W*
137 T12 **Armenia** *off.* Republic of Armenia, *var.* Ajastan, *Arm.* Hayastani Hanrapetut'yun; *prev.* Armenian Soviet Socialist Republic. ◆ *republic* SW Asia
Armenian Soviet Socialist Republic *see* Armenia
Armenia, Republic of *see* Armenia
Armenierstadt *see* Gherla
103 O1 **Armentières** Nord, N France *50°41′N 02°53′E*
40 K14 **Armería** Colima, SW Mexico *18°55′N 103°59′W*
183 T5 **Armidale** New South Wales, SE Australia *30°32′S 151°40′E*
29 P11 **Armour** South Dakota, N USA *43°19′N 98°21′W*
61 B18 **Armstrong** Santa Fe, C Argentina *32°46′S 61°39′W*
11 N16 **Armstrong** British Columbia, SW Canada *50°27′N 119°14′W*
12 D11 **Armstrong** Ontario, S Canada *50°20′N 89°02′W*
29 U11 **Armstrong** Iowa, C USA *43°24′N 94°28′W*
25 S16 **Armstrong** Texas, SW USA *26°55′N 97°47′W*
117 S11 **Armyans'k** *Rus.* Armyansk. Avtonomna Respublika Krym, S Ukraine *46°05′N 33°43′E*
115 H14 **Arnaía** *Cont.* Arnea. Kentrikí Makedonía, N Greece *40°30′N 23°36′E*
121 N2 **Arnaoúti, Akrotíri** *var.* Arnaoútis, Cape Arnaouti. *headland* W Cyprus *35°06′N 32°16′E*
Arnaouti, Cape/Arnaoútis *see* Arnaoúti, Akrotíri
12 L4 **Arnaud** ≈ Québec, E Canada
103 Q8 **Arnay-le-Duc** Côte d'Or, C France *47°08′N 04°27′E*
Arnea *see* Arnaía
105 Q4 **Arnedo** La Rioja, N Spain *42°14′N 02°05′W*
95 I14 **Årnes** Akershus, S Norway *60°07′N 11°28′E*
Årnes *see* Åi Åfjord
26 K9 **Arnett** Oklahoma, C USA *36°08′N 99°46′W*
98 L12 **Arnhem** Gelderland, SE Netherlands *51°59′N 05°54′E*
181 Q2 **Arnhem Land** *physical region* Northern Territory, N Australia
106 F11 **Arno** ≈ C Italy
Arņo *see* Arno Atoll
189 W7 **Arno Atoll** *var.* Arņo. *atoll* Ratak Chain, NE Marshall Islands
182 H8 **Arno Bay** South Australia *33°55′S 136°31′E*
35 Q8 **Arnold** California, W USA *38°15′N 120°19′W*
27 X5 **Arnold** Missouri, C USA *38°25′N 90°22′W*
29 N15 **Arnold** Nebraska, C USA *41°25′N 100°11′W*
109 R10 **Arnoldstein** *Slvn.* Pod Klošter. Kärnten, S Austria *46°34′N 13°43′E*
103 N9 **Arnon** ≈ C France
45 P14 **Arnos Vale** ✈ (Kingstown) Saint Vincent, SE Saint Vincent and the Grenadines *13°08′N 61°13′W*
92 I8 **Arnøya** *Lapp.* Árdni. *island* N Norway
14 L12 **Arnprior** Ontario, SE Canada *45°31′N 76°11′W*
101 G15 **Arnsberg** Nordrhein-Westfalen, W Germany *51°24′N 08°04′E*
101 K16 **Arnstadt** Thüringen, C Germany *50°50′N 10°57′E*
Arnswalde *see* Choszczno
54 K5 **Aroa** Yaracuy, N Venezuela *10°26′N 68°54′W*
83 E21 **Aroab** Karas, SE Namibia *26°47′S 19°40′E*
Ároania *see* Chelmós
191 O6 **Aroa, Pointe** *headland* Moorea, W French Polynesia *17°27′S 149°45′W*
Aroe Islands *see* Aru, Kepulauan
101 H15 **Arolsen** Niedersachsen, C Germany *51°23′N 09°00′E*
106 C7 **Arona** Piemonte, NE Italy *45°45′N 08°33′E*
19 R3 **Aroostook River** ≈ Canada/USA
Arop Island *see* Long Island
38 M12 **Aropuk Lake** ◎ Alaska, USA
191 P4 **Arorae** *atoll* Tungaru, W Kiribati
190 G16 **Arorangi** Rarotonga, S Cook Islands *21°13′S 159°49′W*
108 I9 **Arosa** Graubünden, S Switzerland *46°48′N 09°42′E*
104 F4 **Arousa, Ría de** *estuary* E Atlantic Ocean
184 P8 **Arowhana** ▲ North Island, New Zealand *38°07′S 177°52′E*
137 V12 **Arp'a** *Az.* Arpaçay. ≈ Armenia/Azerbaijan
137 S11 **Arpaçay** Kars, NE Turkey *40°51′N 43°20′E*
Arpaçay *see* Arp'a
Arqalyq *see* Arkalyk
149 N14 **Arra** ≈ SW Pakistan
Arrabona *see* Győr
Arrah *see* Āra
Ar Rahad *see* Er Rahad
139 R9 **Ar Raḩḩālīyah** Al Anbār, C Iraq *32°53′N 43°21′E*
60 Q10 **Arraial do Cabo** Rio de Janeiro, SE Brazil *22°57′S 42°00′W*
104 H11 **Arraiolos** Évora, S Portugal *38°44′N 07°59′W*
139 R8 **Ar Ramādī** *var.* Ramadi, Rumadiya. Al Anbār, SW Iraq *33°27′N 43°19′E*
138 J6 **Ar Rāmī** Ḩimş, C Syria *34°32′N 37°54′E*
Ar Rams *see* Rams
138 H9 **Ar Ramthā** *var.* Ramtha. Irbid, N Jordan *32°34′N 36°00′E*
96 H13 **Arran, Isle of** *island* SW Scotland, United Kingdom
138 L3 **Ar Raqqah** *var.* Rakka; *anc.* Nicephorium. Ar Raqqah, N Syria *35°57′N 39°03′E*
138 L3 **Ar Raqqah** *off.* Muḩāfaẓat al Raqqah, *var.* Raqqah, *Fr.* Rakka. ◈ *governorate* N Syria
103 O2 **Arras** *anc.* Nemetocenna. Pas-de-Calais, N France *50°17′N 02°46′E*
105 P3 **Arrasate** *Cast.* Mondragón. País Vasco, N Spain *43°04′N 02°30′W*
138 G12 **Ar Rashādīyah** Aţ ţafīlah, W Jordan *30°42′N 35°38′E*
138 I5 **Ar Rastān** *var.* Rastâne. Ḩimş, W Syria *34°57′N 36°43′E*
139 X12 **Ar Raţāwī** Al Başrah, E Iraq *30°37′N 47°12′E*
102 L15 **Arrats** ≈ S France
141 N10 **Ar Rawḑah** Makkah, S Saudi Arabia *21°19′N 42°48′E*
141 Q15 **Ar Rawḑah** S Yemen *14°26′N 47°14′E*
142 K11 **Ar Rawḑatayn** *var.* Raudhatain. N Kuwait *29°80′N 47°50′E*
143 N16 **Ar Rayyān** *var.* Al Rayyan. C Qatar *25°18′N 51°29′E*
102 L17 **Arreau** Hautes-Pyrénées, S France *42°55′N 00°21′E*
64 Q11 **Arrecife** *var.* Arrecife de Lanzarote, Puerto Arrecife. Lanzarote, Islas Canarias, NE Atlantic Ocean *28°57′N 13°33′W*
Arrecife de Lanzarote *see* Arrecife
43 P6 **Arrecife Edinburgh** *reef* NE Nicaragua
61 C19 **Arrecifes** Buenos Aires, E Argentina *34°06′S 60°09′W*
102 F6 **Arrée, Monts d'** ▲▲ NW France
Ar Refāʿī *see* Ar Rifāʿī
Arretium *see* Arezzo
Arriaca *see* Guadalajara
109 S9 **Arriach** Kärnten, S Austria *46°43′N 13°52′E*
41 T16 **Arriaga** Chiapas, SE Mexico *16°14′N 93°54′W*
41 N12 **Arriaga** San Luis Potosí, C Mexico *21°55′N 101°23′W*
139 W10 **Ar Rifāʿī** *var.* Ar Refāʿī. Dhī Qār, SE Iraq *31°47′N 46°07′E*
139 V12 **Ar Riḩāb** *salt flat* S Iraq
Arriondas *see* Les Arriondes
141 Q7 **Ar Riyāḑ** *Eng.* Riyadh. ● (Saudi Arabia) Ar Riyāḑ, C Saudi Arabia *24°50′N 46°50′E*
141 O8 **Ar Riyāḑ** *off.* Minţaqat ar Riyāḑ. ◈ *province* C Saudi Arabia
141 S15 **Ar Riyān** S Yemen *14°43′N 49°18′E*
Arrö *see* Ærø
61 H18 **Arroio Grande** Rio Grande do Sul, S Brazil *32°15′S 53°02′W*
102 K15 **Arros** ≈ S France
103 Q9 **Arroux** ≈ C France
25 R5 **Arrowhead, Lake** ⊠ Texas, SW USA
182 L5 **Arrowsmith, Mount** *hill* New South Wales, SE Australia
185 D21 **Arrowtown** Otago, South Island, New Zealand *44°57′S 168°51′E*
61 D17 **Arroyo Barú** Entre Ríos, E Argentina *31°52′S 58°26′W*
104 J10 **Arroyo de la Luz** Extremadura, W Spain *39°28′N 06°36′W*
63 J16 **Arroyo de la Ventana** Río Negro, SE Argentina *41°41′S 66°03′W*
35 P13 **Arroyo Grande** California, W USA *35°07′N 120°35′W*
Ar Ru'ays *see* Ar Ruways
141 R11 **Ar Rubʿ al Khālī** *Eng.* Empty Quarter, Great Sandy Desert. *desert* SW Asia
139 V13 **Ar Ruḑaymah** Al Muthanná, S Iraq *30°20′N 45°26′E*
61 A16 **Arrufó** Santa Fe, C Argentina *30°15′S 61°45′W*
138 I7 **Ar Ruḩaybah** *var.* Ruhaybeh, *Fr.* Rouhaïbé. Rīf Dimashq, W Syria *33°45′N 36°40′E*
139 V15 **Ar Rukhaymīyah** *well* S Iraq
139 U11 **Ar Rumaythah** *var.* Rumaitha. Al Muthanná, S Iraq *31°31′N 45°15′E*
141 X8 **Ar Rustāq** *var.* Rostak, Rustaq. N Oman *23°34′N 57°25′E*
139 N8 **Ar Ruţbah** *var.* Rutba. Al Anbār, SW Iraq *33°03′N 40°16′E*
140 M3 **Ar Rūthīyah** *spring/well* NW Saudi Arabia *31°18′N 41°23′E*
ar-Ruwaida *see* Ar Ruwayḑah
141 O8 **Ar Ruwayḑah** *var.* ar-Ruwaida. Jīzān, C Saudi Arabia *23°48′N 44°44′E*
143 N15 **Ar Ruways** *var.* Al Ruweis, Ar Ru'ays, Ruwais. N Qatar *26°08′N 51°13′E*
143 O17 **Ar Ruways** *var.* Ar Ru'ays. Abū Ẓaby, W United Arab Emirates *24°09′N 52°57′E*
Års *see* Aars
Arsanias *see* Murat Nehri
123 S15 **Arsen'yev** Primorskiy Kray, SE Russian Federation *44°09′N 133°28′E*
155 G19 **Arsikere** Karnātaka, W India *13°20′N 76°13′E′N*
127 R3 **Arsk** Respublika Tatarstan, W Russian Federation *56°07′N 49°54′E*
94 N10 **Årskogen** Gävleborg, C Sweden *62°07′N 17°19′E*
121 O3 **Ársos** C Cyprus *34°51′N 32°46′E*
94 N13 **Årsunda** Gävleborg, C Sweden *60°31′N 16°45′E*
115 C17 **Árta** *anc.* Ambracia. Ípeiros, W Greece *39°08′N 20°59′E*
105 Y9 **Artà** Mallorca, Spain, W Mediterranean Sea *39°42′N 03°20′E*
Arta *see* Árachthos
137 T12 **Artashat** S Armenia *39°57′N 44°34′E*
40 M15 **Arteaga** Michoacán, SW Mexico *18°22′N 102°18′W*
123 S15 **Artem** Primorskiy Kray, SE Russian Federation *43°24′N 132°20′E*
44 C4 **Artemisa** La Habana, W Cuba *22°49′N 82°47′W*
117 W7 **Artemivs'k** Donets'ka Oblast', E Ukraine *48°35′N 37°58′E*
122 K13 **Artemovsk** Krasnoyarskiy Kray, S Russian Federation *54°22′N 93°24′E*
105 U5 **Artesa de Segre** Cataluña, NE Spain *41°54′N 01°03′E*
37 U14 **Artesia** New Mexico, SW USA *32°50′N 104°24′W*
25 Q14 **Artesia Wells** Texas, SW USA *28°13′N 99°18′W*
108 G8 **Arth** Schwyz, C Switzerland *47°05′N 08°39′E*
14 F15 **Arthur** Ontario, S Canada *43°49′N 80°31′W*
30 M14 **Arthur** Illinois, N USA *39°42′N 88°28′W*
28 L14 **Arthur** Nebraska, C USA *41°35′N 101°42′W*
29 Q5 **Arthur** North Dakota, N USA *47°03′N 97°12′W*
185 B21 **Arthur** ≈ South Island, New Zealand
18 B13 **Arthur, Lake** ⊠ Pennsylvania, NE USA
183 N15 **Arthur River** ≈ Tasmania, SE Australia
185 G18 **Arthur's Pass** Canterbury, South Island, New Zealand *42°59′S 171°33′E*
185 G17 **Arthur's Pass** *pass* South Island, New Zealand
44 I3 **Arthur's Town** Cat Island, C The Bahamas *24°34′N 75°39′W*
44 M9 **Artibonite, Rivière de l'** ≈ C Haiti
61 E16 **Artigas** *prev.* San Eugenio, San Eugenio del Cuareim. Artigas, N Uruguay *30°25′S 56°28′W*
61 E16 **Artigas** ◈ *department* N Uruguay
194 H1 **Artigas** *Uruguayan research station* Antarctica *61°57′S 58°23′W*
137 T11 **Art'ik** W Armenia *40°38′N 43°58′E*
187 O16 **Art, Île** *island* Îles Belep, W New Caledonia
103 O2 **Artois** *cultural region* N France
136 L12 **Artova** Tokat, N Turkey *40°04′N 36°17′E*
105 Y9 **Artrutx, Cap d'** *var.* Cabo Dartuch. *cape* Menorca, Spain, W Mediterranean Sea
Artsiz *see* Artsyz
117 N11 **Artsyz** *Rus.* Artsiz. Odes'ka Oblast', SW Ukraine *45°59′N 29°26′E*
158 E7 **Artux** Xinjiang Uygur Zizhiqu, NW China *39°40′N 76°10′E*
137 R11 **Artvin** Artvin, NE Turkey *41°12′N 41°48′E*
137 R11 **Artvın** ◈ *province* NE Turkey
146 G14 **Artyk** Ahal Welaýaty, C Turkmenistan *37°32′N 59°16′E*
79 Q16 **Aru** Orientale, NE Dem. Rep. Congo *02°53′N 30°50′E*
81 E17 **Arua** NW Uganda *03°02′N 30°56′E*
104 I4 **A Rúa de Valdeorras** *var.* La Rúa. Galicia, NW Spain *42°22′N 07°12′W*
Aruângua *see* Luangwa
45 O15 **Aruba** *var.* Oruba. ◇ *Dutch self-governing territory* S West Indies
47 Q4 **Aruba** *island* Aruba, Lesser Antilles
Aru Islands *see* Aru, Kepulauan
171 W15 **Aru, Kepulauan** *Eng.* Aru Islands; *prev.* Aroe Islands. *island group* E Indonesia
153 W10 **Arunāchal Pradesh** *prev.* North East Frontier Agency, North East Frontier Agency of Assam. ◈ *state* NE India
Arun Qi *see* Naji
155 H23 **Aruppukkottai** Tamil Nādu, SE India *09°31′N 78°03′E*
81 I20 **Arusha** Arusha, N Tanzania *03°23′S 36°40′E*
81 I21 **Arusha** ◈ *region* E Tanzania
81 I20 **Arusha** ✈ Arusha, N Tanzania *03°26′S 37°07′E*
54 C9 **Arusí, Punta** *headland* NW Colombia *05°36′N 77°30′W*
155 J23 **Aruvi Aru** ≈ NW Sri Lanka
79 M17 **Aruwimi** *var.* Ituri (upper course). ≈ NE Dem. Rep. Congo
Árva *see* Orava
37 T4 **Arvada** Colorado, C USA *39°48′N 105°06′W*
Arvand Rūd *see* ʿArab, Shaţţ al
162 J8 **Arvayheer** Övörhangay, C Mongolia *46°13′N 102°47′E*
9 O10 **Arviat** *prev.* Eskimo Point. Nunavut, C Canada *61°10′N 94°15′W*
93 I14 **Arvidsjaur** Norrbotten, N Sweden *65°34′N 19°12′E*
95 J15 **Arvika** Värmland, C Sweden *59°41′N 12°38′E*
92 J8 **Årviksand** Troms, N Norway *70°10′N 20°30′E*
35 S13 **Arvin** California, W USA *35°12′N 118°52′W*
163 S8 **Arxan** Nei Mongol Zizhiqu, N China *47.11N 119.58′E*
145 P7 **Arykbalyk** *Kaz.* Aryqbalyq. Severnyy Kazakhstan, N Kazakhstan *53°00′N 68°11′E*
Aryqbalyq *see* Arykbalyk
145 P17 **ʿArys'** *prev.* Arys'. Yuzhnyy Kazakhstan, S Kazakhstan *42°26′N 68°49′E*
Arys' *see* ʿArys'
Arys *see* Orzysz
Arys Köli *see* Arys, Ozero
145 O14 **Arys, Ozero** *Kaz.* Arys Köli. ◎ C Kazakhstan
107 D16 **Arzachena** Sardegna, Italy, C Mediterranean Sea *41°05′N 09°21′E*
127 O4 **Arzamas** Nizhegorodskaya Oblast', W Russian Federation *55°25′N 43°51′E*
141 V13 **Arzāt** S Oman *17°00′N 54°18′E*
104 H3 **Arzúa** Galicia, NW Spain *42°55′N 08°10′W*
111 A16 **Aš** *Ger.* Asch. Karlovarský Kraj, W Czech Republic *50°18′N 12°12′E*
95 H15 **Ås** Akershus, S Norway *59°40′N 10°50′E*
Åsa *see* Asaa
95 H20 **Asaa** *var.* Åsa. Nordjylland, N Denmark *57°07′N 10°24′E*
83 E21 **Asab** Karas, S Namibia *25°29′S 17°59′E*
77 U16 **Asaba** Delta, S Nigeria *06°10′N 06°44′E*
76 J10 **Assaba** *var.* Açâba. ◈ *region* S Mauritania
149 S4 **Asadābād** *var.* Asaḏābād; *prev.* Chaghasarāy. Kunar, E Afghanistan *34°52′N 71°09′E*
Asaḏābād *see* Asadābād
138 K3 **Asad, Buḩayrat al** *Eng.* Lake Assad. ◎ N Syria
63 H20 **Asador, Pampa del** *plain* S Argentina
165 P14 **Asahi** Chiba, Honshū, S Japan *35°43′N 140°38′E*
164 M11 **Asahi** Toyama, Honshū, SW Japan *36°56′N 137°34′E*
165 T13 **Asahi-dake** ▲ Hokkaidō, N Japan *43°42′N 142°50′E*
165 T3 **Asahikawa** Hokkaidō, N Japan *43°46′N 142°23′E*
147 S10 **Asaka** *Rus.* Assake; *prev.* Leninsk. Andijon Viloyati, E Uzbekistan *40°39′N 72°16′E*
77 P17 **Asamankese** SE Ghana *05°47′N 00°41′W*
188 B15 **Asan** W Guam *13°28′N 144°43′E*
188 B15 **Asan Point** *headland* W Guam
153 R15 **Āsānsol** West Bengal, NE India *23°40′N 86°59′E*
80 K12 **Āsayita** Āfar, NE Ethiopia *11°35′S 41°23′E*
171 T12 **Asbakin** Papua Barat, E Indonesia *0°45′S 131°40′E*
Asben *see* Aïr, Massif de l'
15 Q12 **Asbestos** Québec, SE Canada *45°46′N 71°56′W*
29 Y13 **Asbury** Iowa, C USA *42°30′N 90°45′W*
18 K15 **Asbury Park** New Jersey, NE USA *40°13′N 74°00′W*
41 Z12 **Ascención, Bahiá de la** *bay* NW Caribbean Sea
40 I3 **Ascensión** Chihuahua, N Mexico *31°07′N 107°59′W*
Ascension *see* Saint Helena, Ascension and Tristan da Cunha
65 M14 **Ascension Fracture Zone** *tectonic feature* C Atlantic Ocean
Ascension Island ◇ *dependency of St.Helena* C Atlantic Ocean
65 N16 **Ascension Island** *island* C Atlantic Ocean
Asch *see* Aš
109 S3 **Aschach an der Donau** Oberösterreich, N Austria *48°22′N 14°00′E*
101 H18 **Aschaffenburg** Bayern, SW Germany *49°58′N 09°10′E*
101 F14 **Ascheberg** Nordrhein-Westfalen, W Germany *51°46′N 07°36′E*
101 L14 **Aschersleben** Sachsen-Anhalt, C Germany *51°46′N 11°28′E*
106 G12 **Asciano** Toscana, C Italy *43°15′N 11°32′E*
106 J13 **Ascoli Piceno** *anc.* Asculum Picenum. Marche, C Italy *42°51′N 13°34′E*
107 M17 **Ascoli Satriano** *anc.* Asculub, Ausculum Apulum. Puglia, SE Italy *41°13′N 15°32′E*
108 G11 **Ascona** Ticino, S Switzerland *46°10′N 08°45′E*
Asculub *see* Ascoli Satriano
Asculum Picenum *see* Ascoli Piceno
80 L11 **ʿAseb** *var.* Assab, *Amh.* Āseb. SE Eritrea *13°04′N 42°36′E*
95 M20 **Åseda** Kronoberg, S Sweden *57°10′N 15°20′E*
127 T6 **Asekeyevo** Orenburgskaya Oblast', W Russian Federation *53°36′N 52°53′E*
81 J14 **Āsela** *var.* Asella, Aselle, Asselle. Oromīya, C Ethiopia *07°55′N 39°08′E*
93 H15 **Åsele** Västerbotten, N Sweden *64°10′N 17°20′E*
Asella/Aselle *see* Āsela
98 N7 **Assen** Drenthe, NE Netherlands *53°N 06°34′E*
94 K12 **Åsen** Dalarna, C Sweden *61°18′N 13°49′E*
114 J11 **Asenovgrad** *prev.* Stanimaka. Plovdiv, C Bulgaria *42°00′N 24°53′E*
171 O13 **Asera** Sulawesi, C Indonesia *03°24′S 121°42′E*
95 E17 **Åseral** Vest-Agder, S Norway *58°37′N 07°27′E*
118 J3 **Aseri** *var.* Asserien, *Ger.* Asserin. Ida-Virumaa, NE Estonia *59°29′N 26°51′E*
104 G3 **A Serra de Outes** Galicia, NW Spain *42°50′N 08°54′W*
40 J10 **Aserradero** Durango, W Mexico
146 F13 **Aşgabat** *prev.* Ashgabat, Ashkhabad, Poltoratsk. ● (Turkmenistan) Ahal Welaýaty, C Turkmenistan *37°58′N 58°22′E*
146 F13 **Aşgabat** ✈ Ahal Welaýaty, C Turkmenistan *38°06′N 58°10′E*
95 H16 **Åsgårdstrand** Vestfold, S Norway *59°22′N 10°27′E*
Ashara *see* Al ʿAshārah
23 T6 **Ashburn** Georgia, SE USA *31°42′N 83°39′W*
185 G19 **Ashburton** Canterbury, South Island, New Zealand *43°55′S 171°47′E*
185 G19 **Ashburton** ≈ South Island, New Zealand
180 H8 **Ashburton River** ≈ Western Australia
145 V10 **Ashchysu** ≈ E Kazakhstan
10 M16 **Ashcroft** British Columbia, SW Canada *50°41′N 121°17′W*
138 E10 **Ashdod** *anc.* Azotos, *Lat.* Azotus. Central, W Israel *31°48′N 34°38′E*
27 S14 **Ashdown** Arkansas, C USA *33°40′N 94°09′W*
21 T9 **Asheboro** North Carolina, SE USA *35°42′N 79°50′W*
11 X15 **Ashern** Manitoba, S Canada *51°10′N 98°22′W*
21 P10 **Asheville** North Carolina, SE USA *35°36′N 82°33′W*
12 E8 **Asheweig** ≈ Ontario, C Canada
27 V9 **Ash Flat** Arkansas, C USA *36°13′N 91°36′W*
183 T4 **Ashford** New South Wales, SE Australia *29°18′S 151°09′E*
97 P22 **Ashford** SE England, United Kingdom *51°09′N 00°52′E*
36 K11 **Ash Fork** Arizona, SW USA *35°12′N 112°31′W*
Ashgabat *see* Aşgabat
27 T7 **Ash Grove** Missouri, C USA *37°19′N 93°35′W*
165 O12 **Ashikaga** *var.* Asikaga. Tochigi, Honshū, S Japan *36°21′N 139°26′E*
165 Q8 **Ashiro** Iwate, Honshū, C Japan *40°04′N 141°00′E*
164 F15 **Ashizuri-misaki** Shikoku, SW Japan
138 E10 **Ashkelon** *prev.* Ashqelon. Southern, C Israel *31°40′N 34°35′E*
Ashkhabad *see* Aşgabat
23 Q4 **Ashland** Alabama, S USA *33°16′N 85°50′W*
26 K7 **Ashland** Kansas, C USA *37°12′N 99°47′W*
21 P5 **Ashland** Kentucky, S USA *38°28′N 82°40′W*
19 S2 **Ashland** Maine, NE USA *46°36′N 68°24′W*
22 M1 **Ashland** Mississippi, S USA *34°51′N 89°10′W*
27 U4 **Ashland** Missouri, C USA *38°46′N 92°15′W*
29 S15 **Ashland** Nebraska, C USA *41°01′N 96°21′W*
31 T12 **Ashland** Ohio, N USA *40°52′N 82°19′W*
32 G15 **Ashland** Oregon, NW USA *42°11′N 122°42′W*
21 W6 **Ashland** Virginia, NE USA *37°45′N 77°28′W*
30 K3 **Ashland** Wisconsin, N USA *46°34′N 90°54′W*
20 I8 **Ashland City** Tennessee, S USA *36°16′N 87°05′W*
183 S4 **Ashley** New South Wales, SE Australia *29°21′S 149°49′E*
29 O7 **Ashley** North Dakota, N USA *46°00′N 99°22′W*
173 W7 **Ashmore and Cartier Islands** ◇ *Australian external territory* E Indian Ocean
119 I14 **Ashmyany** *Rus.* Oshmyany. Hrodzyenskaya Voblasts', W Belarus *54°24′N 25°57′E*
18 K12 **Ashokan Reservoir** ⊠ New York, NE USA
165 U4 **Ashoro** Hokkaidō, NE Japan *43°16′N 143°33′E*
Ashqelon *see* Ashkelon
Ashraf *see* Behshahr
139 O3 **Ash Shadādah** *var.* Ash Shaddādah, Jisr ash Shadadi, Shaddādī, Shedadi, Tell Shedadi. Al Ḩasakah, NE Syria *36°00′N 40°42′E*
Ash Shaddādah *see* Ash Shadādah
139 Y12 **Ash Shāfī** Al Başrah, E Iraq *30°49′N 47°30′E*
139 R4 **Ash Shakk** *var.* Shaykh. Şalāḩ ad Dīn, C Iraq *35°15′N 43°27′E*
Ash Sham/Ash Shām *see* Rīf Dimashq
139 T10 **Ash Shāmīyah** *var.* Shamiya. Al Qādisīyah, C Iraq *31°56′N 44°37′E*
139 Y13 **Ash Shāmīyah** *var.* Al Bādiyah al Janūbīyah. *desert* S Iraq
139 T11 **Ash Shanāfīyah** *var.* Ash Shināfīyah. Al Qādisīyah, S Iraq *31°35′N 44°38′E*
138 G13 **Ash Sharāh** *var.* Esh Sharā. ▲▲ W Jordan
143 R16 **Ash Shāriqah** *Eng.* Sharjah. Ash Shāriqah, NE United Arab Emirates *25°22′N 55°28′E*
143 R16 **Ash Shāriqah** *var.* Sharjah. ✈ Ash Shāriqah, NE United Arab Emirates *25°19′N 55°37′E*
140 I4 **Ash Sharmah** *var.* Šarma. Tabūk, NW Saudi Arabia *28°02′N 35°16′E*
139 R4 **Ash Sharqāţ** Nīnawýa, NW Iraq *35°31′N 43°15′E*
141 S10 **Ash Sharqīyah** *off.* Al Minţaqah ash Sharqīyah, *Eng.* Eastern Region. ◈ *province* E Saudi Arabia
Ash Sharqīyah *see* Al ʿUbaylah
139 W11 **Ash Shaţrah** *var.* Shatra. Dhī Qār, SE Iraq *31°26′N 46°10′E*
138 G13 **Ash Shawbak** Maʿān, W Jordan *30°32′N 35°34′E*
138 L5 **Ash Shaykh Ibrāhīm** Ḩimş, C Syria *35°03′N 38°50′E*
141 O17 **Ash Shaykh ʿUthmān** SW Yemen *12°53′N 45°00′E*
141 S15 **Ash Shiḩr** SE Yemen *14°45′N 49°24′E*
Ash Shināfīyah *see* Ash Shanāfīyah
141 V12 **Ash Shişar** *var.* Shisur. SW Oman *18°13′N 53°35′E*
139 S13 **Ash Shubrūm** *well* S Iraq
141 R10 **Ash Shuqqān** *desert* E Saudi Arabia
75 O9 **Ash Shuwayrif** *var.* Ash Shwayrif. N Libya *29°54′N 14°16′E*
Ash Shwayrif *see* Ash Shuwayrif
31 U10 **Ashtabula** Ohio, N USA *41°54′N 80°46′W*
29 Q5 **Ashtabula, Lake** ⊠ North Dakota, N USA
137 T12 **Ashtarak** W Armenia *40°18′N 44°22′E*
142 M6 **Āshtīān** *var.* Āshtiyān. Markazī, W Iran *34°23′N 49°55′E*
Āshtiyān *see* Āshtīān
33 R13 **Ashton** Idaho, NW USA *44°04′N 111°27′W*
13 O10 **Ashuanipi Lake** ◎ Newfoundland and Labrador, E Canada
15 P6 **Ashuapmushuan** ≈ Québec, SE Canada
23 Q3 **Ashville** Alabama, S USA *33°50′N 86°15′W*
31 S14 **Ashville** Ohio, N USA *39°43′N 82°57′W*
30 K3 **Ashwabay, Mount** *hill* Wisconsin, N USA
128-129 **Asia** *continent*
171 T11 **Asia, Kepulauan** *island group* E Indonesia
154 N13 **Āsika** Odisha, E India *19°38′N 84°41′E*
Asikaga *see* Ashikaga
93 M18 **Asikkala** *var.* Vääksy. Päijät-Häme, S Finland *61°09′N 25°36′E*
74 G5 **Asilah** N Morocco *35°32′N 06°04′W*
ʿAşī, Nahr al *see* Orontes
107 B16 **Asinara, Isola** *island* W Italy
122 J12 **Asino** Tomskaya Oblast', C Russian Federation *56°56′N 86°02′E*
119 O14 **Asintorf** *Rus.* Osintorf. Vitsyebskaya Voblasts', N Belarus *54°43′N 30°35′E*
119 L17 **Asipovichy** *Rus.* Osipovichi. Mahilyowskaya Voblasts', C Belarus *53°18′N 28°40′E*
141 N12 **ʿAsīr** *off.* Minţaqat ʿAsīr. ◈ *province* SW Saudi Arabia
140 M11 **ʿAsīr** *Eng.* Asir. ▲▲ SW Saudi Arabia
ʿAsīr, Minţaqat *see* ʿAsīr
139 X10 **Askal** Maysān, E Iraq *31°45′N 47°07′E*
137 P13 **Aşkale** Erzurum, NE Turkey *39°56′N 40°39′E*
117 T11 **Askaniya-Nova** Khersons'ka Oblast', S Ukraine *46°27′N 33°54′E*
95 H15 **Asker** Akershus, S Norway *59°52′N 10°26′E*
95 L17 **Askersund** Örebro, C Sweden *58°55′N 14°55′E*
Askī Kalak *see* Eski Kaļak
95 I15 **Askim** Østfold, S Norway *59°15′N 11°10′E*
127 V3 **Askino** Respublika Bashkortostan, W Russian Federation *56°07′N 56°39′E*
115 D14 **Áskio** ▲▲ N Greece
152 L9 **Askot** Uttarakhand, N India *29°44′N 80°20′E*
94 C12 **Askvoll** Sogn Og Fjordane, S Norway *61°21′N 05°04′E*
136 A13 **Aslan Burnu** *headland* W Turkey *38°44′N 26°43′E*
136 L16 **Aslantaş Barajı** ⊠ S Turkey
149 S4 **Asmār** *var.* Bar Kunar. Kunar, E Afghanistan *34°59′N 71°29′E*
Asmara *see* Asmera
80 I9 **Asmera** *var.* Asmara. ● (Eritrea) C Eritrea *15°15′N 38°58′E*
95 L21 **Åsnen** ◎ S Sweden
115 F19 **Asopós** ≈ S Greece
171 W13 **Asori** Papua, E Indonesia *02°37′S 136°06′E*
80 G12 **Āsosa** Bīnishangul Gumuz, W Ethiopia *10°06′N 34°27′E*
32 M10 **Asotin** Washington, NW USA *46°18′N 117°03′W*
Aspadana *see* Eşfahān
Aspang *see* Aspang Markt
109 X6 **Aspang Markt** *var.* Aspang. Niederösterreich, E Austria *47°34′N 16°06′E*
105 S12 **Aspe** Valenciana, E Spain *38°21′N 00°43′W*
37 R5 **Aspen** Colorado, C USA *39°12′N 106°49′W*
25 P6 **Aspermont** Texas, SW USA *33°08′N 100°14′W*
Asphaltites, Lacus *see* Dead Sea
Aspinwall *see* Colón
185 C20 **Aspiring, Mount** ▲ South Island, New Zealand *44°21′S 168°47′E*
115 B16 **Asprókavos, Akrotírio** *headland* Kérkyra, Iónia Nísiá, Greece, C Mediterranean Sea *39°22′N 20°07′E*
Aspropótamos *see* Achelóos
Assab *see* ʿAseb
138 L4 **As Sabkhah** *var.* Sabkha. Ar Raqqah, NE Syria *35°30′N 35°54′E*
139 U6 **As Saʿdīyah** Diyālá, E Iraq *34°11′N 45°09′E*
Assad, Lake *see* Asad, Buḩayrat al
138 I8 **Aş Şafā** ▲ S Syria *33°03′N 37°07′E*
138 I10 **Aş Şafāwī** Al Mafraq, N Jordan *37°12′N 32°30′E*
75 W8 **Aş Şaff** *var.* El Şaff. N Egypt *29°34′N 31°16′E*
139 N2 **Aş Şafīḩ** Al Ḩasakah, N Syria *36°42′N 40°12′E*
Aş Şaḩrā' ash Sharqīyah *see* Sahara el Sharqîya
Assake *see* Asaka
As Salamīyah *see* Salamīyah
141 Q4 **As Sālimī** *var.* Salemy. SW Kuwait *29°07′N 46°41′E*
67 W7 **ʿAssal, Lac** ◎ C Djibouti
75 T7 **As Sallūm** *var.* Salûm. NW Egypt *31°31′N 25°09′E*
139 T13 **As Salmān** Al Muthanná, S Iraq *30°29′N 44°34′E*
138 G10 **As Salţ** *var.* Salt. Al Balqā', NW Jordan *32°03′N 35°44′E*
142 M16 **As Salwā** *var.* Salwa, Salwah. S Qatar *24°44′N 50°52′E*
153 V12 **Assam** ◈ *state* NE India
Assamaka *see* Assamakka
77 T8 **Assamakka** *var.* Assamaka. Agadez, NW Niger *19°24′N 05°53′E*
139 U11 **As Samāwah** *var.* Samawa. Al Muthanná, S Iraq *31°17′N 45°06′E*

◆ Country
● Country Capital
◇ Dependent Territory
○ Dependent Territory Capital
◈ Administrative Regions
✈ International Airport
▲ Mountain
▲▲ Mountain Range
⛰ Volcano
≈ River
◎ Lake
⊠ Reservoir

As Samāwah – Awakino

As Samāwah *see* Al Muthannā
As Saqia al Hamra *see* Saguia al Hamra
138 *J4* **Aş Şā'rān** Ḩamāh, C Syria *35°15′N 37°28′E*
138 *G9* **Aş Şarīḩ** Irbid, N Jordan *32°31′N 35°54′E*
21 *Z5* **Assateague Island** *island* Maryland, NE USA
139 *O6* **As Sayyāl** *var.* Sayyāl. Dayr az Zawr, E Syria *34°37′N 40°52′E*
99 *G18* **Asse** Vlaams Brabant, C Belgium *50°55′N 04°12′E*
99 *D16* **Assebroek** West-Vlaanderen, NW Belgium *51°12′N 03°16′E*
Asselle *see* Āsela
107 *C20* **Assemini** Sardegna, Italy, C Mediterranean Sea *39°16′N 08°58′E*
99 *E16* **Assenede** Oost-Vlaanderen, NW Belgium *51°15′N 03°43′E*
95 *G24* **Assens** Syddtjylland, C Denmark *55°16′N 09°54′N*
Asserien/Asserin *see* Aseri
99 *I21* **Assesse** Namur, SE Belgium *50°22′N 05°01′E*
141 *Y8* **As Sīb** *var.* Seeb. NE Oman *23°40′N 58°03′E*
139 *Z13* **As Sībah** *var.* Sibah. Al Başrah, SE Iraq *30°13′N 47°24′E*
11 *T17* **Assiniboia** Saskatchewan, S Canada *49°39′N 105°59′W*
11 *V15* **Assiniboine** ≈ Manitoba, S Canada
11 *P16* **Assiniboine, Mount** ▲ Alberta/British Columbia, SW Canada *50°54′N 115°43′W*
Assiout *see* Asyūţ
60 *J9* **Assis** São Paulo, S Brazil *22°37′S 50°25′W*
106 *I13* **Assisi** Umbria, C Italy *43°04′N 12°36′E*
Assiut *see* Asyūţ
Assling *see* Jesenice
Assouan *see* Aswān
59 *P14* **Assu** *var.* Açu. Rio Grande do Norte, E Brazil *05°33′S 36°55′W*
Assuan *see* Aswān
142 *K12* **Aş Şubayḩīyah** *var.* Subiyah. S Kuwait *28°55′N 47°57′E*
141 *R16* **As Sufāl** S Yemen *14°06′N 48°42′E*
138 *L5* **As Sukhnah** *var.* Sukhne, *Fr.* Soukhné. Ḩimş, C Syria *34°56′N 38°52′E*
139 *U4* **As Sulaymānīyah** *var.* Sulaimaniya, *Kurd.* Slêmanî. As Sulaymānīyah, NE Iraq *35°32′N 45°27′E*
139 *U4* **As Sulaymānīyah** *off.* Muḩāfa at as Sulaymānīyah, *off. Kurd.* Parêzga-î Slêmanî, *Kurd.* Slêmanî. ⬖ *governorate* N Iraq
as Sulaymānīyah, Muḩāfa at *see* As Sulaymānīyah
141 *P11* **As Sulayyil** Ar Riyāḑ, S Saudi Arabia *20°29′N 45°33′E*
121 *O13* **As Sulţān** N Libya *31°01′N 17°21′E*
141 *Q5* **Aş Şummān** *desert* N Saudi Arabia
141 *Q16* **Aş Şurrah** SW Yemen *13°56′N 46°23′E*
139 *N4* **Aş Şuwār** *var.* Şuwār. Dayr az Zawr, E Syria *35°31′N 40°37′E*
138 *H9* **As Suwaydā'** *var.* El Suweida, Es Suweida, Suweida, *Fr.* Soueida. As Suwaydā', SW Syria *32°43′N 36°33′E*
138 *H9* **As Suwaydā'** *off.* Muḩāfaẓat as Suwaydā', *var.* As Suwaydā, Suwaydá, Suweida. ⬖ *governorate* S Syria
141 *Z9* **As Suwayḩ** NE Oman *22°07′N 59°42′E*
141 *X8* **As Suwayq** *var.* Suwaik. N Oman *23°49′N 57°30′E*
139 *T8* **Aş Şuwayrah** *var.* Suwaira. Wāsiţ, E Iraq *32°57′N 44°47′E*
As Suways *see* Suez
Asta Colonia *see* Asti
Astacus *see* İzmit
115 *M23* **Astakída** *island* SE Greece
145 *Q9* **Astana** *prev.* Akmola, Akmolinsk, Tselinograd, Aqmola. ● (Kazakhstan) Akmola, N Kazakhstan *51°13′N 71°25′E*
142 *M3* **Āstāneh** *var.* Āstāneh-ye Ashrafīyeh. Gīlān, NW Iran *37°17′N 49°58′E*
Āstāneh-ye Ashrafīyeh *see* Āstāneh
Asta Pompeia *see* Asti
137 *Y14* **Astara** S Azerbaijan *38°28′N 48°51′E*
Astarabad *see* Gorgān
99 *L15* **Asten** Noord-Brabant, SE Netherlands *51°24′N 05°45′E*
Asterābād *see* Gorgān
106 *C8* **Asti** *anc.* Asta Colonia, Asta Pompeia, Hasta Colonia, Hasta Pompeia. Piemonte, NW Italy *44°54′N 08°11′E*
Astigi *see* Ecija
Astipálaia *see* Astypálaia
148 *L16* **Astola Island** *island* SW Pakistan
152 *H4* **Astor** Jammu and Kashmir, NW India *35°21′N 74°52′E*
104 *K4* **Astorga** *anc.* Asturica Augusta. Castilla y León, N Spain *42°27′N 06°04′W*
32 *F10* **Astoria** Oregon, NW USA *46°12′N 123°50′W*
0 *F8* **Astoria Fan** *undersea feature* E Pacific Ocean *45°15′N 126°15′W*
95 *J22* **Åstorp** Skåne, S Sweden *56°09′N 12°57′E*
Astrabad *see* Gorgān
127 *Q13* **Astrakhan'** Astrakhanskaya Oblast', SW Russian Federation *46°20′N 48°01′E*
Astrakhan-Bazar *see* Cälilabad
127 *Q11* **Astrakhanskaya Oblast'** ⬖ *province* SW Russian Federation
93 *J15* **Åsträsk** Västerbotten, N Sweden *64°38′N 20°00′E*
Astrida *see* Butare
65 *O22* **Astrid Ridge** *undersea feature* S Atlantic Ocean
187 *P15* **Astrolabe, Récifs de l'** *reef* C New Caledonia
121 *P2* **Astromerítis** N Cyprus *35°09′N 33°02′E*
115 *F20* **Ástros** Pelopónnisos, S Greece *37°24′N 22°43′E*
119 *G16* **Astryna** *Rus.* Ostryna. Hrodzyenskaya Voblasts', W Belarus *53°44′N 24°33′E*
104 *J2* **Asturias** ⬖ *autonomous community* NW Spain
Asturias *see* Oviedo
Asturica Augusta *see* Astorga
115 *L22* **Astypálaia** *var.* Astipálaia, *It.* Stampalia. *island* Kykládes, Greece, Aegean Sea
192 *G16* **Āsuisui, Cape** *headland* Savai'i, W Samoa *13°44′S 172°29′W*
195 *S2* **Asuka** *Japanese research station* Antarctica *71°49′S 23°52′E*
62 *O6* **Asunción** ● (Paraguay) Central, S Paraguay *25°17′S 57°36′W*
62 *O6* **Asunción** ✈ Central, S Paraguay *25°15′S 57°40′W*
188 *K3* **Asuncion Island** *island* N Northern Mariana Islands
42 *E6* **Asunción Mita** Jutiapa, SE Guatemala *14°20′N 89°42′W*
Asunción Nochixtlán *see* Nochixtlán
40 *E3* **Asunción, Río** ≈ NW Mexico
95 *M18* **Åsunden** ◎ S Sweden
118 *K11* **Asvyeya** *Rus.* Osveya. Vitsyebskaya Voblasts', N Belarus *56°00′N 28°05′E*
Aswa *see* Achwa
75 *X11* **Aswān** *var.* Assouan, Assuan, Aswân; *anc.* Syene. SE Egypt *24°03′N 32°59′E*
Aswân *see* Aswān
Aswan Dam *see* Khazzān Aswān
75 *W9* **Asyūţ** *var.* Assiout, Assiut, Asyût, Siut; *anc.* Lycopolis. C Egypt *27°06′N 31°11′E*
Asyût *see* Asyūţ
193 *W15* **Ata** *island* Tongatapu Group, SW Tonga
62 *G8* **Atacama** *off.* Región de Atacama. ⬖ *region* C Chile
Atacama Desert *see* Atacama, Desierto de
62 *H4* **Atacama, Desierto de** *Eng.* Atacama Desert. *desert* N Chile
62 *I6* **Atacama, Puna de** ▲▲ NW Argentina
Atacama, Región de *see* Atacama
62 *I5* **Atacama, Salar de** *salt lake* N Chile
54 *E11* **Ataco** Tolima, C Colombia *03°36′N 75°23′W*
190 *H8* **Atafu Atoll** *island* NW Tokelau
190 *H8* **Atafu Village** Atafu Atoll, NW Tokelau *08°40′S 172°40′W*
74 *K12* **Atakor** ▲▲ SE Algeria
77 *R14* **Atakora, Chaîne de l'** *var.* Atakora Mountains. ▲▲ N Benin
Atakora Mountains *see* Atakora, Chaîne de l'
77 *R16* **Atakpamé** C Togo *07°32′N 01°08′E*
146 *F11* **Atakui** Ahal Welaýaty, C Turkmenistan *40°04′N 58°03′E*
58 *B13* **Atalaia do Norte** Amazonas, N Brazil *04°22′S 70°10′W*
146 *M14* **Atamyrat** *prev.* Kerki. Lebap Welaýaty, E Turkmenistan *37°52′N 65°06′E*
76 *I7* **Aţâr** Adrar, W Mauritania *20°30′N 13°03′W*
162 *G10* **Atas Bogd** ▲ SW Mongolia *43°17′N 96°47′E*
35 *P12* **Atascadero** California, W USA *35°28′N 120°40′W*
25 *S13* **Atascosa River** ≈ Texas, SW USA
145 *R11* **Atasu** Karaganda, C Kazakhstan *48°42′N 71°38′E*
145 *R12* **Atasu** ≈ Karaganda, C Kazakhstan
193 *V15* **Atata** *island* Tongatapu Group, S Tonga
136 *H10* **Atatürk** ✈ (İstanbul) İstanbul, NW Turkey *40°58′N 28°50′E*
137 *N16* **Atatürk Barajī** ⊠ S Turkey
115 *O23* **Atavýros** *prev.* Attavyros. Ródos, Dodekánisa, Aegean Sea *36°10′N 27°50′E*
115 *O23* **Atávyros** *prev.* Attávyros. ▲ Ródos, Dodekánisa, Greece, Aegean Sea *36°10′N 27°50′E*
Atax *see* Aude
80 *G8* **Atbara** *var.* 'Aţbārah. River Nile, NE Sudan *17°42′N 34°E*
80 *H8* **Atbara** *var.* Nahr 'Aţbarah. ≈ Eritrea/Sudan
'Aţbārah/'Aţbarah, Nahr *see* Atbara
145 *P9* **Atbasar** Akmola, N Kazakhstan *51°49′N 68°18′E*
At-Bashi *see* At-Bashy
147 *W9* **At-Bashy** *var.* At-Bashi. Narynskaya Oblast', C Kyrgyzstan *41°07′N 75°48′E*
22 *I10* **Atchafalaya Bay** *bay* Louisiana, S USA
22 *I8* **Atchafalaya River** ≈ Louisiana, S USA
Atchin *see* Aceh
27 *Q3* **Atchison** Kansas, C USA *39°31′N 95°07′W*
77 *P16* **Atebubu** C Ghana *07°47′N 01°00′W*
105 *Q6* **Ateca** Aragón, NE Spain *41°20′N 01°49′W*
40 *K11* **Atengo, Río** ≈ C Mexico
Aternum *see* Pescara
107 *K15* **Atessa** Abruzzo, C Italy *42°03′N 14°25′E*
Ateste *see* Este
99 *E19* **Ath** *var.* Aat. Hainaut, SW Belgium *50°38′N 03°47′E*
11 *Q13* **Athabasca** Alberta, SW Canada *54°44′N 113°15′W*
11 *Q12* **Athabasca** *var.* Athabaska. ≈ Alberta, SW Canada
11 *R10* **Athabasca, Lake** ◎ Alberta/Saskatchewan, SW Canada
Athabaska *see* Athabasca
115 *C16* **Athamánon** ▲▲ C Greece
97 *F17* **Athboy** *Ir.* Baile Átha Buí. E Ireland *53°38′N 06°55′W*
Athenae *see* Athína
97 *C18* **Athenry** *Ir.* Baile Átha an Rí. W Ireland *53°19′N 08°49′W*
23 *P2* **Athens** Alabama, S USA *34°48′N 86°58′W*
23 *T3* **Athens** Georgia, SE USA *33°57′N 83°24′W*
31 *T14* **Athens** Ohio, N USA *39°20′N 82°06′W*
20 *M10* **Athens** Tennessee, S USA *35°27′N 84°38′W*
25 *V7* **Athens** Texas, SW USA *32°12′N 95°51′W*
Athens *see* Athína
115 *B18* **Athéras, Akrotírio** *headland* Kefalloniá, Iónia Nisiá, Greece, C Mediterranean Sea *38°20′N 20°24′E*
181 *W4* **Atherton** Queensland, NE Australia *17°18′S 145°29′E*
81 *I19* **Athi** ≈ S Kenya
121 *Q2* **Athiénou** SE Cyprus *35°01′N 33°31′E*
115 *H19* **Athína** *Eng.* Athens, *prev.* Athínai; *anc.* Athenae. ● (Greece) Attikí, C Greece *37°59′N 23°44′E*
Athínai *see* Athína
139 *S10* **Athīyah** An Najaf, C Iraq *32°01′N 44°04′E*
97 *D18* **Athlone** *Ir.* Baile Átha Luain. C Ireland *53°25′N 07°56′W*
155 *F16* **Athni** Karnātaka, W India *16°43′N 75°04′E*
185 *C23* **Athol** Southland, South Island, New Zealand *45°30′S 168°35′E*
19 *N11* **Athol** Massachusetts, NE USA *42°35′N 72°11′W*
115 *I15* **Áthos** ▲ NE Greece *40°10′N 24°21′E*
Athos, Mount *see* Ágion Óros
Ath Thawrah *see* Madīnat ath Thawrah
141 *P5* **Ath Thumāmī** *spring/well* N Saudi Arabia *27°56′N 45°06′E*
99 *L25* **Athus** Luxembourg, SE Belgium *49°34′N 05°50′E*
97 *E19* **Athy** *Ir.* Baile Átha Í. C Ireland *52°59′N 06°59′W*
78 *I10* **Ati** Batha, C Chad *13°11′N 18°20′E*
81 *F16* **Atiak** NW Uganda *03°14′N 32°05′E*
57 *G17* **Atico** Arequipa, SW Peru *16°13′S 73°13′W*
105 *O6* **Atienza** Castilla-La Mancha, C Spain *41°12′N 02°52′W*
39 *Q6* **Atigun Pass** *pass* Alaska, USA
12 *B12* **Atikokan** Ontario, S Canada *48°45′N 91°38′W*
13 *O9* **Atikonak Lac** ◎ Newfoundland and Labrador, E Canada
42 *C6* **Atitlán, Lago de** ◎ W Guatemala
190 *L16* **Atiu** *island* S Cook Islands
Atjeh *see* Aceh
123 *T9* **Atka** Magadanskaya Oblast', E Russian Federation *60°45′N 151°35′E*
38 *H17* **Atka** Atka Island, Alaska, USA *52°12′N 174°14′W*
38 *H17* **Atka Island** *island* Aleutian Islands, Alaska, USA
127 *O7* **Atkarsk** Saratovskaya Oblast', W Russian Federation *52°15′N 43°48′E*
27 *U11* **Atkins** Arkansas, C USA *35°15′N 92°56′W*
29 *O13* **Atkinson** Nebraska, C USA *42°31′N 98°57′W*
171 *T12* **Atkri** Papua Barat, E Indonesia *01°45′S 130°04′E*
41 *O13* **Atlacomulco** *var.* Atlacomulco de Fabela. México, C Mexico *19°49′N 99°54′W*
Atlacomulco de Fabela *see* Atlacomulco
23 *S3* **Atlanta** *state capital* Georgia, SE USA *33°45′N 84°23′W*
31 *R6* **Atlanta** Michigan, N USA *45°01′N 84°07′W*
25 *X6* **Atlanta** Texas, SW USA *33°06′N 94°09′W*
29 *T15* **Atlantic** Iowa, C USA *41°24′N 95°00′W*
21 *Y10* **Atlantic** North Carolina, SE USA *34°52′N 76°20′W*
23 *W8* **Atlantic Beach** Florida, SE USA *30°19′N 81°24′W*
18 *J17* **Atlantic City** New Jersey, NE USA *39°23′N 74°27′W*
172 *L14* **Atlantic-Indian Basin** *undersea feature* SW Indian Ocean *60°00′S 15°00′E*
172 *K13* **Atlantic-Indian Ridge** *undersea feature* SW Indian Ocean *53°00′S 15°00′E*
54 *E4* **Atlántico** *off.* Departamento del Atlántico. ⬖ *province* NW Colombia
64-65 **Atlantic Ocean** *ocean*
Atlántico, Departamento del *see* Atlántico
42 *K7* **Atlántico Norte, Región Autónoma** *prev.* Zelaya Norte. ⬖ *autonomous region* NE Nicaragua
42 *L10* **Atlántico Sur, Región Autónoma** *prev.* Zelaya Sur. ⬖ *autonomous region* SE Nicaragua
42 *I5* **Atlántida** ⬖ *department* N Honduras
77 *Y15* **Atlantika Mountains** ▲▲ E Nigeria
64 *J10* **Atlantis Fracture Zone** *tectonic feature* NW Atlantic Ocean
74 *H7* **Atlas Mountains** ▲▲ NW Africa
123 *V11* **Atlasova, Ostrov** *island* SE Russian Federation
123 *V10* **Atlasovo** Kamchatskiy Kray, E Russian Federation *55°42′N 159°35′E*
120 *G11* **Atlas Saharien** *var.* Saharan Atlas. ▲▲ Algeria/Morocco
Atlas, Tell *see* Atlas Tellien
120 *H10* **Atlas Tellien** *Eng.* Tell Atlas. ▲▲ N Algeria
10 *I9* **Atlin** British Columbia, W Canada *59°31′N 133°41′W*
10 *I9* **Atlin Lake** ◎ British Columbia, W Canada
41 *P14* **Atlixco** Puebla, S Mexico *18°55′N 98°26′W*
94 *B11* **Atløyna** *island* S Norway
155 *I17* **Ātmakūr** Andhra Pradesh, C India *15°52′N 78°42′E*
23 *O8* **Atmore** Alabama, S USA *31°01′N 87°29′W*
101 *J20* **Atmühl** ≈ S Germany
94 *H11* **Atna** ≈ S Norway
164 *E13* **Atō** Yamaguchi, Honshū, SW Japan *34°24′N 131°42′E*
57 *L21* **Atocha** Potosí, S Bolivia *20°55′S 66°14′W*
27 *P12* **Atoka** Oklahoma, C USA *34°22′N 96°08′W*
27 *O12* **Atoka Lake** *var.* Atoka Reservoir. ⊠ Oklahoma, C USA
Atoka Reservoir *see* Atoka Lake
33 *Q14* **Atomic City** Idaho, NW USA *43°26′N 112°48′W*
40 *L10* **Atotonilco** Zacatecas, C Mexico *24°12′N 102°46′W*
Atotonilco *see* Atotonilco el Alto
40 *M13* **Atotonilco el Alto** *var.* Atotonilco. Jalisco, SW Mexico *20°35′N 102°30′W*
77 *N7* **Atouila, 'Erg** *desert* N Mali
41 *N16* **Atoyac** *var.* Atoyac de Alvarez. Guerrero, S Mexico *17°12′N 100°28′W*
Atoyac de Alvarez *see* Atoyac
41 *P15* **Atoyac, Río** ≈ S Mexico
39 *O5* **Atqasuk** Alaska, USA *70°28′N 157°24′W*
Atrak/Atrak, Rūd-e *see* Etrek
95 *J20* **Ätran** ≈ S Sweden
54 *C7* **Atrato, Río** ≈ NW Colombia
Atrek *see* Etrek
107 *K14* **Atri** Abruzzo, C Italy *42°35′N 13°59′E*
Atria *see* Adria
165 *P9* **Atsumi** Yamagata, Honshū, C Japan *38°38′N 139°36′E*
165 *S3* **Atsuta** Hokkaidō, NE Japan *43°28′N 141°24′E*
143 *Q17* **Aţ Ţaff** *desert* C United Arab Emirates
138 *G12* **Aţ Ţafīlah** *var.* Et Tafila, Tafila. Aţ Ţafīlah, W Jordan *30°52′N 35°36′E*
138 *G12* **Aţ Ţafīlah** *off.* Muḩāfaẓat aţ Ţafīlah. ⬖ *governorate* W Jordan
140 *L10* **Aţ Ţā'if** Makkah, W Saudi Arabia *21°50′N 40°50′E*
Attaleia/Attalia *see* Antalya
23 *Q3* **Attalla** Alabama, S USA *34°01′N 86°05′W*
138 *L2* **At Tall al Abyaḑ** *var.* Tall al Abyaḑ, Tell Abyad, *Fr.* Tell Abiad. Ar Raqqah, N Syria *36°36′N 34°00′E*
At Ta'mīm *see* Kirkūk
138 *L7* **Aţ Ţanf** Ḩimş, S Syria *33°29′N 38°39′E*
163 *N9* **Attanshiree** Dornogovĭ, SE Mongolia *45°36′N 110°30′E*
167 *T10* **Attapu** *var.* Attopeu, Samakhixai. Attapu, S Laos *14°48′N 106°51′E*
139 *S10* **At Ţaqţaqānah** An Najaf, C Iraq *32°03′N 43°54′E*
Attávyros *see* Atávyros
12 *G9* **Attawapiskat** Ontario, C Canada *52°55′N 82°26′W*
12 *F9* **Attawapiskat** ≈ Ontario, S Canada
12 *D9* **Attawapiskat Lake** ◎ Ontario, C Canada
At Taybé *see* Ţayyibah
101 *F16* **Attendorn** Nordrhein-Westfalen, W Germany *51°07′N 07°54′E*
109 *R5* **Attersee** Salzburg, NW Austria *47°55′N 13°31′E*
109 *R5* **Attersee** ◎ N Austria
99 *L24* **Attert** Luxembourg, SE Belgium *49°45′N 05°47′E*
138 *M4* **At Tibnī** *var.* Tibnī. Dayr az Zawr, NE Syria *35°30′N 39°48′E*
31 *N13* **Attica** Indiana, N USA *40°17′N 87°15′W*
18 *E10* **Attica** New York, NE USA *42°51′N 78°13′W*
Attica *see* Attikí
13 *N7* **Attikamagen Lake** ◎ Newfoundland and Labrador, E Canada
115 *H20* **Attikí** *Eng.* Attica. ⬖ *region* C Greece
19 *O12* **Attleboro** Massachusetts, NE USA *41°55′N 71°15′W*
109 *R5* **Attnang** Oberösterreich, N Austria *48°02′N 13°44′E*
149 *U6* **Attock City** Punjab, E Pakistan *33°52′N 72°20′E*
Attopeu *see* Attapu
25 *X8* **Attoyac River** ≈ Texas, SW USA
38 *D16* **Attu** Attu Island, Alaska, USA *52°53′N 173°18′E*
139 *Y12* **Aţ Ţūbah** Al Başrah, E Iraq *30°29′N 47°28′E*
140 *K4* **Aţ Ţubayq** *plain* Jordan/Saudi Arabia
38 *C16* **Attu Island** *island* Aleutian Islands, Alaska, USA
75 *X8* **Aţ Ţūr** *var.* El Ţûr. NE Egypt *28°14′N 33°36′E*
155 *I21* **Āttūr** Tamil Nādu, SE India *11°34′N 78°39′E*
141 *N17* **At Turbah** SW Yemen *12°42′N 43°31′E*
62 *I12* **Atuel, Río** ≈ C Argentina
191 *X7* **Atuona** Hiva Oa, NE French Polynesia *09°47′S 139°03′W*
Aturus *see* Adour
95 *M18* **Åtvidaberg** Östergötland, S Sweden *58°12′N 16°00′E*
35 *P9* **Atwater** California, W USA *37°19′N 120°33′W*
29 *T8* **Atwater** Minnesota, N USA *45°08′N 94°48′W*
26 *I2* **Atwood** Kansas, C USA *39°48′N 101°03′W*
31 *U12* **Atwood Lake** ⊠ Ohio, N USA
127 *P5* **Atyashevo** Respublika Mordoviya, W Russian Federation *54°36′N 46°04′E*
144 *F12* **Atyrau** *prev.* Gur'yev. Atyrau, W Kazakhstan *47°07′N 51°56′E*
144 *E11* **Atyrau** *off.* Atyrauskaya Oblast', *var.Kaz.* Atyraū Oblysy; *prev.* Gur'yevskaya Oblast'. ⬖ *province* W Kazakhstan
Atyraū Oblysy/Atyrauskaya Oblast' *see* Atyrau
Atyrauskaya Oblast' *see* Atyrau
108 *J7* **Au** Vorarlberg, NW Austria *47°19′N 10°01′E*
186 *B4* **Aua Island** *island* NW Papua New Guinea
103 *S16* **Aubagne** *anc.* Albania. Bouches-du-Rhône, SE France *43°17′N 05°35′E*
99 *L25* **Aubange** Luxembourg, SE Belgium *49°35′N 05°49′E*
103 *Q6* **Aube** ⬖ *department* N France
103 *R6* **Aube** ≈ N France
99 *L19* **Aubel** Liège, E Belgium *50°45′N 05°49′E*
103 *Q13* **Aubenas** Ardèche, E France *44°37′N 04°24′E*
103 *O8* **Aubigny-sur-Nère** Cher, C France *47°30′N 02°27′E*
103 *O13* **Aubin** Aveyron, S France *44°30′N 02°18′E*
103 *O13* **Aubrac, Monts d'** ▲▲ S France
36 *J10* **Aubrey Cliffs** *cliff* Arizona, SW USA
23 *R5* **Auburn** Alabama, S USA *32°37′N 85°30′W*
35 *P6* **Auburn** California, W USA *38°53′N 121°03′W*
30 *K14* **Auburn** Illinois, N USA *39°35′N 89°45′W*
31 *Q11* **Auburn** Indiana, N USA *41°22′N 85°03′W*
20 *J7* **Auburn** Kentucky, S USA *36°52′N 86°42′W*
19 *P8* **Auburn** Maine, NE USA *44°05′N 70°15′W*
19 *N11* **Auburn** Massachusetts, NE USA *42°11′N 71°47′W*
29 *S16* **Auburn** Nebraska, C USA *40°23′N 95°50′W*
18 *H10* **Auburn** New York, NE USA *42°55′N 76°31′W*
32 *H8* **Auburn** Washington, NW USA *47°18′N 122°13′W*
103 *N11* **Aubusson** Creuse, C France *45°58′N 02°10′E*
118 *E10* **Auce** *Ger.* Autz. SW Latvia *56°28′N 22°54′E*
102 *L15* **Auch** *Lat.* Augusta Auscorum, Elimberrum. Gers, S France *43°40′N 00°37′E*
77 *U16* **Auchi** Edo, S Nigeria *07°01′N 06°17′E*
23 *T9* **Aucilla River** ≈ Florida/Georgia, SE USA
184 *L6* **Auckland** Auckland, North Island, New Zealand *36°53′S 174°46′E*
184 *K5* **Auckland** *off.* Auckland Region. ⬖ *region* North Island, New Zealand
184 *L6* **Auckland** ✈ Auckland, North Island, New Zealand *37°01′S 174°49′E*
192 *J12* **Auckland Islands** *island group* S New Zealand
Auckland Region *see* Auckland
103 *O16* **Aude** ⬖ *department* S France
103 *N16* **Aude** *anc.* Atax. ≈ S France
Audenarde *see* Oudenaarde
Audern *see* Audru
102 *E6* **Audierne** Finistère, NW France *48°01′N 04°30′W*
102 *E6* **Audierne, Baie d'** *bay* NW France
103 *U7* **Audincourt** Doubs, E France *47°29′N 06°50′E*
118 *G5* **Audru** *Ger.* Audern. Pärnumaa, SW Estonia *58°24′N 24°22′E*
29 *T14* **Audubon** Iowa, C USA *41°44′N 94°56′W*
101 *N17* **Aue** Sachsen, E Germany *50°35′N 12°42′E*
100 *H12* **Aue** ≈ NW Germany
100 *L9* **Auerbach** Bayern, SE Germany *49°41′N 11°41′E*
101 *M17* **Auerbach** Sachsen, E Germany *50°30′N 12°24′E*
108 *I10* **Auererrhein** ≈ SW Switzerland
101 *N17* **Auersberg** ▲ E Germany *50°30′N 12°42′E*
181 *W9* **Augathella** Queensland, E Australia *25°54′S 146°38′E*
31 *Q12* **Auglaize River** ≈ Ohio, N USA
83 *F22* **Augrabies Falls** *waterfall* W South Africa
31 *R7* **Au Gres River** ≈ Michigan, N USA
Augsbourg *see* Augsburg
101 *K22* **Augsburg** *Fr.* Augsbourg; *anc.* Augusta Vindelicorum. Bayern, S Germany *48°22′N 10°54′E*
180 *I14* **Augusta** Western Australia *34°18′S 115°10′E*
107 *L25* **Augusta** *It.* Agosta. Sicilia, Italy, C Mediterranean Sea *37°14′N 15°14′E*
27 *W11* **Augusta** Arkansas, C USA *35°16′N 91°21′W*
23 *V3* **Augusta** Georgia, SE USA *33°29′N 81°58′W*
27 *O6* **Augusta** Kansas, C USA *37°40′N 96°59′W*
19 *Q7* **Augusta** *state capital* Maine, NE USA *44°20′N 69°44′W*
33 *Q8* **Augusta** Montana, NW USA *47°28′N 112°23′W*
Augusta *see* London
Augusta Auscorum *see* Auch
Augusta Emerita *see* Mérida
Augusta Praetoria *see* Aosta
Augusta Suessionum *see* Soissons
Augusta Trajana *see* Stara Zagora
Augusta Treverorum *see* Trier
Augusta Vangionum *see* Worms
Augusta Vindelicorum *see* Augsburg
95 *G24* **Augustenborg** *Ger.* Augustenburg. Syddanmark, SW Denmark *54°57′N 09°53′E*
Augustenburg *see* Augustenborg
39 *Q13* **Augustine Island** *island* Alaska, USA
14 *L9* **Augustines, Lac des** ◎ Québec, SE Canada
Augustobona Tricassium *see* Troyes
Augustodunum *see* Autun
Augustodurum *see* Bayeux
Augustoritum Lemovicensium *see* Limoges
110 *O8* **Augustów** *Rus.* Avgustov. Podlaskie, NE Poland *53°52′N 22°58′E*
Augustow Canal *see* Augustowski, Kanał
110 *O8* **Augustowski, Kanał** *Eng.* Augustow Canal, *Rus.* Avgustovskiy Kanal. *canal* NE Poland
180 *I9* **Augustus, Mount** ▲ Western Australia *24°42′S 117°42′E*
Aujuittuq *see* Grise Fiord
186 *M9* **Auki** Malaita, N Solomon Islands *08°48′S 160°45′E*
21 *W8* **Aulander** North Carolina, SE USA *36°15′N 77°16′W*
180 *L7* **Auld, Lake** *salt lake* C Western Australia
Aulie Ata/Auliye-Ata *see* Taraz
Äülīeköl *see* Auliyekol'
144 *M8* **Auliyekol'** *Kaz.* Äülīeköl; *prev.* Semiozernoye. Kostanay, N Kazakhstan *52°22′N 64°06′E*
106 *E10* **Aulla** Toscana, C Italy *44°15′N 10°00′E*
102 *F6* **Aulne** ≈ NW France
Aulong *see* Ulong
37 *T3* **Ault** Colorado, C USA *40°34′N 104°43′W*
95 *F22* **Aulum** *var.* avlum. Midtjylland, C Denmark *56°16′N 08°48′E*
103 *N3* **Aumale** Seine-Maritime, N France *49°45′N 01°43′E*
Auminzatau, Gory *see* Ovminzatovo Tog'lari
77 *T14* **Auna** Niger, W Nigeria *10°13′N 04°43′E*
166 *L6* **Aunglan** *var.* Allanmyo, Myaydo. Magway, C Myanmar (Burma) *19°25′N 95°13′E*
95 *H21* **Auning** Midtjylland, C Denmark *56°26′N 10°23′E*
192 *K17* **'Aunu'u Island** *island* W American Samoa
83 *E20* **Auob** *var.* Oup. ≈ Namibia/South Africa
93 *K19* **Aura** Varsinais-Suomi, SW Finland *60°37′N 22°35′E*
109 *R5* **Aurach** ≈ N Austria
Aural, Phnom *see* Aôral, Phnum
153 *O14* **Aurangābād** Bihār, N India *24°46′N 84°23′E*
154 *F13* **Aurangābād** Mahārāshtra, C India *19°52′N 75°22′E*
189 *V7* **Aur Atoll** *atoll* E Marshall Islands
102 *G7* **Auray** Morbihan, NW France *47°40′N 02°59′W*
94 *G13* **Aurdal** Oppland, S Norway *60°51′N 09°25′E*
94 *F8* **Aure** Møre og Romsdal, S Norway *63°16′N 08°31′E*
29 *T12* **Aurelia** Iowa, C USA *42°42′N 95°26′W*
Aurelia Aquensis *see* Baden-Baden
Aurelianum *see* Orléans
120 *J10* **Aurès, Massif de l'** ▲▲ NE Algeria
100 *F10* **Aurich** Niedersachsen, NW Germany *53°28′N 07°28′E*
103 *O13* **Aurillac** Cantal, C France *44°56′N 02°26′E*
Aurine, Alpi *see* Zillertaler Alpen
Aurium *see* Ourense
14 *H15* **Aurora** Ontario, S Canada *44°00′N 79°26′W*
55 *S8* **Aurora** NW Guyana *06°46′N 59°45′W*
37 *T4* **Aurora** Colorado, C USA *39°42′N 104°51′W*
30 *M11* **Aurora** Illinois, N USA *41°46′N 88°19′W*
31 *Q15* **Aurora** Indiana, N USA *39°01′N 84°55′W*
29 *W4* **Aurora** Minnesota, N USA *47°31′N 92°14′W*
27 *S8* **Aurora** Missouri, C USA *36°58′N 93°43′W*
29 *P16* **Aurora** Nebraska, C USA *40°52′N 98°00′W*
36 *J5* **Aurora** Utah, W USA *38°55′N 111°55′W*
Aurora *see* San Francisco, Philippines
Aurora *see* Maéwo, Vanuatu
94 *F10* **Aursjøen** ◎ S Norway
94 *I9* **Aursunden** ◎ S Norway
83 *D21* **Aus** Karas, SW Namibia *26°38′S 16°19′E*
Ausa *see* Vic
14 *E16* **Ausable** ≈ Ontario, S Canada
31 *O3* **Au Sable Point** *headland* Michigan, N USA *46°40′N 86°08′W*
31 *S7* **Au Sable Point** *headland* Michigan, N USA *44°19′N 83°20′W*
31 *R6* **Au Sable River** ≈ Michigan, N USA
57 *H16* **Ausangate, Nevado** ▲ C Peru *13°47′S 71°13′W*
Auschwitz *see* Oświęcim
Ausculum Apulum *see* Ascoli Satriano
105 *Q4* **Ausejo** La Rioja, N Spain *42°21′N 02°10′W*
108 *H7* **Ausser Rhoden** ⬖ *canton* NE Switzerland
Aussig *see* Ústí nad Labem
95 *F17* **Aust-Agder** ⬖ *county* S Norway
92 *P2* **Austfonna** *glacier* NE Svalbard
31 *P15* **Austin** Indiana, N USA *38°45′N 85°48′W*
29 *W11* **Austin** Minnesota, N USA *43°40′N 92°58′W*
35 *U5* **Austin** Nevada, W USA *39°30′N 117°05′W*
25 *S10* **Austin** *state capital* Texas, SW USA *30°16′N 97°45′W*
180 *J10* **Austin, Lake** *salt lake* Western Australia
31 *V11* **Austintown** Ohio, N USA *41°06′N 80°45′W*
25 *V9* **Austonio** Texas, SW USA *31°09′N 95°39′W*
Australes, Archipel des *see* Australes, Îles
Australes et Antarctiques Françaises, Terres *see* French Southern and Antarctic Lands
191 *T14* **Australes, Îles** *var.* Archipel des Australes, Îles Tubuai, Tubuai Islands, *Eng.* Austral Islands. *island group* SW French Polynesia
175 *Y11* **Austral Fracture Zone** *tectonic feature* S Pacific Ocean
174 *M8* **Australia** *continent*
181 *O7* **Australia** *off.* Commonwealth of Australia. ◆ *commonwealth republic*
Australia, Commonwealth of *see* Australia
183 *Q12* **Australian Alps** ▲▲ SE Australia
183 *R11* **Australian Capital Territory** *prev.* Federal Capital Territory. ⬖ *territory* SE Australia
Australie, Bassin Nord de l' *see* North Australian Basin
Austral Islands *see* Australes, Îles
Austrava *see* Ostrov
109 *T6* **Austria** *off.* Republic of Austria, *Ger.* Österreich. ◆ *republic* C Europe
Austria, Republic of *see* Austria
92 *K3* **Austurland** ⬖ *region* SE Iceland
92 *G10* **Austvågøya** *island* C Norway
Ausuitoq *see* Grise Fiord
58 *G13* **Autazes** Amazonas, N Brazil *03°37′S 59°08′W*
102 *M16* **Auterive** Haute-Garonne, S France *43°22′N 01°28′E*
Autesiodorum *see* Auxerre
103 *N2* **Authie** ≈ N France
Autissiodorum *see* Auxerre
40 *K14* **Autlán** *var.* Autlán de Navarro. Jalisco, SW Mexico *19°48′N 104°20′W*
Autlán de Navarro *see* Autlán
Autricum *see* Chartres
103 *Q9* **Autun** *anc.* Ædua, Augustodunum. Saône-et-Loire, C France *46°58′N 04°18′E*
Autz *see* Auce
99 *H20* **Auvelais** Namur, S Belgium *50°27′N 04°38′E*
103 *P11* **Auvergne** ⬖ *region* C France
102 *M12* **Auvézère** ≈ W France
103 *P7* **Auxerre** *anc.* Autesiodorum, Autissiodorum. Yonne, C France *47°48′N 03°35′E*
103 *N2* **Auxi-le-Château** Pas-de-Calais, N France *50°14′N 02°06′E*
103 *S8* **Auxonne** Côte d'Or, C France *47°12′N 05°22′E*
55 *P9* **Auyan Tebuy** ▲ SE Venezuela *05°48′N 62°27′W*
103 *O10* **Auzances** Creuse, C France *46°01′N 02°29′E*
27 *U8* **Ava** Missouri, C USA *36°57′N 92°39′W*
142 *M5* **Āvaj** Qazvin, N Iran
95 *C15* **Avaldsnes** Rogaland, S Norway *59°21′N 05°16′E*
103 *Q8* **Avallon** Yonne, C France *47°30′N 03°54′E*
102 *K6* **Avaloirs, Mont des** ▲ NW France *48°27′N 00°11′W*
35 *S16* **Avalon** Santa Catalina Island, California, W USA *33°20′N 118°19′W*
18 *J17* **Avalon** New Jersey, NE USA *39°04′N 74°42′W*
13 *V13* **Avalon Peninsula** *peninsula* Newfoundland and Labrador, E Canada
Avanersuaq *see* Avannaarsua
Avannaarsua *var* Avanersuaq, *Dan.* Nordgrønland. ⬖ *province* N Greenland
60 *K10* **Avaré** São Paulo, S Brazil *23°06′S 48°57′W*
Avaricum *see* Bourges
190 *H16* **Avarua** ○ (Cook Islands) Rarotonga, S Cook Islands *21°12′S 159°46′E*
190 *H16* **Avarua Harbour** *harbor* Rarotonga, S Cook Islands
Avasfelsőfalu *see* Negrești-Oaș
38 *L17* **Avatanak Island** *island* Aleutian Islands, Alaska, USA
190 *B16* **Avatele** S Niue *19°06′S 169°55′E*
190 *H15* **Avatiu Harbour** *harbor* Rarotonga, S Cook Islands
Avdeyevka *see* Avdiyivka
114 *J13* **Ávdira** Anatolikí Makedonía kai Thráki, NE Greece *40°58′N 24°58′E*
117 *X8* **Avdiyivka** *Rus.* Avdeyevka. Donets'ka Oblast', SE Ukraine *48°06′N 37°46′E*
Avdzaga *see* Gurvanbulag
104 *G6* **Ave** ≈ N Portugal
104 *G7* **Aveiro** *anc.* Talabriga. Aveiro, W Portugal *40°38′N 08°40′W*
104 *G7* **Aveiro** ⬖ *district* N Portugal
Avela *see* Ávila
99 *D18* **Avelgem** West-Vlaanderen, W Belgium *50°46′N 03°25′E*
61 *D20* **Avellaneda** Buenos Aires, E Argentina *34°43′S 58°23′W*
107 *L17* **Avellino** *anc.* Abellinum. Campania, S Italy *40°54′N 14°46′E*
35 *Q12* **Avenal** California, W USA *36°00′N 120°07′W*
Avenio *see* Avignon
94 *E8* **Averoya** *island* S Norway
107 *K17* **Aversa** Campania, S Italy *40°58′N 14°13′E*
33 *N9* **Avery** Idaho, NW USA *47°14′N 115°48′W*
25 *W5* **Avery** Texas, SW USA *33°33′N 94°46′W*
Aves, Islas de *see* Las Aves, Islas
Avesnes *see* Avesnes-sur-Helpe
103 *Q2* **Avesnes-sur-Helpe** *var.* Avesnes. Nord, N France *50°08′N 03°57′E*
64 *G12* **Aves Ridge** *undersea feature* SE Caribbean Sea *14°00′N 63°30′W*
95 *M14* **Avesta** Dalarna, C Sweden *60°09′N 16°10′E*
103 *O14* **Aveyron** ⬖ *department* S France
103 *N14* **Aveyron** ≈ S France
107 *J15* **Avezzano** Abruzzo, C Italy *42°02′N 13°26′E*
115 *D16* **Avgó** ▲ C Greece *39°31′N 21°24′E*
Avgustov *see* Augustów
Avgustovskiy Kanal *see* Augustowski, Kanał
96 *J9* **Aviemore** N Scotland, United Kingdom *57°06′N 04°01′W*
185 *F21* **Aviemore, Lake** ◎ South Island, New Zealand
103 *R15* **Avignon** *anc.* Avenio. Vaucluse, SE France *43°57′N 04°49′E*
104 *M7* **Ávila** *var.* Avila; *anc.* Abela, Abula, Abyla, Avela. Castilla y León, C Spain *40°39′N 04°42′W*
104 *L8* **Ávila** ⬖ *province* Castilla y León, C Spain
104 *K2* **Avilés** Asturias, NW Spain *43°33′N 05°55′W*
118 *J4* **Avinurme** *Ger.* Awwinorm. Ida-Virumaa, NE Estonia *58°58′N 26°53′E*
104 *H10* **Avis** Portalegre, C Portugal *39°03′N 07°53′W*
Avlum *see* Aulum
182 *M11* **Avoca** Victoria, SE Australia *37°09′S 143°34′E*
29 *T14* **Avoca** Iowa, C USA *41°27′N 95°20′W*
182 *M11* **Avoca River** ≈ Victoria, SE Australia
107 *L25* **Avola** Sicilia, Italy, C Mediterranean Sea *36°54′N 15°08′E*
18 *F10* **Avon** New York, NE USA *42°53′N 77°41′W*
29 *P12* **Avon** South Dakota, N USA *43°00′N 98°03′W*
97 *M23* **Avon** ≈ S England, United Kingdom
97 *L20* **Avon** ≈ C England, United Kingdom
36 *K13* **Avondale** Arizona, SW USA *33°25′N 112°20′W*
23 *X13* **Avon Park** Florida, SE USA *27°36′N 81°30′W*
102 *J5* **Avranches** Manche, N France *48°42′N 01°21′W*
Avveel *see* Ivalojoki, Finland
186 *M6* **Avuavu** *var.* Kolotambu. Guadalcanal, C Solomon Islands *09°52′S 160°25′E*
103 *O3* **Avure** ≈ N France
Avveel *see* Ivalo, Finland
Avvil *see* Ivalo
77 *O17* **Awaaso** *var.* Awaso. SW Ghana *06°10′N 02°18′W*
141 *X8* **Awābī** *var.* Al 'Awābī. NE Oman *23°20′N 57°35′E*
184 *L9* **Awakino** Waikato, North Island, New Zealand *38°40′S 174°37′E*

◆ Country
● Country Capital
◇ Dependent Territory
○ Dependent Territory Capital
⬖ Administrative Regions
✈ International Airport
▲ Mountain
▲▲ Mountain Range
Volcano
≈ River
◎ Lake
⊠ Reservoir

142 M15 'Awālī C Bahrain 26°07'N 50°33'E
99 K19 Awans Liège, E Belgium 50°39'N 05°30'E
184 I2 Awanui Northland, North Island, New Zealand 35°01'S 173°16'E
148 M14 Awārān Baluchistān, SW Pakistan 26°31'N 65°01'E
81 K16 Awara Plain plain NE Kenya
80 M13 Awarē Sumalē, E Ethiopia 08°12'N 44°09'E
138 M6 'Awārif, Wādī dry watercourse E Syria
185 B20 Awarua Point headland South Island, New Zealand 44°15'S 168°03'E
81 J14 Āwasa Southern Nationalities, S Ethiopia 06°54'N 38°26'E
80 K13 Āwash Afar, NE Ethiopia 08°59'N 40°16'E
80 K12 Āwash var. Hawash. ♒ C Ethiopia
Awaso see Awaaso
158 H7 Awat Xinjiang Uygur Zizhiqu, NW China
185 J15 Awatere ♒ South Island, NZ
75 O10 Awbārī SW Libya 26°35'N 12°46'E
75 N9 Awbārī, Idhān var. Edeyen d'Oubari. desert Algeria/Libya
80 M12 Awdal off. Gobolka Awdal. ◆ N Somalia
80 C13 Aweil Northern Bahr el Ghazal, NW South Sudan 08°42'N 27°20'E
96 H11 Awe, Loch ◎ W Scotland, United Kingdom
77 U16 Awka Anambra, SW Nigeria 06°04'N 07°05'E
39 O6 Awuna River ♒ Alaska, USA
Awwinorm see Avinurme
Axarfjördhur see Öxarfjördhur
103 N17 Axat Aude, S France 42°47'N 02°14'E
99 F16 Axel Zeeland, SW Netherlands 51°16'N 03°55'E
197 P9 Axel Heiberg Island var. Axel Heiberg. island Nunavut, N Canada
Axel Heiburg see Axel Heiberg Island
77 O17 Axim S Ghana 04°53'N 02°14'E
114 F13 Axiós var. Vardar. ♒ Greece/FYR Macedonia see also Vardar
Axiós see Vardar
103 N17 Ax-les-Thermes Ariège, S France 42°43'N 01°49'E
120 D11 Ayachi, Jbel ▲ C Morocco 32°30'N 05°00'W
61 D22 Ayacucho Buenos Aires, E Argentina 37°09'S 58°30'W
57 F15 Ayacucho, S Peru 13°10'S 74°15'W
57 E16 Ayacucho off. Departamento de Ayacucho. ◆ department SW Peru
Ayacucho, Departamento de see Ayacucho
145 W11 Ayagoz var. Ayaguz, Kaz. Ayaköz; prev. Sergiopol. Vostochnyy Kazakhstan, E Kazakhstan 47°54'N 80°25'E
145 V12 Ayagoz var. Ayaguz, Kaz. Ayaköz. ♒ E Kazakhstan
Ayaguz see Ayagoz
Ayakagytma see Oyoqog'itma
158 L10 Ayakkum Hu ◎ NW China
104 H14 Ayamonte Andalucía, S Spain 37°13'N 07°24'W
123 S11 Ayan Khabarovskiy Kray, E Russian Federation 56°27'N 138°09'E
136 D10 Ayancık Sinop, N Turkey 41°56'N 34°35'E
55 S9 Ayangganna Mountain ▲ C Guyana 05°21'N 59°54'W
77 U16 Ayangba Kogi, C Nigeria 07°36'N 07°10'E
123 O7 Ayanka Krasnoyarskiy Kray, E Russian Federation 63°42'N 167°31'E
54 E7 Ayapel Córdoba, NW Colombia 08°16'N 75°10'W
136 H12 Ayaş Ankara, N Turkey 40°02'N 32°21'E
57 I16 Ayaviri Puno, S Peru 14°53'S 70°35'W
Aybak see Aibak
147 N10 Aydarko'l Ko'li Rus. Ozero Aydarkul'. ◎ C Uzbekistan
Aydarkul', Ozero see Aydarko'l Ko'li
21 W10 Ayden North Carolina, SE USA 35°28'N 77°25'W
136 C15 Aydın var. Aïdin; anc. Tralles Aydin. Aydin, SW Turkey 37°51'N 27°51'E
136 C15 Aydın var. Aïdin. ◆ province SW Turkey
136 C15 Aydıncık İçel, S Turkey 36°08'N 33°17'E
136 C15 Aydın Dağları ▲ W Turkey
158 L6 Aydıngkol Hu ◎ NW China
127 X7 Aydyrlinskiy Orenburgskaya Oblast', W Russian Federation 52°03'N 59°54'E
105 S4 Ayerbe Aragón, NE Spain 42°16'N 00°41'W
Ayers Rock see Uluru
96 K8 Ayeyarwady see Ayeyawady
Ayeyarwady see Irrawaddy
168 K8 Ayeyarwady prev. Ayeyarwady. ◆ region SW Myanmar (Burma)
Ayiá see Agiá
Ayia Napa see Agía Nápa
Ayia Phyla see Agía Fylaxis
Ayiássos/Ayiássos see Agiassós
Áyios Evstrátios see Ágios Efstrátios
Áyios Kírikos see Ágios Kírykos
Áyios Nikólaos see Ágios Nikólaos
80 I11 Aykel Āmara, N Ethiopia 12°33'N 37°01'E
123 N9 Aykhal Respublika Sakha (Yakutiya), NE Russian Federation 66°07'N 110°25'E
14 U12 Aylen Lake ◎ Ontario, SE Canada
97 N21 Aylesbury SE England, United Kingdom 51°50'N 00°50'W
15 O6 Ayllón Castilla y León, N Spain 41°25'N 03°23'W

14 F17 Aylmer Ontario, S Canada 42°46'N 80°57'W
14 L12 Aylmer Québec, SE Canada 45°23'N 75°51'W
15 R12 Aylmer, Lac ◎ Québec, SE Canada
8 L9 Aylmer Lake ◎ Northwest Territories, NW Canada
145 V14 Aynabulak Kaz. Aynabulaq. Almaty, SE Kazakhstan 44°37'N 77°59'E
138 K2 'Ayn al 'Arab Kurd. Kobanî. Ḥalab, N Syria 36°55'N 38°21'E
139 V12 'Ayn Ḥamūd Dhī Qār, S Iraq 30°51'N 45°93'E
147 P12 Ayni prev. Varzimanor Ayni. W Tajikistan 39°24'N 68°30'E
140 M10 'Aynīn var. Aynayn. spring/well W Saudi Arabia 20°52'N 41°41'E
21 U12 Aynor South Carolina, SE USA 33°59'N 79°11'W
139 Q7 'Ayn Zāzūh Al Anbār, C Iraq 33°29'N 42°34'E
153 N12 Ayodhya Uttar Pradesh, N India 26°47'N 82°12'E
123 S6 Ayon, Ostrov island NE Russian Federation
105 R11 Ayora Valenciana, E Spain 39°04'N 01°01'W
77 Q11 Ayorou Tillabéri, W Niger 14°45'N 00°54'E
79 E16 Ayos Centre, S Cameroon 03°53'N 12°31'E
76 L5 'Ayoûn 'Abd el Mâlek well N Mauritania
76 K10 'Ayoûn el 'Atroûs var. Aïoun el Atrous, Aïoun el Atroûss. Hodh el Gharbi, SE Mauritania 16°38'N 09°36'W
96 H13 Ayr W Scotland, United Kingdom 55°28'N 04°38'W
96 I13 Ayr ♒ W Scotland, United Kingdom
96 H13 Ayrshire cultural region SW Scotland, United Kingdom
Aysen see Aisén
80 L12 Āysha Sumalē, E Ethiopia 10°36'N 42°31'E
144 L14 Ayteke Bi Kaz. Zhangaqazaly; prev. Novokazalinsk. Kzylorda, SW Kazakhstan
146 K8 Aytim Navoiy Viloyati, N Uzbekistan 42°15'N 62°15'E
181 W4 Ayton Queensland, NE Australia 15°54'S 145°19'E
114 M9 Aytos Burgas, E Bulgaria 42°43'N 27°14'E
171 T11 Ayu, Kepulauan island group E Indonesia
167 U12 A'yun Pa prev. Cheo Reo. Gia Lai, S Vietnam 13°19'N 108°27'E
169 V11 Ayu, Tanjung headland Borneo, N Indonesia 0°25'N 117°34'E
41 P16 Ayutla var. Ayutla de los Libres. Guerrero, S Mexico 16°51'N 99°16'W
40 K13 Ayutla Jalisco, C Mexico 20°07'N 104°18'W
Ayutla de los Libres see Ayutla
167 O11 Ayutthaya var. Phra Nakhon Si Ayutthaya. Phra Nakhon Si Ayutthaya, C Thailand 14°20'N 100°35'E
136 B13 Ayvalık Balıkesir, W Turkey 39°18'N 26°42'E
99 L20 Aywaille Liège, E Belgium 50°28'N 05°40'E
141 R13 'Aywat aş Şay'ar, Wādī seasonal river N Yemen
Azaffal see Azeffâl
141 P6 Az Zilfī Ar Riyāḍ, N Saudi Arabia 26°17'N 44°48'E
139 Y13 Az Zubayr var. Al Zubair. Al Başrah, SE Iraq 30°24'N 47°45'E
Az Zuqur see Jabal Zuqar, Jazīrat

B

187 X15 Ba prev. Mba. Viti Levu, W Fiji 17°35'S 177°40'E
Ba see Da Răng, Sông
171 P17 Baa Pulau Rote, C Indonesia 10°44'S 123°06'E
138 H7 Baalbek var. Ba'labakk; anc. Heliopolis. E Lebanon 34°00'N 36°11'E
108 D7 Baar Zug, N Switzerland 47°12'N 08°32'E
81 L11 Baardheere var. Bardere, It. Bardera. Gedo, SW Somalia 02°13'N 42°19'E
99 I15 Baarle-Hertog Antwerpen, N Belgium 51°26'N 04°56'E
99 I15 Baarle-Nassau Noord-Brabant, S Netherlands 51°27'N 04°56'E
98 J11 Baarn Utrecht, C Netherlands 52°13'N 05°18'E
126 H9 Baatsagaan var. Bayansayr. Bayankhongor, C Mongolia 45°36'N 99°27'E
114 D13 Baba var. Buševa, Gk. Varnoús. ▲ FYR Macedonia/Greece
76 H11 Bababé Brakna, W Mauritania 16°22'N 13°57'W
136 G10 Baba Burnu headland NW Turkey 41°18'N 33°44'E
117 N13 Babadag Tulcea, SE Romania 44°53'N 28°47'E
137 X10 Babadağ Dağı ▲ NE Azerbaijan 41°02'N 48°04'E
95 J17 Babaeski Kırklareli, NW Turkey 41°24'N 27°06'E
58 B7 Babahoyo prev. Bodegas. Los Ríos, C Ecuador 01°53'S 79°31'W
167 T6 Ba Ninh Ha Bắc, N Vietnam 21°10'N 106°04'E
40 J6 Bacoachi Sonora, NW Mexico 30°36'N 110°00'W
171 P6 Bacolod City Negros, C Philippines 10°43'N 122°58'E

64 N2 Azores var. Açores, Ilhas dos Açores, Port. Arquipélago dos Açores. island group Portugal, NE Atlantic Ocean
64 L8 Azores-Biscay Rise undersea feature E Atlantic Ocean 19°00'W 42°40'N
Azotos/Azotus see Ashdod
78 K11 Azoum, Bahr seasonal river SE Chad
126 L12 Azov Rostovskaya Oblast', SW Russian Federation 47°07'N 39°26'E
126 J13 Azov, Sea of Rus. Azovskoye More, Ukr. Azovs'ke More. sea NE Black Sea
Azovs'ke More/Azovskoye More see Azov, Sea of
138 I10 Azraq, Wāḥat al oasis N Jordan
Āzro see Āzrow
74 G6 Azrou C Morocco 33°30'N 05°12'W
149 R5 Āzrow var. Āzro. Lōgar, E Afghanistan 34°11'N 69°39'E
37 P8 Aztec New Mexico, SW USA 36°49'N 107°59'W
36 M13 Aztec Peak ▲ Arizona, SW USA 33°48'N 110°54'W
45 N9 Azua var. Azua de Compostela. S Dominican Republic 18°29'N 70°44'W
Azua de Compostela see Azua
104 K12 Azuaga Extremadura, W Spain 38°16'N 05°40'W
58 B8 Azuay ◆ province W Ecuador
164 C13 Azuchi-Ō-shima island SW Japan
105 O11 Azuer ♒ C Spain
43 S17 Azuero, Península de peninsula S Panama
62 I6 Azufre, Volcán var. Volcán Lastarria. ▲ N Chile 25°16'S 68°45'W
116 J12 Azuga Prahova, SE Romania 45°27'N 25°34'E
61 C22 Azul Buenos Aires, E Argentina 36°46'S 59°50'W
62 I8 Azul, Cerro ▲ NW Argentina 28°28'S 68°43'W
57 E12 Azul, Cordillera ▲ C Peru
165 P11 Azuma-san ▲ Honshū, C Japan 37°44'N 140°05'E
103 U15 Azur, Côte d' coastal region SE France
191 Z3 Azur Lagoon ◎ Kiritimati, E Kiribati
'Azza see Gaza
Az Zāb al Kabīr see Great Zab
138 H7 Az Zabdānī var. Zabadani. Rif Dimashq, W Syria 33°45'N 36°07'E
141 W8 Az Zāhirah desert NW Oman
141 S6 Az Zahrān Eng. Dhahran. Ash Sharqīyah, NE Saudi Arabia 26°18'N 50°02'E
141 R6 Az Zahrān al Khubar var. Dhahran Al Khobar. ✈ Ash Sharqīyah, NE Saudi Arabia 26°28'N 49°42'E
75 W7 Az Zaqāzīq var. Zagazig. N Egypt 30°36'N 31°32'E
138 H10 Az Zarqā' var. Zarqa. Az Zarqā', NW Jordan 32°04'N 36°06'E
138 I11 Az Zarqā' off. Muḥāfazat az Zarqā', var. Zarqa. ◆ governorate N Jordan
75 O7 Az Zāwiyah var. Zawia. NW Libya 32°45'N 12°44'E
141 N15 Az Zaydīyah W Yemen 15°20'N 43°03'E
74 H11 Azzel Matti, Sebkha var. Sebkra Azz el Matti. salt flat C Algeria

Babashy, Gory see Babasy
146 C9 Babasy Rus. Gory Babashy. ▲ W Turkmenistan
168 M13 Babat Sumatera, W Indonesia 02°45'S 101°01'E
Babatag, Khrebet see Bobotog', Tizmasi
81 H21 Babati Manyara, NE Tanzania 04°13'S 35°45'E
124 J13 Babayevo Vologodskaya Oblast', NW Russian Federation 59°23'N 35°52'E
127 Q15 Babayurt Respublika Dagestan, SW Russian Federation 43°38'N 46°49'E
33 P6 Babb Montana, NW USA 48°51'N 113°26'W
29 X4 Babbitt Minnesota, N USA 47°42'N 91°56'W
188 E9 Babeldaob var. Babeldaop, Babelthuap. island N Palau
Babeldaop see Babeldaob
141 N16 Bab el Mandeb strait Gulf of Aden/Red Sea
Babelthuap see Babeldaob
111 K17 Babia Góra var. Babia Hora. ▲ Poland/Slovakia 49°33'N 19°32'E
Babia Hora see Babia Góra
Babian Jiang see Black River
Babichi see Babichy
Babichy Rus. Babichi. Homyel'skaya Voblasts', SE Belarus
169 O13 Babo Papua Barat, E Indonesia 02°33'S 133°25'E
148 K7 Bābol var. Bābul. Māzandarān, N Iran 36°32'N 52°42'E
149 U9 Bābol, Mahallā at Bābil, Babylon, Al Ḥillah see Bābil
149 S6 Bābolsar var. Babolsar, Al Ḥillah. Māzandarān, N Iran 36°43'N 52°38'E
31 S8 Bad Axe Michigan, N USA 43°48'N 83°00'W
101 G16 Bad Berleburg Nordrhein-Westfalen, W Germany 51°03'N 08°24'E
101 L17 Bad Blankenburg Thüringen, C Germany 50°43'N 11°19'E
101 J20 Bad Windsheim Bayern, C Germany 49°30'N 10°25'E
101 J23 Bad Wörishofen Bayern, S Germany 48°00'N 10°36'E
100 G10 Bad Zwischenahn Niedersachsen, NW Germany 53°11'N 08°01'E
104 M13 Baena Andalucía, S Spain 36°37'N 04°20'W
163 V15 Baengnyong-do prev. Paengnyǒng. island NW South Korea
79 X4 Baden var. Baden bei Wien; anc. Aquae Panoniae, Thermae Panonicae. Niederösterreich, NE Austria 48°01'N 16°14'E
101 G21 Baden Aargau, N Switzerland 47°28'N 08°19'E
101 G21 Baden-Baden anc. Aurelia Aquensis. Baden-Württemberg, SW Germany 48°46'N 08°14'E
Baden bei Wien see Baden
101 G22 Baden-Württemberg Fr. Bade-Wurtemberg. ◆ state SW Germany
Bade-Wurtemberg see Baden-Württemberg
Bade-Wurttemberg see Baden-Württemberg
101 H20 Bad Fredrichshall Baden-Württemberg, S Germany 49°13'N 09°15'E
100 P11 Bad Freienwalde Brandenburg, NE Germany 52°47'N 14°04'E
109 Q8 Badgastein var. Gastein. Salzburg, NW Austria 47°07'N 13°09'E
148 L4 Bādghīs ◆ province NW Afghanistan
109 T5 Bad Hall Oberösterreich, N Austria 48°03'N 14°13'E
101 J14 Bad Harzburg Niedersachsen, C Germany 51°52'N 10°34'E
101 I16 Bad Hersfeld Hessen, C Germany 50°52'N 09°42'E
116 K11 Badhoevedorp Noord-Holland, C Netherlands 52°21'N 04°46'E
109 Q8 Bad Hofgastein Salzburg, NW Austria 47°11'N 13°07'E
101 F19 Bad Homburg see Bad Homburg vor der Höhe
101 F19 Bad Homburg vor der Höhe var. Bad Homburg. Hessen, W Germany 50°14'N 08°37'E
101 E17 Bad Honnef Nordrhein-Westfalen, W Germany 50°39'N 07°13'E
101 H20 Bad Hersfeld (duplicate?) ...
21 S10 Badin Sind, SE Pakistan 24°38'N 68°53'E
149 O13 Badin North Carolina, SE USA 35°25'N 80°07'W
79 R6 Bad Ischl Oberösterreich, N Austria 47°43'N 13°38'E
28 J6 Badlands physical region North Dakota/South Dakota, N USA
123 O13 Badarin Respublika Buryatiya, S Russian Federation 52°27'N 113°34'E
61 G17 Bagé Rio Grande do Sul, S Brazil 31°22'S 54°06'W
Bagenalstown see Muine Bheag
153 T16 Bagerhat var. Bagherhat. Khulna, S Bangladesh 22°40'N 89°48'E
Bages et de Sigean, Étang de ◎ S France
33 W17 Baggs Wyoming, C USA 41°02'N 107°39'W
155 F11 Bāgh Madhya Pradesh, C India 22°24'N 74°51'E
139 T8 Baghdad var. Bagdad, Eng. Baghdad. ● Baghdād, C Iraq 33°20'N 44°26'E
139 T8 Baghdād var. Amānat al 'Āşimah. ◆ governorate C Iraq
139 T8 Baghdād ✈ (Baghdād) Baghdad, C Iraq
Baghdād, Muḩāfaz at see Baghdad

111 J24 Bács-Kiskun off. Bács-Kiskun Megye. ◆ county S Hungary
Bács-Kiskun Megye see Bács-Kiskun
Bácsszenttamás see Srbobran
Bácstopolya see Bačka Topola
Bactra see Balkh
Bada see Xilin
155 F21 Badagara var. Vadakara. Kerala, SW India 11°36'N 75°34'E see also Vadakara
162 I13 Badain Jaran Shamo desert N China
104 I11 Badajoz anc. Pax Augusta. Extremadura, W Spain 38°53'N 06°58'W
104 I11 Badajoz ◆ province W Spain
149 S2 Badakhshān ◆ province NE Afghanistan
105 W6 Badalona anc. Baetulo. Cataluña, E Spain 41°27'N 02°15'E
154 O11 Bādāmpāhārh var. Badampahar. Odisha, E India 22°04'N 86°06'E
169 U9 Badas, Kepulauan island group W Indonesia
109 S6 Bad Aussee Salzburg, C Austria 47°35'N 13°44'E
101 G16 Bad Berleburg Nordrhein-Westfalen, W Germany 51°03'N 08°24'E
101 I24 Bad Waldsee Baden-Württemberg, S Germany 47°54'N 09°44'E
35 U11 Badwater Basin depression California, W USA
101 J20 Bad Windsheim Bayern, C Germany 49°30'N 10°25'E
101 J23 Bad Borseck see Borsec
101 G18 Bad Camberg Hessen, W Germany 5°018'N 08°15'E
101 L8 Bad Doberan Mecklenburg-Vorpommern, N Germany 54°06'N 11°54'E
101 N14 Bad Düben Sachsen, E Germany 51°35'N 12°34'E
109 X4 Baden var. Baden bei Wien; anc. Aquae Panoniae, Thermae Panonicae. Niederösterreich, NE Austria 48°01'N 16°14'E
108 F9 Baden Aargau, N Switzerland 47°28'N 08°19'E
105 N13 Baeza Andalucía, S Spain 38°00'N 03°28'E
79 D15 Bafang Ouest, W Cameroon 05°10'N 10°11'E
76 H12 Bafatá C Guinea-Bissau 12°09'N 14°39'W
112 A10 Baderna Istra, NW Croatia 45°12'N 13°45'E
101 H20 Bad Fredrichshall Baden-Württemberg, S Germany 49°13'N 09°15'E
25 T15 Baffin Bay inlet Texas, SW USA
196 M12 Baffin Island island Nunavut, NE Canada
109 Q8 Badgastein var. Gastein. Salzburg, NW Austria 47°07'N 13°09'E
109 Q8 Badger State see Wisconsin
79 E15 Bafia Centre, C Cameroon 04°49'N 11°14'E
77 R14 Bafilo NE Togo 09°22'N 01°14'E
76 J12 Bafing ♒ W Africa
76 J12 Bafoulabé Kayes, W Mali 13°43'N 10°49'W
79 D15 Bafoussam Ouest, W Cameroon 05°31'N 10°25'E
143 R9 Bāfq Yazd, C Iran 31°35'N 55°21'E
136 L10 Bafra Samsun, N Turkey 41°34'N 35°56'E
136 L10 Bafra Burnu headland N Turkey 41°42'N 36°02'E
143 S12 Bāft Kermān, S Iran 29°13'N 56°37'E
79 N18 Bafwabalinga NE Dem. Rep. Congo 0°52'N 26°55'E
79 N18 Bafwaboli Orientale, NE Dem. Rep. Congo 0°06'N 26°08'E
79 N17 Bafwasende Orientale, NE Dem. Rep. Congo 01°00'N 27°09'E
42 K13 Bagaces Guanacaste, NW Costa Rica 10°31'N 85°18'W
153 O12 Bagaha Bihār, N India 27°08'N 84°04'E
155 F16 Bagalkot Karnātaka, W India 16°11'N 75°42'E
81 J22 Bagamoyo Pwani, E Tanzania 06°26'S 38°55'E
168 J9 Bagan Datuk see Bagan Datuk
168 J8 Bagan Datuk var. Bagan Datoh. Perak, Peninsular Malaysia 03°58'N 100°47'E
171 R7 Baganga Mindanao, S Philippines 07°31'N 126°34'E
168 J9 Bagansiapiapi var. Pasirpangarayan. Sumatera, W Indonesia 02°06'N 100°52'E
162 M8 Baganuur var. Nüürst. Töv, C Mongolia 47°44'N 108°22'E
Bagaria see Bagheria
Bagarua Tahoua, W Niger 14°34'N 04°24'E
168 J8 Bagata Bandundu, W Dem. Rep. Congo 03°44'S 17°57'E
123 O13 Bagdad see Baghdad
109 T3 Bad Leonfelden Oberösterreich, N Austria 48°31'N 14°17'E
101 I20 Bad Mergentheim Baden-Württemberg, SW Germany 49°31'N 09°46'E
101 H17 Bad Nauheim Hessen, W Germany 50°22'N 08°45'E
101 E17 Bad Neuenahr-Ahrweiler Rheinland-Pfalz, W Germany 50°33'N 07°07'E
101 H17 Bad Neustadt an der Saale var. Bad Neustadt. Bayern, C Germany 50°21'N 10°13'E
101 H18 Bad Oeynhausen Nordrhein-Westfalen, NW Germany 52°12'N 08°48'E
100 J9 Bad Oldesloe Schleswig-Holstein, N Germany 53°48'N 10°22'E
101 H18 Bad Pyrmont Niedersachsen, C Germany 51°58'N 09°16'E

109 X9 Bad Radkersburg Steiermark, SE Austria 46°40'N 16°02'E
139 V8 Badrah Wāsiṭ, E Iraq 33°06'N 45°58'E
101 N24 Bad Reichenhall Bayern, SE Germany 47°43'N 12°52'E
140 K8 Badr Ḥunayn Al Madīnah, W Saudi Arabia 23°46'N 38°45'E
152 K8 Badrīnāth ▲ N India
30 K4 Bad River ♒ Wisconsin, N USA
101 M24 Bad River ♒ South Dakota, N USA
100 H13 Bad Salzuflen Nordrhein-Westfalen, NW Germany 52°06'N 08°45'E
101 J16 Bad Salzungen Thüringen, C Germany 50°48'N 10°15'E
109 V8 Bad Sankt Leonhard im Lavanttal Kärnten, S Austria 46°55'N 14°51'E
100 K9 Bad Schwartau Schleswig-Holstein, N Germany 53°55'N 10°42'E
101 L24 Bad Tölz Bayern, SE Germany 47°44'N 11°34'E
181 U1 Badu Island island Queensland, NE Australia
109 X5 Bad Vöslau Niederösterreich, NE Austria 47°58'N 16°13'E
101 I24 Bad Waldsee Baden-Württemberg, S Germany 47°54'N 09°44'E
35 U11 Badwater Basin depression California, W USA
76 M13 Bagoé ♒ Ivory Coast/Mali
149 S6 Bagrāmī var. Bagrāmē. Kābōl, E Afghanistan 34°29'N 69°16'E
119 B14 Bagrationovsk Ger. Preussisch Eylau. Kaliningradskaya Oblast', W Russian Federation 54°24'N 20°39'E
Bagrax see Bohu
Bagrax Hu see Bosten Hu
56 C10 Bagua Amazonas, NE Peru 05°37'S 78°36'W
171 O2 Baguio off. Baguio City. Luzon, N Philippines 16°25'N 120°36'E
Baguio City see Baguio
77 V9 Bagzane, Monts ▲ N Niger 17°48'N 08°43'E
Bāḩah, Minṭaqat al see Al Bāḩah
Bahama Islands see Bahamas, The
76 H12 Bahamas, The off. Commonwealth of The Bahamas. ◆ commonwealth republic ♦ N West Indies
0 L13 Bahamas, The var. Bahama Islands. Bahamas. ♦ island group N West Indies
44 H3 Bahamas, The off. Commonwealth of The Bahamas. ♦ commonwealth republic ♦ N West Indies
153 S15 Baharampur var. Berhampore. West Bengal, NE India 24°06'N 88°19'E
146 E12 Baharly var. Bäherden, Rus. Bakharden; prev. Bakherden. Ahal Welayaty, C Turkmenistan 38°30'N 57°32'E
149 U10 Bahāwalnagar Punjab, E Pakistan 30°00'N 73°03'E
149 T11 Bahāwalpur Punjab, E Pakistan 29°25'N 71°40'E
136 L16 Bahçe Osmaniye, S Turkey 37°12'N 36°34'E
160 J8 Ba He ♒ C China
Bäherden see Baharly
59 N16 Bahia off. Estado da Bahia. ♦ state E Brazil
61 B24 Bahía Blanca Buenos Aires, E Argentina 38°43'S 62°19'W
40 L15 Bahía Bufadero Michoacán, SW Mexico
63 J19 Bahía Bustamante Chubut, SE Argentina 45°06'S 66°30'W
40 D5 Bahía de los Ángeles Baja California Norte, NW Mexico 28°58'N 113°34'W
40 C6 Bahía de Tortugas Baja California Sur, NW Mexico 27°42'N 114°54'W
Bahía, Estado da see Bahia
30 J4 Bahía, Islas de la Eng. Bay Islands. island group N Honduras
40 E5 Bahía Kino Sonora, NW Mexico 28°48'N 111°55'W
40 E9 Bahía Tortugas var. Bahía Salina Cruz. Baja California Sur, NW Mexico 24°34'N 112°07'W
54 C8 Bahía Solano var. Ciudad Mutis, Solano. Chocó, W Colombia 06°13'N 77°22'W
80 I11 Bahir Dar var. Bahr Dar, Bahrdar Giyorgis. Āmara, N Ethiopia 11°33'N 37°23'E
141 X8 Bahlā' var. Bahlah, Bahlat. NW Oman 22°58'N 57°16'E
152 M11 Bahraich Uttar Pradesh, N India 27°35'N 81°36'E
143 N12 Bahrain off. Kingdom of Bahrain, Ar. Dawlat al Baḥrayn, Ar. Al Baḥrayn; prev. Tylos, Tyros. ◆ monarchy SW Asia
142 M14 Bahrain ✈ C Bahrain
142 M15 Bahrain, Gulf of gulf Persian Gulf, NW Arabian Sea
Bahrain, State of see Bahrain
138 I7 Baḩrat Mallāḩah ◎ W Syria
Bahr Dar/Bahrdar Giyorgis see Bahir Dar
Bahrein see Bahrain
Baḩr el Azraq see Blue Nile
Bahr el Gebel see Central Equatoria
Bahr el Jebel see Central Equatoria
80 E13 Baḩr ez Zaref ♒ Jonglei, S South Sudan
67 R8 Bahr Kameur ♒ N Central African Republic
Bahr Tabariya, Sea of see Tiberias, Lake
143 W15 Bāhū Kalāt Sīstān va Balūchestān, SE Iran 25°42'N 61°28'E
118 N13 Bahushewsk Rus. Bogushëvsk. Vitsyebskaya Voblasts', NE Belarus 54°51'N 30°11'E
Bai see Tagow Bāy

107 J23 Bagheria var. Bagaria. Sicilia, Italy, C Mediterranean Sea 38°05'N 13°31'E
143 S10 Bāghīn Kermān, C Iran 30°50'N 56°30'E
149 Q3 Baghlān Baghlān, NE Afghanistan 36°11'N 68°44'E
149 Q3 Baghlān var. Baghlān. ◆ province NE Afghanistan
148 M7 Baghlān Helmand, S Afghanistan 31°35'N 64°57'E
29 T4 Bagley Minnesota, N USA 47°31'N 95°24'W
106 H10 Bagnacavallo Emilia-Romagna, C Italy 44°25'N 11°58'E
102 K16 Bagnères-de-Bigorre Hautes-Pyrénées, S France 43°04'N 00°09'E
102 L17 Bagnères-de-Luchon Hautes-Pyrénées, S France 42°46'N 00°34'E
106 F11 Bagni di Lucca Toscana, C Italy 44°01'N 10°38'E
106 H11 Bagno di Romagna Emilia-Romagna, C Italy 43°51'N 11°57'E
103 R14 Bagnols-sur-Cèze Gard, S France 44°10'N 04°37'E
162 M14 Bag Nur ◎ N China
166 L8 Bago var. Pegu. Bago, SW Myanmar (Burma) 17°18'N 96°31'E
171 P6 Bago off. Bago City. Negros, C Philippines 10°30'N 122°49'E
166 L7 Bago var. Pegu. ◆ region S Myanmar (Burma)
Bago City see Bago
76 M13 Bagoé ♒ Ivory Coast/Mali
149 R5 Bagrāmī var. Bagrāmē. Kābōl, E Afghanistan 34°29'N 69°16'E
119 B14 Bagrationovsk Ger. Preussisch Eylau. Kaliningradskaya Oblast', W Russian Federation 54°24'N 20°39'E
56 C10 Bagua Amazonas, NE Peru 05°37'S 78°36'W
171 O2 Baguio off. Baguio City. Luzon, N Philippines 16°25'N 120°36'E
77 V9 Bagzane, Monts ▲ N Niger 17°48'N 08°43'E
76 H12 Bahama Islands see Bahamas, The
44 H3 Bahamas, The off. Commonwealth of The Bahamas. ◆ commonwealth republic ♦ N West Indies
153 S15 Baharampur var. Berhampore. West Bengal, NE India 24°06'N 88°19'E
146 E12 Baharly var. Bäherden, Rus. Bakharden; prev. Bakherden. Ahal Welayaty, C Turkmenistan 38°30'N 57°32'E
149 U10 Bahāwalnagar Punjab, E Pakistan 30°00'N 73°03'E
149 T11 Bahāwalpur Punjab, E Pakistan 29°25'N 71°40'E
136 L16 Bahçe Osmaniye, S Turkey 37°12'N 36°34'E
160 J8 Ba He ♒ C China
59 N16 Bahia off. Estado da Bahia. ♦ state E Brazil
61 B24 Bahía Blanca Buenos Aires, E Argentina 38°43'S 62°19'W

Footer

◆ Country ◇ Dependent Territory ◈ Administrative Regions ▲ Mountain ▲ Volcano ◎ Lake
● Country Capital ○ Dependent Territory Capital ✈ International Airport ▲ Mountain Range ♒ River ▨ Reservoir

Baia de Aramă – Banghāzī

116 G13 **Baia de Aramă** Mehedinţi, SW Romania 45°00′N 22°43′E
116 G11 **Baia de Criş** *Ger.* Altenburg, *Hung.* Körösbánya. Hunedoara, SW Romania 46°10′N 22°41′E
83 A16 **Baia dos Tigres** Namibe, SW Angola 16°36′S 11°44′E
82 A13 **Baia Farta** Benguela, W Angola 12°38′S 13°12′E
116 H9 **Baia Mare** *Ger.* Frauenbach, *Hung.* Nagybánya; *prev.* Neustadt. Maramureş, NW Romania 47°40′N 23°35′E
116 H8 **Baia Sprie** *Ger.* Mittelstadt, *Hung.* Felsőbánya. Maramureş, NW Romania 47°40′N 23°42′E
78 G13 **Baïbokoum** Logone-Oriental, SW Chad 07°46′N 15°43′E
160 F12 **Baicao Ling** ▲▲ SW China
163 U9 **Baicheng** *var.* Pai-ch'eng; *prev.* T'aon-an. Jilin, NE China 45°32′N 122°51′E
158 I6 **Baicheng** *var.* Bay. Xinjiang Uygur Zizhiqu, NW China 41°49′N 81°45′E
116 J13 **Băicoi** Prahova, SE Romania 45°02′N 25°51′E
Baidoa *see* Baydhabo
15 U6 **Baie-Comeau** Québec, SE Canada 49°12′N 68°10′W
15 U6 **Baie-des-Sables** Québec, SE Canada 48°41′N 67°55′W
15 T7 **Baie-des-Bacon** Québec, SE Canada 48°31′N 69°17′W
15 S8 **Baie-des-Rochers** Québec, SE Canada 47°57′N 69°50′W
Baie-du-Poste *see* Mistissini
172 H17 **Baie Lazare** Mahé, NE Seychelles 04°45′S 55°29′E
45 Y5 **Baie-Mahault** Basse Terre, C Guadeloupe 16°16′N 61°35′W
15 R9 **Baie-St-Paul** Québec, SE Canada 47°27′N 70°30′W
15 V5 **Baie-Trinité** Québec, SE Canada 49°25′N 67°20′W
13 T11 **Baie Verte** Newfoundland and Labrador, SE Canada 49°55′N 56°12′W
Baiguan *see* Shangyu
Baihe *see* Erdaobaihe
139 U11 **Bā'ij al Mahdī** Al Muthannā, S Iraq 31°21′N 44°57′E
Baiji *see* Bayjī
Baikal, Lake *see* Baykal, Ozero
Bailādila *see* Kirandul
Baile an Chaistil *see* Ballycastle
Baile an Róba *see* Ballinrobe
Baile an tSratha *see* Ballintra
Baile Átha an Rí *see* Athenry
Baile Átha Buí *see* Athboy
Baile Átha Cliath *see* Dublin
Baile Átha Fhirdhia *see* Ardee
Baile Átha Í *see* Athy
Baile Átha Luain *see* Athlone
Baile Átha Troim *see* Trim
Baile Brigín *see* Balbriggan
Baile Easa Dara *see* Ballysadare
116 I13 **Băile Govora** Vâlcea, SW Romania 45°00′N 24°08′E
116 F13 **Băile Herculane** *Ger.* Herkulesbad, *Hung.* Herkulesfürdő. Caraş-Severin, SW Romania 44°51′N 22°24′E
Baile Locha Riach *see* Loughrea
Baile Mhistéala *see* Mitchelstown
Baile Monaidh *see* Ballymoney
105 N12 **Bailén** Andalucía, S Spain 38°06′N 03°46′W
Baile na hInse *see* Ballynahinch
Baile na Lorgan *see* Castleblayney
Baile na Mainistreach *see* Newtownabbey
Baile Nua na hArda *see* Newtownards
116 I12 **Băile Olăneşti** Vâlcea, SW Romania 45°14′N 24°18′E
116 H14 **Băileşti** Dolj, SW Romania 44°01′N 23°20′E
163 N12 **Bailingmiao** *var.* Darhan Muminggan Lianheqi. Nei Mongol Zizhiqu, N China 41°41′N 110°25′E
58 K11 **Bailique, Ilha** *island* NE Brazil
103 O1 **Bailleul** Nord, N France 50°43′N 02°43′E
78 H12 **Ba Illi** Chari-Baguirmi, SW Chad 10°31′N 16°29′E
159 V12 **Bailong Jiang** ⤳ C China
82 C13 **Bailundo** *Port.* Vila Teixeira da Silva. Huambo, C Angola 12°12′S 15°52′E
159 T13 **Baima** *var.* Sêraitang. Qinghai, C China 32°55′N 100°44′E
Baima *see* Baxoi
186 C8 **Baimuru** Gulf, S Papua New Guinea 07°34′S 144°49′E
158 M16 **Bainang** Xizang Zizhiqu, W China 28°57′N 89°31′E
23 S8 **Bainbridge** Georgia, SE USA 30°54′N 84°33′W
171 O17 **Baing** Pulau Sumba, SE Indonesia 10°09′S 120°34′E
158 M14 **Baingoin** *var.* Pubao. Xizang Zizhiqu, W China 31°22′N 90°00′E
104 G2 **Baio** Galicia, NW Spain 43°08′N 08°58′W
104 G4 **Baiona** Galicia, NW Spain 42°06′N 08°49′W
163 V7 **Baiquan** Heilongjiang, NE China 47°37′N 126°04′E
Bā'ir *see* Bāyir
158 I11 **Bairab Co** ◎ W China
25 Q7 **Baird** Texas, SW USA 32°23′N 99°24′W
39 N7 **Baird Mountains** ▲▲ Alaska, USA
Baireuth *see* Bayreuth
190 H3 **Bairiki** Tarawa, NW Kiribati 01°20′N 173°01′E
Bairin Youqi *see* Daban
Bairin Zuoqi *see* Lindong
Bairkum *see* Bayyrkum
183 P12 **Bairnsdale** Victoria, SE Australia 37°51′S 147°38′E
171 P6 **Bais** Negros, S Philippines 09°36′N 123°07′E
102 L15 **Baïse** *var.* Baise. ⤳ S France
Baise *see* Baïse
163 W11 **Baishan** *prev.* Hunjiang. Jilin, NE China 41°57′N 126°31′E
Baishan *see* Mashan
118 F12 **Baisogala** Šiauliai, C Lithuania 55°38′N 23°44′E
189 Q7 **Baiti** N Nauru 0°30′S 166°55′E
Baitou Shan *see* Paektu-san
104 G13 **Baixo Alentejo** *physical region* S Portugal
64 P5 **Baixo, Ilhéu de** *island* Madeira, Portugal, NE Atlantic Ocean
83 E15 **Baixo Longa** Kuando Kubango, SE Angola 15°39′S 18°39′E
159 V10 **Baiyin** Gansu, C China 36°33′N 104°11′E
160 E8 **Baiyü** *var.* Jianshe. Sichuan, C China 30°37′N 97°15′E
161 N14 **Baiyun** ✈ (Guangzhou) Guangdong, S China 23°12′N 113°19′E
160 K4 **Baiyu Shan** ▲▲ C China
111 J25 **Baja** Bács-Kiskun, S Hungary 46°13′N 18°56′E
40 C4 **Baja California** *Eng.* Lower California. *peninsula* NW Mexico
40 C4 **Baja California Norte** ◈ *state* NW Mexico
40 E9 **Baja California Sur** ◈ *state* NW Mexico
Bājah *see* Béja
Bajan *see* Bayan
191 V16 **Baja, Punta** *headland* Easter Island, Chile, E Pacific Ocean 27°10′S 109°21′W
40 B4 **Baja, Punta** *headland* NW Mexico 29°57′N 115°48′W
55 R5 **Baja, Punta** *headland* NE Venezuela
42 D5 **Baja Verapaz** *off.* Departamento de Baja Verapaz. ◈ *department* C Guatemala
Baja Verapaz, Departamento de *see* Baja Verapaz
171 N16 **Bajawa** *prev.* Badjawa. Flores, S Indonesia 08°46′S 120°59′E
153 S16 **Baj Baj** *prev.* Budge-Budge. West Bengal, E India 22°29′N 88°11′E
141 N15 **Bājil** W Yemen 15°05′N 43°16′E
183 U4 **Bajimba, Mount** ▲ New South Wales, SE Australia 29°19′S 152°04′E
112 K13 **Bajina Bašta** Serbia, W Serbia 43°58′N 19°33′E
153 U14 **Bajitpur** Dhaka, E Bangladesh 24°12′N 90°57′E
112 K8 **Bajmok** Vojvodina, NW Serbia 45°59′N 19°25′E
Bajo Boquete *see* Boquete
113 L17 **Bajram Curri** Kukës, N Albania 42°23′N 20°06′E
79 J14 **Bakala** Ouaka, C Central African Republic 06°03′N 20°31′E
127 T4 **Bakaly** Respublika Bashkortostan, W Russian Federation 55°10′N 53°46′E
Bakan *see* Shimonoseki
145 U14 **Bakanas** *Kaz.* Baqanas. Almaty, SE Kazakhstan 44°50′N 76°13′E
145 V12 **Bakanas** *Kaz.* Baqanas. ⤳ E Kazakhstan
149 R4 **Bākārak** Panjshīr, NE Afghanistan 35°16′N 69°28′E
145 U14 **Bakbakty** *Kaz.* Baqbaqty. Almaty, SE Kazakhstan 44°36′N 76°41′E
122 J12 **Bakchar** Tomskaya Oblast', C Russian Federation 56°58′N 81°59′E
76 I11 **Bakel** E Senegal 14°54′N 12°26′W
35 W13 **Baker** California, W USA 35°15′N 116°04′W
22 J8 **Baker** Louisiana, S USA 30°35′N 91°10′W
33 Y9 **Baker** Montana, NW USA 46°22′N 104°16′W
32 L12 **Baker** Oregon, NW USA 44°46′N 117°50′W
192 L7 **Baker and Howland Islands** ◇ *US unincorporated territory* W Polynesia
36 L12 **Baker Butte** ▲ Arizona, SW USA 34°24′N 111°22′W
39 X15 **Baker Island** *island* Alexander Archipelago, Alaska, USA
9 N9 **Baker Lake** *var.* Qamanittuaq. Nunavut, N Canada 64°20′N 96°10′W
9 N9 **Baker Lake** ◎ Nunavut, N Canada
32 H6 **Baker, Mount** ▲ Washington, NW USA 48°46′N 121°48′W
35 R13 **Bakersfield** California, W USA 35°22′N 119°01′W
24 M9 **Bakersfield** Texas, SW USA 30°54′N 102°21′W
21 P9 **Bakersville** North Carolina, SE USA 36°01′N 82°09′W
Bākhābī *see* Bū Khābī
Bakharden *see* Baharly
Bakhardok *see* Bokurdak
143 U5 **Bākharz, Kuhhā-ye** ▲▲ NE Iran
152 D13 **Bākhāsar** Rājasthān, NW India 24°42′N 71°11′E
Bakhchisaray *see* Bakhchysaray
117 T13 **Bakhchysaray** *Rus.* Bakhchisaray. Avtonomna Respublika Krym, S Ukraine 44°44′N 33°53′E
Bakherden *see* Baharly
117 R3 **Bakhmach** Chernihivs'ka Oblast', N Ukraine 51°10′N 32°48′E
Bākhtarān *see* Kermānshāh
143 Q11 **Bakhtegān, Daryācheh-ye** ◎ C Iran
Bakhty *see* Bakty
137 Z11 **Bakı** *Eng.* Baku. ● (Azerbaijan) E Azerbaijan 40°24′N 49°51′E
80 M12 **Baki** Awdal, N Somalia 10°10′N 43°45′E
137 Z11 **Bakı** ✈ E Azerbaijan 40°26′N 49°55′E
136 C13 **Bakır Çayı** ⤳ W Turkey
92 L1 **Bakkafjörður** Austurland, NE Iceland 66°01′N 14°49′W
92 L1 **Bakkaflói** *sea area* W Norwegian Sea
92 L2 **Bakkagerði** Austurland, NE Iceland 65°32′N 13°46′W
81 I15 **Bako** Southern Nationalities, S Ethiopia 05°45′N 36°39′E
76 L15 **Bako** NW Ivory Coast 09°08′N 07°40′W
Bákó *see* Bacău
111 H23 **Bakony** *Eng.* Bakony Mountains, *Ger.* Bakonywald. ▲▲ W Hungary
Bakony Mountains/Bakonywald *see* Bakony
81 M16 **Bakool** *off.* Gobolka Bakool. ◈ *region* W Somalia
Bakool, Gobolka *see* Bakool
79 L15 **Bakouma** Mbomou, SE Central African Republic 05°42′N 22°43′E
127 N15 **Baksan** Kabardino-Balkarskaya Respublika, SW Russian Federation 43°43′N 43°31′E
119 I16 **Bakshty** Hrodzyenskaya Voblasts', W Belarus 53°56′N 26°11′E
145 X12 **Bakty** *prev.* Bakhty. Vostochnyy Kazakhstan, E Kazakhstan 46°41′N 82°45′E
Baku *see* Bakı
194 K12 **Bakutis Coast** *physical region* Antarctica
Bakwanga *see* Mbuji-Mayi
145 O15 **Bakyrly** Yuzhnyy Kazakhstan, S Kazakhstan 44°30′N 67°41′E
14 H13 **Bala** Ontario, S Canada 45°01′N 79°37′W
136 I13 **Balâ** Ankara, C Turkey 39°34′N 33°07′E
97 J19 **Bala** NW Wales, United Kingdom 52°54′N 03°31′W
170 L7 **Balabac Island** *island* W Philippines
Balabac, Selat *see* Balabac Strait
169 V5 **Balabac Strait** *var.* Selat Balabac. *strait* Malaysia/Philippines
Ba'labakk *see* Baalbek
187 P16 **Balabio, Île** *island* Province Nord, W New Caledonia
116 I14 **Balaci** Teleorman, S Romania 44°21′N 24°55′E
139 S7 **Balad** Şalāḩ ad Dīn, N Iraq 34°00′N 44°07′E
139 U7 **Balad Rūz** Diyālá, E Iraq 33°42′N 45°04′E
154 J11 **Bālāghāt** Madhya Pradesh, C India 21°48′N 80°11′E
155 F14 **Bālāghāt Range** ▲▲ W India
103 X14 **Balagne** *physical region* Corse, France, C Mediterranean Sea
105 U5 **Balaguer** Cataluña, NE Spain 41°48′N 00°48′E
105 S3 **Balaïtous** *var.* Pic de Balaïtous. ▲ France/Spain 42°51′N 00°17′W
Balaïtous, Pic de *see* Balaïtous
Balák *see* Ballangen
127 O3 **Balakhna** Nizhegorodskaya Oblast', W Russian Federation 56°26′N 43°43′E
122 L12 **Balakhta** Krasnoyarskiy Kray, S Russian Federation 55°22′N 91°24′E
182 I9 **Balaklava** South Australia 34°10′S 138°22′E
Balakleya *see* Balakliya
117 V6 **Balakliya** *Rus.* Balakleya. Kharkivs'ka Oblast', E Ukraine 49°27′N 36°53′E
127 Q7 **Balakovo** Saratovskaya Oblast', W Russian Federation 52°03′N 47°47′E
83 P14 **Balama** Cabo Delgado, N Mozambique 13°18′S 38°39′E
169 U6 **Balambangan, Pulau** *island* East Malaysia
Bālā Morghāb *see* Bālā Murghāb
148 L3 **Bālā Murghāb** *prev.* Bālā Morghāb. Laghmān, NW Afghanistan 35°38′N 63°21′E
152 E11 **Bālān** *prev.* Bāhla. Rājasthān, NW India 27°45′N 71°32′E
116 J10 **Bălan** *Hung.* Balánbánya. Harghita, C Romania 46°39′N 25°47′E
Balánbánya *see* Bălan
171 O3 **Balanga** Luzon, N Philippines 14°40′N 120°32′E
154 M12 **Balāngīr** *prev.* Bolangir. Odisha, E India 20°41′N 83°30′E
127 N8 **Balashov** Saratovskaya Oblast', W Russian Federation 51°32′N 43°14′E
Balasore *see* Bāleshwar
111 K21 **Balassagyarmat** Nógrád, N Hungary 48°06′N 19°17′E
29 S10 **Balaton** Minnesota, N USA 44°13′N 95°52′W
111 H24 **Balaton** *var.* Lake Balaton, *Ger.* Plattensee. ◎ W Hungary
111 I23 **Balatonfüred** *var.* Füred. Veszprém, W Hungary 46°59′N 17°53′E
Balaton, Lake *see* Balaton
116 I11 **Bălăuşeri** *Ger.* Bladenmarkt, *Hung.* Balavásár. Mureş, C Romania 46°24′N 24°41′E
Balavásár *see* Bălăuşeri
105 Q11 **Balazote** Castilla-La Mancha, C Spain 38°54′N 02°09′W
Balázsfalva *see* Blaj
119 F14 **Balbieriškis** Kaunas, S Lithuania 54°29′N 23°52′E
186 J7 **Balbi, Mount** ▲ Bougainville Island, NE Papua New Guinea 05°51′S 154°58′E
58 F11 **Balbina, Represa** ⊠ NW Brazil
43 T15 **Balboa** Panamá, C Panama 08°55′N 79°36′W
97 G17 **Balbriggan** *Ir.* Baile Brigín. E Ireland 53°37′N 06°11′W
Balbunar *see* Kubrat
81 N17 **Balcad** Shabeellaha Dhexe, C Somalia 02°19′N 45°19′E
61 D23 **Balcarce** Buenos Aires, E Argentina 37°51′S 58°16′W
11 U16 **Balcarres** Saskatchewan, S Canada 50°49′N 103°31′W
114 O8 **Balchik** Dobrich, NE Bulgaria 43°25′N 28°11′E
185 E24 **Balclutha** Otago, South Island, New Zealand 46°15′S 169°45′E
25 Q12 **Balcones Escarpment** *escarpment* Texas, SW USA
18 F14 **Bald Eagle Creek** ⤳ Pennsylvania, NE USA
Baldenburg *see* Biały Bór
21 V12 **Bald Head Island** *island* North Carolina, SE USA
27 W10 **Bald Knob** Arkansas, C USA 35°18′N 91°34′W
30 K17 **Bald Knob** *hill* Illinois, N USA
Baldohn *see* Baldone
118 G9 **Baldone** *Ger.* Baldohn. W Latvia 56°46′N 24°18′E
22 I9 **Baldwin** Louisiana, S USA 29°50′N 91°32′W
31 P7 **Baldwin** Michigan, N USA 43°54′N 85°50′W
27 Q4 **Baldwin City** Kansas, C USA 38°43′N 95°12′W
39 N8 **Baldwin Peninsula** *headland* Alaska, USA 66°45′N 162°19′W
18 H9 **Baldwinsville** New York, NE USA 43°09′N 76°19′W
23 N2 **Baldwyn** Mississippi, S USA 34°30′N 88°38′W
11 W15 **Baldy Mountain** ▲ Manitoba, S Canada 51°29′N 100°46′W
33 T7 **Baldy Mountain** ▲ Montana, NW USA 48°09′N 109°39′W
37 O13 **Baldy Peak** ▲ Arizona, SW USA 33°56′N 109°37′W
Bâle *see* Basel
Balearic Plain *see* Algerian Basin
Baleares *see* Illes Baleares
105 X11 **Baleares, Islas** *Eng.* Balearic Islands. *island group* Spain, W Mediterranean Sea
Baleares Major *see* Mallorca
Balearic Islands *see* Baleares, Islas
Balearis Minor *see* Menorca
169 S9 **Baleh, Batang** ⤳ East Malaysia
12 J8 **Baleine, Grande Rivière de la** ⤳ Québec, E Canada
12 K7 **Baleine, Petite Rivière de la** ⤳ Québec, SE Canada
12 K7 **Baleine, Petite Rivière de la** ⤳ Québec, NE Canada
13 N6 **Baleine, Rivière à la** ⤳ Québec, E Canada
99 J16 **Balen** Antwerpen, N Belgium 51°12′N 05°12′E
171 O3 **Baler** Luzon, N Philippines 15°47′N 121°30′E
154 P11 **Bāleshwar** *prev.* Balasore. Odisha, E India 21°31′N 86°59′E
77 S12 **Baléyara** Tillabéri, W Niger 13°48′N 02°57′E
127 T1 **Balezino** Udmurtskaya Respublika, NW Russian Federation 57°57′N 53°03′E
42 J4 **Balfate** Colón, N Honduras 15°47′N 86°24′W
11 O17 **Balfour** British Columbia, SW Canada 49°39′N 116°57′W
29 N3 **Balfour** North Dakota, N USA 47°55′N 100°34′W
Balfrush *see* Bābol
114 L9 **Balgarka** ▲ E Bulgaria 42°43′N 26°19′E
122 L14 **Balgazyn** Respublika Tyva, S Russian Federation 50°53′N 95°12′E
11 U16 **Balgonie** Saskatchewan, S Canada 50°30′N 104°12′W
Bālgrad *see* Alba Iulia
81 J19 **Balguda** *spring/well* S Kenya 01°28′S 39°50′E
158 K6 **Balguntay** Xinjiang Uygur Zizhiqu, NW China 42°45′N 86°18′E
141 R16 **Balḩāf** S Yemen 14°02′N 48°16′E
152 F13 **Bāli** Rājasthān, N India 25°10′N 73°20′E
169 U17 **Bali** ◈ *province* S Indonesia
169 T17 **Bali** *island* C Indonesia
111 K16 **Balice** ✈ (Kraków) Małopolskie, S Poland 49°57′N 19°49′E
171 Y14 **Baliem, Sungai** ⤳ Papua, E Indonesia
136 C12 **Balıkesir** Balıkesir, W Turkey 39°38′N 27°52′E
136 C12 **Balıkesir** ◈ *province* NW Turkey
138 L3 **Balīkh, Nahr** ⤳ N Syria
169 V12 **Balikpapan** Borneo, C Indonesia 01°15′S 116°50′E
169 T16 **Bali, Laut** *Eng.* Bali Sea. *sea* C Indonesia
171 N9 **Balimbing** Tawitawi, SW Philippines 05°10′N 120°00′E
186 B8 **Balimo** Western, SW Papua New Guinea 08°00′S 143°00′E
Bálinc *see* Balinţ
101 H23 **Balingen** Baden-Württemberg, SW Germany 48°16′N 08°51′E
116 F11 **Balinţ** *Hung.* Bálinc. Timiş, W Romania 45°48′N 21°54′E
171 O1 **Balintang Channel** *channel* N Philippines
138 K3 **Bālis** Ḩalab, N Syria 36°01′N 38°03′E
Bali Sea *see* Bali, Laut
98 K7 **Balk** Fryslân, N Netherlands 52°54′N 05°34′E
146 B11 **Balkanabat** *Rus.* Nebitdag. Balkan Welaýaty, W Turkmenistan 39°33′N 54°19′E
121 R6 **Balkan Mountains** *Bul./SCr.* Stara Planina. ▲▲ Bulgaria/Serbia
Balkanskiy Welayat *see* Balkan Welaýaty
146 B9 **Balkan Welaýaty** *Rus.* Balkanskiy Velayat. ◈ *province* W Turkmenistan
Balkash *see* Balkhash
145 P8 **Balkashino** Akmola, N Kazakhstan 52°36′N 68°46′E
Balkash, Ozero *see* Balkhash, Ozero
149 O2 **Balkh** *anc.* Bactra. Balkh, N Afghanistan 36°46′N 66°54′E
149 P2 **Balkh** ◈ *province* N Afghanistan
145 T13 **Balkhash** *Kaz.* Balqash, *var.* Balkash. Karaganda, SE Kazakhstan 46°52′N 74°55′E
Balkhash, Lake *see* Balkhash, Ozero
145 T13 **Balkhash, Ozero** *var.* Ozero Balkash, *Eng.* Lake Balkhash, *Kaz.* Balqash. ◎ SE Kazakhstan
Balla Balla *see* Mbalabala
96 H10 **Ballachulish** N Scotland, United Kingdom 56°40′N 05°10′W
180 M12 **Balladonia** Western Australia 32°21′S 123°32′E
97 C16 **Ballaghaderreen** *Ir.* Bealach an Doirín. C Ireland 53°51′N 08°29′W
92 H10 **Ballangen** *Lapp.* Bálák. Nordland, NW Norway 68°18′N 16°48′E
97 H14 **Ballantrae** W Scotland, United Kingdom 55°05′N 05°W
183 N12 **Ballarat** Victoria, SE Australia 37°36′S 143°51′E
180 K11 **Ballard, Lake** *salt lake* Western Australia
Ballari *see* Bellary
76 L11 **Ballé** Koulikoro, W Mali 15°18′N 08°31′W
40 D7 **Ballenas, Bahía de** *bay* NW Mexico
40 D5 **Ballenas, Canal de** *channel* NW Mexico
195 R17 **Balleny Islands** *island group* Antarctica
40 J7 **Balleza** *var.* San Pablo Balleza. Chihuahua, N Mexico 26°55′N 106°21′W
114 M13 **Ballı** Tekirdağ, NW Turkey 40°48′N 27°03′E
153 O13 **Ballia** Uttar Pradesh, N India 25°45′N 84°09′E
183 V4 **Ballina** New South Wales, SE Australia 28°50′S 153°37′E
97 C16 **Ballina** *Ir.* Béal an Átha. W Ireland 54°07′N 09°09′W
97 D16 **Ballinamore** *Ir.* Béal an Átha Móir. NW Ireland 54°03′N 07°47′W
97 D18 **Ballinasloe** *Ir.* Béal Átha na Sluaighe. W Ireland 53°20′N 08°13′W
25 P8 **Ballinger** Texas, SW USA 31°44′N 99°57′W
97 C17 **Ballinrobe** *Ir.* Baile an Róba. W Ireland 53°37′N 09°14′W
97 A21 **Ballinskelligs Bay** *Ir.* Bá na Scealg. *inlet* SW Ireland
97 D15 **Ballintra** *Ir.* Baile an tSratha. NW Ireland 54°35′N 08°07′W
103 T7 **Ballon d'Alsace** ▲ NE France 47°50′N 06°54′E
Ballon de Guebwiller *see* Grand Ballon
113 K21 **Ballsh** *var.* Ballshi. Fier, SW Albania 40°35′N 19°45′E
Ballshi *see* Ballsh
98 K4 **Ballum** Fryslân, N Netherlands 53°27′N 05°40′E
97 F16 **Ballybay** *Ir.* Béal Átha Beithe. N Ireland 54°08′N 06°54′W
97 E14 **Ballybofey** *Ir.* Bealach Féich. NW Ireland 54°49′N 07°47′W
97 G14 **Ballycastle** *Ir.* Baile an Chaistil. N Northern Ireland, United Kingdom 55°12′N 06°14′W
97 G15 **Ballyclare** *Ir.* Bealach Cláir. E Northern Ireland, United Kingdom 54°45′N 06°W
97 E16 **Ballyconnell** *Ir.* Béal Átha Conaill. N Ireland 54°07′N 07°35′W
97 C17 **Ballyhaunis** *Ir.* Beál Átha hAmhnais. W Ireland 53°45′N 08°45′W
97 G14 **Ballymena** *Ir.* An Baile Meánach. NE Northern Ireland, United Kingdom 54°52′N 06°17′W
97 F14 **Ballymoney** *Ir.* Baile Monaidh. NE Northern Ireland, United Kingdom 55°10′N 06°30′W
97 G15 **Ballynahinch** *Ir.* Baile na hInse. SE Northern Ireland, United Kingdom 54°24′N 05°54′W
97 D16 **Ballysadare** *Ir.* Baile Easa Dara. NW Ireland 54°13′N 08°30′W
97 D15 **Ballyshannon** *Ir.* Béal Átha Seanaidh. NW Ireland 54°30′N 08°11′W
63 H19 **Balmaceda** Aisén, S Chile 45°52′S 72°43′W
63 G23 **Balmaceda, Cerro** ▲ S Chile 51°27′S 73°26′W
111 N22 **Balmazújváros** Hajdú-Bihar, E Hungary 47°36′N 21°18′E
108 E10 **Balmhorn** ▲ SW Switzerland 46°27′N 07°41′E
182 L12 **Balmoral** Victoria, SE Australia 37°16′S 141°38′E
24 K9 **Balmorhea** Texas, SW USA 30°58′N 103°44′W
Balneario Claromecó *see* Claromecó
148 M12 **Balochistān** *prev.* Baluchistan, Beluchistan. ◈ *province* SW Pakistan
82 B13 **Balombo** *Port.* Norton de Matos, Vila Norton de Matos. Benguela, W Angola 12°21′S 14°46′E
82 B13 **Balombo** ⤳ W Angola
181 X10 **Balonne River** ⤳ Queensland, E Australia
152 E13 **Bālotra** Rājasthān, N India 25°50′N 72°21′E
145 V14 **Balpyk Bi** *prev.* Kirovskiy, *Kaz.* Kīrov. Almaty, SE Kazakhstan 44°54′N 78°13′E
Balqā'/Balqā', Muḩāfaẓat al *see* Al Balqā'
Balqash *see* Balkhash/Balkhash, Ozero
152 M12 **Balrāmpur** Uttar Pradesh, N India 27°25′N 82°10′E
182 M9 **Balranald** New South Wales, SE Australia 34°39′S 143°33′E
116 H14 **Balş** Olt, S Romania 44°20′N 24°06′E
14 H11 **Balsam Creek** Ontario, S Canada 46°26′N 79°10′W
30 I5 **Balsam Lake** Wisconsin, N USA 45°27′N 92°28′W
14 I14 **Balsam Lake** ◎ Ontario, SE Canada
59 M14 **Balsas** Maranhão, E Brazil 07°30′S 46°W
40 M15 **Balsas, Río** *var.* Río Mexcala. ⤳ S Mexico
43 W16 **Balsas, Río** ⤳ SE Panama
119 O18 **Bal'shavik** *Rus.* Bol'shevik. Homyel'skaya Voblasts', SE Belarus 52°34′N 30°53′E
95 O15 **Bålsta** Uppsala, C Sweden 59°31′N 17°32′E
108 E7 **Balsthal** Solothurn, NW Switzerland 47°18′N 07°42′E
117 O8 **Balta** Odes'ka Oblast', SW Ukraine 47°58′N 29°39′E
105 N5 **Baltanás** Castilla y León, N Spain 41°56′N 04°12′W
61 E16 **Baltasar Brum** Artigas, N Uruguay 30°44′S 57°19′W
116 M9 **Bălţi** *Rus.* Bel'tsy. N Moldova 47°45′N 27°57′E
Baltic Port *see* Paldiski
118 B10 **Baltic Sea** *Ger.* Ostee, *Rus.* Baltiskoye More. *sea* N Europe
21 X3 **Baltimore** Maryland, NE USA 39°17′N 76°37′W
31 T13 **Baltimore** Ohio, N USA 39°48′N 82°33′W
21 X3 **Baltimore-Washington** ✈ Maryland, E USA 39°10′N 76°40′W
Baltischport/Baltiski *see* Paldiski
Baltiskoye More *see* Baltic Sea
119 A14 **Baltiysk** *Ger.* Pillau. Kaliningradskaya Oblast', W Russian Federation 54°39′N 19°54′E
Baltkrievija *see* Belarus
119 H14 **Baltoji Voke** Vilnius, SE Lithuania 54°35′N 25°13′E
Balūchestān va Sīstān *see* Sīstān va Balūchestān
Baluchistan *see* Balochistān
171 P5 **Balud** Masbate, N Philippines 12°03′N 123°12′E
169 T9 **Balui, Batang** ⤳ East Malaysia
153 S13 **Bālurghat** West Bengal, NE India 25°12′N 88°50′E
118 J8 **Balvi** NE Latvia 57°07′N 27°14′E
147 W7 **Balykchy** *Kir.* Ysyk-Köl; *prev.* Issyk-Kul', Rybach'ye. Issyk-Kul'skaya Oblast', NE Kyrgyzstan 42°29′N 76°08′E
56 B7 **Balzar** Guayas, W Ecuador 01°25′S 79°54′W
108 I8 **Balzers** S Liechtenstein 47°04′N 09°32′E
143 T12 **Bam** Kermān, SE Iran 29°08′N 58°27′E
77 Y13 **Bama** Borno, NE Nigeria 11°28′N 13°46′E
76 L12 **Bamako** ● (Mali) Capital District, SW Mali 12°39′N 08°02′W
77 P10 **Bamba** Gao, C Mali 17°03′N 01°19′W
42 M8 **Bambana, Río** ⤳ NE Nicaragua
79 J15 **Bambari** Ouaka, C Central African Republic 05°45′N 20°37′E
181 W5 **Bambaroo** Queensland, NE Australia 19°00′S 146°16′E
101 K19 **Bamberg** Bayern, SE Germany 49°54′N 10°53′E
21 R14 **Bamberg** South Carolina, SE USA 33°16′N 81°02′W
79 M16 **Bambesa** Orientale, N Dem. Rep. Congo 03°25′N 25°43′E
76 G11 **Bambey** W Senegal 14°43′N 16°26′W
79 H16 **Bambio** Sangha-Mbaéré, SW Central African Republic 03°57′N 16°54′E
83 I24 **Bamboesberge** ▲ S South Africa 31°24′S 26°10′E
79 D14 **Bamenda** Nord-Ouest, W Cameroon 05°55′N 10°09′E
10 K17 **Bamfield** Vancouver Island, British Columbia, SW Canada 48°48′N 125°05′W
Bami *see* Bamy
79 J14 **Bamingui** Bamingui-Bangoran, C Central African Republic 07°38′N 20°06′E
78 J13 **Bamingui** ⤳ N Central African Republic
78 J13 **Bamingui-Bangoran** ◈ *prefecture* N Central African Republic
143 V13 **Bampūr** Sīstān va Balūchestān, SE Iran 27°13′N 60°28′E
186 C8 **Bamu** ⤳ SW Papua New Guinea
146 E12 **Bamy** *Rus.* Bami. Ahal Welaýaty, C Turkmenistan 38°42′N 56°47′E
149 P4 **Bāmyān** *prev.* Bāmīān. Bāmyān, NE Afghanistan 34°50′N 67°50′E
149 O4 **Bāmyān** *prev.* Bāmīān. ◈ *province* C Afghanistan
Bán *see* Bánovce nad Bebravou
81 N17 **Banaadir** *off.* Gobolka Banaadir. ◈ *region* S Somalia
Banaadir, Gobolka *see* Banaadir
191 N3 **Banaba** *var.* Ocean Island. *island* Tungaru, W Kiribati
59 O14 **Banabuiú, Açude** ⊠ NE Brazil
57 O19 **Bañados del Izozog** *salt lake* SE Bolivia
97 D18 **Banagher** *Ir.* Beannchar. C Ireland 53°12′N 07°56′W
79 M17 **Banalia** Orientale, N Dem. Rep. Congo 01°33′N 25°23′E
76 L12 **Banamba** Koulikoro, W Mali 13°29′N 07°22′W
40 G4 **Banámichi** Sonora, NW Mexico 30°00′N 110°14′W
181 Y9 **Banana** Queensland, E Australia 24°33′S 150°07′E
191 Z2 **Banana** *prev.* Main Camp. Kiritimati, E Kiribati 02°00′N 157°25′W
59 K16 **Bananal, Ilha do** *island* C Brazil
23 Y12 **Banana River** *lagoon* Florida, SE USA
151 Q22 **Bananga** Andaman and Nicobar Islands, India, NE Indian Ocean 06°57′N 93°54′E
Banaras *see* Vārānasi
114 N13 **Banarlı** Tekirdağ, NW Turkey 41°04′N 27°21′E
152 H12 **Banās** ⤳ N India
75 Z11 **Banās, Rās** *headland* E Egypt 23°55′N 35°47′E
112 N10 **Banatski Karlovac** Vojvodina, NE Serbia 45°03′N 21°02′E
141 P16 **Banā, Wādī** *dry watercourse* SW Yemen
136 E14 **Banaz** Uşak, W Turkey 38°47′N 29°46′E
136 E14 **Banaz Çayı** ⤳ W Turkey
159 P14 **Banbar** *var.* Coka. Xizang Zizhiqu, W China 31°01′N 94°43′E
97 G15 **Banbridge** *Ir.* Droichead na Banna. SE Northern Ireland, United Kingdom 54°21′N 06°16′W
Ban Bua Yai *see* Bua Yai
97 M21 **Banbury** S England, United Kingdom 52°04′N 01°20′W
167 O7 **Ban Chiang Dao** Chiang Mai, NW Thailand 19°22′N 98°59′E
96 K9 **Banchory** NE Scotland, United Kingdom 58°05′N 00°35′W
14 J13 **Bancroft** Ontario, SE Canada 45°03′N 77°52′W
33 R15 **Bancroft** Idaho, NW USA 42°43′N 111°54′W
29 U11 **Bancroft** Iowa, C USA 43°17′N 94°13′W
154 I9 **Bānda** Madhya Pradesh, C India 24°01′N 78°59′E
152 L13 **Bānda** Uttar Pradesh, N India 25°28′N 80°20′E
168 F7 **Banda Aceh** *var.* Banda Atjeh; *prev.* Koetaradja, Kutaradja, Kutaraja. Sumatera, W Indonesia 05°30′N 95°20′E
Banda Atjeh *see* Banda Aceh
171 S14 **Banda, Kepulauan** *island group* E Indonesia
Banda, Laut *see* Banda Sea
77 N17 **Bandama** *var.* Bandama Fleuve. ⤳ S Ivory Coast
77 N15 **Bandama Blanc** ⤳ C Ivory Coast
Bandama Fleuve *see* Bandama
Bandar 'Abbās *see* Bandar-e 'Abbās
153 W16 **Bandarban** Chittagong, SE Bangladesh 22°13′N 92°13′E
80 Q13 **Bandarbeyla** *var.* Bender Beila, Bender Beyla. Bari, NE Somalia 09°28′N 50°48′E
143 R14 **Bandar-e 'Abbās** *var.* Bandar 'Abbās; *prev.* Gombroon. Hormozgān, S Iran 27°11′N 56°11′E
142 M3 **Bandar-e Anzalī** Gīlān, NW Iran 37°26′N 49°29′E
143 N12 **Bandar-e Būshehr** *var.* Būshehr, *Eng.* Bushire. Būshehr, S Iran 28°59′N 50°50′E
143 O13 **Bandar-e Dayyer** *var.* Deyyer. Būshehr, SE Iran 27°50′N 51°55′E
142 M11 **Bandar-e Gonāveh** *var.* Ganāveh; *prev.* Gonāveh. Būshehr, SW Iran 29°33′N 50°39′E
143 T15 **Bandar-e Jāsk** *var.* Jāsk. Hormozgān, SE Iran 25°35′N 58°06′E
143 O13 **Bandar-e Kangān** *var.* Kangān. Būshehr, S Iran 25°50′N 57°30′E
143 R14 **Bandar-e Khamīr** Hormozgān, S Iran 27°00′N 55°50′E
Bandar-e Langeh *see* Bandar-e Lengeh
143 Q14 **Bandar-e Lengeh** *var.* Bandar-e Langeh, Lingeh. Hormozgān, S Iran 26°34′N 54°52′E
142 L10 **Bandar-e Māhshahr** *var.* Māh-Shahr; *prev.* Bandar-e Ma'shūr. Khūzestān, SW Iran 30°34′N 49°10′E
Bandar-e Ma'shūr *see* Bandar-e Māhshahr
143 O14 **Bandar-e Nakhīlū** Hormozgān, S Iran
Bandar-e Shāh *see* Bandar-e Torkaman
143 P4 **Bandar-e Torkaman** *var.* Bandar-e Torkeman, Bandar-e Torkman; *prev.* Bandar-e Shāh. Golestān, N Iran 36°55′N 54°05′E
Bandar-e Torkeman/Bandar-e Torkman *see* Bandar-e Torkaman
Bandar Kassim *see* Boosaaso
168 M15 **Bandar Lampung** *var.* Bandarlampung, Tanjungkarang-Telukbetung; *prev.* Tandjoengkarang, Tanjungkarang, Teloekbetoeng, Telukbetung. Sumatera, W Indonesia 05°28′S 105°16′E
Bandarlampung *see* Bandar Lampung
Bandar Maharani *see* Muar
Bandar Masulipatnam *see* Machilīpatnam
Bandar Penggaram *see* Batu Pahat
169 T7 **Bandar Seri Begawan** *prev.* Brunei Town. ● (Brunei) N Brunei 04°56′N 114°58′E
169 T7 **Bandar Seri Begawan** ✈ N Brunei 04°56′N 114°58′E
171 R15 **Banda Sea** *var.* Laut Banda. *sea* E Indonesia
104 H5 **Bande** Galicia, NW Spain 42°01′N 07°58′W
59 G15 **Bandeirantes** Mato Grosso, W Brazil 09°04′S 57°53′W
59 N20 **Bandeira, Pico da** ▲ SE Brazil 20°25′S 41°45′W
83 K19 **Bandelierkop** Limpopo, NE South Africa 23°21′S 29°46′E
62 L8 **Bandera** Santiago del Estero, N Argentina 28°53′S 62°15′W
25 Q11 **Bandera** Texas, SW USA 29°44′N 99°06′W
40 J13 **Banderas, Bahía de** *bay* W Mexico
77 O11 **Bandiagara** Mopti, C Mali 14°20′N 03°37′W
152 I12 **Bāndīkūi** Rājasthān, N India 27°01′N 76°33′E
136 C11 **Bandırma** *var.* Penderma. Balıkesir, NW Turkey 40°21′N 27°58′E
Bandjarmasin *see* Banjarmasin
Bandoeng *see* Bandung
97 C21 **Bandon** *Ir.* Droicheadna Bandan. SW Ireland 51°44′N 08°44′W
32 E14 **Bandon** Oregon, NW USA 43°07′N 124°24′W
167 R8 **Ban Dong Bang** Nong Khai, E Thailand 18°00′N 104°08′E
167 Q6 **Ban Donkon** Oudômxai, N Laos 20°20′N 101°37′E
172 J14 **Bandrélé** SE Mayotte
79 H20 **Bandundu** *prev.* Banningville. Bandundu, W Dem. Rep. Congo 03°19′S 17°24′E
79 I21 **Bandundu** *off.* Région de Bandundu. ◈ *region* W Dem. Rep. Congo
Bandundu, Région de *see* Bandundu
169 O16 **Bandung** *prev.* Bandoeng. Jawa, C Indonesia 06°47′S 107°28′E
116 L15 **Băneasa** Constanţa, SW Romania 45°56′N 27°55′E
142 J4 **Bāneh** Kordestān, N Iran 35°58′N 45°54′E
44 I7 **Banes** Holguín, E Cuba 20°58′N 75°43′W
11 P16 **Banff** Alberta, SW Canada 51°10′N 115°34′W
96 K8 **Banff** NE Scotland, United Kingdom 57°39′N 02°33′W
96 K8 **Banff** *cultural region* NE Scotland, United Kingdom
Bánffyhunyad *see* Huedin
77 N14 **Banfora** SW Burkina Faso 10°36′N 04°45′W
155 H19 **Bangalore** *var.* Bengalooru, Bengaluru. *state capital* Karnātaka, S India 12°58′N 77°35′E
153 S16 **Bangaon** West Bengal, NE India 23°01′N 88°50′E
79 L15 **Bangassou** Mbomou, SE Central African Republic 04°51′N 22°55′E
186 D7 **Bangeta, Mount** ▲ C Papua New Guinea 06°11′S 147°02′E
171 P12 **Banggai, Kepulauan** *island group* C Indonesia
171 Q12 **Banggai, Pulau** *island* Kepulauan Banggai, N Indonesia
171 X13 **Banggelapa** Papua, E Indonesia 03°47′S 136°53′E
Banggi *see* Banggi, Pulau
169 V6 **Banggi, Pulau** *var.* Banggi. *island* East Malaysia
152 K5 **Banggong Co** *var.* Pangong Tso. ◎ China/India *see also* Pangong Tso
121 P13 **Banghāzī** *Eng.* Bengazi, Benghazi, *It.* Bengasi. NE Libya 32°07′N 20°04′E

◆ Country
● Country Capital
◇ Dependent Territory
○ Dependent Territory Capital
◈ Administrative Regions
✈ International Airport
▲ Mountain
▲▲ Mountain Range
⛰ Volcano
⤳ River
◎ Lake
⊠ Reservoir

BANG HIENG – BASLE

Bang Hieng *see* Xé Banghiang
169 O13 **Bangka-Belitung** *off.* Propinsi Bangka-Belitung. ◐ *province* W Indonesia
169 P11 **Bangkai, Tanjung** *var.* Bankai. *headland* Borneo, N Indonesia 0°21′N 108°53′E
169 S16 **Bangkalan** Pulau Madura, C Indonesia 07°05′S 112°44′E
169 N12 **Bangka, Pulau** *island* W Indonesia
169 N13 **Bangka, Selat** *strait* Sumatera, W Indonesia
169 N13 **Bangka, Selat** *var.* Selat Likupang. *strait* Sulawesi, N Indonesia
168 J11 **Bangkinang** Sumatera, W Indonesia 0°21′N 100°52′E
168 K12 **Bangko** Sumatera, W Indonesia 02°05′S 102°20′E
Bangkok *see* Ao Krung Thep
Bangkok, Bight of *see* Krung Thep, Ao
153 T14 **Bangladesh** *off.* People's Republic of Bangladesh; *prev.* East Pakistan. ◆ *republic* S Asia
Bangladesh, People's Republic of *see* Bangladesh
167 V13 **Ba Ngoi** Khanh Hoa, S Vietnam 11°56′N 109°07′E
Ba Ngoi *see* Cam Ranh
Bangong Co *see* Pangong Tso
97 I18 **Bangor** NW Wales, United Kingdom 53°13′N 04°08′W
97 G15 **Bangor** *Ir.* Beannchar. E Northern Ireland, United Kingdom 54°40′N 05°40′W
19 R6 **Bangor** Maine, NE USA 44°48′N 68°47′W
18 I14 **Bangor** Pennsylvania, NE USA 40°52′N 75°12′W
67 R8 **Bangoran** 〰 S Central African Republic
Bang Phra *see* Trat
Bang Pla Soi *see* Chon Buri
25 Q8 **Bangs** Texas, SW USA 31°43′N 99°07′W
167 N13 **Bang Saphan** *var.* Bang Saphan Yai. Prachuap Khiri Khan, SW Thailand 11°10′N 99°33′E
Bang Saphan Yai *see* Bang Saphan
36 I8 **Bangs, Mount** ▲ Arizona, SW USA 36°47′N 113°51′W
93 E15 **Bangsund** Nord-Trøndelag, C Norway 64°22′N 11°22′E
171 O2 **Bangued** Luzon, N Philippines 17°36′N 120°40′E
79 I15 **Bangui** ● (Central African Republic) Ombella-Mpoko, SW Central African Republic 04°21′N 18°32′E
79 I15 **Bangui** ✈ Ombella-Mpoko, SW Central African Republic 04°19′N 18°34′E
83 N16 **Bangula** Southern, S Malawi 16°38′S 35°04′E
Bangwaketse *see* Southern
82 K12 **Bangweulu, Lake** *var.* Lake Bengweulu. ◎ N Zambia
121 V13 **Banhā** *var.* Benha. N Egypt 30°28′N 31°11′E
Ban Hat Yai *see* Hat Yai
167 Q7 **Ban Hin Heup** Viangchan, C Laos 18°37′N 102°19′E
Ban Houayxay/Ban Houei Sai *see* Houayxay
167 O12 **Ban Hua Hin** *var.* Hua Hin. Prachuap Khiri Khan, SW Thailand 12°34′N 99°58′E
79 L14 **Bani** Haute-Kotto, E Central African Republic 07°06′N 22°51′E
45 O9 **Baní** S Dominican Republic 18°19′N 70°21′W
77 N12 **Bani** 〰 S Mali
Bāmīān *see* Bāmyān
Banias *see* Bāniyās
77 S11 **Bani Bangou** Tillabéri, SW Niger 15°04′N 02°40′E
76 M12 **Banifing** *var.* Ngorolaka. 〰 Burkina Faso/Mali
Banijska Palanka *see* Glina
77 R13 **Banikoara** N Benin 11°18′N 02°26′E
75 W9 **Banī Mazār** *var.* Beni Mazâr. C Egypt 28°29′N 30°48′E
114 K8 **Baniski Lom** 〰 N Bulgaria
21 U7 **Banister River** 〰 Virginia, NE USA
121 V14 **Banī Suwayf** *var.* Beni Suef. N Egypt 29°09′N 31°04′E
75 O8 **Banī Walīd** NW Libya 31°46′N 13°59′E
138 H5 **Bāniyās** *var.* Banias, Baniyas, Paneas. Ṭarṭūs, W Syria 35°12′N 35°57′E
113 K14 **Banja** Serbia, W Serbia 43°33′N 19°35′E
Banjak, Kepulauan *see* Banyak, Kepulauan
112 J12 **Banja Koviljača** Serbia, W Serbia 44°31′N 19°11′E
112 G11 **Banja Luka** ◐ Republika Srpska, NW Bosnia and Herzegovina
169 T13 **Banjarmasin** *prev.* Bandjarmasin. Borneo, C Indonesia 03°22′S 114°33′E
76 F11 **Banjul** *prev.* Bathurst. ● (Gambia) W Gambia 13°26′N 16°43′W
76 F11 **Banjul** ✈ W Gambia 13°18′N 16°39′W
Bank *see* Bankä
137 Y13 **Bankä** *Rus.* Bank. SE Azerbaijan 39°25′N 49°13′E
167 S11 **Ban Kadian** *var.* Ban Kadiene. Champasak, S Laos 14°25′N 105°42′E
Ban Kadiene *see* Ban Kadian
Bankai *see* Bangkai, Tanjung
166 M14 **Ban Kam Phuam** Phangnga, SW Thailand 09°16′N 98°24′E
Ban Kantang *see* Kantang
77 O11 **Bankass** Mopti, S Mali 14°05′N 03°30′W
95 L19 **Bankeryd** Jönköping, S Sweden 57°51′N 14°07′E
83 K16 **Banket** Mashonaland West, N Zimbabwe 17°23′S 30°24′E
167 T11 **Ban Khamphô** Attapu, S Laos 14°35′N 106°18′E
23 O4 **Bankhead Lake** ⊞ Alabama, S USA
77 Q11 **Bankilaré** Tillabéri, SW Niger 14°34′N 00°41′E
Banks, Îles *see* Banks Islands
10 I14 **Banks Island** *island* British Columbia, SW Canada
187 R12 **Banks Islands** *Fr.* Îles Banks. *island group* N Vanuatu
23 U8 **Banks Lake** ◎ Georgia, SE USA
32 K8 **Banks Lake** ⊞ Washington, NW USA
185 I19 **Banks Peninsula** *peninsula* South Island, New Zealand
183 Q15 **Banks Strait** *strait* SW Tasman Sea
Ban Kui Nua *see* Kui Buri
153 R16 **Bānkura** West Bengal, NE India 23°14′N 87°05′E
167 S8 **Ban Lakxao** *var.* Lak Sao. Bolikhamxai, C Laos 18°10′N 104°58′E
167 O16 **Ban Lam Phai** Songkhla, SW Thailand 06°43′N 100°57′E
Ban Mae Sot *see* Mae Sot
Ban Mae Suai *see* Mae Suai
Ban Mak Khaeng *see* Udon Thani
166 M3 **Banmauk** Sagaing, N Myanmar (Burma) 24°26′N 95°54′E
Banmo *see* Bhamo
167 T10 **Ban Mun-Houamuang** S Laos 15°11′N 106°44′E
97 F14 **Bann** *var.* Lower Bann, Upper Bann. 〰 N Northern Ireland, United Kingdom
167 S10 **Ban Nadou** Salavan, S Laos 15°51′N 105°38′E
167 S9 **Ban Nakala** Savannakhét, S Laos 16°14′N 105°09′E
167 Q8 **Ban Nakha** Viangchan, C Laos 18°13′N 102°29′E
167 S9 **Ban Nakham** Khammouan, S Laos 17°10′N 105°25′E
167 P7 **Ban Namoun** Xaignabouli, N Laos 18°40′N 101°34′E
167 O17 **Ban Nang Sata** Yala, SW Thailand 06°15′N 101°13′E
167 N15 **Ban Na San** Surat Thani, SW Thailand 08°53′N 99°17′E
167 R7 **Ban Nasi** Xiangkhoang, N Laos 19°37′N 103°33′E
44 I3 **Bannerman Town** Eleuthera Island, C The Bahamas 24°38′N 76°09′W
35 V15 **Banning** California, W USA 33°55′N 116°52′W
Banningville *see* Bandundu
167 S11 **Ban Nongsim** Champasak, S Laos 14°45′N 106°00′E
149 S7 **Bannu** *prev.* Edwardesabad. Khyber Pakhtunkhwa, NW Pakistan 33°00′N 70°36′E
Bañolas *see* Banyoles
56 C7 **Baños** Tungurahua, C Ecuador 01°26′S 78°24′W
Bánovce *see* Bánovce nad Bebravou
111 I19 **Bánovce nad Bebravou** *var.* Bánovce, *Hung.* Bán. Trenčiansky Kraj, W Slovakia 48°43′N 18°15′E
112 I12 **Banovići** ◐ Federacija Bosne I Hercegovine, E Bosnia and Herzegovina
Banow *see* Andarāb
Ban Pak Phanang *see* Pak Phanang
167 O7 **Ban Pan Nua** Lampang, NW Thailand 18°51′N 99°57′E
167 Q9 **Ban Phai** Khon Kaen, E Thailand 16°00′N 102°42′E
167 Q8 **Ban Phônhông** *var.* Phônhông. C Laos 18°29′N 102°26′E
167 T9 **Ban Phou A Douk** Khammouan, C Laos 17°12′N 106°07′E
167 Q8 **Ban Phu** Uthai Thani, W Thailand
167 O11 **Ban Pong** Ratchaburi, W Thailand 13°49′N 99°53′E
190 I3 **Banraeaba** Tarawa, W Kiribati 01°20′N 173°02′E
167 N10 **Ban Sai Yok** Kanchanaburi, W Thailand 14°24′N 98°54′E
Ban Sattahip/Ban Sattahipp *see* Sattahip
Ban Sichon *see* Sichon
Ban Si Racha *see* Si Racha
111 J19 **Banská Bystrica** *Ger.* Neusohl, *Hung.* Besztercebánya. Banskobystricky Kraj, C Slovakia 48°46′N 19°08′E
111 K20 **Banskobystrický Kraj** ◐ *region* C Slovakia
167 R8 **Ban Sôppheung** Bolikhamxai, C Laos 18°33′N 104°18′E
Ban Sop Prap *see* Sop Prap
152 G15 **Bānswāra** Rājasthān, N India 23°32′N 74°28′E
167 N15 **Ban Ta Khun** Surat Thani, SW Thailand 08°53′N 98°52′E
Ban Takua Pa *see* Takua Pa
167 S8 **Ban Talak** Khammouan, C Laos 17°33′N 105°40′E
77 R15 **Bantè** W Benin 08°25′N 01°58′E
167 Q11 **Bânteay Méan Choăy** *var.* Sisŏphŏn. Bătdâmbâng, NW Cambodia 13°37′N 102°58′E
167 N16 **Banten** *off.* Propinsi Banten. ◐ *province* W Indonesia
Propinsi Banten *see* Banten
167 Q8 **Ban Thabôk** Bolikhamxai, C Laos 18°21′N 103°12′E
167 T9 **Ban Tôp** Savannakhét, S Laos 16°07′N 106°07′E
97 B21 **Bantry** *Ir.* Beanntraí. Cork, SW Ireland 51°41′N 09°27′W
97 A21 **Bantry Bay** *Ir.* Bá Bheanntraí. *bay* SW Ireland
155 F19 **Bantvāl** *var.* Bantwāl. Karnātaka, E India 12°57′N 75°04′E
Bantwāl *see* Bantvāl
114 N9 **Banya** Burgas, E Bulgaria 42°46′N 27°49′E
168 G10 **Banyak, Kepulauan** *prev.* Kepulauan Banjak. *island group* NW Indonesia
105 U8 **Banya, La** *headland* E Spain 40°34′N 00°37′E
79 E14 **Banyo** Adamaoua, NW Cameroon 06°47′N 11°50′E
105 X4 **Banyoles** *var.* Bañolas. Cataluña, NE Spain 42°07′N 02°46′E
167 N16 **Ban Yong Sata** Trang, SW Thailand 07°09′N 99°42′E
195 X14 **Banzare Coast** *physical region* Antarctica
173 Q14 **Banzare Seamounts** *undersea feature* S Indian Ocean
Banzart *see* Bizerte
163 Q12 **Baochang** *var.* Taibus Qi. Nei Mongol Zizhiqu, N China 41°55′N 115°22′E
161 O3 **Baoding** *var.* Pao-ting; *prev.* Tsingyuan. Hebei, E China 38°47′N 115°30′E
Baoebaoe *see* Baubau
Baoi, Oileán *see* Dursey Island
160 J6 **Baoji** *var.* Pao-chi, Paoki. Shaanxi, C China 34°23′N 107°16′E
163 U9 **Baokang** *var.* Hoqin Zuoyi Zhongji. Nei Mongol Zizhiqu, N China 44°08′N 123°18′E
186 L8 **Baolo** Santa Isabel, N Solomon Islands 07°41′S 158°47′E
167 U13 **Bao Lôc** Lâm Đông, S Vietnam 11°33′N 107°48′E
163 Z7 **Baoqing** Heilongjiang, NE China 46°15′N 132°12′E
Baoqing *see* Shaoyang
79 H15 **Baoro** Nana-Mambéré, W Central African Republic 05°40′N 16°00′E
160 E12 **Baoshan** *var.* Pao-shan. Yunnan, SW China 25°05′N 99°07′E
163 N13 **Baotou** *var.* Pao-t'ou, Paotow. Nei Mongol Zizhiqu, N China 40°38′N 109°59′E
76 L14 **Baoulé** 〰 S Mali
76 K12 **Baoulé** 〰 W Mali
Bao Yèn *see* Phô Rang
Ba-Pahalaborwa *see* Phalaborwa
103 O2 **Bapaume** Pas-de-Calais, N France 50°06′N 02°50′E
14 J13 **Baptiste Lake** ◎ Ontario, SE Canada
Bapu *see* Meigu
Baqanas *see* Bakanas
Baqbaqty *see* Bakbakty
159 P14 **Baqên** *var.* Dartang. Xizang Zizhiqu, W China 31°50′N 94°08′E
138 F14 **Bāqir, Jabal** ▲▲ S Jordan
139 T7 **Ba'qūbah** *var.* Qubba. Diyālá, C Iraq 33°45′N 44°40′E
Ba'qūbah *see* Diyālá
62 H5 **Baquedano** Antofagasta, N Chile 23°20′S 69°50′W
Baquerizo Moreno *see* Puerto Baquerizo Moreno
113 J18 **Bar** *It.* Antivari. S Montenegro 42°02′N 19°09′E
116 M6 **Bar** Vinnyts'ka Oblast', C Ukraine 49°05′N 27°40′E
80 E10 **Bara** Northern Kordofan, C Sudan 13°42′N 30°21′E
81 M18 **Baraawe** *It.* Brava. Shabeellaha Hoose, S Somalia 01°10′N 43°59′E
152 M12 **Bāra Banki** Uttar Pradesh, N India 26°56′N 81°11′E
30 L8 **Baraboo** Wisconsin, N USA 43°27′N 89°45′W
30 K8 **Baraboo Range** *hill range* Wisconsin, N USA
Baracaldo *see* San Vicente de Barakaldo
15 Y6 **Barachois** Québec, SE Canada 48°37′N 64°14′W
44 J7 **Baracoa** Guantánamo, E Cuba 20°23′N 74°31′W
61 C19 **Baradero** Buenos Aires, E Argentina 33°50′S 59°30′W
183 R6 **Baradine** New South Wales, SE Australia 30°55′S 149°03′E
Baraf Daja Islands *see* Damar, Kepulauan
Baragarh *see* Bargarh
81 I17 **Baragoi** Samburu, W Kenya 01°39′N 36°46′E
45 N9 **Barahona** SW Dominican Republic 18°13′N 71°07′W
153 W13 **Barail Range** ▲▲ NE India
Baraka *see* Barka
80 G10 **Barakat** Gezira, C Sudan 14°18′N 33°32′E
Barakī *see* Barakī Barak
149 Q6 **Barakī Barak** *var.* Barakī, Baraki Rajan. Lōgar, E Afghanistan 33°58′N 68°58′E
Baraki Rajan *see* Barakī Barak
154 N11 **Bārākot** Odisha, E India 21°35′N 85°00′E
55 S7 **Barama River** 〰 N Guyana
155 E14 **Bārāmati** Mahārāshtra, W India 18°12′N 74°39′E
152 H5 **Bāramūla** Jammu and Kashmir, NW India 34°15′N 74°24′E
119 N14 **Baran'** Vitsyebskaya Voblasts', NE Belarus 54°29′N 30°18′E
152 I14 **Bārān** Rājasthān, N India 25°08′N 76°32′E
Barānān, Shākh-i *see* Beranan, Shax-î
119 I17 **Baranavichy** *Pol.* Baranowicze, *Rus.* Baranovichi. Brestskaya Voblasts', SW Belarus 53°08′N 26°02′E
123 T6 **Baranikha** Chukotskiy Avtonomnyy Okrug, NE Russian Federation 68°29′N 168°13′E
75 Y11 **Baranīs** *var.* Berenice, Mînâ Baranîs. SE Egypt 23°58′N 35°29′E
116 M4 **Baranivka** Zhytomyrs'ka Oblast', N Ukraine 50°16′N 27°40′E
39 W14 **Baranof Island** *island* Alexander Archipelago, Alaska, USA
Baranovichi/Baranowicze *see* Baranavichy
111 N15 **Baranów Sandomierski** Podkarpackie, SE Poland 50°28′N 21°31′E
111 I26 **Baranya** *off.* Baranya Megye. ◐ *county* S Hungary
Baranya Megye *see* Baranya
153 R13 **Barāri** Bihār, NE India 25°31′N 87°23′E
22 L10 **Barataria Bay** *bay* Louisiana, S USA
Barat Daya, Kepulauan *see* Damar, Kepulauan
118 L12 **Baravukha** *Rus.* Borovukha. Vitsyebskaya Voblasts', N Belarus 55°34′N 28°36′E
54 E11 **Baraya** Huila, C Colombia 03°11′N 75°04′W
59 M21 **Barbacena** Minas Gerais, SE Brazil 21°13′S 43°47′W
54 B13 **Barbacoas** Nariño, SW Colombia 01°38′N 78°08′W
54 L6 **Barbacoas** Aragua, N Venezuela 09°29′N 66°58′W
45 Z13 **Barbados** ◆ *commonwealth republic* SE West Indies
47 S3 **Barbados** *island* Barbados
105 U11 **Barbaria, Cap de** *var.* Cabo de Berberia. *headland* Formentera, E Spain 38°39′N 01°24′E
114 N13 **Barbaros** Tekirdağ, NW Turkey 40°55′N 27°28′E
74 A11 **Barbas, Cap** *headland* S Western Sahara 22°14′N 16°45′W
105 T5 **Barbastro** Aragón, NE Spain 42°02′N 00°07′E
104 K16 **Barbate** 〰 SW Spain
104 K16 **Barbate de Franco** Andalucía, S Spain 36°11′N 05°55′W
83 K21 **Barberton** Mpumalanga, NE South Africa 25°48′S 31°03′E
31 U12 **Barberton** Ohio, N USA 41°02′N 81°37′W
102 K12 **Barbezieux-St-Hilaire** Charente, W France 45°28′N 00°09′W
54 G9 **Barbosa** Boyaca, C Colombia 05°57′N 73°37′W
21 N7 **Barbourville** Kentucky, S USA 36°52′N 83°54′W
45 W9 **Barbuda** *island* N Antigua and Barbuda
181 W8 **Barcaldine** Queensland, E Australia 23°33′S 145°21′E
Barcarozsnyó *see* Râşnov
104 I11 **Barcarrota** Extremadura, W Spain 38°31′N 06°51′W
Barcău *see* Berettyó
Barce *see* Al Marj
107 L23 **Barcellona** *var.* Barcellona Pozzo di Gotto. Sicilia, Italy, C Mediterranean Sea 38°10′N 15°15′E
Barcellona Pozzo di Gotto *see* Barcellona
105 W6 **Barcelona** *anc.* Barcino, Barcinona. Cataluña, E Spain 41°25′N 02°10′E
55 N5 **Barcelona** Anzoátegui, NE Venezuela 10°08′N 64°43′W
105 S5 **Barcelona** ◐ *province* Cataluña, NE Spain
105 W6 **Barcelona** ✈ Cataluña, E Spain 41°25′N 02°10′E
103 U14 **Barcelonnette** Alpes-de-Haute-Provence, SE France 44°24′N 06°37′E
58 E12 **Barcelos** Amazonas, N Brazil 0°59′S 62°58′W
104 G5 **Barcelos** Braga, N Portugal 41°32′N 08°37′W
110 I10 **Barcin** *Ger.* Bartschin. Kujawski-pomorskie, C Poland 52°51′N 17°55′E
Barcino/Barcinona *see* Barcelona
Barcoo *see* Cooper Creek
111 H26 **Barcs** Somogy, SW Hungary 45°58′N 17°26′E
137 W11 **Bärdä** *Rus.* Barda. C Azerbaijan 40°25′N 47°07′E
Barda *see* Bärdä
78 H5 **Bardaï** Tibesti, N Chad 21°21′N 16°56′E
139 R2 **Bardaṛash** Dahūk, N Iraq 36°32′N 43°36′E
92 K3 **Bárðarbunga** ▲ C Iceland 64°39′N 17°30′W
92 K2 **Bárðardalur** *valley* C Iceland
139 Q7 **Bardasah** Al Anbār, SW Iraq 34°02′N 42°28′E
153 S16 **Barddhamān** West Bengal, NE India 23°10′N 88°03′E
111 N18 **Bardejov** *Ger.* Bartfeld, *Hung.* Bártfa. Presovský Kraj, E Slovakia 49°17′N 21°18′E
105 R4 **Bárdenas Reales** *physical region* N Spain
Bardera/Bardere *see* Baardheere
Bardesīr *see* Bardsīr
106 E9 **Bardi** Emilia-Romagna, C Italy 44°39′N 09°44′E
106 A8 **Bardonecchia** Piemonte, W Italy 45°04′N 06°40′E
97 H19 **Bardsey Island** *island* NW Wales, United Kingdom
143 S11 **Bardsīr** *var.* Bardesīr, Mashīz. Kermān, C Iran 29°58′N 56°29′E
20 L6 **Bardstown** Kentucky, S USA 37°49′N 85°29′W
Barduli *see* Barletta
20 G7 **Bardwell** Kentucky, S USA 36°52′N 89°01′W
152 K11 **Bareilly** *var.* Bareli. Uttar Pradesh, N India 28°20′N 79°24′E
Bareli *see* Bareilly
Baren *see* Jiashi
98 H13 **Barendrecht** Zuid-Holland, SW Netherlands 51°52′N 04°31′E
102 M3 **Barentin** Seine-Maritime, N France 49°33′N 00°57′E
92 N3 **Barentsburg** Spitsbergen, W Svalbard 78°01′N 14°19′E
Barentsevo More/Barents Havet *see* Barents Sea
92 O3 **Barentsøya** *island* E Svalbard
197 T11 **Barents Plain** *undersea feature* N Barents Sea
125 P3 **Barents Sea** *Nor.* Barents Havet, *Rus.* Barentsevo More. *sea* Arctic Ocean
197 U14 **Barents Trough** *undersea feature* SW Barents Sea 75°00′N 29°00′E
80 I9 **Barentu** W Eritrea 15°08′N 37°35′E
102 J3 **Barfleur** Manche, N France 49°41′N 01°18′W
102 J3 **Barfleur, Pointe de** *headland* N France 49°46′N 01°09′W
Barfrush/Barfurush *see* Bābol
158 H14 **Barga** Xizang Zizhiqu, W China 30°51′N 81°20′E
80 Q12 **Bargaal** *prev.* Baargaal. Bari, NE Somalia 11°12′N 51°04′E
154 M12 **Bargarh** *var.* Baragarh. Odisha, E India 21°25′N 83°35′E
105 N9 **Bargas** Castilla-La Mancha, C Spain 39°56′N 04°00′W
81 I15 **Bargē** Southern Nationalities, S Ethiopia 06°11′N 37°04′E
106 A9 **Barge** Piemonte, NE Italy 44°49′N 07°21′E
153 U16 **Barguna** Barisal, S Bangladesh 22°09′N 90°07′E
Bärguşad *see* Vorotan
123 N13 **Barguzin** Respublika Buryatiya, S Russian Federation 53°37′N 109°47′E
153 O13 **Barhaj** Uttar Pradesh, N India 26°16′N 83°44′E
183 N10 **Barham** New South Wales, SE Australia 35°39′S 144°09′E
152 J12 **Barhan** Uttar Pradesh, N India 27°21′N 78°11′E
19 S7 **Bar Harbor** Mount Desert Island, Maine, NE USA 44°23′N 68°14′W
153 R14 **Barharwa** Jhārkhand, NE India 24°52′N 87°47′E
153 P15 **Barhi** Jhārkhand, N India 24°19′N 85°25′E
107 O17 **Bari** *var.* Bari delle Puglie; *anc.* Barium. Puglia, SE Italy 41°06′N 16°52′E
80 P12 **Bari** *Off.* Gobolka Bari. ◐ *region* NE Somalia
167 T14 **Ba Ria** *var.* Châu Thanh. Ba Ria-Vung Tau, S Vietnam 10°30′N 107°10′E
81 G21 **Bariadi** Simiyu, NE Tanzania 02°48′S 33°59′E
Bāridah *see* Al Bāridah
Bari delle Puglie *see* Bari
Bari, Gobolka *see* Bari
Barikot *see* Barīkowṭ
149 T4 **Barīkowṭ** *var.* Barikot. Kunar, NE Afghanistan 35°18′N 71°36′E
42 C4 **Barillas** *var.* Santa Cruz Barillas. Huehuetenango, NW Guatemala 15°50′N 91°20′W
54 J6 **Barinas** Barinas, W Venezuela 08°36′N 70°15′W
54 I7 **Barinas** *off.* Estado Barinas; *prev.* Zamora. ◐ *state* C Venezuela
Barinas, Estado *see* Barinas
81 H18 **Baringo** ◐ *county* C Kenya
54 I6 **Barinitas** Barinas, NW Venezuela 08°45′N 70°26′W
154 P11 **Bāripada** Odisha, E India 21°56′N 86°43′E
60 K9 **Bariri** São Paulo, S Brazil 22°04′S 48°46′W
75 W11 **Bārīs** *var.* Bârîs. S Egypt 24°28′N 30°39′E
152 G14 **Bari Sādri** Rājasthān, N India 24°25′N 74°28′E
153 U16 **Barisal** Barisal, S Bangladesh 22°41′N 90°20′E
153 U16 **Barisal** ◐ *division* S Bangladesh
168 I10 **Barisan, Pegunungan** ▲▲ Sumatera, W Indonesia
169 T12 **Barito, Sungai** 〰 Borneo, C Indonesia
Barium *see* Bari
Bårjås *see* Porjus
80 I9 **Barka** *var.* Baraka, *Ar.* Khawr Barakah. *seasonal river* Eritrea/Sudan
Barka *see* Al Marj
160 H8 **Barkam** Sichuan, C China 31°56′N 102°22′E
118 J9 **Barkava** C Latvia 56°43′N 26°34′E
10 M15 **Barkerville** British Columbia, SW Canada 53°06′N 121°35′W
14 J12 **Bark Lake** ◎ Ontario, SE Canada
20 H7 **Barkley, Lake** ⊞ Kentucky/Tennessee, S USA
10 K17 **Barkley Sound** *inlet* British Columbia, W Canada
83 J24 **Barkly East** *Afr.* Barkly-Oos. Eastern Cape, SE South Africa 30°58′S 27°35′E
Barkly-Oos *see* Barkly East
181 S4 **Barkly Tableland** *plateau* Northern Territory/Queensland, N Australia
Barkly-Wes *see* Barkly West
83 H22 **Barkly West** *Afr.* Barkly-Wes. Northern Cape, N South Africa 28°32′S 24°32′E
159 O5 **Barkol** *var.* Barkol Kazak Zizhixian. Xinjiang Uygur Zizhiqu, NW China 43°37′N 93°01′E
159 O5 **Barkol Hu** ◎ NW China
Barkol Kazak Zizhixian *see* Barkol
30 J3 **Bark Point** *headland* Wisconsin, N USA 46°53′N 91°11′W
25 P11 **Barksdale** Texas, SW USA 29°43′N 100°03′W
Bar Kunar *see* Asmār
116 L11 **Bârlad** *prev.* Bîrlad. Vaslui, E Romania 46°12′N 27°39′E
116 M11 **Bârlad** *prev.* Bîrlad. 〰 E Romania
76 D9 **Barlavento, Ilhas de** *var.* Windward Islands. *island group* N Cape Verde
103 R5 **Bar-le-Duc** *var.* Bar-sur-Ornain. Meuse, NE France 48°46′N 05°10′E
180 K11 **Barlee, Lake** ◎ Western Australia
180 H8 **Barlee Range** ▲▲ Western Australia
107 N16 **Barletta** *anc.* Barduli. Puglia, SE Italy 41°20′N 16°17′E
110 E10 **Barlinek** *Ger.* Berlinchen. Zachodnio-pomorskie, NW Poland 53°N 15°11′E
27 S11 **Barling** Arkansas, C USA 35°19′N 94°18′W
171 U12 **Barma** Papua Barat, E Indonesia 01°55′S 132°57′E
183 Q9 **Barmedman** New South Wales, SE Australia 34°09′S 147°21′E
Barmen-Elberfeld *see* Wuppertal
152 D12 **Bārmer** Rājasthān, NW India 25°43′N 71°25′E
182 K9 **Barmera** South Australia 34°14′S 140°26′E
97 I19 **Barmouth** NW Wales, United Kingdom 52°44′N 04°04′W
154 F10 **Barnagar** Madhya Pradesh, C India 23°01′N 75°28′E
152 H9 **Barnāla** Punjab, NW India 30°26′N 75°33′E
97 L15 **Barnard Castle** N England, United Kingdom 54°35′N 01°55′W
183 O6 **Barnato** New South Wales, SE Australia 31°39′S 145°01′E
112 I13 **Barnaul** Altayskiy Kray, C Russian Federation 53°21′N 83°45′E
109 V8 **Bärnbach** Steiermark, SE Austria 47°05′N 15°07′E
18 K16 **Barnegat** New Jersey, NE USA 39°43′N 74°12′W
23 S4 **Barnesville** Georgia, SE USA 33°03′N 84°09′W
29 R6 **Barnesville** Minnesota, N USA 46°39′N 96°25′W
31 U13 **Barnesville** Ohio, N USA 39°59′N 81°10′W
98 K11 **Barneveld** *var.* Barnveld. Gelderland, C Netherlands 52°08′N 05°34′E
25 O9 **Barnhart** Texas, SW USA 31°07′N 101°09′W
27 P8 **Barnsdall** Oklahoma, C USA 36°33′N 96°09′W
97 M17 **Barnsley** N England, United Kingdom 53°34′N 01°28′W
19 Q12 **Barnstable** Massachusetts, NE USA 41°42′N 70°17′W
97 I23 **Barnstaple** SW England, United Kingdom 51°05′N 04°04′W
Barnveld *see* Barneveld
21 Q14 **Barnwell** South Carolina, SE USA 33°14′N 81°21′W
77 U15 **Baro** Niger, C Nigeria 08°35′N 06°28′E
67 U8 **Baro** *var.* Baro Wenz. 〰 Ethiopia/Sudan
Baro *see* Baro Wenz
Baroda *see* Vadodara
149 U2 **Baroghil Pass** *var.* Kowtal-e Barowghīl. *pass* Afghanistan/Pakistan
119 Q17 **Baron'ki** *Rus.* Boron'ki. Mahilyowskaya Voblasts', E Belarus 53°09′N 32°08′E
182 J9 **Barossa Valley** *valley* South Australia
Baroui *see* Salisbury
81 H14 **Baro Wenz** *var.* Baro, Nahr Barū. 〰 Ethiopia/Sudan
Baro Wenz *see* Baro
Barowghīl, Kowtal-e *see* Baroghil Pass
153 U12 **Barpeta** Assam, NE India 26°19′N 91°05′E
31 S7 **Barques, Pointe Aux** *headland* Michigan, N USA 44°04′N 82°57′W
54 I5 **Barquisimeto** Lara, NW Venezuela 10°03′N 69°18′W
59 N16 **Barra** Bahia, E Brazil 11°06′S 43°15′W
96 E9 **Barra** *island* NW Scotland, United Kingdom
183 T5 **Barraba** New South Wales, SE Australia 30°24′S 150°37′E
60 L9 **Barra Bonita** São Paulo, S Brazil 22°30′S 48°35′W
64 J12 **Barracuda Fracture Zone** *var.* Fifteen Twenty Fracture Zone. *tectonic feature* SW Atlantic Ocean
64 G11 **Barracuda Ridge** *undersea feature* N Atlantic Ocean
43 N12 **Barra del Colorado** Limón, NE Costa Rica 10°44′N 83°35′W
43 N9 **Barra de Río Grande** Región Autónoma Atlántico Sur, E Nicaragua 12°56′N 83°30′W
82 A11 **Barra do Cuanza** Luanda, NW Angola 09°12′S 13°08′E
60 O9 **Barra do Piraí** Rio de Janeiro, SE Brazil 22°30′S 43°47′W
61 D16 **Barra do Quaraí** Rio Grande do Sul, SE Brazil 31°03′S 58°10′W
59 G14 **Barra do São Manuel** Pará, N Brazil 07°12′S 58°03′W
83 N19 **Barra Falsa, Ponta da** *headland* S Mozambique 22°57′S 35°36′E
96 E10 **Barra Head** *headland* NW Scotland, United Kingdom 56°46′N 07°37′W
60 O9 **Barra Mansa** Rio de Janeiro, SE Brazil 22°25′S 44°03′W
57 D14 **Barranca** Lima, W Peru 10°46′S 77°46′W
54 F8 **Barrancabermeja** Santander, N Colombia 07°01′N 73°51′W
54 H4 **Barrancas** La Guajira, N Colombia 10°59′N 72°46′W
54 J6 **Barrancas** Barinas, NW Venezuela 08°47′N 70°07′W
55 Q6 **Barrancas** Monagas, NE Venezuela 08°45′N 62°12′W
54 F6 **Barranco de Loba** Bolívar, N Colombia 08°56′N 74°07′W
104 I12 **Barrancos** Beja, S Portugal 38°08′N 06°59′W
62 N7 **Barranqueras** Chaco, N Argentina 27°29′S 58°54′W
54 E4 **Barranquilla** Atlántico, N Colombia 10°59′N 74°48′W
83 N20 **Barra, Ponta da** *headland* S Mozambique 23°46′S 35°33′E
105 P11 **Barrax** Castilla-La Mancha, C Spain 39°04′N 02°12′W
19 N11 **Barre** Massachusetts, NE USA 42°24′N 72°06′W
18 M7 **Barre** Vermont, NE USA 44°09′N 72°25′W
59 M17 **Barreiras** Bahia, E Brazil 12°09′S 44°58′W
104 F11 **Barreiro** Setúbal, W Portugal 38°40′N 09°05′W
65 C26 **Barren Island** *island* S Falkland Islands
20 K7 **Barren River Lake** ⊞ Kentucky, S USA
60 L7 **Barretos** São Paulo, S Brazil 20°33′S 48°33′W
11 P14 **Barrhead** Alberta, SW Canada 54°10′N 114°22′W
14 G14 **Barrie** Ontario, S Canada 44°22′N 79°42′W
11 N16 **Barrière** British Columbia, SW Canada 51°10′N 120°06′W
14 H8 **Barrière, Lac** ◎ Québec, SE Canada
182 L6 **Barrier Range** *hill range* New South Wales, SE Australia
42 G3 **Barrier Reef** *reef* E Belize
188 C16 **Barrigada** C Guam 13°27′N 144°48′E
Barrington Island *see* Santa Fe, Isla
183 T7 **Barrington Tops** ▲ New South Wales, SE Australia 32°06′S 151°18′E
183 O4 **Barringun** New South Wales, SE Australia 29°02′S 145°45′E
59 K18 **Barro Alto** Goiás, S Brazil 15°07′S 48°56′W
59 N14 **Barro Duro** Piauí, NE Brazil 05°49′S 42°30′W
30 I5 **Barron** Wisconsin, N USA 45°24′N 91°50′W
14 J12 **Barron** 〰 Ontario, SE Canada
61 H15 **Barros Cassal** Rio Grande do Sul, S Brazil 29°12′S 52°33′W
45 P14 **Barrouallie** Saint Vincent, W Saint Vincent and the Grenadines 13°14′N 61°17′W
39 O4 **Barrow** Alaska, USA 71°17′N 156°47′W
97 E20 **Barrow** *Ir.* An Bhearú. 〰 SE Ireland
181 Q6 **Barrow Creek Roadhouse** Northern Territory, N Australia 21°30′S 133°52′E
97 J16 **Barrow-in-Furness** NW England, United Kingdom 54°07′N 03°14′W
180 G7 **Barrow Island** *island* Western Australia
39 O4 **Barrow, Point** *headland* Alaska, USA 71°23′N 156°28′W
11 V14 **Barrows** Manitoba, S Canada 52°49′N 101°36′W
97 J22 **Barry** S Wales, United Kingdom 51°24′N 03°18′W
14 J12 **Barry's Bay** Ontario, SE Canada 45°30′N 77°41′W
144 K14 **Barsakel'mes, Ostrov** *island* SW Kazakhstan
Baršć Łužyca *see* Forst
147 S14 **Barsem** S Tajikistan 37°36′N 71°43′E
145 V11 **Barshatas** Vostochnyy Kazakhstan, E Kazakhstan 48°13′N 78°33′E
155 F14 **Bārsi** Mahārāshtra, W India 18°14′N 75°42′E
100 I13 **Barsinghausen** Niedersachsen, C Germany 53°19′N 09°30′E
147 X8 **Barskoon** Issyk-Kul'skaya Oblast', E Kyrgyzstan 42°07′N 77°34′E
100 F10 **Barssel** Niedersachsen, NW Germany 53°10′N 07°46′E
35 U14 **Barstow** California, W USA 34°52′N 117°00′W
24 L8 **Barstow** Texas, SW USA 31°27′N 103°23′W
103 R6 **Bar-sur-Aube** Aube, NE France 48°13′N 04°43′E
Bar-sur-Ornain *see* Bar-le-Duc
103 Q6 **Bar-sur-Seine** Aube, N France 48°06′N 04°22′E
147 S13 **Bartang** S Tajikistan 38°06′N 71°48′E
147 T13 **Bartang** 〰 SE Tajikistan
Bartenstein *see* Bartoszyce
Bártfa/Bartfeld *see* Bardejov
100 N7 **Barth** Mecklenburg-Vorpommern, NE Germany 54°21′N 12°43′E
27 W13 **Bartholomew, Bayou** 〰 Arkansas/Louisiana, S USA
55 T8 **Bartica** N Guyana 06°24′N 58°36′W
136 H10 **Bartın** Bartın, NW Turkey 41°37′N 32°20′E
136 H10 **Bartın** ◐ *province* NW Turkey
181 W4 **Bartle Frere** ▲ Queensland, E Australia 17°15′S 145°43′E
27 P8 **Bartlesville** Oklahoma, C USA 36°44′N 95°59′W
29 P14 **Bartlett** Nebraska, C USA 41°51′N 98°32′W
20 E10 **Bartlett** Tennessee, S USA 35°12′N 89°52′W
25 T9 **Bartlett** Texas, SW USA 30°47′N 97°25′W
36 L13 **Bartlett Reservoir** ⊞ Arizona, SW USA
19 N6 **Barton** Vermont, NE USA 44°44′N 72°09′W
110 L7 **Bartoszyce** *Ger.* Bartenstein. Warmińsko-mazurskie, NE Poland 54°16′N 20°49′E
23 W12 **Bartow** Florida, SE USA 27°54′N 81°50′W
Bartschin *see* Barcin
168 J10 **Barumun, Sungai** 〰 Sumatera, W Indonesia
Barū, Nahr *see* Baro Wenz
169 S17 **Barung, Nusa** *island* S Indonesia
168 H9 **Barus** Sumatera, NW Indonesia 02°02′N 98°20′E
162 I9 **Baruunbayan-Ulaan** *var.* Höövör. Övörhangay, C Mongolia 45°10′N 101°19′E
Baruunsuu *see* Tsogttsetsiy
163 P8 **Baruun-Urt** Sühbaatar, E Mongolia 46°40′N 113°17′E
43 P15 **Barú, Volcán** *var.* Volcán de Chiriquí. ⛰ W Panama 08°49′N 82°32′W
99 K21 **Barvaux** Luxembourg, SE Belgium 50°21′N 05°30′E
42 M13 **Barva, Volcán** ⛰ NW Costa Rica 10°07′N 84°08′W
117 W6 **Barvinkove** Kharkivs'ka Oblast', E Ukraine 48°54′N 37°03′E
154 G11 **Barwāh** Madhya Pradesh, C India 22°17′N 76°01′E
Bärwalde Neumark *see* Mieszkowice
154 F11 **Barwāni** Madhya Pradesh, C India 22°02′N 74°56′E
183 P5 **Barwon River** 〰 New South Wales, SE Australia
119 L15 **Barysaw** *Rus.* Borisov. Minskaya Voblasts', NE Belarus 54°14′N 28°30′E
127 Q6 **Barysh** Ul'yanovskaya Oblast', W Russian Federation 53°32′N 47°06′E
117 Q4 **Baryshivka** Kyyivs'ka Oblast', N Ukraine 50°21′N 31°21′E
114 G8 **Barzia** *var.* Bŭrziya. NW Bulgaria
79 J17 **Basankusu** Equateur, NW Dem. Rep. Congo 01°12′N 19°50′E
117 N11 **Basarabeasca** *Rus.* Bessarabka. SE Moldova 46°22′N 28°56′E
116 M14 **Basarabi** Constanţa, SW Romania 44°10′N 28°26′E
40 H6 **Basaseachic** Chihuahua, NW Mexico 28°18′N 108°13′W
Basauri *see* Arizgoiti
61 D18 **Basavilbaso** Entre Ríos, E Argentina 32°23′S 58°55′W
79 F21 **Bas-Congo** *off.* Région du Bas-Congo; *prev.* Bas-Zaïre. ◐ *region* SW Dem. Rep. Congo
108 E6 **Basel** *Eng.* Basle, *Fr.* Bâle. Basel Stadt, NW Switzerland 47°33′N 07°36′E
Baselland *see* Basel Landschaft
108 E7 **Basel Landschaft** *prev.* Baselland. ◐ *canton* NW Switzerland
108 E6 **Basel Stadt** *former canton* Basel. ◐ *canton* NW Switzerland
143 T14 **Bashākerd, Kūhhā-ye** ▲▲ SE Iran
11 Q15 **Bashaw** Alberta, SW Canada 52°40′N 112°53′W
146 K16 **Bashbedeng** Mary Welaýaty, S Turkmenistan 35°44′N 63°07′E
161 T15 **Bashi Channel** *Chin.* Pa-shih Hai-hsia. *channel* Philippines/Taiwan
Bashkiria *see* Bashkortostan, Respublika
122 F11 **Bashkortostan, Respublika** *prev.* Bashkiria. ◐ *autonomous republic* W Russian Federation
127 N6 **Bashmakovo** Penzenskaya Oblast', W Russian Federation 53°13′N 43°00′E
146 J10 **Bashsakarba** Lebap Welaýaty, NE Turkmenistan 40°25′N 62°16′E
117 R9 **Bashtanka** Mykolayivs'ka Oblast', S Ukraine 47°24′N 32°27′E
22 H8 **Basile** Louisiana, S USA 30°29′N 92°36′W
107 M18 **Basilicata** ◐ *region* S Italy
33 V13 **Basin** Wyoming, C USA 44°22′N 108°02′W
97 N22 **Basingstoke** S England, United Kingdom 51°16′N 01°08′W
143 U8 **Başīrān** Khorāsān-e Janūbī, E Iran 31°57′N 59°07′E
112 B10 **Baška** *It.* Bescanuova. Primorje-Gorski Kotar, NW Croatia 44°58′N 14°46′E
137 T15 **Başkale** Van, SE Turkey 38°03′N 43°59′E
14 L10 **Baskatong, Réservoir** ⊞ Québec, SE Canada
137 O14 **Baskil** Elazığ, E Turkey 38°38′N 38°47′E
Basle *see* Basel

◆ Country
● Country Capital
◇ Dependent Territory
○ Dependent Territory Capital
◐ Administrative Regions
✈ International Airport
▲ Mountain
▲▲ Mountain Range
⛰ Volcano
〰 River
◎ Lake
⊞ Reservoir

BĀSODA – BEIJING

154 H9 **Bāsoda** Madhya Pradesh, C India 23°54′N 77°58′E
79 L17 **Basoko** Orientale, N Dem. Rep. Congo 01°14′N 23°26′E
Basque Country, The *see* País Vasco
Basra *see* Al Başrah
Basra *see* Al Başrah
Başrah, Muḩāfa at al *see* Al Başrah
103 U5 **Bas-Rhin** ◈ *department* NE France
Bassam *see* Grand-Bassam
11 Q16 **Bassano** Alberta, SW Canada 50°48′N 112°28′W
106 H7 **Bassano del Grappa** Veneto, NE Italy 45°45′N 11°45′E
77 Q15 **Bassar** *var.* Bassari. NW Togo 09°15′N 00°47′E
Bassari *see* Bassar
172 L9 **Bassas da India** *island group* W Madagascar
108 D7 **Bassecourt** Jura, W Switzerland 47°20′N 07°16′E
Bassein *see* Pathein
79 J15 **Basse-Kotto** ◈ *prefecture* S Central African Republic
102 J5 **Basse-Normandie** *Eng.* Lower Normandy. ◈ *region* N France
45 Q11 **Basse-Pointe** N Martinique 14°52′N 61°07′W
76 H12 **Basse Santa Su** E Gambia 13°18′N 14°10′W
Basse-Saxe *see* Niedersachsen
45 X6 **Basse-Terre** ○ (Guadeloupe) Basse Terre, SW Guadeloupe 16°08′N 61°40′W
45 V10 **Basseterre** ● (Saint Kitts and Nevis) Saint Kitts, Saint Kitts and Nevis 17°16′N 62°45′W
45 X6 **Basse Terre** *island* W Guadeloupe
29 O13 **Bassett** Nebraska, C USA 42°34′N 99°32′W
21 S7 **Bassett** Virginia, NE USA 36°45′N 79°59′W
37 N15 **Bassett Peak** ▲ Arizona, SW USA 32°30′N 110°16′W
76 M10 **Bassikounou** Hodh ech Chargui, SE Mauritania 15°55′N 05°59′W
77 R15 **Bassila** W Benin 08°25′N 01°58′E
Bass, Îlots de *see* Marotiri
31 O11 **Bass Lake** Indiana, N USA 41°12′N 86°35′W
183 O14 **Bass Strait** *strait* SE Australia
100 H11 **Bassum** Niedersachsen, NW Germany 52°52′N 08°44′E
29 X3 **Basswood Lake** ◎ Canada/USA
95 J21 **Båstad** Skåne, S Sweden 56°25′N 12°50′E
Basṭah *see* Beste
153 N12 **Basti** Uttar Pradesh, N India 26°48′N 82°44′E
103 X14 **Bastia** Corse, France, C Mediterranean Sea 42°42′N 09°27′E
99 L23 **Bastogne** Luxembourg, SE Belgium 50°N 05°43′E
22 I5 **Bastrop** Louisiana, S USA 32°46′N 91°54′W
25 T11 **Bastrop** Texas, SW USA 30°07′N 97°21′W
93 J15 **Bastuträsk** Västerbotten, N Sweden 64°47′N 20°05′E
119 J19 **Bastyn'** *Rus.* Bostyn'. Brestskaya Voblasts', SW Belarus 52°23′N 26°45′E
Basuo *see* Dongfang
Basutoland *see* Lesotho
119 O15 **Basya** ≈ E Belarus
Bas-Zaïre *see* Bas-Congo
79 D17 **Bata** NW Equatorial Guinea 01°51′N 09°48′E
79 D17 **Bata** ✈ S Equatorial Guinea 01°55′N 09°48′E
Batae Coritanorum *see* Leicester
123 Q8 **Batagay** Respublika Sakha (Yakutiya), NE Russian Federation 67°36′N 134°44′E
123 P8 **Batagay-Alyta** Respublika Sakha (Yakutiya), NE Russian Federation 67°48′N 130°15′E
112 L10 **Batajnica** Vojvodina, N Serbia 44°55′N 20°17′E
136 H15 **Bataklık Gölü** ◎ S Turkey
114 H11 **Batak, Yazovir** ⊠ SW Bulgaria
152 H7 **Batāla** Punjab, N India 31°48′N 75°12′E
104 F9 **Batalha** Leiria, C Portugal 39°40′N 08°50′W
79 N17 **Batama** Orientale, NE Dem. Rep. Congo 0°54′N 26°25′E
123 Q10 **Batamay** Respublika Sakha (Yakutiya), NE Russian Federation 63°28′N 129°33′E
160 F9 **Batang** *var.* Bazhong. Sichuan, C China 30°04′N 99°10′E
79 I14 **Batangafo** Ouham, NW Central African Republic 07°19′N 18°22′E
171 P8 **Batangas** *off.* Batangas City. Luzon, N Philippines 13°47′N 121°03′E
Batangas City *see* Batangas
Bătania *see* Battonya
171 Q10 **Batan Islands** *island group* N Philippines
60 L8 **Batatais** São Paulo, S Brazil 20°54′S 47°37′W
18 E10 **Batavia** New York, NE USA 43°00′N 78°11′W
Batavia *see* Jakarta
173 T9 **Batavia Seamount** *undersea feature* E Indian Ocean 27°42′S 100°36′E
126 L12 **Bataysk** Rostovskaya Oblast', SW Russian Federation 47°10′N 39°46′E
14 B9 **Batchawana** ≈ Ontario, S Canada
14 B9 **Batchawana Bay** Ontario, S Canada 46°55′N 84°36′W
167 Q12 **Bătdâmbâng** *prev.* Battambang. Bătdâmbâng, NW Cambodia 13°06′N 103°13′E
79 G20 **Batéké, Plateaux** *plateau* S Congo
183 S11 **Batemans Bay** New South Wales, SE Australia 35°45′S 150°09′E
21 Q13 **Batesburg** South Carolina, SE USA 33°54′N 81°33′W
28 K12 **Batesland** South Dakota, N USA 43°05′N 102°07′W
27 V10 **Batesville** Arkansas, C USA 35°45′N 91°39′W
31 Q14 **Batesville** Indiana, N USA 39°18′N 85°13′W
22 L2 **Batesville** Mississippi, S USA 34°18′N 89°56′W
25 Q13 **Batesville** Texas, SW USA 28°56′N 99°38′W
44 L13 **Bath** E Jamaica 17°57′N 76°22′W
97 L22 **Bath** *hist.* Akermanceaster; *anc.* Aquae Calidae, Aquae Solis. SW England, United Kingdom 51°23′N 02°22′W
19 Q8 **Bath** Maine, NE USA 43°54′N 69°49′W
18 F11 **Bath** New York, NE USA 42°20′N 77°16′W
Bath *see* Berkeley Springs
78 I10 **Batha** *off.* Région du Batha. ◈ *region* C Chad
78 I10 **Batha** *seasonal river* C Chad
Batha, Région du *see* Batha
141 Y8 **Baṭhā', Wādī al** *dry watercourse* NE Oman
152 H9 **Bathinda** Punjab, NW India 30°14′N 74°54′E
98 M11 **Bathmen** Overijssel, E Netherlands 52°15′N 06°16′E
45 Z14 **Bathsheba** E Barbados 13°13′N 59°31′W
183 R8 **Bathurst** New South Wales, SE Australia 33°32′S 149°35′E
13 O13 **Bathurst** New Brunswick, SE Canada 47°37′N 65°40′W
Bathurst *see* Banjul
8 H6 **Bathurst, Cape** *headland* Northwest Territories, NW Canada 70°33′N 128°00′W
196 L8 **Bathurst Inlet** Nunavut, N Canada 66°23′N 107°00′W
196 L8 **Bathurst Inlet** *inlet* Nunavut, N Canada
181 N1 **Bathurst Island** *island* Northern Territory, N Australia
197 O9 **Bathurst Island** *island* Parry Islands, Nunavut, N Canada
77 O14 **Batié** SW Burkina Faso 09°53′N 02°53′W
Batinah *see* Al Bāţinah
141 Y9 **Bāṭin, Wādī al** *dry watercourse* SW Asia
15 P9 **Batiscan** ≈ Québec, SE Canada
136 F16 **BatıToroslar** ▲▲ SW Turkey
Batjan *see* Bacan, Pulau
147 R11 **Batken** Batenskaya Oblast', SW Kyrgyzstan 40°03′N 70°50′E
Batken Oblasty *see* Batkenskaya Oblast'
147 Q11 **Batkenskaya Oblast'** *Kir.* Batken Oblasty. ◈ *province* SW Kyrgyzstan
Batlle y Ordóñez *see* José Batlle y Ordóñez
183 Q10 **Batlow** New South Wales, SE Australia 35°32′S 148°09′E
137 Q15 **Batman** *var.* İluh. Batman, SE Turkey 37°52′N 41°06′E
137 Q15 **Batman** ◈ *province* SE Turkey
74 L6 **Batna** NE Algeria 35°34′N 06°10′E
163 O7 **Batnorov** *var.* Dundbürd. Hentiy, E Mongolia 47°55′N 111°37′E
Batoe *see* Batu, Kepulauan
162 K7 **Bat-Öldziy** *var.* Övt. Övörhangay, C Mongolia 46°50′N 102°15′E
Bat-Öldziyt *see* Dzaamar
22 J8 **Baton Rouge** *state capital* Louisiana, S USA 30°28′N 91°09′W
79 G15 **Batouri** Est, E Cameroon 04°26′N 14°27′E
138 G14 **Batrā', Jibāl al** ▲▲ S Jordan
138 G6 **Batroûn** *var.* Al Batrūn. N Lebanon 34°15′N 35°42′E
Batsch *see* Bač
Batsevichi *see* Batsevichy
119 M17 **Batsevichy** *Rus.* Batsevichi. Mahilyowskaya Voblasts', E Belarus 53°24′N 29°14′E
92 M7 **Båtsfjord** Finnmark, N Norway 70°37′N 29°42′E
Batshireet *see* Hentiy
162 L7 **Batsümber** *var.* Mandal. Töv, C Mongolia 48°24′N 106°47′E
Battambang *see* Bătdâmbâng
195 X3 **Batterbee, Cape** *headland* Antarctica
155 L24 **Batticaloa** Eastern Province, E Sri Lanka 07°44′N 81°43′E
99 L19 **Battice** Liège, E Belgium 50°39′N 05°50′E
107 L18 **Battipaglia** Campania, S Italy 40°36′N 14°59′E
11 R15 **Battle** ≈ Alberta/Saskatchewan, SW Canada
Battle Born State *see* Nevada
31 Q10 **Battle Creek** Michigan, N USA 42°20′N 85°10′W
27 T7 **Battlefield** Missouri, C USA 37°07′N 93°22′W
11 S15 **Battleford** Saskatchewan, S Canada 52°45′N 108°20′W
29 S6 **Battle Lake** Minnesota, N USA 46°16′N 95°42′W
35 U3 **Battle Mountain** Nevada, W USA 40°37′N 116°55′W
111 M25 **Battonya** *Rom.* Bătania. Békés, SE Hungary 46°16′N 21°00′E
162 J7 **Battsengel** *var.* Jargalant. Arhangay, C Mongolia 47°46′N 101°56′E
168 D11 **Batu, Kepulauan** *prev.* Batoe. *island group* W Indonesia
137 Q10 **Batumi** W Georgia 41°39′N 41°38′E
168 K10 **Batu Pahat** *prev.* Bandar Penggaram. Johor, Peninsular Malaysia 01°51′N 102°56′E
171 O12 **Baturebe** Sulawesi, N Indonesia 01°43′S 121°43′E
122 J12 **Baturino** Tomskaya Oblast', C Russian Federation 57°46′N 85°08′E
117 R3 **Baturyn** Chernihivs'ka Oblast', N Ukraine 51°20′N 32°54′E
138 F10 **Bat Yam** Tel Aviv, C Israel 32°01′N 34°45′E
127 Q4 **Batyrevo** Chuvashskaya Respublika, W Russian Federation 55°04′N 47°34′E
Batys Qazaqstan Oblysy *see* Zapadnyy Kazakhstan
102 F5 **Batz, Île de** *island* NW France
169 Q10 **Bau** Sarawak, East Malaysia 01°25′N 110°10′E
171 N2 **Bauang** Luzon, N Philippines 16°33′N 120°19′E
171 P14 **Baubau** *var.* Baoebaoe. Pulau Buton, C Indonesia 05°30′S 122°37′E
77 W14 **Bauchi** Bauchi, NE Nigeria 10°18′N 09°46′E
77 W14 **Bauchi** ◈ *state* C Nigeria
102 H7 **Baud** Morbihan, NW France 47°52′N 02°59′W
29 T2 **Baudette** Minnesota, N USA 48°42′N 94°36′W
193 S9 **Bauer Basin** *undersea feature* E Pacific Ocean 10°00′S 101°45′W
187 R14 **Bauer Field** *var.* Port Vila. ✈ (Port-Vila) Éfaté, C Vanuatu 17°42′S 168°21′E
13 T9 **Bauld, Cape** *headland* Newfoundland and Labrador, E Canada 51°35′N 55°22′W
103 T8 **Baume-les-Dames** Doubs, E France 47°22′N 06°20′E
101 I15 **Baunatal** Hessen, C Germany 51°15′N 09°25′E
107 D18 **Baunei** Sardegna, Italy, C Mediterranean Sea 40°04′N 09°36′E
57 M15 **Baures, Río** ≈ N Bolivia
60 K9 **Bauru** São Paulo, S Brazil 22°19′S 49°07′W
Baushar *see* Bawshar
118 G10 **Bauska** *Ger.* Bauske. S Latvia 56°25′N 24°11′E
Bauske *see* Bauska
101 Q15 **Bautzen** *Lus.* Budyšin. Sachsen, E Germany 51°11′N 14°29′E
145 Q16 **Bauyrzhan Momyshuly** *Kaz.* Baūyrzhan Momyshuly; *prev.* Burnoye. Zhambyl, S Kazakhstan 42°36′N 70°46′E
Bauzanum *see* Bolzano
Bavaria *see* Bayern
109 N7 **Bavarian Alps** *Ger.* Bayrische Alpen. ▲▲ Austria/Germany
Bavière *see* Bayern
40 H4 **Bavispe, Río** ≈ NW Mexico
127 T5 **Bavly** Respublika Tatarstan, W Russian Federation 54°20′N 53°21′E
169 P13 **Bawal, Pulau** *island* N Indonesia
169 T12 **Bawan** Borneo, C Indonesia 01°36′S 113°55′E
183 O12 **Baw Baw, Mount** ▲ Victoria, SE Australia 37°49′S 146°16′E
169 S15 **Bawean, Pulau** *island* S Indonesia
75 V9 **Bawīṭī** *var.* Bawîti. N Egypt 28°19′N 28°53′E
Bawîti *see* Bawīṭī
77 Q13 **Bawku** N Ghana 11°00′N 00°12′W
Bawlake *see* Bawlakhe
167 N7 **Bawlakhe** *var.* Bawlake. Kayah State, C Myanmar (Burma) 19°10′N 97°19′E
169 H11 **Bawo Ofuloa** Pulau Tanahmasa, W Indonesia 0°10′S 98°24′E
141 Y8 **Bawshar** *var.* Baushar. NE Oman 23°32′N 58°24′E
Baxian *see* Bazhou
Ba Xian *see* Bazhou
158 M8 **Baxkorgan** Xinjiang Uygur Zizhiqu, W China 39°05′N 90°00′E
23 V6 **Baxley** Georgia, SE USA 31°46′N 82°21′W
159 R15 **Baxoi** *var.* Baima. Xizang Zizhiqu, W China 30°01′N 96°53′E
29 W14 **Baxter** Iowa, C USA 41°49′N 93°09′W
29 U6 **Baxter** Minnesota, N USA 46°21′N 94°18′W
27 R8 **Baxter Springs** Kansas, C USA 37°01′N 94°45′W
81 M17 **Bay** *off.* Gobolka Bay. ◈ *region* SW Somalia
Bay *see* Baicheng
44 H7 **Bayamo** Granma, E Cuba 20°21′N 76°38′W
45 U5 **Bayamón** E Puerto Rico 18°24′N 66°09′W
163 W8 **Bayan** Heilongjiang, NE China 46°05′N 127°24′E
170 L16 **Bayan** *prev.* Bajan. Pulau Lombok, C Indonesia 08°16′S 116°28′E
162 M8 **Bayan** *var.* Maanīt. Töv, C Mongolia 47°14′N 107°34′E
Bayan *see* Hölönbuyr, Dornod, Mongolia
Bayan *see* Ihhet, Dornogovĭ, Mongolia
Bayan *see* Bayan-Uul, Govĭ-Altay, Mongolia
Bayan *see* Bayanhutag, Hentiy, Mongolia
Bayan *see* Bürentogtoh, Hövsgol, Mongolia
152 I12 **Bayāna** Rājasthān, N India 26°55′N 77°18′E
149 N5 **Bāyān, Band-e** ▲▲ C Afghanistan
162 H8 **Bayanbulag** Bayanhongor, C Mongolia 46°46′N 98°07′E
Bayanbulag *see* Ömnödelger
158 J5 **Bayanbulak** Xinjiang Uygur Zizhiqu, W China 43°05′N 84°05′E
162 L7 **Bayanchandmanĭ** *var.* Ihsüüj. Töv, C Mongolia 48°12′N 106°23′E
162 J11 **Bayandalay** *var.* Dalay. Ömnögovĭ, S Mongolia 43°27′N 103°30′E
163 O9 **Bayandelger** *var.* Shireet. Sühbaatar, SE Mongolia 45°33′N 112°19′E
162 I5 **Bayandzürh** *var.* Altraga. Hövsgöl, N Mongolia 50°08′N 98°54′E
Bayan Gol *see* Dengkou, China
Bayangol *see* Bugat, Mongolia
162 I9 **Bayangovĭ** *var.* Örgön. Bayanhongor, C Mongolia 44°43′N 100°23′E
159 R12 **Bayan Har Shan** *var.* Bayan Khar. ▲▲ C China
162 G6 **Bayanhayrhan** *var.* Altanbulag. Dzavhan, N Mongolia 49°16′N 96°22′E
162 I8 **Bayanhongor** Bayanhongor, C Mongolia 46°08′N 100°42′E
162 H9 **Bayanhongor** ◈ *province* C Mongolia
162 K14 **Bayan Hot** *var.* Alxa Zuoqi. Nei Mongol Zizhiqu, N China 38°49′N 105°40′E
163 O8 **Bayanhutag** *var.* Bayan. Hentiy, C Mongolia 47°13′N 110°57′E
163 T9 **Bayan Huxu** *var.* Horqin Zuoyi Zhongqi. Nei Mongol Zizhiqu, N China 45°02′N 121°28′E
Bayan Khar *see* Bayan Har Shan
163 N12 **Bayan Kuang** *prev.* Bayan Obo. Nei Mongol Zizhiqu, N China 41°45′N 109°58′E
168 J7 **Bayan Lepas** ✈ (George Town) Pinang, Peninsular Malaysia 05°18′N 100°15′E
162 I10 **Bayanlig** *var.* Hatansuudal. Bayanhongor, C Mongolia 44°34′N 100°41′E
162 K13 **Bayan Mod** Nei Mongol Zizhiqu, N China 40°45′N 104°29′E
163 N8 **Bayanmönh** *var.* Ulaan-Ereg. Hentiy, E Mongolia 46°50′N 109°39′E
162 L12 **Bayannur** *var.* Linhe. Nei Mongol Zizhiqu, N China 40°46′N 107°27′E
Bayan Nuru *see* Xar Burd
162 E5 **Bayannuur** *var.* Tsul-Ulaan. Bayan-Ölgiy, W Mongolia 48°51′N 91°13′E
Bayan Obo *see* Bayan Kuang
43 V15 **Bayano, Lago** ◎ E Panama
162 C5 **Bayan-Ölgiy** ◈ *province* NW Mongolia
162 H9 **Bayan-Öndör** *var.* Bulgan. Bayanhongor, C Mongolia 44°48′N 98°39′E
162 K8 **Bayan-Öndör** *var.* Bumbat. Övörhangay, C Mongolia 46°30′N 104°08′E
162 L8 **Bayan-Önjüül** *var.* Ihhayrhan. Töv, C Mongolia 46°57′N 105°51′E
163 O7 **Bayan-Ovoo** *var.* Javhlant. Hentiy, E Mongolia 47°46′N 112°06′E
162 L11 **Bayan-Ovoo** *var.* Erdenetsogt. Ömnögovĭ, S Mongolia 42°54′N 106°16′E
Bayansayr *see* Baatsagaan
159 Q9 **Bayan Shan** ▲ C China 37°36′N 96°23′E
162 J9 **Bayanteeg** Övörhangay, C Mongolia 45°39′N 101°30′E
162 G5 **Bayantes** *var.* Altay. Dzavhan, N Mongolia 49°40′N 96°21′E
Bayantöhöm *see* Büren
162 M8 **Bayantsagaan** *var.* Dzogsool. Töv, C Mongolia 46°46′N 107°18′E
163 P7 **Bayantümen** *var.* Tsagaanders. Dornod, NE Mongolia 48°03′N 114°16′E
Bayan-Uhaa *see* Ih-Uul
163 R10 **Bayan Ul** *var.* Xi Ujimqin Qi. Nei Mongol Zizhiqu, N China 44°31′N 117°36′E
Bayan-Ulaan *see* Dzüünbayan-Ulaan
163 O7 **Bayan-Uul** *var.* Javarthushuu. Dornod, NE Mongolia 49°05′N 112°40′E
162 F7 **Bayan-Uul** *var.* Bayan. Govĭ-Altay, W Mongolia 47°05′N 95°13′E
162 M8 **Bayanuur** *var.* Tsul-Ulaan. Töv, C Mongolia 47°44′N 108°22′E
28 J14 **Bayard** Nebraska, C USA 41°45′N 103°19′W
37 P15 **Bayard** New Mexico, SW USA 32°45′N 108°07′W
103 T13 **Bayard, Col** *pass* SE France
Bayasgalant *see* Mönhhaan
136 J12 **Bayat** Çorum, N Turkey 40°34′N 34°07′E
171 P6 **Bayawan** Negros, C Philippines 09°22′N 122°50′E
143 R10 **Bayāẕ** Kermān, C Iran 30°41′N 55°29′E
171 Q6 **Baybay** Leyte, C Philippines 10°41′N 124°49′E
21 X10 **Bayboro** North Carolina, SE USA 35°08′N 76°49′W
137 P12 **Bayburt** Bayburt, NE Turkey 40°15′N 40°16′E
137 P12 **Bayburt** ◈ *province* NE Turkey
31 R8 **Bay City** Michigan, N USA 43°35′N 83°52′W
25 V12 **Bay City** Texas, SW USA 28°59′N 96°00′W
Baydarata Bay *see* Baydaratskaya Guba
122 J7 **Baydaratskaya Guba** *var.* Baydarata Bay. *bay* N Russian Federation
81 M16 **Baydhabo** *var.* Baydhowa, Isha Baydhabo, *It.* Baidoa. Bay, SW Somalia 03°08′N 43°39′E
Baydhowa *see* Baydhabo
101 N21 **Bayerischer Wald** ▲▲ SE Germany
101 K21 **Bayern** *Eng.* Bavaria, *Fr.* Bavière. ◈ *state* SE Germany
147 V9 **Bayetovo** Narynskaya Oblast', C Kyrgyzstan 41°14′N 74°55′E
102 K4 **Bayeux** *anc.* Augustodurum. Calvados, N France 49°16′N 00°42′W
14 E15 **Bayfield** ≈ Ontario, S Canada
145 O15 **Baygekum** *Kaz.* Bäygequm. Kzylorda, S Kazakhstan 44°15′N 66°34′E
Bäygequm *see* Baygekum
136 C14 **Bayındır** İzmir, SW Turkey 38°12′N 27°40′E
138 H12 **Bāyir** *var.* Bā'ir. Ma'ān, S Jordan 30°46′N 36°40′E
Bay Islands *see* Bahía, Islas de la
Bayizhen *see* Nyingchi
139 R5 **Bayjī** *var.* Baiji. Şalāḩ ad Dīn, N Iraq 34°56′N 43°29′E
Baykadam *see* Saudakent
123 N13 **Baykal, Ozero** *Eng.* Lake Baikal. ◎ S Russian Federation
123 M14 **Baykal'sk** Irkutskaya Oblast', S Russian Federation 51°30′N 104°03′E
137 R15 **Baykan** Siirt, SE Turkey 38°08′N 41°43′E
139 S2 **Baykhmah** *var.* Bēkma. Arbīl, E Iraq 36°40′N 44°15′E
123 L11 **Baykit** Krasnoyarskiy Kray, C Russian Federation 61°37′N 96°23′E
145 N12 **Baykonur** *var.* Baykonyr. Karaganda, C Kazakhstan 47°50′N 75°33′E
Baykonur *see* Baykonyr
144 M14 **Baykonyr** *var.* Baykonur, *Kaz.* Bayqongyr; *prev.* Leninsk. Kyzylorda, S Kazakhstan 45°38′N 63°20′E
Baykonyr *see* Baykonur
158 E7 **Baykurt** Xinjiang Uygur Zizhiqu, W China 39°56′N 75°33′E
14 I9 **Bay, Lac** ◎ Québec, SE Canada
127 W6 **Baymak** Respublika Bashkortostan, W Russian Federation 52°34′N 58°20′E
23 O8 **Bay Minette** Alabama, S USA 30°52′N 87°46′W
143 O17 **Baynūnah** *desert* W United Arab Emirates
184 O8 **Bay of Plenty** *off.* Bay of Plenty Region. ◈ *region* North Island, New Zealand
Bay of Plenty Region *see* Bay of Plenty
191 Z3 **Bay of Wrecks** *bay* Kiritimati, E Kiribati
Bayonnaise Rocks *see* Beyonēsu-retsugan
102 I15 **Bayonne** *anc.* Lapurdum. Pyrénées-Atlantiques, SW France 43°30′N 01°28′W
22 H4 **Bayou D'Arbonne Lake** ⊠ Louisiana, S USA
23 N9 **Bayou La Batre** Alabama, S USA 30°24′N 88°15′W
Bayou State *see* Mississippi
Bayqadam *see* Saudakent
Bayqongyr *see* Baykonyr
Bayram-Ali *see* Baýramaly
146 J14 **Baýramaly** *var.* Bayramaly; *prev.* Bayram-Ali. Mary Welaýaty, S Turkmenistan 37°33′N 62°08′E
101 L19 **Bayreuth** *var.* Baireuth. Bayern, SE Germany 49°57′N 11°34′E
Bayrische Alpen *see* Bavarian Alps
Bayrūt *see* Beyrouth
22 L9 **Bay Saint Louis** Mississippi, S USA 30°18′N 89°19′W
Baysān *see* Beit She'an
Bayshint *see* Öndörshireet
14 H13 **Bays, Lake of** ◎ Ontario, S Canada
22 M6 **Bay Springs** Mississippi, S USA 31°58′N 89°17′W
Bay State *see* Massachusetts
Baysun *see* Boysun
14 H13 **Baysville** Ontario, S Canada 45°10′N 79°03′W
141 N15 **Bayt al Faqīh** W Yemen 14°30′N 43°20′E
158 M4 **Baytik Shan** ▲▲ China/Mongolia
Bayt Laḩm *see* Bethlehem
25 W11 **Baytown** Texas, SW USA 29°43′N 94°59′W
169 V11 **Bayur, Tanjung** *headland* Borneo, N Indonesia 0°43′S 117°32′E
121 N14 **Bayy al Kabīr, Wādī** *dry watercourse* NW Libya
145 P17 **Bayyrkum** *Kaz.* Bayyrqum; *prev.* Bairkum. Yuzhnyy Kazakhstan, S Kazakhstan 41°57′N 68°05′E
Bayyrqum *see* Bayyrkum
105 P14 **Baza** Andalucía, S Spain 37°30′N 02°45′W
137 X10 **Bazardüzü Dağı** *Rus.* Gora Bazardyuzyu. ▲ N Azerbaijan 41°13′N 47°50′E
Bazardyuzyu, Gora *see* Bazardüzü Dağı
Bazargic *see* Dobrich
83 N18 **Bazaruto, Ilha do** *island* SE Mozambique
102 K14 **Bazas** Gironde, SW France 44°27′N 00°11′W
105 O14 **Baza, Sierra de** ▲▲ S Spain
160 J8 **Bazhong** *var.* Bazhou. Sichuan, C China 31°55′N 106°44′E
Bazhong *see* Batang
161 P3 **Bazhou** *prev.* Baxian, Ba Xian. Hebei, E China 39°05′N 116°24′E
Bazhou *see* Bazhong
14 M9 **Bazin** ≈ Québec, SE Canada
Bazin *see* Pezinok
139 Q7 **Bāzīyah** Al Anbār, C Iraq 33°50′N 42°41′E
138 H6 **Bcharré** *var.* Bcharreh, Bsharrī, Bsherri. NE Lebanon 34°16′N 36°01′E
Bcharreh *see* Bcharré
28 J5 **Beach** North Dakota, N USA 46°55′N 104°00′W
182 K12 **Beachport** South Australia 37°29′S 140°03′E
97 O23 **Beachy Head** *headland* SE England, United Kingdom 50°44′N 00°16′E
18 K13 **Beacon** New York, NE USA 41°30′N 73°54′W
63 J25 **Beagle Channel** *channel* Argentina/Chile
181 O1 **Beagle Gulf** *gulf* Northern Territory, N Australia
Bealach an Doirín *see* Ballaghaderreen
Bealach Cláir *see* Ballyclare
Bealach Féich *see* Ballybofey
172 J3 **Bealanana** Mahajanga, NE Madagascar 14°33′S 48°44′E
Béal an Átha *see* Ballina
Béal an Átha Móir *see* Ballinamore
Béal an Mhuirhead *see* Belmullet
Béal Átha Beithe *see* Ballybay
Béal Átha Conaill *see* Ballyconnell
Beál Átha hAmhnais *see* Ballyhaunis
Béal Átha na Sluaighe *see* Ballinasloe
Béal Átha Seanaidh *see* Ballyshannon
Bealdovuopmi *see* Peltovuoma
Béal Feirste *see* Belfast
Béal Tairbirt *see* Belturbet
Beanna Boirche *see* Mourne Mountains
Beannchar *see* Banagher, Ireland
Beannchar *see* Bangor, Northern Ireland, UK
Beanntraí *see* Bantry
Bearalváhki *see* Berlevåg
23 N2 **Bear Creek** ≈ Alabama/Mississippi, S USA
30 J13 **Bear Creek** ≈ Illinois, N USA
27 U13 **Bearden** Arkansas, C USA 33°43′N 92°37′W
195 Q10 **Beardmore Glacier** *glacier* Antarctica
30 K13 **Beardstown** Illinois, N USA 40°01′N 90°25′W
28 L14 **Bear Hill** ▲ Nebraska, C USA 41°24′N 101°49′W
Bear Island *see* Bjørnøya
14 H12 **Bear Lake** Ontario, S Canada 45°28′N 79°31′W
36 M1 **Bear Lake** ◎ Idaho/Utah, NW USA
39 U11 **Bear, Mount** ▲ Alaska, USA 61°16′N 141°09′W
102 J16 **Béarn** *cultural region* SW France
194 J11 **Bear Peninsula** *peninsula* Antarctica
152 I7 **Beās** ≈ India/Pakistan
105 P3 **Beasain** País Vasco, N Spain 43°03′N 02°11′W
105 O12 **Beas de Segura** Andalucía, S Spain 38°16′N 02°54′W
45 N10 **Beata, Cabo** *headland* SW Dominican Republic 17°34′N 71°25′W
45 N10 **Beata, Isla** *island* SW Dominican Republic
64 F11 **Beata Ridge** *undersea feature* N Caribbean Sea 16°00′N 72°30′W
29 R17 **Beatrice** Nebraska, C USA 40°14′N 96°43′W
83 L16 **Beatrice** Mashonaland East, NE Zimbabwe 18°15′S 30°55′E
11 N11 **Beatton** ≈ British Columbia, W Canada
11 N11 **Beatton River** British Columbia, W Canada 57°35′N 121°45′W
35 V10 **Beatty** Nevada, W USA 36°53′N 116°44′W
21 N6 **Beattyville** Kentucky, S USA 37°33′N 83°44′W
173 X16 **Beau Bassin** W Mauritius 20°13′S 57°27′E
103 R15 **Beaucaire** Gard, S France 43°49′N 04°37′E
14 I8 **Beauchastel, Lac** ◎ Québec, SE Canada
14 I10 **Beauchêne, Lac** ◎ Québec, SE Canada
183 V3 **Beaudesert** Queensland, E Australia 28°00′S 152°27′E
182 M12 **Beaufort** Victoria, SE Australia 37°27′S 143°24′E
21 X11 **Beaufort** North Carolina, SE USA 34°44′N 76°41′W
21 R15 **Beaufort** South Carolina, SE USA 32°23′N 80°40′W
38 M11 **Beaufort Sea** *sea* Arctic Ocean
Beaufort-Wes *see* Beaufort West
83 G25 **Beaufort West** *Afr.* Beaufort-Wes. Western Cape, SW South Africa 32°21′S 22°35′E
103 N7 **Beaugency** Loiret, C France 47°46′N 01°38′E
19 R1 **Beau Lake** ⊠ Maine, NE USA
96 I8 **Beauly** N Scotland, United Kingdom 57°29′N 04°29′W
99 G21 **Beaumont** Hainaut, S Belgium 50°12′N 04°13′E
185 E23 **Beaumont** Otago, South Island, New Zealand 45°48′S 169°32′E
22 M7 **Beaumont** Mississippi, S USA 31°10′N 88°55′W
25 X10 **Beaumont** Texas, SW USA 30°05′N 94°06′W
102 M15 **Beaumont-de-Lomagne** Tarn-et-Garonne, S France 43°54′N 01°00′E
102 L6 **Beaumont-sur-Sarthe** Sarthe, NW France 48°15′N 00°07′E
103 R8 **Beaune** Côte d'Or, C France 47°02′N 04°50′E
15 R9 **Beaupré** Québec, SE Canada 47°03′N 70°52′W
102 J8 **Beaupréau** Maine-et-Loire, NW France 47°13′N 00°57′W
99 I22 **Beauraing** Namur, SE Belgium 50°07′N 04°57′E
103 R12 **Beaurepaire** Isère, E France 45°20′N 05°03′E
11 Y16 **Beausejour** Manitoba, S Canada 50°04′N 96°30′W
103 N4 **Beauvais** *anc.* Bellovacum, Caesaromagus. Oise, N France 49°27′N 02°04′E
11 S13 **Beauval** Saskatchewan, C Canada 55°10′N 107°37′W
102 I9 **Beauvoir-sur-Mer** Vendée, NW France 46°54′N 02°03′W
39 R8 **Beaver** Alaska, USA 66°22′N 147°31′W
26 J8 **Beaver** Oklahoma, C USA 36°48′N 100°32′W
18 B14 **Beaver** Pennsylvania, NE USA 40°39′N 80°19′W
36 K6 **Beaver** Utah, W USA 38°16′N 112°38′W
10 L9 **Beaver** ≈ British Columbia/Yukon, W Canada
11 S13 **Beaver** ≈ Saskatchewan, C Canada
29 N17 **Beaver City** Nebraska, C USA 40°08′N 99°49′W
10 G6 **Beaver Creek** Yukon, W Canada 62°20′N 140°45′W
31 R14 **Beavercreek** Ohio, N USA 39°42′N 83°58′W
39 S8 **Beaver Creek** ≈ Alaska, USA
26 H3 **Beaver Creek** ≈ Kansas/Nebraska, C USA
28 J5 **Beaver Creek** ≈ Montana/North Dakota, N USA
29 Q14 **Beaver Creek** ≈ Nebraska, C USA
25 Q4 **Beaver Creek** ≈ Texas, SW USA
30 M8 **Beaver Dam** Wisconsin, N USA 43°28′N 88°49′W
30 M8 **Beaver Dam Lake** ◎ Wisconsin, N USA
18 B14 **Beaver Falls** Pennsylvania, NE USA 40°45′N 80°20′W
33 P12 **Beaverhead Mountains** ▲▲ Idaho/Montana, NW USA
33 Q12 **Beaverhead River** ≈ Montana, NW USA
65 A25 **Beaver Island** *island* W Falkland Islands
31 P5 **Beaver Island** *island* Michigan, N USA
27 S9 **Beaver Lake** ⊠ Arkansas, C USA
11 N13 **Beaverlodge** Alberta, W Canada 55°11′N 119°29′W
18 I8 **Beaver River** ≈ New York, NE USA
26 J8 **Beaver River** ≈ Oklahoma, C USA
18 B13 **Beaver River** ≈ Pennsylvania, NE USA
65 A25 **Beaver Settlement** Beaver Island, W Falkland Islands 51°30′S 61°15′W
Beaver State *see* Oregon
14 H14 **Beaverton** Ontario, S Canada 44°24′N 79°07′W
32 G11 **Beaverton** Oregon, NW USA 45°29′N 122°49′W
152 G12 **Beāwar** Rājasthān, N India 26°08′N 74°22′E
Bebas, Dasht-i *see* Bābūs, Dasht-e
60 L8 **Bebedouro** São Paulo, S Brazil 20°58′S 48°28′W
101 I16 **Bebra** Hessen, C Germany 50°59′N 09°46′E
41 W12 **Becal** Campeche, SE Mexico 19°49′N 90°28′W
15 Q11 **Bécancour** ≈ Québec, SE Canada
97 Q19 **Beccles** E England, United Kingdom 52°27′N 01°32′E
112 L9 **Bečej** *Ger.* Altbetsche, *Hung.* Óbecse, Rácz-Becse; *prev.* Magyar-Becse, Stari Bečej. Vojvodina, N Serbia 45°36′N 20°02′E
104 I3 **Becerréa** Galicia, NW Spain 42°51′N 07°10′W
74 H7 **Béchar** *prev.* Colomb-Béchar. W Algeria 31°38′N 02°11′W
39 O14 **Becharof Lake** ◎ Alaska, USA
116 H15 **Bechet** *var.* Bechetu. Dolj, SW Romania 43°45′N 23°57′E
Bechetu *see* Bechet
21 R6 **Beckley** West Virginia, NE USA 37°46′N 81°12′W
101 G14 **Beckum** Nordrhein-Westfalen, W Germany 51°45′N 08°03′E
25 X7 **Beckville** Texas, SW USA 32°14′N 94°27′W
35 X4 **Becky Peak** ▲ Nevada, W USA 39°59′N 114°33′W
116 I9 **Beclean** *Hung.* Bethlen; *prev.* Betlen. Bistriţa-Năsăud, N Romania 47°10′N 24°11′E
Bécs *see* Wien
111 H18 **Bečva** *Ger.* Betschau, *Pol.* Beczwa. ≈ E Czech Republic
Beczwa *see* Bečva
103 P15 **Bédarieux** Hérault, S France 43°37′N 03°10′E
120 B10 **Beddouza, Cap** *headland* W Morocco 32°35′N 09°16′W
80 I13 **Bedelē** Oromīya, C Ethiopia 08°25′N 36°21′E
147 Y8 **Bedel Pass** *Rus.* Pereval Bedel. *pass* China/Kyrgyzstan
Bedel, Pereval *see* Bedel Pass
95 H22 **Beder** Midtjylland, C Denmark 56°03′N 10°13′E
97 N20 **Bedford** E England, United Kingdom 52°08′N 00°29′W
31 O15 **Bedford** Indiana, N USA 38°51′N 86°29′W
29 U16 **Bedford** Iowa, C USA 40°40′N 94°43′W
20 L4 **Bedford** Kentucky, S USA 38°36′N 85°18′W
18 D15 **Bedford** Pennsylvania, NE USA 40°00′N 78°29′W
21 T6 **Bedford** Virginia, NE USA 37°20′N 79°31′W
97 N20 **Bedfordshire** *cultural region* E England, United Kingdom
127 N5 **Bednodem'yanovsk** Penzenskaya Oblast', W Russian Federation 53°55′N 43°14′E
98 N5 **Bedum** Groningen, NE Netherlands 53°18′N 06°36′E
27 V11 **Beebe** Arkansas, C USA 35°04′N 91°52′W
Beechy Group *see* Chichijima-rettō
45 T9 **Beef Island** ✈ (Road Town) Tortola, E British Virgin Islands 18°25′N 64°31′W
Beehive State *see* Utah
99 L18 **Beek** Limburg, SE Netherlands 50°56′N 05°47′E
99 L18 **Beek** ✈ (Maastricht) Limburg, SE Netherlands 50°55′N 05°47′E
99 K14 **Beek-en-Donk** Noord-Brabant, S Netherlands 51°31′N 05°37′E
138 F13 **Be'er Menuha** *prev.* Be'ér Menuḫa. Southern, S Israel 30°22′N 35°09′E
Be'ér Menuḫa *see* Be'er Menuha
99 D16 **Beernem** West-Vlaanderen, NW Belgium 51°09′N 03°18′E
99 I16 **Beerse** Antwerpen, N Belgium 51°20′N 04°52′E
Beersheba *see* Be'er Sheva
138 E11 **Be'er Sheva** *var.* Beersheba, *Ar.* Bir es Saba; *prev.* Be'ér Sheva'. Southern, S Israel 31°15′N 34°47′E
Be'ér Sheva' *see* Be'er Sheva
98 J13 **Beesd** Gelderland, C Netherlands 51°52′N 05°12′E
99 M16 **Beesel** Limburg, SE Netherlands 51°16′N 06°02′E
83 J21 **Beestekraal** North-West, N South Africa 25°21′S 27°40′E
194 J7 **Beethoven Peninsula** *peninsula* Alexander Island, Antarctica
Beetstersweach *see* Beetsterzwaag
98 M6 **Beetsterzwaag** *Fris.* Beetstersweach. Fryslân, N Netherlands 53°03′N 06°04′E
25 S13 **Beeville** Texas, SW USA 28°25′N 97°47′W
79 J18 **Befale** Equateur, NW Dem. Rep. Congo 0°25′N 20°48′E
Befandriana *see* Befandriana Avaratra
172 J3 **Befandriana Avaratra** *var.* Befandriana, Befandriana Nord. Mahajanga, NW Madagascar 15°14′S 48°33′E
Befandriana Nord *see* Befandriana Avaratra
79 K18 **Befori** Equateur, N Dem. Rep. Congo 0°09′N 22°18′E
172 I7 **Befotaka** Fianarantsoa, S Madagascar 23°49′S 47°00′E
183 R11 **Bega** New South Wales, SE Australia 36°43′S 149°50′E
102 G5 **Bégard** Côtes d'Armor, NW France 48°37′N 03°18′W
112 M9 **Begejski Kanal** *canal* Vojvodina, NE Serbia
94 G13 **Begna** ≈ S Norway
Begoml' *see* Byahoml'
Begovat *see* Bekobod
153 Q13 **Begusarai** Bihār, NE India 25°25′N 86°08′E
143 R9 **Behābād** Yazd, C Iran 32°23′N 59°50′E
Behagle *see* Laï
55 Z10 **Béhague, Pointe** *headland* E French Guiana 04°38′N 51°52′W
Behar *see* Bihār
142 M10 **Behbahān** *var.* Behbehān. Khūzestān, SW Iran 30°38′N 50°07′E
Behbehān *see* Behbahān
44 G3 **Behring Point** Andros Island, W The Bahamas 24°28′N 77°44′W
143 P4 **Behshahr** *prev.* Ashraf. Māzandarān, N Iran 36°42′N 53°36′E
163 V6 **Bei'an** Heilongjiang, NE China 48°16′N 126°29′E
Beibunar *see* Sredishte
Beibu Wan *see* Tonkin, Gulf of
Beida *see* Al Bayḑā'
80 H13 **Beigi** Oromīya, C Ethiopia 09°13′N 34°48′E
160 L16 **Beihai** Guangxi Zhuangzu Zizhiqu, S China 21°29′N 109°10′E
159 Q10 **Bei Hulsan Hu** ◎ C China
161 N13 **Bei Jiang** ≈ S China
161 O2 **Beijing** *var.* Pei-ching, *Eng.* Peking; *prev.* Pei-p'ing. ● (China) Beijing Shi, E China 39°58′N 116°23′E
161 P2 **Beijing** ✈ Beijing Shi, N China 39°54′N 116°22′E
Beijing *see* Beijing Shi, China

◆ Country
● Country Capital
◇ Dependent Territory
○ Dependent Territory Capital
◈ Administrative Regions
✈ International Airport
▲ Mountain
▲▲ Mountain Range
⛰ Volcano
≈ River
◎ Lake
⊠ Reservoir

Beijing Shi – Berkåk

161 O2 **Beijing Shi** var. Beijing, Jing, Pei-ching, Eng. Peking; prev. Pei-p'ing. ⬖ municipality E China
76 G8 **Beïla** Trarza, W Mauritania 18°07′N 15°56′W
98 N7 **Beilen** Drenthe, NE Netherlands 52°52′N 06°27′E
160 L15 **Beiliu** var. Lingcheng. Guangxi Zhuangzu Zizhiqu, S China 22°50′N 110°22′E
159 O12 **Beilu He** ≈ W China
Beilul see Beylul
163 U12 **Beining** prev. Beizhen. Liaoning, NE China 41°34′N 121°51′E
96 H8 **Beinn Dearg** ▲ N Scotland, United Kingdom 57°47′N 04°52′W
Beinn MacDuibh see Ben Macdui
160 I12 **Beipan Jiang** ≈ S China
163 T12 **Beipiao** Liaoning, NE China 41°49′N 120°45′E
83 N17 **Beira** Sofala, C Mozambique 19°45′S 34°56′E
83 N17 **Beira** ✈ Sofala, C Mozambique 19°39′S 35°05′E
104 I7 **Beira Alta** former province N Portugal
104 H9 **Beira Baixa** former province C Portugal
104 G8 **Beira Litoral** former province N Portugal
Beirut see Beyrouth
Beisān see Beit She'an
11 Q16 **Beiseker** Alberta, SW Canada 51°20′N 113°34′W
Beitai Ding see Wutai Shan
83 K19 **Beitbridge** Matabeleland South, S Zimbabwe 22°10′S 30°02′E
Beit Lekhem see Bethlehem
138 G9 **Beit She'an** Ar. Baysān, Beisān; anc. Scythopolis, prev. Bet She'an. Northern, N Israel 32°30′N 35°30′E
116 G10 **Beiuş** Hung. Belényes. Bihor, NW Romania 46°40′N 22°21′E
Beizhen see Beining
104 H12 **Beja** anc. Pax Julia. Beja, SE Portugal 38°01′N 07°52′W
74 M5 **Béja** var. Bājah. N Tunisia 36°45′N 09°04′E
104 G13 **Beja** ⬖ district S Portugal
120 I9 **Béjaïa** var. Bejaïa, Fr. Bougie; anc. Saldae. NE Algeria 36°49′N 05°03′E
Bejaïa see Béjaïa
104 K8 **Béjar** Castilla y León, N Spain 40°24′N 05°45′W
Bejraburi see Phetchaburi
Bekaa Valley see El Beqaa
Bekabad see Bekobod
Békás see Bicaz
169 O15 **Bekasi** Jawa, C Indonesia 06°14′S 106°595′E
Bek-Budi see Qarshi
Bekdaş/Bekdash see Garabogaz
147 T10 **Bek-Dzhar** Oshskaya Oblast', SW Kyrgyzstan 40°22′N 73°08′E
111 N24 **Békés** Rom. Bichiş. Békés, SE Hungary 46°45′N 21°09′E
111 M24 **Békés** off. Békés Megye. ⬖ county SE Hungary
111 M24 **Békéscsaba** Rom. Bichiş-Ciaba. Békés, SE Hungary 46°40′N 21°05′E
Békés Megye see Békés
172 H7 **Bekily** Toliara, S Madagascar 24°12′S 45°20′E
165 W4 **Bekkai** var. Betsukai. Hokkaidō, NE Japan 43°23′N 145°07′E
Bēkma see Baykhmah
147 Q11 **Bekobod** Rus. Bekabad; prev. Begovat. Toshkent Viloyati, E Uzbekistan 40°17′N 69°11′E
127 O7 **Bekovo** Penzenskaya Oblast', W Russian Federation 52°27′N 43°41′E
Bel see Beliu
152 M13 **Bela** Uttar Pradesh, N India 25°55′N 82°00′E
149 N15 **Bela** Baluchistān, SW Pakistan 26°12′N 66°20′E
79 F15 **Bélabo** Est, C Cameroon 04°54′N 13°10′E
112 N10 **Bela Crkva** Ger. Weisskirchen, Hung. Fehértemplom. Vojvodina, W Serbia 44°55′N 21°28′E
173 Y16 **Bel Air** var. Rivière Sèche. E Mauritius
104 L12 **Belalcázar** Andalucía, S Spain 38°33′N 05°07′W
113 P15 **Bela Palanka** Serbia, SE Serbia 43°13′N 22°19′E
119 H16 **Belarus** off. Republic of Belarus, var. Belorussia, Latv. Baltkrievija; prev. Belorussian SSR, Rus. Belorusskaya SSR. ◆ republic E Europe
Belarus, Republic of see Belarus
Belau see Palau
59 H21 **Bela Vista** Mato Grosso do Sul, SW Brazil 22°04′S 56°25′W
83 L21 **Bela Vista** Maputo, S Mozambique 26°20′S 32°40′E
168 I8 **Belawan** Sumatera, W Indonesia 03°46′N 98°44′E
Běla Woda see Weisswasser
127 U4 **Belaya** ≈ W Russian Federation
123 R7 **Belaya Gora** Respublika Sakha (Yakutiya), NE Russian Federation 68°25′N 146°12′E
126 M11 **Belaya Kalitiva** Rostovskaya Oblast', SW Russian Federation 48°09′N 40°43′E
125 R14 **Belaya Kholunitsa** Kirovskaya Oblast', NW Russian Federation 58°54′N 50°52′E
Belaya Tserkov' see Bila Tserkva
77 V11 **Belbédji** Zinder, S Niger 14°35′N 08°00′E
111 K14 **Bełchatów** var. Belchatow. Łódzski, C Poland 51°23′N 19°20′E
Belchatow see Bełchatów
Belcher, Îles see Belcher Islands
12 H7 **Belcher Islands** Fr. Îles Belcher. island group Nunavut, SE Canada
105 S6 **Belchite** Aragón, NE Spain 41°18′N 00°45′W
29 O2 **Belcourt** North Dakota, N USA 48°50′N 99°44′W
31 P9 **Belding** Michigan, N USA 43°06′N 85°13′W
127 U5 **Belebey** Respublika Bashkortostan, W Russian Federation 54°04′N 54°13′E
81 N16 **Beledweyne** var. Belet Huen, It. Belet Uen. Hiiraan, C Somalia 04°39′N 45°12′E
146 B10 **Belek** Balkan Welaýaty, W Turkmenistan 39°57′N 53°51′E
58 L12 **Belém** var. Pará. state capital Pará, N Brazil 01°27′S 48°29′W
65 I14 **Belém Ridge** undersea feature C Atlantic Ocean
62 I7 **Belén** Catamarca, NW Argentina 27°36′N 67°00′W
54 G9 **Belén** Boyacá, C Colombia 06°01′N 72°55′W
42 J11 **Belén** Rivas, SW Nicaragua 11°30′N 85°55′W
62 O5 **Belén** Concepción, C Paraguay 23°25′S 57°14′W
61 D16 **Belén** Salto, N Uruguay 30°47′S 57°47′W
37 R12 **Belen** New Mexico, SW USA 34°37′N 106°46′W
61 D20 **Belén de Escobar** Buenos Aires, E Argentina 34°21′S 58°47′W
114 J7 **Belene** Pleven, N Bulgaria 43°39′N 25°09′E
114 J7 **Belene, Ostrov** island N Bulgaria
43 R15 **Belén, Río** ≈ C Panama
Belényes see Beiuş
Embalse de Belesar see Belesar, Encoro de
104 H3 **Belesar, Encoro de** Sp. Embalse de Belesar. ⊠ NW Spain
Belet Huen/Belet Uen see Beledweyne
126 J5 **Belëv** Tul'skaya Oblast', W Russian Federation 53°48′N 36°07′E
19 R7 **Belfast** Maine, NE USA 44°25′N 69°02′W
97 G15 **Belfast** Ir. Béal Feirste. ◐ E Northern Ireland, United Kingdom 54°35′N 05°55′W
97 G15 **Belfast Aldergrove** ✈ E Northern Ireland, United Kingdom 54°37′N 06°11′W
97 G15 **Belfast Lough** Ir. Loch Lao. inlet E Northern Ireland, United Kingdom
28 K5 **Belfield** North Dakota, N USA 46°53′N 103°12′W
103 U7 **Belfort** Territoire-de-Belfort, E France 47°38′N 06°52′E
Belgard see Białogard
155 E17 **Belgaum** Karnātaka, W India 15°52′N 74°30′E
Belgian Congo see Congo (Democratic Republic of)
België/Belgique see Belgium
99 F20 **Belgium** off. Kingdom of Belgium, Dut. België, Fr. Belgique. ◆ monarchy NW Europe
Belgium, Kingdom of see Belgium
126 J8 **Belgorod** Belgorodskaya Oblast', W Russian Federation 50°38′N 36°37′E
Belgorod-Dnestrovskiy see Bilhorod-Dnistrovs'kyy
126 J8 **Belgorodskaya Oblast'** ⬖ province W Russian Federation
Belgrad see Beograd
29 T8 **Belgrade** Minnesota, N USA 45°27′N 94°59′W
33 S11 **Belgrade** Montana, NW USA 45°46′N 111°10′W
Belgrade see Beograd
Belgrano, Cabo see Meredith, Cape
195 N5 **Belgrano II** Argentinian research station Antarctica 77°50′S 35°25′W
21 X9 **Belhaven** North Carolina, SE USA 35°36′N 76°50′W
107 I23 **Belice** anc. Hypsas. ≈ Sicilia, Italy, C Mediterranean Sea
Belice see Belize/Belize City
Beli Drim see Drini i Bardhë
Beligrad see Berat
188 C8 **Beliliou** prev. Peleliu. island S Palau
114 L8 **Beli Lom, Yazovir** ⊠ NE Bulgaria
112 I8 **Beli Manastir** Hung. Pélmonostor; prev. Monostor. Osijek-Baranja, NE Croatia 45°46′N 16°38′E
102 J13 **Bélin-Béliet** Gironde, SW France 44°30′N 00°48′W
79 F17 **Bélinga** Ogooué-Ivindo, NE Gabon 01°05′N 13°12′E
21 S4 **Belington** West Virginia, NE USA 39°01′N 79°57′W
127 O6 **Belinskiy** Penzenskaya Oblast', W Russian Federation 52°58′N 43°25′E
169 N12 **Belinyu** Pulau Bangka, W Indonesia 01°37′S 105°45′E
169 O13 **Belitung, Pulau** island W Indonesia
116 F10 **Beliu** Hung. Bel. Arad, W Romania 46°31′N 21°57′E
114 I9 **Beli Vit** ≈ NW Bulgaria
42 G2 **Belize** Sp. Belice; prev. British Honduras, Colony of Belize. ◆ commonwealth republic Central America
42 F2 **Belize** Sp. Belice. ⬖ district NE Belize
42 G2 **Belize** ≈ Belize/Guatemala
Belize see Belize City
42 G2 **Belize City** var. Belize, Sp. Belice. Belize, NE Belize 17°29′N 88°10′W
42 G2 **Belize City** ✈ Belize, NE Belize 17°31′N 88°15′W
Belize, Colony of see Belize
Beljak see Villach
39 N16 **Belkofski** Alaska, USA 55°07′N 162°04′W
123 O6 **Bel'kovskiy, Ostrov** island Novosibirskiye Ostrova, NE Russian Federation
14 J8 **Bell** ≈ Québec, SE Canada
10 J15 **Bella Bella** British Columbia, SW Canada 52°04′N 128°07′W
102 M10 **Bellac** Haute-Vienne, C France 46°07′N 01°04′E
10 K15 **Bella Coola** British Columbia, SW Canada 52°23′N 126°46′W
106 D6 **Bellagio** Lombardia, N Italy 45°58′N 09°15′E
31 P6 **Bellaire** Michigan, N USA 44°59′N 85°12′W
106 D6 **Bellano** Lombardia, N Italy 46°06′N 09°21′E
155 G17 **Bellary** var. Ballari. Karnātaka, S India 15°11′N 76°54′E
183 S5 **Bellata** New South Wales, SE Australia 29°58′S 149°49′E
61 D16 **Bella Unión** Artigas, N Uruguay 30°18′S 57°35′W
61 C14 **Bella Vista** Corrientes, NE Argentina 28°30′S 59°03′W
62 J7 **Bella Vista** Tucumán, N Argentina 27°05′S 65°19′W
62 P4 **Bella Vista** Amambay, C Paraguay 22°08′S 56°20′W
56 B10 **Bellavista** Cajamarca, N Peru 05°43′S 78°48′W
56 D11 **Bellavista** San Martín, N Peru 07°04′S 76°35′W
183 U6 **Bellbrook** New South Wales, SE Australia 30°48′S 152°32′E
27 V5 **Belle** Missouri, C USA 38°17′N 91°43′W
21 Q5 **Belle** West Virginia, NE USA 38°13′N 81°32′W
31 R13 **Bellefontaine** Ohio, N USA 40°22′N 83°45′W
18 F14 **Bellefonte** Pennsylvania, NE USA 40°54′N 77°43′W
28 J9 **Belle Fourche** South Dakota, N USA 44°40′N 103°50′W
28 J9 **Belle Fourche Reservoir** ⊠ South Dakota, N USA
28 K9 **Belle Fourche River** ≈ South Dakota/Wyoming, N USA
103 S10 **Bellegarde-sur-Valserine** Ain, E France 46°06′N 05°49′E
23 Y14 **Belle Glade** Florida, SE USA 26°40′N 80°40′W
102 G8 **Belle Île** island NW France
13 T9 **Belle Isle** island Belle Isle, Newfoundland and Labrador, E Canada
13 S10 **Belle Isle, Strait of** strait Newfoundland and Labrador, E Canada
Bellenz see Bellinzona
29 W14 **Belle Plaine** Iowa, C USA 41°54′N 92°16′W
29 V9 **Belle Plaine** Minnesota, N USA 44°39′N 93°47′W
14 I9 **Belleterre** Québec, SE Canada 47°24′N 78°40′W
14 J15 **Belleville** Ontario, SE Canada 44°10′N 77°22′W
103 R10 **Belleville** Rhône, E France 46°09′N 04°42′E
30 K15 **Belleville** Illinois, N USA 38°31′N 89°58′W
27 N3 **Belleville** Kansas, C USA 39°51′N 97°38′W
29 Z13 **Bellevue** Iowa, C USA 42°15′N 90°25′W
29 S15 **Bellevue** Nebraska, C USA 41°08′N 95°53′W
31 S11 **Bellevue** Ohio, N USA 41°16′N 82°50′W
25 S5 **Bellevue** Texas, SW USA 33°38′N 98°00′W
32 H8 **Bellevue** Washington, NW USA 47°36′N 122°12′W
55 Y11 **Bellevue de l'Inini, Montagnes** ▲▲ S French Guiana
103 S11 **Belley** Ain, E France 45°46′N 05°41′E
Bellin see Kangirsuk
183 V6 **Bellingen** New South Wales, SE Australia 30°27′S 152°53′E
97 L14 **Bellingham** N England, United Kingdom 55°09′N 02°16′W
32 H6 **Bellingham** Washington, NW USA 48°46′N 122°29′W
Belling Hausen Mulde see Southeast Pacific Basin
194 H2 **Bellingshausen** Russian research station South Shetland Islands, Antarctica 61°57′S 58°23′W
Bellingshausen see Motu One
Bellingshausen Abyssal Plain see Bellingshausen Plain
196 R14 **Bellingshausen Plain** var. Bellingshausen Abyssal Plain. undersea feature SE Pacific Ocean 64°00′S 90°00′W
194 I8 **Bellingshausen Sea** sea Antarctica
98 P6 **Bellingwolde** Groningen, NE Netherlands 53°07′N 07°10′E
108 H11 **Bellinzona** Ger. Bellenz. Ticino, S Switzerland 46°12′N 09°02′E
25 T8 **Bellmead** Texas, SW USA 31°36′N 97°02′W
54 E8 **Bello** Antioquia, W Colombia 06°19′N 75°34′W
61 B21 **Bellocq** Buenos Aires, E Argentina 35°55′S 61°32′W
Bello Horizonte see Belo Horizonte
186 L10 **Bellona** var. Mungiki. island S Solomon Islands
Bellovacum see Beauvais
182 D7 **Bell, Point** headland South Australia 32°13′S 133°08′E
20 F9 **Bells** Tennessee, S USA 35°42′N 89°05′W
25 U5 **Bells** Texas, SW USA 33°36′N 96°24′W
92 N3 **Bellsund** inlet SW Svalbard
106 H6 **Belluno** Veneto, NE Italy 46°08′N 12°13′E
62 L11 **Bell Ville** Córdoba, C Argentina 32°35′S 62°41′W
83 E26 **Bellville** Western Cape, SW South Africa 33°50′S 18°43′E
25 U11 **Bellville** Texas, SW USA 29°57′N 96°15′W
104 L12 **Belméz** Andalucía, S Spain 38°16′N 05°12′W
29 V12 **Belmond** Iowa, C USA 42°51′N 93°36′W
18 E11 **Belmont** New York, NE USA 42°14′N 78°02′W
21 R10 **Belmont** North Carolina, SE USA 35°13′N 81°01′W
59 O18 **Belmonte** Bahia, E Brazil 15°53′S 38°54′W
104 I8 **Belmonte** Castelo Branco, C Portugal 40°21′N 07°20′W
105 P10 **Belmonte** Castilla-La Mancha, C Spain 39°34′N 02°43′W
42 G2 **Belmopan** ● (Belize) Cayo, C Belize 17°13′N 88°48′W
97 B16 **Belmullet** Ir. Béal an Mhuirhead. Mayo, W Ireland 54°14′N 09°59′W
99 E20 **Belœil** Hainaut, SW Belgium 50°33′N 03°45′E
123 R13 **Belogorsk** Amurskaya Oblast', SE Russian Federation 50°53′N 128°24′E
Belogorsk see Bilohirs'k
114 F7 **Belogradchik** Vidin, NW Bulgaria 43°37′N 22°42′E
172 H8 **Beloha** Toliara, S Madagascar 25°09′S 45°04′E
59 M20 **Belo Horizonte** prev. Bello Horizonte. state capital Minas Gerais, SE Brazil 19°54′S 43°54′W
26 M3 **Beloit** Kansas, C USA 39°27′N 98°06′W
30 L9 **Beloit** Wisconsin, N USA 42°31′N 89°01′W
Belokorovichi see Novi Bilokorovychi
124 J8 **Belomorsk** Respublika Kareliya, NW Russian Federation 64°30′N 34°43′E
124 J8 **Belomorsko-Baltiyskiy Kanal** Eng. White Sea-Baltic Canal, White Sea Canal. canal NW Russian Federation
153 V15 **Belonia** Tripura, NE India 23°15′N 91°25′E
Beloozersk see Byelaazyorsk
Belopol'ye see Bilopillya
105 O4 **Belorado** Castilla y León, N Spain 42°25′N 03°11′W
126 L14 **Belorechensk** Krasnodarskiy Kray, SW Russian Federation 44°46′N 39°53′E
127 W5 **Beloretsk** Respublika Bashkortostan, W Russian Federation 53°56′N 58°26′E
Belorussia/Belorussian SSR see Belarus
Belorusskaya Gryada see Byelaruskaya Hrada
Belorusskaya SSR see Belarus
Beloshchel'ye see Nar'yan-Mar
114 N8 **Beloslav** Varna, E Bulgaria 43°13′N 27°42′E
Belostok see Białystok
Belo-sur-Tsiribihina see Belo Tsiribihina
172 H5 **Belo Tsiribihina** var. Belo-sur-Tsiribihina. Toliara, W Madagascar 19°40′S 44°30′E
Belovár see Bjelovar
Belovezhskaya, Pushcha see Białowieska, Puszcza/Byelavyezhskaya, Pushcha
114 H10 **Belovo** Pazardzhik, C Bulgaria 42°10′N 24°01′E
Belovodsk see Bilovods'k
122 H9 **Beloyarskiy** Khanty-Mansiyskiy Avtonomnyy Okrug-Yugra, N Russian Federation 63°40′N 66°31′E
124 K7 **Beloye More** Eng. White Sea. sea NW Russian Federation
124 K13 **Beloye, Ozero** ◎ NW Russian Federation
114 J10 **Belozem** Plovdiv, C Bulgaria 42°11′N 25°00′E
124 K13 **Belozërsk** Vologodskaya Oblast', NW Russian Federation 59°59′N 37°49′E
108 D8 **Belp** Bern, W Switzerland 46°54′N 07°31′E
108 D8 **Belp** ✈ (Bern) Bern, C Switzerland 46°55′N 07°29′E
107 L24 **Belpasso** Sicilia, Italy, C Mediterranean Sea 37°35′N 14°59′E
31 U14 **Belpre** Ohio, N USA 39°14′N 81°34′W
98 M8 **Belterwijde** ◎ N Netherlands
27 R4 **Belton** Missouri, C USA 38°48′N 94°31′W
21 P11 **Belton** South Carolina, SE USA 34°31′N 82°29′W
25 T9 **Belton** Texas, SW USA 31°04′N 97°30′W
25 S9 **Belton Lake** ⊠ Texas, SW USA
Bel'tsy see Bălți
97 E16 **Belturbet** Ir. Béal Tairbirt. Cavan, N Ireland 54°06′N 07°26′W
Beluchistan see Balochistan
145 Z9 **Belukha, Gora** ▲ Kazakhstan/Russian Federation 49°50′N 86°44′E
107 M20 **Belvedere Marittimo** Calabria, SW Italy 39°37′N 15°52′E
30 L10 **Belvidere** Illinois, N USA 42°15′N 88°50′W
18 J14 **Belvidere** New Jersey, NE USA 40°50′N 75°05′W
Bely see Belyy
127 V8 **Belyayevka** Orenburgskaya Oblast', W Russian Federation 51°25′N 56°26′E
Belynichi see Byalynichy
124 H17 **Belyy** var. Bely, Beyj. Tverskaya Oblast', W Russian Federation 55°51′N 32°57′E
126 I6 **Belyye Berega** Bryanskaya Oblast', W Russian Federation 53°11′N 34°42′E
122 J6 **Belyy, Ostrov** island N Russian Federation
122 J11 **Belyy Yar** Tomskaya Oblast', C Russian Federation 58°26′N 84°57′E
100 N13 **Belzig** Brandenburg, NE Germany 52°09′N 12°37′E
22 K4 **Belzoni** Mississippi, S USA 33°10′N 90°29′W
172 H4 **Bemaraha** var. Plateau du Bemaraha. ▲▲ W Madagascar
Bemaraha, Plateau du see Bemaraha
82 B10 **Bembe** Uíge, NW Angola 07°03′S 14°25′E
77 S14 **Bembèrèkè** var. Bimbéréké. N Benin 10°10′N 02°41′E
104 K12 **Bembézar** ≈ SW Spain
104 J3 **Bembibre** Castilla y León, N Spain 42°37′N 06°25′W
29 T4 **Bemidji** Minnesota, N USA 47°27′N 94°53′W
98 L12 **Bemmel** Gelderland, SE Netherlands 51°53′N 05°54′E
171 T13 **Bemu** Pulau Seram, E Indonesia 03°21′S 129°58′E
Benāb see Bonāb
105 T5 **Benabarre** var. Benavarn. Aragón, NE Spain 42°06′N 00°28′E
Benaco see Garda, Lago di
79 L20 **Bena-Dibele** Kasai-Oriental, C Dem. Rep. Congo 04°01′S 22°50′E
105 R9 **Benagéber, Embalse de** E Spain
183 O11 **Benalla** Victoria, SE Australia 36°33′S 146°00′E
104 M14 **Benamejí** Andalucía, S Spain 37°16′N 04°33′W
Benares see Vārānasi
Benavarn see Benabarre
104 F10 **Benavente** Santarém, C Portugal 38°59′N 08°49′W
104 K5 **Benavente** Castilla y León, N Spain 42°N 05°40′W
25 S15 **Benavides** Texas, SW USA 27°36′N 98°24′W
96 F8 **Benbecula** island NW Scotland, United Kingdom
Bencovazzo see Benkovac
32 H13 **Bend** Oregon, NW USA 44°04′N 121°19′W
182 K7 **Benda Range** ▲▲ South Australia
183 T6 **Bendemeer** New South Wales, SE Australia 30°54′S 151°12′E
Bender see Tighina
Bender Beila/Bender Beyla see Bandarbeyla
Bender Cassim/Bender Qaasim see Boosaaso
Bendery see Tighina
183 N11 **Bendigo** Victoria, SE Australia 36°46′S 144°19′E
118 E10 **Bēne** SW Latvia 56°30′N 23°04′E
98 K13 **Beneden-Leeuwen** Gelderland, C Netherlands 51°52′N 05°32′E
101 L24 **Benediktenwand** ▲ S Germany 47°39′N 11°28′E
Benemérita de San Cristóbal see San Cristóbal
77 N12 **Bénéna** Ségou, S Mali 13°04′N 04°20′W
172 I7 **Benenitra** Toliara, S Madagascar 23°25′S 45°06′E
Beneschau see Benešov
Beneški Zaliv see Venice, Gulf of
111 D17 **Benešov** Ger. Beneschau. Středočeský Kraj, W Czech Republic 49°48′N 14°41′E
107 L17 **Benevento** anc. Beneventum, Malventum. Campania, S Italy 41°07′N 14°45′E
Beneventum see Benevento
173 S3 **Bengal, Bay of** bay N Indian Ocean
Bengalooru see Bangalore
Bengaluru see Bangalore
79 M17 **Bengamisa** Orientale, N Dem. Rep. Congo 0°58′N 25°11′E
Bengasi see Banghāzī
Bengazi see Banghāzī
161 P7 **Bengbu** var. Peng-pu. Anhui, E China 32°57′N 117°17′E
32 L9 **Benge** Washington, NW USA 46°55′N 118°01′W
Benghazi see Banghāzī
168 K10 **Bengkalis** Pulau Bengkalis, W Indonesia 01°27′N 102°10′E
168 K10 **Bengkalis, Pulau** island W Indonesia
169 Q10 **Bengkayang** Borneo, C Indonesia 0°45′N 109°28′E
168 K14 **Bengkulu** prev. Bengkoeloe, Benkoelen, Benkulen. Sumatera, W Indonesia 03°46′S 102°16′E
168 K13 **Bengkulu** off. Propinsi Bengkulu; prev. Bengkoeloe, Benkoelen, Benkulen. ⬖ province W Indonesia
Bengkulu, Propinsi see Bengkulu
82 A11 **Bengo** ⬖ province W Angola
95 J16 **Bengtsfors** Västra Götaland, S Sweden 59°03′N 12°14′E
82 B13 **Benguela** var. Benguella. Benguela, W Angola 12°35′S 13°30′E
83 A14 **Benguela** ⬖ province W Angola
Benguella see Benguela
138 F10 **Ben Gurion** ✈ Tel Aviv, C Israel 32°04′N 34°45′E
Bengweulu, Lake see Bangweulu, Lake
Benha see Banhā
192 F6 **Benham Seamount** undersea feature W Philippine Sea 15°48′N 124°15′E
96 H6 **Ben Hope** ▲ N Scotland, United Kingdom 58°25′N 04°36′W
79 P18 **Beni** Nord-Kivu, NE Dem. Rep. Congo 0°31′N 29°30′E
74 H8 **Beni Abbès** W Algeria 30°07′N 02°09′W
105 T8 **Benicarló** Valenciana, E Spain 40°25′N 00°25′E
105 T9 **Benicasim** Cat. Benicàssim. Valenciana, E Spain 40°03′N 00°03′E
Benicàssim see Benicasim
105 T12 **Benidorm** Valenciana, SE Spain 38°33′N 00°09′W
Beni Mazâr see Banī Mazār
121 C11 **Beni-Mellal** C Morocco 32°20′N 06°21′W
77 R14 **Benin** off. Republic of Benin; prev. Dahomey. ◆ republic W Africa
77 S17 **Benin, Bight of** gulf W Africa
77 U16 **Benin City** Edo, SW Nigeria 06°23′N 05°40′E
Benin, Republic of see Benin
57 K16 **Beni, Río** ≈ N Bolivia
120 F10 **Beni Saf** var. Beni-Saf. NW Algeria 35°19′N 01°23′W
Beni-Saf see Beni Saf
Benishangul see Bīnshangul Gumuz
105 T11 **Benissa** Valenciana, E Spain 38°43′N 00°03′E
Beni Suef see Banī Suwayf
11 V15 **Benito** Manitoba, S Canada 51°57′N 101°24′W
Benito see Uolo, Río
61 C23 **Benito Juárez** Buenos Aires, E Argentina 37°43′S 59°48′W
41 P14 **Benito Juárez Internacional** ✈ (México) México, S Mexico 19°24′N 99°02′W
25 P5 **Benjamin** Texas, SW USA 33°35′N 99°49′W
58 B13 **Benjamin Constant** Amazonas, N Brazil 04°22′S 70°02′W
40 F4 **Benjamín Hill** Sonora, NW Mexico 30°13′N 111°08′W
63 F19 **Benjamín, Isla** island Archipiélago de los Chonos, S Chile
164 Q4 **Benkei-misaki** headland Hokkaidō, NE Japan 42°49′N 140°10′E
28 L17 **Benkelman** Nebraska, C USA 40°04′N 101°30′W
96 I7 **Ben Klibreck** ▲ N Scotland, United Kingdom 58°15′N 04°23′W
Benkoelen/Bengkoeloe see Bengkulu
112 D13 **Benkovac** It. Bencovazzo. Zadar, SW Croatia 44°02′N 15°36′E
Benkulen see Bengkulu
96 I11 **Ben Lawers** ▲ C Scotland, United Kingdom 56°33′N 04°13′W
96 J9 **Ben Macdui** var. Beinn MacDuibh. ▲ C Scotland, United Kingdom 57°02′N 03°42′W
96 G11 **Ben More** ▲ W Scotland, United Kingdom 56°26′N 06°00′W
96 I11 **Ben More** ▲ C Scotland, United Kingdom 56°22′N 04°31′W
96 H7 **Ben More Assynt** ▲ N Scotland, United Kingdom 58°09′N 04°51′W
185 E20 **Benmore, Lake** ◎ South Island, New Zealand
98 L12 **Bennekom** Gelderland, SE Netherlands 52°00′N 05°40′E
123 Q5 **Bennetta, Ostrov** island Novosibirskiye Ostrova, NE Russian Federation
21 T11 **Bennettsville** South Carolina, SE USA 34°36′N 79°40′W
96 H10 **Ben Nevis** ▲ N Scotland, United Kingdom 56°80′N 05°00′W
184 M9 **Benneydale** Waikato, North Island, New Zealand 38°31′S 175°22′E
Bennichab see Bennichchâb
76 H8 **Bennichchâb** var. Bennichab. Inchiri, W Mauritania 19°26′N 15°21′W
18 L10 **Bennington** Vermont, NE USA 42°51′N 73°09′W
185 E20 **Ben Ohau Range** ▲▲ South Island, New Zealand
83 J21 **Benoni** Gauteng, NE South Africa 26°04′S 28°18′E
172 J2 **Be, Nosy** var. Nossi-Bé. island NW Madagascar
Bénoué see Benue
42 F2 **Benque Viejo del Carmen** Cayo, W Belize 17°04′N 89°08′W
101 G19 **Bensheim** Hessen, W Germany 49°41′N 08°38′E
37 N16 **Benson** Arizona, SW USA 31°55′N 110°16′W
29 S8 **Benson** Minnesota, N USA 45°19′N 95°36′W
21 U10 **Benson** North Carolina, SE USA 35°22′N 78°33′W
171 N15 **Benteng** Pulau Selayar, C Indonesia 06°07′S 120°28′E
83 A14 **Bentiaba** Namibe, SW Angola 14°18′S 12°27′E
181 T4 **Bentinck Island** island Wellesley Islands, Queensland, N Australia
80 E13 **Bentiu** Unity, N South Sudan 09°14′N 29°49′E
138 G8 **Bent Jbaïl** var. Bint Jubayl. S Lebanon 33°07′N 35°26′E
11 Q15 **Bentley** Alberta, SW Canada 52°27′N 114°02′W
61 I15 **Bento Gonçalves** Rio Grande do Sul, S Brazil 29°12′S 51°34′W
27 U12 **Benton** Arkansas, C USA 34°34′N 92°35′W
30 L16 **Benton** Illinois, N USA 38°00′N 88°55′W
20 H7 **Benton** Kentucky, S USA 36°51′N 88°21′W
22 G5 **Benton** Louisiana, S USA 32°41′N 93°44′W
27 Y7 **Benton** Missouri, C USA 37°05′N 89°34′W
20 M10 **Benton** Tennessee, S USA 35°10′N 84°39′W
31 O10 **Benton Harbor** Michigan, N USA 42°07′N 86°27′W
27 S9 **Bentonville** Arkansas, C USA 36°23′N 94°13′W
77 V16 **Benue** ⬖ state SE Nigeria
78 F13 **Benue** Fr. Bénoué. ≈ Cameroon/Nigeria
163 V12 **Benxi** prev. Pen-ch'i, Penhsihu, Penki. Liaoning, NE China 41°20′N 123°45′E
Benyakoni see Byenyakoni
112 K10 **Beočin** Vojvodina, N Serbia 45°13′N 19°43′E
Beodericsworth see Bury St Edmunds
112 M11 **Beograd** Eng. Belgrade, Ger. Belgrad; anc. Singidunum. ● (Serbia) Serbia, N Serbia 44°48′N 20°27′E
112 L11 **Beograd** Eng. Belgrade. ✈ Serbia, N Serbia 44°45′N 20°21′E
76 M16 **Béoumi** C Ivory Coast 07°40′N 05°34′W
35 V3 **Beowawe** Nevada, W USA 40°33′N 116°31′W
164 E14 **Beppu** Ōita, Kyūshū, SW Japan 33°18′N 131°30′E
187 X15 **Beqa** prev. Mbengga. island W Fiji
45 Y14 **Bequia** island C Saint Vincent and the Grenadines
139 U4 **Beranan, Shax-î** var. Shākh-i Barānān. ▲▲ E Iraq
113 L16 **Berane** prev. Ivangrad. E Montenegro 42°51′N 19°51′E
113 L21 **Berat** var. Berati, SCr. Beligrad. Berat, C Albania 40°43′N 19°58′E
113 L21 **Berat** ⬖ district C Albania
Berătău see Berettyó
Berati see Berat
Beraun see Berounka, Czech Republic
Beraun see Beroun, Czech Republic
171 U13 **Berau, Teluk** var. MacCluer Gulf. bay Papua, E Indonesia
80 G8 **Berber** River Nile, NE Sudan 18°01′N 34°00′E
80 N12 **Berbera** Woqooyi Galbeed, NW Somalia 10°24′N 45°02′E
79 H16 **Berbérati** Mambéré-Kadéï, SW Central African Republic 04°14′N 15°50′E
Berberia, Cabo de see Barbaria, Cap de
55 T9 **Berbice River** ≈ NE Guyana
Berchid see Berrechid
103 N2 **Berck-Plage** Pas-de-Calais, N France 50°24′N 01°35′E
25 T13 **Berclair** Texas, SW USA 28°33′N 97°32′W
117 W10 **Berda** ≈ SE Ukraine
Berdichev see Berdychiv
123 P10 **Berdigestyakh** Respublika Sakha (Yakutiya), NE Russian Federation 62°02′N 127°03′E
122 J12 **Berdsk** Novosibirskaya Oblast', C Russian Federation 54°42′N 82°56′E
117 W10 **Berdyans'k** Rus. Berdyansk; prev. Osipenko. Zaporiz'ka Oblast', SE Ukraine 46°46′N 36°49′E
117 W10 **Berdyans'ka Kosa** spit SE Ukraine
117 V10 **Berdyans'ka Zatoka** gulf S Ukraine
117 N5 **Berdychiv** Rus. Berdichev. Zhytomyrs'ka Oblast', N Ukraine 49°54′N 28°39′E
20 M6 **Berea** Kentucky, S USA 37°34′N 84°18′W
Beregovo/Beregszász see Berehove
116 G8 **Berehove** Cz. Berehovo, Hung. Beregszász, Rus. Beregovo. Zakarpats'ka Oblast', W Ukraine 48°13′N 22°39′E
Berehovo see Berehove
186 D9 **Bereina** Central, S Papua New Guinea 08°29′S 146°30′E
146 C11 **Bereket** prev. Rus. Gazandzhyk, Kazandzhik, Turkm. Gazanjyk. Balkan Welaýaty, W Turkmenistan 39°17′N 55°27′E
45 O12 **Berekua** S Dominica 15°14′N 61°19′W
77 O16 **Berekum** W Ghana 07°27′N 02°35′W
Berenice see Baranīs
11 O14 **Berens** ≈ Manitoba/Ontario, C Canada
11 X14 **Berens River** Manitoba, C Canada 52°22′N 97°00′W
29 R12 **Beresford** South Dakota, N USA 43°02′N 96°45′W
116 J4 **Berestechko** Volyns'ka Oblast', NW Ukraine 50°21′N 25°06′E
116 M11 **Bereşti** Galaţi, E Romania 46°04′N 27°54′E
117 U6 **Berestova** ≈ E Ukraine
Beretău see Berettyó
111 N23 **Berettyó** Rom. Barcău; prev. Berătău, Beretău. ≈ Hungary/Romania
111 N23 **Berettyóújfalu** Hajdú-Bihar, E Hungary 47°15′N 21°33′E
Berëza/Bereza Kartuska see Byaroza
117 Q4 **Berezan'** Kyyivs'ka Oblast', N Ukraine 50°18′N 31°30′E
117 Q10 **Berezanka** Mykolayivs'ka Oblast', S Ukraine 46°51′N 31°24′E
116 J6 **Berezhany** Pol. Brzeżany. Ternopil's'ka Oblast', W Ukraine 49°29′N 25°00′E
Berezina see Byarezina
Berezino see Byerazino
117 P10 **Berezivka** Rus. Berezovka. Odes'ka Oblast', SW Ukraine 47°12′N 30°56′E
117 Q2 **Berezna** Chernihivs'ka Oblast', NE Ukraine 51°35′N 31°50′E
116 L3 **Berezne** Rivnens'ka Oblast', NW Ukraine 51°00′N 26°46′E
117 R9 **Bereznehuvate** Mykolayivs'ka Oblast', S Ukraine 47°18′N 32°52′E
125 N10 **Bereznik** Arkhangel'skaya Oblast', NW Russian Federation 62°50′N 42°40′E
125 U13 **Berezniki** Permskiy Kray, NW Russian Federation 59°26′N 56°49′E
Berëzovka see Byarozawka, Belarus
Berezovka see Berezivka, Ukraine
122 H9 **Berezovo** Khanty-Mansiyskiy Avtonomnyy Okrug-Yugra, N Russian Federation 63°48′N 64°38′E
127 O9 **Berezovskaya** Volgogradskaya Oblast', SW Russian Federation 50°17′N 43°58′E
123 S13 **Berezovyy** Khabarovskiy Kray, E Russian Federation 51°42′N 135°39′E
83 E25 **Berg** ≈ W South Africa
Berg see Berg bei Rohrbach
105 V4 **Berga** Cataluña, NE Spain 42°06′N 01°41′E
95 N20 **Berga** Kalmar, S Sweden 57°13′N 16°03′E
136 B13 **Bergama** İzmir, W Turkey 39°08′N 27°10′E
106 E7 **Bergamo** anc. Bergomum. Lombardia, N Italy 45°42′N 09°40′E
105 P3 **Bergara** País Vasco, N Spain 43°05′N 02°25′W
109 S3 **Berg bei Rohrbach** var. Berg. Oberösterreich, N Austria 48°34′N 14°02′E
100 O6 **Bergen** Mecklenburg-Vorpommern, NE Germany 54°25′N 13°25′E
101 I11 **Bergen** Niedersachsen, NW Germany 52°49′N 09°57′E
98 H8 **Bergen** Noord-Holland, NW Netherlands 52°40′N 04°42′E
94 C13 **Bergen** Hordaland, S Norway 60°24′N 05°19′E
Bergen see Mons
55 W9 **Berg en Dal** Brokopondo, C Suriname 05°15′N 55°20′W
99 G15 **Bergen op Zoom** Noord-Brabant, S Netherlands 51°30′N 04°17′E
102 L13 **Bergerac** Dordogne, SW France 44°51′N 00°30′E
99 J16 **Bergeyk** Noord-Brabant, S Netherlands 51°19′N 05°21′E
101 D16 **Bergheim** Nordrhein-Westfalen, W Germany 50°57′N 06°39′E
55 X10 **Bergi** Sipaliwini, E Suriname 04°36′N 54°24′W
101 E16 **Bergisch Gladbach** Nordrhein-Westfalen, W Germany 50°59′N 07°09′E
101 F14 **Bergkamen** Nordrhein-Westfalen, W Germany 51°32′N 07°41′E
95 N21 **Bergkvara** Kalmar, S Sweden 56°22′N 16°04′E
Bergomum see Bergamo
98 K13 **Bergse Maas** ≈ S Netherlands
95 P15 **Bergshamra** Stockholm, C Sweden 59°37′N 18°40′E
94 N10 **Bergsjö** Gävleborg, C Sweden 62°00′N 17°10′E
93 J14 **Bergsviken** Norrbotten, N Sweden 65°16′N 21°24′E
Bergum see Burgum
98 M6 **Bergumer Meer** ◎ N Netherlands
94 N12 **Bergviken** ◎ C Sweden
168 M11 **Berhala, Selat** strait Sumatera, W Indonesia
Berhampore see Baharampur
99 J17 **Beringen** Limburg, NE Belgium 51°04′N 05°14′E
39 T12 **Bering Glacier** glacier Alaska, USA
Beringov Proliv see Bering Strait
192 L2 **Bering Sea** sea N Pacific Ocean
38 L9 **Bering Strait** Rus. Beringov Proliv. strait Bering Sea/Chukchi Sea
Berislav see Beryslav
105 O15 **Berja** Andalucía, S Spain 36°51′N 02°56′W
94 H9 **Berkåk** Sør-Trøndelag, S Norway 62°50′N 10°01′E

◆ Country
● Country Capital
◇ Dependent Territory
○ Dependent Territory Capital
⬖ Administrative Regions
✈ International Airport
▲ Mountain
▲▲ Mountain Range
Volcano
≈ River
◎ Lake
⊠ Reservoir

◆ Country
● Country Capital
◇ Dependent Territory
○ Dependent Territory Capital
◆ Administrative Regions
✕ International Airport
▲ Mountain
▲ Mountain Range
≈ River
Volcano
⊚ Lake
⊞ Reservoir

BINGA – BOCONÓ

83 J16 **Binga** Matabeleland North, W Zimbabwe 17°40′S 27°22′E
183 T5 **Bingara** New South Wales, SE Australia 29°54′S 150°36′E
101 F18 **Bingen am Rhein** Rheinland-Pfalz, SW Germany 49°58′N 07°54′E
26 M11 **Binger** Oklahoma, C USA 35°19′N 98°19′W
Bingerau see Węgrów
Bin Ghalfān, Jazā'ir see Ḩalāniyāt, Juzur al
19 Q6 **Bingham** Maine, NE USA 45°01′N 69°51′W
18 H11 **Binghamton** New York, NE USA 42°06′N 75°55′W
Bin Ghanīmah, Jabal see Bin Ghunaymah, Jabal
75 P11 **Bin Ghunaymah, Jabal** var. Jabal Bin Ghanīmah. ▲▲ C Libya
139 U3 **Bingird** As Sulaymānīyah, NE Iraq 36°03′N 45°03′E
Bingmei see Congjiang
137 P14 **Bingöl** Bingöl, E Turkey 38°54′N 40°29′E
137 P14 **Bingöl** ◈ province E Turkey
161 R6 **Binhai** var. Dongkan. Jiangsu, E China 34°00′N 119°51′E
167 V11 **Binh Đinh** var. An Nhon. Binh Đinh, C Vietnam 13°53′N 109°07′E
Binh Sơn see Châu Ô
Binimani see Bintimani
168 I8 **Binjai** Sumatera, W Indonesia 03°37′N 98°30′E
183 R6 **Binnaway** New South Wales, SE Australia 31°34′S 149°24′E
108 E6 **Binningen** Basel Landschaft, NW Switzerland 47°32′N 07°35′E
80 H12 **Bīnshangul Gumuz** var. Benishangul. ◈ W Ethiopia
168 J8 **Bintang, Banjaran** ▲▲ Peninsular Malaysia
168 M10 **Bintan, Pulau** island Kepulauan Riau, W Indonesia
76 J14 **Bintimani** var. Binimani. ▲ NE Sierra Leone 09°21′N 11°09′W
Bint Jubayl see Bent Jbaïl
169 S9 **Bintulu** Sarawak, East Malaysia 03°12′N 113°01′E
169 S9 **Bintuni** prev. Steenkool. Papua Barat, E Indonesia 02°03′S 133°45′E
163 W8 **Binxian** prev. Binzhou. Heilongjiang, NE China 45°44′N 127°27′E
160 K14 **Binyang** var. Binzhou. Guangxi Zhuangzu Zizhiqu, S China 23°15′N 108°40′E
161 Q4 **Binzhou** Shandong, E China 37°23′N 118°03′E
Binzhou see Binyang
Binzhou see Binxian
63 G14 **Bío Bío** var. Región del Bío Bío. ◈ region C Chile
Bío Bío, Región del see Bío Bío
63 G14 **Bío Bío, Río** ≈ C Chile
79 C16 **Bioco, Isla de** var. Bioko, Eng. Fernando Po, Sp. Fernando Póo; prev. Macías Nguema Biyogo. island NW Equatorial Guinea
112 D13 **Biograd na Moru** It. Zaravecchia. Zadar, SW Croatia 43°57′N 15°27′E
Bioko see Bioco, Isla de
113 F14 **Biokovo** ▲▲ S Croatia
Biorra see Birr
Bipontium see Zweibrücken
143 W13 **Birag, Kūh-e** ▲▲ SE Iran
75 O10 **Birāk** var. Brak. C Libya 27°32′N 14°17′E
139 S10 **Bi'r al Islām** Karbalā', C Iraq 32°15′N 43°40′E
154 N11 **Biramitrapur** var. Birmitrapur. Odisha, E India 22°24′N 84°42′E
139 T11 **Bi'r an Nişf** An Najaf, S Iraq 31°22′N 44°07′E
78 L12 **Birao** Vakaga, NE Central African Republic 10°14′N 22°49′E
146 J10 **Birata** Rus. Darganata, Dargan-Ata. Lebap Welaýaty, NE Turkmenistan 40°30′N 62°09′E
158 M6 **Biratar Bulak** well NW China
153 R12 **Birātnagar** Eastern, SE Nepal 26°28′N 87°16′E
165 R5 **Biratori** Hokkaidō, NE Japan 42°35′N 142°07′E
39 S8 **Birch Creek** Alaska, USA 66°17′N 145°54′W
38 M11 **Birch Creek** ≈ Alaska, USA
11 T14 **Birch Hills** Saskatchewan, S Canada 52°58′N 105°22′W
182 M10 **Birchip** Victoria, SE Australia 36°01′S 142°55′E
29 X4 **Birch Lake** ◎ Minnesota, N USA
11 Q11 **Birch Mountains** ▲▲ Alberta, W Canada
11 V15 **Birch River** Manitoba, S Canada 52°22′N 101°03′W
44 H12 **Birchs Hill** hill W Jamaica
39 R11 **Birchwood** Alaska, USA 61°24′N 149°28′W
188 I5 **Bird Island** island S Northern Mariana Islands
137 N16 **Birecik** Şanlıurfa, S Turkey 37°03′N 37°59′E
152 M10 **Birendranagar** var. Surkhet. Mid Western, W Nepal 28°35′N 81°36′E
Bir es Saba see Be'er Sheva
74 A12 **Bir-Gandouz** SW Western Sahara 21°35′N 16°27′W
153 P12 **Birganj** Central, C Nepal 27°03′N 84°53′E
81 B14 **Biri** ≈ W South Sudan
Bi'r Ibn Hirmās see Al Bi'r
143 U8 **Birjand** Khorāsān-e Janūbī, E Iran 32°54′N 59°14′E
Birkaland see Pirkanmaa
139 T11 **Birkat Ḩāmid** well S Iraq
95 F18 **Birkeland** Aust-Agder, S Norway 58°18′N 08°13′E
101 E19 **Birkenfeld** Rheinland-Pfalz, SW Germany 49°39′N 07°10′E
97 K18 **Birkenhead** NW England, United Kingdom 53°24′N 03°02′W
109 W7 **Birkfeld** Steiermark, SE Austria 47°21′N 15°40′E
182 A2 **Birksgate Range** ▲▲ South Australia
Birlad see Bârlad
145 S15 **Birlik** var. Novotroickoje, Novotroitskoye; prev. Brlik. Zhambyl, SE Kazakhstan 43°39′N 73°45′E
97 K20 **Birmingham** C England, United Kingdom 52°30′N 01°50′W
23 P4 **Birmingham** Alabama, S USA 33°30′N 86°47′W
97 M20 **Birmingham** ✈ C England, United Kingdom 52°27′N 01°46′W
Birmitrapur see Biramitrapur
Bir Moghrein see Bîr Mogreïn
76 J4 **Bîr Mogreïn** var. Bir Moghrein; prev. Fort-Trinquet. Tiris Zemmour, N Mauritania 25°10′N 11°35′W
191 S4 **Birnie Island** atoll Phoenix Islands, C Kiribati
77 S12 **Birnin Gaouré** var. Birni-Ngaouré. Dosso, SW Niger 12°59′N 03°02′E
Birni-Ngaouré see Birnin Gaouré
77 S12 **Birnin Kebbi** Kebbi, NW Nigeria 12°28′N 04°08′E
77 T12 **Birnin Konni** var. Birni-Nkonni. Tahoua, SW Niger 13°51′N 05°15′E
Birni-Nkonni see Birnin Konni
77 W13 **Birnin Kudu** Jigawa, N Nigeria 11°28′N 09°29′E
123 S16 **Birobidzhan** Yevreyskaya Avtonomnaya Oblast', SE Russian Federation 48°42′N 132°55′E
97 D18 **Birr** var. Parsonstown, Ir. Biorra. C Ireland 53°06′N 07°55′W
183 P4 **Birrie River** ≈ New South Wales/Queensland, SE Australia
108 D7 **Birse** ≈ NW Switzerland
Birsen see Biržai
108 E6 **Birsfelden** Basel Landschaft, NW Switzerland 47°33′N 07°37′E
127 U4 **Birsk** Respublika Bashkortostan, W Russian Federation 55°24′N 55°33′E
119 F14 **Birštonas** Kaunas, C Lithuania 54°37′N 24°00′E
159 P14 **Biru** Xinjiang Uygur Zizhiqu, W China 31°30′N 93°56′E
Biruni see Beruniy
122 L12 **Biryusa** ≈ C Russian Federation
122 L12 **Biryusinsk** Irkutskaya Oblast', C Russian Federation 55°52′N 97°48′E
118 G10 **Biržai** Ger. Birsen. Panevėžys, NE Lithuania 56°12′N 24°47′E
121 P16 **Birżebbuġa** SE Malta 35°50′N 14°32′E
Bisanthe see Tekirdağ
171 R12 **Bisa, Pulau** island Maluku, E Indonesia
37 N17 **Bisbee** Arizona, SW USA 31°27′N 109°55′W
29 O2 **Bisbee** North Dakota, N USA 48°36′N 99°21′W
Biscaia, Baía de see Biscay, Bay of
102 I13 **Biscarrosse et de Parentis, Étang de** ◎ SW France
104 M1 **Biscay, Bay of** Sp. Golfo de Vizcaya, Port. Baía de Biscaia. bay France/Spain
23 Z16 **Biscayne Bay** bay Florida, SE USA
64 M7 **Biscay Plain** undersea feature SE Bay of Biscay 07°15′W 45°00′N
107 N17 **Bisceglie** Puglia, SE Italy 41°14′N 16°31′E
Bischofflack see Škofja Loka
Bischofsburg see Biskupiec
109 Q7 **Bischofshofen** Salzburg, NW Austria 47°25′N 13°13′E
101 P15 **Bischofswerda** Sachsen, E Germany 51°07′N 14°13′E
103 V5 **Bischwiller** Bas-Rhin, NE France 48°46′N 07°52′E
21 T10 **Biscoe** North Carolina, SE USA 35°20′N 79°46′W
194 G5 **Biscoe Islands** island group Antarctica
14 E9 **Biscotasi Lake** ◎ Ontario, S Canada
14 E9 **Biscotasing** Ontario, S Canada 47°16′N 82°04′W
54 J6 **Biscucuy** Portuguesa, NW Venezuela 09°22′N 69°59′W
99 M24 **Bissen** Luxembourg, C Luxembourg 49°47′N 06°04′E
114 K11 **Biser** Haskovo, S Bulgaria 41°52′N 25°59′E
113 D15 **Biševo** It. Busi. island SW Croatia
141 N12 **Bishah, Wādī** dry watercourse C Saudi Arabia
147 U7 **Bishkek** var. Pishpek; prev. Frunze. ● (Kyrgyzstan) Chuyskaya Oblast', N Kyrgyzstan 42°54′N 74°27′E
147 U7 **Bishkek** ✈ Chuyskaya Oblast', N Kyrgyzstan 42°55′N 74°37′E
153 R16 **Bishnupur** West Bengal, NE India 23°05′N 87°20′E
Bisho see Bhisho
35 S9 **Bishop** California, W USA 37°22′N 118°24′W
25 S15 **Bishop** Texas, SW USA 27°36′N 97°49′W
97 L15 **Bishop Auckland** N England, United Kingdom 54°41′N 01°41′W
Bishop's Lynn see King's Lynn
97 O21 **Bishop's Stortford** E England, United Kingdom 51°45′N 00°11′E
21 S12 **Bishopville** South Carolina, SE USA 34°18′N 80°15′W
138 M5 **Bishrī, Jabal** ▲▲ E Syria
163 U4 **Bishui** Heilongjiang, NE China 52°06′N 123°42′E
81 G17 **Bisina, Lake** prev. Lake Salisbury. ◎ E Uganda
Biskara see Biskra
74 L6 **Biskra** var. Beskra, Biskara. NE Algeria 34°51′N 05°44′E
110 M8 **Biskupiec** Ger. Bischofsburg. Warmińsko-Mazurskie, NE Poland 53°52′N 20°57′E
171 R7 **Bislig** Mindanao, S Philippines 08°10′N 126°19′E
27 X6 **Bismarck** Missouri, C USA 37°46′N 90°37′W
28 M5 **Bismarck** state capital North Dakota, N USA 46°49′N 100°47′W
186 D5 **Bismarck Archipelago** island group NE Papua New Guinea
129 Z16 **Bismarck Plate** tectonic feature W Pacific Ocean
186 D7 **Bismarck Range** ▲▲ N Papua New Guinea
186 E6 **Bismarck Sea** sea W Pacific Ocean
137 P15 **Bismil** Diyarbakır, SE Turkey 37°53′N 40°38′E
43 N6 **Bismuna, Laguna** lagoon NE Nicaragua
Bisnulok see Phitsanulok
171 R10 **Bisoa, Tanjung** headland Pulau Halmahera, N Indonesia 02°15′N 127°57′E
28 K7 **Bison** South Dakota, N USA 45°31′N 102°27′W
93 H17 **Bispgården** Jämtland, C Sweden 63°00′N 16°40′E
76 G13 **Bissau** ● (Guinea-Bissau) W Guinea-Bissau 11°52′N 15°39′W
76 G13 **Bissau** ✈ W Guinea-Bissau 11°53′N 15°41′W
Bissojohka see Børselv
76 G13 **Bissorā** W Guinea-Bissau 12°16′N 15°35′W
11 O10 **Bistcho Lake** ◎ Alberta, W Canada
22 G5 **Bistineau, Lake** ⊠ Louisiana, S USA
Bistrica see Ilirska Bistrica
116 I9 **Bistriţa** Ger. Bistritz, Hung. Beszterce; prev. Nösen. Bistriţa-Năsăud, N Romania 47°10′N 24°31′E
116 K10 **Bistriţa** Ger. Bistritz. ≈ NE Romania
116 I9 **Bistriţa-Năsăud** ◈ county N Romania
Bistritz see Bistriţa
Bistritz ober Pernstein see Bystřice nad Pernštejnem
152 L11 **Biswān** Uttar Pradesh, N India 27°30′N 81°00′E
110 M7 **Bisztynek** Warmińsko-Mazurskie, NE Poland 54°05′N 20°53′E
79 E17 **Bitam** Woleu-Ntem, N Gabon 02°05′N 11°30′E
101 D18 **Bitburg** Rheinland-Pfalz, SW Germany 49°58′N 06°31′E
103 U4 **Bitche** Moselle, NE France 49°01′N 07°27′E
78 I11 **Bitkine** Guéra, C Chad 11°59′N 18°13′E
137 R15 **Bitlis** Bitlis, SE Turkey 38°23′N 42°04′E
137 R15 **Bitlis** ◈ province E Turkey
Bitoeng see Bitung
113 N20 **Bitola** Turk. Monastir; prev. Bitolj. S FYR Macedonia 41°01′N 21°22′E
Bitolj see Bitola
107 O17 **Bitonto** anc. Butuntum. Puglia, SE Italy 41°07′N 16°41′E
77 Q13 **Bitou** var. Bittou. SE Burkina Faso 11°19′N 00°17′W
155 C20 **Bitra Island** island Lakshadweep, India, N Indian Ocean
101 M14 **Bitterfeld** Sachsen-Anhalt, E Germany 51°37′N 12°18′E
32 O9 **Bitterroot Range** ▲▲ Idaho/Montana, NW USA
33 P10 **Bitterroot River** ≈ Montana, NW USA
107 D18 **Bitti** Sardegna, Italy, C Mediterranean Sea 40°30′N 09°31′E
Bittou see Bitou
171 Q11 **Bitung** prev. Bitoeng. Sulawesi, C Indonesia 01°28′N 125°13′E
60 I12 **Bituruna** Paraná, S Brazil 26°11′S 51°34′W
77 Y13 **Biu** Borno, E Nigeria 10°35′N 12°13′E
Biumba see Byumba
164 J13 **Biwa-ko** ◎ Honshū, SW Japan
171 X14 **Biwarlaut** Papua, E Indonesia 05°44′S 138°14′E
27 P10 **Bixby** Oklahoma, C USA 35°56′N 95°52′W
122 J13 **Biya** ≈ S Russian Federation
122 J13 **Biysk** Altayskiy Kray, S Russian Federation 52°34′N 85°09′E
164 H13 **Bizen** Okayama, Honshū, SW Japan 34°45′N 134°10′E
Bizerta see Bizerte
120 K10 **Bizerte** Ar. Banzart, Eng. Bizerta. N Tunisia 37°18′N 09°48′E
Bizkaia see Vizcaya
105 O2 **Bizkaia** Cast. Vizcaya. ◈ province País Vasco, N Spain
92 G2 **Bjargtangar** headland W Iceland 63°30′N 24°29′W
Bjärnå see Perniö
95 K22 **Bjärnum** Skåne, S Sweden 56°15′N 13°45′E
93 I16 **Bjästa** Västernorrland, C Sweden 63°12′N 18°30′E
113 I14 **Bjelašnica** ▲ SE Bosnia and Herzegovina 43°13′N 18°18′E
112 C10 **Bjelolasica** ▲ NW Croatia 45°13′N 14°56′E
112 F8 **Bjelovar** Hung. Belovár. Bjelovar-Bilogora, N Croatia 45°54′N 16°49′E
112 F8 **Bjelovar-Bilogora** off. Bjelovarsko-Bilogorska Županija. ◈ province NE Croatia
Bjelovarsko-Bilogorska Županija see Bjelovar-Bilogora
92 H10 **Bjerkvik** Nordland, C Norway 68°31′N 16°08′E
95 G21 **Bjerringbro** Midtjylland, NW Denmark 56°23′N 09°40′E
Bjeshkët e Namuna see North Albanian Alps
95 L14 **Björbo** Dalarna, C Sweden 60°28′N 14°44′E
95 I15 **Bjørkelangen** Akershus, S Norway 59°54′N 11°33′E
95 O14 **Björklinge** Uppsala, C Sweden 60°03′N 17°33′E
93 P14 **Bjørkø-Arholma** Stockholm, C Sweden 59°51′N 19°01′E
93 I14 **Björksele** Västerbotten, N Sweden 64°58′N 18°30′E
93 I16 **Björna** Västernorrland, C Sweden 63°34′N 18°38′E
95 C14 **Bjørnafjorden** fjord S Norway
95 L16 **Björneborg** Värmland, C Sweden 59°13′N 14°15′E
Björneborg see Pori
95 E14 **Bjornesfjorden** ◎ S Norway
92 M9 **Bjørnevatn** Finnmark, N Norway 69°40′N 29°57′E
197 T13 **Bjørnøya** Eng. Bear Island. island N Norway
93 I15 **Bjurholm** Västerbotten, N Sweden 63°57′N 19°16′E
95 J22 **Bjuv** Skåne, S Sweden 56°05′N 12°57′E
76 M12 **Bla** Ségou, W Mali 12°58′N 05°45′W
181 W8 **Blackall** Queensland, E Australia 24°26′S 145°32′E
29 V2 **Black Bay** lake bay Minnesota, N USA
27 N9 **Black Bear Creek** ≈ Oklahoma, C USA
97 K17 **Blackburn** NW England, United Kingdom 53°45′N 02°29′W
39 T11 **Blackburn, Mount** ▲ Alaska, USA 61°43′N 143°25′W
35 N5 **Black Butte Lake** ⊠ California, W USA
194 J5 **Black Coast** physical region Antarctica
11 Q16 **Black Diamond** Alberta, SW Canada 50°42′N 114°09′W
18 K11 **Black Dome** ▲ New York, NE USA 42°16′N 74°07′W
113 L18 **Black Drin** Alb. Lumi i Drinit të Zi, SCr. Crni Drim. ≈ Albania/FYR Macedonia
29 U4 **Blackduck** Minnesota, N USA 47°45′N 94°33′W
12 D6 **Black Duck** ≈ Ontario, C Canada
33 R14 **Blackfoot** Idaho, NW USA 43°11′N 112°20′W
33 P9 **Blackfoot River** ≈ Montana, NW USA
Black Forest see Schwarzwald
28 J10 **Blackhawk** South Dakota, N USA 44°09′N 103°18′W
28 I10 **Black Hills** ▲▲ South Dakota/Wyoming, N USA
11 T10 **Black Lake** ◎ Saskatchewan, C Canada
22 G6 **Black Lake** ⊠ Louisiana, S USA
31 Q5 **Black Lake** ◎ Michigan, N USA
18 I7 **Black Lake** ◎ New York, NE USA
26 F7 **Black Mesa** ▲ Oklahoma, C USA 37°00′N 103°07′W
21 P10 **Black Mountain** North Carolina, SE USA 35°37′N 82°19′W
35 P13 **Black Mountain** ▲ California, W USA 35°22′N 120°21′W
37 Q2 **Black Mountain** ▲ Colorado, C USA 40°47′N 107°23′W
96 K1 **Black Mountains** ▲▲ SE Wales, United Kingdom
36 H10 **Black Mountains** ▲▲ Arizona, SW USA
21 O7 **Black Mountains** ▲▲ Kentucky, E USA 36°54′N 82°53′W
33 Q16 **Black Pine Peak** ▲ Idaho, NW USA 42°07′N 113°07′W
97 K17 **Blackpool** NW England, United Kingdom 53°50′N 03°03′W
37 Q14 **Black Range** ▲▲ New Mexico, SW USA
44 I12 **Black River** W Jamaica 18°02′N 77°52′W
14 J14 **Black River** ≈ Ontario, SE Canada
129 U12 **Black River** Chin. Babian Jiang, Lixian Jiang, Fr. Rivière Noire, Vtn. Sông Đa. ≈ China/Vietnam
44 I12 **Black River** ≈ W Jamaica
39 T7 **Black River** ≈ Alaska, USA
37 N13 **Black River** ≈ Arizona, SW USA
27 X7 **Black River** ≈ Arkansas/Missouri, C USA
22 I7 **Black River** ≈ Louisiana, S USA
31 S8 **Black River** ≈ Michigan, N USA
31 Q5 **Black River** ≈ Michigan, N USA
18 I8 **Black River** ≈ New York, NE USA
21 T13 **Black River** ≈ South Carolina, SE USA
30 J7 **Black River** ≈ Wisconsin, N USA
30 J7 **Black River Falls** Wisconsin, N USA 44°18′N 90°51′W
35 R3 **Black Rock Desert** desert Nevada, W USA
Black Sand Desert see Garagum
21 S7 **Blacksburg** Virginia, NE USA 37°15′N 80°25′W
136 H10 **Black Sea** var. Euxine Sea, Bul. Cherno More, Rom. Marea Neagră, Rus. Chernoye More, Turk. Karadeniz, Ukr. Chorne More. sea Asia/Europe
117 Q10 **Black Sea Lowland** Ukr. Prychornomor'ska Nyzovyna. depression SE Europe
33 S17 **Blacks Fork** ≈ Wyoming, C USA
23 V7 **Blackshear** Georgia, SE USA 31°18′N 82°14′W
23 S6 **Blackshear, Lake** ⊠ Georgia, SE USA
97 A16 **Blacksod Bay** Ir. Cuan an Fhóid Duibh. inlet W Ireland
21 V7 **Blackstone** Virginia, NE USA 37°04′N 78°00′W
77 O14 **Black Volta** var. Borongo, Mouhoun, Moun Hou, Fr. Volta Noire. ≈ W Africa
23 O5 **Black Warrior River** ≈ Alabama, S USA
181 X8 **Blackwater** Queensland, E Australia 23°34′S 148°51′E
97 D20 **Blackwater** Ir. An Abhainn Mhór. ≈ S Ireland
27 T4 **Blackwater River** ≈ Missouri, C USA
21 W7 **Blackwater River** ≈ Virginia, NE USA
Blackwater State see Nebraska
27 N8 **Blackwell** Oklahoma, C USA 36°48′N 97°16′W
25 P7 **Blackwell** Texas, SW USA 32°05′N 100°19′W
99 J15 **Bladel** Noord-Brabant, S Netherlands 51°22′N 05°13′E
Bladenmarkt see Bălăuşeri
114 G11 **Blagoevgrad** prev. Gorna Dzhumaya. Blagoevgrad, W Bulgaria 42°01′N 23°05′E
114 G11 **Blagoevgrad** ◈ province SW Bulgaria
123 Q14 **Blagoveshchensk** Amurskaya Oblast', SE Russian Federation 50°19′N 127°30′E
127 V4 **Blagoveshchensk** Respublika Bashkortostan, W Russian Federation 55°03′N 56°01′E
102 I7 **Blain** Loire-Atlantique, NW France 47°26′N 01°47′W
29 V8 **Blaine** Minnesota, N USA 45°09′N 93°13′W
32 H6 **Blaine** Washington, NW USA 48°59′N 122°45′W
11 T15 **Blaine Lake** Saskatchewan, S Canada 52°49′N 106°48′W
29 S14 **Blair** Nebraska, C USA 41°32′N 96°07′W
96 J10 **Blairgowrie** C Scotland, United Kingdom 56°19′N 03°25′W
18 C15 **Blairsville** Pennsylvania, NE USA 40°25′N 79°12′W
116 H11 **Blaj** Ger. Blasendorf, Hung. Balázsfalva. Alba, SW Romania 46°10′N 23°57′E
64 F9 **Blake-Bahama Ridge** undersea feature W Atlantic Ocean 29°00′N 73°30′W
23 S7 **Blakely** Georgia, SE USA 31°22′N 84°55′W
64 E10 **Blake Plateau** var. Blake Terrace. undersea feature W Atlantic Ocean 31°00′N 79°00′W
30 M1 **Blake Point** headland Michigan, N USA 48°11′N 88°25′W
Blake Terrace see Blake Plateau
61 B24 **Blanca, Bahía** bay E Argentina
56 C12 **Blanca, Cordillera** ▲▲ W Peru
105 T12 **Blanca, Costa** physical region SE Spain
37 S7 **Blanca Peak** ▲ Colorado, C USA 37°34′N 105°29′W
24 I9 **Blanca, Sierra** ▲ Texas, SW USA 31°15′N 105°26′W
120 K9 **Blanc, Cap** headland N Tunisia 37°20′N 09°41′E
Blanc, Cap see Nouâdhibou, Râs
31 R12 **Blanchard River** ≈ Ohio, N USA
182 E8 **Blanche, Cape** headland South Australia 33°03′S 134°10′E
182 J4 **Blanche, Lake** ◎ South Australia
31 R14 **Blanchester** Ohio, N USA 39°17′N 83°59′W
182 I9 **Blanchetown** South Australia 34°21′S 139°36′E
45 U13 **Blanchisseuse** Trinidad, Trinidad and Tobago 10°47′N 61°18′W
103 T11 **Blanc, Mont** It. Monte Bianco. ▲ France/Italy 45°45′N 06°51′E
25 R11 **Blanco** Texas, SW USA 30°06′N 98°25′W
42 K14 **Blanco, Cabo** headland NW Costa Rica 09°34′N 85°06′W
32 D14 **Blanco, Cape** headland Oregon, NW USA 42°49′N 124°33′W
62 H10 **Blanco, Río** ≈ W Argentina
56 F10 **Blanco, Río** ≈ NE Peru
15 O9 **Blanc, Réservoir** ⊠ Québec, SE Canada
21 R7 **Bland** Virginia, NE USA 37°06′N 81°08′W
92 I2 **Blanda** ≈ N Iceland
37 O7 **Blanding** Utah, W USA 37°37′N 109°28′W
105 X5 **Blanes** Cataluña, NE Spain 41°41′N 02°48′E
103 N3 **Blangy-sur-Bresle** Seine-Maritime, N France 49°55′N 01°37′E
111 C18 **Blanice** Ger. Blanitz. ≈ SE Czech Republic
Blanitz see Blanice
99 C16 **Blankenberge** West-Vlaanderen, NW Belgium 51°19′N 03°08′E
101 D17 **Blankenheim** Nordrhein-Westfalen, W Germany 50°26′N 06°41′E
25 R8 **Blanket** Texas, SW USA 31°49′N 98°47′W
55 O3 **Blanquilla, Isla** var. La Blanquilla. island N Venezuela
Blanquilla, La see Blanquilla, Isla
61 F18 **Blanquillo** Durazno, C Uruguay 32°53′S 55°37′W
111 G18 **Blansko** Ger. Blanz. Jihomoravský Kraj, SE Czech Republic 49°22′N 16°39′E
83 N15 **Blantyre** var. Blantyre-Limbe. Southern, S Malawi 15°45′S 35°04′E
83 N15 **Blantyre** ✈ Southern, S Malawi 15°34′S 35°03′E
Blantyre-Limbe see Blantyre
Blanz see Blansko
98 J10 **Blaricum** Noord-Holland, C Netherlands 52°16′N 05°15′E
Blasendorf see Blaj
Blatnitsa see Durankulak
113 F15 **Blato** It. Blatta. Dubrovnik-Neretva, S Croatia 42°57′N 16°47′E
Blatta see Blato
108 E10 **Blatten** Valais, SW Switzerland 46°22′N 08°00′E
101 J20 **Blaufelden** Baden-Württemberg, SW Germany 49°21′N 10°01′E
95 E23 **Blåvands Huk** headland W Denmark 55°33′N 08°04′E
102 G6 **Blavet** ≈ NW France
102 J12 **Blaye** Gironde, SW France 45°08′N 00°40′W
183 R8 **Blayney** New South Wales, SE Australia 33°33′S 149°13′E
65 D25 **Bleaker Island** island SE Falkland Islands
109 T10 **Bled** Ger. Veldes. NW Slovenia 46°23′N 14°06′E
99 D20 **Bléharies** Hainaut, SW Belgium 50°31′N 03°25′E
109 U9 **Bleiburg** Slvn. Pliberk. Kärnten, S Austria 46°36′N 14°49′E
101 L17 **Bleiloch-stausee** ⊠ C Germany
98 H12 **Bleiswijk** Zuid-Holland, W Netherlands 52°01′N 04°32′E
95 L22 **Blekinge** ◈ county S Sweden
14 D17 **Blenheim** Ontario, S Canada 42°20′N 81°59′W
185 K15 **Blenheim** Marlborough, South Island, New Zealand 41°32′S 174°E
99 M15 **Blerick** Limburg, SE Netherlands 51°22′N 06°10′E
Blesae see Blois
25 V13 **Blessing** Texas, SW USA 28°52′N 96°12′W
14 I10 **Bleu, Lac** ◎ Québec, SE Canada
Blibba see Blitta
120 H10 **Blida** var. El Boulaida, El Boulaïda. N Algeria 36°30′N 02°50′E
95 P15 **Blidö** Stockholm, C Sweden 59°37′N 18°55′E
95 K18 **Blidsberg** Västra Götaland, S Sweden 57°55′N 13°30′E
185 A21 **Bligh Sound** sound South Island, New Zealand
187 X14 **Bligh Water** strait NW Fiji
14 D11 **Blind River** Ontario, S Canada 46°12′N 82°59′W
31 R11 **Blissfield** Michigan, N USA 41°49′N 83°51′W
77 R15 **Blitta** prev. Blibba. C Togo 08°19′N 00°59′E
19 O13 **Block Island** island Rhode Island, NE USA
19 O13 **Block Island Sound** sound Rhode Island, NE USA
98 H10 **Bloemendaal** Noord-Holland, W Netherlands 52°23′N 04°39′E
83 H23 **Bloemfontein** var. Mangaung. ● (South Africa-judicial capital) Free State, C South Africa 29°07′S 26°14′E
83 I22 **Bloemhof** North-West, NW South Africa 27°39′S 25°37′E
102 M7 **Blois** anc. Blesae. Loir-et-Cher, C France 47°36′N 01°20′E
98 L8 **Blokzijl** Overijssel, N Netherlands 52°46′N 05°58′E
95 N20 **Blomstermåla** Kalmar, S Sweden 56°58′N 16°19′E
92 I2 **Blönduós** Norðurland Vestra, N Iceland 65°39′N 20°15′W
110 L11 **Błonie** Mazowieckie, C Poland 52°13′N 20°36′E
97 C14 **Bloody Foreland** Ir. Cnoc Fola. headland NW Ireland 55°09′N 08°18′W
31 N15 **Bloomfield** Indiana, N USA 39°01′N 86°58′W
29 X16 **Bloomfield** Iowa, C USA 40°45′N 92°24′W
27 Y8 **Bloomfield** Missouri, C USA 36°54′N 89°58′W
37 P9 **Bloomfield** New Mexico, SW USA 36°42′N 108°00′W
25 U7 **Blooming Grove** Texas, SW USA 32°05′N 96°43′W
29 W10 **Blooming Prairie** Minnesota, N USA 43°52′N 93°03′W
30 L13 **Bloomington** Illinois, N USA 40°28′N 88°59′W
31 O15 **Bloomington** Indiana, N USA 39°10′N 86°31′W
29 V9 **Bloomington** Minnesota, N USA 44°50′N 93°18′W
25 U13 **Bloomington** Texas, SW USA 28°39′N 96°53′W
18 H14 **Bloomsburg** Pennsylvania, NE USA 40°59′N 76°27′W
181 X7 **Bloomsbury** Queensland, NE Australia 20°47′S 148°35′E
169 R16 **Blora** Jawa, C Indonesia 06°55′S 111°29′E
18 G12 **Blossburg** Pennsylvania, NE USA 41°38′N 77°00′W
25 V5 **Blossom** Texas, SW USA 33°39′N 95°23′W
123 T5 **Blossom, Mys** headland Ostrov Vrangelya, NE Russian Federation 70°49′N 178°49′E
23 R8 **Blountstown** Florida, SE USA 30°26′N 85°03′W
21 P8 **Blountville** Tennessee, S USA 36°31′N 82°19′W
21 Q9 **Blowing Rock** North Carolina, SE USA 36°15′N 81°53′W
108 J8 **Bludenz** Vorarlberg, W Austria 47°10′N 09°50′E
36 L6 **Blue Bell Knoll** ▲ Utah, W USA 38°11′N 111°31′W
23 Y12 **Blue Cypress Lake** ◎ Florida, SE USA
29 U11 **Blue Earth** Minnesota, N USA 43°38′N 94°06′W
21 Q7 **Bluefield** Virginia, NE USA 37°15′N 81°16′W
21 R7 **Bluefield** West Virginia, NE USA 37°16′N 81°13′W
43 N10 **Bluefields** Región Autónoma Atlántico Sur, SE Nicaragua 12°01′N 83°47′W
43 N10 **Bluefields, Bahía de** bay W Caribbean Sea
29 Z14 **Blue Grass** Iowa, C USA 41°30′N 90°46′W
Bluegrass State see Kentucky
Blue Hen State see Delaware
19 S7 **Blue Hill** Maine, NE USA 44°25′N 68°36′W
29 P16 **Blue Hill** Nebraska, C USA 40°19′N 98°27′W
30 J5 **Blue Hills** hill range Wisconsin, N USA
34 L3 **Blue Lake** California, W USA 40°52′N 124°00′W
Blue Law State see Connecticut
37 Q6 **Blue Mesa Reservoir** ⊠ Colorado, C USA
27 S12 **Blue Mountain** ▲ Arkansas, C USA 34°42′N 94°04′W
19 O6 **Blue Mountain** ▲ New Hampshire, NE USA 44°48′N 71°26′W
18 K8 **Blue Mountain** ▲ New York, NE USA 43°52′N 74°24′W
18 H15 **Blue Mountain** ridge Pennsylvania, NE USA
44 H10 **Blue Mountain Peak** ▲ E Jamaica 18°02′N 76°34′W
183 S8 **Blue Mountains** ▲▲ New South Wales, SE Australia
32 L11 **Blue Mountains** ▲▲ Oregon/Washington, NW USA
80 G12 **Blue Nile** ◈ state E Sudan
80 H12 **Blue Nile** var. Abai, Bahr el, Azraq, Amh. Ābay Wenz, Ar. An Nīl al Azraq. ≈ Ethiopia/Sudan
8 J7 **Bluenose Lake** ◎ Nunavut, NW Canada
27 O3 **Blue Rapids** Kansas, C USA 39°39′N 96°38′W
23 S1 **Blue Ridge** Georgia, SE USA 34°51′N 84°19′W
17 S11 **Blue Ridge** var. Blue Ridge Mountains. ▲▲ North Carolina/Virginia, USA
23 S1 **Blue Ridge Lake** ⊠ Georgia, SE USA
Blue Ridge Mountains see Blue Ridge
11 N15 **Blue River** British Columbia, SW Canada 52°03′N 119°21′W
27 O12 **Blue River** ≈ Oklahoma, C USA
27 R4 **Blue Springs** Missouri, C USA 39°01′N 94°16′W
21 R6 **Bluestone Lake** ⊠ West Virginia, NE USA
185 C25 **Bluff** Southland, South Island, New Zealand 46°36′S 168°22′E
37 O8 **Bluff** Utah, W USA 37°15′N 109°36′W
21 P8 **Bluff City** Tennessee, S USA 36°28′N 82°15′W
65 E24 **Bluff Cove** East Falkland, Falkland Islands 51°45′S 58°11′W
25 S7 **Bluff Dale** Texas, SW USA 32°18′N 98°01′W
183 N15 **Bluff Hill Point** headland Tasmania, SE Australia 41°03′S 144°35′E
31 Q12 **Bluffton** Indiana, N USA 40°44′N 85°10′W
31 R12 **Bluffton** Ohio, N USA 40°54′N 83°53′W
25 T7 **Blum** Texas, SW USA 32°08′N 97°24′W
101 G24 **Blumberg** Baden-Württemberg, SW Germany 47°48′N 08°31′E
60 K13 **Blumenau** Santa Catarina, S Brazil 26°55′S 49°07′W
29 N9 **Blunt** South Dakota, N USA 44°30′N 99°58′E
32 H15 **Bly** Oregon, NW USA 42°22′N 121°04′W
39 R13 **Blying Sound** sound Alaska, USA
97 M14 **Blyth** N England, United Kingdom 55°07′N 01°30′W
35 Y16 **Blythe** California, W USA 33°35′N 114°36′W
27 Y9 **Blytheville** Arkansas, C USA 35°56′N 89°55′W
117 V7 **Blyznyuky** Kharkivs'ka Oblast', E Ukraine 48°51′N 36°32′E
95 G16 **Bø** Telemark, S Norway 59°24′N 09°04′E
76 I15 **Bo** S Sierra Leone 07°58′N 11°45′W
171 O4 **Boac** Marinduque, N Philippines 13°26′N 121°50′E
42 K10 **Boaco** Boaco, S Nicaragua 12°28′N 85°45′W
42 J10 **Boaco** ◈ department C Nicaragua
79 I15 **Boali** Ombella-Mpoko, SW Central African Republic 04°52′N 18°00′E
Boalsert see Bolsward
31 V12 **Boardman** Ohio, N USA 41°01′N 80°39′W
32 J11 **Boardman** Oregon, NW USA 45°50′N 119°42′W
14 F13 **Boat Lake** ◎ Ontario, S Canada
58 F10 **Boa Vista** state capital Roraima, NW Brazil 02°51′N 60°43′W
76 D9 **Boa Vista** island Ilhas de Barlavento, E Cape Verde
23 Q2 **Boaz** Alabama, S USA 34°12′N 86°10′W
160 L15 **Bobai** Guangxi Zhuangzu Zizhiqu, S China 22°09′N 109°57′E
172 J1 **Bobaomby, Tanjona** Fr. Cap d'Ambre. headland N Madagascar 11°58′S 49°13′E
155 M14 **Bobbili** Andhra Pradesh, E India 18°32′N 83°29′E
106 D9 **Bobbio** Emilia-Romagna, C Italy 44°48′N 09°27′E
14 I14 **Bobcaygeon** Ontario, SE Canada 44°32′N 78°33′W
Bober see Bóbr
103 O5 **Bobigny** Seine-St-Denis, N France 48°55′N 02°27′E
77 N13 **Bobo-Dioulasso** SW Burkina Faso 11°12′N 04°21′W
110 G8 **Bobolice** Ger. Bublitz. Zachodnio-pomorskie, NW Poland 53°56′N 16°37′E
83 J19 **Bobonong** Central, E Botswana 21°58′S 28°26′E
171 R11 **Bobopayo** Pulau Halmahera, E Indonesia 01°07′N 127°26′E
113 J15 **Bobotov Kuk** ▲ N Montenegro 43°06′N 19°00′E
114 G10 **Bobov Dol** var. Bobovdol. Kyustendil, W Bulgaria 42°21′N 23°00′E
Bobovdol see Bobov Dol
119 M15 **Bobr** Minskaya Voblasts', NW Belarus 54°20′N 29°16′E
119 M15 **Bobr** ≈ C Belarus
111 E14 **Bóbr** Eng. Bobrawa, Ger. Bober. ≈ SW Poland
Bobrawa see Bóbr
Bobrik see Bobryk
Bobrinets see Bobrynets'
Bobrka/Bóbrka see Bibrka
126 L8 **Bobrov** Voronezhskaya Oblast', W Russian Federation 51°10′N 40°03′E
117 Q4 **Bobrovytsya** Chernihivs'ka Oblast', N Ukraine 50°43′N 31°24′E
Bobruysk see Babruysk
119 J19 **Bobryk** Rus. Bobrik. ≈ SW Belarus
117 Q8 **Bobrynets'** Rus. Bobrinets. Kirovohrads'ka Oblast', C Ukraine 48°02′N 32°10′E
14 K14 **Bobs Lake** ◎ Ontario, SE Canada
54 I6 **Bobures** Zulia, NW Venezuela 09°15′N 71°10′W
42 H1 **Boca Bacalar Chico** headland N Belize 15°05′N 82°12′W
112 G11 **Bočac** ◈ Republika Srpska, NW Bosnia and Herzegovina
41 R14 **Boca del Río** Veracruz-Llave, S Mexico 19°08′N 96°08′W
55 O4 **Boca de Pozo** Nueva Esparta, NE Venezuela 11°00′N 64°23′W
59 C15 **Boca do Acre** Amazonas, N Brazil 08°45′S 67°23′W
55 N12 **Boca Mavaca** Amazonas, S Venezuela 02°30′N 65°11′W
79 G14 **Bocaranga** Ouham-Pendé, W Central African Republic 07°07′N 15°40′E
23 Z15 **Boca Raton** Florida, SE USA 26°22′N 80°05′W
43 P14 **Bocas del Toro** Bocas del Toro, NW Panama 09°20′N 82°15′W
43 P15 **Bocas del Toro** off. Provincia de Bocas del Toro. ◈ province NW Panama
43 P15 **Bocas del Toro, Archipiélago de** island group NW Panama
Bocas del Toro, Provincia de see Bocas del Toro
42 L7 **Bocay** Jinotega, N Nicaragua 14°19′N 85°08′W
105 N6 **Boceguillas** Castilla y León, N Spain 41°20′N 03°39′W
Bocheykovo see Bacheykava
111 L17 **Bochnia** Małopolskie, SE Poland 49°58′N 20°27′E
99 K16 **Bocholt** Limburg, NE Belgium 51°10′N 05°37′E
101 D14 **Bocholt** Nordrhein-Westfalen, W Germany 51°50′N 06°37′E
101 E15 **Bochum** Nordrhein-Westfalen, W Germany 51°29′N 07°13′E
103 Y15 **Bocognano** Corse, France, C Mediterranean Sea 42°04′N 09°03′E
54 I6 **Boconó** Trujillo, NW Venezuela 09°17′N 70°17′W

◆ Country ● Country Capital ◇ Dependent Territory ○ Dependent Territory Capital ◈ Administrative Regions ✈ International Airport ▲ Mountain ▲▲ Mountain Range ⌂ Volcano ≈ River ◎ Lake ⊠ Reservoir

Bocşa – Borsod-Abaúj-Zemplén Megye

116 F12 Bocşa Ger. Bokschen, Hung. Boksánbánya. Caraş-Severin, SW Romania 45°23′N 21°47′E
79 H15 Boda Lobaye, SW Central African Republic 04°17′N 17°25′E
94 L12 Boda Dalarna, C Sweden 61°00′N 15°15′E
95 O20 Böda Kalmar, S Sweden 57°16′N 17°04′E
95 L19 Bodafors Jönköping, S Sweden 57°30′N 14°40′E
123 O12 Bodaybo Irkutskaya Oblast', E Russian Federation 57°52′N 114°05′E
22 G5 Bodcau, Bayou var. Bodcau Creek. ≈ Louisiana, S USA
Bodcau Creek see Bodcau, Bayou
44 D8 Bodden Town var. Boddentown. Grand Cayman, SW Cayman Islands 19°20′N 81°14′W
Boddentown see Bodden Town
101 K14 Bode ≈ C Germany
34 L7 Bodega Head headland California, W USA 38°16′N 123°04′W
Bodegas see Babahoyo
98 H11 Bodegraven Zuid-Holland, C Netherlands 52°05′N 04°45′E
78 H8 Bodélé depression W Chad
92 J13 Boden Norrbotten, N Sweden 65°50′N 21°44′E
Bodensee see Constance, Lake, C Europe
65 M15 Bode Verde Fracture Zone tectonic feature E Atlantic Ocean
155 H14 Bodhan Telangana, C India 18°40′N 77°51′E
Bodī see Jinst
155 H22 Bodināyakkanūr Tamil Nādu, SE India 10°02′N 77°18′E
108 H10 Bodio Ticino, S Switzerland 46°23′N 08°55′E
Bodjonegoro see Bojonegoro
97 I24 Bodmin SW England, United Kingdom 50°29′N 04°43′W
97 I24 Bodmin Moor moorland SW England, United Kingdom
92 G12 Bodø Nordland, C Norway 67°17′N 14°22′E
59 H20 Bodoquena, Serra da ▲▲ SW Brazil
136 B16 Bodrum Muğla, SW Turkey 37°01′N 27°28′E
Bodzafordulő see Întorsura Buzăului
99 L14 Boekel Noord-Brabant, SE Netherlands 51°35′N 05°42′E
Boeloekoemba see Bulukumba
103 Q11 Boën Loire, E France 45°45′N 04°01′E
79 K18 Boende Equateur, C Dem. Rep. Congo 0°12′S 20°54′E
25 R11 Boerne Texas, SW USA 29°47′N 98°44′W
Boeroe see Buru, Pulau
Boetoeng see Buton, Pulau
22 I5 Boeuf River ≈ Arkansas/Louisiana, S USA
76 H14 Boffa W Guinea 10°12′N 14°02′W
Bó Finne, Inis see Inishbofin
Boga see Bogë
166 L9 Bogale Ayeyawady, SW Myanmar (Burma) 16°16′N 95°21′E
22 L8 Bogalusa Louisiana, S USA 30°47′N 89°51′W
77 Q12 Bogandé C Burkina Faso 13°02′N 00°08′W
79 I15 Bogangolo Ombella-Mpoko, C Central African Republic 05°36′N 18°17′E
183 Q7 Bogan River ≈ New South Wales, SE Australia
25 W5 Bogata Texas, SW USA 33°28′N 95°12′W
111 D14 Bogatynia Ger. Reichenau. Dolnośląskie, SW Poland 50°53′N 14°55′E
136 K13 Boğazlıyan Yozgat, C Turkey 39°13′N 35°17′E
79 J17 Bogbonga Equateur, NW Dem. Rep. Congo 01°36′N 19°24′E
158 J14 Bogcang Zangbo ≈ W China
162 I9 Bogd var. Horiult. Bayanhongor, C Mongolia 45°09′N 100°50′E
162 J10 Bogd var. Hovd. Övörhangay, C Mongolia 44°43′N 102°08′E
158 L5 Bogda Feng ▲ NW China 43°51′N 88°14′E
114 I9 Bogdan ▲ C Bulgaria 42°37′N 24°28′E
113 Q20 Bogdanci SE FYR Macedonia 41°12′N 22°34′E
158 M5 Bogda Shan var. Po-ko-to Shan. ▲▲ NW China
113 K17 Bogë var. Boga. Shkodër, N Albania 42°25′N 19°38′E
Bogeda'er see Wenquan
Bogendorf see Łuków
95 G23 Bogense Syddtjylland, C Denmark 55°34′N 10°06′E
183 T3 Boggabilla New South Wales, SE Australia 28°37′S 150°21′E
183 S6 Boggabri New South Wales, SE Australia 30°44′S 150°00′E
186 D6 Bogia Madang, N Papua New Guinea 04°16′S 144°56′E
97 N23 Bognor Regis SE England, United Kingdom 50°47′N 00°41′W
Bogodukhov see Bohodukhiv
181 V15 Bogong, Mount ▲ Victoria, SE Australia 36°43′S 147°19′E
169 O16 Bogor Dut. Buitenzorg. Jawa, C Indonesia 06°34′S 106°45′E
126 L5 Bogoroditsk Tul'skaya Oblast', W Russian Federation 53°46′N 38°09′E
127 O3 Bogorodsk Nizhegorodskaya Oblast', W Russian Federation 56°06′N 43°29′E
Bogorodskoje see Bogorodskoye
123 S12 Bogorodskoye Khabarovskiy Kray, SE Russian Federation 52°22′N 140°33′E
125 R15 Bogorodskoye var. Bogorodskoje. Kirovskaya Oblast', NW Russian Federation 57°50′N 50°41′E
54 F10 Bogotá prev. Santa Fe, Santa Fe de Bogotá. ● (Colombia) Cundinamarca, C Colombia 04°38′N 74°05′W
153 T14 Bogra Rajshahi, N Bangladesh 24°52′N 89°28′E
Bogschan see Boldu
122 L12 Boguchany Krasnoyarskiy Kray, C Russian Federation 58°20′N 97°20′E
126 M9 Boguchar Voronezhskaya Oblast', W Russian Federation 49°57′N 40°34′E
76 H10 Bogué Brakna, SW Mauritania 16°36′N 14°15′W
22 K8 Bogue Chitto ≈ Louisiana/Mississippi, S USA
Bogushëvsk see Bahushewsk
Boguslav see Bohuslav
44 K12 Bog Walk C Jamaica 18°06′N 77°01′W
161 Q3 Bo Hai var. Gulf of Chihli. gulf NE China
161 R3 Bohai Haixia strait NE China
161 Q3 Bohai Wan bay NE China
111 C17 Bohemia Cz. Čechy, Ger. Böhmen. W Czech Republic
111 B18 Bohemian Forest Cz. Ceský Les, Šumava, Ger. Böhmerwald. ▲▲ C Europe
Bohemian-Moravian Highlands see Českomoravská Vrchovina
77 R16 Bohicon S Benin 07°14′N 02°04′E
109 S11 Bohinjska Bistrica Ger. Wocheiner Feistritz. NW Slovenia 46°16′N 13°55′E
Bohkká see Pokka
Böhmen see Bohemia
Böhmerwald see Bohemian Forest
Böhmisch-Krumau see Český Krumlov
Böhmisch-Leipa see Česká Lípa
Böhmisch-Mährische Höhe see Českomoravská Vrchovina
Böhmisch-Trübau see Česká Třebová
117 U5 Bohodukhiv Rus. Bogodukhov. Kharkivs'ka Oblast', E Ukraine 50°10′N 35°32′E
171 Q6 Bohol island C Philippines
171 Q7 Bohol Sea var. Mindanao Sea. sea S Philippines
116 I7 Bohorodchany Ivano-Frankivs'ka Oblast', W Ukraine 48°46′N 24°31′E
Böhöt see Öndörshil
158 K6 Bohu var. Bagrax. Xinjiang Uygur Zizhiqu, NW China 42°00′N 86°28′E
111 I17 Bohumín Ger. Oderberg; prev. Neuoderberg, Nový Bohumín. Moravskoslezský Kraj, E Czech Republic 49°55′N 18°20′E
117 P6 Bohuslav Rus. Boguslav. Kyyivs'ka Oblast', N Ukraine 49°33′N 30°53′E
58 F11 Boiaçu Roraima, N Brazil 0°27′S 61°46′W
107 K16 Boiano Molise, C Italy 41°28′N 14°28′E
15 R8 Boilleau Québec, SE Canada 48°06′N 70°49′W
59 O17 Boipeba, Ilha de island SE Brazil
104 G3 Boiro Galicia, NW Spain 42°39′N 08°53′W
31 Q5 Bois Blanc Island island Michigan, N USA
29 R7 Bois de Sioux River ≈ Minnesota, N USA
33 N14 Boise var. Boise City. state capital Idaho, NW USA 43°39′N 116°14′W
26 G8 Boise City Oklahoma, C USA 36°44′N 102°31′W
Boise City see Boise
33 N14 Boise River, Middle Fork ≈ Idaho, NW USA
Bois, Lac des see Woods, Lake of the
Bois-le-Duc see 's-Hertogenbosch
11 W17 Boissevain Manitoba, S Canada 49°14′N 100°02′W
15 T7 Boisvert, Pointe au headland Québec, SE Canada 48°34′N 69°07′W
100 K10 Boizenburg Mecklenburg-Vorpommern, N Germany 53°23′N 10°43′E
Bojador see Boujdour
113 K18 Bojana Alb. Bunë. ≈ Albania/Montenegro see also Bunë
Bojana see Bunë, Lumi i
143 S3 Bojnūrd var. Bujnurd. Khorāsān-e Shemālī, N Iran 37°31′N 57°24′E
169 R16 Bojonegoro prev. Bodjonegoro. Jawa, C Indonesia 07°06′S 111°50′E
189 T1 Bokaak Atoll var. Bokak, Taongi. atoll Ratak Chain, NE Marshall Islands
Bokak see Bokaak Atoll
146 K8 Bo'kantov Tog'lari Rus. Gory Bukantau. ▲▲ N Uzbekistan
153 Q15 Bokāro Jhārkhand, N India 23°46′N 85°55′E
79 I18 Bokatola Equateur, NW Dem. Rep. Congo 0°37′S 18°45′E
76 H13 Boké W Guinea 10°56′N 14°18′W
Bokhara see Buxoro
183 Q4 Bokharra River ≈ New South Wales/Queensland, SE Australia
147 X8 Bokonbayevo Kir. Kajisay; prev. Kadzhi-Say. Issyk-Kul'skaya Oblast', NE Kyrgyzstan 42°07′N 76°59′E
78 H11 Bokoro Hadjer-Lamis, W Chad 12°23′N 17°03′E
79 K19 Bokota Equateur, NW Dem. Rep. Congo 0°56′S 22°24′E
167 N13 Bokpyin Taninthayi, S Myanmar (Burma) 11°16′N 98°47′E
Boksánbánya/Bokschen see Bocşa
83 F21 Bokspits Kgalagadi, SW Botswana 26°50′S 20°41′E
79 K18 Bokungu Equateur, C Dem. Rep. Congo 0°41′S 22°19′E
146 F12 Bokurdak Rus. Bakhardok. Ahal Welaýaty, C Turkmenistan 38°51′N 58°34′E
78 G10 Bol Lac, W Chad 13°27′N 14°40′E
76 G13 Bolama SW Guinea-Bissau 11°35′N 15°30′W
Bolangir see Balāngīr
Bolanos see Bolanos, Mount, Guam
Bolaños see Bolaños de Calatrava, Spain
105 N11 Bolaños de Calatrava var. Bolaños. Castilla-La Mancha, C Spain 38°55′N 03°39′W
188 B17 Bolanos, Mount var. Bolanos. ▲ S Guam 13°18′N 144°41′E
40 L12 Bolaños, Río ≈ C Mexico
115 M14 Bolayır Çanakkale, NW Turkey 40°31′N 26°46′E
102 L3 Bolbec Seine-Maritime, N France 49°34′N 00°31′E
116 L13 Boldu var. Bogschan. Buzău, SE Romania 45°18′N 27°15′E
146 H8 Boldumsaz prev. Kalinin, Kalininsk, Porsy. Daşoguz Welaýaty, N Turkmenistan 42°12′N 59°33′E
158 I4 Bole var. Bortala. Xinjiang Uygur Zizhiqu, NW China 44°52′N 82°06′E
77 O15 Bole NW Ghana 09°02′N 02°29′W
79 J19 Boleko Equateur, W Dem. Rep. Congo 01°27′S 19°52′E
111 E14 Bolesławiec Ger. Bunzlau. Dolnośląskie, SW Poland 51°16′N 15°34′E
127 R4 Bolgar prev. Kuybyshev. Respublika Tatarstan, W Russian Federation 54°58′N 49°03′E
77 P14 Bolgatanga N Ghana 10°45′N 00°52′W
Bolgrad see Bolhrad
117 N12 Bolhrad Rus. Bolgrad. Odes'ka Oblast', SW Ukraine 45°42′N 28°35′E
163 Y8 Boli Heilongjiang, NE China 45°45′N 130°32′E
79 I19 Bolia Bandundu, W Dem. Rep. Congo 01°34′S 18°24′E
93 J14 Boliden Västerbotten, N Sweden 64°52′N 20°20′E
171 T13 Bolifar Pulau Seram, E Indonesia 03°08′S 130°34′E
171 N2 Bolinao Luzon, N Philippines 16°22′N 119°52′E
54 C12 Bolívar Cauca, SW Colombia 01°52′N 76°56′W
27 T6 Bolivar Missouri, C USA 37°37′N 93°25′W
20 F10 Bolivar Tennessee, S USA 35°17′N 88°59′W
54 F7 Bolívar off. Departmento de Bolívar. ◈ province N Colombia
56 A13 Bolívar ◈ province C Ecuador
55 N9 Bolívar off. Estado Bolívar. ◈ state SE Venezuela
Bolívar, Departmento de see Bolívar
Bolívar, Estado see Bolívar
25 X12 Bolivar Peninsula headland Texas, SW USA 29°26′N 94°41′W
54 I6 Bolívar, Pico ▲ W Venezuela 08°33′N 71°05′W
57 K17 Bolivia off. Plurinational State of Bolivia; prev. Republic of Bolivia. ◆ republic W South America
Bolivia, Plurinational State of see Bolivia
Bolivia, Republic of see Bolivia
112 O13 Boljevac Serbia, E Serbia 43°50′N 21°57′E
Bolkenhain see Bolków
126 J5 Bolkhov Orlovskaya Oblast', W Russian Federation 53°28′N 36°00′E
111 F14 Bolków Ger. Bolkenhain. Dolnośląskie, SW Poland 50°55′N 15°49′E
182 K3 Bollards Lagoon South Australia 28°58′S 140°52′E
103 R14 Bollène Vaucluse, SE France 44°16′N 04°45′E
94 N12 Bollnäs Gävleborg, C Sweden 61°18′N 16°27′E
181 W10 Bollon Queensland, C Australia 28°07′S 147°28′E
192 L12 Bollons Tablemount undersea feature S Pacific Ocean 49°40′S 176°10′W
93 H17 Bollstabruk Västernorrland, C Sweden 63°00′N 17°41′E
Bolluilos de Par del Condado see Bollullos Par del Condado
104 J14 Bollullos Par del Condado var. Bolluilos de Par del Condado. Andalucía, S Spain 37°20′N 06°32′W
95 K21 Bolmen ◎ S Sweden
137 T10 Bolnisi S Georgia 41°28′N 44°34′E
79 H19 Bolobo Bandundu, W Dem. Rep. Congo 02°10′S 16°17′E
106 G10 Bologna Emilia-Romagna, N Italy 44°30′N 11°20′E
124 I15 Bologoye Tverskaya Oblast', W Russian Federation 57°54′N 34°04′E
79 J18 Bolomba Equateur, NW Dem. Rep. Congo 0°27′N 19°13′E
41 X13 Bolonchén de Rejón var. Bolonchén de Rejón. Campeche, SE Mexico 20°00′N 89°34′W
114 J13 Boloústra, Akrotírio headland NE Greece 40°56′N 24°58′E
167 L8 Bolovén, Phouphiang Fr. Plateau des Bolovens. plateau S Laos
Bolovens, Plateau des see Bolovén, Phouphiang
106 H13 Bolsena Lazio, C Italy 42°39′N 11°59′E
107 G14 Bolsena, Lago di ◎ C Italy
126 B3 Bol'shakovo Ger. Kreuzingen; prev. Gross-Skaisgirren. Kaliningradskaya Oblast', W Russian Federation 54°53′N 21°38′E
Bol'shaya Berëstovitsa see Vyalikaya Byerastavitsa
127 S7 Bol'shaya Chernigovka Samarskaya Oblast', W Russian Federation 52°07′N 50°49′E
127 S7 Bol'shaya Glushitsa Samarskaya Oblast', W Russian Federation 52°22′N 50°29′E
124 J4 Bol'shaya Imandra, Ozero ◎ NW Russian Federation
Bol'shaya Khobda see Kobda
126 M12 Bol'shaya Martynovka Rostovskaya Oblast', SW Russian Federation 47°19′N 41°40′E
122 K12 Bol'shaya Murta Krasnoyarskiy Kray, C Russian Federation 56°51′N 93°10′E
125 V4 Bol'shaya Rogovaya ≈ NW Russian Federation
125 U7 Bol'shaya Synya ≈ NW Russian Federation
145 V9 Bol'shaya Vladimirovka Vostochnyy Kazakhstan, E Kazakhstan 50°53′N 79°29′E
123 V11 Bol'sheretsk Kamchatskiy Kray, E Russian Federation 52°20′N 156°24′E
127 W3 Bol'sheust'ikinskoye Respublika Bashkortostan, W Russian Federation 56°00′N 58°13′E
Bol'shevik see Bal'shavik
122 L5 Bol'shevik, Ostrov island Severnaya Zemlya, N Russian Federation
125 U4 Bol'shezemel'skaya Tundra physical region NW Russian Federation
144 J13 Bol'shiye Barsuki, Peski desert SW Kazakhstan
123 T7 Bol'shoy Anyuy ≈ NE Russian Federation
123 N7 Bol'shoy Begichev, Ostrov island NE Russian Federation
123 S15 Bol'shoye Kamen' Primorskiy Kray, SE Russian Federation 43°06′N 132°21′E
127 O4 Bol'shoye Murashkino Nizhegorodskaya Oblast', W Russian Federation 55°46′N 44°48′E
127 W4 Bol'shoy Iremel' ▲ W Russian Federation 54°31′N 58°47′E
127 R7 Bol'shoy Irgiz ≈ W Russian Federation
123 Q6 Bol'shoy Lyakhovskiy, Ostrov island NE Russian Federation
123 Q11 Bol'shoy Nimnyr Respublika Sakha (Yakutiya), NE Russian Federation 57°55′N 125°34′E
Bol'shoy Rozhan see Vyaliki Rozhan
Bol'shoy Uzen' see Karaozen
40 K6 Bolsón de Mapimí ▲▲ NW Mexico
98 K6 Bolsward Fris. Boalsert. Fryslân, N Netherlands 53°04′N 05°31′E
105 T4 Boltaña Aragón, NE Spain 42°28′N 00°02′E
14 G15 Bolton Ontario, S Canada 43°52′N 79°45′W
97 K17 Bolton prev. Bolton-le-Moors. NW England, United Kingdom 53°35′N 02°26′W
21 V12 Bolton North Carolina, SE USA 34°22′N 78°26′W
Bolton-le-Moors see Bolton
136 G11 Bolu Bolu, NW Turkey 40°45′N 31°38′E
136 G11 Bolu ◈ province NW Turkey
186 G9 Bolubolu Goodenough Island, S Papua New Guinea 09°22′S 150°22′E
92 H1 Bolungarvík Vestfirðir, NW Iceland 66°09′N 23°17′W
159 O10 Boluntay Qinghai, W China 36°30′N 92°11′E
159 P8 Boluozhuanjing, Aksay Kazakzu Zizhixian. Gansu, N China 39°25′N 94°09′E
136 F14 Bolvadin Afyon, W Turkey 38°43′N 31°02′E
114 M10 Bolyarovo prev. Pashkeni. Yambol, E Bulgaria 42°09′N 26°49′E
106 G6 Bolzano Ger. Bozen; anc. Bauzanum. Trentino-Alto Adige, N Italy 46°30′N 11°22′E
79 F22 Boma Bas-Congo, W Dem. Rep. Congo 05°42′S 13°05′E
183 R12 Bombala New South Wales, SE Australia 36°54′S 149°15′E
104 F10 Bombarral Leiria, C Portugal 39°15′N 09°09′W
Bombay see Mumbai
171 U13 Bomberai, Semenanjung cape Papua Barat, E Indonesia
81 F18 Bombo S Uganda 0°36′N 32°33′E
162 I8 Bömbögör var. Dzadgay. Bayanhongor, C Mongolia 46°12′N 99°29′E
79 I17 Bomboma Equateur, NW Dem. Rep. Congo 02°23′N 19°03′E
81 I19 Bomet ◈ county W Kenya
59 I14 Bom Futuro Pará, N Brazil 06°27′S 54°44′W
159 Q15 Bomi var. Bowo, Zhamo. Xizang Zizhiqu, W China 29°43′N 96°12′E
79 N17 Bomili Orientale, NE Dem. Rep. Congo 01°45′N 27°01′E
59 N17 Bom Jesus da Lapa Bahia, E Brazil 13°16′S 43°23′W
60 Q8 Bom Jesus do Itabapoana Rio de Janeiro, SE Brazil 21°07′S 41°43′W
95 C15 Bømlafjorden fjord S Norway
95 B15 Bømlo island S Norway
123 Q12 Bomnak Amurskaya Oblast', SE Russian Federation 54°43′N 128°50′E
79 I17 Bomongo Equateur, NW Dem. Rep. Congo 01°22′N 18°21′E
61 K14 Bom Retiro Santa Catarina, S Brazil 27°45′S 49°31′W
79 L15 Bomu var. Mbomou, Mbomu, M'Bomu. ≈ Central African Republic/Dem. Rep. Congo
142 J3 Bonāb var. Benāb, Bunab. Āžarbāyjān-e Sharqī, N Iran 37°30′N 46°03′E
45 S9 Bonaire ◇ Dutch special municipality S Caribbean Sea
45 Q16 Bonaire island Lesser Antilles
39 U11 Bona, Mount ▲ Alaska, USA 61°22′N 141°45′W
183 Q12 Bonang Victoria, SE Australia 37°13′S 148°43′E
42 L7 Bonanza Región Autónoma Atlántico Norte, NE Nicaragua 13°59′N 84°30′W
37 O4 Bonanza Utah, W USA 40°01′N 109°12′W
45 O9 Bonao C Dominican Republic 18°55′N 70°25′W
180 L3 Bonaparte Archipelago island group Western Australia
32 K6 Bonaparte, Mount ▲ Washington, NW USA 48°47′N 119°07′W
39 N11 Bonasila Dome ▲ Alaska, USA 62°24′N 160°28′W
45 T15 Bonasse Trinidad, Trinidad and Tobago 10°02′N 61°48′W
15 X7 Bonaventure Québec, SE Canada 48°03′N 65°30′W
15 X7 Bonaventure ≈ Québec, SE Canada
13 V11 Bonavista Newfoundland, Newfoundland and Labrador, SE Canada 48°38′N 53°08′W
13 U11 Bonavista Bay inlet NW Atlantic Ocean
79 E19 Bonda Ogooué-Lolo, C Gabon 0°50′S 12°28′E
127 N6 Bondari Tambovskaya Oblast', W Russian Federation 52°58′N 42°02′E
106 G9 Bondeno Emilia-Romagna, C Italy 44°53′N 11°24′E
30 L4 Bond Falls Flowage ◎ Michigan, N USA
79 L16 Bondo Orientale, N Dem. Rep. Congo 03°52′N 23°41′E
171 N17 Bondokodi Pulau Sumba, S Indonesia 09°36′S 119°01′E
77 O15 Bondoukou E Ivory Coast 08°03′N 02°45′W
Bondoukoui/Bondoukuy see Boundoukui
169 T17 Bondowoso Jawa, C Indonesia 07°54′S 113°50′E
33 S14 Bondurant Wyoming, C USA 43°14′N 110°26′W
Bône see Annaba, Algeria
Bone see Watampone, Indonesia
30 I5 Bone Lake ◎ Wisconsin, N USA
171 P14 Bonelipu Pulau Buton, C Indonesia 04°42′S 123°09′E
171 O15 Bonerate, Kepulauan var. Macan. island group C Indonesia
29 O12 Bonesteel South Dakota, N USA 43°01′N 98°55′W
62 I8 Bonete, Cerro ▲ N Argentina 27°58′S 68°22′W
171 O14 Bone, Teluk bay Sulawesi, C Indonesia
108 D6 Bonfol Jura, NW Switzerland 47°28′N 07°08′E
153 U12 Bongaigaon Assam, NE India 26°30′N 90°31′E
79 K17 Bongandanga Equateur, NW Dem. Rep. Congo 01°28′N 21°03′E
78 L13 Bongo, Massif des var. Chaîne des Mongos. ▲▲ NE Central African Republic
78 G12 Bongor Mayo-Kébbi Est, SW Chad 10°18′N 15°20′E
77 N16 Bongouanou E Ivory Coast 06°39′N 04°12′W
167 V11 Bồng Sơn var. Hoai Nhơn. Bình Định, C Vietnam 14°28′N 109°00′E
25 U5 Bonham Texas, SW USA 33°36′N 96°12′W
Bonhard see Bonyhád
103 U6 Bonhomme, Col du pass NE France
103 Y16 Bonifacio Corse, France, C Mediterranean Sea 41°24′N 09°09′E
Bonifacio, Bocche de/Bonifacio, Bouches de see Bonifacio, Strait of
103 Y16 Bonifacio, Strait of Fr. Bouches de Bonifacio, It. Bocche di Bonifacio. strait C Mediterranean Sea
23 Q8 Bonifay Florida, SE USA 30°49′N 85°42′W
Bonin Islands see Ogasawara-shotō
192 H5 Bonin Trench undersea feature NW Pacific Ocean
23 W15 Bonita Springs Florida, SE USA 26°19′N 81°48′W
42 I5 Bonito, Pico ▲ N Honduras 15°33′N 86°55′W
101 E17 Bonn Nordrhein-Westfalen, W Germany 50°44′N 07°06′E
92 H11 Bonnåsjøen Nordland, C Norway 67°35′N 15°39′E
14 J12 Bonnechere Ontario, SE Canada 45°39′N 77°36′W
14 J12 Bonnechere ≈ Ontario, SE Canada
33 N7 Bonners Ferry Idaho, NW USA 48°41′N 116°19′W
27 R4 Bonner Springs Kansas, C USA 39°03′N 94°52′W
102 L6 Bonnétable Sarthe, NW France 48°09′N 00°24′E
27 X6 Bonne Terre Missouri, C USA 37°55′N 90°33′W
10 J5 Bonnet Plume ≈ Yukon, NW Canada
102 M6 Bonneval Eure-et-Loir, C France 48°12′N 01°23′E
103 T10 Bonneville Haute-Savoie, E France 46°05′N 06°25′E
36 J3 Bonneville Salt Flats salt flat Utah, W USA
77 U18 Bonny Rivers, S Nigeria 04°25′N 07°13′E
Bonny, Bight of see Biafra, Bight of
37 W4 Bonny Reservoir ⊠ Colorado, C USA
11 R14 Bonnyville Alberta, SW Canada 54°16′N 110°46′W
107 C18 Bono Sardegna, Italy, C Mediterranean Sea 40°24′N 09°01′E
Bononia see Vidin, Bulgaria
Bononia see Boulogne-sur-Mer, France
107 B18 Bonorva Sardegna, Italy, C Mediterranean Sea 40°27′N 08°46′E
30 M15 Bonpas Creek ≈ Illinois, N USA
190 I3 Bonriki Tarawa, W Kiribati 01°23′N 173°09′E
183 T4 Bonshaw New South Wales, SE Australia 29°06′S 151°15′E
76 I16 Bonthe SW Sierra Leone 07°32′N 12°30′W
171 N2 Bontoc Luzon, N Philippines 17°04′N 120°58′E
25 Y9 Bon Wier Texas, SW USA 30°43′N 93°40′W
111 J25 Bonyhád Ger. Bonhard. Tolna, S Hungary 46°20′N 18°31′E
Bonzabaai see Bonza Bay
83 J25 Bonza Bay Afr. Bonzabaai. Eastern Cape, S South Africa 32°58′S 27°58′E
182 D7 Bookabie South Australia 31°49′S 132°41′E
182 H6 Bookaloo South Australia 31°56′S 137°21′E
37 P5 Book Cliffs cliff Colorado/Utah, W USA
25 P1 Booker Texas, SW USA 36°27′N 100°32′W
76 K15 Boola SE Guinea 08°22′N 08°41′W
183 O8 Booligal New South Wales, SE Australia 33°56′S 144°54′E
99 G17 Boom Antwerpen, N Belgium 51°05′N 04°24′E
43 N6 Boom var. Boon. Región Autónoma Atlántico Norte, NE Nicaragua 14°52′N 83°36′W
183 S3 Boomi New South Wales, SE Australia 28°43′S 149°35′E
Boon see Boom
162 H9 Bööncagaan Nuur ◎ S Mongolia
29 V13 Boone Iowa, C USA 42°04′N 93°52′W
21 Q8 Boone North Carolina, SE USA 36°13′N 81°41′W
27 S11 Booneville Arkansas, C USA 35°09′N 93°57′W
21 N6 Booneville Kentucky, S USA 37°26′N 83°45′W
23 N2 Booneville Mississippi, S USA 34°39′N 88°34′W
21 V3 Boonsboro Maryland, NE USA 39°30′N 77°39′W
34 L6 Boonville California, W USA 38°58′N 123°21′W
31 N16 Boonville Indiana, N USA 38°03′N 87°16′W
27 U4 Boonville Missouri, C USA 38°58′N 92°43′W
18 I9 Boonville New York, NE USA 43°28′N 75°17′W
80 M12 Boorama Awdal, NW Somalia 09°58′N 43°15′E
183 O6 Booroondarra, Mount hill New South Wales, SE Australia
183 N9 Booroorban New South Wales, SE Australia 34°55′S 144°45′E
183 R9 Boorowa New South Wales, SE Australia 34°26′S 148°42′E
99 H17 Boortmeerbeek Vlaams Brabant, C Belgium 50°58′N 04°27′E
80 P11 Boosaaso var. Bandar Kassim, Bender Qaasim, Bosaso, It. Bender Cassim. Bari, N Somalia 11°26′N 49°37′E
19 Q8 Boothbay Harbor Maine, NE USA 43°52′N 69°35′W
Boothia Felix see Boothia Peninsula
9 N6 Boothia, Gulf of gulf Nunavut, NE Canada
9 N6 Boothia Peninsula prev. Boothia Felix. peninsula Nunavut, NE Canada
79 E18 Booué Ogooué-Ivindo, NE Gabon 0°03′S 11°58′E
101 J21 Bopfingen Baden-Württemberg, S Germany 48°51′N 10°21′E
101 F18 Boppard Rheinland-Pfalz, W Germany 50°13′N 07°36′E
62 M4 Boquerón off. Departamento de Boquerón. ◈ department W Paraguay
Boquerón, Departamento de see Boquerón
43 P15 Boquete var. Bajo Boquete. Chiriquí, W Panama 08°45′N 82°26′W
40 J6 Boquilla, Presa de la ⊠ N Mexico
40 L5 Boquillas var. Boquillas del Carmen. Coahuila, NE Mexico 29°10′N 102°55′W
Boquillas del Carmen see Boquillas
112 P12 Bor Serbia, E Serbia 44°05′N 22°07′E
81 F15 Bor Jonglei, E South Sudan 06°12′N 31°33′E
95 L20 Bor Jönköping, S Sweden 57°04′N 14°10′E
136 J15 Bor Niğde, S Turkey 37°49′N 35°00′E
191 S10 Bora-Bora island Îles Sous le Vent, W French Polynesia
167 Q9 Borabu Maha Sarakham, E Thailand 16°01′N 103°06′E
172 K4 Boraha, Nosy island E Madagascar
33 P13 Borah Peak ▲ Idaho, NW USA 44°21′N 113°53′W
145 U16 Boralday prev. Burunday. Almaty, SE Kazakhstan 43°21′N 76°48′E
145 Y11 Boran prev. Buran. Vostochnyy Kazakhstan, E Kazakhstan 48°00′N 85°09′E
144 G13 Borankul prev. Opornyy. Mangistau, SW Kazakhstan 46°09′N 54°32′E
95 J19 Borås Västra Götaland, S Sweden 57°44′N 12°55′E
143 N11 Borāzjān var. Borazjān. Būshehr, S Iran 29°19′N 51°12′E
Borazjān see Borāzjān
58 G13 Borba Amazonas, N Brazil 04°39′S 59°35′W
104 H11 Borba Évora, S Portugal 38°48′N 07°28′W
Borbetomagus see Worms
55 O7 Borbón Bolívar, E Venezuela 07°55′N 64°03′W
59 Q15 Borborema, Planalto da plateau NE Brazil
116 M14 Borcea, Braţul ≈ S Romania
Borchalo see Marneuli
195 R15 Borchgrevink Coast physical region Antarctica
137 Q11 Borçka Artvin, NE Turkey 41°24′N 41°38′E
98 N11 Borculo Gelderland, E Netherlands 52°07′N 06°31′E
182 G10 Borda, Cape headland South Australia 35°45′S 136°34′E
102 K13 Bordeaux anc. Burdigala. Gironde, SW France 44°49′N 00°33′W
11 T15 Borden Saskatchewan, S Canada 52°23′N 107°10′W
14 D8 Borden Lake ◎ Ontario, S Canada
9 N4 Borden Peninsula peninsula Baffin Island, Nunavut, NE Canada
182 K11 Bordertown South Australia 36°21′S 140°48′E
92 H2 Borðeyri Norðurland Vestra, NW Iceland 65°12′N 21°09′W
95 B18 Bordhoy Dan. Bordø. island NE Faroe Islands
106 B11 Bordighera Liguria, NW Italy 43°48′N 07°40′E
74 K5 Bordj-Bou-Arreridj var. Bordj Bou Arreridj, Bordj Bou Arréridj. N Algeria 36°02′N 04°49′E
74 L10 Bordj Omar Driss E Algeria 28°09′N 06°52′E
143 N13 Bord Khūn Hormozgān, S Iran
Bordø see Bordhoy
147 V7 Bordunskiy Chuyskaya Oblast', N Kyrgyzstan 42°37′N 75°31′E
95 M17 Borensberg Östergötland, S Sweden 58°33′N 15°15′E
Borgå see Porvoo
92 H3 Borgarnes Vesturland, W Iceland 64°33′N 21°55′W
93 G14 Børgefjell ▲▲ C Norway
98 O7 Borger Drenthe, NE Netherlands 52°54′N 06°48′E
25 N2 Borger Texas, SW USA 35°40′N 101°24′W
95 N20 Borgholm Kalmar, S Sweden 56°50′N 16°41′E
107 N22 Borgia Calabria, SW Italy 38°48′N 16°28′E
99 J18 Borgloon Limburg, NE Belgium 50°48′N 05°21′E
Borg Massif see Borgmassivet
195 P2 Borgmassivet Eng. Borg Massif. ▲▲ Antarctica
22 L9 Borgne, Lake ◎ Louisiana, S USA
106 C7 Borgomanero Piemonte, NE Italy 45°42′N 08°33′E
106 G10 Borgo Panigale ✈ (Bologna) Emilia-Romagna, N Italy 44°33′N 11°16′E
107 J15 Borgorose Lazio, C Italy 42°10′N 13°15′E
106 A9 Borgo San Dalmazzo Piemonte, N Italy 44°19′N 07°29′E
106 G11 Borgo San Lorenzo Toscana, C Italy 43°58′N 11°22′E
106 C7 Borgosesia Piemonte, NE Italy 45°41′N 08°21′E
106 E9 Borgo Val di Taro Emilia-Romagna, C Italy 44°29′N 09°48′E
106 G6 Borgo Valsugana Trentino-Alto Adige, N Italy 46°04′N 11°31′E
Borhoyn Tal see Dzamīn-Üüd
167 R8 Borikhan var. Borikhane. Bolikhamxai, C Laos 18°36′N 103°43′E
Borikhane see Borikhan
144 G8 Borili prev. Burlin. Zapadnyy Kazakhstan, NW Kazakhstan 51°25′N 52°42′E
Borislav see Boryslav
127 N8 Borisoglebsk Voronezhskaya Oblast', W Russian Federation 51°23′N 42°00′E
Borisov see Barysaw
Borisovgrad see Parvomay
Borispol' see Boryspil'
172 I3 Boriziny prev./Fr. Port-Bergé. Mahajanga, NW Madagascar 15°31′S 47°40′E
105 Q5 Borja Aragón, NE Spain 41°50′N 01°32′W
Borjas Blancas see Les Borges Blanques
137 S10 Borjomi Rus. Borzhomi. C Georgia 41°50′N 43°24′E
118 L12 Borkavichy Rus. Borkovichi. Vitsyebskaya Voblasts', N Belarus 55°40′N 28°20′E
101 H16 Borken Hessen, C Germany 51°01′N 09°16′E
101 E14 Borken Nordrhein-Westfalen, W Germany 51°51′N 06°51′E
92 H10 Borkenes Troms, N Norway 68°46′N 16°10′E
78 H7 Borkou off. Région du Borkou. ◈ region N Chad
Borkou, Région du see Borkou
Borkovichi see Borkavichy
100 E9 Borkum island NW Germany
81 K17 Bor, Lagh var. Lak Bor. dry watercourse NE Kenya
Bor, Lak see Bor, Lagh
95 M14 Borlänge Dalarna, C Sweden 60°29′N 15°25′E
106 C9 Bormida ≈ NW Italy
106 F6 Bormio Lombardia, N Italy 46°27′N 10°24′E
101 M16 Borna Sachsen, E Germany 51°07′N 12°30′E
98 O10 Borne Overijssel, E Netherlands 52°18′N 06°45′E
99 F17 Bornem Antwerpen, N Belgium 51°06′N 04°14′E
169 S10 Borneo island Brunei/Indonesia/Malaysia
101 E16 Bornheim Nordrhein-Westfalen, W Germany 50°46′N 06°58′E
95 L24 Bornholm ◈ county E Denmark
95 L24 Bornholm island E Denmark
77 Y13 Borno ◈ state NE Nigeria
104 K15 Bornos Andalucía, S Spain 36°50′N 05°42′W
162 L7 Bornuur Töv, C Mongolia 48°28′N 106°15′E
117 O4 Borodyanka Kyyivs'ka Oblast', N Ukraine 50°40′N 29°54′E
158 I5 Borohoro Shan ▲▲ NW China
77 O13 Boromo SW Burkina Faso 11°47′N 02°54′W
35 T13 Boron California, W USA 35°00′N 117°42′W
Borongo see Black Volta
Boron'ki see Baron'ki
Borosjenő see Ineu
Borossebes see Sebiş
76 L15 Borotou NW Ivory Coast 08°46′N 07°30′W
117 W6 Borova Kharkivs'ka Oblast', E Ukraine 49°22′N 37°39′E
114 H8 Borovan Vratsa, NW Bulgaria 43°25′N 23°45′E
124 I14 Borovichi Novgorodskaya Oblast', W Russian Federation 58°24′N 33°56′E
Borovlje see Ferlach
114 K8 Borovo Ruse, N Bulgaria 43°28′N 25°46′E
112 J9 Borovo Vukovar-Srijem, NE Croatia 45°22′N 18°57′E
Borovoye see Burabay
126 K4 Borovsk Kaluzhskaya Oblast', W Russian Federation 55°12′N 36°22′E
145 N7 Borovskoye Kostanay, N Kazakhstan 53°48′N 64°17′E
Borovukha see Baravukha
95 L23 Borrby Skåne, S Sweden 55°27′N 14°10′E
105 T9 Borriana var. Burriana. Valenciana, E Spain 39°54′N 00°05′W
181 R3 Borroloola Northern Territory, N Australia 16°09′S 136°18′E
116 F9 Borş Bihor, NW Romania 47°07′N 21°49′E
116 I9 Borşa Hung. Borsa. Maramureş, N Romania 47°40′N 24°37′E
116 J10 Borsec Ger. Bad Borseck, Hung. Borszék. Harghita, C Romania 46°58′N 25°32′E
92 K8 Børselv Lapp. Bissojohka. Finnmark, N Norway 70°18′N 25°35′E
113 L23 Borsh var. Borshi. Vlorë, S Albania 40°04′N 19°51′E
Borshchev see Borshchiv
116 K7 Borshchiv Pol. Borszczów, Rus. Borshchev. Ternopil's'ka Oblast', W Ukraine 48°48′N 26°00′E
Borshi see Borsh
111 L20 Borsod-Abaúj-Zemplén off. Borsod-Abaúj-Zemplén Megye. ◈ county NE Hungary
Borsod-Abaúj-Zemplén Megye see Borsod-Abaúj-Zemplén

228
◆ Country ● Country Capital ◇ Dependent Territory ○ Dependent Territory Capital ◈ Administrative Regions ✈ International Airport ▲ Mountain ▲▲ Mountain Range Volcano ≈ River ◎ Lake ⊠ Reservoir

BORSSELE – BRIDEL

99 *E15* **Borssele** Zeeland, SW Netherlands *51°26′N 03°45′E*
Borszczów *see* Borshchiv
Borszék *see* Borsec
Bortala *see* Bole
103 *O12* **Bort-les-Orgues** Corrèze, C France *45°28′N 02°31′E*
Bor u České Lípy *see* Nový Bor
Bor-Üdzüür *see* Altay
143 *N9* **Borūjen** Chahār MaḤall va Bakhtīārī, C Iran *32°N 51°09′E*
142 *L7* **Borūjerd** ***var.*** Burujird. Lorestān, W Iran *33°55′N 48°46′E*
116 *H6* **Boryslav** ***Pol.*** Borysław, ***Rus.*** Borislav. L'vivs'ka Oblast', NW Ukraine *49°18′N 23°28′E*
Borysław *see* Boryslav
117 *P4* **Boryspil'** ***Rus.*** Borispol'. Kyyivs'ka Oblast', N Ukraine *50°21′N 30°59′E*
117 *P4* **Boryspil'** ***Rus.*** Borispol'. ✈ (Kyyiv) Kyyivs'ka Oblast', N Ukraine *50°21′N 30°46′E*
Borzhomi *see* Borjomi
117 *R3* **Borzna** Chernihivs'ka Oblast', NE Ukraine *51°15′N 32°25′E*
123 *O14* **Borzya** Zabaykal'skiy Kray, S Russian Federation *50°18′N 116°24′E*
107 *B18* **Bosa** Sardegna, Italy, C Mediterranean Sea *40°18′N 08°28′E*
112 *F10* **Bosanska Dubica** ***var.*** Kozarska Dubica. ◈ Republika Srpska, NW Bosnia and Herzegovina
112 *G10* **Bosanska Gradiška** ***var.*** Gradiška. ◈ Republika Srpska, N Bosnia and Herzegovina
112 *F10* **Bosanska Kostajnica** ***var.*** Srpska Kostajnica. ◈ Republika Srpska, NW Bosnia and Herzegovina
112 *E11* **Bosanska Krupa** ***var.*** Krupa, Krupa na Uni. ◈ Federacija Bosne I Hercegovine, NW Bosnia and Herzegovina
112 *H10* **Bosanski Brod** ***var.*** Srpski Brod. ◈ Republika Srpska, N Bosnia and Herzegovina
112 *E10* **Bosanski Novi** ***var.*** Novi Grad. Republika Srpska, NW Bosnia and Herzegovina *45°03′N 16°23′E*
112 *E11* **Bosanski Petrovac** ***var.*** Petrovac. Federacija Bosne I Hercegovine, NW Bosnia and Herzegovina *44°34′N 16°21′E*
112 *I10* **Bosanski Šamac** ***var.*** Šamac. Republika Srpska, N Bosnia and Herzegovina *45°03′N 18°27′E*
112 *E12* **Bosansko Grahovo** ***var.*** Grahovo, Hrvatsko Grahovi. Federacija Bosne I Hercegovine, W Bosnia and Herzegovina *44°10′N 16°22′E*
Bosaso *see* Boosaaso
186 *B7* **Bosavi, Mount** ⛰ W Papua New Guinea *06°33′S 142°50′E*
160 *J14* **Bose** Guangxi Zhuangzu Zizhiqu, S China *23°55′N 106°32′E*
161 *Q5* **Boshan** Shandong, E China *36°32′N 117°47′E*
113 *P16* **Bosilegrad** ***prev.*** Bosiljgrad. Serbia, SE Serbia *42°30′N 22°30′E*
Bosiljgrad *see* Bosilegrad
Bösing *see* Pezinok
98 *H12* **Boskoop** Zuid-Holland, C Netherlands *52°04′N 04°40′E*
111 *G18* **Boskovice** ***Ger.*** Boskowitz. Jihomoravský Kraj, SE Czech Republic *49°30′N 16°39′E*
Boskowitz *see* Boskovice
112 *I10* **Bosna** ~ N Bosnia and Herzegovina
113 *G14* **Bosne I Hercegovine, Federacija** ◈ *republic* Bosnia and Herzegovina
112 *H12* **Bosnia and Herzegovina** ***off.*** Republic of Bosnia and Herzegovina. ◆ *republic* SE Europe
Bosnia and Herzegovina, Republic of *see* Bosnia and Herzegovina
79 *J16* **Bosobolo** Equateur, NW Dem. Rep. Congo *04°11′N 19°55′E*
165 *O14* **Bōsō-hantō** *peninsula* Honshū, S Japan
Bosora *see* Buşrá ash Shām
Bosphorus/Bosporus *see* İstanbul Boğazı
Bosporus Cimmerius *see* Kerch Strait
Bosporus Thracius *see* İstanbul Boğazı
Bosra *see* Buşrá ash Shām
79 *H14* **Bossangoa** Ouham, C Central African Republic *06°32′N 17°25′E*
Bossé Bangou *see* Bossey Bangou
79 *I15* **Bossembélé** Ombella-Mpoko, C Central African Republic *05°13′N 17°39′E*
79 *H15* **Bossentélé** Ouham-Pendé, W Central African Republic *05°36′N 16°37′E*
77 *R12* **Bossey Bangou** ***var.*** Bossé Bangou. Tillabéri, SW Niger *13°22′N 01°18′E*
22 *G5* **Bossier City** Louisiana, S USA *32°31′N 93°43′W*
83 *D20* **Bossiesvlei** Hardap, S Namibia *25°02′S 16°48′E*
77 *Y11* **Bosso** Diffa, SE Niger *13°42′N 13°18′E*
61 *F15* **Bossoroca** Rio Grande do Sul, S Brazil *28°45′S 54°54′W*
158 *J10* **Bostan** Xinjiang Uygur Zizhiqu, W China *41°20′N 83°15′E*
142 *K3* **Bostānābād** Āžarbāyjān-e Sharqī, N Iran *37°52′N 46°51′E*
158 *K6* **Bosten Hu** ***var.*** Bagrax Hu. ◎ NW China
97 *O18* **Boston** ***prev.*** St.Botolph's Town. E England, United Kingdom *52°59′N 00°01′W*
19 *O11* **Boston** *state capital* Massachusetts, NE USA *42°22′N 71°04′W*
146 *I9* **Bo'ston** ***Rus.*** Bustan. Qoraqalpog'iston Respublikasi, W Uzbekistan *41°49′N 60°51′E*
10 *M17* **Boston Bar** British Columbia, SW Canada *49°54′N 121°22′W*
27 *T10* **Boston Mountains** ▲▲ Arkansas, C USA
15 *P8* **Bostonnais** ~ Québec, SE Canada
Bostyn' *see* Bastyn'
112 *J10* **Bosut** ~ E Croatia
154 *C11* **Botād** Gujarāt, W India *22°12′N 71°44′E*
145 *S10* **Botakara** ***Kas.*** Botaqara; ***prev.*** Ul'yanovskiy. Karaganda, C Kazakhstan *50°05′N 73°45′E*
183 *T9* **Botany Bay** *inlet* New South Wales, SE Australia
Botaqara *see* Botakara
83 *G18* **Boteti** ***var.*** Botletle. ~ N Botswana
114 *J9* **Botev** ▲ C Bulgaria *42°45′N 24°57′E*
114 *H9* **Botevgrad** ***prev.*** Orkhaniye. Sofia, W Bulgaria *42°55′N 23°47′E*
93 *J16* **Bothnia, Gulf of** ***Fin.*** Pohjanlahti, ***Swe.*** Bottniska Viken. *gulf* N Baltic Sea
183 *P17* **Bothwell** Tasmania, SE Australia *42°24′S 147°01′E*
104 *H5* **Boticas** Vila Real, N Portugal *41°41′N 07°40′W*
55 *W10* **Boti-Pasi** Sipaliwini, C Suriname *04°15′N 55°27′W*
Botletle *see* Boteti
127 *P16* **Botlikh** Chechenskaya Respublika, SW Russian Federation *42°39′N 46°12′E*
117 *N10* **Botna** ~ E Moldova
116 *I9* **Botoşani** ***Hung.*** Botosány. Botoşani, NE Romania *47°44′N 26°41′E*
116 *K8* **Botoşani** ◈ *county* NE Romania
Botosány *see* Botoşani
147 *P12* **Bototog', Tizmasi** ***Rus.*** Khrebet Babatag. ▲▲ Tajikistan/Uzbekistan
161 *P4* **Botou** ***prev.*** Bozhen. Hebei, E China *38°09′N 116°37′E*
99 *M20* **Botrange** ▲ E Belgium *50°30′N 06°03′E*
107 *O21* **Botricello** Calabria, SW Italy *38°56′N 16°51′E*
83 *I23* **Botshabelo** Free State, C South Africa *29°15′S 26°51′E*
93 *J15* **Botsmark** Västerbotten, N Sweden *64°15′N 20°15′E*
83 *G19* **Botswana** ***off.*** Republic of Botswana. ◆ *republic* S Africa
Botswana, Republic of *see* Botswana
29 *N2* **Bottineau** North Dakota, N USA *48°50′N 100°28′W*
Bottniska Viken *see* Bothnia, Gulf of
60 *L9* **Botucatu** São Paulo, S Brazil *22°52′S 48°30′W*
76 *M16* **Bouaflé** C Ivory Coast *06°59′N 05°45′W*
77 *N16* **Bouaké** ***var.*** Bwake. C Ivory Coast *07°42′N 05°00′W*
79 *G14* **Bouar** Nana-Mambéré, W Central African Republic *05°58′N 15°38′E*
74 *H7* **Bouarfa** NE Morocco *32°33′N 01°54′W*
111 *B19* **Boubín** ▲ SW Czech Republic *49°00′N 13°51′E*
79 *I14* **Bouca** Ouham, W Central African Republic *06°57′N 18°18′E*
15 *T5* **Boucher** ~ Québec, SE Canada
103 *R15* **Bouches-du-Rhône** ◈ *department* SE France
74 *C9* **Bou Craa** ***var.*** Bu Craa. NW Western Sahara *26°32′N 12°52′W*
77 *O9* **Boû Djébéha** *oasis* C Mali
108 *C8* **Boudry** Neuchâtel, W Switzerland *46°57′N 06°46′E*
79 *F21* **Bouenza** ◈ *province* S Congo
186 *J7* **Bougainville** ***off.*** Autonomous Region of Bougainville; ***prev.*** North Solomons. ◈ *autonomous region* Bougainville, NE Papua New Guinea Oceania
Bougainville, Autonomous Region of *see* Bougainville
180 *L2* **Bougainville, Cape** *cape* Western Australia
65 *E24* **Bougainville, Cape** *headland* East Falkland, Falkland Islands *51°18′S 58°28′W*
Bougainville, Détroit de *see* Bougainville Strait
186 *J7* **Bougainville Island** *island* NE Papua New Guinea
186 *I8* **Bougainville Strait** *strait* N Solomon Islands
187 *Q13* **Bougainville Strait** ***Fr.*** Détroit de Bougainville. *strait* C Vanuatu
120 *I9* **Bougaroun, Cap** *headland* NE Algeria *37°07′N 06°18′E*
77 *R8* **Boughessa** Kidal, NE Mali *20°05′N 02°13′E*
Bougie *see* Béjaïa
76 *L13* **Bougouni** Sikasso, SW Mali *11°25′N 07°28′W*
99 *J24* **Bouillon** Luxembourg, SE Belgium *49°47′N 05°04′E*
74 *K5* **Bouira** ***var.*** Bouïra. N Algeria *36°22′N 03°55′E*
74 *D8* **Bou-Izakarn** SW Morocco *29°12′N 09°43′W*
74 *B9* **Boujdour** ***var.*** Bojador. W Western Sahara *26°06′N 14°29′W*
74 *G5* **Boukhalef** ✈ (Tanger) N Morocco *35°45′N 05°53′W*
Boukombé *see* Boukoumbé
77 *R14* **Boukoumbé** ***var.*** Boukombé. C Benin *10°13′N 01°09′E*
76 *G6* **Boû Lanouâr** Dakhlet Nouâdhibou, W Mauritania *21°17′N 16°29′W*
37 *T4* **Boulder** Colorado, C USA *40°02′N 105°18′W*
33 *R10* **Boulder** Montana, NW USA *46°14′N 112°07′W*
35 *X12* **Boulder City** Nevada, W USA *35°58′N 114°49′W*
181 *T7* **Boulia** Queensland, C Australia *23°02′S 139°58′E*
15 *N10* **Boullé** ~ Québec, SE Canada
102 *J9* **Boulogne** ~ NW France
Boulogne *see* Boulogne-sur-Mer
102 *L16* **Boulogne-sur-Gesse** Haute-Garonne, S France *43°18′N 00°38′E*
103 *N1* **Boulogne-sur-Mer** ***var.*** Boulogne; ***anc.*** Bononia, Gesoriacum, Gessoriacum. Pas-de-Calais, N France *50°43′N 01°36′E*
77 *Q12* **Boulsa** C Burkina Faso *12°41′N 00°29′W*
77 *W11* **Boultoum** Zinder, C Niger *14°43′N 10°22′E*
187 *Y14* **Bouma** Taveuni, N Fiji *16°49′S 179°50′W*
79 *G16* **Boumba** ~ SE Cameroon
76 *J9* **Boûmdeïd** ***var.*** Boumdeït. Assaba, S Mauritania *17°26′N 11°21′W*
Boumdeït *see* Boûmdeïd
115 *C17* **Boumistós** ▲ W Greece *38°48′N 20°59′E*
77 *O15* **Bouna** NE Ivory Coast *09°16′N 03°00′W*
19 *P4* **Boundary Bald Mountain** ▲ Maine, NE USA *45°45′N 70°10′W*
35 *S8* **Boundary Peak** ▲ Nevada, W USA *37°50′N 118°21′W*
76 *M14* **Boundiali** N Ivory Coast *09°30′N 06°31′W*
79 *G19* **Boundji** Cuvette, C Congo *01°05′S 15°18′E*
77 *O13* **Boundoukui** ***var.*** Bondoukui, Bondoukuy. W Burkina Faso *11°51′N 03°47′W*
36 *L2* **Bountiful** Utah, W USA *40°53′N 111°52′W*
Bounty Basin *see* Bounty Trough
191 *Q16* **Bounty Bay** *bay* Pitcairn Island, C Pacific Ocean
192 *L12* **Bounty Islands** *island group* S New Zealand
175 *Q13* **Bounty Trough** ***var.*** Bounty Basin. *undersea feature* S Pacific Ocean
187 *P17* **Bourail** Province Sud, C New Caledonia *21°35′S 165°29′E*
27 *V5* **Bourbeuse River** ~ Missouri, C USA
103 *Q9* **Bourbon-Lancy** Saône-et-Loire, C France *46°39′N 03°48′E*
31 *N11* **Bourbonnais** Illinois, N USA *41°08′N 87°52′W*
103 *O10* **Bourbonnais** *cultural region* C France
103 *S7* **Bourbonne-les-Bains** Haute-Marne, N France *48°00′N 05°43′E*
Bourbon Vendée *see* la Roche-sur-Yon
74 *M8* **Bourdj Messaouda** E Algeria *30°18′N 09°19′E*
77 *Q10* **Bourem** Gao, C Mali *16°56′N 00°21′W*
Bourg *see* Bourg-en-Bresse
103 *N11* **Bourganeuf** Creuse, C France *45°57′N 01°47′E*
Bourgas *see* Burgas
Bourge-en-Bresse *see* Bourg-en-Bresse
103 *S10* **Bourg-en-Bresse** ***var.*** Bourg, Bourge-en-Bresse. Ain, E France *46°12′N 05°13′E*
103 *O8* **Bourges** ***anc.*** Avaricum. Cher, C France *47°06′N 02°24′E*
103 *T11* **Bourget, Lac du** ◎ E France
103 *P8* **Bourgogne** ***Eng.*** Burgundy. ◈ *region* E France
103 *S11* **Bourgoin-Jallieu** Isère, E France *45°34′N 05°16′E*
103 *R14* **Bourg-St-Andéol** Ardèche, E France *44°24′N 04°36′E*
103 *U11* **Bourg-St-Maurice** Savoie, E France *45°37′N 06°49′E*
108 *C11* **Bourg St. Pierre** Valais, SW Switzerland *45°54′N 07°10′E*
76 *H8* **Boû Rjeimât** *well* W Mauritania
183 *P5* **Bourke** New South Wales, SE Australia *30°08′S 145°57′E*
97 *M24* **Bournemouth** S England, United Kingdom *50°43′N 01°54′W*
99 *M23* **Bourscheid** Diekirch, NE Luxembourg *49°55′N 06°04′E*
74 *K6* **Bou Saâda** ***var.*** Bou Saada. N Algeria *35°10′N 04°09′E*
36 *I13* **Bouse Wash** ~ Arizona, SW USA
103 *N10* **Boussac** Creuse, C France *46°20′N 02°12′E*
102 *M16* **Boussens** Haute-Garonne, S France *43°11′N 00°58′E*
78 *H12* **Bousso** ***prev.*** Fort-Bretonnet. Chari-Baguirmi, S Chad *10°32′N 16°45′E*
76 *H9* **Boutilimit** Trarza, SW Mauritania *17°33′N 14°42′W*
65 *D21* **Bouvet Island** ◇ *Norwegian dependency* S Atlantic Ocean
77 *U11* **Bouza** Tahoua, SW Niger *14°25′N 06°09′E*
109 *R10* **Bovec** ***Ger.*** Flitsch, ***It.*** Plezzo. NW Slovenia *46°21′N 13°33′E*
98 *J8* **Bovenkarspel** Noord-Holland, NW Netherlands *52°33′N 05°03′E*
29 *V5* **Bovey** Minnesota, N USA *47°18′N 93°25′W*
32 *M9* **Bovill** Idaho, NW USA *46°50′N 116°24′W*
24 *L4* **Bovina** Texas, SW USA *34°30′N 102°52′W*
107 *M17* **Bovino** Puglia, SE Italy *41°14′N 15°19′E*
61 *C17* **Bovril** Entre Ríos, E Argentina *31°24′S 59°25′W*
28 *L2* **Bowbells** North Dakota, N USA *48°48′N 102°15′W*
11 *Q16* **Bow City** Alberta, SW Canada *50°27′N 112°16′W*
29 *O8* **Bowdle** South Dakota, N USA *45°27′N 99°39′W*
181 *X6* **Bowen** Queensland, NE Australia *20°S 148°10′E*
192 *L2* **Bowers Ridge** *undersea feature* S Bering Sea *50°00′N 180°00′W*
25 *S5* **Bowie** Texas, SW USA *33°33′N 97°51′W*
11 *R17* **Bow Island** Alberta, SW Canada *49°53′N 111°24′W*
Bowkān *see* Būkān
20 *J7* **Bowling Green** Kentucky, S USA *37°00′N 86°29′W*
27 *V3* **Bowling Green** Missouri, C USA *39°21′N 91°11′W*
31 *R11* **Bowling Green** Ohio, N USA *41°22′N 83°40′W*
21 *W5* **Bowling Green** Virginia, NE USA *38°02′N 77°22′W*
28 *J6* **Bowman** North Dakota, N USA *46°11′N 103°26′W*
9 *Q7* **Bowman Bay** *bay* NW Atlantic Ocean
194 *I5* **Bowman Coast** *physical region* Antarctica
28 *J7* **Bowman-Haley Lake** ▣ North Dakota, N USA
195 *Z11* **Bowman Island** *island* Antarctica
Bowo *see* Bomi
183 *S9* **Bowral** New South Wales, SE Australia *34°28′S 150°52′E*
186 *E8* **Bowutu Mountains** ▲▲ C Papua New Guinea
83 *I16* **Bowwood** Southern, S Zambia *17°09′S 26°16′E*
28 *I12* **Box Butte Reservoir** ▣ Nebraska, C USA
28 *J10* **Box Elder** South Dakota, N USA *44°06′N 103°04′W*
95 *M18* **Boxholm** Östergötland, S Sweden *58°12′N 15°05′E*
Bo Xian/Boxian *see* Bozhou
161 *Q4* **Boxing** Shandong, E China *37°06′N 118°05′E*
99 *L14* **Boxmeer** Noord-Brabant, SE Netherlands *51°39′N 05°57′E*
99 *J14* **Boxtel** Noord-Brabant, S Netherlands *51°36′N 05°20′E*
136 *J10* **Boyabat** Sinop, N Turkey *41°27′N 34°45′E*
54 *F9* **Boyacá** ***off.*** Departamento de Boyacá. ◈ *province* C Colombia
Boyacá, Departamento de *see* Boyacá
117 *O4* **Boyarka** Kyyivs'ka Oblast', N Ukraine *50°19′N 30°20′E*
22 *H7* **Boyce** Louisiana, S USA *31°23′N 92°40′W*
114 *H8* **Boychinovtsi** Montana, NW Bulgaria *43°28′N 23°20′E*
33 *U11* **Boyd** Montana, NW USA *45°27′N 109°03′W*
25 *S6* **Boyd** Texas, SW USA *33°01′N 97°33′W*
21 *V8* **Boydton** Virginia, NE USA *36°40′N 78°26′W*
Boyer Ahmadī va Kohkīlūyeh *see* Kohgīlūyeh va Bowyer Aḩmad
29 *T13* **Boyer River** ~ Iowa, C USA
21 *W8* **Boykins** Virginia, NE USA *36°35′N 77°11′W*
11 *Q13* **Boyle** Alberta, SW Canada *54°38′N 112°45′W*
97 *D16* **Boyle** ***Ir.*** Mainistirna Búille. C Ireland *53°58′N 08°18′W*
97 *F17* **Boyne** ***Ir.*** An Bhóinn. ~ E Ireland
31 *Q5* **Boyne City** Michigan, N USA *45°13′N 85°00′W*
23 *Z14* **Boynton Beach** Florida, SE USA *26°31′N 80°04′W*
147 *O13* **Boysun** ***Rus.*** Baysun. Surkhondaryo Viloyati, S Uzbekistan *38°14′N 67°08′E*
Bozau *see* Întorsura Buzăului
136 *B12* **Bozcaada** *Island* Çanakkale, NW Turkey
136 *C14* **Boz Dağları** ▲▲ W Turkey
33 *S11* **Bozeman** Montana, NW USA *45°40′N 111°02′W*
Bozen *see* Bolzano
79 *J16* **Bozene** Equateur, NW Dem. Rep. Congo *02°56′N 19°15′E*
Bozhen *see* Botou
161 *P7* **Bozhou** ***var.*** Boxian, Bo Xian. Anhui, E China *33°46′N 115°44′E*
136 *H16* **Bozkır** Konya, S Turkey *37°10′N 32°15′E*
136 *K13* **Bozok Yaylasī** *plateau* C Turkey
79 *H14* **Bozoum** Ouham-Pendé, W Central African Republic *06°17′N 16°26′E*
137 *N16* **Bozova** Sanlıurfa, S Turkey *37°23′N 38°33′E*
Bozrah *see* Buşrá ash Shām
136 *E12* **Bozüyük** Bilecik, NW Turkey *39°55′N 30°02′E*
106 *B9* **Bra** Piemonte, NW Italy *44°42′N 07°51′E*
194 *G4* **Brabant Island** *island* Antarctica
99 *I20* **Brabant Walloon** ◈ *province* C Belgium
113 *F15* **Brač** ***var.*** Brach, ***It.*** Brazza; ***anc.*** Brattia. *island* S Croatia
Bracara Augusta *see* Braga
107 *H15* **Bracciano** Lazio, C Italy *42°04′N 12°12′E*
107 *H14* **Bracciano, Lago di** ◎ C Italy
14 *H13* **Bracebridge** Ontario, S Canada *45°02′N 79°19′W*
Brach *see* Brač
93 *G17* **Bräcke** Jämtland, C Sweden *62°43′N 15°30′E*
25 *P12* **Brackettville** Texas, SW USA *29°19′N 100°27′W*
97 *N22* **Bracknell** S England, United Kingdom *51°26′N 00°46′W*
61 *K14* **Braço do Norte** Santa Catarina, S Brazil *28°16′S 49°11′W*
116 *G11* **Brad** ***Hung.*** Brád. Hunedoara, SW Romania *45°52′N 23°00′E*
107 *N18* **Bradano** ~ S Italy
23 *V13* **Bradenton** Florida, SE USA *27°30′N 82°34′W*
45 *V10* **Brades** ***de facto capital, de jure capital. Previous capital destroyed by volcano in 1995*** Plymouth. ● (Montserrat) SW Montserrat *16°45′N 62°12′W*
14 *H14* **Bradford** Ontario, S Canada *44°09′N 79°34′W*
97 *L17* **Bradford** N England, United Kingdom *53°48′N 01°45′W*
27 *W10* **Bradford** Arkansas, C USA *35°25′N 91°27′W*
18 *D12* **Bradford** Pennsylvania, NE USA *41°57′N 78°38′W*
27 *T15* **Bradley** Arkansas, C USA *33°06′N 93°39′W*
25 *P7* **Bradshaw** Texas, SW USA *32°06′N 99°52′W*
25 *Q9* **Brady** Texas, SW USA *31°08′N 99°22′W*
25 *Q9* **Brady Creek** ~ Texas, SW USA
95 *G22* **Brædstrup** Syddanmark, C Denmark *55°58′N 09°38′E*
96 *J10* **Braemar** NE Scotland, United Kingdom *57°12′N 02°52′W*
116 *K8* **Brăeşti** Botoşani, NW Romania *47°50′N 26°26′E*
104 *G5* **Braga** ***anc.*** Bracara Augusta. Braga, NW Portugal *41°32′N 08°26′W*
104 *G5* **Braga** ◈ *district* N Portugal
116 *J15* **Bragadiru** Teleorman, S Romania *43°43′N 25°32′E*
61 *C20* **Bragado** Buenos Aires, E Argentina *35°10′S 60°29′W*
104 *J5* **Bragança** ***Eng.*** Braganza; ***anc.*** Julio Briga. Bragança, NE Portugal *41°47′N 06°46′W*
104 *I5* **Bragança** ◈ *district* N Portugal
60 *N9* **Bragança Paulista** São Paulo, S Brazil *22°55′S 46°30′W*
Braganza *see* Bragança
Bragin *see* Brahin
29 *V7* **Braham** Minnesota, N USA *45°43′N 93°10′W*
Brahe *see* Brda
Brahestad *see* Raahe
119 *O20* **Brahin** ***Rus.*** Bragin. Homyel'skaya Voblasts', SE Belarus *51°47′N 30°16′E*
153 *U15* **Brahmanbaria** Chittagong, E Bangladesh *23°58′N 91°04′E*
154 *O12* **Brāhmani** ~ E India
154 *N13* **Brahmapur** Odisha, E India *19°21′N 84°51′E*
129 *S10* **Brahmaputra** ***var.*** Padma, Tsangpo, ***Ben.*** Jamuna, ***Chin.*** Yarlung Zangbo Jiang, ***Ind.*** Bramaputra, Dihang, Siang. ~ S Asia
97 *H19* **Braich y Pwll** *headland* NW Wales, United Kingdom *52°47′N 04°46′W*
183 *R10* **Braidwood** New South Wales, SE Australia *35°26′S 149°48′E*
30 *M11* **Braidwood** Illinois, N USA *41°16′N 88°12′W*
116 *M13* **Brăila** Brăila, E Romania *45°17′N 27°57′E*
116 *L13* **Brăila** ◈ *county* SE Romania
99 *G19* **Braine-l'Alleud** Brabant Walloon, C Belgium *50°41′N 04°22′E*
99 *F19* **Braine-le-Comte** Hainaut, SW Belgium *50°37′N 04°08′E*
29 *U6* **Brainerd** Minnesota, N USA *46°22′N 94°10′W*
99 *J19* **Braives** Liège, E Belgium *50°37′N 05°09′E*
83 *H23* **Brak** ~ C South Africa
Brak *see* Birāk
99 *E18* **Brakel** Oost-Vlaanderen, SW Belgium *50°50′N 03°48′E*
98 *J13* **Brakel** Gelderland, C Netherlands *51°49′N 05°05′E*
76 *H9* **Brakna** ◈ *region* S Mauritania
95 *J17* **Brålanda** Västra Götaland, S Sweden *58°32′N 12°18′E*
Bramaputra *see* Brahmaputra
95 *F23* **Bramming** Syddtjylland, W Denmark *55°28′N 08°42′E*
14 *G15* **Brampton** Ontario, S Canada *43°42′N 79°46′W*
100 *F12* **Bramsche** Niedersachsen, NW Germany *52°25′N 07°58′E*
116 *J12* **Bran** ***Ger.*** Törzburg, ***Hung.*** Törcsvár. Braşov, S Romania *45°31′N 25°23′E*
29 *W8* **Branch** Minnesota, N USA *45°29′N 92°57′W*
21 *R14* **Branchville** South Carolina, SE USA *33°15′N 80°49′W*
47 *Y6* **Branco, Cabo** *headland* E Brazil *07°08′S 34°45′W*
58 *F11* **Branco, Rio** ~ N Brazil
108 *J8* **Brand** Vorarlberg, W Austria *47°07′N 09°45′E*
83 *B18* **Brandberg** ▲ NW Namibia *21°20′S 14°22′E*
95 *H14* **Brandbu** Oppland, S Norway *60°24′N 10°30′E*
95 *F22* **Brande** Midtjylland, W Denmark *55°57′N 09°08′E*
Brandebourg *see* Brandenburg
100 *M12* **Brandenburg** ***var.*** Brandenburg an der Havel. Brandenburg, NE Germany *52°25′N 12°34′E*
20 *K5* **Brandenburg** Kentucky, S USA *38°00′N 86°11′W*
100 *N12* **Brandenburg** ***off.*** Freie und Hansestadt Hamburg, ***Fr.*** Brandebourg. ◈ *state* NE Germany
Brandenburg an der Havel *see* Brandenburg
83 *I23* **Brandfort** Free State, C South Africa *28°42′S 26°28′E*
11 *W16* **Brandon** Manitoba, S Canada *49°50′N 99°57′W*
23 *V12* **Brandon** Florida, SE USA *27°56′N 82°17′W*
22 *L6* **Brandon** Mississippi, S USA *32°16′N 90°01′W*
97 *A20* **Brandon Mountain** ***Ir.*** Cnoc Bréanainn. ▲ SW Ireland *52°13′N 10°16′W*
Brandsen *see* Coronel Brandsen
95 *I14* **Brandval** Hedmark, S Norway *60°18′N 12°01′E*
83 *F24* **Brandvlei** Northern Cape, W South Africa *30°27′S 20°29′E*
23 *U9* **Branford** Florida, SE USA *29°57′N 82°54′W*
110 *K7* **Braniewo** ***Ger.*** Braunsberg. Warmińsko-mazurskie, N Poland *54°24′N 19°50′E*
194 *H3* **Bransfield Strait** *strait* Antarctica
37 *U8* **Branson** Colorado, C USA *37°01′N 103°52′W*
27 *T8* **Branson** Missouri, C USA *36°38′N 93°13′W*
14 *G16* **Brantford** Ontario, S Canada *43°09′N 80°17′W*
102 *L12* **Brantôme** Dordogne, SW France *45°21′N 00°37′E*
182 *L12* **Branxholme** Victoria, SE Australia *37°51′S 141°48′E*
Brasil *see* Brazil
59 *C16* **Brasiléia** Acre, W Brazil *10°59′S 48°45′W*
59 *K18* **Brasília** ● (Brazil) Distrito Federal, C Brazil *15°45′S 47°57′W*
Brasil, Repùblica Federativa do *see* Brazil
Braslav *see* Braslaw
118 *J12* **Braslaw** ***Pol.*** Brasław, ***Rus.*** Braslav. Vitsyebskaya Voblasts', N Belarus *55°38′N 27°02′E*
116 *J12* **Braşov** ***Ger.*** Kronstadt, ***Hung.*** Brassó; ***prev.*** Oraşul Stalin. Braşov, C Romania *45°40′N 25°35′E*
116 *I12* **Braşov** ◈ *county* C Romania
77 *U18* **Brass** Bayelsa, S Nigeria *04°19′N 06°21′E*
99 *H16* **Brasschaat** ***var.*** Brasschaet. Antwerpen, N Belgium *51°17′N 04°30′E*
Brasschaet *see* Brasschaat
169 *V8* **Brassey, Banjaran** ***var.*** Brassey Range. ▲▲ East Malaysia
Brassey Range *see* Brassey, Banjaran
Brassó *see* Braşov
23 *T1* **Brasstown Bald** ▲ Georgia, SE USA *34°52′N 83°48′W*
113 *K22* **Brataj** Vlorë, SW Albania *40°18′N 19°37′E*
114 *J10* **Bratan** ***var.*** Morozov. ▲ C Bulgaria *42°31′N 25°08′E*
114 *H10* **Bratia** ***var.*** Bratiya. C Bulgaria *42°36′N 24°08′E*
111 *F21* **Bratislava** ***Ger.*** Pressburg, ***Hung.*** Pozsony. ● (Slovakia) Bratislavský Kraj, W Slovakia *48°10′N 17°10′E*
111 *H21* **Bratislavský Kraj** ◈ *region* W Slovakia
Bratiya *see* Bratia
122 *M12* **Bratsk** Irkutskaya Oblast', C Russian Federation *56°20′N 101°50′E*
117 *Q8* **Brats'ke** Mykolayivs'ka Oblast', S Ukraine *47°52′N 31°34′E*
122 *M13* **Bratskoye Vodokhranilishche** ***Eng.*** Bratsk Reservoir. ▣ S Russian Federation
Bratsk Reservoir *see* Bratskoye Vodokhranilishche
Brattia *see* Brač
94 *D9* **Brattvåg** Møre og Romsdal, S Norway *62°36′N 06°21′E*
112 *K12* **Bratunac** ◈ Republika Srpska, E Bosnia and Herzegovina
114 *J10* **Bratya Daskalovi** ***prev.*** Grozdovo. Stara Zagora, C Bulgaria *42°13′N 25°21′E*
109 *U2* **Braunau** ~ N Austria
Braunau *see* Braunau am Inn
109 *Q4* **Braunau am Inn** ***var.*** Braunau. Oberösterreich, N Austria *48°16′N 13°03′E*
Braunsberg *see* Braniewo
100 *J13* **Braunschweig** ***Eng./Fr.*** Brunswick. Niedersachsen, N Germany *52°16′N 10°32′E*
Brava *see* Baraawe
105 *Y6* **Brava, Costa** *coastal region* NE Spain
43 *V16* **Brava, Punta** *headland* E Panama *08°21′N 78°22′W*
95 *N17* **Bråviken** *inlet* S Sweden
56 *B10* **Bravo, Cerro** ▲ N Peru *05°33′S 79°10′W*
Bravo del Norte, Río/Bravo, Río *see* Grande, Rio
35 *X17* **Brawley** California, W USA *32°58′N 115°31′W*
97 *G18* **Bray** ***Ir.*** Bré. E Ireland *53°12′N 06°06′W*
59 *G16* **Brazil** ***off.*** Federative Republic of Brazil, ***Port.*** Repùblica Federativa do Brasil, ***Sp.*** Brasil; ***prev.*** United States of Brazil. ◆ *federal republic* South America
65 *K15* **Brazil Basin** ***var.*** Brazilian Basin, Brazil'skaya Kotlovina. *undersea feature* W Atlantic Ocean *15°00′S 25°00′W*
Brazil, Federative Republic of *see* Brazil
Brazilian Basin *see* Brazil Basin
Brazilian Highlands *see* Central, Planalto
Brazil'skaya Kotlovina *see* Brazil Basin
Brazil, United States of *see* Brazil
25 *U10* **Brazos River** ~ Texas, SW USA
Brazza *see* Brač
79 *G21* **Brazzaville** ● (Congo) Capital District, S Congo *04°14′S 15°14′E*
79 *G21* **Brazzaville** ✈ Pool, S Congo *04°15′S 15°15′E*
112 *J11* **Brčko** Brčko Distrikt, NE Bosnia and Herzegovina *44°52′N 18°49′E*
112 *J11* **Brčko Distrikt** ◈ *self-governing district* Bosnia and Herzegovina
110 *H8* **Brda** ***Ger.*** Brahe. ~ N Poland
Bré *see* Bray
185 *A23* **Breaksea Sound** *sound* South Island, New Zealand
184 *L4* **Bream Bay** *bay* North Island, New Zealand
184 *L4* **Bream Head** *headland* North Island, New Zealand *35°51′S 174°35′E*
Bréanainn, Cnoc *see* Brandon Mountain
45 *S6* **Brea, Punta** *headland* W Puerto Rico *17°56′N 66°55′W*
22 *I9* **Breaux Bridge** Louisiana, S USA *30°16′N 91°54′W*
116 *J13* **Breaza** Prahova, SE Romania *45°13′N 25°38′E*
169 *P16* **Brebes** Jawa, C Indonesia *06°54′S 109°00′E*
96 *K10* **Brechin** E Scotland, United Kingdom *56°45′N 02°38′W*
99 *H15* **Brecht** Antwerpen, N Belgium *51°21′N 04°33′E*
37 *S4* **Breckenridge** Colorado, C USA *39°28′N 106°02′W*
29 *R6* **Breckenridge** Minnesota, N USA *46°15′N 96°35′W*
25 *R6* **Breckenridge** Texas, SW USA *32°45′N 98°56′W*
97 *J21* **Brecknock** *cultural region* SE Wales, United Kingdom
63 *G25* **Brecknock, Península** *headland* S Chile *54°39′S 71°48′W*
111 *G19* **Břeclav** ***Ger.*** Lundenburg. Jihomoravský Kraj, SE Czech Republic *49°05′N 16°51′E*
97 *J21* **Brecon** E Wales, United Kingdom *51°58′N 03°26′W*
97 *J21* **Brecon Beacons** ▲▲ S Wales, United Kingdom
99 *I14* **Breda** Noord-Brabant, S Netherlands *51°35′N 04°46′E*
95 *K20* **Bredaryd** Jönköping, S Sweden *57°10′N 13°45′E*
83 *F26* **Bredasdorp** Western Cape, SW South Africa *34°32′S 20°02′E*
93 *H16* **Bredbyn** Västernorrland, N Sweden *63°28′N 18°04′E*
122 *F11* **Bredy** Chelyabinskaya Oblast', C Russian Federation *52°23′N 60°24′E*
99 *K17* **Bree** Limburg, NE Belgium *51°08′N 05°36′E*
67 *T15* **Breede** ~ S South Africa
98 *I7* **Breezand** Noord-Holland, NW Netherlands *52°52′N 04°47′E*
113 *P18* **Bregalnica** ~ E FYR Macedonia
108 *I6* **Bregenz** ***anc.*** Brigantium. Vorarlberg, W Austria *47°31′N 09°46′E*
108 *J7* **Bregenzer Wald** ▲▲ W Austria
114 *F6* **Bregovo** Vidin, NW Bulgaria *44°07′N 22°40′E*
102 *H5* **Bréhat, Île de** *island* NW France
92 *H2* **Breiðafjörður** *bay* W Iceland
92 *L3* **Breiðdalsvík** Austurland, E Iceland *64°48′N 14°02′W*
108 *H9* **Breil** ***Ger.*** Brigels. Graubünden, S Switzerland *46°46′N 09°04′E*
92 *J8* **Breivikbotn** Finnmark, N Norway *70°36′N 22°19′E*
94 *I9* **Brekken** Sør-Trøndelag, S Norway *62°39′N 11°49′E*
94 *G7* **Brekstad** Sør-Trøndelag, S Norway *63°42′N 09°40′E*
94 *B10* **Bremangerlandet** *island* S Norway
Brême *see* Bremen
100 *H11* **Bremen** ***Fr.*** Brême. Bremen, NW Germany *53°06′N 08°48′E*
23 *R3* **Bremen** Georgia, SE USA *33°43′N 85°09′W*
31 *O11* **Bremen** Indiana, N USA *41°24′N 86°07′W*
100 *H10* **Bremen** ***off.*** Freie Hansestadt Bremen, ***Fr.*** Brême. ◈ *state* N Germany
100 *G9* **Bremerhaven** Bremen, NW Germany *53°33′N 08°35′E*
Bremersdorp *see* Manzini
32 *G8* **Bremerton** Washington, NW USA *47°34′N 122°37′W*
100 *H10* **Bremervörde** Niedersachsen, NW Germany *53°29′N 09°06′E*
25 *U9* **Bremond** Texas, SW USA *31°10′N 96°40′W*
25 *U10* **Brenham** Texas, SW USA *30°09′N 96°24′W*
108 *M8* **Brenner** Tirol, W Austria *47°10′N 11°51′E*
Brenner, Col du/Brennero, Passo del *see* Brenner Pass
108 *M8* **Brenner Pass** ***var.*** Brenner Sattel, ***Fr.*** Col du Brenner, ***Ger.*** Brennerpass, ***It.*** Passo del Brennero. *pass* Austria/Italy
Brennerpass *see* Brenner Pass
Brenner Sattel *see* Brenner Pass
108 *G10* **Brenno** ~ SW Switzerland
106 *F7* **Breno** Lombardia, N Italy *45°58′N 10°18′E*
23 *O5* **Brent** Alabama, S USA *32°54′N 87°10′W*
106 *H7* **Brenta** ~ NE Italy
97 *P21* **Brentwood** E England, United Kingdom *51°38′N 00°21′E*
18 *L14* **Brentwood** Long Island, New York, NE USA *40°46′N 73°12′W*
106 *F7* **Brescia** ***anc.*** Brixia. Lombardia, N Italy *45°33′N 10°13′E*
99 *D15* **Breskens** Zeeland, SW Netherlands *51°24′N 03°33′E*
Breslau *see* Wrocław
106 *H5* **Bressanone** ***Ger.*** Brixen. Trentino-Alto Adige, N Italy *46°44′N 11°41′E*
96 *M2* **Bressay** *island* NE Scotland, United Kingdom
102 *K9* **Bressuire** Deux-Sèvres, W France *46°50′N 00°29′W*
119 *F20* **Brest** ***Pol.*** Brześć nad Bugiem, ***Rus.*** Brest-Litovsk; ***prev.*** Brześć Litewski. Brestskaya Voblasts', SW Belarus *52°06′N 23°42′E*
102 *F5* **Brest** Finistère, NW France *48°24′N 04°31′W*
Brest-Litovsk *see* Brest
112 *A10* **Brestova** Istra, NW Croatia *45°09′N 14°13′E*
Brestskaya Oblast' *see* Brestskaya Voblasts'
119 *G19* **Brestskaya Voblasts'** ***prev. Rus.*** Brestskaya Oblast'. ◈ *province* SW Belarus
102 *G6* **Bretagne** ***Eng.*** Brittany, ***Lat.*** Britannia Minor. ◈ *region* NW France
116 *G12* **Bretea-Română** ***Hung.*** Oláhbrettye; ***prev.*** Bretea-Romînă. Hunedoara, W Romania *45°39′N 23°00′E*
Bretea-Romînă *see* Bretea-Română
103 *O3* **Breteuil** Oise, N France *49°37′N 02°18′E*
102 *I10* **Breton, Pertuis** *inlet* W France
22 *L10* **Breton Sound** *sound* Louisiana, S USA
184 *K2* **Brett, Cape** *headland* North Island, New Zealand *35°11′S 174°21′E*
101 *G21* **Bretten** Baden-Württemberg, SW Germany *49°01′N 08°42′E*
99 *K15* **Breugel** Noord-Brabant, S Netherlands *51°30′N 05°30′E*
106 *B6* **Breuil-Cervinia** ***It.*** Cervinia. Valle d'Aosta, NW Italy *45°57′N 07°37′E*
98 *I11* **Breukelen** Utrecht, C Netherlands *52°11′N 05°01′E*
21 *P10* **Brevard** North Carolina, SE USA *35°13′N 82°46′W*
38 *L9* **Brevig Mission** Alaska, USA *65°19′N 166°29′W*
95 *G16* **Brevik** Telemark, S Norway *59°05′N 09°42′E*
183 *P5* **Brewarrina** New South Wales, SE Australia *30°01′S 146°50′E*
19 *R6* **Brewer** Maine, NE USA *44°46′N 68°44′W*
29 *T11* **Brewster** Minnesota, N USA *43°43′N 95°28′W*
29 *N14* **Brewster** Nebraska, C USA *41°57′N 99°52′W*
31 *U12* **Brewster** Ohio, N USA *40°42′N 81°36′W*
Brewster, Kap *see* Kangikajik
183 *O8* **Brewster, Lake** ◎ New South Wales, SE Australia
23 *P7* **Brewton** Alabama, S USA *31°06′N 87°04′W*
Brezhnev *see* Naberezhnyye Chelny
109 *W12* **Brežice** ***Ger.*** Rann. E Slovenia *45°54′N 15°35′E*
114 *G9* **Breznik** Pernik, W Bulgaria *42°45′N 22°54′E*
111 *K19* **Brezno** ***Ger.*** Bries, Briesen, ***Hung.*** Breznóbánya; ***prev.*** Brezno nad Hronom. Banskobystrický Kraj, C Slovakia *48°49′N 19°40′E*
Breznóbánya/Brezno nad Hronom *see* Brezno
116 *I12* **Brezoi** Vâlcea, SW Romania *45°18′N 24°15′E*
114 *J10* **Brezovo** ***prev.*** Abrashlare. Plovdiv, C Bulgaria *42°19′N 25°05′E*
79 *K14* **Bria** Haute-Kotto, C Central African Republic *06°30′N 22°00′E*
103 *U13* **Briançon** ***anc.*** Brigantio. Hautes-Alpes, SE France *44°55′N 06°37′E*
36 *K7* **Brian Head** ▲ Utah, W USA *37°40′N 112°49′W*
103 *O7* **Briare** Loiret, C France *47°35′N 02°46′E*
183 *V2* **Bribie Island** *island* Queensland, E Australia
43 *O14* **Bribrí** Limón, E Costa Rica *09°37′N 82°51′W*
116 *L8* **Briceni** ***var.*** Brinceni, ***Rus.*** Brichany. N Moldova *48°21′N 27°02′E*
Bricgstow *see* Bristol
Brichany *see* Briceni
99 *M24* **Bridel** Luxembourg, C Luxembourg *49°40′N 06°03′E*

◆ Country
● Country Capital
◇ Dependent Territory
○ Dependent Territory Capital
◈ Administrative Regions
✈ International Airport
▲ Mountain
▲▲ Mountain Range
⛰ Volcano
~ River
◎ Lake
▣ Reservoir

BRIDGEND – BULL ISLAND

97 J22 **Bridgend** S Wales, United Kingdom *51°30′N 03°37′W*
14 I14 **Bridgenorth** Ontario, SE Canada *44°21′N 78°22′W*
23 Q1 **Bridgeport** Alabama, S USA *34°57′N 85°42′W*
35 R8 **Bridgeport** California, W USA *38°14′N 119°15′W*
18 L13 **Bridgeport** Connecticut, NE USA *41°10′N 73°12′W*
31 N15 **Bridgeport** Illinois, N USA *38°42′N 87°45′W*
28 J14 **Bridgeport** Nebraska, C USA *41°37′N 103°07′W*
25 S6 **Bridgeport** Texas, SW USA *33°12′N 97°45′W*
21 S3 **Bridgeport** West Virginia, NE USA *39°17′N 80°15′W*
25 S5 **Bridgeport, Lake** ▣ Texas, SW USA
33 U11 **Bridger** Montana, NW USA *45°16′N 108°55′W*
18 J17 **Bridgeton** New Jersey, NE USA *39°24′N 75°10′W*
180 J14 **Bridgetown** Western Australia *34°01′S 116°07′E*
45 Y14 **Bridgetown** ● (Barbados) SW Barbados *13°05′N 59°36′W*
183 P17 **Bridgewater** Tasmania, SE Australia *42°47′S 147°15′E*
13 P16 **Bridgewater** Nova Scotia, SE Canada *44°19′N 64°30′W*
19 P12 **Bridgewater** Massachusetts, NE USA *41°59′N 70°58′W*
29 Q11 **Bridgewater** South Dakota, N USA *43°33′N 97°30′W*
21 U5 **Bridgewater** Virginia, NE USA *38°22′N 78°58′W*
19 P8 **Bridgton** Maine, NE USA *44°04′N 70°43′W*
97 K23 **Bridgwater** SW England, United Kingdom *51°08′N 03°00′W*
97 K22 **Bridgwater Bay** *bay* SW England, United Kingdom
97 O16 **Bridlington** E England, United Kingdom *54°05′N 00°12′W*
97 O16 **Bridlington Bay** *bay* E England, United Kingdom
183 P15 **Bridport** Tasmania, SE Australia *41°03′S 147°26′E*
97 K24 **Bridport** S England, United Kingdom *50°44′N 02°43′W*
103 O5 **Brie** *cultural region* N France
Brieg *see* Brzeg
Briel *see* Brielle
98 G12 **Brielle** *var.* Briel, Bril, ***Eng.*** The Brill. Zuid-Holland, SW Netherlands *51°54′N 04°10′E*
108 E9 **Brienz** Bern, C Switzerland *46°45′N 08°00′E*
108 E9 **Brienzer See** ◎ SW Switzerland
Bries/Briesen *see* Brezno
Brietzig *see* Brzesko
103 S4 **Briey** Meurthe-et-Moselle, NE France *49°15′N 05°57′E*
108 E10 **Brig** ***Fr.*** Brigue, ***It.*** Briga. Valais, SW Switzerland *46°19′N 08°E*
Briga *see* Brig
101 G24 **Brigach** ≈ S Germany
18 K17 **Brigantine** New Jersey, NE USA *39°23′N 74°21′W*
Brigantio *see* Briançon
Brigantium *see* Bregenz
Brigels *see* Breil
25 S9 **Briggs** Texas, SW USA *30°52′N 97°55′W*
36 L1 **Brigham City** Utah, W USA *41°30′N 112°00′W*
14 J15 **Brighton** Ontario, SE Canada *44°01′N 77°44′W*
97 O23 **Brighton** SE England, United Kingdom *50°50′N 00°10′W*
37 T4 **Brighton** Colorado, C USA *39°58′N 104°46′W*
30 K15 **Brighton** Illinois, N USA *39°01′N 90°09′W*
103 T16 **Brignoles** Var, SE France *43°25′N 06°03′E*
Brigue *see* Brig
105 O7 **Brihuega** Castilla-La Mancha, C Spain *40°45′N 02°52′W*
112 A10 **Brijuni** ***It.*** Brioni. *island group* NW Croatia
76 G12 **Brikama** W Gambia *13°13′N 16°37′W*
Bril *see* Brielle
Brill, The *see* Brielle
101 G15 **Brilon** Nordrhein-Westfalen, W Germany *51°24′N 08°34′E*
Brinceni *see* Briceni
107 Q18 **Brindisi** *anc.* Brundisium, Brundusium. Puglia, SE Italy *40°39′N 17°55′E*
27 W11 **Brinkley** Arkansas, C USA *34°53′N 91°11′W*
Brioni *see* Brijuni
103 P12 **Brioude** *anc.* Brivas. Haute-Loire, C France *45°18′N 03°23′E*
Briovera *see* St-Lô
183 U2 **Brisbane** *state capital* Queensland, E Australia *27°30′S 153°E*
183 V2 **Brisbane** ✈ Queensland, E Australia *27°30′S 153°00′E*
25 P2 **Briscoe** Texas, SW USA *35°34′N 100°17′W*
106 H10 **Brisighella** Emilia-Romagna, C Italy *44°12′N 11°45′E*
108 G11 **Brissago** Ticino, S Switzerland *46°07′N 08°40′E*
97 K22 **Bristol** *anc.* Bricgstow. SW England, United Kingdom *51°27′N 02°35′W*
18 M12 **Bristol** Connecticut, NE USA *41°40′N 72°56′W*
23 R9 **Bristol** Florida, SE USA *30°25′N 84°58′W*
19 N9 **Bristol** New Hampshire, NE USA *43°34′N 71°42′W*
29 Q8 **Bristol** South Dakota, N USA *45°18′N 97°45′W*
21 P8 **Bristol** Tennessee, S USA *36°36′N 82°11′W*
18 M8 **Bristol** Vermont, NE USA *44°07′N 73°00′W*
39 N14 **Bristol Bay** *bay* Alaska, USA
97 I22 **Bristol Channel** *inlet* England/Wales, United Kingdom
35 W14 **Bristol Lake** ◎ California, W USA
27 P10 **Bristow** Oklahoma, C USA *35°49′N 96°23′W*
86 C10 **Britain** *var.* Great Britain. *island* United Kingdom
Britannia Minor *see* Bretagne
10 L12 **British Columbia** ***Fr.*** Colombie-Britannique. ◈ *province* SW Canada
British Guiana *see* Guyana
British Honduras *see* Belize
173 Q7 **British Indian Ocean Territory** ◇ *UK dependent territory* C Indian Ocean
86 B9 **British Isles** *island group* NW Europe
10 I1 **British Mountains** ▲▲ Yukon, NW Canada
British North Borneo *see* Sabah
British Solomon Islands Protectorate *see* Solomon Islands
45 S8 **British Virgin Islands** *var.* Virgin Islands. ◇ *UK dependent territory* E West Indies
83 J21 **Brits** North-West, N South Africa *25°39′S 27°47′E*
83 H24 **Britstown** Northern Cape, W South Africa *30°36′S 23°30′E*
14 F12 **Britt** Ontario, S Canada *45°46′N 80°34′W*
29 V12 **Britt** Iowa, C USA *43°06′N 93°48′W*
Brittany *see* Bretagne
29 Q7 **Britton** South Dakota, N USA *45°47′N 97°45′W*
Briva Curretia *see* Brive-la-Gaillarde
Briva Isarae *see* Pontoise
Brivas *see* Brioude
Brive *see* Brive-la-Gaillarde
102 M12 **Brive-la-Gaillarde** *prev.* Brive; *anc.* Briva Curretia. Corrèze, C France *45°09′N 01°31′E*
105 O4 **Briviesca** Castilla y León, N Spain *42°33′N 03°19′W*
Brixen *see* Bressanone
Brixia *see* Brescia
Brlik *see* Birlik
Brněnský Kraj *see* Jihomoravský Kraj
111 G18 **Brno** ***Ger.*** Brünn. Jihomoravský Kraj, SE Czech Republic *49°11′N 16°35′E*
96 G7 **Broad Bay** *bay* NW Scotland, United Kingdom
25 X8 **Broaddus** Texas, SW USA *31°18′N 94°16′W*
183 O12 **Broadford** Victoria, SE Australia *37°07′S 145°04′E*
96 G9 **Broadford** N Scotland, United Kingdom *57°14′N 05°54′W*
96 J13 **Broad Law** ▲ S Scotland, United Kingdom *55°30′N 03°22′W*
23 U3 **Broad River** ≈ Georgia, SE USA
21 N8 **Broad River** ≈ North Carolina/South Carolina, SE USA
181 Y8 **Broadsound Range** ▲▲ Queensland, E Australia
33 X11 **Broadus** Montana, NW USA *45°28′N 105°22′W*
21 U4 **Broadway** Virginia, NE USA *38°36′N 78°48′W*
118 E9 **Broceni** SW Latvia *56°41′N 22°31′E*
11 U11 **Brochet** Manitoba, C Canada *57°55′N 101°40′W*
11 U10 **Brochet, Lac** ◎ Manitoba, C Canada
15 S5 **Brochet, Lac au** ◎ Québec, SE Canada
101 K14 **Brocken** ▲ C Germany *51°48′N 10°38′E*
19 O12 **Brockton** Massachusetts, NE USA *42°04′N 71°01′W*
14 L14 **Brockville** Ontario, SE Canada *44°35′N 75°44′W*
18 D13 **Brockway** Pennsylvania, NE USA *41°14′N 78°45′W*
Brod/Bród *see* Slavonski Brod
9 N5 **Brodeur Peninsula** *peninsula* Baffin Island, Nunavut, NE Canada
96 H13 **Brodick** W Scotland, United Kingdom *55°34′N 05°10′W*
Brod na Savi *see* Slavonski Brod
110 K9 **Brodnica** ***Ger.*** Buddenbrock. Kujawski-pomorskie, C Poland *53°15′N 19°23′E*
Brod-Posavina *see* Slavonski Brod-Posavina
Brodsko-Posavska Županija *see* Slavonski Brod-Posavina
116 J5 **Brody** L'vivs'ka Oblast', NW Ukraine *50°05′N 25°08′E*
98 I10 **Broek-in-Waterland** Noord-Holland, C Netherlands *52°27′N 04°59′E*
32 L13 **Brogan** Oregon, NW USA *44°15′N 117°34′W*
110 N10 **Brok** Mazowieckie, C Poland *52°42′N 21°53′E*
27 P9 **Broken Arrow** Oklahoma, C USA *36°03′N 95°47′W*
183 T9 **Broken Bay** *bay* New South Wales, SE Australia
29 N15 **Broken Bow** Nebraska, C USA *41°24′N 99°38′W*
27 R13 **Broken Bow** Oklahoma, C USA *34°01′N 94°44′W*
27 R12 **Broken Bow Lake** ▣ Oklahoma, C USA
182 L6 **Broken Hill** New South Wales, SE Australia *31°58′S 141°27′E*
173 S10 **Broken Ridge** *undersea feature* S Indian Ocean *31°30′S 95°00′E*
186 C6 **Broken Water Bay** *bay* W Bismarck Sea
55 W10 **Brokopondo** Brokopondo, NE Suriname *05°04′N 55°00′W*
55 W10 **Brokopondo** ◈ *district* C Suriname
Bromberg *see* Bydgoszcz
95 L22 **Bromölla** Skåne, S Sweden *56°04′N 14°28′E*
97 L20 **Bromsgrove** W England, United Kingdom *52°20′N 02°03′W*
95 G20 **Brønderslev** Nordjylland, N Denmark *57°16′N 09°58′E*
106 D8 **Broni** Lombardia, N Italy *45°04′N 09°18′E*
10 K11 **Bronlund Peak** ▲ British Columbia, W Canada *57°27′N 126°43′W*
93 F14 **Brønnøysund** Nordland, C Norway *65°38′N 12°15′E*
23 V10 **Bronson** Florida, SE USA *29°25′N 82°38′W*
31 Q11 **Bronson** Michigan, N USA *41°52′N 85°11′W*
25 X8 **Bronson** Texas, SW USA *31°20′N 94°00′W*
107 L24 **Bronte** Sicilia, Italy, C Mediterranean Sea *37°47′N 14°50′E*
25 P8 **Bronte** Texas, SW USA *31°53′N 100°17′W*
25 Y9 **Brookeland** Texas, SW USA *31°05′N 93°57′W*
170 M7 **Brooke's Point** Palawan, W Philippines *08°54′N 117°54′E*
27 T3 **Brookfield** Missouri, C USA *39°46′N 93°04′W*
22 K7 **Brookhaven** Mississippi, S USA *31°34′N 90°26′W*
32 E16 **Brookings** Oregon, NW USA *42°03′N 124°16′W*
29 R10 **Brookings** South Dakota, N USA *44°19′N 96°46′W*
29 W14 **Brooklyn** Iowa, C USA *41°43′N 92°27′W*
29 U8 **Brooklyn Park** Minnesota, N USA *45°06′N 93°18′W*
21 U7 **Brookneal** Virginia, NE USA *37°03′N 78°56′W*
11 R16 **Brooks** Alberta, SW Canada *50°35′N 111°54′W*
25 V11 **Brookshire** Texas, SW USA *29°47′N 95°57′W*
38 L8 **Brooks Mountain** ▲ Alaska, USA *65°31′N 167°24′W*
38 M11 **Brooks Range** ▲▲ Alaska, USA
31 O12 **Brookston** Indiana, N USA *40°34′N 86°53′W*
23 V11 **Brooksville** Florida, SE USA *28°33′N 82°23′W*
23 N4 **Brooksville** Mississippi, S USA *33°13′N 88°34′W*
180 J13 **Brookton** Western Australia *32°24′S 117°04′E*
31 Q14 **Brookville** Indiana, N USA *39°25′N 85°00′W*
18 D13 **Brookville** Pennsylvania, NE USA *41°07′N 79°05′W*
31 Q14 **Brookville Lake** ▣ Indiana, N USA
180 K5 **Broome** Western Australia *17°58′S 122°15′E*
37 S4 **Broomfield** Colorado, C USA *39°55′N 105°05′W*
Broos *see* Orăștie
96 J7 **Brora** N Scotland, United Kingdom *57°59′N 04°00′W*
96 I7 **Brora** ≈ N Scotland, United Kingdom
95 F23 **Brørup** Syddtjylland, W Denmark *55°29′N 09°01′E*
95 L23 **Brösarp** Skåne, S Sweden *55°43′N 14°10′E*
116 J9 **Broșteni** Suceava, NE Romania *47°14′N 25°43′E*
102 M6 **Brou** Eure-et-Loir, C France *48°12′N 01°10′E*
Broucsella *see* Brussel/Bruxelles
Broughton Bay *see* Tongjosŏn-man
Broughton Island *see* Qikiqtarjuaq
138 G7 **Broummâna** C Lebanon *33°53′N 35°39′E*
22 I9 **Broussard** Louisiana, S USA *30°09′N 91°57′W*
98 E13 **Brouwersdam** *dam* SW Netherlands
98 E13 **Brouwershaven** Zeeland, SW Netherlands *51°44′N 03°50′E*
117 P4 **Brovary** Kyyivs'ka Oblast', N Ukraine *50°30′N 30°45′E*
95 G20 **Brovst** Nordjylland, N Denmark *57°06′N 09°32′E*
31 S8 **Brown City** Michigan, N USA *43°12′N 82°50′W*
24 M6 **Brownfield** Texas, SW USA *33°11′N 102°16′W*
33 Q7 **Browning** Montana, NW USA *48°33′N 113°00′W*
33 R6 **Brown, Mount** ▲ Montana, NW USA *48°52′N 111°08′W*
0 M9 **Browns Bank** *undersea feature* NW Atlantic Ocean *42°40′N 66°05′W*
31 O14 **Brownsburg** Indiana, N USA *39°50′N 86°24′W*
18 J16 **Browns Mills** New Jersey, NE USA *39°58′N 74°33′W*
44 J12 **Browns Town** C Jamaica *18°28′N 77°22′W*
31 P15 **Brownstown** Indiana, N USA *38°52′N 86°02′W*
29 R8 **Browns Valley** Minnesota, N USA *45°36′N 96°49′W*
20 K7 **Brownsville** Kentucky, S USA *37°10′N 86°18′W*
20 F9 **Brownsville** Tennessee, S USA *35°35′N 89°15′W*
25 T17 **Brownsville** Texas, SW USA *25°56′N 97°28′W*
55 W10 **Brownsweg** Brokopondo, C Suriname
29 U9 **Brownton** Minnesota, N USA *44°43′N 94°21′W*
19 R5 **Brownville Junction** Maine, NE USA *45°20′N 69°04′W*
25 R8 **Brownwood** Texas, SW USA *31°42′N 98°59′W*
25 R8 **Brownwood Lake** ▣ Texas, SW USA
104 I9 **Brozas** Extremadura, W Spain *39°37′N 06°48′W*
119 M18 **Brozha** Mahilyowskaya Voblasts', E Belarus *52°57′N 29°07′E*
103 O2 **Bruay-en-Artois** Pas-de-Calais, N France *50°31′N 02°30′E*
103 P2 **Bruay-sur-l'Escaut** Nord, N France *50°24′N 03°33′E*
14 F13 **Bruce Peninsula** *peninsula* Ontario, S Canada
20 H9 **Bruceton** Tennessee, S USA *36°02′N 88°14′W*
25 T9 **Bruceville** Texas, SW USA *31°17′N 97°15′W*
101 G21 **Bruchsal** Baden-Württemberg, SW Germany *49°07′N 08°35′E*
109 Q7 **Bruck** Salzburg, NW Austria *47°18′N 12°51′E*
Bruck *see* Bruck an der Mur
109 Y4 **Bruck an der Leitha** Niederösterreich, NE Austria *48°02′N 16°47′E*
109 V7 **Bruck an der Mur** *var.* Bruck. Steiermark, C Austria *47°25′N 15°17′E*
101 M24 **Bruckmühl** Bayern, SE Germany *47°52′N 11°54′E*
168 E7 **Brueuh, Pulau** *island* NW Indonesia
Bruges *see* Brugge
108 F6 **Brugg** Aargau, NW Switzerland *47°29′N 08°13′E*
99 C16 **Brugge** ***Fr.*** Bruges. West-Vlaanderen, NW Belgium *51°13′N 03°14′E*
109 R9 **Bruggen** Kärnten, S Austria *46°46′N 13°13′E*
101 E16 **Brühl** Nordrhein-Westfalen, W Germany *50°50′N 06°55′E*
99 F14 **Bruinisse** Zeeland, SW Netherlands *51°40′N 04°04′E*
169 R9 **Bruit, Pulau** *island* East Malaysia
14 K10 **Brûlé, Lac** ◎ Québec, SE Canada
30 M4 **Brule River** ≈ Michigan/Wisconsin, N USA
99 H23 **Brûly** Namur, S Belgium *49°55′N 04°31′E*
59 N17 **Brumado** Bahia, E Brazil *14°14′S 41°38′W*
98 M11 **Brummen** Gelderland, E Netherlands *52°05′N 06°10′E*
94 H13 **Brumunddal** Hedmark, S Norway *60°54′N 11°00′E*
23 Q6 **Brundidge** Alabama, S USA *31°43′N 85°49′W*
Brundisium/Brundusium *see* Brindisi
33 N15 **Bruneau River** ≈ Idaho, NW USA
Bruneck *see* Brunico
169 T8 **Brunei** *off.* Brunei Darussalam, ***Mal.*** Negara Brunei Darussalam. ◆ *monarchy* SE Asia
169 T7 **Brunei Bay** *var.* Teluk Brunei. *bay* N Brunei
Brunei Darussalam *see* Brunei
Brunei, Teluk *see* Brunei Bay
Brunei Town *see* Bandar Seri Begawan
106 H5 **Brunico** ***Ger.*** Bruneck. Trentino-Alto Adige, N Italy *46°49′N 11°57′E*
Brünn *see* Brno
185 G17 **Brunner, Lake** ◎ South Island, New Zealand
99 M18 **Brunssum** Limburg, SE Netherlands *50°57′N 05°59′E*
23 W7 **Brunswick** Georgia, SE USA *31°09′N 81°30′W*
19 Q8 **Brunswick** Maine, NE USA *43°54′N 69°58′W*
21 V3 **Brunswick** Maryland, NE USA *39°18′N 77°37′W*
27 T3 **Brunswick** Missouri, C USA *39°25′N 93°07′W*
31 T11 **Brunswick** Ohio, N USA *41°14′N 81°50′W*
Brunswick *see* Braunschweig
63 H24 **Brunswick, Península** *headland* S Chile *53°30′S 71°27′W*
111 H17 **Bruntál** ***Ger.*** Freudenthal. Moravskoslezský Kraj, E Czech Republic *50°00′N 17°27′E*
195 N3 **Brunt Ice Shelf** *ice shelf* Antarctica
Brusa *see* Bursa
114 G7 **Brusartsi** Montana, NW Bulgaria *43°39′N 23°04′E*
37 U3 **Brush** Colorado, C USA *40°15′N 103°37′W*
42 M5 **Brus Laguna** Gracias a Dios, E Honduras *15°46′N 84°32′W*
60 K13 **Brusque** Santa Catarina, S Brazil *27°07′S 48°54′W*
Brussa *see* Bursa
99 E18 **Brussel** *var.* Brussels, ***Fr.*** Bruxelles, ***Ger.*** Brüssel; ***anc.*** Broucsella. ● (Belgium) Brussels, C Belgium *50°52′N 04°21′E* ***see also*** Bruxelles
Brussel *see* Bruxelles
Brüssel/Brussels *see* Brussel/Bruxelles
117 O5 **Brusyliv** Zhytomyrs'ka Oblast', N Ukraine *50°16′N 29°31′E*
183 Q12 **Bruthen** Victoria, SE Australia *37°43′S 147°49′E*
Bruttium *see* Calabria
Brüx *see* Most
99 E18 **Bruxelles** *var.* Brussels, ***Dut.*** Brussel, ***Ger.*** Brüssel; ***anc.*** Broucsella. ● Brussels, C Belgium *50°52′N 04°21′E* ***see also*** Brussel
Bruxelles *see* Brussel
54 J7 **Bruzual** Apure, W Venezuela *07°59′N 69°18′W*
31 Q11 **Bryan** Ohio, N USA *41°30′N 84°34′W*
25 U10 **Bryan** Texas, SW USA *30°40′N 96°23′W*
194 J4 **Bryan Coast** *physical region* Antarctica
122 L11 **Bryanka** Krasnoyarskiy Kray, C Russian Federation *59°01′N 93°13′E*
117 Y7 **Bryanka** Luhans'ka Oblast', E Ukraine *48°30′N 38°45′E*
182 J8 **Bryan, Mount** ▲ South Australia *33°25′S 138°59′E*
126 I6 **Bryansk** Bryanskaya Oblast', W Russian Federation *53°16′N 34°07′E*
126 H6 **Bryanskaya Oblast'** ◈ *province* W Russian Federation
194 J5 **Bryant, Cape** *headland* Antarctica
27 U8 **Bryant Creek** ≈ Missouri, C USA
36 K8 **Bryce Canyon** *canyon* Utah, W USA
119 O15 **Bryli** Mahilyowskaya Voblasts', E Belarus *53°54′N 30°33′E*
95 C17 **Bryne** Rogaland, S Norway *58°43′N 05°40′E*
25 R6 **Bryson** Texas, SW USA *33°09′N 98°23′W*
21 N10 **Bryson City** North Carolina, SE USA *35°26′N 83°27′W*
14 K11 **Bryson, Lac** ◎ Québec, SE Canada
126 K13 **Bryukhovetskaya** Krasnodarskiy Kray, SW Russian Federation *45°49′N 38°01′E*
111 H15 **Brzeg** ***Ger.*** Brieg; ***anc.*** Civitas Altae Ripae. Opolskie, S Poland *50°52′N 17°27′E*
111 G14 **Brzeg Dolny** ***Ger.*** Dyhernfurth. Dolnośląskie, SW Poland *51°15′N 16°40′E*
Brześć Litewski/Brześć nad Bugiem *see* Brest
111 L17 **Brzesko** ***Ger.*** Brietzig. Małopolskie, SE Poland *49°59′N 20°34′E*
Brzeżany *see* Berezhany
110 K12 **Brzeziny** Łódzkie, C Poland *51°49′N 19°41′E*
Brzostowica Wielka *see* Vyalikaya Byerastavitsa
111 O17 **Brzozów** Podkarpackie, SE Poland *49°38′N 22°00′E*
Bsharrī/Bsherri *see* Bcharré
187 X14 **Bua** Vanua Levu, N Fiji *16°48′S 178°36′E*
95 J20 **Bua** Halland, S Sweden *57°14′N 12°07′E*
82 M13 **Bua** ≈ C Malawi
Bua *see* Čiovo
81 L18 **Bu'aale** ***It.*** Buale. Jubbada Dhexe, SW Somalia *0°52′N 42°37′E*
Buache, Mount *see* Mutunte, Mount
189 Q8 **Buada Lagoon** *lagoon* Nauru, C Pacific Ocean
186 M8 **Buala** Santa Isabel, E Solomon Islands *08°06′S 159°31′E*
Buale *see* Bu'aale
190 H1 **Buariki** *atoll* Tungaru, W Kiribati
167 Q10 **Bua Yai** *var.* Ban Bua Yai. Nakhon Ratchasima, E Thailand *15°35′N 102°25′E*
75 P8 **Bu'ayrāt al Ḥasūn** *var.* Buwayrāt al Hasūn. C Libya *31°22′N 15°41′E*
76 H13 **Buba** S Guinea-Bissau *11°36′N 14°55′W*
171 P11 **Bubaa** Sulawesi, N Indonesia *0°32′N 122°27′E*
81 D20 **Bubanza** NW Burundi *03°04′S 29°22′E*
83 K18 **Bubi** *prev.* Bubye. ≈ S Zimbabwe
142 L11 **Būbīyan, Jazīrat** *island* E Kuwait
Bublitz *see* Bobolice
Bubye *see* Bubi
187 Y13 **Buca** *prev.* Mbutha. Vanua Levu, N Fiji *16°39′S 179°51′E*
136 F16 **Bucak** Burdur, SW Turkey *37°28′N 30°37′E*
54 G8 **Bucaramanga** Santander, N Colombia *07°08′N 73°10′W*
107 M18 **Buccino** Campania, S Italy *40°37′N 15°25′E*
116 K9 **Bucecea** Botoșani, NE Romania *47°45′N 26°30′E*
116 J6 **Buchach** ***Pol.*** Buczacz. Ternopil's'ka Oblast', W Ukraine *49°04′N 25°23′E*
183 Q12 **Buchan** Victoria, SE Australia *37°26′S 148°11′E*
76 J17 **Buchanan** *prev.* Grand Bassa. SW Liberia *05°53′N 10°03′W*
23 R3 **Buchanan** Georgia, SE USA *33°48′N 85°11′W*
31 O11 **Buchanan** Michigan, N USA *41°49′N 86°21′W*
21 T6 **Buchanan** Virginia, NE USA *37°31′N 79°40′W*
25 R10 **Buchanan Dam** Texas, SW USA *30°42′N 98°24′W*
25 R10 **Buchanan, Lake** ▣ Texas, SW USA
96 L8 **Buchan Ness** *headland* NE Scotland, United Kingdom *57°28′N 01°46′W*
13 T12 **Buchans** Newfoundland and Labrador, SE Canada *48°49′N 56°53′W*
Bucharest *see* București
101 H20 **Buchen** Baden-Württemberg, SW Germany *49°31′N 09°18′E*
100 I10 **Buchholz in der Nordheide** Niedersachsen, NW Germany *53°19′N 09°52′E*
108 F7 **Buchs** Aargau, N Switzerland *47°24′N 08°04′E*
108 I8 **Buchs** Sankt Gallen, NE Switzerland *47°10′N 09°28′E*
100 H13 **Bückeburg** Niedersachsen, NW Germany *52°16′N 09°03′E*
36 K14 **Buckeye** Arizona, SW USA *33°22′N 112°34′W*
Buckeye State *see* Ohio
21 S4 **Buckhannon** West Virginia, NE USA *38°59′N 80°14′W*
25 T9 **Buckholts** Texas, SW USA *30°52′N 97°07′W*
96 K8 **Buckie** NE Scotland, United Kingdom *57°39′N 02°56′W*
14 M12 **Buckingham** Québec, SE Canada *45°35′N 75°25′W*
21 U6 **Buckingham** Virginia, NE USA *37°33′N 78°34′W*
97 N21 **Buckinghamshire** *cultural region* SE England, United Kingdom
39 N8 **Buckland** Alaska, USA *65°58′N 161°07′W*
182 G7 **Buckleboo** South Australia *32°55′S 136°11′E*
26 K7 **Bucklin** Kansas, C USA *37°33′N 99°37′W*
27 T3 **Bucklin** Missouri, C USA *39°46′N 92°53′W*
36 I12 **Buckskin Mountains** ▲▲ Arizona, SW USA
19 R7 **Bucksport** Maine, NE USA *44°34′N 68°46′W*
82 A9 **Buco Zau** Cabinda, NW Angola *04°45′S 12°34′E*
Bu Craa *see* Bou Craa
116 K14 **București** ***Eng.*** Bucharest, ***Ger.*** Bukarest, ***prev.*** Altenburg; ***anc.*** Cetatea Damboviței. ● (Romania) București, S Romania *44°27′N 26°06′E*
31 S12 **Bucyrus** Ohio, N USA *40°47′N 82°57′W*
Buczacz *see* Buchach
94 E9 **Bud** Møre og Romsdal, S Norway *62°55′N 06°55′E*
25 S11 **Buda** Texas, SW USA *30°05′N 97°50′W*
119 O18 **Buda-Kashalyova** ***Rus.*** Buda-Koshelëvo. Homyel'skaya Voblasts', SE Belarus *52°43′N 30°34′E*
Buda-Koshelëvo *see* Buda-Kashalyova
166 L4 **Budalin** Sagaing, C Myanmar (Burma) *22°24′N 95°08′E*
111 J22 **Budapest** *off.* Budapest Főváros, ***SCr.*** Budimpešta. ● (Hungary) Pest, N Hungary *47°30′N 19°03′E*
Budapest Főváros *see* Budapest
92 I2 **Búðardalur** Vesturland, W Iceland *65°07′N 21°45′W*
152 K11 **Budaun** Uttar Pradesh, N India *28°02′N 79°07′E*
141 O9 **Budayyi'ah** *oasis* C Saudi Arabia
195 Y12 **Budd Coast** *physical region* Antarctica
Buddenbrock *see* Brodnica
107 C17 **Budduso** Sardegna, Italy, C Mediterranean Sea *40°37′N 09°19′E*
97 I23 **Bude** SW England, United Kingdom *50°50′N 04°33′W*
22 J7 **Bude** Mississippi, S USA *31°27′N 90°51′W*
Budějovický Kraj *see* Jihočeský Kraj
99 K16 **Budel** Noord-Brabant, SE Netherlands *51°17′N 05°35′E*
100 I8 **Büdelsdorf** Schleswig-Holstein, N Germany *54°20′N 09°40′E*
127 O14 **Budënnovsk** Stavropol'skiy Kray, SW Russian Federation *44°46′N 44°07′E*
116 K14 **Budești** Călărași, SE Romania *44°13′N 26°30′E*
Budge-Budge *see* Baj Baj
Budgewoi *see* Budgewoi Lake
183 T8 **Budgewoi Lake** *var.* Budgewoi. New South Wales, SE Australia *33°14′S 151°34′E*
Budimpešta *see* Budapest
79 J16 **Budjala** Equateur, NW Dem. Rep. Congo *02°39′N 19°42′E*
106 G10 **Budrio** Emilia-Romagna, C Italy *44°33′N 11°34′E*
Budslav *see* Budslaw
119 K14 **Budslaw** ***Rus.*** Budslav. Minskaya Voblasts', N Belarus *54°47′N 27°27′E*
Budua *see* Budva
169 R9 **Budu, Tanjung** *headland* East Malaysia *02°51′N 111°42′E*
113 J17 **Budva** ***It.*** Budua. W Montenegro *42°17′N 18°49′E*
Budweis *see* České Budějovice
Budyšin *see* Bautzen
79 D16 **Buea** Sud-Ouest, SW Cameroon *04°09′N 09°13′E*
103 S13 **Buëch** ≈ SE France
18 J17 **Buena** New Jersey, NE USA *39°30′N 74°55′W*
62 K12 **Buena Esperanza** San Luis, C Argentina *34°45′S 65°15′W*
54 C11 **Buenaventura** Valle del Cauca, W Colombia *03°54′N 77°02′W*
40 I4 **Buenaventura** Chihuahua, N Mexico *29°50′N 107°30′W*
57 M18 **Buena Vista** Santa Cruz, C Bolivia *17°28′S 63°37′W*
40 G10 **Buenavista** Baja California Sur, NW Mexico *23°39′N 109°41′W*
37 S5 **Buena Vista** Colorado, C USA *38°50′N 106°07′W*
23 S5 **Buena Vista** Georgia, SE USA *32°19′N 84°31′W*
21 T6 **Buena Vista** Virginia, NE USA *37°44′N 79°22′W*
44 F5 **Buena Vista, Bahia de** *bay* N Cuba
35 R13 **Buena Vista Lake Bed** ◎ California, W USA
105 P8 **Buendía, Embalse de** ▣ C Spain
63 F16 **Bueno, Río** ≈ S Chile
62 N12 **Buenos Aires** *hist.* Santa Maria del Buen Aire. ● (Argentina) Buenos Aires, E Argentina *34°40′S 58°30′W*
43 O15 **Buenos Aires** Puntarenas, SE Costa Rica *09°10′N 83°23′W*
61 C20 **Buenos Aires** *off.* Provincia de Buenos Aires. ◈ *province* E Argentina
63 H19 **Buenos Aires, Lago** *var.* Lago General Carrera. ◎ Argentina/Chile
Buenos Aires, Provincia de *see* Buenos Aires
54 C13 **Buesaco** Nariño, SW Colombia *01°22′N 77°07′W*
29 U8 **Buffalo** Minnesota, N USA *45°11′N 93°50′W*
26 T6 **Buffalo** Missouri, C USA *37°38′N 93°05′W*
18 D10 **Buffalo** New York, NE USA *42°53′N 78°53′W*
27 K8 **Buffalo** Oklahoma, C USA *36°51′N 99°38′W*
28 J7 **Buffalo** South Dakota, N USA *45°25′N 103°35′W*
25 V8 **Buffalo** Texas, SW USA *31°25′N 96°04′W*
33 W12 **Buffalo** Wyoming, C USA *44°21′N 106°40′W*
29 U11 **Buffalo Center** Iowa, C USA *43°23′N 93°57′W*
24 M3 **Buffalo Lake** ▣ Texas, SW USA
30 K7 **Buffalo Lake** ▣ Wisconsin, N USA
11 S12 **Buffalo Narrows** Saskatchewan, C Canada *55°52′N 108°28′W*
27 U9 **Buffalo River** ≈ Arkansas, C USA
29 R5 **Buffalo River** ≈ Minnesota, N USA
20 I10 **Buffalo River** ≈ Tennessee, S USA
30 J6 **Buffalo River** ≈ Wisconsin, N USA
44 L12 **Buff Bay** E Jamaica *18°18′N 76°40′W*
23 T3 **Buford** Georgia, SE USA *34°07′N 84°00′W*
28 J3 **Buford** North Dakota, N USA *48°00′N 103°58′W*
33 Y17 **Buford** Wyoming, C USA *41°05′N 105°17′W*
116 J14 **Buftea** Ilfov, S Romania *44°34′N 25°58′E*
84 I9 **Bug** ***Bel.*** Zakhodni Buh, ***Eng.*** Western Bug, ***Rus.*** Zapadnyy Bug, ***Ukr.*** Zakhidnyy Buh. ≈ E Europe
54 D11 **Buga** Valle del Cauca, W Colombia *03°53′N 76°17′W*
Buga *see* Dörvöljin
103 O17 **Bugarach, Pic du** ▲ S France *42°52′N 02°23′E*
162 F8 **Bugat** *var.* Bayangol. Govĭ-Altay, SW Mongolia *45°33′N 94°22′E*
146 B12 **Bugdaýly** ***Rus.*** Bugdayly. Balkan Welaýaty, W Turkmenistan *38°42′N 54°14′E*
Bugdayly *see* Bugdaýly
Buggs Island Lake *see* John H. Kerr Reservoir
Bughotu *see* Santa Isabel
171 O14 **Bugingkalo** Sulawesi, C Indonesia *04°49′S 121°42′E*
64 P6 **Bugio** *island* Madeira, Portugal, NE Atlantic Ocean
92 M8 **Bugøynes** Finnmark, N Norway *69°57′N 29°34′E*
125 Q3 **Bugrino** Nenetskiy Avtonomnyy Okrug, NW Russian Federation *68°48′N 49°12′E*
127 T5 **Bugul'ma** Respublika Tatarstan, W Russian Federation *54°31′N 52°45′E*
Bügür *see* Luntai
127 T6 **Buguruslan** Orenburgskaya Oblast', W Russian Federation *53°38′N 52°30′E*
159 R9 **Buh He** ≈ C China
101 F22 **Bühl** Baden-Württemberg, SW Germany *48°42′N 08°07′E*
33 O15 **Buhl** Idaho, NW USA *42°36′N 114°45′W*
116 K10 **Buhuşi** Bacău, E Romania *46°41′N 26°45′E*
Buie d'Istria *see* Buje
97 J20 **Builth Wells** E Wales, United Kingdom *52°07′N 03°28′W*
186 J8 **Buin** Bougainville Island, NE Papua New Guinea *06°52′S 155°42′E*
108 J9 **Buin, Piz** ▲ Austria/Switzerland *46°51′N 10°07′E*
127 Q4 **Buinsk** Chuvashskaya Respublika, W Russian Federation *55°09′N 47°00′E*
127 Q4 **Buinsk** Respublika Tatarstan, W Russian Federation *54°58′N 48°16′E*
163 R8 **Buir Nur** ***Mong.*** Buyr Nuur. ◎ China/Mongolia ***see also*** Buyr Nuur
Buir Nur *see* Buyr Nuur
98 M5 **Buitenpost** ***Fris.*** Bûtenpost. Fryslân, N Netherlands *53°15′N 06°09′E*
Buitenzorg *see* Bogor
83 F19 **Buitepos** Omaheke, E Namibia *22°17′S 19°59′E*
105 N7 **Buitrago del Lozoya** Madrid, C Spain *41°00′N 03°38′W*
Buj *see* Buy
104 M13 **Bujalance** Andalucía, S Spain *37°54′N 04°23′W*
113 O17 **Bujanovac** SE Serbia *42°29′N 21°44′E*
105 S6 **Bujaraloz** Aragón, NE Spain *41°29′N 00°10′W*
112 A9 **Buje** ***It.*** Buie d'Istria. Istra, NW Croatia *45°23′N 13°40′E*
Bujnurd *see* Bojnūrd
81 D21 **Bujumbura** *prev.* Usumbura. ● (Burundi) W Burundi *03°25′S 29°24′E*
81 D20 **Bujumbura** ✈ W Burundi *03°21′S 29°19′E*
186 J6 **Buka** Bougainville, Papua New Guinea *05°24′S 154°40′E*
159 N11 **Buka Daban** *var.* Bukadaban Feng. ▲ C China *36°09′N 90°52′E*
Bukadaban Feng *see* Buka Daban
186 J6 **Buka Island** *island* NE Papua New Guinea
81 F18 **Bukakata** S Uganda *0°18′S 31°57′E*
79 N24 **Bukama** Katanga, SE Dem. Rep. Congo *09°13′S 25°52′E*
142 J4 **Būkān** *var.* Bowkān. Āžarbāyjān-e Gharbī, NW Iran *36°31′N 46°10′E*
Bukantau, Gory *see* Bo'kantov Tog'lari
Bukarest *see* București
79 O19 **Bukavu** *prev.* Costermansville. Sud-Kivu, E Dem. Rep. Congo *02°19′S 28°49′E*
81 F21 **Bukene** Tabora, NW Tanzania *04°15′S 32°51′E*
141 W8 **Bū Khābī** *var.* Bakhābī. NW Oman *23°29′N 56°06′E*
Bukhara *see* Buxoro
Bukharskaya Oblast' *see* Buxoro Viloyati
168 M14 **Bukitkemuning** Sumatera, W Indonesia *04°43′S 104°27′E*
168 I11 **Bukittinggi** *prev.* Fort de Kock. Sumatera, W Indonesia *0°18′S 100°20′E*
111 L21 **Bükk** ▲▲ NE Hungary
81 F19 **Bukoba** Kagera, NW Tanzania *01°19′S 31°49′E*
113 N20 **Bukovo** S FYR Macedonia *40°59′N 21°20′E*
108 G6 **Bülach** Zürich, NW Switzerland *47°31′N 08°30′E*
Būlaevo *see* Bulayevo
Bulag *see* Tünel, Hövsgöl, Mongolia
Bulag *see* Möngönmorĭt, Töv, Mongolia
Bulagiyn Denj *see* Bulgan
183 U7 **Bulahdelah** New South Wales, SE Australia *32°24′S 152°13′E*
171 P4 **Bulan** Luzon, N Philippines *12°40′N 123°55′E*
137 N11 **Bulancak** Giresun, N Turkey *40°57′N 38°14′E*
152 J10 **Bulandshahr** Uttar Pradesh, N India *28°30′N 77°49′E*
137 R14 **Bulanık** Muş, E Turkey *39°04′N 42°16′E*
127 V7 **Bulanovo** Orenburgskaya Oblast', W Russian Federation *52°27′N 55°08′E*
83 J17 **Bulawayo** *var.* Buluwayo. Bulawayo, SW Zimbabwe *20°08′S 28°37′E*
83 J17 **Bulawayo** ✈ Matabeleland North, SW Zimbabwe *20°00′S 28°36′E*
145 Q6 **Bulayevo** ***Kaz.*** Būlaevo. Severnyy Kazakhstan, N Kazakhstan *54°55′N 70°29′E*
136 D15 **Buldan** Denizli, SW Turkey *38°03′N 28°50′E*
154 G12 **Buldāna** Mahārāshtra, C India *20°31′N 76°18′E*
38 E16 **Buldir Island** *island* Aleutian Islands, Alaska, USA
Buldur *see* Burdur
162 I8 **Bulgan** *var.* Bulagiyn Denj. Arhangay, C Mongolia *47°14′N 100°56′E*
162 D7 **Bulgan** *var.* Jargalant. Bayan-Ölgiy, W Mongolia *46°56′N 91°07′E*
162 K6 **Bulgan** Bulgan, N Mongolia *50°31′N 101°30′E*
162 F7 **Bulgan** *var.* Bürenhayrhan. Hovd, W Mongolia *46°04′N 91°34′E*
162 J10 **Bulgan** Ömnögovĭ, S Mongolia *44°07′N 103°28′E*
162 J7 **Bulgan** ◈ *province* N Mongolia
Bulgan *see* Bayan-Öndör, Bayanhongor, C Mongolia
Bulgan *see* Darvi, Hovd, Mongolia
Bulgan *see* Tsagaan-Üür, Hövsgöl, Mongolia
114 H10 **Bulgaria** *off.* Republic of Bulgaria, ***Bul.*** Bŭlgariya; ***prev.*** People's Republic of Bulgaria. ◆ *republic* SE Europe
Bulgaria, People's Republic of *see* Bulgaria
Bulgaria, Republic of *see* Bulgaria
Bŭlgariya *see* Bulgaria
Bŭlgarka *see* Balgarka
171 S11 **Buli** Pulau Halmahera, E Indonesia *0°56′N 128°17′E*
171 S11 **Buli, Teluk** *bay* Pulau Halmahera, E Indonesia
160 J13 **Buliu He** ≈ S China
Bullange *see* Büllingen
Bulla, Ostrov *see* Xärä Zirä Adasi
104 M11 **Bullaque** ≈ C Spain
105 Q13 **Bullas** Murcia, SE Spain *38°02′N 01°40′W*
80 M12 **Bullaxaar** Woqooyi Galbeed, NW Somalia *10°28′N 44°15′E*
108 C9 **Bulle** Fribourg, SW Switzerland *46°37′N 07°04′E*
185 G15 **Buller** ≈ South Island, New Zealand
183 P12 **Buller, Mount** ▲ Victoria, SE Australia *37°10′S 146°31′E*
36 H11 **Bullhead City** Arizona, SW USA *35°07′N 114°32′W*
99 N21 **Büllingen** ***Fr.*** Bullange. Liège, E Belgium *50°23′N 06°15′E*
Bullion State *see* Missouri
21 T14 **Bull Island** *island* South Carolina, SE USA

◆ Country
● Country Capital
◇ Dependent Territory
○ Dependent Territory Capital
◈ Administrative Regions
✈ International Airport
▲ Mountain
▲▲ Mountain Range
Volcano
≈ River
◎ Lake
▣ Reservoir

Bulloo River Overflow – Cadiz Lake

182 M4 Bulloo River Overflow wetland New South Wales, SE Australia
184 M12 Bulls Manawatu-Wanganui, North Island, New Zealand 40°10′S 175°22′E
21 T14 Bulls Bay bay South Carolina, SE USA
27 U9 Bull Shoals Lake ⊠ Arkansas/Missouri, C USA
181 Q2 Bulman Northern Territory, N Australia 13°39′S 134°21′E
162 I6 Bulnayn Nuruu ▲ N Mongolia
171 O11 Bulowa, Gunung ▲ Sulawesi, N Indonesia 0°33′N 123°39′E
Bulqiza see Bulqizë
113 L19 Bulqizë var. Bulqiza. Dibër, C Albania 41°30′N 20°16′E
Bulsar see Valsād
171 N14 Bulukumba prev. Boeloekoemba. Sulawesi, C Indonesia 05°35′S 120°13′E
147 O11 Bulungh'ur Rus. Bulungur; prev. Krasnogvardeysk. Samarqand Viloyati, C Uzbekistan 39°46′N 67°18′E
79 I21 Bulungu Bandundu, SW Dem. Rep. Congo 04°36′S 18°34′E
Bulungur see Bulungh'ur
Buluwayo see Bulawayo
79 K17 Bumba Equateur, N Dem. Rep. Congo 02°14′N 22°25′E
121 R12 Bumbah, Khalīj al gulf N Libya
Bumbat see Bayan-Öndör
81 F19 Bumbire Island island N Tanzania
169 V8 Bum Bun, Pulau island East Malaysia
81 J17 Buna Wajir, NE Kenya 02°40′N 39°34′E
25 Y10 Buna Texas, SW USA 30°25′N 94°00′W
Bunab see Bonāb
Bunai see M'bunai
147 S13 Bunay S Tajikistan 38°29′N 71°41′E
180 I13 Bunbury Western Australia 33°24′S 115°44′E
97 E14 Buncrana Ir. Bun Cranncha. NW Ireland 55°08′N 07°27′W
Bun Cranncha see Buncrana
181 Z9 Bundaberg Queensland, E Australia 24°50′S 152°16′E
183 T5 Bundarra New South Wales, SE Australia 30°12′S 151°06′E
100 G13 Bünde Nordrhein-Westfalen, NW Germany 52°12′N 08°34′E
152 H13 Būndi Rājasthān, N India 25°28′N 75°42′E
Bun Dobhráin see Bundoran
97 D15 Bundoran Ir. Bun Dobhráin. NW Ireland 54°30′N 08°11′W
Bunë see Bojana
113 K18 Bunë, Lumi i SCr. Bojana. ≈ Albania/Montenegro see also Bojana
171 Q8 Bunga ≈ Mindanao, S Philippines
168 I12 Bungalaut, Selat strait W Indonesia
167 R8 Bung Kan Nong Khai, E Thailand 18°19′N 103°39′E
181 N4 Bungle Bungle Range ▲ Western Australia
82 C10 Bungo Uíge, NW Angola 07°30′S 15°24′E
81 G18 Bungoma Western, W Kenya 0°34′N 34°34′E
81 G18 Bungoma ◈ county W Kenya
164 F15 Bungo-suidō strait SW Japan
164 E14 Bungo-Takada Ōita, Kyūshū, SW Japan 33°34′N 131°28′E
100 K8 Bungsberg hill N Germany
Bungur see Bunyu
79 P17 Bunia Orientale, NE Dem. Rep. Congo 01°33′N 30°16′E
35 U6 Bunker Hill ▲ Nevada, W USA 39°16′N 117°06′W
22 I7 Bunkie Louisiana, S USA 30°58′N 92°12′W
23 X10 Bunnell Florida, SE USA 29°28′N 81°15′W
105 S10 Buñol Valenciana, E Spain 39°25′N 00°47′W
98 K11 Bunschoten Utrecht, C Netherlands 52°15′N 05°23′E
136 K14 Bünyan Kayseri, C Turkey 38°51′N 35°50′E
169 W8 Bunyu var. Bungur. Borneo, N Indonesia 03°33′N 117°50′E
169 W8 Bunyu, Pulau island N Indonesia
Bunzlau see Bolesławiec
Buoddobohki see Patoniva
123 P7 Buor-Khaya, Guba bay N Russian Federation
123 P7 Buor-Khaya, Guba bay N Russian Federation
171 Z15 Bupul Papua, E Indonesia 07°24′S 140°57′E
80 P12 Buraan Bari, N Somalia 10°03′N 49°08′E
145 Q7 Burabay prev. Borovoye. Akmola, N Kazakhstan 53°07′N 70°20′E
Buraida see Buraydah
Buraimi see Al Buraymī
Buran see Boran
158 G15 Burang Xizang Zizhiqu, W China 30°28′N 81°13′E
Burao see Burco
138 H8 Burāq Dar'ā, S Syria 33°11′N 36°28′E
141 O6 Buraydah var. Buraida. Al Qaşīm, N Saudi Arabia 26°50′N 44°E
35 S15 Burbank California, W USA 34°10′N 118°25′W
31 N11 Burbank Illinois, N USA 41°45′N 87°48′W
183 Q8 Burcher New South Wales, SE Australia 33°29′S 147°16′E
80 N13 Burco var. Burao, Bur'o. Togdheer, NW Somalia 09°29′N 45°31′E
162 K8 Bürd var. Ongon. Övörhangay, C Mongolia 46°58′N 103°45′E
146 L13 Burdalyk Lebap Welaýaty, E Turkmenistan 38°31′N 64°21′E
181 W6 Burdekin River ≈ Queensland, NE Australia
27 O7 Burden Kansas, C USA 37°18′N 96°45′W
Burdigala see Bordeaux
136 E15 Burdur var. Buldur. Burdur, SW Turkey 37°44′N 30°17′E
136 E15 Burdur var. Buldur. ◈ province SW Turkey
136 E15 Burdur Gölü salt lake SW Turkey
65 H21 Burdwood Bank undersea feature SW Atlantic Ocean
80 I12 Burē Āmara, N Ethiopia 10°43′N 37°09′E
80 H13 Burē Oromīya, C Ethiopia 08°13′N 35°09′E
93 J15 Bureå Västerbotten, N Sweden 64°36′N 21°15′E
162 K7 Büreghangay var. Darhan. Bulgan, C Mongolia 48°07′N 103°54′E
101 G14 Büren Nordrhein-Westfalen, W Germany 51°34′N 08°34′E
162 L8 Büren var. Bayantöhöm. Töv, C Mongolia 46°57′N 105°09′E
162 K6 Bürengiyn Nuruu ▲ N Mongolia
Bürenhayrhan see Bulgan
162 I6 Bürentogtoh var. Bayan. Hövsgol, C Mongolia 49°36′N 99°36′E
149 U10 Būrewāla var. Mandi Būrewāla. Punjab, E Pakistan 30°05′N 72°47′E
92 J9 Burfjord Troms, N Norway 69°55′N 21°54′E
100 L13 Burg var. Burg an der Ihle, Burg bei Magdeburg. Sachsen-Anhalt, C Germany 52°17′N 11°51′E
Burg an der Ihle see Burg
114 N10 Burgas var. Bourgas. Burgas, E Bulgaria 42°31′N 27°30′E
114 M10 Burgas ◈ province E Bulgaria
114 N9 Burgas ✕ Burgas, E Bulgaria 42°35′N 27°33′E
114 N10 Burgaski Zaliv gulf E Bulgaria
114 M10 Burgasko Ezero lagoon E Bulgaria
21 V11 Burgaw North Carolina, SE USA 34°33′N 77°56′W
Burg bei Magdeburg see Burg
108 E8 Burgdorf Bern, NW Switzerland 47°03′N 07°38′E
109 Y7 Burgenland off. Land Burgenland. ◈ state SE Austria
13 S13 Burgeo Newfoundland, Newfoundland and Labrador, SE Canada 47°37′N 57°38′W
83 I24 Burgersdorp Eastern Cape, SE South Africa 31°00′S 26°20′E
83 K20 Burgersfort Mpumalanga, NE South Africa 24°39′S 30°18′E
101 N23 Burghausen Bayern, SE Germany 48°10′N 12°48′E
139 O5 Burghūth, Sabkhat al ◎ E Syria
101 M20 Burglengenfeld Bayern, SE Germany 49°11′N 12°01′E
41 P9 Burgos Tamaulipas, C Mexico 24°55′N 98°46′W
105 N4 Burgos Castilla y León, N Spain 42°21′N 03°41′W
105 N4 Burgos ◈ province Castilla y León, N Spain
Burgstadlberg see Hradiště
95 P20 Burgsvik Gotland, SE Sweden 57°01′N 18°18′E
98 L6 Burgum Dutch. Bergum. Fryslân, N Netherlands 53°12′N 05°59′E
Burgundy see Bourgogne
159 Q11 Burhan Budai Shan ▲ C China
136 B12 Burhaniye Balıkesir, W Turkey 39°29′N 26°59′E
154 G12 Burhānpur Madhya Pradesh, C India 21°18′N 76°14′E
127 W7 Buribay Respublika Bashkortostan, W Russian Federation 51°57′N 58°11′E
43 O17 Burica, Punta headland Costa Rica/Panama 08°02′N 82°53′W
167 Q10 Buriram var. Buri Ram, Puriramya. Buri Ram, E Thailand 15°01′N 103°06′E
Buri Ram see Buriram
105 S10 Burjassot Valenciana, E Spain 39°33′N 00°26′W
81 N16 Burka Giibi Hiiraan, C Somalia 03°52′N 45°07′E
147 X8 Burkan ≈ E Kyrgyzstan
25 R4 Burkburnett Texas, SW USA 34°06′N 98°34′W
29 O12 Burke South Dakota, N USA 43°09′N 99°18′W
10 K15 Burke Channel channel British Columbia, W Canada
194 J10 Burke Island island Antarctica
20 L7 Burkesville Kentucky, S USA 36°48′N 85°21′W
181 T4 Burketown Queensland, NE Australia 17°49′S 139°28′E
25 Q8 Burkett Texas, SW USA 32°01′N 99°17′W
25 Y9 Burkeville Texas, SW USA 30°58′N 93°41′W
21 V7 Burkeville Virginia, NE USA 37°11′N 78°12′W
Burkina see Burkina Faso
77 O12 Burkina Faso off. Burkina Faso; prev. Upper Volta. ◆ republic W Africa
Burkina Faso see Burkina Faso
194 L13 Burks, Cape headland Antarctica
14 H12 Burk's Falls Ontario, S Canada 45°38′N 79°25′W
101 H23 Burladingen Baden-Württemberg, S Germany 48°18′N 09°05′E
25 T7 Burleson Texas, SW USA 32°32′N 97°19′W
33 P15 Burley Idaho, NW USA 42°32′N 113°47′W
Burlin see Borili
14 G16 Burlington Ontario, S Canada 43°19′N 79°48′W
37 W4 Burlington Colorado, C USA 39°17′N 102°17′W
29 Y15 Burlington Iowa, C USA 40°48′N 91°05′W
27 P5 Burlington Kansas, C USA 38°11′N 95°46′W
21 T9 Burlington North Carolina, SE USA 36°05′N 79°27′W
28 M3 Burlington North Dakota, N USA 48°16′N 101°25′W
18 L7 Burlington Vermont, NE USA 44°28′N 73°14′W
30 M9 Burlington Wisconsin, N USA 42°38′N 88°12′W
27 Q1 Burlington Junction Missouri, C USA 40°27′N 95°04′W
Burma see Myanmar (Burma)
10 L17 Burnaby British Columbia, SW Canada 49°16′N 122°58′W
117 O12 Burnas, Ozero ◎ SW Ukraine
25 S10 Burnet Texas, SW USA 30°46′N 98°14′W
35 O3 Burney California, W USA 40°52′N 121°42′W
183 O16 Burnie Tasmania, SE Australia 41°07′S 145°52′E
97 L17 Burnley NW England, United Kingdom 53°48′N 02°14′W
Burnoye see Bauyrzhan Momyshuly
153 R15 Burnpur West Bengal, NE India 23°39′N 86°55′E
32 K14 Burns Oregon, NW USA 43°35′N 119°03′W
26 K11 Burns Flat Oklahoma, C USA 35°21′N 99°10′W
20 M7 Burnside Kentucky, S USA 36°55′N 84°34′W
8 K8 Burnside ≈ Nunavut, NW Canada
32 L15 Burns Junction Oregon, NW USA 42°46′N 117°51′W
10 L13 Burns Lake British Columbia, SW Canada 54°14′N 125°45′W
29 V9 Burnsville Minnesota, N USA 44°49′N 93°14′W
21 P9 Burnsville North Carolina, SE USA 35°56′N 82°18′W
21 R4 Burnsville West Virginia, NE USA 38°50′N 80°39′W
14 I13 Burnt River ≈ Ontario, SE Canada
14 I11 Burntroot Lake ◎ Ontario, SE Canada
11 W12 Burntwood ≈ Manitoba, C Canada
Bur'o see Burco
158 L2 Burqin Xinjiang Uygur Zizhiqu, NW China 47°42′N 86°50′E
182 J8 Burra South Australia 33°41′S 138°54′E
183 S9 Burragorang, Lake ⊠ New South Wales, SE Australia
96 K5 Burray island NE Scotland, United Kingdom
113 L19 Burrel var. Burreli. Dibër, C Albania 41°36′N 20°00′E
Burreli see Burrel
183 R8 Burrendong Reservoir ⊠ New South Wales, SE Australia
183 R5 Burren Junction New South Wales, SE Australia 30°06′S 149°01′E
Burriana see Borriana
183 R10 Burrinjuck Reservoir ⊠ New South Wales, SE Australia
36 J12 Burro Creek ≈ Arizona, SW USA
40 M5 Burro, Serranías del ▲ NW Mexico
62 K7 Burruyacú Tucumán, N Argentina 26°30′S 64°45′W
136 E12 Bursa var. Brussa, prev. Brusa; anc. Prusa. Bursa, NW Turkey 40°12′N 29°04′E
136 D12 Bursa var. Brusa, Brussa. ◈ province NW Turkey
75 Y9 Būr Safājah var. Būr Safājah. E Egypt 26°43′N 33°55′E
Būr Safājah see Būr Safājah
75 W7 Būr Saʿīd var. Port Said. N Egypt 31°17′N 32°18′E
81 O14 Bur Tinle Nugaal, C Somalia 07°50′N 48°01′E
31 Q5 Burt Lake ◎ Michigan, N USA
118 H7 Burtnieks var. Burtnieks Ezers. ◎ N Latvia
Burtnieks Ezers see Burtnieks
31 Q9 Burton Michigan, N USA 43°00′N 84°16′W
Burton on Trent see Burton upon Trent
97 M19 Burton upon Trent var. Burton on Trent, Burton-upon-Trent. C England, United Kingdom 52°48′N 01°36′W
93 J15 Burträsk Västerbotten, N Sweden 64°31′N 20°40′E
Burubaytal see Burybaytal
Burujird see Borūjerd
Burultokay see Fuhai
141 R15 Burūm SE Yemen 14°22′N 48°53′E
Burunday see Boralday
81 D21 Burundi off. Republic of Burundi; prev. Kingdom of Burundi, Urundi. ◆ republic C Africa
Burundi, Kingdom of see Burundi
Burundi, Republic of see Burundi
171 R13 Buru, Pulau prev. Boeroe. island E Indonesia
77 T17 Burutu Delta, S Nigeria 05°18′N 05°32′E
10 G7 Burwash Landing Yukon, W Canada 61°26′N 139°12′W
29 O14 Burwell Nebraska, C USA 41°46′N 99°04′W
97 L17 Bury NW England, United Kingdom 53°36′N 02°17′W
123 N13 Buryatiya, Respublika prev. Buryatskaya ASSR. ◈ autonomous republic S Russian Federation
Buryatskaya ASSR see Buryatiya, Respublika
145 S14 Burybaytal prev. Burubaytal. Zhambyl, SE Kazakhstan 44°56′N 73°59′E
117 S3 Buryn' Sums'ka Oblast', NE Ukraine 51°13′N 33°50′E
97 P20 Bury St Edmunds hist. Beodericsworth. E England, United Kingdom 52°15′N 00°43′E
Būrziya see Barzia
106 D9 Busalla Liguria, NW Italy 44°35′N 08°55′E
163 Z16 Busan off. Pusan-gwangyŏksi, var. Vusan; prev. Pusan, Jap. Fusan. SE South Korea 35°11′N 129°04′E
139 N5 Buşayrah Dayr az Zawr, E Syria 35°10′N 40°25′E
Buševa see Baba
143 N12 Būshehr off. Ostān-e Būshehr. ◈ province SW Iran
Būshehr/Bushire see Bandar-e Būshehr
Būshehr, Ostān-e see Būshehr
25 N2 Bushland Texas, SW USA 35°11′N 102°04′W
30 J12 Bushnell Illinois, N USA 40°33′N 90°30′W
Busi see Biševo
81 H18 Busia Western, W Kenya 0°28′N 34°07′E
81 G18 Busia Eastern, SE Uganda 01°20′N 34°48′E
81 G18 Busia ◈ county W Kenya
Busiasch see Buziaş
79 K16 Businga Equateur, NW Dem. Rep. Congo 03°20′N 20°53′E
79 J18 Busira ≈ NW Dem. Rep. Congo
116 I5 Bus'k Rus. Busk. L'vivs'ka Oblast', W Ukraine 49°59′N 24°34′E
95 E14 Buskerud ◈ county S Norway
113 F14 Buško Jezero ⊠ SW Bosnia and Herzegovina
111 M15 Busko-Zdrój Świętokrzyskie, C Poland 50°28′N 20°44′E
Busra see Al Başrah, Iraq
Buşrá see Buşrá ash Shām, Syria
138 H9 Buşrá ash Shām var. Bosora, Bosra, Bozrah, Buşrá. Dar'ā, S Syria 32°31′N 36°29′E
180 I13 Busselton Western Australia 33°43′S 115°15′E
81 C14 Busseri ≈ W South Sudan
106 E9 Busseto Emilia-Romagna, C Italy 45°00′N 10°06′E
106 A8 Bussoleno Piemonte, NE Italy 45°11′N 07°07′E
Bussora see Al Başrah
41 N7 Bustamante Nuevo León, NE Mexico 26°29′N 100°30′W
63 I23 Bustamante, Punta headland S Argentina 51°35′S 68°58′W
Bustan see Bo'ston
116 J12 Buşteni Prahova, SE Romania 45°23′N 25°32′E
106 D7 Busto Arsizio Lombardia, N Italy 45°37′N 08°51′E
147 Q10 Bŭston Rus. Buston. NW Tajikistan 40°31′N 69°21′E
100 H8 Büsum Schleswig-Holstein, N Germany 54°08′N 08°52′E
98 J10 Bussum Noord-Holland, C Netherlands 52°17′N 05°10′E
79 M16 Buta Orientale, N Dem. Rep. Congo 02°50′N 24°41′E
81 E20 Butare prev. Astrida. S Rwanda 02°39′S 29°44′E
191 O2 Butaritari atoll Tungaru, W Kiribati
Butawal see Butwal
96 H13 Bute cultural region SW Scotland, United Kingdom
162 K6 Büteeliyn Nuruu ▲ N Mongolia
10 L16 Bute Inlet fjord British Columbia, W Canada
96 H12 Bute, Island of island SW Scotland, United Kingdom
79 P18 Butembo Nord-Kivu, NE Dem. Rep. Congo 0°09′N 29°17′E
Bûtenpost see Buitenpost
107 K25 Butera Sicilia, Italy, C Mediterranean Sea 37°12′N 14°12′E
99 M20 Bütgenbach Liège, E Belgium 50°25′N 06°12′E
Butha Qi see Zalantun
166 J5 Buthidaung Rakhine State, W Myanmar (Burma) 20°50′N 92°25′E
61 I16 Butiá Rio Grande do Sul, S Brazil 30°09′S 51°55′W
81 F17 Butiaba NW Uganda 01°48′N 31°21′E
23 N6 Butler Alabama, S USA 32°05′N 88°13′W
23 S5 Butler Georgia, SE USA 32°33′N 84°14′W
31 Q11 Butler Indiana, N USA 41°25′N 84°52′W
27 R5 Butler Missouri, C USA 38°17′N 94°21′W
18 B14 Butler Pennsylvania, NE USA 40°51′N 79°52′W
194 K5 Butler Island island Antarctica
21 U8 Butner North Carolina, SE USA 36°07′N 78°45′W
171 P14 Buton, Pulau var. Pulau Butung; prev. Boetoeng. island C Indonesia
Bütow see Bytów
113 L23 Butrintit, Liqeni i ◎ S Albania
23 N3 Buttahatchee River ≈ Alabama/Mississippi, S USA
33 Q10 Butte Montana, NW USA 46°01′N 112°33′W
29 O12 Butte Nebraska, C USA 42°54′N 98°51′W
168 J7 Butterworth Pinang, Peninsular Malaysia 05°24′N 100°22′E
83 J25 Butterworth var. Gcuwa. Eastern Cape, SE South Africa 32°20′S 28°09′E
13 O3 Button Islands island group Nunavut, NE Canada
35 R13 Buttonwillow California, W USA 35°24′N 119°26′W
171 Q7 Butuan off. Butuan City. Mindanao, S Philippines 08°57′N 125°33′E
Butuan City see Butuan
Butung, Pulau see Buton, Pulau
Butuntum see Bitonto
126 M8 Buturlinovka Voronezhskaya Oblast', W Russian Federation 50°48′N 40°33′E
153 O11 Butwal var. Butawal. Western, C Nepal 27°41′N 83°28′E
101 G17 Butzbach Hessen, W Germany 50°26′N 08°40′E
100 L9 Bützow Mecklenburg-Vorpommern, N Germany 53°49′N 11°58′E
80 N13 Buuhoodle Togdheer, N Somalia 08°18′N 46°15′E
81 N16 Buulobarde var. Buulo Berde. Hiiraan, C Somalia 03°52′N 45°37′E
Buulo Berde see Buulobarde
80 P12 Buuraha Cal Miskaat ▲ NE Somalia
81 L19 Buur Gaabo Jubbada Hoose, S Somalia 01°14′S 41°48′E
99 M22 Buurgplaatz ▲ N Luxembourg 50°09′N 06°02′E
162 H8 Buutsagaan var. Buyant. Bayanhongor, C Mongolia 46°07′N 98°45′E
Buwayrāt al Hasūn see Buʿayrāt al Ḩasūn
146 L11 Buxoro var. Bokhara, Rus. Bukhara. Buxoro Viloyati, C Uzbekistan 39°51′N 64°23′E
146 J11 Buxoro Viloyati Rus. Bukharskaya Oblast'. ◈ province C Uzbekistan
100 I10 Buxtehude Niedersachsen, NW Germany 53°29′N 09°42′E
97 L18 Buxton C England, United Kingdom 53°18′N 01°52′W
124 M14 Buy var. Buj. Kostromskaya Oblast', NW Russian Federation 58°27′N 41°31′E
Buyanbat see Jargalant
162 D6 Buyant Bayan-Ölgiy, W Mongolia 48°31′N 89°36′E
Buyant see Buutsagaan, Bayanhongor, Mongolia
Buyant see Otgon, Dzavhan, Mongolia
Buyant see Galshar, Hentiy, Mongolia
163 N10 Buyant-Uhaa Dornogovĭ, SE Mongolia 44°52′N 110°12′E
162 M7 Buyant Ukha ✕ (Ulaanbaatar) Töv, N Mongolia
127 Q16 Buynaksk Respublika Dagestan, SW Russian Federation 42°53′N 47°03′E
119 L20 Buynavichy Rus. Buynovichi. Homyel'skaya Voblasts', SE Belarus 51°52′N 28°33′E
Buynovichi see Buynavichy
76 L16 Buyo SW Ivory Coast 06°16′N 07°03′W
76 L16 Buyo, Lac de ⊠ W Ivory Coast
163 R7 Buyr Nuur var. Buir Nur. ◎ China/Mongolia see also Buir Nur
Buyr Nuur see Buir Nur
137 T13 Büyükağrı Dağı var. Aghri Dagh, Agri Dagi, Koh I Noh, Masis, Eng. Great Ararat, Mount Ararat. ▲ E Turkey 39°43′N 44°19′E
137 R15 Büyük Çayı ≈ NE Turkey
114 O13 Büyük Çekmece İstanbul, NW Turkey 41°02′N 28°35′E
114 N12 Büyükkarıştıran Kirklareli, NW Turkey 41°17′N 27°33′E
115 L14 Büyükkemikli Burnu cape NW Turkey
136 E15 Büyükmenderes Nehri ≈ SW Turkey
Büyükzap Suyu see Great Zab
102 M9 Buzançais Indre, C France 46°53′N 01°25′E
116 K13 Buzău Buzău, SE Romania 45°08′N 26°51′E
116 K13 Buzău ◈ county SE Romania
116 L12 Buzău ≈ E Romania
75 S11 Buzaymah var. Bzīmah. SE Libya 24°53′N 22°01′E
164 E13 Buzen Fukuoka, Kyūshū, SW Japan 33°37′N 131°06′E
116 F12 Buziaş Ger. Busiasch, Hung. Buziásfürdő; prev. Buziás. Timiş, W Romania 45°38′N 21°36′E
Buziás see Buziaş
Buziásfürdő see Buziaş
83 M18 Búzi, Rio ≈ C Mozambique
117 Q10 Buz'kyy Lyman bay S Ukraine
Büzmeýin see Abadan
127 T6 Buzuluk Orenburgskaya Oblast', W Russian Federation 52°47′N 52°16′E
127 N8 Buzuluk ≈ SW Russian Federation
Buzuluk see Buzylyk
145 O8 Buzylyk prev. Buzuluk. Akmola, C Kazakhstan 51°53′N 66°09′E
19 P12 Buzzards Bay Massachusetts, NE USA 41°45′N 70°37′W
19 P13 Buzzards Bay bay Massachusetts, NE USA
83 G16 Bwabata Caprivi, NE Namibia 17°52′S 22°39′E
186 H10 Bwagaoia Misima Island, SE Papua New Guinea 10°39′S 152°48′E
Bwake see Bouaké
187 R13 Bwatnapne Pentecost, C Vanuatu 15°42′S 168°07′E
119 K14 Byahoml' Rus. Begoml'. Vitsyebskaya Voblasts', N Belarus 54°44′N 28°04′E
114 K8 Byala Ruse, N Bulgaria 43°27′N 25°44′E
114 N9 Byala prev. Ak-Dere. Varna, E Bulgaria 42°52′N 27°53′E
Byala Reka see Erythropótamos
114 H8 Byala Slatina Vratsa, NW Bulgaria 43°28′N 23°56′E
119 N15 Byalynichy Rus. Belynichi. Mahilyowskaya Voblasts', E Belarus 54°00′N 29°42′E
Byan Tumen see Choybalsan
119 L14 Byarezina prev. Byerezino, Rus. Berezina. ≈ C Belarus
119 G19 Byaroza Pol. Bereza Kartuska, Rus. Berëza. Brestskaya Voblasts', SW Belarus 52°32′N 24°59′E
119 H16 Byarozawka Rus. Berëzovka. Hrodzyenskaya Voblasts', W Belarus 53°45′N 25°30′E
Bybles see Jbail
111 O14 Bychawa Lubelskie, SE Poland 51°06′N 22°34′E
Bychikha see Bychykha
118 N11 Bychykha Rus. Bychikha. Vitsyebskaya Voblasts', NE Belarus 55°41′N 29°59′E
111 I14 Byczyna Ger. Pitschen. Opolskie, S Poland 51°06′N 18°13′E
110 I10 Bydgoszcz Ger. Bromberg. Kujawski-pomorskie, C Poland 53°06′N 18°00′E
119 H19 Byelaazyorsk Rus. Beloozersk. Brestskaya Voblasts', SW Belarus 52°28′N 25°10′E
119 I17 Byelaruskaya Hrada Rus. Belorusskaya Gryada. ridge N Belarus
119 G18 Byelavyezhskaya Pushcha Pol. Puszcza Białowieska, Rus. Belovezhskaya Pushcha. forest Belarus/Poland see also Białowieska, Puszcza
Byelavyezhskaya, Pushcha see Białowieska, Puszcza
119 H15 Byenyakoni Rus. Benyakoni. Hrodzyenskaya Voblasts', W Belarus 54°15′N 25°22′E
119 M16 Byerazino Rus. Berezino. Minskaya Voblasts', C Belarus 53°50′N 29°00′E
118 L13 Byerazino Rus. Berezino. Vitsyebskaya Voblasts', N Belarus 54°54′N 28°12′E
Byerezino see Byarezina
118 M13 Byeshankovichy Rus. Beshenkovichi. Vitsyebskaya Voblasts', N Belarus 55°03′N 29°27′E
31 U13 Byesville Ohio, N USA 39°58′N 81°32′W
119 P18 Byesyedz' Rus. Besed'. ≈ SE Belarus
119 H19 Byezdzyezh Rus. Bezdezh. Brestskaya Voblasts', SW Belarus 52°19′N 25°18′E
93 J15 Bygdeå Västerbotten, N Sweden 64°03′N 20°49′E
94 F12 Bygdin ◎ S Norway
93 J15 Bygdsiljum Västerbotten, N Sweden 64°20′N 20°31′E
95 E17 Bygland Aust-Agder, S Norway 58°46′N 07°50′E
95 E17 Byglandsfjord Aust-Agder, S Norway 58°42′N 07°51′E
119 N16 Bykhaw Rus. Bykhov. Mahilyowskaya Voblasts', E Belarus 53°31′N 30°15′E
Bykhov see Bykhaw
127 P9 Bykovo Volgogradskaya Oblast', SW Russian Federation 49°52′N 45°24′E
123 P7 Bykovskiy Respublika Sakha (Yakutiya), NE Russian Federation 71°57′N 129°07′E
195 R12 Byrd Glacier glacier Antarctica
14 K10 Byrd, Lac ◎ Québec, SE Canada
183 P5 Byrock New South Wales, SE Australia 30°40′S 146°24′E
30 L10 Byron Illinois, N USA 42°06′N 89°15′W
183 V4 Byron Bay New South Wales, SE Australia 28°39′S 153°34′E
183 V4 Byron, Cape headland New South Wales, E Australia 28°37′S 153°40′E
63 F21 Byron, Isla island S Chile
Byron Island see Nikunau
65 B24 Byron Sound sound NW Falkland Islands
122 M6 Byrranga, Gory ▲ N Russian Federation
93 J14 Byske Västerbotten, N Sweden 64°58′N 21°10′E
111 K18 Bystrá ▲ N Slovakia 49°10′N 19°49′E
111 F18 Bystřice nad Pernštejnem Ger. Bistritz ober Pernstein. Vysočina, C Czech Republic 49°32′N 16°16′E
Bystrovka see Kemin
111 G16 Bystrzyca Kłodzka Ger. Habelschwerdt. Wałbrzych, SW Poland 50°19′N 16°39′E
111 J18 Bytča Žilinský Kraj, N Slovakia 49°15′N 18°32′E
119 L15 Bytcha Minskaya Voblasts', NE Belarus 54°19′N 28°24′E
Byteń/Byten' see Bytsyen'
111 J16 Bytom Ger. Beuthen. Śląskie, S Poland 50°21′N 18°51′E
110 H7 Bytów Ger. Bütow. Pomorskie, N Poland 54°10′N 17°30′E
119 H18 Bytsyen' Pol. Byteń, Rus. Byten'. Brestskaya Voblasts', SW Belarus 52°53′N 25°30′E
81 E19 Byumba var. Biumba. N Rwanda 01°37′S 30°05′E
Byuzmeyin see Abadan
119 O20 Byval'ki Homyel'skaya Voblasts', SE Belarus 51°51′N 30°38′E
95 O20 Byxelkrok Kalmar, S Sweden 57°18′N 17°01′E
Byzantium see İstanbul
Bzīmah see Buzaymah

C

62 O6 Caacupé Cordillera, S Paraguay 25°23′S 57°05′W
62 P6 Caaguazú off. Departamento de Caaguazú. ◈ department C Paraguay
Caaguazú, Departamento de see Caaguazú
82 C13 Caála var. Kaala, Robert Williams, Port. Vila Robert Williams. Huambo, C Angola 12°51′S 15°33′E
62 P7 Caazapá Caazapá, S Paraguay 26°09′S 56°21′W
62 P7 Caazapá off. Departamento de Caazapá. ◈ department SE Paraguay
Caazapá, Departamento de see Caazapá
81 P15 Cabaad, Raas headland C Somalia 06°13′N 49°01′E
55 N10 Cabadisocaña Amazonas, S Venezuela 04°28′N 64°45′W
44 F5 Cabaiguán Sancti Spíritus, C Cuba 22°04′N 79°32′W
Caballeria, Cabo see Cavalleria, Cap de
37 Q14 Caballo Reservoir ⊠ New Mexico, SW USA
40 L6 Caballos Mesteños, Llano de los plain N Mexico
104 L2 Cabanaquinta Asturias, N Spain 43°10′N 05°37′W
42 B9 Cabañas ◈ department E El Salvador
171 O3 Cabanatuan off. Cabanatuan City. Luzon, N Philippines 15°27′N 120°57′E
Cabanatuan City see Cabanatuan
15 T8 Cabano Québec, SE Canada 47°40′N 68°56′W
104 L11 Cabeza del Buey Extremadura, W Spain 38°44′N 05°13′W
45 V5 Cabezas de San Juan headland E Puerto Rico 18°23′N 65°37′W
105 N2 Cabezón de la Sal Cantabria, N Spain 43°19′N 04°14′W
Cabhán see Cavan
61 B23 Cabildo Buenos Aires, E Argentina 38°28′S 61°50′W
Cabillonum see Chalon-sur-Saône
54 H5 Cabimas Zulia, NW Venezuela 10°26′N 71°27′W
82 A9 Cabinda var. Kabinda. Cabinda, NW Angola 05°34′S 12°12′E
82 A9 Cabinda var. Kabinda. ◈ province NW Angola
33 N7 Cabinet Mountains ▲ Idaho/Montana, NW USA
82 B11 Cabiri Bengo, NW Angola 08°50′S 13°42′E
63 J20 Cabo Blanco Santa Cruz, SE Argentina 47°13′S 65°43′W
82 P13 Cabo Delgado off. Província de Cabo Delgado. ◈ province NE Mozambique
14 L9 Cabonga, Réservoir ⊠ Québec, SE Canada
27 V7 Cabool Missouri, C USA 37°07′N 92°06′W
183 V2 Caboolture Queensland, E Australia 27°05′S 152°50′E
Cabora Bassa, Lake see Cahora Bassa, Albufeira de
40 F3 Caborca Sonora, NW Mexico 30°44′N 112°06′W
Cabo San Lucas see San Lucas
27 V11 Cabot Arkansas, C USA 34°58′N 92°01′W
14 F12 Cabot Head headland Ontario, S Canada 45°13′N 81°17′W
13 R13 Cabot Strait strait E Canada
Cabo Verde, Ilhas do see Cape Verde
104 M14 Cabra Andalucía, S Spain 37°28′N 04°28′W
107 B19 Cabras Sardegna, Italy, C Mediterranean Sea 39°55′N 08°30′E
188 A15 Cabras Island island W Guam
45 O8 Cabrera N Dominican Republic 19°40′N 69°54′W
104 J4 Cabrera ≈ NW Spain
105 X10 Cabrera, Illa de anc. Capraria. island Islas Baleares, Spain, W Mediterranean Sea
105 Q15 Cabrera, Sierra ▲ S Spain
11 S16 Cabri Saskatchewan, S Canada 50°38′N 108°28′W
105 R10 Cabriel ≈ E Spain
54 M7 Cabruta Guárico, C Venezuela 07°39′N 66°19′W
171 N2 Cabugao Luzon, N Philippines 17°55′N 120°29′E
54 G10 Cabuyaro Meta, C Colombia 04°21′N 72°47′W
60 I13 Caçador Santa Catarina, S Brazil 26°47′S 51°00′W
42 G8 Cacaguatique, Cordillera var. Cordillera. ▲ NE El Salvador
112 L13 Čačak Serbia, C Serbia 43°52′N 20°23′E
55 Y10 Cacao NE French Guiana 04°37′N 52°29′W
61 H16 Caçapava do Sul Rio Grande do Sul, S Brazil 30°28′S 53°29′W
21 U3 Cacapon River ≈ West Virginia, NE USA
107 J23 Caccamo Sicilia, Italy, C Mediterranean Sea 37°56′N 13°40′E
107 A17 Caccia, Capo headland Sardegna, Italy, C Mediterranean Sea 40°34′N 08°09′E
146 H15 Çäçe var. Chäche, Rus. Chaacha. Ahal Welaýaty, S Turkmenistan 36°49′N 60°33′E
59 G18 Cáceres Mato Grosso, W Brazil 16°05′S 57°40′W
104 J10 Cáceres Ar. Qazris. Extremadura, W Spain 39°29′N 06°23′W
104 J9 Cáceres ◈ province Extremadura, W Spain
Cachacrou see Scotts Head Village
61 C21 Cacharí Buenos Aires, E Argentina 36°24′S 59°32′W
26 L12 Cache Oklahoma, C USA 34°37′N 98°37′W
10 M16 Cache Creek British Columbia, SW Canada 50°49′N 121°20′W
35 N6 Cache Creek ≈ California, W USA
37 S3 Cache La Poudre River ≈ Colorado, C USA
Cacheo see Cacheu
27 W11 Cache River ≈ Arkansas, C USA
30 L17 Cache River ≈ Illinois, N USA
76 G12 Cacheu var. Cacheo. W Guinea-Bissau 12°12′N 16°10′W
59 I15 Cachimbo Pará, NE Brazil 09°21′S 54°58′W
59 H15 Cachimbo, Serra do ▲ C Brazil
82 D13 Cachingues Bié, C Angola 13°05′S 16°48′E
54 G7 Cáchira Norte de Santander, N Colombia 07°44′N 73°07′W
61 H16 Cachoeira do Sul Rio Grande do Sul, S Brazil 29°58′S 52°54′W
59 O20 Cachoeiro de Itapemirim Espírito Santo, SE Brazil 20°51′S 41°07′W
82 E12 Cacolo Lunda Sul, NE Angola 10°09′S 19°21′E
83 C14 Caconda Huíla, C Angola 13°43′S 15°03′E
82 A9 Cacongo Cabinda, NW Angola 05°13′S 12°08′E
35 U9 Cactus Peak ▲ Nevada, W USA 37°42′N 116°51′W
82 A11 Cacuaco Luanda, NW Angola 08°47′S 13°21′E
83 B14 Cacula Huíla, SW Angola 14°33′S 14°04′E
67 R12 Caculuvar ≈ SW Angola
59 O19 Caçumba, Ilha island SE Brazil
55 N10 Cacuri Amazonas, S Venezuela
81 N17 Cadale Shabeellaha Dhexe, E Somalia 02°48′N 46°19′E
105 X4 Cadaqués Cataluña, NE Spain 42°17′N 03°16′E
111 J18 Čadca Hung. Csaca. Žilinský Kraj, N Slovakia 49°27′N 18°46′E
27 P13 Caddo Oklahoma, C USA 34°07′N 96°15′W
25 R6 Caddo Texas, SW USA 32°42′N 98°40′W
25 X6 Caddo Lake ⊠ Louisiana/Texas, SW USA
27 S12 Caddo Mountains ▲ Arkansas, C USA
41 O8 Cadereyta Nuevo León, NE Mexico 25°35′N 99°54′W
97 J19 Cader Idris ▲ NW Wales, United Kingdom 52°43′N 03°57′W
182 F3 Cadibarrawirracanna, Lake salt lake South Australia
14 I7 Cadillac Québec, SE Canada 48°12′N 78°23′W
11 T17 Cadillac Saskatchewan, S Canada 49°43′N 107°41′W
102 K13 Cadillac Gironde, SW France 44°37′N 00°16′W
31 P7 Cadillac Michigan, N USA 44°15′N 85°23′W
105 V4 Cadí, Toretta de prev. Torre de Cadí. ▲ NE Spain 42°16′N 01°38′E
Torre de Cadí see Cadí, Toretta de
171 P5 Cadiz off. Cadiz City. Negros, C Philippines 10°58′N 123°18′E
104 J15 Cádiz anc. Gades, Gadier, Gadir, Gadire. Andalucía, SW Spain 36°32′N 06°18′W
20 H7 Cadiz Kentucky, S USA 36°52′N 87°50′W
31 U13 Cadiz Ohio, N USA 40°16′N 81°00′W
104 K15 Cádiz ◈ province Andalucía, SW Spain
104 I15 Cádiz, Bahía de bay SW Spain
Cadiz City see Cadiz
104 H15 Cádiz, Golfo de Eng. Gulf of Cadiz. gulf Portugal/Spain
Cadiz, Gulf of see Cádiz, Golfo de
35 X14 Cadiz Lake ◎ California, W USA

◆ Country
● Country Capital
◇ Dependent Territory
○ Dependent Territory Capital
◈ Administrative Regions
✕ International Airport
▲ Mountain
▲ Mountain Range
Volcano
≈ River
◎ Lake
⊠ Reservoir

CADNEY HOMESTEAD – CANTON

182 E2 **Cadney Homestead** South Australia 27°52′S 134°03′E
Cadurcum *see* Cahors
Caecae *see* Xaixai
102 K4 **Caen** Calvados, N France 49°10′N 00°20′W
Caene/Caenepolis *see* Qinā
Caerdydd *see* Cardiff
Caer Glou *see* Gloucester
Caer Gybi *see* Holyhead
Caerleon *see* Chester
Caer Luel *see* Carlisle
97 I18 **Caernarfon** *var.* Caernarvon, Carnarvon. NW Wales, United Kingdom 53°08′N 04°16′W
97 H18 **Caernarfon Bay** *bay* NW Wales, United Kingdom
97 I19 **Caernarvon** *cultural region* NW Wales, United Kingdom
Caernarvon *see* Caernarfon
Caesaraugusta *see* Zaragoza
Caesarea Mazaca *see* Kayseri
Caesarobriga *see* Talavera de la Reina
Caesarodunum *see* Tours
Caesaromagus *see* Beauvais
Caesena *see* Cesena
59 N17 **Caetité** Bahia, E Brazil 14°04′S 42°29′W
62 J6 **Cafayate** Salta, N Argentina 26°02′S 66°00′W
171 O2 **Cagayan** ≈ Luzon, N Philippines
171 Q7 **Cagayan de Oro** *off.* Cagayan de Oro City. Mindanao, S Philippines 08°29′N 124°38′E
Cagayan de Oro City *see* Cagayan de Oro
170 M8 **Cagayan de Tawi Tawi** *island* S Philippines
171 N6 **Cagayan Islands** *island group* C Philippines
31 O14 **Cagles Mill Lake** ⊠ Indiana, N USA
106 I12 **Cagli** Marche, C Italy 43°33′N 12°39′E
107 C20 **Cagliari** *anc.* Caralis. Sardegna, Italy, C Mediterranean Sea 39°15′N 09°06′E
107 C20 **Cagliari, Golfo di** *gulf* Sardegna, Italy, C Mediterranean Sea
103 U15 **Cagnes-sur-Mer** Alpes-Maritimes, SE France 43°40′N 07°09′E
54 L5 **Cagua** Aragua, N Venezuela 10°09′N 67°27′W
171 O1 **Cagua, Mount** ▲ Luzon, N Philippines 18°10′N 122°03′E
54 F13 **Caguán, Río** ≈ SW Colombia
45 U6 **Caguas** E Puerto Rico 18°14′N 66°02′W
146 C9 **Çagyl** *Rus.* Chagyl. Balkan Welaýaty, NW Turkmenistan 40°48′N 55°21′E
23 P5 **Cahaba River** ≈ Alabama, S USA
42 E5 **Cahabón, Río** ≈ C Guatemala
83 B15 **Cahama** Cunene, SW Angola 16°16′S 14°23′E
97 B21 **Caha Mountains** *Ir.* An Cheacha. ▲▲ SW Ireland
97 D20 **Caher** *Ir.* An Cathair. S Ireland 52°21′N 07°58′W
97 A21 **Caherciveen** *Ir.* Cathair Saidhbhín. SW Ireland 51°56′N 10°12′W
30 K15 **Cahokia** Illinois, N USA 38°34′N 90°11′W
83 L15 **Cahora Bassa, Albufeira de** *var.* Lake Cabora Bassa. ⊠ NW Mozambique
97 G20 **Cahore Point** *Ir.* Rinn Chathóir. *headland* SE Ireland 52°33′N 06°11′W
102 M14 **Cahors** *anc.* Cadurcum. Lot, S France 44°26′N 01°27′E
56 D9 **Cahuapanas, Río** ≈ N Peru
116 M12 **Cahul** *Rus.* Kagul. S Moldova 45°53′N 28°13′E
Cahul, Lacul *see* Kahul, Ozero
83 N16 **Caia** Sofala, C Mozambique 17°50′S 35°21′E
59 J19 **Caiapó, Serra do** ▲▲ C Brazil
44 F5 **Caibarién** Villa Clara, C Cuba 22°31′N 79°29′W
55 O5 **Caicara** Monagas, NE Venezuela 09°52′N 63°38′W
54 L5 **Caicara del Orinoco** Bolívar, C Venezuela 07°38′N 66°10′W
59 P14 **Caicó** Rio Grande do Norte, E Brazil 06°25′S 37°04′W
44 M6 **Caicos Islands** *island group* W Turks and Caicos Islands
44 L5 **Caicos Passage** *strait* The Bahamas/Turks and Caicos Islands
161 O9 **Caidian** *prev.* Hanyang. Hubei, C China 30°37′N 114°02′E
Caiffa *see* Hefa
180 M12 **Caiguna** Western Australia 32°14′S 125°33′E
Caillí, Ceann *see* Hag's Head
40 J11 **Caimanero, Laguna del** *var.* Laguna del Camaronero. *lagoon* E Pacific Ocean
117 N10 **Căinari** *Rus.* Kaynary. C Moldova 46°43′N 29°00′E
57 L19 **Caine, Río** ≈ C Bolivia
Caiphas *see* Hefa
195 N4 **Caird Coast** *physical region* Antarctica
96 J9 **Cairn Gorm** ▲ C Scotland, United Kingdom 57°07′N 03°38′W
96 J9 **Cairngorm Mountains** ▲▲ C Scotland, United Kingdom
39 P12 **Cairn Mountain** ▲ Alaska, USA 61°07′N 155°23′W
181 W4 **Cairns** Queensland, NE Australia 16°51′S 145°43′E
121 V13 **Cairo** *var.* El Qâhira, *Ar.* Al Qāhirah. ● (Egypt) N Egypt 30°01′N 31°18′E
23 T8 **Cairo** Georgia, SE USA 30°52′N 84°12′W
30 L17 **Cairo** Illinois, N USA 37°00′N 89°10′W
75 V8 **Cairo** ✈ C Egypt 30°06′N 31°36′E
Caiseal *see* Cashel
Caisleán an Bharraigh *see* Castlebar
Caisleán na Finne *see* Castlefinn
96 J6 **Caithness** *cultural region* N Scotland, United Kingdom
83 D15 **Caiundo** Kuando Kubango, S Angola 15°41′S 17°28′E
56 C11 **Cajamarca** *prev.* Caxamarca. Cajamarca, NW Peru 07°09′S 78°32′W
56 B11 **Cajamarca** *off.* Departamento de Cajamarca. ◈ *department* N Peru
Cajamarca, Departamento de *see* Cajamarca
103 N14 **Cajarc** Lot, S France 44°28′N 01°51′E
42 G6 **Cajón, Represa El** ⊠ NW Honduras
58 N12 **Caju, Ilha do** *island* NE Brazil
159 R10 **Caka Yanhu** ◎ C China
112 E7 **Čakovec** *Ger.* Csakathurn, *Hung.* Csáktornya; *prev. Ger.* Tschakathurn. Medimurje, N Croatia 46°24′N 16°29′E
77 V17 **Calabar** Cross River, S Nigeria 04°56′N 08°25′E
14 K13 **Calabogie** Ontario, SE Canada 45°18′N 76°46′W
54 L6 **Calabozo** Guárico, C Venezuela 08°58′N 67°28′W
107 N20 **Calabria** *anc.* Bruttium. ◈ *region* SW Italy
104 M16 **Calaburra, Punta de** *headland* S Spain 36°30′N 04°38′W
116 G14 **Calafat** Dolj, SW Romania 43°59′N 22°57′E
Calafate *see* El Calafate
105 Q4 **Calahorra** La Rioja, N Spain 42°19′N 01°58′W
103 N1 **Calais** Pas-de-Calais, N France 51°N 01°54′E
19 T5 **Calais** Maine, NE USA 45°09′N 67°15′W
Calais, Pas de *see* Dover, Strait of
Calalen *see* Kallalen
62 H4 **Calama** Antofagasta, N Chile 22°26′S 68°54′W
Calamianes *see* Calamian Group
170 M5 **Calamian Group** *var.* Calamianes. *island group* W Philippines
105 R7 **Calamocha** Aragón, NE Spain 40°54′N 01°18′W
29 N14 **Calamus River** ≈ Nebraska, C USA
116 G12 **Călan** *Ger.* Kalan, *Hung.* Pusztakalán. Hunedoara, SW Romania 45°45′N 22°59′E
105 S7 **Calanda** Aragón, NE Spain 40°56′N 00°15′W
168 F9 **Calang** Sumatera, W Indonesia 04°37′N 95°37′E
171 N4 **Calapan** Mindoro, N Philippines 13°24′N 121°08′E
Călăras *see* Călărași
116 M9 **Călărași** *var.* Călăras, *Rus.* Kalarash. C Moldova 47°19′N 28°13′E
116 L14 **Călărași** Călărași, SE Romania 44°18′N 26°52′E
116 K14 **Călărași** ◈ *county* SE Romania
54 E10 **Calarca** Quindío, W Colombia 04°31′N 75°38′W
105 Q12 **Calasparra** Murcia, SE Spain 38°14′N 01°41′W
107 I23 **Calatafimi** Sicilia, Italy, C Mediterranean Sea 37°54′N 12°52′E
105 Q6 **Calatayud** Aragón, NE Spain 41°21′N 01°39′W
171 O4 **Calauag** Luzon, N Philippines 13°57′N 122°18′E
35 P8 **Calaveras River** ≈ California, W USA
171 N4 **Calavite, Cape** *headland* Mindoro, N Philippines 13°25′N 120°16′E
171 Q8 **Calbayog** *off.* Calbayog City. Samar, C Philippines 12°08′N 124°36′E
Calbayog City *see* Calbayog
22 G9 **Calcasieu Lake** ◎ Louisiana, S USA
22 H8 **Calcasieu River** ≈ Louisiana, S USA
56 B6 **Calceta** Manabí, W Ecuador 0°51′S 80°07′W
61 B16 **Calchaquí** Santa Fe, C Argentina 29°56′S 60°14′W
62 J6 **Calchaquí, Río** ≈ NW Argentina
58 J10 **Calçoene** Amapá, NE Brazil 02°29′N 51°01′W
153 S16 **Calcutta** ✈ West Bengal, N India 22°30′N 88°20′E
Calcutta *see* Kolkata
54 E9 **Caldas** *off.* Departamento de Caldas. ◈ *province* W Colombia
104 F10 **Caldas da Rainha** Leiria, W Portugal 39°24′N 09°08′W
Caldas, Departamento de *see* Caldas
104 G3 **Caldas de Reis** *var.* Caldas de Reyes. Galicia, NW Spain 42°36′N 08°39′W
Caldas de Reyes *see* Caldas de Reis
58 F13 **Caldeirão** Amazonas, NW Brazil 03°18′S 60°22′W
62 G7 **Caldera** Atacama, N Chile 27°05′S 70°48′W
42 L14 **Caldera** Puntarenas, W Costa Rica 09°55′N 84°51′W
105 N10 **Calderina** ▲ C Spain 39°18′N 03°49′W
137 T13 **Çaldıran** Van, E Turkey 39°10′N 43°52′E
32 M14 **Caldwell** Idaho, NW USA 43°39′N 116°41′W
27 N8 **Caldwell** Kansas, C USA 37°01′N 97°36′W
14 G15 **Caledon** Ontario, S Canada 43°51′N 79°58′W
83 I23 **Caledon** *var.* Mohokare. ≈ Lesotho/South Africa
42 G1 **Caledonia** Corozal, N Belize 18°14′N 88°29′W
14 G16 **Caledonia** Ontario, S Canada 43°04′N 79°57′W
29 X11 **Caledonia** Minnesota, N USA 43°37′N 91°30′W
105 X5 **Calella** *var.* Calella de la Costa. Cataluña, NE Spain 41°37′N 02°40′E
Calella de la Costa *see* Calella
23 P4 **Calera** Alabama, S USA 33°06′N 86°45′W
63 I19 **Caleta Olivia** Santa Cruz, SE Argentina 46°21′S 67°37′W
35 X17 **Calexico** California, W USA 32°39′N 115°28′W
97 H16 **Calf of Man** *island* SW Isle of Man
11 Q16 **Calgary** Alberta, SW Canada 51°05′N 114°05′W
11 Q16 **Calgary** ✈ Alberta, SW Canada 51°15′N 114°03′W
37 U5 **Calhan** Colorado, C USA 39°00′N 104°18′W
64 O5 **Calheta** Madeira, Portugal, NE Atlantic Ocean 32°42′N 17°12′W
23 R2 **Calhoun** Georgia, SE USA 34°30′N 84°57′W
20 I6 **Calhoun** Kentucky, S USA 37°32′N 87°15′W
22 M3 **Calhoun City** Mississippi, S USA 33°51′N 89°18′W
21 P12 **Calhoun Falls** South Carolina, SE USA 34°05′N 82°36′W
54 D11 **Cali** Valle del Cauca, W Colombia 03°24′N 76°30′W
27 V9 **Calico Rock** Arkansas, C USA 36°07′N 92°08′W
155 F21 **Calicut** *var.* Kozhikode. Kerala, SW India 11°17′N 75°49′E *see also* Kozhikode
35 Y9 **Caliente** Nevada, W USA 37°37′N 114°30′W
27 U5 **California** Missouri, C USA 38°39′N 92°35′W
18 B15 **California** Pennsylvania, NE USA 40°02′N 79°52′W
35 Q12 **California** *off.* State of California, *also known as* El Dorado, The Golden State. ◈ *state* W USA
35 P11 **California Aqueduct** *aqueduct* California, W USA
35 T13 **California City** California, W USA 35°06′N 117°55′W
40 F6 **California, Golfo de** *Eng.* Gulf of California; *prev.* Sea of Cortez. *gulf* W Mexico
California, Gulf of *see* California, Golfo de
137 Y13 **Cälilabad** *Rus.* Dzhalilabad; *prev.* Astrakhan-Bazar. S Azerbaijan 39°15′N 48°30′E
116 I12 **Călimănești** Vâlcea, SW Romania 45°14′N 24°20′E
116 J9 **Călimani, Munții** ▲▲ N Romania
Calinisc *see* Cupcina
35 X17 **Calipatria** California, W USA 33°07′N 115°30′W
Calisia *see* Kalisz
34 M7 **Calistoga** California, W USA 38°34′N 122°37′W
83 G25 **Calitzdorp** Western Cape, SW South Africa 33°32′S 21°41′E
41 W12 **Calkiní** Campeche, E Mexico 20°21′N 90°03′W
182 K4 **Callabonna Creek** *var.* Tilcha Creek. *seasonal river* New South Wales/South Australia
182 J4 **Callabonna, Lake** ◎ South Australia
102 G5 **Callac** Côtes d'Armor, NW France 48°28′N 03°22′W
35 U5 **Callaghan, Mount** ▲ Nevada, W USA 39°38′N 116°57′W
Callain *see* Callan
97 E19 **Callan** *Ir.* Callain. S Ireland 52°33′N 07°23′W
14 H11 **Callander** Ontario, S Canada 46°14′N 79°21′W
96 I11 **Callander** C Scotland, United Kingdom 56°15′N 04°16′W
98 H7 **Callantsoog** Noord-Holland, NW Netherlands 52°51′N 04°41′E
57 D14 **Callao** Callao, W Peru 12°03′S 77°10′W
57 D15 **Callao** *off.* Departamento del Callao. ◈ *constitutional province* W Peru
Callao, Departamento del *see* Callao
56 F11 **Callaria, Río** ≈ E Peru
Callatis *see* Mangalia
11 Q13 **Calling Lake** Alberta, W Canada 55°12′N 113°07′W
Callosa de Ensarriá *see* Callosa d'En Sarrià
105 T11 **Callosa d'En Sarrià** *var.* Callosa de Ensarriá. Valenciana, E Spain 38°40′N 00°08′W
105 S12 **Callosa de Segura** Valenciana, E Spain 38°07′N 00°53′W
29 X11 **Calmar** Iowa, C USA 43°10′N 91°51′W
Calmar *see* Kalmar
43 R16 **Calobre** Veraguas, C Panama 08°18′N 80°49′W
23 X14 **Caloosahatchee River** ≈ Florida, SE USA
183 V2 **Caloundra** Queensland, E Australia 26°48′S 153°08′E
Calp *see* Calpe
105 T11 **Calpe** *Cat.* Calp. Valenciana, E Spain 38°39′N 00°03′E
41 P14 **Calpulalpan** Tlaxcala, S Mexico 19°36′N 98°26′W
107 K25 **Caltagirone** Sicilia, Italy, C Mediterranean Sea 37°14′N 14°31′E
107 J24 **Caltanissetta** Sicilia, Italy, C Mediterranean Sea 37°30′N 14°01′E
82 E11 **Caluango** Lunda Norte, NE Angola 08°16′S 19°36′E
82 C12 **Calucinga** Bié, W Angola 11°18′S 16°12′E
82 B12 **Calulo** Kwanza Sul, NW Angola 09°58′S 14°56′E
83 B14 **Caluquembe** Huíla, W Angola 13°47′S 14°40′E
80 Q11 **Caluula** Bari, NE Somalia 11°55′N 50°51′E
102 K4 **Calvados** ◈ *department* N France
186 I10 **Calvados Chain, The** *island group* SE Papua New Guinea
25 U9 **Calvert** Texas, SW USA 30°58′N 96°40′W
20 H7 **Calvert City** Kentucky, S USA 37°01′N 88°21′W
103 X14 **Calvi** Corse, France, C Mediterranean Sea 42°34′N 08°44′E
40 L12 **Calvillo** Aguascalientes, C Mexico 21°51′N 102°18′W
83 F24 **Calvinia** Northern Cape, W South Africa 31°25′S 19°47′E
104 K8 **Calvitero** ▲ W Spain 40°16′N 05°48′W
101 G22 **Calw** Baden-Württemberg, SW Germany 48°43′N 08°43′E
Calydon *see* Kalydón
105 N11 **Calzada de Calatrava** Castilla-La Mancha, C Spain 38°42′N 03°46′W
Cama *see* Kama
82 C11 **Camabatela** Kwanza Norte, NW Angola 08°13′S 15°23′E
64 Q5 **Camacha** Porto Santo, Madeira, Portugal, NE Atlantic Ocean 32°41′N 16°52′W
14 L8 **Camachigama, Lac** ◎ Québec, SE Canada
40 M9 **Camacho** Zacatecas, C Mexico 24°23′N 102°20′W
82 D13 **Camacupa** *var.* General Machado, *Port.* Vila General Machado. Bié, C Angola 12°S 17°31′E
54 L7 **Camaguán** Guárico, C Venezuela 08°09′N 67°37′W
44 G6 **Camagüey** *prev.* Puerto Príncipe. Camagüey, C Cuba 21°24′N 77°55′W
44 G5 **Camagüey, Archipiélago de** *island group* C Cuba
40 D5 **Camalli, Sierra de** ▲ NW Mexico 28°21′N 113°26′W
57 G18 **Camana** *var.* Camaná. Arequipa, SW Peru 16°37′S 72°42′W
29 Z14 **Camanche** Iowa, C USA 41°47′N 90°15′W
35 P8 **Camanche Reservoir** ⊠ California, W USA
61 I16 **Camaquã** Rio Grande do Sul, S Brazil 30°50′S 51°47′W
61 H16 **Camaquã, Rio** ≈ S Brazil
64 P6 **Câmara de Lobos** Madeira, Portugal, NE Atlantic Ocean 32°38′N 16°59′W
103 U16 **Camarat, Cap** *headland* SE France 43°12′N 06°42′E
41 O8 **Camargo** Tamaulipas, C Mexico 26°16′N 98°49′W
103 R15 **Camargue** *physical region* SE France
104 F2 **Camariñas** Galicia, NW Spain 43°07′N 09°10′W
Camaronero, Laguna del *see* Caimanero, Laguna del
63 J18 **Camarones** Chubut, S Argentina 44°48′S 65°42′W
63 J18 **Camarones, Bahía** *bay* S Argentina
104 J14 **Camas** Andalucía, S Spain 37°24′N 06°01′W
167 S15 **Ca Mau** *var.* Quan Long. Minh Hai, S Vietnam 09°11′N 105°09′E
82 E11 **Camaxilo** Lunda Norte, NE Angola 08°19′S 18°53′E
104 G3 **Cambados** Galicia, NW Spain 42°31′N 08°49′W
Cambay, Gulf of *see* Khambhāt, Gulf of
Camberia *see* Chambéry
97 N22 **Camberley** SE England, United Kingdom 51°21′N 00°45′W
167 R12 **Cambodia** *off.* Kingdom of Cambodia, *var.* Democratic Kampuchea, Roat Kampuchea, *Cam.* Kampuchea; *prev.* People's Democratic Republic of Kampuchea. ◆ *republic* SE Asia
Cambodia, Kingdom of *see* Cambodia
102 I16 **Cambo-les-Bains** Pyrénées-Atlantiques, SW France 43°22′N 01°24′W
103 P2 **Cambrai** *Flem.* Kambryk, *prev.* Cambray; *anc.* Cameracum. Nord, N France 50°10′N 03°14′E
Cambray *see* Cambrai
104 H2 **Cambre** Galicia, NW Spain 43°18′N 08°21′W
35 O12 **Cambria** California, W USA 35°33′N 121°04′W
97 J20 **Cambrian Mountains** ▲▲ C Wales, United Kingdom
14 G16 **Cambridge** Ontario, S Canada 43°22′N 80°20′W
44 I12 **Cambridge** W Jamaica 18°18′N 77°54′W
184 M8 **Cambridge** Waikato, North Island, New Zealand 37°53′S 175°28′E
97 O20 **Cambridge** *Lat.* Cantabrigia. E England, United Kingdom 52°12′N 00°07′E
32 M12 **Cambridge** Idaho, NW USA 44°34′N 116°42′W
30 K11 **Cambridge** Illinois, N USA 41°18′N 90°11′W
21 Y4 **Cambridge** Maryland, NE USA 38°34′N 76°04′W
19 O11 **Cambridge** Massachusetts, NE USA 42°21′N 71°05′W
29 V7 **Cambridge** Minnesota, N USA 45°34′N 93°13′W
29 N16 **Cambridge** Nebraska, C USA 40°18′N 100°10′W
31 U13 **Cambridge** Ohio, NE USA 40°00′N 81°34′W
8 L7 **Cambridge Bay** *var.* Ikaluktutiak. Victoria Island, Nunavut, NW Canada 68°56′N 105°09′W
97 O20 **Cambridgeshire** *cultural region* E England, United Kingdom
105 U6 **Cambrils** *prev.* Cambrils de Mar. Cataluña, NE Spain 41°06′N 01°02′E
Cambrils de Mar *see* Cambrils
Cambundi-Catembo *see* Nova Gaia
137 N11 **Çam Burnu** *headland* N Turkey 41°07′N 37°48′E
183 S9 **Camden** New South Wales, SE Australia 34°04′S 150°40′E
23 O6 **Camden** Alabama, S USA 31°59′N 87°17′W
27 U14 **Camden** Arkansas, C USA 33°32′N 92°49′W
21 Y3 **Camden** Delaware, NE USA 39°06′N 75°30′W
19 R7 **Camden** Maine, NE USA 44°12′N 69°04′W
18 I16 **Camden** New Jersey, NE USA 39°55′N 75°07′W
18 I9 **Camden** New York, NE USA 43°21′N 75°45′W
21 R12 **Camden** South Carolina, SE USA 34°16′N 80°36′W
20 H8 **Camden** Tennessee, S USA 36°03′N 88°07′W
25 X9 **Camden** Texas, SW USA 30°55′N 94°43′W
39 S5 **Camden Bay** *bay* S Beaufort Sea
27 U6 **Camdenton** Missouri, C USA 38°01′N 92°44′W
Camellia State *see* Alabama
18 M7 **Camels Hump** ▲ Vermont, NE USA 44°18′N 72°50′W
117 N8 **Camenca** *Rus.* Kamenka. N Moldova 48°01′N 28°43′E
Cameracum *see* Cambrai
22 G9 **Cameron** Louisiana, S USA 29°48′N 93°19′W
25 T9 **Cameron** Texas, SW USA 30°51′N 96°58′W
30 J5 **Cameron** Wisconsin, N USA 45°25′N 91°42′W
10 M12 **Cameron** ≈ British Columbia, W Canada
185 A24 **Cameron Mountains** ▲▲ South Island, New Zealand
79 D15 **Cameroon** *off.* Republic of Cameroon, *Fr.* Cameroun. ◆ *republic* W Africa
79 D15 **Cameroon Mountain** ▲ SW Cameroon 04°12′N 09°00′E
Cameroon, Republic of *see* Cameroon
Cameroon Ridge *see* Camerounaise, Dorsale
Cameroun *see* Cameroon
79 E14 **Camerounaise, Dorsale** *Eng.* Cameroon Ridge. *ridge* NW Cameroon
136 B15 **Çamiçi Gölü** ◎ SW Turkey
171 N3 **Camiling** Luzon, N Philippines 15°41′N 120°22′E
23 T7 **Camilla** Georgia, SE USA 31°13′N 84°12′W
104 G5 **Caminha** Viana do Castelo, N Portugal 41°52′N 08°50′W
35 P7 **Camino** California, W USA 38°43′N 120°39′W
107 J24 **Cammarata** Sicilia, Italy, C Mediterranean Sea 37°36′N 13°39′E
42 K10 **Camoapa** Boaco, S Nicaragua 12°25′N 85°30′W
58 O13 **Camocim** Ceará, E Brazil 02°55′S 40°50′W
106 D10 **Camogli** Liguria, NW Italy 44°21′N 09°10′E
181 S5 **Camooweal** Queensland, C Australia 19°57′S 138°14′E
55 Y11 **Camopi** E French Guiana 03°12′N 52°19′W
151 Q22 **Camorta** *island* Nicobar Islands, India, NE Indian Ocean
42 I6 **Campamento** Olancho, C Honduras 14°36′N 86°38′W
61 D19 **Campana** Buenos Aires, E Argentina 34°10′S 58°57′W
63 F21 **Campana, Isla** *island* S Chile
104 K11 **Campanario** Extremadura, W Spain 38°52′N 05°36′W
107 L17 **Campania** *Eng.* Champagne. ◈ *region* S Italy
27 Y8 **Campbell** Missouri, C USA 36°29′N 90°04′W
185 K15 **Campbell, Cape** *headland* South Island, New Zealand 41°44′S 174°16′E
14 J14 **Campbellford** Ontario, SE Canada 44°18′N 77°48′W
31 R13 **Campbell Hill** *hill* Ohio, N USA
192 K13 **Campbell Island** *island* S New Zealand
175 P13 **Campbell Plateau** *undersea feature* SW Pacific Ocean 51°00′S 170°00′E
10 K17 **Campbell River** Vancouver Island, British Columbia, SW Canada 49°59′N 125°18′W
20 L6 **Campbellsville** Kentucky, S USA 37°20′N 85°21′W
13 O13 **Campbellton** New Brunswick, SE Canada 48°00′N 66°41′W
183 S9 **Campbelltown** New South Wales, SE Australia 34°04′S 150°46′E
183 P16 **Campbell Town** Tasmania, SE Australia 41°57′S 147°30′E
96 G13 **Campbeltown** W Scotland, United Kingdom 55°26′N 05°38′W
41 W13 **Campeche** Campeche, SE Mexico 19°47′N 90°29′W
41 W14 **Campeche** ◈ *state* SE Mexico
41 T14 **Campeche, Bahía de** *Eng.* Bay of Campeche. *bay* E Mexico
Campeche, Banco de *see* Campeche Bank
64 C11 **Campeche Bank** *Sp.* Banco de Campeche, Sonda de Campeche. *undersea feature* S Gulf of Mexico 22°00′N 90°00′W
Campeche, Bay of *see* Campeche, Bahía de
Campeche, Sonda de *see* Campeche Bank
44 H7 **Campechuela** Granma, E Cuba 20°15′N 77°17′W
182 M13 **Camperdown** Victoria, SE Australia 38°16′S 143°10′E
167 U6 **Câm Pha** Quang Ninh, N Vietnam 21°04′N 107°20′E
116 H10 **Câmpia Turzii** *Ger.* Jerischmarkt, *Hung.* Aranyosgyéres; *prev.* Cîmpia Turzii, Ghiriș, Gyéres. Cluj, NW Romania 46°33′N 23°53′E
104 K12 **Campillo de Llerena** Extremadura, W Spain 38°30′N 05°48′W
104 L15 **Campillos** Andalucía, S Spain 37°04′N 04°51′W
116 J13 **Câmpina** *prev.* Cîmpina. Prahova, SE Romania 45°08′N 25°44′E
59 Q15 **Campina Grande** Paraíba, E Brazil 07°15′S 35°50′W
60 L9 **Campinas** São Paulo, S Brazil 22°54′S 47°06′W
38 L10 **Camp Kulowiye** Saint Lawrence Island, Alaska, USA 63°15′N 168°45′W
79 D17 **Campo** *var.* Kampo. Sud, SW Cameroon 02°22′N 09°50′E
Campo *see* Ntem
59 N15 **Campo Alegre de Lourdes** Bahia, E Brazil 09°28′S 43°01′W
107 L16 **Campobasso** Molise, C Italy 41°34′N 14°40′E
107 H24 **Campobello di Mazara** Sicilia, Italy, C Mediterranean Sea 37°38′N 12°45′E
Campo Criptana *see* Campo de Criptana
105 O10 **Campo de Criptana** *var.* Campo Criptana. Castilla-La Mancha, C Spain 39°25′N 03°07′W
59 I16 **Campo de Diauarum** *var.* Pôsto Diuarum. Mato Grosso, W Brazil 11°08′S 53°16′W
54 E5 **Campo de la Cruz** Atlántico, N Colombia 10°23′N 74°52′W
105 P11 **Campo de Montiel** *physical region* C Spain
60 H12 **Campo Erê** Santa Catarina, S Brazil 26°24′S 53°04′W
62 L7 **Campo Gallo** Santiago del Estero, N Argentina 26°32′S 62°51′W
59 I20 **Campo Grande** *state capital* Mato Grosso do Sul, SW Brazil 20°24′S 54°35′W
60 K12 **Campo Largo** Paraná, S Brazil 25°27′S 49°29′W
58 N13 **Campo Maior** Piauí, E Brazil 04°50′S 42°12′W
104 I10 **Campo Maior** Portalegre, C Portugal 39°01′N 07°04′W
60 H10 **Campo Mourão** Paraná, S Brazil 24°01′S 52°24′W
59 L17 **Campos Belos** Goiás, S Brazil 13°11′S 46°47′W
60 N9 **Campos do Jordão** São Paulo, S Brazil 22°45′S 45°36′W
60 Q9 **Campos dos Goytacazes** Rio de Janeiro, SE Brazil 21°46′S 41°21′W
60 I13 **Campos Novos** Santa Catarina, S Brazil 27°22′S 51°11′W
59 O14 **Campos Sales** Ceará, E Brazil 07°01′S 40°21′W
25 Q9 **Camp San Saba** Texas, SW USA 30°57′N 99°16′W
21 N6 **Campton** Kentucky, S USA 37°44′N 83°35′W
116 I13 **Câmpulung** *prev.* Câmpulung-Mușcel, Cîmpulung. Argeș, S Romania 45°16′N 25°03′E
116 J9 **Câmpulung Moldovenesc** *var.* Cîmpulung Moldovenesc, *Ger.* Kimpolung, *Hung.* Hosszúmezjő. Suceava, NE Romania 47°32′N 25°34′E
Câmpulung-Mușcel *see* Câmpulung
Campus Stellae *see* Santiago de Compostela
36 L12 **Camp Verde** Arizona, SW USA 34°33′N 111°52′W
25 P11 **Camp Wood** Texas, SW USA 29°40′N 100°00′W
167 V13 **Cam Ranh** *prev.* Ba Ngoi. Khanh Hoa, S Vietnam 11°54′N 109°14′E
11 Q15 **Camrose** Alberta, SW Canada 53°01′N 112°48′W
Camulodunum *see* Colchester
136 B12 **Çan** Çanakkale, NW Turkey 40°03′N 27°03′E
18 L12 **Canaan** Connecticut, NE USA 42°00′N 73°17′W
11 O13 **Canada** ◆ *commonwealth republic* N North America
197 P6 **Canada Basin** *undersea feature* Arctic Ocean 80°00′N 145°00′W
61 B18 **Cañada de Gómez** Santa Fe, C Argentina 32°50′S 61°23′W
197 P6 **Canada Plain** *undersea feature* Arctic Ocean
61 A18 **Cañada Rosquín** Santa Fe, C Argentina 32°04′S 61°35′W
25 P1 **Canadian** Texas, SW USA 35°54′N 100°23′W
16 K12 **Canadian River** ≈ SW USA
8 L12 **Canadian Shield** *physical region* Canada
63 I18 **Cañadón Grande, Sierra** ▲▲ S Argentina
55 P9 **Canaima** Bolívar, SE Venezuela 09°41′N 72°33′W
136 B11 **Çanakkale** *var.* Dardanelli; *prev.* Chanak, Kale Sultanie. Çanakkale, W Turkey 40°09′N 26°25′E
136 B12 **Çanakkale** ◈ *province* NW Turkey
136 B11 **Çanakkale Boğazı** *Eng.* Dardanelles. *strait* NW Turkey
187 Q17 **Canala** Province Nord, C New Caledonia 21°31′S 165°57′E
59 A15 **Canamari** Amazonas, W Brazil 07°37′S 72°33′W
18 G10 **Canandaigua** New York, NE USA 42°52′N 77°14′W
18 F10 **Canandaigua Lake** ◎ New York, NE USA
40 G3 **Cananea** Sonora, NW Mexico 30°59′N 110°20′W
56 B8 **Cañar** ◈ *province* C Ecuador
64 N10 **Canarias, Islas** *Eng.* Canary Islands. ◈ *autonomous community* Spain, NE Atlantic Ocean
Canaries Basin *see* Canary Basin
44 C6 **Canarreos, Archipiélago de los** *island group* W Cuba
Canary Islands *see* Canarias, Islas
66 K3 **Canary Basin** *var.* Canaries Basin, Monaco Basin. *undersea feature* E Atlantic Ocean 30°00′N 25°00′W
42 L13 **Cañas** Guanacaste, NW Costa Rica 10°25′N 85°07′W
18 I10 **Canastota** New York, NE USA 43°04′N 75°45′W
40 K9 **Canatlán** Durango, C Mexico 24°33′N 104°45′W
104 J9 **Cañaveral** Extremadura, W Spain 39°47′N 06°24′W
23 Y11 **Canaveral, Cape** *headland* Florida, SE USA 28°27′N 80°31′W
59 O18 **Canavieiras** Bahia, E Brazil 15°44′S 38°58′W
43 R16 **Cañazas** Veraguas, W Panama 08°25′N 81°10′W
106 H6 **Canazei** Trentino-Alto Adige, N Italy 46°29′N 11°50′E
183 P6 **Canbelego** New South Wales, SE Australia 31°36′S 146°20′E
183 R10 **Canberra** ● (Australia) Australian Capital Territory, SE Australia 35°21′S 149°08′E
183 R10 **Canberra** ✈ Australian Capital Territory, SE Australia 35°19′S 149°12′E
35 P2 **Canby** California, W USA 41°27′N 120°51′W
29 S9 **Canby** Minnesota, N USA 44°42′N 96°17′W
103 N2 **Canche** ≈ N France
102 L13 **Cancon** Lot-et-Garonne, SW France 44°33′N 00°37′E
41 Z11 **Cancún** Quintana Roo, SE Mexico 21°05′N 86°48′W
104 K2 **Candás** Asturias, N Spain 43°35′N 05°45′W
102 J7 **Cande** Maine-et-Loire, NW France 47°33′N 01°03′W
41 W14 **Candelaria** Campeche, SE Mexico 18°10′N 91°00′W
24 J11 **Candelaria** Texas, SW USA 30°05′N 104°40′W
41 W15 **Candelaria, Río** ≈ Guatemala/Mexico
104 L8 **Candeleda** Castilla y León, N Spain 40°10′N 05°14′W
Candia *see* Irákleio
41 P8 **Cándido Aguilar** Tamaulipas, C Mexico 25°30′N 97°57′W
39 N8 **Candle** Alaska, USA 65°54′N 161°55′W
11 T14 **Candle Lake** Saskatchewan, C Canada 53°43′N 105°09′W
18 L13 **Candlewood, Lake** ⊠ Connecticut, NE USA
29 O3 **Cando** North Dakota, N USA 48°29′N 99°12′W
Canea *see* Chaniá
45 O12 **Canefield** ✈ (Roseau) SW Dominica 15°20′N 61°24′W
61 F20 **Canelones** *prev.* Guadalupe. Canelones, S Uruguay 34°32′S 56°17′W
61 E20 **Canelones** ◈ *department* S Uruguay
Canendiyú *see* Canindeyú
63 F14 **Cañete** Bío Bío, C Chile 37°48′S 73°25′W
105 Q9 **Cañete** Castilla-La Mancha, C Spain 40°03′N 01°39′W
Cañete *see* San Vicente de Cañete
27 P8 **Caney** Kansas, C USA 37°00′N 95°56′W
27 P8 **Caney River** ≈ Kansas/Oklahoma, C USA
105 S3 **Canfranc-Estación** Aragón, NE Spain 42°42′N 00°31′W
83 E14 **Cangamba** *Port.* Vila de Aljustrel. Moxico, E Angola 13°40′S 19°47′E
82 C12 **Cangandala** Malanje, NW Angola 09°47′S 16°27′E
104 G4 **Cangas** Galicia, NW Spain 42°16′N 08°46′W
104 J2 **Cangas del Narcea** Asturias, N Spain 43°10′N 06°32′W
Cangas de Onís *see* Cangues d'Onís
161 S11 **Cangnan** *var.* Lingxi. Zhejiang, SE China 27°29′N 120°23′E
82 C10 **Cangola** Uíge, NW Angola 07°54′S 15°57′E
83 E14 **Cangombe** Moxico, E Angola 14°27′S 20°05′E
63 H21 **Cangrejo, Cerro** ▲ S Argentina 49°19′S 72°18′W
61 H17 **Canguçu** Rio Grande do Sul, S Brazil 31°25′S 52°37′W
104 L2 **Cangues d'Onís** *var.* Cangas de Onís. Asturias, N Spain 43°21′N 05°08′W
161 P3 **Cangzhou** Hebei, E China 38°19′N 116°54′E
12 M7 **Caniapiscau** ≈ Québec, E Canada
12 M8 **Caniapiscau, Réservoir de** ⊠ Québec, C Canada
107 J24 **Canicattì** Sicilia, Italy, C Mediterranean Sea 37°22′N 13°51′E
136 L11 **Canik Dağları** ▲▲ N Turkey
105 P14 **Caniles** Andalucía, S Spain 37°24′N 02°41′W
59 B16 **Canindé** Acre, W Brazil 10°55′S 69°45′W
62 P6 **Canindeyú** *var.* Canendiyú, Canindiyú. ◈ *department* E Paraguay
Canindiyú *see* Canindeyú
194 J10 **Canisteo Peninsula** *peninsula* Antarctica
18 F11 **Canisteo River** ≈ New York, NE USA
40 M10 **Cañitas** *var.* Cañitas de Felipe Pescador. Zacatecas, C Mexico 23°35′N 102°39′W
Cañitas de Felipe Pescador *see* Cañitas
105 P15 **Canjáyar** Andalucía, S Spain 37°00′N 02°45′W
136 I12 **Çankırı** *var.* Chankiri; *anc.* Gangra, Germanicopolis. Çankın, N Turkey 40°36′N 33°35′E
136 I11 **Çankırı** *var.* Chankiri. ◈ *province* N Turkey
171 P6 **Canlaon Volcano** ⛰ Negros, C Philippines 10°24′N 123°05′E
11 P16 **Canmore** Alberta, SW Canada 51°07′N 115°18′W
96 F9 **Canna** *island* NW Scotland, United Kingdom
155 F20 **Cannanore** *var.* Kannur, Jagatsinghapur. Kerala, SW India 11°53′N 75°23′E *see also* Kannur
31 O17 **Cannelton** Indiana, N USA 37°54′N 86°44′W
103 U15 **Cannes** Alpes-Maritimes, SE France 43°33′N 06°59′E
39 R5 **Canning River** ≈ Alaska, USA
106 C6 **Cannobio** Piemonte, NE Italy 46°04′N 08°39′E
97 L19 **Cannock** C England, United Kingdom 52°41′N 02°03′W
28 M6 **Cannonball River** ≈ North Dakota, N USA
29 W9 **Cannon Falls** Minnesota, N USA 44°30′N 92°54′W
18 I11 **Cannonsville Reservoir** ⊠ New York, NE USA
183 R12 **Cann River** Victoria, SE Australia 37°34′S 149°11′E
61 I16 **Canoas** Rio Grande do Sul, S Brazil 29°42′S 51°07′W
61 I14 **Canoas, Rio** ≈ S Brazil
14 I12 **Canoe Lake** ◎ Ontario, SE Canada
60 J12 **Canoinhas** Santa Catarina, S Brazil 26°12′S 50°24′W
37 T6 **Canon City** Colorado, C USA 38°25′N 105°14′W
55 P8 **Caño Negro** Bolívar, SE Venezuela
173 X15 **Canonniers Point** *headland* N Mauritius
23 W6 **Canoochee River** ≈ Georgia, SE USA
11 V15 **Canora** Saskatchewan, S Canada 51°38′N 102°28′W
45 Y14 **Canouan** *island* S Saint Vincent and the Grenadines
13 R15 **Canso** Nova Scotia, SE Canada 45°20′N 61°00′W
104 M3 **Cantabria** ◈ *autonomous community* N Spain
104 K3 **Cantábrica, Cordillera** ▲▲ N Spain
Cantabrigia *see* Cambridge
103 O12 **Cantal** ◈ *department* C France
105 N6 **Cantalejo** Castilla y León, N Spain 41°15′N 03°57′W
103 O12 **Cantal, Monts du** ▲▲ C France
104 G8 **Cantanhede** Coimbra, C Portugal 40°21′N 08°37′W
Cantaño *see* Cataño
55 O6 **Cantaura** Anzoátegui, NE Venezuela 09°22′N 64°24′W
116 M11 **Cantemir** *Rus.* Kantemir. S Moldova 46°17′N 28°12′E
97 Q22 **Canterbury** *hist.* Cantwaraburh; *anc.* Durovernum, *Lat.* Cantuaria. SE England, United Kingdom 51°17′N 01°05′E
185 F19 **Canterbury** *off.* Canterbury Region. ◈ *region* South Island, New Zealand
185 H20 **Canterbury Bight** *bight* South Island, New Zealand
185 H19 **Canterbury Plains** *plain* South Island, New Zealand
Canterbury Region *see* Canterbury
167 S14 **Cần Thơ** Cần Thơ, S Vietnam 10°03′N 105°46′E
104 K13 **Cantillana** Andalucía, S Spain 37°34′N 05°48′W
59 N15 **Canto do Buriti** Piauí, NE Brazil 08°07′S 43°00′W
23 S2 **Canton** Georgia, SE USA 34°14′N 84°29′W
30 K12 **Canton** Illinois, N USA 40°33′N 90°02′W
22 L5 **Canton** Mississippi, S USA 32°36′N 90°02′W

◆ Country
● Country Capital
◇ Dependent Territory
○ Dependent Territory Capital
◈ Administrative Regions
✈ International Airport
▲ Mountain
▲▲ Mountain Range
⛰ Volcano
≈ River
◎ Lake
⊠ Reservoir

CANTON – CASTELO DO BODE, BARRAGEM DO

27 V2 **Canton** Missouri, C USA *40°07′N 91°31′W*
18 J7 **Canton** New York, NE USA *44°36′N 75°10′W*
21 O10 **Canton** North Carolina, SE USA *35°31′N 82°50′W*
31 U12 **Canton** Ohio, N USA *40°48′N 81°23′W*
26 L9 **Canton** Oklahoma, C USA *36°03′N 98°35′W*
18 G12 **Canton** Pennsylvania, NE USA *41°38′N 76°49′W*
29 R11 **Canton** South Dakota, N USA *43°19′N 96°33′W*
25 V7 **Canton** Texas, SW USA *32°33′N 95°51′W*
Canton *see* Guangzhou
Canton Island *see* Kanton
26 L9 **Canton Lake** ⊞ Oklahoma, C USA
106 D7 **Cantù** Lombardia, N Italy *45°44′N 09°98′E*
Cantuaria/Cantwaraburh *see* Canterbury
39 R10 **Cantwell** Alaska, USA *63°23′N 148°57′W*
59 O16 **Canudos** Bahia, E Brazil *09°51′S 39°08′W*
47 T7 **Canumã, Rio** ≈ N Brazil
Canusium *see* Puglia, Canosa di
24 G7 **Canutillo** Texas, SW USA *31°53′N 106°34′W*
25 N3 **Canyon** Texas, SW USA *34°58′N 101°56′W*
33 S12 **Canyon** Wyoming, C USA *44°44′N 110°30′W*
32 K13 **Canyon City** Oregon, NW USA *44°24′N 118°58′W*
33 R10 **Canyon Ferry Lake** ◎ Montana, NW USA
25 S11 **Canyon Lake** ⊞ Texas, SW USA
167 T5 **Cao Băng** *var.* Caobang. Cao Băng, N Vietnam *22°40′N 106°16′E*
Caobang *see* Cao Băng
160 J12 **Caodu He** ≈ S China
Caohai *see* Weining
167 S14 **Cao Lanh** Đông Thap, S Vietnam *10°35′N 105°25′E*
82 C11 **Caombo** Malanje, NW Angola *08°42′S 16°33′E*
Caorach, Cuan na g *see* Sheep Haven
Caozhou *see* Heze
171 Q12 **Capalulu** Pulau Mangole, E Indonesia *01°51′S 125°53′E*
54 K8 **Capanaparo, Río** ≈ Colomiba/Venezuela
58 L12 **Capanema** Pará, NE Brazil *01°08′S 47°07′W*
60 L10 **Capão Bonito do Sul** São Paulo, S Brazil *24°01′S 48°23′W*
60 I13 **Capão Doce, Morro do** ▲ S Brazil *26°37′S 51°28′W*
54 I4 **Capatárida** Falcón, N Venezuela *11°11′N 70°37′W*
102 I15 **Capbreton** Landes, SW France *43°40′N 01°25′W*
Cap-Breton, Île du *see* Cape Breton Island
15 W6 **Cap-Chat** Québec, SE Canada *49°04′N 66°43′W*
15 P11 **Cap-de-la-Madeleine** Québec, SE Canada *46°22′N 72°31′W*
103 N13 **Capdenac** Aveyron, S France *44°35′N 02°06′E*
Cap des Palmès *see* Palmas, Cape
183 Q15 **Cape Barren Island** *island* Furneaux Group, Tasmania, SE Australia
65 O18 **Cape Basin** *undersea feature* S Atlantic Ocean *37°00′S 07°00′E*
13 R14 **Cape Breton Island** *Fr.* Île du Cap-Breton. *island* Nova Scotia, SE Canada
23 Y11 **Cape Canaveral** Florida, SE USA *28°24′N 80°36′W*
21 Y6 **Cape Charles** Virginia, NE USA *37°16′N 76°01′W*
77 P17 **Cape Coast** *prev.* Cape Coast Castle. S Ghana *05°10′N 01°13′W*
Cape Coast Castle *see* Cape Coast
19 Q12 **Cape Cod Bay** *bay* Massachusetts, NE USA
23 W15 **Cape Coral** Florida, SE USA *26°33′N 81°57′W*
181 R4 **Cape Crawford Roadhouse** Northern Territory, N Australia *16°39′S 135°44′E*
9 Q7 **Cape Dorset** *var.* Kingait. Baffin Island, Nunavut, NE Canada *76°14′N 76°32′W*
21 N8 **Cape Fear River** ≈ North Carolina, SE USA
27 Y7 **Cape Girardeau** Missouri, C USA *37°19′N 89°31′W*
21 T14 **Cape Island** *island* South Carolina, SE USA
186 A6 **Capella** ▲ NW Papua New Guinea *05°00′S 141°09′E*
98 H12 **Capelle aan den IJssel** Zuid-Holland, SW Netherlands *51°56′N 04°36′E*
83 C15 **Capelongo** Huíla, C Angola *14°45′S 15°02′E*
18 J17 **Cape May** New Jersey, NE USA *38°54′N 74°54′W*
18 J17 **Cape May Court House** New Jersey, NE USA *39°03′N 74°46′W*
Cape Palmas *see* Harper
8 I16 **Cape Parry** Northwest Territories, N Canada *70°10′N 124°33′W*
65 P19 **Cape Rise** *undersea feature* SW Indian Ocean *42°00′S 15°00′E*
Cape Saint Jacques *see* Vung Tau
Capesterre *see* Capesterre-Belle-Eau
45 Y6 **Capesterre-Belle-Eau** *var.* Capesterre. Basse Terre, S Guadeloupe *16°03′N 61°34′W*
83 D26 **Cape Town** *var.* Ekapa, *Afr.* Kaapstad, Kapstad. ● (South Africa-legislative capital) Western Cape, SW South Africa *33°56′S 18°28′E*
83 E26 **Cape Town** ✈ Western Cape, SW South Africa *31°51′S 21°06′E*
76 D9 **Cape Verde** *off.* Republic of Cape Verde, *Port.* Cabo Verde, Ilhas do Cabo Verde. ◆ *republic* E Atlantic Ocean
64 L11 **Cape Verde Basin** *undersea feature* E Atlantic Ocean *15°00′N 30°00′W*
66 K5 **Cape Verde Islands** *island group* E Atlantic Ocean
64 L10 **Cape Verde Plain** *undersea feature* E Atlantic Ocean *23°00′N 26°00′W*
Cape Verde Plateau/Cape Verde Rise *see* Cape Verde Terrace
Cape Verde, Republic of *see* Cape Verde
64 L11 **Cape Verde Terrace** *var.* Cape Verde Plateau, Cape Verde Rise. *undersea feature* E Atlantic Ocean *18°00′N 20°00′W*
181 V2 **Cape York Peninsula** *peninsula* Queensland, N Australia
44 M8 **Cap-Haïtien** *var.* Le Cap. N Haiti *19°44′N 72°12′W*
43 T15 **Capira** Panamá, C Panama *08°48′N 79°51′W*
14 K8 **Capitachouane** ≈ Québec, SE Canada
14 L8 **Capitachouane, Lac** ◎ Québec, SE Canada
37 T13 **Capitan** New Mexico, SW USA *33°33′N 105°34′W*
194 G3 **Capitán Arturo Prat** *Chilean research station* South Shetland Islands, Antarctica *62°24′S 59°42′W*
37 S13 **Capitan Mountains** ▲▲ New Mexico, SW USA
62 M3 **Capitán Pablo Lagerenza** *var.* Mayor Pablo Lagerenza. Chaco, N Paraguay *19°55′S 60°46′W*
37 T13 **Capitan Peak** ▲ New Mexico, SW USA *33°35′N 105°15′W*
188 H5 **Capitol Hill** ● (Northern Mariana Islands-legislative capital) Saipan, S Northern Mariana Islands
60 I9 **Capivara, Represa** ⊞ S Brazil
61 J16 **Capivari** Rio Grande do Sul, S Brazil *30°08′S 50°32′W*
113 H15 **Čapljina** Federicija Bosna I Hercegovina, S Bosnia and Herzegovina *43°07′N 17°42′E*
83 M15 **Capoche** *var.* Kapoche. ≈ Mozambique/Zambia
Capo Delgado, Província de *see* Cabo Delgado
107 K17 **Capodichino** ✈ (Napoli) Campania, S Italy *40°53′N 14°15′E*
Capodistria *see* Koper
106 E12 **Capraia, Isola di** *island* Arcipelago Toscano, C Italy
107 B16 **Caprara, Punta** *var.* Punta dello Scorno. *headland* Isola Asinara, W Italy *41°07′N 08°19′E*
Capraria *see* Cabrera, Illa de
14 F10 **Capreol** Ontario, S Canada *46°43′N 80°56′W*
107 K18 **Capri** Campania, S Italy *40°33′N 14°14′E*
175 S9 **Capricorn Tablemount** *undersea feature* W Pacific Ocean *18°34′S 172°12′W*
107 J18 **Capri, Isola di** *island* S Italy
83 G16 **Caprivi** ⬖ *district* NE Namibia
Caprivi Concession *see* Caprivi Strip
83 F16 **Caprivi Strip** *Ger.* Caprivizipfel; *prev.* Caprivi Concession. *cultural region* NE Namibia
Caprivizipfel *see* Caprivi Strip
25 O5 **Cap Rock Escarpment** *cliffs* Texas, SW USA
15 R10 **Cap-Rouge** Québec, SE Canada *46°45′N 71°18′W*
Cap Saint-Jacques *see* Vung Tau
38 F12 **Captain Cook** Hawaii, USA, C Pacific Ocean *19°30′N 155°55′W*
183 R10 **Captains Flat** New South Wales, SE Australia *35°37′S 149°28′E*
102 K14 **Captieux** Gironde, SW France *44°16′N 00°15′W*
107 K17 **Capua** Campania, S Italy *41°06′N 14°13′E*
54 F14 **Caquetá** *off.* Departamanto del Caquetá. ⬖ *province* S Colombia
Caquetá, Departamanto del *see* Caquetá
54 E13 **Caquetá, Río** *var.* Rio Japurá, Yapurá. ≈ Brazil/Colombia *see also* Japurá, Rio
Caquetá, Río *see* Japurá, Rio
CAR *see* Central African Republic
Cara *see* Kara
57 I16 **Carabaya, Cordillera** ▲▲ E Peru
54 K5 **Carabobo** *off.* Estado Carabobo. ⬖ *state* N Venezuela
Carabobo, Estado *see* Carabobo
116 I14 **Caracal** Olt, S Romania *44°07′N 24°18′E*
58 F10 **Caracaraí** Rondônia, W Brazil *01°47′N 61°11′W*
54 L5 **Caracas** ● (Venezuela) Distrito Federal, N Venezuela *10°29′N 66°54′W*
54 I5 **Carache** Trujillo, N Venezuela *09°40′N 70°15′W*
60 N10 **Caraguatatuba** São Paulo, S Brazil *23°37′S 45°24′W*
48 I7 **Carajás, Serra dos** ▲▲ N Brazil
Caralis *see* Cagliari
54 E9 **Caramanta** Antioquia, W Colombia *05°36′N 75°38′W*
171 P4 **Caramoan** Catanduanes Island, N Philippines *13°47′N 123°49′E*
Caramurat *see* Mihail Kogălniceanu
116 F12 **Caransebeş** *Ger.* Karansebesch, *Hung.* Karánsebes. Caraş-Severin, SW Romania *45°23′N 22°13′E*
Carapella *see* Carapelle
107 M16 **Carapelle** *var.* Carapella. ≈ SE Italy
55 O9 **Carapo** Bolívar, SE Venezuela
13 P13 **Caraquet** New Brunswick, SE Canada *47°48′N 64°59′W*
Caras *see* Caraz
116 F12 **Caraşova** *Hung.* Krassóvár. Caraş-Severin, SW Romania *45°11′N 21°51′E*
116 F12 **Caraş-Severin** ⬖ *county* SW Romania
42 M5 **Caratasca, Laguna de** *lagoon* NE Honduras
58 C13 **Carauari** Amazonas, NW Brazil *04°55′S 66°57′W*
Caravaca *see* Caravaca de la Cruz
105 Q12 **Caravaca de la Cruz** *var.* Caravaca. Murcia, SE Spain *38°06′N 01°51′W*
106 E7 **Caravaggio** Lombardia, N Italy *45°31′N 09°39′E*
107 C18 **Caravai, Passo di** *pass* Sardegna, Italy, C Mediterranean Sea
59 O19 **Caravelas** Bahia, E Brazil *17°45′S 39°15′W*
56 C12 **Caraz** *var.* Caras. Ancash, W Peru *09°03′S 77°47′W*
61 H14 **Carazinho** Rio Grande do Sul, S Brazil *28°16′S 52°46′W*
42 J11 **Carazo** ⬖ *department* SW Nicaragua
Carballiño *see* O Carballiño
104 G2 **Carballo** Galicia, NW Spain *43°13′N 08°41′W*
11 W16 **Carberry** Manitoba, S Canada *49°52′N 99°20′W*
40 F4 **Carbó** Sonora, NW Mexico *29°41′N 111°00′W*
107 C20 **Carbonara, Capo** *headland* Sardegna, Italy, C Mediterranean Sea *39°06′N 09°31′E*
37 Q5 **Carbondale** Colorado, C USA *39°24′N 107°12′W*
30 L17 **Carbondale** Illinois, N USA *37°43′N 89°13′W*
27 Q4 **Carbondale** Kansas, C USA *38°49′N 95°41′W*
18 I13 **Carbondale** Pennsylvania, NE USA *41°34′N 75°30′W*
13 V12 **Carbonear** Newfoundland, Newfoundland and Labrador, SE Canada *47°45′N 53°16′W*
105 Q9 **Carboneras de Guadazaón** *var.* Carboneras de Guadazón. Castilla-La Mancha, C Spain *39°54′N 01°50′W*
Carboneras de Guadazón *see* Carboneras de Guadazaón
23 O3 **Carbon Hill** Alabama, S USA *33°53′N 87°31′W*
107 B20 **Carbonia** *var.* Carbonia Centro. Sardegna, Italy, C Mediterranean Sea *39°11′N 08°31′E*
Carbonia Centro *see* Carbonia
63 I22 **Carbon, Laguna del** *depression* S Argentina
105 S10 **Carcaixent** Valenciana, E Spain *39°08′N 00°28′W*
Carcaso *see* Carcassonne
65 B24 **Carcass Island** *island* NW Falkland Islands
103 O16 **Carcassonne** *anc.* Carcaso. Aude, S France *43°13′N 02°21′E*
105 R12 **Carche** ▲ S Spain *38°24′N 01°11′W*
56 A13 **Carchi** ⬖ *province* N Ecuador
10 I8 **Carcross** Yukon, W Canada *60°11′N 134°41′W*
Cardamomes, Chaîne des *see* Krâvanh, Chuŏr Phnum
155 G22 **Cardamom Hills** ▲▲ SW India
Cardamom Mountains *see* Krâvanh, Chuŏr Phnum
104 M12 **Cardeña** Andalucía, S Spain *38°16′N 04°20′W*
44 D4 **Cárdenas** Matanzas, W Cuba *23°02′N 81°12′W*
41 O11 **Cárdenas** San Luis Potosí, C Mexico *22°03′N 99°30′W*
41 U15 **Cárdenas** Tabasco, SE Mexico *18°00′N 93°21′W*
63 H21 **Cardiel, Lago** ◎ S Argentina
97 K22 **Cardiff** *Wel.* Caerdydd. ◐ S Wales, United Kingdom *51°30′N 03°13′W*
97 J22 **Cardiff-Wales** ✈ S Wales, United Kingdom *51°24′N 03°22′W*
97 I21 **Cardigan** *Wel.* Aberteifi. SW Wales, United Kingdom *52°06′N 04°40′W*
97 I20 **Cardigan** *cultural region* W Wales, United Kingdom
97 I20 **Cardigan Bay** *bay* W Wales, United Kingdom
19 N8 **Cardigan, Mount** ▲ New Hampshire, NE USA *43°39′N 71°52′W*
14 M13 **Cardinal** Ontario, SE Canada *44°48′N 75°22′W*
105 V5 **Cardona** Cataluña, NE Spain *41°55′N 01°41′E*
61 E19 **Cardona** Soriano, SW Uruguay *33°53′S 57°18′W*
105 V4 **Cardoner** ≈ NE Spain
11 Q17 **Cardston** Alberta, SW Canada *49°14′N 113°19′W*
181 W5 **Cardwell** Queensland, NE Australia *18°24′S 146°06′E*
116 G8 **Carei** *Ger.* Gross-Karol, Karol, *Hung.* Nagykároly; *prev.* Careii-Mari. Satu Mare, NW Romania *47°40′N 22°28′E*
Careii-Mari *see* Carei
58 F13 **Careiro** Amazonas, NW Brazil *03°40′S 60°23′W*
102 J4 **Carentan** Manche, N France *49°18′N 01°15′W*
104 M2 **Cares** ≈ N Spain
33 P14 **Carey** Idaho, NW USA *43°17′N 113°58′W*
31 S12 **Carey** Ohio, N USA *40°57′N 83°22′W*
25 P4 **Carey** Texas, SW USA *34°28′N 100°18′W*
180 L11 **Carey, Lake** ◎ Western Australia
173 O8 **Cargados Carajos Bank** *undersea feature* C Indian Ocean
102 G6 **Carhaix-Plouguer** Finistère, NW France *48°16′N 03°35′W*
61 A22 **Carhué** Buenos Aires, E Argentina *37°10′S 62°45′W*
55 O5 **Cariaco** Sucre, NE Venezuela *10°33′N 63°37′W*
107 O20 **Cariati** Calabria, SW Italy *39°30′N 16°57′E*
2 H17 **Caribbean Plate** *tectonic feature*
44 I11 **Caribbean Sea** *sea* W Atlantic Ocean
11 N15 **Cariboo Mountains** ▲▲ British Columbia, SW Canada
11 W9 **Caribou** Manitoba, C Canada *59°27′N 97°43′W*
19 S2 **Caribou** Maine, NE USA *46°51′N 68°00′W*
11 P10 **Caribou Mountains** ▲▲ Alberta, SW Canada
Caribrod *see* Dimitrovgrad
40 I6 **Carichic** Chihuahua, N Mexico *27°57′N 107°01′W*
103 R3 **Carignan** Ardennes, N France *49°38′N 05°08′E*
183 Q5 **Carinda** New South Wales, SE Australia *30°26′S 147°45′E*
105 R6 **Cariñena** Aragón, NE Spain *41°20′N 01°13′W*
107 I23 **Carini** Sicilia, Italy, C Mediterranean Sea *38°06′N 13°09′E*
107 K17 **Carinola** Campania, S Italy *41°14′N 14°03′E*
Carinthi *see* Kärnten
55 O5 **Caripe** Monagas, NE Venezuela *10°13′N 63°30′W*
55 P5 **Caripito** Monagas, NE Venezuela *10°03′N 63°05′W*
15 W7 **Carleton** Québec, SE Canada *48°07′N 66°07′W*
31 S10 **Carleton** Michigan, N USA *42°03′N 83°23′W*
13 O14 **Carleton, Mount** ▲ New Brunswick, SE Canada *47°10′N 66°54′W*
14 L13 **Carleton Place** Ontario, SE Canada *45°08′N 76°09′W*
35 V3 **Carlin** Nevada, W USA *40°40′N 116°09′W*
30 K14 **Carlinville** Illinois, N USA *39°16′N 89°52′W*
97 K14 **Carlisle** *anc.* Caer Luel, Luguvallium, Luguvallum. NW England, United Kingdom *54°54′N 02°55′W*
27 V11 **Carlisle** Arkansas, C USA *34°46′N 91°45′W*
31 N15 **Carlisle** Indiana, N USA *38°57′N 87°23′W*
29 V14 **Carlisle** Iowa, C USA *41°30′N 93°29′W*
21 N5 **Carlisle** Kentucky, S USA *38°19′N 84°02′W*
18 F15 **Carlisle** Pennsylvania, NE USA *40°10′N 77°10′W*
21 Q11 **Carlisle** South Carolina, SE USA *34°35′N 81°27′W*
38 J17 **Carlisle Island** *island* Aleutian Islands, Alaska, USA
27 R7 **Carl Junction** Missouri, C USA *37°10′N 94°34′W*
107 A20 **Carloforte** Sardegna, Italy, C Mediterranean Sea *39°10′N 08°17′E*
Carlopago *see* Karlobag
61 B21 **Carlos Casares** Buenos Aires, E Argentina *35°39′S 61°28′W*
61 E18 **Carlos Reyles** Durazno, C Uruguay *33°00′S 56°30′W*
61 A21 **Carlos Tejedor** Buenos Aires, E Argentina *35°25′S 62°25′W*
97 F19 **Carlow** *Ir.* Ceatharlach. SE Ireland *52°50′N 06°55′W*
97 F19 **Carlow** *Ir.* Cheatharlach. *cultural region* SE Ireland
96 F7 **Carloway** NW Scotland, United Kingdom *58°17′N 06°48′W*
35 U17 **Carlsbad** California, W USA *33°09′N 117°21′W*
37 U15 **Carlsbad** New Mexico, SW USA *32°24′N 104°15′W*
Carlsbad *see* Karlovy Vary
129 N13 **Carlsberg Ridge** *undersea feature* S Arabian Sea *06°00′N 61°00′E*
Carlsruhe *see* Karlsruhe
29 W6 **Carlton** Minnesota, N USA *46°39′N 92°25′W*
11 V17 **Carlyle** Saskatchewan, S Canada *49°39′N 102°18′W*
30 L15 **Carlyle** Illinois, N USA *38°36′N 89°22′W*
30 L15 **Carlyle Lake** ⊞ Illinois, N USA
10 H7 **Carmacks** Yukon, W Canada *62°04′N 136°21′W*
106 B9 **Carmagnola** Piemonte, NW Italy *44°50′N 07°43′E*
11 X16 **Carman** Manitoba, S Canada *49°32′N 97°59′W*
Carmana/Carmania *see* Kermān
97 I21 **Carmarthen** SW Wales, United Kingdom *51°52′N 04°19′W*
97 I21 **Carmarthen** *cultural region* SW Wales, United Kingdom
97 I22 **Carmarthen Bay** *inlet* SW Wales, United Kingdom
103 N14 **Carmaux** Tarn, S France *44°03′N 02°09′E*
35 N11 **Carmel** California, W USA *36°32′N 121°54′W*
31 O13 **Carmel** Indiana, N USA *39°58′N 86°07′W*
18 L13 **Carmel** New York, NE USA *41°25′N 73°40′W*
97 H18 **Carmel Head** *headland* NW Wales, United Kingdom *53°24′N 04°35′W*
42 E2 **Carmelita** Petén, N Guatemala *17°33′N 90°11′W*
61 D19 **Carmelo** Colonia, SW Uruguay *34°00′S 58°20′W*
41 V14 **Carmen** *var.* Ciudad del Carmen. Campeche, SE Mexico *18°38′N 91°50′W*
61 A25 **Carmen de Patagones** Buenos Aires, E Argentina *40°45′S 63°00′W*
40 F8 **Carmen, Isla** *island* NW Mexico
40 M5 **Carmen, Sierra del** ▲▲ NW Mexico
30 M16 **Carmi** Illinois, N USA *38°05′N 88°09′W*
35 O7 **Carmichael** California, W USA *38°36′N 121°21′W*
Carmiel *see* Karmi’el
25 U11 **Carmine** Texas, SW USA *30°07′N 96°40′W*
104 K14 **Carmona** Andalucía, S Spain *37°28′N 05°38′W*
Carmona *see* Uíge
Carnaro *see* Kvarner
180 G9 **Carnarvon** Western Australia *24°57′S 113°38′E*
14 I13 **Carnarvon** Ontario, SE Canada *45°03′N 78°41′W*
83 G24 **Carnarvon** Northern Cape, W South Africa *30°59′S 22°08′E*
Carnarvon *see* Caernarfon
180 K9 **Carnarvon Range** ▲▲ Western Australia
Carn Domhnach *see* Carndonagh
96 E13 **Carndonagh** *Ir.* Carn Domhnach. NW Ireland *55°15′N 07°15′W*
11 V17 **Carnduff** Saskatchewan, S Canada *49°11′N 101°50′W*
26 L11 **Carnegie** Oklahoma, C USA *35°06′N 98°36′W*
180 L9 **Carnegie, Lake** *salt lake* Western Australia
193 U8 **Carnegie Ridge** *undersea feature* E Pacific Ocean *01°00′S 85°00′W*
96 H9 **Carn Eige** ▲ N Scotland, United Kingdom *57°18′N 05°04′W*
182 F5 **Carnes** South Australia *30°12′S 134°31′E*
194 J12 **Carney Island** *island* Antarctica
18 H16 **Carneys Point** New Jersey, NE USA *39°38′N 75°29′W*
Carniche, Alpi *see* Karnische Alpen
151 Q21 **Car Nicobar** *island* Nicobar Islands, India, NE Indian Ocean
79 H15 **Carnot** Mambéré-Kadéï, W Central African Republic *04°58′N 15°55′E*
182 F10 **Carnot, Cape** *headland* South Australia *34°57′S 135°39′E*
96 K11 **Carnoustie** E Scotland, United Kingdom *56°30′N 02°42′W*
97 F20 **Carnsore Point** *Ir.* Ceann an Chairn. *headland* SE Ireland *52°10′N 06°22′W*
8 H7 **Carnwath** ≈ Northwest Territories, NW Canada
31 R8 **Caro** Michigan, N USA *43°29′N 83°24′W*
23 Z15 **Carol City** Florida, SE USA *25°56′N 80°15′W*
59 L14 **Carolina** Maranhão, E Brazil *07°20′S 47°25′W*
45 U5 **Carolina** E Puerto Rico *18°22′N 65°57′W*
21 V12 **Carolina Beach** North Carolina, SE USA *34°02′N 77°53′W*
Caroline Island *see* Millennium Island
189 N15 **Caroline Islands** *island group* C Micronesia
129 Z14 **Caroline Plate** *tectonic feature*
192 H7 **Caroline Ridge** *undersea feature* E Philippine Sea *08°00′N 150°00′E*
Carolopois *see* Châlons-en-Champagne
45 V14 **Caroni Arena Dam** ⊞ Trinidad, Trinidad and Tobago
Caronie, Monti *see* Nebrodi, Monti
55 P7 **Caroní, Río** ≈ E Venezuela
45 U14 **Caroni River** ≈ Trinidad, Trinidad and Tobago
Caronium *see* A Coruña
54 J5 **Carora** Lara, N Venezuela *10°12′N 70°07′W*
86 F12 **Carpathian Mountains** *var.* Carpathians, *Cz./Pol.* Karpaty, *Ger.* Karpaten. ▲▲ E Europe
Carpathians *see* Carpathian Mountains
Carpathos/Carpathus *see* Kárpathos
116 H12 **Carpaţii Meridionalii** *var.* Alpi Transilvaniei, Carpaţii Sudici, *Eng.* South Carpathians, Transylvanian Alps, *Ger.* Südkarpaten, Transsylvanische Alpen, *Hung.* Déli-Kárpátok, Erdélyi-Havasok. ▲▲ C Romania
Carpaţii Sudici *see* Carpaţii Meridionalii
174 L7 **Carpentaria, Gulf of** *gulf* N Australia
Carpentoracte *see* Carpentras
103 R14 **Carpentras** *anc.* Carpentoracte. Vaucluse, SE France *44°03′N 05°03′E*
106 F9 **Carpi** Emilia-Romagna, N Italy *44°47′N 10°53′E*
116 E11 **Cărpiniş** *Hung.* Gyertyámos. Timiş, W Romania *45°46′N 20°53′E*
35 R14 **Carpinteria** California, W USA *34°24′N 119°30′W*
23 S9 **Carrabelle** Florida, SE USA *29°51′N 84°39′W*
Carraig Aonair *see* Fastnet Rock
Carraig Fhearghais *see* Carrickfergus
Carraig Mhachaire Rois *see* Carrickmacross
Carraig na Siúire *see* Carrick-on-Suir
Carrantual *see* Carrauntoohil
106 E10 **Carrara** Toscana, C Italy *44°05′N 10°07′E*
61 F20 **Carrasco** ✈ (Montevideo) Canelones, S Uruguay *34°51′S 56°00′W*
105 P9 **Carrascosa del Campo** Castilla-La Mancha, C Spain *40°02′N 02°35′W*
54 H4 **Carrasquero** Zulia, NW Venezuela *11°00′N 72°01′W*
183 O9 **Carrathool** New South Wales, SE Australia *34°25′S 145°30′E*
Carrauntohil *see* Carrauntoohil
97 B21 **Carrauntoohil** *Ir.* Carrantual, Carrauntohil, Corrán Tuathail. ▲ SW Ireland *51°98′N 09°53′W*
45 Y15 **Carriacou** *island* N Grenada
97 G15 **Carrickfergus** *Ir.* Carraig Fhearghais. NE Northern Ireland, United Kingdom *54°43′N 05°49′W*
97 F16 **Carrickmacross** *Ir.* Carraig Mhachaire Rois. N Ireland *53°58′N 06°43′W*
97 D16 **Carrick-on-Shannon** *Ir.* Cora Droma Rúisc. NW Ireland *53°37′N 08°05′W*
97 E20 **Carrick-on-Suir** *Ir.* Carraig na Siúire. S Ireland *52°21′N 07°25′W*
182 I7 **Carrieton** South Australia *32°27′S 138°33′E*
40 L7 **Carrillo** Chihuahua, N Mexico *25°53′N 103°54′W*
29 O4 **Carrington** North Dakota, N USA *47°27′N 99°07′W*
104 M4 **Carrión** ≈ N Spain
104 M4 **Carrión de los Condes** Castilla y León, N Spain *42°20′N 04°37′W*
25 P13 **Carrizo Springs** Texas, SW USA *28°33′N 99°54′W*
37 S13 **Carrizozo** New Mexico, SW USA *33°38′N 105°52′W*
29 T13 **Carroll** Iowa, C USA *42°04′N 94°52′W*
23 N4 **Carrollton** Alabama, S USA *33°13′N 88°05′W*
23 R3 **Carrollton** Georgia, SE USA *33°33′N 85°04′W*
30 K14 **Carrollton** Illinois, N USA *39°18′N 90°24′W*
20 L4 **Carrollton** Kentucky, S USA *38°41′N 85°09′W*
31 R8 **Carrollton** Michigan, N USA *43°27′N 83°55′W*
27 T3 **Carrollton** Missouri, C USA *39°20′N 93°30′W*
31 U12 **Carrollton** Ohio, N USA *40°34′N 81°05′W*
25 T6 **Carrollton** Texas, SW USA *32°57′N 96°53′W*
11 U14 **Carrot** ≈ Saskatchewan, S Canada
11 U14 **Carrot River** Saskatchewan, C Canada *53°18′N 103°32′W*
18 J7 **Carry Falls Reservoir** ⊞ New York, NE USA
136 L11 **Çarşamba** Samsun, N Turkey *41°13′N 36°43′E*
28 L6 **Carson** North Dakota, N USA *46°26′N 101°34′W*
35 Q6 **Carson City** *state capital* Nevada, W USA *39°10′N 119°46′W*
35 R6 **Carson River** ≈ Nevada, W USA
35 S5 **Carson Sink** *salt flat* Nevada, W USA
11 Q16 **Carstairs** Alberta, SW Canada *51°35′N 114°02′W*
Carstensz, Puntjak *see* Jaya, Puncak
54 E5 **Cartagena** *var.* Cartagena de los Indes. Bolívar, NW Colombia *10°24′N 75°33′W*
105 R13 **Cartagena** *anc.* Carthago Nova. Murcia, SE Spain *37°36′N 00°59′W*
54 E13 **Cartagena de Chaira** Caquetá, S Colombia *01°19′N 74°52′W*
Cartagena de los Indes *see* Cartagena
54 D10 **Cartago** Valle del Cauca, W Colombia *04°45′N 75°55′W*
43 N14 **Cartago** Cartago, C Costa Rica *09°50′N 83°54′W*
42 M14 **Cartago** *off.* Provincia de Cartago. ⬖ *province* C Costa Rica
Cartago, Provincia de *see* Cartago
25 O11 **Carta Valley** Texas, SW USA *29°46′N 100°37′W*
104 F10 **Cartaxo** Santarém, C Portugal *39°10′N 08°47′W*
104 I14 **Cartaya** Andalucía, S Spain *37°16′N 07°09′W*
Carteret Islands *see* Tulun Islands
29 S15 **Carter Lake** Iowa, C USA *41°17′N 95°55′W*
23 S3 **Cartersville** Georgia, SE USA *34°10′N 84°48′W*
185 M14 **Carterton** Wellington, North Island, New Zealand *41°01′S 175°30′E*
30 J13 **Carthage** Illinois, N USA *40°25′N 91°09′W*
22 L5 **Carthage** Mississippi, S USA *32°43′N 89°31′W*
27 R7 **Carthage** Missouri, C USA *37°10′N 94°20′W*
18 I8 **Carthage** New York, NE USA *43°58′N 75°36′W*
21 T10 **Carthage** North Carolina, SE USA *35°21′N 79°27′W*
20 K8 **Carthage** Tennessee, S USA *36°14′N 85°59′W*
25 X7 **Carthage** Texas, SW USA *32°10′N 94°21′W*
74 M5 **Carthage** ✈ (Tunis) N Tunisia *36°51′N 10°12′E*
Carthago Nova *see* Cartagena
14 E10 **Cartier** Ontario, S Canada *46°40′N 81°31′W*
13 S8 **Cartwright** Newfoundland and Labrador, E Canada *53°40′N 57°W*
55 P9 **Caruana de Montaña** Bolívar, SE Venezuela *05°16′N 63°12′W*
59 Q15 **Caruaru** Pernambuco, E Brazil *08°15′S 35°55′W*
55 P5 **Carúpano** Sucre, NE Venezuela *10°39′N 63°14′W*
Carusbur *see* Cherbourg
58 M12 **Carutapera** Maranhão, E Brazil *01°12′S 45°57′W*
27 Y9 **Caruthersville** Missouri, C USA *36°11′N 89°40′W*
103 O1 **Carvin** Pas-de-Calais, N France *50°31′N 03°00′E*
58 E12 **Carvoeiro** Amazonas, NW Brazil *01°24′S 61°59′W*
104 E10 **Carvoeiro, Cabo** *headland* C Portugal *39°19′N 09°27′W*
21 U9 **Cary** North Carolina, SE USA *35°47′N 78°46′W*
182 M3 **Caryapundy Swamp** *wetland* New South Wales/Queensland, SE Australia
65 E24 **Carysfort, Cape** *headland* East Falkland, Falkland Islands *51°26′S 57°50′W*
74 F6 **Casablanca** *Ar.* Dar-el-Beida. NW Morocco *33°39′N 07°31′W*
60 M8 **Casa Branca** São Paulo, S Brazil *21°47′S 47°05′W*
36 L14 **Casa Grande** Arizona, SW USA *32°52′N 111°45′W*
106 C8 **Casale Monferrato** Piemonte, NW Italy *45°08′N 08°27′E*
106 E8 **Casalpusterlengo** Lombardia, N Italy *45°10′N 09°37′E*
54 H10 **Casanare** *off.* Intendencia de Casanare. ⬖ *province* C Colombia
Casanare, Intendencia de *see* Casanare
55 P5 **Casanay** Sucre, NE Venezuela *10°30′N 63°25′W*
24 K11 **Casa Piedra** Texas, SW USA *29°43′N 104°03′W*
107 Q19 **Casarano** Puglia, SE Italy *40°01′N 18°10′E*
42 J11 **Casares** Carazo, W Nicaragua *11°37′N 86°19′W*
105 R10 **Casas Ibáñez** Castilla-La Mancha, C Spain *39°17′N 01°28′W*
61 I14 **Casca** Rio Grande do Sul, S Brazil *28°39′S 51°55′W*
172 I17 **Cascade** Mahé, NE Seychelles *04°39′S 55°29′E*
33 N13 **Cascade** Idaho, NW USA *44°31′N 116°02′W*
29 Y13 **Cascade** Iowa, C USA *42°18′N 91°00′W*
33 R9 **Cascade** Montana, NW USA *47°15′N 111°46′W*
185 B20 **Cascade Point** *headland* South Island, New Zealand *44°00′S 168°23′E*
32 G13 **Cascade Range** ▲▲ Oregon/Washington, NW USA
33 N12 **Cascade Reservoir** ⊞ Idaho, NW USA
0 E8 **Cascadia Basin** *undersea feature* NE Pacific Ocean *47°00′N 127°30′W*
104 E11 **Cascais** Lisboa, C Portugal *38°41′N 09°25′W*
15 W7 **Cascapédia** ≈ Québec, SE Canada
59 I22 **Cascavel** Ceará, E Brazil *04°10′S 38°15′W*
60 G11 **Cascavel** Paraná, S Brazil *24°56′S 53°28′W*
106 I13 **Cascia** Umbria, C Italy *42°45′N 13°01′E*
106 F11 **Cascina** Toscana, C Italy *43°40′N 10°33′E*
19 Q8 **Casco Bay** *bay* Maine, NE USA
194 J7 **Case Island** *island* Antarctica
106 B8 **Caselle** ✈ (Torino) Piemonte, NW Italy *45°06′N 07°41′E*
107 K17 **Caserta** Campania, S Italy *41°05′N 14°20′E*
15 N8 **Casey** Québec, SE Canada *47°50′N 74°09′W*
30 M14 **Casey** Illinois, N USA *39°18′N 87°59′W*
195 Y12 **Casey** *Australian research station* Antarctica *65°58′S 111°04′E*
195 W3 **Casey Bay** *bay* Antarctica
80 Q11 **Caseyr, Raas** *headland* NE Somalia *11°51′S 51°16′E*
97 D20 **Cashel** *Ir.* Caiseal. S Ireland *52°31′N 07°53′W*
54 G6 **Casigua** Zulia, W Venezuela *08°46′N 72°30′W*
61 B19 **Casilda** Santa Fe, C Argentina *33°05′S 61°10′W*
Casim *see* General Toshevo
183 V4 **Casino** New South Wales, SE Australia *28°50′S 153°02′E*
107 J16 **Cassino** *prev.* San Germano; *anc.* Casinum. Lazio, C Italy *41°29′N 13°50′E*
Casinum *see* Cassino
111 E17 **Čáslav** *Ger.* Tschaslau. Střední Čechy, C Czech Republic *49°54′N 15°23′E*
56 C13 **Casma** Ancash, C Peru *09°30′S 78°18′W*
167 S7 **Ca, Sông** ≈ N Vietnam
107 K17 **Casoria** Campania, S Italy *40°54′N 14°28′E*
105 T6 **Caspe** Aragón, NE Spain *41°14′N 00°03′W*
33 X15 **Casper** Wyoming, C USA *42°48′N 106°22′W*
84 M10 **Caspian Depression** *Kaz.* Kaspīy Mangy Oypaty, *Rus.* Prikaspiyskaya Nizmennost’. *depression* Kazakhstan/Russian Federation
130 D10 **Caspian Sea** *Az.* Xäzär Dänizi, *Kaz.* Kaspiy Tengizi, *Per.* Baḩr-e Khazar, Daryā-ye Khazar, *Rus.* Kaspiyskoye More. *inland sea* Asia/Europe
83 L14 **Cassacatiza** Tete, NW Mozambique *14°20′S 32°24′E*
Cassai *see* Kasai
82 F13 **Cassamba** Moxico, E Angola *13°07′S 20°22′E*
107 N20 **Cassano allo Ionio** Calabria, SE Italy *39°46′N 16°16′E*
31 S8 **Cass City** Michigan, N USA *43°36′N 83°10′W*
Cassel *see* Kassel
14 M13 **Casselman** Ontario, SE Canada *45°18′N 75°05′W*
29 R5 **Casselton** North Dakota, N USA *46°53′N 97°10′W*
Cássia *see* Santa Rita de Cassia
10 J9 **Cassiar** British Columbia, W Canada *59°16′N 129°40′W*
10 K10 **Cassiar Mountains** ▲▲ British Columbia, W Canada
83 C15 **Cassinga** Huíla, SW Angola *15°08′S 16°05′E*
29 T4 **Cass Lake** Minnesota, N USA *47°22′N 94°36′W*
29 T4 **Cass Lake** ◎ Minnesota, N USA
31 P10 **Cassopolis** Michigan, N USA *41°56′N 86°00′W*
31 S8 **Cass River** ≈ Michigan, N USA
27 S8 **Cassville** Missouri, C USA *36°42′N 93°52′W*
Castamoni *see* Kastamonu
58 L12 **Castanhal** Pará, NE Brazil *01°16′S 47°55′W*
104 G8 **Castanheira de Pêra** Leiria, C Portugal *40°01′N 08°12′W*
41 N7 **Castaños** Coahuila, NE Mexico *26°48′N 101°26′W*
108 I10 **Castasegna** Graubünden, SE Switzerland *46°21′N 09°30′E*
106 D8 **Casteggio** Lombardia, N Italy *45°02′N 09°10′E*
107 K23 **Castelbuono** Sicilia, Italy, C Mediterranean Sea *37°56′N 14°05′E*
107 K15 **Castel di Sangro** Abruzzo, C Italy *41°46′N 14°03′E*
106 H7 **Castelfranco Veneto** Veneto, NE Italy *45°40′N 11°55′E*
102 K14 **Casteljaloux** Lot-et-Garonne, SW France *44°19′N 00°03′E*
107 L18 **Castellabate** *var.* Santa Maria di Castellabate. Campania, S Italy *40°16′N 14°57′E*
107 I23 **Castellammare del Golfo** Sicilia, Italy, C Mediterranean Sea *38°02′N 12°53′E*
107 H22 **Castellammare, Golfo di** *gulf* Sicilia, Italy, C Mediterranean Sea
103 U15 **Castellane** Alpes-de-Haute-Provence, SE France *43°49′N 06°34′E*
107 O18 **Castellaneta** Puglia, SE Italy *40°38′N 16°57′E*
106 E9 **Castel l’Arquato** Emilia-Romagna, C Italy *44°52′N 09°51′E*
61 E21 **Castelli** Buenos Aires, E Argentina *36°07′S 57°47′W*
105 S8 **Castelló de la Plana** *var.* Castellón de la Plana. ⬖ *province* Valenciana, E Spain
Castelló de la Plana *see* Castellón de la Plana
Castellón *see* Castellón de la Plana
105 T9 **Castellón de la Plana** *var.* Castellón, *Cat.* Castelló de la Plana. Valenciana, E Spain *39°59′N 00°03′W*
Castellón de la Plana *see* Castelló de la Plana
105 S7 **Castellote** Aragón, NE Spain *40°46′N 00°18′W*
103 N16 **Castelnaudary** Aude, S France *43°18′N 01°57′E*
102 L16 **Castelnau-Magnoac** Hautes-Pyrénées, S France *43°18′N 00°30′E*
106 F10 **Castelnovo ne’ Monti** Emilia-Romagna, C Italy *44°26′N 10°24′E*
Castelnuovo *see* Herceg-Novi
104 H9 **Castelo Branco** Castelo Branco, C Portugal *39°50′N 07°30′W*
104 H8 **Castelo Branco** ⬖ *district* C Portugal
104 I10 **Castelo de Vide** Portalegre, C Portugal *39°25′N 07°27′W*
104 G9 **Castelo do Bode, Barragem do** ⊞ C Portugal

◆ Country
● Country Capital
◇ Dependent Territory
○ Dependent Territory Capital
⬖ Administrative Regions
✈ International Airport
▲ Mountain
▲▲ Mountain Range
⛰ Volcano
≈ River
◎ Lake
⊞ Reservoir

CASTEL SAN PIETRO TERME – CHAI BADAN

106 G10 **Castel San Pietro Terme** Emilia-Romagna, C Italy 44°22'N 11°34'E
107 B17 **Castelsardo** Sardegna, Italy, C Mediterranean Sea 40°54'N 08°42'E
102 M14 **Castelsarrasin** Tarn-et-Garonne, S France 44°02'N 01°06'E
107 I24 **Casteltermini** Sicilia, Italy, C Mediterranean Sea 37°33'N 13°38'E
107 H24 **Castelvetrano** Sicilia, Italy, C Mediterranean Sea 37°40'N 12°46'E
182 L12 **Casterton** Victoria, SE Australia 37°37'S 141°22'E
102 J15 **Castets** Landes, SW France 43°55'N 01°08'W
106 H12 **Castiglione del Lago** Umbria, C Italy 43°07'N 12°02'E
106 F13 **Castiglione della Pescaia** Toscana, C Italy 42°46'N 10°53'E
106 F8 **Castiglione delle Stiviere** Lombardia, N Italy 45°24'N 10°31'E
104 M9 **Castilla-La Mancha** ❖ *autonomous community* NE Spain
105 N10 **Castilla Nueva** *cultural region* C Spain
105 N6 **Castilla Vieja** ❖ *cultural region* N Spain
104 L5 **Castilla y León** *var.* Castillia Leon. ❖ *autonomous community* NW Spain
Castillia Leon *see* Castilla y León
Castillo de Locubim *see* Castillo de Locubín
105 N14 **Castillo de Locubín** *var.* Castillo de Locubim. Andalucía, S Spain 37°32'N 03°56'W
102 K13 **Castillon-la-Bataille** Gironde, SW France 44°51'N 00°01'W
63 I19 **Castillo, Pampa del** *plain* S Argentina
61 G19 **Castillos** Rocha, SE Uruguay 34°12'S 53°52'W
97 B16 **Castlebar** *Ir.* Caisleán an Bharraigh. W Ireland 53°52'N 09°17'W
97 F16 **Castleblayney** *Ir.* Baile na Lorgan. N Ireland 54°07'N 06°44'W
45 O11 **Castle Bruce** E Dominica 15°24'N 61°26'W
36 M5 **Castle Dale** Utah, W USA 39°10'N 111°02'W
36 I14 **Castle Dome Peak** ▲ Arizona, SW USA 33°04'N 114°08'W
97 J14 **Castle Douglas** S Scotland, United Kingdom 54°56'N 03°56'W
97 E14 **Castlefinn** *Ir.* Caisleán na Finne. NW Ireland 54°47'N 07°35'W
97 M17 **Castleford** N England, United Kingdom 53°44'N 01°21'W
11 O17 **Castlegar** British Columbia, SW Canada 49°18'N 117°48'W
64 B12 **Castle Harbour** *inlet* Bermuda, NW Atlantic Ocean
21 V12 **Castle Hayne** North Carolina, SE USA 34°23'N 78°07'W
97 B20 **Castleisland** *Ir.* Oileán Ciarraí. SW Ireland 52°12'N 09°30'W
183 N12 **Castlemaine** Victoria, SE Australia 37°06'S 144°13'E
37 R5 **Castle Peak** ▲ Colorado, C USA 39°00'N 106°51'W
33 O13 **Castle Peak** ▲ Idaho, NW USA 44°02'N 114°42'W
184 N13 **Castlepoint** Wellington, North Island, New Zealand 40°54'S 176°13'E
97 D17 **Castlerea** *Ir.* An Caisleán Riabhach. W Ireland 53°45'N 08°32'W
97 G15 **Castlereagh** *Ir.* An Caisleán Riabhach. N Northern Ireland, United Kingdom 54°33'N 05°53'W
183 R6 **Castlereagh River** ≈ New South Wales, SE Australia
37 T5 **Castle Rock** Colorado, C USA 39°22'N 104°51'W
30 K7 **Castle Rock Lake** ⊠ Wisconsin, N USA
65 G25 **Castle Rock Point** *headland* S Saint Helena 16°02'S 05°45'W
97 I16 **Castletown** SE Isle of Man 54°05'N 04°39'W
29 R9 **Castlewood** South Dakota, N USA 44°43'N 97°01'W
11 R15 **Castor** Alberta, SW Canada 52°14'N 111°54'W
14 M13 **Castor** ≈ Ontario, SE Canada
27 X7 **Castor River** ≈ Missouri, C USA
Castra Albiensium *see* Castres
Castra Regina *see* Regensburg
103 N15 **Castres** *anc.* Castra Albiensium. Tarn, S France 43°36'N 02°15'E
98 H9 **Castricum** Noord-Holland, W Netherlands 52°33'N 04°40'E
45 S11 **Castries** ● (Saint Lucia) N Saint Lucia 14°01'N 60°59'W
60 J11 **Castro** Paraná, S Brazil 24°46'S 50°03'W
63 F17 **Castro** Los Lagos, W Chile 42°27'S 73°48'W
104 H7 **Castro Daire** Viseu, N Portugal 40°54'N 07°55'W
104 M13 **Castro del Río** Andalucía, S Spain 37°41'N 04°29'W
Castrogiovanni *see* Enna
104 H14 **Castro Marim** Faro, S Portugal 37°13'N 07°26'W
104 J2 **Castropol** Asturias, N Spain 43°30'N 07°01'W
105 O2 **Castro-Urdiales** *var.* Castro Urdiales. Cantabria, N Spain 43°23'N 03°11'W
104 G13 **Castro Verde** Beja, S Portugal 37°42'N 08°05'W
107 N19 **Castrovillari** Calabria, SW Italy 39°48'N 16°12'E
35 N10 **Castroville** California, W USA 36°46'N 121°46'W
25 R12 **Castroville** Texas, SW USA 29°21'N 98°52'W
104 K11 **Castuera** Extremadura, W Spain 38°44'N 05°33'W
61 F19 **Casupá** Florida, S Uruguay 34°09'S 55°38'W
185 A22 **Caswell Sound** *sound* South Island, New Zealand
137 Q13 **Çat** Erzurum, NE Turkey 39°40'N 41°03'E
42 K6 **Catacamas** Olancho, C Honduras 14°55'N 85°54'W
56 A10 **Catacaos** Piura, NW Peru 05°22'S 80°40'W
22 I7 **Catahoula Lake** ◎ Louisiana, S USA
137 S15 **Çatak** Van, SE Turkey 38°02'N 43°05'E
137 S15 **Çatak Çayi** ≈ SE Turkey
114 O12 **Çatalca** İstanbul, NW Turkey 41°09'N 28°28'E
114 O12 **Çatalca Yarimadasi** *physical region* NW Turkey
62 H6 **Catalina** Antofagasta, N Chile 25°19'S 69°37'W
Catalonia *see* Cataluña
105 U5 **Cataluña** *Cat.* Catalunya, *Eng.* Catalonia. ❖ *autonomous community* N Spain
Catalunya *see* Cataluña
62 I7 **Catamarca** *off.* Provincia de Catamarca. ❖ *province* NW Argentina
Catamarca *see* San Fernando del Valle de Catamarca
Catamarca, Provincia de *see* Catamarca
83 M16 **Catandica** Manica, C Mozambique 18°05'S 33°10'E
171 P4 **Catanduanes Island** *island* N Philippines
60 K8 **Catanduva** São Paulo, S Brazil 21°05'S 49°00'W
107 L24 **Catania** Sicilia, Italy, C Mediterranean Sea 37°31'N 15°04'E
107 M24 **Catania, Golfo di** *gulf* Sicilia, Italy, C Mediterranean Sea
45 U5 **Cataño** *var.* Cantaño. E Puerto Rico 18°26'N 66°06'W
107 O21 **Catanzaro** Calabria, SW Italy 38°53'N 16°36'E
107 O22 **Catanzaro Marina** *var.* Marina di Catanzaro. Calabria, S Italy 38°48'N 16°33'E
Marina di Catanzaro *see* Catanzaro Marina
25 Q14 **Catarina** Texas, SW USA 28°19'N 99°36'W
171 Q5 **Catarman** Samar, C Philippines 12°29'N 124°34'E
105 S10 **Catarroja** Valenciana, E Spain 39°24'N 00°24'W
21 R11 **Catawba River** ≈ North Carolina/South Carolina, SE USA
171 Q5 **Catbalogan** Samar, C Philippines 11°49'N 124°55'E
14 I14 **Catchacoma** Ontario, SE Canada 44°43'N 78°19'W
41 S15 **Catemaco** Veracruz-Llave, SE Mexico 18°28'N 95°10'W
Cathair na Mart *see* Westport
Cathair Saidhbhín *see* Cahersiveen
31 P5 **Cat Head Point** *headland* Michigan, N USA 45°11'N 85°37'W
23 Q2 **Cathedral Caverns** *cave* Alabama, S USA
35 V16 **Cathedral City** California, W USA 33°45'N 116°27'W
24 K10 **Cathedral Mountain** ▲ Texas, SW USA 30°10'N 103°39'W
32 G10 **Cathlamet** Washington, NW USA 46°12'N 123°24'W
76 G13 **Catió** S Guinea-Bissau 11°13'N 15°10'W
55 O10 **Catisimiña** Bolívar, SE Venezuela 04°07'N 63°40'W
44 J3 **Cat Island** *island* C The Bahamas
12 B9 **Cat Lake** Ontario, S Canada 51°47'N 91°52'W
21 P5 **Catlettsburg** Kentucky, S USA 38°24'N 82°37'W
185 D24 **Catlins** ≈ South Island, New Zealand
35 R1 **Catnip Mountain** ▲ Nevada, W USA 41°53'N 119°19'W
41 Z11 **Catoche, Cabo** *headland* SE Mexico 21°36'N 87°04'W
27 P9 **Catoosa** Oklahoma, C USA 36°11'N 95°45'W
41 N10 **Catorce** San Luis Potosí, C Mexico 23°42'N 100°49'W
63 I14 **Catriel** Río Negro, C Argentina 37°54'S 67°52'W
62 K13 **Catriló** La Pampa, C Argentina 36°28'S 63°20'W
58 F11 **Catrimani** Roraima, N Brazil 0°24'N 61°30'W
58 E10 **Catrimani, Rio** ≈ N Brazil
18 K11 **Catskill** New York, NE USA 42°13'N 73°52'W
18 K11 **Catskill Creek** ≈ New York, NE USA
18 J11 **Catskill Mountains** ▲▲ New York, NE USA
18 D11 **Cattaraugus Creek** ≈ New York, NE USA
Cattaro *see* Kotor
Cattaro, Bocche di *see* Kotorska, Boka
107 I24 **Cattolica Eraclea** Sicilia, Italy, C Mediterranean Sea 37°27'N 13°24'E
83 B14 **Catumbela** ≈ W Angola
83 N14 **Catur** Niassa, N Mozambique 13°50'S 35°43'E
82 C10 **Cauale** ≈ NE Angola
171 O2 **Cauayan** Luzon, N Philippines 16°55'N 121°46'E
54 C12 **Cauca** *off.* Departamento del Cauca. ❖ *province* SW Colombia
47 P5 **Cauca** ≈ SE Venezuela
Cauca, Departamento del *see* Cauca
58 P13 **Caucaia** Ceará, E Brazil 03°44'S 38°45'W
54 E7 **Cauca, Río** ≈ N Colombia
54 E7 **Caucasia** Antioquia, NW Colombia 07°59'N 75°13'W
137 Q8 **Caucasus** *Rus.* Kavkaz. ▲▲ Georgia/Russian Federation
62 I10 **Caucete** San Juan, W Argentina 31°38'S 68°16'W
105 R11 **Caudete** Castilla-La Mancha, C Spain 38°42'N 01°00'W
103 P2 **Caudry** Nord, N France 50°07'N 03°24'E
82 D11 **Caungula** Lunda Norte, NE Angola 08°22'S 18°37'E
62 G13 **Cauquenes** Maule, C Chile 35°58'S 72°22'W
55 N8 **Caura, Río** ≈ C Venezuela
15 V7 **Causapscal** Québec, SE Canada 48°22'N 67°14'W
117 N10 **Căuşeni** *Rus.* Kaushany. E Moldova 46°37'N 29°21'E
102 M14 **Caussade** Tarn-et-Garonne, S France 44°10'N 01°31'E
102 K17 **Cauterets** Hautes-Pyrénées, S France 42°53'N 00°08'W
10 J15 **Caution, Cape** *headland* British Columbia, SW Canada 51°10'N 127°43'W
44 H7 **Cauto** ≈ E Cuba
Cauvery *see* Kāveri
102 L3 **Caux, Pays de** *physical region* N France
107 L18 **Cava de' Tirreni** Campania, S Italy 40°42'N 14°42'E
104 G6 **Cávado** ≈ N Portugal
Cavaia *see* Kavajë
103 R15 **Cavaillon** Vaucluse, SE France 43°51'N 05°01'E
103 U16 **Cavalaire-sur-Mer** Var, SE France 43°10'N 06°31'E
106 G6 **Cavalese** *Ger.* Gablös. Trentino-Alto Adige, N Italy 46°18'N 11°29'E
29 Q2 **Cavalier** North Dakota, N USA 48°47'N 97°37'W
76 L17 **Cavalla** *var.* Cavally, Cavally Fleuve. ≈ Ivory Coast/Liberia
105 Y8 **Cavalleria, Cap de** *var.* Cabo Caballeria. *headland* Menorca, Spain, W Mediterranean Sea 40°04'N 04°06'E
184 K2 **Cavalli Islands** *island group* N New Zealand
Cavally/Cavally Fleuve *see* Cavalla
97 E16 **Cavan** *Ir.* Cabhán. N Ireland 54°N 07°21'W
97 E16 **Cavan** *Ir.* An Cabhán. *cultural region* N Ireland
106 H8 **Cavarzere** Veneto, NE Italy 45°08'N 12°05'E
27 W9 **Cave City** Arkansas, C USA 35°56'N 91°33'W
20 K7 **Cave City** Kentucky, S USA 37°08'N 85°57'W
65 M25 **Cave Point** *headland* S Tristan da Cunha
21 N5 **Cave Run Lake** ⊠ Kentucky, S USA
58 K11 **Caviana de Fora, Ilha** *var.* Ilha Caviana. *island* N Brazil
Caviana, Ilha *see* Caviana de Fora, Ilha
113 I16 **Cavtat** *It.* Ragusavecchia. Dubrovnik-Neretva, SE Croatia 42°36'N 18°13'E
Cawnpore *see* Kānpur
Caxamarca *see* Cajamarca
58 A13 **Caxias** Amazonas, W Brazil 04°27'S 71°27'W
58 N13 **Caxias** Maranhão, E Brazil 04°53'S 43°20'W
61 I15 **Caxias do Sul** Rio Grande do Sul, S Brazil 29°14'S 51°10'W
42 J4 **Caxinas, Punta** *headland* N Honduras 16°01'N 86°02'W
82 B11 **Caxito** Bengo, NW Angola 08°34'S 13°38'E
136 F14 **Çay** Afyon, W Turkey 38°35'N 31°01'E
40 L15 **Cayacal, Punta** *var.* Punta Mongrove. *headland* S Mexico 17°55'N 102°09'W
56 C6 **Cayambe** Pichincha, N Ecuador 0°02'N 78°08'W
56 C6 **Cayambe** ▲ N Ecuador 0°00'S 77°58'W
21 R12 **Cayce** South Carolina, SE USA 33°58'N 81°04'W
55 Y10 **Cayenne** ○ (French Guiana) NE French Guiana 04°55'N 52°18'W
55 Y10 **Cayenne** ✈ NE French Guiana 04°55'N 52°18'W
44 K10 **Cayes** *var.* Les Cayes. SW Haiti 18°10'N 73°48'W
45 U6 **Cayey** C Puerto Rico 18°06'N 66°11'W
45 U6 **Cayey, Sierra de** ▲▲ E Puerto Rico
103 N14 **Caylus** Tarn-et-Garonne, S France 44°13'N 01°42'E
44 E8 **Cayman Brac** *island* E Cayman Islands
44 D8 **Cayman Islands** ◇ UK *dependent territory* W West Indies
64 D11 **Cayman Trench** *undersea feature* NW Caribbean Sea 19°00'N 80°00'W
47 O3 **Cayman Trough** *undersea feature* NW Caribbean Sea 18°00'N 81°00'W
80 O13 **Caynabo** Togdheer, N Somalia 08°55'N 46°28'E
42 F3 **Cayo** ❖ *district* SW Belize
Cayo *see* San Ignacio
43 N9 **Cayos Guerrero** *reef* E Nicaragua
43 O9 **Cayos King** *reef* E Nicaragua
44 G4 **Cay Sal** *islet* SW The Bahamas
14 G16 **Cayuga** Ontario, S Canada 42°57'N 79°49'W
25 V8 **Cayuga** Texas, SW USA 31°55'N 95°57'W
18 G10 **Cayuga Lake** ◎ New York, NE USA
104 K13 **Cazalla de la Sierra** Andalucía, S Spain 37°56'N 05°46'W
116 L14 **Căzăneşti** Ialomiţa, SE Romania 44°36'N 27°03'E
102 M16 **Cazères** Haute-Garonne, S France 43°15'N 01°11'E
112 E10 **Cazin** ❖ Federacija Bosne I Hercegovine, NW Bosnia and Herzegovina
82 G13 **Cazombo** Moxico, E Angola 11°54'S 22°56'E
105 O13 **Cazorla** Andalucía, S Spain 37°55'N 03°W
Cazza *see* Sušac
104 L4 **Cea** ≈ NW Spain
Ceadâr-Lunga *see* Ciadîr-Lunga
Ceanannas *see* Kells
Ceann Toirc *see* Kanturk
58 O13 **Ceará** *off.* Estado do Ceará. ❖ *state* C Brazil
Ceará *see* Fortaleza
Ceara Abyssal Plain *see* Ceará Plain
Ceará, Estado do *see* Ceará
59 Q14 **Ceará Mirim** Rio Grande do Norte, E Brazil 05°30'S 35°51'W
64 J13 **Ceará Plain** *var.* Ceara Abyssal Plain. *undersea feature* W Atlantic Ocean 0°00'36°30'W
64 I13 **Ceará Ridge** *undersea feature* C Atlantic Ocean
Ceatharlach *see* Carlow
43 Q17 **Cébaco, Isla** *island* SW Panama
40 K7 **Ceballos** Durango, C Mexico 26°33'N 104°07'W
61 G19 **Cebollatí** Rocha, E Uruguay 33°15'S 53°46'W
61 G19 **Cebollatí, Río** ≈ E Uruguay
105 P5 **Cebollera** ▲ N Spain 42°01'N 02°40'W
104 M8 **Cebreros** Castilla y León, N Spain 40°27'N 04°28'W
171 P6 **Cebu** *off.* Cebu City. Cebu, C Philippines 10°17'N 123°46'E
171 P6 **Cebu** *island* C Philippines
Cebu City *see* Cebu
107 J16 **Ceccano** Lazio, C Italy 41°34'N 13°20'E
Čechy *see* Bohemia
106 F12 **Cecina** Toscana, C Italy 43°19'N 10°31'E
26 K4 **Cedar Bluff Reservoir** ⊠ Kansas, C USA
30 M8 **Cedarburg** Wisconsin, N USA 43°18'N 87°58'W
36 J7 **Cedar City** Utah, W USA 37°40'N 113°03'W
25 T11 **Cedar Creek** Texas, SW USA 30°04'N 97°30'W
28 L7 **Cedar Creek** ≈ North Dakota, N USA
25 U7 **Cedar Creek Reservoir** ⊠ Texas, SW USA
29 W13 **Cedar Falls** Iowa, C USA 42°31'N 92°27'W
31 N8 **Cedar Grove** Wisconsin, N USA 43°31'N 87°48'W
21 Y6 **Cedar Island** *island* Virginia, NE USA
23 U11 **Cedar Key** Cedar Keys, Florida, SE USA 29°08'N 83°03'W
23 U11 **Cedar Keys** *island group* Florida, SE USA
11 V14 **Cedar Lake** ◎ Manitoba, C Canada
14 I11 **Cedar Lake** ◎ Ontario, SE Canada
24 M6 **Cedar Lake** ◎ Texas, SW USA
29 X13 **Cedar Rapids** Iowa, C USA 41°58'N 91°40'W
29 X14 **Cedar River** ≈ Iowa/Minnesota, C USA
29 O14 **Cedar River** ≈ Nebraska, C USA
31 P8 **Cedar Springs** Michigan, N USA 43°13'N 85°33'W
23 R3 **Cedartown** Georgia, SE USA 34°00'N 85°16'W
27 O7 **Cedar Vale** Kansas, C USA 37°06'N 96°30'W
35 Q2 **Cedarville** California, W USA 41°30'N 120°10'W
104 H1 **Cedeira** Galicia, NW Spain 43°40'N 08°03'W
42 H8 **Cedeño** Choluteca, S Honduras 13°10'N 87°25'W
41 N10 **Cedral** San Luis Potosí, C Mexico 23°47'N 100°40'W
42 I6 **Cedros** Francisco Morazán, C Honduras 14°38'N 86°42'W
40 M9 **Cedros** Zacatecas, C Mexico 24°39'N 101°47'W
40 B5 **Cedros, Isla** *island* W Mexico
193 R5 **Cedros Trench** *undersea feature* E Pacific Ocean 27°45'N 115°45'W
182 E7 **Ceduna** South Australia 32°09'S 133°43'E
110 D10 **Cedynia** *Ger.* Zehden. Zachodnio-pomorskie, W Poland 52°54'N 14°15'E
80 P12 **Ceelaayo** Sanaag, N Somalia 11°18'N 49°20'E
81 O16 **Ceel Buur** *It.* El Bur. Galguduud, C Somalia 04°36'N 46°33'E
81 N15 **Ceel Dheere** *var.* Ceel Dher, *It.* El Dere. Galguduud, C Somalia 05°18'N 46°07'E
Ceel Dher *see* Ceel Dheere
80 O12 **Ceerigaabo** *var.* Erigabo, Erigavo. Sanaag, N Somalia 10°34'N 47°22'E
107 J23 **Cefalù** *anc.* Cephaloedium. Sicilia, Italy, C Mediterranean Sea 38°02'N 14°02'E
105 N6 **Cega** ≈ Castilla y León, N Spain
111 K23 **Cegléd** *prev.* Czegléd. Pest, C Hungary 47°10'N 19°47'E
113 N18 **Čegrane** W FYR Macedonia 41°50'N 20°59'E
105 Q13 **Cehegín** Murcia, SE Spain 38°04'N 01°48'W
136 K12 **Çekerek** Yozgat, N Turkey 40°04'N 35°30'E
146 B13 **Çekiçler** *Rus.* Chekishlyar, *Turkm.* Chekichler. Balkan Welaýaty, W Turkmenistan 37°35'N 53°52'E
107 J15 **Celano** Abruzzo, C Italy 42°06'N 13°33'E
104 H4 **Celanova** Galicia, NW Spain 42°09'N 07°58'W
42 F6 **Celaque, Cordillera de** ▲▲ W Honduras
41 N13 **Celaya** Guanajuato, C Mexico 20°32'N 100°48'W
Celebes *see* Sulawesi
192 F7 **Celebes Basin** *undersea feature* SE South China Sea 04°00'N 122°00'E
192 F7 **Celebes Sea** *Ind.* Laut Sulawesi. *sea* Indonesia/Philippines
41 W12 **Celestún** Yucatán, E Mexico 20°50'N 90°22'W
31 Q12 **Celina** Ohio, N USA 40°34'N 84°35'W
20 L8 **Celina** Tennessee, S USA 36°32'N 85°30'W
25 U5 **Celina** Texas, SW USA 33°19'N 96°46'W
112 G11 **Čelinac Donji** Republika Srpska, N Bosnia and Herzegovina 44°43'N 17°19'E
109 V10 **Celje** *Ger.* Cilli. C Slovenia 46°16'N 15°14'E
111 G23 **Celldömölk** Vas, W Hungary 47°16'N 17°10'E
100 J12 **Celle** *var.* Zelle. Niedersachsen, N Germany 52°38'N 10°05'E
99 D19 **Celles** Hainaut, SW Belgium 50°42'N 03°25'E
104 I7 **Celorico da Beira** Guarda, N Portugal 40°38'N 07°24'W
Celovec *see* Klagenfurt
64 M7 **Celtic Sea** *Ir.* An Mhuir Cheilteach. *sea* SW British Isles
64 N7 **Celtic Shelf** *undersea feature* E Atlantic Ocean 07°00'W 49°15'N
114 L13 **Çeltik Gölu** ◎ NW Turkey
146 J17 **Çemenibit** *prev. Rus.* Chemenibit. Mary Welaýaty, S Turkmenistan 35°27'N 62°19'E
113 M14 **Čemerno** ▲▲ C Serbia
105 Q12 **Cenajo, Embalse del** ⊠ S Spain
171 V13 **Cenderawasih, Teluk** *var.* Teluk Irian, Teluk Sarera. *bay* W Pacific Ocean
105 P4 **Cenicero** La Rioja, N Spain 42°29'N 02°38'W
106 E9 **Ceno** ≈ NW Italy
102 K13 **Cenon** Gironde, SW France 44°51'N 00°33'W
14 K13 **Centennial Lake** ◎ Ontario, SE Canada
Centennial State *see* Colorado
37 S7 **Center** Colorado, C USA 37°45'N 106°06'W
29 Q13 **Center** Nebraska, C USA 42°33'N 97°51'W
28 M5 **Center** North Dakota, N USA 47°07'N 101°18'W
25 X8 **Center** Texas, SW USA 31°49'N 94°10'W
29 W8 **Center City** Minnesota, N USA 45°25'N 92°48'W
36 L5 **Centerfield** Utah, W USA 39°07'N 111°49'W
20 K9 **Center Hill Lake** ⊠ Tennessee, S USA
29 X13 **Center Point** Iowa, C USA 42°11'N 91°47'W
25 R11 **Center Point** Texas, SW USA 29°56'N 99°01'W
29 W16 **Centerville** Iowa, C USA 40°43'N 92°51'W
27 W7 **Centerville** Missouri, C USA 37°27'N 91°01'W
29 R12 **Centerville** South Dakota, N USA 43°07'N 96°57'W
20 I9 **Centerville** Tennessee, S USA 35°45'N 87°29'W
25 V9 **Centerville** Texas, SW USA 31°17'N 95°59'W
40 M5 **Centinela, Picacho del** ▲ NE Mexico 29°07'N 102°40'W
106 G9 **Cento** Emilia-Romagna, N Italy 44°43'N 11°16'E
Centrafricaine, République *see* Central African Republic
39 S8 **Central** Alaska, USA 65°34'N 144°48'W
37 P15 **Central** New Mexico, SW USA 32°46'N 108°09'W
83 H18 **Central** ❖ *district* E Botswana
138 E10 **Central** ❖ *district* C Israel
82 M13 **Central** ❖ *region* C Malawi
153 P12 **Central** ❖ *zone* C Nepal
186 E9 **Central** ❖ *province* S Papua New Guinea
63 I21 **Central** ❖ *department* C Paraguay
155 K25 **Central** ❖ *province* C Sri Lanka
83 J14 **Central** ❖ *province* C Zambia
117 P11 **Central** ✈ (Odesa) Odes'ka Oblast', SW Ukraine 46°26'N 30°41'E
Central *see* Centre
Central *see* Rennell and Bellona
79 H14 **Central African Republic** *var.* République Centrafricaine, *abbrev.* CAR; *prev.* Ubangi-Shari, Oubangui-Chari, Territoire de l'Oubangui-Chari. ◆ *republic* C Africa
192 C6 **Central Basin Trough** *undersea feature* W Pacific Ocean 16°45'N 130°00'E
Central Borneo *see* Kalimantan Tengah
149 P12 **Central Brāhui Range** ▲▲ W Pakistan
Central Celebes *see* Sulawesi Tengah
29 Y13 **Central City** Iowa, C USA 42°12'N 91°31'W
20 I6 **Central City** Kentucky, S USA 37°17'N 87°07'W
29 P15 **Central City** Nebraska, C USA 41°04'N 97°59'W
48 D6 **Central, Cordillera** ▲▲ W Bolivia
54 D11 **Central, Cordillera** ▲▲ W Colombia
42 M13 **Central, Cordillera** ▲▲ C Costa Rica
45 N9 **Central, Cordillera** ▲▲ C Dominican Republic
43 R16 **Central, Cordillera** ▲▲ C Panama
45 S6 **Central, Cordillera** ▲▲ Puerto Rico
80 A11 **Central Darfur** ❖ *state* SW Sudan
42 H7 **Central District** *var.* Tegucigalpa. ❖ *district* C Honduras
81 E16 **Central Equatoria** *var.* Bahr el Gebel, Bahr el Jebel. ❖ *state* S South Sudan
Central Finland *see* Keski-Suomi
Central Group *see* Inner Islands
30 L15 **Centralia** Illinois, N USA 38°31'N 89°07'W
27 U4 **Centralia** Missouri, C USA 39°12'N 92°08'W
32 G9 **Centralia** Washington, NW USA 46°43'N 122°57'W
Central Indian Ridge *see* Mid-Indian Ridge
Central Java *see* Jawa Tengah
Central Kalimantan *see* Kalimantan Tengah
148 L14 **Central Makrān Range** ▲▲ W Pakistan
Central Ostrobothnia *see* Keski-Pohjanmaa
192 K7 **Central Pacific Basin** *undersea feature* C Pacific Ocean 05°00'N 175°00'W
59 M19 **Central, Planalto** *var.* Brazilian Highlands. ▲▲ E Brazil
32 F15 **Central Point** Oregon, NW USA 42°22'N 122°55'W
Central Provinces and Berar *see* Madhya Pradesh
186 B6 **Central Range** ▲▲ NW Papua New Guinea
Central Russian Upland *see* Srednerusskaya Vozvyshennost'
Central Siberian Plateau/Central Siberian Uplands *see* Srednesibirskoye Ploskogor'ye
104 K8 **Central, Sistema** ▲▲ C Spain
Central Sulawesi *see* Sulawesi Tengah
35 N3 **Central Valley** California, W USA 40°39'N 122°21'W
35 P8 **Central Valley** *valley* California, W USA
23 Q3 **Centre** Alabama, S USA 34°09'N 85°40'W
79 E15 **Centre** *Eng.* Central. ❖ *province* C Cameroon
102 M8 **Centre** ❖ *region* N France
173 Y16 **Centre de Flacq** E Mauritius 20°12'S 57°43'E
55 Y9 **Centre Spatial Guyanais** *space station* N French Guiana
23 O5 **Centreville** Alabama, S USA 32°58'N 87°08'W
21 X3 **Centreville** Maryland, NE USA 39°03'N 76°04'W
22 J7 **Centreville** Mississippi, S USA 31°05'N 91°04'W
Centum Cellae *see* Civitavecchia
160 M14 **Cenxi** Guangxi Zhuangzu Zizhiqu, S China 22°58'N 111°00'E
Ceos *see* Tziá
Cephaloedium *see* Cefalù
112 I9 **Čepin** *Hung.* Csepén. Osijek-Baranja, E Croatia 45°32'N 18°33'E
Ceram *see* Seram, Pulau
Ceram Sea *see* Seram, Laut
192 G8 **Ceram Trough** *undersea feature* W Pacific Ocean
Cerasus *see* Giresun
36 I10 **Cerbat Mountains** ▲▲ Arizona, SW USA
103 P17 **Cerbère, Cap** *headland* S France 42°28'N 03°15'E
104 F13 **Cercal do Alentejo** Setúbal, S Portugal 37°48'N 08°40'W
111 A18 **Čerchov** *Ger.* Czerkow. ▲ W Czech Republic 49°24'N 12°47'E
103 O13 **Cère** ≈ C France
61 A16 **Ceres** Santa Fe, C Argentina 29°55'S 61°55'W
59 K18 **Ceres** Goiás, C Brazil 15°21'S 49°34'W
Ceresio *see* Lugano, Lago di
103 O17 **Céret** Pyrénées-Orientales, S France 42°30'N 02°44'E
54 E6 **Cereté** Córdoba, NW Colombia 08°54'N 75°51'W
172 I17 **Cerf, Île au** *island* Inner Islands, NE Seychelles
99 G22 **Cerfontaine** Namur, S Belgium 50°08'N 04°25'E
Cergy-Pontoise *see* Pontoise
107 N16 **Cerignola** Puglia, SE Italy 41°17'N 15°53'E
Cerigo *see* Kýthira
103 O9 **Cérilly** Allier, C France 46°38'N 02°51'E
136 I11 **Çerkeş** Çankin, N Turkey 40°51'N 32°52'E
136 D10 **Çerkezköy** Tekirdağ, NW Turkey 41°17'N 28°00'E
109 T12 **Cerknica** *Ger.* Zirknitz. SW Slovenia 45°48'N 14°21'E
109 S11 **Cerkno** W Slovenia 46°07'N 13°58'E
116 F10 **Cermei** *Hung.* Csermő. Arad, W Romania 46°33'N 21°51'E
137 O15 **Çermik** Diyarbakır, SE Turkey 38°09'N 39°27'E
112 I10 **Cerna** Vukovar-Srijem, E Croatia 45°10'N 18°36'E
Cernăuţi *see* Chernivtsi
116 M14 **Cernavodă** Constanţa, SW Romania 44°20'N 28°03'E
103 U7 **Cernay** Haut-Rhin, NE France 47°49'N 07°11'E
Cernice *see* Schwarzach
41 O8 **Cerralvo** Nuevo León, NE Mexico 26°10'N 99°40'W
40 G9 **Cerralvo, Isla** *island* NW Mexico
107 L16 **Cerreto Sannita** Campania, S Italy 41°17'N 14°39'E
113 L20 **Cërrik** *var.* Cerriku. Elbasan, C Albania 41°01'N 19°55'E
Cerriku *see* Cërrik
41 O11 **Cerritos** San Luis Potosí, C Mexico 22°25'N 100°16'W
60 K11 **Cerro Azul** Paraná, S Brazil 24°48'S 49°14'W
61 F18 **Cerro Chato** Treinta y Tres, E Uruguay 33°04'S 55°08'W
61 F19 **Cerro Colorado** Florida, S Uruguay 33°52'S 55°33'W
56 E13 **Cerro de Pasco** Pasco, C Peru 10°43'S 76°15'W
61 G14 **Cêrro Largo** Rio Grande do Sul, S Brazil 28°10'S 54°43'W
61 G18 **Cerro Largo** ❖ *department* NE Uruguay
42 E7 **Cerrón Grande, Embalse** ⊠ N El Salvador
63 I14 **Cerros Colorados, Embalse** ◎ W Argentina
105 V5 **Cervera** Cataluña, NE Spain 41°40'N 01°16'E
104 M3 **Cervera del Pisuerga** Castilla y León, N Spain 42°51'N 04°30'W
105 Q5 **Cervera del Río Alhama** La Rioja, N Spain 42°01'N 01°58'W
107 H15 **Cerveteri** Lazio, C Italy 42°00'N 12°06'E
106 H10 **Cervia** Emilia-Romagna, N Italy 44°14'N 12°22'E
106 J7 **Cervignano del Friuli** Friuli-Venezia Giulia, NE Italy 45°49'N 13°18'E
107 L17 **Cervinara** Campania, S Italy 41°02'N 14°36'E
Cervinia *see* Breuil-Cervinia
106 B6 **Cervino, Monte** *var.* Matterhorn. ▲ Italy/Switzerland 46°00'N 07°39'E
see also Matterhorn
Cervino, Monte *see* Matterhorn
103 Y14 **Cervione** Corse, France, C Mediterranean Sea 42°22'N 09°28'E
104 I1 **Cervo** Galicia, NW Spain 43°39'N 07°25'W
54 F5 **Cesar** *off.* Departamento del Cesar. ❖ *province* N Colombia
Cesar, Departamento del *see* Cesar
106 H10 **Cesena** *anc.* Caesena. Emilia-Romagna, N Italy 44°09'N 12°14'E
106 I10 **Cesenatico** Emilia-Romagna, N Italy 44°12'N 12°24'E
118 H8 **Cēsis** *Ger.* Wenden. C Latvia 57°19'N 25°17'E
111 D15 **Česká Lípa** *Ger.* Böhmisch-Leipa. Liberecký Kraj, N Czech Republic 50°43'N 14°35'E
Česká Republika *see* Czech Republic
111 F17 **Česká Třebová** *Ger.* Böhmisch-Trübau. Pardubický Kraj, C Czech Republic 49°54'N 16°27'E
111 D19 **České Budějovice** *Ger.* Budweis. Jihočeský Kraj, S Czech Republic 48°58'N 14°29'E
111 D19 **České Velenice** Jihočeský Kraj, S Czech Republic 48°49'N 14°57'E
111 E18 **Českomoravská Vrchovina** *var.* Českomoravská Vysočina, *Eng.* Bohemian-Moravian Highlands, *Ger.* Böhmisch-Mährische Höhe. ▲▲ S Czech Republic
Českomoravská Vysočina *see* Českomoravská Vrchovina
111 C19 **Český Krumlov** *var.* Böhmisch-Krumau, *Ger.* Krummau. Jihočeský Kraj, S Czech Republic 48°48'N 14°18'E
Ceský Les *see* Bohemian Forest
112 F8 **Česma** ≈ N Croatia
136 A14 **Çeşme** İzmir, W Turkey 38°19'N 26°20'E
Cess *see* Cestos
183 T8 **Cessnock** New South Wales, SE Australia 32°51'S 151°21'E
76 K17 **Cestos** *var.* Cess. ≈ S Liberia
118 I9 **Cesvaine** E Latvia 56°58'N 26°15'E
116 G14 **Cetate** Dolj, SW Romania 44°06'N 23°01'E
Cetatea Albă *see* Bilhorod-Dnistrovs'kyy
Cetatea Damboviţei *see* Bucureşti
113 J17 **Cetinje** *It.* Cettigne. S Montenegro 42°23'N 18°55'E
107 N20 **Cetraro** Calabria, S Italy 39°30'N 15°59'E
Cette *see* Sète
188 A17 **Cetti Bay** *bay* SW Guam
Cettigne *see* Cetinje
104 L17 **Ceuta** *var.* Sebta. Ceuta, Spain, N Africa 35°53'N 05°19'W
88 C16 **Ceuta** *enclave* Spain, N Africa
106 B9 **Ceva** Piemonte, NE Italy 44°24'N 08°01'E
103 P14 **Cévennes** ▲▲ S France
108 G10 **Cevio** Ticino, S Switzerland 46°18'N 08°36'E
136 K16 **Ceyhan** Adana, S Turkey 37°02'N 35°48'E
136 K17 **Ceyhan Nehri** ≈ S Turkey
137 P17 **Ceylanpınar** Şanlıurfa, SE Turkey 36°53'N 40°02'E
Ceylon *see* Sri Lanka
173 R6 **Ceylon Plain** *undersea feature* N Indian Ocean 04°00'S 82°00'E
Ceyre to the Caribs *see* Marie-Galante
103 Q14 **Cèze** ≈ S France
Chaacha *see* Çäçe
127 P6 **Chaadayevka** Penzenskaya Oblast', W Russian Federation 53°07'N 45°55'E
167 O12 **Cha-Am** Phetchaburi, SW Thailand 12°48'N 99°58'E
143 W15 **Chābahār** *var.* Chāh Bahār, Chahbar. Sīstān va Balūchestān, SE Iran 25°21'N 60°38'E
Chabaricha *see* Khabarikha
61 B19 **Chabas** Santa Fe, C Argentina 33°16'S 61°23'W
103 T10 **Chablais** *physical region* E France
61 B20 **Chacabuco** Buenos Aires, E Argentina 34°40'S 60°27'W
42 K8 **Chachagón, Cerro** ▲ N Nicaragua 13°18'N 85°39'W
56 C10 **Chachapoyas** Amazonas, NW Peru 06°13'S 77°54'W
Chäche *see* Çäçe
119 O18 **Chachersk** *Rus.* Chechersk. Homyel'skaya Voblasts', SE Belarus 52°54'N 30°54'E
119 N16 **Chachevichy** *Rus.* Chechevichi. Mahilyowskaya Voblasts', E Belarus 53°31'N 29°51'E
61 B14 **Chaco** *off.* Provincia de Chaco. ❖ *province* NE Argentina
Chaco *see* Gran Chaco
62 M6 **Chaco Austral** *physical region* N Argentina
62 M3 **Chaco Boreal** *physical region* N Paraguay
62 M6 **Chaco Central** *physical region* N Argentina
39 Y15 **Chacon, Cape** *headland* Prince of Wales Island, Alaska, USA 54°41'N 132°00'W
Chaco, Provincia de *see* Chaco
78 H9 **Chad** *off.* Republic of Chad, *Fr.* Tchad. ◆ *republic* C Africa
122 K14 **Chadan** Respublika Tyva, S Russian Federation 51°16'N 91°25'E
21 U12 **Chadbourn** North Carolina, SE USA 34°19'N 78°49'W
83 L14 **Chadiza** Eastern, E Zambia 14°04'S 32°27'E
67 Q7 **Chad, Lake** *Fr.* Lac Tchad. ◎ C Africa
Chad, Republic of *see* Chad
28 J12 **Chadron** Nebraska, C USA 42°48'N 102°57'W
Chadyr-Lunga *see* Ciadîr-Lunga
163 W14 **Chaeryŏng** SW North Korea 38°22'N 125°35'E
105 P17 **Chafarinas, Islas** *island group* S Spain
27 Y7 **Chaffee** Missouri, C USA 37°10'N 89°39'W
148 L13 **Chāgai Hills** *var.* Chāh Gay. ▲▲ Afghanistan/Pakistan
123 Q11 **Chagda** Respublika Sakha (Yakutiya), NE Russian Federation 58°43'N 130°38'E
Chaghasarāy *see* Asadābād
149 N5 **Chaghcharān** *var.* Chakhcharan, Cheghcheran, Qala Āhangarān. Gōwr, C Afghanistan 34°28'N 65°18'E
103 R9 **Chagny** Saône-et-Loire, C France 46°54'N 04°45'E
173 Q7 **Chagos Archipelago** *var.* Oil Islands. *island group* British Indian Ocean Territory
129 O15 **Chagos Bank** *undersea feature* C Indian Ocean 06°15'S 72°00'E
129 O14 **Chagos-Laccadive Plateau** *undersea feature* N Indian Ocean 03°00'S 73°00'E
173 Q7 **Chagos Trench** *undersea feature* N Indian Ocean 07°00'S 73°30'E
43 T14 **Chagres, Río** ≈ C Panama
45 U14 **Chaguanas** Trinidad, Trinidad and Tobago 10°31'N 61°25'W
54 M6 **Chaguaramas** Guárico, N Venezuela 09°23'N 66°18'W
Chagyl *see* Çagyl
Chahār Maḩāll and Bakhtiyārī *see* Chahār Maḩāl va Bakhtīārī
Chahār Maḩāll va Bakhtīārī, Ostān-e *see* Chahār Maḩāl va Bakhtīārī
142 M9 **Chahār Maḩāll va Bakhtīārī** *off.* Ostān-e Chahār Maḩall va Bakhtīārī, *var.* Chahār Maḩall and Bakhtiyārī. ❖ *province* SW Iran
Chāh Bahār/Chahbar *see* Chābahār
143 V13 **Chāh Derāz** Sīstān va Balūchestān, SE Iran 27°07'N 60°01'E
Chāh Gay *see* Chāgai Hills
167 P10 **Chai Badan** Lop Buri, C Thailand 15°08'N 101°03'E

◆ Country
● Country Capital
◇ Dependent Territory
○ Dependent Territory Capital
❖ Administrative Regions
✈ International Airport
▲ Mountain
▲▲ Mountain Range
Volcano
≈ River
◎ Lake
⊠ Reservoir

◆ Country   ◇ Dependent Territory   ◈ Administrative Regions   ▲ Mountain   ▲ Volcano   ◎ Lake
● Country Capital   ○ Dependent Territory Capital   ✈ International Airport   ▲ Mountain Range   ≈ River   ▦ Reservoir

Chenxi – Chmielnik

160 L11 **Chenxi** *var.* Chenyang. Hunan, S China 28°02′N 110°15′E
Chen Xian/Chenxian/Chen Xiang *see* Chenzhou
Chenyang *see* Chenxi
161 N12 **Chenzhou** *var.* Chenxian, Chen Xian, Chen Xiang. Hunan, S China 25°51′N 113°01′E
163 X15 **Cheonan** *Jap.* Tenan; *prev.* Ch'ŏnan. W South Korea 36°51′N 127°11′E
163 W13 **Cheongju** *prev.* Chŏngju. W North Korea 39°44′N 125°13′E
Cheo Reo *see* A Yun Pa
114 I11 **Chepelare** Smolyan, S Bulgaria 41°44′N 24°41′E
114 I11 **Chepelarska Reka** ≈ S Bulgaria
56 B11 **Chepén** La Libertad, C Peru 07°15′S 79°23′W
62 J10 **Chepes** La Rioja, C Argentina 31°19′S 66°40′W
161 O15 **Chep Lap Kok** ✈ S China 22°23′N 114°11′E
43 U14 **Chepo** Panamá, C Panama 09°09′N 79°03′W
Chepping Wycombe *see* High Wycombe
125 R14 **Cheptsa** ≈ NW Russian Federation
30 K3 **Chequamegon Point** *headland* Wisconsin, N USA 46°42′N 90°45′W
103 O8 **Cher** ◈ *department* C France
102 M8 **Cher** ≈ C France
Cherangani Hills *see* Cherangany Hills
81 H17 **Cherangany Hills** *var.* Cherangani Hills. ▲▲ W Kenya
21 S11 **Cheraw** South Carolina, SE USA 34°42′N 79°52′W
102 I3 **Cherbourg** *anc.* Carusbur. Manche, N France 49°40′N 01°36′W
127 R5 **Cherdakly** Ul'yanovskaya Oblast', W Russian Federation 54°21′N 48°54′E
125 U12 **Cherdyn'** Permskiy Kray, NW Russian Federation 60°21′N 56°39′E
124 J14 **Cherekha** ≈ W Russian Federation
122 M13 **Cheremkhovo** Irkutskaya Oblast', S Russian Federation 53°16′N 102°44′E
Cheren *see* Keren
124 K14 **Cherepovets** Vologodskaya Oblast', NW Russian Federation 59°09′N 37°50′E
125 O11 **Cherevkovo** Arkhangel'skaya Oblast', NW Russian Federation 61°45′N 45°16′E
74 I6 **Chergui, Chott ech** *salt lake* NW Algeria
Cherikov *see* Cherykaw
117 P6 **Cherkas'ka Oblast'** *var.* Cherkasy, *Rus.* Cherkasskaya Oblast'. ◈ *province* C Ukraine
Cherkasskaya Oblast' *see* Cherkas'ka Oblast'
Cherkassy *see* Cherkasy
117 Q6 **Cherkasy** *Rus.* Cherkassy. Cherkas'ka Oblast', C Ukraine 49°26′N 32°05′E
Cherkasy *see* Cherkas'ka Oblast'
126 M15 **Cherkessk** Karachayevo-Cherkesskaya Respublika, SW Russian Federation 44°12′N 42°06′E
122 H12 **Cherlak** Omskaya Oblast', C Russian Federation 54°06′N 74°59′E
122 H12 **Cherlakskoye** Omskaya Oblast', C Russian Federation 53°42′N 74°23′E
125 U13 **Chermoz** Permskiy Kray, NW Russian Federation 58°49′N 56°07′E
Chernavchitsy *see* Charnawchytsy
125 T3 **Chernaya** Nenetskiy Avtonomnyy Okrug, NW Russian Federation 68°36′N 56°34′E
125 T4 **Chernaya** ≈ NW Russian Federation
Chernigov *see* Chernihiv
Chernigovskaya Oblast' *see* Chernihivs'ka Oblast'
117 Q2 **Chernihiv** *Rus.* Chernigov. Chernihivs'ka Oblast', NE Ukraine 51°28′N 31°19′E
Chernihiv *see* Chernihivs'ka Oblast'
117 V9 **Chernihivka** Zaporiz'ka Oblast', SE Ukraine 47°11′N 36°10′E
117 P2 **Chernihivs'ka Oblast'** *var.* Chernihiv, *Rus.* Chernigovskaya Oblast'. ◈ *province* NE Ukraine
114 I9 **Cherni Osŭm** ≈ N Bulgaria
116 J8 **Chernivets'ka Oblast'** *var.* Chernivtsi, *Rus.* Chernovitskaya Oblast'. ◈ *province* W Ukraine
114 I9 **Cherni Vit** ≈ NW Bulgaria
114 G10 **Cherni Vrah** , Cherni Vrŭkh. ▲ W Bulgaria 42°33′N 23°18′E
Cherni Vrŭkh *see* Cherni Vrah
116 K8 **Chernivtsi** *Ger.* Czernowitz, *Rom.* Cernăuţi, *Rus.* Chernovtsy. Chernivets'ka Oblast', W Ukraine 48°18′N 25°55′E
116 M7 **Chernivtsi** Vinnyts'ka Oblast', C Ukraine 48°33′N 28°06′E
Chernivtsi *see* Chernivets'ka Oblast'
Chernobyl' *see* Chornobyl'
Cherno More *see* Black Sea
Chernomorskoye *see* Chornomors'ke
145 T7 **Chernoretsk** *prev.* Chernoretskoye. Pavlodar, NE Kazakhstan 52°51′N 76°37′E
Chernoretskoye *see* Chernoretsk
Chernovitskaya Oblast' *see* Chernivets'ka Oblast'
Chernovtsy *see* Chernivtsi
145 U8 **Chernoye** Pavlodar, NE Kazakhstan 51°40′N 77°33′E
Chernoye More *see* Black Sea
125 U16 **Chernushka** Permskiy Kray, NW Russian Federation 56°30′N 56°07′E
117 N4 **Chernyakhiv** *Rus.* Chernyakhov. Zhytomyrs'ka Oblast', N Ukraine 50°30′N 28°38′E
Chernyakhov *see* Chernyakhiv
119 C14 **Chernyakhovsk** *Ger.* Insterburg. Kaliningradskaya Oblast', W Russian Federation 54°36′N 21°49′E
126 K8 **Chernyanka** Belgorodskaya Oblast', W Russian Federation 50°59′N 37°54′E
125 V5 **Chernyshëva, Gryada** ▲▲ NW Russian Federation
144 J14 **Chernyshëva, Zaliv** *gulf* SW Kazakhstan
123 O10 **Chernyshevskiy** Respublika Sakha (Yakutiya), NE Russian Federation 62°57′N 112°29′E
127 P13 **Chernyye Zemli** *plain* SW Russian Federation
Chërnyy Irtysh *see* Ertix He, China/Kazakhstan
Chërnyy Irtysh *see* Kara Irtysh, Kazakhstan
127 V7 **Chernyy Otrog** Orenburgskaya Oblast', W Russian Federation 52°03′N 56°09′E
29 T12 **Cherokee** Iowa, C USA 42°45′N 95°33′W
26 M8 **Cherokee** Oklahoma, C USA 36°45′N 98°22′W
25 R9 **Cherokee** Texas, SW USA 30°56′N 98°42′W
21 O8 **Cherokee Lake** ⊠ Tennessee, S USA
Cherokees, Lake O' The *see* Grand Lake O' The Cherokees
44 H1 **Cherokee Sound** Great Abaco, N The Bahamas 26°16′N 77°03′W
153 V13 **Cherrapunji** Meghālaya, NE India 25°16′N 91°42′E
28 L9 **Cherry Creek** ≈ South Dakota, N USA
18 J16 **Cherry Hill** New Jersey, NE USA 39°55′N 75°01′W
27 Q7 **Cherryvale** Kansas, C USA 37°16′N 95°33′W
21 Q10 **Cherryville** North Carolina, SE USA 35°22′N 81°22′W
Cherski Range *see* Cherskogo, Khrebet
123 T6 **Cherskiy** Respublika Sakha (Yakutiya), NE Russian Federation 68°45′N 161°15′E
123 R8 **Cherskogo, Khrebet** *var.* Cherski Range. ▲▲ NE Russian Federation
Cherso *see* Cres
126 L10 **Chertkovo** Rostovskaya Oblast', SW Russian Federation 49°22′N 40°10′E
Cherven' *see* Chervyen'
114 H8 **Cherven Bryag** Pleven, N Bulgaria 43°16′N 24°06′E
116 M4 **Chervonoarmiys'k** Zhytomyrs'ka Oblast', N Ukraine 50°27′N 28°15′E
Chervonograd *see* Chervonohrad
116 I4 **Chervonohrad** *Rus.* Chervonograd. L'vivs'ka Oblast', NW Ukraine 50°25′N 24°10′E
117 W6 **Chervonooskil's'ke Vodoskhovyshche** *Rus.* Krasnoosol'skoye Vodokhranilishche. ⊠ NE Ukraine
Chervonoye, Ozero *see* Chyrvonaye, Vozyera
117 S4 **Chervonozavods'ke** Poltavs'ka Oblast', C Ukraine 50°24′N 33°22′E
119 L16 **Chervyen'** *Rus.* Cherven'. Minskaya Voblasts', C Belarus 53°42′N 28°26′E
119 P16 **Cherykaw** *Rus.* Cherikov. Mahilyowskaya Voblasts', E Belarus 53°34′N 31°23′E
31 R9 **Chesaning** Michigan, N USA 43°10′N 84°07′W
21 X5 **Chesapeake Bay** *inlet* NE USA
Chesha Bay *see* Chëshskaya Guba
Cheshevlya *see* Tsyeshawlya
97 K18 **Cheshire** *cultural region* C England, United Kingdom
125 P5 **Chëshskaya Guba** *var.* Archangel Bay, Chesha Bay, Dvina Bay. *bay* NW Russian Federation
14 F14 **Chesley** Ontario, S Canada 44°17′N 81°06′W
21 Q10 **Chesnee** South Carolina, SE USA 35°09′N 81°51′W
97 K18 **Chester** *Wel.* Caerleon, *hist.* Legaceaster, *Lat.* Deva, Devana Castra. C England, United Kingdom 53°12′N 02°54′W
35 O4 **Chester** California, W USA 40°18′N 121°14′W
30 K16 **Chester** Illinois, N USA 37°54′N 89°49′W
33 S7 **Chester** Montana, NW USA 48°30′N 110°59′W
18 I16 **Chester** Pennsylvania, NE USA 39°51′N 75°21′W
21 R1 **Chester** South Carolina, SE USA 34°43′N 81°14′W
25 X9 **Chester** Texas, SW USA 30°55′N 94°36′W
21 W6 **Chester** Virginia, NE USA 37°22′N 77°27′W
21 R11 **Chester** West Virginia, NE USA 40°34′N 80°33′W
97 M18 **Chesterfield** C England, United Kingdom 53°15′N 01°25′W
21 S11 **Chesterfield** South Carolina, SE USA 34°44′N 80°04′W
21 W6 **Chesterfield** Virginia, NE USA 37°22′N 77°31′W
192 J9 **Chesterfield, Îles** *island group* NW New Caledonia
9 O9 **Chesterfield Inlet** Nunavut, NW Canada 63°19′N 90°57′W
9 O9 **Chesterfield Inlet** *inlet* Nunavut, N Canada
21 Y3 **Chester River** ≈ Delaware/Maryland, NE USA
21 X3 **Chestertown** Maryland, NE USA 39°13′N 76°04′W
19 R4 **Chesuncook Lake** ◎ Maine, NE USA
30 J5 **Chetek** Wisconsin, N USA 45°19′N 91°37′W
13 R14 **Chéticamp** Nova Scotia, SE Canada 46°14′N 61°19′W
27 Q8 **Chetopa** Kansas, C USA 37°02′N 95°05′W
41 Y14 **Chetumal** *var.* Payo Obispo. Quintana Roo, SE Mexico 18°32′N 88°16′W
Chetumal, Bahia/Chetumal, Bahía de *see* Chetumal Bay
42 G1 **Chetumal Bay** *var.* Bahia Chetumal, Bahía de Chetumal. *bay* Belize/Mexico
10 M13 **Chetwynd** British Columbia, W Canada 55°42′N 121°36′W
38 M11 **Chevak** Alaska, USA 61°31′N 165°35′W
36 M12 **Chevelon Creek** ≈ Arizona, SW USA
185 J17 **Cheviot** Canterbury, South Island, New Zealand 42°48′S 173°17′E
96 L13 **Cheviot Hills** *hill range* England/Scotland, United Kingdom
96 L13 **Cheviot, The** ▲ NE England, United Kingdom 55°28′N 02°10′W
14 M11 **Chevreuil, Lac du** ◎ Québec, SE Canada
81 I16 **Ch'ew Bahir** *var.* Lake Stefanie. ◎ Ethiopia/Kenya
32 L7 **Chewelah** Washington, NW USA 48°16′N 117°42′W
26 K10 **Cheyenne** Oklahoma, C USA 35°37′N 99°43′W
33 Z17 **Cheyenne** *state capital* Wyoming, C USA 41°08′N 104°46′W
26 L5 **Cheyenne Bottoms** ◎ Kansas, C USA
16 J8 **Cheyenne River** ≈ South Dakota/Wyoming, N USA
37 W5 **Cheyenne Wells** Colorado, C USA 38°49′N 102°21′W
108 C9 **Cheyres** Vaud, W Switzerland 46°48′N 06°48′E
Chezdi-Oşorheiu *see* Târgu Secuiesc
153 P13 **Chhapra** *prev.* Chapra. Bihār, N India 25°50′N 84°42′E
153 V13 **Chhatak** *var.* Chatak. Sylhet, NE Bangladesh 25°02′N 91°33′E
154 J9 **Chhatarpur** Madhya Pradesh, C India 24°54′N 79°35′E
154 N13 **Chhatrapur** *prev.* Chatrapur. Odisha, E India 19°26′N 85°02′E
154 K2 **Chhattīsgarh** ◈ *state* E India
154 L12 **Chhattīsgarh** *plain* C India
154 I11 **Chhindwāra** Madhya Pradesh, C India 22°04′N 78°58′E
153 T12 **Chhukha** SW Bhutan 27°02′N 89°36′E
Chiai *see* Jiayi
Chia-i *see* Jiayi
Chia-mu-ssu *see* Jiamusi
83 B15 **Chiange** *Port.* Vila de Almoster. Huíla, SW Angola 15°44′S 13°54′E
Chiang-hsi *see* Jiangxi
Chiang Kai-shek *see* Taiwan Taoyuan
167 P8 **Chiang Khan** Loei, E Thailand 17°51′N 101°43′E
167 O7 **Chiang Mai** *var.* Chiangmai, Chiengmai, Kiangmai. Chiang Mai, NW Thailand 18°48′N 98°59′E
167 O7 **Chiang Mai** ✈ Chiang Mai, NW Thailand 18°44′N 98°53′E
Chiangmai *see* Chiang Mai
167 O6 **Chiang Rai** *var.* Chianpai, Chienrai, Muang Chiang Rai. Chiang Rai, NW Thailand 19°56′N 99°51′E
Chiang-su *see* Jiangsu
Chianning/Chian-ning *see* Nanjing
Chianpai *see* Chiang Rai
106 G12 **Chianti** *cultural region* C Italy
Chiapa *see* Chiapa de Corzo
41 U16 **Chiapa de Corzo** *var.* Chiapa. Chiapas, SE Mexico 16°42′N 92°59′W
41 V16 **Chiapas** ◈ *state* SE Mexico
106 J12 **Chiaravalle** Marche, C Italy 43°36′N 13°19′E
107 N22 **Chiaravalle Centrale** Calabria, SW Italy 38°40′N 16°25′E
106 E7 **Chiari** Lombardia, N Italy 45°33′N 10°00′E
108 H12 **Chiasso** Ticino, S Switzerland 45°51′N 09°02′E
137 S9 **Chiatura** *prev.* Chiat'ura. C Georgia 42°13′N 43°11′E
Chiat'ura *see* Chiatura
41 P15 **Chiautla** *var.* Chiautla de Tapia. Puebla, S Mexico 18°16′N 98°31′W
Chiautla de Tapia *see* Chiautla
106 D10 **Chiavari** Liguria, NW Italy 44°19′N 09°19′E
106 E6 **Chiavenna** Lombardia, N Italy 46°19′N 09°22′E
Chiayi *see* Jiayi
Chiazza *see* Piazza Armerina
165 O14 **Chiba** *var.* Tiba. Chiba, Honshū, S Japan 35°37′N 140°06′E
165 O13 **Chiba** *off.* Chiba-ken, *var.* Tiba. ◈ *prefecture* Honshū, S Japan
83 M18 **Chibabava** Sofala, C Mozambique 20°17′S 33°39′E
Chiba-ken *see* Chiba
161 O10 **Chibi** *prev.* Puqi. Hubei, C China 29°45′N 113°55′E
83 B15 **Chibia** *Port.* João de Almeida, Vila João de Almeida. Huíla, SW Angola 15°11′S 13°41′E
83 M18 **Chiboma** Sofala, C Mozambique 20°06′S 33°54′E
82 J12 **Chibondo** Luapula, N Zambia 10°42′S 28°42′E
82 K11 **Chibote** Luapula, NE Zambia 09°52′S 29°33′E
12 K12 **Chibougamau** Québec, SE Canada 49°56′N 74°24′W
164 H11 **Chiburi-jima** *island* Oki-shotō, SW Japan
83 M20 **Chibuto** Gaza, S Mozambique 24°40′S 33°33′E
31 N11 **Chicago** Illinois, N USA 41°51′N 87°39′W
31 N11 **Chicago Heights** Illinois, N USA 41°30′N 87°38′W
15 W6 **Chic-Chocs, Monts** *Eng.* Shickshock Mountains. ▲▲ Québec, SE Canada
39 W13 **Chichagof Island** *island* Alexander Archipelago, Alaska, USA
57 K20 **Chichas, Cordillera de** ▲▲ SW Bolivia
41 X12 **Chichén-Itzá, Ruinas** *ruins* Yucatán, SE Mexico
97 N23 **Chichester** SE England, United Kingdom 50°50′N 00°48′W
42 C5 **Chichicastenango** Quiché, W Guatemala 14°55′N 91°06′W
42 I9 **Chichigalpa** Chinandega, NW Nicaragua 12°35′N 87°04′W
Ch'i-ch'i-ha-erh *see* Qiqihar
165 X16 **Chichijima-rettō** *Eng.* Beechy Group. *island group* SE Japan
54 K4 **Chichiriviche** Falcón, N Venezuela 10°58′N 68°17′W
39 R11 **Chickaloon** Alaska, USA 61°48′N 148°27′W
20 L10 **Chickamauga Lake** ⊠ Tennessee, S USA
23 N7 **Chickasawhay River** ≈ Mississippi, S USA
26 M11 **Chickasha** Oklahoma, C USA 35°03′N 97°57′W
39 T9 **Chicken** Alaska, USA 64°04′N 141°56′W
104 J16 **Chiclana de la Frontera** Andalucía, S Spain 36°26′N 06°09′W
56 B11 **Chiclayo** Lambayeque, NW Peru 06°47′S 79°47′W
35 N5 **Chico** California, W USA 39°42′N 121°51′W
83 L15 **Chicoa** Tete, NW Mozambique 15°45′S 32°25′E
83 M20 **Chicomo** Gaza, S Mozambique 24°29′S 34°15′E
18 M11 **Chicopee** Massachusetts, NE USA 42°08′N 72°34′W
63 I19 **Chico, Río** ≈ SE Argentina
63 I21 **Chico, Río** ≈ S Argentina
27 W14 **Chicot, Lake** ◎ Arkansas, C USA
15 R7 **Chicoutimi** Québec, SE Canada 48°24′N 71°04′W
15 Q8 **Chicoutimi** ≈ Québec, SE Canada
83 L19 **Chicualacuala** Gaza, SW Mozambique 22°06′S 31°42′E
83 B14 **Chicuma** Benguela, C Angola 13°33′S 14°41′E
155 J21 **Chidambaram** Tamil Nādu, SE India 11°25′N 79°42′E
196 K13 **Chidley, Cape** *headland* Newfoundland and Labrador, E Canada 60°25′N 64°39′W
101 N24 **Chiemsee** ◎ SE Germany
Chiengmai *see* Chiang Mai
Chienrai *see* Chiang Rai
106 B8 **Chieri** Piemonte, NW Italy 45°01′N 07°49′E
106 F8 **Chiese** ≈ N Italy
107 K14 **Chieti** *var.* Teate. Abruzzo, C Italy 42°22′N 14°10′E
99 E19 **Chièvres** Hainaut, SW Belgium 50°34′N 03°49′E
163 S12 **Chifeng** *var.* Ulanhad. Nei Mongol Zizhiqu, N China 42°17′N 118°56′E
82 F13 **Chifumage** ≈ E Angola
82 M13 **Chifunda** Muchinga, NE Zambia 11°57′S 32°36′E
Chiganak *see* Shyganak
39 P15 **Chiginagak, Mount** ▲ Alaska, USA 57°10′N 157°00′W
Chigirin *see* Chyhyryn
Chigirinskoye Vodokhranilishche *see* Chyhirynskaye Vodaskhovishcha
41 P13 **Chignahuapan** Puebla, S Mexico 19°52′N 98°03′W
39 O15 **Chignik** Alaska, USA 56°18′N 158°24′W
83 M19 **Chigombe** ≈ S Mozambique
54 D7 **Chigorodó** Antioquia, NW Colombia 07°42′N 76°45′W
83 M19 **Chigubo** Gaza, S Mozambique 22°50′S 33°30′E
Chihertey *see* Altay
Chih-fu *see* Yantai
Chihli *see* Hebei
Chihli, Gulf of *see* Bo Hai
40 J6 **Chihuahua** Chihuahua, NW Mexico 28°40′N 106°06′W
40 I6 **Chihuahua** ◈ *state* N Mexico
Chiili *see* Shiyeli
26 M7 **Chikaskia River** ≈ Kansas/Oklahoma, C USA
155 H19 **Chik Ballāpur** Karnātaka, W India 13°28′N 77°42′E
124 G15 **Chikhachevo** Pskovskaya Oblast', W Russian Federation 57°17′N 29°51′E
155 F19 **Chikmagalūr** Karnātaka, W India 13°20′N 75°46′E
129 V7 **Chikoy** ≈ C Russian Federation
83 J15 **Chikumbi** Lusaka, C Zambia 15°11′S 28°20′E
82 M13 **Chikwa** Muchinga, NE Zambia 11°39′S 32°45′E
Chikwana *see* Chikwawa
83 N15 **Chikwawa** *var.* Chikwana. Southern, S Malawi 16°03′S 34°48′E
155 J16 **Chilakalūrupet** Andhra Pradesh, E India 16°09′N 80°13′E
146 L14 **Chilan** Lebap Welaýaty, E Turkmenistan 37°57′N 64°58′E
Chilapa *see* Chilapa de Alvarez
41 P16 **Chilapa de Alvarez** *var.* Chilapa. Guerrero, S Mexico 17°38′N 99°11′W
155 J25 **Chilaw** North Western Province, W Sri Lanka 07°34′N 79°48′E
57 D15 **Chilca** Lima, W Peru 12°35′S 76°41′W
23 Q4 **Childersburg** Alabama, S USA 33°16′N 86°21′W
25 P4 **Childress** Texas, SW USA 34°25′N 100°14′W
63 G14 **Chile** *off.* Republic of Chile. ◆ *republic* SW South America
47 R10 **Chile Basin** *undersea feature* E Pacific Ocean 33°00′S 80°00′W
63 H20 **Chile Chico** Aisén, W Chile 46°34′S 71°44′W
62 I9 **Chilecito** La Rioja, NW Argentina 29°10′S 67°30′W
62 H12 **Chilecito** Mendoza, W Argentina 33°53′S 69°03′W
83 L14 **Chilembwe** Eastern, E Zambia 13°54′S 31°38′E
Chile, Republic of *see* Chile
193 S11 **Chile Rise** *undersea feature* SE Pacific Ocean 40°00′S 90°00′W
117 N13 **Chilia, Braţul** ≈ SE Romania
Chilia-Nouă *see* Kiliya
Chilik *see* Shelek
Chilik *see* Shelek
154 O13 **Chilika Lake** *var.* Chilka Lake. ◎ E India
82 J13 **Chililabombwe** Copperbelt, C Zambia 12°20′S 27°52′E
Chi-lin *see* Jilin
Chilka Lake *see* Chilika Lake
10 H9 **Chilkoot Pass** *pass* British Columbia, W Canada
Chill Ala, Cuan *see* Killala Bay
62 G13 **Chillán** Bío Bío, C Chile 36°37′S 72°10′W
61 C22 **Chillar** Buenos Aires, E Argentina 37°16′S 59°58′W
Chill Chiaráin, Cuan *see* Kilkieran Bay
30 K12 **Chillicothe** Illinois, N USA 40°55′N 89°29′W
27 S3 **Chillicothe** Missouri, C USA 39°47′N 93°33′W
31 S14 **Chillicothe** Ohio, N USA 39°20′N 83°00′W
25 Q4 **Chillicothe** Texas, SW USA 34°15′N 99°31′W
10 M17 **Chilliwack** British Columbia, SW Canada 49°09′N 121°54′W
Chill Mhantáin, Ceann *see* Wicklow Head
Chill Mhantáin, Sléibhte *see* Wicklow Mountains
108 C10 **Chillon** Vaud, W Switzerland 46°24′N 06°56′E
Chil'mamedkum, Peski/Chilmämetgum *see* Çilmämmetgum
63 F17 **Chiloé, Isla de** *var.* Isla Grande de Chiloé. *island* W Chile
32 H15 **Chiloquin** Oregon, NW USA 42°33′N 121°33′W
41 O16 **Chilpancingo** *var.* Chilpancingo de los Bravos. Guerrero, S Mexico 17°33′N 99°30′W
Chilpancingo de los Bravos *see* Chilpancingo
97 N21 **Chiltern Hills** *hill range* S England, United Kingdom
30 M7 **Chilton** Wisconsin, N USA 44°04′N 88°10′W
82 F11 **Chiluage** Lunda Sul, NE Angola 09°32′S 21°48′E
82 N12 **Chilumba** *prev.* Deep Bay. Northern, N Malawi 10°27′S 34°12′E
Chilung *see* Jilong
83 N15 **Chilwa, Lake** *var.* Lago Chirua, Lake Shirwa. ◎ SE Malawi
167 R10 **Chi, Mae Nam** ≈ E Thailand
42 C6 **Chimaltenango** Chimaltenango, C Guatemala 14°40′N 90°48′W
42 A2 **Chimaltenango** *off.* Departamento de Chimaltenango. ◈ *department* S Guatemala
Chimaltenango, Departamento de *see* Chimaltenango
43 V15 **Chimán** Panamá, E Panama 08°42′N 78°35′W
83 M17 **Chimanimani** *prev.* Mandidzudzure, Melsetter. Manicaland, E Zimbabwe 19°48′S 32°52′E
99 G22 **Chimay** Hainaut, S Belgium 50°03′N 04°20′E
37 S10 **Chimayo** New Mexico, SW USA 36°00′N 105°55′W
Chimbay *see* Chimboy
56 A13 **Chimborazo** ◈ *province* C Ecuador
56 C7 **Chimborazo** ⛰ C Ecuador 01°29′S 78°50′W
56 C12 **Chimbote** Ancash, W Peru 09°04′S 78°34′W
146 H7 **Chimboy** *Rus.* Chimbay. Qoraqalpog'iston Respublikasi, NW Uzbekistan 43°03′N 59°52′E
186 D7 **Chimbu** *var.* Simbu. ◈ *province* C Papua New Guinea
54 F6 **Chimichagua** Cesar, N Colombia 09°19′N 73°51′W
Chimishliya *see* Cimişlia
Chimkent *see* Shymkent
Chimkentskaya Oblast' *see* Yuzhnyy Kazakhstan
28 I14 **Chimney Rock** *rock* Nebraska, C USA
83 M17 **Chimoio** Manica, C Mozambique 19°08′S 33°29′E
82 K11 **Chimpembe** Northern, NE Zambia 09°31′S 29°33′E
41 O8 **China** Nuevo León, NE Mexico 25°42′N 99°15′W
156 M9 **China** *off.* People's Republic of China, *Chin.* Chung-hua Jen-min Kung-ho-kuo, Zhonghua Renmin Gongheguo; *prev.* Chinese Empire. ◆ *republic* E Asia
19 Q7 **China Lake** ◎ Maine, NE USA
42 F8 **Chinameca** San Miguel, E El Salvador 13°30′N 88°20′W*
Chi-nan/Chinan *see* Jinan
42 H9 **Chinandega** Chinandega, NW Nicaragua 12°37′N 87°08′W
42 H9 **Chinandega** ◈ *department* NW Nicaragua
China, People's Republic of *see* China
China, Republic of *see* Taiwan
24 J11 **Chinati Mountains** ▲▲ Texas, SW USA
Chinaz *see* Chinoz
57 E15 **Chincha Alta** Ica, SW Peru 13°25′S 76°07′W
11 N11 **Chinchaga** ≈ Alberta, SW Canada
Chin-chiang *see* Quanzhou
Chinchilla *see* Chinchilla de Monte Aragón
105 Q11 **Chinchilla de Monte Aragón** *var.* Chinchilla. Castilla-La Mancha, C Spain 38°56′N 01°44′W
54 D10 **Chinchiná** Caldas, W Colombia 04°59′N 75°37′W
105 O8 **Chinchón** Madrid, C Spain 40°08′N 03°26′W
41 Z14 **Chinchorro, Banco** *island* SE Mexico
Chin-chou/Chinchow *see* Jinzhou
21 Z5 **Chincoteague** Assateague Island, Virginia, NE USA 37°55′N 75°22′W
83 O17 **Chinde** Zambézia, NE Mozambique 18°35′S 36°28′E
Chin-do *see* Jin-do
159 R13 **Chindu** *var.* Chengwen; *prev.* Chuqung. Qinghai, C China 33°19′N 97°08′E
Chindwin *see* Chindwinn
166 M2 **Chindwinn** *var.* Chindwin. ≈ N Myanmar (Burma)
Chinese Empire *see* China
Chinghai *see* Qinghai
Ch'ing Hai *see* Qinghai Hu, China
Chingildi *see* Shengeldi
Chingirlau *see* Shyngghyrlau
82 J13 **Chingola** Copperbelt, C Zambia 12°31′S 27°53′E
Ching-Tao/Ch'ing-tao *see* Qingdao
82 C13 **Chinguar** Huambo, C Angola 12°33′S 16°25′E
76 I7 **Chinguețți** *var.* Chinguetti. Adrar, C Mauritania 20°25′N 12°24′W
Chinhae *see* Jinhae
166 K4 **Chin Hills** ▲▲ W Myanmar (Burma)
83 K16 **Chinhoyi** *prev.* Sinoia. Mashonaland West, N Zimbabwe 17°22′S 30°12′E
Chinhsien *see* Jinzhou
39 Q14 **Chiniak, Cape** *headland* Kodiak Island, Alaska, USA 57°37′N 152°10′W
14 G10 **Chiniguchi Lake** ◎ Ontario, S Canada
149 U8 **Chinīot** Punjab, NE Pakistan 31°40′N 73°00′E
Chinju *see* Jinju
Chinkai *see* Jinhae
78 M13 **Chinko** ≈ E Central African Republic
37 O9 **Chinle** Arizona, SW USA 36°09′N 109°33′W
Chinmen Tao *see* Jinmen Dao
Chinnchâr *see* Shinshār
Chinnereth *see* Tiberias, Lake
164 C12 **Chino** *var.* Tino. Nagano, Honshū, S Japan 36°00′N 138°10′E
102 L8 **Chinon** Indre-et-Loire, C France 47°10′N 00°15′E
33 T7 **Chinook** Montana, NW USA 48°35′N 109°13′W
Chinook State *see* Washington
192 L4 **Chinook Trough** *undersea feature* N Pacific Ocean
36 K11 **Chino Valley** Arizona, SW USA 34°45′N 112°27′W
147 P10 **Chinoz** *Rus.* Chinaz. Toshkent Viloyati, E Uzbekistan 40°58′N 68°46′E
82 L12 **Chinsali** Muchinga, NE Zambia 10°33′S 32°05′E
166 K5 **Chin State** ◈ *state* W Myanmar (Burma)
Chinsura *see* Chunchura
Chin-tō *see* Jin-do
54 E6 **Chinú** Córdoba, NW Colombia 09°07′N 75°25′W
99 K24 **Chiny, Forêt de** *forest* SE Belgium
83 M15 **Chioco** Tete, NW Mozambique 16°22′S 32°50′E
106 H8 **Chioggia** *anc.* Fossa Claudia. Veneto, NE Italy 45°14′N 12°17′E
114 H12 **Chionótrypa** ▲ NE Greece 41°16′N 24°06′E
115 L18 **Chíos** *var.* Hios, Khíos, *It.* Scio, *Turk.* Sakiz-Adasi. Chíos, E Greece 38°23′N 26°07′E
115 K18 **Chíos** *var.* Khíos. *island* E Greece
83 M14 **Chipata** *prev.* Fort Jameson. Eastern, E Zambia 13°40′S 32°42′E
83 C14 **Chipindo** Huíla, C Angola 13°53′S 15°47′E
23 R8 **Chipley** Florida, SE USA 30°46′N 85°32′W
155 D15 **Chiplūn** Mahārāshtra, W India 17°32′N 73°32′E
81 H22 **Chipogolo** Dodoma, C Tanzania 06°52′S 36°03′E
23 R8 **Chipola River** ≈ Florida, SE USA
97 L22 **Chippenham** S England, United Kingdom 51°28′N 02°07′W
30 J6 **Chippewa Falls** Wisconsin, N USA 44°56′N 91°25′W
30 J4 **Chippewa, Lake** ⊠ Wisconsin, N USA
31 Q8 **Chippewa River** ≈ Michigan, N USA
30 I6 **Chippewa River** ≈ Wisconsin, N USA
Chipping Wycombe *see* High Wycombe
114 G8 **Chiprovtsi** Montana, NW Bulgaria 43°23′N 22°53′E
19 T4 **Chiputneticook Lakes** *lakes* Canada/USA
56 D13 **Chiquián** Ancash, W Peru 10°09′S 78°08′W
41 Y11 **Chiquilá** Quintana Roo, SE Mexico 21°25′N 87°20′W
42 E6 **Chiquimula** Chiquimula, SE Guatemala 14°46′N 89°32′W
42 A3 **Chiquimula** *off.* Departamento de Chiquimula. ◈ *department* SE Guatemala
Chiquimula, Departamento de *see* Chiquimula
42 D7 **Chiquimulilla** Santa Rosa, S Guatemala 14°06′N 90°23′W
54 F9 **Chiquinquirá** Boyacá, C Colombia 05°37′N 73°51′W
155 J17 **Chīrāla** Andhra Pradesh, E India 15°49′N 80°21′E
149 N4 **Chīras** Gōwr, N Afghanistan 35°15′N 65°39′E
152 H11 **Chirāwa** Rājasthān, N India 28°12′N 75°42′E
Chirchik *see* Chirchiq
147 Q9 **Chirchiq** *Rus.* Chirchik. Toshkent Viloyati, E Uzbekistan 41°30′N 69°32′E
147 P10 **Chirchiq** ≈ E Uzbekistan
Chire *see* Shire
83 L18 **Chiredzi** Masvingo, SE Zimbabwe 21°00′S 31°38′E
25 X8 **Chireno** Texas, SW USA 31°30′N 94°21′W
77 X7 **Chirfa** Agadez, NE Niger 21°01′N 12°41′E
37 O16 **Chiricahua Mountains** ▲▲ Arizona, SW USA
37 O16 **Chiricahua Peak** ▲ Arizona, SW USA 31°51′N 109°17′W
54 F6 **Chiriguaná** Cesar, N Colombia 09°24′N 73°38′W
39 P15 **Chirikof Island** *island* Alaska, USA
43 P16 **Chiriquí** *off.* Provincia de Chiriquí. ◈ *province* SW Panama
43 P17 **Chiriquí, Golfo de** *Eng.* Chiriqui Gulf. *gulf* SW Panama
43 P15 **Chiriquí Grande** Bocas del Toro, W Panama 08°58′N 82°08′W
Chiriqui Gulf *see* Chiriquí, Golfo de
43 P15 **Chiriquí, Laguna de** *lagoon* NW Panama
Chiriquí, Provincia de *see* Chiriquí
43 O16 **Chiriquí Viejo, Río** ≈ W Panama
Chiriquí, Volcán de *see* Barú, Volcán
83 N15 **Chiromo** Southern, S Malawi 16°32′S 35°07′E
114 J10 **Chirpan** Stara Zagora, C Bulgaria 42°12′N 25°20′E
43 N14 **Chirripó Atlántico, Río** ≈ E Costa Rica
Chirripó, Cerro *see* Chirripó Grande, Cerro
Chirripó del Pacífico, Río *see* Chirripó, Río
43 N14 **Chirripó Grande, Cerro** *var.* Cerro Chirripó. ▲ SE Costa Rica 09°31′N 83°28′W
43 N13 **Chirripó, Río** *var.* Río Chirripó del Pacífico. ≈ NE Costa Rica
Chirua, Lago *see* Chilwa, Lake
83 J15 **Chirundu** Southern, S Zambia 16°03′S 28°50′E
29 W8 **Chisago City** Minnesota, N USA 45°22′N 92°53′W
83 J14 **Chisamba** Central, C Zambia 15°00′S 28°22′E
39 T10 **Chisana** Alaska, USA 62°09′N 142°07′W
82 I13 **Chisasa** North Western, NW Zambia 12°09′S 25°30′E
12 I9 **Chisasibi** *prev.* Fort George. Québec, C Canada 53°50′N 79°01′W
42 D4 **Chisec** Alta Verapaz, C Guatemala 15°50′N 90°18′W
127 U5 **Chishmy** Respublika Bashkortostan, W Russian Federation 54°33′N 55°21′E
29 V4 **Chisholm** Minnesota, N USA 47°29′N 92°52′W
149 U10 **Chishtiān** *var.* Chishtiān Mandi. Punjab, E Pakistan 29°44′N 72°54′E
Chishtiān Mandi *see* Chishtiān
160 I11 **Chishui He** ≈ C China
Chisimaio/Chisimayu *see* Kismaayo
117 N10 **Chişinău** *Rus.* Kishinev. ● (Moldova) C Moldova 47°N 28°51′E
117 N10 **Chişinău** ✈ S Moldova 46°54′N 28°56′E
Chişinău-Criş *see* Chişineu-Criş
116 F10 **Chişineu-Criş** *Hung.* Kisjenő; *prev.* Chişinău-Criş. Arad, W Romania 46°33′N 21°30′E
83 K14 **Chisomo** Central, C Zambia 13°30′S 30°37′E
106 A8 **Chisone** ≈ NW Italy
24 K12 **Chisos Mountains** ▲▲ Texas, SW USA
39 T10 **Chistochina** Alaska, USA 62°34′N 144°39′W
127 R4 **Chistopol'** Respublika Tatarstan, W Russian Federation 55°20′N 50°39′E
145 O8 **Chistopol'ye** Severnyy Kazakhstan, N Kazakhstan 52°37′N 67°14′E
123 O13 **Chita** Zabaykal'skiy Kray, S Russian Federation 52°03′N 113°35′E
83 B16 **Chitado** Cunene, SW Angola 17°16′S 13°54′E
Chitaldroog/Chitaldrug *see* Chitradurga
83 C15 **Chitanda** ≈ S Angola
Chitangwiza *see* Chitungwiza
82 F10 **Chitato** Lunda Norte, NE Angola 07°23′S 20°46′E
83 C14 **Chitembo** Bié, C Angola 13°33′S 16°47′E
39 T11 **Chitina** Alaska, USA 61°31′N 144°26′W
39 T11 **Chitina River** ≈ Alaska, USA
82 M11 **Chitipa** Northern, NW Malawi 09°41′S 33°19′E
165 S4 **Chitose** *var.* Titose. Hokkaidō, NE Japan 42°51′N 141°40′E
155 G18 **Chitradurga** *prev.* Chitaldroog, Chitaldrug. Karnātaka, W India 14°16′N 76°23′E
149 T3 **Chitrāl** Khyber Pakhtunkhwa, NW Pakistan 35°51′N 71°47′E
43 S16 **Chitré** Herrera, S Panama 07°57′N 80°26′W
153 V16 **Chittagong** *Ben.* Chāttagām. Chittagong, SE Bangladesh 22°20′N 91°48′E
153 U16 **Chittagong** ◈ *division* E Bangladesh
153 Q15 **Chittaranjan** West Bengal, NE India 23°52′N 86°40′E
152 G14 **Chittaurgarh** *var.* Chittorgarh. Rājasthān, N India 24°54′N 74°42′E
155 I19 **Chittoor** Andhra Pradesh, E India 13°13′N 79°06′E
Chittorgarh *see* Chittaurgarh
155 G21 **Chittūr** Kerala, SW India 10°42′N 76°46′E
83 K16 **Chitungwiza** *prev.* Chitangwiza. Mashonaland East, NE Zimbabwe 18°S 31°06′E
62 H4 **Chíuchíu** Antofagasta, N Chile 22°13′S 68°34′W
82 F12 **Chiumbe** *var.* Tshiumbe. ≈ Angola/Dem. Rep. Congo
83 F15 **Chiume** Moxico, E Angola 15°08′S 21°09′E
82 K13 **Chiundaponde** Muchinga, NE Zambia 12°14′S 30°40′E
106 H13 **Chiusi** Toscana, C Italy 43°00′N 11°56′E
54 J5 **Chivacoa** Yaracuy, N Venezuela 10°10′N 68°54′W
106 B8 **Chivasso** Piemonte, NW Italy 45°13′N 07°54′E
83 L17 **Chivhu** *prev.* Enkeldoorn. Midlands, C Zimbabwe 19°01′S 30°54′E
61 C20 **Chivilcoy** Buenos Aires, E Argentina 34°55′S 60°00′W
Chiwei Yu *see* Sekibi-sho
82 N12 **Chiweta** Northern, N Malawi 10°36′S 34°09′E
42 D4 **Chixoy, Río** *var.* Río Negro, Río Salinas. ≈ Guatemala/Mexico
82 H13 **Chizela** North Western, NW Zambia 13°11′S 24°59′E
125 O5 **Chizha** Nenetskiy Avtonomnyy Okrug, NW Russian Federation 67°04′N 44°19′E
161 Q9 **Chizhou** *var.* Guichi. Anhui, E China 30°39′N 117°29′E
164 I12 **Chizu** Tottori, Honshū, SW Japan 35°15′N 134°14′E
Chkalov *see* Orenburg
74 J5 **Chlef** *var.* Ech Cheliff, Ech Chleff; *prev.* Al-Asnam, El Asnam, Orléansville. NW Algeria 36°11′N 01°21′E
115 G18 **Chlómo** ▲ C Greece 38°36′N 22°57′E
111 M15 **Chmielnik** Świętokrzyskie, C Poland 50°37′N 20°43′E

◆ Country
● Country Capital
◇ Dependent Territory
○ Dependent Territory Capital
◈ Administrative Regions
✈ International Airport
▲ Mountain
▲▲ Mountain Range
⛰ Volcano
≈ River
◎ Lake
⊠ Reservoir

Chŏâm Khsant – Clavier

167 S11 **Chŏâm Khsant** Preăh Vihéar, N Cambodia 14°13′N 104°56′E
62 G10 **Choapa, Río** *var.* Choapo. ≈ C Chile
Choapas *see* Las Choapas
Choapo *see* Choapa, Río
Choarta *see* Chwarta
67 T13 **Chobe** ≈ N Botswana
14 K8 **Chochocouane** ≈ Québec, SE Canada
110 E13 **Chocianów** *Ger.* Kotzenan. Dolnośląskie, SW Poland 51°23′N 15°55′E
54 C9 **Chocó** *off.* Departamento del Chocó. ◈ *province* W Colombia
Chocó, Departamento del *see* Chocó
35 X16 **Chocolate Mountains** ▲▲ California, W USA
21 W9 **Chocowinity** North Carolina, SE USA 35°33′N 77°03′W
27 N10 **Choctaw** Oklahoma, C USA 35°30′N 97°16′W
23 Q8 **Choctawhatchee Bay** *bay* Florida, SE USA
23 Q8 **Choctawhatchee River** ≈ Florida, SE USA
Chodau *see* Chodov
163 V14 **Ch'o-do** *island* SW North Korea
Chodorów *see* Khodoriv
111 A16 **Chodov** *Ger.* Chodau. Karlovarský Kraj, W Czech Republic 50°15′N 12°45′E
110 G10 **Chodzież** Wielkopolskie, C Poland 53°N 16°55′E
63 J15 **Choele Choel** Río Negro, C Argentina 39°19′S 65°42′W
83 L14 **Chofombo** Tete, NW Mozambique 14°43′S 31°48′E
Chohtan *see* Chauhtan
11 U14 **Choiceland** Saskatchewan, C Canada 53°28′N 104°26′W
186 K8 **Choiseul** ◈ *province* NW Solomon Islands
186 K8 **Choiseul** *var.* Lauru. *island* NW Solomon Islands
63 M23 **Choiseul Sound** *sound* East Falkland, Falkland Islands
40 H7 **Choix** Sinaloa, C Mexico 26°43′N 108°20′W
110 D10 **Chojna** Zachodnio-pomorskie, W Poland 52°56′N 14°25′E
110 H8 **Chojnice** *Ger.* Konitz. Pomorskie, N Poland 53°41′N 17°34′E
111 F14 **Chojnów** *Ger.* Hainau, Haynau. Dolnośląskie, SW Poland 51°16′N 15°55′E
167 Q10 **Chok Chai** Nakhon Ratchasima, C Thailand 14°45′N 102°10′E
80 I12 **Ch'ok'ē** *var.* Choke Mountains. ▲▲ NW Ethiopia
25 R13 **Choke Canyon Lake** ▣ Texas, SW USA
Choke Mountains *see* Ch'ok'ē
Chokpar *see* Shokpar
147 W7 **Chok-Tal** *var.* Choktal. Issyk-Kul'skaya Oblast', E Kyrgyzstan 42°37′N 76°45′E
Choktal *see* Chok-Tal
Chókué *see* Chokwè
123 R7 **Chokurdakh** Respublika Sakha (Yakutiya), NE Russian Federation 70°38′N 148°18′E
83 L20 **Chokwè** *var.* Chókué. Gaza, S Mozambique 24°27′S 32°55′E
188 F8 **Chol** Babeldaob, N Palau
160 E8 **Chola Shan** ▲▲ C China
102 J8 **Cholet** Maine-et-Loire, NW France 47°03′N 00°52′W
63 H17 **Cholila** Chubut, W Argentina 42°33′S 71°28′W
Cholo *see* Thyolo
147 V8 **Cholpon** Narynskaya Oblast', C Kyrgyzstan 42°07′N 75°25′E
147 X7 **Cholpon-Ata** Issyk-Kul'skaya Oblast', E Kyrgyzstan 42°39′N 77°05′E
41 P14 **Cholula** Puebla, S Mexico 19°03′N 98°19′W
42 I8 **Choluteca** Choluteca, S Honduras 13°15′N 87°10′W
42 H8 **Choluteca** ◈ *department* S Honduras
42 H8 **Choluteca, Río** ≈ SW Honduras
83 I15 **Choma** Southern, S Zambia 16°48′S 26°58′E
153 T11 **Chomo Lhari** ▲ NW Bhutan 27°59′N 89°24′E
167 N7 **Chom Thong** Chiang Mai, NW Thailand 18°25′N 98°44′E
111 B15 **Chomutov** *Ger.* Komotau. Ústecký Kraj, NW Czech Republic 50°28′N 13°24′E
123 N11 **Chona** ≈ C Russian Federation
Ch'ŏnan *see* Cheonan
167 P11 **Chon Buri** *prev.* Bang Pla Soi. Chon Buri, S Thailand 13°24′N 100°59′E
56 B6 **Chone** Manabí, W Ecuador 0°44′S 80°04′W
Chong'an *see* Wuyishan
163 W13 **Ch'ŏngch'ŏn-gang** ≈ W North Korea
163 Y11 **Ch'ŏngjin** NE North Korea 41°48′N 129°44′E
Chŏngju *see* Cheongju
161 S8 **Chongming Dao** *island* E China
160 J10 **Chongqing** *var.* Ch'ung-ching, Ch'ung-ch'ing, Chungking, Pahsien, Tchongking, Yuzhou. Chongqing Shi, C China 29°34′N 106°27′E
Chŏngup *see* Chŏnju
161 O10 **Chongyang** *var.* Tiancheng. Hubei, C China 29°35′N 114°03′E
160 J15 **Chongzuo** *prev.* Taiping. Guangxi Zhuangzu Zizhiqu, S China 22°18′N 107°23′E
163 Y16 **Chŏnju** *prev.* Chŏnju, Chŏngup, *Jap.* Seiyu. SW South Korea 35°51′N 127°08′E
Chŏnju *see* Chŏnju
Chŏnju *see* Jeonju
Chonnacht *see* Connaught
Chonogol *see* Erdenetsagaan
63 F19 **Chonos, Archipiélago de los** *island group* S Chile
42 K10 **Chontales** ◈ *department* S Nicaragua
167 T13 **Chơn Thanh** Sông Be, S Vietnam 11°25′N 106°38′E
158 K17 **Cho Oyu** *var.* Qowowuyag. ▲ China/Nepal 28°07′N 86°37′E
116 G7 **Chop** *Cz.* Čop, *Hung.* Csap. Zakarpats'ka Oblast', W Ukraine 48°26′N 22°13′E
21 Y3 **Choptank River** ≈ Maryland, NE USA
115 J22 **Chóra** *prev.* Íos. Íos, Kykládes, Greece, Aegean Sea 36°42′N 25°16′E
115 H25 **Chóra Sfakíon** *var.* Sfákia. Kríti, Greece, E Mediterranean Sea 35°12′N 24°05′E
Chorcaí, Cuan *see* Cork Harbour
43 P15 **Chorcha, Cerro** ▲ W Panama 08°39′N 82°07′W
Chorku *see* Chorkŭh
147 R11 **Chorkŭh** *Rus.* Chorku. N Tajikistan 40°04′N 70°30′E
97 K17 **Chorley** NW England, United Kingdom 53°40′N 02°38′W
Chorne More *see* Black Sea
117 R5 **Chornobay** Cherkas'ka Oblast', C Ukraine 49°40′N 32°20′E
117 O3 **Chornobyl'** *Rus.* Chernobyl'. Kyyivs'ka Oblast', N Ukraine 51°17′N 30°15′E
117 R12 **Chornomors'ke** *Rus.* Chernomorskoye. Avtonomna Respublika Krym, S Ukraine 45°29′N 32°45′E
117 R4 **Chornukhy** Poltavs'ka Oblast', C Ukraine 50°15′N 32°57′E
Chorokh/Chorokhi *see* Çoruh Nehri
110 O9 **Choroszcz** Podlaskie, NE Poland 53°10′N 23°E
116 K6 **Chortkiv** *Rus.* Chortkov. Ternopil's'ka Oblast', W Ukraine 49°01′N 25°46′E
Chortkov *see* Chortkiv
Chorum *see* Çorum
110 M9 **Chorzele** Mazowieckie, C Poland 53°16′N 20°53′E
111 J16 **Chorzów** *Ger.* Königshütte; *prev.* Królewska Huta. Śląskie, S Poland 50°17′N 18°58′E
163 W12 **Ch'osan** N North Korea 40°45′N 125°52′E
Chośebuz *see* Cottbus
Chōsen-kaikyō *see* Korea Strait
164 P14 **Chōshi** *var.* Tyôsi. Chiba, Honshū, S Japan 35°44′N 140°48′E
63 H14 **Chos Malal** Neuquén, W Argentina 37°23′S 70°16′W
Chosŏn-minjujuŭi-inmin-kanghwaguk *see* North Korea
110 E9 **Choszczno** *Ger.* Arnswalde. Zachodnio-pomorskie, NW Poland 53°10′N 15°24′E
153 O15 **Chota Nāgpur** *plateau* N India
33 R8 **Choteau** Montana, NW USA 47°48′N 112°40′W
Chotqol *see* Chatkal
14 M8 **Chouart** ≈ Québec, SE Canada
76 I7 **Choûm** Adrar, C Mauritania 21°19′N 12°59′W
27 Q9 **Chouteau** Oklahoma, C USA 36°11′N 95°20′W
21 X8 **Chowan River** ≈ North Carolina, SE USA
35 Q10 **Chowchilla** California, W USA 37°06′N 120°15′W
163 Q7 **Choybalsan** *var.* Hulstay. Dornod, NE Mongolia 48°25′N 114°56′E
163 P7 **Choybalsan** *prev.* Byan Tumen. Dornod, E Mongolia 48°03′N 114°32′E
162 M9 **Choyr** Govĭ Sumber, C Mongolia 46°20′N 108°21′E
185 I19 **Christchurch** Canterbury, South Island, New Zealand 43°31′S 172°39′E
97 M24 **Christchurch** S England, United Kingdom 50°44′N 01°45′W
185 I18 **Christchurch** ✈ Canterbury, South Island, New Zealand 43°28′S 172°33′E
44 J12 **Christiana** C Jamaica 18°13′N 77°29′W
83 H22 **Christiana** Free State, C South Africa 27°55′S 25°10′E
115 J23 **Christiána** *var.* Christiani. *island* Kykládes, Greece, Aegean Sea
Christiani *see* Christiána
Christiania *see* Oslo
14 G13 **Christian Island** *island* Ontario, S Canada
191 P16 **Christian, Point** *headland* Pitcairn Island, Pitcairn Islands 25°04′S 130°08′E
38 M11 **Christian River** ≈ Alaska, USA
Christiansand *see* Kristiansand
21 S7 **Christiansburg** Virginia, NE USA 37°07′N 80°26′W
95 G23 **Christiansfeld** Syddanmark, SW Denmark 55°21′N 09°30′E
Christianshåb *see* Qasigiannguit
39 X14 **Christian Sound** *inlet* Alaska, USA
45 T9 **Christiansted** Saint Croix, S Virgin Islands (US) 17°43′N 64°42′W
Christiansund *see* Kristiansund
25 R13 **Christine** Texas, SW USA 28°47′N 98°30′W
173 U7 **Christmas Island** ◇ *Australian external territory* E Indian Ocean
129 T17 **Christmas Island** *island* E Indian Ocean
Christmas Island *see* Kiritimati
192 M7 **Christmas Ridge** *undersea feature* C Pacific Ocean
30 L16 **Christopher** Illinois, N USA 37°58′N 89°03′W
25 P9 **Christoval** Texas, SW USA 31°09′N 100°30′W
111 F17 **Chrudim** Pardubický Kraj, C Czech Republic 49°58′N 15°49′E
115 K25 **Chrýsi** *island* SE Greece
121 N2 **Chrysochoú, Kólpos** *var.* Khrysokhou Bay. *bay* E Mediterranean Sea
114 I13 **Chrysoúpoli** *var.* Hrisoupoli; *prev.* Khrisoúpolis. Anatolikí Makedonía kai Thráki, NE Greece 40°59′N 24°42′E
111 K16 **Chrzanów** *var.* Chrzanow, *Ger.* Zaumgarten. Śląskie, S Poland 50°10′N 19°21′E
Chu *see* Shu
42 C5 **Chuacús, Sierra de** ▲▲ W Guatemala
153 S15 **Chuadanga** Khulna, W Bangladesh 23°38′N 88°52′E
Chuan *see* Sichuan
Ch'uan-chou *see* Quanzhou
Chuankou *see* Minhe
39 O11 **Chuathbaluk** Alaska, USA 61°36′N 159°14′W
Chubek *see* Moskva
63 I17 **Chubut** *off.* Provincia de Chubut. ◈ *province* S Argentina
Chubut, Provincia de *see* Chubut
63 I17 **Chubut, Río** ≈ SE Argentina
43 V15 **Chucanti, Cerro** ▲ E Panama 08°48′N 78°27′W
Ch'u-chiang *see* Shaoguan
43 W15 **Chucunaque, Río** ≈ E Panama
Chudin *see* Chudzin
116 M5 **Chudniv** Zhytomyrs'ka Oblast', N Ukraine 50°02′N 28°06′E
124 H13 **Chudovo** Novgorodskaya Oblast', W Russian Federation 59°07′N 31°42′E
Chudskoye Ozero *see* Peipus, Lake
119 J18 **Chudzin** *Rus.* Chudin. Brestskaya Voblasts', SW Belarus 52°44′N 26°59′E
39 Q13 **Chugach Islands** *island group* Alaska, USA
39 S11 **Chugach Mountains** ▲▲ Alaska, USA
164 G12 **Chūgoku-sanchi** ▲▲ Honshū, SW Japan
Chugqênsumdo *see* Jigzhi
Chuguyev *see* Chuhuyiv
117 V5 **Chuhuyiv** *var.* Chuguyev. Kharkivs'ka Oblast', E Ukraine 49°51′N 36°44′E
61 H19 **Chuí** Rio Grande do Sul, S Brazil 33°45′S 53°23′W
Chuí *see* Chuy
Chu-Iliyskiye Gory *see* Gory Shu-Ile
Chukai *see* Cukai
Chukchi Avtonomnyy Okrug *see* Chukotskiy Avtonomnyy Okrug
Chukchi Peninsula *see* Chukotskiy Poluostrov
197 R6 **Chukchi Plain** *undersea feature* Arctic Ocean
197 R6 **Chukchi Plateau** *undersea feature* Arctic Ocean
197 R4 **Chukchi Sea** *Rus.* Chukotskoye More. *sea* Arctic Ocean
125 N14 **Chukhloma** Kostromskaya Oblast', NW Russian Federation 58°42′N 42°39′E
Chukotka *see* Chukotskiy Avtonomnyy Okrug
Chukot Range *see* Anadyrskiy Khrebet
123 V6 **Chukotskiy Avtonomnyy Okrug** *var.* Chukchi Avtonomnyy Okrug, Chukotka. ◈ *autonomous district* NE Russian Federation
123 W5 **Chukotskiy, Mys** *headland* NE Russian Federation 64°15′N 173°03′W
123 V5 **Chukotskiy Poluostrov** *Eng.* Chukchi Peninsula. *peninsula* NE Russian Federation
Chukotskoye More *see* Chukchi Sea
Chukurkak *see* Chuqurqoq
Chulakkurgan *see* Sholakkorgan
35 U17 **Chula Vista** California, W USA 32°38′N 117°04′W
123 Q12 **Chul'man** Respublika Sakha (Yakutiya), NE Russian Federation 56°50′N 124°47′E
56 B9 **Chulucanas** Piura, NW Peru 05°08′S 80°10′W
122 J12 **Chulym** ≈ C Russian Federation
152 K6 **Chumar** Jammu and Kashmir, N India 32°38′N 78°36′E
114 K9 **Chumerna** ▲ C Bulgaria 42°45′N 25°58′E
123 R12 **Chumikan** Khabarovskiy Kray, E Russian Federation 54°41′N 135°12′E
167 Q9 **Chum Phae** Khon Kaen, C Thailand 16°31′N 102°09′E
167 N13 **Chumphon** *var.* Jumporn. Chumphon, SW Thailand 10°30′N 99°11′E
167 O9 **Chumsaeng** *var.* Chum Saeng. Nakhon Sawan, C Thailand 15°52′N 100°18′E
Chum Saeng *see* Chumsaeng
122 L12 **Chuna** ≈ C Russian Federation
161 R9 **Chun'an** *var.* Qiandaohu; *prev.* Pailing. Zhejiang, SE China 29°37′N 118°59′E
Chunan *see* Zhunan
Chuncheng *see* Yangchun
163 Y14 **Chuncheon** *Jap.* Shunsen; *prev.* Ch'unch'ŏn. N South Korea 37°52′N 127°48′E
Ch'unch'ŏn *see* Chuncheon
153 S16 **Chunchura** *prev.* Chinsura. West Bengal, NE India 22°54′N 88°20′E
Chundzha *see* Shonzhy
Ch'ung-ch'ing/Ch'ung-ching *see* Chongqing
Chung-hua Jen-min Kung-ho-kuo *see* China
163 Y15 **Chungju** *Jap.* Chūshū; *prev.* Ch'ungju. C South Korea 36°57′N 127°50′E
Ch'ungju *see* Chungju
Chungking *see* Chongqing
161 T14 **Chungyang Shanmo** *Chin.* Taiwan Shan. ▲▲ C Taiwan
149 V9 **Chūniān** Punjab, E Pakistan 30°57′N 74°01′E
122 L12 **Chunskiy** Irkutskaya Oblast', S Russian Federation 56°10′N 99°15′E
122 M11 **Chunya** ≈ C Russian Federation
124 J6 **Chupa** Respublika Kareliya, NW Russian Federation 66°15′N 33°02′E
125 P8 **Chuprovo** Respublika Komi, NW Russian Federation 64°16′N 46°27′E
57 G17 **Chuquibamba** Arequipa, SW Peru 15°47′S 72°44′W
62 H4 **Chuquicamata** Antofagasta, N Chile 22°20′S 68°56′W
57 L21 **Chuquisaca** ◈ *department* S Bolivia
Chuquisaca *see* Sucre
Chuqung *see* Chindu
146 I8 **Chuqurqoq** *Rus.* Chukurkak. Qoraqalpog'iston Respublikasi, NW Uzbekistan 42°44′N 61°33′E
127 T2 **Chur** Udmurtskaya Respublika, NW Russian Federation 57°06′N 52°57′E
108 I9 **Chur** *Fr.* Coire, *It.* Coira, *Rmsch.* Cuera, Quera; *anc.* Curia Rhaetorum. Graubünden, E Switzerland 46°52′N 09°32′E
123 Q10 **Churapcha** Respublika Sakha (Yakutiya), NE Russian Federation 61°59′N 132°06′E
11 V16 **Churchbridge** Saskatchewan, S Canada 50°55′N 101°53′W
21 O8 **Church Hill** Tennessee, S USA 36°31′N 82°42′W
11 X9 **Churchill** Manitoba, C Canada 58°46′N 94°10′W
11 X10 **Churchill** ≈ Manitoba/Saskatchewan, C Canada
13 P9 **Churchill** ≈ Newfoundland and Labrador, E Canada
11 Y9 **Churchill, Cape** *headland* Manitoba, C Canada 58°42′N 93°12′W
13 P9 **Churchill Falls** Newfoundland and Labrador, E Canada 53°38′N 64°00′W
11 S12 **Churchill Lake** ◎ Saskatchewan, C Canada
19 Q3 **Churchill Lake** ▣ Maine, NE USA
194 I5 **Churchill Peninsula** *peninsula* Antarctica
22 H8 **Church Point** Louisiana, S USA 30°24′N 92°13′W
29 O3 **Churchs Ferry** North Dakota, N USA 48°15′N 99°12′W
146 G12 **Churchuri** Ahal Welaýaty, C Turkmenistan 38°55′N 59°13′E
21 T5 **Churchville** Virginia, NE USA 38°13′N 79°10′W
152 G10 **Chūru** Rājasthān, NW India 28°18′N 75°00′E
54 J4 **Churuguara** Falcón, N Venezuela 10°52′N 69°35′W
167 U11 **Chư Sê** Gia Lai, C Vietnam 13°38′N 108°06′E
144 J12 **Chushkakul, Gory** ▲▲ SW Kazakhstan
Chūshū *see* Chungju
37 O9 **Chuska Mountains** ▲▲ Arizona/New Mexico, SW USA
Chu, Sông *see* Sam, Nam
125 V14 **Chusovoy** Permskiy Kray, NW Russian Federation 58°17′N 57°54′E
147 R10 **Chust** Namangan Viloyati, E Uzbekistan 40°58′N 71°12′E
Chust *see* Khust
15 U6 **Chute-aux-Outardes** Québec, SE Canada 49°07′N 68°25′W
117 U5 **Chutove** Poltavs'ka Oblast', C Ukraine 49°45′N 35°11′E
167 U11 **Chư Ty** *var.* Đưc Cơ. Gia Lai, C Vietnam 13°48′N 107°41′E
189 O15 **Chuuk** *var.* Truk. ◈ *state* C Micronesia
189 P15 **Chuuk Islands** *var.* Hogoley Islands; *prev.* Truk Islands. *island group* Caroline Islands, C Micronesia
Chuvashia *see* Chuvashskaya Respublika
Chuvashiya *see* Chuvashskaya Respublika
127 P4 **Chuvashskaya Respublika** *var.* Chuvashiya, *Eng.* Chuvashia. ◈ *autonomous republic* W Russian Federation
Chuwārtah *see* Chwarta
Chu Xian/Chuxian *see* Chuzhou
160 G13 **Chuxiong** Yunnan, SW China 25°02′N 101°32′E
147 V7 **Chuy** Chuyskaya Oblast', N Kyrgyzstan 42°45′N 75°11′E
61 H19 **Chuy** *var.* Chuí. Rocha, E Uruguay 33°42′S 53°27′W
123 O11 **Chuya** Respublika Sakha (Yakutiya), NE Russian Federation 59°30′N 112°26′E
Chüy Oblasty *see* Chuyskaya Oblast'
147 U8 **Chuyskaya Oblast'** *Kir.* Chüy Oblasty. ◈ *province* N Kyrgyzstan
161 Q7 **Chuzhou** *var.* Chuxian, Chu Xian. Anhui, E China 32°20′N 118°18′E
139 U3 **Chwarta** *Ar.* Juwārtā, *var.* Chwārtā, Choarta, Chuwārtah. As Sulaymānīyah, NE Iraq 35°11′N 45°59′E
Chwārtā *see* Chwarta
119 N16 **Chyhirynskaye Vodaskhovishcha** *Rus.* Chigirinskoye Vodokhranilishche. ▣ E Belarus
117 R6 **Chyhyryn** *Rus.* Chigirin. Cherkas'ka Oblast', N Ukraine 49°03′N 32°40′E
119 J18 **Chyrvonaya Slabada** *Rus.* Krasnaya Slabada, Krasnaya Sloboda. Minskaya Voblasts', S Belarus 52°51′N 27°10′E
119 L19 **Chyrvonaye, Vozyera** *Rus.* Ozero Chervonoye. ◎ SE Belarus
117 N11 **Ciadîr-Lunga** *var.* Ceadâr-Lunga, *Rus.* Chadyr-Lunga. S Moldova 46°03′N 28°50′E
169 P16 **Ciamis** *prev.* Tjiamis. Jawa, C Indonesia 07°20′S 108°21′E
107 I16 **Ciampino** ✈ Lazio, C Italy 41°48′N 12°36′E
169 N16 **Cianjur** *prev.* Tjiandjoer. Jawa, C Indonesia 06°50′S 107°09′E
60 H10 **Cianorte** Paraná, S Brazil 23°42′S 52°31′W
Ciarraí *see* Kerry
112 N13 **Ćićevac** Serbia, E Serbia 43°44′N 21°25′E
187 Z14 **Cicia** *prev.* Thithia. *island* Lau Group, E Fiji
105 P4 **Cidacos** ≈ N Spain
136 I10 **Cide** Kastamonu, N Turkey 41°53′N 33°01′E
110 L10 **Ciechanów** *prev.* Zichenau. Mazowieckie, C Poland 52°53′N 20°37′E
110 O10 **Ciechanowiec** *Ger.* Rudelstadt. Podlaskie, E Poland 52°43′N 22°30′E
110 J10 **Ciechocinek** Kujawskio-pomorskie, C Poland 52°53′N 18°49′E
44 F6 **Ciego de Ávila** Ciego de Ávila, C Cuba 21°50′N 78°44′W
54 F4 **Ciénaga** Magdalena, N Colombia 11°01′N 74°15′W
54 E6 **Ciénaga de Oro** Córdoba, NW Colombia 08°54′N 75°39′W
44 E5 **Cienfuegos** Cienfuegos, C Cuba 22°10′N 80°27′W
104 F4 **Cíes, Illas** *island group* NW Spain
111 P16 **Cieszanów** Podkarpackie, SE Poland 50°15′N 23°09′E
111 J17 **Cieszyn** *Cz.* Těšín, *Ger.* Teschen. Śląskie, S Poland 49°45′N 18°35′E
105 R12 **Cieza** Murcia, SE Spain 38°14′N 01°25′W
136 F13 **Çifteler** Eskişehir, W Turkey 39°25′N 31°00′E
105 P7 **Cifuentes** Castilla-La Mancha, C Spain 40°47′N 02°37′W
Čiganak *see* Shyganak
105 P9 **Cigüela** ≈ C Spain
136 H14 **Cihanbeyli** Konya, C Turkey 38°40′N 32°55′E
136 H14 **Cihanbeyli Yaylası** *plateau* C Turkey
104 L10 **Cíjara, Embalse de** ▣ C Spain
169 P16 **Cikalong** Jawa, S Indonesia 07°46′S 108°13′E
169 N16 **Cikawung** Jawa, S Indonesia 06°49′S 105°29′E
187 Y13 **Cikobia** *prev.* Thikombia. *island* N Fiji
169 P17 **Cilacap** *prev.* Tjilatjap. Jawa, C Indonesia 07°44′S 109°E
173 O16 **Cilaos** C Réunion 21°08′S 55°28′E
137 S11 **Çıldır** Ardahan, NE Turkey 41°08′N 43°08′E
137 S11 **Çıldır Gölü** ◎ NE Turkey
160 M10 **Cili** Hunan, S China 29°24′N 110°59′E
Cilician Gates *see* Gülek Boğazı
121 V10 **Cilicia Trough** *undersea feature* E Mediterranean Sea
Cill Airne *see* Killarney
Cill Chainnigh *see* Kilkenny
Cill Chaoi *see* Kilkee
Cill Choca *see* Kilcock
Cill Dara *see* Kildare
105 N3 **Cilleruelo de Bezana** Castilla y León, N Spain 42°58′N 03°50′W
Cilli *see* Celje
Cill Mhantáin *see* Wicklow
Cill Rois *see* Kilrush
146 C11 **Çilmämmetgum** *Rus.* Peski Chil'mamedkum,, *Turkm.* Chilmämetgum. *desert* Balkan Welaýaty, W Turkmenistan
137 Z11 **Çilov Adası** *Rus.* Ostrov Zhiloy. *island* E Azerbaijan
26 J6 **Cimarron** Kansas, C USA 37°49′N 100°20′W
37 T9 **Cimarron** New Mexico, SW USA 36°30′N 104°55′W
26 M9 **Cimarron River** ≈ Kansas/Oklahoma, C USA
117 N11 **Cimişlia** *Rus.* Chimishliya. S Moldova 46°31′N 28°45′E
Cîmpia Turzii *see* Câmpia Turzii
Cîmpina *see* Câmpina
Cîmpulung *see* Câmpulung
Cîmpulung Moldovenesc *see* Câmpulung Moldovenesc
137 P15 **Çınar** Dıyarbakır, SE Turkey 37°45′N 40°22′E
54 J8 **Cinaruco, Río** ≈ Colombia/Venezuela
Cina Selatan, Laut *see* South China Sea
105 T5 **Cinca** ≈ NE Spain
112 G13 **Cincar** ▲ SW Bosnia and Herzegovina 43°55′N 17°05′E
31 Q15 **Cincinnati** Ohio, N USA 39°04′N 84°34′W
21 M4 **Cincinnati** ✈ Kentucky, S USA 39°03′N 84°39′W
Cinco de Outubro *see* Xá-Muteba
136 C15 **Çine** Aydın, SW Turkey 37°37′N 28°03′E
99 J21 **Ciney** Namur, SE Belgium 50°17′N 05°06′E
104 H6 **Cinfães** Viseu, N Portugal 41°04′N 08°06′W
106 J12 **Cingoli** Marche, C Italy 43°25′N 13°09′E
41 U16 **Cintalapa** *var.* Cintalapa de Figueroa. Chiapas, SE Mexico 16°42′N 93°40′W
Cintalapa de Figueroa *see* Cintalapa
103 X14 **Cinto, Monte** ▲ Corse, France, C Mediterranean Sea 42°22′N 08°57′E
Cintra *see* Sintra
105 Q5 **Cintruénigo** Navarra, N Spain 42°05′N 01°50′W
Cionn tSáile *see* Kinsale
116 K13 **Ciorani** Prahova, SE Romania 44°49′N 26°25′E
113 E14 **Čiovo** *It.* Bua. *island* S Croatia
Cipiúr *see* Kippure
63 I15 **Cipolletti** Río Negro, C Argentina 38°55′S 68°W
120 L7 **Circeo, Capo** *headland* C Italy 41°15′N 13°03′E
39 S8 **Circle** *var.* Circle City. Alaska, USA 65°51′N 144°04′W
33 X8 **Circle** Montana, NW USA 47°25′N 105°32′W
Circle City *see* Circle
31 S14 **Circleville** Ohio, N USA 39°36′N 82°57′W
36 K6 **Circleville** Utah, W USA 38°10′N 112°16′W
169 P16 **Cirebon** *prev.* Tjirebon. Jawa, S Indonesia 06°46′S 108°33′E
97 L21 **Cirencester** *anc.* Corinium, Corinium Dobunorum. C England, United Kingdom 51°44′N 01°59′W
Cirkvenica *see* Crikvenica
107 O20 **Cirò** Calabria, SW Italy 39°22′N 17°02′E
107 O20 **Cirò Marina** Calabria, S Italy 39°21′N 17°07′E
102 K14 **Ciron** ≈ SW France
Cirquenizza *see* Crikvenica
25 R7 **Cisco** Texas, SW USA 32°23′N 98°58′W
116 I12 **Cisnădie** *Ger.* Heltau, *Hung.* Nagydisznód. Sibiu, SW Romania 45°42′N 24°09′E
63 G18 **Cisnes, Río** ≈ S Chile
25 T11 **Cistern** Texas, SW USA 29°46′N 97°12′W
104 L3 **Cistierna** Castilla y León, N Spain 42°47′N 05°08′W
Citharista *see* la Ciotat
Citlaltépetl *see* Orizaba, Volcán Pico de
55 X10 **Citron** NW French Guiana 04°49′N 53°55′W
23 N7 **Citronelle** Alabama, S USA 31°05′N 88°13′W
35 O7 **Citrus Heights** California, W USA 38°42′N 121°18′W
106 H7 **Cittadella** Veneto, NE Italy 45°37′N 11°46′E
106 H13 **Città della Pieve** Umbria, C Italy 42°57′N 12°01′E
106 H12 **Città di Castello** Umbria, C Italy 43°27′N 12°13′E
107 I14 **Cittaducale** Lazio, C Italy 42°24′N 12°55′E
107 N22 **Cittanova** Calabria, SW Italy 38°21′N 16°05′E
Cittavecchia *see* Stari Grad
116 G10 **Ciucea** *Hung.* Csucsa. Cluj, NW Romania 46°58′N 22°50′E
116 M13 **Ciucurova** Tulcea, SE Romania 44°57′N 28°24′E
Ciudad Acuña *see* Villa Acuña
41 N15 **Ciudad Altamirano** Guerrero, S Mexico 18°20′N 100°40′W
42 G7 **Ciudad Barrios** San Miguel, NE El Salvador 13°46′N 88°13′W
54 I7 **Ciudad Bolívar** Barinas, NW Venezuela 08°22′N 70°37′W
55 N7 **Ciudad Bolívar** *prev.* Angostura. Bolívar, E Venezuela 08°08′N 63°31′W
40 K6 **Ciudad Camargo** Chihuahua, N Mexico 27°42′N 105°10′W
40 E8 **Ciudad Constitución** Baja California Sur, NW Mexico 25°09′N 111°43′W
Ciudad Cortés *see* Cortés
41 V17 **Ciudad Cuauhtémoc** Chiapas, SE Mexico 15°38′N 91°59′W
42 J9 **Ciudad Darío** *var.* Dario. Matagalpa, W Nicaragua 12°42′N 86°10′W
Ciudad de Dolores Hidalgo *see* Dolores Hidalgo
42 C6 **Ciudad de Guatemala** *Eng.* Guatemala City; *prev.* Santiago de los Caballeros. ● (Guatemala) Guatemala, C Guatemala 14°38′N 90°29′W
Ciudad del Carmen *see* Carmen
62 Q6 **Ciudad del Este** *prev.* Ciudad Presidente Stroessner, Presidente Stroessner, Puerto Presidente Stroessner. Alto Paraná, SE Paraguay 25°34′S 54°40′W
62 K5 **Ciudad de Libertador General San Martín** *var.* Libertador General San Martín. Jujuy, C Argentina 23°50′S 64°45′W
Ciudad Delicias *see* Delicias
41 O11 **Ciudad del Maíz** San Luis Potosí, C Mexico 22°26′N 99°36′W
Ciudad de México *see* México
54 J7 **Ciudad de Nutrias** Barinas, NW Venezuela 08°03′N 69°17′W
Ciudad de Panama *see* Panamá
55 P7 **Ciudad Guayana** *prev.* San Tomé de Guayana, Santo Tomé de Guayana. Bolívar, NE Venezuela 08°22′N 62°37′W
40 K14 **Ciudad Guzmán** Jalisco, SW Mexico 19°40′N 103°30′W
41 V17 **Ciudad Hidalgo** Chiapas, SE Mexico 14°40′N 92°11′W
41 N14 **Ciudad Hidalgo** Michoacán, SW Mexico 19°40′N 100°34′W
40 J3 **Ciudad Juárez** Chihuahua, N Mexico 31°39′N 106°26′W
40 L8 **Ciudad Lerdo** Durango, C Mexico 25°34′N 103°30′W
41 Q11 **Ciudad Madero** *var.* Villa Cecilia. Tamaulipas, C Mexico 22°18′N 97°56′W
41 P11 **Ciudad Mante** Tamaulipas, C Mexico 22°43′N 99°02′W
42 F2 **Ciudad Melchor de Mencos** *var.* Melchor de Mencos. Petén, NE Guatemala 17°03′N 89°12′W
41 P8 **Ciudad Miguel Alemán** Tamaulipas, C Mexico 26°20′N 98°56′W
Ciudad Mutis *see* Bahía Solano
40 G6 **Ciudad Obregón** Sonora, NW Mexico 27°32′N 109°53′W
54 I5 **Ciudad Ojeda** Zulia, NW Venezuela 10°12′N 71°17′W
55 P7 **Ciudad Piar** Bolívar, E Venezuela 07°25′N 63°19′W
Ciudad Porfirio Díaz *see* Piedras Negras
Ciudad Presidente Stroessner *see* Ciudad del Este
Ciudad Quesada *see* Quesada
105 N11 **Ciudad Real** Castilla-La Mancha, C Spain 38°59′N 03°55′W
105 N11 **Ciudad Real** ◈ *province* Castilla-La Mancha, C Spain
104 J7 **Ciudad-Rodrigo** Castilla y León, N Spain 40°36′N 06°33′W
42 A6 **Ciudad Tecún Umán** San Marcos, SW Guatemala 14°40′N 92°06′W
Ciudad Trujillo *see* Santo Domingo
41 P12 **Ciudad Valles** San Luis Potosí, C Mexico 21°59′N 99°01′W
41 O10 **Ciudad Victoria** Tamaulipas, C Mexico 23°44′N 99°07′W
42 C6 **Ciudad Vieja** Suchitepéquez, S Guatemala 14°30′N 90°46′W
116 L8 **Ciuhuru** *var.* Reuţel. ≈ N Moldova
105 Z8 **Ciutadella** *var.* Ciutadella de Menorca. Menorca, Spain, W Mediterranean Sea 40°N 03°50′E
Ciutadella Ciutadella de Menorca *see* Ciutadella
136 L11 **Civa Burnu** *headland* N Turkey 41°22′N 36°39′E
106 J7 **Cividale del Friuli** Friuli-Venezia Giulia, NE Italy 46°06′N 13°25′E
107 H14 **Civita Castellana** Lazio, C Italy 42°16′N 12°24′E
106 J12 **Civitanova Marche** Marche, C Italy 43°18′N 13°41′E
Civitas Altae Ripae *see* Brzeg
Civitas Carnutum *see* Chartres
Civitas Eburovicum *see* Évreux
Civitas Nemetum *see* Speyer
107 G15 **Civitavecchia** *anc.* Centum Cellae, Trajani Portus. Lazio, C Italy 42°05′N 11°47′E
102 L10 **Civray** Vienne, W France 46°10′N 00°18′E
136 E14 **Çivril** Denizli, W Turkey 38°18′N 29°43′E
161 O5 **Cixian** Hebei, E China 36°19′N 114°22′E
137 R16 **Cizre** Şırnak, SE Turkey 37°21′N 42°11′E
Clacton *see* Clacton-on-Sea
97 Q21 **Clacton-on-Sea** *var.* Clacton. E England, United Kingdom 51°48′N 01°09′E
22 H5 **Claiborne, Lake** ▣ Louisiana, S USA
102 L10 **Clain** ≈ W France
11 Q11 **Claire, Lake** ◎ Alberta, C Canada
25 O6 **Clairemont** Texas, SW USA 33°09′N 100°45′W
34 M3 **Clair Engle Lake** ▣ California, W USA
18 B15 **Clairton** Pennsylvania, NE USA 40°17′N 79°52′W
32 F7 **Clallam Bay** Washington, NW USA 48°13′N 124°16′W
103 P8 **Clamecy** Nièvre, C France 47°28′N 03°30′E
23 P5 **Clanton** Alabama, S USA 32°50′N 86°37′W
61 D17 **Clara** Entre Ríos, E Argentina 31°50′S 58°48′W
97 E18 **Clara** *Ir.* Clóirtheach. C Ireland 53°20′N 07°36′W
29 T9 **Clara City** Minnesota, N USA 44°57′N 95°22′W
61 D23 **Claraz** Buenos Aires, E Argentina 37°56′S 59°18′W
Clár Chlainne Mhuiris *see* Claremorris
182 I8 **Clare** South Australia 33°49′S 138°35′E
97 C19 **Clare** *Ir.* An Clár. *cultural region* W Ireland
97 C18 **Clare** ≈ W Ireland
97 A16 **Clare Island** *Ir.* Cliara. *island* W Ireland
44 J12 **Claremont** C Jamaica 18°23′N 77°11′W
29 W10 **Claremont** Minnesota, N USA 44°01′N 93°00′W
19 N9 **Claremont** New Hampshire, NE USA 43°21′N 72°18′W
27 Q9 **Claremore** Oklahoma, C USA 36°20′N 95°37′W
97 C17 **Claremorris** *Ir.* Clár Chlainne Mhuiris. W Ireland 53°47′N 09°W
185 J16 **Clarence** Canterbury, South Island, New Zealand 42°08′S 173°54′E
185 J16 **Clarence** ≈ South Island, New Zealand
65 F15 **Clarence Bay** *bay* Ascension Island, C Atlantic Ocean
63 H25 **Clarence, Isla** *island* S Chile
194 H2 **Clarence Island** *island* South Shetland Islands, Antarctica
183 V5 **Clarence River** ≈ New South Wales, SE Australia
44 J5 **Clarence Town** Long Island, C The Bahamas 23°03′N 74°57′W
27 W12 **Clarendon** Arkansas, C USA 34°41′N 91°19′W
25 O3 **Clarendon** Texas, SW USA 34°57′N 100°54′W
13 U12 **Clarenville** Newfoundland, Newfoundland and Labrador, SE Canada 48°10′N 54°00′W
11 Q17 **Claresholm** Alberta, SW Canada 50°02′N 113°33′W
29 T16 **Clarinda** Iowa, C USA 40°44′N 95°02′W
55 N5 **Clarines** Anzoátegui, NE Venezuela 09°56′N 65°11′W
29 V12 **Clarion** Iowa, C USA 42°43′N 93°43′W
18 C13 **Clarion** Pennsylvania, NE USA 41°11′N 79°21′W
193 O6 **Clarion Fracture Zone** *tectonic feature* NE Pacific Ocean
18 D13 **Clarion River** ≈ Pennsylvania, NE USA
29 Q9 **Clark** South Dakota, N USA 44°50′N 97°44′W
36 K11 **Clarkdale** Arizona, SW USA 34°46′N 112°03′W
15 W4 **Clarke City** Québec, SE Canada 50°09′N 66°36′W
183 Q15 **Clarke Island** *island* Furneaux Group, Tasmania, SE Australia
181 X6 **Clarke Range** ▲▲ Queensland, E Australia
23 T2 **Clarkesville** Georgia, SE USA 34°36′N 83°31′W
29 S9 **Clarkfield** Minnesota, N USA 44°48′N 95°49′W
33 N7 **Clark Fork** Idaho, NW USA 48°06′N 116°10′W
33 N8 **Clark Fork** ≈ Idaho/Montana, NW USA
21 P13 **Clark Hill Lake** *var.* J.Storm Thurmond Reservoir. ▣ Georgia/South Carolina, SE USA
39 Q12 **Clark, Lake** ◎ Alaska, USA
35 W12 **Clark Mountain** ▲ California, W USA 35°30′N 115°34′W
37 S3 **Clark Peak** ▲ Colorado, C USA 40°36′N 105°57′W
14 D14 **Clark, Point** *headland* Ontario, S Canada 44°04′N 81°45′W
21 S3 **Clarksburg** West Virginia, NE USA 39°16′N 80°22′W
22 K2 **Clarksdale** Mississippi, S USA 34°12′N 90°34′W
33 U12 **Clarks Fork Yellowstone River** ≈ Montana/Wyoming, NW USA
29 R14 **Clarkson** Nebraska, C USA 41°42′N 97°07′W
39 O13 **Clarks Point** Alaska, USA 58°50′N 158°33′W
18 I13 **Clarks Summit** Pennsylvania, NE USA 41°29′N 75°42′W
32 M10 **Clarkston** Washington, NW USA 46°25′N 117°02′W
44 J12 **Clark's Town** C Jamaica 18°25′N 77°32′W
27 T10 **Clarksville** Arkansas, C USA 35°29′N 93°29′W
31 P13 **Clarksville** Indiana, N USA 40°01′N 85°54′W
20 I8 **Clarksville** Tennessee, S USA 36°32′N 87°22′W
25 W5 **Clarksville** Texas, SW USA 33°37′N 95°04′W
21 U8 **Clarksville** Virginia, NE USA 36°36′N 78°36′W
21 U11 **Clarkton** North Carolina, SE USA 34°28′N 78°39′W
61 C24 **Claromecó** *var.* Balneario Claromecó. Buenos Aires, E Argentina 38°51′S 60°01′W
25 N3 **Claude** Texas, SW USA 35°06′N 101°22′W
Clausentum *see* Southampton
171 O1 **Claveria** Luzon, N Philippines 18°36′N 121°04′E
99 J20 **Clavier** Liège, E Belgium 50°27′N 05°21′E

◆ Country
● Country Capital
◇ Dependent Territory
○ Dependent Territory Capital
◈ Administrative Regions
✈ International Airport
▲ Mountain
▲▲ Mountain Range
⛰ Volcano
≈ River
◎ Lake
▣ Reservoir

Claxton – Comoro Basin

23 W6 **Claxton** Georgia, SE USA 32°09′N 81°54′W
21 R4 **Clay** West Virginia, NE USA 38°28′N 81°17′W
27 N3 **Clay Center** Kansas, C USA 39°22′N 97°08′W
29 P16 **Clay Center** Nebraska, C USA 40°31′N 98°03′W
21 Y2 **Claymont** Delaware, NE USA 39°48′N 75°27′W
36 M14 **Claypool** Arizona, SW USA 33°24′N 110°50′W
23 R6 **Clayton** Alabama, S USA 31°52′N 85°27′W
23 T1 **Clayton** Georgia, SE USA 34°52′N 83°24′W
22 J5 **Clayton** Louisiana, S USA 31°43′N 91°32′W
27 X5 **Clayton** Missouri, C USA 38°39′N 90°21′W
37 V9 **Clayton** New Mexico, SW USA 36°27′N 103°12′W
21 V9 **Clayton** North Carolina, SE USA 35°39′N 78°27′W
27 Q12 **Clayton** Oklahoma, C USA 34°35′N 95°21′W
45 V9 **Clayton J. Lloyd** ✈ (The Valley) C Anguilla 18°12′N 63°02′W
182 I4 **Clayton River** *seasonal river* South Australia
21 R7 **Claytor Lake** ▣ Virginia, NE USA
27 P13 **Clear Boggy Creek** ≈ Oklahoma, C USA
97 B22 **Clear, Cape** *var.* The Bill of Cape Clear, *Ir.* Ceann Cléire. *headland* SW Ireland 51°25′N 09°31′W
36 M12 **Clear Creek** ≈ Arizona, SW USA
39 S12 **Cleare, Cape** *headland* Montague Island, Alaska, USA 59°46′N 147°54′W
18 E13 **Clearfield** Pennsylvania, NE USA 41°02′N 78°27′W
36 L2 **Clearfield** Utah, W USA 41°06′N 112°03′W
25 Q6 **Clear Fork Brazos River** ≈ Texas, SW USA
31 T12 **Clear Fork Reservoir** ▣ Ohio, N USA
11 N12 **Clear Hills** ▲ Alberta, SW Canada
34 M6 **Clearlake** California, W USA 38°57′N 122°38′W
29 V12 **Clear Lake** Iowa, C USA 43°07′N 93°27′W
29 R9 **Clear Lake** South Dakota, N USA 44°45′N 96°40′W
34 M6 **Clear Lake** ◎ California, W USA
22 G6 **Clear Lake** ▣ Louisiana, S USA
35 P1 **Clear Lake Reservoir** ▣ California, W USA
11 N16 **Clearwater** British Columbia, SW Canada 51°38′N 120°02′W
23 U12 **Clearwater** Florida, SE USA 27°58′N 82°46′W
11 R12 **Clearwater** ≈ Alberta/Saskatchewan, C Canada
27 W7 **Clearwater Lake** ▣ Missouri, C USA
33 N10 **Clearwater Mountains** ▲ Idaho, NW USA
33 N10 **Clearwater River** ≈ Idaho, NW USA
29 S4 **Clearwater River** ≈ Minnesota, N USA
25 T7 **Cleburne** Texas, SW USA 32°21′N 97°24′W
32 I9 **Cle Elum** Washington, NW USA 47°12′N 120°56′W
97 O17 **Cleethorpes** E England, United Kingdom 53°34′N 00°02′W
Cléire, Ceann *see* Clear, Cape
21 O11 **Clemson** South Carolina, SE USA 34°40′N 82°50′W
21 Q4 **Clendenin** West Virginia, NE USA 38°29′N 81°21′W
26 M9 **Cleo Springs** Oklahoma, C USA 36°25′N 98°25′W
Clerk Island *see* Onotoa
181 X8 **Clermont** Queensland, E Australia 22°47′S 147°41′E
15 S8 **Clermont** Québec, SE Canada 47°41′N 70°15′W
103 O4 **Clermont** Oise, N France 49°23′N 02°26′E
29 X12 **Clermont** Iowa, C USA 43°00′N 91°39′W
103 P11 **Clermont-Ferrand** Puy-de-Dôme, C France 45°47′N 03°05′E
103 Q15 **Clermont-l'Hérault** Hérault, S France 43°37′N 03°25′E
99 M22 **Clervaux** Diekirch, N Luxembourg 50°03′N 06°02′E
106 G6 **Cles** Trentino-Alto Adige, N Italy 46°22′N 11°04′E
182 H8 **Cleve** South Australia 33°43′S 136°30′E
Cleve *see* Kleve
23 T2 **Cleveland** Georgia, SE USA 34°36′N 83°45′W
22 K3 **Cleveland** Mississippi, S USA 33°45′N 90°43′W
31 T11 **Cleveland** Ohio, N USA 41°30′N 81°42′W
27 O9 **Cleveland** Oklahoma, C USA 36°18′N 96°27′W
20 L10 **Cleveland** Tennessee, S USA 35°10′N 84°51′W
25 W10 **Cleveland** Texas, SW USA 30°19′N 95°06′W
31 N7 **Cleveland** Wisconsin, N USA 43°58′N 87°45′W
31 O4 **Cleveland Cliffs Basin** ◎ Michigan, N USA
31 U11 **Cleveland Heights** Ohio, N USA 41°30′N 81°34′W
33 P6 **Cleveland, Mount** ▲ Montana, NW USA 48°55′N 113°51′W
Cleves *see* Kleve
97 B16 **Clew Bay** *Ir.* Cuan Mó. *inlet* W Ireland
23 Y14 **Clewiston** Florida, SE USA 26°45′N 80°55′W
Cliara *see* Clare Island
97 A17 **Clifden** *Ir.* An Clochán. Galway, W Ireland 53°29′N 10°14′W
37 O14 **Clifton** Arizona, SW USA 33°03′N 109°18′W
18 K14 **Clifton** New Jersey, NE USA 40°50′N 74°28′W
25 S8 **Clifton** Texas, SW USA 31°43′N 97°36′W
21 S6 **Clifton Forge** Virginia, NE USA 37°49′N 79°50′W
182 I1 **Clifton Hills** South Australia 27°03′S 138°49′E
11 S17 **Climax** Saskatchewan, S Canada 49°12′N 108°22′W
21 O8 **Clinch River** ≈ Tennessee/Virginia, S USA
25 P12 **Cline** Texas, SW USA 29°14′N 100°07′W
21 N10 **Clingmans Dome** ▲ North Carolina/Tennessee, SE USA 35°33′N 83°30′W
24 H8 **Clint** Texas, SW USA 31°35′N 106°13′W
10 M16 **Clinton** British Columbia, SW Canada 51°06′N 121°31′W
14 E15 **Clinton** Ontario, S Canada 43°36′N 81°33′W
27 U10 **Clinton** Arkansas, C USA 35°34′N 92°28′W
30 L10 **Clinton** Illinois, N USA 40°09′N 88°57′W
29 Z14 **Clinton** Iowa, C USA 41°50′N 90°11′W
20 G7 **Clinton** Kentucky, S USA 36°39′N 89°00′W
22 J8 **Clinton** Louisiana, S USA 30°52′N 91°01′W
19 N11 **Clinton** Massachusetts, NE USA 42°25′N 71°40′W
31 R10 **Clinton** Michigan, N USA 42°04′N 83°58′W
22 K5 **Clinton** Mississippi, S USA 32°22′N 90°22′W
27 S5 **Clinton** Missouri, C USA 38°22′N 93°51′W
21 V10 **Clinton** North Carolina, SE USA 35°00′N 78°19′W
26 L10 **Clinton** Oklahoma, C USA 35°31′N 98°58′W
21 Q12 **Clinton** South Carolina, SE USA 34°28′N 81°52′W
21 M9 **Clinton** Tennessee, S USA 36°07′N 84°08′W
8 L9 **Clinton-Colden Lake** ◎ Northwest Territories, NW Canada
10 H5 **Clinton Creek** Yukon, NW Canada 64°24′N 140°35′W
30 L13 **Clinton Lake** ▣ Illinois, N USA
27 Q4 **Clinton Lake** ▣ Kansas, C USA
21 T11 **Clio** South Carolina, SE USA 34°34′N 79°33′W
193 O7 **Clipperton Fracture Zone** *tectonic feature* E Pacific Ocean
193 Q7 **Clipperton Island** ◇ *French overseas territory* E Pacific Ocean
46 K6 **Clipperton Island** *island* E Pacific Ocean
0 F16 **Clipperton Seamounts** *undersea feature* E Pacific Ocean 08°00′N 111°00′W
102 J8 **Clisson** Loire-Atlantique, NW France 47°06′N 01°19′W
62 K7 **Clodomira** Santiago del Estero, N Argentina 27°35′S 64°14′W
Cloich na Coillte *see* Clonakilty
Clóirtheach *see* Clara
97 C21 **Clonakilty** *Ir.* Cloich na Coillte. SW Ireland 51°37′N 08°54′W
181 T6 **Cloncurry** Queensland, C Australia 20°45′S 140°30′E
97 F18 **Clondalkin** *Ir.* Cluain Dolcáin. E Ireland 53°19′N 06°24′W
97 E16 **Clones** *Ir.* Cluain Eois. N Ireland 54°11′N 07°14′W
97 D20 **Clonmel** *Ir.* Cluain Meala. S Ireland 52°21′N 07°42′W
100 G11 **Cloppenburg** Niedersachsen, NW Germany 52°51′N 08°03′E
29 W6 **Cloquet** Minnesota, N USA 46°43′N 92°27′W
37 S14 **Cloudcroft** New Mexico, SW USA 32°57′N 105°44′W
33 W12 **Cloud Peak** ▲ Wyoming, C USA 44°22′N 107°10′W
185 K14 **Cloudy Bay** *inlet* South Island, New Zealand
21 R10 **Clover** South Carolina, SE USA 35°06′N 81°13′W
34 M6 **Cloverdale** California, W USA 38°49′N 123°03′W
20 J5 **Cloverport** Kentucky, S USA 37°50′N 86°37′W
35 Q10 **Clovis** California, W USA 36°48′N 119°43′W
37 W12 **Clovis** New Mexico, SW USA 34°22′N 103°12′W
14 K13 **Cloyne** Ontario, SE Canada 44°48′N 77°09′W
Cluain Dolcáin *see* Clondalkin
Cluain Eois *see* Clones
Cluainín *see* Manorhamilton
Cluain Meala *see* Clonmel
116 H10 **Cluj** ⬖ *county* NW Romania
Cluj *see* Cluj-Napoca
116 H10 **Cluj-Napoca** *Ger.* Klausenburg, *Hung.* Kolozsvár; *prev.* Cluj. NW Romania 46°47′N 23°36′E
Clunia *see* Feldkirch
103 R10 **Cluny** Saône-et-Loire, C France 46°25′N 04°38′E
103 T10 **Cluses** Haute-Savoie, E France 46°04′N 06°34′E
106 E7 **Clusone** Lombardia, N Italy 45°56′N 10°00′E
25 W12 **Clute** Texas, SW USA 29°01′N 95°24′W
185 D23 **Clutha** ≈ South Island, New Zealand
97 J18 **Clwyd** *cultural region* NE Wales, United Kingdom
185 D22 **Clyde** Otago, South Island, New Zealand 45°12′S 169°21′E
27 N3 **Clyde** Kansas, C USA 39°35′N 97°24′W
29 P2 **Clyde** North Dakota, N USA 48°44′N 98°51′W
31 S11 **Clyde** Ohio, N USA 41°18′N 82°58′W
25 Q7 **Clyde** Texas, SW USA 32°24′N 99°29′W
14 K13 **Clyde** ≈ Ontario, SE Canada
96 J13 **Clyde** ≈ W Scotland, United Kingdom
96 H12 **Clydebank** S Scotland, United Kingdom 55°54′N 04°24′W
96 H13 **Clyde, Firth of** *inlet* S Scotland, United Kingdom
33 S11 **Clyde Park** Montana, NW USA 45°56′N 110°39′W
35 W16 **Coachella** California, W USA 33°38′N 116°10′W
35 W16 **Coachella Canal** *canal* California, W USA
40 I9 **Coacoyole** Durango, C Mexico 24°30′N 106°33′W
25 N7 **Coahoma** Texas, SW USA 32°18′N 101°18′W
10 K8 **Coal** ≈ Yukon, NW Canada
40 L14 **Coalcomán** *var.* Coalcomán de Matamoros. Michoacán, S Mexico 18°49′N 103°13′W
Coalcomán de Matamoros *see* Coalcomán
39 T8 **Coal Creek** Alaska, USA 65°21′N 143°08′W
11 Q17 **Coaldale** Alberta, SW Canada 49°42′N 112°36′W
27 P12 **Coalgate** Oklahoma, C USA 34°33′N 96°15′W
35 P11 **Coalinga** California, W USA 36°08′N 120°21′W
10 L9 **Coal River** British Columbia, W Canada 59°38′N 126°45′W
21 Q6 **Coal River** ≈ West Virginia, NE USA
36 M2 **Coalville** Utah, W USA 40°56′N 111°22′W
58 E13 **Coari** Amazonas, N Brazil 04°08′S 63°07′W
104 I7 **Côa, Rio** ≈ N Portugal
59 D14 **Coari, Rio** ≈ NW Brazil
Coast *see* Pwani
10 G12 **Coast Mountains** *Fr.* Chaîne Côtière. ▲ Canada/USA
16 C7 **Coast Ranges** ▲ W USA
96 I12 **Coatbridge** S Scotland, United Kingdom 55°52′N 04°01′W
42 B6 **Coatepeque** Quezaltenango, SW Guatemala 14°42′N 91°50′W
18 H16 **Coatesville** Pennsylvania, NE USA 39°58′N 75°47′W
15 Q13 **Coaticook** Québec, SE Canada 45°07′N 71°46′W
9 P9 **Coats Island** *island* Nunavut, NE Canada
195 O4 **Coats Land** *physical region* Antarctica
41 T14 **Coatzacoalcos** *var.* Quetzalcoalco; *prev.* Puerto México. Veracruz-Llave, E Mexico 18°06′N 94°26′W
41 S14 **Coatzacoalcos, Río** ≈ SE Mexico
116 M15 **Cobadin** Constanţa, SW Romania 44°05′N 28°13′E
14 H9 **Cobalt** Ontario, S Canada 47°24′N 79°41′W
42 D5 **Cobán** Alta Verapaz, C Guatemala 15°28′N 90°20′W
183 O6 **Cobar** New South Wales, SE Australia 31°31′S 145°51′E
18 F12 **Cobb Hill** ▲ Pennsylvania, NE USA 41°52′N 77°52′W
0 D8 **Cobb Seamount** *undersea feature* E Pacific Ocean 47°00′N 131°00′W
14 K12 **Cobden** Ontario, SE Canada 45°36′N 76°54′W
97 D21 **Cobh** *Ir.* An Cóbh; *prev.* Cove of Cork, Queenstown. SW Ireland 51°51′N 08°17′W
57 J14 **Cobija** Pando, NW Bolivia 11°04′S 68°49′W
Coblence/Coblenz *see* Koblenz
18 J10 **Cobleskill** New York, NE USA 42°40′N 74°29′W
14 I15 **Cobourg** Ontario, SE Canada 43°57′N 78°06′W
181 P1 **Cobourg Peninsula** *headland* Northern Territory, N Australia 11°27′S 132°33′E
183 O10 **Cobram** Victoria, SE Australia 35°56′S 145°36′E
82 N13 **Cóbuè** Niassa, N Mozambique 12°08′S 34°46′E
101 K18 **Coburg** Bayern, SE Germany 50°16′N 10°58′E
19 Q5 **Coburn Mountain** ▲ Maine, NE USA 45°28′N 70°07′W
57 H18 **Cocachacra** Arequipa, SW Peru 17°05′S 71°45′W
59 J17 **Cocalinho** Mato Grosso, W Brazil 14°22′S 51°00′W
Cocanada *see* Kākināda
105 S11 **Cocentaina** Valenciana, E Spain 38°44′N 00°27′W
57 L18 **Cochabamba** *hist.* Oropeza. Cochabamba, C Bolivia 17°23′S 66°10′W
57 K18 **Cochabamba** ⬖ *department* C Bolivia
57 L18 **Cochabamba, Cordillera de** ▲ C Bolivia
101 E18 **Cochem** Rheinland-Pfalz, W Germany 50°09′N 07°09′E
37 R6 **Cochetopa Hills** ▲ Colorado, C USA
155 G22 **Cochin** *var.* Kochchi, Kochi. Kerala, SW India 09°56′N 76°15′E *see also* Kochi
44 D5 **Cochinos, Bahía de** *Eng.* Bay of Pigs. *bay* SE Cuba
37 O16 **Cochise Head** ▲ Arizona, SW USA 32°03′N 109°19′W
23 U5 **Cochran** Georgia, SE USA 32°23′N 83°21′W
11 P16 **Cochrane** Alberta, SW Canada 51°15′N 114°25′W
12 G12 **Cochrane** Ontario, S Canada 49°04′N 81°02′W
63 G20 **Cochrane** Aisén, S Chile 47°16′S 72°33′W
11 U10 **Cochrane** ≈ Manitoba/Saskatchewan, C Canada
Cochrane, Lago *see* Pueyrredón, Lago
Cocibolca *see* Nicaragua, Lago de
Cockade State *see* Maryland
44 M6 **Cockburn Harbour** South Caicos, S Turks and Caicos Islands 21°28′N 71°30′W
14 C11 **Cockburn Island** *island* Ontario, S Canada
44 J3 **Cockburn Town** San Salvador, E The Bahamas 24°01′N 74°31′W
21 X2 **Cockeysville** Maryland, NE USA 39°29′N 76°34′W
181 N12 **Cocklebiddy** Western Australia 32°02′S 125°54′E
44 I12 **Cockpit Country, The** *physical region* W Jamaica
43 S16 **Coclé** *off.* Provincia de Coclé. ⬖ *province* C Panama
43 S15 **Coclé del Norte** Colón, C Panama 09°04′N 80°32′W
Coclé, Provincia de *see* Coclé
23 Y12 **Cocoa** Florida, SE USA 28°23′N 80°44′W
23 Y12 **Cocoa Beach** Florida, SE USA 28°19′N 80°36′W
79 D17 **Cocobeach** Estuaire, NW Gabon 0°59′N 09°34′E
44 G5 **Coco, Cayo** *island* C Cuba
151 Q19 **Coco Channel** *strait* Andaman Sea/Bay of Bengal
173 N6 **Coco-de-Mer Seamounts** *undersea feature* W Indian Ocean 0°30′N 56°00′E
36 K10 **Coconino Plateau** *plain* Arizona, SW USA
43 N6 **Coco, Río** *var.* Río Wanki, Segoviao Wangkí. ≈ Honduras/Nicaragua
173 T7 **Cocos Basin** *undersea feature* E Indian Ocean 05°00′S 94°00′E
188 B17 **Cocos Island** *island* S Guam
Cocos Island Ridge *see* Cocos Ridge
129 S17 **Cocos Islands** *island group* E Indian Ocean
173 T8 **Cocos (Keeling) Islands** ◇ *Australian external territory* E Indian Ocean
0 G15 **Cocos Plate** *tectonic feature*
193 T7 **Cocos Ridge** *var.* Cocos Island Ridge. *undersea feature* E Pacific Ocean 05°30′N 86°00′W
40 K13 **Cocula** Jalisco, SW Mexico 20°22′N 103°50′W
107 D17 **Coda Cavallo, Capo** *headland* Sardegna, Italy, C Mediterranean Sea 40°49′N 09°43′E
58 E13 **Codajás** Amazonas, N Brazil 03°50′S 62°12′W
Codazzi *see* Agustín Codazzi
19 Q12 **Cod, Cape** *headland* Massachusetts, NE USA 41°50′N 69°56′W
185 B25 **Codfish Island** *island* SW New Zealand
106 H9 **Codigoro** Emilia-Romagna, N Italy 44°50′N 12°07′E
13 P5 **Cod Island** *island* Newfoundland and Labrador, E Canada
116 J12 **Codlea** *Ger.* Zeiden, *Hung.* Feketehalom. Braşov, C Romania 45°43′N 25°27′E
58 M13 **Codó** Maranhão, E Brazil 04°28′S 43°51′W
106 E8 **Codogno** Lombardia, N Italy 45°10′N 09°42′E
116 M10 **Codrii** *hill range* C Moldova
45 W9 **Codrington** Barbuda, Antigua and Barbuda 17°43′N 61°49′W
106 J7 **Codroipo** Friuli-Venezia Giulia, NE Italy 45°58′N 13°00′E
28 M12 **Cody** Nebraska, C USA 42°54′N 101°13′W
33 U12 **Cody** Wyoming, C USA 44°31′N 109°04′W
21 P7 **Coeburn** Virginia, NE USA 36°56′N 82°27′W
54 E10 **Coello** Tolima, W Colombia 04°15′N 74°52′W
Coemba *see* Cuemba
181 V2 **Coen** Queensland, NE Australia 14°03′S 143°16′E
101 E14 **Coesfeld** Nordrhein-Westfalen, W Germany 51°55′N 07°10′E
32 M8 **Coeur d'Alene** Idaho, NW USA 47°40′N 116°46′W
32 M8 **Coeur d'Alene Lake** ▣ Idaho, NW USA
98 O8 **Coevorden** Drenthe, NE Netherlands 52°39′N 06°45′E
10 H6 **Coffee Creek** Yukon, W Canada 62°52′N 139°05′W
30 L15 **Coffeen Lake** ▣ Illinois, N USA
22 L3 **Coffeeville** Mississippi, S USA 33°58′N 89°40′W
27 Q8 **Coffeyville** Kansas, C USA 37°02′N 95°37′W
182 G9 **Coffin Bay** South Australia 34°39′S 135°30′E
182 F9 **Coffin Bay Peninsula** *peninsula* South Australia
183 V5 **Coffs Harbour** New South Wales, SE Australia 30°18′S 153°08′E
105 R10 **Cofrentes** Valenciana, E Spain 39°14′N 01°04′W
117 N10 **Cogîlnic** *Ukr.* Kohyl'nyk. ≈ Moldova/Ukraine
102 K11 **Cognac** *anc.* Compniacum. Charente, W France 45°42′N 00°19′W
106 B7 **Cogne** Valle d'Aosta, NW Italy 45°37′N 07°27′E
103 U16 **Cogolin** Var, SE France 43°15′N 06°30′E
105 O7 **Cogolludo** Castilla-La Mancha, C Spain 40°58′N 03°05′W
Cohalm *see* Rupea
92 K8 **Čohkarášša** *var.* Cuokkarašša. ▲ N Norway 69°57′N 24°38′E
Čohkkiras *see* Jukkasjärvi
18 F11 **Cohocton River** ≈ New York, NE USA
18 L10 **Cohoes** New York, NE USA 42°46′N 73°42′W
183 N10 **Cohuna** Victoria, SE Australia 35°51′S 144°15′E
43 P17 **Coiba, Isla de** *island* SW Panama
63 H23 **Coig, Río** ≈ S Argentina
63 G19 **Coihaique** *var.* Coyhaique. Aisén, S Chile 45°32′S 72°00′W
155 G21 **Coimbatore** Tamil Nādu, S India 11°N 76°57′E
104 G8 **Coimbra** *anc.* Conimbria, Conimbriga. Coimbra, W Portugal 40°12′N 08°25′W
104 G8 **Coimbra** ⬖ *district* N Portugal
104 L15 **Coín** Andalucía, S Spain 36°40′N 04°45′W
Coin de Mire *see* Gunner's Quoin
57 J20 **Coipasa, Laguna** ◎ W Bolivia
57 J20 **Coipasa, Salar de** *salt lake* W Bolivia
Coira/Coire *see* Chur
Coirib, Loch *see* Corrib, Lough
54 K6 **Cojedes** *off.* Estado Cojedes. ⬖ *state* N Venezuela
Cojedes, Estado *see* Cojedes
42 F7 **Cojutepeque** Cuscatlán, C El Salvador 13°43′N 88°56′W
33 S16 **Cokeville** Wyoming, C USA 42°03′N 110°55′W
182 M13 **Colac** Victoria, SE Australia 38°22′S 143°38′E
59 O20 **Colatina** Espírito Santo, SE Brazil 19°35′S 40°37′W
27 O13 **Colbert** Oklahoma, C USA 33°51′N 96°30′W
100 L12 **Colbitz-Letzlinger Heide** *heathland* N Germany
26 I3 **Colby** Kansas, C USA 39°24′N 101°04′W
57 H17 **Colca, Río** ≈ SW Peru
97 P21 **Colchester** *hist.* Colneceaste; *anc.* Camulodunum. E England, United Kingdom 51°54′N 00°54′E
19 N13 **Colchester** Connecticut, NE USA 41°34′N 72°17′W
38 M16 **Cold Bay** Alaska, USA 55°11′N 162°43′W
11 R14 **Cold Lake** Alberta, SW Canada 54°26′N 110°16′W
11 R13 **Cold Lake** ◎ Alberta/Saskatchewan, S Canada
29 U8 **Cold Spring** Minnesota, N USA 45°27′N 94°25′W
25 W10 **Coldspring** Texas, SW USA 30°34′N 95°10′W
11 N17 **Coldstream** British Columbia, SW Canada 50°13′N 119°09′W
96 L13 **Coldstream** SE Scotland, United Kingdom 55°39′N 02°19′W
14 H13 **Coldwater** Ontario, S Canada 44°43′N 79°36′W
26 K7 **Coldwater** Kansas, C USA 37°16′N 99°20′W
31 Q10 **Coldwater** Michigan, N USA 41°56′N 85°00′W
25 N1 **Coldwater Creek** ≈ Oklahoma/Texas, SW USA
22 K2 **Coldwater River** ≈ Mississippi, S USA
183 O9 **Coleambally** New South Wales, SE Australia 34°48′S 145°54′E
19 O6 **Colebrook** New Hampshire, NE USA 44°52′N 71°27′W
27 T5 **Cole Camp** Missouri, C USA 38°27′N 93°12′W
39 T6 **Coleen River** ≈ Alaska, USA
11 P17 **Coleman** Alberta, SW Canada 49°36′N 114°26′W
25 Q8 **Coleman** Texas, SW USA 31°50′N 99°27′W
Çölemerik *see* Hakkâri
83 K22 **Colenso** KwaZulu/Natal, E South Africa 28°44′S 29°50′E
182 L12 **Coleraine** Victoria, SE Australia 37°39′S 141°42′E
97 F14 **Coleraine** *Ir.* Cúil Raithin. N Northern Ireland, United Kingdom 55°08′N 06°40′W
185 G18 **Coleridge, Lake** ◎ South Island, New Zealand
83 H24 **Colesberg** Northern Cape, C South Africa 30°41′S 25°08′E
22 H7 **Colfax** Louisiana, S USA 31°31′N 92°42′W
32 L9 **Colfax** Washington, NW USA 46°52′N 117°21′W
30 J6 **Colfax** Wisconsin, N USA 45°00′N 91°44′W
63 I19 **Colhué Huapí, Lago** ◎ S Argentina
45 Z6 **Colibris, Pointe des** *headland* Grande Terre, E Guadeloupe 16°15′N 61°10′W
106 D6 **Colico** Lombardia, N Italy 46°08′N 09°24′E
99 E14 **Colijnsplaat** Zeeland, SW Netherlands 51°36′N 03°47′E
40 L14 **Colima** Colima, S Mexico 19°13′N 103°46′W
40 K14 **Colima** ⬖ *state* SW Mexico
40 L14 **Colima, Nevado de** ♆ C Mexico 19°36′N 103°36′W
59 M14 **Colinas** Maranhão, E Brazil 06°02′S 44°15′W
96 F10 **Coll** *island* W Scotland, United Kingdom
105 N7 **Collado Villalba** *var.* Villalba. Madrid, C Spain 40°38′N 04°00′W
183 R4 **Collarenebri** New South Wales, SE Australia 29°31′S 148°33′E
37 P5 **Collbran** Colorado, C USA 39°14′N 107°57′W
106 G12 **Colle di Val d'Elsa** Toscana, C Italy 43°26′N 11°06′E
39 R9 **College** Alaska, USA 64°49′N 148°06′W
32 K10 **College Place** Washington, NW USA 46°03′N 118°23′W
25 U10 **College Station** Texas, SW USA 30°38′N 96°21′W
183 P4 **Collerina** New South Wales, SE Australia 29°41′S 146°36′E
180 I13 **Collie** Western Australia 33°20′S 116°06′E
180 L4 **Collier Bay** *bay* Western Australia
20 E10 **Collierville** Tennessee, S USA 35°02′N 89°39′W
106 F11 **Collina, Passo della** *pass* C Italy
14 G14 **Collingwood** Ontario, S Canada 44°30′N 80°14′W
184 I13 **Collingwood** Tasman, South Island, New Zealand 40°40′S 172°40′E
22 L7 **Collins** Mississippi, S USA 31°39′N 89°33′W
30 K15 **Collinsville** Illinois, N USA 38°40′N 89°58′W
27 P9 **Collinsville** Oklahoma, C USA 36°21′N 95°50′W
20 H10 **Collinwood** Tennessee, S USA 35°10′N 87°44′W
Collipo *see* Leiria
63 G14 **Collipulli** Araucanía, C Chile 37°55′S 72°30′W
97 D16 **Collooney** *Ir.* Cúil Mhuine. NW Ireland 54°11′N 08°29′W
29 R10 **Colman** South Dakota, N USA 43°58′N 96°48′W
103 U6 **Colmar** *Ger.* Kolmar. Haut-Rhin, NE France 48°05′N 07°21′E
104 M15 **Colmenar** Andalucía, S Spain 36°54′N 04°20′W
Colmenar *see* Colmenar de Oreja
105 O9 **Colmenar de Oreja** *var.* Colmenar. Madrid, C Spain 40°06′N 03°25′W
105 N7 **Colmenar Viejo** Madrid, C Spain 40°39′N 03°46′W
25 X9 **Colmesneil** Texas, SW USA 30°54′N 94°25′W
Cöln *see* Köln
Colneceaste *see* Colchester
59 G15 **Colniza** Mato Grosso, W Brazil 09°16′S 59°25′W
Cologne *see* Köln
42 B6 **Colomba** Quezaltenango, SW Guatemala 14°45′N 91°39′W
Colomb-Béchar *see* Béchar
54 E11 **Colombia** Huila, C Colombia 03°24′N 74°49′W
54 G10 **Colombia** *off.* Republic of Colombia. ◆ *republic* N South America
64 E12 **Colombian Basin** *undersea feature* SW Caribbean Sea 13°00′N 76°00′W
Colombia, Republic of *see* Colombia
Colombie-Britannique *see* British Columbia
15 T6 **Colombier** Québec, SE Canada 48°51′N 68°52′W
155 J25 **Colombo** ● (Sri Lanka) Western Province, W Sri Lanka 06°55′N 79°52′E
155 J25 **Colombo** ✈ Western Province, SW Sri Lanka 06°50′N 79°59′E
29 N11 **Colome** South Dakota, N USA 43°13′N 99°42′W
61 B19 **Colón** Buenos Aires, E Argentina 33°53′S 61°06′W
61 D18 **Colón** Entre Ríos, E Argentina 32°10′S 58°16′W
44 D5 **Colón** Matanzas, C Cuba 22°43′N 80°54′W
43 T14 **Colón** *prev.* Aspinwall. Colón, C Panama 09°04′N 80°33′W
42 K5 **Colón** ⬖ *department* NE Honduras
43 S15 **Colón** *off.* Provincia de Colón. ⬖ *province* N Panama
57 A16 **Colón, Archipiélago de** *var.* Islas de los Galápagos, *Eng.* Galapagos Islands, Tortoise Islands. *island group* Ecuador, E Pacific Ocean
44 K5 **Colonel Hill** Crooked Island, SE The Bahamas 22°43′N 74°12′W
40 C3 **Colonet** Baja California Norte, NW Mexico 31°00′N 116°11′W
40 B3 **Colonet, Cabo** *headland* NW Mexico 30°57′N 116°19′W
188 G14 **Colonia** Yap, W Micronesia 09°29′N 138°06′E
61 D19 **Colonia** ⬖ *department* SW Uruguay
Colonia *see* Kolonia, Micronesia
Colonia *see* Colonia del Sacramento, Uruguay
Colonia Agrippina *see* Köln
61 D20 **Colonia del Sacramento** *var.* Colonia. Colonia, SW Uruguay 34°29′S 57°48′W
62 L8 **Colonia Dora** Santiago del Estero, N Argentina 28°34′S 62°59′W
Colonia Julia Fanestris *see* Fano
21 W5 **Colonial Beach** Virginia, NE USA 38°15′N 76°57′W
21 V6 **Colonial Heights** Virginia, NE USA 37°15′N 77°24′W
Colón, Provincia de *see* Colón
193 S7 **Colón Ridge** *undersea feature* E Pacific Ocean 02°00′N 96°00′W
96 F12 **Colonsay** *island* W Scotland, United Kingdom
57 K22 **Colorada, Laguna** ◎ SW Bolivia
37 R6 **Colorado** *off.* State of Colorado, *also known as* Centennial State, Silver State. ⬖ *state* C USA
63 H22 **Colorado, Cerro** ▲ S Argentina 49°58′S 71°38′W
25 O7 **Colorado City** Texas, SW USA 32°24′N 100°51′W
36 M7 **Colorado Plateau** *plateau* W USA
61 A24 **Colorado, Río** ≈ E Argentina
43 N12 **Colorado, Río** ≈ NE Costa Rica
Colorado, Río *see* Colorado River
16 F12 **Colorado River** *var.* Río Colorado. ≈ Mexico/USA
16 K14 **Colorado River** ≈ Texas, SW USA
35 W15 **Colorado River Aqueduct** *aqueduct* California, W USA
44 A4 **Colorados, Archipiélago de los** *island group* NW Cuba
62 J9 **Colorados, Desagües de los** ◎ W Argentina
37 T5 **Colorado Springs** Colorado, C USA 38°50′N 104°47′W
40 L11 **Colotlán** Jalisco, SW Mexico 22°08′N 103°15′W
57 L19 **Colquechaca** Potosí, C Bolivia 18°40′S 66°00′W
23 S7 **Colquitt** Georgia, SE USA 31°10′N 84°43′W
29 R11 **Colton** South Dakota, N USA 43°47′N 96°55′W
32 M10 **Colton** Washington, NW USA 46°34′N 117°10′W
35 P8 **Columbia** California, W USA 38°01′N 120°22′W
30 K16 **Columbia** Illinois, N USA 38°26′N 90°12′W
20 L7 **Columbia** Kentucky, S USA 37°05′N 85°19′W
22 I6 **Columbia** Louisiana, S USA 32°05′N 92°03′W
21 W3 **Columbia** Maryland, NE USA 39°13′N 76°51′W
22 L7 **Columbia** Mississippi, S USA 31°15′N 89°50′W
27 U4 **Columbia** Missouri, C USA 38°56′N 92°19′W
21 Y9 **Columbia** North Carolina, SE USA 35°55′N 76°15′W
18 G16 **Columbia** Pennsylvania, NE USA 40°01′N 76°30′W
21 Q12 **Columbia** *state capital* South Carolina, SE USA 34°00′N 81°02′W
20 I9 **Columbia** Tennessee, S USA 35°37′N 87°02′W
0 F9 **Columbia** ≈ Canada/USA
32 K9 **Columbia Basin** *basin* Washington, NW USA
197 Q10 **Columbia, Cape** *headland* Ellesmere Island, Nunavut, NE Canada
31 Q12 **Columbia City** Indiana, N USA 41°09′N 85°29′W
21 W3 **Columbia, District of** ⬖ *federal district* NE USA
33 P7 **Columbia Falls** Montana, NW USA 48°22′N 114°10′W
11 O15 **Columbia Icefield** *ice field* Alberta/British Columbia, S Canada
11 O15 **Columbia, Mount** ▲ Alberta/British Columbia, SW Canada 52°07′N 117°30′W
11 N15 **Columbia Mountains** ▲ British Columbia, SW Canada
23 P4 **Columbiana** Alabama, S USA 33°10′N 86°36′W
31 V12 **Columbiana** Ohio, N USA 40°53′N 80°41′W
32 M14 **Columbia Plateau** *plateau* Idaho/Oregon, NW USA
29 P7 **Columbia Road Reservoir** ▣ South Dakota, N USA
65 K16 **Columbia Seamount** *undersea feature* C Atlantic Ocean 20°30′S 32°00′W
83 D25 **Columbine, Cape** *headland* SW South Africa 32°50′S 17°39′E
105 U9 **Columbretes, Illes** *prev.* Islas Columbretes. *island group* E Spain
Columbretes, Islas *see* Columbretes, Illes
23 R5 **Columbus** Georgia, SE USA 32°29′N 84°58′W
31 P14 **Columbus** Indiana, N USA 39°12′N 85°55′W
27 R7 **Columbus** Kansas, C USA 37°09′N 94°52′W
23 N4 **Columbus** Mississippi, S USA 33°30′N 88°25′W
33 U11 **Columbus** Montana, NW USA 45°38′N 109°15′W
29 Q15 **Columbus** Nebraska, C USA 41°25′N 97°22′W
37 Q16 **Columbus** New Mexico, SW USA 31°49′N 107°38′W
21 P10 **Columbus** North Carolina, SE USA 35°15′N 82°09′W
28 K2 **Columbus** North Dakota, N USA 48°52′N 102°47′W
31 S13 **Columbus** *state capital* Ohio, N USA 39°58′N 83°W
25 U11 **Columbus** Texas, SW USA 29°42′N 96°35′W
30 L8 **Columbus** Wisconsin, N USA 43°21′N 89°00′W
31 R12 **Columbus Grove** Ohio, N USA 40°55′N 84°03′W
29 Y15 **Columbus Junction** Iowa, C USA 41°16′N 91°21′W
44 J3 **Columbus Point** *headland* Cat Island, C The Bahamas 24°07′N 75°19′W
35 T8 **Columbus Salt Marsh** *salt marsh* Nevada, W USA
35 N6 **Colusa** California, W USA 39°10′N 122°03′W
32 L7 **Colville** Washington, NW USA 48°33′N 117°54′W
184 M5 **Colville, Cape** *headland* North Island, New Zealand 36°28′S 175°20′E
184 M5 **Colville Channel** *channel* North Island, New Zealand
39 P6 **Colville River** ≈ Alaska, USA
97 J18 **Colwyn Bay** N Wales, United Kingdom 53°18′N 03°43′W
106 H9 **Comacchio** *var.* Commachio; *anc.* Comactium. Emilia-Romagna, N Italy 44°41′N 12°10′E
106 H9 **Comacchio, Valli di** *lagoon* Adriatic Sea, N Mediterranean Sea
Comactium *see* Comacchio
41 V17 **Comalapa** Chiapas, SE Mexico 15°42′N 92°06′W
41 U15 **Comalcalco** Tabasco, SE Mexico 18°16′N 93°05′W
63 H16 **Comallo** Río Negro, SW Argentina 40°58′S 70°13′W
26 M12 **Comanche** Oklahoma, C USA 34°22′N 97°57′W
25 R8 **Comanche** Texas, SW USA 31°55′N 98°36′W
194 H2 **Comandante Ferraz** *Brazilian research station* Antarctica 61°57′S 58°23′W
62 N6 **Comandante Fontana** Formosa, N Argentina 25°19′S 59°42′W
63 I22 **Comandante Luis Peidra Buena** Santa Cruz, S Argentina 50°04′S 68°55′W
59 O18 **Comandatuba** Bahia, SE Brazil 15°13′S 39°00′W
116 K11 **Comăneşti** *Hung.* Kománfalva. Bacău, SW Romania 46°25′N 26°29′E
57 M19 **Comarapa** Santa Cruz, C Bolivia 17°53′S 64°30′W
116 J13 **Comarnic** Prahova, SE Romania 45°18′N 25°37′E
42 H6 **Comayagua** Comayagua, W Honduras 14°30′N 87°39′W
42 H6 **Comayagua** ⬖ *department* W Honduras
42 I6 **Comayagua, Montañas de** ▲ C Honduras
21 R15 **Combahee River** ≈ South Carolina, SE USA
62 G10 **Combarbalá** Coquimbo, C Chile 31°15′S 71°03′W
103 S7 **Combeaufontaine** Haute-Saône, E France 47°43′N 05°52′E
97 G15 **Comber** *Ir.* An Comar. E Northern Ireland, United Kingdom 54°33′N 05°45′W
99 K20 **Comblain-au-Pont** Liège, E Belgium 50°29′N 05°36′E
102 I6 **Combourg** Ille-et-Vilaine, NW France 48°21′N 01°44′W
44 M9 **Comendador** *prev.* Elías Piña. W Dominican Republic 18°53′N 71°42′W
Comer See *see* Como, Lago di
25 R11 **Comfort** Texas, SW USA 29°58′N 98°54′W
153 V15 **Comilla** *Ben.* Kumillā. Chittagong, E Bangladesh 23°28′N 91°10′E
99 B18 **Comines** Hainaut, W Belgium 50°46′N 02°58′E
Comino *see* Kemmuna
107 D18 **Comino, Capo** *headland* Sardegna, Italy, C Mediterranean Sea 40°32′N 09°49′E
107 K25 **Comiso** Sicilia, Italy, C Mediterranean Sea 36°57′N 14°37′E
41 V16 **Comitán** *var.* Comitán de Domínguez. Chiapas, SE Mexico 16°15′N 92°06′W
Comitán de Domínguez *see* Comitán
Commachio *see* Comacchio
Commander Islands *see* Komandorskiye Ostrova
103 O10 **Commentry** Allier, C France 46°18′N 02°46′E
23 T2 **Commerce** Georgia, SE USA 34°12′N 83°27′W
27 R8 **Commerce** Oklahoma, C USA 36°55′N 94°52′W
25 V5 **Commerce** Texas, SW USA 33°16′N 95°52′W
37 T4 **Commerce City** Colorado, C USA 39°48′N 104°54′W
103 S5 **Commercy** Meuse, NE France 48°46′N 05°36′E
55 W9 **Commewijne** *var.* Commewyne. ⬖ *district* NE Suriname
Commewyne *see* Commewijne
15 P8 **Commissaires, Lac des** ◎ Québec, SE Canada
64 A12 **Commissioner's Point** *headland* W Bermuda
9 O7 **Committee Bay** *bay* Nunavut, N Canada
106 D7 **Como** *anc.* Comum. Lombardia, N Italy 45°48′N 09°05′E
63 J19 **Comodoro Rivadavia** Chubut, SE Argentina 45°50′S 67°30′W
106 D6 **Como, Lago di** *var.* Lario, *Eng.* Lake Como, *Ger.* Comer See. ◎ N Italy
Como, Lake *see* Como, Lago di
40 E7 **Comondú** Baja California Sur, NW Mexico 26°01′N 111°50′W
116 F12 **Comorâşte** *Hung.* Komornok. Caraş-Severin, SW Romania 45°13′N 21°34′E
Comores, République Fédérale Islamique des *see* Comoros
155 G24 **Comorin, Cape** *headland* SE India 08°00′N 77°10′E
172 M8 **Comoro Basin** *undersea feature* SW Indian Ocean 14°00′S 44°00′E

◆ Country
● Country Capital
◇ Dependent Territory
○ Dependent Territory Capital
⬖ Administrative Regions
✈ International Airport
▲ Mountain
▲ Mountain Range
♆ Volcano
≈ River
◎ Lake
▣ Reservoir

COMORO ISLANDS – COVINGTON

172 K14 **Comoro Islands** *island group* W Indian Ocean
172 H13 **Comoros** *off.* Federal Islamic Republic of the Comoros, *Fr.* République Fédérale Islamique des Comores. ◆ *republic* W Indian Ocean
Comoros, Federal Islamic Republic of the *see* Comoros
10 L17 **Comox** Vancouver Island, British Columbia, SW Canada 49°40′N 124°55′W
103 O4 **Compiègne** Oise, N France 49°25′N 02°50′E
Complutum *see* Alcalá de Henares
Compniacum *see* Cognac
40 K12 **Compostela** Nayarit, C Mexico 21°12′N 104°52′W
Compostella *see* Santiago de Compostela
60 L11 **Comprida, Ilha** *island* S Brazil
117 N11 **Comrat** *Rus.* Komrat. S Moldova 46°18′N 28°40′E
25 O11 **Comstock** Texas, SW USA 29°39′N 101°10′W
31 P9 **Comstock Park** Michigan, N USA 43°00′N 85°40′W
193 N3 **Comstock Seamount** *undersea feature* N Pacific Ocean 48°15′N 156°55′W
Comum *see* Como
159 N17 **Cona** Xizang Zizhiqu, W China 27°59′N 91°54′E
76 H14 **Conakry** ● (Guinea) SW Guinea 09°31′N 13°43′W
76 H14 **Conakry** ✈ SW Guinea 09°37′N 13°32′W
Conamara *see* Connemara
Conca *see* Cuenca
25 Q12 **Concan** Texas, SW USA 29°27′N 99°43′W
102 F6 **Concarneau** Finistère, NW France 47°53′N 03°55′W
83 O17 **Conceição** Sofala, C Mozambique 18°47′S 36°18′E
59 K15 **Conceição do Araguaia** Pará, NE Brazil 08°15′S 49°15′W
58 F10 **Conceição do Maú** Roraima, W Brazil 03°35′N 59°52′W
61 D14 **Concepción** *var.* Concepcion. Corrientes, NE Argentina 28°25′S 57°54′W
62 J8 **Concepción** Tucumán, N Argentina 27°20′S 65°35′W
57 O17 **Concepción** Santa Cruz, E Bolivia 16°15′S 62°08′W
62 G13 **Concepción** Bío Bío, C Chile 36°47′S 73°01′W
54 E14 **Concepción** Putumayo, S Colombia 00°03′N 75°35′W
62 O5 **Concepción** *var.* Villa Concepción. Concepción, C Paraguay 23°26′S 57°24′W
62 O5 **Concepción** *off.* Departamento de Concepción. ◈ *department* E Paraguay
Concepción *see* La Concepción
Concepción de la Vega *see* La Vega
41 N9 **Concepción del Oro** Zacatecas, C Mexico 24°38′N 101°25′W
61 D18 **Concepción del Uruguay** Entre Ríos, E Argentina 32°30′S 58°15′W
Concepción, Departamento de *see* Concepción
42 K11 **Concepción, Volcán** ✱ SW Nicaragua 11°31′N 85°37′W
44 J4 **Conception Island** *island* C The Bahamas
35 P14 **Conception, Point** *headland* California, W USA 34°27′N 120°28′W
54 H6 **Concha** Zulia, W Venezuela 09°02′N 71°45′W
60 L9 **Conchas** São Paulo, S Brazil 23°00′S 47°58′W
37 U11 **Conchas Dam** New Mexico, SW USA 35°21′N 104°11′W
37 U10 **Conchas Lake** ▣ New Mexico, SW USA
102 M5 **Conches-en-Ouche** Eure, N France 49°00′N 01°00′E
37 N12 **Concho** Arizona, SW USA 34°28′N 109°33′W
40 J5 **Conchos, Río** ≈ NW Mexico
41 O8 **Conchos, Río** ≈ C Mexico
108 C8 **Concise** Vaud, W Switzerland 46°52′N 06°40′E
35 N8 **Concord** California, W USA 37°58′N 122°01′W
19 O9 **Concord** *state capital* New Hampshire, NE USA 43°10′N 71°32′W
21 R10 **Concord** North Carolina, SE USA 35°25′N 80°34′W
61 D17 **Concordia** Entre Ríos, E Argentina 31°25′S 58°W
60 I13 **Concórdia** Santa Catarina, S Brazil 27°14′S 52°01′W
54 D9 **Concordia** Antioquia, W Colombia 06°03′N 75°57′W
40 J10 **Concordia** Sinaloa, C Mexico 23°18′N 106°02′W
57 I19 **Concordia** Tacna, SW Peru 18°12′S 70°19′W
27 N3 **Concordia** Kansas, C USA 39°35′N 97°39′W
27 S4 **Concordia** Missouri, C USA 38°58′N 93°34′W
167 S7 **Con Cuông** Nghê An, N Vietnam 19°02′N 104°54′E
167 T15 **Côn Đao Sơn** *var.* Con Son. *island* S Vietnam
Condate *see* Rennes, Ille-et-Vilaine, France
Condate *see* St-Claude, Jura, France
Condate *see* Montereau-Faut-Yonne, Seine-St-Denis, France
29 P8 **Conde** South Dakota, N USA 45°08′N 98°07′W
42 J8 **Condega** Estelí, NW Nicaragua 13°19′N 86°26′W
103 P2 **Condé-sur-l'Escaut** Nord, N France 50°27′N 03°36′E
102 K5 **Condé-sur-Noireau** Calvados, N France 48°52′N 00°31′W
Condivincum *see* Nantes
183 P8 **Condobolin** New South Wales, SE Australia 33°04′S 147°08′E
102 L15 **Condom** Gers, S France 43°56′N 00°23′E
32 J11 **Condon** Oregon, NW USA 45°15′N 120°10′W
54 D9 **Condoto** Chocó, W Colombia 06°06′N 76°37′W
23 P7 **Conecuh River** ≈ Alabama/Florida, SE USA
106 H7 **Conegliano** Veneto, NE Italy 45°53′N 12°18′E
61 C19 **Conesa** Buenos Aires, E Argentina 33°36′S 60°21′W
14 F15 **Conestogo** ≈ Ontario, S Canada
Confluentes *see* Koblenz
102 L10 **Confolens** Charente, W France 46°00′N 00°40′E
36 J4 **Confusion Range** ▲ Utah, W USA
62 N6 **Confuso, Río** ≈ C Paraguay
21 R12 **Congaree River** ≈ South Carolina, SE USA
Công Hoa Xa Hôi Chu Nghia Viêt Nam *see* Vietnam
160 K12 **Congjiang** *var.* Bingmei. Guizhou, S China 25°48′N 108°55′E
79 G18 **Congo** *off.* Republic of the Congo, *Fr.* Moyen-Congo; *prev.* Middle Congo. ◆ *republic* C Africa
79 K19 **Congo** *off.* Democratic Republic of Congo; *prev.* Zaire, Belgian Congo, Congo (Kinshasa). ◆ *republic* C Africa
67 T11 **Congo** *var.* Kongo, *Fr.* Zaire. ≈ C Africa
Congo *see* Zaire (province) Angola
68 G12 **Congo Basin** *drainage basin* W Dem. Rep. Congo
67 Q11 **Congo Canyon** *var.* Congo Seavalley, Congo Submarine Canyon. *undersea feature* E Atlantic Ocean 06°00′S 11°50′E
Congo Cone *see* Congo Fan
Congo/Congo (Kinshasa) *see* Congo (Democratic Republic of)
65 P15 **Congo Fan** *var.* Congo Cone. *undersea feature* E Atlantic Ocean 06°00′S 09°00′E
Congo Seavalley *see* Congo Canyon
Congo Submarine Canyon *see* Congo Canyon
Coni *see* Cuneo
63 H18 **Cónico, Cerro** ▲ SW Argentina 43°12′S 71°42′W
Conimbria/Conimbriga *see* Coimbra
Conjeeveram *see* Kānchīpuram
11 R13 **Conklin** Alberta, C Canada 55°36′N 111°06′W
24 M1 **Conlen** Texas, SW USA 36°16′N 102°10′W
Con, Loch *see* Conn, Lough
Connacht *see* Connaught
97 B17 **Connaught** *var.* Connacht, *Ir.* Chonnacht, Cúige. *cultural region* W Ireland
31 V10 **Conneaut** Ohio, N USA 41°56′N 80°32′W
18 L13 **Connecticut** *off.* State of Connecticut, *also known as* Blue Law State, Constitution State, Land of Steady Habits, Nutmeg State. ◈ *state* NE USA
19 N8 **Connecticut** ≈ Canada/USA
19 O6 **Connecticut Lakes** *lakes* New Hampshire, NE USA
32 K9 **Connell** Washington, NW USA 46°39′N 118°51′W
97 B17 **Connemara** *Ir.* Conamara. *physical region* W Ireland
31 Q14 **Connersville** Indiana, N USA 39°38′N 85°15′W
97 B16 **Conn, Lough** *Ir.* Loch Con. ◎ W Ireland
35 X6 **Connors Pass** *pass* Nevada, W USA
181 X7 **Connors Range** ▲ Queensland, E Australia
56 E7 **Cononaco, Río** ≈ E Ecuador
29 W13 **Conrad** Iowa, C USA 42°13′N 92°52′W
33 R7 **Conrad** Montana, NW USA 48°10′N 111°58′W
25 W10 **Conroe** Texas, SW USA 30°19′N 95°28′W
25 V10 **Conroe, Lake** ▣ Texas, SW USA
61 C17 **Conscripto Bernardi** Entre Ríos, E Argentina 31°03′S 59°05′W
59 M20 **Conselheiro Lafaiete** Minas Gerais, SE Brazil 20°40′S 43°48′W
Consentia *see* Cosenza
97 L14 **Consett** N England, United Kingdom 54°50′N 01°53′W
44 B5 **Consolación del Sur** Pinar del Río, W Cuba 22°32′N 83°32′W
Con Son *see* Côn Đao Sơn
11 R15 **Consort** Alberta, SW Canada 51°58′N 110°44′W
Constance *see* Konstanz
108 I6 **Constance, Lake** *Ger.* Bodensee. ◎ C Europe
104 G9 **Constância** Santarém, C Portugal 39°29′N 08°22′W
117 N14 **Constanța** *var.* Küstendje, *Eng.* Constanza, *Ger.* Konstanza, *Turk.* Küstence. Constanța, SE Romania 44°09′N 28°37′E
116 L14 **Constanța** ◈ *county* SE Romania
Constantia *see* Coutances
Constantia *see* Konstanz
104 K13 **Constantina** Andalucía, S Spain 37°54′N 05°36′W
74 L5 **Constantine** *var.* Qacentina, *Ar.* Qoussantīna. NE Algeria 36°23′N 06°44′E
39 O14 **Constantine, Cape** *headland* Alaska, USA 58°23′N 158°53′W
Constantinople *see* İstanbul
Constantiola *see* Olteniţa
Constanz *see* Konstanz
Constanza *see* Constanța
62 G13 **Constitución** Maule, C Chile 35°20′S 72°28′W
61 D17 **Constitución** Salto, N Uruguay 31°05′S 57°51′W
Constitution State *see* Connecticut
105 N10 **Consuegra** Castilla-La Mancha, C Spain 39°28′N 03°36′W
181 X9 **Consuelo Peak** ▲ Queensland, E Australia 24°45′S 148°01′E
56 E11 **Contamana** Loreto, N Peru 07°19′S 75°04′W
Contrasto, Colle del *see* Contrasto, Portella del
107 K23 **Contrasto, Portella del** *var.* Colle del Contrasto. *pass* Sicilia, Italy, C Mediterranean Sea
54 G8 **Contratación** Santander, C Colombia 06°18′N 73°27′W
102 M8 **Contres** Loir-et-Cher, C France 47°24′N 01°30′E
107 O17 **Conversano** Puglia, SE Italy 40°58′N 17°07′E
27 U11 **Conway** Arkansas, C USA 35°05′N 92°27′W
19 O8 **Conway** New Hampshire, NE USA 43°58′N 71°05′W
21 U13 **Conway** South Carolina, SE USA 33°51′N 79°04′W
25 N2 **Conway** Texas, SW USA 35°10′N 101°23′W
27 U11 **Conway, Lake** ▣ Arkansas, C USA
27 N7 **Conway Springs** Kansas, C USA 37°23′N 97°38′W
97 J18 **Conwy** N Wales, United Kingdom 53°17′N 03°51′W
23 T3 **Conyers** Georgia, SE USA 33°40′N 84°01′W
Coo *see* Kos
182 F4 **Coober Pedy** South Australia 29°01′S 134°47′E
181 P2 **Cooinda** Northern Territory, N Australia 12°54′S 132°31′E
182 B6 **Cook** South Australia 30°37′S 130°26′E
29 W4 **Cook** Minnesota, N USA 47°51′N 92°41′W
191 N6 **Cook, Baie de** *bay* Moorea, W French Polynesia
10 J16 **Cook, Cape** *headland* Vancouver Island, British Columbia, SW Canada 50°04′N 127°52′W
37 Q15 **Cookes Peak** ▲ New Mexico, SW USA 32°32′N 107°43′W
20 L8 **Cookeville** Tennessee, S USA 361°00′N 85°30′W
175 P9 **Cook Fracture Zone** *tectonic feature* S Pacific Ocean
Cook, Grand Récif de *see* Cook, Récif de
39 Q12 **Cook Inlet** *inlet* Alaska, USA
191 X2 **Cook Island** *island* Line Islands, E Kiribati
190 J14 **Cook Islands** ◇ *self-governing entity in free association with New Zealand* S Pacific Ocean
Cook, Mount *see* Aoraki
187 O15 **Cook, Récif de** *var.* Grand Récif de Cook. *reef* S New Caledonia
14 G14 **Cookstown** Ontario, S Canada 44°12′N 79°39′W
97 F15 **Cookstown** *Ir.* An Chorr Chríochach. C Northern Ireland, United Kingdom 54°39′N 06°45′W
185 K14 **Cook Strait** *var.* Raukawa. *strait* New Zealand
181 W3 **Cooktown** Queensland, NE Australia 15°28′S 145°15′E
183 P6 **Coolabah** New South Wales, SE Australia 31°03′S 146°42′E
182 J11 **Coola Coola Swamp** *wetland* South Australia
183 S7 **Coolah** New South Wales, SE Australia 31°49′S 149°43′E
183 P9 **Coolamon** New South Wales, SE Australia 34°49′S 147°13′E
183 T4 **Coolatai** New South Wales, SE Australia 29°16′S 150°45′E
180 K12 **Coolgardie** Western Australia 31°01′S 121°12′E
36 L14 **Coolidge** Arizona, SW USA 32°58′N 111°29′W
25 U8 **Coolidge** Texas, SW USA 31°45′N 96°39′W
183 Q11 **Cooma** New South Wales, SE Australia 36°16′S 149°09′E
Coomassie *see* Kumasi
183 R6 **Coonabarabran** New South Wales, SE Australia 31°19′S 149°18′E
182 J10 **Coonalpyn** South Australia 35°43′S 139°50′E
183 R6 **Coonamble** New South Wales, SE Australia 30°56′S 148°22′E
Coondapoor *see* Kundāpura
155 G21 **Coonoor** Tamil Nādu, SE India 11°21′N 76°46′E
29 U14 **Coon Rapids** Iowa, C USA 41°52′N 94°40′W
29 V8 **Coon Rapids** Minnesota, N USA 45°12′N 93°18′W
25 V5 **Cooper** Texas, SW USA 33°23′N 95°42′W
181 U9 **Cooper Creek** *var.* Barcoo, Cooper's Creek. *seasonal river* Queensland/South Australia
39 R12 **Cooper Landing** Alaska, USA 60°27′N 149°59′W
21 T14 **Cooper River** ≈ South Carolina, SE USA
Cooper's Creek *see* Cooper Creek
44 H1 **Coopers Town** Great Abaco, N The Bahamas 26°46′N 77°27′W
18 J10 **Cooperstown** New York, NE USA 42°43′N 74°56′W
29 P4 **Cooperstown** North Dakota, N USA 47°26′N 98°07′W
31 P9 **Coopersville** Michigan, N USA 43°03′N 85°55′W
182 D7 **Coorabie** South Australia 31°57′S 132°18′E
23 Q3 **Coosa River** ≈ Alabama/Georgia, S USA
32 E14 **Coos Bay** Oregon, NW USA 43°22′N 124°13′W
183 Q9 **Cootamundra** New South Wales, SE Australia 34°41′S 148°03′E
97 E16 **Cootehill** *Ir.* Muinchille. N Ireland 54°04′N 07°05′W
Čop *see* Chop
57 J17 **Copacabana** La Paz, W Bolivia 16°11′S 69°02′W
63 H14 **Copahué, Volcán** ✱ C Chile 37°56′S 71°04′W
41 U16 **Copainalá** Chiapas, SE Mexico 17°04′N 93°13′W
32 F8 **Copalis Beach** Washington, NW USA 47°05′N 124°11′W
42 F6 **Copán** ◈ *department* W Honduras
Copán *see* Copán Ruinas
25 T14 **Copano Bay** *bay* NW Gulf of Mexico
42 F6 **Copán Ruinas** *var.* Copán. Copán, W Honduras 14°52′N 89°10′W
Copenhagen *see* København
107 Q19 **Copertino** Puglia, SE Italy 40°16′N 18°03′E
62 H7 **Copiapó** Atacama, N Chile 27°17′S 70°25′W
62 G8 **Copiapó, Bahía** *bay* N Chile
62 G7 **Copiapó, Río** ≈ N Chile
114 M12 **Çöpköy** Edirne, NW Turkey 41°14′N 26°51′E
182 I5 **Copley** South Australia 30°36′S 138°26′E
106 H9 **Copparo** Emilia-Romagna, C Italy 44°53′N 11°53′E
55 V10 **Coppename Rivier** *var.* Koppename. ≈ C Suriname
25 S9 **Copperas Cove** Texas, SW USA 31°07′N 97°54′W
82 J13 **Copperbelt** ◈ *province* C Zambia
39 S11 **Copper Center** Alaska, USA 61°57′N 145°21′W
8 K8 **Coppermine** ≈ Northwest Territories/Nunavut, N Canada
Coppermine *see* Kugluktuk
39 T11 **Copper River** ≈ Alaska, USA
Copper State *see* Arizona
116 I11 **Copşa Mică** *Ger.* Kleinkopisch, *Hung.* Kiskapus. Sibiu, C Romania 46°06′N 24°15′E
158 J14 **Coqên** Xizang Zizhiqu, W China 31°13′N 85°12′E
Coquilhatville *see* Mbandaka
32 E14 **Coquille** Oregon, NW USA 43°11′N 124°12′W
62 G9 **Coquimbo** Coquimbo, N Chile 30°S 71°18′W
62 G9 **Coquimbo** *off.* Región de Coquimbo. ◈ *region* C Chile
Coquimbo, Región de *see* Coquimbo
116 I15 **Corabia** Olt, S Romania 43°46′N 24°31′E
57 F17 **Coracora** Ayacucho, SW Peru 15°03′S 73°45′W
Cora Droma Rúisc *see* Carrick-on-Shannon
44 K9 **Corail** SW Haiti 18°34′N 73°53′W
183 V4 **Coraki** New South Wales, SE Australia 29°01′S 153°15′E
180 G8 **Coral Bay** Western Australia 23°02′S 113°51′E
23 Y16 **Coral Gables** Florida, SE USA 25°43′N 80°16′W
9 P8 **Coral Harbour** *var.* Salliq. Southampton Island, Nunavut, NE Canada 64°10′N 83°15′W
192 I9 **Coral Sea** *sea* SW Pacific Ocean
174 M7 **Coral Sea Basin** *undersea feature* N Coral Sea
192 H9 **Coral Sea Islands** ◇ *Australian external territory* SW Pacific Ocean
182 M12 **Corangamite, Lake** ◎ Victoria, SE Australia
Corantijn Rivier *see* Courantyne River
18 B14 **Coraopolis** Pennsylvania, NE USA 40:28′N 80°08′W
107 N17 **Corato** Puglia, SE Italy 41°09′N 16°25′E
103 O17 **Corbières** ▲ S France
103 P8 **Corbigny** Nièvre, C France 47°15′N 03°42′E
21 N7 **Corbin** Kentucky, S USA 36°57′N 84°06′W
104 L14 **Corbones** ≈ SW Spain
Corcaigh *see* Cork
35 R11 **Corcoran** California, W USA 36°06′N 119°33′W
63 G18 **Corcovado, Golfo** *gulf* S Chile
63 G18 **Corcovado, Volcán** ✱ S Chile 43°13′S 72°45′W
104 F3 **Corcubión** Galicia, NW Spain 42°56′N 09°12′W
Corcyra Nigra *see* Korčula
60 Q9 **Cordeiro** Rio de Janeiro, SE Brazil 22°01′S 42°20′W
23 T6 **Cordele** Georgia, SE USA 31°59′N 83°49′W
26 L11 **Cordell** Oklahoma, C USA 35°17′N 98°59′W
103 N14 **Cordes** Tarn, S France 44°03′N 01°57′E
62 O6 **Cordillera** *off.* Departamento de la Cordillera. ◈ *department* C Paraguay
Cordillera *see* Cacaguatique, Cordillera
Cordillera, Departamento de la *see* Cordillera
182 K1 **Cordillo Downs** South Australia 26°44′S 140°37′E
62 K10 **Córdoba** Córdoba, C Argentina 31°25′S 64°11′W
41 R14 **Córdoba** Veracruz-Llave, E Mexico 18°55′N 96°55′W
104 M13 **Córdoba** *var.* Cordoba, *Eng.* Cordova; *anc.* Corduba. Andalucía, SW Spain 37°53′N 04°46′W
62 K11 **Córdoba** *off.* Provincia de Córdoba. ◈ *province* C Argentina
54 D7 **Córdoba** *off.* Departamento de Córdoba. ◈ *province* NW Colombia
104 L13 **Córdoba** ◈ *province* Andalucía, S Spain
Córdoba, Departamento de *see* Córdoba
Córdoba, Provincia de *see* Córdoba
62 K10 **Córdoba, Sierras de** ▲ C Argentina
23 O3 **Cordova** Alabama, S USA 33°45′N 87°10′W
39 S12 **Cordova** Alaska, USA 60°32′N 145°45′W
Cordova/Cordoba *see* Córdoba
Corduba *see* Córdoba
Corentyne River *see* Courantyne River
Corfu *see* Kérkyra
104 J9 **Coria** Extremadura, W Spain 39°59′N 06°32′W
104 J14 **Coria del Río** Andalucía, S Spain 37°18′N 06°04′W
183 S8 **Coricudgy, Mount** ▲ New South Wales, SE Australia 32°49′S 150°28′E
107 N20 **Corigliano Calabro** Calabria, SW Italy 39°36′N 16°32′E
Corinium/Corinium Dobunorum *see* Cirencester
23 N1 **Corinth** Mississippi, S USA 34°56′N 88°29′W
Corinth *see* Kórinthos
Corinth Canal *see* Dióryga Korinthou
Corinth, Gulf of/Corinthiacus Sinus *see* Korinthiakós Kólpos
Corinthus *see* Kórinthos
42 I9 **Corinto** Chinandega, NW Nicaragua 12°29′N 87°14′W
97 C21 **Cork** *Ir.* Corcaigh. S Ireland 51°54′N 08°06′W
97 C21 **Cork** *Ir.* Corcaigh. *cultural region* SW Ireland
97 C21 **Cork** ✈ Cork, SW Ireland 51°52′N 08°25′W
97 D21 **Cork Harbour** *Ir.* Cuan Chorcaí. *inlet* SW Ireland
107 I23 **Corleone** Sicilia, Italy, C Mediterranean Sea 37°49′N 13°18′E
114 N13 **Çorlu** Tekirdağ, NW Turkey 41°11′N 27°48′E
114 N12 **Çorlu Çayı** ≈ NW Turkey
Cormaiore *see* Courmayeur
11 V13 **Cormorant** Manitoba, C Canada 54°12′N 100°33′W
23 T2 **Cornelia** Georgia, SE USA 34°30′N 83°31′W
60 J10 **Cornélio Procópio** Paraná, S Brazil 23°07′S 50°40′W
55 V9 **Corneliskondre** Sipaliwini, N Suriname 05°21′N 56°10′W
30 J5 **Cornell** Wisconsin, N USA 45°09′N 91°10′W
13 S12 **Corner Brook** Newfoundland, Newfoundland and Labrador, E Canada 48°58′N 57°58′W
Corner Rise Seamounts *see* Corner Seamounts
64 I9 **Corner Seamounts** *var.* Corner Rise Seamounts. *undersea feature* NW Atlantic Ocean 35°30′N 51°30′W
116 M9 **Cornești** *Rus.* Korneshty. C Moldova 47°23′N 28°00′E
Corneto *see* Tarquinia
Cornhusker State *see* Nebraska
27 X8 **Corning** Arkansas, C USA 36°26′N 90°35′W
35 N5 **Corning** California, W USA 39°54′N 122°12′W
29 U15 **Corning** Iowa, C USA 40°58′N 94°46′W
18 G11 **Corning** New York, NE USA 42°08′N 77°03′W
Corn Islands *see* Maíz, Islas del
107 J14 **Corno Grande** ▲ C Italy 42°26′N 13°29′E
15 N13 **Cornwall** Ontario, SE Canada 45°02′N 74°45′W
97 H25 **Cornwall** *cultural region* SW England, United Kingdom
97 G25 **Cornwall, Cape** *headland* SW England, United Kingdom 50°11′N 05°39′W
54 J4 **Coro** *prev.* Santa Ana de Coro. Falcón, NW Venezuela 11°27′N 69°41′W
57 J18 **Corocoro** La Paz, W Bolivia 17°10′S 68°28′W
57 K17 **Coroico** La Paz, W Bolivia 16°09′S 67°45′W
184 M5 **Coromandel** Waikato, North Island, New Zealand 36°47′S 175°30′E
155 K20 **Coromandel Coast** *coast* E India
184 M5 **Coromandel Peninsula** *peninsula* North Island, New Zealand
184 M6 **Coromandel Range** ▲ North Island, New Zealand
171 N5 **Coron** Busuanga Island, W Philippines 12°02′N 120°10′E
35 T15 **Corona** California, W USA 33°52′N 117°34′W
37 T12 **Corona** New Mexico, SW USA 34°15′N 105°36′W
11 U17 **Coronach** Saskatchewan, S Canada 49°07′N 105°33′W
35 U17 **Coronado** California, W USA 32°41′N 117°10′W
43 N15 **Coronado, Bahía de** *bay* S Costa Rica
11 R15 **Coronation** Alberta, SW Canada 52°06′N 111°25′W
8 K7 **Coronation Gulf** *gulf* Nunavut, N Canada
194 I1 **Coronation Island** *island* Antarctica
39 X14 **Coronation Island** *island* Alexander Archipelago, Alaska, USA
61 B18 **Coronda** Santa Fe, C Argentina 31°58′S 60°56′W
63 F14 **Coronel** Bío Bío, C Chile 37°01′S 73°08′W
61 D20 **Coronel Brandsen** *var.* Brandsen. Buenos Aires, E Argentina 35°08′S 58°15′W
62 K4 **Coronel Cornejo** Salta, N Argentina 22°46′S 63°49′W
61 B24 **Coronel Dorrego** Buenos Aires, E Argentina 38°38′S 61°15′W
62 P6 **Coronel Oviedo** Caaguazú, SE Paraguay 25°24′S 56°30′W
61 B23 **Coronel Pringles** Buenos Aires, E Argentina 37°56′S 61°25′W
61 B23 **Coronel Suárez** Buenos Aires, E Argentina 37°30′S 61°52′W
61 E22 **Coronel Vidal** Buenos Aires, E Argentina 37°28′S 57°45′W
55 V9 **Coronie** ◈ *district* NW Suriname
57 G17 **Coropuna, Nevado** ▲ S Peru 15°31′S 72°31′W
Çorovoda *see* Çorovodë
113 L22 **Çorovodë** *var.* Çorovoda. Berat, S Albania 40°29′N 20°15′E
183 P11 **Corowa** New South Wales, SE Australia 36°01′S 146°22′E
42 G1 **Corozal** Corozal, N Belize 18°23′N 88°23′W
54 E6 **Corozal** Sucre, NW Colombia 09°18′N 75°19′W
42 G1 **Corozal** ◈ *district* N Belize
25 T14 **Corpus Christi** Texas, SW USA 27°48′N 97°24′W
25 T14 **Corpus Christi Bay** *inlet* Texas, SW USA
25 R14 **Corpus Christi, Lake** ▣ Texas, SW USA
63 F16 **Corral** Los Ríos, C Chile 39°55′S 73°30′W
105 O9 **Corral de Almaguer** Castilla-La Mancha, C Spain 39°45′N 03°10′W
104 K6 **Corrales** Castilla y León, N Spain 41°22′N 05°44′W
37 R11 **Corrales** New Mexico, SW USA 35°11′N 106°37′W
Corrán Tuathail *see* Carrauntoohil
106 F9 **Correggio** Emilia-Romagna, C Italy 44°47′N 10°46′E
59 M16 **Corrente** Piauí, E Brazil 10°29′S 45°11′W
59 I19 **Correntes, Rio** ≈ SW Brazil
103 N12 **Corrèze** ◈ *department* C France
97 C17 **Corrib, Lough** *Ir.* Loch Coirib. ◎ W Ireland
61 C14 **Corrientes** Corrientes, NE Argentina 27°29′S 58°42′W
61 D15 **Corrientes** *off.* Provincia de Corrientes. ◈ *province* NE Argentina
44 A5 **Corrientes, Cabo** *headland* W Cuba 21°48′N 84°30′W
40 I13 **Corrientes, Cabo** *headland* SW Mexico 20°25′N 105°42′W
Corrientes, Provincia de *see* Corrientes
61 C16 **Corrientes, Río** ≈ NE Argentina
56 E8 **Corrientes, Río** ≈ Ecuador/Peru
25 W9 **Corrigan** Texas, SW USA 31°00′N 94°49′W
55 U9 **Corriverton** E Guyana 05°55′N 57°09′W
Corriza *see* Korçë
183 Q11 **Corryong** Victoria, SE Australia 36°14′S 147°54′E
103 F2 **Corse** *Eng.* Corsica. ◈ *region* France, C Mediterranean Sea
101 X13 **Corse** *Eng.* Corsica. *island* France, C Mediterranean Sea
103 Y12 **Corse, Cap** *headland* Corse, France, C Mediterranean Sea 43°01′N 09°25′E
103 X15 **Corse-du-Sud** ◈ *department* Corse, France, C Mediterranean Sea
29 P11 **Corsica** South Dakota, N USA 43°25′N 98°24′W
Corsica *see* Corse
25 U7 **Corsicana** Texas, SW USA 32°05′N 96°27′W
103 Y15 **Corte** Corse, France, C Mediterranean Sea 42°18′N 09°08′E
63 G16 **Corte Alto** Los Lagos, S Chile 40°58′S 73°04′W
104 I13 **Cortegana** Andalucía, S Spain 37°55′N 06°49′W
43 N15 **Cortés** *var.* Ciudad Cortés. Puntarenas, SE Costa Rica 08°59′N 83°32′W
42 G5 **Cortés** ◈ *department* NW Honduras
37 P8 **Cortez** Colorado, C USA 37°22′N 108°36′W
Cortez, Sea of *see* California, Golfo de
106 H6 **Cortina d'Ampezzo** Veneto, NE Italy 46°33′N 12°09′E
18 H11 **Cortland** New York, NE USA 42°34′N 76°09′W
31 V11 **Cortland** Ohio, N USA 41°19′N 80°43′W
106 H12 **Cortona** Toscana, C Italy 43°15′N 12°01′E
76 H13 **Corubal, Rio** ≈ E Guinea-Bissau
104 G10 **Coruche** Santarém, C Portugal 38°58′N 08°31′W
Çoruh *see* Rize
137 R11 **Çoruh Nehri** *Geor.* Chorokh, *Rus.* Chorokhi. ≈ Georgia/Turkey
136 K12 **Çorum** *var.* Chorum. Çorum, N Turkey 40°31′N 34°57′E
136 J12 **Çorum** *var.* Chorum. ◈ *province* N Turkey
59 H19 **Corumbá** Mato Grosso do Sul, S Brazil 19°S 57°35′W
14 D16 **Corunna** Ontario, S Canada 42°49′N 82°25′W
Corunna *see* A Coruña
32 F12 **Corvallis** Oregon, NW USA 44°35′N 122°36′W
64 M1 **Corvo** *var.* Ilha do Corvo. *island* Azores, Portugal, NE Atlantic Ocean
Corvo, Ilha do *see* Corvo
31 O16 **Corydon** Indiana, N USA 38°12′N 86°07′W
29 V16 **Corydon** Iowa, C USA 40°45′N 93°19′W
Cos *see* Kos
40 I9 **Cosalá** Sinaloa, C Mexico 24°25′N 106°39′W
41 R15 **Cosamaloapan** *var.* Cosamaloapan de Carpio. Veracruz-Llave, E Mexico 18°23′N 95°50′W
Cosamaloapan de Carpio *see* Cosamaloapan
107 N21 **Cosenza** *anc.* Consentia. Calabria, SW Italy 39°17′N 16°15′E
31 T13 **Coshocton** Ohio, N USA 40°16′N 81°53′W
42 H9 **Cosigüina, Punta** *headland* NW Nicaragua 12°53′N 87°42′W
29 T9 **Cosmos** Minnesota, N USA 44°56′N 94°42′W
103 O8 **Cosne-Cours-sur-Loire** Nièvre, C France 47°25′N 02°56′E
108 B9 **Cossonay** Vaud, W Switzerland 46°37′N 06°28′E
Cossyra *see* Pantelleria
47 R4 **Costa, Cordillera de la** *var.* Cordillera de Venezuela. ▲ N Venezuela
42 K13 **Costa Rica** *off.* Republic of Costa Rica. ◆ *republic* Central America
Costa Rica, Republic of *see* Costa Rica
43 N15 **Costeña, Fila** ▲ S Costa Rica
Costermansville *see* Bukavu
116 I14 **Costeşti** Argeș, SW Romania 44°40′N 24°53′E
37 S8 **Costilla** New Mexico, SW USA 36°58′N 105°31′W
35 O7 **Cosumnes River** ≈ California, W USA
101 O16 **Coswig** Sachsen, E Germany 51°07′N 13°36′E
101 M14 **Coswig** Sachsen-Anhalt, E Germany 51°53′N 12°26′E
Cosyra *see* Pantelleria
171 Q7 **Cotabato** Mindanao, S Philippines 07°13′N 124°12′E
56 C5 **Cotacachi** ▲ N Ecuador 0°29′N 78°17′W
57 L21 **Cotagaita** Potosí, S Bolivia 20°47′S 65°40′W
103 V15 **Côte d'Azur** *prev.* Nice. ✈ (Nice) Alpes-Maritimes, SE France 43°40′N 07°12′E
Côte d'Ivoire *see* Ivory Coast
Côte d'Ivoire, République de la *see* Ivory Coast
103 R7 **Côte d'Or** ◈ *department* E France
103 R8 **Côte d'Or** *cultural region* C France
Côte Française des Somalis *see* Djibouti
102 J4 **Cotentin** *peninsula* N France
102 G6 **Côtes d'Armor** *prev.* Côtes-du-Nord. ◈ *department* NW France
Côtes-du-Nord *see* Côtes d'Armor
Cothen *see* Köthen
Côtière, Chaîne *see* Coast Mountains
40 M13 **Cotija** *var.* Cotija de la Paz. Michoacán, SW Mexico 19°49′N 102°39′W
Cotija de la Paz *see* Cotija
77 R16 **Cotonou** *var.* Kotonu. S Benin 06°24′N 02°26′E
77 R16 **Cotonou** ✈ S Benin 06°31′N 02°18′E
56 B6 **Cotopaxi** *prev.* León. ◈ *province* C Ecuador
56 C6 **Cotopaxi** ✱ N Ecuador 0°42′S 78°24′W
Cotrone *see* Crotone
97 L21 **Cotswold Hills** *var.* Cotswolds. *hill range* S England, United Kingdom
Cotswolds *see* Cotswold Hills
32 F13 **Cottage Grove** Oregon, NW USA 43°48′N 123°03′W
21 S14 **Cottageville** South Carolina, SE USA 32°55′N 80°28′W
101 P14 **Cottbus** *Lus.* Chośebuz; *prev.* Kottbus. Brandenburg, E Germany 51°42′N 14°22′E
27 U9 **Cotter** Arkansas, C USA 36°16′N 92°30′W
106 A9 **Cottian Alps** *Fr.* Alpes Cottiennes, *It.* Alpi Cozie. ▲ France/Italy
Cottiennes, Alpes *see* Cottian Alps
Cotton State, The *see* Alabama
22 G4 **Cotton Valley** Louisiana, S USA 32°49′N 93°25′W
36 L12 **Cottonwood** Arizona, SW USA 34°43′N 112°00′W
32 M10 **Cottonwood** Idaho, NW USA 46°01′N 116°20′W
29 S9 **Cottonwood** Minnesota, N USA 44°37′N 95°41′W
25 Q7 **Cottonwood** Texas, SW USA 32°12′N 99°14′W
27 O5 **Cottonwood Falls** Kansas, C USA 38°21′N 96°33′W
36 L3 **Cottonwood Heights** Utah, W USA 40°37′N 111°48′W
29 S10 **Cottonwood River** ≈ Minnesota, N USA
45 O9 **Cotuí** C Dominican Republic 19°04′N 70°10′W
25 Q13 **Cotulla** Texas, SW USA 28°27′N 99°15′W
Cotyora *see* Ordu
102 I11 **Coubre, Pointe de la** *headland* W France 45°39′N 01°23′W
18 E12 **Coudersport** Pennsylvania, NE USA 41°45′N 78°00′W
15 S9 **Coudres, Île aux** *island* Québec, SE Canada
182 G11 **Couedic, Cape de** *headland* South Australia 36°04′S 136°43′E
Couentrey *see* Coventry
102 I6 **Couesnon** ≈ NW France
32 H10 **Cougar** Washington, NW USA 46°03′N 122°18′W
102 L10 **Couhé** Vienne, W France 46°18′N 00°10′E
32 K8 **Coulee City** Washington, NW USA 47:36′N 119°18′W
195 Q15 **Coulman Island** *island* Antarctica
103 P5 **Coulommiers** Seine-et-Marne, N France 48°49′N 03°04′E
14 K11 **Coulonge** ≈ Québec, SE Canada
14 K11 **Coulonge Est** ≈ Québec, SE Canada
35 Q9 **Coulterville** California, W USA 37°41′N 120°10′W
38 M9 **Council** Alaska, USA 64°54′N 163°40′W
32 M12 **Council** Idaho, NW USA 44°45′N 116°26′W
29 S15 **Council Bluffs** Iowa, C USA 41°16′N 95°52′W
27 O5 **Council Grove** Kansas, C USA 38°41′N 96°29′W
27 O5 **Council Grove Lake** ▣ Kansas, C USA
32 G7 **Coupeville** Washington, NW USA 48°13′N 122°41′W
55 U12 **Courantyne River** *var.* Corantijn Rivier, Corentyne River. ≈ Guyana/Suriname
99 G21 **Courcelles** Hainaut, S Belgium 50°28′N 04°23′E
108 C7 **Courgenay** Jura, NW Switzerland 47°24′N 07°09′E
126 B2 **Courland Lagoon** *Ger.* Kurisches Haff, *Rus.* Kurskiy Zaliv. *lagoon* Lithuania/Russian Federation
118 B12 **Courland Spit** *Lith.* Kuršių Nerija, *Rus.* Kurshskaya Kosa. *spit* Lithuania/Russian Federation
106 A6 **Courmayeur** *prev.* Cormaiore. Valle d'Aosta, NW Italy 45°48′N 07°00′E
108 D7 **Courroux** Jura, NW Switzerland 47°22′N 07°23′E
10 K17 **Courtenay** Vancouver Island, British Columbia, SW Canada 49°40′N 124°58′W
21 W7 **Courtland** Virginia, NE USA 36°44′N 77°06′W
25 V10 **Courtney** Texas, SW USA 30°16′N 96°04′W
30 J4 **Court Oreilles, Lac** ◎ Wisconsin, N USA
Courtrai *see* Kortrijk
99 H19 **Court-Saint-Étienne** Walloon Brabant, C Belgium 50°38′N 04°34′E
22 G6 **Coushatta** Louisiana, S USA 32°00′N 93°20′W
172 I16 **Cousin** *island* Inner Islands, NE Seychelles
172 I16 **Cousine** *island* Inner Islands, NE Seychelles
102 J4 **Coutances** *anc.* Constantia. Manche, N France 49°04′N 01°27′W
102 K12 **Coutras** Gironde, SW France 45°01′N 00°07′W
45 U14 **Couva** Trinidad, Trinidad and Tobago 10°25′N 61°27′W
108 B8 **Couvet** Neuchâtel, W Switzerland 46°57′N 06°41′E
99 H22 **Couvin** Namur, S Belgium 50°03′N 04°30′E
116 K12 **Covasna** *Ger.* Kowasna, *Hung.* Kovászna. Covasna, E Romania 45°51′N 26°11′E
116 J11 **Covasna** ◈ *county* E Romania
14 E12 **Cove Island** *island* Ontario, S Canada
34 M5 **Covelo** California, W USA 39°46′N 123°16′W
97 M20 **Coventry** *anc.* Couentrey. C England, United Kingdom 52°25′N 01°30′W
Cove of Cork *see* Cobh
21 U5 **Covesville** Virginia, NE USA 37°52′N 78°41′W
104 I8 **Covilhã** Castelo Branco, E Portugal 40°17′N 07°30′W
23 T3 **Covington** Georgia, SE USA 33°34′N 83°52′W
31 N13 **Covington** Indiana, N USA 40°08′N 87°23′W
20 M3 **Covington** Kentucky, S USA 39°04′N 84°30′W

◆ Country
● Country Capital
◇ Dependent Territory
○ Dependent Territory Capital
◈ Administrative Regions
✈ International Airport
▲ Mountain
▲ Mountain Range
✱ Volcano
≈ River
◎ Lake
▣ Reservoir

COVINGTON – DACIA BANK

22 K8 Covington Louisiana, S USA 30°28'N 90°06'W
31 Q13 Covington Ohio, N USA 40°07'N 84°21'W
20 F9 Covington Tennessee, S USA 35°32'N 89°40'W
21 S6 Covington Virginia, NE USA 37°48'N 80°01'W
183 Q8 Cowal, Lake seasonal lake New South Wales, SE Australia
11 W15 Cowan Manitoba, S Canada 51°59'N 100°36'W
18 F12 Cowanesque River ≈ New York/Pennsylvania, NE USA
180 L12 Cowan, Lake ◎ Western Australia
15 P13 Cowansville Québec, SE Canada 45°13'N 72°44'W
182 H8 Cowell South Australia 33°43'S 136°53'E
97 M23 Cowes S England, United Kingdom 50°45'N 01°19'W
27 Q10 Coweta Oklahoma, C USA 35°57'N 95°39'W
0 D6 Cowie Seamount undersea feature NE Pacific Ocean 54°15'N 149°30'W
32 G10 Cowlitz River ≈ Washington, NW USA
21 Q11 Cowpens South Carolina, SE USA 35°01'N 81°48'W
183 R8 Cowra New South Wales, SE Australia 33°50'S 148°45'E
Coxen Hole see Roatán
59 I19 Coxim Mato Grosso do Sul, S Brazil 18°28'S 54°45'W
59 I19 Coxim, Rio ≈ SW Brazil
Coxin Hole see Roatán
153 V17 Cox's Bazar Chittagong, S Bangladesh 21°25'N 91°59'E
76 H14 Coyah Conakry, W Guinea 09°45'N 13°26'W
40 K5 Coyame Chihuahua, N Mexico 29°29'N 105°07'W
24 L9 Coyanosa Draw ≈ Texas, SW USA
Coyhaique see Coihaique
42 C7 Coyolate, Río ≈ S Guatemala
Coyote State, The see South Dakota
40 I10 Coyotitán Sinaloa, C Mexico 23°48'N 106°37'W
41 N15 Coyuca var. Coyuca de Catalán. Guerrero, S Mexico 18°21'N 100°39'W
41 O16 Coyuca var. Coyuca de Benítez. Guerrero, S Mexico 17°01'N 100°08'W
Coyuca de Benítez/Coyuca de Catalán see Coyuca
29 N15 Cozad Nebraska, C USA 40°52'N 99°58'W
158 L14 Cozhê Xizang Zizhiqu, W China 31°53'N 87°51'E
Cozie, Alpi see Cottian Alps
Cozmeni see Kitsman'
40 E3 Cozón, Cerro ▲ NW Mexico 31°16'N 112°29'W
41 Z12 Cozumel Quintana Roo, E Mexico 20°29'N 86°54'W
41 Z12 Cozumel, Isla island SE Mexico
32 K8 Crab Creek ≈ Washington, NW USA
44 H12 Crab Pond Point headland W Jamaica 18°07'N 78°01'W
Cracovia/Cracow see Kraków
83 I25 Cradock Eastern Cape, S South Africa 32°07'S 25°38'E
39 Y14 Craig Prince of Wales Island, Alaska, USA 55°29'N 133°04'W
37 Q3 Craig Colorado, C USA 40°31'N 107°33'W
97 F15 Craigavon C Northern Ireland, United Kingdom 54°28'N 06°25'W
21 T5 Craigsville Virginia, NE USA 38°07'N 79°21'W
101 J21 Crailsheim Baden-Württemberg, S Germany 49°07'N 10°04'E
116 H14 Craiova Dolj, SW Romania 44°19'N 23°49'E
10 K12 Cranberry Junction British Columbia, SW Canada 55°35'N 128°21'W
18 J8 Cranberry Lake ◎ New York, NE USA
11 V13 Cranberry Portage Manitoba, C Canada 54°34'N 101°22'W
11 P17 Cranbrook British Columbia, SW Canada 49°29'N 115°48'W
30 M5 Crandon Wisconsin, N USA 45°34'N 88°54'W
32 K14 Crane Oregon, NW USA 43°24'N 118°35'W
24 M9 Crane Texas, SW USA 31°23'N 102°22'W
Crane see The Crane
25 S8 Cranfills Gap Texas, SW USA 31°46'N 97°49'W
19 O12 Cranston Rhode Island, NE USA 41°46'N 71°26'W
Cranz see Zelenogradsk
59 L15 Craolândia Tocantins, E Brazil 07°17'S 47°23'W
102 J7 Craon Mayenne, NW France 47°52'N 00°57'W
195 V16 Crary, Cape headland Antarctica
Crasna see Kraszna
32 G14 Crater Lake ◎ Oregon, NW USA
33 P14 Craters of the Moon National Monument national park Idaho, NW USA
59 O14 Crateús Ceará, E Brazil 05°10'S 40°39'W
Crathis see Crati
107 N20 Crati anc. Crathis. ≈ S Italy
11 U16 Craven Saskatchewan, S Canada 50°44'N 104°50'W
54 I8 Cravo Norte Arauca, E Colombia 06°17'N 70°15'W
28 J12 Crawford Nebraska, C USA 42°40'N 103°24'W
25 T8 Crawford Texas, SW USA 31°31'N 97°26'W
11 O17 Crawford Bay British Columbia, SW Canada 49°39'N 116°44'W
65 M19 Crawford Seamount undersea feature S Atlantic Ocean 40°30'S 10°00'W
31 O13 Crawfordsville Indiana, N USA 40°02'N 86°52'W
23 S9 Crawfordville Florida, SE USA 30°10'N 84°22'W
97 O23 Crawley SE England, United Kingdom 51°07'N 00°12'W
33 S10 Crazy Mountains ▲▲ Montana, NW USA
11 T11 Cree ≈ Saskatchewan, C Canada
37 R7 Creede Colorado, C USA 37°51'N 106°55'W
40 I6 Creel Chihuahua, N Mexico 27°45'N 107°36'W
11 S11 Cree Lake ◎ Saskatchewan, C Canada
11 V13 Creighton Saskatchewan, C Canada 54°46'N 101°54'W
29 Q13 Creighton Nebraska, C USA 42°28'N 97°54'W
103 O4 Creil Oise, N France 49°16'N 02°29'E
106 E8 Crema Lombardia, N Italy 45°22'N 09°40'E
106 E8 Cremona Lombardia, N Italy 45°08'N 10°02'E
Creole State see Louisiana
112 M10 Crepaja Hung. Cserépalja. Vojvodina, N Serbia 45°02'N 20°36'E
103 O4 Crépy-en-Valois Oise, N France 49°13'N 02°54'E
112 B10 Cres It. Cherso. Primorje-Gorski Kotar, NW Croatia 44°57'N 14°24'E
112 A11 Cres It. Cherso; anc. Crexa. island W Croatia
32 H14 Crescent Oregon, NW USA 43°27'N 121°40'W
34 K1 Crescent City California, W USA 41°45'N 124°14'W
23 W10 Crescent City Florida, SE USA 29°25'N 81°30'W
167 X10 Crescent Group Chin. Yongle Qundao, Viet. Nhom L i Liêm. island group C Paracel Islands
23 W10 Crescent Lake ◎ Florida, SE USA
29 X11 Cresco Iowa, C USA 43°22'N 92°06'W
61 B18 Crespo Entre Ríos, E Argentina 32°05'S 60°20'W
103 R13 Crest Drôme, E France 44°45'N 05°00'E
37 R5 Crested Butte Colorado, C USA 38°52'N 106°59'W
31 S12 Crestline Ohio, N USA 40°47'N 82°44'W
11 O17 Creston British Columbia, SW Canada 49°05'N 116°32'W
29 U15 Creston Iowa, C USA 41°03'N 94°21'W
33 V16 Creston Wyoming, C USA 41°40'N 107°43'W
37 S7 Crestone Peak ▲ Colorado, C USA 37°58'N 105°34'W
23 P8 Crestview Florida, SE USA 30°44'N 86°34'W
121 R10 Cretan Trough undersea feature Aegean Sea, C Mediterranean Sea
29 R16 Crete Nebraska, C USA 40°36'N 96°58'W
Crete see Kríti
103 O5 Créteil Val-de-Marne, N France 48°47'N 02°28'E
Crete, Sea of/Creticum, Mare see Kritikó Pélagos
105 X4 Creus, Cap de headland NE Spain 42°18'N 03°18'E
103 N10 Creuse ◈ department C France
102 L9 Creuse ≈ C France
103 T4 Creutzwald Moselle, NE France 49°13'N 06°41'E
105 S12 Crevillent prev. Crevillente. Valenciana, E Spain 38°15'N 00°48'W
Crevillente see Crevillent
97 L18 Crewe C England, United Kingdom 53°05'N 02°27'W
21 V7 Crewe Virginia, NE USA 37°10'N 78°07'W
Crexa see Cres
43 Q15 Cricamola, Río ≈ NW Panama
61 K14 Criciúma Santa Catarina, S Brazil 28°39'S 49°23'W
96 J11 Crieff C Scotland, United Kingdom 56°22'N 03°49'W
112 B10 Crikvenica It. Cirquenizza; prev. Cirkvenica, Crjkvenica. Primorje-Gorski Kotar, NW Croatia 45°12'N 14°40'E
Crimea/Crimean Oblast see Krym, Avtonomna Respublika
101 M16 Crimmitschau var. Krimmitschau. Sachsen, E Germany 50°48'N 12°23'E
116 G11 Crişcior Hung. Kristyor. Hunedoara, W Romania 46°09'N 22°54'E
21 Y5 Crisfield Maryland, NE USA 37°58'N 75°51'W
31 P3 Crisp Point headland Michigan, N USA 46°45'N 85°15'W
59 L19 Cristalina Goiás, C Brazil 16°43'S 47°37'W
44 J7 Cristal, Sierra del ▲ E Cuba
43 T14 Cristóbal Colón, C Panama 09°18'N 79°52'W
54 F4 Cristóbal Colón, Pico ▲ N Colombia 10°52'N 73°46'W
Cristur/Cristuru Săcuiesc see Cristuru Secuiesc
116 I11 Cristuru Secuiesc prev. Cristur, Cristuru Săcuiesc, Sitaş Cristuru, Ger. Kreutz, Hung. Székelykeresztúr, Szitás-Keresztúr. Harghita, C Romania 46°17'N 25°02'E
116 F10 Crişul Alb var. Weisse Kreisch, Ger. Weisse Körös, Hung. Fehér-Körös. ≈ Hungary/Romania
116 F10 Crişul Negru Ger. Schwarze Körös, Hung. Fekete-Körös. ≈ Hungary/Romania
116 G10 Crişul Repede var. Schnelle Kreisch, Ger. Schnelle Körös, Hung. Sebes-Körös. ≈ Hungary/Romania
117 N10 Criuleni Rus. Kriulyany. C Moldova 47°12'N 29°09'E
Crivadia Vulcanului see Vulcan
Crjkvenica see Crikvenica
113 J17 Crkvice SW Montenegro 42°34'N 18°38'E
113 O17 Crna Gora Alb. Mali i Zi. ▲▲ FYR Macedonia/Serbia
Crna Gora see Montenegro
113 O20 Crna Reka ≈ S FYR Macedonia
Crni Drim see Black Drin
109 V10 Črni vrh ▲ NE Slovenia 46°28'N 15°14'E
109 V13 Črnomelj Ger. Tschernembl. SE Slovenia 45°32'N 15°12'E
97 A17 Croagh Patrick Ir. Cruach Phádraig. ▲ W Ireland 53°45'N 09°39'W
112 D9 Croatia off. Republic of Croatia, Ger. Kroatien, SCr. Hrvatska. ◆ republic SE Europe
Croatia, Republic of see Croatia
Croce, Picco di see Wilde Kreuzspitze
15 P8 Croche ≈ Québec, SE Canada
169 V7 Crocker, Banjaran var. Crocker Range. ▲▲ East Malaysia
Crocker Range see Crocker, Banjaran
25 V9 Crockett Texas, SW USA 31°21'N 95°30'W
67 V14 Crocodile var. Krokodil. ≈ N South Africa
Crocodile see Limpopo
20 I7 Crofton Kentucky, S USA 37°01'N 87°25'W
29 Q12 Crofton Nebraska, C USA 42°43'N 97°30'W
Croia see Krujë
103 R16 Croisette, Cap headland SE France 43°12'N 05°21'E
102 G8 Croisic, Pointe du headland NW France 47°16'N 02°42'W
103 S13 Croix Haute, Col de la pass E France
15 U5 Croix, Pointe à la headland Québec, SE Canada 49°16'N 67°46'W
14 F13 Croker, Cape headland Ontario, S Canada 44°56'N 80°57'W
181 P1 Croker Island island Northern Territory, N Australia
96 I8 Cromarty N Scotland, United Kingdom 57°40'N 04°02'W
99 M21 Crombach Liège, E Belgium 50°14'N 06°07'E
97 Q18 Cromer E England, United Kingdom 52°56'N 01°06'E
185 D22 Cromwell Otago, South Island, New Zealand 45°03'S 169°14'E
185 H16 Cronadun West Coast, South Island, New Zealand 42°03'S 171°52'E
39 O11 Crooked Creek Alaska, USA 61°52'N 158°06'W
44 K5 Crooked Island island SE The Bahamas
44 J5 Crooked Island Passage channel SE The Bahamas
32 I13 Crooked River ≈ Oregon, NW USA
29 R4 Crookston Minnesota, N USA 47°47'N 96°36'W
28 I10 Crooks Tower ▲ South Dakota, N USA 44°09'N 103°55'W
31 T14 Crooksville Ohio, N USA 39°46'N 82°05'W
183 R9 Crookwell New South Wales, SE Australia 34°28'S 149°27'E
14 L14 Crosby Ontario, SE Canada 44°39'N 76°13'W
97 K17 Crosby var. Great Crosby. NW England, United Kingdom 53°30'N 03°02'W
29 U6 Crosby Minnesota, N USA 46°30'N 93°58'W
28 K2 Crosby North Dakota, N USA 48°54'N 103°17'W
25 O5 Crosbyton Texas, SW USA 33°40'N 101°16'W
77 V16 Cross ≈ Cameroon/Nigeria
23 U10 Cross City Florida, SE USA 29°37'N 83°08'W
Crossen see Krosno Odrzańskie
27 V14 Crossett Arkansas, C USA 33°08'N 91°58'W
97 K15 Cross Fell ▲ N England, United Kingdom 54°42'N 02°30'W
11 P16 Crossfield Alberta, SW Canada 51°24'N 114°03'W
21 Q12 Cross Hill South Carolina, SE USA 34°18'N 81°58'W
19 U6 Cross Island island Maine, NE USA
11 X13 Cross Lake Manitoba, C Canada 54°38'N 97°35'W
22 F5 Cross Lake ⊠ Louisiana, S USA
36 I12 Crossman Peak ▲ Arizona, SW USA 34°33'N 114°09'W
25 Q7 Cross Plains Texas, SW USA 32°07'N 99°10'W
77 V17 Cross River ◈ state SE Nigeria
20 L9 Crossville Tennessee, S USA 35°57'N 85°02'W
31 S8 Croswell Michigan, N USA 43°16'N 82°37'W
14 K13 Crotch Lake ◎ Ontario, SE Canada
Croton/Crotona see Crotone
107 O21 Crotone var. Cotrone; anc. Croton, Crotona. Calabria, SW Italy 39°05'N 17°07'E
33 V11 Crow Agency Montana, NW USA 45°35'N 107°28'W
183 U7 Crowdy Head headland New South Wales, SE Australia 31°52'S 152°45'E
25 Q4 Crowell Texas, SW USA 33°59'N 99°45'W
183 O6 Crowl Creek seasonal river New South Wales, SE Australia
22 H9 Crowley Louisiana, S USA 30°11'N 92°21'W
35 S9 Crowley, Lake ⊠ California, W USA
27 X10 Crowleys Ridge hill range Arkansas, C USA
31 N11 Crown Point Indiana, N USA 41°25'N 87°22'W
37 P10 Crownpoint New Mexico, SW USA 35°40'N 108°09'W
33 R10 Crow Peak ▲ Montana, NW USA 46°17'N 111°54'W
11 P17 Crowsnest Pass pass Alberta/British Columbia, SW Canada
29 T6 Crow Wing River ≈ Minnesota, N USA
97 O22 Croydon SE England, United Kingdom 51°21'N 00°06'W
Crozer, Mount see Finkol, Mount
173 P11 Crozet Basin undersea feature S Indian Ocean 39°00'S 60°00'E
173 O12 Crozet Islands island group French Southern and Antarctic Territories
173 N12 Crozet Plateau var. Crozet Plateaus. undersea feature SW Indian Ocean 46°00'S 51°00'E
Crozet Plateaus see Crozet Plateau
102 E6 Crozon Finistère, NW France 48°14'N 04°31'W
Cruacha Dubha, Na see Macgillycuddy's Reeks
Cruach Phádraig see Croagh Patrick
116 M14 Crucea Constanţa, SE Romania 44°30'N 28°18'E
44 E5 Cruces Cienfuegos, C Cuba 22°20'N 80°17'W
107 O20 Crucoli Torretta Calabria, SW Italy 39°26'N 17°03'E
41 P9 Cruillas Tamaulipas, C Mexico 24°43'N 98°26'W
64 K9 Cruiser Tablemount undersea feature E Atlantic Ocean 32°00'N 28°00'W
61 G14 Cruz Alta Rio Grande do Sul, S Brazil 28°38'S 53°38'W
44 G8 Cruz, Cabo headland S Cuba 19°50'N 77°43'W
60 N9 Cruzeiro São Paulo, S Brazil 22°33'S 44°59'W
60 H10 Cruzeiro do Oeste Paraná, S Brazil 23°45'S 53°03'W
59 A15 Cruzeiro do Sul Acre, W Brazil 07°40'S 72°39'W
23 U11 Crystal Bay bay Florida, SE USA NE Gulf of Mexico Atlantic Ocean
182 I8 Crystal Brook South Australia 33°24'S 138°10'E
11 X17 Crystal City Manitoba, S Canada 49°07'N 98°54'W
27 X5 Crystal City Missouri, C USA 38°13'N 90°22'W
25 P13 Crystal City Texas, SW USA 28°43'N 99°51'W
30 M4 Crystal Falls Michigan, N USA 46°06'N 88°20'W
23 Q8 Crystal Lake Florida, SE USA 30°26'N 85°41'W
31 O6 Crystal Lake ◎ Michigan, N USA
23 V11 Crystal River Florida, SE USA 28°54'N 82°35'W
37 Q5 Crystal River ≈ Colorado, C USA
22 K6 Crystal Springs Mississippi, S USA 31°59'N 90°21'W
Csaca see Čadca
Csakathurn/Csáktornya see Čakovec
Csap see Chop
Csepén see Čepin
Cserépalja see Crepaja
Csermő see Cermei
Csíkszereda see Miercurea-Ciuc
111 L24 Csongrád Csongrád, SE Hungary 46°42'N 20°09'E
111 L24 Csongrád off. Csongrád Megye. ◈ county SE Hungary
Csongrád Megye see Csongrád
111 H22 Csorna Győr-Moson-Sopron, NW Hungary 47°37'N 17°14'E
Csucsa see Ciucea
111 G25 Csurgó Somogy, SW Hungary 46°16'N 17°09'E
Csurog see Čurug
54 L5 Cúa Miranda, N Venezuela 10°14'N 66°58'W
82 C11 Cuale Malanje, NW Angola 08°22'S 16°10'E
67 T12 Cuando var. Kwando. ≈ S Africa
Cuando Cubango see Kuando Kubango
83 E16 Cuangar Kuando Kubango, S Angola 17°34'S 18°39'E
82 D11 Cuango Lunda Norte, NE Angola 09°10'S 17°59'E
82 C10 Cuango Uíge, NW Angola 06°20'S 16°42'E
82 C10 Cuango var. Kwango. ≈ Angola/Dem. Rep. Congo see also Kwango
Cuango see Kwango
Cuan, Loch see Strangford Lough
82 C12 Cuanza var. Kwanza. ≈ C Angola
Cuanza Norte see Kwanza Norte
Cuanza Sul see Kwanza Sul
61 E16 Cuareim, Río var. Rio Quaraí. ≈ Brazil/Uruguay see also Quaraí, Rio
Cuareim, Río see Quaraí, Rio
83 D15 Cuatir ≈ S Angola
40 M7 Cuatro Ciénegas var. Cuatro Ciénegas de Carranza. Coahuila, NE Mexico 27°00'N 102°03'W
Cuatro Ciénegas de Carranza see Cuatro Ciénegas
40 I6 Cuauhtémoc Chihuahua, N Mexico 28°22'N 106°52'W
41 P14 Cuautla Morelos, S Mexico 18°48'N 98°56'W
104 H12 Cuba Beja, S Portugal 38°10'N 07°54'W
27 W6 Cuba Missouri, C USA 38°03'N 91°24'W
37 R10 Cuba New Mexico, SW USA 36°01'N 106°57'W
44 E6 Cuba off. Republic of Cuba. ◆ republic W West Indies
47 O2 Cuba island W West Indies
82 B13 Cubal Benguela, W Angola 12°58'S 14°16'E
83 C15 Cubango var. Kuvango, Port. Vila Artur de Paiva, Vila da Ponte. Huíla, SW Angola 14°27'S 16°18'E
83 D16 Cubango var. Kavango, Kavengo, Kubango, Okavango, Okavanggo. ≈ S Africa see also Okavango
Cubango see Okavango
54 H8 Cubará Boyacá, N Colombia 07°01'N 72°07'W
Cuba, Republic of see Cuba
136 I12 Çubuk Ankara, N Turkey 40°13'N 33°02'E
83 D14 Cuchi Kuando Kubango, C Angola 14°40'S 16°58'E
42 C5 Cuchumatanes, Sierra de los ▲▲ W Guatemala
Cuculaya, Rio see Kukalaya, Rio
82 E12 Cucumbi prev. Trás-os-Montes. Lunda Sul, NE Angola 10°13'S 19°04'E
54 G7 Cúcuta var. San José de Cúcuta. Norte de Santander, N Colombia 07°55'N 72°31'W
31 N9 Cudahy Wisconsin, N USA 42°54'N 87°51'W
155 J21 Cuddalore Tamil Nādu, SE India 11°43'N 79°46'E
155 I18 Cuddapah Andhra Pradesh, S India 14°30'N 78°50'E
104 M6 Cuéllar Castilla y León, N Spain 41°24'N 04°19'W
82 D13 Cuemba var. Coemba. Bié, C Angola 12°09'S 18°07'E
56 B8 Cuenca Azuay, S Ecuador 02°54'S 79°W
105 Q9 Cuenca anc. Conca. Castilla-La Mancha, C Spain 40°04'N 02°07'W
105 P9 Cuenca ◈ province Castilla-La Mancha, C Spain
40 L9 Cuencamé var. Cuencamé de Ceniceros. Durango, C Mexico 24°53'N 103°41'W
Cuencamé de Ceniceros see Cuencamé
105 Q8 Cuenca, Serranía de ▲▲ C Spain
Cuera see Chur
105 P5 Cuerda del Pozo, Embalse de la ⊠ N Spain
41 O14 Cuernavaca Morelos, S Mexico 18°57'N 99°15'W
25 T12 Cuero Texas, SW USA 29°06'N 97°19'W
44 I7 Cueto Holguín, E Cuba 20°43'N 75°54'W
41 Q13 Cuetzalán var. Cuetzalán del Progreso. Puebla, S Mexico 20°00'N 97°27'W
Cuetzalán del Progreso see Cuetzalán
105 Q14 Cuevas de Almanzora Andalucía, S Spain 37°19'N 01°52'W
Cuevas de Vinromá see Les Coves de Vinromá
116 H12 Cugir Hung. Kudzsir. Alba, SW Romania 45°48'N 23°25'E
59 H18 Cuiabá prev. Cuyabá. state capital Mato Grosso, SW Brazil 15°32'S 56°05'W
59 H19 Cuiabá, Rio ≈ SW Brazil
41 R15 Cuicatlán var. San Juan Bautista Cuicatlán. Oaxaca, SE Mexico 17°49'N 96°59'W
191 W16 Cuidado, Punta headland Easter Island, Chile, E Pacific Ocean 27°08'S 109°18'W
Cúige see Connaught
Cúige Laighean see Leinster
Cúige Mumhan see Munster
Cuihua see Daguan
98 L13 Cuijck Noord-Brabant, SE Netherlands 51°41'N 05°56'E
Cúil an tSúdaire see Portarlington
42 D7 Cuilapa Santa Rosa, S Guatemala 14°16'N 90°18'W
42 B5 Cuilco, Río ≈ W Guatemala
Cúil Mhuine see Collooney
Cúil Raithin see Coleraine
82 C14 Cuima Huambo, C Angola 13°16'S 15°39'E
83 E16 Cuito var. Kwito. ≈ SE Angola
83 E15 Cuito Cuanavale Kuando Kubango, E Angola 15°01'S 19°07'E
41 N14 Cuitzeo, Lago de ◎ C Mexico
27 W4 Cuivre River ≈ Missouri, C USA
Çuka see Çukë
168 L8 Cukai var. Chukai, Kemaman. Terengganu, Peninsular Malaysia 04°15'N 103°25'E
113 L23 Çukë var. Çuka. Vlorë, S Albania 39°50'N 20°01'E
Cularo see Grenoble
33 Y7 Culbertson Montana, NW USA 48°09'N 104°30'W
28 M16 Culbertson Nebraska, C USA 40°13'N 100°50'W
183 P10 Culcairn New South Wales, SE Australia 35°41'S 147°01'E
45 W5 Culebra var. Dewey. E Puerto Rico 18°19'N 65°17'W
45 W6 Culebra, Isla de island E Puerto Rico
37 T8 Culebra Peak ▲ Colorado, C USA 37°07'N 105°11'W
104 J5 Culebra, Sierra de la ▲▲ NW Spain
98 J12 Culemborg Gelderland, C Netherlands 51°57'N 05°17'E
137 V14 Culfa Rus. Dzhul'fa. SW Azerbaijan 38°58'N 45°37'E
183 P4 Culgoa River ≈ New South Wales/Queensland, SE Australia
40 I9 Culiacán var. Culiacán Rosales, Culiacán-Rosales. Sinaloa, C Mexico 24°48'N 107°25'W
Culiacán-Rosales/Culiacán Rosales see Culiacán
105 P14 Cúllar-Baza Andalucía, S Spain 37°35'N 02°34'W
105 S10 Cullera Valenciana, E Spain 39°10'N 00°15'W
23 P3 Cullman Alabama, S USA 34°10'N 86°50'W
108 B10 Cully Vaud, W Switzerland 46°58'N 06°46'E
Culm see Chełmno
Culmsee see Chełmża
21 V4 Culpeper Virginia, NE USA 38°28'N 78°00'W
185 I17 Culverden Canterbury, South Island, New Zealand 42°46'S 172°51'E
83 H18 Cum var. Xhumo. Central, C Botswana 21°13'S 24°38'E
55 N5 Cumaná Sucre, NE Venezuela 10°29'N 64°12'W
55 O5 Cumanacoa Sucre, NE Venezuela 10°17'N 63°58'W
54 C13 Cumbal, Nevado de elevation S Colombia
21 O7 Cumberland Kentucky, S USA 36°55'N 83°00'W
21 U2 Cumberland Maryland, NE USA 39°40'N 78°47'W
21 V6 Cumberland Virginia, NE USA 37°31'N 78°16'W
187 P12 Cumberland, Cape var. Cape Nahoi. headland Espiritu Santo, N Vanuatu 14°39'S 166°35'E
11 V14 Cumberland House Saskatchewan, C Canada 53°57'N 102°21'W
23 W8 Cumberland Island island Georgia, SE USA
20 L7 Cumberland, Lake ⊠ Kentucky, S USA
9 R5 Cumberland Peninsula peninsula Baffin Island, Nunavut, NE Canada
2 N9 Cumberland Plateau plateau E USA
30 L1 Cumberland Point headland Michigan, N USA 47°51'N 89°14'W
21 O7 Cumberland River ≈ Kentucky/Tennessee, S USA
9 S6 Cumberland Sound inlet Baffin Island, Nunavut, NE Canada
96 I12 Cumbernauld S Scotland, United Kingdom 55°57'N 04°W
97 K15 Cumbria cultural region NW England, United Kingdom
97 K15 Cumbrian Mountains ▲▲ NW England, United Kingdom
23 S2 Cumming Georgia, SE USA 34°12'N 84°08'W
Cummin in Pommern see Kamień Pomorski
182 G9 Cummins South Australia 34°17'S 135°43'E
96 I13 Cumnock W Scotland, United Kingdom 55°32'N 04°28'W
40 G4 Cumpas Sonora, NW Mexico 30°N 109°48'W
136 H16 Çumra Konya, C Turkey 37°34'N 32°38'E
63 G15 Cunco Araucanía, C Chile 38°55'S 72°02'W
54 E9 Cundinamarca off. Departamento de Cundinamarca. ◈ province C Colombia
Cundinamarca, Departamento de see Cundinamarca
41 U15 Cunduacán Tabasco, SE Mexico 18°00'N 93°07'W
83 C16 Cunene ◈ province S Angola
83 A16 Cunene var. Kunene. ≈ Angola/Namibia see also Kunene
Cunene see Kunene
106 A9 Cuneo Fr. Coni. Piemonte, NW Italy 44°23'N 07°32'E
83 E15 Cunjamba Kuando Kubango, E Angola 15°22'S 20°07'E
181 V10 Cunnamulla Queensland, E Australia 28°09'S 145°44'E
Čunusavvon see Junosuando
Cuokkarašša see Čohkarášša
106 B7 Cuorgne Piemonte, NE Italy 45°23'N 07°34'E
96 K11 Cupar E Scotland, United Kingdom 56°19'N 03°01'W
116 L8 Cupcina Rus. Kupchino; prev. Calinisc, Kalinisk. N Moldova 48°07'N 27°22'E
54 C8 Cupica Chocó, W Colombia 06°43'N 77°31'W
54 C8 Cupica, Golfo de gulf W Colombia
112 N13 Ćuprija Serbia, E Serbia 43°57'N 21°21'E
Cura see Villa de Cura
45 S9 Curaçao prev. Dutch West Indies. ◈ Dutch self-governing territory S Caribbean Sea
Curaçao island Lesser Antilles
56 H13 Curanja, Río ≈ E Peru
56 F7 Curaray, Rio ≈ Ecuador/Peru
116 K14 Curcani Călăraşi, SE Romania 44°11'N 26°39'E
182 H4 Curdimurka South Australia 29°27'S 136°56'E
103 P7 Cure ≈ C France
173 Y16 Curepipe C Mauritius 20°19'S 57°31'E
55 R6 Curiapo Delta Amacuro, NE Venezuela 10°03'N 63°05'W
Curia Rhaetorum see Chur
62 G12 Curicó Maule, C Chile 35°00'S 71°15'W
Curieta see Krk
172 I15 Curieuse island Inner Islands, NE Seychelles
59 C16 Curitiba Acre, W Brazil 10°08'S 69°00'W
60 K12 Curitiba prev. Curytiba. state capital Paraná, S Brazil 25°25'S 49°25'W
60 J13 Curitibanos Santa Catarina, S Brazil 27°18'S 50°35'W
183 S6 Curlewis New South Wales, SE Australia 31°09'S 150°18'E
182 J6 Curnamona South Australia 31°39'S 139°35'E
83 A15 Curoca ≈ SW Angola
183 T6 Currabubula New South Wales, SE Australia 31°17'S 150°43'E
59 Q14 Currais Novos Rio Grande do Norte, E Brazil 06°12'S 36°30'W
35 W7 Currant Nevada, W USA 38°43'N 115°27'W
35 W6 Currant Mountain ▲ Nevada, W USA 38°56'N 115°19'W
44 H2 Current Eleuthera Island, C The Bahamas 25°24'N 76°44'W
27 W8 Current River ≈ Arkansas/Missouri, C USA
182 M14 Currie Tasmania, SE Australia 39°59'S 143°51'E
21 Y8 Currituck North Carolina, SE USA 36°29'N 76°02'W
21 Y8 Currituck Sound sound North Carolina, SE USA
39 R11 Curry Alaska, USA 62°36'N 150°00'W
Curtbunar see Tervel
116 I13 Curtea de Argeş var. Curtea-de-Arges. Argeş, S Romania 45°06'N 24°40'E
Curtea-de-Arges see Curtea de Argeş
116 E10 Curtici Ger. Kurtitsch, Hung. Kürtös. Arad, W Romania 46°21'N 21°17'E
104 H2 Curtis Galicia, NW Spain 43°09'N 08°10'W
28 M16 Curtis Nebraska, C USA 40°36'N 100°27'W
183 O14 Curtis Group island group Tasmania, SE Australia
181 Y8 Curtis Island island Queensland, SE Australia
58 K11 Curuá, Ilha do island NE Brazil
47 U7 Curuá, Rio ≈ N Brazil
59 A14 Curuçá, Rio ≈ NW Brazil
112 L9 Čurug Hung. Csurog. Vojvodina, N Serbia 45°30'N 20°02'E
61 D16 Curuzú Cuatiá Corrientes, NE Argentina 29°50'S 58°05'W
59 M19 Curvelo Minas Gerais, SE Brazil 18°45'S 44°27'W
18 E14 Curwensville Pennsylvania, NE USA 40°57'N 78°29'W
30 M3 Curwood, Mount ▲ Michigan, N USA 46°42'N 88°14'W
Curytiba see Curitiba
Curzola see Korčula
42 A10 Cuscatlán ◈ department C El Salvador
57 H15 Cusco var. Cuzco. Cusco, C Peru 13°35'S 72°02'W
57 H15 Cusco off. Departamento de Cusco, var. Cuzco. ◈ department C Peru
Cusco, Departamento de see Cusco
27 O9 Cushing Oklahoma, C USA 36°01'N 96°46'W
25 W8 Cushing Texas, SW USA 31°48'N 94°50'W
40 I6 Cusihuiriachic Chihuahua, N Mexico 28°16'N 106°46'W
103 P10 Cusset Allier, C France 46°08'N 03°27'E
23 S6 Cusseta Georgia, SE USA 32°18'N 84°46'W
28 J10 Custer South Dakota, N USA 43°46'N 103°36'W
Cüstrin see Kostrzyn
33 Q7 Cut Bank Montana, NW USA 48°38'N 112°20'W
Cutch, Gulf of see Kachchh, Gulf of
23 S6 Cuthbert Georgia, SE USA 31°46'N 84°47'W
11 S15 Cut Knife Saskatchewan, S Canada 52°40'N 108°54'W
23 Y16 Cutler Ridge Florida, SE USA 25°34'N 80°21'W
22 K10 Cut Off Louisiana, S USA 29°32'N 90°20'W
63 I15 Cutral-Có Neuquén, C Argentina 38°56'S 69°13'W
107 O21 Cutro Calabria, SW Italy 39°01'N 16°59'E
183 O4 Cuttaburra Channels seasonal river New South Wales, SE Australia
154 O12 Cuttack Odisha, E India 20°28'N 85°53'E
83 C15 Cuvelai Cunene, SW Angola 15°40'S 15°48'E
79 G18 Cuvette var. Région de la Cuvette. ◈ province C Congo
Cuvette, Région de la see Cuvette
173 V9 Cuvier Basin undersea feature E Indian Ocean
173 U9 Cuvier Plateau undersea feature E Indian Ocean
82 B12 Cuvo ≈ W Angola
100 H9 Cuxhaven Niedersachsen, NW Germany 53°51'N 08°43'E
Cuyabá see Cuiabá
Cuyuni, Río see Cuyuni River
55 S8 Cuyuni River var. Río Cuyuni. ≈ Guyana/Venezuela
Cuzco see Cusco
97 K22 Cwmbran Wel. Cwmbrân. SW Wales, United Kingdom 51°39'N 03°W
Cwmbrân see Cwmbran
28 K15 C. W. McConaughy, Lake ⊠ Nebraska, C USA
81 D20 Cyangugu SW Rwanda 02°27'S 29°00'E
110 D11 Cybinka Ger. Ziebingen. Lubuskie, W Poland 52°11'N 14°46'E
Cyclades see Kykládes
Cydonia see Chaniá
Cymru see Wales
20 M5 Cynthiana Kentucky, S USA 38°22'N 84°18'W
11 S17 Cypress Hills ▲▲ Alberta/Saskatchewan, SW Canada
Cypro-Syrian Basin see Cyprus Basin
121 U11 Cyprus off. Republic of Cyprus, Gk. Kypros, Turk. Kıbrıs, Kıbrıs Cumhuriyeti. ◆ republic E Mediterranean Sea
84 L14 Cyprus Gk. Kypros, Turk. Kıbrıs. island E Mediterranean Sea
121 W11 Cyprus Basin var. Cypro-Syrian Basin. undersea feature E Mediterranean Sea 34°00'N 34°00'E
Cyprus, Republic of see Cyprus
Cythera see Kýthira
Cythnos see Kýthnos
110 F9 Czaplinek Ger. Tempelburg. Zachodnio-pomorskie, NW Poland 53°33'N 16°14'E
Czarna Woda see Wda
110 G8 Czarne Pomorskie, N Poland 53°40'N 17°00'E
110 G10 Czarnków Wielkopolskie, C Poland 52°53'N 16°32'E
111 E17 Czech Republic Cz. Česká Republika. ◆ republic C Europe
Czegléd see Cegléd
110 G12 Czempiń Wielkopolskie, C Poland 52°10'N 16°46'E
Czenstochau see Częstochowa
Czerkow see Čerchov
Czernowitz see Chernivtsi
110 I8 Czersk Pomorskie, N Poland 53°48'N 17°58'E
111 J15 Częstochowa Ger. Czenstochau, Tschenstochau, Rus. Chenstokhov. Śląskie, S Poland 50°49'N 19°07'E
110 F10 Człopa Ger. Schloppe. Zachodnio-pomorskie, NW Poland 53°05'N 16°05'E
110 H8 Człuchów Ger. Schlochau. Pomorskie, NW Poland 53°41'N 17°20'E

D

163 V9 Da'an var. Dalai. Jilin, NE China 45°28'N 124°18'E
15 S10 Daaquam Québec, SE Canada 46°36'N 70°03'W
Daawo, Webi see Dawa Wenz
54 I4 Dabajuro Falcón, NW Venezuela 11°00'N 70°41'W
77 N15 Dabakala NE Ivory Coast 08°19'N 04°24'W
163 S11 Daban var. Bairin Youqi. Nei Mongol Zizhiqu, N China 43°33'N 118°40'E
111 K23 Dabas Pest, C Hungary 47°36'N 19°22'E
160 L8 Daba Shan ▲▲ C China
Dabba see Daocheng
140 J5 Dabbāgh, Jabal ▲ NW Saudi Arabia 27°52'N 35°48'E
54 D8 Dabeiba Antioquia, NW Colombia 07°01'N 76°18'W
154 E11 Dabhoi Gujarāt, W India 22°08'N 73°28'E
161 P8 Dabie Shan ▲▲ C China
76 J13 Dabola C Guinea 10°48'N 11°02'W
77 N17 Dabou S Ivory Coast 05°20'N 04°23'W
162 M15 Dabqig prev. Uxin Qi. Nei Mongol Zizhiqu, N China 38°29'N 108°48'E
110 P8 Dąbrowa Białostocka Podlaskie, NE Poland 53°38'N 23°18'E
111 M16 Dąbrowa Tarnowska Małopolskie, S Poland 50°10'N 21°E
119 M20 Dabryn' Rus. Dobryn'. Homyel'skaya Voblasts', SE Belarus 51°46'N 29°12'E
159 P10 Dabsan Hu ◎ C China
161 Q13 Dabu var. Huliao. Guangdong, S China 24°19'N 116°07'E
116 H15 Dăbuleni Dolj, SW Romania 43°48'N 24°05'E
152 G9 Dabwāli Haryāna, NW India 29°56'N 74°40'E
Dacca see Dhaka
101 L23 Dachau Bayern, SE Germany 48°15'N 11°26'E
Dachuan see Dazhou
Dacia Bank see Dacia Seamount

◆ Country
● Country Capital
◇ Dependent Territory
○ Dependent Territory Capital
◈ Administrative Regions
✈ International Airport
▲ Mountain
▲▲ Mountain Range
Volcano
≈ River
◎ Lake
⊠ Reservoir

64 M10 **Dacia Seamount** *var.* Dacia Bank. *undersea feature* E Atlantic Ocean *31°10′N 13°42′W*
37 T3 **Dacono** Colorado, C USA *40°04′N 104°56′W*
Đăc Tô *see* Đăk Tô
Dacura *see* Dákura
23 W12 **Dade City** Florida, SE USA *28°21′N 82°12′W*
152 L10 **Ḍaḍeldhurā** *var.* Dandeldhura. Far Western, W Nepal *29°12′N 80°31′E*
23 Q5 **Dadeville** Alabama, S USA *32°49′N 85°45′W*
Dadong *see* Donggang
103 N15 **Dadou** ~ S France
154 D12 **Dādra and Nagar Haveli** ⬖ *union territory* W India
149 P14 **Dādu** Sind, SE Pakistan *26°42′N 67°48′E*
167 U11 **Da Du Bơloc** Kon Tum, C Vietnam *14°06′N 107°40′E*
160 G9 **Dadu He** ~ C China
163 V15 **Daecheong-do** *prev.* Taechŏng-do. *island* NW South Korea
163 Y16 **Daegu** , *Jap.* Taikyū; *prev.* Taegu. SE South Korea *35°55′N 128°33′E*
163 Y15 **Daejeon** , *Jap.* Taiden; *prev.* Taejŏn. C South Korea *36°20′N 127°28′E*
Daerah Istimewa Aceh *see* Aceh
171 P4 **Daet** Luzon, N Philippines *14°06′N 122°57′E*
160 I11 **Dafang** Guizhou, S China *27°07′N 105°40′E*
Dafeng *see* Shanglin
153 W11 **Dafla Hills** ▲▲ NE India
11 U15 **Dafoe** Saskatchewan, S Canada *51°46′N 104°11′W*
76 G10 **Dagana** N Senegal *16°28′N 15°35′W*
Dagana *see* Massakory, Chad
Dagana *see* Dahana, Tajikistan
Dagcagoin *see* Zoigê
118 K11 **Dagda** SE Latvia *56°06′N 27°36′E*
Dagden *see* Hiiumaa
Dagden-Sund *see* Soela Väin
127 P16 **Dagestan, Respublika** *prev.* Dagestanskaya ASSR, *Eng.* Daghestan. ⬖ *autonomous republic* SW Russian Federation
Dagestanskaya ASSR *see* Dagestan, Respublika
127 R17 **Dagestanskiye Ogni** Respublika Dagestan, SW Russian Federation *42°09′N 48°08′E*
Dagezhen *see* Fengning
185 A23 **Dagg Sound** *sound* South Island, New Zealand
Daghestan *see* Dagestan, Respublika
141 Y8 **Daghmar** NE Oman *23°09′N 59°01′E*
Dağliq Quarabağ *see* Nagorno-Karabakh
Dagö *see* Hiiumaa
54 D11 **Dagua** Valle del Cauca, W Colombia *03°39′N 76°40′W*
160 H11 **Daguan** *var.* Cuihua. Yunnan, SW China *27°42′N 103°51′E*
171 N3 **Dagupan** *off.* Dagupan City. Luzon, N Philippines *16°05′N 120°21′E*
Dagupan City *see* Dagupan
159 N16 **Dagzê** *var.* Dêqên. Xizang Zizhiqu, W China *29°38′N 91°15′E*
147 Q13 **Dahana** *Rus.* Dagana, Dakhana. SW Tajikistan *38°03′N 69°51′E*
163 V10 **Dahei Shan** ▲▲ N China
163 T7 **Da Hinggan Ling** *Eng.* Great Khingan Range. ▲▲ NE China
Dahlac Archipelago *see* Dahlak Archipelago
80 K9 **Dahlak Archipelago** *var.* Dahlac Archipelago. *island group* E Eritrea
23 T2 **Dahlonega** Georgia, SE USA *34°31′N 83°59′W*
101 O14 **Dahme** Brandenburg, E Germany *52°10′N 13°47′E*
100 O13 **Dahme** ~ E Germany
141 O14 **Dahm, Ramlat** *desert* NW Yemen
154 E10 **Dāhod** *prev.* Dohad. Gujarāt, W India *22°48′N 74°18′E*
Dahomey *see* Benin
158 G10 **Dahongliutan** Xinjiang Uygur Zizhiqu, NW China *35°59′N 79°12′E*
Dahra *see* Dara
Dahuaishu *see* Hongtong
139 R2 **Dahūk** *var.* Dohuk, *Kurd.* Dihok, Dihōk. Dahūk, N Iraq *36°52′N 43°01′E*
139 R2 **Dahūk** *off.* Muḥāfa at Dahūk, *var.* Dohuk, *Kurd.* Dihok, *off. Kurd.* Parêzga-î Dihok. ⬖ *governorate* N Iraq
Dahūk, Muḥāfa at *see* Dahūk
116 J15 **Daia** Giurgiu, S Romania *44°00′N 25°59′E*
165 P12 **Daigo** Ibaraki, Honshū, S Japan *36°43′N 140°22′E*
163 O13 **Dai Hai** ◎ N China
Daihoku *see* Taibei
186 M8 **Dai Island** *island* N Solomon Islands
166 M8 **Daik-u** Bago, SW Myanmar (Burma) *17°46′N 96°40′E*
138 H9 **Dā'īl** Dar'ā, S Syria *32°45′N 36°08′E*
167 U12 **Đai Lanh** Khanh Hoa, S Vietnam *12°49′N 109°20′E*
161 Q13 **Daimao Shan** ▲▲ SE China
105 N11 **Daimiel** Castilla-La Mancha, C Spain *39°04′N 03°37′W*
115 F22 **Daimoniá** Pelopónnisos, S Greece *36°38′N 22°54′E*
Dainan *see* Tainan
25 W6 **Daingerfield** Texas, SW USA *33°03′N 94°42′W*
Daingin, Bá an *see* Dingle Bay
159 R13 **Dainkognubma** Xizang Zizhiqu, W China *32°26′N 97°58′E*
164 K14 **Daiō-zaki** *headland* Honshū, SW Japan *34°15′N 136°50′E*
Dairbhre *see* Valencia Island
61 B22 **Daireaux** Buenos Aires, E Argentina *36°34′S 61°40′W*
Dairen *see* Dalian
Dairûṭ *see* Dayrūṭ
25 X10 **Daisetta** Texas, SW USA *30°06′N 94°38′W*
192 G5 **Daitō-jima** *island group* SW Japan
192 G5 **Daitō Ridge** *undersea feature* N Philippine Sea *25°30′N 133°00′E*
161 N3 **Daixian** *var.* Dai Xian, Shangguan. Shanxi, C China *39°10′N 112°57′E*
Dai Xian *see* Daixian
Daiyue *see* Shanyin
161 Q12 **Daiyun Shan** ▲▲ SE China
44 M8 **Dajabón** NW Dominican Republic *19°35′N 71°41′W*
160 G8 **Dajin Chuan** ~ C China
148 J6 **Dak** ◎ W Afghanistan
76 F11 **Dakar** ● (Senegal) W Senegal *14°44′N 17°27′W*
76 F11 **Dakar** ✈ W Senegal *14°42′N 17°27′W*
Đak Glây *see* Đak Glei
167 U10 **Đak Glei** *prev.* Đak Glây. Kon Tum, C Vietnam *15°05′N 107°42′E*
Dakhana *see* Dahana
153 U16 **Dakhin Shahbazpur Island** *island* S Bangladesh
Dakhla *see* Ad Dakhla
76 F7 **Dakhlet Nouâdhibou** ⬖ *region* NW Mauritania
Đak Lap *see* Kiên Đưc
Đak Nông *see* Gia Nghia
77 U11 **Dakoro** Maradi, S Niger *14°29′N 06°45′E*
29 U12 **Dakota City** Iowa, C USA *42°42′N 94°13′W*
29 R13 **Dakota City** Nebraska, C USA *42°25′N 96°25′W*
Đakovica *see* Gjakovë
112 I10 **Đakovo** *var.* Djakovo, *Hung.* Diakovár. Osijek-Baranja, E Croatia *45°18′N 18°24′E*
Dakshin *see* Deccan
167 U11 **Đăk Tô** *var.* Đăc Tô. Kon Tum, C Vietnam *14°35′N 107°55′E*
43 N7 **Dákura** *var.* Dacura. Región Autónoma Atlántico Norte, NE Nicaragua *14°22′N 83°13′W*
95 I14 **Dal** Akershus, S Norway *60°19′N 11°16′E*
82 E12 **Dala** Lunda Sul, E Angola *11°04′S 20°15′E*
108 J8 **Dalaas** Vorarlberg, W Austria *47°08′N 10°03′E*
76 I13 **Dalaba** W Guinea *10°47′N 12°12′W*
Dalai *see* Da'an
162 I12 **Dalain Hob** *var.* Ejin Qi. Nei Mongol Zizhiqu, N China *41°59′N 101°04′E*
Dalai Nor *see* Hulun Nur
163 Q11 **Dalai Nur** *salt lake* N China
Dala-Jarna *see* Järna
95 M14 **Dalälven** ~ C Sweden
136 C16 **Dalaman** Muğla, SW Turkey *36°47′N 28°47′E*
136 C16 **Dalaman** ✈ Muğla, SW Turkey *36°37′N 28°51′E*
136 D16 **Dalaman Çayı** ~ SW Turkey
162 K11 **Dalandzadgad** Ömnögovĭ, S Mongolia *43°35′N 104°23′E*
95 D17 **Dalane** *physical region* S Norway
189 Z2 **Dalap-Uliga-Djarrit** *var.* Delap-Uliga-Darrit, D-U-D. *island group* Ratak Chain, SE Marshall Islands
94 J12 **Dalarna** *prev.* Kopparberg. ⬖ *county* C Sweden
94 L13 **Dalarna** *prev. Eng.* Dalecarlia. *cultural region* C Sweden
95 P16 **Dalarö** Stockholm, C Sweden *59°07′N 18°25′E*
167 U13 **Đa Lat** Lâm Đông, S Vietnam *11°56′N 108°25′E*
Dalay *see* Bayandalay
148 L12 **Dālbandīn** *var.* Dāl Bandin. Baluchistān, SW Pakistan *28°48′N 64°08′E*
95 J17 **Dalbosjön** *lake bay* S Sweden
181 Y10 **Dalby** Queensland, E Australia *27°11′S 151°12′E*
94 D13 **Dale** Hordaland, S Norway *60°35′N 05°48′E*
94 C12 **Dale** Sogn Og Fjordane, S Norway *61°22′N 05°24′E*
32 K12 **Dale** Oregon, NW USA *44°58′N 118°56′W*
25 T11 **Dale** Texas, SW USA *29°56′N 97°34′W*
Dalecarlia *see* Dalarna
21 W4 **Dale City** Virginia, NE USA *38°38′N 77°18′W*
20 L8 **Dale Hollow Lake** ⊠ Kentucky/Tennessee, S USA
98 O8 **Dalen** Drenthe, NE Netherlands *52°42′N 06°45′E*
95 E15 **Dalen** Telemark, S Norway *59°25′N 07°58′E*
166 K14 **Daletme** Chin State, W Myanmar (Burma) *21°44′N 92°48′E*
23 Q7 **Daleville** Alabama, S USA *31°18′N 85°42′W*
98 M9 **Dalfsen** Overijssel, E Netherlands *52°31′N 06°16′E*
114 M8 **Dalgopol** *var.* Dŭlgopol. Varna, E Bulgaria *43°05′N 27°24′E*
24 M1 **Dalhart** Texas, SW USA *36°05′N 102°31′W*
13 O13 **Dalhousie** New Brunswick, SE Canada *48°03′N 66°22′W*
152 I6 **Dalhousie** Himāchal Pradesh, N India *32°32′N 76°01′E*
160 F12 **Dali** *var.* Xiaguan. Yunnan, SW China *25°34′N 100°11′E*
Dali *see* Idálion
163 U14 **Dalian** *var.* Dairen, Dalien, Jay Dairen, Lüda, Ta-lien, *Rus.* Dalny. Liaoning, NE China *38°53′N 121°37′E*
105 O15 **Dalías** Andalucía, S Spain *36°49′N 02°50′W*
Dalien *see* Dalian
Dalijan *see* Delījān
112 J9 **Dalj** *Hung.* Dalja. Osijek-Baranja, E Croatia *45°29′N 19°00′E*
Dalja *see* Dalj
32 F12 **Dallas** Oregon, NW USA *44°56′N 123°20′W*
25 U6 **Dallas** Texas, SW USA *32°47′N 96°48′W*
25 T7 **Dallas-Fort Worth** ✈ Texas, SW USA *32°37′N 97°16′W*
154 K12 **Dalli Rājhara** *var.* Dhalli Rajhara. Chhattīsgarh, C India *20°32′N 81°10′E*
39 X15 **Dall Island** *island* Alexander Archipelago, Alaska, USA
38 M12 **Dall Lake** ◎ Alaska, USA
Dállogilli *see* Korpilombolo
77 S12 **Dallol Bosso** *seasonal river* W Niger
141 U7 **Dalmā** *island* W United Arab Emirates
113 E14 **Dalmacija** *Eng.* Dalmatia, *Ger.* Dalmatien, *It.* Dalmazia. *cultural region* S Croatia
Dalmatia/Dalmatien/Dalmazia *see* Dalmacija
123 S15 **Dal'negorsk** Primorskiy Kray, SE Russian Federation *44°27′N 135°30′E*
Dalny *see* Dalian
76 M16 **Daloa** C Ivory Coast *06°56′N 06°28′W*
160 J11 **Dalou Shan** ▲▲ S China
181 X7 **Dalrymple Lake** ◎ Queensland, E Australia
14 H14 **Dalrymple Lake** ◎ Ontario, S Canada
181 X7 **Dalrymple, Mount** ▲ Queensland, E Australia *21°01′S 148°34′E*
93 K20 **Dalsbruk** *Fin.* Taalintehdas. Varsinais-Suomi, SW Finland *60°02′N 22°31′E*
95 K19 **Dalsjöfors** Västra Götaland, S Sweden *57°43′N 13°05′E*
95 J17 **Dals Långed** *var.* Långed. Västra Götaland, S Sweden *58°54′N 12°20′E*
153 O15 **Dāltenganj** *prev.* Daltonganj. Jhārkhand, N India *24°02′N 84°07′E*
23 R2 **Dalton** Georgia, SE USA *34°46′N 84°58′W*
Daltonganj *see* Dāltenganj
195 X14 **Dalton Iceberg Tongue** *ice feature* Antarctica
92 J1 **Dalvík** Norðurland Eystra, N Iceland *65°58′N 18°31′W*
Dálvvadis *see* Jokkmokk
35 N8 **Daly City** California, W USA *37°44′N 122°27′W*
181 P2 **Daly River** ~ Northern Territory, N Australia
181 Q3 **Daly Waters** Northern Territory, N Australia *16°21′S 133°22′E*
119 F20 **Damachava** *var.* Damachova, *Pol.* Domaczewo, *Rus.* Domachëvo. Brestskaya Voblasts', SW Belarus *51°45′N 23°36′E*
Damachova *see* Damachava
77 W11 **Damagaram Takaya** Zinder, S Niger *14°02′N 09°28′E*
154 D12 **Damān** Damān and Diu, W India *20°25′N 72°58′E*
154 B12 **Damān and Diu** ⬖ *union territory* W India
75 V7 **Damanhûr** *anc.* Hermopolis Parva. N Egypt *31°03′N 30°28′E*
161 O1 **Damaqun Shan** ▲▲ E China
79 I15 **Damara** Ombella-Mpoko, S Central African Republic *05°00′N 18°45′E*
83 D18 **Damaraland** *physical region* C Namibia
171 S15 **Damar, Kepulauan** *var.* Baraf Daja Islands, Kepulauan Barat Daya. *island group* C Indonesia
168 J8 **Damar Laut** Perak, Peninsular Malaysia *04°13′N 100°36′E*
171 S15 **Damar, Pulau** *island* Maluku, E Indonesia
Damas *see* Dimashq
77 Y12 **Damasak** Borno, NE Nigeria *13°10′N 12°40′E*
Damasco *see* Dimashq
21 Q8 **Damascus** Virginia, NE USA *36°37′N 81°46′W*
Damascus *see* Dimashq
77 X13 **Damaturu** Yobe, NE Nigeria *11°44′N 11°58′E*
171 R9 **Damau** Pulau Kaburuang, N Indonesia *03°46′N 126°49′E*
143 O5 **Damāvand, Qolleh-ye** ▲ N Iran *35°56′N 52°08′E*
82 B10 **Damba** Uíge, NW Angola *06°44′S 15°20′E*
114 M12 **Dambaslar** Tekirdağ, NW Turkey *41°13′N 27°13′E*
116 J13 **Dâmbovița** *prev.* Dîmbovița. ⬖ *county* SE Romania
116 J13 **Dâmbovița** *prev.* Dîmbovița. ~ S Romania
173 Y15 **D'Ambre, Île** *island* NE Mauritius
155 K24 **Dambulla** Central Province, C Sri Lanka *07°51′N 80°40′E*
44 K9 **Dame-Marie** SW Haiti *18°36′N 74°26′W*
44 J9 **Dame Marie, Cap** *headland* SW Haiti *18°37′N 74°24′W*
143 Q4 **Dāmghān** Semnān, N Iran *36°13′N 54°22′E*
Damietta *see* Dumyāt
138 G10 **Dāmiyā** Al Balqā', NW Jordan *32°07′N 35°33′E*
146 G11 **Damla** Daşoguz Welaýaty, N Turkmenistan *40°05′N 59°15′E*
Dammām *see* Ad Dammān
100 G12 **Damme** Niedersachsen, NW Germany *52°31′N 08°12′E*
153 R15 **Dāmodar** ~ NE India
154 J9 **Damoh** Madhya Pradesh, C India *23°50′N 79°30′N*
77 P15 **Damongo** NW Ghana *09°05′N 01°49′W*
138 G7 **Damoûr** *var.* Ad Dāmūr. W Lebanon *33°36′N 35°30′E*
171 N11 **Dampal, Teluk** *bay* Sulawesi, C Indonesia
180 H7 **Dampier** Western Australia *20°40′S 116°40′E*
180 H6 **Dampier Archipelago** *island group* Western Australia
141 U14 **Damqawt** *var.* Damqut. E Yemen *16°35′N 52°39′E*
159 O13 **Dam Qu** ~ C China
Damqut *see* Damqawt
167 R13 **Dâmrei, Chuŏr Phmun** *Fr.* Chaîne de l'Éléphant. ▲▲ SW Cambodia
108 C7 **Damvant** Jura, NW Switzerland *47°22′N 06°55′E*
Damwâld *see* Damwoude
98 L5 **Damwoude** *Fris.* Damwâld. Fryslân, N Netherlands *53°18′N 05°59′E*
159 N15 **Damxung** *var.* Gongtang. Xizang Zizhiqu, W China *30°29′N 91°02′E*
80 K11 **Danakil Desert** *var.* Afar Depression, Danakil Plain. *desert* E Africa
Danakil Plain *see* Danakil Desert
35 R8 **Dana, Mount** ▲ California, W USA *37°54′N 119°13′W*
76 L16 **Danané** W Ivory Coast *07°16′N 08°09′W*
167 U10 **Đa Năng** *prev.* Tourane. Quang Nam-Đa Năng, C Vietnam *16°04′N 108°14′E*
160 G9 **Danba** *var.* Zhanggu, *Tib.* Rongzhag. Sichuan, C China *30°54′N 101°49′E*
Danborg *see* Daneborg
18 L13 **Danbury** Connecticut, NE USA *41°21′N 73°27′W*
25 W12 **Danbury** Texas, SW USA *29°13′N 95°20′W*
35 X15 **Danby Lake** ◎ California, W USA
194 H4 **Danco Coast** *physical region* Antarctica
82 B11 **Dande** ~ NW Angola
Dandeldhura *see* Ḍaḍeldhurā
155 E17 **Dandeli** Karnātaka, W India *15°18′N 74°42′E*
183 O12 **Dandenong** Victoria, SE Australia *38°01′S 145°13′E*
163 V13 **Dandong** *var.* Tan-tung; *prev.* An-tung. Liaoning, NE China *40°10′N 124°23′E*
197 Q14 **Daneborg** *var.* Danborg. ⬖ N Greenland
25 V12 **Danevang** Texas, SW USA *29°03′N 96°11′W*
Dänew *see* Galkynyş
Danfeng *see* Shizong
14 L12 **Danford Lake** Québec, SE Canada *45°55′N 76°12′W*
19 T4 **Danforth** Maine, NE USA *45°39′N 67°54′W*
37 P3 **Danforth Hills** ▲▲ Colorado, C USA
Dangara *see* Danghara
159 V12 **Dangchang** Gansu, C China *34°01′N 104°19′E*
159 P8 **Dangchengwan** *var.* Subei, Subei Mongolzu Zizhixian. Gansu, N China *39°33′N 94°50′E*
82 B10 **Dange** Uíge, NW Angola *07°55′S 15°01′E*
Dangerous Archipelago *see* Tuamotu, Îles
83 E26 **Danger Point** *headland* SW South Africa *34°37′S 19°20′E*
147 Q13 **Danghara** *Rus.* Dangara. SW Tajikistan *38°05′N 69°14′E*
159 P8 **Danghe Nanshan** ▲▲ W China
80 I12 **Dangila** *var.* Dānglā. Āmara, NW Ethiopia *11°08′N 36°51′E*
159 P8 **Dangjin Shankou** *pass* N China
Dangla *see* Tanggula Shan, China
Dang La *see* Tanggula Shankou, China
Dānglā *see* Dangila, Ethiopia
Dangme Chu *see* Manās
153 Y11 **Dāngori** Assam, NE India *27°40′N 95°35′E*
Dang Raek, Phanom/Dangrek, Chaîne des *see* Dângrêk, Chuŏr Phnum
167 S11 **Dângrêk, Chuŏr Phnum** *var.* Phanom Dang Raek, Phanom Dong Rak, *Fr.* Chaîne des Dangrek. ▲▲ Cambodia/Thailand
42 G3 **Dangriga** *prev.* Stann Creek. Stann Creek, E Belize *16°59′N 88°13′W*
161 P6 **Dangshan** Anhui, E China *34°22′N 116°21′E*
33 T15 **Daniel** Wyoming, C USA *42°49′N 110°04′W*
83 H22 **Daniëlskuil** Northern Cape, N South Africa *28°11′S 23°33′E*
19 N12 **Danielson** Connecticut, NE USA *41°48′N 71°53′W*
124 M15 **Danilov** Yaroslavskaya Oblast', W Russian Federation *58°11′N 40°11′E*
127 O9 **Danilovka** Volgogradskaya Oblast', SW Russian Federation *50°21′N 44°03′E*
Danish West Indies *see* Virgin Islands (US)
160 L7 **Dan Jiang** ~ C China
160 M7 **Danjiangkou Shuiku** ⊠ C China
141 W8 **Ḍank** *var.* Dhank. NW Oman *23°34′N 56°16′E*
152 J7 **Dankhar** Himāchal Pradesh, N India *32°08′N 78°12′E*
126 L6 **Dankov** Lipetskaya Oblast', W Russian Federation *53°17′N 39°07′E*
42 J7 **Danlí** El Paraíso, S Honduras *14°02′N 86°34′W*
Danmark *see* Denmark
Danmarksstraedet *see* Denmark Strait
95 O14 **Dannemora** Uppsala, C Sweden *60°13′N 17°49′E*
18 L6 **Dannemora** New York, NE USA *44°42′N 73°42′W*
100 K11 **Dannenberg** Niedersachsen, N Germany *53°05′N 11°06′E*
184 N12 **Dannevirke** Manawatu-Wanganui, North Island, New Zealand *40°14′S 176°05′E*
21 U8 **Dan River** ~ Virginia, NE USA
167 P8 **Dan Sai** Loei, C Thailand *17°15′N 101°04′E*
18 F10 **Dansville** New York, NE USA *42°34′N 77°40′W*
Dantzig *see* Gdańsk
86 E12 **Danube** *Bul.* Dunav, *Cz.* Dunaj, *Ger.* Donau, *Hung.* Duna, *Rom.* Dunărea. ~ C Europe
Danubian Plain *see* Dunavska Ravnina
166 L8 **Danubyu** Ayeyawady, SW Myanmar (Burma) *17°15′N 95°35′E*
Danum *see* Doncaster
19 P11 **Danvers** Massachusetts, NE USA *42°34′N 70°54′W*
27 T11 **Danville** Arkansas, C USA *35°03′N 93°22′W*
31 N13 **Danville** Illinois, N USA *40°10′N 87°37′W*
31 O14 **Danville** Indiana, N USA *39°45′N 86°31′W*
29 Y15 **Danville** Iowa, C USA *40°52′N 91°18′W*
20 M6 **Danville** Kentucky, S USA *37°40′N 84°49′W*
18 G14 **Danville** Pennsylvania, NE USA *40°57′N 76°36′W*
21 T6 **Danville** Virginia, NE USA *36°34′N 79°25′W*
Danxian/Dan Xian *see* Danzhou
160 L17 **Danzhou** *prev.* Danxian, Dan Xian, Nada. Hainan, S China *19°31′N 109°31′E*
Danzhou *see* Yichuan
Danzig *see* Gdańsk
Danziger Bucht *see* Danzig, Gulf of
110 J6 **Danzig, Gulf of** *var.* Gulf of Gdańsk, *Ger.* Danziger Bucht, *Pol.* Zakota Gdańska, *Rus.* Gdan'skaya Bukhta. *gulf* N Poland
160 F10 **Daocheng** *var.* Jinzhu, *Tib.* Dabba. Sichuan, C China *29°05′N 100°14′E*
Daojiang *see* Daoxian
Daokou *see* Huaxian
104 H7 **Dão, Rio** ~ N Portugal
Daosa *see* Dausa
77 Y7 **Dao Timmi** Agadez, NE Niger *20°31′N 13°34′E*
160 M13 **Daoxian** *var.* Daojiang. Hunan, S China *25°30′N 111°37′E*
77 Q14 **Dapaong** N Togo *10°52′N 00°12′E*
23 N8 **Daphne** Alabama, S USA *30°36′N 87°54′W*
171 P7 **Dapitan** Mindanao, S Philippines *08°39′N 123°26′E*
159 P9 **Da Qaidam** Qinghai, C China *37°50′N 95°18′E*
163 V8 **Daqing** *var.* Sartu. Heilongjiang, NE China *46°35′N 125°00′E*
163 O13 **Daqing Shan** ▲▲ N China
163 T11 **Daqin Tal** *var.* Naiman Qi. Nei Mongol Zizhiqu, N China *42°51′N 120°41′E*
Daqm *see* Duqm
160 G8 **Da Qu** *var.* Do Qu. ~ C China
139 T5 **Dāqūq** *var.* Tāwūq. Kirkūk, N Iraq *35°08′N 44°27′E*
76 G10 **Dara** *var.* Dahra. NW Senegal *15°20′N 15°28′W*
138 H9 **Dar'ā** *var.* Der'a, *Fr.* Déraa. Dar'ā, SW Syria *32°37′N 36°06′E*
138 H8 **Dar'ā** *off.* Muḥāfaẓat Dar'ā, *var.* Dará, Der'a, Derrá. ⬖ *governorate* S Syria
143 Q12 **Dārāb** Fārs, S Iran *28°52′N 54°25′E*
116 K8 **Darabani** Botoșani, NW Romania *48°10′N 26°39′E*
Daraj *see* Dirj
Dar'ā, Muḥāfaẓat *see* Dar'ā
142 M8 **Dārān** Eşfahān, W Iran *33°00′N 50°27′E*
167 U12 **Da Răng, Sông** *var.* Ba. ~ S Vietnam
Daraut-Kurgan *see* Daroot-Korgon
77 W13 **Darazo** Bauchi, E Nigeria *11°01′N 10°27′E*
139 S3 **Darband** Arbīl, N Iraq *36°15′N 44°17′E*
139 V4 **Darband-i Khān, Sadd** *dam* NE Iraq
139 N1 **Darbāsīyah** *var.* Derbisîye. Al Ḥasakah, N Syria *37°06′N 40°42′E*
118 C11 **Darbėnai** Klaipėda, NW Lithuania *56°02′N 21°16′E*
153 Q13 **Darbhanga** Bihār, N India *26°10′N 85°54′E*
38 M9 **Darby, Cape** *headland* Alaska, USA *64°19′N 162°46′W*
112 I9 **Darda** *Hung.* Dárda. Osijek-Baranja, E Croatia *45°37′N 18°41′E*
Dárda *see* Darda
Dardanë *see* Kamenicë
27 T11 **Dardanelle** Arkansas, C USA *35°11′N 93°09′W*
27 S11 **Dardanelle, Lake** ⊠ Arkansas, C USA
Dardanelles *see* Çanakkale Boğazı
Dardanelli *see* Çanakkale
Dardo *see* Kangding
Dar-el-Beida *see* Casablanca
136 M14 **Darende** Malatya, C Turkey *38°34′N 37°29′E*
81 J22 **Dar es Salaam** Dar es Salaam, E Tanzania *06°51′S 39°18′E*
81 J23 **Dar es Salaam** *off.* Mkoa wa Dar es Salaam. ⬖ *region* E Tanzania
81 J23 **Dar es Salaam** *off.* Mkoa wa Dar es Salaam. ⬖ *region* E Tanzania
81 J22 **Dar es Salaam** ✈ Dar es Salaam, E Tanzania *06°57′S 39°17′E*
Dar es Salaam, Mkoa wa *see* Dar es Salaam
Dar es Salaam, Mkoa wa *see* Dar es Salaam
185 H18 **Darfield** Canterbury, South Island, New Zealand *43°29′S 172°07′E*
106 F7 **Darfo** Lombardia, N Italy *45°54′N 10°12′E*
80 B10 **Darfur** *var.* Darfur Massif. *cultural region* W Sudan
Darfur Massif *see* Darfur
Darganata/Dargan-Ata *see* Birata
143 T3 **Dargaz** *var.* Darreh Gaz; *prev.* Moḥammadābād. Khorāsān-e Razavī, NE Iran *37°28′N 59°08′E*
139 U4 **Dargazayn** As Sulaymānīyah, NE Iraq *35°39′N 45°00′E*
183 P12 **Dargo** Victoria, SE Australia *37°29′S 147°15′E*
162 L6 **Darhan** Darhan Uul, N Mongolia *49°24′N 105°57′E*
163 N8 **Darhan** Hentiy, C Mongolia *46°38′N 109°25′E*
Darhan *see* Büreghangay
Darhan Muminggan Lianheqi *see* Bailingmiao
162 L6 **Darhan Uul** ⬖ *province* N Mongolia
23 W7 **Darien** Georgia, SE USA *31°22′N 81°25′W*
43 W16 **Darién** *off.* Provincia del Darién. ⬖ *province* SE Panama
Darién, Golfo del *see* Darien, Gulf of
43 X14 **Darien, Gulf of** *Sp.* Golfo del Darién. *gulf* S Caribbean Sea
Darien, Isthmus of *see* Panama, Istmo de
Darién, Provincia del *see* Darién
42 K9 **Dariense, Cordillera** ▲▲ C Nicaragua
43 W15 **Darién, Serranía del** ▲▲ Colombia/Panama
163 P10 **Darīganga** *var.* Ovoot. Sühbaatar, SE Mongolia *45°08′N 113°51′E*
Dario *see* Ciudad Darío
Dariorigum *see* Vannes
Dariv *see* Darvi
Darj *see* Dirj
Darjeeling *see* Dārjiling
153 S12 **Dārjiling** *prev.* Darjeeling. West Bengal, NE India *27°00′N 88°13′E*
Darkehnen *see* Ozersk
159 S12 **Darlag** *var.* Gümai. Qinghai, C China *33°43′N 99°42′E*
183 T3 **Darling Downs** *hill range* Queensland, E Australia
28 M2 **Darling, Lake** ⊠ North Dakota, N USA
180 I12 **Darling Range** ▲▲ Western Australia
182 L8 **Darling River** ~ New South Wales, SE Australia
97 M15 **Darlington** N England, United Kingdom *54°31′N 01°34′W*
21 T12 **Darlington** South Carolina, SE USA *34°19′N 79°53′W*
30 K9 **Darlington** Wisconsin, N USA *42°41′N 90°08′W*
110 G7 **Darłowo** Zachodnio-pomorskie, NW Poland *54°24′N 16°21′E*
101 G19 **Darmstadt** Hessen, SW Germany *49°52′N 08°39′E*
75 S7 **Darnah** *var.* Dérna. NE Libya *32°46′N 22°39′E*
103 S6 **Darney** Vosges, NE France *48°06′N 05°58′E*
182 M7 **Darnick** New South Wales, SE Australia *32°52′S 143°38′E*
195 Y6 **Darnley, Cape** *cape* Antarctica
105 R7 **Daroca** Aragón, NE Spain *41°07′N 01°25′W*
147 S11 **Daroot-Korgon** *var.* Daraut-Kurgan. Oshskaya Oblast', SW Kyrgyzstan *39°35′N 72°13′E*
61 A23 **Darragueira** *var.* Darregueira. Buenos Aires, E Argentina *37°40′S 63°12′W*
Darregueira *see* Darragueira
Darreh Gaz *see* Dargaz
142 K7 **Darreh Shahr** *var.* Darreh-ye Shahr. Īlām, W Iran *33°10′N 47°18′E*
Darreh-ye Shahr *see* Darreh Shahr
32 I7 **Darrington** Washington, NW USA *48°15′N 121°36′W*
25 P1 **Darrouzett** Texas, SW USA *36°27′N 100°19′W*
153 S15 **Darsana** *var.* Darshana. Khulna, N Bangladesh *23°32′N 88°49′E*
Darshana *see* Darsana
100 M7 **Darss** *peninsula* NE Germany
100 M7 **Darsser Ort** *headland* NE Germany *54°28′N 12°31′E*
97 J24 **Dart** ~ SW England, United Kingdom
Dartang *see* Baqên
97 P22 **Dartford** SE England, United Kingdom *51°27′N 00°13′E*
182 L12 **Dartmoor** Victoria, SE Australia *37°56′S 141°18′E*
97 I24 **Dartmoor** *moorland* SW England, United Kingdom
13 Q15 **Dartmouth** Nova Scotia, SE Canada *44°40′N 63°35′W*
97 J24 **Dartmouth** SW England, United Kingdom *50°21′N 03°34′W*
15 Y6 **Dartmouth** ~ Québec, SE Canada
183 Q11 **Dartmouth Reservoir** ◎ Victoria, SE Australia
Dartuch, Cabo *see* Artrutx, Cap d'
186 C9 **Daru** Western, SW Papua New Guinea *09°05′S 143°10′E*
112 G9 **Daruvar** *Hung.* Daruvár. Bjelovar-Bilogora, NE Croatia *45°34′N 17°12′E*
Daruvár *see* Daruvar
Darvaza *see* Derweze, Turkmenistan
Darvaza *see* Darvoza, Uzbekistan
Darvazskiy Khrebet *see* Darvoz, Qatorkŭhi
Darvel Bay *see* Lahad Datu, Teluk
Darvel, Teluk *see* Lahad Datu, Teluk
162 F8 **Darvi** *var.* Dariv. Govĭ-Altay, W Mongolia *46°20′N 94°11′E*
162 F7 **Darvi** *var.* Bulgan. Hovd, W Mongolia *46°57′N 93°40′E*
Darvīshān *see* Darwēshān
147 O10 **Darvoza** *Rus.* Darvaza. Jizzax Viloyati, C Uzbekistan *40°59′N 67°19′E*
147 R13 **Darvoz, Qatorkŭhi** *Rus.* Darvazskiy Khrebet. ▲▲ C Tajikistan
148 L9 **Darwēshān** , *var.* Garmser; *prev.* Darvīshān. Helmand, S Afghanistan *31°02′N 64°12′E*
63 J15 **Darwin** Río Negro, S Argentina *39°13′S 65°41′W*
181 O1 **Darwin** *prev.* Palmerston, Port Darwin. *territory capital* Northern Territory, N Australia *12°28′S 130°52′E*
65 D24 **Darwin** *var.* Darwin Settlement. East Falkland, Falkland Islands *51°51′S 58°55′W*
62 H8 **Darwin, Cordillera** ▲▲ N Chile
Darwin Settlement *see* Darwin
57 B17 **Darwin, Volcán** ⛰ Galapagos Islands, Ecuador, E Pacific Ocean *0°12′S 91°17′W*
149 S8 **Darya Khān** Punjab, E Pakistan *31°47′N 71°10′E*
145 O15 **Dar'yalyktakyr, Ravnina** *plain* S Kazakhstan
143 T11 **Dārzīn** Kermān, S Iran *29°11′N 58°09′E*
Dashhowuz *see* Daşoguz
Dashhowuz Welayaty *see* Daşoguz Welaýaty
162 K7 **Dashinchilen** *var.* Süüj. Bulgan, C Mongolia *47°49′N 104°06′E*
119 O16 **Dashkawka** *Rus.* Dashkovka. Mahilyowskaya Voblasts', E Belarus *53°44′N 30°16′E*
Dashkhovuz *see* Daşoguz Welaýaty
Dashkhovuzskiy Velayat *see* Daşoguz Welaýaty
Dashköpri *see* Daşköpri
Dashkovka *see* Dashkawka
148 J15 **Dasht** ~ SW Pakistan
Dasht-i *see* Bābūs, Dasht-e
Dashtidzhum *see* Dashtijum
147 R13 **Dashtijum** *Rus.* Dashtidzhum. SW Tajikistan *38°06′N 70°11′E*
149 W7 **Daska** Punjab, NE Pakistan *32°15′N 74°23′E*
146 J16 **Daşköpri** *var.* Dashköpri, *Rus.* Tashkepri. Mary Welaýaty, S Turkmenistan *36°15′N 62°37′E*
146 H8 **Daşoguz** *Rus.* Dashkhovuz, *Turkm.* Dashhowuz; *prev.* Tashauz. Daşoguz Welaýaty, N Turkmenistan *41°51′N 59°53′E*
146 E9 **Daşoguz Welaýaty** *var.* Dashhowuz Welayaty, *Rus.* Dashkhovuz, Dashkhovuzskiy Velayat. ⬖ *province* N Turkmenistan
Đa, Sông *see* Black River
77 R15 **Dassa** *var.* Dassa-Zoumé. S Benin *07°46′N 02°15′E*
Dassa-Zoumé *see* Dassa
29 U8 **Dassel** Minnesota, N USA *45°06′N 94°18′W*
152 H3 **Dastegil Sar** ▲ N India
136 C16 **Datça** Muğla, SW Turkey *36°46′N 27°40′E*
165 R4 **Date** Hokkaidō, NE Japan *42°28′N 140°51′E*
154 I8 **Datia** *prev.* Duttia. Madhya Pradesh, C India *25°41′N 78°28′E*
Dåtnejaevrie *see* Tunnsjøen
159 T10 **Datong** *var.* Datong Huizu Tuzu Zizhixian, Qiaotou. Qinghai, C China *37°01′N 101°33′E*
161 N2 **Datong** *var.* Tatung, Ta-t'ung. Shanxi, C China *40°09′N 113°17′E*
Datong *see* Tong'an
159 S8 **Datong He** ~ C China
Datong Huizu Tuzu Zizhixian *see* Datong
159 S9 **Datong Shan** ▲▲ C China
169 O10 **Datu, Tanjung** *headland* Indonesia/Malaysia *02°01′N 109°37′E*
Datu, Teluk *see* Lahad Datu, Teluk
Daua *see* Dawa Wenz
172 H16 **Dauban, Mount** ▲ Silhouette, NE Seychelles
149 T7 **Dāūd Khel** Punjab, E Pakistan *32°52′N 71°35′E*
119 G15 **Daugai** Alytus, S Lithuania *54°22′N 24°20′E*
Daugava *see* Western Dvina
118 J11 **Daugavpils** *Ger.* Dünaburg; *prev. Rus.* Dvinsk. SE Latvia *55°53′N 26°34′E*
Dauka *see* Dawkah
Daulatabad *see* Malāyer
101 D18 **Daun** Rheinland-Pfalz, W Germany *50°13′N 06°50′E*
155 E14 **Daund** *prev.* Dhond. Mahārāshtra, W India *18°28′N 74°38′E*
166 M12 **Daung Kyun** *island* S Myanmar (Burma)
11 W15 **Dauphin** Manitoba, S Canada *51°09′N 100°05′W*
103 S13 **Dauphiné** *cultural region* E France
23 N9 **Dauphin Island** *island* Alabama, S USA
11 X15 **Dauphin River** Manitoba, S Canada *51°55′N 98°03′W*
77 V12 **Daura** Katsina, N Nigeria *13°03′N 08°18′E*
152 H12 **Dausa** *prev.* Daosa. Rājasthān, N India *26°51′N 76°21′E*
Dauwa *see* Dawwah
Däväçi *see* Şabran
155 F18 **Dāvangere** Karnātaka, W India *14°30′N 75°52′E*
171 Q8 **Davao** *off.* Davao City. Mindanao, S Philippines *07°06′N 125°36′E*
Davao City *see* Davao
171 Q8 **Davao Gulf** *gulf* Mindanao, S Philippines
15 Q11 **Daveluyville** Québec, SE Canada *46°12′N 72°07′W*
29 Z14 **Davenport** Iowa, C USA *41°31′N 90°35′W*
32 L8 **Davenport** Washington, NW USA *47°39′N 118°09′W*
43 P16 **David** Chiriquí, W Panama *08°26′N 82°26′W*
15 O11 **David** ~ Québec, SE Canada
29 R15 **David City** Nebraska, C USA *41°15′N 97°07′W*
David-Gorodok *see* Davyd-Haradok
11 T16 **Davidson** Saskatchewan, S Canada *51°15′N 105°59′W*
21 R10 **Davidson** North Carolina, SE USA *35°29′N 80°49′W*
26 K12 **Davidson** Oklahoma, C USA *34°15′N 99°06′W*
39 S6 **Davidson Mountains** ▲▲ Alaska, USA
172 M8 **Davie Ridge** *undersea feature* W Indian Ocean *17°10′S 41°45′E*
182 A1 **Davies, Mount** ▲ South Australia *26°14′S 129°14′E*
35 O7 **Davis** California, W USA *38°31′N 121°46′W*
27 N12 **Davis** Oklahoma, C USA *34°30′N 97°07′W*
195 Y7 **Davis** *Australian research station* Antarctica *68°30′S 78°15′E*
194 H3 **Davis Coast** *physical region* Antarctica
18 C16 **Davis, Mount** ▲ Pennsylvania, NE USA *39°47′N 79°10′W*
24 K9 **Davis Mountains** ▲▲ Texas, SW USA
195 Z9 **Davis Sea** *sea* Antarctica
65 O20 **Davis Seamounts** *undersea feature* S Atlantic Ocean
196 M13 **Davis Strait** *strait* Baffin Bay/Labrador Sea
127 U5 **Davlekanovo** Respublika Bashkortostan, W Russian Federation *54°13′N 55°06′E*
108 J9 **Davos** *Rmsch.* Tavau. Graubünden, E Switzerland *46°48′N 09°50′E*
Davvesiida *see* Lebesby
119 J20 **Davyd-Haradok** *Pol.* Dawidgródek, *Rus.* David-Gorodok. Brestskaya Voblasts', SW Belarus *52°03′N 27°13′E*
163 U12 **Dawa** Liaoning, NE China *40°55′N 122°02′E*
141 O11 **Dawāsir, Wādī ad** *dry watercourse* S Saudi Arabia
81 K15 **Dawa Wenz** *var.* Daua, Webi Daawo. ~ E Africa
Dawaymah, Birkat ad *see* Umm al Baqar, Hawr
167 N10 **Dawei** *var.* Tavoy, Htawei. Taninthayi, S Myanmar (Burma) *14°02′N 98°12′E*
119 K14 **Dawhinava** *Rus.* Dolginovo. Minskaya Voblasts', N Belarus *54°39′N 27°29′E*
Dawidgródek *see* Davyd-Haradok
141 V12 **Dawkah** *var.* Dauka. SW Oman *18°32′N 54°03′E*
Dawlat Qaṭar *see* Qatar
24 M3 **Dawn** Texas, SW USA *34°54′N 102°10′W*
Dawo *see* Maqên
140 M11 **Daws** Al BāḤah, SW Saudi Arabia *20°19′N 41°12′E*
10 H5 **Dawson** *var.* Dawson City. Yukon, NW Canada *64°04′N 139°24′W*
23 S6 **Dawson** Georgia, SE USA *31°46′N 84°27′W*
29 S9 **Dawson** Minnesota, N USA *44°55′N 96°03′W*
Dawson City *see* Dawson

◆ Country ● Country Capital ◇ Dependent Territory ○ Dependent Territory Capital ⬖ Administrative Regions ✈ International Airport ▲ Mountain ▲▲ Mountain Range ⛰ Volcano ~ River ◎ Lake ⊠ Reservoir

11 N13 **Dawson Creek** British Columbia, W Canada *55°45′N 120°07′W*
10 H7 **Dawson Range** ▲▲ Yukon, W Canada
181 Y9 **Dawson River** ≈ Queensland, E Australia
10 J15 **Dawsons Landing** British Columbia, SW Canada *51°33′N 127°38′W*
20 I7 **Dawson Springs** Kentucky, S USA *37°10′N 87°41′W*
23 S2 **Dawsonville** Georgia, SE USA *34°28′N 84°07′W*
160 G8 **Dawu** *var.* Xianshui. Sichuan, C China *30°55′N 101°08′E*
Dawu *see* Maqên
Dawukou *see* Huinong
141 Y10 **Dawwah** *var.* Dauwa. W Oman *20°36′N 58°52′E*
102 J15 **Dax** *var.* Ax; *anc.* Aquae Augustae, Aquae Tarbelicae. Landes, SW France *43°43′N 01°03′W*
Daxian *see* Dazhou
Daxiangshan *see* Gangu
Daxue *see* Wencheng
160 G9 **Daxue Shan** ▲▲ C China
Dayan *see* Lijiang
160 G12 **Dayao** *var.* Jinbi. Yunnan, SW China *25°41′N 101°23′E*
Dayishan *see* Gaoyou
Dāykondī *see* Dāykundī
149 O6 **Dāykundī** *prev.* Dāykondī. ◈ *province* C Afghanistan
183 N12 **Daylesford** Victoria, SE Australia *37°24′S 144°07′E*
35 U10 **Daylight Pass** *pass* California, W USA
61 D17 **Daymán, Río** ≈ N Uruguay
Dayong *see* Zhangjiajie
Dayr *see* Ad Dayr
138 G10 **Dayr ʻAllā** *var.* Deir ʻAlla. Al Balqā', N Jordan *32°39′N 36°06′E*
139 N4 **Dayr az Zawr** *var.* Deir ez Zor. Dayr az Zawr, E Syria *35°12′N 40°12′E*
138 M5 **Dayr az Zawr** *off.* Muḩāfaẓat Dayr az Zawr, *var.* Dayr Az-Zor. ◈ *governorate* E Syria
Dayr az Zawr, Muḩāfaẓat *see* Dayr az Zawr
Dayr Az-Zor *see* Dayr az Zawr
75 W9 **Dayrūṭ** *var.* Dairûṭ. C Egypt *27°34′N 30°48′E*
11 Q15 **Daysland** Alberta, SW Canada *52°53′N 112°19′W*
31 R14 **Dayton** Ohio, N USA *39°46′N 84°12′W*
20 L10 **Dayton** Tennessee, S USA *35°30′N 85°01′W*
25 W11 **Dayton** Texas, SW USA *30°03′N 94°53′W*
32 L10 **Dayton** Washington, NW USA *46°19′N 117°58′W*
23 X10 **Daytona Beach** Florida, SE USA *29°12′N 81°03′W*
169 U12 **Dayu** Borneo, C Indonesia *01°59′S 115°04′E*
161 O13 **Dayu Ling** ▲▲ S China
161 R7 **Da Yunhe** *Eng.* Grand Canal. *canal* E China
161 S11 **Dayu Shan** *island* SE China
160 K8 **Dazhou** *prev.* Dachuan, Daxian. Sichuan, C China *31°16′N 107°31′E*
160 J9 **Dazhu** *var.* Zhuyang. Sichuan, C China *30°45′N 107°11′E*
161 T13 **Dazhuoshui** *prev.* Tachoshui. N Taiwan *24°26′N 121°43′E*
160 J9 **Dazu** *var.* Longgang. Chongqing Shi, C China *29°47′N 106°30′E*
83 H24 **De Aar** Northern Cape, C South Africa *30°40′S 24°01′E*
194 K5 **Deacon, Cape** *headland* Antarctica
39 R5 **Deadhorse** Alaska, USA *70°15′N 148°28′W*
33 T12 **Dead Indian Peak** ▲ Wyoming, C USA *44°36′N 109°45′W*
23 R9 **Dead Lake** ◎ Florida, SE USA
44 J4 **Deadman's Cay** Long Island, C The Bahamas *23°09′N 75°06′W*
138 G11 **Dead Sea** *var.* Bahret Lut, Lacus Asphaltites, *Ar.* Al Baḩr al Mayyit, Baḩrat Lūt, *Heb.* Yam HaMelaḥ. *salt lake* Israel/Jordan
28 J9 **Deadwood** South Dakota, N USA *44°22′N 103°43′W*
97 Q22 **Deal** SE England, United Kingdom *51°14′N 01°23′E*
83 I22 **Dealesville** Free State, C South Africa *28°40′S 25°46′E*
161 P10 **De'an** *var.* Puting. Jiangxi, S China *29°24′N 115°46′E*
62 K9 **Deán Funes** Córdoba, C Argentina *30°25′S 64°22′W*
194 L12 **Dean Island** *island* Antarctica
Deanuvuotna *see* Tanafjorden
31 S10 **Dearborn** Michigan, N USA *42°16′N 83°13′W*
27 R3 **Dearborn** Missouri, C USA *39°31′N 94°46′W*
Deargget *see* Tärendö
32 K9 **Deary** Idaho, NW USA *46°46′N 118°33′W*
32 M9 **Deary** Washington, NW USA *46°42′N 116°36′W*
10 J10 **Dease** ≈ British Columbia, W Canada
10 J10 **Dease Lake** British Columbia, W Canada *58°28′N 130°04′W*
35 U11 **Death Valley** California, W USA *36°25′N 116°50′W*
35 U11 **Death Valley** *valley* California, W USA
92 M8 **Deatnu** *Fin.* Tenojoki, *Nor.* Tana. ≈ Finland/Norway *see also* Tana, Tenojoki
Deatnu *see* Tana
102 L4 **Deauville** Calvados, N France *49°21′N 00°06′E*
117 X7 **Debal'tseve** *Rus.* Debal'tsevo. Donets'ka Oblast', SE Ukraine *48°21′N 38°26′E*
Debal'tsevo *see* Debal'tseve
113 M19 **Debar** *Ger.* Dibra, *Turk.* Debre. W FYR Macedonia *41°32′N 20°33′E*
39 O9 **Debauch Mountain** ▲ Alaska, USA *64°31′N 159°52′W*
De Behagle *see* Laï
25 X7 **De Berry** Texas, SW USA *32°18′N 94°09′W*
Debessy *see* Debesy
127 T2 **Debesy** *prev.* Debessy. Udmurtskaya Respublika, NW Russian Federation *57°41′N 53°56′E*
111 N16 **Dębica** Podkarpackie, SE Poland *50°04′N 21°24′E*
De Bildt *see* De Bilt
98 J11 **De Bilt** *var.* De Bildt. Utrecht, C Netherlands *52°06′N 05°11′E*
123 T9 **Debin** Magadanskaya Oblast', E Russian Federation *62°18′N 150°42′E*
110 N13 **Dęblin** *Rus.* Ivangorod. Lubelskie, E Poland *51°34′N 21°50′E*
110 D10 **Dębno** Zachodnio-pomorskie, NW Poland *52°43′N 14°42′E*
39 S10 **Deborah, Mount** ▲ Alaska, USA *63°38′N 147°13′W*
33 N8 **De Borgia** Montana, NW USA *47°23′N 115°24′W*
Debra Birhan *see* Debre Birhan
Debra Marcos *see* Debre Mark'os
Debra Tabor *see* Debre Tabor
Debre *see* Debar
80 J13 **Debre Birhan** *var.* Debra Birhan. Āmara, N Ethiopia *09°45′N 39°40′E*
111 N22 **Debrecen** *Ger.* Debreczin, *Rom.* Debreţin; *prev.* Debreczen. Hajdú-Bihar, E Hungary *47°32′N 21°38′E*
Debreczen/Debreczin *see* Debrecen
80 I12 **Debre Mark'os** *var.* Debra Marcos. Āmara, N Ethiopia *10°18′N 37°48′E*
113 N19 **Debrešte** SW FYR Macedonia *41°29′N 21°20′E*
80 J11 **Debre Tabor** *var.* Debra Tabor. Āmara, N Ethiopia *11°46′N 38°06′E*
Debreţin *see* Debrecen
80 J13 **Debre Zeyt** Oromīya, C Ethiopia *08°41′N 39°00′E*
113 L16 **Deçan** *Serb.* Dečane; *prev.* Dečani. W Kosovo *42°33′N 20°18′E*
Dečane *see* Deçan
Dečani *see* Deçan
23 P2 **Decatur** Alabama, S USA *34°36′N 86°58′W*
23 S3 **Decatur** Georgia, SE USA *33°46′N 84°18′W*
30 L13 **Decatur** Illinois, N USA *39°50′N 88°57′W*
31 Q12 **Decatur** Indiana, N USA *40°40′N 84°57′W*
22 M5 **Decatur** Mississippi, S USA *32°26′N 89°06′W*
29 S14 **Decatur** Nebraska, C USA *42°00′N 96°19′W*
25 S6 **Decatur** Texas, SW USA *33°14′N 97°35′W*
20 H9 **Decaturville** Tennessee, S USA *35°35′N 88°08′W*
103 O13 **Decazeville** Aveyron, S France *44°34′N 02°18′E*
155 H17 **Deccan** *Hind.* Dakshin. *plateau* C India
14 J8 **Decelles, Réservoir** ▣ Québec, SE Canada
12 K2 **Déception** Québec, NE Canada *62°06′N 74°36′W*
160 G11 **Dechang** *var.* Dezhou. Sichuan, C China *27°24′N 102°09′E*
111 C15 **Děčín** *Ger.* Tetschen. Ústecký Kraj, NW Czech Republic *50°48′N 14°15′E*
103 P9 **Decize** Nièvre, C France *46°51′N 03°25′E*
98 I6 **De Cocksdorp** Noord-Holland, NW Netherlands *53°09′N 04°52′E*
29 X11 **Decorah** Iowa, C USA *43°18′N 91°47′W*
Dedeagaç/Dedeagach *see* Alexandroúpoli
188 C15 **Dededo** N Guam *13°30′N 144°51′E*
98 N9 **Dedemsvaart** Overijssel, E Netherlands *52°36′N 06°28′E*
19 O11 **Dedham** Massachusetts, NE USA *42°14′N 71°10′W*
63 H19 **Dedo, Cerro** ▲ SW Argentina *44°46′S 71°48′W*
77 O13 **Dédougou** W Burkina Faso *12°29′N 03°25′W*
124 G15 **Dedovichi** Pskovskaya Oblast', W Russian Federation *57°31′N 29°53′E*
Dedu *see* Wudalianchi
155 J24 **Deduru Oya** ≈ W Sri Lanka
83 N14 **Dedza** Central, S Malawi *14°20′S 34°24′E*
83 N14 **Dedza Mountain** ▲ C Malawi *14°22′S 34°16′E*
96 K9 **Dee** ≈ NE Scotland, United Kingdom
97 J19 **Dee** *Wel.* Afon Dyfrdwy. ≈ England/Wales, United Kingdom
Deep Bay *see* Chilumba
21 T3 **Deep Creek Lake** ▣ Maryland, NE USA
36 J4 **Deep Creek Range** ▲▲ Utah, W USA
27 P10 **Deep Fork River** ≈ Oklahoma, C USA
14 J11 **Deep River** Ontario, SE Canada *46°04′N 77°29′W*
21 T10 **Deep River** ≈ North Carolina, SE USA
183 U4 **Deepwater** New South Wales, SE Australia *29°27′S 151°52′E*
31 S14 **Deer Creek Lake** ◎ Ohio, N USA
23 Z15 **Deerfield Beach** Florida, SE USA *26°19′N 80°06′W*
39 N8 **Deering** Alaska, USA *66°04′N 162°43′W*
38 M16 **Deer Island** *island* Alaska, USA
19 S7 **Deer Isle** *island* Maine, NE USA
13 S11 **Deer Lake** Newfoundland and Labrador, SE Canada *49°11′N 57°27′N*
99 D18 **Deerlijk** West-Vlaanderen, W Belgium *50°52′N 03°21′E*
33 Q10 **Deer Lodge** Montana, NW USA *46°24′N 112°43′W*
32 L8 **Deer Park** Washington, NW USA *47°55′N 117°28′W*
29 U5 **Deer River** Minnesota, N USA *47°19′N 93°47′W*
Deés *see* Dej
Defeng *see* Liping
31 R11 **Defiance** Ohio, N USA *41°17′N 84°21′W*
23 Q8 **De Funiak Springs** Florida, SE USA *30°43′N 86°07′W*
95 L23 **Degeberga** Skåne, S Sweden *55°48′N 14°06′E*
104 H12 **Degebe, Ribeira** ≈ S Portugal
80 M13 **Degeh Bur** Sumalē, E Ethiopia *08°08′N 43°35′E*
15 U9 **Dégelis** Québec, SE Canada *47°30′N 68°38′W*
77 U17 **Degema** Rivers, S Nigeria *04°46′N 06°47′E*
95 L16 **Degerfors** Örebro, C Sweden *59°14′N 14°26′E*
193 R14 **De Gerlache Seamounts** *undersea feature* SE Pacific Ocean
101 N21 **Deggendorf** Bayern, SE Germany *48°50′N 12°58′E*
80 I11 **Degoma** Āmara, N Ethiopia *12°22′N 37°36′E*
De Gordyk *see* Gorredijk
27 T12 **De Gray Lake** ▣ Arkansas, C USA
180 J6 **De Grey River** ≈ Western Australia
126 M10 **Degtevo** Rostovskaya Oblast', SW Russian Federation *49°12′N 40°39′E*
Dehbārez *see* Rūdān
Deh Bīd *see* Şafāshahr
142 M10 **Deh Dasht** Kohkīlūyeh va Būyer AḤmad, SW Iran *30°49′N 50°36′E*
75 N8 **Dehibat** SE Tunisia *31°58′N 10°43′E*
Dehli *see* Delhi
142 K8 **Dehlorān** Īlām, W Iran *32°41′N 47°18′E*
147 N13 **Dehqonobod** *Rus.* Dekhkanabad. Qashqadaryo Viloyati, S Uzbekistan *38°24′N 66°31′E*
152 J9 **Dehra Dūn** Uttararanchal, N India *30°19′N 78°04′E*
153 O14 **Dehri** Bihār, N India *24°55′N 84°11′E*
Deh Shū *see* Dīshū
163 W9 **Dehui** Jilin, NE China *44°23′N 125°42′E*
99 D17 **Deinze** Oost-Vlaanderen, NW Belgium *50°59′N 03°32′E*
Deir ʻAlla *see* Dayr ʻAllā
Deir ez Zor *see* Dayr az Zawr
Deirgeirt, Loch *see* Derg, Lough
116 H9 **Dej** *Hung.* Dés; *prev.* Deés. Cluj, NW Romania *47°08′N 23°55′E*
95 K15 **Deje** Värmland, C Sweden *59°35′N 13°29′E*
171 Y15 **De Jongs, Tanjung** *headland* Papua, SE Indonesia *06°56′S 138°32′E*
De Jouwer *see* Joure
30 M10 **De Kalb** Illinois, N USA *41°55′N 88°45′W*
22 M5 **De Kalb** Mississippi, S USA *32°46′N 88°39′W*
25 W5 **De Kalb** Texas, SW USA *33°30′N 94°34′W*
83 G18 **Dekar** *var.* D'Kar. Ghanzi, NW Botswana *21°31′S 21°55′E*
Dekéleia *see* Dhekélia
79 K20 **Dekese** Kasai-Occidental, C Dem. Rep. Congo *03°28′S 21°24′E*
Dekhkanabad *see* Dehqonobod
79 I14 **Dékoa** Kémo, C Central African Republic *06°17′N 19°07′E*
98 H6 **De Koog** Noord-Holland, NW Netherlands *53°06′N 04°43′E*
30 M9 **Delafield** Wisconsin, N USA *43°03′N 88°22′W*
61 C23 **De La Garma** Buenos Aires, E Argentina *37°58′S 60°25′W*
14 K10 **Delahey, Lac** ◎ Québec, SE Canada
80 E11 **Delami** Southern Kordofan, C Sudan *11°51′N 30°30′E*
23 X11 **De Land** Florida, SE USA *29°01′N 81°18′W*
35 R12 **Delano** California, W USA *35°46′N 119°15′W*
29 V8 **Delano** Minnesota, N USA *45°03′N 93°46′W*
36 K6 **Delano Peak** ▲ Utah, W USA *38°22′N 112°21′W*
Delap-Uliga-Darrit *see* Dalap-Uliga-Djarrit
Delārām *see* Dilārām
38 F17 **Delarof Islands** *island group* Aleutian Islands, Alaska, USA
30 M9 **Delavan** Wisconsin, N USA *42°37′N 88°37′W*
31 S13 **Delaware** Ohio, N USA *40°18′N 83°06′W*
18 I17 **Delaware** *off.* State of Delaware, ***also known as*** Blue Hen State, Diamond State, First State. ◈ *state* NE USA
18 I17 **Delaware Bay** *bay* NE USA
24 J8 **Delaware Mountains** ▲▲ Texas, SW USA
18 I12 **Delaware River** ≈ NE USA
27 Q3 **Delaware River** ≈ Kansas, C USA
18 J14 **Delaware Water Gap** *valley* New Jersey/Pennsylvania, NE USA
101 G14 **Delbrück** Nordrhein-Westfalen, W Germany *51°46′N 08°34′E*
11 Q15 **Delburne** Alberta, SW Canada *52°09′N 113°11′W*
172 M12 **Del Cano Rise** *undersea feature* SW Indian Ocean *45°15′S 44°15′E*
113 Q18 **Delčevo** NE FYR Macedonia *41°57′N 22°45′E*
Delcommune, Lac *see* Nzilo, Lac
98 O10 **Delden** Overijssel, E Netherlands *52°16′N 06°41′E*
183 R12 **Delegate** New South Wales, SE Australia *37°04′S 148°57′E*
De Lemmer *see* Lemmer
108 D7 **Delémont** *Ger.* Delsberg. Jura, NW Switzerland *47°22′N 07°21′E*
25 R7 **De Leon** Texas, SW USA *32°06′N 98°33′W*
115 F18 **Delfoí** Stereá Elláda, C Greece *38°28′N 22°31′E*
98 G12 **Delft** Zuid-Holland, W Netherlands *52°01′N 04°22′E*
155 J23 **Delft** *island* NW Sri Lanka
98 O5 **Delfzijl** Groningen, NE Netherlands *53°20′N 06°55′E*
0 E9 **Delgada Fan** *undersea feature* NE Pacific Ocean *39°15′N 126°00′W*
42 F7 **Delgado** San Salvador, SW El Salvador *13°44′N 89°07′W*
82 Q12 **Delgado, Cabo** *headland* N Mozambique *10°41′S 40°40′E*
162 G8 **Delger** *var.* Taygan. Govĭ-Altay, C Mongolia *46°20′N 97°22′E*
163 O9 **Delgereh** *var.* Hongor. Dornogovĭ, SE Mongolia *45°49′N 111°20′E*
162 J8 **Delgerhaan** *var.* Hujirt. Töv, C Mongolia *46°41′N 104°40′E*
162 K9 **Delgerhangay** *var.* Hashaat. Dundgovĭ, C Mongolia *45°09′N 104°51′E*
162 L9 **Delgertsogt** *var.* Amardalay. Dundgovĭ, C Mongolia *46°09′N 106°24′E*
80 E6 **Delgo** Northern, N Sudan *20°08′N 30°35′E*
159 R10 **Delhi** *var.* Delingha. Qinghai, C China *37°19′N 97°22′E*
152 I10 **Delhi** *var.* Dehli, *Hind.* Dilli, *hist.* Shahjahanabad. *union territory capital* Delhi, N India *28°40′N 77°11′E*
22 J5 **Delhi** Louisiana, S USA *32°28′N 91°29′W*
18 J11 **Delhi** New York, NE USA *42°16′N 74°55′W*
152 I10 **Delhi** ◈ *union territory* NW India
136 J17 **Deli Burnu** *headland* S Turkey *36°43′N 34°55′E*
136 J12 **Delice Çayı** ≈ C Turkey
55 X10 **Délices** C French Guiana *04°45′N 53°45′W*
40 J6 **Delicias** *var.* Ciudad Delicias. Chihuahua, N Mexico *28°09′N 105°22′W*
143 N7 **Delījān** *var.* Dalijan, Dilijan. Markazī, W Iran *34°02′N 50°39′E*
112 P12 **Deli Jovan** ▲▲ E Serbia
Déli-Kárpátok *see* Carpaţii Meridionalii
8 I8 **Délįne** *prev.* Fort Franklin. Northwest Territories, NW Canada *65°10′N 123°30′W*
Delingha *see* Delhi
15 Q7 **Delisle** Québec, SE Canada *48°39′N 71°42′W*
11 T15 **Delisle** Saskatchewan, S Canada *51°54′N 107°01′W*
101 M15 **Delitzsch** Sachsen, E Germany *51°31′N 12°19′E*
33 Q12 **Dell** Montana, NW USA *44°41′N 112°42′W*
24 I7 **Dell City** Texas, SW USA *31°56′N 105°12′W*
103 U7 **Delle** Territoire-de-Belfort, E France *47°30′N 07°00′E*
36 J9 **Dellenbaugh, Mount** ▲ Arizona, SW USA *36°06′N 113°32′W*
29 R11 **Dell Rapids** South Dakota, N USA *43°50′N 96°42′W*
21 Y4 **Delmar** Maryland, NE USA *38°26′N 75°32′W*
18 K11 **Delmar** New York, NE USA *42°37′N 73°49′W*
100 G11 **Delmenhorst** Niedersachsen, NW Germany *53°03′N 08°38′E*
112 C9 **Delnice** Primorje-Gorski Kotar, NW Croatia *45°24′N 14°49′E*
37 R7 **Del Norte** Colorado, C USA *37°40′N 106°21′W*
39 N6 **De Long Mountains** ▲▲ Alaska, USA
183 P16 **Deloraine** Tasmania, SE Australia *41°34′S 146°43′E*
11 W17 **Deloraine** Manitoba, S Canada *49°12′N 100°28′W*
31 O12 **Delphi** Indiana, N USA *40°35′N 86°40′W*
31 Q12 **Delphos** Ohio, N USA *40°49′N 84°20′W*
23 Z15 **Delray Beach** Florida, SE USA *26°27′N 80°04′W*
25 O12 **Del Rio** Texas, SW USA *29°23′N 100°56′W*
Delsberg *see* Delémont
94 N11 **Delsbo** Gävleborg, C Sweden *61°49′N 16°34′E*
37 P6 **Delta** Colorado, C USA *38°44′N 108°04′W*
36 K5 **Delta** Utah, W USA *39°21′N 112°34′W*
77 T17 **Delta** ◈ *state* S Nigeria
55 Q6 **Delta Amacuro** *off.* Territorio Delta Amacuro. ◈ *federal district* NE Venezuela
Delta Amacuro, Territorio *see* Delta Amacuro
39 S9 **Delta Junction** Alaska, USA *64°02′N 145°43′W*
23 X11 **Deltona** Florida, SE USA *28°54′N 81°15′W*
183 T5 **Delungra** New South Wales, SE Australia *29°40′S 150°49′E*
162 D6 **Delüün** *var.* Rashaant. Bayan-Ölgiy, W Mongolia *47°48′N 90°45′E*
154 C12 **Delvāda** Gujarāt, W India *20°46′N 71°02′E*
61 B21 **Del Valle** Buenos Aires, E Argentina *35°55′S 60°42′W*
Delvina *see* Delvinë
115 C15 **Delvináki** *var.* Dhelvinákion; *prev.* Pogónion. Ípeiros, W Greece *39°57′N 20°28′E*
113 L23 **Delvinë** *var.* Delvina, *It.* Delvino. Vlorë, S Albania *39°56′N 20°07′E*
Delvino *see* Delvinë
116 I7 **Delyatyn** Ivano-Frankivs'ka Oblast', W Ukraine *48°32′N 24°38′E*
127 U5 **Dëma** ≈ W Russian Federation
105 O5 **Demanda, Sierra de la** ▲▲ W Spain
39 T5 **Demarcation Point** *headland* Alaska, USA *69°40′N 141°19′W*
79 K21 **Demba** Kasai-Occidental, C Dem. Rep. Congo *05°24′S 22°16′E*
172 H13 **Dembéni** Grande Comore, NW Comoros *11°50′S 43°25′E*
79 M15 **Dembia** Mbomou, SE Central African Republic *05°08′N 24°25′E*
Dembidollo *see* Dembī Dolo
80 H13 **Dembī Dolo** *var.* Dembidollo. Oromīya, C Ethiopia *08°33′N 34°49′E*
152 K6 **Demchok** *var.* Dêmqog. China/India *32°30′N 79°42′E* *see also* Dêmqog
152 L6 **Demchok** *var.* Dêmqog. *disputed region* China/India *see also* Dêmqog
98 I12 **De Meern** Utrecht, C Netherlands *52°06′N 05°00′E*
99 I17 **Demer** ≈ C Belgium
64 H12 **Demerara Plain** *undersea feature* W Atlantic Ocean *10°00′N 48°00′W*
64 H12 **Demerara Plateau** *undersea feature* W Atlantic Ocean
55 T9 **Demerara River** ≈ NE Guyana
126 H3 **Demidov** Smolenskaya Oblast', W Russian Federation *55°15′N 31°30′E*
37 Q15 **Deming** New Mexico, SW USA *32°17′N 107°46′W*
32 H6 **Deming** Washington, NW USA *48°49′N 122°13′W*
58 E10 **Demini, Rio** ≈ NW Brazil
136 D13 **Demirci** Manisa, W Turkey *39°03′N 28°40′E*
113 P19 **Demir Kapija** *prev.* Železna Vrata. SE FYR Macedonia *41°25′N 22°15′E*
114 N11 **Demirköy** Kırklareli, NW Turkey *41°48′N 27°49′E*
100 N9 **Demmin** Mecklenburg-Vorpommern, NE Germany *53°53′N 13°03′E*
23 O5 **Demopolis** Alabama, S USA *32°31′N 87°50′W*
31 N11 **Demotte** Indiana, N USA *41°13′N 87°07′W*
158 F13 **Dêmqog** *var.* Demchok. China/India *32°36′N 79°29′E* *see also* Demchok
152 L6 **Dêmqog** *var.* Demchok. *disputed region* China/India *see also* Demchok
171 Y13 **Demta** Papua, E Indonesia *02°19′S 140°06′E*
121 K11 **Dem'yanka** ≈ C Russian Federation
124 H15 **Demyansk** Novgorodskaya Oblast', W Russian Federation *57°39′N 32°31′E*
122 H10 **Dem'yanskoye** Tyumenskaya Oblast', C Russian Federation *59°39′N 69°15′E*
103 P2 **Denain** Nord, N France *50°19′N 03°24′E*
39 S10 **Denali** Alaska, USA *63°08′N 147°33′W*
Denali *see* McKinley, Mount
81 M14 **Denan** Sumalē, E Ethiopia *06°40′N 43°31′E*
Denau *see* Denov
97 J18 **Denbigh** *Wel.* Dinbych. NE Wales, United Kingdom *53°11′N 03°25′W*
97 J18 **Denbigh** *cultural region* N Wales, United Kingdom
98 I6 **Den Burg** Noord-Holland, NW Netherlands *53°03′N 04°47′E*
99 F18 **Dender** *Fr.* Dendre. ≈ W Belgium
99 F18 **Denderleeuw** Oost-Vlaanderen, NW Belgium *50°53′N 04°05′E*
99 F17 **Dendermonde** *Fr.* Termonde. Oost-Vlaanderen, NW Belgium *51°02′N 04°08′E*
Dendre *see* Dender
194 I9 **Dendtler Island** *island* Antarctica
98 P10 **Denekamp** Overijssel, E Netherlands *52°23′N 07°E*
77 W12 **Dengas** Zinder, S Niger *13°15′N 09°43′E*
Dêngka *see* Têwo
Dêngkagoin *see* Têwo
162 L13 **Dengkou** *var.* Bayan Gol. Nei Mongol Zizhiqu, N China *40°15′N 106°59′E*
159 Q14 **Dêngqên** *var.* Gyamotang. Xizang Zizhiqu, W China *31°28′N 95°28′E*
Deng Xian *see* Dengzhou
160 M7 **Dengzhou** *prev.* Deng Xian. Henan, C China *32°48′N 111°59′E*
Dengzhou *see* Penglai
Den Haag *see* 's-Gravenhage
180 H10 **Denham** Western Australia *25°56′S 113°35′E*
98 N9 **Den Ham** Overijssel, E Netherlands *52°30′N 06°31′E*
44 J12 **Denham, Mount** ▲ C Jamaica *18°13′N 77°33′W*
22 J8 **Denham Springs** Louisiana, S USA *30°29′N 90°57′W*
98 I7 **Den Helder** Noord-Holland, NW Netherlands *52°54′N 04°45′E*
105 T11 **Dénia** Valenciana, E Spain *38°51′N 00°07′E*
189 Q8 **Denig** W Nauru
183 N10 **Deniliquin** New South Wales, SE Australia *35°33′S 144°58′E*
29 T14 **Denison** Iowa, C USA *42°00′N 95°22′W*
25 U5 **Denison** Texas, SW USA *33°45′N 96°32′W*
144 L8 **Denisovka** *prev.* Ordzhonikidze. Kostanay, N Kazakhstan *52°27′N 61°42′E*
136 D15 **Denizli** Denizli, SW Turkey *37°46′N 29°05′E*
136 D15 **Denizli** ◈ *province* SW Turkey
Denjong *see* Sikkim
183 S7 **Denman** New South Wales, SE Australia *32°24′S 150°43′E*
195 Y10 **Denman Glacier** *glacier* Antarctica
21 R14 **Denmark** South Carolina, SE USA *33°19′N 81°08′W*
95 G23 **Denmark** *off.* Kingdom of Denmark, *Dan.* Danmark; *anc.* Hafnia. ◆ *monarchy* N Europe
Denmark, Kingdom of *see* Denmark
92 H1 **Denmark Strait** *var.* Danmarksstraedet. *strait* Greenland/Iceland
45 T11 **Dennery** E Saint Lucia *13°55′N 60°53′W*
98 I7 **Den Oever** Noord-Holland, NW Netherlands *52°56′N 05°01′E*
147 O13 **Denov** *Rus.* Denau. Surkhondaryo Viloyati, S Uzbekistan *38°20′N 67°48′E*
169 U17 **Denpasar** *prev.* Paloe. Bali, C Indonesia *08°40′S 115°14′E*
116 E12 **Denta** Timiş, W Romania *45°20′N 21°15′E*
21 Y3 **Denton** Maryland, NE USA *38°53′N 75°50′W*
25 T6 **Denton** Texas, SW USA *33°11′N 97°08′W*
186 G9 **D'Entrecasteaux Islands** *island group* SE Papua New Guinea
37 T4 **Denver** *state capital* Colorado, C USA *39°45′N 105°W*
37 T4 **Denver** ✈ Colorado, C USA *39°57′N 104°38′W*
24 L6 **Denver City** Texas, SW USA *32°57′N 102°49′W*
152 J9 **Deoband** Uttar Pradesh, N India *29°41′N 77°40′E*
Deoghar *see* Devghar
163 X15 **Deokjeok-gundo** *prev.* Tŏkchŏk-kundo. *island group* NW South Korea
154 E13 **Deolāli** Mahārāshtra, W India *19°55′N 73°49′E*
154 I10 **Deori** Madhya Pradesh, C India *23°09′N 78°39′E*
153 O12 **Deoria** Uttar Pradesh, N India *26°31′N 83°48′E*
99 A17 **De Panne** West-Vlaanderen, W Belgium *51°06′N 02°35′E*
Departamento del Quindío *see* Quindío
Departamento de Narino, *see* Nariño
54 M5 **Dependencia Federal** *off.* Territorio Dependencia Federal. ◈ *federal dependency* N Venezuela
Dependencia Federal, Territorio *see* Dependencia Federal
30 M7 **De Pere** Wisconsin, N USA *44°26′N 88°03′W*
18 D10 **Depew** New York, NE USA *42°54′N 78°41′W*
99 E17 **De Pinte** Oost-Vlaanderen, NW Belgium *51°00′N 03°37′E*
25 V5 **Deport** Texas, SW USA *33°31′N 95°19′W*
123 Q8 **Deputatskiy** Respublika Sakha (Yakutiya), NE Russian Federation *69°18′N 139°48′E*
Dêqên *see* Dagzê
27 S13 **De Queen** Arkansas, C USA *34°02′N 94°20′W*
22 G8 **De Quincy** Louisiana, S USA *30°27′N 93°25′W*
81 J20 **Dera** *spring/well* S Kenya *02°39′S 39°52′E*
Der'a/Derá/Déraa *see* Darʻā
149 S10 **Dera Ghāzi Khān** *var.* Dera Ghāzikhān. Punjab, C Pakistan *30°01′N 70°37′E*
Dera Ghāzikhān *see* Dera Ghāzi Khān
149 S8 **Dera Ismāīl Khān** Khyber Pakhtunkhwa, C Pakistan *31°51′N 70°56′E*
Đeravica *see* Gjeravicë
116 L6 **Derazhnya** Khmel'nyts'ka Oblast', W Ukraine *49°16′N 27°24′E*
127 R17 **Derbent** Respublika Dagestan, SW Russian Federation *42°01′N 48°16′E*
147 N13 **Derbent** Surkhondaryo Viloyati, S Uzbekistan *38°15′N 66°59′E*
Derbisīye *see* Darbāsīyah
79 M15 **Derbissaka** Mbomou, SE Central African Republic *05°43′N 24°48′E*
180 L4 **Derby** Western Australia *17°18′S 123°37′E*
97 M19 **Derby** C England, United Kingdom *52°55′N 01°30′W*
27 N7 **Derby** Kansas, C USA *37°33′N 97°16′W*
97 L18 **Derbyshire** *cultural region* C England, United Kingdom
112 O11 **Đerdap** *physical region* E Serbia
Derelí *see* Gónnoi
162 L9 **Deren** *var.* Tsant. Dundgovĭ, C Mongolia *46°16′N 106°55′E*
171 W13 **Derew** ≈ Papua, E Indonesia
127 R8 **Dergachi** Saratovskaya Oblast', W Russian Federation *51°15′N 48°58′E*
Dergachi *see* Derhachi
97 C19 **Derg, Lough** *Ir.* Loch Deirgeirt. ◎ W Ireland
117 V5 **Derhachi** *Rus.* Dergachi. Kharkivs'ka Oblast', E Ukraine *50°09′N 36°11′E*
22 G8 **De Ridder** Louisiana, S USA *30°51′N 93°18′W*
137 P16 **Derik** Mardin, SE Turkey *37°22′N 40°16′E*
83 E20 **Derm** Hardap, C Namibia *23°38′S 18°12′E*
Dermentobe *see* Diirmentobe
27 W14 **Dermott** Arkansas, C USA *33°31′N 91°26′W*
Dérna *see* Darnah
Dernberg, Cape *see* Dolphin Head
22 J11 **Dernieres, Isles** *island group* Louisiana, S USA
102 I4 **Déroute, Passage de la** *strait* Channel Islands/France
Derrá *see* Darʻā
81 O16 **Derri** *prev.* Dirri. Galguduud, C Somalia *04°15′N 46°31′E*
Derry *see* Londonderry
Dertona *see* Tortona
Dertosa *see* Tortosa
80 H8 **Derudeb** Red Sea, NE Sudan *17°31′N 36°07′E*
112 H10 **Derventa** Republika Srpska, N Bosnia and Herzegovina *44°57′N 17°55′E*
183 O16 **Derwent Bridge** Tasmania, SE Australia *42°10′S 146°13′E*
183 O17 **Derwent, River** ≈ Tasmania, SE Australia
146 F10 **Derweze** *Rus.* Darvaza. Ahal Welaýaty, C Turkmenistan *40°10′N 58°27′E*
Deržavinsk *see* Derzhavinsk
145 O9 **Derzhavinsk** *var.* Deržavinsk. ◈ Akmola, C Kazakhstan
Dés *see* Dej
57 J18 **Desaguadero** Puno, S Peru *16°35′S 69°05′W*
57 J18 **Desaguadero, Río** ≈ Bolivia/Peru
191 W9 **Désappointement, Îles du** *island group* Îles Tuamotu, C French Polynesia
27 W11 **Des Arc** Arkansas, C USA *34°58′N 91°30′W*
14 C10 **Desbarats** Ontario, S Canada *46°20′N 83°52′W*
62 H13 **Descabezado Grande, Volcán** ⩚ C Chile *35°34′S 70°40′W*
102 L9 **Descartes** Indre-et-Loire, C France *46°58′N 00°40′E*
11 T13 **Deschambault Lake** ◎ Saskatchewan, C Canada
Deschnaer Koppe *see* Velká Deštná
32 I11 **Deschutes River** ≈ Oregon, NW USA
80 J12 **Desē** *var.* Desse, *It.* Dessie. Āmara, N Ethiopia *11°02′N 39°39′E*
63 I20 **Deseado, Río** ≈ S Argentina
106 F8 **Desenzano del Garda** Lombardia, N Italy *45°28′N 10°31′E*
36 K3 **Deseret Peak** ▲ Utah, W USA *40°27′N 112°37′W*
64 P6 **Deserta Grande** *island* Madeira, Portugal, NE Atlantic Ocean
64 P6 **Desertas, Ilhas** *island group* Madeira, Portugal, NE Atlantic Ocean
35 X16 **Desert Center** California, W USA *33°42′N 115°22′W*
35 V15 **Desert Hot Springs** California, W USA *33°57′N 116°33′W*
14 K10 **Désert, Lac** ◎ Québec, SE Canada
36 J2 **Desert Peak** ▲ Utah, W USA *41°03′N 113°22′W*
31 R11 **Deshler** Ohio, N USA *41°12′N 83°55′W*
Deshu *see* Dīshū
Desiderii Fanum *see* St-Dizier
106 D7 **Desio** Lombardia, N Italy *45°37′N 09°12′E*
115 E15 **Deskáti** *var.* Dheskáti. Dytikí Makedonía, N Greece *39°55′N 21°49′E*
28 L2 **Des Lacs River** ≈ North Dakota, N USA
27 X6 **Desloge** Missouri, C USA *37°52′N 90°31′W*
11 Q12 **Desmarais** Alberta, W Canada *55°58′N 113°56′W*
29 Q10 **De Smet** South Dakota, N USA *44°23′N 97°33′W*
29 V14 **Des Moines** *state capital* Iowa, C USA *41°36′N 93°37′W*
17 N9 **Des Moines River** ≈ C USA
117 P4 **Desna** ≈ Russian Federation/Ukraine
116 G14 **Desnăţui** ≈ S Romania
63 F24 **Desolación, Isla** *island* S Chile
29 V14 **De Soto** Iowa, C USA *41°31′N 94°00′W*
23 Q4 **De Soto Falls** *waterfall* Alabama, S USA
83 I25 **Despatch** Eastern Cape, S South Africa *33°48′S 25°28′E*
105 N12 **Despeñaperros, Desfiladero de** *pass* S Spain
31 N10 **Des Plaines** Illinois, N USA *42°01′N 87°52′W*
115 J21 **Despotikó** *island* Kykládes, Greece, Aegean Sea
112 N12 **Despotovac** Serbia, E Serbia *44°06′N 21°25′E*
101 M14 **Dessau** Sachsen-Anhalt, E Germany *51°51′N 12°15′E*
Desse *see* Desē
99 J16 **Dessel** Antwerpen, N Belgium *51°15′N 05°07′E*
Dessie *see* Desē
Destêrro *see* Florianópolis
23 P9 **Destin** Florida, SE USA *30°23′N 86°30′W*
Deštná *see* Velká Deštná
193 T10 **Desventurados, Islas de los** *island group* W Chile
103 N1 **Desvres** Pas-de-Calais, N France *50°41′N 01°48′E*
116 E12 **Deta** *Ger.* Detta. Timiş, W Romania *45°24′N 21°14′E*
101 H14 **Detmold** Nordrhein-Westfalen, W Germany *51°55′N 08°52′E*
31 S10 **Detroit** Michigan, N USA *42°20′N 83°03′W*
25 W5 **Detroit** Texas, SW USA *33°39′N 95°16′W*
31 S10 **Detroit** ≈ Canada/USA
29 S6 **Detroit Lakes** Minnesota, N USA *46°49′N 95°49′W*
31 S10 **Detroit Metropolitan** ✈ Michigan, N USA *42°12′N 83°16′W*
Detta *see* Deta
167 S10 **Det Udom** Ubon Ratchathani, E Thailand *14°54′N 105°03′E*
111 K20 **Detva** *Hung.* Gyeva. Bankobýstrický Kraj, C Slovakia *48°35′N 19°25′E*
154 G13 **Deūlgaon Rāja** Mahārāshtra, C India *20°04′N 76°08′E*
99 L15 **Deurne** Noord-Brabant, SE Netherlands *51°28′N 05°47′E*
99 H16 **Deurne** ✈ (Antwerpen) Antwerpen, N Belgium *51°10′N 04°28′E*
Deutsch-Brod *see* Havlíčkův Brod
Deutschendorf *see* Poprad
Deutsch-Eylau *see* Iława
109 Y6 **Deutschkreutz** Burgenland, E Austria *47°37′N 16°37′E*
Deutsch Krone *see* Wałcz
Deutschland/Deutschland, Bundesrepublik *see* Germany
109 V9 **Deutschlandsberg** Steiermark, SE Austria *46°52′N 15°13′E*
Deutsch-Südwestafrika *see* Namibia
109 Y3 **Deutsch-Wagram** Niederösterreich, E Austria *48°19′N 16°33′E*
Deux-Ponts *see* Zweibrücken
14 I11 **Deux Rivieres** Ontario, SE Canada *46°13′N 78°16′W*
102 K9 **Deux-Sèvres** ◈ *department* W France
116 G11 **Deva** *Ger.* Diemrich, *Hung.* Déva. Hunedoara, W Romania *45°55′N 22°55′E*
Déva *see* Deva
Deva *see* Chester
Devana *see* Aberdeen
Devana Castra *see* Chester
Đevđelija *see* Gevgelija
136 L12 **Deveci Dağları** ▲▲ N Turkey
137 P15 **Devegeçidi Barajī** ▣ SE Turkey
136 K15 **Develi** Kayseri, C Turkey *38°22′N 35°28′E*
98 M11 **Deventer** Overijssel, E Netherlands *52°15′N 06°10′E*
15 O10 **Devenyns, Lac** ◎ Québec, SE Canada
96 K8 **Deveron** ≈ NE Scotland, United Kingdom
153 R14 **Devghar** *prev.* Deoghar. Jhārkhand, NE India
27 R10 **Devil's Den** *plateau* Arkansas, C USA
35 R7 **Devils Gate** *pass* California, W USA
30 J2 **Devils Island** *island* Apostle Islands, Wisconsin, N USA
Devil's Island *see* Diable, Île du
29 P3 **Devils Lake** North Dakota, N USA *48°08′N 98°50′W*
31 R10 **Devils Lake** ◎ Michigan, N USA
29 O3 **Devils Lake** ◎ North Dakota, N USA
35 W13 **Devils Playground** *desert* California, W USA
25 O11 **Devils River** ≈ Texas, SW USA
33 Y12 **Devils Tower** ▲ Wyoming, C USA *44°33′N 104°45′W*
114 I11 **Devin** *prev.* Dovlen. Smolyan, S Bulgaria *41°45′N 24°24′E*
25 R12 **Devine** Texas, SW USA *29°08′N 98°54′W*
152 H13 **Devli** Rājasthān, N India *25°47′N 75°23′E*
Devne *see* Devnya
114 N8 **Devnya** *prev.* Devne. Varna, E Bulgaria *43°15′N 27°35′E*
31 U14 **Devola** Ohio, N USA *39°28′N 81°28′W*
Devoll *see* Devollit, Lumi i

◆ Country
● Country Capital
◇ Dependent Territory
○ Dependent Territory Capital
◈ Administrative Regions
✈ International Airport
▲ Mountain
▲▲ Mountain Range
⩚ Volcano
≈ River
◎ Lake
▣ Reservoir

◆ Country ● Country Capital ◇ Dependent Territory ○ Dependent Territory Capital ◆ Administrative Regions ✕ International Airport ▲ Mountain ▲ Mountain Range ☒ Volcano ☑ River ◎ Lake ☒ Reservoir

DOKUCHAYEVSK – DUCHCOV

Dokuchayevsk see Dokuchayevs'k
Dolak, Pulau see Yos Sudarso, Pulau
29 P9 Doland South Dakota, N USA 44°51′N 98°06′W
63 J18 Dolavón Chaco, S Argentina 43°16′S 65°44′W
15 P6 Dolbeau Québec, SE Canada 48°52′N 72°15′W
15 P6 Dolbeau-Mistassini Québec, SE Canada 48°54′N 72°13′W
102 I5 Dol-de-Bretagne Ille-et-Vilaine, NW France 48°33′N 01°45′W
64 J13 Doldrums Fracture Zone tectonic feature W Atlantic Ocean
103 S8 Dôle Jura, E France 47°05′N 05°30′E
97 J19 Dolgellau NW Wales, United Kingdom 52°45′N 03°54′W
Dolginovo see Dawhinava
Dolgi, Ostrov see Dolgiy, Ostrov
125 U2 Dolgiy, Ostrov var. Ostrov Dolgi. island NW Russian Federation
162 J9 Dölgöön Övörhangay, C Mongolia 45°57′N 103°14′E
107 C20 Dolianova Sardegna, Italy, C Mediterranean Sea 39°23′N 09°08′E
Dolina see Dolyna
123 T13 Dolinsk Ostrov Sakhalin, Sakhalinskaya Oblast', SE Russian Federation 47°20′N 142°52′E
Dolinskaya see Dolyns'ka
79 F21 Dolisie prev. Loubomo. Niari, S Congo 04°12′S 12°41′E
116 G14 Dolj ◐ county SW Romania
98 P5 Dollard bay NW Germany
194 J5 Dolleman Island island Antarctica
114 K8 Dolna Oryahovitsa var. Dolna Oryakhovits. Veliko Tarnovo, N Bulgaria 43°09′N 25°44′E
Dolna Oryakhovits see Dolna Oryahovitsa
114 N9 Dolni Chiflik Varna, E Bulgaria 42°59′N 27°43′E
114 I8 Dolni Dabnik var. Dolni Dŭbnik. Pleven, N Bulgaria 43°24′N 24°25′E
Dolni Dŭbnik see Dolni Dabnik
114 F8 Dolni Lom Vidin, NW Bulgaria 43°31′N 22°46′E
Dolnja Lendava see Lendava
129 F14 Dolnośląskie ◐ province SW Poland
111 K18 Dolný Kubín Hung. Alsókubin. Žilinský Kraj, N Slovakia 49°12′N 19°17′E
106 H8 Dolo Veneto, NE Italy 45°25′N 12°06′E
Dolomites/Dolomiti see Dolomitiche, Alpi
106 H6 Dolomitiche, Alpi var. Dolomiti, Eng. Dolomites. ▲▲ NE Italy
Dolonnur see Duolun
Doloon see Tsogt-Ovoo
61 E21 Dolores Buenos Aires, E Argentina 36°21′S 57°39′W
42 E3 Dolores Petén, N Guatemala 16°33′N 89°26′W
171 Q5 Dolores Samar, C Philippines 12°01′N 125°27′E
105 S12 Dolores Valenciana, E Spain 38°09′N 00°45′W
61 D19 Dolores Soriano, SW Uruguay 33°34′S 58°15′W
41 N12 Dolores Hidalgo var. Ciudad de Dolores Hidalgo. Guanajuato, C Mexico 21°10′N 100°55′W
8 J7 Dolphin and Union Strait strait Northwest Territories/ Nunavut, N Canada
65 D23 Dolphin, Cape headland East Falkland, Falkland Islands 51°15′S 58°57′W
44 H12 Dolphin Head hill W Jamaica
83 B21 Dolphin Head var. Cape Dernberg. headland SW Namibia 25°33′S 14°36′E
110 G12 Dolsk Ger. Dolzig. Weilkopolskie, C Poland 51°59′N 17°03′E
167 S8 Đô Lương Nghê An, N Vietnam 18°51′N 105°19′E
116 I6 Dolyna Rus. Dolina. Ivano-Frankivs'ka Oblast', W Ukraine 48°58′N 24°01′E
117 R8 Dolyns'ka Rus. Dolinskaya. Kirovohrads'ka Oblast', S Ukraine 48°06′N 32°46′E
Dolzig see Dolsk
Domachëvo/Domaczewo see Damachava
117 P9 Domanivka Mykolayivs'ka Oblast', S Ukraine 47°40′N 30°56′E
153 S13 Domar Rajshahi, N Bangladesh 26°08′N 88°57′E
108 I9 Domat/Ems Graubünden, SE Switzerland 46°50′N 09°28′E
111 A18 Domažlice Ger. Taus. Plzeňský Kraj, W Czech Republic 49°26′N 12°54′E
127 X8 Dombarovskiy Orenburgskaya Oblast', W Russian Federation 50°53′N 59°18′E
94 G10 Dombås Oppland, S Norway 62°04′N 09°07′E
83 M17 Dombe Manica, C Mozambique 19°59′S 33°24′E
82 A13 Dombe Grande Benguela, C Angola 12°57′S 13°07′E
103 R10 Dombes physical region E France
111 I25 Dombóvár Tolna, S Hungary 46°24′N 18°09′E
99 D14 Domburg Zeeland, SW Netherlands 51°34′N 03°30′E
58 L13 Dom Eliseu Pará, NE Brazil 04°02′S 47°31′W
Domel Island see Letsôk-aw Kyun
103 O11 Dôme, Puy de ▲ C France 45°46′N 03°00′E
36 H13 Dome Rock Mountains ▲▲ Arizona, SW USA
Domesnes, Cape see Kolkasrags
62 G8 Domeyko Atacama, N Chile 28°58′S 70°54′W
62 H5 Domeyko, Cordillera ▲▲ N Chile
102 K5 Domfront Orne, N France 48°35′N 00°39′W
171 X13 Dom, Gunung ▲ Papua, E Indonesia 02°41′S 137°00′E
45 X11 Dominica off. Commonwealth of Dominica. ◆ republic E West Indies
47 S3 Dominica island Dominica
Dominica Channel see Martinique Passage
Dominica, Commonwealth of see Dominica
43 N15 Dominical Puntarenas, SE Costa Rica 09°16′N 83°52′W
45 O8 Dominican Republic ◆ republic C West Indies
45 X11 Dominica Passage passage E Caribbean Sea
99 K14 Dommel ≈ S Netherlands
81 O14 Domo Somalē, E Ethiopia 07°53′N 46°55′E
126 L4 Domodedovo ✈ (Moskva) Moskovskaya Oblast', W Russian Federation 55°19′N 37°55′E
106 C6 Domodossola Piemonte, NE Italy 46°07′N 08°20′E
115 F17 Domokós var. Dhomokós. Stereá Elláda, C Greece 39°07′N 22°18′E
172 I14 Domoni Anjouan, SE Comoros 12°15′S 44°39′E
61 G16 Dom Pedrito Rio Grande do Sul, S Brazil 31°00′S 54°40′W
Dompoe see Dompu
170 M16 Dompu prev. Dompoe. Sumbawa, C Indonesia 08°30′S 118°28′E
Domschale see Domžale
62 H13 Domuyo, Volcán ⛰ W Argentina 36°36′S 70°22′W
109 U11 Domžale Ger. Domschale. C Slovenia 46°09′N 14°33′E
127 O10 Don var. Duna, Tanais. ≈ SW Russian Federation
96 K9 Don ≈ NE Scotland, United Kingdom
182 M11 Donald Victoria, SE Australia 36°27′S 143°03′E
22 J9 Donaldsonville Louisiana, S USA 30°06′N 90°59′W
23 S8 Donalsonville Georgia, SE USA 31°02′N 84°52′W
Donau see Danube
101 G23 Donaueschingen Baden-Württemberg, SW Germany 47°57′N 08°30′E
101 K22 Donaumoos wetland S Germany
101 K22 Donauwörth Bayern, S Germany 48°43′N 10°46′E
109 U7 Donawitz Steiermark, SE Austria 47°23′N 15°00′E
117 X7 Donbass industrial region Russian Federation/Ukraine
104 K11 Don Benito Extremadura, W Spain 38°57′N 05°52′W
97 M17 Doncaster anc. Danum. N England, United Kingdom 53°32′N 01°07′W
44 K12 Don Christophers Point headland C Jamaica 18°19′N 76°48′W
55 V9 Donderkamp Sipaliwini, NW Suriname 05°18′N 56°22′W
82 B12 Dondo Kwanza Norte, NW Angola 09°40′S 14°24′E
171 O12 Dondo Sulawesi, N Indonesia 0°54′S 121°33′E
83 N17 Dondo Sofala, C Mozambique 19°41′S 34°45′E
155 K26 Dondra Head headland S Sri Lanka 05°57′N 80°33′E
Dondușani see Dondușeni
116 M8 Dondușeni var. Dondușani, Rus. Dondyushany. N Moldova 48°13′N 27°38′E
Dondyushany see Dondușeni
97 D15 Donegal Ir. Dún na nGall. Donegal, NW Ireland 54°39′N 08°06′W
97 D14 Donegal Ir. Dún na nGall. cultural region NW Ireland
97 C15 Donegal Bay Ir. Bá Dhún na nGall. bay NW Ireland
84 K10 Donets ≈ Russian Federation/Ukraine
117 X8 Donets'k Rus. Donetsk; prev. Stalino. Donets'ka Oblast', E Ukraine 47°58′N 37°50′E
117 W8 Donets'k ✈ Donets'ka Oblast', E Ukraine 48°03′N 37°44′E
Donets'k see Donets'ka Oblast'
117 W8 Donets'ka Oblast' var. Donets'k, Rus. Donetskaya Oblast'; prev. Rus. Stalins'kaya Oblast'. ◐ province SE Ukraine
Donetskaya Oblast' see Donets'ka Oblast'
67 P8 Donga ≈ Cameroon/Nigeria
157 O13 Dongchuan Yunnan, SW China 26°09′N 103°10′E
161 Q14 Dongchuan Dao prev. Donghan Dao. island SE China
Dong Dao see Lincoln Island
99 I14 Dongen Noord-Brabant, S Netherlands 51°38′N 04°56′E
160 K17 Dongfang var. Basuo. Hainan, S China 19°05′N 108°40′E
163 Z7 Dongfanghong Heilongjiang, NE China 46°13′N 133°13′E
163 W11 Dongfeng Jilin, NE China 42°39′N 125°33′E
171 N12 Donggala Sulawesi, C Indonesia 0°40′S 119°44′E
163 V13 Donggang var. Dadong; prev. Donggou. Liaoning, NE China 39°52′N 124°08′E
Donggou see Donggang
161 O14 Dongguan Guangdong, S China 23°03′N 113°43′E
167 T9 Đông Ha Quang Tri, C Vietnam 16°45′N 107°10′E
163 Y14 Donghae prev. Tonghae. NE South Korea 37°26′N 129°09′E
Dong Hai see East China Sea
160 M16 Donghai Dao island S China
Donghe see Omon Gol
Donghe see Wangcang
167 T9 Đông Hơi Quang Binh, C Vietnam 17°32′N 106°35′E
Donghua see Huating
Donghuang see Xishui
108 H10 Dongio Ticino, S Switzerland 46°27′N 08°58′E
160 L11 Dongkou Hunan, S China 27°06′N 110°35′E
167 S8 Đông Lê Quang Binh, C Vietnam 17°52′N 105°49′E
Dongliao see Liaoyuan
Dong-nai see Đông Nai, Sông
167 U13 Đông Nai, Sông var. Dong-nai, Dong Noi, Donnai. ≈ S Vietnam
161 N14 Dongnan Qiuling plateau SE China
163 Y9 Dongning Heilongjiang, NE China 44°01′N 131°03′E
Dong Noi see Đông Nai, Sông
83 C14 Dongo Huíla, C Angola 14°35′S 15°51′E
80 E7 Dongola var. Donqola, Dunqulah. Northern, N Sudan 19°10′N 30°27′E
79 I17 Dongou Likouala, NE Congo 02°05′N 18°E
Đông Phu see Đông Xoai
Dongping see Anhua
Dong Rak, Phanom see Dângrêk, Chuŏr Phnum
Dongshan Dao see Dongchuan Dao
Dongsha Qundao see Tungsha Tao
Dongsheng see Ordos
161 S13 Dongshi Jap. Tōsei; prev. Tungshih. N Taiwan 24°13′N 120°54′E
161 R7 Dongtai Jiangsu, E China 32°50′N 120°22′E
161 N10 Dongting Hu var. Tung-t'ing Hu. ◎ S China
161 P10 Dongxiang var. Xiaogang. Jiangxi, S China 28°16′N 116°32′E
167 T13 Đông Xoai var. Đông Phu. Sông Be, S Vietnam 11°31′N 106°55′E
161 Q4 Dongying Shandong, E China 37°27′N 118°01′E
27 X8 Doniphan Missouri, C USA 36°39′N 90°51′W
Donja Łužica see Niederlausitz
10 G7 Donjek ≈ Yukon, W Canada
112 E11 Donji Lapac Lika-Senj, W Croatia 44°33′N 15°58′E
112 H8 Donji Miholjac Osijek-Baranja, NE Croatia 45°45′N 18°10′E
112 P12 Donji Milanovac Serbia, E Serbia 44°27′N 22°06′E
112 G12 Donji Vakuf var. Srbobran. ◐ Federacija Bosne I Hercegovine, C Bosnia and Herzegovina
98 M6 Donkerbroek Fryslân, N Netherlands 52°58′N 05°15′E
167 P11 Don Muang ✈ (Krung Thep) Nonthaburi, C Thailand 13°51′N 100°40′E
25 S17 Donna Texas, SW USA 26°10′N 98°03′W
15 Q10 Donnacona Québec, SE Canada 46°41′N 71°46′W
Donnai see Đông Nai, Sông
29 Y16 Donnellson Iowa, C USA 40°38′N 91°33′W
11 O13 Donnelly Alberta, W Canada 55°42′N 117°06′W
35 P6 Donner Pass Pass California, W USA
101 F19 Donnersberg ▲ W Germany 49°37′N 07°54′E
Donoso see Miguel de la Borda
105 P2 Donostia-San Sebastián País Vasco, N Spain 43°19′N 01°59′W
115 K21 Donoúsa var. Donoússa. island Kykládes, Greece, Aegean Sea
Donoússa see Donoúsa
35 P8 Don Pedro Reservoir ▣ California, W USA
Donqola see Dongola
126 L5 Donskoy Tul'skaya Oblast', W Russian Federation 54°02′N 38°27′E
81 L16 Doolow Sumalē, E Ethiopia 04°10′N 42°05′E
39 Q7 Doonerak, Mount ▲ Alaska, USA 67°54′N 150°33′W
98 J12 Doorn Utrecht, C Netherlands 52°02′N 05°21′E
Doornik see Tournai
31 N6 Door Peninsula peninsula Wisconsin, N USA
80 P13 Dooxo Nugaaleed var. Nogal Valley. valley E Somalia
Do Qu see Da Qu
106 B7 Dora Baltea anc. Duria Major. ≈ NW Italy
180 K7 Dora, Lake salt lake Western Australia
106 A8 Dora Riparia anc. Duria Minor. ≈ NW Italy
Dorbiljin see Emin
Dorbod/Dorbod Mongolzu Zizhixian see Taikang
Dorbod Mongolzu Zizhixian see Taikang
113 N18 Dorče Petrov var. Djorče Petrov, Gorče Petrov. N Macedonia 42°01′N 21°21′E
14 F16 Dorchester Ontario, S Canada 43°00′N 81°04′W
97 L24 Dorchester anc. Durnovaria. S England, United Kingdom 50°43′N 02°26′W
9 P7 Dorchester, Cape headland Baffin Island, Nunavut, NE Canada 65°25′N 77°25′W
83 D19 Dordabis Khomas, C Namibia 22°57′S 17°39′E
102 L12 Dordogne ◐ department SW France
103 N12 Dordogne ≈ W France
98 H13 Dordrecht var. Dordt, Dort. Zuid-Holland, SW Netherlands 51°48′N 04°40′E
Dordt see Dordrecht
103 P11 Dore ≈ C France
11 S13 Doré Lake Saskatchewan, C Canada 54°37′N 107°36′W
103 O12 Dore, Monts ▲▲ C France
101 M23 Dorfen Bayern, SE Germany 48°16′N 12°06′E
107 D18 Dorgali Sardegna, Italy, C Mediterranean Sea 40°18′N 09°34′E
159 N11 Dorgê Co var. Elsen Nur. ◎ C China
162 E6 Dörgön var. Seer. Hovd, W Mongolia 48°18′N 92°37′E
162 F7 Dörgön Nuur ◎ NW Mongolia
77 Q12 Dori N Burkina Faso 14°03′N 00°02′W
83 E24 Doring ≈ S South Africa
101 E16 Dormagen Nordrhein-Westfalen, W Germany 51°06′N 06°49′E
103 P4 Dormans Marne, N France 49°03′N 03°44′E
108 E6 Dornach Solothurn, NW Switzerland 47°29′N 07°37′E
Dorna Watra see Vatra Dornei
108 J7 Dornbirn Vorarlberg, W Austria 47°25′N 09°46′E
96 I7 Dornoch N Scotland, United Kingdom 57°52′N 04°01′W
96 J7 Dornoch Firth inlet N Scotland, United Kingdom
163 P7 Dornod ◐ province E Mongolia
163 N10 Dornogovĭ ◐ province SE Mongolia
77 P10 Doro Tombouctou, S Mali 16°07′N 00°57′W
116 L14 Dorobanţu Călăraşi, S Romania 44°15′N 26°55′E
111 J22 Dorog Komárom-Esztergom, N Hungary 47°43′N 18°42′E
126 I4 Dorogobuzh Smolenskaya Oblast', W Russian Federation 54°56′N 33°16′E
116 K8 Dorohoi Botoşani, NE Romania 47°57′N 26°24′E
93 H15 Dorotea Västerbotten, N Sweden 64°17′N 16°30′E
Dorpat see Tartu
180 G10 Dorre Island island Western Australia
183 U5 Dorrigo New South Wales, SE Australia 30°22′S 152°43′E
35 N1 Dorris California, W USA 41°58′N 121°54′W
14 H13 Dorset Ontario, SE Canada 45°12′N 78°52′W
97 K23 Dorset cultural region S England, United Kingdom
101 E14 Dorsten Nordrhein-Westfalen, W Germany 51°38′N 06°58′E
Dort see Dordrecht
101 F15 Dortmund Nordrhein-Westfalen, W Germany 51°31′N 07°28′E
100 F12 Dortmund-Ems-Kanal canal W Germany
136 L17 Dörtyol Hatay, S Turkey 36°51′N 36°11′E
Do Rūd see Dow Rūd
79 O15 Doruma Orientale, N Dem. Rep. Congo 04°35′N 27°43′E
15 O12 Dorval ✈ (Montréal) Québec, SE Canada 45°27′N 73°46′W
162 F7 Dörvöljin var. Buga. Dzavhan, W Mongolia 47°42′N 94°53′E
45 T5 Dos Bocas, Lago ◎ C Puerto Rico
104 K14 Dos Hermanas Andalucía, S Spain 37°16′N 05°55′W
Dospad Dagh see Rhodope Mountains
35 P10 Dos Palos California, W USA 37°00′N 120°39′W
114 I11 Dospat Smolyan, S Bulgaria 41°39′N 24°10′E
114 H11 Dospat, Yazovir ▣ SW Bulgaria
100 M11 Dosse ≈ NE Germany
77 S12 Dosso Dosso, SW Niger 13°03′N 03°10′E
77 S12 Dosso ◐ department SW Niger
144 G12 Dossor Atyrau, SW Kazakhstan 47°31′N 53°01′E
147 O10 Do'stlik Jizzax Viloyati, C Uzbekistan 40°37′N 67°59′E
147 V9 Dostuk Narynskaya Oblast', C Kyrgyzstan 41°19′N 75°40′E
145 X13 Dostyk prev. Druzhba. Almaty, SE Kazakhstan 45°15′N 82°29′E
23 R7 Dothan Alabama, S USA 31°13′N 85°23′W
39 T9 Dot Lake Alaska, USA 63°39′N 144°10′W
118 F12 Dotnuva Kaunas, C Lithuania 55°23′N 23°53′E
99 D19 Dottignies Hainaut, W Belgium 50°43′N 03°16′E
103 P2 Douai prev. Douay; anc. Duacum. Nord, N France 50°22′N 03°04′E
15 L9 Douaire, Lac ◎ Québec, SE Canada
79 D16 Douala var. Duala. Littoral, W Cameroon 04°04′N 09°43′E
79 D16 Douala ✈ Littoral, W Cameroon 03°57′N 09°48′E
102 F6 Douarnenez Finistère, NW France 48°05′N 04°20′W
102 E6 Douarnenez, Baie de bay NW France
Douay see Douai
25 O6 Double Mountain Fork Brazos River ≈ Texas, SW USA
23 O3 Double Springs Alabama, S USA 34°09′N 87°24′W
103 T8 Doubs ◐ department E France
108 C8 Doubs ≈ France/ Switzerland
185 A22 Doubtful Sound sound South Island, New Zealand
184 J2 Doubtless Bay bay North Island, New Zealand
25 X9 Doucette Texas, SW USA 30°48′N 94°25′W
102 K8 Doué-la-Fontaine Maine-et-Loire, NW France 47°12′N 00°16′W
77 O11 Douentza Mopti, S Mali 14°59′N 02°57′W
65 E24 Douglas East Falkland, Falkland Islands
97 I16 Douglas ○ (Isle of Man) E Isle of Man 54°09′N 04°28′W
83 H23 Douglas Northern Cape, C South Africa 29°04′S 23°47′E
39 X13 Douglas Alexander Archipelago, Alaska, USA 58°12′N 134°18′W
37 O17 Douglas Arizona, SW USA 31°20′N 109°32′W
23 U7 Douglas Georgia, SE USA 31°30′N 82°51′W
33 Y15 Douglas Wyoming, C USA 42°48′N 105°23′W
21 O7 Douglas Cape headland Alaska, N USA 64°59′N 166°41′W
10 J14 Douglas Channel channel British Columbia, W Canada
182 G3 Douglas Creek seasonal river South Australia
31 P5 Douglas Lake ◎ Michigan, N USA
21 O9 Douglas Lake ▣ Tennessee, S USA
39 Q13 Douglas, Mount ▲ Alaska, USA 58°51′N 153°31′W
194 I6 Douglas Range ▲▲ Alexander Island, Antarctica
161 S14 Douliu prev. Touliu. C Taiwan 23°44′N 120°27′E
103 O2 Doullens Somme, N France 50°09′N 02°21′E
Douma see Dūmā
79 F15 Doumé Est, E Cameroon 04°14′N 13°27′E
99 E21 Dour Hainaut, S Belgium 50°24′N 03°47′E
59 K18 Dourada, Serra ▲▲ S Brazil
59 I21 Dourados Mato Grosso do Sul, S Brazil 22°09′S 54°52′W
103 N5 Dourdan Essonne, N France 48°33′N 01°58′E
104 I6 Douro Sp. Duero. ≈ Portugal/Spain see also Duero
Douro see Duero
104 G6 Douro Litoral former province N Portugal
Douvres see Dover
102 K15 Douze ≈ SW France
183 P17 Dover Tasmania, SE Australia 43°19′S 147°01′E
97 Q22 Dover Fr. Douvres, Lat. Dubris Portus. SE England, United Kingdom 51°08′N 01°19′E
21 Y3 Dover state capital Delaware, NE USA 39°09′N 75°31′W
19 P9 Dover New Hampshire, NE USA 43°10′N 70°50′W
18 J14 Dover New Jersey, NE USA 40°51′N 74°33′W
31 U12 Dover Ohio, N USA 40°31′N 81°28′W
20 H8 Dover Tennessee, S USA 36°30′N 87°50′W
97 Q23 Dover, Strait of var. Straits of Dover, Fr. Pas de Calais. strait England, United Kingdom/France
Dover, Straits of see Dover, Strait of
Dovlen see Devin
94 G11 Dovre Oppland, S Norway 61°59′N 09°16′E
94 G10 Dovrefjell plateau S Norway
Dovsk see Dowsk
83 M14 Dowa Central, C Malawi 13°40′S 33°55′E
31 O10 Dowagiac Michigan, N USA 41°58′N 86°06′W
Dow Gonbadān see Do Gonbadān
148 M2 Dowlatābād Fāryāb, N Afghanistan 36°30′N 64°51′E
97 G16 Down cultural region SE Northern Ireland, United Kingdom
33 R16 Downey Idaho, NW USA 42°25′N 112°06′W
35 P5 Downieville California, W USA 39°34′N 120°49′W
97 G16 Downpatrick Ir. Dún Pádraig. SE Northern Ireland, United Kingdom 54°20′N 05°43′W
26 M3 Downs Kansas, C USA 39°30′N 98°33′W
18 J12 Downsville New York, NE USA 42°03′N 74°59′W
142 L7 Dow Rūd var. Do Rūd, Durud. Lorestān, W Iran 33°28′N 49°04′E
29 V12 Dows Iowa, C USA 42°39′N 93°30′W
119 O17 Dowsk Rus. Dovsk. Homyel'skaya Voblasts', SE Belarus 53°09′N 30°28′E
35 Q4 Doyle California, W USA 40°00′N 120°06′W
18 I15 Doylestown Pennsylvania, NE USA 40°18′N 75°08′W
Doyransko, Ezero see Dojran, Lake
114 I8 Doyrentsi Lovech, N Bulgaria 43°13′N 24°46′E
164 G11 Dōzen island Oki-shotō, SW Japan
14 K9 Dozois, Réservoir ▣ Québec, SE Canada
74 D9 Drâa seasonal river S Morocco
Drâa, Hammada du see Dra, Hamada du
Drabble see José Enrique Rodó
117 Q5 Drabiv Cherkas'ka Oblast', C Ukraine 49°57′N 32°10′E
Drable see José Enrique Rodó
103 S13 Drac ≈ E France
Drač/Draç see Durrës
60 I8 Dracena São Paulo, S Brazil 21°27′S 51°30′W
98 M6 Drachten Fryslân, N Netherlands 53°07′N 06°06′E
92 H11 Drag Lapp. Ájluokta. Nordland, C Norway 68°02′N 16°E
116 L14 Dragalina Călăraşi, SE Romania 44°26′N 27°19′E
116 I14 Draganesti-Olt Olt, S Romania 44°06′N 25°00′E
116 J14 Drăgănești-Vlașca Teleorman, S Romania 44°05′N 25°39′E
116 I13 Drăgășani Vâlcea, SW Romania 44°40′N 24°16′E
114 G9 Dragoman Sofia, W Bulgaria 42°32′N 22°56′E
115 L25 Dragonáda island SE Greece
Dragonera, Isla see Sa Dragonera
45 T14 Dragon's Mouths, The strait Trinidad and Tobago/ Venezuela
95 J23 Dragør Sjælland, E Denmark 55°36′N 12°42′E
114 F10 Dragovishtitsa Kyustendil, W Bulgaria 42°22′N 22°39′E
103 U15 Draguignan Var, SE France 43°31′N 06°31′E
74 E9 Dra, Hamada du var. Hammada du Drâa, Haut Plateau du Dra. plateau W Algeria
Dra, Haut Plateau du see Dra, Hamada du
119 H19 Drahichyn Pol. Drohiczyn Poleski, Rus. Drogichin. Brestskaya Voblasts', SW Belarus 52°11′N 25°10′E
29 N4 Drake North Dakota, N USA 47°54′N 100°23′W
83 K23 Drakensberg ▲▲ Lesotho/ South Africa
194 F3 Drake Passage passage Atlantic Ocean/Pacific Ocean
114 L8 Dralfa Tŭrgovishte, N Bulgaria 43°17′N 26°25′E
114 I12 Dráma var. Dhráma. Anatolikí Makedonía kai Thráki, NE Greece 41°09′N 24°10′E
Dramburg see Drawsko Pomorskie
95 H15 Drammen Buskerud, S Norway 59°44′N 10°12′E
95 H15 Drammensfjorden fjord S Norway
92 H1 Drangajökull ▲ NW Iceland 66°13′N 22°18′W
95 F16 Drangedal Telemark, S Norway 59°05′N 09°05′E
92 I2 Drangsnes Vestfirðir, NW Iceland 65°42′N 21°27′W
Drann see Dravinja
109 T10 Drau var. Drava, Eng. Drave, Hung. Dráva. ≈ C Europe see also Drava
Drau see Drava
84 I11 Drava var. Drau, Eng. Drave, Hung. Dráva. ≈ C Europe see also Drau
Dráva/Drave see Drau/ Drava
109 W10 Dravinja Ger. Drann. ≈ NE Slovenia
109 V9 Dravograd Ger. Unterdrauburg; prev. Spodnji Dravograd. N Slovenia 46°36′N 15°00′E
110 F10 Drawa ≈ NW Poland
110 F9 Drawno Zachodnio-pomorskie, NW Poland 53°12′N 15°44′E
110 F9 Drawsko Pomorskie Ger. Dramburg. Zachodnio-pomorskie, NW Poland 53°32′N 15°48′E
29 R3 Drayton North Dakota, N USA 48°34′N 97°10′W
11 P14 Drayton Valley Alberta, SW Canada 53°15′N 115°00′W
186 B6 Dreikikir East Sepik, NW Papua New Guinea 03°42′S 142°46′E
Dreikirchen see Teiuş
98 N7 Drenthe ◐ province NE Netherlands
115 H15 Drépano, Akrotírio var. Akrotírio Dhrepanon. headland N Greece 39°56′N 23°57′E
Drepanum see Trapani
14 D17 Dresden Ontario, S Canada 42°34′N 82°09′W
101 O16 Dresden Sachsen, E Germany 51°03′N 13°43′E
20 G8 Dresden Tennessee, S USA 36°17′N 88°42′W
118 M11 Dretun' Vitsyebskaya Voblasts', N Belarus 55°41′N 29°13′E
102 M5 Dreux anc. Drocae, Durocasses. Eure-et-Loir, C France 48°44′N 01°23′E
94 I11 Drevsjø Hedmark, S Norway 61°52′N 12°01′E
22 K3 Drew Mississippi, S USA 33°48′N 90°31′W
110 F10 Drezdenko Ger. Driesen. Lubuskie, W Poland 52°51′N 15°50′E
98 J12 Driebergen var. Driebergen-Rijsenburg. Utrecht, C Netherlands 52°03′N 05°17′E
Driebergen-Rijsenburg see Driebergen
Driesen see Drezdenko
97 N16 Driffield E England, United Kingdom 54°00′N 00°28′W
65 D25 Driftwood Point headland East Falkland, Falkland Islands 52°15′S 59°00′W
33 S14 Driggs Idaho, NW USA 43°44′N 111°06′W
Drin see Drinit, Lumi i
112 K12 Drina ≈ Bosnia and Herzegovina/Serbia
Drin, Gulf of see Drinit, Gjiri i
113 M16 Drini i Bardhë Serb. Beli Drim. ≈ Albania/Serbia
113 K18 Drinit, Gjiri i var. Pellg i Drinit, Eng. Gulf of Drin. gulf NW Albania
113 L17 Drinit, Lumi i var. Drin. ≈ NW Albania
Drinit, Pellg i see Drinit, Gjiri i
Drinit të Zi, Lumi i see Black Drin
113 L22 Dríno var. Drino, Drínos Pótamos, Alb. Lumi i Drinos. ≈ Albania/Greece
Drinos, Lumi i/Drínos Pótamos see Dríno
25 S11 Dripping Springs Texas, SW USA 30°11′N 98°04′W
25 S15 Driscoll Texas, SW USA 27°40′N 97°45′W
22 H5 Driskill Mountain ▲ Louisiana, S USA 32°25′N 92°54′W
Drissa see Drysa
94 G10 Driva ≈ S Norway
112 E13 Drniš It. Šibenik-Knin. Šibenik-Knin, S Croatia 43°51′N 16°10′E
95 H15 Drøbak Akershus, S Norway 59°40′N 10°40′E
116 G13 Drobeta-Turnu Severin prev. Turnu Severin. Mehedinți, SW Romania 44°39′N 22°40′E
Drocae see Dreux
116 M8 Drochia Rus. Drokiya. N Moldova 48°02′N 27°46′E
97 F17 Drogheda Ir. Droichead Átha. NE Ireland 53°43′N 06°21′W
Drogichin see Drahichyn
Drogobych see Drohobych
Drohiczyn Poleski see Drahichyn
116 H6 Drohobych Pol. Drohobycz, Rus. Drogobych. L'vivs'ka Oblast', NW Ukraine 49°22′N 23°33′E
Drohobycz see Drohobych
Droichead Átha see Drogheda
Droicheadna Bandan see Bandon
Droichead na Banna see Banbridge
Droim Mór see Dromore
Drokiya see Drochia
103 R13 Drôme ◐ department E France
103 S13 Drôme ≈ E France
97 G15 Dromore Ir. Droim Mór. SE Northern Ireland, United Kingdom 54°25′N 06°09′W
106 A9 Dronero Piemonte, NE Italy 44°28′N 07°25′E
102 L12 Dronne ≈ SW France
195 T3 Dronning Fabiolafjella var. Mount Victor. ▲ Antarctica 72°49′S 33°01′E
195 Q3 Dronning Maud Land physical region Antarctica
98 K6 Dronrijp Fris. Dronryp. Fryslân, N Netherlands 53°12′N 05°37′E
Dronryp see Dronrijp
98 L9 Dronten Flevoland, C Netherlands 52°31′N 05°41′E
Drontheim see Trondheim
102 L13 Dropt ≈ SW France
149 T4 Drosh Khyber Pakhtunkhwa, NW Pakistan 35°33′N 71°48′E
Drossen see Ośno Lubuskie
Drug see Durg
Drujba see Pitnak
118 I12 Drūkšiai ◎ NE Lithuania
Druk-yul see Bhutan
11 Q16 Drumheller Alberta, SW Canada 51°28′N 112°42′W
33 Q10 Drummond Montana, NW USA 46°39′N 113°12′W
31 R4 Drummond Island island Michigan, N USA
Drummond Island see Tabiteuea
21 X7 Drummond, Lake ◎ Virginia, NE USA
15 P12 Drummondville Québec, SE Canada 45°52′N 72°28′W
39 T11 Drum, Mount ▲ Alaska, USA 62°11′N 144°37′W
27 O9 Drumright Oklahoma, C USA 35°59′N 96°36′W
99 J14 Drunen Noord-Brabant, S Netherlands 51°41′N 05°08′E
Druskieniki see Druskininkai
119 F15 Druskininkai Pol. Druskienniki. Alytus, S Lithuania 54°00′N 24°00′E
98 K13 Druten Gelderland, SE Netherlands 51°53′N 05°37′E
118 K11 Druya Vitsyebskaya Voblasts', NW Belarus 55°47′N 27°27′E
117 S2 Druzhba Sums'ka Oblast', NE Ukraine 52°01′N 33°56′E
Druzhba see Dostyk, Kazakhstan
Druzhba see Pitnak, Uzbekistan
123 R7 Druzhina Respublika Sakha (Yakutiya), NE Russian Federation 68°01′N 144°58′E
117 X7 Druzhkivka Donets'ka Oblast', E Ukraine 48°38′N 37°31′E
112 E12 Drvar Federacija Bosne I Hercegovine, W Bosnia and Hercegovina 44°21′N 16°24′E
113 G15 Drvenik Split-Dalmacija, SE Croatia 43°10′N 17°13′E
114 K9 Dryanovo Gabrovo, N Bulgaria 42°58′N 25°28′E
26 G7 Dry Cimarron River ≈ Kansas/Oklahoma, C USA
12 B11 Dryden Ontario, C Canada 49°48′N 92°48′W
24 M11 Dryden Texas, SW USA 30°01′N 102°06′W
195 Q14 Drygalski Ice Tongue ice feature Antarctica
118 L11 Drysa Rus. Drissa. ≈ N Belarus
23 V17 Dry Tortugas island Florida, SE USA
79 D15 Dschang Ouest, W Cameroon 05°28′N 10°02′E
54 J5 Duaca Lara, N Venezuela 10°22′N 69°08′W
Duacum see Douai
Duala see Douala
45 N9 Duarte, Pico ▲ C Dominican Republic 19°02′N 70°57′W
140 J5 Dubā Tabūk, NW Saudi Arabia 27°26′N 35°42′E
Dubai see Dubayy
117 N9 Dubăsari Rus. Dubossary. NE Moldova 47°16′N 29°07′E
117 N9 Dubăsari Reservoir ▣ NE Moldova
8 M10 Dubawnt ≈ Nunavut, NW Canada
8 L9 Dubawnt Lake ◎ Northwest Territories/Nunavut, N Canada
30 L6 Du Bay, Lake ▣ Wisconsin, N USA
141 U7 Dubayy Eng. Dubai. Dubayy, NE United Arab Emirates 25°11′N 55°18′E
141 W7 Dubayy Eng. Dubai. ✈ Dubayy, NE United Arab Emirates 25°15′N 55°22′E
183 R7 Dubbo New South Wales, SE Australia 32°16′S 148°41′E
108 G7 Dübendorf Zürich, NW Switzerland 47°23′N 08°37′E
97 F18 Dublin Ir. Baile Átha Cliath; anc. Eblana. ● (Ireland) Dublin, E Ireland 53°20′N 06°15′W
23 U5 Dublin Georgia, SE USA 32°32′N 82°54′W
25 R7 Dublin Texas, SW USA 32°05′N 98°20′W
97 G18 Dublin Ir. Baile Átha Cliath; anc. Eblana. cultural region E Ireland
97 G18 Dublin Airport ✈ Dublin, E Ireland 53°25′N 06°18′W
189 V12 Dublon var. Tonoas. island Chuuk Islands, C Micronesia
126 K2 Dubna Moskovskaya Oblast', W Russian Federation 56°45′N 37°09′E
111 G19 Dubňany Ger. Dubnian. Jihomoravský Kraj, SE Czech Republic 48°54′N 17°00′E
Dubnian see Dubňany
111 I19 Dubnica nad Váhom Hung. Máriatölgyes; prev. Dubnicz. Trenčiansky Kraj, W Slovakia 48°58′N 18°10′E
Dubnicz see Dubnica nad Váhom
116 K4 Dubno Rivnens'ka Oblast', NW Ukraine 50°28′N 25°40′E
33 R13 Dubois Idaho, NW USA 44°10′N 112°13′W
18 D13 Du Bois Pennsylvania, NE USA 41°07′N 78°45′W
33 T14 Dubois Wyoming, C USA 43°31′N 109°38′W
Dubossary see Dubăsari
127 O10 Dubovka Volgogradskaya Oblast', SW Russian Federation 49°10′N 48°49′E
76 H14 Dubréka SW Guinea 09°48′N 13°31′W
14 B7 Dubreuilville Ontario, S Canada 48°21′N 84°31′W
Dubris Portus see Dover
119 L20 Dubrova Homyel'skaya Voblasts', SE Belarus 51°47′N 28°13′E
Dubrovačko-Neretvanska Županija see Dubrovnik-Neretva
126 I5 Dubrovka Bryanskaya Oblast', W Russian Federation 53°44′N 33°27′E
113 H16 Dubrovnik It. Ragusa. Dubrovnik-Neretva, SE Croatia 42°40′N 18°06′E
113 I16 Dubrovnik ✈ Dubrovnik-Neretva, SE Croatia 42°34′N 18°17′E
113 F16 Dubrovnik-Neretva off. Dubrovačko-Neretvanska Županija. ◐ province SE Croatia
Dubrovno see Dubrowna
116 L2 Dubrovytsya Rivnens'ka Oblast', NW Ukraine 51°34′N 26°37′E
119 O14 Dubrowna Rus. Dubrovno. Vitsyebskaya Voblasts', N Belarus 54°35′N 30°41′E
29 Z13 Dubuque Iowa, C USA 42°30′N 90°40′W
118 E12 Dubysa ≈ C Lithuania
Đưc Cơ see Chư Ty
191 V12 Duc de Gloucester, Îles du Eng. Duke of Gloucester Islands. island group C French Polynesia
111 C15 Duchcov Ger. Dux. Ústecký Kraj, NW Czech Republic 50°37′N 13°45′E

◆ Country ● Country Capital ◇ Dependent Territory ○ Dependent Territory Capital ◐ Administrative Regions ✈ International Airport ▲ Mountain ▲▲ Mountain Range ⛰ Volcano ≈ River ◎ Lake ▣ Reservoir

37 N3 **Duchesne** Utah, W USA 40°09′N 110°24′W
191 P17 **Ducie Island** *atoll* E Pitcairn Group of Islands
11 W15 **Duck Bay** Manitoba, S Canada 52°11′N 100°08′W
23 X17 **Duck Key** *island* Florida Keys, Florida, SE USA
11 T14 **Duck Lake** Saskatchewan, S Canada 52°52′N 106°12′W
11 V15 **Duck Mountain** ▲ Manitoba, S Canada
20 I9 **Duck River** ♣ Tennessee, S USA
20 M10 **Ducktown** Tennessee, S USA 35°01′N 84°24′W
167 Q12 **Đức Phô** Quang Ngai, C Vietnam 14°56′N 108°55′E
Đức Tho *see* Lin Camh
Đức Trong *see* Liên Nghia
D-U-D *see* Dalap-Uliga-Djarrit
153 N15 **Dūddhinagar** *var.* Dũdhi. Uttar Pradesh, N India 24°09′N 83°16′E
99 M25 **Dudelange** *var.* Forge du Sud, *Ger.* Dudelingen. Luxembourg, S Luxembourg 49°28′N 06°05′E
Dudelingen *see* Dudelange
101 J15 **Duderstadt** Niedersachsen, C Germany 51°31′N 10°16′E
Dũdhi *see* Dũddhinagar
122 K8 **Dudinka** Krasnoyarskiy Kray, N Russian Federation 69°27′N 86°13′E
97 L20 **Dudley** C England, United Kingdom 52°30′N 02°05′W
154 G13 **Dudna** ♣ C India
76 L16 **Duékoué** W Ivory Coast 05°50′N 05°22′W
104 M5 **Dueñas** Castilla y León, N Spain 41°52′N 04°33′W
104 K4 **Duerna** ♣ NW Spain
105 O6 **Duero Port.** Douro. ♣ Portugal/Spain *see also* Douro
Duero *see* Douro
Duesseldorf *see* Düsseldorf
21 P12 **Due West** South Carolina, SE USA 34°19′N 82°23′W
195 P11 **Dufek Coast** *physical region* Antarctica
99 H17 **Duffel** Antwerpen, C Belgium 51°06′N 04°30′E
35 S2 **Duffer Peak** ▲ Nevada, W USA 41°40′N 118°45′W
187 Q9 **Duff Islands** *island group* E Solomon Islands
Dufour, Pizzo/Dufour, Punta *see* Dufour Spitze
108 E12 **Dufour Spitze** *It.* Pizzo Dufour, Punta Dufour. ▲ Italy/Switzerland 45°54′N 07°51′E
112 D9 **Duga Resa** Karlovac, C Croatia 45°25′N 15°30′E
22 H5 **Dugdemona River** ♣ Louisiana, S USA
154 J12 **Duggipar** Mahārāshtra, C India 21°06′N 80°01′E
112 B13 **Dugi Otok** *var.* Isola Grossa, *It.* Isola Lunga. *island* W Croatia
113 F14 **Dugopolje** Split-Dalmacija, S Croatia 43°35′N 16°35′E
160 L8 **Du Hé** ♣ C China
54 M11 **Duida, Cerro** ▲ S Venezuela 03°21′N 65°43′W
Duinekerke *see* Dunkerque
101 E14 **Duisburg** *prev.* Duisburg-Hamborn. Nordrhein-Westfalen, W Germany 51°25′N 06°47′E
Duisburg-Hamborn *see* Duisburg
99 F14 **Duiveland** *island* SW Netherlands
98 M12 **Duiven** Gelderland, E Netherlands 51°57′N 06°02′E
139 W10 **Dujaylah, Hawr ad** ♦ S Iraq
160 H9 **Dujiangyan** *var.* Guanxian, Guan Xian. Sichuan, C China 31°01′N 103°40′E
81 L18 **Dujuuma** Shabeellaha Hoose, S Somalia 01°04′N 42°37′E
139 T3 **Dūkān** *Ar.* Dūkān, *var.* Dokan. As Sulaymānīyah, E Iraq 35°55′N 44°58′E
Dūkān *see* Dūkan
39 Z14 **Duke Island** *island* Alexander Archipelago, Alaska, USA
Dukelský Priesmy/Dukelský Průsmyk *see* Dukla Pass
Duke of Gloucester Islands *see* Duc de Gloucester, Îles du
81 F14 **Duk Faiwil** Jonglei, E South Sudan 07°30′N 31°27′E
141 T7 **Dukhān** C Qatar 25°29′N 50°48′E
Dukhan Heights *see* Dukhān, Jabal
143 N16 **Dukhān, Jabal** *var.* Dukhan Heights. *hill range* S Qatar
127 Q7 **Dukhovnitskoye** Saratovskaya Oblast', W Russian Federation 52°31′N 48°32′E
126 H4 **Dukhovshchina** Smolenskaya Oblast', W Russian Federation 55°15′N 32°22′E
Dukielska, Przełęcz *see* Dukla Pass
111 N17 **Dukla** Podkarpackie, SE Poland 49°33′N 21°40′E
Duklai Hág *see* Dukla Pass
111 N18 **Dukla Pass** *Cz.* Dukelský Průsmyk, *Ger.* Dukla-Pass, *Hung.* Duklai Hág, *Pol.* Przełęcz Dukielska, *Slvk.* Dukelský Priesmy. *pass* Poland/Slovakia
Dukla-Pass *see* Dukla Pass
Dukou *see* Panzhihua
118 G12 **Dūkštas** Utena, E Lithuania 55°32′N 26°21′E
Dulaan *see* Herlenbayan-Ulaan
159 R10 **Dulan** *var.* Qagan Us. Qinghai, C China 36°11′N 97°51′E
37 R6 **Dulce** New Mexico, SW USA 36°55′N 107°00′W
43 N13 **Dulce, Golfo** *gulf* S Costa Rica
Dulce, Golfo *see* Izabal, Lago de
42 K6 **Dulce Nombre de Culmí** Olancho, C Honduras 15°09′N 85°33′W
63 C20 **Dulce, Río** ♣ C Argentina
123 Q9 **Dulgalakh** ♣ NE Russian Federation
153 V14 **Dullabchara** Assam, NE India 24°25′N 92°27′E
20 D11 **Dulles** ✈ (Washington DC) Virginia, NE USA 39°00′N 77°27′W

101 E14 **Dülmen** Nordrhein-Westfalen, W Germany 51°51′N 07°17′E
114 M7 **Dulovo** Silistra, NE Bulgaria 43°51′N 27°10′E
29 W5 **Duluth** Minnesota, N USA 46°47′N 92°06′W
138 H7 **Dûmâ** *Fr.* Douma. Rif Dimashq, SW Syria 33°33′N 36°24′E
171 O8 **Dumagasa Point** *headland* Mindanao, S Philippines 07°01′N 121°54′E
171 P6 **Dumaguete** *var.* Dumaguete City. Negros, C Philippines 09°16′N 123°17′E
Dumaguete City *see* Dumaguete
168 J10 **Dumai** Sumatera, W Indonesia 01°39′N 101°28′E
183 R7 **Dumaresq River** ♣ New South Wales/Queensland, SE Australia
27 W13 **Dumas** Arkansas, C USA 33°53′N 91°29′W
25 N1 **Dumas** Texas, SW USA 35°51′N 101°57′W
138 I7 **Dumayr** Rif Dimashq, SW Syria 33°36′N 36°28′E
96 I12 **Dumbarton** *cultural region* C Scotland, United Kingdom 55°57′N 04°35′W
96 I12 **Dumbarton** C Scotland, United Kingdom 55°57′N 04°35′W
187 Q17 **Dumbéa** Province Sud, S New Caledonia 22°11′S 166°27′E
111 K19 **Ďumbier** *Ger.* Djumbir, *Hung.* Gyömbér. ▲ C Slovakia 48°54′N 19°36′E
116 I11 **Dumbrăveni** *Ger.* Elisabethstadt, *Hung.* Erzsébetváros; *prev.* Ebesfalva, Eppeschdorf, Ibaşfalău. Sibiu, C Romania 46°14′N 24°34′E
116 L12 **Dumbrăveni** Vrancea, E Romania 45°31′N 27°09′E
97 J14 **Dumfries** S Scotland, United Kingdom 55°04′N 03°37′W
97 J14 **Dumfries** *cultural region* SW Scotland, United Kingdom
153 R15 **Dumka** Jhārkhand, NE India 24°17′N 87°15′E
15 P13 **Dumham** Québec, SE Canada 45°08′N 72°48′W
Dümmer *see* Dümmersee
100 G12 **Dümmersee** *var.* Dümmer. ◎ NW Germany
14 J11 **Dumoine** ♣ Québec, SE Canada
14 J10 **Dumoine, Lac** ◎ Québec, SE Canada
195 V16 **Dumont d'Urville** *French research station* Antarctica 66°24′S 139°38′E
195 W15 **Dumont d'Urville Sea** *sea* S Pacific Ocean
14 K11 **Dumont, Lac** ◎ Québec, SE Canada
75 W7 **Dumyât** *var.* Dumyât, *Eng.* Damietta. N Egypt 31°26′N 31°48′E
Dún *see* Doon
111 J24 **Dunaföldvár** Tolna, C Hungary 46°48′N 18°55′E
Dunaj *see* Wien, Austria
Dunaj *see* Danube, C Europe
111 L18 **Dunajec** ♣ S Poland
111 H21 **Dunajská Streda** *Hung.* Dunaszerdahely. Trnavský Kraj, W Slovakia 48°N 17°28′E
Dunántúl *see* Transdanubia
Dunapentele *see* Dunaújváros
116 M13 **Dunărea** *Ger.* Donau, *Eng.* Danube, *Rus.* Dunay. ♣ SE Romania
117 N13 **Dunării, Delta** *delta* SE Romania
111 J23 **Dunaújváros** *prev.* Dunapentele, Sztálinváros. Fejér, C Hungary 46°59′N 18°55′E
Dunav *see* Danube
114 J8 **Dunavska Ravnina** *Eng.* Danubian Plain. *lowlands* N Bulgaria
114 G7 **Dunavtsi** Vidin, NW Bulgaria 43°54′N 22°49′E
112 S15 **Dunay** Primorskiy Kray, SE Russian Federation 42°54′N 132°25′E
Dunay *see* Dunayivtsi
116 L7 **Dunayivtsi** *Rus.* Dunayivtsy. Khmel'nyts'ka Oblast', NW Ukraine 48°56′N 26°50′E
185 F22 **Dunback** Otago, South Island, New Zealand 45°22′S 170°37′E
10 L17 **Duncan** Vancouver Island, British Columbia, SW Canada 48°46′N 123°10′W
27 O13 **Duncan** Oklahoma, C USA 34°30′N 97°57′W
26 M12 **Duncan** Oklahoma, C USA 34°30′N 97°57′W
185 D21 **Duncan Mountains** ▲ South Island, New Zealand
151 Q20 **Duncan Passage** *strait* Andaman Sea/Bay of Bengal
96 K6 **Duncansby Head** *headland* N Scotland, United Kingdom 58°37′N 03°01′W
14 L14 **Dunchurch** Ontario, S Canada 45°36′N 79°54′W
118 D12 **Dundaga** NW Latvia 57°29′N 22°19′E
14 G14 **Dundalk** Ontario, S Canada 44°11′N 80°22′W
97 F16 **Dundalk** *Ir.* Dún Dealgan. Louth, NE Ireland 54°N 06°25′W
21 X3 **Dundalk** Maryland, NE USA 39°15′N 76°31′W
97 F16 **Dundalk Bay** *Ir.* Cuan Dún Dealgan. *bay* NE Ireland
23 G16 **Dundas** Ontario, S Canada 43°16′N 79°55′W
180 L12 **Dundas, Lake** *salt lake* Western Australia
Dundbürd *see* Batnorov
51 N13 **Dún Dealgan** *see* Dundalk 45°01′N 74°27′W
14 G14 **Dundonald** Ontario, S Canada 44°11′N 80°22′W
96 K11 **Dundee** E Scotland, United Kingdom 56°28′N 03°03′W
31 R10 **Dundee** Michigan, N USA 41°57′N 83°39′W
29 R5 **Dundee** Minnesota, SW USA 43°43′N 95°33′W
194 H3 **Dundee Island** *island* Antarctica
162 L9 **Dundgovi** ♦ province C Mongolia
97 F16 **Dundrum Bay** *Ir.* Cuan Dhún Droma. *inlet* NW Irish Sea

11 T15 **Dundurn** Saskatchewan, S Canada 51°43′N 106°22′W
Dund-Us *see* Hovd
Dund-Us *see* Hovd
185 F23 **Dunedin** Otago, South Island, New Zealand 45°52′S 170°31′E
183 R7 **Dunedoo** New South Wales, SE Australia 32°04′S 149°23′E
97 D14 **Dunfanaghy** *Ir.* Dún Fionnachaidh. NW Ireland 55°11′N 07°59′W
96 J12 **Dunfermline** C Scotland, United Kingdom 56°04′N 03°29′W
Dún Fionnachaidh *see* Dunfanaghy
149 V10 **Dunga Bunga** Punjab, E Pakistan 29°51′N 73°19′E
97 F15 **Dungannon** *Ir.* Dún Geanainn. C Northern Ireland, United Kingdom 54°31′N 06°46′W
152 F15 **Dūngarpur** Rājasthān, N India 23°50′N 73°43′E
97 E21 **Dungarvan** *Ir.* Dún Garbháin. Waterford, S Ireland 52°05′N 07°37′W
101 N21 **Dungau** *cultural region* SE Germany
Dún Geanainn *see* Dungannon
97 P23 **Dungeness** *headland* SE England, United Kingdom 50°55′N 00°58′E
63 J25 **Dungeness, Punta** *headland* S Argentina 52°25′S 68°25′W
97 E16 **Dunglow** *see* Dunglow
97 D14 **Dunglow** *var.* Dungloe, *Ir.* An Clochán Liath. Donegal, NW Ireland 54°57′N 08°22′W
183 T7 **Dungog** New South Wales, SE Australia 32°24′S 151°45′E
79 O16 **Dungu** Orientale, NE Dem. Rep. Congo 03°40′N 28°32′E
168 L8 **Dungun** *var.* Kuala Dungun. Terengganu, Peninsular Malaysia 04°47′N 103°26′E
80 H6 **Dungūnab** Red Sea, NE Sudan 21°10′N 37°09′E
15 P13 **Dunham** Québec, SE Canada 45°08′N 72°48′W
Dunheved *see* Launceston
Dunholme *see* Durham
163 X10 **Dunhua** Jilin, NE China 43°22′N 128°12′E
159 P8 **Dunhuang** Gansu, N China 40°10′N 94°40′E
182 L12 **Dunkeld** Victoria, SE Australia 37°41′S 142°19′E
103 O1 **Dunkerque** *Eng.* Dunkirk, *Flem.* Duinekerke; *prev.* Dunquerque. Nord, N France 51°06′N 02°34′E
97 K23 **Dunkery Beacon** ▲ SW England, United Kingdom 51°10′N 03°36′W
18 C11 **Dunkirk** New York, NE USA 42°28′N 79°19′W
77 P17 **Dunkwa** SW Ghana 05°59′N 01°45′W
97 G18 **Dún Laoghaire** *Eng.* Dunleary; *prev.* Kingstown. E Ireland 53°17′N 06°08′W
29 S14 **Dunlap** Iowa, C USA 41°51′N 95°36′W
20 J9 **Dunlap** Tennessee, S USA 35°22′N 85°23′W
Dunleary *see* Dún Laoghaire
Dún Mánmhaí *see* Dunmanway
97 B21 **Dunmanway** *Ir.* Dún Mánmhaí. Cork, SW Ireland 51°43′N 09°07′W
18 H3 **Dunmore** Pennsylvania, NE USA 41°25′N 75°37′W
21 U10 **Dunn** North Carolina, SE USA 35°19′N 78°36′W
Dún na nGall *see* Donegal
13 U13 **Dunnellon** Florida, SE USA 29°03′N 82°27′W
96 J6 **Dunnet Head** *headland* N Scotland, United Kingdom 58°40′N 03°27′W
29 N14 **Dunning** Nebraska, C USA 41°49′N 100°04′W
65 B24 **Dunnose Head Settlement** West Falkland, Falkland Islands 51°24′S 60°29′W
14 G17 **Dunnville** Ontario, S Canada 42°54′N 79°36′W
Dún Pádraig *see* Downpatrick
79 L17 **Dunqulah** *see* Dongola
185 E22 **Dunsandel** Otago, South Island, New Zealand 43°40′S 172°10′E
97 N21 **Dunstable** *Lat.* Durocobrivae. E England, United Kingdom 51°53′N 00°32′W
185 D21 **Dunstan Mountains** ▲ South Island, New Zealand
103 O9 **Dun-sur-Auron** Cher, C France 46°56′N 02°40′E
185 F21 **Dunrobin** Canterbury, South Island, New Zealand 44°52′S 170°40′E
149 T10 **Dunyāpur** Punjab, E Pakistan 29°48′N 71°48′E
163 U5 **Duobukur He** ♣ NE China
163 R12 **Duolun** *var.* Dolonnur. Nei Mongol Zizhiqu, N China 42°11′N 116°30′E
97 R16 **Dương Đông** Kiên Giang, S Vietnam 10°15′N 103°58′E
38 L8 **Dupree** South Dakota, N USA 45°03′N 101°36′W
33 Q7 **Dupuyer** Montana, NW USA 48°12′N 112°30′W
141 Y11 **Duqm** *var.* Daqm. E Oman 19°42′N 57°40′E
118 N4 **Durack Range** ▲ Western Australia
67 V16 **Du Toit Fracture Zone** *tectonic feature* SW Indian Ocean
25 U8 **Durance** ♣ SE France
31 R9 **Durand** Michigan, N USA 42°54′N 83°58′W
30 J7 **Durand** Wisconsin, N USA 41°57′N 83°59′W
40 K10 **Durango** *var.* Victoria de Durango. Durango, W Mexico 24°01′N 104°38′W
37 Q8 **Durango** Colorado, C USA 37°13′N 107°51′W
40 J9 **Durango** ♦ *state* C Mexico

114 O7 **Durankulak** *Rom.* Răcari; *prev.* Blatnitsa, Duranulac. Dobrich, NE Bulgaria 43°41′N 28°31′E
22 L4 **Durant** Mississippi, S USA 33°04′N 89°51′W
27 P13 **Durant** Oklahoma, C USA 33°59′N 96°24′W
105 N6 **Duratón** ♣ N Spain
61 E19 **Durazno** *var.* San Pedro de Durazno. Durazno, C Uruguay 33°22′S 56°31′W
61 E19 **Durazno** ♦ *department* C Uruguay
Durazzo *see* Durrës
83 K23 **Durban** *var.* Port Natal. KwaZulu/Natal, E South Africa 29°51′S 31°E
83 K23 **Durban ✈** KwaZulu/Natal, E South Africa 29°55′S 31°01′E
118 C9 **Durbe** *Ger.* Durben. W Latvia 56°34′N 21°22′E
Durben *see* Durbe
99 H19 **Durbuy** Luxembourg, SE Belgium 50°21′N 05°27′E
105 N15 **Dúrcal** Andalucía, S Spain 37°00′N 03°24′W
128 F8 **Đurđevac** *Ger.* Sankt Georgen, *Hung.* Szentgyörgy; *prev.* Djurdjevac, Gjurgjevac. Koprivnica-Križevci, N Croatia 46°02′N 17°03′E
113 K15 **Đurđevica Tara** N Montenegro 43°09′N 19°18′E
97 L24 **Durdle Door** *natural arch* S England, United Kingdom
158 L3 **Düre** Xinjiang Uygur Zizhiqu, W China 46°30′N 88°26′E
101 D16 **Düren** *anc.* Marcodurum. Nordrhein-Westfalen, W Germany 50°48′N 06°30′E
154 K12 **Durg** *prev.* Drug. Chhattisgarh, C India 21°12′N 81°20′E
153 U13 **Durgapur** Dhaka, N Bangladesh 25°10′N 90°41′E
153 R15 **Durgapur** West Bengal, NE India 23°30′N 87°20′E
14 F14 **Durham** Ontario, S Canada 44°10′N 80°48′W
97 M14 **Durham** *hist.* Dunholme. N England, United Kingdom 54°47′N 01°34′W
21 U9 **Durham** North Carolina, SE USA 36°N 78°54′W
97 L15 **Durham** *cultural region* N England, United Kingdom
168 J10 **Duri** Sumatera, W Indonesia 01°13′N 101°12′E
Duria Major *see* Dora Baltea
Duria Minor *see* Dora Riparia
Durlas *see* Thurles
141 P8 **Durmā** Ar Riyāḍ, C Saudi Arabia 24°37′N 46°06′E
113 J15 **Durmitor** ▲ N Montenegro
96 H6 **Durness** N Scotland, United Kingdom 58°34′N 04°46′W
109 Y3 **Dürnkrut** Niederösterreich, E Austria 48°28′N 16°50′E
Durnovaria *see* Dorchester
Durobrivae *see* Rochester
Durocasses *see* Dreux
Durocortorum *see* Reims
Durostorum *see* Silistra
Duroverum *see* Canterbury
113 K20 **Durrës** *var.* Durrësi, Dursi, *It.* Durazzo, *SCr.* Draç, *Turk.* Draç. Durrës, W Albania 41°19′N 19°27′E
113 K19 **Durrës** ♦ *district* W Albania
Durrësi *see* Durrës
97 A21 **Dursey Island** *Ir.* Oileán Baoi. *island* SW Ireland
Dursi *see* Durrës
Duru *see* Wuchuan
Durud *see* Dow Rūd
114 P12 **Durusu** Istanbul, NW Turkey 41°19′N 28°41′E
114 O12 **Durusu Gölü** ◎ NW Turkey
138 I9 **Durūz, Jabal ad** ▲ SW Syria 37°00′N 32°32′E
184 K13 **D'Urville Island** *island* C New Zealand
171 X12 **D'Urville, Tanjung** *headland* Papua, E Indonesia 01°26′S 137°52′E
115 H18 **Dýstos, Límni** *var.* Límni Dístos. ◎ Évvoia, C Greece
115 D18 **Dytikí Elláda** *Eng.* West, *var.* Dytikí Ellás. ♦ *region* C Greece
111 I11 **Dytikí Makedonía** *Eng.* Macedonia West. ♦ *region* N Greece
160 L2 **Duss** SE Scotland, United Kingdom 55°46′N 02°13′W
147 P13 **Dushanbe** *var.* Dyushambe; *prev.* Stalinabad, *Taj.* Stalinobod. ● (Tajikistan) W Tajikistan 38°35′N 68°44′E
147 P13 **Dushanbe ✈** W Tajikistan 38°31′N 68°49′E
137 T9 **Dusheti** *prev.* Dushet'i. E Georgia 42°07′N 44°44′E
Dushet'i *see* Dusheti
18 H13 **Dushore** Pennsylvania, NE USA 41°30′N 76°23′W
185 A23 **Dusky Sound** *sound* South Island, New Zealand
101 E15 **Düsseldorf** *var.* Duesseldorf. Nordrhein-Westfalen, W Germany 51°14′N 06°49′E
Dūstī *Rus.* Dusti. SW Tajikistan 37°22′N 68°41′E
194 I9 **Dustin Island** *island* Antarctica
Dutch East Indies *see* Indonesia
Dutch Guiana *see* Suriname
38 L17 **Dutch Harbor** Unalaska Island, Alaska, USA 53°51′N 166°33′W
Dutch Mount ▲ Utah, W USA 40°11′N 113°56′W
Dutch New Guinea *see* Papua
Dutch West Indies *see* Curaçao
163 U10 **Düüreg** *see* Ögiynuur
141 Y11 **Duqm** *var.* Daqm. E Oman 19°42′N 57°40′E
79 V13 **Dutse** Jigawa, N Nigeria 11°43′N 09°25′E
154 E17 **Dutton, Lake** ◎ S Australia 32°37′N 133°07′E
147 T10 **Dutton, Mount** ▲ Utah, W USA 38°00′N 112°10′W

83 J25 **Dutywa** *prev.* Idutywa. Eastern Cape, SE South Africa 32°06′S 28°20′E *see also* Idutywa
162 E7 **Duut** Hovd, W Mongolia 47°28′N 91°52′E
14 K11 **Duval, Lac** ◎ Québec, SE Canada
127 W3 **Duvan** Respublika Bashkortostan, W Russian Federation 55°42′N 57°56′E
138 L9 **Duwayhilat Satiḥ ar Ruwayshid** *seasonal river* SE Jordan
Dux *see* Duchcov
160 J13 **Duyang Shan** ▲ S China
167 T14 **Duyên Hai** Tra Vinh, S Vietnam
160 K12 **Duyun** Guizhou, S China 26°16′N 107°29′E
136 G11 **Düzce** Düzce, NW Turkey 40°51′N 31°09′E
136 K14 **Düzce** ♦ *province* NW Turkey
Duzdab *see* Zāhedān
Duzenkyr, Khrebet *see* Duzkyr, Khrebet
146 I16 **Duzkyr, Khrebet** *prev.* Khrebet Duzenkyr. ▲ S Turkmenistan
114 K8 **Dve Mogili** Ruse, N Bulgaria 43°35′N 25°51′E
124 L7 **Dvina Bay** *see* Chëshskaya Guba
118 K8 **Dvinsk** *see* Daugavpils
112 E10 **Dvor** Sisak-Moslavina, C Croatia 45°05′N 16°22′E
117 W5 **Dvorichna** Kharkiv'ska Oblast', E Ukraine 49°52′N 37°43′E
111 F16 **Dvůr Králové nad Labem** *Ger.* Königinhof an der Elbe. Královéhradecký Kraj, N Czech Republic 50°27′N 15°50′E
123 P8 **Dzhugdzhur, Khrebet** ▲ E Russian Federation
Dzhul'fa *see* Culfa
154 A10 **Dwārka** Gujarāt, W India 22°14′N 68°58′E
30 M4 **Dwight** Illinois, N USA 41°05′N 88°25′W
98 N8 **Dwingeloo** Drenthe, NE Netherlands 52°49′N 06°20′E
33 N10 **Dworshak Reservoir** ◎ Idaho, NW USA
Dyal *see* Dihang
Dyanev *see* Galkynyş
Dyatlovo *see* Dzyatlava
186 G5 **Dyaul Island** *var.* Djaul, Dyal. *island* NE Papua New Guinea
20 L9 **Dyer** Tennessee, S USA 36°04′N 88°59′W
41 X11 **Dzilam de Bravo** Yucatán, E Mexico 21°24′N 88°54′W
118 L12 **Dzisna** *Rus.* Disna. Vitsyebskaya Voblasts', N Belarus 55°33′N 28°13′E
118 K12 **Dzisna** *Lith.* Dysna, *Rus.* ♣ Belarus/Lithuania
119 G20 **Dzivin** *Rus.* Divin. Brestskaya Voblasts', SW Belarus 51°58′N 24°33′E
115 M15 **Dzmitravichy** *Rus.* Dmitrovichi. Minskaya Voblasts', C Belarus 53°58′N 29°14′E
111 E19 **Dyje** *var.* Thaya. ♣ Austria/Czech Republic *see also* Thaya
Dyje *see* Thaya
117 T5 **Dykan'ka** Poltavs'ka Oblast', C Ukraine 49°49′N 34°33′E
127 N16 **Dykhtau** ▲ SW Russian Federation 43°01′N 42°56′E
111 A16 **Dýleň** *Ger.* Tillenberg. ▲ NW Czech Republic 49°58′N 12°31′E
110 K9 **Dylewska Góra** ▲ N Poland 53°33′N 19°57′E
117 O4 **Dymer** Kyyivs'ka Oblast', N Ukraine 50°50′N 30°20′E
117 W7 **Dymytrov** *Rus.* Dimitrov. Donets'ka Oblast', E Ukraine 48°18′N 37°20′E
111 O17 **Dynów** Podkarpackie, SE Poland 49°49′N 22°14′E
29 X13 **Dysart** Iowa, C USA 42°10′N 92°18′W
Dysna *see* Dzisna
115 H18 **Dýstos, Límni** *var.* Límni Dístos. ◎ Évvoia, C Greece
115 D18 **Dytikí Elláda** *var.* Dytikí Ellás ♦ *region* C Greece
115 C14 **Dytikí Makedonía** ♦ *region* N Greece
Dyurment'yube *see* Diirmentobe
127 U4 **Dyurtyuli** Respublika Bashkortostan, W Russian Federation 55°31′N 54°49′E
Dyushambe *see* Dushanbe
162 K7 **Dzaamar** *var.* Bat-Öldziyt. Töv, C Mongolia 48°10′N 104°47′E
137 T9 **Dzaanhushuu** *var.* Ihtamir. Arhangay, C Mongolia 13°09′N 108°14′E
162 H8 **Dzag** Bayanhongor, C Mongolia 46°54′N 99°11′E
163 O11 **Dzamïn-Üüd** *var.* Borhoyn Tal. Dornogovĭ, SE Mongolia 43°43′N 111°53′E
172 J14 **Dzaoudzi** E Mayotte 12°48′S 45°18′E
162 G7 **Dzavhan** ♦ *province* NW Mongolia
162 G7 **Dzavhan Gol** ♣ NW Mongolia
162 G6 **Dzavhanmandal** *var.* Nuga. Dzavhan, W Mongolia 53°51′N 96°57′E
162 E7 **Dzereg** *var.* Altanteel. Hovd, W Mongolia 47°08′N 92°50′E
127 O3 **Dzerzhinsk** Nizhegorodskaya Oblast', W Russian Federation 56°N 43°22′E
Dzerzhinsk *see* Romaniv
Dzerzhinsk *see* Romaniv
Dzerzhinskaya *see* Nar'yan-Mar
Dzerzhinskoye *see* Tokzhaylau
Dzerzhinskoye *see* Tokzhaylau
117 V13 **Dutsan Wai** *var.* Dutsen Wai. Kaduna, C Nigeria 10°49′N 08°15′E
79 W13 **Dutse Jigawa**, N Nigeria 11°43′N 09°25′E
Duttia *see* Datia
147 T10 **Dzhalagash** *see* Zhalagash
147 T10 **Dzhalal-Abad** *Kir.* Jalal-Abad. Dzhalal-Abadskaya Oblast', W Kyrgyzstan 40°56′N 73°00′E

147 S9 **Dzhalal-Abadskaya Oblast'** *Kir.* Jalal-Abad Oblasty. ♦ *province* W Kyrgyzstan
Dzhalilabad *see* Cälilabad
161 Q12 **Dzhambeyty** *see* Zhympity
Dzhambul *see* Taraz
Dzhambulskaya Oblast' *see* Zhambyl
144 D9 **Dzhanibek**, *Kaz.* Zhänibek. Zapadnyy Kazakhstan, W Kazakhstan 49°27′N 46°51′E
Dzhankel'dy *see* Jongeldi
117 T12 **Dzhankoy** Avtonomna Respublika Krym, S Ukraine 45°40′N 34°27′E
Dzhansugurov *see* Zhansugirov
147 R9 **Dzhany-Bazar** *var.* Yangibazar. Dzhalal-Abadskaya Oblast', W Kyrgyzstan 41°40′N 70°49′E
123 P8 **Dzhardzhan** Respublika Sakha (Yakutiya), NE Russian Federation 68°47′N 123°51′E
117 S11 **Dzharylhats'ka Zatoka** *gulf* S Ukraine
Dzhayilgan *see* Jayilgan
Dzhebel *see* Jebel
144 T14 **Dzhelandy** SE Tajikistan 37°34′N 72°35′E
147 Y7 **Dzhergatal'** *Kir.* Jyrgalan. Issyk-Kul'skaya Oblast', NE Kyrgyzstan 42°37′N 78°56′E
Dzhetysay *see* Zhetysay
Dzhezkazgan *see* Zhezkazgan
146 J12 **Dzhigirbent** *see* Jigerbent
Dzhirgatal' *see* Jirgatol
146 K14 **Dzhizak** *see* Jizzax
Dzhizakskaya Oblast' *see* Jizzax Viloyati
123 P8 **Dzhugdzhur, Khrebet** ▲ E Russian Federation
Dzhul'fa *see* Culfa

E

E *see* Hubei
Éadan Doire *see* Edenderry
10 U12 **Ea Đrăng** *var.* Ea H'leo. Đắc Lắc, S Vietnam 13°09′N 108°14′E
37 W6 **Eads** Colorado, C USA 38°28′N 102°46′W
37 O13 **Eagar** Arizona, SW USA 34°06′N 109°17′W
39 T8 **Eagle** Alaska, USA 64°47′N 141°12′W
13 S8 **Eagle** ♣ Newfoundland and Labrador, E Canada
10 J13 **Eagle** ♣ Yukon, NW Canada
29 T7 **Eagle Bend** Minnesota, N USA 46°10′N 95°00′W
38 M8 **Eagle Butte** South Dakota, N USA 45°00′N 101°13′W
19 R2 **Eagle Grove** Iowa, C USA 42°39′N 93°54′W
19 R2 **Eagle Lake** Maine, NE USA 46°20′N 69°22′W
25 U11 **Eagle Lake** Texas, SW USA 29°35′N 96°19′W
2 A11 **Eagle Lake** ◎ Ontario, S Canada
35 P3 **Eagle Lake** ◎ California, W USA
19 R3 **Eagle Lake** ◎ Maine, NE USA
29 Y3 **Eagle Mountain** ▲ Minnesota, N USA
5 T6 **Eagle Mountain Lake** ◎ Texas, SW USA
37 S9 **Eagle Nest Lake** ◎ New Mexico, SW USA
25 P13 **Eagle Pass** Texas, SW USA 28°43′N 100°30′W
65 C25 **Eagle Passage** *passage* SW Atlantic Ocean
35 R8 **Eagle Peak** ▲ California, W USA 41°16′N 120°12′W
37 P13 **Eagle Peak** ▲ New Mexico, SW USA

10 I4 **Eagle Plain** Yukon, NW Canada 66°23′N 136°42′W
32 G15 **Eagle Point** Oregon, NW USA 42°28′N 122°48′W
186 P10 **Eagle Point** ▲ SE Papua New Guinea 10°31′S 149°53′E
39 R11 **Eagle River** Alaska, USA 61°18′N 149°38′W
30 M2 **Eagle River** Michigan, N USA 47°24′N 88°18′W
30 L4 **Eagle River** Wisconsin, N USA 45°55′N 89°15′W
21 S6 **Eagle Rock** Virginia, NE USA 37°40′N 79°46′W
36 J13 **Eagletail Mountains** ▲ Arizona, SW USA
167 U12 **Ea Kar** Đắc Lắc, S Vietnam 12°47′N 108°26′E
Eanjum *see* Anjum
Eanodat *see* Enontekiö
12 B10 **Ear Falls** Ontario, C Canada 50°38′N 93°13′W
27 X10 **Earle** Arkansas, C USA 35°16′N 90°28′W
35 R12 **Earlimart** California, W USA 35°53′N 119°17′W
20 I6 **Earlington** Kentucky, S USA 37°16′N 87°30′W
14 H13 **Earlton** Ontario, S Canada 47°41′N 79°46′W
29 T13 **Early** Iowa, C USA 42°27′N 95°09′W
96 J11 **Earn** ♣ N Scotland, United Kingdom
185 C21 **Earnslaw, Mount** ▲ South Island, New Zealand 44°34′S 168°26′E
24 M4 **Earth** Texas, SW USA 34°13′N 102°25′W
21 P11 **Easley** South Carolina, SE USA 34°49′N 82°36′W
East *see* Est
97 P19 **East Anglia** *physical region* E England, United Kingdom
15 Q12 **East Angus** Québec, SE Canada 45°29′N 71°39′W
195 V8 **East Antarctica** *prev.* Greater Antarctica. *physical region* Antarctica
18 E10 **East Aurora** New York, NE USA 42°46′N 78°36′W
181 T8 **East Australian Basin** *see* Tasman Basin
East Azerbaijan *see* Āzarbāyjān-e Sharqī
64 L9 **East Azores Fracture Zone** *var.* Açores Fracture Zone. *tectonic feature* E Atlantic Ocean
22 M11 **East Bay** *bay* Louisiana, S USA
25 V9 **East Bernard** Texas, SW USA 29°32′N 96°04′W
29 V8 **East Bethel** Minnesota, N USA 45°24′N 93°14′W
East Borneo *see* Kalimantan Timur
97 P23 **Eastbourne** SE England, United Kingdom 50°46′N 00°16′E
15 R11 **East-Broughton** Québec, SE Canada 46°14′N 71°05′W
44 M6 **East Caicos** *island* E Turks and Caicos Islands
184 R7 **East Cape** *headland* North Island, New Zealand 37°30′S 178°31′E
174 M4 **East Caroline Basin** *undersea feature* SW Pacific Ocean 04°00′N 146°00′E
192 P4 **East China Sea** *Chin.* Dong Hai. *sea* W Pacific Ocean
97 P19 **East Dereham** E England, United Kingdom 52°41′N 00°55′E
30 J9 **East Dubuque** Illinois, N USA 42°29′N 90°38′W
11 S17 **East Saskatchewan** Canada
193 S10 **Easter Fracture Zone** *tectonic feature* E Pacific Ocean
153 G23 **Easter Island** *see* Pascua, Isla de
155 K25 **Eastern** ♦ *province* E Sri Lanka
82 L13 **Eastern** ♦ *province* E Zambia
83 H24 **Eastern Cape** ♦ *Afr.* Oos-Kaap. ♦ *province* SE South Africa
Eastern Cape Province *see* Eastern Cape
80 C12 **Eastern Darfur** ♦ *state* SW Sudan
Eastern Desert *see* Sahara el Sharqiya
81 F15 **Eastern Equatoria** ♦ *state* SE South Sudan
Eastern Euphrates *see* Murat Nehri
155 I21 **Eastern Ghats** ▲ SE India
186 E7 **Eastern Highlands** ♦ province C Papua New Guinea
Eastern Region *see* Ash Sharqīyah
Eastern Sayans *see* Vostochnyy Sayan
Eastern Scheldt *see* Oosterschelde
Eastern Sierra Madre *see* Madre Oriental, Sierra
Eastern Transvaal *see* Mpumalanga
11 W14 **Easterville** Manitoba, C Canada 53°06′N 99°53′W
Easterwälde *see* Oosterwolde
63 M23 **East Falkland** *var.* Isla Soledad. *island* E Falkland Islands
19 P12 **East Falmouth** Massachusetts, NE USA 41°34′N 70°31′W
East Fayu *see* Fayu
East Flanders *see* Oost-Vlaanderen
39 S6 **East Fork Chandalar River** ♣ Alaska, USA
29 U12 **East Fork Des Moines River** ♣ Iowa/Minnesota, C USA
East Frisian Islands *see* Ostfriesische Inseln
18 K10 **East Glenville** New York, NE USA 42°53′N 73°55′W
29 R4 **East Grand Forks** Minnesota, N USA 47°55′N 97°00′W
97 O23 **East Grinstead** SE England, United Kingdom 51°08′N 00°01′W
19 N12 **East Hartford** Connecticut, NE USA 41°46′N 72°36′W
18 M13 **East Haven** Connecticut, NE USA 41°16′N 72°52′W

◆ Country   ◇ Dependent Territory   ◈ Administrative Regions   ▲ Mountain   ⊠ Volcano
● Country Capital   ○ Dependent Territory Capital   ✕ International Airport   ▲ Mountain Range   ♣ River   ◎ Lake   ▨ Reservoir

245

◆ Country · ● Country Capital · ◇ Dependent Territory · ○ Dependent Territory Capital · ◆ Administrative Regions · ▲ Mountain Range · ▲ Mountain · ✈ International Airport · ☈ Volcano · ☞ River · ◉ Lake · ▣ Reservoir

El Higo – Erdenetsagaan

El Higo *see* Higos
114 *L10* **Elhovo** *var.* Elkhovo; *prev.* Kizilagach. Yambol, E Bulgaria 42°10′N 26°34′E
171 *T16* **Eliase** Pulau Selaru, E Indonesia 08°16′S 130°49′E
Elías Piña *see* Comendador
25 *R6* **Eliasville** Texas, SW USA 32°55′N 98°46′W
Elichpur *see* Achalpur
37 *V13* **Elida** New Mexico, SW USA 33°57′N 103°39′W
115 *F18* **Elikónas** ▲ C Greece
67 *T10* **Elila** ⚡ W Dem. Rep. Congo
39 *N9* **Elim** Alaska, USA 64°37′N 162°15′W
Elimberrum *see* Auch
Eliocroca *see* Lorca
61 *B16* **Elisa** Santa Fe, C Argentina 30°42′S 61°04′W
Elisabethstedt *see* Dumbrăveni
Élisabethville *see* Lubumbashi
127 *O13* **Elista** Respublika Kalmykiya, SW Russian Federation 46°18′N 44°09′E
182 *I9* **Elizabeth** South Australia 34°44′S 138°39′E
21 *Q3* **Elizabeth** West Virginia, NE USA 39°04′N 81°24′W
19 *Q9* **Elizabeth, Cape** *headland* Maine, NE USA 43°34′N 70°12′W
21 *Y8* **Elizabeth City** North Carolina, SE USA 36°18′N 76°16′W
21 *P8* **Elizabethton** Tennessee, S USA 36°22′N 82°15′W
30 *M17* **Elizabethtown** Illinois, N USA 37°24′N 88°21′W
20 *K6* **Elizabethtown** Kentucky, S USA 37°41′N 85°51′W
18 *L7* **Elizabethtown** New York, NE USA 44°13′N 73°38′W
21 *U11* **Elizabethtown** North Carolina, SE USA 34°36′N 78°36′W
18 *G15* **Elizabethtown** Pennsylvania, NE USA 40°08′N 76°36′W
74 *E6* **El-Jadida** *prev.* Mazagan. W Morocco 33°15′N 08°27′W
El Jafr *see* Jafr, Qāʿ al
80 *F11* **El Jebelein** White Nile, E Sudan 12°38′N 32°51′E
110 *N8* **Ełk** *Ger.* Lyck. Warmińsko-mazurskie, NE Poland 53°51′N 22°20′E
110 *O8* **Ełk** ⚡ NE Poland
29 *Y12* **Elkader** Iowa, C USA 42°51′N 91°24′W
80 *G9* **El Kamlin** Gezira, C Sudan 15°03′N 33°11′E
33 *N11* **Elk City** Idaho, NW USA 45°50′N 115°28′W
26 *K10* **Elk City** Oklahoma, C USA 35°24′N 99°24′W
27 *P7* **Elk City Lake** ⊠ Kansas, C USA
34 *M5* **Elk Creek** California, W USA 39°34′N 122°34′W
28 *J10* **Elk Creek** ⚡ South Dakota, N USA
74 *M5* **El Kef** *var.* Al Kāf, Le Kef. NW Tunisia 36°13′N 08°44′E
74 *F7* **El Kelâa Srarhna** *var.* Kal al Sraghna. C Morocco 32°05′N 07°20′W
El Kerak *see* Al Karak
11 *P17* **Elkford** British Columbia, SW Canada 49°58′N 114°57′W
El Khalil *see* Hebron
80 *E7* **El Khandaq** Northern, N Sudan 18°34′N 30°34′E
El Khârga *see* Al Khārijah
31 *P11* **Elkhart** Indiana, N USA 41°40′N 85°58′W
26 *H7* **Elkhart** Kansas, C USA 37°00′N 101°51′W
25 *V8* **Elkhart** Texas, SW USA 31°37′N 95°34′W
30 *M7* **Elkhart Lake** ◎ Wisconsin, N USA
El Khartûm *see* Khartoum
37 *Q3* **Elkhead Mountains** ▲ Colorado, C USA
18 *I12* **Elk Hill** ▲ Pennsylvania, NE USA 41°42′N 75°33′W
138 *G8* **El Khiyam** *var.* Al Khiyām, Khiam. S Lebanon 33°12′N 35°42′E
29 *S15* **Elkhorn** Nebraska, C USA 41°17′N 96°13′W
30 *M9* **Elkhorn** Wisconsin, N USA 42°40′N 88°34′W
29 *R14* **Elkhorn River** ⚡ Nebraska, C USA
127 *O16* **El'khotovo** Respublika Severnaya Osetiya, SW Russian Federation 43°18′N 44°17′E
Elkhovo *see* Elhovo
21 *R8* **Elkin** North Carolina, SE USA 36°14′N 80°51′W
21 *S4* **Elkins** West Virginia, NE USA 38°56′N 79°53′W
195 *X3* **Elkins, Mount** ▲ Antarctica 66°25′S 53°54′E
14 *G8* **Elk Lake** Ontario, S Canada 47°44′N 80°19′W
31 *P6* **Elk Lake** ◎ Michigan, N USA
18 *F12* **Elkland** Pennsylvania, NE USA 41°59′N 77°16′W
35 *W3* **Elko** Nevada, W USA 40°48′N 115°46′W
11 *R14* **Elk Point** Alberta, SW Canada 53°52′N 110°49′W
29 *R12* **Elk Point** South Dakota, N USA 42°42′N 96°37′W
29 *V8* **Elk River** Minnesota, N USA 45°18′N 93°34′W
20 *J10* **Elk River** ⚡ Alabama/ Tennessee, S USA
21 *R4* **Elk River** ⚡ West Virginia, NE USA
20 *I7* **Elkton** Kentucky, S USA 36°49′N 87°11′W
21 *Y2* **Elkton** Maryland, NE USA 39°37′N 75°50′W
29 *R10* **Elkton** South Dakota, N USA 44°14′N 96°28′W
20 *I10* **Elkton** Tennessee, S USA 35°01′N 86°51′W
21 *U5* **Elkton** Virginia, NE USA 38°22′N 78°35′W
El Kuneitra *see* Al Qunayṭirah
81 *L15* **El Kure** Somali, E Ethiopia 05°37′N 42°05′E
80 *D12* **El Lagowa** Western Kordofan, C Sudan 11°23′N 29°10′E
39 *S12* **Ellamar** Alaska, USA 60°54′N 146°37′W
Ellás *see* Greece
23 *S6* **Ellaville** Georgia, SE USA 32°14′N 84°18′W
197 *P9* **Ellef Ringnes Island** *island* Nunavut, N Canada
29 *V10* **Ellendale** Minnesota, N USA 43°53′N 93°19′W
29 *P7* **Ellendale** North Dakota, N USA 45°57′N 98°33′W
36 *M6* **Ellen, Mount** ▲ Utah, W USA 38°06′N 110°48′W
32 *I9* **Ellensburg** Washington, NW USA 47°00′N 120°34′W
18 *K12* **Ellenville** New York, NE USA 41°43′N 74°24′W
Ellep *see* Lib
21 *T10* **Ellerbe** North Carolina, SE USA 35°04′N 79°45′W
197 *P10* **Ellesmere Island** *island* Queen Elizabeth Islands, Nunavut, N Canada
185 *H19* **Ellesmere, Lake** ◎ South Island, New Zealand
97 *K18* **Ellesmere Port** C England, United Kingdom 53°17′N 02°54′W
31 *O14* **Ellettsville** Indiana, N USA 39°13′N 86°37′W
99 *E19* **Ellezelles** Hainaut, SW Belgium 50°44′N 03°40′E
8 *L7* **Ellice** ⚡ Nunavut, NE Canada
Ellice Islands *see* Tuvalu
Ellichpur *see* Achalpur
21 *W3* **Ellicott City** Maryland, NE USA 39°16′N 76°48′W
23 *S2* **Ellijay** Georgia, SE USA 34°42′N 84°28′W
27 *W7* **Ellington** Missouri, C USA 37°14′N 90°58′W
26 *L5* **Ellinwood** Kansas, C USA 38°21′N 98°34′W
83 *J24* **Elliot** Eastern Cape, SE South Africa 31°20′S 27°51′E
14 *D10* **Elliot Lake** Ontario, S Canada 46°24′N 82°38′W
181 *X6* **Elliot, Mount** ▲ Queensland, E Australia 19°36′S 147°02′E
21 *T5* **Elliott Knob** ▲ Virginia, NE USA 38°10′N 79°18′W
26 *K4* **Ellis** Kansas, C USA 38°56′N 99°33′W
182 *F8* **Elliston** South Australia 33°40′S 134°56′E
22 *M7* **Ellisville** Mississippi, S USA 31°36′N 89°12′W
105 *V5* **El Llobregat** ⚡ NE Spain
96 *L9* **Ellon** NE Scotland, United Kingdom 57°22′N 02°06′W
Ellore *see* Elūru
21 *S13* **Elloree** South Carolina, SE USA 33°34′N 80°37′W
26 *M4* **Ellsworth** Kansas, C USA 38°45′N 98°15′W
19 *S7* **Ellsworth** Maine, NE USA 44°32′N 68°25′W
30 *I6* **Ellsworth** Wisconsin, N USA 44°44′N 92°29′W
26 *M11* **Ellsworth, Lake** ⊠ Oklahoma, C USA
194 *K9* **Ellsworth Land** *physical region* Antarctica
194 *L9* **Ellsworth Mountains** ▲ Antarctica
101 *J21* **Ellwangen** Baden-Württemberg, S Germany 48°58′N 10°07′E
18 *B14* **Ellwood City** Pennsylvania, NE USA 40°49′N 80°15′W
108 *H8* **Elm** Glarus, NE Switzerland 46°55′N 09°09′E
32 *G9* **Elma** Washington, NW USA 47°00′N 123°24′W
121 *V13* **El Maḥalla el Kubra** *var.* Al Maḥallah al Kubrá, Mahalla el Kubra. N Egypt 30°59′N 31°10′E
74 *E9* **El Mahbas** *var.* Mahbés. SW Western Sahara 27°26′N 09°09′W
63 *H17* **El Maitén** Chubut, W Argentina 42°03′S 71°10′W
136 *E16* **Elmalı** Antalya, SW Turkey 36°43′N 29°19′E
80 *G10* **El Manaqil** Gezira, C Sudan 14°12′N 33°01′E
54 *M12* **El Mango** Amazonas, S Venezuela 01°55′N 66°35′W
El Manṣûra *see* Al Manṣūrah
55 *P8* **El Manteco** Bolívar, E Venezuela 07°27′N 62°32′W
29 *O16* **Elm Creek** Nebraska, C USA 40°43′N 99°22′W
El Mediyya *see* Médéa
77 *V9* **Elméki** Agadez, C Niger 17°52′N 08°07′E
108 *K7* **Elmen** Tirol, W Austria 47°22′N 10°34′E
18 *I16* **Elmer** New Jersey, NE USA 39°34′N 75°09′W
138 *G6* **El Mina** *var.* Al Mīnā'. N Lebanon 34°28′N 35°49′E
El Minya *see* Al Minyā
14 *F15* **Elmira** Ontario, S Canada 43°35′N 80°34′W
18 *G11* **Elmira** New York, NE USA 42°06′N 76°50′W
36 *K13* **El Mirage** Arizona, SW USA 33°36′N 112°19′W
29 *O7* **Elm Lake** ⊠ South Dakota, N USA
El Moján *see* San Rafael
105 *N7* **El Molar** Madrid, C Spain 40°43′N 03°35′W
76 *L7* **El Mrâyer** *well* C Mauritania
76 *L5* **El Mreiti** *well* N Mauritania
76 *L8* **El Mreyyé** *desert* E Mauritania
29 *P8* **Elm River** ⚡ North Dakota/ South Dakota, N USA
100 *I9* **Elmshorn** Schleswig-Holstein, N Germany 53°45′N 09°39′E
80 *D12* **El Muglad** Western Kordofan, C Sudan 11°02′N 27°44′E
El Muwaqqar *see* Al Muwaqqar
14 *G14* **Elmvale** Ontario, S Canada 44°34′N 79°53′W
30 *K12* **Elmwood** Illinois, N USA 40°46′N 89°58′W
26 *J8* **Elmwood** Oklahoma, C USA 36°37′N 100°31′W
103 *P17* **Elne** *anc.* Illiberis. Pyrénées-Orientales, S France 42°36′N 02°58′E
54 *F11* **El Nevado, Cerro** *elevation* C Colombia
171 *N5* **El Nido** Palawan, W Philippines 11°10′N 119°25′E
62 *I12* **El Nihuil** Mendoza, W Argentina 34°58′S 68°40′W
75 *W7* **El Nouzha** ✈ (Alexandria) N Egypt 31°06′N 29°58′E
80 *E10* **El Obeid** *var.* Al Obayyid, Al Ubayyiḍ. Northern Kordofan, C Sudan 13°11′N 30°10′E
41 *O13* **El Oro** México, S Mexico 19°51′N 100°07′W
56 *B8* **El Oro** ◆ *province* SW Ecuador
61 *B19* **Elortondo** Santa Fe, C Argentina 33°42′S 61°37′W
54 *J8* **Elorza** Apure, C Venezuela 07°02′N 69°30′W
El Ouâdi *see* El Oued
74 *L7* **El Oued** *var.* Al Oued, El Ouâdi, El Wad. NE Algeria 33°20′N 06°53′E
36 *L15* **Eloy** Arizona, SW USA 32°47′N 111°33′W
55 *Q7* **El Palmar** Bolívar, E Venezuela 08°01′N 61°53′W
40 *K8* **El Palmito** Durango, W Mexico 25°40′N 104°59′W
55 *P7* **El Pao** Bolívar, E Venezuela 08°03′N 62°40′W
54 *K5* **El Pao** Cojedes, N Venezuela 09°40′N 68°08′W
42 *J7* **El Paraíso** El Paraíso, S Honduras 13°51′N 86°31′W
42 *I7* **El Paraíso** ◆ *department* SE Honduras
30 *L12* **El Paso** Illinois, N USA 40°44′N 89°01′W
24 *G8* **El Paso** Texas, SW USA 31°45′N 106°30′W
24 *G8* **El Paso** ✈ Texas, SW USA 31°48′N 106°24′W
105 *U7* **El Perelló** Cataluña, NE Spain 40°53′N 00°43′E
55 *P5* **El Pilar** Sucre, NE Venezuela 10°31′N 63°12′W
42 *F7* **El Pital, Cerro** ▲ El Salvador/Honduras 14°19′N 89°06′W
35 *Q9* **El Portal** California, W USA 37°40′N 119°46′W
40 *J3* **El Porvenir** Chihuahua, N Mexico 31°15′N 105°48′W
43 *U14* **El Porvenir** Kuna Yala, N Panama 09°33′N 78°56′W
105 *W6* **El Prat de Llobregat** Cataluña, NE Spain 41°20′N 02°05′E
42 *H5* **El Progreso** Yoro, NW Honduras 15°25′N 87°49′W
42 *A2* **El Progreso** *off.* Departamento de El Progreso. ◆ *department* C Guatemala
El Progreso *see* Guastatoya
El Progreso, Departamento de *see* El Progreso
104 *L9* **El Puente del Arzobispo** Castilla-La Mancha, C Spain 39°48′N 05°10′W
104 *J15* **El Puerto de Santa María** Andalucía, S Spain 36°36′N 06°13′W
62 *I8* **El Puesto** Catamarca, NW Argentina 27°57′S 67°37′W
El Qâhira *see* Cairo
El Qaṣr *see* Al Qaṣr
El Qatrani *see* Al Qaṭrānah
40 *I10* **El Quelite** Sinaloa, C Mexico 23°37′N 106°26′W
62 *G9* **Elqui, Río** ⚡ N Chile
El Q'unayṭirah *see* Al Qunayṭirah
El Quneitra *see* Al Qunayṭirah
El Quseir *see* Al Quṣayr
El Quweira *see* Al Quwayrah
141 *O15* **El-Rahaba** ✈ (Ṣan'ā') W Yemen 15°28′N 44°12′E
42 *M10* **El Rama** Región Autónoma Atlántico Sur, SE Nicaragua 12°09′N 84°15′W
43 *W16* **El Real** *var.* El Real de Santa María. Darién, SE Panama 08°06′N 77°42′W
El Real de Santa María *see* El Real
26 *M10* **El Reno** Oklahoma, C USA 35°32′N 95°57′W
40 *K9* **El Rodeo** Durango, C Mexico 25°12′N 104°35′W
104 *J13* **El Ronquillo** Andalucía, S Spain 37°43′N 06°09′W
11 *S16* **Elrose** Saskatchewan, S Canada 51°07′N 107°59′W
30 *K8* **Elroy** Wisconsin, N USA 43°43′N 90°16′W
25 *S17* **Elsa** Texas, SW USA 26°17′N 97°59′W
El Ṣaff *see* Aṣ Ṣaff
40 *J10* **El Salto** Durango, C Mexico 23°47′N 105°22′W
42 *D8* **El Salvador** *off.* Republica de El Salvador. ◆ *republic* Central America
El Salvador, Republica de *see* El Salvador
54 *K7* **El Samán de Apure** Apure, C Venezuela 07°54′N 68°44′W
14 *D7* **Elsas** Ontario, S Canada 48°31′N 82°53′W
40 *F3* **El Sásabe** *var.* Aduana del Sásabe. Sonora, NW Mexico 31°27′N 111°31′W
Elsass *see* Alsace
40 *J5* **El Sáuz** Chihuahua, N Mexico 29°03′N 106°15′W
27 *W4* **Elsberry** Missouri, C USA 39°10′N 90°46′W
45 *P9* **El Seibo** *var.* Santa Cruz de El Seibo, Santa Cruz del Seibo. E Dominican Republic 18°45′N 69°04′W
42 *B7* **El Semillero Barra Nahualate** Escuintla, SW Guatemala 14°01′N 91°28′W
Elsene *see* Ixelles
Elsen Nur *see* Dorgê Co
36 *L6* **Elsinore** Utah, W USA 38°40′N 112°09′W
Elsinore *see* Helsingør
99 *L18* **Elsloo** Limburg, SE Netherlands 50°57′N 05°46′E
60 *G13* **El Soberbio** Misiones, NE Argentina 27°15′S 54°05′W
55 *N6* **El Socorro** Guárico, C Venezuela 09°00′N 65°42′W
54 *L6* **El Sombrero** Guárico, N Venezuela 09°25′N 67°06′W
98 *L10* **Elspeet** Gelderland, E Netherlands 52°19′N 05°47′E
98 *L12* **Elst** Gelderland, E Netherlands 51°55′N 05°51′E
101 *O15* **Elsterwerda** Brandenburg, E Germany 51°27′N 13°32′E
40 *J4* **El Sueco** Chihuahua, N Mexico 29°53′N 106°24′W
El Suweida *see* As Suwaydā'
El Suweis *see* Suez
54 *D12* **El Tambo** Cauca, SW Colombia 02°25′N 76°50′W
175 *T13* **Eltanin Fracture Zone** *tectonic feature* SE Pacific Ocean
105 *X5* **El Ter** ⚡ NE Spain
184 *K11* **Eltham** Taranaki, North Island, New Zealand 39°26′S 174°25′E
55 *O6* **El Tigre** Anzoátegui, NE Venezuela 08°55′N 64°15′W
El Tigrito *see* San José de Guanipa
54 *J5* **El Tocuyo** Lara, N Venezuela 09°48′N 69°51′W
127 *Q10* **El'ton** Volgogradskaya Oblast', SW Russian Federation 49°07′N 46°50′E
32 *K10* **Eltopia** Washington, NW USA 46°33′N 118°59′W
El Toro *see* Mare de Déu del Toro
61 *A18* **El Trébol** Santa Fe, C Argentina 32°12′S 61°40′W
40 *J13* **El Tuito** Jalisco, SW Mexico 20°19′N 105°22′W
El Ṭûr *see* Aṭ Ṭūr
161 *S15* **Eluan Bi** *Eng.* Cape Olwanpi; *prev.* Oluan Pi. *headland* S Taiwan 21°57′N 120°48′E
155 *K16* **Elūru** *prev.* Ellore. Andhra Pradesh, E India 16°45′N 81°10′E
118 *H13* **Elva** *Ger.* Elwa. Tartumaa, SE Estonia 58°13′N 26°25′E
37 *R9* **El Vado Reservoir** ⊠ New Mexico, SW USA
43 *S15* **El Valle** Coclé, C Panama 08°39′N 80°08′W
104 *I11* **Elvas** Portalegre, C Portugal 38°53′N 07°10′W
54 *K7* **El Venado** Apure, C Venezuela 07°25′N 68°46′W
105 *V6* **El Vendrell** Cataluña, NE Spain 41°13′N 01°32′E
94 *I13* **Elverum** Hedmark, S Norway 60°54′N 11°33′E
42 *I9* **El Viejo** Chinandega, NW Nicaragua 12°39′N 87°11′W
54 *G7* **El Viejo, Cerro** ▲ C Colombia 07°31′N 72°56′W
54 *H6* **El Vigía** Mérida, NW Venezuela 08°38′N 71°39′W
105 *Q4* **El Villar de Arnedo** La Rioja, N Spain 42°19′N 02°05′W
59 *A14* **Elvira** Amazonas, W Brazil 07°12′S 69°56′W
Elwa *see* Elva
El Wad *see* El Oued
81 *K17* **El Wak** Mandera, NE Kenya 02°46′N 40°57′E
33 *R7* **Elwell, Lake** ⊠ Montana, NW USA
31 *P13* **Elwood** Indiana, N USA 40°16′N 85°50′W
27 *R3* **Elwood** Kansas, C USA 39°43′N 94°52′W
29 *N16* **Elwood** Nebraska, C USA 40°35′N 99°51′W
Elx *see* Elche
97 *O20* **Ely** E England, United Kingdom 52°24′N 00°15′E
29 *X4* **Ely** Minnesota, N USA 47°54′N 91°52′W
35 *X6* **Ely** Nevada, W USA 39°15′N 114°53′W
El Yopal *see* Yopal
31 *T11* **Elyria** Ohio, N USA 41°22′N 82°06′W
45 *S9* **El Yunque** ▲ E Puerto Rico 18°15′N 65°46′W
101 *F23* **Elz** ⚡ SW Germany
187 *R14* **Emae** *island* Shepherd Islands, C Vanuatu
118 *I5* **Emajõgi** *Ger.* Embach. ⚡ SE Estonia
Emāmrūd *see* Shāhrūd
Emām Ṣāḥeb *see* Imām Ṣāḥib
Emam Saheb *see* Imām Ṣāḥib
Emāmshahr *see* Shāhrūd
95 *M20* **Emån** ⚡ S Sweden
144 *J11* **Emba** *Kaz.* Embi. Aktyubinsk, W Kazakhstan 48°50′N 58°10′E
Emba *see* Zhem
Embach *see* Emajõgi
62 *K5* **Embarcación** Salta, N Argentina 23°15′S 64°05′W
30 *M15* **Embarras River** ⚡ Illinois, N USA
Embi *see* Emba
81 *I19* **Embu** Embu, C Kenya 0°32′N 37°28′E
81 *I19* **Embu** ◆ *county* C Kenya
100 *E10* **Emden** Niedersachsen, NW Germany 53°22′N 07°12′E
160 *H9* **Emei Shan** ▲ Sichuan, C China 29°32′N 103°21′E
29 *Q4* **Emerado** North Dakota, N USA 47°55′N 97°21′W
181 *X8* **Emerald** Queensland, E Australia 23°33′S 148°11′E
Emerald Isle *see* Montserrat
57 *J15* **Emero, Río** ⚡ W Bolivia
11 *Y17* **Emerson** Manitoba, S Canada 49°01′N 97°07′W
29 *T15* **Emerson** Iowa, C USA 41°00′N 95°22′W
29 *R13* **Emerson** Nebraska, C USA 42°16′N 96°43′W
36 *M5* **Emery** Utah, W USA 38°54′N 111°16′W
Emesa *see* Ḥimṣ
136 *E13* **Emet** Kütahya, W Turkey 39°22′N 29°15′E
186 *B8* **Emeti** Western, SW Papua New Guinea 07°54′S 143°18′E
35 *V3* **Emigrant Pass** *pass* Nevada, W USA
78 *I6* **Emi Koussi** ▲ N Chad 19°52′N 18°34′E
Emilia *see* Emilia-Romagna
41 *V15* **Emiliano Zapata** Chiapas, SE Mexico 17°42′N 91°46′W
106 *E9* **Emilia-Romagna** *prev.* Emilia; *anc.* Æmilia. ◆ *region* N Italy
158 *J3* **Emin** *var.* Dorbiljin. Xinjiang Uygur Zizhiqu, NW China 46°30′N 83°42′E
149 *W8* **Eminābād** Punjab, E Pakistan 32°02′N 73°51′E
21 *L5* **Eminence** Kentucky, S USA 38°22′N 85°10′W
27 *W7* **Eminence** Missouri, C USA 37°09′N 91°22′W
114 *N9* **Emine, Nos** *headland* E Bulgaria 42°43′N 27°53′E
158 *I3* **Emin He** ⚡ NW China
186 *G4* **Emirau Island** *island* N Papua New Guinea
136 *F13* **Emirdağ** Afyon, W Turkey 39°01′N 31°09′E
83 *K22* **eMkhondo** *prev.* Piet Retief. Mpumalanga, E South Africa 27°00′S 30°49′E *see also* Piet Retief
95 *M21* **Emmaboda** Kalmar, S Sweden 56°36′N 15°30′E
118 *E5* **Emmaste** Hiiumaa, W Estonia 58°43′N 22°36′E
18 *I15* **Emmaus** Pennsylvania, NE USA 40°32′N 75°28′W
183 *U4* **Emmaville** New South Wales, SE Australia 29°26′S 151°38′E
98 *L8* **Emmeloord** Flevoland, N Netherlands 52°43′N 05°46′E
98 *O8* **Emmen** Drenthe, NE Netherlands 52°48′N 06°57′E
108 *F8* **Emmen** Luzern, C Switzerland 47°03′N 08°14′E
101 *F23* **Emmendingen** Baden-Württemberg, SW Germany 48°07′N 07°51′E
98 *P8* **Emmer-Compascuum** Drenthe, NE Netherlands 52°47′N 07°03′E
101 *D14* **Emmerich** Nordrhein-Westfalen, W Germany 51°49′N 06°16′E
29 *U12* **Emmetsburg** Iowa, C USA 43°06′N 94°40′W
32 *M14* **Emmett** Idaho, NW USA 43°52′N 116°30′W
38 *M10* **Emmonak** Alaska, USA 62°46′N 164°31′W
Emona *see* Ljubljana
Emonti *see* East London
24 *L12* **Emory Peak** ▲ Texas, SW USA 29°15′N 103°18′W
40 *F6* **Empalme** Sonora, NW Mexico 27°57′N 110°49′W
83 *L23* **Empangeni** KwaZulu/Natal, E South Africa 28°45′S 31°54′E
61 *C14* **Empedrado** Corrientes, NE Argentina 27°59′S 58°47′W
192 *K3* **Emperor Seamounts** *undersea feature* NW Pacific Ocean 42°00′N 170°00′E
192 *L3* **Emperor Trough** *undersea feature* N Pacific Ocean
35 *R4* **Empire** Nevada, W USA 40°26′N 119°21′W
Empire State of the South *see* Georgia
Emplawas *see* Amplawas
106 *F11* **Empoli** Toscana, C Italy 43°43′N 10°57′E
27 *P5* **Emporia** Kansas, C USA 38°24′N 96°10′W
21 *W7* **Emporia** Virginia, NE USA 36°42′N 77°33′W
18 *E13* **Emporium** Pennsylvania, NE USA 41°31′N 78°14′W
Empty Quarter *see* Ar Rub' al Khālī
100 *E10* **Ems** *Dut.* Eems. ⚡ NW Germany
100 *F13* **Emsdetten** Nordrhein-Westfalen, NW Germany 52°11′N 07°32′E
Ems-Hunte Canal *see* Küstenkanal
100 *F10* **Ems-Jade-Kanal** *canal* NW Germany
100 *F11* **Emsland** *cultural region* NW Germany
182 *D3* **Emu Junction** South Australia 28°39′S 132°13′E
163 *T3* **Emur He** ⚡ NE China
55 *R8* **Enachu Landing** NW Guyana 06°09′N 60°02′W
93 *F16* **Enafors** Jämtland, C Sweden 63°17′N 12°24′E
94 *N11* **Enånger** Gävleborg, C Sweden 61°30′N 17°10′E
96 *G7* **Enard Bay** *bay* NW Scotland, United Kingdom
Enareträsk *see* Inarijärvi
171 *X14* **Enarotali** Papua, E Indonesia 03°55′S 136°21′E
En 'Avedat *see* Ein Avdat
165 *T2* **Enbetsu** Hokkaidō, NE Japan 44°44′N 141°47′E
61 *H16* **Encantadas, Serra das** ▲ S Brazil
40 *E7* **Encantado, Cerro** ▲ NW Mexico 26°46′N 112°33′W
62 *P7* **Encarnación** Itapúa, S Paraguay 27°20′S 55°50′W
40 *M12* **Encarnación de Díaz** Jalisco, SW Mexico 21°33′N 102°13′W
77 *O17* **Enchi** SW Ghana 05°53′N 02°48′W
25 *Q14* **Encinal** Texas, SW USA 28°02′N 99°21′W
35 *U17* **Encinitas** California, W USA 33°02′N 117°17′W
25 *S16* **Encino** Texas, SW USA 26°58′N 98°06′W
54 *H6* **Encontrados** Zulia, NW Venezuela 09°04′N 72°16′W
182 *I10* **Encounter Bay** *inlet* South Australia
61 *F15* **Encruzilhada** Rio Grande do Sul, S Brazil 28°58′S 55°31′W
61 *H16* **Encruzilhada do Sul** Rio Grande do Sul, S Brazil 30°30′S 52°32′W
111 *M20* **Encs** Borsod-Abaúj-Zemplén, NE Hungary 48°21′N 21°09′E
193 *P3* **Endeavour Seamount** *undersea feature* N Pacific Ocean 48°15′N 129°04′W
181 *V1* **Endeavour Strait** *strait* Queensland, NE Australia
171 *O16* **Endeh** Flores, S Indonesia 08°48′S 121°37′E
95 *G23* **Endelave** *island* C Denmark
191 *T4* **Enderbury Island** *atoll* Phoenix Islands, C Kiribati
11 *N16* **Enderby** British Columbia, SW Canada 50°34′N 119°09′W
195 *W4* **Enderby Land** *physical region* Antarctica
173 *N14* **Enderby Plain** *undersea feature* S Indian Ocean
29 *Q6* **Enderlin** North Dakota, N USA 46°37′N 97°36′W
Endersdorf *see* Jędrzejów
28 *K16* **Enders Reservoir** ⊠ Nebraska, C USA
18 *H11* **Endicott** New York, NE USA 42°06′N 76°03′W
39 *P7* **Endicott Mountains** ▲ Alaska, USA
118 *I5* **Endla Raba** *wetland* C Estonia
117 *T9* **Enerhodar** Zaporiz'ka Oblast', SE Ukraine 47°30′N 34°40′E
57 *F14* **Ene, Río** ⚡ C Peru
189 *N4* **Enewetak Atoll** *var.* Ānewetak, Eniwetok. *atoll* Ralik Chain, W Marshall Islands
114 *L13* **Enez** Edirne, NW Turkey 40°44′N 26°05′E
21 *W8* **Enfield** North Carolina, SE USA 36°10′N 77°40′W
186 *B7* **Enga** ◆ *province* W Papua New Guinea
45 *Q9* **Engaño, Cabo** *headland* E Dominican Republic 18°36′N 68°19′W
164 *U3* **Engaru** Hokkaidō, NE Japan 44°06′N 143°30′E
108 *F9* **Engelberg** Unterwalden, C Switzerland 46°51′N 08°25′E
21 *Y9* **Engelhard** North Carolina, SE USA 35°30′N 76°00′W
127 *P8* **Engel's** Saratovskaya Oblast', W Russian Federation 51°27′N 46°09′E
101 *G24* **Engen** Baden-Württemberg, SW Germany 47°52′N 08°46′E
Engeten *see* Aiud
168 *K15* **Enggano, Pulau** *island* W Indonesia
80 *J8* **Enghershatu** ▲ N Eritrea 16°41′N 38°21′E
99 *F19* **Enghien** *Dut.* Edingen. Hainaut, SW Belgium 50°42′N 04°03′E
27 *V12* **England** Arkansas, C USA 34°32′N 91°58′W
97 *M20* **England** *Lat.* Anglia. ◇ *national region* England, United Kingdom
14 *H8* **Englehart** Ontario, S Canada 47°50′N 79°52′W
37 *T4* **Englewood** Colorado, C USA 39°39′N 104°59′W
31 *O16* **English** Indiana, N USA 38°20′N 86°28′W
39 *Q13* **English Bay** Alaska, USA 59°21′N 151°55′W
English Bazar *see* Ingrāj Bāzār
97 *N25* **English Channel** *var.* The Channel, *Fr.* la Manche. *channel* NW Europe
194 *J7* **English Coast** *physical region* Antarctica
105 *S11* **Enguera** Valenciana, E Spain 38°58′N 00°42′W
118 *E8* **Engure** W Latvia 57°09′N 23°13′E
118 *E8* **Engures Ezers** ◎ NW Latvia
137 *R9* **Enguri** *Rus.* Inguri. ⚡ NW Georgia
Engyum *see* Gangi
26 *M9* **Enid** Oklahoma, C USA 36°25′N 97°53′W
22 *L3* **Enid Lake** ⊠ Mississippi, S USA
189 *Y2* **Enigu** *island* Ratak Chain, SE Marshall Islands
Enikale Strait *see* Kerch Strait
147 *Z8* **Enil'chek** Issyk-Kul'skaya Oblast', E Kyrgyzstan 42°04′N 79°01′E
115 *F17* **Enipéfs** ⚡ C Greece
165 *S4* **Eniwa** Hokkaidō, NE Japan 42°53′N 141°14′E
Eniwetok *see* Enewetak Atoll
Enjiang *see* Yongfeng
Enkeldoorn *see* Chivhu
123 *S11* **Enken, Mys** *prev.* Mys Enkan. *headland* NE Russian Federation 58°29′N 141°27′E
Enken, Mys *see* Enken, Mys
98 *J8* **Enkhuizen** Noord-Holland, NW Netherlands 52°34′N 05°03′E
109 *Q4* **Enknach** ⚡ N Austria
95 *N15* **Enköping** Uppsala, C Sweden 59°38′N 17°05′E
107 *K24* **Enna** *var.* Castrogiovanni, Henna. Sicilia, Italy, C Mediterranean Sea 37°34′N 14°16′E
80 *D11* **En Nahud** Western Kordofan, C Sudan 12°41′N 28°28′E
138 *F8* **En Nâqoûra** *var.* An Nāqūrah. SW Lebanon 33°06′N 33°30′E
En Nazira *see* Natzrat
78 *K8* **Ennedi** *plateau* NE Chad
78 *J7* **Ennedi-Ouest** ◆ *region* N Chad
Ennedi-Ouest, Région de l' *see* Ennedi-Ouest
101 *E15* **Ennepetal** Nordrhein-Westfalen, W Germany 51°18′N 07°23′E
183 *P4* **Enngonia** New South Wales, SE Australia 29°19′S 145°52′E
97 *C19* **Ennis** *Ir.* Inis. Clare, W Ireland 52°50′N 08°59′W
33 *R11* **Ennis** Montana, NW USA 45°21′N 111°45′W
25 *U7* **Ennis** Texas, SW USA 32°19′N 96°37′W
97 *F20* **Enniscorthy** *Ir.* Inis Córthaidh. SE Ireland 52°30′N 06°34′W
97 *E15* **Enniskillen** *var.* Inniskilling, *Ir.* Inis Ceithleann. SW Northern Ireland, United Kingdom 54°21′N 07°38′W
97 *B19* **Ennistimon** *Ir.* Inis Díomáin. Clare, W Ireland 52°57′N 09°17′W
109 *T4* **Enns** Oberösterreich, N Austria 48°13′N 14°28′E
109 *T4* **Enns** ⚡ C Austria
93 *O16* **Eno** Pohjois-Karjala, SE Finland 62°45′N 30°15′E
24 *M5* **Enochs** Texas, SW USA 33°51′N 102°46′W
93 *N17* **Enonkoski** Etelä-Savo, E Finland 62°04′N 28°53′E
92 *K10* **Enontekiö** *Lapp.* Eanodat. Lappi, N Finland 68°25′N 23°40′E
21 *Q11* **Enoree** South Carolina, SE USA 34°39′N 81°58′W
21 *P11* **Enoree River** ⚡ South Carolina, SE USA
18 *M6* **Enosburg Falls** Vermont, NE USA 44°54′N 72°50′W
171 *N13* **Enrekang** Sulawesi, C Indonesia 03°33′S 119°46′E
45 *N10* **Enriquillo** SW Dominican Republic 17°57′N 71°13′W
45 *N9* **Enriquillo, Lago** ◎ SW Dominican Republic
98 *L9* **Ens** Flevoland, N Netherlands 52°39′N 05°49′E
98 *P11* **Enschede** Overijssel, E Netherlands 52°13′N 06°55′E
40 *B2* **Ensenada** Baja California Norte, NW Mexico 31°52′N 116°32′W
101 *E20* **Ensheim** ✈ (Saarbrücken) Saarland, W Germany 49°13′N 07°09′E
160 *L9* **Enshi** Hubei, C China 30°16′N 109°26′E
164 *L14* **Enshū-nada** *gulf* SW Japan
23 *O8* **Ensley** Florida, SE USA 30°31′N 87°16′W
Enso *see* Svetogorsk
81 *F18* **Entebbe** S Uganda 0°07′N 32°30′E
81 *F18* **Entebbe** ✈ C Uganda 0°04′N 32°29′E
101 *M18* **Entenbühl** ▲ Czech Republic/Germany 50°09′N 12°10′E
98 *N10* **Enter** Overijssel, E Netherlands 52°19′N 06°34′E
23 *Q7* **Enterprise** Alabama, S USA 31°19′N 85°50′W
32 *L11* **Enterprise** Oregon, NW USA 45°25′N 117°18′W
36 *J7* **Enterprise** Utah, W USA 37°33′N 113°42′W
32 *J8* **Entiat** Washington, NW USA 47°39′N 120°15′W
105 *P15* **Entinas, Punta de las** *headland* S Spain 36°40′N 02°44′W
108 *F8* **Entlebuch** Luzern, W Switzerland 47°02′N 08°04′E
108 *F8* **Entlebuch** *valley* C Switzerland
63 *I22* **Entrada, Punta** *headland* S Argentina
103 *O13* **Entraygues-sur-Truyère** Aveyron, S France 44°39′N 02°36′E
187 *O14* **Entrecasteaux, Récifs d'** *reef* N New Caledonia
61 *C17* **Entre Ríos** *off.* Provincia de Entre Ríos. ◆ *province* NE Argentina
42 *K7* **Entre Ríos, Cordillera** ▲ Honduras/Nicaragua
Entre Ríos, Provincia de *see* Entre Ríos
104 *G9* **Entroncamento** Santarém, C Portugal 39°28′N 08°28′W
77 *V16* **Enugu** Enugu, S Nigeria 06°24′N 07°24′E
77 *U16* **Enugu** ◆ *state* SE Nigeria
123 *V5* **Enurmino** Chukotskiy Avtonomnyy Okrug, NE Russian Federation 66°46′N 171°40′W
54 *E9* **Envigado** Antioquia, W Colombia 06°09′N 75°38′W
59 *B15* **Envira** Amazonas, W Brazil 07°12′S 69°59′W
Enyélé *see* Enyellé
79 *I16* **Enyellé** *var.* Enyélé. Likouala, NE Congo 02°49′N 18°02′E
101 *H21* **Enz** ⚡ SW Germany
165 *N13* **Enzan** *var.* Kōshū. Yamanashi, Honshū, S Japan 35°44′N 138°43′E
104 *I2* **Eo** ⚡ NW Spain
Eochaill *see* Youghal
Eochaille, Cuan *see* Youghal Bay
107 *K22* **Eolie, Isole** *var.* Isole Lipari, *Eng.* Aeolian Islands, Lipari Islands. *island group* S Italy
189 *U12* **Eot** *island* Chuuk, C Micronesia
Epáno Archánes/Epáno Arkhánai *see* Archánes
115 *G14* **Epanomí** Kentrikí Makedonía, N Greece 40°25′N 22°57′E
98 *M10* **Epe** Gelderland, E Netherlands 52°21′N 05°59′E
77 *S16* **Epe** Lagos, S Nigeria 06°37′N 04°01′E
79 *I17* **Epéna** Likouala, NE Congo 01°28′N 17°29′E
Eperies/Eperjes *see* Prešov
103 *Q4* **Épernay** *anc.* Sparnacum. Marne, N France 49°02′N 03°58′E
36 *L5* **Ephraim** Utah, W USA 39°21′N 111°35′W
18 *H15* **Ephrata** Pennsylvania, NE USA 40°09′N 76°08′W
32 *J8* **Ephrata** Washington, NW USA 47°19′N 119°33′W
187 *R14* **Epi** *var.* Épi. *island* C Vanuatu
Épi *see* Epi
105 *R6* **Épila** Aragón, NE Spain 41°34′N 01°19′W
103 *T6* **Épinal** Vosges, NE France 48°10′N 06°28′E
Epiphania *see* Ḥamāh
Epirus *see* Ípeiros
121 *P3* **Episkopí** SW Cyprus 34°37′N 32°53′E
Episkopi Bay *see* Episkopí, Kólpos
121 *P3* **Episkopí, Kólpos** *var.* Episkopi Bay. *bay* SE Cyprus
Epoon *see* Ebon Atoll
Eporedia *see* Ivrea
Eppeschdorf *see* Dumbrăveni
101 *H21* **Eppingen** Baden-Württemberg, SW Germany 49°09′N 08°54′E
83 *E18* **Epukiro** Omaheke, E Namibia 21°40′S 19°09′E
29 *Y13* **Epworth** Iowa, C USA 42°27′N 90°55′W
143 *O10* **Eqlīd** *var.* Iqlīd. Fārs, C Iran 30°54′N 52°40′E
Equality State *see* Wyoming
79 *J18* **Equateur** *off.* Région de l' Equateur. ◆ *region* N Dem. Rep. Congo
Equateur, Région de l' *see* Equateur
151 *K22* **Equatorial Channel** *channel* S Maldives
79 *B17* **Equatorial Guinea** *off.* Equatorial Guinea, Republic of. ◆ *republic* C Africa
Equatorial Guinea, Republic of *see* Equatorial Guinea
121 *V11* **Eratosthenes Tablemount** *undersea feature* E Mediterranean Sea 33°48′N 32°53′E
Erautini *see* Johannesburg
136 *L12* **Erbaa** Tokat, N Turkey 40°42′N 36°37′E
101 *E19* **Erbeskopf** ▲ W Germany 49°43′N 07°04′E
139 *U4* **Erbet** *Ar.* 'Arbat, *var.* Arbat. As Sulaymānīyah, NE Iraq 35°26′N 45°34′E
Erbil *see* Arbil
121 *P2* **Ercan** ✈ (Nicosia) N Cyprus 35°07′N 33°30′E
Ercegnovi *see* Herceg-Novi
137 *T14* **Erçek Gölü** ◎ E Turkey
137 *S14* **Erciş** Van, E Turkey 39°02′N 43°21′E
136 *K14* **Erciyes Dağı** *anc.* Argaeus. ▲ C Turkey 38°32′N 35°28′E
111 *J22* **Érd** *Ger.* Hanselbeck. Pest, C Hungary 47°22′N 18°56′E
163 *X11* **Erdaobaihe** *prev.* Baihe. Jilin, NE China 42°24′N 128°09′E
159 *O12* **Erdaogou** Qinghai, W China 34°30′N 92°50′E
163 *X11* **Erdao Jiang** ⚡ NE China
Erdát-Sângeorz *see* Sângeorgiu de Pădure
136 *C11* **Erdek** Balıkesir, NW Turkey 40°24′N 27°47′E
Erdély *see* Transylvania
Erdélyi-Havasok *see* Carpații Meridionalii
136 *J17* **Erdemli** İçel, S Turkey 36°35′N 34°19′E
163 *O10* **Erdene** *var.* Ulaan-Uul. Dornogovĭ, SE Mongolia 44°21′N 111°06′E
162 *H9* **Erdene** *var.* Sangiyn Dalay. Govĭ-Altay, C Mongolia 45°12′N 97°51′E
162 *E6* **Erdenebüren** *var.* Har-Us. Hovd, W Mongolia 48°30′N 91°25′E
162 *K9* **Erdenedalay** *var.* Sangiyn Dalay. Dundgovĭ, C Mongolia 45°59′N 104°58′E
162 *G7* **Erdenehayrhan** *var.* Altan. Dzavhan, W Mongolia 48°05′N 95°48′E
162 *J7* **Erdenemandal** *var.* Öldziyt. Arhangay, C Mongolia 48°30′N 101°25′E
162 *K6* **Erdenet** Orhon, N Mongolia 49°01′N 104°07′E
163 *Q9* **Erdenetsagaan** *var.* Chonogol. Sühbaatar, E Mongolia 45°55′N 115°19′E

◆ Country ● Country Capital ◇ Dependent Territory ○ Dependent Territory Capital ◆ Administrative Regions ✈ International Airport ▲ Mountain ▲ Mountain Range ⛰ Volcano ⚡ River ◎ Lake ⊠ Reservoir
247

ERDENETSOGT – FADIFFOLU

162 I8 **Erdenetsogt** Bayanhongor, C Mongolia 46°27′N 100°53′E
Erdenetsogt see Bayan-Ovoo
78 K7 **Erdi** plateau NE Chad
78 L7 **Erdi Ma** desert NE Chad
101 M23 **Erding** Bayern, SE Germany 48°18′N 11°54′E
Erdőszáda see Ardusat
Erdőszentgyörgy see Sângeorgiu de Pădure
102 I7 **Erdre** ≈ NW France
195 R13 **Erebus, Mount** 🌋 Ross Island, Antarctica 78°11′S 165°09′E
61 H14 **Erechim** Rio Grande do Sul, S Brazil 27°35′S 52°15′W
163 O7 **Ereen Davaanï Nuruu** ⛰ NE Mongolia
163 Q6 **Ereentsav** Dornod, NE Mongolia 49°51′N 115°41′E
136 I16 **Ereğli** Konya, S Turkey 37°30′N 34°02′E
115 A15 **Ereíkoussa** island Iónia Nisiá, Greece, C Mediterranean Sea
163 O11 **Erenhot** var. Erlian. Nei Mongol Zizhiqu, NE China 43°35′N 112°E
104 M6 **Eresma** ≈ N Spain
115 K17 **Eresós** var. Eressós. Lésvos, E Greece 39°11′N 25°57′E
Eressós see Eresós
Ereymentaū see Yereymentau
99 K21 **Érezée** Luxembourg, SE Belgium 50°16′N 05°34′E
74 G7 **Erfoud** SE Morocco 31°29′N 04°18′W
101 D16 **Erft** ≈ W Germany
101 K16 **Erfurt** Thüringen, C Germany 50°59′N 11°02′E
137 P15 **Ergani** Dıyarbakır, SE Turkey 38°17′N 39°44′E
Ergel see Hatanbulag
Ergene Çayı see Ergene Irmağı
136 C10 **Ergene Irmağı** var. Ergene Çayı. ≈ NW Turkey
118 I9 **Ērgli** C Latvia 56°55′N 25°38′E
78 H11 **Erguig, Bahr** ≈ SW Chad
163 S5 **Ergun** var. Labudalin; prev. Ergun Youqi. Nei Mongol Zizhiqu, N China 50°13′N 120°09′E
Ergun see Gegan Gol
Ergun He see Argun
Ergun Youqi see Ergun
Ergun Zuoqi see Gegen Gol
160 F12 **Er Hai** ⊠ SW China
104 K4 **Ería** ≈ NW Spain
80 H8 **Eriba** Kassala, NE Sudan 16°37′N 36°04′E
96 I6 **Eriboll, Loch** inlet NW Scotland, United Kingdom
65 Q18 **Erica Seamount** undersea feature SW Indian Ocean 38°15′S 14°30′E
107 H23 **Erice** Sicilia, Italy, C Mediterranean Sea 38°02′N 12°35′E
104 E10 **Ericeira** Lisboa, C Portugal 38°58′N 09°25′W
96 H10 **Ericht, Loch** ◎ C Scotland, United Kingdom
26 J11 **Erick** Oklahoma, C USA 35°13′N 99°52′W
18 B11 **Erie** Pennsylvania, NE USA 42°07′N 80°04′W
18 E9 **Erie Canal** canal New York, NE USA
Érié, Lac see Erie, Lake
31 T10 **Erie, Lake** Fr. Lac Érié. ◎ Canada/USA
Erigabo see Ceerigaabo
77 N8 **'Erîgât** desert N Mali
Erigavo see Ceerigaabo
92 P2 **Erik Eriksenstretet** strait E Svalbard
11 X15 **Eriksdale** Manitoba, S Canada 50°52′N 98°07′W
189 V6 **Erikub Atoll** var. Ādkup. atoll Ratak Chain, C Marshall Islands
102 G4 **Er, Îles d'** Island group NW France
Erimanthos see Erýmanthos
165 T6 **Erimo** Hokkaidō, NE Japan 42°01′N 143°07′E
165 T6 **Erimo-misaki** headland Hokkaidō, NE Japan 41°57′N 143°12′E
20 H8 **Erin** Tennessee, S USA 36°19′N 87°42′W
96 E9 **Eriskay** island NW Scotland, United Kingdom
Erithraí see Erythrés
80 I9 **Eritrea** off. State of Eritrea, Ērtra. ◆ transitional government E Africa
Eritrea, State of see Eritrea
Erivan see Yerevan
101 D16 **Erkelenz** Nordrhein-Westfalen, W Germany 51°04′N 06°19′E
95 P15 **Erken** ◎ C Sweden
101 K19 **Erlangen** Bayern, S Germany 49°36′N 11°E
160 G9 **Erlang Shan** ▲ C China 29°56′N 102°24′E
Erlau see Eger
109 V5 **Erlauf** ≈ NE Austria
181 Q8 **Erldunda Roadhouse** Northern Territory, N Australia 25°13′S 133°13′E
Erlian see Erenhot
27 T15 **Erling, Lake** ⊠ Arkansas, USA
109 O8 **Erlsbach** Tirol, W Austria 46°54′N 12°15′E
Ermak see Aksu
98 K10 **Ermelo** Gelderland, C Netherlands 52°18′N 05°38′E
83 K21 **Ermelo** Mpumalanga, NE South Africa 26°32′S 29°59′E
136 H17 **Ermenek** Karaman, S Turkey 36°38′N 32°55′E
Érmihályfalva see Valea lui Mihai
115 G20 **Ermióni** Pelopónnisos, S Greece 37°24′N 23°15′E
115 J20 **Ermoúpoli** var. Hermoupolis; prev. Ermoúpolis. Sýros, Kykládes, Greece, Aegean Sea 37°26′N 24°55′E
Ermoúpolis see Ermoúpoli
Ernabella see Pukatja
155 G22 **Ernākulam** Kerala, SW India 10°04′N 76°18′E
102 J6 **Ernée** Mayenne, NW France 48°18′N 00°54′W
61 H14 **Ernestina, Barragem** ⊠ S Brazil
54 E4 **Ernesto Cortissoz** ✈ (Barranquilla) Atlántico, N Colombia
155 H21 **Erode** Tamil Nādu, SE India 11°21′N 77°43′E
Eroj see Iroj
83 C19 **Erongo** ◈ district W Namibia
99 F21 **Erquelinnes** Hainaut, S Belgium 50°18′N 04°08′E
74 G7 **Er-Rachidia** var. Ksar al Soule. E Morocco 31°58′N 04°22′W
80 E11 **Er Rahad** var. Ar Rahad. Northern Kordofan, C Sudan 12°43′N 30°39′E
Er Ramle see Ramla
83 O15 **Errego** Zambézia, NE Mozambique 16°02′S 37°11′E
105 Q2 **Errenteria** Cast. Rentería. País Vasco, N Spain 43°17′N 01°54′W
Er Rif/Er Riff see Rif
97 D14 **Errigal Mountain** Ir. An Earagail. ▲ N Ireland 55°03′N 08°09′W
97 A15 **Erris Head** Ir. Ceann Iorrais. headland W Ireland 54°18′N 10°01′W
187 S15 **Erromango** island S Vanuatu
Error Guyot see Error Tablemount
173 O4 **Error Tablemount** var. Error Guyot. undersea feature W Indian Ocean 10°20′N 56°05′E
80 G11 **Er Roseires** Blue Nile, E Sudan 11°52′N 34°23′E
Erseka see Ersekë
113 M22 **Ersekë** var. Erseka, Kolonjë. Korçë, SE Albania 40°19′N 20°39′E
Érsekújvár see Nové Zámky
29 S4 **Erskine** Minnesota, N USA 47°42′N 96°00′W
103 V6 **Erstein** Bas-Rhin, NE France 48°24′N 07°39′E
108 G9 **Erstfeld** Uri, C Switzerland 46°49′N 08°41′E
158 M3 **Ertai** Xinjiang Uygur Zizhiqu, NW China 46°04′N 90°06′E
126 M7 **Ertil'** Voronezhskaya Oblast', W Russian Federation 51°51′N 40°46′E
Ertis see Irtysh, C Asia
Ertis see Irtyshsk, Kazakhstan
158 K2 **Ertix He** Rus. Chërnyy Irtysh. ≈ China/Kazakhstan
Ērtra see Eritrea
21 P9 **Erwin** North Carolina, SE USA 35°19′N 78°40′W
Erydropótamos see Erythropótamos
115 E19 **Erýmanthos** var. Erimanthos. ▲ S Greece 37°57′N 21°51′E
115 G19 **Erythrés** prev. Erithraí. Stereá Elláda, C Greece 38°18′N 23°20′E
114 L12 **Erythropótamos** Bul. Byala Reka, var. Erydropótamos. ≈ Bulgaria/Greece
160 F12 **Eryuan** var. Yuhu. Yunnan, SW China 26°09′N 100°01′E
109 U6 **Erzbach** ≈ W Austria
Erzerum see Erzurum
101 N17 **Erzgebirge** Cz. Krušné Hory, Eng. Ore Mountains. ⛰ Czech Republic/Germany see also Krušné Hory
Erzgebirge see Krušné Hory
122 L14 **Erzin** Respublika Tyva, S Russian Federation 50°17′N 95°03′E
137 O13 **Erzincan** var. Erzinjan. Erzincan, E Turkey 39°44′N 39°30′E
137 N13 **Erzincan** var. Erzinjan. ◈ province NE Turkey
Erzinjan see Erzincan
Erzsébetváros see Dumbrăveni
137 Q13 **Erzurum** prev. Erzerum. Erzurum, NE Turkey 39°57′N 41°17′E
137 Q12 **Erzurum** prev. Erzerum. ◈ province NE Turkey
186 G9 **Esa'ala** Normanby Island, SE Papua New Guinea 09°45′S 150°47′E
165 T2 **Esashi** Hokkaidō, NE Japan 44°57′N 142°32′E
165 Q9 **Esashi** var. Esasi. Iwate, Honshū, C Japan 39°13′N 141°11′E
165 Q5 **Esasho** Hokkaidō, N Japan
Esasi see Esashi
95 F23 **Esbjerg** Syddtjylland, W Denmark 55°28′N 08°28′E
Esbo see Espoo
36 L7 **Escalante** Utah, W USA 37°46′N 111°36′W
36 M7 **Escalante River** ≈ Utah, W USA
14 L12 **Escalier, Réservoir l'** ⊠ Québec, SE Canada
40 K7 **Escalón** Chihuahua, N Mexico 26°43′N 104°20′W
104 M8 **Escalona** Castilla-La Mancha, C Spain 40°10′N 04°24′W
23 O8 **Escambia River** ≈ Florida, SE USA
31 N5 **Escanaba** Michigan, N USA 45°45′N 87°03′W
31 O4 **Escanaba River** ≈ Michigan, N USA
105 R8 **Escandón, Puerto de** pass E Spain
41 W14 **Escárcega** Campeche, SE Mexico 18°33′N 90°41′W
171 O1 **Escarpada Point** headland Luzon, N Philippines 18°28′N 122°10′E
23 N8 **Escatawpa River** ≈ Alabama/Mississippi, S USA
103 P2 **Escaut** ≈ N France
Escaut see Scheldt
99 M25 **Esch-sur-Alzette** Luxembourg, S Luxembourg 49°30′N 05°59′E
101 J15 **Eschwege** Hessen, C Germany 51°10′N 10°03′E
101 D16 **Eschweiler** Nordrhein-Westfalen, W Germany 50°49′N 06°16′E
Esclaves, Grand Lac des see Great Slave Lake
45 O8 **Escocesa, Bahía** bay N Dominican Republic
43 W15 **Escocés, Punta** headland NE Panama 08°50′N 77°37′W
35 U17 **Escondido** California, W USA 33°07′N 117°05′W
42 M10 **Escondido, Río** ≈ SE Nicaragua
15 S7 **Escoumins, Rivière des** ≈ Québec, SE Canada
37 O13 **Escudilla Mountain** ▲ Arizona, SW USA 33°57′N 109°07′W
40 J11 **Escuinapa** var. Escuinapa de Hidalgo. Sinaloa, C Mexico 22°50′N 105°46′W
Escuinapa de Hidalgo see Escuinapa
42 C6 **Escuintla** Escuintla, S Guatemala 14°17′N 90°46′W
41 V17 **Escuintla** Chiapas, SE Mexico 15°20′N 92°40′W
42 A2 **Escuintla** off. Departamento de Escuintla. ◈ department S Guatemala
Escuintla, Departamento de see Escuintla
15 W7 **Escuminac** ≈ Québec, SE Canada
79 D16 **Eséka** Centre, SW Cameroon 03°40′N 10°48′E
136 I12 **Esenboğa** ✈ (Ankara) Ankara, C Turkey 40°05′N 33°01′E
136 D17 **Eşen Çayı** ≈ SW Turkey
146 B13 **Esenguly** Rus. Gasan-Kuli. Balkan Welaýaty, W Turkmenistan 37°29′N 53°57′E
105 T4 **Ésera** ≈ NE Spain
143 N8 **Eşfahān** Eng. Isfahan; anc. Aspadana. Eşfahān, C Iran 32°41′N 51°41′E
143 O7 **Eşfahān** off. Ostān-e Eşfahān. ◈ province C Iran
Eşfahān, Ostān-e see Eşfahān
105 N5 **Esgueva** ≈ N Spain
Eshkamesh see Ishkamish
Eshkāshem see Ishkāshim
83 L23 **Eshowe** KwaZulu/Natal, E South Africa 28°53′S 31°28′E
143 T5 **'Eshqābād** Khorāsān-e Razavī, NE Iran 36°00′N 59°01′E
Esh Sham see Rif Dimashq
Esh Sharā see Ash Sharāh
Esik see Yesik
Esil see Yesil'
Esil see Ishim, Kazakhstan/Russian Federation
183 V2 **Esk** Queensland, E Australia 27°15′S 152°23′E
184 O11 **Eskdale** Hawke's Bay, North Island, New Zealand 39°24′S 176°51′E
Eski Dzhumaya see Targovishte
92 L2 **Eskifjörður** Austurland, E Iceland 65°04′N 14°01′W
139 S3 **Eski Kaļak** var. Askī Kalak, Kalak. Arbīl, N Iraq 36°16′N 43°40′E
95 N16 **Eskilstuna** Södermanland, C Sweden 59°22′N 16°31′E
8 H6 **Eskimo Lakes** ◎ Northwest Territories, NW Canada
9 O10 **Eskimo Point** headland Nunavut, E Canada 61°19′N 93°49′W
Eskimo Point see Arviat
139 Q2 **Eski Mosul** Nīnawá, N Iraq 36°31′N 42°45′E
Eski-Nookat see Nookat
136 F12 **Eskişehir** var. Eskishehr. Eskişehir, W Turkey 39°46′N 30°30′E
136 F13 **Eskişehir** var. Eski shehr. ◈ province NW Turkey
Eskishehr see Eskişehir
104 K5 **Esla** ≈ NW Spain
Eslāmābād see Eslāmābād-e Gharb
142 J6 **Eslāmābād-e Gharb** var. Eslāmābād; prev. Harunabad, Shāhābād. Kermānshāhān, W Iran 34°08′N 46°35′E
148 J4 **Eslām Qal'eh** Pash. Islam Qala. Herāt, W Afghanistan 34°41′N 61°03′E
95 K23 **Eslöv** Skåne, S Sweden 55°50′N 13°20′E
143 S12 **Esmā'īlābād** Kermān, S Iran 28°48′N 56°39′E
143 U8 **Esmā'īlābād** Khorāsān-e Jonūbī, E Iran 35°20′N 60°30′E
136 D14 **Eşme** Uşak, W Turkey 38°26′N 28°59′E
44 G6 **Esmeralda** Camagüey, E Cuba 21°51′N 78°10′W
63 F21 **Esmeralda, Isla** island S Chile
56 B5 **Esmeraldas** Esmeraldas, N Ecuador 0°55′N 79°40′W
56 B5 **Esmeraldas** ◈ province NW Ecuador
Esna see Isnā
14 B6 **Esnagi Lake** ◎ Ontario, S Canada
143 V14 **Espakeh** Sīstān va Balūchestān, SE Iran 26°54′N 60°09′E
103 O13 **Espalion** Aveyron, S France 44°31′N 02°45′E
España see Spain
14 E11 **Espanola** Ontario, S Canada 46°15′N 81°46′W
37 S10 **Espanola** New Mexico, SW USA 35°59′N 106°04′W
57 C18 **Española, Isla** var. Hood Island. island Galapagos Islands, Ecuador, E Pacific Ocean
104 M13 **Espejo** Andalucía, S Spain 37°40′N 04°34′W
94 C13 **Espeland** Hordaland, S Norway 60°22′N 05°27′E
100 G12 **Espelkamp** Nordrhein-Westfalen, NW Germany 52°22′N 08°37′E
38 M8 **Espenberg, Cape** headland Alaska, USA 66°33′N 163°36′W
180 L13 **Esperance** Western Australia 33°49′S 121°52′E
186 L9 **Esperance, Cape** headland Guadacanal, C Solomon Islands 09°09′S 159°38′E
57 P18 **Esperancita** Santa Cruz, E Bolivia
61 B17 **Esperanza** Santa Fe, C Argentina 31°29′S 61°00′W
40 G6 **Esperanza** Sonora, NW Mexico 27°37′N 109°51′W
24 H9 **Esperanza** Texas, SW USA 31°09′N 105°40′W
194 H3 **Esperanza** Argentinian research station Antarctica 63°29′S 56°53′W
104 E12 **Espichel, Cabo** headland S Portugal 38°25′N 09°15′W
54 E10 **Espinal** Tolima, C Colombia 04°08′N 74°53′W
48 K10 **Espinhaço, Serra do** ⛰ SE Brazil
104 G6 **Espinho** Aveiro, N Portugal 41°01′N 08°38′W
59 N18 **Espinosa** Minas Gerais, SE Brazil 14°58′S 42°49′W
103 O15 **Espinouse** ⛰ S France
60 Q8 **Espírito Santo** off. Estado do Espírito Santo. ◈ state E Brazil
Espírito Santo, Estado do see Espírito Santo
187 P13 **Espiritu Santo** var. Santo. island W Vanuatu
41 Z13 **Espíritu Santo, Bahía del** bay SE Mexico
40 F9 **Espíritu Santo, Isla del** island NW Mexico
41 Y12 **Espita** Yucatán, SE Mexico 21°00′N 88°17′W
15 Y7 **Espoir, Cap d'** headland Québec, SE Canada 48°24′N 64°21′W
Esponsede/Esponsende see Esposende
93 L20 **Espoo** Swe. Esbo. Uusimaa, S Finland 60°10′N 24°42′E
104 G5 **Esposende** var. Esponsede, Esponsende. Braga, N Portugal 41°32′N 08°47′W
83 M18 **Espungabera** Manica, SW Mozambique 20°29′S 32°48′E
63 H17 **Esquel** Chubut, SW Argentina 42°55′S 71°20′W
10 L17 **Esquimalt** Vancouver Island, British Columbia, SW Canada 48°26′N 123°27′W
61 C16 **Esquina** Corrientes, NE Argentina 30°00′S 59°30′W
42 E6 **Esquipulas** Chiquimula, SE Guatemala 14°36′N 89°22′W
42 K9 **Esquipulas** Matagalpa, C Nicaragua 12°30′N 85°55′W
94 I8 **Essandsjøen** ◎ S Norway
74 E7 **Essaouira** prev. Mogador. W Morocco 31°33′N 09°40′W
Esseg see Osijek
Es Semara see Smara
99 G15 **Essen** Antwerpen, N Belgium 51°28′N 04°28′E
101 E15 **Essen** var. Essen an der Ruhr. Nordrhein-Westfalen, W Germany 51°28′N 07°01′E
Essen an der Ruhr see Essen
74 I5 **Es Senia** ✈ (Oran) NW Algeria 35°34′N 00°42′W
55 T8 **Essequibo Islands** island group N Guyana
55 T11 **Essequibo River** ≈ C Guyana
14 C18 **Essex** Ontario, S Canada 42°10′N 82°50′W
29 T16 **Essex** Iowa, C USA 40°49′N 95°18′W
97 P21 **Essex** cultural region E England, United Kingdom
31 R8 **Essexville** Michigan, N USA 43°37′N 83°50′W
101 H22 **Esslingen** var. Esslingen am Neckar. Baden-Württemberg, SW Germany 48°45′N 09°19′E
Esslingen am Neckar see Esslingen
103 N6 **Essonne** ◈ department N France
Es Suweida see As Suwaydā'
79 F16 **Est** Eng. East. ◈ province SE Cameroon
104 I1 **Estaca de Bares, Punta de** point NW Spain
24 M5 **Estacado, Llano** plain New Mexico/Texas, SW USA
41 P12 **Estación Tamuín** San Luis Potosí, C Mexico 22°00′N 98°44′W
63 K25 **Estados, Isla de los** prev. Eng. Staten Island. island S Argentina
Estado Vargas see Vargas
143 P12 **Eşţahbān** Fārs, S Iran
14 F11 **Estaire** Ontario, S Canada 46°19′N 80°47′W
59 P16 **Estância** Sergipe, E Brazil 11°15′S 37°28′W
37 S12 **Estancia** New Mexico, SW USA 34°45′N 106°03′W
104 G7 **Estarreja** Aveiro, N Portugal 40°45′N 08°35′W
102 M17 **Estats, Pica d'** Sp. Pico d'Estats. ▲ France/Spain 42°39′N 01°24′E
Estats, Pico d' see Estats, Pica d'
83 K23 **Estcourt** KwaZulu/Natal, E South Africa 29°00′S 29°53′E
106 H8 **Este** anc. Ateste. Veneto, NE Italy 45°14′N 11°40′E
42 J9 **Esteli** Estelí, NW Nicaragua 13°05′N 86°21′W
42 J9 **Esteli** ◈ department NW Nicaragua
105 Q4 **Estella** Bas. Lizarra. Navarra, N Spain 42°41′N 02°02′W
29 R9 **Estelline** South Dakota, N USA 44°34′N 96°54′W
25 P4 **Estelline** Texas, SW USA 34°33′N 100°26′W
104 L14 **Estepa** Andalucía, S Spain 37°17′N 04°52′W
104 L16 **Estepona** Andalucía, S Spain 36°26′N 05°09′W
39 R9 **Ester** Alaska, USA 64°49′N 148°03′W
11 V16 **Esterhazy** Saskatchewan, S Canada 50°40′N 102°02′W
37 S3 **Estes Park** Colorado, C USA 40°22′N 105°31′W
11 V17 **Estevan** Saskatchewan, S Canada 49°07′N 103°05′W
29 T11 **Estherville** Iowa, C USA 43°24′N 94°49′W
21 R15 **Estill** South Carolina, SE USA 32°45′N 81°14′W
103 Q6 **Estissac** Aube, N France 48°17′N 03°51′E
15 T9 **Est, Lac de l'** ◎ Québec, SE Canada
Estland see Estonia
11 S16 **Eston** Saskatchewan, S Canada 51°09′N 108°42′W
118 G5 **Estonia** off. Republic of Estonia, Est. Eesti Vabariik, Ger. Estland, Latv. Igaunija; prev. Estonian SSR, Rus. Estonskaya SSR. ◆ republic NE Europe
Estonian SSR see Estonia
Estonia, Republic of see Estonia
Estonskaya SSR see Estonia
104 E11 **Estoril** Lisboa, W Portugal 38°42′N 09°23′W
59 L14 **Estreito** Maranhão, E Brazil 06°34′S 47°22′W
104 I8 **Estrela, Serra da** ⛰ C Portugal
40 D3 **Estrella, Punta** headland NW Mexico 30°53′N 114°45′W
104 F10 **Estremadura** cultural and historical region W Portugal
Estremadura see Extremadura
104 H11 **Estremoz** Évora, S Portugal 38°50′N 07°35′W
79 D18 **Estuaire** off. Province de l'Estuaire, var. L'Estuaire. ◈ province NW Gabon
Estuaire, Province de l' see Estuaire
Eszék see Osijek
111 I22 **Esztergom** Ger. Gran; anc. Strigonium. Komárom-Esztergom, N Hungary 47°47′N 18°44′E
152 K11 **Etah** Uttar Pradesh, N India 27°33′N 78°39′E
189 R17 **Etal Atoll** atoll Mortlock Islands, C Micronesia
99 K24 **Étalle** Luxembourg, SE Belgium 49°41′N 05°36′E
103 N6 **Étampes** Essonne, N France 48°26′N 02°10′E
182 J1 **Etamunbanie, Lake** salt lake South Australia
103 N1 **Étaples** Pas-de-Calais, N France 50°31′N 01°39′E
152 K12 **Etāwah** Uttar Pradesh, N India 26°46′N 79°01′E
15 R10 **Etchemin** ≈ Québec, SE Canada
Etchmiadzin see Vagharshapat
40 G7 **Etchojoa** Sonora, NW Mexico 26°54′N 109°37′W
93 L19 **Etelä-Karjala** Swe. Södra Karelen, Eng. South Karelia. ◈ region S Finland
93 K17 **Etelä-Pohjanmaa** Swe. Södra Österbotten, Eng. South Ostrobothnia. ◈ region W Finland
93 M18 **Etelä-Savo** Swe. Södra Savolax. ◈ region SE Finland
83 B16 **Etengua** Kunene, NW Namibia 17°24′S 13°05′E
99 K25 **Éthe** Luxembourg, SE Belgium 49°34′N 05°32′E
11 W15 **Ethelbert** Manitoba, S Canada 51°30′N 100°22′W
80 H12 **Ethiopia** off. Federal Democratic Republic of Ethiopia; prev. Abyssinia, People's Democratic Republic of Ethiopia. ◆ republic E Africa
Ethiopia, Federal Democratic Republic of see Ethiopia
80 I13 **Ethiopian Highlands** var. Ethiopian Plateau. plateau N Ethiopia
Ethiopian Plateau see Ethiopian Highlands
Ethiopia, People's Democratic Republic of see Ethiopia
34 M2 **Etna** California, W USA 41°25′N 122°53′W
18 B14 **Etna** Pennsylvania, NE USA 40°29′N 79°55′W
94 G12 **Etna** ≈ S Norway
107 L24 **Etna, Monte** Eng. Mount Etna. 🌋 Sicilia, Italy, C Mediterranean Sea 37°46′N 15°00′E
Etna, Mount see Etna, Monte
95 C15 **Etne** Hordaland, S Norway 59°40′N 05°55′E
Etoliko see Aitolikó
39 Y14 **Etolin Island** island Alexander Archipelago, Alaska, USA
38 L12 **Etolin Strait** strait Alaska, USA
83 C17 **Etosha Pan** salt lake N Namibia
79 G18 **Etoumbi** Cuvette Ouest, NW Congo 0°01′N 14°57′E
20 M10 **Etowah** Tennessee, S USA 35°19′N 84°31′W
23 S2 **Etowah River** ≈ Georgia, SE USA
146 B13 **Etrek** var. Gyzyletrek, Rus. Kizyl-Atrek. Balkan Welaýaty, W Turkmenistan 37°40′N 54°44′E
146 C13 **Etrek** Per. Rūd-e Atrak, Rus. Atrak, Atrek. ≈ Iran/Turkmenistan
102 L3 **Étretat** Seine-Maritime, N France 49°46′N 00°23′E
114 H9 **Etropole** Sofia, W Bulgaria 42°50′N 24°00′E
Etsch see Adige
Et Tafila see Aţ Ţafīlah
99 M23 **Ettelbrück** Diekirch, C Luxembourg 49°51′N 06°06′E
189 V12 **Etten** atoll Chuuk Islands, C Micronesia
99 H14 **Etten-Leur** Noord-Brabant, S Netherlands 51°34′N 04°37′E
76 G7 **Et Tîdra** var. Ile Tîdra. island Dakhlet Nouâdhibou, NW Mauritania
101 G21 **Ettlingen** Baden-Württemberg, SW Germany 48°57′N 08°25′E
102 M2 **Eu** Seine-Maritime, N France 50°01′N 01°24′E
193 W16 **'Eua** prev. Middleburg Island. island Tongatapu Group, SE Tonga
193 W15 **Eua Iki** island Tongatapu Group, S Tonga
Euboea see Évvoia
181 O12 **Eucla** Western Australia 31°41′S 128°51′E
31 U11 **Euclid** Ohio, N USA 41°34′N 81°33′W
27 W14 **Eudora** Arkansas, C USA 33°06′N 91°15′W
27 Q4 **Eudora** Kansas, C USA 38°56′N 95°06′W
182 J9 **Eudunda** South Australia 34°11′S 139°03′E
23 R6 **Eufaula** Alabama, S USA 31°53′N 85°09′W
27 Q11 **Eufaula** Oklahoma, C USA 35°16′N 95°36′W
27 Q11 **Eufaula Lake** var. Eufaula Reservoir. ⊠ Oklahoma, C USA
Eufaula Reservoir see Eufaula Lake
32 F13 **Eugene** Oregon, NW USA 44°03′N 123°05′W
40 B6 **Eugenia, Punta** headland NW Mexico 27°48′N 115°03′W
183 Q8 **Eugowra** New South Wales, SE Australia 33°28′S 148°21′E
104 I2 **Eume** ≈ NW Spain
104 H2 **Eume, Encoro do** ⊠ NW Spain
Eumolpias see Plovdiv
59 O18 **Eunápolis** Bahia, SE Brazil 16°20′S 39°36′W
22 H8 **Eunice** Louisiana, S USA 30°29′N 92°25′W
37 W15 **Eunice** New Mexico, SW USA 32°26′N 103°09′W
99 M19 **Eupen** Liège, E Belgium 50°38′N 06°02′E
130 B10 **Euphrates** Ar. Al-Furāt, Turk. Firat Nehri. ≈ SW Asia
138 L3 **Euphrates Dam** dam N Syria
22 M4 **Eupora** Mississippi, S USA 33°32′N 89°16′W
93 K19 **Eura** Satakunta, SW Finland 61°07′N 22°12′E
93 K19 **Eurajoki** Satakunta, SW Finland 61°13′N 21°45′E
0-1 **Eurasian Plate** tectonic feature
102 L4 **Eure** ◈ department N France
102 M4 **Eure** ≈ N France
102 M6 **Eure-et-Loir** ◈ department C France
34 K3 **Eureka** California, W USA 40°47′N 124°12′W
27 P6 **Eureka** Kansas, C USA 37°51′N 96°17′W
33 O6 **Eureka** Montana, NW USA 48°52′N 115°03′W
35 V5 **Eureka** Nevada, W USA 39°31′N 115°58′W
29 O7 **Eureka** South Dakota, N USA 45°46′N 99°37′W
36 L4 **Eureka** Utah, W USA 39°57′N 112°07′W
32 K10 **Eureka** Washington, NW USA 46°21′N 118°41′W
27 S9 **Eureka Springs** Arkansas, C USA 36°25′N 93°45′W
182 K6 **Eurinilla Creek** seasonal river South Australia
183 O11 **Euroa** Victoria, SE Australia 36°46′S 145°35′E
172 M9 **Europa, Île** island W Madagascar
104 L3 **Europa, Picos de** ⛰ N Spain
104 L16 **Europa Point** headland S Gibraltar 36°07′N 05°20′W
84-85 **Europe** continent
98 F12 **Europoort** Zuid-Holland, W Netherlands 51°59′N 04°08′E
Euskadi see País Vasco
101 D17 **Euskirchen** Nordrhein-Westfalen, W Germany 50°40′N 06°47′E
23 W11 **Eustis** Florida, SE USA 28°51′N 81°41′W
182 M9 **Euston** New South Wales, SE Australia 34°34′S 142°45′E
23 N5 **Eutaw** Alabama, S USA 32°50′N 87°53′W
100 K8 **Eutin** Schleswig-Holstein, N Germany 54°08′N 10°38′E
10 K14 **Eutsuk Lake** ◎ British Columbia, SW Canada
Euxine Sea see Black Sea
83 C16 **Evale** Cunene, SW Angola 16°36′S 15°46′E
37 T3 **Evans** Colorado, C USA 40°22′N 104°41′W
11 P14 **Evansburg** Alberta, SW Canada 53°34′N 114°57′W
29 X13 **Evansdale** Iowa, C USA 42°28′N 92°16′W
183 V4 **Evans Head** New South Wales, SE Australia 29°07′S 153°27′E
12 J11 **Evans, Lac** ◎ Québec, SE Canada
37 S5 **Evans, Mount** ▲ Colorado, C USA 39°15′N 106°10′W
9 Q6 **Evans Strait** strait Nunavut, N Canada
31 N10 **Evanston** Illinois, N USA 42°02′N 87°41′W
33 S17 **Evanston** Wyoming, C USA 41°16′N 110°57′W
14 D11 **Evansville** Manitoulin Island, Ontario, S Canada 45°48′N 82°34′W
31 N16 **Evansville** Indiana, N USA 37°58′N 87°33′W
30 L9 **Evansville** Wisconsin, N USA 42°46′N 89°16′W
25 S8 **Evant** Texas, SW USA 31°28′N 98°09′W
143 P13 **Evaz** Fārs, S Iran 27°48′N 53°58′E
29 W4 **Eveleth** Minnesota, N USA 47°27′N 92°32′W
182 E3 **Evelyn Creek** seasonal river South Australia
181 Q2 **Evelyn, Mount** ▲ Northern Territory, N Australia 13°28′S 132°50′E
122 K10 **Evenkiyskiy Avtonomnyy Okrug** ◈ autonomous district Krasnoyarskiy Kray, N Russian Federation
183 R13 **Everard, Cape** headland Victoria, SE Australia 37°48′S 149°21′E
182 F6 **Everard, Lake** salt lake South Australia
182 C2 **Everard Ranges** ⛰ South Australia
153 R11 **Everest, Mount** Chin. Qomolangma Feng, Nep. Sagarmāthā. ▲ China/Nepal 27°59′N 86°57′E
18 E15 **Everett** Pennsylvania, NE USA 40°00′N 78°22′W
32 H7 **Everett** Washington, NW USA 47°59′N 122°12′W
99 E17 **Evergem** Oost-Vlaanderen, NW Belgium 51°07′N 03°43′E
23 X16 **Everglades City** Florida, SE USA 25°51′N 81°22′W
23 Y16 **Everglades, The** wetland Florida, SE USA
23 P7 **Evergreen** Alabama, S USA 31°25′N 86°55′W
37 T4 **Evergreen** Colorado, C USA 39°37′N 105°19′W
Evergreen State see Washington
97 L21 **Evesham** C England, United Kingdom 52°06′N 01°57′W
103 T10 **Évian-les-Bains** Haute-Savoie, E France 46°22′N 06°34′E
93 K16 **Evijärvi** Etelä-Pohjanmaa, W Finland 63°22′N 23°30′E
79 D17 **Evinayong** var. Ebinayon, Evinayoung. C Equatorial Guinea 01°28′N 10°17′E
Evinayoung see Evinayong
115 E18 **Évinos** ≈ C Greece
95 E17 **Evje** Aust-Agder, S Norway 58°35′N 07°49′E
Evmolpia see Plovdiv
104 H11 **Évora** anc. Ebora, Lat. Liberalitas Julia. Évora, C Portugal 38°34′N 07°54′W
104 G11 **Évora** ◈ district S Portugal
102 M4 **Évreux** anc. Civitas Eburovicum. Eure, N France 49°02′N 01°10′E
102 K6 **Évron** Mayenne, NW France 48°10′N 00°24′W
114 L13 **Évros** Bul. Maritsa, Turk. Meriç; anc. Hebrus. ≈ SE Europe see also Maritsa/Meriç
Évros see Meriç
Évros see Maritsa
115 F21 **Evrótas** ≈ S Greece
103 O5 **Évry** Essonne, N France 48°38′N 02°34′E
25 O8 **E. V. Spence Reservoir** ⊠ Texas, SW USA
115 I18 **Évvoia** Lat. Euboea. island C Greece
38 D9 **'Ewa Beach** var. Ewa Beach. O'ahu, Hawaii, USA, C Pacific Ocean 21°19′N 158°00′W
Ewa Beach see 'Ewa Beach
32 L9 **Ewan** Washington, NW USA 47°06′N 117°46′W
44 K12 **Ewarton** C Jamaica 18°11′N 77°06′W
81 J18 **Ewaso Ng'iro** var. Nyiro. ≈ C Kenya
29 P13 **Ewing** Nebraska, C USA 42°13′N 98°20′W
194 J5 **Ewing Island** island Antarctica
65 P17 **Ewing Seamount** undersea feature E Atlantic Ocean 23°20′S 08°45′E
158 L6 **Ewirgol** Xinjiang Uygur Zizhiqu, W China 42°56′N 87°39′E
79 G19 **Ewo** Cuvette, W Congo 0°55′S 14°49′E
27 S3 **Excelsior Springs** Missouri, C USA 39°20′N 94°13′W
97 J23 **Exe** ≈ SW England, United Kingdom
194 L12 **Executive Committee Range** ⛰ Antarctica
14 E16 **Exeter** Ontario, S Canada 43°19′N 81°26′W
97 J24 **Exeter** anc. Isca Damnoniorum. SW England, United Kingdom 50°43′N 03°31′W
35 R11 **Exeter** California, W USA 36°17′N 119°08′W
19 P10 **Exeter** New Hampshire, NE USA 42°57′N 70°55′W
Exin see Kcynia
29 T14 **Exira** Iowa, C USA 41°36′N 94°55′W
97 J23 **Exmoor** moorland SW England, United Kingdom
21 Y6 **Exmore** Virginia, NE USA 37°31′N 75°48′W
180 G8 **Exmouth** Western Australia 22°01′S 114°06′E
97 J24 **Exmouth** SW England, United Kingdom 50°36′N 03°25′W
180 G8 **Exmouth Gulf** gulf Western Australia
173 V8 **Exmouth Plateau** undersea feature E Indian Ocean 19°00′S 116°00′E
83 K23 **eXobho** prev. Ixopo. KwaZulu/Natal, E South Africa 30°10′S 30°05′E
115 J20 **Exompourgo** ancient monument Tínos, Kykládes, Greece, Aegean Sea
104 I10 **Extremadura** var. Estremadura. ◈ autonomous community W Spain
78 F12 **Extrême-Nord** Eng. Extreme North. ◈ province N Cameroon
Extreme North see Extrême-Nord
44 I3 **Exuma Cays** islets C The Bahamas
44 I3 **Exuma Sound** sound C The Bahamas
81 H20 **Eyasi, Lake** ◎ N Tanzania
95 F17 **Eydehavn** Aust-Agder, S Norway 58°31′N 08°53′E
96 L12 **Eyemouth** SE Scotland, United Kingdom 55°52′N 02°07′W
96 G7 **Eye Peninsula** peninsula NW Scotland, United Kingdom
92 J4 **Eyjafjallajökull** ▲ S Iceland 63°37′N 19°37′W
80 Q13 **Eyl** It. Eil. Nugaal, E Somalia 08°03′N 49°49′E
103 N11 **Eymoutiers** Haute-Vienne, C France 45°45′N 01°43′E
Eyo (lower course) see Uolo, Río
29 X10 **Eyota** Minnesota, N USA 44°00′N 92°13′W
182 H2 **Eyre Basin, Lake** salt lake South Australia
182 I1 **Eyre Creek** seasonal river Northern Territory/South Australia
174 L9 **Eyre, Lake** salt lake South Australia
185 C22 **Eyre Mountains** ⛰ South Island, New Zealand
182 H3 **Eyre North, Lake** salt lake South Australia
182 G7 **Eyre Peninsula** peninsula South Australia
182 H4 **Eyre South, Lake** salt lake South Australia
95 B18 **Eysturoy** Dan. Østerø. island N Faroe Islands
61 D20 **Ezeiza** ✈ (Buenos Aires) Buenos Aires, E Argentina 34°49′S 58°30′W
Ezeres see Ezeriş
116 F12 **Ezeriş** Hung. Ezeres. Caraş-Severin, W Romania 45°21′N 21°55′E
161 O9 **Ezhou** prev. Echeng. Hubei, C China 30°23′N 114°52′E
125 R11 **Ezhva** Respublika Komi, NW Russian Federation 61°45′N 50°43′E
136 B12 **Ezine** Çanakkale, NW Turkey 39°46′N 26°20′E
Ezo see Hokkaidō
Ezra/Ezraa see Izra'

F

191 P7 **Faaa** Tahiti, W French Polynesia 17°32′S 149°36′W
191 P7 **Faaa** ✈ (Papeete) Tahiti, W French Polynesia 17°31′S 149°36′W
95 H24 **Faaborg** var. Fåborg. Syddtjylland, C Denmark 55°06′N 10°10′E
151 K19 **Faadhippolhu Atoll** var. Fadiffolu, Lhaviyani Atoll. atoll N Maldives
191 U10 **Faaite** atoll Îles Tuamotu, C French Polynesia
191 Q8 **Faaone** Tahiti, W French Polynesia 17°39′S 149°18′W
24 H8 **Fabens** Texas, SW USA 31°30′N 106°09′W
94 H12 **Fåberg** Oppland, S Norway 61°15′N 10°21′E
Fåborg see Faaborg
106 I12 **Fabriano** Marche, C Italy 43°20′N 12°54′E
145 U16 **Fabrichnoye** prev. Fabrichnyy. Almaty, SE Kazakhstan 43°12′N 76°19′E
Fabrichnyy see Fabrichnoye
54 F10 **Facatativá** Cundinamarca, C Colombia 04°49′N 74°22′W
77 X9 **Fachi** Agadez, C Niger 18°01′N 11°36′E
188 B16 **Facpi Point** headland W Guam
18 I13 **Factoryville** Pennsylvania, NE USA 41°34′N 75°45′W
78 K8 **Fada** Ennedi-Ouest, E Chad 17°14′N 21°32′E
77 Q13 **Fada-Ngourma** E Burkina Faso 12°05′N 00°26′E
123 N6 **Faddeya, Zaliv** bay N Russian Federation
123 Q5 **Faddeyevskiy, Poluostrov** island Novosibirskiye Ostrova, NE Russian Federation
141 W12 **Fadhi** S Oman 17°54′N 55°30′E
Fadiffolu see Faadhippolhu Atoll

◆ Country ● Country Capital ◇ Dependent Territory ○ Dependent Territory Capital ◈ Administrative Regions ✈ International Airport ▲ Mountain ⛰ Mountain Range 🌋 Volcano ≈ River ◎ Lake ⊠ Reservoir

Faenza – Fierza

106 H10 **Faenza** anc. Faventia. Emilia-Romagna, N Italy 44°17′N 11°53′E
Faeroe-Iceland Ridge see Faroe-Iceland Ridge
Faeroe Islands see Faroe Islands
Færoerne see Faroe Islands
Faeroe-Shetland Trough see Faroe-Shetland Trough
104 H6 **Fafe** Braga, N Portugal 41°27′N 08°11′W
80 K13 **Fafen Shet'** ∽ E Ethiopia
193 V15 **Fafo** island Tongatapu Group, S Tonga
192 I16 **Fagaloa Bay** bay Upolu, E Samoa
192 H15 **Fagamālo** Savai'i, N Samoa 13°27′S 172°22′W
116 I12 **Făgăraş** Ger. Fogarasch, Hung. Fogaras. Braşov, C Romania 45°50′N 24°59′E
191 W10 **Fagatau** prev. Fangatau. atoll Îles Tuamotu, C French Polynesia
191 X12 **Fagataufa** prev. Fangataufa. island Îles Tuamotu, SE French Polynesia
95 M20 **Fagerhult** Kalmar, S Sweden 57°07′N 15°40′E
94 G13 **Fagernes** Oppland, S Norway 60°59′N 09°14′E
92 I9 **Fagernes** Troms, N Norway 69°31′N 19°16′E
95 M14 **Fagersta** Västmanland, C Sweden 59°59′N 15°49′E
77 W13 **Faggo** var. Foggo. Bauchi, N Nigeria 11°22′N 09°55′E
Faghman see Fughmah
Fagibina, Lake see Faguibine, Lac
63 J25 **Fagnano, Lago** ⊗ S Argentina
99 G22 **Fagne** hill range S Belgium
77 N10 **Faguibine, Lac** var. Lake Fagibina. ⊗ NW Mali
Fahaheel see Al Fuḩayḩīl
Fahlun see Falun
143 U12 **Fahraj** Kermān, SE Iran 29°00′N 59°00′E
64 P5 **Faial** Madeira, Portugal, NE Atlantic Ocean 32°47′N 16°53′W
64 N2 **Faial** var. Ilha do Faial. island Azores, Portugal, NE Atlantic Ocean
Faial, Ilha do see Faial
108 G10 **Faido** Ticino, S Switzerland 46°30′N 08°48′E
Faifo see Hôi An
Failaka Island see Faylakah
190 G12 **Faioa, Île** island N Wallis and Futuna
181 W8 **Fairbairn Reservoir** ⊠ Queensland, E Australia
39 R9 **Fairbanks** Alaska, USA 64°48′N 147°47′W
21 U12 **Fair Bluff** North Carolina, SE USA 34°18′N 79°02′W
31 R14 **Fairborn** Ohio, N USA 39°48′N 84°03′W
23 S3 **Fairburn** Georgia, SE USA 33°34′N 84°34′W
30 M12 **Fairbury** Illinois, N USA 40°45′N 88°30′W
29 Q17 **Fairbury** Nebraska, C USA 40°08′N 97°10′W
29 T9 **Fairfax** Minnesota, N USA 44°31′N 94°43′W
27 O8 **Fairfax** Oklahoma, C USA 36°34′N 96°42′W
21 R14 **Fairfax** South Carolina, SE USA 32°57′N 81°14′W
35 N8 **Fairfield** California, W USA 38°14′N 122°03′W
33 O14 **Fairfield** Idaho, NW USA 43°20′N 114°45′W
30 M16 **Fairfield** Illinois, N USA 38°22′N 88°23′W
29 X15 **Fairfield** Iowa, C USA 41°00′N 91°57′W
33 R8 **Fairfield** Montana, NW USA 47°36′N 111°59′W
31 Q14 **Fairfield** Ohio, N USA 39°21′N 84°34′W
25 U8 **Fairfield** Texas, SW USA 31°43′N 96°10′W
27 T7 **Fair Grove** Missouri, C USA 37°22′N 93°09′W
19 P12 **Fairhaven** Massachusetts, NE USA 41°38′N 70°51′W
23 N8 **Fairhope** Alabama, S USA 30°31′N 87°54′W
96 L4 **Fair Isle** island NE Scotland, United Kingdom
185 F20 **Fairlie** Canterbury, South Island, New Zealand 44°06′S 170°50′E
29 U11 **Fairmont** Minnesota, N USA 43°40′N 94°27′W
29 Q16 **Fairmont** Nebraska, C USA 40°37′N 97°36′W
21 S3 **Fairmont** West Virginia, NE USA 39°28′N 80°08′W
31 P13 **Fairmount** Indiana, N USA 40°25′N 85°39′W
18 H10 **Fairmount** New York, NE USA 43°03′N 76°14′W
29 R7 **Fairmount** North Dakota, N USA 46°02′N 96°36′W
37 S5 **Fairplay** Colorado, C USA 39°13′N 106°00′W
18 F9 **Fairport** New York, NE USA 43°06′N 77°26′W
11 O12 **Fairview** Alberta, W Canada 56°03′N 118°28′W
26 L9 **Fairview** Oklahoma, C USA 36°16′N 98°29′W
36 L4 **Fairview** Utah, W USA 39°37′N 111°26′W
35 T6 **Fairview Peak** ▲ Nevada, W USA 39°13′N 118°09′W
188 H14 **Fais** atoll Caroline Islands, W Micronesia
149 U8 **Faisalābād** prev. Lyallpur. Punjab, NE Pakistan 31°26′N 73°06′E
Faisaliya see Fayşalīyah
28 L8 **Faith** South Dakota, N USA 45°01′N 102°02′W
153 N12 **Faizābād** Uttar Pradesh, N India 26°46′N 82°08′E
Faizabad/Faizābād see Feizābād
45 S9 **Fajardo** E Puerto Rico 18°20′N 65°39′W
139 R9 **Fajj, Wādī al** dry watercourse S Iraq
140 K4 **Fajr, Bi'r** well NW Saudi Arabia
191 W10 **Fakahina** atoll Îles Tuamotu, C French Polynesia
190 L10 **Fakaofo Atoll** island SE Tokelau
191 U10 **Fakarava** atoll Îles Tuamotu, C French Polynesia
127 T2 **Fakel** Udmurtskaya Respublika, NW Russian Federation 57°35′N 53°00′E
97 P19 **Fakenham** E England, United Kingdom 52°50′N 00°51′E
171 U13 **Fakfak** Papua Barat, E Indonesia 02°55′S 132°17′E
153 T12 **Fakīragrām** Assam, NE India 26°22′N 90°15′E
114 M10 **Fakiyska Reka** ∽ SE Bulgaria
95 J24 **Fakse** Sjælland, SE Denmark 55°16′N 12°08′E
95 J24 **Fakse Bugt** bay SE Denmark
95 J24 **Fakse Ladeplads** Sjælland, SE Denmark 55°14′N 12°11′E
163 V11 **Faku** Liaoning, NE China 42°30′N 123°27′E
76 J14 **Falaba** N Sierra Leone 09°54′N 11°22′W
102 K5 **Falaise** Calvados, N France 48°52′N 00°12′W
114 H12 **Falakró** ▲▲ NE Greece
189 T12 **Falalu** island Chuuk, C Micronesia
166 L4 **Falam** Chin State, W Myanmar (Burma) 22°58′N 93°45′E
143 N8 **Falāvarjān** Eşfahān, C Iran 32°33′N 51°28′E
116 M11 **Fălciu** Vaslui, E Romania 46°19′N 28°10′E
54 I4 **Falcón** off. Estado Falcón. ◆ state NW Venezuela
106 J12 **Falconara Marittima** Marche, C Italy 43°37′N 13°23′E
Falcone, Capo del see Falcone, Punta del
107 A16 **Falcone, Punta del** var. Capo del Falcone. headland Sardegna, Italy, C Mediterranean Sea 40°57′N 8°12′E
Falcón, Estado see Falcón
11 Y16 **Falcon Lake** Manitoba, S Canada 49°44′N 95°18′W
Falcon Lake see Falcón, Presa/Falcon Reservoir
41 O7 **Falcón, Presa** var. Falcon Lake, Falcon Reservoir. ⊠ Mexico/USA see also Falcon Reservoir
Falcón, Presa see Falcon Reservoir
25 Q16 **Falcon Reservoir** var. Falcon Lake, Presa Falcón. ⊠ Mexico/USA see also Falcón, Presa
Falcon Reservoir see Falcón, Presa
190 L10 **Fale** island Fakaofo Atoll, SE Tokelau
192 F15 **Faleālupo** Savai'i, NW Samoa 13°30′S 172°46′W
190 B10 **Falefatu** island Funafuti Atoll, C Tuvalu
192 G15 **Fālelima** Savai'i, NW Samoa 13°30′S 172°41′W
95 N18 **Falerum** Östergötland, S Sweden 58°07′N 16°15′E
Faleshty see Făleşti
116 M9 **Făleşti** Rus. Faleshty. NW Moldova 47°33′N 27°43′E
25 S15 **Falfurrias** Texas, SW USA 27°17′N 98°10′W
11 O13 **Falher** Alberta, W Canada 55°45′N 117°18′W
Falkenau an der Eger see Sokolov
95 J21 **Falkenberg** Halland, S Sweden 56°55′N 12°30′E
Falkenberg see Niemodlin
Falkenburg in Pommern see Złocieniec
100 N12 **Falkensee** Brandenburg, NE Germany 52°34′N 13°04′E
96 J12 **Falkirk** C Scotland, United Kingdom 56°N 03°48′W
65 I20 **Falkland Escarpment** undersea feature SW Atlantic Ocean 50°00′S 45°00′W
63 K24 **Falkland Islands** var. Falklands, Islas Malvinas. ◇ UK dependent territory SW Atlantic Ocean
47 W14 **Falkland Islands** island group SW Atlantic Ocean
65 I20 **Falkland Plateau** var. Argentine Rise. undersea feature SW Atlantic Ocean 51°00′S 50°00′W
Falklands see Falkland Islands
63 M23 **Falkland Sound** var. Estrecho de San Carlos. strait C Falkland Islands
Falknov nad Ohří see Sokolov
115 H21 **Falkonéra** island S Greece
95 K18 **Falköping** Västra Götaland, S Sweden 58°10′N 13°31′E
139 U8 **Fallāh** Wāsiţ, E Iraq 32°58′N 45°09′E
35 U16 **Fallbrook** California, W USA 33°22′N 117°15′W
189 U12 **Falleallej Pass** passage Chuuk Islands, C Micronesia
93 J14 **Fällfors** Västerbotten, N Sweden 65°07′N 20°46′E
194 I6 **Fallières Coast** physical region Antarctica
100 I11 **Fallingbostel** Niedersachsen, NW Germany 52°52′N 09°42′E
33 X9 **Fallon** Montana, NW USA 46°49′N 105°07′W
35 S5 **Fallon** Nevada, W USA 39°29′N 118°47′W
19 O12 **Fall River** Massachusetts, NE USA 41°42′N 71°09′W
27 P6 **Fall River Lake** ⊠ Kansas, C USA
35 O3 **Fall River Mills** California, W USA 41°00′N 121°28′W
21 W4 **Falls Church** Virginia, NE USA 38°53′N 77°11′W
29 S17 **Falls City** Nebraska, C USA 40°03′N 95°36′W
25 S12 **Falls City** Texas, SW USA 28°58′N 98°01′W
Falluja see Al Fallūjah
77 S12 **Falmey** Dosso, SW Niger 12°29′N 02°58′E
45 W10 **Falmouth** Antigua, Antigua and Barbuda 17°02′N 61°47′W
44 J11 **Falmouth** W Jamaica 18°28′N 77°39′W
97 H25 **Falmouth** SW England, United Kingdom 50°08′N 05°04′W
20 M4 **Falmouth** Kentucky, S USA 38°40′N 84°20′W
19 P13 **Falmouth** Massachusetts, NE USA 41°31′N 70°36′W
21 W5 **Falmouth** Virginia, NE USA 38°19′N 77°28′W
189 U12 **Falos** island Chuuk, C Micronesia
83 E26 **False Bay** Afr. Valsbaai. bay SW South Africa
155 K17 **False Divi Point** headland E India 15°46′N 80°43′E
38 M16 **False Pass** Unimak Island, Alaska, USA 54°52′N 163°15′W
154 P12 **False Point** headland E India 20°23′N 86°52′E
105 U6 **Falset** Cataluña, NE Spain 41°08′N 00°49′E
95 I25 **Falster** island SE Denmark
116 K9 **Fălticeni** Hung. Falticsén. Suceava, NE Romania 47°27′N 26°20′E
Falticsén see Fălticeni
94 M13 **Falun** var. Fahlun. Kopparberg, C Sweden 60°36′N 15°36′E
Famagusta see Gazimağusa
Famagusta Bay see Gazimağusa Körfezi
62 I8 **Famatina** La Rioja, NW Argentina 28°58′S 67°46′W
99 J21 **Famenne** physical region SE Belgium
77 X15 **Fan** ∽ E Nigeria
76 M12 **Fana** Koulikoro, SW Mali 12°45′N 06°55′W
115 K19 **Fána** ancient harbor Chíos, SE Greece
189 V13 **Fanan** island Chuuk, C Micronesia
189 U12 **Fanapanges** island Chuuk, C Micronesia
115 L20 **Fanári, Akrotírio** headland Ikaría, Dodekánisa, Greece, Aegean Sea 37°40′N 26°21′E
45 Q13 **Fancy** Saint Vincent, Saint Vincent and the Grenadines 13°22′N 61°10′W
172 I5 **Fandriana** Fianarantsoa, SE Madagascar 20°14′S 47°21′E
167 O6 **Fang** Chiang Mai, NW Thailand 19°56′N 99°14′E
80 E13 **Fangak** Jonglei, E South Sudan 09°05′N 30°52′E
Fangatau see Fagatau
Fangataufa see Fagataufa
193 V15 **Fanga Uta** bay S Tonga
161 N7 **Fangcheng** Henan, C China 33°18′N 113°03′E
Fangcheng see Fangchenggang
160 K15 **Fangchenggang** var. Fangcheng Gezu Zizhixian; prev. Fangcheng. Guangxi Zhuangzu Zizhiqu, S China 21°49′N 108°21′E
Fangcheng Gezu Zizhixian see Fangchenggang
161 S15 **Fangshan** S Taiwan 22°19′N 120°41′E
163 X8 **Fangzheng** Heilongjiang, NE China 45°50′N 128°50′E
Fani see Fanit, Lumi i
119 K16 **Fanipal'** Rus. Fanipol'. Minskaya Voblasts', C Belarus 53°45′N 27°20′E
Fanipol' see Fanipal'
113 D22 **Fanit, Lumi i** var. Fani. ∽ N Albania
25 T13 **Fannin** Texas, SW USA 28°41′N 97°13′W
Fanning Island see Tabuaeran
94 G8 **Fannrem** Sør-Trøndelag, S Norway 63°16′N 09°48′E
106 I11 **Fano** anc. Colonia Julia Fanestris, Fanum Fortunae. Marche, C Italy 43°50′N 13°E
95 E23 **Fanø** island W Denmark
167 R5 **Fan Si Pan** ▲ N Vietnam 22°18′N 103°36′E
Fanum Fortunae see Fano
Fao see Al Fāw
141 W7 **Faq'** var. Al Faqa. Dubayy, E United Arab Emirates 24°42′N 55°37′E
Farab see Farap
185 G16 **Faraday, Mount** ▲ South Island, New Zealand 42°01′S 171°37′E
79 P16 **Faradje** Orientale, NE Dem. Rep. Congo 03°45′N 29°43′E
Faradofay see Tôlañaro
172 I7 **Farafangana** Fianarantsoa, SE Madagascar 22°50′S 47°50′E
148 J7 **Farāh** var. Farah, Fararud. Farāh, W Afghanistan 32°22′N 62°07′E
148 K7 **Farāh** ◆ province W Afghanistan
148 J7 **Farāh Rūd** ∽ W Afghanistan
188 K7 **Farallon de Medinilla** island C Northern Mariana Islands
188 J2 **Farallon de Pajaros** var. Uracas. island N Northern Mariana Islands
76 J14 **Faranah** Haute-Guinée, S Guinea 10°02′N 10°44′W
146 K12 **Farap** Rus. Farab. Lebap Welaýaty, NE Turkmenistan 39°15′N 63°32′E
Fararud see Farāh
140 M13 **Farasān, Jazā'ir** island group SW Saudi Arabia
172 I5 **Faratsiho** Antananarivo, C Madagascar 19°24′S 46°57′E
188 K15 **Faraulep Atoll** atoll Caroline Islands, C Micronesia
99 H20 **Farciennes** Hainaut, S Belgium 50°26′N 04°33′E
105 O14 **Fardes** ∽ S Spain
191 S10 **Fare** Huahine, W French Polynesia 16°42′S 151°01′W
97 M23 **Fareham** S England, United Kingdom 50°51′N 01°10′W
39 P11 **Farewell** Alaska, USA 62°35′N 153°59′W
184 H13 **Farewell, Cape** headland South Island, New Zealand 40°30′S 172°39′E
Farewell, Cape see Nunap Isua
184 I13 **Farewell Spit** spit South Island, New Zealand
95 I17 **Färgelanda** Västra Götaland, S Sweden 58°34′N 11°59′E
Farghona, Wodii/ Farghona Valley see Fergana Valley
Farghona Wodiysi see Fergana Valley
23 V8 **Fargo** Georgia, SE USA 30°42′N 82°33′W
29 R5 **Fargo** North Dakota, N USA 46°53′N 96°47′W
147 S10 **Farg'ona** Rus. Fergana; prev. Novyy Margilan. Farg'ona Viloyati, E Uzbekistan 40°28′N 71°44′E
147 R10 **Farg'ona Viloyati** Rus. Ferganskaya Oblast'. ◆ province E Uzbekistan
29 V10 **Faribault** Minnesota, N USA 44°18′N 93°16′W
152 J11 **Farīdābād** Haryāna, N India 28°26′N 77°19′E
152 H8 **Farīdkot** Punjab, NW India 30°42′N 74°47′E
153 T15 **Faridpur** Dhaka, C Bangladesh 23°29′N 89°50′E
121 P14 **Fārigh, Wādī al** ∽ N Libya
172 I4 **Farihy Alaotra** ⊗ C Madagascar
94 M11 **Färila** Gävleborg, C Sweden 61°48′N 15°55′E
104 E9 **Farilhões** island C Portugal
76 G12 **Farim** NW Guinea-Bissau 12°30′N 15°09′W
Farish see Forish
141 T11 **Fāris, Qalamat** well SE Saudi Arabia
95 N21 **Färjestaden** Kalmar, S Sweden 56°38′N 16°30′E
149 R2 **Farkhār** Takhār, NE Afghanistan 36°39′N 69°43′E
147 Q14 **Farkhor** Rus. Parkhar. SW Tajikistan 37°32′N 69°22′E
116 F12 **Fârliug** prev. Fîrliug, Hung. Furluk. Caraş-Severin, SW Romania 45°21′N 21°55′E
115 M21 **Farmakonísi** island Dodekánisa, Greece, Aegean Sea
30 M13 **Farmer City** Illinois, N USA 40°14′N 88°38′W
31 N14 **Farmersburg** Indiana, N USA 39°14′N 87°21′W
25 U6 **Farmersville** Texas, SW USA 33°09′N 96°21′W
22 H5 **Farmerville** Louisiana, S USA 32°46′N 92°24′W
29 X16 **Farmington** Iowa, C USA 40°37′N 91°43′W
19 Q6 **Farmington** Maine, NE USA 44°40′N 70°09′W
29 V9 **Farmington** Minnesota, N USA 44°39′N 93°09′W
27 X6 **Farmington** Missouri, C USA 37°46′N 90°26′W
19 O9 **Farmington** New Hampshire, NE USA 43°23′N 71°04′W
37 P9 **Farmington** New Mexico, SW USA 36°44′N 108°13′W
36 L2 **Farmington** Utah, W USA 40°58′N 111°53′W
21 W9 **Farmville** North Carolina, SE USA 35°37′N 77°36′W
21 U6 **Farmville** Virginia, NE USA 37°17′N 78°25′W
97 N22 **Farnborough** S England, United Kingdom 51°17′N 00°46′W
97 N22 **Farnham** S England, United Kingdom 51°13′N 00°49′W
10 J7 **Faro** Yukon, W Canada 62°15′N 133°30′W
104 G14 **Faro** Faro, S Portugal 37°01′N 07°56′W
95 Q18 **Fårö** Gotland, SE Sweden 57°55′N 19°10′E
104 G14 **Faro** ◆ district S Portugal
78 F13 **Faro** ∽ Cameroon/Nigeria
104 G14 **Faro** ✈ Faro, S Portugal 37°02′N 08°01′W
64 M5 **Faroe-Iceland Ridge** var. Faeroe-Iceland Ridge. undersea ridge NW Norwegian Sea
86 C8 **Faroe Islands** var. Faeroe Islands. island group N Atlantic Ocean
64 M5 **Faroe Islands** var. Faeroe Islands, Dan. Færoerne, Faer. Føroyar. Self-governing territory of Denmark N Atlantic Ocean
64 N6 **Faroe-Shetland Trough** var. Faeroe-Shetland Trough. trough NE Atlantic Ocean
Faro, Punta del see Peloro, Capo
95 Q18 **Fårösund** Gotland, SE Sweden 57°51′N 19°02′E
173 N7 **Farquhar Group** island group S Seychelles
18 B13 **Farrell** Pennsylvania, NE USA 41°12′N 80°28′W
152 K11 **Farrukhābād** Uttar Pradesh, N India 27°24′N 79°34′E
143 P11 **Fārs** off. Ostān-e Fārs; anc. Persis. ◆ province S Iran
115 F16 **Fársala** Thessalía, C Greece 39°17′N 22°23′E
143 R4 **Fārsīān** Golestán, N Iran 37°14′N 55°58′E
Fars, Khalīj-e see Persian Gulf
95 G21 **Farsø** Nordjylland, N Denmark 56°47′N 09°21′E
Fārs, Ostān-e see Fārs
95 D18 **Farsund** Vest-Agder, S Norway 58°05′N 06°49′E
141 U14 **Fartak, Ra's** headland E Yemen 15°34′N 52°19′E
60 H13 **Fartura, Serra da** ▲▲ S Brazil
Farvel, Kap see Nunap Isua
24 L4 **Farwell** Texas, SW USA 34°23′N 103°03′W
194 J9 **Farwell Island** island Antarctica
152 L9 **Far Western** ◆ zone W Nepal
148 M3 **Fāryāb** ◆ province N Afghanistan
143 P12 **Fasā** Fārs, S Iran 28°55′N 53°39′E
141 U12 **Fasad, Ramlat** desert SW Oman
107 P17 **Fasano** Puglia, SE Italy 40°50′N 17°21′E
92 L3 **Fáskrúðsfjörður** Austurland, E Iceland 64°55′N 14°01′W
117 O5 **Fastiv** Rus. Fastov. Kyyivs'ka Oblast', NW Ukraine 50°08′N 29°59′E
97 B22 **Fastnet Rock** Ir. Carraig Aonair. island SW Ireland
Fastov see Fastiv
190 C9 **Fatato** island Funafuti Atoll, C Tuvalu
152 K12 **Fatehgarh** Uttar Pradesh, N India 27°22′N 79°38′E
149 U6 **Fatehjang** Punjab, E Pakistan 33°33′N 72°43′E
152 G11 **Fatehpur** Rājasthān, N India 27°59′N 74°58′E
152 L13 **Fatehpur** Uttar Pradesh, N India 25°56′N 80°55′E
126 J7 **Fatezh** Kurskaya Oblast', W Russian Federation 52°01′N 35°51′E
76 G11 **Fatick** W Senegal 14°19′N 16°27′W
104 G9 **Fátima** Santarém, W Portugal 39°37′N 08°39′W
136 M11 **Fatsa** Ordu, N Turkey 41°02′N 37°31′E
Fatshan see Foshan
190 D12 **Fatua, Pointe** var. Pointe Nord. headland Île Futuna, S Wallis and Futuna
191 X7 **Fatu Hiva** island Îles Marquises, NE French Polynesia
Fatunda see Fatundu
79 H21 **Fatundu** var. Fatunda. Bandundu, W Dem. Rep. Congo 04°08′S 17°13′E
29 O8 **Faulkton** South Dakota, N USA 45°02′N 99°07′W
116 L13 **Făurei** prev. Filimon Sîrbu. Brăila, SE Romania 45°05′N 27°15′E
92 G12 **Fauske** Nordland, C Norway 67°15′N 15°27′E
11 P13 **Faust** Alberta, W Canada 55°19′N 115°33′W
99 L23 **Fauvillers** Luxembourg, SE Belgium 49°52′N 05°40′E
107 J24 **Favara** Sicilia, Italy, C Mediterranean Sea 37°19′N 13°40′E
Faventia see Faenza
107 G23 **Favignana, Isola** island Isole Egadi, S Italy
12 D8 **Fawn** ∽ Ontario, SE Canada
Faxa Bay see Faxaflói
92 H3 **Faxaflói** Eng. Faxa Bay. bay W Iceland
78 I7 **Faya** prev. Faya-Largeau, Largeau. Borkou, N Chad 17°58′N 19°06′E
Faya-Largeau see Faya
187 Q16 **Fayaoué** Province des Îles Loyauté, C New Caledonia 20°41′S 166°31′E
138 M5 **Fayḑāt** hill range E Syria
23 O3 **Fayette** Alabama, S USA 33°40′N 87°49′W
29 X12 **Fayette** Iowa, C USA 42°50′N 91°48′W
22 J6 **Fayette** Mississippi, S USA 31°42′N 91°03′W
27 U4 **Fayette** Missouri, C USA 39°09′N 92°40′W
27 S9 **Fayetteville** Arkansas, C USA 36°04′N 94°10′W
21 U10 **Fayetteville** North Carolina, SE USA 35°03′N 78°53′W
20 J10 **Fayetteville** Tennessee, S USA 35°08′N 86°33′W
25 U11 **Fayetteville** Texas, SW USA 29°52′N 96°40′W
21 R5 **Fayetteville** West Virginia, NE USA 38°03′N 81°09′W
141 R4 **Faylakah** var. Failaka Island. island E Kuwait
139 T10 **Fayşalīyah** var. Faisaliya. Al Qādisīyah, S Iraq 31°48′N 44°36′E
189 P15 **Fayu** var. East Fayu. island Hall Islands, C Micronesia
152 G8 **Fāzilka** Punjab, NW India 30°26′N 74°04′E
Fdérick see Fdérik
76 I6 **Fdérik** var. Fdérick, Fr. Fort Gouraud. Tiris Zemmour, NW Mauritania 22°40′N 12°41′W
Feabhail, Loch see Foyle, Lough
97 B20 **Feale** ∽ SW Ireland
21 V12 **Fear, Cape** headland Bald Head Island, North Carolina, SE USA 33°50′N 77°57′W
35 O6 **Feather River** ∽ California, W USA
185 M14 **Featherston** Wellington, North Island, New Zealand 41°07′S 175°28′E
102 L3 **Fécamp** Seine-Maritime, N France 49°45′N 00°22′E
Fédala see Mohammedia
61 D17 **Federación** Entre Ríos, E Argentina 31°00′S 57°55′W
61 D17 **Federal** Entre Ríos, E Argentina 30°55′S 58°45′W
77 T15 **Federal Capital District** ◆ capital territory C Nigeria
Federal Capital Territory see Australian Capital Territory
Federal District see Distrito Federal
21 Y4 **Federalsburg** Maryland, NE USA 38°41′N 75°46′W
74 M6 **Fedjaj, Chott el** var. Chott el Fejaj, Shaţţ al Fijāj. salt lake C Tunisia
94 B13 **Fedje** island S Norway
144 M7 **Fedorovka** Kostanay, N Kazakhstan 51°12′N 52°00′E
127 U6 **Fedorovka** Respublika Bashkortostan, W Russian Federation 53°09′N 55°07′E
Fëdory see Fyadory
117 U11 **Fedotova Kosa** spit SE Ukraine
189 V13 **Fefan** atoll Chuuk Islands, C Micronesia
111 O21 **Fehérgyarmat** Szabolcs-Szatmár-Bereg, E Hungary 47°59′N 22°29′E
Fehér-Körös see Crişul Alb
Fehértemplom see Bela Crkva
Fehérvölgy see Albac
100 L7 **Fehmarn** island N Germany
95 H25 **Fehmarn Belt** Dan. Femern Bælt, Ger. Fehmarnbelt. strait Denmark /Germany see also Femern Bælt
Fehmarnbelt see Fehmarn Belt/Femer Bælt
109 X8 **Fehring** Steiermark, SE Austria 46°56′N 16°00′E
59 B15 **Feijó** Acre, W Brazil 08°07′S 70°27′W
184 M12 **Feilding** Manawatu-Wanganui, North Island, New Zealand 40°15′S 175°34′E
Feira see Feira de Santana
59 O17 **Feira de Santana** var. Feira. Bahia, E Brazil 12°17′S 38°53′W
109 X7 **Feistritz** ∽ SE Austria
Feistritz see Ilirska Bistrica
161 P8 **Feixi** var. Shangpai; prev. Shangpaihe. Anhui, E China 31°40′N 117°08′E
149 S2 **Feizābād** var. Faizabad, Faizābād, Feyzābād, Fyzabad; prev. Feyzābād. Badakhshān, NE Afghanistan 37°06′N 70°34′E
Fejaj, Chott el see Fedjaj, Chott el
111 I23 **Fejér** off. Fejér Megye. ◆ county W Hungary
Fejér Megye see Fejér
95 I24 **Fejø** island SE Denmark
136 K15 **Feke** Adana, S Turkey 37°49′N 35°55′E
Feketehalom see Codlea
Fekete-Körös see Crişul Negru
105 Y9 **Felanitx** Mallorca, Spain, W Mediterranean Sea 39°28′N 03°08′E
109 T3 **Feldaist** ∽ N Austria
109 W8 **Feldbach** Steiermark, SE Austria 46°58′N 15°53′E
101 F24 **Feldberg** ▲ SW Germany 47°52′N 08°01′E
116 J12 **Feldioara** Ger. Marienburg, Hung. Földvár. Braşov, C Romania 45°49′N 25°36′E
108 I7 **Feldkirch** anc. Clunia. Vorarlberg, W Austria 47°15′N 09°38′E
109 S9 **Feldkirchen in Kärnten** Slvn. Trg. Kärnten, S Austria 46°42′N 14°01′E
Félegyháza see Kiskunfélegyháza
192 H16 **Feleolo** ✈ (Āpia) Upolu, C Samoa 13°49′S 171°59′W
104 H6 **Felgueiras** Porto, N Portugal 41°22′N 08°12′W
Felicitas Julia see Lisboa
172 J16 **Félicité** island Inner Islands, NE Seychelles
151 K20 **Felidhu Atoll** atoll C Maldives
41 Y13 **Felipe Carrillo Puerto** Quintana Roo, SE Mexico 19°34′N 88°02′W
97 Q21 **Felixstowe** E England, United Kingdom 51°58′N 01°20′E
103 N11 **Felletin** Creuse, C France 45°53′N 02°12′E
Fellin see Viljandi
Felsőbánya see Baia Sprie
Felsőmuzslya see Mužlja
Felsővisó see Vişeu de Sus
35 N10 **Felton** California, W USA 37°03′N 122°04′W
106 H7 **Feltre** Veneto, NE Italy 46°01′N 11°55′E
95 H25 **Femer Bælt** Dan. Fehmarn Belt, Ger. Fehmarnbelt. strait Denmark/Germany see also Fehmarn Belt
95 I24 **Femø** island SE Denmark
94 I10 **Femunden** ⊗ S Norway
104 H2 **Fene** Galicia, NW Spain 43°28′N 08°10′W
14 I14 **Fenelon Falls** Ontario, SE Canada 44°34′N 78°43′W
189 U13 **Feneppi** atoll Chuuk Islands, C Micronesia
137 O11 **Fener Burnu** headland N Turkey 41°07′N 39°26′E
Fénérive see Fenoarivo Atsinanana
115 J14 **Fengári** ▲ Samothráki, E Greece 40°27′N 25°37′E
163 V13 **Fengcheng** var. Feng-cheng, Fenghwangcheng. Liaoning, NE China 40°28′N 124°01′E
Fengcheng see Lianjiang
Feng-cheng see Fengcheng
160 K11 **Fenggang** var. Longquan. Guizhou, S China 27°57′N 107°42′E
161 O7 **Feng He** ∽ C China
161 S9 **Fenghua** Zhejiang, SE China 29°40′N 121°25′E
Fenghwangcheng see Fengcheng
Fengjiaba see Wangcang
160 L9 **Fengjie** var. Yong'an. Sichuan, C China 31°03′N 109°31′E
160 M14 **Fengkai** var. Jiangkou. Guangdong, S China 23°26′N 111°28′E
161 T13 **Fenglin** Jap. Hōrin. C Taiwan 23°52′N 121°30′E
161 P1 **Fengning** prev. Dagezhen. Hebei, E China 41°12′N 116°37′E
160 E13 **Fengqing** var. Fengshan. Yunnan, SW China 24°38′N 99°54′E
161 O6 **Fengqiu** Henan, C China 35°02′N 114°24′E
161 Q2 **Fengrun** Hebei, E China 39°50′N 118°10′E
Fengshan see Luoyuan, Fujian, China
Fengshan see Fengqing, Yunnan, China
163 T4 **Fengshui Shan** ▲ NE China 52°20′N 123°22′E
161 P14 **Fengshun** Guangdong, S China 23°51′N 116°11′E
Fengtien see Liaoning, China
Fengtien see Shenyang, China
160 J7 **Fengxian** var. Feng Xian; prev. Shuangshipu. Shaanxi, C China 33°50′N 106°33′E
Feng Xian see Fengxian
Fengxiang see Luobei
Fengyizhen see Maoxian
163 P13 **Fengzhen** Nei Mongol Zizhiqu, N China 40°25′N 113°09′E
160 M6 **Fen He** ∽ C China
153 V15 **Feni** Chittagong, E Bangladesh 23°00′N 91°24′E
186 I6 **Feni Islands** island group NE Papua New Guinea
38 H17 **Fenimore Pass** strait Aleutian Islands, Alaska, USA
84 B9 **Feni Ridge** undersea feature N Atlantic Ocean 53°45′N 18°00′W
Fennern see Vändra
30 J9 **Fennimore** Wisconsin, N USA 42°58′N 90°39′W
172 J4 **Fenoarivo Atsinanana** prev./Fr. Fénérive. Toamasina, E Madagascar 20°52′S 46°52′E
95 I24 **Fensmark** Sjælland, SE Denmark 55°17′N 11°48′E
97 O19 **Fens, The** wetland E England, United Kingdom
31 R9 **Fenton** Michigan, N USA 42°48′N 83°42′W
190 K10 **Fenua Fala** island SE Tokelau
190 F12 **Fenuafo'ou, Île** island E Wallis and Futuna
190 L10 **Fenua Loa** island Fakaofo Atoll, E Tokelau
160 M4 **Fenyang** Shanxi, C China 37°14′N 111°40′E
117 U13 **Feodosiya** var. Kefe, It. Kaffa; anc. Theodosia. Avtonomna Respublika Krym, S Ukraine 45°03′N 35°24′E
94 I10 **Feragen** ⊗ S Norway
74 L5 **Fer, Cap de** headland NE Algeria 37°05′N 07°10′E
31 O16 **Ferdinand** Indiana, N USA 38°13′N 86°51′W
Ferdinand see Montana, Bulgaria
Ferdinand see Mihail Kogălniceanu, Romania
Ferdinandsberg see Oţelu Roşu
143 T7 **Ferdows** var. Firdaus; prev. Tūn. Khorāsān-e Razavī, E Iran 34°00′N 58°09′E
103 Q5 **Fère-Champenoise** Marne, N France 48°45′N 03°59′E
Ferencz-József Csúcs see Gerlachovský štít
107 J16 **Ferentino** Lazio, C Italy 41°40′N 13°16′E
114 L13 **Féres** Anatolikí Makedonía kai Thrákí, NE Greece 40°54′N 26°12′E
Fergana see Farg'ona
147 S10 **Fergana Valley** var. Farghona Valley, Rus. Ferganskaya Dolina, Taj. Wodii Farghona, Uzb. Farghona Wodiysi. basin Tajikistan/Uzbekistan
Ferganskaya Dolina see Fergana Valley
Ferganskaya Oblast' see Farg'ona Viloyati
147 U9 **Ferganskiy Khrebet** ▲▲ C Kyrgyzstan
14 F15 **Fergus** Ontario, S Canada 43°42′N 80°22′W
29 S6 **Fergus Falls** Minnesota, N USA 46°15′N 96°02′W
186 G9 **Fergusson Island** var. Kaluwawa. island SE Papua New Guinea
111 K22 **Ferihegy** ✈ (Budapest) Budapest, C Hungary 47°25′N 19°13′E
113 N17 **Ferizaj** Serb. Uroševac. C Kosovo 42°23′N 21°09′E
77 N14 **Ferkessédougou** N Ivory Coast 09°36′N 05°12′W
109 T10 **Ferlach** Slvn. Borovlje. Kärnten, S Austria 46°31′N 14°18′E
97 E16 **Fermanagh** cultural region SW Northern Ireland, United Kingdom
106 J13 **Fermo** anc. Firmum Picenum. Marche, C Italy 43°09′N 13°44′E
104 J6 **Fermoselle** Castilla y León, N Spain 41°19′N 06°24′W
97 D20 **Fermoy** Ir. Mainistir Fhear Maí. SW Ireland 52°08′N 08°16′W
23 W8 **Fernandina Beach** Amelia Island, Florida, SE USA 30°40′N 81°27′W
57 A17 **Fernandina, Isla** var. Narborough Island. island Galapagos Islands, Ecuador, E Pacific Ocean
47 Y5 **Fernando de Noronha** island E Brazil
Fernando Po/Fernando Póo see Bioco, Isla de
60 J7 **Fernandópolis** São Paulo, S Brazil 20°18′S 50°13′W
104 M13 **Fernán Núñez** Andalucía, S Spain 37°40′N 04°44′W
83 Q14 **Fernão Veloso, Baia de** bay NE Mozambique
34 K3 **Ferndale** California, W USA 40°34′N 124°16′W
32 H6 **Ferndale** Washington, NW USA 48°51′N 122°35′W
11 P17 **Fernie** British Columbia, SW Canada 49°30′N 115°00′W
35 R5 **Fernley** Nevada, W USA 39°35′N 119°15′W
Ferozepore see Firozpur
107 N18 **Ferrandina** Basilicata, S Italy 40°30′N 16°25′E
106 G9 **Ferrara** anc. Forum Alieni. Emilia-Romagna, N Italy 44°50′N 11°36′E
120 F9 **Ferrat, Cap** headland NW Algeria 35°52′N 00°24′W
107 D20 **Ferrato, Capo** headland Sardegna, Italy, C Mediterranean Sea 39°18′N 09°37′E
104 G12 **Ferreira do Alentejo** Beja, S Portugal 38°04′N 08°06′W
56 B11 **Ferreñafe** Lambayeque, W Peru 06°42′S 79°45′W
108 C12 **Ferret** Valais, SW Switzerland 45°57′N 07°04′E
102 I13 **Ferret, Cap** headland W France 44°37′N 01°15′W
22 I6 **Ferriday** Louisiana, S USA 31°37′N 91°33′W
Ferro see Hierro
107 D16 **Ferro, Capo** headland Sardegna, Italy, C Mediterranean Sea 41°09′N 09°31′E
104 H2 **Ferrol** var. El Ferrol; prev. El Ferrol del Caudillo. Galicia, NW Spain 43°29′N 08°14′W
56 B12 **Ferrol, Península de** peninsula W Peru
36 M5 **Ferron** Utah, W USA 39°05′N 111°07′W
21 S7 **Ferrum** Virginia, NE USA 36°54′N 80°01′W
23 O8 **Ferry Pass** Florida, SE USA 30°30′N 87°12′W
Ferryville see Menzel Bourguiba
29 S4 **Fertile** Minnesota, N USA 47°32′N 96°16′W
Fertő see Neusiedler See
Ferwerd see Ferwert
98 L5 **Ferwert** Dutch. Ferwerd. Fryslân, N Netherlands 53°21′N 05°47′E
74 G6 **Fès** Eng. Fez. N Morocco 34°06′N 04°57′W
79 I22 **Feshi** Bandundu, SW Dem. Rep. Congo 06°08′S 18°12′E
29 O4 **Fessenden** North Dakota, N USA 47°36′N 99°37′W
Festenberg see Twardogóra
27 X5 **Festus** Missouri, C USA 38°13′N 90°24′W
116 M14 **Feteşti** Ialomiţa, SE Romania 44°22′N 27°51′E
136 D17 **Fethiye** Muğla, SW Turkey 36°37′N 29°08′E
96 M1 **Fetlar** island NE Scotland, United Kingdom
95 I15 **Fetsund** Akershus, S Norway 59°55′N 11°03′E
12 L5 **Feuilles, Lac aux** ⊗ Québec, E Canada
12 L5 **Feuilles, Rivière aux** ∽ Québec, E Canada
99 M23 **Feulen** Diekirch, C Luxembourg 49°52′N 06°03′E
103 Q11 **Feurs** Loire, E France 45°44′N 04°14′E
95 F18 **Fevik** Aust-Agder, S Norway 58°22′N 08°40′E
123 R13 **Fevral'sk** Amurskaya Oblast', SE Russian Federation 52°25′N 131°06′E
Feyzābād see Feizābād
Feyzābād see Feizābād
Fez see Fès
97 J19 **Ffestiniog** NW Wales, United Kingdom 52°55′N 03°54′W
Fhóid Duibh, Cuan an see Blacksod Bay
62 I8 **Fiambalá** Catamarca, NW Argentina 27°45′S 67°37′W
172 I6 **Fianarantsoa** Fianarantsoa, C Madagascar 21°27′S 47°05′E
172 H6 **Fianarantsoa** ◆ province SE Madagascar
78 G12 **Fianga** Mayo-Kébbi Est, SW Chad 09°57′N 15°09′E
Ficce see Fichē
80 J12 **Fichē** It. Ficce. Oromīya, C Ethiopia 09°48′N 38°43′E
101 N17 **Fichtelberg** ▲ Czech Republic/Germany 50°26′N 12°57′E
101 M18 **Fichtelgebirge** ▲▲ SE Germany
101 M19 **Fichtelnaab** ∽ SE Germany
106 E9 **Fidenza** Emilia-Romagna, N Italy 44°52′N 10°04′E
113 K21 **Fier** var. Fieri. Fier, SW Albania 40°44′N 19°34′E
113 K21 **Fier** ◆ district W Albania
Fieri see Fier
Fierza see Fierzë

◆ Country
● Country Capital
◇ Dependent Territory
○ Dependent Territory Capital
◆ Administrative Regions
✈ International Airport
▲ Mountain
▲▲ Mountain Range
Volcano
∽ River
⊗ Lake
⊠ Reservoir

Fierzë – Four Corners

113 L17 **Fierzë** *var.* Fierza. Shkodër, N Albania 42°15′N 20°02′E
113 L17 **Fierzës, Liqeni i** ◎ N Albania
108 F10 **Fiesch** Valais, SW Switzerland 46°25′N 08°09′E
106 G11 **Fiesole** Toscana, C Italy 43°50′N 11°18′E
138 G12 **Fīfah** Aţ Ţafīlah, W Jordan 30°55′N 35°25′E
96 K11 **Fife** *var.* Kingdom of Fife. *cultural region* E Scotland, United Kingdom
Fife, Kingdom of *see* Fife
96 K11 **Fife Ness** *headland* E Scotland, United Kingdom 56°16′N 02°35′W
Fifteen Twenty Fracture Zone *see* Barracuda Fracture Zone
103 N13 **Figeac** Lot, S France 44°37′N 02°01′E
95 N19 **Figeholm** Kalmar, SE Sweden 57°12′N 16°34′E
Figig *see* Figuig
83 J18 **Figtree** Matabeleland South, SW Zimbabwe 20°24′S 28°21′E
104 F8 **Figueira da Foz** Coimbra, W Portugal 40°09′N 08°51′W
105 X4 **Figueres** Cataluña, E Spain 42°16′N 02°57′E
74 H7 **Figuig** *var.* Figig. E Morocco 32°09′N 01°13′W
Fijājj, Shaţţ al *see* Fedjaj, Chott el
187 Y15 **Fiji** *off.* Republic of Fiji, *prev.* Sovereign Democratic Republic of Fiji, *prev.* Republic of the Fiji Islands, *Fij.* Viti. ◆ *republic* SW Pacific Ocean
192 K9 **Fiji** *island group* SW Pacific Ocean
Fiji Islands, Republic of the *see* Fiji
175 Q8 **Fiji Plate** *tectonic feature*
Fiji, Republic of *see* Fiji
Fiji, Sovereign Democratic Republic of *see* Fiji
105 P14 **Filabres, Sierra de los** ▲▲ SE Spain
83 K18 **Filabusi** Matabeleland South, S Zimbabwe 20°34′S 29°20′E
42 K13 **Filadelfia** Guanacaste, W Costa Rica 10°28′N 85°33′W
111 K20 **Fil'akovo** *Hung.* Fülek. Banskobýstricky Kraj, C Slovakia 48°15′N 19°53′E
195 N5 **Filchner Ice Shelf** *ice shelf* Antarctica
14 J11 **Fildegrand** ≈ Québec, SE Canada
33 O15 **Filer** Idaho, NW USA 42°34′N 114°36′W
Filevo *see* Varbitsa
116 H14 **Filiaşi** Dolj, SW Romania 44°32′N 23°31′E
115 B16 **Filiátes** Ípeiros, W Greece 39°38′N 20°16′E
115 D21 **Filiatrá** Pelopónnisos, S Greece 37°09′N 21°35′E
107 K22 **Filicudi, Isola** *island* Isole Eolie, S Italy
141 Y10 **Filim** E Oman 20°37′N 58°11′E
Filimon Sîrbu *see* Făurei
77 S11 **Filingué** Tillabéri, W Niger 14°21′N 03°22′E
Filiouri *see* Líssos
114 I13 **Fílippoi** *anc.* Philippi. *site of ancient city* Anatolikí Makedonía kai Thráki, NE Greece
95 L15 **Filipstad** Värmland, C Sweden 59°44′N 14°10′E
108 I9 **Filisur** Graubünden, S Switzerland 46°40′N 09°43′E
94 E12 **Fillefjell** ▲▲ S Norway
35 R14 **Fillmore** California, W USA 34°23′N 118°56′W
36 K5 **Fillmore** Utah, W USA 38°57′N 112°19′W
14 J10 **Fils, Lac du** ◎ Québec, SE Canada
Filyos Çayı *see* Yenice Çayı
Fimbul Ice Shelf *see* Fimbulisen
195 Q2 **Fimbutheimen** *physical region* Antarctica
106 G9 **Finale Emilia** Emilia-Romagna, C Italy 44°50′N 11°17′E
106 C10 **Finale Ligure** Liguria, NW Italy 44°11′N 08°22′E
105 P14 **Fiñana** Andalucía, S Spain 37°09′N 02°47′W
21 S6 **Fincastle** Virginia, NE USA 37°30′N 79°54′W
99 M25 **Findel** ✈ (Luxembourg) Luxembourg, C Luxembourg 49°39′N 06°16′E
96 J9 **Findhorn** ≈ N Scotland, United Kingdom
31 R12 **Findlay** Ohio, N USA 41°02′N 83°40′W
18 G11 **Finger Lakes** ◎ New York, NE USA
83 L14 **Fingoè** Tete, NW Mozambique 15°10′S 31°51′E
136 E17 **Finike** Antalya, SW Turkey 36°18′N 30°08′E
102 F6 **Finistère** ⬖ *department* NW France
186 D7 **Finisterre Range** ▲▲ N Papua New Guinea
181 Q8 **Finke** Northern Territory, N Australia 25°37′S 134°35′E
109 S10 **Finkenstein** Kärnten, S Austria 46°34′N 13°53′E
189 Y15 **Finkol, Mount** *var.* Mount Crozer. ▲ Kosrae, E Micronesia 05°18′N 163°00′E
93 L17 **Finland** *off.* Republic of Finland, *Fin.* Suomen Tasavalta, Suomi. ◆ *republic* N Europe
124 F12 **Finland, Gulf of** *Est.* Soome Laht, *Fin.* Suomenlahti, *Ger.* Finnischer Meerbusen, *Rus.* Finskiy Zaliv, *Swe.* Finska Viken. *gulf* E Baltic Sea
Finland, Republic of *see* Finland
10 L11 **Finlay** ≈ British Columbia, W Canada
183 O10 **Finley** New South Wales, SE Australia 35°41′S 145°33′E
29 Q4 **Finley** North Dakota, N USA 47°30′N 97°50′W
Finnischer Meerbusen *see* Finland, Gulf of
92 K9 **Finnmark** ⬖ *county* N Norway
92 K9 **Finnmarksvidda** *physical region* N Norway
92 I9 **Finnsnes** Troms, N Norway 69°16′N 18°00′E
186 E7 **Finschhafen** Morobe, C Papua New Guinea 06°35′S 147°51′E
94 E13 **Finse** Hordaland, S Norway 60°35′N 07°33′E
Finska Viken/Finskiy Zaliv *see* Finland, Gulf of
95 M17 **Finspång** Östergötland, S Sweden 58°42′N 15°45′E
108 F10 **Finsteraarhorn** ▲ Switzerland 46°33′N 08°07′E
101 O14 **Finsterwalde** Brandenburg, E Germany 51°38′N 13°43′E
185 A23 **Fiordland** *physical region* South Island, New Zealand
106 E9 **Fiorenzuola d'Arda** Emilia-Romagna, C Italy 44°57′N 09°53′E
Firat Nehri *see* Euphrates
Firdaus *see* Ferdows
18 M14 **Fire Island** *island* New York, NE USA
106 G11 **Firenze** *Eng.* Florence; *anc.* Florentia. Toscana, C Italy 43°47′N 11°15′E
106 G10 **Firenzuola** Toscana, C Italy 44°07′N 11°22′E
14 C6 **Fire River** Ontario, S Canada 48°46′N 83°34′W
Fîrliug *see* Fârliug
61 B19 **Firmat** Santa Fe, C Argentina 33°29′S 61°29′W
103 Q12 **Firminy** Loire, E France 45°22′N 04°18′E
Firmum Picenum *see* Fermo
152 J12 **Fīrozābād** Uttar Pradesh, N India 27°09′N 78°24′E
152 G8 **Fīrozpur** *var.* Ferozepore. Punjab, NW India 30°55′N 74°38′E
First State *see* Delaware
143 O12 **Fīrūzābād** Fārs, S Iran 28°51′N 52°35′E
Fischamend *see* Fischamend Markt
109 Y4 **Fischamend Markt** *var.* Fischamend. Niederösterreich, NE Austria 48°08′N 16°37′E
109 W6 **Fischbacher Alpen** ▲▲ E Austria
Fischhausen *see* Primorsk
83 D21 **Fish** *var.* Vis. ≈ S Namibia
83 F24 **Fish** *Afr.* Vis. ≈ SW South Africa
11 X15 **Fisher Branch** Manitoba, S Canada 51°09′N 97°34′W
11 X15 **Fisher River** Manitoba, S Canada 51°25′N 97°23′W
19 N13 **Fishers Island** *island* New York, NE USA
37 U8 **Fishers Peak** ▲ Colorado, C USA 37°06′N 104°27′W
9 P9 **Fisher Strait** *strait* Nunavut, N Canada
97 H21 **Fishguard** *Wel.* Abergwaun. SW Wales, United Kingdom 51°59′N 04°49′W
19 R2 **Fish River Lake** ◎ Maine, NE USA
194 K6 **Fiske, Cape** *headland* Antarctica 74°27′S 60°28′W
103 P4 **Fismes** Marne, N France 49°19′N 03°41′E
104 F3 **Fisterra, Cabo** *headland* NW Spain 42°53′N 09°16′W
19 N11 **Fitchburg** Massachusetts, NE USA 42°34′N 71°48′W
96 L3 **Fitful Head** *headland* NE Scotland, United Kingdom 59°57′N 01°24′W
95 C14 **Fitjar** Hordaland, S Norway 59°55′N 05°19′E
192 H16 **Fito, Mauga** ▲ Upolu, C Samoa 13°57′S 171°42′W
23 U6 **Fitzgerald** Georgia, SE USA 31°42′N 83°15′W
180 M5 **Fitzroy Crossing** Western Australia 18°10′S 125°40′E
63 G21 **Fitzroy, Monte** *var.* Cerro Chaltel. ▲ S Argentina 49°18′S 73°06′W
181 Y8 **Fitzroy River** ≈ Queensland, E Australia
180 L5 **Fitzroy River** ≈ Western Australia
14 E12 **Fitzwilliam Island** *island* Ontario, S Canada
107 J15 **Fiuggi** Lazio, C Italy 41°47′N 13°16′E
Fiume *see* Rijeka
107 H15 **Fiumicino** Lazio, C Italy 41°46′N 12°13′E
Fiumicino *see* Leonardo da Vinci
106 E10 **Fivizzano** Toscana, C Italy 44°13′N 10°06′E
79 O21 **Fizi** Sud-Kivu, E Dem. Rep. Congo 04°15′S 28°57′E
Fizuli *see* Füzuli
92 I11 **Fjällåsen** Norrbotten, N Sweden 67°31′N 20°08′E
95 G20 **Fjerritslev** Nordjylland, N Denmark 57°06′N 09°17′E
F.J.S. *see* Franz Josef Strauss
95 L16 **Fjugesta** Örebro, C Sweden 59°10′N 14°50′E
Fladstrand *see* Frederikshavn
37 V5 **Flagler** Colorado, C USA 39°17′N 103°04′W
23 X10 **Flagler Beach** Florida, SE USA 29°28′N 81°07′W
36 L11 **Flagstaff** Arizona, SW USA 35°12′N 111°39′W
65 H24 **Flagstaff Bay** *bay* N Saint Helena, C Atlantic Ocean
19 P5 **Flagstaff Lake** ◎ Maine, NE USA
94 E13 **Flåm** Sogn Og Fjordane, S Norway 60°51′N 07°06′E
15 O8 **Flamand** ≈ Québec, SE Canada
30 J5 **Flambeau River** ≈ Wisconsin, N USA
97 O16 **Flamborough Head** *headland* E England, United Kingdom 54°06′N 00°03′W
100 N13 **Fläming** *hill range* NE Germany
16 H8 **Flaming Gorge Reservoir** ▣ Utah/Wyoming, NW USA
Flanders *see* Vlaanderen
Flandre *see* Vlaanderen
29 R10 **Flandreau** South Dakota, N USA 44°03′N 96°36′W
96 D6 **Flannan Isles** *island group* NW Scotland, United Kingdom
28 M6 **Flasher** North Dakota, N USA 46°25′N 101°12′W
93 G15 **Flåsjön** ◎ N Sweden
39 O11 **Flat** Alaska, USA 62°27′N 158°00′W
92 H1 **Flateyri** Vestfirðir, NW Iceland 66°03′N 23°28′W
33 P8 **Flathead Lake** ◎ Montana, NW USA
173 Y15 **Flat Island** *Fr.* Île Plate. *island* N Mauritius
25 T11 **Flatonia** Texas, SW USA 29°41′N 97°06′W
185 M14 **Flat Point** *headland* North Island, New Zealand 41°12′S 176°03′E
27 X6 **Flat River** Missouri, C USA 37°51′N 90°31′W
31 P8 **Flat River** ≈ Michigan, N USA
31 P14 **Flatrock River** ≈ Indiana, N USA
32 E6 **Flattery, Cape** *headland* Washington, NW USA 48°22′N 124°43′W
64 B12 **Flatts Village** *var.* The Flatts Village. C Bermuda 32°19′N 64°44′W
108 H7 **Flawil** Sankt Gallen, NE Switzerland 47°25′N 09°12′E
97 N22 **Fleet** S England, United Kingdom 51°16′N 00°50′W
97 K16 **Fleetwood** NW England, United Kingdom 53°55′N 03°02′W
18 H15 **Fleetwood** Pennsylvania, NE USA 40°27′N 75°49′W
95 D18 **Flekkefjord** Vest-Agder, S Norway 58°17′N 06°40′E
21 N5 **Flemingsburg** Kentucky, S USA 38°26′N 83°43′W
18 J15 **Flemington** New Jersey, NE USA 40°30′N 74°51′W
64 I7 **Flemish Cap** *undersea feature* NW Atlantic Ocean 47°00′N 45°00′W
95 N16 **Flen** Södermanland, C Sweden 59°04′N 16°39′E
100 I6 **Flensburg** Schleswig-Holstein, N Germany 54°47′N 09°26′E
100 J6 **Flensburger Förde** *inlet* Denmark/Germany
102 K5 **Flers** Orne, N France 48°45′N 00°34′W
95 C14 **Flesland** ✈ (Bergen) Hordaland, S Norway 60°18′N 05°15′E
Flessingue *see* Vlissingen
21 P10 **Fletcher** North Carolina, SE USA 35°24′N 82°29′W
31 R6 **Fletcher Pond** ◎ Michigan, N USA
102 L15 **Fleurance** Gers, S France 43°50′N 00°39′E
108 B8 **Fleurier** Neuchâtel, W Switzerland 46°55′N 06°37′E
99 H20 **Fleurus** Hainaut, S Belgium 50°28′N 04°33′E
103 N7 **Fleury-les-Aubrais** Loiret, C France 47°55′N 01°55′E
98 K10 **Flevoland** ⬖ *province* C Netherlands
Flickertail State *see* North Dakota
108 H9 **Flims** Glarus, NE Switzerland 46°50′N 09°16′E
182 F8 **Flinders Island** *island* Investigator Group, South Australia
183 P14 **Flinders Island** *island* Furneaux Group, Tasmania, SE Australia
182 I6 **Flinders Ranges** ▲▲ South Australia
181 U5 **Flinders River** ≈ Queensland, NE Australia
11 V13 **Flin Flon** Manitoba, C Canada 54°47′N 101°51′W
97 K18 **Flint** NE Wales, United Kingdom 53°15′N 03°10′W
31 R9 **Flint** Michigan, N USA 43°01′N 83°41′W
97 J18 **Flint** *cultural region* NE Wales, United Kingdom
27 O7 **Flint Hills** *hill range* Kansas, C USA
191 Y6 **Flint Island** *island* Line Islands, E Kiribati
23 S4 **Flint River** ≈ Georgia, SE USA
31 R9 **Flint River** ≈ Michigan, N USA
189 X12 **Flipper Point** *headland* C Wake Island 19°18′N 166°37′E
94 I13 **Flisa** Hedmark, S Norway 60°36′N 12°02′E
94 J13 **Flisa** ≈ S Norway
122 J5 **Flissingskiy, Mys** *headland* Novaya Zemlya, NW Russian Federation 76°43′N 69°01′E
Flitsch *see* Bovec
105 U6 **Flix** Cataluña, NE Spain 41°13′N 00°32′E
95 J19 **Floda** Västra Götaland, S Sweden 57°47′N 12°20′E
101 O16 **Flöha** ≈ E Germany
25 O4 **Flomot** Texas, SW USA 34°13′N 100°58′W
29 V5 **Floodwood** Minnesota, N USA 46°55′N 92°55′W
30 M15 **Flora** Illinois, N USA 38°40′N 88°29′W
103 P14 **Florac** Lozère, S France 44°18′N 03°35′E
23 Q8 **Florala** Alabama, S USA 31°00′N 86°19′W
103 S4 **Florange** Moselle, NE France 49°21′N 06°06′E
Floreana, Isla *see* Santa María, Isla
23 O2 **Florence** Alabama, S USA 34°48′N 87°40′W
36 L14 **Florence** Arizona, SW USA 33°01′N 111°23′W
37 T6 **Florence** Colorado, C USA 38°20′N 105°06′W
27 O5 **Florence** Kansas, C USA 38°13′N 96°56′W
20 M4 **Florence** Kentucky, S USA 39°00′N 84°37′W
32 E13 **Florence** Oregon, NW USA 43°58′N 124°06′W
21 T12 **Florence** South Carolina, SE USA 34°12′N 79°44′W
25 S9 **Florence** Texas, SW USA 30°50′N 97°47′W
Florence *see* Firenze
54 E13 **Florencia** Caquetá, S Colombia 01°37′N 75°37′W
99 H21 **Florennes** Namur, S Belgium 50°15′N 04°36′E
Florentia *see* Firenze
63 J18 **Florentino Ameghino, Embalse** ◎ S Argentina
99 J24 **Florenville** Luxembourg, SE Belgium 49°42′N 05°19′E
42 E3 **Flores** Petén, N Guatemala 16°55′N 89°56′W
61 E19 **Flores** ⬖ *department* S Uruguay
171 O16 **Flores** *island* Nusa Tenggara, C Indonesia
64 M1 **Flores** *island* Azores, Portugal, NE Atlantic Ocean
Floreshty *see* Floreşti
Flores, Lago de *see* Petén Itzá, Lago
Flores, Laut *see* Flores Sea
171 N15 **Flores Sea** *Ind.* Laut Flores. *sea* C Indonesia
116 M8 **Floreşti** *Rus.* Floreshty. N Moldova 47°52′N 28°19′E
25 S12 **Floresville** Texas, SW USA 29°09′N 98°10′W
59 N14 **Floriano** Piauí, E Brazil 06°45′S 43°W
61 K14 **Florianópolis** *prev.* Destêrro. *state capital* Santa Catarina, S Brazil 27°35′S 48°32′W
44 G6 **Florida** Camagüey, C Cuba 21°32′N 78°14′W
61 F19 **Florida** Florida, S Uruguay 34°04′S 56°14′W
61 F19 **Florida** ⬖ *department* S Uruguay
23 U9 **Florida** *off.* State of Florida, *also known as* Peninsular State, Sunshine State. ⬖ *state* SE USA
23 Y17 **Florida Bay** *bay* Florida, SE USA
54 G8 **Floridablanca** Santander, N Colombia 07°04′N 73°06′W
23 Y17 **Florida Keys** *island group* Florida, SE USA
37 Q16 **Florida Mountains** ▲▲ New Mexico, SW USA
64 D10 **Florida, Straits of** *strait* Atlantic Ocean/Gulf of Mexico
114 D13 **Flórina** *var.* Phlórina. Dytikí Makedonía, N Greece 40°48′N 21°26′E
27 X4 **Florissant** Missouri, C USA 38°48′N 90°20′W
94 C11 **Florø** Sogn Og Fjordane, S Norway 61°36′N 05°04′E
115 L22 **Floúda, Akrotírio** *headland* Astypálaia, Kykládes, Greece, Aegean Sea 36°38′N 26°23′E
21 S7 **Floyd** Virginia, NE USA 36°55′N 80°22′W
25 N4 **Floydada** Texas, SW USA 33°58′N 101°20′W
Flüela Wisshorn *see* Weisshorn
98 K7 **Fluessen** ◎ N Netherlands
105 S5 **Flúmen** ≈ NE Spain
107 C20 **Flumendosa** ≈ Sardegna, Italy, C Mediterranean Sea
31 R9 **Flushing** Michigan, N USA 43°03′N 83°51′W
Flushing *see* Vlissingen
25 O6 **Fluvanna** Texas, SW USA 32°54′N 101°06′W
186 B8 **Fly** ≈ Indonesia/Papua New Guinea
194 I10 **Flying Fish, Cape** *headland* Thurston Island, Antarctica 72°00′S 102°25′W
Flylân *see* Vlieland
Fly River *see* Western
193 Y15 **Foa** *island* Ha'apai Group, C Tonga
11 U15 **Foam Lake** Saskatchewan, S Canada 51°38′N 103°31′W
113 J14 **Foča** *var.* Srbinje. SE Bosnia and Herzegovina 43°32′N 18°46′E
116 L12 **Focşani** Vrancea, E Romania 45°45′N 27°13′E
Fogaras/Fogarasch *see* Făgăraş
107 M16 **Foggia** Puglia, SE Italy 41°28′N 15°31′E
Foggo *see* Faggo
76 D10 **Fogo** *island* Ilhas de Sotavento, SW Cape Verde
13 U11 **Fogo Island** *island* Newfoundland and Labrador, E Canada
109 U7 **Fohnsdorf** Steiermark, SE Austria 47°13′N 14°40′E
100 G7 **Föhr** *island* NW Germany
104 F14 **Fóia** ▲ S Portugal 37°19′N 08°39′W
14 I10 **Foins, Lac aux** ◎ Québec, SE Canada
103 N17 **Foix** Ariège, S France 42°58′N 01°39′E
126 I5 **Fokino** Bryanskaya Oblast', W Russian Federation 53°22′N 34°22′E
123 S15 **Fokino** Primorskiy Kray, W Russian Federation 42°58′N 132°25′E
Fola, Cnoc *see* Bloody Foreland
94 E13 **Folarskardnuten** ▲ S Norway 60°34′N 07°18′E
92 G11 **Folda** *prev.* Foldafjorden. *fjord* C Norway
Foldafjorden *see* Folda
93 F14 **Foldereid** Nord-Trøndelag, C Norway 64°58′N 12°09′E
Földvár *see* Feldioara
115 J22 **Folégandros** *island* Kykládes, Greece, Aegean Sea
23 O9 **Foley** Alabama, S USA 30°24′N 87°40′W
29 U7 **Foley** Minnesota, N USA 45°39′N 93°54′W
14 E7 **Foleyet** Ontario, S Canada 48°15′N 82°26′W
95 D14 **Folgefonna** *glacier* S Norway
106 I13 **Foligno** Umbria, C Italy 42°58′N 12°40′E
97 Q23 **Folkestone** SE England, United Kingdom 51°05′N 01°11′E
23 W8 **Folkston** Georgia, SE USA 30°49′N 82°00′W
94 H10 **Folldal** Hedmark, S Norway 62°08′N 10°00′E
25 P1 **Follett** Texas, SW USA 36°25′N 100°08′W
106 F13 **Follonica** Toscana, C Italy 42°55′N 10°45′E
21 T15 **Folly Beach** South Carolina, SE USA 32°39′N 79°56′W
35 O7 **Folsom** California, W USA 38°40′N 121°11′W
116 M12 **Folteşti** Galaţi, E Romania 45°45′N 28°00′E
172 H14 **Fomboni** Mohéli, S Comoros 12°18′S 43°46′E
18 K10 **Fonda** New York, NE USA 42°57′N 74°24′W
11 S10 **Fond-du-Lac** Saskatchewan, C Canada 59°20′N 107°09′W
30 M8 **Fond du Lac** Wisconsin, N USA 43°48′N 88°27′W
11 T10 **Fond-du-Lac** ≈ Saskatchewan, C Canada
190 G8 **Fongafale** *atoll* C Tuvalu
107 C18 **Fonni** Sardegna, Italy, C Mediterranean Sea 40°07′N 09°17′E
189 V12 **Fono** *island* Chuuk, C Micronesia
54 G4 **Fonseca** La Guajira, N Colombia 10°53′N 72°51′W
Fonseca, Golfo de *see* Fonseca, Gulf of
42 H8 **Fonseca, Gulf of** *Sp.* Golfo de Fonseca. *gulf* C Central America
103 O6 **Fontainebleau** Seine-et-Marne, N France 48°24′N 02°42′E
63 G19 **Fontana, Lago** ◎ W Argentina
21 N10 **Fontana Lake** ▣ North Carolina, SE USA
107 L24 **Fontanarossa** ✈ (Catania) Sicilia, Italy, C Mediterranean Sea 37°28′N 15°04′E
11 N11 **Fontas** ≈ British Columbia, W Canada
58 D12 **Fonte Boa** Amazonas, N Brazil 02°32′S 66°01′W
102 J10 **Fontenay-le-Comte** Vendée, NW France 46°28′N 00°48′W
33 T16 **Fontenelle Reservoir** ▣ Wyoming, C USA
193 Y14 **Fonualei** *island* Vava'u Group, N Tonga
111 H24 **Fonyód** Somogy, W Hungary 46°43′N 17°32′E
Foochow *see* Fuzhou
39 Q10 **Foraker, Mount** ▲ Alaska, USA 62°57′N 151°24′W
187 R14 **Forari** Éfaté, C Vanuatu 17°42′S 168°33′E
103 U4 **Forbach** Moselle, NE France 49°11′N 06°54′E
183 Q8 **Forbes** New South Wales, SE Australia 33°24′S 148°00′E
77 T17 **Forcados** Delta, S Nigeria 05°16′N 05°25′E
103 S14 **Forcalquier** Alpes-de-Haute-Provence, SE France 43°57′N 05°46′E
101 K19 **Forchheim** Bayern, SE Germany 49°43′N 11°07′E
35 R13 **Ford City** California, W USA 35°09′N 119°27′W
94 D11 **Førde** Sogn Og Fjordane, S Norway 61°27′N 05°51′E
31 N4 **Ford River** ≈ Michigan, N USA
183 O4 **Fords Bridge** New South Wales, SE Australia 29°44′S 145°25′E
20 J6 **Fordsville** Kentucky, S USA 37°36′N 86°39′W
27 U13 **Fordyce** Arkansas, C USA 33°49′N 92°23′W
76 I14 **Forécariah** SW Guinea 09°28′N 13°06′W
197 O14 **Forel, Mont** ▲ SE Greenland 66°55′N 36°45′W
11 R17 **Foremost** Alberta, SW Canada 49°30′N 111°34′W
14 D16 **Forest** Ontario, S Canada 43°05′N 82°00′W
22 L5 **Forest** Mississippi, S USA 32°22′N 89°30′W
31 S12 **Forest** Ohio, N USA 40°47′N 83°26′W
29 V11 **Forest City** Iowa, C USA 43°15′N 93°38′W
21 Q10 **Forest City** North Carolina, SE USA 35°19′N 81°52′W
32 G11 **Forest Grove** Oregon, NW USA 45°31′N 123°06′W
183 P17 **Forestier Peninsula** *peninsula* Tasmania, SE Australia
29 V8 **Forest Lake** Minnesota, N USA 45°16′N 92°59′W
23 S3 **Forest Park** Georgia, SE USA 33°37′N 84°22′W
29 Q3 **Forest River** ≈ North Dakota, N USA
15 T6 **Forestville** Québec, SE Canada 48°45′N 69°04′W
103 Q11 **Forez, Monts du** ▲▲ C France
96 K10 **Forfar** E Scotland, United Kingdom 56°38′N 02°54′W
26 J8 **Forgan** Oklahoma, C USA 36°54′N 100°32′W
Forge du Sud *see* Dudelange
101 J24 **Forggensee** ◎ S Germany
147 N10 **Forish** *Rus.* Farish. Jizzax Viloyati, C Uzbekistan 40°33′N 66°52′E
20 F9 **Forked Deer River** ≈ Tennessee, S USA
32 F7 **Forks** Washington, NW USA 47°57′N 124°22′W
92 N2 **Forlandsundet** *sound* W Svalbard
106 H10 **Forlì** *anc.* Forum Livii. Emilia-Romagna, N Italy 44°14′N 12°02′E
29 Q7 **Forman** North Dakota, N USA 46°07′N 97°39′W
97 K17 **Formby** NW England, United Kingdom 53°34′N 03°05′W
105 V11 **Formentera** *anc.* Ophiusa, *Lat.* Frumentum. *island* Islas Baleares, Spain, W Mediterranean Sea
Formentor, Cabo de *see* Formentor, Cap de
105 Y9 **Formentor, Cap de** *var.* Cabo de Formentor, Cape Formentor. *headland* Mallorca, Spain, W Mediterranean Sea 39°57′N 03°12′E
Formentor, Cape *see* Formentor, Cap de
107 J16 **Formia** Lazio, C Italy 41°16′N 13°37′E
62 O7 **Formosa** Formosa, NE Argentina 26°07′S 58°14′W
62 M6 **Formosa** *off.* Provincia de Formosa. ⬖ *province* NE Argentina
Formosa/Formo'sa *see* Taiwan
Formosa, Provincia de *see* Formosa
59 I17 **Formosa, Serra** ▲▲ C Brazil
Formosa Strait *see* Taiwan Strait
95 H21 **Fornæs** *headland* C Denmark 56°26′N 10°57′E
25 U6 **Forney** Texas, SW USA 32°45′N 96°28′W
106 E9 **Fornovo di Taro** Emilia-Romagna, C Italy 44°42′N 10°07′E
117 T14 **Foros** Avtonomna Respublika Krym, S Ukraine 44°24′N 33°47′E
Føroyar *see* Faroe Islands
96 J8 **Forres** NE Scotland, United Kingdom 57°32′N 03°38′W
27 X11 **Forrest City** Arkansas, C USA 35°01′N 90°48′W
39 Y15 **Forrester Island** *island* Alexander Archipelago, Alaska, USA
25 N7 **Forsan** Texas, SW USA 32°06′N 101°22′W
181 V5 **Forsayth** Queensland, NE Australia 18°31′S 143°37′E
95 L19 **Forserum** Jönköping, S Sweden 57°42′N 14°28′E
95 K15 **Forshaga** Värmland, C Sweden 59°33′N 13°29′E
93 L19 **Forssa** Kanta-Häme, S Finland 60°49′N 23°40′E
101 Q14 **Forst** *Lus.* Baršć Łužyca. Brandenburg, E Germany 51°43′N 14°38′E
183 U7 **Forster-Tuncurry** New South Wales, SE Australia 32°11′S 152°30′E
23 T4 **Forsyth** Georgia, SE USA 33°00′N 83°57′W
27 T8 **Forsyth** Missouri, C USA 36°41′N 93°07′W
33 W10 **Forsyth** Montana, NW USA 46°16′N 106°40′W
149 U11 **Fort Abbās** Punjab, E Pakistan 29°12′N 73°00′E
12 G10 **Fort Albany** Ontario, C Canada 52°15′N 81°35′W
56 L13 **Fortaleza** Pando, N Bolivia 09°48′S 65°29′W
58 P13 **Fortaleza** *prev.* Ceará. *state capital* Ceará, NE Brazil 03°45′S 38°35′W
59 D16 **Fortaleza** Rondônia, W Brazil 08°45′S 64°06′W
56 C13 **Fortaleza, Río** ≈ W Peru
Fort-Archambault *see* Sarh
21 U3 **Fort Ashby** West Virginia, NE USA 39°30′N 78°46′W
96 I9 **Fort Augustus** N Scotland, United Kingdom 57°14′N 04°38′W
Fort-Bayard *see* Zhanjiang
33 S8 **Fort Benton** Montana, NW USA 47°49′N 110°40′W
35 Q1 **Fort Bidwell** California, W USA 41°50′N 120°07′W
34 L5 **Fort Bragg** California, W USA 39°25′N 123°48′W
31 N16 **Fort Branch** Indiana, N USA 38°15′N 87°34′W
Fort-Bretonnet *see* Bousso
33 T17 **Fort Bridger** Wyoming, C USA 41°18′N 110°19′W
Fort-Cappolani *see* Tidjikja
Fort-Carnot *see* Ikongo
Fort Charlet *see* Djanet
Fort-Chimo *see* Kuujjuaq
11 R10 **Fort Chipewyan** Alberta, C Canada 58°42′N 111°08′W
Fort Cobb Lake *see* Fort Cobb Reservoir
26 L11 **Fort Cobb Reservoir** *var.* Fort Cobb Lake. ▣ Oklahoma, C USA
37 T3 **Fort Collins** Colorado, C USA 40°35′N 105°05′W
14 K12 **Fort-Coulonge** Québec, SE Canada 45°50′N 76°45′W
Fort-Crampel *see* Kaga Bandoro
Fort-Dauphin *see* Tôlañaro
24 K10 **Fort Davis** Texas, SW USA 30°35′N 103°54′W
37 O10 **Fort Defiance** Arizona, SW USA 35°44′N 109°04′W
45 Q12 **Fort-de-France** *prev.* Fort-Royal. ○ (Martinique) W Martinique 14°36′N 61°05′W
45 P12 **Fort-de-France, Baie de** *bay* W Martinique
Fort de Kock *see* Bukittinggi
23 P6 **Fort Deposit** Alabama, S USA 31°58′N 86°34′W
29 U13 **Fort Dodge** Iowa, C USA 42°30′N 94°10′W
13 S10 **Forteau** Québec, E Canada 51°30′N 56°55′W
106 E11 **Forte dei Marmi** Toscana, C Italy 43°59′N 10°10′E
14 H17 **Fort Erie** Ontario, S Canada 42°55′N 78°56′W
180 H7 **Fortescue River** ≈ Western Australia
19 S2 **Fort Fairfield** Maine, NE USA 46°45′N 67°51′W
Fort-Foureau *see* Kousséri
12 A11 **Fort Frances** Ontario, S Canada 48°37′N 93°23′W
Fort Franklin *see* Délı̨ne
23 R7 **Fort Gaines** Georgia, SE USA 31°36′N 85°03′W
37 T8 **Fort Garland** Colorado, C USA 37°22′N 105°24′W
21 P5 **Fort Gay** West Virginia, NE USA 38°06′N 82°35′W
Fort George *see* Chisasibi
Fort George *see* La Grande Rivière
27 Q10 **Fort Gibson** Oklahoma, C USA 35°48′N 95°15′W
27 Q9 **Fort Gibson Lake** ▣ Oklahoma, C USA
8 H7 **Fort Good Hope** *var.* Rádeyilikóé. Northwest Territories, NW Canada 66°16′N 128°37′W
23 V4 **Fort Gordon** Georgia, SE USA 33°25′N 82°09′W
Fort Gouraud *see* Fdérik
96 I11 **Forth** ≈ C Scotland, United Kingdom
Fort Hall *see* Murang'a
24 H8 **Fort Hancock** Texas, SW USA 31°18′N 105°49′W
Fort Hertz *see* Putao
96 K12 **Forth, Firth of** *estuary* E Scotland, United Kingdom
14 L14 **Forthton** Ontario, SE Canada 44°43′N 75°31′W
14 M8 **Fortier** ≈ Québec, SE Canada
Fortín General Eugenio Garay *see* General Eugenio A. Garay
Fort Jameson *see* Chipata
Fort Johnston *see* Mangochi
19 R1 **Fort Kent** Maine, NE USA 47°15′N 68°33′W
Fort-Lamy *see* N'Djaména
23 Z15 **Fort Lauderdale** Florida, SE USA 26°07′N 80°09′W
21 R11 **Fort Lawn** South Carolina, SE USA 34°43′N 80°46′W
8 H10 **Fort Liard** *var.* Liard. Northwest Territories, W Canada 60°14′N 123°28′W
44 M8 **Fort-Liberté** NE Haiti 19°42′N 71°51′W
21 N9 **Fort Loudoun Lake** ▣ Tennessee, S USA
37 T3 **Fort Lupton** Colorado, C USA 40°04′N 104°48′W
11 R12 **Fort MacKay** Alberta, C Canada 57°12′N 111°41′W
11 Q17 **Fort Macleod** *var.* MacLeod. Alberta, SW Canada 49°44′N 113°24′W
29 Y16 **Fort Madison** Iowa, C USA 40°37′N 91°15′W
Fort Manning *see* Mchinji
25 P9 **Fort McKavett** Texas, SW USA 30°50′N 100°07′W
11 R12 **Fort McMurray** Alberta, C Canada 56°44′N 111°23′W
8 G7 **Fort McPherson** *var.* McPherson. Northwest Territories, NW Canada 67°29′N 134°50′W
21 R11 **Fort Mill** South Carolina, SE USA 35°00′N 80°57′W
Fort-Millot *see* Ngouri
37 U3 **Fort Morgan** Colorado, C USA 40°14′N 103°48′W
23 W14 **Fort Myers** Florida, SE USA 26°39′N 81°52′W
23 W15 **Fort Myers Beach** Florida, SE USA 26°27′N 81°57′W
10 M10 **Fort Nelson** British Columbia, W Canada 58°48′N 122°44′W
10 M10 **Fort Nelson** ≈ British Columbia, W Canada
Fort Norman *see* Tulita
23 Q2 **Fort Payne** Alabama, S USA 34°23′N 85°43′W
33 W7 **Fort Peck** Montana, NW USA 48°00′N 106°28′W
33 V8 **Fort Peck Lake** ▣ Montana, NW USA
23 Y13 **Fort Pierce** Florida, SE USA 27°28′N 80°20′W
29 N10 **Fort Pierre** South Dakota, N USA 44°21′N 100°22′W
81 E18 **Fort Portal** SW Uganda 0°39′N 30°17′E
8 J10 **Fort Providence** *var.* Providence. Northwest Territories, W Canada 61°21′N 117°39′W
11 U16 **Fort Qu'Appelle** Saskatchewan, S Canada 50°50′N 103°52′W
Fort-Repoux *see* Akjoujt
8 K10 **Fort Resolution** *var.* Resolution. Northwest Territories, W Canada 61°10′N 113°39′W
33 T13 **Fortress Mountain** ▲ Wyoming, C USA 44°20′N 109°51′W
Fort Rosebery *see* Mansa
Fort Rousset *see* Owando
Fort-Royal *see* Fort-de-France
Fort Rupert *see* Waskaganish
8 H13 **Fort St. James** British Columbia, SW Canada 54°26′N 124°15′W
11 N12 **Fort St. John** British Columbia, W Canada 56°16′N 120°52′W
Fort Sandeman *see* Zhob
11 Q14 **Fort Saskatchewan** Alberta, SW Canada 53°42′N 113°12′W
27 R6 **Fort Scott** Kansas, C USA 37°52′N 94°43′W
12 E6 **Fort Severn** Ontario, C Canada 56°N 87°40′W
31 R12 **Fort Shawnee** Ohio, N USA 40°41′N 84°08′W
144 E14 **Fort-Shevchenko** Mangistau, W Kazakhstan 44°29′N 50°16′E
Fort-Sibut *see* Sibut
8 I10 **Fort Simpson** *var.* Simpson. Northwest Territories, W Canada 61°52′N 121°23′W
8 I10 **Fort Smith** Northwest Territories, W Canada 60°01′N 111°55′W
27 R10 **Fort Smith** Arkansas, C USA 35°23′N 94°24′W
37 T13 **Fort Stanton** New Mexico, SW USA 33°28′N 105°31′W
24 L9 **Fort Stockton** Texas, SW USA 30°54′N 102°54′W
37 U12 **Fort Sumner** New Mexico, SW USA 34°28′N 104°15′W
26 K8 **Fort Supply** Oklahoma, C USA 36°34′N 99°34′W
26 K8 **Fort Supply Lake** ▣ Oklahoma, C USA
29 O10 **Fort Thompson** South Dakota, N USA 44°01′N 99°22′W
Fort-Trinquet *see* Bîr Mogreïn
105 R12 **Fortuna** Murcia, SE Spain 38°11′N 01°07′W
34 K3 **Fortuna** California, W USA 40°35′N 124°07′W
28 J2 **Fortuna** North Dakota, N USA 48°53′N 103°46′W
23 T5 **Fort Valley** Georgia, SE USA 32°33′N 83°53′W
11 P11 **Fort Vermilion** Alberta, W Canada 58°22′N 115°59′W
Fort Victoria *see* Masvingo
31 P13 **Fortville** Indiana, N USA 39°55′N 85°51′W
23 P9 **Fort Walton Beach** Florida, SE USA 30°24′N 86°37′W
31 P12 **Fort Wayne** Indiana, N USA 41°08′N 85°08′W
96 H10 **Fort William** N Scotland, United Kingdom 56°49′N 05°07′W
25 T6 **Fort Worth** Texas, SW USA 32°44′N 97°19′W
28 M7 **Fort Yates** North Dakota, N USA 46°05′N 100°37′W
39 S7 **Fort Yukon** Alaska, USA 66°35′N 145°05′W
Forum Alieni *see* Ferrara
Forum Julii *see* Fréjus
Forum Livii *see* Forlì
143 Q15 **Forūr-e Bozorg, Jazīreh-ye** *island* S Iran
94 H7 **Fosen** *physical region* S Norway
161 N14 **Foshan** *var.* Fatshan, Fo-shan, Namhoi. Guangdong, S China 23°03′N 113°08′E
Fo-shan *see* Foshan
194 J6 **Fossil Bluff** *UK research station* Antarctica 71°30′S 68°30′W
Fossa Claudia *see* Chioggia
106 B9 **Fossano** Piemonte, NW Italy 44°33′N 07°43′E
99 H21 **Fosses-la-Ville** Namur, S Belgium 50°24′N 04°42′E
32 J12 **Fossil** Oregon, NW USA 45°01′N 120°14′W
Foss Lake *see* Foss Reservoir
106 I11 **Fossombrone** Marche, C Italy 43°42′N 12°48′E
26 K10 **Foss Reservoir** *var.* Foss Lake. ▣ Oklahoma, C USA
29 S4 **Fosston** Minnesota, N USA 47°34′N 95°45′W
183 O13 **Foster** Victoria, SE Australia 38°40′S 146°15′E
11 T12 **Foster Lakes** ◎ Saskatchewan, C Canada
31 S12 **Fostoria** Ohio, N USA 41°09′N 83°25′W
79 D19 **Fougamou** Ngounié, C Gabon 01°16′S 10°30′E
102 J6 **Fougères** Ille-et-Vilaine, NW France 48°21′N 01°12′W
Fou-hsin *see* Fuxin
27 S14 **Fouke** Arkansas, C USA 33°15′N 93°53′W
96 K2 **Foula** *island* NE Scotland, United Kingdom
65 D24 **Foul Bay** *bay* East Falkland, Falkland Islands
97 P21 **Foulness Island** *island* SE England, United Kingdom
185 F15 **Foulwind, Cape** *headland* South Island, New Zealand 41°45′S 171°28′E
79 E15 **Foumban** Ouest, NW Cameroon 05°43′N 10°50′E
172 H13 **Foumbouni** Grande Comore, NW Comoros 11°49′S 43°30′E
195 N8 **Foundation Ice Stream** *glacier* Antarctica
37 T6 **Fountain** Colorado, C USA 38°40′N 104°42′W
36 L4 **Fountain Green** Utah, W USA 39°37′N 111°37′W
21 P11 **Fountain Inn** South Carolina, SE USA 34°41′N 82°12′W
27 S11 **Fourche LaFave River** ≈ Arkansas, C USA
33 Z13 **Four Corners** Wyoming, C USA 44°04′N 104°08′W

◆ Country
● Country Capital
◇ Dependent Territory
○ Dependent Territory Capital
⬖ Administrative Regions
✈ International Airport
▲ Mountain
▲▲ Mountain Range
Volcano
≈ River
◎ Lake
▣ Reservoir

FOURMIES – GÄDDEDE

103 Q2 **Fourmies** Nord, N France 50°01′N 04°03′E
38 J17 **Four Mountains, Islands of** island group Aleutian Islands, Alaska, USA
173 P17 **Fournaise, Piton de la** ⛰ SE Réunion 21°14′S 55°43′E
14 J8 **Fournière, Lac** ◎ Québec, SE Canada
115 L20 **Foúrnoi** island Dodekánisa, Greece, Aegean Sea
64 K13 **Four North Fracture Zone** tectonic feature W Atlantic Ocean
Fouron-Saint-Martin see Sint-Martens-Voeren
30 L3 **Fourteen Mile Point** headland Michigan, N USA 46°59′N 89°07′W
Fou-shan see Fushun
76 I13 **Fouta Djallon** var. Futa Jallon. ▲▲ W Guinea
185 C25 **Foveaux Strait** strait S New Zealand
35 Q11 **Fowler** California, W USA 36°35′N 119°40′W
37 U6 **Fowler** Colorado, C USA 38°07′N 104°01′W
31 N12 **Fowler** Indiana, N USA 40°36′N 87°20′W
182 D7 **Fowlers Bay** bay South Australia
25 R13 **Fowlerton** Texas, SW USA 28°27′N 98°48′W
142 M3 **Fowman** var. Fuman, Fumen. Gīlān, NW Iran 37°15′N 49°19′E
65 C25 **Fox Bay East** West Falkland, Falkland Islands
65 C25 **Fox Bay West** West Falkland, Falkland Islands
14 J14 **Foxboro** Ontario, SE Canada 44°16′N 77°23′W
11 O14 **Fox Creek** Alberta, W Canada 54°25′N 116°57′W
64 G5 **Foxe Basin** sea Nunavut, N Canada
64 G5 **Foxe Channel** channel Nunavut, N Canada
95 I16 **Foxen** ◎ C Sweden
9 Q7 **Foxe Peninsula** peninsula Baffin Island, Nunavut, NE Canada
185 E19 **Fox Glacier** West Coast, South Island, New Zealand 43°28′S 170°00′E
38 L17 **Fox Islands** island Aleutian Islands, Alaska, USA
30 M10 **Fox Lake** Illinois, N USA 42°24′N 88°10′W
9 V12 **Fox Mine** Manitoba, C Canada 56°36′N 101°48′W
35 R3 **Fox Mountain** ▲ Nevada, W USA 41°01′N 119°30′W
65 E25 **Fox Point** headland East Falkland, Falkland Islands 51°55′S 58°24′W
30 M11 **Fox River** ∽ Illinois/ Wisconsin, N USA
30 L7 **Fox River** ∽ Wisconsin, N USA
184 L13 **Foxton** Manawatu-Wanganui, North Island, New Zealand 40°27′S 175°18′E
11 S16 **Fox Valley** Saskatchewan, S Canada 50°29′N 109°29′W
11 W16 **Foxwarren** Manitoba, S Canada 50°30′N 101°09′W
97 E14 **Foyle, Lough** Ir. Loch Feabhail. inlet N Ireland
194 H5 **Foyn Coast** physical region Antarctica
104 I2 **Foz** Galicia, NW Spain 43°33′N 07°16′W
60 I12 **Foz do Areia, Represa de** ⊠ S Brazil
59 A16 **Foz do Breu** Acre, W Brazil 09°21′S 72°41′W
83 A16 **Foz do Cunene** Namibe, SW Angola 17°11′S 11°52′E
60 G12 **Foz do Iguaçu** Paraná, S Brazil 25°33′S 54°31′W
58 C12 **Foz do Mamoriá** Amazonas, NW Brazil 02°28′S 66°06′W
105 T6 **Fraga** Aragón, NE Spain 41°32′N 00°21′E
44 F5 **Fragoso, Cayo** island C Cuba
61 G18 **Fraile Muerto** Cerro Largo, NE Uruguay 32°30′S 54°30′W
99 H21 **Fraire** Namur, S Belgium 50°16′N 04°30′E
99 L21 **Fraiture, Baraque de** hill SE Belgium
Frakštát see Hlohovec
Fram Basin see Amundsen Basin
99 F20 **Frameries** Hainaut, S Belgium 50°25′N 03°41′E
19 O11 **Framingham** Massachusetts, NE USA 42°15′N 71°24′W
60 L7 **Franca** São Paulo, S Brazil 20°33′S 47°27′W
187 O15 **Français, Récif des** reef W New Caledonia
107 K14 **Francavilla al Mare** Abruzzo, C Italy 42°25′N 14°16′E
107 P18 **Francavilla Fontana** Puglia, SE Italy 40°32′N 17°35′E
102 M8 **France** off. French Republic, It./Sp. Francia; prev. Gaul, Gaule, Lat. Gallia. ♦ republic W Europe
45 O8 **Francés Viejo, Cabo** headland NE Dominican Republic 19°39′N 69°57′W
79 F19 **Franceville** var. Massoukou, Masuku. Haut-Ogooué, E Gabon 01°40′S 13°31′E
79 F19 **Franceville** ✈ Haut-Ogooué, E Gabon 01°38′S 13°24′E
Francfort see Frankfurt am Main
103 T8 **Franche-Comté** ◈ region E France
Francia see France
29 O11 **Francis Case, Lake** ⊠ South Dakota, N USA
60 H12 **Francisco Beltrão** Paraná, S Brazil 26°05′S 53°04′W
Francisco I. Madero see Villa Madero
61 A21 **Francisco Madero** Buenos Aires, E Argentina 35°52′S 62°03′W
42 H6 **Francisco Morazán** prev. Tegucigalpa. ◈ department C Honduras
83 J18 **Francistown** North East, NE Botswana 21°08′S 27°31′E
Franconian Forest see Frankenwald
Franconian Jura see Fränkische Alb
98 K6 **Franeker** Fris. Frentsjer. Fryslân, N Netherlands 53°11′N 05°33′E
Frankenalb see Fränkische Alb
101 H16 **Frankenberg** Hessen, C Germany 51°04′N 08°49′E
101 J20 **Frankenhöhe** hill range C Germany
31 R8 **Frankenmuth** Michigan, N USA 43°19′N 83°44′W
101 F20 **Frankenstein** hill W Germany
Frankenstein/Frankenstein in Schlesien see Ząbkowice Śląskie
101 G20 **Frankenthal** Rheinland-Pfalz, W Germany 49°32′N 08°22′E
101 L18 **Frankenwald** Eng. Franconian Forest. ▲▲ C Germany
44 J12 **Frankfield** C Jamaica 18°08′N 77°22′W
14 J14 **Frankford** Ontario, SE Canada 44°12′N 77°36′W
31 O13 **Frankfort** Indiana, N USA 40°16′N 86°30′W
27 O3 **Frankfort** Kansas, C USA 39°42′N 96°25′W
20 L5 **Frankfort** state capital Kentucky, S USA 38°12′N 84°52′W
Frankfort on the Main see Frankfurt am Main
Frankfurt see Frankfurt am Main, Germany
Frankfurt see Słubice, Poland
101 G18 **Frankfurt am Main** var. Frankfurt, Fr. Francfort; prev. Eng. Frankfort on the Main. Hessen, SW Germany 50°07′N 08°41′E
100 Q12 **Frankfurt an der Oder** Brandenburg, E Germany 52°20′N 14°32′E
101 L21 **Fränkische Alb** var. Frankenalb, Eng. Franconian Jura. ▲▲ S Germany
101 I18 **Fränkische Saale** ∽ C Germany
101 L19 **Fränkische Schweiz** hill range C Germany
23 R4 **Franklin** Georgia, SE USA 33°15′N 85°06′W
31 P14 **Franklin** Indiana, N USA 39°29′N 86°02′W
20 J7 **Franklin** Kentucky, S USA 36°42′N 86°35′W
22 I9 **Franklin** Louisiana, S USA 29°48′N 91°30′W
29 O17 **Franklin** Nebraska, C USA 40°06′N 98°57′W
21 N10 **Franklin** North Carolina, SE USA 35°12′N 83°23′W
18 C13 **Franklin** Pennsylvania, NE USA 41°24′N 79°49′W
20 J9 **Franklin** Tennessee, S USA 35°55′N 86°52′W
25 U9 **Franklin** Texas, SW USA 31°02′N 96°30′W
21 X7 **Franklin** Virginia, NE USA 36°41′N 76°58′W
21 T4 **Franklin** West Virginia, NE USA 38°39′N 79°21′W
30 M9 **Franklin** Wisconsin, N USA 42°53′N 88°00′W
8 I6 **Franklin Bay** inlet Northwest Territories, N Canada
32 K7 **Franklin D. Roosevelt Lake** ⊠ Washington, NW USA
35 W4 **Franklin Lake** ◎ Nevada, W USA
185 B22 **Franklin Mountains** ▲▲ South Island, New Zealand
39 R5 **Franklin Mountains** ▲▲ Alaska, USA
39 N4 **Franklin, Point** headland Alaska, USA 70°54′N 158°48′W
183 O17 **Franklin River** ∽ Tasmania, SE Australia
22 K8 **Franklinton** Louisiana, S USA 30°51′N 90°09′W
21 U9 **Franklinton** North Carolina, SE USA 36°06′N 78°27′W
Frankstadt see Frenštát pod Radhoštěm
25 V7 **Frankston** Texas, SW USA 32°03′N 95°30′W
33 U12 **Frannie** Wyoming, C USA 44°57′N 108°37′W
15 U5 **Franquelin** Québec, SE Canada 49°17′N 67°52′W
15 U5 **Franquelin** ∽ Québec, SE Canada
83 C18 **Fransfontein** Kunene, NW Namibia 20°12′S 15°01′E
93 H17 **Fränsta** Västernorrland, C Sweden 62°30′N 16°06′E
122 J3 **Frantsa-Iosifa, Zemlya** Eng. Franz Josef Land. island group N Russian Federation
185 E18 **Franz Josef Glacier** West Coast, South Island, New Zealand 43°22′S 170°11′E
Franz Josef Land see Frantsa-Iosifa, Zemlya
Franz-Josef Spitze see Gerlachovský štít
101 L23 **Franz Josef Strauss** abbrev. F.J.S.. ✈ (München) Bayern, SE Germany 48°07′N 11°43′E
107 A19 **Frasca, Capo della** headland Sardegna, Italy, C Mediterranean Sea 39°46′N 08°27′E
107 I15 **Frascati** Lazio, C Italy 41°48′N 12°41′E
11 N14 **Fraser** ∽ British Columbia, SW Canada
83 G24 **Fraserburg** Western Cape, SW South Africa 31°55′S 21°31′E
96 L8 **Fraserburgh** NE Scotland, United Kingdom 57°42′N 02°20′W
181 Z9 **Fraser Island** var. Great Sandy Island. island Queensland, E Australia
10 L14 **Fraser Lake** British Columbia, SW Canada 54°00′N 124°45′W
10 L15 **Fraser Plateau** plateau British Columbia, SW Canada
184 P10 **Frasertown** Hawke's Bay, North Island, New Zealand 38°58′S 177°25′E
99 E19 **Frasnes-lez-Buissenal** Hainaut, SW Belgium 50°40′N 03°37′E
108 I7 **Frastanz** Vorarlberg, NW Austria 47°13′N 09°38′E
14 B8 **Frater** Ontario, S Canada 47°19′N 84°28′W
Frauenbach see Baia Mare
Frauenburg see Saldus, Latvia
Frauenburg see Frombork, Poland
108 H6 **Frauenfeld** Thurgau, NE Switzerland 47°34′N 08°54′E
109 Z5 **Frauenkirchen** Burgenland, E Austria 47°50′N 16°57′E
61 D19 **Fray Bentos** Río Negro, W Uruguay 33°09′S 58°14′W
61 F19 **Fray Marcos** Florida, S Uruguay 34°13′S 55°43′W
29 S6 **Frazee** Minnesota, N USA 46°42′N 95°40′W
104 M5 **Frechilla** Castilla y León, N Spain 42°08′N 04°50′W
30 I4 **Frederic** Wisconsin, N USA 45°42′N 92°30′W
95 G23 **Fredericia** Syddanmark, C Denmark 55°34′N 09°47′E
21 W3 **Frederick** Maryland, NE USA 39°25′N 77°25′W
26 L12 **Frederick** Oklahoma, C USA 34°24′N 99°03′W
29 P7 **Frederick** South Dakota, N USA 45°49′N 98°31′W
29 X12 **Fredericksburg** Iowa, C USA 42°58′N 92°12′W
25 R10 **Fredericksburg** Texas, SW USA 30°17′N 98°52′W
21 W5 **Fredericksburg** Virginia, NE USA 38°16′N 77°27′W
39 X13 **Frederick Sound** sound Alaska, USA
27 X6 **Fredericktown** Missouri, C USA 37°33′N 90°19′W
60 H13 **Frederico Westphalen** Rio Grande do Sul, S Brazil 27°22′S 53°20′W
13 O15 **Fredericton** province capital New Brunswick, SE Canada 45°57′N 66°40′W
Frederiksborgs Amt see Hovedstaden
Frederikshåb see Paamiut
95 H19 **Frederikshavn** prev. Fladstrand. Nordjylland, N Denmark 57°28′N 10°33′E
95 J22 **Frederikssund** Hovedstaden, E Denmark 55°51′N 12°05′E
45 T9 **Frederiksted** Saint Croix, S Virgin Islands (US) 17°41′N 64°51′W
95 J22 **Frederiksværk** var. Frederiksværk og Hanehoved. Hovedstaden, E Denmark 55°58′N 12°02′E
Frederiksværk og Hanehoved see Frederiksværk
54 E9 **Fredonia** Antioquia, W Colombia 05°57′N 75°42′W
36 K8 **Fredonia** Arizona, SW USA 36°57′N 112°31′W
27 P7 **Fredonia** Kansas, C USA 37°32′N 95°50′W
18 C11 **Fredonia** New York, NE USA 42°26′N 79°19′W
35 P4 **Fredonyer Pass** pass California, W USA
93 I15 **Fredrika** Västerbotten, N Sweden 64°03′N 18°25′E
95 L14 **Fredriksberg** Dalarna, C Sweden 60°07′N 14°23′E
Fredrikshald see Halden
Fredrikshamn see Hamina
95 H16 **Fredrikstad** Østfold, S Norway 59°12′N 10°57′E
30 K16 **Freeburg** Illinois, N USA 38°25′N 89°54′W
18 K15 **Freehold** New Jersey, NE USA 40°14′N 74°14′W
18 H14 **Freeland** Pennsylvania, NE USA 41°01′N 75°54′W
182 J5 **Freeling Heights** ▲ South Australia 30°09′S 139°24′E
35 Q7 **Freel Peak** ▲ California, W USA 38°52′N 119°52′W
11 Z9 **Freels, Cape** headland Newfoundland and Labrador, E Canada 49°16′N 53°30′W
29 Q11 **Freeman** South Dakota, N USA 43°21′N 97°26′W
44 G1 **Freeport** Grand Bahama Island, N The Bahamas 26°28′N 78°43′W
30 L10 **Freeport** Illinois, N USA 42°18′N 89°37′W
25 W12 **Freeport** Texas, SW USA 28°57′N 95°21′W
44 G1 **Freeport** ✈ Grand Bahama Island, N The Bahamas 26°31′N 78°48′W
25 R14 **Freer** Texas, SW USA 27°52′N 98°37′W
83 I22 **Free State** off. Free State Province; prev. Orange Free State, Afr. Oranje Vrystaat. ◈ province C South Africa
Free State see Maryland
Free State Province see Free State
76 G15 **Freetown** ● (Sierra Leone) W Sierra Leone 08°27′N 13°16′W
172 J16 **Frégate** island Inner Islands, NE Seychelles
104 J12 **Fregenal de la Sierra** Extremadura, W Spain 38°10′N 06°39′W
182 C2 **Fregon** South Australia 26°44′S 132°03′E
102 H5 **Fréhel, Cap** headland NW France 48°41′N 02°21′W
94 F8 **Frei** Møre og Romsdal, S Norway 63°02′N 07°47′E
101 O16 **Freiberg** Sachsen, E Germany 50°55′N 13°21′E
101 O16 **Freiberger Mulde** ∽ E Germany
Freiburg see Freiburg im Breisgau, Germany
Freiburg see Fribourg, Switzerland
101 F23 **Freiburg im Breisgau** var. Freiburg, Fr. Fribourg-en-Brisgau. Baden-Württemberg, SW Germany 48°N 07°52′E
Freiburg in Schlesien see Świebodzice
Freie Hansestadt Bremen see Bremen
Freie und Hansestadt Hamburg see Brandenburg
101 L22 **Freising** Bayern, SE Germany 48°24′N 11°45′E
109 T3 **Freistadt** Oberösterreich, N Austria 48°31′N 14°31′E
Freistadtl see Hlohovec
101 O16 **Freital** Sachsen, E Germany 51°00′N 13°40′E
Freiwaldau see Jeseník
104 J6 **Freixo de Espada à Cinta** Bragança, N Portugal 41°05′N 06°49′W
103 U15 **Fréjus** anc. Forum Julii. Var, SE France 43°26′N 06°44′E
180 I13 **Fremantle** Western Australia 32°07′S 115°44′E
35 N9 **Fremont** California, W USA 37°34′N 122°01′W
31 Q11 **Fremont** Indiana, N USA 41°43′N 84°54′W
29 W15 **Fremont** Iowa, C USA 41°12′N 92°26′W
31 P8 **Fremont** Michigan, N USA 43°28′N 85°56′W
29 R15 **Fremont** Nebraska, C USA 41°25′N 96°30′W
31 S11 **Fremont** Ohio, N USA 41°21′N 83°08′W
33 T14 **Fremont Peak** ▲ Wyoming, C USA 43°07′N 109°37′W
36 M6 **Fremont River** ∽ Utah, W USA
21 O9 **French Broad River** ∽ Tennessee, S USA
21 N5 **Frenchburg** Kentucky, S USA 37°58′N 83°37′W
18 C12 **French Creek** ∽ Pennsylvania, NE USA
32 K15 **Frenchglen** Oregon, NW USA 42°49′N 118°55′W
55 Y10 **French Guiana** var. Guiana, Guyane. ◇ French overseas department N South America
French Guinea see Guinea
31 O15 **French Lick** Indiana, N USA 38°33′N 86°37′W
185 J14 **French Pass** Marlborough, South Island, New Zealand 40°57′S 173°49′E
191 T11 **French Polynesia** ◇ French overseas territory S Pacific Ocean
French Republic see France
14 F11 **French River** ∽ Ontario, S Canada
French Somaliland see Djibouti
173 P12 **French Southern and Antarctic Lands** prev. French Southern and Antarctic Territories, Fr. Terres Australes et Antarctiques Françaises. ◈ French overseas territory S Indian Ocean
French Southern and Antarctic Territories see French Southern and Antarctic Lands
French Sudan see Mali
French Territory of the Afars and Issas see Djibouti
French Togoland see Togo
74 J6 **Frenda** NW Algeria 35°04′N 01°03′E
111 I18 **Frenštát pod Radhoštěm** Ger. Frankstadt. Moravskoslezský Kraj, E Czech Republic 49°33′N 18°10′E
Frentsjer see Franeker
76 M17 **Fresco** S Ivory Coast 05°03′N 05°31′W
195 U16 **Freshfield, Cape** headland Antarctica
40 L10 **Fresnillo** var. Fresnillo de González Echeverría. Zacatecas, C Mexico 23°11′N 102°53′W
Fresnillo de González Echeverría see Fresnillo
35 Q10 **Fresno** California, W USA 36°45′N 119°48′W
Freu, Cabo del see Freu, Cap des
105 Y9 **Freu, Cap des** var. Cabo del Freu. cape Mallorca, Spain, W Mediterranean Sea
101 G22 **Freudenstadt** Baden-Württemberg, SW Germany 48°28′N 08°25′E
Freudenthal see Bruntál
183 Q17 **Freycinet Peninsula** peninsula Tasmania, SE Australia
76 H14 **Fria** W Guinea 10°27′N 13°38′W
83 A17 **Fria, Cape** headland NW Namibia 18°32′S 12°00′E
35 Q10 **Friant** California, W USA 36°56′N 119°44′W
62 K8 **Frías** Catamarca, N Argentina 28°41′S 65°00′W
108 D9 **Fribourg** Ger. Freiburg. Fribourg, W Switzerland 46°50′N 07°10′E
108 C9 **Fribourg** Ger. Freiburg. ◈ canton W Switzerland
Fribourg-en-Brisgau see Freiburg im Breisgau
32 G7 **Friday Harbor** San Juan Islands, Washington, NW USA 48°31′N 123°01′W
Friedau see Ormož
101 K23 **Friedberg** Bayern, S Germany 48°21′N 10°58′E
101 H18 **Friedberg** Hessen, W Germany 50°19′N 08°46′E
Friedeberg Neumark see Strzelce Krajeńskie
Friedek-Mistek see Frýdek-Místek
Friedland see Pravdinsk
101 I24 **Friedrichshafen** Baden-Württemberg, S Germany 47°39′N 09°29′E
Friedrichstadt see Jaunjelgava
29 Q16 **Friend** Nebraska, C USA 40°37′N 97°16′W
Friendly Islands see Tonga
55 V9 **Friendship** Coronie, N Suriname 05°50′N 56°16′W
30 L7 **Friendship** Wisconsin, N USA 43°58′N 89°48′W
109 T8 **Friesach** Kärnten, S Austria 46°58′N 14°24′E
Friesche Eilanden see Frisian Islands
101 F22 **Friesenheim** Baden-Württemberg, SW Germany 48°27′N 07°56′E
Friesische Inseln see Frisian Islands
Friesland see Fryslân
60 Q10 **Frio, Cabo** headland SE Brazil 23°01′S 41°59′W
24 M3 **Friona** Texas, SW USA 34°38′N 102°43′W
42 L12 **Frío, Río** ∽ N Costa Rica
25 R13 **Frio River** ∽ Texas, SW USA
99 M25 **Frisange** Luxembourg, S Luxembourg 49°31′N 06°12′E
Frisches Haff see Vistula Lagoon
36 J6 **Frisco Peak** ▲ Utah, W USA 38°31′N 113°17′W
84 F9 **Frisian Islands** Dut. Friesche Eilanden, Ger. Friesische Inseln. island group N Europe
18 L12 **Frissell, Mount** ▲ Connecticut, NE USA 42°01′N 73°25′W
95 J19 **Fristad** Västra Götaland, S Sweden 57°50′N 13°01′E
25 N2 **Fritch** Texas, SW USA 35°38′N 101°36′W
95 J19 **Fritsla** Västra Götaland, S Sweden 57°33′N 12°47′E
101 H16 **Fritzlar** Hessen, C Germany 51°09′N 09°16′E
106 H6 **Friuli-Venezia Giulia** ◈ region NE Italy
196 L13 **Frobisher Bay** inlet Baffin Island, Nunavut, NE Canada
Frobisher Bay see Iqaluit
11 S12 **Frobisher Lake** ◎ Saskatchewan, C Canada
94 G7 **Frohavet** sound C Norway
Frohenbruck see Veselí nad Lužnicí
109 V7 **Frohnleiten** Steiermark, SE Austria 47°17′N 15°20′E
99 G22 **Froidchapelle** Hainaut, S Belgium 50°10′N 04°18′E
127 O9 **Frolovo** Volgogradskaya Oblast', SW Russian Federation 49°46′N 43°38′E
110 K7 **Frombork** Ger. Frauenburg. Warmińsko-Mazurskie, NE Poland 54°21′N 19°40′E
97 L22 **Frome** SW England, United Kingdom 51°15′N 02°22′W
182 I4 **Frome Creek** seasonal river South Australia
182 J6 **Frome Downs** South Australia 31°17′S 139°48′E
182 J5 **Frome, Lake** salt lake South Australia
Fronicken see Wronki
104 H10 **Fronteira** Portalegre, C Portugal 39°03′N 07°39′W
40 M7 **Frontera** Coahuila, NE Mexico 26°55′N 101°27′W
41 U14 **Frontera** Tabasco, SE Mexico 18°32′N 92°39′W
40 G3 **Fronteras** Sonora, NW Mexico 30°51′N 109°33′W
103 Q16 **Frontignan** Hérault, S France 43°27′N 03°45′E
54 D8 **Frontino** Antioquia, NW Colombia 06°46′N 76°10′W
21 V4 **Front Royal** Virginia, NE USA 38°56′N 78°13′W
107 J16 **Frosinone** anc. Frusino. Lazio, C Italy 41°38′N 13°22′E
107 K16 **Frosolone** Molise, C Italy 41°34′N 14°25′E
25 U7 **Frost** Texas, SW USA 32°04′N 96°48′W
21 U2 **Frostburg** Maryland, NE USA 39°39′N 78°55′W
23 X13 **Frostproof** Florida, SE USA 27°45′N 81°31′W
Frostviken see Kvarnbergsvattnet
95 M15 **Frövi** Örebro, C Sweden 59°28′N 15°24′E
94 F7 **Frøya** island W Norway
37 P5 **Fruita** Colorado, C USA 39°10′N 108°42′W
28 J9 **Fruitdale** South Dakota, N USA 44°39′N 103°38′W
23 W11 **Fruitland Park** Florida, SE USA 28°51′N 81°54′W
Frumentum see Formentera
147 S11 **Frunze** Batkenskaya Oblast', SW Kyrgyzstan 40°07′N 71°40′E
Frunze see Bishkek
117 O9 **Frunzivka** Odes'ka Oblast', SW Ukraine 47°19′N 29°46′E
Frusino see Frosinone
108 E9 **Frutigen** Bern, W Switzerland 46°35′N 07°38′E
111 I17 **Frýdek-Místek** Ger. Friedek-Mistek. Moravskoslezský Kraj, E Czech Republic 49°40′N 18°22′E
98 K6 **Fryslân** prev. Friesland. ◈ province N Netherlands
193 V16 **Fua'amotu** Tongatapu, S Tonga 21°15′S 175°08′W
190 A9 **Fuafatu** island Funafuti Atoll, C Tuvalu
190 A9 **Fuagea** island Funafuti Atoll, C Tuvalu
190 B8 **Fualifeke** atoll C Tuvalu
190 A8 **Fualopa** island Funafuti Atoll, C Tuvalu
151 K22 **Fuammulah** var. Fuammulah, Gnaviyani. atoll S Maldives
Fuammulah see Fuammulah
161 R11 **Fu'an** Fujian, SE China 27°11′N 119°42′E
Fu-chien see Fujian
Fu-chou see Fuzhou
164 G13 **Fuchū** var. Hutyû. Hiroshima, Honshū, SW Japan 34°35′N 133°12′E
160 M13 **Fuchuan** var. Fuyang. Guangxi Zhuangzu Zizhiqu, S China 24°56′N 111°15′E
161 R9 **Fuchun Jiang** var. Tsien Tang. ∽ SE China
165 R8 **Fudai** Iwate, Honshū, C Japan 39°59′N 141°50′E
161 S11 **Fuding** var. Tongshan. Fujian, SE China 27°21′N 120°10′E
81 J20 **Fudua** spring/well S Kenya 02°13′S 39°43′E
104 M16 **Fuengirola** Andalucía, S Spain 36°32′N 04°38′W
104 J12 **Fuente de Cantos** Extremadura, W Spain 38°15′N 06°18′W
104 J11 **Fuente del Maestre** Extremadura, W Spain 38°31′N 06°26′W
104 L12 **Fuente Obejuna** Andalucía, S Spain 38°15′N 05°25′W
104 L6 **Fuentesaúco** Castilla y León, N Spain 41°14′N 05°30′W
62 O3 **Fuerte Olimpo** var. Olimpo. Alto Paraguay, NE Paraguay 21°02′S 57°51′W
40 H8 **Fuerte, Río** ∽ C Mexico
64 Q11 **Fuerteventura** island Islas Canarias, Spain, NE Atlantic Ocean
141 S14 **Fughmah** var. Faghman, Fugma. C Yemen 16°08′N 49°23′E
92 M2 **Fuglehuken** headland W Svalbard 78°54′N 10°30′E
Fuglø see Fugloy
95 B18 **Fugloy** Dan. Fuglø. island NE Faroe Islands
197 T15 **Fugløya Bank** undersea feature E Norwegian Sea 71°00′N 19°20′E
Fugma see Fughmah
160 E11 **Fugong** Yunnan, SW China 27°00′N 98°48′E
81 K16 **Fugugo** spring/well NE Kenya 03°19′N 39°39′E
158 L2 **Fuhai** var. Burultokay. Xinjiang Uygur Zizhiqu, NW China 47°15′N 87°39′E
161 P10 **Fu He** ∽ S China
Fuhkien see Fujian
100 J9 **Fuhlsbüttel** ✈ (Hamburg) Hamburg, N Germany 53°37′N 09°57′E
101 L14 **Fuhne** ∽ C Germany
Fu-hsin see Fuxin
Fujairah see Al Fujayrah
164 M14 **Fuji** var. Huzi. Shizuoka, Honshū, S Japan 35°08′N 138°39′E
161 Q12 **Fujian** var. Fu-chien, Fuhkien, Fukien, Min, Fujian Sheng. ◈ province SE China
160 I9 **Fu Jiang** ∽ C China
Fujian Sheng see Fujian
164 M14 **Fujieda** var. Huzieda. Shizuoka, Honshū, S Japan 34°54′N 138°15′E
Fuji, Mount/Fujiyama see Fuji-san
163 Y7 **Fujin** Heilongjiang, NE China 47°12′N 132°01′E
164 M13 **Fujinomiya** var. Huzinomiya. Shizuoka, Honshū, S Japan 35°16′N 138°33′E
164 N13 **Fuji-san** var. Fujiyama, Eng. Mount Fuji. ▲ Honshū, SE Japan 35°23′N 138°44′E
165 N14 **Fujisawa** var. Huzisawa. Kanagawa, Honshū, S Japan 35°22′N 139°29′E
165 T3 **Fukagawa** var. Hukagawa. Hokkaidō, NE Japan 43°44′N 142°03′E
158 L5 **Fukang** Xinjiang Uygur Zizhiqu, W China 44°07′N 87°55′E
165 P7 **Fukaura** Aomori, Honshū, C Japan 40°38′N 139°55′E
193 W15 **Fukave** island Tongatapu Group, S Tonga
Fukien see Fujian
164 J13 **Fukuchiyama** var. Hukutiyama. Kyōto, Honshū, SW Japan 35°19′N 135°08′E
Fukue see Gotō
164 A13 **Fukue-jima** island Gotō-rettō, SW Japan
164 K12 **Fukui** var. Hukui. Fukui, Honshū, SW Japan 36°03′N 136°12′E
164 K12 **Fukui** off. Fukui-ken, var. Hukui. ◈ prefecture Honshū, SW Japan
Fukui-ken see Fukui
164 D13 **Fukuoka** var. Hukuoka, hist. Najima. Fukuoka, Kyūshū, SW Japan 33°36′N 130°24′E
164 D13 **Fukuoka** off. Fukuoka-ken, var. Hukuoka. ◈ prefecture Kyūshū, SW Japan
Fukuoka-ken see Fukuoka
165 Q6 **Fukushima** Hokkaidō, NE Japan 41°27′N 140°14′E
165 Q12 **Fukushima** off. Fukushima-ken, var. Hukusima. ◈ prefecture Honshū, C Japan
Fukushima-ken see Fukushima
164 G13 **Fukuyama** var. Hukuyama. Hiroshima, Honshū, SW Japan 34°29′N 133°21′E
76 G13 **Fulacunda** C Guinea-Bissau 11°44′N 15°03′W
129 P8 **Fūlādī, Kūh-e** ▲ E Afghanistan 34°38′N 67°32′E
187 Z15 **Fulaga** island Lau Group, E Fiji
101 I17 **Fulda** Hessen, C Germany 50°33′N 09°41′E
29 S10 **Fulda** Minnesota, N USA 43°52′N 95°36′W
101 I16 **Fulda** ∽ C Germany
Fülek see Fil'akovo
Fuli see Jixian
Fulin see Hanyuan
160 K10 **Fuling** Chongqing Shi, C China 29°45′N 107°23′E
35 T15 **Fullerton** California, SE USA 33°53′N 117°55′W
29 P15 **Fullerton** Nebraska, C USA 41°21′N 97°58′W
108 M8 **Fulpmes** Tirol, W Austria 47°11′N 11°22′E
20 G8 **Fulton** Kentucky, S USA 36°31′N 88°52′W
23 N2 **Fulton** Mississippi, S USA 34°16′N 88°24′W
27 V4 **Fulton** Missouri, C USA 38°50′N 91°57′W
18 H9 **Fulton** New York, NE USA 43°18′N 76°22′W
Fuman/Fumen see Fowman
103 R3 **Fumay** Ardennes, N France 49°58′N 04°42′E
102 M13 **Fumel** Lot-et-Garonne, SW France 44°31′N 00°58′E
190 B10 **Funafara** atoll C Tuvalu
190 C9 **Funafuti** ✈ Funafuti Atoll, C Tuvalu 08°30′S 179°12′E
Funafuti see Vaiaku
190 C9 **Funafuti Atoll** ● (Tuvalu) C Tuvalu 08°31′N 179°11′E
190 C9 **Funafuti Atoll** atoll C Tuvalu
Funan see Fusui
190 B9 **Funangongo** atoll C Tuvalu
93 F17 **Funäsdalen** Jämtland, C Sweden 62°33′N 12°33′E
64 O6 **Funchal** Madeira, Portugal, NE Atlantic Ocean 32°40′N 16°55′W
64 P5 **Funchal** ✈ Madeira, Portugal, NE Atlantic Ocean 32°38′N 16°53′W
54 F5 **Fundación** Magdalena, N Colombia 10°31′N 74°09′W
104 I8 **Fundão** var. Fundão. Castelo Branco, C Portugal 40°08′N 07°30′W
Fundão see Fundão
13 O16 **Fundy, Bay of** bay Canada/USA
Fünen see Fyn
56 C13 **Fúnes** Nariño, SW Colombia 0°59′N 77°27′W
Fünfkirchen see Pécs
83 M19 **Funhalouro** Inhambane, S Mozambique 23°04′S 34°24′E
161 R6 **Funing** Jiangsu, E China 33°43′N 119°47′E
160 I14 **Funing** var. Xinhua. Yunnan, SW China 23°39′N 105°41′E
160 M7 **Funiu Shan** ▲▲ C China
77 U13 **Funtua** Katsina, N Nigeria 11°31′N 07°19′E
161 R12 **Fuqing** Fujian, SE China 25°40′N 119°23′E
83 M14 **Furancungo** Tete, NW Mozambique 14°51′S 33°39′E
116 J15 **Furculeşti** Teleorman, S Romania 43°51′N 25°07′E
Füred see Balatonfüred
165 W4 **Füren-ko** ◎ Hokkaidō, NE Japan
Fürg see Doborji
Furluk see Fârliug
Fürmanov/Furmanovka see Moyynkum
Furmanovo see Zhalpaktal
59 L20 **Furnas, Represa de** ⊠ SE Brazil
183 Q14 **Furneaux Group** island group Tasmania, SE Australia
Furnes see Veurne
160 J10 **Furong Jiang** ∽ S China
138 I5 **Furqlus** Ḥimṣ, W Syria 34°40′N 37°02′E
100 F12 **Fürstenau** Niedersachsen, NW Germany 52°30′N 07°40′E
109 X8 **Fürstenfeld** Steiermark, SE Austria 47°03′N 16°05′E
101 L23 **Fürstenfeldbruck** Bayern, S Germany 48°10′N 11°16′E
100 P12 **Fürstenwalde** Brandenburg, NE Germany 52°22′N 14°04′E
101 K20 **Fürth** Bayern, S Germany 49°29′N 10°59′E
109 W3 **Furth bei Göttweig** Niederösterreich, NW Austria 48°22′N 15°33′E
165 R3 **Furubira** Hokkaidō, NE Japan 43°14′N 140°38′E
94 L12 **Furudal** Dalarna, C Sweden 61°10′N 15°07′E
164 L12 **Furukawa** var. Hida. Gifu, Honshū, SW Japan 36°13′N 137°11′E
165 Q10 **Furukawa** var. Hurukawa, Ōsaki. Miyagi, Honshū, C Japan 38°36′N 140°57′E
54 F10 **Fusagasugá** Cundinamarca, C Colombia 04°22′N 74°21′W
Fusan see Busan
Fushë-Arëzi/Fushë-Arrësi see Fushë-Arrëz
113 L18 **Fushë-Arrëz** var. Fushë-Arëzi, Fushë-Arrësi. Shkodër, N Albania 42°05′N 20°01′E
113 N16 **Fushë Kosovë** Serb. Kosovo Polje. C Kosovo 42°40′N 21°07′E
Fushë-Kruja see Fushë-Krujë
113 K19 **Fushë-Krujë** var. Fushë-Kruja. Durrës, C Albania 41°30′N 19°43′E
163 V12 **Fushun** var. Fou-shan, Fu-shun. Liaoning, NE China 41°50′N 123°54′E
Fu-shun see Fushun
Fusin see Fuxin
108 G10 **Fusio** Ticino, S Switzerland 46°27′N 08°39′E
163 X11 **Fusong** Jilin, NE China 42°20′N 127°17′E
101 K24 **Füssen** Bayern, S Germany 47°34′N 10°43′E
160 K15 **Fusui** var. Xinning; prev. Funan. Guangxi Zhuangzu Zizhiqu, S China 22°39′N 107°49′E
Futa Jallon see Fouta Djallon
63 G18 **Futaleufú** Los Lagos, S Chile 43°14′S 71°50′W
112 K10 **Futog** Vojvodina, NW Serbia 45°15′N 19°43′E
165 O14 **Futtsu** var. Huttu. Chiba, Honshū, S Japan 35°11′N 139°52′E
187 S15 **Futuna** island S Vanuatu
190 D12 **Futuna, Île** island S Wallis and Futuna
161 Q11 **Futun Xi** ∽ SE China
160 L5 **Fuxian** var. Fu Xian. Shaanxi, C China 36°03′N 109°19′E
Fuxian see Wafangdian
Fu Xian see Fuxian
160 G13 **Fuxian Hu** ⊠ SW China
163 U12 **Fuxin** var. Fou-hsin, Fu-hsin, Fusin. Liaoning, NE China 41°59′N 121°40′E
Fuxing see Wangmo
161 P7 **Fuyang** Anhui, E China 32°52′N 115°51′E
Fuyang see Fuchuan
161 O4 **Fuyang He** ∽ E China
163 U7 **Fuyu** Heilongjiang, NE China 47°48′N 124°26′E
163 Z6 **Fuyuan** Heilongjiang, NE China 48°20′N 134°22′E
Fuyu/Fu-yü see Songyuan
158 M3 **Fuyun** var. Koktokay. Xinjiang Uygur Zizhiqu, NW China 46°58′N 89°30′E
111 L22 **Füzesabony** Heves, E Hungary 47°46′N 20°25′E
161 R12 **Fuzhou** var. Foochow, Fu-chou. province capital Fujian, SE China 26°09′N 119°17′E
161 P11 **Fuzhou** prev. Linchuan. Jiangxi, S China 27°58′N 116°20′E
137 W13 **Füzuli** Rus. Fizuli. SW Azerbaijan 39°33′N 47°09′E
119 I20 **Fyadory** Rus. Fëdory. Brestskaya Voblasts', SW Belarus 51°57′N 26°24′E
95 G23 **Fyn** Ger. Fünen. island C Denmark
96 H12 **Fyne, Loch** inlet W Scotland, United Kingdom
95 E16 **Fyresvatnet** ◎ S Norway
FYR Macedonia/FYROM see Macedonia, FYR
Fyzabad see Feizābād

G

Gaafu Alifu Atoll see North Huvadhu Atoll
81 O14 **Gaalkacyo** var. Galka'yo, It. Galcaio. Mudug, C Somalia 06°42′N 47°24′E
146 J11 **Gabakly** Rus. Kabakly. Lebap Welaýaty, NE Turkmenistan 39°45′N 62°30′E
114 H8 **Gabare** Vratsa, NW Bulgaria 43°20′N 23°57′E
102 K15 **Gabas** ∽ SW France
Gabasumdo see Tongde
35 T7 **Gabbs** Nevada, W USA 38°51′N 117°55′W
82 B12 **Gabela** Kwanza Sul, W Angola 10°50′S 14°21′E
Gaberones see Gaborone
189 X14 **Gabert** island Caroline Islands, E Micronesia
74 M7 **Gabès** var. Qābis. E Tunisia 33°53′N 10°03′E
74 M6 **Gabès, Golfe de** Ar. Khalīj Qābis. gulf E Tunisia
Gablonz an der Neisse see Jablonec nad Nisou
Gablös see Cavalese
79 E18 **Gabon** off. Gabonese Republic. ♦ republic C Africa
Gabonese Republic see Gabon
83 I20 **Gaborone** prev. Gaberones. ● (Botswana) South East, SE Botswana 24°42′S 25°50′E
83 I20 **Gaborone** ✈ South East, SE Botswana 24°45′S 25°49′E
104 K8 **Gabriel y Galán, Embalse de** ⊠ W Spain
143 U15 **Gābrīk, Rūd-e** ∽ SE Iran
114 J9 **Gabrovo** Gabrovo, N Bulgaria 42°54′N 25°19′E
114 J9 **Gabrovo** ◈ province N Bulgaria
76 H12 **Gabú** prev. Nova Lamego. E Guinea-Bissau 12°16′N 14°09′W
29 O6 **Gackle** North Dakota, N USA 46°34′N 99°07′W
113 I15 **Gacko** Republika Srpska, S Bosnia and Herzegovina 43°08′N 18°29′E
155 F17 **Gadag** Karnātaka, W India 15°25′N 75°37′E
93 G15 **Gäddede** Jämtland, C Sweden 64°30′N 14°15′E

♦ Country
● Country Capital
◇ Dependent Territory
○ Dependent Territory Capital
◈ Administrative Regions
✈ International Airport
▲ Mountain
▲▲ Mountain Range
⛰ Volcano
∽ River
◎ Lake
⊠ Reservoir

GADÊ – GEDAREF

159 S12 **Gadê** var. Kequ; prev. Pagqên. Qinghai, C China 33°56′N 99°49′E
Gades/Gadier/Gadir/Gadire see Cádiz
105 P15 **Gádor, Sierra de** ▲▲ S Spain
149 S15 **Gadra** Sind, SE Pakistan 25°39′N 70°28′E
23 Q3 **Gadsden** Alabama, S USA 34°00′N 86°00′W
36 H15 **Gadsden** Arizona, SW USA 32°33′N 114°45′W
Gadyach see Hadyach
124 J3 **Gadzhiyevo** Murmanskaya Oblast', NW Russian Federation 69°16′N 33°20′E
79 H15 **Gadzi** Mambéré-Kadéï, SW Central African Republic 04°46′N 16°42′E
116 J13 **Găeşti** Dâmboviţa, S Romania 44°42′N 25°19′E
107 J17 **Gaeta** Lazio, C Italy 41°12′N 13°35′E
107 J17 **Gaeta, Golfo di** var. Gulf of Gaeta. gulf C Italy
Gaeta, Gulf of see Gaeta, Golfo di
188 L14 **Gaferut** atoll Caroline Islands, W Micronesia
21 Q10 **Gaffney** South Carolina, SE USA 35°03′N 81°40′W
Gäfle see Gävle
Gäfleborg see Gävleborg
74 M6 **Gafsa** var. Qafşah. W Tunisia 34°25′N 08°52′E
Gafurov see Ghafurov
126 J3 **Gagarin** prev. Gzhatsk. Smolenskaya Oblast', W Russian Federation 53°33′N 35°00′E
147 O10 **Gagarin** Jizzax Viloyati, C Uzbekistan 40°40′N 68°04′E
116 M12 **Găgăuzia** ◈ cultural region S Moldavia
101 G21 **Gaggenau** Baden-Württemberg, SW Germany 48°48′N 08°19′E
188 F16 **Gagil Tamil** var. Gagil-Tomil. island Caroline Islands, W Micronesia
Gagil-Tomil see Gagil Tamil
127 O4 **Gagino** Nizhegorodskaya Oblast', W Russian Federation 55°18′N 45°01′E
107 Q19 **Gagliano del Capo** Puglia, SE Italy 39°49′N 18°22′E
94 L13 **Gagnef** Dalarna, C Sweden 60°34′N 15°04′E
76 M17 **Gagnoa** C Ivory Coast 06°11′N 05°56′W
13 N10 **Gagnon** Québec, E Canada 51°56′N 68°16′W
Gago Coutinho see Lumbala N'Guimbo
137 P8 **Gagra** NW Georgia 43°17′N 40°18′E
31 S13 **Gahanna** Ohio, N USA 40°01′N 82°52′W
143 R13 **Gahkom** Hormozgān, S Iran 28°14′N 55°48′E
Gahnpa see Ganta
57 Q19 **Gaíba, Laguna** ◎ E Bolivia
153 T13 **Gaibandha** var. Gaibanda. Rajshahi, NW Bangladesh 25°21′N 89°36′E
Gaibhlte, Cnoc Mór na n see Galtymore Mountain
109 R9 **Gail** ≈ S Austria
101 I21 **Gaildorf** Baden-Württemberg, S Germany 48°41′N 10°08′E
103 N15 **Gaillac** var. Gaillac-sur-Tarn. Tarn, S France 43°54′N 01°54′E
Gaillac-sur-Tarn see Gaillac
Gaillimh see Galway
Gaillimhe, Cuan na see Galway Bay
109 Q9 **Gailtaler Alpen** ▲▲ S Austria
63 J17 **Gaimán** Chaco, S Argentina 43°15′S 65°30′W
20 K8 **Gainesboro** Tennessee, S USA 36°20′N 85°41′W
23 V10 **Gainesville** Florida, SE USA 29°39′N 82°19′W
23 T2 **Gainesville** Georgia, SE USA 34°18′N 83°49′W
27 U8 **Gainesville** Missouri, C USA 36°37′N 92°28′W
25 T5 **Gainesville** Texas, SW USA 33°39′N 97°08′W
109 X5 **Gainfarn** Niederösterreich, NE Austria 47°59′N 16°11′E
97 N18 **Gainsborough** E England, United Kingdom 53°24′N 00°48′W
182 G6 **Gairdner, Lake** salt lake South Australia
Gaissane see Gáissát
92 L8 **Gáissát** var. Gaissane. ▲▲ N Norway
43 T15 **Gaital, Cerro** ▲ C Panama 08°37′N 80°04′W
21 W3 **Gaithersburg** Maryland, NE USA 39°08′N 77°13′W
163 U13 **Gaizhou** Liaoning, NE China 40°24′N 122°17′E
Gaizina Kalns see Gaiziņkalns
118 H7 **Gaiziņkalns** var. Gaizina Kalns. ▲ E Latvia 56°51′N 25°58′E
Gajac see Villeneuve-sur-Lot
197 T10 **Gakkel Ridge** var. Arctic Mid Oceanic Ridge; prev. Nansen Cordillera. seamount range Arctic Ocean
39 S10 **Gakona** Alaska, USA 62°21′N 145°16′W
158 M16 **Gala** Xizang Zizhiqu, China 28°17′N 89°21′E
Galaassiya see Galaosiyo
114 K10 **Galabovo** , Gŭlŭbovo. Stara Zagora, C Bulgaria 42°08′N 25°51′E
Ğalāğıl see Jalājil
Galam, Pulau see Gelam, Pulau
62 J6 **Galán, Cerro** ▲ NW Argentina 25°54′S 66°45′W
111 H21 **Galanta** Hung. Galánta. Trnavský Kraj, W Slovakia 48°12′N 17°45′E
146 L11 **Galaosiyo** Rus. Galaassiya. Buxoro Viloyati, C Uzbekistan 39°53′N 64°25′E
57 B17 **Galápagos** off. Provincia de Galápagos. ◈ province W Ecuador, E Pacific Ocean
193 P8 **Galapagos Fracture Zone** tectonic feature E Pacific Ocean
Galapagos Islands see Colón, Archipiélago de
Galápagos, Islas de los see Colón, Archipiélago de
Galápagos, Provincia de see Galápagos
193 S9 **Galapagos Rise** undersea feature E Pacific Ocean 15°00′S 97°00′W
96 K13 **Galashiels** SE Scotland, United Kingdom 55°37′N 02°49′W
116 M12 **Galaţi** Ger. Galatz. Galaţi, E Romania 45°27′N 28°00′E
116 L12 **Galaţi** ◈ county E Romania
107 Q19 **Galatina** Puglia, SE Italy 40°10′N 18°10′E
107 Q19 **Galatone** Puglia, SE Italy 40°09′N 18°05′E
Galatz see Galaţi
21 R8 **Galax** Virginia, NE USA 36°40′N 88°56′W
146 J16 **Galaýmor** Rus. Kala-i-Mor. Mary Welaýaty, S Turkmenistan 35°40′N 62°28′E
Galcaio see Gaalkacyo
64 P11 **Gáldar** Gran Canaria, Islas Canarias, NE Atlantic Ocean 28°09′N 15°40′W
94 F11 **Galdhøpiggen** ▲ S Norway 61°30′N 08°08′E
40 I4 **Galeana** Chihuahua, N Mexico 30°08′N 107°38′W
41 O9 **Galeana** Nuevo León, NE Mexico 24°45′N 99°59′W
60 P9 **Galeão** ✕ (Rio de Janeiro) Rio de Janeiro, SE Brazil 22°48′S 43°16′W
171 R10 **Galela** Pulau Halmahera, E Indonesia 01°52′N 127°48′E
39 O9 **Galena** Alaska, USA 64°43′N 156°55′W
30 K10 **Galena** Illinois, N USA 42°25′N 90°25′W
27 R7 **Galena** Kansas, C USA 37°04′N 94°38′W
27 T8 **Galena** Missouri, C USA 36°45′N 93°30′W
45 V15 **Galeota Point** headland Trinidad, Trinidad and Tobago 10°07′N 60°59′W
105 P13 **Galera** Andalucía, S Spain 37°45′N 02°33′W
45 Y16 **Galera Point** headland Trinidad, Trinidad and Tobago 10°49′N 60°54′W
56 A5 **Galera, Punta** headland NW Ecuador 0°49′N 80°03′W
30 K12 **Galesburg** Illinois, N USA 40°57′N 90°22′W
30 J7 **Galesville** Wisconsin, N USA 44°04′N 91°21′W
18 F12 **Galeton** Pennsylvania, NE USA 41°43′N 77°38′W
116 H9 **Gâlgău** Hung. Galgó; prev. Gîlgău. Sălaj, NW Romania 47°17′N 23°43′E
Galgó see Gâlgău
Galgóc see Hlohovec
81 N15 **Galguduud** off. Gobolka Galguduud. ◈ region E Somalia
Galguduud, Gobolka see Galguduud
137 Q9 **Gali** W Georgia 42°40′N 41°39′E
125 N14 **Galich** Kostromskaya Oblast', NW Russian Federation 58°21′N 42°21′E
114 H7 **Galiche** Vratsa, NW Bulgaria 43°36′N 23°53′E
104 H3 **Galicia** anc. Gallaecia. ◈ autonomous community NW Spain
64 M8 **Galicia Bank** undersea feature E Atlantic Ocean 11°45′W 42°40′N
Galilee see HaGalil
181 W7 **Galilee, Lake** ◎ Queensland, NE Australia
Galilee, Sea of see Tiberias, Lake
106 E11 **Galileo Galilei** ✕ (Pisa) Toscana, C Italy 43°40′N 10°22′E
31 S12 **Galion** Ohio, N USA 40°43′N 82°47′W
Galka'yo see Gaalkacyo
146 K12 **Galkynyş** prev. Rus. Deynau, Dyanev, Turkm. Dänew. Lebap Welaýaty, NE Turkmenistan 39°16′N 63°10′E
80 H11 **Gallabat** Gedaref, E Sudan 12°57′N 36°10′E
Gallaecia see Galicia
147 O11 **G'allaorol** Jizzax Viloyati, C Uzbekistan 40°01′N 67°30′E
106 C7 **Gallarate** Lombardia, NW Italy 45°39′N 08°47′E
27 S2 **Gallatin** Missouri, C USA 39°54′N 93°57′W
20 J8 **Gallatin** Tennessee, S USA 36°22′N 86°28′W
33 R11 **Gallatin Peak** ▲ Montana, NW USA 45°22′N 111°21′W
33 R12 **Gallatin River** ≈ Montana/Wyoming, NW USA
155 J26 **Galle** prev. Point de Galle. Southern Province, SW Sri Lanka 06°04′N 80°12′E
105 S5 **Gállego** ≈ NE Spain
193 Q8 **Gallego Rise** undersea feature E Pacific Ocean 02°00′S 115°00′W
Gallegos see Río Gallegos
63 H23 **Gallegos, Río** ≈ Argentina/Chile
Gallia see France
22 K10 **Galliano** Louisiana, S USA 29°26′N 90°18′W
114 G13 **Gallikós** ≈ N Greece
37 S12 **Gallinas Peak** ▲ New Mexico, SW USA 34°14′N 105°47′W
54 H3 **Gallinas, Punta** headland NE Colombia 12°27′N 71°44′W
37 T11 **Gallinas River** ≈ New Mexico, SW USA
107 Q19 **Gallipoli** Puglia, SE Italy 40°08′N 18°E
Gallipoli see Gelibolu
Gallipoli Peninsula see Gelibolu Yarımadası
31 T15 **Gallipolis** Ohio, N USA 38°49′N 82°14′W
92 J12 **Gällivare** Lapp. Váhtjer. Norrbotten, N Sweden 67°08′N 20°39′E
109 T4 **Gallneukirchen** Oberösterreich, N Austria 48°22′N 14°25′E
93 G17 **Gällö** Jämtland, C Sweden 62°57′N 15°15′E
105 Q7 **Gallo** ≈ C Spain
107 I23 **Gallo, Capo** headland Sicilia, Italy, C Mediterranean Sea 38°13′N 13°18′E
37 P13 **Gallo Mountains** ▲▲ New Mexico, SW USA
18 G8 **Galloo Island** island New York, NE USA
97 H15 **Galloway, Mull of** headland S Scotland, United Kingdom 54°37′N 04°54′W
37 P10 **Gallup** New Mexico, SW USA 35°32′N 108°45′W
105 R5 **Gallur** Aragón, NE Spain 41°51′N 01°21′W
Gâlma see Guelma
163 N9 **Galshar** var. Buyant. Hentiy, C Mongolia 46°15′N 110°50′E
162 I6 **Galt** var. Ider. Hövsgöl, C Mongolia 48°45′N 99°52′E
35 O8 **Galt** California, W USA 38°13′N 121°19′W
74 C10 **Galtat-Zemmour** C Western Sahara 25°07′N 12°21′W
95 G22 **Galten** Midtjylland, C Denmark 56°09′N 09°54′E
Gålto see Kultsjön
97 D20 **Galtymore Mountain** Ir. Cnoc Mór na nGaibhlte. ▲ S Ireland 52°21′N 08°09′W
97 D20 **Galty Mountains** Ir. Na Gaibhlte. ▲▲ S Ireland
30 K11 **Galva** Illinois, N USA 41°10′N 90°02′W
25 X12 **Galveston** Texas, SW USA 29°17′N 94°48′W
25 W11 **Galveston Bay** inlet Texas, SW USA
25 W12 **Galveston Island** island Texas, SW USA
61 B18 **Gálvez** Santa Fe, C Argentina 32°03′S 61°14′W
97 C18 **Galway** Ir. Gaillimh. W Ireland 53°16′N 09°03′W
97 B18 **Galway** Ir. Gaillimh. cultural region W Ireland
97 B18 **Galway Bay** Ir. Cuan na Gaillimhe. bay W Ireland
83 F18 **Gam** Otjozondjupa, NE Namibia 20°10′S 20°51′E
164 L14 **Gamagōri** Aichi, Honshū, SW Japan 34°49′N 137°15′E
54 F7 **Gamarra** Cesar, N Colombia 08°21′N 73°46′W
Gámas see Kaamanen
158 L17 **Gamba** Xizang Zizhiqu, W China 28°13′N 88°32′E
Gamba see Zamtang
77 P14 **Gambaga** NE Ghana 10°32′N 00°28′W
80 G13 **Gambēla** Gambēla Hizboch, W Ethiopia 08°09′N 34°15′E
81 H14 **Gambēla Hizboch** ◈ federal region W Ethiopia
38 K10 **Gambell** Saint Lawrence Island, Alaska, USA 63°44′N 171°41′W
76 E12 **Gambia** off. Republic of The Gambia, The Gambia. ◆ republic W Africa
76 I12 **Gambia** Fr. Gambie. ≈ W Africa
64 K12 **Gambia Plain** undersea feature E Atlantic Ocean
Gambia, Republic of The see Gambia
Gambia, The see Gambia
Gambie see Gambia
31 T13 **Gambier** Ohio, N USA 40°22′N 82°24′W
191 Y13 **Gambier, Îles** island group E French Polynesia
182 G10 **Gambier Islands** island group South Australia
79 H19 **Gamboma** Plateaux, E Congo 01°53′S 15°51′E
79 G16 **Gamboula** Mambéré-Kadéï, SW Central African Republic 04°09′N 15°12′E
37 P10 **Gamerco** New Mexico, SW USA 35°34′N 108°45′W
137 V12 **Gamış Dağı** ▲ W Azerbaijan 40°18′N 46°15′E
Gamlakarleby see Kokkola
95 N18 **Gamleby** Kalmar, S Sweden 57°54′N 16°25′E
Gammelstad see Gammelstaden
93 J14 **Gammelstaden** var. Gammelstad. Norrbotten, N Sweden 65°38′N 22°05′E
Gammouda see Sidi Bouzid
155 J25 **Gampaha** Western Province, W Sri Lanka 07°05′N 80°00′E
155 K25 **Gampola** Central Province, C Sri Lanka 07°10′N 80°34′E
167 S5 **Gâm, Sông** ≈ N Vietnam
92 L7 **Gamvik** Finnmark, N Norway 71°04′N 28°08′E
150 H13 **Gan** Addu Atoll, C Maldives
Gan see Gansu, China
Gan see Jiangxi, China
Ganaane see Juba
37 O10 **Ganado** Arizona, SW USA 35°42′N 109°31′W
25 U12 **Ganado** Texas, SW USA 29°02′N 96°30′W
14 L14 **Gananoque** Ontario, SE Canada 44°21′N 76°11′W
Ganāveh see Bandar-e Gonāveh
137 V11 **Gäncä** Rus. Gyandzha; prev. Kirovabad, Yelisavetpol. W Azerbaijan 40°42′N 46°23′E
Ganchi see Ghonchí
Gand see Gent
82 B13 **Ganda** var. Mariano Machado, Port. Vila Mariano Machado. Benguela, W Angola 13°02′S 14°40′E
79 L22 **Gandajika** Kasai-Oriental, S Dem. Rep. Congo 06°42′S 24°01′E
153 O12 **Gandak** Nep. Nārāyāni. ≈ India/Nepal
13 U11 **Gander** Newfoundland and Labrador, SE Canada 48°56′N 54°33′W
13 U11 **Gander** ✕ Newfoundland and Labrador, E Canada 49°03′N 54°49′W
100 G11 **Ganderkesee** Niedersachsen, NW Germany 53°01′N 08°33′E
105 T7 **Gandesa** Cataluña, NE Spain 41°03′N 00°26′E
154 B10 **Gāndhīdhām** Gujarāt, W India 23°08′N 70°05′E
154 D10 **Gāndhīnagar** state capital Gujarāt, W India 23°12′N 72°37′E
154 F9 **Gāndhi Sāgar** ◎ C India
105 T11 **Gandia** prev. Gandía. Valenciana, E Spain 38°59′N 00°11′W
Gandía see Gandia
159 O10 **Gang** Qinghai, W China
152 G9 **Gangānagar** Rājasthān, NW India 29°54′N 73°56′E
152 I12 **Gangāpur** Rājasthān, N India 26°30′N 76°49′E
153 S17 **Ganga Sāgar** West Bengal, NE India 21°39′N 88°05′E
Gangavathi see Gangāwati
155 G17 **Gangāwati** var. Gangavathi. Karnātaka, C India 15°26′N 76°35′E
159 S9 **Gangca** var. Shaliuhe. Qinghai, C China 37°21′N 100°09′E
158 H14 **Gangdisê Shan** Eng. Kailas Range. ▲▲ W China
103 Q15 **Ganges** Hérault, S France 43°57′N 03°42′E
153 P13 **Ganges** Ben. Padma. ≈ Bangladesh/India see also Padma
Ganges see Padma
Ganges Cone see Ganges Fan
173 S3 **Ganges Fan** var. Ganges Cone. undersea feature N Bay of Bengal 12°00′N 87°00′E
153 U17 **Ganges, Mouths of the** delta Bangladesh/India
107 K23 **Gangi** anc. Engyum. Sicilia, Italy, C Mediterranean Sea 37°48′N 14°13′E
163 Y14 **Gangneung** Jap. Kōryō; prev. Kangnŭng. NE South Korea 37°47′N 128°51′E
152 K8 **Gangotri** Uttarakhand, N India 30°56′N 79°02′E
Gangra see Çankırı
153 S11 **Gangtok** state capital Sikkim, N India 27°20′N 88°39′E
159 W11 **Gangu** var. Daxiangshan. Gansu, C China 34°38′N 105°18′E
163 U5 **Gan He** ≈ NE China
171 S12 **Gani** Pulau Halmahera, E Indonesia 0°45′S 128°13′E
161 O12 **Gan Jiang** ≈ S China
163 U11 **Ganjig** var. Horqin Zuoyi Houqi. Nei Mongol Zizhiqu, N China 42°53′N 122°22′E
146 H15 **Gannaly** Ahal Welaýaty, S Turkmenistan 37°02′N 60°43′E
163 U7 **Gannan** Heilongjiang, NE China 47°58′N 123°36′E
103 P10 **Gannat** Allier, C France 46°06′N 03°12′E
33 T14 **Gannett Peak** ▲ Wyoming, C USA 43°10′N 109°39′W
29 O10 **Gannvalley** South Dakota, N USA 44°01′N 98°59′W
Ganqu see Lhünzhub
109 Y3 **Gänserndorf** Niederösterreich, NE Austria 48°21′N 16°43′E
Gansos, Lago dos see Goose Lake
159 T9 **Gansu** var. Gan, Gansu Sheng, Kansu. ◈ province N China
Gansu Sheng see Gansu
76 K16 **Ganta** var. Gahnpa. NE Liberia 07°15′N 08°59′W
182 H11 **Gantheaume, Cape** headland South Australia 36°04′S 137°28′E
Gantsevichi see Hantsavichy
161 Q6 **Ganyu** var. Qingkou. Jiangsu, E China 34°52′N 119°11′E
144 D12 **Ganyushkino** Atyrau, SW Kazakhstan 46°38′N 49°12′E
161 O12 **Ganzhou** Jiangxi, S China 25°51′N 114°59′E
Ganzhou see Zhangye
77 Q10 **Gao** Gao, E Mali 16°16′N 00°03′E
77 R10 **Gao** ◈ region SE Mali
161 O10 **Gao'an** Jiangxi, S China 28°24′N 115°27′E
Gaocheng see Litang
Gaoleshan see Xianfeng
161 R5 **Gaomi** Shandong, C China 36°23′N 119°45′E
161 N5 **Gaoping** Shanxi, C China 35°51′N 112°55′E
159 S8 **Gaotai** Gansu, N China 39°22′N 99°44′E
Gaoth Dobhair see Gweedore
77 O14 **Gaoua** SW Burkina Faso 10°18′N 03°12′W
76 I13 **Gaoual** N Guinea 11°44′N 13°14′W
161 S14 **Gaoxiong** var. Kaohsiung, Jap. Takao, Takow. S Taiwan 22°36′N 120°17′E
161 S14 **Gaoxiong** ✕ S Taiwan 22°26′N 120°32′E
161 R7 **Gaoyou** var. Dayishan. Jiangsu, E China 32°48′N 119°26′E
161 R7 **Gaoyou Hu** ◎ E China
160 M15 **Gaozhou** Guangdong, S China 21°56′N 110°49′E
103 T13 **Gap** anc. Vapincum. Hautes-Alpes, SE France 44°33′N 06°05′E
146 E9 **Gaplaňgyr Platosy** Rus. Plato Kaplangky. ridge Turkmenistan/Uzbekistan
156 G13 **Gar** var. Shiquanhe. Xizang Zizhiqu, W China 32°31′N 80°04′E
Gar see Gar Xincun
Garabekevyul see Garabekewül
146 L13 **Garabekewül** Rus. Garabekevyul, Karabekaul. Lebap Welaýaty, E Turkmenistan 38°31′N 64°04′E
146 K15 **Garabil Belentligi** Rus. Vozvyshennost' Karabil'. ▲▲ S Turkmenistan
146 A8 **Garabogaz** Rus. Bekdash. Balkan Welaýaty, NW Turkmenistan 41°33′N 52°33′E
146 B9 **Garabogaz Aylagy** Rus. Zaliv Kara-Bogaz-Gol. bay NW Turkmenistan
146 A9 **Garabogazköl** Rus. Kara-Bogaz-Gol. Balkan Welaýaty, NW Turkmenistan 41°03′N 52°52′E
43 V16 **Garachiné** Darién, SE Panama 08°03′N 78°22′W
43 V16 **Garachiné, Punta** headland SE Panama 08°05′N 78°23′W
146 K12 **Garagan** Rus. Karagan. Ahal Welaýaty, C Turkmenistan 38°16′N 57°34′E
54 G10 **Garagoa** Boyacá, C Colombia 05°05′N 73°20′W
146 A11 **Garagöl'** Rus. Karagel'. Balkan Welaýaty, W Turkmenistan 39°24′N 53°13′E
146 F12 **Garagum** var. Garagumy, Qara Qum, Eng. Black Sand Desert, Kara Kum; prev. Peski Karakumy. desert C Turkmenistan
146 E12 **Garagum Kanaly** var. Kara Kum Canal, Rus. Karagumskiy Kanal, Karakumskiy Kanal. canal C Turkmenistan
Garagumy see Garagum
183 S4 **Garah** New South Wales, SE Australia 29°07′S 149°37′E
64 O11 **Garajonay** ▲ Gomera, Islas Canarias, NE Atlantic Ocean 28°07′N 17°14′W
76 L13 **Garalo** Sikasso, SW Mali 10°58′N 07°26′W
Garam see Hron, Slovakia
146 L14 **Garamätnyýaz** Rus. Karamet-Niyaz. Lebap Welaýaty, E Turkmenistan 37°45′N 64°28′E
Garamszentkereszt see Žiar nad Hronom
77 Q13 **Garango** S Burkina Faso 11°45′N 00°30′W
59 Q15 **Garanhuns** Pernambuco, E Brazil 08°53′S 36°28′W
188 H5 **Garapan** Saipan, S Northern Mariana Islands 15°12′S 145°43′E
Gárasavvon see Karesuando
Gárassavon see Kaaresuvanto
78 J13 **Garba** Bamingui-Bangoran, N Central African Republic 09°09′N 20°24′E
Garba see Jiulong
81 L16 **Garbahaarrey** It. Garba Harre. Gedo, SW Somalia 03°14′N 42°18′E
Garba Harre see Garbahaarrey
81 J18 **Garba Tula** Isiolo, C Kenya 0°31′N 38°35′E
27 N9 **Garber** Oklahoma, C USA 36°26′N 97°35′W
34 L4 **Garberville** California, W USA 40°07′N 123°48′W
Garbo see Lhozhag
100 I12 **Garbsen** Niedersachsen, N Germany 52°25′N 09°36′E
60 K9 **Garça** São Paulo, S Brazil 22°14′S 49°36′W
104 L10 **García de Solá, Embalse de** ⊠ C Spain
103 Q14 **Gard** ◈ department S France
103 Q14 **Gard** ≈ S France
106 F7 **Garda, Lago di** var. Benaco, Eng. Lake Garda, Ger. Gardasee. ◎ NE Italy
Garda, Lake see Garda, Lago di
Gardan Divāl see Gardan Dīwāl
149 Q5 **Gardan Dīwāl** var. Gardan Divāl. Wardak, C Afghanistan 34°30′N 68°15′E
103 S15 **Gardanne** Bouches-du-Rhône, SE France 43°27′N 05°28′E
Gardasee see Garda, Lago di
100 L12 **Gardelegen** Sachsen-Anhalt, C Germany 52°31′N 11°25′E
14 B10 **Garden** ≈ Ontario, S Canada
23 X6 **Garden City** Georgia, SE USA 32°06′N 81°09′W
26 I6 **Garden City** Kansas, C USA 37°57′N 100°54′W
27 S5 **Garden City** Missouri, C USA 38°34′N 94°12′W
25 N8 **Garden City** Texas, SW USA 31°51′N 101°30′W
23 P3 **Gardendale** Alabama, S USA 33°39′N 86°48′W
31 P5 **Garden Island** island Michigan, N USA
22 M11 **Garden Island Bay** bay Louisiana, S USA
31 O5 **Garden Peninsula** peninsula Michigan, N USA
Garden State, The see New Jersey
95 I14 **Gardermoen** Akershus, S Norway 60°10′N 11°04′E
95 I14 **Gardermoen** ✕ (Oslo) Akershus, S Norway 60°12′N 11°05′E
Gardeyz/Gardez see Gardēz
149 Q6 **Gardēz** var. Gardeyz, Gordiaz; prev. Gardīz. Paktiyā, E Afghanistan 33°35′N 69°14′E
93 G14 **Gardiken** ◎ N Sweden
19 Q7 **Gardiner** Maine, NE USA 44°13′N 69°46′W
33 S12 **Gardiner** Montana, NW USA 45°02′N 110°42′W
19 N13 **Gardiners Island** island New York, NE USA
Gardner Island see Nikumaroro
19 T6 **Gardner Lake** ◎ Maine, NE USA
35 Q6 **Gardnerville** Nevada, W USA 38°55′N 119°44′W
Gardo see Qardho
106 F7 **Gardone Val Trompia** Lombardia, N Italy 45°40′N 10°11′E
Garegegasnjárga see Karigasniemi
38 F17 **Gareloi Island** island Aleutian Islands, Alaska, USA
Gares see Puente la Reina
106 B10 **Garessio** Piemonte, NE Italy 44°14′N 08°01′E
32 M9 **Garfield** Washington, NW USA 47°00′N 117°07′W
31 U11 **Garfield Heights** Ohio, N USA 41°25′N 81°36′W
Gargaliani see Gargaliánoi
115 D21 **Gargaliánoi** var. Gargaliani. Pelopónnisos, S Greece 37°04′N 21°38′E
107 N15 **Gargàno, Promontorio del** headland SE Italy 41°51′N 16°11′E
108 J8 **Gargellen** Graubünden, W Switzerland 46°57′N 09°55′E
93 I14 **Gargnäs** Västerbotten, N Sweden 65°19′N 18°00′E
118 C11 **Gargždai** Klaipėda, W Lithuania 55°42′N 21°24′E
154 J13 **Garhchiroli** Mahārāshtra, C India 20°14′N 79°58′E
153 O15 **Garhwa** Jhārkhand, N India 24°07′N 83°52′E
171 V13 **Gariau** Papua Barat, E Indonesia 03°43′S 134°54′E
83 E24 **Garies** Northern Cape, W South Africa 30°30′S 18°00′E
107 K17 **Garigliano** ≈ C Italy
81 K19 **Garissa** Garissa, E Kenya 0°27′S 39°39′E
81 K19 **Garissa** ◈ county SE Kenya
21 V11 **Garland** North Carolina, SE USA 34°45′N 78°25′W
25 T6 **Garland** Texas, SW USA 32°50′N 96°37′W
36 L1 **Garland** Utah, W USA 41°43′N 112°07′W
106 D8 **Garlasco** Lombardia, N Italy 45°12′N 08°59′E
119 F14 **Garliava** Kaunas, S Lithuania 54°49′N 23°52′E
Garm see Gharm
142 M9 **Garm, Āb-e** var. Rūd-e Khersān. ≈ SW Iran
Garmik see Germik
101 K25 **Garmisch-Partenkirchen** Bayern, S Germany 47°30′N 11°05′E
143 O5 **Garmsār** prev. Qishlaq. Semnān, N Iran 35°18′N 52°22′E
Garmser see Darwēshān
29 V12 **Garner** Iowa, C USA 43°06′N 93°36′W
21 U9 **Garner** North Carolina, SE USA 35°42′N 78°36′W
27 Q5 **Garnett** Kansas, C USA 38°16′N 95°15′W
99 M25 **Garnich** Luxembourg, SW Luxembourg 49°37′N 05°57′E
182 M8 **Garnpung, Lake** salt lake New South Wales, SE Australia
Garoe see Garoowe
Garoet see Garut
153 U13 **Gāro Hills** hill range NE India
102 K13 **Garonne** anc. Garumna. ≈ S France
80 P13 **Garoowe** var. Garoe. Nugaal, N Somalia 08°24′N 48°29′E
78 F12 **Garoua** var. Garua. Nord, N Cameroon 09°17′N 13°22′E
79 G14 **Garoua Boulaï** Est, E Cameroon 05°54′N 14°33′E
77 O10 **Garou, Lac** ◎ C Mali
95 L16 **Garphyttan** Örebro, C Sweden 59°18′N 14°54′E
29 R11 **Garretson** South Dakota, N USA 43°43′N 96°30′W
31 Q11 **Garrett** Indiana, N USA 41°21′N 85°08′W
33 Q10 **Garrison** Montana, NW USA 46°32′N 112°46′W
28 M4 **Garrison** North Dakota, N USA 47°36′N 101°25′W
25 X8 **Garrison** Texas, SW USA 31°49′N 94°29′W
28 L4 **Garrison Dam** dam North Dakota, N USA
104 J9 **Garrovillas** Extremadura, W Spain 39°43′N 06°33′W
Garrygala see Magtymguly
8 L8 **Garry Lake** ◎ Nunavut, N Canada
Gars see Gars am Kamp
109 W3 **Gars am Kamp** var. Gars. Niederösterreich, NE Austria 48°35′N 15°40′E
81 K20 **Garsen** Tana River, S Kenya 02°16′S 40°07′E
Garshy see Garşy
14 F10 **Garson** Ontario, S Canada 46°33′N 80°51′W
109 T5 **Garsten** Oberösterreich, N Austria 48°00′N 14°24′E
146 A9 **Garşy** var. Garshy, Rus. Karshi. Balkan Welaýaty, NW Turkmenistan 40°45′N 52°50′E
Gartar see Qianning
102 M10 **Gartempe** ≈ C France
Gartog see Markam
Garua see Garoua
83 D21 **Garub** Karas, SW Namibia 26°33′S 16°00′E
Garumna see Garonne
169 P16 **Garut** prev. Garoet. Jawa, C Indonesia 07°15′S 107°55′E
185 C20 **Garvie Mountains** ▲▲ South Island, New Zealand
110 N12 **Garwolin** Mazowieckie, E Poland 51°54′N 21°36′E
25 U12 **Garwood** Texas, SW USA 29°25′N 96°26′W
158 G13 **Gar Xincun** prev. Gar. Xizang Zizhiqu, W China 32°04′N 80°01′E
31 N11 **Gary** Indiana, N USA 41°36′N 87°21′W
25 X7 **Gary** Texas, SW USA 32°00′N 94°21′W
158 G13 **Garyarsa** Xizang Zizhiqu, SW China 31°44′N 80°20′E
158 G13 **Gar Zangbo** ≈ W China
160 F8 **Garzê** Sichuan, C China 31°40′N 99°58′E
54 E12 **Garzón** Huila, S Colombia 02°14′N 75°37′W
Gasan-Kuli see Esenguly
31 P13 **Gas City** Indiana, N USA 40°29′N 85°36′W
102 K15 **Gascogne** Eng. Gascony. cultural region S France
26 V5 **Gasconade River** ≈ Missouri, C USA
Gascony see Gascogne
180 H9 **Gascoyne Junction** Western Australia 25°06′S 115°10′E
173 V8 **Gascoyne Plain** undersea feature E Indian Ocean
180 H9 **Gascoyne River** ≈ Western Australia
192 J11 **Gascoyne Tablemount** undersea feature N Tasman Sea 36°30′S 156°30′E
67 U6 **Gash** var. Nahr al Qāsh. ≈ W Sudan
149 X3 **Gasherbrum** ▲ NE Pakistan 35°39′N 76°34′E
Gas Hu see Gas Hure Hu
77 X12 **Gashua** Yobe, NE Nigeria 12°55′N 11°10′E
159 N9 **Gas Hure Hu** var. Gas Hu. ◎ C China
Gásluokta see Kjøpsvik
186 G7 **Gasmata** New Britain, E Papua New Guinea 06°12′S 150°25′E
23 V14 **Gasparilla Island** island Florida, SE USA
169 O13 **Gaspar, Selat** strait W Indonesia
15 Y6 **Gaspé** Québec, SE Canada 48°50′N 64°33′W
15 Z6 **Gaspé, Cap de** headland Québec, SE Canada 48°45′N 64°10′W
15 X6 **Gaspé, Péninsule de** var. Péninsule de la Gaspésie. peninsula Québec, SE Canada
Gaspésie, Péninsule de la see Gaspé, Péninsule de
77 W15 **Gassol** Taraba, E Nigeria 08°28′N 10°24′E
Gastein see Badgastein
21 R10 **Gastonia** North Carolina, SE USA 35°14′N 81°12′W
21 V8 **Gaston, Lake** ⊠ North Carolina/Virginia, SE USA
115 D19 **Gastoúni** Dytikí Elláda, S Greece 37°51′N 21°15′E
63 I17 **Gastre** Chubut, S Argentina 42°20′S 69°10′W
Gat see Ghāt
105 P15 **Gata, Cabo de** cape S Spain
Gata, Cape see Gátas, Akrotíri
105 T11 **Gata de Gorgos** Valenciana, E Spain 38°45′N 00°06′E
116 E12 **Gătaia** Ger. Gataja, Hung. Gátalja; prev. Gáttája. Timiş, W Romania 45°24′N 21°26′E
Gataja/Gátalja see Gătaia
121 P3 **Gátas, Akrotíri** var. Cape Gata. headland S Cyprus 34°34′N 33°03′E
104 J8 **Gata, Sierra de** ▲▲ W Spain
124 G13 **Gatchina** Leningradskaya Oblast', NW Russian Federation 59°34′N 30°06′E
21 P8 **Gate City** Virginia, NE USA 36°38′N 82°37′W
97 M14 **Gateshead** NE England, United Kingdom 54°57′N 01°37′W
21 X8 **Gatesville** North Carolina, SE USA 36°24′N 76°46′W
25 S8 **Gatesville** Texas, SW USA 31°26′N 97°46′W
14 L12 **Gatineau** Québec, SE Canada 45°29′N 75°40′W
14 L11 **Gatineau** ≈ Ontario/Québec, SE Canada
21 N9 **Gatlinburg** Tennessee, S USA 35°42′N 83°30′W
Gatooma see Kadoma
Gáttája see Gătaia
43 T14 **Gatún, Lago** ⊠ C Panama
59 N14 **Gaturiano** Piauí, NE Brazil 06°53′S 41°45′W
97 O22 **Gatwick** ✕ (London) SE England, United Kingdom 51°10′N 00°12′W
187 Y14 **Gau** prev. Ngau. island C Fiji
187 R12 **Gaua** var. Santa Maria. Island Banks Island, N Vanuatu
104 L16 **Gaucín** Andalucía, S Spain 36°31′N 05°19′W
Gauhāti see Guwāhāti
118 I8 **Gauja** Ger. Aa. ≈ Estonia/Latvia
118 I7 **Gaujiena** NE Latvia 57°31′N 26°24′E
94 H9 **Gauldalen** valley S Norway
21 R5 **Gauley River** ≈ West Virginia, NE USA
Gaul/Gaule see France
99 D19 **Gaurain-Ramecroix** Hainaut, SW Belgium 50°35′N 03°31′E
95 F15 **Gaustatoppen** ▲ S Norway 59°50′N 08°39′E
83 J21 **Gauteng** off. Gauteng Province; prev. Pretoria-Witwatersrand-Vereeniging. ◈ province NE South Africa
Gauteng see Johannesburg, South Africa
Gauteng see Germiston, South Africa
Gauteng Province see Gauteng
137 U11 **Gavarr** prev. Kamo. C Armenia 40°21′N 45°07′E
143 P14 **Gāvbandī** Hormozgān, S Iran 27°07′N 53°21′E
115 H25 **Gavdopoúla** island SE Greece
115 H26 **Gávdos** island SE Greece
102 K16 **Gave de Pau** var. Gave-de-Pay. ≈ SW France
Gave-de-Pay see Gave de Pau
102 J16 **Gave d'Oloron** ≈ SW France
99 E18 **Gavere** Oost-Vlaanderen, NW Belgium 50°56′N 03°41′E
94 N13 **Gävle** var. Gäfle; prev. Gefle. Gävleborg, C Sweden 60°41′N 17°09′E
94 M11 **Gävleborg** var. Gäfleborg, Gefleborg. ◈ county C Sweden
94 O13 **Gävlebukten** bay C Sweden
124 L16 **Gavrilov-Yam** Yaroslavskaya Oblast', W Russian Federation 57°19′N 39°52′E
182 I9 **Gawler** South Australia 34°38′S 138°44′E
182 G7 **Gawler Ranges** hill range South Australia
Gawso see Goaso
162 H11 **Gaxun Nur** ◎ N China
153 P14 **Gaya** Bihār, N India 24°48′N 85°E
77 S13 **Gaya** Dosso, SW Niger 11°52′N 03°28′E
Gaya see Kyjov
31 Q6 **Gaylord** Michigan, N USA 45°01′N 84°40′W
29 U9 **Gaylord** Minnesota, N USA 44°33′N 94°13′W
181 Y9 **Gayndah** Queensland, E Australia 25°37′S 151°31′E
125 T12 **Gayny** Komi-Permyatskiy Okrug, NW Russian Federation 60°19′N 54°15′E
Gaysin see Haysyn
Gayvorno see Hayvoron
138 E11 **Gaza** Ar. Ghazzah, Heb. 'Azza. NE Gaza Strip 31°30′N 34°E
83 L20 **Gaza** off. Província de Gaza. ◈ province SW Mozambique
Gaz-Achak see Gazojak
147 Q9 **G'azalkent** Rus. Gazalkent. Toshkent Viloyati, E Uzbekistan 41°30′N 69°46′E
Gazalkent see G'azalkent
Gazandzhyk/Gazanjyk see Bereket
77 V12 **Gazaoua** Maradi, S Niger 13°28′N 07°54′E
Gaza, Província de see Gaza
138 E11 **Gaza Strip** Ar. Qita Ghazzah. disputed region SW Asia
136 M16 **Gaziantep** var. Gazi Antep; prev. Aintab, Antep. Gaziantep, S Turkey 37°04′N 37°21′E
136 M17 **Gaziantep** var. Gazi Antep. ◈ province S Turkey
Gazi Antep see Gaziantep
114 M13 **Gaziköy** Tekirdağ, NW Turkey 40°45′N 27°18′E
121 Q2 **Gazimağusa** var. Famagusta, Gk. Ammóchostos. E Cyprus 35°07′N 33°57′E
121 Q2 **Gazimağusa Körfezi** var. Famagusta Bay, Gk. Kólpos Ammóchostos. bay E Cyprus
146 K11 **Gazli** Buxoro Viloyati, C Uzbekistan 40°09′N 63°28′E
146 I9 **Gazojak** Rus. Gaz-Achak. Lebap Welaýaty, NE Turkmenistan 41°12′N 61°24′E
79 K15 **Gbadolite** Equateur, NW Dem. Rep. Congo 04°14′N 20°59′E
76 K16 **Gbanga** var. Gbarnga. N Liberia 07°02′N 09°30′W
Gbarnga see Gbanga
77 S14 **Gbéroubouè** var. Béroubouay. N Benin 10°35′N 02°47′E
77 W16 **Gboko** Benue, S Nigeria 07°21′N 08°57′E
Gcuwa see Butterworth
110 J7 **Gdańsk** Fr. Dantzig, Ger. Danzig. Pomorskie, N Poland 54°22′N 18°35′E
Gdan'skaya Bukhta/Gdańsk, Gulf of see Danzig, Gulf of
Gdańska, Zakota see Danzig, Gulf of
Gdingen see Gdynia
124 F13 **Gdov** Pskovskaya Oblast', W Russian Federation 58°43′N 27°51′E
110 I6 **Gdynia** Ger. Gdingen. Pomorskie, N Poland 54°31′N 18°30′E
26 M10 **Geary** Oklahoma, C USA 35°37′N 98°19′W
Geavvú see Kevo
76 H12 **Gêba, Rio** ≈ C Guinea-Bissau
136 E11 **Gebze** Kocaeli, NW Turkey 40°48′N 29°26′E
80 H10 **Gedaref** var. Al Qaḑārif, El Gedaref. Gedaref, E Sudan 14°03′N 35°24′E
80 H10 **Gedaref** ◈ state E Sudan

◆ Country
● Country Capital
◇ Dependent Territory
○ Dependent Territory Capital
◈ Administrative Regions
✕ International Airport
▲ Mountain
▲▲ Mountain Range
⛰ Volcano
≈ River
◎ Lake
⊠ Reservoir

GEDID RAS EL FIL – GLÁMA

80 B11 **Gedid Ras el Fil** Southern Darfur, W Sudan 12°45′N 25°45′E
99 I23 **Gedinne** Namur, SE Belgium 49°57′N 04°55′E
136 E13 **Gediz** Kütahya, W Turkey 39°04′N 29°25′E
136 C14 **Gediz Nehri** ≈ W Turkey
81 N14 **Gedlegubē** Sumalē, E Ethiopia 06°53′N 45°08′E
81 L17 **Gedo** *off.* Gobolka Gedo. ⬖ *region* SW Somalia
Gedo, Gobolka *see* Gedo
95 I25 **Gedser** Sjælland, SE Denmark 54°34′N 11°57′E
99 I16 **Geel** *var.* Gheel. Antwerpen, N Belgium 51°10′N 04°59′E
183 N13 **Geelong** Victoria, SE Australia 38°10′S 144°21′E
Ge'e'mu *see* Golmud
99 I14 **Geertruidenberg** Noord-Brabant, S Netherlands 51°43′N 04°52′E
100 H10 **Geeste** ≈ NW Germany
100 J10 **Geesthacht** Schleswig-Holstein, N Germany 53°25′N 10°22′E
183 P17 **Geeveston** Tasmania, SE Australia 43°12′S 146°54′E
Gefle *see* Gävle
Gefleborg *see* Gävleborg
163 S5 **Gegan Gol** *prev.* Ergun, Gen He, Zuoqi. NE China
163 T5 **Gegen Gol** *prev.* Ergun Zuoqi, Genhe. Nei Mongol Zizhiqu, N China 50°48′N 121°30′E
158 G13 **Gê'gyai** Xizang Zizhiqu, W China 32°29′N 81°04′E
77 X12 **Geidam** Yobe, NE Nigeria 12°52′N 11°55′E
11 T11 **Geikie** ≈ Saskatchewan, C Canada
94 F13 **Geilo** Buskerud, S Norway 60°32′N 08°13′E
94 E10 **Geiranger** Møre og Romsdal, S Norway 62°07′N 07°12′E
101 I22 **Geislingen** *var.* Geislingen an der Steige. Baden-Württemberg, SW Germany 48°37′N 09°50′E
Geislingen an der Steige *see* Geislingen
81 F20 **Geita** Geita, NW Tanzania 02°52′S 32°12′E
81 F21 **Geita** *off.* Mkoa wa Geita. ⬖ *region* N Tanzania
Geita, Mkoa wa *see* Geita
95 G15 **Geithus** Buskerud, S Norway 59°56′N 09°58′E
160 H14 **Gejiu** *var.* Kochiu. Yunnan, S China 23°22′N 103°07′E
Gëkdepe *see* Gökdepe
146 E9 **Geklengkui, Solonchak** *var.* Solonchak Goklenkuy. *salt marsh* NW Turkmenistan
81 D14 **Gel** ≈ C South Sudan
107 K25 **Gela** *prev.* Terranova di Sicilia. Sicilia, Italy, C Mediterranean Sea 37°05′N 14°15′E
81 N14 **Geladī** SE Ethiopia 06°58′N 46°24′E
169 P13 **Gelam, Pulau** *var.* Pulau Galam. *island* N Indonesia
Gelaozu Miaozu Zhizhixian *see* Wuchuan
98 L11 **Gelderland** *prev. Eng.* Guelders. ⬖ *province* E Netherlands
98 J13 **Geldermalsen** Gelderland, C Netherlands 51°53′N 05°17′E
101 D14 **Geldern** Nordrhein-Westfalen, W Germany 51°31′N 06°19′E
99 K15 **Geldrop** Noord-Brabant, SE Netherlands 51°25′N 05°34′E
99 L17 **Geleen** Limburg, SE Netherlands 50°57′N 05°49′E
126 K14 **Gelendzhik** Krasnodarskiy Kray, SW Russian Federation 44°34′N 38°06′E
Gelib *see* Jilib
136 B11 **Gelibolu** *Eng.* Gallipoli. Çanakkale, NW Turkey 40°25′N 26°41′E
115 L14 **Gelibolu Yarımadası** *Eng.* Gallipoli Peninsula. *peninsula* NW Turkey
81 O14 **Gellinsor** Galguduud, C Somalia 06°25′N 46°44′E
101 H18 **Gelnhausen** Hessen, C Germany 50°12′N 09°12′E
101 E14 **Gelsenkirchen** Nordrhein-Westfalen, W Germany 51°30′N 07°05′E
83 C20 **Geluk** Hardap, SW Namibia 24°35′S 15°48′E
99 H20 **Gembloux** Namur, Belgium 50°34′N 04°42′E
79 J16 **Gemena** Equateur, NW Dem. Rep. Congo 03°13′N 19°49′E
99 L14 **Gemert** Noord-Brabant, SE Netherlands 51°33′N 05°41′E
136 E11 **Gemlik** Bursa, NW Turkey 40°26′N 29°10′E
Gem of the Mountains *see* Idaho
106 J6 **Gemona del Friuli** Friuli-Venezia Giulia, NE Italy 46°18′N 13°12′E
Gem State *see* Idaho
Genalē Wenz *see* Juba
169 R10 **Genali, Danau** ◎ Borneo, N Indonesia
99 G19 **Genappe** Walloon Brabant, C Belgium 50°39′N 04°27′E
137 P14 **Genç** Bingöl, E Turkey 38°44′N 40°35′E
Genck *see* Genk
98 M9 **Genemuiden** Overijssel, E Netherlands 52°38′N 06°03′E
63 K14 **General Acha** La Pampa, C Argentina 37°25′S 64°38′W
61 C21 **General Alvear** Buenos Aires, E Argentina 36°03′S 60°01′W
62 I12 **General Alvear** Mendoza, W Argentina 34°59′S 67°40′W
61 B20 **General Arenales** Buenos Aires, E Argentina 34°21′S 61°20′W
61 D21 **General Belgrano** Buenos Aires, E Argentina 35°47′S 58°30′W
194 H3 **General Bernardo O'Higgins** *Chilean research station* Antarctica 63°09′S 57°13′W
41 O8 **General Bravo** Nuevo León, NE Mexico 25°47′N 99°04′W
62 M7 **General Capdevila** Chaco, N Argentina 27°25′S 61°30′W
General Carrera, Lago *see* Buenos Aires, Lago
41 N9 **General Cepeda** Coahuila, NE Mexico 25°18′N 101°24′W
63 K15 **General Conesa** Río Negro, E Argentina 40°06′S 64°26′W
61 G18 **General Enrique Martínez** Treinta y Tres, E Uruguay 33°13′S 53°47′W
62 L3 **General Eugenio A. Garay** *var.* Fortín General Eugenio Garay; *prev.* Yrendagüé. Nueva Asunción, NW Paraguay 20°30′S 61°56′W
61 C18 **General Galarza** Entre Ríos, E Argentina 32°43′S 59°24′W
61 E22 **General Guido** Buenos Aires, E Argentina 36°36′S 57°45′W
General José F.Uriburu *see* Zárate
61 E22 **General Juan Madariaga** Buenos Aires, E Argentina 37°00′S 57°09′W
41 O16 **General Juan N Alvarez** ✈ (Acapulco) Guerrero, S Mexico 16°47′N 99°47′W
61 B22 **General La Madrid** Buenos Aires, E Argentina 37°17′S 61°20′W
61 E21 **General Lavalle** Buenos Aires, E Argentina 36°25′S 56°56′W
General Machado *see* Camacupa
62 I8 **General Manuel Belgrano, Cerro** ▲ W Argentina 29°05′S 67°05′W
41 O8 **General Mariano Escobero** ✈ (Monterrey) Nuevo León, NE Mexico 25°47′N 100°00′W
61 B20 **General O'Brien** Buenos Aires, E Argentina 34°54′S 60°45′W
62 K13 **General Pico** La Pampa, C Argentina 35°43′S 63°45′W
62 M7 **General Pinedo** Chaco, N Argentina 27°17′S 61°20′W
61 B20 **General Pinto** Buenos Aires, E Argentina 34°45′S 61°50′W
61 E22 **General Pirán** Buenos Aires, E Argentina 37°16′S 57°46′W
43 N15 **General, Río** ≈ S Costa Rica
63 I15 **General Roca** Río Negro, C Argentina 39°00′S 67°35′W
171 Q8 **General Santos** *off.* General Santos City. Mindanao, S Philippines 06°10′N 125°10′E
General Santos City *see* General Santos
41 O9 **General Terán** Nuevo León, NE Mexico 25°18′N 99°40′W
114 N7 **General Toshevo** *Rom.* I.G.Duca; *prev.* Casim, Kasimköj. Dobrich, NE Bulgaria 43°43′N 28°04′E
61 B20 **General Viamonte** Buenos Aires, E Argentina 35°01′S 61°00′W
61 A20 **General Villegas** Buenos Aires, E Argentina 35°02′S 63°01′W
Gênes *see* Genova
18 E11 **Genesee River** ≈ New York/Pennsylvania, NE USA
30 K11 **Geneseo** Illinois, N USA 41°27′N 90°08′W
18 F10 **Geneseo** New York, NE USA 42°48′N 77°46′W
57 L14 **Geneshuaya, Río** ≈ N Bolivia
23 Q8 **Geneva** Alabama, S USA 31°01′N 85°51′W
30 M10 **Geneva** Illinois, N USA 41°53′N 88°18′W
29 Q16 **Geneva** Nebraska, C USA 40°31′N 97°36′W
18 G10 **Geneva** New York, NE USA 42°52′N 76°58′W
31 U10 **Geneva** Ohio, NE USA 41°48′N 80°53′W
Geneva *see* Genève
108 B10 **Geneva, Lake** *Fr.* Lac de Genève, Lac Léman, le Léman, *Ger.* Genfer See. ◎ France/Switzerland
108 A10 **Genève** *Eng.* Geneva, *Ger.* Genf, *It.* Ginevra. Genève, SW Switzerland 46°13′N 06°09′E
108 A11 **Genève** *Eng.* Geneva, *Ger.* Genf, *It.* Ginevra. ⬖ *canton* SW Switzerland
108 A10 **Genève** *var.* Geneva. ✈ Vaud, SW Switzerland 46°13′N 06°06′E
Genève, Lac de *see* Geneva, Lake
Genf *see* Genève
Genfer See *see* Geneva, Lake
Gen He *see* Gegan Gol
Genhe *see* Gegen Gol
Genichesk *see* Heniches'k
104 L14 **Genil** ≈ S Spain
99 K18 **Genk** *var.* Genck. Limburg, NE Belgium 50°58′N 05°30′E
164 C13 **Genkai-nada** *gulf* Kyūshū, SW Japan
107 C19 **Gennargentu, Monti del** ▲▲ Sardegna, Italy, C Mediterranean Sea
99 M14 **Gennep** Limburg, SE Netherlands 51°43′N 05°58′E
30 M10 **Genoa** Illinois, N USA 42°06′N 88°41′W
29 Q15 **Genoa** Nebraska, C USA 41°27′N 97°43′W
Genoa *see* Genova
Genoa, Gulf of *see* Genova, Golfo di
106 D10 **Genova** *Eng.* Genoa, *Fr.* Gênes; *anc.* Genua. Liguria, NW Italy 44°28′N 09°E
106 D10 **Genova, Golfo di** *Eng.* Gulf of Genoa. *gulf* NW Italy
57 C17 **Genovesa, Isla** *var.* Tower Island. *island* Galapagos Islands, Ecuador, E Pacific Ocean
Genshū *see* Wonju
99 E17 **Gent** *Eng.* Ghent, *Fr.* Gand. Oost-Vlaanderen, NW Belgium 51°02′N 03°42′E
169 N16 **Genteng** Jawa, C Indonesia 07°21′S 106°20′E
100 M12 **Genthin** Sachsen-Anhalt, E Germany 52°24′N 12°10′E
27 R9 **Gentry** Arkansas, C USA 36°16′N 94°28′W
Genua *see* Genova
107 I15 **Genzano di Roma** Lazio, C Italy 41°42′N 12°42′E
163 Y17 **Geogeum-do** *prev.* Kōgŭm-do. *island* S South Korea
163 Z16 **Geogeum-do** *Jap.* Kyōsai-tō; *prev.* Kōje-do. *island* S South Korea
Geokchay *see* Göyçay
Geok-Tepe *see* Gökdepe
122 I3 **Georga, Zemlya** *Eng.* George Land. *island* Zemlya Frantsa-Iosifa, N Russian Federation
83 G26 **George** Western Cape, S South Africa 33°57′S 22°28′E
29 S11 **George** Iowa, C USA 43°20′N 96°00′W
13 O5 **George** ≈ Newfoundland and Labrador/Québec, E Canada
George F L Charles *see* Vigie
65 C25 **George Island** *island* S Falkland Islands
183 R10 **George, Lake** ◎ New South Wales, SE Australia
81 E18 **George, Lake** ◎ SW Uganda
23 W10 **George, Lake** ◎ Florida, SE USA
18 L8 **George, Lake** ⊠ New York, NE USA
George Land *see* Georga, Zemlya
Georgenburg *see* Jurbarkas
George River *see* Kangiqsualujjuaq
64 G8 **Georges Bank** *undersea feature* W Atlantic Ocean 41°15′N 67°30′W
185 A21 **George Sound** *sound* South Island, New Zealand
65 F15 **Georgetown** ○ (Ascension Island) NW Ascension Island 17°56′S 14°25′W
181 V5 **Georgetown** Queensland, NE Australia 18°17′S 143°37′E
183 P15 **George Town** Tasmania, SE Australia 41°04′S 146°48′E
44 D8 **George Town** *var.* Georgetown. ○ (Cayman Islands) Grand Cayman, SW Cayman Islands 19°16′N 81°23′W
76 H12 **Georgetown** E Gambia 13°33′N 14°49′W
55 T8 **Georgetown** ● (Guyana) N Guyana 06°46′N 58°10′W
168 I7 **George Town** *var.* Penang, Pinang. Pinang, Peninsular Malaysia 05°28′N 100°20′E
45 Y14 **Georgetown** Saint Vincent, Saint Vincent and the Grenadines 13°19′N 61°09′W
44 I4 **George Town** Great Exuma Island, C The Bahamas 23°28′N 75°47′W
21 Y4 **Georgetown** Delaware, NE USA 38°39′N 75°22′W
23 R6 **Georgetown** Georgia, SE USA 31°52′N 85°04′W
20 M5 **Georgetown** Kentucky, S USA 38°13′N 84°30′W
21 T13 **Georgetown** South Carolina, SE USA 33°23′N 79°18′W
25 S10 **Georgetown** Texas, SW USA 30°39′N 97°42′W
55 T8 **Georgetown** ✈ N Guyana 06°46′N 58°10′W
Georgetown *see* George Town
195 U16 **George V Coast** *physical region* Antarctica
194 J7 **George VI Ice Shelf** *ice shelf* Antarctica
194 J6 **George VI Sound** *sound* Antarctica
195 T15 **George V Land** *physical region* Antarctica
25 S14 **George West** Texas, SW USA 28°21′N 98°08′W
137 R9 **Georgia** *off.* Republic of Georgia, *Geor.* Sak'art'velo, *Rus.* Gruzinskaya SSR, Gruziya. ◆ *republic* SW Asia
23 S5 **Georgia** *off.* State of Georgia, *also known as* Empire State of the South, Peach State. ⬖ *state* SE USA
14 F12 **Georgian Bay** *lake bay* Ontario, S Canada
Georgia, Republic of *see* Georgia
10 L17 **Georgia, Strait of** *strait* British Columbia, W Canada
Georgi Dimitrov *see* Kostenets
Georgi Dimitrov, Yazovir *see* Koprinka, Yazovir
Georgiu-Dezh *see* Liski
145 W10 **Georgiyevka** Vostochnyy Kazakhstan, E Kazakhstan 49°19′N 81°35′E
Georgiyevka *see* Korday
127 N15 **Georgiyevsk** Stavropol'skiy Kray, SW Russian Federation 44°07′N 43°22′E
100 G13 **Georgsmarienhütte** Niedersachsen, NW Germany 52°13′N 08°02′E
195 O1 **Georg von Neumayer** *German research station* Antarctica 70°41′S 08°18′W
101 M16 **Gera** Thüringen, E Germany 50°51′N 12°13′E
101 K16 **Gera** ≈ C Germany
99 E19 **Geraardsbergen** Oost-Vlaanderen, SW Belgium 50°47′N 03°53′E
115 F21 **Geráki** Pelopónnisos, S Greece 36°56′N 22°46′E
27 W5 **Gerald** Missouri, C USA 38°24′N 91°20′W
47 V8 **Geral de Goiás, Serra** ▲▲ E Brazil
185 G20 **Geraldine** Canterbury, South Island, New Zealand 44°06′S 171°14′E
180 H11 **Geraldton** Western Australia 28°48′S 114°40′E
12 E11 **Geraldton** Ontario, S Canada 49°44′N 86°59′W
60 J12 **Geral, Serra** ▲▲ S Brazil
103 U6 **Gérardmer** Vosges, NE France 48°05′N 06°54′E
Gerasa *see* Jarash
Gerdauen *see* Zheleznodorozhnyy
39 Q11 **Gerdine, Mount** ▲ Alaska, USA 61°40′N 152°21′W
136 H11 **Gerede** Bolu, N Turkey 40°48′N 32°13′E
136 H11 **Gerede Çayı** ≈ N Turkey
148 M8 **Gereshk** Helmand, SW Afghanistan 31°50′N 64°32′E
101 L24 **Geretsried** Bayern, S Germany 47°51′N 11°28′E
105 P14 **Gérgal** Andalucía, S Spain 37°07′N 02°34′W
28 I14 **Gering** Nebraska, C USA 41°49′N 103°39′W
35 R3 **Gerlach** Nevada, W USA 40°38′N 119°21′W
Gerlachfalvi Csúcs/Gerlachovka *see* Gerlachovský štít
111 L18 **Gerlachovský štít** *var.* Gerlachovka, *Ger.* Gerlsdorfer Spitze, *Hung.* Gerlachfalvi Csúcs; *prev.* Stalinov Štít, *Ger.* Franz-Josef Spitze, *Hung.* Ferencz-József Csúcs. ▲ N Slovakia 49°12′N 20°09′E
108 E8 **Gerlafingen** Solothurn, NW Switzerland 47°10′N 07°35′E
Gerlsdorfer Spitze *see* Gerlachovský štít
Germak *see* Germik
German East Africa *see* Tanzania
Germanicopolis *see* Çankırı
Germanicum, Mare/German Ocean *see* North Sea
Germanovichi *see* Hyermanavichy
German Southwest Africa *see* Namibia
20 E10 **Germantown** Tennessee, S USA 35°06′N 89°51′W
101 I15 **Germany** *off.* Federal Republic of Germany, Bundesrepublik Deutschland, *Ger.* Deutschland. ◆ *federal republic* N Europe
Germany, Federal Republic of *see* Germany
101 L23 **Germering** Bayern, SE Germany 48°07′N 11°22′E
139 V3 **Germik** *Ar.* Garmik, *var.* Germak. As Sulaymānīyah, E Iraq 35°49′N 46°09′E
83 J21 **Germiston** *var.* Gauteng. Gauteng, NE South Africa 26°15′S 28°10′E
Gernika *see* Gernika-Lumo
105 P2 **Gernika-Lumo** *var.* Gernika, Guernica, Guernica y Lumo. País Vasco, N Spain 43°19′N 02°40′W
164 L12 **Gero** Gifu, Honshū, SW Japan 35°48′N 137°15′E
115 F22 **Geroliménas** Pelopónnisos, S Greece 36°28′N 22°25′E
Gerona *see* Girona
99 H21 **Gerpinnes** Hainaut, S Belgium 50°20′N 04°38′E
102 L15 **Gers** ⬖ *department* S France
102 L14 **Gers** ≈ S France
Gerunda *see* Girona
158 I13 **Gêrzê** *var.* Luring. Xizang Zizhiqu, W China 32°19′N 84°05′E
136 K10 **Gerze** Sinop, N Turkey 41°48′N 35°13′E
Gesoriacum *see* Boulogne-sur-Mer
Gessoriacum *see* Boulogne-sur-Mer
99 J21 **Gesves** Namur, SE Belgium 50°24′N 05°04′E
93 J20 **Geta** Åland, SW Finland 60°22′N 19°49′E
105 N8 **Getafe** Madrid, C Spain 40°18′N 03°44′W
95 J21 **Getinge** Halland, S Sweden 56°46′N 12°42′E
18 F16 **Gettysburg** Pennsylvania, NE USA 39°49′N 77°13′W
29 N8 **Gettysburg** South Dakota, N USA 45°00′N 99°57′W
194 K12 **Getz Ice Shelf** *ice shelf* Antarctica
137 S15 **Gevaş** Van, SE Turkey 38°16′N 43°05′E
Gevgeli *see* Gevgelija
113 Q20 **Gevgelija** *var.* Đevđelija, Djevdjelija, *Turk.* Gevgeli. SE Macedonia 41°09′N 22°30′E
103 T10 **Gex** Ain, E France 46°21′N 06°02′E
92 I3 **Geysir** *physical region* SW Iceland
136 F11 **Geyve** Sakarya, NW Turkey 40°32′N 30°18′E
80 G10 **Gezira** ⬖ *state* E Sudan
109 V3 **Gföhl** Niederösterreich, N Austria 48°30′N 15°27′E
83 H22 **Ghaap Plateau** *Afr.* Ghaapplato. *plateau* C South Africa
Ghaapplato *see* Ghaap Plateau
Ghaba *see* Al Ghābah
138 J8 **Ghāb, Tall** ▲ SE Syria 33°09′N 37°48′E
139 Q9 **Ghadaf, Wādī al** *dry watercourse* C Iraq
Ghadamès *see* Ghadāmis
74 M9 **Ghadāmis** *var.* Ghadamès, Rhadames. W Libya 30°08′N 09°30′E
141 Y10 **Ghadan** E Oman 20°20′N 57°58′E
75 O10 **Ghaddūwah** C Libya 26°36′N 14°26′E
147 Q11 **Ghafurov** *Rus.* Gafurov; *prev.* Sovetabad. NW Tajikistan 40°13′N 69°42′E
153 N12 **Ghāghara** ≈ S Asia
149 P13 **Ghaibi Dero** Sind, SE Pakistan 27°35′N 67°42′E
141 Y10 **Ghalat** E Oman 21°06′N 58°51′E
139 W11 **Ghamūkah, Hawr** ◎ S Iraq
77 P15 **Ghana** *off.* Republic of Ghana. ◆ *republic* W Africa
141 X12 **Ghānah** *spring/well* S Oman 18°35′N 56°34′E
Ghanongga *see* Ranongga
Ghansi/Ghansiland *see* Ghanzi
83 F18 **Ghanzi** *var.* Khanzi. Ghanzi, W Botswana 21°39′S 21°38′E
83 G19 **Ghanzi** *var.* Ghansi, Ghansiland, Khanzi. ⬖ *district* C Botswana
67 T14 **Ghanzi** *var.* Knanzi. ≈ Botswana/South Africa
Ghap'an *see* Kapan
138 F13 **Gharandal** Al'Aqabah, SW Jordan 30°12′N 35°18′E
139 U14 **Gharbīyah, Sha'īb al** ≈ S Iraq
Gharbt, Jabal al *see* Liban, Jebel
74 K7 **Ghardaïa** N Algeria 32°30′N 03°44′E
147 R12 **Gharm** *Rus.* Garm. C Tajikistan 39°03′N 70°25′E
149 P17 **Gharo** Sind, SE Pakistan 24°44′N 67°35′E
139 W10 **Gharrāf, Shaṭṭ al** ≈ S Iraq
Gharvān *see* Gharyān
75 O7 **Gharyān** *var.* Gharvān. NW Libya 32°10′N 13°01′E
74 M11 **Ghāt** *var.* Gat. SW Libya 24°58′N 10°11′E
Ghawdex *see* Gozo
141 U8 **Ghayathi** Abū Ẓaby, W United Arab Emirates 23°51′N 53°01′E
Ghazāl, Baḥr al *see* Ghazal, Bahr el
78 H9 **Ghazal, Bahr el** *var.* Soro. *seasonal river* C Chad
80 E13 **Ghazal, Bahr el** *var.* Baḥr al Ghazāl. ≈ N South Sudan
74 H6 **Ghazaouet** NW Algeria 35°08′N 01°50′W
152 J10 **Ghāziābād** Uttar Pradesh, N India 28°42′N 77°28′E
153 O13 **Ghāzīpur** Uttar Pradesh, N India 25°36′N 83°36′E
149 Q6 **Ghaznī** *var.* Ghazni. Ghaznī, E Afghanistan 33°31′N 68°24′E
149 P7 **Ghaznī** ⬖ *province* SE Afghanistan
Ghazzah *see* Gaza
Gheel *see* Geel
Ghelîzâne *see* Relizane
Ghent *see* Gent
Gheorghe Brațul *see* Sfântu Gheorghe, Brațul
Gheorghe Gheorghiu-Dej *see* Onești
116 J10 **Gheorgheni** *prev.* Gheorghieni, Sîn-Miclăuș, *Ger.* Niklasmarkt, *Hung.* Gyergyószentmiklós. Harghita, C Romania 46°43′N 25°36′E
Gheorghieni *see* Gheorgheni
116 H10 **Gherla** *Ger.* Neuschliss, *Hung.* Szamosújvár; *prev.* Armenierstadt. Cluj, NW Romania 47°02′N 23°55′E
Gheweifat *see* Ghuwayfāt
Ghilan *see* Gīlān
107 C18 **Ghilarza** Sardegna, Italy, C Mediterranean Sea 40°09′N 08°50′E
Ghilizane *see* Relizane
Ghimbi *see* Gīmbī
Ghiriș *see* Câmpia Turzii
103 Y15 **Ghisonaccia** Corse, France, C Mediterranean Sea 42°00′N 09°25′E
147 Q11 **Ghonchí** *Rus.* Ganchi. NW Tajikistan 39°57′N 69°10′E
153 T13 **Ghoraghat** Rajshahi, NW Bangladesh 25°18′N 89°20′E
148 J5 **Ghōriān** *prev.* Ghūriān. Herāt, W Afghanistan 34°20′N 61°26′E
149 R13 **Ghotki** Sind, SE Pakistan 28°00′N 69°21′E
Ghowr *see* Gōwr
147 T13 **Ghŭdara** *var.* Gudara, *Rus.* Kudara. SE Tajikistan 38°28′N 72°39′E
153 R13 **Ghugri** ≈ N India
147 S14 **Ghund** *Rus.* Gunt. ≈ SE Tajikistan
Ghurābīyah, Sha'īb al *see* Gharbīyah, Sha'īb al
Ghurdaqah *see* Al Ghurdaqah
Ghūriān *see* Ghōriān
141 T8 **Ghuwayfāt** *var.* Gheweifat. Abū Ẓaby, W United Arab Emirates 24°06′N 51°40′E
121 O14 **Ghuzayyil, Sabkhat** *salt lake* N Libya
126 J3 **Ghzatsk** Smolenskaya Oblast', W Russian Federation 55°33′N 35°00′E
115 G17 **Giáltra** Évvoia, C Greece 38°21′N 22°58′E
Giamame *see* Jamaame
167 U13 **Gia Nghia** *var.* Đak Nông. Đăc Lăc, S Vietnam 11°58′N 107°42′E
114 F13 **Giannitsá** *var.* Yiannitsá. Kentrikí Makedonía, N Greece 40°49′N 22°24′E
107 F14 **Giannutri, Isola di** *island* Archipelago Toscano, C Italy
96 F13 **Giant's Causeway** *Ir.* Clochán an Aifir. *lava flow* N Northern Ireland, United Kingdom
167 S15 **Gia Rai** Minh Hai, S Vietnam 09°14′N 105°28′E
107 L24 **Giarre** Sicilia, Italy, C Mediterranean Sea 37°44′N 15°12′E
44 I7 **Gibara** Holguín, E Cuba 21°09′N 76°11′W
29 O16 **Gibbon** Nebraska, C USA 40°45′N 98°50′W
32 K11 **Gibbon** Oregon, NW USA 45°40′N 118°22′W
33 P11 **Gibbonsville** Idaho, NW USA 45°33′N 113°55′W
64 A13 **Gibbs Hill** *hill* S Bermuda
92 I9 **Gibostad** Troms, N Norway 69°21′N 18°01′E
104 I14 **Gibraleón** Andalucía, S Spain 37°23′N 06°58′W
104 L16 **Gibraltar** ○ (Gibraltar) S Gibraltar 36°08′N 05°21′W
104 L16 **Gibraltar** ◇ *UK dependent territory* SW Europe
Gibraltar, Détroit de/Gibraltar, Estrecho de *see* Gibraltar, Strait of
104 J17 **Gibraltar, Strait of** *Fr.* Détroit de Gibraltar, *Sp.* Estrecho de Gibraltar. *strait* Atlantic Ocean/Mediterranean Sea
31 S11 **Gibsonburg** Ohio, N USA 41°22′N 83°19′W
30 M13 **Gibson City** Illinois, N USA 40°27′N 88°24′W
180 L8 **Gibson Desert** *desert* Western Australia
10 L17 **Gibsons** British Columbia, SW Canada 49°24′N 123°32′W
149 N12 **Gīdār** Baluchistān, SW Pakistan 28°16′N 66°00′E
155 I17 **Giddalūr** Andhra Pradesh, E India 15°24′N 78°54′E
25 U10 **Giddings** Texas, SW USA 30°12′N 96°59′W
27 Y8 **Gideon** Missouri, C USA 36°27′N 89°55′W
81 I15 **Gīdolē** Southern Nationalities, S Ethiopia 05°31′N 37°26′E
Giebnegáisi *see* Kebnekaise
118 H13 **Giedraičiai** Utena, E Lithuania 55°05′N 25°16′E
103 O7 **Gien** Loiret, C France 47°40′N 02°37′E
101 G17 **Giessen** Hessen, W Germany 50°35′N 08°41′E
98 O6 **Gieten** Drenthe, NE Netherlands 53°00′N 06°43′E
23 Y13 **Gifford** Florida, SE USA 27°40′N 80°24′W
9 O5 **Gifford** ≈ Baffin Island, Nunavut, NE Canada
100 J12 **Gifhorn** Niedersachsen, N Germany 52°28′N 10°33′E
11 P13 **Gift Lake** Alberta, W Canada 55°49′N 115°57′W
Giftokastro *see* Tsigansko Gradishte
164 L13 **Gifu** *var.* Gihu. Gifu, Honshū, SW Japan 35°24′N 136°46′E
164 K13 **Gifu** *off.* Gifu-ken, *var.* Gihu. ⬖ *prefecture* Honshū, SW Japan
Gifu-ken *see* Gifu
126 M13 **Gigant** Rostovskaya Oblast', SW Russian Federation 46°29′N 41°18′E
40 E8 **Giganta, Sierra de la** ▲▲ NW Mexico
54 E12 **Gigante** Huila, S Colombia 02°24′N 75°34′W
114 I7 **Gigen** Pleven, N Bulgaria 43°40′N 24°31′E
Giggiga *see* Jijiga
96 G12 **Gigha Island** *island* SW Scotland, United Kingdom
107 E14 **Giglio, Isola del** *island* Archipelago Toscano, C Italy
Gihu *see* Gifu
146 L11 **G'ijduvon** *Rus.* Gizhduvon. Buxoro Viloyati, C Uzbekistan 40°06′N 64°38′E
104 L2 **Gijón** *var.* Xixón. Asturias, NW Spain 43°32′N 05°40′W
81 D20 **Gikongoro** SW Rwanda 02°30′S 29°32′E
36 K14 **Gila Bend** Arizona, SW USA 32°57′N 112°43′W
36 J14 **Gila Bend Mountains** ▲▲ Arizona, SW USA
36 I15 **Gila Mountains** ▲▲ Arizona, SW USA
37 N14 **Gila Mountains** ▲▲ Arizona, SW USA
142 M4 **Gīlān** *off.* Ostān-e Gīlān, *var.* Ghilan, Guilan. ⬖ *province* NW Iran
Gīlān, Ostān-e *see* Gīlān
36 L14 **Gila River** ≈ Arizona, SW USA
29 W4 **Gilbert** Minnesota, N USA 47°29′N 92°27′W
Gilbert Islands *see* Tungaru
10 L16 **Gilbert, Mount** ▲ British Columbia, SW Canada 50°49′N 124°03′W
181 U4 **Gilbert River** ≈ Queensland, NE Australia
0 C6 **Gilbert Seamounts** *undersea feature* NE Pacific Ocean 52°50′N 150°10′W
33 S7 **Gildford** Montana, NW USA 48°34′N 110°21′W
83 P15 **Gilé** Zambézia, NE Mozambique 16°10′S 28°17′E
30 K4 **Gile Flowage** ◎ Wisconsin, N USA
182 G7 **Giles, Lake** *salt lake* South Australia
Gilf Kebir Plateau *see* Haḍabat al Jilf al Kabīr
183 R6 **Gilgandra** New South Wales, SE Australia 31°43′S 148°39′E
Gilgău *see* Gâlgău
81 I19 **Gilgil** Nakuru, SW Kenya 0°29′S 36°19′E
183 S4 **Gil Gil Creek** ≈ New South Wales, SE Australia
149 V3 **Gilgit** Jammu and Kashmir, NE Pakistan 35°54′N 74°20′E
149 V3 **Gilgit** ≈ N Pakistan
11 X11 **Gillam** Manitoba, C Canada 56°25′N 94°45′W
95 J22 **Gilleleje** Hovedstaden, E Denmark 56°05′N 12°17′E
30 K14 **Gillespie** Illinois, N USA 39°07′N 89°49′W
27 W13 **Gillett** Arkansas, C USA 34°07′N 91°22′W
33 X12 **Gillette** Wyoming, C USA 44°17′N 105°30′W
97 P22 **Gillingham** SE England, United Kingdom 51°24′N 00°33′E
195 X6 **Gillock Island** *island* Antarctica
173 O16 **Gillot** ✈ (St-Denis) N Réunion 20°53′S 55°31′E
65 H25 **Gill Point** *headland* E Saint Helena 15°59′S 05°38′W
30 M12 **Gilman** Illinois, N USA 40°44′N 87°58′W
25 W6 **Gilmer** Texas, SW USA 32°44′N 94°58′W
Gilolo *see* Halmahera, Pulau
81 G14 **Gīlo Wenz** ≈ SW Ethiopia
35 O10 **Gilroy** California, W USA 37°00′N 121°34′W
123 Q12 **Gilyuy** ≈ SE Russian Federation
99 I14 **Gilze** Noord-Brabant, S Netherlands 51°33′N 04°56′E
165 R16 **Gima** Okinawa, Kume-jima, SW Japan
80 H13 **Gīmbī** *It.* Ghimbi. Oromīya, C Ethiopia 09°13′N 35°39′E
163 Y15 **Gimcheon** *prev.* Kimch'ŏn. C South Korea 36°08′N 128°06′E
163 Z16 **Gimhae** *prev.* Kim Hae. ✈ (Busan) SE South Korea 35°10′N 128°57′E
45 T12 **Gimie, Mount** ▲ C Saint Lucia 13°51′N 61°00′W
11 X16 **Gimli** Manitoba, S Canada 50°39′N 97°00′W
Gimma *see* Jīma
95 O14 **Gimo** Uppsala, C Sweden 60°11′N 18°12′E
102 L15 **Gimone** ≈ S France
Gimpoe *see* Gimpu
171 N12 **Gimpu** *prev.* Gimpoe. Sulawesi, C Indonesia 01°38′S 120°00′E
182 F5 **Gina** South Australia 29°56′S 134°33′E
Ginevra *see* Genève
99 J19 **Gingelom** Limburg, NE Belgium 50°46′N 05°09′E
180 I12 **Gingin** Western Australia 31°22′S 115°51′E
171 Q7 **Gingoog** Mindanao, S Philippines 08°47′N 125°05′E
81 K14 **Gīnīr** Oromīya, C Ethiopia 07°12′N 40°43′E
Giohar *see* Jawhar
107 O17 **Gióia del Colle** Puglia, SE Italy 40°47′N 16°56′E
107 M22 **Gioia, Golfo di** *gulf* S Italy
Giona *see* Gkióna
115 I16 **Gioúra** *island* Vóreies Sporádes, Greece, Aegean Sea
107 O17 **Giovinazzo** Puglia, SE Italy 41°11′N 16°40′E
Gipeswic *see* Ipswich
105 P3 **Gipuzkoa** *Cast.* guipuzcoa. ⬖ *province* País Vasco, N Spain
Giran *see* Ilan
30 K14 **Girard** Illinois, N USA 39°27′N 89°46′W
27 R7 **Girard** Kansas, C USA 37°30′N 94°50′W
25 O6 **Girard** Texas, SW USA 33°18′N 100°38′W
54 E10 **Girardot** Cundinamarca, C Colombia 04°19′N 74°47′W
172 M7 **Giraud Seamount** *undersea feature* SW Indian Ocean 09°57′S 46°55′E
83 A15 **Giraul** ≈ SW Angola
96 L9 **Girdle Ness** *headland* NE Scotland, United Kingdom 57°09′N 02°04′W
137 N11 **Giresun** *var.* Kerasunt; *anc.* Cerasus, Pharnacia. Giresun, NE Turkey 40°55′N 38°35′E
137 N12 **Giresun** *var.* Kerasunt. ⬖ *province* NE Turkey
137 N12 **Giresun Dağları** ▲▲ N Turkey
Girga *see* Jirjā
Girgeh *see* Jirjā
Girgenti *see* Agrigento
153 Q15 **Girīdīh** Jhārkhand, NE India 24°10′N 86°20′E
183 P6 **Girilambone** New South Wales, SE Australia 31°19′S 146°57′E
Girin *see* Jilin
121 W10 **Girne** *Gk.* Kerýneia, Kyrenia. N Cyprus 35°20′N 33°20′E
Giron *see* Kiruna
105 X5 **Girona** *var.* Gerona; *anc.* Gerunda. Cataluña, NE Spain 41°59′N 02°49′E
105 W5 **Girona** *var.* Gerona. ⬖ *province* Cataluña, NE Spain
102 J12 **Gironde** ⬖ *department* SW France
102 J11 **Gironde** *estuary* SW France
105 V5 **Gironella** Cataluña, NE Spain 42°02′N 01°53′E
103 N15 **Girou** ≈ S France
97 H14 **Girvan** W Scotland, United Kingdom 55°14′N 04°53′W
24 M9 **Girvin** Texas, SW USA 31°05′N 102°24′W
184 Q9 **Gisborne** Gisborne, North Island, New Zealand 38°41′S 178°01′E
184 P9 **Gisborne** *off.* Gisborne District. ⬖ *unitary authority* North Island, New Zealand
Gisborne District *see* Gisborne
Giseifu *see* Uijeongbu
Gisenye *see* Gisenyi
81 D19 **Gisenyi** *var.* Gisenye. NW Rwanda 01°42′S 29°18′E
95 K20 **Gislaved** Jönköping, S Sweden 57°19′N 13°30′E
103 N4 **Gisors** Eure, N France 49°18′N 01°46′E
Gissar *see* Hisor
147 P12 **Gissar Range** *Rus.* Gissarskiy Khrebet. ▲▲ Tajikistan/Uzbekistan
Gissarskiy Khrebet *see* Gissar Range
99 B16 **Gistel** West-Vlaanderen, W Belgium 51°09′N 02°58′E
108 F9 **Giswil** Obwalden, C Switzerland 46°49′N 08°11′E
115 B16 **Gitánes** *ancient monument* Ípeiros, W Greece
81 E20 **Gitarama** C Rwanda 02°05′S 29°45′E
81 E20 **Gitega** C Burundi 03°20′S 29°56′E
Githio *see* Gýtheio
108 H11 **Giubiasco** Ticino, S Switzerland 46°11′N 09°01′E
106 K13 **Giulianova** Abruzzi, C Italy 42°45′N 13°58′E
Giulie, Alpi *see* Julian Alps
Giumri *see* Gyumri
116 M13 **Giurgeni** Ialomița, SE Romania 44°45′N 27°48′E
116 J15 **Giurgiu** Giurgiu, S Romania 43°54′N 25°58′E
116 J14 **Giurgiu** ⬖ *county* SE Romania
95 F22 **Give** Syddanmark, C Denmark 55°51′N 09°15′E
103 R2 **Givet** Ardennes, N France 50°08′N 04°50′E
103 R11 **Givors** Rhône, E France 45°35′N 04°47′E
83 K19 **Giyani** Limpopo, NE South Africa 23°20′S 30°37′E
80 I13 **Giyon** Oromīya, C Ethiopia 08°31′N 37°56′E
75 W8 **Giza** *var.* Al Jīzah, El Gîza, Gizeh. N Egypt 30°01′N 31°13′E
75 V8 **Giza, Pyramids of** *ancient monument* N Egypt
Gizhduvon *see* G'ijduvon
123 U8 **Gizhiga** Magadanskaya Oblast', E Russian Federation 61°58′N 160°16′E
123 T9 **Gizhiginskaya Guba** *bay* E Russian Federation
186 K8 **Gizo** Gizo, NW Solomon Islands 08°03′S 156°49′E
110 N7 **Giżycko** *Ger.* Lötzen. Warmińsko-Mazurskie, NE Poland 54°03′N 21°48′E
Gízymałów *see* Hrymayliv
113 M17 **Gjakovë** *Serb.* Đakovica. W Kosovo 42°22′N 20°30′E
94 F12 **Gjende** ◎ S Norway
113 L16 **Gjeravicë** *Serb.* Đeravica. ▲ S Serbia 42°33′N 20°08′E
95 F17 **Gjerstad** Aust-Agder, S Norway 58°54′N 09°03′E
113 O17 **Gjilan** *Serb.* Gnjilane. E Kosovo 42°27′N 21°28′E
Gjinokastër *see* Gjirokastër
113 L23 **Gjirokastër** *var.* Gjirokastra; *prev.* Gjinokastër, *Gk.* Argyrokastron, *It.* Argirocastro. Gjirokastër, S Albania 40°04′N 20°09′E
113 L22 **Gjirokastër** ⬖ *district* S Albania
Gjirokastra *see* Gjirokastër
9 N7 **Gjoa Haven** *var.* Uqsuqtuuq. King William Island, Nunavut, NW Canada 68°38′N 95°57′W
94 H13 **Gjøvik** Oppland, S Norway 60°47′N 10°40′E
113 J22 **Gjuhëzës, Kepi i** *headland* SW Albania 40°25′N 19°19′E
Gjurgjevac *see* Đurđevac
115 E18 **Gkióna** *var.* Giona. ▲▲ C Greece
121 R3 **Gkréko, Akrotíri** *var.* Cape Greco, Pidálion. *cape* E Cyprus
99 I18 **Glabbeek-Zuurbemde** Vlaams Brabant, C Belgium 50°54′N 04°58′E
13 R14 **Glace Bay** Cape Breton Island, Nova Scotia, SE Canada 46°12′N 59°57′W
11 O16 **Glacier** British Columbia, SW Canada 51°12′N 117°33′W
39 W12 **Glacier Bay** *inlet* Alaska, USA
32 I7 **Glacier Peak** ▲ Washington, NW USA 48°06′N 121°06′W
159 N13 **Gladaindong Feng** ▲ C China 33°24′N 91°00′E
21 Q7 **Glade Spring** Virginia, NE USA 36°47′N 81°46′W
43 W7 **Gladewater** Texas, SW USA 32°32′N 94°57′W
181 Y8 **Gladstone** Queensland, E Australia 23°52′S 151°16′E
182 I8 **Gladstone** South Australia 33°16′S 138°21′E
11 X16 **Gladstone** Manitoba, S Canada 50°12′N 98°56′W
31 O5 **Gladstone** Michigan, N USA 45°51′N 87°01′W
27 R4 **Gladstone** Missouri, C USA 39°12′N 94°33′W
31 Q7 **Gladwin** Michigan, N USA 43°58′N 84°29′W
95 J15 **Glafsfjorden** ◎ C Sweden
92 H2 **Gláma** *physical region* NW Iceland
94 I12 **Gláma** *var.* Glommen. ≈ S Norway

◆ Country
● Country Capital
◇ Dependent Territory
○ Dependent Territory Capital
⬖ Administrative Regions
✈ International Airport
▲ Mountain
▲▲ Mountain Range
Volcano
≈ River
◎ Lake
⊠ Reservoir

GLAMOČ – GOURIN

112 F13 **Glamoč** Federacija Bosne I Hercegovine, NE Boznia and Herzegovina 44°01′N 16°51′E
97 J22 **Glamorgan** *cultural region* S Wales, United Kingdom
95 G24 **Glamsbjerg** Syddtjylland, C Denmark 55°17′N 10°07′E
171 Q8 **Glan** Mindanao, S Philippines 05°49′N 125°11′E
109 T9 **Glan** ≈ SE Austria
101 F19 **Glan** ≈ W Germany
95 M17 **Glan** ◎ S Sweden
Glaris *see* Glarus
108 H9 **Glarner Alpen** *Eng.* Glarus Alps. ▲▲ E Switzerland
108 H8 **Glarus** Glarus, E Switzerland 47°03′N 09°04′E
108 H9 **Glarus** *Fr.* Glaris. ◈ *canton* C Switzerland
Glarus Alps *see* Glarner Alpen
27 N3 **Glasco** Kansas, C USA 39°21′N 97°50′W
96 I12 **Glasgow** S Scotland, United Kingdom 55°53′N 04°15′W
20 K7 **Glasgow** Kentucky, S USA 37°00′N 85°54′W
27 T4 **Glasgow** Missouri, C USA 39°13′N 92°51′W
33 W7 **Glasgow** Montana, NW USA 48°12′N 106°37′W
21 T6 **Glasgow** Virginia, NE USA 37°37′N 79°27′W
96 I12 **Glasgow** ✈ W Scotland, United Kingdom 55°52′N 04°27′W
11 S14 **Glaslyn** Saskatchewan, S Canada 53°20′N 108°18′W
18 I16 **Glassboro** New Jersey, NE USA 39°40′N 75°05′W
24 L10 **Glass Mountains** ▲▲ Texas, SW USA
97 K23 **Glastonbury** SW England, United Kingdom 51°09′N 02°43′W
Glatz *see* Kłodzko
101 N16 **Glauchau** Sachsen, E Germany 50°48′N 12°32′E
Glavn'a Morava *see* Velika Morava
Glavnik *see* Gllamnik
127 T1 **Glazov** Udmurtskaya Respublika, NW Russian Federation 58°06′N 52°38′E
Głda *see* Gwda
109 U8 **Gleinalpe** ▲▲ SE Austria
109 W8 **Gleisdorf** Steiermark, SE Austria 47°07′N 15°43′E
Gleiwitz *see* Gliwice
39 S11 **Glenallen** Alaska, USA 62°06′N 145°33′W
102 F7 **Glénan, Îles** *island group* NW France
185 G21 **Glenavy** Canterbury, South Island, New Zealand 44°53′S 171°04′E
10 H5 **Glenboyle** Yukon, NW Canada 63°55′N 138°43′W
21 X3 **Glen Burnie** Maryland, NE USA 39°09′N 76°37′W
36 L8 **Glen Canyon** *canyon* Utah, W USA
36 L8 **Glen Canyon Dam** *dam* Arizona, SW USA
30 K15 **Glen Carbon** Illinois, N USA 38°45′N 89°58′W
14 E17 **Glencoe** Ontario, S Canada 42°44′N 81°42′W
83 K22 **Glencoe** KwaZulu/Natal, E South Africa 28°10′S 30°15′E
29 U9 **Glencoe** Minnesota, N USA 44°46′N 94°09′W
96 H10 **Glen Coe** *valley* N Scotland, United Kingdom
36 K13 **Glendale** Arizona, SW USA 33°32′N 112°11′W
35 S15 **Glendale** California, W USA 34°09′N 118°20′W
182 G5 **Glendambo** South Australia 30°59′S 135°45′E
33 Y8 **Glendive** Montana, NW USA 47°08′N 104°42′W
33 Y15 **Glendo** Wyoming, C USA 42°27′N 105°01′W
55 S10 **Glendor Mountains** C Guyana
182 K12 **Glenelg River** ≈ South Australia/Victoria, SE Australia
29 P4 **Glenfield** North Dakota, N USA 47°25′N 98°33′W
25 V12 **Glen Flora** Texas, SW USA 29°22′N 96°12′W
181 P7 **Glen Helen** Northern Territory, N Australia 23°45′S 132°46′E
183 U5 **Glen Innes** New South Wales, SE Australia 29°42′S 151°45′E
31 P6 **Glen Lake** ◎ Michigan, N USA
10 I7 **Glenlyon Peak** ▲ Yukon, W Canada 62°32′N 134°51′W
37 N16 **Glenn, Mount** ▲ Arizona, SW USA 31°55′N 110°00′W
33 N15 **Glenns Ferry** Idaho, NW USA 42°57′N 115°18′W
23 W6 **Glennville** Georgia, SE USA 31°56′N 81°55′W
10 J10 **Glenora** British Columbia, W Canada 57°52′N 131°16′W
182 M11 **Glenorchy** Victoria, SE Australia 36°56′S 142°39′E
183 V5 **Glenreagh** New South Wales, SE Australia 30°04′S 153°00′E
33 X15 **Glenrock** Wyoming, C USA 42°51′N 105°52′W
96 K11 **Glenrothes** E Scotland, United Kingdom 56°11′N 03°09′W
18 L9 **Glens Falls** New York, NE USA 43°18′N 73°38′W
97 D14 **Glenties** *Ir.* Na Gleannta. Donegal, NW Ireland 54°47′N 08°17′W
28 L5 **Glen Ullin** North Dakota, N USA 46°49′N 101°49′W
21 R4 **Glenville** West Virginia, NE USA 38°57′N 80°51′W
27 T12 **Glenwood** Arkansas, C USA 34°19′N 93°33′W
29 S15 **Glenwood** Iowa, C USA 41°03′N 95°44′W
29 T7 **Glenwood** Minnesota, N USA 45°39′N 95°23′W
36 L5 **Glenwood** Utah, W USA 38°45′N 111°59′W
30 I5 **Glenwood City** Wisconsin, N USA 45°04′N 92°11′W
37 Q4 **Glenwood Springs** Colorado, C USA 39°33′N 107°21′W
108 F10 **Gletsch** Valais, S Switzerland 46°34′N 08°21′E
Glevum *see* Gloucester
29 U14 **Glidden** Iowa, C USA 42°03′N 94°43′W
112 E9 **Glina** *var.* Banijska Palanka. Sisak-Moslavina, NE Croatia 45°19′N 16°07′E
94 F11 **Glittertind** ▲ S Norway 61°24′N 08°19′E
111 J16 **Gliwice** *Ger.* Gleiwitz. Śląskie, S Poland 50°19′N 18°49′E
113 N16 **Gllamnik** *Serb.* Glavnik. N Kosovo 42°53′N 21°10′E
36 M14 **Globe** Arizona, SW USA 33°24′N 110°47′W
Globino *see* Hlobyne
108 L9 **Glockturm** ▲ SW Austria 46°54′N 10°42′E
116 L9 **Glodeni** *Rus.* Glodyany. N Moldova 47°47′N 27°33′E
109 S9 **Glödnitz** Kärnten, S Austria 46°57′N 14°03′E
Glodyany *see* Glodeni
Glogau *see* Głogów
109 W6 **Gloggnitz** Niederösterreich, E Austria 47°41′N 15°57′E
110 F13 **Głogów** *Ger.* Glogau, Glogow. Dolnośląskie, SW Poland 51°40′N 16°04′E
Glogow *see* Głogów
111 I16 **Głogówek** *Ger.* Oberglogau. Opolskie, S Poland 50°21′N 17°51′E
92 G12 **Glomfjord** Nordland, C Norway 66°49′N 14°00′E
Glomma *see* Gláma
Glommen *see* Gláma
93 I14 **Glommersträsk** Norrbotten, N Sweden 65°17′N 19°40′E
172 I1 **Glorieuses, Îles** *Eng.* Glorioso Islands. *island* (to France) N Madagascar
Glorioso Islands *see* Glorieuses, Îles
65 C25 **Glorious Hill** *hill* East Falkland, Falkland Islands
38 J12 **Glory of Russia Cape** *headland* Saint Matthew Island, Alaska, USA 60°36′N 172°57′W
22 J7 **Gloster** Mississippi, S USA 31°12′N 91°01′W
183 U7 **Gloucester** New South Wales, SE Australia 32°01′S 152°00′E
186 F7 **Gloucester** New Britain, E Papua New Guinea 05°30′S 148°30′E
97 L21 **Gloucester** *hist.* Caer Glou, *Lat.* Glevum. C England, United Kingdom 51°53′N 02°14′W
19 P10 **Gloucester** Massachusetts, NE USA 42°36′N 70°36′W
21 X6 **Gloucester** Virginia, NE USA 37°26′N 76°33′W
97 K21 **Gloucestershire** *cultural region* C England, United Kingdom
31 T14 **Glouster** Ohio, N USA 39°30′N 82°04′W
42 H3 **Glovers Reef** *reef* E Belize
18 K10 **Gloversville** New York, NE USA 43°03′N 74°20′W
110 K12 **Głowno** Łódź, C Poland 51°58′N 19°43′E
111 H16 **Głubczyce** *Ger.* Leobschütz. Opolskie, S Poland 50°13′N 17°50′E
126 L11 **Glubokiy** Rostovskaya Oblast', SW Russian Federation 48°34′N 40°16′E
145 W9 **Glubokoye** Vostochnyy Kazakhstan, E Kazakhstan 50°08′N 82°16′E
Glubokoye *see* Hlybokaye
111 H16 **Głuchołazy** *Ger.* Ziegenhais. Opolskie, S Poland 50°20′N 17°23′E
100 I9 **Glückstadt** Schleswig-Holstein, N Germany 53°47′N 09°26′E
Glukhov *see* Hlukhiv
Glushkevichi *see* Hlushkavichy
Glusk/Glussk *see* Hlusk
Glybokaya *see* Hlyboka
95 F21 **Glyngøre** Midtjylland, NW Denmark 56°45′N 08°55′E
127 Q9 **Gmelinka** Volgogradskaya Oblast', SW Russian Federation 50°50′N 46°51′E
109 R8 **Gmünd** Kärnten, S Austria 46°56′N 13°32′E
109 U2 **Gmünd** Niederösterreich, N Austria 48°47′N 14°59′E
Gmünd *see* Schwäbisch Gmünd
109 S5 **Gmunden** Oberösterreich, N Austria 47°56′N 13°48′E
Gmundner See *see* Traunsee
94 N10 **Gnarp** Gävleborg, C Sweden 62°03′N 17°19′E
109 W8 **Gnas** Steiermark, SE Austria 46°53′N 15°48′E
Gnaviyani *see* Fuammulah
Gnesen *see* Gniezno
95 O16 **Gnesta** Södermanland, C Sweden 59°05′N 17°20′E
110 H11 **Gniezno** *Ger.* Gnesen. Weilkopolskie, C Poland 52°33′N 17°35′E
Gnjilane *see* Gjilan
95 K20 **Gnosjö** Jönköping, S Sweden 57°22′N 13°44′E
155 E17 **Goa** *prev.* Old Goa, Vela Goa, Velha Goa. Goa, W India 15°31′N 73°56′E
155 E17 **Goa** *var.* Old Goa. ◈ *state* W India
Goabddális *see* Kåbdalis
42 H7 **Goascorán, Río** ≈ El Salvador/Honduras
77 O16 **Goaso** *var.* Gawso. W Ghana 06°49′N 02°27′W
81 K14 **Goba** Oromīya, C Ethiopia 07°02′N 39°58′E
83 C20 **Gobabed** Erongo, W Namibia 23°36′S 15°03′E
83 E19 **Gobabis** Omaheke, E Namibia 22°25′S 18°58′E
Gobannium *see* Abergavenny
64 M7 **Goban Spur** *undersea feature* NW Atlantic Ocean
63 H21 **Gobernador Gregores** Santa Cruz, S Argentina 48°43′S 70°21′W
61 F14 **Gobernador Ingeniero Virasoro** Corrientes, NE Argentina 28°06′S 56°00′W
162 L12 **Gobi** *desert* China/Mongolia
164 I14 **Gobō** Wakayama, Honshū, SW Japan 33°52′N 135°09′E
Gobolka Awdal *see* Awdal
Gobolka Sool *see* Sool
101 D14 **Goch** Nordrhein-Westfalen, W Germany 51°41′N 06°10′E
83 E20 **Gochas** Hardap, S Namibia 24°54′S 18°43′E
155 I14 **Godāvari** *var.* Godavari. ≈ C India
Godavari *see* Godāvari
155 L16 **Godāvari, Mouths of the** *delta* E India
15 V5 **Godbout** Québec, SE Canada 49°19′N 67°37′W
15 U5 **Godbout** ≈ Québec, SE Canada
15 U5 **Godbout Est** ≈ Québec, SE Canada
27 N6 **Goddard** Kansas, C USA 37°39′N 97°34′W
14 E15 **Goderich** Ontario, S Canada 43°43′N 81°43′W
Godhavn *see* Qeqertarsuaq
154 E10 **Godhra** Gujarāt, W India 22°49′N 73°40′E
Göding *see* Hodonín
111 K22 **Gödöllő** Pest, N Hungary 47°36′N 19°20′E
62 H11 **Godoy Cruz** Mendoza, W Argentina 32°59′S 68°49′W
11 Y11 **Gods** ≈ Manitoba, C Canada
11 X13 **Gods Lake** ◎ Manitoba, C Canada
11 Y13 **Gods Lake Narrows** Manitoba, C Canada 54°29′N 94°21′W
Godthaab/Godthåb *see* Nuuk
Godwin Austen, Mount *see* K2
Goede Hoop, Kaap de *see* Good Hope, Cape of
Goedgegun *see* Nhlangano
Goeie Hoop, Kaap die *see* Good Hope, Cape of
13 O7 **Goélands, Lac aux** ◎ Québec, SE Canada
98 E13 **Goeree** *island* SW Netherlands
99 F15 **Goes** Zeeland, SW Netherlands 51°30′N 03°55′E
Goettingen *see* Göttingen
19 O10 **Goffstown** New Hampshire, NE USA 43°01′N 71°34′W
14 E8 **Gogama** Ontario, S Canada 47°42′N 81°44′W
30 L3 **Gogebic, Lake** ◎ Michigan, N USA
30 K3 **Gogebic Range** *hill range* Michigan/Wisconsin, N USA
Gogi Lerr *see* Gogi, Mount
137 V13 **Gogi, Mount** *Arm.* Gogi Lerr, *Az.* Küküdağ. ▲ Armenia/Azerbaijan 39°33′N 45°35′E
124 F12 **Gogland, Ostrov** *island* NW Russian Federation
111 I15 **Gogolin** Opolskie, S Poland 50°28′N 18°04′E
Gogonou *see* Gogounou
77 S14 **Gogounou** *var.* Gogonou. N Benin 10°50′N 02°50′E
152 I10 **Gohāna** Haryāna, N India 29°06′N 76°43′E
59 K18 **Goianésia** Goiás, C Brazil 15°21′S 49°02′W
59 K18 **Goiânia** *prev.* Goyania. *state capital* Goiás, C Brazil 16°43′S 49°18′W
59 K18 **Goiás** Goiás, C Brazil 15°57′S 50°07′W
59 J18 **Goiás** *off.* Estado de Goiás; *prev.* Goiaz, Goyaz. ◈ *state* C Brazil
Goiás, Estado de *see* Goiás
Goiaz *see* Goiás
Goidhoo Atoll *see* Horsburgh Atoll
159 R14 **Goinsargoin** Xizang Zizhiqu, W China 31°56′N 98°04′E
60 H10 **Goio-Erê** Paraná, SW Brazil 24°08′S 53°07′W
99 I15 **Goirle** Noord-Brabant, S Netherlands 51°31′N 05°04′E
104 H8 **Góis** Coimbra, N Portugal 40°10′N 08°06′W
165 Q8 **Gojōme** Akita, Honshū, NW Japan 39°55′N 140°07′E
149 U9 **Gojra** Punjab, E Pakistan 31°10′N 72°43′E
136 A11 **Gökçeada** *var.* Imroz Adasī, *Gk.* Imbros. *island* NW Turkey
Gökçeada *see* İmroz
146 F13 **Gökdepe** *Rus.* Gëkdepe, Geok-Tepe. Ahal Welaýaty, C Turkmenistan 38°05′N 58°08′E
136 I10 **Gökırmak** ≈ N Turkey
Goklenkuy, Solonchak *see* Geklengkui, Solonchak
136 C16 **Gökova Körfezi** *gulf* SW Turkey
136 K15 **Göksu** ≈ S Turkey
136 L15 **Göksun** Kahramanmaraş, C Turkey 38°03′N 36°30′E
136 I17 **Göksu Nehri** ≈ S Turkey
83 J16 **Gokwe** Midlands, NW Zimbabwe 18°13′S 28°55′E
94 F13 **Gol** Buskerud, S Norway 60°42′N 08°57′E
153 X12 **Golāghāt** Assam, NE India 26°31′N 93°54′E
110 H10 **Gołańcz** Wielkopolskie, C Poland 52°57′N 17°17′E
138 G8 **Golan Heights** *Ar.* Al Jawlān, *Heb.* HaGolan. ▲▲ SW Syria
Golārā *see* Ārān-va-Bidgol
Golaya Pristan *see* Hola Prystan'
143 T11 **Golbāf** Kermān, C Iran 29°51′N 57°44′E
136 M15 **Gölbaşı** Adıyaman, S Turkey 37°46′N 37°40′E
109 P9 **Gölbner** ▲ SW Austria 46°51′N 12°31′E
30 M17 **Golconda** Illinois, N USA 37°20′N 88°29′W
35 T3 **Golconda** Nevada, W USA 40°56′N 117°29′W
136 E11 **Gölcük** Kocaeli, NW Turkey 40°44′N 29°50′E
108 I7 **Goldach** Sankt Gallen, NE Switzerland 47°28′N 09°28′E
110 N7 **Gołdap** *Ger.* Goldap. Warmińsko-Mazurskie, NE Poland 54°19′N 22°23′E
32 E15 **Gold Beach** Oregon, NW USA 42°25′N 124°27′W
Goldberg *see* Złotoryja
183 V3 **Gold Coast** *cultural region* Queensland, E Australia
68 D12 **Gold Coast** *coastal region* S Ghana
39 R10 **Gold Creek** Alaska, USA 62°48′N 149°40′W
11 O16 **Golden** British Columbia, SW Canada 51°19′N 116°58′W
37 T4 **Golden** Colorado, C USA 39°40′N 105°12′W
184 I13 **Golden Bay** *bay* South Island, New Zealand
27 R7 **Golden City** Missouri, C USA 37°23′N 94°05′W
32 I11 **Goldendale** Washington, NW USA 45°49′N 120°49′W
Goldener Tisch *see* Zlatý Stôl
44 L13 **Golden Grove** E Jamaica 17°56′N 76°17′W
14 J12 **Golden Lake** ◎ Ontario, SE Canada
22 K10 **Golden Meadow** Louisiana, S USA 29°22′N 90°15′W
Golden Sands *see* Zlatni Pyasatsi
Golden State, The *see* California
83 K16 **Golden Valley** Mashonaland West, N Zimbabwe 18°11′S 29°50′E
35 U9 **Goldfield** Nevada, W USA 37°42′N 117°15′W
Goldingen *see* Kuldīga
Goldmarkt *see* Zlatna
10 K17 **Gold River** Vancouver Island, British Columbia, SW Canada 49°41′N 126°05′W
21 V10 **Goldsboro** North Carolina, SE USA 35°23′N 78°00′W
24 M8 **Goldsmith** Texas, SW USA 31°58′N 102°36′W
25 R8 **Goldthwaite** Texas, SW USA 31°28′N 98°35′W
137 R11 **Göle** Ardahan, NE Turkey 40°47′N 42°36′E
Golema Ada *see* Ostrovo
114 H9 **Golema Planina** ▲▲ W Bulgaria
114 F9 **Golemi Vrah** *var.* Golemi Vrŭkh. ▲ W Bulgaria 42°41′N 22°38′E
Golemi Vrŭkh *see* Golemi Vrah
110 D8 **Goleniów** *Ger.* Gollnow. Zachodnio-pomorskie, NW Poland 53°34′N 14°48′E
143 R3 **Golestān** *off.* Ostān-e Golestān. ◈ *province* N Iran
Golestān, Ostān-e *see* Golestān
35 Q14 **Goleta** California, W USA 34°26′N 119°50′W
43 O16 **Golfito** Puntarenas, SE Costa Rica 08°42′N 83°10′W
25 T13 **Goliad** Texas, SW USA 28°40′N 97°26′W
113 L14 **Golija** ▲▲ SW Serbia
Golinka *see* Gongbo'gyamda
113 O16 **Goljak** ▲▲ SE Serbia
136 M12 **Gölköy** Ordu, N Turkey 40°42′N 37°37′E
Gollel *see* Lavumisa
109 X3 **Göllersbach** ≈ NE Austria
Gollnow *see* Goleniów
Golmo *see* Golmud
159 P10 **Golmud** *var.* Ge'e'mu, Golmo, *Chin.* Ko-erh-mu. Qinghai, C China 36°23′N 94°56′E
103 Y14 **Golo** ≈ Corse, France, C Mediterranean Sea
Golovanevsk *see* Holovanivs'k
Golovchin *see* Halowchyn
39 N9 **Golovin** Alaska, USA 64°33′N 162°54′W
142 M7 **Golpāyegān** *var.* Gulpaigan. Eşfahān, W Iran 33°23′N 50°18′E
Golshan *see* Ţabas
Gol'shany *see* Hal'shany
96 J7 **Golspie** N Scotland, United Kingdom 57°59′N 03°56′W
112 O11 **Golubac** Serbia, NE Serbia 44°38′N 21°36′E
110 J9 **Golub-Dobrzyń** Kujawski-pomorskie, C Poland 53°07′N 19°03′E
145 S7 **Golubovka** Pavlodar, N Kazakhstan 53°07′N 74°11′E
82 B11 **Golungo Alto** Kwanza Norte, NW Angola 09°10′S 14°45′E
114 M8 **Golyama Kamchia** *var.* Golyama Kamchiya. ≈ E Bulgaria
Golyama Kamchiya *see* Golyama Kamchia
114 L8 **Golyama Reka** ≈ N Bulgaria
Golyama Syutka *see* Golyama Syutkya
114 H11 **Golyama Syutkya** *var.* Golyama Syutka. ▲ SW Bulgaria 41°55′N 24°03′E
114 I12 **Golyam Perelik** ▲ S Bulgaria 41°37′N 24°34′E
114 I11 **Golyam Persenk** ▲ S Bulgaria 41°50′N 24°33′E
79 P19 **Goma** Nord-Kivu, NE Dem. Rep. Congo 01°36′S 29°08′E
153 N13 **Gomati** *var.* Gumti. ≈ N India
77 X14 **Gombe** Gombe, E Nigeria 10°19′N 11°02′E
67 U10 **Gombe** *var.* Igombe. ≈ E Tanzania
77 Y14 **Gombi** Adamawa, E Nigeria 10°07′N 12°45′E
Gombroon *see* Bandar-e 'Abbās
Gomel' *see* Homyel'
Gomel'skaya Oblast' *see* Homyel'skaya Voblasts'
64 N11 **Gomera** *island* Islas Canarias, Spain, NE Atlantic Ocean
40 I5 **Gómez Farías** Chihuahua, N Mexico 29°25′N 107°46′W
40 L8 **Gómez Palacio** Durango, C Mexico 25°39′N 103°30′W
158 J13 **Gomo** Xizang Zizhiqu, W China 33°37′N 86°40′E
143 T6 **Gonābād** *var.* Gunabad. Khorāsān-e Razavī, NE Iran 36°30′N 59°18′E
44 L8 **Gonaïves** *var.* Les Gonaïves. N Haiti 19°26′N 72°41′W
123 Q12 **Gonam** ≈ NE Russian Federation
44 L9 **Gonâve, Canal de la** *var.* Canal de Sud. *channel* N Caribbean Sea
44 K9 **Gonâve, Golfe de la** *gulf* N Caribbean Sea
Gonāveh *see* Bandar-e Gonāveh
44 K9 **Gonâve, Île de la** *island* C Haiti
Gonbadān *see* Do Gonbadān
143 Q3 **Gonbad-e Kāvūs** *var.* Gunbad-i-Qawus. Golestán, N Iran 37°15′N 55°11′E
152 M12 **Gonda** Uttar Pradesh, N India 27°08′N 81°58′E
Gondar *see* Gonder
80 I11 **Gonder** *var.* Gondar. Āmara, NW Ethiopia 12°36′N 37°27′E
78 J13 **Gondey** Moyen-Chari, S Chad 09°07′N 19°10′E
154 J12 **Gondia** Mahārāshtra, C India 21°27′N 80°12′E
104 G6 **Gondomar** Porto, NW Portugal 41°10′N 08°35′W
136 C12 **Gönen** Balıkesir, W Turkey 40°06′N 27°39′E
136 C12 **Gönen Çayı** ≈ NW Turkey
159 O15 **Gongbo'gyamda** *var.* Golinka. Xizang Zizhiqu, W China 30°03′N 93°10′E
Gongchang *see* Longxi
159 N16 **Gonggar** *var.* Gyixong. Xizang Zizhiqu, W China 29°18′N 90°56′E
160 G9 **Gongga Shan** ▲ C China 29°50′N 101°55′E
159 T10 **Gonghe** *var.* Qabqa. Qinghai, C China 36°20′N 100°46′E
Gongjiang *see* Yudu
158 I5 **Gongliu** *var.* Tokkuztara. Xinjiang Uygur Zizhiqu, NW China 43°29′N 82°16′E
77 W14 **Gongola** ≈ E Nigeria
Gongoleh State *see* Jonglei
183 P5 **Gongolgon** New South Wales, SE Australia 30°19′S 146°57′E
159 Q6 **Gongpoquan** Gansu, N China 41°45′N 100°27′E
Gongquan *see* Gongxian
Gongtang *see* Damxung
160 I10 **Gongxian** *var.* Gongquan, Gong Xian. Sichuan, C China 28°25′N 104°51′E
Gong Xian *see* Gongxian
157 V10 **Gongzhuling** *prev.* Huaide. Jilin, NE China 43°30′N 124°48′E
159 S14 **Gonjo** Xizang Zizhiqu, W China 30°51′N 98°16′E
107 B20 **Gonnesa** Sardegna, Italy, C Mediterranean Sea 39°15′N 08°27′E
Gonni/Gónnos *see* Gónnoi
115 F15 **Gónnoi** *var.* Gonni, Gónnos; *prev.* Derelí. Thessalía, C Greece 39°52′N 22°27′E
Gōnoura *see* Iki
35 O11 **Gonzales** California, W USA 36°30′N 121°26′W
22 J9 **Gonzales** Louisiana, S USA 30°14′N 90°55′W
25 T12 **Gonzales** Texas, SW USA 29°31′N 97°29′W
41 P11 **González** Tamaulipas, C Mexico 22°50′N 98°25′W
21 V6 **Goochland** Virginia, NE USA 37°42′N 77°54′W
195 X14 **Goodenough, Cape** *headland* Antarctica 66°15′S 126°35′E
186 F9 **Goodenough Island** *var.* Morata. *island* SE Papua New Guinea
39 N8 **Goodhope Bay** *bay* Alaska, USA
83 D26 **Good Hope, Cape of** *Afr.* Kaap de Goede Hoop, Kaap die Goeie Hoop. *headland* SW South Africa 34°19′S 18°25′E
10 K10 **Good Hope Lake** British Columbia, W Canada 59°15′N 129°18′W
83 E23 **Goodhouse** Northern Cape, W South Africa 28°54′S 18°13′E
33 O15 **Gooding** Idaho, NW USA 42°56′N 114°42′W
26 H3 **Goodland** Kansas, C USA 39°20′N 101°43′W
173 Y15 **Goodlands** NW Mauritius 20°02′S 57°39′E
20 J8 **Goodlettsville** Tennessee, S USA 36°19′N 86°42′W
39 N13 **Goodnews** Alaska, USA 59°07′N 161°35′W
25 O3 **Goodnight** Texas, SW USA 35°00′N 101°07′W
183 Q4 **Goodooga** New South Wales, SE Australia 29°09′S 147°30′E
29 N4 **Goodrich** North Dakota, N USA 47°24′N 100°07′W
25 W10 **Goodrich** Texas, SW USA 30°36′N 94°57′W
29 X10 **Goodview** Minnesota, N USA 44°04′N 91°42′W
26 H8 **Goodwell** Oklahoma, C USA 36°36′N 101°38′W
97 N17 **Goole** E England, United Kingdom 53°43′N 00°46′W
183 O8 **Goolgowi** New South Wales, SE Australia 34°00′S 145°43′E
182 I10 **Goolwa** South Australia 35°31′S 138°43′E
181 Y11 **Goondiwindi** Queensland, E Australia 28°33′S 150°22′E
98 O11 **Goor** Overijssel, E Netherlands 52°13′N 06°33′E
Goose Bay *see* Happy Valley-Goose Bay
33 V13 **Gooseberry Creek** ≈ Wyoming, C USA
21 S14 **Goose Creek** South Carolina, SE USA 32°58′N 80°01′W
63 M23 **Goose Green** *var.* Prado del Ganso. East Falkland, Falkland Islands 51°52′S 59°W
16 D8 **Goose Lake** *var.* Lago dos Gansos. ◎ California/Oregon, W USA
29 Q4 **Goose River** ≈ North Dakota, N USA
153 T16 **Gopalganj** Dhaka, S Bangladesh 23°00′N 89°48′E
153 O12 **Gopālganj** Bihār, N India 26°28′N 84°26′E
Gopher State *see* Minnesota
101 I22 **Göppingen** Baden-Württemberg, SW Germany 48°42′N 09°39′E
110 G13 **Góra** *Ger.* Guhrau. Dolnośląskie, SW Poland 51°40′N 16°03′E
110 M12 **Góra Kalwaria** Mazowieckie, C Poland 52°00′N 21°14′E
153 O12 **Gorakhpur** Uttar Pradesh, N India 26°45′N 83°23′E
Gora Kyuren *see* Kürendag
Gorany *see* Harany
113 J14 **Goražde** Federacija Bosne I Hercegovine, SE Bosnia and Herzegovina 43°39′N 18°58′E
Gorbovichi *see* Harbavichy
Gorče Petrov *see* Đorče Petrov
0 E9 **Gorda Ridges** *undersea feature* NE Pacific Ocean 41°30′N 128°00′W
Gordiaz *see* Gardēz
78 K12 **Gordil** Vakaga, N Central African Republic 09°37′N 21°42′E
23 U5 **Gordon** Georgia, SE USA 32°52′N 83°19′W
28 K12 **Gordon** Nebraska, C USA 42°48′N 102°12′W
25 R7 **Gordon** Texas, SW USA 32°32′N 98°21′W
28 L13 **Gordon Creek** ≈ Nebraska, C USA
63 I25 **Gordon, Isla** *island* S Chile
183 O17 **Gordon, Lake** ◎ Tasmania, SE Australia
183 O17 **Gordon River** ≈ Tasmania, SE Australia
21 V5 **Gordonsville** Virginia, NE USA 38°08′N 78°11′W
78 H13 **Goré** Logone-Oriental, S Chad 07°55′N 16°38′E
80 H13 **Gorē** Oromīya, C Ethiopia 08°08′N 35°33′E
185 D24 **Gore** Southland, South Island, New Zealand 46°06′S 168°58′E
14 D11 **Gore Bay** Manitoulin Island, Ontario, S Canada 45°54′N 82°28′W
25 Q5 **Goree** Texas, SW USA 33°28′N 99°31′W
137 O11 **Görele** Giresun, NE Turkey 41°00′N 39°00′E
19 N6 **Gore Mountain** ▲ Vermont, NE USA 44°55′N 71°47′W
39 R13 **Gore Point** *headland* Alaska, USA 59°12′N 150°57′W
37 R4 **Gore Range** ▲▲ Colorado, C USA
97 F19 **Gorey** *Ir.* Guaire. Wexford, SE Ireland 52°40′N 06°18′W
143 R12 **Gorgāb** Kermān, S Iran
143 Q4 **Gorgān** *var.* Astarabad, Astrabad, Gurgan, *prev.* Asterābād; *anc.* Hyrcania. Golestán, N Iran 36°53′N 54°28′E
143 Q4 **Gorgān, Rūd-e** ≈ N Iran
76 I10 **Gorgol** ◈ *region* S Mauritania
106 D12 **Gorgona, Isola di** *island* Archipelago Toscano, C Italy
19 P8 **Gorham** Maine, NE USA 43°41′N 70°27′W
137 T10 **Gori** C Georgia 42°00′N 44°07′E
98 I13 **Gorinchem** *var.* Gorkum. Zuid-Holland, C Netherlands 51°50′N 04°59′E
137 V13 **Goris** SE Armenia 39°31′N 46°20′E
124 K16 **Goritsy** Tverskaya Oblast', W Russian Federation 57°09′N 36°44′E
106 J7 **Gorizia** *Ger.* Görz. Friuli-Venezia Giulia, NE Italy 45°57′N 13°37′E
116 G13 **Gorj** ◈ *county* SW Romania
109 W12 **Gorjanci** *var.* Uskočke Planine, Žumberak, Žumberačko Gorje, *Ger.* Uskokengebirge; *prev.* Sichelburger Gerbirge. ▲ Croatia/Slovenia Europe *see also* Žumberačko Gorje
Görkau *see* Jirkov
Gorki *see* Horki
Gor'kiy *see* Nizhniy Novgorod
Gorkum *see* Gorinchem
95 I23 **Gørlev** Sjælland, E Denmark 55°33′N 11°14′E
111 M17 **Gorlice** Małopolskie, S Poland 49°40′N 21°09′E
101 Q15 **Görlitz** Sachsen, E Germany 51°09′N 14°58′E
Görlitz *see* Zgorzelec
Gorlovka *see* Horlivka
25 R7 **Gorman** Texas, SW USA 32°12′N 98°40′W
21 T3 **Gormania** West Virginia, NE USA 39°16′N 79°18′W
Gorna Dzhumaya *see* Blagoevgrad
114 K8 **Gorna Oryahovitsa** , Gorna Oryakhovitsa. Veliko Tŭrnovo, N Bulgaria 43°07′N 25°40′E
Gorna Oryakhovitsa *see* Gorna Oryahovitsa
114 J8 **Gorna Studena** Veliko Tŭrnovo, N Bulgaria 43°26′N 25°21′E
Gornja Mužlja *see* Mužlja
109 X9 **Gornja Radgona** *Ger.* Oberradkersburg. NE Slovenia 46°39′N 16°00′E
112 M13 **Gornji Milanovac** Serbia, C Serbia 44°01′N 20°26′E
112 G13 **Gornji Vakuf** *var.* Uskoplje. Federacija Bosne I Hercegovine, SW Bosnia and Herzegovina 43°55′N 17°34′E
122 J13 **Gorno-Altaysk** Respublika Altay, S Russian Federation 51°59′N 85°56′E
Gorno-Altayskaya Respublika *see* Altay, Respublika
123 N12 **Gorno-Chuyskiy** Irkutskaya Oblast', C Russian Federation 57°33′N 111°38′E
125 V14 **Gornozavodsk** Permskiy Kray, NW Russian Federation 58°21′N 58°24′E
125 V14 **Gornozavodsk** Ostrov Sakhalin, Sakhalinskaya Oblast', SE Russian Federation 46°34′N 141°52′E
122 I13 **Gornyak** Altayskiy Kray, S Russian Federation 50°58′N 81°24′E
123 O14 **Gornyy** Chitunskaya Oblast', Russian Federation 51°42′N 114°16′E
127 R8 **Gornyy** Saratovskaya Oblast', W Russian Federation 51°42′N 48°26′E
Gornyy Altay *see* Altay, Respublika
127 O10 **Gornyy Balykley** Volgogradskaya Oblast', SW Russian Federation 49°37′N 45°03′E
80 I13 **Goroch'an** ▲ W Ethiopia 09°09′N 37°16′E
116 J7 **Gorodenka** *var.* Horodenka. Ivano-Frankivs'ka Oblast', W Ukraine 48°41′N 25°28′E
127 O3 **Gorodets** Nizhegorodskaya Oblast', W Russian Federation 56°36′N 43°27′E
Gorodets *see* Haradzyets
Gorodeya *see* Haradzyeya
127 P6 **Gorodishche** Penzenskaya Oblast', W Russian Federation 53°17′N 45°39′E
Gorodishche *see* Horodyshche
Gorodnya *see* Horodnya
Gorodok *see* Haradok
Gorodok/Gorodok Yagellonski *see* Horodok
126 M13 **Gorodovikovsk** Respublika Kalmykiya, SW Russian Federation 46°07′N 41°56′E
186 D7 **Goroka** Eastern Highlands, C Papua New Guinea 06°02′S 145°22′E
Gorokhov *see* Horokhiv
127 N3 **Gorokhovets** Vladimirskaya Oblast', W Russian Federation 56°12′N 42°40′E
77 Q11 **Gorom-Gorom** NE Burkina Faso 14°27′N 00°14′W
171 U13 **Gorong, Kepulauan** *island group* E Indonesia
83 M17 **Gorongosa** Sofala, C Mozambique 18°40′S 34°03′E
171 P11 **Gorontalo** Sulawesi, C Indonesia 0°33′N 123°05′E
171 O11 **Gorontalo** *off.* Propinsi Gorontalo. ◈ *province* N Indonesia
Propinsi Gorontalo *see* Gorontalo
Gorontalo, Teluk *see* Tomini, Teluk
110 L7 **Górowo Iławeckie** *Ger.* Landsberg. Warmińsko-Mazurskie, NE Poland 54°18′N 20°30′E
98 M7 **Gorredijk** *Fris.* De Gordyk. Fryslân, N Netherlands 53°00′N 06°04′E
84 C14 **Gorringe Ridge** *undersea feature* E Atlantic Ocean 36°40′N 11°35′W
98 M11 **Gorssel** Gelderland, E Netherlands 52°12′N 06°13′E
109 T8 **Görtschitz** ≈ S Austria
Goryn *see* Horyn'
145 S15 **Gory Shu-Ile** *Kaz.* Shū-Ile Taūlary; *prev.* Chu-Iliyskiye Gory. ▲▲ S Kazakhstan
Görz *see* Gorizia
110 E10 **Gorzów Wielkopolski** *Ger.* Landsberg, Landsberg an der Warthe. Lubuskie, W Poland 52°44′N 15°12′E
146 B10 **Goşabo** *var.* Goshoba, *Rus.* Koshoba. Balkan Welaýaty, NW Turkmenistan 40°28′N 54°11′E
108 G9 **Göschenen** Uri, C Switzerland 46°40′N 08°36′E
165 O11 **Gosen** Niigata, Honshū, C Japan 37°45′N 139°11′E
163 Y13 **Goseong** *prev.* Kosŏng. SE North Korea 38°41′N 128°14′E
183 T8 **Gosford** New South Wales, SE Australia 33°25′S 151°18′E
31 P11 **Goshen** Indiana, N USA 41°34′N 85°49′W
18 K13 **Goshen** New York, NE USA 41°24′N 74°17′W
Goshoba *see* Goşabo
Goshoba *see* Goşoba
165 Q7 **Goshogawara** *var.* Gosyogawara. Aomori, Honshū, C Japan 40°47′N 140°24′E
Goshquduq Qum *see* Tosquduq Qumlari
101 J14 **Goslar** Niedersachsen, C Germany 51°55′N 10°25′E
27 Y9 **Gosnell** Arkansas, C USA 35°57′N 89°58′W
146 B10 **Goşoba** *var.* Goshoba, *Rus.* Koshoba. Balkanskiy Velayat, NW Turkmenistan 40°28′N 54°11′E
112 C11 **Gospić** Lika-Senj, C Croatia 44°32′N 15°21′E
97 N23 **Gosport** S England, United Kingdom 50°48′N 01°08′W
94 D9 **Gossa** *island* S Norway
108 H7 **Gossau** Sankt Gallen, NE Switzerland 47°25′N 09°16′E
99 G20 **Gosselies** *var.* Goss'lies. Hainaut, S Belgium 50°28′N 04°26′E
77 P10 **Gossi** Tombouctou, C Mali 15°44′N 01°19′W
Goss'lies *see* Gosselies
113 N18 **Gostivar** W FYR Macedonia 41°48′N 20°55′E
Gostomel' *see* Hostomel'
110 G12 **Gostyń** *var.* Gostyn. Wielkopolskie, C Poland 51°52′N 17°00′E
Gostyn *see* Gostyń
110 K11 **Gostynin** Mazowiesckie, C Poland 52°25′N 19°27′E
Gosyogawara *see* Goshogawara
95 J18 **Göta Älv** ≈ S Sweden
95 N17 **Göta kanal** *canal* S Sweden
95 K18 **Götaland** *cultural region* S Sweden
95 H17 **Göteborg** *Eng.* Gothenburg. Västra Götaland, S Sweden 57°43′N 11°58′E
77 X16 **Gotel Mountains** ▲▲ E Nigeria
95 K17 **Götene** Västra Götaland, S Sweden 58°32′N 13°29′E
Gotera *see* San Francisco
101 K16 **Gotha** Thüringen, C Germany 50°57′N 10°43′E
29 N15 **Gothenburg** Nebraska, C USA 40°57′N 100°09′W
Gothenburg *see* Göteborg
77 R12 **Gothèye** Tillabéri, SW Niger 13°52′N 01°27′E
Gothland *see* Gotland
95 P19 **Gotland** *var.* Gothland, Gottland. ◈ *county* SE Sweden
95 O18 **Gotland** *island* SE Sweden
164 A14 **Gotō** *var.* Hukue; *prev.* Fukue. Nagasaki, Fukue-jima, SW Japan 32°41′N 128°52′E
164 B13 **Gotō-rettō** *island group* SW Japan
114 H12 **Gotse Delchev** *prev.* Nevrokop. Blagoevgrad, SW Bulgaria 41°33′N 23°42′E
95 P17 **Gotska Sandön** *island* SE Sweden
101 I15 **Göttingen** *var.* Goettingen. Niedersachsen, C Germany 51°33′N 09°55′E
Gottland *see* Gotland
93 I16 **Gottne** Västernorrland, C Sweden 63°27′N 18°25′E
Gottschee *see* Kočevje
Gottwaldov *see* Zlín
146 B11 **Goturdepe** *Rus.* Koturdepe. Balkan Welaýaty, W Turkmenistan 39°32′N 53°39′E
108 I7 **Götzis** Vorarlberg, NW Austria 47°21′N 09°40′E
98 H12 **Gouda** Zuid-Holland, C Netherlands 52°01′N 04°42′E
76 I11 **Goudiri** *var.* Goudiry. E Senegal 14°12′N 12°41′W
Goudiry *see* Goudiri
77 X12 **Goudoumaria** Diffa, S Niger 13°28′N 11°15′E
15 R9 **Gouffre, Rivière du** ≈ Québec, SE Canada
65 M19 **Gough Fracture Zone** *tectonic feature* S Atlantic Ocean
65 M19 **Gough Island** *island* Tristan da Cunha, S Atlantic Ocean
15 N8 **Gouin, Réservoir** ⊞ Québec, SE Canada
14 B10 **Goulais River** Ontario, S Canada 46°41′N 84°22′W
183 R9 **Goulburn** New South Wales, SE Australia 34°45′S 149°44′E
183 O11 **Goulburn River** ≈ Victoria, SE Australia
195 O10 **Gould Coast** *physical region* Antarctica
Goulimime *see* Guelmime
114 F13 **Gouménissa** Kentrikí Makedonía, N Greece 40°56′N 22°27′E
77 O10 **Goundam** Tombouctou, NW Mali 16°27′N 03°39′W
78 H12 **Goundi** Mandoul, S Chad 09°22′N 17°21′E
78 G12 **Gounou-Gaya** Mayo-Kébbi Est, SW Chad 09°37′N 15°30′E
Gourci *see* Goursi
Gourcy *see* Goursi
102 M13 **Gourdon** Lot, S France 44°45′N 01°22′E
77 W11 **Gouré** Zinder, SE Niger 13°59′N 10°16′E
102 G6 **Gourin** Morbihan, NW France 48°07′N 03°37′W

◆ Country
● Country Capital
◇ Dependent Territory
○ Dependent Territory Capital
◈ Administrative Regions
✈ International Airport
▲ Mountain
▲▲ Mountain Range
Volcano
≈ River
◎ Lake
⊞ Reservoir

GOURMA-RHAROUS – GRENAA

77 P10 **Gourma-Rharous** Tombouctou, C Mali 16°54′N 01°55′W
103 N4 **Gournay-en-Bray** Seine-Maritime, N France 49°29′N 01°42′E
78 J6 **Gouro** Ennedi-Ouest, N Chad 19°26′N 19°36′E
77 O12 **Goursi** *var.* Gourci, Gourcy. NW Burkina Faso 13°13′N 02°20′W
104 H8 **Gouveia** Guarda, N Portugal 40°29′N 07°35′W
18 I7 **Gouverneur** New York, NE USA 44°20′N 75°27′W
99 L21 **Gouvy** Luxembourg, E Belgium 50°10′N 05°55′E
45 R14 **Gouyave** *var.* Charlotte Town. NW Grenada 12°10′N 61°44′W
Goverla, Gora *see* Hoverla, Hora
59 N20 **Governador Valadares** Minas Gerais, SE Brazil 18°51′S 41°57′W
171 R8 **Governor Generoso** Mindanao, S Philippines 06°36′N 126°06′E
44 I2 **Governor's Harbour** Eleuthera Island, C The Bahamas 25°11′N 76°15′W
162 F9 **Govĭ-Altay** ◈ *province* SW Mongolia
162 I10 **Govĭ Altayn Nuruu** ▲▲ S Mongolia
154 L9 **Govind Ballabh Pant Sāgar** ⊠ C India
152 I7 **Govind Sāgar** ⊠ NE India
162 M8 **Govĭ-Sumber** ◈ *province* C Mongolia
Govurdak *see* Magdanly
18 D11 **Gowanda** New York, NE USA 42°25′N 78°55′W
148 J10 **Gowd-e Zereh, Dasht-e** *var.* Guad-i-Zirreh. *marsh* SW Afghanistan
14 F8 **Gowganda** Ontario, S Canada 47°41′N 80°46′W
14 G8 **Gowganda Lake** ◎ Ontario, S Canada
148 M5 **Gōwr** *prev.* Ghowr. ◈ *province* C Afghanistan
29 U13 **Gowrie** Iowa, C USA 42°16′N 94°17′W
Gowurdak *see* Magdanly
61 C15 **Goya** Corrientes, NE Argentina 29°10′S 59°15′W
Goyania *see* Goiânia
Goyaz *see* Goiás
137 X11 **Göyçay** *Rus.* Geokchay. C Azerbaijan 40°38′N 47°44′E
137 V11 **Göygöl** ; *prev.* Xanlar. NW Azerbaijan 40°37′N 46°18′E
146 D10 **Goymat** *Rus.* Koymat. Balkan Welaýaty, NW Turkmenistan 40°23′N 55°45′E
146 D10 **Goymatdag, Gory** *Rus.* Gory Koymatdag. *hill range* Balkan Welaýaty, NW Turkmenistan
136 F12 **Göynük** Bolu, NW Turkey 40°24′N 30°45′E
165 R9 **Goyō-san** ▲ Honshū, C Japan 39°12′N 141°40′E
78 K11 **Goz Beïda** Sila, SE Chad 12°06′N 21°22′E
146 M10 **G'ozg'on** *Rus.* Gazgan. Navoiy Viloyati, C Uzbekistan 40°36′N 65°29′E
158 H11 **Gozha Co** ◎ W China
121 O15 **Gozo** *var.* Ghawdex. *island* N Malta
80 H9 **Göz Regeb** Kassala, NE Sudan 16°03′N 35°33′E
83 H25 **Graaff-Reinet** Eastern Cape, S South Africa 32°15′S 24°32′E
Graasten *see* Gråsten
76 L17 **Grabo** SW Ivory Coast 04°57′N 07°30′W
112 P11 **Grabovica** Serbia, E Serbia 44°30′N 22°29′E
110 I13 **Grabów nad Prosną** Wielkopolskie, C Poland 51°30′N 18°06′E
108 I8 **Grabs** Sankt Gallen, NE Switzerland 47°10′N 09°27′E
112 D12 **Gračac** Zadar, SW Croatia 44°18′N 15°52′E
112 I11 **Gračanica** Federacija Bosne I Hercegovine, NE Bosnia and Herzegovina 44°41′N 18°20′E
14 L11 **Gracefield** Québec, SE Canada 46°06′N 76°03′W
99 K19 **Grâce-Hollogne** Liège, E Belgium 50°38′N 05°30′E
23 R8 **Graceville** Florida, SE USA 30°57′N 85°31′W
29 R8 **Graceville** Minnesota, N USA 45°34′N 96°25′W
42 G6 **Gracias** Lempira, W Honduras 14°35′N 88°35′W
Gracias *see* Lempira
42 L5 **Gracias a Dios** ◈ *department* E Honduras
43 O6 **Gracias a Dios, Cabo de** *headland* Honduras/Nicaragua 15°00′N 83°10′W
64 O2 **Graciosa** *var.* Ilha Graciosa. *island* Azores, Portugal, NE Atlantic Ocean
64 Q11 **Graciosa** *island* Islas Canarias, Spain, NE Atlantic Ocean
Graciosa, Ilha *see* Graciosa
112 I11 **Gradačac** Federacija Bosne I Hercegovine, N Bosnia and Herzegovina 44°51′N 18°24′E
59 J15 **Gradaús, Serra dos** ▲▲ C Brazil
104 L3 **Gradefes** Castilla y León, N Spain 42°37′N 05°14′W
Gradiška *see* Bosanska Gradiška
Gradizhsk *see* Hradyz'k
106 J7 **Grado** Friuli-Venezia Giulia, NE Italy 45°41′N 13°24′E
Grado *see* Grau
113 P19 **Gradsko** C FYR Macedonia 41°34′N 21°56′E
37 V11 **Grady** New Mexico, SW USA 34°49′N 103°19′W
Grad Zagreb *see* Zagreb
29 T12 **Graettinger** Iowa, C USA 43°14′N 94°45′W
101 M23 **Grafing** Bayern, SE Germany 48°01′N 11°57′E
25 S6 **Graford** Texas, SW USA 32°56′N 98°15′W
183 V5 **Grafton** New South Wales, SE Australia 29°41′S 152°55′E
29 Q3 **Grafton** North Dakota, N USA 48°24′N 97°24′W
21 S3 **Grafton** West Virginia, NE USA 39°21′N 80°03′W
21 T9 **Graham** North Carolina, SE USA 36°05′N 79°25′W
25 R6 **Graham** Texas, SW USA 33°07′N 98°36′W
Graham Bell Island *see* Greem-Bell, Ostrov
10 I13 **Graham Island** *island* Queen Charlotte Islands, British Columbia, SW Canada
19 S6 **Graham Lake** ◎ Maine, NE USA
194 H4 **Graham Land** *physical region* Antarctica
37 N15 **Graham, Mount** ▲ Arizona, SW USA 32°42′N 109°52′W
Grahamstad *see* Grahamstown
83 I25 **Grahamstown** *Afr.* Grahamstad. Eastern Cape, S South Africa 33°18′S 26°32′E
Grahovo *see* Bosansko Grahovo
68 C12 **Grain Coast** *coastal region* S Liberia
169 S17 **Grajagan, Teluk** *bay* Jawa, S Indonesia
59 L14 **Grajaú** Maranhão, E Brazil 05°50′S 45°12′W
58 M13 **Grajaú, Rio** ∽ NE Brazil
110 O8 **Grajewo** Podlaskie, NE Poland 53°38′N 22°26′E
95 F24 **Gram** Syddanmark, SW Denmark 55°18′N 09°03′E
103 N13 **Gramat** Lot, S France 44°45′N 01°45′E
22 H5 **Grambling** Louisiana, S USA 32°31′N 92°43′W
115 C14 **Grámmos** ▲▲ Albania/Greece
96 I9 **Grampian Mountains** ▲▲ C Scotland, United Kingdom
182 L12 **Grampians, The** ▲▲ Victoria, SE Australia
98 O9 **Gramsbergen** Overijssel, E Netherlands 52°37′N 06°39′E
113 L21 **Gramsh** *var.* Gramshi. Elbasan, C Albania 40°52′N 20°12′E
Gramshi *see* Gramsh
Gran *see* Esztergom, Hungary
Gran *see* Hron
54 F11 **Granada** Meta, C Colombia 03°33′N 73°44′W
42 J10 **Granada** Granada, SW Nicaragua 11°55′N 85°58′W
105 N14 **Granada** Andalucía, S Spain 37°13′N 03°41′W
37 W6 **Granada** Colorado, C USA 38°00′N 102°18′W
42 J11 **Granada** ◈ *department* SW Nicaragua
105 N14 **Granada** ◈ *province* Andalucía, S Spain
63 I21 **Gran Antiplanicie Central** *plain* S Argentina
97 E17 **Granard** *Ir.* Gránard. C Ireland 53°47′N 07°30′W
Gránard *see* Granard
63 J20 **Gran Bajo** *basin* S Argentina
63 J15 **Gran Bajo del Gualicho** *basin* E Argentina
63 I21 **Gran Bajo de San Julián** *basin* SE Argentina
25 S7 **Granbury** Texas, SW USA 32°27′N 97°47′W
15 P12 **Granby** Québec, SE Canada 45°23′N 72°44′W
27 S8 **Granby** Missouri, C USA 36°55′N 94°14′W
37 S3 **Granby, Lake** ⊠ Colorado, C USA
64 O12 **Gran Canaria** *var.* Grand Canary. *island* Islas Canarias, Spain, NE Atlantic Ocean
47 T11 **Gran Chaco** *var.* Chaco. *lowland plain* South America
45 R14 **Grand Anse** SW Grenada 12°01′N 61°45′W
Grand-Anse *see* Portsmouth
44 G1 **Grand Bahama Island** *island* N The Bahamas
Grand Balé *see* Tui
103 U7 **Grand Ballon** *Ger.* Ballon de Guebwiller. ▲ NE France 47°53′N 07°06′E
13 T13 **Grand Bank** Newfoundland, Newfoundland and Labrador, SE Canada 47°06′N 55°48′W
64 I7 **Grand Banks of Newfoundland** *undersea feature* NW Atlantic Ocean 45°00′N 40°00′W
Grand Bassa *see* Buchanan
77 N17 **Grand-Bassam** *var.* Bassam. SE Ivory Coast 05°14′N 03°45′W
14 E16 **Grand Bend** Ontario, S Canada 43°17′N 81°46′W
76 L17 **Grand-Bérébi** *var.* Grand-Béréby. SW Ivory Coast 04°38′N 06°55′W
Grand-Béréby *see* Grand-Bérébi
45 X11 **Grand-Bourg** Marie-Galante, SE Guadeloupe 15°53′N 61°19′W
44 M6 **Grand Caicos** *var.* Middle Caicos. *island* C Turks and Caicos Islands
14 K12 **Grand Calumet, Île du** *island* Québec, SE Canada
97 E18 **Grand Canal** *Ir.* An Chanáil Mhór. *canal* C Ireland
Grand Canal *see* Da Yunhe
Grand Canary *see* Gran Canaria
36 K10 **Grand Canyon** Arizona, SW USA 36°01′N 112°10′W
36 J9 **Grand Canyon** *canyon* Arizona, SW USA
Grand Canyon State *see* Arizona
44 D8 **Grand Cayman** *island* SW Cayman Islands
11 R14 **Grand Centre** Alberta, SW Canada 54°25′N 110°13′W
76 L17 **Grand Cess** SE Liberia 04°36′N 08°12′W
108 D12 **Grand Combin** ▲ S Switzerland 45°58′N 07°27′E
32 K8 **Grand Coulee** Washington, NW USA 47°56′N 119°00′W
32 J8 **Grand Coulee** *valley* Washington, NW USA
45 X5 **Grand Cul-de-Sac Marin** *bay* N Guadeloupe
Grand Duchy of Luxembourg *see* Luxembourg
63 I22 **Grande, Bahía** *bay* S Argentina
11 N14 **Grande Cache** Alberta, W Canada 53°53′N 119°07′W
103 U12 **Grande Casse** ▲ E France 45°22′N 06°50′E
Grande Comore *see* Ngazidja
61 G18 **Grande, Cuchilla** *hill range* E Uruguay
45 S5 **Grande de Añasco, Río** ∽ W Puerto Rico
Grande de Chiloé, Isla *see* Chiloé, Isla de
58 J12 **Grande de Gurupá, Ilha** *river island* NE Brazil
57 K21 **Grande de Lipez, Río** ∽ SW Bolivia
45 U6 **Grande de Loíza, Río** ∽ E Puerto Rico
45 T5 **Grande de Manatí, Río** ∽ C Puerto Rico
42 L9 **Grande de Matagalpa, Río** ∽ C Nicaragua
40 K12 **Grande de Santiago, Río** *var.* Santiago. ∽ C Mexico
43 O15 **Grande de Térraba, Río** *var.* Río Térraba. ∽ SE Costa Rica
12 J9 **Grande Deux, Réservoir la** ⊠ Québec, E Canada
60 O10 **Grande, Ilha** *island* SE Brazil
11 O13 **Grande Prairie** Alberta, W Canada 55°10′N 118°52′W
74 I8 **Grand Erg Occidental** *desert* W Algeria
74 L9 **Grand Erg Oriental** *desert* Algeria/Tunisia
57 M18 **Grande, Río** ∽ C Bolivia
59 J20 **Grande, Rio** ∽ S Brazil
2 F15 **Grande, Rio** *var.* Río Bravo, *Sp.* Río Bravo del Norte, Bravo del Norte. ∽ Mexico/USA
15 Y7 **Grande-Rivière** Québec, SE Canada 48°27′N 64°37′W
15 Y6 **Grande Rivière** ∽ Québec, SE Canada
44 M8 **Grande-Rivière-du-Nord** N Haiti 19°36′N 72°10′W
62 K9 **Grande, Salina** *var.* Gran Salitral. *salt lake* C Argentina
15 S7 **Grandes-Bergeronnes** Québec, SE Canada 48°16′N 69°32′W
47 W6 **Grande, Serra** ▲▲ W Brazil
40 K4 **Grande, Sierra** ▲▲ N Mexico
103 S12 **Grandes Rousses** ▲▲ E France
63 K17 **Grandes, Salinas** *salt lake* E Argentina
45 Y5 **Grande Terre** *island* E West Indies
15 X5 **Grande-Vallée** Québec, SE Canada 49°14′N 65°08′W
45 Y5 **Grande Vigie, Pointe de la** *headland* Grande Terre, N Guadeloupe 16°31′N 61°27′W
13 N14 **Grand Falls** New Brunswick, SE Canada 47°02′N 67°46′W
13 T11 **Grand Falls** Newfoundland, Newfoundland and Labrador, SE Canada 48°57′N 55°48′W
24 L9 **Grandfalls** Texas, SW USA 31°20′N 102°51′W
21 P9 **Grandfather Mountain** ▲ North Carolina, SE USA 36°06′N 81°48′W
26 L13 **Grandfield** Oklahoma, C USA 34°15′N 98°40′W
11 N17 **Grand Forks** British Columbia, SW Canada 49°02′N 118°30′W
29 R4 **Grand Forks** North Dakota, N USA 47°54′N 97°03′W
31 O9 **Grand Haven** Michigan, N USA 43°03′N 86°13′W
Grandichi *see* Hrandzichy
29 P15 **Grand Island** Nebraska, C USA 40°55′N 98°20′W
31 O3 **Grand Island** *island* Michigan, N USA
22 K10 **Grand Isle** Louisiana, S USA 29°12′N 90°00′W
65 A23 **Grand Jason** *island* Jason Islands, NW Falkland Islands
37 P5 **Grand Junction** Colorado, C USA 39°03′N 108°33′W
20 F10 **Grand Junction** Tennessee, S USA 35°03′N 89°11′W
14 J9 **Grand-Lac-Victoria** Québec, SE Canada 47°33′N 77°28′W
14 J9 **Grand lac Victoria** ◎ Québec, SE Canada
77 N17 **Grand-Lahou** *var.* Grand Lahu. S Ivory Coast 05°09′N 05°01′W
Grand Lahu *see* Grand-Lahou
37 S3 **Grand Lake** Colorado, C USA 40°15′N 105°49′W
13 S11 **Grand Lake** ◎ Newfoundland and Labrador, E Canada
22 G9 **Grand Lake** ◎ Louisiana, S USA
31 R5 **Grand Lake** ◎ Michigan, N USA
31 Q13 **Grand Lake** ⊠ Ohio, N USA
27 R9 **Grand Lake O' The Cherokees** *var.* Lake O' The Cherokees. ⊠ Oklahoma, C USA
31 Q9 **Grand Ledge** Michigan, N USA 42°45′N 84°45′W
102 I8 **Grand-Lieu, Lac de** ◎ NW France
19 U6 **Grand Manan Channel** *channel* Canada/USA
13 O15 **Grand Manan Island** *island* New Brunswick, SE Canada
29 Y4 **Grand Marais** Minnesota, N USA 47°45′N 90°19′W
15 P10 **Grand-Mère** Québec, SE Canada 46°36′N 72°41′W
37 P5 **Grand Mesa** ▲▲ Colorado, C USA
108 C10 **Grand Muveran** ▲ W Switzerland 46°16′N 07°12′E
104 G12 **Grândola** Setúbal, S Portugal 38°10′N 08°34′W
Grand Paradis *see* Gran Paradiso
187 O15 **Grand Passage** *passage* N New Caledonia
77 R16 **Grand-Popo** S Benin 06°19′N 01°50′E
29 Z3 **Grand Portage** Minnesota, N USA 48°00′N 89°36′W
25 T6 **Grand Prairie** Texas, SW USA 32°45′N 97°00′W
11 W14 **Grand Rapids** Manitoba, C Canada 53°12′N 99°19′W
31 P9 **Grand Rapids** Michigan, N USA 42°57′N 86°40′W
29 V5 **Grand Rapids** Minnesota, N USA 47°14′N 93°31′W
14 L10 **Grand-Remous** Québec, SE Canada 46°36′N 75°53′W
14 F15 **Grand River** ∽ Ontario, S Canada
31 P9 **Grand River** ∽ Michigan, N USA
27 T3 **Grand River** ∽ Missouri, C USA
28 M7 **Grand River** ∽ South Dakota, N USA
45 Q11 **Grand' Rivière** N Martinique 14°52′N 61°11′W
32 F11 **Grand Ronde** Oregon, NW USA 45°03′N 123°43′W
32 L11 **Grand Ronde River** ∽ Oregon/Washington, NW USA
Grand-Saint-Bernard, Col du *see* Great Saint Bernard Pass
25 V6 **Grand Saline** Texas, SW USA 32°40′N 95°42′W
55 X10 **Grand-Santi** W French Guiana 04°19′N 54°24′W
Grandsee *see* Grandson
172 J16 **Grand Sœur** *island* Les Sœurs, NE Seychelles
108 B9 **Grandson** *prev.* Grandsee. Vaud, W Switzerland 46°49′N 06°39′E
33 S14 **Grand Teton** ▲ Wyoming, C USA 43°44′N 110°48′W
31 P5 **Grand Traverse Bay** *lake bay* Michigan, N USA
45 N6 **Grand Turk** ○ (Turks and Caicos Islands) Grand Turk Island, S Turks and Caicos Islands 21°24′N 71°08′W
45 N6 **Grand Turk Island** *island* SE Turks and Caicos Islands
103 S13 **Grand Veymont** ▲ E France 44°51′N 05°32′E
11 W15 **Grandview** Manitoba, S Canada 51°11′N 100°41′W
27 R4 **Grandview** Missouri, C USA 38°53′N 94°31′W
36 I10 **Grand Wash Cliffs** *cliff* Arizona, SW USA
14 J8 **Granet, Lac** ◎ Québec, SE Canada
95 L14 **Grangärde** Dalarna, C Sweden 60°15′N 15°00′E
44 H12 **Grange Hill** W Jamaica 18°19′N 78°11′W
96 J12 **Grangemouth** C Scotland, United Kingdom 56°01′N 03°44′W
25 T10 **Granger** Texas, SW USA 30°43′N 97°26′W
32 J10 **Granger** Washington, NW USA 46°20′N 120°11′W
33 T17 **Granger** Wyoming, C USA 41°37′N 109°58′W
Granges *see* Grenchen
95 L14 **Grängesberg** Dalarna, C Sweden 60°06′N 15°00′E
33 N11 **Grangeville** Idaho, NW USA 45°55′N 116°07′W
10 K13 **Granisle** British Columbia, SW Canada 54°55′N 126°14′W
30 K15 **Granite City** Illinois, N USA 38°42′N 90°09′W
29 S9 **Granite Falls** Minnesota, N USA 44°48′N 95°33′W
21 Q9 **Granite Falls** North Carolina, SE USA 35°48′N 81°25′W
36 K12 **Granite Mountain** ▲ Arizona, SW USA 34°38′N 112°34′W
33 T12 **Granite Peak** ▲ Montana, NW USA 45°09′N 109°48′W
35 T2 **Granite Peak** ▲ Nevada, W USA 41°40′N 117°35′W
36 J3 **Granite Peak** ▲ Utah, W USA 40°09′N 113°18′W
Granite State *see* New Hampshire
107 H24 **Granitola, Capo** *headland* Sicilia, Italy, C Mediterranean Sea 37°33′N 12°39′E
185 H15 **Granity** West Coast, South Island, New Zealand 41°37′S 171°53′E
Gran Lago *see* Nicaragua, Lago de
63 J18 **Gran Laguna Salada** ◎ S Argentina
Gran Malvina *see* West Falkland
95 L18 **Gränna** Jönköping, S Sweden 58°02′N 14°30′E
105 W5 **Granollers** *var.* Granollérs. Cataluña, NE Spain 41°37′N 02°18′E
Granollérs *see* Granollers
106 A7 **Gran Paradiso** *Fr.* Grand Paradis. ▲ NW Italy 45°31′N 07°13′E
Gran Pilastro *see* Hochfeiler
Gran Salitral *see* Grande, Salina
Gran San Bernardo, Passo di *see* Great Saint Bernard Pass
Gran Santiago *see* Santiago
107 J14 **Gran Sasso d'Italia** ▲▲ C Italy
100 N11 **Gransee** Brandenburg, NE Germany 53°00′N 13°10′E
28 L15 **Grant** Nebraska, C USA 40°50′N 101°43′W
27 R1 **Grant City** Missouri, C USA 40°29′N 94°25′W
97 N19 **Grantham** E England, United Kingdom 52°55′N 00°39′W
65 D24 **Grantham Sound** *sound* East Falkland, Falkland Islands
194 K13 **Grant Island** *island* Antarctica
45 Z14 **Grantley Adams** ✈ (Bridgetown) SE Barbados 13°04′N 59°29′W
35 S7 **Grant, Mount** ▲ Nevada, W USA 38°34′N 118°47′W
96 J9 **Grantown-on-Spey** N Scotland, United Kingdom 57°11′N 03°53′W
35 W8 **Grant Range** ▲▲ Nevada, W USA
37 Q11 **Grants** New Mexico, SW USA 35°09′N 107°50′W
30 I4 **Grantsburg** Wisconsin, N USA 45°47′N 92°40′W
32 F15 **Grants Pass** Oregon, NW USA 42°26′N 123°20′W
36 K3 **Grantsville** Utah, W USA 40°36′N 112°27′W
21 R4 **Grantsville** West Virginia, NE USA 38°55′N 81°07′W
102 I5 **Granville** Manche, N France 48°50′N 01°35′W
11 V12 **Granville Lake** ◎ Manitoba, C Canada
25 V8 **Grapeland** Texas, SW USA 31°29′N 95°28′W
25 T6 **Grapevine** Texas, SW USA 32°55′N 97°04′W
83 K20 **Graskop** Mpumalanga, NE South Africa 24°58′S 30°49′E
95 P14 **Gräsö** Uppsala, C Sweden 60°22′N 18°30′E
93 I19 **Gräsö** *island* C Sweden
103 U15 **Grasse** Alpes-Maritimes, SE France 43°42′N 06°52′E
18 E14 **Grassflat** Pennsylvania, NE USA 41°00′N 78°04′W
33 U9 **Grassrange** Montana, NW USA 47°02′N 108°48′W
18 J6 **Grass River** ∽ New York, NE USA
35 P6 **Grass Valley** California, W USA 39°12′N 121°04′W
183 N14 **Grassy** Tasmania, SE Australia 40°03′S 144°04′E
28 K4 **Grassy Butte** North Dakota, N USA 47°20′N 103°13′W
21 R5 **Grassy Knob** ▲ West Virginia, NE USA 38°04′N 80°31′W
95 G24 **Gråsten** *var.* Graasten. Syddanmark, SW Denmark 54°55′N 09°37′E
95 J18 **Grästorp** Västra Götaland, S Sweden 58°20′N 12°40′E
Gratianopolis *see* Grenoble
109 V8 **Gratwein** Steiermark, SE Austria 47°08′N 15°20′E
Gratz *see* Graz
104 K2 **Grau** *var.* Grado. Asturias, N Spain 43°23′N 06°04′W
108 I9 **Graubünden** *Fr.* Grisons, *It.* Grigioni. ◈ *canton* SE Switzerland
Graudenz *see* Grudziądz
103 N15 **Graulhet** Tarn, S France 43°45′N 01°58′E
105 T4 **Graus** Aragón, NE Spain 42°11′N 00°21′E
61 I16 **Gravataí** Rio Grande do Sul, S Brazil 29°55′S 51°00′W
98 L13 **Grave** Noord-Brabant, SE Netherlands 51°45′N 05°45′E
11 T17 **Gravelbourg** Saskatchewan, S Canada 49°53′N 106°33′W
103 N1 **Gravelines** Nord, N France 51°00′N 02°07′E
Graven *see* Grez-Doiceau
14 H13 **Gravenhurst** Ontario, S Canada 44°55′N 79°22′W
33 O10 **Grave Peak** ▲ Idaho, NW USA 46°24′N 114°43′W
102 I11 **Grave, Pointe de** *headland* W France 45°33′N 01°24′W
183 S4 **Gravesend** New South Wales, SE Australia 29°37′S 150°15′E
97 P22 **Gravesend** SE England, United Kingdom 51°27′N 00°24′E
107 N17 **Gravina in Puglia** Puglia, SE Italy 40°48′N 16°25′E
103 S8 **Gray** Haute-Saône, E France 47°28′N 05°34′E
23 T4 **Gray** Georgia, SE USA 33°00′N 83°31′W
195 V16 **Gray, Cape** *headland* Antarctica 67°30′N 143°30′E
32 F9 **Grayland** Washington, NW USA 46°46′N 124°07′W
39 N10 **Grayling** Alaska, USA 62°55′N 160°07′W
31 Q6 **Grayling** Michigan, N USA 44°40′N 84°43′W
32 F9 **Grays Harbor** *inlet* Washington, NW USA
21 O5 **Grayson** Kentucky, S USA 38°21′N 82°59′W
37 S4 **Grays Peak** ▲ Colorado, C USA 39°37′N 105°49′W
30 M16 **Grayville** Illinois, N USA 38°15′N 87°59′W
109 V8 **Graz** *prev.* Gratz. Steiermark, SE Austria 47°05′N 15°23′E
104 L15 **Grazalema** Andalucía, S Spain 36°46′N 05°23′W
113 P15 **Grdelica** Serbia, SE Serbia 42°54′N 22°05′E
44 H1 **Great Abaco** *var.* Abaco Island. *island* N The Bahamas
Great Admiralty Island *see* Manus Island
Great Alfold *see* Great Hungarian Plain
Great Ararat *see* Büyükağrı Dağı
181 U8 **Great Artesian Basin** *lowlands* Queensland, C Australia
181 O12 **Great Australian Bight** *bight* S Australia
64 E11 **Great Bahama Bank** *undersea feature* E Gulf of Mexico 23°15′N 78°00′W
184 M4 **Great Barrier Island** *island* N New Zealand
181 X4 **Great Barrier Reef** *reef* Queensland, NE Australia
18 L11 **Great Barrington** Massachusetts, NE USA 42°11′N 73°20′W
0 F10 **Great Basin** *basin* W USA
8 I8 **Great Bear Lake** *Fr.* Grand Lac de l'Ours. ◎ Northwest Territories, NW Canada
Great Belt *see* Storebælt
26 L5 **Great Bend** Kansas, C USA 38°22′N 98°47′W
Great Bermuda *see* Bermuda
97 A20 **Great Blasket Island** *Ir.* An Blascaod Mór. *island* SW Ireland
Great Britain *see* Britain
151 Q23 **Great Channel** *channel* Andaman Sea/Indian Ocean
166 J10 **Great Coco Island** *island* SW Myanmar (Burma)
Great Crosby *see* Crosby
21 X7 **Great Dismal Swamp** *wetland* North Carolina/Virginia, SE USA
33 V16 **Great Divide Basin** *basin* Wyoming, C USA
181 W7 **Great Dividing Range** ▲▲ NE Australia
14 D12 **Great Duck Island** *island* Ontario, S Canada
Great Elder Reservoir *see* Waconda Lake
44 G8 **Greater Antilles** *island group* West Indies
129 V16 **Greater Sunda Islands** *var.* Sunda Islands. *island group* Indonesia
184 I1 **Great Exhibition Bay** *inlet* North Island, New Zealand
44 H4 **Great Exuma Island** *island* C The Bahamas
33 R8 **Great Falls** Montana, NW USA 47°30′N 111°18′W
21 R11 **Great Falls** South Carolina, SE USA 34°34′N 80°54′W
84 F9 **Great Fisher Bank** *undersea feature* C North Sea 57°00′N 04°00′E
Great Glen *see* Mor, Glen
Great Grimsby *see* Grimsby
44 I4 **Great Guana Cay** *island* C The Bahamas
64 I5 **Great Hellefiske Bank** *undersea feature* N Atlantic Ocean
111 L24 **Great Hungarian Plain** *var.* Great Alfold, Plain of Hungary, *Hung.* Alföld. *plain* SE Europe
44 L7 **Great Inagua** *var.* Inagua Islands. *island* S The Bahamas
Great Indian Desert *see* Thar Desert
83 G25 **Great Karoo** *var.* Great Karroo, High Veld, *Afr.* Groot Karoo, Hoë Karoo. *plateau region* S South Africa
Great Karroo *see* Great Karoo
Great Kei *see* Nciba
Great Khingan Range *see* Da Hinggan Ling
14 E11 **Great La Cloche Island** *island* Ontario, S Canada
183 P16 **Great Lake** ◎ Tasmania, SE Australia
Great Lake *see* Tônlé Sap
11 R15 **Great Lakes** *lakes* Ontario, Canada/USA
Great Lakes State *see* Michigan
97 L20 **Great Malvern** W England, United Kingdom 52°07′N 02°19′W
184 M5 **Great Mercury Island** *island* N New Zealand
Great Meteor Seamount *see* Great Meteor Tablemount
64 K10 **Great Meteor Tablemount** *var.* Great Meteor Seamount. *undersea feature* E Atlantic Ocean 30°00′N 28°30′W
31 Q14 **Great Miami River** ∽ Ohio, N USA
151 Q24 **Great Nicobar** *island* Nicobar Islands, India, NE Indian Ocean
97 O19 **Great Ouse** *var.* Ouse. ∽ E England, United Kingdom
183 Q17 **Great Oyster Bay** *bay* Tasmania, SE Australia
44 I13 **Great Pedro Bluff** *headland* W Jamaica 17°51′N 77°44′W
21 T12 **Great Pee Dee River** ∽ North Carolina/South Carolina, SE USA
129 W9 **Great Plain of China** *plain* E China
0 F12 **Great Plains** *var.* High Plains. *plains* Canada/USA
37 W6 **Great Plains Reservoirs** ⊠ Colorado, C USA
19 Q13 **Great Point** *headland* Nantucket Island, Massachusetts, NE USA 41°23′N 70°03′W
68 I13 **Great Rift Valley** *var.* Rift Valley. *depression* Asia/Africa
81 I23 **Great Ruaha** ∽ S Tanzania
18 K10 **Great Sacandaga Lake** ⊠ New York, NE USA
108 C12 **Great Saint Bernard Pass** *Fr.* Col du Grand-Saint-Bernard, *It.* Passo del Gran San Bernardo. *pass* Italy/Switzerland
44 F1 **Great Sale Cay** *island* N The Bahamas
Great Salt Desert *see* Kavīr, Dasht-e
36 K1 **Great Salt Lake** *salt lake* Utah, W USA
36 J3 **Great Salt Lake Desert** *plain* Utah, W USA
26 M8 **Great Salt Plains Lake** ⊠ Oklahoma, C USA
75 T9 **Great Sand Sea** *desert* Egypt/Libya
180 L6 **Great Sandy Desert** *desert* Western Australia
Great Sandy Desert *see* Ar Rub' al Khālī
Great Sandy Island *see* Fraser Island
187 Y13 **Great Sea Reef** *reef* Vanua Levu, N Fiji
38 H17 **Great Sitkin Island** *island* Aleutian Islands, Alaska, USA
8 J10 **Great Slave Lake** *Fr.* Grand Lac des Esclaves. ◎ Northwest Territories, NW Canada
21 O10 **Great Smoky Mountains** ▲▲ North Carolina/Tennessee, SE USA
10 L11 **Great Snow Mountain** ▲ British Columbia, W Canada 57°22′N 124°08′W
Great Socialist People's Libyan Arab Jamahiriya *see* Libya
64 A12 **Great Sound** *sound* Bermuda, NW Atlantic Ocean
180 M10 **Great Victoria Desert** *desert* South Australia/Western Australia
194 H2 **Great Wall** *Chinese research station* South Shetland Islands, Antarctica 61°57′S 58°23′W
19 T7 **Great Wass Island** *island* Maine, NE USA
97 Q19 **Great Yarmouth** *var.* Yarmouth. E England, United Kingdom 52°37′N 01°44′E
139 S1 **Great Zab** *Ar.* Az Zāb al Kabīr, *Kurd.* Zē-i Bādīnān, *Turk.* Büyükzap Suyu. ∽ Iraq/Turkey
95 I17 **Grebbestad** Västra Götaland, S Sweden 58°42′N 11°15′E
Grebenka *see* Hrebinka
42 M13 **Grecia** Alajuela, C Costa Rica 10°04′N 84°19′W
61 E18 **Greco** Río Negro, W Uruguay 32°49′S 57°03′W
Greco, Cape *see* Gkréko, Akrotíri
104 L8 **Gredos, Sierra de** ▲▲ W Spain
18 F9 **Greece** New York, NE USA 43°12′N 77°41′W
115 E17 **Greece** *off.* Hellenic Republic, *Gk.* Ellás; *anc.* Hellas. ◆ *republic* SE Europe
Greece Central *see* Stereá Elláda
Greece West *see* Dytikí Elláda
37 T3 **Greeley** Colorado, C USA 40°21′N 104°41′W
29 P14 **Greeley** Nebraska, C USA 41°33′N 98°31′W
122 K3 **Greem-Bell, Ostrov** *Eng.* Graham Bell Island. *island* Zemlya Frantsa-Iosifa, N Russian Federation
30 M6 **Green Bay** Wisconsin, N USA 44°32′N 88°W
31 N6 **Green Bay** *lake bay* Michigan/Wisconsin, N USA
21 S5 **Greenbrier River** ∽ West Virginia, NE USA
29 S2 **Greenbush** Minnesota, N USA 48°42′N 96°10′W
183 R12 **Green Cape** *headland* New South Wales, SE Australia 37°15′S 150°03′E
31 O14 **Greencastle** Indiana, N USA 39°38′N 86°51′W
18 F16 **Greencastle** Pennsylvania, NE USA 39°47′N 77°43′W
27 T2 **Green City** Missouri, C USA 40°16′N 92°57′W
21 O9 **Greeneville** Tennessee, S USA 36°10′N 82°50′W
35 O11 **Greenfield** California, W USA 36°19′N 121°15′W
31 P14 **Greenfield** Indiana, N USA 39°47′N 85°46′W
29 U15 **Greenfield** Iowa, C USA 41°18′N 94°27′W
18 M11 **Greenfield** Massachusetts, NE USA 42°33′N 72°34′W
27 S7 **Greenfield** Missouri, C USA 37°25′N 93°50′W
31 S14 **Greenfield** Ohio, N USA 39°21′N 83°22′W
20 G8 **Greenfield** Tennessee, S USA 36°09′N 88°48′W
30 M9 **Greenfield** Wisconsin, N USA 42°55′N 87°59′W
27 T9 **Green Forest** Arkansas, C USA 36°19′N 93°24′W
37 T7 **Greenhorn Mountain** ▲ Colorado, C USA 37°50′N 104°59′W
Green Island *see* Lü Dao
186 I6 **Green Islands** *var.* Nissan Islands. *island group* NE Papua New Guinea
11 S14 **Green Lake** Saskatchewan, C Canada 54°15′N 107°51′W
30 L8 **Green Lake** ◎ Wisconsin, N USA
197 O14 **Greenland** *Dan.* Grønland, *Inuit* Kalaallit Nunaat. ◇ *Danish self-governing territory* NE North America
84 D4 **Greenland** *island* NE North America
197 R13 **Greenland Plain** *undersea feature* N Greenland Sea
197 R14 **Greenland Sea** *sea* Arctic Ocean
37 R4 **Green Mountain Reservoir** ⊠ Colorado, C USA
18 M8 **Green Mountains** ▲▲ Vermont, NE USA
Green Mountain State *see* Vermont
96 H12 **Greenock** W Scotland, United Kingdom 55°57′N 04°45′W
39 T5 **Greenough, Mount** ▲ Alaska, USA 69°15′N 141°37′W
186 A6 **Green River** West Sepik, NW Papua New Guinea 03°54′S 141°08′E
37 N5 **Green River** Utah, W USA 39°00′N 110°07′W
33 U17 **Green River** Wyoming, C USA 41°33′N 109°27′W
16 H9 **Green River** ∽ W USA
30 K11 **Green River** ∽ Illinois, N USA
20 J7 **Green River** ∽ Kentucky, C USA
28 K5 **Green River** ∽ North Dakota, N USA
37 N6 **Green River** ∽ Utah, W USA
33 T16 **Green River** ∽ Wyoming, C USA
20 L7 **Green River Lake** ⊠ Kentucky, S USA
23 O5 **Greensboro** Alabama, S USA 32°42′N 87°36′W
23 U3 **Greensboro** Georgia, SE USA 33°34′N 83°10′W
21 T9 **Greensboro** North Carolina, SE USA 36°04′N 79°48′W
31 P14 **Greensburg** Indiana, N USA 39°20′N 85°28′W
26 K6 **Greensburg** Kansas, C USA 37°36′N 99°17′W
20 L7 **Greensburg** Kentucky, S USA 37°14′N 85°30′W
18 C15 **Greensburg** Pennsylvania, NE USA 40°18′N 79°32′W
37 O13 **Greens Peak** ▲ Arizona, SW USA 34°06′N 109°34′W
21 V12 **Green Swamp** *wetland* North Carolina, SE USA
21 O4 **Greenup** Kentucky, S USA 38°34′N 82°49′W
36 M16 **Green Valley** Arizona, SW USA 31°49′N 111°00′W
76 K17 **Greenville** *var.* Sino, Sinoe. SE Liberia 05°01′N 09°03′W
23 P6 **Greenville** Alabama, S USA 31°49′N 86°37′W
23 T8 **Greenville** Florida, SE USA 30°28′N 83°37′W
23 S4 **Greenville** Georgia, SE USA 33°03′N 84°42′W
30 L15 **Greenville** Illinois, N USA 38°53′N 89°24′W
20 I7 **Greenville** Kentucky, S USA 37°11′N 87°11′W
19 Q5 **Greenville** Maine, NE USA 45°26′N 69°36′W
31 P9 **Greenville** Michigan, N USA 43°10′N 85°15′W
22 J4 **Greenville** Mississippi, S USA 33°24′N 91°03′W
21 W9 **Greenville** North Carolina, SE USA 35°36′N 77°23′W
31 Q13 **Greenville** Ohio, N USA 40°06′N 84°37′W
19 O12 **Greenville** Rhode Island, NE USA 41°52′N 71°33′W
21 P11 **Greenville** South Carolina, SE USA 34°51′N 82°24′W
25 U6 **Greenville** Texas, SW USA 33°09′N 96°07′W
31 T12 **Greenwich** Ohio, N USA 41°01′N 82°31′W
27 S11 **Greenwood** Arkansas, C USA 35°13′N 94°15′W
31 O14 **Greenwood** Indiana, N USA 39°38′N 86°06′W
22 K4 **Greenwood** Mississippi, S USA 33°30′N 90°11′W
21 P12 **Greenwood** South Carolina, SE USA 34°11′N 82°10′W
21 Q12 **Greenwood, Lake** ⊠ South Carolina, SE USA
21 P11 **Greer** South Carolina, SE USA 34°56′N 82°13′W
27 V10 **Greers Ferry Lake** ⊠ Arkansas, C USA
27 S13 **Greeson, Lake** ⊠ Arkansas, C USA
45 N8 **Gregorio Luperón** ✈ N Dominican Republic 19°43′N 70°43′W
29 O12 **Gregory** South Dakota, N USA 43°11′N 99°26′W
182 J3 **Gregory, Lake** *salt lake* South Australia
180 J9 **Gregory Lake** ◎ Western Australia
181 V5 **Gregory Range** ▲▲ Queensland, E Australia
Greifenberg/Greifenberg in Pommern *see* Gryfice
Greifenhagen *see* Gryfino
100 O8 **Greifswald** Mecklenburg-Vorpommern, NE Germany 54°04′N 13°24′E
100 O8 **Greifswalder Bodden** *bay* NE Germany
109 U4 **Grein** Oberösterreich, N Austria 48°14′N 14°50′E
101 M17 **Greiz** Thüringen, C Germany 50°40′N 12°11′E
125 V14 **Gremyachinsk** Permskiy Kray, NW Russian Federation 58°33′N 57°52′E
Grenå *see* Grenaa
95 H21 **Grenaa** *var.* Grenå. Midtjylland, C Denmark 56°25′N 10°53′E

◆ Country
● Country Capital
◇ Dependent Territory
○ Dependent Territory Capital
◈ Administrative Regions
✈ International Airport
▲ Mountain
▲▲ Mountain Range
Volcano
∽ River
◎ Lake
⊠ Reservoir

Grenada – Gulin

22 L3 **Grenada** Mississippi, S USA 33°46′N 89°48′W
45 W15 **Grenada** ◆ *commonwealth republic* SE West Indies
47 S4 **Grenada** *island* Grenada
47 R4 **Grenada Basin** *undersea feature* W Atlantic Ocean 13°30′N 62°00′W
22 L3 **Grenada Lake** ⊠ Mississippi, S USA
45 Y14 **Grenadines, The** *island group* Grenada/St Vincent and the Grenadines
108 D7 **Grenchen** *Fr.* Granges. Solothurn, NW Switzerland 47°13′N 07°24′E
183 Q9 **Grenfell** New South Wales, SE Australia 33°54′S 148°09′E
11 V16 **Grenfell** Saskatchewan, S Canada 50°24′N 102°56′W
92 J1 **Grenivík** Norðurland Eystra, N Iceland 65°57′N 18°10′W
103 S12 **Grenoble** *anc.* Cularo, Gratianopolis. Isère, E France 45°11′N 05°42′E
28 J2 **Grenora** North Dakota, N USA 48°36′N 103°57′W
92 N8 **Grense-Jakobselv** Finnmark, N Norway 69°46′N 30°39′E
45 S14 **Grenville** E Grenada 12°07′N 61°37′W
32 G11 **Gresham** Oregon, NW USA 45°30′N 122°25′W
Gresk *see* Hresk
106 B7 **Gressoney-St-Jean** Valle d'Aosta, NW Italy 45°48′N 07°49′E
22 K9 **Gretna** Louisiana, S USA 29°54′N 90°03′W
21 T7 **Gretna** Virginia, NE USA 36°57′N 79°21′W
98 F13 **Grevelingen** *inlet* S North Sea
100 F13 **Greven** Nordrhein-Westfalen, NW Germany 52°07′N 07°38′E
115 D15 **Grevená** Dytikí Makedonía, N Greece 40°05′N 21°26′E
101 D16 **Grevenbroich** Nordrhein-Westfalen, W Germany 51°06′N 06°34′E
99 N24 **Grevenmacher** Grevenmacher, E Luxembourg 49°41′N 06°27′E
99 M24 **Grevenmacher** ◈ *district* E Luxembourg
100 K9 **Grevesmühlen** Mecklenburg-Vorpommern, N Germany 53°52′N 11°12′E
185 H16 **Grey** ≈ South Island, New Zealand
33 V12 **Greybull** Wyoming, C USA 44°29′N 108°03′W
33 U13 **Greybull River** ≈ Wyoming, C USA
65 A24 **Grey Channel** *sound* Falkland Islands
Greyerzer See *see* Gruyère, Lac de la
13 T10 **Grey Islands** *island group* Newfoundland and Labrador, E Canada
18 L10 **Greylock, Mount** ▲ Massachusetts, NE USA 42°38′N 73°09′W
185 G17 **Greymouth** West Coast, South Island, New Zealand 42°29′S 171°14′E
181 U10 **Grey Range** ▲▲ New South Wales/Queensland, E Australia
97 G18 **Greystones** *Ir.* Na Clocha Liatha. E Ireland 53°08′N 06°05′W
185 M14 **Greytown** Wellington, North Island, New Zealand 41°04′S 175°29′E
83 K23 **Greytown** KwaZulu/Natal, E South Africa 29°04′S 30°35′E
Greytown *see* San Juan del Norte
99 H19 **Grez-Doiceau** *Dut.* Graven. Walloon Brabant, C Belgium 50°43′N 04°41′E
115 J19 **Griá, Akrotírio** *headland* Ándros, Kykládes, Greece, Aegean Sea 37°54′N 24°57′E
127 N8 **Gribanovskiy** Voronezhskaya Oblast', W Russian Federation 51°27′N 41°53′E
78 I13 **Gribingui** ≈ N Central African Republic
35 O6 **Gridley** California, W USA 39°21′N 121°41′W
83 G23 **Griekwastad** *var.* Griquatown. Northern Cape, C South Africa 28°50′S 23°16′E
23 S4 **Griffin** Georgia, SE USA 33°15′N 84°17′W
183 O9 **Griffith** New South Wales, SE Australia 34°18′S 146°04′E
14 F13 **Griffith Island** *island* Ontario, S Canada
21 W10 **Grifton** North Carolina, SE USA 35°22′N 77°26′W
Grigioni *see* Graubünden
119 H14 **Grigiškes** Vilnius, SE Lithuania 54°42′N 25°00′E
117 N10 **Grigoriopol** C Moldova 47°09′N 29°18′E
147 X7 **Grigor'yevka** Issyk-Kul'skaya Oblast', E Kyrgyzstan 42°43′N 77°27′E
193 U8 **Grijalva Ridge** *undersea feature* E Pacific Ocean
41 U15 **Grijalva, Río** *var.* Tabasco. ≈ Guatemala/Mexico
98 N5 **Grijpskerk** Groningen, NE Netherlands 53°15′N 06°18′E
83 C22 **Grillenthal** Karas, SW Namibia 26°55′S 15°24′E
79 J15 **Grimari** Ouaka, C Central African Republic 05°44′N 20°02′E
Grimaylov *see* Hrymayliv
99 G18 **Grimbergen** Vlaams Brabant, C Belgium 50°56′N 04°22′E
183 N15 **Grim, Cape** *headland* Tasmania, SE Australia 40°42′S 144°42′E
100 N8 **Grimmen** Mecklenburg-Vorpommern, NE Germany 54°06′N 13°03′E
14 G16 **Grimsby** Ontario, S Canada 43°12′N 79°35′W
97 O17 **Grimsby** *prev.* Great Grimsby. E England, United Kingdom 53°35′N 00°05′W
92 J1 **Grímsey** *var.* Grimsey. *island* N Iceland
Grimsey *see* Grímsey
11 O12 **Grimshaw** Alberta, W Canada 56°11′N 117°37′W
95 F18 **Grimstad** Aust-Agder, S Norway 58°20′N 08°35′E
92 H4 **Grindavík** Suðurnes, W Iceland 65°57′N 18°10′W
108 F9 **Grindelwald** Bern, S Switzerland 46°37′N 08°04′E
95 F23 **Grindsted** Syddtjylland, W Denmark 55°46′N 08°56′E
29 W14 **Grinnell** Iowa, C USA 41°44′N 92°43′W
109 U10 **Grintovec** ▲ N Slovenia 46°21′N 14°31′E
Griquatown *see* Griekwastad
9 N4 **Grise Fiord** *var* Aujuittuq. Northwest Territories, Ellesmere Island, N Canada 76°10′N 83°15′W
182 H1 **Griselda, Lake** *salt lake* South Australia
Grisons *see* Graubünden
95 P14 **Grisslehamn** Stockholm, C Sweden 60°04′N 18°50′E
29 T15 **Griswold** Iowa, C USA 41°14′N 95°08′W
102 M1 **Griz Nez, Cap** *headland* N France 50°51′N 01°34′E
112 P13 **Grljan** Serbia, E Serbia 43°52′N 22°18′E
112 E11 **Grmeč** ▲▲ NW Bosnia and Herzegovina
99 H16 **Grobbendonk** Antwerpen, N Belgium 51°12′N 04°41′E
Grobin *see* Grobiņa
118 C10 **Grobiņa** *Ger.* Grobin. W Latvia 56°32′N 21°12′E
83 K20 **Groblersdal** Mpumalanga, NE South Africa 25°15′S 29°25′E
83 G23 **Groblershoop** Northern Cape, W South Africa 28°51′S 22°01′E
Gródek Jagielloński *see* Horodok
109 Q6 **Grödig** Salzburg, W Austria 47°42′N 13°06′E
111 H15 **Grodków** Opolskie, S Poland 50°42′N 17°23′E
Grodnenskaya Oblast' *see* Hrodzyenskaya Voblasts'
Grodno *see* Hrodna
110 L12 **Grodzisk Mazowiecki** Mazowieckie, C Poland 52°09′N 20°38′E
110 F12 **Grodzisk Wielkopolski** Wielkopolskie, C Poland 52°13′N 16°21′E
Grodzyanka *see* Hradzyanka
98 O12 **Groenlo** Gelderland, E Netherlands 52°02′N 06°36′E
83 E22 **Groenrivier** Karas, SE Namibia 27°27′S 18°52′E
25 U8 **Groesbeck** Texas, SW USA 31°31′N 96°35′W
98 L13 **Groesbeek** Gelderland, SE Netherlands 51°47′N 05°56′E
102 G7 **Groix, Îles de** *island group* NW France
110 M12 **Grójec** Mazowieckie, C Poland 51°51′N 20°52′E
65 K15 **Gröll Seamount** *undersea feature* C Atlantic Ocean 12°54′S 33°24′W
100 E13 **Gronau** *var.* Gronau in Westfalen. Nordrhein-Westfalen, NW Germany 52°13′N 07°02′E
Gronau in Westfalen *see* Gronau
93 F15 **Grong** Nord-Trøndelag, C Norway 64°29′N 12°19′E
95 N22 **Grönhögen** Kalmar, S Sweden 56°16′N 16°09′E
98 N5 **Groningen** Groningen, NE Netherlands 53°13′N 06°35′E
55 W9 **Groningen** Saramacca, N Suriname 05°45′N 55°31′W
98 N5 **Groningen** ◈ *province* NE Netherlands
Grønland *see* Greenland
108 H11 **Grono** Graubünden, S Switzerland 46°15′N 09°07′E
95 M20 **Grönskåra** Kalmar, S Sweden 57°04′N 15°45′E
25 O2 **Groom** Texas, SW USA 35°12′N 101°06′W
35 W9 **Groom Lake** ◎ Nevada, W USA
83 H25 **Groot** ≈ S South Africa
181 S2 **Groote Eylandt** *island* Northern Territory, N Australia
98 M6 **Grootegast** Groningen, NE Netherlands 53°11′N 06°12′E
83 D17 **Grootfontein** Otjozondjupa, N Namibia 19°32′S 18°05′E
83 E22 **Groot Karasberge** ▲▲ S Namibia
Groot Karoo *see* Great Karoo
Groot-Kei *see* Nciba
15 V6 **Grosses-Roches** Québec, SE Canada 48°55′N 67°06′W
109 V2 **Gross-Siegharts** Niederösterreich, N Austria 48°48′N 15°25′E
45 T10 **Gros Islet** N Saint Lucia 14°04′N 60°57′W
44 L8 **Gros-Morne** NW Haiti 19°45′N 72°46′W
13 S11 **Gros Morne** ▲ Newfoundland, Newfoundland and Labrador, E Canada 49°38′N 57°45′W
103 R9 **Grosne** ≈ C France
45 S12 **Gros Piton** ▲ SW Saint Lucia 13°48′N 61°04′W
Grossa, Isola *see* Dugi Otok
Grossbetschkerek *see* Zrenjanin
Grosse Isper *see* Grosse Ysper
Grosse Kokel *see* Târnava Mare
101 M21 **Grosse Laaber** *var.* Grosse Laber. ≈ SE Germany
Grosse Laber *see* Grosse Laaber
Grosse Morava *see* Velika Morava
101 O15 **Grossenhain** Sachsen, E Germany 51°18′N 13°31′E
109 Y4 **Grossenzersdorf** Niederösterreich, NE Austria 48°12′N 16°33′E
101 O21 **Grosser Arber** ▲ SE Germany 49°07′N 13°10′E
101 K17 **Grosser Beerberg** ▲ C Germany 50°39′N 10°45′E
101 G18 **Grosser Feldberg** ▲ W Germany 50°13′N 08°28′E
109 O8 **Grosser Löffler** *It.* Monte Lovello. ▲ Austria/Italy 47°02′N 11°56′E
109 N8 **Grosser Möseler** *var.* Mesule. ▲ Austria/Italy 47°01′N 11°52′E
100 J8 **Grosser Plöner See** ◎ N Germany
101 O21 **Grosser Rachel** ▲ SE Germany 48°59′N 13°23′E
Grosser Sund *see* Suur Väin
109 P8 **Grosses Wiesbachhorn** *var.* Wiesbachhorn. ▲ W Austria 47°09′N 12°44′E
106 F13 **Grosseto** Toscana, C Italy 42°45′N 11°07′E
101 M22 **Grosse Vils** ≈ SE Germany
109 U4 **Grosse Ysper** *var.* Grosse Isper. ≈ N Austria
101 G19 **Gross-Gerau** Hessen, W Germany 49°55′N 08°28′E
109 U3 **Gross Gerungs** Niederösterreich, N Austria 48°33′N 14°58′E
109 P8 **Grossglockner** ▲ W Austria 47°05′N 12°39′E
Grosskanizsa *see* Nagykanizsa
Gross-Karol *see* Carei
Grosskikinda *see* Kikinda
109 W9 **Grossklein** Steiermark, SE Austria 46°43′N 15°24′E
Grosskoppe *see* Velká Deštná
Grossmeseritsch *see* Velké Meziříčí
Grossmichel *see* Michalovce
101 H19 **Grossostheim** Bayern, C Germany 49°54′N 09°03′E
109 X7 **Grosspetersdorf** Burgenland, SE Austria 47°15′N 16°19′E
109 T5 **Grossraming** Oberösterreich, C Austria 47°54′N 14°34′E
101 P14 **Grossräschen** Brandenburg, E Germany 51°34′N 14°00′E
Grossrauschenbach *see* Revúca
Gross-Sankt-Johannis *see* Suure-Jaani
Gross-Schlatten *see* Abrud
Gross-Skaisgirren *see* Bol'shakovo
Gross-Steffelsdorf *see* Rimavská Sobota
Gross Strehlitz *see* Strzelce Opolskie
109 O8 **Grossvenediger** ▲ W Austria 47°07′N 12°19′E
Grosswardein *see* Oradea
Gross Wartenberg *see* Syców
109 U11 **Grosuplje** C Slovenia 46°00′N 14°36′E
99 H17 **Grote Nete** ≈ N Belgium
94 E10 **Grotli** Oppland, S Norway 62°02′N 07°36′E
19 N13 **Groton** Connecticut, NE USA 41°20′N 72°03′W
29 P8 **Groton** South Dakota, N USA 45°27′N 98°06′W
107 P18 **Grottaglie** Puglia, SE Italy 40°32′N 17°26′E
107 L17 **Grottaminarda** Campania, S Italy 41°04′N 15°02′E
106 K13 **Grottammare** Marche, C Italy 43°00′N 13°52′E
21 U5 **Grottoes** Virginia, NE USA 38°16′N 78°49′W
98 L6 **Grou** *Dutch.* Grouw. Fryslân, N Netherlands 53°07′N 05°51′E
13 N10 **Groulx, Monts** ▲▲ Québec, E Canada
14 E7 **Groundhog** ≈ Ontario, S Canada
36 J1 **Grouse Creek** Utah, W USA 41°41′N 113°52′W
36 J1 **Grouse Creek Mountains** ▲▲ Utah, W USA
Grouw *see* Grou
27 R8 **Grove** Oklahoma, C USA 36°35′N 94°46′W
31 S13 **Grove City** Ohio, N USA 39°52′N 83°05′W
18 B13 **Grove City** Pennsylvania, NE USA 41°09′N 80°02′W
23 O6 **Grove Hill** Alabama, S USA 31°42′N 87°46′W
33 S15 **Grover** Wyoming, C USA 42°48′N 110°57′W
35 P13 **Grover City** California, W USA 35°08′N 120°37′W
25 Y11 **Groves** Texas, SW USA 29°57′N 93°55′W
19 O7 **Groveton** New Hampshire, NE USA 44°35′N 71°28′W
25 W9 **Groveton** Texas, SW USA 31°04′N 95°08′W
36 J15 **Growler Mountains** ▲▲ Arizona, SW USA
Grozdovo *see* Bratya Daskalovi
127 P16 **Groznyy** Chechenskaya Respublika, SW Russian Federation 43°20′N 45°43′E
Grubeshov *see* Hrubieszów
112 G9 **Grubišno Polje** Bjelovar-Bilogora, NE Croatia 45°42′N 17°09′E
Grudovo *see* Sredets
110 J9 **Grudziądz** *Ger.* Graudenz. Kujawsko-pomorskie, C Poland 53°29′N 18°45′E
25 R17 **Grulla** *var.* La Grulla. Texas, SW USA 26°15′N 98°37′W
40 K14 **Grullo** Jalisco, SW Mexico 19°45′N 104°15′W
67 V10 **Grumeti** ≈ N Tanzania
95 K16 **Grums** Värmland, C Sweden 59°22′N 13°11′E
109 S5 **Grünau im Almtal** Oberösterreich, N Austria 47°51′N 13°56′E
101 H17 **Grünberg** Hessen, W Germany 50°36′N 08°57′E
Grünberg/Grünberg in Schlesien *see* Zielona Góra
92 H3 **Grundarfjörður** Vestfirðir, W Iceland 64°55′N 23°15′W
21 P7 **Grundy** Virginia, NE USA 37°17′N 82°06′W
29 W13 **Grundy Center** Iowa, C USA 42°21′N 92°46′W
Grüneberg *see* Zielona Góra
25 N1 **Gruver** Texas, SW USA 36°16′N 101°24′W
108 C9 **Gruyère, Lac de la** *Ger.* Greyerzer See. ◎ SW Switzerland
108 C9 **Gruyères** Fribourg, W Switzerland 46°34′N 07°04′E
118 E11 **Gruzdžiai** Šiauliai, N Lithuania 56°06′N 23°15′E
Gruzinskaya SSR/Gruziya *see* Georgia
Gryada Akkyr *see* Akgyr Erezi
126 L7 **Gryazi** Lipetskaya Oblast', W Russian Federation 52°27′N 39°56′E
124 M14 **Gryazovets** Vologodskaya Oblast', NW Russian Federation 58°52′N 40°12′E
111 M17 **Grybów** Małopolskie, SE Poland 49°35′N 20°54′E
94 M13 **Grycksbo** Dalarna, C Sweden 60°40′N 15°30′E
110 E8 **Gryfice** *Ger.* Greifenberg, Greifenberg in Pommern. Zachodnio-pomorskie, NW Poland 53°55′N 15°11′E
110 D9 **Gryfino** *Ger.* Greifenhagen. Zachodnio-pomorskie, NW Poland 53°15′N 14°30′E
92 H9 **Gryllefjord** Troms, N Norway 69°21′N 17°07′E
95 L15 **Grythyttan** Örebro, C Sweden 59°52′N 14°31′E
108 D10 **Gstaad** Bern, W Switzerland 46°30′N 07°16′E
43 P14 **Guabito** Bocas del Toro, NW Panama 09°30′N 82°35′W
44 G7 **Guacanayabo, Golfo de** *gulf* S Cuba
40 I7 **Guachochi** Chihuahua, N Mexico
104 J11 **Guadajira** ≈ SW Spain
104 M13 **Guadajoz** ≈ S Spain
40 L13 **Guadalajara** Jalisco, C Mexico 20°43′N 103°24′W
105 O8 **Guadalajara** *Ar.* Wad Al-Hajarah; *anc.* Arriaca. Castilla-La Mancha, C Spain 40°37′N 03°10′W
105 O7 **Guadalajara** ◈ *province* Castilla-La Mancha, C Spain
104 K12 **Guadalcanal** Andalucía, S Spain 38°06′N 05°49′W
186 L10 **Guadalcanal** *off.* Guadalcanal Province. ◈ *province* C Solomon Islands
186 M9 **Guadalcanal** *island* C Solomon Islands
Guadalcanal Province *see* Guadalcanal
105 O12 **Guadalén** ≈ S Spain
105 R13 **Guadalentín** ≈ SE Spain
104 K15 **Guadalete** ≈ SW Spain
105 O13 **Guadalimar** ≈ S Spain
105 P12 **Guadalmena** ≈ S Spain
104 L11 **Guadalmez** ≈ W Spain
105 S7 **Guadalope** ≈ E Spain
104 K13 **Guadalquivir** ≈ W Spain
104 J14 **Guadalquivir, Marismas del** *var.* Las Marismas. *wetland* SW Spain
40 M11 **Guadalupe** Zacatecas, C Mexico 22°47′N 102°30′W
57 E16 **Guadalupe** Ica, W Peru 13°59′S 75°49′W
104 L10 **Guadalupe** Extremadura, W Spain 39°26′N 05°18′W
36 L14 **Guadalupe** Arizona, SW USA 33°20′N 111°57′W
35 P13 **Guadalupe** California, W USA 34°55′N 120°34′W
Guadalupe *see* Canelones
40 J3 **Guadalupe Bravos** Chihuahua, N Mexico 31°22′N 106°04′W
40 A4 **Guadalupe, Isla** *island* NW Mexico
37 U15 **Guadalupe Mountains** ▲▲ New Mexico/Texas, SW USA
24 J8 **Guadalupe Peak** ▲ Texas, SW USA 31°53′N 104°51′W
25 R11 **Guadalupe River** ≈ SW USA
104 K10 **Guadalupe, Sierra de** ▲▲ W Spain
40 K9 **Guadalupe Victoria** Durango, C Mexico 24°30′N 104°08′W
40 I8 **Guadalupe y Calvo** Chihuahua, N Mexico 26°04′N 106°58′W
105 N7 **Guadarrama** Madrid, C Spain 40°40′N 04°06′W
105 N7 **Guadarrama** ≈ C Spain
104 M7 **Guadarrama, Puerto de** *pass* C Spain
105 N9 **Guadarrama, Sierra de** ▲▲ C Spain
105 Q9 **Guadazaón** ≈ C Spain
45 X10 **Guadeloupe** ◇ *French overseas department* E West Indies
47 S3 **Guadeloupe** *island group* E West Indies
45 W10 **Guadeloupe Passage** *passage* E Caribbean Sea
104 H13 **Guadiana** ≈ Portugal/Spain
105 O13 **Guadiana Menor** ≈ S Spain
105 Q8 **Guadiela** ≈ C Spain
105 O14 **Guadix** Andalucía, S Spain 37°19′N 03°08′W
Guad-i-Zirreh *see* Gowd-e Zereh, Dasht-e
193 T12 **Guafo Fracture Zone** *tectonic feature* SE Pacific Ocean
63 F18 **Guafo, Isla** *island* S Chile
42 I6 **Guaimaca** Francisco Morazán, C Honduras 14°34′N 86°49′W
54 J12 **Guainía** *off.* Comisaría del Guainía. ◈ *province* E Colombia
Guainía, Comisaría del *see* Guainía
54 K12 **Guainía, Río** ≈ Colombia/Venezuela
55 O9 **Guaiquinima, Cerro** *elevation* SE Venezuela
60 G10 **Guaíra** Paraná, S Brazil 24°05′S 54°15′W
60 L7 **Guaíra** São Paulo, S Brazil 20°17′S 48°21′W
62 O7 **Guairá** *off.* Departamento del Guairá. ◈ *department* S Paraguay
Guairá, Departamento del *see* Guairá
Guaire *see* Gorey
63 F18 **Guaiteca, Isla** *island* S Chile
44 G6 **Guajaba, Cayo** *headland* C Cuba 21°50′N 77°33′W
59 D16 **Guajará-Mirim** Rondônia, W Brazil 10°50′S 65°21′W
Guajira *see* La Guajira
Guajira, Departamento de La *see* La Guajira
54 H3 **Guajira, Península de la** *peninsula* N Colombia
42 J6 **Gualaco** Olancho, C Honduras 15°00′N 86°03′W
34 L7 **Gualala** California, W USA 38°45′N 123°33′W
42 E5 **Gualán** Zacapa, C Guatemala 15°06′N 89°22′W
61 C19 **Gualeguay** Entre Ríos, E Argentina 33°09′S 59°20′W
61 D18 **Gualeguaychú** Entre Ríos, E Argentina 33°03′S 58°31′W
61 C18 **Gualeguay, Río** ≈ E Argentina
63 K16 **Gualicho, Salina del** *salt lake* E Argentina
188 B15 **Guam** ◇ *US unincorporated territory* W Pacific Ocean
63 F19 **Guamblin, Isla** *island* Archipiélago de los Chonos, S Chile
61 A22 **Guaminí** Buenos Aires, E Argentina 37°01′S 62°28′W
40 H8 **Guamúchil** Sinaloa, C Mexico 25°23′N 108°01′W
54 H4 **Guana** *var.* Misión de Guana. Zulia, NW Venezuela 11°07′N 72°17′W
44 C4 **Guanabacoa** La Habana, W Cuba 23°02′N 82°12′W
42 K13 **Guanacaste** *off.* Provincia de Guanacaste. ◈ *province* NW Costa Rica
42 K12 **Guanacaste, Cordillera de** ▲▲ NW Costa Rica
Guanacaste, Provincia de *see* Guanacaste
40 J8 **Guanaceví** Durango, C Mexico 25°55′N 105°51′W
44 A5 **Guanahacabibes, Golfo de** *gulf* W Cuba
42 K4 **Guanaja, Isla de** *island* Islas de la Bahía, N Honduras
44 C4 **Guanajay** La Habana, W Cuba 22°56′N 82°42′W
41 N12 **Guanajuato** Guanajuato, C Mexico 21°N 101°19′W
40 M12 **Guanajuato** ◈ *state* C Mexico
54 J6 **Guanare** Portuguesa, N Venezuela 09°04′N 69°45′W
54 K7 **Guanare, Río** ≈ W Venezuela
54 J6 **Guanarito** Portuguesa, NW Venezuela 08°43′N 69°12′W
160 M3 **Guancen Shan** ▲▲ C China
62 I9 **Guandacol** La Rioja, W Argentina 29°32′S 68°37′W
44 A5 **Guane** Pinar del Río, W Cuba 22°12′N 84°05′W
161 N14 **Guangdong** *var.* Guangdong Sheng, Kuang-tung, Kwangtung, Yue. ◈ *province* S China
Guangdong Sheng *see* Guangdong
Guanghua *see* Laohekou
Guangju *see* Gwangju
161 Q9 **Guangming Ding** ▲ Anhui, China 30°06′N 118°04′E
160 I13 **Guangnan** *var.* Liancheng. Yunnan, SW China 24°07′N 104°54′E
161 N8 **Guangshui** *prev.* Yinshan. Hubei, C China 31°41′N 113°53′E
Guangxi *see* Guangxi Zhuangzu Zizhiqu
160 K14 **Guangxi Zhuangzu Zizhiqu** *var.* Guangxi, Gui, Kuang-hsi, Kwangsi, *Eng.* Kwangsi Chuang Autonomous Region. ◈ *autonomous region* S China
160 J8 **Guangyuan** *var.* Kuang-yuan, Kwangyuan. Sichuan, C China 32°27′N 105°49′E
161 N14 **Guangzhou** *var.* Kuang-chou, Kwangchow, *Eng.* Canton. *province capital* Guangdong, S China 23°11′N 113°19′E
59 N19 **Guanhães** Minas Gerais, SE Brazil 18°46′S 42°58′W
160 I12 **Guanling** *var.* Guanling Bouyeizu Miaozu Zizhixian. Guizhou, S China 26°00′N 105°40′E
Guanling Bouyeizu Miaozu Zizhixian *see* Guanling
55 N5 **Guanta** Anzoátegui, NE Venezuela 10°15′N 64°38′W
44 J8 **Guantánamo** Guantánamo, SE Cuba 20°06′N 75°16′W
44 J8 **Guantánamo, Bahia de** *Eng.* Guantanamo Bay. *US military base* SE Cuba 20°06′N 75°16′W
Guantanamo Bay *see* Guantánamo, Bahia de
Guanxian/Guan Xian *see* Dujiangyan
161 Q6 **Guanyun** *var.* Yishan. Jiangsu, E China 34°18′N 119°14′E
54 C12 **Guapí** Cauca, SW Colombia 02°36′N 77°54′W
43 N13 **Guápiles** Limón, NE Costa Rica 10°15′N 83°46′W
61 I15 **Guaporé** Rio Grande do Sul, S Brazil 28°55′S 51°53′W
47 S8 **Guaporé, Rio** *var.* Río Iténez. ≈ Bolivia/Brazil *see also* Río Iténez
Guaporé, Rio *see* Iténez, Río
56 B7 **Guaranda** Bolívar, C Ecuador 01°35′S 78°59′W
60 H11 **Guaraniaçu** Paraná, S Brazil 25°05′S 52°52′W
59 O20 **Guarapari** Espírito Santo, SE Brazil 20°39′S 40°31′W
60 I12 **Guarapuava** Paraná, S Brazil 25°22′S 51°28′W
60 J8 **Guararapes** São Paulo, S Brazil 21°16′S 50°37′W
105 S4 **Guara, Sierra de** ▲▲ NE Spain
60 N10 **Guaratinguetá** São Paulo, S Brazil 22°44′S 45°16′W
104 I7 **Guarda** Guarda, N Portugal 40°32′N 07°17′W
104 I7 **Guarda** ◈ *district* N Portugal
Guardak *see* Magdanly
104 M3 **Guardo** Castilla y León, N Spain 42°48′N 04°50′W
104 K11 **Guareña** Extremadura, W Spain 38°51′N 06°06′W
60 J11 **Guaricana, Pico** ▲ S Brazil 25°13′S 48°50′W
54 L6 **Guárico** *off.* Estado Guárico. ◈ *state* N Venezuela
Guárico, Estado *see* Guárico
44 J7 **Guarico, Punta** *headland* E Cuba 20°36′N 74°43′W
54 L7 **Guárico, Río** ≈ C Venezuela
60 M10 **Guarujá** São Paulo, SE Brazil 23°50′S 46°27′W
61 L22 **Guarulhos** ✈ (São Paulo) São Paulo, S Brazil 23°23′S 46°32′W
43 R17 **Guarumal** Veraguas, S Panama 07°48′N 81°15′W
Guasapa *see* Guasopa
40 H8 **Guasave** Sinaloa, C Mexico 25°33′N 108°29′W
54 I8 **Guasdualito** Apure, C Venezuela 07°15′N 70°40′W
55 Q7 **Guasipati** Bolívar, E Venezuela 07°28′N 61°58′W
186 I9 **Guasopa** *var.* Guasapa. Woodlark Island, SE Papua New Guinea 09°12′S 152°58′E
106 F9 **Guastalla** Emilia-Romagna, C Italy 44°54′N 10°38′E
42 D6 **Guastatoya** *var.* El Progreso. El Progreso, C Guatemala 14°51′N 90°01′W
42 D5 **Guatemala** *off.* Republic of Guatemala. ◆ *republic* Central America
42 A2 **Guatemala** *off.* Departamento de Guatemala. ◈ *department* S Guatemala
193 S7 **Guatemala Basin** *undersea feature* E Pacific Ocean 11°00′N 95°00′W
Guatemala City *see* Ciudad de Guatemala
Guatemala, Departamento de *see* Guatemala
Guatemala, Republic of *see* Guatemala
45 V14 **Guatuaro Point** *headland* Trinidad, Trinidad and Tobago 10°19′N 60°58′W
186 B8 **Guavi** ≈ SW Papua New Guinea
54 G13 **Guaviare** *off.* Comisaría Guaviare. ◈ *province* S Colombia
Guaviare, Comisaría *see* Guaviare
54 J11 **Guaviare, Río** ≈ E Colombia
61 E15 **Guaviravi** Corrientes, NE Argentina 29°20′S 56°50′W
54 G12 **Guayabero, Río** ≈ SW Colombia
45 U6 **Guayama** E Puerto Rico 17°59′N 66°07′W
42 J7 **Guayambre, Río** ≈ S Honduras
Guayanas, Macizo de las *see* Guiana Highlands
45 V6 **Guayanés, Punta** *headland* E Puerto Rico 18°03′N 65°48′W
42 J6 **Guayape, Río** ≈ C Honduras
56 B7 **Guayaquil** *var.* Santiago de Guayaquil. Guayas, SW Ecuador 02°13′S 79°54′W
Guayaquil *see* Simon Bolivar
56 A8 **Guayaquil, Golfo de** *var.* Gulf of Guayaquil. *gulf* SW Ecuador
Guayaquil, Gulf of *see* Guayaquil, Golfo de
56 A7 **Guayas** ◈ *province* W Ecuador
62 N7 **Guaycurú, Río** ≈ NE Argentina
40 F6 **Guaymas** Sonora, NW Mexico 27°56′N 110°54′W
45 U5 **Guaynabo** E Puerto Rico 18°19′N 66°05′W
80 H12 **Guba** Bīnishangul Gumuz, W Ethiopia 11°11′N 35°21′E
146 H8 **Gubadag** *Turkm.* Tel'man; *prev.* Tel'mansk. Daşoguz Welaýaty, N Turkmenistan 42°07′N 59°55′E
125 V13 **Gubakha** Permskiy Kray, NW Russian Federation 58°52′N 57°35′E
106 I12 **Gubbio** Umbria, C Italy 43°27′N 12°34′E
100 Q13 **Guben** *var.* Wilhelm-Pieck-Stadt. Brandenburg, E Germany 51°59′N 14°42′E
Guben *see* Gubin
110 D12 **Gubin** *Ger.* Guben. Lubuskie, W Poland 51°59′N 14°43′E
126 K8 **Gubkin** Belgorodskaya Oblast', W Russian Federation 51°16′N 37°32′E
162 J9 **Guchin-Us** *var.* Arguut. Övörhangay, C Mongolia 45°27′N 102°25′E
Gudara *see* Ghŭdara
105 S8 **Gúdar, Sierra de** ▲▲ E Spain
137 P8 **Gudauta** *prev.* Gudaut'a. NW Georgia 43°07′N 40°35′E
Gudaut'a *see* Gudauta
94 G12 **Gudbrandsdalen** *valley* S Norway
95 G21 **Gudenå** *var.* Gudenaa. ≈ C Denmark
Gudenaa *see* Gudenå
127 P16 **Gudermes** Chechenskaya Respublika, SW Russian Federation 43°23′N 46°06′E
155 J18 **Gūdūr** Andhra Pradesh, E India 14°10′N 79°51′E
146 B13 **Gudurolum** Balkan Welaýaty, W Turkmenistan 37°28′N 54°30′E
94 D13 **Gudvangen** Sogn Og Fjordane, S Norway 60°54′N 06°49′E
103 U7 **Guebwiller** Haut-Rhin, NE France 47°55′N 07°13′E
Guéckédou *see* Guékédou
14 K8 **Guéguen, Lac** ◎ Québec, SE Canada
76 J15 **Guékédou** *var.* Guéckédou. Guinée-Forestière, S Guinea 08°33′N 10°08′W
41 R16 **Guelatao** Oaxaca, SE Mexico
Guelders *see* Gelderland
78 G11 **Guélengdeng** Mayo-Kébbi Est, W Chad 10°55′N 15°31′E
74 L5 **Guelma** *var.* Gâlma. NE Algeria 36°29′N 07°25′E
74 D8 **Guelmime** *var.* Goulimime. SW Morocco 28°59′N 10°10′W
14 G15 **Guelph** Ontario, S Canada 43°34′N 80°16′W
102 I7 **Gueméné-Penfao** Loire-Atlantique, NW France 47°37′N 01°49′W
102 I7 **Guer** Morbihan, NW France 47°54′N 02°07′W
78 I11 **Guéra** *off.* Région du Guéra. ◈ *region* S Chad
102 H8 **Guérande** Loire-Atlantique, NW France 47°20′N 02°25′W
Guéra, Région du *see* Guéra
78 K9 **Guéréda** Wadi Fira, E Chad 14°30′N 22°05′E
103 N10 **Guéret** Creuse, C France 46°10′N 01°52′E
Guernica/Guernica y Lumo *see* Gernika-Lumo
33 Z15 **Guernsey** Wyoming, C USA 42°16′N 104°44′W
97 K25 **Guernsey** ◇ *British Crown Dependency* Channel Islands, NW Europe
97 K25 **Guernsey** *island* Channel Islands, NW Europe
76 J10 **Guérou** Assaba, S Mauritania 16°48′N 11°40′W
25 R16 **Guerra** Texas, SW USA 26°54′N 98°53′W
41 O15 **Guerrero** ◈ *state* S Mexico
40 D6 **Guerrero Negro** Baja California Sur, NW Mexico 27°56′N 114°04′W
103 P9 **Gueugnon** Saône-et-Loire, C France 46°36′N 04°03′E
76 M17 **Guéyo** S Ivory Coast 05°25′N 06°04′W
107 L15 **Guglionesi** Molise, C Italy 41°54′N 14°54′E
188 K5 **Guguan** *island* C Northern Mariana Islands
Guhrau *see* Góra
Gui *see* Guangxi Zhuangzu Zizhiqu
Guiana *see* French Guiana
47 V4 **Guiana Basin** *undersea feature* W Atlantic Ocean 11°00′N 52°00′W
48 G6 **Guiana Highlands** *var.* Macizo de las Guayanas. ▲▲ N South America
Guiba *see* Juba
102 I7 **Guichen** Ille-et-Vilaine, NW France 47°57′N 01°47′W
Guichi *see* Chizhou
61 E18 **Guichón** Paysandú, W Uruguay 32°30′S 57°13′W
77 U12 **Guidan-Roumji** Maradi, S Niger 13°40′N 06°41′E
Guidder *see* Guider
159 T10 **Guide** *var.* Heyin. Qinghai, C China 36°06′N 101°25′E
78 F12 **Guider** *var.* Guidder. Nord, N Cameroon 09°55′N 13°59′E
76 I11 **Guidimaka** ◈ *region* S Mauritania
77 W12 **Guidimouni** Zinder, S Niger 13°40′N 09°31′E
76 G10 **Guier, Lac de** *var.* Lac de Guiers. ◎ N Senegal
Guiers, Lac de *see* Guier, Lac de
160 L14 **Guigang** *var.* Guixian, Gui Xian. Guangxi Zhuangzu Zizhiqu, S China 23°06′N 109°36′E
76 L16 **Guiglo** W Ivory Coast 06°33′N 07°29′W
54 L5 **Güigüe** Carabobo, N Venezuela 10°05′N 67°48′W
83 M20 **Guijá** Gaza, S Mozambique 24°31′S 33°02′E
42 E7 **Güija, Lago de** ◎ El Salvador/Guatemala
160 L14 **Gui Jiang** *var.* Gui Shui. ≈ S China
104 K8 **Guijuelo** Castilla y León, N Spain 40°34′N 05°40′W
Guilan *see* Gīlān
97 N22 **Guildford** SE England, United Kingdom 51°14′N 00°35′W
19 R5 **Guildford** Maine, NE USA 45°10′N 69°22′W
19 O7 **Guildhall** Vermont, NE USA 44°34′N 71°36′W
103 R13 **Guilherand** Ardèche, E France 44°57′N 04°49′E
160 L13 **Guilin** *var.* Kuei-lin, Kweilin. Guangxi Zhuangzu Zizhiqu, S China 25°15′N 110°16′E
12 J6 **Guillaume-Delisle, Lac** ◎ Québec, NE Canada
103 U13 **Guillestre** Hautes-Alpes, SE France 44°41′N 06°39′E
104 H6 **Guimarães** *var.* Guimaráes. Braga, N Portugal 41°26′N 08°19′W
Guimaráes *see* Guimarães
58 D11 **Guimarães Rosas, Pico** ▲ NW Brazil
23 N3 **Guin** Alabama, S USA 33°58′N 87°54′W
Güina *see* Wina
76 I14 **Guinea** *off.* Republic of Guinea, *var.* Guinée; *prev.* French Guinea, People's Revolutionary Republic of Guinea. ◆ *republic* W Africa
64 N13 **Guinea Basin** *undersea feature* E Atlantic Ocean 0°00′N 05°00′W
76 E12 **Guinea-Bissau** *off.* Republic of Guinea-Bissau, *Fr.* Guinée-Bissau, *Port.* Guiné-Bissau; *prev.* Portuguese Guinea. ◆ *republic* W Africa
Guinea-Bissau, Republic of *see* Guinea-Bissau
66 K7 **Guinea Fracture Zone** *tectonic feature* E Atlantic Ocean
64 O13 **Guinea, Gulf of** *Fr.* Golfe de Guinée. *gulf* E Atlantic Ocean
Guinea, People's Revolutionary Republic of *see* Guinea
Guinea, Republic of *see* Guinea
Guiné-Bissau *see* Guinea-Bissau
Guinée *see* Guinea
Guinée-Bissau *see* Guinea-Bissau
Guinée, Golfe de *see* Guinea, Gulf of
44 C4 **Güines** La Habana, W Cuba 22°50′N 82°02′W
102 G5 **Guingamp** Côtes d'Armor, NW France 48°34′N 03°09′W
Guipúzcoa *see* Gipuzkoa
44 C5 **Güira de Melena** La Habana, W Cuba 22°47′N 82°33′W
74 G8 **Guir, Hamada du** *desert* Algeria/Morocco
55 P5 **Güiria** Sucre, NE Venezuela 10°37′N 62°21′W
Gui Shui *see* Gui Jiang
104 H2 **Guitiriz** Galicia, NW Spain 43°10′N 07°52′W
77 N17 **Guitri** S Ivory Coast 05°31′N 05°14′W
171 Q5 **Guiuan** Samar, C Philippines 11°02′N 125°45′E
Gui Xian/Guixian *see* Guigang
160 J12 **Guiyang** *var.* Kuei-Yang, Kuei-yang, Kueyang, Kweiyang; *prev.* Kweichu. *province capital* Guizhou, S China 26°33′N 106°45′E
160 J12 **Guizhou** *var.* Guizhou Sheng, Kuei-chou, Kweichow, Qian. ◈ *province* S China
Guizhou Sheng *see* Guizhou
102 J13 **Gujan-Mestras** Gironde, SW France 44°39′N 01°04′W
154 B10 **Gujarāt** *var.* Gujerat. ◈ *state* W India
149 V6 **Gūjar Khān** Punjab, E Pakistan 33°19′N 73°23′E
Gujerat *see* Gujarāt
149 V7 **Gujrānwāla** Punjab, NE Pakistan 32°11′N 74°09′E
149 V7 **Gujrāt** Punjab, E Pakistan 32°34′N 74°04′E
146 B8 **Gulandag** *Rus.* Gory Kulandag. ▲▲ Balkan Welaýaty, W Turkmenistan
159 U9 **Gulang** Gansu, C China 37°31′N 102°55′E
183 R6 **Gulargambone** New South Wales, SE Australia 31°19′S 148°31′E
155 G15 **Gulbarga** Karnātaka, C India 17°22′N 76°47′E
118 J8 **Gulbene** *Ger.* Alt-Schwanenburg. NE Latvia 57°10′N 26°44′E
147 U10 **Gul'cha** *Kir.* Gülchö. Oshskaya Oblast', SW Kyrgyzstan 40°16′N 73°27′E
Gülchö *see* Gul'cha
173 T10 **Gulden Draak Seamount** *undersea feature* E Indian Ocean 33°45′S 101°00′E
136 J16 **Gülek Boğazı** *var.* Cilician Gates. *pass* S Turkey
186 D8 **Gulf** ◈ *province* S Papua New Guinea
23 O9 **Gulf Breeze** Florida, SE USA 30°21′N 87°09′W
Gulf of Liaotung *see* Liaodong Wan
23 V13 **Gulfport** Florida, SE USA 27°45′N 82°42′W
22 M9 **Gulfport** Mississippi, S USA 30°22′N 89°06′W
23 O9 **Gulf Shores** Alabama, S USA 30°15′N 87°40′W
183 R7 **Gulgong** New South Wales, SE Australia 32°22′S 149°31′E
160 I11 **Gulin** Sichuan, C China 28°06′N 105°47′E

◆ Country
● Country Capital
◇ Dependent Territory
○ Dependent Territory Capital
◈ Administrative Regions
✈ International Airport
▲ Mountain
▲▲ Mountain Range
⛰ Volcano
≈ River
◎ Lake
⊠ Reservoir

Gulir – Halmahera, Pulau

171 U14 Gulir Pulau Kasiui, E Indonesia 04°27′S 131°41′E
Gulistan see Guliston
147 P10 Guliston Rus. Gulistan. Sirdaryo Viloyati, E Uzbekistan 40°29′N 68°46′E
163 T6 Guliya Shan ▲ NE China 49°42′N 122°22′E
Gulja see Yining
39 S11 Gulkana Alaska, USA 62°17′N 145°25′W
11 S17 Gull Lake Saskatchewan, S Canada 50°05′N 108°30′W
31 P10 Gull Lake ◎ Michigan, N USA
29 T6 Gull Lake ◎ Minnesota, N USA
95 L16 Gullspång Västra Götaland, S Sweden 58°58′N 14°04′E
136 B15 Güllük Körfezi prev. Akbük Limanı. bay W Turkey
152 H5 Gulmarg Jammu and Kashmir, NW India 34°04′N 74°25′E
Gulpaigan see Golpāyegān
99 L18 Gulpen Limburg, SE Netherlands 50°48′N 05°53′E
Gul'shad see Gul'shat
145 S13 Gul'shat var. Gul'shad. Karaganda, E Kazakhstan 46°37′N 74°22′E
81 F17 Gulu N Uganda 02°46′N 32°21′E
Gülübovo see Galabovo
114 I7 Gulyantsi Pleven, N Bulgaria 43°37′N 24°40′E
Gulyaypole see Hulyaypole
Guma see Pishan
Gümai see Darlag
79 K16 Gumba Equateur, NW Dem. Rep. Congo 02°58′N 21°23′E
Gumbinnen see Gusev
81 H24 Gumbiro Ruvuma, S Tanzania 10°19′S 35°40′E
146 B11 Gumdag prev. Kum-Dag. Balkan Welaýaty, W Turkmenistan 39°13′N 54°35′E
77 W12 Gumel Jigawa, N Nigeria 12°37′N 09°23′E
105 N5 Gumiel de Hizán Castilla y León, N Spain 41°46′N 03°42′W
153 P16 Gumla Jhārkhand, N India 23°03′N 84°36′E
Gumma see Gunma
101 F16 Gummersbach Nordrhein-Westfalen, W Germany 51°01′N 07°34′E
77 T13 Gummi Zamfara, NW Nigeria 12°07′N 05°07′E
Gumpolds see Humpolec
Gumti see Gomati
Gümülcine/Gümüljina see Komotiní
Gümüşane see Gümüşhane
137 O12 Gümüşhane var. Gümüşane, Gumushkhane. Gümüşhane, NE Turkey 40°31′N 39°27′E
137 O12 Gümüşhane var. Gümüşane, Gumushkhane. ◈ province NE Turkey
Gumushkhane see Gümüşhane
171 V14 Gumzai Pulau Kola, E Indonesia 05°27′S 134°38′E
154 H9 Guna Madhya Pradesh, C India 24°39′N 77°18′E
Gunabad see Gonābād
Gunan see Qijiang
Gunbad-i-Qawus see Gonbad-e Kāvūs
183 O9 Gunbar New South Wales, SE Australia 34°03′S 145°32′E
183 O9 Gun Creek seasonal river New South Wales, SE Australia
183 Q10 Gundagai New South Wales, SE Australia 35°06′S 148°03′E
79 K17 Gundji Equateur, N Dem. Rep. Congo 02°03′N 21°31′E
155 G20 Gundlupet Karnātaka, W India 11°48′N 76°42′E
136 G16 Gündoğmuş Antalya, S Turkey 36°50′N 32°07′E
137 O14 Güney Doğu Toroslar ▲▲ SE Turkey
79 J21 Gungu Bandundu, SW Dem. Rep. Congo 05°43′S 19°20′E
127 P17 Gunib Respublika Dagestan, SW Russian Federation 42°24′N 46°55′E
112 J11 Gunja Vukovar-Srijem, E Croatia 44°53′N 18°51′E
31 P9 Gun Lake ◎ Michigan, N USA
165 N12 Gunma off. Gunma-ken, var. Gumma. ◈ prefecture Honshū, S Japan
Gunma-ken see Gunma
197 P15 Gunnbjørn Fjeld var. Gunnbjörns Bjerge. ▲ C Greenland 69°03′N 29°36′W
Gunnbjörns Bjerge see Gunnbjørn Fjeld
183 S6 Gunnedah New South Wales, SE Australia 30°59′S 150°15′E
173 Y15 Gunner's Quoin var. Coin de Mire. island N Mauritius
37 R6 Gunnison Colorado, C USA 38°33′N 106°55′W
36 L5 Gunnison Utah, W USA 39°09′N 111°49′W
37 P5 Gunnison River ≈ Colorado, C USA
21 X2 Gunpowder River ≈ Maryland, NE USA
Güns see Kőszeg
163 X16 Gunsan var. Gunsan, Jap. Gunzan; prev. Kunsan. W South Korea 35°58′N 126°42′E
Gunsan see Gunsan
109 S4 Gunskirchen Oberösterreich, N Austria 48°07′N 13°54′E
Gunt see Ghund
155 H17 Guntakal Andhra Pradesh, C India 15°11′N 77°24′E
23 Q2 Guntersville Alabama, S USA 34°21′N 86°17′W
23 Q2 Guntersville Lake ⊠ Alabama, S USA
109 X4 Guntramsdorf Niederösterreich, E Austria 48°03′N 16°19′E
155 J16 Guntūr var. Guntur. Andhra Pradesh, SE India 16°20′N 80°27′E
168 H10 Gunungsitoli Pulau Nias, W Indonesia 01°11′N 97°35′E
155 M14 Gunupur Odisha, E India 19°04′N 83°52′E
101 J23 Günz ≈ S Germany
Gunzan see Gunsan
101 J22 Günzburg Bayern, S Germany 48°26′N 10°18′E
101 K21 Gunzenhausen Bayern, S Germany 49°07′N 10°45′E
Guolüezhen see Lingbao
Guovdageaidnu see Kautokeino
161 P7 Guoyang Anhui, E China 33°30′N 116°12′E
116 G11 Gurahonț Hung. Honctő. Arad, W Romania 46°16′N 22°21′E
Gurahumora see Gura Humorului
116 K9 Gura Humorului Ger. Gurahumora. Suceava, NE Romania 47°31′N 26°00′E
146 H8 Gurbansoltan Eje prev. Ýylanly, Rus. Il'yaly. Daşoguz Welaýaty, N Turkmenistan 41°57′N 59°42′E
158 K4 Gurbantünggüt Shamo desert W China
152 H7 Gurdāspur Punjab, N India 32°04′N 75°28′E
27 T13 Gurdon Arkansas, C USA 33°55′N 93°09′W
Gurdzhaani see Gurjaani
Gurgan see Gorgān
152 I10 Gurgaon Haryāna, N India 28°27′N 77°01′E
59 M15 Gurguéia, Rio ≈ NE Brazil
55 Q7 Guri, Embalse de ⊠ E Venezuela
137 V10 Gurjaani Rus. Gurdzhaani. E Georgia 41°42′N 45°47′E
109 T8 Gurk Kärnten, S Austria 46°52′N 14°17′E
109 T9 Gurk Slvn. Krka. ≈ S Austria
Gurkfeld see Krško
114 K9 Gurkovo prev. Kolupchii. Stara Zagora, C Bulgaria 42°42′N 25°46′E
109 S9 Gurktaler Alpen ▲ S Austria
146 H8 Gurlan Rus. Gurlen. Xorazm Viloyati, W Uzbekistan 41°54′N 60°18′E
Gurlen see Gurlan
83 M16 Guro Manica, C Mozambique 17°28′S 33°18′E
136 M14 Gürün Sivas, C Turkey 38°44′N 37°15′E
59 K16 Gurupi Tocantins, C Brazil 11°44′S 49°01′W
58 L12 Gurupi, Rio ≈ NE Brazil
152 E14 Guru Sikhar ▲ NW India 24°45′N 72°51′E
162 H8 Gurvanbulag var. Höviyn Am. Bayanhongor, C Mongolia 47°08′N 98°41′E
162 K7 Gurvanbulag var. Avdzaga. Bulgan, C Mongolia 47°43′N 103°30′E
162 I11 Gurvantes var. Urt. Ömnögovĭ, S Mongolia 43°16′N 101°00′E
Gur'yev/Gur'yevskaya Oblast' see Atyrau
77 U13 Gusau Zamfara, NW Nigeria 12°18′N 06°27′E
126 C3 Gusev Ger. Gumbinnen. Kaliningradskaya Oblast', W Russian Federation 54°36′N 22°14′E
146 J17 Gushgy Rus. Kushka. ≈ Mary Welaýaty, S Turkmenistan
Gushiago see Gushiegu
77 Q14 Gushiegu var. Gushiago. NE Ghana 09°54′N 00°12′W
165 S17 Gushikawa Okinawa, Okinawa, SW Japan 26°21′N 127°50′E
113 L16 Gusinje E Montenegro 42°34′N 19°51′E
126 M4 Gus'-Khrustal'nyy Vladimirskaya Oblast', W Russian Federation 55°39′N 40°42′E
107 B19 Guspini Sardegna, Italy, C Mediterranean Sea 39°33′N 08°39′E
109 X8 Güssing Burgenland, SE Austria 47°03′N 16°19′E
109 V6 Gusswerk Steiermark, E Austria 47°43′N 15°18′E
92 O2 Gustav Adolf Land physical region NE Svalbard
195 X5 Gustav Bull Mountains ▲▲ Antarctica
39 W13 Gustavus Alaska, USA 58°24′N 135°44′W
92 O1 Gustav V Land physical region NE Svalbard
35 P9 Gustine California, W USA 37°14′N 121°00′W
25 R8 Gustine Texas, SW USA 31°51′N 98°24′W
100 M9 Güstrow Mecklenburg-Vorpommern, NE Germany 53°48′N 12°12′E
95 N18 Gusum Östergötland, S Sweden 58°15′N 16°30′E
Guta/Gúta see Kolárovo
Gutenstein see Ravne na Koroškem
101 G14 Gütersloh Nordrhein-Westfalen, W Germany 51°54′N 08°23′E
27 N10 Guthrie Oklahoma, C USA 35°53′N 97°26′W
25 P5 Guthrie Texas, SW USA 33°38′N 100°21′W
29 U14 Guthrie Center Iowa, C USA 41°40′N 94°30′W
41 Q13 Gutiérrez Zamora Veracruz-Llave, E Mexico 20°29′N 97°07′W
Gutta see Kolárovo
29 Y12 Guttenberg Iowa, C USA 42°47′N 91°06′W
Guttentag see Dobrodzień
Guttstadt see Dobre Miasto
162 G8 Guulin Govĭ-Altay, C Mongolia 46°33′N 97°21′E
153 V12 Guwāhāti prev. Gauhāti. Assam, NE India 26°09′N 91°42′E
139 R3 Guwēr var. Al Kuwayr, Al Quwayr, Quwair. Arbīl, N Iraq 36°03′N 43°30′E
146 A10 Guwlumaýak Rus. Kuuli-Mayak. Balkan Welaýaty, NW Turkmenistan 40°14′N 52°43′E
55 R9 Guyana off. Co-operative Republic of Guyana; prev. British Guiana. ◆ republic N South America
Guyana, Co-operative Republic of see Guyana
21 P5 Guyandotte River ≈ West Virginia, NE USA
Guyane see French Guiana
Guyi see Sanjiang
26 H8 Guymon Oklahoma, C USA 36°42′N 101°30′W
146 K12 Guynuk Lebap Welaýaty, NE Turkmenistan 39°18′N 63°00′E
Guyong see Jiangle
21 O9 Guyot, Mount ▲ North Carolina/Tennessee, SE USA 35°42′N 83°15′W
183 U5 Guyra New South Wales, SE Australia 30°13′S 151°42′E
159 W10 Guyuan Ningxia, N China 35°57′N 106°13′E
Guzar see G'uzor
121 P2 Güzelyurt Gk. Kólpos Mórfu, Morphou. W Cyprus 35°12′N 33°E
121 N2 Güzelyurt Körfezi var. Morfou Bay, Morphou Bay, Gk. Kólpos Mórfou. bay W Cyprus
Guzhou see Rongjiang
40 I3 Guzmán Chihuahua, N Mexico 31°13′N 107°27′W
147 N13 G'uzor Rus. Guzar. Qashqadaryo Viloyati, S Uzbekistan 38°41′N 66°12′E
119 B14 Gvardeysk Ger. Tapaiu. Kaliningradskaya Oblast', W Russian Federation 54°39′N 21°02′E
Gvardeyskoye see Hvardiys'ke
183 R5 Gwabegar New South Wales, SE Australia 30°34′S 148°58′E
148 J16 Gwādar var. Gwadur. Baluchistān, SW Pakistan 25°09′N 62°21′E
148 J16 Gwādar East Bay bay SW Pakistan
148 J16 Gwādar West Bay bay SW Pakistan
Gwadur see Gwādar
83 J17 Gwai Matabeleland North, W Zimbabwe 19°17′S 27°37′E
154 I7 Gwalior Madhya Pradesh, C India 26°16′N 78°12′E
83 J18 Gwanda Matabeleland South, SW Zimbabwe 20°56′S 29°E
79 N15 Gwane Orientale, N Dem. Rep. Congo 04°40′N 25°51′E
163 X16 Gwangju off. Kwangju-gwangyŏksi, var. Guangju, Kwangchu, Jap. Kōshū; prev. Kwangju. SW South Korea 35°09′N 126°53′E
83 I17 Gwayi ≈ W Zimbabwe
110 G8 Gwda var. Głda, Ger. Küddow. ≈ NW Poland
97 C14 Gweebarra Bay Ir. Béal an Bheara. inlet W Ireland
97 D14 Gweedore Ir. Gaoth Dobhair. Donegal, NW Ireland 55°03′N 08°14′W
Gwelo see Gweru
97 K21 Gwent cultural region S Wales, United Kingdom
83 K17 Gweru prev. Gwelo. Midlands, C Zimbabwe 19°27′S 29°49′E
29 Q7 Gwinner North Dakota, N USA 46°10′N 97°42′W
77 Y13 Gwoza Borno, NE Nigeria 11°07′N 13°40′E
Gwy see Wye
183 R4 Gwydir River ≈ New South Wales, SE Australia
97 I19 Gwynedd var. Gwyneth. cultural region NW Wales, United Kingdom
Gwyneth see Gwynedd
159 O16 Gyaca var. Ngarrab. Xizang Zizhiqu, W China 29°06′N 92°37′E
Gya'gya see Saga
Gyaijêpozhanggê see Zhidoi
Gyaisi see Jiulong
115 M22 Gyalí var. Yialí. island Dodekánisa, Greece, Aegean Sea
Gyamotang see Dêngqên
Gyandzha see Gäncä
Gyangkar see Dinggyê
158 M16 Gyangzê Xizang Zizhiqu, W China 28°50′N 89°38′E
158 L14 Gyaring Co ◎ W China
159 Q12 Gyaring Hu ◎ C China
115 I20 Gyáros var. Yioúra. island Kykládes, Greece, Aegean Sea
122 J7 Gyda Yamalo-Nenetskiy Avtonomnyy Okrug, N Russian Federation 70°55′N 78°34′E
122 J7 Gydanskiy Poluostrov Eng. Gyda Peninsula. peninsula N Russian Federation
Gyda Peninsula see Gydanskiy Poluostrov
Gyêgu see Yushu
163 W15 Gyeonggi-man prev. Kyŏnggi-man. bay NW South Korea
163 Z16 Gyeongju Jap. Keishū; prev. Kyŏngju. SE South Korea 35°49′N 129°09′E
Gyéres see Câmpia Turzii
Gyergyószentmiklós see Gheorgheni
Gyergyótölgyes see Tulgheș
Gyertyámos see Cărpiniș
Gyeva see Detva
Gyigang see Zayü
Gyixong see Gonggar
95 I23 Gyldenløveshøy hill range C Denmark
181 Z10 Gympie Queensland, E Australia 26°05′S 152°40′E
166 L7 Gyobingauk Bago, SW Myanmar (Burma) 18°14′N 95°39′E
111 M23 Gyomaendrőd Békés, SE Hungary 46°56′N 20°50′E
Gyömbér see Ďumbier
111 L22 Gyöngyös Heves, NE Hungary 47°44′N 19°49′E
111 H22 Győr Ger. Raab, Lat. Arrabona. Győr-Moson-Sopron, NW Hungary 47°41′N 17°40′E
111 G22 Győr-Moson-Sopron off. Győr-Moson-Sopron Megye. ◈ county NW Hungary
Győr-Moson-Sopron Megye see Győr-Moson-Sopron
11 X15 Gypsumville Manitoba, S Canada 51°47′N 98°38′W
12 M4 Gyrfalcon Islands island group Northwest Territories, NE Canada
95 N14 Gysinge Gävleborg, C Sweden 60°16′N 16°55′E
115 F22 Gýtheio var. Githio; prev. Ýithion. Pelopónnisos, S Greece 36°46′N 22°34′E
146 L13 Gyuichbirleshik Lebap Welaýaty, E Turkmenistan 38°10′N 64°33′E
111 N24 Gyula Rom. Jula. Békés, SE Hungary 46°39′N 21°17′E
Gyulafehérvár see Alba Iulia
Gyulovo see Roza
137 T11 Gyumri var. Giumri, Rus. Kumayri; prev. Aleksandropol', Leninakan. W Armenia 40°48′N 43°51′E
146 D13 Gyunuzyndag, Gora ▲ Balkan Welaýaty, W Turkmenistan 38°15′N 56°25′E
146 J15 Gyzylbaydak Rus. Krasnoye Znamya. Mary Welaýaty, S Turkmenistan 36°51′N 62°24′E
Gyzyletrek see Etrek
146 D10 Gyzylgaýa Rus. Kizyl-Kaya. Balkan Welaýaty, NW Turkmenistan 40°37′N 55°15′E
146 A10 Gyzylsuw Rus. Kizyl-Su. Balkan Welaýaty, W Turkmenistan 39°49′N 53°00′E
Gyzyrlabat see Serdar
Gzhatsk see Gagarin

H

153 T12 Ha W Bhutan 27°17′N 89°22′E
Haabai see Ha'apai Group
99 H17 Haacht Vlaams Brabant, C Belgium 50°58′N 04°38′E
109 T4 Haag Niederösterreich, NE Austria 48°07′N 14°32′E
194 L8 Haag Nunataks ▲ Antarctica
92 N2 Haakon VII Land physical region NW Svalbard
98 O11 Haaksbergen Overijssel, E Netherlands 52°09′N 06°45′E
99 E14 Haamstede Zeeland, SW Netherlands 51°43′N 03°45′E
193 Y15 Ha'ano island Ha'apai Group, C Tonga
193 Y15 Ha'apai Group var. Haabai. island group C Tonga
93 L15 Haapajärvi Pohjois-Pohjanmaa, C Finland 63°45′N 25°20′E
93 L17 Haapamäki Pirkanmaa, C Finland 62°11′N 24°32′E
93 L15 Haapavesi Pohjois-Pohjanmaa, C Finland 64°09′N 25°25′E
191 N7 Haapiti Moorea, W French Polynesia 17°33′S 149°52′W
118 F4 Haapsalu Ger. Hapsal. Läänemaa, W Estonia 58°58′N 23°32′E
Ha'Arava see 'Arabah, Wādī al
95 G24 Haarby var. Hårby. Syddtjylland, C Denmark 55°13′N 10°07′E
98 H10 Haarlem prev. Harlem. Noord-Holland, W Netherlands 52°23′N 04°39′E
185 D19 Haast West Coast, South Island, New Zealand 43°53′S 169°02′E
185 C20 Haast ≈ South Island, New Zealand
185 D20 Haast Pass pass South Island, New Zealand
193 W16 Ha'atua 'Eau, E Tonga 21°23′S 174°57′W
149 P15 Hab ≈ SW Pakistan
141 W7 Haba var. Al Haba. Dubayy, NE United Arab Emirates 25°01′N 55°37′E
158 K2 Habahe var. Kaba. Xinjiang Uygur Zizhiqu, NW China 48°04′N 86°20′E
141 U13 Ḩabarūt var. Habrut. SW Oman 17°19′N 52°45′E
81 J18 Habaswein Isiolo, NE Kenya 01°01′N 39°27′E
99 L24 Habay-la-Neuve Luxembourg, SE Belgium 49°43′N 05°38′E
139 S8 Ḩabbānīyah, Buḩayrat ◎ C Iraq
Habelschwerdt see Bystrzyca Kłodzka
153 V14 Habiganj Sylhet, NE Bangladesh 24°23′N 91°25′E
163 Q12 Habirag Nei Mongol Zizhiqu, N China 42°18′N 115°40′E
95 L19 Habo Västra Götaland, S Sweden 57°55′N 14°05′E
123 V14 Habomai Islands island group Kuril'skiye Ostrova, SE Russian Federation
165 S2 Haboro Hokkaidō, NE Japan 44°19′N 141°42′E
153 S16 Habra West Bengal, NE India 22°39′N 88°17′E
Habrut see Ḩabarūt
143 P17 Ḩabshān Abū Ẓaby, C United Arab Emirates 23°51′N 53°34′E
54 E14 Hacha Putumayo, S Colombia 0°02′S 75°30′W
165 X13 Hachijō-jima island Izu-shotō, SE Japan
164 L12 Hachiman Gifu, Honshū, SW Japan 35°46′N 136°57′E
165 P7 Hachimori Akita, Honshū, C Japan 40°22′N 139°59′E
165 R7 Hachinohe Aomori, Honshū, C Japan 40°30′N 141°29′E
165 X13 Hachiōji Tōkyō, Hachijō-jima, SE Japan 35°40′N 139°20′E
137 Y12 Hacıqabal prev. Qazımämmäd. SE Azerbaijan 40°03′N 48°56′E
93 G17 Hackås Jämtland, C Sweden 62°55′N 14°31′E
18 K14 Hackensack New Jersey, NE USA 40°51′N 73°57′W
75 U12 Hadabat al Jilf al Kabīr var. Gilf Kebir Plateau. plateau SW Egypt
Hadama see Nazrēt
141 W13 Ḩadbaram S Oman 17°27′N 55°13′E
139 U13 Ḩaddānīyah well S Iraq
96 K12 Haddington SE Scotland, United Kingdom 55°59′N 02°46′W
141 Z8 Ḩadd, Ra's al headland NE Oman 22°28′N 59°58′E
Haded see Xadeed
77 W12 Hadejia Jigawa, N Nigeria 12°22′N 10°02′E
77 W12 Hadejia ≈ N Nigeria
138 F9 Hadera var. Khadera; prev. Ḩadera. Haifa, N Israel 32°26′N 34°55′E
Ḩadera see Hadera
Hadersleben see Haderslev
95 G24 Haderslev Ger. Hadersleben. Syddanmark, SW Denmark 55°15′N 09°30′E
151 J21 Hadhdhunmathi Atoll atoll S Maldives
141 W17 Hadībūh Suquṭrā, SE Yemen 12°38′N 54°05′E
158 K9 Hadilik Xinjiang Uygur Zizhiqu, W China 37°51′N 86°10′E
136 H16 Hadım Konya, S Turkey 36°58′N 32°27′E
140 K7 Hadīyah Al Madīnah, W Saudi Arabia 25°36′N 38°31′E
78 H11 Hadjer-Lamis off. Région du Hadjer-Lamis. ◈ region SW Chad
Hadjer-Lamis, Région du see Hadjer-Lamis
8 L5 Hadley Bay bay Victoria Island, Nunavut, N Canada
167 S6 Ha Đông var. Hadong. Ha Tây, N Vietnam 20°58′N 105°46′E
Hadong see Ha Đông
Hadramaut see Ḩaḑramawt
141 R15 Ḩaḑramawt Eng. Hadramaut. ▲▲ S Yemen
Hadria see Adria
Hadrianopolis see Edirne
Hadria Picena see Apricena
95 G22 Hadsten Midtjylland, C Denmark 56°19′N 10°03′E
95 G21 Hadsund Nordjylland, N Denmark 56°43′N 10°08′E
117 S4 Hadyach Rus. Gadyach. Poltavs'ka Oblast', NE Ukraine 50°21′N 34°00′E
114 N9 Hadzhiyska Reka var. Khadzhiyska Reka. ≈ E Bulgaria
112 I13 Hadžići Federacija Bosne I Hercegovine, SE Bosnia and Herzegovina 43°49′N 18°12′E
163 W14 Haeju S North Korea 38°04′N 125°40′E
Haerbin/Haerhpin/Ha-erh-pin see Harbin
141 P5 Ḩafar al Bāţin Ash Sharqīyah, N Saudi Arabia 28°25′N 45°59′E
11 T15 Hafford Saskatchewan, S Canada 52°43′N 107°19′W
136 M13 Hafik Sivas, N Turkey 39°53′N 37°24′E
149 V8 Hāfizābād Punjab, E Pakistan 32°03′N 73°42′E
92 H4 Hafnarfjörður Höfuðborgarsvæðið, W Iceland 64°03′N 21°57′W
Hafnia see Denmark
Hafnia see København
Hafren see Severn
Hafun see Xaafuun
Hafun, Ras see Xaafuun, Raas
80 G10 Hag 'Abdullah Sinnar, E Sudan 13°59′N 33°35′E
81 K18 Hagadera Garissa, E Kenya 0°06′N 40°23′E
138 G8 HaGalil Eng. Galilee. ▲ N Israel
14 G10 Hagar Ontario, S Canada 46°27′N 80°22′W
155 G18 Hagari var. Vedāvati. ≈ W India
188 B16 Hagåtña, var. Agaña. ○ (Guam) NW Guam 13°27′N 144°45′E
100 M13 Hagelberg hill NE Germany
39 N14 Hagemeister Island island Alaska, USA
101 F15 Hagen Nordrhein-Westfalen, W Germany 51°22′N 07°27′E
100 K10 Hagenow Mecklenburg-Vorpommern, N Germany 53°27′N 11°10′E
10 K15 Hagensborg British Columbia, SW Canada 52°24′N 126°24′W
80 I13 Hägere Hiywet var. Agere Hiywet, Ambo. Oromīya, C Ethiopia 09°00′N 37°55′E
33 O15 Hagerman Idaho, NW USA 42°48′N 114°53′W
37 U14 Hagerman New Mexico, SW USA 33°07′N 104°19′W
21 V2 Hagerstown Maryland, NE USA 39°39′N 77°44′W
14 G16 Hagersville Ontario, S Canada 42°58′N 80°03′W
102 J15 Hagetmau Landes, SW France 43°40′N 00°36′W
95 K14 Hagfors Värmland, C Sweden 60°03′N 13°45′E
93 G16 Häggenås Jämtland, C Sweden 63°24′N 14°53′E
164 E12 Hagi Yamaguchi, Honshū, SW Japan 34°25′N 131°22′E
167 S5 Ha Giang Ha Giang, N Vietnam 22°50′N 104°58′E
Hagios Evstrátios see Ágios Efstrátios
HaGolan see Golan Heights
103 T4 Hagondange Moselle, NE France 49°16′N 06°06′E
97 B18 Hag's Head Ir. Ceann Caillí. headland W Ireland 52°56′N 09°29′W
102 I3 Hague, Cap de la headland N France 49°43′N 01°56′W
103 V5 Haguenau Bas-Rhin, NE France 48°49′N 07°47′E
165 X16 Hahajima-rettō island group SE Japan
15 R8 Hā Hā, Lac ◎ Québec, SE Canada
172 H13 Hahaya ✈ (Moroni) Grande Comore, NW Comoros
22 K9 Hahnville Louisiana, S USA 29°58′N 90°24′W
83 E22 Haib Karas, S Namibia 28°12′S 18°19′E
Haibak see Aibak
149 N15 Haibo ≈ SW Pakistan
Haibowan see Wuhai
163 U12 Haicheng Liaoning, NE China 40°53′N 122°45′E
Haicheng see Haifeng
Haicheng see Haiyuan
Haida see Nový Bor
Haidarabad see Hyderābād
Haidenschaft see Ajdovščina
167 T6 Hai Dương Hai Hưng, N Vietnam 20°56′N 106°21′E
138 F9 Haifa ◈ district NW Israel
Haifa see Hefa
Haifa, Bay of see Mifrats Hefa
161 P14 Haifeng var. Haicheng. Guangdong, S China 22°56′N 115°19′E
Haifong see Hai Phong
161 P3 Hai He ≈ E China
Haikang see Leizhou
160 L17 Haikou var. Hai-k'ou, Hoihow, Fr. Hoï-Hao. province capital Hainan, S China 20°10′N 110°17′E
Hai-k'ou see Haikou
140 M6 Ḩā'il Ḩā'il, NW Saudi Arabia 27°N 42°50′E
141 N5 Ḩā'il off. Mintaqah Ḩā'il. ◈ province N Saudi Arabia
163 S6 Hailar He ≈ NE China
33 P14 Hailey Idaho, NW USA 43°31′N 114°18′W
14 H9 Haileybury Ontario, S Canada 47°27′N 79°39′W
163 X9 Hailin Heilongjiang, NE China 44°37′N 129°24′E
Ḩā'il, Minţaqah see Ḩā'il
Hailong see Meihekou
93 K14 Hailuoto Swe. Karlö. island W Finland
Haima see Haymā'
Haimen see Taizhou
160 M17 Hainan var. Hainan Sheng, Qiong. ◈ province S China
160 K17 Hainan Dao island S China
Hainan Sheng see Hainan
Hainan Strait see Qiongzhou Haixia
Hainasch see Ainaži
Hainau see Chojnów
99 E20 Hainaut ◈ province SW Belgium
Hainburg see Hainburg an der Donau
109 Z4 Hainburg an der Donau var. Hainburg. Niederösterreich, NE Austria 48°09′N 16°57′E
39 W12 Haines Alaska, USA 59°13′N 135°27′W
32 L12 Haines Oregon, NW USA 44°53′N 117°56′W
23 W12 Haines City Florida, SE USA 28°06′N 81°37′W
10 H8 Haines Junction Yukon, W Canada 60°45′N 137°30′W
109 W4 Hainfeld Niederösterreich, NE Austria 48°03′N 15°47′E
101 N16 Hainichen Sachsen, E Germany 50°58′N 13°08′E
Hai Ninh see Mong Cai
167 T6 Hai Phong var. Haifong, Haiphong. N Vietnam 20°50′N 106°41′E
Haiphong see Hai Phong
161 S12 Haitan Dao island SE China
44 K8 Haiti off. Republic of Haiti. ◆ republic C West Indies
Haiti, Republic of see Haiti
35 T11 Haiwee Reservoir ⊠ California, W USA
80 I7 Haiya Red Sea, NE Sudan 18°17′N 36°21′E
159 T10 Haiyan var. Sanjiaocheng. Qinghai, W China 36°55′N 100°54′E
160 M13 Haiyang Shan ▲▲ S China
159 V10 Haiyuan Ningxia, N China 36°32′N 105°31′E
Hajda see Nový Bor
111 M22 Hajdú-Bihar off. Hajdú-Bihar Megye. ◈ county E Hungary
Hajdú-Bihar Megye see Hajdú-Bihar
111 N22 Hajdúböszörmény Hajdú-Bihar, E Hungary 47°39′N 21°32′E
111 N22 Hajdúhadház Hajdú-Bihar, E Hungary 47°40′N 21°40′E
111 N21 Hajdúnánás Hajdú-Bihar, E Hungary 47°50′N 21°26′E
111 N22 Hajdúszoboszló Hajdú-Bihar, E Hungary 47°27′N 21°24′E
142 I3 Ḩājī Ebrāhīm, Kūh-e ▲ Iran/Iraq 36°53′N 44°56′E
165 O9 Hajiki-zaki headland Sado, C Japan 38°19′N 138°28′E
Hajine see Abū Ḩardān
153 P13 Hājīpur Bihār, N India 25°41′N 85°13′E
141 N14 Ḩajjah W Yemen 15°43′N 43°33′E
139 U11 Ḩajjama Al Muthannā, S Iraq 31°24′N 45°20′E
143 R12 Ḩājjīābād Hormozgān, C Iran
139 U14 Ḩājj, Thaqb al well S Iraq
113 L16 Hajla ▲▲ E Montenegro
110 P10 Hajnówka Ger. Hermhausen. Podlaskie, NE Poland 52°45′N 23°32′E
Haka see Hakha
Hakapehi see Punaauia
Hakâri see Hakkâri
138 F12 HaKatan, HaMakhtesh prev. HaMakhtesh HaQatan. ▲ S Israel
166 K4 Hakha var. Haka. Chin State, W Myanmar (Burma) 22°42′N 93°41′E
137 T16 Hakkâri var. Çölemerik, Hakâri. Hakkâri, SE Turkey 37°36′N 43°45′E
137 T16 Hakkâri var. Hakkari. ◈ province SE Turkey
Hakkari see Hakkâri
92 J12 Hakkas Norrbotten, N Sweden 66°53′N 21°36′E
164 J14 Hakken-zan ▲ Honshū, SW Japan 34°11′N 135°57′E
165 R7 Hakkōda-san ▲ Honshū, C Japan 40°40′N 140°49′E
165 T2 Hako-dake ▲ Hokkaidō, NE Japan 44°40′N 142°22′E
165 R5 Hakodate Hokkaidō, NE Japan 41°46′N 140°43′E
164 L11 Hakui Ishikawa, Honshū, SW Japan 36°55′N 136°46′E
190 B16 Hakupu SE Niue 19°06′S 169°50′E
164 L12 Haku-san ▲ Honshū, SW Japan 36°07′N 136°45′E
Hakusan see Mattō
Hal see Halle
149 Q15 Hāla Sind, SE Pakistan 25°47′N 68°28′E
138 J3 Ḩalab Eng. Aleppo, Fr. Alep; anc. Beroea. Ḩalab, NW Syria 36°14′N 37°10′E
138 J3 Ḩalab off. Muḩāfaz̧at Ḩalab, var. Aleppo, Halab. ◈ governorate NW Syria
138 J3 Ḩalab ✈ Ḩalab, NW Syria 36°12′N 37°10′E
Halab see Ḩalab
141 O8 Ḩalabān var. Halibān. Ar Riyāḑ, C Saudi Arabia 23°29′N 44°20′E
Ḩalabja see Helebce
Ḩalabjah see Helebce
Ḩalab, Muḩāfaz̧at see Ḩalab
146 L13 Halaç Rus. Khalach. Lebap Welaýaty, E Turkmenistan 38°05′N 64°46′E
190 A16 Halagigie Point headland W Niue
75 Z11 Halaib SE Egypt 22°10′N 36°33′E
76 Z11 Hala'ib Triangle ◈ disputed region S Egypt / N Sudan
190 H11 Halalo Île Uvea, N Wallis and Futuna 13°21′S 176°11′W
167 U10 Ha Lam Quang Nam-Da Nang, C Vietnam 15°42′N 108°24′E
Halandri see Chalándri
141 X13 Ḩalānīyāt, Juzur al var. Jazā'ir Bin Ghalfān, Eng. Kuria Muria Islands. island group S Oman
141 W13 Ḩalānīyāt, Khalīj al Eng. Kuria Muria Bay. bay S Oman
Halas see Kiskunhalas
38 G11 Hālawa var. Halawa. Hawaii, USA, C Pacific Ocean 20°13′N 155°46′W
Halawa see Hālawa
38 F9 Hālawa, Cape var. Cape Halawa. headland Molokaʻi, Hawaiʻi, USA 21°09′N 156°43′W
Cape Halawa see Hālawa, Cape
Halban see Tsetserleg
101 K14 Halberstadt Sachsen-Anhalt, C Germany 51°54′N 11°04′E
184 M12 Halcombe Manawatu-Wanganui, North Island, New Zealand 40°09′S 175°30′E
95 I16 Halden prev. Fredrikshald. Østfold, S Norway 59°08′N 11°20′E
100 L13 Haldensleben Sachsen-Anhalt, C Germany 52°18′N 11°25′E
Háldi see Halti
153 S17 Haldia West Bengal, NE India 22°04′N 88°02′E
152 K10 Haldwāni Uttarakhand, N India 29°13′N 79°31′E
163 P9 Haldzan var. Hatavch. Sühbaatar, E Mongolia 46°10′N 112°57′E
163 P9 Haldzan Sühbaatar, E Mongolia 46°10′N 112°57′E
38 F10 Haleakalā var. Haleakala. crater Maui, Hawaiʻi, USA
Haleakala see Haleakalā
25 N4 Hale Center Texas, SW USA 34°03′N 101°50′W
99 J18 Halen Limburg, NE Belgium 50°35′N 05°08′E
23 O2 Haleyville Alabama, S USA 34°13′N 87°37′W
77 O17 Half Assini SW Ghana 05°03′N 02°57′W
35 R8 Half Dome ▲ California, W USA 37°46′N 119°27′W
185 C25 Halfmoon Bay var. Oban. Stewart Island, Southland, New Zealand 46°53′S 168°08′E
182 E5 Half Moon Lake salt lake South Australia
163 R7 Halhgol Dornod, E Mongolia 47°57′N 118°07′E
163 S8 Halhgol var. Tsagaannuur. Dornod, E Mongolia 47°30′N 118°45′E
Haliacmon see Aliákmonas
Halibān see Ḩalabān
14 I13 Haliburton Ontario, SE Canada 45°03′N 78°20′W
14 I12 Haliburton Highlands var. Madawaska Highlands. hill range Ontario, SE Canada
13 Q15 Halifax province capital Nova Scotia, SE Canada 44°38′N 63°35′W
97 L17 Halifax N England, United Kingdom 53°44′N 01°52′W
21 W8 Halifax North Carolina, SE USA 36°19′N 77°37′W
21 U7 Halifax Virginia, NE USA 36°46′N 78°55′W
13 Q15 Halifax ✈ Nova Scotia, SE Canada 44°33′N 63°48′W
143 T13 Halīl Rūd seasonal river SE Iran
138 I6 Ḩalīmah ▲ Lebanon/Syria 34°12′N 36°37′E
162 G8 Haliun Govĭ-Altay, W Mongolia 45°55′N 96°06′E
118 I3 Haljala Ger. Halljal. Lääne-Virumaa, N Estonia 59°25′N 26°18′E
39 Q4 Halkett, Cape headland Alaska, USA 70°48′N 152°11′W
Halkida see Chalkida
96 J6 Halkirk N Scotland, United Kingdom 58°30′N 03°29′W
15 X7 Hall ≈ Québec, SE Canada
Hall see Schwäbisch Hall
93 H15 Hålla Västerbotten, N Sweden 63°56′N 17°20′E
96 J6 Halladale ≈ N Scotland, United Kingdom
95 J21 Halland ◈ county S Sweden
23 Z15 Hallandale Florida, SE USA 25°58′N 80°09′W
95 K22 Hallandsås physical region S Sweden
9 P6 Hall Beach var Sanirajak. Nunavut, N Canada 68°10′N 81°56′W
99 G19 Halle Fr. Hal. Vlaams Brabant, C Belgium 50°44′N 04°14′E
101 M15 Halle var. Halle an der Saale. Sachsen-Anhalt, C Germany 51°28′N 11°58′E
Halle an der Saale see Halle
35 W3 Halleck Nevada, W USA 40°57′N 115°27′W
95 L15 Hällefors Örebro, C Sweden 59°46′N 14°30′E
95 N16 Hälleforsnäs Södermanland, C Sweden 59°10′N 16°30′E
109 Q6 Hallein Salzburg, N Austria 47°41′N 13°06′E
101 L15 Halle-Neustadt Sachsen-Anhalt, C Germany 51°29′N 11°54′E
25 U12 Hallettsville Texas, SW USA 29°27′N 96°57′W
195 N4 Halley UK research station Antarctica 75°42′S 26°30′W
28 L4 Halliday North Dakota, N USA 47°19′N 102°19′W
37 S2 Halligan Reservoir ⊠ Colorado, C USA
100 G7 Halligen island group N Germany
94 G13 Hallingdal valley S Norway
38 J12 Hall Island island Alaska, USA
Hall Island see Maiana
189 P15 Hall Islands island group C Micronesia
118 H6 Halliste ≈ S Estonia
Halljal see Haljala
93 I15 Hällnäs Västerbotten, N Sweden 64°20′N 19°41′E
29 R2 Hallock Minnesota, N USA 48°47′N 96°56′W
9 S6 Hall Peninsula peninsula Baffin Island, Nunavut, NE Canada
20 F9 Halls Tennessee, S USA 35°52′N 89°24′W
95 M16 Hallsberg Örebro, C Sweden 59°05′N 15°07′E
181 N5 Halls Creek Western Australia 18°17′S 127°39′E
182 L12 Halls Gap Victoria, SE Australia 37°09′S 142°30′E
95 N15 Hallstahammar Västmanland, C Sweden 59°37′N 16°13′E
109 R6 Hallstatt Salzburg, W Austria 47°32′N 13°39′E
109 R6 Hallstätter See ◎ C Austria
95 P14 Hallstavik Stockholm, C Sweden 60°12′N 18°45′E
25 X7 Hallsville Texas, SW USA 32°31′N 94°30′W
103 P1 Halluin Nord, N France 50°46′N 03°07′E
171 S12 Halmahera, Laut Eng. Halmahera Sea. sea E Indonesia
171 R11 Halmahera, Pulau prev. Djailolo, Gilolo, Jailolo. island E Indonesia

◆ Country ● Country Capital ◇ Dependent Territory ○ Dependent Territory Capital ◈ Administrative Regions ✈ International Airport ▲ Mountain ▲▲ Mountain Range Volcano ≈ River ◎ Lake ⊠ Reservoir

HALMAHERA SEA – HAUT, ISLE AU

Halmahera Sea see Halmahera, Laut
95 J21 **Halmstad** Halland, S Sweden 56°41'N 12°49'E
167 T6 **Ha Long** prev. Hông Gai, var. Hon Gai, Hongay. Quang Ninh, N Vietnam 20°57'N 107°06'E
119 N15 **Halowchyn** Rus. Golovchin. Mahilyowskaya Voblasts', E Belarus 54°04'N 29°55'E
95 H20 **Hals** Nordjylland, N Denmark 57°00'N 10°19'E
94 F8 **Halsa** Møre og Romsdal, S Norway 63°04'N 08°13'E
119 I15 **Hal'shany** Rus. Gol'shany. Hrodzyenskaya Voblasts', W Belarus 54°15'N 26°01'E
Hälsingborg see Helsingborg
29 R5 **Halstad** Minnesota, N USA 47°21'N 96°49'W
27 N6 **Halstead** Kansas, C USA 38°00'N 97°30'W
99 G15 **Halsteren** Noord-Brabant, S Netherlands 51°32'N 04°16'E
93 L16 **Halsua** Keski-Pohjanmaa, W Finland 63°28'N 24°10'E
101 E14 **Haltern** Nordrhein-Westfalen, W Germany 51°45'N 07°10'E
92 J9 **Halti** var. Haltiatunturi, Lapp. Hálditšohkka. ▲ Finland/Norway 69°18'N 21°19'E
Haltiatunturi see Halti
116 J6 **Halych** Ivano-Frankivs'ka Oblast', W Ukraine 49°08'N 24°44'E
Halycus see Platani
103 P3 **Ham** Somme, N France 49°46'N 03°03'E
Hama see Ḩamāh
164 F12 **Hamada** Shimane, Honshū, SW Japan 34°54'N 132°07'E
142 L6 **Hamadān** anc. Ecbatana. Hamadān, W Iran 34°51'N 48°31'E
142 L6 **Hamadān** off. Ostān-e Hamadān. ❖ province W Iran
Hamadān, Ostān-e see Hamadān
138 I5 **Ḩamāh** var. Hama; anc. Epiphania, Bibl. Hamath. Ḩamāh, W Syria 35°09'N 36°44'E
138 I5 **Ḩamāh** off. Muḩāfaẓat Ḩamāh, var. Hama. ❖ governorate C Syria
Ḩamāh, Muḩāfaẓat see Ḩamāh
1665 S3 **Hamamasu** Hokkaidō, NE Japan 43°37'N 141°24'E
164 L14 **Hamamatsu** var. Hamamatu. Shizuoka, Honshū, S Japan 34°43'N 137°46'E
Hamamatu see Hamamatsu
165 W14 **Hamanaka** Hokkaidō, NE Japan 43°05'N 145°05'E
164 L14 **Hamana-ko** ◎ Honshū, S Japan
94 I13 **Hamar** prev. Storhammer. Hedmark, S Norway 60°57'N 10°55'E
141 U10 **Ḩamārīr al Kidan, Qalamat** well E Saudi Arabia
164 I12 **Hamasaka** Hyōgo, Honshū, SW Japan 35°37'N 134°27'E
Hamath see Ḩamāh
165 T1 **Hamatonbetsu** Hokkaidō, NE Japan 45°07'N 142°21'E
155 K26 **Hambantota** Southern Province, SE Sri Lanka 06°07'N 81°07'E
Hambourg see Hamburg
100 J9 **Hamburg** Hamburg, N Germany 53°33'N 10°03'E
27 V14 **Hamburg** Arkansas, C USA 33°13'N 91°50'W
29 S16 **Hamburg** Iowa, C USA 40°36'N 95°39'W
18 D10 **Hamburg** New York, NE USA 42°40'N 78°49'W
100 I10 **Hamburg** Fr. Hambourg. ❖ state N Germany
148 K5 **Hamdam Āb, Dasht-e** Pash. Dasht-i Hamdamab. ▲▲ W Afghanistan
Hamdamab, Dasht-i see Hamdam Āb, Dasht-e
18 M13 **Hamden** Connecticut, NE USA 41°23'N 72°55'W
140 K6 **Ḩamḑ, Wādī al** dry watercourse W Saudi Arabia
93 K18 **Hämeenkyrö** Pirkanmaa, W Finland 61°39'N 23°10'E
93 L19 **Hämeenlinna** Swe. Tavastehus. Kanta-Häme, S Finland 61°N 24°25'E
HaMela h, Yam see Dead Sea
Hamelin see Hameln
100 I13 **Hameln** Eng. Hamelin. Niedersachsen, N Germany 52°07'N 09°22'E
180 I8 **Hamersley Range** ▲▲ Western Australia
163 Y12 **Hamgyŏng-sanmaek** ▲▲ N North Korea
163 X13 **Hamhŭng** C North Korea 39°53'N 127°31'E
159 O6 **Hami** var. Ha-mi, Uigh. Kumul, Qomul. Xinjiang Uygur Zizhiqu, NW China 42°48'N 93°27'E
Ha-mi see Hami
139 X10 **Ḩāmid Amīn** Maysān, E Iraq 32°06'N 46°53'E
141 W11 **Hamīdān, Khawr** oasis SE Saudi Arabia
114 L12 **Hamidiye** Edirne, NW Turkey 41°09'N 26°40'E
182 L12 **Hamilton** Victoria, SE Australia 37°45'S 142°04'E
64 B12 **Hamilton** ○ (Bermuda) C Bermuda 32°18'N 64°48'W
14 G16 **Hamilton** Ontario, S Canada 43°15'N 79°50'W
184 M7 **Hamilton** Waikato, North Island, New Zealand 37°49'S 175°16'E
96 I12 **Hamilton** S Scotland, United Kingdom 55°47'N 04°03'W
23 N3 **Hamilton** Alabama, S USA 34°08'N 87°59'W
38 M10 **Hamilton** Alaska, USA 62°54'N 163°53'W
30 J13 **Hamilton** Illinois, N USA 40°24'N 91°20'W
27 S3 **Hamilton** Missouri, C USA 39°44'N 94°00'W
33 P10 **Hamilton** Montana, NW USA 46°15'N 114°09'W
25 S8 **Hamilton** Texas, SW USA 31°42'N 98°08'W
14 G16 **Hamilton** ✈ Ontario, SE Canada 43°12'N 79°54'W
64 I6 **Hamilton Bank** undersea feature SE Labrador Sea
182 E1 **Hamilton Creek** seasonal river South Australia
13 R8 **Hamilton Inlet** inlet Newfoundland and Labrador, E Canada
27 T12 **Hamilton, Lake** ⊠ Arkansas, C USA
35 W6 **Hamilton, Mount** ▲ Nevada, W USA 39°15'N 115°30'W
75 S8 **Ḩamīm, Wādī al** ≈ NE Libya
93 N19 **Hamina** Swe. Fredrikshamn. Kymenlaakso, S Finland 60°33'N 27°15'E
11 W16 **Hamiota** Manitoba, S Canada 50°13'N 100°37'W
152 L13 **Hamīrpur** Uttar Pradesh, N India 25°57'N 80°08'E
Hamis Musait see Khamīs Mushayṭ
21 T11 **Hamlet** North Carolina, SE USA 34°52'N 79°41'W
25 P6 **Hamlin** Texas, SW USA 32°52'N 100°07'W
21 P5 **Hamlin** West Virginia, NE USA 38°16'N 82°07'W
31 O7 **Hamlin Lake** ◎ Michigan, N USA
101 F14 **Hamm** var. Hamm in Westfalen. Nordrhein-Westfalen, W Germany 51°39'N 07°49'E
Ḩammāmāt, Khalīj al see Hammamet, Golfe de
75 N5 **Hammamet, Golfe de** Ar. Khalīj al Ḩammāmāt. gulf NE Tunisia
139 R3 **Ḩammān al 'Alīl** Nīnawé, N Iraq 36°07'N 43°15'E
139 X12 **Ḩammār, Hawr al** ◎ SE Iraq
93 J20 **Hammarland** Åland, SW Finland 60°13'N 19°45'E
93 H16 **Hammarstrand** Jämtland, C Sweden 63°07'N 16°27'E
93 O17 **Hammaslahti** Pohjois-Karjala, SE Finland 62°26'N 29°58'E
99 F17 **Hamme** Oost-Vlaanderen, NW Belgium 51°06'N 04°08'E
100 H10 **Hamme** ≈ NW Germany
95 G22 **Hammel** Midtjylland, C Denmark 56°15'N 09°53'E
101 I18 **Hammelburg** Bayern, C Germany 50°06'N 09°50'E
99 H18 **Hamme-Mille** Walloon Brabant, C Belgium 50°48'N 04°42'E
100 H10 **Hamme-Oste-Kanal** canal NW Germany
93 G16 **Hammerdal** Jämtland, C Sweden 63°34'N 15°19'E
92 K8 **Hammerfest** Finnmark, N Norway 70°40'N 23°44'E
101 D14 **Hamminkeln** Nordrhein-Westfalen, W Germany 51°43'N 06°36'E
Hamm in Westfalen see Hamm
26 K10 **Hammon** Oklahoma, C USA 35°37'N 99°22'W
31 N11 **Hammond** Indiana, N USA 41°35'N 87°30'W
22 K8 **Hammond** Louisiana, S USA 30°30'N 90°27'W
99 K20 **Hamoir** Liège, E Belgium 50°28'N 05°35'E
99 J21 **Hamois** Namur, SE Belgium 50°21'N 05°09'E
99 K16 **Hamont** Limburg, NE Belgium 51°15'N 05°33'E
185 F22 **Hampden** Otago, South Island, New Zealand 45°18'S 170°49'E
19 R6 **Hampden** Maine, NE USA 44°44'N 68°51'W
97 M23 **Hampshire** cultural region S England, United Kingdom
13 O15 **Hampton** New Brunswick, SE Canada 45°30'N 65°50'W
27 U14 **Hampton** Arkansas, C USA 33°33'N 92°28'W
29 V12 **Hampton** Iowa, C USA 42°44'N 93°12'W
19 P10 **Hampton** New Hampshire, NE USA 42°55'N 70°48'W
21 R14 **Hampton** South Carolina, SE USA 32°52'N 81°06'W
21 P8 **Hampton** Tennessee, S USA 36°16'N 82°10'W
21 X7 **Hampton** Virginia, NE USA 37°02'N 76°23'W
94 L11 **Hamra** Gävleborg, C Sweden 61°40'N 15°00'E
80 D10 **Hamrat esh Sheikh** Northern Kordofan, C Sudan 14°38'N 27°56'E
139 S5 **Ḩamrīn, Jabal** ▲▲ N Iraq
121 P16 **Ḩamrun** C Malta 35°53'N 14°28'E
Ham Thuân Nam see Thuân Nam
Hāmūn, Daryācheh-ye see Şāberī, Hāmūn-e/Sīstān, Daryācheh-ye
Hamwih see Southampton
38 G10 **Hāna** var. Hana. Maui, Hawaii, USA, C Pacific Ocean 20°45'N 155°59'W
Hana see Hāna
21 S14 **Hanahan** South Carolina, SE USA 32°55'N 80°01'W
38 B8 **Hanalei** Kaua'i, Hawaii, USA, C Pacific Ocean 22°12'N 159°30'W
165 Q9 **Hanamaki** Iwate, Honshū, C Japan 39°25'N 141°04'E
38 F10 **Hanamanioa, Cape** headland Maui, Hawai'i, USA 20°34'N 156°22'W
190 B16 **Hanan** ✈ (Alofi) SW Niue
101 H18 **Hanau** Hessen, W Germany 50°06'N 08°56'E
162 M11 **Hanbogd** var. Ih Bulag. Ömnögovĭ, S Mongolia 43°04'N 107°42'E
8 L9 **Hanbury** ≈ Northwest Territories, NW Canada
Hâncești see Hîncești
10 M15 **Hanceville** British Columbia, SW Canada 51°54'N 122°56'W
23 P3 **Hanceville** Alabama, S USA 34°03'N 86°46'W
Hancewicze see Hantsavichy
160 L6 **Hancheng** Shaanxi, C China 35°22'N 110°27'E
21 V2 **Hancock** Maryland, NE USA 39°42'N 78°10'W
30 M3 **Hancock** Michigan, N USA 47°07'N 88°34'W
29 S8 **Hancock** Minnesota, N USA 45°30'N 95°47'W
18 I12 **Hancock** New York, NE USA 41°57'N 75°15'W
80 Q12 **Handa** Bari, NE Somalia 10°35'N 51°09'E
161 O5 **Handan** var. Han-tan. Hebei, E China 36°35'N 114°28'E
95 P16 **Handen** Stockholm, C Sweden 59°12'N 18°09'E
81 J22 **Handeni** Tanga, E Tanzania 05°25'S 38°04'E
37 Q7 **Handies Peak** ▲ Colorado, C USA 37°54'N 107°30'W
111 J19 **Handlová** Ger. Krickerhäu, Hung. Nyitrabánya; prev. Kriegerhaj. Trenčiansky Kraj, C Slovakia 48°45'N 18°45'E
165 O13 **Haneda** ✈ (Tōkyō) Tōkyō, Honshū, S Japan 35°33'N 139°45'E
138 F13 **HaNegev** Eng. Negev. desert S Israel
Hanfeng see Kaixian
35 Q11 **Hanford** California, W USA 36°19'N 119°39'W
191 V16 **Hanga Roa** Easter Island, Chile, E Pacific Ocean 27°09'S 109°26'W
162 I7 **Hangay** var. Hunt. Arhangay, C Mongolia 47°49'N 99°24'E
162 H7 **Hangayn Nuruu** ▲▲ C Mongolia
Hang-chou/Hangchow see Hangzhou
95 K20 **Hånger** Jönköping, S Sweden 57°06'N 13°58'E
Hangō see Hanko
161 R9 **Hangzhou** var. Hang-chou, Hangchow. province capital Zhejiang, SE China 30°18'N 120°07'E
162 J4 **Hanh** var. Turt. Hövsgöl, N Mongolia 51°30'N 100°40'E
162 F5 **Hanhöhiy Uul** ▲▲ NW Mongolia
162 K10 **Hanhongor** var. Ögöömör. Ömnögovĭ, S Mongolia 43°47'N 104°31'E
146 I14 **Hanhowuz** Rus. Khauz-Khan. Ahal Welaýaty, S Turkmenistan 37°15'N 61°12'E
146 I14 **Hanhowuz Suw Howdany** Rus. Khauzkhanskoye Vodoranilishche. ⊠ S Turkmenistan
137 P15 **Hani** Diyarbakır, SE Turkey 38°26'N 40°23'E
Hania see Chaniá
141 R11 **Ḩanīsh al Kabīr, Jazīrat al** island SW Yemen
Hanka, Lake see Khanka, Lake
93 M17 **Hankasalmi** Keski-Suomi, C Finland 62°25'N 26°27'E
29 R7 **Hankinson** North Dakota, N USA 46°04'N 96°54'W
93 K20 **Hanko** Swe. Hangö. Uusimaa, SW Finland 59°50'N 23°E
Han-kou/Han-k'ou/Hankow see Wuhan
36 M6 **Hanksville** Utah, W USA 38°21'N 110°43'W
152 K6 **Hanle** Jammu and Kashmir, NW India 32°46'N 79°01'E
185 I17 **Hanmer Springs** Canterbury, South Island, New Zealand 42°31'S 172°49'E
11 R16 **Hanna** Alberta, SW Canada 51°38'N 111°56'W
27 V3 **Hannibal** Missouri, C USA 39°42'N 91°23'W
180 M3 **Hann, Mount** ▲ Western Australia 15°53'S 125°46'E
100 I12 **Hannover** Eng. Hanover. Niedersachsen, NW Germany 52°23'N 09°43'E
99 J19 **Hannut** Liège, C Belgium 50°40'N 05°05'E
95 L22 **Hanöbukten** bay S Sweden
167 T6 **Ha Nôi** Eng. Hanoi, Fr. Hanoï. ● (Vietnam) N Vietnam 21°01'N 105°52'E
14 F14 **Hanover** Ontario, S Canada 44°10'N 81°03'W
31 P15 **Hanover** Indiana, N USA 38°42'N 85°28'W
18 G16 **Hanover** Pennsylvania, NE USA 39°46'N 76°57'W
21 W6 **Hanover** Virginia, NE USA 37°44'N 77°21'W
Hanover see Hannover
63 G23 **Hanover, Isla** island S Chile
Hanselbeck see Érd
195 X5 **Hansen Mountains** ▲▲ Antarctica
160 M8 **Han Shui** ≈ C China
152 H10 **Hānsi** Haryāna, NW India 29°06'N 76°01'E
95 F20 **Hanstholm** Midtjylland, NW Denmark 57°05'N 08°39'E
Han-tan see Handan
158 H6 **Hantengri Feng** var. Pik Khan-Tengri. ▲ China/Kazakhstan 42°17'N 80°11'E see also Khan-Tengri, Pik
119 I18 **Hantsavichy** Pol. Hancewicze, Rus. Gantsevichi. Brestskaya Voblasts', SW Belarus 52°45'N 26°27'E
9 Q6 **Hantzsch** ≈ Baffin Island, Nunavut, NE Canada
152 G9 **Hanumāngarh** Rājasthān, NW India 29°33'N 74°21'E
183 O9 **Hanwood** New South Wales, SE Australia 34°19'S 146°03'E
Hanyang see Wuhan
Hanyang see Caidian
160 H10 **Hanyuan** var. Fulin. Sichuan, C China 29°29'N 102°45'E
Hanyuan see Xihe
160 J7 **Hanzhong** Shaanxi, C China 33°12'N 107°E
191 W11 **Hao** atoll Îles Tuamotu, C French Polynesia
153 S16 **Hāora** prev. Howrah. West Bengal, NE India 22°35'N 88°20'E
78 K8 **Haouach, Ouadi** dry watercourse E Chad
92 K13 **Haparanda** Norrbotten, N Sweden 65°49'N 24°05'E
25 N3 **Happy** Texas, SW USA 34°44'N 101°51'W
34 M1 **Happy Camp** California, W USA 41°48'N 123°24'W
13 Q9 **Happy Valley-Goose Bay** prev. Goose Bay. Newfoundland and Labrador, E Canada 53°19'N 60°24'W
Hapsal see Haapsalu
152 J10 **Hāpur** Uttar Pradesh, N India 28°43'N 77°47'E
HaQatan, HaMakhtesh see HaKatan, HaMakhtesh
140 I4 **Ḩaql** Tabūk, NW Saudi Arabia 29°18'N 34°58'E
171 U14 **Har** Pulau Kai Besar, E Indonesia 05°21'S 133°09'E
Haraat see Tsagaandelger
141 R8 **Ḩaraḑ** var. Haradh. Ash Sharqīyah, E Saudi Arabia 24°08'N 49°02'E
Haradh see Ḩaraḑ
118 N12 **Haradok** Rus. Gorodok. Vitsyebskaya Voblasts', N Belarus 55°28'N 30°00'E
92 J13 **Harads** Norrbotten, N Sweden 66°04'N 21°04'E
119 G19 **Haradzyets** Rus. Gorodets. Brestskaya Voblasts', SW Belarus 52°12'N 24°40'E
119 J17 **Haradzyeya** Rus. Gorodeya. Minskaya Voblasts', C Belarus 53°19'N 26°32'E
191 V10 **Haraiki** atoll Îles Tuamotu, C French Polynesia
165 Q11 **Haramachi** Fukushima, Honshū, E Japan 37°40'N 140°55'E
118 M12 **Harany** Rus. Gorany. Vitsyebskaya Voblasts', N Belarus 55°25'N 29°03'E
83 L16 **Harare** prev. Salisbury. ● (Zimbabwe) Mashonaland East, NE Zimbabwe 17°47'S 31°04'E
83 L16 **Harare** ✈ Mashonaland East, NE Zimbabwe 17°51'S 31°06'E
78 J10 **Haraz-Djombo** Batha, C Chad 14°10'N 19°35'E
119 O16 **Harbavichy** Rus. Gorbovichi. Mahilyowskaya Voblasts', E Belarus 53°49'N 30°42'E
76 J16 **Harbel** W Liberia 06°19'N 10°20'W
163 W8 **Harbin** var. Haerbin, Ha-erh-pin, Kharbin; prev. Haerhpin, Pingkiang, Pinkiang. province capital Heilongjiang, NE China 45°45'N 126°41'E
31 S7 **Harbor Beach** Michigan, N USA 43°50'N 82°39'W
13 T13 **Harbour Breton** Newfoundland, Newfoundland and Labrador, E Canada 47°29'N 55°50'W
65 D25 **Harbours, Bay of** bay East Falkland, Falkland Islands
Hårby see Haarby
36 I13 **Harcuvar Mountains** ▲▲ Arizona, SW USA
108 I7 **Hard** Vorarlberg, NW Austria 47°29'N 09°42'E
154 H11 **Harda Khas** Madhya Pradesh, C India 22°22'N 77°06'E
95 D14 **Hardanger** physical region S Norway
95 D14 **Hardangerfjorden** fjord S Norway
94 E13 **Hardangerjøkulen** glacier S Norway
95 E14 **Hardangervidda** plateau S Norway
83 D20 **Hardap** ❖ district S Namibia
21 R15 **Hardeeville** South Carolina, SE USA 32°18'N 81°04'W
Hardegarijp see Hurdegaryp
98 O9 **Hardenberg** Overijssel, E Netherlands 52°34'N 06°38'E
183 Q9 **Harden-Murrumburrah** New South Wales, SE Australia 34°33'S 148°22'E
98 K10 **Harderwijk** Gelderland, C Netherlands 52°21'N 05°37'E
30 J14 **Hardin** Illinois, N USA 39°10'N 90°38'W
33 V11 **Hardin** Montana, NW USA 45°44'N 107°35'W
23 R5 **Harding, Lake** ⊠ Alabama/Georgia, SE USA
20 J6 **Hardinsburg** Kentucky, S USA 37°46'N 86°29'W
98 I13 **Hardinxveld-Giessendam** Zuid-Holland, C Netherlands 51°52'N 04°49'E
11 R15 **Hardisty** Alberta, SW Canada 52°42'N 111°22'W
152 L12 **Hardoi** Uttar Pradesh, N India 27°23'N 80°06'E
Hardwar see Haridwār
23 U4 **Hardwick** Georgia, SE USA 33°03'N 83°13'W
27 W9 **Hardy** Arkansas, C USA 36°19'N 91°29'W
94 D10 **Hareid** Møre og Romsdal, S Norway 62°22'N 06°02'E
8 H7 **Hare Indian** ≈ Northwest Territories, NW Canada
99 D18 **Harelbeke** var. Harlebeke. West-Vlaanderen, W Belgium 50°51'N 03°19'E
Harem see Ḩārim
100 E11 **Haren** Niedersachsen, NW Germany 52°47'N 07°16'E
98 N6 **Haren** Groningen, NE Netherlands 53°10'N 06°37'E
80 L13 **Hārer** E Ethiopia 09°17'N 42°19'E
95 P14 **Harg** Uppsala, C Sweden 60°13'N 18°25'E
Hargeisa see Hargeysa
80 M13 **Hargeysa** var. Hargeisa. Woqooyi Galbeed, NW Somalia 09°32'N 44°07'E
116 J10 **Harghita** ❖ county NE Romania
25 S17 **Hargill** Texas, SW USA 26°26'N 98°00'W
162 J8 **Harhorin** Övörhangay, C Mongolia 47°13'N 102°48'E
159 Q9 **Har Hu** ◎ C China
Hariana see Haryāna
141 P15 **Ḩarīb** W Yemen 15°08'N 45°35'E
168 M12 **Hari, Batang** prev. Djambi. ≈ Sumatera, W Indonesia
152 J9 **Haridwār** prev. Hardwar. Uttarakhand, N India 29°58'N 78°09'E
155 F18 **Harihar** Karnātaka, W India 14°33'N 75°44'E
185 F18 **Harihari** West Coast, South Island, New Zealand 43°09'S 170°35'E
138 I3 **Ḩārim** var. Harem. Idlib, W Syria 36°30'N 36°30'E
98 F13 **Haringvliet** channel SW Netherlands
98 F13 **Haringvlietdam** dam SW Netherlands
149 U5 **Haripur** Khyber Pakhtunkhwa, NW Pakistan 34°00'N 73°01'E
148 J4 **Harīrūd** var. Tedzhen, Turkm. Tejen. ≈ Afghanistan/Iran see also Tejen
Harīrūd see Tejen
94 J11 **Härjahågnen** Swe. Härjahågnen, var. Härjehågna. ▲ Norway/Sweden 61°43'N 12°07'E
Härjahågnen see Østrehogna
Härjåhågnen see Härjåhågnen
93 K18 **Harjavalta** Satakunta, W Finland 61°19'N 22°10'E
Härjehågna see Østrehogna
Härjehågna see Härjåhågnen
118 G4 **Harjumaa** var. Harju Maakond. ❖ province NW Estonia
Harju Maakond see Harjumaa
21 X11 **Harkers Island** North Carolina, SE USA 34°42'N 76°33'W
Harki see Hurkê
29 T14 **Harlan** Iowa, C USA 41°40'N 95°19'W
21 O7 **Harlan** Kentucky, S USA 36°50'N 83°19'W
29 N17 **Harlan County Lake** ⊠ Nebraska, C USA
116 L9 **Hârlău** var. Hîrlău. Iași, NE Romania 47°26'N 26°54'E
Harlebeke see Harelbeke
33 U7 **Harlem** Montana, NW USA 48°31'N 108°46'W
Harlem see Haarlem
95 G22 **Harlev** Midtjylland, C Denmark 56°08'N 10°00'E
98 K6 **Harlingen** Fris. Harns. Fryslân, N Netherlands 53°10'N 05°25'E
25 T17 **Harlingen** Texas, SW USA 26°12'N 97°43'W
97 O21 **Harlow** E England, United Kingdom 51°47'N 00°07'E
33 T10 **Harlowton** Montana, NW USA 46°26'N 109°49'W
94 N11 **Harmånger** Gävleborg, C Sweden 61°55'N 17°19'E
114 K11 **Harmanli** var. Kharmanli. Haskovo, S Bulgaria 41°56'N 25°54'E
114 K11 **Harmanliyska Reka** var. Kharmanliyska Reka. ≈ S Bulgaria
98 I11 **Harmelen** Utrecht, C Netherlands 52°06'N 04°58'E
29 X11 **Harmony** Minnesota, N USA 43°33'N 92°00'W
32 J14 **Harney Basin** basin Oregon, NW USA
32 J14 **Harney Lake** ◎ Oregon, NW USA
28 J10 **Harney Peak** ▲ South Dakota, N USA 43°52'N 103°31'W
93 H17 **Härnösand** var. Hernösand. Västernorrland, C Sweden 62°37'N 17°55'E
Harns see Harlingen
105 P4 **Haro** La Rioja, N Spain 42°34'N 02°52'W
40 F6 **Haro, Cabo** headland NW Mexico 27°50'N 110°55'W
94 D9 **Harøy** island S Norway
97 N21 **Harpenden** E England, United Kingdom 51°49'N 00°22'W
76 L18 **Harper** var. Cape Palmas. NE Liberia 04°25'N 07°43'W
26 M7 **Harper** Kansas, C USA 37°17'N 98°01'W
32 L13 **Harper** Oregon, NW USA 43°51'N 117°37'W
25 Q10 **Harper** Texas, SW USA 30°18'N 99°18'W
35 U13 **Harper Lake** salt flat California, W USA
39 T9 **Harper, Mount** ▲ Alaska, USA 64°18'N 143°54'W
95 J21 **Harplinge** Halland, S Sweden 56°45'N 12°45'E
36 J13 **Harquahala Mountains** ▲▲ Arizona, SW USA
141 T15 **Ḩarrah** SE Yemen 15°02'N 50°23'E
12 H11 **Harricana** ≈ Québec, SE Canada
20 M9 **Harriman** Tennessee, S USA 35°57'N 84°33'W
13 R11 **Harrington Harbour** Québec, E Canada 50°34'N 59°29'W
64 B12 **Harrington Sound** bay Bermuda, NW Atlantic Ocean
96 F8 **Harris** physical region NW Scotland, United Kingdom
27 X10 **Harrisburg** Arkansas, C USA 35°33'N 90°43'W
30 M17 **Harrisburg** Illinois, N USA 37°44'N 88°32'W
28 I14 **Harrisburg** Nebraska, C USA 41°33'N 103°46'W
32 F12 **Harrisburg** Oregon, NW USA 44°16'N 123°10'W
18 G15 **Harrisburg** state capital Pennsylvania, NE USA 40°16'N 76°53'W
182 F6 **Harris, Lake** ◎ South Australia
23 W11 **Harris, Lake** ◎ Florida, SE USA
83 K22 **Harrismith** Free State, E South Africa 28°16'S 29°08'E
27 T9 **Harrison** Arkansas, C USA 36°13'N 93°07'W
31 Q7 **Harrison** Michigan, N USA 44°02'N 84°46'W
28 I12 **Harrison** Nebraska, C USA 42°42'N 103°53'W
39 Q5 **Harrison Bay** inlet Alaska, USA
22 I6 **Harrisonburg** Louisiana, S USA 31°44'N 91°51'W
21 U4 **Harrisonburg** Virginia, NE USA 38°27'N 78°54'W
13 R7 **Harrison, Cape** headland Newfoundland and Labrador, E Canada 54°55'N 57°48'W
27 R5 **Harrisonville** Missouri, C USA 38°40'N 94°21'W
Harris Ridge see Lomonosov Ridge
192 M3 **Harris Seamount** undersea feature N Pacific Ocean 46°09'N 161°25'W
96 F8 **Harris, Sound of** strait NW Scotland, United Kingdom
31 R6 **Harrisville** Michigan, N USA 44°41'N 83°19'W
21 R3 **Harrisville** West Virginia, NE USA 39°13'N 81°04'W
20 M6 **Harrodsburg** Kentucky, S USA 37°45'N 84°51'W
97 M16 **Harrogate** N England, United Kingdom 54°N 01°33'W
25 Q4 **Harrold** Texas, SW USA 34°05'N 99°02'W
27 S5 **Harry S. Truman Reservoir** ⊠ Missouri, C USA
100 G13 **Harsewinkel** Nordrhein-Westfalen, W Germany 51°58'N 08°13'E
116 M14 **Hârşova** prev. Hîrşova. Constanţa, SE Romania 44°41'N 27°56'E
92 H10 **Harstad** Troms, N Norway 68°48'N 16°31'E
31 O8 **Hart** Michigan, N USA 43°43'N 86°22'W
24 M4 **Hart** Texas, SW USA 34°23'N 102°07'W
10 I5 **Hart** ≈ Yukon, NW Canada
83 F23 **Hartbees** ≈ C South Africa
109 X7 **Hartberg** Steiermark, SE Austria 47°18'N 15°58'E
182 I10 **Hart, Cape** headland South Australia 35°54'S 138°03'E
95 E14 **Hårteigen** ▲ S Norway 60°11'N 07°01'E
23 Q7 **Hartford** Alabama, S USA 31°06'N 85°42'W
27 R11 **Hartford** Arkansas, C USA 35°01'N 94°22'W
18 M12 **Hartford** state capital Connecticut, NE USA 41°46'N 72°41'W
20 J6 **Hartford** Kentucky, S USA 37°26'N 86°57'W
31 P10 **Hartford** Michigan, N USA 42°12'N 85°54'W
29 R11 **Hartford** South Dakota, N USA 43°37'N 96°56'W
30 M8 **Hartford** Wisconsin, N USA 43°19'N 88°25'W
31 P13 **Hartford City** Indiana, N USA 40°27'N 85°22'W
29 Q13 **Hartington** Nebraska, C USA 42°37'N 97°15'W
13 N14 **Hartland** New Brunswick, SE Canada 46°18'N 67°31'W
97 H23 **Hartland Point** headland SW England, United Kingdom 51°01'N 04°33'W
97 M15 **Hartlepool** N England, United Kingdom 54°41'N 01°13'W
29 T12 **Hartley** Iowa, C USA 43°10'N 95°28'W
24 M1 **Hartley** Texas, SW USA 35°52'N 102°24'W
32 J15 **Hart Mountain** ▲ Oregon, NW USA 42°24'N 119°46'W
173 U10 **Hartog Ridge** undersea feature W Indian Ocean
93 M18 **Hartola** Päijät-Häme, S Finland 61°34'N 26°04'E
67 U14 **Harts** var. Hartz. ≈ N South Africa
23 P2 **Hartselle** Alabama, S USA 34°26'N 86°56'W
23 S3 **Hartsfield Atlanta** ✈ Georgia, SE USA 33°38'N 84°24'W
27 Q11 **Hartshorne** Oklahoma, C USA 34°51'N 95°33'W
21 S12 **Hartsville** South Carolina, SE USA 34°22'N 80°04'W
20 K8 **Hartsville** Tennessee, S USA 36°23'N 86°11'W
27 U7 **Hartville** Missouri, C USA 37°15'N 92°30'W
23 U2 **Hartwell** Georgia, SE USA 34°21'N 82°55'W
21 O11 **Hartwell Lake** ⊠ Georgia/South Carolina, SE USA
Hartz see Harts
Harunabad see Eslāmābād-e Gharb
Har-Us see Erdenebüren
162 F6 **Har Us Gol** ◎ Hovd, W Mongolia
162 E6 **Har Us Nuur** ◎ NW Mongolia
30 M10 **Harvard** Illinois, N USA 42°25'N 88°36'W
29 P16 **Harvard** Nebraska, C USA 40°37'N 98°06'W
37 R5 **Harvard, Mount** ▲ Colorado, C USA 38°55'N 106°19'W
31 N11 **Harvey** Illinois, N USA 41°36'N 87°39'W
29 N4 **Harvey** North Dakota, N USA 47°43'N 99°55'W
97 Q21 **Harwich** E England, United Kingdom 51°56'N 01°16'E
152 H10 **Haryāna** var. Hariana. ❖ state N India
141 Y9 **Ḩaryān, Ṭawī al** spring/well NE Oman 21°56'N 58°33'E
101 J14 **Harz** ▲▲ C Germany
Hasakah, Muḩāfaẓat al see Al Ḩasakah
165 Q9 **Hasama** Miyagi, Honshū, C Japan 38°42'N 141°09'E
137 Y13 **Häsänabad** prev. 26 Bakinskikh Komissarov. SE Azerbaijan 39°18'N 49°13'E
136 J15 **Hasan Dağı** ▲ C Turkey 38°09'N 34°15'E
139 T9 **Ḩasan Ibn Ḩassūn** An Najaf, C Iraq 32°24'N 44°13'E
149 R6 **Ḩasan Khēl** var. Ahmad Khel. Paktīyā, SE Afghanistan 33°46'N 69°37'E
100 F12 **Hase** ≈ NW Germany
Haselberg see Krasnoznamensk
100 F12 **Haselünne** Niedersachsen, NW Germany 52°40'N 07°28'E
Hashaat see Delgerhangay
Hashemite Kingdom of Jordan see Jordan
139 V8 **Ḩāshimah** Wāsiţ, E Iraq 33°22'N 45°56'E
142 K3 **Hashtrūd** var. Azaran. Āżarbāyjān-e Khāvarī, N Iran 37°34'N 47°10'E
141 W13 **Ḩāsik** S Oman 17°22'N 55°18'E
149 U10 **Hāsilpur** Punjab, E Pakistan 29°42'N 72°40'E
27 Q10 **Haskell** Oklahoma, C USA 35°49'N 95°40'W
25 Q6 **Haskell** Texas, SW USA 33°10'N 99°45'W
114 K11 **Haskovo** var. Khaskovo. Haskovo, S Bulgaria 41°56'N 25°35'E
114 K11 **Haskovo** var. Khaskovo. ❖ province S Bulgaria
114 M11 **Hasköy** Edirne, NW Turkey 41°37'N 26°51'E
95 L24 **Hasle** Bornholm, E Denmark 55°12'N 14°43'E
97 N23 **Haslemere** SE England, United Kingdom 51°06'N 00°45'W
102 I16 **Hasparren** Pyrénées-Atlantiques, SW France 43°23'N 01°18'W
Hassakeh see Al Ḩasakah
155 G19 **Hassan** Karnātaka, W India 13°01'N 76°03'E
36 J13 **Hassayampa River** ≈ Arizona, SW USA
101 J18 **Hassberge** hill range C Germany
94 N10 **Hassela** Gävleborg, C Sweden 62°06'N 16°45'E
99 J18 **Hasselt** Limburg, NE Belgium 50°56'N 05°20'E
98 M9 **Hasselt** Overijssel, E Netherlands 52°36'N 06°06'E
Hassetché see Al Ḩasakah
101 J18 **Hassfurt** Bayern, C Germany 50°02'N 10°32'E
74 L9 **Hassi Bel Guebbour** E Algeria 28°41'N 06°29'E
74 L8 **Hassi Messaoud** E Algeria 31°41'N 06°10'E
95 K22 **Hässleholm** Skåne, S Sweden 56°09'N 13°45'E
Hasta Colonia/Hasta Pompeia see Asti
183 O13 **Hastings** Victoria, SE Australia 38°18'S 145°12'E
184 O11 **Hastings** Hawke's Bay, North Island, New Zealand 39°39'S 176°51'E
97 P23 **Hastings** SE England, United Kingdom 50°51'N 00°36'E
31 P9 **Hastings** Michigan, N USA 42°38'N 85°17'W
29 W9 **Hastings** Minnesota, N USA 44°44'N 92°51'W
29 P16 **Hastings** Nebraska, C USA 40°35'N 98°23'W
95 K22 **Hästveda** Skåne, S Sweden 56°16'N 13°55'E
92 J8 **Hasvik** Finnmark, N Norway 70°29'N 22°08'E
37 V6 **Haswell** Colorado, C USA 38°27'N 103°09'W
163 N11 **Hatanbulag** var. Ergel. Dornogovĭ, SE Mongolia 43°10'N 109°13'E
Hatansuudal see Bayanlig
Hatavch see Haldzan
136 K17 **Hatay** ❖ province S Turkey
37 R15 **Hatch** New Mexico, SW USA 32°40'N 107°10'W
36 K7 **Hatch** Utah, W USA 37°39'N 112°25'W
20 F9 **Hatchie River** ≈ Tennessee, S USA
116 G12 **Haţeg** Ger. Wallenthal, Hung. Hátszeg; prev. Hatzeg, Hötzing. Hunedoara, SW Romania 45°35'N 22°57'E
165 O17 **Hateruma-jima** island Yaeyama-shotō, SW Japan
183 N8 **Hatfield** New South Wales, SE Australia 33°54'S 143°43'E
162 I5 **Hatgal** Hövsgöl, N Mongolia 50°24'N 100°12'E
153 V16 **Hathazari** Chittagong, SE Bangladesh 22°30'N 91°46'E
141 T13 **Hathūt, Hişā'** oasis NE Yemen
167 R14 **Ha Tiên** Kiên Giang, S Vietnam 10°24'N 104°30'E
167 T8 **Ha Tinh** Ha Tinh, N Vietnam 18°21'N 105°55'E
Ḩatira, Haré see Hatira, Harei
138 F12 **Hatira, Harei** prev. Haré Ḩatira. hill range S Israel
167 R6 **Hat Lot** var. Mai Son. Son La, N Vietnam 21°07'N 104°10'E
45 P16 **Hato Airport** ✈ (Willemstad) Curaçao 12°10'N 68°56'W
54 H9 **Hato Corozal** Casanare, C Colombia 06°08'N 71°45'W
Hato del Volcán see Volcán
45 P9 **Hato Mayor** E Dominican Republic 18°49'N 69°16'W
Hatra see Al Ḩaḑr
Hatria see Adria
Hátszeg see Haţeg
143 R16 **Ḩattā** Dubayy, NE United Arab Emirates 24°50'N 56°06'E
182 L9 **Hattah** Victoria, SE Australia 34°49'S 142°18'E
98 M9 **Hattem** Gelderland, E Netherlands 52°29'N 06°04'E
21 Z10 **Hatteras** Hatteras Island, North Carolina, SE USA 35°13'N 75°39'W
21 R10 **Hatteras, Cape** headland North Carolina, SE USA 35°29'N 75°33'W
21 Z9 **Hatteras Island** island North Carolina, SE USA
64 F10 **Hatteras Plain** undersea feature W Atlantic Ocean 31°00'N 71°00'W
93 G14 **Hattfjelldal** Troms, N Norway 65°37'N 13°58'E
22 M7 **Hattiesburg** Mississippi, S USA 31°20'N 89°17'W
29 Q4 **Hatton** North Dakota, N USA 47°38'N 97°27'W
Hatton Bank see Hatton Ridge
64 L6 **Hatton Ridge** var. Hatton Bank. undersea feature N Atlantic Ocean 59°00'N 17°30'W
191 W6 **Hatutu** island Îles Marquises, NE French Polynesia
111 K22 **Hatvan** Heves, NE Hungary 47°40'N 19°39'E
167 O16 **Hat Yai** var. Ban Hat Yai. Songkhla, SW Thailand 07°01'N 100°27'E
Hatzeg see Haţeg
Hatzfeld see Jimbolia
80 N13 **Haud** plateau Ethiopia/Somalia
95 D18 **Hauge** Rogaland, S Norway 58°20'N 06°17'E
95 C15 **Haugesund** Rogaland, S Norway 59°24'N 05°17'E
109 X2 **Haugsdorf** Niederösterreich, NE Austria 48°41'N 16°04'E
184 M9 **Hauhungaroa Range** ▲▲ North Island, New Zealand
95 E15 **Haukeligrend** Telemark, S Norway 59°45'N 07°33'E
93 L14 **Haukipudas** Pohjois-Pohjanmaa, C Finland 65°11'N 25°21'E
93 M17 **Haukivesi** ◎ SE Finland
93 M17 **Haukivuori** Etelä-Savo, E Finland 62°02'N 27°11'E
Hauptkanal see Havelländ Grosse
187 N10 **Hauraha** Makira-Ulawa, SE Solomon Islands 10°47'S 162°00'E
184 L5 **Hauraki Gulf** gulf North Island, New Zealand
185 B24 **Hauroko, Lake** ◎ South Island, New Zealand
167 S14 **Hâu, Sông** ≈ S Vietnam
92 N12 **Hautajärvi** Lappi, NE Finland 66°30'N 29°01'E
74 F7 **Haut Atlas** Eng. High Atlas. ▲▲ C Morocco
79 M17 **Haut-Congo** off. Région du Haut-Congo; prev. Haut-Zaïre. ❖ region NE Dem. Rep. Congo
103 Y14 **Haute-Corse** ❖ department Corse, France, C Mediterranean Sea
102 L16 **Haute-Garonne** ❖ department S France
79 K14 **Haute-Kotto** ❖ prefecture E Central African Republic
103 P12 **Haute-Loire** ❖ department C France
103 R6 **Haute-Marne** ❖ department N France
102 M3 **Haute-Normandie** ❖ region N France
15 U6 **Hauterive** Québec, SE Canada 49°11'N 68°16'W
103 T13 **Hautes-Alpes** ❖ department SE France
103 S7 **Haute-Saône** ❖ department E France
103 T10 **Haute-Savoie** ❖ department E France
99 M20 **Hautes Fagnes** Ger. Hohes Venn. ▲▲ E Belgium
102 K16 **Hautes-Pyrénées** ❖ department S France
99 L23 **Haute Sûre, Lac de la** ⊠ NW Luxembourg
102 M11 **Haute-Vienne** ❖ department C France
19 S8 **Haut, Isle au** island Maine, NE USA

◆ Country
● Country Capital
◇ Dependent Territory
○ Dependent Territory Capital
❖ Administrative Regions
✈ International Airport
▲ Mountain
▲▲ Mountain Range
Volcano
≈ River
◎ Lake
⊠ Reservoir

Haut-Mbomou ◇ *prefecture* SE Central African Republic — 79 M14
Hautmont Nord, N France 50°15′N 03°55′E — 103 Q2
Haut-Ogooué *off.* Province du Haut-Ogooué, *var.* Le Haut-Ogooué. ◇ *province* SE Gabon — 79 F19
Haut-Ogooué, Le *see* Haut-Ogooué
Haut-Ogooué, Province du *see* Haut-Ogooué
Haut-Rhin ◆ *department* NE France — 103 U7
Hauts Plateaux *plateau* Algeria/Morocco — 74 I6
Haut-Zaïre *see* Haut-Congo
Hau'ula *var.* Hauula. O'ahu, Hawaii, USA, C Pacific Ocean 21°36′N 157°54′W — 38 D9
Hauula *see* Hau'ula
Hauzenberg Bayern, SE Germany 48°39′N 13°37′E — 101 O22
Havana Illinois, N USA 40°18′N 90°03′W — 30 K13
Havana *see* La Habana
Havant S England, United Kingdom 50°51′N 00°59′W — 97 N23
Havasu, Lake ◎ Arizona/California, W USA — 35 Y14
Havdrup Sjælland, E Denmark 55°33′N 12°08′E — 95 J23
Havel ◢ NE Germany — 100 N10
Havelange Namur, SE Belgium 50°23′N 05°14′E — 99 J21
Havelberg Sachsen-Anhalt, NE Germany 52°49′N 12°05′E — 100 M11
Havelián Khyber Pakhtunkhwa, NW Pakistan 34°05′N 73°14′E — 149 U5
Havelländ Grosse *var.* Hauptkanal. *canal* NE Germany — 100 N12
Havelock Ontario, SE Canada 44°22′N 77°57′W — 14 J14
Havelock Marlborough, South Island, New Zealand 41°17′S 173°46′E — 185 J14
Havelock North Carolina, SE USA 34°52′N 76°54′W — 21 X11
Havelock North Hawke's Bay, North Island, New Zealand 39°40′S 176°53′E — 184 O11
Havelte Drenthe, NE Netherlands 52°46′N 06°14′E — 98 M8
Haven Kansas, C USA 37°54′N 97°46′W — 27 N6
Haverfordwest SW Wales, United Kingdom 51°50′N 04°57′W — 97 H21
Haverhill E England, United Kingdom 52°05′N 00°26′E — 97 P20
Haverhill Massachusetts, NE USA 42°46′N 71°02′W — 19 O10
Haverö Västernorrland, C Sweden 62°25′N 15°04′E — 93 G17
Havířov Moravskoslezský Kraj, E Czech Republic 49°47′N 18°30′E — 111 I17
Havlíčkův Brod *Ger.* Deutsch-Brod; *prev.* Německý Brod. Vysočina, C Czech Republic 49°38′N 15°46′E — 111 E17
Havøysund Finnmark, N Norway 70°59′N 24°39′E — 92 K7
Havré Hainaut, S Belgium — 99 P20
Havre Montana, NW USA 48°33′N 109°41′W — 33 T7
Havre-St-Pierre Québec, E Canada 50°16′N 63°36′W — 13 P11
Havsa Edirne, NW Turkey 41°32′N 26°49′E — 136 B10
Hawai'i *off.* State of Hawai'i, *also known as* Aloha State, Paradise of the Pacific, *var.* Hawaii. ◆ *state* USA, C Pacific Ocean — 38 D8
Hawai'i *var.* Hawaii. *island* Hawaii, USA, C Pacific Ocean — 38 G12
Hawaiian Islands *prev.* Sandwich Islands. *island group* Hawaii, USA — 192 M5
Hawaiian Ridge *undersea feature* N Pacific Ocean 24°00′N 165°00′W — 192 L5
Hawaiian Trough *undersea feature* N Pacific Ocean — 193 N6
Hawarden Iowa, C USA 43°00′N 96°29′W — 29 R12
Hawash *see* Āwash
Hawban al Gharbiyah Al Anbār, C Iraq 34°24′N 42°06′E — 139 P6
Hawea, Lake ◎ South Island, New Zealand — 185 D21
Hawera Taranaki, North Island, New Zealand 39°36′S 174°16′E — 184 K11
Hawesville Kentucky, S USA 37°53′N 86°47′W — 21 J5
Hawi Hawaii, USA, C Pacific Ocean 20°14′N 155°50′W — 38 G11
Hāwī *var.* Hawi. Hawaii, USA, C Pacific Ocean 20°13′N 155°49′E — 38 G11
Hawi *see* Hāwī
Hawick SE Scotland, United Kingdom 55°24′N 02°49′W — 96 K5
Ḩawījah Kirkūk, C Iraq 35°15′N 43°54′E — 139 S4
Hawizah, Hawr al ◎ S Iraq — 35 Y10
Hawkdun Range ▲ South Island, New Zealand — 185 E21
Hawke Bay *bay* North Island, New Zealand — 184 P10
Hawker South Australia 31°54′S 138°25′E — 182 I6
Hawke's Bay *off.* Hawkes Bay Region. ◇ *region* North Island, New Zealand — 184 N11
Hawkes Bay *bay* SE Pakistan — 149 O16
Hawkes Bay Region *see* Hawke's Bay
Hawkesbury Ontario, SE Canada 45°36′N 74°38′W — 15 N12
Hawkinsville Georgia, SE USA — 23 T5
Hawk Junction Ontario, S Canada 48°05′N 84°34′W — 14 B7
Haw Knob ▲ North Carolina/Tennessee, SE USA 35°18′N 84°01′W — 21 N10
Hawksbill Mountain ▲ North Carolina, SE USA 35°54′N 81°53′W — 21 Q9
Hawk Springs Wyoming, C USA 41°48′N 104°17′W — Z16
Hawler *see* Arbīl
Hawley Minnesota, N USA 46°53′N 96°18′W — 29 S5
Hawley Texas, SW USA 32°36′N 99°47′W — 25 P7
Ḩawrā' C Yemen — 141 R14
Hawrān, Wadi *dry watercourse* W Iraq — 139 P7
Haw River ◢ North Carolina, SE USA — 21 T9

Hawshqūrah Diyālá, E Iraq 34°34′N 45°33′E — 139 U5
Hawthorne Nevada, W USA 38°30′N 118°38′W — 35 S7
Haxtun Colorado, C USA 40°36′N 102°38′W — 37 W3
Hay New South Wales, SE Australia 34°31′S 144°51′E — 183 N9
Hay ◢ W Canada — 11 O10
Haya Pulau Seram, E Indonesia 03°22′S 129°31′E — 171 S13
Hayachine-san ▲ Honshū, C Japan 39°31′N 141°28′E — 165 R9
Hayange Moselle, NE France 49°19′N 06°04′E — 103 S4
Ha'Yarden *see* Jordan
Hayastani Hanrapetut'yun *see* Armenia
Hayasu-seto *see* Hōyo-kaikyō
Haycock Alaska, USA 65°12′N 161°10′W — 39 Q3
Hayden Arizona, SW USA 33°00′N 110°46′W — 36 M14
Hayden Colorado, C USA 40°29′N 107°15′W — 37 Q3
Hayes South Dakota, N USA 44°20′N 101°01′W — 28 M10
Hayes ◢ Manitoba, C Canada — 9 X13
Hayes ◢ Nunavut, NE Canada — 11 P12
Hayes Center Nebraska, C USA 40°30′N 101°02′W — 28 M16
Hayes, Mount ▲ Alaska, USA 63°36′N 146°43′W — 39 S10
Hayesville North Carolina, SE USA 35°03′N 83°49′W — 21 N11
Hayford Peak ▲ Nevada, W USA 36°40′N 115°10′W — 35 X10
Hayfork California, W USA 40°33′N 123°10′W — 34 M3
Hayir, Qasr al *see* Ḩayr al Gharbī, Qasr al
Haylaastay *see* Sühbaatar
Hay Lake ◎ Ontario, SE Canada — 14 I12
Haymā' *var.* Haima. C Oman 19°59′N 56°20′E — 141 X11
Haymana Ankara, C Turkey 39°26′N 32°30′E — 136 H13
Ḩaymūr, Jabal ▲ W Syria — 138 J7
Haynesville Louisiana, S USA 32°57′N 93°08′W — 22 G4
Hayneville Alabama, S USA 32°13′N 86°34′W — 23 P6
Hayrabolu Tekirdağ, NW Turkey 41°14′N 27°04′E — 114 M12
Hayrabolu Deresi ◢ NW Turkey — 136 C10
Ḩayr al Gharbī, Qasr al *var.* Qasr al Hayir, Qasr al Hir *var.* Qasr al Hayr. *ruins* Ḩimş, C Syria — 138 J6
Ḩayr ash Sharqī, Qasr al *var.* Qasr al Hayr ash Sharqī. *ruins* Ḩimş, C Syria — 138 L5
Hayrhan *var.* Uubulan. Arhangay, C Mongolia 48°33′N 101°58′E — 162 J7
Hayrhandulaan *var.* Mardzad. Övörhangay, C Mongolia 45°58′N 102°06′E — 162 J9
Hay River Northwest Territories, W Canada 60°51′N 115°42′W — 8 J10
Hays Kansas, C USA 38°53′N 99°20′W — 26 K4
Hays Stacks Nebraska, C USA — 25 V8
Haystack, The ▲ NE Saint Helena 15°55′S 05°40′W — 65 H25
Haysville Kansas, C USA 37°34′N 97°21′W — 27 N7
Haysyn *Rus.* Gaysin. Vinnyts'ka Oblast', C Ukraine 48°49′N 29°26′E — 117 O7
Hayti Missouri, C USA 36°13′N 89°45′W — 27 Y9
Hayti South Dakota, N USA 44°40′N 97°13′W — 29 Q9
Hayvoron *Rus.* Gayvoron. Kirovohrads'ka Oblast', C Ukraine 48°20′N 29°52′E — 117 N9
Hayward California, W USA 37°40′N 122°07′W — 35 N9
Hayward Wisconsin, N USA 46°02′N 91°26′W — 30 J4
Haywards Heath SE England, United Kingdom 51°N 00°06′W — 97 O23
Hazar *prev. Rus.* Cheleken. Balkan Welaýaty, W Turkmenistan 39°26′N 53°07′E — 146 A11
Ḩazārān, Kūh-e *var.* Kūh-e ā Hazr. ▲ SE Iran 26°55′N 57°15′E — 143 S11
Hazard Kentucky, S USA 37°14′N 83°11′W — 21 Q7
Hazar Gölü ◎ C Turkey — 137 O15
Hazārībāg *var.* Hazārībāgh. Jhārkhand, N India 24°00′N 85°23′E — 153 P15
Hazārībāgh *see* Hazārībāg
Hazebrouck Nord, N France 50°43′N 02°32′E — 103 O1
Hazel Green Wisconsin, N USA 42°33′N 90°26′W — 30 K9
Hazel Holme Bank *undersea feature* S Pacific Ocean — 192 K9
Hazelton British Columbia, SW Canada 55°15′N 127°38′W — 10 K13
Hazelton North Dakota, N USA 46°27′N 100°17′W — 29 N6
Hazen Nevada, W USA 39°33′N 119°02′W — 35 R5
Hazen North Dakota, N USA 47°18′N 101°37′W — 28 L5
Hazen Bay *bay* E Bering Sea — 28 L12
Hazen, Lake ◎ Nunavut, N Canada — 9 N1
Hazim, Bi'r *well* C Iraq — 139 S5
Hazlehurst Georgia, SE USA 31°51′N 82°35′W — 23 V6
Hazlehurst Mississippi, S USA 31°51′N 90°24′W — 22 K6
Hazlet New Jersey, NE USA 40°24′N 74°10′W — 18 I14
Hazorasp *Rus.* Khazarasp. Xorazm Viloyati, W Uzbekistan 41°21′N 61°01′E — 146 I9
Hazratishoh, Qatorkŭhi *var.* Khrebet Khazretishi, *Rus.* Khrebet Khozretishi. ▲ S Tajikistan — 147 R13
Hazr, Kūh-e ā *see* Ḩazārān, Kūh-e
Hazro Punjab, E Pakistan 33°55′N 72°33′E — 149 U6
Headland Alabama, S USA 31°21′N 85°20′W — 23 R7
Head of Bight *headland* South Australia 31°33′S 131°15′E — 182 C6
Headquarters Idaho, NW USA 46°38′N 115°52′W — 35 N10
Healdsburg California, W USA 38°36′N 122°52′W — 34 M7

Healdton Oklahoma, C USA 34°13′N 97°29′W — 27 N13
Healesville Victoria, SE Australia 37°41′S 145°31′E — 183 O12
Healy Alaska, USA 63°51′N 148°58′W — 39 R10
Heard and McDonald Islands ◇ *Australian external territory* S Indian Ocean — 173 R13
Heard Island *island* Heard and McDonald Islands, S Indian Ocean — 173 R13
Hearne Texas, SW USA 30°52′N 96°35′W — 25 U9
Hearst Ontario, S Canada 49°42′N 83°40′W — 12 C12
Hearst Island *island* Antarctica — 194 J5
Heart of Dixie *see* Alabama
Heart River ◢ North Dakota, N USA — 31 T13
Heathcote Victoria, SE Australia 36°57′S 144°43′E — 183 N11
Heathrow ✈ (London) SE England, United Kingdom 51°27′N 00°27′W — 97 N22
Heathsville Virginia, NE USA 37°55′N 76°29′W — 21 X5
Heavener Oklahoma, C USA 34°53′N 94°36′W — 27 R11
Hebbronville Texas, SW USA 27°19′N 98°41′W — 25 R15
Hebei *var.* Hebei Sheng, Hopeh, Hopei, Ji; *prev.* Chihli. ◇ *province* NE China — 163 Q13
Hebei Sheng *see* Hebei
Heber City Utah, W USA 40°31′N 111°25′W — 36 M3
Heber Springs Arkansas, C USA 35°30′N 92°07′W — 27 V10
Hebi Henan, C China 35°57′N 114°08′E — 161 N5
Hebo Oregon, NW USA 45°10′N 123°55′W — 35 P11
Hebrides, Sea of the *sea* NW Scotland, United Kingdom — 96 F9
Hebron Newfoundland and Labrador, E Canada 58°15′N 62°45′W — 13 P5
Hebron Indiana, N USA 41°19′N 87°12′W — 31 N16
Hebron Nebraska, C USA 40°10′N 97°35′W — 29 Q17
Hebron North Dakota, N USA 46°54′N 102°00′W — 28 L5
Hebron *var.* Al Khalīl, El Khalīl, *Heb.* Hevron; *anc.* Kiriath-Arba. S West Bank 31°30′N 35°E — 138 F11
Hebrus *see* Évros/Maritsa/Meriç
Heby Västmanland, C Sweden 59°56′N 16°53′E — 95 N14
Hecate Strait *strait* British Columbia, W Canada — 10 I14
Hecelchakán Campeche, SE Mexico 20°09′N 90°04′W — 41 W12
Hechi *var.* Jinchengjiang. Guangxi Zhuangzu Zizhiqu, S China 24°39′N 108°02′E — 160 K13
Hechingen Baden-Württemberg, S Germany 48°20′N 08°58′E — 101 H23
Hechtel Limburg, NE Belgium 51°07′N 05°24′E — 99 I17
Hechuan *var.* He, Heyang. Chongqing Shi, C China 30°02′N 106°16′E — 160 J9
Hecla South Dakota, N USA 45°52′N 98°09′W — 29 P7
Hecla, Cape *headland* Nunavut, N Canada 82°00′N 64°00′W — 9 N1
Hector Minnesota, N USA 44°44′N 94°43′W — 29 T9
Hede Jämtland, C Sweden 62°25′N 13°33′E — 93 F17
Hede *see* Sheyang
Hedemora Dalarna, C Sweden 60°18′N 15°58′E — 95 M14
Hedenäset *Finn.* Hietaniemi. Norrbotten, N Sweden 66°12′N 23°40′E — 95 K13
Hedensted Syddanmark, C Denmark 55°47′N 09°43′E — 95 G23
Hedesunda Gävleborg, C Sweden 60°25′N 17°00′E — 95 N14
Hedesundafjärden ◎ C Sweden — 95 N14
Hedley Texas, SW USA 34°52′N 100°39′W — 25 Q5
Hedo-misaki *headland* Okinawa, SW Japan 26°55′N 128°15′E — 165 T16
Hedrick Iowa, C USA 41°10′N 92°18′W — 29 X15
Heel Limburg, SE Netherlands 51°12′N 06°01′E — 98 L16
Heel Point *point* Wake Island — 189 Y12
Heemskerk Noord-Holland, W Netherlands 52°31′N 04°40′E — 98 H9
Heerde Gelderland, E Netherlands 52°24′N 06°02′E — 98 M10
Heerenveen *Fris.* It Hearrenfean. Fryslân, N Netherlands 52°57′N 05°55′E — L7
Heerhugowaard Noord-Holland, NW Netherlands 52°40′N 04°50′E — 98 I8
Heer Land *physical region* S Svalbard — 92 I4
Heerlen Limburg, SE Netherlands 50°55′N 06°E — 99 M18
Heerwegen *see* Polkowice
Heesch Noord-Brabant, S Netherlands 51°44′N 05°32′E — K13
Heeze Noord-Brabant, SE Netherlands 51°23′N 05°35′E — K15
Hefa Haifa, *hist.* Caiffa, Caiphas; *anc.* Sycaminum. Haifa, N Israel 32°49′N 34°59′E — 138 F8
Hefa, Mifraz *see* Mifrats Hefa
Hefei *var.* Hofei, *hist.* Luchow. *province capital* Anhui, E China 31°51′N 117°20′E — 161 Q8
Heflin Alabama, S USA 33°39′N 85°35′W — 94 E10
Hegang Heilongjiang, NE China 47°18′N 130°16′E — 163 X7
Hegura-jima *island* SW Japan — 164 L10
Hei *see* Heilongjiang
Heide Schleswig-Holstein, N Germany 54°13′N 09°06′E — 100 H8
Heidelberg Baden-Württemberg, SW Germany 49°24′N 08°41′E — 101 G20
Heidelberg Gauteng, NE South Africa 26°31′S 28°21′E — 83 J21
Heidelberg Mississippi, S USA 31°53′N 88°59′W — 22 M6

Heidenheim *see* Heidenheim an der Brenz
Heidenheim an der Brenz *var.* Heidenheim. S Germany 48°41′N 10°09′E — 101 J22
Heidenreichstein Niederösterreich, N Austria 48°53′N 15°07′E — 109 U2
Heigun-tō *var.* Heguri-jima. *island* SW Japan — 164 F17
Heihe *prev.* Ai-hun. Heilongjiang, NE China 50°13′N 127°29′E — 163 W5
Hei He ◢ China — 162 S8
Heilbron Free State, N South Africa 27°17′S 27°57′E — 83 J22
Heilbronn Baden-Württemberg, SW Germany 49°08′N 09°14′E — 101 H21
Heiligenbeil *see* Mamonovo
Heiligenblut Tirol, W Austria 49°17′N 14°10′E — 109 Q8
Heiligenhafen Schleswig-Holstein, N Germany 54°22′N 10°57′E — 100 K7
Heiligenkreuz *see* Žiar nad Hronom
Heiligenstadt Thüringen, C Germany 51°22′N 10°09′E — 111 J15
Heilongjiang *var.* Hei, Heilongjiang Sheng, Hei-lung-chiang, Heilungkiang. ◇ *province* NE China — 163 W8
Heilong Jiang *see* Amur
Heilongjiang Sheng *see* Heilongjiang
Heiloo Noord-Holland, NW Netherlands 52°36′N 04°44′E — 98 H9
Heilsberg *see* Lidzbark Warmiński
Heishan Liaoning, NE China 41°43′N 122°12′E — 163 U12
Heishui Sichuan, C China — 160 H7
Heist-op-den-Berg Antwerpen, C Belgium 51°04′N 04°43′E — 99 H17
Heitō *see* Pingdong
Heitske Papua, E Indonesia 07°00′S 138°45′E — 171 X15
Hejanah *var.* Al Hijānah
Hejaz *see* Al Hijāz
Hejiang Sichuan — 35 X11
Hejiayan *see* Lüeyang — 158 K6
Hekimhan Malatya, C Turkey 38°50′N 37°56′E — 137 N14
Hekla ▲ S Iceland 64°S 19°42′W — 92 J4
Hekou *see* Yanshan, Jiangxi, China
Hekou *see* Yajiang, Sichuan, China
Hel *Ger.* Hela. Pomorskie, N Poland 54°35′N 18°48′E — 110 J6
Hela ◆ *province* W Papua New Guinea — 186 B6
Helagsfjället ▲ C Sweden 62°52′N 12°31′E — 93 F17
Helan *var.* Xigang. Ningxia, N China 38°33′N 106°21′E — 159 W8
Helan Shan ▲ N China — 162 K14
Helden Limburg, SE Netherlands 51°20′N 06°00′E — 98 M16
Ḩelebce *Ar.* Ḩalabjah, *var.* Halabja. As Sulaymānīyah, NE Iraq 35°11′N 45°59′E — 139 V4
Helena Arkansas, C USA 34°32′N 90°34′W — 27 X12
Helena *state capital* Montana, NW USA 46°36′N 112°02′W — 33 R10
Helena Alabama, S USA 33°17′N 86°51′W — 23 R3
Helensburgh W Scotland, United Kingdom 56°00′N 04°45′W — 96 H12
Helensville Auckland, North Island, New Zealand 36°42′S 174°26′E — 184 K5
Helgasjön ◎ S Sweden — 95 G8
Helgoland *Eng.* Heligoland. *island* NW Germany — 100 G8
Helgoland Bay *see* Helgoländer Bucht
Helgoländer Bucht *var.* Helgoland Bay, Heligoland Bight. *bay* NW Germany — 100 G8
Heligoland *see* Helgoland
Heligoland Bight *see* Helgoländer Bucht
Heliopolis *see* Baalbek
Hella Suðurland, SW Iceland 63°51′N 20°24′W — 21 I4
Hellas *see* Greece
Ḩelleh, Rūd-e ◢ S Iran — 143 N11
Hellendoorn Overijssel, E Netherlands 52°24′N 06°27′E — 98 I8
Hellenic Republic *see* Greece
Hellenic Trough *undersea feature* Aegean Sea, C Mediterranean Sea — 121 Q10
Hellesylt Møre og Romsdal, S Norway 62°06′N 06°51′E — 94 E10
Hellevoetsluis Zuid-Holland, SW Netherlands 51°49′N 04°08′E — 98 F13
Hellín Castilla-La Mancha, C Spain 38°31′N 01°43′W — 105 Q11
Hellinikon ✈ (Athína) Attikí, C Greece 37°55′N 23°43′E — 115 H19
Hells Canyon *valley* Idaho/Oregon, NW USA — 32 M12
Helmand, Daryā-ye ◢ Afghanistan/Iran, *see also* Hīrmand, Rūd-e — 148 K10
Helmand, Daryā-ye *see* Hīrmand, Rūd-e

Helme ◢ C Germany — 101 K15
Helmond Noord-Brabant, S Netherlands 51°29′N 05°41′E — 99 L15
Helmsdale N Scotland, United Kingdom 58°06′N 03°36′W — 96 J7
Helmstedt Niedersachsen, N Germany 52°14′N 11°01′E — 100 K13
Helong Jilin, NE China 42°32′N 129°00′E — 163 Y10
Helper Utah, W USA 39°40′N 110°52′W — 36 M4
Helpter Berge *hill* NE Germany — 100 O10
Helsingborg *prev.* Hälsingborg. Skåne, S Sweden 56°N 12°48′E — 95 J22
Helsingfors *see* Helsinki
Helsingør *Eng.* Elsinore. Hovedstaden, E Denmark 56°03′N 12°38′E — 95 J22
Helsinki *Swe.* Helsingfors. ● (Finland) Uusimaa, S Finland 60°18′N 24°58′E — 93 M20
Helston SW England, United Kingdom 50°04′N 05°17′W — 97 H25
Heltau *see* Cisnădie
Heradsvötn ◢ Iceland — 92 J2
Herakleion *see* Irákleio
Herāt *var.* Herat; *anc.* Aria. Herāt, W Afghanistan 34°23′N 62°11′E — 148 K5
Herāt ◆ *province* W Afghanistan — 148 J5
Hérault ◆ *department* S France — 103 P14
Hérault ◢ S France — 103 P15
Herbert Saskatchewan, S Canada 50°24′N 107°09′W — 11 T16
Herbert Otago, South Island, New Zealand 45°14′S 170°48′E — 185 F22
Herbert Island *island* Aleutian Islands, Alaska, USA — 38 J17
Herbertville Québec, SE Canada 48°23′N 71°42′W — 15 Q7
Herborn Hessen, W Germany 50°40′N 08°18′E — 101 G17
Herceg-Novi *It.* Castelnuovo; *prev.* Ercegnovi. SW Montenegro 42°28′N 18°35′E — 113 I17
Herchmer Manitoba, C Canada 57°25′N 94°12′W — 11 X10
Hercules Bay *bay* E Papua New Guinea — 186 E8
Heredia Heredia, C Costa Rica 10°N 84°06′W — 42 M13
Heredia *off.* Provincia de Heredia. ◆ *province* N Costa Rica — 42 M12
Heredia, Provincia de *see* Heredia
Hereford W England, United Kingdom 52°04′N 02°43′W — 97 K21
Hereford Texas, SW USA 34°49′N 102°25′W — 24 M3
Herefordshire *cultural region* W England, United Kingdom — 97 K21
Herent Vlaams Brabant, C Belgium 50°54′N 04°40′E — 99 H18
Herentals *var.* Herenthals. Antwerpen, N Belgium 51°11′N 04°50′E — 99 I16
Herenthals *see* Herentals
Herenthout Antwerpen, N Belgium 51°09′N 04°45′E — 99 H17
Herfølge Sjælland, E Denmark 55°25′N 12°09′E — 99 G23
Herford Nordrhein-Westfalen, NW Germany 52°07′N 08°41′E — 100 G13
Herington Kansas, C USA 38°37′N 96°55′W — 27 O5
Herisau *Fr.* Hérisau. Ausser Rhoden, NE Switzerland 47°23′N 09°17′E — 108 H7
Hérisau *see* Herisau
Hendorābi, Jazireh-ye *island* S Iran — 143 O14
Hendrik Top *var.* Hendriktopp. *elevation* C Suriname — 55 V10
Hendriktopp *see* Hendrik Top
Hendu Kosh *see* Hindu Kush
Heney, Lac ◎ Québec, SE Canada — 14 L12
Hengchow *see* Hengyang
Hengch'un ⬩ S Taiwan 22°09′N 120°43′E — 161 S15
Hengduan Shan ▲ SW China — 159 R16
Hengelo Gelderland, E Netherlands 52°03′N 06°19′E — 98 N12
Hengelo Overijssel, E Netherlands 52°16′N 06°46′E — 98 O10
Hengshan Hunan, S China 27°17′N 112°51′E — 161 N13
Hengshui Hebei, E China 37°42′N 115°39′E — 161 O4
Hengyang *var.* Hengnan, Heng-yang; *prev.* Henchow. Hunan, S China 26°55′N 112°34′E — 161 N13
Heniches'k *Rus.* Genichesk. Khersons'ka Oblast', S Ukraine 46°10′N 34°49′E — 117 U11
Henlopen, Cape *headland* Delaware, NE USA 38°48′N 75°06′W — 21 Z4
Hennan Gävleborg, C Sweden 62°01′N 15°55′E — 94 M10
Hennebont Morbihan, NW France 47°49′N 03°16′W — 102 G7
Hennigsdorf *var.* Hennigsdorf bei Berlin. Brandenburg, NE Germany 52°37′N 13°15′E — 100 N12
Hennigsdorf bei Berlin *see* Hennigsdorf
Hennessey Oklahoma, C USA 36°07′N 97°53′W — 26 M9
Henneberg ▲ S Germany — 101 I21
Henrietta Texas, SW USA 33°49′N 98°13′W — 27 Y7
Henrique de Carvalho *see* Saurimo
Henry Illinois, N USA 41°06′N 89°22′W — 30 L11
Henrád *Ger.* Kundert. ◢ Hungary/Slovakia — 111 N20
Hernández Entre Ríos, E Argentina 32°21′S 60°02′W — 61 C18
Henry Ice Rise *ice cap* Antarctica — 194 M7

Henry Kater, Cape *headland* Baffin Island, Nunavut, NE Canada 69°09′N 66°45′W — 9 R5
Henrys Fork ◢ Idaho, NW USA — 33 R13
Hensall Ontario, S Canada 43°25′N 81°28′W — 14 E15
Henstedt-Ulzburg Schleswig-Holstein, N Germany 53°45′N 09°59′E — 100 J9
Hentiy *var.* Batshireet, Eg. ⬩ N Mongolia — 163 N7
Hentiyn Nuruu ▲ N Mongolia — 162 M7
Henty New South Wales, SE Australia 35°33′S 147°03′E — 183 P10
Henza Hordaland, S Norway 59°06′N 09°34′E — 95 G16
Herreid South Dakota, N USA 45°49′N 100°04′W — 21 N7
Herrenberg Baden-Württemberg, SW Germany 48°36′N 08°52′E — 101 H22
Herrera Andalucía, S Spain 37°22′N 04°50′W — 104 L14
Herrera *off.* Provincia de Herrera. ◆ *province* C Panama — 43 R17
Herrera del Duque Extremadura, W Spain 39°10′N 05°03′W — 104 L14
Herrera de Pisuerga Castilla y León, N Spain 42°35′N 04°20′W — 104 M4
Herrera, Provincia de *see* Herrera
Herrero, Punta *headland* SE Mexico 19°16′S 87°28′W — 41 Z13
Herrick Tasmania, SE Australia 41°07′S 147°53′E — 183 P16
Herrin Illinois, N USA 37°48′N 89°01′W — 30 L17
Herrington Lake ◎ Kentucky, S USA — 20 M6
Herrljunga Västra Götaland, S Sweden 58°05′N 13°02′E — 95 K18
Hers ◢ S France — 103 N16
Herschel Island *island* Yukon, NW Canada — 10 I1
Herselt Antwerpen, C Belgium 51°04′N 04°53′E — 99 I17
Hershey Pennsylvania, NE USA 40°17′N 76°38′W — 18 G15
Herstal *Fr.* Hérstal. Liège, E Belgium 50°40′N 05°38′E — 99 K19
Hertford E England, United Kingdom 51°48′N 00°05′W — 103 N16
Hertford North Carolina, SE USA 36°11′N 76°28′W — 21 X8
Hertfordshire *cultural region* E England, United Kingdom — 97 O21
Hervey Bay Queensland, E Australia 25°13′S 152°48′E — 181 Z9
Herzberg Brandenburg, E Germany 51°42′N 13°15′E — 101 O14
Herzele Oost-Vlaanderen, NW Belgium 50°53′N 03°52′E — 99 E18
Herzogenaurach Bayern, SE Germany 49°34′N 10°52′E — 101 K20
Herzogenburg Niederösterreich, NE Austria 48°18′N 15°43′E — 109 W4
Herzogenbusch *see* 's-Hertogenbosch
Hesdin Pas-de-Calais, N France 50°22′N 02°02′E — 103 N2
Heshan Guangxi Zhuangzu Zizhiqu, S China 23°45′N 108°58′E — 160 K14
Heshui *var.* Xihuachi. Gansu, C China 35°42′N 108°06′E — 159 X10
Hespérange Luxembourg, SE Luxembourg — W25
Hesperia California, W USA 34°25′N 117°17′W — 35 U14
Hesperus Mountain ▲ Colorado, C USA 37°17′N 108°05′W — 37 P7
Hess ◢ Yukon, NW Canada — 10 J6
Hesse *see* Hessen
Hesselberg ▲ S Germany — 101 I21
Hesselo *island* E Denmark — 95 I22
Hessen *Eng./Fr.* Hesse. ◆ *state* C Germany — 101 H17
Hess Tablemount *undersea feature* N Pacific Ocean 17°49′N 174°15′W — 192 L6
Hesston Kansas, C USA 38°08′N 97°25′W — 27 N6
Hestkjøltoppen ▲ C Norway 64°21′N 13°57′E — 93 G15
Heswall NW England, United Kingdom 53°20′N 03°06′W — 97 K18
Hetauda Central, C Nepal 27°25′N 85°02′E — 153 P12
Hettinger North Dakota, N USA 46°00′N 102°38′W — 28 K7
Hettstedt Sachsen-Anhalt, C Germany 51°39′N 11°31′E — 101 L14
Heuglin, Kapp *headland* W Svalbard 78°15′N 22°49′E — 92 P3
Heuksan-jedo *var.* Hüksan-chedo. *island group* SW South Korea — W17
Heuru Makira-Ulawa, SE Solomon Islands 10°13′S 161°37′E — 187 N10
Heusden Limburg, NE Belgium 51°02′N 05°17′E — 99 I13
Heusden Noord-Brabant, S Netherlands 51°44′N 05°08′E — 98 J13
Hève, Cap de la *headland* N France 49°28′N 00°04′E — 102 K3
Heverlee Vlaams Brabant, C Belgium 50°51′N 04°42′E — 99 H18
Heves Heves, NE Hungary 47°36′N 20°17′E — 111 L22
Heves *off.* Heves Megye. ◆ *county* NE Hungary — 111 L22
Heves Megye *see* Heves
Hevron *see* Hebron
Hewanorra ✈ (Saint Lucia) S Saint Lucia 13°41′N 60°57′W — 45 Y13
Hewlêr *see* Arbīl
Hewlêr, Parêzga-i *see* Arbīl
Hexian *see* Hezhou
Heyang Shaanxi, C China 35°14′N 110°02′E — 163 L6
Heyang *see* Hechuan
Heydebrech *see* Kędzierzyn-Kozle
Heydekrug *see* Šilutė
Heysham NW England, United Kingdom 54°03′N 02°53′W — 97 K16
Heyuan *var.* Yuancheng. Guangdong, S China 23°41′N 114°45′E — 161 N14
Heywood Victoria, SE Australia 38°09′S 141°38′E — 182 L12
Heywood Islands *island group* Western Australia — 180 K3
Heze *var.* Caozhou. Shandong, E China 35°16′N 115°27′E — 161 O6

◆ Country ◇ Dependent Territory ◈ Administrative Regions ▲ Mountain 🌋 Volcano ◎ Lake
● Country Capital ○ Dependent Territory Capital ✈ International Airport ▲ Mountain Range ◢ River ▨ Reservoir

Hezheng – Hongwan

159 U11 **Hezheng** Gansu, C China 35°29′N 103°36′E
160 M13 **Hezhou** *var.* Babu; *prev.* Hexian. Guangxi Zhuangzu Zizhiqu, S China 24°33′N 11°30′E
159 U11 **Hezuo** Gansu, C China 34°55′N 102°49′E
23 Z16 **Hialeah** Florida, SE USA 25°51′N 80°16′W
27 Q3 **Hiawatha** Kansas, C USA 39°51′N 95°34′W
36 M4 **Hiawatha** Utah, W USA 39°28′N 111°00′W
29 V4 **Hibbing** Minnesota, N USA 47°24′N 92°55′W
183 N17 **Hibbs, Point** *headland* Tasmania, SE Australia 42°37′S 145°15′E
Hibernia *see* Ireland
20 F8 **Hickman** Kentucky, S USA 36°33′N 89°11′W
21 Q9 **Hickory** North Carolina, SE USA 35°44′N 81°20′W
21 Q9 **Hickory, Lake** ◎ North Carolina, SE USA
184 Q7 **Hicks Bay** Gisborne, North Island, New Zealand 37°36′S 178°18′E
25 S8 **Hico** Texas, SW USA 31°58′N 98°01′W
Hida *see* Furukawa
165 T4 **Hidaka** Hokkaidō, NE Japan 42°53′N 142°24′E
164 I12 **Hidaka** Hyōgo, Honshū, SW Japan 35°27′N 134°43′E
165 T5 **Hidaka-sanmyaku** ▲▲ Hokkaidō, NE Japan
41 O6 **Hidalgo** *var.* Villa Hidalgo. Coahuila, NE Mexico 27°46′N 99°54′W
41 N8 **Hidalgo** Nuevo León, NE Mexico 29°59′N 100°27′W
41 O10 **Hidalgo** Tamaulipas, C Mexico 24°16′N 99°28′W
41 O13 **Hidalgo** ◈ *state* C Mexico
40 J7 **Hidalgo del Parral** *var.* Parral. Chihuahua, N Mexico 26°58′N 105°40′W
100 N7 **Hiddensee** *island* NE Germany
80 G6 **Hidiglib, Wadi** ≈ NE Sudan
109 U6 **Hieflau** Salzburg, E Austria 47°36′N 14°34′E
187 P16 **Hienghène** Province Nord, C New Caledonia 20°43′S 164°54′E
Hierosolyma *see* Jerusalem
64 N12 **Hierro** *var.* Ferro. *island* Islas Canarias, Spain, NE Atlantic Ocean
Hietaniemi *see* Hedenäset
164 G13 **Higashi-Hiroshima** *var.* Higasihirosima. Hiroshima, Honshū, SW Japan 34°27′N 132°43′E
164 C12 **Higashi-suidō** *strait* SW Japan
Higasihirosima *see* Higashi-Hiroshima
25 P1 **Higgins** Texas, SW USA 36°06′N 100°01′W
31 P7 **Higgins Lake** ◎ Michigan, N USA
27 S4 **Higginsville** Missouri, C USA 39°04′N 93°43′W
High Atlas *see* Haut Atlas
30 M5 **High Falls Reservoir** ▣ Wisconsin, N USA
44 K12 **Highgate** C Jamaica 18°16′N 76°53′W
25 X11 **High Island** Texas, SW USA 29°35′N 94°24′W
31 O5 **High Island** *island* Michigan, N USA
30 K15 **Highland** Illinois, N USA 38°44′N 89°40′W
31 N10 **Highland Park** Illinois, N USA 42°10′N 87°48′W
21 O10 **Highlands** North Carolina, SE USA 35°04′N 83°10′W
11 O11 **High Level** Alberta, W Canada 58°31′N 117°08′W
29 O9 **Highmore** South Dakota, N USA 44°29′N 99°26′W
171 N3 **High Peak** ▲ Luzon, N Philippines 15°28′N 120°07′E
High Plains *see* Great Plains
21 S9 **High Point** North Carolina, SE USA 35°58′N 80°00′W
18 J13 **High Point** *hill* New Jersey, NE USA
11 P13 **High Prairie** Alberta, W Canada 55°27′N 116°28′W
11 Q16 **High River** Alberta, SW Canada 50°35′N 113°50′W
21 S9 **High Rock Lake** ▣ North Carolina, SE USA
23 V9 **High Springs** Florida, SE USA 29°49′N 82°36′W
High Veld *see* Great Karoo
97 J24 **High Willhays** ▲ SW England, United Kingdom 50°39′N 03°58′W
97 N22 **High Wycombe** *prev.* Chepping Wycombe, Chipping Wycombe. SE England, United Kingdom 51°38′N 00°46′W
41 P12 **Higos** *var.* El Higo. Veracruz-Llave, E Mexico 21°48′N 98°25′W
102 I16 **Higuer, Cap** *headland* NE Spain 43°23′N 01°46′W
45 R5 **Higüero, Punta** *headland* W Puerto Rico 18°21′N 67°15′W
45 P9 **Higüey** *var.* Salvaleón de Higüey. E Dominican Republic 18°40′N 68°43′W
190 G11 **Hihifo** ✈ (Matāʻutu) Île Uvea, N Wallis and Futuna
81 N16 **Hiiraan** *off.* Gobolka Hiiraan. ◈ *region* C Somalia
Hiiraan, Gobolka *see* Hiiraan
118 E4 **Hiiumaa** *var.* Hiiumaa Maakond. ◈ *province* W Estonia
118 D4 **Hiiumaa** *Ger.* Dagden, *Swe.* Dagö. *island* W Estonia
Hiiumaa Maakond *see* Hiiumaa
Hijanah *see* Al Hījānah
105 S6 **Híjar** Aragón, NE Spain 41°10′N 00°27′W
191 V10 **Hikueru** *atoll* Îles Tuamotu, C French Polynesia
184 K3 **Hikurangi** Northland, North Island, New Zealand 35°37′S 174°16′E
184 Q8 **Hikurangi** ▲ North Island, New Zealand 37°55′S 177°59′E
192 L11 **Hikurangi Trench** *var.* Hikurangi Trough. *undersea feature* SW Pacific Ocean
Hikurangi Trough *see* Hikurangi Trench
190 B15 **Hikutavake** NW Niue
121 Q12 **Hilāl, Ra's al** *headland* N Libya 32°55′N 22°09′E
61 A24 **Hilario Ascasubi** Buenos Aires, E Argentina 39°22′S 62°39′W
101 K17 **Hildburghausen** Thüringen, C Germany 50°26′N 10°44′E
101 E15 **Hilden** Nordrhein-Westfalen, W Germany 51°12′N 06°56′E
100 I13 **Hildesheim** Niedersachsen, N Germany 52°09′N 09°57′E
33 T9 **Hilger** Montana, NW USA 47°15′N 109°18′W
153 S13 **Hili** *var.* Hilli. Rajshahi, NW Bangladesh 25°16′N 89°04′E
Hilla *see* Al Ḥillah
45 O14 **Hillaby, Mount** ▲ N Barbados 13°12′N 59°34′W
Ḥillah, Al *see* Bābil
95 K19 **Hillared** Västra Götaland, S Sweden 57°37′N 13°10′E
195 R12 **Hillary Coast** *physical region* Antarctica
42 G2 **Hill Bank** Orange Walk, N Belize 17°36′N 88°43′W
33 O14 **Hill City** Idaho, NW USA 43°18′N 115°03′W
26 K3 **Hill City** Kansas, C USA 39°23′N 99°51′W
29 V5 **Hill City** Minnesota, N USA 46°59′N 93°36′W
28 J10 **Hill City** South Dakota, N USA 43°54′N 103°33′W
65 C24 **Hill Cove Settlement** West Falkland, Falkland Islands
98 H10 **Hillegom** Zuid-Holland, W Netherlands 52°18′N 04°35′E
95 J22 **Hillerød** Hovedstaden, E Denmark 55°56′N 12°19′E
36 M7 **Hillers, Mount** ▲ Utah, W USA 37°53′N 110°42′W
Hilli *see* Hili
29 R11 **Hills** Minnesota, N USA 43°31′N 96°21′W
30 L14 **Hillsboro** Illinois, N USA 39°09′N 89°29′W
27 N5 **Hillsboro** Kansas, C USA 38°21′N 97°12′W
27 X5 **Hillsboro** Missouri, C USA 38°13′N 90°33′W
19 N10 **Hillsboro** New Hampshire, NE USA 43°06′N 71°52′W
37 Q14 **Hillsboro** New Mexico, SW USA 32°55′N 107°33′W
29 R4 **Hillsboro** North Dakota, N USA 47°25′N 97°03′W
31 R14 **Hillsboro** Ohio, N USA 39°12′N 83°36′W
32 G11 **Hillsboro** Oregon, NW USA 45°32′N 122°59′W
25 T8 **Hillsboro** Texas, SW USA 32°01′N 97°08′W
30 K8 **Hillsboro** Wisconsin, N USA 43°40′N 90°21′W
23 Y14 **Hillsboro Canal** *canal* Florida, SE USA
45 Y15 **Hillsborough** Carriacou, N Grenada 12°28′N 61°28′W
97 G15 **Hillsborough** E Northern Ireland, United Kingdom 54°27′N 06°06′W
21 U9 **Hillsborough** North Carolina, SE USA 36°04′N 79°06′W
31 Q10 **Hillsdale** Michigan, N USA 41°55′N 84°37′W
183 O8 **Hillston** New South Wales, SE Australia 33°30′S 145°33′E
21 R7 **Hillsville** Virginia, NE USA 36°46′N 80°44′W
96 L2 **Hillswick** NE Scotland, United Kingdom 60°28′N 01°37′W
Hill Tippera *see* Tripura
38 H11 **Hilo** Hawaii, USA, C Pacific Ocean 19°42′N 155°04′W
18 F9 **Hilton** New York, NE USA 43°17′N 77°47′W
14 C10 **Hilton Beach** Ontario, S Canada 46°14′N 83°51′W
21 R16 **Hilton Head Island** South Carolina, SE USA 32°13′N 80°45′W
21 R16 **Hilton Head Island** *island* South Carolina, SE USA
99 J15 **Hilvarenbeek** Noord-Brabant, S Netherlands 51°29′N 05°08′E
98 J11 **Hilversum** Noord-Holland, C Netherlands 52°14′N 05°10′E
75 W8 **Hilwān** *var.* Helwân, Hilwân, Hulwan, Hulwân. N Egypt 29°51′N 31°20′E
Hilwân *see* Hilwān
152 J7 **Himāchal Pradesh** ◈ *state* NW India
Himalaya/Himalaya Shan *see* Himalayas
152 M9 **Himalayas** *var.* Himalaya, *Chin.* Himalaya Shan. ▲▲ S Asia
171 P6 **Himamaylan** Negros, C Philippines 10°04′N 122°52′E
93 K15 **Himanka** Pohjois-Pohjanmaa, W Finland 64°04′N 23°40′E
Himara *see* Himarë
113 L23 **Himarë** *var.* Himara. Vlorë, S Albania 40°06′N 19°45′E
138 M2 **Ḥimār, Wādī al** *dry watercourse* N Syria
154 D9 **Himatnagar** Gujarāt, W India 23°38′N 73°02′E
109 Y4 **Himberg** Niederösterreich, E Austria 48°05′N 16°27′E
164 I13 **Himeji** *var.* Himezi. Hyōgo, Honshū, SW Japan 34°47′N 134°32′E
164 E14 **Hime-jima** *island* SW Japan
Himezi *see* Himeji
164 L13 **Himi** Toyama, Honshū, SW Japan 36°54′N 136°59′E
109 S9 **Himmelberg** Kärnten, S Austria 46°45′N 14°01′E
138 I5 **Ḥimṣ** *var.* Homs; *anc.* Emesa. Ḥimṣ, C Syria 34°44′N 36°43′E
138 K6 **Ḥimṣ** *off.* Muḥāfazat Hims, *var.* Homs. ◈ *governorate* C Syria
138 I5 **Ḥimṣ, Buḥayrat** *var.* Buhayrat Qattinah. ◎ W Syria
171 R7 **Hinatuan** Mindanao, S Philippines 08°21′N 126°19′E
117 N10 **Hînceşti** *var.* Hânceşti; *prev.* Kotovsk. C Moldova 46°48′N 28°33′E
44 M9 **Hinche** C Haiti 19°07′N 72°00′W
181 X5 **Hinchinbrook Island** *island* Queensland, NE Australia
39 S12 **Hinchinbrook Island** *island* Alaska, USA
97 M19 **Hinckley** C England, United Kingdom 52°33′N 01°21′W
29 V7 **Hinckley** Minnesota, N USA 46°01′N 92°57′W
36 K5 **Hinckley** Utah, W USA 39°21′N 112°39′W
18 J9 **Hinckley Reservoir** ▣ New York, NE USA
152 I12 **Hindaun** Rājasthān, N India 26°44′N 77°02′E
Hindenburg/Hindenburg in Oberschlesien *see* Zabrze
Hindiya *see* Al Hindīyah
21 O6 **Hindman** Kentucky, S USA 37°20′N 82°58′W
182 L10 **Hindmarsh, Lake** ◎ Victoria, SE Australia
185 G19 **Hinds** Canterbury, South Island, New Zealand 44°01′S 171°33′E
185 G19 **Hinds** ≈ South Island, New Zealand
95 H23 **Hindsholm** *island* C Denmark
149 S4 **Hindu Kush** *Per.* Hendū Kosh. ▲▲ Afghanistan/Pakistan
155 H19 **Hindupur** Andhra Pradesh, E India 13°46′N 77°33′E
11 O12 **Hines Creek** Alberta, W Canada 56°14′N 118°36′W
23 W6 **Hinesville** Georgia, SE USA 31°51′N 81°36′W
154 I12 **Hinganghāt** Mahārāshtra, C India 20°32′N 78°52′E
149 N15 **Hingol** ≈ SW Pakistan
154 H13 **Hingoli** Mahārāshtra, C India 19°45′N 77°08′E
137 R13 **Hınıs** Erzurum, E Turkey 39°22′N 41°44′E
92 O2 **Hinlopenstretet** *strait* N Svalbard
92 G10 **Hinnøya** *Lapp.* Iinnasuolu. *island* C Norway
108 H10 **Hinterrhein** ≈ SW Switzerland
166 L8 **Hinthada** *var.* Henzada. Ayeyawady, SW Myanmar (Burma) 17°36′N 95°26′E
11 O14 **Hinton** Alberta, SW Canada 53°24′N 117°35′W
26 M10 **Hinton** Oklahoma, C USA 35°28′N 98°21′W
21 R6 **Hinton** West Virginia, NE USA 37°42′N 80°54′W
Hios *see* Chíos
41 N8 **Hipólito** Coahuila, NE Mexico 25°42′N 101°22′W
Hipponium *see* Vibo Valentia
164 B13 **Hirado** Nagasaki, Hirado-shima, SW Japan 33°22′N 129°31′E
164 B13 **Hirado-shima** *island* SW Japan
165 P16 **Hirakubo-saki** *headland* Ishigaki-jima, SW Japan 24°36′N 124°19′E
154 M11 **Hīrākud Reservoir** ▣ E India
Hir al Gharbi, Qasr al *see* Ḥayr al Gharbī, Qaṣr al
165 Q16 **Hirara** Okinawa, Miyako-jima, SW Japan 24°48′N 125°17′E
Hir Ash Sharqi, Qasr al *see* Ḥayr ash Sharqī, Qaṣr al
164 G12 **Hirata** Shimane, Honshū, SW Japan 35°25′N 132°45′E
136 I13 **Hirfanlı Barajı** ▣ C Turkey
155 G18 **Hiriyūr** Karnātaka, W India 13°58′N 76°33′E
Hîrlău *see* Hârlău
148 K10 **Hīrmand, Rūd-e** *var.* Daryā-ye Helmand. ≈ Afghanistan/Iran *see also* Helmand, Daryā-ye
Hīrmand, Rūd-e *see* Helmand, Daryā-ye
Hirmil *see* Hermel
165 T5 **Hiroo** Hokkaidō, NE Japan 42°16′N 143°16′E
165 Q7 **Hirosaki** Aomori, Honshū, C Japan 40°34′N 140°28′E
164 G13 **Hiroshima** *var.* Hirosima. Hiroshima, Honshū, SW Japan 34°23′N 132°26′E
164 F13 **Hiroshima** *off.* Hiroshima-ken, *var.* Hirosima. ◈ *prefecture* Honshū, SW Japan
Hiroshima-ken *see* Hiroshima
Hirosima *see* Hiroshima
Hirschberg/Hirschberg im Riesengebirge/Hirschberg in Schlesien *see* Jelenia Góra
103 Q3 **Hirson** Aisne, N France 49°56′N 04°05′E
Hîrşova *see* Hârşova
95 G19 **Hirtshals** Nordjylland, N Denmark 57°34′N 09°58′E
152 H10 **Hisār** Haryāna, NW India 29°10′N 75°45′E
114 I10 **Hisarya** *var.* Khisarya. Plovdiv, C Bulgaria 42°33′N 24°43′E
162 K7 **Hishig Öndör** *var.* Maanĭt. Bulgan, C Mongolia 48°17′N 103°29′E
186 E9 **Hisiu** Central, SW Papua New Guinea 09°20′S 146°48′E
147 P13 **Hisor** *Rus.* Gissar. W Tajikistan 38°34′N 68°29′E
Hispalis *see* Sevilla
Hispana/Hispania *see* Spain
44 M7 **Hispaniola** *island* Dominion Republic/Haiti
64 F11 **Hispaniola Basin** *var.* Hispaniola Trough. *undersea feature* SW Atlantic Ocean
Hispaniola Trough *see* Hispaniola Basin
Histonium *see* Vasto
139 R7 **Hīt** Al Anbār, SW Iraq 33°38′N 42°50′E
165 P14 **Hita** Ōita, Kyūshū, SW Japan 33°19′N 130°55′E
165 P12 **Hitachi** *var.* Hitati. Ibaraki, Honshū, S Japan 36°40′N 140°42′E
165 P12 **Hitachiōta** Ibaraki, Honshū, S Japan 36°32′N 140°31′E
Hitati *see* Hitachi
97 O21 **Hitchin** E England, United Kingdom 51°57′N 00°17′W
191 Q7 **Hitiaa** Tahiti, W French Polynesia 17°35′S 149°17′W
164 D15 **Hitoyoshi** *var.* Hitoyosi. Kumamoto, Kyūshū, SW Japan 32°12′N 130°48′E
Hitoyosi *see* Hitoyoshi
94 F7 **Hitra** *prev.* Hitteren. *island* S Norway
Hitteren *see* Hitra
187 Q11 **Hiu** *island* Torres Islands, N Vanuatu
165 O11 **Hiuchiga-take** ▲ Honshū, C Japan 36°57′N 139°18′E
191 X7 **Hiva Oa** *island* Îles Marquises, N French Polynesia
20 M10 **Hiwassee Lake** ▣ North Carolina, SE USA
20 M10 **Hiwassee River** ≈ SE USA
95 H20 **Hjallerup** Nordjylland, N Denmark 57°10′N 10°10′E
95 M16 **Hjälmaren** *Eng.* Lake Hjalmar. ◎ C Sweden
Hjalmar, Lake *see* Hjälmaren
95 C14 **Hjellestad** Hordaland, S Norway 60°15′N 05°13′E
95 D16 **Hjelmeland** Rogaland, S Norway 59°12′N 06°07′E
94 G10 **Hjerkinn** Oppland, S Norway 62°13′N 09°37′E
95 L18 **Hjo** Västra Götaland, S Sweden 58°18′N 14°17′E
95 G19 **Hjørring** Nordjylland, N Denmark 57°28′N 09°59′E
167 O1 **Hkakabo Razi** ▲ Myanmar (Burma)/China 28°17′N 97°28′E
166 M2 **Hkamti** *var.* Singkaling Hkamti. Sagaing, N Myanmar (Burma) 26°00′N 95°43′E
167 N1 **Hkring Bum** ▲ N Myanmar (Burma) 27°05′N 97°16′E
83 L21 **Hlathikulu** *var.* Hlatikulu. S Swaziland 26°58′S 31°19′E
Hlatikulu *see* Hlathikulu
Hliboka *see* Hlyboka
111 F17 **Hlinsko** *var.* Hlinsko v Čechách. Pardubický Kraj, C Czech Republic 49°46′N 15°54′E
Hlinsko v Čechách *see* Hlinsko
117 S6 **Hlobyne** *Rus.* Globino. Poltavs'ka Oblast', NE Ukraine 49°24′N 33°16′E
111 H20 **Hlohovec** *Ger.* Freistadtl, *Hung.* Galgóc; *prev.* Frakštát. Trnavský Kraj, W Slovakia 48°26′N 17°49′E
83 J23 **Hlotse** *var.* Leribe. NW Lesotho 28°55′S 28°01′E
111 I17 **Hlučín** *Ger.* Hultschin, *Pol.* Hulczyn. Moravskoslezský Kraj, E Czech Republic 49°54′N 18°11′E
117 S2 **Hlukhiv** *Rus.* Glukhov. Sums'ka Oblast', NE Ukraine 51°40′N 33°53′E
119 K21 **Hlushkavichy** *Rus.* Glushkevichi. Homyel'skaya Voblasts', SE Belarus 51°34′N 27°47′E
119 L18 **Hlusk** *Rus.* Glusk, Glussk. Mahilyowskaya Voblasts', E Belarus 52°54′N 28°41′E
116 K8 **Hlyboka** *Ger.* Hliboka, *Rus.* Glybokaya. Chernivets'ka Oblast', W Ukraine 48°04′N 25°56′E
118 K13 **Hlybokaye** *Rus.* Glubokoye. Vitsyebskaya Voblasts', N Belarus 55°08′N 27°41′E
77 Q16 **Ho** SE Ghana 06°36′N 00°28′E
167 S6 **Hoa Binh** Hoa Binh, N Vietnam 20°49′N 105°20′E
83 E20 **Hoachanas** Hardap, C Namibia 23°55′S 18°04′E
Hoai Nhơn *see* Bông Sơn
167 T8 **Hoa Lac** Quang Binh, C Vietnam 17°54′N 106°24′E
167 S5 **Hoang Liên Sơn** ▲▲ N Vietnam
Hoang Sa, Quân Đao *see* Paracel Islands
83 B17 **Hoanib** ≈ NW Namibia
33 S15 **Hoback Peak** ▲ Wyoming, C USA 43°04′N 110°34′W
183 P17 **Hobart** *prev.* Hobarton, Hobart Town. *state capital* Tasmania, SE Australia 42°54′S 147°18′E
26 L11 **Hobart** Oklahoma, C USA 35°03′N 99°04′W
183 P17 **Hobart** ✈ Tasmania, SE Australia 42°52′S 147°28′E
Hobarton/Hobart Town *see* Hobart
37 W14 **Hobbs** New Mexico, SW USA 32°42′N 103°08′W
194 L12 **Hobbs Coast** *physical region* Antarctica
23 Z14 **Hobe Sound** Florida, SE USA 27°03′N 80°08′W
Hobicaurikány *see* Uricani
54 E12 **Hobo** Huila, S Colombia 02°34′N 75°28′W
99 G16 **Hoboken** Antwerpen, N Belgium 51°12′N 04°22′E
158 K3 **Hoboksar** *var.* Hoboksar Mongol Zizhixian. Xinjiang Uygur Zizhiqu, NW China 46°48′N 85°42′E
Hoboksar Mongol Zizhixian *see* Hoboksar
95 G21 **Hobro** Nordjylland, N Denmark 56°39′N 09°51′E
21 X10 **Hobucken** North Carolina, SE USA 35°15′N 76°31′W
95 O20 **Hoburgen** *headland* SE Sweden 56°54′N 18°07′E
81 P15 **Hobyo** *It.* Obbia. Mudug, E Somalia 05°16′N 48°24′E
109 R8 **Hochalmspitze** ▲ SW Austria 47°00′N 13°19′E
109 Q4 **Hochburg** Oberösterreich, N Austria 48°10′N 12°57′E
108 F8 **Hochdorf** Luzern, N Switzerland 47°10′N 08°16′E
109 N8 **Hochfeiler** *It.* Gran Pilastro. ▲ Austria/Italy 46°59′N 11°42′E
167 T14 **Hô Chi Minh** *var.* Ho Chi Minh City; *prev.* Saigon. S Vietnam 10°46′N 106°43′E
Ho Chi Minh City *see* Hô Chi Minh
108 I7 **Höchst** Vorarlberg, NW Austria 47°28′N 09°40′E
Höchstadt *see* Höchstadt an der Aisch
101 K19 **Höchstadt an der Aisch** *var.* Höchstadt. Bayern, C Germany 49°43′N 10°48′E
108 L9 **Hochwilde** *It.* L'Altissima. ▲ Austria/Italy 46°45′N 11°00′E
109 S7 **Hochwildstelle** ▲ C Austria 47°21′N 13°53′E
31 T14 **Hocking River** ≈ Ohio, N USA
Hoctúm *see* Hoctún
41 X12 **Hoctún** *var.* Hoctúm. Yucatán, E Mexico 20°48′N 89°14′W
Hodeida *see* Al Ḥudaydah
20 K6 **Hodgenville** Kentucky, S USA 37°34′N 85°45′W
11 T17 **Hodgeville** Saskatchewan, S Canada 50°06′N 106°55′W
76 L9 **Hodh ech Chargui** ◈ *region* E Mauritania
Hodh el Garbi *see* Hodh el Gharbi
76 J10 **Hodh el Gharbi** *var.* Hodh el Garbi. ◈ *region* S Mauritania
111 L25 **Hódmezővásárhely** Csongrád, SE Hungary 46°27′N 20°18′E
74 J6 **Hodna, Chott El** *var.* Chott el-Hodna, *Ar.* Shatt al-Hodna. *salt lake* N Algeria
Hodna, Chott el-/Hodna, Shatt al- *see* Hodna, Chott El
111 G19 **Hodonín** *Ger.* Göding. Jihomoravský Kraj, SE Czech Republic 48°52′N 17°07′E
Hödrögö *see* Nömrög
Hodság/Hodschag *see* Odžaci
39 R7 **Hodzana River** ≈ Alaska, USA
Hoei *see* Huy
99 H19 **Hoeilaart** Vlaams Brabant, C Belgium 50°46′N 04°28′E
Hoë Karoo *see* Great Karoo
98 F12 **Hoek van Holland** *Eng.* Hook of Holland. Zuid-Holland, W Netherlands 52°00′N 04°07′E
98 L11 **Hoenderloo** Gelderland, E Netherlands 52°08′N 05°46′E
99 L18 **Hoensbroek** Limburg, SE Netherlands 50°55′N 05°55′E
163 Y11 **Hoeryŏng** NE North Korea 42°23′N 129°46′E
99 K18 **Hoeselt** Limburg, NE Belgium 50°50′N 05°30′E
98 K11 **Hoevelaken** Gelderland, C Netherlands 52°10′N 05°27′E
Hoey *see* Huy
101 M18 **Hof** Bayern, SE Germany 50°19′N 11°55′E
Höfdhakaupstadhur *see* Skagaströnd
Hofei *see* Hefei
101 G18 **Hofheim am Taunus** Hessen, W Germany 50°04′N 08°27′E
Hofmarkt *see* Odorheiu Secuiesc
92 L3 **Höfn** Austurland, SE Iceland 64°14′N 15°17′W
94 N13 **Hofors** Gävleborg, C Sweden 60°33′N 16°21′E
92 J6 **Hofsjökull** *glacier* C Iceland
92 J1 **Hofsós** Norðurland Vestra, N Iceland 65°54′N 19°25′W
164 E13 **Hōfu** Yamaguchi, Honshū, SW Japan 34°01′N 131°34′E
92 I4 **Höfuðborgarsvæðið** SW Iceland
Hofuf *see* Al Hufūf
95 J22 **Höganäs** Skåne, S Sweden 56°11′N 12°39′E
183 P14 **Hogan Group** *island group* Tasmania, SE Australia
23 R4 **Hogansville** Georgia, SE USA 33°10′N 84°55′W
39 P8 **Hogatza River** ≈ Alaska, USA
28 I14 **Hogback Mountain** ▲ Nebraska, C USA 41°40′N 103°44′W
95 G14 **Høgevarde** ▲ S Norway 60°19′N 09°27′E
Högfors *see* Karkkila
31 P5 **Hog Island** *island* Michigan, N USA
21 Y6 **Hog Island** *island* Virginia, NE USA
Hogoley Islands *see* Chuuk Islands
95 N20 **Högsby** Kalmar, S Sweden 57°10′N 16°03′E
36 K1 **Hogup Mountains** ▲▲ Utah, W USA
101 E17 **Hohe Acht** ▲ W Germany 50°23′N 07°00′E
Hohenelbe *see* Vrchlabí
108 I7 **Hohenems** Vorarlberg, W Austria 47°23′N 09°43′E
Hohenmauth *see* Vysoké Mýto
Hohensalza *see* Inowrocław
Hohenstadt *see* Zábřeh
Hohenstein in Ostpreussen *see* Olsztynek
20 I9 **Hohenwald** Tennessee, S USA 35°33′N 87°31′W
101 L17 **Hohenwarte-Stausee** ▣ C Germany
Hohes Venn *see* Hautes Fagnes
109 Q8 **Hohe Tauern** ▲▲ W Austria
163 O13 **Hohhot** *var.* Huhehot, Huhohaote, *Mong.* Kukukhoto; *prev.* Kweisui, Kwesui. Nei Mongol Zizhiqu, N China 40°49′N 111°37′E
162 F7 **Hohmorĭt** *var.* Sayn-Ust. Govĭ-Altay, W Mongolia 47°23′N 94°19′E
103 U6 **Hohneck** ▲ NE France 48°04′N 07°01′E
77 Q16 **Hohoe** E Ghana 07°08′N 00°32′E
164 E12 **Hōhoku** Yamaguchi, Honshū, SW Japan 34°15′N 130°56′E
159 O11 **Hoh Sai Hu** ◎ C China
159 N11 **Hoh Xil Hu** ◎ C China
158 L11 **Hoh Xil Shan** ▲▲ W China
167 U10 **Hôi An** *prev.* Faifo. Quang Nam-Đa Năng, C Vietnam 15°54′N 108°19′E
Hoï-Hao/Hoihow *see* Haikou
159 S11 **Hoika** *prev.* Heka. Qinghai, W China 35°49′N 99°50′E
81 F17 **Hoima** W Uganda 01°25′N 31°22′E
26 L5 **Hoisington** Kansas, C USA 38°31′N 98°46′W
146 D12 **Hojagala** *Rus.* Khodzhakala. Balkan Welaýaty, W Turkmenistan 38°46′N 56°14′E
146 M13 **Hojambaz** *Rus.* Khodzhambas. Lebap Welaýaty, E Turkmenistan 38°11′N 64°33′E
95 H23 **Højby** Syddtjylland, C Denmark 55°20′N 10°27′E
95 F24 **Højer** Syddanmark, SW Denmark 54°57′N 08°43′E
164 E14 **Hōjō** *var.* Hôzyô. Ehime, Shikoku, SW Japan 33°58′N 132°47′E
184 J3 **Hokianga Harbour** *inlet* SE Tasman Sea
185 F17 **Hokitika** West Coast, South Island, New Zealand 42°44′S 170°58′E
165 U4 **Hokkai-dō** ◈ *territory* Hokkaidō, NE Japan
165 T3 **Hokkaidō** *prev.* Ezo, Yeso, Yezo. *island* NE Japan
95 G15 **Hokksund** Buskerud, S Norway 59°45′N 09°55′E
143 S4 **Hokmābād** Khorāsān-e Razavī, N Iran 36°37′N 57°34′E
Hokō *see* Pohang
Hoko-guntō/Hoko-shotō *see* Penghu Liedao
Hoktemberyan *see* Armavir
94 F13 **Hol** Buskerud, S Norway 60°36′N 08°18′E
81 K19 **Hola** Tana River, SE Kenya 01°06′S 40°01′E
117 R11 **Hola Prystan'** *Rus.* Golaya Pristan. Khersons'ka Oblast', S Ukraine 46°31′N 32°31′E
95 I23 **Holbæk** Sjælland, E Denmark 55°42′N 11°42′E
Holboo *see* Santmargats
183 P10 **Holbrook** New South Wales, SE Australia 35°45′S 147°18′E
37 N11 **Holbrook** Arizona, SW USA 34°54′N 110°09′W
27 S5 **Holden** Missouri, C USA 38°42′N 93°59′W
36 K5 **Holden** Utah, W USA 39°06′N 112°16′W
27 O11 **Holdenville** Oklahoma, C USA 35°05′N 96°25′W
29 O16 **Holdrege** Nebraska, C USA 40°28′N 99°28′W
35 X3 **Hole in the Mountain Peak** ▲ Nevada, W USA 40°54′N 115°06′W
155 G20 **Hole Narsipur** Karnātaka, W India 12°46′N 76°14′E
111 H18 **Holešov** *Ger.* Holleschau. Zlínský Kraj, E Czech Republic 49°20′N 17°35′E
45 N14 **Holetown** *prev.* Jamestown. W Barbados 13°11′N 59°38′W
31 Q12 **Holgate** Ohio, N USA 41°12′N 84°06′W
44 I7 **Holguín** Holguín, SE Cuba 20°51′N 76°16′W
23 V12 **Holiday** Florida, SE USA 28°11′N 82°44′W
163 S9 **Holin Gol** *prev.* Hulingol. Nei Mongol Zizhiqu, N China 45°36′N 119°54′E
39 O12 **Holitna River** ≈ Alaska, USA
94 J13 **Höljes** Värmland, C Sweden 60°54′N 12°34′E
109 X3 **Hollabrunn** Niederösterreich, NE Austria 48°33′N 16°06′E
36 L3 **Holladay** Utah, W USA 40°39′N 111°49′W
11 X16 **Holland** Manitoba, S Canada 49°36′N 98°52′W
31 O9 **Holland** Michigan, N USA 42°47′N 86°06′W
25 T9 **Holland** Texas, SW USA 30°52′N 97°24′W
Holland *see* Netherlands
22 K4 **Hollandale** Mississippi, S USA 33°10′N 90°51′W
Hollandia *see* Jayapura
Hollandsch Diep *see* Hollands Diep
99 H14 **Hollands Diep** *var.* Hollandsch Diep. *channel* SW Netherlands
Holleschau *see* Holešov
25 R5 **Holliday** Texas, SW USA 33°49′N 98°41′W
18 E15 **Hollidaysburg** Pennsylvania, NE USA 40°24′N 78°22′W
21 S6 **Hollins** Virginia, NE USA 37°20′N 79°56′W
26 J12 **Hollis** Oklahoma, C USA 34°42′N 99°56′W
35 O10 **Hollister** California, W USA 36°51′N 121°25′W
27 T8 **Hollister** Missouri, C USA 36°37′N 93°13′W
93 M19 **Hollola** Päijät-Häme, S Finland 61°N 25°32′E
98 K4 **Hollum** Fryslân, N Netherlands 53°27′N 05°38′E
95 J23 **Höllviken** *prev.* Höllviksnäs. Skåne, S Sweden 55°25′N 12°57′E
Höllviksnäs *see* Höllviken
37 W6 **Holly** Colorado, C USA 38°03′N 102°07′W
31 R9 **Holly** Michigan, N USA 42°47′N 83°37′W
21 S14 **Holly Hill** South Carolina, SE USA 33°19′N 80°24′W
21 W11 **Holly Ridge** North Carolina, SE USA 34°31′N 77°31′W
22 L1 **Holly Springs** Mississippi, S USA 34°47′N 89°25′W
23 Z15 **Hollywood** Florida, SE USA 26°00′N 80°09′W
8 J6 **Holman** Victoria Island, Northwest Territories, N Canada 70°42′N 117°45′W
92 I2 **Hólmavík** Vestfirðir, NW Iceland 65°42′N 21°43′W
30 J7 **Holmen** Wisconsin, N USA 43°57′N 91°14′W
23 R8 **Holmes Creek** ≈ Alabama/Florida, SE USA
95 H16 **Holmestrand** Vestfold, S Norway 59°29′N 10°20′E
93 J16 **Holmön** *island* N Sweden
95 E22 **Holmsland Klit** *beach* W Denmark
93 J16 **Holmsund** Västerbotten, N Sweden 63°42′N 20°26′E
95 Q18 **Holmudden** *headland* SE Sweden 57°59′N 19°14′E
138 F10 **Holon** *var.* Kholon; *prev.* Ḥolon. Tel Aviv, C Israel 32°01′N 34°46′E
Ḥolon *see* Holon
163 P7 **Hölönbuyr** *var.* Bayan. Dornod, E Mongolia 47°56′N 112°58′E
117 P8 **Holovanivs'k** *Rus.* Golovanevsk. Kirovohrads'ka Oblast', C Ukraine 48°21′N 30°26′E
95 F21 **Holstebro** Midtjylland, W Denmark 56°22′N 08°38′E
95 F23 **Holsted** Syddtjylland, W Denmark 55°30′N 08°54′E
29 T13 **Holstein** Iowa, C USA 42°29′N 95°32′W
Holsteinborg/Holsteinsborg/Holstenborg/Holstensborg *see* Sisimiut
21 O8 **Holston River** ≈ Tennessee, S USA
31 Q9 **Holt** Michigan, N USA 42°38′N 84°31′W
98 N10 **Holten** Overijssel, E Netherlands 52°16′N 06°25′E
27 P3 **Holton** Kansas, C USA 39°27′N 95°44′W
27 U5 **Holts Summit** Missouri, C USA 38°38′N 92°07′W
35 X17 **Holtville** California, W USA 32°48′N 115°22′W
98 L5 **Holwerd** *Fris.* Holwert. Fryslân, N Netherlands 53°22′N 05°51′E
Holwert *see* Holwerd
39 O11 **Holy Cross** Alaska, USA 62°12′N 159°46′W
37 R4 **Holy Cross, Mount Of The** ▲ Colorado, C USA 39°28′N 106°28′W
97 I18 **Holyhead** *Wel.* Caer Gybi. NW Wales, United Kingdom 53°19′N 04°38′W
97 H18 **Holy Island** *island* NW Wales, United Kingdom
96 L12 **Holy Island** *island* NE England, United Kingdom
37 W3 **Holyoke** Colorado, C USA 40°31′N 102°18′W
18 M11 **Holyoke** Massachusetts, NE USA 42°12′N 72°37′W
101 I14 **Holzminden** Niedersachsen, C Germany 51°49′N 09°27′E
81 G19 **Homa Bay** Homa Bay, W Kenya 0°31′S 34°30′E
81 G19 **Homa Bay** ◈ *county* W Kenya
Homāyūnshahr *see* Khomeynīshahr
77 P11 **Hombori** Mopti, S Mali 15°13′N 01°39′W
101 E20 **Homburg** Saarland, SW Germany 49°20′N 07°20′E
9 R5 **Home Bay** *bay* Baffin Bay, Nunavut, NE Canada
Homenau *see* Humenné
39 Q13 **Homer** Alaska, USA 59°38′N 151°33′W
22 H4 **Homer** Louisiana, S USA 32°47′N 93°03′W
18 H10 **Homer** New York, NE USA 42°38′N 76°10′W
23 V7 **Homerville** Georgia, SE USA 31°02′N 82°45′W
23 Y16 **Homestead** Florida, SE USA 25°28′N 80°28′W
27 O9 **Hominy** Oklahoma, C USA 36°24′N 96°24′W
94 H8 **Hommelvik** Sør-Trøndelag, S Norway 63°24′N 10°48′E
95 C16 **Hommersåk** Rogaland, S Norway 58°55′N 05°51′E
155 H15 **Homnābād** Karnātaka, C India 17°46′N 77°08′E
22 J7 **Homochitto River** ≈ Mississippi, S USA
83 N20 **Homoíne** Inhambane, SE Mozambique 23°51′S 35°04′E
112 O12 **Homoljske Planine** ▲▲ E Serbia
Homonna *see* Humenné
Homs *see* Al Khums, Libya
Homs *see* Ḥimṣ
119 P19 **Homyel'** *Rus.* Gomel'. Homyel'skaya Voblasts', SE Belarus 52°25′N 31°E
118 L12 **Homyel'** Vitsyebskaya Voblasts', N Belarus 55°20′N 28°52′E
119 L19 **Homyel'skaya Voblasts'** *Rus.* Gomel'skaya Oblast'. ◈ *province* SE Belarus
Honan *see* Henan, China
Honan *see* Luoyang, China
164 U4 **Honbetsu** Hokkaidō, NE Japan 43°09′N 143°46′E
Honctő *see* Gurahonţ
54 E9 **Honda** Tolima, C Colombia 05°12′N 74°45′W
83 D24 **Hondeklip** *Afr.* Hondeklipbaai. Northern Cape, W South Africa 30°15′S 17°17′E
Hondeklipbaai *see* Hondeklip
11 Q13 **Hondo** Alberta, W Canada 54°43′N 113°14′W
25 Q12 **Hondo** Texas, SW USA 29°21′N 99°09′W
42 G1 **Hondo** ≈ Central America
Hondo *see* Honshū
Hondo *see* Amakusa
42 G6 **Honduras** *off.* Republic of Honduras. ◆ *republic* Central America
Honduras, Golfo de *see* Honduras, Gulf of
42 H4 **Honduras, Gulf of** *Sp.* Golfo de Honduras. *gulf* W Caribbean Sea
Honduras, Republic of *see* Honduras
11 V12 **Hone** Manitoba, C Canada 56°13′N 101°12′W
21 P12 **Honea Path** South Carolina, SE USA 34°27′N 82°23′W
95 H14 **Hønefoss** Buskerud, S Norway 60°10′N 10°15′E
31 S12 **Honey Creek** ≈ Ohio, N USA
25 V5 **Honey Grove** Texas, SW USA 33°34′N 95°54′W
35 Q4 **Honey Lake** ◎ California, W USA
102 L4 **Honfleur** Calvados, N France 49°25′N 00°14′E
Hon Gai *see* Ha Long
161 O8 **Hong'an** *prev.* Huang'an. Hubei, C China 31°20′N 114°43′E
Hongay *see* Ha Long
Hông Gai *see* Ha Long
161 O15 **Honghai Wan** *bay* N South China Sea
Hông Hà, Sông *see* Red River
161 O7 **Hong He** ≈ C China
161 N9 **Hong Hu** ◎ C China
160 L11 **Hongjiang** Hunan, S China 27°09′N 109°58′E
Hongjiang *see* Wangcang
161 O15 **Hong Kong** *off.* Hong Kong Special Administrative Region, *var.* Hong Kong S.A.R., *Chin.* Xianggang Tebie Xingzhengqu. S China 22°17′N 114°09′E
Hong Kong S.A.R. *see* Hong Kong
Hong Kong Special Administrative Region *see* Hong Kong
160 L4 **Hongliu He** ≈ C China
160 L4 **Hongliu He** ≈ C China
159 P8 **Hongliuwan** *var.* Aksay, Aksay Kazakzu Zizhixian. Gansu, N China 39°25′N 94°09′E
159 P7 **Hongliuyuan** Gansu, N China 41°02′N 95°24′E
Hongor *see* Delgereh
161 S8 **Hongqiao** ✈ (Shanghai) Shanghai Shi, E China 31°28′N 121°08′E
160 K14 **Hongshui He** ≈ S China
160 M5 **Hongtong** *var.* Dahuaishu. Shanxi, C China 36°30′N 111°42′E
164 J15 **Hongū** Wakayama, Honshū, SW Japan 33°50′N 135°42′E
Honguedo, Détroit d' *see* Honguedo Passage
15 Y5 **Honguedo Passage** *var.* Honguedo Strait, *Fr.* Détroit d'Honguedo. *strait* Québec, E Canada
Honguedo Strait *see* Honguedo Passage
Hongwan *see* Hongwansi

◆ Country
● Country Capital
◇ Dependent Territory
○ Dependent Territory Capital
◈ Administrative Regions
✈ International Airport
▲ Mountain
▲▲ Mountain Range
Ѫ Volcano
≈ River
◎ Lake
▣ Reservoir

HONGWANSI – HUIXIAN

159 S8 **Hongwansi** *var.* Sunan, Sunan Yugurzu Zizhixian; *prev.* Hongwan. Gansu, N China 38°55′N 99°29′E
163 X13 **Hongwŏn** E North Korea 40°03′N 127°54′E
160 H7 **Hongyuan** *var.* Qiongxi; *prev.* Hurama. Sichuan, C China 32°49′N 102°40′E
161 Q7 **Hongze Hu** *var.* Hung-tse Hu. ◎ E China
186 L9 **Honiara** ● (Solomon Islands) Guadalcanal, C Solomon Islands 09°27′S 159°56′E
165 P8 **Honjō** *var.* Honzyô, Yurihonjô. Akita, Honshū, C Japan 39°23′N 140°03′E
93 K18 **Honkajoki** Satakunta, SW Finland 62°00′N 22°15′E
92 K7 **Honningsvåg** Finnmark, N Norway 70°58′N 25°59′E
95 I19 **Hönö** Västra Götaland, S Sweden 57°42′N 11°39′E
38 G11 **Honoka'a** Hawaii, USA, C Pacific Ocean 20°04′N 155°27′W
38 G11 **Honoka'a** *var.* Honokaa. Hawaii, USA, C Pacific Ocean 20°04′N 155°27′W
Honokaa *see* Honoka'a
38 D9 **Honolulu** *state capital* O'ahu, Hawaii, USA, C Pacific Ocean 21°18′N 157°52′W
38 H11 **Honomú** *var.* Honomu. Hawaii, USA, C Pacific Ocean 19°51′N 155°06′W
105 P10 **Honrubia** Castilla-La Mancha, C Spain 39°36′N 02°17′W
164 M12 **Honshū** *var.* Hondo, Honsyû. *island* SW Japan
Honsyû *see* Honshū
Honte *see* Westerschelde
Honzyô *see* Honjō
8 K8 **Hood** ≈ Nunavut, NW Canada
Hood Island *see* Española, Isla
32 H11 **Hood, Mount** ▲ Oregon, NW USA 45°22′N 121°41′W
32 H11 **Hood River** Oregon, NW USA 45°44′N 121°31′W
98 H10 **Hoofddorp** Noord-Holland, W Netherlands 52°18′N 04°41′E
99 G15 **Hoogerheide** Noord-Brabant, S Netherlands 51°25′N 04°19′E
98 N8 **Hoogeveen** Drenthe, NE Netherlands 52°44′N 06°30′E
98 O6 **Hoogezand-Sappemeer** Groningen, NE Netherlands 53°10′N 06°47′E
98 J8 **Hoogkarspel** Noord-Holland, NW Netherlands 52°42′N 04°59′E
98 N5 **Hoogkerk** Groningen, NE Netherlands 53°13′N 06°30′E
98 G13 **Hoogvliet** Zuid-Holland, SW Netherlands 51°51′N 04°23′E
26 I8 **Hooker** Oklahoma, C USA 36°51′N 101°12′W
97 E21 **Hook Head** *Ir.* Rinn Duáin. *headland* SE Ireland 52°07′N 06°55′W
Hook of Holland *see* Hoek van Holland
Hoolt *see* Tögrög
39 W13 **Hoonah** Chichagof Island, Alaska, USA 58°05′N 135°21′W
38 L11 **Hooper Bay** Alaska, USA 61°31′N 166°06′W
31 N13 **Hoopeston** Illinois, N USA 40°28′N 87°40′W
95 K22 **Höör** Skåne, S Sweden 55°55′N 13°33′E
98 I9 **Hoorn** Noord-Holland, NW Netherlands 52°38′N 05°04′E
18 L10 **Hoosic River** ≈ New York, NE USA
Hoosier State *see* Indiana
35 Y11 **Hoover Dam** *dam* Arizona/Nevada, W USA
Höövör *see* Baruunbayan-Ulaan
137 Q11 **Hopa** Artvin, NE Turkey 41°23′N 41°28′E
18 J14 **Hopatcong** New Jersey, NE USA 40°55′N 74°39′W
10 M17 **Hope** British Columbia, SW Canada 49°21′N 121°28′W
39 R12 **Hope** Alaska, USA 60°55′N 149°38′W
27 T14 **Hope** Arkansas, C USA 33°40′N 93°36′W
31 P14 **Hope** Indiana, N USA 39°18′N 85°46′W
29 Q5 **Hope** North Dakota, N USA 47°18′N 97°42′W
13 Q7 **Hopedale** Newfoundland and Labrador, NE Canada 55°26′N 60°14′W
Hopeh/Hopei *see* Hebei
180 K13 **Hope, Lake** *salt lake* Western Australia
41 X13 **Hopelchén** Campeche, SE Mexico 19°46′N 89°50′W
21 U11 **Hope Mills** North Carolina, SE USA 34°58′N 78°57′W
183 O7 **Hope, Mount** New South Wales, SE Australia 32°49′S 145°55′E
92 P4 **Hopen** *island* SE Svalbard
197 Q4 **Hope, Point** *headland* Alaska, USA
12 M3 **Hopes Advance, Cap** *cape* Québec, NE Canada
182 L10 **Hopetoun** Victoria, SE Australia 35°46′S 142°23′E
83 H23 **Hopetown** Northern Cape, W South Africa 29°37′S 24°05′E
21 W6 **Hopewell** Virginia, NE USA 37°16′N 77°15′W
109 O7 **Hopfgarten im Brixental** Tirol, W Austria 47°28′N 12°14′E
181 N8 **Hopkins Lake** *salt lake* Western Australia
182 M12 **Hopkins River** ≈ Victoria, SE Australia
20 I7 **Hopkinsville** Kentucky, S USA 36°50′N 87°30′W
34 M6 **Hopland** California, W USA 38°58′N 123°09′W
95 G24 **Hoptrup** Syddanmark, SW Denmark 55°09′N 09°27′E
Hoqin Zuoyi Zhongji *see* Baokang
32 F9 **Hoquiam** Washington, NW USA 46°58′N 123°53′W
29 R6 **Horace** North Dakota, N USA 46°44′N 96°54′W
117 T14 **Hora Roman-Kosh** ▲ S Ukraine 44°37′N 34°13′E
137 R12 **Horasan** Erzurum, NE Turkey 40°03′N 42°10′E
101 G22 **Horb am Neckar** Baden-Württemberg, S Germany 48°27′N 08°42′E
95 K23 **Hörby** Skåne, S Sweden 55°51′N 13°42′E
43 P16 **Horconcitos** Chiriquí, W Panama 08°20′N 82°10′W
95 C14 **Hordaland** ◆ *county* S Norway
116 H13 **Horezu** Vâlcea, SW Romania 45°06′N 24°00′E
108 G7 **Horgen** Zürich, N Switzerland 47°16′N 08°36′E
Horgo *see* Tariat
Hôrin *see* Fenglin
163 O13 **Horinger** Nei Mongol Zizhiqu, N China 40°23′N 111°48′E
Horiult *see* Bogd
11 U17 **Horizon** Saskatchewan, S Canada 49°33′N 105°05′W
192 K9 **Horizon Bank** *undersea feature* S Pacific Ocean
192 L10 **Horizon Deep** *undersea feature* W Pacific Ocean
95 L14 **Hörken** Örebro, S Sweden 60°03′N 14°55′E
119 O15 **Horki** *Rus.* Gorki. Mahilyowskaya Voblasts', E Belarus 54°18′N 31°E
195 O10 **Horlick Mountains** ▲ Antarctica
117 X7 **Horlivka** *Rom.* Adâncata, *Rus.* Gorlovka. Donets'ka Oblast', E Ukraine 48°19′N 38°04′E
143 V11 **Hormak** Sīstān va Balūchestān, SE Iran 30°00′N 60°50′E
143 R13 **Hormozgān** *off.* Ostān-e Hormozgān. ◆ *province* S Iran
Hormozgān, Ostān-e *see* Hormozgān
Hormoz, Tangeh-ye *see* Hormuz, Strait of
141 W6 **Hormuz, Strait of** *var.* Strait of Ormuz, *Per.* Tangeh-ye Hormoz. *strait* Iran/Oman
109 W2 **Horn** Niederösterreich, NE Austria 48°40′N 15°40′E
95 M18 **Horn** Östergötland, S Sweden 57°54′N 15°49′E
8 J9 **Horn** ≈ Northwest Territories, NW Canada
Hornád *see* Hernád
8 I6 **Hornaday** ≈ Northwest Territories, NW Canada
92 H13 **Hornavan** ◎ N Sweden
65 C24 **Hornby Mountains** *hill range* West Falkland, Falkland Islands
Horn, Cape *see* Hornos, Cabo de
97 O18 **Horncastle** E England, United Kingdom 53°12′N 00°07′W
95 N14 **Horndal** Dalarna, C Sweden 60°16′N 16°25′E
93 I16 **Hörnefors** Västerbotten, N Sweden 63°37′N 19°54′E
18 F11 **Hornell** New York, NE USA 42°19′N 77°38′W
Horné Nové Mesto *see* Kysucké Nové Mesto
12 F12 **Hornepayne** Ontario, S Canada 49°14′N 84°48′W
94 D10 **Hornindalsvatnet** ◎ S Norway
101 G22 **Hornisgrinde** ▲ SW Germany 48°37′N 08°13′E
22 M9 **Horn Island** *island* Mississippi, S USA
Hornja Łužica *see* Oberlausitz
63 J26 **Hornos, Cabo de** *Eng.* Cape Horn. *headland* S Chile 55°52′S 67°00′W
117 S10 **Hornostayivka** Khersons'ka Oblast', S Ukraine 47°00′N 33°42′E
183 T9 **Hornsby** New South Wales, SE Australia 33°44′S 151°08′E
97 O16 **Hornsea** E England, United Kingdom 53°54′N 00°10′W
94 O11 **Hornslandet** *peninsula* C Sweden
95 H22 **Hornslet** Midtjylland, C Denmark 56°19′N 10°20′E
92 O4 **Hornsundtind** ▲ S Svalbard 76°54′N 16°07′E
Horochów *see* Horokhiv
Horodenka *see* Gorodenka
117 Q2 **Horodnya** *Rus.* Gorodnya. Chernihivs'ka Oblast', NE Ukraine 51°54′N 31°30′E
116 K6 **Horodok** Khmel'nyts'ka Oblast', W Ukraine 49°10′N 26°34′E
116 H5 **Horodok** *Pol.* Gródek Jagielloński, *Rus.* Gorodok, Gorodok Yagellonski. L'vivs'ka Oblast', NW Ukraine 49°48′N 23°39′E
117 Q6 **Horodyshche** *Rus.* Gorodishche. Cherkas'ka Oblast', C Ukraine 49°19′N 31°27′E
165 T3 **Horokanai** Hokkaidō, NE Japan 44°02′N 142°08′E
116 J4 **Horokhiv** *Pol.* Horochów, *Rus.* Gorokhov. Volyns'ka Oblast', NW Ukraine 50°31′N 24°50′E
165 T4 **Horoshiri-dake** *var.* Horosiri Dake. ▲ Hokkaidō, N Japan 42°43′N 142°41′E
Horosiri Dake *see* Horoshiri-dake
111 C17 **Hořovice** *Ger.* Horowitz. Střední Čechy, W Czech Republic 49°49′N 13°53′E
Horowitz *see* Hořovice
Horqin Zuoyi Houqi *see* Ganjig
Horqin Zuoyi Zhongqi *see* Bayan Huxu
62 O5 **Horqueta** Concepción, C Paraguay 23°24′S 56°53′W
55 O12 **Horqueta Minas** Amazonas, S Venezuela 02°20′N 63°32′W
95 J20 **Horred** Västra Götaland, S Sweden 57°22′N 12°25′E
151 J19 **Horsburgh Atoll** *var.* Goidhoo Atoll. *atoll* N Maldives
20 K7 **Horse Cave** Kentucky, S USA 37°10′N 85°54′W
37 V6 **Horse Creek** ≈ Colorado, C USA
27 S6 **Horse Creek** ≈ Missouri, C USA
18 G11 **Horseheads** New York, NE USA 42°10′N 76°49′W
37 P13 **Horse Mount** ▲ New Mexico, SW USA 33°58′N 108°10′W
95 G22 **Horsens** Syddanmark, C Denmark 55°53′N 09°53′E
65 F25 **Horse Pasture Point** *headland* W Saint Helena 15°57′S 05°46′W
33 N13 **Horseshoe Bend** Idaho, NW USA 43°55′N 116°11′W
36 L13 **Horseshoe Reservoir** ⊠ Arizona, SW USA
64 M9 **Horseshoe Seamounts** *undersea feature* E Atlantic Ocean 36°30′N 15°00′W
182 L11 **Horsham** Victoria, SE Australia 36°44′S 142°13′E
97 O23 **Horsham** SE England, United Kingdom 51°01′N 00°21′W
99 M15 **Horst** Limburg, SE Netherlands 51°30′N 06°05′E
64 N2 **Horta** Faial, Azores, Portugal, NE Atlantic Ocean 38°32′N 28°39′W
105 S12 **Horta, Cap de l'** *Cast.* Cabo Huertas. *headland* SE Spain 38°21′N 00°25′W
95 H16 **Horten** Vestfold, S Norway 59°25′N 10°25′E
111 M23 **Hortobágy-Berettyó** ≈ E Hungary
27 Q3 **Horton** Kansas, C USA 39°39′N 95°31′W
8 I7 **Horton** ≈ Northwest Territories, NW Canada
95 I23 **Hørve** Sjælland, E Denmark 55°46′N 11°28′E
95 L22 **Hörvik** Blekinge, S Sweden 56°01′N 14°45′E
Horvot Haluza *see* Horvot Halutsa
14 E7 **Horwood Lake** ◎ Ontario, S Canada
116 K4 **Horyn'** *Rus.* Goryn. ≈ NW Ukraine
81 I14 **Hosa'ina** *var.* Hosseina, *It.* Hosanna. Southern Nationalities, S Ethiopia 07°38′N 37°58′E
Hosanna *see* Hosa'ina
101 H18 **Hösbach** Bayern, C Germany 50°00′N 09°12′E
Hose Mountains *see* Hose, Pegunungan
169 T9 **Hose, Pegunungan** *var.* Hose Mountains. ▲ East Malaysia
148 L15 **Hoshāb** Baluchistān, SW Pakistan 26°01′N 63°51′E
154 H10 **Hoshangābād** Madhya Pradesh, C India 22°44′N 77°45′E
116 L4 **Hoshcha** Rivnens'ka Oblast', NW Ukraine 50°37′N 26°38′E
152 I7 **Hoshiārpur** Punjab, NW India 31°30′N 75°59′E
Höshööt *see* Öldziyt
99 M23 **Hosingen** Diekirch, NE Luxembourg 50°01′N 06°05′E
186 G7 **Hoskins** New Britain, E Papua New Guinea 05°28′S 150°25′E
155 G17 **Hospet** Karnātaka, C India 15°16′N 76°20′E
104 K4 **Hospital de Orbigo** Castilla y León, N Spain 42°27′N 05°53′W
Hospitalet *see* L'Hospitalet de Llobregat
92 N13 **Hossa** Kainuu, E Finland 65°28′N 29°36′E
Hosseina *see* Hosa'ina
Hosszúmezjő *see* Câmpulung Moldovenesc
63 I25 **Hoste, Isla** *island* S Chile
117 O4 **Hostomel'** *Rus.* Gostomel'. Kyyivs'ka Oblast', N Ukraine 50°41′N 30°15′E
155 H20 **Hosūr** Tamil Nādu, SE India 12°45′N 77°51′E
167 N8 **Hot** Chiang Mai, NW Thailand 18°07′N 98°34′E
158 G10 **Hotan** *var.* Khotan, *Chin.* Ho-t'ien. Xinjiang Uygur Zizhiqu, NW China 37°10′N 79°51′E
158 H9 **Hotan He** ≈ NW China
83 G22 **Hotazel** Northern Cape, N South Africa 27°12′S 22°58′E
37 Q5 **Hotchkiss** Colorado, C USA 38°47′N 107°43′W
35 V7 **Hot Creek Range** ▲ Nevada, W USA
Hote *see* Hoti
171 T13 **Hoti** *var.* Hote. Pulau Seram, E Indonesia 02°58′S 130°19′E
Ho-t'ien *see* Hotan
Hotin *see* Khotyn
93 H15 **Hoting** Jämtland, C Sweden 64°07′N 16°14′E
162 L14 **Hotong Qagan Nur** ◎ N China
162 J8 **Hotont** Arhangay, C Mongolia 47°21′N 102°27′E
27 T12 **Hot Springs** Arkansas, C USA 34°31′N 93°03′W
28 J11 **Hot Springs** South Dakota, N USA 43°26′N 103°29′W
21 S5 **Hot Springs** Virginia, NE USA 38°00′N 79°50′W
35 Q4 **Hot Springs Peak** ▲ California, W USA 40°23′N 120°06′W
27 T12 **Hot Springs Village** Arkansas, C USA 34°39′N 93°03′W
Hotspur Bank *see* Hotspur Seamount
65 J16 **Hotspur Seamount** *var.* Hotspur Bank. *undersea feature* C Atlantic Ocean 18°00′S 35°00′W
8 J8 **Hottah Lake** ◎ Northwest Territories, NW Canada
44 K9 **Hotte, Massif de la** ▲ SW Haiti
99 K21 **Hotton** Luxembourg, SE Belgium 50°18′N 05°25′E
Hötzing *see* Hațeg
187 P17 **Houaïlou** Province Nord, C New Caledonia 21°17′S 165°37′E
74 K5 **Houari Boumediène** ✈ (Alger) N Algeria 36°38′N 03°15′E
167 P6 **Houayxay** *var.* Ban Houayxay. Bokèo, N Laos 20°17′N 100°27′E
103 N5 **Houdan** Yvelines, N France 48°48′N 01°36′E
99 F20 **Houdeng-Goegnies** *var.* Houdeng-Goegnies. Hainaut, S Belgium 50°29′N 04°10′E
102 K14 **Houeillès** Lot-et-Garonne, SW France 44°15′N 00°02′E
99 L22 **Houffalize** Luxembourg, SE Belgium 50°08′N 05°47′E
30 M3 **Houghton** Michigan, N USA 47°07′N 88°34′W
31 Q7 **Houghton Lake** Michigan, N USA 44°18′N 84°45′W
31 Q7 **Houghton Lake** ◎ Michigan, N USA
19 T3 **Houlton** Maine, NE USA 46°09′N 67°50′W
160 M5 **Houma** Shanxi, C China 35°36′N 11°23′E
193 U16 **Houma** Tongatapu, S Tonga 21°18′S 174°55′W
22 J10 **Houma** Louisiana, S USA 29°35′N 90°44′W
196 V16 **Houma Taloa** *headland* Tongatapu, S Tonga 21°16′S 175°08′W
77 O13 **Houndé** SW Burkina Faso 11°34′N 03°31′W
102 J12 **Hourtin-Carcans, Lac d'** ◎ SW France
36 J5 **House Range** ▲ Utah, W USA
10 K13 **Houston** British Columbia, SW Canada 54°24′N 126°39′W
39 R11 **Houston** Alaska, USA 61°37′N 149°50′W
29 X10 **Houston** Minnesota, N USA 43°45′N 91°34′W
22 M3 **Houston** Mississippi, S USA 33°54′N 89°00′W
27 V7 **Houston** Missouri, C USA 37°19′N 91°59′W
25 W11 **Houston** Texas, SW USA 29°46′N 95°22′W
25 W11 **Houston** ✈ Texas, SW USA 30°03′N 95°18′W
98 J12 **Houten** Utrecht, C Netherlands 52°02′N 05°10′E
99 K17 **Houthalen** Limburg, NE Belgium 51°02′N 05°22′E
99 I22 **Houyet** Namur, SE Belgium 50°10′N 05°00′E
95 H22 **Hov** Midtjylland, C Denmark 55°54′N 10°13′E
95 L17 **Hova** Västra Götaland, S Sweden 58°52′N 14°13′E
162 E6 **Hovd** *var.* Dund-Us. Hovd, W Mongolia 48°06′N 91°22′E
162 J10 **Hovd** *var.* Dund-Us. Hovd, W Mongolia 48°06′N 91°22′E
162 E6 **Hovd** *var.* Khovd, Kobdo; *prev.* Jirgalanta. Hovd, W Mongolia 47°59′N 91°41′E
162 E7 **Hovd** ◆ *province* W Mongolia
Hovd *see* Bogd
162 C5 **Hovd Gol** ≈ NW Mongolia
97 O23 **Hove** SE England, United Kingdom 50°49′N 00°11′W
95 I22 **Hovedstaden** *off.* Frederiksborgs Amt. ◆ *county* E Denmark
29 N8 **Hoven** South Dakota, N USA 45°12′N 99°47′W
116 I8 **Hoverla, Hora** *Rus.* Gora Goverla. ▲ W Ukraine 48°09′N 24°30′E
Höviyn Am *see* Gurvanbulag
95 M21 **Hovmantorp** Kronoberg, S Sweden 56°47′N 15°08′E
163 N11 **Hövsgöl** Dornogovĭ, SE Mongolia 43°35′N 109°40′E
162 I5 **Hövsgöl** ◆ *province* N Mongolia
Hovsgol, Lake *see* Hövsgöl Nuur
162 J5 **Hövsgöl Nuur** *var.* Lake Hovsgol. ◎ N Mongolia
78 L9 **Howa, Ouadi** *var.* Wādi Howar. ≈ Chad/Sudan *see also* Howar, Wādi
Howa, Ouadi *see* Howar, Wādi
27 P7 **Howard** Kansas, C USA 37°27′N 96°16′W
29 Q10 **Howard** South Dakota, N USA 43°58′N 97°31′W
25 N10 **Howard Draw** *valley* Texas, SW USA
29 U8 **Howard Lake** Minnesota, N USA 45°03′N 94°03′W
80 B8 **Howar, Wādi** *var.* Ouadi Howa. ≈ Chad/Sudan *see also* Howa, Ouadi
Howar, Wādi *see* Howa, Ouadi
25 U5 **Howe** Texas, SW USA 33°29′N 96°38′W
183 R12 **Howe, Cape** *headland* New South Wales/Victoria, SE Australia 37°30′S 149°58′E
31 R9 **Howell** Michigan, N USA 42°36′N 83°55′W
28 L9 **Howes** South Dakota, N USA 44°34′N 102°03′W
83 K23 **Howick** KwaZulu/Natal, E South Africa 29°29′S 30°13′E
Howrah *see* Hāora
167 T9 **Hô Xa** *prev.* Vinh Linh. Quang Tri, C Vietnam 17°02′N 107°03′E
27 W9 **Hoxie** Arkansas, C USA 36°03′N 90°58′W
26 J3 **Hoxie** Kansas, C USA 39°21′N 100°27′W
101 I14 **Höxter** Nordrhein-Westfalen, W Germany 51°46′N 09°22′E
158 K6 **Hoxud** *var.* Tewulike. Xinjiang Uygur Zizhiqu, NW China 42°18′N 86°51′E
96 J5 **Hoy** *island* N Scotland, United Kingdom
43 S17 **Hoya, Cerro** ▲ S Panama 07°22′N 80°38′W
94 D12 **Høyanger** Sogn Og Fjordane, S Norway 61°13′N 06°05′E
101 P15 **Hoyerswerda** *Lus.* Wojerecy. Sachsen, E Germany 51°27′N 14°18′E
164 R14 **Hōyo-kaikyō** *var.* Hayasui-seto. *strait* SW Japan
104 J8 **Hoyos** Extremadura, W Spain 40°10′N 06°43′W
29 W4 **Hoyt Lakes** Minnesota, N USA 47°31′N 92°08′W
87 V2 **Hoyvík** Streymoy, N Faroe Islands
137 O14 **Hozat** Tunceli, E Turkey 39°09′N 39°13′E
Hôzyô *see* Hōjō
167 N8 **Hpa-an** *var.* Pa-an. Kayin State, S Myanmar (Burma) 16°51′N 97°37′E
167 N8 **Hpapun** *var.* Papun. Kayin State, S Myanmar (Burma) 18°05′N 97°26′E
167 N7 **Hpasawng** *var.* Pasawng. Kayah State, C Myanmar (Burma) 18°50′N 97°16′E
Hpyu *see* Phyu
111 F16 **Hradec Králové** *Ger.* Königgrätz. Královéhradecký Kraj, N Czech Republic 50°13′N 15°50′E
Hradecký Kraj *see* Královéhradecký Kraj
111 B16 **Hradiště** *Ger.* Burgstadlberg. ▲ NW Czech Republic 50°12′N 13°04′E
117 R6 **Hradyz'k** *Rus.* Gradizhsk. Poltavs'ka Oblast', NE Ukraine 49°14′N 33°07′E
119 M16 **Hradzyanka** *Rus.* Grodzyanka. Mahilyowskaya Voblasts', E Belarus 53°33′N 28°45′E
119 F16 **Hrandzichy** *Rus.* Grandichi. Hrodzyenskaya Voblasts', W Belarus 53°43′N 23°49′E
111 H18 **Hranice** *Ger.* Mährisch-Weisskirchen. Olomoucký Kraj, E Czech Republic 49°34′N 17°45′E
112 I13 **Hrasnica** Federacija Bosna I Hercegovina, SE Bosnia and Herzegovina 43°48′N 18°19′E
109 V11 **Hrastnik** C Slovenia 46°09′N 15°08′E
137 U12 **Hrazdan** *Rus.* Razdan. C Armenia 40°30′N 44°50′E
137 T12 **Hrazdan** *var.* Zanga, *Rus.* Razdan. ≈ C Armenia
117 R5 **Hrebinka** *Rus.* Grebenka. Poltavs'ka Oblast', NE Ukraine 50°08′N 32°27′E
119 K17 **Hresk** *Rus.* Gresk. Minskaya Voblasts', C Belarus 53°10′N 27°29′E
Hrisoupoli *see* Chrysoúpoli
119 F16 **Hrodna** *Rus.* Grodno. Hrodzyenskaya Voblasts', W Belarus 53°40′N 23°50′E
119 F16 **Hrodzyenskaya Voblasts'** *Rus.* Grodnenskaya Oblast'. ◆ *province* W Belarus
111 J21 **Hron** *Ger.* Gran, *Hung.* Garam. ≈ C Slovakia
111 Q14 **Hrubieszów** *Rus.* Grubeshov. Lubelskie, E Poland 50°49′N 23°53′E
112 F13 **Hrvace** Split-Dalmacija, SE Croatia 43°46′N 16°35′E
Hrvatska *see* Croatia
112 F10 **Hrvatska Kostanjnica** *var.* Kostajnica. Sisak-Moslavina, C Croatia 45°14′N 16°35′E
Hrvatsko Grahovi *see* Bosansko Grahovo
116 K6 **Hrymayliv** *Pol.* Gízymałów, *Rus.* Grimaylov. Ternopil's'ka Oblast', W Ukraine 49°18′N 26°02′E
167 N4 **Hseni** *var.* Hsenwi. Shan State, E Myanmar (Burma) 23°20′N 97°59′E
Hsenwi *see* Hseni
Hsia-men *see* Xiamen
Hsiang-t'an *see* Xiangtan
Hsi Chiang *see* Xi Jiang
167 N6 **Hsihseng** Shan State, C Myanmar (Burma) 20°07′N 97°17′E
Hsinchu *see* Xinzhu
Hsing-K'ai Hu *see* Khanka, Lake
Hsi-ning/Hsining *see* Xining
Hsinking *see* Changchun
Hsin-yang *see* Xinyang
Hsinying *see* Xinying
167 N4 **Hsipaw** Shan State, C Myanmar (Burma) 22°32′N 97°12′E
Hsu-chou *see* Xuzhou
Hsüeh Shan *see* Xue Shan
Htawei *see* Dawei
Hu *see* Shanghai Shi
83 B18 **Huab** ≈ W Namibia
57 M21 **Huacaya** Chuquisaca, S Bolivia 20°45′S 63°42′W
57 J19 **Huachacalla** Oruro, SW Bolivia 18°47′S 68°23′W
159 X9 **Huachi** *var.* Rouyuan, Rouyuanchengzi. Gansu, C China 36°24′N 107°58′E
57 N16 **Huachi, Laguna** ◎ N Bolivia
57 D14 **Huacho** Lima, W Peru 11°05′S 77°36′W
163 Y7 **Huachuan** Heilongjiang, NE China 46°59′N 130°21′E
163 P12 **Huade** Nei Mongol Zizhiqu, N China 41°52′N 113°58′E
163 W10 **Huadian** Jilin, NE China 42°59′N 126°38′E
56 E13 **Huagaruncho, Cordillera** ▲ C Peru
Hua Hin *see* Ban Hua Hin
191 S10 **Huahine** *island* Îles Sous le Vent, W French Polynesia
Huahua, Rio *see* Wawa, Río
167 R8 **Huai** ≈ E Thailand
161 Q7 **Huai'an** *var.* Qingjiang. Jiangsu, E China 33°33′N 119°03′E
161 P6 **Huaibei** Anhui, E China 34°00′N 116°48′E
Huaide *see* Gongzhuling
157 T10 **Huai He** ≈ C China
160 L11 **Huaihua** Hunan, S China 27°36′N 109°57′E
161 N14 **Huaiji** Guangdong, S China 23°54′N 112°12′E
161 O2 **Huailai** *var.* Shacheng. Hebei, E China 40°22′N 115°34′E
161 P7 **Huainan** *var.* Huai-nan, Hwainan. Anhui, E China 32°37′N 116°57′E
Huai-nan *see* Huainan
161 N2 **Huairen** *var.* Yunzhong. Shanxi, C China 35°28′N 110°29′E
161 O7 **Huaiyang** Henan, C China 33°44′N 114°55′E
161 Q7 **Huaiyin** Jiangsu, E China 33°31′N 119°03′E
167 N16 **Huai Yot** Trang, SW Thailand 07°45′N 99°36′E
41 Q15 **Huajuapan** *var.* Huajuapan de León. Oaxaca, SE Mexico 17°50′N 97°48′W
Huajuapan de León *see* Huajuapan
41 O9 **Hualahuises** Nuevo León, NE Mexico 24°56′N 99°42′W
36 I11 **Hualapai Mountains** ▲ Arizona, SW USA
36 I11 **Hualapai Peak** ▲ Arizona, SW USA 35°04′N 113°54′W
62 J7 **Hualfín** Catamarca, N Argentina 27°15′S 66°53′W
161 T13 **Hualian** *var.* Hualien, Hwalien, *Jap.* Karen. C Taiwan 23°58′N 121°35′E
Hualien *see* Hualian
56 E10 **Huallaga, Río** ≈ N Peru
56 C11 **Huamachuco** La Libertad, C Peru 07°50′S 78°01′W
41 Q14 **Huamantla** Tlaxcala, S Mexico 19°18′N 97°57′W
82 C13 **Huambo** *Port.* Nova Lisboa. Huambo, C Angola 12°48′S 15°45′E
82 B13 **Huambo** ◆ *province* C Angola
41 P15 **Huamuxtitlán** Guerrero, S Mexico 17°49′N 98°34′W
163 Y8 **Huanan** Heilongjiang, NE China 46°21′N 130°43′E
63 H17 **Huancache, Sierra** ▲ SW Argentina
57 I17 **Huancané** Puno, SE Peru 15°10′S 69°44′W
57 F16 **Huancapi** Ayacucho, C Peru 13°40′S 74°05′W
57 E15 **Huancavelica** Huancavelica, SW Peru 12°45′S 75°03′W
57 E15 **Huancavelica** *off.* Departamento de Huancavelica. ◆ *department* W Peru
Huancavelica, Departamento de *see* Huancavelica
57 E14 **Huancayo** Junín, C Peru 12°03′S 75°14′W
57 K20 **Huanchaca, Cerro** ▲ S Bolivia 20°12′S 66°35′W
Huancheng *see* Huanxian
56 C12 **Huandoy, Nevado** ▲ W Peru 08°48′S 77°33′W
Huang'an *see* Hong'an
161 O8 **Huangchuan** Henan, C China 32°00′N 115°02′E
161 O9 **Huanggang** Hubei, C China 30°27′N 114°48′E
Huang Hai *see* Yellow Sea
157 Q8 **Huang He** *var.* Yellow River. ≈ C China
Huanghe *see* Madoi
161 Q4 **Huanghe Kou** *delta* E China
Huangheyan *see* Madoi
160 L5 **Huangling** Shaanxi, C China 35°40′N 109°14′E
161 O9 **Huangpi** Hubei, C China 30°53′N 114°22′E
163 P13 **Huangqi Hai** ◎ N China
161 Q9 **Huangshan** *var.* Tunxi. Anhui, E China 29°43′N 118°20′E
161 O9 **Huangshi** *var.* Huang-shih, Hwangshih. Hubei, C China 30°14′N 115°E
Huang-shih *see* Huangshi
160 L5 **Huangtu Gaoyuan** *plateau* C China
61 B22 **Huanguelén** Buenos Aires, E Argentina 37°02′S 61°57′W
161 S10 **Huangyan** Zhejiang, SE China 28°39′N 121°19′E
159 T10 **Huangyuan** Qinghai, C China 36°40′N 101°12′E
159 T10 **Huangzhong** *var.* Lushar. Qinghai, C China 36°31′N 101°32′E
163 W12 **Huanren** *var.* Huanren Manzu Zizhixian. Liaoning, NE China 41°16′N 125°25′E
Huanren Manzu Zizhixian *see* Huanren
57 F15 **Huanta** Ayacucho, C Peru 12°54′S 74°13′W
56 E13 **Huánuco** Huánuco, C Peru 09°58′S 76°16′W
56 D13 **Huánuco** *off.* Departamento de Huánuco. ◆ *department* C Peru
Huánuco, Departamento de *see* Huánuco
57 K19 **Huanuni** Oruro, W Bolivia 18°15′S 66°48′W
159 X9 **Huanxian** *var.* Huancheng. Gansu, C China 36°30′N 107°20′E
161 S12 **Huaping Yu** *prev.* Huap'ing Yu. *island* N Taiwan
62 H3 **Huara** Tarapacá, N Chile 19°59′S 69°42′W
57 D14 **Huaral** Lima, W Peru 11°31′S 77°10′W
Huarás *see* Huaraz
56 D13 **Huaraz** *var.* Huarás. Ancash, W Peru 09°31′S 77°32′W
57 I16 **Huari Huari, Río** ≈ S Peru
56 C13 **Huarmey** Ancash, W Peru 10°03′S 78°08′W
40 H4 **Huásabas** Sonora, NW Mexico 29°47′N 109°18′W
56 D8 **Huasaga, Río** ≈ Ecuador/Peru
167 O15 **Hua Sai** Nakhon Si Thammarat, SW Thailand 08°02′N 100°18′E
56 D12 **Huascarán, Nevado** ▲ W Peru 09°01′S 77°27′W
62 G8 **Huasco** Atacama, N Chile 28°30′S 71°15′W
62 G8 **Huasco, Río** ≈ C Chile
159 S11 **Huashixia** Qinghai, W China
40 G7 **Huatabampo** Sonora, NW Mexico 26°49′N 109°40′W
159 W10 **Huating** *var.* Donghua. Gansu, C China 35°13′N 106°39′E
167 S7 **Huatt, Phou** ▲ N Vietnam 19°45′N 104°48′E
41 Q14 **Huatusco** *var.* Huatusco de Chicuellar. Veracruz-Llave, C Mexico 19°13′N 96°57′W
Huatusco de Chicuellar *see* Huatusco
41 P13 **Huauchinango** Puebla, S Mexico 20°11′N 98°04′W
Huaunta *see* Wounta
41 R15 **Huautla** *var.* Huautla de Jiménez. Oaxaca, SE Mexico 18°10′N 96°51′W
Huautla de Jiménez *see* Huautla
161 O5 **Huaxian** *var.* Daokou, Hua Xian. Henan, C China 35°33′N 114°30′E
Hua Xian *see* Huaxian
Huazangsi *see* Tianzhu
29 V13 **Hubbard** Iowa, C USA 42°18′N 93°18′W
25 U8 **Hubbard** Texas, SW USA 31°52′N 96°43′W
25 Q6 **Hubbard Creek Lake** ◎ Texas, SW USA
31 R6 **Hubbard Lake** ◎ Michigan, N USA
160 M9 **Hubei** *var.* E, Hubei Sheng, Hupeh, Hupei. ◆ *province* C China
Hubei Sheng *see* Hubei
109 P8 **Huben** Tirol, W Austria 46°55′N 12°35′E
31 R13 **Huber Heights** Ohio, N USA 39°50′N 84°07′W
155 F17 **Hubli** Karnātaka, SW India 15°20′N 75°14′E
163 X12 **Huch'ang** N North Korea 41°25′N 127°04′E
97 M18 **Hucknall** C England, United Kingdom 53°02′N 01°11′W
97 L17 **Huddersfield** N England, United Kingdom 53°39′N 01°47′W
95 O16 **Huddinge** Stockholm, C Sweden 59°15′N 17°57′E
94 N11 **Hudiksvall** Gävleborg, C Sweden 61°45′N 17°12′E
29 W13 **Hudson** Iowa, C USA 42°24′N 92°27′W
19 O11 **Hudson** Massachusetts, NE USA 42°24′N 71°34′W
31 Q11 **Hudson** Michigan, N USA 41°51′N 84°21′W
30 H6 **Hudson** Wisconsin, N USA 44°59′N 92°45′W
11 V14 **Hudson Bay** Saskatchewan, S Canada 52°51′N 102°23′W
12 G6 **Hudson Bay** *bay* NE Canada
195 T16 **Hudson, Cape** *headland* Antarctica 68°15′S 154°00′E
Hudson, Détroit d' *see* Hudson Strait
27 Q9 **Hudson, Lake** ⊠ Oklahoma, C USA
18 K9 **Hudson River** ≈ New Jersey/New York, NE USA
10 M12 **Hudson's Hope** British Columbia, W Canada 56°03′N 121°59′W
12 L2 **Hudson Strait** *Fr.* Détroit d'Hudson. *strait* Northwest Territories/Québec, NE Canada
Ḩudūd ash Shamālīyah, Minṭaqat al *see* Al Ḩudūd ash Shamālīyah
Hudur *see* Xuddur
167 U9 **Huê** Th,a Thiên-Huê, C Vietnam 16°28′N 107°35′E
104 J7 **Huebra** ≈ W Spain
24 H8 **Hueco Mountains** ▲ Texas, SW USA
116 G10 **Huedin** *Hung.* Bánffyhunyad. Cluj, NW Romania 46°52′N 23°02′E
40 J10 **Huehuento, Cerro** ▲ C Mexico 24°04′N 105°42′W
42 B5 **Huehuetenango** Huehuetenango, W Guatemala 15°19′N 91°26′W
42 B4 **Huehuetenango** *off.* Departamento de Huehuetenango. ◆ *department* W Guatemala
Huehuetenango, Departamento de *see* Huehuetenango
40 L11 **Huejuquilla** Jalisco, SW Mexico 22°40′N 103°52′W
41 P12 **Huejutla** *var.* Huejutla de Reyes. Hidalgo, C Mexico 21°10′N 98°25′W
Huejutla de Reyes *see* Huejutla
102 G6 **Huelgoat** Finistère, NW France 48°22′N 03°45′W
105 O13 **Huelma** Andalucía, S Spain 37°39′N 03°28′W
104 I14 **Huelva** *anc.* Onuba. Andalucía, SW Spain 37°15′N 06°56′W
104 I13 **Huelva** ◆ *province* Andalucía, SW Spain
104 J13 **Huelva** ≈ SW Spain
105 Q14 **Huercal-Overa** Andalucía, S Spain 37°23′N 01°56′W
37 Q9 **Huerfano Mountain** ▲ New Mexico, SW USA 36°25′N 107°50′W
37 T7 **Huerfano River** ≈ Colorado, C USA
Huertas, Cabo *see* Horta, Cap de l'
105 R6 **Huerva** ≈ N Spain
105 S4 **Huesca** *anc.* Osca. Aragón, NE Spain 42°08′N 00°25′W
105 T4 **Huesca** ◆ *province* Aragón, NE Spain
105 P13 **Huéscar** Andalucía, S Spain 37°39′N 02°32′W
41 N15 **Huetamo** *var.* Huetamo de Núñez. Michoacán, SW Mexico 18°36′N 100°54′W
Huetamo de Núñez *see* Huetamo
105 P8 **Huete** Castilla-La Mancha, C Spain 40°09′N 02°42′W
23 P4 **Hueytown** Alabama, S USA 33°27′N 87°00′W
28 L16 **Hugh Butler Lake** ⊠ Nebraska, C USA
181 V6 **Hughenden** Queensland, NE Australia 20°57′S 144°16′E
182 A6 **Hughes** South Australia 30°41′S 129°31′E
39 P8 **Hughes** Alaska, USA 66°03′N 154°15′W
27 X11 **Hughes** Arkansas, C USA 34°57′N 90°28′W
25 W6 **Hughes Springs** Texas, SW USA 33°00′N 94°37′W
37 V5 **Hugo** Colorado, C USA 39°08′N 103°28′W
27 Q13 **Hugo** Oklahoma, C USA 34°01′N 95°31′W
27 Q13 **Hugo Lake** ⊠ Oklahoma, C USA
26 H7 **Hugoton** Kansas, C USA 37°11′N 101°22′W
Huhehot/Huhohaote *see* Hohhot
Huhttán *see* Kvikkjokk
161 R13 **Hui'an** *var.* Luocheng. Fujian, SE China 25°06′N 118°45′E
184 O9 **Huiarau Range** ▲ North Island, New Zealand
83 D22 **Huib-Hoch Plateau** *plateau* S Namibia
41 O13 **Huichapán** Hidalgo, C Mexico 20°24′N 99°40′W
Huicheng *see* Shexian
163 W13 **Hŭich'ŏn** C North Korea 40°09′N 126°17′E
83 B15 **Huíla** ◆ *province* SW Angola
54 E12 **Huila** *off.* Departamento del Huila. ◆ *province* S Colombia
Huila, Departamento del *see* Huila
54 D11 **Huila, Nevado del** *elevation* C Colombia
83 B15 **Huíla Plateau** *plateau* S Angola
160 G12 **Huili** Sichuan, C China 26°39′N 102°13′E
161 P4 **Huimin** Shandong, E China 37°29′N 117°30′E
163 W11 **Huinan** *var.* Chaoyang. Jilin, NE China 42°40′N 126°03′E
62 K12 **Huinca Renancó** Córdoba, C Argentina 34°51′S 64°22′W
159 V10 **Huining** *var.* Huishi. Gansu, C China 35°42′N 105°01′E
159 W8 **Huinong** *var.* Dawukou. Ningxia, N China 39°04′N 106°22′E
Huishi *see* Huining
160 J12 **Huishui** *var.* Heping. Guizhou, S China 26°07′N 106°39′E
102 L6 **Huisne** ≈ NW France
98 L12 **Huissen** Gelderland, SE Netherlands 51°57′N 05°57′E
159 N11 **Huiten Nur** ◎ C China
93 K19 **Huittinen** Satakunta, SW Finland 61°11′N 22°40′E
41 O15 **Huitzuco** *var.* Huitzuco de los Figueroa. Guerrero, S Mexico 18°18′N 99°22′W
Huitzuco de los Figueroa *see* Huitzuco
159 W11 **Huixian** *var.* Hui Xian. Gansu, C China 33°48′N 106°02′E

◆ Country ● Country Capital ◇ Dependent Territory ○ Dependent Territory Capital ◆ Administrative Regions ✈ International Airport ▲ Mountain ▲ Mountain Range ☒ Volcano ≈ River ◎ Lake ⊠ Reservoir

HUI XIAN – ILIO POINT

Hui Xian see Huixian
41 V17 **Huixtla** Chiapas, SE Mexico 15°09′N 92°30′W
160 H12 **Huize** var. Zhongping. Yunnan, SW China 26°28′N 103°18′E
98 J10 **Huizen** Noord-Holland, C Netherlands 52°17′N 05°15′E
161 O14 **Huizhou** Guangdong, S China 23°02′N 114°28′E
162 J6 **Hujirt** Arhangay, C Mongolia 48°49′N 101°20′E
Hujirt see Tsetserleg, Övörhangay, Mongolia
Hujirt see Delgerhaan, Töv, Mongolia
Hukagawa see Fukagawa
Hŭksan-gundo see Heuksan-jedo
Hukue see Gotō
Hukui see Fukui
83 G20 **Hukuntsi** Kgalagadi, SW Botswana 23°59′S 21°44′E
Hukuoka see Fukuoka
Hukusima see Fukushima
Hukutiyama see Fukuchiyama
Hukuyama see Fukuyama
163 W8 **Hulan** Heilongjiang, NE China 45°59′N 126°37′E
163 W8 **Hulan He** ≈ NE China
31 Q4 **Hulbert Lake** ◎ Michigan, N USA
Hulczyn see Hlučín
Huliao see Dabu
163 Z8 **Hulin** Heilongjiang, NE China 45°48′N 133°06′E
Hulingol see Holin Gol
14 L12 **Hull** Québec, SE Canada 45°26′N 75°45′W
29 S12 **Hull** Iowa, C USA 43°11′N 96°07′W
Hull see Kingston upon Hull
Hull Island see Orona
99 F16 **Hulst** Zeeland, SW Netherlands 51°17′N 04°03′E
Hulstay see Choybalsan
Hultschin see Hlučín
95 M19 **Hultsfred** Kalmar, S Sweden 57°30′N 15°50′E
163 T13 **Huludao** prev. Jinxi, Lianshan. Liaoning, NE China 40°46′N 120°47′E
Hulun see Hulun Buir
163 S6 **Hulun Buir** var. Hailar; prev. Hulun. Nei Mongol Zizhiqu, N China 49°15′N 119°41′E
Hu-lun Ch'ih see Hulun Nur
163 Q6 **Hulun Nur** var. Hu-lun Ch'ih; prev. Dalai Nor. ◎ NE China
Hulwan/Hulwān see Hilwān
117 V8 **Hulyaypole** Rus. Gulyaypole. Zaporiz'ka Oblast', SE Ukraine 47°41′N 36°10′E
163 V4 **Huma** Heilongjiang, NE China 51°40′N 126°38′E
45 V6 **Humacao** E Puerto Rico 18°09′N 65°50′W
163 U4 **Huma He** ≈ NE China
62 J5 **Humahuaca** Jujuy, N Argentina 23°13′S 65°20′W
59 E14 **Humaitá** Amazonas, N Brazil 07°33′S 63°01′W
62 N7 **Humaitá** Ñeembucú, S Paraguay 27°02′S 58°31′W
83 H26 **Humansdorp** Eastern Cape, S South Africa 34°01′S 24°45′E
27 S6 **Humansville** Missouri, C USA 37°47′N 93°34′W
40 I8 **Humaya, Río** ≈ C Mexico
83 C16 **Humbe** Cunene, SW Angola 16°37′S 14°52′E
97 N17 **Humber** estuary E England, United Kingdom
97 N17 **Humberside** cultural region E England, United Kingdom
Humberto see Umberto
25 W11 **Humble** Texas, SW USA 29°58′N 95°15′W
11 U15 **Humboldt** Saskatchewan, S Canada 52°13′N 105°09′W
29 U12 **Humboldt** Iowa, C USA 42°42′N 94°13′W
27 Q6 **Humboldt** Kansas, C USA 37°48′N 95°26′W
29 S17 **Humboldt** Nebraska, C USA 40°09′N 95°56′W
35 S3 **Humboldt** Nevada, W USA 40°36′N 118°15′W
20 G9 **Humboldt** Tennessee, S USA 35°49′N 88°55′W
34 K3 **Humboldt Bay** bay California, W USA
35 S4 **Humboldt Lake** ◎ Nevada, W USA
35 S4 **Humboldt River** ≈ Nevada, W USA
35 T5 **Humboldt Salt Marsh** wetland Nevada, W USA
183 P11 **Hume, Lake** ◎ New South Wales/Victoria, SE Australia
111 N19 **Humenné** Ger. Homenau, Hung. Homonna. Prešovský Kraj, E Slovakia 48°57′N 21°54′E
29 V15 **Humeston** Iowa, C USA 40°51′N 93°30′W
54 J5 **Humocaro Bajo** Lara, N Venezuela 09°41′N 70°00′W
29 Q14 **Humphrey** Nebraska, C USA 41°38′N 97°29′W
35 S9 **Humphreys, Mount** ▲ California, W USA 37°11′N 118°39′W
36 L11 **Humphreys Peak** ▲ Arizona, SW USA 35°18′N 111°40′W
111 E17 **Humpolec** Ger. Gumpolds, Humpoletz. Vysočina, C Czech Republic 49°33′N 15°23′E
Humpoletz see Humpolec
93 K19 **Humppila** Kanta-Häme, S Finland 60°54′N 23°21′E
32 F8 **Humptulips** Washington, NW USA 47°13′N 123°57′W
42 H7 **Humuya, Río** ≈ W Honduras
75 P9 **Hūn** N Libya 29°06′N 15°56′E
92 I1 **Húnaflói** bay NW Iceland
160 M11 **Hunan** var. Hunan Sheng, Xiang. ◈ province S China
Hunan Sheng see Hunan
163 Y10 **Hunchun** Jilin, NE China 42°51′N 130°21′E
95 I22 **Hundested** Hovedstaden, E Denmark 55°58′N 11°53′E
Hundred Mile House see 100 Mile House
116 G12 **Hunedoara** Ger. Eisenmarkt, Hung. Vajdahunyad. Hunedoara, SW Romania 45°45′N 22°54′E
116 G12 **Hunedoara** ◈ county W Romania
101 I17 **Hünfeld** Hessen, C Germany 50°41′N 09°46′E
Hungarian People's Republic see Hungary
111 H23 **Hungary** off. Republic of Hungary, Ger. Ungarn, Hung. Magyarország, Rom. Ungaria, SCr. Mađarska, Ukr. Uhorshchyna; prev. Hungarian People's Republic. ◆ republic C Europe
Hungary, Plain of see Great Hungarian Plain
Hungary, Republic of see Hungary
Hungiy see Urgamal
163 X13 **Hŭngnam** E North Korea 39°50′N 127°36′E
33 P8 **Hungry Horse Reservoir** ▣ Montana, NW USA
Hungt'ou see Lan Yu
Hung-tse Hu see Hongze Hu
167 T6 **Hưng Yên** Hai Hung, N Vietnam 20°38′N 106°05′E
Hunjiang see Baishan
95 I18 **Hunnebostrand** Västra Götaland, S Sweden 58°26′N 11°19′E
101 E19 **Hunsrück** ▲▲ W Germany
97 P18 **Hunstanton** E England, United Kingdom 52°57′N 00°27′E
155 G20 **Hunsūr** Karnātaka, E India 12°18′N 76°15′E
Hunt see Hangay
100 G12 **Hunte** ≈ NW Germany
29 Q5 **Hunter** North Dakota, N USA 47°10′N 97°11′W
25 S11 **Hunter** Texas, SW USA 29°47′N 98°01′W
185 D20 **Hunter** ≈ South Island, New Zealand
183 N15 **Hunter Island** island Tasmania, SE Australia
18 K11 **Hunter Mountain** ▲ New York, NE USA 42°10′N 74°13′W
185 B23 **Hunter Mountains** ▲▲ South Island, New Zealand
183 S7 **Hunter River** ≈ New South Wales, SE Australia
32 L7 **Hunters** Washington, NW USA 48°07′N 118°13′W
185 F20 **Hunters Hills, The** hill range South Island, New Zealand
184 M12 **Hunterville** Manawatu-Wanganui, North Island, New Zealand 39°55′S 175°34′E
31 N16 **Huntingburg** Indiana, N USA 38°18′N 86°57′W
97 O20 **Huntingdon** E England, United Kingdom 52°20′N 00°12′W
18 E15 **Huntingdon** Pennsylvania, NE USA 40°28′N 78°00′W
20 G9 **Huntingdon** Tennessee, S USA 36°00′N 88°25′W
97 O20 **Huntingdonshire** cultural region C England, United Kingdom
31 P12 **Huntington** Indiana, N USA 40°52′N 85°30′W
32 L13 **Huntington** Oregon, NW USA 44°22′N 117°18′W
25 X9 **Huntington** Texas, SW USA 31°16′N 94°34′W
36 M5 **Huntington** Utah, W USA 39°19′N 110°57′W
21 P5 **Huntington** West Virginia, NE USA 38°25′N 82°27′W
35 T16 **Huntington Beach** California, W USA 33°39′N 118°00′W
35 W4 **Huntington Creek** ≈ Nevada, W USA
184 L7 **Huntly** Waikato, North Island, New Zealand 37°34′S 175°09′E
96 K8 **Huntly** NE Scotland, United Kingdom 57°25′N 02°48′W
10 K8 **Hunt, Mount** ▲ Yukon, NW Canada 61°29′N 129°10′W
14 H12 **Huntsville** Ontario, S Canada 45°20′N 79°14′W
23 P2 **Huntsville** Alabama, S USA 34°44′N 86°35′W
27 S9 **Huntsville** Arkansas, C USA 36°04′N 93°46′W
27 U3 **Huntsville** Missouri, C USA 39°27′N 92°31′W
20 M8 **Huntsville** Tennessee, S USA 36°25′N 84°30′W
25 V10 **Huntsville** Texas, SW USA 30°43′N 95°34′W
36 L2 **Huntsville** Utah, W USA 41°16′N 111°47′W
41 W12 **Hunucmá** Yucatán, SE Mexico 20°59′N 89°55′W
149 W3 **Hunza** ≈ NE Pakistan
Hunza see Karīmābād
Hunze see Oostermoers Vaart
158 H4 **Huocheng** var. Shuiding. Xinjiang Uygur Zizhiqu, NW China 44°03′N 80°49′E
161 N6 **Huojia** Henan, C China 35°14′N 113°38′E
186 N14 **Huon** reef N New Caledonia
186 E7 **Huon Peninsula** headland C Papua New Guinea 06°24′S 147°50′E
Huoshao Dao see Lü Dao
Huoshao Tao see Lan Yu
Hupeh/Hupei see Hubei
Hurama see Hongyuan
95 H14 **Hurdalsjøen** prev. Hurdalssjøen. ◎ S Norway
Hurdalssjøen see Hurdalsjøen
14 E13 **Hurd, Cape** headland Ontario, S Canada 45°12′N 81°43′W
98 L5 **Hurdegaryp** Dutch. Hardegarijp. Fryslân, N Netherlands 53°13′N 05°57′E
29 N4 **Hurdsfield** North Dakota, N USA 47°24′N 99°55′W
Hüremt see Sayhan, Bulgan, Mongolia
Hüremt see Taragt, Övörhangay, Mongolia
Hurghada see Al Ghurdaqah
67 V9 **Huri Hills** ▲ NW Kenya
Hürkay see Hurkë
139 S1 **Hurkë** Ar. Hürkay, var. Harki. Dahūk, N Iraq 37°03′N 43°39′E
37 P15 **Hurley** New Mexico, SW USA 32°42′N 108°07′W
30 K4 **Hurley** Wisconsin, N USA 46°25′N 90°15′W
21 Y4 **Hurlock** Maryland, NE USA 38°37′N 75°51′W
162 K11 **Hürmen** var. Tsoohor. Ömnögovĭ, S Mongolia 43°15′N 104°04′E
29 P10 **Huron** South Dakota, N USA 44°19′N 98°13′W
31 S6 **Huron, Lake** ◎ Canada/USA
31 N3 **Huron Mountains** hill range Michigan, N USA
36 J8 **Hurricane** Utah, W USA 37°10′N 113°18′W
21 P5 **Hurricane** West Virginia, NE USA 38°25′N 82°01′W
36 J8 **Hurricane Cliffs** cliff Arizona, SW USA
23 V6 **Hurricane Creek** ≈ Georgia, SE USA
94 E12 **Hurrungane** ▲ S Norway 61°25′N 07°48′E
101 E16 **Hürth** Nordrhein-Westfalen, W Germany 50°52′N 06°49′E
Hurukawa see Furukawa
185 I17 **Hurunui** ≈ South Island, New Zealand
95 F21 **Hurup** Midtjylland, NW Denmark 56°46′N 08°26′E
117 T14 **Hurzuf** Avtonomna Respublika Krym, S Ukraine 44°33′N 34°18′E
Huş see Huşi
95 B19 **Húsavík** Dan. Husevig. Sandoy, C Faroe Islands 61°24′N 06°38′W
92 K1 **Húsavík** Norðurland Eystra, NE Iceland 66°03′N 17°20′W
Husevig see Húsavík
116 M10 **Huşi** var. Huş. Vaslui, E Romania 46°40′N 28°05′E
95 L19 **Huskvarna** Jönköping, S Sweden 57°47′N 14°15′E
39 P8 **Huslia** Alaska, USA 65°42′N 156°24′W
Husn see Al Ḩuşn
95 C15 **Husnes** Hordaland, S Norway 59°52′N 05°46′E
94 D8 **Hustadvika** sea area S Norway
Husté see Khust
100 H7 **Husum** Schleswig-Holstein, N Germany 54°29′N 09°04′E
93 I16 **Husum** Västernorrland, C Sweden 63°21′N 19°12′E
116 K6 **Husyatyn** Ternopil's'ka Oblast', W Ukraine 49°04′N 26°10′E
Huszt see Khust
Hutag see Hutag-Öndör
162 K6 **Hutag-Öndör** var. Hutag. Bulgan, N Mongolia 49°22′N 102°50′E
26 M6 **Hutchinson** Kansas, C USA 38°03′N 97°56′W
29 U9 **Hutchinson** Minnesota, N USA 44°53′N 94°22′W
23 Y13 **Hutchinson Island** island Florida, SE USA
36 L11 **Hutch Mountain** ▲ Arizona, SW USA 34°49′N 111°22′W
141 O14 **Ḩūth** NW Yemen 16°14′N 44°E
186 I7 **Hutjena** Buka Island, NE Papua New Guinea 05°19′S 154°40′E
109 T8 **Hüttenberg** Kärnten, S Austria 46°58′N 14°33′E
25 T10 **Hutto** Texas, SW USA 30°32′N 97°33′W
Huttu see Futtsu
108 E8 **Huttwil** Bern, W Switzerland 47°06′N 07°48′E
158 K5 **Hutubi** Xinjiang Uygur Zizhiqu, NW China 44°10′N 86°51′E
161 N4 **Hutuo He** ≈ C China
Hutyû see Fuchū
185 E20 **Huxley, Mount** ▲ South Island, New Zealand 44°02′S 169°42′E
99 J20 **Huy** Dut. Hoei, Hoey. Liège, E Belgium 50°32′N 05°14′E
161 R8 **Huzhou** var. Wuxing. Zhejiang, SE China 30°52′N 120°06′E
Huzi see Fuji
Huzieda see Fujieda
Huzinomiya see Fujinomiya
Huzisawa see Fujisawa
92 I2 **Hvammstangi** Norðurland Vestra, N Iceland 65°22′N 20°54′W
92 K4 **Hvannadalshnúkur** var. Hvannadalshnukur. ⛰ S Iceland 64°01′N 16°39′W
Hvannadalshnukur see Hvannadalshnúkur
113 E15 **Hvar** It. Lesina. Split-Dalmacija, S Croatia 43°10′N 16°27′E
113 F15 **Hvar** It. Lesina; anc. Pharus. island S Croatia
117 T13 **Hvardiys'ke** Rus. Gvardeyskoye. Avtonomna Respublika Krym, S Ukraine 45°08′N 34°01′E
92 I4 **Hveragerði** Suðurland, SW Iceland 64°00′N 21°13′W
95 E22 **Hvide Sande** Midtjylland, W Denmark 56°00′N 08°08′E
92 I3 **Hvítá** ≈ C Iceland
95 G15 **Hvittingfoss** Buskerud, S Norway 59°28′N 10°00′E
92 I4 **Hvolsvöllur** Suðurland, SW Iceland 63°44′N 20°12′W
Hwach'ŏn-chŏsuji see Paro-ho
Hwainan see Huainan
Hwalien see Hualian
83 I16 **Hwange** prev. Wankie. Matabeleland North, W Zimbabwe 18°18′S 26°31′E
Hwang-Hae see Yellow Sea
Hwangshih see Huangshi
83 L17 **Hwedza** Mashonaland East, E Zimbabwe 18°35′S 31°35′E
63 G20 **Hyades, Cerro** ▲ S Chile 46°57′S 73°09′W
162 K6 **Hyalganat** var. Selenge. Bulgan, N Mongolia 49°34′N 104°18′E
19 Q12 **Hyannis** Massachusetts, NE USA 41°38′N 70°15′W
28 L13 **Hyannis** Nebraska, C USA 42°00′N 101°45′W
162 F6 **Hyargas Nuur** ◎ NW Mongolia
Hybla/Hybla Major see Paternò
39 Y14 **Hydaburg** Prince of Wales Island, Alaska, USA 55°10′N 132°44′W
185 F22 **Hyde** Otago, South Island, New Zealand 45°17′S 170°17′E
21 O7 **Hyden** Kentucky, S USA 37°08′N 83°23′W
18 K12 **Hyde Park** New York, NE USA 41°46′N 73°52′W
39 Z14 **Hyder** Alaska, USA 55°55′N 130°01′W
155 I15 **Hyderābād** var. Haidarabad. state capital Telangana/Andhra Pradesh, C India 17°22′N 78°26′E
149 Q16 **Hyderābād** var. Haidarabad. Sind, SE Pakistan 25°26′N 68°22′E
103 T16 **Hyères** Var, SE France 43°07′N 06°08′E
103 T16 **Hyères, Îles d'** island group S France
118 K12 **Hyermanavichy** Rus. Germanovichi. Vitsyebskaya Voblasts', N Belarus 55°24′N 27°48′E
163 X12 **Hyesan** NE North Korea 41°18′N 128°13′E
10 K8 **Hyland** ≈ Yukon, NW Canada
95 K20 **Hyltebruk** Halland, S Sweden 57°N 13°14′E
18 D16 **Hyndman** Pennsylvania, NE USA 39°49′N 78°42′W
33 P14 **Hyndman Peak** ▲ Idaho, NW USA 43°45′N 114°07′W
164 I13 **Hyōgo** off. Hyōgo-ken. ◈ prefecture Honshū, SW Japan
Hyōgo-ken see Hyōgo
Hypsas see Belice
Hyrcania see Gorgān
36 L1 **Hyrum** Utah, W USA 41°37′N 111°51′W
93 N14 **Hyrynsalmi** Kainuu, C Finland 64°41′N 28°30′E
33 V10 **Hysham** Montana, NW USA 46°16′N 107°14′W
11 N13 **Hythe** Alberta, W Canada 55°18′N 119°44′W
97 Q23 **Hythe** SE England, United Kingdom 51°05′N 01°04′E
Hyvinge see Hyvinkää
93 L19 **Hyvinkää** Swe. Hyvinge. Uusimaa, S Finland 60°37′N 24°50′E

I

116 J9 **Iacobeni** Ger. Jakobeny. Suceava, NE Romania 47°24′N 25°20′E
Iader see Zadar
172 I7 **Iakora** Fianarantsoa, SE Madagascar 23°04′S 46°40′E
116 K14 **Ialomiţa** var. Jalomitsa. ◈ county SE Romania
116 K14 **Ialomiţa** ≈ SE Romania
117 N10 **Ialoveni** Rus. Yaloveny. C Moldova 46°57′N 28°47′E
117 N11 **Ialpug** var. Ialpugul Mare, Rus. Yalpug. ≈ Moldova/Ukraine
Ialpugul Mare see Ialpug
23 T8 **Iamonia, Lake** ◎ Florida, SE USA
116 L13 **Ianca** Brăila, SE Romania 45°06′N 27°29′E
116 M10 **Iaşi** Ger. Jassy. Iaşi, NE Romania 47°08′N 27°38′E
116 L9 **Iaşi** Ger. Jassy, Yassy. ◈ county NE Romania
114 J13 **Íasmos** Anatolikí Makedonía kai Thráki, NE Greece 41°07′N 25°12′E
22 H6 **Iatt, Lake** ◎ Louisiana, S USA
58 B11 **Iauretê** Amazonas, NW Brazil 0°37′N 69°10′W
171 N3 **Iba** Luzon, N Philippines 15°25′N 119°55′E
77 S16 **Ibadan** Oyo, SW Nigeria 07°22′N 04°01′E
54 E10 **Ibagué** Tolima, C Colombia 04°27′N 75°14′W
60 J10 **Ibaiti** Paraná, S Brazil 23°49′S 50°15′W
36 J4 **Ibapah Peak** ▲ Utah, W USA 39°51′N 113°55′W
Ibar see Ibër
165 P13 **Ibaraki** off. Ibaraki-ken. ◈ prefecture Honshū, S Japan
Ibaraki-ken see Ibaraki
56 C5 **Ibarra** var. San Miguel de Ibarra. Imbabura, N Ecuador 0°23′S 78°08′W
Ibaşfalău see Dumbrăveni
141 O16 **Ibb** W Yemen 13°55′N 44°10′E
100 F13 **Ibbenbüren** Nordrhein-Westfalen, NW Germany 52°17′N 07°43′E
79 H16 **Ibenga** ≈ N Congo
113 M15 **Ibër** Serb. Ibar. ≈ C Serbia
57 I14 **Iberia** Madre de Dios, E Peru 11°21′S 69°36′W
Iberia see Spain
66 M1 **Iberian Basin** undersea feature E Atlantic Ocean 39°00′N 16°00′W
Iberian Mountains see Ibérico, Sistema
84 D12 **Iberian Peninsula** physical region Portugal/Spain
64 M8 **Iberian Plain** undersea feature E Atlantic Ocean 13°30′W 43°45′N
Ibérica, Cordillera see Ibérico, Sistema
105 P6 **Ibérico, Sistema** var. Cordillera Ibérica, Eng. Iberian Mountains. ▲▲ NE Spain
12 K7 **Iberville Lac d'** ◎ Québec, NE Canada
77 T14 **Ibeto** Niger, W Nigeria 10°30′N 05°07′E
77 W15 **Ibi** Taraba, C Nigeria 08°13′N 09°46′E
105 S11 **Ibi** Valenciana, E Spain 38°38′N 00°35′W
59 L20 **Ibiá** Minas Gerais, SE Brazil 19°30′S 46°31′W
61 F15 **Ibicuí, Rio** ≈ S Brazil
61 C19 **Ibicuy** Entre Ríos, E Argentina 33°44′S 59°10′W
61 F16 **Ibirapuitã** ≈ S Brazil
105 V10 **Ibiza** var. Iviza, Cast. Eivissa; anc. Ebusus. island Islas Baleares, Spain, W Mediterranean Sea
Ibiza see Eivissa
138 J4 **Ibn Wardān, Qaşr** ruins Ḩamāh, C Syria
Ibo see Sassandra
188 E9 **Ibobang** Babeldaob, N Palau
171 V13 **Ibonma** Papua Barat, E Indonesia 03°27′S 133°30′E
59 N17 **Ibotirama** Bahia, E Brazil 12°13′S 43°12′W
141 Y8 **Ibrā** NE Oman 22°45′N 58°30′E
127 Q4 **Ibresi** Chuvashskaya Respublika, W Russian Federation 55°22′N 47°04′E
141 X8 **ʿIbrī** NW Oman 23°12′N 56°28′E
164 C16 **Ibusuki** Kagoshima, Kyūshū, SW Japan 31°15′N 130°40′E
57 E16 **Ica** Ica, SW Peru 14°02′S 75°48′W
57 E16 **Ica** off. Departamento de Ica. ◈ department SW Peru
Ica, Departamento de see Ica
58 C11 **Içana** Amazonas, NW Brazil 0°22′N 67°25′W
Içaria see Ikaria
58 B13 **Içá, Rio** var. Río Putumayo. ≈ NW South America see also Putumayo, Río
Içá, Rio see Putumayo, Río
136 I17 **İçel** prev. Ichili; prev. Mersin. ◈ province S Turkey
İçel see Mersin
92 I3 **Iceland** off. Republic of Iceland, Dan. Island, Icel. Ísland. ◆ republic N Atlantic Ocean
86 B6 **Iceland** island N Atlantic Ocean
64 L5 **Iceland Basin** undersea feature N Atlantic Ocean 61°00′N 19°00′W
Icelandic Plateau see Iceland Plateau
197 Q15 **Iceland Plateau** var. Icelandic Plateau. undersea feature S Greenland Sea 12°00′W 69°30′N
Iceland, Republic of see Iceland
155 E16 **Ichalkaranji** Mahārāshtra, W India 16°42′N 74°28′E
164 D15 **Ichifusa-yama** ▲ Kyūshū, SW Japan 32°18′N 131°05′E
Ichili see İçel
164 K13 **Ichinomiya** var. Itinomiya. Aichi, Honshū, SW Japan 35°18′N 136°48′E
165 Q9 **Ichinoseki** var. Itinoseki. Iwate, Honshū, C Japan 38°56′N 141°08′E
117 R3 **Ichnya** Chernihivs'ka Oblast', NE Ukraine 50°52′N 32°24′E
57 L17 **Ichoa, Río** ≈ C Bolivia
Iconium see Konya
Iculisma see Angoulême
39 U12 **Icy Bay** inlet Alaska, USA
39 N5 **Icy Cape** headland Alaska, USA 70°19′N 161°52′W
39 W13 **Icy Strait** strait Alaska, USA
27 R13 **Idabel** Oklahoma, C USA 33°54′N 94°50′W
29 T13 **Ida Grove** Iowa, C USA 42°21′N 95°28′W
77 U16 **Idah** Kogi, S Nigeria 07°06′N 06°45′E
33 N13 **Idaho** off. State of Idaho, also known as Gem of the Mountains, Gem State. ◈ state NW USA
33 N14 **Idaho City** Idaho, NW USA 43°48′N 115°51′W
33 R14 **Idaho Falls** Idaho, NW USA 43°28′N 112°01′W
121 P2 **Idálion** var. Dali, Dhali. C Cyprus 35°00′N 33°25′E
25 N5 **Idalou** Texas, SW USA 33°40′N 101°40′W
104 I9 **Idanha-a-Nova** Castelo Branco, C Portugal 39°55′N 07°15′W
101 E19 **Idar-Oberstein** Rheinland-Pfalz, SW Germany 49°43′N 07°19′E
118 J3 **Ida-Virumaa** var. Ida-Viru Maakond. ◈ province NE Estonia
Ida-Viru Maakond see Ida-Virumaa
124 J8 **Idel'** Respublika Kareliya, NW Russian Federation 64°08′N 34°12′E
79 C15 **Idenao** Sud-Ouest, SW Cameroon 04°04′N 09°01′E
Idensalmi see Iisalmi
162 I6 **Ider** var. Dzuunmod. Hövsgöl, C Mongolia 48°09′N 97°22′E
Ider see Galt
75 X10 **Idfū** var. Edfu. SE Egypt 24°55′N 32°52′E
Ídhi Óros see Ídi
Ídhra see Ýdra
80 L10 **ʿĪdī** var. Ed. SE Eritrea 13°54′N 41°39′E
168 H7 **Idi** Sumatera, W Indonesia 05°00′N 98°00′E
115 I25 **Ídi** var. Ídhi Óros. ▲ Kríti, Greece, E Mediterranean Sea
Idi Amin, Lac see Edward, Lake
106 G10 **Idice** ≈ N Italy
76 G9 **Idini** Trarza, W Mauritania 17°58′N 15°40′W
79 J21 **Idiofa** Bandundu, SW Dem. Rep. Congo 05°00′S 19°38′E
39 O10 **Iditarod River** ≈ Alaska, USA
Idjavuotna see Ifjord
95 M14 **Idkerberget** Dalarna, C Sweden 60°22′N 15°15′E
138 I3 **Idlib** Idlib, NW Syria 35°57′N 36°38′E
138 I4 **Idlib** off. Muḩāfaz̧at Idlib. ◈ governorate NW Syria
Idlib, Muḩāfaz̧at see Idlib
Idra see Ýdra
94 J11 **Idre** Dalarna, C Sweden 61°52′N 12°45′E
Idria see Idrija
109 S11 **Idrija** It. Idria. W Slovenia 46°00′N 14°59′E
101 G18 **Idstein** Hessen, W Germany 50°10′N 08°16′E
Idutywa see Dutywa
Idzhevan see Ijevan
118 G9 **Iecava** S Latvia 56°36′N 24°10′E
165 T16 **Ie-jima** var. Ii-shima. island Nansei-shotō, SW Japan
99 B18 **Ieper** Fr. Ypres. West-Vlaanderen, W Belgium 50°51′N 02°53′E
115 K25 **Ierápetra** Kríti, Greece, E Mediterranean Sea 35°00′N 25°45′E
115 G22 **Iérax, Akrotírio** headland S Greece 36°45′N 23°06′E
Ierisós see Ierissós
115 H14 **Ierissós** var. Ierisós. Kentrikí Makedonía, N Greece 40°24′N 23°53′E
116 I11 **Iernut** Hung. Radnót. Mureş, C Romania 46°27′N 24°15′E
106 J12 **Iesi** var. Jesi. Marche, C Italy 43°33′N 13°16′E
92 K9 **Iešjavri** ◎ N Norway
Iesolo see Jesolo
188 K16 **Ifalik Atoll** atoll Caroline Islands, C Micronesia
172 I6 **Ifanadiana** Fianarantsoa, SE Madagascar 21°19′S 47°39′E
77 T16 **Ife** Osun, SW Nigeria 07°25′N 04°31′E
77 V8 **Iferouâne** Agadez, N Niger 19°05′N 08°24′E
Iferten see Yverdon
92 L8 **Ifjord** Lapp. Idjavuotna. Finnmark, N Norway 70°27′N 27°06′E
77 R8 **Ifôghas, Adrar des** var. Adrar des Iforas. ▲▲ NE Mali
Iforas, Adrar des see Ifôghas, Adrar des
182 D6 **Ifould lake** salt lake South Australia
74 G6 **Ifrane** C Morocco 33°31′N 05°09′W
171 S11 **Iga** Pulau Halmahera, E Indonesia 01°23′N 128°17′E
81 G18 **Iganga** SE Uganda 0°34′N 33°27′E
60 L7 **Igarapava** São Paulo, S Brazil 20°01′S 47°46′W
122 K9 **Igarka** Krasnoyarskiy Kray, N Russian Federation 67°31′N 86°33′E
Igaunija see Estonia
137 T12 **Iğdır** ◈ province NE Turkey
I.G.Duca see General Toshevo
Igel see Jihlava
94 N11 **Iggesund** Gävleborg, C Sweden 61°38′N 17°04′E
39 P7 **Igikpak, Mount** ▲ Alaska, USA 67°28′N 154°55′W
39 P13 **Igiugig** Alaska, USA 59°19′N 155°53′W
Iglau/Iglawa/Igława see Jihlava
107 B20 **Iglesias** Sardegna, Italy, C Mediterranean Sea 39°20′N 08°34′E
127 V4 **Iglino** Respublika Bashkortostan, W Russian Federation 54°51′N 56°29′E
Igló see Spišská Nová Ves
9 O6 **Igloolik** Nunavut, N Canada 69°24′N 81°55′W
12 B11 **Ignace** Ontario, S Canada 49°26′N 91°40′W
118 I12 **Ignalina** Utena, E Lithuania 55°20′N 26°10′E
127 Q5 **Ignatovka** Ul'yanovskaya Oblast', W Russian Federation 53°56′N 47°40′E
124 K12 **Ignatovo** Vologodskaya Oblast', NW Russian Federation 60°47′N 37°51′E
114 N11 **İğneada** Kırklareli, NW Turkey 41°54′N 27°58′E
121 S7 **İğneada Burnu** headland NW Turkey 41°54′N 28°03′E
Igombe see Gombe
115 B16 **Igoumenítsa** Ípeiros, W Greece 39°30′N 20°16′E
127 T2 **Igra** Udmurtskaya Respublika, NW Russian Federation 57°30′N 53°01′E
122 H9 **Igrim** Khanty-Mansiyskiy Avtonomnyy Okrug-Yugra, N Russian Federation 63°09′N 64°33′E
60 G12 **Iguaçu, Rio** Sp. Río Iguazú. ≈ Argentina/Brazil see also Iguazú, Río
Iguaçu, Rio see Iguazú, Río
59 I22 **Iguaçu, Salto do** Sp. Cataratas del Iguazú; prev. Victoria Falls. waterfall Argentina/Brazil see also Iguazú, Cataratas del
Iguaçu, Salto do see Iguazú, Cataratas del
41 O15 **Iguala** var. Iguala de la Independencia. Guerrero, S Mexico 18°21′N 99°31′W
105 V5 **Igualada** Cataluña, NE Spain 41°35′N 01°37′E
Iguala de la Independencia see Iguala
60 G12 **Iguazú, Cataratas del** Port. Salto do Iguaçu; prev. Victoria Falls. waterfall Argentina/Brazil see also Iguaçu, Salto do
Iguazú, Cataratas del see Iguaçu, Salto do
62 Q6 **Iguazú, Río** var. Río Iguaçu. ≈ Argentina/Brazil see also Iguaçu, Rio
Iguazú, Río see Iguaçu, Rio
79 D19 **Iguéla** prev. Iguéla. Ogooué-Maritime, SW Gabon 02°00′S 09°23′E
Iguéla see Iguéla
Iguid, Erg see Iguîdi, 'Erg
67 M5 **Iguîdi, 'Erg** var. Erg Iguid. desert Algeria/Mauritania
172 K2 **Iharaña** prev. Vohémar. Antsiranana, NE Madagascar 13°22′S 50°00′E
151 K18 **Ihavandhippolhu Atoll** var. Ihavandiffulu Atoll. atoll N Maldives
Ihavandiffulu Atoll see Ihavandhippolhu Atoll
Ih Bulag see Hanbogd
165 T16 **Iheya-jima** island Nansei-shotō, SW Japan
Ihhayrhan see Bayan-Önjüül
163 N9 **Ihhet** var. Bayan. Dornogovĭ, SE Mongolia 46°15′N 110°16′E
172 I6 **Ihosy** Fianarantsoa, S Madagascar 22°23′S 46°09′E
Ihsüüj see Bayanchandmanĭ
162 I7 **Ihtamir** var. Dzaanhushuu. Arhangay, C Mongolia 47°36′N 101°06′E
114 H10 **Ihtiman** Sofia, W Bulgaria 42°26′N 23°49′E
162 H6 **Ih-Uul** var. Bayan-Uhaa. Dzavhan, C Mongolia 48°41′N 98°46′E
162 J6 **Ih-Uul** var. Selenge. Hövsgöl, N Mongolia 49°25′N 101°30′E
93 L14 **Ii** Pohjois-Pohjanmaa, C Finland 65°18′N 25°23′E
164 M13 **Iida** Nagano, Honshū, S Japan 35°32′N 137°48′E
93 M14 **Iijoki** ≈ C Finland
Iinnasuolu see Hinnøya
118 J4 **Iisaku** Ger. Isaak. Ida-Virumaa, NE Estonia 59°06′N 27°19′E
93 M16 **Iisalmi** var. Idensalmi. Pohjois-Savo, C Finland 63°32′N 27°10′E
Ii-shima see Ie-jima
165 N11 **Iiyama** Nagano, Honshū, S Japan 36°52′N 138°22′E
77 S16 **Ijebu-Ode** Ogun, SW Nigeria 06°46′N 03°57′E
137 U11 **Ijevan** Rus. Idzhevan. N Armenia 40°53′N 45°07′E
98 H9 **IJmuiden** Noord-Holland, W Netherlands 52°28′N 04°38′E
98 M12 **IJssel** var. Yssel. ≈ Netherlands
98 J8 **IJsselmeer** prev. Zuider Zee. ◎ N Netherlands
98 L9 **IJsselmuiden** Overijssel, E Netherlands 52°34′N 05°55′E
98 I12 **IJsselstein** Utrecht, C Netherlands 52°01′N 05°02′E
61 G14 **Ijuí** Rio Grande do Sul, S Brazil 28°23′S 53°55′W
61 G14 **Ijuí, Rio** ≈ S Brazil
189 R8 **Ijuw** NE Nauru 0°30′S 166°57′E
99 E16 **IJzendijke** Zeeland, SW Netherlands 51°20′N 03°36′E
99 A18 **IJzer** ≈ W Belgium
Ikaahuk see Sachs Harbour
93 K18 **Ikaalinen** Pirkanmaa, W Finland 61°46′N 23°05′E
172 I6 **Ikalamavony** Fianarantsoa, SE Madagascar 21°10′S 46°35′E
Ikaluktutiak see Cambridge Bay
185 G16 **Ikamatua** West Coast, South Island, New Zealand 42°16′S 171°42′E
145 P16 **Ikan** prev. Staroikan. Yuzhnyy Kazakhstan, S Kazakhstan 43°09′N 68°34′E
77 U16 **Ikare** Ondo, SW Nigeria 07°36′N 05°52′E
115 L20 **Ikaría** var. Kariot, Nicaria, Nikaria; anc. Icaria. island Dodekánisa, Greece, Aegean Sea
95 F22 **Ikast** Midtjylland, W Denmark 56°09′N 09°10′E
184 O9 **Ikawhenua Range** ▲▲ North Island, New Zealand
165 U4 **Ikeda** Hokkaidō, NE Japan 42°54′N 143°25′E
164 H14 **Ikeda** Tokushima, Shikoku, SW Japan 34°00′N 133°47′E
77 S16 **Ikeja** Lagos, SW Nigeria 06°36′N 03°16′E
79 L19 **Ikela** Equateur, C Dem. Rep. Congo 01°11′S 23°16′E
164 C13 **Iki** prev. Gōnoura. Nagasaki, Iki, SW Japan 33°44′N 129°41′E
164 C13 **Iki** island SW Japan
127 O13 **Iki Burul** Respublika Kalmykiya, SW Russian Federation 45°48′N 44°44′E
137 P11 **İkizdere** Rize, NE Turkey 40°47′N 40°34′E
39 P14 **Ikolik, Cape** headland Kodiak Island, Alaska, USA 57°12′N 154°46′W
77 V17 **Ikom** Cross River, SE Nigeria 05°57′N 08°43′E
172 I6 **Ikongo** prev. Fort-Carnot. Fianarantsoa, SE Madagascar 21°52′S 47°27′E
39 P5 **Ikpikpuk River** ≈ Alaska, USA
190 H1 **Iku** prev. Lone Tree Islet. atoll Tungaru, W Kiribati
164 I12 **Ikuno** Hyōgo, Honshū, SW Japan 35°13′N 134°48′E
190 H16 **Ikurangi** ▲ Rarotonga, S Cook Islands 21°12′S 159°45′W
171 X14 **Ilaga** Papua, E Indonesia 03°54′S 137°30′E
171 O2 **Ilagan** Luzon, N Philippines 17°08′N 121°54′E
142 J7 **Īlām** var. Elam. Īlām, W Iran 33°37′N 46°27′E
153 R12 **Ilām** Eastern, E Nepal 26°52′N 87°58′E
142 J8 **Īlām** off. Ostān-e Īlām. ◈ province W Iran
Īlām, Ostān-e see Īlām
161 T13 **Ilan** Jap. Giran. N Taiwan 24°45′N 121°44′E
146 G9 **Ilanly Obvodnitel'nyy Kanal** canal N Turkmenistan
122 L12 **Ilanskiy** Krasnoyarskiy Kray, S Russian Federation 56°16′N 95°59′E
108 H9 **Ilanz** Graubünden, S Switzerland 46°46′N 09°10′E
77 S16 **Ilaro** Ogun, SW Nigeria 06°52′N 03°01′E
57 I17 **Ilave** Puno, S Peru 16°07′S 69°40′W
110 K8 **Iława** Ger. Deutsch-Eylau. Warmińsko-Mazurskie, NE Poland 53°36′N 19°35′E
121 P16 **Il-Bajja ta' Marsaxlokk** var. Marsaxlokk Bay. bay SE Malta
123 P10 **Ilbenge** Respublika Sakha (Yakutiya), NE Russian Federation 62°52′N 124°13′E
Ile see Ile/Ili He
Ile see Ili He
11 S13 **Île-à-la-Crosse** Saskatchewan, C Canada 55°29′N 108°00′W
79 J21 **Ilebo** prev. Port-Francqui. Kasai-Occidental, W Dem. Rep. Congo 04°19′S 20°32′E
103 N5 **Île-de-France** ◈ region N France
Ilek see Yelek
Ilerda see Lleida
77 T16 **Ilesha** Osun, SW Nigeria 07°35′N 04°49′E
187 Q16 **Îles Loyauté, Province des** ◈ province E New Caledonia
11 X12 **Ilford** Manitoba, C Canada 56°02′N 95°48′W
116 K14 **Ilfov** ◈ county S Romania
97 I23 **Ilfracombe** SW England, United Kingdom 51°12′N 04°10′W
136 I11 **Ilgaz Dağları** ▲▲ N Turkey
136 G15 **Ilgın** Konya, W Turkey 38°16′N 31°57′E
60 I7 **Ilha Solteira** São Paulo, S Brazil 20°28′S 51°19′W
104 G7 **Ílhavo** Aveiro, N Portugal 40°36′N 08°40′W
59 O18 **Ilhéus** Bahia, E Brazil 14°50′S 39°06′W
129 R7 **Ili** var. Ile, Chin. Ili He, Rus. Reka Ili. ≈ China/Kazakhstan see also Ili He
116 G11 **Ilia** Hung. Marosillye. Hunedoara, SW Romania 45°57′N 22°40′E
39 P13 **Iliamna** Alaska, USA 59°42′N 154°49′W
39 P13 **Iliamna Lake** ◎ Alaska, USA
137 N13 **İliç** Erzincan, C Turkey 39°27′N 38°34′E
Il'ichevsk see Şărur, Azerbaijan
Il'ichevsk see Illichivs'k, Ukraine
Ilici see Elche
37 V2 **Iliff** Colorado, C USA 40°46′N 103°04′W
171 Q7 **Iligan** off. Iligan City. Mindanao, S Philippines 08°12′N 124°16′E
171 Q7 **Iligan Bay** bay S Philippines
Iligan City see Iligan
158 I5 **Ili He** var. Ili, Kaz. Ile, Rus. Reka Ili. ≈ China/Kazakhstan see also Ili
Ili He see Ili
56 C6 **Iliniza** ▲ N Ecuador 0°37′S 78°41′W
Ilinski see Il'inskiy
125 U14 **Il'inskiy** Permskiy Kray, NW Russian Federation 58°33′N 55°31′E
125 U14 **Il'inskiy** Ostrov Sakhalin, Sakhalinskaya Oblast', SE Russian Federation 47°59′N 142°14′E
18 I10 **Ilion** New York, NE USA 43°01′N 75°02′W
38 E9 **ʻĪlio Point** var. Ilio Point. headland Molokaʻi, Hawaiʻi, USA 21°13′N 157°15′W
Ilio Point see ʻĪlio Point

◆ Country ● Country Capital ◇ Dependent Territory ○ Dependent Territory Capital ◈ Administrative Regions ✈ International Airport ▲ Mountain ▲▲ Mountain Range ⛰ Volcano ≈ River ◎ Lake ▣ Reservoir

ILIRSKA BISTRICA – ÍSAFJARÐARDJÚP

109 T13 **Ilirska Bistrica** prev. Bistrica, Ger. Feistritz, Illyrisch-Feistritz, It. Villa del Nevoso. SW Slovenia 45°34′N 14°12′E
137 Q16 **Ilısu Barajı** ⊠ SE Turkey
155 G17 **Ilkal** Karnātaka, C India 15°59′N 76°08′E
97 M19 **Ilkeston** C England, United Kingdom 52°59′N 01°18′W
121 O16 **Il-Kullana** headland SW Malta 35°49′N 14°26′E
108 J8 **Ill** 〰 W Austria
103 U6 **Ill** 〰 NE France
62 G10 **Illapel** Coquimbo, C Chile 31°40′S 71°13′W
Illaue Fartak Trench see Alula-Fartak Trench
182 C2 **Illbillee, Mount** ▲ South Australia 27°01′S 132°13′E
102 I6 **Ille-et-Vilaine** ◈ department NW France
77 T11 **Illéla** Tahoua, SW Niger 14°25′N 05°10′E
101 J24 **Iller** 〰 S Germany
101 J23 **Illertissen** Bayern, S Germany 48°13′N 10°08′E
105 X9 **Illes Baleares** ◈ autonomous community E Spain
105 N8 **Illescas** Castilla-La Mancha, C Spain 40°08′N 03°51′W
Ille-sur-la-Têt see Ille-sur-Têt
103 O17 **Ille-sur-Têt** var. Ille-sur-la-Têt. Pyrénées-Orientales, S France 42°40′N 02°37′E
Illiberis see Elne
117 P11 **Illichivs'k** Rus. Il'ichevsk. Odes'ka Oblast', SW Ukraine 46°18′N 30°36′E
Illicis see Elche
102 M6 **Illiers-Combray** Eure-et-Loir, C France 48°18′N 01°15′E
30 K12 **Illinois** off. State of Illinois, also known as Prairie State, Sucker State. ◈ state C USA
30 J13 **Illinois River** 〰 Illinois, N USA
117 N6 **Illintsi** Vinnyts'ka Oblast', C Ukraine 49°07′N 29°13′E
Illiturgis see Andújar
74 M10 **Illizi** SE Algeria 26°30′N 08°28′E
27 Y7 **Illmo** Missouri, C USA 37°13′N 89°30′W
Illurco see Lorca
Illuro see Mataró
Illyrisch-Feistritz see Ilirska Bistrica
101 K16 **Ilm** 〰 C Germany
101 K17 **Ilmenau** Thüringen, C Germany 50°40′N 10°55′E
124 H14 **Il'men', Ozero** ◎ NW Russian Federation
57 H18 **Ilo** Moquegua, SW Peru 17°42′S 71°20′W
171 O6 **Iloilo** off. Iloilo City. Panay Island, C Philippines 10°42′N 122°34′E
Iloilo City see Iloilo
112 K10 **Ilok** Hung. Újlak. Vojvodina, NW Serbia 45°12′N 19°22′E
93 O16 **Ilomantsi** Pohjois-Karjala, SE Finland 62°40′N 30°55′E
42 F8 **Ilopango, Lago de** volcanic lake C El Salvador
77 T15 **Ilorin** Kwara, W Nigeria 08°32′N 04°35′E
117 X8 **Ilovays'k** Rus. Ilovaysk. Donets'ka Oblast', SE Ukraine 47°55′N 38°14′E
Ilovaysk see Ilovays'k
127 O10 **Ilovlya** Volgogradskaya Oblast', SW Russian Federation 49°45′N 44°19′E
127 O10 **Ilovlya** 〰 SW Russian Federation
121 N15 **Il-Ponta ta' San Dimitri** var. Ras San Dimitri, San Dimitri Point. headland Gozo, NW Malta 36°04′N 14°12′E
126 K14 **Il'skiy** Krasnodarskiy Kray, SW Russian Federation 44°52′N 38°26′E
182 B2 **Iltur** South Australia 27°33′S 130°31′E
171 Y13 **Ilugwa** Papua, E Indonesia 03°42′S 139°09′E
İluh see Batman
118 I11 **Ilūkste** SE Latvia 55°58′N 26°21′E
196 N13 **Ilulissat** Qaasuitsup, C Greenland 68°13′N 51°06′W
171 Y13 **Ilur** Pulau Gorong, E Indonesia 04°00′S 131°25′E
32 F10 **Ilwaco** Washington, NW USA 46°19′N 124°03′W
Il'yaly see Gurbansoltan Eje
İlyasbaba Burnu see Tekke Burnu
125 U9 **Ilych** 〰 NW Russian Federation
101 O21 **Ilz** 〰 SE Germany
111 M14 **Iłża** Radom, SE Poland 51°09′N 21°15′E
164 G13 **Imabari** var. Imaharu. Ehime, Shikoku, SW Japan 34°04′N 132°59′E
Imaharu see Imabari
165 O12 **Imaichi** var. Imaiti. Tochigi, Honshū, S Japan 36°43′N 139°41′E
Imaiti see Imaichi
164 K12 **Imajō** Fukui, Honshū, SW Japan 35°45′N 136°10′E
139 R9 **Imām Ibn Hāshim** Karbalā', C Iraq 32°46′N 43°21′E
149 Q2 **Imām Şāḩib** var. Emam Saheb, Hazarat Imam; prev. Emām Şāḩeb. Kunduz, NE Afghanistan 37°11′N 68°55′E
139 T11 **Imān 'Abd Allāh** Al Qādisīyah, S Iraq 31°36′N 44°34′E
164 F15 **Imano-yama** ▲ Shikoku, SW Japan 32°51′N 132°48′E
164 C13 **Imari** Saga, Kyūshū, SW Japan 33°18′N 129°51′E
Imarssuak Mid-Ocean Seachannel see Imarssuak Seachannel
64 J6 **Imarssuak Seachannel** var. Imarssuak Mid-Ocean Seachannel. channel N Atlantic Ocean
93 N18 **Imatra** Etelä-Karjala, SE Finland 61°14′N 28°50′E
164 K13 **Imazu** Shiga, Honshū, SW Japan 35°25′N 136°00′E
56 C6 **Imbabura** ◈ province N Ecuador
55 R9 **Imbaimadai** W Guyana 05°44′N 60°23′W
61 K14 **Imbituba** Santa Catarina, S Brazil 28°15′S 48°44′W
27 W9 **Imboden** Arkansas, C USA 36°12′N 91°10′W
Imbros see Gökçeada
Imeni 26 Bakinskikh Komissarov see Uzboý
125 N13 **Imeni Babushkina** Vologodskaya Oblast', NW Russian Federation 59°40′N 43°04′E
126 J7 **Imeni Karla Libknekhta** Kurskaya Oblast', W Russian Federation 51°36′N 35°28′E
Imeni Mollanepesa see Mollanepes Adyndaky
Imeni S. A. Niyazova see S. A.Nyýazow Adyndaky
Imeni Sverdlova Rudnik see Sverdlovs'k
188 E9 **Imeong** Babeldaob, N Palau
81 L14 **Īmī** Sumalē, E Ethiopia 06°27′N 42°10′E
115 M21 **Imia** Turk. Kardak. island Dodekánisa, Greece, Aegean Sea
Imishli see İmişli
137 X12 **İmişli** Rus. Imishli. C Azerbaijan 39°54′N 48°04′E
163 X14 **Imjin-gang** 〰 North Korea/South Korea
35 S3 **Imlay** Nevada, W USA 40°39′N 118°10′W
31 S9 **Imlay City** Michigan, N USA 43°01′N 83°04′W
23 X15 **Immokalee** Florida, SE USA 26°24′N 81°25′W
77 U17 **Imo** ◈ state SE Nigeria
106 G10 **Imola** Emilia-Romagna, N Italy 44°22′N 11°43′E
186 A5 **Imonda** West Sepik, NW Papua New Guinea 03°21′S 141°10′E
Imoschi see Imotski
113 G14 **Imotski** It. Imoschi. Split-Dalmacija, SE Croatia 43°28′N 17°13′E
59 L14 **Imperatriz** Maranhão, NE Brazil 05°32′S 47°28′W
106 B10 **Imperia** Liguria, NW Italy 43°53′N 08°03′E
57 E15 **Imperial** Lima, W Peru 13°04′S 76°21′W
35 X17 **Imperial** California, W USA 32°51′N 115°34′W
28 L16 **Imperial** Nebraska, C USA 40°30′N 101°37′W
24 M9 **Imperial** Texas, SW USA 31°15′N 102°40′W
35 Y17 **Imperial Dam** dam California, W USA
79 I17 **Impfondo** Likouala, NE Congo 01°37′N 18°04′E
153 X14 **Imphāl** state capital Manipur, NE India 24°47′N 93°55′E
103 P9 **Imphy** Nièvre, C France 46°55′N 03°16′E
106 G11 **Impruneta** Toscana, C Italy 43°42′N 11°16′E
115 K15 **İmroz** var. Gökçeada. Çanakkale, NW Turkey 40°06′N 25°50′E
Imroz Adası see Gökçeada
108 L7 **Imst** Tirol, W Austria 47°14′N 10°45′E
40 F3 **Imuris** Sonora, NW Mexico 30°48′N 110°52′W
164 M13 **Ina** Nagano, Honshū, S Japan 35°52′N 137°58′E
65 M18 **Inaccessible Island** island W Tristan da Cunha
115 F20 **Ínachos** 〰 S Greece
188 H6 **I Naftan, Puntan** headland Saipan, S Northern Mariana Islands
Inagua Islands see Little Inagua
Inagua Islands see Great Inagua
185 H15 **Inangahua** West Coast, South Island, New Zealand 41°51′S 171°58′E
57 I14 **Iñapari** Madre de Dios, E Peru 11°00′S 69°34′W
188 B17 **Inarajan** SE Guam 13°16′N 144°45′E
92 L10 **Inari** Lapp. Anár, Aanaar. Lappi, N Finland 68°54′N 27°06′E
92 L10 **Inarijärvi** Lapp. Aanaarjävri, Swe. Enareträsk. ◎ N Finland
92 L9 **Inarijoki** Lapp. Anárjohka. 〰 Finland/Norway
Inău see Ineu
165 P11 **Inawashiro-ko** var. Inawasiro Ko. ◎ Honshū, C Japan
Inawasiro Ko see Inawashiro-ko
105 X9 **Inca** Mallorca, Spain, W Mediterranean Sea 39°43′N 02°54′E
62 H7 **Inca de Oro** Atacama, N Chile 26°45′S 69°54′W
115 J15 **İnce Burnu** cape NW Turkey
136 K9 **İnce Burnu** headland N Turkey 42°06′N 34°57′E
136 I17 **İncekum Burnu** headland S Turkey 36°13′N 33°57′E
163 X15 **Incheon** , Jap. Jinsen; prev. Chemulpo, Inch'ŏn. NW South Korea 37°27′N 126°41′E
161 X15 **Incheon** ✈ (Seoul) NW South Korea 37°37′N 126°42′E
76 G7 **Inchiri** ◈ region NW Mauritania
Inch'ŏn see Incheon
83 M17 **Inchope** Manica, C Mozambique 19°09′S 33°54′E
Incoronata see Kornat
103 Y15 **Incudine, Monte** ▲ Corse, France, C Mediterranean Sea 41°52′N 09°13′E
60 M10 **Indaiatuba** São Paulo, S Brazil 23°03′S 47°14′W
93 H17 **Indal** Västernorrland, C Sweden 62°36′N 17°06′E
93 H17 **Indalsälven** 〰 C Sweden
40 K8 **Inde** Durango, C Mexico 25°55′N 105°10′W
Indefatigable Island see Santa Cruz, Isla
35 S10 **Independence** California, W USA 36°48′N 118°14′W
29 X13 **Independence** Iowa, C USA 42°28′N 91°42′W
27 P7 **Independence** Kansas, C USA 37°13′N 95°43′W
20 M4 **Independence** Kentucky, S USA 38°56′N 84°32′W
27 R4 **Independence** Missouri, C USA 39°04′N 94°27′W
21 R8 **Independence** Virginia, NE USA 36°38′N 81°11′W
30 J7 **Independence** Wisconsin, N USA 44°21′N 91°25′W
197 R12 **Independence Fjord** fjord N Greenland
Independence Island see Malden Island
35 W2 **Independence Mountains** ▲▲ Nevada, W USA
57 K18 **Independencia** Cochabamba, C Bolivia 17°08′S 66°52′W
57 E16 **Independencia, Bahía de la** bay W Peru
Independencia, Monte see Adam, Mount
116 M12 **Independenţa** Galaţi, SE Romania 45°29′N 27°45′E
Inderagiri see Indragiri, Sungai
144 F11 **Inderbor** prev. Inderborskiy. Atyrau, W Kazakhstan 48°35′N 51°45′E
Inderborskiy see Inderbor
151 I14 **India** off. Republic of India, var. Indian Union, Union of India, Hind. Bhārat. ◆ republic S Asia
India see Inđija
18 D14 **Indiana** Pennsylvania, NE USA 40°37′N 79°09′W
31 N13 **Indiana** off. State of Indiana, also known as Hoosier State. ◈ state N USA
31 O14 **Indianapolis** state capital Indiana, N USA 39°46′N 86°09′W
11 O10 **Indian Cabins** Alberta, W Canada 59°51′N 117°06′W
42 G1 **Indian Church** Orange Walk, N Belize 17°47′N 88°39′W
Indian Desert see Thar Desert
11 U16 **Indian Head** Saskatchewan, S Canada 50°32′N 103°41′W
31 O4 **Indian Lake** ◎ Michigan, N USA
18 K9 **Indian Lake** ◎ New York, NE USA
31 R13 **Indian Lake** ⊠ Ohio, N USA
172-173 **Indian Ocean** ocean
29 V15 **Indianola** Iowa, C USA 41°21′N 93°33′W
22 K4 **Indianola** Mississippi, S USA 33°27′N 90°39′W
36 J6 **Indian Peak** ▲ Utah, W USA 38°18′N 113°52′W
23 Y13 **Indian River** lagoon Florida, SE USA
35 W10 **Indian Springs** Nevada, W USA 36°33′N 115°40′W
23 Y14 **Indiantown** Florida, SE USA 27°01′N 80°29′W
Indian Union see India
59 K19 **Indiara** Goiás, S Brazil 17°12′S 50°09′W
India, Republic of see India
India, Union of see India
125 Q4 **Indiga** Nenetskiy Avtonomnyy Okrug, NW Russian Federation 67°40′N 49°01′E
123 R9 **Indigirka** 〰 NE Russian Federation
112 L10 **Inđija** Hung. India; prev. Indjija. Vojvodina, N Serbia 45°03′N 20°04′E
35 V16 **Indio** California, W USA 33°42′N 116°13′W
42 M12 **Indio, Río** 〰 SE Nicaragua
152 I10 **Indira Gandhi** ✈ (Delhi) Delhi, N India
151 Q23 **Indira Point** headland Andaman and Nicobar Island, India, NE Indian Ocean 6°54′N 93°54′E
Indjija see Inđija
129 Q13 **Indo-Australian Plate** tectonic feature
173 N11 **Indomed Fracture Zone** tectonic feature SW Indian Ocean
170 L12 **Indonesia** off. Republic of Indonesia, Ind. Republik Indonesia; prev. Dutch East Indies, Netherlands East Indies, United States of Indonesia. ◆ republic SE Asia
Indonesian Borneo see Kalimantan
Indonesia, Republic of see Indonesia
Indonesia, Republik see Indonesia
Indonesia, United States of see Indonesia
154 G10 **Indore** Madhya Pradesh, C India 22°42′N 75°51′E
168 L11 **Indragiri, Sungai** var. Batang Kuantan, Inderagiri. 〰 Sumatera, W Indonesia
Indramajoe/Indramaju see Indramayu
169 P15 **Indramayu** prev. Indramajoe, Indramaju. Jawa, C Indonesia 06°22′S 108°20′E
155 K14 **Indrāvati** 〰 S India
103 N9 **Indre** ◈ department C France
102 M8 **Indre** 〰 C France
94 D13 **Indre Ålvik** Hordaland, S Norway 60°26′N 06°27′E
102 L8 **Indre-et-Loire** ◈ department C France
Indreville see Châteauroux
152 G3 **Indus** Chin. Yindu He; prev. Yin-tu Ho. 〰 S Asia
Indus Cone see Indus Fan
173 P3 **Indus Fan** var. Indus Cone. undersea feature N Arabian Sea 16°00′N 66°30′E
149 P17 **Indus, Mouths of the** delta S Pakistan
83 I24 **Indwe** Eastern Cape, SE South Africa 31°28′S 27°20′E
136 I10 **İnebolu** Kastamonu, N Turkey 41°57′N 33°45′E
77 P8 **I-n-Échaï** oasis C Mali
114 M13 **İnecik** Tekirdağ, NW Turkey 40°55′N 27°16′E
136 E12 **İnegöl** Bursa, NW Turkey 40°06′N 29°31′E
Inessa see Biancavilla
116 F10 **Ineu** Hung. Borosjenő; prev. Inău. Arad, W Romania 46°26′N 21°51′E
Ineul/Ineu, Vîrful see Ineu, Vârful
116 J9 **Ineu, Vârful** var. Ineul; prev. Vîrful Ineu. ▲ N Romania 47°31′N 24°52′E
21 P6 **Inez** Kentucky, S USA 37°53′N 82°33′W
74 E8 **Inezgane** ✈ (Agadir) W Morocco 30°35′N 09°27′W
41 T17 **Inferior, Laguna** lagoon S Mexico
40 M15 **Infiernillo, Presa del** ⊠ S Mexico
Infiesto see L'Infiestu
93 L20 **Ingå** Fin. Inkoo. Uusimaa, S Finland 60°01′N 24°05′E
77 U10 **Ingal** var. I-n-Gall. Agadez, C Niger 16°52′N 06°57′E
I-n-Gall see Ingal
99 C18 **Ingelmunster** West-Vlaanderen, W Belgium 50°12′N 03°15′E
79 I18 **Ingende** Equateur, W Dem. Rep. Congo 0°15′S 18°58′E
62 L5 **Ingeniero Guillermo Nueva Juárez** Formosa, N Argentina 23°55′S 61°50′W
63 H16 **Ingeniero Jacobacci** Río Negro, C Argentina 41°18′S 69°35′W
14 F16 **Ingersoll** Ontario, S Canada 43°03′N 80°53′W
Ingettolgoy see Selenge
181 W5 **Ingham** Queensland, NE Australia 18°35′S 146°12′E
146 M11 **Ingichka** Samarqand Viloyati, C Uzbekistan 39°46′N 65°56′E
97 L16 **Ingleborough** ▲ N England, United Kingdom 54°07′N 02°22′W
25 T14 **Ingleside** Texas, SW USA 27°52′N 97°12′W
184 K10 **Inglewood** Taranaki, North Island, New Zealand 39°07′S 174°13′E
35 S15 **Inglewood** California, W USA 33°57′N 118°21′W
101 L21 **Ingolstadt** Bayern, S Germany 48°46′N 11°26′E
33 V9 **Ingomar** Montana, NW USA 46°34′N 107°21′W
13 R14 **Ingonish Beach** Cape Breton Island, Nova Scotia, SE Canada 46°42′N 60°22′W
153 S14 **Ingrāj Bāzār** prev. English Bazar. West Bengal, NE India 25°00′N 88°10′E
25 Q11 **Ingram** Texas, SW USA 30°04′N 99°14′W
195 X7 **Ingrid Christensen Coast** physical region Antarctica
74 K14 **I-n-Guezzam** S Algeria 19°35′N 05°49′E
Ingulets see Inhulets'
Inguri see Enguri
Ingushetia/Ingushetiya, Respublika see Ingushetiya, Respublika
127 O15 **Ingushetiya, Respublika** var. Respublika Ingushetiya, Eng. Ingushetia. ◈ autonomous republic SW Russian Federation
83 N20 **Inhambane** Inhambane, SE Mozambique 23°52′S 35°31′E
83 M20 **Inhambane** off. Província de Inhambane. ◈ province S Mozambique
Inhambane, Província de see Inhambane
83 N17 **Inhaminga** Sofala, C Mozambique 18°24′S 35°00′E
83 N20 **Inharrime** Inhambane, SE Mozambique 24°29′S 35°01′E
83 M18 **Inhassoro** Inhambane, E Mozambique 21°32′S 35°13′E
117 S9 **Inhulets'** Rus. Ingulets. Dnipropetrovs'ka Oblast', E Ukraine 47°43′N 33°16′E
117 R10 **Inhulets'** Rus. Ingulets. 〰 S Ukraine
105 Q10 **Iniesta** Castilla-La Mancha, C Spain 39°27′N 01°45′W
I-ning see Yining
54 K11 **Inírida, Río** 〰 E Colombia
Inis see Ennis
Inis Ceithleann see Enniskillen
Inis Córthaidh see Enniscorthy
Inis Díomáin see Ennistimon
97 A17 **Inishbofin** Ir. Inis Bó Finne. island W Ireland
97 B18 **Inisheer** var. Inishere, Ir. Inis Oírr. island W Ireland
Inishere see Inisheer
97 B18 **Inishmaan** Ir. Inis Meáin. island W Ireland
97 A18 **Inishmore** Ir. Árainn. island W Ireland
96 E13 **Inishtrahull** Ir. Inis Trá Tholl. island NW Ireland
97 A17 **Inishturk** Ir. Inis Toirc. island W Ireland
Inkoo see Ingå
185 J16 **Inland Kaikoura Range** ▲▲ South Island, New Zealand
Inland Sea see Seto-naikai
21 P11 **Inman** South Carolina, SE USA 35°03′N 82°05′W
108 L7 **Inn** 〰 C Europe
197 O11 **Innaanganeq** var. Kap York. headland NW Greenland 75°54′N 66°27′W
182 K2 **Innamincka** South Australia 27°47′S 140°45′E
92 G12 **Inndyr** Nordland, C Norway 67°01′N 14°00′E
42 G3 **Inner Channel** inlet SE Belize
96 F11 **Inner Hebrides** island group W Scotland, United Kingdom
172 H15 **Inner Islands** var. Central Group. island group NE Seychelles
Inner Mongolia/Inner Mongolian Autonomous Region see Nei Mongol Zizhiqu
109 I7 **Inner Rhoden** former canton Appenzell. ◈ NW Switzerland
96 G8 **Inner Sound** strait NW Scotland, United Kingdom
100 J13 **Innerste** 〰 C Germany
181 W5 **Innisfail** Queensland, NE Australia 17°29′S 146°03′E
11 Q15 **Innisfail** Alberta, SW Canada 52°01′N 113°59′W
Inniskilling see Enniskillen
39 O11 **Innoko River** 〰 Alaska, USA
Innsbruch see Innsbruck
108 M7 **Innsbruck** var. Innsbruch. Tirol, W Austria 47°17′N 11°25′E
79 I19 **Inongo** Bandundu, W Dem. Rep. Congo 01°55′S 18°20′E
Inoucdjouac see Inukjuak
Inowrazlaw see Inowrocław
110 I10 **Inowrocław** Ger. Hohensalza; prev. Inowrazlaw. Kujawski-pomorskie, C Poland 52°47′N 18°15′E
57 K18 **Inquisivi** La Paz, W Bolivia 16°55′S 67°10′W
Inrin see Yuanlin
77 O8 **I-n-Sâkâne, 'Erg** desert N Mali
77 O8 **I-n-Sâkâne, 'Erg** desert N Mali
74 J10 **I-n-Salah** var. In Salah. C Algeria 27°11′N 02°31′E
127 O5 **Insar** Respublika Mordoviya, W Russian Federation 53°52′N 44°26′E
189 X15 **Insiaf** Kosrae, E Micronesia
94 L13 **Insjön** Dalarna, C Sweden 60°40′N 15°05′E
Insterburg see Chernyakhovsk
Insula see Lille
116 L13 **Însurăţei** Brăila, SE Romania 44°55′N 27°40′E
125 V6 **Inta** Respublika Komi, NW Russian Federation 66°00′N 60°10′E
77 R9 **I-n-Tebezas** Kidal, E Mali 17°58′N 01°51′E
Interamna see Teramo
Interamna Nahars see Terni
28 L11 **Interior** South Dakota, N USA 43°42′N 101°57′W
108 E9 **Interlaken** Bern, SW Switzerland 46°41′N 07°51′E
29 V2 **International Falls** Minnesota, N USA 48°38′N 93°26′W
167 O7 **Inthanon, Doi** ▲ NW Thailand 18°33′N 98°29′E
42 G7 **Intibucá** ◈ department SW Honduras
42 G8 **Intipucá** La Unión, SE El Salvador 13°10′N 88°03′W
61 B15 **Intiyaco** Santa Fe, C Argentina 28°43′S 60°04′W
116 K12 **Întorsura Buzăului** Ger. Bozau, Hung. Bodzafordulő. Covasna, E Romania 45°40′N 26°02′E
22 H9 **Intracoastal Waterway** inland waterway system Louisiana, S USA
25 V13 **Intracoastal Waterway** inland waterway system Texas, SW USA
108 G11 **Intragna** Ticino, S Switzerland 46°12′N 08°42′E
165 P14 **Inubō-zaki** headland Honshū, S Japan 35°42′N 140°51′E
164 E14 **Inukai** Ōita, Kyūshū, SW Japan 33°05′N 131°37′E
12 I5 **Inukjuak** var. Inoucdjouac; prev. Port Harrison. Québec, NE Canada 58°28′N 77°58′W
63 I24 **Inútil, Bahía** bay S Chile
Inuuvik see Inuvik
11 R8 **Inuvik** var. Inuuvik. Northwest Territories, NW Canada 68°25′N 133°35′W
164 L13 **Inuyama** Aichi, Honshū, SW Japan 35°23′N 136°56′E
56 G13 **Inuya, Río** 〰 E Peru
125 U13 **In'va** 〰 NW Russian Federation
96 H11 **Inveraray** W Scotland, United Kingdom 56°13′N 05°05′W
185 C24 **Invercargill** Southland, South Island, New Zealand 46°25′S 168°22′E
183 T5 **Inverell** New South Wales, SE Australia 29°46′S 151°10′E
96 I8 **Invergordon** N Scotland, United Kingdom 57°42′N 04°02′W
11 P16 **Invermere** British Columbia, SW Canada 50°30′N 116°00′W
13 R14 **Inverness** Cape Breton Island, Nova Scotia, SE Canada 46°14′N 61°19′W
96 I8 **Inverness** N Scotland, United Kingdom 57°27′N 04°15′W
23 V11 **Inverness** Florida, SE USA 28°50′N 82°19′W
96 I9 **Inverness** cultural region NW Scotland, United Kingdom
96 K9 **Inverurie** NE Scotland, United Kingdom 57°14′N 02°14′W
182 F8 **Investigator Group** island group South Australia
173 T7 **Investigator Ridge** undersea feature E Indian Ocean 11°30′S 98°10′E
182 H10 **Investigator Strait** strait South Australia
29 R11 **Inwood** Iowa, C USA 43°16′N 96°25′W
123 S10 **Inya** 〰 E Russian Federation
Inyanga see Nyanga
83 M16 **Inyangani** ▲ NE Zimbabwe 18°22′S 32°57′E
83 J17 **Inyathi** Matabeleland North, SW Zimbabwe 19°39′S 28°54′E
35 T12 **Inyokern** California, W USA 35°37′N 117°48′W
35 T10 **Inyo Mountains** ▲▲ California, W USA
127 P6 **Inza** Ul'yanovskaya Oblast', W Russian Federation 53°51′N 46°21′E
127 W5 **Inzer** Respublika Bashkortostan, W Russian Federation 54°11′N 57°37′E
127 N7 **Inzhavino** Tambovskaya Oblast', W Russian Federation 52°18′N 42°28′E
115 C16 **Ioánnina** var. Janina, Yannina. Ípeiros, W Greece 39°39′N 20°52′E
164 B17 **Iō-jima** var. Iwojima. island Nansei-shotō, SW Japan
124 L4 **Iokan'ga** 〰 NW Russian Federation
27 Q6 **Iola** Kansas, C USA 37°55′N 95°24′W
Iolcus see Iolkós
115 G16 **Iolkós** anc. Iolcus. site of ancient city Thessalía, C Greece
Iolotan' see Ýolöten
83 A16 **Iona** Namibe, SW Angola 16°54′S 12°39′E
96 F11 **Iona** island W Scotland, United Kingdom
116 M15 **Ion Corvin** Constanţa, SE Romania 44°07′N 27°50′E
35 P7 **Ione** California, W USA 38°21′N 120°55′W
116 I13 **Ioneşti** Vâlcea, SW Romania 44°52′N 24°12′E
31 Q9 **Ionia** Michigan, N USA 42°59′N 85°04′W
Ionia Basin see Ionian Basin
121 O10 **Ionian Basin** var. Ionia Basin. undersea feature Ionian Sea, C Mediterranean Sea 36°00′N 20°00′E
115 B17 **Iónia Nisiá** var. Iónioi Nísoi, Eng. Ionian Islands. island group W Greece
Ionian Islands see Iónia Nisiá/Iónioi Nísoi
121 O10 **Ionian Sea** Gk. Iónio Pélagos, It. Mar Ionio. sea C Mediterranean Sea
115 B17 **Iónioi Nísoi** Eng. Ionian Islands. ◈ region W Greece
Iónioi Nísoi see Iónia Nisiá
Ionio, Mar/Iónio Pélagos see Ionian Sea
Iordan see Yordon
137 U10 **Iori** var. Qabırrı. 〰 Azerbaijan/Georgia
Iorrais, Ceann see Erris Head
115 J22 **Íos** var. Nio. island Kykládes, Greece, Aegean Sea
Íos see Chóra
165 U15 **Io-Tori-shima** prev. Tori-shima. island Izu-shotō, SE Japan
115 I20 **Ioulís** prev. Kéa. Tziá, Kykládes, Greece, Aegean Sea 37°40′N 24°19′E
22 G9 **Iowa** Louisiana, S USA 30°12′N 93°00′W
29 V13 **Iowa** off. State of Iowa, also known as Hawkeye State. ◈ state C USA
29 Y14 **Iowa City** Iowa, C USA 41°40′N 91°32′W
29 V13 **Iowa Falls** Iowa, C USA 42°31′N 93°15′W
25 R4 **Iowa Park** Texas, SW USA 33°57′N 98°40′W
29 Y14 **Iowa River** 〰 Iowa, C USA
119 M19 **Ipa** 〰 SE Belarus
59 N20 **Ipatinga** Minas Gerais, SE Brazil 19°32′S 42°30′W
127 N13 **Ipatovo** Stavropol'skiy Kray, SW Russian Federation 45°40′N 42°51′E
115 C16 **Ípeiros** Eng. Epirus. ◈ region W Greece
111 J21 **Ipel'** var. Ipoly, Ger. Eipel. 〰 Hungary/Slovakia
54 C13 **Ipiales** Nariño, SW Colombia 0°52′N 77°38′W
189 V14 **Ipis** atoll Chuuk Islands, C Micronesia
59 A14 **Ipixuna** Amazonas, W Brazil 06°57′S 71°42′W
168 J8 **Ipoh** Perak, Peninsular Malaysia 04°36′N 101°02′E
Ipoly see Ipel'
187 S15 **Ipota** Erromango, S Vanuatu 18°54′S 169°19′E
79 K14 **Ippy** Ouaka, C Central African Republic 06°17′N 21°13′E
114 L13 **İpsala** Edirne, NW Turkey 40°56′N 26°23′E
Ipsario see Ypsário
183 V3 **Ipswich** Queensland, E Australia 27°28′S 152°40′E
97 Q20 **Ipswich** hist. Gipeswic. E England, United Kingdom 52°05′N 01°08′E
29 O8 **Ipswich** South Dakota, N USA 45°24′N 99°00′W
Iput' see Iputs'
119 P18 **Iputs'** Rus. Iput'. 〰 Belarus/Russian Federation
9 R7 **Iqaluit** prev. Frobisher Bay. province capital Baffin Island, Nunavut, NE Canada 63°44′N 68°28′W
Iqlid see Eqlīd
62 G3 **Iquique** Tarapacá, N Chile 20°15′S 70°08′W
56 G8 **Iquitos** Loreto, N Peru 03°51′S 73°13′W
25 N9 **Iraan** Texas, SW USA 30°52′N 101°52′W
79 K14 **Ira Banda** Haute-Kotto, E Central African Republic 05°57′N 22°05′E
165 P16 **Irabu-jima** island Miyako-shotō, SW Japan
55 Y9 **Iracoubo** N French Guiana 05°28′N 53°15′W
60 H13 **Iraí** Rio Grande do Sul, S Brazil 27°15′S 53°17′W
114 G12 **Irákleia** Kentrikí Makedonía, N Greece 41°09′N 23°16′E
115 J21 **Irákleia** island Kykládes, Greece, Aegean Sea
115 J25 **Irákleio** var. Herakleion, Eng. Candia; prev. Iráklion. Kríti, Greece, E Mediterranean Sea 35°20′N 25°08′E
115 F15 **Irákleio** anc. Heracleum. castle Kentrikí Makedonía, N Greece
115 J25 **Irákleio** ✈ Kríti, Greece, E Mediterranean Sea 35°20′N 25°10′E
Iráklion see Irákleio
143 O7 **Iran** off. Islamic Republic of Iran; prev. Persia. ◆ republic SW Asia
58 F13 **Iranduba** Amazonas, NW Brazil 03°19′S 60°09′W
85 P13 **Iranian Plate** tectonic feature
143 Q9 **Iranian Plateau** var. Plateau of Iran. plateau N Iran
Iran, Islamic Republic of see Iran
Iran Mountains see Iran, Pegunungan
169 U9 **Iran, Pegunungan** var. Iran Mountains. ▲▲ Indonesia/Malaysia
Iran, Plateau of see Iranian Plateau
143 W13 **Īrānshahr** Sīstān va Balūchestān, SE Iran 27°14′N 60°40′E
55 P5 **Irapa** Sucre, NE Venezuela 10°37′N 62°35′W
41 N13 **Irapuato** Guanajuato, C Mexico 20°40′N 101°23′W
139 R7 **Iraq** off. Republic of Iraq, Ar. 'Irāq. ◆ republic SW Asia
'Irāq see Iraq
Iraq, Republic of see Iraq
60 J12 **Irati** Paraná, S Brazil 25°25′S 50°38′W
105 R3 **Irati** 〰 N Spain
125 T8 **Irayël'** Respublika Komi, NW Russian Federation 64°28′N 55°20′E
43 N13 **Irazú, Volcán** 🌋 C Costa Rica 09°57′N 83°52′W
Irbenskiy Zaliv/Irbes Šaurums see Irbe Strait
118 D7 **Irbe Strait** Est. Kura Kurk, Latv. Irbes Šaurums, Irbenskiy Zaliv; prev. Est. Irbe Väin. strait Estonia/Latvia
Irbe Väin see Irbe Strait
138 G9 **Irbid** Irbid, N Jordan 32°33′N 35°51′E
138 G9 **Irbid** off. Muḩāfaẓat Irbid. ◈ governorate N Jordan
Irbid, Muḩāfaẓat see Irbid
Irbil see Arbīl
Irbīl see Arbīl
109 S6 **Irdning** Steiermark, SE Austria 47°29′N 14°04′E
79 I18 **Irebu** Equateur, W Dem. Rep. Congo 0°32′S 17°44′E
97 D17 **Ireland** off. Republic of Ireland, Ir. Éire. ◆ republic NW Europe
84 C9 **Ireland** Lat. Hibernia. island Ireland/United Kingdom
64 A12 **Ireland Island North** island W Bermuda
64 A12 **Ireland Island South** island W Bermuda
Ireland, Republic of see Ireland
125 V15 **Iren'** 〰 NW Russian Federation
185 A22 **Irene, Mount** ▲ South Island, New Zealand 45°04′S 167°24′E
Irgalem see Yirga 'Alem
Irgiz see Yrghyz
Irian see New Guinea
Irian Barat see Papua
Irian Jaya see Papua
Irian Jaya Barat see Papua Barat
Irian, Teluk see Cenderawasih, Teluk
78 K9 **Iriba** Wadi Fira, NE Chad 15°10′N 22°11′E
127 X7 **Iriklinskoye Vodokhranilishche** ⊠ W Russian Federation
81 H23 **Iringa** Iringa, C Tanzania 07°49′S 35°39′E
81 H23 **Iringa** ◈ region S Tanzania
165 O16 **Iriomote-jima** island Sakishima-shotō, SW Japan
42 L4 **Iriona** Colón, NE Honduras 15°55′N 85°10′W
47 U7 **Iriri** 〰 N Brazil
58 I13 **Iriri, Rio** 〰 C Brazil
Iris see Yeşilırmak
35 W9 **Irish, Mount** ▲ Nevada, W USA 37°39′N 115°22′W
97 H17 **Irish Sea** Ir. Muir Éireann. sea C British Isles
139 U12 **Irjal ash Shaykhīyah** Al Muthanná, S Iraq 30°49′N 44°58′E
147 U11 **Irkeshtam** Oshskaya Oblast', SW Kyrgyzstan 39°39′N 73°49′E
122 M13 **Irkutsk** Irkutskaya Oblast', S Russian Federation 52°18′N 104°15′E
122 M12 **Irkutskaya Oblast'** ◈ province S Russian Federation
Irlir, Gora see Irlir Tog'i
146 K8 **Irlir Tog'i** var. Gora Irlir. ▲ N Uzbekistan 42°43′N 63°24′E
Irminger Basin see Reykjanes Basin
21 R12 **Irmo** South Carolina, SE USA 34°05′N 81°10′W
102 E6 **Iroise** sea NW France
189 X2 **Iroj** var. Eroj. island Ratak Chain, SE Marshall Islands
182 H7 **Iron Baron** South Australia 33°01′S 137°13′E
14 C10 **Iron Bridge** Ontario, S Canada 46°16′N 83°12′W
20 H10 **Iron City** Tennessee, S USA 35°01′N 87°34′W
14 I13 **Irondale** 〰 Ontario, SE Canada
182 H7 **Iron Knob** South Australia 32°46′S 137°08′E
30 M5 **Iron Mountain** Michigan, N USA 45°51′N 88°03′W
30 M4 **Iron River** Michigan, N USA 46°05′N 88°38′W
30 J3 **Iron River** Wisconsin, N USA 46°34′N 91°22′W
27 X6 **Ironton** Missouri, C USA 37°37′N 90°40′W
31 S15 **Ironton** Ohio, N USA 38°32′N 82°40′W
30 K4 **Ironwood** Michigan, N USA 46°27′N 90°10′W
12 H12 **Iroquois Falls** Ontario, S Canada 48°47′N 80°41′W
31 N12 **Iroquois River** 〰 Illinois/Indiana, N USA
164 M15 **Irō-zaki** headland Honshū, S Japan 34°36′N 138°49′E
Irpen' see Irpin'
117 O4 **Irpin'** Rus. Irpen'. Kyyivs'ka Oblast', N Ukraine 50°31′N 30°16′E
117 O4 **Irpin'** Rus. Irpen'. 〰 N Ukraine
141 Q16 **'Irqah** SW Yemen 13°42′N 47°21′E
166 L6 **Irrawaddy** var. Ayeyarwady. 〰 W Myanmar (Burma)
Irrawaddy see Ayeyawady
166 K8 **Irrawaddy, Mouths of the** delta SW Myanmar (Burma)
117 N4 **Irsha** 〰 N Ukraine
116 H7 **Irshava** Zakarpats'ka Oblast', W Ukraine 48°19′N 23°03′E
107 N18 **Irsina** Basilicata, S Italy 40°42′N 16°18′E
Irtish see Yertis
Irtysh see Yertis
Irtyshsk see Yertis
79 P17 **Irumu** Orientale, E Dem. Rep. Congo 01°27′N 29°52′E
105 Q2 **Irun** Cast. Irún. País Vasco, N Spain 43°20′N 01°48′W
Irún see Irun
Iruña see Pamplona
105 Q3 **Irurtzun** Navarra, N Spain 42°55′N 01°50′W
96 I13 **Irvine** W Scotland, United Kingdom 55°37′N 04°40′W
21 N6 **Irvine** Kentucky, S USA 37°42′N 83°59′W
25 T6 **Irving** Texas, SW USA 32°47′N 96°57′W
20 K5 **Irvington** Kentucky, S USA 37°52′N 86°16′W
164 C15 **Isa** prev. Ōkuchi, Ōkuti. Kagoshima, Kyūshū, SW Japan 32°04′N 130°36′E
Isaak see Iisaku
28 L8 **Isabel** South Dakota, N USA 45°21′N 101°25′W
186 L8 **Isabel** off. Isabel Province. ◈ province N Solomon Islands
171 O8 **Isabela** Basilan Island, SW Philippines 06°41′N 122°00′E
45 S5 **Isabela** W Puerto Rico 18°30′N 67°02′W
45 N8 **Isabela, Cabo** headland NW Dominican Republic 19°54′N 71°03′W
57 A18 **Isabela, Isla** var. Albemarle Island. island Galapagos Islands, Ecuador, E Pacific Ocean
40 I12 **Isabela, Isla** island C Mexico
42 K9 **Isabella, Cordillera** ▲▲ NW Nicaragua
35 S12 **Isabella Lake** ⊠ California, W USA
31 N2 **Isabelle, Point** headland Michigan, N USA 47°20′N 87°56′W
Isabel Province see Isabel
Isabel Segunda see Vieques
116 M13 **Isaccea** Tulcea, E Romania 45°16′N 28°28′E
92 H1 **Ísafjarðardjúp** inlet NW Iceland

◆ Country ● Country Capital ◇ Dependent Territory ○ Dependent Territory Capital ◈ Administrative Regions ✈ International Airport ▲ Mountain ▲▲ Mountain Range 🌋 Volcano 〰 River ◎ Lake ⊠ Reservoir

ÍSAFJARÐARDJÚP – JAMAARE

92 *H1* **Ísafjarðardjúp** *inlet* NW Iceland
92 *H1* **Ísafjörður** Vestfirðir, NW Iceland 66°04′N 23°09′W
164 *C14* **Isahaya** Nagasaki, Kyūshū, SW Japan 32°51′N 130°02′E
149 *S7* **Īsa Khel** Punjab, E Pakistan 32°39′N 71°20′E
172 *H7* **Isalo** *var.* Massif de L'Isalo. ▲▲ SW Madagascar
79 *K20* **Isandja** Kasai-Occidental, C Dem. Rep. Congo 03°03′S 21°57′E
187 *R15* **Isangel** Tanna, S Vanuatu 19°34′S 169°17′E
79 *M18* **Isangi** Orientale, C Dem. Rep. Congo 0°46′N 24°15′E
101 *L24* **Isar** ≈ Austria/Germany
101 *M23* **Isar-Kanal** *canal* SE Germany
Isbarta *see* Isparta
Isca Damnoniorum *see* Exeter
107 *K18* **Ischia** *var.* Isola d'Ischia; *anc.* Aenaria. Campania, S Italy 40°44′N 13°57′E
107 *J18* **Ischia, Isola d'** *island* S Italy
54 *B12* **Iscuandé** *var.* Santa Bárbara. Nariño, SW Colombia 02°32′N 78°00′W
164 *K14* **Ise** Mie, Honshū, SW Japan 34°29′N 136°43′E
100 *J12* **Ise** ≈ N Germany
95 *I23* **Isefjord** *fjord* E Denmark
Iseghem *see* Izegem
192 *M14* **Iselin Seamount** *undersea feature* S Pacific Ocean 72°30′S 179°00′W
Isenhof *see* Püssi
106 *E7* **Iseo** Lombardia, N Italy 45°40′N 10°03′E
103 *U12* **Iseran, Col de l'** *pass* E France
103 *S11* **Isère** ⬖ *department* E France
103 *S12* **Isère** ≈ E France
101 *F15* **Iserlohn** Nordrhein-Westfalen, W Germany 51°23′N 07°42′E
107 *K16* **Isernia** *var.* Æsernia. Molise, C Italy 41°35′N 14°14′E
165 *N12* **Isesaki** Gunma, Honshū, S Japan 36°19′N 139°11′E
129 *Q5* **Iset'** ≈ C Russian Federation
77 *S15* **Iseyin** Oyo, W Nigeria 07°56′N 03°33′E
Isfahan *see* Eşfahān
147 *Q11* **Isfana** Batkenskaya Oblast', SW Kyrgyzstan 39°51′N 69°31′E
147 *R11* **Isfara** N Tajikistan 40°06′N 70°34′E
149 *O4* **Isfi Maidān** Gōwr, N Afghanistan 35°09′N 66°16′E
92 *O3* **Isfjorden** *fjord* W Svalbard
Isgender *see* Kul'mach
Isha Baydhabo *see* Baydhabo
125 *V11* **Isherim, Gora** ▲ NW Russian Federation 61°06′N 59°09′E
127 *Q5* **Isheyevka** Ul'yanovskaya Oblast', W Russian Federation 54°27′N 48°18′E
165 *P16* **Ishigaki** Okinawa, Ishigaki-jima, SW Japan 24°20′N 124°09′E
165 *P16* **Ishigaki-jima** *island* Sakishima-shotō, SW Japan
165 *R3* **Ishikari-wan** *bay* Hokkaidō, NE Japan
165 *S16* **Ishikawa** *var.* Isikawa. Okinawa, Okinawa, SW Japan 26°25′N 127°47′E
164 *K11* **Ishikawa** *off.* Ishikawa-ken, *var.* Isikawa. ⬖ *prefecture* Honshū, SW Japan
Ishikawa-ken *see* Ishikawa
122 *H11* **Ishim** Tyumenskaya Oblast', C Russian Federation 56°13′N 69°25′E
127 *V6* **Ishimbay** Respublika Bashkortostan, W Russian Federation 53°21′N 56°03′E
145 *O9* **Ishimskoye** Akmola, C Kazakhstan 51°23′N 67°07′E
165 *Q10* **Ishinomaki** *var.* Isinomaki. Miyagi, Honshū, C Japan 38°26′N 141°17′E
165 *P13* **Ishioka** *var.* Isioka. Ibaraki, Honshū, S Japan 36°11′N 140°16′E
149 *Q3* **Ishkamish** *prev.* Eshkamesh. Takhār, NE Afghanistan 36°25′N 69°11′E
149 *T2* **Ishkāshim** *prev.* Eshkāshem. Badakhshān, NE Afghanistan 36°43′N 71°34′E
Ishkashim *see* Ishkoshim
Ishkashimskiy Khrebet *see* Ishkoshim, Qatorkŭhi
147 *S15* **Ishkoshim** *Rus.* Ishkashim. S Tajikistan 36°46′N 71°35′E
147 *S15* **Ishkoshim, Qatorkŭhi** *Rus.* Ishkashimskiy Khrebet. ▲▲ SE Tajikistan
31 *N4* **Ishpeming** Michigan, N USA 46°29′N 87°40′W
147 *N11* **Ishtixon** *Rus.* Ishtykhan. Samarqand Viloyati, C Uzbekistan 39°59′N 66°28′E
Ishtykhan *see* Ishtixon
Ishurdi *see* Iswardi
61 *G17* **Isidoro Noblia** Cerro Largo, NE Uruguay 31°58′S 54°09′W
102 *J4* **Isigny-sur-Mer** Calvados, N France 49°20′N 01°06′W
Isikawa *see* Ishikawa
136 *C11* **Işıklar Dağı** ▲▲ NW Turkey
107 *C19* **Isili** Sardegna, Italy, C Mediterranean Sea 39°46′N 09°06′E
122 *H12* **Isil'kul'** Omskaya Oblast', C Russian Federation 54°52′N 71°07′E
Isinomaki *see* Ishinomaki
Isioka *see* Ishioka
81 *I18* **Isiolo** Isiolo, C Kenya 0°20′N 37°36′E
81 *J18* **Isiolo** ⬖ *county* C Kenya
79 *O16* **Isiro** Orientale, NE Dem. Rep. Congo 02°51′N 27°47′E
92 *P2* **Isispynten** *headland* NE Svalbard 79°51′N 26°44′E
123 *P11* **Isit** Respublika Sakha (Yakutiya), NE Russian Federation 60°53′N 125°32′E
149 *O2* **Iskabad Canal** *canal* N Afghanistan
147 *Q9* **Iskandar** *Rus.* Iskander. Toshkent Viloyati, E Uzbekistan 41°32′N 69°46′E
Iskander *see* Iskandar
114 *G10* **Iskar** *var.* Iskăr, Iskŭr. ≈ NW Bulgaria
Iskăr *see* Iskar
114 *H10* **Iskar, Yazovir** *var.* Yazovir Iskŭr; *prev.* Yazovir Stalin. ⊠ W Bulgaria
121 *Q2* **İskele** *var.* Trikomo, *Gk.* Tríkomon. E Cyprus 35°16′N 33°54′E
136 *K17* **İskenderun** *Eng.* Alexandretta. Hatay, S Turkey 36°34′N 36°10′E
138 *H2* **İskenderun Körfezi** *Eng.* Gulf of Alexandretta. *gulf* S Turkey
136 *J11* **İskilip** Çorum, N Turkey 40°45′N 34°28′E
Iski-Nauket *see* Nookat
114 *J11* **Iskra** *prev.* Popovo. Haskovo, S Bulgaria 41°55′N 25°12′E
Iskŭr *see* Iskar
Iskŭr, Yazovir *see* Iskar, Yazovir
41 *S15* **Isla** Veracruz-Llave, SE Mexico 18°01′N 95°30′W
119 *J15* **Islach** *Rus.* Isloch'. ≈ C Belarus
104 *H14* **Isla Cristina** Andalucía, S Spain 37°12′N 07°20′W
Isla de León *see* San Fernando
149 *U6* **Islāmābād** ● (Pakistan) Federal Capital Territory Islāmābād, NE Pakistan 33°40′N 73°08′E
149 *V6* **Islāmābād** ✈ Federal Capital Territory Islāmābād, NE Pakistan 33°40′N 73°08′E
Islamabad *see* Anantnāg
149 *R17* **Islāmkot** Sind, SE Pakistan 24°37′N 70°04′E
23 *Y17* **Islamorada** Florida Keys, Florida, SE USA 24°55′N 80°37′W
153 *P14* **Islāmpur** Bihār, N India 25°09′N 85°13′E
Islam Qala *see* Eslām Qal'eh
18 *K16* **Island Beach** *spit* New Jersey, NE USA
19 *S4* **Island Falls** Maine, NE USA 45°59′N 68°16′W
Island/Ísland *see* Iceland
182 *H6* **Island Lagoon** ◎ South Australia
11 *Y13* **Island Lake** ◎ Manitoba, C Canada
29 *W5* **Island Lake Reservoir** ⊠ Minnesota, N USA
33 *R13* **Island Park** Idaho, NW USA 44°27′N 111°21′W
19 *N6* **Island Pond** Vermont, NE USA 44°48′N 71°51′W
184 *K2* **Islands, Bay of** *inlet* North Island, New Zealand
103 *R7* **Is-sur-Tille** Côte d'Or, C France 47°34′N 05°03′E
42 *J3* **Islas de la Bahía** ⬖ *department* N Honduras
65 *L20* **Islas Orcadas Rise** *undersea feature* S Atlantic Ocean
96 *F12* **Islay** *island* SW Scotland, United Kingdom
116 *I15* **Islaz** Teleorman, S Romania 43°44′N 24°45′E
29 *V7* **Isle** Minnesota, N USA 46°08′N 93°28′W
102 *M12* **Isle** ≈ W France
97 *I16* **Isle of Man** ◇ *British Crown Dependency* NW Europe
97 *I16* **Isle of Man** *island* NW Europe
21 *X7* **Isle of Wight** Virginia, NE USA 36°54′N 76°41′W
97 *M24* **Isle of Wight** *cultural region* S England, United Kingdom
191 *Y3* **Isles Lagoon** ◎ Kiritimati, E Kiribati
37 *R11* **Isleta Pueblo** New Mexico, SW USA 34°54′N 106°40′W
Isloch' *see* Islach
61 *E19* **Ismael Cortinas** Flores, S Uruguay 33°57′S 57°05′W
Ismailia *see* Al Ismā'īlīya
Ismâ'ilîya *see* Al Ismā'īlīya
Ismailly *see* İsmayıllı
137 *X11* **İsmayıllı** *Rus.* Ismailly. N Azerbaijan 40°47′N 48°09′E
Ismid *see* İzmit
147 *S12* **Ismoili Somoní, Qullai** *prev.* Qullai Kommunizm. ▲ E Tajikistan
75 *X10* **Isnā** *var.* Esna. SE Egypt 25°16′N 32°30′E
93 *K18* **Isojoki** Etelä-Pohjanmaa, W Finland 62°07′N 22°00′E
82 *M12* **Isoka** Muchinga, NE Zambia 10°08′S 32°43′E
Isola d'Ischia *see* Ischia
Isola d'Istria *see* Izola
Isonzo *see* Soča
15 *U4* **Isoukustouc** ≈ Québec, SE Canada
136 *F15* **Isparta** *var.* Isbarta. Isparta, SW Turkey 37°46′N 30°32′E
136 *F15* **Isparta** *var.* Isbarta. ⬖ *province* SW Turkey
114 *M7* **Isperih** *var.* Isperikh; *prev.* Kemanlar. Razgrad, N Bulgaria 43°43′N 26°49′E
Isperikh *see* Isperih
107 *L26* **Ispica** Sicilia, Italy, C Mediterranean Sea 36°47′N 14°55′E
148 *J14* **Ispikān** Baluchistān, SW Pakistan 26°21′N 62°15′E
137 *Q12* **İspir** Erzurum, NE Turkey 40°29′N 41°02′E
138 *E12* **Israel** *off.* State of Israel, *var.* Medinat Israel, *Heb.* Yisrael, Yisra'el. ◆ *republic* SW Asia
Israel, State of *see* Israel
Issa *see* Vis
55 *S9* **Issano** C Guyana 05°49′N 59°28′W
76 *M16* **Issia** SW Ivory Coast 06°33′N 06°33′W
Issiq Köl *see* Issyk-Kul', Ozero
103 *P11* **Issoire** Puy-de-Dôme, C France 45°33′N 03°15′E
103 *N9* **Issoudun** *anc.* Uxellodunum. Indre, C France 46°57′N 01°59′E
81 *H22* **Issuna** Singida, C Tanzania 05°24′S 34°48′E
Issyk *see* Yesik
Issyk-Kul' *see* Balykchy
147 *X7* **Issyk-Kul', Ozero** *var.* Issiq Köl, *Kir.* Ysyk-Köl. ◎ E Kyrgyzstan
147 *X7* **Issyk-Kul'skaya Oblast'** *Kir.* Ysyk-Köl Oblasty. ⬖ *province* E Kyrgyzstan
149 *Q7* **Īstādeh-ye Moqor, Āb-e-** *var.* Āb-i-Istāda. ◎ SE Afghanistan
136 *D11* **İstanbul** *Bul.* Tsarigrad, *Eng.* Istanbul, *prev.* Constantinople; *anc.* Byzantium. Istanbul, NW Turkey 41°02′N 28°57′E
114 *P12* **İstanbul** ⬖ *province* NW Turkey
114 *P12* **İstanbul Boğazı** *var.* Bosporus Thracius, *Eng.* Bosphorus, Bosporus, *Turk.* Karadeniz Boğazi. *strait* NW Turkey
Istarska Županija *see* Istra
115 *G19* **Isthmía** Pelopónnisos, S Greece 37°55′N 23°02′E
115 *G17* **Istiaía** Évvoia, C Greece 38°57′N 23°09′E
54 *D9* **Istmina** Chocó, W Colombia 05°09′N 76°42′W
23 *W13* **Istokpoga, Lake** ◎ Florida, SE USA
112 *A9* **Istra** *off.* Istarska Županija. ⬖ *province* NW Croatia
112 *I10* **Istra** *Eng.* Istria, *Ger.* Istrien. *cultural region* NW Croatia
103 *R15* **Istres** Bouches-du-Rhône, SE France 43°30′N 04°59′E
Istria/Istrien *see* Istra
153 *T15* **Iswardi** *var.* Ishurdi. Rajshahi, W Bangladesh 24°10′N 89°04′E
127 *V7* **Isyangulovo** Respublika Bashkortostan, W Russian Federation 52°10′N 56°38′E
62 *O6* **Itá** Central, S Paraguay 25°29′S 57°21′W
59 *O17* **Itaberaba** Bahia, E Brazil 12°34′S 40°21′W
59 *M20* **Itabira** *prev.* Presidente Vargas. Minas Gerais, SE Brazil 19°39′S 43°14′W
59 *O18* **Itabuna** Bahia, E Brazil 14°48′S 39°18′W
59 *J18* **Itacaiu** Mato Grosso, S Brazil 14°49′S 51°21′W
58 *G12* **Itacoatiara** Amazonas, N Brazil 03°06′S 58°22′W
54 *D9* **Itagüí** Antioquia, W Colombia 06°12′N 75°40′W
60 *D13* **Itá Ibaté** Corrientes, NE Argentina 27°27′S 57°24′W
60 *G11* **Itaipú, Represa de** ⊠ Brazil/Paraguay
58 *H13* **Itaituba** Pará, NE Brazil 04°15′S 55°56′W
60 *K13* **Itajaí** Santa Catarina, S Brazil 26°50′S 48°39′W
Italia/Italiana, Republica/Italian Republic, The *see* Italy
Italian Somaliland *see* Somalia
25 *T7* **Italy** Texas, SW USA 32°10′N 96°52′W
106 *G12* **Italy** *off.* The Italian Republic, *It.* Italia, Repubblica Italiana. ◆ *republic* S Europe
59 *O19* **Itamaraju** Bahia, E Brazil 16°58′S 39°32′W
59 *C14* **Itamarati** Amazonas, W Brazil 06°13′S 68°17′W
59 *M19* **Itambé, Pico de** ▲ SE Brazil 18°23′S 43°21′W
164 *J13* **Itami** ✈ (Ōsaka) Ōsaka, Honshū, SW Japan 34°47′N 135°24′E
115 *H15* **Ítamos** ▲ N Greece 40°06′N 23°51′E
153 *W11* **Itānagar** *state capital* Arunāchal Pradesh, NE India 27°02′N 93°38′E
Itany *see* Litani
59 *N19* **Itaobim** Minas Gerais, SE Brazil 16°34′S 41°27′W
59 *P15* **Itaparica, Represa de** ⊠ E Brazil
58 *M13* **Itapecuru-Mirim** Maranhão, E Brazil 03°24′S 44°20′W
60 *Q8* **Itaperuna** Rio de Janeiro, SE Brazil 21°14′S 41°51′W
59 *O18* **Itapetinga** Bahia, E Brazil 15°17′S 40°16′W
60 *L10* **Itapetininga** São Paulo, S Brazil 23°36′S 48°07′W
60 *K10* **Itapeva** São Paulo, S Brazil 23°58′S 48°54′W
47 *W6* **Itapicuru, Rio** ≈ NE Brazil
58 *O13* **Itapipoca** Ceará, E Brazil 03°29′S 39°35′W
60 *M9* **Itapira** São Paulo, S Brazil 22°25′S 46°46′W
60 *K8* **Itápolis** São Paulo, S Brazil 21°36′S 48°43′W
60 *K10* **Itaporanga** São Paulo, S Brazil 23°43′S 49°28′W
62 *P7* **Itapúa** *off.* Departamento de Itapúa. ⬖ *department* SE Paraguay
Itapúa, Departamento de *see* Itapúa
59 *E15* **Itapuã do Oeste** Rondônia, W Brazil 09°21′S 63°07′W
61 *E15* **Itaqui** Rio Grande do Sul, S Brazil 29°10′S 56°28′W
60 *K10* **Itararé** São Paulo, S Brazil 24°07′S 49°16′W
60 *K10* **Itararé, Rio** ≈ S Brazil
154 *H11* **Itārsi** Madhya Pradesh, C India 22°39′N 77°48′E
25 *T7* **Itasca** Texas, SW USA 32°09′N 97°09′W
Itassi *see* Vieille Case
60 *D13* **Itatí** Corrientes, NE Argentina 27°16′S 58°15′W
60 *K10* **Itatinga** São Paulo, S Brazil 23°08′S 48°36′W
115 *F18* **Itéas, Kólpos** *gulf* C Greece
57 *N15* **Iténez, Río** *var.* Rio Guaporé. ≈ Bolivia/Brazil *see also* Rio Guaporé
Iténez, Río *see* Guaporé, Rio
54 *H11* **Iteviate, Río** ≈ C Colombia
100 *I13* **Ith** *hill range* C Germany
31 *Q8* **Ithaca** Michigan, N USA 43°17′N 84°36′W
18 *H11* **Ithaca** New York, NE USA 42°26′N 76°30′W
115 *C18* **Itháki** *island* Iónia Nísiá, Greece, C Mediterranean Sea
Itháki *see* Vathy
It Hearrenfean *see* Heerenveen
79 *L17* **Itimbiri** ≈ N Dem. Rep. Congo
Itinomiya *see* Ichinomiya
Itinoseki *see* Ichinoseki
39 *Q5* **Itkilik River** ≈ Alaska, USA
164 *M11* **Itoigawa** Niigata, Honshū, C Japan 37°02′N 137°53′E
15 *R6* **Itomamo, Lac** ◎ Québec, SE Canada
165 *S17* **Itoman** Okinawa, SW Japan 26°05′N 127°40′E
102 *M5* **Iton** ≈ N France
57 *M16* **Itonamas Río** ≈ NE Bolivia
Itoupé, Mont *see* Sommet Tabulaire
Itseqqortoormiit *see* Ittoqqortoormiit
22 *K4* **Itta Bena** Mississippi, S USA 33°30′N 90°19′W
107 *B17* **Ittiri** Sardegna, Italy, C Mediterranean Sea 40°36′N 08°34′E
197 *Q14* **Ittoqqortoormiit** *var.* Itseqqortoormiit, *Dan.* Scoresbysund, *Eng.* Scoresby Sound. Sermersooq, C Greenland 70°33′N 21°52′W
60 *M10* **Itu** São Paulo, S Brazil 23°17′S 47°16′W
54 *D8* **Ituango** Antioquia, NW Colombia 07°07′N 75°46′W
59 *A14* **Ituí, Rio** ≈ NW Brazil
79 *O20* **Itula** Sud-Kivu, E Dem. Rep. Congo 03°30′S 27°50′E
59 *K19* **Itumbiara** Goiás, C Brazil 18°25′S 49°15′W
55 *T9* **Ituni** E Guyana 05°24′N 58°18′W
41 *X13* **Iturbide** Campeche, SE Mexico 19°41′N 89°29′W
Ituri *see* Aruwimi
123 *V13* **Iturup, Ostrov** *island* Kuril'skiye Ostrova, SE Russian Federation
60 *L7* **Ituverava** São Paulo, S Brazil 20°22′S 47°48′W
59 *C15* **Ituxi, Rio** ≈ W Brazil
61 *E14* **Ituzaingó** Corrientes, NE Argentina 27°34′S 56°44′W
101 *K18* **Itz** ≈ C Germany
100 *I9* **Itzehoe** Schleswig-Holstein, N Germany 53°56′N 09°31′E
23 *N2* **Iuka** Mississippi, S USA 34°48′N 88°11′W
60 *I11* **Ivaiporã** Paraná, S Brazil 24°16′S 51°46′W
60 *I11* **Ivaí, Rio** ≈ S Brazil
92 *L10* **Ivalo** *Lapp.* Avveel, Avvil. Lappi, N Finland 68°34′N 27°29′E
92 *L10* **Ivalojoki** *Lapp.* Avveel. ≈ N Finland
119 *H20* **Ivanava** *Pol.* Janów, Janów Poleski, *Rus.* Ivanovo. Brestskaya Voblasts', SW Belarus 52°09′N 25°32′E
79 *F18* **Ivando** *var.* Djidji. ≈ Congo/Gabon
Ivangorod *see* Dęblin
Ivangrad *see* Berane
183 *N7* **Ivanhoe** New South Wales, SE Australia 32°55′S 144°21′E
29 *S9* **Ivanhoe** Minnesota, N USA 44°27′N 96°15′W
14 *D8* **Ivanhoe** ≈ Ontario, S Canada
112 *E8* **Ivanić-Grad** Sisak-Moslavina, N Croatia 45°43′N 16°23′E
117 *T10* **Ivanivka** Khersons'ka Oblast', S Ukraine 46°43′N 34°28′E
117 *P10* **Ivanivka** Odes'ka Oblast', SW Ukraine 46°57′N 30°26′E
113 *L14* **Ivanjica** Serbia, C Serbia 43°36′N 20°14′E
112 *G11* **Ivanjska** *var.* Potkozarje. Republika Srpska, NW Bosnia and Herzegovina 44°54′N 17°04′E
111 *H21* **Ivanka** ✈ (Bratislava) Bratislavský Kraj, W Slovakia 48°10′N 17°13′E
117 *O3* **Ivankiv** *Rus.* Ivankov. Kyyivs'ka Oblast', N Ukraine 50°55′N 29°53′E
Ivankov *see* Ivankiv
39 *O15* **Ivanof Bay** Alaska, USA 55°55′N 159°28′W
116 *J7* **Ivano-Frankivs'k** *Ger.* Stanislau, *Pol.* Stanisławów, *Rus.* Ivano-Frankovsk; *prev.* Stanislav. Ivano-Frankivs'ka Oblast', W Ukraine 48°55′N 24°45′E
Ivano-Frankivs'k *see* Ivano-Frankivs'ka Oblast'
116 *I7* **Ivano-Frankivs'ka Oblast'** *var.* Ivano-Frankivs'k, *Rus.* Ivano-Frankovskaya Oblast'; *prev.* Stanislavskaya Oblast'. ⬖ *province* W Ukraine
Ivano-Frankovsk *see* Ivano-Frankivs'k
Ivano-Frankovskaya Oblast' *see* Ivano-Frankivs'ka Oblast'
124 *M16* **Ivanovo** Ivanovskaya Oblast', W Russian Federation 57°02′N 40°58′E
Ivanovo *see* Ivanava
124 *M16* **Ivanovskaya Oblast'** ⬖ *province* W Russian Federation
35 *X12* **Ivanpah Lake** ◎ California, W USA
112 *E7* **Ivanščica** ▲▲ NE Croatia
127 *R7* **Ivanteyevka** Saratovskaya Oblast', W Russian Federation 52°13′N 49°06′E
Ivantsevichi/Ivatsevichi *see* Ivatsevichy
116 *I4* **Ivanychi** Volyns'ka Oblast', NW Ukraine 50°37′N 24°22′E
119 *H18* **Ivatsevichy** *Pol.* Iwacewicze, *Rus.* Ivantsevichi, Ivatsevichi. Brestskaya Voblasts', SW Belarus 52°43′N 25°21′E
114 *L12* **Ivaylovgrad** Haskovo, S Bulgaria 41°32′N 26°06′E
114 *K11* **Ivaylovgrad, Yazovir** ⊠ S Bulgaria
122 *G9* **Ivdel'** Sverdlovskaya Oblast', C Russian Federation 60°42′N 60°07′E
Ivenets *see* Ivyanyets
116 *L12* **Iveşti** Galaţi, E Romania 45°27′N 28°00′E
Ivgovuotna *see* Lyngen
Ivigtut *see* Ivittuut
59 *I21* **Ivinheima** Mato Grosso do Sul, SW Brazil 22°16′S 53°52′W
196 *M15* **Ivittuut** *var.* Ivigtut. Sermersooq, S Greenland 61°12′N 48°10′W
Iviza *see* Eivissa\Ibiza
172 *I6* **Ivohibe** Fianarantsoa, SE Madagascar 22°28′S 46°53′E
Ivoire, Côte d' *see* Ivory Coast
76 *L15* **Ivory Coast** *off.* Republic of the Ivory Coast, *Fr.* Côte d'Ivoire, République de la Côte d'Ivoire. ◆ *republic* W Africa
68 *C12* **Ivory Coast** *Fr.* Côte d'Ivoire. *coastal region* S Ivory Coast
Ivory Coast, Republic of the *see* Ivory Coast
95 *L22* **Ivösjön** ◎ S Sweden
106 *B7* **Ivrea** *anc.* Eporedia. Piemonte, NW Italy 45°28′N 07°52′E
12 *J2* **Ivujivik** Québec, NE Canada 62°26′N 77°49′W
119 *J16* **Ivyanyets** *Rus.* Ivenets. Minskaya Voblasts', C Belarus 53°53′N 26°45′E
Iv'ye *see* Iwye
Iwacewicze *see* Ivatsevichy
165 *R8* **Iwaizumi** Iwate, Honshū, NE Japan 39°48′N 141°46′E
165 *P12* **Iwaki** Fukushima, Honshū, N Japan 37°01′N 140°52′E
164 *F13* **Iwakuni** Yamaguchi, Honshū, SW Japan 34°08′N 132°06′E
165 *S4* **Iwamizawa** Hokkaidō, NE Japan 43°12′N 141°47′E
165 *R4* **Iwanai** Hokkaidō, NE Japan 42°51′N 140°21′E
165 *Q10* **Iwanuma** Miyagi, Honshū, C Japan 38°06′N 140°51′E
164 *L14* **Iwata** Shizuoka, Honshū, S Japan 34°42′N 137°51′E
165 *R8* **Iwate** Iwate, Honshū, N Japan 40°03′N 141°12′E
165 *R8* **Iwate** *off.* Iwate-ken. ⬖ *prefecture* Honshū, C Japan
Iwate-ken *see* Iwate
77 *S16* **Iwo** Oyo, SW Nigeria 07°21′N 03°58′E
Iwojima *see* Iō-jima
119 *I16* **Iwye** *Pol.* Iwje, *Rus.* Iv'ye. Hrodzyenskaya Voblasts', W Belarus 53°56′N 25°46′E
42 *C4* **Ixcán, Río** ≈ Guatemala/Mexico
99 *G18* **Ixelles** *Dut.* Elsene. Brussels, C Belgium 50°49′N 04°21′E
57 *J16* **Ixiamas** La Paz, NW Bolivia 13°45′S 68°10′W
41 *O13* **Ixmiquilpan** *var.* Ixmiquilpán. Hidalgo, C Mexico 20°30′N 99°15′W
Ixmiquilpán *see* Ixmiquilpan
Ixopo *see* eXobho
Ixtaccíhuatal, Volcán *see* Iztaccíhuati, Volcán
40 *M16* **Ixtapa** Guerrero, S Mexico 17°38′N 101°29′W
41 *S16* **Ixtepec** Oaxaca, SE Mexico 16°32′N 95°03′W
40 *K12* **Ixtlán** *var.* Ixtlán del Río. Nayarit, C Mexico 21°02′N 104°21′W
Ixtlán del Río *see* Ixtlán
122 *H11* **Iyevlevo** Tyumenskaya Oblast', C Russian Federation 57°36′N 67°20′E
164 *F14* **Iyo** Ehime, Shikoku, SW Japan 33°43′N 132°42′E
164 *E14* **Iyo-nada** *sea* S Japan
42 *E4* **Izabal** *off.* Departamento de Izabal. ⬖ *department* E Guatemala
Izabal, Departamento de *see* Izabal
42 *F5* **Izabal, Lago de** *prev.* Golfo Dulce. ◎ E Guatemala
143 *O9* **Īzad Khvāst** Fārs, C Iran 31°31′N 52°09′E
41 *X12* **Izamal** Yucatán, SE Mexico 20°58′N 89°00′W
127 *Q16* **Izberbash** Respublika Dagestan, SW Russian Federation 42°32′N 47°51′E
99 *C18* **Izegem** *prev.* Iseghem. West-Vlaanderen, W Belgium 50°55′N 03°13′E
142 *M9* **Īzeh** Khūzestān, SW Iran 31°48′N 49°49′E
165 *T16* **Izena-jima** *island* Nansei-shotō, SW Japan
114 *N10* **Izgrev** Burgas, E Bulgaria 42°09′N 27°49′E
127 *T2* **Izhevsk** *prev.* Ustinov. Udmurtskaya Respublika, NW Russian Federation 56°48′N 53°12′E
125 *S7* **Izhma** Respublika Komi, NW Russian Federation 64°56′N 53°52′E
125 *S7* **Izhma** ≈ NW Russian Federation
141 *X8* **Izkī** NE Oman 22°45′N 57°36′E
Izmail *see* Izmayil
117 *N13* **Izmayil** *Rus.* Izmail. Odes'ka Oblast', SW Ukraine 45°19′N 28°49′E
136 *B14* **İzmir** *prev.* Smyrna. İzmir, W Turkey 38°25′N 27°10′E
136 *C14* **İzmir** *prev.* Smyrna. ⬖ *province* W Turkey
136 *E11* **İzmit** *var.* Ismid; *anc.* Astacus. Kocaeli, NW Turkey 40°47′N 29°55′E
104 *M14* **Iznájar** Andalucía, S Spain 37°17′N 04°16′W
104 *M14* **Iznajar, Embalse de** ⊠ S Spain
105 *N14* **Iznalloz** Andalucía, S Spain 37°23′N 03°31′W
136 *E11* **İznik** Bursa, NW Turkey 40°27′N 29°43′E
136 *E12* **İznik Gölü** ◎ NW Turkey
126 *M14* **Izobil'nyy** Stavropol'skiy Kray, SW Russian Federation 45°22′N 41°40′E
109 *S13* **Izola** *It.* Isola d'Istria. SW Slovenia 45°31′N 13°40′E
138 *H9* **Izra'** *var.* Ezra, Ezraa. Dar'ā, S Syria 32°52′N 36°15′E
41 *P14* **Iztaccíhuati, Volcán** *var.* Volcán Ixtaccíhuatal. 🌋 S Mexico 19°07′N 98°37′W
42 *C7* **Iztapa** Escuintla, SE Guatemala 13°58′N 90°42′W
Izúcar de Matamoros *see* Matamoros
165 *N14* **Izu-hantō** *peninsula* Honshū, S Japan
Izuhara *see* Tsushima
164 *J14* **Izumiōtsu** Ōsaka, Honshū, SW Japan 34°29′N 135°25′E
164 *I14* **Izumisano** Ōsaka, Honshū, SW Japan 34°23′N 135°18′E
164 *G12* **Izumo** Shimane, Honshū, SW Japan 35°22′N 132°46′E
192 *H5* **Izu Trench** *undersea feature* NW Pacific Ocean
122 *K6* **Izvestiy TsIK, Ostrova** *island* N Russian Federation
114 *G10* **Izvor** Pernik, W Bulgaria 42°27′N 22°53′E
116 *L5* **Izyaslav** Khmel'nyts'ka Oblast', W Ukraine 50°08′N 26°53′E
117 *W6* **Izyum** Kharkivs'ka Oblast', E Ukraine 49°12′N 37°19′E

J

93 *M18* **Jaala** Kymenlaakso, S Finland 61°04′N 26°30′E
140 *J5* **Jabal ash Shifā** *desert* NW Saudi Arabia
141 *U8* **Jabal aẓ Ẓannah** *var.* Jebel Dhanna. Abū Ẓaby, W United Arab Emirates 24°10′N 52°36′E
138 *E11* **Jabalīya** *var.* Jabāliyah. NE Gaza Strip 31°32′N 34°29′E
Jabāliyah *see* Jabalīya
105 *N11* **Jabalón** ≈ C Spain
154 *J10* **Jabalpur** *prev.* Jubbulpore. Madhya Pradesh, C India 23°10′N 79°59′E
141 *N15* **Jabal Zuqar, Jazīrat** *var.* Az Zuqur. *island* SW Yemen
Jabar *see* Jawa Barat
Jabat *see* Jabwot
138 *J3* **Jabbūl, Sabkhat al** *sabkha* NW Syria
181 *P1* **Jabiru** Northern Territory, N Australia 12°44′S 132°48′E
138 *H4* **Jablah** *var.* Jeble, *Fr.* Djéblé. Al Lādhiqīyah, W Syria 35°00′N 36°00′E
112 *C11* **Jablanac** Lika-Senj, W Croatia 44°43′N 14°54′E
113 *H14* **Jablanica** Federacija Bosne I Hercegovine, SW Bosnia and Herzegovina 43°39′N 17°43′E
113 *M20* **Jablanica** *Alb.* Mali i Jabllanicës, *var.* Malet e Jabllanicës. ▲ Albania/FYR Macedonia *see also* Jabllanicës, Mali i
Jabllanicës, Malet e *see* Jablanica/Jabllanicës, Mali i
113 *M20* **Jabllanicës, Mali i**, *Mac.* Jablanica. ▲ Albania/FYR Macedonia *see also* Jablanica
111 *E15* **Jablonec nad Nisou** *Ger.* Gablonz an der Neisse. Liberecký Kraj, N Czech Republic 50°44′N 15°10′E
Jabłonków/Jablunkau *see* Jablunkov
110 *J9* **Jabłonowo Pomorskie** Kujawski-pomorskie, C Poland 53°24′N 19°08′E
111 *J17* **Jablunkov** *Ger.* Jablunkau, *Pol.* Jabłonków. Moravskoslezský Kraj, E Czech Republic 49°35′N 18°46′E
59 *Q15* **Jaboatão** Pernambuco, E Brazil 08°05′S 35°W
60 *L8* **Jaboticabal** São Paulo, S Brazil 21°15′S 48°17′W
189 *U7* **Jabwot** *var.* Jabat, Jebat, Jōwat. *island* Ralik Chain, S Marshall Islands
105 *S4* **Jaca** Aragón, NE Spain 42°34′N 00°33′W
42 *B4* **Jacaltenango** Huehuetenango, W Guatemala 15°39′N 91°46′W
59 *G14* **Jacaré-a-Canga** Pará, NE Brazil 05°59′S 57°32′W
60 *N10* **Jacareí** São Paulo, S Brazil 23°18′S 45°55′W
59 *I18* **Jaciara** Mato Grosso, W Brazil 15°59′S 54°57′W
59 *E15* **Jaciparaná** Rondônia, W Brazil 09°20′S 64°28′W
19 *P5* **Jackman** Maine, NE USA 45°35′N 70°14′W
35 *X1* **Jackpot** Nevada, W USA 41°57′N 114°41′W
20 *M8* **Jacksboro** Tennessee, S USA 36°19′N 84°11′W
25 *S6* **Jacksboro** Texas, SW USA 33°13′N 98°11′W
23 *N7* **Jackson** Alabama, S USA 31°30′N 87°53′W
35 *P7* **Jackson** California, W USA 38°19′N 120°46′W
23 *T4* **Jackson** Georgia, SE USA 33°17′N 83°58′W
21 *O6* **Jackson** Kentucky, S USA 37°32′N 83°24′W
22 *J8* **Jackson** Louisiana, S USA 30°50′N 91°13′W
31 *Q10* **Jackson** Michigan, N USA 42°15′N 84°24′W
29 *T11* **Jackson** Minnesota, N USA 43°38′N 95°00′W
22 *K5* **Jackson** *state capital* Mississippi, S USA 32°19′N 90°12′W
27 *Y7* **Jackson** Missouri, C USA 37°23′N 89°40′W
21 *W8* **Jackson** North Carolina, SE USA 36°24′N 77°25′W
31 *T15* **Jackson** Ohio, NE USA 39°03′N 82°40′W
20 *G9* **Jackson** Tennessee, S USA 35°37′N 88°50′W
33 *S14* **Jackson** Wyoming, C USA 43°28′N 110°45′W
185 *C19* **Jackson Bay** *bay* South Island, New Zealand
186 *E9* **Jackson Field** ✈ (Port Moresby) Central/National Capital District, S Papua New Guinea 09°28′S 147°12′E
185 *C20* **Jackson Head** *headland* South Island, New Zealand 43°57′S 168°38′E
23 *S8* **Jackson, Lake** ◎ Florida, SE USA
33 *S13* **Jackson Lake** ◎ Wyoming, C USA
194 *J6* **Jackson, Mount** ▲ Antarctica 71°43′S 63°45′W
37 *U3* **Jackson Reservoir** ◎ Colorado, C USA
23 *Q3* **Jacksonville** Alabama, S USA 33°48′N 85°45′W
27 *V11* **Jacksonville** Arkansas, C USA 34°52′N 92°08′W
23 *W8* **Jacksonville** Florida, SE USA 30°20′N 81°39′W
30 *K14* **Jacksonville** Illinois, N USA 39°43′N 90°13′W
21 *W11* **Jacksonville** North Carolina, SE USA 34°45′N 77°26′W
25 *W7* **Jacksonville** Texas, SW USA 31°57′N 95°16′W
23 *X9* **Jacksonville Beach** Florida, SE USA 30°17′N 81°23′W
44 *L9* **Jacmel** *var.* Jaquemel. S Haiti 18°13′N 72°33′W
Jacob *see* Nkayi
149 *Q12* **Jacobābād** Sind, SE Pakistan 28°16′N 68°30′E
55 *T11* **Jacobs Ladder Falls** *waterfall* S Guyana
45 *O11* **Jaco, Pointe** *headland* N Dominica 15°38′N 61°25′W
15 *Q9* **Jacques-Cartier** ≈ Québec, SE Canada
13 *P11* **Jacques-Cartier, Détroit de** *var.* Jacques-Cartier Passage. *strait* Gulf of St. Lawrence/St. Lawrence River, Canada
15 *W6* **Jacques-Cartier, Mont** ▲ Québec, SE Canada 48°58′N 66°00′W
Jacques-Cartier Passage *see* Jacques-Cartier, Détroit de
61 *H16* **Jacuí, Rio** ≈ S Brazil
60 *L11* **Jacupiranga** São Paulo, S Brazil 24°42′S 48°00′W
100 *G10* **Jade** ≈ NW Germany
100 *G10* **Jadebusen** *bay* NW Germany
Jadotville *see* Likasi
Jadransko More/Jadransko Morje *see* Adriatic Sea
105 *O7* **Jadraque** Castilla-La Mancha, C Spain 40°55′N 02°55′W
95 *I22* **Jægerspris** Hovedstaden, E Denmark 55°52′N 11°59′E
56 *C10* **Jaén** Cajamarca, N Peru 05°45′S 78°51′W
105 *N13* **Jaén** Andalucía, SW Spain 37°46′N 03°48′W
105 *N13* **Jaén** ⬖ *province* Andalucía, S Spain
95 *C17* **Jæren** *physical region* S Norway
155 *J23* **Jaffna** Northern Province, N Sri Lanka 09°42′N 80°03′E
155 *K23* **Jaffna Lagoon** *lagoon* N Sri Lanka
19 *N10* **Jaffrey** New Hampshire, NE USA 42°46′N 72°00′W
138 *H13* **Jafr, Qā' al** *var.* El Jafr. *salt pan* S Jordan
152 *J9* **Jagādhri** Haryāna, N India 30°11′N 77°18′E
118 *H4* **Jägala** *var.* Jägala Jōgi, *Ger.* Jaggowal. ≈ NW Estonia
Jägala Jōgi *see* Jägala
Jagannath *see* Puri
155 *L14* **Jagdalpur** Chhattīsgarh, C India 19°07′N 82°04′E
163 *U5* **Jagdaqi** Nei Mongol Zizhiqu, N China 50°26′N 124°03′E
Jägerndorf *see* Krnov
Jaggowal *see* Jägala
139 *O2* **Jaghjaghah, Nahr** ≈ N Syria
112 *N13* **Jagodina** *prev.* Svetozarevo. Serbia, C Serbia 43°59′N 21°15′E
112 *K12* **Jagodnja** ▲▲ W Serbia
101 *I20* **Jagst** ≈ SW Germany
155 *I14* **Jagtiāl** Telangana, C India 18°49′N 78°53′E
61 *H18* **Jaguarão** Rio Grande do Sul, S Brazil 32°30′S 53°25′W
61 *H18* **Jaguarão, Rio** *var.* Río Yaguarón. ≈ Brazil/Uruguay
60 *K11* **Jaguariaíva** Paraná, S Brazil 24°15′S 49°44′W
44 *D5* **Jagüey Grande** Matanzas, W Cuba 22°31′N 81°07′W
153 *P14* **Jahānābād** Bihār, N India 25°13′N 84°59′E
Jahra *see* Al Jahrā'
143 *P12* **Jahrom** *var.* Jahrum. Fārs, S Iran 28°35′N 53°32′E
Jahrum *see* Jahrom
Jailolo *see* Halmahera, Pulau
Jainat *see* Chai Nat
Jainti *see* Jayanti
152 *H12* **Jaipur** *prev.* Jeypore. *state capital* Rājasthān, N India 26°54′N 75°47′E
153 *T14* **Jaipurhat** *var.* Joypurhat. Rajshahi, NW Bangladesh 25°04′N 89°06′E
152 *D11* **Jaisalmer** Rājasthān, NW India 26°55′N 70°56′E
154 *O12* **Jājapur** *var.* Jajpur, Panikoilli. Odisha, E India 18°54′N 82°36′E
143 *R4* **Jājarm** Khorāsān-e Shemālī, NE Iran 36°58′N 56°26′E
112 *G12* **Jajce** Federacija Bosne I Hercegovine, W Bosnia and Herzegovina 44°20′N 17°16′E
Jajı *see* 'Alī Khēl
Jajpur *see* Jājapur
83 *D17* **Jakalsberg** Otjozondjupa, N Namibia 19°23′S 17°28′E
169 *O15* **Jakarta** *prev.* Djakarta, *Dut.* Batavia. ● (Indonesia) Jawa, C Indonesia 06°08′S 106°45′E
10 *I8* **Jakes Corner** Yukon, W Canada 60°18′N 134°00′W
152 *H9* **Jākhal** Haryāna, NW India 29°46′N 75°51′E
Jakobeny *see* Iacobeni
93 *K16* **Jakobstad** *Fin.* Pietarsaari. Österbotten, W Finland 63°41′N 22°40′E
Jakobstadt *see* Jēkabpils
113 *O18* **Jakupica** ▲▲ C FYR Macedonia
37 *W15* **Jal** New Mexico, SW USA 32°07′N 103°10′W
141 *P7* **Jalājil** *var.* Ğalāğıl. Ar Riyāḍ, C Saudi Arabia 25°43′N 45°22′E
149 *S5* **Jalālābād** *var.* Jalalabad, Jelalabad. Nangarhār, E Afghanistan 34°26′N 70°28′E
Jalal-Abad *see* Dzhalal-Abad, Dzhalal-Abadskaya Oblast', Kyrgyzstan
Jalal-Abad Oblasty *see* Dzhalal-Abadskaya Oblast'
149 *V7* **Jalālpur** Punjab, E Pakistan 32°39′N 74°11′E
149 *T11* **Jalālpur Pīrwāla** Punjab, E Pakistan 29°30′N 71°20′E
152 *H8* **Jalandhar** *prev.* Jullundur. Punjab, N India 31°20′N 75°37′E
42 *J7* **Jalán, Río** ≈ S Honduras
42 *E6* **Jalapa** Jalapa, C Guatemala 14°39′N 89°59′W
42 *J7* **Jalapa** Nueva Segovia, NW Nicaragua 13°56′N 86°11′W
42 *A3* **Jalapa** *off.* Departamento de Jalapa. ⬖ *department* SE Guatemala
Jalapa, Departamento de *see* Jalapa
42 *E6* **Jalapa, Río** ≈ SE Guatemala
143 *X13* **Jālaq** Sīstān va Balūchestān, SE Iran
93 *K17* **Jalasjärvi** Etelä-Pohjanmaa, W Finland 62°30′N 22°50′E
149 *O8* **Jaldak** Zābul, SE Afghanistan 32°00′N 66°45′E
60 *J7* **Jales** São Paulo, S Brazil 20°15′S 50°34′W
154 *P11* **Jaleshwar** *var.* Jaleswar. Odisha, NE India 21°51′N 87°15′E
Jaleswar *see* Jaleshwar
154 *F12* **Jalgaon** Mahārāshtra, C India 21°01′N 75°34′E
139 *W12* **Jalībah** Dhī Qār, S Iraq 30°37′N 46°31′E
139 *W13* **Jalīb Shahāb** Al Muthanná, S Iraq 30°26′N 46°09′E
77 *X15* **Jalingo** Taraba, E Nigeria 08°54′N 11°22′E
40 *K13* **Jalisco** ⬖ *state* SW Mexico
154 *G13* **Jālna** Mahārāshtra, W India 19°50′N 75°53′N
Jalomitsa *see* Ialomiţa
105 *R5* **Jalón** ≈ N Spain
152 *E13* **Jālor** Rājasthān, N India 25°21′N 72°43′E
112 *K11* **Jalovik** Serbia, W Serbia 44°37′N 19°48′E
40 *L12* **Jalpa** Zacatecas, C Mexico 21°40′N 103°W
153 *S12* **Jalpāiguri** West Bengal, NE India 26°43′N 88°24′E
41 *O12* **Jalpán** *var.* Jalpan. Querétaro de Arteaga, C Mexico 21°13′N 99°28′W
Jalpan *see* Jalpán
67 *P2* **Jalta** *island* N Tunisia
75 *S9* **Jālū** *var.* Jūlā. NE Libya 29°02′N 21°33′E
189 *U8* **Jaluit Atoll** *var.* Jālwōj. *atoll* Ralik Chain, S Marshall Islands
Jālwōj *see* Jaluit Atoll
81 *L18* **Jamaame** *It.* Giamame; *prev.* Margherita. Jubbada Hoose, S Somalia 0°00′N 42°43′E
77 *W13* **Jamaare** ≈ NE Nigeria

◆ Country
● Country Capital
◇ Dependent Territory
○ Dependent Territory Capital
⬖ Administrative Regions
✈ International Airport
▲ Mountain
▲▲ Mountain Range
🌋 Volcano
≈ River
◎ Lake
⊠ Reservoir

JAMAICA – JINZHOU

44 G9 **Jamaica** ◆ commonwealth republic W West Indies
47 P3 **Jamaica** island W West Indies
44 I9 **Jamaica Channel** channel Haiti/Jamaica
153 T14 **Jamalpur** Dhaka, N Bangladesh 24°54′N 89°57′E
153 Q14 **Jamālpur** Bihār, NE India 25°19′N 86°30′E
168 L9 **Jamaluang** var. Jemaluang. Johor, Peninsular Malaysia 02°15′N 103°50′E
59 I14 **Jamanxim, Rio** ≈ C Brazil
56 B8 **Jambeli, Canal de** channel S Ecuador
99 I20 **Jambes** Namur, SE Belgium 50°26′N 04°51′E
168 L12 **Jambi** var. Telanaipura; prev. Djambi. Sumatera, W Indonesia 01°34′S 103°37′E
168 K12 **Jambi** off. Propinsi Jambi, var. Djambi. ◐ province W Indonesia
Jambi, Propinsi see Jambi
Jamdena see Yamdena, Pulau
12 H8 **James Bay** bay Ontario/Québec, E Canada
63 F19 **James, Isla** island Archipiélago de los Chonos, S Chile
181 Q8 **James Ranges** ▲▲ Northern Territory, C Australia
29 P8 **James River** ≈ North Dakota/South Dakota, N USA
21 X7 **James River** ≈ Virginia, NE USA
194 H4 **James Ross Island** island Antarctica
182 I8 **Jamestown** South Australia 33°13′S 138°36′E
65 G25 **Jamestown** ○ (Saint Helena) NW Saint Helena 15°56′S 05°44′W
35 P8 **Jamestown** California, W USA 37°57′N 120°25′W
20 L7 **Jamestown** Kentucky, S USA 36°58′N 85°03′W
18 D11 **Jamestown** New York, NE USA 42°05′N 79°15′W
29 P5 **Jamestown** North Dakota, N USA 46°54′N 98°42′W
20 L8 **Jamestown** Tennessee, S USA 36°24′N 84°58′W
Jamestown see Holetown
15 N10 **Jamet** ≈ Québec, SE Canada
41 Q17 **Jamiltepec** var. Santiago Jamiltepec. Oaxaca, SE Mexico 16°18′N 97°51′W
95 F20 **Jammerbugten** bay Skagerrak, E North Sea
152 H6 **Jammu** prev. Jummoo. state capital Jammu and Kashmir, NW India 32°43′N 74°54′E
152 I5 **Jammu and Kashmir** var. Jammu-Kashmir, Kashmir. ◐ state NW India
149 V4 **Jammu and Kashmir** disputed region India/Pakistan
Jammu-Kashmir see Jammu and Kashmir
154 B10 **Jāmnagar** prev. Navanagar. Gujarāt, W India 22°28′N 70°06′E
149 S11 **Jāmpur** Punjab, E Pakistan 29°38′N 70°40′E
93 L18 **Jämsä** Keski-Suomi, C Finland 61°51′N 25°10′E
93 L18 **Jämsänkoski** Keski-Suomi, C Finland 61°54′N 25°10′E
153 Q16 **Jamshedpur** Jhārkhand, NE India 22°47′N 86°12′E
94 K9 **Jämtland** ◐ county C Sweden
153 Q14 **Jamūi** Bihār, NE India 24°57′N 86°14′E
Jamuna see Brahmaputra
153 T14 **Jamuna Nadi** ≈ N Bangladesh
Jamundá see Nhamundá, Rio
54 D11 **Jamundí** Valle del Cauca, SW Colombia 03°16′N 76°31′W
153 Q12 **Janakpur** Central, C Nepal 26°45′N 85°55′E
59 N18 **Janaúba** Minas Gerais, SE Brazil 15°47′S 43°16′W
58 K11 **Janaucu, Ilha** island NE Brazil
143 Q7 **Jandaq** Eşfahān, C Iran 34°04′N 54°26′E
64 Q11 **Jandia, Punta de** headland Fuerteventura, Islas Canarias, Spain, NE Atlantic Ocean 28°03′N 14°32′W
59 B14 **Jandiatuba, Rio** ≈ NW Brazil
105 N12 **Jándula** ≈ S Spain
29 V10 **Janesville** Minnesota, N USA 44°07′N 93°43′W
30 L9 **Janesville** Wisconsin, N USA 42°42′N 89°02′W
83 N20 **Jangamo** Inhambane, SE Mozambique 24°04′S 35°25′E
155 J14 **Jangaon** Telangana, C India 18°47′N 79°25′E
153 S14 **Jangīpur** West Bengal, NE India 24°31′N 88°03′E
Janina see Ioánnina
Janischken see Joniškis
112 J11 **Janja** NE Bosnia and Herzegovina 44°40′N 19°15′E
Jankovac see Jánoshalma
197 Q15 **Jan Mayen** ◇ constituent part of Norway N Atlantic Ocean
84 D5 **Jan Mayen** island N Atlantic Ocean
197 R15 **Jan Mayen Fracture Zone** tectonic feature Greenland Sea/Norwegian Sea
197 R15 **Jan Mayen Ridge** undersea feature Greenland Sea/Norwegian Sea 69°00′N 08°00′W
40 H3 **Janos** Chihuahua, N Mexico 30°50′N 108°10′W
111 K25 **Jánoshalma** SCr. Jankovac. Bács-Kiskun, S Hungary 46°19′N 19°16′E
Janów see Ivanava, Belarus
110 H10 **Janowiec Wielkopolski** Ger. Janowitz. Kujawski-pomorskie, C Poland 52°47′N 17°30′E
Janowitz see Janowiec Wielkopolski
Janow/Janów see Jonava, Lithuania
111 O15 **Janów Lubelski** Lubelski, E Poland 50°42′N 22°24′E
Janów Poleski see Ivanava
83 H25 **Jansenville** Eastern Cape, S South Africa 32°56′S 24°40′E
59 M18 **Januária** Minas Gerais, SE Brazil 15°28′S 44°23′W
Janūbīyah, Al Bādiyah al see Ash Shāmīyah
102 I7 **Janzé** Ille-et-Vilaine, NW France 47°55′N 01°28′W
154 F10 **Jaora** Madhya Pradesh, C India 23°40′N 75°10′E
131 Y9 **Japan** var. Nippon, Jap. Nihon. ◆ monarchy E Asia
129 Y9 **Japan** island group E Asia
192 H4 **Japan Basin** undersea feature N Sea of Japan 40°00′N 135°00′E
129 Y8 **Japan, Sea of** var. East Sea, Rus. Yapanskoye More. sea NW Pacific Ocean see also East Sea
192 H4 **Japan Trench** undersea feature NW Pacific Ocean 37°00′N 143°00′E
Japen see Yapen, Pulau
59 A15 **Japiim** var. Máncio Lima. Acre, W Brazil 08°00′S 73°39′W
58 D12 **Japurá** Amazonas, N Brazil 01°43′S 66°14′W
58 C12 **Japurá, Rio** var. Río Caquetá, Yapurá. ≈ Brazil/Colombia see also Caquetá, Río
Japurá, Rio see Caquetá, Río
43 W17 **Jaqué** Darién, SE Panama 07°31′N 78°09′W
Jaquemel see Jacmel
Jarablos see Jarābulus
138 K2 **Jarābulus** var. Jarablos, Jerablus, Fr. Djérablous. Ḩalab, N Syria 36°51′N 38°02′E
60 K13 **Jaraguá do Sul** Santa Catarina, S Brazil 26°29′S 49°07′W
104 K9 **Jaraicejo** Extremadura, W Spain 39°40′N 05°49′W
104 K9 **Jaraíz de la Vera** Extremadura, W Spain 40°04′N 05°45′W
105 O7 **Jarama** ≈ C Spain
63 J20 **Jaramillo** Santa Cruz, SE Argentina 47°10′S 67°07′W
Jarandilla de la Vega see Jarandilla de la Vera
104 K8 **Jarandilla de la Vera** var. Jarandilla de la Vega. Extremadura, W Spain 40°08′N 05°39′W
149 V9 **Jarānwāla** Punjab, E Pakistan 31°20′N 73°26′E
138 G9 **Jarash** var. Jerash; anc. Gerasa. Jarash, NW Jordan 32°17′N 35°54′E
138 G8 **Jarash** off. Muḩāfa at Jarash. ◐ governorate N Jordan
Jarash, Muḩāfa at see Jarash
Jarbah, Jazīrat see Jerba, Île de
94 N13 **Järbo** Gävleborg, C Sweden 60°43′N 16°40′E
Jardan see Yordon
44 F7 **Jardines de la Reina, Archipiélago de los** island group C Cuba
Jargalant see Battsengel
162 I8 **Jargalant** Bayanhongor, C Mongolia 47°14′N 99°43′E
162 K6 **Jargalant** Bulgan, N Mongolia 49°09′N 104°19′E
162 G7 **Jargalant** var. Buyanbat. Govĭ-Altay, W Mongolia 47°00′N 95°57′E
162 I6 **Jargalant** var. Orgil. Hövsgöl, C Mongolia 48°31′N 99°19′E
Jargalant see Battsengel
Jargalant see Bulgan, Bayan-Ölgiy, Mongolia
Jargalant see Biger, Govĭ-Altay, Mongolia
Jarid, Shaṭṭ al see Jerid, Chott el
58 I11 **Jari, Rio** var. Jary. ≈ N Brazil
141 N7 **Jarīr, Wādī al** dry watercourse C Saudi Arabia
Jarja see Yur'ya
94 L13 **Järna** var. Dala-Jarna. Dalarna, C Sweden 60°31′N 14°22′E
95 O16 **Järna** Stockholm, C Sweden 59°05′N 17°35′E
102 K11 **Jarnac** Charente, W France 45°41′N 00°10′W
110 H12 **Jarocin** Wielkopolskie, C Poland 51°59′N 17°30′E
111 F16 **Jaroměř** Ger. Jermer. Královéhradecký Kraj, N Czech Republic 50°22′N 15°55′E
Jaroslau see Jarosław
111 O16 **Jarosław** Ger. Jaroslau, Rus. Yaroslav. Podkarpackie, SE Poland 50°01′N 22°41′E
93 F16 **Järpen** Jämtland, C Sweden 63°21′N 13°30′E
147 O14 **Jarqo'rg'on** Rus. Dzharkurgan. Surkhondaryo Viloyati, S Uzbekistan 37°31′N 67°20′E
139 P2 **Jarrah, Wadi** dry watercourse NE Syria
Jars, Plain of see Xiangkhoang, Plateau de
162 K14 **Jartai Yanchi** ◎ N China
59 E16 **Jaru** Rondônia, W Brazil 10°24′S 62°45′W
Jarud Qi see Lubei
118 I4 **Järva-Jaani** Ger. Sankt-Johannis. Järvamaa, N Estonia 59°03′N 25°54′E
118 G5 **Järvakandi** Ger. Jerwakant. Raplamaa, NW Estonia 58°45′N 24°49′E
118 H4 **Järvamaa** var. Järva Maakond. ◐ province N Estonia
Järva Maakond see Järvamaa
93 L19 **Järvenpää** Uusimaa, S Finland 60°29′N 25°06′E
14 G17 **Jarvis** Ontario, S Canada 42°53′N 80°06′W
177 R8 **Jarvis Island** ◇ US unincorporated territory C Pacific Ocean
94 M11 **Järvsö** Gävleborg, C Sweden 61°43′N 16°25′E
Jary see Jari, Rio
112 M9 **Jaša Tomić** Vojvodina, NE Serbia 45°27′N 20°51′E
112 D12 **Jasenice** Zadar, SW Croatia 44°15′N 15°33′E
138 I11 **Jashshat al 'Adlah, Wādī al** dry watercourse C Jordan
77 Q16 **Jasikan** E Ghana 07°24′N 00°28′E
Jāsk see Bandar-e Jāsk
146 F6 **Jasliq** Rus. Zhaslyk. Qoraqalpog'iston Respublikasi, NW Uzbekistan 43°57′N 57°30′E
111 N17 **Jasło** Podkarpackie, SE Poland 49°45′N 21°28′E
11 U16 **Jasmin** Saskatchewan, S Canada 51°11′N 103°34′W
65 A23 **Jason Islands** island group NW Falkland Islands
194 I4 **Jason Peninsula** peninsula Antarctica
31 N15 **Jasonville** Indiana, N USA 39°09′N 87°12′W
11 O15 **Jasper** Alberta, SW Canada 52°55′N 118°05′W
14 L13 **Jasper** Ontario, SE Canada 44°50′N 75°57′W
23 O3 **Jasper** Alabama, S USA 33°49′N 87°16′W
27 T9 **Jasper** Arkansas, C USA 36°00′N 93°11′W
23 U8 **Jasper** Florida, SE USA 30°31′N 82°57′W
31 N16 **Jasper** Indiana, N USA 38°22′N 86°57′W
29 R11 **Jasper** Minnesota, N USA 43°51′N 96°24′W
27 S7 **Jasper** Missouri, C USA 37°20′N 94°18′W
20 K10 **Jasper** Tennessee, S USA 35°04′N 85°36′W
25 Y9 **Jasper** Texas, SW USA 30°55′N 94°00′W
11 O15 **Jasper National Park** national park Alberta/British Columbia, SW Canada
Jassy see Iaşi
113 N14 **Jastrebac** ▲▲ SE Serbia
112 D9 **Jastrebarsko** Zagreb, N Croatia 45°40′N 15°40′E
Jastrow see Jastrowie
110 G9 **Jastrowie** Ger. Jastrow. Wielkopolskie, C Poland 53°25′N 16°48′E
111 J17 **Jastrzębie-Zdrój** Śląskie, S Poland 49°58′N 18°34′E
111 L22 **Jászapáti** Jász-Nagykun-Szolnok, E Hungary 47°30′N 20°10′E
111 L22 **Jászberény** Jász-Nagykun-Szolnok, E Hungary 47°30′N 19°56′E
111 L23 **Jász-Nagykun-Szolnok** off. Jász-Nagykun-Szolnok Megye. ◐ county E Hungary
Jász-Nagykun-Szolnok Megye see Jász-Nagykun-Szolnok
59 J19 **Jataí** Goiás, C Brazil 17°58′S 51°45′W
58 G12 **Jatapu, Serra do** ▲▲ N Brazil
41 W16 **Jatate, Río** ≈ SE Mexico
Jateng see Jawa Tengah
149 P17 **Jāti** Sind, SE Pakistan 24°20′N 68°18′E
44 F6 **Jatibonico** Sancti Spíritus, C Cuba 21°56′N 79°11′W
169 O16 **Jatiluhur, Danau** ◎ Jawa, S Indonesia
Jatim see Jawa Timur
Jativa see Xàtiva
149 S11 **Jatoi** prev. Jattoi. Punjab, E Pakistan 29°29′N 70°58′E
Jattoi see Jatoi
60 L9 **Jaú** São Paulo, S Brazil 22°11′S 48°35′W
58 F11 **Jauaperi, Rio** ≈ N Brazil
99 I19 **Jauche** Walloon Brabant, C Belgium 50°42′N 04°55′E
Jauer see Jawor
Jauf see Al Jawf
149 U7 **Jauharābād** Punjab, E Pakistan 32°16′N 72°17′E
57 E14 **Jauja** Junín, C Peru 11°48′S 75°30′W
41 O10 **Jaumave** Tamaulipas, C Mexico 23°28′N 99°22′W
118 H10 **Jaunjelgava** Ger. Friedrichstadt. S Latvia 56°38′N 25°03′E
Jaunlatgale see Pytalovo
118 I8 **Jaunpiebalga** NE Latvia 57°10′N 26°02′E
118 E9 **Jaunpils** C Latvia 56°45′N 23°03′E
153 N13 **Jaunpur** Uttar Pradesh, N India 25°44′N 82°41′E
29 N8 **Java** South Dakota, N USA 45°29′N 99°54′W
Java see Jawa
105 R9 **Javalambre** ▲ E Spain 40°02′N 01°06′W
173 V7 **Java Ridge** undersea feature E Indian Ocean
59 A14 **Javari, Rio** var. Yavarí. ≈ Brazil/Peru
Javarthushuu see Bayan-Uul
169 Q15 **Java Sea** Ind. Laut Jawa. sea W Indonesia
173 U7 **Java Trench** var. Sunda Trench. undersea feature E Indian Ocean
143 Q10 **Javazm** var. Jowzam. Kermān, C Iran 30°31′N 55°01′E
105 T11 **Jávea** Cat. Xàbia. Valenciana, E Spain 38°48′N 00°10′E
Javhlant see Bayan-Ovoo
63 G20 **Javier, Isla** island S Chile
113 L14 **Javor** ▲▲ Bosnia and Herzegovina/Serbia
111 K20 **Javorie** Hung. Jávoros. ▲ S Slovakia 48°26′N 19°16′E
Jávoros see Javorie
93 J14 **Jävre** Norrbotten, N Sweden 65°07′N 21°31′E
192 E8 **Jawa** Eng. Java; prev. Djawa. island C Indonesia
169 O16 **Jawa Barat** off. Propinsi Jawa Barat, var. Jabar, Eng. West Java. ◐ province S Indonesia
Jawa Barat, Propinsi see Jawa Barat
Jawa, Laut see Java Sea
139 R3 **Jawān** Nīnawá, NW Iraq 35°57′N 43°03′E
169 P16 **Jawa Tengah** off. Propinsi Jawa Tengah, var. Jateng, Eng. Central Java. ◐ province S Indonesia
Jawa Tengah, Propinsi see Jawa Tengah
169 R16 **Jawa Timur** off. Propinsi Jawa Timur, var. Jatim, Eng. East Java. ◐ province S Indonesia
Jawa Timur, Propinsi see Jawa Timur
81 N17 **Jawhar** var. Jowhar, It. Giohar. Shabeellaha Dhexe, S Somalia 02°37′N 45°30′E
111 F14 **Jawor** Ger. Jauer. Dolnośląskie, SW Poland 51°01′N 16°11′E
Jaworów see Yavoriv
111 J16 **Jaworzno** Śląskie, S Poland 50°13′N 19°11′E
Jaxartes see Syr Darya
27 R9 **Jay** Oklahoma, C USA 36°25′N 94°49′W
Jayabum see Chaiyaphum
Jayanath see Chai Nat
153 T12 **Jayanti** prev. Jainti. West Bengal, NE India 26°45′N 89°40′E
171 X14 **Jaya, Puncak** prev. Puntjak Carstensz, Puntjak Sukarno. ▲ Papua, E Indonesia 04°00′S 137°10′E
171 Z13 **Jayapura** var. Djajapura, Dut. Hollandia; prev. Kotabaru, Sukarnapura. Papua, E Indonesia 02°37′S 140°39′E
Jay Dairen see Dalian
Jayhawker State see Kansas
147 S12 **Jayilgan** Rus. Dzhailgan, Dzhayilgan. C Tajikistan 39°17′N 71°32′E
155 L14 **Jaypur** var. Jeypore, Jeypur. Odisha, E India 18°54′N 82°36′E
25 O6 **Jayton** Texas, SW USA 33°16′N 100°35′W
143 U13 **Jaz Mūrīān, Hāmūn-e** ◎ SE Iran
138 M4 **Jazrah** Ar Raqqah, C Syria 35°56′N 39°02′E
138 G6 **Jbaïl** var. Jebeil, Jubayl, Jubeil; anc. Biblical Gebal, Bybles. W Lebanon 34°00′N 35°45′E
25 O7 **J. B. Thomas, Lake** ⊠ Texas, SW USA
Jdaïdé see Judaydah
35 X12 **Jean** Nevada, W USA 35°45′N 115°20′W
22 I9 **Jeanerette** Louisiana, S USA 29°54′N 91°39′W
44 L8 **Jean-Rabel** NW Haiti 19°48′N 73°05′W
143 T12 **Jebāl Bārez, Kūh-e** ▲▲ SE Iran
Jebat see Jabwot
77 T15 **Jebba** Kwara, W Nigeria 09°04′N 04°50′E
Jebeil see Jbaïl
116 E12 **Jebel** Hung. Széphely; prev. Hung. Zsebely. Timiş, W Romania 45°33′N 21°14′E
146 B11 **Jebel** Rus. Dzhebel. Balkan Welaýaty, W Turkmenistan 39°42′N 54°10′E
Jebel, Bahr el see White Nile
Jebel Dhanna see Jabal az Zannah
Jeble see Jablah
163 Y15 **Jecheon** Jap. Teisen; prev. Chech'ŏn. N South Korea 37°06′N 128°15′E
96 K13 **Jedburgh** SE Scotland, United Kingdom 55°29′N 02°34′W
Jedda see Jiddah
111 L15 **Jędrzejów** Ger. Endersdorf. Świętokrzyskie, C Poland 50°39′N 20°18′E
100 K12 **Jeetze** var. Jeetzel. ≈ C Germany
Jeetzel see Jeetze
29 U14 **Jefferson** Iowa, C USA 42°01′N 94°22′W
21 Q8 **Jefferson** North Carolina, SE USA 36°24′N 81°33′W
25 X6 **Jefferson** Texas, SW USA 32°45′N 94°21′W
30 M9 **Jefferson** Wisconsin, N USA 43°01′N 88°48′W
27 U5 **Jefferson City** state capital Missouri, C USA 38°33′N 92°13′W
33 R10 **Jefferson City** Montana, NW USA 46°24′N 112°01′W
21 N9 **Jefferson City** Tennessee, S USA 36°07′N 83°29′W
35 U7 **Jefferson, Mount** ▲ Nevada, W USA 38°49′N 116°54′W
32 H12 **Jefferson, Mount** ▲ Oregon, NW USA 44°40′N 121°48′W
20 L5 **Jeffersontown** Kentucky, S USA 38°11′N 85°33′W
31 P16 **Jeffersonville** Indiana, N USA 38°16′N 85°45′W
33 V15 **Jeffrey City** Wyoming, C USA 42°29′N 107°49′W
77 T13 **Jega** Kebbi, NW Nigeria 12°15′N 04°21′E
Jehol see Chengde
163 X17 **Jeju** Jap. Saishū; prev. Cheju. S South Korea 33°31′N 126°34′E
163 Y17 **Jeju-do** Jap. Saishū; prev. Cheju-do, Quelpart. island S South Korea
163 Y17 **Jeju-haehyeop** Eng. Cheju Strait; prev. Cheju-haehyeop. strait S South Korea
62 P5 **Jejui-Guazú, Río** ≈ E Paraguay
118 I10 **Jēkabpils** Ger. Jakobstadt. S Latvia 56°30′N 25°56′E
23 W7 **Jekyll Island** island Georgia, SE USA
169 R13 **Jelai, Sungai** ≈ Borneo, N Indonesia
Jelalabad see Jalālābād
111 H14 **Jelcz-Laskowice** Dolnośląskie, SW Poland 51°01′N 17°24′E
111 E14 **Jelenia Góra** Ger. Hirschberg, Hirschberg im Riesengebirge, Hirschberg im Riesengebirge, Hirschberg in Schlesien. Dolnośląskie, SW Poland 50°54′N 15°48′E
153 S11 **Jelep La** pass N India
118 F9 **Jelgava** Ger. Mitau. C Latvia 56°38′N 23°47′E
112 L13 **Jelica** ▲▲ C Serbia
20 M8 **Jellico** Tennessee, S USA 36°33′N 84°06′W
95 G23 **Jelling** Syddanmark, C Denmark 55°45′N 09°24′E
169 N9 **Jemaja, Pulau** island W Indonesia
Jemaluang see Jamaluang
99 E20 **Jemappes** Hainaut, S Belgium 50°27′N 03°53′E
169 S17 **Jember** prev. Djember. Jawa, C Indonesia 08°07′S 113°45′E
99 I20 **Jemeppe-sur-Sambre** Namur, S Belgium 50°27′N 04°41′E
37 R10 **Jemez Pueblo** New Mexico, SW USA 35°36′N 106°43′W
158 K2 **Jeminay** var. Tuotiereke. Xinjiang Uygur Zizhiqu, NW China 47°28′N 85°49′E
189 U5 **Jemo Island** atoll Ratak Chain, C Marshall Islands
169 U11 **Jempang, Danau** ◎ Borneo, N Indonesia
101 L16 **Jena** Thüringen, C Germany 50°56′N 11°35′E
22 I6 **Jena** Louisiana, S USA 31°40′N 92°07′W
108 I8 **Jenaz** Graubünden, SE Switzerland 46°56′N 09°43′E
109 N7 **Jenbach** Tirol, W Austria 47°24′N 11°47′E
171 N15 **Jeneponto** prev. Djeneponto. Sulawesi, C Indonesia 05°41′S 119°42′E
138 F9 **Jenin** N West Bank 32°28′N 35°17′E
21 P7 **Jenkins** Kentucky, S USA 37°10′N 82°37′W
27 P9 **Jenks** Oklahoma, C USA 36°01′N 95°58′W
Jenné see Djenné
109 X8 **Jennersdorf** Burgenland, SE Austria 46°57′N 16°08′E
22 H9 **Jennings** Louisiana, S USA 30°13′N 92°39′W
11 N7 **Jenny Lind Island** island Nunavut, N Canada
23 Y13 **Jensen Beach** Florida, SE USA 27°15′N 80°13′W
9 P6 **Jens Munk Island** island Nunavut, NE Canada
163 Y15 **Jeonju** Jap. Zenshū; prev. Chŏnju. SW South Korea 35°51′N 127°08′E
59 O17 **Jequié** Bahia, E Brazil 13°52′S 40°06′W
59 O18 **Jequitinhonha, Rio** ≈ E Brazil
Jerablus see Jarābulus
74 H6 **Jerada** NE Morocco 34°16′N 02°07′W
Jerash see Jarash
75 N7 **Jerba, Île de** var. Djerba, Jazīrat Jarbah. island E Tunisia
44 K9 **Jérémie** SW Haiti 18°39′N 74°11′W
Jeréz see Jeréz de García Salinas, Mexico
Jerez see Jeréz de la Frontera, Spain
40 L11 **Jerez de García Salinas** var. Jeréz. Zacatecas, C Mexico 22°40′N 103°00′W
104 J15 **Jerez de la Frontera** var. Jerez; prev. Xeres. Andalucía, SW Spain 36°41′N 06°08′W
104 I12 **Jerez de los Caballeros** Extremadura, W Spain 38°20′N 06°45′W
Jergucati see Jorgucat
138 G10 **Jericho** Ar. Arīḩā, Heb. Yeriḩo. E West Bank 31°51′N 35°27′E
74 M7 **Jerid, Chott el** var. Shaṭṭ al Jarīd. salt lake SW Tunisia
183 O10 **Jerilderie** New South Wales, SE Australia 35°24′S 145°43′E
Jerischmarkt see Câmpia Turzii
92 K11 **Jerisjärvi** ◎ NW Finland
Jermentau see Yereymentau
Jermer see Jaroměř
36 K11 **Jerome** Arizona, SW USA 34°45′N 112°06′W
33 O15 **Jerome** Idaho, NW USA 42°43′N 114°31′W
97 K26 **Jersey** ◇ British Crown Dependency Channel Islands, NW Europe
97 L26 **Jersey** island Channel Islands, NW Europe
18 K14 **Jersey City** New Jersey, NE USA 40°42′N 74°01′W
18 F13 **Jersey Shore** Pennsylvania, NE USA 41°12′N 77°13′W
30 K14 **Jerseyville** Illinois, N USA 39°07′N 90°19′W
104 K8 **Jerte** ≈ W Spain
138 F10 **Jerusalem** Ar. Al Quds, Al Quds ash Sharīf, Heb. Yerushalayim; anc. Hierosolyma. ● (Israel) Jerusalem, NE Israel 31°47′N 35°13′E
138 G10 **Jerusalem** ◐ district E Israel
183 S10 **Jervis Bay** New South Wales, SE Australia 35°09′S 150°42′E
183 S10 **Jervis Bay Territory** ◇ territory SE Australia
Jerwakant see Järvakandi
109 S10 **Jesenice** Ger. Assling. NW Slovenia 46°26′N 14°01′E
111 H16 **Jeseník** Ger. Freiwaldau. Olomoucký Kraj, E Czech Republic 50°14′N 17°12′E
Jesi see Iesi
106 I8 **Jesolo** var. Iesolo. Veneto, NE Italy 45°32′N 12°37′E
Jesselton see Kota Kinabalu
95 I14 **Jessheim** Akershus, S Norway 60°07′N 11°10′E
153 T15 **Jessore** Khulna, W Bangladesh 23°10′N 89°12′E
23 W6 **Jesup** Georgia, SE USA 31°36′N 81°54′W
41 S15 **Jesús Carranza** Veracruz-Llave, SE Mexico 17°30′N 95°01′W
62 K10 **Jesús María** Córdoba, C Argentina 30°59′S 64°05′W
26 K6 **Jetmore** Kansas, C USA 38°05′N 99°55′W
103 Q2 **Jeumont** Nord, N France 50°18′N 04°06′E
95 H14 **Jevnaker** Oppland, S Norway 60°15′N 10°25′E
Jewe see Jõhvi
25 V9 **Jewett** Texas, SW USA 31°21′N 96°08′W
19 N12 **Jewett City** Connecticut, NE USA 41°36′N 71°58′W
Jeypore see Jaipur, Rājasthān, India
Jeypore/Jeypur see Jaypur, Orissa, India
113 L17 **Jezercës, Maja e** ▲ N Albania 42°27′N 19°49′E
111 B18 **Jezerní Hora** ▲ SW Czech Republic 49°10′N 13°11′E
154 F10 **Jhābua** Madhya Pradesh, C India 22°44′N 74°37′E
152 H14 **Jhālāwār** Rājasthān, N India 24°37′N 76°12′E
Jhang/Jhang Sadar see Jhang Sadr
149 U9 **Jhang Sadr** var. Jhang, Jhang Sadar. Punjab, NE Pakistan 31°16′N 72°19′E
152 J13 **Jhānsi** Uttar Pradesh, N India 25°27′N 78°34′E
153 O16 **Jhārkhand** ◐ state NE India
154 M11 **Jhārsuguda** Odisha, E India 21°56′N 84°04′E
149 V7 **Jhelum** Punjab, NE Pakistan 32°55′N 73°42′E
129 P9 **Jhelum** ≈ E Pakistan
Jhenaida see Jhenaidaha
153 T15 **Jhenaidaha** var. Jhenaida, Jhenida. Khulna, Bangladesh 23°32′N 89°09′E
Jhenaidaha see Jhenida
153 T15 **Jhenida** var. Jhenaidaha. Dhaka, W Bangladesh 23°34′N 89°39′E
jhenida see Jhenaidaha
149 P16 **Jhimpir** Sind, SE Pakistan 25°00′N 68°01′E
Jhind see Jīnd
149 R16 **Jhudo** Sind, SE Pakistan 24°58′N 69°18′E
Jhumra see Chak Jhumra
152 H11 **Jhunjhunūn** Rājasthān, N India 28°05′N 75°30′E
Ji see Hebei, China
Ji see Jilin, China
Jiading see Xinfeng
153 S14 **Jiāganj** West Bengal, NE India 24°18′N 88°07′E
Jiahe see Qionghai
160 J7 **Jialing Jiang** ≈ C China
163 Y7 **Jiamusi** var. Chia-mu-ssu, Kiamusze. Heilongjiang, NE China 46°46′N 130°19′E
161 O11 **Ji'an** Jiangxi, S China 27°08′N 115°00′E
163 W12 **Ji'an** Jilin, NE China 41°04′N 126°07′E
163 T13 **Jianchang** Liaoning, NE China 40°48′N 119°51′E
Jianchang see Nancheng
Jiancheng see Jianyang
160 F11 **Jianchuan** var. Jinhuan. Yunnan, SW China 26°28′N 99°49′E
158 M4 **Jiangjunmiao** Xinjiang Uygur Zizhiqu, W China 44°42′N 90°06′E
160 K11 **Jiangkou** var. Shuangjiang. Guizhou, S China 27°46′N 108°53′E
Jiangkou see Fengkai
161 Q12 **Jiangle** var. Guyong. Fujian, SE China 26°44′N 117°26′E
161 N15 **Jiangmen** Guangdong, S China 22°35′N 113°02′E
Jiangna see Yanshan
161 Q10 **Jiangshan** Zhejiang, SE China 28°41′N 118°33′E
161 Q7 **Jiangsu** var. Chiang-su, Jiangsu Sheng, Kiangsu, Su. ◐ province E China
Jiangsu see Nanjing
Jiangsu Sheng see Jiangsu
161 O11 **Jiangxi** var. Chiang-hsi, Gan, Jiangxi Sheng, Kiangsi. ◐ province S China
Jiangxi Sheng see Jiangxi
160 I8 **Jiangyou** prev. Zhongba. Sichuan, C China 31°52′N 104°52′E
161 N9 **Jianli** var. Rongcheng. Hubei, C China 29°51′N 112°50′E
161 Q11 **Jian'ou** Fujian, SE China 27°04′N 118°20′E
163 S12 **Jianping** var. Yebaishou. Liaoning, NE China 41°13′N 119°37′E
Jianshe see Baiyü
160 L9 **Jianshi** var. Yezhou. Hubei, C China 30°37′N 109°42′E
Jiantang see Xamgyi'nyilha
129 V11 **Jian Xi** ≈ SE China
161 Q11 **Jianyang** Fujian, SE China 27°20′N 118°01′E
160 I9 **Jianyang** var. Jiancheng. Sichuan, C China 30°22′N 104°31′E
163 X10 **Jiaohe** Jilin, NE China 43°41′N 127°20′E
Jiaojiang see Taizhou
Jiaoxian see Jiaozhou
161 R5 **Jiaozhou** prev. Jiaoxian. Shandong, E China 36°17′N 120°00′E
161 N6 **Jiaozuo** Henan, C China 35°14′N 113°13′E
Jiashan see Mingguang
158 F8 **Jiashi** var. Baren, Payzawat. Xinjiang Uygur Zizhiqu, NW China 39°27′N 76°45′E
154 L9 **Jiāwān** Madhya Pradesh, C India 24°20′N 82°17′E
161 S9 **Jiaxing** Zhejiang, SE China 30°44′N 120°46′E
161 S14 **Jiayi** var. Chia-i, Chiai, Chiayi, Kiayi, Jap. Kagi. C Taiwan 23°29′N 120°27′E
163 X6 **Jiayin** var. Chaoyang. Heilongjiang, NE China 48°51′N 130°24′E
159 R8 **Jiayuguan** Gansu, N China 39°47′N 98°14′E
Jibhalanta see Uliastay
138 M4 **Jibli** Ar Raqqah, C Syria 35°49′N 39°23′E
116 H9 **Jibou** Hung. Zsibó. Sălaj, NW Romania 47°15′N 23°17′E
141 Z9 **Jibsh, Ra's al** headland E Oman 21°20′N 59°23′E
Jibuti see Djibouti
111 E15 **Jičín** Ger. Jitschin. Královéhradecký Kraj, N Czech Republic 50°27′N 15°20′E
140 K10 **Jiddah** Eng. Jedda. (Saudi Arabia) Makkah, W Saudi Arabia 21°34′N 39°13′E
141 W11 **Jiddat al Ḩarāsīs** desert C Oman
Jiesjavrre see Iešjavri
160 M4 **Jiexiu** Shanxi, C China 37°00′N 111°50′E
161 P14 **Jieyang** Guangdong, S China 23°32′N 116°20′E
119 F14 **Jieznas** Kaunas, S Lithuania 54°37′N 24°10′E
Jifa', Bi'r see Jif'īyah, Bi'r
141 P15 **Jif'īyah, Bi'r** var. Bi'r Jifa'. well C Yemen
77 W13 **Jigawa** ◐ state N Nigeria
146 J10 **Jigerbent** Rus. Dzhigirbent. Lebap Welaýaty, NE Turkmenistan 40°44′N 61°56′E
44 I7 **Jiguaní** Granma, E Cuba 20°24′N 76°26′W
159 T12 **Jigzhi** var. Chugqênsumdo. Qinghai, C China 33°23′N 101°25′E
Jih-k'a-tse see Xigazê
111 E18 **Jihlava** Ger. Iglau, Pol. Igława. Vysocina, S Czech Republic 49°22′N 15°36′E
111 E18 **Jihlava** var. Igel, Ger. Iglawa. ≈ Vysočina, C Czech Republic
Jihlavský Kraj see Vysočina
111 C18 **Jihočeský Kraj** prev. Budějovický Kraj. ◐ region Czech Republic
111 G19 **Jihomoravský Kraj** prev. Brněnský Kraj. ◐ region SE Czech Republic
74 L5 **Jijel** var. Djidjel; prev. Djidjelli. NE Algeria 36°50′N 05°43′E
116 L9 **Jijia** ≈ N Romania
80 L13 **Jijiga** It. Giggiga. Sumalē, E Ethiopia 09°21′N 42°53′E
105 S12 **Jijona** var. Xixona. Valenciana, E Spain 38°34′N 00°29′W
81 L18 **Jilib** It. Gelib. Jubbada Dhexe, S Somalia 0°18′N 42°48′E
163 W10 **Jilin** var. Chi-lin, Girin, Kirin; prev. Yungki, Yunki. Jilin, NE China 43°46′N 126°32′E
163 W10 **Jilin** var. Chi-lin, Girin, Ji, Jilin Sheng, Kirin. ◐ province NE China
163 W11 **Jilin Hada Ling** ▲▲ NE China
Jilin Sheng see Jilin
163 S4 **Jiliu He** ≈ NE China
105 Q6 **Jiloca** ≈ N Spain
161 T12 **Jilong** var. Keelung, Jap. Kirun, Kirun'; prev. Chilung, prev. Sp. Santissima Trinidad. N Taiwan 25°10′N 121°43′E
81 I14 **Jīma** var. Jimma, It. Gimma. Oromīya, C Ethiopia 07°42′N 36°51′E
44 M9 **Jimaní** W Dominican Republic 18°29′N 71°49′W
116 E11 **Jimbolia** Ger. Hatzfeld, Hung. Zsombolya. Timiş, W Romania 45°47′N 20°43′E
104 K16 **Jimena de la Frontera** Andalucía, S Spain 36°27′N 05°28′W
40 K7 **Jiménez** Chihuahua, N Mexico 27°09′N 104°54′W
41 N5 **Jiménez** Coahuila, NE Mexico 29°05′N 100°40′W
41 P9 **Jiménez** var. Santander Jiménez. Tamaulipas, C Mexico 24°11′N 98°29′W
40 L10 **Jiménez del Teul** Zacatecas, C Mexico 23°13′N 103°46′W
77 Y14 **Jimeta** Adamawa, E Nigeria 09°16′N 12°25′E
Jimma see Jīma
158 M5 **Jimsar** Xinjiang Uygur Zizhiqu, NW China 44°05′N 88°48′E
18 I14 **Jim Thorpe** Pennsylvania, NE USA 40°51′N 75°43′W
Jin see Shanxi
Jin see Tianjin Shi
161 P5 **Jinan** var. Chinan, Chi-nan, Tsinan. province capital Shandong, E China 36°43′N 116°58′E
Jin'an see Songpan
Jinbi see Dayao
159 T8 **Jinchang** Gansu, N China 38°31′N 102°07′E
161 N5 **Jincheng** Shanxi, C China 35°30′N 112°52′E
Jincheng see Wuding
Jinchengjiang see Hechi
152 I9 **Jīnd** prev. Jhind. Haryāna, NW India 29°29′N 76°22′E
183 Q11 **Jindabyne** New South Wales, SE Australia 36°28′S 148°36′E
163 X17 **Jin-do** Jap. Chin-tō; prev. Chin-do. island SW South Korea
111 O18 **Jindřichův Hradec** Ger. Neuhaus. Jihočeský Kraj, S Czech Republic 49°09′N 15°01′E
Jing see Beijing Shi
Jing see Jinghe, China
159 X10 **Jingchuan** Gansu, C China 35°20′N 107°45′E
161 Q10 **Jingdezhen** Jiangxi, S China 29°18′N 117°18′E
161 O12 **Jinggangshan** Jiangxi, S China 26°36′N 114°11′E
161 P3 **Jinghai** Tianjin Shi, E China 38°53′N 116°45′E
158 I4 **Jinghe** var. Jing. Xinjiang Uygur Zizhiqu, NW China 44°35′N 82°55′E
160 K6 **Jing He** ≈ C China
160 F15 **Jinghong** var. Yunjinghong. Yunnan, SW China 22°03′N 100°56′E
160 M9 **Jingmen** Hubei, C China 30°58′N 112°09′E
163 X10 **Jingpo Hu** ◎ NE China
160 M8 **Jing Shan** ▲ C China
159 V9 **Jingtai** var. Yitiaoshan. Gansu, C China 37°12′N 104°06′E
160 J14 **Jingxi** var. Xinjing. Guangxi Zhuangzu Zizhiqu, S China 23°10′N 106°22′E
Jing Xian see Jingzhou, Hunan, China
163 W11 **Jingyu** Jilin, NE China 42°23′N 126°48′E
159 V10 **Jingyuan** var. Wulan. Gansu, C China 36°35′N 104°40′E
160 M9 **Jingzhou** prev. Shashi, Shashih, Shasi. Hubei, C China 30°21′N 112°09′E
160 L12 **Jingzhou** var. Jing Xian, Jingzhou Miaouzu Dongzu Zizhixian, Quyang. Hunan, S China 26°35′N 109°40′E
Jingzhou Miaouzu Dongzu Zizhixian see Jingzhou
163 Z16 **Jinhae** Jap. Chinkai; prev. Chinhae. S South Korea 35°06′N 128°48′E
Jinhe see Jinping
161 R10 **Jinhua** Zhejiang, SE China 29°15′N 119°36′E
Jinhuan see Jianchuan
161 P5 **Jining** Shandong, E China 35°25′N 116°35′E
Jining see Ulan Qab
81 G18 **Jinja** S Uganda 0°27′N 33°14′E
161 R13 **Jinjiang** var. Qingyang. Fujian, SE China 24°53′N 118°36′E
161 O11 **Jin Jiang** ≈ S China
Jinjiang see Chengmai
163 Y16 **Jinju** prev. Chinju, Jap. Shinshū. S South Korea 35°12′N 128°06′E
171 V15 **Jin, Kepulauan** island group E Indonesia
161 R13 **Jinmen Dao** var. Chinmen Tao, Quemoy. island W Taiwan
42 J9 **Jinotega** Jinotega, NW Nicaragua 13°03′N 85°59′W
42 K7 **Jinotega** ◐ department N Nicaragua
42 J11 **Jinotepe** Carazo, SW Nicaragua 11°50′N 86°10′W
160 L13 **Jinping** var. Sanjiang. Guizhou, S China 26°42′N 109°13′E
160 H14 **Jinping** var. Jinhe. Yunnan, SW China 22°47′N 103°12′E
Jinsen see Incheon
160 J11 **Jinsha** Guizhou, S China 27°24′N 106°16′E
160 M10 **Jinshi** Hunan, S China 29°42′N 111°46′E
Jinshi see Xinning
162 I9 **Jinst** var. Bodĭ. Bayanhongor, C Mongolia 45°25′N 100°33′E
159 R7 **Jinta** Gansu, N China 40°01′N 98°57′E
161 Q12 **Jin Xi** ≈ SE China
Jinxi see Huludao
Jinxian see Jinzhou
161 P6 **Jinxiang** Shandong, E China 35°08′N 116°19′E
161 P8 **Jinzhai** var. Meishan. Anhui, E China 31°42′N 115°47′E
161 N4 **Jinzhong** var. Yuci. Shanxi, C China 37°34′N 112°45′E
163 T12 **Jinzhou** var. Chin-chou, Chinchow; prev. Chinhsien. Liaoning, NE China 41°07′N 121°06′E

◆ Country
● Country Capital
◇ Dependent Territory
○ Dependent Territory Capital
◐ Administrative Regions
✈ International Airport
▲ Mountain
▲▲ Mountain Range
Volcano
≈ River
◎ Lake
⊠ Reservoir

JINZHOU – KÂHTA

163 U14 Jinzhou prev. Jinxian. Liaoning, NE China 39°04′N 121°45′E
Jinzhu see Daocheng
138 H12 Jinz, Qā' al ◎ C Jordan
47 S8 Jiparaná, Rio ≈ W Brazil
56 A7 Jipijapa Manabí, W Ecuador 01°23′S 80°35′W
42 F8 Jiquilisco Usulután, S El Salvador 13°19′N 88°35′W
Jirgalanta see Hovd
147 S12 Jirgatol Rus. Dzhirgatal'. C Tajikistan 39°13′N 71°09′E
75 X10 Jirjā var. Girga, Girgeh, Jirjā. C Egypt 26°17′N 31°58′E
Jirjā see Jirjā
111 B15 Jirkov Ger. Görkau. Ústecký Kraj, NW Czech Republic 50°30′N 13°27′E
143 T12 Jiroft var. Sabzawaran, Sabzvārān. Kermān, SE Iran 28°40′N 57°40′E
81 P14 Jirriiban Mudug, E Somalia 07°15′N 48°55′E
160 L11 Jishou Hunan, S China 28°20′N 109°43′E
Jisr ash Shadadi see Ash Shadādah
116 I14 Jitaru Olt, S Romania 44°27′N 24°32′E
Jitschin see Jičín
116 H14 Jiu Ger. Schil, Schyl, Hung. Zsil, Zsily. ≈ S Romania
161 R11 Jiufeng Shan ▲ SE China
161 P9 Jiujiang Jiangxi, S China 29°45′N 115°59′E
161 O10 Jiuling Shan ▲ S China
160 G10 Jiulong var. Garba, Tib. Gyaisi. Sichuan, C China 29°00′N 101°30′E
161 Q13 Jiulong Jiang ≈ SE China
161 Q12 Jiulong Xi ≈ SE China
159 R8 Jiuquan var. Suzhou. Gansu, N China 39°47′N 98°30′E
160 K17 Jiusuo Hainan, S China 18°25′N 109°55′E
163 W10 Jiutai Jilin, NE China 44°01′N 125°51′E
160 K13 Jiuwan Dashan ▲ S China
160 I7 Jiuzhaigou var. Nongle; prev. Nanping. Sichuan, C China 33°25′N 104°05′E
186 C7 Jiwaka ◆ province C Papua New Guinea
148 I16 Jīwani Baluchistān, SW Pakistan 25°05′N 61°46′E
163 Y8 Jixi Heilongjiang, NE China 45°17′N 131°01′E
163 Y7 Jixian var. Fuli. Heilongjiang, NE China 46°38′N 131°04′E
160 M5 Jixian var. Ji Xian. Shanxi, C China 36°15′N 110°41′E
Ji Xian see Jixian
Jiza see Al Jīzah
141 N13 Jīzān var. Qīzān. Jīzān, SW Saudi Arabia 17°50′N 42°50′E
141 N13 Jīzān var. Minṭaqat Jīzān. ◆ province SW Saudi Arabia
Jīzān, Minṭaqat see Jīzān
140 K6 Jizl, Wādī al dry watercourse W Saudi Arabia
164 H12 Jizō-zaki headland Honshū, SW Japan 35°34′N 133°16′E
141 U14 Jiz', Wādī al dry watercourse E Yemen
147 O11 Jizzax Rus. Dzhizak. Jizzax Viloyati, C Uzbekistan 40°08′N 67°47′E
147 N10 Jizzax Viloyati Rus. Dzhizakskaya Oblast'. ◆ province C Uzbekistan
60 I13 Joaçaba Santa Catarina, S Brazil 27°08′S 51°30′W
Joal see Joal-Fadiout
76 F11 Joal-Fadiout prev. Joal. W Senegal 14°09′N 16°50′W
76 E10 João Barrosa Boa Vista, E Cape Verde 16°01′N 22°44′W
João Belo see Xai-Xai
João de Almeida see Chibia
59 Q15 João Pessoa prev. Paraíba. state capital Paraíba, E Brazil 07°06′S 34°53′W
25 X7 Joaquin Texas, SW USA 31°58′N 94°03′W
62 K6 Joaquín V. González Salta, N Argentina 25°06′S 64°07′W
Joazeiro see Juazeiro
Job'urg see Johannesburg
109 O7 Jochberger Ache ≈ W Austria
Jo-ch'iang see Ruoqiang
92 K12 Jock Norrbotten, N Sweden 66°40′N 22°45′E
42 I5 Jocón Yoro, N Honduras 15°17′N 86°55′W
105 O13 Jódar Andalucía, S Spain 37°51′N 03°18′W
152 F12 Jodhpur Rājasthān, NW India 26°17′N 73°02′E
99 I19 Jodoigne Walloon Brabant, C Belgium 50°43′N 04°52′E
93 O16 Joensuu Pohjois-Karjala, SE Finland 62°36′N 29°45′E
37 W4 Joes Colorado, C USA 39°36′N 102°40′W
191 Z3 Joe's Hill hill Kiritimati, NE Kiribati
165 N11 Jōetsu var. Zyôetu. Niigata, Honshū, C Japan 37°09′N 138°13′E
83 M18 Jofane Inhambane, S Mozambique 21°16′S 34°21′E
153 R12 Jogbani Bihār, NE India 26°23′N 87°16′E
118 I5 Jõgeva Ger. Laisholm. Jõgevamaa, E Estonia 58°48′N 26°28′E
118 I4 Jõgevamaa off. Jõgeva Maakond. ◆ province E Estonia
Jõgeva Maakond see Jõgevamaa
155 E18 Jog Falls Waterfall Karnātaka, W India
143 S4 Joghatāy Khorāsān-e Razavī, NE Iran 36°34′N 57°00′E
153 U12 Jogīghopa Assam, NE India 26°14′N 90°35′E
152 I7 Jogindarnagar Himāchal Pradesh, N India 31°51′N 76°47′E
Jogjakarta see Yogyakarta
164 L11 Jōhana Toyama, Honshū, SW Japan 36°30′N 136°53′E
83 J21 Johannesburg var. Egoli, Erautini, Gauteng, abbrev. Job'urg. Gauteng, NE South Africa 26°10′S 28°02′E
35 T13 Johannesburg California, W USA 35°20′N 117°37′W
Johannisburg see Pisz
149 P14 Johi Sind, SE Pakistan 26°46′N 67°28′E
55 T13 Johi Village S Guyana 01°48′N 58°33′W
45 W10 John A. Osborne ✈ (Plymouth) E Montserrat 16°45′N 62°09′W
32 K13 John Day Oregon, NW USA 44°25′N 118°57′W
32 I11 John Day River ≈ Oregon, NW USA
18 L14 John F Kennedy ✈ (New York) Long Island, New York, NE USA 40°39′N 73°45′W
21 V8 John H. Kerr Reservoir var. Buggs Island Lake, Kerr Lake. ▣ North Carolina/Virginia, SE USA
37 V6 John Martin Reservoir ▣ Colorado, C USA
96 K6 John o'Groats N Scotland, United Kingdom 58°38′N 03°03′W
27 P5 John Redmond Reservoir ▣ Kansas, C USA
39 Q7 John River ≈ Alaska, USA
26 H6 Johnson Kansas, C USA 37°33′N 101°46′W
18 M7 Johnson Vermont, NE USA 44°39′N 72°40′W
18 D13 Johnsonburg Pennsylvania, NE USA 41°28′N 78°37′W
18 H11 Johnson City New York, NE USA 42°06′N 75°54′W
21 P8 Johnson City Tennessee, S USA 36°18′N 82°21′W
25 R10 Johnson City Texas, SW USA 30°17′N 98°27′W
35 S12 Johnsondale California, W USA 35°58′N 118°32′W
10 I8 Johnsons Crossing Yukon, W Canada 60°30′N 133°15′W
21 T13 Johnsonville South Carolina, SE USA 33°50′N 79°26′W
21 Q13 Johnston South Carolina, SE USA 33°49′N 81°48′W
192 M6 Johnston Atoll ◇ US unincorporated territory C Pacific Ocean
175 Q3 Johnston Atoll atoll C Pacific Ocean
30 L17 Johnston City Illinois, N USA 37°49′N 88°55′W
180 K12 Johnston, Lake salt lake Western Australia
31 S13 Johnstown Ohio, N USA 40°08′N 82°39′W
18 D15 Johnstown Pennsylvania, NE USA 40°20′N 78°56′W
168 L10 Johor var. Johore. ◆ state Peninsular Malaysia
Johor Baharu see Johor Bahru
168 K10 Johor Bahru var. Johor Baharu, Johore Bahru. Johor, Peninsular Malaysia 01°29′N 103°44′E
Johore see Johor
Johore Bahru see Johor Bahru
118 K3 Jõhvi Ger. Jewe. Ida-Virumaa, NE Estonia 59°21′N 27°25′E
103 P7 Joigny Yonne, C France 47°58′N 03°24′E
Joinvile see Joinville
60 K12 Joinville var. Joinvile. Santa Catarina, S Brazil 26°20′S 48°55′W
103 R6 Joinville Haute-Marne, N France 48°26′N 05°07′E
194 H3 Joinville Island island Antarctica
41 O15 Jojutla var. Jojutla de Juárez. Morelos, S Mexico 18°38′N 99°10′W
Jojutla de Juárez see Jojutla
92 I12 Jokkmokk Lapp. Dálvvadis. Norrbotten, N Sweden 66°35′N 19°57′E
92 L2 Jökuldalur E Iceland
92 K2 Jökulsá á Fjöllum ≈ NE Iceland
Jokyakarta see Yogyakarta
30 M11 Joliet Illinois, N USA 41°33′N 88°05′W
15 O11 Joliette Québec, SE Canada 46°02′N 73°27′W
171 O8 Jolo Jolo Island, SW Philippines 06°02′N 121°00′E
94 D11 Jølstervatnet ◎ S Norway
169 S16 Jombang prev. Djombang. Jawa, S Indonesia 07°33′S 112°14′E
159 R14 Jomda Xizang Zizhiqu, W China 31°26′N 98°09′E
118 G13 Jonava Ger. Janow, Pol. Janów. Kaunas, C Lithuania 55°05′N 24°19′E
146 L11 Jondor Rus. Zhondor. Buxoro Viloyati, C Uzbekistan 39°46′N 64°11′E
159 V11 Jonê var. Liulin. Gansu, C China 34°36′N 103°39′E
27 X9 Jonesboro Arkansas, C USA 35°50′N 90°42′W
23 S4 Jonesboro Georgia, SE USA 33°31′N 84°21′W
30 L17 Jonesboro Illinois, N USA 37°25′N 89°19′W
22 H5 Jonesboro Louisiana, S USA 32°14′N 92°43′W
21 P8 Jonesboro Tennessee, S USA 36°17′N 82°28′W
19 T6 Jonesport Maine, NE USA 44°33′N 67°35′W
0 J4 Jones Sound channel Nunavut, N Canada
22 I6 Jonesville Louisiana, S USA 31°37′N 91°49′W
31 Q10 Jonesville Michigan, N USA 41°58′N 84°39′W
21 Q11 Jonesville South Carolina, SE USA 34°49′N 81°38′W
146 K10 Jongeldi Rus. Dzhankel'dy. Buxoro Viloyati, C Uzbekistan 40°50′N 63°16′E
81 F14 Jonglei Jonglei, E South Sudan 06°54′N 31°19′E
81 F14 Jonglei var. Gongoleh State. ◆ state E South Sudan
81 F14 Jonglei Canal canal E South Sudan
118 F11 Joniškėlis Panevėžys, N Lithuania 56°02′N 24°10′E
118 F10 Joniškis Ger. Janischken. Šiaulai, N Lithuania 56°15′N 23°36′E
95 L19 Jönköping Jönköping, S Sweden 57°45′N 14°10′E
95 K20 Jönköping ◆ county S Sweden
15 Q7 Jonquière Québec, SE Canada 48°25′N 71°16′W
41 V15 Jonuta Tabasco, SE Mexico 18°04′N 92°03′W
102 K12 Jonzac Charente-Maritime, W France 45°26′N 00°25′W
27 R7 Joplin Missouri, C USA 37°04′N 94°31′W
33 W8 Jordan Montana, NW USA 47°18′N 106°54′W
138 H12 Jordan off. Hashemite Kingdom of Jordan, Ar. Al Mamlaka al Urduniya al Hashemiyah, Al Urdunn; prev. Transjordan. ◆ monarchy SW Asia
138 G9 Jordan Ar. Urdunn, Heb. HaYarden. ≈ SW Asia
Jordan Lake see B. Everett Jordan Reservoir
111 K17 Jordanów Małopolskie, S Poland 49°39′N 19°51′E
32 M15 Jordan Valley Oregon, NW USA 42°59′N 117°03′W
138 G9 Jordan Valley valley N Israel
57 D15 Jorge Chávez Internacional var. Lima. ✈ (Lima) Lima, W Peru 12°07′S 77°01′W
113 L23 Jorgucat var. Jergucati, Jorgucati. Gjirokastër, S Albania 39°57′N 20°14′E
Jorgucati see Jorgucat
153 X12 Jorhāt Assam, NE India 26°45′N 94°09′E
93 J14 Jörn Västerbotten, N Sweden 65°03′N 20°04′E
37 R14 Jornada Del Muerto valley New Mexico, SW USA
93 N17 Joroinen Etelä-Savo, E Finland 62°11′N 27°50′E
95 C16 Jørpeland Rogaland, S Norway 59°01′N 06°04′E
77 W14 Jos Plateau, C Nigeria 09°59′N 08°57′E
171 Q8 Jose Abad Santos var. Trinidad. Mindanao, S Philippines 05°51′N 125°35′E
61 F19 José Batlle y Ordóñez var. Batlle y Ordóñez. Florida, C Uruguay 33°28′S 55°08′W
63 H18 José de San Martín Chubut, S Argentina 44°04′S 70°29′W
61 E19 José Enrique Rodó var. Rodó, José E.Rodo; prev. Drabble, Drable. Soriano, SW Uruguay 33°43′S 57°33′W
José E.Rodo see José Enrique Rodó
Josefsdorf see Žabalj
44 C4 José Martí ✈ (La Habana) Cuidad de La Habana, N Cuba 23°03′N 82°29′W
61 F19 José Pedro Varela var. José P.Varela. Lavalleja, S Uruguay 33°30′S 54°28′W
181 N2 Joseph Bonaparte Gulf gulf N Australia
37 N11 Joseph City Arizona, SW USA 34°56′N 110°18′W
13 O9 Joseph, Lake ◎ Newfoundland and Labrador, E Canada
14 G13 Joseph, Lake ◎ Ontario, S Canada
186 C6 Josephstaal Madang, N Papua New Guinea 04°42′S 144°55′E
José P.Varela see José Pedro Varela
59 J14 José Rodrigues Pará, N Brazil 05°45′S 51°20′W
152 K9 Joshimath Uttarakhand, N India 30°33′N 79°35′E
25 T7 Joshua Texas, SW USA 32°27′N 97°23′W
35 V15 Joshua Tree California, W USA 34°07′N 116°18′W
77 V14 Jos Plateau plateau C Nigeria
102 H6 Josselin Morbihan, NW France 47°57′N 02°35′W
Jos Sudarso see Yos Sudarso, Pulau
94 E11 Jostedalsbreen glacier S Norway
94 F12 Jotunheimen ▲ S Norway
138 G7 Joûnié var. Juniyah. W Lebanon 33°54′N 33°36′E
25 R13 Jourdanton Texas, SW USA 28°55′N 98°34′W
98 L7 Joure Fris. De Jouwer. Fryslân, N Netherlands 52°58′N 05°48′E
93 M18 Joutsa Keski-Suomi, C Finland 61°46′N 26°09′E
93 N18 Joutseno Etelä-Karjala, SE Finland 61°06′N 28°30′E
92 M12 Joutsijärvi Lappi, NE Finland 66°40′N 28°00′E
108 A9 Joux, Lac de ◎ W Switzerland
Jovakān see Jowkān
44 D5 Jovellanos Matanzas, W Cuba 22°49′N 81°11′W
153 V13 Jowai Meghālaya, NE India 25°25′N 92°21′E
Jowat see Jabwot
Jowhar see Jawhar
143 O12 Jowkān var. Jovakān. Fārs, S Iran
Jowzam see Javazm
149 N2 Jowzjān ◆ province N Afghanistan
Joypurhat see Jaipurhat
Józseffalva see Žabalj
J.Storm Thurmond Reservoir see Clark Hill Lake
45 T6 Juana Díaz C Puerto Rico 18°03′N 66°30′W
40 L9 Juan Aldama Zacatecas, C Mexico 24°20′N 103°23′W
0 E9 Juan de Fuca Plate tectonic feature
32 F7 Juan de Fuca, Strait of strait Canada/USA
Juan Fernandez Islands see Juan Fernández, Islas
193 S11 Juan Fernández, Islas Eng. Juan Fernandez Islands. island group W Chile
55 O4 Juangriego Nueva Esparta, NE Venezuela 11°06′N 63°59′W
56 D11 Juanjuí var. Juanjuy. San Martín, N Peru 07°10′S 76°44′W
Juanjuy see Juanjuí
93 N16 Juankoski Pohjois-Savo, C Finland 63°01′N 28°24′E
Juan Lacaze see Juan L. Lacaze
61 E20 Juan L. Lacaze var. Juan Lacaze, Puerto Sauce; prev. Sauce. Colonia, SW Uruguay 34°26′S 57°25′W
62 L5 Juan Solá Salta, N Argentina 23°30′S 62°42′W
63 F21 Juan Stuven, Isla island S Chile
59 H16 Juará Mato Grosso, W Brazil 11°10′S 57°28′W
41 N7 Juárez var. Villa Juárez. Coahuila, NE Mexico 27°39′N 100°43′W
40 C2 Juárez, Sierra de ▲ NW Mexico
59 O15 Juazeiro prev. Joazeiro. Bahia, E Brazil 09°25′S 40°30′W
59 P14 Juazeiro do Norte Ceará, E Brazil 07°10′S 39°18′W
81 F15 Juba var. Jūbā. ● Central Equatoria, S South Sudan 04°50′N 31°35′E
81 L17 Juba Amh. Genalē Wenz, It. Guiba, Som. Ganaane, Webi Jubba. ≈ Ethiopia/Somalia
194 H2 Jubany Argentinian research station Antarctica 61°57′S 58°23′W
Jubayl see Jbail
81 L18 Jubbada Dhexe off. Gobolka Jubbada Dhexe. ◆ region SW Somalia
Jubbada Dhexe, Gobolka see Jubbada Dhexe
81 K18 Jubbada Hoose ◆ region SW Somalia
Jubba, Webi see Juba
Jubbulpore see Jabalpur
Jubeil see Jbail
74 B9 Juby, Cap headland SW Morocco 27°58′N 12°56′W
105 R10 Júcar var. Jucar. ≈ C Spain
40 L12 Juchipila Zacatecas, C Mexico 21°25′N 103°06′W
41 S16 Juchitán var. Juchitán de Zaragosa. Oaxaca, SE Mexico 16°27′N 95°W
Juchitán de Zaragosa see Juchitán
138 G11 Judaea cultural region Israel/West Bank
138 F11 Judaean Hills Heb. Haré Yehuda. hill range E Israel
138 H8 Judaydah Fr. Jdaïdé. Rif Dimashq, W Syria 33°17′N 36°15′E
139 P11 Judayyidat Hāmir Al Anbār, S Iraq 31°50′N 41°50′E
109 U8 Judenburg Steiermark, C Austria 47°09′N 14°43′E
33 T8 Judith River ≈ Montana, NW USA
27 V11 Judsonia Arkansas, C USA 35°16′N 91°38′W
141 P14 Jufrah, Wādī al dry watercourse NW Yemen
Jugar see Sêrxü
Jugoslavija see Serbia
42 K10 Juigalpa Chontales, S Nicaragua 12°04′N 85°21′W
100 E9 Juist island NW Germany
Juisui see Ruisui
59 M21 Juiz de Fora Minas Gerais, SE Brazil 21°47′S 43°23′W
62 J5 Jujuy off. Provincia de Jujuy. ◆ province N Argentina
Jujuy see San Salvador de Jujuy
Jujuy, Provincia de see Jujuy
92 J11 Jukkasjärvi Lapp. Čohkkiras. Norrbotten, N Sweden 67°52′N 20°39′E
Jula see Gyula, Hungary
Jūlā see Jālū, Libya
37 W2 Julesburg Colorado, C USA 40°59′N 102°15′W
Julia Beterrae see Béziers
57 I17 Juliaca Puno, SE Peru 15°32′S 70°10′W
181 U6 Julia Creek Queensland, C Australia 20°40′S 141°49′E
35 V17 Julian California, W USA 33°04′N 116°36′W
98 H7 Julianadorp Noord-Holland, NW Netherlands 52°53′N 04°43′E
109 S11 Julian Alps Ger. Julische Alpen, It. Alpi Giulie, Slvn. Julijske Alpe. ▲ Italy/Slovenia
55 V11 Juliana Top ▲ C Suriname 03°39′N 56°36′W
Julianehåb see Qaqortoq
Julijske Alpe see Julian Alps
40 J6 Julimes Chihuahua, N Mexico 28°29′N 105°21′W
Juliobriga see Bragança
Juliobriga see Logroño
61 G15 Júlio de Castilhos Rio Grande do Sul, S Brazil 29°14′S 53°42′W
Juliomagus see Angers
Julische Alpen see Julian Alps
Jullundur see Jalandhar
147 N11 Juma Rus. Dzhuma. Samarqand Viloyati, C Uzbekistan 39°43′N 66°37′E
161 O3 Juma He ≈ E China
81 L18 Jumboo prev. Jumbo. Jubbada Hoose, S Somalia 0°12′S 42°34′E
Jumbo see Jumboo
35 Y11 Jumbo Peak ▲ Nevada, W USA 36°12′N 114°09′W
105 R12 Jumilla Murcia, SE Spain 38°28′N 01°19′W
153 N10 Jumla Mid Western, NW Nepal 29°22′N 82°13′E
Jummoo see Jammu
Jumna see Yamuna
Jumporn see Chumphon
30 K5 Jump River ≈ Wisconsin, N USA
154 B11 Jūnāgadh var. Junagarh. Gujarāt, W India 21°32′N 70°32′E
Junagarh see Jūnāgadh
161 Q6 Junan var. Shizilu. Shandong, E China 35°11′N 118°47′E
62 G11 Juncal, Cerro ▲ C Chile 33°03′S 70°02′W
25 Q10 Junction Texas, SW USA 30°31′N 99°48′W
36 K6 Junction Utah, W USA 38°14′N 112°13′W
27 O4 Junction City Kansas, C USA 39°02′N 96°51′W
32 F13 Junction City Oregon, NW USA 44°13′N 123°12′W
60 M10 Jundiaí São Paulo, S Brazil 23°10′S 46°54′W
39 X12 Juneau state capital Alaska, USA 58°13′N 134°11′W
30 M8 Juneau Wisconsin, N USA 43°23′N 88°42′W
105 U6 Juneda Cataluña, NE Spain 41°33′N 00°49′E
183 Q9 Junee New South Wales, SE Australia 34°51′S 147°33′E
35 R8 June Lake California, W USA 37°46′N 119°04′W
Jungbunzlau see Mladá Boleslav
158 L4 Junggar Pendi Eng. Dzungarian Basin. basin NW China
99 N24 Junglinster Grevenmacher, C Luxembourg 49°43′N 06°15′E
18 F14 Juniata River ≈ Pennsylvania, NE USA
61 B20 Junín Buenos Aires, E Argentina 34°36′S 61°02′W
57 E14 Junín Junín, C Peru 11°11′S 76°00′W
57 F14 Junín off. Departamento de Junín. ◆ department C Peru
63 H15 Junín de los Andes Neuquén, W Argentina 39°57′S 71°05′W
Junín, Departamento de see Junín
57 D14 Junín, Lago de ◎ C Peru
Junīyah see Joûnié
Junkseylon see Phuket
160 I11 Junlian Sichuan, C China 28°11′N 104°31′E
25 O11 Juno Texas, SW USA 30°09′N 101°07′W
92 J11 Junosuando Lapp. Čunusavvon. Norrbotten, N Sweden 67°25′N 22°28′E
93 H16 Junsele Västernorrland, C Sweden 63°42′N 16°54′E
32 L14 Juntura Oregon, NW USA 43°43′N 118°05′W
93 N14 Juntusranta Kainuu, E Finland 65°12′N 29°30′E
118 H11 Juodupė Panevėžys, NE Lithuania 56°07′N 25°37′E
119 H14 Juozapinės Kalnas ▲ SE Lithuania 54°29′N 25°27′E
99 K19 Juprelle Liège, E Belgium 50°43′N 05°31′E
80 D13 Jur ≈ W South Sudan
103 S9 Jura ◆ department E France
108 C7 Jura ◆ canton NW Switzerland
108 B8 Jura var. Jura Mountains. ▲ France/Switzerland
96 G12 Jura island SW Scotland, United Kingdom
Juraciszki see Yuratsishki
54 C8 Jurado Chocó, NW Colombia 07°07′N 77°45′W
Jura Mountains see Jura
96 G12 Jura, Sound of strait W Scotland, United Kingdom
139 V15 Juraybīyāt, Bi'r well S Iraq
118 E13 Jurbarkas Ger. Georgenburg, Jurburg. Tauragė, W Lithuania 55°04′N 22°45′E
99 F20 Jurbise Hainaut, SW Belgium 50°33′N 03°54′E
Jurburg see Jurbarkas
118 F9 Jūrmala C Latvia 56°57′N 23°42′E
58 D13 Juruá Amazonas, NW Brazil 03°08′S 65°59′W
48 F7 Juruá, Rio var. Río Yuruá. ≈ Brazil/Peru
59 G16 Juruena Mato Grosso, W Brazil 10°32′S 58°38′W
59 G16 Juruena, Rio ≈ W Brazil
165 Q6 Jūsan-ko ◎ Honshū, C Japan
25 O6 Justiceburg Texas, SW USA 32°57′N 101°07′W
Justinianopolis see Kirşehir
62 K11 Justo Daract San Luis, C Argentina 33°52′S 65°12′W
59 C14 Jutaí Amazonas, W Brazil 05°10′S 68°45′W
58 C13 Jutaí, Rio ≈ NW Brazil
100 N13 Jüterbog Brandenburg, E Germany 51°58′N 13°06′E
42 E6 Jutiapa Jutiapa, S Guatemala 14°18′N 89°52′W
42 A3 Jutiapa off. Departamento de Jutiapa. ◆ department SE Guatemala
Jutiapa, Departamento de see Jutiapa
42 J6 Juticalpa Olancho, C Honduras 14°39′N 86°12′W
82 I13 Jutila North Western, NW Zambia 12°33′S 26°09′E
Jutland see Jylland
84 F8 Jutland Bank undersea feature SE North Sea 56°50′N 07°20′E
93 N16 Juuka Pohjois-Karjala, E Finland 63°12′N 29°17′E
93 N17 Juva Etelä-Savo, E Finland 61°55′N 27°54′E
Juvavum see Salzburg
44 A6 Juventud, Isla de la var. Isla de Pinos, Eng. Isle of Youth; prev. The Isle of the Pines. island W Cuba
Juwārtā see Chwarta
Juwārtā see Chemchemal
161 Q5 Juxian var. Chengyang, Ju Xian. Shandong, E China 35°33′N 118°45′E
Ju Xian see Juxian
161 P6 Juye Shandong, E China 35°26′N 116°04′E
113 O15 Južna Morava Ger. Südliche Morava. ≈ SE Serbia
83 H20 Jwaneng Southern, S Botswana 24°35′S 24°45′E
95 I23 Jyderup Sjælland, E Denmark 55°40′N 11°25′E
95 F22 Jylland Eng. Jutland. peninsula W Denmark
Jyrgalan see Dzhergalan
93 M17 Jyväskylä Keski-Suomi, C Finland 62°08′N 25°47′E

K

38 D9 Ka'a'awa var. Kaaawa. O'ahu, Hawaii, USA, C Pacific Ocean 21°33′N 157°47′W
Kaaawa see Ka'a'awa
81 G16 Kaabong NE Uganda 03°30′N 34°08′E
Kaaden see Kadaň
Kaafu Atoll see Male' Atoll
55 V9 Kaaimanston Sipaliwini, N Suriname 05°06′N 56°04′W
Kaakhka see Kaka
Kaala see Caála
187 O16 Kaala-Gomen Province Nord, W New Caledonia 20°40′S 164°24′E
92 L9 Kaamanen Lapp. Gámas. Lappi, N Finland 69°05′N 27°16′E
Kaapstad see Cape Town
Kaarasjoki see Karasjok
Kaaresuanto see Karesuando
92 J10 Kaaresuvanto Lapp. Gárassavon. Lappi, N Finland 68°28′N 22°29′E
93 K19 Kaarina Varsinais-Suomi, SW Finland 60°24′N 22°25′E
99 I14 Kaatsheuvel Noord-Brabant, S Netherlands 51°39′N 05°02′E
93 N16 Kaavi Pohjois-Savo, C Finland 62°58′N 28°30′E
Kaba see Habahe
171 O14 Kabaena, Pulau island C Indonesia
Kabakly see Gabakly
76 J14 Kabala N Sierra Leone 09°40′N 11°36′W
81 E19 Kabale SW Uganda 01°15′S 29°58′E
55 U10 Kabalebo Rivier ≈ W Suriname
79 N22 Kabalo Katanga, SE Dem. Rep. Congo 06°02′S 26°55′E
79 O21 Kabambare Maniema, E Dem. Rep. Congo 04°40′S 27°41′E
145 W13 Kabanbay Kaz. Qabanbay; prev. Andreyevka, Kaz. Andreevka. Almaty, SE Kazakhstan 45°50′N 80°34′E
145 Q9 Kabanbay Batyr prev. Rozhdestvenka. Akmola, C Kazakhstan 50°51′N 71°25′E
187 Y15 Kabara prev. Kambara. island Lau Group, E Fiji
Kabardino-Balkaria see Kabardino-Balkarskaya Respublika
126 M15 Kabardino-Balkarskaya Respublika Eng. Kabardino-Balkaria. ◆ autonomous republic SW Russian Federation
79 O19 Kabare Sud-Kivu, E Dem. Rep. Congo 02°13′S 28°40′E
171 T11 Kabarei Papua Barat, E Indonesia 0°01′S 130°58′E
171 P7 Kabasalan Mindanao, S Philippines 07°46′N 122°49′E
77 U15 Kabba Kogi, S Nigeria 07°48′N 06°07′E
92 I13 Kåbdalis Lapp. Goabddális. Norrbotten, N Sweden 66°08′N 20°03′E
138 M6 Kabd aş Şārim hill range E Syria
14 B7 Kabenung Lake ◎ Ontario, S Canada
29 W3 Kabetogama Lake ◎ Minnesota, N USA
Kabia, Pulau see Kabin, Pulau
79 M22 Kabinda Kasai-Oriental, SE Dem. Rep. Congo 06°09′S 24°29′E
Kabinda see Cabinda
171 O15 Kabin, Pulau var. Pulau Kabia. island W Indonesia
171 P16 Kabir Pulau Pantar, S Indonesia 08°15′S 124°12′E
149 T10 Kabīrwāla Punjab, E Pakistan 30°24′N 71°51′E
114 M9 Kableshkovo Burgas, E Bulgaria 42°65′N 27°34′E
78 I13 Kabo Ouham, NW Central African Republic 07°43′N 18°38′E
Kābol see Kābul
83 H14 Kabompo North Western, W Zambia 13°36′S 24°10′E
83 H14 Kabompo ≈ W Zambia
79 M22 Kabongo Katanga, SE Dem. Rep. Congo 07°20′S 25°34′E
120 K11 Kaboudia, Rass headland E Tunisia 35°13′N 11°09′E
124 J14 Kabozha Novgorodskaya Oblast', W Russian Federation 58°48′N 35°00′E
Kabūd Gonbad see Kalāt
142 L5 Kabūd Rāhang Hamadān, W Iran 35°12′N 48°44′E
82 L12 Kabuko Muchinga, NE Zambia 11°31′S 31°16′E
149 Q5 Kābul prev. Kābol. ● (Afghanistan) Kābul, E Afghanistan 34°34′N 69°08′E
149 Q5 Kābul ; prev. Kābol. ◆ province E Afghanistan
149 Q5 Kābul ✈ Kābul, E Afghanistan 34°31′N 69°11′E
149 R5 Kabul var. Daryā-ye Kābul. ≈ Afghanistan/Pakistan see also Kābul, Daryā-ye
Kabul see Kābul, Daryā-ye
149 S5 Kābul, Daryā-ye var. Kabul. ≈ Afghanistan/Pakistan see also Kabul
Kābul, Daryā-ye see Kabul
79 O25 Kabunda Katanga, SE Dem. Rep. Congo 12°21′S 29°14′E
171 R9 Kaburuang, Pulau island Kepulauan Talaud, N Indonesia
80 G8 Kabushiya River Nile, NE Sudan 16°54′N 33°41′E
83 J14 Kabwe Central, C Zambia 14°29′S 28°25′E
186 E7 Kabwum Morobe, C Papua New Guinea 06°04′S 147°09′E
113 N17 Kaçanik Serb. Kačanik. S Kosovo 42°13′N 21°16′E
Kačanik see Kaçanik
118 F13 Kačerginė Kaunas, C Lithuania 54°55′N 23°40′E
117 S13 Kacha Avtonomna Respublika Krym, S Ukraine 44°46′N 33°33′E
154 A10 Kachchh, Gulf of var. Gulf of Cutch, Gulf of Kutch. gulf W India
154 I11 Kachchhīdhāna Madhya Pradesh, C India 21°33′N 78°54′E
149 Q11 Kachchh, Rann of var. Rann of Kachh, Rann of Kutch. salt marsh India/Pakistan
39 Q13 Kachemak Bay bay Alaska, USA
Kachh, Rann of see Kachchh, Rann of
77 V14 Kachia Kaduna, C Nigeria 09°52′N 08°00′E
167 N2 Kachin State ◆ state N Myanmar (Burma)
Kachiry see Kashyr
137 Q11 Kaçkar Dağları ▲ NE Turkey
155 C21 Kadamatt Island island Lakshadweep, India, N Indian Ocean
111 B15 Kadaň Ger. Kaaden. Ústecký Kraj, NW Czech Republic 50°24′N 13°16′E
1667 N11 Kadan Kyun prev. King Island. island Mergui Archipelago, S Myanmar (Burma)
187 X15 Kadavu prev. Kandavu. island S Fiji
187 X15 Kadavu Passage channel S Fiji
79 G16 Kadéï ≈ Cameroon/Central African Republic
Kadhimain see Al Kāẓimīyah
Kadijica see Kadiytsa
114 M13 Kadıköy Barajı ▣ NW Turkey
182 I8 Kadina South Australia 33°59′S 137°43′E
136 H15 Kadınhanı Konya, C Turkey 38°15′N 32°14′E
76 M14 Kadiolo Sikasso, S Mali 10°30′N 05°43′W
136 L16 Kadirli Osmaniye, S Turkey 37°22′N 36°05′E
114 G11 Kadiytsa Mac. Kadijica. ▲ Bulgaria/FYR Macedonia 41°48′N 22°58′E
28 L10 Kadoka South Dakota, N USA 43°49′N 101°30′W
127 N5 Kadom Ryazanskaya Oblast', W Russian Federation 54°35′N 42°27′E
83 K16 Kadoma prev. Gatooma. Mashonaland West, C Zimbabwe 18°22′S 29°55′E
80 E12 Kadugli Southern Kordofan, S Sudan 11°N 29°44′E
77 V14 Kaduna Kaduna, C Nigeria 10°32′N 07°26′E
77 V14 Kaduna ◆ state C Nigeria
77 V15 Kaduna ≈ N Nigeria
124 K14 Kaduy Vologodskaya Oblast', NW Russian Federation 59°10′N 37°11′E
154 E13 Kadwa ≈ W India
123 S9 Kadykchan Magadanskaya Oblast', E Russian Federation 62°54′N 146°53′E
Kadzharan see K'ajaran
125 T7 Kadzherom Respublika Komi, NW Russian Federation 64°42′N 55°51′E
Kadzhi-Say see Bokonbayevo
76 I10 Kaédi Gorgol, S Mauritania 16°12′N 13°32′W
78 G12 Kaélé Extrême-Nord, N Cameroon 10°05′N 14°28′E
38 C9 Ka'ena Point var. Kaena Point. headland O'ahu, Hawai'i, USA 21°34′N 158°16′W
184 J2 Kaeo Northland, North Island, New Zealand 35°03′S 173°40′E
163 X14 Kaesŏng var. Kaesŏng-si. S North Korea 37°58′N 126°31′E
Kaesŏng-si see Kaesŏng
Kaewieng see Kavieng
79 L24 Kafakumba Shaba, S Dem. Rep. Congo 09°39′S 23°43′E
Kafan see Kapan
77 V14 Kafanchan Kaduna, C Nigeria 09°32′N 08°18′E
Kaffa see Feodosiya
76 G11 Kaffrine C Senegal 14°07′N 15°27′W
Kafiréas, Akrotírio see Ntóro, Kávo
115 I19 Kafiréos, Stenó strait Évvoia/Kykládes, Greece, Aegean Sea
Kafirnigan see Kofarnihon
Kafo see Kafu
75 W7 Kafr ash Shaykh var. Kafrel Sheik, Kafr el Sheikh. N Egypt 31°07′N 30°56′E
Kafr el Sheikh see Kafr ash Shaykh
81 F17 Kafu var. Kafo. ≈ W Uganda
83 J15 Kafue Lusaka, SE Zambia 15°44′S 28°10′E
83 I14 Kafue ≈ C Zambia
67 T13 Kafue Flats plain C Zambia
164 K12 Kaga Ishikawa, Honshū, SW Japan 36°18′N 136°19′E
79 J14 Kaga Bandoro prev. Fort-Crampel. Nana-Grébizi, C Central African Republic 06°54′N 19°10′E
81 E18 Kagadi W Uganda 0°57′N 30°52′E
38 H17 Kagalaska Island island Aleutian Islands, Alaska, USA
Kagan see Kogon
Kaganovichabad see Kolkhozobod
Kagarlyk see Kaharlyk
164 H14 Kagawa off. Kagawa-ken. ◆ prefecture Shikoku, SW Japan
Kagawa-ken see Kagawa
154 J13 Kagaznagar Telangana, C India 19°25′N 79°30′E
93 J14 Kåge Västerbotten, N Sweden 64°49′N 21°00′E
81 E19 Kagera var. Ziwa Magharibi, Eng. West Lake. ◆ region NW Tanzania
81 E19 Kagera var. Akagera. ≈ Rwanda/Tanzania see also Akagera
76 L5 Kâgheț var. Karet. physical region N Mauritania
Kagi see Jiayi
137 S12 Kağızman Kars, NE Turkey 40°08′N 43°07′E
188 I6 Kagman Point headland Saipan, S Northern Mariana Islands
164 C16 Kagoshima var. Kagosima. Kagoshima, Kyūshū, SW Japan 31°37′N 130°33′E
164 C16 Kagoshima off. Kagoshima-ken, var. Kagosima. ◆ prefecture Kyūshū, SW Japan
Kagoshima-ken see Kagoshima
Kagosima see Kagoshima
Kagul see Cahul
Kagul, Ozero see Kahul, Ozero
38 B8 Kahala Point headland Kaua'i, Hawai'i, USA 22°08′N 159°17′W
81 F21 Kahama Shinyanga, NW Tanzania 03°48′S 32°36′E
117 P5 Kaharlyk Rus. Kagarlyk. Kyyivs'ka Oblast', N Ukraine 49°50′N 30°50′E
169 T13 Kahayan, Sungai ≈ Borneo, C Indonesia
79 I22 Kahemba Bandundu, SW Dem. Rep. Congo 07°20′S 19°00′E
185 A23 Kaherekoau Mountains ▲ South Island, New Zealand
143 W14 Kahīrī var. Kūhīrī. Sīstān va Balūchestān, SE Iran 26°55′N 61°04′E
101 L16 Kahla Thüringen, C Germany 50°49′N 11°33′E
101 G15 Kahler Asten ▲ W Germany 51°11′N 08°32′E
149 Q4 Kahmard, Daryā-ye prev. Darya-i-surkhab. ≈ NE Afghanistan
143 T13 Kahnūj Kermān, SE Iran 28°N 57°41′E
27 V1 Kahoka Missouri, C USA 40°24′N 91°44′W
38 E10 Kaho'olawe var. Kahoolawe. island Hawai'i, USA, C Pacific Ocean
Kahoolawe see Kaho'olawe
136 M16 Kahramanmaraş var. Kahraman Maraş, Maraş, Marash. Kahramanmaraş, S Turkey 37°34′N 36°54′E
136 L15 Kahramanmaraş var. Kahraman Maraş, Maraş, Marash. ◆ province C Turkey
Kahraman Maraş see Kahramanmaraş
Kahror/Kahror Pakka see Kahror Pakka
149 T11 Kahror Pakka var. Kahror, Koror Pacca. Punjab, E Pakistan 29°38′N 71°59′E
137 N15 Kâhta Adıyaman, S Turkey 37°48′N 38°35′E

◆ Country
● Country Capital
◇ Dependent Territory
○ Dependent Territory Capital
◆ Administrative Regions
✈ International Airport
▲ Mountain
▲ Mountain Range
Volcano
≈ River
◎ Lake
▣ Reservoir

KAHUKU – KANEVSKAYA

38 D8 Kahuku O'ahu, Hawaii, USA, C Pacific Ocean 21°40′N 157°57′W
38 D8 Kahuku Point headland O'ahu, Hawai'i, USA 21°42′N 157°59′W
116 M12 Kahul, Ozero var. Lacul Cahul, Rus. Ozero Kagul. ◎ Moldova/Ukraine
143 V11 Kahūrak Sīstān va Balūchestān, SE Iran 29°25′N 59°38′E
184 G13 Kahurangi Point headland South Island, New Zealand 40°41′S 171°57′E
149 V6 Kahūta Punjab, E Pakistan 33°38′N 73°27′E
77 S14 Kaiama Kwara, W Nigeria 09°37′N 03°58′E
186 D7 Kaiapit Morobe, C Papua New Guinea 06°12′S 146°09′E
185 I18 Kaiapoi Canterbury, South Island, New Zealand 43°23′S 172°40′E
36 K9 Kaibab Plateau plain Arizona, SW USA
171 U14 Kai Besar, Pulau island Kepulauan Kai, E Indonesia
36 L9 Kaibito Plateau plain Arizona, SW USA
158 K6 Kaidu He var. Karaxahar. 〜 NW China
55 S10 Kaieteur Falls waterfall C Guyana
161 O6 Kaifeng Henan, C China 34°47′N 114°20′E
184 J3 Kaihu Northland, North Island, New Zealand 35°47′S 173°39′E
Kaihua see Wenshan
171 U14 Kai Kecil, Pulau island Kepulauan Kai, E Indonesia
169 U16 Kai, Kepulauan prev. Kei Islands. island group Maluku, SE Indonesia
184 J3 Kaikohe Northland, North Island, New Zealand 35°25′S 173°48′E
185 J16 Kaikoura Canterbury, South Island, New Zealand 42°22′S 173°40′E
185 J16 Kaikoura Peninsula peninsula South Island, New Zealand
Kailas Range see Gangdisê Shan
160 K12 Kaili Guizhou, S China 26°34′N 107°58′E
38 F10 Kailua Maui, Hawaii, USA, C Pacific Ocean 20°53′N 156°13′W
Kailua see Kalaoa
38 G11 Kailua-Kona var. Kona. Hawaii, USA, C Pacific Ocean 19°43′N 155°58′W
186 B7 Kaim 〜 W Papua New Guinea
171 X14 Kaima Papua, E Indonesia 05°36′S 138°39′E
184 M7 Kaimai Range ▲▲ North Island, New Zealand
114 E13 Kaïmaktsalán var. Kajmakčalan. ▲ Greece/FYR Macedonia 40°57′N 21°48′E see also Kajmakčalan
Kaïmaktsalán see Kajmakčalan
185 C20 Kaimanawa Mountains ▲▲ North Island, New Zealand
118 E4 Käina Ger. Keinis; prev. Keina. Hiiumaa, W Estonia 58°50′N 22°49′E
109 V7 Kainach 〜 SE Austria
164 I14 Kainan Tokushima, Shikoku, SW Japan 33°36′N 134°20′E
164 H15 Kainan Wakayama, Honshū, SW Japan 34°09′N 135°12′E
147 U7 Kaindy Kir. Kayyngdy. Chuyskaya Oblast', N Kyrgyzstan 42°48′N 73°39′E
77 T14 Kainji Dam dam W Nigeria
Kainji Lake see Kainji Reservoir
77 T14 Kainji Reservoir var. Kainji Lake. ▣ W Nigeria
186 D8 Kaintiba var. Kamina. Gulf, S Papua New Guinea 07°29′S 146°04′E
92 K12 Kainulasjärvi Norrbotten, N Sweden 67°00′N 22°31′E
93 M14 Kainuu Swe. Kajanaland. ◈ region N Finland
184 K5 Kaipara Harbour harbor North Island, New Zealand
152 I10 Kairāna Uttar Pradesh, N India 29°24′N 77°10′E
74 M6 Kairouan var. Al Qayrawān. E Tunisia 35°46′N 10°11′E
Kaisaria see Kayseri
101 F20 Kaiserslautern Rheinland-Pfalz, SW Germany 49°27′N 07°46′E
118 G13 Kaišiadorys Kaunas, S Lithuania 54°51′N 24°27′E
184 I2 Kaitaia Northland, North Island, New Zealand 35°07′S 173°13′E
185 E24 Kaitangata Otago, South Island, New Zealand 46°18′S 169°52′E
152 I9 Kaithal Haryāna, NW India 29°47′N 76°26′E
Kaitong see Tongyu
169 N13 Kait, Tanjung headland Sumatera, W Indonesia 03°13′S 106°03′E
38 E9 Kaiwi Channel channel Hawai'i, USA, C Pacific Ocean
160 K9 Kaixian var. Hanfeng. Sichuan, C China 31°13′N 108°25′E
163 V11 Kaiyuan var. K'ai-yüan. Liaoning, NE China 42°33′N 124°04′E
160 H14 Kaiyuan Yunnan, SW China 23°42′N 103°14′E
K'ai-yüan see Kaiyuan
39 O9 Kaiyuh Mountains ▲▲ Alaska, USA
93 M15 Kajaani Swe. Kajana. Kainuu, C Finland 64°17′N 27°46′E
149 N7 Kajakī, Band-e ◎ C Afghanistan
Kajan see Kayan, Sungai
Kajana see Kajaani
Kajanaland see Kainuu
137 V13 K'ajaran Rus. Kadzharan. SE Armenia 39°10′N 46°09′E
81 I19 Kajiado Kajiado, S Kenya 01°51′S 36°48′E
81 I20 Kajiado ◈ county S Kenya
Kajisay see Bokonbayevo
113 O20 Kajmakčalan var. Kaïmaktsalán. ▲ S FYR Macedonia 40°57′N 21°48′E see also Kaïmaktsalán
Kajmakčalan see Kaïmaktsalán
Kajnar see Kaynar
149 N6 Kajrān Dāykundī, C Afghanistan 33°12′N 65°28′E
149 N5 Kaj Rūd 〜 C Afghanistan
146 G14 Kaka Rus. Kaakhka. Ahal Welaýaty, S Turkmenistan 37°20′N 59°37′E
12 C12 Kakabeka Falls Ontario, S Canada 48°24′N 89°40′W
83 F23 Kakamas Northern Cape, W South Africa 28°45′S 20°33′E
81 H18 Kakamega Kakamega, W Kenya 0°17′N 34°47′E
81 H18 Kakamega ◈ county W Kenya
112 H13 Kakanj Federacija Bosne I Hercegovine, C Bosnia and Herzegovina 44°06′N 18°07′E
185 F22 Kakanui Mountains ▲▲ South Island, New Zealand
184 K11 Kakaramea Taranaki, North Island, New Zealand 39°42′S 174°27′E
76 J16 Kakata C Liberia 06°35′N 10°19′W
184 M11 Kakatahi Manawatu-Wanganui, North Island, New Zealand 39°40′S 175°20′E
113 M23 Kakavi Gjirokastër, S Albania 39°55′N 20°19′E
147 O14 Kakaydi Surkhondaryo Viloyati, S Uzbekistan 37°37′N 67°30′E
164 F13 Kake Hiroshima, Honshū, SW Japan 34°37′N 132°17′E
39 X13 Kake Kupreanof Island, Alaska, USA 56°58′N 133°57′W
171 P14 Kakea Pulau Wowoni, C Indonesia 04°09′S 123°06′E
164 M14 Kakegawa Shizuoka, Honshū, S Japan 34°47′N 138°02′E
165 V16 Kakeroma-jima Kagoshima, SW Japan
143 T6 Kākhak Khorāsān-e Razavī, E Iran
118 L11 Kakhanavichy Rus. Kokhanovichi. Vitsyebskaya Voblasts', N Belarus 55°52′N 28°08′E
39 P13 Kakhonak Alaska, USA 59°26′N 154°48′W
117 S10 Kakhovka Khersons'ka Oblast', S Ukraine 46°40′N 33°30′E
117 U9 Kakhovs'ke Vodoskhovyshche Rus. Kakhovskoye Vodokhranilishche. ▣ SE Ukraine
Kakhovskoye Vodokhranilishche see Kakhovs'ke Vodoskhovyshche
117 T11 Kakhovs'kyy Kanal canal S Ukraine
Kakia see Khakhea
155 L16 Kākināda prev. Cocanada. Andhra Pradesh, E India 16°56′N 82°13′E
Käkisalmi see Priozersk
164 I13 Kakogawa Hyōgo, Honshū, SW Japan 34°49′N 134°52′E
81 F18 Kakoge C Uganda 01°03′N 32°30′E
145 O7 Kak, Ozero ◎ N Kazakhstan
Ka-Krem see Malyy Yenisey
Kakshaal-Too, Khrebet see Kokshaal-Tau
39 S5 Kaktovik Alaska, USA 70°08′N 143°37′W
165 Q11 Kakuda Miyagi, Honshū, C Japan 37°59′N 140°48′E
165 Q8 Kakunodate Akita, Honshū, C Japan 39°37′N 140°35′E
Kalaallit Nunaat see Greenland
149 T7 Kālābāgh Punjab, E Pakistan 33°00′N 71°35′E
171 Q16 Kalabahi Pulau Alor, S Indonesia 08°14′S 124°32′E
188 I5 Kalabera Saipan, S Northern Mariana Islands
83 G14 Kalabo Western, W Zambia 15°00′S 22°37′E
126 M9 Kalach Voronezhskaya Oblast', W Russian Federation 50°24′N 41°00′E
127 N10 Kalach-na-Donu Volgogradskaya Oblast', SW Russian Federation 48°45′N 43°29′E
166 K5 Kaladan 〜 W Myanmar (Burma)
14 K14 Kaladar Ontario, SE Canada 44°38′N 77°06′W
38 G13 Ka Lae var. South Cape, South Point. headland Hawai'i, USA, C Pacific Ocean 18°54′N 155°40′W
83 G19 Kalahari Desert desert Southern Africa
38 B8 Kalāheo var. Kalaheo. Kaua'i, Hawaii, USA, C Pacific Ocean 21°55′N 159°31′W
Kalaheo see Kalāheo
Kalaikhum see Qal'aikhum
Kala-i-Mor see Galaýmor
93 K15 Kalajoki Pohjois-Pohjanmaa, W Finland 64°15′N 24°E
Kalak see Eski Kaļak
Kal al Sraghna see El Kelâa Srarhna
32 G10 Kalama Washington, NW USA 46°00′N 122°50′W
Kalámai see Kalámata
115 G14 Kalamariá Kentrikí Makedonía, N Greece 40°37′N 22°58′E
115 C15 Kalamás var. Thiamis; prev. Thýamis. 〜 W Greece
115 E21 Kalámata prev. Kalámai. Pelopónnisos, S Greece 37°02′N 22°07′E
148 L16 Kalamat, Khor Eng. Kalmat Lagoon. lagoon SW Pakistan
148 L16 Kalamat, Khor Eng. Kalmat Lagoon. lagoon SW Pakistan
31 P10 Kalamazoo Michigan, N USA 42°17′N 85°35′W
31 P9 Kalamazoo River 〜 Michigan, N USA
Kalambaka see Kalampáka
117 S13 Kalamits'ka Zatoka Rus. Kalamitskiy Zaliv. gulf S Ukraine
Kalamitskiy Zaliv see Kalamits'ka Zatoka
115 H18 Kálamos Attikí, C Greece 38°16′N 23°51′E
115 C18 Kálamos island Iónioi Nísia, Greece, C Mediterranean Sea
115 D15 Kalampáka var. Kalambaka. Thessalía, C Greece 39°43′N 21°36′E
Kalan see Cãlan, Romania
Kalan see Tunceli, Turkey
117 S11 Kalanchak Khersons'ka Oblast', S Ukraine 46°14′N 33°19′E
38 G11 Kalaoa var. Kailua. Hawaii, USA, C Pacific Ocean 19°43′N 155°59′W
171 O15 Kalaotoa, Pulau island W Indonesia
155 J24 Kala Oya 〜 NW Sri Lanka
Kalarash see Călăraşi
93 H17 Kälarne Jämtland, C Sweden 63°00′N 16°10′E
143 V15 Kalar Rūd 〜 SE Iran
169 R9 Kalasin var. Muang Kalasin. Kalasin, E Thailand 16°29′N 103°31′E
143 U4 Kalāt var. Kabūd Gonbad. Khorāsān-e Razavī, NE Iran 37°02′N 59°46′E
149 O11 Kālat var. Kelat, Khelat. Baluchistān, SW Pakistan 29°01′N 66°38′E
115 J14 Kalathriá, Ákrotírio headland Samothráki, NE Greece 40°24′N 25°34′E
193 W147 Kalau island Tongatapu Group, SE Tonga
38 E9 Kalaupapa Moloka'i, Hawaii, USA, C Pacific Ocean 21°11′N 156°59′W
127 N13 Kalaus 〜 SW Russian Federation
Kalávrita see Kalávryta
115 E19 Kalávryta var. Kalávrita. Dytikí Elláda, S Greece 38°02′N 22°06′E
141 Y10 Kalbān W Oman 20°19′N 58°40′E
Kalbar see Kalimantan Barat
180 H11 Kalbarri Western Australia 27°43′S 114°08′E
Kalbinskiy Khrebet see Khrebet Kalba
144 G10 Kaldygayty 〜 W Kazakhstan
136 I12 Kalecik Ankara, N Turkey 40°08′N 33°27′E
79 O19 Kalehe Sud-Kivu, E Dem. Rep. Congo 02°05′S 28°52′E
79 P22 Kalemie prev. Albertville. Katanga, SE Dem. Rep. Congo 05°55′S 29°09′E
166 L4 Kalemyo Sagaing, W Myanmar (Burma) 23°11′N 94°03′E
82 H12 Kalene Hill North Western, NW Zambia 11°10′S 24°12′E
167 T11 Kalêng prev. Phumĭ Kalêng. Stœng Trêng, NE Cambodia 13°57′N 106°17′E
Kale Sultanie see Çanakkale
124 I7 Kalevala Respublika Kareliya, NW Russian Federation 65°12′N 31°16′E
166 L4 Kalewa Sagaing, C Myanmar (Burma) 23°15′N 94°19′E
Kalgan see Zhangjiakou
39 Q12 Kalgin Island island Alaska, USA
180 L12 Kalgoorlie Western Australia 30°51′S 121°27′E
Kali see Sārda
115 E17 Kaliakoúda ▲ C Greece 38°47′N 21°42′E
114 O8 Kaliakra, Nos headland NE Bulgaria 43°22′N 28°28′E
115 F19 Kaliánoi Pelopónnisos, S Greece 37°55′N 22°28′E
115 N24 Kalí Límni ▲ Kárpathos, SE Greece 35°34′N 27°08′E
79 N20 Kalima Maniema, E Dem. Rep. Congo 02°34′S 26°27′E
169 S11 Kalimantan Eng. Indonesian Borneo. ◈ geopolitical region Borneo, C Indonesia
169 Q11 Kalimantan Barat off. Propinsi Kalimantan Berat, var. Kalbar, Eng. West Borneo, West Kalimantan. ◈ province N Indonesia
Kalimantan Barat, Propinsi see Kalimantan Barat
169 T13 Kalimantan Selatan off. Propinsi Kalimantan Selatan, var. Kalsel, Eng. South Borneo, South Kalimantan. ◈ province N Indonesia
Kalimantan Selatan, Propinsi see Kalimantan Selatan
169 R12 Kalimantan Tengah off. Propinsi Kalimantan Tengah, var. Kalteng, Eng. Central Borneo, Central Kalimantan. ◈ province N Indonesia
Kalimantan Tengah, Propinsi see Kalimantan Tengah
169 U10 Kalimantan Timur off. Propinsi Kalimantan Timur, var. Kaltim, Eng. East Borneo, East Kalimantan. ◈ province N Indonesia
Kalimantan Timur, Propinsi see Kalimantan Timur
169 V9 Kalimantan Utara off. Propinsi Kalimantan Utara, var. Kaltara, Eng. North Kalimantan. ◈ province N Indonesia
Kalimantan Utara, Propinsi see Kalimantan Utara
Kálimnos see Kálymnos
153 S12 Kālimpang West Bengal, NE India 27°02′N 88°34′E
Kalinin see Tver'
Kalinin see Boldumsaz
Kalininabad see Kalininobod
126 B3 Kaliningrad Kaliningradskaya Oblast', W Russian Federation 54°48′N 21°33′E
Kaliningrad see Kaliningradskaya Oblast'
126 A3 Kaliningradskaya Oblast' var. Kaliningrad. ◈ province and enclave W Russian Federation
Kalinino see Tashir
147 P14 Kalininobod Rus. Kalininabad. SW Tajikistan 37°49′N 68°55′E
127 O8 Kalininsk Saratovskaya Oblast', W Russian Federation 51°31′N 44°25′E
Kalininsk see Boldumsaz
Kalinisk see Cupcina
119 M19 Kalinkavichy Rus. Kalinkovichi. Homyel'skaya Voblasts', SE Belarus 52°08′N 29°19′E
Kalinkovichi see Kalinkavichy
81 G18 Kaliro SE Uganda 0°54′N 33°30′E
Kalisch/Kalish see Kalisz
33 O7 Kalispell Montana, NW USA 48°12′N 114°18′W
110 I13 Kalisz Ger. Kalisch, Rus. Kalish; anc. Calisia. Wielkopolskie, C Poland 51°46′N 18°04′E
110 F9 Kalisz Pomorski Ger. Kallies. Zachodnio-pomorskie, NW Poland 53°55′N 15°55′E
126 M10 Kalitva 〜 SW Russian Federation
81 F21 Kaliua Tabora, C Tanzania 05°03′S 31°48′E
92 K13 Kalix Norrbotten, N Sweden 65°51′N 23°14′E
92 K12 Kalixälven 〜 N Sweden
92 J11 Kalixfors Norrbotten, N Sweden 67°45′N 20°20′E
145 T8 Kalkaman Kaz. Qalqaman. Pavlodar, NE Kazakhstan 51°57′N 75°58′E
Kalkandelen see Tetovo
181 O4 Kalkarindji Northern Territory, N Australia 17°32′S 130°40′E
31 P6 Kalkaska Michigan, N USA 44°44′N 85°11′W
93 F16 Kall Jämtland, C Sweden 63°31′N 13°16′E
189 X2 Kallalen var. Calalen. island Ratak Chain, SE Marshall Islands
118 J5 Kallaste Ger. Krasnogor. Tartumaa, SE Estonia 58°40′N 27°12′E
93 N16 Kallavesi ◎ SE Finland
115 F17 Kallídromo ▲▲ C Greece
Kallies see Kalisz Pomorski
95 M22 Kallinge Blekinge, S Sweden 56°14′N 15°17′E
115 L16 Kalloní Lésvos, E Greece 39°14′N 26°16′E
93 F16 Kallsjön ◎ C Sweden
95 N21 Kalmar var. Calmar. Kalmar, S Sweden 56°40′N 16°22′E
95 M19 Kalmar var. Calmar. ◈ county S Sweden
95 N20 Kalmarsund strait S Sweden
Kalmat Lagoon see Kalamat, Khor
117 X9 Kal'mius 〜 E Ukraine
99 H15 Kalmthout Antwerpen, N Belgium 51°24′N 04°27′E
Kalmykia/Kalmykiya-Khal'mg Tangch, Respublika see Kalmykiya, Respublika
127 O12 Kalmykiya, Respublika var. Respublika Kalmykiya-Khal'mg Tangch, Eng. Kalmykia; prev. Kalmytskaya ASSR. ◈ autonomous republic SW Russian Federation
Kalmytskaya ASSR see Kalmykiya, Respublika
118 F9 Kalnciems C Latvia 56°46′N 23°37′E
114 L10 Kalnitsa 〜 SE Bulgaria
111 J24 Kalocsa Bács-Kiskun, S Hungary 46°31′N 19°00′E
114 J9 Kalofer Plovdiv, C Bulgaria 42°36′N 25°00′E
38 E10 Kalohi Channel channel C Pacific Ocean
83 I16 Kalomo Southern, S Zambia 17°02′S 26°29′E
29 X14 Kalona Iowa, C USA 41°28′N 91°42′W
115 K22 Kalotási, Akrotírio cape Amorgós, Kykládes, Greece, Aegean Sea
152 J8 Kalpa Himāchal Pradesh, N India 31°33′N 78°16′E
115 C15 Kalpáki Ípeiros, W Greece 39°53′N 20°38′E
155 C22 Kalpeni Island island Lakshadweep, India, N Indian Ocean
152 K13 Kālpi Uttar Pradesh, N India 26°07′N 79°44′E
158 G7 Kalpin Xinjiang Uygur Zizhiqu, NW China 40°35′N 78°52′E
149 P16 Kalri Lake ◎ SE Pakistan
Kalsel see Kalimantan Selatan
143 R5 Kāl Shūr 〜 N Iran
39 N11 Kalskag Alaska, USA 61°32′N 160°15′W
Kalsø see Kalsoy
95 B18 Kalsoy Dan. Kalsø. island N Faroe Islands
39 O9 Kaltag Alaska, USA 64°19′N 158°43′W
Kaltara see Kalimantan Utara
108 H7 Kaltbrunn Sankt Gallen, NE Switzerland 47°11′N 09°00′E
Kaltdorf see Pruszków
Kalteng see Kalimantan Tengah
Kaltim see Kalimantan Timur
77 X14 Kaltungo Gombe, E Nigeria 09°49′N 11°22′E
126 K4 Kaluga Kaluzhskaya Oblast', W Russian Federation 54°31′N 36°16′E
155 J26 Kalu Ganga 〜 S Sri Lanka
82 J13 Kalulushi Copperbelt, C Zambia 12°50′S 28°03′E
180 M2 Kalumburu Western Australia 14°11′S 126°40′E
95 H23 Kalundborg Sjælland, E Denmark 55°42′N 11°06′E
82 K11 Kalungwishi 〜 N Zambia
149 T8 Kalūr Kot Punjab, E Pakistan 32°08′N 71°20′E
116 I6 Kalush Pol. Kałusz. Ivano-Frankivs'ka Oblast', W Ukraine 49°02′N 24°20′E
Kałusz see Kalush
110 N11 Kałuszyn Mazowieckie, C Poland 52°12′N 21°43′E
155 J26 Kalutara Western Province, SW Sri Lanka 06°35′N 79°59′E
Kaluwawa see Fergusson Island
126 I5 Kaluzhskaya Oblast' ◈ province W Russian Federation
119 E14 Kalvarija Pol. Kalwaria. Marijampolė, S Lithuania 54°25′N 23°13′E
93 K15 Kälviä Keski-Pohjanmaa, W Finland 63°53′N 23°30′E
109 U6 Kalwang Steiermark, E Austria 47°25′N 14°48′E
Kalwaria see Kalvarija
154 D13 Kalyān Mahārāshtra, W India 19°17′N 73°11′E
124 K16 Kalyazin Tverskaya Oblast', W Russian Federation 57°15′N 37°53′E
115 D18 Kalydon anc. Calydon. site of ancient city Dytikí Elláda, C Greece
115 M21 Kálymnos var. Kálimnos. Kálymnos, Dodekánisa, Greece, Aegean Sea 36°57′N 26°59′E
115 M21 Kálymnos var. Kálimnos. island Dodekánisa, Greece, Aegean Sea
117 O5 Kalynivka Kyyivs'ka Oblast', N Ukraine 50°14′N 30°16′E
117 N6 Kalynivka Vinnyts'ka Oblast', C Ukraine 49°27′N 28°32′E
145 W15 Kalzhat prev. Kol'zhat. Almaty, SE Kazakhstan 43°29′N 80°37′E
42 M10 Kama var. Cama. Región Autónoma Atlántico Sur, SE Nicaragua 12°06′N 83°55′W
165 R9 Kamaishi var. Kamaisi. Iwate, Honshū, C Japan 39°18′N 141°52′E
Kamaisi see Kamaishi
118 H13 Kamajai Utena, E Lithuania 55°49′N 25°30′E
Kamajai see Toliejai
149 U9 Kamālia Punjab, NE Pakistan 30°44′N 72°39′E
83 I14 Kamalondo North Western, NW Zambia 13°42′S 25°38′E
136 I13 Kaman Kirşehir, C Turkey 39°22′N 33°43′E
79 O20 Kamanyola Sud-Kivu, E Dem. Rep. Congo 02°54′S 29°04′E
141 N14 Kamarān island W Yemen
55 R9 Kamarang W Guyana 05°49′N 60°38′W
Kāmāreddi/Kamareddy see Rāmāreddi
Kama Reservoir see Kamskoye Vodokhranilishche
148 K13 Kamarod Baluchistān, SW Pakistan 27°34′N 63°36′E
171 P14 Kamaru Pulau Buton, C Indonesia 05°10′S 123°03′E
77 S13 Kamba Kebbi, NW Nigeria 11°50′N 03°44′E
Kambaeng Petch see Kamphaeng Phet
180 L12 Kambalda Western Australia 31°15′S 121°33′E
149 P13 Kambar var. Qambar. Sind, SE Pakistan 27°35′N 68°03′E
Kambara see Kabara
76 I14 Kambia W Sierra Leone 09°09′N 12°53′W
Kambos see Kámpos
79 N25 Kambove Katanga, SE Dem. Rep. Congo 10°50′S 26°39′E
Kambryk see Cambrai
123 V10 Kamchatka 〜 E Russian Federation
Kamchatka see Kamchatka, Poluostrov
Kamchatka Basin see Komandorskaya Basin
123 U10 Kamchatka, Poluostrov Eng. Kamchatka. peninsula E Russian Federation
123 V10 Kamchatskiy Kray ◈ province E Russian Federation
123 V10 Kamchatskiy Zaliv gulf E Russian Federation
114 N9 Kamchia var. Kamchiya. 〜 E Bulgaria
114 L9 Kamchia, Yazovir var. Yazovir Kamchiya. ◎ E Bulgaria
Kamchiya see Kamchia
Kamchiya, Yazovir see Kamchia, Yazovir
149 T4 Kāmdēsh var. Kamdesh; prev. Kāmdeysh. Nūrestān, E Afghanistan 35°25′N 71°26′E
Kamdesh see Kāmdēsh
Kāmdeysh see Kāmdēsh
Kamen' see Kamyen'
Kamenets see Kamyanyets
Kamenets-Podol'skaya Oblast' see Khmel'nyts'ka Oblast'
Kamenets-Podol'skiy see Kam"yanets'-Podil's'kyy
113 Q18 Kamenica NE Macedonia 42°03′N 22°34′E
113 O16 Kamenicë var. Dardanë, Serb. Kosovska Kamenica. E Kosovo 42°37′N 21°33′E
112 A11 Kamenjak, Rt headland NW Croatia
125 O6 Kamenka Arkhangel'skaya Oblast', NW Russian Federation 65°55′N 44°01′E
126 6 Kamenka Penzenskaya Oblast', W Russian Federation 53°12′N 44°00′E
127 L8 Kamenka Voronezhskaya Oblast', W Russian Federation 50°44′N 39°31′E
Kamenka see Taskala
Kamenka see Camenca
Kamenka see Kam"yanka
Kamenka-Bugskaya see Kam"yanka-Buz'ka
Kamenka Dneprovskaya see Kam"yanka-Dniprovs'ka
Kamen Kashirskiy see Kamin'-Kashyrs'kyy
Kamenka-Strumilov see Kam"yanka-Buz'ka
126 L15 Kamennomostskiy Respublika Adygeya, SW Russian Federation 44°13′N 40°12′E
126 L11 Kamenolomni Rostovskaya Oblast', SW Russian Federation 47°36′N 40°18′E
127 P8 Kamenskiy Saratovskaya Oblast', W Russian Federation 50°56′N 45°32′E
Kamenskoye see Romaniv
126 L11 Kamensk-Shakhtinskiy Rostovskaya Oblast', SW Russian Federation 48°18′N 40°16′E
101 P15 Kamenz Sachsen, E Germany 51°15′N 14°06′E
164 J13 Kameoka Kyōto, Honshū, SW Japan 35°02′N 135°35′E
126 M3 Kameshkovo Vladimirskaya Oblast', W Russian Federation 56°21′N 41°01′E
164 C11 Kami-Agata Nagasaki, Tsushima, SW Japan 34°40′N 129°27′E
33 N10 Kamiah Idaho, NW USA 46°13′N 116°01′W
Kamień Koszyrski see Kamin'-Kashyrs'kyy
110 H9 Kamień Krajeński Ger. Kamin in Westpreussen. Kujawski-pomorskie, C Poland 53°31′N 17°31′E
111 F15 Kamienna Góra Ger. Landeshut, Landeshut in Schlesien. Dolnośląskie, SW Poland 50°48′N 16°00′E
110 D8 Kamień Pomorski Ger. Cummin in Pommern. Zachodnio-pomorskie, NW Poland 53°57′N 14°44′E
165 R5 Kamiiso Hokkaidō, NE Japan 41°50′N 140°38′E
79 L22 Kamiji Kasai-Oriental, S Dem. Rep. Congo 06°39′S 23°27′E
165 T3 Kamikawa Hokkaidō, NE Japan 43°51′N 142°47′E
164 B15 Kami-Koshiki-jima island SW Japan
79 M23 Kamina Katanga, S Dem. Rep. Congo 08°42′S 25°01′E
Kamina see Kaintiba
42 C6 Kaminaljuyú ruins Guatemala, C Guatemala
Kamin in Westpreussen see Kamień Krajeński
116 J2 Kamin'-Kashyrs'kyy Pol. Kamień Koszyrski, Rus. Kamen Kashirskiy. Volyns'ka Oblast', NW Ukraine 51°39′N 24°59′E
Kaminka Strumiłowa see Kam"yanka-Buz'ka
165 Q5 Kaminokuni Hokkaidō, NE Japan 41°48′N 140°05′E
165 P10 Kaminoyama Yamagata, Honshū, C Japan 38°10′N 140°16′E
39 Q13 Kamishak Bay bay Alaska, USA
165 U4 Kamishihoro Hokkaidō, NE Japan 43°14′N 143°18′E
Kamishli see Al Qāmishlī
Kamissar see Kamsar
164 C11 Kami-Tsushima Nagasaki, Tsushima, SW Japan 34°40′N 129°27′E
79 O20 Kamituga Sud-Kivu, E Dem. Rep. Congo 03°07′S 28°10′E
164 B17 Kamiyaku Kagoshima, Yaku-shima, SW Japan 30°24′N 130°32′E
11 N16 Kamloops British Columbia, SW Canada 50°39′N 120°24′W
107 G25 Kamma Sicilia, Italy, C Mediterranean Sea 36°46′N 12°03′E
192 K4 Kammu Seamount undersea feature N Pacific Ocean 32°09′N 173°00′E
109 U11 Kamnik Ger. Stein. C Slovenia 46°13′N 14°34′E
Kamniške Alpe see Kamniško-Savinjske Alpe
109 T10 Kamniško-Savinjske Alpe var. Kamniške Alpe, Sanntaler Alpen, Ger. Steiner Alpen. ▲▲ N Slovenia
Kamo see Gavarr
165 R3 Kamoenai var. Kamuenai. Hokkaidō, NE Japan 43°07′N 140°25′E
165 O14 Kamogawa Chiba, Honshū, S Japan 35°05′N 140°04′E
149 W8 Kāmoke Punjab, E Pakistan 31°58′N 74°15′E
82 L13 Kamoto Eastern, E Zambia 13°16′S 32°04′E
109 V3 Kamp 〜 N Austria
81 F18 Kampala ● (Uganda) S Uganda 0°20′N 32°28′E
168 K11 Kampar, Sungai 〜 Sumatera, W Indonesia
98 L9 Kampen Overijssel, E Netherlands 52°33′N 05°55′E
79 N20 Kampene Maniema, E Dem. Rep. Congo 03°35′S 26°40′E
29 Q9 Kampeska, Lake ◎ South Dakota, N USA
167 O9 Kamphaeng Phet var. Kambaeng Petch. Kamphaeng Phet, W Thailand 16°28′N 99°31′E
Kampo see Campo, Cameroon
Kampo see Ntem, Cameroon/Equatorial Guinea
167 S12 Kâmpóng Cham prev. Kompong Cham. Kâmpóng Cham, C Cambodia 12°N 105°27′E
167 R12 Kâmpóng Chhnăng prev. Kompong. Kâmpóng Chhnăng, C Cambodia 12°15′N 104°40′E
167 R12 Kâmpóng Khleăng prev. Kompong Kleang. Siĕmréab, NW Cambodia 13°04′N 104°07′E
Kâmpóng Saôm see Sihanoukville
167 R13 Kâmpóng Spoe prev. Kompong Speu. Kâmpóng Spœ, S Cambodia 11°28′N 104°29′E
167 S12 Kâmpóng Thum var. Trâpeăng Vêng. Kâmpóng Thum, C Cambodia 12°37′N 104°58′E
167 S12 Kâmpóng Trâbêk prev. Phumĭ Kâmpóng Trâbêk, Phum Kompong Trabek. Kâmpóng Thum, C Cambodia 13°06′N 105°16′E
121 O2 Kámpos var. Kambos. NW Cyprus 35°03′N 32°44′E
167 R14 Kâmpôt Kâmpôt, SW Cambodia 10°37′N 104°11′E
Kamptee see Kāmthi
77 O14 Kampti SW Burkina Faso 10°07′N 03°22′W
Kampuchea see Cambodia
Kampuchea, Democratic see Cambodia
Kampuchea, People's Democratic Republic of see Cambodia
169 Q9 Kampung Sirik Sarawak, East Malaysia 02°42′N 111°28′E
11 V15 Kamsack Saskatchewan, S Canada 51°34′N 101°51′W
76 H13 Kamsar var. Kamissar. Guinée-Maritime, W Guinea 10°33′N 14°34′W
127 R4 Kamskoye Ust'ye Respublika Tatarstan, W Russian Federation 55°13′N 49°11′E
125 U14 Kamskoye Vodokhranilishche var. Kama Reservoir. ▣ NW Russian Federation
154 I12 Kāmthi prev. Kamptee. Mahārāshtra, C India 21°19′N 79°11′E
Kamuela see Waimea
Kamuenai see Kamoenai
165 T5 Kamui-dake ⛰ Hokkaidō, NE Japan 42°24′N 142°57′E
165 R3 Kamui-misaki headland Hokkaidō, NE Japan 43°20′N 140°20′E
43 O15 Kámuk, Cerro ▲ SE Costa Rica 09°15′N 83°01′W
116 K7 Kam"yanets'-Podil's'kyy Rus. Kamenets-Podol'skiy. Khmel'nyts'ka Oblast', W Ukraine 48°43′N 26°36′E
117 Q6 Kam"yanka Rus. Kamenka. Cherkas'ka Oblast', C Ukraine 49°03′N 32°06′E
116 I5 Kam"yanka-Buz'ka prev. Kamenka-Strumilov, Pol. Kaminka Strumiłowa, Rus. Kamenka-Bugskaya. L'vivs'ka Oblast', NW Ukraine 50°04′N 24°21′E
117 T9 Kam"yanka-Dniprovs'ka Rus. Kamenka Dneprovskaya. Zaporiz'ka Oblast', SE Ukraine 47°28′N 34°24′E
119 F19 Kamyanyets Rus. Kamenets. Brestskaya Voblasts', SW Belarus 52°24′N 23°49′E
118 M13 Kamyen' Rus. Kamen'. Vitsyebskaya Voblasts', N Belarus 55°01′N 28°53′E
127 P9 Kamyshin Volgogradskaya Oblast', SW Russian Federation 50°07′N 45°20′E
127 Q13 Kamyzyak Astrakhanskaya Oblast', SW Russian Federation 46°07′N 48°03′E
12 K8 Kanaaupscow 〜 Québec, C Canada
36 K8 Kanab Utah, W USA 37°03′N 112°31′W
36 K9 Kanab Creek 〜 Arizona/Utah, SW USA
187 Y14 Kanacea prev. Kanathea. Taveuni, N Fiji 16°59′S 179°54′E
38 G17 Kanaga Island island Aleutian Islands, Alaska, USA
38 G17 Kanaga Volcano ⛰ Kanaga Island, Alaska, USA 51°55′N 177°09′W
164 N14 Kanagawa off. Kanagawa-ken. ◈ prefecture Honshū, S Japan
Kanagawa-ken see Kanagawa
13 Q8 Kanairiktok 〜 Newfoundland and Labrador, E Canada
Kanaky see New Caledonia
79 K22 Kananga prev. Luluabourg. Kasai-Occidental, S Dem. Rep. Congo 05°51′S 22°22′E
Kanara see Karnātaka
36 J7 Kanarraville Utah, W USA 37°32′N 113°10′W
127 Q4 Kanash Chuvashskaya Respublika, W Russian Federation 55°30′N 47°27′E
Kanathea see Kanacea
21 Q4 Kanawha River 〜 West Virginia, NE USA
164 L13 Kanayama Gifu, Honshū, SW Japan 35°46′N 137°15′E
164 L11 Kanazawa Ishikawa, Honshū, SW Japan 36°35′N 136°40′E
166 M4 Kanbalu Sagaing, C Myanmar (Burma) 23°10′N 95°31′E
166 L8 Kanbe Yangon, SW Myanmar (Burma) 16°40′N 96°01′E
167 O11 Kanchanaburi Kanchanaburi, W Thailand 14°02′N 99°32′E
Kanchanjańghā/Kānchenjunga see Kangchenjunga
155 J19 Kānchīpuram prev. Conjeeveram. Tamil Nādu, SE India 12°50′N 79°44′E
149 N8 Kandahār Per. Qandahār. Kandahār, S Afghanistan 31°36′N 65°48′E
149 N9 Kandahār Per. Qandahār. ◈ province SE Afghanistan
167 S13 Kândal var. Ta Khmau. Kândal, S Cambodia 11°30′N 104°59′E
Kandalakša see Kandalaksha
124 I5 Kandalaksha var. Kandalakša, Fin. Kantalahti. Murmanskaya Oblast', NW Russian Federation 67°09′N 32°14′E
Kandalaksha Gulf/Kandalakshskaya Guba see Kandalakshskiy Zaliv
124 K6 Kandalakshskiy Zaliv var. Kandalakshskaya Guba, Eng. Kandalaksha Gulf. bay NW Russian Federation
83 G17 Kandalengoti var. Kandalengoti. Ngamiland, NW Botswana 19°25′S 22°12′E
Kandalengoti see Kandalengoti
169 U13 Kandangan Borneo, C Indonesia 02°50′S 115°15′E
Kandau see Kandava
118 E8 Kandava Ger. Kandau. W Latvia 57°02′N 22°48′E
Kandavu see Kadavu
77 R14 Kandé var. Kanté. NE Togo 09°55′N 01°01′E
101 F23 Kandel ▲ SW Germany 48°03′N 08°00′E
186 C7 Kandep Enga, W Papua New Guinea 05°54′S 143°34′E
149 R12 Kandh Kot Sind, SE Pakistan 28°15′N 69°18′E
77 S13 Kandi N Benin 11°05′N 02°59′E
149 P14 Kandiāro Sind, SE Pakistan 27°02′N 68°16′E
136 F11 Kandıra Kocaeli, NW Turkey 41°05′N 30°08′E
183 S8 Kandos New South Wales, SE Australia 32°52′S 149°58′E
148 M16 Kandrāch var. Kanrach. Baluchistān, SW Pakistan 25°26′N 65°28′E
172 I4 Kandreho Mahajanga, C Madagascar 17°27′S 46°06′E
186 F7 Kandrian New Britain, E Papua New Guinea 06°14′S 149°32′E
Kandukur see Kondūkūr
155 K25 Kandy Central Province, C Sri Lanka 07°17′N 80°40′E
144 I10 Kandyagash Kaz. Qandyaghash; prev. Oktyab'rsk. Aktyubinsk, W Kazakhstan 49°25′N 57°24′E
18 D12 Kane Pennsylvania, NE USA 41°39′N 78°47′W
64 I11 Kane Fracture Zone tectonic feature NW Atlantic Ocean
Kanëka see Kanëvka
78 G9 Kanem off. Région du Kanem. ◈ region W Chad
Kanem, Région du see Kanem
38 D9 Kāne'ohe var. Kaneohe. O'ahu, Hawaii, USA, C Pacific Ocean 21°25′N 157°48′W
Kanestron, Akrotírio see Palioúri, Akrotírio
Kanëv see Kaniv
124 M5 Kanëvka var. Kanëka. Murmanskaya Oblast', NW Russian Federation 67°07′N 39°43′E
126 K13 Kanevskaya Krasnodarskiy Kray, SW Russian Federation 46°07′N 38°57′E

◆ Country ● Country Capital ◇ Dependent Territory ○ Dependent Territory Capital ◈ Administrative Regions ✈ International Airport ▲ Mountain ▲▲ Mountain Range ⛰ Volcano 〜 River ◎ Lake ▣ Reservoir

KANEVSKOYE VODOKHRANILISHCHE – KĀSGANJ

Kanevskoye Vodokhranilishche *see* Kanivs'ke Vodoskhovyshche
165 *P9* **Kaneyama** Yamagata, Honshū, C Japan *38°54′N 140°20′E*
83 *G20* **Kang** Kgalagadi, C Botswana *23°41′S 22°50′E*
76 *L13* **Kangaba** Koulikoro, SW Mali *11°57′N 08°24′W*
136 *M13* **Kangal** Sivas, C Turkey *39°15′N 37°23′E*
Kangān *see* Bandar-e Kangān
168 *J6* **Kangar** Perlis, Peninsular Malaysia *06°28′N 100°10′E*
76 *L13* **Kangaré** Sikasso, S Mali *11°39′N 08°10′W*
182 *F10* **Kangaroo Island** *island* South Australia
93 *M17* **Kangasniemi** Etelä-Savo, E Finland *61°58′N 26°37′E*
142 *K6* **Kangāvar** *var.* Kangāwar. Kermānshāhān, W Iran *34°29′N 47°55′E*
Kangāwar *see* Kangāvar
153 *S11* **Kangchenjunga** *var.* Kānchenjunga, *Nep.* Kanchanjaṅghā. ▲ NE India *27°36′N 88°06′E*
160 *G9* **Kangding** *var.* Lucheng, *Tib.* Dardo. Sichuan, C China *30°03′N 101°56′E*
169 *U16* **Kangean, Kepulauan** *island group* S Indonesia
169 *T16* **Kangean, Pulau** *island* Kepulauan Kangean, S Indonesia
67 *U8* **Kangen** *var.* Kengen. ≈ E South Sudan
197 *N14* **Kangerlussuaq** *Dan.* Sondre Strømfjord. ✕ Qeqqata, W Greenland *66°59′N 50°28′E*
197 *Q15* **Kangertittivaq** *Dan.* Scoresby Sund. *fjord* E Greenland
167 *O2* **Kangfang** Kachin State, N Myanmar (Burma) *26°09′N 98°36′E*
163 *X12* **Kanggye** N North Korea *40°58′N 126°37′E*
197 *P15* **Kangikajik** *var.* Kap Brewster. *headland* E Greenland *70°10′N 22°00′W*
13 *N5* **Kangiqsualujjuaq** *prev.* George River, Port-Nouveau-Québec. Québec, E Canada *58°35′N 65°59′W*
12 *L2* **Kangiqsujuaq** *prev.* Maricourt, Wakeham Bay. Québec, NE Canada *61°35′N 72°00′W*
12 *M4* **Kangirsuk** *prev.* Bellin, Payne. Québec, E Canada *60°00′N 70°01′W*
Kangle *see* Wanzai
158 *M16* **Kangmar** Xizang Zizhiqu, W China *28°34′N 89°40′E*
Kangnŭng *see* Gangneung
79 *D18* **Kango** Estuaire, NW Gabon *0°17′N 10°00′E*
152 *I7* **Kāngra** Himāchal Pradesh, NW India *32°04′N 76°16′E*
153 *Q16* **Kangsabati Reservoir** ⊠ N India
159 *O17* **Kangto** ▲ China/India *27°54′N 92°33′E*
159 *W12* **Kangxian** *var.* Kang Xian, Zuitai, Zuitaizi. Gansu, C China *33°21′N 105°40′E*
Kang Xian *see* Kangxian
76 *M15* **Kani** NW Ivory Coast *08°29′N 06°36′W*
166 *L4* **Kani** Sagaing, C Myanmar (Burma) *22°24′N 94°55′E*
79 *M23* **Kaniama** Katanga, S Dem. Rep. Congo *07°32′S 24°11′E*
Kanibadam *see* Konibodom
169 *V6* **Kanibongan** Sabah, East Malaysia *06°40′N 117°12′E*
185 *F17* **Kaniere** West Coast, South Island, New Zealand *42°45′S 171°00′E*
185 *G17* **Kaniere, Lake** ◎ South Island, New Zealand
188 *E17* **Kanifaay** Yap, W Micronesia
125 *O4* **Kanin Kamen'** ▲▲ NW Russian Federation
125 *N3* **Kanin Nos** Nenetskiy Avtonomnyy Okrug, NW Russian Federation *68°38′N 43°19′E*
125 *N3* **Kanin Nos, Mys** *cape* NW Russian Federation
125 *O5* **Kanin, Poluostrov** *peninsula* NW Russian Federation
139 *V8* **Kānī Sakht** Wāsiṭ, E Iraq *33°19′N 46°04′E*
139 *T3* **Kanî Slêman** *Ar.* Kānī Sulaymān. Arbīl, N Iraq *35°54′N 44°35′E*
Kānī Sulaymān *see* Kanî Slêman
165 *Q6* **Kanita** Aomori, Honshū, C Japan *41°04′N 140°36′E*
117 *Q5* **Kaniv** *Rus.* Kanëv. Cherkas'ka Oblast', C Ukraine *49°46′N 31°28′E*
182 *K11* **Kaniva** Victoria, SE Australia *36°25′S 141°13′E*
117 *Q5* **Kanivs'ke Vodoskhovyshche** *Rus.* Kanevskoye Vodokhranilishche. ⊠ C Ukraine
112 *L8* **Kanjiža** *Ger.* Altkanischa, *Hung.* Magyarkanizsa, Ókanizsa; *prev.* Stara Kanjiža. Vojvodina, N Serbia *46°03′N 20°03′E*
93 *K18* **Kankaanpää** Satakunta, SW Finland *61°47′N 22°25′E*
30 *M12* **Kankakee** Illinois, N USA *41°07′N 87°51′W*
31 *O11* **Kankakee River** ≈ Illinois/Indiana, N USA
76 *K14* **Kankan** E Guinea *10°25′N 09°19′W*
154 *K13* **Kānker** Chhattīsgarh, C India *20°19′N 81°29′E*
76 *J10* **Kankossa** Assaba, S Mauritania *15°54′N 11°31′W*
169 *N12* **Kanmaw Kyun** *var.* Kisseraing, Kithareng. *island* Mergui Archipelago, S Myanmar (Burma)
164 *F12* **Kanmuri-yama** ▲ Kyūshū, SW Japan *34°28′N 132°03′E*
21 *R10* **Kannapolis** North Carolina, SE USA *35°30′N 80°36′W*
93 *L16* **Kannonkoski** Keski-Suomi, C Finland *62°59′N 25°20′E*
93 *K15* **Kannus** Keski-Pohjanmaa, W Finland *63°55′N 23°55′E*
77 *V13* **Kano** Kano, N Nigeria *11°56′N 08°31′E*
77 *V13* **Kano** ⬖ *state* N Nigeria
77 *V13* **Kano** ✕ Kano, N Nigeria *11°56′N 08°26′E*
164 *G14* **Kan'onji** *var.* Kanonzi. Kagawa, Shikoku, SW Japan *34°08′N 133°38′E*
Kanonzi *see* Kan'onji
26 *M5* **Kanopolis Lake** ⊠ Kansas, C USA
36 *K5* **Kanosh** Utah, W USA *38°48′N 112°26′W*
169 *R9* **Kanowit** Sarawak, East Malaysia *02°03′N 112°15′E*
164 *C16* **Kanoya** Kagoshima, Kyūshū, SW Japan *31°22′N 130°50′E*
152 *L13* **Kānpur** *Eng.* Cawnpore. Uttar Pradesh, N India *26°28′N 80°21′E*
Kanrach *see* Kandrāch
164 *I14* **Kansai** ✕ (Ōsaka) Ōsaka, Honshū, SW Japan *34°25′N 135°13′E*
27 *R9* **Kansas** Oklahoma, C USA *36°14′N 94°46′W*
26 *L5* **Kansas** *off.* State of Kansas, *also known as* Jayhawker State, Sunflower State. ⬖ *state* C USA
27 *R4* **Kansas City** Kansas, C USA *39°07′N 94°38′W*
27 *R4* **Kansas City** Missouri, C USA *39°06′N 94°35′W*
27 *R3* **Kansas City** ✕ Missouri, C USA *39°18′N 94°45′W*
27 *P4* **Kansas River** ≈ Kansas, C USA
122 *L14* **Kansk** Krasnoyarskiy Kray, S Russian Federation *56°11′N 95°32′E*
Kansu *see* Gansu
147 *V7* **Kant** Chuyskaya Oblast', N Kyrgyzstan *42°54′N 74°47′E*
93 *L19* **Kanta-Häme** *Swe.* Egentliga Tavastland. ⬖ *region* S Finland
Kantalahti *see* Kandalaksha
167 *N16* **Kantang** *var.* Ban Kantang. Trang, SW Thailand *07°25′N 99°30′E*
115 *H25* **Kántanos** Kríti, Greece, E Mediterranean Sea *35°20′N 23°42′E*
77 *R12* **Kantchari** E Burkina Faso *12°47′N 01°37′E*
Kanté *see* Kandé
Kantemir *see* Cantemir
126 *L9* **Kantemirovka** Voronezhskaya Oblast', W Russian Federation *49°44′N 39°53′E*
167 *R11* **Kantharalak** Si Sa Ket, E Thailand *14°32′N 104°37′E*
Kantipur *see* Kathmandu
39 *Q9* **Kantishna River** ≈ Alaska, USA
191 *S3* **Kanton** *var.* Abariringa, Canton Island; *prev.* Mary Island. *atoll* Phoenix Islands, C Kiribati
97 *C20* **Kanturk** *Ir.* Ceann Toirc. Cork, SW Ireland *52°12′N 08°54′W*
55 *T11* **Kanuku Mountains** ▲▲ S Guyana
165 *O12* **Kanuma** Tochigi, Honshū, S Japan *36°34′N 139°44′E*
83 *H20* **Kanye** Southern, SE Botswana *24°55′S 25°14′E*
83 *H17* **Kanyu** North-West, C Botswana *20°04′S 24°36′E*
166 *M7* **Kanyutkwin** Bago, C Myanmar (Burma) *18°19′N 96°30′E*
79 *M24* **Kanzenze** Katanga, SE Dem. Rep. Congo *10°33′S 25°28′E*
193 *Y15* **Kao** *island* Kotu Group, W Tonga
167 *Q13* **Kaôh Kŏng** *var.* Krŏng Kaôh Kŏng. Kaôh Kông, SW Cambodia *11°37′N 102°59′E*
Kaohsiung *see* Gaoxiong
Kaokaona *see* Kirakira
83 *B17* **Kaoko Veld** ▲▲ N Namibia
76 *G11* **Kaolack** *var.* Kaolak. W Senegal *14°09′N 16°08′W*
Kaolak *see* Kaolack
Kaolan *see* Lanzhou
186 *M8* **Kaolo** San Jorge, N Solomon Islands *08°24′S 159°35′E*
83 *H14* **Kaoma** Western, W Zambia *14°50′S 24°48′E*
38 *B8* **Kapa'a** *var.* Kapaa. Kaua'i, Hawaii, USA, C Pacific Ocean *22°04′N 159°19′W*
Kapaa *see* Kapa'a
113 *J16* **Kapa Moračka** ▲ C Montenegro *42°53′N 19°01′E*
137 *V13* **Kapan** *Rus.* Kafan; *prev.* Ghap'an. SE Armenia *39°13′N 46°25′E*
82 *L13* **Kapandashila** Muchinga, NE Zambia *12°43′S 31°00′E*
79 *L23* **Kapanga** Katanga, S Dem. Rep. Congo *08°22′S 22°37′E*
Kapchagay *see* Kapshagay
Kapchagayskoye Vodokhranilishche *see* Vodokhranilishche Kapshagay
99 *F15* **Kapelle** Zeeland, SW Netherlands *51°29′N 03°58′E*
99 *G16* **Kapellen** Antwerpen, N Belgium *51°19′N 04°25′E*
95 *P15* **Kapellskär** Stockholm, C Sweden *59°43′N 19°03′E*
81 *H18* **Kapenguria** West Pokit, W Kenya *01°14′N 35°08′E*
109 *V6* **Kapfenberg** Steiermark, C Austria *47°27′N 15°18′E*
83 *J14* **Kapiri Mposhi** Central, C Zambia *13°59′S 28°40′E*
149 *R4* **Kāpīsā** ⬖ *province* E Afghanistan
12 *G10* **Kapiskau** ≈ Ontario, C Canada
184 *K13* **Kapiti Island** *island* C New Zealand
78 *K9* **Kapka, Massif du** ▲▲ E Chad
Kaplamada *see* Kaubalatmada, Gunung
22 *H9* **Kaplan** Louisiana, S USA *30°00′N 92°16′W*
Kaplangky, Plato *see* Gaplaňgyr Platosy
111 *D19* **Kaplice** *Ger.* Kaplitz. Jihočeský Kraj, S Czech Republic *48°42′N 14°27′E*
Kaplitz *see* Kaplice
Kapoche *see* Capoche
171 *T12* **Kapocol** Papua Barat, E Indonesia *01°59′S 130°11′E*
167 *N14* **Kapoe** Ranong, SW Thailand *09°33′N 98°37′E*
Kapoeas *see* Kapuas, Sungai
81 *G15* **Kapoeta** Eastern Equatoria, SE South Sudan *04°50′N 33°35′E*
111 *I25* **Kapos** ≈ S Hungary
111 *H25* **Kaposvár** Somogy, SW Hungary *46°23′N 17°54′E*
94 *H13* **Kapp** Oppland, S Norway *60°42′N 10°49′E*
100 *I7* **Kappeln** Schleswig-Holstein, N Germany *54°41′N 09°56′E*
Kaproncza *see* Koprivnica
109 *P7* **Kaprun** Salzburg, C Austria *47°15′N 12°48′E*
145 *U15* **Kapshagay** *prev.* Kapchagay. Almaty, SE Kazakhstan *43°52′N 77°05′E*
Kapstad *see* Cape Town
Kapsukas *see* Marijampolė
171 *Y13* **Kaptiau** Papua, E Indonesia *02°23′S 139°51′E*
119 *L19* **Kaptsevichy** *Rus.* Koptsevichi. Homyel'skaya Voblasts', SE Belarus *52°14′N 28°19′E*
Kapuas Hulu, Banjaran/Kapuas Hulu, Pegunungan *see* Kapuas Mountains
169 *S10* **Kapuas Mountains** *Ind.* Banjaran Kapuas Hulu, Pegunungan Kapuas Hulu. ▲▲ Indonesia/Malaysia
169 *P11* **Kapuas, Sungai** ≈ Borneo, N Indonesia
169 *T13* **Kapuas, Sungai** *prev.* Kapoeas. ≈ Borneo, C Indonesia
182 *J9* **Kapunda** South Australia *34°23′S 138°51′E*
152 *H8* **Kapūrthala** Punjab, N India *31°20′N 75°26′E*
12 *G12* **Kapuskasing** Ontario, S Canada *49°25′N 82°26′W*
14 *D6* **Kapuskasing** ≈ Ontario, S Canada
127 *P11* **Kapustin Yar** Astrakhanskaya Oblast', SW Russian Federation *48°36′N 45°49′E*
82 *K11* **Kaputa** Northern, NE Zambia *08°28′S 29°41′E*
111 *G22* **Kapuvár** Győr-Moson-Sopron, NW Hungary *47°35′N 17°01′E*
Kapydzhik, Gora *see* Qazangödağ
119 *J17* **Kapyl'** *Rus.* Kopyl'. Minskaya Voblasts', C Belarus *53°09′N 27°05′E*
43 *N9* **Kara** *var.* Cara. Región Autónoma Atlántico Sur, E Nicaragua *12°50′N 83°35′W*
77 *R14* **Kara** *var.* Lama-Kara. NE Togo *09°33′N 01°12′E*
77 *Q14* **Kara** ≈ N Togo
147 *U7* **Kara-Balta** Chuyskaya Oblast', N Kyrgyzstan *42°51′N 73°51′E*
144 *L7* **Karabalyk** *var.* Komsomolets, *Kaz.* Komsomol. Kostanay, N Kazakhstan *53°47′N 61°58′E*
144 *G11* **Karabau** *Kaz.* Qarabaū. Atyrau, W Kazakhstan *48°29′N 53°05′E*
146 *E7* **Karabaur', Uval** *Kaz.* Korabavur Pastligi, *Uzb.* Qorabowur Kirlari. *physical region* Kazakhstan/Uzbekistan
Karabekaul *see* Garabekewül
Karabil', Vozvyshennost' *see* Garabil Belentligi
Kara-Bogaz-Gol *see* Garabogazköl
Kara-Bogaz-Gol, Zaliv *see* Garabogaz Aylagy
145 *R15* **Karaboget** *Kaz.* Qaraböget. Zhambyl, S Kazakhstan *44°36′N 72°03′E*
136 *H11* **Karabük** Karabük, NW Turkey *41°12′N 32°36′E*
136 *H11* **Karabük** ⬖ *province* NW Turkey
122 *L12* **Karabula** Krasnoyarskiy Kray, C Russian Federation *58°01′N 97°17′E*
145 *V14* **Karabulak** *Kaz.* Qarabulaq. Taldykorgan, SE Kazakhstan *44°53′N 78°29′E*
145 *Y11* **Karabulak** *Kaz.* Qarabulaq. Vostochnyy Kazakhstan, E Kazakhstan *47°34′N 84°40′E*
145 *Q17* **Karabulak** *Kaz.* Qarabulaq. Yuzhnyy Kazakhstan, S Kazakhstan *42°31′N 69°47′E*
Karabura *see* Yumin
136 *C17* **Kara Burnu** *headland* SW Turkey *36°34′N 28°00′E*
144 *K10* **Karabutak** *Kaz.* Qarabutaq. Aktyubinsk, W Kazakhstan *49°55′N 60°05′E*
136 *D12* **Karacabey** Bursa, NW Turkey *40°14′N 28°22′E*
114 *O12* **Karacaköy** İstanbul, NW Turkey *41°24′N 28°21′E*
114 *M12* **Karacaoğlan** Kırklareli, NW Turkey *41°30′N 27°06′E*
Karachay-Cherkessia *see* Karachayevo-Cherkesskaya Respublika
126 *L15* **Karachayevo-Cherkesskaya Respublika** *Eng.* Karachay-Cherkessia. ⬖ *autonomous republic* SW Russian Federation
126 *M15* **Karachayevsk** Karachayevo-Cherkesskaya Respublika, SW Russian Federation *43°43′N 41°53′E*
126 *J6* **Karachev** Bryanskaya Oblast', W Russian Federation *53°07′N 35°56′E*
149 *O16* **Karāchi** Sind, SE Pakistan *24°51′N 67°02′E*
149 *O16* **Karāchi** ✕ Sind, S Pakistan *24°51′N 67°02′E*
Karácsonkő *see* Piatra-Neamţ
155 *E15* **Karād** Mahārāshtra, W India *17°19′N 74°15′E*
136 *H16* **Karadağ** ▲ S Turkey *37°00′N 33°00′E*
147 *T10* **Karadar'ya** *Uzb.* Qoradaryo. ≈ Kyrgyzstan/Uzbekistan
Karadeniz *see* Black Sea
Karadeniz Boğazi *see* İstanbul Boğazı
146 *B13* **Karadepe** Balkan Welaýaty, W Turkmenistan *38°04′N 54°01′E*
Karadzhar *see* Qorajar
Karaferiye *see* Véroia
Karagan *see* Garagan
Karaganda *see* Karagandy
Karaganda *see* Karagandy
Karagandinskaya Oblast' *see* Karagandy
145 *R10* **Karagandy** *Kaz.* Qaraghandy; *prev.* Karaganda. Karaganda, C Kazakhstan *49°53′N 73°07′E*
145 *R10* **Karagandy** *off.* Karagandinskaya Oblast', *Kaz.* Qaraghandy Oblysy; *prev.* Karaganda. ⬖ *province* C Kazakhstan
145 *T10* **Karagayly** *Kaz.* Qaraghayly. Karaganda, C Kazakhstan *49°25′N 75°31′E*
Karagel' *see* Garagöl'
123 *U9* **Karaginskiy, Ostrov** *island* E Russian Federation
197 *T1* **Karaginskiy Zaliv** *bay* E Russian Federation
137 *P13* **Karagöl Dağları** ▲▲ NE Turkey
Karagumskiy Kanal *see* Garagum Kanaly
114 *L13* **Karahisar** Edirne, NW Turkey *40°47′N 26°34′E*
127 *V3* **Karaidel'** Respublika Bashkortostan, W Russian Federation *55°50′N 56°55′E*
114 *L13* **Karaidemir Barajı** ⊠ NW Turkey
155 *J21* **Kāraikāl** Puducherry, SE India *10°58′N 79°50′E*
155 *I22* **Kāraikkudi** Tamil Nādu, SE India *10°04′N 78°46′E*
Kara Irtysh *see* Kara Yertis
143 *N5* **Karaj** Alborz, N Iran *35°44′N 51°26′E*
168 *K8* **Karak** Pahang, Peninsular Malaysia *03°24′N 101°59′E*
Karak *see* Al Karak
147 *T11* **Kara-Kabak** Oshskaya Oblast', SW Kyrgyzstan *39°40′N 72°45′E*
Kara-Kala *see* Magtymguly
Karakala *see* Oqqal'a
Karakalpakstan, Respublika *see* Qoraqalpog'iston Respublikasi
Karakalpakya *see* Qoraqalpog'iston
Karakax *see* Moyu
158 *G10* **Karakax He** ≈ NW China
121 *X8* **Karakaya Barajı** ⊠ C Turkey
171 *Q9* **Karakelong, Pulau** *island* N Indonesia
Karakılısse *see* Ağrı
Karak, Muḥāfaẓat al *see* Al Karak
147 *X8* **Karakol** *var.* Karakolka. Issyk-Kul'skaya Oblast', NE Kyrgyzstan *41°30′N 77°18′E*
147 *Y7* **Karakol** *prev.* Przheval'sk. Issyk-Kul'skaya Oblast', NE Kyrgyzstan *42°32′N 78°21′E*
Kara-Köl *see* Kara-Kul'
Karakolka *see* Karakol
149 *W2* **Karakoram Highway** *road* China/Pakistan
149 *Z3* **Karakoram Pass** *Chin.* Karakoram Shankou. *pass* C Asia
152 *I3* **Karakoram Range** ▲▲ C Asia
Karakoram Shankou *see* Karakoram Pass
Karaköse *see* Ağrı
145 *P14* **Karakoyyn, Ozero** *Kaz.* Qaraqoyyn. ◎ C Kazakhstan
83 *F19* **Karakubis** Ghanzi, W Botswana *22°03′S 20°36′E*
147 *T9* **Kara-Kul'** *Kir.* Kara-Köl. Dzhalal-Abadskaya Oblast', W Kyrgyzstan *40°35′N 73°36′E*
Karakul' *see* Qarokŭl, Tajikistan
Karakul' *see* Qorako'l, Uzbekistan
147 *U10* **Kara-Kul'dzha** Oshskaya Oblast', SW Kyrgyzstan *40°32′N 73°50′E*
127 *T3* **Karakulino** Udmurtskaya Respublika, NW Russian Federation *56°02′N 53°45′E*
Karakul', Ozero *see* Qarokŭl
Kara Kum *see* Garagum
Kara Kum Canal/Karakumskiy Kanal *see* Garagum Kanaly
Karakumy, Peski *see* Garagum
83 *E17* **Karakuwisa** Okavango, NE Namibia *18°56′S 19°40′E*
122 *M13* **Karam** Irkutskaya Oblast', S Russian Federation *55°07′N 107°21′E*
Karamai *see* Karamay
169 *T14* **Karamain, Pulau** *island* N Indonesia
136 *I16* **Karaman** Karaman, S Turkey *37°11′N 33°13′E*
136 *H16* **Karaman** ⬖ *province* S Turkey
114 *M8* **Karamandere** ≈ NE Bulgaria
158 *J4* **Karamay** *var.* Karamai, Kelamayi; *prev. Chin.* K'o-la-ma-i. Xinjiang Uygur Zizhiqu, NW China *45°33′N 84°45′E*
169 *U14* **Karambu** Borneo, N Indonesia *03°48′S 116°06′E*
185 *H14* **Karamea** West Coast, South Island, New Zealand *41°15′S 172°07′E*
185 *H14* **Karamea** ≈ South Island, New Zealand
185 *G15* **Karamea Bight** *gulf* South Island, New Zealand
Karamet-Niyaz *see* Garamätnyýaz
158 *K10* **Karamiran He** ≈ NW China
147 *S11* **Karamyk** Oshskaya Oblast', SW Kyrgyzstan *39°28′N 71°45′E*
169 *U17* **Karangasem** Bali, S Indonesia *08°24′S 115°40′E*
154 *H12* **Kāranja** Mahārāshtra, C India *20°30′N 77°26′E*
Karanpur *see* Karanpura
152 *F9* **Karanpura** *var.* Karanpur. Rājasthān, NW India *29°46′N 73°30′E*
Karánsebes/Karansebesch *see* Caransebeş
145 *T14* **Karaoy** *Kaz.* Qaraoy. Almaty, SE Kazakhstan *45°52′N 74°44′E*
144 *E10* **Karaozen** *Kaz.* Ülkenözen; *prev.* Bol'shoy Uzen'. ≈ Kazakhstan/Russian Federation
114 *N7* **Karapelit** *Rom.* Stejarul. Dobrich, NE Bulgaria *43°40′N 27°33′E*
136 *I15* **Karapınar** Konya, C Turkey *37°43′N 33°34′E*
83 *D22* **Karas** ⬖ *district* S Namibia
147 *Y8* **Kara-Say** Issyk-Kul'skaya Oblast', NE Kyrgyzstan *41°34′N 77°56′E*
83 *E22* **Karasburg** Karas, S Namibia *27°59′S 18°46′E*
Kara Sea *see* Karskoye More
92 *K9* **Kárášjohka** *var.* Karašjokka. ≈ N Norway
Kárášjohka *see* Karasjok
92 *L9* **Karasjok** *Fin.* Kaarasjoki, *Lapp.* Kárášjohka. Finnmark, N Norway *69°27′N 25°28′E*
Karašjokka *see* Kárášjohka
Kara Strait *see* Karskiye Vorota, Proliv
145 *N8* **Karasu** *Kaz.* Qarasū. Kostanay, N Kazakhstan *52°44′N 65°29′E*
136 *F11* **Karasu** Sakarya, NW Turkey *41°07′N 30°37′E*
Kara Su *see* Mesta/Néstos
Karasubazar *see* Bilohirs'k
122 *I12* **Karasuk** Novosibirskaya Oblast', C Russian Federation *53°41′N 78°04′E*
145 *U13* **Karatal** *Kaz.* Qaratal. ≈ SE Kazakhstan
136 *K17* **Karataş** Adana, S Turkey *36°32′N 35°22′E*
145 *Q16* **Karatau** *Kaz.* Qaratau. Zhambyl, S Kazakhstan *43°09′N 70°28′E*
Karatau *see* Karatau, Khrebet
145 *P16* **Karatau, Khrebet** *var.* Karatau, *Kaz.* Qaratau. ▲▲ S Kazakhstan
144 *G13* **Karaton** *Kaz.* Qaraton. Atyrau, W Kazakhstan *46°33′N 53°31′E*
164 *C13* **Karatsu** *var.* Karatu. Saga, Kyūshū, SW Japan *33°28′N 129°48′E*
Karatu *see* Karatsu
122 *K8* **Karaul** Krasnoyarskiy Kray, N Russian Federation *70°07′N 83°12′E*
Karaulbazar *see* Qorowulbozor
Karauzyak *see* Qorao'zak
115 *D16* **Karáva** ▲ C Greece *39°19′N 21°33′E*
Karavanke *see* Karawanken
115 *F22* **Karavás** Kýthira, S Greece *36°21′N 22°57′E*
113 *J20* **Karavastasë, Laguna e** *var.* Kënet' e Karavastas, Kravasta Lagoon. *lagoon* W Albania
Karavastas, Kënet' e *see* Karavastasë, Laguna e
118 *I5* **Käravere** Tartumaa, E Estonia *58°25′N 26°29′E*
115 *L23* **Karavonísia** *island* Kykládes, Greece, Aegean Sea
169 *O15* **Karawang** *prev.* Krawang. Jawa, C Indonesia *06°13′S 107°16′E*
109 *T10* **Karawanken** *Slvn.* Karavanke. ▲▲ Austria/Serbia
Karaxahar *see* Kaidu He
137 *R13* **Karayazı** Erzurum, NE Turkey *39°40′N 42°09′E*
145 *Y11* **Kara Yertis** *prev.* Kara Irtysh *Rus.* Chërnyy Irtysh. ≈ NE Kazakhstan
145 *Q12* **Karazhal** *Kaz.* Qarazhal. Karaganda, C Kazakhstan *48°002′N 70°52′E*
139 *S9* **Karbalā'** *var.* Kerbala, Kerbela. Karbalā', S Iraq *32°37′N 44°03′E*
139 *S9* **Karbalā'** *off.* Muḥāfa at Karbalā'. ⬖ *governorate* S Iraq
Karbalā', Muḥāfa at *see* Karbalā'
94 *L11* **Kårböle** Gävleborg, C Sweden *61°59′N 15°16′E*
111 *M23* **Karcag** Jász-Nagykun-Szolnok, E Hungary *47°22′N 20°51′E*
Kardak *see* Imia
114 *N7* **Kardam** Dobrich, NE Bulgaria *43°45′N 28°06′E*
Kardamila *see* Kardámyla
115 *L18* **Kardámyla** *var.* Kardamila, Kardhámila. Chíos, E Greece *38°33′N 26°04′E*
Kardeljevo *see* Ploče
Kardh *see* Qardho
Kardhámila *see* Kardámyla
Kardhítsa *see* Kardítsa
115 *E16* **Kardítsa** *var.* Kardhítsa. Thessalía, C Greece *39°22′N 21°56′E*
118 *E4* **Kärdla** *Ger.* Kertel. Hiiumaa, W Estonia *59°00′N 22°42′E*
Kărdžali *see* Kardzhali
114 *J11* **Kardzhali** *var.* Kŭrdzhali, Kărdžali, Kirdzhali. Kardzhali, S Bulgaria *41°39′N 25°23′E*
114 *K11* **Kardzhali** *var* Kŭrdzhali. ⬖ *province* S Bulgaria
114 *J11* **Kardzhali, Yazovir** *var.* Yazovir Kŭrdzhali. ⊠ S Bulgaria
Karelia *see* Kareliya, Respublika
119 *I16* **Karelichy** *Pol.* Korelicze, *Rus.* Korelichi. Hrodzyenskaya Voblasts', W Belarus *53°34′N 26°08′E*
124 *I10* **Kareliya, Respublika** *prev.* Karel'skaya ASSR, *Eng.* Karelia. ⬖ *autonomous republic* NW Russian Federation
Karel'skaya ASSR *see* Kareliya, Respublika
81 *E22* **Karema** Katavi, W Tanzania *06°50′S 30°25′E*
Karen *see* Hualian
83 *I14* **Karenda** Central, C Zambia *14°42′S 26°52′E*
Karen State *see* Kayin State
92 *J10* **Karesuando** *Fin.* Kaaresuanto, *Lapp.* Gárasavvon. Norrbotten, N Sweden *68°25′N 22°28′E*
Karet *see* Kâgheţ
Kareyz-e-Elyās/Kārez Iliās *see* Kārīz-e Elyās
122 *J11* **Kargasok** Tomskaya Oblast', C Russian Federation *59°01′N 80°34′E*
122 *I12* **Kargat** Novosibirskaya Oblast', C Russian Federation *55°07′N 80°19′E*
136 *J11* **Kargı** Çorum, N Turkey *41°09′N 34°32′E*
152 *I5* **Kargil** Jammu and Kashmir, NW India *34°34′N 76°06′E*
Kargilik *see* Yecheng
124 *L11* **Kargopol'** Arkhangel'skaya Oblast', NW Russian Federation *61°30′N 38°53′E*
110 *F12* **Kargowa** *Ger.* Unruhstadt. Lubuskie, W Poland *52°05′N 15°50′E*
77 *X13* **Kari** Bauchi, E Nigeria *11°13′N 10°34′E*
83 *J15* **Kariba** Mashonaland West, N Zimbabwe *16°29′S 28°48′E*
83 *J16* **Kariba, Lake** ⊠ Zambia/Zimbabwe
165 *Q4* **Kariba-yama** ▲ Hokkaidō, NE Japan *42°36′N 139°55′E*
83 *C19* **Karibib** Erongo, C Namibia *21°56′S 15°51′E*
Karies *see* Karyés
92 *L9* **Karigasniemi** *Lapp.* Garegegasnjárga. Lappi, N Finland *69°24′N 25°52′E*
184 *J2* **Karikari, Cape** *headland* North Island, New Zealand *34°47′S 173°24′E*
149 *W3* **Karīmābād** *prev.* Hunza. Jammu and Kashmir, NE Pakistan *36°23′N 74°43′E*
169 *P12* **Karimata, Kepulauan** *island group* N Indonesia
169 *P12* **Karimata, Pulau** *island* Kepulauan Karimata, N Indonesia
169 *O11* **Karimata, Selat** *strait* W Indonesia
155 *I14* **Karīmnagar** Telangana, C India *18°28′N 79°09′E*
186 *C7* **Karimui** Chimbu, C Papua New Guinea *06°19′S 144°48′E*
169 *Q15* **Karimunjawa, Pulau** *island* S Indonesia
80 *N12* **Karin** Woqooyi Galbeed, N Somalia *10°48′N 45°46′E*
Kariot *see* Ikaría
93 *L20* **Karis** *Fin.* Karjaa. Uusimaa, SW Finland *60°05′N 23°39′E*
Káristos *see* Kárystos
148 *J4* **Kārīz-e Elyās** *var.* Kareyz-e-Elyās, Kārez Iliās. Herāt, NW Afghanistan *35°26′N 61°24′E*
Karjaa *see* Karis
Karkaralinsk *see* Karkaraly
145 *T10* **Karkaraly** *prev.* Karkaralinsk. Karaganda, E Kazakhstan *49°31′N 75°53′E*
186 *D6* **Karkar Island** *island* N Papua New Guinea
143 *N7* **Karkas, Kūh-e** ▲▲ C Iran
142 *K8* **Karkheh, Rūd-e** ≈ SW Iran
115 *L20* **Karkinágri** *var.* Karkinagrio. Ikaría, Dodekánisa, Greece, Aegean Sea *37°31′N 26°01′E*
Karkinagrio *see* Karkinágri
117 *R12* **Karkinits'ka Zatoka** *Rus.* Karkinitskiy Zaliv. *gulf* S Ukraine
Karkinitskiy Zaliv *see* Karkinits'ka Zatoka
93 *L19* **Karkkila** *Swe.* Högfors. Uusimaa, S Finland *60°32′N 24°10′E*
93 *M19* **Kärkölä** Päijät-Häme, S Finland *60°52′N 25°17′E*
182 *G9* **Karkoo** South Australia *34°03′S 135°45′E*
Karkūk *see* Āltūn Kūbrî
118 *D5* **Kärla** *Ger.* Kergel. Saaremaa, W Estonia *58°20′N 22°15′E*
Karleby *see* Kokkola
110 *F7* **Karlino** *Ger.* Körlin an der Persante. Zachodinio-pomorskie, NW Poland *54°02′N 15°52′E*
137 *Q13* **Karlıova** Bingöl, E Turkey *39°16′N 41°01′E*
117 *U6* **Karlivka** Poltavs'ka Oblast', C Ukraine *49°27′N 35°08′E*
Karl-Marx-Stadt *see* Chemnitz
Karlö *see* Hailuoto
112 *C11* **Karlobag** *It.* Carlopago. Lika-Senj, W Croatia *44°31′N 15°06′E*
112 *D9* **Karlovac** *Ger.* Karlstadt, *Hung.* Károlyváros. Karlovac, C Croatia *45°29′N 15°31′E*
112 *C10* **Karlovac** *off.* Karlovačka Županija. ⬖ *province* C Croatia
Karlovačka Županija *see* Karlovac
111 *A16* **Karlovarský Kraj** ⬖ W Czech Republic
115 *M19* **Karlovási** *var.* Néon Karlovásion, Néon Karlovasi. Sámos, Dodekánisa, Greece, Aegean Sea *37°47′N 26°40′E*
114 *J9* **Karlovo** *prev.* Levskigrad. Plovdiv, C Bulgaria *42°38′N 24°49′E*
111 *A16* **Karlovy Vary** *Ger.* Karlsbad; *prev. Eng.* Carlsbad. Karlovarský Kraj, W Czech Republic *50°13′N 12°51′E*
Karlsbad *see* Karlovy Vary
95 *L17* **Karlsborg** Västra Götaland, S Sweden *58°32′N 14°32′E*
Karlsburg *see* Alba Iulia
95 *L22* **Karlshamn** Blekinge, S Sweden *56°10′N 14°50′E*
95 *L16* **Karlskoga** Örebro, C Sweden *59°19′N 14°33′E*
95 *M22* **Karlskrona** Blekinge, S Sweden *56°11′N 15°39′E*
101 *G21* **Karlsruhe** *var.* Carlsruhe. Baden-Württemberg, SW Germany *49°01′N 08°24′E*
95 *K16* **Karlstad** Värmland, C Sweden *59°22′N 13°36′E*
29 *R3* **Karlstad** Minnesota, N USA *48°34′N 96°31′W*
101 *I18* **Karlstadt** Bayern, C Germany *49°58′N 09°46′E*
Karlstadt *see* Karlovac
39 *Q14* **Karluk** Kodiak Island, Alaska, USA *57°34′N 154°27′W*
Karluk *see* Qarluq
119 *O17* **Karma** *Rus.* Korma. Homyel'skaya Voblasts', SE Belarus *53°07′N 30°48′E*
155 *F14* **Karmāla** Mahārāshtra, W India *18°26′N 75°08′E*
146 *M11* **Karmana** Navoiy Viloyati, C Uzbekistan *40°09′N 65°18′E*
138 *G8* **Karmi'el** *var.* Carmiel. Northern, N Israel *32°55′N 35°18′E*
95 *B16* **Karmøy** *island* S Norway
152 *I9* **Karnāl** Haryāna, N India *29°41′N 76°58′E*
153 *W15* **Karnaphuli Reservoir** ⊠ NE India
155 *F17* **Karnātaka** *var.* Kanara; *prev.* Maisur, Mysore. ⬖ *state* W India
25 *S13* **Karnes City** Texas, SW USA *28°54′N 97°55′W*
109 *P9* **Karnische Alpen** *It.* Alpi Carniche. ▲▲ Austria/Italy
114 *M9* **Karnobat** Burgas, E Bulgaria *42°39′N 26°59′E*
109 *Q9* **Kärnten** *off.* Land Kärten, *Eng.* Carinthi, *Slvn.* Koroška. ⬖ *state* S Austria
Karnul *see* Kurnool
83 *K16* **Karoi** Mashonaland West, N Zimbabwe *16°50′S 29°40′E*
Karol *see* Carei
Károly-Fehérvár *see* Alba Iulia
Károlyváros *see* Karlovac
82 *M12* **Karonga** Northern, N Malawi *09°54′S 33°55′E*
147 *W10* **Karool-Dëbë** *Kas.* Karool-Döbö; *prev.* Karool-Tëbë. Narynskaya Oblast', C Kyrgyzstan *40°33′N 75°52′E*
Karool-Döbö *see* Karool-Dëbë
Karool-Tëbë *see* Karool-Dëbë
182 *J9* **Karoonda** South Australia *35°04′S 139°58′E*
149 *S9* **Karor** *var.* Koror Lāl Esan. Punjab, E Pakistan *31°15′N 70°58′E*
Karosa *see* Karossa
171 *N12* **Karossa** *var.* Karosa. Sulawesi, C Indonesia *01°38′S 119°21′E*
Karpaten *see* Carpathian Mountains
115 *L22* **Karpáthio Pélagos** *sea* Dodekánisa, Greece, Aegean Sea
115 *N24* **Kárpathos** Kárpathos, SE Greece *35°30′N 27°13′E*
115 *N24* **Kárpathos** *It.* Scarpanto; *anc.* Carpathos, Carpathus. *island* SE Greece
Karpathos Strait *see* Karpathou, Stenó
115 *N24* **Karpathou, Stenó** *var.* Karpathos Strait, Scarpanto Strait. *strait* Dodekánisa, Greece, Aegean Sea
Karpaty *see* Carpathian Mountains
115 *E17* **Karpenísi** *prev.* Karpenísion. Stereá Elláda, C Greece *38°55′N 21°46′E*
Karpenísion *see* Karpenísi
Karpilovka *see* Aktsyabrski
125 *O8* **Karpogory** Arkhangel'skaya Oblast', NW Russian Federation *64°01′N 44°22′E*
180 *I7* **Karratha** Western Australia *20°44′S 116°52′E*
137 *S12* **Kars** *var.* Qars. Kars, NE Turkey *40°35′N 43°05′E*
137 *S12* **Kars** *var.* Qars. ⬖ *province* NE Turkey
145 *O12* **Karsakpay** *Kaz.* Qarsaqbay. Karaganda, C Kazakhstan *47°51′N 66°42′E*
93 *L15* **Kärsämäki** Pohjois-Pohjanmaa, C Finland *63°58′N 25°49′E*
Karsau *see* Kārsava
118 *K9* **Kārsava** *Ger.* Karsau; *prev. Rus.* Korsovka. E Latvia *56°46′N 27°39′E*
Karshi *see* Garşy, Turkmenistan
Karshi *see* Qarshi, Uzbekistan
Karshinskaya Step *see* Qarshi Cho'li
Karshinskiy Kanal *see* Qarshi Kanali
84 *I5* **Karskiye Vorota, Proliv** *Eng.* Kara Strait. *strait* N Russian Federation
122 *J6* **Karskoye More** *Eng.* Kara Sea. *sea* Arctic Ocean
93 *L17* **Karstula** Keski-Suomi, C Finland *62°52′N 24°48′E*
127 *Q5* **Karsun** Ul'yanovskaya Oblast', W Russian Federation *54°12′N 47°00′E*
122 *F11* **Kartaly** Chelyabinskaya Oblast', C Russian Federation *53°02′N 60°42′E*
18 *E13* **Karthaus** Pennsylvania, NE USA *41°06′N 78°03′W*
110 *I7* **Kartuzy** Pomorskie, NW Poland *54°21′N 18°11′E*
165 *R8* **Karumai** Iwate, Honshū, C Japan *40°19′N 141°27′E*
181 *U4* **Karumba** Queensland, NE Australia *17°31′S 140°51′E*
142 *L10* **Kārūn** *var.* Rūd-e Kārūn. ≈ SW Iran
92 *K13* **Karungi** Norrbotten, N Sweden *66°03′N 23°55′E*
92 *K13* **Karunki** Lappi, N Finland *66°01′N 24°06′E*
Kārūn, Rūd-e *see* Kārūn
155 *H21* **Karūr** Tamil Nādu, SE India *10°58′N 78°03′E*
93 *K17* **Karvia** Satakunta, SW Finland *62°07′N 22°34′E*
111 *J17* **Karviná** *Ger.* Karwin, *Pol.* Karwina; *prev.* Nová Karvinná. Moravskoslezský Krajj, E Czech Republic *49°50′N 18°30′E*
155 *E17* **Kārwār** Karnātaka, W India *14°50′N 74°09′E*
108 *M7* **Karwendelgebirge** ▲▲ Austria/Germany
Karwin/Karwina *see* Karviná
115 *I14* **Karyés** *var.* Karies. Ágion Óros, N Greece *40°15′N 24°15′E*
115 *I19* **Kárystos** *var.* Káristos. Évvoia, C Greece *38°01′N 24°25′E*
136 *E17* **Kaş** Antalya, SW Turkey *36°12′N 29°38′E*
39 *Y14* **Kasaan** Prince of Wales Island, Alaska, USA *55°32′N 132°24′W*
164 *I13* **Kasai** Hyōgo, Honshū, SW Japan *34°56′N 134°49′E*
79 *K21* **Kasai** *var.* Cassai, Kassai. ≈ Angola/Dem. Rep. Congo
79 *K22* **Kasai-Occidental** *off.* Région Kasai Occidental. ⬖ *region* S Dem. Rep. Congo
Kasai Occidental, Région *see* Kasai-Occidental
79 *L21* **Kasai-Oriental** *off.* Région Kasai Oriental. ⬖ *region* C Dem. Rep. Congo
Kasai Oriental, Région *see* Kasai-Oriental
79 *L24* **Kasaji** Katanga, S Dem. Rep. Congo *10°22′S 23°29′E*
82 *L12* **Kasama** Northern, N Zambia *10°14′S 31°12′E*
Kasan *see* Koson
83 *H16* **Kasane** North-West, NE Botswana *17°48′S 20°06′E*
81 *E23* **Kasanga** Rukwa, W Tanzania *08°27′S 31°10′E*
79 *G21* **Kasangulu** Bas-Congo, W Dem. Rep. Congo *04°33′S 15°12′E*
Kasansay *see* Kosonsoy
155 *E20* **Kāsaragod** Kerala, SW India *12°30′N 74°59′E*
Kasargen *see* Kasari
118 *P13* **Kasari** *var.* Kasari Jõgi, *Ger.* Kasargen. ≈ W Estonia
Kasari Jõgi *see* Kasari
8 *L11* **Kasba Lake** ◎ Northwest Territories, Nunavut N Canada
Kaschau *see* Košice
Kaseda *see* Minamisatsuma
83 *I14* **Kasempa** North Western, NW Zambia *13°27′S 25°49′E*
79 *O24* **Kasenga** Katanga, SE Dem. Rep. Congo *10°22′S 28°37′E*
79 *P17* **Kasenye** *var.* Kasenyi. Orientale, NE Dem. Rep. Congo *01°23′N 30°25′E*
Kasenyi *see* Kasenye
79 *O19* **Kasese** Maniema, E Dem. Rep. Congo *01°36′S 27°31′E*
81 *E18* **Kasese** SW Uganda *0°10′N 30°06′E*
152 *J11* **Kāsganj** Uttar Pradesh, N India *27°48′N 78°38′E*

◆ Country
● Country Capital
◇ Dependent Territory
○ Dependent Territory Capital
⬖ Administrative Regions
✕ International Airport
▲ Mountain
▲▲ Mountain Range
Volcano
≈ River
◎ Lake
⊠ Reservoir

KASHAF RŪD – KENYA, MOUNT

143 U4 **Kashaf Rūd** ≈ NE Iran
143 N7 **Kāshān** Eşfahān, C Iran 33°57′N 51°31′E
126 M10 **Kashary** Rostovskaya Oblast', SW Russian Federation 49°02′N 40°58′E
39 O12 **Kashegelok** Alaska, USA 60°57′N 157°46′W
Kashgar *see* Kashi
158 E7 **Kashi** *Chin.* Kaxgar, K'o-shih, *Uigh.* Kashgar. Xinjiang Uygur Zizhiqu, NW China 39°32′N 75°58′E
164 J14 **Kashihara** *var.* Kasihara. Nara, Honshū, SW Japan 34°28′N 135°46′E
165 P13 **Kashima-nada** *gulf* S Japan
124 K15 **Kashin** Tverskaya Oblast', W Russian Federation 57°20′N 37°34′E
152 K10 **Kāshīpur** Uttarakhand, N India 29°13′N 78°58′E
126 L4 **Kashira** Moskovskaya Oblast', W Russian Federation 54°53′N 38°13′E
165 N11 **Kashiwazaki** *var.* Kasiwazaki. Niigata, Honshū, C Japan 37°22′N 138°33′E
Kashkadar'inskaya Oblast' *see* Qashqadaryo Viloyati
143 T5 **Kāshmar** *var.* Turshiz; *prev.* Solţānābād, Torshiz. Khorāsān, NE Iran 35°13′N 58°25′E
Kashmir *see* Jammu and Kashmir
149 R12 **Kashmor** Sind, SE Pakistan 28°24′N 69°42′E
149 S5 **Kashmūnd Ghar** *Eng.* Kashmund Range. ▲▲ E Afghanistan
Kashmund Range *see* Kashmūnd Ghar
145 T7 **Kashyr** *prev.* Kachiry. Pavlodar, NE Kazakhstan 53°07′N 76°08′E
Kasi *see* Vārānasi
153 O12 **Kasia** Uttar Pradesh, N India 26°45′N 83°55′E
39 N12 **Kasigluk** Alaska, USA 60°54′N 162°31′W
Kasihara *see* Kashihara
39 R12 **Kasilof** Alaska, USA 60°20′N 151°16′W
Kasimköj *see* General Toshevo
126 M4 **Kasimov** Ryazanskaya Oblast', W Russian Federation 54°59′N 41°22′E
79 P18 **Kasindi** Nord-Kivu, E Dem. Rep. Congo 0°03′N 29°43′E
82 M12 **Kasitu** ≈ N Malawi
Kasiwazaki *see* Kashiwazaki
30 L14 **Kaskaskia River** ≈ Illinois, N USA
93 J17 **Kaskinen** *Swe.* Kaskö. Österbotten, W Finland 62°23′N 21°10′E
Kaskö *see* Kaskinen
Kas Kong *see* Kŏng, Kaôh
11 O17 **Kaslo** British Columbia, SW Canada 49°54′N 116°57′W
Käsmark *see* Kežmarok
169 T12 **Kasongan** Borneo, C Indonesia 02°01′S 113°21′E
79 N21 **Kasongo** Maniema, E Dem. Rep. Congo 04°22′S 26°42′E
79 H22 **Kasongo-Lunda** Bandundu, SW Dem. Rep. Congo 06°30′S 16°51′E
115 M24 **Kásos** *island* S Greece
Kasos Strait *see* Kásou, Stenó
115 M25 **Kásou, Stenó** *var.* Kasos Strait. *strait* Dodekánisos/Kríti, Greece, Aegean Sea
137 T10 **K'asp'i** *prev.* Kaspi. C Georgia 41°54′N 44°25′E
Kaspi *see* K'asp'i
114 M8 **Kaspichan** Shumen, NE Bulgaria 43°18′N 27°09′E
Kaspīy Mangy Oypaty *see* Caspian Depression
127 Q16 **Kaspiysk** Respublika Dagestan, SW Russian Federation 42°52′N 47°40′E
Kaspiyskiy *see* Lagan'
Kaspiyskoye More/Kaspiy Tengizi *see* Caspian Sea
Kassa *see* Košice
Kassai *see* Kasai
80 I9 **Kassala** Kassala, E Sudan 15°24′N 36°25′E
80 H9 **Kassala** ◈ *state* NE Sudan
115 G15 **Kassándra** *prev.* Pallíni; *anc.* Pallene. *peninsula* NE Greece
115 G15 **Kassándra** *headland* N Greece 39°58′N 23°22′E
115 H15 **Kassándras, Kólpos** *var.* Kólpos Toronaíos. *gulf* N Greece
139 Y11 **Kassārah** Maysān, E Iraq 31°21′N 47°25′E
101 I15 **Kassel** *prev.* Cassel. Hessen, C Germany 51°19′N 09°30′E
74 M6 **Kasserine** *var.* Al Qaşrayn. W Tunisia 35°15′N 08°52′E
14 J14 **Kasshabog Lake** ◎ Ontario, SE Canada
139 O5 **Kassīr, Sabkhat al** ◎ E Syria
29 W10 **Kasson** Minnesota, N USA 44°00′N 92°42′W
115 C17 **Kassópeia** *Var.* Kassópi. *site of ancient city* Ípeiros, W Greece
Kassópi *see* Kassópeia
136 I11 **Kastamonu** *var.* Castamoni, Kastamuni. Kastamonu, N Turkey 41°22′N 33°47′E
136 I10 **Kastamonu** *var.* Kastamuni. ◈ *province* N Turkey
Kastamuni *see* Kastamonu
Kastaneá *see* Kastaniá
115 E14 **Kastaniá** *prev.* Kastaneá. Kentrikí Makedonía, N Greece 40°25′N 22°09′E
Kastélli *see* Kíssamos
Kastellórizon *see* Megísti
115 N24 **Kástelo, Akrotírio** *prev.* Akrotírio Kastállou. *headland* Kárpathos, SE Greece 35°24′N 27°08′E
95 N21 **Kastlösa** Kalmar, S Sweden 56°25′N 16°25′E
115 D14 **Kastoría** Dytikí Makedonía, N Greece 40°33′N 21°15′E
126 K7 **Kastornoye** Kurskaya Oblast', W Russian Federation 51°49′N 38°07′E
115 I21 **Kástro** Sífnos, Kykládes, Greece, Aegean Sea 36°58′N 24°45′E
95 J23 **Kastrup** ✈ (København) København, E Denmark 55°36′N 12°39′E
119 Q17 **Kastsyukovichy** *Rus.* Kostyukovichi. Mahilyowskaya Voblasts', E Belarus 53°20′N 32°03′E
119 O18 **Kastsyukowka** *Rus.* Kostyukovka. Homyel'skaya Voblasts', SE Belarus 52°32′N 30°54′E
164 D13 **Kasuga** Fukuoka, Kyūshū, SW Japan 33°31′N 130°27′E
164 L13 **Kasugai** Aichi, Honshū, SW Japan 35°15′N 136°57′E
81 E21 **Kasulu** Kigoma, W Tanzania 04°33′S 30°06′E
164 I12 **Kasumi** Hyōgo, Honshū, SW Japan 35°36′N 134°37′E
127 R17 **Kasumkent** Respublika Dagestan, SW Russian Federation 41°39′N 48°09′E
82 M13 **Kasungu** Central, C Malawi 13°04′S 33°29′E
149 W9 **Kasūr** Punjab, E Pakistan 31°07′N 74°30′E
83 G15 **Kataba** Western, W Zambia 15°28′S 23°25′E
19 R4 **Katahdin, Mount** ▲ Maine, NE USA 45°55′N 68°52′W
79 M20 **Katako-Kombe** Kasai-Oriental, C Dem. Rep. Congo 03°24′S 24°25′E
39 T12 **Katalla** Alaska, USA 60°12′N 144°31′W
Katana *see* Qaţanā
79 L24 **Katanga** *off.* Région du Katanga; *prev.* Shaba. ◈ *region* SE Dem. Rep. Congo
122 M11 **Katanga** ≈ C Russian Federation
Katanga, Région du *see* Katanga
154 J11 **Katāngi** Madhya Pradesh, C India 21°46′N 79°50′E
180 J13 **Katanning** Western Australia 33°45′S 117°33′E
181 P8 **Kata Tjuṯa** *var.* Mount Olga. ▲ Northern Territory, C Australia 25°20′S 130°47′E
81 F23 **Katavi** *off.* Mkoa wa Katavi. ◈ *region* SW Tanzania
Katawaz *see* Zarghūn Shahr
151 Q22 **Katchall Island** *island* Nicobar Islands, India, NE Indian Ocean
115 F14 **Kateríni** Kentrikí Makedonía, N Greece 40°15′N 22°30′E
117 P7 **Katerynopil'** Cherkas'ka Oblast', C Ukraine 49°00′N 30°59′E
166 M3 **Katha** Sagaing, N Myanmar (Burma) 24°11′N 96°20′E
181 P2 **Katherine** Northern Territory, N Australia 14°29′S 132°20′E
154 B11 **Kāthiāwār Peninsula** *peninsula* W India
153 P11 **Kathmandu** *prev.* Kantipur. ● (Nepal) Central, C Nepal 27°46′N 85°17′E
152 H7 **Kathua** Jammu and Kashmir, NW India 32°23′N 75°34′E
76 L12 **Kati** Koulikoro, SW Mali 12°41′N 08°04′W
153 R13 **Katihār** Bihār, NE India 25°33′N 87°34′E
184 N7 **Katikati** Bay of Plenty, North Island, New Zealand 37°34′S 175°55′E
83 H16 **Katima Mulilo** Caprivi, NE Namibia 17°31′S 24°20′E
77 N15 **Katiola** C Ivory Coast 08°11′N 05°04′W
191 V10 **Katiu** *atoll* Îles Tuamotu, C French Polynesia
92 J5 **Katla** ⛰ S Iceland 63°38′N 19°03′W
117 N12 **Katlabuh, Ozero** ◎ SW Ukraine
39 P14 **Katmai, Mount** ▲ Alaska, USA 58°16′N 154°57′W
154 J9 **Katni** Madhya Pradesh, C India 23°47′N 80°29′E
115 D19 **Káto Achaïa** *var.* Kato Ahaia, Káto Akhaía. Dytikí Elláda, S Greece 38°08′N 21°33′E
Kato Ahaia/Káto Akhaía *see* Káto Achaïa
121 P2 **Kato Lakatámeia** *var.* Kato Lakatamia. C Cyprus 35°07′N 33°20′E
Kato Lakatamia *see* Kato Lakatámeia
79 N22 **Katompi** Katanga, SE Dem. Rep. Congo 06°10′S 26°19′E
83 K14 **Katondwe** Lusaka, C Zambia 15°08′S 30°10′E
114 H12 **Káto Nevrokópi** *prev.* Káto Nevrokópion. Anatolikí Makedonía kai Thráki, NE Greece 41°21′N 23°51′E
Káto Nevrokópion *see* Káto Nevrokópi
81 E18 **Katonga** ≈ S Uganda
115 F15 **Káto Ólympos** ▲▲ C Greece
115 D17 **Katoúna** Dytikí Elláda, C Greece 38°47′N 21°07′E
115 E19 **Káto Vlasiá** Dytikí Makedonía, S Greece 38°02′N 21°54′E
111 J16 **Katowice** *Ger.* Kattowitz. Śląskie, S Poland 50°15′N 19°01′E
153 S15 **Kātoya** West Bengal, NE India 23°39′N 88°11′E
136 E16 **Katrançık Dağı** ▲▲ SW Turkey
95 N16 **Katrineholm** Södermanland, C Sweden 58°59′N 16°15′E
96 I11 **Katrine, Loch** ◎ C Scotland, United Kingdom
77 V12 **Katsina** Katsina, N Nigeria 12°59′N 07°33′E
77 U12 **Katsina** ◈ *state* N Nigeria
67 P8 **Katsina Ala** ≈ S Nigeria
164 C13 **Katsumoto** Nagasaki, Iki, SW Japan 33°49′N 129°42′E
165 P13 **Katsuta** *var.* Katuta. Ibaraki, Honshū, S Japan 36°24′N 140°32′E
165 O14 **Katsuura** *var.* Katuura. Chiba, Honshū, S Japan 35°09′N 140°16′E
164 K12 **Katsuyama** *var.* Katuyama. Fukui, Honshū, SW Japan 36°00′N 136°30′E
164 H12 **Katsuyama** Okayama, Honshū, SW Japan 35°06′N 133°43′E
Kattakurgan *see* Kattaqo'rg'on
147 N11 **Kattaqo'rg'on** *Rus.* Kattakurgan. Samarqand Viloyati, C Uzbekistan 39°56′N 66°11′E
115 O23 **Kattavía** Ródos, Dodekánisa, Greece, Aegean Sea 35°56′N 27°47′E
95 I21 **Kattegat** *Dan.* Kattegatt. *strait* N Europe
Kattegatt *see* Kattegat
95 P19 **Katthammarsvik** Gotland, SE Sweden 57°27′N 18°54′E
Kattowitz *see* Katowice
122 J13 **Katun'** ≈ S Russian Federation
Katuta *see* Katsuta
Katuura *see* Katsuura
Katuyama *see* Katsuyama
Katwijk *see* Katwijk aan Zee
98 G11 **Katwijk aan Zee** *var.* Katwijk. Zuid-Holland, W Netherlands 59°12′N 04°24′E
38 B8 **Kaua'i** *var.* Kauai. *island* Hawaiian Islands, Hawai'i, USA, C Pacific Ocean
Kauai *see* Kaua'i
38 C8 **Kaua'i Channel** *var.* Kauai Channel. *channel* Hawai'i, USA, C Pacific Ocean
Kauai Channel *see* Kaua'i Channel
171 R13 **Kaubalatmada, Gunung** *var.* Kaplamada. ▲ Pulau Buru, E Indonesia 03°16′S 126°17′E
191 U10 **Kauehi** *atoll* Îles Tuamotu, C French Polynesia
Kauen *see* Kaunas
101 K24 **Kaufbeuren** Bayern, S Germany 47°53′N 10°37′E
25 U7 **Kaufman** Texas, SW USA 32°35′N 96°18′W
101 I15 **Kaufungen** Hessen, C Germany 51°16′N 09°39′E
93 K17 **Kauhajoki** Etelä-Pohjanmaa, W Finland 62°26′N 22°10′E
93 K16 **Kauhava** Etelä-Pohjanmaa, W Finland 63°06′N 23°08′E
30 M7 **Kaukauna** Wisconsin, N USA 44°18′N 88°18′W
92 L11 **Kaukonen** Lappi, N Finland 67°28′N 24°49′E
38 A8 **Kaulakahi Channel** *channel* Hawai'i, USA, C Pacific Ocean
38 E9 **Kaunakakai** Moloka'i, Hawaii, USA, C Pacific Ocean 21°05′N 157°01′W
38 F12 **Kaunā Point** *var.* Kauna Point. *headland* Hawai'i, USA, C Pacific Ocean 19°02′N 155°52′W
Kauna Point *see* Kaunā Point
118 F13 **Kaunas** *Ger.* Kauen, *Pol.* Kowno; *prev. Rus.* Kovno. Kaunas, C Lithuania 54°54′N 23°57′E
118 F13 **Kaunas** ◈ *province* C Lithuania
186 C6 **Kaup** East Sepik, NW Papua New Guinea 03°50′S 144°01′E
77 U12 **Kaura Namoda** Zamfara, NW Nigeria 12°43′N 06°17′E
Kaushany *see* Căuşeni
93 K16 **Kaustinen** Keski-Pohjanmaa, W Finland 63°33′N 23°40′E
99 M23 **Kautenbach** Diekirch, NE Luxembourg 49°57′N 06°01′E
92 K10 **Kautokeino** *Lapp.* Guovdageaidnu. Finnmark, N Norway 69°N 23°01′E
Kavadar *see* Kavadarci
113 P19 **Kavadarci** *Turk.* Kavadar. C Macedonia 41°25′N 22°00′E
Kavaja *see* Kavajë
113 K20 **Kavajë** *It.* Cavaia, Kavaja. Tiranë, W Albania 41°11′N 19°33′E
114 M13 **Kavak Çayı** ≈ NW Turkey
Kavakli *see* Topolovgrad
114 I13 **Kavála** *prev.* Kaválla. Anatolikí Makedonía kai Thráki, NE Greece 40°57′N 24°26′E
114 I13 **Kaválas, Kólpos** *gulf* Aegean Sea, NE Mediterranean Sea
155 J17 **Kāvali** Andhra Pradesh, E India 15°05′N 80°02′E
Kaválla *see* Kavála
Kavango *see* Cubango/Okavango
155 C21 **Kavaratti** Lakshadweep, SW India 10°33′N 72°38′E
114 O8 **Kavarna** Dobrich, NE Bulgaria 43°27′N 28°21′E
118 G12 **Kavarskas** Utena, E Lithuania 55°27′N 24°55′E
76 I13 **Kavendou** ▲ C Guinea 10°49′N 12°14′W
Kavengo *see* Cubango/Okavango
155 F20 **Kāveri** *var.* Cauvery. ≈ S India
186 G5 **Kavieng** *var.* Kaewieng. New Ireland, NE Papua New Guinea 04°13′S 152°11′E
83 H16 **Kavimba** North-West, NE Botswana 18°02′S 24°38′E
83 I15 **Kavingu** Southern, S Zambia 15°39′S 26°03′E
143 Q6 **Kavīr, Dasht-e** *var.* Great Salt Desert. *salt pan* N Iran
Kavirondo Gulf *see* Winam Gulf
Kavkaz *see* Caucasus
95 K23 **Kävlinge** Skåne, S Sweden 55°47′N 13°05′E
82 G12 **Kavungo** Moxico, E Angola 11°31′S 22°59′E
165 Q8 **Kawabe** Akita, Honshū, C Japan 39°39′N 140°14′E
165 R9 **Kawai** Iwate, Honshū, C Japan 39°36′N 141°40′E
38 A8 **Kawaihoa Point** *headland* Ni'ihau, Hawai'i, USA, C Pacific Ocean 21°47′N 160°12′W
184 K3 **Kawakawa** Northland, North Island, New Zealand 35°23′S 174°06′E
82 I13 **Kawama** North Western, NW Zambia 13°04′S 25°59′E
82 K11 **Kawambwa** Luapula, N Zambia 09°45′S 29°10′E
154 K11 **Kawardha** Chhattīsgarh, C India 21°59′N 81°12′E
14 I14 **Kawartha Lakes** ◎ Ontario, SE Canada
165 O13 **Kawasaki** Kanagawa, Honshū, S Japan 35°32′N 139°41′E
171 R12 **Kawassi** Pulau Obi, E Indonesia 01°32′S 127°25′E
165 R6 **Kawauchi** Aomori, Honshū, C Japan 41°11′N 141°00′E
184 L5 **Kawau Island** *island* N New Zealand
184 N10 **Kaweka Range** ▲▲ North Island, New Zealand
Kawelecht *see* Puhja
184 O8 **Kawerau** Bay of Plenty, North Island, New Zealand 38°06′S 176°43′E
184 L8 **Kawhia** Waikato, North Island, New Zealand 38°04′S 174°49′E
184 K8 **Kawhia Harbour** *inlet* North Island, New Zealand
35 V8 **Kawich Peak** ▲ Nevada, W USA 38°00′N 116°27′W
35 V9 **Kawich Range** ▲▲ Nevada, W USA
14 G12 **Kawigamog Lake** ◎ Ontario, S Canada
171 P9 **Kawio, Kepulauan** *island group* N Indonesia
167 N9 **Kawkareik** Kayin State, S Myanmar (Burma) 16°33′N 98°18′E
27 O8 **Kaw Lake** ▣ Oklahoma, C USA
166 M3 **Kawlin** Sagaing, N Myanmar (Burma) 23°48′N 95°41′E
Kawm Umbū *see* Kom Ombo
Kawthule State *see* Kayin State
Kaxgar *see* Kashi
158 D7 **Kaxgar He** ≈ NW China
158 J5 **Kax He** ≈ NW China
77 P12 **Kaya** C Burkina Faso 13°04′N 01°09′W
167 N6 **Kayah State** ◈ *state* C Myanmar (Burma)
39 T12 **Kayak Island** *island* Alaska, USA
114 M11 **Kayalıköy Barajı** ▣ NW Turkey
166 M8 **Kayan** Yangon, SW Myanmar (Burma) 16°54′N 96°35′E
Kayangel Islands *see* Ngcheangel
155 G23 **Kāyankulam** Kerala, SW India 09°10′N 76°31′E
169 V9 **Kayan, Sungai** *prev.* Kajan. ≈ Borneo, C Indonesia
144 F14 **Kaydak, Sor** *salt flat* SW Kazakhstan
Kaydanovo *see* Dzyarzhynsk
37 N9 **Kayenta** Arizona, SW USA 36°43′N 110°15′W
76 J11 **Kayes** Kayes, W Mali 14°26′N 11°22′W
76 J11 **Kayes** ◈ *region* SW Mali
167 N8 **Kayin State** *var.* Kawthule State, Karen State. ◈ *state* S Myanmar (Burma)
145 U10 **Kaynar** *Kaz.* Qaynar, *var.* Kajnar. Vostochnyy Kazakhstan, E Kazakhstan 49°13′N 77°27′E
Kaynary *see* Căinari
83 H15 **Kayoya** Western, W Zambia 16°13′S 24°09′E
Kayrakkum *see* Qayroqqum
Kayrakkumskoye Vodokhranilishche *see* Qayroqqum, Obanbori
136 K14 **Kayseri** *var.* Kaisaria; *anc.* Caesarea Mazaca, Mazaca. Kayseri, C Turkey 38°42′N 35°28′E
136 K14 **Kayseri** *var.* Kaisaria. ◈ *province* C Turkey
36 L2 **Kaysville** Utah, W USA 41°10′N 111°55′W
Kayyngdy *see* Kaindy
14 L11 **Kazabazua** Québec, SE Canada 45°58′N 76°00′W
14 L12 **Kazabazua** ≈ Québec, SE Canada
123 Q7 **Kazach'ye** Respublika Sakha (Yakutiya), NE Russian Federation 70°38′N 135°54′E
Kazakdar'ya *see* Qozoqdaryo
146 E9 **Kazakhlyshor, Solonchak** *var.* Solonchak Shorkazakhly. *salt marsh* NW Turkmenistan
Kazakhskaya SSR/Kazakh Soviet Socialist Republic *see* Kazakhstan
144 L12 **Kazakhstan** *off.* Republic of Kazakhstan, *var.* Kazakstan, *Kaz.* Qazaqstan, Qazaqstan Respublikasy; *prev.* Kazakh Soviet Socialist Republic, *Rus.* Kazakhskaya SSR. ◆ *republic* C Asia
Kazakhstan, Republic of *see* Kazakhstan
Kazakh Uplands *see* Saryarka
Kazakstan *see* Kazakhstan
Kazalinsk *see* Kazaly
144 L14 **Kazaly** *prev.* Kazalinsk. Kzyl-Orda, S Kazakhstan 45°45′N 62°01′E
127 R4 **Kazan'** Respublika Tatarstan, W Russian Federation 55°43′N 49°07′E
8 M10 **Kazan** ≈ Nunavut, NW Canada
127 R4 **Kazan'** ✈ Respublika Tatarstan, W Russian Federation 55°46′N 49°21′E
Kazandzhik *see* Bereket
117 R8 **Kazanka** Mykolayivs'ka Oblast', S Ukraine 47°49′N 32°50′E
Kazanketken *see* Qozonketkan
114 J9 **Kazanlak** *var.* Kazanlŭk; *prev.* Kazanlik. Stara Zagora, C Bulgaria 42°38′N 25°24′E
Kazanlik *see* Kazanlak
Kazanlŭk *see* Kazanlak
165 Y16 **Kazan-rettō** *Eng.* Volcano Islands. *island group* SE Japan
Kazantip, Mys *see* Kazantyp, Mys
117 V12 **Kazantyp, Mys** *prev.* Mys Kazantip. *headland* S Ukraine 45°27′N 35°50′E
147 U9 **Kazarman** Narynskaya Oblast', C Kyrgyzstan 41°21′N 74°03′E
Kazatin *see* Kozyatyn
Kazbegi *see* Q'azbegi
Kazbegi *see* Kazbek
137 T9 **Kazbek** *var.* Kazbegi, *Geor.* Mqinvartsveri. ▲ N Georgia 42°43′N 44°28′E
82 M13 **Kazembe** Eastern, NE Zambia 12°06′S 32°45′E
143 N11 **Kāzerūn** Fārs, S Iran 29°41′N 51°38′E
125 R12 **Kazhym** Respublika Komi, NW Russian Federation 60°19′N 51°26′E
Kazi Ahmad *see* Qāzi Ahmad
136 H16 **Kazımkarabekir** Karaman, S Turkey 37°13′N 32°58′E
111 M20 **Kazincbarcika** Borsod-Abaúj-Zemplén, NE Hungary 48°15′N 20°40′E
119 H17 **Kazlowshchyna** *Pol.* Kozlowszczyzna, *Rus.* Kozlovshchina. Hrodzyenskaya Voblasts', W Belarus 53°19′N 25°18′E
119 E14 **Kazlų Rūda** Marijampolė, S Lithuania 54°45′N 23°28′E
144 E9 **Kaztalovka** Zapadnyy Kazakhstan, W Kazakhstan 49°45′N 48°42′E
79 K22 **Kazumba** Kasai-Occidental, S Dem. Rep. Congo 06°25′S 22°02′E
165 Q8 **Kazuno** Akita, Honshū, C Japan 40°14′N 140°48′E
Kazvin *see* Qazvīn
118 J12 **Kazyany** *Rus.* Koz'yany. Vitsyebskaya Voblasts', NW Belarus 55°18′N 26°52′E
122 H9 **Kazym** ≈ N Russian Federation
110 H10 **Kcynia** *Ger.* Exin. Kujawsko-pomorskie, C Poland 53°00′N 17°29′E
Kéa *see* Tziá
Kéa *see* Ioulís
38 H11 **Kea'au** *var.* Keaau. Hawaii, USA, C Pacific Ocean 19°36′N 155°01′W
Keaau *see* Kea'au
38 F11 **Keāhole Point** *var.* Keahole Point. *headland* Hawai'i, USA, C Pacific Ocean 19°43′N 156°03′W
38 G12 **Kealakekua** Hawaii, USA, C Pacific Ocean 19°31′N 155°56′W
38 H11 **Kea, Mauna** ▲ Hawai'i, USA 19°50′N 155°30′W
37 N10 **Keams** Arizona, SW USA 35°47′N 110°09′W
Kéamu *see* Aneityum
29 O16 **Kearney** Nebraska, C USA 40°42′N 99°06′W
36 L3 **Kearns** Utah, W USA 40°39′N 112°00′W
115 H20 **Kéas, Stenó** *strait* SE Greece
137 O14 **Keban Barajı** *dam* C Turkey
137 O14 **Keban Barajı** ▣ C Turkey
77 S13 **Kebbi** ◈ *state* NW Nigeria
76 G10 **Kébémèr** NW Senegal 15°24′N 16°25′W
74 M7 **Kebili** *var.* Qibilī. C Tunisia 33°42′N 09°06′E
138 H4 **Kebir, Nahr el** ≈ NW Syria
80 A10 **Kebkabiya** Northern Darfur, W Sudan 13°39′N 24°05′E
92 I11 **Kebnekaise** *Lapp.* Giebnegáisi. ▲ N Sweden 68°01′N 18°24′E
81 M14 **K'ebrī Dehar** Sumalē, E Ethiopia 06°43′N 44°15′E
148 K15 **Kech** ≈ SW Pakistan
10 K10 **Kechika** ≈ British Columbia, W Canada
111 K23 **Kecskemét** Bács-Kiskun, C Hungary 46°54′N 19°42′E
168 J6 **Kedah** ◈ *state* Peninsular Malaysia
118 F12 **Kėdainiai** Kaunas, C Lithuania 55°19′N 24°00′E
152 K8 **Kedārnāth** Uttarakhand, N India 30°44′N 79°03′E
Kedder *see* Kehra
13 N13 **Kedgwick** New Brunswick, SE Canada 47°38′N 67°21′W
169 R16 **Kediri** Jawa, C Indonesia 07°45′S 112°01′E
171 Y13 **Kedir Sarmi** Papua, E Indonesia 02°00′S 139°01′E
163 V7 **Kedong** Heilongjiang, NE China 48°00′N 126°15′E
76 I12 **Kédougou** SE Senegal 12°35′N 12°09′W
122 I11 **Kedrovyy** Tomskaya Oblast', C Russian Federation 57°31′N 79°45′E
111 H16 **Kędzierzyn-Kozle** *Ger.* Heydebrech. Opolskie, S Poland 50°20′N 18°12′E
8 H8 **Keele** ≈ Northwest Territories, NW Canada
10 K6 **Keele Peak** ▲ Yukon, NW Canada 63°31′N 130°21′W
Keelung *see* Jilong
19 N10 **Keene** New Hampshire, NE USA 42°56′N 72°14′W
99 H17 **Keerbergen** Vlaams Brabant, C Belgium 51°01′N 04°39′E
83 E21 **Keetmanshoop** Karas, S Namibia 26°36′S 18°08′E
12 A11 **Keewatin** Ontario, S Canada 49°47′N 94°30′W
29 V4 **Keewatin** Minnesota, N USA 47°24′N 93°04′W
Kefallinía *see* Kefallonía
115 B18 **Kefallonía** *var.* Kefallinía. *island* Iónia Nisiá, Greece, C Mediterranean Sea
115 M22 **Kéfalos** Kos, Dodekánisa, Greece, Aegean Sea 36°44′N 26°58′E
171 Q17 **Kefamenanu** Timor, C Indonesia 09°31′S 124°29′E
Kefar Sava *see* Kfar Sava
Kefe *see* Feodosiya
77 V15 **Keffi** Nassarawa, C Nigeria 08°52′N 07°54′E
92 H4 **Keflavík** Suðurnes, W Iceland 64°01′N 22°35′W
92 H4 **Keflavik** ✈ (Reykjavík) Suðurnes, W Iceland 63°58′N 22°37′W
Kegalee *see* Kegalla
155 J25 **Kegalla** *var.* Kegalee, Kegalle. Sabaragamuwa Province, C Sri Lanka 07°14′N 80°21′E
Kegalle *see* Kegalla
Kegayli *see* Kegeyli
Kegel *see* Keila
145 W16 **Kegen** Almaty, SE Kazakhstan 42°58′N 79°12′E
146 H7 **Kegeyli** *prev.* Kegayli. Qoraqalpog'iston Respublikasi, W Uzbekistan 42°46′N 59°49′E
101 F22 **Kehl** Baden-Württemberg, SW Germany 48°34′N 07°49′E
118 H3 **Kehra** *Ger.* Kedder. Harjumaa, NW Estonia 59°19′N 25°21′E
117 U6 **Kehychivka** Kharkivs'ka Oblast', E Ukraine 49°18′N 35°46′E
97 L17 **Keighley** N England, United Kingdom 53°51′N 01°58′W
Kei Islands *see* Kai, Kepulauan
Keijō *see* Seoul
118 G3 **Keila** *Ger.* Kegel. Harjumaa, NW Estonia 59°18′N 24°29′E
Keilberg *see* Klínovec
83 F23 **Keimoes** Northern Cape, W South Africa 28°41′S 20°59′E
Keina/Keinis *see* Käina
Keishū *see* Gyeongju
77 T11 **Keïta** Tahoua, C Niger 14°45′N 05°40′E
78 J12 **Kéita, Bahr** *var.* Doka. ≈ S Chad
182 K10 **Keith** South Australia 36°01′S 140°22′E
96 K8 **Keith** NE Scotland, United Kingdom 57°33′N 02°57′W
26 K3 **Keith Sebelius Lake** ▣ Kansas, C USA
32 G11 **Keizer** Oregon, NW USA 44°59′N 123°01′W
38 A8 **Kekaha** Kaua'i, Hawaii, USA, C Pacific Ocean 21°58′N 159°43′W
147 U10 **Këk-Art** *prev.* Alaykel', Alay-Kuu. Oshskaya Oblast', SW Kyrgyzstan 40°16′N 74°21′E
147 W10 **Këk-Aygyr** *var.* Keyaygyr. Narynskaya Oblast', C Kyrgyzstan 40°42′N 75°37′E
147 V9 **Këk-Dzhar** Narynskaya Oblast', C Kyrgyzstan 41°28′N 74°48′E
14 L8 **Kekek** ≈ Québec, SE Canada
185 K15 **Kekerengu** Canterbury, South Island, New Zealand 41°55′S 174°05′E
111 L21 **Kékes** ▲ N Hungary 47°53′N 19°59′E
171 P17 **Kekneno, Gunung** ▲ Timor, S Indonesia
147 S9 **Këk-Tash** *Kir.* Kök-Tash. Dzhalal-Abadskaya Oblast', W Kyrgyzstan 41°08′N 72°25′E
81 M15 **K'elafo** Sumalē, E Ethiopia 05°36′N 44°12′E
169 U10 **Kelai, Sungai** ≈ Borneo, N Indonesia
Kelamayi *see* Karamay
Kelang *see* Klang
168 K7 **Kelantan** ◈ *state* Peninsular Malaysia
Kelantan *see* Kelantan, Sungai
168 K7 **Kelantan, Sungai** *var.* Kelantan. ≈ Peninsular Malaysia
Kelat *see* Kālat
Këlcyra *see* Këlcyrë
113 L22 **Këlcyrë** *var.* Këlcyra. Gjirokastër, S Albania 40°19′N 20°10′E
Kelifskiy Uzboy *see* Kelif Uzboýy
146 L14 **Kelif Uzboýy** *Rus.* Kelifskiy Uzboy. *salt marsh* E Turkmenistan
137 O12 **Kelkit** Gümüşhane, NE Turkey 40°07′N 39°28′E
136 M12 **Kelkit Çayı** ≈ N Turkey
79 G18 **Kéllé** Cuvette-Quest, W Congo 0°04′S 14°33′E
77 W11 **Kellé** Zinder, S Niger 14°10′N 10°10′E
145 P7 **Kellerovka** Severnyy Kazakhstan, N Kazakhstan 53°51′N 69°15′E
8 I5 **Kellett, Cape** *headland* Banks Island, Northwest Territories, NW Canada 71°57′N 125°55′W
31 S11 **Kelleys Island** *island* Ohio, N USA
33 N8 **Kellogg** Idaho, NW USA 47°30′N 116°07′W
92 M12 **Kelloselkä** Lappi, N Finland 66°56′N 28°52′E
97 F17 **Kells** *Ir.* Ceanannas. Meath, E Ireland 53°44′N 06°53′W
118 E12 **Kelmė** Šiaulai, C Lithuania 55°39′N 22°57′E
99 M19 **Kelmis** *var.* La Calamine. Liège, E Belgium 50°43′N 06°01′E
78 H12 **Kélo** Tandjilé, SW Chad 09°21′N 15°50′E
83 I14 **Kelongwa** North Western, NW Zambia 13°41′S 26°19′E
11 N17 **Kelowna** British Columbia, SW Canada 49°50′N 119°29′W
11 X12 **Kelsey** Manitoba, C Canada 56°02′N 96°31′W
34 M6 **Kelseyville** California, W USA 38°58′N 122°51′W
96 K13 **Kelso** SE Scotland, United Kingdom 55°36′N 02°27′W
32 G10 **Kelso** Washington, NW USA 46°09′N 122°54′W
195 W15 **Keltie, Cape** *headland* Antarctica
Keltsy *see* Kielce
168 L9 **Keluang** *var.* Kluang. Johor, Peninsular Malaysia 02°01′N 103°18′E
168 M11 **Kelume** Pulau Lingga, W Indonesia 0°12′S 104°27′E
11 U15 **Kelvington** Saskatchewan, S Canada 52°10′N 103°30′W
124 J7 **Kem'** Respublika Kareliya, NW Russian Federation 64°55′N 34°18′E
124 I7 **Kem'** ≈ NW Russian Federation
137 O13 **Kemah** Erzincan, E Turkey 39°35′N 39°02′E
137 N13 **Kemaliye** Erzincan, C Turkey 39°16′N 38°29′E
Kemaman *see* Cukai
Kemanlar *see* Isperih
10 K14 **Kemano** British Columbia, SW Canada 53°39′N 127°58′W
Kemarat *see* Khemmarat
171 P12 **Kembani** Pulau Peleng, N Indonesia 01°32′S 122°57′E
136 F17 **Kemer** Antalya, SW Turkey 36°39′N 30°33′E
122 J12 **Kemerovo** *prev.* Shcheglovsk. Kemerovskaya Oblast', C Russian Federation 55°25′N 86°05′E
122 K12 **Kemerovskaya Oblast'** ◈ *province* S Russian Federation
92 L13 **Kemi** Lappi, NW Finland 65°46′N 24°34′E
92 M12 **Kemijärvi** *Swe.* Kemiträsk. Lappi, N Finland 66°41′N 27°24′E
92 M12 **Kemijärvi** ◎ N Finland
92 L13 **Kemijoki** ≈ NW Finland
147 V7 **Kemin** *prev.* Bystrovka. Chuyskaya Oblast', N Kyrgyzstan
92 L13 **Keminmaa** Lappi, NW Finland 65°49′N 24°34′E
Kemins Island *see* Nikumaroro
Kemiö *see* Kimito
Kemiträsk *see* Kemijärvi
127 P5 **Kemlya** Respublika Mordoviya, W Russian Federation 54°42′N 45°16′E
99 B18 **Kemmel** West-Vlaanderen, W Belgium 50°42′N 02°51′E
33 S16 **Kemmerer** Wyoming, C USA 41°47′N 110°32′W
121 O15 **Kemmuna** *var.* Comino. *island* C Malta
79 I14 **Kémo** ◈ *prefecture* S Central African Republic
25 U7 **Kemp** Texas, SW USA 32°26′N 96°13′W
93 L14 **Kempele** Pohjois-Pohjanmaa, C Finland 64°56′N 25°26′E
101 D15 **Kempen** Nordrhein-Westfalen, W Germany 51°22′N 06°25′E
25 Q5 **Kemp, Lake** ◎ Texas, SW USA
195 W5 **Kemp Land** *physical region* Antarctica
25 S9 **Kempner** Texas, SW USA 31°03′N 98°01′W
44 H3 **Kemp's Bay** Andros Island, W The Bahamas 24°02′N 77°32′W
183 U6 **Kempsey** New South Wales, SE Australia 31°05′S 152°50′E
101 J24 **Kempten** Bayern, S Germany 47°44′N 10°19′E
15 N9 **Kempt, Lac** ◎ Québec, SE Canada
183 P17 **Kempton** Tasmania, SE Australia 42°34′S 147°13′E
154 J9 **Ken** ≈ C India
39 R12 **Kenai** Alaska, USA 60°33′N 151°15′W
0 D5 **Kenai Mountains** ▲▲ Alaska, USA
39 R12 **Kenai Peninsula** *peninsula* Alaska, USA
21 V11 **Kenansville** North Carolina, SE USA 34°57′N 77°54′W
146 A10 **Kenar** *prev. Rus.* Ufra. Balkan Welaýaty, NW Turkmenistan 40°00′N 53°05′E
121 U13 **Kenâyis, Râs el-** *headland* N Egypt 31°13′N 27°53′E
97 K16 **Kendal** NW England, United Kingdom 54°20′N 02°45′W
23 Y16 **Kendall** Florida, SE USA 25°39′N 80°18′W
9 O8 **Kendall, Cape** *headland* Nunavut, E Canada 63°31′N 87°09′W
18 J15 **Kendall Park** New Jersey, NE USA 40°25′N 74°33′W
31 Q11 **Kendallville** Indiana, N USA 41°24′N 85°10′W
171 P14 **Kendari** Sulawesi, C Indonesia 03°57′S 122°36′E
169 Q13 **Kendawangan** Borneo, C Indonesia 02°32′S 110°13′E
Kendrāpāra *see* Kendrāparha
154 O12 **Kendrāparha** *var.* Kendrāpāra. Odisha, E India 20°29′N 86°25′E
154 O11 **Kendujhargarh** *prev.* Keonjihargarh. Odisha, E India 21°38′N 85°40′E
25 S13 **Kenedy** Texas, SW USA 28°49′N 97°51′W
76 J15 **Kenema** SE Sierra Leone 07°55′N 11°12′W
29 P16 **Kenesaw** Nebraska, C USA 40°37′N 98°39′W
Këneurgench *see* Köneürgenç
79 H21 **Kenge** Bandundu, SW Dem. Rep. Congo 04°52′S 16°59′E
Kengen *see* Kangen
167 O5 **Kengtung** Shan State, E Myanmar (Burma) 21°18′N 99°36′E
83 F23 **Kenhardt** Northern Cape, W South Africa 29°19′S 21°08′E
76 J12 **Kéniéba** Kayes, W Mali 12°47′N 11°16′W
Kenimekh *see* Konimex
169 U7 **Keningau** Sabah, East Malaysia 05°21′N 116°11′E
74 F6 **Kénitra** *prev.* Port-Lyautey. NW Morocco 34°20′N 06°29′W
21 V9 **Kenly** North Carolina, SE USA 35°39′N 78°16′W
97 B21 **Kenmare** *Ir.* Neidín. S Ireland 51°53′N 09°35′W
28 L2 **Kenmare** North Dakota, N USA 48°40′N 102°04′W
97 A21 **Kenmare River** *Ir.* An Ribhéar. *inlet* NE Atlantic Ocean
18 D10 **Kenmore** New York, NE USA 42°58′N 78°52′W
25 W8 **Kennard** Texas, SW USA 31°21′N 95°10′W
29 N10 **Kennebec** South Dakota, N USA 43°53′N 99°51′W
19 Q7 **Kennebec River** ≈ Maine, NE USA
19 P9 **Kennebunk** Maine, NE USA 43°22′N 70°33′W
39 R13 **Kennedy Entrance** *strait* Alaska, USA
166 L3 **Kennedy Peak** ▲ W Myanmar (Burma) 23°18′N 93°52′E
22 K9 **Kenner** Louisiana, S USA 29°57′N 90°15′W
180 I8 **Kenneth Range** ▲▲ Western Australia
27 Y9 **Kennett** Missouri, C USA 36°15′N 90°04′W
18 I16 **Kennett Square** Pennsylvania, NE USA 39°50′N 75°40′W
32 K10 **Kennewick** Washington, NW USA 46°12′N 119°08′W
12 E11 **Kenogami** ≈ Ontario, S Canada
15 Q7 **Kénogami, Lac** ◎ Québec, SE Canada
14 G8 **Kenogami Lake** Ontario, S Canada 48°04′N 80°10′W
14 F7 **Kenogamissi Lake** ◎ Ontario, S Canada
10 I6 **Keno Hill** Yukon, NW Canada 63°54′N 135°18′W
12 A11 **Kenora** Ontario, S Canada 49°47′N 94°26′W
31 N9 **Kenosha** Wisconsin, N USA 42°34′N 87°50′W
13 P14 **Kensington** Prince Edward Island, SE Canada 46°26′N 63°39′W
26 L3 **Kensington** Kansas, C USA 39°46′N 99°01′W
32 I11 **Kent** Oregon, NW USA 45°14′N 120°43′W
24 J9 **Kent** Texas, SW USA 31°03′N 104°13′W
32 H8 **Kent** Washington, NW USA 47°22′N 122°13′W
97 P22 **Kent** *cultural region* SE England, United Kingdom
145 P16 **Kentau** Yuzhnyy Kazakhstan, S Kazakhstan 43°28′N 68°41′E
183 P14 **Kent Group** *island group* Tasmania, SE Australia
31 N12 **Kentland** Indiana, N USA 40°46′N 87°26′W
31 R12 **Kenton** Ohio, N USA 40°39′N 83°36′W
8 K7 **Kent Peninsula** *peninsula* Nunavut, N Canada
115 F14 **Kentrikí Makedonía** *Eng.* Macedonia Central. ◈ *region* N Greece
20 J6 **Kentucky** *off.* Commonwealth of Kentucky, *also known as* Bluegrass State. ◈ *state* C USA
20 H8 **Kentucky Lake** ▣ Kentucky/Tennessee, S USA
13 P15 **Kentville** Nova Scotia, SE Canada 45°04′N 64°30′W
22 K8 **Kentwood** Louisiana, S USA 30°56′N 90°30′W
31 P9 **Kentwood** Michigan, N USA 42°52′N 85°33′W
81 H17 **Kenya** *off.* Republic of Kenya. ◆ *republic* E Africa
Kenya, Mount *see* Kirinyaga

◆ Country
● Country Capital
◇ Dependent Territory
○ Dependent Territory Capital
◈ Administrative Regions
✈ International Airport
▲ Mountain
▲▲ Mountain Range
⛰ Volcano
≈ River
◎ Lake
▣ Reservoir

KENYA, REPUBLIC OF – KICKAPOO RIVER

Kenya, Republic of *see* Kenya
168 L7 **Kenyir, Tasik** *var.* Tasek Kenyir. ◎ Peninsular Malaysia
29 W10 **Kenyon** Minnesota, N USA 44°16′N 92°59′W
29 Y16 **Keokuk** Iowa, C USA 40°24′N 91°22′W
Keonjihargarh *see* Kendujhargarh
Kéos *see* Tziá
29 X16 **Keosauqua** Iowa, C USA 40°43′N 91°58′W
29 X15 **Keota** Iowa, C USA 41°21′N 91°57′W
21 O11 **Keowee, Lake** ▣ South Carolina, SE USA
124 I7 **Kepa** *var.* Kepe. Respublika Kareliya, NW Russian Federation 65°09′N 32°15′E
Kepe *see* Kepa
189 O13 **Kepirohi Falls** *waterfall* Pohnpei, E Micronesia
185 B22 **Kepler Mountains** ▲▲ South Island, New Zealand
111 I14 **Kępno** Wielkopolskie, C Poland 51°17′N 17°57′E
65 C24 **Keppel Island** *island* N Falkland Islands
Keppel Island *see* Niuatoputapu
65 C23 **Keppel Sound** *sound* N Falkland Islands
Kepri *see* Kepulauan Riau
136 D12 **Kepsut** Balıkesir, NW Turkey 39°41′N 28°09′E
168 M11 **Kepulauan Riau** *off.* Propinsi Kepulauan Riau, *var.* Kepri. ◆ *province* NW Indonesia
Kequ *see* Gadê
171 V13 **Kerai** Papua Barat, E Indonesia 03°53′S 134°30′E
Kerak *see* Al Karak
155 F22 **Kerala** ◆ *state* S India
165 R16 **Kerama-rettō** *island group* SW Japan
183 N10 **Kerang** Victoria, SE Australia 35°46′S 144°01′E
Kerasunt *see* Giresun
115 H19 **Keratéa** *var.* Keratea. Attikí, C Greece 37°48′N 23°58′E
Keratea *see* Keratéa
93 M19 **Kerava** *Swe.* Kervo. Uusimaa, S Finland 60°25′N 25°10′E
Kerbala/Kerbela *see* Karbalā'
32 F15 **Kerby** Oregon, NW USA 42°10′N 123°39′W
117 W12 **Kerch** *Rus.* Kerch'. Avtonomna Respublika Krym, SE Ukraine 45°22′N 36°30′E
Kerch' *see* Kerch
Kerchens'ka Protska/Kerchenskiy Proliv *see* Kerch Strait
117 V13 **Kerchens'kyy Pivostriv** *peninsula* S Ukraine
121 V4 **Kerch Strait** *var.* Bosporus Cimmerius, Enikale Strait, *Rus.* Kerchenskiy Proliv, *Ukr.* Kerchens'ka Protska. *strait* Black Sea/Sea of Azov
Kerdilio *see* Kerdýlio
114 H13 **Kerdýlio** *var.* Kerdilio. ▲ N Greece 40°46′N 23°37′E
186 D8 **Kerema** Gulf, S Papua New Guinea 07°59′S 145°46′E
Keremitlik *see* Lyulyakovo
136 I9 **Kerempe Burnu** *headland* N Turkey 42°01′N 33°20′E
80 J9 **Keren** *var.* Cheren. C Eritrea 15°45′N 38°22′E
25 U7 **Kerens** Texas, SW USA 32°07′N 96°13′W
184 M6 **Kerepehi** Waikato, North Island, New Zealand 37°18′S 175°33′E
145 P10 **Kerey, Ozero** ◎ C Kazakhstan
Kergel *see* Kärla
173 Q12 **Kerguelen** *island* C French Southern and Antarctic Territories
173 Q13 **Kerguelen Plateau** *undersea feature* S Indian Ocean
115 C20 **Kerí** Zákynthos, Iónia Nisiá, Greece, C Mediterranean Sea 37°40′N 20°48′E
81 H19 **Kericho** Kericho, W Kenya 0°22′S 35°19′E
81 H19 **Kericho** ◆ *county* W Kenya
184 K2 **Kerikeri** Northland, North Island, New Zealand 35°14′S 173°58′E
93 O17 **Kerimäki** Etelä-Savo, E Finland 61°56′N 29°18′E
168 K12 **Kerinci, Gunung** ▲ Sumatera, W Indonesia 02°00′S 101°40′E
Keriya *see* Yutian
158 H9 **Keriya He** ≋ NW China
98 J9 **Kerkbuurt** Noord-Holland, C Netherlands 52°29′N 05°08′E
98 J13 **Kerkdriel** Gelderland, C Netherlands 51°46′N 05°21′E
75 N6 **Kerkenah, Îles de** *var.* Kerkenna Islands, *Ar.* Juzur Qarqannah. *island group* E Tunisia
Kerkenna Islands *see* Kerkenah, Îles de
115 M20 **Kerketévs** ▲ Sámos, Dodekánisa, Greece, Aegean Sea 37°44′N 26°39′E
29 T8 **Kerkhoven** Minnesota, N USA 45°12′N 95°18′W
Kerki *see* Atamyrat
Kerkichi *see* Kerkiçi
146 M14 **Kerkiçi** *Rus.* Kerkichi. Lebap Welaýaty, E Turkmenistan 37°46′N 65°18′E
115 F16 **Kerkíneo** *prehistoric site* Thessalía, C Greece
114 G12 **Kerkíni, Límni** *var.* Límni Kerkinitis. ◎ N Greece
Kerkinitis Límni *see* Kerkíni, Límni
Kérkira *see* Kérkyra
99 M18 **Kerkrade** Limburg, SE Netherlands 50°53′N 06°04′E
Kerkuk *see* Āltūn Kūbrī
115 B16 **Kérkyra** *var.* Kérkira, *Eng.* Corfu. Kérkyra, Iónia Nisiá, Greece, C Mediterranean Sea 39°37′N 19°56′E
115 B16 **Kerkyra** ✈ Kérkyra, Iónia Nisiá, Greece, C Mediterranean Sea 39°36′N 19°55′E
115 A16 **Kérkyra** *var.* Kérkira, *Eng.* Corfu. *island* Iónia Nisiá, Greece, C Mediterranean Sea
192 K10 **Kermadec Islands** *island group* New Zealand, SW Pacific Ocean
175 R10 **Kermadec Ridge** *undersea feature* SW Pacific Ocean 30°30′S 178°30′W
175 R11 **Kermadec Trench** *undersea feature* SW Pacific Ocean
143 S10 **Kermān** *var.* Kirman; *anc.* Carmana. Kermān, C Iran 30°18′N 57°05′E
143 R11 **Kermān** *off.* Ostān-e Kermān, *var.* Kirman; *anc.* Carmania. ◆ *province* SE Iran
143 U12 **Kermān, Bīābān-e** *desert* SE Iran
Kermān, Ostān-e *see* Kermān
142 K6 **Kermānshāh** *var.* Qahremānshahr; *prev.* Bākhtarān. Kermānshāhān, W Iran 34°19′N 47°04′E
143 Q9 **Kermānshāh** Yazd, C Iran 34°19′N 47°04′E
142 J6 **Kermānshāh** *off.* Ostān-e Kermānshāh; *prev.* Bākhtarān. ◆ *province* W Iran
Kermānshāh, Ostān-e *see* Kermānshāh
114 L10 **Kermen** Sliven, C Bulgaria 42°30′N 26°12′E
24 L8 **Kermit** Texas, SW USA 31°49′N 103°07′W
21 P6 **Kermit** West Virginia, NE USA 37°51′N 82°24′W
21 S9 **Kernersville** North Carolina, SE USA 36°12′N 80°13′W
35 S12 **Kern River** ≋ California, W USA
35 S12 **Kernville** California, W USA 35°44′N 118°25′W
115 K21 **Kéros** *island* Kykládes, Greece, Aegean Sea
76 K14 **Kérouané** SE Guinea 09°16′N 09°00′W
101 D16 **Kerpen** Nordrhein-Westfalen, W Germany 50°51′N 06°40′E
146 I11 **Kerpichli** Lebap Welaýaty, NE Turkmenistan 40°12′N 61°09′E
24 M1 **Kerrick** Texas, SW USA 36°29′N 102°14′W
Kerr Lake *see* John H. Kerr Reservoir
11 S15 **Kerrobert** Saskatchewan, S Canada 51°56′N 109°09′W
25 Q11 **Kerrville** Texas, SW USA 30°03′N 99°09′W
97 B20 **Kerry** *Ir.* Ciarraí. *cultural region* SW Ireland
21 S11 **Kershaw** South Carolina, SE USA 34°33′N 80°34′W
Kertel *see* Kärdla
95 H23 **Kerteminde** Syddtjylland, C Denmark 55°27′N 10°40′E
163 Q7 **Kerulen** *Chin.* Herlen He, *Mong.* Herlen Gol. ≋ China/Mongolia
Kervo *see* Kerava
Kerýneia *see* Girne
12 H11 **Kesagami Lake** ◎ Ontario, SE Canada
93 O17 **Kesälahti** Pohjois-Karjala, SE Finland 61°54′N 29°49′E
136 B11 **Keşan** Edirne, NW Turkey 40°52′N 26°37′E
165 R9 **Kesennuma** Miyagi, Honshū, C Japan 38°55′N 141°35′E
163 V7 **Keshan** Heilongjiang, NE China 48°00′N 125°46′E
30 M6 **Keshena** Wisconsin, N USA 44°54′N 88°37′W
136 I13 **Keskin** Kırıkkale, C Turkey 39°41′N 33°36′E
93 K16 **Keski-Pohjanmaa** *Swe.* Mellersta Österbotten, *Eng.* centralostrobothnia. ◆ *region* W Finland
93 M17 **Keski-Suomi** *Swe.* Mellersta Finland, *Eng.* Central Finland. ◆ *region* C Finland
Késmárk *see* Kežmarok
124 I6 **Kesten'ga** *var.* Kest Enga. Respublika Kareliya, NW Russian Federation 65°53′N 31°47′E
Kest Enga *see* Kesten'ga
98 K12 **Kesteren** Gelderland, C Netherlands 51°55′N 05°34′E
14 H14 **Keswick** Ontario, S Canada 44°15′N 79°26′W
97 K15 **Keswick** NW England, United Kingdom 54°30′N 03°04′W
111 H24 **Keszthely** Zala, SW Hungary 46°47′N 17°16′E
122 K11 **Ket'** ≋ C Russian Federation
77 R17 **Keta** SE Ghana 05°55′N 00°59′E
169 Q12 **Ketapang** Borneo, C Indonesia 01°50′S 109°59′E
127 O12 **Ketchenery** *prev.* Sovetskoye. Respublika Kalmykiya, SW Russian Federation 47°18′N 44°31′E
39 Y14 **Ketchikan** Revillagigedo Island, Alaska, USA 55°21′N 131°39′W
33 O14 **Ketchum** Idaho, NW USA 43°40′N 114°24′W
Kete/Kete Krakye *see* Kete-Krachi
77 Q15 **Kete-Krachi** *var.* Kete, Kete Krakye. E Ghana 07°50′N 00°03′W
98 L9 **Ketelmeer** *channel* E Netherlands
149 P17 **Keti Bandar** Sind, SE Pakistan 23°55′N 67°31′E
77 S16 **Kétou** SE Benin 07°25′N 02°36′E
110 M7 **Kętrzyn** *Ger.* Rastenburg. Warmińsko-Mazurskie, NE Poland 54°05′N 21°24′E
97 N20 **Kettering** C England, United Kingdom 52°24′N 00°44′W
31 R14 **Kettering** Ohio, N USA 39°41′N 84°10′W
18 F13 **Kettle Creek** ≋ Pennsylvania, NE USA
32 L7 **Kettle Falls** Washington, NW USA 48°36′N 118°03′W
14 D16 **Kettle Point** *headland* Ontario, S Canada 43°12′N 82°01′W
29 V6 **Kettle River** ≋ Minnesota, N USA
186 B7 **Ketu** ≋ W Papua New Guinea
18 G10 **Keuka Lake** ◎ New York, NE USA
Keupriya *see* Primorsko
93 L17 **Keuruu** Keski-Suomi, C Finland 62°15′N 24°34′E
Kevevára *see* Kovin
92 L9 **Kevo** *Lapp.* Geavvú. Lappi, N Finland 69°42′N 27°08′E
44 M6 **Kew** North Caicos, N Turks and Caicos Islands 21°52′N 71°57′W
30 K11 **Kewanee** Illinois, N USA 41°15′N 89°55′W
31 N7 **Kewaunee** Wisconsin, N USA 44°27′N 87°31′W
30 M3 **Keweenaw Bay** ◎ Michigan, N USA
31 N2 **Keweenaw Peninsula** *peninsula* Michigan, N USA
31 N2 **Keweenaw Point** *peninsula* Michigan, N USA
29 N12 **Keya Paha River** ≋ Nebraska/South Dakota, N USA
Keyaygyr *see* Këk-Aygyr
23 Z16 **Key Biscayne** Florida, SE USA 25°41′N 80°09′W
26 G8 **Keyes** Oklahoma, C USA 36°48′N 102°15′W
23 Y17 **Key Largo** Key Largo, Florida, SE USA 25°06′N 80°25′W
21 U3 **Keyser** West Virginia, NE USA 39°25′N 78°59′W
27 O9 **Keystone Lake** ▣ Oklahoma, C USA
36 L16 **Keystone Peak** ▲ Arizona, SW USA 31°52′N 111°12′W
Keystone State *see* Pennsylvania
21 U7 **Keysville** Virginia, NE USA 37°02′N 78°28′W
27 T3 **Keytesville** Missouri, C USA 39°25′N 92°56′W
23 W17 **Key West** Florida Keys, Florida, SE USA 24°34′N 81°48′W
127 T1 **Kez** Udmurtskaya Respublika, NW Russian Federation 57°55′N 53°42′E
Kezdivásárhely *see* Târgu Secuiesc
122 M12 **Kezhma** Krasnoyarskiy Kray, C Russian Federation 58°57′N 101°00′E
111 L18 **Kežmarok** *Ger.* Käsmark, *Hung.* Késmárk. Prešovský Kraj, E Slovakia 49°09′N 20°25′E
Kfar Saba *see* Kfar Sava
138 F10 **Kfar Sava** *var.* Kfar Saba; *prev.* Kefar Sava. Central, C Israel 32°11′N 34°58′E
83 F20 **Kgalagadi** ◆ *district* SW Botswana
83 I20 **Kgatleng** ◆ *district* SE Botswana
188 F8 **Kgkeklau** Babeldaob, N Palau
125 R6 **Khabarikha** *var.* Chabaricha. Respublika Komi, NW Russian Federation 65°52′N 52°19′E
123 S14 **Khabarovsk** Khabarovskiy Kray, SE Russian Federation 48°32′N 135°08′E
123 R11 **Khabarovskiy Kray** ◆ *territory* E Russian Federation
141 W7 **Khabb** Abū Ẓaby, E United Arab Emirates 24°39′N 55°43′E
Khabour, Nahr al *see* Khābūr, Nahr al
Khabura *see* Al Khābūrah
139 N2 **Khābūr, Nahr al** *var.* Nahr al Khabour. ≋ Syria/Turkey
Khachmas *see* Xaçmaz
80 B12 **Khadari** ≋ W Sudan
Khadera *see* Hadera
141 X12 **Khādhil** *var.* Khudal. SE Oman 18°48′N 56°48′E
155 E14 **Khadki** *prev.* Kirkee. Mahārāshtra, W India 18°34′N 73°52′E
126 L14 **Khadyzhensk** Krasnodarskiy Kray, SW Russian Federation 44°26′N 39°31′E
Khadzhiyska Reka *see* Hadzhiyska Reka
117 P10 **Khadzhybeys'kyy Lyman** ◎ SW Ukraine
138 K3 **Khafsah** Ḥalab, N Syria 36°16′N 38°03′E
152 M13 **Khāga** Uttar Pradesh, N India 25°47′N 81°05′E
153 Q13 **Khagaria** Bihār, NE India 25°31′N 86°27′E
149 Q13 **Khairpur** Sind, SE Pakistan 27°30′N 68°50′E
122 K13 **Khakasiya, Respublika** *prev.* Khakasskaya Avtonomnaya Oblast', *Eng.* Khakassia. ◆ *autonomous republic* C Russian Federation
Khakassia/Khakasskaya Avtonomnaya Oblast' *see* Khakasiya, Respublika
167 N9 **Kha Khaeng, Khao** ▲ W Thailand 16°13′N 99°03′E
83 G20 **Khakhea** *var.* Kakia. Southern, S Botswana 24°41′S 23°29′E
Khalach *see* Halaç
Khalándrion *see* Chalándri
75 T7 **Khalīj as Sallūm** *Ar.* Gulf of Salûm. *gulf* Egypt/Libya
75 X8 **Khalīj as Suways** *var.* Suez, Gulf of. *gulf* NE Egypt
127 W7 **Khalilovo** Orenburgskaya Oblast', W Russian Federation 51°25′N 58°13′E
Khalkabad *see* Xalqobod
142 L3 **Khalkhāl** *prev.* Herowābād. Ardabīl, NW Iran 37°36′N 48°36′E
Khalkidhikí *see* Chalkidikí
Khalkís *see* Chalkída
125 W3 **Khal'mer-Yu** Respublika Komi, NW Russian Federation 68°00′N 64°45′E
119 M14 **Khalopyenichy** *Rus.* Kholopenichi. Minskaya Voblasts', NE Belarus 54°31′N 28°58′E
Khalturin *see* Orlov
141 Y10 **Khalūf** *var.* Al Khaluf. E Oman 20°27′N 57°59′E
154 K10 **Khamaria** Madhya Pradesh, C India 23°07′N 80°54′E
154 D11 **Khambhāt** Gujarāt, W India 22°19′N 72°39′E
154 C12 **Khambhāt, Gulf of** *Eng.* Gulf of Cambay. *gulf* W India
167 U10 **Khâm Đuc** *var.* Phươc Sơn. Quang Nam-Đa Năng, C Vietnam 15°28′N 107°49′E
154 G12 **Khāmgaon** Mahārāshtra, C India 20°41′N 76°34′E
141 O14 **Khamir** *var.* Khamr. W Yemen 16°N 43°56′E
141 N12 **Khamīs Mushayṭ** *var.* Hamīs Musait. 'Asīr, SW Saudi Arabia 18°19′N 42°41′E
155 J16 **Khammam** Telangana, India 17°16′N 80°13′E
123 P10 **Khampa** Respublika Sakha (Yakutiya), NE Russian Federation 63°43′N 123°02′E
Khamr *see* Khamir
83 C19 **Khan** ≋ W Namibia
149 Q2 **Khānābād** Kunduz, NE Afghanistan 36°42′N 69°08′E
Khân Abou Châmâte/Khan Abou Ech Cham *see* Khān Abū Shāmāt
138 I7 **Khān Abū Shāmāt** *var.* Khân Abou Châmâte, Khan Abou Ech Cham. Rīf Dimashq, W Syria 33°43′N 36°56′E
Khān al Baghdādī *see* Al Baghdādī
Khān al Maḥāwīl *see* Al Maḥāwīl
139 T7 **Khān al Mashāhidah** Baghdād, C Iraq 33°40′N 44°15′E
139 T10 **Khān al Muṣallá** An Najaf, S Iraq 32°09′N 44°20′E
139 U6 **Khānaqīn** Diyālá, E Iraq 34°22′N 45°22′E
139 T11 **Khān ar Ruḥbah** An Najaf, S Iraq 31°42′N 44°18′E
139 P2 **Khān as Sūr** Nīnawá, N Iraq 36°28′N 41°36′E
139 T8 **Khān Āzād** Baghdād, C Iraq 33°08′N 44°21′E
154 N13 **Khandaparha** *prev.* Khandpara. Odisha, E India 20°15′N 85°11′E
Khandpara *see* Khandaparha
149 T2 **Khandūd** *var.* Khandud, Wakhan. Badakhshān, NE Afghanistan 36°57′N 72°19′E
Khandud *see* Khandūd
154 G11 **Khandwa** Madhya Pradesh, C India 21°49′N 76°23′E
123 R10 **Khandyga** Respublika Sakha (Yakutiya), NE Russian Federation 62°39′N 135°30′E
149 T10 **Khānewāl** Punjab, NE Pakistan 30°18′N 71°56′E
149 S10 **Khāngarh** Punjab, E Pakistan 29°57′N 71°14′E
Khanh Hung *see* Soc Trăng
Khaniá *see* Chaniá
Khanka *see* Xonqa
163 Z8 **Khanka, Lake** *var.* Hsing-K'ai Hu, Lake Hanka, *Chin.* Xingkai Hu, *Rus.* Ozero Khanka. ◎ China/Russian Federation
Khanka, Ozero *see* Khanka, Lake
Khankendi *see* Xankändi
123 O9 **Khannya** ≋ NE Russian Federation
144 D10 **Khan Ordasy** *prev.* Urda. Zapadnyy Kazakhstan, W Kazakhstan 48°52′N 47°31′E
149 S12 **Khānpur** Punjab, E Pakistan 28°31′N 70°30′E
138 I4 **Khān Shaykhūn** *var.* Khan Sheikhun. Idlib, NW Syria 35°27′N 36°38′E
Khan Sheikhun *see* Khān Shaykhūn
Khanshyngghys *see* Khrebet Khanshyngys
145 S15 **Khantau** Zhambyl, S Kazakhstan 44°13′N 73°47′E
145 W16 **Khan Tengri, Pik** ▲ SE Kazakhstan 42°17′N 80°11′E
Khan-Tengri, Pik *see* Hantengri Feng
Khanthabouli *see* Savannakhét
127 V8 **Khanty-Mansiysk** *prev.* Ostyako-Voguls'k. Khanty-Mansiyskiy Avtonomnyy Okrug-Yugra, C Russian Federation 61°01′N 69°E
125 V8 **Khanty-Mansiyskiy Avtonomnyy Okrug-Yugra** ◆ *autonomous district* C Russian Federation
139 R4 **Khānūqah** Nīnawýé, C Iraq 35°25′N 43°15′E
138 E11 **Khān Yūnis** *var.* Khān Yūnus. S Gaza Strip 31°21′N 34°18′E
Khān Yūnus *see* Khān Yūnis
Khanzi *see* Ghanzi
Khān Zūr *see* Xan Sûr
167 N10 **Khao Laem Reservoir** ▣ W Thailand
123 O14 **Khapcheranga** Zabaykal'skiy Kray, S Russian Federation 49°46′N 112°21′E
127 Q12 **Kharabali** Astrakhanskaya Oblast', SW Russian Federation 47°28′N 47°14′E
153 R16 **Kharagpur** West Bengal, NE India 22°30′N 87°19′E
139 V11 **Kharā'ib 'Abd al Karīm** Al Muthanná, S Iraq 31°07′N 45°33′E
143 Q8 **Kharānaq** Yazd, C Iran 31°54′N 54°21′E
Kharbin *see* Harbin
Kharchi *see* Mārwār
146 H13 **Khardzhagaz** Ahal Welaýaty, C Turkmenistan 37°54′N 60°10′E
154 F11 **Khargon** Madhya Pradesh, C India 21°49′N 75°39′E
149 V7 **Khāriān** Punjab, NE Pakistan 32°52′N 73°52′E
117 V5 **Kharkiv** *Rus.* Khar'kov. Kharkivs'ka Oblast', NE Ukraine 50°N 36°14′E
117 V5 **Kharkiv** ✈ Kharkivs'ka Oblast', E Ukraine 49°54′N 36°20′E
Kharkiv *see* Kharkivs'ka Oblast'
117 U5 **Kharkivs'ka Oblast'** *var.* Kharkiv, *Rus.* Khar'kovskaya Oblast'. ◆ *province* E Ukraine
Khar'kov *see* Kharkiv
Khar'kovskaya Oblast' *see* Kharkivs'ka Oblast'
124 L3 **Kharlovka** Murmanskaya Oblast', NW Russian Federation 68°47′N 37°09′E
Kharmanli *see* Harmanli
Kharmanliyska Reka *see* Harmanliyska Reka
124 M13 **Kharovsk** Vologodskaya Oblast', NW Russian Federation 59°57′N 40°05′E
80 F9 **Khartoum** *var.* El Khartûm, Khartum. ● (Sudan) Khartoum, C Sudan 15°33′N 32°32′E
80 F9 **Khartoum** ◆ *state* NE Sudan
80 F9 **Khartoum** ✈ Khartoum, C Sudan 15°36′N 32°37′E
80 F9 **Khartoum North** Khartoum, C Sudan 15°38′N 32°33′E
117 X8 **Khartsyz'k** *Rus.* Khartsyzsk. Donets'ka Oblast', SE Ukraine 48°01′N 38°10′E
117 X8 **Khartsyz'k** Donets'ka Oblast', E Ukraine 48°01′N 38°10′E
Khartsyzsk *see* Khartsyz'k
Khartum *see* Khartoum
Kharwazawk *see* Xêrzok
Khasab *see* Al Khaṣab
123 S15 **Khasan** Primorskiy Kray, SE Russian Federation 42°24′N 130°45′E
127 P16 **Khasavyurt** Respublika Dagestan, SW Russian Federation 43°16′N 46°33′E
143 W12 **Khāsh** *prev.* Vāsht. Sīstān va Balūchestān, SE Iran 28°15′N 61°11′E
148 K8 **Khāsh, Dasht-e** *Eng.* Khash Desert. *desert* SW Afghanistan
Khash Desert *see* Khāsh, Dasht-e
Khashim Al Qirba/Khashm al Qirbah *see* Khashm el Girba
80 H9 **Khashm el Girba** *var.* Khashim Al Qirba, Khashm al Qirbah. Kassala, E Sudan 15°00′N 35°59′E
138 G14 **Khashsh, Jabal al** ▲▲ S Jordan
137 S10 **Khashuri** C Georgia 41°59′N 43°36′E
153 V13 **Khāsi Hills** *hill range* NE India
Khaskovo *see* Haskovo
Khaskovo *see* Haskovo
122 M7 **Khatanga** ≋ N Russian Federation
Khatanga, Gulf of *see* Khatangskiy Zaliv
123 N7 **Khatangskiy Zaliv** *var.* Gulf of Khatanga. *bay* N Russian Federation
141 W7 **Khatmat al Malāḥah** N Oman 24°58′N 56°22′E
143 S16 **Khaṭmat al Malāḥah** Ash Shāriqah, E United Arab Emirates
123 V7 **Khatyrka** Chukotskiy Avtonomnyy Okrug, NE Russian Federation 62°03′N 175°09′E
Khauz-Khan *see* Hanhowuz
Khauzkhanskoye Vodoranilishche *see* Hanhowuz Suw Howdany
Khavaling *see* Khovaling
Khavast *see* Xovos
139 W10 **Khawrah, Nahr al** ≋ S Iraq
Khawr Barakah *see* Barka
141 W7 **Khawr Fakkān** *var.* Khor Fakkan. Ash Shāriqah, NE United Arab Emirates 25°22′N 56°19′E
140 L6 **Khaybar** Al Madīnah, NW Saudi Arabia 25°53′N 39°16′E
Khaybar, Kowtal-e *see* Khyber Pass
147 S11 **Khaydarkan** *var.* Khaydarken. Batkenskaya Oblast', SW Kyrgyzstan 39°56′N 71°17′E
Khaydarken *see* Khaydarkan
125 U2 **Khaypudyrskaya Guba** *bay* NW Russian Federation
Khayrūzuk *see* Xêrzok
Khạzar, Baḥr-e/Khazar, Daryā-ye *see* Caspian Sea
Khazarosp *see* Hazorasp
Khazretishi, Khrebet *see* Hazratishoh, Qatorkŭhi
75 X11 **Khazzān Aswān** *var.* Aswan Dam. *dam* SE Egypt
Khelat *see* Kālat
74 F6 **Khemisset** NW Morocco 33°52′N 06°04′W
167 R10 **Khemmarat** *var.* Kemarat. Ubon Ratchathani, E Thailand 16°03′N 105°11′E
74 L6 **Khenchela** *var.* Khenchla. NE Algeria 35°22′N 07°09′E
Khenchla *see* Khenchela
74 G7 **Khénifra** C Morocco 32°59′N 05°37′W
Khersān, Rūd-e *see* Garm, Āb-e
117 R10 **Kherson** Khersons'ka Oblast', S Ukraine 46°39′N 32°38′E
Kherson *see* Khersons'ka Oblast'
117 S14 **Khersones, Mys** *Rus.* Mys Khersonesskiy. *headland* S Ukraine 44°34′N 33°24′E
Khersonesskiy, Mys *see* Khersones, Mys
117 R10 **Khersons'ka Oblast'** *var.* Kherson, *Rus.* Khersonskaya Oblast'. ◆ *province* S Ukraine
Khersonskaya Oblast' *see* Khersons'ka Oblast'
122 L8 **Kheta** ≋ N Russian Federation
149 U7 **Khewra** Punjab, E Pakistan 32°41′N 73°04′E
Khiam *see* El Khiyam
124 J4 **Khibiny** ▲▲ NW Russian Federation
126 K3 **Khimki** Moskovskaya Oblast', W Russian Federation 55°57′N 37°48′E
147 S12 **Khingov** *Rus.* Obi-Khingou. ≋ C Tajikistan
Khíos *see* Chíos
149 R15 **Khipro** Sind, SE Pakistan 25°50′N 69°24′E
139 S10 **Khirr, Wādī al** *dry watercourse* S Iraq
Khisarya *see* Hisarya
Khiva/Khiwa *see* Xiva
167 N9 **Khlong Khlung** Kamphaeng Phet, W Thailand 16°15′N 99°41′E
167 N15 **Khlong Thom** Krabi, SW Thailand 07°55′N 99°09′E
167 P12 **Khlung** Chantaburi, S Thailand 12°25′N 102°12′E
Khmel'nik *see* Khmil'nyk
Khmel'nitskaya Oblast' *see* Khmel'nyts'ka Oblast'
Khmel'nitskiy *see* Khmel 'nyts'kyy
116 K5 **Khmel'nyts'ka Oblast'** *var.* Khmel'nyts'kyy, *Rus.* Khmel'nitskaya Oblast'; *prev.* Kamenets-Podol'skaya Oblast'. ◆ *province* NW Ukraine
116 L6 **Khmel 'nyts'kyy** *Rus.* Khmel'nitskiy; *prev.* Proskurov. Khmel'nyts'ka Oblast', W Ukraine 49°24′N 26°59′E
Khmel'nyts'kyy *see* Khmel'nyts'ka Oblast'
116 M6 **Khmil'nyk** *Rus.* Khmel'nik. Vinnyts'ka Oblast', C Ukraine 49°36′N 27°59′E
Khobda *see* Kobda
137 R9 **Khobi** W Georgia 42°20′N 41°54′E
119 P15 **Khodasy** *Rus.* Khodosy. Mahilyowskaya Voblasts', E Belarus 53°56′N 31°29′E
116 I6 **Khodoriv** *Pol.* Chodorów, *Rus.* Khodorov. L'vivs'ka Oblast', NW Ukraine 49°20′N 24°19′E
Khodorov *see* Khodoriv
Khodosy *see* Khodasy
Khodzhakala *see* Hojagala
Khodzhambas *see* Hojambaz
Khodzhent *see* Khujand
Khodzheyli *see* Xo'jayli
Khoi *see* Khvoy
Khojend *see* Khujand
Khokand *see* Qo'qon
126 L8 **Khokhol'skiy** Voronezhskaya Oblast', W Russian Federation 51°33′N 38°43′E
167 P10 **Khok Samrong** Lop Buri, C Thailand 15°03′N 100°44′E
124 H15 **Kholm** Novgorodskaya Oblast', W Russian Federation 57°10′N 31°06′E
Kholm *see* Khulm
Kholm *see* Chełm
Kholmech' *see* Kholmyech
123 T13 **Kholmsk** Ostrov Sakhalin, Sakhalinskaya Oblast', SE Russian Federation 46°57′N 142°10′E
119 O19 **Kholmyech** *Rus.* Kholmech'. Homyel'skaya Voblasts', SE Belarus 52°09′N 30°37′E
Kholon *see* Holon
Kholopenichi *see* Khalopyenichy
83 D19 **Khomas** ◆ *district* C Namibia
83 D19 **Khomas Hochland** *var.* Khomasplato. *plateau* C Namibia
Khomasplato *see* Khomas Hochland
Khomein *see* Khomeyn
142 M7 **Khomeyn** *var.* Khomein, Khumain. Markazī, W Iran 33°38′N 50°03′E
143 N8 **Khomeynīshahr** *prev.* Homāyūnshahr. Eṣfahān, C Iran 32°42′N 51°28′E
Khoms *see* Al Khums
Khong Sedone *see* Muang Khôngxédôn
167 Q9 **Khon Kaen** *var.* Muang Khon Kaen. Khon Kaen, E Thailand 16°25′N 102°50′E
Khonqa *see* Xonqa
167 Q9 **Khon San** Khon Kaen, E Thailand 16°40′N 101°51′E
123 R8 **Khonuu** Respublika Sakha (Yakutiya), NE Russian Federation 66°24′N 143°15′E
127 N8 **Khopër** *var.* Khoper. ≋ SW Russian Federation
Khoper *see* Khopër
123 S14 **Khor** Khabarovskiy Kray, SE Russian Federation 47°44′N 134°48′E
143 U9 **Khorāsān-e Jonūbī** *off.* Ostan-e Khorāsān-e Jonūbī. ◆ *province* E Iran
143 U6 **Khorāsān-e Raẕavī** *off.* Ostān-e Khorāsān-e Raẕavī, *var.* Khorassan, Khurasan. ◆ *province* NE Iran
Khorāsān-e Raẕavī, Ostān-e *see* Khorāsān-e Raẕavī
143 S3 **Khorāsān-e Shomālī** *off.* Ostan-e Khorāsān-e Shomālī. ◆ *province* NE Iran
Khorassan *see* Khorāsān-e Raẕavī
Khorat *see* Nakhon Ratchasima
154 O13 **Khordha** *prev.* Khurda. Odisha, E India 20°10′N 85°42′E
125 U4 **Khorey-Ver** Nenetskiy Avtonomnyy Okrug, NW Russian Federation 67°25′N 58°05′E
Khorezmskaya Oblast' *see* Xorazm Viloyati
Khor Fakkan *see* Khawr Fakkān
Khorgos *see* Korgas
123 N13 **Khorinsk** Respublika Buryatiya, S Russian Federation 52°13′N 109°52′E
83 C18 **Khorixas** Kunene, NW Namibia 20°23′S 14°55′E
141 O17 **Khormaksar** *var.* Aden. ✈ ('Adan) SW Yemen 12°56′N 45°00′E
Khormal *see* Xurmał
Khormuj *see* Khvormūj
Khorog *see* Khorugh
117 S5 **Khorol** Poltavs'ka Oblast', NE Ukraine 49°49′N 33°17′E
142 L7 **Khorramābād** *var.* Khurramabad. Lorestān, W Iran 33°29′N 48°21′E
143 R9 **Khorramdasht** Kermān, C Iran 31°41′N 56°10′E
142 K10 **Khorramshahr** *var.* Khurramshahr, Muhammerah; *prev.* Mohammerah. Khūzestān, SW Iran 30°30′N 48°09′E
147 S14 **Khorugh** *Rus.* Khorog. S Tajikistan 37°30′N 71°31′E
Khorvot Khalutsa *see* Horvot Halutsa
127 Q12 **Khosheutovo** Astrakhanskaya Oblast', SW Russian Federation 47°04′N 47°49′E
149 R6 **Khōst** *prev.* Khowst. Khōst, E Afghanistan 33°22′N 69°57′E
149 S6 **Khōst** *prev.* Khowst. ◆ *province* E Afghanistan
Khotan *see* Hotan
Khotimsk *see* Khotsimsk
Khotin *see* Khotyn
119 R16 **Khotsimsk** *Rus.* Khotimsk. Mahilyowskaya Voblasts', E Belarus 53°24′N 32°35′E
116 K7 **Khotyn** *Rom.* Hotin, *Rus.* Khotin. Chernivets'ka Oblast', W Ukraine 48°29′N 26°30′E
74 F7 **Khouribga** C Morocco 32°55′N 06°51′W
147 Q13 **Khovaling** *Rus.* Khavaling. SW Tajikistan 38°22′N 69°54′E
Khovd *see* Hovd
Khowst *see* Khōst
Khoy *see* Khvoy
119 N20 **Khoyniki** Homyel'skaya Voblasts', SE Belarus 51°54′N 29°58′E
Khozretishi, Khrebet *see* Hazratishoh, Qatorkŭhi
145 V11 **Khrebet Khanshyngys** *Kas.* Khanshyngghys; *prev.* Khrebet Kanchingiz. ▲▲ E Kazakhstan
145 X10 **Khrebet Kalba** *Kaz.* Qalba Zhotasy; *prev.* Kalbinskiy Khrebet. ▲▲ E Kazakhstan
Khrebet Kanchingiz *see* Khrebet Khanshyngys
Khrebet Ketmen *see* Khrebet Uzynkara
145 Y10 **Khrebet Naryn** *Kaz.* Naryn Zhotasy; *prev.* Narymskiy Khrebet. ▲▲ E Kazakhstan
145 W16 **Khrebet Uzynkara** *prev.* Khrebet Ketmen. ▲▲ SE Kazakhstan
Khrisoúpolis *see* Chrysoúpoli
144 J10 **Khromtau** *Kaz.* Khromtaū. Aktyubinsk, W Kazakhstan 50°14′N 58°22′E
Khromtaū *see* Khromtau
Khrysokhou Bay *see* Chrysochoú, Kólpos
117 O7 **Khrystynivka** Cherkas'ka Oblast', C Ukraine 48°49′N 29°55′E
167 R10 **Khuang Nai** Ubon Ratchathani, E Thailand 15°22′N 104°33′E
Khudal *see* Khādhil
Khudat *see* Xudat
149 W9 **Khudiān** Punjab, E Pakistan 30°59′N 74°19′E
Khudzhand *see* Khujand
83 G21 **Khuis** Kgalagadi, SW Botswana 26°37′S 21°50′E
147 Q11 **Khujand** *var.* Khodzhent, Khojend, *Rus.* Khudzhand; *prev.* Leninabad, *Taj.* Leninobod. N Tajikistan 40°17′N 69°37′E
167 R11 **Khukhan** Si Sa Ket, E Thailand 14°38′N 104°12′E
149 P2 **Khulm** *var.* Tashqurghan; *prev.* Kholm. Balkh, N Afghanistan 36°42′N 67°41′E
153 T16 **Khulna** Khulna, SW Bangladesh 22°48′N 89°32′E
153 T16 **Khulna** ◆ *division* SW Bangladesh
Khumain *see* Khomeyn
Khums *see* Al Khums
149 W2 **Khunjerāb Pass** *pass* China/Pakistan
Khūnjerāb Pass *see* Kunjirap Daban
153 P16 **Khunti** Jhārkhand, N India 23°02′N 85°19′E
167 N7 **Khun Yuam** Mae Hong Son, NW Thailand 18°54′N 97°54′E
Khurais *see* Khurayṣ
Khurasan *see* Khorāsān-e Raẕavī
141 R7 **Khurayṣ** *var.* Khurais. Ash Sharqīyah, C Saudi Arabia 25°06′N 48°03′E
Khurda *see* Khordha
152 J11 **Khurja** Uttar Pradesh, N India 28°15′N 77°51′E
Khūrmāl *see* Xurmał
Khurramabad *see* Khorramābād
Khurramshahr *see* Khorramshahr
149 U7 **Khushāb** Punjab, NE Pakistan 32°16′N 72°18′E
116 H8 **Khust** *var.* Husté, *Cz.* Chust, *Hung.* Huszt. Zakarpats'ka Oblast', W Ukraine 48°11′N 23°19′E
80 D11 **Khuwei** Western Kordofan, C Sudan 13°02′N 29°13′E
149 O13 **Khuzdār** Baluchistān, SW Pakistan 27°49′N 66°39′E
142 L9 **Khūzestān** *off.* Ostān-e Khūzestān, *var.* Khuzistan, *prev.* Arabistan; *anc.* Susiana. ◆ *province* SW Iran
Khūzestān, Ostān-e *see* Khūzestān
Khuzistan *see* Khūzestān
Khvājeh Ghār *see* Khwājeh Ghār
127 Q7 **Khvalynsk** Saratovskaya Oblast', W Russian Federation 52°30′N 48°06′E
143 N12 **Khvormūj** *var.* Khormuj. Būshehr, S Iran 28°32′N 51°22′E
142 I2 **Khvoy** *var.* Khoi, Khoy. Āẕarbāyjān-e Bākhtarī, NW Iran 38°36′N 45°04′E
Khwajaghar/Khwaja-i-Ghar *see* Khwājeh Ghār
149 R2 **Khwājeh Ghār** *var.* Khwajaghar, Khwaja-i-Ghar; *prev.* Khvājeh Ghār. Takhār, NE Afghanistan 37°08′N 69°24′E
149 U4 **Khyber Pakhtunkhwa** *prev.* North-West Frontier Province. ◆ *province* NW Pakistan
149 S5 **Khyber Pass** *var.* Kowtal-e Khaybar. *pass* Afghanistan/Pakistan
186 L8 **Kia** Santa Isabel, N Solomon Islands 07°34′S 158°31′E
183 S10 **Kiama** New South Wales, SE Australia 34°41′S 150°49′E
79 O22 **Kiambi** Katanga, SE Dem. Rep. Congo 07°15′S 28°01′E
81 I19 **Kiambu** ◆ *county* C Kenya
27 Q12 **Kiamichi Mountains** ▲▲ Oklahoma, C USA
27 Q12 **Kiamichi River** ≋ Oklahoma, C USA
14 M10 **Kiamika, Réservoir** ▣ Québec, SE Canada
Kiamusze *see* Jiamusi
39 N7 **Kiana** Alaska, USA 66°58′N 160°25′W
Kiangmai *see* Chiang Mai
Kiang-ning *see* Nanjing
Kiangsi *see* Jiangxi
Kiangsu *see* Jiangsu
93 M14 **Kiantajärvi** ◎ E Finland
115 F19 **Kiáto** *prev.* Kiáton. Pelopónnisos, S Greece 38°01′N 22°45′E
Kiáton *see* Kiáto
Kiayi *see* Jiayi
95 F22 **Kibæk** Midtjylland, W Denmark 56°03′N 08°52′E
67 T9 **Kibali** *var.* Uele (upper course). ≋ NE Dem. Rep. Congo
79 E20 **Kibangou** Niari, SW Congo 03°27′S 12°21′E
Kibarty *see* Kybartai
92 M8 **Kiberg** Finnmark, N Norway 70°17′N 30°47′E
79 N20 **Kibombo** Maniema, E Dem. Rep. Congo 03°52′S 25°59′E
81 E20 **Kibondo** Kigoma, NW Tanzania 03°34′S 30°41′E
81 J15 **Kibre Mengist** *var.* Adola. Oromīya, C Ethiopia 05°50′N 39°06′E
Kıbrıs *see* Cyprus
Kıbrıs/Kıbrıs Cumhuriyeti *see* Cyprus
81 E20 **Kibungo** *var.* Kibungu. SE Rwanda 02°09′S 30°30′E
Kibungu *see* Kibungo
113 N19 **Kičevo** SW FYR Macedonia 41°31′N 20°57′E
125 P13 **Kichmengskiy Gorodok** Vologodskaya Oblast', NW Russian Federation 60°00′N 45°52′E
30 J8 **Kickapoo River** ≋ Wisconsin, N USA

◆ Country
● Country Capital
◇ Dependent Territory
○ Dependent Territory Capital
◆ Administrative Regions
✈ International Airport
▲ Mountain
▲▲ Mountain Range
Volcano
≋ River
◎ Lake
▣ Reservoir

Kicking Horse Pass – Kizlyar

11 P16 Kicking Horse Pass pass Alberta/British Columbia, SW Canada
77 R9 Kidal Kidal, C Mali 18°22′N 01°21′E
77 Q8 Kidal ❖ region NE Mali
171 Q7 Kidapawan Mindanao, S Philippines 07°02′N 125°04′E
97 L20 Kidderminster C England, United Kingdom 52°23′N 02°14′W
76 I11 Kidira E Senegal 14°28′N 12°13′W
184 O11 Kidnappers, Cape headland North Island, New Zealand 41°13′S 175°15′E
100 J8 Kiel Schleswig-Holstein, N Germany 54°21′N 10°05′E
111 L15 Kielce Rus. Keltsy. Świętokrzyskie, C Poland 50°53′N 20°39′E
100 K7 Kieler Bucht bay N Germany
100 J7 Kieler Förde inlet N Germany
167 U13 Kiến Đức var. Đak Lap. Đắc Lắc, S Vietnam 11°59′N 107°30′E
79 N24 Kienge Katanga, SE Dem. Rep. Congo 10°33′S 27°33′E
100 Q12 Kietz Brandenburg, NE Germany 52°33′N 14°36′E
Kiev see Kyyiv
Kiev Reservoir see Kyyivs'ke Vodoskhovyshche
76 J10 Kiffa Assaba, S Mauritania 16°38′N 11°23′W
115 H19 Kifisiá Attikí, C Greece 38°04′N 23°49′E
115 F18 Kifisós ≈ C Greece
139 U5 Kifrī At Ta'mím, N Iraq 34°44′N 44°58′E
81 D20 Kigali ● (Rwanda) C Rwanda 01°59′S 30°02′E
81 E20 Kigali ✈ C Rwanda 01°43′S 30°01′E
137 P13 Kiğı Bingöl, E Turkey 39°19′N 40°20′E
81 E21 Kigoma Kigoma, W Tanzania 04°52′S 29°36′E
81 E21 Kigoma ❖ region W Tanzania
38 F10 Kīhei var. Kihei. Maui, Hawaii, USA, C Pacific Ocean 20°47′N 156°28′W
93 K17 Kihniö Pirkanmaa, W Finland 62°11′N 23°10′E
118 F6 Kihnu var. Kihnu Saar, Ger. Kühnö. island SW Estonia
Kihnu Saar see Kihnu
38 A8 Kii Landing Ni'ihau, Hawaii, USA, C Pacific Ocean 21°58′N 160°03′W
93 L14 Kiiminki Pohjois-Pohjanmaa, C Finland 65°05′N 25°47′E
164 J14 Kii-Nagashima var. Nagashima. Mie, Honshū, SW Japan 34°10′N 136°18′E
164 J14 Kii-sanchi ▲▲ Honshū, SW Japan
92 L11 Kiistala Lappi, N Finland 67°52′N 25°19′E
164 I15 Kii-suidō strait S Japan
165 V16 Kikai-shima island Nansei-shotō, SW Japan
112 M8 Kikinda Ger. Grosskikinda, Hung. Nagykikinda; prev. Velika Kikinda. Vojvodina, N Serbia 45°48′N 20°29′E
Kikládhes see Kykládes
165 Q5 Kikonai Hokkaidō, NE Japan 41°40′N 140°25′E
186 C8 Kikori Gulf, S Papua New Guinea 07°25′S 144°13′E
186 C8 Kikori ≈ W Papua New Guinea
165 O14 Kikuchi var. Kikuti. Kumamoto, Kyūshū, SW Japan 33°00′N 130°49′E
Kikuti see Kikuchi
127 N8 Kikvidze Volgogradskaya Oblast', SW Russian Federation 50°47′N 42°58′E
14 I10 Kikwissi, Lac ◎ Québec, SE Canada
79 I21 Kikwit Bandundu, W Dem. Rep. Congo 05°S 18°53′E
95 K15 Kil Värmland, C Sweden 59°30′N 13°20′E
94 N12 Kilafors Gävleborg, C Sweden 61°13′N 16°34′E
38 B8 Kilauea Kaua'i, Hawaii, USA, C Pacific Ocean 22°12′N 159°24′W
38 H12 Kilauea Caldera var. Kilauea Caldera. crater Hawai'i, USA, C Pacific Ocean
Kilauea Caldera see Kilauea Caldera
109 V4 Kilb Niederösterreich, C Austria 48°06′N 15°21′E
39 O12 Kilbuck Mountains ▲▲ Alaska, USA
163 Y12 Kilchu NE North Korea 40°58′N 129°22′E
97 F18 Kilcock Ir. Cill Choca. Kildare, E Ireland 53°25′N 06°40′W
183 V2 Kilcoy Queensland, E Australia 26°58′S 152°30′E
97 F18 Kildare Ir. Cill Dara. E Ireland 53°10′N 06°55′W
97 F18 Kildare Ir. Cill Dara. cultural region E Ireland
124 K2 Kil'din, Ostrov island NW Russian Federation
25 W7 Kilgore Texas, SW USA 32°23′N 94°52′W
Kilien Mountains see Qilian Shan
114 K9 Kilifarevo Veliko Tŭrnovo, N Bulgaria 43°00′N 25°36′E
81 K20 Kilifi Kilifi, SE Kenya 03°37′S 39°50′E
81 J21 Kilifi ❖ county SE Kenya
189 U9 Kili Island var. Kōle. island Ralik Chain, S Marshall Islands
149 V2 Kilik Pass pass Afghanistan/China
Kilimane see Quelimane
81 I21 Kilimanjaro ❖ region E Tanzania
81 I20 Kilimanjaro var. Uhuru Peak. ☒ NE Tanzania 03°01′S 37°14′E
Kilimbangara see Kolombangara
Kilinailau Islands see Tulun Islands
81 K23 Kilindoni Pwani, E Tanzania 07°56′S 39°40′E
118 H6 Kilingi-Nõmme Ger. Kurkund. Pärnumaa, SW Estonia 58°07′N 24°00′E
136 M17 Kilis Kilis, S Turkey 36°43′N 37°07′E
136 M16 Kilis ❖ province S Turkey
117 N12 Kiliya Rom. Chilia-Nouă. Odes'ka Oblast', SW Ukraine 45°30′N 29°16′E
97 B19 Kilkee Ir. Cill Chaoi. Clare, W Ireland 52°41′N 09°38′W
97 E19 Kilkenny Ir. Cill Chainnigh. Kilkenny, S Ireland 52°39′N 07°15′W
97 E19 Kilkenny Ir. Cill Chainnigh. cultural region S Ireland
97 B18 Kilkieran Bay Ir. Cuan Chill Chiaráin. bay W Ireland
114 G13 Kilkís Kentrikí Makedonía, N Greece 40°59′N 22°55′E
97 C15 Killala Bay Ir. Cuan Chill Ala. inlet NW Ireland
11 R15 Killam Alberta, SW Canada 52°45′N 111°46′W
183 U3 Killarney Queensland, E Australia 28°18′S 152°15′E
11 W17 Killarney Manitoba, S Canada 49°12′N 99°40′W
14 E11 Killarney Ontario, S Canada 45°58′N 81°27′W
97 B20 Killarney Ir. Cill Airne. Kerry, SW Ireland 52°03′N 09°30′W
28 K4 Killdeer North Dakota, N USA 47°21′N 102°45′W
28 J4 Killdeer Mountains ▲▲ North Dakota, N USA
45 V15 Killdeer River ≈ Trinidad, Trinidad and Tobago
25 S9 Killeen Texas, SW USA 31°07′N 97°44′W
39 P6 Killik River ≈ Alaska, USA
11 T7 Killinek Island island Nunavut, NE Canada
Killini see Kyllíni
115 C19 Killínis, Akrotírio headland S Greece 37°55′N 21°07′E
97 D15 Killybegs Ir. Na Cealla Beaga. NW Ireland 54°38′N 08°27′W
Kilmain see Quelimane
96 I13 Kilmarnock W Scotland, United Kingdom 55°37′N 04°30′W
21 X6 Kilmarnock Virginia, NE USA 37°42′N 76°22′W
125 S16 Kil'mez' Kirovskaya Oblast', NW Russian Federation 56°55′N 51°03′E
127 S2 Kil'mez' Udmurtskaya Respublika, NW Russian Federation 57°04′N 51°22′E
125 R16 Kil'mez' ≈ NW Russian Federation
67 V11 Kilombero ≈ S Tanzania
92 J10 Kilpisjärvi Lappi, N Finland 69°03′N 20°50′E
97 B19 Kilrush Ir. Cill Rois. Clare, W Ireland 52°39′N 09°29′W
79 O24 Kilwa Katanga, SE Dem. Rep. Congo 09°22′S 28°19′E
Kilwa see Kilwa Kivinje
81 J24 Kilwa Kivinje var. Kilwa. Lindi, SE Tanzania 08°45′S 39°21′E
81 J24 Kilwa Masoko Lindi, SE Tanzania 08°55′S 39°31′E
171 T13 Kilwo Pulau Seram, E Indonesia 03°36′S 130°48′E
114 P12 Kilyos İstanbul, NW Turkey 41°15′N 29°01′E
37 V8 Kim Colorado, C USA 37°12′N 103°22′W
145 O9 Kima prev. Kiyma. Akmola, C Kazakhstan 51°37′N 67°31′E
169 U7 Kimanis, Teluk bay Sabah, East Malaysia
182 H8 Kimba South Australia 33°09′S 136°26′E
28 I15 Kimball Nebraska, C USA 41°16′N 103°40′W
29 O11 Kimball South Dakota, N USA 43°45′N 98°57′W
79 I21 Kimbao Bandundu, SW Dem. Rep. Congo 05°27′S 17°40′E
186 F7 Kimbe New Britain, E Papua New Guinea 05°36′S 150°10′E
186 G7 Kimbe Bay inlet New Britain, E Papua New Guinea
11 P17 Kimberley British Columbia, SW Canada 49°40′N 115°58′W
83 H23 Kimberley Northern Cape, C South Africa 28°45′S 24°46′E
180 M4 Kimberley Plateau plateau Western Australia
33 P15 Kimberly Idaho, NW USA 42°31′N 114°21′W
163 Y12 Kimch'aek prev. Sŏngjin. E North Korea 40°42′N 129°13′E
Kimch'ŏn see Gimcheon
Kim Hae see Gimhae
Kími see Kými
93 K20 Kimito Swe. Kemiö. Varsinais-Suomi, SW Finland 60°10′N 22°45′E
9 R7 Kimmirut prev. Lake Harbour. Baffin Island, Nunavut, NE Canada 62°48′N 69°49′W
165 R4 Kimobetsu Hokkaidō, NE Japan 42°47′N 140°55′E
115 I21 Kímolos island Kykládes, Greece, Aegean Sea
115 I21 Kimolou Sífnou, Stenó strait Kykládes, Greece, Aegean Sea
126 L5 Kimovsk Tul'skaya Oblast', W Russian Federation 53°59′N 38°34′E
Kimpolung see Câmpulung Moldovenesc
124 K16 Kimry Tverskaya Oblast', W Russian Federation 56°52′N 37°21′E
79 H21 Kimvula Bas-Congo, SW Dem. Rep. Congo 05°44′S 15°58′E
169 U6 Kinabalu, Gunung ▲ East Malaysia 05°52′N 116°08′E
Kinabatangan see Kinabatangan, Sungai
169 V7 Kinabatangan, Sungai var. Kinabatangan. ≈ East Malaysia
115 L21 Kinaros island Kykládes, Greece, Aegean Sea
11 O15 Kinbasket Lake ⊠ British Columbia, SW Canada
96 I7 Kinbrace N Scotland, United Kingdom 58°16′N 02°59′W
14 E14 Kincardine Ontario, S Canada 44°11′N 81°38′W
96 K10 Kincardine cultural region E Scotland, United Kingdom
79 K21 Kinda Kasai-Occidental, SE Dem. Rep. Congo 04°48′S 21°50′E
79 M24 Kinda Katanga, SE Dem. Rep. Congo 09°20′S 25°06′E
166 L3 Kindat Sagaing, N Myanmar (Burma) 23°42′N 94°29′E
109 V6 Kindberg Steiermark, C Austria 47°31′N 15°27′E
22 H8 Kinder Louisiana, S USA 30°29′N 92°51′W
98 H13 Kinderdijk Zuid-Holland, SW Netherlands 51°52′N 04°37′E
97 M17 Kinder Scout ▲ C England, United Kingdom 53°25′N 01°52′W
11 S16 Kindersley Saskatchewan, S Canada 51°29′N 109°08′W
76 I14 Kindia Guinée-Maritime, SW Guinea 10°12′N 12°26′W
64 B11 Kindley Field air base E Bermuda
29 R6 Kindred North Dakota, N USA 46°39′N 97°01′W
79 N20 Kindu prev. Kindu-Port-Empain. Maniema, C Dem. Rep. Congo 02°57′S 25°54′E
Kindu-Port-Empain see Kindu
127 S6 Kinel' Samarskaya Oblast', W Russian Federation 53°14′N 50°40′E
125 N15 Kineshma Ivanovskaya Oblast', W Russian Federation 57°28′N 42°08′E
King see King William's Town
140 K10 King Abdul Aziz ✈ (Makkah) Makkah, W Saudi Arabia 21°44′N 39°08′E
Kingait see Cape Dorset
21 X6 King and Queen Court House Virginia, NE USA 37°40′N 76°49′W
King Charles Islands see Kong Karls Land
King Christian IX Land see Kong Christian IX Land
King Christian X Land see Kong Christian X Land
35 O11 King City California, W USA 36°12′N 121°09′W
27 R2 King City Missouri, C USA 40°03′N 94°31′W
38 M16 King Cove Alaska, USA 55°03′N 162°19′W
26 M10 Kingfisher Oklahoma, C USA 35°53′N 97°56′W
King Frederik VI Coast see Kong Frederik VI Kyst
King Frederik VIII Land see Kong Frederik VIII Land
65 B24 King George Bay bay West Falkland, Falkland Islands
194 G3 King George Island var. King George Land. island South Shetland Islands, Antarctica
12 I6 King George Islands island group Northwest Territories, C Canada
King George Land see King George Island
124 G13 Kingisepp Leningradskaya Oblast', NW Russian Federation 59°23′N 28°37′E
183 N14 King Island island Tasmania, SE Australia
10 J15 King Island island British Columbia, SW Canada
King Island see Kadan Kyun
Kingissepp see Kuressaare
141 Q7 King Khalid ✈ (Ar Riyāḍ) Ar Riyāḍ, C Saudi Arabia 25°00′N 46°40′E
35 S2 King Lear Peak ▲ Nevada, W USA 41°13′N 118°30′W
195 Y8 King Leopold and Queen Astrid Land physical region Antarctica
180 M4 King Leopold Ranges ▲▲ Western Australia
36 I11 Kingman Arizona, SW USA 35°12′N 114°02′W
26 M6 Kingman Kansas, C USA 37°39′N 98°07′W
192 L7 Kingman Reef ◇ US unincorporated territory C Pacific Ocean
79 N20 Kingombe Maniema, E Dem. Rep. Congo 02°37′S 26°39′E
182 F5 Kingoonya South Australia 30°56′S 135°20′E
194 J10 King Peninsula peninsula Antarctica
39 P13 King Salmon Alaska, USA 58°41′N 156°39′W
35 Q6 Kings Beach California, W USA 39°13′N 120°02′W
35 R11 Kingsburg California, W USA 36°30′N 119°33′W
182 I10 Kingscote South Australia 35°41′S 137°36′E
King's County see Offaly
194 H2 King Sejong South Korean research station Antarctica 61°57′S 58°23′W
183 T9 Kingsford Smith ✈ (Sydney) New South Wales, SE Australia 33°58′S 151°09′E
11 P17 Kingsgate British Columbia, SW Canada 48°58′N 116°09′W
23 W8 Kingsland Georgia, SE USA 30°48′N 81°41′W
29 S13 Kingsley Iowa, C USA 42°35′N 95°58′W
97 O19 King's Lynn var. Bishop's Lynn, Kings Lynn, Lynn, Lynn Regis. E England, United Kingdom 52°45′N 00°24′E
Kings Lynn see King's Lynn
21 Q10 Kings Mountain North Carolina, SE USA 35°15′N 81°20′W
180 K4 King Sound sound Western Australia
37 N2 Kings Peak ▲ Utah, W USA 40°43′N 110°27′W
21 O8 Kingsport Tennessee, S USA 36°32′N 82°33′W
35 R11 Kings River ≈ California, W USA
183 P17 Kingston Tasmania, SE Australia 42°57′S 147°18′E
14 K14 Kingston Ontario, SE Canada 44°14′N 76°30′W
44 K13 Kingston ● (Jamaica) E Jamaica 17°58′N 76°48′W
185 C22 Kingston Otago, South Island, New Zealand 45°20′S 168°45′E
19 P12 Kingston Massachusetts, NE USA 41°59′N 70°43′W
27 S3 Kingston Missouri, C USA 39°36′N 94°02′W
18 K12 Kingston New York, NE USA 41°55′N 74°00′W
31 S14 Kingston Ohio, N USA 39°28′N 82°54′W
19 O13 Kingston Rhode Island, NE USA 41°28′N 71°31′W
20 M9 Kingston Tennessee, S USA 35°52′N 84°30′W
35 W12 Kingston Peak ▲ California, W USA 35°43′N 115°54′W
182 J11 Kingston Southeast South Australia 36°51′S 139°53′E
97 N17 Kingston upon Hull var. Hull. E England, United Kingdom 53°45′N 00°20′W
97 N22 Kingston upon Thames SE England, United Kingdom 51°26′N 00°18′W
45 P14 Kingstown ● (Saint Vincent and the Grenadines) Saint Vincent, Saint Vincent and the Grenadines 13°09′N 61°14′W
Kingstown see Dún Laoghaire
21 T13 Kingstree South Carolina, SE USA 33°40′N 79°50′W
64 L8 Kings Trough undersea feature E Atlantic Ocean 22°00′W 43°48′N
14 C18 Kingsville Ontario, S Canada 42°03′N 82°43′W
25 S15 Kingsville Texas, SW USA 27°32′N 97°53′W
21 W6 King William Virginia, NE USA 37°42′N 77°03′W
9 N7 King William Island island Nunavut, N Canada
83 I25 King William's Town var. King, Kingwilliamstown. Eastern Cape, S South Africa 32°53′S 27°24′E
Kingwilliamstown see King William's Town
21 T3 Kingwood West Virginia, NE USA 39°27′N 79°43′W
136 C13 Kınık İzmir, W Turkey 39°05′N 27°25′E
79 G21 Kinkala Pool, S Congo 04°18′S 14°49′E
165 R10 Kinka-san headland Honshū, C Japan 38°17′N 141°34′E
184 M8 Kinleith Waikato, North Island, New Zealand 38°16′S 175°53′E
95 J19 Kinna Västra Götaland, S Sweden 57°32′N 12°42′E
96 L8 Kinnaird Head var. Kinnairds Head. headland NE Scotland, United Kingdom 58°39′N 03°22′W
Kinnairds Head see Kinnaird Head
95 K20 Kinnared Halland, S Sweden 57°01′N 13°04′E
92 L7 Kinnarodden headland N Norway 71°07′N 27°40′E
Kinneret, Yam see Tiberias, Lake
155 K24 Kinniyai Eastern Province, NE Sri Lanka 08°30′N 81°11′E
93 L16 Kinnula Keski-Suomi, C Finland 63°24′N 25°E
14 I8 Kinojévis ≈ Québec, SE Canada
164 I14 Kino-kawa ≈ Honshū, SW Japan
11 U11 Kinoosao Saskatchewan, C Canada 57°06′N 101°02′W
99 L17 Kinrooi Limburg, NE Belgium 51°09′N 05°48′E
96 J11 Kinross C Scotland, United Kingdom 56°14′N 03°27′W
96 J11 Kinross cultural region C Scotland, United Kingdom
97 C21 Kinsale Ir. Cionn tSáile. Cork, SW Ireland 51°42′N 08°32′W
95 D14 Kinsarvik Hordaland, S Norway 60°22′N 06°43′E
79 G21 Kinshasa prev. Léopoldville. ● Kinshasa, W Dem. Rep. Congo 04°21′S 15°16′E
79 G21 Kinshasa off. Ville de Kinshasa, var. Kinshasa City. ❖ region (Dem. Rep. Congo) SW Dem. Rep. Congo
79 G21 Kinshasa ✈ Kinshasa, SW Dem. Rep. Congo 04°23′S 15°30′E
Kinshasa City see Kinshasa
117 U9 Kins'ka ≈ SE Ukraine
26 K6 Kinsley Kansas, C USA 37°55′N 99°26′W
21 W10 Kinston North Carolina, SE USA 35°16′N 77°35′W
77 P15 Kintampo W Ghana 08°06′N 01°40′W
182 B1 Kintore, Mount ▲ South Australia 26°30′S 130°24′E
96 G13 Kintyre peninsula W Scotland, United Kingdom
96 G13 Kintyre, Mull of headland W Scotland, United Kingdom 55°16′N 05°46′W
166 M4 Kin-u Sagaing, C Myanmar (Burma) 22°47′N 95°36′E
12 G8 Kinushseo ≈ Ontario, C Canada
11 P13 Kinuso Alberta, W Canada 55°19′N 115°23′W
154 I13 Kinwat Mahārāshtra, C India 19°37′N 78°12′E
81 F16 Kinyeti ▲ S South Sudan 03°56′N 32°52′E
101 I17 Kinzig ≈ SW Germany
Kioga, Lake see Kyoga, Lake
26 M8 Kiowa Kansas, C USA 37°01′N 98°29′W
27 P12 Kiowa Oklahoma, C USA 34°43′N 95°54′W
Kiparissia see Kyparissía
14 H10 Kipawa, Lac ◎ Québec, SE Canada
81 G24 Kipengere Range ▲▲ SW Tanzania
81 E23 Kipili Rukwa, W Tanzania 07°30′S 30°39′E
81 K20 Kipini Tana River, SE Kenya 02°30′S 40°30′E
11 V16 Kipling Saskatchewan, S Canada 50°04′N 102°45′W
38 M13 Kipnuk Alaska, USA 59°56′N 164°02′W
97 F18 Kippure Ir. Cipiúr. ▲ E Ireland 53°10′N 06°22′W
79 N25 Kipushi Katanga, SE Dem. Rep. Congo 11°45′S 27°20′E
187 N10 Kirakira var. Kaokaona. Makira-Ulawa, SE Solomon Islands 10°28′S 161°54′E
155 K14 Kirandul var. Bailādila. Chhattīsgarh, C India 18°46′N 81°18′E
155 I21 Kīranūr Tamil Nādu, SE India 11°37′N 79°10′E
119 N21 Kiraw Rus. Kirovo. Homyel'skaya Voblasts', SE Belarus 51°30′N 29°25′E
119 M17 Kirawsk Rus. Kirovsk; prev. Startsy. Mahilyowskaya Voblasts', E Belarus 53°16′N 29°29′E
118 F5 Kirbla Läänemaa, W Estonia 58°45′N 23°57′E
25 Y9 Kirbyville Texas, SW USA 30°39′N 93°53′W
114 M12 Kircasalih Edirne, NW Turkey 41°24′N 26°48′E
109 W8 Kirchbach var. Kirchbach in Steiermark. Steiermark, SE Austria 46°55′N 15°40′E
Kirchbach in Steiermark see Kirchbach
108 H7 Kirchberg Sankt Gallen, NE Switzerland 47°24′N 09°03′E
109 S5 Kirchdorf an der Krems Oberösterreich, N Austria 47°55′N 14°08′E
Kirchheim see Kirchheim unter Teck
101 I22 Kirchheim unter Teck var. Kirchheim. Baden-Württemberg, SW Germany 48°39′N 09°28′E
139 T1 Kirdī Kawrāw, Qimmat var. Sar-i Kōrāwa. NE Iraq 37°08′N 44°39′E
Kirdzhali see Kardzhali
123 N13 Kirenga ≈ S Russian Federation
123 N12 Kirensk Irkutskaya Oblast', C Russian Federation 57°37′N 107°54′E
Kirghizia see Kyrgyzstan
145 S16 Kirghiz Range Rus. Kirgizskiy Khrebet; prev. Alexander Range. ▲▲ Kazakhstan/Kyrgyzstan
Kirghiz SSR see Kyrgyzstan
Kirghiz Steppe see Saryarka
Kirgizskaya SSR see Kyrgyzstan
Kirgizskiy Khrebet see Kirghiz Range
79 I19 Kiri Bandundu, W Dem. Rep. Congo 01°29′S 19°00′E
Kiriath-Arba see Hebron
191 R3 Kiribati off. Republic of Kiribati. ◆ republic C Pacific Ocean
Kiribati, Republic of see Kiribati
136 L17 Kırıkhan Hatay, S Turkey 36°30′N 36°20′E
136 I13 Kırıkkale Kırıkkale, C Turkey 39°50′N 33°31′E
136 C10 Kırıkkale ❖ province C Turkey
124 L13 Kirillov Vologodskaya Oblast', NW Russian Federation 59°52′N 38°24′E
Kirin see Jilin
81 I19 Kirinyaga ❖ county C Kenya
81 I18 Kirinyaga prev. Mount Kenya. ☒ Kirinyaga, C Kenya 0°02′S 37°19′E
124 H13 Kirishi var. Kirisi. Leningradskaya Oblast', NW Russian Federation 59°28′N 32°02′E
164 C16 Kirishima-yama ▲ Kyūshū, SW Japan 31°58′N 130°51′E
Kirisi see Kirishi
191 Y2 Kiritimati ✈ Kiritimati, E Kiribati 02°00′N 157°30′W
191 Y2 Kiritimati prev. Christmas Island. atoll Line Islands, E Kiribati
186 G9 Kiriwina Island Eng. Trobriand Island. island SE Papua New Guinea
186 G9 Kiriwina Islands var. Trobriand Islands. island group S Papua New Guinea
96 K12 Kirkcaldy E Scotland, United Kingdom 56°07′N 03°10′W
97 I14 Kirkcudbright S Scotland, United Kingdom 54°50′N 04°03′W
97 I14 Kirkcudbright cultural region S Scotland, United Kingdom
Kirkee see Khadki
95 I14 Kirkenær Hedmark, S Norway 60°27′N 12°04′E
92 M8 Kirkenes Fin. Kirkkoniemi. Finnmark, N Norway 69°43′N 30°02′E
92 J4 Kirkjubæjarklaustur Suðurland, S Iceland 63°46′N 18°03′W
Kirk-Kilissa see Kırklareli
Kirkkoniemi see Kirkenes
93 L20 Kirkkonummi Swe. Kyrkslätt. Uusimaa, S Finland 60°06′N 24°20′E
14 G7 Kirkland Lake Ontario, S Canada 48°10′N 80°02′W
136 C9 Kırklareli prev. Kirk-Kilissa. Kırklareli, NW Turkey 41°45′N 27°12′E
136 I13 Kırklareli ❖ province NW Turkey
185 F20 Kirkliston Range ▲▲ South Island, New Zealand
14 D10 Kirkpatrick Lake ◎ Ontario, S Canada
195 Q11 Kirkpatrick, Mount ▲ Antarctica 84°37′S 164°36′E
27 U2 Kirksville Missouri, C USA 40°12′N 92°35′W
139 S4 Kirkūk off. Muḥāfa at at Kirkūk; prev. At Ta'mīm. ❖ governorate NE Iraq
Kirkūk see Āltūn Kübrī
Kirkūk, Muḥāfa at at see Kirkūk
139 U7 Kir Kush Diyālá, E Iraq 33°42′N 45°15′E
96 K5 Kirkwall NE Scotland, United Kingdom 58°59′N 02°58′W
83 H25 Kirkwood Eastern Cape, S South Africa 33°23′S 25°19′E
27 X5 Kirkwood Missouri, C USA 38°35′N 90°24′W
Kirman see Kermān
Kir Moab/Kir of Moab see Al Karak
126 I5 Kirov Kaluzhskaya Oblast', W Russian Federation 54°02′N 34°17′E
125 R14 Kirov prev. Vyatka. Kirovskaya Oblast', NW Russian Federation 58°35′N 49°39′E
Kirov see Balpyk Bi/Ust'yevoye
Kirovabad see Gäncä
Kirovabad see Panj, Tajikistan
Kirovakan see Vanadzor
Kirov/Kirova see Kopbirlik, Kazakhstan
Kirovo see Kiraw, Belarus
Kirovo see Beshariq, Uzbekistan
125 R14 Kirovo-Chepetsk Kirovskaya Oblast', NW Russian Federation 58°33′N 50°06′E
Kirovogradskaya Oblast'/Kirovohrad see Kirovohrads'ka Oblast'
117 R7 Kirovohrad Rus. Kirovograd; prev. Kirovo, Yelizavetgrad, Zinov'yevsk. Kirovohrads'ka Oblast', C Ukraine 48°30′N 31°17′E
117 P7 Kirovohrads'ka Oblast' var. Kirovohrad, Rus. Kirovogradskaya Oblast'. ❖ province C Ukraine
Kirovo/Kirovograd see Kirovohrad
124 J4 Kirovsk Murmanskaya Oblast', NW Russian Federation 67°37′N 33°38′E
117 X7 Kirovs'k Luhans'ka Oblast', E Ukraine 48°40′N 38°39′E
Kirovsk see Kirawsk, Belarus
Kirovsk see Babadaýhan, Turkmenistan
122 E9 Kirovskaya Oblast' ❖ province NW Russian Federation
117 U13 Kirovs'ke Rus. Kirovskoye. Avtonomna Respublika Krym, S Ukraine 45°13′N 35°12′E
117 X8 Kirovs'ke Donets'ka Oblast', E Ukraine 48°12′N 38°20′E
Kirovskiy see Balpyk Bi
Kirovskiy see Ust'yevoye
Kirovskoye see Kyzyl-Adyr
Kirovskoye see Kirovs'ke
146 E11 Kirpili Ahal Welaýaty, C Turkmenistan 39°31′N 57°13′E
96 K10 Kirriemuir E Scotland, United Kingdom 56°38′N 03°01′W
125 S13 Kirs Kirovskaya Oblast', NW Russian Federation 59°22′N 52°20′E
127 N7 Kirsanov Tambovskaya Oblast', W Russian Federation 52°40′N 42°48′E
136 J14 Kırşehir anc. Justinianopolis. Kırşehir, C Turkey 39°09′N 34°08′E
136 I13 Kırşehir ❖ province C Turkey
149 P4 Kīrthar Range ▲▲ S Pakistan
37 P9 Kirtland New Mexico, SW USA 36°43′N 108°21′W
92 J11 Kiruna Lapp. Giron. Norrbotten, N Sweden 67°50′N 20°16′E
79 M18 Kirundu Orientale, NE Dem. Rep. Congo 0°45′S 25°28′E
Kirun/Kirun' see Jilong
26 L3 Kirwin Reservoir ⊠ Kansas, C USA
127 Q4 Kirya Chuvashskaya Respublika, W Russian Federation 55°04′N 46°50′E
138 G8 Kiryat Shmona prev. Qiryat Shemona. Northern, N Israel 33°13′N 35°35′E
95 M18 Kisa Östergötland, S Sweden 58°N 15°39′E
165 P9 Kisakata Akita, Honshū, C Japan 39°12′N 139°55′E
Kisalföld see Little Alföld
79 L18 Kisangani prev. Stanleyville. Orientale, NE Dem. Rep. Congo 0°30′N 25°14′E
39 N12 Kisaralik River ≈ Alaska, USA
165 O14 Kisarazu Chiba, Honshū, S Japan 35°23′N 139°57′E
111 I22 Kisbér Komárom-Esztergom, NW Hungary 47°30′N 18°00′E
11 V17 Kisbey Saskatchewan, S Canada 49°41′N 102°39′W
122 J13 Kiselevsk Kemerovskaya Oblast', S Russian Federation 54°00′N 86°38′E
153 R13 Kishanganj Bihār, NE India 26°06′N 87°57′E
152 G12 Kishangarh Rājasthān, N India 26°33′N 74°52′E
Kishegyes see Mali Iđoš
77 S15 Kishi Oyo, W Nigeria 09°01′N 03°53′E
Kishinev see Chişinău
Kishiözen see Saryozen
164 I14 Kishiwada var. Kisiwada. Ōsaka, Honshū, SW Japan 34°28′N 135°22′E
143 P14 Kīsh, Jazīreh-ye var. Qey. island S Iran
145 R7 Kishkenekol' prev. Kzyltu, Kaz. Qyzyltū. Kokshetau, N Kazakhstan 53°39′N 72°22′E
138 G9 Kishon, Nahal prev. Naḥal Qishon. ≈ N Israel
152 I6 Kishtwār Jammu and Kashmir, NW India 33°20′N 75°49′E
81 H19 Kisii Kisii, SW Kenya 0°40′S 34°47′E
81 H19 Kisii ❖ county W Kenya
81 J23 Kisiju Pwani, E Tanzania 07°25′S 39°20′E
Kisiwada see Kishiwada
Kisjenő see Chişineu-Criş
38 E17 Kiska Island island Aleutian Islands, Alaska, USA
Kiskapus see Copşa Mică
111 M22 Kiskőrei-víztároló ⊠ E Hungary
Kis-Küküllo see Târnava Mică
111 L24 Kiskunfélegyháza var. Félegyháza. Bács-Kiskun, C Hungary 46°42′N 19°52′E
111 K25 Kiskunhalas var. Halas. Bács-Kiskun, S Hungary 46°26′N 19°29′E
111 K24 Kiskunmajsa Bács-Kiskun, S Hungary 46°30′N 19°46′E
127 N15 Kislovodsk Stavropol'skiy Kray, SW Russian Federation 43°55′N 42°45′E
81 L18 Kismaayo var. Chisimayu, Kismayu, It. Chisimaio. Jubbada Hoose, S Somalia 0°05′S 42°35′E
Kismayu see Kismaayo
164 M13 Kiso-sanmyaku ▲▲ Honshū, S Japan
115 H24 Kíssamos prev. Kastélli. Kriti, Greece, E Mediterranean Sea 35°30′N 23°39′E
Kisseraing see Kanmaw Kyun
76 K14 Kissidougou Guinée-Forestière, S Guinea 09°15′N 10°08′W
23 X12 Kissimmee Florida, SE USA 28°17′N 81°24′W
23 X12 Kissimmee, Lake ◎ Florida, SE USA
23 X13 Kissimmee River ≈ Florida, SE USA
11 V13 Kississing Lake ◎ Manitoba, C Canada
111 L24 Kistelek Csongrád, SE Hungary 46°27′N 19°58′E
Kistna see Krishna
111 M23 Kisújszállás Jász-Nagykun-Szolnok, E Hungary 47°14′N 20°45′E
164 G12 Kisuki var. Unnan. Shimane, Honshū, SW Japan 35°25′N 133°15′E
81 H18 Kisumu prev. Port Florence. Kisumu, W Kenya 0°02′N 34°42′E
81 H18 Kisumu ❖ county W Kenya
Kisutzaneustadtl see Kysucké Nové Mesto
111 O20 Kisvárda Ger. Kleinwardein. Szabolcs-Szatmár-Bereg, E Hungary 48°13′N 22°03′E
81 J24 Kiswere Lindi, SE Tanzania 09°24′S 39°37′E
Kiszucaújhely see Kysucké Nové Mesto
76 K12 Kita Kayes, W Mali 13°00′N 09°28′W
Kitaa ❖ province W Greenland
Kita-Akita see Takanosu
Kitab see Kitob
165 Q4 Kitahiyama Hokkaidō, NE Japan 42°25′N 139°55′E
165 P12 Kitaibaraki Ibaraki, Honshū, S Japan 36°46′N 140°45′E
165 X16 Kita-Iō-jima Eng. San Alessandro. island SE Japan
165 Q9 Kitakami Iwate, Honshū, C Japan 39°18′N 141°05′E
165 P11 Kitakata Fukushima, Honshū, C Japan 37°38′N 139°52′E
164 D13 Kitakyūshū var. Kitakyûsyû. Fukuoka, Kyūshū, SW Japan 33°51′N 130°49′E
Kitakyûsyû see Kitakyūshū
81 H18 Kitale Trans Nzoia, W Kenya 01°01′N 35°01′E
165 U3 Kitami Hokkaidō, NE Japan 43°52′N 143°51′E
165 T2 Kitami-sanchi ▲▲ Hokkaidō, NE Japan
37 W5 Kit Carson Colorado, C USA 38°45′N 102°47′W
180 M12 Kitchener Western Australia 31°03′S 124°00′E
14 F16 Kitchener Ontario, S Canada 43°28′N 80°27′W
93 O17 Kitee Pohjois-Karjala, SE Finland 62°06′N 30°09′E
81 G16 Kitgum N Uganda 03°17′N 32°54′E
Kithareng see Kanmaw Kyun
Kíthira see Kýthira
Kíthnos see Kýthnos
8 L8 Kitikmeot ❖ cultural region Nunavut, N Canada
10 J13 Kitimat British Columbia, SW Canada 54°05′N 128°38′W
92 L11 Kitinen ≈ N Finland
147 N12 Kitob Rus. Kitab. Qashqadaryo Viloyati, S Uzbekistan 39°06′N 66°47′E
116 K7 Kitsman' Ger. Kotzman, Rom. Cozmeni, Rus. Kitsman. Chernivets'ka Oblast', W Ukraine 48°30′N 25°50′E
164 E14 Kitsuki var. Kituki. Ōita, Kyūshū, SW Japan 33°24′N 131°36′E
18 C14 Kittanning Pennsylvania, NE USA 40°48′N 79°28′W
19 P10 Kittery Maine, NE USA 43°05′N 70°44′W
92 L11 Kittilä Lappi, N Finland 67°39′N 24°53′E
109 Z4 Kittsee Burgenland, E Austria 48°06′N 17°03′E
81 J19 Kitui Kitui, S Kenya 01°25′S 38°00′E
81 J20 Kitui ❖ county S Kenya
Kituki see Kitsuki
81 G22 Kitunda Tabora, C Tanzania 06°47′S 33°13′E
10 K13 Kitwanga British Columbia, SW Canada 55°07′N 128°03′W
82 J13 Kitwe var. Kitwe-Nkana. Copperbelt, C Zambia 12°48′S 28°13′E
Kitwe-Nkana see Kitwe
109 O7 Kitzbühel Tirol, W Austria 47°27′N 12°23′E
109 O7 Kitzbüheler Alpen ▲▲ W Austria
101 J19 Kitzingen Bayern, SE Germany 49°45′N 10°11′E
153 Q14 Kiul Bihār, NE India 25°10′N 86°06′E
186 A7 Kiunga Western, SW Papua New Guinea 06°10′S 141°15′E
93 M16 Kiuruvesi Pohjois-Savo, C Finland 63°38′N 26°40′E
38 M7 Kivalina Alaska, USA 67°44′N 164°32′W
9 O10 Kivalliq ❖ cultural region Nunavut, N Canada
92 L13 Kivalo ridge C Finland
116 J3 Kivertsi Pol. Kiwerce, Rus. Kivertsy. Volyns'ka Oblast', NW Ukraine 50°50′N 25°31′E
Kivertsy see Kivertsi
93 L16 Kivijärvi Keski-Suomi, C Finland 63°09′N 25°04′E
95 L23 Kivik Skåne, S Sweden 55°40′N 14°15′E
118 J3 Kiviõli Ida-Virumaa, NE Estonia 59°20′N 27°00′E
Kivu, Lac see Kivu, Lake
67 U10 Kivu, Lake Fr. Lac Kivu. ◎ Rwanda/Dem. Rep. Congo
186 C9 Kiwai Island island SW Papua New Guinea
39 N8 Kiwalik Alaska, USA 66°01′N 161°50′W
Kiwerce see Kivertsi
Kiyev see Kyyiv
145 R10 Kiyevka Karaganda, C Kazakhstan 50°15′N 71°33′E
Kiyevskaya Oblast' see Kyyivs'ka Oblast'
Kiyevskoye Vodokhranilishche see Kyyivs'ke Vodoskhovyshche
136 D10 Kıyıköy Kırklareli, NW Turkey 41°37′N 28°07′E
Kiyma see Kima
125 V13 Kizel Permskiy Kray, NW Russian Federation 58°59′N 57°37′E
125 O12 Kizema var. Kizëma. Arkhangel'skaya Oblast', NW Russian Federation 61°06′N 44°51′E
Kizëma see Kizema
Kizilagach see Elhovo
136 H12 Kızılcahamam Ankara, N Turkey 40°28′N 32°37′E
136 J10 Kızıl Irmak ≈ C Turkey
Kizil Kum see Kyzyl Kum
137 P16 Kızıltepe Mardin, SE Turkey 37°12′N 40°36′E
Ki Zil Uzen see Qezel Owzan, Rūd-e
127 Q16 Kizilyurt Respublika Dagestan, SW Russian Federation 43°13′N 46°54′E
127 Q15 Kizlyar Respublika Dagestan, SW Russian Federation 43°51′N 46°39′E

◆ Country ● Country Capital ◇ Dependent Territory ○ Dependent Territory Capital ❖ Administrative Regions ✈ International Airport ▲ Mountain ▲▲ Mountain Range ☒ Volcano ≈ River ◎ Lake ⊠ Reservoir

KIZNER – KONKOURÉ

127 *S3* **Kizner** Udmurtskaya Respublika, NW Russian Federation *56°19′N 51°37′E*

Kizyl-Arvat *see* Serdar

Kizyl-Atrek *see* Etrek

Kizyl-Kaya *see* Gyzylgaýa

Kizyl-Su *see* Gyzylsuw

95 *H16* **Kjerkøy** *island* S Norway

Kjølen *see* Kölen

92 *L7* **Kjøllefjord** Finnmark, N Norway *70°55′N 27°19′E*

92 *H11* **Kjøpsvik** *Lapp.* Gásluokta. Nordland, C Norway *68°06′N 16°21′E*

169 *N12* **Klabat, Teluk** *bay* Pulau Bangka, W Indonesia

112 *I12* **Kladanj** ◈ Fedederacija Bosna I Hercegovina, E Bosnia and Herzegovina

171 *X16* **Kladar** Papua, E Indonesia *08°14′S 137°46′E*

111 *C16* **Kladno** Středočeský, NW Czech Republic *50°10′N 14°05′E*

112 *P11* **Kladovo** Serbia, E Serbia *44°37′N 22°36′E*

167 *P12* **Klaeng** Rayong, S Thailand *12°48′N 101°41′E*

109 *T9* **Klagenfurt** *Slvn.* Celovec. Kärnten, S Austria *46°38′N 14°20′E*

118 *B11* **Klaipėda** *Ger.* Memel. Klaipeda, NW Lithuania *55°42′N 21°09′E*

118 *C11* **Klaipėda** ◈ *province* W Lithuania

Klaksvig *see* Klaksvík

95 *B18* **Klaksvík** *Dan.* Klaksvig. Faroe Islands *62°13′N 06°34′W*

34 *L2* **Klamath** California, W USA *41°31′N 124°02′W*

32 *H16* **Klamath Falls** Oregon, NW USA *42°14′N 121°47′W*

34 *M1* **Klamath Mountains** ▲▲ California/Oregon, W USA

34 *L2* **Klamath River** ≈ California/Oregon, W USA

168 *K9* **Klang** *var.* Kelang; *prev.* Port Swettenham. Selangor, Peninsular Malaysia *03°02′N 101°27′E*

94 *J13* **Klarälven** ≈ Norway/Sweden

111 *B15* **Klášterec nad Ohří** *Ger.* Klösterle an der Eger. Ústecky Kraj, NW Czech Republic *50°24′N 13°10′E*

111 *B18* **Klatovy** *Ger.* Klattau. Plzeňský Kraj, W Czech Republic *49°24′N 13°16′E*

Klattau *see* Klatovy

Klausenburg *see* Cluj-Napoca

39 *Y14* **Klawock** Prince of Wales Island, Alaska, USA *55°33′N 133°06′W*

98 *P8* **Klazienaveen** Drenthe, NE Netherlands *52°43′N 07°E*

Kleck *see* Klyetsk

110 *H11* **Kłecko** Weilkopolskie, C Poland *52°37′N 17°27′E*

110 *I11* **Kleczew** Wielkopolskie, C Poland *52°22′N 18°12′E*

10 *L15* **Kleena Kleene** British Columbia, SW Canada *51°55′N 124°54′W*

83 *D20* **Klein Aub** Hardap, C Namibia *23°48′S 16°39′E*

Kleine Donau *see* Mosoni-Duna

101 *O14* **Kleine Elster** ≈ E Germany

Kleine Kokel *see* Târnava Mică

99 *I16* **Kleine Nete** ≈ N Belgium

Kleines Ungarisches Tiefland *see* Little Alföld

83 *E22* **Klein Karas** Karas, S Namibia *27°36′S 18°05′E*

Kleinkopisch *see* Copşa Mică

Klein-Marien *see* Väike-Maarja

Kleinschlatten *see* Zlatna

83 *D23* **Kleinsee** Northern Cape, W South Africa *29°43′S 17°03′E*

Kleinwardein *see* Kisvárda

115 *C16* **Kleisoúra** Ípeiros, W Greece *39°21′N 20°52′E*

95 *C17* **Klepp** Rogaland, S Norway *58°46′N 05°39′E*

83 *I22* **Klerksdorp** North-West, N South Africa *26°52′S 26°39′E*

126 *I5* **Kletnya** Bryanskaya Oblast', W Russian Federation *53°25′N 32°58′E*

Kletsk *see* Klyetsk

101 *D14* **Kleve** *Eng.* Cleves, *Fr.* Clèves; *prev.* Cleve. Nordrhein-Westfalen, W Germany *51°47′N 06°11′E*

113 *J16* **Kličevo** C Montenegro *42°45′N 18°58′E*

119 *M16* **Klichaw** *Rus.* Klichev. Mahilyowskaya Voblasts', E Belarus *53°29′N 29°21′E*

Klichev *see* Klichaw

119 *Q16* **Klimavichy** *Rus.* Klimovichi. Mahilyowskaya Voblasts', E Belarus *53°37′N 31°58′E*

114 *M7* **Kliment** Shumen, NE Bulgaria *43°37′N 27°00′E*

Klimovichi *see* Klimavichy

93 *G14* **Klimpfjäll** Västerbotten, N Sweden *65°05′N 14°50′E*

126 *K3* **Klin** Moskovskaya Oblast', W Russian Federation *56°19′N 36°45′E*

Klina *see* Klinë

113 *M16* **Klinë** *Serb.* Klina. W Kosovo *42°38′N 20°35′E*

111 *B15* **Klínovec** *Ger.* Keilberg. ▲ NW Czech Republic *50°23′N 12°57′E*

95 *P19* **Klintehamn** Gotland, SE Sweden *57°22′N 18°15′E*

127 *R8* **Klintsovka** Saratovskaya Oblast', W Russian Federation *51°42′N 49°17′E*

126 *H6* **Klintsy** Bryanskaya Oblast', W Russian Federation *52°46′N 32°21′E*

95 *K22* **Klippan** Skåne, S Sweden *56°08′N 13°10′E*

92 *G13* **Klippen** Västerbotten, N Sweden *65°50′N 15°07′E*

121 *P2* **Klírou** W Cyprus *35°01′N 33°11′E*

114 *I9* **Klisura** Plovdiv, C Bulgaria *42°40′N 24°28′E*

95 *F20* **Klitmøller** Midtjylland, NW Denmark *57°01′N 08°29′E*

112 *F11* **Ključ** Federacija Bosne I Hercegovine, NW Bosnia and Herzegovina *44°32′N 16°46′E*

111 *J14* **Kłobuck** Śląskie, S Poland *50°56′N 18°55′E*

110 *J11* **Kłodawa** Wielkopolskie, C Poland *52°14′N 18°55′E*

111 *G16* **Kłodzko** *Ger.* Glatz. Dolnośląskie, SW Poland *50°27′N 16°37′E*

95 *I14* **Kløfta** Akershus, S Norway *60°04′N 11°06′E*

112 *P12* **Klokočevac** Serbia, E Serbia *44°19′N 22°11′E*

118 *G3* **Klooga** *Ger.* Lodensee. Harjumaa, NW Estonia *59°18′N 24°10′E*

99 *F15* **Kloosterzande** Zeeland, SW Netherlands *51°22′N 04°01′E*

113 *L19* **Klos** *var.* Klosi. Dibër, C Albania *41°30′N 20°07′E*

Klosi *see* Klos

Klösterle an der Eger *see* Klášterec nad Ohří

109 *X3* **Klosterneuburg** Niederösterreich, NE Austria *48°19′N 16°20′E*

108 *J9* **Klosters** Graubünden, SE Switzerland *46°54′N 09°52′E*

108 *G7* **Kloten** Zürich, N Switzerland *47°27′N 08°35′E*

108 *G7* **Kloten** ✕ (Zürich) Zürich, N Switzerland *47°25′N 08°36′E*

100 *K12* **Klötze** Sachsen-Anhalt, C Germany *52°37′N 11°09′E*

12 *K3* **Klotz, Lac** ◎ Québec, NE Canada

101 *O15* **Klotzsche** ✕ (Dresden) Sachsen, E Germany *51°06′N 13°44′E*

10 *H7* **Kluane Lake** ◎ Yukon, W Canada

Kluang *see* Keluang

111 *I14* **Kluczbork** *Ger.* Kreuzburg, Kreuzburg in Oberschlesien. Opolskie, S Poland *50°59′N 18°13′E*

39 *W12* **Klukwan** Alaska, USA *59°24′N 135°49′W*

Klyastitsy *see* Klyastsitsy

118 *L11* **Klyastsitsy** *Rus.* Klyastitsy. Vitsyebskaya Voblasts', N Belarus *55°53′N 28°36′E*

127 *T5* **Klyavlino** Samarskaya Oblast', W Russian Federation *54°21′N 52°12′E*

84 *K9* **Klyaz'in** ≈ W Russian Federation

127 *N3* **Klyaz'ma** ≈ W Russian Federation

119 *J17* **Klyetsk** *Pol.* Kleck, *Rus.* Kletsk. Minskaya Voblasts', SW Belarus *53°04′N 26°38′E*

147 *S8* **Klyuchevka** Talasskaya Oblast', NW Kyrgyzstan *42°34′N 71°45′E*

123 *V10* **Klyuchevskaya Sopka, Vulkan** ⋀ E Russian Federation *56°03′N 160°38′E*

95 *D17* **Knaben** Vest-Agder, S Norway *58°46′N 07°04′E*

95 *K21* **Knäred** Halland, S Sweden *56°30′N 13°21′E*

97 *M16* **Knaresborough** N England, United Kingdom *54°01′N 01°35′W*

114 *H8* **Knezha** Vratsa, NW Bulgaria *43°29′N 24°04′E*

25 *O9* **Knickerbocker** Texas, SW USA *31°18′N 100°35′W*

28 *K5* **Knife River** ≈ North Dakota, N USA

10 *K16* **Knight Inlet** *inlet* British Columbia, W Canada

39 *S12* **Knight Island** *island* Alaska, USA

97 *K20* **Knighton** E Wales, United Kingdom *52°20′N 03°01′W*

35 *O7* **Knights Landing** California, W USA *38°47′N 121°43′W*

112 *E13* **Knin** Šibenik-Knin, S Croatia *44°03′N 16°12′E*

25 *Q12* **Knippal** Texas, SW USA

109 *U7* **Knittelfeld** Steiermark, C Austria *47°14′N 14°50′E*

95 *O15* **Knivsta** Uppsala, C Sweden *59°43′N 17°49′E*

113 *P14* **Knjaževac** Serbia, E Serbia *43°34′N 22°16′E*

27 *S4* **Knob Noster** Missouri, C USA *38°47′N 93°33′W*

99 *D15* **Knokke-Heist** West-Vlaanderen, NW Belgium *51°21′N 03°19′E*

95 *H20* **Knøsen** *hill* N Denmark

Knosós *see* Knossos

115 *J25* **Knossos** *Gk.* Knosós. *prehistoric site* Kríti, Greece, E Mediterranean Sea

25 *N7* **Knott** Texas, SW USA *32°21′N 101°35′W*

194 *K5* **Knowles, Cape** *headland* Antarctica *71°45′S 60°20′W*

31 *O11* **Knox** Indiana, N USA *41°17′N 86°37′W*

29 *O3* **Knox** North Dakota, N USA *48°19′N 99°43′W*

18 *C13* **Knox** Pennsylvania, NE USA *41°13′N 79°33′W*

189 *X8* **Knox Atoll** *var.* Ņadikdik, Narikrik. *atoll* Ratak Chain, SE Marshall Islands

10 *H13* **Knox, Cape** *headland* Graham Island, British Columbia, SW Canada *54°05′N 133°02′W*

25 *P5* **Knox City** Texas, SW USA *33°25′N 99°49′W*

195 *Y11* **Knox Coast** *physical region* Antarctica

31 *T12* **Knox Lake** ⊞ Ohio, N USA

23 *T5* **Knoxville** Georgia, SE USA *32°44′N 83°58′W*

30 *K12* **Knoxville** Illinois, N USA *40°54′N 90°16′W*

29 *W15* **Knoxville** Iowa, C USA *41°19′N 93°06′W*

21 *N9* **Knoxville** Tennessee, S USA *35°58′N 83°55′W*

197 *P11* **Knud Rasmussen Land** *physical region* N Greenland

Knüll *see* Knüllgebirge

101 *I16* **Knüllgebirge** *var.* Knüll. ▲▲ C Germany

124 *I5* **Knyazhegubskoye Vodokhranilishche** ◎ NW Russian Federation

Knyazhevo *see* Sredishte

Knyazhitsy *see* Knyazhytsy

119 *O15* **Knyazhytsy** *Rus.* Knyazhitsy. Mahilyowskaya Voblasts', E Belarus *54°10′N 30°28′E*

83 *G26* **Knysna** Western Cape, SW South Africa *34°03′S 23°03′E*

Koartac *see* Quaqtaq

169 *N13* **Koba** Pulau Bangka, W Indonesia *02°30′S 106°26′E*

Kobanî *see* 'Ayn al 'Arab

164 *D16* **Kobayashi** *var.* Kobayasi. Miyazaki, Kyūshū, SW Japan *32°00′N 130°58′E*

Kobayasi *see* Kobayashi

144 *I10* **Kobda** *prev.* Khobda, Novoalekseyevka. Aktyubinsk, W Kazakhstan *50°09′N 55°39′E*

144 *H9* **Kobda** *Kaz.* Ülkenqobda; *prev.* Bol'shaya Khobda. ≈ Kazakhstan/Russian Federation

Kobdo *see* Hovd

164 *I13* **Kōbe** Hyōgo, Honshū, SW Japan *34°40′N 135°10′E*

Kobelyaki *see* Kobelyaky

117 *T6* **Kobelyaky** *Rus.* Kobelyaki. Poltavs'ka Oblast', NE Ukraine *49°10′N 34°13′E*

95 *J22* **København** *Eng.* Copenhagen; *anc.* Hafnia. ● (Denmark) Sjælland, København, E Denmark *55°43′N 12°34′E*

76 *K10* **Kobenni** Hodh el Gharbi, S Mauritania *15°58′N 09°24′W*

171 *T13* **Kobi** Pulau Seram, E Indonesia *02°56′S 129°53′E*

101 *F17* **Koblenz** *prev.* Coblenz, *Fr.* Coblence; *anc.* Confluentes. Rheinland-Pfalz, W Germany *50°21′N 07°36′E*

108 *F6* **Koblenz** Aargau, N Switzerland *47°34′N 08°16′E*

Kobrin *see* Kobryn

171 *V15* **Kobroor, Pulau** *island* Kepulauan Aru, E Indonesia

119 *G19* **Kobryn** *Rus.* Kobrin. Brestskaya Voblasts', SW Belarus *52°13′N 24°21′E*

39 *O7* **Kobuk** Alaska, USA *66°54′N 156°52′W*

39 *O7* **Kobuk River** ≈ Alaska, USA

137 *Q10* **Kobuleti** *prev.* K'obulet'i. W Georgia *41°47′N 41°47′E*

K'obulet'i *see* Kobuleti

123 *P10* **Kobyay** Respublika Sakha (Yakutiya), NE Russian Federation *63°36′N 126°33′E*

136 *E11* **Kocaeli** ◈ *province* NW Turkey

113 *P18* **Kočani** NE FYR Macedonia *41°55′N 22°25′E*

112 *K12* **Koceljevo** Serbia, W Serbia *44°28′N 19°49′E*

109 *U12* **Kočevje** *Ger.* Gottschee. S Slovenia *45°41′N 14°48′E*

153 *T12* **Koch Bihār** West Bengal, NE India *26°19′N 89°26′E*

Kochchi *see* Cochin/Kochi

122 *M9* **Kochechum** ≈ N Russian Federation

101 *I20* **Kocher** ≈ SW Germany

125 *T13* **Kochevo** Komi-Permyatskiy Okrug, NW Russian Federation *59°37′N 54°16′E*

164 *G14* **Kōchi** *var.* Kôti. Kōchi, Shikoku, SW Japan *33°31′N 133°30′E*

164 *G14* **Kōchi** *off.* Kōchi-ken, *var.* Kôti. ◈ *prefecture* Shikoku, SW Japan

Kōchi-ken *see* Kōchi

Kochiu *see* Gejiu

Kochkor *see* Kochkorka

147 *V8* **Kochkorka** *Kir.* Kochkor. Narynskaya Oblast', C Kyrgyzstan *42°09′N 75°42′E*

125 *V5* **Kochmes** Respublika Komi, NW Russian Federation *66°10′N 60°46′E*

127 *P15* **Kochubey** Respublika Dagestan, SW Russian Federation *44°25′N 46°33′E*

115 *I17* **Kochýlas** ▲ Skńyros, Vóreies Sporádes, Greece, Aegean Sea *38°50′N 24°35′E*

110 *O13* **Kock** Lubelskie, E Poland *51°39′N 22°26′E*

81 *J19* **Kodacho** *spring/well* S Kenya *01°52′S 39°22′E*

155 *K24* **Koddiyar Bay** *bay* NE Sri Lanka

39 *Q14* **Kodiak** Kodiak Island, Alaska, USA *57°47′N 152°24′W*

39 *Q14* **Kodiak Island** *island* Alaska, USA

154 *B12* **Kodīnār** Gujarāt, W India *20°44′N 70°46′E*

124 *M9* **Kodino** Arkhangel'skaya Oblast', NW Russian Federation *63°36′N 39°54′E*

122 *M12* **Kodinsk** Krasnoyarskiy Kray, C Russian Federation *58°37′N 99°18′E*

80 *F12* **Kodok** Upper Nile, NE South Sudan *09°51′N 32°07′E*

117 *N8* **Kodyma** Odes'ka Oblast', SW Ukraine *48°05′N 29°09′E*

Koedoes *see* Kudus

99 *B17* **Koekelare** West-Vlaanderen, W Belgium *51°07′N 02°58′E*

Koeln *see* Köln

Koepang *see* Kupang

Ko-erh-mu *see* Golmud

99 *J17* **Koersel** Limburg, NE Belgium *51°04′N 05°17′E*

83 *E21* **Koës** Karas, SE Namibia *25°59′S 19°08′E*

Koetai *see* Mahakam, Sungai

Koetaradja *see* Banda Aceh

36 *I14* **Kofa Mountains** ▲▲ Arizona, SW USA

171 *Y15* **Kofarau** Papua, E Indonesia *07°29′S 140°28′E*

147 *P13* **Kofarnihon** *Rus.* Kofarnikhon; *prev.* Ordzhonikidzeabad, *Taj.* Orjonikidzeobod, Yangi-Bazar. W Tajikistan *38°32′N 68°56′E*

147 *P14* **Kofarnihon** *Rus.* Kafirnigan. ≈ SW Tajikistan

Kofarnikhon *see* Kofarnihon

114 *M11* **Kofçaz** Kırklareli, NW Turkey *41°58′N 27°12′E*

115 *J25* **Kófinas** ▲ Kríti, Greece, E Mediterranean Sea *34°58′N 25°03′E*

121 *P3* **Kofínou** *var.* Kophinou. S Cyprus *34°49′N 33°24′E*

109 *V8* **Köflach** Steiermark, SE Austria *47°04′N 15°04′E*

77 *Q17* **Koforidua** SE Ghana *06°01′N 00°12′W*

164 *H12* **Kōfu** Tottori, Honshū, SW Japan *35°16′N 133°31′E*

164 *M13* **Kōfu** *var.* Kôhu. Yamanashi, Honshū, S Japan *35°41′N 138°33′E*

81 *F22* **Koga** Tabora, C Tanzania *06°08′S 32°20′E*

Kogălniceanu *see* Mihail Kogălniceanu

13 *P6* **Kogaluk** ≈ Newfoundland and Labrador, E Canada

12 *J4* **Kogaluk, Rivière** ≈ Québec, NE Canada

145 *V14* **Kogaly** *Kaz.* Qoghaly; *prev.* Kugaly. Almaty, SE Kazakhstan *44°30′N 78°40′E*

122 *I10* **Kogalym** Khanty-Mansiyskiy Avtonomnyy Okrug-Yugra, C Russian Federation *62°13′N 74°34′E*

95 *J23* **Køge** Sjælland, E Denmark *55°28′N 12°12′E*

95 *J23* **Køge Bugt** *bay* E Denmark

77 *U16* **Kogi** ◈ *state* C Nigeria

146 *L11* **Kogon** *Rus.* Kagan. Buxoro Viloyati, C Uzbekistan *39°47′N 64°29′E*

Kŏgŭm-do *see* Geogeum-do

Kőhalom *see* Rupea

149 *T6* **Kohāt** Khyber Pakhtunkhwa, NW Pakistan *33°37′N 71°30′E*

142 *L10* **Kohgīlūyeh va Bowyer Aḥmad** *off.* Ostān-e Kohgīlūyeh va Bowyer Aḥmad, *var.* Boyer Ahmadī va Kohkīlūyeh. ◈ *province* SW Iran

Kohgīlūyeh va Bowyer Aḥmad, Ostān-e *see* Kohgīlūyeh va Bowyer Aḥmad

118 *G4* **Kohila** *Ger.* Koil. Raplamaa, NW Estonia *59°09′N 24°45′E*

153 *X13* **Kohīma** *state capital* Nāgāland, E India *25°40′N 94°08′E*

Koh I Noh *see* Büyükağrı Dağı

Kohsān *see* Kūhestān

118 *J3* **Kohtla-Järve** Ida-Virumaa, NE Estonia *59°22′N 27°21′E*

Kôhu *see* Kōfu

Kohyl'nyk *see* Cogîlnic

165 *N11* **Koide** Niigata, Honshū, C Japan *37°13′N 138°58′E*

10 *G7* **Koidern** Yukon, W Canada *61°55′N 140°22′W*

76 *J15* **Koidu** E Sierra Leone *08°40′N 11°01′W*

118 *I4* **Koigi** Järvamaa, C Estonia *58°51′N 25°45′E*

Koil *see* Kohila

172 *H13* **Koimbani** Grande Comore, NW Comoros *11°37′S 43°23′E*

93 *O16* **Koitere** ◎ E Finland

Koivisto *see* Primorsk

Kŏje-do *see* Geogeum-do

80 *J13* **K'ok'a Hāyk'** ◎ C Ethiopia

Kokand *see* Qo'qon

182 *F6* **Kokatha** South Australia *31°17′S 135°16′E*

146 *M10* **Ko'kcha** *Rus.* Kokcha. Buxoro Viloyati, C Uzbekistan *40°30′N 64°58′E*

Kokcha *see* Ko'kcha

Kokchetav *see* Kokshetau

93 *K18* **Kokemäenjoki** ≈ SW Finland

171 *W14* **Kokenau** *var.* Kokonau. Papua, E Indonesia *04°38′S 136°24′E*

83 *E22* **Kokerboom** Karas, SE Namibia *28°11′S 19°25′E*

119 *N14* **Kokhanava** *Rus.* Kokhanovo. Vitsyebskaya Voblasts', NE Belarus *54°28′N 29°59′E*

Kokhanovichi *see* Kakhanavichy

Kokhanovo *see* Kokhanava

Kök-Janggak *see* Kok-Yangak

93 *K16* **Kokkola** *Swe.* Karleby; *prev. Swe.* Gamlakarleby. Keski-Pohjanmaa, W Finland *63°50′N 23°10′E*

158 *L3* **Kok Kuduk** *spring/well* N China *46°03′N 87°34′E*

118 *H9* **Koknese** C Latvia *56°38′N 25°27′E*

77 *T13* **Koko** Kebbi, W Nigeria *11°25′N 04°33′E*

186 *E9* **Kokoda** Northern, S Papua New Guinea *08°52′S 147°44′E*

76 *K12* **Kokofata** Kayes, W Mali *12°48′N 09°56′W*

39 *N6* **Kokolik River** ≈ Alaska, USA

31 *O13* **Kokomo** Indiana, N USA *40°29′N 86°07′W*

Kokonau *see* Kokenau

Koko Nor *see* Qinghai, China

Koko Nor *see* Qinghai Hu, China

186 *H6* **Kokopo** *var.* Kopopo; *prev.* Herbertshöhe. New Britain, E Papua New Guinea *04°18′S 152°17′E*

Kokpekti *see* Kokpekty

Kokpekti *see* Kokpekty

145 *X10* **Kokpekty** *prev.* Kokpekti. Vostochnyy Kazakhstan, E Kazakhstan *48°47′N 82°28′E*

145 *X11* **Kokpekty** *prev.* Kokpekti. ≈ E Kazakhstan

39 *P9* **Kokrines** Alaska, USA *64°58′N 154°42′W*

39 *P9* **Kokrines Hills** ▲▲ Alaska, USA

145 *P17* **Koksaray** Yuzhnyy Kazakhstan, S Kazakhstan *42°34′N 68°06′E*

147 *X9* **Kokshaal-Tau** *Rus.* Khrebet Kakshaal-Too. ▲▲ China/Kyrgyzstan

145 *P7* **Kokshetau** *Kaz.* Kökshetaū; *prev.* Kokchetav. Kokshetau, N Kazakhstan *53°18′N 69°25′E*

Kökshetaū *see* Kokshetau

99 *A17* **Koksijde** West-Vlaanderen, W Belgium *51°07′N 02°40′E*

12 *M5* **Koksoak** ≈ Québec, E Canada

83 *K24* **Kokstad** KwaZulu/Natal, E South Africa *30°23′S 29°23′E*

145 *V14* **Koksu** *Kaz.* Rūdnīchnyy. Almaty, SE Kazakhstan *44°39′N 78°57′E*

145 *W15* **Koktal** *Kaz.* Köktal. Almaty, SE Kazakhstan *44°05′N 79°44′E*

145 *Q12* **Koktas** ≈ C Kazakhstan

Kök-Tash *see* Këk-Tash

Koktokay *see* Fuyun

147 *T9* **Kok-Yangak** *Kir.* Kök-Janggak. Dzhalal-Abadskaya Oblast', W Kyrgyzstan *41°02′N 73°11′E*

158 *F9* **Kokyar** Xinjiang Uygur Zizhiqu, W China *37°41′N 77°15′E*

149 *O13* **Kolāchi** *var.* Kulachi. ≈ SW Pakistan

76 *J15* **Kolahun** N Liberia *08°24′N 10°02′W*

171 *O14* **Kolaka** Sulawesi, C Indonesia *04°04′S 121°38′E*

K'o-la-ma-i *see* Karamay

Kola Peninsula *see* Kol'skiy Poluostrov

155 *H19* **Kolār** Karnātaka, E India *13°10′N 78°10′E*

155 *H19* **Kolār Gold Fields** Karnātaka, E India *12°56′N 78°16′E*

92 *K11* **Kolari** Lappi, NW Finland *67°20′N 23°51′E*

111 *I21* **Kolárovo** *Ger.* Gutta; *prev.* Guta, *Hung.* Gúta. Nitriansky Kraj, SW Slovakia *47°54′N 18°01′E*

113 *K16* **Kolašin** E Montenegro *42°49′N 19°32′E*

Kolāyat *see* Shir Kolāyat

95 *N15* **Kolbäck** Västmanland, C Sweden *59°33′N 16°15′E*

Kolbcha *see* Kowbcha

197 *Q15* **Kolbeinsey Ridge** *undersea feature* Denmark Strait/Norwegian Sea *69°00′N 17°30′W*

Kolberg *see* Kołobrzeg

95 *H15* **Kolbotn** Akershus, S Norway *62°15′N 10°24′E*

111 *N16* **Kolbuszowa** Podkarpackie, SE Poland *50°12′N 22°07′E*

126 *L3* **Kol'chugino** Vladimirskaya Oblast', W Russian Federation *56°19′N 39°24′E*

76 *H12* **Kolda** S Senegal *12°58′N 14°58′W*

95 *G23* **Kolding** Syddanmark, C Denmark *55°29′N 09°30′E*

79 *K20* **Kole** Kasai-Oriental, SW Dem. Rep. Congo *03°30′S 22°28′E*

79 *M17* **Kole** Orientale, N Dem. Rep. Congo *02°08′N 25°25′E*

Kōle *see* Kili Island

84 *F6* **Kölen** *Nor.* Kjølen. ▲▲ Norway/Sweden

Kolepom, Pulau *see* Yos Sudarso, Pulau

118 *H3* **Kolga Laht** *Ger.* Kolko-Wiek. *bay* N Estonia

125 *Q3* **Kolguyev, Ostrov** *island* NW Russian Federation

155 *E16* **Kolhāpur** Mahārāshtra, SW India *16°42′N 74°13′E*

151 *K21* **Kolhumadulu** *var.* Thaa Atoll. *atoll* S Maldives

93 *O16* **Koli** *var.* Kolinkylä. Pohjois-Karjala, E Finland *63°06′N 29°46′E*

39 *O13* **Koliganek** Alaska, USA *59°43′N 157°16′W*

111 *E16* **Kolín** *Ger.* Kolin. Střední Čechy, C Czech Republic *50°02′N 15°10′E*

Kolinkylä *see* Koli

190 *E12* **Koliu** Île Futuna, W Wallis and Futuna

118 *E7* **Kolka** NW Latvia *57°44′N 22°34′E*

118 *E7* **Kolkasrags** *prev. Eng.* Cape Domesnes. *headland* NW Latvia *57°45′N 22°35′E*

153 *S16* **Kolkata** *prev.* Calcutta. *state capital* West Bengal, NE India *22°30′N 88°20′E*

Kolkhozabad *see* Kolkhozobod

147 *P14* **Kolkhozobod** *Rus.* Kolkhozabad; *prev.* Kaganovichabad, Tugalan. SW Tajikistan *37°33′N 68°34′E*

Kolki/Kołki *see* Kolky

Kolko-Wiek *see* Kolga Laht

116 *K3* **Kolky** *Pol.* Kołki, *Rus.* Kolki. Volyns'ka Oblast', NW Ukraine *51°05′N 25°40′E*

155 *G20* **Kollegāl** Karnātaka, W India *12°08′N 77°06′E*

98 *M5* **Kollum** Fryslân, N Netherlands *53°17′N 06°09′E*

Kolmar *see* Colmar

101 *E16* **Köln** *var.* Koeln, *Eng./Fr.* Cologne, *prev.* Cöln; *anc.* Colonia Agrippina, Oppidum Ubiorum. Nordrhein-Westfalen, W Germany *50°57′N 06°57′E*

110 *N9* **Kolno** Podlaskie, NE Poland *53°24′N 21°57′E*

110 *J12* **Koło** Wielkopolskie, C Poland *52°11′N 18°39′E*

38 *B8* **Kōloa** *var.* Koloa. Kaua'i, Hawaii, USA, C Pacific Ocean *21°54′N 159°28′W*

Koloa *see* Kōloa

110 *E7* **Kołobrzeg** *Ger.* Kolberg. Zachodnio-pomorskie, NW Poland *54°11′N 15°34′E*

126 *H4* **Kolodnya** Smolenskaya Oblast', W Russian Federation *54°57′N 32°22′E*

190 *E13* **Kolofau, Mont** ⋀ Île Alofi, S Wallis and Futuna *14°21′S 178°02′W*

125 *O14* **Kologriv** Kostromskaya Oblast', NW Russian Federation *58°49′N 44°22′E*

76 *L12* **Kolokani** Koulikoro, W Mali *13°35′N 08°01′W*

77 *N13* **Koloko** W Burkina Faso *11°06′N 05°18′W*

186 *K8* **Kolombangara** *var.* Kilimbangara, Nduke. *island* New Georgia Islands, NW Solomon Islands

Kolomea *see* Kolomyya

126 *L4* **Kolomna** Moskovskaya Oblast', W Russian Federation *55°03′N 38°52′E*

116 *J7* **Kolomyya** *Ger.* Kolomea. Ivano-Frankivs'ka Oblast', W Ukraine *48°31′N 25°00′E*

76 *M13* **Kolondiéba** Sikasso, SW Mali *11°04′N 06°55′W*

193 *V15* **Kolonga** Tongatapu, S Tonga *21°07′S 175°05′W*

189 *U16* **Kolonia** *var.* Colonia. Pohnpei, E Micronesia *06°57′N 158°12′E*

Kolonja *see* Kolonjë

113 *K21* **Kolonjë** *var.* Kolonja. Fier, C Albania *40°49′N 19°37′E*

Kolonjë *see* Ersekë

Kolosjoki *see* Nikel'

Kolotambu *see* Avuavu

193 *U15* **Kolovai** Tongatapu, S Tonga *21°05′S 175°20′W*

Kolozsvár *see* Cluj-Napoca

112 *C9* **Kolpa** *Ger.* Kulpa, *SCr.* Kupa. ≈ Croatia/Slovenia

122 *J11* **Kolpashevo** Tomskaya Oblast', C Russian Federation *58°21′N 82°44′E*

124 *H13* **Kolpino** Leningradskaya Oblast', NW Russian Federation *59°44′N 30°39′E*

100 *M10* **Kölpinsee** ◎ NE Germany

Kólpos Mórfu *see* Güzelyurt

146 *K8* **Ko'lquduq** *Rus.* Kulkuduk. Navoiy Viloyati, N Uzbekistan *42°36′N 63°24′E*

124 *K5* **Kol'skiy Poluostrov** *Eng.* Kola Peninsula. *peninsula* NW Russian Federation

127 *T6* **Koltubanovskiy** Orenburgskaya Oblast', W Russian Federation *53°00′N 52°00′E*

112 *L11* **Kolubara** ≈ C Serbia

Kolupchii *see* Gurkovo

110 *K13* **Koluszki** Łódźskie, C Poland *51°44′N 19°50′E*

125 *T6* **Kolva** ≈ NW Russian Federation

93 *E14* **Kolvereid** Nord-Trøndelag, W Norway *64°47′N 11°22′E*

79 *M24* **Kolwezi** Katanga, S Dem. Rep. Congo *10°43′S 25°29′E*

123 *S7* **Kolyma** ≈ NE Russian Federation

Kolyma Lowland *see* Kolymskaya Nizmennost'

Kolyma Range/Kolymskiy, Khrebet *see* Kolymskoye Nagor'ye

123 *S7* **Kolymskaya Nizmennost'** *Eng.* Kolyma Lowland. *lowlands* NE Russian Federation

123 *S7* **Kolymskoye** Respublika Sakha (Yakutiya), NE Russian Federation *68°42′N 158°46′E*

123 *U8* **Kolymskoye Nagor'ye** *var.* Khrebet Kolymskiy, *Eng.* Kolyma Range. ▲▲ E Russian Federation

123 *V5* **Kolyuchinskaya Guba** *bay* NE Russian Federation

Kol'zhat *see* Kalzhat

114 *G8* **Kom** ▲ NW Bulgaria *43°10′N 23°02′E*

80 *I13* **Koma** Oromīya, C Ethiopia *08°19′N 36°48′E*

77 *X12* **Komadugu Gana** ≈ NE Nigeria

164 *M13* **Komagane** Nagano, Honshū, S Japan *35°44′N 137°54′E*

79 *P17* **Komanda** Orientale, NE Dem. Rep. Congo *01°25′N 29°43′E*

197 *U1* **Komandorskaya Basin** *var.* Kamchatka Basin. *undersea feature* SW Bering Sea *57°00′N 168°00′E*

125 *Pp9* **Komandorskiye Ostrova** *Eng.* Commander Islands. *island group* E Russian Federation

Kománfalva *see* Comăneşti

111 *I22* **Komárno** *Ger.* Komorn, *Hung.* Komárom. Nitriansky Kraj, SW Slovakia *47°46′N 18°07′E*

111 *I22* **Komárom** Komárom-Esztergom, NW Hungary *47°43′N 18°06′E*

Komárom *see* Komárno

111 *I22* **Komárom-Esztergom** *off.* Komárom-Esztergom Megye. ◈ *county* N Hungary

Komárom-Esztergom Megye *see* Komárom-Esztergom

164 *K11* **Komatsu** *var.* Komatu. Ishikawa, Honshū, SW Japan *36°25′N 136°27′E*

Komatu *see* Komatsu

83 *D17* **Kombat** Otjozondjupa, N Namibia *19°42′S 17°45′E*

Kombissiguiri *see* Kombissiri

77 *P13* **Kombissiri** *var.* Kombissiguiri. C Burkina Faso *12°01′N 01°27′W*

188 *E10* **Komebail Lagoon** *lagoon* N Palau

81 *F20* **Kome Island** *island* N Tanzania

Komeyo *see* Wandai

117 *P10* **Kominternivs'ke** Odes'ka Oblast', SW Ukraine *46°52′N 30°56′E*

125 *R8* **Komi, Respublika** ◈ *autonomous republic* NW Russian Federation

111 *I25* **Komló** Baranya, SW Hungary *46°11′N 18°15′E*

Kommunarsk *see* Alchevs'k

Kommunizm, Qullai *see* Ismoili Somoní, Qullai

186 *B7* **Komo** Hela, W Papua New Guinea *06°06′S 142°52′E*

170 *M16* **Komodo, Pulau** *island* Nusa Tenggara, S Indonesia

77 *N15* **Komoé** *var.* Komoé Fleuve. ≈ E Ivory Coast

Komoé Fleuve *see* Komoé

75 *X11* **Kom Ombo** *var.* Kôm Ombo, Kawm Umbū. SE Egypt *24°26′N 32°57′E*

79 *F20* **Komono** Lékoumou, SW Congo *03°15′S 13°14′E*

171 *Y16* **Komoran** Papua, E Indonesia *08°14′S 138°51′E*

171 *Y16* **Komoran, Pulau** *island* E Indonesia

Komorn *see* Komárno

Komornok *see* Comorâşte

Komosolabad *see* Komsomolobod

Komotau *see* Chomutov

114 *K13* **Komotiní** *var.* Gümüljina, *Turk.* Gümülcine. Anatolikí Makedonía kai Thráki, NE Greece *41°07′N 25°27′E*

113 *K16* **Komovi** ▲▲ E Montenegro

117 *R8* **Kompaniyivka** Kirovohrads'ka Oblast', C Ukraine *48°16′N 32°12′E*

Kompong *see* Kâmpóng Chhnăng

Kompong Cham *see* Kâmpóng Cham

Kompong Kleang *see* Kâmpóng Khleăng

Kompong Som *see* Sihanoukville

Kompong Speu *see* Kâmpóng Spoe

Komrat *see* Comrat

144 *G12* **Komsomol** *prev.* Komsomol'skiy. Atyrau, W Kazakhstan *47°18′N 53°37′E*

125 *W4* **Komsomol** *prev.* Komsomol'skiy. Respublika Komi, NW Russian Federation *67°33′N 64°00′E*

122 *K14* **Komsomolets, Ostrov** *island* Severnaya Zemlya, N Russian Federation

144 *F13* **Komsomolets, Zaliv** *lake gulf* SW Kazakhstan

Komsomol/Komsomolets *see* Karabalyk, Kostanay, Kazakhstan

147 *Q12* **Komsomolobod** *Rus.* Komosolabad. C Tajikistan *38°51′N 69°54′E*

124 *M16* **Komsomol'sk** Ivanovskaya Oblast', W Russian Federation *56°58′N 40°15′E*

117 *S6* **Komsomol's'k** Poltavs'ka Oblast', C Ukraine *49°01′N 33°37′E*

146 *M11* **Komsomol'sk** Navoiy Viloyati, N Uzbekistan *40°14′N 65°10′E*

Komsomol'skiy *see* Komsomol

Komsomol'skiy *see* Komsomol

123 *S13* **Komsomol'sk-na-Amure** Khabarovskiy Kray, SE Russian Federation *50°32′N 136°59′E*

Komsomol'sk-na-Ustyurte *see* Kubla-Ustyurt

144 *K10* **Komsomol'skoye** Aktyubinsk, NW Kazakhstan

127 *Q8* **Komsomol'skoye** Saratovskaya Oblast', W Russian Federation *50°45′N 47°00′E*

145 *P10* **Kon** ≈ C Kazakhstan

Kona *see* Kailua-Kona

124 *K16* **Konakovo** Tverskaya Oblast', W Russian Federation *56°42′N 36°44′E*

Konar *see* Kunar

143 *V15* **Konārak** Sīstān va Balūchestān, SE Iran *25°26′N 60°23′E*

Konarhā *see* Kunar

27 *O11* **Konawa** Oklahoma, C USA *34°57′N 96°45′W*

122 *H10* **Konda** ≈ C Russian Federation

154 *L13* **Kondagaon** Chhattīsgarh, C India *19°38′N 81°41′E*

14 *K10* **Kondiaronk, Lac** ◎ Québec, SE Canada

180 *J13* **Kondinin** Western Australia *32°31′S 118°15′E*

81 *H21* **Kondoa** Dodoma, C Tanzania *04°54′S 35°46′E*

127 *P6* **Kondol'** Penzenskaya Oblast', W Russian Federation *52°49′N 45°03′E*

114 *N10* **Kondolovo** Burgas, E Bulgaria *42°07′N 27°43′E*

171 *Z16* **Kondomirat** Papua, E Indonesia *08°57′S 140°55′E*

124 *J10* **Kondopoga** Respublika Kareliya, NW Russian Federation *62°13′N 34°17′E*

Kondoz *see* Kunduz

155 *J17* **Kondukūr** *var.* Kandukur. Andhra Pradesh, E India *15°17′N 79°49′E*

Kondūz *see* Kunduz

187 *P16* **Koné** Province Nord, W New Caledonia *21°04′S 164°51′E*

146 *E13* **Könekesir** Balkan Welaýaty, W Turkmenistan *38°16′N 56°51′E*

146 *G8* **Köneürgenç** *var.* Köneürgench, *Rus.* Këneurgench; *prev.* Kunya-Urgench. Daşoguz Welaýaty, N Turkmenistan *42°21′N 59°09′E*

77 *N15* **Kong** N Ivory Coast *09°10′N 04°33′W*

39 *S5* **Kongakut River** ≈ Alaska, USA

197 *O14* **Kong Christian IX Land** *Eng.* King Christian IX Land. *physical region* SE Greenland

197 *P13* **Kong Christian X Land** *Eng.* King Christian X Land. *physical region* E Greenland

197 *N13* **Kong Frederik IX Land** *physical region* SW Greenland

197 *Q12* **Kong Frederik VIII Land** *Eng.* King Frederik VIII Land. *physical region* NE Greenland

197 *N15* **Kong Frederik VI Kyst** *Eng.* King Frederik VI Coast. *physical region* SE Greenland

167 *P13* **Kŏng, Kaôh** *prev.* Kas Kong. *island* SW Cambodia

92 *P2* **Kong Karls Land** *Eng.* King Charles Islands. *island group* SE Svalbard

81 *G14* **Kong Kong** ≈ E South Sudan

Kongo *see* Congo (river)

83 *G16* **Kongola** Caprivi, NE Namibia *17°47′S 23°24′E*

79 *N21* **Kongolo** Katanga, E Dem. Rep. Congo *05°20′S 26°58′E*

81 *F14* **Kongor** Jonglei, E South Sudan *07°09′N 31°44′E*

197 *Q14* **Kong Oscar Fjord** *fjord* E Greenland

77 *P12* **Kongoussi** N Burkina Faso *13°19′N 01°31′W*

95 *G15* **Kongsberg** Buskerud, S Norway *59°39′N 09°38′E*

92 *Q2* **Kongsøya** *island* Kong Karls Land, E Svalbard

95 *I14* **Kongsvinger** Hedmark, S Norway *60°10′N 12°00′E*

Kongtong *see* Pingliang

167 *T11* **Kông, Tônlé** *var.* Xê Kong. ≈ Cambodia/Laos

158 *E8* **Kongur Shan** ▲ NW China *38°39′N 75°21′E*

81 *I22* **Kongwa** Dodoma, C Tanzania *06°13′S 36°28′E*

Kong, Xê *see* Kông, Tônlé

Konia *see* Konya

147 *R11* **Konibodom** *Rus.* Kanibadam. N Tajikistan *40°16′N 70°20′E*

111 *K15* **Koniecpol nad Pilicą** Śląskie, S Poland *50°47′N 19°45′E*

Konieh *see* Konya

Königgrätz *see* Hradec Králové

Königinhof an der Elbe *see* Dvůr Králové nad Labem

101 *K23* **Königsbrunn** Bayern, S Germany *48°16′N 10°52′E*

101 *O24* **Königssee** ◎ SE Germany

Königshütte *see* Chorzów

109 *S8* **Königstuhl** ▲ S Austria *46°57′N 13°47′E*

109 *U3* **Königswiesen** Oberösterreich, N Austria *48°25′N 14°48′E*

101 *E17* **Königswinter** Nordrhein-Westfalen, W Germany *50°40′N 07°12′E*

146 *M11* **Konimex** *Rus.* Kenimekh. Navoiy Viloyati, N Uzbekistan *40°14′N 65°10′E*

110 *I12* **Konin** *Ger.* Kuhnau. Weilkopolskie, C Poland *52°13′N 18°17′E*

Koninkrijk der Nederlanden *see* Netherlands

113 *L24* **Konispol** *var.* Konispoli. Vlorë, S Albania *39°40′N 20°10′E*

Konispoli *see* Konispol

115 *C15* **Kónitsa** Ípeiros, W Greece *40°04′N 20°48′E*

Konitz *see* Chojnice

108 *D8* **Köniz** Bern, W Switzerland *46°56′N 07°25′E*

113 *H14* **Konjic** ◈ Federacija Bosne I Hercegovine, S Bosnia and Herzegovina

92 *J10* **Könkämäälven** ≈ Finland/Sweden

155 *D14* **Konkan** *plain* W India

83 *D22* **Konkiep** ≈ S Namibia

76 *I14* **Konkouré** ≈ W Guinea

◆ Country
● Country Capital
◇ Dependent Territory
○ Dependent Territory Capital
◈ Administrative Regions
✕ International Airport
▲ Mountain
▲▲ Mountain Range
⋀ Volcano
≈ River
◎ Lake
⊞ Reservoir

KONNA – KRASNOLESNYY

77 O11 **Konna** Mopti, S Mali 14°58′N 03°49′W
186 H6 **Konogaiang, Mount** ▲ New Ireland, NE Papua New Guinea 04°05′S 152°43′E
186 H5 **Konogogo** New Ireland, NE Papua New Guinea 03°25′S 152°09′E
108 E9 **Konolfingen** Bern, W Switzerland 46°53′N 07°36′E
77 P16 **Konongo** C Ghana 06°39′N 01°06′W
186 H5 **Konos** New Ireland, NE Papua New Guinea 03°09′S 151°47′E
124 M12 **Konosha** Arkhangel'skaya Oblast', NW Russian Federation 60°58′N 40°09′E
117 R3 **Konotop** Sums'ka Oblast', NE Ukraine 51°15′N 33°14′E
158 L7 **Konqi He** ≈ NW China
111 L14 **Końskie** Świętokrzyskie, C Poland 51°12′N 20°23′E
Konstantinovka *see* Kostyantynivka
126 M11 **Konstantinovsk** Rostovskaya Oblast', SW Russian Federation 47°37′N 41°07′E
101 H24 **Konstanz** *var.* Constanz, *Eng.* Constance, *hist.* Kostnitz; *anc.* Constantia. Baden-Württemberg, S Germany 47°40′N 09°10′E
Konstanza *see* Constanța
77 T14 **Kontagora** Niger, W Nigeria 10°25′N 05°29′E
78 E13 **Kontcha** Nord, N Cameroon 08°00′N 12°13′E
99 G17 **Kontich** Antwerpen, N Belgium 51°08′N 04°27′E
93 O16 **Kontiolahti** Pohjois-Karjala, SE Finland 62°46′N 29°51′E
93 M15 **Kontiomäki** Kainuu, C Finland 64°20′N 28°09′E
167 U11 **Kon Tum** *var.* Kontum. Kon Tum, C Vietnam 14°23′N 108°00′E
Kontum *see* Kon Tum
Konur *see* Sulakyurt
136 H15 **Konya** *var.* Konieh, *prev.* Konia; *anc.* Iconium. Konya, C Turkey 37°51′N 32°30′E
136 H15 **Konya** *var.* Konia, Konieh. ◆ *province* C Turkey
151 E15 **Konya Reservoir** *prev.* Shivāji Sāgar. ⊠ W India
145 T13 **Konyrat** *var.* Kounradskiy, *Kaz.* Qongyrat. Karaganda, SE Kazakhstan 46°57′N 75°01′E
145 W15 **Konyrolen** Almaty, SE Kazakhstan 44°16′N 79°18′E
81 I19 **Konza** Kajiado, S Kenya 01°44′S 37°07′E
98 I9 **Koog aan den Zaan** Noord-Holland, C Netherlands 52°28′N 04°49′E
182 E7 **Koonibba** South Australia 31°55′S 133°23′E
31 O11 **Koontz Lake** Indiana, N USA 41°25′N 86°24′W
171 U12 **Koor** Papua Barat, E Indonesia 0°21′S 132°28′E
183 R9 **Koorawatha** New South Wales, SE Australia 34°03′S 148°33′E
118 J5 **Koosa** Tartumaa, E Estonia 58°31′N 27°06′E
33 N7 **Kootenai** *var.* Kootenay. ≈ Canada/USA *see also* Kootenay
Kootenai *see* Kootenay
11 P17 **Kootenay** *var.* Kootenai. ≈ Canada/USA *see also* Kootenai
Kootenay *see* Kootenai
83 F24 **Kootjieskolk** Northern Cape, W South Africa 31°16′S 20°21′E
113 M15 **Kopaonik** ▲▲ S Serbia
Kopar *see* Koper
92 K1 **Kópasker** Norðurland Eystra, N Iceland 66°15′N 16°23′W
92 H4 **Kópavogur** Höfuðborgarsvæðið, W Iceland 64°06′N 21°47′W
145 V14 **Kopbirlik** *prev.* Kirov, Kirova. Almaty, SE Kazakhstan 46°24′N 77°16′E
109 S13 **Koper** *It.* Capodistria; *prev.* Kopar. SW Slovenia 45°32′N 13°43′E
95 C16 **Kopervik** Rogaland, S Norway 59°17′N 05°20′E
Köpetdag Gershi/Kopetdag, Khrebet *see* Koppeh Dāgh
Kophinou *see* Kofinou
182 G8 **Kopi** South Australia 33°24′S 135°40′E
153 W12 **Kopili** ≈ NE India
95 M15 **Köping** Västmanland, C Sweden 59°31′N 16°00′E
113 K17 **Koplik** *var.* Kopliku. Shkodër, NW Albania 42°12′N 19°26′E
Kopliku *see* Koplik
Kopopo *see* Kokopo
94 I11 **Koppang** Hedmark, S Norway 61°34′N 11°04′E
Kopparberg *see* Dalarna
143 S3 **Koppeh Dāgh** *Rus.* Khrebet Kopetdag, *Turkm.* Köpetdag Gershi. ▲▲ Iran/Turkmenistan
Koppename *see* Coppename Rivier
95 J15 **Koppom** Värmland, C Sweden 59°42′N 12°07′E
Kopreinitz *see* Koprivnica
114 K9 **Koprinka, Yazovir** *prev.* Yazovir Georgi Dimitrov. ⊠ C Bulgaria
112 F7 **Koprivnica** *Ger.* Kopreinitz, *Hung.* Kaproncza. Koprivnica-Križevci, N Croatia 46°10′N 16°49′E
112 F8 **Koprivnica-Križevci** *off.* Koprivničko-Križevačka Županija. ◆ *province* N Croatia
111 I17 **Kopřivnice** *Ger.* Nesselsdorf. Moravskoslezský Kraj, E Czech Republic 49°36′N 18°09′E
Koprivničko-Križevačka Županija *see* Koprivnica-Križevci
Köprülü *see* Veles
Koptsevichi *see* Kaptsevichy
Kopyl' *see* Kapyl'
119 O14 **Kopys'** Vitsyebskaya Voblasts', NE Belarus 54°19′N 30°18′E
113 M18 **Korab** ▲ Albania/FYR Macedonia 41°48′N 20°33′E
Korabavur Pastligi *see* Karabaur', Uval
124 M5 **Korabel'noye** Murmanskaya Oblast', NW Russian Federation 67°00′N 41°10′E
81 M14 **K'orahē** Sumalē, E Ethiopia 06°36′N 44°21′E
115 L16 **Kórakas, Akrotírio** *cape* Lésvos, E Greece
112 D9 **Korana** ≈ C Croatia
155 L14 **Korāput** Odisha, E India 18°48′N 82°41′E
Korat *see* Nakhon Ratchasima
167 Q9 **Korat Plateau** *plateau* E Thailand
Kōrāwa, Sar-I *see* Kirdī Kawrāw, Qimmat
154 L11 **Korba** Chhattīsgarh, C India 22°25′N 82°43′E
101 H15 **Korbach** Hessen, C Germany 51°16′N 08°52′E
Korça *see* Korçë
113 M21 **Korçë** *var.* Korça, *Gk.* Korytsa, *It.* Corriza; *prev.* Koritsa. Korçë, SE Albania 40°38′N 20°47′E
113 M21 **Korçë** ◆ *district* SE Albania
113 G15 **Korčula** *It.* Curzola. Dubrovnik-Neretva, S Croatia 42°57′N 17°08′E
113 F15 **Korčula** *It.* Curzola; *anc.* Corcyra Nigra. *island* S Croatia
113 F15 **Korčulanski Kanal** *channel* S Croatia
145 T6 **Korday** *prev.* Georgiyevka. Zhambyl, SE Kazakhstan 43°03′N 74°43′E
142 J5 **Kordestān** *off.* Ostān-e Kordestān, *var.* Kurdestan. ◆ *province* W Iran
Kordestān, Ostān-e *see* Kordestān
143 P4 **Kord Küy** *var.* Kurd Kui. Golestān, N Iran 36°49′N 54°05′E
165 V13 **Korea Bay** *bay* China/North Korea
Korea, Democratic People's Republic of *see* North Korea
171 T15 **Koreare** Pulau Yamdena, E Indonesia 07°33′S 131°13′E
Korea, Republic of *see* South Korea
163 Z17 **Korea Strait** *Jap.* Chōsen-kaikyō, *Kor.* Taehan-haehyŏp. *channel* Japan/South Korea
Korelichi/Korelicze *see* Karelichy
80 J11 **Korem** Tigrai, N Ethiopia 12°32′N 39°29′E
77 U11 **Korén Adoua** ≈ C Niger
126 I7 **Korenevo** Kurskaya Oblast', W Russian Federation 51°21′N 34°53′E
126 L13 **Korenovsk** Krasnodarskiy Kray, SW Russian Federation 45°28′N 39°25′E
116 L4 **Korets'** *Pol.* Korzec, *Rus.* Korets. Rivnens'ka Oblast', NW Ukraine 50°38′N 27°12′E
Korets *see* Korets'
194 L7 **Korff Ice Rise** *ice cap* Antarctica
145 Q10 **Korgalzhyn** *var.* Kurgal'dzhino, Kurgal'dzhinsky, *Kaz.* Qorghalzhyn. Akmola, C Kazakhstan 50°33′N 69°58′E
145 W15 **Korgas**; *prev.* Khorgos. Almaty, SE Kazakhstan 44°13′N 80°22′E
92 G13 **Korgen** Troms, N Norway 66°04′N 13°51′E
147 R9 **Korgon-Dēbē** Dzhalal-Abadskaya Oblast', W Kyrgyzstan 41°51′N 70°52′E
76 M14 **Korhogo** N Ivory Coast 09°29′N 05°39′W
115 F19 **Korinthiakós Kólpos** *Eng.* Gulf of Corinth; *anc.* Corinthiacus Sinus. *gulf* C Greece
115 F19 **Kórinthos** *anc.* Corinthus *Eng.* Corinth. Pelopónnisos, S Greece 37°56′N 22°55′E
113 M18 **Koritnik** ▲ S Serbia 42°06′N 20°34′E
Koritsa *see* Korçë
165 P11 **Kōriyama** Fukushima, Honshū, C Japan 37°25′N 140°20′E
136 E16 **Korkuteli** Antalya, SW Turkey 37°07′N 30°11′E
158 K6 **Korla** *Chin.* K'u-erh-lo. Xinjiang Uygur Zizhiqu, NW China 41°48′N 86°10′E
122 J10 **Korliki** Khanty-Mansiyskiy Avtonomnyy Okrug-Yugra, C Russian Federation 61°28′N 82°12′E
Körlin an der Persante *see* Karlino
Korma *see* Karma
14 D8 **Kormak** Ontario, S Canada 47°38′N 83°00′W
Kormakiti, Akrotíri/Kormakiti, Cape/Kormakítis *see* Koruçam Burnu
111 G23 **Körmend** Vas, W Hungary 47°02′N 16°35′E
139 S5 **Kōrmōr** Şalāḩ ad Dīn, E Iraq 35°06′N 44°47′E
112 C13 **Kornat** *It.* Incoronata. *island* W Croatia
Korneshty *see* Cornești
109 X3 **Korneuburg** Niederösterreich, NE Austria 48°22′N 16°20′E
145 P7 **Korneyevka** Severnyy Kazakhstan, N Kazakhstan 54°01′N 68°30′E
95 I17 **Kornsjø** Østfold, S Norway 58°55′N 11°40′E
77 O11 **Koro** Mopti, S Mali 14°05′N 03°06′W
1873 Y14 **Koro** *island* C Fiji
186 B7 **Koroba** Hela, W Papua New Guinea 05°46′S 142°48′E
126 K8 **Korocha** Belgorodskaya Oblast', W Russian Federation 50°49′N 37°08′E
136 H12 **Köroğlu Dağları** ▲▲ C Turkey
183 V6 **Korogoro Point** *headland* New South Wales, SE Australia 31°03′S 153°04′E
81 J21 **Korogwe** Tanga, E Tanzania 05°10′S 38°30′E
182 L13 **Koroit** Victoria, SE Australia 38°17′S 142°22′E
187 X15 **Korolevu** Viti Levu, W Fiji 18°12′S 177°44′E
190 I17 **Koromiri** *island* S Cook Islands
171 Q8 **Koronadal** Mindanao, S Philippines 06°23′N 124°54′E
114 G13 **Korónia, Límni** *var.* Límni Korónia. ◎ N Greece
115 E22 **Koróni** Pelopónnisos, S Greece 36°47′N 21°57′E
Korónia, Límni *see* Korōneia, Límni
110 I9 **Koronowo** *Ger.* Krone an der Brahe. Kujawski-pomorskie, C Poland 53°18′N 17°56′E
117 R2 **Korop** Chernihivs'ka Oblast', N Ukraine 51°35′N 32°57′E
115 H19 **Koropí** Attikí, C Greece 37°54′N 23°52′E
188 C8 **Koror** (Palau) Oreor, N Palau 07°21′N 134°28′E
Koror *see* Oreor
Koror Lāl Esan *see* Karor
Koror Pacca *see* Kahror Pakka
111 L23 **Körös** ≈ E Hungary
Kőrös *see* Križevci
Kőrösbánya *see* Baia de Criș
187 Y14 **Koro Sea** *sea* C Fiji
Koroška *see* Kärnten
117 N3 **Korosten'** Zhytomyrs'ka Oblast', NW Ukraine 50°56′N 28°39′E
Korostyshev *see* Korostyshiv
117 N4 **Korostyshiv** *Rus.* Korostyshev. Zhytomyrs'ka Oblast', N Ukraine 50°18′N 29°05′E
125 V3 **Korotaikha** ≈ NW Russian Federation
122 J9 **Korotchayevo** Yamalo-Nenetskiy Avtonomnyy Okrug, N Russian Federation 66°00′N 78°11′E
78 I8 **Koro Toro** Borkou, N Chad 16°01′N 18°27′E
39 N16 **Korovin Island** *island* Shumagin Islands, Alaska, USA
187 X14 **Korovou** Viti Levu, W Fiji 17°48′S 178°32′E
93 M17 **Korpilahti** Keski-Suomi, C Finland 62°02′N 25°34′E
92 K12 **Korpilombolo** *Lapp.* Dállogilli. Norrbotten, N Sweden 66°51′N 23°00′E
123 T13 **Korsakov** Ostrov Sakhalin, Sakhalinskaya Oblast', SE Russian Federation 46°41′N 142°45′E
93 J16 **Korsholm** *Fin.* Mustasaari. Österbotten, W Finland 63°05′N 21°43′E
95 I23 **Korsør** Sjælland, E Denmark 55°19′N 11°09′E
Korsovka *see* Kārsava
117 P6 **Korsun'-Shevchenkivs'kyy** *Rus.* Korsun'-Shevchenkovskiy. Cherkas'ka Oblast', C Ukraine 49°26′N 31°15′E
Korsun'-Shevchenkovskiy *see* Korsun'-Shevchenkivs'kyy
99 C17 **Kortemark** West-Vlaanderen, W Belgium 51°03′N 03°03′E
99 H18 **Kortenberg** Vlaams Brabant, C Belgium 50°53′N 04°33′E
99 K18 **Kortessem** Limburg, NE Belgium 50°52′N 05°22′E
99 E14 **Kortgene** Zeeland, SW Netherlands 51°34′N 03°48′E
80 F8 **Korti** Northern, N Sudan 18°06′N 31°33′E
99 C18 **Kortrijk** *Fr.* Courtrai. West-Vlaanderen, W Belgium 50°50′N 03°17′E
121 O2 **Koruçam Burnu** *var.* Cape Kormakiti, Kormakítis, *Gk.* Akrotíri Kormakíti. *headland* N Cyprus 35°24′N 32°55′E
183 O13 **Korumburra** Victoria, SE Australia 38°27′S 145°48′E
Koryak Range *see* Koryakskoye Nagor'ye
Koryakskiy Khrebet *see* Koryakskoye Nagor'ye
123 V8 **Koryakskiy Okrug** ◆ *autonomous district* E Russian Federation
123 V7 **Koryakskoye Nagor'ye** *var.* Koryakskiy Khrebet, *Eng.* Koryak Range. ▲▲ NE Russian Federation
125 P11 **Koryazhma** Arkhangel'skaya Oblast', NW Russian Federation 61°16′N 47°07′E
Kōryō *see* Gangneung
Korytsa *see* Korçë
117 Q2 **Koryukivka** Chernihivs'ka Oblast', N Ukraine 51°45′N 32°16′E
Korzec *see* Korets'
115 N21 **Kos** Kos, Dodekánisa, Greece, Aegean Sea 36°53′N 27°19′E
115 M21 **Kos** *It.* Coo; *anc.* Cos. *island* Dodekánisa, Greece, Aegean Sea
125 T12 **Kosa** Komi-Permyatskiy Okrug, NW Russian Federation 59°55′N 54°54′E
125 T13 **Kosa** ≈ NW Russian Federation
164 B12 **Kō-saki** *headland* Nagasaki, Tsushima, SW Japan 34°06′N 129°13′E
163 X13 **Kosan** SE North Korea 38°50′N 127°26′E
119 H18 **Kosava** *Rus.* Kosovo. Brestskaya Voblasts', SW Belarus 52°45′N 25°16′E
Kosch *see* Kose
Koschagyl *see* Kosshagyl
110 G12 **Kościan** *Ger.* Kosten. Wielkopolskie, C Poland 52°05′N 16°38′E
110 I7 **Kościerzyna** Pomorskie, NW Poland 54°07′N 17°55′E
22 L4 **Kosciusko** Mississippi, S USA 33°03′N 89°35′W
Kosciusko, Mount *see* Kosciuszko, Mount
183 R11 **Kosciuszko, Mount** *prev.* Mount Kosciusko. ▲ New South Wales, SE Australia 36°28′S 148°15′E
118 H4 **Kose** *Ger.* Kosch. Harjumaa, NW Estonia 59°11′N 25°10′E
25 U9 **Kosse** Texas, SW USA 31°16′N 96°38′W
114 G6 **Koshava** Vidin, NW Bulgaria 44°03′N 23°00′E
147 U9 **Kosh-Dēbē** *var.* Koshtebë. Narynskaya Oblast', C Kyrgyzstan 41°03′N 74°08′E
K'o-shih *see* Kashi
164 B12 **Koshikijima-rettō** *var.* Kosikizima Rettô. *island group* SW Japan
145 W13 **Koshkarkol', Ozero** ◎ SE Kazakhstan
30 L9 **Koshkonong, Lake** ◎ Wisconsin, N USA
Koshoba *see* Goşoba
Koshoba *see* Goşoba
164 M12 **Kōshoku** *var.* Kôsyoku. Nagano, Honshū, S Japan 36°33′N 138°09′E
Koshtebë *see* Kosh-Dēbē
Kōshū *see* Enzan
Kōshū *see* Gwangju
111 N19 **Košice** *Ger.* Kaschau, *Hung.* Kassa. Košický Kraj, E Slovakia 48°44′N 21°15′E
111 M20 **Košický Kraj** ◆ *region* E Slovakia
Kosikizima Rettô *see* Koshikijima-rettō
153 R12 **Kosi Reservoir** ⊠ E Nepal
116 J8 **Kosiv** Ivano-Frankivs'ka Oblast', W Ukraine 48°19′N 25°04′E
145 O11 **Koskol'** *Kaz.* Qoskōl. Karaganda, C Kazakhstan 49°32′N 67°08′E
125 Q9 **Koslan** Respublika Komi, NW Russian Federation 63°27′N 48°52′E
Köslin *see* Koszalin
146 M12 **Koson** *Rus.* Kasan. Qashqadaryo Viloyati, S Uzbekistan 39°04′N 65°35′E
Kosōng *see* Goseong
147 S9 **Kosonsoy** *Rus.* Kasansay. Namangan Viloyati, E Uzbekistan 41°15′N 71°28′E
113 M16 **Kosovo** *prev.* Autonomous Province of Kosovo and Metohija. ◆ *republic* SE Europe
Kosovo *see* Kosava
Kosovo and Metohija, Autonomous Province of *see* Kosovo
Kosovo Polje *see* Fushë Kosovë
Kosovska Kamenica *see* Kamenicë
Kosovska Mitrovica *see* Mitrovicë
189 X17 **Kosrae** ◆ *state* E Micronesia
189 Y14 **Kosrae** *prev.* Kusaie. *island* Caroline Islands, E Micronesia
109 P6 **Kössen** Tirol, W Austria 47°40′N 12°24′E
144 G12 **Kosshagyl** *prev.* Koschagyl, *Kaz.* Qosshaghyl. Atyrau, W Kazakhstan 46°52′N 53°46′E
76 M16 **Kossou, Lac de** ◎ C Ivory Coast
Kossukavak *see* Krumovgrad
Kostajnica *see* Hrvatska Kostajnica
Kostamus *see* Kostomuksha
144 M7 **Kostanay** *var.* Kustanay, *Kaz.* Qostanay. Kostanay, N Kazakhstan 53°16′N 63°34′E
144 L8 **Kostanay** *var.* Kostanayskaya Oblast', *Kaz.* Qostanay Oblysy. ◆ *province* N Kazakhstan
Kostanayskaya Oblast' *see* Kostanay
Kosten *see* Kościan
114 H10 **Kostenets** *prev.* Georgi Dimitrov. Sofia, W Bulgaria 42°15′N 23°48′E
80 F10 **Kosti** White Nile, C Sudan 13°11′N 32°38′E
Kostnitz *see* Konstanz
124 H7 **Kostomuksha** *Fin.* Kostamus. Respublika Kareliya, NW Russian Federation 64°33′N 30°28′E
116 K3 **Kostopil'** *Rus.* Kostopol'. Rivnens'ka Oblast', NW Ukraine 50°20′N 26°29′E
Kostopol' *see* Kostopil'
124 M15 **Kostroma** Kostromskaya Oblast', NW Russian Federation 57°46′N 41°E
125 N14 **Kostroma** ≈ NW Russian Federation
125 N14 **Kostromskaya Oblast'** ◆ *province* NW Russian Federation
110 D11 **Kostrzyn** *Ger.* Cüstrin, Küstrin. Lubuskie, W Poland 52°35′N 14°40′E
110 H11 **Kostrzyn** Wielkopolskie, C Poland 52°23′N 17°13′E
117 X7 **Kostyantynivka** *Rus.* Konstantinovka. Donets'ka Oblast', SE Ukraine 48°33′N 37°45′E
Kostyukovichi *see* Kastsyukovichy
Kostyukovka *see* Kastsyukowka
Kōsyoku *see* Kōshoku
125 U6 **Kos'yu** Respublika Komi, NW Russian Federation 65°39′N 59°01′E
125 U6 **Kos'yu** ≈ NW Russian Federation
110 F7 **Koszalin** *Ger.* Köslin. Zachodnio-pomorskie, NW Poland 54°12′N 16°10′E
111 F22 **Kőszeg** *Ger.* Güns. Vas, W Hungary 47°23′N 16°33′E
152 H13 **Kota** *prev.* Kotah. Rājasthān, N India 25°14′N 75°52′E
Kota Baharu *see* Kota Bharu
Kota Bahru *see* Kota Bharu
169 U13 **Kotabaru** Pulau Laut, C Indonesia 03°15′S 116°15′E
168 K12 **Kota Baru** Sumatera, W Indonesia 01°07′S 101°43′E
Kotabaru *see* Jayapura
168 K6 **Kota Bharu** *var.* Kota Baharu, Kota Bahru. Kelantan, Peninsular Malaysia 06°07′N 102°15′E
Kotaboemi *see* Kotabumi
168 M14 **Kotabumi** *prev.* Kotaboemi. Sumatera, W Indonesia 04°50′S 104°54′E
149 S10 **Kot Addu** Punjab, E Pakistan 30°28′N 70°58′E
Kotah *see* Kota
169 U7 **Kota Kinabalu** *prev.* Jesselton. Sabah, East Malaysia 05°59′N 116°04′E
169 U7 **Kota Kinabalu** ✕ Sabah, East Malaysia 05°59′N 116°04′E
92 M12 **Kotala** Lappi, N Finland 67°01′N 29°00′E
Kotamobagoe *see* Kotamobagu
171 Q11 **Kotamobagu** *prev.* Kotamobagoe. Sulawesi, C Indonesia 0°46′N 124°21′E
155 L14 **Kotapad** *var.* Kotapārh. Odisha, E India 19°10′N 82°23′E
Kotapārh *see* Kotapad
166 N17 **Ko Ta Ru Tao** *island* SW Thailand
169 R13 **Kotawaringin, Teluk** *bay* Borneo, C Indonesia
149 Q13 **Kot Diji** Sind, SE Pakistan 27°16′N 68°44′E
152 K9 **Kotdwāra** Uttarakhand, N India 29°44′N 78°33′E
125 Q14 **Kotel'nich** Kirovskaya Oblast', NW Russian Federation 58°19′N 48°12′E
127 N12 **Kotel'nikovo** Volgogradskaya Oblast', SW Russian Federation 47°37′N 43°07′E
123 Q6 **Kotel'nyy, Ostrov** *island* Novosibirskiye Ostrova, N Russian Federation
117 T5 **Kotel'va** Poltavs'ka Oblast', C Ukraine 50°04′N 34°46′E
101 M14 **Köthen** *var.* Cöthen. Sachsen-Anhalt, C Germany 51°46′N 11°59′E
Kōti *see* Kōchi
81 G17 **Kotido** NE Uganda 03°03′N 34°07′E
93 N19 **Kotka** Kymenlaakso, S Finland 60°28′N 26°55′E
125 P11 **Kotlas** Arkhangel'skaya Oblast', NW Russian Federation 61°14′N 46°43′E
38 M10 **Kotlik** Alaska, USA 63°01′N 163°33′W
77 Q17 **Kotoka** ✕ (Accra) S Ghana 05°41′N 00°10′W
Kotonu *see* Cotonou
113 J17 **Kotor** *It.* Cattaro. SW Montenegro 42°25′N 18°47′E
Kotor *see* Kotoriba
112 F7 **Kotoriba** *Hung.* Kotor. Medimurje, N Croatia 46°20′N 16°47′E
113 I17 **Kotorska, Boka** *It.* Bocche di Cattaro. *bay* SW Montenegro
112 H11 **Kotorsko** ◆ Republika Srpska, N Bosnia and Herzegovina
112 G11 **Kotor Varoš** ◆ Republika Srpska, N Bosnia and Herzegovina
Koto Sho/Kotosho *see* Lan Yu
126 M7 **Kotovsk** Tambovskaya Oblast', W Russian Federation 52°39′N 41°31′E
117 O9 **Kotovs'k** *Rus.* Kotovsk. Odes'ka Oblast', SW Ukraine 47°42′N 29°30′E
Kotovsk *see* Hîncești
119 G16 **Kotra** ≈ W Belarus
149 P16 **Kotri** Sind, SE Pakistan 25°22′N 68°18′E
109 Q9 **Kötschach** Kärnten, S Austria 46°41′N 12°57′E
155 K15 **Kottagūdem** Telangana, E India 17°36′N 80°40′E
155 F21 **Kottappadi** Kerala, SW India 11°38′N 76°03′E
155 G23 **Kottayam** Kerala, SW India 09°34′N 76°31′E
Kottbus *see* Cottbus
79 K15 **Kotto** ≈ Central African Republic/Dem. Rep. Congo
193 X15 **Kotu Group** *island group* W Tonga
Koturdepe *see* Goturdepe
122 M9 **Kotuy** ≈ N Russian Federation
83 M16 **Kotwa** Mashonaland East, NE Zimbabwe 16°58′S 32°46′E
39 N7 **Kotzebue** Alaska, USA 66°54′N 162°36′W
38 M7 **Kotzebue Sound** *inlet* Alaska, USA
Kotzenan *see* Chocianów
Kotzman *see* Kitsman'
77 R14 **Kouandé** NW Benin 10°20′N 01°42′E
79 J15 **Kouango** Ouaka, S Central African Republic 05°00′N 20°01′E
77 O13 **Koudougou** C Burkina Faso 12°15′N 02°23′W
98 K7 **Koudum** Fryslân, N Netherlands 52°55′N 05°26′E
115 L25 **Koufonísi** *island* SE Greece
115 K21 **Koufonísi** *island* Kykládes, Greece, Aegean Sea
38 M8 **Kougarok Mountain** ▲ Alaska, USA 65°41′N 165°29′W
79 E21 **Kouilou** ◆ *province* SW Congo
79 E20 **Kouilou** ≈ S Congo
167 Q11 **Koŭk Kduŏch** *prev.* Phumĭ Koŭk Kduŏch. Bătdâmbâng, NW Cambodia 13°16′N 103°08′E
121 O3 **Koúklia** SW Cyprus 34°42′N 32°35′E
79 E19 **Koulamoutou** Ogooué-Lolo, C Gabon 01°07′S 12°27′E
76 L12 **Koulikoro** Koulikoro, SW Mali 12°55′N 07°31′W
76 L11 **Koulikoro** ◆ *region* SW Mali
187 P16 **Koumac** Province Nord, W New Caledonia 20°34′S 164°18′E
165 N12 **Koumi** Nagano, Honshū, S Japan 36°06′N 138°27′E
78 I13 **Koumra** Mandoul, S Chad 08°56′N 17°32′E
Kounadougou *see* Koundougou
76 M15 **Kounahiri** C Ivory Coast 07°47′N 05°51′W
76 I12 **Koundâra** Moyenne-Guinée, NW Guinea 12°28′N 13°15′W
77 N13 **Koundougou** *var.* Kounadougou. C Burkina Faso 11°43′N 04°40′W
76 H11 **Koungheul** C Senegal 14°00′N 14°48′W
Kounradskiy *see* Konyrat
25 X10 **Kountze** Texas, SW USA 30°22′N 94°20′W
77 Q13 **Koupéla** C Burkina Faso 12°07′N 00°21′W
77 N13 **Kouri** Sikasso, SW Mali 12°09′N 04°46′W
55 Y9 **Kourou** N French Guiana 05°08′N 52°37′W
114 J12 **Kouroú** ≈ NE Greece
76 K14 **Kouroussa** C Guinea 10°40′N 09°50′W
Kousseir *see* Al Quşayr
78 G11 **Kousséri** *prev.* Fort-Foureau. Extrême-Nord, NE Cameroon 12°05′N 14°56′E
Kouteifé *see* Al Quţayfah
76 M13 **Koutiala** Sikasso, S Mali 12°20′N 05°23′W
76 M14 **Kouto** NW Ivory Coast 09°51′N 06°25′W
93 M19 **Kouvola** Kymenlaakso, S Finland 60°50′N 26°48′E
79 G18 **Kouyou** ≈ C Congo
112 M10 **Kovačica** *Hung.* Antafalva; *prev.* Kovacsicza. Vojvodina, N Serbia 45°08′N 20°36′E
Kovacsicza *see* Kovačica
Kővárhosszúfalu *see* Satulung
Kovászna *see* Covasna
124 I4 **Kovdor** Murmanskaya Oblast', NW Russian Federation 67°32′N 30°27′E
116 J3 **Kovel'** *Pol.* Kowel. Volyns'ka Oblast', NW Ukraine 51°14′N 24°43′E
112 M11 **Kovin** *Hung.* Kevevára; *prev.* Temes-Kubin. Vojvodina, NE Serbia 44°45′N 20°59′E
Kovno *see* Kaunas
127 N3 **Kovrov** Vladimirskaya Oblast', W Russian Federation 56°24′N 41°21′E
127 O5 **Kovylkino** Respublika Mordoviya, W Russian Federation 54°03′N 43°52′E
110 J11 **Kowal** Kujawsko-pomorskie, C Poland 52°31′N 19°09′E
110 J9 **Kowalewo Pomorskie** *Ger.* Schönsee. Kujawsko-pomorskie, N Poland 53°07′N 18°48′E
Kowasna *see* Covasna
119 M16 **Kowbcha** *Rus.* Kolbcha. Mahilyowskaya Voblasts', E Belarus 53°39′N 29°14′E
Koweit *see* Kuwait
Kowel *see* Kovel'
185 F17 **Kowhitirangi** West Coast, South Island, New Zealand 42°54′S 171°01′E
161 O15 **Kowloon** Hong Kong, S China
Kowno *see* Kaunas
159 N7 **Kox Kuduk** *well* NW China
136 D16 **Köyceğiz** Muğla, SW Turkey 36°57′N 28°40′E
125 N6 **Koyda** Arkhangel'skaya Oblast', NW Russian Federation 66°22′N 42°42′E
139 T3 **Koye** *Ar.* Kūysanjaq, *var.* Koi Sanjaq. Arbīl, N Iraq 36°05′N 44°38′E
Koymat *see* Goymat
Koymatdag, Gory *see* Goymatdag, Gory
151 E15 **Koyna Reservoir** ⊠ W India
165 P9 **Koyoshi-gawa** ≈ Honshū, C Japan
Koi Sanjaq *see* Koye
Koytash *see* Qo'ytosh
146 M14 **Köýtendag** *prev. Rus.* Charshanga, Charshangy, *Turkm.* Charshangngy. Lebap Welaýaty, E Turkmenistan 37°31′N 65°58′E
39 N9 **Koyuk** Alaska, USA 64°55′N 161°09′W
39 N9 **Koyuk River** ≈ Alaska, USA
39 O9 **Koyukuk** Alaska, USA 64°52′N 157°42′W
39 O9 **Koyukuk River** ≈ Alaska, USA
136 J13 **Kozaklı** Nevşehir, C Turkey 39°12′N 34°48′E
136 K16 **Kozan** Adana, S Turkey 37°27′N 35°47′E
115 E14 **Kozáni** Dytikí Makedonía, N Greece 40°19′N 21°48′E
112 F10 **Kozara** ▲▲ NW Bosnia and Herzegovina
Kozarska Dubica *see* Bosanska Dubica
117 P3 **Kozelets'** *Rus.* Kozelets. Chernihivs'ka Oblast', NE Ukraine 50°54′N 31°09′E
Kozelets *see* Kozelets'
117 S6 **Kozel'shchyna** Poltavs'ka Oblast', C Ukraine 49°13′N 33°49′E
126 J5 **Kozel'sk** Kaluzhskaya Oblast', W Russian Federation 54°04′N 35°51′E
151 F21 **Kozhikode** *var.* Calicut. Kerala, SW India 11°17′N 75°49′E *see also* Calicut
Kozhimiz, Gora *see* Kozhymiz, Gora
124 L9 **Kozhozero, Ozero** ◎ NW Russian Federation
125 T7 **Kozhva** Respublika Komi, NW Russian Federation 65°06′N 57°00′E
125 T7 **Kozhva** ≈ NW Russian Federation
125 U6 **Kozhym** Respublika Komi, NW Russian Federation 65°43′N 59°25′E
125 V9 **Kozhymiz, Gora** *prev.* Gora Kozhimiz. ▲ NW Russian Federation 63°13′N 58°54′E
110 N13 **Kozienice** Mazowieckie, C Poland 51°35′N 21°31′E
109 S13 **Kozina** SW Slovenia 45°36′N 13°56′E
114 H7 **Kozloduy** Vratsa, NW Bulgaria 43°48′N 23°42′E
127 Q3 **Kozlovka** Chuvashskaya Respublika, W Russian Federation 55°53′N 48°07′E
Kozlovshchina/Kozłowszczyzna *see* Kazlowshchyna
127 P3 **Koz'modem'yansk** Respublika Mariy El, W Russian Federation 56°19′N 46°33′E
116 J6 **Kozova** Ternopil's'ka Oblast', W Ukraine 49°25′N 25°09′E
113 P20 **Kožuf** ▲▲ Greece/Macedonia
165 N15 **Kōzu-shima** *island* E Japan
Koz'yany *see* Kazyany
117 N5 **Kozyatyn** *Rus.* Kazatin. Vinnyts'ka Oblast', C Ukraine 49°41′N 28°49′E
77 Q16 **Kpalimé** *var.* Palimé. SW Togo 06°54′N 00°38′E
77 Q16 **Kpandu** E Ghana 07°00′N 00°18′E
99 F15 **Krabbendijke** Zeeland, SW Netherlands 51°25′N 04°07′E
167 N15 **Krabi** *var.* Muang Krabi. Krabi, SW Thailand 08°04′N 98°52′E
167 N13 **Kra Buri** Ranong, SW Thailand 10°25′N 98°48′E
167 S12 **Krâchéh** *prev.* Kratie. Krâchéh, E Cambodia 12°29′N 106°01′E
95 G17 **Kragerø** Telemark, S Norway 58°54′N 09°25′E
112 M13 **Kragujevac** Serbia, C Serbia 44°01′N 20°55′E
Krainburg *see* Kranj
166 N13 **Kra, Isthmus of** *isthmus* Malaysia/Thailand
112 D12 **Krajina** *cultural region* SW Croatia
Krakatau, Pulau *see* Rakata, Pulau
Krakau *see* Kraków
111 L16 **Kraków** *Eng.* Cracow, *Ger.* Krakau; *anc.* Cracovia. Małopolskie, S Poland 50°03′N 19°58′E
100 L9 **Krakower See** ◎ NE Germany
167 Q11 **Krâlănh** Siĕmréab, NW Cambodia 13°35′N 103°27′E
45 Q16 **Kralendijk** ○ Bonaire 12°07′N 68°13′W
112 B10 **Kraljevica** *It.* Porto Re. Primorje-Gorski Kotar, NW Croatia 45°15′N 14°36′E
112 M13 **Kraljevo** *prev.* Rankovićevo. Serbia, C Serbia 43°44′N 20°40′E
111 E16 **Královehradecký Kraj** *prev.* Hradecký Kraj. ◆ *region* N Czech Republic
Kralup an der Moldau *see* Kralupy nad Vltavou
111 C16 **Kralupy nad Vltavou** *Ger.* Kralup an der Moldau. Středočeský Kraj, NW Czech Republic 50°15′N 14°20′E
117 W7 **Kramators'k** *Rus.* Kramatorsk. Donets'ka Oblast', SE Ukraine 48°43′N 37°34′E
Kramatorsk *see* Kramators'k
93 H17 **Kramfors** Västernorrland, C Sweden 62°55′N 17°50′E
Kranéa *see* Kraniá
108 M7 **Kranebitten** ✕ (Innsbruck) Tirol, W Austria 47°18′N 11°21′E
115 D15 **Kraniá** *var.* Kranéa. Dytikí Makedonía, N Greece 39°54′N 21°21′E
115 G20 **Kranídi** Pelopónnisos, S Greece 37°21′N 23°09′E
109 T11 **Kranj** *Ger.* Krainburg. NW Slovenia 46°17′N 14°16′E
115 F16 **Krannón** *battleground* Thessalía, C Greece
Kranz *see* Zelenogradsk
112 D7 **Krapina** Krapina-Zagorje, N Croatia 46°12′N 15°52′E
112 E8 **Krapina** ≈ N Croatia
112 D8 **Krapina-Zagorje** *off.* Krapinsko-Zagorska Županija. ◆ *province* N Croatia
114 L7 **Krapinets** ≈ NE Bulgaria
Krapinsko-Zagorska Županija *see* Krapina-Zagorje
111 I15 **Krapkowice** *Ger.* Krappitz. Opolskie, SW Poland 50°29′N 17°56′E
Krappitz *see* Krapkowice
125 O12 **Krasavino** Vologodskaya Oblast', NW Russian Federation 60°56′N 46°27′E
122 H6 **Krasino** Novaya Zemlya, Arkhangel'skaya Oblast', N Russian Federation 70°45′N 54°16′E
123 S15 **Kraskino** Primorskiy Kray, SE Russian Federation 42°40′N 130°51′E
118 J11 **Krāslava** SE Latvia 55°56′N 27°08′E
119 M14 **Krasnaluki** *Rus.* Krasnoluki. Vitsyebskaya Voblasts', N Belarus 54°37′N 28°50′E
119 P17 **Krasnapollye** *Rus.* Krasnopol'ye. Mahilyowskaya Voblasts', E Belarus 53°20′N 31°24′E
126 L15 **Krasnaya Polyana** Krasnodarskiy Kray, SW Russian Federation 43°40′N 40°13′E
Krasnaya Slabada / Krasnaya Sloboda *see* Chyrvonaya Slabada
119 J15 **Krasnaye** *Rus.* Krasnoye. Minskaya Voblasts', C Belarus 54°14′N 27°05′E
111 O14 **Kraśnik** *Ger.* Kratznick. Lubelskie, E Poland 50°56′N 22°14′E
117 O9 **Krasni Okny** Odes'ka Oblast', SW Ukraine 47°33′N 29°28′E
127 P8 **Krasnoarmeysk** Saratovskaya Oblast', W Russian Federation 51°02′N 45°42′E
Krasnoarmeysk *see* Tayynsha
Krasnoarmeysk *see* Krasnoarmiys'k/Tayynsha
123 T6 **Krasnoarmeyskiy** Chukotskiy Avtonomnyy Okrug, NE Russian Federation 69°30′N 171°44′E
117 W7 **Krasnoarmiys'k** *Rus.* Krasnoarmeysk. Donets'ka Oblast', SE Ukraine 48°17′N 37°14′E
125 P11 **Krasnoborsk** Arkhangel'skaya Oblast', NW Russian Federation 61°31′N 45°57′E
126 K14 **Krasnodar** *prev.* Ekaterinodar, Yekaterinodar. Krasnodarskiy Kray, SW Russian Federation 45°06′N 39°01′E
126 K13 **Krasnodarskiy Kray** ◆ *territory* SW Russian Federation
117 Z7 **Krasnodon** Luhans'ka Oblast', E Ukraine 48°17′N 39°44′E
Krasnogor *see* Kallaste
127 T2 **Krasnogorskoye** Udmurtskaya Respublika, NW Russian Federation 57°42′N 52°29′E
Krasnograd *see* Krasnohrad
Krasnogvardeysk *see* Bulung'ur
126 M13 **Krasnogvardeyskoye** Stavropol'skiy Kray, SW Russian Federation 45°49′N 41°31′E
Krasnogvardeyskoye *see* Krasnohvardiys'ke
117 U6 **Krasnohrad** *Rus.* Krasnograd. Kharkivs'ka Oblast', E Ukraine 49°22′N 35°28′E
117 S12 **Krasnohvardiys'ke** *Rus.* Krasnogvardeyskoye. Avtonomna Respublika Krym, S Ukraine 45°30′N 34°19′E
123 P14 **Krasnokamensk** Zabaykal'skiy Kray, S Russian Federation 50°03′N 118°01′E
125 U14 **Krasnokamsk** Permskiy Kray, W Russian Federation 58°08′N 55°48′E
127 U8 **Krasnokholm** Orenburgskaya Oblast', W Russian Federation 51°34′N 54°11′E
117 U5 **Krasnokuts'k** *Rus.* Krasnokutsk. Kharkivs'ka Oblast', E Ukraine 50°01′N 35°03′E
Krasnokutsk *see* Krasnokuts'k
Krasnoluki *see* Krasnaluki
126 L7 **Krasnolesnyy** Voronezhskaya Oblast', W Russian Federation 51°53′N 39°37′E

◆ Country ● Country Capital ◇ Dependent Territory ○ Dependent Territory Capital ◆ Administrative Regions ✕ International Airport ▲ Mountain ▲▲ Mountain Range Volcano ≈ River ◎ Lake ⊠ Reservoir
273

KRASNOLUKI – KUPYANSK

Krasnoluki see Krasnaluki
Krasnoosol'skoye Vodokhranilishche see Chervonooskil's'ke Vodoskhovyshche
117 S11 **Krasnoperekops'k** *Rus.* Krasnoperekopsk. Avtonomna Respublika Krym, S Ukraine 45°56′N 33°47′E
Krasnoperekopsk see Krasnoperekops'k
117 U4 **Krasnopillya** Sums'ka Oblast', NE Ukraine 50°46′N 35°17′E
Krasnopol'ye see Krasnapollye
124 L5 **Krasnoshchel'ye** Murmanskaya Oblast', NW Russian Federation 67°22′N 37°03′E
127 O5 **Krasnoslobodsk** Respublika Mordoviya, W Russian Federation 54°24′N 43°51′E
127 T2 **Krasnoslobodsk** Volgogradskaya Oblast', SW Russian Federation 48°41′N 44°34′E
Krasnostav see Krasnystaw
127 V5 **Krasnousol'skiy** Respublika Bashkortostan, W Russian Federation 53°55′N 56°22′E
125 U12 **Krasnovishersk** Permskiy Kray, NW Russian Federation 60°22′N 57°04′E
Krasnovodsk see Türkmenbaşy
Krasnovodskiy Zaliv see Türkmenbasy Aylagy
146 B10 **Krasnovodskoye Plato** *Turkm.* Krasnowodsk Platosy. *plateau* NW Turkmenistan
Krasnowodsk Aylagy see Türkmenbasy Aylagy
Krasnowodsk Platosy see Krasnovodskoye Plato
122 K12 **Krasnoyarsk** Krasnoyarskiy Kray, S Russian Federation 56°05′N 92°46′E
127 X7 **Krasnoyarskiy** Orenburgskaya Oblast', W Russian Federation 51°56′N 59°54′E
122 K11 **Krasnoyarskiy Kray** ◆ *territory* C Russian Federation
Krasnoye see Krasnaye
Krasnoye Znamya see Gyzylbaydak
125 R11 **Krasnozatonskiy** Respublika Komi, NW Russian Federation 61°39′N 51°00′E
118 D13 **Krasnoznamensk** *prev.* Lasdehnen, *Ger.* Haselberg. Kaliningradskaya Oblast', W Russian Federation 54°57′N 22°28′E
126 K3 **Krasnoznamensk** Moskovskaya Oblast', W Russian Federation 55°40′N 37°05′E
117 R11 **Krasnoznam"yans'kyy Kanal** *canal* S Ukraine
111 P14 **Krasnystaw** *Rus.* Krasnostav. Lubelskie, SE Poland 51°N 23°10′E
126 H4 **Krasnyy** Smolenskaya Oblast', W Russian Federation 54°36′N 31°27′E
127 P2 **Krasnyye Baki** Nizhegorodskaya Oblast', W Russian Federation 57°07′N 45°12′E
127 Q13 **Krasnyye Barrikady** Astrakhanskaya Oblast', SW Russian Federation 46°14′N 47°48′E
124 K15 **Krasnyy Kholm** Tverskaya Oblast', W Russian Federation 58°04′N 37°05′E
127 Q8 **Krasnyy Kut** Saratovskaya Oblast', W Russian Federation 50°54′N 46°58′E
Krasnyy Liman see Krasnyy Lyman
117 Y7 **Krasnyy Luch** *prev.* Krindachevka. Luhans'ka Oblast', E Ukraine 48°09′N 38°52′E
117 X6 **Krasnyy Lyman** *Rus.* Krasnyy Liman. Donets'ka Oblast', SE Ukraine 49°00′N 37°50′E
127 R3 **Krasnyy Steklovar** Respublika Mariy El, W Russian Federation 56°14′N 48°49′E
127 P8 **Krasnyy Tekstil'shchik** Saratovskaya Oblast', W Russian Federation 51°35′N 45°49′E
127 R13 **Krasnyy Yar** Astrakhanskaya Oblast', SW Russian Federation 46°33′N 48°21′E
Krassóvár see Caraşova
116 L5 **Krasyliv** Khmel'nyts'ka Oblast', W Ukraine 49°38′N 26°59′E
111 O21 **Kraszna** *Rom.* Crasna. ≈ Hungary/Romania
Kratie see Krâchéh
113 P17 **Kratovo** NE FYR Macedonia 42°04′N 22°08′E
Kratznick see Kraśnik
171 Y13 **Krau** Papua, E Indonesia 03°15′S 140°07′E
167 Q13 **Krâvanh, Chuŏr Phnum** *Eng.* Cardamom Mountains, *Fr.* Chaîne des Cardamomes. ▲ W Cambodia
Kravasta Lagoon see Karavastasë, Laguna e
Krawang see Karawang
127 Q15 **Kraynovka** Respublika Dagestan, SW Russian Federation 43°58′N 47°24′E
118 D12 **Kražiai** Šiaulai, C Lithuania 55°36′N 22°41′E
27 P11 **Krebs** Oklahoma, C USA 34°55′N 95°43′W
101 D15 **Krefeld** Nordrhein-Westfalen, W Germany 51°20′N 06°34′E
Kreisstadt see Krosno Odrzańskie
115 D17 **Kremastón, Technití Límni** ◎ C Greece
Kremenchug see Kremenchuk
Kremenchugskoye Vodokhranilishche/Kremenchuk Reservoir see Kremenchuts'ke Vodoskhovyshche
117 S6 **Kremenchuk** *Rus.* Kremenchug. Poltavs'ka Oblast', NE Ukraine 49°04′N 33°27′E
117 R6 **Kremenchuts'ke Vodoskhovyshche** *Eng.* Kremenchuk Reservoir, *Rus.* Kremenchugskoye Vodokhranilishche. ◎ C Ukraine
116 K5 **Kremenets'** *Pol.* Krzemieniec, *Rus.* Kremenets. Ternopil's'ka Oblast', W Ukraine 50°06′N 25°43′E
Kremennaya see Kreminna
117 X6 **Kreminna** *Rus.* Kremennaya. Luhans'ka Oblast', E Ukraine 49°03′N 38°15′E
37 R4 **Kremmling** Colorado, C USA 40°03′N 106°23′W
109 V3 **Krems** ≈ NE Austria
Krems see Krems an der Donau
109 W3 **Krems an der Donau** *var.* Krems. Niederösterreich, N Austria 48°25′N 15°36′E
Kremsier see Kroměříž
109 S4 **Kremsmünster** Oberösterreich, N Austria 48°04′N 14°08′E
38 M17 **Krenitzin Islands** *island* Aleutian Islands, Alaska, USA
Kresena see Kresna
114 G11 **Kresna** *var.* Kresena. Blagoevgrad, SW Bulgaria 41°43′N 23°10′E
112 O12 **Krespoljin** Serbia, E Serbia 44°22′N 21°36′E
25 N4 **Kress** Texas, SW USA 34°21′N 101°43′W
123 V6 **Kresta, Zaliv** *bay* E Russian Federation
115 D20 **Kréstena** *prev.* Selinoús. Dytikí Elláda, S Greece 37°36′N 21°36′E
124 H14 **Krestsy** Novgorodskaya Oblast', W Russian Federation 58°15′N 32°28′E
Kretikon Delagos see Kritikó Pélagos
118 C11 **Kretinga** *Ger.* Krottingen. Klaipėda, NW Lithuania 55°53′N 21°13′E
Kreutz see Cristuru Secuiesc
Kreuz see Križevci, Croatia
Kreuz see Risti, Estonia
Kreuzburg/Kreuzburg in Oberschlesien see Kluczbork
Kreuzingen see Bol'shakovo
108 H6 **Kreuzlingen** Thurgau, NE Switzerland 47°38′N 09°12′E
101 K25 **Kreuzspitze** ▲ S Germany 47°30′N 10°55′E
101 F16 **Kreuztal** Nordrhein-Westfalen, W Germany 50°58′N 08°00′E
119 I15 **Kreva** *Rus.* Krevo. Hrodzyenskaya Voblasts', W Belarus 54°19′N 26°17′E
Krevo see Kreva
Kría Vrísi see Krýa Vrýsi
79 D16 **Kribi** Sud, SW Cameroon 02°53′N 09°57′E
Krichëv see Krychaw
Krickerhäu/Kriegerhaj see Handlová
109 W6 **Krieglach** Steiermark, E Austria 47°33′N 15°37′E
108 F8 **Kriens** Luzern, W Switzerland 47°03′N 08°17′E
Krievija see Russian Federation
Krimmitschau see Crimmitschau
98 H12 **Krimpen aan den IJssel** Zuid-Holland, SW Netherlands 51°56′N 04°39′E
Krindachevka see Krasnyy Luch
115 G25 **Kríos, Akrotírio** *headland* Kríti, Greece, E Mediterranean Sea 35°17′N 23°31′E
155 J16 **Krishna** *prev.* Kistna. ≈ C India
155 H20 **Krishnagiri** Tamil Nādu, SE India 12°33′N 78°11′E
155 K17 **Krishna, Mouths of the** *delta* SE India
153 S15 **Krishnanagar** West Bengal, N India 23°22′N 88°32′E
155 G20 **Krishnarājāsāgara** *var.* Paradip. ◎ W India
95 N19 **Kristdala** Kalmar, S Sweden 57°24′N 16°12′E
Kristiania see Oslo
95 E18 **Kristiansand** *var.* Christiansand. Vest-Agder, S Norway 58°08′N 07°52′E
95 L22 **Kristianstad** Skåne, S Sweden 56°02′N 14°10′E
94 F8 **Kristiansund** *var.* Christiansund. Møre og Romsdal, S Norway 63°07′N 07°45′E
Kristiinankaupunki see Kristinestad
93 I14 **Kristineberg** Västerbotten, N Sweden 65°07′N 18°36′E
95 L16 **Kristinehamn** Värmland, C Sweden 59°17′N 14°09′E
93 J17 **Kristinestad** *Fin.* Kristiinankaupunki. Österbotten, W Finland 62°15′N 21°24′E
Kristyor see Crişcior
115 J25 **Kríti** *Eng.* Crete. ◆ *region* Greece, Aegean Sea
115 J24 **Kríti** *Eng.* Crete. *island* Greece, Aegean Sea
115 J23 **Kritikó Pélagos** *var.* Kretikon Delagos, *Eng.* Sea of Crete; *anc.* Mare Creticum. *sea* Greece, Aegean Sea
Kriulyany see Criuleni
112 I12 **Krivaja** ≈ NE Bosnia and Herzegovina
Krivaja see Mali Idoš
113 P17 **Kriva Palanka** *Turk.* Eğri Palanka. NE Macedonia 42°13′N 22°19′E
Krivichi see Kryvichy
114 H8 **Krivodol** Vratsa, NW Bulgaria 43°23′N 23°30′E
126 M10 **Krivorozh'ye** Rostovskaya Oblast', SW Russian Federation 48°51′N 40°49′E
Krivoshin see Kryvoshyn
Krivoy Rog see Kryvyy Rih
112 F7 **Križevci** *Ger.* Kreuz, *Hung.* Kőrös. Varaždin, NE Croatia 46°02′N 16°32′E
112 B10 **Krk** *It.* Veglia. Primorje-Gorski Kotar, NW Croatia 45°01′N 14°36′E
112 B10 **Krk** *It.* Veglia; *anc.* Curieta. *island* NW Croatia
109 V12 **Krka** ≈ SE Slovenia
Krka see Gurk
109 R11 **Krn** ▲ NW Slovenia 46°15′N 13°37′E
111 H16 **Krnov** *Ger.* Jägerndorf. Moravskoslezský Kraj, E Czech Republic 50°05′N 17°40′E
Kroatien see Croatia
95 G14 **Krøderen** Buskerud, S Norway 60°06′N 09°48′E
95 G14 **Krøderen** ◎ S Norway
Kroi see Krui
95 N17 **Krokek** Östergötland, S Sweden 58°40′N 16°25′E
Krokodil see Crocodile
93 G16 **Krokom** Jämtland, C Sweden 63°20′N 14°30′E
117 S2 **Krolevets'** *Rus.* Krolevets. Sums'ka Oblast', NE Ukraine 51°34′N 33°24′E
Krolevets' see Krolevets'
Królewska Huta see Chorzów
111 H18 **Kroměříž** *Ger.* Kremsier. Zlínský Kraj, E Czech Republic 49°18′N 17°24′E
98 I9 **Krommenie** Noord-Holland, C Netherlands 52°30′N 04°46′E
126 J6 **Kromy** Orlovskaya Oblast', W Russian Federation 52°41′N 35°45′E
101 L18 **Kronach** Bayern, E Germany 50°14′N 11°19′E
Krone an der Brahe see Koronowo
Krŏng Kaôh Kŏng see Kaôh Kŏng
95 K21 **Kronoberg** ◆ *county* S Sweden
123 V10 **Kronotskiy Zaliv** *bay* E Russian Federation
195 O2 **Kronprinsesse Märtha Kyst** *physical region* Antarctica
195 V3 **Kronprins Olav Kyst** *physical region* Antarctica
124 G12 **Kronshtadt** Leningradskaya Oblast', NW Russian Federation 60°01′N 29°42′E
Kronstadt see Braşov
83 I22 **Kroonstad** Free State, C South Africa 27°40′S 27°15′E
123 O12 **Kropotkin** Irkutskaya Oblast', C Russian Federation 58°30′N 115°21′E
126 L14 **Kropotkin** Krasnodarskiy Kray, SW Russian Federation 45°29′N 40°31′E
110 J11 **Krośniewice** Łódzkie, C Poland 52°14′N 19°10′E
111 N17 **Krosno** *Ger.* Krossen. Podkarpackie, SE Poland 49°40′N 21°46′E
110 E12 **Krosno Odrzańskie** *Ger.* Crossen, Kreisstadt. Lubuskie, W Poland 52°02′N 15°06′E
Krossen see Krosno
Krotoschin see Krotoszyn
110 H13 **Krotoszyn** *Ger.* Krotoschin. Wielkopolskie, C Poland 51°43′N 17°24′E
Krottingen see Kretinga
Krousón see Krousónas
115 J25 **Krousónas** *prev.* Krousón, Kroussón. Kríti, Greece, E Mediterranean Sea 35°14′N 24°59′E
Kroussón see Krousónas
Krraba see Krrabë
113 L20 **Krrabë** *var.* Krraba. Tiranë, C Albania 41°15′N 19°56′E
113 L17 **Krrabit, Mali i** ▲ N Albania
109 W12 **Krško** *Ger.* Gurkfeld; *prev.* Videm-Krško. E Slovenia 45°57′N 15°31′E
83 K19 **Kruger National Park** *national park* Northern, N South Africa
83 J21 **Krugersdorp** Gauteng, NE South Africa 26°06′S 27°46′E
38 D16 **Krugloi Point** *headland* Agattu Island, Alaska, USA 52°30′N 173°46′E
Kruglyye see Kruhlaye
119 N15 **Kruhlaye** *Rus.* Krugloye. Mahilyowskaya Voblasts', E Belarus 54°15′N 29°48′E
168 L15 **Krui** *var.* Kroi. Sumatera, SW Indonesia 05°11′S 103°55′E
99 G16 **Kruibeke** Oost-Vlaanderen, N Belgium 51°10′N 04°18′E
83 G25 **Kruidfontein** Western Cape, SW South Africa 32°50′S 21°59′E
99 F15 **Kruiningen** Zeeland, SW Netherlands 51°28′N 04°01′E
Kruja see Krujë
113 L19 **Krujë** *var.* Kruja, *It.* Croia. Durrës, C Albania 41°30′N 19°48′E
Krulevshchina/Krulewshchyna see Krulyewshchyzna
118 K13 **Krulyewshchyzna** *Rus.* Krulevshchina, Krulewshchyna. Vitsyebskaya Voblasts', N Belarus 55°02′N 27°45′E
25 T6 **Krum** Texas, SW USA 33°15′N 97°14′W
101 J23 **Krumbach** Bayern, S Germany 48°12′N 10°21′E
113 M17 **Krumë** Kukës, NE Albania 42°11′N 20°25′E
Krummau see Český Krumlov
114 K12 **Krumovgrad** *prev.* Kossukavak. Yambol, E Bulgaria 41°27′N 25°40′E
114 K12 **Krumovitsa** ≈ S Bulgaria
114 L10 **Krumovo** Yambol, E Bulgaria 42°16′N 26°25′E
167 O11 **Krung Thep, Ao** *var.* Bight of Bangkok. *bay* S Thailand
Krung Thep Mahanakhon see Ao Krung Thep
Krupa/Krupa na Uni see Bosanska Krupa
119 M15 **Krupki** Minskaya Voblasts', C Belarus 54°19′N 29°08′E
95 G24 **Kruså** *var.* Krusaa. Syddanmark, SW Denmark 54°50′N 09°25′E
Krusaa see Kruså
113 N14 **Kruševac** Serbia, C Serbia 43°37′N 21°20′E
113 N19 **Kruševo** SW FYR Macedonia 41°22′N 21°15′E
111 A15 **Krušné Hory** *Eng.* Ore Mountains, *Ger.* Erzgebirge. ▲ Czech Republic/Germany *see also* Erzgebirge
Krušné Hory see Erzgebirge
39 W13 **Kruzof Island** *island* Alexander Archipelago, Alaska, USA
114 F13 **Krýa Vrýsi** *var.* Kría Vrísi. Kentrikí Makedonía, N Greece 40°41′N 22°18′E
119 P16 **Krychaw** *Rus.* Krichëv. Mahilyowskaya Voblasts', E Belarus 53°42′N 31°43′E
64 K11 **Krylov Seamount** *undersea feature* E Atlantic Ocean 17°35′N 30°07′W
Krym see Krym, Avtonomna Respublika
117 S13 **Krym, Avtonomna Respublika** *var.* Krym, *Eng.* Crimea, Crimean Oblast; *prev. Rus.* Krymskaya ASSR, Krymskaya Oblast'. ◆ *province* SE Ukraine
126 K14 **Krymsk** Krasnodarskiy Kray, SW Russian Federation 44°56′N 38°02′E
Krymskaya ASSR/Krymskaya Oblast' see Krym, Avtonomna Respublika
117 T13 **Kryms'ki Hory** ▲ S Ukraine
117 T13 **Kryms'kyy Pivostriv** *peninsula* S Ukraine
111 M18 **Krynica** *Ger.* Tannenhof. Małopolskie, S Poland 49°25′N 20°56′E
117 P8 **Kryve Ozero** Odes'ka Oblast', SW Ukraine 47°54′N 30°19′E
119 K14 **Kryvichy** *Rus.* Krivichi. Minskaya Voblasts', C Belarus 54°43′N 27°17′E
119 I18 **Kryvoshyn** *Rus.* Krivoshin. Brestskaya Voblasts', SW Belarus 52°52′N 26°08′E
117 S8 **Kryvyy Rih** *Rus.* Krivoy Rog. Dnipropetrovs'ka Oblast', SE Ukraine 47°53′N 33°24′E
117 N8 **Kryzhopil'** Vinnyts'ka Oblast', C Ukraine 48°22′N 28°51′E
Krzemieniec see Kremenets'
111 J14 **Krzepice** Śląskie, S Poland 50°58′N 18°42′E
110 F10 **Krzyż Wielkopolskie** Wielkopolskie, W Poland 52°52′N 16°03′E
Ksar al Kabir see Ksar-el-Kebir
Ksar al Soule see Er-Rachidia
74 J5 **Ksar El Boukhari** N Algeria 35°55′N 02°47′E
74 G5 **Ksar-el-Kebir** *var.* Alcázar, Ksar al Kabir, Ksar-el-Kébir, *Ar.* Al-Kasr al-Kebir, Al-Qsar al-Kbir, *Sp.* Alcazarquivir. NW Morocco 35°04′N 05°56′W
Ksar-el-Kébir see Ksar-el-Kebir
110 H12 **Książ Wielkopolski** *Ger.* Xions. Weilkopolskie, W Poland 52°03′N 17°10′E
127 O3 **Kstovo** Nizhegorodskaya Oblast', W Russian Federation 56°07′N 44°12′E
169 T8 **Kuala Belait** W Brunei 04°48′N 114°12′E
Kuala Dungun see Dungun
169 S10 **Kualakeriau** Borneo, C Indonesia
169 S12 **Kualakuayan** Borneo, C Indonesia 02°01′S 112°35′E
168 K8 **Kuala Lipis** Pahang, Peninsular Malaysia 04°11′N 102°00′E
168 K9 **Kuala Lumpur** ● (Malaysia) Kuala Lumpur, Peninsular Malaysia 03°08′N 101°42′E
168 K9 **Kuala Lumpur International** ✈ Selangor, Peninsular Malaysia 02°51′N 101°45′E
Kuala Pelabohan Kelang see Pelabuhan Klang
169 U7 **Kuala Penyu** Sabah, East Malaysia 05°37′N 115°36′E
38 E9 **Kualapu'u** *var.* Kualapuu. Moloka'i, Hawaii, USA, C Pacific Ocean 21°09′N 157°02′W
Kualapuu see Kualapu'u
168 L7 **Kuala Terengganu** *var.* Kuala Trengganu. Terengganu, Peninsular Malaysia 05°20′N 103°07′E
168 L11 **Kualatungkal** Sumatera, W Indonesia 0°49′S 103°22′E
171 P11 **Kuandang** Sulawesi, N Indonesia 0°50′N 122°55′E
163 V12 **Kuandian** *var.* Kuandian Manzu Zizhixian. Liaoning, NE China 40°41′N 124°46′E
Kuandian Manzu Zizhixian see Kuandian
83 E15 **Kuando Kubango** *prev.* Cuando Cubango. ◆ *province* SE Angola
Kuang-chou see Guangzhou
Kuang-hsi see Guangxi Zhuangzu Zizhiqu
Kuang-tung see Guangdong
Kuang-yuan see Guangyuan
Kuantan, Batang see Indragiri, Sungai
Kuanzhou see Qingjian
Kuba see Quba
Kubango see Cubango/Okavango
141 X8 **Kubārah** NW Oman 23°03′N 56°52′E
93 H16 **Kubbe** Västernorrland, C Sweden 63°31′N 18°04′E
80 A11 **Kubbum** Southern Darfur, W Sudan 11°47′N 23°47′E
124 L13 **Kubenskoye, Ozero** ◎ NW Russian Federation
146 G6 **Kubla-Ustyurt** *Rus.* Komsomol'sk-na-Ustyurte. Qoraqalpog'iston Respublikasi, NW Uzbekistan 44°06′N 58°14′E
164 G15 **Kubokawa** Kōchi, Shikoku, SW Japan 33°22′N 133°14′E
114 L7 **Kubrat** *prev.* Balbunar. Razgrad, N Bulgaria 43°48′N 26°31′E
112 O13 **Kučajske Planine** ▲ E Serbia
165 T1 **Kuccharo-ko** ◎ Hokkaidō, N Japan
112 O11 **Kučevo** Serbia, NE Serbia 44°29′N 21°42′E
Kuchan see Qūchān
169 Q10 **Kuching** *prev.* Sarawak. Sarawak, East Malaysia 01°32′N 110°20′E
169 Q10 **Kuching** ✈ Sarawak, East Malaysia 01°32′N 110°20′E
164 B17 **Kuchinoerabu-jima** *island* Nansei-shotō, SW Japan
Kuchinotsu see Minamishimabara
109 Q6 **Kuchl** Salzburg, NW Austria 47°37′N 13°12′E
148 L9 **Kūchnay Darwēshān** *prev.* Kūchnay Darweyshān. Helmand, S Afghanistan 31°02′N 64°10′E
Kūchnay Darweyshān see Kūchnay Darwēshān
Kuchurgan see Kuchurhan
117 O9 **Kuchurhan** *Rus.* Kuchurgan. ≈ NE Ukraine
Kuçova see Kuçovë
113 L21 **Kuçovë** *var.* Kuçova; *prev.* Qyteti Stalin. Berat, C Albania 40°48′N 19°55′E
136 D11 **Küçük Çekmece** İstanbul, NW Turkey 41°01′N 28°47′E
164 F14 **Kudamatsu** *var.* Kudamatu. Yamaguchi, Honshū, SW Japan 34°00′N 131°53′E
Kudamatu see Kudamatsu
Kudara see Ghŭdara
169 V6 **Kudat** Sabah, East Malaysia 06°54′N 116°47′E
Küddow see Gwda
155 G17 **Kūdligi** Karnātaka, W India 14°58′N 76°24′E
Kudowa see Kudowa-Zdrój
111 F16 **Kudowa-Zdrój** *Ger.* Kudowa. Wałbrzych, SW Poland 50°27′N 16°20′E
117 P9 **Kudryavtsivka** Mykolayivs'ka Oblast', S Ukraine 47°18′N 31°02′E
169 R16 **Kudus** *prev.* Koedoes. Jawa, C Indonesia 06°46′S 110°48′E
125 T13 **Kudymkar** Permskiy Kray, NW Russian Federation 59°01′N 54°40′E
Kudzsir see Cugir
Kuei-chou see Guizhou
Kuei-lin see Guilin
Kuei-Yang/Kuei-yang see Guiyang
K'u-erh-lo see Korla
Kueyang see Guiyang
Kufa see Al Kūfah
136 E14 **Küfiçayı** ≈ C Turkey
109 O6 **Kufstein** Tirol, W Austria 47°36′N 12°10′E
9 N7 **Kugaaruk** *prev.* Pelly Bay. Nunavut, N Canada 68°38′N 89°45′W
Kugaly see Kogaly
8 K8 **Kugluktuk** *var.* Qurlurtuuq; *prev.* Coppermine. Nunavut, NW Canada 67°49′N 115°12′W
143 Y13 **Kūhak** Sīstān va Balūchestān, SE Iran 27°10′N 63°15′E
143 R9 **Kūhbonān** Kermān, C Iran 31°23′N 56°16′E
148 J5 **Kūhestān** *var.* Kohsān. Herāt, W Afghanistan 34°40′N 61°11′E
Kūhīrī see Kahīrī
93 N15 **Kuhmo** Kainuu, E Finland 64°04′N 29°34′E
93 L18 **Kuhmoinen** Keski-Suomi, C Finland 61°32′N 25°09′E
Kuhnau see Konin
Kühnö see Kihnu
143 O8 **Kūhpāyeh** Eşfahān, C Iran 32°42′N 52°25′E
167 O12 **Kui Buri** *var.* Ban Kui Nua. Prachuap Khiri Khan, SW Thailand 12°10′N 99°49′E
Kuibyshev see Kuybyshevskoye Vodokhranilishche
82 D13 **Kuito** *Port.* Silva Porto. Bié, C Angola 12°21′S 16°55′E
39 X14 **Kuiu Island** *island* Alexander Archipelago, Alaska, USA
92 L13 **Kuivaniemi** Pohjois-Pohjanmaa, C Finland 65°34′N 25°13′E
196 M15 **Kujalleo** ◆ *municipality* S Greenland
Kujalleo, Kommune see Kujalleo
77 V14 **Kujama** Kaduna, C Nigeria 10°27′N 07°39′E
110 I10 **Kujawsko-pomorskie** ◆ *province* C Poland
165 R8 **Kuji** *var.* Kuzi. Iwate, Honshū, C Japan 40°12′N 141°47′E
Kujto, Ozero see Yushkozerskoye Vodokhranilishche
Kujū-renzan see Kujū-san
164 D14 **Kujū-san** *var.* Kujū-renzan. ▲ Kyūshū, SW Japan 33°07′N 131°13′E
43 N7 **Kukalaya, Rio** *var.* Rio Cuculaya, Rio Kukulaya. ≈ NE Nicaragua
113 O16 **Kukavica** *var.* Vlajna. ▲ SE Serbia 42°46′N 21°58′E
113 M18 **Kukës** *var.* Kukësi. Kukës, NE Albania 42°03′N 20°25′E
113 L18 **Kukës** ◆ *district* NE Albania
Kukësi see Kukës
186 D8 **Kukipi** Gulf, S Papua New Guinea 08°11′S 146°09′E
127 S3 **Kukmor** Respublika Tatarstan, W Russian Federation 56°11′N 50°56′E
Kukong see Shaoguan
39 N6 **Kukpowruk River** ≈ Alaska, USA
38 M6 **Kukpuk River** ≈ Alaska, USA
Kükürtdağ see Gogi, Mount
Kukukhoto see Hohhot
Kukulaya, Rio see Kukalaya, Rio
189 W12 **Kuku Point** *headland* NW Wake Island 19°19′N 166°36′E
146 G11 **Kukurtli** Ahal Welaýaty, C Turkmenistan 39°58′N 58°47′E
Kül see Kūl, Rūd-e
114 F7 **Kula** Vidin, NW Bulgaria 43°55′N 22°32′E
112 K9 **Kula** Vojvodina, NW Serbia 45°37′N 19°31′E
136 D14 **Kula** Manisa, W Turkey 38°33′N 28°38′E
149 S8 **Kulāchi** Khyber Pakhtunkhwa, NW Pakistan 31°58′N 70°30′E
Kulachi see Kolāchi
Kulagino see Yesbol
168 L10 **Kulai** Johor, Peninsular Malaysia 01°41′N 103°33′E
114 M7 **Kulak** ≈ NE Bulgaria
153 T11 **Kula Kangri** *var.* Kulhakangri. ▲ Bhutan/China 28°06′N 90°19′E
144 E13 **Kulaly, Ostrov** *island* SW Kazakhstan
145 S16 **Kulan** *Kaz.* Qulan; *prev.* Lugovoy, Lugovoye. Zhambyl, S Kazakhstan 42°54′N 72°45′E
147 V9 **Kulanak** Narynskaya Oblast', C Kyrgyzstan 41°18′N 75°38′E
Gory Kulandag see Gulandag
153 V14 **Kulaura** Sylhet, NE Bangladesh 24°32′N 92°02′E
118 D9 **Kuldīga** *Ger.* Goldingen. W Latvia 56°57′N 21°59′E
Kuldja see Yining
Kul'dzhuktau, Gory see QuljuqtovTog'lari
127 N4 **Kulebaki** Nizhegorodskaya Oblast', W Russian Federation 55°25′N 42°31′E
112 E11 **Kulen Vakuf** *var.* Spasovo. ◆ Federacija Bosne I Hercegovine, NW Bosnia and Herzegovina
181 Q9 **Kulgera Roadhouse** Northern Territory, N Australia 25°49′S 133°30′E
Kulhakangri see Kula Kangri
127 T1 **Kuliga** Udmurtskaya Respublika, NW Russian Federation 58°14′N 53°49′E
Kulkuduk see Ko'lquduq
118 G4 **Kullamaa** Läänemaa, W Estonia 58°52′N 24°05′E
197 O12 **Kullorsuaq** *var.* Kuvdlorssuak. ◆ Qaasuitsup, C Greenland
29 O6 **Kulm** North Dakota, N USA 46°18′N 98°57′W
Kulm see Chełmno
146 D12 **Kul'mach** *prev. Turkm.* Isgender. Balkan Welaýaty, W Turkmenistan 39°04′N 55°49′E
101 L18 **Kulmbach** Bayern, SE Germany 50°07′N 11°27′E
Kulmsee see Chełmża
147 Q14 **Kŭlob** *Rus.* Kulyab. SW Tajikistan 37°55′N 68°46′E
92 M13 **Kuloharju** Lappi, N Finland 65°51′N 28°10′E
125 N7 **Kuloy** Arkhangel'skaya Oblast', NW Russian Federation 64°55′N 43°35′E
125 N7 **Kuloy** ≈ NW Russian Federation
137 Q14 **Kulp** Diyarbakır, SE Turkey 38°32′N 41°01′E
Kulpa see Kolpa
77 P14 **Kulpawn** ≈ N Ghana
143 R13 **Kūl, Rūd-e** *var.* Kül. ≈ S Iran
144 G12 **Kul'sary** *Kaz.* Qulsary. Atyrau, W Kazakhstan 46°59′N 54°02′E
153 R15 **Kulti** West Bengal, NE India 23°45′N 86°50′E
93 G14 **Kultsjön** *Lapp.* Gålto. ◎ N Sweden
136 I14 **Kulu** Konya, W Turkey 39°06′N 33°02′E
123 S9 **Kulu** ≈ E Russian Federation
122 I13 **Kulunda** Altayskiy Kray, S Russian Federation 52°33′N 79°04′E
Kulunda Steppe see Ravnina Kulyndy
Kulundinskaya Ravnina see Ravnina Kulyndy
182 M9 **Kulwin** Victoria, SE Australia 35°04′S 142°37′E
Kulyab see Kŭlob
117 Q3 **Kulykivka** Chernihivs'ka Oblast', N Ukraine 51°23′N 31°39′E
Kum see Qom
164 F14 **Kuma** Ehime, Shikoku, SW Japan 33°36′N 132°53′E
127 P14 **Kuma** ≈ SW Russian Federation
165 O12 **Kumagaya** Saitama, Honshū, S Japan 36°09′N 139°22′E
165 Q5 **Kumaishi** Hokkaidō, NE Japan 42°08′N 139°57′E
169 R13 **Kumai, Teluk** *bay* Borneo, C Indonesia
127 Y7 **Kumak** Orenburgskaya Oblast', W Russian Federation 51°16′N 60°06′E
164 C14 **Kumamoto** Kumamoto, Kyūshū, SW Japan 32°49′N 130°41′E
164 D15 **Kumamoto** *off.* Kumamoto-ken. ◆ *prefecture* Kyūshū, SW Japan
Kumamoto-ken see Kumamoto
164 J15 **Kumano** Mie, Honshū, SW Japan 33°54′N 136°08′E
Kumanova see Kumanovo
113 O17 **Kumanovo** *Turk.* Kumanova. N Macedonia 42°08′N 21°43′E
185 G17 **Kumara** West Coast, South Island, New Zealand 42°39′S 171°12′E
180 J8 **Kumarina Roadhouse** Western Australia 24°46′S 119°39′E
153 T15 **Kumarkhali** Khulna, W Bangladesh 23°54′N 89°16′E
77 P16 **Kumasi** *prev.* Coomassie. C Ghana 06°41′N 01°40′W
Kumayri see Gyumri
79 D15 **Kumba** Sud-Ouest, W Cameroon 04°39′N 09°26′E
114 N13 **Kumbağ** Tekirdağ, NW Turkey 40°51′N 27°26′E
155 J21 **Kumbakonam** Tamil Nādu, SE India 10°59′N 79°24′E
Kum-Dag see Gumdag
165 R16 **Kume-jima** *island* Nansei-shotō, SW Japan
127 V6 **Kumertau** Respublika Bashkortostan, W Russian Federation 52°48′N 55°48′E
Kumillā see Comilla
35 R4 **Kumiva Peak** ▲ Nevada, W USA 40°24′N 119°16′W
159 N7 **Kum Kuduk** Xinjiang Uygur Zizhiqu, W China 40°15′N 91°55′E
159 N8 **Kum Kuduk** *well* NW China
Kumkurgan see Qumqo'rg'on
95 M16 **Kumla** Örebro, C Sweden 59°08′N 15°09′E
136 E17 **Kumluca** Antalya, SW Turkey 36°23′N 30°17′E
100 N9 **Kummerower See** ◎ NE Germany
77 X14 **Kumo** Gombe, E Nigeria 10°03′N 11°14′E
145 O13 **Kumola** ≈ C Kazakhstan
167 N1 **Kumon Range** ▲ N Myanmar (Burma)
83 F22 **Kums** Karas, SE Namibia 28°07′S 19°40′E
155 E18 **Kumta** Karnātaka, W India 14°25′N 74°24′E
38 H12 **Kumukahi, Cape** *headland* Hawai'i, USA, C Pacific Ocean 19°31′N 154°48′W
127 Q17 **Kumukh** Respublika Dagestan, SW Russian Federation 42°10′N 47°07′E
Kumul see Hami
158 L6 **Kümüx** Xinjiang Uygur Zizhiqu, W China 42°10′N 88°16′E
127 N9 **Kumylzhenskaya** Volgogradskaya Oblast', SW Russian Federation 49°54′N 42°35′E
141 W6 **Kumzār** N Oman 26°19′N 56°26′E
43 W15 **Kuna de Wargandí** ◆ *special territory* NE Panama
149 S4 **Kunar** *Per.* Konarhā; *prev.* Konar. ◆ *province* E Afghanistan
Kunashiri see Kunashir, Ostrov
123 U14 **Kunashir, Ostrov** *var.* Kunashiri. *island* Kuril'skiye Ostrova, SE Russian Federation
43 V14 **Kuna Yala** *prev.* San Blas. ◆ *special territory* NE Panama
118 I3 **Kunda** Lääne-Virumaa, NE Estonia 59°31′N 26°33′E
152 M13 **Kunda** Uttar Pradesh, N India 25°43′N 81°31′E
155 E19 **Kundāpura** *var.* Coondapoor. Karnātaka, W India 13°39′N 74°41′E
79 O24 **Kundelungu, Monts** ▲ S Dem. Rep. Congo
Kundert see Hernád
186 D7 **Kundiawa** Chimbu, W Papua New Guinea 06°00′S 144°57′E
Kundla see Sāvarkundla
Kunduk, Ozero see Sasyk, Ozero
Kunduk, Ozero Sasyk see Sasyk, Ozero
168 L10 **Kundur, Pulau** *island* W Indonesia
149 Q2 **Kunduz** *var.* Kondūz, Qondūz; *prev.* Kondoz, Kundūz. Kunduz, NE Afghanistan 36°49′N 68°50′E
149 Q2 **Kunduz** *prev.* Kondoz. ◆ *province* NE Afghanistan
Kunduz/Kundūz see Kunduz
Kuneitra see Al Qunayţirah
83 B18 **Kunene** ◆ *district* NE Namibia
83 A16 **Kunene** *var.* Cunene. ≈ Angola/Namibia *see also* Cunene
Kunene see Cunene
Künes see Xinyuan
158 J5 **Künes He** ≈ NW China
95 I19 **Kungälv** Västra Götaland, S Sweden 57°54′N 12°00′E
147 W7 **Kungei Ala-Tau** *Rus.* Khrebet Kyungëy Ala-Too, *Kir.* Küngöy Ala-Too. ▲ Kazakhstan/Kyrgyzstan
Küngöy Ala-Too see Kungei Ala-Tau
Kungrad see Qo'ng'irot
95 J19 **Kungsbacka** Halland, S Sweden 57°30′N 12°05′E
95 I18 **Kungshamn** Västra Götaland, S Sweden 58°21′N 11°15′E
95 M16 **Kungsör** Västmanland, C Sweden 59°25′N 16°05′E
79 J16 **Kungu** Equateur, NW Dem. Rep. Congo 02°47′N 19°12′E
125 V15 **Kungur** Permskiy Kray, NW Russian Federation 57°24′N 56°56′E
166 L9 **Kungyangon** Yangon, SW Myanmar (Burma) 16°27′N 96°00′E
111 M22 **Kunhegyes** Jász-Nagykun-Szolnok, E Hungary 47°22′N 20°36′E
167 O5 **Kunhing** Shan State, E Myanmar (Burma) 21°17′N 98°26′E
158 D9 **Kunjirap Daban** *var.* Khūnjerāb Pass. *pass* China/Pakistan *see also* Khūnjerāb Pass
Kunjirap Daban see Khunjerāb Pass
Kunlun Mountains see Kunlun Shan
158 H10 **Kunlun Shan** *Eng.* Kunlun Mountains. ▲ NW China
159 P11 **Kunlun Shankou** *pass* C China
160 G13 **Kunming** *var.* K'un-ming; *prev.* Yunnan. *province capital* Yunnan, SW China 25°04′N 102°41′E
K'un-ming see Kunming
Kunø see Kunoy
95 B18 **Kunoy** *Dan.* Kunø. *island* N Faroe Islands
Kunsan see Gunsan
111 L24 **Kunszentmárton** Jász-Nagykun-Szolnok, E Hungary 46°50′N 20°19′E
111 J23 **Kunszentmiklós** Bács-Kiskun, C Hungary 47°00′N 19°07′E
181 N3 **Kununurra** Western Australia 15°50′S 128°44′E
Kunyang see Pingyang
Kunya-Urgench see Köneürgenç
Kunyé see Pins, Île des
169 T11 **Kunyi** Borneo, C Indonesia 03°23′S 119°20′E
101 I20 **Künzelsau** Baden-Württemberg, S Germany 49°22′N 09°43′E
161 S10 **Kuocang Shan** ▲ SE China
Kuolajarvi see Kuoloyarvi
124 H5 **Kuoloyarvi** *Finn.* Kuolajärvi, *var.* Luolajarvi. Murmanskaya Oblast', NW Russian Federation 66°58′N 29°13′E
93 N16 **Kuopio** Pohjois-Savo, C Finland 62°54′N 27°41′E
93 K17 **Kuortane** Etelä-Pohjanmaa, W Finland 62°48′N 23°30′E
93 M18 **Kuortti** Etelä-Savo, E Finland 61°25′N 26°25′E
Kupa see Kolpa
171 P17 **Kupang** *prev.* Koepang. Timor, C Indonesia 10°13′S 123°38′E
39 Q5 **Kuparuk River** ≈ Alaska, USA
Kupchino see Cupcina
186 E9 **Kupiano** Central, S Papua New Guinea 10°06′S 148°12′E
180 M4 **Kupingarri** Western Australia 16°46′S 125°57′E
122 I12 **Kupino** Novosibirskaya Oblast', C Russian Federation 54°22′N 77°09′E
118 H11 **Kupiškis** Panevėžys, NE Lithuania 55°51′N 24°58′E
114 L13 **Küplü** Edirne, NW Turkey 41°06′N 26°23′E
39 X13 **Kupreanof Island** *island* Alexander Archipelago, Alaska, USA
39 O16 **Kupreanof Point** *headland* Alaska, USA 55°34′N 159°36′W
112 G13 **Kupres** ◆ Federacija Bosne I Hercegovine, SW Bosnia and Herzegovina
117 W5 **Kup"yans'k** *Rus.* Kupyansk. Kharkivs'ka Oblast', E Ukraine 49°42′N 37°36′E
Kupyansk see Kup"yans'k

◆ Country ● Country Capital ◇ Dependent Territory ○ Dependent Territory Capital ◆ Administrative Regions ✈ International Airport ▲ Mountain ▲ Mountain Range ⛰ Volcano ≈ River ◎ Lake ▣ Reservoir

Column 1

117 W5 **Kup"yans'k-Vuzlovyy**
Kharkivs'ka Oblast', E Ukraine
49°40´N 37°41´E

158 I6 **Kuqa** Xinjiang Uygur
Zizhiqu, NW China
41°43´N 82°58´E

Kür see Kura

137 W11 **Kura** Az. Kür, Geor.
Mtkvari, Turk. Kura Nehri.
⟳ SW Asia

55 R8 **Kuracki** NW Guyana
06°52´N 60°13´W

Kura Kurk see Irbe Strait

147 Q10 **Kurama Range** Rus.
Kuraminskiy Khrebet.
▲ Tajikistan/Uzbekistan

Kuraminskiy Khrebet see
Kurama Range

Kura Nehri see Kura

119 J14 **Kuranets** Rus. Kurenets.
C Belarus 54°33´N 26°57´E

164 H13 **Kurashiki** var. Kurasiki.
Okayama, Honshū, SW Japan
34°35´N 133°44´E

154 L10 **Kurasia** Chhattisgarh,
C India 23°11´N 82°16´E

Kurasiki see Kurashiki

164 H12 **Kurayoshi** var. Kurayosi.
Tottori, Honshū, SW Japan
35°27´N 133°52´E

Kurayosi see Kurayoshi

163 X6 **Kurbin He** ⟳ NE China

Kurchum see Kurshim

Kurchum see Kurshim

137 X11 **Kürdämir**
Kyurdamir. C Azerbaijan
40°21´N 48°08´E

Kurdestan see Kordestān

139 S1 **Kurdistan** cultural region
SW Asia

Kurd Kui see Kord Kūy

155 F15 **Kurduvādi** Mahārāshtra,
W India 18°06´N 75°31´E

Kürdzhali see Kardzhali

Kürdzhali see Kardzhali

Kürdzhali, Yazovir see
Kardzhali, Yazovir

164 F13 **Kure** Hiroshima,
Honshū, SW Japan
34°15´N 132°33´E

192 K5 **Kure Atoll** var. Ocean
atoll Hawaiian Islands,
Hawaii, USA

136 D10 **Küre Dağları** ▲ N Turkey

146 C11 **Kürendag** Rus. Gora
Kyuren. ▲ W Turkmenistan
39°05´N 55°09´E

Kurenets see Kuranets

118 E6 **Kuressaare** Ger. Arensburg;
prev. Kingissepp. Saaremaa,
W Estonia 58°17´N 22°29´E

122 K9 **Kureyka** Krasnoyarskiy
Kray, N Russian Federation
66°22´N 87°21´E

122 K9 **Kureyka** ⟳ N Russian
Federation

Kurgal'dzhino/
Kurgal'dzhinsky see
Korgalzhyn

122 G11 **Kurgan** Kurganskaya
Oblast', C Russian Federation
55°30´N 65°20´E

126 L14 **Kurganinsk** Krasnodarskiy
Kray, SW Russian Federation
44°52´N 40°36´E

122 G11 **Kurganskaya Oblast'**
◆ province C Russian
Federation

Kurgan–Tyube see
Qürghonteppa

191 O2 **Kuria** prev. Woodle Island.
island Tungaru, W Kiribati

Kuria Muria Bay see
Ḩalāniyāt, Khalīj al

Kuria Muria Islands see
Ḩalāniyāt, Juzur al

153 T13 **Kurigram** Rajshahi,
N Bangladesh 25°49´N 89°39´E

93 K17 **Kurikka** Etelä-Pohjanmaa,
W Finland 62°36´N 22°25´E

192 J3 **Kuril Basin** var. Kurile
Basin. undersea basin
NW Pacific Ocean

Kurile Basin see Kuril Basin

Kurile Islands see Kuril'skiye
Ostrova

Kurile-Kamchatka
Depression see Kuril-
Kamchatka Trench

Kurile Trench see Kuril-
Kamchatka Trench

Kurile Islands see Kuril'skiye
Ostrova

192 J3 **Kuril-Kamchatka Trench**
var. Kurile-Kamchatka
Depression, Kurile Trench.
trench NW Pacific Ocean

127 Q9 **Kurilovka** Saratovskaya
Oblast', W Russian Federation
50°39´N 48°02´E

123 U13 **Kuril'sk** Jap. Shana.
Kuril'skiye Ostrova,
Sakhalinskaya Oblast',
SE Russian Federation
45°10´N 147°51´E

122 G11 **Kuril'skiye Ostrova** Eng.
Kuril Islands, Kurile Islands.
island group SE Russian
Federation

42 M9 **Kurinwás, Río**
⟳ E Nicaragua

Kurishes Haff see Courland
Lagoon

Kurkund see Kilingi-Nõmme

126 M4 **Kurlovskiy** Vladimirskaya
Oblast', W Russian Federation
55°25´N 40°39´E

80 G12 **Kurmuk** Blue Nile, SE Sudan
10°36´N 34°16´E

Kurna see Al Qurnah

155 F17 **Kurnool** var. Karnul.
Andhra Pradesh, S India
15°51´N 78°01´E

164 M11 **Kurobe** Toyama, Honshū,
SW Japan 36°55´N 137°24´E

165 Q7 **Kuroisi** var. Kuroisi.
Aomori, Honshū, NE Japan
40°37´N 140°34´E

Kuroisi see Kuroishi

165 Q4 **Kuroiso** Tochigi, Honshū,
S Japan 36°58´N 140°02´E

164 B17 **Kuro-shima** island SW Japan

185 F21 **Kurow** Canterbury, South
Island, New Zealand
44°44´S 170°28´E

127 N15 **Kursavka** Stavropol'skiy
Kray, SW Russian Federation
44°28´N 42°32´E

118 E11 **Kuršėnai** Šiauliai,
N Lithuania 56°00´N 22°56´E

145 X10 **Kurshim** var. Kurchum.
Vostochnyy Kazakhstan,
E Kazakhstan 48°35´N 83°42´E

145 Y10 **Kurshim** var. Kurchum.
⟳ E Kazakhstan

**Kurshskaya Kosa/Kuršiu̇
Nerija** see Courland Spit

Column 2

126 J7 **Kursk** Kurskaya Oblast',
W Russian Federation
51°44´N 36°47´E

126 I7 **Kurskaya Oblast'**
◇ province W Russian
Federation

Kurskiy Zaliv see Courland
Lagoon

137 R15 **Kurtalan** Siirt, SE Turkey
37°58´N 41°36´E

Kurtbunar see Tervel

Kurtitsch/Kürtös see Curtici

72 M5 **Kuujjuaq** prev. Fort-
Chimo. Québec, E Canada
58°10´N 68°15´W

12 I7 **Kuujjuarapik** Québec,
C Canada 55°07´N 78°09´W

12 I7 **Kuujjuarapik** prev.
Poste-de-la-Baleine. Québec,
NE Canada 55°13´N 77°54´W

93 N13 **Kuusamo** Pohjois-
Pohjanmaa, E Finland
65°57´N 29°15´E

93 M16 **Kuusankoski** Kymenlaakso,
S Finland 60°51´N 26°54´E

127 W7 **Kuvandyk** Orenburgskaya
Oblast', W Russian Federation
51°27´N 57°18´E

79 H20 **Kuvango** see Cubango

Kuvasay see Quvasoy

Kuvdlorssuak see
Kullorsuaq

124 I16 **Kuvshinovo** Tverskaya
Oblast', W Russian Federation
57°03´N 34°09´E

141 Q4 **Kuwait** off. State of Kuwait,
var. Dawlat al Kuwait, Koweit,
Kuweit. ◆ monarchy SW Asia

Kuwait Bay see Kuwayt, Jūn
al

Kuwait City see Al Kuwayt

Kuwait, Dawlat al see
Kuwait

Kuwait, State of see Kuwait

Kuwajleen see Kwajalein
Atoll

164 K13 **Kuwana** Mie, Honshū,
SW Japan 35°04´N 136°40´E

139 X9 **Kuwayt** Maysān, E Iraq
32°26´N 47°12´E

141 Q4 **Kuwayt, Jūn al** var. Kuwait
Bay. bay E Kuwait

Kuweit see Kuwait

117 P10 **Kuyal'nyts'kyy Lyman**
⟳ SW Ukraine

122 I12 **Kuybyshev** Novosibirskaya
Oblast', C Russian Federation
55°28´N 77°55´E

Kuybyshev see Bolgar,
Respublika Tatarstan, Russian
Federation

Kuybyshev see Samara

127 W9 **Kuybysheve** Rus.
Kuybyshevo. Zaporiz'ka
Oblast', SE Ukraine
47°20´N 36°41´E

Kuybyshevo see Kuybysheve

Kuybyshev Reservoir
see Kuybyshevskoye
Vodokhranilishche

Kuybyshevskaya Oblast'
see Samarskaya Oblast'

Kuybyshevskiy see
Novoishimskiy

127 R4 **Kuybyshevskoye**
Vodokhranilishche var.
Kuibyshev, Eng. Kuybyshev
Reservoir. ⚿ W Russian
Federation

127 U4 **Kushnarenkovo** Respublika
Bashkortostan, W Russian
Federation 55°07´N 55°24´E

Kushrabat see Qo'shrabot

Kushtia see Kustia

Kusima see Kushima

Kusiro see Kushiro

38 M13 **Kuskokwim Bay** bay Alaska,
USA

39 P11 **Kuskokwim Mountains**
▲ Alaska, USA

39 N12 **Kuskokwim River**
⟳ Alaska, USA

145 N8 **Kusmuryn** Kaz. Qusmuryn;
prev. Kushmurun. Kostanay,
N Kazakhstan 52°27´N 64°31´E

145 N8 **Kusmuryn, Ozero** Kaz.
Qusmuryn; prev. Ozero
Kushmurun. ⊚ N Kazakhstan

108 G7 **Küsnacht** Zürich,
N Switzerland 47°19´N 08°34´E

165 V4 **Kussharo-ko** var. Kussyaro.
⊚ Hokkaidō, NE Japan

108 F8 **Küssnacht am Rigi**
var. Küssnacht. Schwyz,
C Switzerland 47°03´N 08°25´E

Kussyaro see Kussharo-ko

Kustanay see Kostanay

Küstence/Küstendje see
Constanţa

100 F11 **Küstenkanal** var.
Ems-Hunte Canal. canal
NW Germany

153 T15 **Kustia** var. Kushtia. Khulna,
W Bangladesh 23°54´N 89°07´E

Küstrin see Kostrzyn

171 R11 **Kusu** Pulau Halmahera,
E Indonesia 01°51´N 127°41´E

170 L16 **Kuta** Pulau Lombok,
S Indonesia 08°53´S 116°15´E

139 T4 **Kutabān** Kirkūk, N Iraq
35°21´N 44°45´E

136 E13 **Kütahya** prev. Kutaia.
Kütahya, W Turkey
39°25´N 29°56´E

136 E13 **Kütahya** var. Kutaia.
◇ province W Turkey

137 R9 **Kutaisi** var. K'ut'aisi.
W Georgia 42°16´N 42°42´E

K'ut'aisi see Kutaisi

139 Q5 **Kūt al 'Amārah** see Al Kūt

139 Q5 **Kut al Hai/Kūt al Ḩayy** see
Al Ḩayy

139 Q5 **Kut al Imara** see Al Kūt

123 Q11 **Kutana** Respublika Sakha
(Yakutiya), NE Russian
Federation 59°13´N 131°43´E

Kutaradja/Kutaraja see
Banda Aceh

165 R4 **Kutchan** Hokkaidō,
NE Japan 42°54´N 140°46´E

Kutch, Gulf of see Kachchh,
Gulf of

Kutch, Rann of see Kachchh,
Rann of

112 F9 **Kutina** Sisak-Moslavina,
NE Croatia 45°29´N 16°45´E

112 H9 **Kutjevo** Požega-Slavonija,
NE Croatia 45°26´N 17°54´E

111 E17 **Kutná Hora** Ger.
Kuttenberg. Středočeský
Kraj, C Czech Republic
49°57´N 15°18´E

Column 3

110 K12 **Kutno** Łódzkie, C Poland
52°14´N 19°23´E

Kuttenberg see Kutná Hora

79 I20 **Kutu** Bandundu, W Dem.
Rep. Congo 02°42´S 18°10´E

153 V17 **Kutubdia Island** island
SE Bangladesh

80 B10 **Kutum** Northern Darfur,
W Sudan 14°10´N 24°40´E

147 Y7 **Kuturgu** Issyk-Kul'skaya
Oblast', E Kyrgyzstan
42°45´N 78°04´E

12 I7 **Kuujjuaq** prev. Fort-
Chimo. Québec, E Canada
58°10´N 68°15´W

55 T5 **Kwara** ◇ state SW Nigeria

83 K17 **KwaZulu/Natal** off.
KwaZulu/Natal Province;
prev. Natal. ◇ province
E South Africa

83 K17 **KwaZulu/Natal Province**
see KwaZulu/Natal

92 N13 **Kuusamo** Pohjois-
Pohjanmaa, E Finland
65°57´N 29°15´E

110 I6 **Kuulei-Mayak** see
Guwlumaýak

92 N13 **Kuulsemägi** ▲ S Estonia

127 W7 **Kuvandyk** Orenburgskaya

83 G20 **Kweneng** ◇ district
S Botswana

39 N12 **Kwethluk** Alaska, USA
60°48´N 161°26´W

39 N12 **Kwethluk River** ⟳ Alaska,
USA

110 J8 **Kwidzyń** Ger. Marienwerder.
Pomorskie, N Poland
53°44´N 18°55´E

38 M13 **Kwigillingok** Alaska, USA
59°52´N 163°08´W

186 E9 **Kwikila** Central,
S Papua New Guinea
09°51´S 147°43´E

79 I20 **Kwilu** ⟳ W Dem. Rep.
Congo

Kwito see Cuíto

171 U12 **Kwoka, Gunung**
▲ Papua Barat, E Indonesia
0°33´S 132°25´E

78 I12 **Kyabé** Moyen-Chari, S Chad
09°28´N 18°54´E

183 O11 **Kyabram** Victoria, SE
Australia 36°21´S 145°05´E

166 M9 **Kyaikkami** prev. Amherst.
Mon State, S Myanmar
(Burma) 16°03´N 97°36´E

166 L9 **Kyaiklat** Ayeyarwady,
SW Myanmar (Burma)
16°25´N 95°42´E

166 M8 **Kyaikto** Mon State,
S Myanmar (Burma)
17°16´N 97°01´E

123 N14 **Kyakhta** Respublika
Buryatiya, S Russian
Federation 50°25´N 106°13´E

182 G8 **Kyancutta** South Australia
33°09´S 135°34´E

167 T8 **Ky Anh** Ha Tinh, N Vietnam
18°05´N 106°16´E

166 L5 **Kyaukpadaung** Mandalay,
C Myanmar (Burma)
20°50´N 95°08´E

166 M5 **Kyaukpyu** see Kyaunkpyu

166 M5 **Kyaukse** Mandalay,
C Myanmar (Burma)
21°33´N 96°06´E

166 L8 **Kyaunggon** Ayeyarwady,
SW Myanmar (Burma)
17°04´N 95°12´E

166 J6 **Kyaunkpyu** var. Kyaukpyu.
Rakhine State, W Myanmar
(Burma) 19°27´N 93°33´E

119 E14 **Kybartai** Pol. Kibarty.
Marijampolė, S Lithuania
54°37´N 22°44´E

152 I7 **Kyelang** Himāchal Pradesh,
NW India 32°33´N 77°03´E

111 G19 **Kyjov** Ger. Gaya.
Jihomoravský Kraj, SE Czech
Republic 49°00´N 17°07´E

115 J21 **Kykládes** var. Kikládhes.
island group SE Greece

25 S11 **Kyle** Texas, SW USA
29°59´N 97°52´W

96 G9 **Kyle of Lochalsh**
N Scotland, United Kingdom
57°18´N 05°39´W

101 D19 **Kyll** ⟳ W Germany

115 F19 **Kyllíni** var. Killini.
⟳ S Greece

93 N20 **Kymenlaakso** Swe.
Kymmenedalen. ◇ region
S Finland

Kymmenedalen see
Kymenlaakso

115 I18 **Kými** prev. Kími. Évvoia,
C Greece 38°38´N 24°06´E

94 G11 **Kymijoki** ⟳ S Finland

115 H18 **Kýmis, Akrotírio**
headland Évvoia, C Greece
38°39´N 24°08´E

Kyoto see Kyōto

183 N12 **Kyneton** Victoria,
SE Australia 37°14´S 144°28´E

81 G17 **Kyoga, Lake** var. Lake
Kioga. ⊚ C Uganda

183 V4 **Kyogle** New South Wales,
SE Australia 28°35´S 153°00´E

164 H12 **Kyōga-misaki** headland
Honshū, SW Japan
35°46´N 135°13´E

Kyonggi-man see
Gyeonggi-man

Kyongju see Gyeongju

Kyŏngsŏng see Seoul

Kyŏsai-tō see Geogeum-do

81 I19 **Kyotera** S Uganda
0°38´S 31°32´E

164 J13 **Kyōto** Kyōto, Honshū,
SW Japan 35°01´N 135°46´E

164 J13 **Kyōto** off. Kyōto-fu, var.
Kyoto. ◇ urban prefecture
Honshū, SW Japan

Kyōto-fu/Kyōto Hu see
Kyōto

115 D21 **Kyparissía** var. Kiparissía.
Pelopónnisos, S Greece
37°15´N 21°40´E

115 D20 **Kyparissiakós Kólpos** gulf
S Greece

115 G22 **Kypárissos** S Greece

104 I11 **Kyperounta** Kyperoúnta

104 I11 **Kyperoúnta** var.
Kyperounta. C Cyprus
34°57´N 33°02´E

104 O7 **Kypros** see Cyprus

115 G20 **Kyrá Panagía** island Vóreies
Sporádes, Greece, Aegean Sea

Kyrenia see Girne

104 P9 **Kyrenia Mountains** see
Beşparmak Dağları

Kyrgyz Republic see
Kyrgyzstan

Column 4

79 H20 **Kwango** Port. Cuango.
⟳ Angola/Dem. Rep. Congo
see also Cuango

Kwango see Cuango

**Kwangsi/Kwangsi Chuang
Autonomous Region** see
Guangxi Zhuangzu Zizhiqu

Kwangtung see Guangdong

Kwangyuan see Guangyuan

81 F17 **Kwania, Lake** ⊚ C Uganda

82 B11 **Kwanza** see Cuanza

82 B12 **Kwanza Norte** prev. Cuanza
Norte. ◇ province
NE Angola

82 B12 **Kwanza Sul** prev. Cuanza
Sul. ◇ province NE Angola

55 T5 **Kwara** ◇ state SW Nigeria

83 K17 **KwaZulu/Natal** off.
KwaZulu/Natal Province;
prev. Natal. ◇ province
E South Africa

KwaZulu/Natal Province
see KwaZulu/Natal

Kweichow see Guizhou

Kweichu see Guiyang

Kweilin see Guilin

Kweisui see Hohhot

Kweiyang see Guiyang

83 K17 **Kwekwe** prev. Que Que.
Midlands, C Zimbabwe
18°56´S 29°49´E

83 G20 **Kweneng** ◇ district
S Botswana

Kwesui see Hohhot

39 N12 **Kwethluk** Alaska, USA
60°48´N 161°26´W

39 N12 **Kwethluk River** ⟳ Alaska,
USA

110 J8 **Kwidzyń** Ger. Marienwerder.
Pomorskie, N Poland
53°44´N 18°55´E

38 M13 **Kwigillingok** Alaska, USA
59°52´N 163°08´W

186 E9 **Kwikila** Central,
S Papua New Guinea
09°51´S 147°43´E

79 I20 **Kwilu** ⟳ W Dem. Rep.
Congo

Kwito see Cuíto

171 U12 **Kwoka, Gunung**
▲ Papua Barat, E Indonesia
0°33´S 132°25´E

78 I12 **Kyabé** Moyen-Chari, S Chad
09°28´N 18°54´E

183 N12 **Kyneton** Victoria,
SE Australia 37°14´S 144°28´E

167 I7 **Kyelang** see Kyelang

111 P16 **Kyjov**

183 N12 **Kyela** Njombe, S Tanzania

79 G15 **Kyarnbergsvattnet** var.
Frostviken. ⊚ N Sweden

183 N12 **Kyneton** Victoria,
SE Australia 37°14´S 144°28´E

81 G17 **Kyoga, Lake**

Column 5 (147 U9 etc.)

147 U9 **Kyrgyzstan** off. Kyrgyz
Republic, var. Kirghizia; prev.
Kirgizskaya SSR, Kirghiz
SSR, Respublika Kyrgyzstan.
◆ republic C Asia

Kyrgyzstan, Republic of see
Kyrgyzstan

138 F11 **Kyriat Gat** prev. Qiryat
Gat. Southern, C Israel

100 M11 **Kyritz** Brandenburg,
NE Germany 52°56´N 12°24´E

94 G8 **Kyrksæterøra** Sør-Trøndelag,
S Norway 63°17´N 09°06´E

Kyrkslätt see Kirkkonummi

125 U8 **Kyrta** Respublika Komi,
NW Russian Federation
64°03´N 57°41´E

111 J18 **Kysucké Nové Mesto**
prev. Horné Nové Mesto,
Ger. Kischütz-Neustadtl,
Obernestadtl, Hung.
Kiszucaújhely. Žilinský Kraj,
N Slovakia 49°18´N 18°48´E

117 N12 **Kytay, Ozero** ⊚ SW Ukraine

115 F23 **Kýthira** var. Kíthira,
It. Cerigo, Lat. Cythera.
Kýythira, S Greece
41°39´N 26°30´E

115 F23 **Kýthira** var. Kíthira, It.
Cerigo, Lat. Cythera. island
S Greece

115 I20 **Kýthnos** Kýnthnos,
Kykládes, Greece, Aegean Sea
37°24´N 24°28´E

115 I20 **Kýthnos** var. Kíthnos,
Thermiá, It. Termia; anc.
Cythnos. island Kykládes,
Greece, Aegean Sea

115 I20 **Kýthnou, Stenó** strait
Kykládes, Greece, Aegean Sea

164 D15 **Kyūshū** var. Kyûsyû. island
SW Japan

192 H6 **Kyushu-Palau Ridge** var.
Kyusu-Palau Ridge. undersea
feature W Pacific Ocean
20°00´N 136°00´E

114 F10 **Kyustendil** anc. Pautalia.
W Bulgaria
42°17´N 22°42´E

114 G11 **Kyustendil** ◇ province
W Bulgaria

Kyûsyû see Kyūshū

Kyusu-Palau Ridge see
Kyushu-Palau Ridge

123 P8 **Kyusyur** Respublika Sakha
(Yakutiya), NE Russian
Federation 70°36´N 127°19´E

183 P10 **Kywong** New South
Wales, SE Australia
34°59´S 146°42´E

117 P4 **Kyyiv** Eng. Kiev, Rus. Kiyev.
● (Ukraine) Kyyivs'ka Oblast',
N Ukraine 50°26´N 30°32´E

117 O4 **Kyyivs'ka Oblast'** var.
Kyyiv, Rus. Kiyevskaya
Oblast'. ◇ province N Ukraine

117 P3 **Kyyivs'ke
Vodoskhovyshche**
Eng. Kiev Reservoir,
Rus. Kiyevskoye
Vodokhranilishche.
⚿ N Ukraine

93 L16 **Kyyjärvi** Keski-Suomi,
C Finland 63°02´N 24°34´E

122 K14 **Kyzyl** Respublika Tyva,
C Russian Federation
51°42´N 94°28´E

147 S8 **Kyzyl-Adyr** var.
Kirovskoye. Talasskaya
Oblast', NW Kyrgyzstan
42°37´N 71°34´E

145 V14 **Kyzylagash** Kaz.
Qyzylaghash. Almaty,
SE Kazakhstan 45°18´N 78°45´E

146 C13 **Kyzylbair** Balkan
Welaýaty, W Turkmenistan
38°13´N 55°38´E

122 K14 **Kyzyl Kum** var. Kizil Kum,
Qizil Qum, Uzb. Qizilqum.
desert Kazakhstan/Uzbekistan

145 N15 **Kyzylorda** var. Kyzyl-Orda,
Qizil Orda, Qyzylorda;
prev. Kzylorda, Perovsk.
Kyzylorda, S Kazakhstan
44°54´N 65°31´E

145 N15 **Kyzylorda** off.
Qyzylordinskaya Oblast', Kaz.
Qyzylorda Oblïsy. ◇ province
S Kazakhstan

144 L14 **Kyzylorda** off.
Qyzylordinskaya Oblast'
see Kyzylorda

Kyzylrabat see Qizilrabot

Kyzylsu see Kyzyl-Suu

147 X7 **Kyzyl-Suu** prev. Pokrovka.
Issyk-Kul'skaya Oblast',
NE Kyrgyzstan 42°20´N 77°55´E

147 S12 **Kyzyl-Suu** var. Kyzylsu.
⟳ Kyrgyzstan/Tajikistan

147 X8 **Kyzyl-Tuu** Issyk-Kul'skaya
Oblast', E Kyrgyzstan
42°06´N 76°54´E

145 Q12 **Kyzylzhar** Kaz. Qyzylzhar.
Karaganda, C Kazakhstan
48°22´N 70°00´E

108 C9 **Kzyl-Orda** see Kyzylorda

Kzylorda see Kyzylorda

Kzyltu see Kishkenekol'

Column 6 (L section)

L

109 X2 **Laa an der Thaya**
Niederösterreich, NE Austria
48°44´N 16°23´E

63 K15 **La Adela** La Pampa,
E Argentina 38°57´S 64°02´W

Laagen see Numedalslågen

115 D21 **Laakirchen** Oberösterreich,
N Austria 47°59´N 13°49´E

104 I11 **La Albuera** Extremadura,
W Spain 38°43´N 06°49´W

105 O7 **La Alcarria** physical region
C Spain

104 L13 **La Algaba** Andalucía, S Spain
37°27´N 06°01´W

105 N14 **La Almarcha** Castilla-
La Mancha, C Spain
39°41´N 02°22´W

Column 7

105 R6 **La Almunia de Doña
Godina** Aragón, NE Spain
41°28´N 01°23´W

41 N5 **La Amistad, Presa** ⚿
NW Mexico

118 F4 **Läänemaa** var. Lääne
Maakond. ◇ province
NW Estonia

118 I3 **Lääne-Virumaa** off. Lääne-
Viru Maakond. ◇ province
NE Estonia

Lääne Maakond see
Läänemaa

Lääne-Viru Maakond see
Lääne-Virumaa

62 J9 **La Antigua, Salina** salt lake
W Argentina

99 E17 **Laarne** Oost-Vlaanderen,
NW Belgium 51°03´N 03°50´E

80 O13 **Laas Caanood** Sool,
N Somalia 08°33´N 47°44´E

41 O9 **La Ascensión** Nuevo León,
NE Mexico 24°25´N 99°55´W

80 N12 **Laas Dhaareed** Togdheer,
N Somalia 10°12´N 46°09´E

55 O4 **La Asunción** Nueva Esparta,
NE Venezuela 11°06´N 63°28´W

Laatokka see Ladozhskoye,
Ozero

100 I13 **Laatzen** Niedersachsen,
N Germany 52°19´N 09°46´E

38 E9 **La'au Point** var. Laau
Point. headland Molokai,
Hawai'i, USA 21°06´N 157°18´W

42 D6 **La Aurora** × (Ciudad de
Guatemala) Guatemala,
C Guatemala 14°34´N 90°30´W

74 C9 **Laâyoune** var. Aaiún.
● (Western Sahara)
NW Western Sahara
27°10´N 13°11´W

126 L14 **Laba** ⟳ SW Russian
Federation

40 M6 **La Babia** Coahuila,
NE Mexico 28°30´N 102°00´W

15 R7 **La Baie** Québec, SE Canada
48°20´N 70°54´W

171 P16 **Labala** Pulau Lomblen,
S Indonesia 08°30´S 123°27´E

62 K8 **La Banda** Santiago del
Estero, N Argentina
27°44´S 64°14´W

104 K4 **La Bañeza** Castilla y León,
N Spain 42°19´N 05°55´W

167 T11 **Labäng** prev. Phumĭ Labăng.
Rôtânôkiri, NE Cambodia
13°51´N 107°01´E

40 M13 **La Barca** Jalisco, SW Mexico
20°20´N 102°33´W

40 K14 **La Barra de Navidad**
Jalisco, C Mexico
19°11´N 104°43´W

32 G9 **Lacey** Washington, NW USA
47°01´N 122°49´W

103 P12 **la Chaise-Dieu** Haute-Loire,
C France 45°19´N 03°41´E

14 G13 **Lachans** Kentriki
Makedonia, N Greece
40°57´N 23°15´E

124 L11 **Lacha, Ozero** ⊚ NW Russian
Federation

103 O8 **la Charité-sur-Loire** Nièvre,
C France 47°10´N 03°02´E

103 N9 **la Châtre** Indre, C France
46°35´N 01°59´E

108 C8 **La Chaux-de-Fonds**
Neuchâtel, W Switzerland
47°07´N 06°51´E

108 C8 **Lachen** Schwyz,
C Switzerland 47°12´N 08°51´E

183 Q8 **Lachlan River** ⟳ New
South Wales, SE Australia

43 T15 **La Chorrera** Panamá,
C Panama 08°51´N 79°46´W

15 V7 **Lac-Humqui** Québec, SE
Canada 48°18´N 67°28´W

15 R12 **Lac-Mégantic** var.
Mégantic. Québec, SE Canada
45°34´N 70°52´W

Lacobriga see Lagos

40 G5 **La Colorada** Sonora,
NW Mexico 28°49´N 110°32´W

11 Q16 **La Lacombe** Alberta,
SW Canada 52°30´N 113°42´W

30 L12 **Lacon** Illinois, N USA
41°01´N 89°24´W

43 P16 **La Concepción** Chiriquí,
W Panama 08°31´N 82°37´W

54 H5 **La Concepción**
Zulia, NW Venezuela
10°48´N 71°46´W

107 C19 **Laconi** Sardegna, Italy,
C Mediterranean Sea
39°52´N 09°02´E

19 O9 **Laconia** New Hampshire,
NE USA 43°31´N 71°28´W

61 H19 **La Coronilla** Rocha,
E Uruguay 33°44´S 53°31´W

104 H2 **La Coruña** see A Coruña

102 J11 **la Courtine** Creuse, C France
45°43´N 02°16´E

59 D14 **Lábrea** Amazonas, N Brazil

45 U15 **La Brea** Trinidad, Trinidad
and Tobago 10°14´N 61°37´W

102 K14 **Labrit** Landes, SW France

108 C9 **La Broye** ⟳ SW Switzerland

168 I7 **Labuan** Malay. Victoria.
05°20´N 115°14´E

168 I7 **Labuan** ◆ federal territory
East Malaysia

168 I7 **Labuan, Pulau** var.
Labuan. island East Malaysia

169 S16 **Labuanbajo** Flores,
S Indonesia 08°33´S 119°55´E

171 N16 **Labuhanbajo** Flores,

104 I11 **La Albuera**

168 J9 **Labuhanbilik** Sumatera,
W Indonesia

168 G8 **Labuhanhaji** Sumatera,
W Indonesia

169 V7 **Labuk, Sungai** var. Labuk.
⟳ East Malaysia

Column 8

169 W6 **Labuk, Teluk** var. Labuk
Bay, Telukan Labuk. bay
S Sulu Sea
Labuk, Telukan see Labuk,
Teluk

166 K9 **Labutta** Ayeyarwady,
SW Myanmar (Burma)
16°08´N 94°45´E

122 I8 **Labytnangi** Yamalo-
Nenetskiy Avtonomnyy
Okrug, N Russian Federation

113 K19 **Laç** var. Laci. Lezhë,
C Albania 41°37´N 19°37´E

78 F10 **Lac** off. Région du Lac.
◇ region W Chad

57 K19 **Lacajahuira, Río**
⟳ W Bolivia

62 G11 **La Calera** Valparaíso, C Chile
32°47´S 71°16´W

13 P11 **Lac-Allard** Québec, E Canada
50°37´N 63°26´W

104 L13 **La Campana** Andalucía,
S Spain 37°35´N 05°25´W

102 J12 **Lacanau** Gironde, SW France
44°59´N 01°04´W

42 C2 **Lacandón, Sierra del**
▲ Guatemala/Mexico
la Cañada see A Cañiza

41 W16 **Lacantún, Río** ⟳ SE Mexico

103 Q3 **la Capelle** Aisne, N France
49°58´N 03°55´E

112 K10 **Lačarak** Vojvodina,
NW Serbia 45°00´N 19°34´E

62 L11 **La Carlota** Córdoba,
C Argentina 33°30´S 63°15´W

104 L13 **La Carlota** Andalucía,
S Spain 37°40´N 04°56´W

105 N12 **La Carolina** Castilla-
La Mancha, S Spain
38°15´N 03°37´W

103 O15 **Lacaune** Tarn, S France
43°42´N 02°42´E

15 P7 **Lac-Bouchette**
Québec, SE Canada
48°14´N 72°11´W

**Laccadive Islands/
Laccadive Minicoy and
Amindivi Islands, the** see
Lakshadweep

11 Y16 **Lac du Bonnet**
S Canada 50°13´N 96°04´W

30 L4 **Lac du Flambeau** Wisconsin,
N USA 45°58´N 89°51´W

15 P8 **Lac-Édouard** Québec,
SE Canada 47°39´N 72°16´W

42 I4 **La Ceiba** Atlántida,
N Honduras 15°45´N 86°29´W

54 E9 **La Ceja** Antioquia,
W Colombia 06°02´N 75°30´W

182 J11 **Lacepede Bay** bay South
Australia

103 O11 **la Courtine** Creuse, C France
45°43´N 02°16´E

103 O11 **la Courtine**

59 D14 **Lábrea**

102 J16 **Lacq** Pyrénées-Atlantiques,
SW France 43°25´N 00°38´W

15 P9 **La Croche** Québec,
SE Canada 47°37´N 72°42´W

X3 **La Croix, Lac** ⊚ Canada/USA

26 K5 **La Crosse** Kansas, C USA
38°31´N 99°18´W

21 V7 **La Crosse** Virginia, NE USA
36°41´N 78°03´W

30 J7 **La Crosse** Wisconsin, N USA
43°46´N 91°12´W

42 K12 **La Cruz** Guanacaste,
NW Costa Rica
11°05´N 85°39´W

40 J10 **La Cruz** Sinaloa, W Mexico
23°53´N 106°53´W

61 F19 **La Cruz** Florida, S Uruguay
33°56´S 56°15´W

42 M9 **La Cruz de Río Grande**
Región Autónoma
Atlántico Sur, E Nicaragua
13°04´N 84°12´W

54 J4 **La Cruz de Taratara** Falcón,
N Venezuela 11°04´N 69°42´W

15 Q10 **Lac-St-Charles**
Québec, SE Canada
46°57´N 71°23´W

40 M6 **La Cuesta** Coahuila,
NE Mexico 28°45´N 102°00´W

◆ Country   ◇ Dependent Territory   ◆ Administrative Regions   ▲ Mountain   ⚐ Volcano   ⊚ Lake
● Country Capital   ○ Dependent Territory Capital   ✕ International Airport   ▲ Mountain Range   ⟳ River   ⊟ Reservoir

275

◆ Country ◇ Dependent Territory ◆ Administrative Regions ▲ Mountain ▲ Volcano ◎ Lake
● Country Capital ○ Dependent Territory Capital ✕ International Airport ▲▲ Mountain Range ⬥ River ⊡ Reservoir

LANSDALE – LEBANON

18 I15 **Lansdale** Pennsylvania, NE USA 40°14′N 75°13′W
14 L14 **Lansdowne** Ontario, SE Canada 44°25′N 76°00′W
152 K9 **Lansdowne** Uttarakhand, N India 29°50′N 78°42′E
30 M3 **L'Anse** Michigan, N USA 46°45′N 88°27′W
15 S7 **L'Anse-St-Jean** Québec, SE Canada 48°14′N 70°13′W
29 Y11 **Lansing** Iowa, C USA 43°22′N 91°11′W
27 R4 **Lansing** Kansas, C USA 39°15′N 94°54′W
31 Q9 **Lansing** *state capital* Michigan, N USA 42°44′N 84°33′W
Länsi-Suomi ◈ *province* W Finland
92 J12 **Lansjärv** Norrbotten, N Sweden 66°39′N 22°10′E
111 G17 **Lanškroun** *Ger.* Landskron. Pardubický Kraj, E Czech Republic 49°55′N 16°38′E
167 N16 **Lanta, Ko** *island* S Thailand
161 O15 **Lantau Island** *Cant.* Tai Yue Shan, *Chin.* Landao. *island* Hong Kong, S China
Lantian *see* Lianyuan
Lan-ts'ang Chiang *see* Mekong
Lantung, Gulf of *see* Liaodong Wan
171 O11 **Lanu** Sulawesi, N Indonesia 01°00′N 121°33′E
107 D19 **Lanusei** Sardegna, Italy, C Mediterranean Sea 39°55′N 09°31′E
102 H7 **Lanvaux, Landes de** *physical region* NW France
163 W8 **Lanxi** Heilongjiang, NE China 46°18′N 126°15′E
161 R10 **Lanxi** Zhejiang, SE China 29°12′N 119°27′E
La Nyanga *see* Nyanga
161 T15 **Lan Yu** *var.* Huoshao Tao, Hungt'ou, Lan Hsü, Lanyü, *Eng.* Orchid Island; *prev.* Kotosho, Koto Sho, Lan Yü. *island* SE Taiwan
Lanyü *see* Lan Yu
64 P11 **Lanzarote** *island* Islas Canarias, Spain, NE Atlantic Ocean
159 V10 **Lanzhou** *var.* Lan-chou, Lanchow, Lan-chow; *prev.* Kaolan. *province capital* Gansu, C China 36°01′N 103°52′E
106 B8 **Lanzo Torinese** Piemonte, NE Italy 45°18′N 07°26′E
171 O11 **Laoag** Luzon, N Philippines 18°11′N 120°34′E
171 Q5 **Laoang** Samar, C Philippines 12°29′N 125°01′E
167 R5 **Lao Cai** Lao Cai, N Vietnam 22°29′N 104°00′E
Laodicea/Laodicea ad Mare *see* Al Lādhiqīyah
Laoet *see* Laut, Pulau
163 T11 **Laoha He** ≈ NE China
160 M8 **Laohekou** *var.* Guanghua. Hubei, C China 32°20′N 111°42′E
Laoi, An *see* Lee
97 E19 **Laois** *prev.* Leix, Queen's County. *cultural region* C Ireland
163 W12 **Lao Ling** ▲▲ N China
64 Q11 **La Oliva** *var.* Oliva. Fuerteventura, Islas Canarias, Spain, NE Atlantic Ocean 28°36′N 13°53′W
Lao, Loch *see* Belfast Lough
Laolong *see* Longchuan
Lao Mangnai *see* Mangnai
103 P3 **Laon** *var.* la Laon; *anc.* Laudunum. Aisne, N France 49°34′N 03°37′E
Lao People's Democratic Republic *see* Laos
54 M3 **La Orchila, Isla** *island* N Venezuela
64 O11 **La Orotava** Tenerife, Islas Canarias, Spain, NE Atlantic Ocean 28°23′N 16°32′W
57 E14 **La Oroya** Junín, C Peru 11°36′S 75°54′W
167 Q7 **Laos** *off.* Lao People's Democratic Republic. ◆ *republic* SE Asia
161 R5 **Laoshan Wan** *bay* E China
163 Y10 **Laoye Ling** ▲▲ NE China
60 J12 **Lapa** Paraná, S Brazil 25°46′S 49°44′W
103 P10 **Lapalisse** Allier, C France 46°13′N 03°39′E
54 F9 **La Palma** Cundinamarca, C Colombia 05°23′N 74°24′W
42 F7 **La Palma** Chalatenango, N El Salvador 14°19′N 89°10′W
43 W16 **La Palma** Darién, SE Panama 08°24′N 78°09′W
64 N11 **La Palma** *island* Islas Canarias, Spain, NE Atlantic Ocean
104 J14 **La Palma del Condado** Andalucía, S Spain 37°23′N 06°33′W
61 F18 **La Paloma** Durazno, C Uruguay 32°54′S 55°36′W
61 G20 **La Paloma** Rocha, E Uruguay 34°37′S 54°08′W
61 A21 **La Pampa** *off.* Provincia de La Pampa. ◈ *province* C Argentina
La Pampa, Provincia de *see* La Pampa
55 P8 **La Paragua** Bolívar, E Venezuela 06°53′N 63°16′W
119 O16 **Lapatsichy** *Rus.* Lopatichi. Mahilyowskaya Voblasts', E Belarus 53°34′N 30°53′E
61 C16 **La Paz** Entre Ríos, E Argentina 30°45′S 59°36′W
62 I11 **La Paz** Mendoza, C Argentina 33°30′S 67°36′W
57 J18 **La Paz** *var.* La Paz de Ayacucho. ● (Bolivia-seat of government) La Paz, W Bolivia 16°30′S 68°13′W
42 H6 **La Paz** La Paz, SW Honduras 14°20′N 87°40′W
40 F9 **La Paz** Baja California Sur, NW Mexico 24°07′N 110°18′W
61 F20 **La Paz** Canelones, S Uruguay 34°46′S 56°13′W
57 J16 **La Paz** ◈ *department* W Bolivia
42 B9 **La Paz** ◈ *department* S El Salvador
42 G7 **La Paz** ◈ *department* SW Honduras
La Paz *see* El Alto, Bolivia
La Paz *see* Robles, Colombia
La Paz *see* La Paz Centro
40 F9 **La Paz, Bahía de** *bay* NW Mexico
42 I10 **La Paz Centro** *var.* La Paz. León, W Nicaragua 12°20′N 86°41′W
La Paz de Ayacucho *see* La Paz
54 J15 **La Pedrera** Amazonas, SE Colombia 01°19′S 69°31′W
31 S9 **Lapeer** Michigan, N USA 43°03′N 83°19′W
40 K6 **La Perla** Chihuahua, N Mexico 28°18′N 104°34′W
165 T1 **La Pérouse Strait** *Jap.* Sōya-kaikyō, *Rus.* Proliv Laperuza. *strait* Japan/Russian Federation
63 I14 **La Perra, Salitral de** *salt lake* C Argentina
Laperuza, Proliv *see* La Pérouse Strait
41 Q10 **La Pesca** Tamaulipas, C Mexico 23°49′N 97°45′W
40 M13 **La Piedad Cavadas** Michoacán, C Mexico 20°20′N 102°01′W
Lapines *see* Lafnitz
93 M16 **Lapinlahti** Pohjois-Savo, C Finland 63°21′N 27°25′E
Lápithos *see* Lapta
22 K9 **Laplace** Louisiana, S USA 30°04′N 90°28′W
45 X12 **La Plaine** SE Dominica 15°20′N 61°15′W
173 P16 **La Plaine-des-Palmistes** C Réunion
92 K11 **Lapland** *Fin.* Lappi, *Swe.* Lappland. *cultural region* N Europe
Lapland *see* Lappi
28 M8 **La Plant** South Dakota, N USA 45°06′N 100°40′W
61 D20 **La Plata** Buenos Aires, E Argentina 34°56′S 57°55′W
54 D12 **La Plata** Huila, SW Colombia 02°33′N 75°55′W
21 W4 **La Plata** Maryland, NE USA 38°32′N 76°59′W
La Plata *see* Sucre
45 U6 **la Plata, Río de** ≈ C Puerto Rico
105 W4 **La Pobla de Lillet** Cataluña, NE Spain 42°15′N 01°57′E
105 U4 **La Pobla de Segur** Cataluña, NE Spain 42°15′N 00°58′E
15 S9 **La Pocatière** Québec, SE Canada 47°21′N 70°04′W
104 K2 **La Pola** *prev.* Pola de Lena. Asturias, N Spain 43°10′N 05°49′W
104 L3 **La Pola de Gordón** Castilla y León, N Spain 42°50′N 05°38′W
104 L2 **La Pola Siero** *prev.* Pola de Siero. Asturias, N Spain 43°24′N 05°39′W
31 O11 **La Porte** Indiana, N USA 41°36′N 86°43′W
18 H13 **Laporte** Pennsylvania, NE USA 41°25′N 76°29′W
29 X13 **La Porte City** Iowa, C USA 42°19′N 92°11′W
62 J8 **La Posta** Catamarca, C Argentina 27°59′S 65°32′W
40 E8 **La Poza Grande** Baja California Sur, NW Mexico 25°50′N 112°00′W
93 K16 **Lappajärvi** Etelä-Pohjanmaa, W Finland 63°13′N 23°40′E
93 L16 **Lappajärvi** ◎ W Finland
93 N18 **Lappeenranta** *Swe.* Villmanstrand. Etelä-Karjala, SE Finland 61°04′N 28°15′E
93 J17 **Lappfjärd** *Fin.* Lapväärtti. Österbotten, W Finland 62°14′N 21°30′E
92 L12 **Lappi** *Swe.* Lappland, *Eng.* Lapland. ◈ *region* N Finland
Lappi/Lappland *see* Lapland
Lappo *see* Lapua
61 C23 **Laprida** Buenos Aires, E Argentina 37°34′S 60°45′W
25 P13 **La Pryor** Texas, SW USA 28°56′N 99°51′W
136 B11 **Lâpseki** Çanakkale, NW Turkey 40°22′N 26°42′E
121 P2 **Lapta** *Gk.* Lápithos. NW Cyprus 35°20′N 33°11′E
Laptev Sea *see* Laptevykh, More
122 N6 **Laptevykh, More** *Eng.* Laptev Sea. *sea* Arctic Ocean
93 K16 **Lapua** *Swe.* Lappo. Etelä-Pohjanmaa, W Finland 62°57′N 23°E
105 P3 **La Puebla de Arganzón** País Vasco, N Spain 42°45′N 02°49′W
104 L14 **La Puebla de Cazalla** Andalucía, S Spain 37°14′N 05°18′W
104 M9 **La Puebla de Montalbán** Castilla-La Mancha, C Spain 39°52′N 04°22′W
54 I6 **La Puerta** Trujillo, NW Venezuela 09°08′N 70°46′W
Lapurdum *see* Bayonne
40 E7 **La Purísima** Baja California Sur, NW Mexico 26°10′N 112°05′W
Lapväärtti *see* Lappfjärd
110 O10 **Łapy** Podlaskie, NE Poland 53°N 22°54′E
80 D6 **Laqiya Arba'in** Northern, NW Sudan 20°01′N 28°01′E
62 J4 **La Quiaca** Jujuy, N Argentina 22°12′S 65°36′W
107 J14 **L'Aquila** *var.* Aquila, Aquila degli Abruzzi. Abruzzo, C Italy 42°21′N 13°24′E
143 Q13 **Lār** Fārs, S Iran 27°42′N 54°19′E
54 J5 **Lara** *off.* Estado Lara. ◈ *state* NW Venezuela
104 G2 **Laracha** Galicia, NW Spain 43°14′N 08°34′W
74 G5 **Larache** *var.* al Araïch, El Araïch, *prev.* El Araïche; *anc.* Lixus. NW Morocco 35°12′N 06°10′W
Lara, Estado *see* Lara
103 T14 **Laragne-Montéglin** Hautes-Alpes, SE France 44°21′N 05°46′E
104 M13 **La Rambla** Andalucía, S Spain 37°37′N 04°44′W
33 Y17 **Laramie** Wyoming, C USA 41°18′N 105°35′W
33 X15 **Laramie Mountains** ▲▲ Wyoming, C USA
33 Y16 **Laramie River** ≈ Wyoming, C USA
60 H12 **Laranjeiras do Sul** Paraná, S Brazil 25°23′S 52°23′W
Larantoeka *see* Larantuka
171 P16 **Larantuka** *prev.* Larantoeka. Flores, C Indonesia 08°20′S 123°00′E
171 U15 **Larat** Pulau Larat, E Indonesia 07°07′S 131°46′E
171 U15 **Larat, Pulau** *island* Kepulauan Tanimbar, E Indonesia
95 P19 **Lärbro** Gotland, SE Sweden 57°46′N 18°49′E
106 A9 **Larche, Col de** *pass* France/Italy
14 H8 **Larder Lake** Ontario, S Canada 48°06′N 79°44′W
105 O2 **Laredo** Cantabria, N Spain 43°23′N 03°22′W
25 Q15 **Laredo** Texas, SW USA 27°30′N 99°30′W
40 H9 **La Reforma** Sinaloa, W Mexico 25°05′N 108°03′W
98 N11 **Laren** Gelderland, E Netherlands 52°12′N 06°22′E
98 J11 **Laren** Noord-Holland, C Netherlands 52°15′N 05°13′E
102 K13 **la Réole** Gironde, SW France 44°34′N 00°00′W
La Réunion *see* Réunion
Largeau *see* Faya
103 U13 **l'Argentière-la-Bessée** Hautes-Alpes, SE France 44°49′N 06°34′E
149 O4 **Lar Gerd** *var.* Largird. Balkh, N Afghanistan 35°36′N 66°48′E
Largird *see* Lar Gerd
23 V12 **Largo** Florida, SE USA 27°55′N 82°47′W
37 Q9 **Largo, Canon** *valley* New Mexico, SW USA
44 D6 **Largo, Cayo** *island* W Cuba
23 Z17 **Largo, Key** *island* Florida Keys, Florida, SE USA
96 H12 **Largs** W Scotland, United Kingdom 55°48′N 04°50′W
102 I16 **la Rhune** *var.* Larrún. ▲ France/Spain 43°19′N 01°36′W *see also* Larrún
la Rhune *see* Larrún
la Riege *see* Ariège
29 Q4 **Larimore** North Dakota, N USA 47°54′N 97°37′W
107 L15 **Larino** Molise, C Italy 41°46′N 14°50′E
Lario *see* Como, Lago di
62 J9 **La Rioja** La Rioja, NW Argentina 29°26′S 66°50′W
62 I9 **La Rioja** *off.* Provincia de La Rioja. ◈ *province* NW Argentina
105 O4 **La Rioja** ◈ *autonomous community* N Spain
La Rioja, Provincia de *see* La Rioja
115 F16 **Lárisa** *var.* Larissa. Thessalía, C Greece 39°38′N 22°27′E
Larissa *see* Lárisa
149 Q13 **Lārkāna** *var.* Larkhana. Sind, SE Pakistan 27°32′N 68°18′E
Larkhana *see* Lārkāna
Larnaca *see* Lárnaka
121 Q3 **Lárnaka** *var.* Larnaca, Larnax. SE Cyprus 34°55′N 33°39′E
121 Q3 **Lárnaka** ✈ SE Cyprus 34°52′N 33°38′E
Larnax *see* Lárnaka
97 G14 **Larne** *Ir.* Latharna. E Northern Ireland, United Kingdom 54°51′N 05°49′W
26 L5 **Larned** Kansas, C USA 38°12′N 99°05′W
104 L3 **La Robla** Castilla y León, N Spain 42°48′N 05°37′W
104 J10 **La Roca de la Sierra** Extremadura, W Spain 39°06′N 06°41′W
99 K22 **La Roche-en-Ardenne** Luxembourg, SE Belgium 50°11′N 05°35′E
102 L11 **la Rochefoucauld** Charente, W France 45°43′N 00°23′E
102 J10 **la Rochelle** *anc.* Rupella. Charente-Maritime, W France 46°09′N 01°07′W
102 I9 **la Roche-sur-Yon** *prev.* Bourbon Vendée, Napoléon-Vendée. Vendée, NW France 46°40′N 01°26′W
105 Q10 **La Roda** Castilla-La Mancha, C Spain 39°13′N 02°10′W
104 L14 **La Roda de Andalucía** Andalucía, S Spain 37°12′N 04°45′W
45 P9 **La Romana** E Dominican Republic 18°25′N 69°00′W
11 T13 **La Ronge** Saskatchewan, C Canada 55°07′N 105°18′W
11 U13 **La Ronge, Lac** ◎ Saskatchewan, C Canada
22 K10 **Larose** Louisiana, S USA 29°34′N 90°22′W
42 M7 **La Rosita** Región Autónoma Atlántico Norte, NE Nicaragua 13°55′N 84°23′W
181 Q3 **Larrimah** Northern Territory, N Australia 15°30′S 133°12′E
62 N11 **Larroque** Entre Ríos, E Argentina 33°05′S 59°06′W
105 Q2 **Larrún** *Fr.* la Rhune. ▲ France/Spain 43°18′N 01°35′W *see also* la Rhune
Larrún *see* la Rhune
195 X6 **Lars Christensen Coast** *physical region* Antarctica
39 Q14 **Larsen Bay** Kodiak Island, Alaska, USA 57°32′N 153°58′W
194 I5 **Larsen Ice Shelf** *ice shelf* Antarctica
8 M6 **Larsen Sound** *sound* Nunavut, N Canada
La Rúa *see* A Rúa de Valdeorras
102 K16 **Laruns** Pyrénées-Atlantiques, SW France 43°00′N 00°25′W
95 G16 **Larvik** Vestfold, S Norway 59°04′N 10°02′E
La-sa *see* Lhasa
171 S13 **Lasahata** Pulau Seram, E Indonesia 02°52′S 128°27′E
Lasahau *see* Lasihao
37 O6 **La Sal** Utah, W USA 38°19′N 109°14′W
14 C17 **La Salle** Ontario, S Canada 42°13′N 83°05′W
30 L11 **La Salle** Illinois, N USA 41°19′N 89°06′W
45 O9 **Las Americas** ✈ (Santo Domingo) S Dominican Republic 18°24′N 69°38′W
37 V6 **Las Animas** Colorado, C USA 38°04′N 103°13′W
108 D10 **La Sarine** *var.* Sarine. ≈ SW Switzerland
108 B9 **La Sarraz** Vaud, W Switzerland 46°40′N 06°32′E
12 H12 **La Sarre** Québec, SE Canada 48°49′N 79°12′W
54 L3 **Las Aves, Islas** *var.* Islas de Aves. *island group* N Venezuela
55 N7 **Las Bonitas** Bolívar, C Venezuela 07°50′N 65°40′W
104 K15 **Las Cabezas de San Juan** Andalucía, S Spain 36°59′N 05°56′W
61 G19 **Lascano** Rocha, E Uruguay 33°40′S 54°12′W
62 I5 **Lascar, Volcán** ⛰ N Chile 23°22′S 67°33′W
41 T15 **Las Choapas** *var.* Choapas. Veracruz-Llave, SE Mexico 17°51′N 94°00′W
37 R15 **Las Cruces** New Mexico, SW USA 32°19′N 106°49′W
Lasdehnen *see* Krasnoznamensk
La Selle *see* Selle, Pic de la
62 G9 **La Serena** Coquimbo, C Chile 29°54′S 71°18′W
104 K11 **La Serena** *physical region* W Spain
105 V4 **La Seu d'Urgell** *var.* La Seu d'Urgell; *prev.* Seo de Urgel. Cataluña, NE Spain 42°22′N 01°27′E
La Seu d'Urgell *see* La Seu d'Urgell
103 T16 **la Seyne-sur-Mer** Var, SE France 43°07′N 05°53′E
61 D21 **Las Flores** Buenos Aires, E Argentina 36°03′S 59°08′W
62 H9 **Las Flores** San Juan, W Argentina 30°14′S 69°10′W
11 S14 **Lashburn** Saskatchewan, S Canada 53°09′N 109°37′W
62 I11 **Las Heras** Mendoza, W Argentina 32°48′S 68°50′W
148 M8 **Lashkar Gāh** *var.* Lash-Kar-Gar'. Helmand, S Afghanistan 31°35′N 64°21′E
Lash-Kar-Gar' *see* Lashkar Gāh
171 P14 **Lasihao** *var.* Lasahau. Pulau Muna, C Indonesia 05°01′S 122°23′E
107 N21 **La Sila** ▲▲ SW Italy
63 H23 **La Silueta, Cerro** ▲ S Chile 52°22′S 72°09′W
42 L9 **La Sirena** Región Autónoma Atlántico Sur, E Nicaragua 12°59′N 84°35′W
110 J13 **Łask** Łódzkie, C Poland 51°36′N 19°06′E
109 V11 **Laško** *Ger.* Tüffer. C Slovenia 46°08′N 15°13′E
63 H14 **Las Lajas** Neuquén, W Argentina 38°31′S 70°22′W
63 H15 **Las Lajas, Cerro** ▲ W Argentina 38°49′S 70°42′W
62 M6 **Las Lomitas** Formosa, N Argentina 24°45′S 60°35′W
41 V16 **Las Margaritas** Chiapas, SE Mexico 16°15′N 91°58′W
Las Marismas *see* Guadalquivir, Marismas del
54 M6 **Las Mercedes** Guárico, N Venezuela 09°08′N 66°27′W
42 F6 **Las Minas, Cerro** ▲ W Honduras 14°33′N 88°41′W
105 O11 **La Solana** Castilla-La Mancha, C Spain 38°56′N 03°14′W
45 Q14 **La Soufrière** ⛰ Saint Vincent, Saint Vincent and the Grenadines 13°20′N 61°11′W
102 M10 **la Souterraine** Creuse, C France 46°15′N 01°28′E
62 N7 **Las Palmas** Chaco, N Argentina 27°08′S 58°45′W
43 Q16 **Las Palmas** Veraguas, W Panama 08°09′N 81°28′W
64 P12 **Las Palmas** *var.* Las Palmas de Gran Canaria. Gran Canaria, Islas Canarias, Spain, NE Atlantic Ocean 28°08′N 15°27′W
64 P12 **Las Palmas** ◈ *province* Islas Canarias, Spain, NE Atlantic Ocean
64 Q12 **Las Palmas** ✈ Gran Canaria, Islas Canarias, Spain, NE Atlantic Ocean
Las Palmas de Gran Canaria *see* Las Palmas
40 D6 **Las Palomas** Baja California Norte, W Mexico 31°44′N 107°37′W
105 P10 **Las Pedroñeras** Castilla-La Mancha, C Spain 39°27′N 02°41′W
106 E10 **La Spezia** Liguria, NW Italy 44°08′N 09°50′E
61 F20 **Las Piedras** Canelones, S Uruguay 34°42′S 56°14′W
63 J18 **Las Plumas** Chubut, S Argentina 43°46′S 67°15′W
61 B18 **Las Rosas** Santa Fe, C Argentina 32°27′S 61°30′W
Lassa *see* Lhasa
35 O4 **Lassen Peak** ⛰ California, W USA 40°27′N 121°28′W
194 K6 **Lassiter Coast** *physical region* Antarctica
109 V9 **Lassnitz** ≈ SE Austria
15 O12 **L'Assomption** Québec, SE Canada 45°48′N 73°27′W
15 N11 **L'Assomption** ≈ Québec, SE Canada
43 S17 **Las Tablas** Los Santos, S Panama 07°45′N 80°17′W
Lastarria, Volcán *see* Azufre, Volcán
37 V4 **Last Chance** Colorado, C USA 39°41′N 103°34′W
Last Frontier, The *see* Alaska
11 U16 **Last Mountain Lake** ◎ Saskatchewan, S Canada
62 H9 **Las Tórtolas, Cerro** ▲ W Argentina 29°57′S 69°49′W
61 C14 **Las Toscas** Santa Fe, C Argentina 28°22′S 59°20′W
79 F19 **Lastoursville** Ogooué-Lolo, E Gabon 0°50′S 12°43′E
113 F16 **Lastovo** *It.* Lagosta. *island* SW Croatia
113 F16 **Lastovski Kanal** *channel* SW Croatia
40 E6 **Las Tres Vírgenes, Volcán** ⛰ NW Mexico 27°27′N 112°34′W
40 F4 **Las Trincheras** Sonora, NW Mexico 30°21′N 111°27′W
55 N8 **Las Trincheras** Bolívar, E Venezuela 06°57′N 64°49′W
44 H7 **Las Tunas** *var.* Victoria de las Tunas. Las Tunas, E Cuba 20°58′N 76°59′W
La Suisse *see* Switzerland
40 I5 **Las Varas** Chihuahua, N Mexico 29°35′N 108°01′W
40 J12 **Las Varas** Nayarit, C Mexico 21°12′N 105°10′W
62 L10 **Las Varillas** Córdoba, E Argentina 31°54′S 62°45′W
35 X11 **Las Vegas** Nevada, W USA 36°09′N 115°10′W
37 T10 **Las Vegas** New Mexico, SW USA 35°35′N 105°15′W
187 P10 **Lata** Nendö, Solomon Islands 10°45′S 165°43′E
13 R10 **La Tabatière** Québec, E Canada 50°51′N 58°59′W
56 C6 **Latacunga** Cotopaxi, C Ecuador 0°58′S 78°36′W
194 I7 **Latady Island** *island* Antarctica
54 E14 **La Tagua** Putumayo, S Colombia 0°05′S 74°39′W
Latakia *see* Al Lādhiqīyah
92 J10 **Lätäseno** ≈ NW Finland
14 H9 **Latchford** Ontario, S Canada 47°20′N 79°45′W
14 J13 **Latchford Bridge** Ontario, SE Canada 45°16′N 77°29′W
193 Y14 **Late** *island* Vava'u Group, N Tonga
153 P15 **Lātehār** Jhārkhand, N India 23°48′N 84°28′E
15 R7 **Laterrière** Québec, SE Canada 48°17′N 71°10′W
102 J13 **la Teste** Gironde, SW France 44°38′N 01°04′W
25 V8 **Latexo** Texas, SW USA 31°24′N 95°28′W
18 L10 **Latham** New York, NE USA 42°45′N 73°45′W
Latharna *see* Larne
108 B9 **La Thielle** *var.* Thièle. ≈ W Switzerland
27 R3 **Lathrop** Missouri, C USA 39°33′N 94°19′W
107 I16 **Latina** *prev.* Littoria. Lazio, C Italy 41°28′N 12°53′E
41 R14 **La Tinaja** Veracruz-Llave, S Mexico
106 J7 **Latisana** Friuli-Venezia Giulia, NE Italy 45°47′N 13°01′E
Latium *see* Lazio
115 K25 **Lató** *site of ancient city* Kríti, Greece, E Mediterranean Sea
187 Q17 **La Tontouta** ✈ (Noumea) Province Sud, S New Caledonia 22°06′S 166°12′E
55 N4 **La Tortuga, Isla** *var.* Isla Tortuga. *island* N Venezuela
108 C10 **La Tour-de-Peilz** *var.* La Tour de Peilz. Vaud, SW Switzerland 46°28′N 06°52′E
La Tour de Peilz *see* La Tour-de-Peilz
103 S11 **la Tour-du-Pin** Isère, E France 45°34′N 05°25′E
102 J11 **la Tremblade** Charente-Maritime, W France 45°45′N 01°07′W
102 L10 **la Trimouille** Vienne, W France 46°27′N 01°02′E
42 J9 **La Trinidad** Estelí, NW Nicaragua 12°57′N 86°15′W
41 V16 **La Trinitaria** Chiapas, SE Mexico 16°02′N 92°00′W
45 Q11 **la Trinité** E Martinique 14°44′N 60°58′W
15 U7 **La Trinité-des-Monts** Québec, SE Canada 48°07′N 68°31′W
18 C15 **Latrobe** Pennsylvania, NE USA 40°18′N 79°19′W
183 P13 **La Trobe River** ≈ Victoria, SE Australia
Lattakia/Lattaquié *see* Al Lādhiqīyah
171 S13 **Latu** Pulau Seram, E Indonesia 03°24′S 128°37′E
15 P9 **La Tuque** Québec, SE Canada 47°26′N 72°47′W
155 G14 **Lātūr** Mahārāshtra, C India 18°24′N 76°34′E
118 G8 **Latvia** *off.* Republic of Latvia, *Ger.* Lettland, *Latv.* Latvija, Latvijas Republika; *prev.* Latvian SSR, *Rus.* Latviyskaya SSR. ◆ *republic* NE Europe
Latvian SSR/Latvija/Latvijas Republika/Latviyskaya SSR *see* Latvia
Latvia, Republic of *see* Latvia
186 H7 **Lau** New Britain, E Papua New Guinea 05°46′S 151°21′E
175 R9 **Lau Basin** *undersea feature* S Pacific Ocean
101 O15 **Lauchhammer** Brandenburg, E Germany 51°30′N 13°48′E
105 O3 **Laudio** *var.* Llodio. País Vasco, N Spain 43°08′N 02°59′W
Laudunum *see* Laon
Laudus *see* St-Lô
Lauenburg/Lauenburg in Pommern *see* Lębork
101 L20 **Lauf an der Pegnitz** Bayern, SE Germany 49°31′N 11°16′E
108 D7 **Laufen** Basel, NW Switzerland 47°26′N 07°31′E
109 P5 **Lauffen** Salzburg, NW Austria 47°54′N 12°57′E
92 I2 **Laugarbakki** Norðurland Vestra, N Iceland 65°18′N 20°51′W
92 I4 **Laugarvatn** Suðurland, SW Iceland 64°09′N 20°43′W
31 O3 **Laughing Fish Point** *headland* Michigan, N USA 46°31′N 87°01′W
187 Z14 **Lau Group** *island group* E Fiji
Lauis *see* Lugano
93 M17 **Laukaa** Keski-Suomi, C Finland 62°27′N 25°58′E
118 D12 **Laukuva** Tauragė, W Lithuania 55°37′N 22°12′E
Laun *see* Louny
183 P16 **Launceston** Tasmania, SE Australia 41°25′S 147°07′E
97 I24 **Launceston** *anc.* Dunheved. SW England, United Kingdom 50°38′N 04°21′W
54 C13 **La Unión** Nariño, SW Colombia 01°35′N 77°09′W
42 H8 **La Unión** La Unión, SE El Salvador 13°20′N 87°50′W
42 I6 **La Unión** Olancho, C Honduras 15°02′N 86°40′W
40 M15 **La Unión** Guerrero, S Mexico 17°58′N 101°49′W
41 Y14 **La Unión** Quintana Roo, E Mexico 18°00′N 101°48′W
105 S13 **La Unión** Murcia, SE Spain 37°37′N 00°54′W
54 L7 **La Unión** Barinas, C Venezuela 08°15′N 67°46′W
42 B10 **La Unión** ◈ *department* E El Salvador
38 H11 **Laupāhoehoe** *var.* Laupahoehoe. Hawaii, USA, C Pacific Ocean 20°00′N 155°15′W
Laupahoehoe *see* Laupāhoehoe
101 I23 **Laupheim** Baden-Württemberg, S Germany 48°13′N 09°54′E
181 W3 **Laura** Queensland, NE Australia 15°37′S 144°34′E
189 X2 **Laura** *atoll* Majuro Atoll, SE Marshall Islands
Laurana *see* Lovran
54 L8 **La Urbana** Bolívar, C Venezuela 07°05′N 66°58′W
21 Y4 **Laurel** Delaware, NE USA 38°33′N 75°34′W
23 V14 **Laurel** Florida, SE USA 27°07′N 82°27′W
21 W3 **Laurel** Maryland, NE USA 39°06′N 76°51′W
22 M6 **Laurel** Mississippi, S USA 31°41′N 89°10′W
33 U11 **Laurel** Montana, NW USA 45°40′N 108°46′W
29 R13 **Laurel** Nebraska, C USA 42°25′N 97°04′W
18 H15 **Laureldale** Pennsylvania, NE USA 40°24′N 75°52′W
18 C16 **Laurel Hill** *ridge* Pennsylvania, NE USA
29 T12 **Laurens** Iowa, C USA 42°51′N 94°51′W
21 P11 **Laurens** South Carolina, SE USA 34°29′N 82°01′W
Laurentian Highlands *see* Laurentian Mountains
15 P10 **Laurentian Mountains** *var.* Laurentian Highlands, *Fr.* Les Laurentides. *plateau* Newfoundland and Labrador/Québec, Canada
15 O12 **Laurentides** Québec, SE Canada 45°51′N 73°49′W
Laurentides, Les *see* Laurentian Mountains
107 M19 **Lauria** Basilicata, S Italy 40°03′N 15°50′E
194 I1 **Laurie Island** *island* Antarctica
21 T11 **Laurinburg** North Carolina, SE USA 34°46′N 79°29′W
30 M2 **Laurium** Michigan, N USA 47°14′N 88°26′W
Lauru *see* Choiseul
108 B9 **Lausanne** *It.* Losanna. Vaud, SW Switzerland 46°32′N 06°39′E
101 Q16 **Lausche** *var.* Luže. ▲ Czech Republic/Germany 50°52′N 14°39′E *see also* Luže
Lausche *see* Luže
101 Q16 **Lausitzer Bergland** *var.* Lausitzer Gebirge, *Cz.* Gory Lužycké, Lužické Hory', *Eng.* Lusatian Mountains. ▲▲ E Germany
Lausitzer Gebirge *see* Lausitzer Bergland
Lausitzer Neisse *see* Neisse
103 T12 **Lautaret, Col du** *pass* SE France
63 G15 **Lautaro** Araucanía, C Chile 38°30′S 71°30′W
101 F21 **Lauter** ≈ W Germany
108 I7 **Lauterach** Vorarlberg, NW Austria 47°29′N 09°44′E
101 I17 **Lauterbach** Hessen, C Germany 50°37′N 09°24′E
108 E9 **Lauterbrunnen** Bern, C Switzerland 46°36′N 07°52′E
169 U14 **Laut Kecil, Kepulauan** *island group* N Indonesia
187 X14 **Lautoka** Viti Levu, W Fiji 17°36′S 177°28′E
169 O8 **Laut, Pulau** *prev.* Laoet. *island* Borneo, C Indonesia
169 V14 **Laut, Pulau** *island* Kepulauan Natuna, W Indonesia
169 U13 **Laut, Selat** *strait* Borneo, C Indonesia
168 H8 **Laut Tawar, Danau** ◎ Sumatera, NW Indonesia
189 V14 **Lauvergne Island** *island* Chuuk, C Micronesia
98 M5 **Lauwers Meer** ◎ N Netherlands
98 M4 **Lauwersoog** Groningen, NE Netherlands 53°25′N 06°14′E
102 M14 **Lauzerte** Tarn-et-Garonne, S France 44°15′N 01°08′E
25 U13 **Lavaca Bay** *bay* Texas, SW USA
25 U12 **Lavaca River** ≈ Texas, SW USA
15 O12 **Laval** Québec, SE Canada 45°32′N 73°44′W
102 J6 **Laval** Mayenne, NW France 48°04′N 00°46′W
15 T6 **Laval** ≈ Québec, SE Canada
105 S9 **La Vall d'Uixó** *var.* Vall D'Uxó. Valenciana, E Spain 39°49′N 00°15′W
61 F19 **Lavalleja** ◈ *department* S Uruguay
15 O12 **Lavaltrie** Québec, SE Canada 45°56′N 73°14′W
186 M10 **Lavanggu** Rennell, S Solomon Islands 11°39′S 160°13′E
143 O14 **Lāvān, Jazīreh-ye** *island* S Iran
109 U8 **Lavant** ≈ S Austria
118 G5 **Lavassaare** *Ger.* Lawassaar. Pärnumaa, SW Estonia 58°29′N 24°22′E
104 L3 **La Vecilla de Curueño** Castilla y León, N Spain 42°51′N 05°24′W
45 N8 **La Vega** *var.* Concepción de la Vega. C Dominican Republic 19°15′N 70°33′W
La Vela *see* La Vela de Coro
54 J4 **La Vela de Coro** *var.* La Vela. Falcón, N Venezuela 11°30′N 69°33′W
103 N17 **Lavelanet** Ariège, S France 42°56′N 01°50′E
107 M17 **Lavello** Basilicata, S Italy 41°03′N 15°48′E
36 J8 **La Verkin** Utah, W USA 37°12′N 113°16′W
26 J8 **Laverne** Oklahoma, C USA 36°42′N 99°53′W
25 S12 **La Vernia** Texas, SW USA 29°19′N 98°07′W
93 K18 **Lavia** Satakunta, SW Finland 61°36′N 22°34′E
14 I12 **Lavieille, Lake** ◎ Ontario, SE Canada
94 C12 **Lavik** Sogn Og Fjordane, S Norway 61°06′N 05°25′E
La Vila Joíosa *see* Villajoyosa
33 U10 **Lavina** Montana, NW USA 46°18′N 108°55′W
194 H5 **Lavoisier Island** *island* Antarctica
23 U2 **Lavonia** Georgia, SE USA 34°26′N 83°06′W
103 R13 **la Voulte-sur-Rhône** Ardèche, E France 44°49′N 04°46′E
123 W5 **Lavrentiya** Chukotskiy Avtonomnyy Okrug, NE Russian Federation 65°33′N 171°12′W
115 H20 **Lávrio** *prev.* Lávrion. Attikí, C Greece 37°43′N 24°03′E
Lávrion *see* Lávrio
83 L22 **Lavumisa** *prev.* Gollel. SE Swaziland 27°18′S 31°55′E
149 T4 **Lawarai Pass** *pass* N Pakistan
Lawassaar *see* Lavassaare
141 P16 **Lawdar** SW Yemen 13°49′N 45°55′E
25 Q7 **Lawn** Texas, SW USA 32°07′N 99°45′W
195 Y4 **Law Promontory** *headland* Antarctica
77 O14 **Lawra** NW Ghana 10°40′N 02°49′W
185 E23 **Lawrence** Otago, South Island, New Zealand 45°53′S 169°43′E
31 P14 **Lawrence** Indiana, N USA 39°49′N 86°01′W
27 Q4 **Lawrence** Kansas, C USA 38°58′N 95°15′W
19 O10 **Lawrence** Massachusetts, NE USA 42°42′N 71°09′W
20 L5 **Lawrenceburg** Kentucky, S USA 38°02′N 84°53′W
20 I10 **Lawrenceburg** Tennessee, S USA 35°16′N 87°20′W
23 T3 **Lawrenceville** Georgia, SE USA 33°57′N 83°59′W
31 N15 **Lawrenceville** Illinois, N USA 38°43′N 87°40′W
21 V7 **Lawrenceville** Virginia, NE USA 36°45′N 77°50′W
27 S3 **Lawson** Missouri, C USA 39°26′N 94°12′W
26 L12 **Lawton** Oklahoma, C USA 34°35′N 98°20′W
140 I4 **Lawz, Jabal al** ▲ NW Saudi Arabia 28°45′N 35°20′E
95 L16 **Laxå** Örebro, C Sweden 59°00′N 14°37′E
125 T5 **Laya** ≈ NW Russian Federation
57 I19 **La Yarada** Tacna, SW Peru 18°14′S 70°30′W
141 S15 **Layjūn** C Yemen 15°27′N 49°16′E
141 Q9 **Laylā** *var.* Laila. Ar Riyāḑ, C Saudi Arabia 22°14′N 46°40′E
23 P4 **Lay Lake** ▣ Alabama, S USA
45 P14 **Layou** Saint Vincent, Saint Vincent and the Grenadines 13°11′N 61°16′W
La Youne *see* El Ayoun
192 L5 **Laysan Island** *island* Hawaiian Islands, Hawai'i, USA
36 L2 **Layton** Utah, W USA 41°03′N 112°00′W
34 L5 **Laytonville** California, W USA 39°39′N 123°30′W
172 H17 **Lazare, Pointe** *headland* Mahé, NE Seychelles 04°46′S 55°28′E
123 T12 **Lazarev** Khabarovskiy Kray, SE Russian Federation 52°11′N 141°18′E
112 L12 **Lazarevac** Serbia, C Serbia 44°25′N 20°17′E
65 N22 **Lazarev Sea** *sea* Antarctica
40 M15 **Lázaro Cárdenas** Michoacán, SW Mexico 17°56′N 102°13′W
119 F15 **Lazdijai** Alytus, S Lithuania 54°13′N 23°33′E
107 H15 **Lazio** *anc.* Latium. ◈ *region* C Italy
111 A16 **Lázně Kynžvart** *Ger.* Bad Königswart. Karlovarský Kraj, W Czech Republic 50°00′N 12°40′E
Lazovsk *see* Sîngerei
167 R12 **Leach** Poŭthĭsăt, W Cambodia 12°19′N 103°45′E
27 X9 **Leachville** Arkansas, C USA 35°56′N 90°15′W
28 I9 **Lead** South Dakota, N USA 44°21′N 103°45′W
11 S16 **Leader** Saskatchewan, S Canada 50°55′N 109°31′W
19 S6 **Lead Mountain** ▲ Maine, NE USA 44°53′N 68°07′W
37 R5 **Leadville** Colorado, C USA 39°15′N 106°17′W
11 V12 **Leaf Rapids** Manitoba, C Canada 56°30′N 100°02′W
22 M7 **Leaf River** ≈ Mississippi, S USA
25 W11 **League City** Texas, SW USA 29°30′N 95°05′W
92 K8 **Leaibevuotna** *Nor.* Olderfjord. Finnmark, N Norway 70°29′N 24°58′E
23 N7 **Leakesville** Mississippi, S USA 31°09′N 88°33′W
25 Q11 **Leakey** Texas, SW USA 29°44′N 99°48′W
Leal *see* Lihula
83 G15 **Lealui** Western, W Zambia 15°12′S 22°59′E
Leamhcán *see* Lucan
14 C18 **Leamington** Ontario, S Canada 42°03′N 82°35′W
Leamington/Leamington Spa *see* Royal Leamington Spa
Leammi *see* Lemmenjoki
25 S10 **Leander** Texas, SW USA 30°34′N 97°51′W
60 F13 **Leandro N. Alem** Misiones, NE Argentina 27°34′S 55°15′W
97 A20 **Leane, Lough** *Ir.* Loch Léin. ◎ SW Ireland
180 G8 **Learmonth** Western Australia 22°17′S 114°03′E
Leau *see* Zoutleeuw
L'Eau d'Heure *see* Plate Taille, Lac de la
190 D12 **Leava** Île Futuna, S Wallis and Futuna
Leavdnja *see* Lakselv
27 R3 **Leavenworth** Kansas, C USA 39°19′N 94°55′W
32 I8 **Leavenworth** Washington, NW USA 47°36′N 120°39′W
92 L8 **Leavvajohka** *var.* Levajok. Finnmark, N Norway 69°57′N 26°18′E
27 R4 **Leawood** Kansas, C USA 38°57′N 94°37′W
110 H6 **Łeba** *Ger.* Leba. Pomorskie, N Poland 54°45′N 17°32′E
110 I6 **Łeba** *Ger.* Leba. ≈ N Poland
Leba *see* Łeba
101 D20 **Lebach** Saarland, SW Germany 49°25′N 06°54′E
Łeba, Jezioro *see* Łebsko, Jezioro
171 P8 **Lebak** Mindanao, S Philippines 06°28′N 124°03′E
Lebanese Republic *see* Lebanon
31 O13 **Lebanon** Indiana, N USA 40°03′N 86°28′W
20 L6 **Lebanon** Kentucky, S USA 37°33′N 85°15′W
27 U6 **Lebanon** Missouri, C USA 37°40′N 92°40′W
19 N9 **Lebanon** New Hampshire, NE USA 43°40′N 72°15′W
32 G12 **Lebanon** Oregon, NW USA 44°32′N 122°54′W
18 H15 **Lebanon** Pennsylvania, NE USA 40°20′N 76°24′W
20 J8 **Lebanon** Tennessee, S USA 36°11′N 86°19′W
21 P7 **Lebanon** Virginia, NE USA 36°52′N 82°07′W
138 G6 **Lebanon** *off.* Lebanese Republic, *Ar.* Al Lubnān, *Fr.* Liban. ◆ *republic* SW Asia

◆ Country ● Country Capital ◇ Dependent Territory ○ Dependent Territory Capital ◈ Administrative Regions ✈ International Airport ▲ Mountain ▲▲ Mountain Range ⛰ Volcano ≈ River ◎ Lake ▣ Reservoir

LEBANON JUNCTION – LEXINGTON

20 K6 **Lebanon Junction** Kentucky, S USA 37°49′N 85°43′W
Lebanon, Mount see Liban, Jebel
146 J10 **Lebap** Lebapskiy Velayat, NE Turkmenistan 41°04′N 61°49′E
Lebapskiy Velayat see Lebap Welaýaty
146 J11 **Lebap Welaýaty** *Rus.* Lebapskiy Velayat; *prev. Rus.* Chardzhevskaya Oblast, *Turkm.* Chärjew Oblasty. ◆ *province* E Turkmenistan
Lebasee see Łebsko, Jezioro
99 F17 **Lebbeke** Oost-Vlaanderen, NW Belgium 51°00′N 04°08′E
35 S14 **Lebec** California, W USA 34°51′N 118°52′W
Lebedin see Lebedyn
123 Q11 **Lebedinyy** Respublika Sakha (Yakutiya), NE Russian Federation 58°23′N 125°24′E
126 L6 **Lebedyan'** Lipetskaya Oblast', W Russian Federation 53°00′N 39°11′E
117 T4 **Lebedyn** *Rus.* Lebedin. Sums'ka Oblast', NE Ukraine 50°36′N 34°30′E
12 I12 **Lebel-sur-Quévillon** Québec, SE Canada 49°01′N 76°56′W
92 L8 **Lebesby** *Lapp.* Davvesiida. Finnmark, N Norway 70°31′N 27°00′E
102 M9 **le Blanc** Indre, C France 46°38′N 01°04′E
79 L15 **Lebo** Orientale, N Dem. Rep. Congo 04°30′N 23°58′E
27 P5 **Lebo** Kansas, C USA 38°22′N 95°50′W
110 H6 **Lębork** *var.* Lębórk, *Ger.* Lauenburg, Lauenburg in Pommern. Pomorskie, N Poland 54°32′N 17°43′E
103 O17 **le Boulou** Pyrénées-Orientales, S France 42°32′N 02°50′E
108 A9 **Le Brassus** Vaud, W Switzerland 46°35′N 06°14′E
104 J15 **Lebrija** Andalucía, S Spain 36°55′N 06°04′W
110 G6 **Łebsko, Jezioro** *Ger.* Lebasee; *prev.* Jezioro Łeba. ◎ N Poland
63 F14 **Lebu** Bío Bío, C Chile 37°38′S 73°43′W
Lebyazh'ye see Akku
104 F6 **Leça da Palmeira** Porto, N Portugal 41°12′N 08°43′W
103 U15 **le Cannet** Alpes-Maritimes, SE France 43°35′N 07°E
Le Cap see Cap-Haïtien
103 P2 **le Cateau-Cambrésis** Nord, N France 50°05′N 03°32′E
107 Q18 **Lecce** Puglia, SE Italy 40°23′N 18°11′E
106 D7 **Lecco** Lombardia, N Italy 45°51′N 09°23′E
29 V10 **Le Center** Minnesota, N USA 44°23′N 93°43′W
108 J7 **Lech** Vorarlberg, W Austria 47°14′N 10°10′E
101 K22 **Lech** ≈ Austria/Germany
115 D19 **Lechainá** *var.* Lehena, Lekhainá. Dytikí Elláda, S Greece 37°57′N 21°16′E
102 J11 **le Château d'Oléron** Charente-Maritime, W France 45°53′N 01°12′W
103 R3 **le Chesne** Ardennes, N France 49°33′N 04°42′E
103 R13 **le Cheylard** Ardèche, E France 44°55′N 04°27′E
108 K7 **Lechtaler Alpen** ▲▲ W Austria
100 H6 **Leck** Schleswig-Holstein, N Germany 54°45′N 09°00′E
14 L9 **Lecointre, Lac** ◎ Québec, SE Canada
22 H7 **Lecompte** Louisiana, S USA 31°05′N 92°24′W
103 Q9 **le Creusot** Saône-et-Loire, C France 46°48′N 04°27′E
Lecumberri see Lekunberri
110 P13 **Łęczna** Lubelskie, E Poland 51°20′N 22°52′E
110 J12 **Łęczyca** *Ger.* Lentschiza, *Rus.* Lenchitsa. Łódzkie, C Poland 52°04′N 19°10′E
100 F10 **Leda** ≈ NW Germany
109 Y9 **Ledava** ≈ NE Slovenia
99 F17 **Lede** Oost-Vlaanderen, NW Belgium 50°58′N 03°59′E
104 K6 **Ledesma** Castilla y León, N Spain 41°05′N 06°00′W
45 Q12 **le Diamant** SW Martinique 14°29′N 61°02′W
172 J16 **Le Digue** *island* Inner Islands, NE Seychelles
103 Q10 **le Donjon** Allier, C France 46°19′N 03°50′E
102 M10 **le Dorat** Haute-Vienne, C France 46°14′N 01°05′E
Ledo Salinarius see Lons-le-Saunier
11 Q14 **Leduc** Alberta, SW Canada 53°17′N 113°30′W
123 V7 **Ledyanaya, Gora** ▲ E Russian Federation 61°51′N 171°03′E
97 C21 **Lee** *Ir.* An Laoi. ≈ SW Ireland
29 U5 **Leech Lake** ◎ Minnesota, N USA
26 K10 **Leedey** Oklahoma, C USA 35°54′N 99°21′W
97 M17 **Leeds** N England, United Kingdom 53°50′N 01°35′W
23 P4 **Leeds** Alabama, S USA 33°33′N 86°32′W
29 O3 **Leeds** North Dakota, N USA 48°19′N 99°43′W
98 N6 **Leek** Groningen, NE Netherlands 53°10′N 06°24′E
99 K15 **Leende** Noord-Brabant, SE Netherlands 51°21′N 05°34′E
100 F10 **Leer** Niedersachsen, NW Germany 53°14′N 07°26′E
98 J13 **Leerdam** Zuid-Holland, C Netherlands 51°54′N 05°06′E
98 K12 **Leersum** Utrecht, C Netherlands 52°01′N 05°26′E
23 W11 **Leesburg** Florida, SE USA 28°48′N 81°52′W
21 V3 **Leesburg** Virginia, NE USA 39°09′N 77°34′W
27 R4 **Lees Summit** Missouri, C USA 38°55′N 94°21′W
22 G7 **Leesville** Louisiana, S USA 31°08′N 93°15′W
25 S12 **Leesville** Texas, SW USA 29°22′N 97°45′W
31 U13 **Leesville Lake** ⊠ Ohio, N USA
Leesville Lake see Smith Mountain Lake
183 P9 **Leeton** New South Wales, SE Australia 34°33′S 146°24′E
98 L6 **Leeuwarden** *Fris.* Ljouwert. Fryslân, N Netherlands 53°15′N 05°48′E
180 I14 **Leeuwin, Cape** *headland* Western Australia 34°18′S 115°03′E
35 R8 **Lee Vining** California, W USA 37°57′N 119°07′W
45 V8 **Leeward Islands** *island group* E West Indies
Leeward Islands see Sotavento, Ilhas de
Leeward Islands see Vent, Îles Sous le
79 G20 **Léfini** ≈ SE Congo
115 C17 **Lefkáda** *prev.* Levkás. Lefkáda, Iónia Nisiá, Greece, C Mediterranean Sea 38°50′N 20°42′E
115 B17 **Lefkáda** *It.* Santa Maura, *prev.* Levkás; *anc.* Leucas. *island* Iónia Nisiá, Greece, C Mediterranean Sea
115 H25 **Lefká Óri** ▲▲ Kríti, Greece, E Mediterranean Sea
115 B16 **Lefkímmi** *var.* Levkímmi. Kérkyra, Iónia Nisiá, Greece, C Mediterranean Sea 39°26′N 20°05′E
Lefkosia/Lefkoşa see Nicosia
25 O2 **Lefors** Texas, SW USA 35°26′N 100°48′W
45 R12 **le François** E Martinique 14°36′N 60°59′W
180 L12 **Lefroy, Lake** *salt lake* Western Australia
Legaceaster see Chester
105 N8 **Leganés** Madrid, C Spain 40°20′N 03°46′W
Legaspi see Legazpi City
Leghorn see Livorno
110 M11 **Legionowo** Mazowieckie, C Poland 52°25′N 20°56′E
99 K24 **Léglise** Luxembourg, SE Belgium 49°48′N 05°31′E
106 G8 **Legnago** Lombardia, NE Italy 45°13′N 11°18′E
106 D7 **Legnano** Veneto, NE Italy 45°36′N 08°54′E
111 F14 **Legnica** *Ger.* Liegnitz. Dolnośląskie, SW Poland 51°12′N 16°11′E
35 Q9 **Le Grand** California, W USA 37°12′N 120°15′W
103 Q15 **le Grau-du-Roi** Gard, S France 43°32′N 04°10′E
183 U3 **Legume** New South Wales, SE Australia 28°24′S 152°20′E
102 L4 **le Havre** *Eng.* Havre; *prev.* le Havre-de-Grâce. Seine-Maritime, N France 49°30′N 00°06′E
le Havre-de-Grâce see le Havre
Lehena see Lechainá
36 L3 **Lehi** Utah, W USA 40°23′N 111°51′W
18 I14 **Lehighton** Pennsylvania, NE USA 40°49′N 75°42′W
29 O6 **Lehr** North Dakota, N USA 46°15′N 99°21′W
38 A8 **Lehua Island** *island* Hawaiian Islands, Hawai'i, USA
149 S9 **Leiāh** Punjab, NE Pakistan 30°59′N 70°58′E
109 W9 **Leibnitz** Steiermark, SE Austria 46°48′N 15°33′E
97 M19 **Leicester** *Lat.* Batae Coritanorum. C England, United Kingdom 52°38′N 01°05′W
97 M19 **Leicestershire** *cultural region* C England, United Kingdom
Leicheng see Leizhou
98 H11 **Leiden** *prev.* Leyden; *anc.* Lugdunum Batavorum. Zuid-Holland, W Netherlands 52°09′N 04°30′E
98 H11 **Leiderdorp** Zuid-Holland, W Netherlands 52°08′N 04°32′E
98 G11 **Leidschendam** Zuid-Holland, W Netherlands 52°05′N 04°24′E
99 D18 **Leie** *Fr.* Lys. ≈ Belgium/France
Leifear see Lifford
184 L4 **Leigh** Auckland, North Island, New Zealand 36°17′S 174°48′E
97 K17 **Leigh** NW England, United Kingdom 53°30′N 02°33′W
182 I5 **Leigh Creek** South Australia 30°27′S 138°23′E
23 O2 **Leighton** Alabama, S USA 34°42′N 87°31′W
97 M21 **Leighton Buzzard** E England, United Kingdom 51°55′N 00°41′W
Léim an Bhradáin see Leixlip
Léim An Mhadaidh see Limavady
Léime, Ceann see Loop Head, Ireland
Léime, Ceann see Slyne Head, Ireland
101 G20 **Leimen** Baden-Württemberg, SW Germany 49°21′N 08°40′E
100 I13 **Leine** ≈ NW Germany
101 J15 **Leinefelde** Thüringen, C Germany 51°22′N 10°19′E
Léin, Loch see Leane, Lough
97 D19 **Leinster** *Ir.* Cúige Laighean. *cultural region* E Ireland
97 F19 **Leinster, Mount** *Ir.* Stua Laighean. ▲ SE Ireland 52°36′N 06°45′W
119 F15 **Leipalingis** Alytus, S Lithuania 54°05′N 23°52′E
92 J12 **Leipojärvi** Norrbotten, N Sweden 67°03′N 21°15′E
31 R12 **Leipsic** Ohio, N USA 41°06′N 83°58′W
Leipsic see Leipzig
115 M20 **Leipsoí** *island* Dodekánisa, Greece, Aegean Sea
101 M15 **Leipzig** *Pol.* Lipsk, *hist.* Leipsic; *anc.* Lipsia. Sachsen, E Germany 51°19′N 12°24′E
101 M15 **Leipzig Halle** ✕ Sachsen, E Germany 51°26′N 12°14′E
104 G9 **Leiria** *anc.* Collipo. Leiria, C Portugal 39°45′N 08°49′W
104 F9 **Leiria** ◆ *district* C Portugal
95 C15 **Leirvik** Hordaland, S Norway 59°49′N 05°27′E
118 E5 **Leisi** *Ger.* Laisberg. Saaremaa, W Estonia 58°33′N 22°41′E
104 J3 **Leitariegos, Puerto de** *pass* NW Spain
20 J6 **Leitchfield** Kentucky, S USA 37°28′N 86°19′W
109 Y5 **Leitha** *Hung.* Lajta. ≈ Austria/Hungary
Leitir Ceanainn see Letterkenny
Leitmeritz see Litoměřice
Leitomischl see Litomyšl
97 D16 **Leitrim** *Ir.* Liatroim. *cultural region* NW Ireland
Leix see Laois
97 F18 **Leixlip** *Eng.* Salmon Leap, *Ir.* Léim an Bhradáin. Kildare, E Ireland 53°23′N 06°32′W
64 N8 **Leixões** Porto, N Portugal 41°11′N 08°41′W
161 N12 **Leiyang** Hunan, S China 26°23′N 112°49′E
160 L16 **Leizhou** *var.* Haikang, Leicheng. Guangdong, S China 20°54′N 110°05′E
160 L16 **Leizhou Bandao** *var.* Luichow Peninsula. *peninsula* S China
98 H13 **Lek** ≈ SW Netherlands
114 I13 **Lekánis** ▲▲ NE Greece
172 H13 **Le Kartala** ▲ Grande Comore, NW Comoros
Le Kef see El Kef
79 G20 **Lékéti, Monts de la** ▲▲ S Congo
Lekhainá see Lechainá
92 G11 **Leknes** Nordland, C Norway 68°07′N 13°36′E
79 F20 **Lékoumou** ◆ *province* SW Congo
94 L13 **Leksand** Dalarna, C Sweden 60°44′N 15°E
124 H8 **Leksozero, Ozero** ◎ NW Russian Federation
105 Q3 **Lekunberri** *var.* Lecumberri. Navarra, N Spain 43°00′N 01°54′W
171 S11 **Lelai, Tanjung** *headland* Pulau Halmahera, N Indonesia 01°32′N 128°43′E
45 Q12 **le Lamentin** *var.* Lamentin. C Martinique 14°37′N 61°01′W
31 P6 **Leland** Michigan, N USA 45°01′N 85°44′W
22 J4 **Leland** Mississippi, S USA 33°24′N 90°54′W
95 J16 **Lelång** *var.* Lelången. ◎ S Sweden
Lelången see Lelång
Lel'chitsy see Lyel'chytsy
le Léman see Geneva, Lake
Leli see Tianlin
25 O3 **Lelia Lake** Texas, SW USA 34°52′N 100°42′W
113 I14 **Lelija** ▲ SE Bosnia and Herzegovina 43°25′N 18°31′E
108 C8 **Le Locle** Neuchâtel, W Switzerland 47°04′N 06°45′E
189 Y14 **Lelu** Kosrae, E Micronesia
Lelu see Lelu Island
189 Y14 **Lelu Island** *var.* Lelu. *island* Kosrae, E Micronesia
55 W9 **Lelydorp** Wanica, N Suriname 05°36′N 55°04′W
98 K9 **Lelystad** Flevoland, C Netherlands 52°30′N 05°26′E
63 K25 **Le Maire, Estrecho de** *strait* S Argentina
168 L10 **Lemang** Pulau Rangsang, W Indonesia 01°04′N 102°44′E
186 I7 **Lemankoa** Buka Island, NE Papua New Guinea 05°06′S 154°23′E
Léman, Lac see Geneva, Lake
102 L6 **le Mans** Sarthe, NW France 48°N 00°12′E
29 S12 **Le Mars** Iowa, C USA 42°47′N 96°10′W
109 S3 **Lembach im Mühlkreis** Oberösterreich, N Austria 48°28′N 13°53′E
101 G23 **Lemberg** ▲ SW Germany 48°09′N 08°47′E
Lemberg see L'viv
Lemdiyya see Médéa
121 P3 **Lemesós** *var.* Limassol. SW Cyprus 34°41′N 33°02′E
100 H13 **Lemgo** Nordrhein-Westfalen, W Germany 52°02′N 08°54′E
33 P13 **Lemhi Range** ▲▲ Idaho, NW USA
9 S6 **Lemieux Islands** *island group* Nunavut, NE Canada
171 O11 **Lemito** Sulawesi, N Indonesia 0°34′N 121°31′E
92 L10 **Lemmenjoki** *Lapp.* Leammi. ≈ NE Finland
98 L7 **Lemmer** *Fris.* De Lemmer. Fryslân, N Netherlands 52°50′N 05°43′E
28 L7 **Lemmon** South Dakota, N USA 45°54′N 102°08′W
36 M15 **Lemmon, Mount** ▲ Arizona, SW USA 32°26′N 110°47′W
Lemnos see Límnos
31 O14 **Lemon, Lake** ◎ Indiana, N USA
102 J5 **le Mont St-Michel** *castle* Manche, N France
35 Q11 **Lemoore** California, W USA 36°16′N 119°48′W
189 T13 **Lemotol Bay** *bay* Chuuk Islands, C Micronesia
45 Y5 **le Moule** *var.* Moule. Grande Terre, NE Guadeloupe 16°20′N 61°21′W
Lemovices see Limoges
Le Moyen-Ogooué see Moyen-Ogooué
12 M6 **le Moyne, Lac** ◎ Québec, E Canada
93 L18 **Lempäälä** Pirkanmaa, W Finland 61°14′N 23°47′E
42 E7 **Lempa, Río** ≈ Central America
42 F7 **Lempira** *prev.* Gracias. ◆ *department* SW Honduras
Lemsalu see Limbaži
107 N17 **Le Murge** ▲▲ SE Italy
125 V6 **Lemva** ≈ NW Russian Federation
95 F21 **Lemvig** Midtjylland, W Denmark 56°31′N 08°19′E
166 K8 **Lemyethna** Ayeyawady, SW Myanmar (Burma) 17°36′N 95°08′E
30 K10 **Lena** Illinois, N USA 42°22′N 89°49′W
129 V4 **Lena** ≈ NE Russian Federation
173 N13 **Lena Tablemount** *undersea feature* S Indian Ocean 51°06′S 56°54′E
Lenchitsa see Łęczyca
59 N17 **Lençóis** Bahia, E Brazil 12°36′S 41°24′W
60 K9 **Lençóis Paulista** São Paulo, S Brazil 22°35′S 48°51′W
109 Y9 **Lendava** *Hung.* Lendva, *Ger.* Unterlimbach; *prev.* Dolnja Lendava. NE Slovenia 46°33′N 16°27′E
83 F20 **Lendepas** Hardap, SE Namibia 24°41′S 19°58′E
124 H9 **Lendery** *Finn.* Lentiira. Respublika Kareliya, NW Russian Federation 63°20′N 31°18′E
Lendum see Lens
Lendva see Lendava
27 R4 **Lenexa** Kansas, C USA 38°57′N 94°43′W
109 Q5 **Lengau** Oberösterreich, N Austria 48°01′N 13°17′E
145 Q17 **Lenger** Yuzhnyy Kazakhstan, S Kazakhstan 42°10′N 69°54′E
Lenghu see Lenghuzhen
159 O9 **Lenghuzhen** *var.* Lenghu. Qinghai, C China 38°50′N 93°25′E
159 T9 **Lenglong Ling** ▲ N China
108 D7 **Lengnau** Bern, W Switzerland 47°12′N 07°23′E
95 M20 **Lenhovda** Kronoberg, S Sweden 57°00′N 15°16′E
Lenīn see Uzynkol', Kazakhstan
Lenin see Akdepe, Turkmenistan
Leninabad see Khujand
Leninakan see Gyumri
Lenina, Pik see Lenin Peak
117 V12 **Lenine** *Rus.* Lenino. Avtonomna Respublika Krym, S Ukraine 45°18′N 35°47′E
Leningor see Leninogorsk
Leningrad see Sankt-Peterburg
Leningrad see Mu'minobod
126 L13 **Leningradskaya** Krasnodarskiy Kray, SW Russian Federation 46°19′N 39°23′E
195 S16 **Leningradskaya** *Russian research station* Antarctica 69°30′S 159°51′E
124 H12 **Leningradskaya Oblast'** ◆ *province* NW Russian Federation
Leningradskiy see Mu'minobod
Lenino see Lyenina, Belarus
Lenino see Lenine, Ukraine
Leninobod see Khujand
145 X9 **Leninogorsk** *Kaz.* Leningor. Vostochnyy Kazakhstan, E Kazakhstan 50°20′N 83°34′E
147 T12 **Lenin Peak** *Rus.* Pik Lenina, *Taj.* Qullai Lenin. ▲ Kyrgyzstan/Tajikistan 39°20′N 72°50′E
147 S8 **Leninpol'** Talasskaya Oblast', NW Kyrgyzstan 42°29′N 71°54′E
Lenin, Qullai see Lenin Peak
127 P11 **Leninsk** Volgogradskaya Oblast', SW Russian Federation 48°41′N 45°18′E
Leninsk see Baykonyr, Kazakhstan
Leninsk see Akdepe, Turkmenistan
Leninsk see Asaka, Uzbekistan
145 T8 **Leninskiy** Pavlodar, E Kazakhstan 52°13′N 76°50′E
122 I13 **Leninsk-Kuznetskiy** Kemerovskaya Oblast', S Russian Federation 54°42′N 86°16′E
125 P15 **Leninskoje** Kirovskaya Oblast', NW Russian Federation 58°19′N 47°03′E
Leninskoye see Uzynkol'
Lenin-Turkmenski see Türkmenabat
Leninváros see Tiszaújváros
Lenkoran' see Länkäran
101 F15 **Lenne** ≈ W Germany
101 G16 **Lennestadt** Nordrhein-Westfalen, W Germany 51°07′N 08°04′E
29 R11 **Lennox** South Dakota, N USA 43°21′N 96°53′W
63 J25 **Lennox, Isla** *Eng.* Lennox Island. *island* S Chile
Lennox Island see Lennox, Isla
21 Q9 **Lenoir** North Carolina, SE USA 35°56′N 81°31′W
20 M9 **Lenoir City** Tennessee, S USA 35°48′N 84°15′W
108 C7 **Le Noirmont** Jura, NW Switzerland 47°14′N 06°57′E
14 L9 **Lenôtre, Lac** ◎ Québec, SE Canada
29 U15 **Lenox** Iowa, C USA 40°52′N 94°33′W
103 O2 **Lens** *anc.* Lendum, Lentium. Pas-de-Calais, N France 50°26′N 02°50′E
123 O11 **Lensk** Respublika Sakha (Yakutiya), NE Russian Federation 60°43′N 115°16′E
111 F24 **Lenti** Zala, SW Hungary 46°38′N 16°30′E
Lentia see Linz
93 N14 **Lentiira** Kainuu, E Finland 64°22′N 29°52′E
Lentiira see Lendery
107 L25 **Lentini** *anc.* Leontini. Sicilia, Italy, C Mediterranean Sea 37°17′N 15°00′E
Lentium see Lens
Lentschiza see Łęczyca
93 N15 **Lentua** ◎ E Finland
119 H14 **Lentvaris** *Pol.* Landwarów. Vilnius, SE Lithuania 54°39′N 24°58′E
108 F7 **Lenzburg** Aargau, N Switzerland 47°23′N 08°11′E
109 R5 **Lenzing** Oberösterreich, N Austria 47°58′N 13°34′E
77 P13 **Léo** SW Burkina Faso 11°07′N 02°08′W
109 V7 **Leoben** Steiermark, C Austria 47°23′N 15°06′E
Leobschütz see Głubczyce
44 L9 **Léogâne** S Haiti 18°32′N 72°37′W
171 O11 **Leok** Sulawesi, N Indonesia 01°10′N 121°20′E
29 O7 **Leola** South Dakota, N USA 45°41′N 98°58′W
97 K20 **Leominster** W England, United Kingdom 52°09′N 02°18′W
19 N11 **Leominster** Massachusetts, NE USA 42°29′N 71°43′W
102 I15 **Léon** Landes, SW France 43°54′N 01°17′W
40 M12 **León** *var.* León de los Aldamas. Guanajuato, C Mexico 21°05′N 101°43′W
42 I10 **León** León, NW Nicaragua 12°24′N 86°52′W
104 L4 **León** Castilla y León, NW Spain 42°34′N 05°34′W
29 V16 **Leon** Iowa, C USA 40°44′N 93°45′W
42 I9 **León** ◆ *department* W Nicaragua
104 K4 **León** ◆ *province* Castilla y León, NW Spain
León see Cotopaxi
25 V9 **Leona** Texas, SW USA 31°09′N 95°58′W
25 U5 **Leonard** Texas, SW USA 33°22′N 96°15′W
107 H15 **Leonardo da Vinci** *prev.* Fiumicino. ✕ (Roma) Lazio, C Italy 41°48′N 12°15′E
21 X5 **Leonardtown** Maryland, NE USA 38°17′N 76°35′W
25 Q13 **Leona River** ≈ Texas, SW USA
41 Z11 **Leona Vicario** Quintana Roo, SE Mexico 20°57′N 87°06′W
101 H21 **Leonberg** Baden-Württemberg, SW Germany 48°48′N 09°01′E
62 M3 **León, Cerro** ▲ NW Paraguay 20°21′S 60°16′W
León de los Aldamas see León
109 T4 **Leonding** Oberösterreich, N Austria 48°17′N 14°15′E
107 I14 **Leonessa** Lazio, C Italy 42°36′N 12°56′E
107 K24 **Leonforte** Sicilia, Italy, C Mediterranean Sea 37°38′N 14°23′E
183 O13 **Leongatha** Victoria, SE Australia 38°30′S 145°56′E
Leonídi see Leonídio
115 F21 **Leonídio** *var.* Leonídi. Pelopónnisos, S Greece 37°11′N 22°50′E
104 J4 **León, Montes de** ▲▲ NW Spain
180 K11 **Leonora** Western Australia 28°52′S 121°16′E
25 S8 **Leon River** ≈ Texas, SW USA
Leontini see Lentini
Léopold II, Lac see Mai-Ndombe, Lac
99 J17 **Leopoldsburg** Limburg, NE Belgium 51°07′N 05°16′E
Léopoldville see Kinshasa
26 I5 **Leoti** Kansas, C USA 38°28′N 101°22′W
116 M11 **Leova** *Rus.* Leovo. SW Moldova 46°31′N 28°16′E
Leovo see Leova
102 G8 **le Palais** Morbihan, NW France 47°20′N 03°08′W
27 X10 **Lepanto** Arkansas, C USA 35°34′N 90°21′W
169 N13 **Lepar, Pulau** *island* W Indonesia
104 I14 **Lepe** Andalucía, S Spain 37°15′N 07°12′W
Lepel' see Lyepyel'
83 E25 **Lepelle** *var.* Elefantes; *prev.* Olifants. ≈ SW South Africa
83 I20 **Lephepe** *var.* Lephephe. Kweneng, SE Botswana 23°20′S 25°50′E
Lephephe see Lephepe
161 Q10 **Leping** Jiangxi, S China 28°57′N 117°07′E
Lépontiennes, Alpes/Lepontine, Alpi see Lepontine Alps
108 G10 **Lepontine Alps** *Fr.* Alpes Lépontiennes, *It.* Alpi Lepontine. ▲▲ Italy/Switzerland
79 G20 **Le Pool** ◆ *province* S Congo
173 O16 **Le Port** NW Réunion
103 N1 **le Portel** Pas-de-Calais, N France 50°42′N 01°35′E
93 N17 **Leppävirta** Pohjois-Savo, C Finland 62°30′N 27°50′E
45 Q11 **le Prêcheur** NW Martinique 14°48′N 61°14′W
145 V13 **Lepsi** *prev.* Lepsy. Taldykorgan, SE Kazakhstan 46°14′N 78°56′E
145 V13 **Lepsi** *prev.* Lepsy. ≈ SE Kazakhstan
Lepsy see Lepsi
Le Puglie see Puglia
103 Q12 **le Puy** *prev.* le Puy-en-Velay, *hist.* Anicium, Podium Anicensis. Haute-Loire, C France 45°03′N 03°53′E
le Puy-en-Velay see le Puy
45 X11 **le Raizet** *var.* Le Raizet. ✕ (Pointe-à-Pitre) Grande Terre, C Guadeloupe 16°16′N 61°31′W
107 J24 **Lercara Friddi** Sicilia, Italy, C Mediterranean Sea 37°45′N 13°37′E
78 G12 **Léré** Mayo-Kébbi Ouest, SW Chad 09°41′N 14°17′E
Leribe see Hlotse
106 E10 **Lerici** Liguria, NW Italy 44°06′N 09°53′E
54 I14 **Lérida** Vaupés, SE Colombia 0°01′S 70°28′W
105 U5 **Lérida** ◆ *province* Cataluña, NE Spain
Lérida see Lleida
105 N5 **Lerma** Castilla y León, N Spain 42°02′N 03°46′W
40 M13 **Lerma, Río** ≈ C Mexico
Lerna see Lérni
115 F20 **Lérni** *var.* Lerna. *prehistoric site* Pelopónnisos, S Greece
45 R11 **le Robert** E Martinique 14°41′N 60°57′W
115 M21 **Léros** *island* Dodekánisa, Greece, Aegean Sea
30 L13 **Le Roy** Illinois, N USA 40°21′N 88°45′W
27 Q6 **Le Roy** Kansas, C USA 38°04′N 95°37′W
29 W11 **Le Roy** Minnesota, N USA 43°30′N 92°30′W
18 E10 **Le Roy** New York, NE USA 42°58′N 77°58′W
Lerrnayin Gharabakh see Nagorno-Karabakh
95 J19 **Lerum** Västra Götaland, S Sweden 57°46′N 12°12′E
96 M2 **Lerwick** NE Scotland, United Kingdom 60°09′N 01°09′W
45 Y6 **les Abymes** *var.* Abymes. Grande Terre, C Guadeloupe 16°16′N 61°31′W
les Albères see Albères, Chaîne des
102 M4 **les Andelys** Eure, N France 49°15′N 01°26′E
45 Q12 **les Anses-d'Arlets** SW Martinique 14°29′N 61°05′W
104 L2 **Les Arriondes** *prev.* Arriondas. Asturias, N Spain 43°23′N 05°11′W
105 U6 **Les Borges Blanques** *var.* Borjas Blancas. Cataluña, NE Spain 41°31′N 00°52′E
Lesbos see Lésvos
Les Cayes see Cayes
31 Q4 **Les Cheneaux Islands** *island* Michigan, N USA
105 T8 **Les Coves de Vinromá** *Cast.* Cuevas de Vinromá. Valenciana, E Spain 40°18′N 00°07′E
103 T12 **les Écrins** ▲ E France 44°54′N 06°25′E
108 C10 **Le Sépey** Vaud, W Switzerland 46°21′N 07°04′E
15 T7 **Les Escoumins** Québec, SE Canada 48°21′N 69°26′W
Les Gonaïves see Gonaïves
160 H9 **Leshan** Sichuan, C China 29°42′N 103°43′E
108 D11 **Les Haudères** Valais, SW Switzerland 46°02′N 07°27′E
102 J9 **les Herbiers** Vendée, NW France 46°52′N 01°01′W
Lesh/Leshi see Lezhë
125 O8 **Leshukonskoye** Arkhangel'skaya Oblast', NW Russian Federation 64°54′N 45°48′E
Lesina see Hvar
107 M15 **Lesina, Lago di** ◎ SE Italy
114 K13 **Lesítse** ▲▲ NE Greece
94 G10 **Lesja** Oppland, S Norway 62°07′N 08°56′E
95 L15 **Lesjöfors** Värmland, C Sweden 59°57′N 14°12′E
111 O18 **Lesko** Podkarpackie, SE Poland 49°28′N 22°19′E
113 O15 **Leskovac** Serbia, SE Serbia 43°00′N 21°58′E
113 M22 **Leskovik** *var.* Leskoviku. Korçë, S Albania 40°09′N 20°39′E
Leskoviku see Leskovik
33 P14 **Leslie** Idaho, NW USA 43°51′N 113°28′W
31 Q10 **Leslie** Michigan, N USA 42°27′N 84°25′W
Leśna/Lesnaya see Lyasnaya
Lesnaya see Lyasnaya
102 F5 **Lesneven** Finistère, NW France 48°35′N 04°19′W
112 J11 **Lešnica** Serbia, W Serbia 44°40′N 19°18′E
125 S13 **Lesnoy** Kirovskaya Oblast', NW Russian Federation 59°49′N 52°07′E
122 G10 **Lesnoy** Sverdlovskaya Oblast', C Russian Federation 58°40′N 59°48′E
122 K12 **Lesosibirsk** Krasnoyarskiy Kray, C Russian Federation 58°13′N 92°23′E
83 J23 **Lesotho** *off.* Kingdom of Lesotho; *prev.* Basutoland. ◆ *monarchy* S Africa
Lesotho, Kingdom of see Lesotho
102 J12 **Lesparre-Médoc** Gironde, SW France 45°18′N 00°57′W
108 C8 **Les Ponts-de-Martel** Neuchâtel, W Switzerland 47°00′N 06°45′E
103 P1 **Lesquin** ✕ Nord, N France 50°34′N 3°07′E
102 I9 **les Sables-d'Olonne** Vendée, NW France 46°30′N 01°47′W
109 S7 **Lessach** *var.* Lessachbach. ≈ E Austria
Lessachbach see Lessach
45 W11 **les Saintes** *var.* Îles des Saintes. *island group* S Guadeloupe
74 L5 **Les Salines** ✕ (Annaba) NE Algeria 36°45′N 07°57′E
99 J22 **Lesse** ≈ SE Belgium
95 M21 **Lessebo** Kronoberg, S Sweden 56°45′N 15°19′E
Lesser Antarctica see West Antarctica
45 P15 **Lesser Antilles** *island group* E West Indies
137 T10 **Lesser Caucasus** *Rus.* Malyy Kavkaz. ▲▲ SW Asia
Lesser Khingan Range see Xiao Hinggan Ling
11 P13 **Lesser Slave Lake** ◎ Alberta, W Canada
99 E19 **Lessines** Hainaut, SW Belgium 50°43′N 03°50′E
103 R16 **les Stes-Maries-de-la-Mer** Bouches-du-Rhône, SE France 43°27′N 04°26′E
14 G15 **Lester B. Pearson** *var.* Toronto. ✕ (Toronto) Ontario, S Canada 43°59′N 81°30′W
29 U9 **Lester Prairie** Minnesota, N USA 44°52′N 94°02′W
93 L16 **Lestijärvi** Keski-Pohjanmaa, W Finland 63°29′N 24°41′E
L'Estuaire see Estuaire
29 U9 **Le Sueur** Minnesota, N USA 44°27′N 93°53′W
108 B8 **Les Verrières** Neuchâtel, W Switzerland 46°54′N 06°29′E
115 L17 **Lésvos** *anc.* Lesbos. *island* E Greece
110 G12 **Leszno** *Ger.* Lissa. Wielkopolskie, C Poland 51°51′N 16°35′E
83 L20 **Letaba** Northern, NE South Africa 23°44′S 31°29′E
173 P17 **Le Tampon** SW Réunion
97 O21 **Letchworth** E England, United Kingdom 51°58′N 00°14′W
111 G25 **Letenye** Zala, SW Hungary 46°25′N 16°44′E
11 Q17 **Lethbridge** Alberta, SW Canada 49°43′N 112°48′W
55 S11 **Lethem** S Guyana 03°24′N 59°45′W
83 H18 **Letiahau** ≈ W Botswana
54 J18 **Leticia** Amazonas, S Colombia 04°09′S 69°57′W
171 S16 **Leti, Kepulauan** *island group* E Indonesia
83 I18 **Letlhakane** Central, C Botswana 21°26′S 25°36′E
83 H20 **Letlhakeng** Kweneng, SE Botswana 24°05′S 25°03′E
114 J8 **Letnitsa** Lovech, N Bulgaria 43°19′N 25°02′E
103 N1 **le Touquet-Paris-Plage** Pas-de-Calais, N France 50°31′N 01°36′E
166 L8 **Letpadan** Bago, SW Myanmar (Burma) 17°46′N 95°45′E
166 K6 **Letpan** Rakhine State, W Myanmar (Burma) 19°22′N 94°11′E
102 M2 **le Tréport** Seine-Maritime, N France 50°03′N 01°21′E
166 M12 **Letsôk-aw Kyun** *var.* Letsutan Island; *prev.* Domel Island. *island* Mergui Archipelago, S Myanmar (Burma)
Letsutan Island see Letsôk-aw Kyun
97 E14 **Letterkenny** *Ir.* Leitir Ceanainn. Donegal, NW Ireland 54°57′N 07°44′W
Lettland see Latvia
116 M6 **Letychiv** Khmel'nyts'ka Oblast', W Ukraine 49°24′N 27°39′E
Lëtzebuerg see Luxembourg
116 H14 **Leu** Dolj, SW Romania 44°10′N 24°01′E
Leucas see Lefkáda
103 P17 **Leucate** Aude, S France 42°55′N 03°03′E
103 P17 **Leucate, Étang de** ◎ S France
108 E10 **Leuk** Valais, SW Switzerland 46°18′N 07°46′E
108 E10 **Leukerbad** Valais, SW Switzerland 46°22′N 07°47′E
Leusden see Leusden-Centrum
98 K11 **Leusden-Centrum** *var.* Leusden. Utrecht, C Netherlands 52°08′N 05°25′E
Leutensdorf see Litvínov
Leutschau see Levoča
99 H18 **Leuven** *Fr.* Louvain, *Ger.* Löwen. Vlaams Brabant, C Belgium 50°53′N 04°42′E
99 I20 **Leuze** Namur, C Belgium 50°33′N 04°55′E
Leuze see Leuze-en-Hainaut
99 E19 **Leuze-en-Hainaut** *var.* Leuze. Hainaut, SW Belgium 50°36′N 03°37′E
Léva see Levice
Levádia see Livádeia
Levajok see Leavvajohka
36 L4 **Levan** Utah, W USA 39°33′N 111°51′W
93 E16 **Levanger** Nord-Trøndelag, C Norway 63°45′N 11°18′E
106 D10 **Levanto** Liguria, W Italy 44°12′N 09°33′E
107 H23 **Levanzo, Isola di** *island* Isole Egadi, S Italy
127 Q17 **Levashi** Respublika Dagestan, SW Russian Federation 42°27′N 47°19′E
24 M5 **Levelland** Texas, SW USA 33°35′N 102°23′W
39 P13 **Levelock** Alaska, USA 59°07′N 156°51′W
101 E16 **Leverkusen** Nordrhein-Westfalen, W Germany 51°02′N 07°E
111 J21 **Levice** *Ger.* Lewentz, *Hung.* Léva, Lewenz. Nitriansky Kraj, SW Slovakia 48°14′N 18°38′E
106 G6 **Levico Terme** Trentino-Alto Adige, N Italy 46°02′N 11°19′E
115 E20 **Levídi** Pelopónnisos, S Greece 37°39′N 22°13′E
103 P14 **le Vigan** Gard, S France 43°00′N 03°36′E
184 L13 **Levin** Manawatu-Wanganui, North Island, New Zealand 40°38′S 175°17′E
15 R10 **Lévis** *var.* Levis. Québec, SE Canada 46°47′N 71°12′W
Levis see Lévis
21 P6 **Levisa Fork** ≈ Kentucky/Virginia, S USA
115 L21 **Levítha** *island* Kykládes, Greece, Aegean Sea
18 L14 **Levittown** Long Island, New York, NE USA 40°42′N 73°29′W
18 J15 **Levittown** Pennsylvania, NE USA 40°09′N 74°50′W
Levkás see Lefkáda
Levkímmi see Lefkímmi
111 L19 **Levoča** *Ger.* Leutschau, *Hung.* Locse. Prešovský Kraj, E Slovakia 49°01′N 20°34′E
Lévrier, Baie du see Nouâdhibou, Dakhlet
103 N9 **Levroux** Indre, C France 47°00′N 01°37′E
114 J8 **Levski** Pleven, N Bulgaria 43°21′N 25°11′E
Levskigrad see Karlovo
126 L6 **Lev Tolstoy** Lipetskaya Oblast', W Russian Federation 53°12′N 39°28′E
187 X14 **Levuka** Ovalau, C Fiji 17°42′S 178°50′E
166 L6 **Lewe** Mandalay, C Myanmar (Burma) 19°40′N 96°04′E
Lewentz/Lewenz see Levice
97 O23 **Lewes** SE England, United Kingdom 50°52′N 00°01′E
21 Z4 **Lewes** Delaware, NE USA 38°46′N 75°08′W
29 Q12 **Lewis And Clark Lake** ⊠ Nebraska/South Dakota, N USA
18 G14 **Lewisburg** Pennsylvania, NE USA 40°57′N 76°52′W
20 J10 **Lewisburg** Tennessee, S USA 35°29′N 86°49′W
21 S6 **Lewisburg** West Virginia, NE USA 37°49′N 80°28′W
96 F6 **Lewis, Butt of** *headland* NW Scotland, United Kingdom 58°31′N 06°18′W
96 F7 **Lewis, Isle of** *island* NW Scotland, United Kingdom
35 U4 **Lewis, Mount** ▲ Nevada, W USA 40°22′N 116°50′W
185 H16 **Lewis Pass** *pass* South Island, New Zealand
33 P7 **Lewis Range** ▲▲ Montana, NW USA
23 O3 **Lewis Smith Lake** ⊠ Alabama, S USA
32 M10 **Lewiston** Idaho, NW USA 46°25′N 117°01′W
19 P7 **Lewiston** Maine, NE USA 44°08′N 70°14′W
29 X10 **Lewiston** Minnesota, N USA 43°58′N 91°52′W
18 D9 **Lewiston** New York, NE USA 43°10′N 79°02′W
36 L1 **Lewiston** Utah, W USA 41°58′N 111°52′W
30 K13 **Lewistown** Illinois, N USA 40°23′N 90°09′W
33 T9 **Lewistown** Montana, NW USA 47°04′N 109°26′W
27 T14 **Lewisville** Arkansas, C USA 33°21′N 93°38′W
25 T6 **Lewisville** Texas, SW USA 33°00′N 96°57′W
25 T6 **Lewisville, Lake** ⊠ Texas, SW USA
Le Woleu-Ntem see Woleu-Ntem
23 U3 **Lexington** Georgia, SE USA 33°51′N 83°04′W
20 M5 **Lexington** Kentucky, S USA 38°03′N 84°30′W
22 L4 **Lexington** Mississippi, S USA 33°06′N 90°03′W
27 S4 **Lexington** Missouri, C USA 39°11′N 93°52′W
29 N16 **Lexington** Nebraska, C USA 40°46′N 99°44′W
20 S9 **Lexington** North Carolina, SE USA 35°49′N 80°15′W
27 N11 **Lexington** Oklahoma, C USA 35°00′N 97°20′W
21 R12 **Lexington** South Carolina, SE USA 33°59′N 81°15′W
20 G9 **Lexington** Tennessee, S USA 35°39′N 88°24′W
25 T10 **Lexington** Texas, SW USA 30°25′N 97°00′W

◆ Country
● Country Capital
◇ Dependent Territory
○ Dependent Territory Capital
◆ Administrative Regions
✕ International Airport
▲ Mountain
▲▲ Mountain Range
⛰ Volcano
≈ River
◎ Lake
⊠ Reservoir

LEXINGTON – LITTLE CURRENT

21 T6 **Lexington** Virginia, NE USA 37°47′N 79°27′W
21 X5 **Lexington Park** Maryland, NE USA 38°16′N 76°27′W
Leyden *see* Leiden
102 J14 **Leyre** ≈ SW France
171 Q5 **Leyte** *island* C Philippines
171 Q6 **Leyte Gulf** *gulf* E Philippines
111 O16 **Leżajsk** Podkarpackie, SE Poland 50°15′N 22°25′E
Lezha *see* Lezhë
113 K18 **Lezhë** *var.* Lezha; *prev.* Lesh, Leshi. Lezhë, NW Albania 41°46′N 19°40′E
113 K18 **Lezhë** ◈ *district* NW Albania
103 O16 **Lézignan-Corbières** Aude, S France 43°12′N 02°46′E
126 J7 **L'gov** Kurskaya Oblast', W Russian Federation 51°38′N 35°17′E
159 P15 **Lhari** Xizang Zizhiqu, W China 30°34′N 93°40′E
159 N16 **Lhasa** *var.* La-sa, Lassa. Xizang Zizhiqu, W China 29°41′N 91°10′E
159 O15 **Lhasa He** ≈ W China
Lhaviyani Atoll *see* Faadhippolhu Atoll
158 K16 **Lhazê** *var.* Quxar. Xizang Zizhiqu, W China 29°07′N 87°32′E
158 K14 **Lhazhong** Xizang Zizhiqu, W China 31°58′N 86°43′E
168 H7 **Lhoksukon** Sumatera, W Indonesia 05°04′N 97°19′E
159 Q15 **Lhorong** *var.* Zito. Xizang Zizhiqu, W China 30°51′N 95°41′E
105 W6 **L'Hospitalet de Llobregat** *var.* Hospitalet. Cataluña, NE Spain 41°21′N 02°06′E
153 R11 **Lhotse** ▲ China/Nepal 28°00′N 86°55′E
159 N17 **Lhozhag** *var.* Garbo. Xizang Zizhiqu, W China 28°21′N 90°47′E
159 O16 **Lhünzê** *var.* Xingba. Xizang Zizhiqu, W China 28°25′N 92°30′E
159 N15 **Lhünzhub** *var.* Ganqu. Xizang Zizhiqu, W China 30°14′N 91°20′E
167 N8 **Li** Lamphun, NW Thailand 17°46′N 98°54′E
115 L21 **Liádi** *var.* Livádi. *island* Kykládes, Greece, Aegean Sea
161 P12 **Liancheng** *var.* Lianfeng. Fujian, SE China 25°47′N 116°42′E
Liancheng *see* Lianjiang, Guangdong, China
Liancheng *see* Qinglong, Guizhou, China
Liancheng *see* Guangnan, Yunnan, China
Lianfeng *see* Liancheng
160 K9 **Liangping** *var.* Liangshan. Sichuan, C China 30°40′N 107°46′E
Liangshan *see* Liangping
Liangzhou *see* Wuwei
161 O9 **Liangzi Hu** ▣ C China
161 R12 **Lianjiang** *var.* Fengcheng. Fujian, SE China 26°14′N 119°33′E
160 L15 **Lianjiang** *var.* Liancheng. Guangdong, S China 21°41′N 110°12′E
Lianjiang *see* Xingguo
161 O13 **Lianping** *var.* Yuanshan. Guangdong, S China 24°18′N 114°27′E
Lianshan *see* Huludao
Lian Xian *see* Lianzhou
160 M11 **Lianyuan** *prev.* Lantian. Hunan, S China 27°51′N 111°44′E
161 Q6 **Lianyungang** *var.* Xinpu. Jiangsu, E China 34°38′N 119°12′E
161 N13 **Lianzhou** *var.* Linxian; *prev.* Lian Xian. Guangdong, S China 24°48′N 112°26′E
Lianzhou *see* Hepu
Liao *see* Liaoning
161 P5 **Liaocheng** Shandong, E China 36°31′N 115°59′E
163 U13 **Liaodong Bandao** *var.* Liaotung Peninsula. *peninsula* NE China
163 T13 **Liaodong Wan** *Eng.* Gulf of Lantung, Gulf of Liaotung. *gulf* NE China
163 U11 **Liao He** ≈ NE China
163 U12 **Liaoning** *var.* Liao, Liaoning Sheng, Shengking, *hist.* Fengtien, Shenking. ◈ *province* NE China
Liaoning Sheng *see* Liaoning
Liaotung Peninsula *see* Liaodong Bandao
163 V12 **Liaoyang** *var.* Liao-yang. Liaoning, NE China 41°16′N 123°12′E
Liao-yang *see* Liaoyang
163 V11 **Liaoyuan** *var.* Dongliao, Shuang-liao, *Jap.* Chengchiatun. Jilin, NE China 42°52′N 125°09′E
163 U12 **Liaozhong** Liaoning, NE China 41°33′N 122°54′E
Liaqatabad *see* Piplān
10 M10 **Liard** ≈ W Canada
Liard *see* Fort Liard
10 L10 **Liard River** British Columbia, W Canada 59°23′N 126°05′W
149 O15 **Liāri** Baluchistān, SW Pakistan 25°43′N 66°28′E
Liatroim *see* Leitrim
189 S6 **Lib** *var.* Ellep. *island* Ralik Chain, C Marshall Islands
Liban *see* Lebanon
138 H6 **Liban, Jebel** *Ar.* Jabal al Gharbt, Jabal Lubnān, *Eng.* Mount Lebanon. ▲▲ C Lebanon
Libau *see* Liepāja
33 N7 **Libby** Montana, NW USA 48°25′N 115°33′W
79 I16 **Libenge** Equateur, NW Dem. Rep. Congo 03°39′N 18°39′E
26 I7 **Liberal** Kansas, C USA 37°03′N 100°56′W
27 R7 **Liberal** Missouri, C USA 37°33′N 94°31′W
Liberalitas Julia *see* Évora
111 D15 **Liberec** *Ger.* Reichenberg. Liberecký Kraj, N Czech Republic 50°45′N 15°05′E
111 D15 **Liberecky Kraj** ◈ *region* N Czech Republic
42 K12 **Liberia** Guanacaste, NW Costa Rica 10°36′N 85°26′W
76 K17 **Liberia** *off.* Republic of Liberia. ◆ *republic* W Africa
Liberia, Republic of *see* Liberia
61 D16 **Libertad** Corrientes, NE Argentina 30°01′S 57°51′W
61 E20 **Libertad** San José, S Uruguay 34°38′S 56°39′W
54 I7 **Libertad** Barinas, NW Venezuela 08°21′N 69°39′W
54 K6 **Libertad** Cojedes, N Venezuela 09°15′N 68°30′W
62 G12 **Libertador** *off.* Región del Libertador General Bernardo O'Higgins. ◈ *region* C Chile
Libertador General Bernardo O'Higgins, Región del *see* Libertador
Libertador General San Martín *see* Ciudad de Libertador General San Martín
20 L6 **Liberty** Kentucky, S USA 37°19′N 84°58′W
22 J7 **Liberty** Mississippi, S USA 31°09′N 90°49′W
27 R4 **Liberty** Missouri, C USA 39°15′N 94°22′W
18 J12 **Liberty** New York, NE USA 41°48′N 74°45′W
21 T9 **Liberty** North Carolina, SE USA 35°49′N 79°34′W
Libian Desert *see* Libyan Desert
99 J23 **Libin** Luxembourg, SE Belgium 50°01′N 05°13′E
Lībīyah, Aş Şahrā' al *see* Libyan Desert
160 K13 **Libo** *var.* Yuping. Guizhou, S China 25°28′N 107°52′E
Libohova *see* Libohovë
113 L23 **Libohovë** *var.* Libohova. Gjirokastër, S Albania 40°03′N 20°13′E
81 K18 **Liboi** Wajir, E Kenya 0°23′N 40°55′E
102 K13 **Libourne** Gironde, SW France 44°55′N 00°14′W
99 K23 **Libramont** Luxembourg, SE Belgium 49°55′N 05°21′E
113 M20 **Librazhd** *var.* Librazhdi. Elbasan, E Albania 41°10′N 20°22′E
Librazhdi *see* Librazhd
79 C18 **Libreville** ● (Gabon) Estuaire, NW Gabon 0°25′N 09°29′E
75 P10 **Libya** *off.* Great Socialist People's Libyan Arab Jamahiriya, *Ar.* Al Jamāhīrīyah al 'Arabīyah al Lībīyah ash Sha'bīyah al Ishtirākīy; *prev.* Libyan Arab Republic. ◆ *Islamic state* N Africa
Libyan Arab Republic *see* Libya
75 T11 **Libyan Desert** *var.* Libian Desert, *Ar.* Aş Şahrā' al Lībīyah. *desert* N Africa
75 T8 **Libyan Plateau** *var.* Aḍ Ḍiffah. *plateau* Egypt/Libya
62 G12 **Licantén** Maule, C Chile 35°00′S 72°00′W
107 J25 **Licata** *anc.* Phintias. Sicilia, Italy, C Mediterranean Sea 37°07′N 13°57′E
137 P14 **Lice** Dıyarbakır, SE Turkey 38°29′N 40°39′E
Licheng *see* Lipu
97 L19 **Lichfield** C England, United Kingdom 52°42′N 01°48′W
83 N14 **Lichinga** Niassa, N Mozambique 13°19′S 35°13′E
109 V3 **Lichtenau** Niederösterreich, N Austria 48°29′N 15°24′E
83 I21 **Lichtenburg** North-West, N South Africa 26°09′S 26°11′E
101 K18 **Lichtenfels** Bayern, SE Germany 50°09′N 11°04′E
98 O12 **Lichtenvoorde** Gelderland, E Netherlands 51°59′N 06°34′E
Lichtenwald *see* Sevnica
99 C17 **Lichtervelde** West-Vlaanderen, W Belgium 51°02′N 03°09′E
160 L9 **Lichuan** Hubei, C China 30°20′N 108°56′E
27 V7 **Licking** Missouri, C USA 37°30′N 91°51′W
20 M4 **Licking River** ≈ Kentucky, S USA
112 C11 **Lički Osik** Lika-Senj, C Croatia 44°36′N 15°24′E
Ličko-Senjska Županija *see* Lika-Senj
107 K19 **Licosa, Punta** *headland* S Italy 40°15′N 14°54′E
119 H16 **Lida** Hrodzyenskaya Voblasts', W Belarus 53°53′N 25°20′E
93 H17 **Liden** Västernorrland, C Sweden 62°43′N 16°49′E
29 R7 **Lidgerwood** North Dakota, N USA 46°04′N 97°09′W
Lidhorikíon *see* Lidoríki
95 K21 **Lidhult** Kronoberg, S Sweden 56°49′N 13°25′E
95 P16 **Lidingö** Stockholm, C Sweden 59°22′N 18°10′E
95 K17 **Lidköping** Västra Götaland, S Sweden 58°30′N 13°10′E
Lido di Iesolo *see* Lido di Jesolo
106 I8 **Lido di Jesolo** *var.* Lido di Iesolo. Veneto, NE Italy 45°30′N 12°37′E
107 H15 **Lido di Ostia** Lazio, C Italy 41°42′N 12°19′E
Lidokhorikion *see* Lidoríki
115 E18 **Lidoríki** *prev.* Lidhorikíon, Lidokhorikion. Stereá Ellàda, C Greece 38°32′N 22°12′E
110 K9 **Lidzbark** Warmińsko-Mazurskie, NE Poland 53°15′N 19°49′E
110 L7 **Lidzbark Warmiński** *Ger.* Heilsberg. Olsztyn, N Poland 54°08′N 20°35′E
109 U3 **Liebenau** Oberösterreich, N Austria 48°33′N 14°48′E
181 P7 **Liebig, Mount** ▲ Northern Territory, C Australia 23°19′S 131°30′E
109 V8 **Lieboch** Steiermark, SE Austria 47°00′N 15°21′E
108 I8 **Liechtenstein** *off.* Principality of Liechtenstein. ◆ *principality* C Europe
Liechtenstein, Principality of *see* Liechtenstein
99 F18 **Liedekerke** Vlaams Brabant, C Belgium 50°51′N 04°05′E
99 K19 **Liège** *Dut.* Luik, *Ger.* Lüttich. Liège, E Belgium 50°38′N 05°35′E
99 K20 **Liège** *Dut.* Luik. ◈ *province* E Belgium
Liegnitz *see* Legnica
93 O16 **Lieksa** Pohjois-Karjala, E Finland 63°20′N 30°E
118 F10 **Lielupe** ≈ Latvia/Lithuania
118 G9 **Lielvārde** C Latvia 56°45′N 24°48′E
167 U13 **Liên Hương** *var.* Tuy Phong. Binh Thuân, S Vietnam 11°13′N 108°40′E
167 U13 **Liên Nghia** *var.* Liên Nghia *var.* Đưc Trong. Lâm Đông, S Vietnam 11°45′N 108°24′E
Liên Nghia *see* Liên Nghia
109 P9 **Lienz** Tirol, W Austria 46°50′N 12°45′E
118 B10 **Liepāja** *Ger.* Libau. W Latvia 56°32′N 21°02′E
99 H17 **Lier** *Fr.* Lierre. Antwerpen, N Belgium 51°08′N 04°35′E
95 H15 **Lierbyen** Buskerud, S Norway 59°57′N 10°14′E
99 L21 **Lierneux** Liège, E Belgium 50°12′N 05°51′E
Lierre *see* Lier
101 D18 **Lieser** ≈ W Germany
109 U7 **Liesing** ≈ E Austria
108 E6 **Liestal** Basel Landschaft, N Switzerland 47°29′N 07°43′E
Lietuva *see* Lithuania
Lievenhof *see* Līvāni
103 O2 **Liévin** Pas-de-Calais, N France 50°25′N 02°48′E
14 M9 **Lièvre, Rivière du** ≈ Québec, SE Canada
109 T6 **Liezen** Steiermark, C Austria 47°34′N 14°12′E
97 E14 **Lifford** *Ir.* Leifear. Donegal, NW Ireland 54°50′N 07°29′W
187 Q16 **Lifou** *island* Îles Loyauté, E New Caledonia
193 Y15 **Lifuka** *island* Ha'apai Group, C Tonga
171 P4 **Ligao** Luzon, N Philippines 13°16′N 123°30′E
Liger *see* Loire
42 H2 **Lighthouse Reef** *reef* E Belize
183 Q4 **Lightning Ridge** New South Wales, SE Australia 29°29′S 148°00′E
103 N9 **Lignières** Cher, C France 46°45′N 02°10′E
103 S5 **Ligny-en-Barrois** Meuse, NE France 48°42′N 05°22′E
83 P15 **Ligonha** ≈ NE Mozambique
31 P11 **Ligonier** Indiana, N USA 41°25′N 85°33′W
81 I25 **Ligunga** Ruvuma, S Tanzania 10°51′S 37°10′E
106 D9 **Ligure, Appennino** *Eng.* Ligurian Mountains. ▲▲ NW Italy
Ligure, Mar *see* Ligurian Sea
106 C9 **Liguria** ◈ *region* NW Italy
Ligurian Mountains *see* Ligure, Appennino
120 K6 **Ligurian Sea** *Fr.* Mer Ligurienne, *It.* Mar Ligure. *sea* N Mediterranean Sea
Ligurienne, Mer *see* Ligurian Sea
186 H5 **Lihir Group** *island group* NE Papua New Guinea
38 B8 **Līhu'e** *var.* Lihue. Kaua'i, Hawaii, USA 21°59′N 159°23′W
Lihue *see* Līhu'e
118 F5 **Lihula** *Ger.* Leal. Läänemaa, W Estonia 58°44′N 23°49′E
124 I2 **Liinakhamari** *var.* Linacmamari. Murmanskaya Oblast', NW Russian Federation 69°40′N 31°27′E
Liivi Laht *see* Riga, Gulf of
160 F11 **Lijiang** *var.* Dayan, Lijiang Naxizu Zizhixian. Yunnan, SW China 26°52′N 100°10′E
Lijiang Naxizu Zizhixian *see* Lijiang
112 C11 **Lika-Senj** *off.* Ličko-Senjska Županija. ◈ *province* W Croatia
79 N25 **Likasi** *prev.* Jadotville. Shaba, SE Dem. Rep. Congo 11°02′S 26°51′E
79 L16 **Likati** Orientale, N Dem. Rep. Congo 03°28′N 23°45′E
10 M15 **Likely** British Columbia, SW Canada 52°40′N 121°34′W
153 Y11 **Likhapāni** Assam, NE India 27°19′N 95°54′E
124 J16 **Likhoslavl'** Tverskaya Oblast', W Russian Federation 57°08′N 35°27′E
189 U5 **Likiep Atoll** *atoll* Ratak Chain, C Marshall Islands
95 D18 **Liknes** Vest-Agder, S Norway 58°20′N 06°58′E
79 H16 **Likouala** ◈ *province* NE Congo
79 H18 **Likouala** ≈ N Congo
79 H18 **Likouala aux Herbes** ≈ E Congo
190 B16 **Liku** E Niue 19°02′S 169°47′E
Likupang, Selat *see* Bangka, Selat
27 Y8 **Lilbourn** Missouri, C USA 36°35′N 89°37′W
103 X14 **l'Île-Rousse** Corse, France, C Mediterranean Sea 42°39′N 08°59′E
109 W5 **Lilienfeld** Niederösterreich, NE Austria 48°01′N 15°36′E
161 N11 **Liling** Hunan, S China 27°42′N 113°49′E
95 J18 **Lilla Edet** Västra Götaland, S Sweden 58°08′N 12°08′E
103 P1 **Lille** *var.* l'Isle, *Dut.* Rijssel, *Flem.* Ryssel, *prev.* Lisle; *anc.* Insula. Nord, N France 50°38′N 03°04′E
95 G24 **Lillebælt** *var.* Lille Bælt, *Eng.* Little Belt. *strait* S Denmark
Lille Bælt *see* Lillebælt
102 L3 **Lillebonne** Seine-Maritime, N France 49°30′N 00°34′E
94 H12 **Lillehammer** Oppland, S Norway 61°07′N 10°28′E
103 O1 **Lillers** Pas-de-Calais, N France 50°34′N 02°26′E
95 F18 **Lillesand** Aust-Agder, S Norway 58°15′N 08°23′E
95 I15 **Lillestrøm** Akershus, S Norway 59°58′N 11°05′E
93 F18 **Lillhärdal** Jämtland, C Sweden 61°51′N 14°04′E
21 U10 **Lillington** North Carolina, SE USA 35°25′N 78°50′W
105 O9 **Lillo** Castilla-La Mancha, C Spain 39°43′N 03°19′W
10 M16 **Lillooet** British Columbia, SW Canada 50°41′N 121°59′W
83 M14 **Lilongwe** ● (Malawi) Central , W Malawi 13°58′S 33°48′E
83 M14 **Lilongwe** ✈ Central, W Malawi 13°46′S 33°44′E
83 M14 **Lilongwe** ≈ W Malawi
171 P7 **Liloy** Mindanao, S Philippines 08°04′N 122°42′E
Lilybaeum *see* Marsala
182 J7 **Lilydale** South Australia 32°57′S 140°00′E
183 P16 **Lilydale** Tasmania, SE Australia 41°17′S 147°13′E
113 J14 **Lim** ≈ SE Europe
57 D15 **Lima** ● (Peru) Lima, W Peru 12°06′S 78°W
94 K13 **Lima** Dalarna, C Sweden 60°55′N 13°19′E
31 R12 **Lima** Ohio, NE USA 40°43′N 84°06′W
57 D14 **Lima** ◈ *department* W Peru
Lima *see* Jorge Chávez Internacional
137 Y13 **Liman** *prev.* Port-İliç. SE Azerbaijan 38°54′N 48°49′E
111 L17 **Limanowa** Małopolskie, S Poland 49°43′N 20°25′E
104 G5 **Lima, Rio** *Sp.* Limia. ≈ Portugal/Spain *see also* Limia
Lima, Rio *see* Limia
168 M11 **Limas** Pulau Sebangka, W Indonesia 0°09′N 104°31′E
Limassol *see* Lemesós
97 F14 **Limavady** *Ir.* Léim An Mhadaidh. NW Northern Ireland, United Kingdom 55°03′N 06°57′W
63 J14 **Limay Mahuida** La Pampa, C Argentina 37°09′S 66°40′W
63 H15 **Limay, Río** ≈ W Argentina
101 N16 **Limbach-Oberfrohna** Sachsen, E Germany 50°52′N 12°46′E
81 F22 **Limba Limba** ≈ C Tanzania
107 C17 **Limbara, Monte** ▲ Sardegna, Italy, C Mediterranean Sea 40°50′N 09°10′E
118 G7 **Limbaži** *Est.* Lemsalu. N Latvia 57°33′N 24°46′E
44 M8 **Limbé** N Haiti 19°44′N 72°25′W
99 L19 **Limbourg** Liège, E Belgium 50°37′N 05°56′E
99 K17 **Limburg** ◈ *province* NE Belgium
99 L16 **Limburg** ◈ *province* SE Netherlands
101 F17 **Limburg an der Lahn** Hessen, W Germany 50°22′N 08°04′E
94 K13 **Limedsforsen** Dalarna, C Sweden 60°52′N 13°25′E
60 L9 **Limeira** São Paulo, S Brazil 22°34′S 47°25′W
97 C19 **Limerick** *Ir.* Luimneach. Limerick, SW Ireland 52°40′N 08°38′W
97 C20 **Limerick** *Ir.* Luimneach. *cultural region* SW Ireland
19 S2 **Limestone** Maine, NE USA 46°52′N 67°49′W
25 U9 **Limestone, Lake** ◎ Texas, SW USA
39 P12 **Lime Village** Alaska, USA 61°21′N 155°26′W
95 F20 **Limfjorden** *fjord* N Denmark
95 J23 **Limhamn** Skåne, S Sweden 55°34′N 12°57′E
104 H5 **Limia** *Port.* Rio Lima. ≈ Portugal/Spain *see also* Lima,Rio
Limia *see* Lima, Rio
93 L14 **Liminka** Pohjois-Pohjanmaa, C Finland 64°48′N 25°19′E
Limín Vathéos *see* Sámos
115 G17 **Límni** Évvoia, C Greece 38°46′N 23°20′E
115 J15 **Límnos** *anc.* Lemnos. *island* E Greece
102 M11 **Limoges** *anc.* Augustoritum Lemovicensium, Lemovices. Haute-Vienne, C France 45°51′N 01°16′E
43 O13 **Limón** *var.* Puerto Limón. Limón, E Costa Rica 09°59′N 83°02′W
42 K4 **Limón** Colón, NE Honduras 15°50′N 85°31′W
37 U5 **Limon** Colorado, C USA 39°15′N 103°41′W
43 N13 **Limón** *off.* Provincia de Limón. ◈ *province* E Costa Rica
106 A10 **Limone Piemonte** Piemonte, NE Italy 44°12′N 07°37′E
Limones *see* Valdéz
Limón, Provincia de *see* Limón
Limonum *see* Poitiers
103 N11 **Limousin** ◈ *region* C France
103 N16 **Limoux** Aude, S France 43°03′N 02°13′E
83 J20 **Limpopo** *off.* Limpopo Province; *prev.* Northern, Northern Transvaal. ◈ *province* NE South Africa
83 L19 **Limpopo** *var.* Crocodile. ≈ S Africa
Limpopo Province *see* Limpopo
160 K17 **Limu Ling** ▲▲ S China
113 M20 **Lin** *var.* Lini. Elbasan, E Albania 41°03′N 20°37′E
Linacmamari *see* Liinakhamari
62 G13 **Linares** Maule, C Chile 35°50′S 71°37′W
54 C13 **Linares** Nariño, SW Colombia 01°24′N 77°30′W
41 O9 **Linares** Nuevo León, NE Mexico 24°54′N 99°38′W
105 N12 **Linares** Andalucía, S Spain 38°05′N 03°38′W
107 G15 **Linaro, Capo** *headland* C Italy 42°01′N 11°49′E
106 D8 **Linate** ✈ (Milano) Lombardia, N Italy 45°27′N 09°18′E
167 T8 **Lin Camh** *prev.* Đưc Tho. Ha Tinh, N Vietnam 18°30′N 105°36′E
160 F13 **Lincang** Yunnan, SW China 23°55′N 100°03′E
Lincheng *see* Lingao
Linchuan *see* Fuzhou
61 B20 **Lincoln** Buenos Aires, E Argentina 34°54′S 61°30′W
185 H19 **Lincoln** Canterbury, South Island, New Zealand 43°37′S 172°30′E
97 N18 **Lincoln** *anc.* Lindum, Lindum Colonia. E England, United Kingdom 53°14′N 00°33′W
35 O6 **Lincoln** California, W USA 38°52′N 121°18′W
30 L13 **Lincoln** Illinois, N USA 40°09′N 89°21′W
26 M4 **Lincoln** Kansas, C USA 39°03′N 98°09′W
19 S5 **Lincoln** Maine, NE USA 45°22′N 68°30′W
27 T5 **Lincoln** Missouri, C USA 38°23′N 93°19′W
29 R16 **Lincoln** *state capital* Nebraska, C USA 40°46′N 96°43′W
32 F11 **Lincoln City** Oregon, NW USA 44°57′N 124°01′W
167 X10 **Lincoln Island** *Chin.* Dong Dao, *Viet.* Đao Linh Côn. *island* E Paracel Islands
197 Q11 **Lincoln Sea** *sea* Arctic Ocean
97 N18 **Lincolnshire** *cultural region* E England, United Kingdom
21 R10 **Lincolnton** North Carolina, SE USA 35°27′N 81°16′W
25 V7 **Lindale** Texas, SW USA 32°31′N 95°24′W
101 I25 **Lindau** *var.* Lindau am Bodensee. Bayern, S Germany 47°33′N 09°41′E
Lindau am Bodensee *see* Lindau
123 P9 **Linde** ≈ NE Russian Federation
55 T9 **Linden** E Guyana 05°58′N 58°12′W
23 O6 **Linden** Alabama, S USA 32°18′N 87°48′W
20 H9 **Linden** Tennessee, S USA 35°38′N 87°50′W
25 X6 **Linden** Texas, SW USA 33°01′N 94°22′W
44 H2 **Linden Pindling** ✈ New Providence, C The Bahamas 25°00′N 77°26′W
18 J16 **Lindenwold** New Jersey, NE USA 39°47′N 74°58′W
95 M15 **Lindesberg** Örebro, C Sweden 59°36′N 15°15′E
95 D18 **Lindesnes** *headland* S Norway 57°58′N 07°03′E
Líndhos *see* Líndos
81 K24 **Lindi** Lindi, SE Tanzania 10°S 39°41′E
81 J24 **Lindi** ◈ *region* SE Tanzania
79 N17 **Lindi** ≈ NE Dem. Rep. Congo
163 V7 **Lindian** Heilongjiang, NE China 47°15′N 124°51′E
185 E21 **Lindis Pass** *pass* South Island, New Zealand
83 J22 **Lindley** Free State, C South Africa 27°52′S 27°55′E
95 J19 **Lindome** Västra Götaland, S Sweden 57°34′N 12°05′E
163 S10 **Lindong** *var.* Bairin Zuoqi. Nei Mongol Zizhiqu, N China 43°59′N 119°24′E
115 O23 **Líndos** *var.* Líndhos. Ródos, Dodekánisa, Greece, Aegean Sea 36°05′N 28°05′E
14 I14 **Lindsay** Ontario, SE Canada 44°21′N 78°44′W
35 R11 **Lindsay** California, W USA 36°11′N 119°06′W
33 X8 **Lindsay** Montana, NW USA 47°13′N 105°10′W
27 N11 **Lindsay** Oklahoma, C USA 34°50′N 97°37′W
27 N5 **Lindsborg** Kansas, C USA 38°34′N 97°39′W
95 N21 **Lindsdal** Kalmar, S Sweden 56°44′N 16°18′E
Lindum/Lindum Colonia *see* Lincoln
191 W3 **Line Islands** *island group* E Kiribati
Linëvo *see* Linova
160 M5 **Linfen** *var.* Lin-fen. Shanxi, C China 36°08′N 111°34′E
Lin-fen *see* Linfen
104 L2 **L'Infiestu** *prev.* Infiesto. Asturias, N Spain 43°21′N 05°21′W
155 F18 **Linganamakki Reservoir** ▣ SW India
160 L17 **Lingao** *var.* Lincheng. Hainan, S China 19°44′N 109°23′E
171 N3 **Lingayen** Luzon, N Philippines 16°00′N 120°12′E
160 M6 **Lingbao** *var.* Guolüezhen. Henan, C China 34°34′N 110°50′E
94 N12 **Lingbo** Gävleborg, C Sweden 61°04′N 16°45′E
Lingcheng *see* Lingshan, Guangxi, China
Lingcheng *see* Beiliu,Guangxi, China
Lingeh *see* Bandar-e Lengeh
100 E12 **Lingen** *var.* Lingen an der Ems. Niedersachsen, NW Germany 52°31′N 07°19′E
Lingen an der Ems *see* Lingen
168 M11 **Lingga, Kepulauan** *island group* W Indonesia
168 L11 **Lingga, Pulau** *island* Kepulauan Lingga, W Indonesia
14 J14 **Lingham Lake** ◎ Ontario, SE Canada
94 M13 **Linghed** Dalarna, C Sweden 60°48′N 15°55′E
33 Z15 **Lingle** Wyoming, C USA 42°07′N 104°21′W
18 G15 **Linglestown** Pennsylvania, NE USA 40°20′N 76°46′W
160 M12 **Lingling** *prev.* Yongzhou, Zhishan. Hunan, S China 26°13′N 111°36′E
79 K18 **Lingomo 11** Equateur, NW Dem. Rep. Congo 0°42′N 21°59′E
160 L15 **Lingshan** *var.* Lingcheng. Guangxi Zhuangzu Zizhiqu, S China 22°28′N 109°19′E
160 L17 **Lingshui** *var.* Lingshui Lizu Zizhixian. Hainan, S China 18°35′N 110°03′E
Lingshui Lizu Zizhixian *see* Lingshui
155 G16 **Lingsugūr** Karnātaka, C India 16°13′N 76°33′E
107 L23 **Linguaglossa** Sicilia, Italy, C Mediterranean Sea 37°51′N 15°06′E
76 H10 **Linguère** N Senegal 15°24′N 15°06′W
159 W8 **Lingwu** Ningxia, N China 38°04′N 106°21′E
Lingxi *see* Yongshun, Hunan, China
Lingxi *see* Cangnan, Zhejiang, China
Lingxian/Ling Xian *see* Yanling
163 S12 **Lingyuan** Liaoning, NE China 41°09′N 119°24′E
163 U4 **Linhai** Heilongjiang, NE China 51°30′N 124°18′E
161 S10 **Linhai** *var.* Taizhou. Zhejiang, SE China 28°54′N 121°08′E
59 O20 **Linhares** Espírito Santo, SE Brazil 19°22′S 40°04′W
Linh Côn, Đao *see* Lincoln Island
Linhe *see* Bayannur
Lini *see* Lin
Līnik, Chiyā-ē *see* Linkî, Chiya-î
Linjiang *see* Shanghang
139 S1 **Linkî, Chiya-î** *Ar.* Jabal Linkī, *var.* Chiyā-ē Līnik, Chiyā-î Linkî Kurezûr. ▲▲ N Iraq
Linkī, Jabal *see* Linkî, Chiya-î
95 M18 **Linköping** Östergötland, S Sweden 58°25′N 15°37′E
163 Y8 **Linkou** Heilongjiang, NE China 45°18′N 130°17′E
118 F11 **Linkuva** Šiaulai, N Lithuania 56°06′N 23°58′E
27 V5 **Linn** Missouri, C USA 38°29′N 91°51′W
25 S16 **Linn** Texas, SW USA 26°32′N 98°06′W
27 T2 **Linneus** Missouri, C USA 39°53′N 93°10′W
96 H10 **Linnhe, Loch** *inlet* W Scotland, United Kingdom
119 G19 **Linova** *Rus.* Linëvo. Brestskaya Voblasts', SW Belarus 52°29′N 24°30′E
161 O5 **Linqing** Shandong, E China 36°51′N 115°42′E
60 K8 **Lins** São Paulo, S Brazil 21°40′S 49°44′W
93 F17 **Linsell** Jämtland, C Sweden 62°10′N 14°E
160 J9 **Linshui** Sichuan, C China 30°24′N 106°54′E
44 K12 **Linstead** C Jamaica 18°08′N 77°02′W
159 U11 **Lintan** Gansu, N China 34°43′N 101°27′E
159 V11 **Lintao** *var.* Taoyang. Gansu, C China 35°23′N 103°54′E
15 S12 **Lintère** ≈ Québec, SE Canada
108 H8 **Linth** ≈ NW Switzerland
108 H8 **Linthal** Glarus, NE Switzerland 46°59′N 08°57′E
31 N15 **Linton** Indiana, N USA 39°01′N 87°10′W
29 N6 **Linton** North Dakota, N USA 46°16′N 100°13′W
163 R11 **Linxi** Nei Mongol Zizhiqu, N China 43°29′N 117°59′E
159 U11 **Linxia** *var.* Linxia Huizu Zizhizhou. Gansu, C China 35°34′N 103°08′E
Linxia Huizu Zizhizhou *see* Linxia
Linxian *see* Lianzhou
161 P4 **Linyi** *var.* Yishi. Shandong, E China 37°12′N 116°54′E
161 Q6 **Linyi** Shandong, S China 37°06′N 118°18′E
160 M6 **Linyi** Shanxi, C China 35°10′N 118°18′E
109 T4 **Linz** *anc.* Lentia. Oberösterreich, N Austria 48°19′N 14°18′E
159 S8 **Linze** *var.* Shahe; *prev.* Shahepu. Gansu, N China 39°06′N 100°03′E
44 J13 **Lionel Town** C Jamaica 17°49′N 77°14′W
103 Q16 **Lion, Golfe du** *Eng.* Gulf of Lion, Gulf of Lions; *anc.* Sinus Gallicus. *gulf* S France
Lion, Gulf of/Lions, Gulf of *see* Lion, Golfe du
83 K16 **Lions Den** Mashonaland West, N Zimbabwe 17°16′S 30°00′E
14 F13 **Lion's Head** Ontario, S Canada 44°59′N 81°16′W
Lios Ceannúir, Bá *see* Liscannor Bay
Lios Mór *see* Lismore
Lios na gCearrbhach *see* Lisburn
Lios Tuathail *see* Listowel
79 G17 **Liouesso** Sangha, N Congo 01°02′N 15°43′E
Liozno *see* Lyozna
171 O4 **Lipa** *off.* Lipa City. Luzon, N Philippines 13°57′N 121°10′E
Lipa City *see* Lipa
25 S7 **Lipan** Texas, SW USA 32°31′N 98°03′W
Lipari Islands/Lipari, Isole *see* Eolie, Isole
107 L22 **Lipari, Isola** *island* Isole Eolie, S Italy
116 L8 **Lipcani** *Rus.* Lipkany. N Moldova 48°16′N 26°47′E
93 N17 **Liperi** Pohjois-Karjala, SE Finland 62°33′N 29°29′E
126 L7 **Lipetsk** Lipetskaya Oblast', W Russian Federation 52°37′N 39°38′E
126 K6 **Lipetskaya Oblast'** ◈ *province* W Russian Federation
57 K22 **Lipez, Cordillera de** ▲▲ SW Bolivia
110 E10 **Lipiany** *Ger.* Lippehne. Zachodnio-pomorskie, W Poland 53°00′N 14°58′E
112 G9 **Lipik** Požega-Slavonija, NE Croatia 45°24′N 17°08′E
124 L12 **Lipin Bor** Vologodskaya Oblast', NW Russian Federation 60°12′N 38°04′E
160 L12 **Liping** *var.* Defeng. Guizhou, S China 26°16′N 109°08′E
Lipkany *see* Lipcani
119 H15 **Lipnishki** Hrodzyenskaya Voblasts', W Belarus 54°00′N 25°37′E
110 J10 **Lipno** Kujawsko-pomorskie, C Poland 52°52′N 19°11′E
116 F11 **Lipova** *Hung.* Lippa. Arad, W Romania 46°05′N 21°42′E
Lipovets *see* Lypovets'
Lippa *see* Lipova
101 E14 **Lippe** ≈ W Germany
Lippehne *see* Lipiany
101 G14 **Lippstadt** Nordrhein-Westfalen, W Germany 51°41′N 08°20′E
25 P1 **Lipscomb** Texas, SW USA 36°14′N 100°16′W
Lipsia/Lipsk *see* Leipzig
Liptau-Sankt-Nikolaus/Liptószentmiklós *see* Liptovský Mikuláš
111 K19 **Liptovský Mikuláš** *Ger.* Liptau-Sankt-Nikolaus, *Hung.* Liptószentmiklós. Zílinský Kraj, N Slovakia 49°06′N 19°36′E
183 O13 **Liptrap, Cape** *headland* Victoria, SE Australia 38°55′S 145°58′E
160 L13 **Lipu** *var.* Licheng. Guangxi Zhuangzu Zizhiqu, S China 24°25′N 110°15′E
81 G17 **Lira** N Uganda 02°15′N 32°55′E
57 F15 **Lircay** Huancavelica, C Peru 12°59′S 74°44′W
107 K17 **Liri** ≈ C Italy
144 M8 **Lisakovsk** Kostanay, NW Kazakhstan 52°32′N 62°32′E
79 K17 **Lisala** Equateur, N Dem. Rep. Congo 02°10′N 21°29′E
104 F11 **Lisboa** *Eng.* Lisbon; *anc.* Felicitas Julia, Olisipo. ● (Portugal) Lisboa, W Portugal 38°44′N 09°08′W
104 F10 **Lisboa** *Eng.* Lisbon. ◈ *district* C Portugal
19 N7 **Lisbon** New Hampshire, NE USA 44°11′N 71°52′W
29 Q6 **Lisbon** North Dakota, N USA 46°27′N 97°42′W
Lisbon *see* Lisboa
19 Q8 **Lisbon Falls** Maine, NE USA 44°00′N 70°03′W
97 G15 **Lisburn** *Ir.* Lios na gCearrbhach. E Northern Ireland, United Kingdom 54°31′N 06°03′W
38 L6 **Lisburne, Cape** *headland* Alaska, USA 68°52′N 166°13′W
97 B19 **Liscannor Bay** *Ir.* Bá Lios Ceannúir. *inlet* W Ireland
113 Q18 **Lisec** ▲ E FYR Macedonia 41°46′N 22°30′E
160 F13 **Lishe Jiang** ≈ SW China
Lishi *see* Lüliang
163 V10 **Lishu** Jilin, NE China 43°25′N 124°19′E
161 R10 **Lishui** Zhejiang, SE China 28°27′N 119°25′E
192 L5 **Lisianski Island** *island* Hawaiian Islands, Hawai'i, USA
Lisichansk *see* Lysychans'k
102 L4 **Lisieux** *anc.* Noviomagus. Calvados, N France 49°09′N 00°13′E
126 L8 **Liski** *prev.* Georgiu-Dezh. Voronezhskaya Oblast', W Russian Federation 51°00′N 39°36′E
103 N4 **l'Isle-Adam** Val-d'Oise, N France 49°07′N 02°13′E
Lisle/l'Isle *see* Lille
103 R15 **l'Isle-sur-la-Sorgue** Vaucluse, SE France 43°55′N 05°03′E
15 S9 **L'Islet** Québec, SE Canada 47°07′N 70°18′W
183 V4 **Lismore** New South Wales, SE Australia 28°48′S 153°12′E
182 M12 **Lismore** Victoria, SE Australia 37°59′S 143°18′E
97 D20 **Lismore** *Ir.* Lios Mór. S Ireland 52°10′N 07°10′W
Lissa *see* Vis, Croatia
Lissa *see* Leszno, Poland
98 H11 **Lisse** Zuid-Holland, W Netherlands 52°15′N 04°33′E
114 K13 **Líssos** *var.* Filiouri. ≈ NE Greece
95 D18 **Lista** *peninsula* S Norway
95 D18 **Listafjorden** *fjord* S Norway
195 R13 **Lister, Mount** ▲ Antarctica 78°12′S 161°46′E
126 M8 **Listopadovka** Voronezhskaya Oblast', W Russian Federation 51°54′N 41°08′E
14 F15 **Listowel** Ontario, S Canada 43°44′N 80°57′W
97 B20 **Listowel** *Ir.* Lios Tuathail. Kerry, SW Ireland 52°27′N 09°29′W
160 L14 **Litang** Guangxi Zhuangzu Zizhiqu, S China 23°09′N 109°08′E
160 F9 **Litang** *var.* Gaocheng. Sichuan, C China 30°03′N 100°12′E
160 F10 **Litang Qu** ≈ C China
55 X12 **Litani** *var.* Itany. ≈ French Guiana/Suriname
138 G8 **Līṭānī, Nahr el** *var.* Nahr al Litant. ≈ C Lebanon
Litant, Nahr al *see* Līṭānī, Nahr el
Litauen *see* Lithuania
30 K14 **Litchfield** Illinois, N USA 39°17′N 89°52′W
29 U8 **Litchfield** Minnesota, N USA 45°09′N 94°31′W
36 K13 **Litchfield Park** Arizona, SW USA 33°29′N 112°21′W
183 S8 **Lithgow** New South Wales, SE Australia 33°30′S 150°09′E
115 I26 **Líthino, Akrotírio** *headland* Kríti, Greece, E Mediterranean Sea 34°55′N 24°43′E
118 D12 **Lithuania** *off.* Republic of Lithuania, *Ger.* Litauen, *Lith.* Lietuva, *Pol.* Litwa, *Rus.* Litva; *prev.* Lithuanian SSR, *Rus.* Litovskaya SSR. ◆ *republic* NE Europe
Lithuanian SSR *see* Lithuania
Lithuania, Republic of *see* Lithuania
109 U11 **Litija** *Ger.* Littai. C Slovenia 46°03′N 14°50′E
18 H15 **Lititz** Pennsylvania, NE USA 40°09′N 76°18′W
115 F15 **Litóchoro** *var.* Litohoro, Litókhoron. Kentrikí Makedonía, N Greece 40°06′N 22°30′E
Litohoro/Litókhoron *see* Litóchoro
111 C15 **Litoměřice** *Ger.* Leitmeritz. Ústecký Kraj, NW Czech Republic 50°33′N 14°10′E
111 F17 **Litomyšl** *Ger.* Leitomischl. Pardubický Kraj, C Czech Republic 49°54′N 16°18′E
111 G17 **Litovel** *Ger.* Littau. Olomoucký Kraj, E Czech Republic 49°42′N 17°05′E
123 S13 **Litovko** Khabarovskiy Kray, SE Russian Federation 49°22′N 135°10′E
Litovskaya SSR *see* Lithuania
Littai *see* Litija
Littau *see* Litovel
44 G1 **Little Abaco** *var.* Abaco Island. *island* N The Bahamas
111 I21 **Little Alföld** *Ger.* Kleines Ungarisches Tiefland, *Hung.* Kisalföld, *Slvk.* Podunajská Rovina. *plain* Hungary/Slovakia
151 Q20 **Little Andaman** *island* Andaman Islands, India, NE Indian Ocean
26 M5 **Little Arkansas River** ≈ Kansas, C USA
184 L4 **Little Barrier Island** *island* N New Zealand
Little Belt *see* Lillebælt
38 M11 **Little Black River** ≈ Alaska, USA
27 O2 **Little Blue River** ≈ Kansas/Nebraska, C USA
44 D8 **Little Cayman** *island* E Cayman Islands
11 X11 **Little Churchill** ≈ Manitoba, C Canada
166 J10 **Little Coco Island** *island* SW Myanmar (Burma)
36 L10 **Little Colorado River** ≈ Arizona, SW USA
14 E11 **Little Current** Manitoulin Island, Ontario, S Canada 45°57′N 81°56′W

◆ Country
● Country Capital
◇ Dependent Territory
○ Dependent Territory Capital
◈ Administrative Regions
✈ International Airport
▲ Mountain
▲▲ Mountain Range
Volcano
≈ River
◎ Lake
▣ Reservoir

LORRAINE – LUODING

103 T5 **Lorraine** ◈ *region* NE France
Lorungau *see* Lorengau
94 L11 **Los** Gävleborg, C Sweden 61°43′N 15°15′E
35 P14 **Los Alamos** California, W USA 34°44′N 120°16′W
37 S10 **Los Alamos** New Mexico, SW USA 35°52′N 106°17′W
42 F5 **Los Amates** Izabal, E Guatemala 15°14′N 89°06′W
63 G14 **Los Ángeles** Bío Bío, C Chile 37°30′S 72°18′W
35 S15 **Los Angeles** California, W USA 34°03′N 118°15′W
35 S15 **Los Angeles** ✈ California, W USA 33°54′N 118°24′W
35 T13 **Los Angeles Aqueduct** *aqueduct* California, W USA
Losanna *see* Lausanne
63 H20 **Los Antiguos** Santa Cruz, SW Argentina 46°36′S 71°31′W
189 Q16 **Losap Atoll** *atoll* C Micronesia
35 P10 **Los Banos** California, W USA 37°00′N 120°39′W
104 K16 **Los Barrios** Andalucía, S Spain 36°11′N 05°30′W
62 L5 **Los Blancos** Salta, N Argentina 23°36′S 62°35′W
42 L12 **Los Chiles** Alajuela, NW Costa Rica 11°00′N 84°42′W
105 O2 **Los Corrales de Buelna** Cantabria, N Spain 43°15′N 04°04′W
25 T17 **Los Fresnos** Texas, SW USA 26°03′N 97°28′W
35 N9 **Los Gatos** California, W USA 37°13′N 121°58′W
127 P10 **Loshchina** Volgogradskaya Oblast', SW Russian Federation 48°58′N 46°14′E
110 O11 **Łosice** Mazowieckie, C Poland 52°13′N 22°42′E
112 B11 **Lošinj** *Ger.* Lussin, *It.* Lussino. *island* W Croatia
Los Jardines *see* Ngetik Atoll
63 G15 **Los Lagos** Los Ríos, C Chile 39°50′S 72°50′W
63 F17 **Los Lagos** *off.* Región de los Lagos. ◈ *region* C Chile
los Lagos, Región de *see* Los Lagos
Loslau *see* Wodzisław Śląski
64 N11 **Los Llanos de Aridane** *var.* Los Llanos de Aridane. La Palma, Islas Canarias, Spain, NE Atlantic Ocean 28°39′N 17°54′W
Los Llanos de Aridane *see* Los Llanos de Aridane
37 R11 **Los Lunas** New Mexico, SW USA 34°48′N 106°43′W
63 I16 **Los Menucos** Río Negro, C Argentina 40°52′S 68°07′W
40 H8 **Los Mochis** Sinaloa, C Mexico 25°48′N 108°58′W
35 N4 **Los Molinos** California, W USA 40°00′N 122°05′W
104 M9 **Los Navalmorales** Castilla-La Mancha, C Spain 39°43′N 04°38′W
25 S15 **Los Olmos Creek** ≈ Texas, SW USA
Losonc/Losontz *see* Lučenec
167 S5 **Lô, Sông** *var.* Panlong Jiang. ≈ China/Vietnam
44 B5 **Los Palacios** Pinar del Río, W Cuba 22°35′N 83°16′W
104 K14 **Los Palacios y Villafranca** Andalucía, S Spain 37°10′N 05°55′W
37 R12 **Los Pinos Mountains** ▲▲ New Mexico, SW USA
37 R11 **Los Ranchos de Albuquerque** New Mexico, SW USA 35°09′N 106°37′W
40 M14 **Los Reyes** Michoacán, SW Mexico 19°36′N 102°29′W
63 G15 **Los Ríos** ◈ *region* C Chile
56 B7 **Los Ríos** ◈ *province* C Ecuador
64 O11 **Los Rodeos** ✈ (Santa Cruz de Tenerife) Tenerife, Islas Canarias, Spain, NE Atlantic Ocean 28°27′N 16°20′W
54 L4 **Los Roques, Islas** *island group* N Venezuela
43 S17 **Los Santos** Los Santos, S Panama 07°56′N 80°23′W
43 S17 **Los Santos** *off.* Provincia de Los Santos. ◈ *province* S Panama
Los Santos *see* Los Santos de Maimona
104 J12 **Los Santos de Maimona** *var.* Los Santos. Extremadura, W Spain 38°27′N 06°22′W
Los Santos, Provincia de *see* Los Santos
98 P10 **Losser** Overijssel, E Netherlands 52°16′N 06°25′E
96 J8 **Lossiemouth** NE Scotland, United Kingdom 57°43′N 03°18′W
61 B14 **Los Tábanos** Santa Fe, C Argentina 28°27′S 59°57′W
54 J4 **Los Taques** Falcón, N Venezuela 11°50′N 70°16′W
14 G11 **Lost Channel** Ontario, S Canada 45°54′N 80°20′W
54 L5 **Los Teques** Miranda, N Venezuela 10°25′N 67°01′W
35 Q12 **Lost Hills** California, W USA 35°35′N 119°40′W
36 I7 **Lost Peak** ▲ Utah, W USA 37°30′N 113°57′W
33 P11 **Lost Trail Pass** *pass* Montana, NW USA
186 G9 **Losuia** Kiriwina Island, SE Papua New Guinea 08°29′S 151°03′E
62 G10 **Los Vilos** Coquimbo, C Chile 31°56′S 71°35′W
105 N10 **Los Yébenes** Castilla-La Mancha, C Spain 39°35′N 03°52′W
103 N13 **Lot** ◈ *department* S France
103 N13 **Lot** ≈ S France
63 F14 **Lota** Bío Bío, C Chile 37°07′S 73°10′W
81 G15 **Lotagipi Swamp** *wetland* Kenya/Sudan
102 K14 **Lot-et-Garonne** ◈ *department* SW France
83 K21 **Lothair** Mpumalanga, NE South Africa 26°23′S 30°26′E
33 R7 **Lothair** Montana, NW USA 48°28′N 111°15′W
79 L20 **Loto** Kasai-Oriental, C Dem. Rep. Congo 02°48′S 22°30′E
108 E10 **Lötschbergtunnel** *tunnel* Valais, SW Switzerland
25 T9 **Lott** Texas, SW USA 31°12′N 97°02′W
124 H3 **Lotta** *var.* Lutto. ≈ Finland/Russian Federation
184 Q7 **Lottin Point** *headland* North Island, New Zealand 37°26′S 178°07′E
Lötzen *see* Giżycko
Loualaba *see* Lualaba
167 P6 **Louangnamtha** *var.* Luong Nam Tha. Louang Namtha, N Laos 20°55′N 101°24′E
167 Q7 **Louangphabang** *var.* Louangphrabang, Luang Prabang. Louangphabang, N Laos 19°51′N 102°08′E
Louangphrabang *see* Louangphabang
194 H5 **Loubet Coast** *physical region* Antarctica
Loubomo *see* Dolisie
Louch *see* Loukhi
102 H6 **Loudéac** Côtes d'Armor, NW France 48°11′N 02°45′W
160 M11 **Loudi** Hunan, S China 27°51′N 111°59′E
79 F21 **Loudima** Bouenza, S Congo 04°06′S 13°05′E
20 M9 **Loudon** Tennessee, S USA 35°43′N 84°19′W
31 T12 **Loudonville** Ohio, N USA 40°38′N 82°13′W
102 L8 **Loudun** Vienne, W France 47°01′N 00°05′E
102 K7 **Loué** Sarthe, NW France 48°00′N 00°14′W
76 G10 **Louga** NW Senegal 15°36′N 16°15′W
97 M19 **Loughborough** C England, United Kingdom 52°47′N 01°11′W
97 C18 **Loughrea** *Ir.* Baile Locha Riach. Galway, W Ireland 53°12′N 08°34′W
103 S9 **Louhans** Saône-et-Loire, C France 46°38′N 05°12′E
21 P5 **Louisa** Kentucky, S USA 38°06′N 82°37′W
21 V5 **Louisa** Virginia, NE USA 38°02′N 78°00′W
21 V9 **Louisburg** North Carolina, SE USA 36°05′N 78°18′W
25 U12 **Louise** Texas, SW USA 29°07′N 96°22′W
15 P11 **Louiseville** Québec, SE Canada 46°15′N 72°54′W
27 W3 **Louisiana** Missouri, C USA 39°25′N 91°03′W
22 G8 **Louisiana** *off.* State of Louisiana, *also known as* Creole State, Pelican State. ◈ *state* S USA
83 K19 **Louis Trichardt** *prev.* Makhado. Northern, NE South Africa 23°01′S 29°43′E
23 V4 **Louisville** Georgia, SE USA 33°00′N 82°24′W
30 M15 **Louisville** Illinois, N USA 38°46′N 88°32′W
20 K5 **Louisville** Kentucky, S USA 38°15′N 85°46′W
22 M4 **Louisville** Mississippi, S USA 33°07′N 89°03′W
29 S15 **Louisville** Nebraska, C USA 41°00′N 96°09′W
192 L11 **Louisville Ridge** *undersea feature* S Pacific Ocean
124 J6 **Loukhi** *var.* Louch. Respublika Kareliya, NW Russian Federation 66°05′N 33°04′E
79 H19 **Loukoléla** Cuvette, E Congo 01°04′S 17°10′E
104 G14 **Loulé** Faro, S Portugal 37°08′N 08°02′W
111 C16 **Louny** *Ger.* Laun. Ústecký Kraj, NW Czech Republic 50°22′N 13°50′E
29 O15 **Loup City** Nebraska, C USA 41°16′N 98°58′W
29 P15 **Loup River** ≈ Nebraska, C USA
15 S9 **Loup, Rivière du** ≈ Québec, SE Canada
12 K7 **Loups Marins, Lacs des** ◎ Québec, NE Canada
102 K16 **Lourdes** Hautes-Pyrénées, S France 43°06′N 00°03′W
Lourenço Marques *see* Maputo
104 F11 **Loures** Lisboa, C Portugal 38°50′N 09°10′W
104 F10 **Lourinhã** Lisboa, C Portugal 39°14′N 09°19′W
115 C16 **Loúros** ≈ W Greece
104 G8 **Lousã** Coimbra, N Portugal 40°07′N 08°15′W
Loushanguan *see* Tongzi
160 M10 **Lou Shui** ≈ C China
183 O5 **Louth** New South Wales, SE Australia 30°34′S 145°07′E
97 O18 **Louth** E England, United Kingdom 53°19′N 00°00′E
97 F17 **Louth** *Ir.* Lú. *cultural region* NE Ireland
115 H15 **Loutrá** Kentrikí Makedonía, N Greece 39°55′N 23°37′E
115 G19 **Loutráki** Pelopónnisos, S Greece 37°55′N 22°55′E
Louvain *see* Leuven
99 H19 **Louvain-la Neuve** Walloon Brabant, C Belgium 50°39′N 04°36′E
14 J8 **Louvicourt** Québec, SE Canada 48°04′N 77°22′W
102 M4 **Louviers** Eure, N France 49°13′N 01°11′E
30 K14 **Lou Yaeger, Lake** ⊡ Illinois, N USA
93 J15 **Lövånger** Västerbotten, N Sweden 64°22′N 21°19′E
124 I14 **Lovat'** ≈ NW Russian Federation
113 J17 **Lovćen** ▲ SW Montenegro 42°22′N 18°49′E
114 I8 **Lovech** Lovech, N Bulgaria 43°08′N 24°45′E
114 I9 **Lovech** ◈ *province* N Bulgaria
25 V9 **Lovelady** Texas, SW USA 31°07′N 95°27′W
37 T3 **Loveland** Colorado, C USA 40°24′N 105°04′W
33 U12 **Lovell** Wyoming, C USA 44°50′N 108°23′W
Lovello, Monte *see* Grosser Löffler
35 S4 **Lovelock** Nevada, W USA 40°11′N 118°30′W
106 E7 **Lovere** Lombardia, N Italy 45°51′N 10°06′E
30 L10 **Loves Park** Illinois, N USA 42°19′N 89°03′W
26 M2 **Lovewell Reservoir** ⊡ Kansas, C USA
93 M19 **Loviisa** *Swe.* Lovisa. Uusimaa, S Finland 60°27′N 26°15′E
37 V15 **Loving** New Mexico, SW USA 32°17′N 104°06′W
21 U6 **Lovingston** Virginia, NE USA 37°46′N 78°54′W
37 V14 **Lovington** New Mexico, SW USA 32°56′N 103°21′W
Lovisa *see* Loviisa
111 C15 **Lovosice** *Ger.* Lobositz. Ústecký Kraj, NW Czech Republic 50°30′N 14°02′E
124 K4 **Lovozero** Murmanskaya Oblast', NW Russian Federation 68°00′N 35°03′E
124 K4 **Lovozero, Ozero** ◎ NW Russian Federation
112 B9 **Lovran** *It.* Laurana. Primorje-Gorski Kotar, NW Croatia 45°16′N 14°15′E
116 E11 **Lovrin** *Ger.* Lowrin. Timiş, W Romania 45°58′N 20°49′E
82 E10 **Lóvua** Lunda Norte, NE Angola 07°21′S 20°09′E
82 G12 **Lóvua** Moxico, E Angola 11°33′S 23°35′E
65 D25 **Low Bay** *bay* East Falkland, Falkland Islands
9 P9 **Low, Cape** *headland* Nunavut, E Canada 63°05′N 85°27′W
33 N10 **Lowell** Idaho, NW USA 46°07′N 115°36′W
19 O10 **Lowell** Massachusetts, NE USA 42°38′N 71°19′W
Löwen *see* Leuven
Löwenberg in Schlesien *see* Lwówek Śląski
Lower Austria *see* Niederösterreich
Lower Bann *see* Bann
Lower California *see* Baja California
Lower Danube *see* Niederösterreich
185 L14 **Lower Hutt** Wellington, North Island, New Zealand 41°13′S 174°51′E
39 N11 **Lower Kalskag** Alaska, USA 61°30′N 160°28′W
35 O1 **Lower Klamath Lake** ◎ California, W USA
35 Q2 **Lower Lake** ◎ California/Nevada, W USA
97 E15 **Lower Lough Erne** ◎ SW Northern Ireland, United Kingdom
Lower Lusatia *see* Niederlausitz
Lower Normandy *see* Basse-Normandie
10 K9 **Lower Post** British Columbia, W Canada 59°53′N 128°19′W
29 T4 **Lower Red Lake** ◎ Minnesota, N USA
Lower Rhine *see* Neder Rijn
Lower Saxony *see* Niedersachsen
97 Q19 **Lowestoft** E England, United Kingdom 52°29′N 01°45′E
Lowgar *see* Lōgar
182 H7 **Low Hill** South Australia 32°17′S 136°46′E
110 K12 **Łowicz** Łódzkie, C Poland 52°06′N 19°55′E
33 N13 **Lowman** Idaho, NW USA 44°04′N 115°37′W
149 P8 **Lowrah** *var.* Lora. ≈ SE Afghanistan
Lowrin *see* Lovrin
183 N17 **Low Rocky Point** *headland* Tasmania, SE Australia 42°59′S 145°28′E
18 I8 **Lowville** New York, NE USA 43°47′N 75°29′W
Loxa *see* Loksa
182 K9 **Loxton** South Australia 34°30′S 140°36′E
81 G21 **Loya** Tabora, C Tanzania 04°57′S 33°53′E
30 K6 **Loyal** Wisconsin, N USA 44°45′N 90°30′W
18 G13 **Loyalsock Creek** ≈ Pennsylvania, NE USA
35 Q5 **Loyalton** California, W USA 39°39′N 120°16′W
Lo-yang *see* Luoyang
187 Q16 **Loyauté, Îles** *island group* S New Caledonia
Loyev *see* Loyew
119 O20 **Loyew** *Rus.* Loyev. Homyel'skaya Voblasts', SE Belarus 51°56′N 30°48′E
125 S13 **Loyno** Kirovskaya Oblast', NW Russian Federation 59°44′N 52°42′E
103 P13 **Lozère** ◈ *department* S France
103 Q14 **Lozère, Mont** ▲ S France 44°27′N 03°44′E
112 J11 **Loznica** Serbia, W Serbia 44°32′N 19°13′E
114 L8 **Loznitsa** Razgrad, N Bulgaria 43°22′N 26°36′E
117 V7 **Lozova** *Rus.* Lozovaya. Kharkivs'ka Oblast', E Ukraine 48°54′N 36°23′E
Lozovaya *see* Lozova
105 N7 **Lozoyuela** Madrid, C Spain 40°55′N 03°36′W
Lu *see* Shandong, China
Lú *see* Louth, Ireland
82 F12 **Luacano** Moxico, E Angola 11°19′S 21°30′E
79 N21 **Lualaba** *Fr.* Loualaba. ≈ SE Dem. Rep. Congo
83 H14 **Luampa** Western, NW Zambia 15°02′S 24°27′E
83 H15 **Luampa Kuta** Western, W Zambia 15°22′S 24°40′E
161 P8 **Lu'an** Anhui, E China 31°46′N 116°31′E
Luanco *see* Lluanco
82 A11 **Luanda** *var.* Loanda, *Port.* São Paulo de Loanda. ● (Angola) Luanda, NW Angola 08°48′S 13°17′E
82 A11 **Luanda** ◈ *province* (Angola) NW Angola
82 A11 **Luanda** ✈ Luanda, NW Angola 08°49′S 13°16′E
82 D10 **Luando** ≈ C Angola
Luang *see* Tapi, Mae Nam
83 G14 **Luanginga** *var.* Luanguinga. ≈ Angola/Zambia
167 N15 **Luang, Khao** ▲ SW Thailand 08°21′N 99°46′E
Luang Prabang *see* Louangphabang
167 P8 **Luang Prabang Range** *Th.* Thiukhaoluang Phrahang. ▲▲ Laos/Thailand
167 N16 **Luang, Thale** *lagoon* S Thailand
Luangua, Rio *see* Luangwa
82 E11 **Luangue** ≈ NE Angola
Luanguinga *see* Luanginga
83 K15 **Luangwa** *var.* Aruângua. Lusaka, C Zambia 15°36′S 30°27′E
83 K14 **Luangwa** *var.* Aruângua, Rio Luangua. ≈ Mozambique/Zambia
161 Q2 **Luan He** ≈ E China
190 G11 **Luaniva, Île** *island* E Wallis and Futuna
161 P2 **Luanping** *var.* Anjiangying. Hebei, E China 40°55′N 117°19′E
82 J13 **Luanshya** Copperbelt, C Zambia 13°09′S 28°24′E
62 K13 **Luan Toro** La Pampa, C Argentina 36°14′S 64°15′W
161 Q2 **Luanxian** *var.* Luan Xian. Hebei, E China 39°46′N 118°46′E
Luan Xian *see* Luanxian
82 J12 **Luapula** ◈ *province* N Zambia
79 O25 **Luapula** ≈ Dem. Rep. Congo/Zambia
104 J2 **Luarca** Asturias, N Spain 43°33′N 06°31′W
169 R10 **Luar, Danau** ◎ Borneo, N Indonesia
79 L25 **Luashi** Katanga, S Dem. Rep. Congo 10°54′S 23°55′E
82 G12 **Luau** *Port.* Vila Teixeira de Sousa. Moxico, NE Angola 10°42′S 22°12′E
79 C16 **Luba** *prev.* San Carlos. Isla de Bioco, NW Equatorial Guinea 03°26′N 08°36′E
42 F4 **Lubaantun** *ruins* Toledo, S Belize
111 P16 **Lubaczów** *var.* Lúbaczów. Podkarpackie, SE Poland 50°10′N 23°08′E
Lubale *see* Lubalo
82 E11 **Lubalo** Lunda Norte, NE Angola 09°02′S 19°11′E
82 E11 **Lubalo** *var.* Lubale. ≈ Angola/Dem. Rep. Congo
118 J9 **Lubāna** E Latvia 56°55′N 29°43′E
Lubānas Ezers *see* Lubāns
171 N4 **Lubang Island** *island* N Philippines
83 B15 **Lubango** *Port.* Sá da Bandeira. Huíla, SW Angola 14°55′S 13°33′E
118 J9 **Lubāns** *var.* Lubānas Ezers. ◎ E Latvia
79 M21 **Lubao** Kasai-Oriental, C Dem. Rep. Congo 05°21′S 25°42′E
110 O13 **Lubartów** *Ger.* Qumälisch. Lublin, E Poland 51°29′N 22°38′E
100 G13 **Lübbecke** Nordrhein-Westfalen, NW Germany 52°18′N 08°37′E
100 O13 **Lübben** Brandenburg, E Germany 51°56′N 13°52′E
101 P14 **Lübbenau** Brandenburg, E Germany 51°52′N 13°57′E
25 N5 **Lubbock** Texas, SW USA 33°35′N 101°51′W
Lubcz *see* Lyubcha
19 U6 **Lubec** Maine, NE USA 44°49′N 67°00′W
100 K9 **Lübeck** Schleswig-Holstein, N Germany 53°52′N 10°41′E
100 K8 **Lübecker Bucht** *bay* N Germany
79 M21 **Lubefu** Kasai-Oriental, C Dem. Rep. Congo 04°43′S 24°25′E
163 T10 **Lubei** *var.* Jarud Qi. Nei Mongol Zizhiqu, N China 44°25′N 121°12′E
111 O14 **Lubelska, Wyżyna** *plateau* SE Poland
110 O13 **Lubelskie** ◈ *province* E Poland
Lubembe *see* Luembe
Lüben *see* Lubin
144 H9 **Lubenka** Zapadnyy Kazakhstan, W Kazakhstan 50°27′N 54°07′E
79 P18 **Lubero** Nord-Kivu, E Dem. Rep. Congo 0°10′S 29°12′E
79 L22 **Lubi** ≈ S Dem. Rep. Congo
Lubiana *see* Ljubljana
110 J11 **Lubień Kujawski** Kujawsko-pomorskie, C Poland 52°25′N 19°10′E
67 T11 **Lubilandji** ≈ S Dem. Rep. Congo
110 F13 **Lubin** *Ger.* Lüben. Dolnośląskie, SW Poland 51°23′N 16°12′E
111 O14 **Lublin** *Rus.* Lyublin. Lubelskie, E Poland 51°15′N 22°33′E
111 J15 **Lubliniec** Śląskie, S Poland 50°41′N 18°41′E
Lubnān, Jabal *see* Liban, Jebel
117 R5 **Lubny** Poltavs'ka Oblast', NE Ukraine 50°00′N 33°00′E
Luboml *see* Lyuboml'
110 G11 **Luboń** *Ger.* Peterhof. Wielkolpolskie, C Poland 52°23′N 16°54′E
Luboten *see* Ljuboten
110 D12 **Lubsko** *Ger.* Sommerfeld. Lubuskie, W Poland 51°47′N 14°57′E
79 N24 **Lubudi** Katanga, SE Dem. Rep. Congo 09°57′S 25°59′E
168 L13 **Lubuklinggau** Sumatera, W Indonesia 03°10′S 102°52′E
79 N25 **Lubumbashi** *prev.* Élisabethville. Shaba, SE Dem. Rep. Congo 11°40′S 27°31′E
83 I14 **Lubungu** Central, C Zambia 14°28′S 26°30′E
110 E12 **Lubuskie** ◈ *province* W Poland
Lubuskie *see* Dobiegniew
79 N18 **Lubutu** Maniema, E Dem. Rep. Congo 0°48′S 26°39′E
Luca *see* Lucca
82 C11 **Lucala** ≈ W Angola
14 E16 **Lucan** Ontario, S Canada 43°10′N 81°22′W
97 F18 **Lucan** *Ir.* Leamhcán. Dublin, E Ireland 53°22′N 06°27′W
Lucanian Mountains *see* Lucano, Appennino
107 M18 **Lucano, Appennino** *Eng.* Lucanian Mountains. ▲▲ S Italy
82 F11 **Lucapa** *var.* Lukapa. Lunda Norte, NE Angola 08°24′S 20°42′E
29 V15 **Lucas** Iowa, C USA 41°01′N 93°26′W
61 C18 **Lucas González** Entre Ríos, E Argentina 32°25′S 59°33′W
65 C25 **Lucas Point** *headland* West Falkland, Falkland Islands 52°10′S 60°22′W
31 S15 **Lucasville** Ohio, N USA 38°52′N 83°00′W
106 F11 **Lucca** *anc.* Luca. Toscana, C Italy 43°50′N 10°30′E
44 H12 **Lucea** W Jamaica 18°26′N 78°11′W
97 H15 **Luce Bay** *inlet* SW Scotland, United Kingdom
22 M8 **Lucedale** Mississippi, S USA 30°55′N 88°35′W
171 O4 **Lucena** *off.* Lucena City. Luzon, N Philippines 13°57′N 121°38′E
104 M14 **Lucena** Andalucía, S Spain 37°25′N 04°29′W
Lucena City *see* Lucena
105 S8 **Lucena del Cid** Valenciana, E Spain 40°07′N 00°15′W
111 D15 **Lučenec** *Ger.* Losontz, *Hung.* Losonc. Banskobystrický Kraj, C Slovakia 48°21′N 19°37′E
Lucentum *see* Alicante
107 M16 **Lucera** Puglia, SE Italy 41°30′N 15°19′E
Lucerna/Lucerne *see* Luzern
Lucerne, Lake of *see* Vierwaldstätter See
40 J4 **Lucero** Chihuahua, N Mexico 30°50′N 106°30′W
123 S14 **Luchegorsk** Primorskiy Kray, SE Russian Federation 46°26′N 134°10′E
105 Q13 **Luchena** ≈ SE Spain
Lucheng *see* Kangding
82 N13 **Lucheringo** *var.* Luchulingo. ≈ N Mozambique
Luchesa *see* Luchosa
Luchin *see* Luchyn
118 N13 **Luchosa** *Rus.* Luchesa. ≈ N Belarus
100 K11 **Lüchow** Mecklenburg-Vorpommern, N Germany 52°57′N 11°10′E
Luchow *see* Hefei
Luchulingo *see* Lucheringo
119 N17 **Luchyn** *Rus.* Luchin. Homyel'skaya Voblasts', SE Belarus 53°01′N 30°01′E
55 U11 **Lucie Rivier** ≈ W Suriname
182 K11 **Lucindale** South Australia 36°57′S 140°20′E
83 A14 **Lucira** Namibe, SW Angola 13°51′S 12°35′E
Łuck *see* Luts'k
101 O14 **Luckau** Brandenburg, E Germany 51°50′N 13°42′E
100 N13 **Luckenwalde** Brandenburg, E Germany 52°05′N 13°11′E
14 E15 **Lucknow** Ontario, S Canada 43°58′N 81°30′W
152 L12 **Lucknow** *var.* Lakhnau. *state capital* Uttar Pradesh, N India 26°50′N 80°54′E
102 J10 **Luçon** Vendée, NW France 46°27′N 01°10′W
44 I7 **Lucrecia, Cabo** *headland* E Cuba 21°00′N 75°34′W
82 F13 **Lucusse** Moxico, E Angola 12°32′S 20°46′E
Lüda *see* Dalian
114 M9 **Luda Kamchia** *var.* Luda Kamchiya. ≈ E Bulgaria
Luda Kamchiya *see* Luda Kamchia
161 T14 **Lü Dao** *var.* Huoshao Dao, Lütao, *Eng.* Green Island; *prev.* Lü Tao. *island* SE Taiwan
Ludasch *see* Luduş
114 I10 **Luda Yana** ≈ C Bulgaria
112 F7 **Ludbreg** Varaždin, N Croatia 46°15′N 16°36′E
29 P7 **Ludden** North Dakota, N USA 45°58′N 98°07′W
101 F15 **Lüdenscheid** Nordrhein-Westfalen, W Germany 51°13′N 07°38′E
83 C21 **Lüderitz** *prev.* Angra Pequena. Karas, SW Namibia 26°38′S 15°10′E
152 H8 **Ludhiāna** Punjab, N India 30°56′N 75°52′E
31 O7 **Ludington** Michigan, N USA 43°58′N 86°27′W
97 K20 **Ludlow** W England, United Kingdom 52°20′N 02°28′W
35 W14 **Ludlow** California, W USA 34°43′N 116°07′W
28 J7 **Ludlow** South Dakota, N USA 45°48′N 103°21′W
18 M9 **Ludlow** Vermont, NE USA 43°24′N 72°39′W
114 L7 **Ludogorie** *physical region* NE Bulgaria
23 W6 **Ludowici** Georgia, SE USA 31°42′N 81°44′W
Ludsan *see* Ludza
116 I10 **Luduş** *Ger.* Ludasch, *Hung.* Marosludas. Mureş, C Romania 46°28′N 24°05′E
95 M14 **Ludvika** Dalarna, C Sweden 60°08′N 15°14′E
101 H21 **Ludwigsburg** Baden-Württemberg, SW Germany 48°54′N 09°12′E
100 O13 **Ludwigsfelde** Brandenburg, NE Germany 52°17′N 13°15′E
101 G20 **Ludwigshafen** *var.* Ludwigshafen am Rhein. Rheinland-Pfalz, W Germany 49°29′N 08°24′E
Ludwigshafen am Rhein *see* Ludwigshafen
101 L20 **Ludwigskanal** *canal* SE Germany
100 L10 **Ludwigslust** Mecklenburg-Vorpommern, N Germany 53°19′N 11°29′E
118 K10 **Ludza** *Ger.* Ludsan. E Latvia 56°32′N 27°41′E
79 K21 **Luebo** Kasai-Occidental, SW Dem. Rep. Congo 05°19′S 21°21′E
25 Q6 **Lueders** Texas, SW USA 32°46′N 99°38′W
79 N20 **Lueki** Maniema, C Dem. Rep. Congo 03°25′S 25°50′E
82 F10 **Luembe** *var.* Lubembe. ≈ Angola/Dem. Rep. Congo
82 E13 **Luena** *var.* Lwena, *Port.* Luso. Moxico, E Angola 11°47′S 19°52′E
79 M24 **Luena** Katanga, SE Dem. Rep. Congo 09°28′S 25°45′E
82 K12 **Luena** Northern, NE Zambia 10°40′S 30°21′E
83 F13 **Luena** ≈ E Angola
83 F16 **Luengue** ≈ SE Angola
67 V13 **Luenha** ≈ W Mozambique
83 G15 **Lueti** ≈ Angola/Zambia
160 J7 **Lüeyang** *var.* Hejiayan. Shaanxi, C China 33°12′N 106°31′E
161 P14 **Lufeng** Guangdong, S China 22°59′N 115°40′E
79 N24 **Lufira** ≈ SE Dem. Rep. Congo
79 N25 **Lufira, Lac de Retenue de la** *var.* Lac Tshangalele. ◎ SE Dem. Rep. Congo
25 W8 **Lufkin** Texas, SW USA 31°21′N 94°47′W
82 L11 **Lufubu** ≈ N Zambia
124 G14 **Luga** Leningradskaya Oblast', NW Russian Federation 58°43′N 29°46′E
124 G13 **Luga** ≈ NW Russian Federation
Luganer See *see* Lugano, Lago di
108 H11 **Lugano** *Ger.* Lauis. Ticino, S Switzerland 46°01′N 08°57′E
108 H12 **Lugano, Lago di** *var.* Ceresio, *Ger.* Luganer See. ◎ S Switzerland
Lugansk *see* Luhans'k
187 Q13 **Luganville** Espiritu Santo, C Vanuatu 15°31′S 167°12′E
Lugdunum *see* Lyon
Lugdunum Batavorum *see* Leiden
83 O15 **Lugela** Zambézia, NE Mozambique 16°27′S 36°47′E
83 O16 **Lugela** ≈ C Mozambique
82 P13 **Lugenda, Rio** ≈ N Mozambique
Luggarus *see* Locarno
Lugh Ganana *see* Luuq
97 G19 **Lugnaquillia Mountain** *Ir.* Log na Coille. ▲ E Ireland 52°58′N 06°27′W
106 H10 **Lugo** Emilia-Romagna, N Italy 44°25′N 11°53′E
104 I3 **Lugo** *anc.* Lugus Augusti. Galicia, NW Spain 43°N 07°33′W
104 I3 **Lugo** ◈ *province* Galicia, NW Spain
21 R12 **Lugoff** South Carolina, SE USA 34°13′N 80°41′W
116 F12 **Lugoj** *Ger.* Lugosch, *Hung.* Lugos. Timiş, W Romania 45°41′N 21°56′E
Lugos/Lugosch *see* Lugoj
Lugovoy/Lugovoye *see* Kulan
158 I13 **Lugu** Xizang Zizhiqu, W China 33°26′N 84°10′E
Lugus Augusti *see* Lugo
Luguvallium/Luguvallum *see* Carlisle
117 Y7 **Luhans'k** *Rus.* Lugansk; *prev.* Voroshilovgrad. Luhans'ka Oblast', E Ukraine 48°32′N 39°21′E
117 Y7 **Luhans'k** ✈ Luhans'ka Oblast', E Ukraine 48°25′N 39°24′E
Luhans'k *see* Luhans'ka Oblast'
117 X6 **Luhans'ka Oblast'** *var.* Luhans'k; *prev.* Voroshilovgrad, *Rus.* Voroshilovgradskaya Oblast'. ◈ *province* E Ukraine
161 Q7 **Luhe** Jiangsu, E China 32°20′N 118°52′E
171 S13 **Luhu** Pulau Seram, E Indonesia 03°20′S 127°58′E
Luhua *see* Heishui
160 G8 **Luhuo** *var.* Xindu, *Tib.* Zhaggo. Sichuan, C China 31°18′N 100°39′E
116 M3 **Luhyny** Zhytomyrs'ka Oblast', N Ukraine 51°06′N 28°24′E
83 G15 **Lui** ≈ W Zambia
83 G16 **Luiana** ≈ SE Angola
83 L15 **Luia, Rio** *var.* Ruya. ≈ Mozambique/Zimbabwe
Luichow Peninsula *see* Leizhou Bandao
Luik *see* Liège
82 C13 **Luimbale** Huambo, C Angola 12°15′S 15°19′E
Luimneach *see* Limerick
106 D6 **Luino** Lombardia, N Italy 46°00′N 08°45′E
82 E13 **Luio** ≈ E Angola
92 L11 **Luiro** ≈ NE Finland
79 N25 **Luishia** Katanga, SE Dem. Rep. Congo 11°18′S 27°08′E
59 M19 **Luislândia do Oeste** Minas Gerais, SE Brazil 17°59′S 45°35′W
40 K5 **Luis L. León, Presa** ⊡ N Mexico
45 U5 **Luis Muñoz Marín** *var.* Luis Muñoz Marin. ✈ NE Puerto Rico 18°27′N 66°05′W
Luis Muñoz Marin *see* Luis Muñoz Marín
195 N5 **Luitpold Coast** *physical region* Antarctica
79 K22 **Luiza** Kasai-Occidental, S Dem. Rep. Congo 07°11′S 22°27′E
61 D20 **Luján** Buenos Aires, E Argentina 34°34′S 59°07′W
79 N24 **Lukafu** Katanga, SE Dem. Rep. Congo 10°28′S 27°32′E
Lukapa *see* Lucapa
112 I11 **Lukavac** ◈ Federacija Bosne I Hercegovine, NE Bosnia and Herzegovina
79 I20 **Lukenie** ≈ C Dem. Rep. Congo
Lŭki *see* Laki
79 H19 **Lukolela** Equateur, W Dem. Rep. Congo 01°10′S 17°11′E
Lukoml'skaye, Vozyera *see* Lukomskaye, Vozyera
Lukoml'skoye, Ozero *see* Lukomskaye, Vozyera
119 M14 **Lukomskaye, Vozyera** *Rus.* Ozero Lukoml'skoye; *prev.* Vozyera Lukoml'skaye. ◎ N Belarus
114 I8 **Lukovit** Lovech, N Bulgaria 43°11′N 24°10′E
110 O12 **Łuków** *Ger.* Bogendorf. Lubelskie, E Poland 51°57′N 22°22′E
127 O4 **Lukoyanov** Nizhegorodskaya Oblast', W Russian Federation 55°02′N 44°26′E
Lukransar *see* Lūnkaransar
79 N22 **Lukuga** ≈ SE Dem. Rep. Congo
79 F21 **Lukula** Bas-Congo, SW Dem. Rep. Congo 05°23′S 12°57′E
83 G14 **Lukulu** Western, NW Zambia 14°24′S 23°12′E
189 R17 **Lukunor Atoll** *atoll* Mortlock Islands, C Micronesia
82 J12 **Lukwesa** Luapula, NE Zambia 10°03′S 28°42′E
93 K14 **Luleå** Norrbotten, N Sweden 65°35′N 22°10′E
92 J13 **Luleälven** ≈ N Sweden
136 C10 **Lüleburgaz** Kırklareli, NW Turkey 41°25′N 27°22′E
160 M4 **Lüliang** *var.* Lishi. Shanxi, C China 37°27′N 111°05′E
160 M4 **Lüliang Shan** ▲▲ C China
79 O21 **Lulimba** Maniema, E Dem. Rep. Congo 04°42′S 28°38′E
22 K9 **Luling** Louisiana, S USA 29°55′N 90°22′W
25 T11 **Luling** Texas, SW USA 29°40′N 97°39′W
79 I18 **Lulonga** ≈ NW Dem. Rep. Congo
79 K22 **Lulua** ≈ S Dem. Rep. Congo
Luluabourg *see* Kananga
192 L17 **Luma** Ta'ū, E American Samoa 14°15′S 169°30′W
169 S17 **Lumajang** Jawa, C Indonesia 08°06′S 113°13′E
158 G12 **Lumajangdong Co** ◎ W China
82 G13 **Lumbala Kaquengue** Moxico, E Angola 12°40′S 22°34′E
83 F14 **Lumbala N'Guimbo** *var.* Nguimbo, Gago Coutinho, *Port.* Vila Gago Coutinho. Moxico, E Angola 14°08′S 21°25′E
21 T11 **Lumber River** ≈ North Carolina/South Carolina, SE USA
Lumber State *see* Maine
22 L8 **Lumberton** Mississippi, S USA 31°00′N 89°27′W
21 U11 **Lumberton** North Carolina, SE USA 34°37′N 79°00′W
105 R4 **Lumbier** Navarra, N Spain 42°39′N 01°19′W
83 Q15 **Lumbo** Nampula, NE Mozambique 15°S 40°40′E
124 M4 **Lumbovka** Murmanskaya Oblast', NW Russian Federation 67°41′N 40°31′E
104 J7 **Lumbrales** Castilla y León, N Spain 40°57′N 06°43′W
153 W13 **Lumding** Assam, NE India 25°46′N 93°10′E
82 F12 **Lumege** *var.* Lumeje. Moxico, E Angola 11°30′S 20°57′E
Lumeje *see* Lumege
99 J17 **Lummen** Limburg, NE Belgium 50°58′N 05°12′E
93 J20 **Lumparland** Åland, SW Finland 60°06′N 20°15′E
167 T11 **Lumphăt** *prev.* Lomphat. Rôtânôkiri, NE Cambodia 13°30′N
11 U16 **Lumsden** Saskatchewan, S Canada 50°39′N 104°52′W
185 C23 **Lumsden** Southland, South Island, New Zealand 45°43′S 168°26′E
169 N14 **Lumut, Tanjung** *headland* Sumatera, W Indonesia 03°47′S 105°55′E
157 P4 **Lün** Töv, C Mongolia 47°51′N 105°11′E
116 I13 **Lunca Corbului** Argeş, S Romania 44°41′N 24°46′E
95 K23 **Lund** Skåne, S Sweden 55°42′N 13°10′E
35 X6 **Lund** Nevada, W USA 38°50′N 115°00′W
82 D11 **Lunda Norte** ◈ *province* NE Angola
82 E12 **Lunda Sul** ◈ *province* NE Angola
82 M13 **Lundazi** Eastern, NE Zambia 12°19′S 33°11′E
95 G16 **Lunde** Telemark, S Norway 61°31′N 06°38′E
Lundenburg *see* Břeclav
95 C17 **Lundevatnet** ◎ S Norway
Lundi *see* Runde
97 I23 **Lundy** *island* SW England, United Kingdom
100 J10 **Lüneburg** Niedersachsen, N Germany 53°15′N 10°25′E
100 J11 **Lüneburger Heide** *heathland* NW Germany
103 Q15 **Lunel** Hérault, S France 43°40′N 04°08′E
101 F14 **Lünen** Nordrhein-Westfalen, W Germany 51°37′N 07°31′E
13 P16 **Lunenburg** Nova Scotia, SE Canada 44°23′N 64°21′W
21 V7 **Lunenburg** Virginia, NE USA 36°56′N 78°15′W
103 T5 **Lunéville** Meurthe-et-Moselle, NE France 48°35′N 06°30′E
83 I14 **Lunga** ≈ C Zambia
Lunga, Isola *see* Dugi Otok
158 H12 **Lungdo** Xizang Zizhiqu, W China 33°45′N 82°09′E
158 I14 **Lunggar** Xizang Zizhiqu, W China 31°10′N 84°01′E
76 I15 **Lungi** ✈ (Freetown) W Sierra Leone 08°36′N 13°10′W
Lungkiang *see* Qiqihar
Lungleh *see* Lunglei
153 W15 **Lunglei** *prev.* Lungleh. Mizoram, NE India 22°55′N 92°49′E
158 L15 **Lungsang** Xizang Zizhiqu, W China 29°50′N 88°27′E
82 E13 **Lungué-Bungo** *var.* Lungwebungu. ≈ Angola/Zambia *see also* Lungwebungu
Lungué-Bungo *see* Lungwebungu
83 G14 **Lungwebungu** *var.* Lungué-Bungo. ≈ Angola/Zambia *see also* Lungué-Bungo
Lungwebungu *see* Lungué-Bungo
152 F12 **Lūni** Rājasthān, N India 26°03′N 73°00′E
152 F12 **Lūni** ≈ N India
Luninets *see* Luninyets
35 S7 **Luning** Nevada, W USA 38°29′N 118°10′W
Łuniniec *see* Luninyets
127 P6 **Lunino** Penzenskaya Oblast', W Russian Federation 53°35′N 45°12′E
119 J19 **Luninyets** *Pol.* Łuniniec, *Rus.* Luninets. Brestskaya Voblasts', SW Belarus 52°15′N 26°48′E
152 F10 **Lūnkaransar** *var.* Lookransar, Lukransar. Rājasthān, NW India 28°32′N 73°50′E
119 G17 **Lunna** *Pol.* Łunna, *Rus.* Lunno. Hrodzyenskaya Voblasts', W Belarus 53°27′N 24°16′E
Łunna *see* Lunna
Lunno *see* Lunna
76 I15 **Lunsar** W Sierra Leone 08°41′N 12°32′W
83 K14 **Lunsemfwa** ≈ C Zambia
158 J6 **Luntai** *var.* Bügür. Xinjiang Uygur Zizhiqu, NW China 41°48′N 84°14′E
98 K11 **Lunteren** Gelderland, C Netherlands 52°05′N 05°38′E
109 U5 **Lunz am See** Niederösterreich, C Austria 47°54′N 15°01′E
163 Y7 **Luobei** *var.* Fengxiang. Heilongjiang, NE China 47°35′N 130°50′E
Luocheng *see* Hui'an, Fujian, China
Luocheng *see* Luoding, Guangdong, China
160 J13 **Luodian** *var.* Longping. Guizhou, S China 25°25′N 106°49′E
160 M15 **Luoding** *var.* Luocheng. Guangdong, S China 22°44′N 111°28′E

◆ Country
● Country Capital
◇ Dependent Territory
○ Dependent Territory Capital
◈ Administrative Regions
✈ International Airport
▲ Mountain
▲▲ Mountain Range
Volcano
≈ River
◎ Lake
⊡ Reservoir

LUOHE – MADURA, SELAT

161 N7 Luohe Henan, C China 33°37′N 114°00′E
160 M6 Luo He C China
160 L5 Luo He C China
L i Liêm, Nhom see Crescent Group
Luolajarvi see Kuoloyarvi
Luong Nam Tha see Louangnamtha
160 L13 Luoqing Jiang S China
161 O8 Luoshan Henan, C China 32°12′N 114°30′E
161 O12 Luoxiao Shan ▲▲ S China
161 N6 Luoyang var. Honan, Lo-yang. Henan, C China 34°41′N 112°25′E
161 R12 Luoyuan var. Fengshan. Fujian, SE China 26°29′N 119°32′E
79 F21 Luozi Bas-Congo, W Dem. Rep. Congo 04°57′S 14°08′E
83 J17 Lupane Matabeleland North, W Zimbabwe 18°54′S 27°44′E
160 I12 Lupanshui var. Liupanshui; prev. Shuicheng. Guizhou, S China 26°38′N 104°49′E
169 R10 Lupar, Batang East Malaysia
Lupatia see Altamura
116 G12 Lupeni Hung. Lupény. Hunedoara, SW Romania 45°20′N 23°10′E
Lupény see Lupeni
82 N13 Lupiliche Niassa, N Mozambique 11°36′S 35°15′E
83 E14 Lupire Kuando Kubango, E Angola 14°39′S 19°39′E
79 L22 Luputa Kasai-Oriental, S Dem. Rep. Congo 07°07′S 23°43′E
121 P16 Luqa ✈ (Valletta) S Malta 35°53′N 14°27′E
159 U11 Luqu var. Ma'ai. Gansu, C China 34°34′N 102°27′E
45 U5 Luquillo, Sierra de ▲▲ E Puerto Rico
26 L4 Luray Kansas, C USA 39°06′N 98°41′W
21 U4 Luray Virginia, NE USA 38°40′N 78°28′W
103 T7 Lure Haute-Saône, E France 47°42′N 06°30′E
82 D11 Luremo Lunda Norte, NE Angola 08°32′S 17°55′E
97 F15 Lurgan Ir. An Lorgain. S Northern Ireland, United Kingdom 54°28′N 06°20′W
57 K18 Luribay La Paz, W Bolivia 17°05′S 67°37′W
Luring see Gêrzê
83 Q14 Lúrio Nampula, NE Mozambique 13°32′S 40°34′E
83 P14 Lúrio, Rio NE Mozambique
Luristan see Lorestān
Lurka see Lorca
83 J15 Lusaka ● (Zambia) Lusaka, SE Zambia 15°24′S 28°17′E
83 J15 Lusaka ◈ province C Zambia
83 J15 Lusaka ✈ Lusaka, C Zambia 15°10′S 28°22′E
79 L21 Lusambo Kasai-Oriental, C Dem. Rep. Congo 04°59′S 23°26′E
186 F8 Lusancay Islands and Reefs island group SE Papua New Guinea
79 I21 Lusanga Bandundu, SW Dem. Rep. Congo 04°55′S 18°40′E
79 N21 Lusangi Maniema, E Dem. Rep. Congo 04°39′S 27°10′E
Lusatian Mountains see Lausitzer Bergland
Lushar see Huangzhong
Lushnja see Lushnjë
113 K21 Lushnjë var. Lushnja. Fier, C Albania 40°54′N 19°43′E
81 J21 Lushoto Tanga, E Tanzania 04°48′S 38°20′E
102 L10 Lusignan Vienne, W France 46°25′N 00°06′E
33 Z15 Lusk Wyoming, C USA 42°45′N 104°27′W
Luso see Luena
102 L10 Lussac-les-Châteaux Vienne, W France 46°23′N 00°44′E
Lussin/Lussino see Lošinj
Lussinpiccolo see Mali Lošinj
108 I7 Lustenau Vorarlberg, W Austria 47°26′N 09°42′E
Lütao see Lü Dao
Lü Tao see Lü Dao
Lūt, Baḩrat/Lut, Bahret see Dead Sea
22 K9 Lutcher Louisiana, S USA 30°02′N 90°42′W
143 T9 Lūt, Dasht-e var. Kavīr-e Lūt. desert E Iran
83 F14 Lutembo Moxico, E Angola 13°30′S 21°21′E
Lutetia/Lutetia Parisiorum see Paris
Luteva see Lodève
14 G15 Luther Lake ◎ Ontario, S Canada
186 K8 Luti Choiseul, NW Solomon Islands 07°13′S 157°01′E
Lūt, Kavīr-e see Lūt, Dasht-e
97 N21 Luton E England, United Kingdom 51°53′N 00°25′W
97 N21 Luton ✈ (London) SE England, United Kingdom 51°54′N 00°24′W
108 B10 Lutry Vaud, SW Switzerland 46°31′N 06°32′E
8 K10 Lutselk'e prev. Snowdrift. Northwest Territories, W Canada 62°24′N 110°42′W
8 K10 Lutselk'e var. Snowdrift. Northwest Territories, NW Canada
29 Y4 Lutsen Minnesota, N USA 47°39′N 90°37′W
116 J4 Luts'k Pol. Łuck, Rus. Lutsk. Volyns'ka Oblast', NW Ukraine 50°45′N 25°23′E
Lutsk see Luts'k
Luttenberg see Ljutomer
Lüttich see Liège
83 G25 Luttig Western Cape, SW South Africa 32°33′S 22°13′E
Lutto see Lotta
82 E13 Lutuai Moxico, E Angola 12°38′S 20°06′E
117 Y7 Lutuhyne Luhans'ka Oblast', E Ukraine 48°24′N 39°12′E
171 V14 Lutur, Pulau island Kepulauan Aru, E Indonesia
23 V12 Lutz Florida, SE USA 28°09′N 82°27′W
Lutzow-Holm Bay see Lützow Holmbukta
195 V2 Lützow Holmbukta var. Lutzow-Holm Bay. bay Antarctica
81 L16 Luuq It. Lugh Ganana. Gedo, SW Somalia 03°42′N 42°34′E
92 M12 Luusua Lappi, NE Finland 66°28′N 27°16′E
23 Q6 Luverne Alabama, S USA 31°43′N 86°15′W
29 S11 Luverne Minnesota, N USA 43°39′N 96°12′W
79 O22 Luvua SE Dem. Rep. Congo
82 F13 Luvuei Moxico, E Angola 13°08′S 21°09′E
81 H24 Luwego S Tanzania
82 K12 Luwingu Northern, NE Zambia 10°13′S 29°58′E
171 P12 Luwuk prev. Loewoek. Sulawesi, C Indonesia 0°56′S 122°47′E
23 N3 Luxapallila Creek Alabama/Mississippi, S USA
99 M25 Luxembourg ● (Luxembourg) Luxembourg, S Luxembourg 49°37′N 06°08′E
99 M25 Luxembourg off. Grand Duchy of Luxembourg, var. Lëtzebuerg, Luxemburg. ◆ monarchy NW Europe
99 J23 Luxembourg ◈ province SE Belgium
99 L24 Luxembourg ◈ district S Luxembourg
31 N6 Luxemburg Wisconsin, N USA 44°32′N 87°42′W
Luxemburg see Luxembourg
103 U7 Luxeuil-les-Bains Haute-Saône, E France 47°49′N 06°22′E
160 E13 Luxi prev. Mangshi. Yunnan, SW China 24°27′N 98°31′E
82 E10 Luxico Angola/Dem. Rep. Congo
75 X10 Luxor Ar. Al Uqṣur. E Egypt 25°39′N 32°39′E
75 X10 Luxor ✈ C Egypt 25°39′N 32°48′E
160 M4 Luya Shan ▲▲ C China
102 J15 Luy de Béarn SW France
102 J15 Luy de France SW France
125 P12 Luza Kirovskaya Oblast', NW Russian Federation 60°38′N 47°13′E
125 Q12 Luza NW Russian Federation
104 I16 Luz, Costa de la coastal region SW Spain
111 K20 Luže var. Lausche. ▲ Czech Republic/Germany 50°51′N 14°40′E see also Lausche
Luže see Lausche
108 F8 Luzern Fr. Lucerne, It. Lucerna. Luzern, C Switzerland 47°03′N 08°17′E
108 E8 Luzern Fr. Lucerne. ◈ canton C Switzerland
160 L13 Luzhai Guangxi Zhuangzu Zizhiqu, S China 24°31′N 109°46′E
118 K12 Luzhki Vitsyebskaya Voblasts', N Belarus 55°21′N 27°52′E
160 I10 Luzhou Sichuan, C China 28°55′N 105°25′E
Lužická Nisa see Neisse
Lužické Hory' see Lausitzer Bergland
Lužnice see Lainsitz
171 O2 Luzon island N Philippines
171 N1 Luzon Strait strait Philippines/Taiwan
Lužyckie, Gory see Lausitzer Bergland
116 I5 L'viv Ger. Lemberg, Pol. Lwów, Rus. L'vov. L'vivs'ka Oblast', W Ukraine 49°49′N 24°05′E
L'viv see L'vivs'ka Oblast'
116 I4 L'vivs'ka Oblast' var. L'viv, Rus. L'vovskaya Oblast'. ◈ province NW Ukraine
L'vov see L'viv
L'vovskaya Oblast' see L'vivs'ka Oblast'
Lwena see Luena
Lwów see L'viv
110 F11 Lwówek Ger. Neustadt bei Pinne. Wielkopolskie, C Poland 52°27′N 16°10′E
111 E14 Lwówek Śląski Ger. Löwenberg in Schlesien. Jelenia Góra, SW Poland 51°06′N 15°35′E
119 I18 Lyakhavichy Rus. Lyakhovichi. Brestskaya Voblasts', SW Belarus 53°02′N 26°16′E
Lyakhovichi see Lyakhavichy
185 B22 Lyall, Mount ▲ South Island, New Zealand 45°14′S 167°31′E
Lyallpur see Faisalābād
Lyangar see Langar
124 H11 Lyaskelya Respublika Kareliya, NW Russian Federation 61°42′N 31°06′E
119 I18 Lyasnaya Rus. Lesnaya. Brestskaya Voblasts', SW Belarus 52°59′N 25°46′E
119 F19 Lyasnaya Pol. Leśna, Rus. Lesnaya. SW Belarus
124 H15 Lychkovo Novgorodskaya Oblast', W Russian Federation 57°55′N 32°24′E
Lyck see Ełk
93 I15 Lycksele Västerbotten, N Sweden 64°34′N 18°40′E
18 G13 Lycoming Creek Pennsylvania, NE USA
Lycopolis see Asyūṭ
195 N3 Lyddan Island island Antarctica
Lydenburg see Mashishing
119 L20 Lyel'chytsy Rus. Lel'chitsy. Homyel'skaya Voblasts', SE Belarus 51°47′N 28°20′E
119 P14 Lyenina Rus. Lenino. Mahilyowskaya Voblasts', E Belarus 54°25′N 31°08′E
118 L13 Lyepyel' Rus. Lepel'. Vitsyebskaya Voblasts', N Belarus 54°48′N 28°44′E
25 S17 Lyford Texas, SW USA 26°24′N 97°47′W
95 E17 Lygna S Norway
18 G14 Lykens Pennsylvania, NE USA 40°33′N 76°42′W
115 E21 Lykódimo ▲ S Greece 36°56′N 21°49′E
97 K24 Lyme Bay bay S England, United Kingdom
97 K24 Lyme Regis S England, United Kingdom 50°44′N 02°56′W
110 L7 Łyna Ger. Alle. N Poland
29 P12 Lynch Nebraska, C USA 42°49′N 98°27′W
20 J10 Lynchburg Tennessee, S USA 35°17′N 86°22′W
21 T6 Lynchburg Virginia, NE USA 37°24′N 79°09′W
21 T12 Lynches River South Carolina, SE USA
32 H6 Lynden Washington, NW USA 48°57′N 122°27′W
182 I5 Lyndhurst South Australia 30°19′S 138°20′E
27 Q5 Lyndon Kansas, C USA 38°37′N 95°40′W
19 N7 Lyndonville Vermont, NE USA 44°31′N 71°58′W
95 D18 Lyngdal Vest-Agder, S Norway 58°10′N 07°08′E
92 I9 Lyngen Lapp. Ivgovuotna. inlet Arctic Ocean
95 G17 Lyngør Aust-Agder, S Norway 58°38′N 09°05′E
92 I9 Lyngseidet Troms, N Norway 69°36′N 20°07′E
19 P11 Lynn Massachusetts, NE USA 42°28′N 70°57′W
Lynn see King's Lynn
23 R9 Lynn Haven Florida, SE USA 30°15′N 85°39′W
11 V11 Lynn Lake Manitoba, C Canada 56°51′N 101°01′W
Lynn Regis see King's Lynn
118 I13 Lyntupy Vitsyebskaya Voblasts', NW Belarus 55°03′N 26°19′E
103 R11 Lyon Eng. Lyons; anc. Lugdunum. Rhône, E France 45°46′N 04°50′E
8 I6 Lyon, Cape headland Northwest Territories, NW Canada 69°47′N 123°10′W
18 K6 Lyon Mountain ▲ New York, NE USA 44°42′N 73°52′W
103 Q11 Lyonnais, Monts du ▲▲ C France
65 N25 Lyon Point headland SE Tristan da Cunha 37°06′S 12°13′W
182 E5 Lyons South Australia 30°40′S 133°50′E
37 T3 Lyons Colorado, C USA 40°13′N 105°16′W
23 V6 Lyons Georgia, SE USA 32°12′N 82°19′W
26 M5 Lyons Kansas, C USA 38°22′N 98°13′W
29 R14 Lyons Nebraska, C USA 41°56′N 96°28′W
18 G10 Lyons New York, NE USA 43°03′N 76°58′W
Lyons see Lyon
118 O13 Lyozna Rus. Liozno. Vitsyebskaya Voblasts', NE Belarus 55°02′N 30°48′E
117 S4 Lypova Dolyna Sums'ka Oblast', NE Ukraine 50°36′N 33°50′E
117 N6 Lypovets' Rus. Lipovets. Vinnyts'ka Oblast', C Ukraine 49°13′N 29°06′E
Lys see Leie
111 I18 Lysá Hora ▲ E Czech Republic 49°31′N 18°27′E
95 D16 Lysefjorden fjord S Norway
95 I18 Lysekil Västra Götaland, S Sweden 58°16′N 11°26′E
Lýsi see Akdoğan
33 V14 Lysite Wyoming, C USA 43°16′N 107°42′W
127 P3 Lyskovo Nizhegorodskaya Oblast', W Russian Federation 56°04′N 45°01′E
108 D8 Lyss Bern, W Switzerland 47°04′N 07°19′E
95 H22 Lystrup Midtjylland, C Denmark 56°14′N 10°14′E
125 V14 Lys'va Permskiy Kray, NW Russian Federation 58°04′N 57°48′E
117 P6 Lysyanka Cherkas'ka Oblast', C Ukraine 49°15′N 30°50′E
117 X6 Lysychans'k Rus. Lisichansk. Luhans'ka Oblast', E Ukraine 48°52′N 38°27′E
97 K17 Lytham St Anne's NW England, United Kingdom 53°45′N 03°01′W
185 I19 Lyttelton South Island, New Zealand 43°35′S 172°44′E
10 M17 Lytton British Columbia, SW Canada 50°12′N 121°34′W
119 L18 Lyuban' Minskaya Voblasts', S Belarus 52°48′N 28°00′E
119 L18 Lyubanskaye Vodaskhovishcha Rus. Lyubanskoye Vodokhranilishche. ⊠ C Belarus
Lyubanskoye Vodokhranilishche see Lyubanskaye Vodaskhovishcha
116 M5 Lyubar Zhytomyrs'ka Oblast', N Ukraine 49°54′N 27°48′E
Lyubashëvka see Lyubashivka
117 O8 Lyubashivka Rus. Lyubashëvka. Odes'ka Oblast', SW Ukraine 47°49′N 30°18′E
119 I16 Lyubcha Pol. Lubcz. Hrodzyenskaya Voblasts', W Belarus 53°45′N 26°04′E
126 L4 Lyubertsy Moskovskaya Oblast', W Russian Federation 55°37′N 38°02′E
116 K2 Lyubeshiv Volyns'ka Oblast', NW Ukraine 51°46′N 25°33′E
124 M14 Lyubim Yaroslavskaya Oblast', NW Russian Federation 58°21′N 40°46′E
114 K11 Lyubimets Haskovo, S Bulgaria 41°51′N 26°03′E
Lyublin see Lublin
116 I3 Lyuboml' Pol. Luboml. Volyns'ka Oblast', NW Ukraine 51°12′N 24°01′E
Lyubotin see Lyubotyn
117 U5 Lyubotyn Rus. Lyubotin. Kharkivs'ka Oblast', E Ukraine 49°57′N 35°57′E
126 I5 Lyudinovo Kaluzhskaya Oblast', W Russian Federation 53°52′N 34°28′E
127 T2 Lyuk Udmurtskaya Respublika, NW Russian Federation 56°55′N 52°45′E
114 M9 Lyulyakovo prev. Keremitlik. Burgas, E Bulgaria 42°53′N 27°05′E
119 I18 Lyusina Rus. Lyusino. Brestskaya Voblasts', SW Belarus 52°38′N 26°31′E
Lyusino see Lyusina

M

138 G9 Ma'ād Irbid, N Jordan 32°37′N 35°36′E
Ma'ai see Luqu
Maalahti see Malax
Maale see Male'
138 G13 Ma'ān Ma'ān, SW Jordan 30°11′N 35°45′E
138 H13 Ma'ān off. Muḩāfaẓat Ma'ān, var. Ma'an, Ma'ān. ◈ governorate S Jordan
93 M16 Maaninka Pohjois-Savo, C Finland 63°10′N 27°19′E
Maanīt see Bayan, Töv, Mongolia
Maanīt see Hishig Öndör, Bulgan, Mongolia
Ma'ān, Muḩāfaẓat see Ma'ān
93 N15 Maanselkä Kainuu, C Finland 63°54′N 28°28′E
161 Q8 Ma'anshan Anhui, E China 31°45′N 118°32′E
188 F16 Maap island Caroline Islands, W Micronesia
118 H3 Maardu Ger. Maart. Harjumaa, NW Estonia 59°28′N 24°56′E
Ma'aret-en-Nu'man see Ma'arrat an Nu'mān
99 K16 Maarheeze Noord-Brabant, SE Netherlands 51°19′N 05°37′E
Maarianhamina see Mariehamn
138 I4 Ma'arrat an Nu'mān var. Ma'aret-en-Nu'man, Fr. Maarret enn Naamâne. Idlib, NW Syria 35°40′N 36°40′E
Maarret enn Naamâne see Ma'arrat an Nu'mān
98 I11 Maarssen Utrecht, C Netherlands 52°08′N 05°03′E
Maart see Maardu
99 L17 Maas Fr. Meuse. W Europe see also Meuse
Maas see Meuse
99 M15 Maasbree Limburg, SE Netherlands 51°22′N 06°03′E
99 L17 Maaseik prev. Maeseyck. Limburg, NE Belgium 51°05′N 05°48′E
171 Q6 Maasin Leyte, C Philippines 10°10′N 124°55′E
99 L17 Maasmechelen Limburg, NE Belgium 50°58′N 05°42′E
98 G12 Maassluis Zuid-Holland, SW Netherlands 51°55′N 04°15′E
99 L18 Maastricht var. Maestricht; anc. Traiectum ad Mosam, Traiectum Tungorum. Limburg, SE Netherlands 50°51′N 05°42′E
183 N18 Maatsuyker Group island group Tasmania, SE Australia
Maba see Qujiang
83 L20 Mabalane Gaza, S Mozambique 23°43′S 32°37′E
25 V7 Mabank Texas, SW USA 32°22′N 96°06′W
97 O18 Mablethorpe E England, United Kingdom 53°21′N 00°14′E
171 V12 Maboi Papua Barat, E Indonesia 01°00′S 134°02′E
83 M19 Mabote Inhambane, S Mozambique 22°03′S 34°09′E
32 J10 Mabton Washington, NW USA 46°13′N 120°00′W
83 H20 Mabutsane Southern, S Botswana 24°24′S 23°34′E
63 G19 Macá, Cerro ▲ S Chile 45°07′S 73°11′W
60 Q9 Macaé Rio de Janeiro, SE Brazil 22°21′S 41°48′W
82 N13 Macaloge Niassa, N Mozambique 12°27′S 35°25′E
Macan see Bonerate, Kepulauan
161 N15 Macao off. Macao Special Administrative Region, var. Macao S.A.R., Chin. Aomen Tebie Xingzhengqu, Port. Região Administrativa Especial de Macau. Guangdong, SE China 22°06′N 113°30′E
104 H9 Mação Santarém, C Portugal 39°33′N 08°00′W
Macao S.A.R. see Macao
Macao Special Administrative Region see Macao
58 J11 Macapá state capital Amapá, N Brazil 0°04′N 51°04′W
43 S17 Macaracas Los Santos, S Panama 07°46′N 80°31′W
55 P6 Macare, Caño NE Venezuela
55 Q6 Macareo, Caño NE Venezuela
Macarsca see Makarska
182 L12 Macarthur Victoria, SE Australia 38°04′S 142°02′E
MacArthur see Ormoc
56 C7 Macas Morona Santiago, SE Ecuador 02°22′S 78°08′W
Macassar see Makassar
59 Q14 Macau Rio Grande do Norte, E Brazil 05°05′S 36°37′W
Macău see Makó, Hungary
Macau, Região Administrativa Especial de see Macao
65 E24 Macbride Head headland East Falkland, Falkland Islands 51°25′S 57°55′W
23 V9 Macclenny Florida, SE USA 30°16′N 82°07′W
97 L18 Macclesfield C England, United Kingdom 53°16′N 02°07′W
192 F6 Macclesfield Bank undersea feature N South China Sea 15°50′N 114°20′E
MacCluer Gulf see Berau, Teluk
181 N7 Macdonald, Lake salt lake Western Australia
181 Q7 Macdonnell Ranges ▲▲ Northern Territory, C Australia
96 K8 Macduff NE Scotland, United Kingdom 57°40′N 02°29′W
104 I6 Macedo de Cavaleiros Bragança, N Portugal 41°31′N 06°57′W
Macedonia see Macedonia, FYR
Macedonia Central see Kentrikí Makedonía
Macedonia East and Thrace see Anatolikí Makedonía kai Thráki
113 O19 Macedonia, FYR off. the Former Yugoslav Republic of Macedonia, var. Macedonia, Mac. Makedonija, abbrev. FYR Macedonia, FYROM. ◆ republic SE Europe
Macedonia, the Former Yugoslav Republic of see Macedonia, FYR
Macedonia West see Dytikí Makedonía
59 Q16 Maceió state capital Alagoas, E Brazil 09°40′S 35°44′W
76 K15 Macenta SE Guinea 08°31′N 09°32′W
106 J12 Macerata Marche, C Italy 43°18′N 13°27′E
11 S11 MacFarlane Saskatchewan, C Canada
182 H7 Macfarlane, Lake var. Lake Mcfarlane. ◎ South Australia
Macgillicuddy's Reeks Mountains see Macgillycuddy's Reeks
97 B21 Macgillycuddy's Reeks var. Macgillicuddy's Reeks Mountains, Ir. Na Cruacha Dubha. ▲▲ SW Ireland
11 X16 MacGregor Manitoba, S Canada 49°58′N 98°49′W
149 O10 Mach Baluchistān, SW Pakistan 29°52′N 67°20′E
56 C6 Machachi Pichincha, C Ecuador 0°33′S 78°34′W
83 M19 Machaila Gaza, S Mozambique 22°15′S 32°57′E
Machaire Fíolta see Magherafelt
Machaire Rátha see Maghera
81 I19 Machakos Machakos, S Kenya 01°31′S 37°16′E
81 I20 Machakos ◈ county C Kenya
56 B8 Machala El Oro, SW Ecuador 03°20′S 79°57′W
83 J19 Machaneng Central, SE Botswana 23°12′S 27°30′E
83 M18 Machanga Sofala, E Mozambique 20°56′S 35°04′E
80 G13 Machar Marshes wetland SE Sudan
102 I8 Machecoul Loire-Atlantique, NW France 46°59′N 01°51′W
161 O8 Macheng Hubei, C China 31°10′N 115°00′E
155 J16 Mācherla Andhra Pradesh, C India 16°29′N 79°25′E
153 O11 Māchhāpuchhre ▲ C Nepal 28°30′N 83°57′E
19 T6 Machias Maine, NE USA 44°44′N 67°28′W
19 T6 Machias River Maine, NE USA
19 R3 Machias River Maine, NE USA
64 P5 Machico Madeira, Portugal, NE Atlantic Ocean 32°43′N 16°47′W
155 K16 Machilīpatnam var. Bandar Masulipatnam. Andhra Pradesh, E India 16°12′N 81°11′E
54 G5 Machiques Zulia, NW Venezuela 10°04′N 72°37′W
57 G15 Machu Picchu Cusco, C Peru 13°08′S 72°30′W
83 M20 Macia var. Vila de Macia. Gaza, S Mozambique 25°02′S 33°08′E
Macías Nguema Biyogo see Bioco, Isla de
116 M13 Măcin Tulcea, SE Romania 45°15′N 28°09′E
183 T4 Macintyre River New South Wales/Queensland, SE Australia
181 Y7 Mackay Queensland, NE Australia 21°10′S 149°10′E
181 O7 Mackay, Lake salt lake Northern Territory/Western Australia
10 M13 Mackenzie British Columbia, W Canada 55°18′N 123°09′W
8 I9 Mackenzie Northwest Territories, NW Canada
195 Y6 Mackenzie Bay bay Antarctica
10 J1 Mackenzie Bay bay NW Canada
2 D9 Mackenzie Delta delta Northwest Territories, NW Canada
197 P8 Mackenzie King Island island Queen Elizabeth Islands, Northwest Territories, N Canada
8 H8 Mackenzie Mountains ▲▲ Northwest Territories, NW Canada
31 Q5 Mackinac, Straits of ◎ Michigan, N USA
194 K5 Mackintosh, Cape headland Antarctica 72°52′S 60°00′W
11 R15 Macklin Saskatchewan, S Canada 52°19′N 109°51′W
183 V6 Macksville New South Wales, SE Australia 30°39′S 152°54′E
183 V5 Maclean New South Wales, SE Australia 29°30′S 153°15′E
83 J24 Maclear Eastern Cape, SE South Africa 31°05′S 28°22′E
183 U6 Macleay River New South Wales, SE Australia
MacLeod see Fort Macleod
180 G9 Macleod, Lake ◎ Western Australia
10 I6 Macmillan Yukon, NW Canada
30 J12 Macomb Illinois, N USA 40°27′N 90°40′W
107 B18 Macomer Sardegna, Italy, C Mediterranean Sea 40°15′N 08°47′E
82 Q13 Macomia Cabo Delgado, NE Mozambique 12°15′S 40°06′E
103 R10 Mâcon anc. Matisco, Matisco Ædourum. Saône-et-Loire, C France 46°19′N 04°49′E
23 T5 Macon Georgia, SE USA 32°49′N 83°41′W
23 N4 Macon Mississippi, S USA 33°06′N 88°33′W
27 U3 Macon Missouri, C USA 39°44′N 92°27′W
22 J6 Macon, Bayou Arkansas/Louisiana, S USA
82 G13 Macondo Moxico, E Angola 12°31′S 23°45′E
83 M16 Macossa Manica, C Mozambique 17°51′S 33°54′E
11 T12 Macoun Lake ◎ Saskatchewan, C Canada
30 K14 Macoupin Creek Illinois, N USA
Macouria see Tonate
83 N18 Macovane Inhambane, SE Mozambique 21°30′S 35°07′E
183 N17 Macquarie Harbour inlet Tasmania, SE Australia
192 J13 Macquarie Island island New Zealand, SW Pacific Ocean
183 T8 Macquarie, Lake lagoon New South Wales, SE Australia
183 Q6 Macquarie Marshes wetland New South Wales, SE Australia
175 O13 Macquarie Ridge undersea feature SW Pacific Ocean 57°00′S 159°00′E
183 Q6 Macquarie River New South Wales, SE Australia
183 P17 Macquarie River Tasmania, SE Australia
195 V5 Mac. Robertson Land physical region Antarctica
97 C21 Macroom Ir. Maigh Chromtha. Cork, SW Ireland 51°54′N 08°57′W
42 G5 Macuelizo Santa Bárbara, NW Honduras 15°21′N 88°31′W
182 G2 Macumba River South Australia
57 I16 Macusani Puno, S Peru 14°05′S 70°24′W
56 E8 Macusari, Río N Peru
41 U15 Macuspana Tabasco, SE Mexico 17°43′N 92°36′W
138 G10 Mādabā var. Ma'dabā, Madeba; anc. Medeba. Mādabā, NW Jordan 31°44′N 35°48′E
138 G11 Mādabā off. Muḩāfaẓat Mādabā, var. Ma'dabā. ◈ governorate C Jordan
Ma'dabā see Mādabā
Ma'dabā see Mādabā
Mādabā, Muḩāfaẓat see Mādabā
172 G2 Madagascar off. Democratic Republic of Madagascar, Malg. Madagasikara; prev. Malagasy Republic. ◆ republic W Indian Ocean
172 I5 Madagascar island W Indian Ocean
128 L17 Madagascar Basin undersea feature W Indian Ocean 27°00′S 53°00′E
Madagascar, Democratic Republic of see Madagascar
128 L16 Madagascar Plain undersea feature W Indian Ocean 19°00′S 52°00′E
67 Y14 Madagascar Plateau var. Madagascar Ridge, Madagascar Rise, Rus. Madagaskarskiy Khrebet. undersea feature W Indian Ocean 30°00′S 45°00′E
Madagascar Rise/Madagascar Ridge see Madagascar Plateau
Madagasikara see Madagascar
Madagaskarskiy Khrebet see Madagascar Plateau
64 N2 Madalena Pico, Azores, Portugal, NE Atlantic Ocean 38°32′N 28°15′W
77 Y6 Madama Agadez, NE Niger 21°54′N 13°43′E
114 J12 Madan Smolyan, S Bulgaria 41°29′N 24°56′E
155 I19 Madanapalle Andhra Pradesh, E India 13°33′N 78°31′E
186 D7 Madang Madang, N Papua New Guinea 05°14′S 145°45′E
186 C6 Madang ◈ province N Papua New Guinea
146 G7 Madaniyat Rus. Madeniyet. Qoraqalpog'iston Respublikasi, W Uzbekistan 42°48′N 59°00′E
Madanīyīn see Médenine
77 U11 Madaoua Tahoua, SW Niger 14°06′N 06°01′E
Madaras see Vtáčnik
153 U15 Madaripur Dhaka, C Bangladesh 23°09′N 90°11′E
77 U12 Madarounfa Maradi, S Niger 13°16′N 07°07′E
Mađarska see Hungary
146 B13 Madau Balkan Welaýaty, W Turkmenistan 38°11′N 54°46′E
186 H9 Madau Island island SE Papua New Guinea
19 S1 Madawaska Maine, NE USA 47°19′N 68°19′W
14 J13 Madawaska Ontario, SE Canada
Madawaska Highlands see Haliburton Highlands
166 M4 Madaya Mandalay, C Myanmar (Burma) 22°12′N 96°05′E
107 K17 Maddaloni Campania, S Italy 41°03′N 14°23′E
29 O3 Maddock North Dakota, N USA 47°57′N 99°31′W
99 I14 Made Noord-Brabant, S Netherlands 51°41′N 04°48′E
Madeba see Mādabā
64 L9 Madeira var. Ilha de Madeira. island Madeira, Portugal, NE Atlantic Ocean
Madeira, Ilha de see Madeira
64 O5 Madeira Islands Port. Região Autoónoma da Madeira. ◈ autonomous region Madeira, Portugal, NE Atlantic Ocean
64 L9 Madeira Plain undersea feature E Atlantic Ocean
Madeira, Região Autoónoma da see Madeira Islands
64 L9 Madeira Ridge undersea feature E Atlantic Ocean 35°30′N 15°45′W
59 F14 Madeira, Rio var. Río Madera. Bolivia/Brazil see also Madera, Río
Madeira, Rio see Madera, Río
101 J25 Mädelegabel ▲ Austria/Germany 47°18′N 10°19′E
15 X6 Madeleine Québec, SE Canada
15 X5 Madeleine, Cap de la headland Québec, SE Canada 49°13′N 65°20′W
13 Q13 Madeleine, Îles de la Eng. Magdalen Islands. island group Québec, E Canada
29 U10 Madelia Minnesota, N USA 44°03′N 94°26′W
35 P3 Madeline California, W USA 41°02′N 120°28′W
30 K3 Madeline Island island Apostle Islands, Wisconsin, N USA
137 O15 Maden Elazığ, SE Turkey 38°24′N 39°42′E
145 V12 Madeniyet Vostochnyy Kazakhstan, E Kazakhstan 47°51′N 78°37′E
Madeniyet see Madaniyat
40 H5 Madera Chihuahua, N Mexico 29°10′N 108°10′W
35 Q10 Madera California, W USA 36°57′N 120°02′W
56 L13 Madera, Río Port. Rio Madeira. Bolivia/Brazil see also Madeira, Rio
Madera, Río see Madeira, Rio
106 D6 Madesimo Lombardia, N Italy 46°20′N 09°26′E
141 O14 Madhāb, Wādī dry watercourse NW Yemen
153 R13 Madhepura prev. Madhipure. Bihār, NE India 25°56′N 86°48′E
Madhipure see Madhepura
153 Q13 Madhubani Bihār, N India 26°21′N 86°05′E
153 Q15 Madhupur Jhārkhand, NE India 24°17′N 86°38′E
154 I10 Madhya Pradesh prev. Central Provinces and Berar. ◈ state C India
57 K15 Madidi, Río W Bolivia
155 F20 Madikeri prev. Mercara. Karnātaka, W India 12°29′N 75°40′E
27 O13 Madill Oklahoma, C USA 34°06′N 96°46′W
79 G21 Madimba Bas-Congo, SW Dem. Rep. Congo 04°58′S 15°08′E
138 M4 Ma'dīn Ar Raqqah, C Syria 35°45′N 39°36′E
Madīnah, Minṭaqat al see Al Madīnah
76 M14 Madinani NW Ivory Coast 09°37′N 06°57′W
141 O17 Madīnat ash Sha'b prev. Al Ittiḩād. SW Yemen 12°52′N 44°55′E
138 K3 Madīnat ath Thawrah var. Ath Thawrah. Ar Raqqah, N Syria 35°36′N 39°00′E
173 O6 Madingley Rise undersea feature W Indian Ocean
79 E21 Madingo-Kayes Kouilou, S Congo 04°27′S 11°43′E
79 F21 Madingou Bouenza, S Congo 04°10′S 13°33′E
Madioen see Madiun
23 U8 Madison Florida, SE USA 30°27′N 83°24′W
23 T3 Madison Georgia, SE USA 33°37′N 83°28′W
31 P15 Madison Indiana, N USA 38°44′N 85°22′W
27 P6 Madison Kansas, C USA 45°01′N 96°11′W
19 Q6 Madison Maine, NE USA 44°48′N 69°52′W
29 S9 Madison Minnesota, N USA 45°00′N 96°12′W
22 K5 Madison Mississippi, S USA 32°27′N 90°07′W
29 Q14 Madison Nebraska, C USA 41°49′N 97°27′W
29 R10 Madison South Dakota, N USA 44°00′N 97°06′W
21 V5 Madison Virginia, NE USA 38°23′N 78°16′W
21 Q5 Madison West Virginia, NE USA 38°03′N 81°50′W
30 L9 Madison state capital Wisconsin, N USA 43°04′N 89°22′W
21 T6 Madison Heights Virginia, NE USA 37°25′N 79°07′W
20 I6 Madisonville Kentucky, S USA 37°20′N 87°30′W
20 M10 Madisonville Tennessee, S USA 35°31′N 84°21′W
25 V9 Madisonville Texas, SW USA 30°58′N 95°56′W
Madisonville see Taiohae
169 R16 Madiun prev. Madioen. Jawa, C Indonesia 07°37′S 111°33′E
Madjene see Majene
14 J14 Madoc Ontario, SE Canada 44°31′N 77°27′W
Madoera see Madura, Pulau
81 J18 Mado Gashi Garissa, E Kenya 0°40′N 39°09′E
159 R11 Madoi var. Huanghe; prev. Huangheyan. Qinghai, C China 34°53′N 98°07′E
189 O13 Madolenihmw Pohnpei, E Micronesia
118 I9 Madona Ger. Modohn. E Latvia 56°51′N 26°10′E
107 J23 Madonie ▲▲ Sicilia, Italy, C Mediterranean Sea
141 Y11 Madrakah, Ra's headland E Oman 18°56′N 57°54′E
32 I12 Madras Oregon, NW USA 44°39′N 121°08′W
Madras see Chennai
Madras see Tamil Nādu
57 H14 Madre de Dios off. Departamento de Madre de Dios. ◈ department E Peru
Madre de Dios, Departamento de see Madre de Dios
63 F22 Madre de Dios, Isla island S Chile
57 J14 Madre de Dios, Río Bolivia/Peru
0 H15 Madre del Sur, Sierra ▲▲ S Mexico
41 Q9 Madre, Laguna lagoon NE Mexico
25 T16 Madre, Laguna lagoon Texas, SW USA
37 Q12 Madre Mount ▲ New Mexico, SW USA 34°18′N 107°54′W
0 H13 Madre Occidental, Sierra var. Western Sierra Madre. ▲▲ C Mexico
0 H13 Madre Oriental, Sierra var. Eastern Sierra Madre. ▲▲ C Mexico
41 U17 Madre, Sierra var. Sierra de Soconusco. ▲▲ Guatemala/Mexico
37 R2 Madre, Sierra ▲▲ Colorado/Wyoming, C USA
105 N8 Madrid ● (Spain) Madrid, C Spain 40°25′N 03°43′W
29 V14 Madrid Iowa, C USA 41°52′N 93°49′W
105 N7 Madrid ◈ autonomous community C Spain
105 N10 Madridejos Castilla-La Mancha, C Spain 39°29′N 03°32′W
104 L7 Madrigal de las Altas Torres Castilla y León, N Spain 41°05′N 05°00′W
104 K10 Madrigalejo Extremadura, W Spain 39°08′N 05°36′W
34 L3 Mad River California, W USA
42 J8 Madriz ◈ department NW Nicaragua
104 K10 Madroñera Extremadura, W Spain 39°25′N 05°46′W
181 N12 Madura Western Australia 31°52′S 127°01′E
Madura see Madurai
155 H22 Madurai prev. Madura, Mathurai. Tamil Nādu, S India 09°55′N 78°07′E
169 S16 Madura, Pulau prev. Madoera. island C Indonesia
169 S16 Madura, Selat strait C Indonesia

282
◆ Country ● Country Capital ◇ Dependent Territory ○ Dependent Territory Capital ◈ Administrative Regions ✈ International Airport ▲ Mountain ▲▲ Mountain Range Volcano River ◎ Lake ⊠ Reservoir

Madzhalis – Malko Sharkovo, Yazovir

127 *Q17* **Madzhalis** Respublika Dagestan, SW Russian Federation 42°12′N 47°46′E
114 *K12* **Madzharovo** Haskovo, S Bulgaria 41°36′N 25°52′E
83 *M14* **Madzimoyo** Eastern, E Zambia 13°42′S 32°34′E
165 *O12* **Maebashi** *var.* Maebasi, Mayebashi. Gunma, Honshū, S Japan 36°24′N 139°02′E
Maebasi *see* Maebashi
167 *O6* **Mae Chan** Chiang Rai, NW Thailand 20°13′N 99°52′E
167 *N7* **Mae Hong Son** *var.* Maehongson, Muai To. Mae Hong Son, NW Thailand 19°16′N 97°56′E
Maehongson *see* Mae Hong Son
Mae Nam Khong *see* Mekong
167 *Q7* **Mae Nam Nan** ≈ NW Thailand
167 *O10* **Mae Nam Tha Chin** ≈ W Thailand
167 *P7* **Mae Nam Yom** ≈ W Thailand
37 *O3* **Maeser** Utah, W USA 40°28′N 109°35′W
Maeseyck *see* Maaseik
167 *N9* **Mae Sot** *var.* Ban Mae Sot. Tak, W Thailand 16°44′N 98°32′E
44 *H8* **Maestra, Sierra** ▲▲ E Cuba
Maestricht *see* Maastricht
167 *O7* **Mae Suai** *var.* Ban Mae Suai. Chiang Rai, NW Thailand 19°43′N 99°30′E
167 *O7* **Mae Tho, Doi** ▲ NW Thailand 18°56′N 99°20′E
172 *I4* **Maevatanana** Mahajanga, C Madagascar 16°57′S 46°50′E
187 *R13* **Maéwo** *prev.* Aurora. *island* C Vanuatu
171 *S11* **Mafa** Pulau Halmahera, E Indonesia 0°01′N 127°50′E
83 *I23* **Mafeteng** W Lesotho 29°48′S 27°15′E
99 *J21* **Maffe** Namur, SE Belgium 50°21′N 05°19′E
183 *P12* **Maffra** Victoria, SE Australia 37°59′S 147°03′E
81 *K23* **Mafia** *island* E Tanzania
81 *J23* **Mafia Channel** *sea waterway* E Tanzania
83 *I21* **Mafikeng** *off.* Mahikeng. North-West, N South Africa 25°53′S 25°39′E
60 *J12* **Mafra** Santa Catarina, S Brazil 26°08′S 49°47′W
104 *F10* **Mafra** Lisboa, C Portugal 38°57′N 09°19′W
143 *Q17* **Mafraq** Abū Ẓaby, C United Arab Emirates 24°21′N 54°33′E
Mafraq/Muḩāfaẓat al Mafraq *see* Al Mafraq
123 *T10* **Magadan** Magadanskaya Oblast', E Russian Federation 59°38′N 150°50′E
123 *T9* **Magadanskaya Oblast'** ⬖ *province* E Russian Federation
108 *G11* **Magadino** Ticino, S Switzerland 46°09′N 08°50′E
63 *G23* **Magallanes** *var.* Región de Magallanes y de la Antártica Chilena. ⬖ *region* S Chile
Magallanes *see* Punta Arenas
Magallanes, Estrecho de *see* Magellan, Strait of
Magallanes y de la Antártica Chilena, Región de *see* Magallanes
14 *I10* **Maganasipi, Lac** ◎ Québec, SE Canada
54 *F6* **Magangué** Bolívar, N Colombia 09°14′N 74°46′W
191 *Y13* **Magareva** *var.* Mangareva. *island* Îles Tuamotu, SE French Polynesia
77 *V12* **Magaria** Zinder, S Niger 13°00′N 08°55′E
186 *F10* **Magarida** Central, SW Papua New Guinea 10°10′S 149°21′E
171 *O2* **Magat** ≈ Luzon, N Philippines
27 *T11* **Magazine Mountain** ▲ Arkansas, C USA 35°10′N 93°38′W
76 *I15* **Magburaka** C Sierra Leone 08°44′N 11°57′W
123 *Q13* **Magdagachi** Amurskaya Oblast', SE Russian Federation 53°25′N 125°41′E
62 *O12* **Magdalena** Buenos Aires, E Argentina 35°05′S 57°30′W
57 *M15* **Magdalena** El Beni, N Bolivia 13°22′S 64°07′W
40 *F4* **Magdalena** Sonora, NW Mexico 30°38′N 110°59′W
37 *Q13* **Magdalena** New Mexico, SW USA 34°07′N 107°14′W
54 *F5* **Magdalena** *off.* Departamento del Magdalena. ⬖ *province* N Colombia
40 *E9* **Magdalena, Bahía** *bay* W Mexico
Magdalena, Departamento del *see* Magdalena
63 *G19* **Magdalena, Isla** *island* Archipiélago de los Chonos, S Chile
40 *D8* **Magdalena, Isla** *island* NW Mexico
47 *P6* **Magdalena, Río** ≈ C Colombia
40 *F4* **Magdalena, Río** ≈ NW Mexico
Magdalen Islands *see* Madeleine, Îles de la
147 *N14* **Magdanly** *Rus.* Govurdak; *prev.* gowurdak, Guardak. Lebap Welaýaty, E Turkmenistan 37°50′N 66°06′E
100 *L13* **Magdeburg** Sachsen-Anhalt, C Germany 52°08′N 11°39′E
22 *L6* **Magee** Mississippi, S USA 31°52′N 89°43′W
169 *Q16* **Magelang** Jawa, C Indonesia 07°28′S 110°11′E
192 *K7* **Magellan Rise** *undersea feature* C Pacific Ocean
63 *H24* **Magellan, Strait of** *Sp.* Estrecho de Magallanes. *strait* Argentina/Chile
106 *D7* **Magenta** Lombardia, NW Italy 45°28′N 08°52′E
Magerøy *see* Magerøya
92 *K7* **Magerøya** *var.* Magerøy, *Lapp.* Máhkarávju. *island* N Norway
164 *C17* **Mage-shima** *island* Nansei-shotō, SW Japan
108 *G11* **Maggia** Ticino, S Switzerland 46°15′N 08°42′E
108 *G10* **Maggia** ≈ SW Switzerland
Maggiore, Lago *see* Maggiore, Lake
106 *C6* **Maggiore, Lake** *It.* Lago Maggiore. ◎ Italy/Switzerland
44 *I12* **Maggotty** W Jamaica 18°09′N 77°46′W
76 *I10* **Maghama** Gorgol, S Mauritania 15°31′N 12°50′W
97 *F14* **Maghera** *Ir.* Machaire Rátha. C Northern Ireland, United Kingdom 54°51′N 06°40′W
97 *F15* **Magherafelt** *Ir.* Machaire Fíolta. C Northern Ireland, United Kingdom 54°45′N 06°36′W
188 *H6* **Magicienne Bay** *bay* Saipan, S Northern Mariana Islands
105 *O13* **Magina** ▲ S Spain 37°43′N 03°24′W
81 *H24* **Magingo** Ruvuma, S Tanzania 09°57′S 35°23′E
112 *H11* **Maglaj** ⬖ Federacija Bosne I Hercegovine, N Bosnia and Herzegovina
107 *Q19* **Maglie** Puglia, SE Italy 40°07′N 18°18′E
114 *K10* **Maglizh** *var.* Mŭglizh. Stara Zagora, C Bulgaria 42°36′N 25°32′E
36 *L2* **Magna** Utah, W USA 40°42′N 112°06′W
Magnesia *see* Manisa
14 *G12* **Magnetawan** ≈ Ontario, S Canada
27 *T14* **Magnolia** Arkansas, C USA 33°17′N 93°16′W
22 *K7* **Magnolia** Mississippi, S USA 31°08′N 90°27′W
25 *V10* **Magnolia** Texas, SW USA 30°12′N 95°46′W
Magnolia State *see* Mississippi
95 *J15* **Magnor** Hedmark, S Norway 59°57′N 12°14′E
187 *Y14* **Mago** *prev.* Mango. *island* Lau Group, E Fiji
83 *L15* **Màgoé** Tete, NW Mozambique 15°50′S 31°42′E
15 *Q13* **Magog** Québec, SE Canada 45°16′N 72°09′W
83 *J15* **Magoye** Southern, S Zambia 16°00′S 27°34′E
41 *Q12* **Magozal** Veracruz-Llave, C Mexico 21°33′N 97°57′W
14 *B7* **Magpie** ≈ Ontario, S Canada
11 *Q17* **Magrath** Alberta, SW Canada 49°27′N 112°52′W
105 *R10* **Magre** ≈ Valenciana, E Spain
76 *I9* **Magṭa' Laḥjar** *var.* Magta Lahjar, Magta' Lahjar, Magtá Lahjar. Brakna, SW Mauritania 17°27′N 13°07′W
146 *D12* **Magtymguly** *prev.* Garrygala, *Rus.* Kara-Kala. Balkan Welaýaty, W Turkmenistan 38°27′N 56°15′E
83 *L20* **Magude** Maputo, S Mozambique 25°02′S 32°40′E
77 *Y12* **Magumeri** Borno, NE Nigeria 12°07′N 12°48′E
189 *O14* **Magur Islands** *island group* Caroline Islands, C Micronesia
166 *L6* **Magway** *var.* Magwe. Magway, W Myanmar (Burma) 20°08′N 94°55′E
166 *L6* **Magway** *var.* Magwe. ⬖ *region* C Myanmar (Burma)
Magwe *see* Magway
Magyar-Becse *see* Bečej
Magyarkanizsa *see* Kanjiža
Magyarország *see* Hungary
Magyarzsombor *see* Zimbor
142 *J4* **Mahābād** *var.* Mehabad; *prev.* Sāūjbulāgh. Āžarbāyjān-e Gharbī, NW Iran 36°44′N 45°44′E
172 *H5* **Mahabo** Toliara, W Madagascar 20°22′S 44°39′E
Maha Chai *see* Samut Sakhon
155 *D14* **Mahād** Mahārāshtra, W India 18°04′N 73°21′E
81 *N17* **Mahadday Weyne** Shabeellaha Dhexe, C Somalia 02°55′N 45°30′E
79 *Q17* **Mahagi** Orientale, NE Dem. Rep. Congo 02°16′N 30°59′E
Mahāil *see* Muhāyil
172 *I4* **Mahajamba** *seasonal river* NW Madagascar
152 *G10* **Mahājan** Rājasthān, NW India 28°47′N 73°50′E
172 *I3* **Mahajanga** *var.* Majunga. Mahajanga, NW Madagascar 15°40′S 46°20′E
172 *I3* **Mahajanga** ⬖ *province* W Madagascar
172 *I3* **Mahajanga** ✈ Mahajanga, NW Madagascar 15°40′S 46°20′E
169 *U10* **Mahakam, Sungai** *var.* Koetai, Kutai. ≈ Borneo, C Indonesia
83 *I19* **Mahalapye** *var.* Mahalatswe. Central, SE Botswana 23°02′S 26°53′E
Mahalatswe *see* Mahalapye
Mahalla el Kubra *see* El Maḥalla el Kubra
171 *O13* **Mahalona** Sulawesi, C Indonesia 02°37′S 121°26′E
Mahameru *see* Semeru, Gunung
143 *S11* **Māhān** Kermān, E Iran 30°00′N 57°00′E
154 *N12* **Mahanādi** ≈ E India
172 *J5* **Mahanoro** Toamasina, E Madagascar 19°53′S 48°48′E
153 *P13* **Mahārājganj** Bihār, N India 26°07′N 84°31′E
154 *G13* **Mahārāshtra** ⬖ *state* W India
172 *I4* **Mahavavy** *seasonal river* N Madagascar
155 *K24* **Mahaweli Ganga** ≈ C Sri Lanka
Mahbés *see* El Mahbas
155 *J15* **Mahbūbābād** Telangana, E India 17°35′N 80°00′E
155 *H16* **Mahbūbnagar** Telangana, C India 16°46′N 78°01′E
140 *M8* **Mahd adh Dhahab** Al Madīnah, W Saudi Arabia 23°33′N 40°56′E
55 *S9* **Mahdia** C Guyana 05°16′N 59°08′W
75 *N6* **Mahdia** *var.* Al Mahdīyah, Mehdia. NE Tunisia 35°14′N 11°06′E
155 *F20* **Mahe** *Fr.* Mahé; *prev.* Mayyali. Puducherry, SW India 11°41′N 75°31′E
172 *I16* **Mahé** ✈ Mahé, NE Seychelles 04°37′S 55°27′E
172 *H16* **Mahé** *island* Inner Islands, NE Seychelles
Mahé *see* Mahe
173 *Y17* **Mahebourg** SE Mauritius 20°24′S 57°42′E
152 *I11* **Mahendragarh** *prev.* Mohendergarh. Haryāna, N India 28°17′N 76°14′E
152 *L10* **Mahendranagar** Far Western, W Nepal 28°58′N 80°13′E
81 *I23* **Mahenge** Morogoro, SE Tanzania 08°41′S 36°41′E
185 *F22* **Maheno** Otago, South Island, New Zealand 45°10′S 170°51′E
154 *D9* **Mahesāna** Gujarāt, W India 23°37′N 72°28′E
154 *F11* **Maheshwar** Madhya Pradesh, C India 22°11′N 75°40′E
153 *V17* **Maheskhali Island** *var.* Maiskhal Island. *island* SE Bangladesh
151 *F14* **Mahi** ≈ N India
184 *Q10* **Mahia Peninsula** *peninsula* North Island, New Zealand
Mahikeng *see* Mafikeng
119 *O16* **Mahilyow** *Rus.* Mogilëv. Mahilyowskaya Voblasts', E Belarus 53°55′N 30°23′E
119 *M16* **Mahilyowskaya Voblasts'** *Rus.* Mogilëvskaya Oblast'. ⬖ *province* E Belarus
191 *P7* **Mahina** Tahiti, W French Polynesia 17°29′S 149°27′W
185 *E23* **Mahinerangi, Lake** ◎ South Island, New Zealand
Máhkarávju *see* Magerøya
83 *L22* **Mahlabatini** KwaZulu/Natal, E South Africa 28°15′S 31°28′E
166 *L5* **Mahlaing** Mandalay, C Myanmar (Burma) 21°03′N 95°44′E
109 *X8* **Mahldorf** Steiermark, SE Austria 46°54′N 15°55′E
Mahmūd-e 'Erāqī *see* Maḥmūd-e Rāqī
149 *R4* **Maḥmūd-e Rāqī** *var.* Mahmūd-e 'Erāqī. Kāpīsā, NE Afghanistan 35°01′N 69°20′E
Mahmudiya *see* Al Maḥmūdīyah
29 *S5* **Mahnomen** Minnesota, N USA 47°19′N 95°58′W
152 *K14* **Mahoba** Uttar Pradesh, N India 25°18′N 79°53′E
Mahón *see* Maó
18 *D14* **Mahoning Creek Lake** ⊠ Pennsylvania, NE USA
105 *Q10* **Mahora** Castilla-La Mancha, C Spain 39°13′N 01°44′W
Mähren *see* Moravia
Mährisch-Budwitz *see* Moravské Budějovice
Mährisch-Kromau *see* Moravský Krumlov
Mährisch-Neustadt *see* Uničov
Mährisch-Schönberg *see* Šumperk
Mährisch-Trübau *see* Moravská Třebová
Mährisch-Weisskirchen *see* Hranice
Māh-Shahr *see* Bandar-e Māhshahr
79 *N19* **Mahulu** Maniema, E Dem. Rep. Congo 01°04′S 27°10′E
154 *C12* **Mahuva** Gujarāt, W India 21°06′N 71°46′E
114 *N11* **Mahya Dağı** ▲ NW Turkey 41°47′N 27°34′E
105 *T6* **Maials** *var.* Mayals. Cataluña, NE Spain 41°22′N 00°30′E
191 *O2* **Maiana** *prev.* Hall Island. *atoll* Tungaru, W Kiribati
191 *S11* **Maiao** *var.* Tapuaemanu, Tubuai-Manu. *island* Îles du Vent, W French Polynesia
54 *H4* **Maicao** La Guajira, N Colombia 11°23′N 72°16′W
Mai Ceu/Mai Chio *see* Maych'ew
103 *U8* **Maîche** Doubs, E France 47°15′N 06°43′E
149 *Q5* **Maīdān Shahr** *var.* Maydān Shahr; *prev.* Meydān Shahr. Wardak, E Afghanistan 34°27′N 68°48′E
97 *N22* **Maidenhead** S England, United Kingdom 51°32′N 00°44′W
11 *S15* **Maidstone** Saskatchewan, S Canada 53°06′N 109°21′W
97 *P22* **Maidstone** SE England, United Kingdom 51°17′N 00°31′E
77 *Y13* **Maiduguri** Borno, NE Nigeria 11°51′N 13°10′E
108 *I8* **Maienfeld** Sankt Gallen, NE Switzerland 47°01′N 09°30′E
116 *J12* **Măieruş** *Hung.* Szászmagyarós. Braşov, C Romania 45°55′N 25°30′E
Maigh Chromtha *see* Macroom
Maigh Eo *see* Mayo
55 *N9* **Maigualida, Sierra** ▲▲ S Venezuela
154 *K9* **Maihar** Madhya Pradesh, C India 24°18′N 80°46′E
154 *K11* **Maikala Range** ▲▲ C India
67 *T10* **Maiko** ≈ W Dem. Rep. Congo
Mailand *see* Milano
152 *L11* **Mailāni** Uttar Pradesh, N India 28°17′N 80°20′E
149 *U10* **Māilsi** Punjab, E Pakistan 29°46′N 72°15′E
147 *R8* **Maimak** Talasskaya Oblast', NW Kyrgyzstan 42°40′N 71°12′E
Maimāna *see* Maīmanah
148 *M3* **Maīmanah** *var.* Maimāna, Maymana; *prev.* Meymaneh. Fāryāb, NW Afghanistan 35°57′N 64°48′E
171 *V13* **Maimawa** Papua Barat, E Indonesia 03°21′S 133°36′E
Maimuna *see* Al Maymūnah
101 *G18* **Main** ≈ C Germany
115 *F22* **Maïna** *ancient monument* Pelopónnisos, S Greece
115 *E20* **Maínalo** ▲▲ S Greece
101 *L22* **Mainburg** Bayern, SE Germany 48°40′N 11°48′E
Main Camp *see* Banana
14 *E12* **Main Channel** *lake channel* Ontario, S Canada
79 *I20* **Mai-Ndombe, Lac** *prev.* Lac Léopold II. ◎ W Dem. Rep. Congo
101 *K20* **Main-Donau-Kanal** *canal* SE Germany
19 *R6* **Maine** *off.* State of Maine, *also known as* Lumber State, Pine Tree State. ⬖ *state* NE USA
102 *K6* **Maine** *cultural region* NW France
102 *J7* **Maine-et-Loire** ⬖ *department* NW France
19 *Q9* **Maine, Gulf of** *gulf* NE USA
77 *X12* **Maïné-Soroa** Diffa, SE Niger 13°14′N 12°00′E
167 *N2* **Maingkwan** *var.* Mungkawn. Kachin State, N Myanmar (Burma) 26°20′N 96°37′E
Main Island *see* Bermuda
Mainistir Fhear Maí *see* Fermoy
Mainistirna Búille *see* Boyle
Mainistir na Corann *see* Midleton
Mainistir na Féile *see* Abbeyfeale
96 *J5* **Mainland** *island* N Scotland, United Kingdom
96 *L2* **Mainland** *island* NE Scotland, United Kingdom
159 *P16* **Mainling** *var.* Tungdor. Xizang Zizhiqu, W China 29°12′N 94°06′E
152 *K12* **Mainpuri** Uttar Pradesh, N India 27°14′N 79°01′E
103 *N5* **Maintenon** Eure-et-Loir, C France 48°35′N 01°34′E
172 *H4* **Maintirano** Mahajanga, W Madagascar 18°01′S 44°03′E
93 *M15* **Mainua** Kainuu, C Finland 64°05′N 27°28′E
101 *G18* **Mainz** *Fr.* Mayence. Rheinland-Pfalz, SW Germany 50°00′N 08°16′E
76 *I9* **Maio** *var.* Vila do Maio. Maio, S Cape Verde 15°07′N 23°12′W
76 *E10* **Maio** *var.* Mayo. *island* Ilhas de Sotavento, SE Cape Verde
62 *G12* **Maipo, Río** ≈ C Chile
62 *H12* **Maipo, Volcán** ⛰ W Argentina 34°09′S 69°51′W
61 *E22* **Maipú** Buenos Aires, E Argentina 36°52′S 57°52′W
62 *I11* **Maipú** Mendoza, E Argentina 33°00′S 68°46′W
62 *H11* **Maipú** Santiago, C Chile 33°30′S 70°52′W
106 *A9* **Maira** ≈ NW Italy
108 *I10* **Maira** *It.* Mera. ≈ Italy/Switzerland
153 *V12* **Mairābari** Assam, NE India 26°28′N 92°22′E
44 *K7* **Maisí** Guantánamo, E Cuba 20°13′N 74°08′W
118 *H13* **Maišiagala** Vilnius, SE Lithuania 54°52′N 25°03′E
Maiskhal Island *see* Maheskhali Island
167 *N13* **Mai Sombun** Chumphon, SW Thailand 10°49′N 99°13′E
Mai Son *see* Hat Lot
Maisur *see* Karnātaka, India
Maisur *see* Mysore, India
183 *T8* **Maitland** New South Wales, SE Australia 32°33′S 151°33′E
182 *I9* **Maitland** South Australia 34°21′S 137°42′E
14 *F15* **Maitland** ≈ Ontario, S Canada
195 *R1* **Maitri** *Indian research station* Antarctica 70°03′S 08°59′E
159 *N15* **Maizhokunggar** Xizang Zizhiqu, W China 29°50′N 91°40′E
43 *O10* **Maíz, Islas del** *var.* Corn Islands. *island group* SE Nicaragua
164 *J12* **Maizuru** Kyōto, Honshū, SW Japan 35°30′N 135°20′E
54 *F6* **Majagual** Sucre, N Colombia 08°36′N 74°39′W
41 *Z13* **Majahual** Quintana Roo, E Mexico 18°43′N 87°43′W
Mājeej *see* Mejit Island
171 *N13* **Majene** *prev.* Madjene. Sulawesi, C Indonesia 03°33′S 118°59′E
43 *V15* **Majé, Serranía de** ▲▲ E Panama
112 *I11* **Majevica** ▲▲ NE Bosnia and Herzegovina
81 *H15* **Majī** Southern Nationalities, S Ethiopia 06°11′N 35°32′E
141 *X7* **Majīs** NW Oman 24°25′N 56°34′E
Majorca *see* Mallorca
105 *X9* **Major, Puig** ▲ Mallorca, Spain, W Mediterranean Sea 39°50′N 2°50′E
Mājro *see* Majuro Atoll
Majunga *see* Mahajanga
189 *Y3* **Majuro** ✈ Majuro Atoll, SE Marshall Islands 07°05′N 171°08′E
189 *Y2* **Majuro Atoll** *var.* Mājro. *atoll* Ratak Chain, SE Marshall Islands
189 *X2* **Majuro Lagoon** *lagoon* Majuro Atoll, SE Marshall Islands
76 *H11* **Maka** C Senegal 13°40′N 14°12′W
79 *F20* **Makabana** Niari, SW Congo 03°28′S 12°36′E
38 *D9* **Mākaha** *var.* Makaha. O'ahu, Hawaii, USA, C Pacific Ocean 21°28′N 158°13′W
38 *B8* **Makahū'ena Point** *var.* Makahuena Point. *headland* Kaua'i, Hawai'i, USA 21°52′N 159°28′W
38 *D9* **Makakilo City** O'ahu, Hawaii, USA, C Pacific Ocean 21°21′N 158°05′W
83 *H18* **Makalamabedi** Central, C Botswana 20°19′S 23°51′E
Makale *see* Mek'elē
158 *K17* **Makalu** *Chin.* Makaru Shan. ▲ China/Nepal 27°53′N 87°09′E
81 *G23* **Makampi** Mbeya, S Tanzania 08°00′S 33°17′E
Makanchi *see* Makanshy
145 *X12* **Makanshy** *prev.* Makanchi. Vostochnyy Kazakhstan, E Kazakhstan 46°47′N 82°00′E
42 *M8* **Makantaka** Región Autónoma Atlántico Norte, NE Nicaragua 13°13′N 84°04′W
190 *B16* **Makapu Point** *headland* W Niue 18°59′S 169°56′E
185 *C24* **Makarewa** Southland, South Island, New Zealand 46°17′S 168°16′E
117 *O4* **Makariv** Kyyivs'ka Oblast', N Ukraine 50°28′N 29°49′E
185 *D20* **Makarora** ≈ South Island, New Zealand
123 *T13* **Makarov** Ostrov Sakhalin, Sakhalinskaya Oblast', SE Russian Federation 48°24′N 142°37′E
197 *R9* **Makarov Basin** *undersea feature* Arctic Ocean
192 *I5* **Makarov Seamount** *undersea feature* W Pacific Ocean 29°30′N 153°30′E
113 *F15* **Makarska** *It.* Macarsca. Split-Dalmacija, SE Croatia 43°18′N 17°00′E
Makaru Shan *see* Makalu
125 *O15* **Makar'yev** Kostromskaya Oblast', NW Russian Federation 57°52′N 43°46′E
82 *L11* **Makasa** Northern, NE Zambia 09°42′S 31°54′E
Makasar *see* Makassar
Makasar, Selat *see* Makassar Straits
170 *M14* **Makassar** *var.* Macassar, Makasar; *prev.* Ujungpandang. Sulawesi, C Indonesia 05°09′S 119°28′E
192 *F7* **Makassar Straits** *Ind.* Makasar Selat. *strait* C Indonesia
144 *G12* **Makat** *Kaz.* Maqat. Atyrau, SW Kazakhstan 47°40′N 53°28′E
191 *T10* **Makatea** *island* Îles Tuamotu, C French Polynesia
139 *U7* **Makātū** Diyālá, E Iraq 33°55′N 45°25′E
172 *H6* **Makay** *var.* Massif du Makay. ▲▲ SW Madagascar
Makay, Massif du *see* Makay
114 *J12* **Makaza** *pass* Bulgaria/Greece
Makedonija *see* Macedonia, FYR
190 *B16* **Makefu** W Niue 18°59′S 169°55′W
191 *V10* **Makemo** *atoll* Îles Tuamotu, C French Polynesia
76 *I15* **Makeni** C Sierra Leone 08°57′N 12°02′W
Makenzen *see* Orlyak
Makeyevka *see* Makiyivka
127 *Q16* **Makhachkala** *prev.* Petrovsk-Port. Respublika Dagestan, SW Russian Federation 42°58′N 47°30′E
Makhado *see* Louis Trichardt
144 *F11* **Makhambet** Atyrau, W Kazakhstan 47°35′N 51°35′E
Makharadze *see* Ozurgeti
139 *W13* **Makhfar al Buşayyah** Al Muthanná, S Iraq 30°09′N 46°09′E
Makhmūr *see* Mexmûr
138 *I11* **Makhrūq, Wadi al** *dry watercourse* E Jordan
139 *R4* **Makḥūl, Jabal** ▲▲ C Iraq
141 *R13* **Makhyah, Wādī** *dry watercourse* N Yemen
171 *V13* **Maki** Papua Barat, E Indonesia 03°00′S 134°10′E
185 *G21* **Makikihi** Canterbury, South Island, New Zealand 44°36′S 171°09′E
191 *O2* **Makin** *prev.* Pitt Island. *atoll* Tungaru, W Kiribati
81 *I20* **Makindu** Makueni, S Kenya 02°15′S 37°49′E
145 *Q8* **Makinsk** Akmola, N Kazakhstan 52°40′N 70°28′E
Makira *see* Makira-Ulawa
Makira *see* San Cristobal
187 *N10* **Makira-Ulawa** *prev.* Makira. ⬖ *province* SE Solomon Islands
117 *X8* **Makiyivka** *Rus.* Makeyevka; *prev.* Dmitriyevsk. Donets'ka Oblast', E Ukraine 47°57′N 37°47′E
140 *L10* **Makkah** *Eng.* Mecca. Makkah, W Saudi Arabia 21°28′N 39°50′E
140 *M10* **Makkah** *var.* Minṭaqat Makkah. ⬖ *province* W Saudi Arabia
Makkah, Minṭaqat *see* Makkah
13 *R7* **Makkovik** Newfoundland and Labrador, NE Canada 55°06′N 59°07′W
98 *K6* **Makkum** Fryslân, N Netherlands 53°03′N 05°25′E
111 *M25* **Makó** *Rom.* Macău. Csongrád, SE Hungary 46°14′N 20°28′E
Mako *see* Makung
14 *G9* **Makobe Lake** ◎ Ontario, S Canada
79 *F18* **Makokou** Ogooué-Ivindo, NE Gabon 0°38′N 12°47′E
81 *G23* **Makongolosi** Mbeya, S Tanzania 08°24′S 33°09′E
81 *E19* **Makota** SW Uganda 0°37′S 30°12′E
79 *G18* **Makoua** Cuvette, C Congo 0°01′S 15°40′E
110 *M10* **Maków Mazowiecki** Mazowieckie, C Poland 52°51′N 21°06′E
111 *K17* **Maków Podhalański** Małopolskie, S Poland 49°43′N 19°40′E
143 *V14* **Makran** *cultural region* Iran/Pakistan
152 *G12* **Makrāna** Rājasthān, N India 27°02′N 74°44′E
143 *U15* **Makran Coast** *coastal region* SE Iran
119 *F20* **Makrany** *Rus.* Mokrany. Brestskaya Voblasts', SW Belarus 51°50′N 24°15′E
Makrinoros *see* Makrynóros
115 *H20* **Makrónisosi** *island* Kykládes, Greece, Aegean Sea
115 *D17* **Makrynóros** *var.* Makrinoros. ▲▲ C Greece
115 *G19* **Makryplági** ▲ C Greece 38°00′N 23°06′E
Maksamaa *see* Maxmo
Maksatha *see* Maksatikha
Maksaticha *see* Maksatikha
124 *J15* **Maksatikha** *var.* Maksatha, Maksaticha. Tverskaya Oblast', W Russian Federation 57°49′N 35°46′E
154 *G10* **Maksi** Madhya Pradesh, C India 23°20′N 76°36′E
142 *I1* **Mākū** Āžarbāyjān-e Gharbī, NW Iran 39°20′N 44°50′E
81 *I20* **Makueni** ⬖ *county* S Kenya
153 *Y11* **Mākum** Assam, NE India 27°28′N 95°28′E
Makun *see* Makung
161 *R14* **Makung** *prev.* Mako, Makun. W Taiwan 23°35′N 119°35′E
164 *B16* **Makurazaki** Kagoshima, Kyūshū, SW Japan 31°16′N 130°18′E
77 *V15* **Makurdi** Benue, C Nigeria 07°42′N 08°36′E
38 *L17* **Makushin Volcano** ▲ Unalaska Island, Alaska, USA 53°53′N 166°55′W
83 *K16* **Makwiro** Mashonaland West, N Zimbabwe 17°58′S 30°25′E
57 *D15* **Mala** Lima, W Peru 12°40′S 76°36′W
93 *I14* **Malå** Västerbotten, N Sweden 65°12′N 18°45′E
Mala *see* Mallow, Ireland
Mala *see* Malaita, Solomon Islands
171 *P8* **Malabang** Mindanao, S Philippines
155 *E21* **Malabār Coast** *coast* SW India
79 *C16* **Malabo** *prev.* Santa Isabel. ● (Equatorial Guinea) Isla de Bioco, NW Equatorial Guinea 03°43′N 08°52′E
79 *C16* **Malabo** ✈ Isla de Bioco, N Equatorial Guinea 03°44′N 08°51′E
Malaca *see* Málaga
Malacca *see* Melaka
168 *I7* **Malacca, Strait of** *Ind.* Selat Malaka. *strait* Indonesia/Malaysia
Malacka *see* Malacky
111 *G20* **Malacky** *Hung.* Malacka. Bratislavský Kraj, W Slovakia 48°26′N 17°01′E
33 *R16* **Malad City** Idaho, NW USA 42°10′N 112°16′W
117 *Q4* **Mala Divytsya** Chernihivs'ka Oblast', N Ukraine 50°40′N 32°13′E
119 *J15* **Maladzyechna** *Pol.* Molodeczno, *Rus.* Molodechno. Minskaya Voblasts', C Belarus 54°19′N 26°51′E
190 *D12* **Malaee** Île Futuna, N Wallis and Futuna
190 *G12* **Mala'etoli** Île Uvea, E Wallis and Futuna
54 *G8* **Málaga** Santander, C Colombia 06°44′N 72°45′W
104 *M15* **Málaga** *anc.* Malaca. Andalucía, S Spain 36°43′N 04°25′W
37 *V15* **Malaga** New Mexico, SW USA 32°10′N 104°04′W
104 *L15* **Málaga** ⬖ *province* Andalucía, S Spain
104 *M15* **Málaga** ✈ Andalucía, S Spain 36°38′N 04°36′W
Malagasy Republic *see* Madagascar
105 *N10* **Malagón** Castilla-La Mancha, C Spain 39°10′N 03°51′W
97 *G18* **Malahide** *Ir.* Mullach Íde. Dublin, E Ireland 53°27′N 06°09′W
187 *N9* **Malaita** *off.* Malaita Province. ⬖ *province* N Solomon Islands
187 *N8* **Malaita** *var.* Mala. *island* N Solomon Islands
Malaita Province *see* Malaita
80 *F13* **Malakal** Upper Nile, NE South Sudan 09°31′N 31°40′E
112 *C10* **Mala Kapela** ▲▲ NW Croatia
25 *V7* **Malakoff** Texas, SW USA 32°10′N 96°00′W
Malakula *see* Malekula
149 *V7* **Malakwāl** *var.* Mālikwāla. Punjab, E Pakistan 32°32′N 73°18′E
186 *E7* **Malalamai** Madang, W Papua New Guinea 05°49′S 146°44′E
171 *O13* **Malamala** Sulawesi, C Indonesia 03°21′S 120°58′E
169 *S17* **Malang** Jawa, C Indonesia 07°59′S 112°45′E
83 *O14* **Malanga** Niassa, N Mozambique 13°27′S 36°05′E
Malange *see* Malanje
92 *I9* **Malangen** *sound* N Norway
82 *C11* **Malanje** *var.* Malange. Malanje, NW Angola 09°34′S 16°25′E
82 *C11* **Malanje** *var.* Malange. ⬖ *province* N Angola
148 *M16* **Malān, Rās** *cape* SW Pakistan
77 *S13* **Malanville** NE Benin 11°50′N 03°23′E
Malapane *see* Ozimek
155 *F21* **Malappuram** Kerala, SW India 11°00′N 76°02′E
43 *T17* **Mala, Punta** *headland* S Panama 07°28′N 79°58′W
148 *L15* **Malar** Baluchistān, SW Pakistan 26°19′N 64°55′E
95 *N16* **Mälaren** ◎ C Sweden
62 *H13* **Malargüe** Mendoza, W Argentina 35°32′S 69°35′W
14 *J8* **Malartic** Québec, SE Canada 48°09′N 78°09′W
119 *F20* **Malaryta** *Pol.* Maloryta, *Rus.* Malorita. Brestskaya Voblasts', SW Belarus 51°47′N 24°05′E
63 *J19* **Malaspina** Chubut, SE Argentina 44°56′S 66°52′W
10 *G8* **Malaspina Glacier** *glacier* Yukon, W Canada
39 *U12* **Malaspina Glacier** *glacier* Alaska, USA
137 *N15* **Malatya** *anc.* Melitene. Malatya, SE Turkey 38°22′N 38°18′E
136 *M14* **Malatya** ⬖ *province* C Turkey
Malavate *see* Pédima
117 *Q7* **Mala Vyska** *Rus.* Malaya Viska. Kirovohrads'ka Oblast', S Ukraine 48°37′N 31°36′E
83 *M14* **Malawi** *off.* Republic of Malaŵi; *prev.* Nyasaland, Nyasaland Protectorate. ◆ *republic* S Africa
Malawi, Lake *see* Nyasa, Lake
Malaŵi, Republic of *see* Malawi
93 *J17* **Malax** *Fin.* Maalahti. Österbotten, W Finland 62°55′N 21°30′E
124 *H14* **Malaya Vishera** Novgorodskaya Oblast', W Russian Federation 58°52′N 32°12′E
Malaya Viska *see* Mala Vyska
171 *Q7* **Malaybalay** Mindanao, S Philippines 08°10′N 125°08′E
142 *L6* **Malāyer** *prev.* Daulatabad. Hamadān, W Iran 34°20′N 48°47′E
168 *J7* **Malay Peninsula** *peninsula* Malaysia/Thailand
168 *L7* **Malaysia** *off.* Malaysia, *var.* Federation of Malaysia; *prev.* the separate territories of Federation of Malaya, Sarawak and Sabah (North Borneo) and Singapore. ◆ *monarchy* SE Asia
Malaysia, Federation of *see* Malaysia
137 *R14* **Malazgirt** Muş, E Turkey 39°09′N 42°30′E
15 *R8* **Malbaie** ≈ Québec, SE Canada
77 *T12* **Malbaza** Tahoua, S Niger 13°57′N 05°32′E
110 *J7* **Malbork** *Ger.* Marienburg, Marienburg in Westpreussen. Pomorskie, N Poland 54°01′N 19°03′E
100 *N9* **Malchin** Mecklenburg-Vorpommern, N Germany 53°43′N 12°46′E
100 *M9* **Malchiner See** ◎ NE Germany
99 *D16* **Maldegem** Oost-Vlaanderen, NW Belgium 51°12′N 03°27′E
98 *L13* **Malden** Gelderland, SE Netherlands 51°47′N 05°51′E
19 *O11* **Malden** Massachusetts, NE USA 42°25′N 71°04′W
27 *Y8* **Malden** Missouri, C USA 36°33′N 89°58′W
191 *X4* **Malden Island** *prev.* Independence Island. *atoll* E Kiribati
173 *Q6* **Maldives** *off.* Maldivian Divehi, Republic of Maldives. ◆ *republic* N Indian Ocean
Maldives, Republic of *see* Maldives
Maldivian Divehi *see* Maldives
97 *P21* **Maldon** E England, United Kingdom 51°44′N 00°40′E
61 *G20* **Maldonado** Maldonado, S Uruguay 34°57′S 54°59′W
61 *G20* **Maldonado** ⬖ *department* S Uruguay
41 *P17* **Maldonado, Punta** *headland* S Mexico 16°18′N 98°31′W
106 *G6* **Malè** Trentino-Alto Adige, N Italy 46°21′N 10°51′E
151 *K19* **Male'** *Div.* Maale. ● (Maldives) Male' Atoll, C Maldives 04°10′N 73°29′E
76 *K13* **Maléa** *var.* Maléya. NE Guinea 11°46′N 09°43′W
Maléas, Ákra *see* Agriliá, Akrotírio
115 *G22* **Maléas, Akrotírio** *headland* S Greece 36°25′N 23°11′E
151 *K19* **Male' Atoll** *var.* Kaafu Atoll. *atoll* C Maldives
Malebo, Pool *see* Stanley Pool
154 *E12* **Mālegaon** Mahārāshtra, W India 20°33′N 74°32′E
81 *F15* **Malek** Jonglei, E South Sudan 06°04′N 31°36′E
187 *Q13* **Malekula** *var.* Malakula; *prev.* Mallicolo. *island* W Vanuatu
189 *Y15* **Malem** Kosrae, E Micronesia 05°16′N 163°01′E
83 *O15* **Malema** Nampula, N Mozambique 14°57′S 37°28′E
79 *N23* **Malemba-Nkulu** Katanga, SE Dem. Rep. Congo 08°01′S 26°48′E
139 *T1* **Małe Mela** *Ar.* Marī Mīlā, *var.* Mārī Mīlah. Arbīl, E Iraq 36°58′N 44°42′E
124 *K9* **Malen'ga** Respublika Kareliya, NW Russian Federation 63°50′N 36°21′E
95 *M20* **Målerås** Kalmar, S Sweden 56°55′N 15°34′E
103 *O6* **Malesherbes** Loiret, C France 48°18′N 02°25′E
115 *G18* **Malesína** Stereá Elláda, E Greece 38°37′N 23°15′E
Maléya *see* Maléa
127 *O15* **Malgobek** Respublika Ingushetiya, SW Russian Federation 43°34′N 44°34′E
105 *X5* **Malgrat de Mar** Cataluña, NE Spain 41°39′N 02°45′E
80 *C9* **Malha** Northern Darfur, W Sudan 15°07′N 26°00′E
139 *Q5* **Malḩah** *var.* Malḩāt. Şalāḩ ad Dīn, C Iraq 34°44′N 42°41′E
Malḩāt *see* Malḩah
32 *K14* **Malheur Lake** ◎ Oregon, NW USA
32 *L14* **Malheur River** ≈ Oregon, NW USA
76 *I13* **Mali** NW Guinea 12°08′N 12°29′W
77 *O9* **Mali** *off.* Republic of Mali, *Fr.* République du Mali; *prev.* French Sudan, Sudanese Republic. ◆ *republic* W Africa
171 *Q16* **Maliana** W East Timor 08°57′S 125°25′E
167 *O2* **Mali Hka** ≈ N Myanmar (Burma)
Mali Idjoš *see* Mali Iđoš
112 *K8* **Mali Iđoš** *var.* Mali Idjoš, *Hung.* Kishegyes; *prev.* Krivaja. Vojvodina, N Serbia 45°43′N 19°40′E
113 *M18* **Mali i Sharrit** *Serb.* Šar Planina. ▲▲ FYR Macedonia/Serbia
Mali i Zi *see* Crna Gora
112 *K9* **Mali Kanal** *canal* N Serbia
171 *P12* **Maliku** Sulawesi, N Indonesia 0°36′S 123°13′E
Malik, Wadi al *see* Milk, Wadi el
Mālikwāla *see* Malakwāl
167 *N11* **Mali Kyun** *var.* Tavoy Island. *island* Mergui Archipelago, S Myanmar (Burma)
95 *M19* **Målilla** Kalmar, S Sweden 57°24′N 15°49′E
112 *B11* **Mali Lošinj** *It.* Lussinpiccolo. Primorje-Gorski Kotar, W Croatia 44°31′N 14°28′E
Malin *see* Malyn
171 *P7* **Malindang, Mount** ▲ Mindanao, S Philippines 08°12′N 123°37′E
81 *K20* **Malindi** Kilifi, SE Kenya 03°14′S 40°05′E
Malines *see* Mechelen
96 *E13* **Malin Head** *Ir.* Cionn Mhálanna. *headland* NW Ireland 55°37′N 07°37′W
171 *O11* **Malino, Gunung** ▲ Sulawesi, N Indonesia 0°44′N 120°45′E
113 *M21* **Maliq** *var.* Maliqi. Korçë, SE Albania 40°45′N 20°45′E
Maliqi *see* Maliq
Mali, Republic of *see* Mali
Mali, République du *see* Mali
171 *Q8* **Malita** Mindanao, S Philippines 06°13′N 125°39′E
Maljovica *see* Malyovitsa
154 *G12* **Malkāpur** Mahārāshtra, C India 20°52′N 76°18′E
136 *B10* **Malkara** Tekirdağ, NW Turkey 40°54′N 26°54′E
119 *J19* **Mal'kavichy** *Rus.* Mal'kovichi. Brestskaya Voblasts', SW Belarus 52°31′N 26°36′E
Malkiye *see* Al Mālikīyah
114 *L11* **Malko Sharkovo, Yazovir** ⊠ SE Bulgaria

◆ Country
● Country Capital
◇ Dependent Territory
○ Dependent Territory Capital
⬖ Administrative Regions
✈ International Airport
▲ Mountain
▲▲ Mountain Range
⛰ Volcano
≈ River
◎ Lake
⊠ Reservoir

MALKO TARNOVO – MAPLETON

114 *N11* **Malko Tarnovo** *var.* Malko Tŭrnovo. Burgas, E Bulgaria 42°00′N 27°33′E
Malko Tŭrnovo *see* Malko Tarnovo
Mal'kovichi *see* Mal'kavichy
183 *R12* **Mallacoota** Victoria, SE Australia 37°34′S 149°45′E
96 *G10* **Mallaig** N Scotland, United Kingdom 57°04′N 05°48′W
182 *I9* **Mallala** South Australia 34°29′S 138°30′E
75 *W9* **Mallawī** *var.* Mallawi. C Egypt 27°44′N 30°50′E
Mallawi *see* Mallawī
105 *R5* **Mallén** Aragón, NE Spain 41°53′N 01°25′W
106 *F5* **Malles Venosta** *Ger.* Mals im Vinschgau. Trentino-Alto Adige, N Italy 46°40′N 10°37′E
Mallicolo *see* Malekula
109 *Q8* **Mallnitz** Salzburg, S Austria 46°58′N 13°09′E
105 *W9* **Mallorca** *Eng.* Majorca; *anc.* Baleares Major. *island* Islas Baleares, Spain, W Mediterranean Sea
97 *C20* **Mallow** *Ir.* Mala. SW Ireland 52°08′N 08°39′W
93 *E15* **Malm** Nord-Trøndelag, C Norway 64°04′N 11°12′E
95 *L19* **Malmbäck** Jönköping, S Sweden 57°34′N 14°30′E
92 *J12* **Malmberget** *Lapp.* Malmivaara. Norrbotten, N Sweden 67°09′N 20°39′E
99 *M20* **Malmédy** Liège, E Belgium 50°26′N 06°02′E
83 *E25* **Malmesbury** Western Cape, SW South Africa 33°28′S 18°43′E
Malmivaara *see* Malmberget
95 *N16* **Malmköping** Södermanland, C Sweden 59°08′N 16°49′E
95 *K23* **Malmö** Skåne, S Sweden 55°36′N 13°E
95 *K23* **Malmo** ✈ Skåne, S Sweden 55°33′N 13°23′E
45 *Q16* **Malmok** *headland* N Bonaire 12°16′N 68°21′W
95 *M18* **Malmslätt** Östergötland, S Sweden 58°25′N 15°30′E
125 *R16* **Malmyzh** Kirovskaya Oblast', NW Russian Federation 56°30′N 50°37′E
187 *Q13* **Malo** *island* W Vanuatu
126 *J7* **Maloarkhangel'sk** Orlovskaya Oblast', W Russian Federation 52°25′N 36°37′E
Maļoeļap *see* Maloelap Atoll
189 *V6* **Maloelap Atoll** *var.* Maļoeļap. *atoll* E Marshall Islands
Maloenda *see* Malunda
108 *I10* **Maloja** Graubünden, S Switzerland 46°25′N 09°42′E
82 *L12* **Malole** Northern, NE Zambia 10°05′S 31°37′E
171 *O3* **Malolos** Luzon, N Philippines 14°51′N 120°49′E
18 *K6* **Malone** New York, NE USA 44°51′N 74°18′W
79 *K25* **Malonga** Katanga, S Dem. Rep. Congo 10°26′S 23°10′E
111 *L17* **Małopolskie** ◈ *province* SE Poland
Malorita/Maloryta *see* Malaryta
124 *K9* **Maloshuyka** Arkhangel'skaya Oblast', NW Russian Federation 63°43′N 37°20′E
Mal'ovitsa *see* Malyovitsa
145 *V15* **Malovodnoye** Almaty, SE Kazakhstan 43°31′N 77°42′E
94 *C10* **Måløy** Sogn Og Fjordane, S Norway 61°57′N 05°06′E
126 *K4* **Maloyaroslavets** Kaluzhskaya Oblast', W Russian Federation 55°03′N 36°31′E
122 *G7* **Malozemel'skaya Tundra** *physical region* NW Russian Federation
104 *J10* **Malpartida de Cáceres** Extremadura, W Spain 39°26′N 06°30′W
104 *K9* **Malpartida de Plasencia** Extremadura, W Spain 39°59′N 06°03′W
106 *C7* **Malpensa** ✈ (Milano) Lombardia, N Italy 45°41′N 08°40′E
76 *J6* **Malqteïr** *desert* N Mauritania
Mals im Vinschgau *see* Malles Venosta
118 *J10* **Malta** SE Latvia 56°19′N 27°11′E
33 *V7* **Malta** Montana, NW USA 48°21′N 107°52′W
120 *M11* **Malta** *off.* Republic of Malta. ◆ *republic* C Mediterranean Sea
109 *R8* **Malta** *var.* Maltabach. ≈ S Austria
120 *M11* **Malta** *island* Malta, C Mediterranean Sea
Maltabach *see* Malta
Malta, Canale di *see* Malta Channel
120 *M11* **Malta Channel** *It.* Canale di Malta. *strait* Italy/Malta
83 *D20* **Maltahöhe** Hardap, SW Namibia 24°50′S 17°00′E
Malta, Republic of *see* Malta
97 *N16* **Malton** N England, United Kingdom 54°07′N 00°50′W
171 *R13* **Maluku** *off.* Propinsi Maluku, *Dut.* Molukken, *Eng.* Moluccas. ◈ *province* E Indonesia
171 *R13* **Maluku** *Dut.* Molukken, *Eng.* Moluccas; *prev.* Spice Islands. *island group* E Indonesia
Maluku, Laut *see* Molucca Sea
Maluku, Propinsi *see* Maluku
171 *R11* **Maluku Utara** *off.* Propinsi Maluku Utara. ◈ *province* E Indonesia
Maluku Utara, Propinsi *see* Maluku Utara
77 *V13* **Malumfashi** Katsina, N Nigeria 11°51′N 07°39′E
171 *N13* **Malunda** *prev.* Maloenda. Sulawesi, C Indonesia 02°58′S 118°52′E
94 *K13* **Malung** Dalarna, C Sweden 60°40′N 13°45′E
94 *K13* **Malungsfors** Dalarna, C Sweden 60°43′N 13°34′E
186 *M8* **Maluu** *var.* Malu'u. Malaita, N Solomon Islands 08°22′S 160°39′E
Malu'u *see* Maluu
155 *D16* **Mālvan** Mahārāshtra, W India 16°05′N 73°28′E
Malventum *see* Benevento
27 *U12* **Malvern** Arkansas, C USA 34°21′N 92°50′W
29 *S15* **Malvern** Iowa, C USA 40°59′N 95°36′W
44 *I13* **Malvern** ▲ W Jamaica 17°59′N 77°42′W
Malvina, Isla Gran *see* West Falkland
Malvinas, Islas *see* Falkland Islands
117 *N4* **Malyn** *Rus.* Malin. Zhytomyrs'ka Oblast', N Ukraine 50°46′N 29°14′E
114 *G10* **Malyovitsa** *var.* Maljovica, Mal'ovitsa. ▲ W Bulgaria 42°12′N 23°19′E
127 *O11* **Malyye Derbety** Respublika Kalmykiya, SW Russian Federation 47°57′N 44°39′E
Malyy Kavkaz *see* Lesser Caucasus
123 *Q6* **Malyy Lyakhovskiy, Ostrov** *island* NE Russian Federation
Malyy Pamir *see* Little Pamir
122 *N5* **Malyy Taymyr, Ostrov** *island* Severnaya Zemlya, N Russian Federation
Malyy Uzen' *see* Saryozen
122 *L14* **Malyy Yenisey** *var.* Ka-Krem. ≈ S Russian Federation
127 *S3* **Mamadysh** Respublika Tatarstan, W Russian Federation 55°46′N 51°22′E
117 *N14* **Mamaia** Constanţa, E Romania 44°13′N 28°37′E
187 *W14* **Mamanuca Group** *island group* Yasawa Group, W Fiji
146 *L13* **Mamash** Lebap Welaýaty, E Turkmenistan 38°24′N 64°12′E
79 *O17* **Mambasa** Orientale, NE Dem. Rep. Congo 01°20′N 29°05′E
171 *X13* **Mamberamo, Sungai** ≈ Papua, E Indonesia
79 *G15* **Mambéré** ≈ SW Central African Republic
79 *G15* **Mambéré-Kadéï** ◈ *prefecture* SW Central African Republic
Mambij *see* Manbij
79 *H18* **Mambili** ≈ W Congo
83 *N18* **Mambone** *var.* Nova Mambone. Inhambane, E Mozambique 20°59′S 35°04′E
171 *O4* **Mamburao** Mindoro, N Philippines 13°16′N 120°36′E
172 *I16* **Mamelles** *island* Inner Islands, NE Seychelles
99 *M25* **Mamer** Luxembourg, SW Luxembourg 49°37′N 06°01′E
102 *L6* **Mamers** Sarthe, NW France 48°21′N 00°22′E
79 *D15* **Mamfe** Sud-Ouest, W Cameroon 05°46′N 09°18′E
145 *P6* **Mamlyutka** Severnyy Kazakhstan, N Kazakhstan 54°54′N 68°36′E
36 *M15* **Mammoth** Arizona, SW USA 32°43′N 110°38′W
33 *S12* **Mammoth Hot Springs** Wyoming, C USA 44°57′N 110°40′W
Mamoedjoe *see* Mamuju
119 *A14* **Mamonovo** *Ger.* Heiligenbeil. Kaliningradskaya Oblast', W Russian Federation 54°28′N 19°57′E
57 *L14* **Mamoré, Rio** ≈ Bolivia/Brazil
76 *I14* **Mamou** W Guinea 10°24′N 12°05′W
22 *H8* **Mamou** Louisiana, S USA 30°37′N 92°25′W
172 *I14* **Mamoudzou** ○ (Mayotte) C Mayotte 12°48′S 45°E
172 *I3* **Mampikony** Mahajanga, N Madagascar 16°03′S 47°39′E
77 *P16* **Mampong** C Ghana 07°06′N 01°20′W
110 *M7* **Mamry, Jezioro** *Ger.* Mauersee. ◎ NE Poland
171 *N13* **Mamuju** *prev.* Mamoedjoe. Sulawesi, S Indonesia 02°41′S 118°55′E
83 *F19* **Mamuno** Ghanzi, W Botswana 22°15′S 20°02′E
113 *K19* **Mamuras** *var.* Mamurasi, Mamurras. Lezhë, C Albania 41°34′N 19°42′E
Mamurasi/Mamurras *see* Mamuras
76 *L16* **Man** W Ivory Coast 07°24′N 07°33′W
55 *X9* **Mana** NW French Guiana 05°40′N 53°49′W
56 *A6* **Manabí** ◈ *province* W Ecuador
42 *G4* **Manabique, Punta** *var.* Cabo Tres Puntas. *headland* E Guatemala 15°57′N 88°37′W
54 *G11* **Manacacías, Río** ≈ C Colombia
58 *F13* **Manacapuru** Amazonas, N Brazil 03°16′S 60°37′W
105 *Y9* **Manacor** Mallorca, Spain, W Mediterranean Sea 39°35′N 03°12′E
171 *Q11* **Manado** *prev.* Menado. Sulawesi, C Indonesia 01°32′N 124°55′E
188 *H5* **Managaha** *island* S Northern Mariana Islands
99 *G20* **Manage** Hainaut, S Belgium 50°30′N 04°14′E
42 *J10* **Managua** ● (Nicaragua) Managua, W Nicaragua 12°08′N 86°15′W
42 *J10* **Managua** ◈ *department* W Nicaragua
42 *J10* **Managua** ✈ Managua, W Nicaragua 12°07′N 86°11′W
42 *J10* **Managua, Lago de** *var.* Xolotlán. ◎ W Nicaragua
Manaḥ *see* Bilād Manaḥ
18 *K16* **Manahawkin** New Jersey, NE USA 39°39′N 74°12′W
184 *K11* **Manaia** Taranaki, North Island, New Zealand 39°33′S 174°07′E
172 *J6* **Manakara** Fianarantsoa, SE Madagascar 22°09′S 48°E
152 *J7* **Manāli** Himāchal Pradesh, NW India 32°12′N 77°06′E
Ma, Nam *see* Sông Ma
Manama *see* Al Manāmah
186 *D6* **Manam Island** *island* N Papua New Guinea
67 *Y13* **Mananara Avaratra** ≈ SE Madagascar
182 *M9* **Manangatang** Victoria, SE Australia 35°04′S 142°53′E
172 *J6* **Mananjary** Fianarantsoa, SE Madagascar 21°13′S 48°20′E
76 *L14* **Manankoro** Sikasso, SW Mali 10°33′N 07°25′W
76 *J12* **Manantali, Lac de** ▣ W Mali
Manáos *see* Manaus
185 *B23* **Manapouri** Southland, South Island, New Zealand 45°33′S 167°38′E
185 *B23* **Manapouri, Lake** ◎ South Island, New Zealand
58 *F13* **Manaquiri** Amazonas, NW Brazil 03°27′S 60°37′W
Manar *see* Mannar
158 *K5* **Manas** Xinjiang Uygur Zizhiqu, NW China 44°16′N 86°12′E
153 *U12* **Manās** *var.* Dangme Chu. ≈ Bhutan/India
153 *P10* **Manāsalu** *var.* Manaslu. ▲ C Nepal 28°33′N 84°33′E
147 *R8* **Manas, Gora** ▲ Kyrgyzstan/Uzbekistan 42°17′N 71°04′E
158 *K3* **Manas Hu** ◎ NW China
Manaslu *see* Manāsalu
37 *S8* **Manassa** Colorado, C USA 37°10′N 105°56′W
21 *W4* **Manassas** Virginia, NE USA 38°45′N 77°28′W
45 *T5* **Manatí** C Puerto Rico 18°26′N 66°29′W
186 *E8* **Manau** Northern, S Papua New Guinea 08°02′S 148°00′E
54 *H4* **Manaure** La Guajira, N Colombia 11°46′N 72°28′W
58 *F12* **Manaus** *prev.* Manáos. *state capital* Amazonas, NW Brazil 03°06′S 60°W
136 *G17* **Manavgat** Antalya, SW Turkey 36°47′N 31°28′E
184 *M13* **Manawatu** ≈ North Island, New Zealand
184 *L11* **Manawatu-Wanganui** *off.* Manawatu-Wanganui Region. ◈ *region* North Island, New Zealand
Manawatu-Wanganui Region *see* Manawatu-Wanganui
171 *R7* **Manay** Mindanao, S Philippines 07°12′N 126°29′E
138 *K2* **Manbij** *var.* Mambij, *Fr.* Membidj. Ḥalab, N Syria 36°32′N 37°55′E
105 *N13* **Mancha Real** Andalucía, S Spain 37°47′N 03°37′W
102 *I4* **Manche** ◈ *department* N France
97 *L17* **Manchester** *Lat.* Mancunium. NW England, United Kingdom 53°30′N 02°15′W
23 *S5* **Manchester** Georgia, SE USA 32°51′N 84°37′W
29 *Y13* **Manchester** Iowa, C USA 42°28′N 91°27′W
21 *N7* **Manchester** Kentucky, S USA 37°09′N 83°46′W
19 *O10* **Manchester** New Hampshire, NE USA 42°59′N 71°26′W
20 *K10* **Manchester** Tennessee, S USA 35°28′N 86°05′W
18 *M9* **Manchester** Vermont, NE USA 43°09′N 73°03′W
97 *L18* **Manchester** ✈ NW England, United Kingdom 53°21′N 02°16′W
149 *P15* **Manchhar Lake** ◎ SE Pakistan
Man-chou-li *see* Manzhouli
129 *X7* **Manchurian Plain** *plain* NE China
Máncio Lima *see* Japiim
Mancunium *see* Manchester
148 *J15* **Mand** Baluchistān, SW Pakistan 26°06′N 61°58′E
Mand *see* Mand, Rūd-e
81 *H25* **Manda** Njombe, SW Tanzania 10°30′S 34°37′E
172 *H6* **Mandabe** Toliara, W Madagascar 21°02′S 44°56′E
162 *M10* **Mandah** *var.* Töhöm. Dornogovĭ, SE Mongolia 44°25′N 108°18′E
95 *E18* **Mandal** Vest-Agder, S Norway 58°02′N 07°30′E
Mandal *see* Arbulag, Hövsgöl, Mongolia
Mandal *see* Batsümber, Töv, Mongolia
166 *L5* **Mandalay** Mandalay, C Myanmar (Burma) 21°57′N 96°04′E
166 *M6* **Mandalay** ◈ *region* C Myanmar (Burma)
162 *L9* **Mandalgovĭ** Dundgovĭ, C Mongolia 45°47′N 106°18′E
139 *V7* **Mandalī** Diyālá, E Iraq 33°43′N 45°33′E
162 *K10* **Mandal-Ovoo** *var.* Sharhulsan. Ömnögovĭ, S Mongolia 44°43′N 104°06′E
95 *E18* **Mandalselva** ≈ S Norway
163 *P11* **Mandalt** *var.* Sonid Zuoqi. Nei Mongol Zizhiqu, N China 43°49′N 113°36′E
28 *M5* **Mandan** North Dakota, N USA 46°49′N 100°53′W
Mandargiri Hill *see* Mandār Hill
153 *R14* **Mandār Hill** *prev.* Mandargiri Hill. Bihār, NE India 24°51′N 87°03′E
170 *M13* **Mandar, Teluk** *bay* Sulawesi, C Indonesia
107 *C19* **Mandas** Sardegna, Italy, C Mediterranean Sea 39°40′N 09°07′E
Mandasor *see* Mandsaur
81 *L16* **Mandera** Mandera, NE Kenya 03°56′N 41°53′E
81 *K17* **Mandera** ◈ *county* NE Kenya
33 *V13* **Manderson** Wyoming, C USA 44°13′N 107°55′W
44 *J12* **Mandeville** C Jamaica 18°02′N 77°31′W
22 *K9* **Mandeville** Louisiana, S USA 30°21′N 90°04′W
152 *I7* **Mandi** Himāchal Pradesh, NW India 31°40′N 76°59′E
76 *K14* **Mandiana** E Guinea 10°37′N 08°39′W
Mandi Būrewāla *see* Būrewāla
Mandidzudzure *see* Chimanimani
83 *M15* **Mandié** Manica, NW Mozambique 16°27′S 33°28′E
83 *N14* **Mandimba** Niassa, N Mozambique 14°21′S 35°40′E
57 *Q19* **Mandioré, Laguna** ◎ E Bolivia
154 *J10* **Mandla** Madhya Pradesh, C India 22°36′N 80°23′E
83 *M20* **Mandlakazi** *var.* Manjacaze. Gaza, S Mozambique 24°47′S 33°50′E
95 *E24* **Mandø** *var.* Manø. *island* W Denmark
Mandoúdhion/Mandoudi *see* Mantoúdi
78 *I13* **Mandoul** *off.* Région du Mandoul. ◈ *region* S Chad
Mandoul, Région du *see* Mandoul
115 *G19* **Mándra** Attikí, C Greece 38°04′N 23°30′E
172 *I7* **Mandrare** ≈ S Madagascar
114 *M10* **Mandra, Yazovir** *salt lake* SE Bulgaria
107 *L23* **Mandrazzi, Portella** *pass* Sicilia, Italy, C Mediterranean Sea
172 *J3* **Mandritsara** Mahajanga, N Madagascar 15°49′S 48°50′E
143 *O13* **Mand, Rūd-e** *var.* Mand. ≈ S Iran
154 *F9* **Mandsaur** *prev.* Mandasor. Madhya Pradesh, C India 24°03′N 75°10′E
154 *F11* **Māndu** Madhya Pradesh, C India 22°22′N 75°24′E
169 *W8* **Mandul, Pulau** *island* N Indonesia
83 *G15* **Mandundu** Western, W Zambia 16°34′S 22°18′E
180 *I13* **Mandurah** Western Australia 32°31′S 115°41′E
107 *P18* **Manduria** Puglia, SE Italy 40°24′N 17°38′E
155 *G20* **Mandya** Karnātaka, C India 12°34′N 76°55′E
77 *P12* **Mané** C Burkina Faso 12°59′N 01°21′W
106 *E8* **Manerbio** Lombardia, NW Italy 45°22′N 10°09′E
Manevichi *see* Manevychi
116 *K3* **Manevychi** *Pol.* Maniewicze, *Rus.* Manevichi. Volyns'ka Oblast', NW Ukraine 51°18′N 25°29′E
107 *N16* **Manfredonia** Puglia, SE Italy 41°38′N 15°54′E
107 *N16* **Manfredonia, Golfo di** *gulf* Adriatic Sea, N Mediterranean Sea
77 *P13* **Manga** C Burkina Faso 11°41′N 01°04′W
59 *L16* **Mangabeiras, Chapada das** ▲▲ E Brazil
79 *J20* **Mangai** Bandundu, W Dem. Rep. Congo 03°58′S 19°32′E
190 *L17* **Mangaia** *island group* S Cook Islands
184 *M9* **Mangakino** Waikato, North Island, New Zealand 38°23′S 175°47′E
116 *M15* **Mangalia** *anc.* Callatis. Constanţa, SE Romania 43°48′N 28°35′E
78 *J11* **Mangalmé** Guéra, SE Chad 12°26′N 19°37′E
155 *E19* **Mangalore** Karnātaka, W India 12°54′N 74°51′E
Mangareva *see* Magareva
83 *I23* **Mangaung** Free State, C South Africa 29°10′S 26°19′E
Mangaung *see* Bloemfontein
154 *K9* **Mangawān** Madhya Pradesh, C India 24°39′N 81°33′E
184 *M11* **Mangaweka** Manawatu-Wanganui, North Island, New Zealand 39°49′S 175°47′E
184 *N11* **Mangaweka** ▲ North Island, New Zealand 39°51′S 176°06′E
79 *P17* **Mangbwalu** Orientale, NE Dem. Rep. Congo 02°06′N 30°04′E
139 *R1* **Mangêsh** *Ar.* Māngīsh, *var.* Mangish. Dahūk, N Iraq 37°03′N 43°04′E
101 *L24* **Mangfall** ≈ SE Germany
169 *P13* **Manggar** Pulau Belitung, W Indonesia 02°52′S 108°13′E
Mangghystaū Üstirti' *see* Mangystau, Plato
166 *M2* **Mangin Range** ▲▲ N Myanmar (Burma)
Māngīsh *see* Mangêsh
Mangish *see* Mangêsh
Mangistau *see* Mangystau
146 *H8* **Mang'it** *Rus.* Mangit. Qoraqalpog'iston Respublikasi, W Uzbekistan 42°06′N 60°02′E
Mangit *see* Mang'it
54 *A13* **Manglares, Cabo** *headland* SW Colombia 01°36′N 79°02′W
149 *V6* **Mangla Reservoir** ▣ NE Pakistan
159 *N9* **Mangnai** *var.* Lao Mangnai. Qinghai, C China 37°52′N 91°45′E
Mango *see* Mago, Fiji
Mango *see* Sansanné-Mango, Togo
Mangoche *see* Mangochi
83 *N14* **Mangochi** *var.* Mangoche; *prev.* Fort Johnston. Southern, SE Malawi 14°30′S 35°15′E
77 *N14* **Mangodara** SW Burkina Faso 09°49′N 04°22′W
172 *H6* **Mangoky** ≈ W Madagascar
171 *Q12* **Mangole, Pulau** *island* Kepulauan Sula, E Indonesia
184 *J2* **Mangonui** Northland, North Island, New Zealand 35°00′S 173°30′E
Mangqystaū Oblysy *see* Mangystau
Mangqystaū Shyghanaghy *see* Mangystau Zaliv
Mangshi *see* Luxi
104 *H7* **Mangualde** Viseu, N Portugal 40°36′N 07°46′W
61 *H18* **Mangueira, Lagoa** ◎ S Brazil
77 *X6* **Manguéni, Plateau du** ▲▲ NE Niger
163 *T4* **Mangui** Nei Mongol Zizhiqu, N China 52°02′N 122°13′E
26 *K11* **Mangum** Oklahoma, C USA 34°52′N 99°30′W
79 *O18* **Manguredjipa** Nord-Kivu, E Dem. Rep. Congo 0°28′N 28°33′E
83 *L16* **Mangwendi** Mashonaland East, E Zimbabwe 18°22′S 31°24′E
Mangyshlak, Plato *see* Mangystau, Plato
Mangyshlakskiy Zaliv *see* Mangystau Zaliv
Mangyshlaskaya *see* Mangystau
144 *F15* **Mangystau** *Kaz.* Mangqystaū Oblysy *prev.* Mangistau; *prev.* Mangyshlaskaya. ◈ *province* SW Kazakhstan
144 *F15* **Mangystau, Plato** *plateau* SW Kazakhstan
144 *E14* **Mangystau Zaliv** *Kaz.* Mangqystaū Shyghanaghy; *prev.* Mangyshlakskiy Zaliv. *gulf* SW Kazakhstan
162 *E7* **Manhan** *var.* Tögrög. Hovd, W Mongolia 47°24′N 92°06′E
Manhan *see* Alag-Erdene
27 *O4* **Manhattan** Kansas, C USA 39°11′N 96°35′W
99 *L21* **Manhay** Luxembourg, SE Belgium 50°13′N 05°43′E
83 *L21* **Manhiça** *prev.* Vila de Manhiça. Maputo, S Mozambique 25°25′S 32°49′E
83 *L21* **Manhoca** Maputo, S Mozambique 26°49′S 32°36′E
59 *N20* **Manhuaçu** Minas Gerais, SE Brazil 20°16′S 42°01′W
117 *W9* **Manhush** *prev.* Pershotravneve. Donets'ka Oblast', E Ukraine 47°03′N 37°20′E
54 *H10* **Maní** Casanare, C Colombia 04°50′N 72°15′W
143 *R11* **Mānī** Kermān, C Iran
83 *M17* **Manica** *var.* Vila de Manica. Manica, W Mozambique 18°56′S 32°52′E
83 *M17* **Manica** *off.* Província de Manica. ◈ *province* W Mozambique
83 *L17* **Manicaland** ◈ *province* E Zimbabwe
Manica, Província de *see* Manica
15 *U5* **Manic Deux, Réservoir** ▣ Québec, SE Canada
Manich *see* Manych
59 *F14* **Manicoré** Amazonas, N Brazil 05°48′S 61°16′W
13 *N11* **Manicouagan** Québec, SE Canada 50°40′N 68°46′W
13 *N11* **Manicouagan** ≈ Québec, SE Canada
15 *U6* **Manicouagan, Péninsule de** *peninsula* Québec, SE Canada
13 *N11* **Manicouagan, Réservoir** ◎ Québec, E Canada
15 *T4* **Manic Trois, Réservoir** ▣ Québec, SE Canada
79 *M20* **Maniema** *off.* Région du Maniema. ◈ *region* E Dem. Rep. Congo
Maniema, Région du *see* Maniema
Maniewicze *see* Manevychi
160 *F8* **Maniganggo** Sichuan, C China 32°01′N 99°04′E
11 *Y15* **Manigotagan** Manitoba, S Canada 51°06′N 96°18′W
153 *R13* **Manihāri** Bihār, N India 25°21′N 87°37′E
191 *U9* **Manihi** *island* Îles Tuamotu, C French Polynesia
190 *L13* **Manihiki** *atoll* N Cook Islands
175 *U8* **Manihiki Plateau** *undersea feature* C Pacific Ocean
196 *M14* **Maniitsoq** *var.* Manîtsoq, *Dan.* Sukkertoppen. ◈ Qeqqata, S Greenland
153 *T15* **Manikganj** Dhaka, C Bangladesh 23°52′N 90°00′E
152 *M14* **Mānikpur** Uttar Pradesh, N India 25°04′N 81°06′E
171 *N4* **Manila** *off.* City of Manila. ● (Philippines) Luzon, N Philippines 14°34′N 120°59′E
27 *Y9* **Manila** Arkansas, C USA 35°52′N 90°10′W
Manila, City of *see* Manila
189 *N16* **Manila Reef** *reef* W Micronesia
183 *T6* **Manilla** New South Wales, SE Australia 30°44′S 150°43′E
192 *P6* **Maniloa** *island* Tongatapu Group, S Tonga
123 *U8* **Manily** Krasnoyarskiy Kray, E Russian Federation 62°33′N 165°03′E
171 *V12* **Manim, Pulau** *island* E Indonesia
168 *I11* **Maninjau, Danau** ◎ Sumatera, W Indonesia
153 *W13* **Manipur** ◈ *state* NE India
153 *X14* **Manipur Hills** *hill range* E India
136 *C14* **Manisa** *var.* Manissa, *prev.* Saruhan; *anc.* Magnesia. Manisa, W Turkey 38°36′N 27°29′E
136 *C13* **Manisa** *var.* Manissa. ◈ *province* W Turkey
Manissa *see* Manisa
31 *O7* **Manistee** Michigan, N USA 44°14′N 86°19′W
31 *P7* **Manistee River** ≈ Michigan, N USA
31 *O4* **Manistique** Michigan, N USA 45°57′N 86°15′W
31 *P4* **Manistique Lake** ◎ Michigan, N USA
11 *W13* **Manitoba** ◈ *province* S Canada
11 *X16* **Manitoba, Lake** ◎ Manitoba, S Canada
11 *X17* **Manitou** Manitoba, S Canada 49°12′N 98°28′W
31 *N2* **Manitou Island** *island* Michigan, N USA
14 *H11* **Manitou Lake** ◎ Ontario, SE Canada
12 *G15* **Manitoulin Island** *island* Ontario, S Canada
37 *T5* **Manitou Springs** Colorado, C USA 38°51′N 104°56′W
14 *G12* **Manitouwabing Lake** ◎ Ontario, S Canada
12 *E12* **Manitouwadge** Ontario, S Canada 49°14′N 85°51′W
12 *G15* **Manitowaning** Manitoulin Island, Ontario, S Canada 45°44′N 81°50′W
14 *B7* **Manitowik Lake** ◎ Ontario, S Canada
31 *N7* **Manitowoc** Wisconsin, N USA 44°04′N 87°40′W
Manîtsoq *see* Maniitsoq
139 *O7* **Mānī', Wādī al** *dry watercourse* W Iraq
12 *J14* **Maniwaki** Québec, SE Canada 46°22′N 75°58′W
171 *W13* **Maniwori** Papua, E Indonesia 02°49′S 136°00′E
54 *E10* **Manizales** Caldas, W Colombia 05°03′N 73°52′W
112 *F11* **Manjača** ▲▲ NW Bosnia and Herzegovina
Manjacaze *see* Mandlakazi
180 *J14* **Manjimup** Western Australia 34°18′S 116°14′E
109 *V4* **Mank** Niederösterreich, C Austria 48°06′N 15°13′E
79 *I17* **Mankanza** Equateur, NW Dem. Rep. Congo 01°40′N 19°08′E
153 *N12* **Mankāpur** Uttar Pradesh, N India 27°03′N 82°12′E
26 *M3* **Mankato** Kansas, C USA 39°48′N 98°13′W
29 *U10* **Mankato** Minnesota, N USA 44°10′N 94°00′W
117 *O7* **Man'kivka** Cherkas'ka Oblast', C Ukraine 48°58′N 30°10′E
76 *M15* **Mankono** C Ivory Coast 08°01′N 06°09′W
11 *T17* **Mankota** Saskatchewan, S Canada 49°25′N 107°05′W
155 *K23* **Mankulam** Northern Province, N Sri Lanka 09°07′N 80°27′E
162 *L10* **Manlay** *var.* Üydzen. Ömnögovĭ, S Mongolia 44°08′N 106°48′E
39 *Q9* **Manley Hot Springs** Alaska, USA 65°00′N 150°37′W
18 *H10* **Manlius** New York, NE USA 43°00′N 75°58′W
105 *W5* **Manlleu** Cataluña, NE Spain 41°59′N 02°17′E
29 *V11* **Manly** Iowa, C USA 43°17′N 93°12′W
154 *E13* **Manmād** Mahārāshtra, W India 20°15′N 74°29′E
182 *J7* **Mannahill** South Australia 32°29′S 139°58′E
155 *J23* **Mannar** *var.* Manar. Northern Province, NW Sri Lanka 09°01′N 79°53′E
155 *I24* **Mannar, Gulf of** *gulf* India/Sri Lanka
155 *J23* **Mannar Island** *island* N Sri Lanka
Mannersdorf *see* Mannersdorf am Leithagebirge
109 *Y5* **Mannersdorf am Leithagebirge** *var.* Mannersdorf. Niederösterreich, E Austria 47°59′N 16°36′E
109 *Y6* **Mannersdorf an der Rabnitz** Burgenland, E Austria 47°25′N 16°32′E
101 *G20* **Mannheim** Baden-Württemberg, SW Germany 49°29′N 08°29′E
11 *O12* **Manning** Alberta, W Canada 56°53′N 117°39′W
29 *T14* **Manning** Iowa, C USA 41°54′N 95°03′W
28 *K5* **Manning** North Dakota, N USA 47°15′N 102°48′W
21 *S13* **Manning** South Carolina, SE USA 33°42′N 80°12′W
191 *Y2* **Manning, Cape** *headland* Kiritimati, NE Kiribati 02°02′N 157°26′W
21 *S3* **Mannington** West Virginia, NE USA 39°31′N 80°20′W
182 *A1* **Mann Ranges** ▲▲ South Australia
107 *C19* **Mannu** ≈ Sardegna, Italy, C Mediterranean Sea
11 *R14* **Mannville** Alberta, SW Canada 53°19′N 111°08′W
76 *J15* **Mano** ≈ Liberia/Sierra Leone
Manø *see* Mandø
61 *F15* **Manoel Viana** Rio Grande do Sul, S Brazil 29°33′S 55°28′W
39 *O13* **Manokotak** Alaska, USA 59°00′N 158°58′W
171 *V12* **Manokwari** Papua Barat, E Indonesia 0°53′S 134°05′S
79 *N22* **Manono** Shaba, SE Dem. Rep. Congo 07°18′S 27°25′E
25 *T10* **Manor** Texas, SW USA 30°20′N 97°33′W
97 *D16* **Manorhamilton** *Ir.* Cluainín. Leitrim, NW Ireland 54°18′N 08°10′W
103 *S15* **Manosque** Alpes-de-Haute-Provence, SE France 43°50′N 05°47′E
12 *L11* **Manouane, Lac** ◎ Québec, SE Canada
163 *W12* **Manp'o** *var.* Manp'ojin. NW North Korea 41°10′N 126°24′E
Manp'ojin *see* Manp'o
191 *T4* **Manra** *prev.* Sydney Island. *atoll* Phoenix Islands, C Kiribati
105 *V5* **Manresa** Cataluña, NE Spain 41°43′N 01°50′E
152 *H9* **Mānsa** Punjab, NW India 30°00′N 75°25′E
82 *J12* **Mansa** *prev.* Fort Rosebery. Luapula, N Zambia 11°14′S 28°55′E
76 *G12* **Mansa Konko** C Gambia 13°26′N 15°29′W
15 *Q11* **Manseau** Québec, SE Canada 46°23′N 71°59′W
149 *U5* **Mānsehra** Khyber Pakhtunkhwa, NW Pakistan 34°23′N 73°18′E
9 *Q9* **Mansel Island** *island* Nunavut, NE Canada
183 *O12* **Mansfield** Victoria, SE Australia 37°04′S 146°06′E
97 *M18* **Mansfield** C England, United Kingdom 53°09′N 01°11′W
27 *S11* **Mansfield** Arkansas, C USA 35°03′N 94°15′W
22 *G6* **Mansfield** Louisiana, S USA 32°02′N 93°42′W
19 *O12* **Mansfield** Massachusetts, NE USA 42°00′N 71°11′W
31 *T12* **Mansfield** Ohio, N USA 40°45′N 82°31′W
18 *G12* **Mansfield** Pennsylvania, NE USA 41°46′N 77°02′W
18 *M7* **Mansfield, Mount** ▲ Vermont, NE USA 44°31′N 72°49′W
59 *M16* **Mansidão** Bahia, E Brazil 10°46′S 44°04′W
102 *L11* **Mansle** Charente, W France 45°52′N 00°11′E
76 *G12* **Mansôa** C Guinea-Bissau 12°08′N 15°18′W
47 *V8* **Manso, Rio** ≈ C Brazil
Manṣûra *see* Al Manṣūrah
Mansurabad *see* Mehrān, Rūd-e
56 *A6* **Manta** Manabí, W Ecuador 0°59′S 80°44′W
56 *A6* **Manta, Bahía de** *bay* W Ecuador
57 *F14* **Mantaro, Río** ≈ C Peru
35 *O8* **Manteca** California, W USA 37°48′N 121°13′W
54 *J7* **Mantecal** Apure, C Venezuela 07°34′N 69°07′W
31 *N11* **Manteno** Illinois, N USA 41°15′N 87°49′W
21 *Y9* **Manteo** Roanoke Island, North Carolina, SE USA 35°54′N 75°42′W
Mantes-Gassicourt *see* Mantes-la-Jolie
103 *N5* **Mantes-la-Jolie** *prev.* Mantes-Gassicourt, Mantes-sur-Seine; *anc.* Medunta. Yvelines, N France 48°59′N 01°43′E
Mantes-sur-Seine *see* Mantes-la-Jolie
36 *L5* **Manti** Utah, W USA 39°16′N 111°38′W
Mantinea *see* Mantíneia
115 *F20* **Mantíneia** *anc.* Mantinea. *site of ancient city* Pelopónnisos, S Greece
59 *M21* **Mantiqueira, Serra da** ▲▲ S Brazil
29 *W10* **Mantorville** Minnesota, N USA 44°04′N 92°45′W
115 *G17* **Mantoúdi** *var.* Mandoudi; *prev.* Mandoúdhion. Évvoia, C Greece 38°47′N 23°29′E
Mantoue *see* Mantova
106 *F8* **Mantova** *Eng.* Mantua, *Fr.* Mantoue. Lombardia, NW Italy 45°10′N 10°47′E
93 *M19* **Mäntsälä** Uusimaa, S Finland 60°38′N 25°21′E
93 *L17* **Mänttä** Pirkanmaa, W Finland 62°00′N 24°36′E
Mantua *see* Mantova
125 *O14* **Manturovo** Kostromskaya Oblast', NW Russian Federation 58°19′N 44°42′E
93 *M18* **Mäntyharju** Etelä-Savo, SE Finland 61°25′N 26°53′E
92 *M13* **Mäntyjärvi** Lappi, N Finland 66°00′N 27°35′E
190 *L16* **Manuae** *island* S Cook Islands
191 *Q10* **Manuae** *atoll* Îles Sous le Vent, W French Polynesia
192 *L16* **Manu'a Islands** *island group* E American Samoa
40 *L5* **Manuel Benavides** Chihuahua, N Mexico 29°07′N 103°52′W
61 *D21* **Manuel J. Cobo** Buenos Aires, E Argentina 35°49′S 57°54′W
58 *M12* **Manuel Luís, Recife** *reef* E Brazil
59 *I14* **Manuel Zinho** Pará, N Brazil 07°21′S 54°47′W
191 *V11* **Manuhagi** *prev.* Manuhangi. *atoll* Îles Tuamotu, C French Polynesia
Manuhangi *see* Manuhagi
185 *E22* **Manuherikia** ≈ South Island, New Zealand
171 *P13* **Manui, Pulau** *island* N Indonesia
Manukau *see* Manurewa
184 *L6* **Manukau Harbour** *harbor* North Island, New Zealand
191 *Z2* **Manulu Lagoon** ◎ Kiritimati, E Kiribati
182 *J7* **Manunda Creek** *seasonal river* South Australia
57 *K15* **Manupari, Río** ≈ N Bolivia
184 *L6* **Manurewa** *var.* Manukau. Auckland, North Island, New Zealand 37°01′S 174°55′E
57 *K15* **Manurimi, Río** ≈ NW Bolivia
186 *D5* **Manus** ◈ *province* N Papua New Guinea
186 *D5* **Manus Island** *var.* Great Admiralty Island. *island* N Papua New Guinea
171 *T16* **Manuwui** Pulau Babar, E Indonesia 07°47′S 129°39′E
29 *Q3* **Manvel** North Dakota, N USA 48°07′N 97°15′W
33 *Z14* **Manville** Wyoming, C USA 42°45′N 104°38′W
22 *G6* **Many** Louisiana, S USA 31°34′N 93°28′W
81 *H21* **Manyara, Lake** ◎ NE Tanzania
126 *L12* **Manych** *var.* Manich. ≈ SW Russian Federation
83 *H14* **Manyinga** North Western, NW Zambia 13°28′S 24°18′E
105 *O11* **Manzanares** Castilla-La Mancha, C Spain 39°N 03°23′W
44 *H7* **Manzanillo** Granma, E Cuba 20°21′N 77°07′W
40 *K14* **Manzanillo** Colima, SW Mexico 19°00′N 104°19′W
40 *K14* **Manzanillo, Bahía** *bay* SW Mexico
37 *S11* **Manzano Mountains** ▲▲ New Mexico, SW USA
37 *R12* **Manzano Peak** ▲ New Mexico, SW USA 34°35′N 106°27′W
163 *R6* **Manzhouli** *var.* Man-chou-li. Nei Mongol Zizhiqu, N China 49°36′N 117°28′E
Manzil Bū Ruqaybah *see* Menzel Bourguiba
139 *X9* **Manzilīyah** Maysān, E Iraq 32°26′N 47°01′E
83 *L21* **Manzini** *prev.* Bremersdorp. C Swaziland 26°30′S 31°22′E
83 *L21* **Manzini** ✈ (Mbabane) C Swaziland 26°36′S 31°25′E
78 *G10* **Mao** Kanem, W Chad 14°06′N 15°17′E
45 *N8* **Mao** NW Dominican Republic 19°37′N 71°04′W
105 *Z9* **Maó** *Cast.* Mahón, *Eng.* Port Mahon; *anc.* Portus Magonis. Menorca, Spain, W Mediterranean Sea 39°54′N 04°15′E
Maoemere *see* Maumere
159 *W9* **Maojing** Gansu, N China 36°26′N 106°36′E
171 *Y14* **Maoke, Pegunungan** *Dut.* Sneeuw-gebergte, *Eng.* Snow Mountains. ▲▲ Papua, E Indonesia
Maol Réidh, Caoc *see* Mweelrea
160 *M15* **Maoming** Guangdong, S China 21°46′N 110°51′E
160 *H8* **Maoxian** *var.* Mao Xian; *prev.* Fengyizhen. Sichuan, C China 31°42′N 103°48′E
Mao Xian *see* Maoxian
83 *L19* **Mapai** Gaza, SW Mozambique 22°52′S 32°00′E
158 *H15* **Mapam Yumco** ◎ W China
83 *I15* **Mapanza** Southern, S Zambia 16°16′S 26°54′E
54 *J4* **Mapararí** Falcón, N Venezuela 10°52′N 69°27′W
41 *U17* **Mapastepec** Chiapas, SE Mexico 15°28′N 93°00′W
169 *V9* **Mapat, Pulau** *island* N Indonesia
171 *Y15* **Mapi** Papua, E Indonesia 07°02′S 139°24′E
171 *V11* **Mapia, Kepulauan** *island group* E Indonesia
40 *L8* **Mapimí** Durango, C Mexico 25°50′N 103°50′W
83 *N19* **Mapinhane** Inhambane, SE Mozambique 22°14′S 35°07′E
55 *N7* **Mapire** Monagas, NE Venezuela 07°48′N 64°40′W
11 *S17* **Maple Creek** Saskatchewan, S Canada 49°55′N 109°28′W
31 *Q9* **Maple River** ≈ Michigan, N USA
29 *P7* **Maple River** ≈ North Dakota/South Dakota, N USA
29 *S13* **Mapleton** Iowa, C USA 42°10′N 95°47′W
29 *U10* **Mapleton** Minnesota, N USA 43°55′N 93°57′W
29 *R5* **Mapleton** North Dakota, N USA 46°51′N 97°04′W
32 *F13* **Mapleton** Oregon, NW USA 44°01′N 123°56′W

◆ Country
● Country Capital
◇ Dependent Territory
○ Dependent Territory Capital
◈ Administrative Regions
✈ International Airport
▲ Mountain
▲▲ Mountain Range
Ѧ Volcano
≈ River
◎ Lake
▣ Reservoir

Mapleton – Marysvale

36 L3 **Mapleton** Utah, W USA 40°07′N 111°37′W
192 K5 **Mapmaker Seamounts** *undersea feature* N Pacific Ocean 25°00′N 165°00′E
186 B6 **Maprik** East Sepik, NW Papua New Guinea 03°38′S 143°02′E
83 L21 **Maputo** *prev.* Lourenço Marques. ● (Mozambique) Maputo, S Mozambique 25°58′S 32°35′E
83 L21 **Maputo** ◆ *province* S Mozambique
67 V14 **Maputo** ~ S Mozambique
83 L21 **Maputo** ✈ Maputo, S Mozambique 25°47′S 32°36′E
Maqat *see* Makat
113 M19 **Maqellarë** Dibër, C Albania 41°36′N 20°29′E
159 S12 **Maqên** *var.* Dawo; *prev.* Dawu. Qinghai, C China 34°32′N 100°17′E
159 S11 **Maqên Kangri** ▲ C China 34°44′N 99°25′E
141 X7 **Maqiz al Kurbá** N Oman 24°13′N 56°48′E
159 U12 **Maqu** *var.* Nyinma. Gansu, C China 34°02′N 102°00′E
104 M9 **Maqueda** Castilla-La Mancha, C Spain 40°04′N 04°22′W
82 B9 **Maquela do Zombo** Uíge, NW Angola 06°06′S 15°12′E
63 I16 **Maquinchao** Río Negro, C Argentina 41°19′S 68°47′W
29 Z13 **Maquoketa** Iowa, C USA 42°03′N 90°42′W
29 Y13 **Maquoketa River** ~ Iowa, C USA
14 F13 **Mar** Ontario, S Canada 44°48′N 81°12′W
95 F14 **Mår** ⊛ S Norway
81 G19 **Mara** ◆ *region* N Tanzania
58 D12 **Maraã** Amazonas, NW Brazil 01°48′S 65°21′W
191 P8 **Maraa** Tahiti, W French Polynesia 17°44′S 149°34′W
191 O8 **Maraa, Pointe** *headland* Tahiti, W French Polynesia 17°44′S 149°34′W
59 K14 **Marabá** Pará, NE Brazil 05°23′S 49°10′W
54 H5 **Maracaibo** Zulia, NW Venezuela 10°40′N 71°39′W
Maracaibo, Gulf of *see* Venezuela, Golfo de
54 H5 **Maracaibo, Lago de** *var.* Lake Maracaibo. *inlet* NW Venezuela
Maracaibo, Lake *see* Maracaibo, Lago de
58 K10 **Maracá, Ilha de** *island* NE Brazil
59 H20 **Maracaju, Serra de** ▲▲ S Brazil
58 I11 **Maracanaquará, Planalto** ▲▲ NE Brazil
54 L5 **Maracay** Aragua, N Venezuela 10°15′N 67°36′W
Marada *see* Marādah
75 R9 **Marādah** *var.* Marada. N Libya 29°16′N 19°29′E
77 U12 **Maradi** Maradi, S Niger 13°30′N 07°05′E
77 U11 **Maradi** ◆ *department* S Niger
81 E21 **Maragarazi** *var.* Muragarazi. ~ Burundi/Tanzania
Maragha *see* Marāgheh
142 J3 **Marāgheh** *var.* Maragha. Āžarbāyjān-e Khāvarī, NW Iran 37°21′N 46°14′E
141 P7 **Marāh** *var.* Marrāt. Ar Riyāḑ, C Saudi Arabia 25°04′N 45°30′E
55 N11 **Marahuaca, Cerro** ▲ S Venezuela 03°37′N 65°25′W
27 R5 **Marais des Cygnes River** ~ Kansas/Missouri, C USA
58 L11 **Marajó, Baía de** *bay* N Brazil
59 K12 **Marajó, Ilha de** *island* N Brazil
191 O2 **Marakei** *atoll* Tungaru, W Kiribati
Marakesh *see* Marrakech
81 I18 **Maralal** Samburu, C Kenya 01°05′N 36°42′E
83 G21 **Maralaleng** Kgalagadi, S Botswana 25°42′S 22°39′E
145 U8 **Maraldy, Ozero** ⊛ NE Kazakhstan
182 C5 **Maralinga** South Australia 30°16′S 131°35′E
Máramarossziget *see* Sighetu Marmației
187 N9 **Maramasike** *var.* Small Malaita. *island* N Solomon Islands
Maramba *see* Livingstone
194 H3 **Marambio** *Argentinian research station* Antarctica 64°22′S 57°18′W
116 H9 **Maramureș** ◆ *county* NW Romania
36 L15 **Marana** Arizona, SW USA 32°24′N 111°12′W
105 P7 **Maranchón** Castilla-La Mancha, C Spain 41°02′N 02°11′W
142 J2 **Marand** *var.* Merend. Āžarbāyjān-e Sharqī, NW Iran 38°25′N 45°40′E
Marandellas *see* Marondera
58 L13 **Maranhão** *off.* Estado do Maranhão. ◆ *state* E Brazil
104 H10 **Maranhão, Barragem do** ⊠ C Portugal
Maranhão, Estado do *see* Maranhão
149 O11 **Mārān, Koh-i** ▲ SW Pakistan 29°24′N 66°50′E
106 J7 **Marano, Laguna di** *lagoon* NE Italy
56 E9 **Marañón, Río** ~ N Peru
102 J10 **Marans** Charente-Maritime, W France 46°19′N 00°58′W
83 M20 **Marão** Inhambane, S Mozambique 24°15′S 34°09′E
185 B23 **Mararoa** ~ South Island, New Zealand
Maraş/Marash *see* Kahramanmaraş
107 M19 **Maratea** Basilicata, S Italy 39°57′N 15°44′E
104 G11 **Marateca** Setúbal, S Portugal 38°34′N 08°40′W
115 B20 **Marathiá, Akrotírio** *headland* Zákynthos, Iónia Nisiá, Greece, C Mediterranean Sea 37°39′N 20°49′E
12 E12 **Marathon** Ontario, S Canada 48°44′N 86°23′W
23 Y17 **Marathon** Florida Keys, Florida, SE USA 24°42′N 81°05′W
24 L10 **Marathon** Texas, SW USA 30°10′N 103°14′W
Marathón *see* Marathónas
115 H19 **Marathónas** *prev.* Marathón. Attikí, C Greece 38°09′N 23°57′E
169 W9 **Maratua, Pulau** *island* N Indonesia
59 O18 **Maraú** Bahia, SE Brazil 14°07′S 39°02′W
143 R3 **Marāveh Tappeh** Golestán, N Iran 37°53′N 55°57′E
24 L11 **Maravillas Creek** ~ Texas, SW USA
186 D8 **Marawaka** Eastern Highlands, C Papua New Guinea 06°56′S 145°54′E
171 Q7 **Marawi** Mindanao, S Philippines 07°59′N 124°16′E
Märäzä *see* Qobustan
Marbat *see* Mirbāț
104 L16 **Marbella** Andalucía, S Spain 36°31′N 04°50′W
180 J7 **Marble Bar** Western Australia 21°13′S 119°48′E
36 L9 **Marble Canyon** *canyon* Arizona, SW USA
25 S10 **Marble Falls** Texas, SW USA 30°34′N 98°16′W
27 Y7 **Marble Hill** Missouri, C USA 37°18′N 89°58′W
33 T15 **Marbleton** Wyoming, C USA 42°31′N 110°06′W
Marburg *see* Marburg an der Lahn, Germany
Marburg *see* Maribor, Slovenia
101 H16 **Marburg an der Lahn** *hist.* Marburg. Hessen, W Germany 50°49′N 08°46′E
111 H23 **Marcal** ~ W Hungary
42 G7 **Marcala** La Paz, SW Honduras 14°11′N 88°00′W
111 H24 **Marcali** Somogy, SW Hungary 46°33′N 17°29′E
83 A16 **Marca, Ponta da** *headland* SW Angola 16°31′S 11°42′E
59 I16 **Marcelândia** Mato Grosso, W Brazil 11°18′S 54°49′W
27 T3 **Marceline** Missouri, C USA 39°42′N 92°57′W
60 I13 **Marcelino Ramos** Rio Grande do Sul, S Brazil 27°31′S 51°57′W
55 Y12 **Marcel, Mont** ▲ S French Guiana 02°32′N 53°00′W
97 O19 **March** E England, United Kingdom 52°37′N 00°13′E
109 Z3 **March** *var.* Morava. ~ C Europe *see also* Morava
March *see* Morava
106 I12 **Marche** *Eng.* Marches. ◆ *region* C Italy
103 N11 **Marche** *cultural region* C France
99 J21 **Marche-en-Famenne** Luxembourg, SE Belgium 50°13′N 05°21′E
104 K14 **Marchena** Andalucía, S Spain 37°20′N 05°24′W
57 B17 **Marchena, Isla** *var.* Bindloe Island. *island* Galapagos Islands, Ecuador, E Pacific Ocean
Marches *see* Marche
99 J20 **Marchin** Liège, E Belgium 50°30′N 05°17′E
181 S1 **Marchinbar Island** *island* Wessel Islands, Northern Territory, N Australia
62 L9 **Mar Chiquita, Laguna** ⊛ C Argentina
103 Q10 **Marcigny** Saône-et-Loire, C France 46°16′N 04°04′E
23 W16 **Marco** Florida, SE USA 25°56′N 81°43′W
Marcodurum *see* Düren
59 O15 **Marcolândia** Pernambuco, E Brazil 07°21′S 40°40′W
106 I8 **Marco Polo** ✈ (Venezia) Veneto, NE Italy 45°30′N 12°21′E
Marcounda *see* Markounda
Marcq *see* Mark
116 M8 **Mărculești** *Rus.* Markuleshty. N Moldova 47°54′N 28°14′E
29 S12 **Marcus** Iowa, C USA 42°49′N 95°48′W
39 S11 **Marcus Baker, Mount** ▲ Alaska, USA 61°26′N 147°45′W
192 I5 **Marcus Island** *var.* Minami Tori Shima. *island* E Japan
18 K8 **Marcy, Mount** ▲ New York, NE USA 44°06′N 73°55′W
149 T5 **Mardān** Khyber Pakhtunkhwa, N Pakistan 34°14′N 71°59′E
63 N14 **Mar del Plata** Buenos Aires, E Argentina 38°S 57°32′W
137 Q16 **Mardin** Mardin, SE Turkey 37°19′N 40°43′E
137 Q16 **Mardin** ◆ *province* SE Turkey
137 Q16 **Mardin Dağları** ▲▲ SE Turkey
Mardzad *see* Hayrhandulaan
187 R17 **Maré** *island* Îles Loyauté, E New Caledonia
Marea Neagră *see* Black Sea
105 Z8 **Mare de Déu del Toro** *var.* El Toro. ▲ Menorca, Spain, W Mediterranean Sea 39°59′N 04°06′E
181 W4 **Mareeba** Queensland, NE Australia 17°03′S 145°30′E
96 G8 **Maree, Loch** ⊛ N Scotland, United Kingdom
Mareeq *see* Mereeg
Marek *see* Dupnitsa
76 J11 **Maréna** Kayes, W Mali 14°36′N 10°57′W
190 I2 **Marenanuka** *atoll* Tungaru, W Kiribati
29 X14 **Marengo** Iowa, C USA 41°48′N 92°04′W
102 J11 **Marennes** Charente-Maritime, W France 45°47′N 01°04′W
107 G23 **Marettimo, Isola** *island* Isole Egadi, S Italy
24 K10 **Marfa** Texas, SW USA 30°19′N 104°03′W
57 P17 **Marfil, Laguna** ⊛ E Bolivia
Marganets *see* Marhanets'
25 Q4 **Margaret** Texas, SW USA 34°00′N 99°38′W
180 I14 **Margaret River** Western Australia 33°58′S 115°10′E
186 C7 **Margarima** Hela, W Papua New Guinea 06°00′S 143°23′E
55 N4 **Margarita, Isla de** *island* N Venezuela
115 I25 **Margarítes** Kríti, Greece, E Mediterranean Sea 35°19′N 24°40′E
97 Q22 **Margate** *prev.* Mergate. SE England, United Kingdom 51°24′N 01°24′E
23 Z15 **Margate** Florida, SE USA 26°14′N 80°12′W
Margelan *see* Marg'ilon
103 P13 **Margeride, Montagnes de la** ▲▲ C France
Margherita *see* Jamaame
107 N16 **Margherita di Savoia** Puglia, SE Italy 41°23′N 16°09′E
Margherita, Lake *see* Ābaya Hāyk'
81 E18 **Margherita Peak** *Fr.* Pic Marguerite. ▲ Uganda/Dem. Rep. Congo 0°28′N 29°58′E
149 O4 **Marghī** Bāmyān, N Afghanistan 35°10′N 66°26′E
116 G9 **Marghita** *Hung.* Margitta. Bihor, NW Romania 47°20′N 22°20′E
Margilan *see* Marg'ilon
147 S10 **Marg'ilon** *var.* Margelan, *Rus.* Margilan. Farg'ona Viloyati, E Uzbekistan 40°29′N 71°43′E
116 K8 **Marginea** Suceava, NE Romania 47°49′N 25°47′E
Margitta *see* Marghita
148 K9 **Māṛgow, Dasht-e** *desert* SW Afghanistan
99 L18 **Margraten** Limburg, SE Netherlands 50°49′N 05°49′E
10 M15 **Marguerite** British Columbia, SW Canada 52°17′N 122°10′W
15 V3 **Marguerite** ~ Quebec, SE Canada
194 I6 **Marguerite Bay** *bay* Antarctica
Marguerite, Pic *see* Margherita Peak
117 T9 **Marhanets'** *Rus.* Marganets. Dnipropetrovs'ka Oblast', E Ukraine 47°35′N 34°37′E
186 B9 **Mari** Western, SW Papua New Guinea 09°10′S 141°39′E
191 Y12 **Maria** *atoll* Groupe Actéon, SE French Polynesia
191 R12 **Maria** *island* Îles Australes, SW French Polynesia
40 I12 **María Cleofas, Isla** *island* C Mexico
62 H4 **María Elena** *var.* Oficina María Elena. Antofagasta, N Chile 22°18′S 69°40′W
95 G21 **Mariager** Midtjylland, C Denmark 56°39′N 09°59′E
61 C22 **María Ignacia** Buenos Aires, E Argentina 37°24′S 59°30′W
183 P17 **Maria Island** *island* Tasmania, SE Australia
40 H12 **María Madre, Isla** *island* C Mexico
40 I12 **María Magdalena, Isla** *island* C Mexico
192 H6 **Mariana Islands** *island group* Guam/Northern Mariana Islands
175 N3 **Mariana Trench** *var.* Challenger Deep. *undersea feature* W Pacific Ocean 15°00′N 147°30′E
153 X12 **Mariāni** Assam, NE India 26°39′N 94°18′E
27 X11 **Marianna** Arkansas, C USA 34°46′N 90°49′W
23 R8 **Marianna** Florida, SE USA 30°46′N 85°13′W
172 J16 **Marianne** *island* Inner Islands, NE Seychelles
95 M19 **Mariannelund** Jönköping, S Sweden 57°37′N 15°33′E
61 D15 **Mariano I. Loza** Corrientes, NE Argentina 29°22′S 58°12′W
Mariano Machado *see* Ganda
111 A16 **Mariánské Lázně** *Ger.* Marienbad. Karlovarský Kraj, W Czech Republic 49°57′N 12°43′E
Máriaradna *see* Radna
33 S7 **Marias River** ~ Montana, NW USA
Maria-Theresiopel *see* Subotica
Máriatölgyes *see* Dubnica nad Váhom
184 H1 **Maria van Diemen, Cape** *headland* North Island, New Zealand 34°27′S 172°38′E
109 V5 **Mariazell** Steiermark, E Austria 47°47′N 15°20′E
141 P15 **Ma'rib** W Yemen 15°28′N 45°25′E
95 I25 **Maribo** Sjælland, S Denmark 54°47′N 11°30′E
109 W9 **Maribor** *Ger.* Marburg. NE Slovenia 46°34′N 15°40′E
Marica *see* Maritsa
35 R13 **Maricopa** California, W USA 35°03′N 119°24′W
Maricourt *see* Kangiqsujuaq
81 D15 **Maridi** Western Equatoria, SW South Sudan 04°55′N 29°30′E
194 M11 **Marie Byrd Land** *physical region* Antarctica
193 P14 **Marie Byrd Seamount** *undersea feature* N Amundsen Sea 70°00′S 118°00′W
45 X11 **Marie-Galante** *var.* Ceyre to the Caribs. *island* SE Guadeloupe
45 Y6 **Marie-Galante, Canal de** *channel* S Guadeloupe
93 J20 **Mariehamn** *Fin.* Maarianhamina. Åland, SW Finland 60°05′N 19°55′E
44 C4 **Mariel** La Habana, W Cuba 23°02′N 82°44′W
99 H22 **Mariembourg** Namur, S Belgium 50°07′N 04°30′E
Marienbad *see* Mariánské Lázně
Marienburg *see* Alūksne, Latvia
Marienburg *see* Malbork, Poland
Marienburg *see* Feldioara, Romania
Marienburg in Westpreussen *see* Malbork
Marienhausen *see* Viļaka
83 D20 **Mariental** Hardap, SW Namibia 24°35′S 17°56′E
18 D13 **Marienville** Pennsylvania, NE USA 41°27′N 79°07′W
Marienwerder *see* Kwidzyń
58 C12 **Marié, Rio** ~ NW Brazil
95 K17 **Mariestad** Västra Götaland, S Sweden 58°42′N 13°50′E
23 S3 **Marietta** Georgia, SE USA 33°57′N 84°34′W
31 U14 **Marietta** Ohio, N USA 39°25′N 81°27′W
27 N13 **Marietta** Oklahoma, C USA 33°57′N 97°08′W
81 H18 **Marigat** Baringo, W Kenya 0°29′S 35°59′E
103 S16 **Marignane** Bouches-du-Rhône, SE France 43°25′N 05°12′E
Marignano *see* Melegnano
45 O11 **Marigot** NE Dominica 15°32′N 61°18′W
122 K12 **Mariinsk** Kemerovskaya Oblast', S Russian Federation 56°13′N 87°27′E
127 Q3 **Mariinskiy Posad** Respublika Mariy El, W Russian Federation 56°07′N 47°44′E
119 E14 **Marijampolė** *prev.* Kapsukas. Marijampolė, S Lithuania 54°33′N 23°21′E
Marikostenovo *see* Marikostinovo
114 G12 **Marikostinovo** *prev.* Marikostenovo. Blagoevgrad, SW Bulgaria 41°25′N 23°21′E
60 J9 **Marília** São Paulo, S Brazil 22°13′S 49°58′W
82 D11 **Marimba** Malanje, NW Angola 08°18′S 16°58′E
Marī Mīlā *see* Małe Mela
Mārī Mīlah *see* Małe Mela
104 G4 **Marín** Galicia, NW Spain 42°23′N 08°43′W
35 N10 **Marina** California, W USA 36°40′N 121°48′W
Mar'ina Gorka *see* Mar"ina Horka
119 L17 **Mar"ina Horka** *Rus.* Mar'ina Gorka. Minskaya Voblasts', C Belarus 53°31′N 28°09′E
171 O4 **Marinduque** *island* C Philippines
31 S9 **Marine City** Michigan, N USA 42°43′N 82°29′W
31 N6 **Marinette** Wisconsin, N USA 45°06′N 87°38′W
60 I10 **Maringá** Paraná, S Brazil 23°26′S 51°55′W
83 N16 **Maringuè** Sofala, C Mozambique 17°57′S 34°23′E
104 F9 **Marinha Grande** Leiria, C Portugal 39°45′N 08°55′W
107 I15 **Marino** Lazio, C Italy 41°46′N 12°40′E
59 A15 **Mário Lobão** Acre, W Brazil 08°21′S 72°58′W
23 O5 **Marion** Alabama, S USA 32°37′N 87°19′W
27 Y11 **Marion** Arkansas, C USA 35°12′N 90°12′W
30 L17 **Marion** Illinois, N USA 37°43′N 88°55′W
31 P13 **Marion** Indiana, N USA 40°32′N 85°40′W
29 X13 **Marion** Iowa, C USA 42°01′N 91°36′W
27 O5 **Marion** Kansas, C USA 38°22′N 97°02′W
20 H6 **Marion** Kentucky, S USA 37°19′N 88°06′W
21 P9 **Marion** North Carolina, SE USA 35°43′N 82°00′W
31 S12 **Marion** Ohio, N USA 40°35′N 83°08′W
21 T12 **Marion** South Carolina, SE USA 34°11′N 79°23′W
21 Q7 **Marion** Virginia, NE USA 36°51′N 81°30′W
27 O5 **Marion Lake** ⊠ Kansas, C USA
21 S13 **Marion, Lake** ⊠ South Carolina, SE USA
27 S8 **Marionville** Missouri, C USA 37°00′N 93°38′W
55 N7 **Maripa** Bolívar, E Venezuela 07°27′N 65°10′W
55 X11 **Maripasoula** W French Guiana 03°43′N 54°04′W
35 Q9 **Mariposa** California, W USA 37°28′N 119°59′W
61 G19 **Mariscala** Lavalleja, S Uruguay 34°03′S 54°47′W
62 M4 **Mariscal Estigarribia** Boquerón, NW Paraguay 22°03′S 60°39′W
56 C6 **Mariscal Sucre** *var.* Quito. ✈ (Quito) Pichincha, C Ecuador 0°21′S 78°37′W
30 K16 **Marissa** Illinois, N USA 38°15′N 89°45′W
103 U14 **Maritime Alps** *Fr.* Alpes Maritimes, *It.* Alpi Marittime. ▲▲ France/Italy
Maritimes, Alpes *see* Maritime Alps
Maritime Territory *see* Primorskiy Kray
114 K11 **Maritsa** *var.* Marica, *Gk.* Évros, *Turk.* Meriç; *anc.* Hebrus. ~ SW Europe *see also* Évros/Meriç
Maritsa *see* Simeonovgrad, Bulgaria
Marittime, Alpi *see* Maritime Alps
Maritzburg *see* Pietermaritzburg
117 X9 **Mariupol'** *prev.* Zhdanov. Donets'ka Oblast', SE Ukraine 47°06′N 37°34′E
55 Q6 **Mariusa, Caño** ~ NE Venezuela
142 J5 **Marīvān** *prev.* Dezh Shāhpūr. Kordestān, W Iran 35°30′N 46°09′E
127 R3 **Mariyets** Respublika Mariy El, W Russian Federation 56°31′N 49°48′E
118 G4 **Märjamaa** *Ger.* Merjama. Raplamaa, NW Estonia 58°54′N 24°21′E
99 I15 **Mark** *Fr.* Marcq. ~ Belgium/Netherlands
81 N17 **Marka** *var.* Merca. Shabeellaha Hoose, S Somalia 01°43′N 44°45′E
145 Z10 **Markakol', Ozero** *Kaz.* Marqaköl. ⊛ E Kazakhstan
76 M12 **Markala** Ségou, W Mali 13°38′N 06°07′W
159 S15 **Markam** *var.* Gartog. Xizang Zizhiqu, W China 29°40′N 98°33′E
95 K21 **Markaryd** Kronoberg, S Sweden 56°26′N 13°35′E
142 L7 **Markazī** *off.* Ostān-e Markazī. ◆ *province* W Iran
Markazī, Ostān-e *see* Markazī
14 F14 **Markdale** Ontario, S Canada 44°19′N 80°37′W
27 X10 **Marked Tree** Arkansas, C USA 35°31′N 90°25′W
98 N11 **Markelo** Overijssel, E Netherlands 52°15′N 06°30′E
98 J9 **Markermeer** ⊛ C Netherlands
97 N20 **Market Harborough** C England, United Kingdom 52°30′N 00°57′W
97 N18 **Market Rasen** E England, United Kingdom 53°23′N 00°21′W
123 O10 **Markha** ~ NE Russian Federation
12 H16 **Markham** Ontario, S Canada 43°54′N 79°16′W
25 V12 **Markham** Texas, SW USA 28°57′N 96°04′W
186 E7 **Markham** ~ C Papua New Guinea
195 Q11 **Markham, Mount** ▲ Antarctica 82°58′S 163°30′E
110 M11 **Marki** Mazowieckie, C Poland 52°20′N 21°07′E
158 F8 **Markit** Xinjiang Uygur Zizhiqu, NW China 38°55′N 77°40′E
117 Y5 **Markivka** *Rus.* Markovka. Luhans'ka Oblast', E Ukraine 49°34′N 39°35′E
35 Q7 **Markleeville** California, W USA 38°41′N 119°46′W
98 L8 **Marknesse** Flevoland, N Netherlands 52°44′N 05°54′E
79 H14 **Markounda** *var.* Marcounda. Ouham, NW Central African Republic 07°38′N 17°00′E
Markovka *see* Markivka
123 U7 **Markovo** Chukotskiy Avtonomnyy Okrug, NE Russian Federation 64°43′N 170°13′E
127 P8 **Marks** Saratovskaya Oblast', W Russian Federation 51°40′N 46°44′E
22 K2 **Marks** Mississippi, S USA 34°15′N 90°16′W
22 I7 **Marksville** Louisiana, S USA 31°07′N 92°04′W
101 I19 **Marktheidenfeld** Bayern, C Germany 49°50′N 09°36′E
101 J24 **Marktoberdorf** Bayern, S Germany 47°45′N 10°36′E
101 M18 **Marktredwitz** Bayern, E Germany 50°N 12°04′E
Markt-Übelbach *see* Übelbach
27 V3 **Mark Twain Lake** ⊠ Missouri, C USA
Markuleshty *see* Mărculești
101 E14 **Marl** Nordrhein-Westfalen, W Germany 51°38′N 07°06′E
182 E2 **Marla** South Australia 27°19′S 133°35′E
181 Y8 **Marlborough** Queensland, E Australia 22°55′S 150°07′E
97 M22 **Marlborough** S England, United Kingdom 51°25′N 01°45′W
185 I15 **Marlborough** *off.* Marlborough District. ◆ *unitary authority* South Island, New Zealand
Marlborough District *see* Marlborough
103 P3 **Marle** Aisne, N France 49°44′N 03°47′E
31 S8 **Marlette** Michigan, N USA 43°20′N 83°05′W
25 T9 **Marlin** Texas, SW USA 31°20′N 96°55′W
21 S5 **Marlinton** West Virginia, NE USA 38°14′N 80°06′W
26 M12 **Marlow** Oklahoma, C USA 34°39′N 97°57′W
155 E17 **Marmagao** Goa, W India 15°26′N 73°50′E
Marmanda *see* Marmande
102 L13 **Marmande** *anc.* Marmanda. Lot-et-Garonne, SW France 44°30′N 00°10′E
136 C11 **Marmara** Balıkesir, NW Turkey 40°36′N 27°34′E
136 D11 **Marmara Denizi** *Eng.* Sea of Marmara. *sea* NW Turkey
114 N13 **Marmaraereğlisi** Tekirdağ, NW Turkey 40°59′N 27°57′E
Marmara, Sea of *see* Marmara Denizi
136 C16 **Marmaris** Muğla, SW Turkey 36°52′N 28°17′E
28 J6 **Marmarth** North Dakota, N USA 46°17′N 103°55′W
21 Q5 **Marmet** West Virginia, NE USA 38°12′N 81°31′W
106 H5 **Marmolada, Monte** ▲ N Italy 46°36′N 11°58′E
104 M13 **Marmolejo** Andalucía, S Spain 38°03′N 04°10′W
14 J14 **Marmora** Ontario, SE Canada 44°29′N 77°40′W
39 Q14 **Marmot Bay** *bay* Alaska, USA
103 Q4 **Marne** ◆ *department* N France
103 Q4 **Marne** ~ N France
137 U10 **Marneuli** *prev.* Borchalo, Sarvani. S Georgia 41°28′N 44°45′E
78 I13 **Maro** Moyen-Chari, S Chad 08°25′N 18°46′E
54 L12 **Maroa** Amazonas, S Venezuela 02°40′N 67°33′W
172 J3 **Maroantsetra** Toamasina, NE Madagascar 15°23′S 49°44′E
191 W11 **Marokau** *atoll* Îles Tuamotu, C French Polynesia
172 J5 **Marolambo** Toamasina, E Madagascar 20°03′S 48°08′E
172 J2 **Maromokotro** ▲ N Madagascar
83 L16 **Marondera** *prev.* Marandellas. Mashonaland East, NE Zimbabwe 18°11′S 31°33′E
55 X9 **Maroni** *Dut.* Marowijne. ~ French Guiana/Suriname
183 V2 **Maroochydore-Mooloolaba** Queensland, E Australia 26°36′S 153°04′E
171 N14 **Maros** Sulawesi, C Indonesia 04°59′S 119°35′E
116 H11 **Maros** *var.* Mureș, Mureșul, *Ger.* Marosch, Mieresch. ~ Hungary/Romania *see also* Mureś
Marosch *see* Maros/Mureș
Maroshevíz *see* Toplița
Marosillye *see* Ilia
Marosludas *see* Luduș
Marosújvár/Marosújvárakna *see* Ocna Mureș
Marosvásárhely *see* Târgu Mureș
191 V14 **Marotiri** *var.* Îlots de Bass, Morotiri. *island group* Îles Australes, SW French Polynesia
78 G12 **Maroua** Extrême-Nord, N Cameroon 10°35′N 14°20′E
55 X12 **Marouini Rivier** ~ SE Suriname
172 I3 **Marovoay** Mahajanga, NW Madagascar 16°05′S 46°40′E
55 W9 **Marowijne** ◆ *district* NE Suriname
Marowijne *see* Maroni
Marqakö1 *see* Markakol', Ozero
193 P8 **Marquesas Fracture Zone** *tectonic feature* E Pacific Ocean
Marquesas Islands *see* Marquises, Îles
23 W17 **Marquesas Keys** *island group* Florida, SE USA
29 Y12 **Marquette** Iowa, C USA 43°02′N 91°10′W
31 N3 **Marquette** Michigan, N USA 46°32′N 87°24′W
103 N1 **Marquise** Pas-de-Calais, N France 50°49′N 01°42′E
191 X7 **Marquises, Îles** *Eng.* Marquesas Islands. *island group* N French Polynesia
183 Q6 **Marra Creek** ~ New South Wales, SE Australia
80 B10 **Marra Hills** *plateau* W Sudan
80 B11 **Marra, Jebel** ▲ W Sudan 12°59′N 24°16′E
74 E7 **Marrakech** *var.* Marakesh, *Eng.* Marrakesh; *prev.* Morocco. W Morocco 31°39′N 07°58′W
Marrakesh *see* Marrakech
Marrāt *see* Marāh
183 N15 **Marrawah** Tasmania, SE Australia 40°56′S 144°41′E
182 I4 **Marree** South Australia 29°40′S 138°06′E
81 L17 **Marrehan** ▲▲ SW Somalia
83 N17 **Marromeu** Sofala, C Mozambique 18°18′S 35°58′E
104 J17 **Marroquí, Punta** *headland* SW Spain 36°01′N 05°39′W
183 N8 **Marrowie Creek** *seasonal river* New South Wales, SE Australia
83 O14 **Marrupa** Niassa, N Mozambique 13°10′S 37°30′E
182 D1 **Marryat** South Australia 26°22′S 133°22′E
75 Y10 **Marsā al 'Alam** *var.* Marsa 'Alam. SE Egypt 25°03′N 33°44′E
Marsa 'Alam *see* Marsā al 'Alam
75 R8 **Marsá al Burayqah** *var.* Al Burayqah. N Libya 30°21′N 19°37′E
81 J17 **Marsabit** Marsabit, N Kenya 02°20′N 37°59′E
81 I17 **Marsabit** ◆ *county* N Kenya
107 H23 **Marsala** *anc.* Lilybaeum. Sicilia, Italy, C Mediterranean Sea 37°48′N 12°26′E
75 U7 **Marsā Maṭroūḥ** *var.* Maṭrûḥ; *anc.* Paraetonium. NW Egypt 31°21′N 27°15′E
Marsaxlokk Bay *see* Il-Bajja ta' Marsaxlokk
65 G15 **Mars Bay** *bay* Ascension Island, C Atlantic Ocean
101 H15 **Marsberg** Nordrhein-Westfalen, W Germany 51°28′N 08°51′E
11 R15 **Marsden** Saskatchewan, S Canada 52°50′N 109°45′W
98 H7 **Marsdiep** *strait* NW Netherlands
103 R16 **Marseille** *Eng.* Marseilles; *anc.* Massilia. Bouches-du-Rhône, SE France 43°19′N 05°22′E
Marseille-Marigrane *see* Provence
30 M11 **Marseilles** Illinois, N USA 41°19′N 88°42′W
Marseilles *see* Marseille
76 J16 **Marshall** W Liberia 06°10′N 10°23′W
39 N11 **Marshall** Alaska, USA 61°52′N 162°04′W
27 U9 **Marshall** Arkansas, C USA 35°54′N 92°40′W
31 N14 **Marshall** Illinois, N USA 39°23′N 87°41′W
31 Q10 **Marshall** Michigan, N USA 42°16′N 84°57′W
29 S9 **Marshall** Minnesota, N USA 44°26′N 95°48′W
27 T4 **Marshall** Missouri, C USA 39°07′N 93°12′W
21 O9 **Marshall** North Carolina, SE USA 35°48′N 82°43′W
25 X6 **Marshall** Texas, SW USA 32°33′N 94°22′W
189 S4 **Marshall Islands** *off.* Republic of the Marshall Islands. ◆ *republic* W Pacific Ocean
175 Q3 **Marshall Islands** *island group* W Pacific Ocean
Marshall Islands, Republic of the *see* Marshall Islands
192 K6 **Marshall Seamounts** *undersea feature* SW Pacific Ocean 10°00′N 165°00′E
29 W13 **Marshalltown** Iowa, C USA 42°01′N 92°54′W
19 P12 **Marshfield** Massachusetts, NE USA 42°04′N 70°40′W
27 T7 **Marshfield** Missouri, C USA 37°20′N 92°55′W
30 K6 **Marshfield** Wisconsin, N USA 44°41′N 90°12′W
44 H1 **Marsh Harbour** Great Abaco, W The Bahamas 26°31′N 77°03′W
19 S3 **Mars Hill** Maine, NE USA 46°31′N 67°51′W
21 P9 **Mars Hill** North Carolina, SE USA 35°49′N 82°33′W
22 H10 **Marsh Island** *island* Louisiana, S USA
21 S11 **Marshville** North Carolina, SE USA 34°59′N 80°22′W
15 W5 **Marsoui** Québec, SE Canada 49°12′N 65°58′W
15 R8 **Mars, Rivière à** ~ Québec, SE Canada
95 O15 **Märsta** Stockholm, C Sweden 59°37′N 17°52′E
95 H24 **Marstal** Syddtjylland, C Denmark 54°52′N 10°32′E
95 I19 **Marstrand** Västra Götaland, S Sweden 57°54′N 11°31′E
25 U8 **Mart** Texas, SW USA 31°32′N 96°49′W
Martaban *see* Mottama
Martaban, Gulf of *see* Mottama, Gulf of
107 Q19 **Martano** Puglia, SE Italy 40°12′N 18°19′E
Martapoera *see* Martapura
169 T13 **Martapura** *prev.* Martapoera. Borneo, C Indonesia 03°25′S 114°51′E
99 L23 **Martelange** Luxembourg, SE Belgium 49°50′N 05°43′E
114 L7 **Marten** Ruse, N Bulgaria 43°57′N 26°08′E
14 H10 **Marten River** Ontario, S Canada 46°43′N 79°45′W
11 T15 **Martensville** Saskatchewan, S Canada 52°15′N 106°42′W
Marteskirch *see* Târnăveni
Martes Tolosane *see* Martres-Tolosane
115 K25 **Mártha** Kríti, Greece, E Mediterranean Sea 35°03′N 25°22′E
183 Q6 **Marthaguy Creek** ~ New South Wales, SE Australia
19 P13 **Martha's Vineyard** *island* Massachusetts, NE USA
108 C11 **Martigny** Valais, SW Switzerland 46°06′N 07°04′E
103 R16 **Martigues** Bouches-du-Rhône, SE France 43°24′N 05°03′E
111 J19 **Martin** *Ger.* Sankt Martin, *Hung.* Turócszentmárton; *prev.* Turčianský Svätý Martin. Žilinský Kraj, N Slovakia 49°03′N 18°54′E
28 L11 **Martin** South Dakota, N USA 43°10′N 101°43′W
20 G8 **Martin** Tennessee, S USA 36°20′N 88°51′W
105 S7 **Martín** ~ E Spain
107 P18 **Martina Franca** Puglia, SE Italy 40°42′N 17°20′E
185 M14 **Martinborough** Wellington, North Island, New Zealand 41°12′S 175°28′E
25 S11 **Martindale** Texas, SW USA 29°49′N 97°49′W
35 N8 **Martinez** California, W USA 38°00′N 122°12′W
23 V3 **Martinez** Georgia, SE USA 33°31′N 82°04′W
41 Q13 **Martínez de La Torre** Veracruz-Llave, E Mexico 20°05′N 97°02′W
45 Y12 **Martinique** ◇ *French overseas department* E West Indies
1 O15 **Martinique** *island* E West Indies
Martinique Channel *see* Martinique Passage
45 X12 **Martinique Passage** *var.* Dominica Channel, Martinique Channel. *channel* Dominica/Martinique
23 Q5 **Martin Lake** ⊠ Alabama, S USA
115 G18 **Martíno** *prev.* Martínon. Stereá Elláda, C Greece 38°34′N 23°13′E
Martínon *see* Martíno
194 J11 **Martin Peninsula** *peninsula* Antarctica
39 S5 **Martin Point** *headland* Alaska, USA 70°06′N 143°04′W
109 V3 **Martinsberg** Niederösterreich, NE Austria 48°23′N 15°09′E
21 V3 **Martinsburg** West Virginia, NE USA 39°28′N 77°59′W
31 V13 **Martins Ferry** Ohio, N USA 40°06′N 80°43′W
Martinskirch *see* Târnăveni
31 O14 **Martinsville** Indiana, N USA 39°25′N 86°25′W
21 S8 **Martinsville** Virginia, NE USA 36°43′N 79°53′W
65 K16 **Martin Vaz, Ilhas** *island group* E Brazil
144 I9 **Martok** *prev.* Martuk. Aktyubinsk, NW Kazakhstan 50°45′N 56°30′E
184 M12 **Marton** Manawatu-Wanganui, North Island, New Zealand 40°05′S 175°22′E
105 N13 **Martos** Andalucía, S Spain 37°44′N 03°58′W
102 M16 **Martres-Tolosane** *var.* Martes Tolosane. Haute-Garonne, S France 43°13′N 01°00′E
92 M11 **Martti** Lappi, NE Finland 67°28′N 28°20′E
Martuk *see* Martok
137 U12 **Martuni** E Armenia 40°07′N 45°20′E
58 L11 **Marudá** Pará, E Brazil 05°25′S 49°04′W
169 V6 **Marudu, Teluk** *bay* East Malaysia
149 O8 **Ma'rūf** Kandahār, SE Afghanistan 31°34′N 67°06′E
164 H13 **Marugame** Kagawa, Shikoku, SW Japan 34°17′N 133°46′E
185 H16 **Maruia** ~ South Island, New Zealand
98 M6 **Marum** Groningen, NE Netherlands 53°07′N 06°16′E
187 R13 **Marum, Mount** ▲ Ambrym, C Vanuatu 16°15′S 168°07′E
79 P23 **Marungu** ▲▲ SE Dem. Rep. Congo
191 Y12 **Marutea** *atoll* Groupe Actéon, C French Polynesia
143 O11 **Marv Dasht** *var.* Mervdasht. Fārs, S Iran 29°50′N 52°40′E
103 P13 **Marvejols** Lozère, S France 44°35′N 03°16′E
27 X12 **Marvell** Arkansas, C USA 34°33′N 90°52′W
36 L6 **Marvine, Mount** ▲ Utah, W USA 38°40′N 111°38′W
139 Q7 **Marwānīyah** Al Anbār, C Iraq 33°58′N 42°31′E
152 F13 **Mārwār** *var.* Kharchi, Marwar Junction. Rājasthān, N India 25°41′N 73°42′E
Marwar Junction *see* Mārwār
11 R14 **Marwayne** Alberta, SW Canada 53°30′N 110°25′W
146 I14 **Mary** *prev.* Merv. Mary Welaýaty, S Turkmenistan 37°25′N 61°48′E
Mary *see* Mary Welaýaty
181 Z9 **Maryborough** Queensland, E Australia 25°32′S 152°36′E
182 M11 **Maryborough** Victoria, SE Australia 37°05′S 143°47′E
Maryborough *see* Port Laoise
83 G23 **Marydale** Northern Cape, W South Africa 29°25′S 22°06′E
117 W8 **Mar"yinka** Donets'ka Oblast', E Ukraine 47°57′N 37°27′E
Mary Island *see* Kanton
21 W4 **Maryland** *off.* State of Maryland, *also known as* America in Miniature, Cockade State, Free State, Old Line State. ◆ *state* NE USA
Maryland, State of *see* Maryland
25 P7 **Maryneal** Texas, SW USA 32°12′N 100°25′W
97 J15 **Maryport** NW England, United Kingdom 54°45′N 03°28′W
13 U13 **Marystown** Newfoundland, Newfoundland and Labrador, SE Canada 47°10′N 55°10′W
36 K6 **Marysvale** Utah, W USA 38°26′N 112°14′W

◆ Country
● Country Capital
◇ Dependent Territory
○ Dependent Territory Capital
◆ Administrative Regions
✈ International Airport
▲ Mountain
▲▲ Mountain Range
Volcano
~ River
⊛ Lake
⊠ Reservoir

MARYSVILLE – MBOUT

35 O6 **Marysville** California, W USA 39°07′N 121°35′W
27 O3 **Marysville** Kansas, C USA 39°48′N 96°37′W
31 S13 **Marysville** Michigan, N USA 42°54′N 82°29′W
31 S9 **Marysville** Ohio, NE USA 40°13′N 83°22′W
32 H7 **Marysville** Washington, NW USA 48°03′N 122°10′W
27 R2 **Maryville** Missouri, C USA 40°20′N 94°53′W
21 N9 **Maryville** Tennessee, S USA 35°45′N 83°59′W
146 I15 **Mary Welaýaty** *var.* Mary, *Rus.* Maryyskiy Velayat. ◐ *province* S Turkmenistan
Maryyskiy Velayat *see* Mary Welaýaty
Marzūq *see* Murzuq
42 J11 **Masachapa** *var.* Puerto Masachapa. Managua, W Nicaragua 11°47′N 86°31′W
81 G19 **Masai Mara National Reserve** *reserve* SW Kenya
81 I21 **Masai Steppe** *grassland* NW Tanzania
81 F19 **Masaka** SW Uganda 0°20′S 31°46′E
169 T15 **Masalembo Besar, Pulau** *island* S Indonesia
137 Y13 **Masallı** *Rus.* Masally. S Azerbaijan 39°03′N 48°39′E
Masally *see* Masallı
171 N13 **Masamba** Sulawesi, C Indonesia 02°33′S 120°20′E
Masampo *see* Masan
163 Y16 **Masan** *prev.* Masampo. S South Korea 35°11′N 128°36′E
Masandam Peninsula *see* Musandam Peninsula
81 J25 **Masasi** Mtwara, SE Tanzania 10°43′S 38°48′E
Masawa/Massawa *see* Mits'iwa
42 J10 **Masaya** Masaya, W Nicaragua 11°59′N 86°06′W
42 J10 **Masaya** ◐ *department* W Nicaragua
171 P5 **Masbate** Masbate, N Philippines 12°21′N 123°34′E
171 P5 **Masbate** *island* C Philippines
74 I6 **Mascara** *var.* Mouaskar. NW Algeria 35°20′N 00°09′E
173 O7 **Mascarene Basin** *undersea feature* W Indian Ocean 15°00′S 56°00′E
173 O9 **Mascarene Islands** *island group* W Indian Ocean
173 N9 **Mascarene Plain** *undersea feature* W Indian Ocean 19°00′S 52°00′E
173 O7 **Mascarene Plateau** *undersea feature* W Indian Ocean 10°00′S 60°00′E
194 H5 **Mascart, Cape** *headland* Adelaide Island, Antarctica
62 J10 **Mascasín, Salinas de** *salt lake* C Argentina
40 K13 **Mascota** Jalisco, C Mexico 20°31′N 104°46′W
15 O12 **Mascouche** Québec, SE Canada 45°46′N 73°37′W
124 J9 **Masel'gskaya** Respublika Kareliya, NW Russian Federation 63°09′N 34°22′E
83 J23 **Maseru** ● (Lesotho) W Lesotho 29°21′S 27°35′E
83 J23 **Maseru** ✈ W Lesotho 29°27′S 27°37′E
Mashaba *see* Mashava
160 K14 **Mashan** *var.* Baishan. Guangxi Zhuangzu Zizhiqu, S China 23°40′N 108°10′E
83 K17 **Mashava** *prev.* Mashaba. Masvingo, SE Zimbabwe 20°03′S 30°29′E
143 U4 **Mashhad** *var.* Meshed. Khorāsān-e Razavī, NE Iran 36°16′N 59°34′E
165 S3 **Mashike** Hokkaidō, NE Japan 43°51′N 141°30′E
83 K20 **Mashishing** *prev.* Lydenburg. Mpumalanga, NE South Africa 25°10′S 30°29′E
Mashīz *see* Bardsīr
149 N14 **Mashkai** ≈ SW Pakistan
143 X13 **Māshkel** *var.* Rūd-i Māshkel, Rūd-e Māshkīd. ≈ Iran/Pakistan
148 K12 **Māshkel, Hāmūn-i** *salt marsh* SW Pakistan
Māshkel, Rūd-i/Māshkīd, Rūd-e *see* Māshkel
83 K15 **Mashonaland Central** ◐ *province* N Zimbabwe
83 K16 **Mashonaland East** ◐ *province* NE Zimbabwe
83 J16 **Mashonaland West** ◐ *province* NW Zimbabwe
Mashtagi *see* Maştağa
141 S14 **Masīlah, Wādī al** *dry watercourse* SE Yemen
79 I21 **Masi-Manimba** Bandundu, SW Dem. Rep. Congo 04°47′S 17°54′E
81 F17 **Masindi** W Uganda 01°41′N 31°45′E
81 I19 **Masinga Reservoir** ⊞ S Kenya
Masīra *see* Maşīrah, Jazīrat
Masira, Gulf of *see* Maşīrah, Khalīj
141 Y10 **Maşīrah, Jazīrat** *var.* Masīra. *island* E Oman
141 Y10 **Maşīrah, Khalīj** *var.* Gulf of Masira. *bay* E Oman
Masis *see* Büyükağrı Dağı
79 O19 **Masisi** Nord-Kivu, E Dem. Rep. Congo 01°25′S 28°50′E
Masjed-e Soleymān *see* Masjed Soleymān
142 L9 **Masjed Soleymān** *var.* Masjed-e Soleymān, Masjid-i Sulaiman. Khūzestān, SW Iran 31°59′N 49°18′E
Masjid-i Sulaiman *see* Masjed Soleymān
Maskat *see* Masqaţ
139 Q7 **Maskhān** Al Anbār, C Iraq 33°41′N 42°46′E
141 X8 **Maskin** *var.* Miskin. NW Oman 23°28′N 56°46′E
97 B17 **Mask, Lough** *Ir.* Loch Measca. ◎ W Ireland
114 N10 **Maslen Nos** *headland* E Bulgaria 42°19′N 27°47′E
172 K3 **Masoala, Tanjona** *headland* NE Madagascar 15°59′N 50°13′E
Masohi *see* Amahai
31 Q9 **Mason** Michigan, N USA 42°33′N 84°25′W
31 R14 **Mason** Ohio, N USA 39°21′N 84°18′W
25 Q10 **Mason** Texas, SW USA 30°45′N 99°15′W
21 P4 **Mason** West Virginia, NE USA 39°01′N 82°01′W
185 B25 **Mason Bay** *bay* Stewart Island, New Zealand
30 K13 **Mason City** Illinois, N USA 40°12′N 89°42′W
29 V12 **Mason City** Iowa, C USA 43°09′N 93°12′W
18 B16 **Masontown** Pennsylvania, NE USA 39°49′N 79°53′W
141 Y8 **Masqaţ** *var.* Maskat, *Eng.* Muscat. ● (Oman) NE Oman 23°35′N 58°36′E
106 E10 **Massa** Toscana, C Italy 44°02′N 10°07′E
18 M11 **Massachusetts** *off.* Commonwealth of Massachusetts, *also known as* Bay State, Old Bay State, Old Colony State. ◐ *state* NE USA
19 P11 **Massachusetts Bay** *bay* Massachusetts, NE USA
35 R2 **Massacre Lake** ◎ Nevada, W USA
107 O18 **Massafra** Puglia, SE Italy 40°35′N 17°08′E
108 G11 **Massagno** Ticino, S Switzerland 46°01′N 08°55′E
78 G11 **Massaguet** Hadjer-Lamis, W Chad 12°28′N 15°26′E
Massakori *see* Massakory
78 G10 **Massakory** *var.* Massakori; *prev.* Dagana. Hadjer-Lamis, W Chad 13°02′N 15°43′E
78 H11 **Massalassef** Hadjer-Lamis, SW Chad 11°37′N 17°09′E
106 F13 **Massa Marittima** Toscana, C Italy 43°03′N 10°55′E
82 B11 **Massangano** Kwanza Norte, NW Angola 09°40′S 14°13′E
83 M18 **Massangena** Gaza, S Mozambique 21°34′S 32°57′E
80 K9 **Massawa Channel** *channel* E Eritrea
18 J6 **Massena** New York, NE USA 44°55′N 74°53′W
78 H11 **Massenya** Chari-Baguirmi, SW Chad 11°21′N 16°09′E
10 I13 **Masset** Graham Island, British Columbia, SW Canada 54°00′N 132°09′W
102 L16 **Masseube** Gers, S France 43°26′N 00°33′E
14 E11 **Massey** Ontario, S Canada 46°13′N 82°06′W
103 P12 **Massiac** Cantal, C France 45°16′N 03°13′E
103 P12 **Massif Central** *plateau* C France
Massif de L'Isalo *see* Isalo
Massilia *see* Marseille
31 U12 **Massillon** Ohio, N USA 40°48′N 81°31′W
77 N12 **Massina** Ségou, W Mali 13°58′N 05°24′W
83 N19 **Massinga** Inhambane, SE Mozambique 23°20′S 35°25′E
83 L20 **Massingir** Gaza, SW Mozambique 23°51′S 31°58′E
195 Z10 **Masson Island** *island* Antarctica
Massoukou *see* Franceville
137 Z11 **Maştağa** *Rus.* Mashtagi, Mastaga. E Azerbaijan 40°31′N 50°01′E
Mastanli *see* Momchilgrad
184 M13 **Masterton** Wellington, North Island, New Zealand 40°56′S 175°40′E
18 M14 **Mastic** Long Island, New York, NE USA 40°48′N 72°50′W
149 O10 **Mastung** Baluchistān, SW Pakistan 29°44′N 66°56′E
119 J20 **Mastva** *Rus.* Mostva. ≈ SW Belarus
119 G17 **Masty** *Rus.* Mosty. Hrodzyenskaya Voblasts', W Belarus 53°25′N 24°32′E
164 F12 **Masuda** Shimane, Honshū, SW Japan 34°40′N 131°50′E
92 J11 **Masugnsbyn** Norrbotten, N Sweden 67°28′N 22°01′E
Masuku *see* Franceville
83 K17 **Masvingo** *prev.* Fort Victoria, Nyanda, Victoria. Masvingo, SE Zimbabwe 20°05′S 30°50′E
83 K18 **Masvingo** *prev.* Victoria. ◐ *province* SE Zimbabwe
138 H5 **Maşyāf** *Fr.* Misiaf. Ḥamāh, C Syria 35°04′N 36°21′E
110 E9 **Maszewo** Zachodniopomorskie, NW Poland 53°29′N 15°01′E
83 I17 **Matabeleland North** ◐ *province* W Zimbabwe
83 J18 **Matabeleland South** ◐ *province* S Zimbabwe
82 O13 **Mataca** Niassa, N Mozambique 12°27′S 36°13′E
14 G8 **Matachewan** Ontario, S Canada 47°58′N 80°37′W
163 Q8 **Matad** *var.* Dzüünbulag. Dornod, E Mongolia 46°48′N 115°21′E
79 F22 **Matadi** Bas-Congo, W Dem. Rep. Congo 05°49′S 13°31′E
25 O4 **Matador** Texas, SW USA 34°01′N 100°50′W
42 J9 **Matagalpa** Matagalpa, C Nicaragua 12°53′N 85°56′W
42 K9 **Matagalpa** ◐ *department* W Nicaragua
12 I12 **Matagami** Québec, S Canada 49°47′N 77°38′W
25 U13 **Matagorda** Texas, SW USA 28°40′N 96°57′W
25 U13 **Matagorda Bay** *inlet* Texas, SW USA
25 U14 **Matagorda Island** *island* Texas, SW USA
25 V13 **Matagorda Peninsula** *headland* Texas, SW USA 28°34′N 96°01′W
191 Q8 **Mataiea** Tahiti, W French Polynesia 17°46′S 149°25′W
191 T9 **Mataiva** *atoll* Îles Tuamotu, C French Polynesia
183 O7 **Matakana** New South Wales, SE Australia 32°59′S 145°53′E
184 N7 **Matakana Island** *island* NE New Zealand
83 C15 **Matala** Huíla, SW Angola 14°45′S 15°02′E
190 G12 **Matala'a Pointe** *headland* Île Uvea, N Wallis and Futuna 13°20′S 176°08′W
155 K25 **Matale** Central Province, C Sri Lanka 07°29′N 80°38′E
190 E12 **Matalesina, Pointe** *headland* Île Alofi, W Wallis and Futuna
76 I10 **Matam** NE Senegal 15°40′N 13°18′W
184 M8 **Matamata** Waikato, North Island, New Zealand 37°49′S 175°45′E
77 V12 **Matamey** Zinder, S Niger 13°27′N 08°27′E
40 L8 **Matamoros** Coahuila, NE Mexico 25°34′N 103°13′W
41 P15 **Matamoros** *var.* Izúcar de Matamoros. Puebla, S Mexico 18°38′N 98°30′W
41 Q8 **Matamoros** Tamaulipas, C Mexico 25°50′N 97°31′W
75 S13 **Ma'ţan as Sārah** SE Libya 21°45′N 21°55′E
82 J12 **Matanda** Luapula, N Zambia 11°24′S 28°25′E
81 J24 **Matandu** ≈ S Tanzania
15 V6 **Matane** Québec, SE Canada 48°50′N 67°31′W
15 V6 **Matane** ≈ Québec, SE Canada
77 S12 **Matankari** Dosso, SW Niger 13°39′N 04°03′E
39 R11 **Matanuska River** ≈ Alaska, USA
54 G7 **Matanza** Santander, N Colombia 07°22′N 73°02′W
44 D4 **Matanzas** Matanzas, NW Cuba 23°N 81°32′W
15 V7 **Matapédia** ≈ Québec, SE Canada
15 V6 **Matapédia, Lac** ◎ Québec, SE Canada
190 B17 **Mata Point** *headland* SE Niue 19°07′S 169°51′E
190 D12 **Matapu, Pointe** *headland* Île Futuna, W Wallis and Futuna
62 G12 **Mataquito, Río** ≈ C Chile
155 K26 **Matara** Southern Province, S Sri Lanka 05°58′N 80°33′E
115 D18 **Matarágka** *var.* Mataránga. Dytikí Elláda, C Greece 38°32′N 21°28′E
171 K16 **Mataram** Pulau Lombok, C Indonesia 08°36′S 116°07′E
Mataránga *see* Matarágka
181 Q3 **Mataranka** Northern Territory, N Australia 14°55′S 133°03′E
105 W6 **Mataró** *anc.* Illuro. Cataluña, E Spain 41°32′N 02°27′E
184 O8 **Matata** Bay of Plenty, North Island, New Zealand 37°54′S 176°45′E
192 K16 **Matātula, Cape** *headland* Tutuila, W American Samoa 14°15′S 170°35′W
185 D24 **Mataura** Southland, South Island, New Zealand 46°12′S 168°53′E
185 D24 **Mataura** ≈ South Island, New Zealand
Mata Uta *see* Matā'utu
192 H16 **Matāutu** Upolu, C Samoa 13°57′S 171°55′W
190 G11 **Matā'utu** *var.* Mata Uta. ○ (Wallis and Futuna) Île Uvea, Wallis and Futuna 13°22′S 176°12′W
190 G12 **Matā'utu, Baie de** *bay* Île Uvea, Wallis and Futuna
191 P7 **Matavai, Baie de** *bay* Tahiti, W French Polynesia
190 I16 **Matavera** Rarotonga, S Cook Islands 21°13′S 159°44′W
191 V16 **Mataveri** Easter Island, Chile, E Pacific Ocean 27°10′S 109°27′W
191 V17 **Mataveri** ✈ (Easter Island) Easter Island, Chile, E Pacific Ocean 27°10′S 109°27′W
184 P9 **Matawai** Gisborne, North Island, New Zealand 38°23′S 177°31′E
15 O10 **Matawin** ≈ Québec, SE Canada
145 V13 **Matay** Almaty, SE Kazakhstan 45°53′N 78°45′E
14 K8 **Matchi-Manitou, Lac** ◎ Québec, SE Canada
41 O10 **Matehuala** San Luis Potosí, C Mexico 23°40′N 100°40′W
45 V13 **Matelot** Trinidad, Trinidad and Tobago 10°48′N 61°06′W
83 M15 **Matenge** Tete, NW Mozambique 15°22′S 33°47′E
107 O18 **Matera** Basilicata, S Italy 40°39′N 16°35′E
111 O21 **Mátészalka** Szabolcs-Szatmár-Bereg, E Hungary 47°58′N 22°17′E
93 H17 **Matfors** Västernorrland, C Sweden 62°21′N 17°02′E
102 K11 **Matha** Charente-Maritime, W France 45°50′N 00°13′W
0 F15 **Mathematicians Seamounts** *undersea feature* E Pacific Ocean 15°00′N 111°00′W
21 X6 **Mathews** Virginia, NE USA 37°26′N 76°20′W
25 S14 **Mathis** Texas, SW USA 28°05′N 97°49′W
152 J11 **Mathura** *prev.* Muttra. Uttar Pradesh, N India 27°30′N 77°42′E
Mathurai *see* Madurai
171 R7 **Mati** Mindanao, S Philippines 06°58′N 126°11′E
Matianus *see* Orūmīyeh, Daryācheh-ye
Matiara *see* Matiāri
149 Q15 **Matiāri** *var.* Matiara. Sind, SE Pakistan 25°38′N 68°29′E
41 S16 **Matías Romero** Oaxaca, SE Mexico 16°53′N 95°02′W
43 O13 **Matina** Limón, E Costa Rica 10°06′N 83°18′W
14 D10 **Matinenda Lake** ◎ Ontario, S Canada
19 R8 **Matinicus Island** *island* Maine, NE USA
Matisco/Matisco Ædourum *see* Mâcon
113 K19 **Matit, Lumi i** ≈ NW Albania
149 Q16 **Mātli** Sind, SE Pakistan 25°06′N 68°37′E
97 M18 **Matlock** C England, United Kingdom 53°08′N 01°32′W
59 F18 **Mato Grosso** *prev.* Vila Bela da Santissima Trindade. Mato Grosso, W Brazil 14°53′S 59°58′W
59 G17 **Mato Grosso** *off.* Estado de Mato Grosso; *prev.* Matto Grosso. ◐ *state* W Brazil
60 H8 **Mato Grosso do Sul** *off.* Estado de Mato Grosso do Sul. ◐ *state* S Brazil
Mato Grosso do Sul, Estado de *see* Mato Grosso do Sul
Mato Grosso, Estado de *see* Mato Grosso
59 I18 **Mato Grosso, Planalto de** *plateau* C Brazil
83 L21 **Matola** Maputo, S Mozambique 25°57′S 32°27′E
104 G6 **Matosinhos** *prev.* Matozinhos. Porto, NW Portugal 41°11′N 08°42′W
Matou *see* Pingguo
55 Z10 **Matoury** NE French Guiana 04°49′N 52°17′W
Matozinhos *see* Matosinhos
111 L21 **Mátra** ▲▲ N Hungary
141 Y8 **Maţraḩ** *var.* Mutrah. NE Oman 23°35′N 58°31′E
116 L12 **Mătraşeşti** Vrancea, E Romania 45°53′N 27°14′E
108 M8 **Matrei am Brenner** Tirol, W Austria 47°09′N 11°28′E
109 P8 **Matrei in Osttirol** Tirol, W Austria 47°04′N 12°31′E
76 I15 **Matru** SW Sierra Leone 07°37′N 12°08′W
Maţrûḩ *see* Marsā Maţroūḩ
165 U16 **Matsubara** *var.* Matubara. Kagoshima, Tokuno-shima, SW Japan 32°58′N 129°56′E
161 S12 **Matsu Dao** *var.* Mazu Tao; *prev.* Matsu Tao. *island* NW Taiwan
164 G12 **Matsue** *var.* Matsuye, Matue. Shimane, Honshū, SW Japan 35°27′N 133°04′E
165 Q6 **Matsumae** Hokkaidō, NE Japan 41°27′N 140°04′E
164 M12 **Matsumoto** *var.* Matumoto. Nagano, Honshū, S Japan 36°18′N 137°58′E
164 K14 **Matsusaka** *var.* Matsuzaka, Matusaka. Mie, Honshū, SW Japan 34°33′N 136°31′E
Matsu Tao *see* Matsu Dao
Matsutō *see* Mattō
164 F14 **Matsuyama** *var.* Matuyama. Ehime, Shikoku, SW Japan 33°50′N 132°47′E
Matsuye *see* Matsue
Matsuzaka *see* Matsusaka
164 M14 **Matsuzaki** Shizuoka, Honshū, S Japan 34°43′N 138°45′E
14 F8 **Mattagami** ≈ Ontario, S Canada
14 F8 **Mattagami Lake** ◎ Ontario, S Canada
62 K12 **Mattaldi** Córdoba, C Argentina 34°26′S 64°14′W
21 Y9 **Mattamuskeet, Lake** ◎ North Carolina, SE USA
21 W6 **Mattaponi River** ≈ Virginia, NE USA
14 I11 **Mattawa** Ontario, SE Canada 46°19′N 78°42′W
14 I11 **Mattawa** ≈ Ontario, SE Canada
19 S5 **Mattawamkeag** Maine, NE USA 45°30′N 68°20′W
19 S4 **Mattawamkeag Lake** ◎ Maine, NE USA
108 D11 **Matterhorn** *It.* Monte Cervino. ▲ Italy/Switzerland 45°58′N 07°36′E *see also* Cervino, Monte
32 L12 **Matterhorn** *var.* Sacajawea Peak. ▲ Oregon, NW USA 45°12′N 117°18′W
35 W1 **Matterhorn** ▲ Nevada, W USA 41°48′N 115°22′W
Matterhorn *see* Cervino, Monte
35 R8 **Matterhorn Peak** ▲ California, W USA 38°06′N 119°19′W
109 Y5 **Mattersburg** Burgenland, E Austria 47°45′N 16°24′E
108 E11 **Matter Vispa** ≈ S Switzerland
55 R7 **Matthews Ridge** N Guyana 07°30′N 60°07′W
44 K7 **Matthew Town** Great Inagua, S The Bahamas 20°56′N 73°41′W
109 Q4 **Mattighofen** Oberösterreich, NW Austria 48°07′N 13°09′E
107 N16 **Mattinata** Puglia, SE Italy 41°41′N 16°01′E
141 T9 **Maţţī, Sabkhat** *salt flat* Saudi Arabia/United Arab Emirates
18 M14 **Mattituck** Long Island, New York, NE USA 40°59′N 72°31′W
164 L11 **Mattō** *var.* Hakusan, Matsutō. Ishikawa, Honshū, SW Japan 36°31′N 136°34′E
Matto Grosso *see* Mato Grosso
30 M14 **Mattoon** Illinois, N USA 39°28′N 88°22′W
57 L16 **Mattos, Río** ≈ C Bolivia
Mattu *see* Metu
169 R9 **Matu** Sarawak, East Malaysia 02°39′N 111°31′E
Matubara *see* Matsubara
57 E14 **Matucana** Lima, W Peru 11°54′S 76°25′W
Matue *see* Matsue
187 Y15 **Matuku** *island* S Fiji
112 B9 **Matulji** Primorje-Gorski Kotar, NW Croatia 45°21′N 14°18′E
Matumoto *see* Matsumoto
55 P5 **Maturín** Monagas, NE Venezuela 09°45′N 63°10′W
Matusaka *see* Matsusaka
Matuyama *see* Matsuyama
126 K11 **Matveyev Kurgan** Rostovskaya Oblast', SW Russian Federation 47°31′N 38°55′E
127 O8 **Matyshevo** Volgogradskaya Oblast', SW Russian Federation 50°53′N 44°09′E
153 O13 **Mau** *var.* Maunāth Bhanjan. Uttar Pradesh, N India 25°57′N 83°33′E
83 O14 **Maúa** Niassa, N Mozambique 13°53′S 37°10′E
102 M17 **Maubermé, Pic de** *var.* Tuc de Moubermé, *Sp.* Pico Mauberme; *prev.* Tuc de Maubermé. ▲ France/Spain 42°48′N 00°54′E *see also* Moubermé, Tuc de
Maubermé, Pic de *see* Moubermé, Tuc de
Mauberme, Pico *see* Maubermé, Pic de/Moubermé, Tuc de
Maubermé, Tuc de *see* Maubermé, Pic de/Moubermé, Tuc de
103 Q2 **Maubeuge** Nord, N France 50°16′N 04°00′E
166 L8 **Maubin** Ayeyawady, SW Myanmar (Burma) 16°44′N 95°37′E
152 L13 **Maudaha** Uttar Pradesh, N India 25°41′N 80°07′E
183 N9 **Maude** New South Wales, SE Australia 34°30′S 144°20′E
195 P3 **Maudheimvidda** *physical region* Antarctica
65 N22 **Maud Rise** *undersea feature* S Atlantic Ocean
109 Q4 **Mauerkirchen** Oberösterreich, NW Austria 48°11′N 13°08′E
Mauersee *see* Mamry, Jezioro
188 K2 **Maug Islands** *island group* N Northern Mariana Islands
103 Q15 **Mauguio** Hérault, S France 43°37′N 04°01′E
193 N5 **Maui** *island* Hawai'i, USA, C Pacific Ocean
190 M16 **Mauke** *atoll* S Cook Islands
62 G13 **Maule** *var.* Región del Maule. ◐ *region* C Chile
102 J9 **Mauléon** Deux-Sèvres, W France 46°55′N 00°45′W
102 J16 **Mauléon-Licharre** Pyrénées-Atlantiques, SW France 43°14′N 00°51′W
Maule, Región del *see* Maule
62 G13 **Maule, Río** ≈ C Chile
63 G17 **Maullín** Los Lagos, S Chile 41°38′S 73°35′W
31 R11 **Maumee** Ohio, N USA 41°34′N 83°40′W
31 Q12 **Maumee River** ≈ Indiana/Ohio, N USA
27 U11 **Maumelle** Arkansas, C USA 34°51′N 92°24′W
27 T11 **Maumelle, Lake** ⊞ Arkansas, C USA
171 O16 **Maumere** *prev.* Maoemere. Flores, S Indonesia 08°35′S 122°13′E
83 G17 **Maun** North-West, C Botswana 20°01′S 23°28′E
Maunāth Bhanjan *see* Mau
Maunawai *see* Waimea
190 H16 **Maungaroa** ▲ Rarotonga, S Cook Islands 21°13′S 159°48′W
184 K3 **Maungatapere** Northland, North Island, New Zealand 35°46′S 174°10′E
184 K4 **Maungaturoto** Northland, North Island, New Zealand 36°06′S 174°21′E
166 J5 **Maungdaw** *var.* Zullapara. Rakhine State, W Myanmar (Burma) 20°51′N 92°23′E
191 R10 **Maupiti** *var.* Maurua. *island* Îles Sous le Vent, W French Polynesia
152 K14 **Mau Rānīpur** Uttar Pradesh, N India 25°14′N 79°07′E
22 K9 **Maurepas, Lake** ◎ Louisiana, S USA
103 T16 **Maures** ▲▲ SE France
103 O12 **Mauriac** Cantal, C France 45°13′N 02°21′E
Maurice *see* Mauritius
65 J20 **Maurice Ewing Bank** *undersea feature* SW Atlantic Ocean 51°00′S 43°00′W
182 C4 **Maurice, Lake** *salt lake* South Australia
18 I17 **Maurice River** ≈ New Jersey, NE USA
25 Y10 **Mauriceville** Texas, SW USA 30°13′N 93°52′W
98 K12 **Maurik** Gelderland, C Netherlands 51°57′N 05°25′E
76 H8 **Mauritania** *off.* Islamic Republic of Mauritania, *Ar.* Mūrītānīyah. ◆ *republic* W Africa
Mauritania, Islamic Republic of *see* Mauritania
173 W15 **Mauritius** *off.* Republic of Mauritius, *Fr.* Maurice. ◆ *republic* W Indian Ocean
128 M17 **Mauritius** *island* W Indian Ocean
Mauritius, Republic of *see* Mauritius
173 N9 **Mauritius Trench** *undersea feature* W Indian Ocean
102 H6 **Mauron** Morbihan, NW France 48°06′N 02°16′W
103 N13 **Maurs** Cantal, C France 44°45′N 02°12′E
Maurua *see* Maupiti
Maury Mid-Ocean Channel *see* Maury Seachannel
64 L6 **Maury Seachannel** *var.* Maury Mid-Ocean Channel. *undersea feature* N Atlantic Ocean 56°33′N 24°00′W
30 K8 **Mauston** Wisconsin, N USA 43°46′N 90°06′W
109 R8 **Mauterndorf** Salzburg, NW Austria 47°09′N 13°39′E
109 T4 **Mauthausen** Oberösterreich, N Austria 48°13′N 14°30′E
109 Q9 **Mauthen** Kärnten, S Austria 46°39′N 12°58′E
83 F15 **Mavinga** Kuando Kubango, SE Angola 15°44′S 20°21′E
83 M17 **Mavita** Manica, W Mozambique 19°31′S 33°09′E
115 K22 **Mavrópetra, Akrotírio** *headland* Santoríni, Kykládes, Greece, Aegean Sea 36°28′N 25°22′E
115 F16 **Mavrovoúni** ▲ C Greece 39°37′N 22°45′E
184 Q8 **Mawhai Point** *headland* North Island, New Zealand 38°08′S 178°24′E
166 L3 **Mawlaik** Sagaing, C Myanmar (Burma) 23°40′N 94°26′E
Mawlamyaing *see* Mawlamyine
166 M9 **Mawlamyine** *var.* Mawlamyaing, Moulmein. Mon State, S Myanmar (Burma) 16°30′N 97°39′E
166 L8 **Mawlamyinegyunn** *var.* Moulmeingyun. Ayeyawady, SW Myanmar (Burma) 16°24′N 95°15′E
141 N14 **Mawr, Wādī** *dry watercourse* NW Yemen
Mawşil, Al *see* Nīnawá
195 X5 **Mawson** *Australian research station* Antarctica 67°24′S 63°16′E
195 X5 **Mawson Coast** *physical region* Antarctica
28 M4 **Max** North Dakota, N USA 47°48′N 101°18′W
41 W12 **Maxcanú** Yucatán, SE Mexico 20°35′N 90°00′W
Maxesibebi *see* Mount Ayliff
109 Q5 **Maxglan** ✈ (Salzburg) Salzburg, W Austria 47°46′N 13°00′E
93 K16 **Maxmo** *Fin.* Maksamaa. Österbotten, W Finland 63°13′N 22°04′E
21 T11 **Maxton** North Carolina, SE USA 34°47′N 79°34′W
25 R8 **May** Texas, SW USA 31°58′N 98°54′W
186 B6 **May** ≈ NW Papua New Guinea
123 R10 **Maya** ≈ E Russian Federation
151 Q19 **Māyābandar** Andaman and Nicobar Islands, India, E Indian Ocean 12°43′N 92°52′E
Mayadin *see* Al Mayādīn
44 L5 **Mayaguana** *island* SE The Bahamas
44 L5 **Mayaguana Passage** *passage* SE The Bahamas
45 S6 **Mayagüez** W Puerto Rico 18°12′N 67°08′W
45 R6 **Mayagüez, Bahía de** *bay* W Puerto Rico
Mayals *see* Maials
79 G20 **Mayama** Pool, SE Congo 03°50′S 14°52′E
37 V8 **Maya, Mesa De** ▲ Colorado, C USA 37°06′N 103°30′W
143 R4 **Mayamey** Semnān, N Iran 36°50′N 55°50′E
42 F3 **Maya Mountains** *Sp.* Montañas Mayas. ▲▲ Belize/Guatemala
44 I7 **Mayarí** Holguín, E Cuba 20°41′N 75°42′W
Mayas, Montañas *see* Maya Mountains
18 I17 **May, Cape** *headland* New Jersey, NE USA 38°55′N 74°57′W
80 J11 **Maych'ew** *var.* Mai Chio, *It.* Mai Ceu. Tigray, N Ethiopia 12°55′N 39°30′E
138 I2 **Maydān Ikbiz** Ḥalab, N Syria 36°51′N 36°40′E
Maydān Shahr *see* Maīdān Shahr
80 O12 **Maydh** Sanaag, N Somalia 10°57′N 47°07′E
Maydī *see* Mīdī
Mayebashi *see* Maebashi
Mayence *see* Mainz
102 K6 **Mayenne** Mayenne, NW France 48°18′N 00°37′W
102 J6 **Mayenne** ◐ *department* NW France
102 J7 **Mayenne** ≈ N France
36 K12 **Mayer** Arizona, SW USA 34°25′N 112°15′W
22 J4 **Mayersville** Mississippi, S USA 32°54′N 91°04′W
11 P14 **Mayerthorpe** Alberta, SW Canada 53°59′N 115°06′W
21 S12 **Mayesville** South Carolina, SE USA 34°00′N 80°10′W
185 G19 **Mayfield** Canterbury, South Island, New Zealand 43°50′S 171°24′E
33 N14 **Mayfield** Idaho, NW USA 43°24′N 115°56′W
20 G7 **Mayfield** Kentucky, S USA 36°45′N 88°40′W
36 L5 **Mayfield** Utah, W USA 39°06′N 111°42′W
Mayhan *see* Sant
37 T14 **Mayhill** New Mexico, SW USA 32°52′N 105°28′W
Maykain *see* Maykayyn
145 T9 **Maykayyn** *prev.* Maykain *Kaz.* Mayqayyng. Pavlodar, NE Kazakhstan 51°27′N 75°52′E
126 L14 **Maykop** Respublika Adygeya, SW Russian Federation 44°36′N 40°07′E
Maylibash *see* Maylybas
Mayli-Say *see* Mayluu-Suu
147 T9 **Mayluu-Suu** *prev.* Mayli-Say, *Kir.* Mayly-Say. Dzhalal-Abadskaya Oblast', W Kyrgyzstan 41°16′N 72°27′E
144 L14 **Maylybas** *prev.* Maylibash. Kzylorda, S Kazakhstan 45°51′N 62°37′E
Mayly-Say *see* Mayluu-Suu
Maymana *see* Maīmanah
Maymyo *see* Pyin-Oo-Lwin
123 V7 **Mayn** ≈ NE Russian Federation
127 Q5 **Mayna** Ul'yanovskaya Oblast', W Russian Federation 54°04′N 47°20′E
21 N8 **Maynardville** Tennessee, S USA 36°15′N 83°48′W
14 J13 **Maynooth** Ontario, SE Canada 45°14′N 77°54′W
10 I6 **Mayo** Yukon, NW Canada 63°37′N 135°48′W
23 U9 **Mayo** Florida, SE USA 30°03′N 83°10′W
97 B16 **Mayo** *Ir.* Maigh Eo. *cultural region* W Ireland
Mayo *see* Maio
78 G12 **Mayo-Kébbi Est** *off.* Région du Mayo-Kébbi Est. ◐ *region* SW Chad
Mayo-Kébbi Est, Région du *see* Mayo-Kébbi Est
78 G13 **Mayo-Kébbi Ouest** *off.* Région du mayo-Kébbi Ouest. ◐ *region* SW Chad
Mayo-Kébbi Ouest, Région du *see* Mayo-Kébbi Ouest
79 F19 **Mayoko** Niari, SW Congo 02°19′S 12°47′E
171 P4 **Mayon Volcano** ✶ Luzon, N Philippines 13°15′N 123°41′E
61 A24 **Mayor Buratovich** Buenos Aires, E Argentina 39°15′S 62°35′W
104 L4 **Mayorga** Castilla y León, N Spain 42°10′N 05°16′W
184 N6 **Mayor Island** *island* NE New Zealand
Mayor Pablo Lagerenza *see* Capitán Pablo Lagerenza
173 I14 **Mayotte** ◇ *French overseas department* E Africa
Mayoumba *see* Mayumba
44 J13 **May Pen** C Jamaica 17°58′N 77°15′W
Mayqayyng *see* Maykayyn
171 O1 **Mayraira Point** *headland* Luzon, N Philippines 18°36′N 120°47′E
109 N8 **Mayrhofen** Tirol, W Austria 47°09′N 11°52′E
186 A6 **May River** East Sepik, NW Papua New Guinea 04°24′S 141°52′E
139 Y10 **Maysān** *off.* Muḩāfa at Maysān, *var.* Al 'Amārah, Mīsān. ◐ *governorate* SE Iraq
Maysān, Muḩāfa at *see* Maysān
123 R13 **Mayskiy** Amurskaya Oblast', SE Russian Federation 52°13′N 129°30′E
127 O15 **Mayskiy** Kabardino-Balkarskaya Respublika, SW Russian Federation 43°37′N 44°04′E
145 U9 **Mayskoye** Pavlodar, NE Kazakhstan 50°55′N 78°11′E
18 J17 **Mays Landing** New Jersey, NE USA 39°27′N 74°44′W
21 N4 **Maysville** Kentucky, S USA 38°38′N 83°46′W
27 R2 **Maysville** Missouri, C USA 39°53′N 94°21′W
79 D20 **Mayumba** *var.* Mayoumba. Nyanga, S Gabon 03°23′S 10°38′E
31 S8 **Mayville** Michigan, N USA 43°18′N 83°16′W
18 C11 **Mayville** New York, NE USA 42°15′N 79°32′W
29 Q4 **Mayville** North Dakota, N USA 47°27′N 97°17′W
Mayyali *see* Mahe
Mayyit, Al Baḩr al *see* Dead Sea
83 J15 **Mazabuka** Southern, S Zambia 15°52′S 27°46′E
Mazaca *see* Kayseri
Mazagan *see* El-Jadida
32 J7 **Mazama** Washington, NW USA 48°34′N 120°26′W
103 O15 **Mazamet** Tarn, S France 43°30′N 02°21′E
143 O4 **Māzandarān** *off.* Ostān-e Māzandarān. ◐ *province* N Iran
Māzandarān, Ostān-e *see* Māzandarān
156 F7 **Mazar** Xinjiang Uygur Zizhiqu, NW China 36°28′N 77°00′E
107 H24 **Mazara del Vallo** Sicilia, Italy, C Mediterranean Sea 37°39′N 12°36′E
149 O2 **Mazār-e Sharīf** *var.* Mazār-i Sharif. Balkh, N Afghanistan 36°44′N 67°06′E
Mazār-i Sharif *see* Mazār-e Sharīf
105 R13 **Mazarrón** Murcia, SE Spain 37°36′N 01°19′W
105 R14 **Mazarrón, Golfo de** *gulf* SE Spain
55 S9 **Mazaruni River** ≈ N Guyana
42 B6 **Mazatenango** Suchitepéquez, SW Guatemala 14°31′N 91°30′W
40 I10 **Mazatlán** Sinaloa, C Mexico 23°15′N 106°24′W
36 L12 **Mazatzal Mountains** ▲▲ Arizona, SW USA
118 D10 **Mažeikiai** Telšiai, NW Lithuania 56°19′N 22°22′E
118 D7 **Mazirbe** NW Latvia 57°39′N 22°16′E
40 G5 **Mazocahui** Sonora, NW Mexico 29°32′N 110°09′W
57 I18 **Mazocruz** Puno, S Peru 16°41′S 69°42′W
Mazoe, Rio *see* Mazowe
79 N21 **Mazomeno** Maniema, E Dem. Rep. Congo 04°54′S 27°13′E
159 Q6 **Mazong Shan** ▲ N China 41°40′N 97°10′E
83 L16 **Mazowe** *var.* Rio Mazoe. ≈ Mozambique/Zimbabwe
110 M11 **Mazowieckie** ◐ *province* C Poland
Mazra'a *see* Al Mazra'ah
138 G6 **Mazraat Kfar Debiâne** C Lebanon 34°00′N 35°51′E
118 H7 **Mazsalaca** *Est.* Väike-Salatsi, *Ger.* Salisburg. N Latvia 57°52′N 25°03′E
110 L9 **Mazury** *physical region* NE Poland
Mazu Tao *see* Matsu Dao
119 M20 **Mazyr** *Rus.* Mozyr'. Homyel'skaya Voblasts', SE Belarus 52°04′N 29°15′E
107 K25 **Mazzarino** Sicilia, Italy, C Mediterranean Sea 37°18′N 14°13′E
Mba *see* Ba
83 L21 **Mbabane** ● (Swaziland) NW Swaziland 26°24′S 31°13′E
Mbacké *see* Mbaké
77 N16 **Mbahiakro** E Ivory Coast 07°33′N 04°19′W
79 I16 **Mbaïki** *var.* M'Baiki. Lobaye, SW Central African Republic 03°52′N 17°58′E
M'Baiki *see* Mbaïki
79 F14 **Mbakaou, Lac de** ◎ C Cameroon
76 G11 **Mbaké** *var.* Mbacké. W Senegal 14°47′N 15°54′W
82 L11 **Mbala** *prev.* Abercorn. Northern, NE Zambia 08°50′S 31°23′E
83 J18 **Mbalabala** *prev.* Balla Balla. Matabeleland South, SW Zimbabwe 20°27′S 29°03′E
81 G18 **Mbale** E Uganda 01°04′N 34°12′E
79 E16 **Mbalmayo** *var.* M'Balmayo. Centre, S Cameroon 03°30′N 11°31′E
M'Balmayo *see* Mbalmayo
81 H25 **Mbamba Bay** Ruvuma, S Tanzania 11°15′S 34°44′E
79 I18 **Mbandaka** *prev.* Coquilhatville. Equateur, NW Dem. Rep. Congo 0°07′N 18°12′E
82 B9 **M'banza Kongo** Zaire Province, NW Angola 06°11′S 14°16′E
79 G21 **Mbanza-Ngungu** Bas-Congo, W Dem. Rep. Congo 05°19′S 14°45′E
67 V11 **Mbarangandu** ≈ E Tanzania
81 E19 **Mbarara** SW Uganda 0°36′S 30°40′E
79 L15 **Mbari** ≈ SE Central African Republic
81 I24 **Mbarika Mountains** ▲▲ S Tanzania
78 F13 **Mbé** Nord, N Cameroon 07°51′N 13°36′E
81 J24 **Mbemkuru** *var.* Mbwemkuru. ≈ S Tanzania
Mbengga *see* Beqa
172 H13 **Mbéni** Grande Comore, NW Comoros
83 K18 **Mberengwa** Midlands, S Zimbabwe 20°29′S 29°55′E
81 G24 **Mbeya** Mbeya, SW Tanzania 08°54′S 33°29′E
81 G23 **Mbeya** ◐ *region* S Tanzania
83 J24 **Mbhashe** *prev.* Mbashe. ≈ S South Africa
79 E19 **Mbigou** Ngounié, C Gabon 01°54′S 12°00′E
Mbilua *see* Vella Lavella
79 F19 **Mbinda** Niari, SW Congo 02°07′S 12°52′E
79 D17 **Mbini** W Equatorial Guinea 01°34′N 09°39′E
Mbini *see* Uolo, Río
83 L18 **Mbizi** Masvingo, SE Zimbabwe 21°23′S 30°54′E
81 G23 **Mbogo** Mbeya, W Tanzania 07°24′S 33°26′E
79 N15 **Mboki** Haut-Mbomou, SE Central African Republic 05°18′N 25°52′E
83 K21 **Mbombela** *prev.* Nelspruit. Mpumalanga, NE South Africa 25°28′S 30°58′E *see also* Nelspruit
79 G18 **Mbomo** Cuvette, NW Congo 0°25′N 14°42′E
79 L15 **Mbomou** ◐ *prefecture* SE Central African Republic
Mbomou/M'Bomu/Mbomu *see* Bomu
76 F11 **Mbour** W Senegal 14°22′N 16°54′W
76 I10 **Mbout** Gorgol, S Mauritania 16°02′N 12°38′W

◆ Country
● Country Capital
◇ Dependent Territory
○ Dependent Territory Capital
◐ Administrative Regions
✈ International Airport
▲ Mountain
▲▲ Mountain Range
✶ Volcano
≈ River
◎ Lake
⊞ Reservoir

Mbrès – Mercara

79 J14 **Mbrès** var. Mbrés. Nana-Grébizi, C Central African Republic 06°40′N 19°46′E
Mbrés see Mbrès
79 L22 **Mbuji-Mayi** prev. Bakwanga. Kasai-Oriental, S Dem. Rep. Congo 06°05′S 23°30′E
81 H21 **Mbulu** Manyara, N Tanzania 03°45′S 35°33′E
186 E5 **M'bunai** var. Bunai. Manus Island, N Papua New Guinea 02°08′S 147°13′E
62 N8 **Mburucuyá** Corrientes, NE Argentina 28°03′S 58°15′W
Mbutha see Buca
Mbwemkuru see Mbemkuru
81 G21 **Mbwikwe** Singida, C Tanzania 05°19′S 34°09′E
13 O15 **McAdam** New Brunswick, SE Canada 45°34′N 67°20′W
25 O5 **McAdoo** Texas, SW USA 33°41′N 100°58′W
35 V2 **McAfee Peak** ▲ Nevada, W USA 41°31′N 115°57′W
27 P11 **McAlester** Oklahoma, C USA 34°56′N 95°46′W
25 S17 **McAllen** Texas, SW USA 26°12′N 98°14′W
21 S11 **McBee** South Carolina, SE USA 34°30′N 80°12′W
11 N14 **McBride** British Columbia, SW Canada 53°21′N 120°19′W
24 M9 **McCamey** Texas, SW USA 31°08′N 102°13′W
33 R15 **McCammon** Idaho, NW USA 42°38′N 112°10′W
35 X11 **McCarran** ✈ (Las Vegas) Nevada, W USA 36°04′N 115°07′W
39 T11 **McCarthy** Alaska, USA 61°25′N 142°55′W
30 M5 **McCaslin Mountain** hill Wisconsin, N USA
25 O2 **McClellan Creek** ≈ Texas, SW USA
21 T14 **McClellanville** South Carolina, SE USA 33°07′N 79°27′W
195 R12 **McClintock, Mount** ▲ Antarctica 80°09′S 156°42′E
35 N2 **McCloud** California, W USA 41°15′N 122°09′W
35 N3 **McCloud River** ≈ California, W USA
35 Q9 **McClure, Lake** ⊠ California, W USA
197 O8 **McClure Strait** strait Northwest Territories, N Canada
29 N4 **McClusky** North Dakota, N USA 47°27′N 100°25′W
21 T11 **McColl** South Carolina, SE USA 34°40′N 79°33′W
22 K7 **McComb** Mississippi, S USA 31°14′N 90°27′W
18 E16 **McConnellsburg** Pennsylvania, NE USA 39°56′N 78°00′W
31 T14 **McConnelsville** Ohio, N USA 39°39′N 81°51′W
28 M17 **McCook** Nebraska, C USA 40°12′N 100°38′W
21 P13 **McCormick** South Carolina, SE USA 33°55′N 82°19′W
11 W16 **McCreary** Manitoba, S Canada 50°48′N 99°34′W
27 W11 **McCrory** Arkansas, C USA 35°15′N 91°12′W
25 T10 **McDade** Texas, SW USA 30°15′N 97°15′W
23 O8 **McDavid** Florida, SE USA 30°51′N 87°18′W
35 T1 **McDermitt** Nevada, W USA 41°57′N 117°43′W
23 S4 **McDonough** Georgia, SE USA 33°27′N 84°09′W
36 L12 **McDowell Mountains** ▲▲ Arizona, SW USA
20 H8 **McEwen** Tennessee, S USA 36°06′N 87°37′W
35 R12 **McFarland** California, W USA 35°41′N 119°14′W
Mcfarlane, Lake see Macfarlane, Lake
27 P12 **McGee Creek Lake** ⊠ Oklahoma, C USA
27 W13 **McGehee** Arkansas, C USA 33°37′N 91°24′W
35 X5 **McGill** Nevada, W USA 39°24′N 114°46′W
14 K11 **McGillivray, Lac** ◎ Québec, SE Canada
39 P10 **McGrath** Alaska, USA 62°57′N 155°36′W
25 T8 **McGregor** Texas, SW USA 31°26′N 97°24′W
33 N11 **McGuire, Mount** ▲ Idaho, NW USA 45°10′N 114°36′W
83 M14 **Mchinji** prev. Fort Manning. Central, W Malawi 13°48′S 32°55′E
28 M7 **McIntosh** South Dakota, N USA 45°56′N 101°21′W
9 S7 **McKeand** ≈ Baffin Island, Nunavut, NE Canada
191 R4 **McKean Island** island Phoenix Islands, C Kiribati
30 J13 **McKee Creek** ≈ Illinois, N USA
18 C15 **Mckeesport** Pennsylvania, NE USA 40°18′N 79°48′W
21 V7 **McKenney** Virginia, NE USA 36°57′N 77°42′W
20 G8 **McKenzie** Tennessee, S USA 36°07′N 88°31′W
185 B20 **McKerrow, Lake** ◎ South Island, New Zealand
39 Q10 **McKinley, Mount** var. Denali. ▲ Alaska, USA 63°04′N 151°00′W
39 R10 **McKinley Park** Alaska, USA 63°42′N 149°01′W
34 K3 **McKinleyville** California, W USA 40°56′N 124°06′W
25 U6 **McKinney** Texas, SW USA 33°14′N 96°37′W
26 I5 **McKinney, Lake** ⊠ Kansas, C USA
28 M7 **McLaughlin** South Dakota, N USA 45°48′N 100°48′W
25 O2 **McLean** Texas, SW USA 35°13′N 100°36′W
30 M16 **Mcleansboro** Illinois, N USA 38°05′N 88°32′W
11 O13 **McLennan** Alberta, W Canada 55°42′N 116°50′W
14 L9 **McLennan, Lac** ◎ Québec, SE Canada
10 M13 **McLeod Lake** British Columbia, W Canada 55°03′N 123°02′W
8 L6 **M'Clintock Channel** channel Nunavut, N Canada
27 N10 **McLoud** Oklahoma, C USA 35°26′N 97°05′W
32 G15 **McLoughlin, Mount** ▲ Oregon, NW USA 42°27′N 122°18′W
37 U15 **McMillan, Lake** ⊠ New Mexico, SW USA
32 G11 **McMinnville** Oregon, NW USA 45°14′N 123°12′W
20 K9 **McMinnville** Tennessee, S USA 35°40′N 85°49′W
195 R13 **McMurdo** US research station Antarctica 77°40′S 167°16′E
37 N13 **McNary** Arizona, SW USA 34°04′N 109°51′W
24 H9 **McNary** Texas, SW USA 31°15′N 105°46′W
27 N5 **McPherson** Kansas, C USA 38°22′N 97°41′W
McPherson see Fort McPherson
23 U6 **McRae** Georgia, SE USA 32°04′N 82°54′W
29 P4 **McVille** North Dakota, N USA 47°45′N 98°10′W
83 J25 **Mdantsane** Eastern Cape, SE South Africa 32°55′S 27°39′E
167 T6 **Me** Ninh Binh, N Vietnam 20°21′N 105°49′E
26 J7 **Meade** Kansas, C USA 37°17′N 100°21′W
39 O5 **Meade River** ≈ Alaska, USA
35 Y11 **Mead, Lake** ⊠ Arizona/Nevada, W USA
24 M5 **Meadow** Texas, SW USA 33°20′N 102°12′W
11 S14 **Meadow Lake** Saskatchewan, C Canada 54°90′N 108°30′W
35 Y10 **Meadow Valley Wash** ≈ Nevada, W USA
22 J7 **Meadville** Mississippi, S USA 31°28′N 90°51′W
18 B12 **Meadville** Pennsylvania, NE USA 41°38′N 80°09′W
14 F14 **Meaford** Ontario, S Canada 44°35′N 80°35′W
Meáin, Inis see Inishmaan
104 G8 **Mealhada** Aveiro, N Portugal 40°22′N 08°27′W
13 R8 **Mealy Mountains** ▲▲ Newfoundland and Labrador, E Canada
11 O10 **Meander River** Alberta, W Canada 59°02′N 117°42′W
32 E11 **Meares, Cape** headland Oregon, NW USA 45°29′N 123°59′W
47 V6 **Mearim, Rio** ≈ NE Brazil
Measca, Loch see Mask, Lough
97 F17 **Meath** Ir. An Mhí. cultural region E Ireland
11 T14 **Meath Park** Saskatchewan, S Canada 53°25′N 105°18′W
103 O5 **Meaux** Seine-et-Marne, N France 48°47′N 02°54′E
21 T9 **Mebane** North Carolina, SE USA 36°06′N 79°16′W
171 U12 **Mebo, Gunung** ▲ Papua Barat, E Indonesia 01°10′S 133°53′E
94 I8 **Mebonden** Sør-Trøndelag, S Norway 63°13′N 11°00′E
82 A10 **Mebridege** ≈ NW Angola
35 W16 **Mecca** California, W USA 33°34′N 116°04′W
Mecca see Makkah
29 Y14 **Mechanicsville** Iowa, C USA 41°54′N 91°15′W
18 L10 **Mechanicville** New York, NE USA 42°54′N 73°41′W
99 H17 **Mechelen** Eng. Mechlin, Fr. Malines. Antwerpen, C Belgium 51°02′N 04°29′E
188 C8 **Mecherchar** var. Eil Malk. island Palau Islands, Palau
101 D17 **Mechernich** Nordrhein-Westfalen, W Germany 50°36′N 06°39′E
126 L12 **Mechetinskaya** Rostovskaya Oblast', SW Russian Federation 46°46′N 40°30′E
114 J11 **Mechka** ≈ S Bulgaria
Mechlin see Mechelen
61 D23 **Mechongué** Buenos Aires, E Argentina 38°09′S 58°13′W
115 L14 **Mecidiye** Edirne, NW Turkey 40°39′N 26°33′E
101 I24 **Meckenbeuren** Baden-Württemberg, S Germany 47°42′N 09°34′E
100 L8 **Mecklenburger Bucht** bay N Germany
100 M10 **Mecklenburgische Seenplatte** wetland NE Germany
100 L9 **Mecklenburg-Vorpommern** ◆ state NE Germany
83 Q15 **Meconta** Nampula, NE Mozambique 15°01′S 39°52′E
111 I25 **Mecsek** ▲▲ SW Hungary
83 P14 **Mecubúri** ≈ N Mozambique
83 Q14 **Mecúfi** Cabo Delgado, NE Mozambique 13°20′S 40°32′E
82 O13 **Mecula** Niassa, N Mozambique 12°03′S 37°37′E
168 I8 **Medan** Sumatera, E Indonesia 03°35′N 98°39′E
61 A24 **Médanos** var. Medanos. Buenos Aires, E Argentina 38°52′S 62°45′W
61 C19 **Médanos** Entre Ríos, E Argentina 33°28′S 59°07′W
155 K24 **Medawachchiya** North Central Province, N Sri Lanka 08°32′N 80°30′E
106 C8 **Mede** Lombardia, N Italy 45°06′N 08°43′E
74 J5 **Médéa** var. El Mediyya, Lemdiyya. N Algeria 36°15′N 02°48′E
Medeba see Mādabā
54 E8 **Medellín** Antioquia, NW Colombia 06°15′N 75°36′W
100 H9 **Medem** ≈ NW Germany
98 J8 **Medemblik** Noord-Holland, NW Netherlands 52°47′N 05°06′E
75 N7 **Médenine** var. Madanīyīn. SE Tunisia 33°23′N 10°30′E
76 G9 **Mederdra** Trarza, SW Mauritania 16°56′N 15°40′W
Medeshamstede see Peterborough
42 F4 **Medesto Mendez** Izabal, NE Guatemala 15°54′N 89°13′W
19 O11 **Medford** Massachusetts, NE USA 42°25′N 71°08′W
27 N8 **Medford** Oklahoma, C USA 36°48′N 97°45′W
32 G15 **Medford** Oregon, NW USA 42°20′N 122°52′W
30 K6 **Medford** Wisconsin, N USA 45°08′N 90°22′W
39 P10 **Medfra** Alaska, USA 63°06′N 154°42′W
116 M14 **Medgidia** Constanța, SE Romania 44°15′N 28°16′E
Medgyes see Mediaş
43 O5 **Media Luna, Arrecifes de la** reef E Honduras
60 G11 **Medianeira** Paraná, S Brazil 25°15′S 54°07′W
29 Y15 **Mediapolis** Iowa, C USA 41°00′N 91°09′W
116 I11 **Mediaş** Ger. Mediasch, Hung. Medgyes. Sibiu, C Romania 46°10′N 24°20′E
41 S15 **Medias Aguas** Veracruz-Llave, SE Mexico 17°40′N 95°02′W
Mediasch see Mediaş
106 G10 **Medicina** Emilia-Romagna, C Italy 44°29′N 11°41′E
33 X16 **Medicine Bow** Wyoming, C USA 41°52′N 106°11′W
37 S2 **Medicine Bow Mountains** ▲▲ Colorado/Wyoming, C USA
33 X16 **Medicine Bow River** ≈ Wyoming, C USA
11 R17 **Medicine Hat** Alberta, SW Canada 50°03′N 110°41′W
26 L7 **Medicine Lodge** Kansas, C USA 37°18′N 98°35′W
26 L7 **Medicine Lodge River** ≈ Kansas/Oklahoma, C USA
112 E7 **Međimurje** off. Međimurska Županija. ◆ province N Croatia
Međimurska Županija see Međimurje
54 G10 **Medina** Cundinamarca, C Colombia 04°31′N 73°21′W
18 E9 **Medina** New York, NE USA 43°13′N 78°23′W
29 O5 **Medina** North Dakota, N USA 46°53′N 99°18′W
31 T11 **Medina** Ohio, N USA 41°08′N 81°51′W
25 Q11 **Medina** Texas, SW USA 29°46′N 99°14′W
Medina see Al Madīnah
105 P6 **Medinaceli** Castilla y León, N Spain 41°10′N 02°26′W
104 L6 **Medina del Campo** Castilla y León, N Spain 41°18′N 04°55′W
104 L5 **Medina de Ríoseco** Castilla y León, N Spain 41°53′N 05°03′W
Médina Gonassé see Médina Gounas
76 H12 **Médina Gounas** var. Médina Gonassé. S Senegal 13°06′N 13°49′W
25 S12 **Medina River** ≈ Texas, SW USA
104 K16 **Medina Sidonia** Andalucía, S Spain 36°28′N 05°55′W
Medinat Israel see Israel
119 H14 **Medininkai** Vilnius, SE Lithuania 54°32′N 25°40′E
153 R16 **Medinīpur** West Bengal, NE India 22°25′N 87°24′E
Mediolanum see Saintes, France
Mediolanum see Milano, Italy
Mediomatrica see Metz
121 Q11 **Mediterranean Ridge** undersea feature C Mediterranean Sea 34°00′N 23°00′E
121 O16 **Mediterranean Sea** Fr. Mer Méditerranée. sea Africa/Asia/Europe
Méditerranée, Mer see Mediterranean Sea
79 N17 **Medje** Orientale, NE Dem. Rep. Congo 02°27′N 27°14′E
Medjerda, Oued see Mejerda
114 G7 **Medkovets** Montana, NW Bulgaria 43°39′N 23°22′E
93 J15 **Medle** Västerbotten, N Sweden 64°45′N 20°45′E
127 W7 **Mednogorsk** Orenburgskaya Oblast', W Russian Federation 51°24′N 57°37′E
123 W9 **Mednyy, Ostrov** island E Russian Federation
102 J12 **Médoc** cultural region SW France
159 Q16 **Mêdog** Xizang Zizhiqu, W China 29°26′N 95°26′E
28 J5 **Medora** North Dakota, N USA 46°56′N 103°40′W
79 E17 **Médouneu** Woleu-Ntem, N Gabon 0°58′N 10°50′E
106 I7 **Meduna** ≈ NE Italy
Medunta see Mantes-la-Jolie
Medvedica see Medveditsa
124 J16 **Medveditsa** var. Medvedica. ≈ W Russian Federation
127 O9 **Medveditsa** ≈ SW Russian Federation
112 E8 **Medvednica** ▲▲ NE Croatia
125 R15 **Medvedok** Kirovskaya Oblast', NW Russian Federation 57°23′N 50°01′E
123 S6 **Medvezh'i, Ostrova** island group NE Russian Federation
124 J9 **Medvezh'yegorsk** Respublika Kareliya, NW Russian Federation 62°56′N 34°26′E
109 T11 **Medvode** Ger. Zwischenwässern. NW Slovenia 46°09′N 14°21′E
126 J4 **Medyn'** Kaluzhskaya Oblast', W Russian Federation 54°59′N 35°52′E
180 J10 **Meekatharra** Western Australia 26°37′S 118°35′E
37 Q4 **Meeker** Colorado, C USA 40°02′N 107°54′W
13 T12 **Meelpaeg Lake** ◎ Newfoundland, Newfoundland and Labrador, E Canada
Meemu Atoll see Mulakatholhu
Meenen see Menen
101 M16 **Meerane** Sachsen, E Germany 50°50′N 12°28′E
101 D15 **Meerbusch** Nordrhein-Westfalen, W Germany 51°19′N 06°43′E
98 I12 **Meerkerk** Zuid-Holland, C Netherlands 51°55′N 05°00′E
99 L18 **Meerssen** var. Mersen. Limburg, SE Netherlands 50°53′N 05°45′E
152 J10 **Meerut** Uttar Pradesh, N India 29°01′N 77°41′E
33 U13 **Meeteetse** Wyoming, C USA 44°10′N 108°53′W
99 K17 **Meeuwen** Limburg, NE Belgium 51°04′N 05°36′E
81 J16 **Mēga** Oromīya, C Ethiopia 04°03′N 38°15′E
81 J16 **Mēga Escarpment** escarpment S Ethiopia
Megála Kalívia see Megála Kalývia
115 E16 **Megála Kalývia** var. Megála Kalívia. Thessalía, C Greece 39°30′N 21°48′E
115 H14 **Megáli Panagiá** var. Megáli Panayía. Kentrikí Makedonía, N Greece 40°24′N 23°42′E
Megáli Panayía see Megáli Panagiá
Megáli Préspa, Límni see Prespa, Lake
114 K12 **Megálo Livádi** ▲ Bulgaria/Greece 41°18′N 25°51′E
115 E20 **Megalópoli** prev. Megalópolis. Pelopónnisos, S Greece 37°24′N 22°08′E
Megalópolis see Megalópoli
171 U12 **Megamo** Papua Barat, E Indonesia 0°55′S 131°46′E
115 C18 **Meganísi** island Iónia Nisiá, Greece, C Mediterranean Sea
Meganom, Mys see Mehanom, Mys
Mégantic see Lac-Mégantic
15 R12 **Mégantic, Mont** ▲ Québec, SE Canada 45°27′N 71°09′W
115 G19 **Mégara** Attikí, C Greece 38°00′N 23°20′E
25 R5 **Megargel** Texas, SW USA 33°27′N 98°55′W
98 K13 **Megen** Noord-Brabant, S Netherlands 51°49′N 05°34′E
153 U13 **Meghālaya** ◆ state NE India
153 U16 **Meghna Nadi** ≈ S Bangladesh
137 V14 **Meghri** Rus. Megri. SE Armenia 38°57′N 46°15′E
115 Q23 **Megísti** var. Kastellórizon. island SE Greece
Megri see Meghri
Mehabad see Mahābād
116 F13 **Mehadia** Hung. Mehádia. Caraş-Severin, SW Romania 44°53′N 22°20′E
Mehádia see Mehadia
92 L7 **Mehamn** Finnmark, N Norway 71°01′N 27°46′E
117 U13 **Mehanom, Mys** Rus. Mys Meganom. headland S Ukraine 44°48′N 35°04′E
149 P14 **Mehar** Sind, SE Pakistan 27°12′N 67°51′E
180 J8 **Meharry, Mount** ▲ Western Australia 23°17′S 118°48′E
Mehdia see Mahdia
116 G14 **Mehedinți** ◆ county SW Romania
153 S15 **Meherpur** Khulna, W Bangladesh 23°47′N 88°40′E
21 W8 **Meherrin River** ≈ North Carolina/Virginia, SE USA
Meheso see Mi'eso
191 T11 **Mehetia** island Îles du Vent, W French Polynesia
118 K6 **Mehikoorma** Tartumaa, E Estonia 58°14′N 27°29′E
Me Hka see Nmai Hka
143 N5 **Mehrabad** ✈ (Tehrān) Tehrān, N Iran 35°46′N 51°07′E
142 J7 **Mehrān** Īlām, W Iran 33°07′N 46°10′E
143 Q14 **Mehrān, Rūd-e** prev. Mansurabad. ≈ W Iran
143 Q9 **Mehrīz** Yazd, C Iran 31°32′N 54°28′E
149 R5 **Mehtar Lām** var. Mehtarlām, Meterlam, Metharlam, Metharlam. Laghmān, E Afghanistan 34°39′N 70°10′E
Mehtarlām see Mehtar Lām
103 N8 **Mehun-sur-Yèvre** Cher, C France 47°09′N 02°15′E
79 G14 **Meiganga** Adamaoua, NE Cameroon 06°31′N 14°07′E
160 H10 **Meigu** var. Bapu. Sichuan, C China 28°16′N 103°20′E
163 W11 **Meihekou** var. Hailong. Jilin, NE China 42°31′N 125°40′E
99 L15 **Meijel** Limburg, SE Netherlands 51°22′N 05°52′E
Meijiang see Ningdu
166 M5 **Meiktila** Mandalay, C Myanmar (Burma) 20°53′N 95°54′E
Meilbhe, Loch see Melvin, Lough
108 G7 **Meilen** Zürich, N Switzerland 47°17′N 08°39′E
Meilu see Wuchuan
101 J17 **Meiningen** Thüringen, C Germany 50°34′N 10°25′E
108 F9 **Meiringen** Bern, S Switzerland 46°42′N 08°13′E
Meishan see Jinzhai
101 O15 **Meissen** Ger. Meißen. Sachsen, E Germany 51°10′N 13°28′E
Meißen see Meissen
101 I15 **Meissner** ▲ C Germany 51°13′N 09°52′E
99 K25 **Meix-devant-Virton** Luxembourg, SE Belgium 49°36′N 05°27′E
Meixian see Meizhou
Meixing see Xinjin
161 P13 **Meizhou** var. Meixian, Mei Xian. Guangdong, S China 24°21′N 116°05′E
67 P2 **Mejerda** var. Oued Medjerda, Wādī Majardah. ≈ Algeria/Tunisia see also Medjerda, Oued
42 F7 **Mejicanos** San Salvador, C El Salvador 13°50′N 89°13′W
Méjico see Mexico
62 G5 **Mejillones** Antofagasta, N Chile 23°03′S 70°25′W
189 V5 **Mejit Island** var. Mājeej. island Ratak Chain, NE Marshall Islands
79 F17 **Mékambo** Ogooué-Ivindo, NE Gabon 01°03′N 13°50′E
80 J10 **Mek'elē** var. Makale. Tigray, N Ethiopia 13°36′N 39°29′E
74 I10 **Mekerrhane, Sebkha** var. Sebkha Meqerghane, Sebkra Mekerrhane. salt flat C Algeria
Mekerrhane, Sebkra see Mekerrhane, Sebkha
76 G10 **Mékhé** NW Senegal 15°02′N 16°40′W
146 G14 **Mekhinli** Ahal Welaýaty, C Turkmenistan 37°28′N 59°20′E
15 P9 **Mékinac, Lac** ◎ Québec, SE Canada
Meklong see Samut Songkhram
74 G6 **Meknès** N Morocco 33°54′N 05°27′W
129 U12 **Mekong** var. Lan-ts'ang Chiang, Cam. Mékôngk, Chin. Lancang Jiang, Lao. Mènam Khong, Nam Khong, Th. Mae Nam Khong, Tib. Dza Chu, Vtn. Sông Tiên Giang. ≈ SE Asia
Mékôngk see Mekong
167 T15 **Mekong, Mouths of the** delta S Vietnam
38 L12 **Mekoryuk** Nunivak Island, Alaska, USA 60°23′N 166°11′W
77 R14 **Mékrou** ≈ N Benin
168 K9 **Melaka** var. Malacca. Melaka, Peninsular Malaysia 02°14′N 102°14′E
168 L9 **Melaka** var. Malacca. ◆ state Peninsular Malaysia
Melaka, Selat see Malacca, Strait of
175 O6 **Melanesia** island group W Pacific Ocean
175 P5 **Melanesian Basin** undersea feature W Pacific Ocean 0°05′N 160°35′E
171 R9 **Melanguane** Pulau Karakelang, N Indonesia 04°02′N 126°43′E
169 R11 **Melawi, Sungai** ≈ Borneo, N Indonesia
183 N12 **Melbourne** state capital Victoria, SE Australia 37°51′S 144°56′E
27 V9 **Melbourne** Arkansas, C USA 36°04′N 91°54′W
23 Y12 **Melbourne** Florida, SE USA 28°04′N 80°36′W
29 W14 **Melbourne** Iowa, C USA 41°57′N 93°07′W
92 G10 **Melbu** Nordland, C Norway 68°31′N 14°50′E
Melchor de Mencos see Ciudad Melchor de Mencos
63 F19 **Melchor, Isla** island Archipiélago de los Chonos, S Chile
40 M9 **Melchor Ocampo** Zacatecas, C Mexico 24°45′N 101°38′W
14 C11 **Meldrum Bay** Manitoulin Island, Ontario, S Canada 45°55′N 83°06′W
Meleda see Mljet
106 D8 **Melegnano** prev. Marignano. Lombardia, N Italy 45°22′N 09°19′E
188 F9 **Melekeok** ● Babeldaob, N Palau 07°30′N 134°37′E
112 L9 **Melenci** Hung. Melencze. Vojvodina, N Serbia 45°32′N 20°18′E
Melencze see Melenci
127 N4 **Melenki** Vladimirskaya Oblast', W Russian Federation 55°21′N 41°37′E
127 V6 **Meleuz** Respublika Bashkortostan, W Russian Federation 52°55′N 55°54′E
12 L6 **Mélèzes, Rivière aux** ≈ Québec, C Canada
78 I11 **Melfi** Guéra, S Chad 11°05′N 17°57′E
107 M17 **Melfi** Basilicata, S Italy 41°00′N 15°33′E
11 U14 **Melfort** Saskatchewan, S Canada 52°52′N 104°38′W
104 H4 **Melgaço** Viana do Castelo, N Portugal 42°07′N 08°15′W
105 N4 **Melgar de Fernamental** Castilla y León, N Spain 42°24′N 04°15′W
74 L6 **Melghir, Chott** var. Chott Melrhir. salt lake E Algeria
94 H8 **Melhus** Sør-Trøndelag, S Norway 63°17′N 10°18′E
104 H3 **Melide** Galicia, NW Spain 42°54′N 08°01′W
Meligalá see Meligalás
115 E21 **Meligalás** prev. Meligalá. Pelopónnisos, S Greece 37°13′N 21°58′E
60 L12 **Mel, Ilha do** island S Brazil
120 E10 **Melilla** anc. Rusaddir, Russadir. Melilla, Spain, N Africa 35°18′N 02°56′W
71 N1 **Melilla** enclave Spain, N Africa
63 G18 **Melimoyu, Monte** ▲ S Chile 44°06′S 72°49′W
169 V11 **Melintang, Danau** ◎ Borneo, N Indonesia
117 U7 **Melioratyvne** Dnipropetrovs'ka Oblast', E Ukraine 48°35′N 35°18′E
62 G11 **Melipilla** Santiago, C Chile 33°42′S 71°15′W
115 I25 **Mélissa, Akrotírio** headland Kríti, Greece, E Mediterranean Sea 35°06′N 24°33′E
11 W17 **Melita** Manitoba, S Canada 49°16′N 100°59′W
Melita see Mljet
Melitene see Malatya
107 M23 **Melito di Porto Salvo** Calabria, SW Italy 37°55′N 15°48′E
117 U10 **Melitopol'** Zaporiz'ka Oblast', SE Ukraine 46°49′N 35°23′E
109 V4 **Melk** Niederösterreich, NE Austria 48°14′N 15°21′E
95 K15 **Mellan-Fryken** ◎ C Sweden
99 E17 **Melle** Oost-Vlaanderen, NW Belgium 51°N 03°48′E
100 G13 **Melle** Niedersachsen, NW Germany 52°12′N 08°19′E
Mellersta Finland see Keski-Suomi
Mellersta Österbotten see Keski-Pohjanmaa
95 J17 **Mellerud** Västra Götaland, S Sweden 58°42′N 12°27′E
102 K10 **Melle-sur-Bretonne** Deux-Sèvres, W France 46°13′N 00°07′W
29 P8 **Mellette** South Dakota, N USA 45°07′N 98°29′W
121 O15 **Mellieħa** E Malta 35°58′N 14°21′E
80 B10 **Mellit** Northern Darfur, W Sudan 14°07′N 25°34′E
75 N7 **Mellita** ✈ SE Tunisia 33°47′N 10°51′E
63 G21 **Mellizo Sur, Cerro** ▲ S Chile 48°27′S 73°10′W
100 G9 **Mellum** island NW Germany
83 L22 **Melmoth** KwaZulu/Natal, E South Africa 28°35′S 31°25′E
111 D16 **Mělník** Ger. Melnik. Středočeský Kraj, NW Czech Republic 50°21′N 14°30′E
122 J12 **Mel'nikovo** Tomskaya Oblast', C Russian Federation 56°35′N 84°11′E
61 G18 **Melo** Cerro Largo, NE Uruguay 32°22′S 54°10′W
Melodunum see Melun
Melrhir, Chott see Melghir, Chott
183 P7 **Melrose** New South Wales, SE Australia 32°41′S 146°58′E
182 I7 **Melrose** South Australia 32°52′S 138°16′E
29 T7 **Melrose** Minnesota, N USA 45°40′N 94°46′W
33 Q11 **Melrose** Montana, NW USA 45°33′N 112°41′W
37 V12 **Melrose** New Mexico, SW USA 34°25′N 103°37′W
108 I8 **Mels** Sankt Gallen, NE Switzerland 47°03′N 09°26′E
Melsetter see Chimanimani
33 V9 **Melstone** Montana, NW USA 46°37′N 107°49′W
101 I16 **Melsungen** Hessen, C Germany 51°08′N 09°33′E
92 L12 **Meltaus** Lappi, NW Finland 66°54′N 25°18′E
97 N19 **Melton Mowbray** C England, United Kingdom 52°46′N 01°04′W
82 Q13 **Meluco** Cabo Delgado, NE Mozambique 12°31′S 39°39′E
103 O5 **Melun** anc. Melodunum. Seine-et-Marne, N France 48°32′N 02°40′E
80 F12 **Melut** Upper Nile, NE South Sudan 10°27′N 32°13′E
27 P5 **Melvern Lake** ⊠ Kansas, C USA
11 V16 **Melville** Saskatchewan, S Canada 50°57′N 102°49′W
Melville Bay/Melville Bugt see Qimusseriarsuaq
45 O11 **Melville Hall** ✈ (Dominica) NE Dominica 15°33′N 61°19′W
181 O1 **Melville Island** island Northern Territory, N Australia
197 O8 **Melville Island** island Parry Islands, Northwest Territories, NW Canada
11 W9 **Melville, Lake** ◎ Newfoundland and Labrador, E Canada
9 O7 **Melville Peninsula** peninsula Nunavut, NE Canada
Melville Sound see Viscount Melville Sound
25 Q9 **Melvin** Texas, SW USA 31°12′N 99°34′W
97 D15 **Melvin, Lough** Ir. Loch Meilbhe. ◎ S Northern Ireland, United Kingdom/Ireland
169 S12 **Memala** Borneo, C Indonesia 01°44′S 112°36′E
113 L22 **Memaliaj** Gjirokastër, S Albania 40°21′N 19°56′E
83 Q14 **Memba** Nampula, NE Mozambique 14°07′S 40°33′E
83 Q14 **Memba, Baia de** inlet NE Mozambique
Membidj see Manbij
Memel see Neman, NE Europe
Memel see Klaipeda, Lithuania
101 J23 **Memmingen** Bayern, S Germany 47°59′N 10°11′E
27 U1 **Memphis** Missouri, C USA 40°28′N 92°11′W
20 E10 **Memphis** Tennessee, S USA 35°09′N 90°03′W
25 P3 **Memphis** Texas, SW USA 34°43′N 100°34′W
20 E10 **Memphis** ✈ Tennessee, S USA 35°02′N 89°57′W
15 Q13 **Memphrémagog, Lac** var. Lake Memphremagog. ◎ Canada/USA see also Lake Memphremagog
19 N6 **Memphremagog, Lake** var. Lac Memphrémagog. ◎ Canada/USA see also Memphrémagog, Lac
117 Q2 **Mena** Chernihivs'ka Oblast', NE Ukraine 51°30′N 32°15′E
27 S12 **Mena** Arkansas, C USA 34°40′N 94°15′W
Menaam see Menaldum
Menado see Manado
106 D6 **Menaggio** Lombardia, N Italy 46°03′N 09°14′E
29 T6 **Menahga** Minnesota, N USA 46°45′N 95°06′W
77 R10 **Ménaka** Goa, E Mali 15°55′N 02°25′E
98 K5 **Menaldum** Fris. Menaam. Fryslân, N Netherlands 53°14′N 05°38′E
Mènam Khong see Mekong
74 E7 **Menara** ✈ (Marrakech) C Morocco 31°36′N 08°00′W
25 Q9 **Menard** Texas, SW USA 30°56′N 99°48′W
193 Q12 **Menard Fracture Zone** tectonic feature E Pacific Ocean
30 M7 **Menasha** Wisconsin, N USA 44°13′N 88°25′W
Mencezi Garagum see Merkezi Garagumy
193 U9 **Mendaña Fracture Zone** tectonic feature E Pacific Ocean
169 S13 **Mendawai, Sungai** ≈ Borneo, C Indonesia
103 P13 **Mende** anc. Mimatum. Lozère, S France 44°32′N 03°30′E
81 J14 **Mendebo** ▲▲ C Ethiopia
80 J9 **Mendefera** prev. Adi Ugri. S Eritrea 14°53′N 38°51′E
197 S7 **Mendeleyev Ridge** undersea feature Arctic Ocean
127 T3 **Mendeleyevsk** Respublika Tatarstan, W Russian Federation 55°54′N 52°19′E
101 F15 **Menden** Nordrhein-Westfalen, W Germany 51°26′N 07°47′E
22 L6 **Mendenhall** Mississippi, S USA 31°57′N 89°52′W
38 L13 **Mendenhall, Cape** headland Nunivak Island, Alaska, USA 59°45′N 166°10′W
41 P9 **Méndez** var. Villa de Méndez. Tamaulipas, C Mexico 25°06′N 98°32′W
80 H13 **Mendī** Oromīya C Ethiopia 09°43′N 35°07′E
186 C7 **Mendi** Southern Highlands, W Papua New Guinea 06°13′S 143°39′E
97 K22 **Mendip Hills** var. Mendips. hill range S England, United Kingdom
Mendips see Mendip Hills
34 L6 **Mendocino** California, W USA 39°18′N 123°48′W
34 J3 **Mendocino, Cape** headland California, W USA 40°26′N 124°24′W
0 B8 **Mendocino Fracture Zone** tectonic feature NE Pacific Ocean
35 P10 **Mendota** California, W USA 36°44′N 120°24′W
30 L11 **Mendota** Illinois, N USA 41°32′N 89°04′W
30 K8 **Mendota, Lake** ◎ Wisconsin, N USA
62 I11 **Mendoza** Mendoza, W Argentina 33°00′S 68°47′W
62 I12 **Mendoza** off. Provincia de Mendoza. ◆ province W Argentina
Mendoza, Provincia de see Mendoza
108 H12 **Mendrisio** Ticino, S Switzerland 45°53′N 08°59′E
168 L10 **Mendung** Pulau Mendol, W Indonesia 0°33′N 103°09′E
54 I5 **Mene de Mauroa** Falcón, NW Venezuela 10°39′N 71°04′W
54 I5 **Mene Grande** Zulia, NW Venezuela 09°51′N 70°57′W
136 B14 **Menemen** İzmir, W Turkey 38°34′N 27°03′E
99 C18 **Menen** var. Meenen, Fr. Menin. West-Vlaanderen, W Belgium 50°48′N 03°07′E
163 Q8 **Menengiyn Tal** plain E Mongolia
189 R9 **Meneng Point** headland SW Nauru 0°33′S 166°57′E
92 L10 **Menesjärvi** Lapp. Menešjávri. Lappi, N Finland 68°39′N 26°22′E
Menešjávri see Menesjärvi
107 I24 **Menfi** Sicilia, Italy, C Mediterranean Sea 37°36′N 12°59′E
161 P7 **Mengcheng** Anhui, E China 33°15′N 116°33′E
160 F15 **Menghai** Yunnan, SW China 22°02′N 100°18′E
160 F15 **Mengla** Yunnan, SW China 21°30′N 101°33′E
65 F24 **Menguera Point** headland East Falkland, Falkland Islands
160 M13 **Mengzhu Ling** ▲▲ S China
160 H14 **Mengzi** Yunnan, SW China 23°20′N 103°32′E
114 H13 **Meníkio** var. Menoíkio. ▲ NE Greece 40°50′N 12°40′E
Menin see Menen
182 L7 **Menindee** New South Wales, SE Australia 32°24′S 142°25′E
182 L7 **Menindee Lake** ◎ New South Wales, SE Australia
182 J10 **Meningie** South Australia 35°43′S 139°20′E
103 O5 **Mennecy** Essonne, N France 48°34′N 02°25′E
29 Q12 **Menno** South Dakota, N USA 43°14′N 97°34′W
114 H13 **Menoíkio** ▲ NE Greece
Menoíkio see Meníkio
31 N5 **Menominee** Michigan, N USA 45°06′N 87°36′W
30 M5 **Menominee River** ≈ Michigan/Wisconsin, N USA
30 M8 **Menomonee Falls** Wisconsin, N USA 43°11′N 88°09′W
30 I6 **Menomonie** Wisconsin, N USA 44°52′N 91°55′W
83 D14 **Menongue** var. Vila Serpa Pinto, Port. Serpa Pinto. Kuando Kubango, C Angola 14°38′S 17°39′E
120 H8 **Menorca** Eng. Minorca; anc. Balearis Minor. island Islas Baleares, Spain, W Mediterranean Sea
105 S13 **Menor, Mar** lagoon SE Spain
39 S10 **Mentasta Lake** ◎ Alaska, USA
39 S10 **Mentasta Mountains** ▲▲ Alaska, USA
168 I13 **Mentawai, Kepulauan** island group W Indonesia
168 I12 **Mentawai, Selat** strait W Indonesia
168 M12 **Mentok** Pulau Bangka, W Indonesia 02°01′S 105°10′E
103 V15 **Menton** It. Mentone. Alpes-Maritimes, SE France 43°47′N 07°30′E
24 K8 **Mentone** Texas, SW USA 31°42′N 103°36′W
Mentone see Menton
31 U11 **Mentor** Ohio, N USA 41°40′N 81°20′W
169 U10 **Menyapa, Gunung** ▲ Borneo, N Indonesia 01°04′N 116°01′E
159 T9 **Menyuan** var. Menyuan Huizu Zizhixian. Qinghai, C China 37°27′N 101°33′E
Menyuan Huizu Zizhixian see Menyuan
74 M5 **Menzel Bourguiba** var. Manzil Bū Ruqaybah; prev. Ferryville. N Tunisia 37°09′N 09°51′E
136 M15 **Menzelet Barajı** ⊠ C Turkey
127 T4 **Menzelinsk** Respublika Tatarstan, W Russian Federation 55°44′N 53°00′E
180 K11 **Menzies** Western Australia 29°42′S 121°04′E
195 V6 **Menzies, Mount** ▲ Antarctica 73°32′S 61°02′E
40 J6 **Meoqui** Chihuahua, N Mexico 28°18′N 105°30′W
83 N14 **Meponda** Niassa, NE Mozambique 13°20′S 34°53′E
98 M8 **Meppel** Drenthe, NE Netherlands 52°42′N 06°12′E
100 E12 **Meppen** Niedersachsen, NW Germany 52°42′N 07°18′E
Meqerghane, Sebkha see Mekerrhane, Sebkha
105 T6 **Mequinenza, Embalse de** ⊠ NE Spain
30 M8 **Mequon** Wisconsin, N USA 43°13′N 87°57′W
Mera see Maíra
182 D3 **Meramangye, Lake** salt lake South Australia
27 W5 **Meramec River** ≈ Missouri, C USA
Meran see Merano
168 K13 **Merangin** ≈ Sumatera, W Indonesia
106 G5 **Merano** Ger. Meran. Trentino-Alto Adige, N Italy 46°40′N 11°10′E
168 K8 **Merapuh Lama** Pahang, Peninsular Malaysia 04°37′N 101°58′E
106 D7 **Merate** Lombardia, N Italy 45°42′N 09°26′E
169 U13 **Meratus, Pegunungan** ▲▲ Borneo, N Indonesia
171 Y16 **Merauke, Sungai** ≈ Papua, E Indonesia
182 L9 **Merbein** Victoria, SE Australia 34°11′S 142°03′E
99 F21 **Merbes-le-Château** Hainaut, S Belgium 50°19′N 04°09′E
Merca see Marka
54 C13 **Mercaderes** Cauca, SW Colombia 01°46′N 77°09′W
Mercara see Madikeri

◆ Country
● Country Capital
◇ Dependent Territory
○ Dependent Territory Capital
◆ Administrative Regions
✈ International Airport
▲ Mountain
▲▲ Mountain Range
Volcano
≈ River
◎ Lake
⊠ Reservoir
287

MERCED – MILLER, MOUNT

35 P9 **Merced** California, W USA *37°17′N 120°30′W*
61 C20 **Mercedes** Buenos Aires, E Argentina *34°42′S 59°30′W*
61 D15 **Mercedes** Corrientes, NE Argentina *29°09′S 58°05′W*
61 D19 **Mercedes** Soriano, SW Uruguay *33°16′S 58°01′W*
25 S17 **Mercedes** Texas, SW USA *26°09′N 97°54′W*
Mercedes *see* Villa Mercedes
35 R9 **Merced Peak** ▲ California, W USA *37°34′N 119°30′W*
35 P9 **Merced River** ≈ California, W USA
18 B13 **Mercer** Pennsylvania, NE USA *41°14′N 80°14′W*
99 G18 **Merchtem** Vlaams Brabant, C Belgium *50°57′N 04°14′E*
15 O13 **Mercier** Québec, SE Canada *45°15′N 73°45′W*
25 Q9 **Mercury** Texas, SW USA *31°23′N 99°09′W*
184 M5 **Mercury Islands** *island group* N New Zealand
19 O9 **Meredith** New Hampshire, NE USA *43°36′N 71°28′W*
65 B25 **Meredith, Cape** *var.* Cabo Belgrano. *headland* West Falkland, Falkland Islands *52°15′S 60°40′W*
37 V6 **Meredith, Lake** ◎ Colorado, C USA
25 N2 **Meredith, Lake** ⊞ Texas, SW USA
81 O16 **Mereeg** *var.* Mareeq, *It.* Meregh. Galguduud, E Somalia *03°47′N 47°19′E*
117 V5 **Merefa** Kharkivs'ka Oblast', E Ukraine *49°49′N 36°05′E*
Meregh *see* Mereeg
99 E17 **Merelbeke** Oost-Vlaanderen, NW Belgium *51°00′N 03°45′E*
Merend *see* Marand
167 T12 **Méreuch** Môndól Kiri, E Cambodia *13°01′N 107°26′E*
Mergate *see* Margate
Mergui *see* Myeik
Mergui Archipelago *see* Myeik Archipelago
114 L12 **Meriç** Edirne, NW Turkey *41°12′N 26°24′E*
114 L12 **Meriç** *Bul.* Maritsa, *Gk.* Évros; *anc.* Hebrus. ≈ SE Europe *see also* Évros/Maritsa
41 X12 **Mérida** Yucatán, SW Mexico *20°58′N 89°35′W*
104 J11 **Mérida** *anc.* Augusta Emerita. Extremadura, W Spain *38°55′N 06°20′W*
54 I6 **Mérida** Mérida, W Venezuela *08°36′N 71°08′W*
54 H7 **Mérida** *off.* Estado Mérida. ◈ *state* W Venezuela
Mérida, Estado *see* Mérida
18 M13 **Meriden** Connecticut, NE USA *41°32′N 72°48′W*
22 M5 **Meridian** Mississippi, S USA *32°24′N 88°43′W*
25 S8 **Meridian** Texas, SW USA *31°56′N 97°40′W*
102 J13 **Mérignac** Gironde, SW France *44°50′N 00°40′W*
102 J13 **Mérignac** ✈ (Bordeaux) Gironde, SW France *44°51′N 00°44′W*
93 J18 **Merikarvia** Satakunta, SW Finland *61°51′N 21°30′E*
183 R12 **Merimbula** New South Wales, SE Australia *36°52′S 149°51′E*
182 L9 **Meringur** Victoria, SE Australia *34°26′S 141°19′E*
Merín, Laguna *see* Mirim Lagoon
97 I19 **Merioneth** *cultural region* W Wales, United Kingdom
188 A11 **Merir** *island* Palau Islands, N Palau
188 B17 **Merizo** SW Guam *13°15′N 144°40′E*
Merjama *see* Märjamaa
Merke *see* Merki
25 P7 **Merkel** Texas, SW USA *32°28′N 100°00′W*
146 E12 **Merkezi Garagumy** *var.* Mencezi Garagum, *Rus.* Tsentral'nyye Nizmennyye Garagumy. *desert* C Turkmenistan
145 S16 **Merki** *prev.* Merke. Zhambyl, S Kazakhstan *42°48′N 73°10′E*
119 F15 **Merkinė** Alytus, S Lithuania *54°09′N 24°11′E*
99 G16 **Merksem** Antwerpen, N Belgium *51°17′N 04°26′E*
99 I15 **Merksplas** Antwerpen, N Belgium *51°22′N 04°54′E*
Merkulovichi *see* Myerkulavichy
119 G15 **Merkys** ≈ S Lithuania
32 F15 **Merlin** Oregon, NW USA *42°34′N 123°23′W*
61 C20 **Merlo** Buenos Aires, E Argentina *34°39′S 58°45′W*
138 G8 **Meron, Harei** *prev.* Haré Meron. ▲ N Israel *35°06′N 33°00′E*
74 K6 **Merouane, Chott** *salt lake* NE Algeria
80 F7 **Merowe** Northern, N Sudan *18°29′N 31°49′E*
180 J12 **Merredin** Western Australia *31°31′S 118°18′E*
97 I14 **Merrick** ▲ S Scotland, United Kingdom *55°09′N 04°28′W*
32 H16 **Merrill** Oregon, NW USA *42°00′N 121°37′W*
30 L5 **Merrill** Wisconsin, N USA *45°12′N 89°43′W*
31 N11 **Merrillville** Indiana, N USA *41°28′N 87°19′W*
19 O10 **Merrimack River** ≈ Massachusetts/New Hampshire, NE USA
28 L12 **Merriman** Nebraska, C USA *42°54′N 101°42′W*
11 N17 **Merritt** British Columbia, SW Canada *50°09′N 120°49′W*
23 Y12 **Merritt Island** Florida, SE USA *28°21′N 80°42′W*
23 Y11 **Merritt Island** *island* Florida, SE USA
28 M12 **Merritt Reservoir** ⊞ Nebraska, C USA
183 S7 **Merriwa** New South Wales, SE Australia *32°09′S 150°24′E*
183 O8 **Merriwagga** New South Wales, SE Australia *33°51′S 145°38′E*
22 G8 **Merryville** Louisiana, S USA *30°45′N 93°32′W*
80 K9 **Mersa Fat'ma** E Eritrea *14°52′N 40°16′E*
102 M7 **Mer St-Aubin** Loir-et-Cher, C France *47°42′N 01°31′E*
99 M24 **Mersch** Luxembourg, C Luxembourg *49°45′N 06°06′E*
101 M15 **Merseburg** Sachsen-Anhalt, C Germany *51°22′N 12°00′E*
Mersen *see* Meerssen
97 K18 **Mersey** ≈ NW England, United Kingdom
136 J17 **Mersin** *var.* İçel. İçel, S Turkey *36°50′N 34°39′E*
Mersin *see* İçel
168 L9 **Mersing** Johor, Peninsular Malaysia *02°25′N 103°50′E*
118 E8 **Mērsrags** NW Latvia *57°21′N 23°05′E*
Merta *see* Merta City
152 G12 **Merta City** *var.* Merta. Rājasthān, N India *26°40′N 74°04′E*
152 F12 **Merta Road** Rājasthān, N India *26°42′N 73°54′E*
97 J21 **Merthyr Tydfil** S Wales, United Kingdom *51°46′N 03°23′W*
104 H13 **Mértola** Beja, S Portugal *37°38′N 07°40′W*
144 G14 **Mertvyy Kultuk, Sor** *salt flat* SW Kazakhstan
195 V16 **Mertz Glacier** *glacier* Antarctica
99 M24 **Mertzig** Diekirch, C Luxembourg *49°50′N 06°00′E*
25 O9 **Mertzon** Texas, SW USA *31°16′N 100°50′W*
103 N4 **Méru** Oise, N France *49°15′N 02°07′E*
81 I18 **Meru** Meru, C Kenya *0°03′N 37°38′E*
81 I19 **Meru** ◈ *county* C Kenya
81 I20 **Meru, Mount** ▲ NE Tanzania *03°12′S 36°45′E*
Merv *see* Mary
Mervdasht *see* Marv Dasht
136 K11 **Merzifon** Amasya, N Turkey *40°52′N 35°28′E*
101 D20 **Merzig** Saarland, SW Germany *49°27′N 06°39′E*
36 L14 **Mesa** Arizona, SW USA *33°25′N 111°49′W*
29 V4 **Mesabi Range** ▲▲ Minnesota, N USA
54 H6 **Mesa Bolívar** Mérida, NW Venezuela *08°30′N 71°38′W*
107 Q18 **Mesagne** Puglia, SE Italy *40°33′N 17°49′E*
39 P12 **Mesa Mountain** ▲ Alaska, USA *60°26′N 155°14′W*
115 J25 **Mesará** *lowland* Kríti, Greece, E Mediterranean Sea
37 S14 **Mescalero** New Mexico, SW USA *33°09′N 105°46′W*
101 G15 **Meschede** Nordrhein-Westfalen, W Germany *51°21′N 08°16′E*
137 Q12 **Mescit Dağları** ▲▲ NE Turkey
189 V13 **Mesegon** *island* Chuuk, C Micronesia
Meseritz *see* Międzyrzecz
54 F11 **Mesetas** Meta, C Colombia *03°14′N 74°09′W*
Meshchera Lowland *see* Meshcherskaya Nizmennost'
Meshcherskaya Nizina *see* Meshcherskaya Nizmennost'
126 M4 **Meshcherskaya Nizmennost'** *var.* Meshcherskaya Nizina, *Eng.* Meshchera Lowland. *basin* W Russian Federation
126 J5 **Meshchovsk** Kaluzhskaya Oblast', W Russian Federation *54°21′N 35°23′E*
125 R9 **Meshchura** Respublika Komi, NW Russian Federation *63°18′N 50°56′E*
Meshed *see* Mashhad
Meshed-i-Sar *see* Bābolsar
80 E13 **Meshra'er Req** Warap, W South Sudan *08°30′N 29°27′E*
37 R15 **Mesilla** New Mexico, SW USA *32°15′N 106°49′W*
108 H10 **Mesocco** *Ger.* Misox. Ticino, S Switzerland *46°18′N 09°13′E*
115 D18 **Mesolóngi** *prev.* Mesolóngion. Dytikí Elláda, W Greece *38°21′N 21°26′E*
Mesolóngion *see* Mesolóngi
14 E8 **Mesomikenda Lake** ◎ Ontario, S Canada
61 D15 **Mesopotamia** *var.* Mesopotamia Argentina. *physical region* NE Argentina
Mesopotamia Argentina *see* Mesopotamia
35 Y10 **Mesquite** Nevada, W USA *36°47′N 114°04′W*
82 Q13 **Messalo, Rio** *var.* Mualo. ≈ NE Mozambique
Messana/Messene *see* Messina
99 L25 **Messancy** Luxembourg, SE Belgium *49°36′N 05°49′E*
107 M23 **Messina** *var.* Messana, Messene; *anc.* Zancle. Sicilia, Italy, C Mediterranean Sea *38°12′N 15°33′E*
Messina *see* Musina
Messina, Strait of *see* Messina, Stretto di
107 M23 **Messina, Stretto di** *Eng.* Strait of Messina. *strait* SW Italy
115 E21 **Messíni** Pelopónnisos, S Greece *37°03′N 22°00′E*
115 E21 **Messinía** *peninsula* S Greece
115 E22 **Messiniakós Kólpos** *gulf* S Greece
122 J8 **Messoyakha** ≈ N Russian Federation
114 H11 **Mesta** *Gk.* Néstos, *Turk.* Kara Su. ≈ Bulgaria/Greece *see also* Néstos
Mesta *see* Néstos
Mestghanem *see* Mostaganem
137 R8 **Mest'ia** *prev.* Mestia, *var.* Mestiya. N Georgia *43°03′N 42°50′E*
Mestia *see* Mest'ia
Mestiya *see* Mest'ia
115 K18 **Mestón, Akrotírio** *cape* Chíos, E Greece
106 H8 **Mestre** Veneto, NE Italy *45°30′N 12°14′E*
59 M16 **Mestre, Espigão** ▲▲ E Brazil
169 N14 **Mesuji** ≈ Sumatera, W Indonesia
Mesule *see* Grosser Möseler
10 J10 **Meszah Peak** ▲ British Columbia, W Canada *58°31′N 131°28′W*
54 G11 **Meta** *off.* Departamento del Meta. ◈ *province* C Colombia
15 Q8 **Metabetchouane** ≈ Québec, SE Canada
Meta, Departamento del *see* Meta
9 S7 **Meta Incognita Peninsula** *peninsula* Baffin Island, Nunavut, NE Canada
22 K9 **Metairie** Louisiana, S USA *29°58′N 90°09′W*
32 M6 **Metaline Falls** Washington, NW USA *48°51′N 117°21′W*
62 K6 **Metán** Salta, N Argentina *25°29′S 64°57′W*
82 N13 **Metangula** Niassa, N Mozambique *12°41′S 34°50′E*
42 E7 **Metapán** Santa Ana, NW El Salvador *14°20′N 89°28′W*
54 K9 **Meta, Río** ≈ Colombia/Venezuela
106 I11 **Metauro** ≈ C Italy
80 H11 **Metema** Āmara, N Ethiopia *12°53′N 36°10′E*
115 D15 **Metéora** *religious building* Thessalía, C Greece
65 O20 **Meteor Rise** *undersea feature* SW Indian Ocean *46°00′S 05°30′E*
186 G5 **Meteran** New Hanover, NE Papua New Guinea *02°40′S 150°12′E*
Meterlam *see* Mehtar Lām
115 G20 **Methanon** *peninsula* S Greece
Methariam/Metharlam *see* Mehtar Lām
32 J6 **Methow River** ≈ Washington, NW USA
19 O10 **Methuen** Massachusetts, NE USA *42°43′N 71°10′W*
185 G19 **Methven** Canterbury, South Island, New Zealand *43°37′S 171°38′E*
Metis *see* Metz
113 G15 **Metković** Dubrovnik-Neretva, SE Croatia *43°02′N 17°37′E*
39 Y14 **Metlakatla** Annette Island, Alaska, USA *55°07′N 131°34′W*
109 V13 **Metlika** *Ger.* Möttling. SE Slovenia *45°38′N 15°18′E*
109 T8 **Metnitz** Kärnten, S Austria *46°58′N 14°09′E*
27 W12 **Meto, Bayou** ≈ Arkansas, C USA
168 M15 **Metro** Sumatera, W Indonesia *05°05′S 105°20′E*
30 M17 **Metropolis** Illinois, N USA *37°09′N 88°43′W*
Metropolitan *see* Santiago
35 N8 **Metropolitan Oakland** ✈ California, W USA *37°42′N 122°13′W*
115 D15 **Métsovo** *prev.* Métsovon. Ípeiros, C Greece *39°47′N 21°12′E*
Métsovon *see* Métsovo
23 V5 **Metter** Georgia, SE USA *32°24′N 82°03′W*
99 H21 **Mettet** Namur, S Belgium *50°19′N 04°43′E*
101 D20 **Mettlach** Saarland, SW Germany *49°28′N 06°37′E*
Mettu *see* Metu
80 H13 **Metu** *var.* Mattu, Mettu. Oromīya, C Ethiopia *08°18′N 35°39′E*
138 G8 **Metula** *prev.* Metulla. Northern, N Israel *33°16′N 35°35′E*
169 T10 **Metulang** Borneo, N Indonesia *01°28′N 114°40′E*
Metulla *see* Metula
103 T4 **Metz** *anc.* Divodurum Mediomatricum, Mediomatrica, Metis. Moselle, NE France *49°07′N 06°09′E*
101 H22 **Metzingen** Baden-Württemberg, S Germany *48°31′N 09°16′E*
168 G8 **Meulaboh** Sumatera, W Indonesia *04°10′N 96°09′E*
99 D18 **Meulebeke** West-Vlaanderen, W Belgium *50°57′N 03°18′E*
103 U6 **Meurthe** ≈ NE France
103 S5 **Meurthe-et-Moselle** ◈ *department* NE France
103 S4 **Meuse** ◈ *department* NE France
84 F10 **Meuse** *Dut.* Maas. ≈ W Europe *see also* Maas
Meuse *see* Maas
Mexcala, Río *see* Balsas, Río
25 U8 **Mexia** Texas, SW USA *31°40′N 96°28′W*
58 K11 **Mexiana, Ilha** *island* NE Brazil
40 C1 **Mexicali** Baja California Norte, NW Mexico *32°34′N 115°26′W*
Mexicanos, Estados Unidos *see* Mexico
41 O14 **México** *var.* Ciudad de México, *Eng.* Mexico City. ● (Mexico) México, C Mexico *19°26′N 99°08′W*
27 V4 **Mexico** Missouri, C USA *19°10′N 99°04′W*
18 H9 **Mexico** New York, NE USA *43°27′N 76°14′W*
40 L7 **Mexico** *off.* United Mexican States, *var.* Méjico, México, *Sp.* Estados Unidos Mexicanos. ◆ *federal republic* N Central America
41 O13 **México** ◈ *state* S Mexico
México *see* Mexico
0 J13 **Mexico Basin** *var.* Sigsbee Deep. *undersea feature* C Gulf of Mexico *25°00′N 92°00′W*
Mexico City *see* México
México, Golfo de *see* Mexico, Gulf of
44 B4 **Mexico, Gulf of** *Sp.* Golfo de México. *gulf* W Atlantic Ocean
139 R4 **Mexmûr** *Ar.* Makhmūr. Arbīl, N Iraq *35°47′N 43°32′E*
Meyadine *see* Al Mayādīn
Meydān Shahr *see* Maīdān Shahr
39 Y14 **Meyers Chuck** Etolin Island, Alaska, USA *55°44′N 132°15′W*
Meymaneh *see* Maīmanah
143 N7 **Meymeh** Eşfahān, C Iran *33°29′N 51°09′E*
123 V7 **Meynypil'gyno** Chukotskiy Avtonomnyy Okrug, NE Russian Federation *62°33′N 177°00′E*
108 A10 **Meyrin** Genève, SW Switzerland *46°14′N 06°05′E*
166 L7 **Mezaligon** Ayeyawady, SW Myanmar (Burma) *17°53′N 95°12′E*
41 O15 **Mezcala** Guerrero, S Mexico *17°55′N 99°34′W*
114 H8 **Mezdra** Vratsa, NW Bulgaria *43°08′N 23°42′E*
103 P16 **Mèze** Hérault, S France *43°26′N 03°37′E*
125 O6 **Mezen'** Arkhangel'skaya Oblast', NW Russian Federation *65°54′N 44°10′E*
125 P8 **Mezen'** ≈ NW Russian Federation
Mezen, Bay of *see* Mezenskaya Guba
103 Q13 **Mézenc, Mont** ▲ C France *44°57′N 04°15′E*
125 O8 **Mezenskaya Guba** *var.* Bay of Mezen. *bay* NW Russian Federation
Mezha *see* Myazha
122 H6 **Mezhdusharskiy, Ostrov** *island* Novaya Zemlya, N Russian Federation
Mezhëvo *see* Myezhava
Mezhgor'ye *see* Mizhhir"ya
117 V8 **Mezhova** Dnipropetrovs'ka Oblast', E Ukraine *48°15′N 36°44′E*
10 J12 **Meziadin Junction** British Columbia, W Canada *56°06′N 129°15′W*
111 G16 **Mezileské Sedlo** *var.* Przełęcz Międzyleska. *pass* Czech Republic/Poland
102 L14 **Mézin** Lot-et-Garonne, SW France *44°03′N 00°16′E*
111 M24 **Mezőberény** Békés, SE Hungary *46°49′N 21°00′E*
111 M25 **Mezőhegyes** Békés, SE Hungary *46°20′N 20°48′E*
111 M25 **Mezőkovácsháza** Békés, SE Hungary *46°24′N 20°52′E*
111 M21 **Mezőkövesd** Borsod-Abaúj-Zemplén, NE Hungary *47°49′N 20°32′E*
Mezőtelegd *see* Tileagd
111 M23 **Mezőtúr** Jász-Nagykun-Szolnok, E Hungary *47°N 20°37′E*
40 K10 **Mezquital** Durango, C Mexico *23°31′N 104°19′W*
106 G6 **Mezzolombardo** Trentino-Alto Adige, N Italy *46°13′N 11°08′E*
82 L13 **Mfuwe** Muchinga, N Zambia *13°00′S 31°45′E*
121 O15 **Mġarr** Gozo, N Malta *36°01′N 14°18′E*
126 H6 **Mglin** Bryanskaya Oblast', W Russian Federation *53°01′N 32°54′E*
Mhálanna, Cionn *see* Malin Head
154 G10 **Mhow** Madhya Pradesh, C India *22°32′N 75°49′E*
Miadzioł Nowy *see* Myadzyel
171 O6 **Miagao** Panay Island, C Philippines *10°40′N 122°15′E*
41 R17 **Miahuatlán** *var.* Miahuatlán de Porfirio Díaz. Oaxaca, SE Mexico *16°21′N 96°36′W*
Miahuatlán de Porfirio Díaz *see* Miahuatlán
104 K10 **Miajadas** Extremadura, W Spain *39°10′N 05°54′W*
Miajlar *see* Myājlār
36 M14 **Miami** Arizona, SW USA *33°23′N 110°53′W*
23 Z16 **Miami** Florida, SE USA *25°46′N 80°12′W*
27 R8 **Miami** Oklahoma, C USA *36°53′N 94°54′W*
25 O2 **Miami** Texas, SW USA *35°42′N 100°37′W*
23 Z16 **Miami** ✈ Florida, SE USA *25°47′N 80°16′W*
23 Z16 **Miami Beach** Florida, SE USA *25°47′N 80°08′W*
23 Y15 **Miami Canal** *canal* Florida, SE USA
31 R14 **Miamisburg** Ohio, N USA *39°38′N 84°17′W*
149 U10 **Miān Channūn** Punjab, E Pakistan *30°24′N 72°27′E*
142 J4 **Mīāndowāb** *var.* Mianduab, Miyāndoāb. Āzarbāyjān-e Gharbī, NW Iran *36°57′N 46°06′E*
172 H5 **Miandrivazo** Toliara, C Madagascar *19°31′S 45°29′E*
Mianduab *see* Mīāndowāb
142 K3 **Mīāneh** *var.* Miyāneh. Āzarbāyjān-e Sharqī, NW Iran *37°23′N 47°45′E*
161 T12 **Mianhua Yu** *prev.* Mienhua Yü. *island* N Taiwan
149 O16 **Miāni Hōr** *lagoon* S Pakistan
160 G10 **Mianning** Sichuan, C China *28°34′N 102°12′E*
149 T7 **Miānwāli** Punjab, NE Pakistan *32°32′N 71°33′E*
160 J7 **Mianxian** *var.* Mian Xian. Shaanxi, C China *33°12′N 106°36′E*
Mian Xian *see* Mianxian
160 I8 **Mianyang** Sichuan, C China *31°29′N 104°43′E*
Mianyang *see* Xiantao
161 R3 **Miaodao Qundao** *island group* E China
161 S13 **Miaoli** N Taiwan *24°33′N 120°48′E*
122 F11 **Miass** Chelyabinskaya Oblast', C Russian Federation *55°00′N 59°55′E*
110 G8 **Miastko** *Ger.* Rummelsburg in Pommern. Pomorskie, N Poland *54°N 16°58′E*
Miava *see* Myjava
11 O15 **Mica Creek** British Columbia, SW Canada *51°58′N 118°29′W*
160 J7 **Micang Shan** ▲▲ C China
Mi Chai *see* Nong Khai
111 O19 **Michalovce** *Ger.* Grossmichel, *Hung.* Nagymihály. Košický Kraj, E Slovakia *48°46′N 21°55′E*
99 M20 **Michel, Baraque** *hill* E Belgium
39 S5 **Michelson, Mount** ▲ Alaska, USA *69°19′N 144°16′W*
45 P9 **Miches** E Dominican Republic *18°59′N 69°03′W*
30 M4 **Michigamme, Lake** ◎ Michigan, N USA
30 M4 **Michigamme Reservoir** ⊞ Michigan, N USA
31 N4 **Michigamme River** ≈ Michigan, N USA
31 O7 **Michigan** *off.* State of Michigan, *also known as* Great Lakes State, Lake State, Wolverine State. ◈ *state* N USA
31 O11 **Michigan City** Indiana, N USA *41°43′N 86°52′W*
31 O8 **Michigan, Lake** ◎ N USA
31 P2 **Michipicoten Bay** *lake bay* Ontario, N Canada
14 A8 **Michipicoten Island** *island* Ontario, S Canada
14 B7 **Michipicoten River** Ontario, S Canada *47°56′N 84°48′W*
Michurin *see* Tsarevo
126 M6 **Michurinsk** Tambovskaya Oblast', W Russian Federation *52°56′N 40°31′E*
Mico, Punta/Mico, Punto *see* Monkey Point
42 L10 **Mico, Río** ≈ SE Nicaragua
45 T12 **Micoud** SE Saint Lucia *13°49′N 60°54′W*
189 N16 **Micronesia** *off.* Federated States of Micronesia. ◆ *federation* W Pacific Ocean
175 P4 **Micronesia** *island group* W Pacific Ocean
Micronesia, Federated States of *see* Micronesia
169 O9 **Midai, Pulau** *island* Kepulauan Natuna, W Indonesia
Mid-Atlantic Cordillera *see* Mid-Atlantic Ridge
65 M17 **Mid-Atlantic Ridge** *var.* Mid-Atlantic Cordillera, Mid-Atlantic Rise, Mid-Atlantic Swell. *undersea feature* Atlantic Ocean *0°00′N 20°00′W*
Mid-Atlantic Rise/Mid-Atlantic Swell *see* Mid-Atlantic Ridge
99 E15 **Middelburg** Zeeland, SW Netherlands *51°30′N 03°36′E*
83 H24 **Middelburg** Eastern Cape, S South Africa *31°28′S 25°01′E*
83 K21 **Middelburg** Mpumalanga, NE South Africa *25°47′S 29°28′E*
95 G23 **Middelfart** Syddtjylland, C Denmark *55°30′N 09°44′E*
98 G13 **Middelharnis** Zuid-Holland, SW Netherlands *51°45′N 04°10′E*
99 B16 **Middelkerke** West-Vlaanderen, W Belgium *51°12′N 02°51′E*
98 I9 **Middenbeemster** Noord-Holland, C Netherlands *52°33′N 04°55′E*
98 I8 **Middenmeer** Noord-Holland, NW Netherlands *52°48′N 04°58′E*
35 Q2 **Middle Alkali Lake** ◎ California, W USA
193 S6 **Middle America Trench** *undersea feature* E Pacific Ocean *15°00′N 95°00′W*
151 P19 **Middle Andaman** *island* Andaman Islands, India, NE Indian Ocean
Middle Atlas *see* Moyen Atlas
21 R3 **Middlebourne** West Virginia, NE USA *39°30′N 80°53′W*
23 W9 **Middleburg** Florida, SE USA *30°03′N 81°55′W*
Middleburg Island *see* 'Eua
Middle Caicos *see* Grand Caicos
25 N8 **Middle Concho River** ≈ Texas, SW USA
Middle Congo *see* Congo (Republic of)
39 R6 **Middle Fork Chandalar River** ≈ Alaska, USA
39 Q7 **Middle Fork Koyukuk River** ≈ Alaska, USA
33 O12 **Middle Fork Salmon River** ≈ Idaho, NW USA
11 T15 **Middle Lake** Saskatchewan, S Canada *52°31′N 105°16′W*
28 L13 **Middle Loup River** ≈ Nebraska, C USA
185 E22 **Middlemarch** Otago, South Island, New Zealand *45°30′S 170°07′E*
31 T15 **Middleport** Ohio, N USA *39°00′N 82°03′W*
29 U14 **Middle Raccoon River** ≈ Iowa, C USA
29 R3 **Middle River** ≈ Minnesota, N USA
21 N8 **Middlesboro** Kentucky, S USA *36°37′N 83°42′W*
97 M15 **Middlesbrough** N England, United Kingdom *54°35′N 01°14′W*
42 G3 **Middlesex** Stann Creek, C Belize *17°00′N 88°31′W*
97 N22 **Middlesex** *cultural region* SE England, United Kingdom
13 P15 **Middleton** Nova Scotia, SE Canada *44°56′N 65°04′W*
20 F10 **Middleton** Tennessee, S USA *35°05′N 88°57′W*
30 L9 **Middleton** Wisconsin, N USA *43°06′N 89°30′W*
39 S13 **Middleton Island** *island* Alaska, USA
34 M7 **Middletown** California, W USA *38°44′N 122°39′W*
21 Y2 **Middletown** Delaware, NE USA *39°25′N 75°39′W*
18 K15 **Middletown** New Jersey, NE USA *40°23′N 74°08′W*
18 K13 **Middletown** New York, NE USA *41°27′N 74°25′W*
31 R14 **Middletown** Ohio, N USA *39°33′N 84°19′W*
18 G15 **Middletown** Pennsylvania, NE USA *40°11′N 76°42′W*
141 N14 **Mīdī** *var.* Maydī. NW Yemen *16°18′N 42°51′E*
103 O16 **Midi, Canal du** *canal* S France
102 K17 **Midi de Bigorre, Pic du** ▲ S France *42°57′N 00°08′E*
102 K17 **Midi d'Ossau, Pic du** ▲ SW France *42°51′N 00°27′W*
173 R7 **Mid-Indian Basin** *undersea feature* N Indian Ocean *10°00′S 80°00′E*
173 P7 **Mid-Indian Ridge** *var.* Central Indian Ridge. *undersea feature* C Indian Ocean *12°00′S 66°00′E*
103 N14 **Midi-Pyrénées** ◈ *region* S France
25 N8 **Midkiff** Texas, SW USA *31°35′N 101°51′W*
14 G13 **Midland** Ontario, S Canada *44°45′N 79°53′W*
31 R8 **Midland** Michigan, N USA *43°37′N 84°15′W*
28 M10 **Midland** South Dakota, N USA *44°04′N 101°07′W*
24 M8 **Midland** Texas, SW USA *32°N 102°05′W*
83 K17 **Midlands** ◈ *province* C Zimbabwe
97 D21 **Midleton** *Ir.* Mainistir na Corann. SW Ireland *51°55′N 08°10′W*
25 T7 **Midlothian** Texas, SW USA *32°28′N 96°59′W*
96 K12 **Midlothian** *cultural region* S Scotland, United Kingdom
172 I7 **Midongy Atsimo** Fianarantsoa, S Madagascar *21°58′S 47°47′E*
102 K15 **Midou** ≈ SW France
192 J6 **Mid-Pacific Mountains** *var.* Mid-Pacific Seamounts. *undersea feature* NW Pacific Ocean *20°00′N 178°00′W*
Mid-Pacific Seamounts *see* Mid-Pacific Mountains
171 Q7 **Midsayap** Mindanao, S Philippines *07°12′N 124°31′E*
95 F21 **Midtjylland** ◈ *county* NW Denmark
36 L3 **Midway** Utah, W USA *40°30′N 111°28′W*
192 L5 **Midway Islands** ◇ *US unincorporated territory* C Pacific Ocean
33 X14 **Midwest** Wyoming, C USA *43°24′N 106°15′W*
27 N10 **Midwest City** Oklahoma, C USA *35°28′N 98°24′W*
152 M10 **Mid Western** ◈ *zone* W Nepal
98 P5 **Midwolda** Groningen, NE Netherlands *53°12′N 07°00′E*
137 Q16 **Midyat** Mardin, SE Turkey *37°25′N 41°20′E*
114 F8 **Midzhur** *SCr.* Midžor. ▲ Bulgaria/Serbia *43°24′N 22°41′E* *see also* Midžor
Midzhur *see* Midžor
113 Q14 **Midžor** *Bul.* Midzhur. ▲ Bulgaria/Serbia *43°24′N 22°40′E* *see also* Midzhur
Midžor *see* Midzhur
164 K14 **Mie** *off.* Mie-ken. ◈ *prefecture* Honshū, SW Japan
111 L16 **Miechów** Małopolskie, S Poland *50°21′N 20°01′E*
110 F11 **Międzychód** *Ger.* Mitteldorf. Wielkopolskie, C Poland *52°36′N 15°53′E*
Międzyleska, Przełęcz *see* Mezileské Sedlo
110 O12 **Międzyrzec Podlaski** Lubelskie, E Poland *52°N 22°47′E*
110 E11 **Międzyrzecz** *Ger.* Meseritz. Lubuskie, W Poland *52°26′N 15°33′E*
Mie-ken *see* Mie
102 L16 **Miélan** Gers, S France *43°26′N 00°19′E*
111 N16 **Mielec** Podkarpackie, SE Poland *50°18′N 21°27′E*
95 L21 **Mien** ◎ S Sweden
Mienhua Yü *see* Mianhua Yu
41 O8 **Mier** Tamaulipas, C Mexico *26°28′N 99°10′W*
116 J11 **Miercurea-Ciuc** *Ger.* Szeklerburg, *Hung.* Csíkszereda. Harghita, C Romania *46°24′N 25°48′E*
Mieresch *see* Maros/Mureș
104 K2 **Mieres del Camín** *var.* Mieres del Camino. Asturias, NW Spain *43°15′N 05°46′W*
Mieres del Camino *see* Mieres del Camín
99 K15 **Mierlo** Noord-Brabant, SE Netherlands *51°27′N 05°37′E*
41 O10 **Mier y Noriega** Nuevo León, NE Mexico *23°24′N 100°06′W*
Mies *see* Stříbro
80 K13 **Mī'ēso** *var.* Meheso, Miesso. Oromīya, C Ethiopia *09°13′N 40°47′E*
Miesso *see* Mī'ēso
110 D10 **Mieszkowice** *Ger.* Bärwalde Neumark. Zachodnio-pomorskie, W Poland *52°45′N 14°24′E*
18 G14 **Mifflinburg** Pennsylvania, NE USA *40°55′N 77°03′W*
18 F14 **Mifflintown** Pennsylvania, NE USA *40°34′N 77°24′W*
138 F8 **Mifrats Hefa** *Eng.* Bay of Haifa; *prev.* MifrazHefa. *bay* N Israel
81 G19 **Migori** ◈ *county* W Kenya
41 R15 **Miguel Alemán, Presa** ⊞ SE Mexico
40 L9 **Miguel Asua** *var.* Miguel Auza. Zacatecas, C Mexico *24°17′N 103°29′W*
Miguel Auza *see* Miguel Asua
43 S15 **Miguel de la Borda** *var.* Donoso. Colón, C Panama *09°09′N 80°20′W*
41 N13 **Miguel Hidalgo** ✈ (Guadalajara) Jalisco, SW Mexico *20°52′N 101°09′W*
40 H7 **Miguel Hidalgo, Presa** ⊞ W Mexico
116 J14 **Mihăilești** Giurgiu, S Romania *44°20′N 25°54′E*
116 M14 **Mihail Kogălniceanu** *var.* Kogălniceanu; *prev.* Caramurat, Ferdinand. Constanța, SE Romania *44°22′N 28°27′E*
117 N14 **Mihai Viteazu** Constanța, SE Romania *44°37′N 28°41′E*
136 G12 **Mihalıççık** Eskişehir, NW Turkey *39°52′N 31°30′E*
164 G13 **Mihara** Hiroshima, Honshū, SW Japan *34°24′N 133°04′E*
165 N14 **Mihara-yama** 🌋 Miyako-jima, SE Japan *34°43′N 139°23′E*
105 S8 **Mijares** ≈ E Spain
98 I11 **Mijdrecht** Utrecht, C Netherlands *52°12′N 04°52′E*
165 S4 **Mikasa** Hokkaidō, NE Japan *43°15′N 141°57′E*
Mikashevichi *see* Mikashevichy
119 K19 **Mikashevichy** *Pol.* Mikaszewicze, *Rus.* Mikashevichi. Brestskaya Voblasts', SW Belarus *52°13′N 27°28′E*
Mikaszewicze *see* Mikashevichy
126 L5 **Mikhaylov** Ryazanskaya Oblast', W Russian Federation *54°12′N 39°03′E*
Mikhaylovgrad *see* Montana
195 Z8 **Mikhaylov Island** *island* Antarctica
145 T6 **Mikhaylovka** Pavlodar, N Kazakhstan *53°49′N 76°31′E*
127 N9 **Mikhaylovka** Volgogradskaya Oblast', SW Russian Federation *50°06′N 43°17′E*
Mikhaylovka *see* Mykhaylivka
81 K24 **Mikindani** Mtwara, SE Tanzania *10°16′S 40°05′E*
93 N18 **Mikkeli** *Swe.* Sankt Michel. Etelä-Savo, SE Finland *61°41′N 27°14′E*
110 M8 **Mikołajki** *Ger.* Nikolaiken. Warmińsko-Mazurskie, NE Poland *53°49′N 21°31′E*
Míkonos *see* Mýkonos
114 I9 **Mikre** Lovech, N Bulgaria *43°01′N 24°31′E*
114 C13 **Mikrí Préspa, Límni** ◎ N Greece
125 P4 **Mikulkin, Mys** *headland* NW Russian Federation *67°50′N 46°36′E*
81 I23 **Mikumi** Morogoro, SE Tanzania *07°22′S 37°00′E*
125 R10 **Mikun'** Respublika Komi, NW Russian Federation *62°20′N 50°02′E*
164 K13 **Mikuni** Fukui, Honshū, SW Japan *36°12′N 136°09′E*
165 X13 **Mikura-jima** *island* E Japan
29 V7 **Milaca** Minnesota, N USA *45°45′N 93°40′W*
62 J10 **Milagro** La Rioja, C Argentina *31°00′S 66°01′W*
56 B7 **Milagro** Guayas, SW Ecuador *02°11′S 79°36′W*
31 P4 **Milakokia Lake** ◎ Michigan, N USA
30 J1 **Milan** Illinois, N USA *41°27′N 90°33′W*
31 R10 **Milan** Michigan, N USA *42°05′N 83°40′W*
27 T2 **Milan** Missouri, C USA *40°12′N 93°08′W*
37 Q11 **Milan** New Mexico, SW USA *35°10′N 107°53′W*
20 G9 **Milan** Tennessee, S USA *35°55′N 88°45′W*
Milan *see* Milano
95 F15 **Miland** Telemark, S Norway *59°57′N 08°48′E*
83 N15 **Milange** Zambézia, NE Mozambique *16°09′S 35°44′E*
106 D8 **Milano** *Eng.* Milan, *Ger.* Mailand; *anc.* Mediolanum. Lombardia, N Italy *45°28′N 09°10′E*
25 U10 **Milano** Texas, SW USA *30°42′N 96°51′W*
136 C15 **Milas** Muğla, SW Turkey *37°17′N 27°46′E*
119 K21 **Milashavichy** *Rus.* Milashevichi. Homyel'skaya Voblasts', SE Belarus *51°39′N 27°56′E*
Milashevichi *see* Milashavichy
119 I18 **Milavidy** *Rus.* Milovidy. Brestskaya Voblasts', SW Belarus *52°54′N 25°51′E*
107 L23 **Milazzo** *anc.* Mylae. Sicilia, Italy, C Mediterranean Sea *38°13′N 15°15′E*
29 R8 **Milbank** South Dakota, N USA *45°12′N 96°36′W*
19 T7 **Milbridge** Maine, NE USA *44°31′N 67°55′W*
100 L11 **Milde** ≈ C Germany
14 F14 **Mildmay** Ontario, S Canada *44°03′N 81°07′W*
182 L9 **Mildura** Victoria, SE Australia *34°13′S 142°09′E*
137 X12 **Mil Düzü** *Rus.* Mil'skaya Ravnina, Mil'skaya Step'. *physical region* C Azerbaijan
160 H13 **Mile** *var.* Miyang. Yunnan, SW China *24°28′N 103°26′E*
Mile *see* Mili Atoll
181 Y10 **Miles** Queensland, E Australia *26°41′S 150°15′E*
25 P8 **Miles** Texas, SW USA *31°36′N 100°10′W*
33 X9 **Miles City** Montana, NW USA *46°24′N 105°48′W*
11 U17 **Milestone** Saskatchewan, S Canada *50°00′N 104°24′W*
107 N22 **Mileto** Calabria, SW Italy *38°35′N 16°03′E*
107 K16 **Miletto, Monte** ▲ C Italy *41°28′N 14°21′E*
18 M13 **Milford** Connecticut, NE USA *41°12′N 73°01′W*
21 Y3 **Milford** *var.* Milford City. Delaware, NE USA *38°54′N 75°25′W*
29 T11 **Milford** Iowa, C USA *43°19′N 95°09′W*
19 S6 **Milford** Maine, NE USA *44°57′N 68°37′W*
29 R16 **Milford** Nebraska, C USA *40°46′N 97°03′W*
19 O10 **Milford** New Hampshire, NE USA *42°49′N 71°38′W*
18 J13 **Milford** Pennsylvania, NE USA *41°20′N 74°48′W*
25 T7 **Milford** Texas, SW USA *32°07′N 96°57′W*
36 K6 **Milford** Utah, W USA *38°22′N 112°57′W*
Milford *see* Milford Haven
Milford City *see* Milford
97 H21 **Milford Haven** *prev.* Milford. SW Wales, United Kingdom *51°44′N 05°02′W*
27 O4 **Milford Lake** ⊞ Kansas, C USA
185 B21 **Milford Sound** Southland, South Island, New Zealand *44°41′S 167°57′E*
185 B21 **Milford Sound** *inlet* South Island, New Zealand
Milhau *see* Millau
Milḩ, Baḩr al *see* Razāzah, Buḩayrat ar
139 T10 **Milḩ, Wādī al** *dry watercourse* S Iraq
189 W8 **Mili Atoll** *var.* Mile. *atoll* Ratak Chain, SE Marshall Islands
110 H13 **Milicz** Dolnośląskie, SW Poland *51°32′N 17°15′E*
107 L25 **Militello in Val di Catania** Sicilia, Italy, C Mediterranean Sea *37°17′N 14°47′E*
11 R17 **Milk River** Alberta, SW Canada *49°10′N 112°06′W*
44 J13 **Milk River** ≈ C Jamaica
33 W7 **Milk River** ≈ Montana, NW USA
80 D9 **Milk, Wadi el** *var.* Wadi al Malik. ≈ C Sudan
99 L14 **Mill** Noord-Brabant, SE Netherlands *51°42′N 05°46′E*
103 P14 **Millau** *var.* Milhau; *anc.* Æmilianum. Aveyron, S France *44°06′N 03°05′E*
14 I14 **Millbrook** Ontario, SE Canada *44°09′N 78°26′W*
23 U4 **Milledgeville** Georgia, SE USA *33°04′N 83°13′W*
12 C12 **Mille Lacs, Lac des** ◎ Ontario, S Canada
29 V6 **Mille Lacs Lake** ◎ Minnesota, N USA
23 V4 **Millen** Georgia, SE USA *32°50′N 81°56′W*
191 Y5 **Millennium Island** *prev.* Caroline Island, Thornton Island. *atoll* Line Islands, E Kiribati
29 O9 **Miller** South Dakota, N USA *44°31′N 98°59′W*
30 K5 **Miller Dam Flowage** ◎ Wisconsin, N USA
39 U12 **Miller, Mount** ▲ Alaska, USA *60°29′N 142°16′W*

◆ Country
● Country Capital
◇ Dependent Territory
○ Dependent Territory Capital
◈ Administrative Regions
✈ International Airport
▲ Mountain
▲▲ Mountain Range
🌋 Volcano
≈ River
◎ Lake
⊞ Reservoir

MILLEROVO – MOGILËVSKAYA OBLAST'

126 L10 **Millerovo** Rostovskaya Oblast', SW Russian Federation 48°57′N 40°26′E
37 N17 **Miller Peak** ▲ Arizona, SW USA 31°23′N 110°17′W
31 T12 **Millersburg** Ohio, N USA 40°33′N 81°55′W
18 G15 **Millersburg** Pennsylvania, NE USA 40°31′N 76°56′W
185 D23 **Millers Flat** Otago, South Island, New Zealand 45°42′S 169°25′E
25 Q8 **Millersview** Texas, SW USA 31°26′N 99°44′W
106 B10 **Millesimo** Piemonte, NE Italy 44°24′N 08°09′E
12 C12 **Milles Lacs, Lac des** ◎ Ontario, SW Canada
25 Q13 **Millett** Texas, SW USA 28°33′N 99°10′W
103 N11 **Millevaches, Plateau de** plateau C France
182 K12 **Millicent** South Australia 37°29′S 140°01′E
98 M13 **Millingen aan den Rijn** Gelderland, SE Netherlands 51°52′N 06°02′E
20 E10 **Millington** Tennessee, S USA 35°20′N 89°54′W
19 R4 **Millinocket** Maine, NE USA 45°38′N 68°45′W
19 R4 **Millinocket Lake** ◎ Maine, NE USA
195 Z11 **Mill Island** island Antarctica
183 T3 **Millmerran** Queensland, E Australia 27°53′S 151°15′E
109 R9 **Millstatt** Kärnten, S Austria 46°45′N 13°36′E
97 B19 **Milltown Malbay** Ir. Sráid na Cathrach. W Ireland 52°51′N 09°23′W
18 J17 **Millville** New Jersey, NE USA 39°24′N 75°01′W
27 S13 **Millwood Lake** ▣ Arkansas, C USA
Milne Bank see Milne Seamounts
186 G10 **Milne Bay** ◈ province SE Papua New Guinea
64 J8 **Milne Seamounts** var. Milne Bank. undersea feature N Atlantic Ocean
29 Q6 **Milnor** North Dakota, N USA 46°15′N 97°27′W
19 R5 **Milo** Maine, NE USA 45°15′N 69°01′W
115 I22 **Mílos** island Kykládes, Greece, Aegean Sea
Mílos see Pláka
110 H11 **Miłosław** Wielkopolskie, C Poland 52°13′N 17°28′E
113 K19 **Milot** var. Miloti. Lezhë, C Albania 41°42′N 19°43′E
Miloti see Milot
117 Z5 **Milove** Luhans'ka Oblast', E Ukraine 49°22′N 40°09′E
Milovidy see Milavidy
182 L4 **Milparinka** New South Wales, SE Australia 29°48′S 141°57′E
35 N9 **Milpitas** California, W USA 37°25′N 121°54′W
Mil'skaya Ravnina/ Mil'skaya Step' see Mil Düzü
14 G15 **Milton** Ontario, S Canada 43°31′N 79°53′W
185 E24 **Milton** Otago, South Island, New Zealand 46°08′S 169°59′E
21 Y4 **Milton** Delaware, NE USA 38°48′N 75°21′W
23 P8 **Milton** Florida, SE USA 30°37′N 87°02′W
18 G14 **Milton** Pennsylvania, NE USA 41°01′N 76°49′W
18 L7 **Milton** Vermont, NE USA 44°37′N 73°04′W
32 K11 **Milton-Freewater** Oregon, NW USA 45°54′N 118°24′W
97 N21 **Milton Keynes** SE England, United Kingdom 52°N 00°43′W
27 N3 **Miltonvale** Kansas, C USA 39°21′N 97°27′W
161 N10 **Miluo** Hunan, S China 28°52′N 113°00′E
30 M9 **Milwaukee** Wisconsin, N USA 43°03′N 87°56′W
Milyang see Miryang
Mimatum see Mende
37 Q15 **Mimbres Mountains** ▲▲ New Mexico, SW USA
182 D2 **Mimili** South Australia 27°01′S 132°33′E
102 J14 **Mimizan** Landes, SW France 44°12′N 01°12′W
Mimmaya see Minmaya
79 E19 **Mimongo** Ngounié, C Gabon 01°36′S 11°44′E
Min see Fujian
35 T7 **Mina** Nevada, W USA 38°23′N 118°07′W
143 S14 **Mīnāb** Hormozgān, SE Iran 27°08′N 57°02′E
Mina Baranīs see Baranīs
149 R9 **Mīna Bāzār** Balūchistān, SW Pakistan 30°58′N 69°11′E
Minami-Awaji see Nandan
165 X17 **Minami-Iō-jima** Eng. San Augustine. island SE Japan
165 R5 **Minami-Kayabe** Hokkaidō, NE Japan 41°54′N 140°58′E
164 B16 **Minamisatsuma** var. Kaseda. Kagoshima, Kyūshū, SW Japan 31°25′N 130°17′E
164 C14 **Minamishimabara** var. Kuchinotsu. Nagasaki, Kyūshū, SW Japan 32°36′N 130°11′E
164 C17 **Minamitane** Kagoshima, Tanega-shima, SW Japan 30°23′N 130°54′E
Minami Tori Shima see Marcus Island
Min'an see Longshan
62 J4 **Mina Pirquitas** Jujuy, NW Argentina 22°48′S 66°24′W
173 O3 **Mīnā' Qābūs** NE Oman
61 F19 **Minas** Lavalleja, S Uruguay 34°20′S 55°15′W
13 P15 **Minas Basin** bay Nova Scotia, SE Canada
61 F17 **Minas de Corrales** Rivera, NE Uruguay 31°35′S 55°20′W
44 A5 **Minas de Matahambre** Pinar del Río, W Cuba 22°34′N 83°57′W
104 J13 **Minas de Ríotinto** Andalucía, S Spain 37°40′N 06°36′W
60 K7 **Minas Gerais** off. Estado de Minas Gerais. ◈ state E Brazil
Minas Gerais, Estado de see Minas Gerais
42 E5 **Minas, Sierra de las** ▲▲ E Guatemala
41 T15 **Minatitlán** Veracruz-Llave, E Mexico 17°59′N 94°32′W
166 L6 **Minbu** Magway, W Myanmar (Burma) 20°09′N 94°52′E
149 V10 **Minchinābād** Punjab, E Pakistan 30°10′N 73°40′E
63 G17 **Minchinmávida, Volcán** ⛰ S Chile 42°51′S 72°23′W
96 G7 **Minch, The** var. North Minch. strait NW Scotland, United Kingdom
106 F8 **Mincio** anc. Mincius. ≈ N Italy
Mincius see Mincio
26 M11 **Minco** Oklahoma, C USA 35°18′N 97°56′W
171 Q7 **Mindanao** island S Philippines
Mindanao Sea see Bohol Sea
101 J23 **Mindel** ≈ S Germany
101 J23 **Mindelheim** Bayern, S Germany 48°03′N 10°30′E
Mindello see Mindelo
76 C9 **Mindelo** var. Mindello; prev. Porto Grande. São Vicente, N Cape Verde 16°54′N 25°01′W
14 I13 **Minden** Ontario, SE Canada 44°54′N 78°41′W
100 H13 **Minden** anc. Minthun. Nordrhein-Westfalen, NW Germany 52°18′N 08°55′E
22 G5 **Minden** Louisiana, S USA 32°37′N 93°17′W
29 O16 **Minden** Nebraska, C USA 40°30′N 98°57′W
35 Q6 **Minden** Nevada, W USA 38°58′N 119°47′W
182 L8 **Mindona Lake** seasonal lake New South Wales, SE Australia
171 O4 **Mindoro** island N Philippines
171 N5 **Mindoro Strait** strait W Philippines
97 J23 **Minehead** SW England, United Kingdom 51°13′N 03°29′W
97 E21 **Mine Head** Ir. Mionn Ard. headland S Ireland 51°58′N 07°36′W
59 J19 **Mineiros** Goiás, C Brazil 17°34′S 52°33′W
25 V6 **Mineola** Texas, SW USA 32°39′N 95°29′W
25 S13 **Mineral** Texas, SW USA 28°32′N 97°54′W
127 N15 **Mineral'nyye Vody** Stavropol'skiy Kray, SW Russian Federation 44°13′N 43°06′E
30 K9 **Mineral Point** Wisconsin, N USA 42°54′N 90°09′W
25 S6 **Mineral Wells** Texas, SW USA 32°48′N 98°06′W
36 K6 **Minersville** Utah, W USA 38°12′N 112°56′W
31 U12 **Minerva** Ohio, N USA 40°43′N 81°06′W
107 N17 **Minervino Murge** Puglia, SE Italy 41°06′N 16°05′E
103 O16 **Minervois** physical region S France
158 I10 **Minfeng** var. Niya. Xinjiang Uygur Zizhiqu, NW China 37°07′N 82°43′E
79 O25 **Minga** Katanga, SE Dem. Rep. Congo 11°06′S 27°57′E
137 W11 **Mingäçevir** Rus. Mingechaur, Mingechevir. C Azerbaijan 40°46′N 47°02′E
137 W11 **Mingäçevir Su Anbarı** Rus. Mingechaurskoye Vodokhranilishche, Mingechevirskoye Vodokhranilishche. ▣ NW Azerbaijan
166 L8 **Mingaladon** ✈ (Yangon) Yangon, SW Myanmar (Burma) 16°55′N 96°11′E
13 P11 **Mingan** Québec, E Canada 50°19′N 64°02′W
Mingãora see Saīdu
146 K8 **Mingbuloq** Rus. Mynbulak. Navoiy Viloyati, N Uzbekistan 42°18′N 62°53′E
146 K9 **Mingbuloq Botig'i** Rus. Vpadina Mynbulak. depression N Uzbekistan
Mingechaur/Mingechevir see Mingäçevir
Mingechaurskoye Vodokhranilishche/ Mingechevirskoye Vodokhranilishche see Mingäçevir Su Anbarı
161 Q7 **Mingguang** prev. Jiashan. Anhui, SE China 32°45′N 117°59′E
166 L4 **Mingin** Sagaing, C Myanmar (Burma) 22°51′N 94°30′E
105 Q10 **Minglanilla** Castilla-La Mancha, C Spain 39°32′N 01°36′W
31 V13 **Mingo Junction** Ohio, N USA 40°19′N 80°36′W
Mingora see Saīdu
163 V7 **Mingshui** Heilongjiang, NE China 47°10′N 125°53′E
Mingtekl Daban see Mintaka Pass
Mingu see Zhenfeng
83 Q14 **Minguri** Nampula, NE Mozambique 14°30′S 40°37′E
Mingzhou see Suide
159 U10 **Minhe** var. Chuankou; prev. Minhe Huizu Tuzu Zizhixian, Shangchuankou. Qinghai, C China 36°21′N 102°40′E
Minhe Huizu Tuzu Zizhixian see Minhe
166 L6 **Minhla** Magway, W Myanmar (Burma) 19°58′N 95°03′E
167 S14 **Minh Lương** Kiên Giang, S Vietnam 09°52′N 105°10′E
104 G5 **Minho** former province N Portugal
104 G5 **Minho, Rio** Sp. Miño. ≈ Portugal/Spain see also Miño
Minho, Rio see Miño
155 C24 **Minicoy Island** island SW India
33 P15 **Minidoka** Idaho, NW USA 42°45′N 113°29′W
118 C11 **Minija** ≈ W Lithuania
180 G9 **Minilya** Western Australia 23°45′S 114°03′E
14 E8 **Minisinakwa Lake** ◎ Ontario, S Canada
45 T12 **Ministre Point** headland S Saint Lucia 13°42′N 60°57′W
11 V15 **Minitonas** Manitoba, S Canada 52°07′N 101°02′W
Minius see Miño
186 D7 **Minj** Jiwaka, Papua New Guinea 05°55′S 144°37′E
161 R12 **Min Jiang** ≈ SE China
160 H10 **Min Jiang** ≈ C China
182 H9 **Minlaton** South Australia 34°52′S 137°33′E
159 S9 **Minle** Gansu, N China
165 Q6 **Minmaya** var. Mimmaya. Aomori, Honshū, C Japan 41°10′N 140°24′E
77 U14 **Minna** Niger, C Nigeria 09°33′N 06°33′E
165 P16 **Minna-jima** island Sakishima-shotō, SW Japan
27 N4 **Minneapolis** Kansas, C USA 39°08′N 97°43′W
29 U9 **Minneapolis** Minnesota, N USA 44°59′N 93°16′W
29 V8 **Minneapolis-Saint Paul** ✈ Minnesota, N USA 44°53′N 93°13′W
11 W16 **Minnedosa** Manitoba, S Canada 50°14′N 99°50′W
26 J7 **Minneola** Kansas, C USA 37°26′N 100°00′W
29 S7 **Minnesota** off. State of Minnesota, also known as Gopher State, New England of the West, North Star State. ◈ state N USA
29 S9 **Minnesota River** ≈ Minnesota/South Dakota, N USA
29 V9 **Minnetonka** Minnesota, N USA 44°55′N 93°28′W
29 O3 **Minnewaukan** North Dakota, N USA 48°04′N 99°14′W
182 F7 **Minnipa** South Australia 32°52′S 135°07′E
104 G5 **Miño** var. Mino, Minius, Port. Rio Minho. ≈ Portugal/Spain see also Minho, Rio
Miño see Minho, Rio
30 L4 **Minocqua** Wisconsin, N USA 45°53′N 89°42′W
30 L12 **Minonk** Illinois, N USA 40°54′N 89°01′W
Minorca see Menorca
28 M3 **Minot** North Dakota, N USA 48°15′N 101°19′W
159 U8 **Minqin** Gansu, N China 38°35′N 103°07′E
119 J16 **Minsk** ● (Belarus) Horad Minsk, C Belarus 53°52′N 27°34′E
Minsk-2 see Minsk National
Minskaya Oblast' see Minskaya Voblasts'
119 K16 **Minskaya Voblasts'** prev. Rus. Minskaya Oblast'. ◈ province C Belarus
Minskaya Vozvyshennost' see Minskaye Wzvyshsha
119 J16 **Minskaye Wzvyshsha** Rus. Minskaya Vozvyshennost'. ▲▲ C Belarus
Minsk, Gorod see Minsk, Horad
119 J16 **Minsk, Horad** Russ. Gorod Minsk. ◈ province C Belarus
110 N12 **Mińsk Mazowiecki** var. Nowo-Minsk. Mazowieckie, C Poland 52°10′N 21°31′E
119 L16 **Minsk National** prev. Minsk-2. ✈ Minskaya Voblasts', C Belarus 53°52′N 27°58′E
31 Q13 **Minster** Ohio, N USA 40°23′N 84°22′W
79 F15 **Minta** Centre, C Cameroon 04°34′N 12°54′E
149 W2 **Mintaka Pass** Chin. Mingtekl Daban. pass China/Pakistan
115 D20 **Mínthi** ▲▲ S Greece
Minthun see Minden
13 O14 **Minto** New Brunswick, SE Canada 46°05′N 66°05′W
10 H6 **Minto** Yukon, W Canada 62°33′N 136°45′W
39 R9 **Minto** Alaska, USA 65°07′N 149°22′W
29 Q3 **Minto** North Dakota, N USA 48°17′N 97°22′W
12 K6 **Minto, Lac** ◎ Québec, C Canada
195 R16 **Minto, Mount** ▲ Antarctica 71°38′S 169°11′E
11 U17 **Minton** Saskatchewan, S Canada 49°12′N 104°33′W
189 R15 **Minto Reef** atoll Caroline Islands, C Micronesia
37 R4 **Minturn** Colorado, C USA 39°34′N 106°21′W
107 J16 **Minturno** Lazio, C Italy 41°15′N 13°47′E
122 K13 **Minusinsk** Krasnoyarskiy Kray, S Russian Federation 53°37′N 91°49′E
108 G11 **Minusio** Ticino, S Switzerland 46°11′N 08°47′E
79 E17 **Minvoul** Woleu-Ntem, N Gabon 02°08′N 12°12′E
141 R13 **Minwakh** N Yemen 16°55′N 48°04′E
159 V11 **Minxian** var. Min Xian, Minyang. Gansu, C China 34°20′N 104°09′E
Min Xian see Minxian
Minya see Al Minyā
Minyang see Minxian
31 R6 **Mio** Michigan, N USA 44°40′N 84°09′W
Mionn Ard see Mine Head
Miory see Myory
158 L5 **Miquan** Xinjiang Uygur Zizhiqu, NW China 44°04′N 87°40′E
119 I17 **Mir** Hrodzyenskaya Voblasts', W Belarus 53°25′N 26°28′E
106 H8 **Mira** Veneto, NE Italy 45°25′N 12°07′E
15 O12 **Mirabel** var. Montreal. ✈ (Montréal) Québec, SE Canada 45°27′N 73°47′W
60 Q8 **Miracema** Rio de Janeiro, SE Brazil 21°24′S 42°10′W
54 G9 **Miraflores** Boyacá, C Colombia 05°07′N 73°09′W
40 G10 **Miraflores** Baja California Sur, NW Mexico 23°24′N 109°45′W
44 L9 **Miragoâne** S Haiti 18°25′N 73°07′W
155 E16 **Miraj** Mahārāshtra, W India 16°51′N 74°42′E
61 E23 **Miramar** Buenos Aires, E Argentina 38°15′S 57°50′W
103 R15 **Miramas** Bouches-du-Rhône, SE France 43°33′N 05°02′E
102 K12 **Mirambeau** Charente-Maritime, W France 45°23′N 00°33′W
102 L13 **Miramont-de-Guyenne** Lot-et-Garonne, SW France 44°34′N 00°20′E
115 L25 **Mirampéllou Kólpos** gulf Kríti, Greece, E Mediterranean Sea
158 L8 **Miran** Xinjiang Uygur Zizhiqu, NW China 39°13′N 88°58′E
54 M5 **Miranda** off. Estado Miranda. ◈ state N Venezuela
Miranda de Corvo see Miranda do Corvo
105 O3 **Miranda de Ebro** La Rioja, N Spain 42°41′N 02°57′W
104 G8 **Miranda do Corvo** var. Miranda de Corvo. Coimbra, N Portugal 40°05′N 08°20′W
104 J6 **Miranda do Douro** Bragança, N Portugal 41°30′N 06°16′W
Miranda, Estado see Miranda
102 L15 **Mirande** Gers, S France 43°31′N 00°25′E
104 I6 **Mirandela** Bragança, N Portugal 41°28′N 07°10′W
25 R15 **Mirando City** Texas, SW USA 27°24′N 99°00′W
106 G9 **Mirandola** Emilia-Romagna, N Italy 44°52′N 11°04′E
60 I8 **Mirandópolis** São Paulo, S Brazil 21°10′S 51°03′W
104 G13 **Mira, Rio** ≈ S Portugal
60 K8 **Mirassol** São Paulo, S Brazil 20°50′S 49°30′W
104 J3 **Miravalles** ▲ NW Spain 42°52′N 06°45′W
42 L12 **Miravalles, Volcán** ⛰ NW Costa Rica 10°43′N 85°07′W
141 W13 **Mirbāṭ** var. Marbat. S Oman 17°03′N 54°44′E
44 M9 **Mirebalais** C Haiti 18°51′N 72°08′W
103 T6 **Mirecourt** Vosges, NE France 48°19′N 06°04′E
103 N16 **Mirepoix** Ariège, S France 43°05′N 01°51′E
Mirgorod see Myrhorod
139 W10 **Mīr Ḥājī Khalīl** Wāsiţ, E Iraq 32°11′N 46°19′E
169 T8 **Miri** Sarawak, East Malaysia 04°23′N 113°59′E
77 W12 **Miria** Zinder, S Niger 13°39′N 09°15′E
182 F5 **Mirikata** South Australia 29°56′S 135°13′E
54 K4 **Mirimire** Falcón, N Venezuela 11°14′N 68°39′W
61 H18 **Mirim Lagoon** var. Lake Mirim, Sp. Laguna Merín. lagoon Brazil/Uruguay
Mirim, Lake see Mirim Lagoon
Mírina see Mýrina
172 H14 **Miringoni** Mohéli, S Comoros 12°17′S 93°39′E
143 W11 **Mīrjāveh** Sīstān va Balūchestān, SE Iran 29°04′N 61°24′E
195 Z9 **Mirny** Russian research station Antarctica 66°25′S 93°09′E
124 M10 **Mirnyy** Arkhangel'skaya Oblast', NW Russian Federation 62°50′N 40°20′E
123 O10 **Mirnyy** Respublika Sakha (Yakutiya), NE Russian Federation 62°30′N 113°58′E
Mironovka see Myronivka
110 F9 **Mirosławiec** Zachodnio-pomorskie, NW Poland 53°21′N 16°04′E
Mirovo see Vrattsa
100 N10 **Mirow** Mecklenburg-Vorpommern, N Germany 53°16′N 12°48′E
152 G6 **Mirpur** Jammu and Kashmir, NW India 33°06′N 73°49′E
Mirpur see New Mirpur
149 P17 **Mīrpur Batoro** Sind, SE Pakistan 24°40′N 68°15′E
149 Q16 **Mīrpur Khās** Sind, SE Pakistan 25°31′N 69°01′E
149 P17 **Mīrpur Sakro** Sind, SE Pakistan 24°32′N 67°38′E
143 T14 **Mīr Shahdād** Hormozgān, S Iran 26°15′N 58°29′E
Mirtoan Sea see Mirtóo Pélagos
115 G21 **Mirtóo Pélagos** Eng. Mirtoan Sea; anc. Myrtoum Mare. sea S Greece
163 Z16 **Miryang** var. Milyang, Jap. Mitsuō. SE South Korea 35°30′N 128°46′E
Mirzachirla see Murzechirla
164 E14 **Misaki** Ehime, Shikoku, SW Japan 33°22′N 132°04′E
Misān see Maysān
41 Q13 **Misantla** Veracruz-Llave, E Mexico 19°54′N 96°51′W
165 R7 **Misawa** Aomori, Honshū, C Japan 40°42′N 141°26′E
57 G14 **Mishagua, Río** ≈ C Peru
163 Z8 **Mishan** Heilongjiang, NE China 45°30′N 131°53′E
31 O11 **Mishawaka** Indiana, N USA 41°40′N 86°10′W
39 N6 **Misheguk Mountain** ▲ Alaska, USA 68°13′N 161°11′W
165 N14 **Mishima** var. Misima. Shizuoka, Honshū, S Japan 35°08′N 138°54′E
164 E12 **Mi-shima** island SW Japan
127 V4 **Mishkino** Respublika Bashkortostan, W Russian Federation 55°31′N 55°57′E
153 Y10 **Mishmi Hills** hill range NE India
161 N11 **Mi Shui** ≈ S China
Misiaf see Maşyāf
107 J23 **Misilmeri** Sicilia, Italy, C Mediterranean Sea 38°03′N 13°27′E
Misima see Mishima
14 C7 **Missinaibi Lake** ◎ Ontario, S Canada
Misión de Guana see Guana
60 F13 **Misiones** off. Provincia de Misiones. ◈ province NE Argentina
62 P8 **Misiones** off. Departamento de las Misiones. ◈ department S Paraguay
Misiones, Departamento de las see Misiones
Misiones, Provincia de see Misiones
Misión San Fernando see San Fernando
Miskin see Maskin
Miskito Coast see La Mosquitia
43 O7 **Miskitos, Cayos** island group NE Nicaragua
111 M21 **Miskolc** Borsod-Abaúj-Zemplén, NE Hungary 48°05′N 20°46′E
171 T12 **Misool, Pulau** island Papua Barat, E Indonesia
Misox see Mesocco
29 Y3 **Misquah Hills** hill range Minnesota, N USA
75 P7 **Mişrātah** var. Misurata. NW Libya 32°23′N 15°06′E
75 P7 **Mişrātah, Rās** headland N Libya 32°22′N 15°16′E
14 C7 **Missanabie** Ontario, S Canada 48°18′N 84°04′W
58 E10 **Missão Catrimani** Roraima, N Brazil 01°26′N 62°05′W
14 D6 **Missinaibi** ≈ Ontario, S Canada
11 T13 **Missinipe** Saskatchewan, C Canada 55°36′N 104°45′W
28 M11 **Mission** South Dakota, N USA 43°16′N 100°38′W
25 S17 **Mission** Texas, SW USA 26°13′N 98°19′W
12 F10 **Missisa Lake** ◎ Ontario, C Canada
18 M6 **Missisquoi Bay** lake bay Canada/USA
14 C10 **Mississagi** ≈ Ontario, S Canada
14 G15 **Mississauga** Ontario, S Canada 43°38′N 79°36′W
31 P12 **Mississinewa Lake** ▣ Indiana, N USA
31 P12 **Mississinewa River** ≈ Indiana/Ohio, N USA
22 K4 **Mississippi** off. State of Mississippi, also known as Bayou State, Magnolia State. ◈ state SE USA
14 K13 **Mississippi** ≈ Ontario, SE Canada
47 N1 **Mississippi Fan** undersea feature N Gulf of Mexico 26°45′N 88°30′W
14 L13 **Mississippi Lake** ◎ Ontario, SE Canada
22 M10 **Mississippi Delta** delta Louisiana, S USA
0 J11 **Mississippi River** ≈ C USA
22 M9 **Mississippi Sound** sound Alabama/Mississippi, S USA
33 P9 **Missoula** Montana, NW USA 46°54′N 114°03′W
27 T5 **Missouri** off. State of Missouri, also known as Bullion State, Show Me State. ◈ state C USA
25 V11 **Missouri City** Texas, SW USA 29°37′N 95°32′W
0 J10 **Missouri River** ≈ C USA
15 Q6 **Mistassibi** ≈ Québec, SE Canada
15 P6 **Mistassini** ≈ Québec, SE Canada
12 J11 **Mistassini, Lac** ◎ Québec, SE Canada
109 Y3 **Mistelbach an der Zaya** Niederösterreich, NE Austria 48°34′N 16°33′E
107 L24 **Misterbianco** Sicilia, Italy, C Mediterranean Sea 37°31′N 15°01′E
95 N19 **Misterhult** Kalmar, S Sweden 57°28′N 16°34′E
12 K11 **Mistissini** var. Baie-du-Poste. Québec, SE Canada 50°20′N 73°50′W
57 H17 **Misti, Volcán** ⛰ S Peru 16°20′S 71°22′W
Mistras see Mystrás
107 K23 **Mistretta** anc. Amestratus. Sicilia, Italy, C Mediterranean Sea 37°56′N 14°22′E
164 F12 **Misumi** Shimane, Honshū, SW Japan 34°47′N 132°00′E
Misurata see Mişrātah
83 O14 **Mitande** Niassa, N Mozambique 14°06′S 36°03′E
40 J13 **Mita, Punta de** headland C Mexico 20°46′N 105°31′W
55 W12 **Mitaraka, Massif du** ▲ NE South America 02°18′N 54°31′W
Mitau see Jelgava
181 X9 **Mitchell** Queensland, E Australia 26°29′S 148°00′E
14 E15 **Mitchell** Ontario, S Canada 43°28′N 81°11′W
28 I13 **Mitchell** Nebraska, C USA 41°56′N 103°48′W
32 J12 **Mitchell** Oregon, NW USA 44°34′N 120°09′W
29 P11 **Mitchell** South Dakota, N USA 43°42′N 98°01′W
23 P5 **Mitchell Lake** ▣ Alabama, S USA
31 P7 **Mitchell, Lake** ◎ Michigan, N USA
21 P9 **Mitchell, Mount** ▲ North Carolina, SE USA 35°46′N 82°16′W
181 V3 **Mitchell River** ≈ Queensland, NE Australia
97 D20 **Mitchelstown** Ir. Baile Mhistéala. SW Ireland 52°20′N 08°16′W
14 M9 **Mitchinamécus, Lac** ◎ Québec, SE Canada
Mitèmboni see Mitemele, Río
79 D17 **Mitemele, Río** var. Mitèmboni, Temboni, Utamboni. ≈ S Equatorial Guinea
149 S12 **Mithān Kot** Punjab, E Pakistan 28°53′N 70°25′E
149 T7 **Mitha Tiwāna** Punjab, E Pakistan 32°18′N 72°08′E
149 R17 **Mithi** Sind, SE Pakistan 24°43′N 69°53′E
Míthimna see Míthymna
Mi Tho see My Tho
115 L16 **Míthymna** var. Míthimna. Lésvos, E Greece 39°20′N 26°12′E
190 L16 **Mitiaro** island S Cook Islands
Mitilíni see Mytilíni
15 U7 **Mitis** ≈ Québec, SE Canada
41 R16 **Mitla** Oaxaca, SE Mexico 16°56′N 96°19′W
165 P13 **Mito** Ibaraki, Honshū, S Japan 36°21′N 140°26′E
92 N2 **Mitra, Kapp** headland W Svalbard 79°07′N 11°11′E
184 M13 **Mitre** ▲ North Island, New Zealand 40°46′S 175°27′E
185 B21 **Mitre Peak** ▲ South Island, New Zealand 44°37′S 167°45′E
39 O15 **Mitrofania Island** island Alaska, USA
Mitrovica/Mitrovicë see Kosovska Mitrovica, Serbia
Mitrovica/Mitrowitz see Sremska Mitrovica, Serbia
113 M16 **Mitrovicë** Serb. Mitrovica, Kosovska Mitrovica, Titova Mitrovica. N Kosovo 42°54′N 20°52′E
172 H12 **Mitsamiouli** Grande Comore, NW Comoros 11°22′S 43°19′E
172 I3 **Mitsinjo** Mahajanga, NW Madagascar 16°00′S 45°52′E
80 J9 **Mits'iwa** var. Masawa, Massawa. E Eritrea 15°37′N 39°27′E
172 H13 **Mitsoudjé** Grande Comore, NW Comoros
138 F12 **Mitspe Ramon** prev. Mizpe Ramon. Southern, S Israel 30°38′N 34°47′E
165 T5 **Mitsuishi** Hokkaidō, NE Japan 42°12′N 142°40′E
165 O11 **Mitsuke** var. Mituke. Niigata, Honshū, C Japan 37°30′N 138°54′E
Mitsuō see Miryang
164 C12 **Mitsushima** Nagasaki, Tsushima, SW Japan 34°16′N 129°18′E
100 G12 **Mittellandkanal** canal NW Germany
108 J7 **Mittelberg** Vorarlberg, NW Austria 47°19′N 10°09′E
Mitteldorf see Międzychód
Mittelstadt see Baia Sprie
Mitterburg see Pazin
109 P7 **Mittersill** Salzburg, NW Austria 47°16′N 12°27′E
Mittimatalik see Pond Inlet
101 N16 **Mittweida** Sachsen, E Germany 50°59′N 12°57′E
54 J13 **Mitú** Vaupés, SE Colombia 01°07′N 70°05′W
Mituke see Mitsuke
Mitumba, Chaîne des/ Mitumba Range see Mitumba, Monts
79 O22 **Mitumba, Monts** var. Chaîne des Mitumba, Mitumba Range. ▲▲ E Dem. Rep. Congo
79 N23 **Mitwaba** Katanga, SE Dem. Rep. Congo 08°37′S 27°20′E
79 E18 **Mitzic** Woleu-Ntem, N Gabon 0°48′N 11°30′E
82 K11 **Miueru Wantipa, Lake** ◎ N Zambia
165 N14 **Miura** Kanagawa, Honshū, S Japan 35°08′N 139°37′E
165 Q10 **Miyagi** off. Miyagi-ken. ◈ prefecture Honshū, C Japan
Miyagi-ken see Miyagi
138 M7 **Miyāh, Wādī al** dry watercourse E Syria
165 X13 **Miyake** Tōkyō, Miyako-jima, SE Japan 34°35′N 135°33′E
165 Q16 **Miyake-jima** island Sakishima-shotō, SW Japan
165 R8 **Miyako** Iwate, Honshū, C Japan 39°39′N 141°57′E
164 D16 **Miyakonojō** var. Miyakonzyô. Miyazaki, Kyūshū, SW Japan 31°42′N 131°04′E
Miyakonzyô see Miyakonojō
165 Q16 **Miyako-shotō** island group SW Japan
144 G11 **Miyaly** Atyrau, W Kazakhstan 48°52′N 53°55′E
Miyāndoāb see Mīāndowāb
Miyāneh see Mīāneh
Miyang see Mile
164 D16 **Miyazaki** Miyazaki, Kyūshū, SW Japan 31°55′N 131°24′E
164 D16 **Miyazaki** off. Miyazaki-ken. ◈ prefecture Kyūshū, SW Japan
Miyazaki-ken see Miyazaki
164 J12 **Miyazu** Kyōto, Honshū, SW Japan 35°33′N 135°12′E
Miyory see Myory
164 G12 **Miyoshi** var. Miyosi. Hiroshima, Honshū, SW Japan 34°48′N 132°51′E
Miyosi see Miyoshi
Miza see Mizë
81 H14 **Mīzan Teferī** Southern Nationalities, S Ethiopia 06°57′N 35°30′E
Mizda see Mizdah
75 O8 **Mizdah** var. Mizda. NW Libya 31°26′N 12°59′E
113 K20 **Mizë** var. Miza. Fier, W Albania 40°58′N 19°32′E
97 A22 **Mizen Head** Ir. Carn Uí Néid. headland SW Ireland 51°26′N 09°50′W
116 H7 **Mizhhir"ya** Rus. Mezhgor'ye. Zakarpats'ka Oblast', W Ukraine 48°30′N 23°30′E
160 L4 **Mizhi** Shaanxi, C China 37°50′N 110°03′E
114 H7 **Mizia** var. Miziya. Vratsa, NW Bulgaria 43°42′N 23°52′E
116 K13 **Mizil** Prahova, SE Romania 45°00′N 26°29′E
Miziya see Mizia
153 W15 **Mizo Hills** hill range E India
153 W15 **Mizoram** ◈ state NE India
Mizpe Ramon see Mitspe Ramon
57 L19 **Mizque** Cochabamba, C Bolivia 17°57′S 65°10′W
57 M19 **Mizque, Río** ≈ C Bolivia
165 Q9 **Mizusawa** var. Ōshū. Iwate, Honshū, C Japan 39°10′N 141°07′E
95 M18 **Mjölby** Östergötland, S Sweden 58°19′N 15°10′E
95 G15 **Mjøndalen** Buskerud, S Norway 59°45′N 09°58′E
95 J19 **Mjörn** ◎ S Sweden
94 I13 **Mjøsa** var. Mjøsen. ◎ S Norway
Mjøsen see Mjøsa
81 G21 **Mkalama** Singida, C Tanzania 04°09′S 34°35′E
80 K13 **Mkata** ≈ C Tanzania
83 K14 **Mkushi** Central, C Zambia 13°40′S 29°26′E
83 L22 **Mkuze** KwaZulu/Natal, E South Africa 27°37′S 32°03′E
81 J22 **Mkwaja** Tanga, E Tanzania 05°42′S 38°48′E
111 D16 **Mladá Boleslav** Ger. Jungbunzlau. Středočeský Kraj, N Czech Republic 50°26′N 14°55′E
112 M12 **Mladenovac** Serbia, C Serbia 44°27′N 20°42′E
114 L11 **Mladinovo** Haskovo, S Bulgaria 41°57′N 26°13′E
113 O17 **Mlado Nagoričane** N FYR Macedonia 42°11′N 21°49′E
112 N12 **Mlava** ≈ E Serbia
110 L9 **Mława** Mazowieckie, C Poland 53°07′N 20°23′E
113 G16 **Mljet** It. Meleda; anc. Melita. island S Croatia
167 S11 **Mlu Prey** prev. Phumĭ Mlu Prey. Preăh Vihéar, N Cambodia 13°48′N 105°16′E
116 K4 **Mlyniv** Rivnens'ka Oblast', NW Ukraine 50°31′N 25°36′E
83 I21 **Mmabatho** North-West, N South Africa 25°51′S 25°37′E
83 I19 **Mmashoro** Central, E Botswana 21°56′S 26°39′E
44 J7 **Moa** Holguín, E Cuba 20°42′N 74°57′W
76 J15 **Moa** ≈ Guinea/Sierra Leone
37 O6 **Moab** Utah, W USA 38°35′N 109°34′W
181 V1 **Moa Island** island Queensland, NE Australia
187 Y15 **Moala** island S Fiji
83 L21 **Moamba** Maputo, SW Mozambique 25°35′S 32°13′E
79 F19 **Moanda** var. Mouanda. Haut-Ogooué, SE Gabon 01°31′S 13°07′E
83 M15 **Moatize** Tete, NW Mozambique 16°04′S 33°43′E
79 P22 **Moba** Katanga, E Dem. Rep. Congo 07°03′S 29°52′E
Mobay see Montego Bay
79 K15 **Mobaye** Basse-Kotto, S Central African Republic 04°19′N 21°17′E
79 K15 **Mobayi-Mbongo** Equateur, NW Dem. Rep. Congo 04°21′N 21°10′E
25 P2 **Mobeetie** Texas, SW USA 35°33′N 100°25′W
27 U3 **Moberly** Missouri, C USA 39°25′N 92°26′W
23 N8 **Mobile** Alabama, S USA 30°42′N 88°03′W
23 N9 **Mobile Bay** bay Alabama, S USA
23 N8 **Mobile River** ≈ Alabama, S USA
29 N8 **Mobridge** South Dakota, N USA 45°32′N 100°25′W
Mobutu Sese Seko, Lac see Albert, Lake
45 N8 **Moca** N Dominican Republic 19°26′N 70°33′W
83 Q15 **Moçambique** Nampula, NE Mozambique 15°00′S 40°44′E
Moçâmedes see Namibe
167 S6 **Môc Châu** Sơn La, N Vietnam 20°49′N 104°38′E
187 Z15 **Moce** island Lau Group, E Fiji
Mocha see Al Mukhā
193 T11 **Mocha Fracture Zone** tectonic feature SE Pacific Ocean
63 F14 **Mocha, Isla** island C Chile
56 C12 **Moche, Río** ≈ W Peru
167 S14 **Môc Hoa** Long An, S Vietnam 10°46′N 105°56′E
83 I20 **Mochudi** Kgatleng, SE Botswana 24°25′S 26°07′E
82 Q13 **Mocímboa da Praia** var. Vila de Mocímboa da Praia. Cabo Delgado, N Mozambique 11°17′S 40°21′E
94 L13 **Mockfjärd** Dalarna, C Sweden 60°30′N 14°57′E
21 R9 **Mocksville** North Carolina, SE USA 35°53′N 80°33′W
32 F8 **Moclips** Washington, NW USA 47°11′N 124°13′W
82 C13 **Môco** var. Morro de Môco. ▲ W Angola 12°36′S 15°09′E
54 D13 **Mocoa** Putumayo, SW Colombia 01°07′N 76°38′W
60 M8 **Mococa** São Paulo, S Brazil 21°30′S 47°00′W
Môco, Morro de see Môco
40 H8 **Mocorito** Sinaloa, C Mexico 25°24′N 107°55′W
40 J4 **Moctezuma** Chihuahua, N Mexico 30°10′N 106°28′W
41 N11 **Moctezuma** San Luis Potosí, C Mexico 22°46′N 101°06′W
40 G4 **Moctezuma** Sonora, NW Mexico 29°50′N 109°40′W
41 P12 **Moctezuma, Río** ≈ C Mexico
Mó, Cuan see Clew Bay
83 O16 **Mocuba** Zambézia, NE Mozambique 16°50′S 37°02′E
103 U12 **Modane** Savoie, E France 45°14′N 06°41′E
106 F9 **Modena** anc. Mutina. Emilia-Romagna, N Italy 44°39′N 10°55′E
36 I7 **Modena** Utah, W USA 37°46′N 113°54′W
35 O9 **Modesto** California, W USA 37°38′N 121°02′W
107 L25 **Modica** anc. Motyca. Sicilia, Italy, C Mediterranean Sea 36°52′N 14°45′E
83 J20 **Modimolle** prev. Nylstroom. Limpopo, NE South Africa 24°39′S 28°23′E
79 K17 **Modjamboli** Equateur, N Dem. Rep. Congo 02°27′N 22°03′E
109 X4 **Mödling** Niederösterreich, NE Austria 48°06′N 16°18′E
Modohn see Madona
Modot see Tsenhermandal
171 V14 **Modowi** Papua Barat, E Indonesia 04°05′S 134°39′E
112 I12 **Modračko Jezero** ▣ NE Bosnia and Herzegovina
112 I10 **Modriča** Republika Srpska, N Bosnia and Herzegovina 44°57′N 18°17′E
183 O13 **Moe** Victoria, SE Australia 38°11′S 146°18′E
Moearatewe see Muaratewe
Moei, Mae Nam see Thaungyin
94 H13 **Moelv** Hedmark, S Norway 60°55′N 10°47′E
92 I10 **Moen** Troms, N Norway 69°08′N 18°35′E
Möen see Møn, Denmark
Moen see Weno, Micronesia
Moena see Muna, Pulau
36 M10 **Moenkopi Wash** ≈ Arizona, SW USA
185 F22 **Moeraki Point** headland South Island, New Zealand 45°23′S 170°52′E
99 F16 **Moerbeke** Oost-Vlaanderen, NW Belgium 51°11′N 03°57′E
99 H14 **Moerdijk** Noord-Brabant, S Netherlands 51°42′N 04°37′E
Moero, Lac see Mweru, Lake
101 D15 **Moers** var. Mörs. Nordrhein-Westfalen, W Germany 51°27′N 06°36′E
Moesi see Musi, Air
Moeskroen see Mouscron
96 J13 **Moffat** S Scotland, United Kingdom 55°29′N 03°36′W
185 C22 **Moffat Peak** ▲ South Island, New Zealand 44°57′S 168°10′E
79 N19 **Moga** Sud-Kivu, E Dem. Rep. Congo 02°16′S 26°54′E
152 H8 **Moga** Punjab, N India 30°49′N 75°13′E
Mogadiscio/Mogadishu see Muqdisho
Mogador see Essaouira
104 J6 **Mogadouro** Bragança, N Portugal 41°20′N 06°43′W
167 N2 **Mogaung** Kachin State, N Myanmar (Burma) 25°20′N 96°54′E
110 L13 **Mogielnica** Mazowieckie, C Poland 51°40′N 20°42′E
Mogilëv see Mahilyow
Mogilev-Podol'skiy see Mohyliv-Podil's'kyy
Mogilëvskaya Oblast' see Mahilyowskaya Voblasts'

◆ Country ● Country Capital ◇ Dependent Territory ○ Dependent Territory Capital ◈ Administrative Regions ✈ International Airport ▲ Mountain ▲▲ Mountain Range ⛰ Volcano ≈ River ◎ Lake ▣ Reservoir
289

MOGILNO – MONTROSE

110 I11 **Mogilno** Kujawsko-pomorskie, C Poland 52°39′N 17°58′E
83 Q15 **Mogincual** Nampula, NE Mozambique 15°33′S 40°28′E
114 E13 **Moglenítsas** ≈ N Greece
106 H8 **Mogliano Veneto** Veneto, NE Italy 45°34′N 12°14′E
113 M21 **Moglicë** Korçë, SE Albania 40°43′N 20°22′E
123 O13 **Mogocha** Zabaykal'skiy Kray, S Russian Federation 53°39′N 119°47′E
122 J11 **Mogochin** Tomskaya Oblast', C Russian Federation 57°42′N 83°24′E
80 F13 **Mogogh** Jonglei, E South Sudan 08°26′N 31°19′E
171 U12 **Mogoi** Papua Barat, E Indonesia 01°44′S 133°13′E
166 M4 **Mogok** Mandalay, C Myanmar (Burma) 22°55′N 96°29′E
37 P14 **Mogollon Mountains** ▲▲ New Mexico, SW USA
36 M12 **Mogollon Rim** *cliff* Arizona, SW USA
61 E23 **Mogotes, Punta** *headland* E Argentina 38°03′S 57°31′W
42 J8 **Mogotón** ▲ NW Nicaragua 13°45′N 86°22′W
104 I14 **Moguer** Andalucía, S Spain 37°15′N 06°52′W
111 J26 **Mohács** Baranya, SW Hungary 46°N 18°40′E
185 C20 **Mohaka** ≈ North Island, New Zealand
28 M2 **Mohall** North Dakota, N USA 48°45′N 101°30′W
Moḩammadābād *see* Dargaz
143 U12 **Moḩammadābād-e Rīgān** Kermān, SE Iran 28°39′N 59°01′E
74 F6 **Mohammedia** *prev.* Fédala. NW Morocco 33°46′N 07°16′W
74 F6 **Mohammed V** ✈ (Casablanca) W Morocco 33°07′N 08°28′W
Mohammerah *see* Khorramshahr
36 H10 **Mohave, Lake** ▣ Arizona/Nevada, W USA
36 I12 **Mohave Mountains** ▲▲ Arizona, SW USA
36 I15 **Mohawk Mountains** ▲▲ Arizona, SW USA
18 J10 **Mohawk River** ≈ New York, NE USA
163 T3 **Mohe** *var.* Xilinji. Heilongjiang, NE China 53°01′N 122°26′E
95 L20 **Moheda** Kronoberg, S Sweden 57°00′N 14°34′E
Mohéli *see* Mwali
Mohendergarh *see* Mahendragarh
38 K12 **Mohican, Cape** *headland* Nunivak Island, Alaska, USA 60°12′N 167°25′W
Mohn *see* Muhu
101 G15 **Möhne** ≈ W Germany
101 G15 **Möhne-Stausee** ▣ W Germany
92 P2 **Mohn, Kapp** *headland* NW Svalbard 79°26′N 25°44′E
197 S14 **Mohns Ridge** *undersea feature* Greenland Sea/Norwegian Sea 72°30′N 05°00′E
57 I17 **Moho** Puno, SE Peru 15°21′S 69°32′W
Mohokare *see* Caledon
95 L17 **Moholm** Västra Götaland, S Sweden 58°37′N 14°04′E
36 J11 **Mohon Peak** ▲ Arizona, SW USA 34°55′N 113°07′W
81 J23 **Mohoro** Pwani, E Tanzania 08°09′S 39°10′E
Mohra *see* Moravice
Mohrungen *see* Morąg
116 M7 **Mohyliv-Podil's'kyy** *Rus.* Mogilev-Podol'skiy. Vinnyts'ka Oblast', C Ukraine 48°29′N 27°49′E
95 D17 **Moi** Rogaland, S Norway 58°27′N 06°32′E
Moili *see* Mwali
116 K11 **Moinești** *Hung.* Mojnest. Bacău, E Romania 46°27′N 26°31′E
Móinteach Mílic *see* Mountmellick
14 J14 **Moira** ≈ Ontario, SE Canada
92 G13 **Mo i Rana** Nordland, C Norway 66°19′N 14°10′E
153 X14 **Moirāng** Manipur, NE India 24°29′N 93°45′E
115 J25 **Moíres** Kríti, Greece, E Mediterranean Sea 35°03′N 24°51′E
118 H6 **Mõisaküla** *Ger.* Moiseküll. Viljandimaa, S Estonia 58°05′N 25°12′E
Moiseküll *see* Mõisaküla
15 W4 **Moisie** Québec, E Canada 50°12′N 66°06′W
15 W3 **Moisie** ≈ Québec, SE Canada
102 M14 **Moissac** Tarn-et-Garonne, S France 44°07′N 01°05′E
78 I13 **Moïssala** Mandoul, S Chad 08°21′N 17°46′E
55 O7 **Moitaco** Bolívar, E Venezuela 08°00′N 64°22′W
95 P15 **Möja** Stockholm, C Sweden 59°25′N 18°55′E
105 Q14 **Mojácar** Andalucía, S Spain 37°09′N 01°50′W
35 T13 **Mojave** California, W USA 35°03′N 118°10′W
35 V13 **Mojave Desert** *plain* California, W USA
35 V13 **Mojave River** ≈ California, W USA
60 L9 **Moji-Mirim** *var.* Moji-Mirim. São Paulo, S Brazil 22°26′S 46°55′W
Moji-Mirim *see* Moji-Mirim
113 K15 **Mojkovac** E Montenegro 42°57′N 19°34′E
Mojnest *see* Moinești
Mōka *see* Mooka
153 Q13 **Mokāma** *prev.* Mokameh, Mukama. Bihār, N India 25°24′N 85°55′E
79 O25 **Mokambo** Katanga, SE Dem. Rep. Congo 12°23′S 28°21′E
Mokameh *see* Mokāma
38 D9 **Mōkapu Point** *var.* Mokapu Point. *headland* O'ahu, Hawai'i, USA 21°27′N 157°43′W
184 L9 **Mokau** Waikato, North Island, New Zealand 38°42′S 174°37′E
184 L9 **Mokau** ≈ North Island, New Zealand
35 P7 **Mokelumne River** ≈ California, W USA
83 J23 **Mokhotlong** NE Lesotho 29°19′S 29°06′E
Mokil Atoll *see* Mwokil Atoll
95 N14 **Möklinta** Västmanland, C Sweden 60°04′N 16°34′E
Mokna *see* Mokra Gora
184 L4 **Mokohinau Islands** *island group* N New Zealand
153 X12 **Mokokchūng** Nāgāland, NE India 26°20′N 94°30′E
78 F12 **Mokolo** Extrême-Nord, N Cameroon 10°49′N 13°54′E
83 J20 **Mokopane** *prev.* Potgietersrus. Limpopo, NE South Africa 24°09′S 28°58′E
185 D24 **Mokoreta** ≈ South Island, New Zealand
163 X17 **Mokpo** *Jap.* Moppo; *prev.* Mokp'o. SW South Korea 34°50′N 126°26′E
Mokp'o *see* Mokpo
113 L16 **Mokra Gora** *Alb.* Mokna. ▲▲ S Serbia
Mokrany *see* Makrany
127 O5 **Moksha** ≈ W Russian Federation
143 X12 **Mok Sukhteh-ye Pāyīn** Sīstān va Balūchestān, SE Iran
Moktama *see* Mottama
77 T14 **Mokwa** Niger, W Nigeria 09°19′N 05°01′E
99 J16 **Mol** *prev.* Moll. Antwerpen, N Belgium 51°11′N 05°07′E
107 O17 **Mola di Bari** Puglia, SE Italy 41°03′N 17°05′E
Molai *see* Moláoi
41 P13 **Molango** Hidalgo, C Mexico 20°48′N 98°44′W
115 F22 **Moláoi** *var.* Molai. Pelopónnisos, S Greece 36°48′N 22°51′E
41 Z12 **Molas del Norte, Punta** *var.* Punta Molas. *headland* SE Mexico 20°34′N 86°43′W
Molas, Punta *see* Molas del Norte, Punta
105 R11 **Molatón** ▲ C Spain 38°58′N 01°19′W
97 K18 **Mold** NE Wales, United Kingdom 53°10′N 03°08′W
Moldau *see* Vltava, Czech Republic
Moldau *see* Moldova
Moldavia *see* Moldova
Moldavian SSR/Moldavskaya SSR *see* Moldova
94 E9 **Molde** Møre og Romsdal, S Norway 62°44′N 07°08′E
Moldotau, Khrebet *see* Moldo-Too, Khrebet
147 V9 **Moldo-Too, Khrebet** *prev.* Khrebet Moldotau. ▲▲ C Kyrgyzstan
116 L9 **Moldova** *off.* Republic of Moldova, *var.* Moldavia; *prev.* Moldavian SSR, *Rus.* Moldavskaya SSR. ◆ *republic* SE Europe
116 K9 **Moldova** *Eng.* Moldavia, *Ger.* Moldau. *former province* NE Romania
116 K9 **Moldova** ≈ N Romania
116 F13 **Moldova Nouă** *Ger.* Neumoldowa, *Hung.* Újmoldova. Caraș-Severin, SW Romania 44°45′N 21°39′E
Moldova, Republic of *see* Moldova
116 F13 **Moldova Veche** *Ger.* Altmoldowa, *Hung.* Ómoldova. Caraș-Severin, SW Romania 44°45′N 21°13′E
Moldoveanul *see* Vârful Moldoveanu
83 I20 **Molepolole** Kweneng, SE Botswana 24°25′S 25°30′E
44 L8 **Môle-St-Nicolas** NW Haiti 19°46′N 73°19′W
118 H13 **Molėtai** Utena, E Lithuania 55°14′N 25°25′E
107 O17 **Molfetta** Puglia, SE Italy 41°12′N 16°35′E
171 P11 **Molibagu** Sulawesi, N Indonesia 0°25′N 123°57′E
62 G12 **Molina** Maule, C Chile 35°06′S 71°18′W
105 Q7 **Molina de Aragón** Castilla-La Mancha, C Spain 40°50′N 01°54′W
105 R13 **Molina de Segura** Murcia, SE Spain 38°03′N 01°11′W
30 J11 **Moline** Illinois, N USA 41°30′N 90°31′W
27 P7 **Moline** Kansas, C USA 37°21′N 96°18′W
79 P23 **Moliro** Katanga, SE Dem. Rep. Congo 08°11′S 30°31′E
107 K16 **Molise** ◈ *region* S Italy
95 K15 **Molkom** Värmland, C Sweden 59°36′N 13°43′E
109 Q9 **Möll** ≈ S Austria
Moll *see* Mol
146 I14 **Mollanepes Adyndaky** *Rus.* Imeni Mollanepesa. Mary Welaýaty, S Turkmenistan 37°36′N 61°54′E
95 J22 **Mölle** Skåne, S Sweden 56°15′N 12°19′E
57 H18 **Mollendo** Arequipa, SW Peru 17°02′S 72°01′W
105 U5 **Mollerussa** Cataluña, NE Spain 41°37′N 00°53′E
108 H8 **Mollis** Glarus, NE Switzerland 47°05′N 09°03′E
95 J19 **Mölndal** Västra Götaland, S Sweden 57°39′N 12°05′E
95 J19 **Mölnlycke** Västra Götaland, S Sweden 57°39′N 12°09′E
117 U9 **Molochans'k** *Rus.* Molochansk. Zaporiz'ka Oblast', SE Ukraine 47°10′N 35°38′E
117 U10 **Molochna** *Rus.* Molochnaya. ≈ S Ukraine
Molochnaya *see* Molochna
117 U10 **Molochnyy Lyman** *bay* N Black Sea
Molodechno/Molodeczno *see* Maladzyechna
195 V3 **Molodezhnaya** *Russian research station* Antarctica 67°33′S 46°12′E
124 J14 **Mologa** ≈ NW Russian Federation
38 E9 **Moloka'i** *var.* Molokai. *island* Hawaiian Islands, Hawai'i, USA
175 X3 **Molokai Fracture Zone** *tectonic feature* NE Pacific Ocean
124 K15 **Molokovo** Tverskaya Oblast', W Russian Federation 58°10′N 36°43′E
125 Q14 **Moloma** ≈ NW Russian Federation
183 R8 **Molong** New South Wales, SE Australia 33°07′S 148°52′E
83 H21 **Molopo** *seasonal river* Botswana/South Africa
115 F17 **Mólos** Stereá Elláda, C Greece 38°48′N 22°39′E
171 O11 **Molosipat** Sulawesi, N Indonesia 0°28′N 121°08′E
Molotov *see* Severodvinsk, Arkhangel'skaya Oblast', Russian Federation
Molotov *see* Perm', Permskaya Oblast', Russian Federation
79 G17 **Moloundou** Est, SE Cameroon 02°03′N 15°14′E
103 U5 **Molsheim** Bas-Rhin, NE France 48°33′N 07°30′E
11 X13 **Molson Lake** ◎ Manitoba, C Canada
Moluccas *see* Maluku
171 Q12 **Molucca Sea** *Ind.* Laut Maluku. *sea* E Indonesia
Molukken *see* Maluku
83 O15 **Molumbo** Zambézia, N Mozambique 15°33′S 36°19′E
171 T15 **Molu, Pulau** *island* Maluku, E Indonesia
83 P16 **Moma** Nampula, NE Mozambique 16°42′S 39°12′E
171 X14 **Momats** ≈ Papua, E Indonesia
42 J11 **Mombacho, Volcán** ⛰ SW Nicaragua 11°49′N 85°58′W
81 K21 **Mombasa** Mombasa, SE Kenya 04°04′N 39°40′E
81 K21 **Mombasa** ◈ *county* SE Kenya
81 J21 **Mombasa** ✈ Mombasa, SE Kenya 04°01′S 39°31′E
Mombetsu *see* Monbetsu
114 J12 **Momchilgrad** *prev.* Mastanli. Kardzhali, S Bulgaria 41°33′N 25°25′E
99 F23 **Momignies** Hainaut, S Belgium 50°02′N 04°10′E
54 E6 **Momil** Córdoba, NW Colombia 09°15′N 75°40′W
42 I10 **Momotombo, Volcán** ⛰ W Nicaragua 12°25′N 86°33′W
56 B5 **Mompiche, Ensenada de** *bay* NW Ecuador
79 K18 **Mompono** Equateur, NW Dem. Rep. Congo 0°11′N 21°31′E
54 F6 **Mompós** Bolívar, NW Colombia 09°15′N 74°29′W
95 J24 **Møn** *prev.* Möen. *island* SE Denmark
36 L4 **Mona** Utah, W USA 39°49′N 111°52′W
Mona, Canal de la *see* Mona Passage
96 E8 **Monach Islands** *island group* NW Scotland, United Kingdom
103 V14 **Monaco** *var.* Monaco-Ville; *anc.* Monoecus. ● (Monaco) S Monaco 43°46′N 07°23′E
103 V14 **Monaco** *off.* Principality of Monaco. ◆ *monarchy* W Europe
Monaco *see* München
Monaco Basin *see* Canary Basin
Monaco, Principality of *see* Monaco
Monaco-Ville *see* Monaco
96 I9 **Monadhliath Mountains** ▲▲ N Scotland, United Kingdom
55 O6 **Monagas** *off.* Estado Monagas. ◈ *state* NE Venezuela
Monagas, Estado *see* Monagas
97 F16 **Monaghan** *Ir.* Muineachán. Monaghan, N Ireland 54°15′N 06°58′W
97 E16 **Monaghan** *Ir.* Muineachán. *cultural region* N Ireland
43 S16 **Monagrillo** Herrera, S Panama 08°00′N 80°28′W
24 L8 **Monahans** Texas, SW USA 31°35′N 102°54′W
45 Q9 **Mona, Isla** *island* W Puerto Rico
45 Q9 **Mona Passage** *Sp.* Canal de la Mona. *channel* Dominican Republic/Puerto Rico
43 O14 **Mona, Punta** *headland* E Costa Rica 09°44′N 82°48′W
155 K25 **Monaragala** Uva Province, SE Sri Lanka 06°52′N 81°22′E
33 S9 **Monarch** Montana, NW USA 47°04′N 110°51′W
10 H14 **Monarch Mountain** ▲ British Columbia, SW Canada 51°59′N 125°56′W
Monasterio *see* Monesterio
Monasterzyska *see* Monastyrys'ka
Monastir *see* Bitola
Monastyriska *see* Monastyrys'ka
117 O7 **Monastyryshche** Cherkas'ka Oblast', C Ukraine 48°59′N 29°47′E
116 J6 **Monastyrys'ka** *Pol.* Monasterzyska, *Rus.* Monastyriska. Ternopil's'ka Oblast', W Ukraine 49°05′N 25°10′E
79 E15 **Monatélé** Centre, SW Cameroon 04°16′N 11°12′E
165 U2 **Monbetsu** *var.* Mombetsu, Monbetu. Hokkaidō, NE Japan 44°23′N 143°22′E
Monbetu *see* Monbetsu
106 B8 **Moncalieri** Piemonte, NW Italy 45°N 07°41′E
104 G4 **Monção** Viana do Castelo, N Portugal 42°03′N 08°29′W
105 Q5 **Moncayo** ▲ N Spain 41°43′N 01°51′W
105 Q5 **Moncayo, Sierra del** ▲▲ N Spain
124 J4 **Monchegorsk** Murmanskaya Oblast', NW Russian Federation 67°56′N 32°47′E
101 D15 **Mönchengladbach** *prev.* München-Gladbach. Nordrhein-Westfalen, W Germany 51°12′N 06°35′E
104 F14 **Monchique** Faro, S Portugal 37°19′N 08°33′W
104 G14 **Monchique, Serra de** ▲▲ S Portugal
21 S14 **Moncks Corner** South Carolina, SE USA 33°12′N 80°00′W
41 N7 **Monclova** Coahuila, NE Mexico 26°55′N 101°25′W
Moncorvo *see* Torre de Moncorvo
13 P14 **Moncton** New Brunswick, SE Canada 46°04′N 64°50′W
104 F8 **Mondego, Cabo** *headland* N Portugal 40°10′N 08°58′W
104 G8 **Mondego, Rio** ≈ N Portugal
104 I2 **Mondoñedo** Galicia, NW Spain 43°25′N 07°22′W
99 N25 **Mondorf-les-Bains** Grevenmacher, SE Luxembourg 49°30′N 06°16′E
102 M7 **Mondoubleau** Loir-et-Cher, C France 48°00′N 00°49′E
106 B9 **Mondovì** Piemonte, NW Italy 44°23′N 07°56′E
30 J6 **Mondovi** Wisconsin, N USA 44°34′N 91°40′W
Mondragón *see* Arrasate
107 J17 **Mondragone** Campania, S Italy 41°07′N 13°53′E
109 R5 **Mondsee** ◎ N Austria
115 G22 **Monemvasiá** *var.* Monemvasía. Pelopónnisos, S Greece 36°22′N 23°03′E
18 B15 **Monessen** Pennsylvania, NE USA 40°07′N 79°51′W
104 J12 **Monesterio** *var.* Monasterio. Extremadura, W Spain 38°05′N 06°16′W
14 L8 **Monet** Québec, SE Canada 48°09′N 75°37′W
27 S8 **Monett** Missouri, C USA 36°55′N 93°55′W
27 X9 **Monette** Arkansas, C USA 35°53′N 90°20′W
14 G11 **Monetville** Ontario, S Canada 46°08′N 80°24′W
106 J7 **Monfalcone** Friuli-Venezia Giulia, NE Italy 45°49′N 13°32′E
104 H10 **Monforte** Portalegre, C Portugal 39°03′N 07°26′W
104 I4 **Monforte de Lemos** Galicia, NW Spain 42°32′N 07°30′W
79 L16 **Monga** Orientale, N Dem. Rep. Congo 04°12′N 22°49′E
81 I24 **Monga** Lindi, SE Tanzania 09°05′S 37°51′E
81 F15 **Mongalla** Central Equatoria, S South Sudan 05°12′N 31°42′E
153 U11 **Mongar** E Bhutan 27°16′N 91°07′E
167 U6 **Mong Cai** *var.* Hai Ninh. Quang Ninh, N Vietnam 21°33′N 107°56′E
180 I11 **Mongers Lake** *salt lake* Western Australia
186 K8 **Mongga** Kolombangara, NW Solomon Islands 07°51′S 157°00′E
167 O6 **Möng Hpayak** Shan State, E Myanmar (Burma) 20°56′N 100°00′E
Monghyr *see* Munger
106 B10 **Mongioie** ▲ NW Italy 44°13′N 07°46′E
153 T16 **Mongla** *var.* Mungla. Khulna, S Bangladesh 22°18′N 89°34′E
188 C15 **Mongmong** C Guam
167 N6 **Möng Nai** Shan State, E Myanmar (Burma) 20°28′N 97°51′E
78 I11 **Mongo** Guéra, C Chad 12°12′N 18°40′E
76 I14 **Mongo** ≈ N Sierra Leone
163 I8 **Mongolia** *Mong.* Mongol Uls. ◆ *republic* E Asia
129 V8 **Mongolia, Plateau of** *plateau* E Mongolia
Mongolküre *see* Zhaosu
Mongol Uls *see* Mongolia
79 E17 **Mongomo** E Equatorial Guinea 01°39′N 11°18′E
162 M7 **Möngönmorīt** *var.* Bulag. Töv, C Mongolia 48°09′N 108°33′E
77 Y12 **Mongonu** *var.* Monguno. Borno, NE Nigeria 12°42′N 13°37′E
Mongora *see* Saīdu
78 K11 **Mongororo** Sila, SE Chad 12°03′N 22°26′E
Mongos, Chaîne des *see* Bongo, Massif des
79 I16 **Mongoumba** Lobaye, SW Central African Republic 03°39′N 18°30′E
Mongrove, Punta *see* Cayacal, Punta
83 G15 **Mongu** Western, W Zambia 15°13′S 23°09′E
76 I10 **Mônguel** Gorgol, SW Mauritania 16°25′N 13°08′W
Monguno *see* Mongonu
167 N4 **Möng Yai** Shan State, E Myanmar (Burma) 22°25′N 98°02′E
167 O5 **Möng Yang** Shan State, E Myanmar (Burma) 21°52′N 99°31′E
167 N3 **Möng Yu** Shan State, E Myanmar (Burma) 24°00′N 97°57′E
Mönhbulag *see* Yösöndzüyl
163 O8 **Mönhhaan** *var.* Bayasgalant. Sühbaatar, E Mongolia 46°55′N 112°11′E
162 E7 **Mönhhayrhan** *var.* Tsenher. Hovd, W Mongolia 47°07′N 92°04′E
Mönh Sarīdag *see* Munku-Sardyk, Gora
186 P9 **Moni** ≈ S Papau New Guinea
115 I15 **Moní Megístis Lávras** *monastery* Kentrikí Makedonía, N Greece
115 F18 **Moní Osíou Loúkas** *monastery* Stereá Elláda, C Greece
54 F9 **Moniquirá** Boyacá, C Colombia 05°54′N 73°35′W
103 Q12 **Monistrol-sur-Loire** Haute-Loire, C France 45°19′N 04°12′E
35 V7 **Monitor Range** ▲▲ Nevada, W USA
115 I14 **Moní Vatopedíou** *monastery* Kentrikí Makedonía, N Greece
Monkchester *see* Newcastle upon Tyne
83 N14 **Monkey Bay** Southern, SE Malawi 14°09′S 34°53′E
43 N11 **Monkey Point** *var.* Punta Mico, Punte Mono, Punto Mico. *headland* SE Nicaragua 11°37′N 83°39′W
Monkey River *see* Monkey River Town
42 G3 **Monkey River Town** *var.* Monkey River. Toledo, SE Belize 16°22′N 88°29′W
14 M13 **Monkland** Ontario, SE Canada 45°11′N 74°51′W
79 J19 **Monkoto** Equateur, NW Dem. Rep. Congo 01°39′S 20°41′E
97 K21 **Monmouth** *Wel.* Trefynwy. SE Wales, United Kingdom 51°50′N 02°43′W
30 J12 **Monmouth** Illinois, N USA 40°54′N 90°39′W
32 F12 **Monmouth** Oregon, NW USA 44°51′N 123°13′W
97 K21 **Monmouth** *cultural region* SE Wales, United Kingdom
98 I10 **Monnickendam** Noord-Holland, C Netherlands 52°28′N 05°02′E
77 R15 **Mono** ≈ C Togo
Monoecus *see* Monaco
35 R8 **Mono Lake** ◎ California, W USA
115 O23 **Monólithos** Ródos, Dodekánisa, Greece, Aegean Sea 36°08′N 27°45′E
19 Q12 **Monomoy Island** *island* Massachusetts, NE USA
31 O12 **Monon** Indiana, N USA 40°52′N 86°54′W
29 Y12 **Monona** Iowa, C USA 43°03′N 91°23′W
30 L9 **Monona** Wisconsin, N USA 43°03′N 89°18′W
18 B15 **Monongahela** Pennsylvania, NE USA 40°10′N 79°54′W
18 B16 **Monongahela River** ≈ NE USA
107 P17 **Monopoli** Puglia, SE Italy 40°57′N 17°18′E
Mono, Punte *see* Monkey Point
111 K23 **Monor** Pest, C Hungary 47°21′N 18°27′E
Monostor *see* Beli Manastir
78 K8 **Monou** Ennedi-Ouest, NE Chad 16°22′N 22°15′E
105 S12 **Monóvar** *Cat.* Monòver. Valenciana, E Spain 38°26′N 00°50′W
Monòver *see* Monóvar
105 R7 **Monreal del Campo** Aragón, NE Spain 40°47′N 01°20′W
107 I23 **Monreale** Sicilia, Italy, C Mediterranean Sea 38°05′N 13°17′E
23 T3 **Monroe** Georgia, SE USA 33°47′N 83°42′W
29 W14 **Monroe** Iowa, C USA 41°31′N 93°06′W
22 I5 **Monroe** Louisiana, S USA 32°32′N 92°06′W
31 S10 **Monroe** Michigan, N USA 41°55′N 83°24′W
18 K13 **Monroe** New York, NE USA 41°18′N 74°09′W
21 S11 **Monroe** North Carolina, SE USA 35°00′N 80°35′W
36 L6 **Monroe** Utah, W USA 38°37′N 112°07′W
32 H7 **Monroe** Washington, NW USA 47°51′N 121°58′W
30 L9 **Monroe** Wisconsin, N USA 42°35′N 89°39′W
27 V3 **Monroe City** Missouri, C USA 39°39′N 91°43′W
31 O15 **Monroe Lake** ▣ Indiana, N USA
23 O7 **Monroeville** Alabama, S USA 31°31′N 87°19′W
18 C15 **Monroeville** Pennsylvania, NE USA 40°24′N 79°44′W
76 J16 **Monrovia** ● (Liberia) W Liberia 06°18′N 10°48′W
76 J16 **Monrovia** ✈ W Liberia 06°22′N 10°50′W
105 T7 **Monroyo** Aragón, NE Spain 40°47′N 00°03′W
99 F20 **Mons** *Dut.* Bergen. Hainaut, S Belgium 50°28′N 03°58′E
104 I8 **Monsanto** Castelo Branco, C Portugal 40°02′N 07°07′W
106 H8 **Monselice** Veneto, NE Italy 45°15′N 11°47′E
166 M9 **Mon State** ◈ *state* S Myanmar (Burma)
98 G12 **Monster** Zuid-Holland, W Netherlands 52°01′N 04°10′E
95 N20 **Mönsterås** Kalmar, S Sweden 57°03′N 16°27′E
101 F17 **Montabaur** Rheinland-Pfalz, W Germany 50°25′N 07°48′E
106 G8 **Montagnana** Veneto, NE Italy 45°14′N 11°31′E
35 N1 **Montague** California, W USA 41°43′N 122°31′W
25 S5 **Montague** Texas, SW USA 33°40′N 97°44′W
183 S11 **Montague Island** *island* New South Wales, SE Australia
39 S12 **Montague Island** *island* Alaska, USA
39 S13 **Montague Strait** *strait* N Gulf of Alaska
102 J8 **Montaigu** Vendée, NW France 46°58′N 01°18′W
Montaigu *see* Scherpenheuvel
105 S7 **Montalbán** Aragón, NE Spain 40°49′N 00°48′W
106 G13 **Montalcino** Toscana, C Italy 43°01′N 11°34′E
104 H5 **Montalegre** Vila Real, N Portugal 41°49′N 07°48′W
114 G8 **Montana** *prev.* Ferdinand, Mikhaylovgrad. Montana, NW Bulgaria 43°25′N 23°14′E
108 D10 **Montana** Valais, SW Switzerland 46°23′N 07°29′E
39 R11 **Montana** Alaska, USA 62°06′N 150°03′W
114 G8 **Montana** ◈ *province* NW Bulgaria
33 T9 **Montana** *off.* State of Montana, *also known as* Mountain State, Treasure State. ◈ *state* NW USA
104 J10 **Montánchez** Extremadura, W Spain 39°15′N 06°07′W
Montañita *see* La Montañita
15 Q8 **Mont-Apica** Québec, SE Canada 47°57′N 71°24′W
104 G10 **Montargil** Portalegre, C Portugal 39°05′N 08°10′W
104 G10 **Montargil, Barragem de** ▣ C Portugal
103 O7 **Montargis** Loiret, C France 48°N 02°44′E
103 O4 **Montataire** Oise, N France 49°16′N 02°24′E
102 M14 **Montauban** Tarn-et-Garonne, S France 44°01′N 01°20′E
19 N14 **Montauk** Long Island, New York, NE USA 41°01′N 71°58′W
19 N14 **Montauk Point** *headland* Long Island, New York, NE USA 41°04′N 71°51′W
103 Q7 **Montbard** Côte d'Or, C France 47°35′N 04°25′E
103 U7 **Montbéliard** Doubs, E France 47°31′N 06°49′E
25 W11 **Mont Belvieu** Texas, SW USA 29°51′N 94°53′W
105 U6 **Montblanc** *prev.* Montblanch. Cataluña, NE Spain 41°23′N 01°10′E
Montblanch *see* Montblanc
103 Q11 **Montbrison** Loire, E France 45°37′N 04°04′E
Montcalm, Lake *see* Dogai Coring
103 Q9 **Montceau-les-Mines** Saône-et-Loire, C France 46°40′N 04°19′E
103 U12 **Mont Cenis, Col du** *pass* E France
102 K15 **Mont-de-Marsan** Landes, SW France 43°54′N 00°30′W
103 O3 **Montdidier** Somme, N France 49°39′N 02°35′E
187 Q17 **Mont-Dore** Province Sud, S New Caledonia 22°18′S 166°34′E
20 K10 **Monteagle** Tennessee, S USA 35°15′N 85°47′W
57 M20 **Monteagudo** Chuquisaca, S Bolivia 19°48′S 63°57′W
41 R16 **Monte Albán** *ruins* Oaxaca, S Mexico
105 R11 **Montealegre del Castillo** Castilla-La Mancha, C Spain 38°48′N 01°18′W
59 N18 **Monte Azul** Minas Gerais, SE Brazil 15°13′S 42°53′W
14 M12 **Montebello** Québec, SE Canada 45°40′N 74°56′W
106 H7 **Montebelluna** Veneto, NE Italy 45°46′N 12°03′E
60 G13 **Montecarlo** Misiones, NE Argentina 26°38′S 54°45′W
61 D16 **Monte Caseros** Corrientes, NE Argentina 30°15′S 57°39′W
60 J13 **Monte Castelo** Santa Catarina, S Brazil 26°34′S 50°12′W
106 F11 **Montecatini Terme** Toscana, C Italy 43°53′N 10°46′E
42 H7 **Montecillos, Cordillera de** ▲▲ W Honduras
62 I12 **Monte Comén** Mendoza, W Argentina 34°35′S 67°53′W
44 M8 **Monte Cristi** *var.* San Fernando de Monte Cristi. NW Dominican Republic 19°52′N 71°39′W
58 C13 **Monte Cristo** Amazonas, W Brazil 03°14′S 68°00′W
107 E14 **Montecristo, Isola di** *island* Archipelago Toscano, C Italy
Monte Croce Carnico, Passo di *see* Plöcken Pass
58 J12 **Monte Dourado** Pará, NE Brazil 0°48′S 52°32′W
40 L11 **Monte Escobedo** Zacatecas, C Mexico 22°19′N 103°30′W
106 I13 **Montefalco** Umbria, C Italy 42°54′N 12°40′E
107 H14 **Montefiascone** Lazio, C Italy 42°33′N 12°01′E
105 N14 **Montefrío** Andalucía, S Spain 37°19′N 04°00′W
44 I11 **Montego Bay** *var.* Mobay. W Jamaica 18°28′N 77°55′W
Montego Bay *see* Sangster
104 J8 **Montehermoso** Extremadura, W Spain 40°05′N 06°20′W
104 F10 **Montejunto, Serra de** ▲ C Portugal 39°10′N 09°01′W
Monteleone di Calabria *see* Vibo Valentia
54 E7 **Montelíbano** Córdoba, NW Colombia 08°02′N 75°29′W
103 R13 **Montélimar** *anc.* Acunum Acusio, Montilium Adhemari. Drôme, E France 44°33′N 04°45′E
104 K15 **Montellano** Andalucía, S Spain 37°00′N 05°34′W
35 Y2 **Montello** Nevada, W USA 41°18′N 114°10′W
30 L8 **Montello** Wisconsin, N USA 43°47′N 89°20′W
63 J18 **Montemayor, Meseta de** *plain* SE Argentina
41 O9 **Montemorelos** Nuevo León, NE Mexico 25°10′N 99°52′W
104 G11 **Montemor-o-Novo** Évora, S Portugal 38°38′N 08°13′W
104 G8 **Montemor-o-Velho** *var.* Montemor-o-Vélho. Coimbra, N Portugal 40°11′N 08°41′W
Montemor-o-Vélho *see* Montemor-o-Velho
104 H7 **Montemuro, Serra de** ▲ N Portugal 40°59′N 07°59′W
102 K12 **Montendre** Charente-Maritime, W France 45°17′N 00°24′W
61 I15 **Montenegro** Rio Grande do Sul, S Brazil 29°40′S 51°32′W
113 J16 **Montenegro** *Serb.* Crna Gora. ◆ *republic* SW Europe
62 G10 **Monte Patria** Coquimbo, N Chile 30°40′S 71°00′W
45 O9 **Monte Plata** E Dominican Republic 18°50′N 69°47′W
83 P14 **Montepuez** Cabo Delgado, N Mozambique 13°09′S 39°00′E
83 P14 **Montepuez** ≈ N Mozambique
106 G13 **Montepulciano** Toscana, C Italy 43°02′N 11°51′E
62 L6 **Monte Quemado** Santiago del Estero, N Argentina 25°48′S 62°52′W
103 O6 **Montereau-Faut-Yonne** *anc.* Condate. Seine-St-Denis, N France 48°23′N 02°57′E
35 N11 **Monterey** California, W USA 36°36′N 121°53′W
20 L9 **Monterey** Tennessee, S USA 36°09′N 85°16′W
21 T5 **Monterey** Virginia, NE USA 38°24′N 79°36′W
Monterey *see* Monterrey
35 N10 **Monterey Bay** *bay* California, W USA
54 D6 **Montería** Córdoba, NW Colombia 08°45′N 75°54′W
57 N18 **Montero** Santa Cruz, C Bolivia 17°20′S 63°15′W
62 J7 **Monteros** Tucumán, C Argentina 27°12′S 65°30′W
104 I5 **Monterrei** Galicia, NW Spain 41°56′N 07°27′W
41 O8 **Monterrey** *var.* Monterey. Nuevo León, NE Mexico 25°41′N 100°16′W
32 F9 **Montesano** Washington, NW USA 46°58′N 123°37′W
107 M19 **Montesano sulla Marcellana** Campania, S Italy 40°15′N 15°41′E
107 N16 **Monte Sant' Angelo** Puglia, SE Italy 41°43′N 15°58′E
59 O16 **Monte Santo** Bahia, E Brazil 10°25′S 39°18′W
107 D18 **Monte Santu, Capo di** *headland* Sardegna, Italy, C Mediterranean Sea 40°05′N 09°43′E
59 M19 **Montes Claros** Minas Gerais, SE Brazil 16°45′S 43°52′W
107 K14 **Montesilvano Marina** Abruzzo, C Italy 42°28′N 14°07′E
23 P4 **Montevallo** Alabama, S USA 33°06′N 86°51′W
106 G12 **Montevarchi** Toscana, C Italy 43°32′N 11°34′E
61 F20 **Montevideo** ● (Uruguay) Montevideo, S Uruguay 34°55′S 56°10′W
29 S9 **Montevideo** Minnesota, N USA 44°56′N 95°43′W
37 S7 **Monte Vista** Colorado, C USA 37°33′N 106°08′W
23 T5 **Montezuma** Georgia, SE USA 32°18′N 84°01′W
29 W14 **Montezuma** Iowa, C USA 41°35′N 92°31′W
26 J6 **Montezuma** Kansas, C USA 37°33′N 100°25′W
103 U12 **Montgenèvre, Col de** *pass* France/Italy
97 K20 **Montgomery** E Wales, United Kingdom 52°38′N 03°05′W
23 Q5 **Montgomery** *state capital* Alabama, S USA 32°22′N 86°18′W
29 V9 **Montgomery** Minnesota, N USA 44°26′N 93°34′W
18 G13 **Montgomery** Pennsylvania, NE USA 41°08′N 76°52′W
21 Q5 **Montgomery** West Virginia, NE USA 38°07′N 81°19′W
97 K19 **Montgomery** *cultural region* E Wales, United Kingdom
Montgomery *see* Sāhīwāl
27 V4 **Montgomery City** Missouri, C USA 38°57′N 91°27′W
35 S8 **Montgomery Pass** *pass* Nevada, W USA
102 K12 **Montguyon** Charente-Maritime, W France 45°12′N 00°13′W
108 C10 **Monthey** Valais, SW Switzerland 46°15′N 06°56′E
27 V13 **Monticello** Arkansas, C USA 33°38′N 91°49′W
23 T4 **Monticello** Florida, SE USA 30°33′N 83°52′W
23 T8 **Monticello** Georgia, SE USA 33°18′N 83°40′W
30 M13 **Monticello** Illinois, N USA 40°01′N 88°34′W
31 O12 **Monticello** Indiana, N USA 40°45′N 86°46′W
29 Y13 **Monticello** Iowa, C USA 42°14′N 91°11′W
20 L7 **Monticello** Kentucky, S USA 36°50′N 84°50′W
29 V8 **Monticello** Minnesota, N USA 45°19′N 93°45′W
22 K7 **Monticello** Mississippi, S USA 31°33′N 90°06′W
27 V2 **Monticello** Missouri, C USA 40°07′N 91°42′W
18 J12 **Monticello** New York, NE USA 41°39′N 74°41′W
37 O7 **Monticello** Utah, W USA 37°52′N 109°20′W
106 F8 **Montichiari** Lombardia, N Italy 45°24′N 10°27′E
102 M12 **Montignac** Dordogne, SW France 45°24′N 00°54′E
99 G21 **Montignies-le-Tilleul** *var.* Montigny-le-Tilleul. Hainaut, S Belgium 50°22′N 04°23′E
14 J8 **Montigny, Lac de** ◎ Québec, SE Canada
103 S6 **Montigny-le-Roi** Haute-Marne, N France 48°02′N 05°28′E
Montigny-le-Tilleul *see* Montignies-le-Tilleul
43 R16 **Montijo** Veraguas, S Panama 07°59′N 80°58′W
104 F11 **Montijo** Setúbal, W Portugal 38°42′N 08°59′W
104 J11 **Montijo** Extremadura, W Spain 38°55′N 06°38′W
Montilium Adhemari *see* Montélimar
104 M13 **Montilla** Andalucía, S Spain 37°36′N 04°39′W
102 L3 **Montivilliers** Seine-Maritime, N France 49°31′N 00°10′E
15 U7 **Mont-Joli** Québec, SE Canada 48°36′N 68°14′W
14 M10 **Mont-Laurier** Québec, SE Canada 46°33′N 75°31′W
15 X5 **Mont-Louis** Québec, SE Canada 49°15′N 65°46′W
103 N17 **Mont-Louis** *var.* Mont Louis. Pyrénées-Orientales, S France 42°30′N 02°08′E
103 O10 **Montluçon** Allier, C France 46°21′N 02°37′E
15 R10 **Montmagny** Québec, SE Canada 47°00′N 70°31′W
103 S3 **Montmédy** Meuse, NE France 49°31′N 05°21′E
103 P5 **Montmirail** Marne, N France 48°53′N 03°31′E
15 R9 **Montmorency** ≈ Québec, SE Canada
102 M10 **Montmorillon** Vienne, W France 46°26′N 00°52′E
107 J14 **Montorio al Vomano** Abruzzo, C Italy 42°31′N 13°39′E
104 M13 **Montoro** Andalucía, S Spain 38°00′N 04°21′W
33 S16 **Montpelier** Idaho, NW USA 42°19′N 111°18′W
29 P6 **Montpelier** North Dakota, N USA 46°40′N 98°34′W
18 M7 **Montpelier** *state capital* Vermont, NE USA 44°16′N 72°32′W
103 Q15 **Montpellier** Hérault, S France 43°37′N 03°52′E
102 L12 **Montpon-Ménestérol** Dordogne, SW France 45°01′N 00°10′E
12 K15 **Montréal** *Eng.* Montreal. Québec, SE Canada 45°30′N 73°36′W
14 G8 **Montreal** ≈ Ontario, S Canada
Montreal *see* Mirabel
11 T14 **Montreal Lake** ◎ Saskatchewan, C Canada
14 B9 **Montreal River** Ontario, S Canada 47°13′N 84°36′W
103 N2 **Montreuil** Pas-de-Calais, N France 50°28′N 01°46′E
102 K8 **Montreuil-Bellay** Maine-et-Loire, NW France 47°07′N 00°10′W
108 C10 **Montreux** Vaud, SW Switzerland 46°27′N 06°55′E
108 B9 **Montricher** Vaud, W Switzerland 46°37′N 06°24′E
96 K10 **Montrose** E Scotland, United Kingdom 56°43′N 02°29′W
27 W14 **Montrose** Arkansas, C USA 33°18′N 91°29′W
37 Q6 **Montrose** Colorado, C USA 38°29′N 107°53′W

◆ Country
● Country Capital
◇ Dependent Territory
○ Dependent Territory Capital
◈ Administrative Regions
✈ International Airport
▲ Mountain
▲▲ Mountain Range
⛰ Volcano
≈ River
◎ Lake
▣ Reservoir

MONTROSE – MOYALĒ

29 Y16 **Montrose** Iowa, C USA 40°31′N 91°24′W
18 H12 **Montrose** Pennsylvania, NE USA 41°49′N 75°53′W
21 X5 **Montross** Virginia, NE USA 38°04′N 76°51′W
15 O12 **Mont-St-Hilaire** Québec, SE Canada 45°34′N 73°10′W
103 S3 **Mont-St-Martin** Meurthe-et-Moselle, NE France 49°31′N 05°51′E
45 V10 **Montserrat** *var.* Emerald Isle. ◇ *UK dependent territory* E West Indies
105 V5 **Montserrat** ▲ NE Spain 41°39′N 01°44′E
104 M7 **Montuenga** Castilla y León, N Spain 41°04′N 04°38′W
99 M19 **Montzen** Liège, E Belgium 50°42′N 05°59′E
37 N8 **Monument Valley** *valley* Arizona/Utah, SW USA
166 L4 **Monywa** Sagaing, C Myanmar (Burma) 22°05′N 95°12′E
106 D7 **Monza** Lombardia, N Italy 45°35′N 09°16′E
83 J15 **Monze** Southern, S Zambia 16°20′S 27°29′E
105 T5 **Monzón** Aragón, NE Spain 41°54′N 00°12′E
25 T9 **Moody** Texas, SW USA 31°18′N 97°21′W
98 L13 **Mook** Limburg, SE Netherlands 51°45′N 05°52′E
165 O12 **Mooka** *var.* Mōka. Tochigi, Honshū, S Japan 36°27′N 139°59′E
182 K3 **Moomba** South Australia 28°07′S 140°12′E
14 G13 **Moon** ≈ Ontario, S Canada
Moon *see* Muhu
181 Y10 **Moonie** Queensland, E Australia 27°46′S 150°22′E
193 O5 **Moonless Mountains** *undersea feature* E Pacific Ocean 30°40′N 140°00′W
182 L13 **Moonlight Head** *headland* Victoria, SE Australia 38°47′S 143°12′E
Moon-Sund *see* Väinameri
182 H8 **Moonta** South Australia 34°03′S 137°36′E
Moor *see* Mór
180 I12 **Moora** Western Australia 30°23′S 116°05′E
98 H12 **Moordrecht** Zuid-Holland, C Netherlands 51°59′N 04°40′E
33 T9 **Moore** Montana, NW USA 47°00′N 109°40′W
27 N11 **Moore** Oklahoma, C USA 35°21′N 97°30′W
25 R12 **Moore** Texas, SW USA 29°03′N 99°01′W
191 S10 **Moorea** *island* Îles du Vent, W French Polynesia
21 U3 **Moorefield** West Virginia, NE USA 39°04′N 78°59′W
23 X14 **Moore Haven** Florida, SE USA 26°49′N 81°05′W
180 J11 **Moore, Lake** ◎ Western Australia
19 N7 **Moore Reservoir** ⊠ New Hampshire/Vermont, NE USA
44 G1 **Moores Island** *island* N The Bahamas
21 R10 **Mooresville** North Carolina, SE USA 35°34′N 80°48′W
29 R5 **Moorhead** Minnesota, N USA 46°51′N 96°44′W
22 K4 **Moorhead** Mississippi, S USA 33°27′N 90°30′W
99 F18 **Moorsel** Oost-Vlaanderen, C Belgium 50°58′N 04°06′E
99 C18 **Moorslede** West-Vlaanderen, W Belgium 50°53′N 03°03′E
18 L8 **Moosalamoo, Mount** ▲ Vermont, NE USA 43°55′N 73°03′W
101 M22 **Moosburg in der Isar** Bayern, SE Germany 48°28′N 11°55′E
33 S14 **Moose** Wyoming, C USA 43°38′N 110°42′W
12 H11 **Moose** ≈ Ontario, S Canada
12 H10 **Moose Factory** Ontario, S Canada 51°16′N 80°32′W
19 Q4 **Moosehead Lake** ◎ Maine, NE USA
11 U16 **Moose Jaw** Saskatchewan, S Canada 50°23′N 105°35′W
11 V14 **Moose Lake** Manitoba, C Canada 53°42′N 100°22′W
29 W6 **Moose Lake** Minnesota, N USA 46°28′N 92°46′W
19 P6 **Mooselookmeguntic Lake** ◎ Maine, NE USA
39 R12 **Moose Pass** Alaska, USA 60°28′N 149°21′W
19 P5 **Moose River** ≈ Maine, NE USA
18 J9 **Moose River** ≈ New York, NE USA
11 V16 **Moosomin** Saskatchewan, S Canada 50°09′N 101°41′W
12 H10 **Moosonee** Ontario, SE Canada 51°18′N 80°40′W
19 N12 **Moosup** Connecticut, NE USA 41°42′N 71°51′W
83 N16 **Mopeia** Zambézia, NE Mozambique 17°59′S 35°43′E
83 H18 **Mopipi** Central, C Botswana 21°07′S 24°55′E
Moppo *see* Mokpo
77 N11 **Mopti** Mopti, C Mali 14°30′N 04°15′W
77 O11 **Mopti** ◈ *region* S Mali
57 H18 **Moquegua** Moquegua, SE Peru 17°07′S 70°55′W
57 H18 **Moquegua** *off.* Departamento de Moquegua. ◈ *department* S Peru
Moquegua, Departamento de *see* Moquegua
111 I23 **Mór** *Ger.* Moor. Fejér, C Hungary 47°21′N 18°12′E
78 G11 **Mora** Extrême-Nord, N Cameroon 11°02′N 14°07′E
104 G11 **Mora** Évora, S Portugal 38°56′N 08°10′W
105 N9 **Mora** Castilla-La Mancha, C Spain 39°40′N 03°46′W
94 L12 **Mora** Dalarna, C Sweden 61°N 14°30′E
29 V7 **Mora** Minnesota, N USA 45°52′N 93°18′W
37 T10 **Mora** New Mexico, SW USA 35°56′N 105°16′W
113 J17 **Morača** ≈ S Montenegro
152 K10 **Morādābād** Uttar Pradesh, N India 28°50′N 78°45′E
105 U6 **Móra d'Ebre** *var.* Mora de Ebro. Cataluña, NE Spain 41°05′N 00°38′E
Mora de Ebro *see* Móra d'Ebre
105 S8 **Mora de Rubielos** Aragón, NE Spain 40°15′N 00°45′W
172 H4 **Morafenobe** Mahajanga, W Madagascar 17°49′S 44°54′E
110 K8 **Morąg** *Ger.* Mohrungen. Warmińsko-Mazurskie, N Poland 53°55′N 19°56′E
111 L25 **Mórahalom** Csongrád, S Hungary 46°14′N 19°52′E
105 N11 **Moral de Calatrava** Castilla-La Mancha, C Spain 38°50′N 03°34′W
63 G19 **Moraleda, Canal** *strait* SE Pacific Ocean
54 J3 **Morales** Bolívar, N Colombia 08°17′N 73°52′W
54 D12 **Morales** Cauca, SW Colombia 02°46′N 76°44′W
42 F5 **Morales** Izabal, E Guatemala 15°28′N 88°46′W
172 J5 **Moramanga** Toamasina, E Madagascar 18°57′S 48°13′E
27 Q6 **Moran** Kansas, C USA 37°55′N 95°10′W
25 Q7 **Moran** Texas, SW USA 32°33′N 99°10′W
181 X7 **Moranbah** Queensland, NE Australia 22°01′S 148°08′E
44 L13 **Morant Bay** E Jamaica 17°53′N 76°25′W
96 G10 **Morar, Loch** ◎ N Scotland, United Kingdom
Morata *see* Goodenough Island
105 Q12 **Moratalla** Murcia, SE Spain 38°11′N 01°53′W
108 C8 **Morat, Lac de** *Ger.* Murtensee. ◎ W Switzerland
84 I11 **Morava** *var.* March. ≈ C Europe *see also* March
Morava *see* March
Morava *see* Moravia, Czech Republic
Morava *see* Velika Morava, Serbia
29 W15 **Moravia** Iowa, C USA 40°53′N 92°49′W
111 F18 **Moravia** *Cz.* Morava, *Ger.* Mähren. *cultural region* E Czech Republic
111 H17 **Moravice** *Ger.* Mohra. ≈ NE Czech Republic
116 E12 **Moravița** Timiș, SW Romania 45°15′N 21°17′E
111 G17 **Moravská Třebová** *Ger.* Mährisch-Trübau. Pardubický Kraj, C Czech Republic 49°47′N 16°40′E
111 E19 **Moravské Budějovice** *Ger.* Mährisch-Budwitz. Vysočina, C Czech Republic 49°03′N 15°48′E
111 H17 **Moravskoslezský Kraj** *prev.* Ostravský Kraj. ◈ *region* E Czech Republic
111 F19 **Moravský Krumlov** *Ger.* Mährisch-Kromau. Jihomoravský Kraj, SE Czech Republic 48°58′N 16°30′E
96 J8 **Moray** *cultural region* N Scotland, United Kingdom
96 J8 **Moray Firth** *inlet* N Scotland, United Kingdom
42 B10 **Morazán** ◈ *department* NE El Salvador
154 C10 **Morbi** Gujarāt, W India 22°51′N 70°49′E
102 G7 **Morbihan** ◈ *department* NW France
Mörbisch *see* Mörbisch am See
109 Y5 **Mörbisch am See** *var.* Mörbisch. Burgenland, E Austria 47°43′N 16°40′E
95 N21 **Mörbylånga** Kalmar, S Sweden 56°31′N 16°25′E
102 J14 **Morcenx** Landes, SW France 44°04′N 00°55′W
Morcheh Khort *see* Mūrcheh Khvort
163 T5 **Mordaga** Nei Mongol Zizhiqu, N China 51°15′N 120°47′E
11 X17 **Morden** Manitoba, S Canada 49°12′N 98°05′W
Mordovia *see* Mordoviya, Respublika
127 N5 **Mordoviya, Respublika** *prev.* Mordovskaya ASSR, *Eng.* Mordovia, Mordvinia. ◈ *autonomous republic* W Russian Federation
126 M7 **Mordovo** Tambovskaya Oblast', W Russian Federation 52°05′N 40°49′E
Mordovskaya ASSR/Mordvinia *see* Mordoviya, Respublika
Morea *see* Pelopónnisos
28 K8 **Moreau River** ≈ South Dakota, N USA
97 K16 **Morecambe** NW England, United Kingdom 54°04′N 02°53′W
97 K16 **Morecambe Bay** *inlet* NW England, United Kingdom
183 S4 **Moree** New South Wales, SE Australia 29°29′S 149°53′E
21 N5 **Morehead** Kentucky, S USA 38°11′N 83°27′W
21 X11 **Morehead City** North Carolina, SE USA 34°43′N 76°43′W
27 Y8 **Morehouse** Missouri, C USA 36°51′N 89°41′W
108 E10 **Mörel** Valais, SW Switzerland 46°22′N 08°03′E
54 D13 **Morelia** Caquetá, S Colombia 01°30′N 75°43′W
41 N14 **Morelia** Michoacán, S Mexico 19°40′N 101°11′W
105 T7 **Morella** Valenciana, E Spain 40°37′N 00°06′W
40 I7 **Morelos** Chihuahua, N Mexico 26°37′N 107°37′W
41 O15 **Morelos** ◈ *state* S Mexico
154 H7 **Morena** Madhya Pradesh, C India 26°30′N 78°04′E
104 L12 **Morena, Sierra** ▲▲ S Spain
37 O14 **Morenci** Arizona, SW USA 33°05′N 109°21′W
31 R11 **Morenci** Michigan, N USA 41°43′N 84°13′W
116 J13 **Moreni** Dâmbovița, S Romania 44°59′N 25°39′E
94 D9 **Møre og Romsdal** ◈ *county* S Norway
10 I14 **Moresby Island** *island* Queen Charlotte Islands, British Columbia, SW Canada
183 W2 **Moreton Island** *island* Queensland, E Australia
103 O3 **Moreuil** Somme, N France 49°47′N 02°28′E
35 V7 **Morey Peak** ▲ Nevada, W USA 38°40′N 116°16′W
125 U4 **More-Yu** ≈ NW Russian Federation
103 T9 **Morez** Jura, E France 46°33′N 06°01′E
Morfou Bay/Mórfou, Kólpos *see* Güzelyurt Körfezi
182 J8 **Morgan** South Australia 34°02′S 139°39′E
23 S7 **Morgan** Georgia, SE USA 31°31′N 84°34′W
25 S8 **Morgan** Texas, SW USA 32°01′N 97°36′W
22 J10 **Morgan City** Louisiana, S USA 29°42′N 91°12′W
20 H6 **Morganfield** Kentucky, S USA 37°41′N 87°55′W
35 O10 **Morgan Hill** California, W USA 37°05′N 121°38′W
21 Q9 **Morganton** North Carolina, SE USA 35°44′N 81°43′W
20 J7 **Morgantown** Kentucky, S USA 37°12′N 86°42′W
21 S2 **Morgantown** West Virginia, NE USA 39°38′N 79°57′W
108 B10 **Morges** Vaud, SW Switzerland 46°31′N 06°30′E
Morghāb, Daryā-ye *see* Murgap
Morghāb, Daryā-ye *see* Murghāb, Daryā-ye
96 I9 **Mor, Glen** *var.* Glen Albyn, Great Glen. *valley* N Scotland, United Kingdom
103 T5 **Morhange** Moselle, NE France 48°56′N 06°37′E
158 M5 **Mori** *var.* Mori Kazak Zizhixian. Xinjiang Uygur Zizhiqu, NW China 43°48′N 90°21′E
165 R5 **Mori** Hokkaidō, NE Japan 42°04′N 140°36′E
35 Y6 **Moriah, Mount** ▲ Nevada, W USA 39°16′N 114°10′W
37 S11 **Moriarty** New Mexico, SW USA 34°59′N 106°03′W
54 J12 **Morichal** Guaviare, E Colombia 02°09′N 70°35′W
Mori Kazak Zizhixian *see* Mori
Morin Dawa Daurzu Zizhiqi *see* Nirji
11 Q14 **Morinville** Alberta, SW Canada 53°48′N 113°38′W
165 R8 **Morioka** Iwate, Honshū, C Japan 39°42′N 141°08′E
183 T8 **Morisset** New South Wales, SE Australia 33°07′S 151°32′E
165 Q8 **Moriyoshi-zan** ▲ Honshū, C Japan 39°58′N 140°32′E
92 K13 **Morjärv** Norrbotten, N Sweden 66°03′N 22°45′E
127 R3 **Morki** Respublika Mariy El, W Russian Federation 56°27′N 49°01′E
123 N10 **Morkoka** ≈ NE Russian Federation
102 F5 **Morlaix** Finistère, NW France 48°35′N 03°50′W
95 M20 **Mörlunda** Kalmar, S Sweden 57°19′N 15°52′E
107 N19 **Mormanno** Calabria, SW Italy 39°54′N 15°58′E
36 L11 **Mormon Lake** ◎ Arizona, SW USA
35 Y10 **Mormon Peak** ▲ Nevada, W USA 36°59′N 114°25′W
Mormon State *see* Utah
45 Y5 **Morne-à-l'Eau** Grande Terre, N Guadeloupe 16°20′N 61°31′W
29 Y15 **Morning Sun** Iowa, C USA 41°06′N 91°15′W
193 S12 **Mornington Abyssal Plain** *undersea feature* SE Pacific Ocean 50°00′S 90°00′W
63 F22 **Mornington, Isla** *island* S Chile
181 T4 **Mornington Island** *island* Wellesley Islands, Queensland, N Australia
115 E18 **Mórnos** ≈ C Greece
149 P14 **Moro** Sind, SE Pakistan 26°36′N 67°59′E
32 I11 **Moro** Oregon, NW USA 45°30′N 120°46′W
186 E8 **Morobe** Morobe, C Papua New Guinea 07°46′S 147°35′E
186 E8 **Morobe** ◈ *province* C Papua New Guinea
31 N12 **Morocco** Indiana, N USA 40°57′N 87°27′W
74 E8 **Morocco** *off.* Kingdom of Morocco, *Ar.* Al Mamlakah. ◆ *monarchy* N Africa
Morocco *see* Marrakech
Morocco, Kingdom of *see* Morocco
81 I22 **Morogoro** Morogoro, E Tanzania 06°49′S 37°40′E
81 H24 **Morogoro** ◈ *region* SE Tanzania
171 Q7 **Moro Gulf** *gulf* S Philippines
41 N13 **Moroleón** Guanajuato, C Mexico 20°00′N 101°13′W
172 H6 **Morombe** Toliara, W Madagascar 21°47′S 43°21′E
44 G5 **Morón** Ciego de Ávila, C Cuba 22°08′N 78°39′W
163 N8 **Mörön** Hentiy, C Mongolia 47°21′N 110°21′E
162 I6 **Mörön** Hövsgöl, N Mongolia 49°39′N 100°08′E
54 K5 **Morón** Carabobo, N Venezuela 10°29′N 68°11′W
Morón *see* Morón de la Frontera
56 D8 **Morona, Río** ≈ N Peru
56 C8 **Morona Santiago** ◈ *province* E Ecuador
172 H5 **Morondava** Toliara, W Madagascar 20°19′S 44°17′E
104 K14 **Morón de la Frontera** *var.* Morón. Andalucía, S Spain 37°07′N 05°27′W
172 G13 **Moroni** ● (Comoros) Grande Comore, NW Comoros 11°41′S 43°16′E
171 S10 **Morotai, Pulau** *island* Maluku, E Indonesia
Morotiri *see* Marotiri
81 H17 **Moroto** NE Uganda 02°32′N 34°41′E
Morozov *see* Bratan
126 M11 **Morozovsk** Rostovskaya Oblast', SW Russian Federation 48°21′N 41°54′E
97 L14 **Morpeth** N England, United Kingdom 55°10′N 01°41′W
Morphou *see* Güzelyurt
Morphou Bay *see* Güzelyurt Körfezi
28 I13 **Morrill** Nebraska, C USA 41°57′N 103°55′W
27 U11 **Morrilton** Arkansas, C USA 35°09′N 92°45′W
11 Q16 **Morrin** Alberta, SW Canada 51°40′N 112°45′W
184 M7 **Morrinsville** Waikato, North Island, New Zealand 37°41′S 175°32′E
11 X16 **Morris** Manitoba, S Canada 49°22′N 94°21′W
30 M11 **Morris** Illinois, N USA 41°21′N 88°25′W
29 S8 **Morris** Minnesota, N USA 45°32′N 95°53′W
14 M13 **Morrisburg** Ontario, SE Canada 44°55′N 75°07′W
197 R11 **Morris Jesup, Kap** *headland* N Greenland 83°33′N 32°40′W
182 B1 **Morris, Mount** ▲ South Australia 26°04′S 131°03′E
30 K10 **Morrison** Illinois, N USA 41°48′N 89°58′W
36 K13 **Morristown** Arizona, SW USA 33°48′N 112°34′W
18 J14 **Morristown** New Jersey, NE USA 40°48′N 74°29′W
21 O8 **Morristown** Tennessee, S USA 36°13′N 83°18′W
42 L11 **Morrito** Río San Juan, SW Nicaragua 11°37′N 85°05′W
35 P13 **Morro Bay** California, W USA 35°21′N 120°51′W
95 L22 **Mörrum** Blekinge, S Sweden 56°11′N 14°45′E
83 N16 **Morrumbala** Zambézia, NE Mozambique 17°17′S 35°35′E
83 N20 **Morrumbene** Inhambane, SE Mozambique 23°41′S 35°25′E
95 F21 **Mors** *island* NW Denmark
Mörs *see* Moers
25 N1 **Morse** Texas, SW USA 36°03′N 101°28′W
127 N6 **Morshansk** Tambovskaya Oblast', W Russian Federation 53°27′N 41°46′E
102 L5 **Mortagne-au-Perche** Orne, N France 48°32′N 00°31′E
102 J8 **Mortagne-sur-Sèvre** Vendée, NW France 47°00′N 00°57′W
104 G8 **Mortágua** Viseu, N Portugal 40°24′N 08°14′W
102 J5 **Mortain** Manche, N France 48°39′N 00°51′W
106 C8 **Mortara** Lombardia, N Italy 45°15′N 08°44′E
59 J17 **Mortes, Rio das** ≈ C Brazil
182 M12 **Mortlake** Victoria, SE Australia 38°06′S 142°48′E
Mortlock Group *see* Takuu Islands
189 Q17 **Mortlock Islands** *prev.* Nomoi Islands. *island group* C Micronesia
29 T9 **Morton** Minnesota, N USA 44°33′N 94°58′W
22 L5 **Morton** Mississippi, S USA 32°21′N 89°39′W
24 M5 **Morton** Texas, SW USA 33°40′N 102°45′W
32 H9 **Morton** Washington, NW USA 46°33′N 122°16′W
0 D7 **Morton Seamount** *undersea feature* NE Pacific Ocean 50°15′N 142°45′W
45 U15 **Moruga** Trinidad, Trinidad and Tobago 10°04′N 61°16′W
183 P9 **Morundah** New South Wales, SE Australia 34°57′S 146°18′E
191 X12 **Moruroa** *var.* Mururoa. *atoll* Îles Tuamotu, SE French Polynesia
183 S11 **Moruya** New South Wales, SE Australia 35°55′S 150°04′E
103 Q8 **Morvan** *physical region* C France
185 G21 **Morven** Canterbury, South Island, New Zealand 44°51′S 171°07′E
183 O13 **Morwell** Victoria, SE Australia 38°14′S 146°25′E
125 N6 **Morzhovets, Ostrov** *island* NW Russian Federation
126 J4 **Mosal'sk** Kaluzhskaya Oblast', W Russian Federation 54°30′N 34°55′E
101 H20 **Mosbach** Baden-Württemberg, SW Germany 49°21′N 09°06′E
95 E18 **Mosby** Vest-Agder, S Norway 58°12′N 07°55′E
33 V9 **Mosby** Montana, NW USA 46°58′N 107°53′W
32 M9 **Moscow** Idaho, NW USA 46°43′N 117°00′W
20 F10 **Moscow** Tennessee, S USA 35°04′N 89°27′W
Moscow *see* Moskva
101 D19 **Mosel** *Fr.* Moselle. ≈ W Europe *see also* Moselle
Mosel *see* Moselle
103 T4 **Moselle** ◈ *department* NE France
103 T6 **Moselle** *Ger.* Mosel. ≈ W Europe *see also* Mosel
Moselle *see* Mosel
32 K9 **Moses Lake** ◎ Washington, NW USA
83 I18 **Mosetse** Central, E Botswana 20°40′S 26°38′E
92 H4 **Mosfellsbær** Höfuðborgarsvæðið, SW Iceland 65°09′N 21°43′W
183 N8 **Mossgiel** New South Wales, SE Australia 33°16′S 144°34′E
185 F23 **Mosgiel** Otago, South Island, New Zealand 45°51′S 170°22′E
124 M11 **Mosha** ≈ NW Russian Federation
81 I20 **Moshi** Kilimanjaro, NE Tanzania 03°21′S 37°19′E
110 G12 **Mosina** Wielkolpolskie, C Poland 52°15′N 16°50′E
30 L6 **Mosinee** Wisconsin, N USA 44°45′N 89°39′W
92 F13 **Mosjøen** Nordland, C Norway 65°49′N 13°12′E
123 S12 **Moskal'vo** Ostrov Sakhalin, Sakhalinskaya Oblast', SE Russian Federation 53°36′N 142°31′E
92 I13 **Moskosel** Norrbotten, N Sweden 65°52′N 19°30′E
126 K4 **Moskovskaya Oblast'** ◈ *province* W Russian Federation
Moskovskiy *see* Moskva
126 J3 **Moskva** *Eng.* Moscow. ● (Russian Federation) Gorod Moskva, W Russian Federation 55°45′N 37°42′E
147 Q14 **Moskva** *Rus.* Moskovskiy; *prev.* Chubek. SW Tajikistan 37°41′N 69°33′E
126 L4 **Moskva** ≈ W Russian Federation
83 I20 **Mosomane** Kgatleng, SE Botswana 24°04′S 26°15′E
Moson and Magyaróvár *see* Mosonmagyaróvár
111 H21 **Mosoni-Duna** *Ger.* Kleine Donau. ≈ NW Hungary
111 H21 **Mosonmagyaróvár** *Ger.* Wieselburg-Ungarisch-Altenburg; *prev.* Moson and Magyaróvár, *Ger.* Wieselburg and Ungarisch-Altenburg. Győr-Moson-Sopron, NW Hungary 47°52′N 17°15′E
Mospino *see* Mospyne
117 X8 **Mospyne** *Rus.* Mospino. Donets'ka Oblast', E Ukraine 47°53′N 38°03′E
54 B12 **Mosquera** Nariño, SW Colombia 02°32′N 78°24′W
37 U10 **Mosquero** New Mexico, SW USA 35°46′N 103°57′W
Mosquito Coast *see* La Mosquitia
31 U11 **Mosquito Creek Lake** ⊠ Ohio, N USA
Mosquito Gulf *see* Mosquitos, Golfo de los
23 X11 **Mosquito Lagoon** *wetland* Florida, SE USA
43 N10 **Mosquito, Punta** *headland* E Nicaragua 12°18′N 83°38′W
43 W14 **Mosquito, Punta** *headland* NE Panama 09°06′N 77°52′W
43 Q15 **Mosquitos, Golfo de los** *Eng.* Mosquito Gulf. *gulf* N Panama
95 H16 **Moss** Østfold, S Norway 59°25′N 10°40′E
Mossâmedes *see* Namibe
22 G8 **Moss Bluff** Louisiana, S USA 30°18′N 93°11′W
185 C23 **Mossburn** Southland, South Island, New Zealand 45°40′S 168°15′E
83 G26 **Mosselbaai** *var.* Mosselbai, *Eng.* Mossel Bay. Western Cape, SW South Africa 34°11′S 22°08′E
Mosselbai/Mossel Bay *see* Mosselbaai
79 F20 **Mossendjo** Niari, SW Congo 02°57′S 12°40′E
101 H22 **Mössingen** Baden-Württemberg, S Germany 48°22′N 09°01′E
181 W4 **Mossman** Queensland, NE Australia 16°34′S 145°27′E
59 P14 **Mossoró** Rio Grande do Norte, NE Brazil 05°11′S 37°20′W
23 N9 **Moss Point** Mississippi, S USA 30°24′N 88°31′W
183 S9 **Moss Vale** New South Wales, SE Australia 34°33′S 150°20′E
32 G9 **Mossyrock** Washington, NW USA 46°32′N 122°30′W
111 B15 **Most** *Ger.* Brüx. Ústecký Kraj, NW Czech Republic 50°30′N 13°37′E
162 E7 **Möst** *var.* Ulaantolgoy. Hovd, W Mongolia 46°39′N 92°50′E
121 P16 **Mosta** *var.* Musta. C Malta 35°54′N 14°25′E
74 I5 **Mostaganem** *var.* Mestghanem. NW Algeria 35°54′N 00°05′E
113 H14 **Mostar** Federacija Bosne I Hercegovine, S Bosnia and Herzegovina 43°21′N 17°47′E
61 J17 **Mostardas** Rio Grande do Sul, S Brazil 31°02′S 50°51′W
116 K14 **Mostiștea** ≈ S Romania
Mostva *see* Mastva
Mosty *see* Masty
116 H5 **Mostys'ka** L'vivs'ka Oblast', W Ukraine 49°47′N 23°09′E
Mosul *see* Al Mawșil
95 F15 **Møsvatnet** ◎ S Norway
80 J12 **Mot'a** Āmara, N Ethiopia 11°03′N 38°03′E
79 H16 **Motaba** ≈ N Congo
105 O10 **Mota del Cuervo** Castilla-La Mancha, C Spain 39°30′N 02°52′W
104 L5 **Mota del Marqués** Castilla y León, N Spain 41°38′N 05°11′W
42 F5 **Motagua, Río** ≈ Guatemala/Honduras
119 H19 **Motal'** Brestskaya Voblasts', SW Belarus 52°13′N 25°34′E
95 L17 **Motala** Östergötland, S Sweden 58°34′N 15°05′E
191 X7 **Motane** *island* Îles Marquises, NE French Polynesia
152 K13 **Moth** Uttar Pradesh, N India 25°44′N 78°56′E
Mother of Presidents/Mother of States *see* Virginia
96 I12 **Motherwell** C Scotland, United Kingdom 55°48′N 04°W
153 P12 **Motīhāri** Bihār, N India 26°40′N 84°55′E
105 Q10 **Motilla del Palancar** Castilla-La Mancha, C Spain 39°34′N 01°55′W
184 N7 **Motiti Island** *island* NE New Zealand
65 E25 **Motley Island** *island* SE Falkland Islands
83 J19 **Motloutse** ≈ E Botswana
41 V17 **Motozintla de Mendoza** Chiapas, SE Mexico 15°21′N 92°14′W
105 N15 **Motril** Andalucía, S Spain 36°45′N 03°30′W
116 G13 **Motru** Gorj, SW Romania 44°49′N 22°56′E
165 Q4 **Motsuta-misaki** *headland* Hokkaidō, NE Japan 42°36′N 139°48′E
28 L6 **Mott** North Dakota, N USA 46°21′N 102°17′W
166 M9 **Mottama** *var.* Martaban, Moktama. Mon State, S Myanmar (Burma) 16°32′N 97°35′E
166 L9 **Mottama, Gulf of** *var.* Gulf of Martaban. *gulf* S Myanmar (Burma)
Möttling *see* Metlika
107 O18 **Mottola** Puglia, SE Italy 40°38′N 17°02′E
184 P8 **Motu** ≈ North Island, New Zealand
185 I14 **Motueka** Tasman, South Island, New Zealand 41°08′S 173°00′E
185 I14 **Motueka** ≈ South Island, New Zealand
Motu Iti *see* Tupai
41 X12 **Motul** *var.* Motul de Felipe Carrillo Puerto. Yucatán, SE Mexico 21°06′N 89°17′W
Motul de Felipe Carrillo Puerto *see* Motul
191 U17 **Motu Nui** *island* Easter Island, Chile, E Pacific Ocean
191 Q10 **Motu One** *var.* Bellingshausen. *atoll* Îles Sous le Vent, W French Polynesia
190 I16 **Motutapu** *island* E Cook Islands
193 V15 **Motu Tapu** *island* Tongatapu Group, S Tonga
184 L5 **Motutapu Island** *island* N New Zealand
Motyca *see* Modica
Mouanda *see* Moanda
Mouaskar *see* Mascara
105 U3 **Moubermé, Tuc de** *Fr.* Pic de Maubermé, *Sp.* Pico Mauberme; *prev.* Tuc de Maubermé. ▲ France/Spain 42°48′N 00°57′E *see also* Maubermé, Pic de
Moubermé, Tuc de *see* Maubermé, Pic de
45 N7 **Mouchoir Passage** *passage* SE Turks and Caicos Islands
76 I9 **Moudjéria** Tagant, SW Mauritania 17°52′N 12°20′W
108 C9 **Moudon** Vaud, W Switzerland 46°41′N 06°49′E
Mouhoun *see* Black Volta
79 E19 **Mouila** Ngounié, C Gabon 01°50′S 11°02′E
79 K14 **Mouka** Haute-Kotto, C Central African Republic 07°12′N 21°52′E
Moukden *see* Shenyang
183 N10 **Moulamein** New South Wales, SE Australia 35°06′S 144°03′E
Moulamein Creek *see* Billabong Creek
74 F6 **Moulay-Bousselham** NW Morocco 35°00′N 06°22′W
Moule *see* le Moule
80 M11 **Moulhoulé** N Djibouti 12°34′N 43°06′E
103 P9 **Moulins** Allier, C France 46°34′N 03°20′E
Moulmein *see* Mawlamyine
Moulmeingyun *see* Mawlamyinegyunn
74 G6 **Moulouya** *var.* Mulucha, Muluya, Mulwiya. *seasonal river* NE Morocco
23 O2 **Moulton** Alabama, S USA 34°28′N 87°18′W
29 W16 **Moulton** Iowa, C USA 40°41′N 92°40′W
25 T11 **Moulton** Texas, SW USA 29°34′N 97°08′W
23 T7 **Moultrie** Georgia, SE USA 31°10′N 83°47′W
21 S14 **Moultrie, Lake** ⊠ South Carolina, SE USA
22 K3 **Mound Bayou** Mississippi, S USA 33°52′N 90°43′W
30 L17 **Mound City** Illinois, N USA 37°06′N 89°09′W
27 R6 **Mound City** Kansas, C USA 38°07′N 94°49′W
27 Q2 **Mound City** Missouri, C USA 40°07′N 95°13′W
29 N7 **Mound City** South Dakota, N USA 45°44′N 100°03′W
78 H13 **Moundou** Logone-Occidental, SW Chad 08°35′N 16°01′E
27 P10 **Mounds** Oklahoma, C USA 35°52′N 96°03′W
21 R2 **Moundsville** West Virginia, NE USA 39°54′N 80°44′W
167 R11 **Moŭng** *prev.* Phumĭ Moŭng. Siĕmréab, NW Cambodia 13°45′N 103°35′E
167 Q12 **Moŭng Roessei** Bătdâmbâng, W Cambodia 12°47′N 103°28′E
Moun Hou *see* Black Volta
8 H8 **Mountain** ≈ Northwest Territories, NW Canada
37 S12 **Mountainair** New Mexico, SW USA 34°31′N 106°14′W
35 V1 **Mountain City** Nevada, W USA 41°48′N 115°58′W
21 Q8 **Mountain City** Tennessee, S USA 36°28′N 81°48′W
27 U7 **Mountain Grove** Missouri, C USA 37°07′N 92°15′W
27 U9 **Mountain Home** Arkansas, C USA 36°19′N 92°24′W
33 N15 **Mountain Home** Idaho, NW USA 43°07′N 115°42′W
25 Q11 **Mountain Home** Texas, SW USA 30°11′N 99°19′W
29 W4 **Mountain Iron** Minnesota, N USA 47°31′N 92°37′W
29 T10 **Mountain Lake** Minnesota, N USA 43°57′N 94°54′W
23 S3 **Mountain Park** Georgia, SE USA 34°04′N 84°24′W
35 W12 **Mountain Pass** *pass* California, W USA
27 T12 **Mountain Pine** Arkansas, C USA 34°34′N 93°10′W
39 Y14 **Mountain Point** Annette Island, Alaska, USA 55°17′N 131°31′W
Mountain State *see* Montana
Mountain State *see* West Virginia
27 V7 **Mountain View** Arkansas, C USA 35°52′N 92°07′W
38 H12 **Mountain View** Hawaii, USA, C Pacific Ocean 19°32′N 155°03′W
27 V10 **Mountain View** Missouri, C USA 37°00′N 91°42′W
38 M11 **Mountain Village** Alaska, USA 62°06′N 163°42′W
21 R8 **Mount Airy** North Carolina, SE USA 36°30′N 80°36′W
83 K24 **Mount Ayliff** *Xh.* Maxesibebi. Eastern Cape, SE South Africa 30°48′S 29°23′E
29 U16 **Mount Ayr** Iowa, C USA 40°42′N 94°14′W
182 J9 **Mount Barker** South Australia 35°06′S 138°52′E
180 J14 **Mount Barker** Western Australia 34°42′S 117°40′E
183 P11 **Mount Beauty** Victoria, SE Australia 36°47′S 147°12′E
14 E16 **Mount Brydges** Ontario, S Canada 42°54′N 81°29′W
31 N16 **Mount Carmel** Illinois, N USA 38°23′N 87°46′W
30 K10 **Mount Carroll** Illinois, N USA 42°05′N 89°59′W
31 S9 **Mount Clemens** Michigan, N USA 42°36′N 82°52′W
185 E19 **Mount Cook** Canterbury, South Island, New Zealand 43°47′S 170°06′E
83 L16 **Mount Darwin** Mashonaland Central, NE Zimbabwe 16°45′S 31°39′E
19 S7 **Mount Desert Island** *island* Maine, NE USA
23 W11 **Mount Dora** Florida, SE USA 28°48′N 81°38′W
182 G5 **Mount Eba** South Australia 30°11′S 135°40′E
25 W8 **Mount Enterprise** Texas, SW USA 31°53′N 94°40′W
182 J4 **Mount Fitton** South Australia 29°55′S 139°26′E
83 J24 **Mount Fletcher** Eastern Cape, SE South Africa 30°41′S 28°30′E
14 F15 **Mount Forest** Ontario, S Canada 43°58′N 80°44′W
182 K12 **Mount Gambier** South Australia 37°47′S 140°49′E
181 W5 **Mount Garnet** Queensland, NE Australia 17°41′S 145°07′E
21 P6 **Mount Gay** West Virginia, NE USA 37°49′N 82°00′W
31 S12 **Mount Gilead** Ohio, N USA 40°33′N 82°49′W
186 C7 **Mount Hagen** Western Highlands, C Papua New Guinea 05°54′S 144°13′E
18 J16 **Mount Holly** New Jersey, NE USA 39°59′N 74°46′W
21 R10 **Mount Holly** North Carolina, SE USA 35°18′N 81°01′W
27 T12 **Mount Ida** Arkansas, C USA 34°32′N 93°38′W
181 T6 **Mount Isa** Queensland, C Australia 20°48′S 139°32′E
21 U4 **Mount Jackson** Virginia, NE USA 38°45′N 78°38′W
18 D12 **Mount Jewett** Pennsylvania, NE USA 41°43′N 78°37′W
18 L13 **Mount Kisco** New York, NE USA 41°12′N 73°42′W
18 B15 **Mount Lebanon** Pennsylvania, NE USA 40°21′N 80°03′W
182 J8 **Mount Lofty Ranges** ▲▲ South Australia
180 J10 **Mount Magnet** Western Australia 28°09′S 117°52′E
184 N7 **Mount Maunganui** Bay of Plenty, North Island, New Zealand 37°39′S 176°11′E
97 E18 **Mountmellick** *Ir.* Móinteach Mílic. Laois, C Ireland 53°07′N 07°20′W
30 L10 **Mount Morris** Illinois, N USA 42°03′N 89°25′W
31 R9 **Mount Morris** Michigan, N USA 43°07′N 83°42′W
18 F10 **Mount Morris** New York, NE USA 42°43′N 77°51′W
18 B16 **Mount Morris** Pennsylvania, NE USA 39°43′N 80°06′W
30 K15 **Mount Olive** Illinois, N USA 39°04′N 89°43′W
21 V10 **Mount Olive** North Carolina, SE USA 35°12′N 78°03′W
21 N4 **Mount Olivet** Kentucky, S USA 38°32′N 84°01′W
29 Y15 **Mount Pleasant** Iowa, C USA 40°57′N 91°33′W
31 Q8 **Mount Pleasant** Michigan, N USA 43°36′N 84°46′W
18 C15 **Mount Pleasant** Pennsylvania, NE USA 40°07′N 79°33′W
21 T14 **Mount Pleasant** South Carolina, SE USA 32°47′N 79°51′W
20 I9 **Mount Pleasant** Tennessee, S USA 35°32′N 87°11′W
25 W6 **Mount Pleasant** Texas, SW USA 33°10′N 94°49′W
36 L4 **Mount Pleasant** Utah, W USA 39°33′N 111°27′W
63 N23 **Mount Pleasant** ✈ (Stanley) East Falkland, Falkland Islands
97 G25 **Mount's Bay** *inlet* SW England, United Kingdom
35 N2 **Mount Shasta** California, W USA 41°18′N 122°19′W
30 J13 **Mount Sterling** Illinois, N USA 39°59′N 90°44′W
21 N5 **Mount Sterling** Kentucky, S USA 38°03′N 83°56′W
18 E15 **Mount Union** Pennsylvania, NE USA 40°21′N 77°51′W
23 V6 **Mount Vernon** Georgia, SE USA 32°10′N 82°35′W
30 L16 **Mount Vernon** Illinois, N USA 38°19′N 88°54′W
20 M6 **Mount Vernon** Kentucky, S USA 37°20′N 84°20′W
27 S7 **Mount Vernon** Missouri, C USA 37°05′N 93°49′W
31 T13 **Mount Vernon** Ohio, N USA 40°23′N 82°29′W
32 K13 **Mount Vernon** Oregon, NW USA 44°22′N 119°07′W
25 W6 **Mount Vernon** Texas, SW USA 33°11′N 95°13′W
32 H7 **Mount Vernon** Washington, NW USA 48°25′N 122°19′W
20 L5 **Mount Washington** Kentucky, S USA 38°03′N 85°33′W
182 F8 **Mount Wedge** South Australia 33°29′S 135°08′E
30 L14 **Mount Zion** Illinois, N USA 39°46′N 88°52′W
181 Y9 **Moura** Queensland, NE Australia 24°34′S 149°57′E
58 F12 **Moura** Amazonas, NW Brazil 01°32′S 61°43′W
104 H12 **Moura** Beja, S Portugal 38°08′N 07°27′W
104 I12 **Mourão** Évora, S Portugal 38°22′N 07°20′W
76 L11 **Mourdiah** Koulikoro, W Mali 14°28′N 07°31′W
78 K7 **Mourdi, Dépression du** *desert lowland* Chad/Sudan
102 J16 **Mourenx** Pyrénées-Atlantiques, SW France 43°24′N 00°37′W
Mourgana *see* Mourgkána
115 C15 **Mourgkána** *var.* Mourgana. ▲ Albania/Greece 39°48′N 20°24′E
97 G16 **Mourne Mountains** *Ir.* Beanna Boirche. ▲▲ SE Northern Ireland, United Kingdom
115 I15 **Moúrtzeflos, Akrotírio** *headland* Límnos, E Greece 40°00′N 25°02′E
99 C19 **Mouscron** *Dut.* Moeskroen. Hainaut, W Belgium 50°44′N 03°14′E
Mouse River *see* Souris River
78 H10 **Moussoro** Bahr el Gazel, W Chad 13°41′N 16°31′E
103 T11 **Moûtiers** Savoie, E France 45°28′N 06°31′E
172 J14 **Moutsamudou** *var.* Moutsamudou. Anjouan, SE Comoros 12°10′S 44°25′E
Moutsamudou *see* Moutsamudou
74 K11 **Mouydir, Monts de** ▲▲ S Algeria
79 F20 **Mouyondzi** Bouenza, S Congo 03°58′S 13°57′E
115 E16 **Mouzáki** *prev.* Mouzákion. Thessalía, C Greece 39°25′N 21°40′E
Mouzákion *see* Mouzáki
29 S13 **Moville** Iowa, C USA 42°30′N 96°04′W
82 E13 **Moxico** ◈ *province* E Angola
172 I14 **Moya** Anjouan, SE Comoros 12°18′S 44°27′E
40 L12 **Moyahua** Zacatecas, C Mexico 21°18′N 103°09′W
81 J16 **Moyalē** Oromīya, C Ethiopia 03°34′N 38°58′E

◆ Country
● Country Capital
◇ Dependent Territory
○ Dependent Territory Capital
◈ Administrative Regions
✈ International Airport
▲ Mountain
▲▲ Mountain Range
⛰ Volcano
≈ River
◎ Lake
⊠ Reservoir

76 *I15* **Moyamba** W Sierra Leone 08°04′N 12°30′W
74 *G7* **Moyen Atlas** *Eng.* Middle Atlas. ⛰ N Morocco
78 *H13* **Moyen-Chari** *off.* Région du Moyen-Chari. ◈ *region* S Chad
Moyen-Chari, Région du *see* Moyen-Chari
Moyen-Congo *see* Congo (Republic of)
83 *J24* **Moyeni** *var.* Quthing. SW Lesotho 30°25′S 27°43′E
79 *D18* **Moyen-Ogooué** *off.* Province du Moyen-Ogooué, *var.* Le Moyen-Ogooué. ◈ *province* C Gabon
Moyen-Ogooué, Province du *see* Moyen-Ogooué
103 *S4* **Moyeuvre-Grande** Moselle, NE France 49°15′N 06°03′E
33 *N7* **Moyie Springs** Idaho, NW USA 48°43′N 116°15′W
146 *G6* **Mo'ynoq** *Rus.* Muynak. Qoraqalpog'iston Respublikasi, NW Uzbekistan 43°45′N 59°03′E
81 *F16* **Moyo** NW Uganda 03°38′N 31°43′E
56 *D10* **Moyobamba** San Martín, NW Peru 06°04′S 76°56′W
78 *H10* **Moyto** Hadjer-Lamis, W Chad 12°35′N 16°33′E
158 *G9* **Moyu** *var.* Karakax. Xinjiang Uygur Zizhiqu, NW China 37°16′N 79°39′E
122 *M9* **Moyyero** ≈ N Russian Federation
145 *S15* **Moyynkum** *var.* Furmanovka, *Kaz.* Fūrmanov. Zhambyl, S Kazakhstan 44°15′N 72°55′E
145 *Q15* **Moyynkum, Peski** *Kaz.* Moyynqum. *desert* S Kazakhstan
Moyynqum *see* Moyynkum, Peski
145 *S12* **Moyynty** Karaganda, C Kazakhstan 47°10′N 73°24′E
145 *S12* **Moyynty** ≈ Karaganda, C Kazakhstan
Mozambika, Lakandranon' i *see* Mozambique Channel
83 *M18* **Mozambique** *off.* Republic of Mozambique; *prev.* People's Republic of Mozambique, Portuguese East Africa. ◆ *republic* S Africa
Mozambique Basin *see* Natal Basin
Mozambique, Canal de *see* Mozambique Channel
83 *P17* **Mozambique Channel** *Fr.* Canal de Mozambique, *Mal.* Lakandranon' i Mozambika. *strait* W Indian Ocean
172 *L11* **Mozambique Escarpment** *var.* Mozambique Scarp. *undersea feature* SW Indian Ocean 33°00′S 36°30′E
Mozambique, People's Republic of *see* Mozambique
172 *L10* **Mozambique Plateau** *var.* Mozambique Rise. *undersea feature* SW Indian Ocean 32°00′S 35°00′E
Mozambique, Republic of *see* Mozambique
Mozambique Rise *see* Mozambique Plateau
Mozambique Scarp *see* Mozambique Escarpment
127 *O15* **Mozdok** Respublika Severnaya Osetiya, SW Russian Federation 43°48′N 44°42′E
57 *K17* **Mozetenes, Serranías de** ⛰ C Bolivia
126 *J4* **Mozhaysk** Moskovskaya Oblast', W Russian Federation 55°31′N 36°01′E
127 *T3* **Mozhga** Udmurtskaya Respublika, NW Russian Federation 56°24′N 52°13′E
Mozyr' *see* Mazyr
79 *P22* **Mpala** Katanga, E Dem. Rep. Congo 06°43′S 29°28′E
79 *G19* **Mpama** ≈ C Congo
81 *E22* **Mpanda** Katavi, W Tanzania 06°21′S 31°01′E
82 *L11* **Mpande** Northern, NE Zambia 09°13′S 31°42′E
83 *J18* **Mphoengs** Matabeleland South, SW Zimbabwe 21°04′S 27°56′E
81 *F18* **Mpigi** S Uganda 0°14′N 32°19′E
82 *L13* **Mpika** Muchinga, NE Zambia 11°50′S 31°30′E
83 *J14* **Mpima** Central, C Zambia 14°25′S 28°34′E
82 *J13* **Mpongwe** Copperbelt, C Zambia 13°25′S 28°13′E
82 *K11* **Mporokoso** Northern, N Zambia 09°22′S 30°06′E
79 *H20* **Mpouya** Plateaux, SE Congo 02°38′S 16°13′E
77 *P16* **Mpraeso** C Ghana 06°36′N 00°43′W
82 *L11* **Mpulungu** Northern, N Zambia 08°50′S 31°06′E
83 *K21* **Mpumalanga** *prev.* Eastern Transvaal, *Afr.* Oos-Transvaal. ◈ *province* NE South Africa
83 *D16* **Mpungu** Okavango, N Namibia 17°36′S 18°16′E
81 *I22* **Mpwapwa** Dodoma, C Tanzania 06°21′S 36°29′E
Mqinvartsveri *see* Kazbek
110 *M8* **Mrągowo** *Ger.* Sensburg. Warmińsko-Mazurskie, NE Poland 53°53′N 21°19′E
127 *V6* **Mrakovo** Respublika Bashkortostan, W Russian Federation 52°43′N 56°36′E
172 *I13* **Mramani** Anjouan, E Comoros 12°18′N 44°39′E
166 *K5* **Mrauk-oo** *var.* Mrauk U, Myohaung. Rakhine State, W Myanmar (Burma) 20°35′N 93°12′E
Mrauk U *see* Mrauk-oo
112 *F12* **Mrkonjić Grad** ◈ Republika Srpska, W Bosnia and Herzegovina
110 *H9* **Mrocza** Kujawsko-pomorskie, C Poland 53°15′N 17°38′E
124 *I14* **Msta** ≈ NW Russian Federation
Mstislavl' *see* Mstsislaw
119 *P15* **Mstsislaw** *Rus.* Mstislavl'. Mahilyowskaya Voblasts', E Belarus 54°01′N 31°43′E
83 *J24* **Mthatha** *prev.* Umtata. Eastern Cape, SE South Africa 31°33′S 28°47′E *see also* Umtata
Mtkvari *see* Kura
Mtoko *see* Mutoko
126 *K6* **Mtsensk** Orlovskaya Oblast', W Russian Federation 53°17′N 36°34′E
81 *K24* **Mtwara** Mtwara, SE Tanzania 10°17′S 40°11′E
81 *J25* **Mtwara** ◈ *region* SE Tanzania
104 *G14* **Mu** ▲ S Portugal 37°24′N 08°04′W
193 *V15* **Mu'a** Tongatapu, S Tonga 21°11′S 175°07′W
Muai To *see* Mae Hong Son
83 *P16* **Mualama** Zambézia, NE Mozambique 16°51′S 38°21′E
Mualo *see* Messalo, Rio
79 *E22* **Muanda** Bas-Congo, SW Dem. Rep. Congo 05°53′S 12°17′E
Muang Chiang Rai *see* Chiang Rai
167 *R6* **Muang Ham** Houaphan, N Laos 20°19′N 104°00′E
167 *S8* **Muang Hinboun** Khammouan, C Laos 17°37′N 104°37′E
Muang Kalasin *see* Kalasin
Muang Khammouan *see* Thakhèk
167 *S11* **Muang Không** Champasak, S Laos 14°08′N 105°48′E
167 *S10* **Muang Khôngxédôn** *var.* Khong Sedone. Salavan, S Laos 15°34′N 105°46′E
Muang Khon Kaen *see* Khon Kaen
167 *Q6* **Muang Khoua** Phôngsali, N Laos 21°07′N 102°31′E
Muang Krabi *see* Krabi
Muang Lampang *see* Lampang
Muang Lamphun *see* Lamphun
Muang Loei *see* Loei
Muang Lom Sak *see* Lom Sak
Muang Nakhon Sawan *see* Nakhon Sawan
167 *Q6* **Muang Namo** Oudômxai, N Laos 20°58′N 101°46′E
Muang Nan *see* Nan
167 *Q6* **Muang Ngoy** Louangphabang, N Laos 20°43′N 102°42′E
167 *Q5* **Muang Ou Tai** Phôngsali, N Laos 22°06′N 101°59′E
Muang Pak Lay *see* Pak Lay
Muang Pakxan *see* Pakxan
167 *T10* **Muang Pakxong** Champasak, S Laos 15°10′N 106°17′E
167 *S9* **Muang Phalan** *var.* Muang Phalane. Savannakhét, S Laos 16°40′N 105°33′E
Muang Phalane *see* Muang Phalan
Muang Phan *see* Phan
Muang Phayao *see* Phayao
Muang Phichit *see* Phichit
167 *T9* **Muang Phin** Savannakhét, S Laos 16°31′N 106°01′E
Muang Phitsanulok *see* Phitsanulok
Muang Phrae *see* Phrae
Muang Roi Et *see* Roi Et
Muang Sakon Nakhon *see* Sakon Nakhon
Muang Samut Prakan *see* Samut Prakan
167 *P6* **Muang Sing** Louang Namtha, N Laos 21°12′N 101°09′E
Muang Ubon *see* Ubon Ratchathani
Muang Uthai Thani *see* Uthai Thani
167 *P7* **Muang Vangviang** Viangchan, C Laos 18°53′N 102°27′E
Muang Xaignabouri *see* Xaignabouli
Muang Xay *see* Oudômxai
167 *S9* **Muang Xépôn** *var.* Sepone. Savannakhét, S Laos 16°40′N 106°15′E
168 *K10* **Muar** *var.* Bandar Maharani. Johor, Peninsular Malaysia 02°01′N 102°35′E
168 *J9* **Muara** Sumatera, W Indonesia 02°18′N 98°54′E
168 *L13* **Muarabeliti** Sumatera, W Indonesia 03°13′S 103°00′E
168 *K12* **Muarabungo** Sumatera, W Indonesia 01°28′S 102°06′E
168 *L13* **Muaraenim** Sumatera, W Indonesia 03°40′S 103°48′E
169 *T11* **Muarajuloi** Borneo, C Indonesia 0°12′S 114°03′E
169 *U12* **Muarakaman** Borneo, C Indonesia 0°09′S 116°43′E
168 *H12* **Muarasigep** Pulau Siberut, W Indonesia 01°01′S 98°48′E
168 *L12* **Muaratembesi** Sumatera, W Indonesia 01°40′S 103°08′E
169 *T12* **Muaratewe** *var.* Muarateweh; *prev.* Moearatewe. Borneo, C Indonesia 0°58′S 114°52′E
Muarateweh *see* Muaratewe
169 *U10* **Muarawahau** Borneo, N Indonesia 01°03′N 116°48′E
138 *G13* **Mubārak, Jabal** ▲ S Jordan 29°19′N 35°13′E
153 *N13* **Mubārakpur** Uttar Pradesh, N India 26°05′N 83°19′E
Mubarek *see* Muborak
81 *F18* **Mubende** SW Uganda 0°35′N 31°24′E
77 *Y14* **Mubi** Adamawa, NE Nigeria 10°15′N 13°18′E
146 *M12* **Muborak** *Rus.* Mubarek. Qashqadaryo Viloyati, S Uzbekistan 39°17′N 65°10′E
171 *U12* **Mubrani** Papua Barat, E Indonesia 0°42′S 133°25′E
82 *L12* **Muchinga** ◈ *province* NE Zambia
67 *U12* **Muchinga Escarpment** *escarpment* NE Zambia
127 *N7* **Muchkapskiy** Tambovskaya Oblast', W Russian Federation 51°51′N 42°25′E
96 *G10* **Muck** *island* W Scotland, United Kingdom
82 *Q13* **Mucojo** Cabo Delgado, N Mozambique 12°05′S 40°30′E
82 *F12* **Muconda** Lunda Sul, NE Angola 10°37′S 21°19′E
54 *I10* **Muco, Río** ≈ E Colombia
83 *O16* **Mucubela** Zambézia, NE Mozambique 16°51′S 37°48′E
42 *J5* **Mucupina, Monte** ▲ N Honduras 15°07′N 86°36′W
136 *J14* **Mucur** Kirşehir, C Turkey 39°05′N 34°25′E
143 *U8* **Mūd** Khorāsān-e Janūbī, E Iran 32°41′N 59°30′E
163 *Y9* **Mudanjiang** *var.* Mu-tan-chiang. Heilongjiang, NE China 44°33′N 129°40′E
163 *Y9* **Mudan Jiang** ≈ NE China
136 *D11* **Mudanya** Bursa, NW Turkey 40°23′N 28°53′E
28 *K8* **Mud Butte** South Dakota, N USA 45°00′N 102°51′W
155 *G16* **Muddebihāl** Karnātaka, C India 16°26′N 76°07′E
27 *P12* **Muddy Boggy Creek** ≈ Oklahoma, C USA
36 *M6* **Muddy Creek** ≈ Utah, W USA
37 *V7* **Muddy Creek Reservoir** ⊠ Colorado, C USA
33 *W15* **Muddy Gap** Wyoming, C USA 42°21′N 107°27′W
35 *Y11* **Muddy Peak** ▲ Nevada, W USA 36°17′N 114°40′W
183 *R7* **Mudgee** New South Wales, SE Australia 32°37′S 149°36′E
29 *S3* **Mud Lake** ◎ Minnesota, N USA
29 *P7* **Mud Lake Reservoir** ⊠ South Dakota, N USA
167 *N9* **Mudon** Mon State, S Myanmar (Burma) 16°17′N 97°40′E
81 *O14* **Mudug** *off.* Gobolka Mudug. ◈ *region* NE Somalia
81 *O14* **Mudug** *var.* Mudugh. *plain* N Somalia
Mudug, Gobolka *see* Mudug
Mudugh *see* Mudug
83 *Q15* **Muecate** Nampula, NE Mozambique 14°56′S 39°38′E
82 *Q13* **Mueda** Cabo Delgado, NE Mozambique 11°40′S 39°31′E
42 *L10* **Muelle de los Bueyes** Región Autónoma Atlántico Sur, SE Nicaragua 12°03′N 84°34′W
Muenchen *see* München
83 *M14* **Muende** Tete, NW Mozambique 14°22′S 33°00′E
25 *T5* **Muenster** Texas, SW USA 33°39′N 97°22′W
Muenster *see* Münster
43 *O6* **Muerto, Cayo** *reef* NE Nicaragua
41 *T17* **Muerto, Mar** *lagoon* SE Mexico
64 *F11* **Muertos Trough** *undersea feature* N Caribbean Sea
83 *H14* **Mufaya Kuta** Western, NW Zambia 14°30′S 24°18′E
82 *J13* **Mufulira** Copperbelt, C Zambia 12°33′S 28°16′E
161 *O10* **Mufu Shan** ⛰ C China
Mugalla *see* Yutian
Mugalzhar Taūlary *see* Mugodzhary, Gory
137 *Y12* **Muğan Düzü** *Rus.* Muganskaya Ravnina, Muganskaya Step'. *physical region* S Azerbaijan
Muganskaya Ravnina/ Muganskaya Step' *see* Muğan Düzü
106 *K8* **Múggia** Friuli-Venezia Giulia, NE Italy 45°36′N 13°48′E
153 *N14* **Mughal Sarāi** Uttar Pradesh, N India 25°18′N 83°07′E
Mughla *see* Muğla
141 *W11* **Mughshin** *var.* Muqshin. S Oman 19°26′N 54°38′E
147 *S12* **Mughsu** *Rus.* Muksu. ≈ C Tajikistan
164 *H14* **Mugi** Tokushima, Shikoku, SW Japan 33°39′N 134°24′E
136 *C16* **Muğla** *var.* Mughla. Muğla, SW Turkey 37°13′N 28°22′E
136 *C16* **Muğla** *var.* Mughla. ◈ *province* SW Turkey
Müglizh *see* Maglizh
144 *J11* **Mugodzhary, Gory** *Kaz.* Mugalzhar Taūlary. ⛰ W Kazakhstan
83 *O15* **Mugulama** Zambézia, NE Mozambique 16°01′S 37°33′E
Muḥāfazat Hims *see* Ḥimş
139 *U9* **Muḥammad** Wāsiţ, E Iraq 32°46′N 45°14′E
139 *R8* **Muḥammadīyah** Al Anbār, C Iraq 33°22′N 42°48′E
80 *I6* **Muhammad Qol** Red Sea, NE Sudan 20°53′N 37°09′E
75 *Y9* **Muḥammad, Rās** *headland* E Egypt 27°45′N 34°18′E
Muhammerah *see* Khorramshahr
140 *M12* **Muḥāyil** *var.* Mahāil. 'Asīr, SW Saudi Arabia 18°34′N 42°01′E
139 *O7* **Muḥaywīr** Al Anbār, W Iraq 33°35′N 41°06′E
101 *H21* **Mühlacker** Baden-Württemberg, SW Germany 48°57′N 08°51′E
Mühlbach *see* Sebeş
Mühldorf *see* Mühldorf am Inn
101 *N23* **Mühldorf am Inn** *var.* Mühldorf. Bayern, SE Germany 48°14′N 12°32′E
101 *J15* **Mühlhausen** *var.* Mühlhausen in Thüringen. Thüringen, C Germany 51°13′N 10°28′E
Mühlhausen in Thüringen *see* Mühlhausen
195 *Q2* **Mühlig-Hofmanfjella** *Eng.* Mühlig-Hofmann Mountains. ⛰ Antarctica
93 *L14* **Muhos** Pohjois-Pohjanmaa, C Finland 64°48′N 26°00′E
138 *K6* **Muḥ, Sabkhat al** ◎ C Syria
118 *E5* **Muhu** *Ger.* Mohn, Moon. *island* W Estonia
81 *F19* **Muhutwe** Kagera, NW Tanzania 01°31′S 31°41′E
Muhu Väin *see* Väinameri
98 *J10* **Muiden** Noord-Holland, C Netherlands 52°19′N 05°04′E
193 *W15* **Mui Hopohoponga** *headland* Tongatapu, S Tonga 21°09′S 175°02′W
Muinchille *see* Cootehill
Muineachán *see* Monaghan
97 *F19* **Muine Bheag** *Eng.* Bagenalstown. Carlow, SE Ireland 52°42′N 06°57′W
56 *B5* **Muisne** Esmeraldas, NW Ecuador 0°35′N 79°58′W
83 *P14* **Muite** Nampula, NE Mozambique 14°02′S 39°06′E
41 *Z11* **Mujeres, Isla** *island* E Mexico
116 *G7* **Mukacheve** *Hung.* Munkács, *Rus.* Mukachevo. Zakarpats'ka Oblast', W Ukraine 48°27′N 22°45′E
Mukachevo *see* Mukacheve
169 *R9* **Mukah** Sarawak, East Malaysia 02°56′N 112°02′E
Mukalla *see* Al Mukallā
Mukama *see* Mokama
Mukāshafa/Mukashshafah *see* Mukayshīfah
139 *S6* **Mukayshīfah** *var.* Mukāshafa, Mukashshafah. Şalāḥ ad Dīn, N Iraq 34°24′N 43°44′E
167 *R9* **Mukdahan** Mukdahan, E Thailand 16°31′N 104°43′E
Mukden *see* Shenyang
165 *Y15* **Mukojima-rettō** *Eng.* Parry group. *island group* SE Japan
146 *M14* **Mukry** Lebap Welaýaty, E Turkmenistan 37°39′N 65°37′E
Muksu *see* Mughsu
Muktagacha *see* Muktagachha
153 *U14* **Muktagachha** *var.* Muktagacha. N Bangladesh 24°46′N 90°16′E
82 *K13* **Mukuku** Central, C Zambia 12°10′S 29°50′E
82 *K11* **Mukupa Kaoma** Northern, NE Zambia 09°55′S 30°19′E
81 *I18* **Mukutan** Baringo, W Kenya 01°06′N 36°16′E
83 *F16* **Mukwe** Caprivi, NE Namibia 18°01′S 21°24′E
105 *R13* **Mula** Murcia, SE Spain 38°02′N 01°29′W
151 *K20* **Mulakatholhu** *var.* Meemu Atoll, Mulaku Atoll. *atoll* C Maldives
Mulaku Atoll *see* Mulakatholhu
83 *J15* **Mulalika** Lusaka, C Zambia 15°37′S 28°48′E
163 *X8* **Mulan** Heilongjiang, NE China 45°57′N 128°00′E
83 *N15* **Mulanje** *var.* Mlanje. Southern, S Malawi 16°05′S 35°29′E
40 *H5* **Mulatos** Sonora, NW Mexico 28°42′N 108°44′W
23 *P3* **Mulberry Fork** ≈ Alabama, S USA
39 *P12* **Mulchatna River** ≈ Alaska, USA
125 *W4* **Mul'da** Respublika Komi, NW Russian Federation 67°29′N 63°55′E
101 *M14* **Mulde** ≈ E Germany
27 *R10* **Muldrow** Oklahoma, C USA 35°25′N 94°34′W
40 *E7* **Mulegé** Baja California Sur, NW Mexico 26°54′N 112°00′W
108 *I10* **Mulegns** Graubünden, S Switzerland 46°30′N 09°36′E
79 *M21* **Mulenda** Kasai-Oriental, C Dem. Rep. Congo 04°19′S 24°55′E
24 *M4* **Muleshoe** Texas, SW USA 34°13′N 102°43′W
83 *O15* **Mulevala** Zambézia, NE Mozambique 16°25′S 37°35′E
183 *P5* **Mulgoa Creek** *seasonal river* New South Wales, SE Australia
105 *O15* **Mulhacén** *var.* Cerro de Mulhacén. ▲ S Spain 37°07′N 03°11′W
Mulhacén, Cerro de *see* Mulhacén
Mülhausen *see* Mulhouse
101 *E24* **Müllheim** Baden-Württemberg, SW Germany 47°48′N 07°37′E
101 *E15* **Mülheim** *var.* Mulheim an der Ruhr. Nordrhein-Westfalen, W Germany 51°25′N 06°50′E
Mulheim an der Ruhr *see* Mülheim
103 *U7* **Mulhouse** *Ger.* Mülhausen. Haut-Rhin, NE France 47°45′N 07°20′E
160 *G11* **Muli** *var.* Qiaowa, Muli Zangzu Zizhixian. Sichuan, C China 27°49′N 101°10′E
171 *X15* **Muli** *channel* Papua, E Indonesia
Mülig-Hofmann Mountains *see* Mühlig-Hofmanfjella
163 *Y9* **Muling** Heilongjiang, NE China 44°54′N 130°35′E
Muli Zangzu Zizhixian *see* Muli
Mullach Íde *see* Malahide
Mullaitivu *see* Mullaittivu
155 *K23* **Mullaittivu** *var.* Mullaitivu. Northern Province, N Sri Lanka 09°15′N 80°48′E
33 *N8* **Mullan** Idaho, NW USA 47°28′N 115°48′W
28 *M13* **Mullen** Nebraska, C USA 42°02′N 101°01′W
183 *Q6* **Mullengudgery** New South Wales, SE Australia 31°42′S 147°24′E
21 *Q6* **Mullens** West Virginia, NE USA 37°34′N 81°22′W
Müller-gerbergte *see* Muller, Pegunungan
169 *T10* **Muller, Pegunungan** *Dut.* Müller-gerbergte. ⛰ Borneo, C Indonesia
31 *Q5* **Mullett Lake** ◎ Michigan, N USA
18 *J16* **Mullica River** ≈ New Jersey, NE USA
25 *R8* **Mullin** Texas, SW USA 31°33′N 98°40′W
97 *E17* **Mullingar** *Ir.* An Muileann gCearr. C Ireland 53°32′N 07°20′W
21 *T12* **Mullins** South Carolina, SE USA 34°12′N 79°15′W
96 *G11* **Mull, Isle of** *island* W Scotland, United Kingdom
127 *R5* **Mullovka** Ul'yanovskaya Oblast', W Russian Federation 54°13′N 49°19′E
95 *K19* **Mullsjö** Västra Götaland, S Sweden 57°56′N 13°55′E
183 *V4* **Mullumbimby** New South Wales, SE Australia 28°34′S 153°28′E
83 *H15* **Mulobezi** Western, SW Zambia 16°48′S 25°11′E
83 *C15* **Mulondo** Huíla, SW Angola 15°41′S 15°09′E
83 *G15* **Mulonga Plain** *plain* W Zambia
79 *N23* **Mulongo** Katanga, SE Dem. Rep. Congo 07°44′S 26°57′E
149 *T10* **Multān** Punjab, E Pakistan 30°12′N 71°30′E
93 *L17* **Multia** Keski-Suomi, C Finland 62°27′N 24°49′E
Mulucha *see* Moulouya
83 *J14* **Mulungushi** Central, C Zambia 14°15′S 28°27′E
83 *K14* **Mulungwe** Central, C Zambia 13°57′S 29°51′E
Muluya *see* Moulouya
27 *N7* **Mulvane** Kansas, C USA 37°28′N 97°14′W
183 *O10* **Mulwala** New South Wales, SE Australia 35°59′S 146°00′E
Mulwiya *see* Moulouya
182 *K6* **Mulyungarie** South Australia 31°29′S 140°45′E
154 *D13* **Mumbai** *prev.* Bombay. *state capital* Mahārāshtra, W India 18°56′N 72°51′E
154 *D13* **Mumbai** ✈ Mahārāshtra, W India 19°10′N 72°51′E
83 *D14* **Mumbué** Bié, C Angola 13°52′S 17°15′E
186 *E8* **Mumeng** Morobe, C Papua New Guinea 06°57′S 146°34′E
171 *V12* **Mumi** Papua Barat, E Indonesia 01°33′S 134°09′E
Muminabad/Mŭ'minobod *see* Mu'minobod
147 *Q13* **Mu'minobod** *Rus.* Leningradskiy, Muminabad; *prev.* Leningrad. SW Tajikistan 38°03′N 69°50′E
127 *Q13* **Mumra** Astrakhanskaya Oblast', SW Russian Federation 45°46′N 47°46′E
41 *X12* **Muna** Yucatán, SE Mexico 20°29′N 89°41′W
123 *O9* **Muna** ≈ NE Russian Federation
152 *C12* **Munābāo** Rājasthān, NW India 25°46′N 70°19′E
Munamägi *see* Suur Munamägi
171 *O14* **Muna, Pulau** *prev.* Moena. *island* C Indonesia
101 *L18* **Münchberg** Bayern, E Germany 50°10′N 11°50′E
101 *L23* **München** *var.* Muenchen, *Eng.* Munich, *It.* Monaco. Bayern, SE Germany 48°09′N 11°34′E
München-Gladbach *see* Mönchengladbach
108 *E6* **Münchenstein** Basel Landschaft, NW Switzerland 47°31′N 07°38′E
10 *L10* **Muncho Lake** British Columbia, W Canada 58°52′N 125°40′W
31 *P13* **Muncie** Indiana, N USA 40°11′N 85°22′W
18 *G13* **Muncy** Pennsylvania, NE USA 41°10′N 76°46′W
11 *Q14* **Mundare** Alberta, SW Canada 53°34′N 112°20′W
25 *Q5* **Munday** Texas, SW USA 33°27′N 99°37′W
31 *N10* **Mundelein** Illinois, N USA 42°15′N 88°00′W
101 *I15* **Münden** Niedersachsen, C Germany 52°16′N 08°54′E
105 *Q12* **Mundo** ≈ S Spain
82 *B12* **Munenga** Kwanza Sul, NW Angola 10°03′S 14°40′E
105 *P11* **Munera** Castilla-La Mancha, C Spain 39°03′N 02°29′W
20 *E9* **Munford** Tennessee, S USA 35°27′N 89°49′W
20 *K7* **Munfordville** Kentucky, S USA 37°17′N 85°55′W
182 *D5* **Mungala** South Australia 30°36′S 132°57′E
83 *M16* **Mungári** Manica, C Mozambique 17°09′S 33°33′E
79 *O16* **Mungbere** Orientale, NE Dem. Rep. Congo 02°38′N 28°30′E
153 *Q13* **Munger** *prev.* Monghyr. Bihār, NE India 25°23′N 86°28′E
182 *I2* **Mungeranie** South Australia 28°02′S 138°42′E
Mu Nggava *see* Rennell
169 *O10* **Mungguresak, Tanjung** *headland* Borneo, N Indonesia 01°57′N 109°19′E
Mungiki *see* Bellona
183 *R4* **Mungindi** New South Wales, SE Australia 28°59′S 149°00′E
Mungkawn *see* Maingkwan
Mungla *see* Mongla
82 *C13* **Mungo** Huambo, W Angola 11°49′S 16°16′E
188 *F16* **Munguuy Bay** *bay* Yap, W Micronesia
82 *E13* **Muhango** Bié, C Angola 12°12′S 18°34′E
Munich *see* München
105 *S7* **Muniesa** Aragón, NE Spain 41°02′N 00°49′W
31 *O4* **Munising** Michigan, N USA 46°24′N 86°39′W
Munkács *see* Mukacheve
95 *I17* **Munkedal** Västra Götaland, S Sweden 58°28′N 11°38′E
95 *K15* **Munkfors** Värmland, C Sweden 59°50′N 13°35′E
122 *M14* **Munku-Sardyk, Gora** *var.* Mönh Sarĭdag. ▲ Mongolia/Russian Federation 51°45′N 100°22′E
99 *E18* **Munkzwalm** Oost-Vlaanderen, NW Belgium 50°53′N 03°44′E
167 *R10* **Mun, Mae Nam** ≈ E Thailand
153 *U15* **Munshiganj** Dhaka, C Bangladesh 23°32′N 90°32′E
108 *D8* **Münsingen** Bern, W Switzerland 46°53′N 07°34′E
103 *U6* **Munster** Haut-Rhin, NE France 48°03′N 07°09′E
100 *J11* **Münster** Niedersachsen, NW Germany 52°59′N 10°07′E
100 *F13* **Münster** *var.* Muenster, Münster in Westfalen. Nordrhein-Westfalen, W Germany 51°58′N 07°38′E
108 *F10* **Münster** Valais, S Switzerland 46°31′N 08°18′E
97 *B20* **Munster** *Ir.* Cúige Mumhan. *cultural region* S Ireland
Münsterberg in Schlesien *see* Ziębice
Münster in Westfalen *see* Münster
100 *E13* **Münsterland** *cultural region* NW Germany
100 *F13* **Münster-Osnabruck** ✈ Nordrhein-Westfalen, NW Germany 52°08′N 07°41′E
Muntafiq, Al *see* Dhī Qār
31 *R4* **Munuscong Lake** ◎ Michigan, N USA
83 *K17* **Munyati** ≈ C Zimbabwe
109 *R3* **Münzkirchen** Oberösterreich, N Austria 48°29′N 13°37′E
92 *K11* **Muodoslompolo** Norrbotten, N Sweden 67°57′N 23°31′E
92 *M13* **Muojärvi** ◎ NE Finland
167 *S6* **Muong Khên** Hoa Binh, N Vietnam 20°34′N 105°18′E
Muong Sai *see* Oudômxai
167 *Q7* **Muong Xiang Ngeun** *var.* Xieng Ngeun. Louangphabang, N Laos 19°43′N 102°09′E
92 *K11* **Muonio** Lappi, N Finland 67°58′N 23°40′E
Muonioälv/Muoniojoki *see* Muonionjoki
92 *K11* **Muonionjoki** *var.* Muoniojoki, *Swe.* Muonioälv. ≈ Finland/Sweden
Muorjek *see* Murjek
83 *N17* **Mupa** ≈ C Mozambique
83 *E16* **Mupini** Okavango, NE Namibia 17°55′S 19°34′E
80 *F8* **Muqaddam, Wadi** ≈ N Sudan
138 *K9* **Muqāţ** Al Mafraq, E Jordan 32°28′N 38°04′E
81 *N17* **Muqdisho** *Eng.* Mogadishu, *It.* Mogadiscio. ● (Somalia) Banaadir, S Somalia 02°06′N 45°27′E
81 *N17* **Muqdisho** ✈ Banaadir, E Somalia 01°58′N 45°18′E
Muqshin *see* Mughshin
109 *T8* **Mur** *SCr.* Mura. ≈ C Europe
109 *X9* **Mura** ≈ NE Slovenia
Mura *see* Mur
137 *T14* **Muradiye** Van, E Turkey 39°N 43°44′E
Muragarazi *see* Maragarazi
165 *O10* **Murakami** Niigata, Honshū, C Japan 38°13′N 139°28′E
63 *G22* **Murallón, Cerro** ▲ S Argentina 49°49′S 73°25′W
81 *E20* **Muramvya** C Burundi 03°18′S 29°41′E
81 *I19* **Murang'a** *prev.* Fort Hall. Murang'a, SW Kenya 0°43′S 37°10′E
81 *I19* **Murang'a** ◈ *county* C Kenya
81 *H16* **Murangering** Turkana, NW Kenya 03°48′N 35°29′E
Murapara *see* Murupara
140 *M5* **Murār, Bi'r al** *well* NW Saudi Arabia
125 *Q13* **Murashi** Kirovskaya Oblast', NW Russian Federation 59°27′N 48°02′E
103 *O12* **Murat** Cantal, C France 45°07′N 02°52′E
114 *N12* **Muratlı** Tekirdağ, NW Turkey 41°12′N 27°30′E
137 *R14* **Murat Nehri** *var.* Eastern Euphrates; *anc.* Arsanias. ≈ NE Turkey
107 *D20* **Muravera** Sardegna, Italy, C Mediterranean Sea 39°24′N 09°34′E
165 *P10* **Murayama** Yamagata, Honshū, C Japan 38°29′N 140°21′E
121 *R13* **Muraysah, Ra's al** *headland* N Libya 31°58′N 25°00′E
104 *I6* **Murça** Vila Real, N Portugal 41°24′N 07°28′W
80 *Q11* **Murcanyo** Bari, NE Somalia 11°39′N 50°27′E
143 *N8* **Mürcheh Khvort** *var.* Morcheh Khort. Eşfahān, C Iran 33°07′N 51°26′E
185 *H15* **Murchison** Tasman, South Island, New Zealand 41°48′S 172°19′E
185 *B22* **Murchison Mountains** ⛰ South Island, New Zealand
180 *I10* **Murchison River** ≈ Western Australia
105 *R13* **Murcia** Murcia, SE Spain 37°59′N 01°08′W
105 *Q13* **Murcia** ◈ *autonomous community* SE Spain
103 *O13* **Mur-de-Barrez** Aveyron, S France 44°48′N 02°39′E
182 *G8* **Murdinga** South Australia 33°46′S 135°46′E
28 *M10* **Murdo** South Dakota, N USA 43°53′N 100°42′W
15 *X6* **Murdochville** Québec, SE Canada 48°57′N 65°30′W
109 *W9* **Mureck** Steiermark, SE Austria 46°42′N 15°46′E
114 *M13* **Mürefte** Tekirdağ, NW Turkey 40°40′N 27°15′E
116 *I10* **Mureş** ◈ *county* N Romania
84 *J11* **Mureş** ≈ Hungary/Romania
Mureş *see* Maros
Mureşul *see* Maros/Mureş
102 *M16* **Muret** Haute-Garonne, S France 43°28′N 01°19′E
27 *T13* **Murfreesboro** Arkansas, C USA 34°04′N 93°42′W
21 *W8* **Murfreesboro** North Carolina, SE USA 36°26′N 77°06′W
20 *J9* **Murfreesboro** Tennessee, S USA 35°50′N 86°25′W
Murgab *see* Morghāb, Daryā-ye/Murgap
Murgab *see* Murghob
146 *I14* **Murgap** Mary Welaýaty, S Turkmenistan 37°19′N 61°48′E
146 *J16* **Murgap** *var.* Deryasy Murgap, Murghab, *Pash.* Daryā-ye Morghāb, *Rus.* Murgab. ≈ Afghanistan/Turkmenistan *see also* Morghāb, Daryā-ye
Murgap *see* Morghāb, Daryā-ye
Murgap, Deryasy *see* Morghāb, Daryā-ye/Murgap
114 *H9* **Murgash** ▲ W Bulgaria 42°51′N 23°58′E
Murghab *see* Morghāb, Daryā-ye/Murgap
148 *M4* **Murghāb, Daryā-ye** *Rus.* Murgab, Murghab, *Turk.* Murgap, Deryasy Murgap; *prev.* Morghāb, Daryā-ye. ≈ Afghanistan/Turkmenistan *see also* Murgap
147 *U13* **Murghob** *Rus.* Murgab. SE Tajikistan 38°11′N 74°E
147 *U13* **Murghob** *Rus.* Murgab. ≈ SE Tajikistan
181 *Z10* **Murgon** Queensland, E Australia 26°08′S 152°04′E
190 *I16* **Muri** Rarotonga, S Cook Islands 21°15′S 159°44′W
108 *F7* **Muri** Aargau, W Switzerland 47°16′N 08°21′E
108 *D8* **Muri** *var.* Muri bei Bern. Bern, W Switzerland 46°55′N 07°30′E
104 *K3* **Murias de Paredes** Castilla y León, N Spain 42°51′N 06°11′W
Muri bei Bern *see* Muri
82 *F11* **Muriege** Lunda Sul, NE Angola 09°55′S 21°12′E
189 *P14* **Murilo Atoll** *atoll* Hall Islands, C Micronesia
Mūrītāniyah *see* Mauritania
100 *N10* **Müritz** *var.* Müritzee. ◎ NE Germany
Müritzee *see* Müritz
100 *L10* **Müritz-Elde-Wasserstrasse** *canal* N Germany
184 *K6* **Muriwai Beach** Auckland, North Island, New Zealand 36°56′S 174°28′E
92 *J13* **Murjek** *Lapp.* Muorjek. Norrbotten, N Sweden 66°27′N 20°54′E
124 *J3* **Murmansk** Murmanskaya Oblast', NW Russian Federation 68°59′N 33°08′E
124 *I4* **Murmanskaya Oblast'** ◈ *province* NW Russian Federation
197 *V14* **Murmansk Rise** *undersea feature* SW Barents Sea 71°00′N 37°00′E
124 *J3* **Murmashi** Murmanskaya Oblast', NW Russian Federation 68°49′N 32°43′E
126 *M5* **Murmino** Ryazanskaya Oblast', W Russian Federation 54°31′N 40°01′E
101 *K24* **Murnau** Bayern, SE Germany 47°41′N 11°12′E
103 *X16* **Muro, Capo di** *headland* Corse, France, C Mediterranean Sea 41°45′N 08°40′E
107 *M18* **Muro Lucano** Basilicata, S Italy 40°48′N 15°33′E
127 *N4* **Murom** Vladimirskaya Oblast', W Russian Federation 55°33′N 42°03′E
122 *I11* **Muromtsevo** Omskaya Oblast', C Russian Federation 56°18′N 75°15′E
165 *R5* **Muroran** Hokkaidō, NE Japan 42°20′N 140°58′E
104 *G3* **Muros** Galicia, NW Spain 42°47′N 09°04′W
104 *F3* **Muros e Noia, Ría de** *estuary* NW Spain
164 *H15* **Muroto** Kōchi, Shikoku, SW Japan 33°16′N 134°10′E
164 *H15* **Muroto-zaki** Shikoku, SW Japan
116 *L7* **Murovani Kurylivtsi** Vinnyts'ka Oblast', C Ukraine 48°43′N 27°31′E
110 *G11* **Murowana Goślina** Wielkolpolskie, W Poland 52°33′N 16°59′E
32 *M14* **Murphy** Idaho, NW USA 43°14′N 116°36′W
21 *N10* **Murphy** North Carolina, SE USA 35°05′N 84°02′W
35 *P8* **Murphys** California, W USA 38°07′N 120°27′W
30 *L17* **Murphysboro** Illinois, N USA 37°45′N 89°20′W
29 *V15* **Murray** Iowa, C USA 41°03′N 93°56′W
20 *H8* **Murray** Kentucky, S USA 36°35′N 88°20′W
182 *J10* **Murray Bridge** South Australia 35°10′S 139°17′E
175 *X2* **Murray Fracture Zone** *tectonic feature* NE Pacific Ocean
192 *H11* **Murray, Lake** ◎ SW Papua New Guinea
21 *P12* **Murray, Lake** ⊠ South Carolina, SE USA
10 *K8* **Murray, Mount** ▲ Yukon, NW Canada 60°49′N 128°57′W
Murray Range *see* Murray Ridge
173 *O3* **Murray Ridge** *var.* Murray Range. *undersea feature* N Arabian Sea 21°45′N 61°50′E
183 *N10* **Murray River** ≈ SE Australia
182 *K10* **Murrayville** Victoria, SE Australia 35°17′S 141°12′E
149 *U5* **Murree** Punjab, E Pakistan 33°55′N 73°26′E
101 *I21* **Murrhardt** Baden-Württemberg, S Germany 49°00′N 09°34′E
183 *O9* **Murrumbidgee River** ≈ New South Wales, SE Australia
83 *P15* **Murrupula** Nampula, NE Mozambique 15°26′S 38°46′E
183 *T7* **Murrurundi** New South Wales, SE Australia 31°47′S 150°51′E
109 *X9* **Murska Sobota** *Ger.* Olsnitz. NE Slovenia 46°41′N 16°09′E
154 *G12* **Murtajāpur** *prev.* Murtazapur. Mahārāshtra, C India 20°43′N 77°28′E
77 *S16* **Murtala Muhammed** ✈ (Lagos) Ogun, SW Nigeria 06°31′N 03°12′E
Murtazapur *see* Murtajāpur
108 *C8* **Murten** Neuchâtel, W Switzerland 46°55′N 07°06′E
Murtensee *see* Morat, Lac de
182 *L11* **Murtoa** Victoria, SE Australia 36°39′S 142°27′E
92 *N13* **Murtovaara** Pohjois-Pohjanmaa, E Finland 65°40′N 29°25′E
Murua Island *see* Woodlark Island
155 *D14* **Murud** Mahārāshtra, W India 18°27′N 72°56′E
184 *O9* **Murupara** *var.* Murapara. Bay of Plenty, North Island, New Zealand 38°27′S 176°41′E
Mururoa *see* Moruroa
Murviedro *see* Sagunto
154 *J9* **Murwāra** Madhya Pradesh, N India 23°50′N 80°23′E
183 *V4* **Murwillumbah** New South Wales, SE Australia 28°20′S 153°24′E
146 *H11* **Murzechirla** *prev.* Mirzachirla. Ahal Welaýaty, C Turkmenistan 39°33′N 60°02′E
Murzuk *see* Murzuq
75 *O11* **Murzuq** *var.* Marzūq, Murzuk. SW Libya 25°55′N 13°55′E
Murzuq, Edeyin *see* Murzuq, Idhān
75 *N11* **Murzuq, Ḥammādat** *plateau* W Libya
75 *O11* **Murzuq, Idhān** *var.* Edeyin Murzuq. *desert* SW Libya
109 *W6* **Mürzzuschlag** Steiermark, E Austria 47°35′N 15°41′E
137 *Q14* **Muş** *var.* Mush. Muş, E Turkey 38°45′N 41°30′E
137 *Q14* **Muş** *var.* Mush. ◈ *province* E Turkey
118 *G11* **Mūša** ≈ Latvia/Lithuania
186 *F9* **Musa** ≈ S Papua New Guinea
Mūsa, Gebel *see* Mūsá, Jabal
Musaiyib *see* Al Musayyib
75 *X8* **Mūsá, Jabal** *var.* Gebel Mûsa. ▲ NE Egypt 28°33′N 33°51′E
149 *R9* **Mūsā Khel** *var.* Mūsa Khel Bāzār. Baluchistān, SW Pakistan 30°53′N 69°52′E
Mūsa Khel Bāzār *see* Mūsā Khel
139 *Z13* **Mūsá, Khowr-e** *bay* Iraq/Kuwait
114 *H10* **Musala** ▲ W Bulgaria 42°12′N 23°36′E
168 *H10* **Musala, Pulau** *island* W Indonesia

◆ Country ● Country Capital ◇ Dependent Territory ○ Dependent Territory Capital ◈ Administrative Regions ✈ International Airport ▲ Mountain ⛰ Mountain Range Volcano ≈ River ◎ Lake ⊠ Reservoir

MUSALE – NAMPULA

83 I15 Musale Southern, S Zambia 15°27′S 26°50′E
141 Y9 Muşalla NE Oman 22°20′N 58°03′E
141 W6 Musandam Peninsula Ar. Masandam Peninsula. peninsula N Oman
Musay'id see Umm Sa'id
Muscat see Masqaţ
Muscat and Oman see Oman
29 Y14 Muscatine Iowa, C USA 41°25′N 91°03′W
Muscat Sīb Airport see Seeb
31 O15 Muscatuck River ≈ Indiana, N USA
30 K8 Muscoda Wisconsin, N USA 43°11′N 90°27′W
185 F19 Musgrave, Mount ▲ South Island, New Zealand 43°48′S 170°43′E
181 P9 Musgrave Ranges ▲▲ South Australia
Mush see Muş
138 H12 Mushayyish, Qaşr al castle Ma'ān, C Jordan
79 H20 Mushie Bandundu, W Dem. Rep. Congo 03°00′S 16°55′E
168 M13 Musi, Air prev. Moesi. ≈ Sumatera, W Indonesia
192 M4 Musicians Seamounts undersea feature N Pacific Ocean
83 K19 Musina prev. Messina. Limpopo, NE South Africa 22°18′S 30°02′E
54 D8 Musinga, Alto ▲ NW Colombia 06°49′N 76°24′W
29 T2 Muskeg Bay lake bay Minnesota, N USA
31 O8 Muskegon Michigan, N USA 43°13′N 86°15′W
31 O8 Muskegon Heights Michigan, N USA 43°12′N 86°14′W
31 P8 Muskegon River ≈ Michigan, N USA
31 T14 Muskingum River ≈ Ohio, N USA
95 P16 Muskö Stockholm, C Sweden 58°58′N 18°10′E
Muskogean see Tallahassee
27 Q10 Muskogee Oklahoma, C USA 35°45′N 95°21′W
14 H13 Muskoka, Lake ◎ Ontario, S Canada
80 H8 Musmar Red Sea, NE Sudan 18°13′N 35°40′E
83 K14 Musofu Central, C Zambia 13°31′S 29°02′E
81 G19 Musoma Mara, N Tanzania 01°31′S 33°49′E
82 L13 Musoro Central, C Zambia 13°21′S 31°04′E
186 F4 Mussau Island island NE Papua New Guinea
98 P7 Musselkanaal Groningen, NE Netherlands 52°55′N 07°01′E
33 V9 Musselshell River ≈ Montana, NW USA
82 C12 Mussende Kwanza Sul, NW Angola 10°33′S 16°02′E
102 L12 Mussidan Dordogne, SW France 45°03′N 00°22′E
99 L25 Musson Luxembourg, SE Belgium 49°33′N 05°42′E
152 J9 Mussoorie Uttarakhand, N India 30°26′N 78°04′E
Musta see Mosta
152 M13 Mustafābād Uttar Pradesh, N India 25°54′N 81°17′E
136 D12 Mustafakemalpaşa Bursa, NW Turkey 40°03′N 28°25′E
Mustafa-Pasha see Svilengrad
81 M15 Mustahīl Sumalē, E Ethiopia 05°18′N 44°34′E
24 M7 Mustang Draw valley Texas, SW USA
25 T14 Mustang Island island Texas, SW USA
Mustasaari see Korsholm
Mustér see Disentis
63 I19 Musters, Lago ◎ S Argentina
45 Y14 Mustique island C Saint Vincent and the Grenadines
118 I6 Mustla Viljandimaa, S Estonia 58°12′N 25°50′E
118 J4 Mustvee Ger. Tschorna. Jõgevamaa, E Estonia 58°51′N 26°59′E
42 L9 Musún, Cerro ▲ NE Nicaragua 13°01′N 85°02′W
183 T7 Muswellbrook New South Wales, SE Australia 32°17′S 150°55′E
111 M18 Muszyna Małopolskie, SE Poland 49°21′N 20°54′E
75 V10 Mūţ var. Mut. C Egypt 25°28′N 28°58′E
136 I17 Mut İçel, S Turkey 36°38′N 33°27′E
109 V9 Muta N Slovenia 46°37′N 15°09′E
190 B15 Mutalau N Niue 18°56′S 169°50′E
Mu-tan-chiang see Mudanjiang
82 I13 Mutanda North Western, NW Zambia 12°24′S 26°13′E
59 O17 Mutá, Ponta do headland E Brazil 13°54′S 38°54′W
83 L17 Mutare var. Mutari; prev. Umtali. Manicaland, E Zimbabwe 18°55′S 32°36′E
Mutari see Mutare
54 D8 Mutatá Antioquia, NW Colombia 07°16′N 76°32′W
Muthannā, Muḩāfa at al see Al Muthannā
Mutina see Modena
83 L16 Mutoko prev. Mtoko. Mashonaland East, NE Zimbabwe 17°24′S 32°13′E
81 J20 Mutomo Kitui, S Kenya 01°50′S 38°13′E
Mutrah see Maţraḩ
79 M24 Mutshatsha Katanga, S Dem. Rep. Congo 10°40′S 24°26′E
165 R6 Mutsu var. Mutu. Aomori, Honshū, N Japan 41°18′N 141°11′E
165 R6 Mutsu-wan bay N Japan
108 E6 Muttenz Basel Landschaft, NW Switzerland 47°31′N 07°39′E
185 A26 Muttonbird Islands island group SW New Zealand
Muttra see Mathura
Mutu see Mutsu
83 O15 Mutuáli Nampula, N Mozambique 14°51′S 37°01′E
82 D13 Mutumbo Bié, C Angola 13°10′S 17°22′E
189 Y14 Mutunte, Mount var. Mount Buache. ▲ Kosrae, E Micronesia 05°21′N 163°00′E
155 K24 Mutur Eastern Province, E Sri Lanka 08°27′N 81°15′E
92 L13 Muurola Lappi, NW Finland 66°22′N 25°20′E
162 M14 Mu Us Shadi var. Ordos Desert; prev. Mu Us Shamo. desert N China
Mu Us Shamo see Mu Us Shadi
82 B11 Muxima Bengo, NW Angola 09°33′S 13°58′E
124 I8 Muyezerskiy Respublika Kareliya, NW Russian Federation 63°54′N 32°00′E
81 E20 Muyinga NE Burundi 02°54′S 30°19′E
42 K9 Muy Muy Matagalpa, C Nicaragua 12°43′N 85°35′W
Muynak see Mo'ynoq
79 N22 Muyumba Katanga, SE Dem. Rep. Congo 07°13′S 27°02′E
149 V5 Muzaffarābād Jammu and Kashmir, NE Pakistan 34°23′N 73°34′E
149 S10 Muzaffargarh Punjab, E Pakistan 30°04′N 71°15′E
152 J9 Muzaffarnagar Uttar Pradesh, N India 29°28′N 77°42′E
153 P13 Muzaffarpur Bihār, N India 26°07′N 85°23′E
158 H6 Muzat He ≈ W China
83 L15 Muze Tete, NW Mozambique 15°05′S 31°16′E
122 H8 Muzhi Yamalo-Nenetskiy Avtonomnyy Okrug, N Russian Federation 65°25′N 64°28′E
102 H7 Muzillac Morbihan, NW France 47°34′N 02°30′W
Muzkol, Khrebet see Muzqŭl, Qatorkŭhi
112 L9 Mužlja Hung. Felsőmuzslya; prev. Gornja Mužlja. Vojvodina, N Serbia 45°21′N 20°25′E
54 F9 Muzo Boyacá, C Colombia 05°34′N 74°07′W
83 J15 Muzoka Southern, S Zambia 16°39′S 27°18′E
39 Y15 Muzon, Cape headland Dall Island, Alaska, USA 54°39′N 132°41′W
40 M6 Múzquiz Coahuila, NE Mexico 27°54′N 101°30′W
147 U13 Muzqŭl, Qatorkŭhi Rus. Khrebet Muzkol. ▲▲ SE Tajikistan
Muztag see Muztag Feng
158 D8 Muztagata ▲ NW China 38°16′N 75°03′E
158 K10 Muztag Feng var. Muztag. ▲ W China 36°26′N 87°15′E
83 K17 Mvuma prev. Umvuma. Midlands, C Zimbabwe 19°17′S 30°32′E
172 H13 Mwali var. Moili, Fr. Mohéli. island S Comoros
82 L13 Mwanya Eastern, E Zambia 12°40′S 32°15′E
79 N23 Mwanza Katanga, SE Dem. Rep. Congo 07°49′S 26°49′E
81 F20 Mwanza ◈ region N Tanzania
81 G20 Mwanza ◈ state capital Mwanza, NW Tanzania
82 M13 Mwase Lundazi Eastern, E Zambia 12°26′S 33°20′E
97 B17 Mweelrea Ir. Caoc Maol Réidh. ▲ W Ireland 53°37′N 09°47′W
79 K21 Mweka Kasai-Occidental, C Dem. Rep. Congo 04°52′S 21°38′E
82 K12 Mwenda Luapula, N Zambia 10°30′S 30°21′E
79 L22 Mwene-Ditu Kasai-Oriental, S Dem. Rep. Congo 07°06′S 23°34′E
83 L18 Mwenezi ≈ S Zimbabwe
79 O20 Mwenga Sud-Kivu, E Dem. Rep. Congo 03°00′S 28°28′E
82 K11 Mweru, Lake var. Lac Moero. ◎ Dem. Rep. Congo/Zambia
82 H13 Mwinilunga North Western, NW Zambia 11°44′S 24°24′E
189 V16 Mwokil Atoll prev. Mokil Atoll. atoll Caroline Islands, E Micronesia
Myadel' see Myadzyel
118 J13 Myadzyel Pol. Miadzioł Nowy, Rus. Myadel'. Minskaya Voblasts', N Belarus 54°51′N 26°51′E
152 C12 Myājlār var. Miajlar. Rājasthān, NW India 26°16′N 70°21′E
123 T9 Myakit Magadanskaya Oblast', E Russian Federation 61°23′N 151°58′E
23 W13 Myakka River ≈ Florida, SE USA
124 L14 Myaksa Vologodskaya Oblast', NW Russian Federation 58°54′N 38°15′E
183 U8 Myall Lake ◎ New South Wales, SE Australia
166 L7 Myanaung Ayeyawady, SW Myanmar (Burma) 18°17′N 95°19′E
166 M4 Myanmar (Burma) off. Republic of the Union of Myanmar; prev. Union of Myanmar, var. Burma. ◆ transitional democracy SE Asia
Myanmar, Republic of the Union of see Myanmar (Burma)
Myanmar, Union of see Myanmar (Burma)
166 K8 Myaungmya Ayeyawady, SW Myanmar (Burma) 16°33′N 94°55′E
Myaydo see Aunglan
118 N11 Myazha Rus. Mezha. Vitsyebskaya Voblasts', NE Belarus 55°41′N 30°25′E
167 N12 Myeik var. Mergui. Tanintharyi, S Myanmar (Burma) 12°26′N 98°34′E
166 M12 Myeik Archipelago var. Mergui Archipelago. island group S Myanmar (Burma)
119 O18 Myerkulavichy Rus. Merkulovichi. Homyel'skaya Voblasts', SE Belarus 52°58′N 30°36′E
119 N14 Myezhava Rus. Mezhëvo. Vitsyebskaya Voblasts', NE Belarus 54°38′N 30°20′E
Myggenaes see Mykines
166 L5 Myingyan Mandalay, C Myanmar (Burma) 21°25′N 95°20′E
167 N12 Myitkyina Kachin State, N Myanmar (Burma) 25°24′N 97°25′E
166 M5 Myittha Mandalay, C Myanmar (Burma) 21°21′N 96°06′E
111 H19 Myjava Hung. Miava. Trenčiansky Kraj, W Slovakia 48°45′N 17°35′E
Myjeldino see Myyëldino
117 U9 Mykhaylivka Rus. Mikhaylovka. Zaporiz'ka Oblast', SE Ukraine 47°16′N 35°14′E
95 A18 Mykines Dan. Myggenaes. island W Faroe Islands
116 I5 Mykolayiv L'vivs'ka Oblast', W Ukraine 49°34′N 23°58′E
117 Q10 Mykolayiv Rus. Nikolayev. Mykolayivs'ka Oblast', S Ukraine 46°58′N 31°59′E
117 Q10 Mykolayiv ✈ Mykolayivs'ka Oblast', S Ukraine 47°02′N 31°54′E
Mykolayiv see Mykolayivs'ka Oblast'
117 S13 Mykolayivka Avtonomna Respublika Krym, S Ukraine 44°58′N 33°37′E
117 P9 Mykolayivka Odes'ka Oblast', SW Ukraine 47°34′N 30°48′E
117 P9 Mykolayivs'ka Oblast' var. Mykolayiv, Rus. Nikolayevskaya Oblast'. ◈ province S Ukraine
115 J20 Mýkonos Mýkonos, Kykládes, Greece, Aegean Sea 37°27′N 25°20′E
115 K20 Mýkonos var. Míkonos. island Kykládes, Greece, Aegean Sea
125 R7 Myla Respublika Komi, NW Russian Federation 65°24′N 50°51′E
Mylae see Milazzo
93 M19 Myllykoski Kymenlaakso, S Finland 60°45′N 26°52′E
153 U14 Mymensing var. Mymensingh. Dhaka, N Bangladesh 24°45′N 90°23′E
Mymensingh see Mymensing
93 K19 Mynämäki Varsinais-Suomi, SW Finland 60°41′N 22°00′E
145 S14 Mynaral Kaz. Myngaral. Zhambyl, S Kazakhstan 45°25′N 73°37′E
Mynbulak see Mingbuloq
Mynbulak, Vpadina see Mingbuloq Botig'I
Myngaral see Mynaral
Myohaung see Mrauk-oo
163 W13 Myohyang-sanmaek ▲▲ C North Korea
164 M11 Myōkō-san ▲ Honshū, S Japan 36°54′N 138°05′E
83 J15 Myooye Central, C Zambia 15°11′S 27°10′E
118 K12 Myory prev. Miyory, Rus. Miory. Vitsyebskaya Voblasts', N Belarus 55°39′N 27°39′E
92 J4 Mýrdalsjökull glacier S Iceland
92 G10 Myre Nordland, C Norway 68°54′N 15°04′E
117 S5 Myrhorod Rus. Mirgorod. Poltavs'ka Oblast', NE Ukraine 49°58′N 33°37′E
115 J15 Mýrina var. Mírina. Límnos, SE Greece 39°52′N 25°04′E
117 P5 Myronivka Rus. Mironovka. Kyyivs'ka Oblast', N Ukraine 49°40′N 30°59′E
21 U13 Myrtle Beach South Carolina, SE USA 33°41′N 78°53′W
32 F14 Myrtle Creek Oregon, NW USA 43°01′N 123°19′W
183 P11 Myrtleford Victoria, SE Australia 36°34′S 146°45′E
32 E14 Myrtle Point Oregon, NW USA 43°04′N 124°08′W
115 K25 Mýrtos Kríti, Greece, E Mediterranean Sea 35°00′N 25°34′E
Myrtoum Mare see Mirtóo Pélagos
93 G17 Myrviken Jämtland, C Sweden 62°59′N 14°19′E
95 I15 Mysen Østfold, S Norway 59°33′N 11°20′E
124 L15 Myshkin Yaroslavskaya Oblast', NW Russian Federation 57°47′N 38°28′E
111 K17 Myślenice Małopolskie, S Poland 49°50′N 19°55′E
110 D10 Myślibórz Zachodnio-pomorskie, NW Poland 52°55′N 14°51′E
155 G20 Mysore var. Maisur. Karnātaka, W India 12°18′N 76°37′E
Mysore see Karnātaka
115 F21 Mystrás var. Mistras. Pelopónnisos, S Greece 37°03′N 22°22′E
111 K15 Myszków Śląskie, S Poland 50°36′N 19°20′E
167 T14 My Tho var. Mi Tho. Tiên Giang, S Vietnam 10°21′N 106°21′E
Mytilene see Mytilíni
115 L17 Mytilíni var. Mitilíni; anc. Mytilene. Lésvos, E Greece 39°06′N 26°33′E
126 K3 Mytishchi Moskovskaya Oblast', W Russian Federation 56°00′N 37°51′E
37 N3 Myton Utah, W USA 40°11′N 110°03′W
92 K2 Mývatn ◎ C Iceland
125 T11 Myyëldino var. Myjeldino. Respublika Komi, NW Russian Federation 61°46′N 54°48′E
82 M13 Mzimba Northern, NW Malawi 11°56′S 33°36′E
82 M12 Mzuzu Northern, N Malawi 11°23′S 34°03′E

N

101 M19 Naab ≈ SE Germany
98 G12 Naaldwijk Zuid-Holland, W Netherlands 52°00′N 04°13′E
38 G12 Nā'ālehu var. Naalehu. Hawaii, USA, C Pacific Ocean 19°04′N 155°36′W
93 K19 Naantali Swe. Nådendal. Varsinais-Suomi, SW Finland 60°28′N 22°05′E
98 J10 Naarden Noord-Holland, C Netherlands 52°18′N 05°10′E
109 U4 Naarn ≈ N Austria
97 F18 Naas Ir. An Nás, Nás na Ríogh. Kildare, C Ireland 53°13′N 06°39′W
92 M9 Näätämöjoki Lapp. Njävdám. ≈ NE Finland
83 E23 Nababeep var. Nababiep. Northern Cape, W South Africa 29°36′S 17°46′E
Nababiep see Nababeep
Nabadwip see Navadwip
164 J14 Nabari Mie, Honshū, SW Japan 34°37′N 136°05′E
Nabatié see Nabatîyé
138 G8 Nabatîyé var. An Nabatīyah at Taḩtā, Nabatié, Nabatiyet et Tahta. SW Lebanon 33°18′N 35°36′E
Nabatiyet et Tahta see Nabatîyé
187 X14 Nabavatu Vanua Levu, N Fiji 16°35′S 178°55′E
190 I2 Nabeina island Tungaru, W Kiribati
127 T4 Naberezhnyye Chelny prev. Brezhnev. Respublika Tatarstan, W Russian Federation 55°43′N 52°21′E
39 T10 Nabesna Alaska, USA 62°22′N 143°00′W
39 T10 Nabesna River ≈ Alaska, USA
75 N5 Nabeul var. Nābul. NE Tunisia 36°32′N 10°45′E
152 I9 Nābha Punjab, NW India 30°22′N 76°12′E
171 W13 Nabire Papua, E Indonesia 03°23′S 135°31′E
141 O15 Nabī Shu'ayb, Jabal an ▲ W Yemen 15°24′N 44°04′E
138 F10 Nablus var. Nābulus, Heb. Shekhem; anc. Neapolis, Bibl. Shechem. N West Bank 32°13′N 35°16′E
187 X14 Nabouwalu Vanua Levu, N Fiji 17°00′S 178°43′E
Nābul see Nabeul
Nābulus see Nablus
187 Y13 Nabuna Vanua Levu, N Fiji 16°13′S 179°46′E
83 Q14 Nacala Nampula, NE Mozambique 14°30′S 40°37′E
42 H8 Nacaome Valle, S Honduras 13°30′N 87°31′W
Na Cealla Beaga see Killybegs
Na-Ch'ii see Nagqu
164 J15 Nachikatsuura var. Nachi-Katsuura. Wakayama, Honshū, SE Japan 33°37′N 135°54′E
Nachi-Katsuura see Nachikatsuura
81 J24 Nachingwea Lindi, SE Tanzania 10°21′S 38°46′E
111 F16 Náchod Královéhradecký Kraj, N Czech Republic 50°26′N 16°10′E
Na Clocha Liatha see Greystones
40 G3 Naco Sonora, NW Mexico 31°16′N 109°56′W
25 X8 Nacogdoches Texas, SW USA 31°36′N 94°40′W
40 G4 Nacozari de García Sonora, NW Mexico 30°27′N 109°43′W
Nada see Danzhou
77 O14 Nadawli NW Ghana 10°30′N 02°40′W
104 I3 Nadela Galicia, NW Spain 42°58′N 07°33′W
Nådendal see Naantali
144 M7 Nadezhdinka prev. Nadezhdinskiy. Kostanay, N Kazakhstan 53°46′N 63°44′E
Nadezhdinskiy see Nadezhdinka
Nadgan see Nadqān, Qalamat
187 W14 Nadi prev. Nandi. Viti Levu, W Fiji 17°47′S 177°32′E
187 X14 Nadi prev. Nandi. ✈ Viti Levu, W Fiji 17°46′S 177°28′E
154 D10 Nadiād Gujarāt, W India 22°42′N 72°55′E
Nadikdik see Knox Atoll
116 E11 Nădlac Ger. Nadlak, Hung. Nagylak. Arad, W Romania 46°10′N 20°47′E
Nadlak see Nădlac
74 H6 Nador prev. Villa Nador. NE Morocco 35°10′N 05°22′W
141 S9 Nadqān, Qalamat var. Nadgan. well E Saudi Arabia
111 N22 Nádudvar Hajdú-Bihar, E Hungary 47°36′N 21°09′E
121 O15 Nadur Gozo, N Malta 36°03′N 14°18′E
187 X13 Naduri prev. Nanduri. Vanua Levu, N Fiji 16°26′S 179°08′E
116 I7 Nadvirna Pol. Nadwórna, Rus. Nadvornaya. Ivano-Frankivs'ka Oblast', W Ukraine 48°27′N 24°30′E
124 J8 Nadvoitsy Respublika Kareliya, NW Russian Federation 63°53′N 34°17′E
Nadvornaya/Nadwórna see Nadvirna
122 I9 Nadym Yamalo-Nenetskiy Avtonomnyy Okrug, N Russian Federation 65°25′N 72°40′E
122 I9 Nadym ≈ C Russian Federation
186 E7 Nadzab Morobe, C Papua New Guinea 06°36′S 146°46′E
95 C17 Nærbø Rogaland, S Norway 58°40′N 05°39′E
95 I24 Næstved Sjælland, SE Denmark 55°12′N 11°47′E
77 X13 Nafada Gombe, E Nigeria 11°02′N 11°18′E
108 H8 Näfels Glarus, NE Switzerland 47°06′N 09°04′E
115 E18 Náfpaktos var. Návpaktos. Dytikí Elláda, C Greece 38°23′N 21°50′E
115 F20 Náfplio prev. Návplion. Pelopónnisos, S Greece 37°34′N 22°50′E
139 U6 Naft Khāneh Diyālá, E Iraq 34°01′N 45°26′E
149 N13 Nāg Baluchistān, SW Pakistan 27°43′N 65°31′E
171 P4 Naga off. Naga City; prev. Nueva Caceres. Luzon, N Philippines 13°36′N 123°10′E
Nagaarzê see Nagarzê
Naga City see Naga
12 F11 Nagagami ≈ Ontario, S Canada
164 F14 Nagahama Ehime, Shikoku, SW Japan 33°36′N 132°29′E
153 X12 Nāga Hills ▲▲ NE India
165 P10 Nagai Yamagata, Honshū, C Japan 38°08′N 140°00′E
Na Gaibhlte see Galty Mountains
39 N16 Nagai Island island Shumagin Islands, Alaska, USA
153 X12 Nāgāland ◈ state NE India
164 M11 Nagano Nagano, Honshū, S Japan 36°39′N 138°11′E
164 M12 Nagano off. Nagano-ken. ◈ prefecture Honshū, S Japan
Nagano-ken see Nagano
165 N11 Nagaoka Niigata, Honshū, C Japan 37°26′N 138°48′E
153 W12 Nagaon prev. Nowgong. Assam, NE India 26°21′N 92°41′E
155 J21 Nāgappattinam var. Negapatam, Negapattinam. Tamil Nādu, SE India 10°45′N 79°50′E
Nagara Nayok see Nakhon Nayok
Nagara Panom see Nakhon Phanom
Nagara Pathom see Nakhon Pathom
Nagara Sridharmaraj see Nakhon Si Thammarat
Nagara Svarga see Nakhon Sawan
155 H16 Nāgārjuna Sāgar ⊠ E India
42 I10 Nagarote León, SW Nicaragua 12°15′N 86°35′W
158 M16 Nagarzê var. Nagaarzê. Xizang Zizhiqu, W China 28°57′N 90°26′E
164 C14 Nagasaki Nagasaki, Kyūshū, SW Japan 32°45′N 129°52′E
164 C14 Nagasaki off. Nagasaki-ken. ◈ prefecture Kyūshū, SW Japan
Nagasaki-ken see Nagasaki
Nagashima see Kii-Nagashima
164 E12 Nagato Yamaguchi, Honshū, SW Japan 34°22′N 131°10′E
152 F11 Nāgaur Rājasthān, NW India 27°12′N 73°48′E
154 F10 Nāgda Madhya Pradesh, C India 23°30′N 75°29′E
98 L8 Nagele Flevoland, N Netherlands 52°39′N 05°43′E
155 H24 Nāgercoil Tamil Nādu, SE India 08°11′N 77°30′E
153 X12 Nāginimāra Nāgāland, NE India 26°44′N 94°51′E
Na Gleannta see Glenties
165 T16 Nago Okinawa, Okinawa, SW Japan 26°36′N 127°59′E
154 K9 Nāgod Madhya Pradesh, C India 24°34′N 80°34′E
155 J26 Nagoda Southern Province, S Sri Lanka 06°13′N 80°13′E
101 G22 Nagold Baden-Württemberg, SW Germany 48°33′N 08°43′E
137 V12 Nagorno-Karabakh var. Nagorno- Karabakhskaya Avtonomnaya Oblast, Arm. Lerrnayin Gharabakh, Az. Dağliq Quarabağ, Rus. Nagornyy Karabakh. former autonomous region SW Azerbaijan
Nagorno- Karabakhskaya Avtonomnaya Oblast see Nagorno-Karabakh
123 Q12 Nagornyy Respublika Sakha (Yakutiya), NE Russian Federation 55°53′N 124°58′E
Nagornyy Karabakh see Nagorno-Karabakh
125 R13 Nagorsk Kirovskaya Oblast', NW Russian Federation 59°18′N 50°49′E
164 K13 Nagoya Aichi, Honshū, SW Japan 35°10′N 136°53′E
154 I12 Nāgpur Mahārāshtra, C India 21°09′N 79°06′E
156 K10 Nagqu Chin. Na-Ch'ii; prev. Hei-ho. Xizang Zizhiqu, W China 31°30′N 91°57′E
152 J8 Nāg Tibba Range ▲▲ N India
45 O8 Nagua NE Dominican Republic 19°25′N 69°49′W
111 H25 Nagyatád Somogy, SW Hungary 46°15′N 17°25′E
Nagybánya see Baia Mare
Nagybecskerek see Zrenjanin
Nagydisznód see Cisnădie
Nagyenyed see Aiud
111 N21 Nagykálló Szabolcs-Szatmár-Bereg, E Hungary 47°50′N 21°47′E
111 G25 Nagykanizsa Ger. Grosskanizsa. Zala, SW Hungary 46°27′N 17°E
Nagykároly see Carei
111 K22 Nagykáta Pest, C Hungary 47°25′N 19°45′E
Nagykikinda see Kikinda
111 K23 Nagykőrös Pest, C Hungary 47°01′N 19°46′E
Nagy-Küküllő see Târnava Mare
Nagylak see Nădlac
Nagymihály see Michalovce
Nagyrőce see Revúca
Nagysomkút see Şomcuta Mare
Nagysurány see Šurany
Nagyszalonta see Salonta
Nagyszeben see Sibiu
Nagyszentmiklós see Sânnicolau Mare
Nagyszöllös see Vynohradiv
Nagyszombat see Trnava
Nagytapolcsány see Topoľčany
Nagyvárad see Oradea
165 S17 Naha Okinawa, Okinawa, SW Japan 26°10′N 127°40′E
152 J8 Nāhan Himāchal Pradesh, NW India 30°33′N 77°18′E
Nahang, Rūd-e see Nihing
138 F8 Nahariya prev. Nahariyya. Northern, N Israel 33°01′N 35°05′E
Nahariyya see Nahariya
142 L6 Nahāvand var. Nehavend. Hamadān, W Iran 34°13′N 48°21′E
101 F19 Nahe ≈ SW Germany
Na H-Iarmhidhe see Westmeath
189 O13 Nahnalaud ▲ Pohnpei, E Micronesia
Nahoi, Cape see Cumberland, Cape
Nahtavárr see Nattavaara
63 H16 Nahuel Huapí, Lago ◎ W Argentina
23 W7 Nahunta Georgia, SE USA 31°11′N 81°58′W
40 J6 Naica Chihuahua, N Mexico 27°53′N 105°30′W
11 U15 Naicam Saskatchewan, S Canada 52°26′N 104°30′W
Naiman Qi see Daqin Tal
158 M4 Naimin Bulak spring NW China
13 P6 Nain Newfoundland and Labrador, NE Canada 56°33′N 61°46′W
143 P8 Nā'īn Eşfahān, C Iran 32°52′N 53°05′E
152 K10 Naini Tāl Uttarakhand, N India 29°22′N 79°26′E
154 J11 Nainpur Madhya Pradesh, C India 22°26′N 80°10′E
96 J8 Nairn N Scotland, United Kingdom 57°36′N 03°51′W
96 I8 Nairn cultural region NE Scotland, United Kingdom
81 I19 Nairobi ● (Kenya) Nairobi City, S Kenya 01°17′S 36°50′E
81 I19 Nairobi ✈ Nairobi City, S Kenya 01°21′S 37°01′E
81 I19 Nairobi City ◈ county Nairobi City, Kenya, Africa
82 P13 Nairoto Cabo Delgado, NE Mozambique 12°22′S 39°05′E
118 G3 Naissaar island N Estonia
Naissus see Niš
187 Z14 Naitaba var. Naitauba; prev. Naitamba. island Lau Group, E Fiji
Naitamba/Naitauba see Naitaba
81 I19 Naivasha Nakuru, SW Kenya 0°44′S 36°26′E
81 H19 Naivasha, Lake ◎ SW Kenya
Najaf see An Najaf
143 N8 Najafābād var. Nejafabad. Eşfahān, C Iran 32°38′N 51°23′E
Najaf, Muḩāfa at an see An Najaf
141 N7 Najd var. Nejd. cultural region C Saudi Arabia
105 O4 Nájera La Rioja, N Spain 42°25′N 02°45′W
105 P4 Najerilla ≈ N Spain
163 U7 Naji var. Arun Qi. Nei Mongol Zizhiqu, N China 48°05′N 123°28′E
152 J9 Najībābād Uttar Pradesh, N India 29°37′N 78°19′E
Najima see Fukuoka
163 Y11 Najin NE North Korea 42°13′N 130°16′E
139 T9 Najm al Ḩassūn Bābil, C Iraq 32°24′N 44°13′E
141 O13 Najrān var. Abā as Su'ūd. Najrān, S Saudi Arabia 17°31′N 44°09′E
141 P12 Najrān var. Minţaqat al Najrān. ◈ province S Saudi Arabia
Najrān, Minţaqat al see Najrān
165 T2 Nakagawa Hokkaidō, NE Japan 44°49′N 142°04′E
38 F9 Nākālele Point var. Nakalele Point. headland Maui, Hawai'i, USA 21°01′N 156°35′W
164 D13 Nakama Fukuoka, Kyūshū, SW Japan 33°53′N 130°48′E
Nakambé see White Volta
Nakamti see Nek'emtē
164 F15 Nakamura var. Shimanto. Kōchi, Shikoku, SW Japan 33°00′N 132°55′E
186 H7 Nakanai Mountains ▲▲ New Britain, E Papua New Guinea
164 H11 Nakano-shima island Okishotō, SW Japan
165 Q6 Nakasato Aomori, Honshū, C Japan 40°58′N 140°26′E
165 T5 Nakasatsunai Hokkaidō, NE Japan 42°42′N 143°09′E
165 W4 Nakashibetsu Hokkaidō, NE Japan 43°31′N 144°58′E
81 F18 Nakasongola C Uganda 01°19′N 32°28′E
165 T1 Nakatonbetsu Hokkaidō, NE Japan 44°58′N 142°18′E
164 L13 Nakatsugawa var. Nakatugawa. Gifu, Honshū, SW Japan 35°30′N 137°29′E
Nakatugawa see Nakatsugawa
Nakdong see Nakdong-gang
163 Y15 Nakdong-gang var. Nakdong, Jap. Rakutō-kō; prev. Naktong-gang. ≈ C South Korea
Nakel see Nakło nad Notecią
Nakfa see Nak'fa
80 J8 Nak'fa var. Nakfa. N Eritrea 16°38′N 38°26′E
Nakhichevan' see Naxçıvan
123 S15 Nakhodka Primorskiy Kray, SE Russian Federation 42°46′N 132°48′E
122 J8 Nakhodka Yamalo-Nenetskiy Avtonomnyy Okrug, N Russian Federation 67°48′N 77°48′E
Nakhon Navok see Nakhon Nayok
167 P11 Nakhon Nayok var. Nagara Nayok, Nakhon Navok. Nakhon Nayok, C Thailand 14°15′N 101°12′E
167 O11 Nakhon Pathom var. Nagara Pathom, Nakorn Pathom. Nakhon Pathom, W Thailand 13°49′N 100°06′E
167 R8 Nakhon Phanom var. Nagara Panom. Nakhon Phanom, E Thailand 17°22′N 104°46′E
167 Q10 Nakhon Ratchasima var. Khorat, Korat. Nakhon Ratchasima, E Thailand 15°N 102°06′E
167 O9 Nakhon Sawan var. Muang Nakhon Sawan, Nagara Svarga. Nakhon Sawan, W Thailand 15°42′N 100°06′E
167 N15 Nakhon Si Thammarat var. Nagara Sridharmaraj, Nakhon Sithammaraj. Nakhon Si Thammarat, SW Thailand 08°24′N 99°58′E
Nakhon Sithammaraj see Nakhon Si Thammarat
139 Y11 Nakhrash Al Başrah, SE Iraq 31°13′N 47°24′E
10 I9 Nakina British Columbia, W Canada 59°12′N 132°48′W
110 H9 Nakło nad Notecią Ger. Nakel. Kujawsko-pomorskie, C Poland 53°08′N 17°35′E
39 P13 Naknek Alaska, USA 58°45′N 157°01′W
152 H8 Nakodar Punjab, NW India 31°06′N 75°31′E
82 M11 Nakonde Muchinga, NE Zambia 09°22′S 32°47′E
Nakorn Pathom see Nakhon Pathom
95 H24 Nakskov Sjælland, SE Denmark 54°50′N 11°10′E
Naktong-gang see Nakdong-gang
81 H18 Nakuru Nakuru, SW Kenya 0°16′S 36°04′E
81 I19 Nakuru ◈ county W Kenya
81 H19 Nakuru, Lake ◎ Nakuru, C Kenya
11 O17 Nakusp British Columbia, SW Canada 50°14′N 117°48′W
149 N15 Nāl ≈ W Pakistan
162 M7 Nalayh Töv, C Mongolia 47°48′N 107°17′E
153 V12 Nalbāri Assam, NE India 26°36′N 91°49′E
63 G19 Nalcayec, Isla island Archipiélago de los Chonos, S Chile
127 N15 Nal'chik Kabardino-Balkarskaya Respublika, SW Russian Federation 43°30′N 43°39′E
155 I16 Nalgonda Telangana, C India 17°04′N 79°15′E
153 S14 Nalhāti West Bengal, NE India 24°19′N 87°53′E
153 U14 Nalitabari Dhaka, N Bangladesh 25°06′N 90°11′E
155 I17 Nallamala Hills ▲▲ E India
136 G12 Nallıhan Ankara, NW Turkey 40°12′N 31°22′E
104 K2 Nalón ≈ NW Spain
167 N3 Nalong Kachin State, N Myanmar (Burma) 24°42′N 97°27′E
75 N8 Nālūt NW Libya 31°52′N 10°59′E
171 T14 Nama Pulau Manawoka, E Indonesia 04°07′S 131°22′E
189 Q16 Nama island C Micronesia
83 O16 Namacurra Zambézia, NE Mozambique 17°31′S 37°03′E
188 F9 Namai Bay bay Babeldaob, N Palau
29 W2 Namakan Lake ◎ Canada/USA
143 O6 Namak, Daryācheh-ye marsh N Iran
143 T6 Namak, Kavīr-e salt pan NE Iran
167 O6 Namaklwe Shan State, E Myanmar (Burma) 19°45′N 99°01′E
Namakzār, Kowl-e/Namakzār, Daryācheh-ye see Namakzar
148 J5 Namakzar Pash. Daryācheh-ye Namakzār, Kowl-e Namakzār. marsh Afghanistan/Iran
171 V15 Namalau Pulau Jursian, E Indonesia 05°50′S 134°43′E
81 I20 Namanga Kajiado, S Kenya 02°33′S 36°48′E
147 S10 Namangan Namangan Viloyati, E Uzbekistan 40°59′N 71°34′E
Namanganskaya Oblast' see Namangan Viloyati
147 R10 Namangan Viloyati Rus. Namanganskaya Oblast'. ◈ province E Uzbekistan
83 Q14 Namapa Nampula, NE Mozambique 13°43′S 39°48′E
83 C21 Namaqualand physical region S Namibia
81 G18 Namasagali C Uganda 01°02′N 32°58′E
186 H6 Namatanai New Ireland, NE Papua New Guinea 03°40′S 152°26′E
83 I14 Nambala Central, C Zambia 15°04′S 26°56′E
81 J23 Nambanje Lindi, SE Tanzania 08°37′S 38°21′E
183 V2 Nambour Queensland, E Australia 26°40′S 152°52′E
183 V6 Nambucca Heads New South Wales, SE Australia 30°37′S 153°00′E
159 N15 Nam Co ◎ W China
167 R5 Nâm Cum Lai Châu, N Vietnam 22°37′N 103°12′E
Namdik see Namorik Atoll
167 T6 Nam Định Nam Ha, N Vietnam 20°25′N 106°12′E
99 I20 Namêche Namur, SE Belgium 50°29′N 05°02′E
30 J4 Namekagon Lake ◎ Wisconsin, N USA
188 F10 Namekakl Passage passage Babeldaob, N Palau
Namen see Namur
83 P15 Nametil Nampula, NE Mozambique 15°46′S 39°21′E
163 X14 Nam-gang ≈ C North Korea
163 Y16 Nam-gang ≈ S South Korea
163 Y17 Namhae-do Jap. Nankai-tō. island S South Korea
Namhoi see Foshan
83 C19 Namib Desert desert W Namibia
83 A15 Namibe Port. Moçâmedes, Mossâmedes. Namibe, SW Angola 15°10′S 12°09′E
83 A15 Namibe ◈ province SE Angola
83 C18 Namibia off. Republic of Namibia, var. South West Africa, Afr. Suidwes-Afrika, Ger. Deutsch-Südwestafrika; prev. German Southwest Africa, South-West Africa. ◆ republic S Africa
65 O17 Namibia Plain undersea feature S Atlantic Ocean
Namibia, Republic of see Namibia
165 Q11 Namie Fukushima, Honshū, C Japan 37°29′N 140°58′E
165 Q7 Namioka Aomori, Honshū, C Japan 40°43′N 140°34′E
40 I5 Namiquipa Chihuahua, N Mexico 29°15′N 107°25′W
159 P15 Namjagbarwa Feng ▲ W China 29°39′N 95°00′E
Namka see Doilungdêqên
Nam Khong see Mekong
171 R13 Namlea Pulau Buru, E Indonesia 03°12′S 127°06′E
158 L16 Namling Xizang Zizhiqu, W China 29°40′N 88°58′E
Namnetes see Nantes
167 R8 Nam Ngum ≈ C Laos
Namo see Namu Atoll
183 R5 Namoi River ≈ New South Wales, SE Australia
189 Q17 Namoluk Atoll atoll Mortlock Islands, C Micronesia
189 O15 Namonuito Atoll atoll Caroline Islands, C Micronesia
189 T9 Namorik Atoll var. Namdik. atoll Ralik Chain, S Marshall Islands
167 Q6 Nam Ou ≈ N Laos
32 M14 Nampa Idaho, NW USA 43°32′N 116°33′W
76 M11 Nampala Ségou, W Mali 15°21′N 05°32′W
163 W14 Namp'o SW North Korea 38°46′N 125°25′E
83 P15 Nampula Nampula, NE Mozambique 15°09′S 39°14′E

◆ Country ● Country Capital ◇ Dependent Territory ○ Dependent Territory Capital ◈ Administrative Regions ✈ International Airport ▲ Mountain ▲▲ Mountain Range Volcano ≈ River ◎ Lake ⊠ Reservoir

83 P15 **Nampula** *off.* Província de Nampula. ◆ *province* NE Mozambique
Nampula, Província de *see* Nampula
163 W13 **Namsan-ni** NW North Korea 40°25′N 125°01′E
Namslau *see* Namysłów
93 E15 **Namsos** Nord-Trøndelag, C Norway 64°28′N 11°31′E
93 F14 **Namsskogan** Nord-Trøndelag, C Norway 64°57′N 13°04′E
167 O6 **Nam Teng** ≈ E Myanmar (Burma)
167 P6 **Nam Tha** ≈ N Laos
123 Q10 **Namtsy** Respublika Sakha (Yakutiya), NE Russian Federation 62°42′N 129°30′E
167 N4 **Namtu** Shan State, E Myanmar (Burma) 23°04′N 97°26′E
10 J15 **Namu** British Columbia, SW Canada 51°46′N 127°49′W
189 T7 **Namu Atoll** *var.* Namo. *atoll* Ralik Chain, C Marshall Islands
187 Y15 **Namuka-i-lau** *island* Lau Group, E Fiji
83 O15 **Namuli, Mont** ▲ NE Mozambique 15°15′S 37°33′E
83 P14 **Namuno** Cabo Delgado, N Mozambique 13°39′S 38°50′E
99 I20 **Namur** *Dut.* Namen. Namur, SE Belgium 50°28′N 04°52′E
99 H21 **Namur** *Dut.* Namen. ◆ *province* S Belgium
83 D17 **Namutoni** Kunene, N Namibia 18°49′S 16°55′E
163 Y16 **Namwon** *Jap.* Nangen; *prev.* Namwŏn. S South Korea 35°24′N 127°20′E
Namwŏn *see* Namwon
111 H14 **Namysłów** *Ger.* Namslau. Opole, SW Poland 51°03′N 17°41′E
167 P7 **Nan** *var.* Muang Nan. Nan, NW Thailand 18°47′N 100°50′E
79 G15 **Nana** ≈ W Central African Republic
165 R5 **Nanae** Hokkaidō, NE Japan 41°55′N 140°40′E
79 I14 **Nana-Grébizi** ◆ *prefecture* N Central African Republic
10 L17 **Nanaimo** Vancouver Island, British Columbia, SW Canada 49°08′N 123°58′W
38 C9 **Nānākuli** *var.* Nanakuli. O'ahu, Hawaii, USA, C Pacific Ocean 21°23′N 158°09′W
79 G15 **Nana-Mambéré** ◆ *prefecture* W Central African Republic
161 R13 **Nan'an** Fujian, SE China 24°57′N 118°22′E
183 U2 **Nanango** Queensland, E Australia 26°42′S 151°58′E
164 L11 **Nanao** Ishikawa, Honshū, SW Japan 37°03′N 136°58′E
161 Q14 **Nan'ao Dao** *island* S China
164 L10 **Nanatsu-shima** *island* SW Japan
56 F8 **Nanay, Río** ≈ NE Peru
160 J8 **Nanbu** Sichuan, C China 31°19′N 106°02′E
163 X7 **Nancha** Heilongjiang, NE China 47°09′N 129°17′E
161 P10 **Nanchang** *var.* Nan-ch'ang, Nanch'ang-hsien. *province capital* Jiangxi, S China 28°38′N 115°58′E
Nan-ch'ang *see* Nanchang
Nanch'ang-hsien *see* Nanchang
161 P11 **Nancheng** *var.* Jianchang. Jiangxi, S China 27°37′N 116°37′E
Nan-ching *see* Nanjing
160 J9 **Nanchong** Sichuan, C China 30°47′N 106°03′E
160 J10 **Nanchuan** Chongqing Shi, C China 29°06′N 107°13′E
103 T5 **Nancy** Meurthe-et-Moselle, NE France 48°40′N 06°11′E
185 A22 **Nancy Sound** *sound* South Island, New Zealand
152 L9 **Nanda Devi** ▲ NW India 30°27′N 80°00′E
42 J11 **Nandaime** Granada, SW Nicaragua 11°45′N 86°02′W
160 K13 **Nandan** *var.* Minami-Awaji. Guangxi Zhuangzu Zizhiqu, S China 25°03′N 107°31′E
155 H14 **Nānded** Mahārāshtra, C India 19°11′N 77°21′E
183 S5 **Nandewar Range** ▲▲ New South Wales, SE Australia
81 H18 **Nandi** ◆ *county* W Kenya
Nandi *see* Nadi
160 E13 **Nanding He** ≈ China/Vietnam
Nándorhgy *see* Oţelu Roşu
154 E11 **Nandurbār** Mahārāshtra, W India 21°22′N 74°18′E
Nanduri *see* Naduri
155 I17 **Nandyāl** Andhra Pradesh, E India 15°30′N 78°28′E
161 P11 **Nanfeng** *var.* Qincheng. Jiangxi, S China 27°15′N 116°16′E
Nang *see* Nangxian
79 E15 **Nanga Eboko** Centre, C Cameroon 04°38′N 12°21′E
Nangah Serawai *see* Nangaserawai
149 W4 **Nanga Parbat** ▲ India/Pakistan 35°15′N 74°36′E
169 R11 **Nangapinoh** Borneo, C Indonesia 0°21′S 111°44′E
149 R5 **Nangarhār** ◆ *province* E Afghanistan
169 S11 **Nangaserawai** *var.* Nangah Serawai. Borneo, C Indonesia 0°20′S 112°26′E
169 Q12 **Nangatayap** Borneo, C Indonesia 01°30′S 110°33′E
Nangen *see* Namwon
103 P5 **Nangis** Seine-et-Marne, N France 48°36′N 03°02′E
163 X13 **Nangnim-sanmaek** ▲▲ C North Korea
161 O4 **Nangong** Hebei, E China 37°22′N 115°20′E
159 Q14 **Nangqên** *var.* Xangda. Qinghai, C China 32°05′N 96°28′E
167 Q10 **Nang Rong** Buri Ram, E Thailand 14°37′N 102°48′E
159 O16 **Nangxian** *var.* Nang. Xizang Zizhiqu, W China 29°04′N 93°03′E
Nan Hai *see* South China Sea
160 L8 **Nan He** ≈ C China
160 F12 **Nanhua** *var.* Longchuan. Yunnan, SW China 25°15′N 101°15′E
Naniwa *see* Ōsaka
155 G20 **Nanjangūd** Karnātaka, W India 12°07′N 76°40′E
161 Q8 **Nanjing** *var.* Nan-ching, Nanking; *prev.* Chianning, Chian-ning, Kiang-ning, Jiangsu. *province capital* Jiangsu, E China 32°03′N 118°47′E
Nankai-tō *see* Namhae-do
161 O12 **Nankang** *var.* Rongjiang. Jiangxi, S China 25°42′N 114°45′E
Nanking *see* Nanjing
161 N13 **Nan Ling** ▲▲ S China
160 L15 **Nanliu Jiang** ≈ S China
189 P13 **Nan Madol** *ruins* Temwen Island, E Micronesia
160 K15 **Nanning** *var.* Nan-ning; *prev.* Yung-ning. Guangxi Zhuangzu Zizhiqu, S China 22°50′N 108°19′E
Nan-ning *see* Nanning
196 M15 **Nanortalik** Kujalleo, S Greenland 60°08′N 45°14′W
Nanouki *see* Aranuka
160 H13 **Nanpan Jiang** ≈ S China
152 M11 **Nānpāra** Uttar Pradesh, N India 27°51′N 81°30′E
161 Q12 **Nanping** *var.* Nan-p'ing; *prev.* Yenping. Fujian, SE China 26°40′N 118°07′E
Nan-p'ing *see* Nanping
Nanping *see* Jiuzhaigou
Nanpu *see* Pucheng
161 R12 **Nanri Dao** *island* SE China
165 S16 **Nansei-shotō** *Eng.* Ryukyu Islands. *island group* SW Japan
Nansei Syotō Trench *see* Ryukyu Trench
197 T10 **Nansen Basin** *undersea feature* Arctic Ocean
Nansen Cordillera *see* Gakkel Ridge
129 T9 **Nan Shan** ▲▲ C China
Nansha Qundao *see* Spratly Islands
12 K3 **Nantais, Lac** ◎ Québec, NE Canada
103 N5 **Nanterre** Hauts-de-Seine, N France 48°53′N 02°13′E
102 I8 **Nantes** *Bret.* Naoned; *anc.* Condivincum, Namnetes. Loire-Atlantique, NW France 47°12′N 01°32′W
14 G17 **Nanticoke** Ontario, S Canada 42°49′N 80°04′W
18 H13 **Nanticoke** Pennsylvania, NE USA 41°12′N 76°00′W
21 Y4 **Nanticoke River** ≈ Delaware/Maryland, NE USA
11 Q17 **Nanton** Alberta, SW Canada 50°21′N 113°47′W
161 S8 **Nantong** Jiangsu, E China 32°00′N 120°52′E
161 S13 **Nantou** *prev.* Nant'ou. W Taiwan 23°54′N 120°51′E
Nant'ou *see* Nantou
103 S10 **Nantua** Ain, E France 46°10′N 05°34′E
19 Q13 **Nantucket** Nantucket Island, Massachusetts, NE USA 41°15′N 70°05′W
19 Q13 **Nantucket Island** *island* Massachusetts, NE USA
19 Q13 **Nantucket Sound** *sound* Massachusetts, NE USA
82 P13 **Nantulo** Cabo Delgado, N Mozambique 12°30′S 39°03′E
189 O12 **Nanuh** Pohnpei, E Micronesia
190 D6 **Nanumaga** *var.* Nanumanga. *atoll* NW Tuvalu
Nanumanga *see* Nanumaga
190 D5 **Nanumea Atoll** *atoll* NW Tuvalu
59 O19 **Nanuque** Minas Gerais, SE Brazil 17°49′S 40°21′W
171 R10 **Nanusa, Kepulauan** *island group* N Indonesia
Nanwei Dao *see* Spratly Island
163 U4 **Nanweng He** ≈ NE China
160 I10 **Nanxi** Sichuan, C China 28°54′N 104°59′E
161 N10 **Nanxian** *var.* Nan Xian, Nanzhou. Hunan, S China 29°23′N 112°18′E
Nan Xian *see* Nanxian
161 N7 **Nanyang** *var.* Nan-yang. Henan, C China 32°59′N 112°29′E
Nan-yang *see* Nanyang
161 P6 **Nanyang Hu** ◎ E China
165 P10 **Nan'yō** Yamagata, Honshū, C Japan 38°04′N 140°06′E
81 I18 **Nanyuki** Laikipia, C Kenya 0°01′N 37°05′E
160 M8 **Nanzhang** Hubei, C China 31°47′N 111°48′E
Nanzhou *see* Nanxian
105 T11 **Nao, Cabo de La** *headland* E Spain 38°43′N 00°13′E
12 M9 **Naococane, Lac** ◎ Québec, E Canada
153 S14 **Naogaon** Rajshahi, NW Bangladesh 24°49′N 88°59′E
Naokot *see* Naukot
187 R13 **Naone** Maewo, C Vanuatu 15°03′S 168°06′E
Naoned *see* Nantes
115 E14 **Náousa** Kentrikí Makedonía, N Greece 40°38′N 22°04′E
35 N8 **Napa** California, W USA 38°15′N 122°17′W
39 O11 **Napaimiut** Alaska, USA 61°32′N 158°46′W
39 N12 **Napakiak** Alaska, USA 60°42′N 161°57′W
122 J7 **Napalkovo** Yamalo-Nenetskiy Avtonomnyy Okrug, N Russian Federation 70°06′N 73°43′E
12 I16 **Napanee** Ontario, SE Canada 44°13′N 76°57′W
39 N12 **Napaskiak** Alaska, USA 60°42′N 161°46′W
167 S5 **Na Phác** Cao Bằng, N Vietnam 22°24′N 105°54′E
184 O11 **Napier** Hawke's Bay, North Island, New Zealand 39°30′S 176°55′E
195 X3 **Napier Mountains** ▲▲ Antarctica
15 O13 **Napierville** Québec, SE Canada 45°12′N 73°25′W
23 W15 **Naples** Florida, SE USA 26°08′N 81°48′W
25 W5 **Naples** Texas, SW USA 33°12′N 94°40′W
Naples *see* Napoli
160 I14 **Napo** Guangxi Zhuangzu Zizhiqu, S China 23°21′N 105°47′E
56 C6 **Napo** ◆ *province* NE Ecuador
29 O6 **Napoleon** North Dakota, N USA 46°30′N 99°46′W
31 R11 **Napoleon** Ohio, N USA 41°23′N 84°07′W
Napoléon-Vendée *see* la Roche-sur-Yon
22 J9 **Napoleonville** Louisiana, S USA 29°55′N 91°01′W
107 K17 **Napoli** *Eng.* Naples, *Ger.* Neapel; *anc.* Neapolis. Campania, S Italy 40°52′N 14°15′E
107 J18 **Napoli, Golfo di** *gulf* S Italy
57 F7 **Napo, Río** ≈ Ecuador/Peru
191 W9 **Napuka** *island* Îles Tuamotu, C French Polynesia
142 J3 **Naqadeh** Āžarbāyjān-e Bākhtarī, NW Iran 36°57′N 45°24′E
139 U6 **Naqnah** Diyālá, E Iraq 34°13′N 45°33′E
Nar *see* Nera
164 J14 **Nara** Nara, Honshū, SW Japan 34°41′N 135°49′E
76 L11 **Nara** Koulikoro, W Mali 15°04′N 07°19′W
149 R14 **Nāra Canal** *irrigation canal* S Pakistan
182 K11 **Naracoorte** South Australia 37°02′S 140°45′E
183 P8 **Naradhan** New South Wales, SE Australia 33°37′S 146°19′E
Naradhivas *see* Narathiwat
56 B8 **Naranjal** Guayas, W Ecuador 02°43′S 79°38′W
57 Q19 **Naranjos** Santa Cruz, E Bolivia
41 Q12 **Naranjos** Veracruz-Llave, E Mexico 21°21′N 97°41′W
159 Q6 **Naran Sebstein Bulag** *spring* NW China
164 B14 **Narao** Nagasaki, Nakadōri-jima, SW Japan 32°40′N 129°03′E
155 J16 **Narasaraopet** Andhra Pradesh, E India 16°16′N 80°06′E
158 J5 **Narat** Xinjiang Uygur Zizhiqu, W China 43°20′N 84°02′E
167 P17 **Narathiwat** *var.* Naradhivas. Narathiwat, SW Thailand 06°25′N 101°48′E
37 V10 **Nara Visa** New Mexico, SW USA 35°35′N 103°06′W
Nārāyāni *see* Gandak
Narbada *see* Narmada
Narbo Martius *see* Narbonne
103 P16 **Narbonne** *anc.* Narbo Martius. Aude, S France 43°11′N 03°E
Narborough Island *see* Fernandina, Isla
104 J2 **Narcea** ≈ NW Spain
152 J9 **Narendranagar** Uttarakhand, N India 30°10′N 78°21′E
Nares Abyssal Plain *see* Nares Plain
64 G11 **Nares Plain** *var.* Nares Abyssal Plain. *undersea feature* NW Atlantic Ocean 23°30′N 63°00′W
Nares Stræde *see* Nares Strait
197 P10 **Nares Strait** *Dan.* Nares Stræde. *strait* Canada/Greenland
110 O9 **Narew** ≈ E Poland
155 F17 **Nargund** Karnātaka, W India 15°43′N 75°23′E
83 D20 **Narib** Hardap, S Namibia 24°11′S 17°46′E
Narikrik *see* Knox Atoll
Narin Gol *see* Omon Gol
54 B13 **Nariño** *off.* Departamento de Nariño. ◆ *province* SW Colombia
165 P13 **Narita** Chiba, Honshū, S Japan 35°46′N 140°20′E
165 P13 **Narita** ✈ (Tōkyō) Chiba, Honshū, S Japan 35°45′N 140°23′E
Nariya *see* An Nu'ayrīyah
162 F5 **Nariyn Gol** ≈ Mongolia/Russian Federation
162 J8 **Nariynteel** *var.* Tsagaan-Ovoo. Övörhangay, C Mongolia 45°57′N 101°25′E
152 J8 **Narkanda** Himāchal Pradesh, NW India 31°14′N 77°27′E
92 L13 **Narkaus** Lappi, NW Finland 66°13′N 26°09′E
154 E11 **Narmada** *var.* Narbada. ≈ C India
152 H11 **Narnaul** *var.* Nārnaul. Haryāna, N India 28°04′N 76°10′E
107 I14 **Narni** Umbria, C Italy 42°31′N 12°31′E
107 J24 **Naro** Sicilia, Italy, C Mediterranean Sea 37°18′N 13°48′E
Narodichi *see* Narodychi
125 V7 **Narodnaya, Gora** ▲ NW Russian Federation 65°04′N 60°12′E
117 N3 **Narodychi** *Rus.* Narodichi. Zhytomyrs'ka Oblast', N Ukraine 51°11′N 29°01′E
126 J4 **Naro-Fominsk** Moskovskaya Oblast', W Russian Federation 55°25′N 36°41′E
81 H19 **Narok** Narok, SW Kenya 01°04′S 35°54′E
81 H20 **Narok** ◆ *county* SW Kenya
104 H2 **Narón** Galicia, NW Spain 43°31′N 08°08′W
183 S11 **Narooma** New South Wales, SE Australia 36°16′S 150°08′E
Narova *see* Narva
Narowlya *see* Narowlya
149 W8 **Nārowāl** Punjab, E Pakistan 32°04′N 74°54′E
119 N20 **Narowlya** *Rus.* Narovlya. Homyel'skaya Voblasts', SE Belarus 51°48′N 29°30′E
93 J17 **Närpes** *Fin.* Närpiö. Österbotten, W Finland 62°28′N 21°19′E
Närpiö *see* Närpes
183 S5 **Narrabri** New South Wales, SE Australia 30°21′S 149°48′E
183 P9 **Narrandera** New South Wales, SE Australia 34°46′S 146°32′E
183 Q4 **Narran Lake** ◎ New South Wales, SE Australia
183 Q4 **Narran River** ≈ New South Wales/Queensland, SE Australia
180 J13 **Narrogin** Western Australia 32°53′S 117°17′E
183 Q7 **Narromine** New South Wales, SE Australia 32°16′S 148°15′E
21 R6 **Narrows** Virginia, NE USA 37°19′N 80°48′W
196 M15 **Narsarsuaq** ✈ Kujalleo, S Greenland 61°07′N 45°03′W
154 I10 **Narsimhapur** Madhya Pradesh, C India 22°58′N 79°15′E
153 U15 **Narsingdi** *var.* Narsinghdi. Dhaka, C Bangladesh 23°56′N 90°40′E
Narsinghdi *see* Narsingdi
154 H9 **Narsinghgarh** Madhya Pradesh, C India 23°42′N 77°08′E
163 Q11 **Nart** Nei Mongol Zizhiqu, N China 42°54′N 115°55′E
Nartës, Gjol i/Nartës, Laguna e *see* Nartës, Liqeni i
113 J22 **Nartës, Liqeni i** *var.* Gjol i Nartës, Laguna e Nartës. ◎ SW Albania
115 F17 **Nartháki** ▲ C Greece 39°12′N 22°24′E
127 O15 **Nartkala** Kabardino-Balkarskaya Respublika, SW Russian Federation 43°34′N 43°55′E
118 K3 **Narva** Ida-Virumaa, NE Estonia 59°23′N 28°12′E
118 K4 **Narva** *prev.* Narova. ≈ Estonia/Russian Federation
118 J3 **Narva Bay** *Est.* Narva Laht, *Ger.* Narwa-Bucht, *Rus.* Narvskiy Zaliv. *bay* Estonia/Russian Federation
Narva Laht *see* Narva Bay
124 F13 **Narva Reservoir** *Est.* Narva Veehoidla, *Rus.* Narvskoye Vodokhranilishche. ⊠ Estonia/Russian Federation
Narva Veehoidla *see* Narva Reservoir
92 H10 **Narvik** Nordland, C Norway 68°26′N 17°24′E
Narvskiy Zaliv *see* Narva Bay
Narvskoye Vodokhranilishche *see* Narva Reservoir
Narwa-Bucht *see* Narva Bay
152 I9 **Narwāna** Haryāna, NW India 29°36′N 76°11′E
125 R4 **Nar'yan-Mar** *prev.* Beloshchel'ye, Dzerzhinskiy. Nenetskiy Avtonomnyy Okrug, NW Russian Federation 67°38′N 53°E
122 J12 **Narym** Tomskaya Oblast', C Russian Federation 58°59′N 81°20′E
Narymskiy Khrebet *see* Khrebet Naryn
147 W9 **Naryn** Narynskaya Oblast', C Kyrgyzstan 41°24′N 76°E
147 U8 **Naryn** ≈ Kyrgyzstan/Uzbekistan
145 W16 **Narynkol** *Kaz.* Naryynqol. Almaty, SE Kazakhstan 42°45′N 80°12′E
Naryn Oblasty *see* Narynskaya Oblast'
Naryynqol *see* Narynkol
147 V9 **Narynskaya Oblast'** *Kir.* Naryn Oblasty. ◆ *province* C Kyrgyzstan
Naryn Zhotasy *see* Khrebet Naryn
126 J6 **Naryshkino** Orlovskaya Oblast', W Russian Federation 53°00′N 35°41′E
95 L14 **Nås** Dalarna, C Sweden 60°28′N 14°30′E
10 J11 **Nass** ≈ British Columbia, SW Canada
92 G13 **Nasafjellet** *Lapp.* Násávárre. ▲ C Norway 66°29′N 15°23′E
93 H16 **Näsåker** Västernorrland, C Sweden 63°27′N 16°55′E
187 Y14 **Nasau** Koro, C Fiji 17°20′S 179°26′E
190 J13 **Nassau** *island* N Cook Islands
116 I9 **Năsăud** *Ger.* Nussdorf, *Hung.* Naszód. Bistrița-Năsăud, N Romania 47°16′N 24°24′E
Násávárre *see* Nasafjellet
103 P13 **Nasbinals** Lozère, S France 44°40′N 03°03′E
Na Sceirí *see* Skerries
Nase *see* Naze
185 E22 **Naseby** Otago, South Island, New Zealand 45°02′S 170°09′E
143 R10 **Nāşerīyeh** Kermān, C Iran
25 X5 **Nash** Texas, SW USA 33°26′N 94°04′W
154 E13 **Nāshik** *prev.* Nāsik. Mahārāshtra, W India 20°05′N 73°48′E
56 E7 **Nashiño, Río** ≈ Ecuador/Peru
29 W12 **Nashua** Iowa, C USA 42°57′N 92°32′W
33 W7 **Nashua** Montana, NW USA 48°06′N 106°16′W
19 O10 **Nashua** New Hampshire, NE USA 42°45′N 71°26′W
27 S13 **Nashville** Arkansas, C USA 33°57′N 93°50′W
23 U7 **Nashville** Georgia, SE USA 31°12′N 83°15′W
30 L16 **Nashville** Illinois, N USA 38°20′N 89°22′W
31 O14 **Nashville** Indiana, N USA 39°13′N 86°15′W
21 V9 **Nashville** North Carolina, SE USA 35°58′N 78°00′W
20 J8 **Nashville** *state capital* Tennessee, S USA 36°11′N 86°48′W
20 J9 **Nashville** ✈ Tennessee, S USA 36°06′N 86°44′W
64 H10 **Nashville Seamount** *undersea feature* NW Atlantic Ocean 30°00′N 57°20′W
112 H9 **Našice** Osijek-Baranja, E Croatia 45°29′N 18°05′E
110 M11 **Nasielsk** Mazowieckie, C Poland 52°33′N 20°46′E
93 K18 **Näsijärvi** ◎ SW Finland
Nāsik *see* Nāshik
80 G13 **Nasir** Upper Nile, NE South Sudan 08°37′N 33°06′E
148 K15 **Nasīrābād** Baluchistān, SW Pakistan 26°15′N 62°32′E
Nasir, Buhayrat/Nâşir, Buḩeiret *see* Nasser, Lake
Nāsiri *see* Ahvāz
Nasiriya *see* An Nāşirīyah
Nāşirīyah, An *see* Dhī Qār
Nás na Ríogh *see* Naas
107 L23 **Naso** Sicilia, Italy, C Mediterranean Sea 38°07′N 14°46′E
Nasratabad *see* Zābol
77 V15 **Nassarawa** Nassarawa, C Nigeria 08°33′N 07°42′E
44 H2 **Nassau** ● (The Bahamas) New Providence, N The Bahamas 25°03′N 77°21′W
23 W8 **Nassau Sound** *sound* Florida, SE USA
108 L7 **Nassereith** Tirol, W Austria 47°19′N 10°51′E
80 F5 **Nasser, Lake** *var.* Buhayrat Nasir, Buḩayrat Nāşir, Buheiret Nâşir. ◎ Egypt/Sudan
95 L19 **Nässjö** Jönköping, S Sweden 57°39′N 14°40′E
99 K22 **Nassogne** Luxembourg, SE Belgium 50°08′N 05°19′E
12 J6 **Nastapoka Islands** *island group* Northwest Territories, C Canada
93 M19 **Nastola** Päijät-Häme, S Finland 60°57′N 25°56′E
171 O4 **Nasugbu** Luzon, N Philippines 14°03′N 120°39′E
94 N11 **Näsviken** Gävleborg, C Sweden 61°46′N 16°55′E
Naszód *see* Năsăud
83 I17 **Nata** Central, NE Botswana 20°11′S 26°10′E
54 E11 **Natagaima** Tolima, C Colombia 03°38′N 75°07′W
59 Q14 **Natal** *state capital* Rio Grande do Norte, E Brazil 05°46′S 35°15′W
168 I11 **Natal** Sumatera, N Indonesia 0°32′N 99°07′E
Natal *see* KwaZulu/Natal
173 L10 **Natal Basin** *var.* Mozambique Basin. *undersea feature* W Indian Ocean 30°00′S 40°00′E
25 R12 **Natalia** Texas, SW USA 29°11′N 98°51′W
67 W15 **Natal Valley** *undersea feature* SW Indian Ocean 31°00′S 33°15′E
Natanya *see* Netanya
143 O7 **Naţanz** Eşfahān, C Iran 33°31′N 51°55′E
13 Q11 **Natashquan** Québec, E Canada 50°10′N 61°50′W
13 Q10 **Natashquan** ≈ Newfoundland and Labrador/Québec, E Canada
22 J7 **Natchez** Mississippi, S USA 31°34′N 91°24′W
22 G6 **Natchitoches** Louisiana, S USA 31°45′N 93°05′W
108 E10 **Naters** Valais, S Switzerland 46°22′N 08°00′E
Nathanya *see* Netanya
92 O3 **Nathorst Land** *physical region* W Svalbard
186 E9 **National Capital District** ◆ *province* S Papua New Guinea
35 U17 **National City** California, W USA 32°40′N 117°06′W
184 M10 **National Park** Manawatu-Wanganui, North Island, New Zealand 39°11′S 175°22′E
77 R14 **Natitingou** NW Benin 10°21′N 01°26′E
40 B5 **Natividad, Isla** *island* NW Mexico
165 Q10 **Natori** Miyagi, Honshū, C Japan 38°12′N 140°51′E
18 C14 **Natrona Heights** Pennsylvania, NE USA 40°37′N 79°42′W
81 H20 **Natron, Lake** ◎ Kenya/Tanzania
Natsrat *see* Natzrat
166 L7 **Nattalin** Bago, C Myanmar (Burma) 18°25′N 95°34′E
92 J12 **Nattavaara** *Lapp.* Nahtavárr. Norrbotten, N Sweden 66°45′N 20°58′E
109 S3 **Natternbach** Oberösterreich, N Austria 48°26′N 13°44′E
95 M22 **Nättraby** Blekinge, S Sweden 56°12′N 15°30′E
169 P10 **Natuna Besar, Pulau** *island* Kepulauan Natuna, W Indonesia
Natuna Islands *see* Natuna, Kepulauan
169 O9 **Natuna, Kepulauan** *var.* Natuna Islands. *island group* W Indonesia
169 N9 **Natuna, Laut** *Eng.* Natuna Sea. *sea* W Indonesia
Natuna Sea *see* Natuna, Laut
21 N6 **Natural Bridge** *tourist site* Kentucky, C USA
173 V11 **Naturaliste Fracture Zone** *tectonic feature* E Indian Ocean
174 J10 **Naturaliste Plateau** *undersea feature* E Indian Ocean
138 G9 **Natzrat** *var.* Natsrat, *Ar.* En Nazira, *Eng.* Nazareth; *prev.* Nazerat. Northern, N Israel 32°42′N 35°18′E
Nau *see* Nov
103 O14 **Naucelle** Aveyron, S France 44°10′N 02°19′E
83 D20 **Nauchas** Hardap, C Namibia 23°40′S 16°19′E
108 K9 **Nauders** Tirol, W Austria 46°52′N 10°31′E
Naugard *see* Nowogard
118 F12 **Naujamiestis** Panevėžys, C Lithuania 55°42′N 24°10′E
118 E10 **Naujoji Akmenė** Šiauliai, NW Lithuania 56°20′N 22°57′E
149 R16 **Naukot** *var.* Naokot. Sind, SE Pakistan 24°52′N 69°27′E
101 L16 **Naumburg** *var.* Naumburg an der Saale. Sachsen-Anhalt, C Germany 51°09′N 11°48′E
Naumburg am Queis *see* Nowogrodziec
Naumburg an der Saale *see* Naumburg
191 W15 **Naunau** *ancient monument* Easter Island, Chile, E Pacific Ocean
138 G10 **Nā'ūr** 'Ammān, W Jordan 31°52′N 35°50′E
189 Q8 **Nauru** *off.* Republic of Nauru; *prev.* Pleasant Island. ◆ *republic* W Pacific Ocean
175 P5 **Nauru** *island* W Pacific Ocean
189 Q9 **Nauru International** ✈ S Nauru
Nauru, Republic of *see* Nauru
Nausari *see* Navsāri
19 Q12 **Nauset Beach** *beach* Massachusetts, NE USA
Naushahra *see* Nowshera
149 P14 **Naushahro Fīroz** Sind, SE Pakistan 26°51′N 68°11′E
Naushara *see* Nowshera
187 X14 **Nausori** Viti Levu, W Fiji 18°01′S 178°31′E
56 F9 **Nauta** Loreto, N Peru 04°31′S 73°36′W
153 O12 **Nautanwa** Uttar Pradesh, N India 27°26′N 83°25′E
41 R13 **Nautla** Veracruz-Llave, E Mexico 20°15′N 96°45′W
Nauzad *see* Now Zād
41 N6 **Nava** Coahuila, NE Mexico 28°28′N 100°45′W
Navabad *see* Navobod
104 L6 **Nava del Rey** Castilla y León, N Spain 41°19′N 05°04′W
153 S15 **Navadwīp** *prev.* Nabadwip. West Bengal, NE India 23°24′N 88°23′E
104 M9 **Navahermosa** Castilla-La Mancha, C Spain 39°39′N 04°25′W
119 I16 **Navahrudak** *Pol.* Nowogródek, *Rus.* Novogrudok. Hrodzyenskaya Voblasts', W Belarus 53°36′N 25°50′E
119 I16 **Navahrudskaye Wzvyshsha** *Rus.* Novogrudskaya Vozvyshennost'. ▲ W Belarus
36 M8 **Navajo Mount** ▲ Utah, W USA 37°00′N 110°52′W
37 Q9 **Navajo Reservoir** ⊠ New Mexico, SW USA
104 K9 **Navalmoral de la Mata** Extremadura, W Spain 39°54′N 05°33′W
104 K10 **Navalvillar de Pelea** Extremadura, W Spain 39°05′N 05°27′W
97 F17 **Navan** *Ir.* An Uaimh. E Ireland 53°39′N 06°41′W
Navanagar *see* Jāmnagar
118 L12 **Navapolatsk** *Rus.* Novopolotsk. Vitsyebskaya Voblasts', N Belarus 55°34′N 28°35′E
149 P6 **Nāvar, Dasht-e** *Pash.* Dasht-i-Nawar. *desert* C Afghanistan
123 W6 **Navarin, Mys** *headland* NE Russian Federation 62°18′N 179°06′E
63 I25 **Navarino, Isla** *island* S Chile
105 Q4 **Navarra** *Eng./Fr.* Navarre. ◆ *autonomous community* N Spain
Navarre *see* Navarra
105 P4 **Navarrete** La Rioja, N Spain 42°26′N 02°34′W
61 C20 **Navarro** Buenos Aires, E Argentina 35°00′S 59°15′W
105 O12 **Navas de San Juan** Andalucía, S Spain 38°11′N 03°19′W
25 V10 **Navasota** Texas, SW USA 30°23′N 96°05′W
25 U9 **Navasota River** ≈ Texas, SW USA
44 I9 **Navassa Island** ◇ *US unincorporated territory* C West Indies
119 L19 **Navasyolki** *Rus.* Novosëlki. Homyel'skaya Voblasts', SE Belarus 52°24′N 28°33′E
119 H17 **Navayel'nya** *Pol.* Nowojelnia, *Rus.* Novoyel'nya. Hrodzyenskaya Voblasts', W Belarus 53°28′N 25°35′E
171 Y13 **Naver** Papua, E Indonesia 03°27′S 139°45′E
98 H15 **Navesti** ≈ C Estonia
104 J2 **Navia** Asturias, N Spain 43°33′N 06°43′W
104 J2 **Navia** ≈ NW Spain
59 I21 **Naviraí** Mato Grosso do Sul, SW Brazil 23°01′S 54°09′W
126 I6 **Navlya** Bryanskaya Oblast', W Russian Federation 52°47′N 34°28′E
187 X13 **Navoalevu** Vanua Levu, N Fiji 16°22′S 179°28′E
147 P13 **Navobod** *Rus.* Navabad. W Tajikistan 38°37′N 68°42′E
147 R12 **Navobod** *Rus.* Navabad, Novabad. C Tajikistan 39°00′N 70°06′E
Navoi *see* Navoiy
146 M11 **Navoiy** *Rus.* Navoi. Navoiy Viloyati, C Uzbekistan 40°05′N 65°23′E
Navoiyskaya Oblast' *see* Navoiy Viloyati
146 K8 **Navoiy Viloyati** *Rus.* Navoiyskaya Oblast'. ◆ *province* N Uzbekistan
40 G7 **Navojoa** Sonora, NW Mexico 27°04′N 109°28′W
Navolat *see* Navolato
40 H9 **Navolato** *var.* Navolat. Sinaloa, C Mexico 24°46′N 107°42′W
187 Q13 **Navonda** Ambae, C Vanuatu 15°21′S 167°58′E
Návpaktos *see* Náfpaktos
Návplion *see* Náfplio
77 P14 **Navrongo** N Ghana 10°51′N 01°03′W
154 D12 **Navsāri** *var.* Nausari. Gujarāt, W India 20°55′N 72°55′E
187 X15 **Navua** Viti Levu, W Fiji 18°15′S 178°10′E
138 H8 **Nawá** Dar'ā, S Syria 32°53′N 36°03′E
Nawabashah *see* Nawābshāh
153 S14 **Nawabganj** Rajshahi, NW Bangladesh 24°35′N 88°21′E
153 S14 **Nawābganj** Uttar Pradesh, N India 26°52′N 82°09′E
149 Q15 **Nawābshāh** *var.* Nawabashah. Sind, S Pakistan 26°15′N 68°26′E
153 P14 **Nawāda** Bihār, N India 24°54′N 85°33′E
152 H11 **Nawalgarh** Rājasthān, N India 27°48′N 75°21′E
Nawāl, Sabkhat an *see* Noual, Sebkhet en
Nawar, Dasht-i- *see* Nāvar, Dasht-e
167 N4 **Nawnghkio** *var.* Nawngkio. Shan State, C Myanmar (Burma) 22°17′N 96°50′E
Nawngkio *see* Nawnghkio
137 U13 **Naxçıvan** *Rus.* Nakhichevan'. SW Azerbaijan 39°14′N 45°24′E
160 I10 **Naxi** Sichuan, C China 28°50′N 105°20′E
115 K21 **Náxos** *var.* Naxos. Náxos, Kykládes, Greece, Aegean Sea 36°07′N 25°24′E
115 K21 **Náxos** *island* Kykládes, Greece, Aegean Sea
40 J11 **Nayarit** ◆ *state* C Mexico
187 Y14 **Nayau** *island* Lau Group, E Fiji
143 S8 **Nāy Band** Yazd, E Iran 32°26′N 57°30′E
165 T2 **Nayoro** Hokkaidō, NE Japan 44°22′N 142°27′E
166 M7 **Nay Pyi Taw** ◆ *union territory* C Myanmar (Burma)
104 F9 **Nazaré** *var.* Nazare. Leiria, C Portugal 39°36′N 09°04′W
Nazare *see* Nazaré
24 M4 **Nazareth** Texas, SW USA 34°32′N 102°06′W
Nazareth *see* Natzrat
173 O8 **Nazareth Bank** *undersea feature* W Indian Ocean
40 K9 **Nazas** Durango, C Mexico 25°15′N 104°06′W
57 F16 **Nazca** Ica, S Peru 14°53′S 74°54′W
0 L17 **Nazca Plate** *tectonic feature*
193 U9 **Nazca Ridge** *undersea feature* E Pacific Ocean 22°00′S 82°00′W
165 V15 **Naze** *var.* Nase. Kagoshima, Amami-ōshima, SW Japan 28°21′N 129°30′E
Nazerat *see* Natzrat
137 R14 **Nazik Gölü** ◎ E Turkey
136 C15 **Nazilli** Aydın, SW Turkey 37°55′N 28°20′E
137 P14 **Nazimiye** Tunceli, E Turkey 39°12′N 39°51′E
Nazinon *see* Red Volta
10 L15 **Nazko** British Columbia, SW Canada 52°57′N 123°44′W
127 O16 **Nazran'** Respublika Ingushetiya, SW Russian Federation 43°14′N 44°47′E
80 J13 **Nazrēt** *var.* Adama, Hadama. Oromīya, C Ethiopia 08°31′N 39°20′E
Nazwāh *see* Nizwa
82 J13 **Nchanga** Copperbelt, C Zambia 12°30′S 27°53′E
82 J11 **Nchelenge** Luapula, N Zambia 09°20′S 28°50′E
Ncheu *see* Ntcheu
83 J25 **Nciba** *Eng.* Great Kei; *prev.* Groot-Kei. ≈ S South Africa
Ndaghamcha, Sebkra de *see* Te-n-Dghâmcha, Sebkhet
81 G21 **Ndala** Tabora, C Tanzania 04°45′S 33°15′E
82 B11 **N'Dalatando** *Port.* Salazar, Vila Salazar. Kwanza Norte, NW Angola 09°19′S 14°48′E
77 S14 **Ndali** C Benin 09°50′N 02°46′E
81 E18 **Ndeke** SW Uganda 0°11′S 30°04′E
78 J13 **Ndélé** Bamingui-Bangoran, N Central African Republic 08°24′N 20°41′E
79 E19 **Ndendé** Ngounié, S Gabon 02°21′S 11°20′E
79 E20 **Ndindi** Nyanga, S Gabon 03°47′S 11°06′E
78 G11 **N'Djaména** *var.* Ndjamena; *prev.* Fort-Lamy. ● (Chad) Ville de N'Djaména, W Chad 12°08′N 15°02′E
78 G11 **N'Djaména** ✈ Ville de N'Djaména, W Chad 12°09′N 15°00′E
Ndjamena *see* N'Djaména
N'Djaména, Région de la Ville de *see* N'Djaména, Ville de
78 G11 **N'Djaména, Ville de** ◆ *region* SW Chad
79 D18 **Ndjolé** Moyen-Ogooué, W Gabon 0°07′S 10°45′E
82 J13 **Ndola** Copperbelt, C Zambia 12°59′S 28°35′E
Ndrhamcha, Sebkha de *see* Te-n-Dghâmcha, Sebkhet
79 L15 **Ndu** Orientale, N Dem. Rep. Congo 04°36′N 22°49′E
81 H21 **Nduguti** Singida, C Tanzania 04°19′S 34°40′E
186 M9 **Nduindui** Guadalcanal, C Solomon Islands 09°46′S 159°54′E
Nduke *see* Kolombangara
Ndzouani *see* Nzwani
115 F16 **Néa Anchíalos** *var.* Nea Anhialos, Néa Ankhíalos. Thessalía, C Greece 39°16′N 22°49′E
Nea Anhialos/Néa Ankhíalos *see* Néa Anchíalos
115 H18 **Néa Artáki** Évvoia, C Greece 38°31′N 23°39′E
97 F15 **Neagh, Lough** ◎ E Northern Ireland, United Kingdom
32 F7 **Neah Bay** Washington, NW USA 48°21′N 124°39′W
115 J22 **Nea Kaméni** *island* Kykládes, Greece, Aegean Sea
181 O8 **Neale, Lake** ◎ Northern Territory, C Australia
182 G2 **Neales River** *seasonal river* South Australia
115 G14 **Néa Moudanía** *var.* Néa Moudhaniá. Kentrikí Makedonía, N Greece 40°14′N 23°17′E
Néa Moudhaniá *see* Néa Moudanía
116 K10 **Neamț** ◆ *county* NE Romania
Neapel *see* Napoli
115 D14 **Neápoli** *prev.* Neápolis. Dytikí Makedonía, N Greece 40°19′N 21°23′E
115 K25 **Neápoli** Kríti, Greece, E Mediterranean Sea 35°15′N 25°37′E
115 G22 **Neápoli** Pelopónnisos, S Greece 36°29′N 23°05′E
Neápolis *see* Neápoli, Greece
Neapolis *see* Napoli, Italy
Neapolis *see* Nablus, West Bank
38 D16 **Near Islands** *island group* Aleutian Islands, Alaska, USA
97 J21 **Neath** S Wales, United Kingdom 51°40′N 03°48′W
114 H13 **Néa Zíchni** *var.* Néa Zíkhni; *prev.* Néa Zíkhna. Kentrikí Makedonía, NE Greece 41°02′N 23°50′E
Néa Zíkhna/Néa Zíkhni *see* Néa Zíchni
42 C5 **Nebaj** Quiché, W Guatemala 15°25′N 91°05′W
77 P13 **Nebbou** S Burkina Faso 11°22′N 01°49′W
Nebitdag *see* Balkanabat
54 M13 **Neblina, Pico da** ▲ NW Brazil 0°49′N 66°31′W
124 I13 **Nebolchi** Novgorodskaya Oblast', W Russian Federation 59°08′N 33°19′E
36 L4 **Nebo, Mount** ▲ Utah, W USA 39°47′N 111°46′W
28 L14 **Nebraska** *off.* State of Nebraska, *also known as* Blackwater State, Cornhusker State, Tree Planters State. ◆ *state* C USA
29 S16 **Nebraska City** Nebraska, C USA 40°38′N 95°52′W
107 K23 **Nebrodi, Monti** *var.* Monti Caronie. ▲▲ Sicilia, Italy, C Mediterranean Sea
10 L14 **Nechako** ≈ British Columbia, SW Canada
29 Q2 **Neche** North Dakota, N USA 48°57′N 97°33′W
25 V8 **Neches** Texas, SW USA 31°51′N 95°28′W
25 W8 **Neches River** ≈ Texas, SW USA
101 H20 **Neckar** ≈ SW Germany
101 H20 **Neckarsulm** Baden-Württemberg, SW Germany 49°12′N 09°13′E
192 L5 **Necker Island** *island* C British Virgin Islands
175 U3 **Necker Ridge** *undersea feature* N Pacific Ocean
61 D23 **Necochea** Buenos Aires, E Argentina 38°34′S 58°42′W
104 H2 **Neda** Galicia, NW Spain 43°29′N 08°09′W
115 E20 **Néda** *var.* Nédas. ≈ S Greece
Nédas *see* Néda
114 J12 **Nedelino** Smolyan, S Bulgaria 41°27′N 25°05′E
25 Y11 **Nederland** Texas, SW USA 29°58′N 93°59′W
Nederland *see* Netherlands
98 K12 **Neder Rijn** *Eng.* Lower Rhine. ≈ C Netherlands

◆ Country
● Country Capital
◇ Dependent Territory
○ Dependent Territory Capital
◆ Administrative Regions
✈ International Airport
▲ Mountain
▲▲ Mountain Range
Volcano
≈ River
◎ Lake
⊠ Reservoir

NEDERWEERT – NEW TOWN

99 L16 **Nederweert** Limburg, SE Netherlands *51°17′N 05°45′E*
95 G16 **Nedre Tokke** ◎ S Norway
Nedrigaylov *see* Nedryhayliv
117 S3 **Nedryhayliv** *Rus.* Nedrigaylov. Sums'ka Oblast', NE Ukraine *50°51′N 33°54′E*
98 O11 **Neede** Gelderland, E Netherlands *52°08′N 06°36′E*
33 T13 **Needle Mountain** ▲ Wyoming, C USA *44°03′N 109°33′W*
35 Y14 **Needles** California, W USA *34°50′N 114°37′W*
97 M24 **Needles, The** *rocks* S England, United Kingdom
62 O7 **Ñeembucú** *off.* Departamento de Ñeembucú. ◈ *department* SW Paraguay
Ñeembucú, Departamento de *see* Ñeembucú
30 M7 **Neenah** Wisconsin, N USA *44°09′N 88°26′W*
11 W16 **Neepawa** Manitoba, S Canada *50°14′N 99°29′W*
99 K16 **Neerpelt** Limburg, NE Belgium *51°13′N 05°26′E*
74 M6 **Nefta** ✈ W Tunisia *34°03′N 08°05′E*
126 L15 **Neftegorsk** Krasnodarskiy Kray, SW Russian Federation *44°21′N 39°40′E*
127 U3 **Neftekamsk** Respublika Bashkortostan, W Russian Federation *56°07′N 54°13′E*
127 O14 **Neftekumsk** Stavropol'skiy Kray, SW Russian Federation *44°45′N 45°00′E*
Neftezavodsk *see* Seýdi
82 C10 **Negage** *var.* N'Gage. Uíge, NW Angola *07°47′S 15°27′E*
Negapatam/Negapattinam *see* Nāgappattinam
169 T17 **Negara** Bali, Indonesia *08°21′S 114°35′E*
169 T13 **Negara** Borneo, C Indonesia *02°40′S 115°05′E*
Negara Brunei Darussalam *see* Brunei
31 N4 **Negaunee** Michigan, N USA *46°30′N 87°36′W*
81 J15 **Negēlē** *var.* Negelli, *It.* Neghelli. Oromīya, C Ethiopia *05°13′N 39°43′E*
Negelli *see* Negēlē
Negeri Pahang Darul Makmur *see* Pahang
Negeri Selangor Darul Ehsan *see* Selangor
168 K9 **Negeri Sembilan** *var.* Negri Sembilan. ◈ *state* Peninsular Malaysia
92 P3 **Negerpynten** *headland* S Svalbard *77°15′N 22°40′E*
Negev *see* HaNegev
Neghelli *see* Negēlē
116 I12 **Negoiu** *var.* Negoiul. ▲ S Romania *45°34′N 24°34′E*
Negoiul *see* Negoiu
82 P13 **Negomane** *var.* Negomano. Cabo Delgado, N Mozambique *11°22′S 38°32′E*
Negomano *see* Negomane
155 J25 **Negombo** Western Province, SW Sri Lanka *07°13′N 79°51′E*
191 W11 **Negonego** *prev.* Nengonengo. *atoll* Îles Tuamotu, C French Polynesia
Negoreloye *see* Nyeharelaye
112 P12 **Negotin** Serbia, E Serbia *44°14′N 22°31′E*
113 P19 **Negotino** C Macedonia *41°29′N 22°04′E*
56 A10 **Negra, Punto** *headland* NW Peru *06°03′S 81°08′W*
104 G3 **Negreira** Galicia, NW Spain *42°54′N 08°46′W*
116 L10 **Negrești** Vaslui, E Romania *46°50′N 27°28′E*
Negrești *see* Negrești-Oaș
116 H8 **Negrești-Oaș** *Hung.* Avasfelsőfalu; *prev.* Negrești. Satu Mare, NE Romania *47°56′N 23°22′E*
44 H12 **Negril** W Jamaica *18°16′N 78°21′W*
Negri Sembilan *see* Negeri Sembilan
63 K15 **Negro, Río** ≈ E Argentina
62 N7 **Negro, Río** ≈ NE Argentina
57 N17 **Negro, Río** ≈ E Bolivia
48 F6 **Negro, Rio** ≈ N South America
61 E18 **Negro, Río** ≈ Brazil/ Uruguay
62 O5 **Negro, Río** ≈ C Paraguay
Negro, Río *see* Chixoy, Río, Guatemala/Mexico
Negro, Río *see* Sico Tinto, Río, Honduras
171 P6 **Negros** *island* C Philippines
116 M15 **Negru Vodă** Constanța, SE Romania *43°49′N 28°12′E*
13 P13 **Neguac** New Brunswick, SE Canada *47°16′N 65°04′W*
14 B7 **Negwazu, Lake** ◎ Ontario, S Canada
Négyfalu *see* Săcele
32 F10 **Nehalem** Oregon, NW USA *45°42′N 123°55′W*
32 F10 **Nehalem River** ≈ Oregon, NW USA
Nehavend *see* Nahāvand
143 V9 **Nehbandān** Khorāsān-e Jonūbī, E Iran *31°00′N 60°00′E*
163 V6 **Nehe** Heilongjiang, NE China *48°28′N 124°52′E*
193 Y14 **Neiafu** 'Uta Vava'u, N Tonga *18°36′S 173°58′W*
45 N9 **Neiba** *var.* Neyba. SW Dominican Republic *18°31′N 71°25′W*
Néid, Carn Uí *see* Mizen Head
92 M9 **Neiden** Finnmark, N Norway *69°41′N 29°23′E*
Neidín *see* Kenmare
Néifinn *see* Nephin
103 S10 **Neige, Crêt de la** ▲ E France *46°18′N 05°58′E*
173 O16 **Neiges, Piton des** ▲ C Réunion *21°05′S 55°28′E*
15 R9 **Neiges, Rivière des** ≈ Québec, SE Canada
160 I10 **Neijiang** Sichuan, C China *29°32′N 105°03′E*
30 K6 **Neillsville** Wisconsin, N USA *44°34′N 90°36′W*
Nei Monggol Zizhiqu/Nei Mongol *see* Nei Mongol Zizhiqu
163 Q10 **Nei Mongol Gaoyuan** *plateau* NE China
163 O12 **Nei Mongol Zizhiqu** *var.* Nei Mongol, *Eng.* Inner Mongolia, Inner Mongolian Autonomous Region; *prev.* Nei Monggol Zizhiqu. ◈ *autonomous region* N China
161 O4 **Neiqiu** Hebei, E China *37°22′N 114°34′E*
Neiriz *see* Neyrīz
101 Q16 **Neisse** *Pol.* Nisa *Cz.* Lužická Nisa, *Ger.* Lausitzer Neisse, Nysa Łużycka. ≈ C Europe
Neisse *see* Nysa
54 E11 **Neiva** Huila, S Colombia *02°58′N 75°15′W*
160 M7 **Neixiang** Henan, C China *33°08′N 111°50′E*
Nejafabad *see* Najafābād
11 V9 **Nejanilini Lake** ◎ Manitoba, C Canada
Nejd *see* Najd
80 I13 **Nek'emtē** *var.* Lakemti, Nakamti. Oromīya, C Ethiopia *09°06′N 36°31′E*
126 M9 **Nekhayevskaya** Volgogradskaya Oblast', SW Russian Federation *50°25′N 41°44′E*
30 K7 **Nekoosa** Wisconsin, N USA *44°19′N 89°54′W*
Neksø Bornholm *see* Nexø
104 H7 **Nelas** Viseu, N Portugal *40°32′N 07°52′W*
124 H16 **Nelidovo** Tverskaya Oblast', W Russian Federation *56°13′N 32°45′E*
29 P13 **Neligh** Nebraska, C USA *42°07′N 98°01′W*
123 R11 **Nel'kan** Khabarovskiy Kray, E Russian Federation *57°44′N 136°09′E*
92 M10 **Nellim** *var.* Nellimö, *Lapp.* Njellim. Lappi, N Finland *68°49′N 28°18′E*
Nellimö *see* Nellim
155 J18 **Nellore** Andhra Pradesh, E India *14°29′N 80°E*
61 B17 **Nelson** Santa Fe, C Argentina *31°16′S 60°45′W*
11 O17 **Nelson** British Columbia, SW Canada *49°29′N 117°17′W*
185 I14 **Nelson** Nelson, South Island, New Zealand *41°17′S 173°17′E*
97 L17 **Nelson** NW England, United Kingdom *53°51′N 02°13′W*
29 P17 **Nelson** Nebraska, C USA *40°12′N 98°04′W*
185 J14 **Nelson** ◈ *unitary authority* South Island, New Zealand
11 X12 **Nelson** ≈ Manitoba, C Canada
183 U8 **Nelson Bay** New South Wales, SE Australia *32°48′S 152°10′E*
182 K13 **Nelson, Cape** *headland* Victoria, SE Australia *38°25′S 141°33′E*
63 G23 **Nelson, Estrecho** *strait* SE Pacific Ocean
11 W12 **Nelson House** Manitoba, C Canada *55°49′N 98°51′W*
30 J4 **Nelson Lake** ◎ Wisconsin, N USA
31 T14 **Nelsonville** Ohio, N USA *39°27′N 82°13′W*
27 S2 **Nelsoon River** ≈ Iowa/ Missouri, C USA
83 K21 **Nelspruit** Mpumalanga, NE South Africa *25°28′S 30°58′E*
Nelspruit *see* Mbombela
76 L10 **Néma** Hodh ech Chargui, SE Mauritania *16°32′N 07°12′W*
118 D13 **Neman** *Ger.* Ragnit. Kaliningradskaya Oblast', W Russian Federation *55°01′N 22°00′E*
84 I9 **Neman** *Bel.* Nyoman, *Ger.* Memel, *Lith.* Nemunas, *Pol.* Niemen. ≈ NE Europe
Nemausus *see* Nîmes
115 F19 **Neméa** Pelopónnisos, S Greece *37°49′N 22°40′E*
Německý Brod *see* Havlíčkův Brod
14 D7 **Nemegosenda** ≈ Ontario, S Canada
14 D8 **Nemegosenda Lake** ◎ Ontario, S Canada
119 H14 **Nemencinė** Vilnius, SE Lithuania *54°50′N 25°29′E*
Nemetocenna *see* Arras
Nemirov *see* Nemyriv
103 O6 **Nemours** Seine-et-Marne, N France *48°16′N 02°41′E*
Nemunas *see* Neman
165 W4 **Nemuro** Hokkaidō, NE Japan *43°20′N 145°35′E*
165 W4 **Nemuro-hantō** *peninsula* Hokkaidō, NE Japan
165 W3 **Nemuro-kaikyō** *strait* Japan/Russian Federation
165 W4 **Nemuro-wan** *bay* N Japan
116 H5 **Nemyriv** *Rus.* Nemirov. L'vivs'ka Oblast', NW Ukraine *50°08′N 23°28′E*
117 N7 **Nemyriv** *Rus.* Nemirov. Vinnyts'ka Oblast', C Ukraine *48°58′N 28°50′E*
97 D19 **Nenagh** *Ir.* An tAonach. Tipperary, C Ireland *52°52′N 08°12′W*
39 R9 **Nenana** Alaska, USA *64°33′N 149°05′W*
39 R9 **Nenana River** ≈ Alaska, USA
187 P10 **Nendö** *var.* Swallow Island. *island* Santa Cruz Islands, E Solomon Islands
97 O19 **Nene** ≈ E England, United Kingdom
125 R4 **Nenetskiy Avtonomnyy Okrug** ◈ *autonomous district* Arkhangel'skaya Oblast', NW Russian Federation
Nengonengo *see* Negonego
163 V6 **Nenjiang** Heilongjiang, NE China *49°11′N 125°18′E*
163 U6 **Nen Jiang** *var.* Nonni. ≈ NE China
189 P16 **Neoch** *atoll* Caroline Islands, C Micronesia
115 D18 **Neochóri** Dytikí Elláda, C Greece *38°23′N 21°14′E*
27 Q7 **Neodesha** Kansas, C USA *37°25′N 95°40′W*
29 S14 **Neola** Iowa, C USA *41°27′N 95°40′W*
115 E16 **Néo Monastíri** *var.* Néon Monastiri. Thessalía, C Greece *39°22′N 21°55′E*
Néon Karlovasi/Néon Karlovásion *see* Karlovási
Néon Monastiri *see* Néo Monastíri
27 R8 **Neosho** Missouri, C USA *36°53′N 94°24′W*
27 Q7 **Neosho River** ≈ Kansas/ Oklahoma, C USA
123 N12 **Nepa** ≈ C Russian Federation
153 N10 **Nepal** *off.* Nepal. ◆ *monarchy* S Asia
Nepal *see* Nepal
152 M11 **Nepālganj** Mid Western, SW Nepal *28°04′N 81°37′E*
14 L13 **Nepean** Ontario, SE Canada *45°19′N 75°54′W*
36 L4 **Nephi** Utah, W USA *39°43′N 111°50′W*
97 B16 **Nephin** *Ir.* Néifinn. ▲ W Ireland *54°00′N 09°21′W*
67 T9 **Nepoko** ≈ NE Dem. Rep. Congo
18 K15 **Neptune** New Jersey, NE USA *40°10′N 74°03′W*
182 G10 **Neptune Islands** *island group* South Australia
107 I14 **Nera** *anc.* Nar. ≈ C Italy
102 L14 **Nérac** Lot-et-Garonne, SW France *44°08′N 00°21′E*
111 D16 **Neratovice** *Ger.* Neratowitz. Středocesky Kraj, C Czech Republic *50°16′N 14°31′E*
Neratowitz *see* Neratovice
123 O13 **Nercha** ≈ S Russian Federation
123 O13 **Nerchinsk** Zabaykal'skiy Kray, S Russian Federation *52°01′N 116°25′E*
123 P14 **Nerchinskiy Zavod** Zabaykal'skiy Kray, S Russian Federation *51°13′N 119°25′E*
124 M15 **Nerekhta** Kostromskaya Oblast', NW Russian Federation *57°27′N 40°33′E*
118 H10 **Nereta** S Latvia *56°12′N 25°18′E*
106 K13 **Nereto** Abruzzo, C Italy *42°49′N 13°50′E*
113 H15 **Neretva** ≈ Bosnia and Herzegovina/Croatia
115 C17 **Nerikós** *ruins* Lefkáda, Iónia Nísiá, Greece, C Mediterranean Sea
83 F15 **Neriquinha** Kuando Kubango, SE Angola *15°44′S 21°34′E*
118 I13 **Neris** *Bel.* Viliya, *Pol.* Wilia; *prev. Pol.* Wilja. ≈ Belarus/ Lithuania
Neris *see* Viliya
105 N15 **Nerja** Andalucía, S Spain *36°45′N 03°35′W*
124 L16 **Nerl'** ≈ W Russian Federation
105 P12 **Nerpio** Castilla-La Mancha, C Spain *38°08′N 02°18′W*
104 J13 **Nerva** Andalucía, S Spain *37°40′N 06°31′W*
98 L4 **Nes** Fryslân, N Netherlands *53°28′N 05°46′E*
94 G13 **Nesbyen** Buskerud, S Norway *60°36′N 09°35′E*
114 M9 **Nesebar** *var.* Nesebŭr. Burgas, E Bulgaria *42°40′N 27°43′E*
Nesebŭr *see* Nesebar
Neshcherdo, Ozero *see* Nyeshcharda, Vozyera
92 L2 **Neskaupstaður** Austurland, E Iceland *65°08′N 13°45′W*
92 F13 **Nesna** Nordland, C Norway *66°11′N 12°54′E*
26 K5 **Ness City** Kansas, C USA *38°27′N 99°54′W*
Nesselsdorf *see* Kopřivnice
108 H7 **Nesslau** Sankt Gallen, NE Switzerland *47°13′N 09°12′E*
96 I9 **Ness, Loch** ◎ N Scotland, United Kingdom
Nesterov *see* Zhovkva
114 I12 **Néstos** *Bul.* Mesta, *Turk.* Kara Su. ≈ Bulgaria/Greece *see also* Mesta
Néstos *see* Mesta
95 C14 **Nesttun** Hordaland, S Norway *60°19′N 05°16′E*
Nesvizh *see* Nyasvizh
138 F9 **Netanya** *var.* Natanya, Nathanya. Central, C Israel *32°20′N 34°51′E*
98 I9 **Netherlands** *off.* Kingdom of the Netherlands, *var.* Holland, *Dut.* Koninkrijk der Nederlanden, Nederland. ◆ *monarchy* NW Europe
Netherlands East Indies *see* Indonesia
Netherlands Guiana *see* Suriname
Netherlands, Kingdom of the *see* Netherlands
Netherlands New Guinea *see* Papua
116 L4 **Netishyn** Khmel'nyts'ka Oblast', W Ukraine *50°20′N 26°38′E*
138 E11 **Netivot** Southern, S Israel *31°26′N 34°36′E*
107 O21 **Neto** ≈ S Italy
9 Q6 **Nettilling Lake** ◎ Baffin Island, Nunavut, N Canada
29 V3 **Nett Lake** ◎ Minnesota, N USA
107 I16 **Nettuno** Lazio, C Italy *41°27′N 12°40′E*
Netum *see* Noto
41 U16 **Netzahualcóyotl, Presa** ⊠ SE Mexico
Netze *see* Noteć
Neu Amerika *see* Puławy
Neubetsche *see* Novi Bečej
Neubidschow *see* Nový Bydžov
Neubistritz *see* Nová Bystřice
100 N9 **Neubrandenburg** Mecklenburg-Vorpommern, NE Germany *53°33′N 13°16′E*
101 K22 **Neuburg an der Donau** Bayern, S Germany *48°43′N 11°10′E*
108 C8 **Neuchâtel** *Ger.* Neuenburg. Neuchâtel, W Switzerland *46°59′N 06°55′E*
108 C8 **Neuchâtel** *Ger.* Neuenburg. ◈ *canton* W Switzerland
108 C8 **Neuchâtel, Lac de** *Ger.* Neuenburger See. ◎ W Switzerland
Neudorf *see* Spišská Nová Ves
100 L10 **Neue Elde** *canal* N Germany
Neuenburg *see* Neuchâtel
Neuenburg an der Elbe *see* Nymburk
Neuenburger See *see* Neuchâtel, Lac de
108 F7 **Neuenhof** Aargau, N Switzerland *47°27′N 08°17′E*
100 H11 **Neuenland** ✈ (Bremen) Bremen, NW Germany *53°03′N 08°46′E*
Neuenstadt *see* La Neuveville
101 C18 **Neuerburg** Rheinland-Pfalz, W Germany *50°01′N 06°13′E*
99 K24 **Neufchâteau** Luxembourg, SE Belgium *49°50′N 05°26′E*
103 S6 **Neufchâteau** Vosges, NE France *48°21′N 05°42′E*
102 M3 **Neufchâtel-en-Bray** Seine-Maritime, N France *49°44′N 01°26′E*
109 S3 **Neufelden** Oberösterreich, N Austria *48°27′N 14°01′E*
Neugradisk *see* Nova Gradiška
Neuhaus *see* Jindřichův Hradec
Neuhäusel *see* Nové Zámky
108 G6 **Neuhausen** *var.* Neuhausen am Rheinfall. Schaffhausen, N Switzerland *47°24′N 08°37′E*
Neuhausen am Rheinfall *see* Neuhausen
101 I17 **Neuhof** Hessen, C Germany *50°26′N 09°34′E*
Neuhof *see* Zgierz
Neukuhren *see* Pionerskiy
Neu-Langenburg *see* Tukuyu
109 W4 **Neulengbach** Niederösterreich, NE Austria *48°10′N 15°53′E*
113 G15 **Neum** Federacija Bosne I Hercegovine, S Bosnia and Herzegovina *42°57′N 17°33′E*
Neumark *see* Nowy Targ, Małopolskie, Poland
Neumark *see* Nowe Miasto Lubawskie, Warmińsko-Mazurskie, Poland
Neumarkt *see* Neumarkt im Hausruckkreis, Oberösterreich, Austria
Neumarkt *see* Neumarkt am Wallersee, Salzburg, Austria
Neumarkt *see* Sroda Śląska, Dolnośląskie, Poland
Neumarkt *see* Târgu Secuiesc, Covasna, Romania
Neumarkt *see* Târgu Mureș
109 Q5 **Neumarkt am Wallersee** *var.* Neumarkt. Salzburg, NW Austria *47°55′N 13°16′E*
109 R4 **Neumarkt im Hausruckkreis** *var.* Neumarkt. Oberösterreich, N Austria *48°16′N 13°40′E*
101 L20 **Neumarkt in der Oberpfalz** Bayern, SE Germany *49°16′N 11°28′E*
Neumarktl *see* Tržič
Neumoldowa *see* Moldova Nouă
100 J8 **Neumünster** Schleswig-Holstein, N Germany *54°04′N 09°59′E*
109 X5 **Neunkirchen** *var.* Neunkirchen am Steinfeld. Niederösterreich, E Austria *47°44′N 16°05′E*
101 E20 **Neunkirchen** Saarland, SW Germany *49°21′N 07°11′E*
Neunkirchen am Steinfeld *see* Neunkirchen
Neuoderberg *see* Bohumín
63 I15 **Neuquén** Neuquén, SE Argentina *39°03′S 68°36′W*
63 H14 **Neuquén** *off.* Provincia de Neuquén. ◈ *province* W Argentina
Neuquén, Provincia de *see* Neuquén
63 H14 **Neuquén, Río** ≈ W Argentina
Neurode *see* Nowa Ruda
100 N11 **Neuruppin** Brandenburg, NE Germany *52°56′N 12°49′E*
Neusalz an der Oder *see* Nowa Sól
Neu Sandec *see* Nowy Sącz
101 K22 **Neusäss** Bayern, S Germany *48°24′N 10°49′E*
Neusatz *see* Novi Sad
Neuschliss *see* Gherla
21 N8 **Neuse River** ≈ North Carolina, SE USA
109 Z5 **Neusiedl am See** Burgenland, E Austria *47°58′N 16°51′E*
111 G22 **Neusiedler See** *Hung.* Fertő. ◎ Austria/Hungary
Neusohl *see* Banská Bystrica
101 D15 **Neuss** *anc.* Novaesium, Novesium. Nordrhein-Westfalen, W Germany *51°12′N 06°42′E*
Neuss *see* Nyon
Neustadt *see* Neustadt bei Coburg, Bayern, Germany
Neustadt *see* Neustadt an der Aisch, Bayern, Germany
Neustadt *see* Prudnik, Opole, Poland
Neustadt *see* Baia Mare, Maramureș, Romania
100 I12 **Neustadt am Rübenberge** Niedersachsen, N Germany *52°30′N 09°28′E*
101 J19 **Neustadt an der Aisch** *var.* Neustadt. Bayern, C Germany *49°34′N 10°36′E*
Neustadt an der Haardt *see* Neustadt an der Weinstrasse
101 F20 **Neustadt an der Weinstrasse** *prev.* Neustadt an der Haardt, *hist.* Niewenstat; *anc.* Nova Civitas. Rheinland-Pfalz, SW Germany *49°21′N 08°09′E*
101 K18 **Neustadt bei Coburg** *var.* Neustadt. Bayern, C Germany *50°19′N 11°06′E*
Neustadt bei Pinne *see* Lwówek
Neustadt in Oberschlesien *see* Prudnik
Neustadtl *see* Novo mesto
Neustadtl in Mähren *see* Nové Město na Moravě
Neustettin *see* Szczecinek
108 M8 **Neustift im Stubaital** *var.* Stubaital. Tirol, W Austria *47°07′N 11°26′E*
100 N10 **Neustrelitz** Mecklenburg-Vorpommern, NE Germany *53°22′N 13°05′E*
Neutitschein *see* Nový Jičín
Neutra *see* Nitra
101 J22 **Neu-Ulm** Bayern, S Germany *48°23′N 10°02′E*
Neuveville *see* La Neuveville
103 N12 **Neuvic** Corrèze, C France *45°23′N 02°16′E*
Neuwarp *see* Nowe Warpno
100 G9 **Neuwerk** *island* NW Germany
101 E17 **Neuwied** Rheinland-Pfalz, W Germany *50°26′N 07°28′E*
Neuzen *see* Terneuzen
124 H12 **Neva** ≈ NW Russian Federation
29 V14 **Nevada** Iowa, C USA *42°01′N 93°27′W*
27 R6 **Nevada** Missouri, C USA *37°51′N 94°22′W*
35 R5 **Nevada** *off.* State of Nevada, *also known as* Battle Born State, Sagebrush State, Silver State. ◈ *state* W USA
35 P6 **Nevada City** California, W USA *39°15′N 121°02′W*
105 O14 **NevadaSierra** ▲▲ S Spain
35 P6 **Nevada, Sierra** ▲▲ W USA
62 I13 **Nevado, Sierra del** ▲▲ W Argentina
124 G16 **Nevel'** Pskovskaya Oblast', W Russian Federation *56°01′N 29°54′E*
123 T14 **Nevel'sk** Ostrov Sakhalin, Sakhalinskaya Oblast', SE Russian Federation *46°41′N 141°54′E*
123 Q13 **Never** Amurskaya Oblast', SE Russian Federation *53°58′N 124°04′E*
127 Q6 **Neverkino** Penzenskaya Oblast', W Russian Federation *52°53′N 46°46′E*
103 P9 **Nevers** *anc.* Noviodunum. Nièvre, C France *47°N 03°09′E*
18 J12 **Neversink River** ≈ New York, NE USA
183 Q6 **Nevertire** New South Wales, SE Australia *31°52′S 147°42′E*
113 H15 **Nevesinje** ◈ Republika Srpska, S Bosnia and Herzegovina
118 G12 **Nevežis** ≈ C Lithuania
138 F11 **Neve Zohar** *prev.* Newé Zohar. Southern, E Israel *31°07′N 35°23′E*
126 M14 **Nevinnomyssk** Stavropol'skiy Kray, SW Russian Federation *44°39′N 41°57′E*
45 W10 **Nevis** *island* Saint Kitts and Nevis
Nevoso, Monte *see* Veliki Snežnik
Nevrokop *see* Gotse Delchev
136 J14 **Nevșehir** *var.* Nevshehr. Nevșehir, C Turkey *38°38′N 34°43′E*
136 J14 **Nevșehir** *var.* Nevshehr. ◈ *province* C Turkey
Nevshehr *see* Nevșehir
122 G10 **Nev'yansk** Sverdlovskaya Oblast', C Russian Federation *57°26′N 60°15′E*
81 J25 **Newala** Mtwara, SE Tanzania *10°59′S 39°18′E*
31 P16 **New Albany** Indiana, N USA *38°17′N 85°50′W*
22 M2 **New Albany** Mississippi, S USA *34°29′N 89°00′W*
29 Y11 **New Albin** Iowa, C USA *43°30′N 91°17′W*
55 U8 **New Amsterdam** E Guyana *06°17′N 57°31′W*
183 Q4 **New Angledool** New South Wales, SE Australia *29°06′S 147°54′E*
21 Y2 **Newark** Delaware, NE USA *39°42′N 75°45′W*
18 K14 **Newark** New Jersey, NE USA *40°42′N 74°12′W*
18 G10 **Newark** New York, NE USA *43°01′N 77°04′W*
31 T13 **Newark** Ohio, N USA *40°03′N 82°24′W*
Newark *see* Newark-on-Trent
35 W5 **Newark Lake** ◎ Nevada, W USA
97 N18 **Newark-on-Trent** *var.* Newark. C England, United Kingdom *53°05′N 00°49′W*
22 M7 **New Augusta** Mississippi, S USA *31°12′N 89°03′W*
19 P12 **New Bedford** Massachusetts, NE USA *41°38′N 70°55′W*
32 G11 **Newberg** Oregon, NW USA *45°18′N 122°58′W*
21 X10 **New Bern** North Carolina, SE USA *35°05′N 77°04′W*
20 F8 **Newbern** Tennessee, S USA *36°06′N 89°15′W*
31 P4 **Newberry** Michigan, N USA *46°21′N 85°30′W*
21 Q12 **Newberry** South Carolina, SE USA *34°17′N 81°39′W*
18 F15 **New Bloomfield** Pennsylvania, NE USA *40°24′N 77°08′W*
25 X5 **New Boston** Texas, SW USA *33°27′N 94°25′W*
25 S11 **New Braunfels** Texas, SW USA *29°43′N 98°09′W*
31 Q13 **New Bremen** Ohio, N USA *40°26′N 84°22′W*
97 F18 **Newbridge** *Ir.* An Droichead Nua. Kildare, C Ireland *53°11′N 06°48′W*
18 B14 **New Brighton** Pennsylvania, NE USA *40°44′N 80°18′W*
18 M12 **New Britain** Connecticut, NE USA *41°37′N 72°45′W*
186 G7 **New Britain** *island* E Papua New Guinea
192 I8 **New Britain Trench** *undersea feature* W Pacific Ocean
18 J15 **New Brunswick** New Jersey, NE USA *40°29′N 74°27′W*
15 V8 **New Brunswick** *Fr.* Nouveau-Brunswick. ◈ *province* SE Canada
18 K13 **Newburgh** New York, NE USA *41°30′N 74°00′W*
97 M22 **Newbury** S England, United Kingdom *51°25′N 01°20′W*
19 P10 **Newburyport** Massachusetts, NE USA *42°49′N 70°53′W*
77 T14 **New Bussa** Niger, W Nigeria *09°50′N 04°32′E*
187 O17 **New Caledonia** *var.* Kanaky, *Fr.* Nouvelle-Calédonie. ◇ *French self-governing territory of special status* SW Pacific Ocean
187 O15 **New Caledonia** *island* SW Pacific Ocean
175 O10 **New Caledonia Basin** *undersea feature* W Pacific Ocean
183 T8 **Newcastle** New South Wales, SE Australia *32°55′S 151°46′E*
13 O14 **Newcastle** New Brunswick, SE Canada *47°01′N 65°36′W*
14 I15 **Newcastle** Ontario, SE Canada *43°55′N 78°35′W*
83 K22 **Newcastle** KwaZulu/Natal, E South Africa *27°45′S 29°55′E*
97 G16 **Newcastle** *Ir.* An Caisleán Nua. SE Northern Ireland, United Kingdom *54°12′N 05°54′W*
31 P13 **New Castle** Indiana, N USA *39°56′N 85°21′W*
20 L5 **New Castle** Kentucky, S USA *38°28′N 85°10′W*
27 N11 **Newcastle** Oklahoma, C USA *35°15′N 97°36′W*
18 B13 **New Castle** Pennsylvania, NE USA *41°00′N 80°22′W*
25 R6 **Newcastle** Texas, SW USA *33°11′N 98°44′W*
36 J7 **Newcastle** Utah, W USA *37°40′N 113°31′W*
21 S6 **New Castle** Virginia, NE USA *37°31′N 80°09′W*
33 Z13 **Newcastle** Wyoming, C USA *43°52′N 104°14′W*
97 L14 **Newcastle** ✈ NE England, United Kingdom *55°03′N 01°42′W*
Newcastle *see* Newcastle upon Tyne
97 L18 **Newcastle-under-Lyme** C England, United Kingdom *53°N 02°14′W*
97 M14 **Newcastle upon Tyne** *var.* Newcastle, *hist.* Monkchester, *Lat.* Pons Aelii. NE England, United Kingdom *54°59′N 01°35′W*
181 Q4 **Newcastle Waters** Northern Territory, N Australia *17°20′S 133°26′E*
Newchwang *see* Yingkou
18 K13 **New City** New York, NE USA *41°08′N 73°57′W*
31 U13 **Newcomerstown** Ohio, N USA *40°16′N 81°36′W*
18 G15 **New Cumberland** Pennsylvania, NE USA *40°13′N 76°52′W*
21 R1 **New Cumberland** West Virginia, NE USA *40°30′N 80°35′W*
152 I10 **New Delhi** ● (India) Delhi, N India *28°35′N 77°15′E*
11 O17 **New Denver** British Columbia, SW Canada *49°58′N 117°21′W*
28 J9 **Newell** South Dakota, N USA *44°42′N 103°25′W*
21 Q13 **New Ellenton** South Carolina, SE USA *33°25′N 81°41′W*
22 J6 **Newellton** Louisiana, S USA *32°04′N 91°14′W*
28 K6 **New England** North Dakota, N USA *46°32′N 102°52′W*
19 P8 **New England** *cultural region* NE USA
New England of the West *see* Minnesota
183 U5 **New England Range** ▲▲ New South Wales, SE Australia
64 G9 **New England Seamounts** *var.* Bermuda-New England Seamount Arc. *undersea feature* W Atlantic Ocean *38°00′N 61°00′W*
38 M14 **Newenham, Cape** *headland* Alaska, USA *58°39′N 162°10′W*
Newé Zohar *see* Neve Zohar
18 D9 **Newfane** New York, NE USA *43°16′N 78°40′W*
97 M23 **New Forest** *physical region* S England, United Kingdom
13 T12 **Newfoundland** *Fr.* Terre-Neuve. *island* Newfoundland and Labrador, SE Canada
13 R9 **Newfoundland and Labrador** *Fr.* Terre Neuve. ◈ *province* E Canada
65 J8 **Newfoundland Basin** *undersea feature* NW Atlantic Ocean *45°00′N 40°00′W*
64 I8 **Newfoundland Ridge** *undersea feature* NW Atlantic Ocean
64 J8 **Newfoundland Seamounts** *undersea feature* N Sargasso Sea
18 G16 **New Freedom** Pennsylvania, NE USA *39°43′N 76°41′W*
186 K9 **New Georgia** *island* New Georgia Islands, NW Solomon Islands
186 K8 **New Georgia Islands** *island group* NW Solomon Islands
186 L8 **New Georgia Sound** *var.* The Slot. *sound* E Solomon Sea
30 L9 **New Glarus** Wisconsin, N USA *42°50′N 89°38′W*
13 Q15 **New Glasgow** Nova Scotia, SE Canada *45°36′N 62°38′W*
New Goa *see* Panaji
186 A6 **New Guinea** *Dut.* Nieuw Guinea, *Ind.* Irian. *island* Indonesia/Papua New Guinea
192 H8 **New Guinea Trench** *undersea feature* SW Pacific Ocean
32 I6 **Newhalem** Washington, NW USA *48°40′N 121°18′W*
39 P13 **Newhalen** Alaska, USA *59°43′N 154°54′W*
29 X13 **Newhall** Iowa, C USA *42°00′N 91°58′W*
14 F16 **New Hamburg** Ontario, S Canada *43°24′N 80°37′W*
19 N9 **New Hampshire** *off.* State of New Hampshire, *also known as* Granite State. ◈ *state* NE USA
29 W12 **New Hampton** Iowa, C USA *43°03′N 92°19′W*
186 G5 **New Hanover** *island* NE Papua New Guinea
97 P23 **Newhaven** SE England, United Kingdom *50°48′N 00°00′E*
18 M13 **New Haven** Connecticut, NE USA *41°18′N 72°55′W*
31 Q12 **New Haven** Indiana, N USA *41°02′N 84°59′W*
27 W5 **New Haven** Missouri, C USA *38°34′N 91°15′W*
10 K13 **New Hazelton** British Columbia, SW Canada *55°15′N 127°30′W*
New Hebrides *see* Vanuatu
175 P9 **New Hebrides Trench** *undersea feature* N Coral Sea
18 H15 **New Holland** Pennsylvania, NE USA *40°06′N 76°05′W*
22 I9 **New Iberia** Louisiana, S USA *30°00′N 91°51′W*
186 G5 **New Ireland** ◈ *province* NE Papua New Guinea
186 G5 **New Ireland** *island* NE Papua New Guinea
65 A24 **New Island** *island* W Falkland Islands
18 J15 **New Jersey** *off.* State of New Jersey, *also known as* The Garden State. ◈ *state* NE USA
18 C14 **New Kensington** Pennsylvania, NE USA *40°33′N 79°45′W*
21 W6 **New Kent** Virginia, NE USA *37°32′N 76°59′W*
27 O8 **Newkirk** Oklahoma, C USA *36°54′N 97°03′W*
21 Q9 **Newland** North Carolina, SE USA *36°04′N 81°50′W*
28 L6 **New Leipzig** North Dakota, N USA *46°21′N 101°54′W*
14 H9 **New Liskeard** Ontario, S Canada *47°31′N 79°41′W*
22 G7 **Newllano** Louisiana, S USA *31°06′N 93°16′W*
19 N13 **New London** Connecticut, NE USA *41°21′N 72°04′W*
29 Y15 **New London** Iowa, C USA *40°55′N 91°24′W*
29 T8 **New London** Minnesota, N USA *45°18′N 94°56′W*
27 V3 **New London** Missouri, C USA *39°34′N 91°24′W*
30 M7 **New London** Wisconsin, N USA *44°25′N 88°44′W*
27 Y8 **New Madrid** Missouri, C USA *36°34′N 89°32′W*
180 J8 **Newman** Western Australia *23°18′S 119°45′E*
194 M13 **Newman Island** *island* Antarctica
14 H15 **Newmarket** Ontario, S Canada *44°03′N 79°27′W*
97 P20 **Newmarket** E England, United Kingdom *52°18′N 00°28′E*
19 P10 **Newmarket** New Hampshire, NE USA *43°04′N 70°53′W*
21 U4 **New Market** Virginia, NE USA *38°39′N 78°40′W*
21 R2 **New Martinsville** West Virginia, NE USA *39°39′N 80°52′W*
31 U14 **New Matamoras** Ohio, N USA *39°32′N 81°04′W*
32 M12 **New Meadows** Idaho, NW USA *44°57′N 116°16′W*
26 R12 **New Mexico** *off.* State of New Mexico, *also known as* Land of Enchantment, Sunshine State. ◈ *state* SW USA
149 V6 **New Mīrpur** *var.* Mīrpur. Sind, SE Pakistan *33°11′N 73°46′E*
151 N15 **New Moore Island** *Island* E India
23 S4 **Newnan** Georgia, SE USA *33°22′N 84°48′W*
183 P17 **New Norfolk** Tasmania, SE Australia *42°46′S 147°02′E*
22 K9 **New Orleans** Louisiana, S USA *30°00′N 90°01′W*
22 K9 **New Orleans** ✈ Louisiana, S USA *29°59′N 90°17′W*
18 K12 **New Paltz** New York, NE USA *41°44′N 74°04′W*
31 U12 **New Philadelphia** Ohio, N USA *40°29′N 81°27′W*
184 K10 **New Plymouth** Taranaki, North Island, New Zealand *39°04′S 174°06′E*
97 M24 **Newport** S England, United Kingdom *50°42′N 01°18′W*
97 K22 **Newport** SE Wales, United Kingdom *51°35′N 03°W*
27 W10 **Newport** Arkansas, C USA *35°36′N 91°16′W*
31 N13 **Newport** Indiana, N USA *39°52′N 87°24′W*
20 M3 **Newport** Kentucky, S USA *39°05′N 84°27′W*
29 W9 **Newport** Minnesota, N USA *44°52′N 93°00′W*
32 F12 **Newport** Oregon, NW USA *44°39′N 124°04′W*
19 O13 **Newport** Rhode Island, NE USA *41°29′N 71°17′W*
21 O9 **Newport** Tennessee, S USA *35°58′N 83°13′W*
19 N6 **Newport** Vermont, NE USA *44°56′N 72°13′W*
32 M7 **Newport** Washington, NW USA *48°08′N 117°05′W*
21 X7 **Newport News** Virginia, NE USA *36°59′N 76°26′W*
97 N20 **Newport Pagnell** SE England, United Kingdom *52°05′N 00°44′W*
23 U12 **New Port Richey** Florida, SE USA *28°14′N 82°42′W*
29 V9 **New Prague** Minnesota, N USA *44°32′N 93°34′W*
44 H3 **New Providence** *island* N The Bahamas
97 I20 **New Quay** SW Wales, United Kingdom *52°13′N 04°22′W*
97 H24 **Newquay** SW England, United Kingdom *50°27′N 05°03′W*
29 V10 **New Richland** Minnesota, N USA *43°53′N 93°29′W*
15 X7 **New-Richmond** Québec, SE Canada *48°12′N 65°52′W*
31 R15 **New Richmond** Ohio, N USA *38°57′N 84°16′W*
30 I5 **New Richmond** Wisconsin, N USA *45°09′N 92°31′W*
42 G1 **New River** ≈ N Belize
55 T12 **New River** ≈ SE Guyana
21 R6 **New River** ≈ West Virginia, NE USA
42 G1 **New River Lagoon** ◎ N Belize
22 J8 **New Roads** Louisiana, S USA *30°42′N 91°26′W*
18 L14 **New Rochelle** New York, NE USA *40°55′N 73°44′W*
29 O4 **New Rockford** North Dakota, N USA *47°40′N 99°08′W*
97 P23 **New Romney** SE England, United Kingdom *50°58′N 00°56′E*
97 F20 **New Ross** *Ir.* Ros Mhic Thriúin. Wexford, SE Ireland *52°24′N 06°56′W*
97 F16 **Newry** *Ir.* An tIúr. SE Northern Ireland, United Kingdom *54°11′N 06°20′W*
28 M5 **New Salem** North Dakota, N USA *46°51′N 101°24′W*
New Sarum *see* Salisbury
29 W14 **New Sharon** Iowa, C USA *41°28′N 92°39′W*
New Siberian Islands *see* Novosibirskiye Ostrova
23 X11 **New Smyrna Beach** Florida, SE USA *29°01′N 80°55′W*
183 O7 **New South Wales** ◈ *state* SE Australia
39 O13 **New Stuyahok** Alaska, USA *59°27′N 157°18′W*
21 N8 **New Tazewell** Tennessee, S USA *36°26′N 83°36′W*
152 K9 **New Tehri** *prev.* Tehri. Uttarakhand, N India *30°12′N 78°29′E*
38 M12 **Newtok** Alaska, USA *60°56′N 164°37′W*
23 S7 **Newton** Georgia, SE USA *31°18′N 84°20′W*
29 W14 **Newton** Iowa, C USA *41°42′N 93°03′W*
27 N6 **Newton** Kansas, C USA *38°02′N 97°22′W*
19 O11 **Newton** Massachusetts, NE USA *42°19′N 71°10′W*
22 M5 **Newton** Mississippi, S USA *32°19′N 89°09′W*
18 J14 **Newton** New Jersey, NE USA *41°03′N 74°45′W*
21 R9 **Newton** North Carolina, SE USA *35°42′N 81°14′W*
25 Y9 **Newton** Texas, SW USA *30°51′N 93°45′W*
97 J24 **Newton Abbot** SW England, United Kingdom *50°33′N 03°34′W*
96 K13 **Newton St Boswells** SE Scotland, United Kingdom *55°34′N 02°40′W*
97 I14 **Newton Stewart** S Scotland, United Kingdom *54°58′N 04°30′W*
92 O2 **Newtontoppen** ▲ C Svalbard *78°57′N 17°34′E*
97 J20 **Newtown** E Wales, United Kingdom *52°32′N 03°19′W*
28 K3 **New Town** North Dakota, N USA *47°58′N 102°30′W*

◆ Country
● Country Capital
◇ Dependent Territory
○ Dependent Territory Capital
◈ Administrative Regions
✈ International Airport
▲ Mountain
▲▲ Mountain Range
Volcano
≈ River
◎ Lake
⊠ Reservoir

NEWTOWNABBEY – NOMAD

97 G15 **Newtownabbey** Ir. Baile na Mainistreach. E Northern Ireland, United Kingdom 54°40′N 05°57′W
97 G15 **Newtownards** Ir. Baile Nua na hArda. SE Northern Ireland, United Kingdom 54°36′N 05°41′W
29 U10 **New Ulm** Minnesota, N USA 44°20′N 94°28′W
28 K10 **New Underwood** South Dakota, N USA 44°05′N 102°46′W
25 V10 **New Waverly** Texas, SW USA 30°32′N 95°28′W
18 K14 **New York** New York, NE USA 40°45′N 73°57′W
18 G10 **New York** ◐ state NE USA
35 X13 **New York Mountains** ▲▲ California, W USA
184 K12 **New Zealand** ◆ commonwealth republic SW Pacific Ocean
95 M24 **Nexø** var. Neksø Bornholm. E Denmark 55°04′N 15°09′E
125 O15 **Neya** Kostromskaya Oblast', NW Russian Federation 58°19′N 43°51′E
Neyba see Neiba
143 Q12 **Neyrīz** var. Neiriz, Niriz. Fārs, S Iran 29°14′N 54°18′E
143 T4 **Neyshābūr** var. Nishapur. Khorāsān-Razavī, NE Iran 36°15′N 58°47′E
155 J21 **Neyveli** Tamil Nādu, SE India 11°36′N 79°26′E
Nezhin see Nizhyn
33 N10 **Nezperce** Idaho, NW USA 46°14′N 116°15′W
22 H8 **Nezpique, Bayou** ≈ Louisiana, S USA
77 Y13 **Ngadda** ≈ NE Nigeria
N'Gage see Negage
185 G16 **Ngahere** West Coast, South Island, New Zealand 42°22′S 171°29′E
77 Z12 **Ngala** Borno, NE Nigeria 12°19′N 14°11′E
158 K16 **Ngamring** Xizang Zizhiqu, W China 29°16′N 87°10′E
81 K19 **Ngangerabeli Plain** plain SE Kenya
158 I14 **Ngangla Ringco** ◎ W China
158 H13 **Nganglong Kangri** ▲ W China 32°55′N 81°00′E
158 K15 **Ngangzê Co** ◎ W China
79 F14 **Ngaoundéré** var. N'Gaoundéré. Adamaoua, N Cameroon 07°20′N 13°35′E
N'Gaoundéré see Ngaoundéré
81 E20 **Ngara** Kagera, NW Tanzania 02°30′S 30°40′E
188 F8 **Ngardmau Bay** bay Babeldaob, N Palau
188 F7 **Ngaregur** island Palau Islands, N Palau
Ngarrab see Gyaca
184 L7 **Ngaruawahia** Waikato, North Island, New Zealand 37°41′S 175°10′E
184 N11 **Ngaruroro** ≈ North Island, New Zealand
190 I16 **Ngatangiia** Rarotonga, S Cook Islands 21°14′S 159°44′W
184 M6 **Ngatea** Waikato, North Island, New Zealand 37°16′S 175°29′E
166 L8 **Ngathainggyaung** Ayeyawady, SW Myanmar (Burma) 17°22′N 95°04′E
Ngatik see Ngetik Atoll
Ngau see Gau
Ngawa see Aba
172 G12 **Ngazidja** Fr. Grande Comore, var. Njazidja. island NW Comoros
188 C7 **Ngcheangel** var. Kayangel Islands. island Palau Islands, N Palau
188 E10 **Ngchemiangel** Babeldaob, N Palau
188 C8 **Ngeaur** var. Angaur. island Palau Islands, S Palau
188 E10 **Ngerkeai** Babeldaob, N Palau
188 F9 **Ngermechau** Babeldaob, N Palau 07°35′N 134°39′E
188 C8 **Ngeruktabel** prev. Urukthapel. island Palau Islands, S Palau
188 F8 **Ngetbong** Babeldaob, N Palau 07°37′N 134°35′E
189 T17 **Ngetik Atoll** var. Ngatik; prev. Los Jardines. atoll Caroline Islands, E Micronesia
188 E10 **Ngetkip** Babeldaob, N Palau
Nghia Dan see Thai Hoa
N'Giva see Ondjiva
79 G20 **Ngo** Plateaux, SE Congo 02°28′S 15°43′E
167 S7 **Ngoc Lăc** Thanh Hoa, N Vietnam 20°06′N 105°21′E
79 G17 **Ngoko** ≈ Cameroon/Congo
81 H19 **Ngorengore** Narok, SW Kenya 01°01′S 35°26′E
159 Q11 **Ngoring Hu** ◎ C China
Ngorolaka see Banifing
81 H20 **Ngorongoro Crater** crater N Tanzania
79 D19 **Ngounié** off. Province de la Ngounié, var. La Ngounié. ◐ province S Gabon
79 D19 **Ngounié** ≈ Congo/Gabon
Ngounié, Province de la see Ngounié
78 H10 **Ngoura** var. NGoura. Hadjer-Lamis, W Chad 12°52′N 16°27′E
NGoura see Ngoura
78 G10 **Ngouri** var. NGouri; prev. Fort-Millot. Lac, W Chad 13°42′N 15°19′E
NGouri see Ngouri
77 Y10 **Ngourti** Diffa, E Niger 15°22′N 13°13′E
77 Y11 **Nguigmi** var. N'Guigmi. Diffa, SE Niger 14°17′N 13°07′E
N'Guigmi see Nguigmi
Nguimbo see Lumbala N'Guimbo
188 F15 **Ngulu Atoll** atoll Caroline Islands, W Micronesia
187 R14 **Nguna** island C Vanuatu
N'Gunza see Sumbe
169 U17 **Ngurah Rai** ✕ (Bali) Bali, S Indonesia 8°40′S 115°14′E
77 W12 **Nguru** Yobe, NE Nigeria 12°55′N 10°31′E
Ngwaketze see Southern
83 I16 **Ngweze** ≈ S Zambia
83 M17 **Nhamatanda** Sofala, C Mozambique 19°16′S 34°10′E
58 G12 **Nhamundá, Rio** var. Jamundá, Yamundá. ≈ N Brazil
60 J7 **Nhandeara** São Paulo, S Brazil 20°40′S 50°03′W
82 D12 **Nharêa** var. N'Harea, Nhareia. Bié, W Angola 11°38′S 16°58′E
N'Harea see Nharêa
Nhareia see Nharêa
167 V12 **Nha Trang** Khanh Hoa, S Vietnam 12°15′N 109°10′E
182 L11 **Nhill** Victoria, SE Australia 36°21′S 141°38′E
83 L22 **Nhlangano** prev. Goedgegun. SW Swaziland 27°06′S 31°12′E
181 S1 **Nhulunbuy** Northern Territory, N Australia 12°16′S 136°46′E
77 N10 **Niafounké** Tombouctou, W Mali 15°54′N 03°58′W
31 N5 **Niagara** Wisconsin, N USA 45°45′N 87°57′W
14 H16 **Niagara** ≈ Ontario, S Canada
14 G15 **Niagara Escarpment** hill range Ontario, S Canada
14 H16 **Niagara Falls** Ontario, S Canada 43°05′N 79°06′W
18 D9 **Niagara Falls** New York, NE USA 43°06′N 79°04′W
14 H16 **Niagara Falls** waterfall Canada/USA
76 K12 **Niagassola** var. Nyagassola. Haute-Guinée, NE Guinea 12°24′N 09°03′W
77 R12 **Niamey** ● (Niger) Niamey, SW Niger 13°28′N 02°03′E
77 R12 **Niamey** ✕ Niamey, SW Niger 13°28′N 02°14′E
77 R14 **Niamtougou** N Togo 09°50′N 01°08′E
79 O16 **Niangara** Orientale, NE Dem. Rep. Congo 03°45′N 27°54′E
77 O10 **Niangay, Lac** ◎ E Mali
77 N14 **Niangoloko** SW Burkina Faso 10°15′N 04°53′W
27 U6 **Niangua River** ≈ Missouri, C USA
79 O17 **Nia-Nia** Orientale, NE Dem. Rep. Congo 01°26′N 27°38′E
19 N13 **Niantic** Connecticut, NE USA 41°19′N 72°11′W
163 U7 **Nianzishan** Heilongjiang, NE China 47°31′N 122°53′E
79 E20 **Niari** ◐ province SW Congo
168 H10 **Nias, Pulau** island W Indonesia
82 O13 **Niassa** off. Província do Niassa. ◐ province N Mozambique
Niassa, Província do see Niassa
191 U10 **Niau** island Îles Tuamotu, C French Polynesia
95 G20 **Nibe** Nordjylland, N Denmark 56°59′N 09°39′E
189 Q8 **Nibok** N Nauru 0°31′S 166°55′E
118 C10 **Nīca** W Latvia 56°21′N 21°03′E
Nicaea see Nice
42 J9 **Nicaragua** off. Republic of Nicaragua. ◆ republic Central America
42 K11 **Nicaragua, Lago de** var. Cocibolca, Gran Lago, Eng. Lake Nicaragua. ◎ S Nicaragua
Nicaragua, Lake see Nicaragua, Lago de
64 D11 **Nicaraguan Rise** undersea feature NW Caribbean Sea 16°00′N 80°00′W
Nicaragua, Republic of see Nicaragua
Nicaria see Ikaría
107 N21 **Nicastro** Calabria, SW Italy 38°59′N 16°20′E
103 V15 **Nice** It. Nizza; anc. Nicaea. Alpes-Maritimes, SE France 43°43′N 07°13′E
Nice see Côte d'Azur
Nicephorium see Ar Raqqah
12 M9 **Nichicun, Lac** ◎ Québec, E Canada
164 D16 **Nichinan** var. Nitinan. Miyazaki, Kyūshū, SW Japan 31°36′N 131°23′E
44 E4 **Nicholas Channel** channel N Cuba
Nicholas II Land see Severnaya Zemlya
149 U2 **Nicholas Range** Pash. Selselehye Kuhe Vākhān, Taj. Qatorkŭhi Vakhon. ▲▲ Afghanistan/Tajikistan
20 M6 **Nicholasville** Kentucky, S USA 37°52′N 84°34′W
44 G2 **Nicholls Town** Andros Island, NW The Bahamas 25°07′N 78°01′W
21 U12 **Nichols** South Carolina, SE USA 34°13′N 79°09′W
55 U9 **Nickerie** ◐ district NW Suriname
55 V9 **Nickerie Rivier** ≈ NW Suriname
151 P22 **Nicobar Islands** island group India, E Indian Ocean
116 L9 **Nicolae Bălcescu** Botoşani, NE Romania 47°33′N 26°52′E
15 P11 **Nicolet** Québec, SE Canada 46°13′N 72°37′W
15 Q12 **Nicolet** ≈ Québec, SE Canada
31 Q4 **Nicolet, Lake** ◎ Michigan, N USA
29 U10 **Nicollet** Minnesota, N USA 44°16′N 94°11′W
61 F19 **Nico Pérez** Florida, S Uruguay 33°30′S 55°10′W
Nicopolis see Nikopol, Bulgaria
Nicopolis see Nikópoli, Greece
121 P2 **Nicosia** Gk. Lefkosía, Turk. Lefkoşa. ● (Cyprus) C Cyprus 35°10′N 33°23′E
107 K24 **Nicosia** Sicilia, Italy, C Mediterranean Sea 37°45′N 14°24′E
107 N22 **Nicotera** Calabria, SW Italy 38°33′N 15°55′E
42 K13 **Nicoya** Guanacaste, W Costa Rica 10°09′N 85°26′W
42 L14 **Nicoya, Golfo de** gulf W Costa Rica
42 L14 **Nicoya, Península de** peninsula NW Costa Rica
Nictheroy see Niterói
118 B12 **Nida** Ger. Nidden. Klaipėda, SW Lithuania 55°20′N 21°00′E
111 L15 **Nida** ≈ S Poland
Nidaros see Trondheim
108 D8 **Nidau** Bern, W Switzerland 47°07′N 07°15′E
101 H17 **Nidda** ≈ W Germany
Nidden see Nida
95 F17 **Nidelva** ≈ S Norway
108 F9 **Nidwalden** ◐ canton C Switzerland
110 L9 **Nidzica** Ger. Niedenburg. Warmińsko-Mazurskie, NE Poland 53°22′N 20°27′E
100 H6 **Niebüll** Schleswig-Holstein, N Germany 54°47′N 08°51′E
Niedenburg see Nidzica
99 N25 **Niederanven** Luxembourg, C Luxembourg 49°39′N 06°15′E
103 V4 **Niederbronn-les-Bains** Bas-Rhin, NE France 48°57′N 07°37′E
Niederdonau see Niederösterreich
109 S7 **Niedere Tauern** ▲▲ C Austria
101 P14 **Niederlausitz** Eng. Lower Lusatia, Lus. Donja Łužica. physical region E Germany
109 U5 **Niederösterreich** off. Land Niederösterreich, Eng. Lower Austria, Ger. Niederdonau; prev. Lower Danube. ◐ state NE Austria
Niederösterreich, Land see Niederösterreich
100 G12 **Niedersachsen** Eng. Lower Saxony, Fr. Basse-Saxe. ◐ state NW Germany
79 D17 **Niefang** var. Sevilla de Niefang. NW Equatorial Guinea 01°52′N 10°12′E
83 G23 **Niekerkshoop** Northern Cape, W South Africa 29°21′S 22°49′E
99 G17 **Niel** Antwerpen, N Belgium 51°07′N 04°20′E
Niélé see Niellé
76 M14 **Niellé** var. Niélé. N Ivory Coast 10°12′N 05°38′W
79 O22 **Niemba** Katanga, E Dem. Rep. Congo 05°58′S 28°24′E
111 G15 **Niemcza** Ger. Nimptsch. Dolnośląskie, SW Poland 50°45′N 16°52′E
Niemen see Neman
92 J13 **Niemisel** Norrbotten, N Sweden 66°00′N 22°00′E
111 H15 **Niemodlin** Ger. Falkenberg. Opolskie, SW Poland 50°37′N 17°45′E
76 M13 **Niéna** Sikasso, SW Mali 11°24′N 06°20′W
100 H12 **Nienburg** Niedersachsen, N Germany 52°37′N 09°12′E
100 N13 **Nieplitz** ≈ NE Germany
111 L16 **Niepołomice** Malopolskie, S Poland 50°02′N 20°12′E
101 D14 **Niers** ≈ Germany/Netherlands
101 Q15 **Niesky** Lus. Niska. Sachsen, E Germany 51°16′N 14°49′E
Nieśwież see Nyasvizh
Nieuport see Nieuwpoort
98 O8 **Nieuw-Amsterdam** Drenthe, NE Netherlands 52°43′N 06°52′E
55 W9 **Nieuw Amsterdam** Commewijne, NE Suriname 05°53′N 55°05′W
99 M14 **Nieuw-Bergen** Limburg, SE Netherlands 51°36′N 06°04′E
98 O7 **Nieuw-Buinen** Drenthe, NE Netherlands 52°57′N 06°55′E
98 J12 **Nieuwegein** Utrecht, C Netherlands 52°03′N 05°06′E
98 P6 **Nieuwe Pekela** Groningen, NE Netherlands 53°04′N 06°58′E
98 P5 **Nieuweschans** Groningen, NE Netherlands 53°10′N 07°10′E
Nieuw Guinea see New Guinea
98 I11 **Nieuwkoop** Zuid-Holland, C Netherlands 52°09′N 04°46′E
98 M9 **Nieuwleusen** Overijssel, S Netherlands 52°34′N 06°16′E
98 J11 **Nieuw-Loosdrecht** Noord-Holland, C Netherlands 52°12′N 05°08′E
55 U9 **Nieuw Nickerie** Nickerie, NW Suriname 05°56′N 57°W
98 P5 **Nieuwolda** Groningen, NE Netherlands 53°15′N 06°58′E
99 B17 **Nieuwpoort** var. Nieuport. West-Vlaanderen , W Belgium 51°08′N 02°45′E
99 G14 **Nieuw-Vossemeer** Noord-Brabant, S Netherlands 51°34′N 04°13′E
98 P7 **Nieuw-Weerdinge** Drenthe, NE Netherlands 52°51′N 07°00′E
40 L10 **Nieves** Zacatecas, C Mexico 24°00′N 102°57′W
64 O11 **Nieves, Pico de las** ▲ Gran Canaria, Islas Canarias, Spain, NE Atlantic Ocean 27°58′N 15°34′W
103 P8 **Nièvre** ◐ department C France
Niewenstat see Neustadt an der Weinstrasse
136 J15 **Niğde** Niğde, C Turkey 37°58′N 34°42′E
136 J15 **Niğde** ◐ province C Turkey
83 J21 **Nigel** Gauteng, NE South Africa 26°25′S 28°28′E
77 V10 **Niger** off. Republic of Niger. ◆ republic W Africa
77 T14 **Niger** ◐ state C Nigeria
67 P8 **Niger** ≈ W Africa
Niger Cone see Niger Fan
67 P9 **Niger Delta** delta S Nigeria
67 P9 **Niger Fan** var. Niger Cone. undersea feature E Atlantic Ocean 04°15′N 05°00′E
77 T13 **Nigeria** off. Federal Republic of Nigeria. ◆ federal republic W Africa
Nigeria, Federal Republic of see Nigeria
77 T17 **Niger, Mouths of the** delta S Nigeria
Niger, Republic of see Niger
185 C24 **Nightcaps** Southland, South Island, New Zealand 45°58′S 168°03′E
14 F7 **Night Hawk Lake** ◎ Ontario, S Canada
65 M19 **Nightingale Island** island S Tristan da Cunha, S Atlantic Ocean
38 M12 **Nightmute** Alaska, USA 60°28′N 164°43′W
114 G13 **Nigríta** Kentrikí Makedonía, NE Greece 40°54′N 23°29′E
148 J15 **Nihing** Per. Rūd-e Nahang. ≈ Iran/Pakistan
191 V10 **Nihiru** atoll Îles Tuamotu, C French Polynesia
Nihommatsu see Nihonmatsu
Nihon see Japan
165 P11 **Nihonmatsu** var. Nihommatsu, Nihonmatu. Fukushima, Honshū, C Japan 37°34′N 140°25′E
Nihonmatu see Nihonmatsu
62 I12 **Nihuil, Embalse del** ◎ W Argentina
165 O10 **Niigata** Niigata, Honshū, C Japan 37°55′N 139°01′E
165 O11 **Niigata** off. Niigata-ken. ◐ prefecture Honshū, C Japan
Niigata-ken see Niigata
165 G14 **Niihama** Ehime, Shikoku, SW Japan 33°57′N 133°15′E
38 A8 **Ni'ihau** var. Niihau. island Hawai'i, USA, C Pacific Ocean
165 X12 **Nii-jima** island E Japan
165 H12 **Niimi** Okayama, Honshū, SW Japan 35°00′N 133°27′E
165 O10 **Niitsu** var. Niitu. Niigata, Honshū, C Japan 37°48′N 139°09′E
Niitu see Niitsu
105 P15 **Níjar** Andalucía, S Spain 36°57′N 02°13′W
98 K11 **Nijkerk** Gelderland, C Netherlands 52°13′N 05°30′E
99 H16 **Nijlen** Antwerpen, N Belgium 51°10′N 04°40′E
98 L13 **Nijmegen** Ger. Nimwegen; anc. Noviomagus. Gelderland, SE Netherlands 51°50′N 05°52′E
98 N10 **Nijverdal** Overijssel, E Netherlands 52°22′N 06°28′E
190 G16 **Nikao** Rarotonga, S Cook Islands
Nikaria see Ikaría
124 I2 **Nikel'** Finn. Kolosjoki. Murmanskaya Oblast', NW Russian Federation 69°25′N 30°12′E
171 Q17 **Nikiniki** Timor, S Indonesia 10°S 124°30′E
129 Q15 **Nikitin Seamount** undersea feature E Indian Ocean 05°48′S 84°48′E
77 S14 **Nikki** E Benin 09°55′N 03°12′E
Niklasmarkt see Gheorgheni
39 P10 **Nikolai** Alaska, USA 63°00′N 154°22′W
Nikolaiken see Mikołajki
Nikolainkaupunki see Vaasa
Nikolayev see Mykolayiv
145 O6 **Nikolayevka** Severnyy Kazakhstan, N Kazakhstan
Nikolayevka see Zhetigen
127 P9 **Nikolayevsk** Volgogradskaya Oblast', SW Russian Federation 50°03′N 45°30′E
Nikolayevskaya Oblast' see Mykolayivs'ka Oblast'
123 S12 **Nikolayevsk-na-Amure** Khabarovskiy Kray, SE Russian Federation 53°04′N 140°39′E
127 P6 **Nikol'sk** Penzenskaya Oblast', W Russian Federation 53°46′N 46°03′E
125 O13 **Nikol'sk** Vologodskaya Oblast', NW Russian Federation 59°35′N 45°31′E
Nikol'sk see Ussuriysk
38 K17 **Nikolski** Umnak Island, Alaska, USA 52°56′N 168°52′W
Nikol'skiy see Satpayev
127 V7 **Nikol'skoye** Orenburgskaya Oblast', W Russian Federation 52°01′N 55°48′E
Nikol'sk-Ussuriyskiy see Ussuriysk
114 J7 **Nikopol** anc. Nicopolis. Pleven, N Bulgaria 43°43′N 24°55′E
117 S9 **Nikopol'** Dnipropetrovs'ka Oblast', SE Ukraine 47°34′N 34°23′E
115 C17 **Nikópoli** anc. Nicopolis. site of ancient city Ípeiros, W Greece
136 M12 **Niksar** Tokat, N Turkey 40°35′N 36°59′E
143 V14 **Nīkshahr** Sīstān va Balūchestān, SE Iran 26°15′N 60°10′E
113 J16 **Nikšić** C Montenegro 42°47′N 18°56′E
191 R4 **Nikumaroro** ; prev. Gardner Island. atoll Phoenix Islands, C Kiribati
191 P3 **Nikunau** var. Nukunau; prev. Byron Island. atoll Tungaru, W Kiribati
155 G21 **Nilambūr** Kerala, SW India 11°17′N 76°15′E
35 X16 **Niland** California, W USA 33°14′N 115°31′W
80 G8 **Nile** former province NW Uganda
67 T3 **Nile** Ar. Nahr an Nīl. ≈ N Africa
75 W7 **Nile Delta** delta N Egypt
67 T3 **Nile Fan** undersea feature E Mediterranean Sea 33°00′N 31°00′E
31 O11 **Niles** Michigan, N USA 41°49′N 86°15′W
31 V11 **Niles** Ohio, N USA 41°10′N 80°46′W
155 F20 **Nīleswaram** Kerala, SW India 12°18′N 75°07′E
14 K10 **Nilgaut, Lac** ◎ Québec, SE Canada
149 O6 **Nīlī** Dāykundī, C Afghanistan 33°43′N 66°07′E
158 I5 **Nilka** Xinjiang Uygur Zizhiqu, NW China 43°46′N 82°33′E
Nīl, Nahr an see Nile
93 N16 **Nilsiä** Pohjois-Savo, C Finland 63°13′N 28°00′E
154 F9 **Nīmach** Madhya Pradesh, C India 24°27′N 74°56′E
152 G14 **Nimbāhera** Rājasthān, N India 24°38′N 74°45′E
76 L15 **Nimba, Monts** var. Nimba Mountains. ▲▲ W Africa
Nimba Mountains see Nimba, Monts
Nimburg see Nymburk
103 Q15 **Nîmes** anc. Nemausus, Nismes. Gard, S France 43°49′N 04°20′E
152 H11 **Nīm ka Thāna** Rājasthān, N India 27°42′N 75°50′E
183 R11 **Nimmitabel** New South Wales, SE Australia 36°34′S 149°18′E
Nimptsch see Niemcza
195 R11 **Nimrod Glacier** glacier Antarctica
148 K8 **Nīmrōz** var. Nimroze; prev. Chakhānsūr, Nīmrūz. ◐ province SW Afghanistan
Nimroze see Nīmrōz
Nīmrūz see Nīmrōz
81 F16 **Nimule** Eastern Equatoria, S South Sudan 03°35′N 32°03′E
Nimwegen see Nijmegen
139 Q3 **Nīnawá** off. Muḩāfa at Nīnawá, var. Al Mawşil, Nineveh. ◐ governorate NW Iraq
Nīnawá, Muḩāfa at see Nīnawá
155 C23 **Nine Degree Channel** channel India/Maldives
18 G9 **Ninemile Point** headland New York, NE USA 43°31′N 76°22′W
173 S8 **Ninetyeast Ridge** undersea feature E Indian Ocean 04°00′S 90°00′E
183 P13 **Ninety Mile Beach** beach Victoria, SE Australia
184 I2 **Ninety Mile Beach** beach North Island, New Zealand
21 P12 **Ninety Six** South Carolina, SE USA 34°10′N 82°01′W
Nineveh see Nīnawá
163 Y9 **Ning'an** Heilongjiang, NE China 44°20′N 129°28′E
161 S9 **Ningbo** var. Ning-po, Yin-hsien; prev. Ninghsien. Zhejiang, SE China 29°54′N 121°33′E
161 U12 **Ningde** Fujian, SE China 26°48′N 119°33′E
161 P12 **Ningdu** var. Meijiang. Jiangxi, S China 26°28′N 115°53′E
Ning'er see Pu'er
186 A7 **Ningerum** Western, SW Papua New Guinea 05°40′S 141°10′E
161 R9 **Ningguo** Anhui, E China 30°33′N 118°58′E
161 S9 **Ninghai** Zhejiang, SE China 29°18′N 121°26′E
Ning-hsia see Ningxia
Ninghsien see Ningbo
160 J15 **Ningming** var. Chengzhong. Guangxi Zhuangzu Zizhiqu, S China 22°07′N 106°43′E
160 H11 **Ningnan** var. Pisha. Sichuan, C China 26°59′N 102°49′E
Ning-po see Ningbo
Ningsia/Ningsia Hui/Ningsia Hui Autonomous Region see Ningxia
160 J5 **Ningxia** off. Ningxia Huizu Zizhiqu, var. Ning-hsia, Ningsia, Eng. Ningsia Hui, Ningsia Hui Autonomous Region. ◐ autonomous region N China
Ningxia Huizu Zizhiqu see Ningxia
159 X10 **Ningxian** var. Xinning. Gansu, N China 35°30′N 108°05′E
167 T7 **Ninh Binh** Ninh Binh, N Vietnam 20°14′N 106°00′E
167 V12 **Ninh Hoa** Khanh Hoa, S Vietnam 12°28′N 109°07′E
186 C4 **Ninigo Group** island group N Papua New Guinea
39 Q12 **Ninilchik** Alaska, USA 60°03′N 151°40′W
27 N7 **Ninnescah River** ≈ Kansas, C USA
195 U16 **Ninnis Glacier** glacier Antarctica
165 R8 **Ninohe** Iwate, Honshū, C Japan 40°16′N 141°18′E
99 F18 **Ninove** Oost-Vlaanderen, C Belgium 50°50′N 04°02′E
171 O4 **Ninoy Aquino** ✕ (Manila) Luzon, N Philippines 14°26′N 121°00′E
Nio see Íos
29 P12 **Niobrara** Nebraska, C USA 42°43′N 97°59′W
28 M12 **Niobrara River** ≈ Nebraska/Wyoming, C USA
79 I20 **Nioki** Bandundu, W Dem. Rep. Congo 02°44′S 17°42′E
76 M11 **Niono** Ségou, C Mali 14°18′N 05°59′W
76 K11 **Nioro** var. Nioro du Sahel. Kayes, W Mali 15°13′N 09°39′W
76 G11 **Nioro du Rip** SW Senegal 13°44′N 15°48′W
Nioro du Sahel see Nioro
102 K10 **Niort** Deux-Sèvres, W France 46°21′N 00°25′W
172 H14 **Nioumachoua** Mohéli, S Comoros 12°21′S 43°43′E
186 C7 **Nipa** Southern Highlands, W Papua New Guinea 06°11′S 143°27′E
11 U14 **Nipawin** Saskatchewan, S Canada 52°23′N 104°01′W
12 D12 **Nipigon** Ontario, S Canada 49°02′N 88°15′W
12 D11 **Nipigon, Lake** ◎ Ontario, S Canada
11 S13 **Nipin** ≈ Saskatchewan, C Canada
14 G11 **Nipissing, Lake** ◎ Ontario, S Canada
35 P13 **Nipomo** California, W USA 35°02′N 120°28′W
Nippon see Japan
138 K6 **Niqniqīyah, Jabal an** ▲▲ C Syria
62 I9 **Niquivil** San Juan, W Argentina 30°25′S 68°42′W
171 Y13 **Nirabotong** Papua, E Indonesia 02°35′S 140°08′E
Niriz see Neyrīz
163 U7 **Nirji** var. Morin Dawa Daurzu Zizhiqi. Nei Mongol Zizhiqu, N China 48°21′N 124°32′E
155 I14 **Nirmal** Telangana, C India 19°04′N 78°21′E
153 Q13 **Nirmāli** Bihār, NE India 26°18′N 86°35′E
113 O14 **Niš** Eng. Nish, Ger. Nisch; anc. Naissus. Serbia, SE Serbia 43°21′N 21°53′E
104 H9 **Nisa** Portalegre, C Portugal 39°31′N 07°39′W
Nisa see Neisse
141 P4 **Nişab** Al Ḩudūd ash Shamālīyah, N Saudi Arabia 29°11′N 44°43′E
141 Q15 **Nişāb** var. Anşāb. SW Yemen 14°24′N 46°47′E
113 P14 **Nišava** Bul. Nishava. ≈ Bulgaria/Serbia see also Nishava
Nišava see Nishava
107 K25 **Niscemi** Sicilia, Italy, C Mediterranean Sea 37°09′N 14°23′E
Nisch/Nish see Niš
165 R4 **Niseko** Hokkaidō, NE Japan 42°50′N 140°43′E
Nishapur see Neyshābūr
114 G9 **Nishava** var. Nišava. ≈ Bulgaria/Serbia see also Nišava
Nishava see Nišava
118 L11 **Nishcha** ≈ N Belarus
165 C17 **Nishinoomote** Kagoshima, Tanega-shima, SW Japan 30°42′N 130°59′E
165 X15 **Nishino-shima** Eng. Rosario. island Ogasawara-shotō, SE Japan
165 I13 **Nishiwaki** var. Nisiwaki. Hyōgo, Honshū, SW Japan 34°59′N 134°58′E
141 U14 **Nishtūn** SE Yemen 15°47′N 52°08′E
Nisibin see Nusaybin
Nisiros see Nísyros
Nisiwaki see Nishiwaki
Niska see Niesky
113 O14 **Niška Banja** Serbia, SE Serbia 43°18′N 22°01′E
12 D6 **Niskibi** ≈ Ontario, C Canada
111 O15 **Nisko** Podkrapackie, SE Poland 50°31′N 22°09′E
10 H7 **Nisling** ≈ Yukon, W Canada
99 H22 **Nismes** Namur, S Belgium 50°04′N 04°31′E
Nismes see Nîmes
116 M10 **Nisporeni** Rus. Nisporeny. W Moldova 47°04′N 28°10′E
Nisporeny see Nisporeni
95 K20 **Nissan** ≈ S Sweden
Nissan Islands see Green Islands
95 F16 **Nisser** ◎ S Norway
95 E21 **Nissum Bredning** inlet NW Denmark
29 U6 **Nisswa** Minnesota, N USA 46°31′N 94°17′W
Nistru see Dniester
115 M22 **Nísyros** var. Nisiros. island Dodekánisa, Greece, Aegean Sea
118 H8 **Nītaure** C Latvia 57°05′N 25°12′E
60 P10 **Niterói** prev. Nictheroy. Rio de Janeiro, SE Brazil 22°54′S 43°06′W
14 F16 **Nith** ≈ Ontario, S Canada
96 J13 **Nith** ≈ S Scotland, United Kingdom
Nitinan see Nichinan
111 I21 **Nitra** Ger. Neutra, Hung. Nyitra. Nitriansky Kraj, SW Slovakia 48°20′N 18°05′E
111 I20 **Nitra** Ger. Neutra, Hung. Nyitra. ≈ W Slovakia
111 I21 **Nitriansky Kraj** ◐ region SW Slovakia
21 Q5 **Nitro** West Virginia, NE USA 38°24′N 81°50′W
Niuatobutabu see Niuatoputapu
193 X13 **Niuatoputapu** var. Niuatobutabu; prev. Keppel Island. island N Tonga
193 U15 **Niu'Aunofa** headland Tongatapu, S Tonga 21°03′S 175°19′W
Niuchwang see Yingkou
190 B16 **Niue** ◇ self-governing territory in free association with New Zealand S Pacific Ocean
190 F10 **Niulakita** var. Nurakita. atoll S Tuvalu
190 E6 **Niutao** atoll NW Tuvalu
93 L15 **Nivala** Pohjois-Pohjanmaa, C Finland 63°56′N 25°00′E
102 I15 **Nive** ≈ SW France
99 G19 **Nivelles** Walloon Brabant, C Belgium 50°36′N 04°04′E
103 P8 **Nivernais** cultural region C France
15 N8 **Niverville, Lac** ◎ Québec, SE Canada
27 T7 **Nixa** Missouri, C USA 37°02′N 93°17′W
35 R5 **Nixon** Nevada, W USA 39°48′N 119°24′W
25 S12 **Nixon** Texas, SW USA 29°16′N 97°45′W
Niya see Minfeng
Niyazov see Nyýazow
155 H14 **Nizāmābād** Telangana, C India 18°40′N 78°05′E
155 H15 **Nizām Sāgar** ◎ C India
125 N16 **Nizhegorodskaya Oblast'** ◐ province W Russian Federation
Nizhnegorskiy see Nyzhn'ohirs'kyy
127 S4 **Nizhnekamsk** Respublika Tatarstan, W Russian Federation 55°36′N 51°45′E
127 U3 **Nizhnekamskoye Vodokhranilishche** ⊠ W Russian Federation
123 S14 **Nizhneleninskoye** Yevreyskaya Avtonomnaya Oblast', SE Russian Federation 47°50′N 132°30′E
122 L13 **Nizhneudinsk** Irkutskaya Oblast', S Russian Federation 54°48′N 98°51′E
122 I10 **Nizhnevartovsk** Khanty-Mansiyskiy Avtonomnyy Okrug-Yugra, C Russian Federation 60°57′N 76°40′E
123 Q7 **Nizhneyansk** Respublika Sakha (Yakutiya), NE Russian Federation 71°25′N 135°59′E
127 Q11 **Nizhniy Baskunchak** Astrakhanskaya Oblast', SW Russian Federation 48°15′N 46°49′E
127 O6 **Nizhniy Lomov** Penzenskaya Oblast', W Russian Federation 53°32′N 43°39′E
127 P3 **Nizhniy Novgorod** prev. Gor'kiy. Nizhegorodskaya Oblast', W Russian Federation 56°17′N 44°E
125 T8 **Nizhniy Odes** Respublika Komi, NW Russian Federation 63°42′N 54°59′E
Nizhniy Pyandzh see Panji Poyon
122 G10 **Nizhniy Tagil** Sverdlovskaya Oblast', C Russian Federation 57°57′N 59°51′E
125 T9 **Nizhnyaya-Omra** Respublika Komi, NW Russian Federation 62°46′N 55°54′E
125 P5 **Nizhnyaya Pësha** Nenetskiy Avtonomnyy Okrug, NW Russian Federation 66°54′N 47°37′E
117 Q3 **Nizhyn** Rus. Nezhin. Chernihivs'ka Oblast', NE Ukraine 51°03′N 31°54′E
136 M17 **Nizip** Gaziantep, S Turkey 37°02′N 37°47′E
141 X8 **Nizwa** var. Nazwāh. NE Oman 23°00′N 57°50′E
Nizza see Nice
106 C9 **Nizza Monferrato** Piemonte, NE Italy 44°47′N 08°22′E
Njávdám see Näätämöjoki
Njazidja see Ngazidja
Njellim see Nellim
81 H24 **Njombe** Njombe, S Tanzania 09°20′S 34°47′E
81 H25 **Njombe** off. Mkoa wa Njombe. ◐ region S Tanzania
81 G23 **Njombe** ≈ C Tanzania
Njombe, Mkoa wa see Njombe
Njuk, Ozero see Nyuk, Ozero
Njuksenica see Nyuksenitsa
92 I10 **Njunis** ▲ N Norway 68°47′N 19°24′E
Njurunda see Njurundabommen
93 H17 **Njurundabommen** prev. Njurunda. Västernorrland, C Sweden 62°15′N 17°24′E
94 N11 **Njutånger** Gävleborg, C Sweden 61°37′N 17°04′E
79 D14 **Nkambe** Nord-Ouest, NW Cameroon 06°35′N 10°44′E
Nkata Bay see Nkhata Bay
79 F21 **Nkayi** prev. Jacob. Bouenza, S Congo 04°11′S 13°17′E
83 J17 **Nkayi** Matabeleland North, W Zimbabwe 19°00′S 28°54′E
82 N13 **Nkhata Bay** var. Nkata Bay. Northern, N Malawi 11°37′S 34°20′E
81 E22 **Nkonde** Kigoma, N Tanzania 06°16′S 30°17′E
79 D15 **Nkongsamba** var. N'Kongsamba. Littoral, W Cameroon 04°59′N 09°53′E
N'Kongsamba see Nkongsamba
83 E16 **Nkurenkuru** Okavango, N Namibia 17°38′S 18°39′E
77 Q15 **Nkwanta** E Ghana 08°18′N 00°27′E
167 O2 **Nmai Hka** var. Me Hka. ≈ N Myanmar (Burma)
Noardwâlde see Noordwolde
39 N7 **Noatak** Alaska, USA 67°34′N 162°58′W
39 N7 **Noatak River** ≈ Alaska, USA
164 E15 **Nobeoka** Miyazaki, Kyūshū, SW Japan 32°34′N 131°37′E
27 N11 **Noble** Oklahoma, C USA 35°08′N 97°23′W
31 P13 **Noblesville** Indiana, N USA 40°03′N 86°00′W
165 R5 **Noboribetsu** var. Noboribetu. Hokkaidō, NE Japan 42°27′N 141°08′E
Noboribetu see Noboribetsu
59 H18 **Nobres** Mato Grosso, W Brazil 14°44′S 56°15′W
107 N21 **Nocera Terinese** Calabria, S Italy 39°03′N 16°10′E
41 Q16 **Nochixtlán** var. Asunción Nochixtlán. Oaxaca, SE Mexico 17°29′N 97°17′W
25 S5 **Nocona** Texas, SW USA 33°47′N 97°43′W
63 K21 **Nodales, Bahía de los** bay S Argentina
27 Q2 **Nodaway River** ≈ Iowa/Missouri, C USA
27 R8 **Noel** Missouri, C USA 36°33′N 94°29′W
40 H3 **Nogales** Chihuahua, NW Mexico 18°50′N 97°12′W
40 F3 **Nogales** Sonora, NW Mexico 31°17′N 110°53′W
36 M17 **Nogales** Arizona, SW USA 31°20′N 110°55′W
Nogal Valley see Dooxo Nugaaleed
102 K15 **Nogaro** Gers, S France 43°46′N 00°01′W
110 J7 **Nogat** ≈ N Poland
164 D12 **Nōgata** Fukuoka, Kyūshū, SW Japan 33°46′N 130°42′E
127 P15 **Nogayskaya Step'** steppe SW Russian Federation
102 M6 **Nogent-le-Rotrou** Eure-et-Loir, C France 48°19′N 00°50′E
103 O4 **Nogent-sur-Oise** Oise, N France 49°16′N 02°28′E
103 P6 **Nogent-sur-Seine** Aube, N France 48°30′N 03°31′E
122 L10 **Noginsk** Krasnoyarskiy Kray, N Russian Federation 64°28′N 91°09′E
126 L3 **Noginsk** Moskovskaya Oblast', W Russian Federation 55°51′N 38°23′E
123 T12 **Nogliki** Ostrov Sakhalin, Sakhalinskaya Oblast', SE Russian Federation 51°44′N 143°14′E
164 K12 **Nōgōhaku-san** ▲ Honshū, SW Japan 35°46′N 136°30′E
162 D5 **Nogoonnuur** Bayan-Ölgiy, NW Mongolia 49°31′N 89°48′E
61 C18 **Nogoyá** Entre Ríos, E Argentina 32°25′S 59°50′W
111 K21 **Nógrád** off. Nógrád Megye. ◐ county N Hungary
Nógrád Megye see Nógrád
105 U5 **Noguera Pallaresa** ≈ NE Spain
105 U4 **Noguera Ribagorçana** ≈ NE Spain
101 E19 **Nohfelden** Saarland, SW Germany 49°35′N 07°08′E
38 A8 **Nohili Point** headland Kaua'i, Hawai'i, USA 22°03′N 159°48′W
104 G3 **Noia** Galicia, NW Spain 42°48′N 08°52′W
103 N16 **Noire, Montagne** ▲▲ S France
14 J10 **Noire, Rivière** ≈ Québec, SE Canada
15 P12 **Noire, Rivière** ≈ Québec, SE Canada
Noire, Rivi`ere see Black River
102 G6 **Noires, Montagnes** ▲▲ NW France
102 H8 **Noirmoutier-en-l'Île** Vendée, NW France 47°00′N 02°15′W
102 H8 **Noirmoutier, Île de** island NW France
187 Q10 **Noka** Nendö, E Solomon Islands 10°42′S 165°57′E
83 G17 **Nokaneng** North West, NW Botswana 19°40′S 22°12′E
93 L18 **Nokia** Pirkanmaa, W Finland 61°29′N 23°30′E
148 K11 **Nok Kundi** Baluchistān, SW Pakistan 28°49′N 62°39′E
30 L14 **Nokomis** Illinois, N USA 39°18′N 89°17′W
30 K5 **Nokomis, Lake** ⊠ Wisconsin, N USA
78 G9 **Nokou** Kanem, W Chad 14°36′N 14°45′E
187 Q12 **Nokuku** Espiritu Santo, W Vanuatu 14°56′S 166°34′E
95 J18 **Nol** Västra Götaland, S Sweden 57°55′N 12°03′E
79 H16 **Nola** Sangha-Mbaéré, SW Central African Republic 03°29′N 16°05′E
25 P7 **Nolan** Texas, SW USA 32°15′N 100°15′W
125 R15 **Nolinsk** Kirovskaya Oblast', NW Russian Federation 57°35′N 49°54′E
Nolsø see Nólsoy
95 B19 **Nólsoy** Dan. Nolsø. island E Faroe Islands
186 B7 **Nomad** Western, SW Papau New Guinea 06°11′S 142°13′E

◆ Country
● Country Capital
◇ Dependent Territory
○ Dependent Territory Capital
◐ Administrative Regions
✕ International Airport
▲ Mountain
▲▲ Mountain Range
△ Volcano
≈ River
◎ Lake
⊠ Reservoir

NOMA-ZAKI – NOUPOORT

164 *B16* **Noma-zaki** Kyūshū, SW Japan
40 *K10* **Nombre de Dios** Durango, C Mexico *23°51′N 104°14′W*
42 *I5* **Nombre de Dios, Cordillera** ▲▲ N Honduras
38 *M9* **Nome** Alaska, USA *64°30′N 165°24′W*
29 *Q6* **Nome** North Dakota, N USA *46°39′N 97°49′W*
38 *M9* **Nome, Cape** *headland* Alaska, USA *64°25′N 165°00′W*
162 *K11* **Nomgon** *var.* Sangiyn Dalay. Ömnögovĭ, S Mongolia *42°50′N 105°04′E*
14 *M11* **Nominingue, Lac** ◎ Québec, SE Canada
Nomoi Islands *see* Mortlock Islands
164 *B16* **Nomo-zaki** *headland* Kyūshū, SW Japan *32°34′N 129°45′E*
162 *G6* **Nömrög** *var.* Hödrögö. Dzavhan, N Mongolia *48°51′N 96°48′E*
193 *X15* **Nomuka** *island* Nomuka Group, C Tonga
193 *X15* **Nomuka Group** *island group* W Tonga
189 *Q15* **Nomwin Atoll** *atoll* Hall Islands, C Micronesia
8 *L10* **Nonacho Lake** ◎ Northwest Territories, NW Canada
Nondaburi *see* Nonthaburi
39 *P12* **Nondalton** Alaska, USA *59°58′N 154°51′W*
163 *V10* **Nong'an** Jilin, NE China *44°25′N 125°10′E*
169 *P10* **Nong Bua Khok** Nakhon Ratchasima, C Thailand *15°23′N 101°51′E*
167 *Q9* **Nong Bua Lamphu** Udon Thani, E Thailand *17°11′N 102°27′E*
167 *R7* **Nông Hèt** Xiangkhoang, N Laos *19°27′N 104°02′E*
Nongkaya *see* Nong Khai
167 *Q8* **Nong Khai** *var.* Mi Chai, Nongkaya. Nong Khai, E Thailand *17°52′N 102°44′E*
Nongle *see* Jiuzhaigou
167 *N14* **Nong Met** Surat Thani, SW Thailand *09°27′N 99°09′E*
83 *L22* **Nongoma** KwaZulu/Natal, E South Africa *27°54′S 31°40′E*
167 *P9* **Nong Phai** Phetchabun, C Thailand *15°58′N 101°02′E*
153 *U13* **Nongstoin** Meghālaya, NE India *25°24′N 91°19′E*
83 *C19* **Nonidas** Erongo, N Namibia *22°36′S 14°40′E*
Nonni *see* Nen Jiang
40 *I7* **Nonoava** Chihuahua, N Mexico *27°24′N 106°18′W*
191 *O3* **Nonouti** *prev.* Sydenham Island. *atoll* Tungaru, W Kiribati
167 *O11* **Nonthaburi** *var.* Nondaburi, Nontha Buri. Nonthaburi, C Thailand *13°48′N 100°11′E*
Nontha Buri *see* Nonthaburi
102 *L11* **Nontron** Dordogne, SW France *45°34′N 00°41′E*
147 *T10* **Nookat** *var.* Iski-Nauket; *prev.* Eski-Nookat. Oshskaya Oblast', SW Kyrgyzstan *40°18′N 72°29′E*
181 *P1* **Noonamah** Northern Territory, N Australia *12°46′S 131°08′E*
28 *K2* **Noonan** North Dakota, N USA *48°51′N 102°57′W*
Noonu *see* South Miladhunmadulu Atoll
99 *E14* **Noord-Beveland** *var.* North Beveland. *island* SW Netherlands
99 *J14* **Noord-Brabant** *Eng.* North Brabant. ◈ *province* S Netherlands
98 *H7* **Noorder Haaks** *spit* NW Netherlands
98 *H9* **Noord-Holland** *Eng.* North Holland. ◈ *province* NW Netherlands
Noordhollandsch Kanaal *see* Noordhollands Kanaal
98 *H8* **Noordhollands Kanaal** *var.* Noordhollandsch Kanaal. *canal* NW Netherlands
Noord-Kaap *see* Northern Cape
98 *L8* **Noordoostpolder** *island* N Netherlands
45 *P16* **Noordpunt** *headland* N Curaçao *12°21′N 69°08′W*
98 *I8* **Noord-Scharwoude** Noord-Holland, NW Netherlands *52°42′N 04°48′E*
Noordwes *see* North-West
98 *G11* **Noordwijk aan Zee** Zuid-Holland, W Netherlands *52°15′N 04°25′E*
98 *H11* **Noordwijkerhout** Zuid-Holland, W Netherlands *52°16′N 04°30′E*
98 *M7* **Noordwolde** *Fris.* Noardwâlde. Fryslân, N Netherlands *52°54′N 06°10′E*
Noordzee *see* North Sea
98 *H10* **Noordzee-Kanaal** *canal* NW Netherlands
93 *K18* **Noormarkku** *Swe.* Norrmark. Satakunta, SW Finland *61°35′N 21°54′E*
39 *N8* **Noorvik** Alaska, USA *66°50′N 161°01′W*
10 *J17* **Nootka Sound** *inlet* British Columbia, W Canada
82 *A9* **Nóqui** Zaire Province, NW Angola *05°54′S 13°30′E*
95 *L15* **Nora** Örebro, C Sweden *59°31′N 15°02′E*
147 *Q13* **Norak** *Rus.* Nurek. W Tajikistan *38°23′N 69°14′E*
13 *I13* **Noranda** Québec, SE Canada *48°16′N 79°03′W*
29 *W12* **Nora Springs** Iowa, C USA *43°08′N 93°00′W*
95 *M14* **Norberg** Västmanland, C Sweden *60°04′N 15°56′E*
14 *K13* **Norcan Lake** ◎ Ontario, SE Canada
197 *R12* **Nord** N Greenland *81°38′N 12°51′W*
78 *F13* **Nord** *Eng.* North. ◈ *province* N Cameroon
103 *P2* **Nord** ◈ *department* N France
92 *P1* **Nordaustlandet** *island* NE Svalbard
95 *G24* **Nordborg** *Ger.* Nordburg. Syddanmark, SW Denmark *55°04′N 09°41′E*
Nordburg *see* Nordborg
95 *F23* **Nordby** Syddtjylland, W Denmark *55°27′N 08°25′E*
11 *P15* **Nordegg** Alberta, SW Canada *52°27′N 116°06′W*
100 *E9* **Norden** Niedersachsen, NW Germany *53°36′N 07°12′E*
100 *G10* **Nordenham** Niedersachsen, NW Germany *53°30′N 08°29′E*
122 *M6* **Nordenshel'da, Arkhipelag** *island group* N Russian Federation
92 *O3* **Nordenskiold Land** *physical region* W Svalbard
100 *E9* **Norderney** *island* NW Germany
100 *J9* **Norderstedt** Schleswig-Holstein, N Germany *53°42′N 09°59′E*
94 *D11* **Nordfjord** *fjord* S Norway
94 *C11* **Nordfjord** *physical region* S Norway
94 *D11* **Nordfjordeid** Sogn og Fjordane, S Norway *61°54′N 06°E*
92 *G11* **Nordfold** Nordland, C Norway *67°48′N 15°16′E*
Nordfriesische Inseln *see* North Frisian Islands
100 *H7* **Nordfriesland** *cultural region* N Germany
Nordgrønland *see* Avannaarsua
101 *K15* **Nordhausen** Thüringen, C Germany *51°31′N 10°48′E*
25 *T13* **Nordheim** Texas, SW USA *28°55′N 97°36′W*
94 *C13* **Nordhordland** *physical region* S Norway
100 *E12* **Nordhorn** Niedersachsen, NW Germany *52°26′N 07°04′E*
172 *H16* **Nord, Île du** *island* Inner Islands, NE Seychelles
95 *F20* **Nordjylland** ◈ *county* N Denmark
92 *K7* **Nordkapp** *Eng.* North Cape. *headland* N Norway *25°47′E 71°10′N*
92 *O1* **Nordkapp** *headland* N Svalbard *80°31′N 19°58′E*
79 *N19* **Nord-Kivu** *off.* Région du Nord Kivu. ◈ *region* E Dem. Rep. Congo
Nord Kivu, Région du *see* Nord-Kivu
92 *G12* **Nordland** ◈ *county* C Norway
101 *J21* **Nördlingen** Bayern, S Germany *48°49′N 10°28′E*
93 *I16* **Nordmaling** Västerbotten, N Sweden *63°35′N 19°30′E*
95 *K15* **Nordmark** Värmland, C Sweden *59°52′N 14°04′E*
Nord, Mer du *see* North Sea
94 *F8* **Nordmøre** *physical region* S Norway
100 *I8* **Nord-Ostee-Kanal** *canal* N Germany
0 *J3* **Nordøstrundingen** *cape* NE Greenland
79 *D14* **Nord-Ouest** *Eng.* North-West. ◈ *province* NW Cameroon
Nord-Ouest, Territoires du *see* Northwest Territories
103 *N2* **Nord-Pas-de-Calais** ◈ *region* N France
101 *F19* **Nordpfälzer Bergland** ▲▲ W Germany
Nord, Pointe *see* Fatua, Pointe
187 *P16* **Nord, Province** ◈ *province* C New Caledonia
101 *D14* **Nordrhein-Westfalen** *Eng.* North Rhine-Westphalia, *Fr.* Rhénanie du Nord-Westphalie. ◈ *state* W Germany
Nordsee/Nordsjøen/Nordsøen *see* North Sea
100 *H7* **Nordstrand** *island* N Germany
93 *E15* **Nord-Trøndelag** ◈ *county* C Norway
92 *I1* **Norðurfjörður** Vestfirðir, NW Iceland *66°01′N 21°33′W*
92 *J1* **Norðurland Eystra** ◈ *region* N Iceland
92 *I2* **Norðurland Vestra** ◈ *region* N Iceland
97 *E19* **Nore** *Ir.* An Fheoir. ≈ S Ireland
29 *Q14* **Norfolk** Nebraska, C USA *42°01′N 97°25′W*
21 *X7* **Norfolk** Virginia, NE USA *36°51′N 76°17′W*
97 *P19* **Norfolk** *cultural region* E England, United Kingdom
192 *K10* **Norfolk Island** ◈ *Australian self-governing territory* SW Pacific Ocean
175 *P9* **Norfolk Ridge** *undersea feature* W Pacific Ocean
27 *U8* **Norfork Lake** ⊠ Arkansas/Missouri, C USA
98 *N6* **Norg** Drenthe, NE Netherlands *53°04′N 06°28′E*
Norge *see* Norway
95 *D14* **Norheimsund** Hordaland, S Norway *60°22′N 06°09′E*
25 *S16* **Norias** Texas, SW USA *26°47′N 97°45′W*
164 *L12* **Norikura-dake** ▲ Honshū, S Japan *36°06′N 137°33′E*
122 *K8* **Noril'sk** Krasnoyarskiy Kray, N Russian Federation *69°21′N 88°02′E*
14 *I13* **Norland** Ontario, SE Canada *44°46′N 78°48′W*
21 *V8* **Norlina** North Carolina, SE USA *36°26′N 78°11′W*
30 *L13* **Normal** Illinois, N USA *40°30′N 88°59′W*
27 *N11* **Norman** Oklahoma, C USA *35°13′N 97°27′W*
Norman *see* Tulita
186 *G9* **Normanby Island** *island* SE Papua New Guinea
Normandes, Îles *see* Channel Islands
58 *G9* **Normandia** Roraima, N Brazil *03°57′N 59°39′W*
102 *L5* **Normandie** *Eng.* Normandy. *cultural region* N France
102 *J5* **Normandie, Collines de** *hill range* NW France
Normandy *see* Normandie
25 *V9* **Normangee** Texas, SW USA *31°01′N 96°06′W*
21 *Q10* **Norman, Lake** ⊠ North Carolina, SE USA
44 *K13* **Norman Manley** ✈ (Kingston) E Jamaica *17°55′N 76°46′W*
181 *U5* **Norman River** ≈ Queensland, NE Australia
181 *U4* **Normanton** Queensland, NE Australia *17°49′S 141°08′E*
8 *I8* **Norman Wells** Northwest Territories, NW Canada *65°18′N 126°42′W*
12 *H12* **Normétal** Québec, S Canada *48°59′N 79°23′W*
163 *O7* **Norovlin** *var.* Uldz. Hentiy, NE Mongolia *48°47′N 112°01′E*
11 *V15* **Norquay** Saskatchewan, S Canada *51°51′N 102°04′W*
Norra Karelen *see* Pohjois-Karjala
93 *G15* **Norråker** Jämtland, C Sweden *64°25′N 15°40′E*
94 *N12* **Norrala** Gävleborg, C Sweden *61°22′N 17°04′E*
Norra Ny *see* Stöllet
Norra Österbotten *see* Pohjois-Pohjanmaa
Norra Savolax *see* Pohjois-Savo
92 *G13* **Norra Storjfjället** ▲ N Sweden *65°57′N 15°15′E*
92 *I13* **Norrbotten** ◈ *county* N Sweden
94 *N11* **Norrdellen** ◎ C Sweden
95 *G23* **Nørre Aaby** *var.* Nørre Åby. Syddtjylland, C Denmark *55°28′N 09°53′E*
Nørre Åby *see* Nørre Aaby
95 *I24* **Nørre Alslev** Sjælland, SE Denmark *54°54′N 11°53′E*
95 *E23* **Nørre Nebel** Syddtjylland, W Denmark *55°45′N 08°16′E*
95 *G20* **Nørresundby** Nordjylland, N Denmark *57°05′N 09°55′E*
21 *N8* **Norris Lake** ⊠ Tennessee, S USA
18 *I15* **Norristown** Pennsylvania, NE USA *40°07′N 75°20′W*
95 *N17* **Norrköping** Östergötland, S Sweden *58°35′N 16°10′E*
Norrmark *see* Noormarkku
94 *N13* **Norrsundet** Gävleborg, C Sweden *60°55′N 17°09′E*
95 *P15* **Norrtälje** Stockholm, C Sweden *59°46′N 18°42′E*
180 *L12* **Norseman** Western Australia *32°16′S 121°46′E*
93 *I14* **Norsjö** Västerbotten, N Sweden *64°55′N 19°30′E*
95 *G16* **Norsjø** ◎ S Norway
123 *R13* **Norsk** Amurskaya Oblast', SE Russian Federation *52°20′N 129°57′E*
Norske Havet *see* Norwegian Sea
187 *Q13* **Norsup** Malekula, C Vanuatu *16°05′S 167°24′E*
191 *V15* **Norte, Cabo** *headland* Easter Island, Chile, E Pacific Ocean *27°03′S 109°24′W*
54 *F7* **Norte de Santander** *off.* Departamento de Norte de Santander. ◈ *province* N Colombia
Norte de Santander, Departamento de *see* Norte de Santander
61 *E21* **Norte, Punta** *headland* E Argentina *36°17′S 56°46′W*
21 *R13* **North** South Carolina, SE USA *33°37′N 81°06′W*
North *see* Nord
18 *L10* **North Adams** Massachusetts, NE USA *42°40′N 73°06′W*
113 *L17* **North Albanian Alps** *Alb.* Bjeshkët e Namuna, *SCr.* Prokletije. ▲▲ SE Europe
97 *M15* **Northallerton** N England, United Kingdom *54°20′N 01°26′W*
180 *J12* **Northam** Western Australia *31°40′S 116°40′E*
83 *J20* **Northam** Northern, N South Africa *24°56′S 27°18′E*
1 **North America** *continent*
1 *N12* **North American Basin** *undersea feature* W Sargasso Sea *30°00′N 60°00′W*
0 *C5* **North American Plate** *tectonic feature*
18 *M11* **North Amherst** Massachusetts, NE USA *42°24′N 72°31′W*
97 *N20* **Northampton** C England, United Kingdom *52°14′N 00°54′W*
97 *M20* **Northamptonshire** *cultural region* C England, United Kingdom
151 *P18* **North Andaman** *island* Andaman Islands, India, NE Indian Ocean
65 *D25* **North Arm** East Falkland, Falkland Islands *52°06′S 59°21′W*
21 *Q13* **North Augusta** South Carolina, SE USA *33°30′N 81°58′W*
173 *W8* **North Australian Basin** *Fr.* Bassin Nord de l' Australie. *undersea feature* E Indian Ocean
31 *R11* **North Baltimore** Ohio, N USA *41°10′N 83°40′W*
11 *T15* **North Battleford** Saskatchewan, S Canada *52°47′N 108°19′W*
14 *H11* **North Bay** Ontario, S Canada *46°20′N 79°28′W*
12 *H6* **North Belcher Islands** *island group* Belcher Islands, Nunavut, C Canada
29 *R15* **North Bend** Nebraska, C USA *41°27′N 96°46′W*
32 *E14* **North Bend** Oregon, NW USA *43°24′N 124°13′W*
96 *K12* **North Berwick** SE Scotland, United Kingdom *56°04′N 02°44′W*
North Beveland *see* Noord-Beveland
North Borneo *see* Sabah
183 *P5* **North Bourke** New South Wales, SE Australia *30°03′S 145°56′E*
North Brabant *see* Noord-Brabant
182 *F2* **North Branch Neales** *seasonal river* South Australia
44 *M6* **North Caicos** *island* NW Turks and Caicos Islands
26 *L10* **North Canadian River** ≈ Oklahoma, C USA
31 *U12* **North Canton** Ohio, N USA *40°52′N 81°24′W*
13 *R13* **North, Cape** *headland* Cape Breton Island, Nova Scotia, SE Canada *47°06′N 60°24′W*
184 *I1* **North Cape** *headland* North Island, New Zealand *34°23′S 173°02′E*
186 *G5* **North Cape** *headland* New Ireland, NE Papua New Guinea *02°33′S 150°48′E*
North Cape *see* Nordkapp
18 *J17* **North Cape May** New Jersey, NE USA *38°59′N 74°55′W*
12 *C9* **North Caribou Lake** ◎ Ontario, C Canada
21 *U10* **North Carolina** *off.* State of North Carolina, ***also known as*** Old North State, Tar Heel State, Turpentine State. ◈ *state* SE USA
North Celebes *see* Sulawesi Utara
155 *J24* **North Central** ◈ *province* N Sri Lanka
31 *S4* **North Channel** *lake channel* Canada/USA
97 *G14* **North Channel** *strait* Northern Ireland/Scotland, United Kingdom
21 *S14* **North Charleston** South Carolina, SE USA *32°53′N 79°59′W*
31 *N10* **North Chicago** Illinois, N USA *42°19′N 87°50′W*
195 *Y10* **Northcliffe Glacier** *glacier* Antarctica
31 *Q14* **North College Hill** Ohio, N USA *39°13′N 84°33′W*
25 *O8* **North Concho River** ≈ Texas, SW USA
19 *O8* **North Conway** New Hampshire, NE USA *44°03′N 71°06′W*
27 *V14* **North Crossett** Arkansas, C USA *33°10′N 91°56′W*
28 *L4* **North Dakota** *off.* State of North Dakota, ***also known as*** Flickertail State, Peace Garden State, Sioux State. ◈ *state* N USA
North Devon Island *see* Devon Island
97 *O22* **North Downs** *hill range* SE England, United Kingdom
18 *C11* **North East** Pennsylvania, NE USA *42°13′N 79°49′W*
83 *I18* **North East** ◈ *district* NE Botswana
65 *G15* **North East Bay** *bay* Ascension Island, C Atlantic Ocean
38 *L10* **Northeast Cape** *headland* Saint Lawrence Island, Alaska, USA *63°16′N 168°50′W*
North East Frontier Agency/North East Frontier Agency of Assam *see* Arunāchal Pradesh
65 *E25* **North East Island** *island* E Falkland Islands
189 *V11* **Northeast Island** *island* Chuuk, C Micronesia
44 *L12* **North East Point** *headland* E Jamaica *18°09′N 76°19′W*
191 *Z2* **Northeast Point** *headland* Kiritimati, E Kiribati *10°23′S 105°45′E*
44 *L6* **Northeast Point** *headland* Great Inagua, S The Bahamas *21°18′N 73°01′W*
44 *K5* **Northeast Point** *headland* Acklins Island, SE The Bahamas *22°43′N 73°50′W*
44 *H2* **Northeast Providence Channel** *channel* N The Bahamas
101 *J14* **Northeim** Niedersachsen, C Germany *51°42′N 10°E*
29 *X14* **North English** Iowa, C USA *41°30′N 92°04′W*
138 *G8* **Northern** ◈ *district* N Israel
82 *M12* **Northern** ◈ *region* N Malawi
186 *F8* **Northern** *var.* Oro. ◈ *province* S Papua New Guinea
155 *J23* **Northern** ◈ *province* N Sri Lanka
80 *D7* **Northern** ◈ *state* N Sudan
82 *K12* **Northern** ◈ *province* NE Zambia
Northern *see* Limpopo
80 *B13* **Northern Bahr el Ghazal** ◈ *state* NW South Sudan
Northern Border Region *see* Al Ḥudūd ash Shamālīyah
83 *F24* **Northern Cape** *off.* Northern Cape Province, *Afr.* Noord-Kaap. ◈ *province* W South Africa
Northern Cape Province *see* Northern Cape
190 *K14* **Northern Cook Islands** *island group* N Cook Islands
80 *B8* **Northern Darfur** ◈ *state* NW Sudan
Northern Dvina *see* Severnaya Dvina
97 *F14* **Northern Ireland** *var.* The Six Counties. *cultural region* Northern Ireland, United Kingdom
97 *F14* **Northern Ireland** *var.* The Six Counties. ◈ *political division* Northern Ireland, United Kingdom
80 *D9* **Northern Kordofan** ◈ *state* C Sudan
187 *Z14* **Northern Lau Group** *island group* Lau Group, NE Fiji
188 *K3* **Northern Mariana Islands** ◇ *US commonwealth territory* W Pacific Ocean
Northern Rhodesia *see* Zambia
Northern Sporades *see* Vóreies Sporádes
182 *D1* **Northern Territory** ◈ *territory* N Australia
Northern Transvaal *see* Limpopo
Northern Ural Hills *see* Severnyye Uvaly
84 *I9* **North European Plain** *plain* N Europe
27 *V2* **North Fabius River** ≈ Missouri, C USA
65 *D24* **North Falkland Sound** *sound* N Falkland Islands
29 *V9* **Northfield** Minnesota, N USA *44°27′N 93°10′W*
19 *O9* **Northfield** New Hampshire, NE USA *43°26′N 71°34′W*
175 *Q8* **North Fiji Basin** *undersea feature* N Coral Sea
97 *Q22* **North Foreland** *headland* SE England, United Kingdom *51°22′N 01°26′E*
35 *P6* **North Fork American River** ≈ California, W USA
39 *R7* **North Fork Chandalar River** ≈ Alaska, USA
28 *K7* **North Fork Grand River** ≈ North Dakota/South Dakota, N USA
21 *O6* **North Fork Kentucky River** ≈ Kentucky, S USA
39 *Q7* **North Fork Koyukuk River** ≈ Alaska, USA
39 *Q10* **North Fork Kuskokwim River** ≈ Alaska, USA
26 *K11* **North Fork Red River** ≈ Oklahoma/Texas, SW USA
26 *K3* **North Fork Solomon River** ≈ Kansas, C USA
23 *W14* **North Fort Myers** Florida, SE USA *26°40′N 81°52′W*
31 *P5* **North Fox Island** *island* Michigan, N USA
100 *G6* **North Frisian Islands** *var.* Nordfriesische Inseln. *island group* N Germany
197 *N9* **North Geomagnetic Pole** *pole* Arctic Ocean
18 *M13* **North Haven** Connecticut, NE USA *41°25′N 72°51′W*
184 *J5* **North Head** *headland* North Island, New Zealand *36°23′S 174°01′E*
18 *L6* **North Hero** Vermont, NE USA *44°49′N 73°14′W*
35 *O7* **North Highlands** California, W USA *38°40′N 121°25′W*
North Holland *see* Noord-Holland
81 *I16* **North Horr** Marsabit, N Kenya *03°17′N 37°08′E*
151 *K21* **North Huvadhu Atoll** *var.* Gaafu Alifu Atoll. *atoll* S Maldives
65 *A24* **North Island** *island* W Falkland Islands
184 *N9* **North Island** *island* N New Zealand
21 *U14* **North Island** *island* South Carolina, SE USA
31 *O11* **North Judson** Indiana, N USA *41°12′N 86°44′W*
North Kalimantan *see* Kalimantan Utara
North Karelia *see* Pohjois-Karjala
North Kazakhstan *see* Severnyy Kazakhstan
31 *V10* **North Kingsville** Ohio, N USA *41°54′N 80°41′W*
163 *Y13* **North Korea** *off.* Democratic People's Republic of Korea, *Kor.* Chosŏn-minjujuŭi-inmin-kanghwaguk. ◆ *republic* E Asia
153 *X11* **North Lakhimpur** Assam, NE India *27°10′N 94°00′E*
184 *J3* **Northland** *off.* Northland Region. ◈ *region* North Island, New Zealand
192 *K11* **Northland Plateau** *undersea feature* S Pacific Ocean
Northland Region *see* Northland
35 *X11* **North Las Vegas** Nevada, W USA *36°12′N 115°07′W*
31 *O11* **North Liberty** Indiana, N USA *41°36′N 86°22′W*
29 *X14* **North Liberty** Iowa, C USA *41°45′N 91°36′W*
27 *V12* **North Little Rock** Arkansas, C USA *34°46′N 92°15′W*
28 *M13* **North Loup River** ≈ Nebraska, C USA
151 *K18* **North Maalhosmadulu Atoll** *var.* North Malosmadulu Atoll, Raa Atoll. *atoll* N Maldives
31 *U10* **North Madison** Ohio, N USA *41°48′N 81°03′W*
North Malosmadulu Atoll *see* North Maalhosmadulu Atoll
31 *P12* **North Manchester** Indiana, N USA *41°00′N 85°45′W*
31 *P6* **North Manitou Island** *island* Michigan, N USA
29 *U10* **North Mankato** Minnesota, N USA *44°11′N 94°03′W*
23 *Z15* **North Miami** Florida, SE USA *25°54′N 80°11′W*
151 *K18* **North Miladhunmadulu Atoll** *var.* Shaviyani Atoll. *atoll* N Maldives
North Minch *see* Minch, The
23 *W15* **North Naples** Florida, SE USA *26°13′N 81°47′W*
175 *P8* **North New Hebrides Trench** *undersea feature* N Coral Sea
23 *Y15* **North New River Canal** ≈ Florida, SE USA
151 *K20* **North Nilandhe Atoll** *atoll* C Maldives
36 *L2* **North Ogden** Utah, W USA *41°18′N 111°57′W*
North Ossetia *see* Severnaya Osetiya-Alaniya, Respublika
North Ostrobothnia *see* Pohjois-Pohjanmaa
35 *S10* **North Palisade** ▲ California, W USA *37°06′N 118°31′W*
189 *U11* **North Pass** *passage* Chuuk Islands, C Micronesia
28 *M15* **North Platte** Nebraska, C USA *41°07′N 100°46′W*
33 *X17* **North Platte River** ≈ C USA
65 *G14* **North Point** *headland* Ascension Island, C Atlantic Ocean
172 *I16* **North Point** *headland* Mahé, NE Seychelles *04°23′S 55°28′E*
31 *R5* **North Point** *headland* Michigan, N USA *45°21′N 83°30′W*
31 *S6* **North Point** *headland* Michigan, N USA *45°01′N 83°16′W*
39 *S9* **North Pole** Alaska, USA *64°42′N 147°09′W*
197 *R9* **North Pole** *pole* Arctic Ocean
23 *O4* **Northport** Alabama, S USA *33°13′N 87°34′W*
23 *W14* **North Port** Florida, SE USA *27°03′N 82°15′W*
32 *L6* **Northport** Washington, NW USA *48°54′N 117°48′W*
32 *L12* **North Powder** Oregon, NW USA *45°00′N 117°56′W*
29 *U13* **North Raccoon River** ≈ Iowa, C USA
North Rhine-Westphalia *see* Nordrhein-Westfalen
97 *M16* **North Riding** *cultural region* N England, United Kingdom
96 *G5* **North Rona** *island* NW Scotland, United Kingdom
96 *K4* **North Ronaldsay** *island* NE Scotland, United Kingdom
36 *L2* **North Salt Lake** Utah, W USA *40°51′N 111°54′W*
11 *P15* **North Saskatchewan** ≈ Alberta/Saskatchewan, S Canada
35 *X5* **North Schell Peak** ▲ Nevada, W USA *39°25′N 114°34′W*
North Scotia Ridge *see* South Georgia Ridge
86 *D10* **North Sea** *Dan.* Nordsøen, *Dut.* Noordzee, *Fr.* Mer du Nord, *Ger.* Nordsee, *Nor.* Nordsjøen; *prev.* German Ocean, *Lat.* Mare Germanicum. *sea* NW Europe
35 *T6* **North Shoshone Peak** ▲ Nevada, W USA *39°08′N 117°28′W*
North Siberian Lowland/North Siberian Plain *see* Severo-Sibirskaya Nizmennost'
29 *R13* **North Sioux City** South Dakota, N USA *42°31′N 96°28′W*
North Solomons *see* Bougainville
96 *K4* **North Sound, The** *sound* N Scotland, United Kingdom
183 *T4* **North Star** New South Wales, SE Australia *28°55′S 150°25′E*
North Star State *see* Minnesota
183 *V3* **North Stradbroke Island** *island* Queensland, E Australia
North Sulawesi *see* Sulawesi Utara
North Sumatra *see* Sumatera Utara
14 *D17* **North Sydenham** ≈ Ontario, S Canada
18 *H9* **North Syracuse** New York, NE USA *43°07′N 76°07′W*
184 *K9* **North Taranaki Bight** *gulf* North Island, New Zealand
12 *H9* **North Twin Island** *island* Nunavut, C Canada
96 *E8* **North Uist** *island* NW Scotland, United Kingdom
97 *L14* **Northumberland** *cultural region* N England, United Kingdom
181 *Y7* **Northumberland Isles** *island group* Queensland, NE Australia
13 *Q14* **Northumberland Strait** *strait* SE Canada
32 *G14* **North Umpqua River** ≈ Oregon, NW USA
45 *Q13* **North Union** Saint Vincent , Saint Vincent and the Grenadines *13°15′N 61°07′W*
10 *L17* **North Vancouver** British Columbia, SW Canada *49°21′N 123°05′W*
18 *K9* **Northville** New York, NE USA *43°13′N 74°08′W*
97 *Q19* **North Walsham** E England, United Kingdom *52°49′N 01°22′E*
39 *T10* **Northway** Alaska, USA *62°57′N 141°56′W*
83 *G17* **North-West** ◈ *district* NW Botswana
83 *G21* **North-West** *off.* North-West Province, *Afr.* Noordwes. ◈ *province* N South Africa
North-West *see* Nord-Ouest
64 *I6* **Northwest Atlantic Mid-Ocean Canyon** *undersea feature* N Atlantic Ocean
180 *G8* **North West Cape** *cape* Western Australia
38 *J9* **Northwest Cape** *headland* Saint Lawrence Island, Alaska, USA *63°46′N 171°45′W*
155 *J24* **North Western** ◈ *province* W Sri Lanka
82 *H13* **North Western** ◈ *province* W Zambia
North-West Frontier Province *see* Khyber Pakhtunkhwa
96 *H8* **North West Highlands** ▲▲ N Scotland, United Kingdom
192 *J4* **Northwest Pacific Basin** *undersea feature* NW Pacific Ocean *40°00′N 150°00′E*
191 *Y2* **Northwest Point** *headland* Kiritimati, E Kiribati *10°25′S 105°35′E*
44 *G1* **Northwest Providence Channel** *channel* N The Bahamas
North-West Province *see* North-West
13 *Q8* **North West River** Newfoundland and Labrador, E Canada *53°30′N 60°10′W*
8 *J9* **Northwest Territories** *Fr.* Territoires du Nord-Ouest. ◈ *territory* NW Canada
97 *K18* **Northwich** C England, United Kingdom *53°16′N 02°32′W*
25 *Q5* **North Wichita River** ≈ Texas, SW USA
18 *J17* **North Wildwood** New Jersey, NE USA *39°00′N 74°45′W*
21 *R9* **North Wilkesboro** North Carolina, SE USA *36°09′N 81°09′W*
19 *P8* **North Windham** Maine, NE USA *43°51′N 70°25′W*
197 *Q6* **Northwind Plain** *undersea feature* Arctic Ocean
29 *V11* **Northwood** Iowa, C USA *43°26′N 93°13′W*
29 *Q4* **Northwood** North Dakota, N USA *47°43′N 97°34′W*
97 *M15* **North York Moors** *moorland* N England, United Kingdom
25 *V9* **North Zulch** Texas, SW USA *30°54′N 96°06′W*
26 *K2* **Norton** Kansas, C USA *39°51′N 99°53′W*
31 *S13* **Norton** Ohio, N USA *40°25′N 83°04′W*
21 *P7* **Norton** Virginia, NE USA *36°56′N 82°37′W*
39 *N9* **Norton Bay** *bay* Alaska, USA
Norton de Matos *see* Balombo
31 *O9* **Norton Shores** Michigan, N USA *43°10′N 86°15′W*
38 *M10* **Norton Sound** *inlet* Alaska, USA
27 *Q3* **Nortonville** Kansas, C USA *39°25′N 95°19′W*
102 *I8* **Nort-sur-Erdre** Loire-Atlantique, NW France *47°27′N 01°30′W*
195 *N2* **Norvegia, Cape** *headland* Antarctica *71°16′S 12°25′W*
18 *L13* **Norwalk** Connecticut, NE USA *41°08′N 73°28′W*
29 *V14* **Norwalk** Iowa, C USA *41°28′N 93°40′W*
31 *S11* **Norwalk** Ohio, N USA *41°14′N 82°37′W*
19 *P7* **Norway** Maine, NE USA *44°13′N 70°30′W*
31 *N5* **Norway** Michigan, N USA *45°47′N 87°54′W*
93 *E17* **Norway** *off.* Kingdom of Norway, *Nor.* Norge. ◆ *monarchy* N Europe
11 *X13* **Norway House** Manitoba, C Canada *53°59′N 97°50′W*
Norway, Kingdom of *see* Norway
197 *R16* **Norwegian Basin** *undersea feature* NW Norwegian Sea *68°00′N 02°00′W*
84 *D6* **Norwegian Sea** *var.* Norske Havet. *sea* NE Atlantic Ocean
197 *S17* **Norwegian Trench** *undersea feature* NE North Sea *59°00′N 04°30′E*
14 *F16* **Norwich** Ontario, S Canada *42°57′N 80°37′W*
97 *Q19* **Norwich** E England, United Kingdom *52°38′N 01°18′E*
19 *N13* **Norwich** Connecticut, NE USA *41°30′N 72°02′W*
18 *I11* **Norwich** New York, NE USA *42°31′N 75°31′W*
29 *U9* **Norwood** Minnesota, N USA *44°46′N 93°55′W*
31 *Q15* **Norwood** Ohio, N USA *39°07′N 84°27′W*
14 *H11* **Nosbonsing, Lake** ◎ Ontario, S Canada
Nösen *see* Bistriţa
165 *T1* **Noshappu-misaki** *headland* Hokkaidō, NE Japan *45°26′N 141°38′E*
165 *P7* **Noshiro** *var.* Nosiro; *prev.* Noshirominato. Akita, Honshū, C Japan *40°11′N 140°02′E*
Noshirominato/Nosiro *see* Noshiro
117 *Q3* **Nosivka** *Rus.* Nosovka. Chernihivs'ka Oblast', NE Ukraine *50°55′N 31°37′E*
67 *T14* **Nosop** *var.* Nossob, Nossop. ≈ Botswana/Namibia
83 *E20* **Nossop** ≈ E Namibia
125 *S4* **Nosovaya** Nenetskiy Avtonomnyy Okrug, NW Russian Federation *68°12′N 54°33′E*
Nosovka *see* Nosivka
143 *V11* **Noşratābād** Sīstān va Balūchestān, E Iran *29°53′N 59°57′E*
95 *J18* **Nossebro** Västra Götaland, S Sweden *58°12′N 12°42′E*
96 *K6* **Noss Head** *headland* N Scotland, United Kingdom *58°29′N 03°03′W*
Nossi-Bé *see* Be, Nosy
Nossob/Nossop *see* Nosop
172 *J2* **Nosy Be** ✈ Antsirănana, N Madagascar *23°36′S 47°36′E*
172 *J6* **Nosy Varika** Fianarantsoa, SE Madagascar *20°36′S 48°31′E*
14 *L10* **Notawassi** ≈ Québec, SE Canada
14 *M9* **Notawassi, Lac** ◎ Québec, SE Canada
36 *J5* **Notch Peak** ▲ Utah, W USA *39°08′N 113°24′W*
110 *G10* **Noteć** *Ger.* Netze. ≈ NW Poland
Nóties Sporádes *see* Dodekánisa
115 *J22* **Nótion Aigaíon** *Eng.* Aegean South. ◈ *region* E Greece
115 *H18* **Nótios Evvoïkós Kólpos** *gulf* E Greece
115 *B16* **Nótio Stenó Kérkyras** *strait* W Greece
107 *L25* **Noto** *anc.* Netum. Sicilia, Italy, C Mediterranean Sea *36°53′N 15°05′E*
164 *M10* **Noto** Ishikawa, Honshū, SW Japan *37°18′N 137°11′E*
95 *G15* **Notodden** Telemark, S Norway *59°35′N 09°18′E*
107 *L25* **Noto, Golfo di** *gulf* Sicilia, Italy, C Mediterranean Sea
164 *L10* **Noto-hantō** *peninsula* Honshū, SW Japan
164 *L11* **Noto-jima** *island* SW Japan
13 *T11* **Notre Dame Bay** *bay* Newfoundland, Newfoundland and Labrador, E Canada
15 *P6* **Notre-Dame-de-Lorette** Québec, SE Canada *49°05′N 72°24′W*
14 *L11* **Notre-Dame-de-Pontmain** Québec, SE Canada *46°18′N 75°37′W*
15 *T8* **Notre-Dame-du-Lac** Québec, SE Canada *47°36′N 68°48′W*
15 *Q6* **Notre-Dame-du-Rosaire** Québec, SE Canada *48°48′N 71°27′W*
15 *U8* **Notre-Dame, Monts** ▲▲ Québec, S Canada
77 *R16* **Notsé** S Togo *06°59′N 01°12′E*
14 *G14* **Nottawasaga** ≈ Ontario, S Canada
14 *G14* **Nottawasaga Bay** *lake bay* Ontario, S Canada
12 *I11* **Nottaway** ≈ Québec, SE Canada
23 *S1* **Nottely Lake** ⊠ Georgia, SE USA
95 *H16* **Nøtterøy** *island* S Norway
97 *M19* **Nottingham** C England, United Kingdom *52°58′N 01°10′W*
9 *E14* **Nottingham Island** *island* Nunavut, NE Canada
97 *N18* **Nottinghamshire** *cultural region* C England, United Kingdom
21 *V7* **Nottoway** Virginia, NE USA *37°07′N 78°03′W*
21 *V7* **Nottoway River** ≈ Virginia, NE USA
76 *G7* **Nouâdhibou** *prev.* Port-Étienne. Dakhlet Nouâdhibou, W Mauritania *20°54′N 17°01′W*
76 *G7* **Nouâdhibou** ✈ Dakhlet Nouâdhibou, W Mauritania *20°59′N 17°02′W*
76 *F7* **Nouâdhibou, Dakhlet** *prev.* Baie du Lévrier. *bay* W Mauritania
76 *F7* **Nouâdhibou, Râs** *prev.* Cap Blanc. *headland* NW Mauritania *20°48′N 17°03′W*
76 *G9* **Nouakchott** ● (Mauritania) Nouakchott District, SW Mauritania *18°09′N 15°58′W*
76 *G9* **Nouakchott** ✈ Trarza, SW Mauritania *18°18′N 15°54′W*
120 *J11* **Noual, Sebkhet en** *var.* Sabkhat an Nawāl. *salt flat* C Tunisia
76 *G8* **Nouâmghâr** *var.* Nouamrhar. Dakhlet Nouâdhibou, W Mauritania *19°22′N 16°31′W*
Nouamrhar *see* Nouâmghâr
Nouă Sulița *see* Novoselytsya
187 *Q17* **Nouméa** ○ (New Caledonia) Province Sud, S New Caledonia *22°13′S 166°29′E*
79 *E15* **Noun** ≈ C Cameroon
77 *N12* **Nouna** W Burkina Faso *12°44′N 03°54′W*
83 *H24* **Noupoort** Northern Cape, C South Africa *31°11′S 24°57′E*

◆ Country
● Country Capital
◇ Dependent Territory
○ Dependent Territory Capital
◈ Administrative Regions
✈ International Airport
▲ Mountain
▲▲ Mountain Range
Volcano
≈ River
◎ Lake
⊠ Reservoir

NOUVEAU-BRUNSWICK – NZÉRÉKORÉ

Nouveau-Brunswick *see* New Brunswick
Nouveau-Comptoir *see* Wemindji
15 *T4* **Nouvel, Lacs** ◎ Québec, SE Canada
15 *W7* **Nouvelle** Québec, SE Canada 48°07′N 66°16′W
15 *W7* **Nouvelle** ≈ Québec, SE Canada
Nouvelle-Calédonie *see* New Caledonia
Nouvelle Écosse *see* Nova Scotia
103 *R3* **Nouzonville** Ardennes, N France 49°49′N 04°45′E
147 *Q11* **Nov** *Rus.* Nau. NW Tajikistan 40°10′N 69°16′E
59 *I21* **Nova Alvorada** Mato Grosso do Sul, SW Brazil 21°25′S 54°19′W
Novabad *see* Navobod
111 *D19* **Nová Bystřice** *Ger.* Neubistritz. Jihočeský Kraj, S Czech Republic 49°N 15°05′E
116 *H13* **Novaci** Gorj, SW Romania 45°07′N 23°37′E
Nova Civitas *see* Neustadt an der Weinstrasse
Novaesium *see* Neuss
60 *H10* **Nova Esperança** Paraná, S Brazil 23°09′S 52°13′W
106 *H11* **Novafeltria** Marche, C Italy 43°54′N 12°18′E
60 *Q9* **Nova Friburgo** Rio de Janeiro, SE Brazil 22°16′S 42°34′W
82 *D12* **Nova Gaia** *var.* Cambundi-Catembo. Malanje, NE Angola 10°09′S 17°31′E
109 *S12* **Nova Gorica** W Slovenia 45°57′N 13°40′E
112 *G10* **Nova Gradiška** *Ger.* Neugradisk, *Hung.* Újgradiska. Brod-Posavina, NE Croatia 45°15′N 17°23′E
60 *K7* **Nova Granada** São Paulo, S Brazil 20°33′S 49°19′W
60 *O10* **Nova Iguaçu** Rio de Janeiro, SE Brazil 22°31′S 44°05′W
117 *S10* **Nova Kakhovka** *Rus.* Novaya Kakhovka. Khersons'ka Oblast', SE Ukraine 46°45′N 33°20′E
Nová Karvinná *see* Karviná
Nova Lamego *see* Gabú
Nova Lisboa *see* Huambo
112 *C11* **Novalja** Lika-Senj, W Croatia 44°33′N 14°53′E
119 *M14* **Novalukoml'** *Rus.* Novolukoml'. Vitsyebskaya Voblasts', N Belarus 54°40′N 29°09′E
Nova Mambone *see* Mambone
83 *P16* **Nova Nabúri** Zambézia, NE Mozambique 16°47′S 38°55′E
117 *Q9* **Nova Odesa** *var.* Novaya Odessa. Mykolayivs'ka Oblast', S Ukraine 47°19′N 31°45′E
60 *H10* **Nova Olímpia** Paraná, S Brazil 23°28′S 53°12′W
61 *I15* **Nova Prata** Rio Grande do Sul, S Brazil 28°45′S 51°37′W
14 *H12* **Novar** Ontario, S Canada 45°26′N 79°14′W
106 *C7* **Novara** *anc.* Novaria. Piemonte, NW Italy 45°27′N 08°36′E
Novaria *see* Novara
13 *P15* **Nova Scotia** *Fr.* Nouvelle Écosse. ◈ *province* SE Canada
0 *M9* **Nova Scotia** *physical region* SE Canada
34 *M8* **Novato** California, W USA 38°06′N 122°35′W
192 *M7* **Nova Trough** *undersea feature* W Pacific Ocean
116 *L7* **Nova Ushytsya** Khmel'nyts'ka Oblast', W Ukraine 48°50′N 27°16′E
83 *M17* **Nova Vanduzi** Manica, C Mozambique 18°54′S 33°18′E
117 *U5* **Nova Vodolaha** *Rus.* Novaya Vodolaga. Kharkivs'ka Oblast', E Ukraine 49°43′N 35°49′E
123 *O12* **Novaya Chara** Zabaykal'skiy Kray, S Russian Federation 56°45′N 117°58′E
122 *M12* **Novaya Igirma** Irkutskaya Oblast', C Russian Federation 57°08′N 103°52′E
Novaya Kakhovka *see* Nova Kakhovka
Novaya Kazanka *see* Zhanakazan
124 *I12* **Novaya Ladoga** Leningradskaya Oblast', NW Russian Federation 60°03′N 32°15′E
127 *R5* **Novaya Malykla** Ul'yanovskaya Oblast', W Russian Federation 54°13′N 49°55′E
Novaya Odessa *see* Nova Odesa
123 *Q5* **Novaya Sibir', Ostrov** *island* Novosibirskiye Ostrova, NE Russian Federation
Novaya Vodolaga *see* Nova Vodolaha
122 *I6* **Novaya Zemlya** *island group* N Russian Federation
Novaya Zemlya Trough *see* East Novaya Zemlya Trough
114 *K10* **Nova Zagora** Sliven, C Bulgaria 42°29′N 26°00′E
105 *S12* **Novelda** Valenciana, E Spain 38°24′N 00°45′W
111 *H19* **Nové Mesto nad Váhom** *Ger.* Waagneustadtl, *Hung.* Vágújhely. Trenčiansky Kraj, W Slovakia 48°46′N 17°50′E
111 *F17* **Nové Město na Moravě** *Ger.* Neustadtl in Mähren. Vysočina, C Czech Republic 49°34′N 16°05′E
Novesium *see* Neuss
111 *I21* **Nové Zámky** *Ger.* Neuhäusel, *Hung.* Érsekújvár. Nitriansky Kraj, SW Slovakia 49°00′N 18°10′E
Novgorod *see* Velikiy Novgorod
Novgorod-Severskiy *see* Novhorod-Sivers'kyy
122 *C7* **Novgorodskaya Oblast'** ◈ *province* W Russian Federation
117 *R8* **Novhorodka** Kirovohrads'ka Oblast', C Ukraine 48°21′N 32°38′E
117 *R2* **Novhorod-Sivers'kyy** *Rus.* Novgorod-Severskiy. Chernihivs'ka Oblast', NE Ukraine 52°00′N 33°15′E
31 *R10* **Novi** Michigan, N USA 42°28′N 83°28′W
Novi *see* Novi Vinodolski
112 *L9* **Novi Bečej** *prev.* Új-Becse, Vološinovo, *Ger.* Neubetsche, *Hung.* Törökbecse. Vojvodina, N Serbia 45°36′N 20°09′E
116 *M3* **Novi Bilokorovychi** *Rus.* Belokorovichi; *prev.* Bilokorovychi. Zhytomyrs'ka Oblast', N Ukraine 51°07′N 28°02′E
25 *Q8* **Novice** Texas, SW USA 32°00′N 99°38′W
112 *A9* **Novigrad** Istra, NW Croatia 45°19′N 13°33′E
Novi Grad *see* Bosanski Novi
114 *G9* **Novi Iskar** Sofia Grad, W Bulgaria 42°46′N 23°19′E
106 *C9* **Novi Ligure** Piemonte, NW Italy 44°46′N 08°47′E
99 *L22* **Noville** Luxembourg, SE Belgium 50°04′N 05°46′E
194 *I10* **Noville Peninsula** *peninsula* Thurston Island, Antarctica
Noviodunum *see* Soissons, Aisne, France
Noviodunum *see* Nevers, Nièvre, France
Noviodunum *see* Nyon, Vaud, Switzerland
Noviomagus *see* Lisieux, Calvados, France
Noviomagus *see* Nijmegen, Netherlands
114 *M8* **Novi Pazar** Shumen, NE Bulgaria 43°20′N 27°12′E
113 *M15* **Novi Pazar** *Turk.* Yenipazar. Serbia, S Serbia 43°09′N 20°31′E
112 *K10* **Novi Sad** *Ger.* Neusatz, *Hung.* Újvidék. Vojvodina, N Serbia 45°16′N 19°49′E
117 *T6* **Novi Sanzhary** Poltavs'ka Oblast', C Ukraine 49°21′N 34°18′E
112 *H12* **Novi Travnik** *prev.* Pučarevo. Federacija Bosne I Hercegovine, C Bosnia and Herzegovina 44°12′N 17°39′E
112 *B10* **Novi Vinodolski** *var.* Novi. Primorje-Gorski Kotar, NW Croatia 45°08′N 14°46′E
58 *F12* **Novo Airão** Amazonas, N Brazil 02°06′S 61°20′W
Novoalekseyevka *see* Kobda
127 *N9* **Novoanninskiy** Volgogradskaya Oblast', SW Russian Federation 50°31′N 42°43′E
58 *F13* **Novo Aripuanã** Amazonas, N Brazil 05°05′S 60°20′W
117 *P7* **Novoarkhangel's'k** Kirovohrads'ka Oblast', C Ukraine 48°39′N 30°48′E
117 *Y6* **Novoaydar** Luhans'ka Oblast', E Ukraine 49°00′N 39°00′E
117 *X9* **Novoazovs'k** *Rus.* Novoazovsk. Donets'ka Oblast', E Ukraine 47°07′N 38°06′E
123 *R14* **Novobureyskiy** Amurskaya Oblast', SE Russian Federation 49°42′N 129°46′E
127 *Q3* **Novocheboksarsk** Chuvashskaya Respublika, W Russian Federation 56°07′N 47°33′E
127 *R5* **Novocheremshansk** Ul'yanovskaya Oblast', W Russian Federation 54°23′N 50°08′E
126 *L12* **Novocherkassk** Rostovskaya Oblast', SW Russian Federation 47°23′N 40°E
127 *R6* **Novodevich'ye** Samarskaya Oblast', W Russian Federation 53°33′N 48°51′E
124 *M8* **Novodvinsk** Arkhangel'skaya Oblast', NW Russian Federation 64°22′N 40°49′E
Novograd-Volynskiy *see* Novohrad-Volyns'kyy
Novogrudok *see* Navahrudak
Novogrudskaya Vozvyshennost' *see* Navahrudskaye Wzvyshsha
61 *I15* **Novo Hamburgo** Rio Grande do Sul, S Brazil 29°42′S 51°07′W
59 *H16* **Novo Horizonte** Mato Grosso, W Brazil 11°19′S 57°11′W
60 *K8* **Novo Horizonte** São Paulo, S Brazil 21°27′S 49°14′W
116 *M4* **Novohrad-Volyns'kyy** *Rus.* Novograd-Volynskiy. Zhytomyrs'ka Oblast', N Ukraine 50°34′N 27°32′E
145 *O7* **Novoishimskiy** *prev.* Kuybyshevskiy. Severnyy Kazakhstan, N Kazakhstan 53°15′N 66°51′E
Novokazalinsk *see* Ayteke Bi
126 *M8* **Novokhopersk** Voronezhskaya Oblast', W Russian Federation 51°09′N 41°34′E
127 *R6* **Novokuybyshevsk** Samarskaya Oblast', W Russian Federation 53°06′N 49°56′E
122 *J13* **Novokuznetsk** *prev.* Stalinsk. Kemerovskaya Oblast', S Russian Federation 53°45′N 87°12′E
195 *R1* **Novolazarevskaya** *Russian research station* Antarctica 70°42′S 11°31′E
Novolukoml' *see* Novalukoml'
109 *V12* **Novo mesto** *Ger.* Rudolfswert; *prev. Ger.* Neustadtl. SE Slovenia 45°49′N 15°09′E
126 *K15* **Novomikhaylovskiy** Krasnodarskiy Kray, SW Russian Federation 44°18′N 38°49′E
112 *L8* **Novo Miloševo** Vojvodina, N Serbia 45°43′N 20°20′E
Novomirgorod *see* Novomyrhorod
126 *L5* **Novomoskovsk** Tul'skaya Oblast', W Russian Federation 54°05′N 38°23′E
117 *U7* **Novomoskovs'k** *Rus.* Novomoskovsk. Dnipropetrovs'ka Oblast', E Ukraine 48°38′N 35°15′E
117 *V8* **Novomykolayivka** Zaporiz'ka Oblast', SE Ukraine 47°58′N 35°54′E
117 *Q7* **Novomyrhorod** *Rus.* Novomirgorod. Kirovohrads'ka Oblast', S Ukraine 48°46′N 31°39′E
127 *N8* **Novonikolayevskiy** Volgogradskaya Oblast', SW Russian Federation 50°55′N 42°24′E
127 *P10* **Novonikol'skoye** Volgogradskaya Oblast', SW Russian Federation 49°23′N 45°06′E
127 *X7* **Novoorsk** Orenburgskaya Oblast', W Russian Federation 51°21′N 59°03′E
126 *M13* **Novopokrovskaya** Krasnodarskiy Kray, SW Russian Federation 45°58′N 40°43′E
Novopolotsk *see* Navapolatsk
117 *Y5* **Novopskov** Luhans'ka Oblast', E Ukraine 49°33′N 39°07′E
Novoradomsk *see* Radomsko
Novo Redondo *see* Sumbe
127 *R8* **Novorepnoye** Saratovskaya Oblast', W Russian Federation 51°04′N 48°34′E
126 *K14* **Novorossiysk** Krasnodarskiy Kray, SW Russian Federation 44°50′N 37°38′E
Novorossiyskiy/Novorossiyskoye *see* Akzhar
124 *F15* **Novorzhev** Pskovskaya Oblast', W Russian Federation 57°01′N 29°19′E
Novoselitsa *see* Novoselytsya
117 *S12* **Novoselivs'ke** Avtonomna Respublika Krym, S Ukraine 45°26′N 33°37′E
Novosëlki *see* Navasyolki
114 *G6* **Novo Selo** Vidin, NW Bulgaria 44°08′N 22°48′E
113 *M14* **Novo Selo** Serbia, C Serbia 43°39′N 20°54′E
116 *K8* **Novoselytsya** *Rom.* Nouă Suliţa, *Rus.* Novoselitsa. Chernivets'ka Oblast', W Ukraine 48°14′N 26°18′E
127 *U7* **Novosergiyevka** Orenburgskaya Oblast', W Russian Federation 52°04′N 53°40′E
126 *L11* **Novoshakhtinsk** Rostovskaya Oblast', SW Russian Federation 47°48′N 39°51′E
122 *J12* **Novosibirsk** Novosibirskaya Oblast', C Russian Federation 55°04′N 83°05′E
122 *J12* **Novosibirskaya Oblast'** ◈ *province* C Russian Federation
122 *M4* **Novosibirskiye Ostrova** *Eng.* New Siberian Islands. *island group* N Russian Federation
126 *K6* **Novosil'** Orlovskaya Oblast', W Russian Federation 53°00′N 37°59′E
124 *G16* **Novosokol'niki** Pskovskaya Oblast', W Russian Federation 56°21′N 30°07′E
127 *Q6* **Novospasskoye** Ul'yanovskaya Oblast', W Russian Federation 53°08′N 47°48′E
Novotroickoje *see* Birlik
127 *X8* **Novotroitsk** Orenburgskaya Oblast', W Russian Federation 51°10′N 58°18′E
Novotroitskoye *see* Brlik, Kazakhstan
Novotroitskoye *see* Novotroyits'ke, Ukraine
117 *T11* **Novotroyits'ke** *Rus.* Novotroitskoye. Khersons'ka Oblast', S Ukraine 46°21′N 34°21′E
Novoukrainka *see* Novoukrayinka
117 *Q8* **Novoukrayinka** *Rus.* Novoukrainka. Kirovohrads'ka Oblast', C Ukraine 48°19′N 31°33′E
127 *Q5* **Novoul'yanovsk** Ul'yanovskaya Oblast', W Russian Federation 54°10′N 48°19′E
127 *W8* **Novoural'sk** Orenburgskaya Oblast', W Russian Federation 51°19′N 56°57′E
Novo-Urgench *see* Urganch
116 *I4* **Novovolyns'k** *Rus.* Novovolynsk. Volyns'ka Oblast', NW Ukraine 50°46′N 24°09′E
117 *S9* **Novovorontsovka** Khersons'ka Oblast', S Ukraine 47°28′N 33°55′E
147 *Y7* **Novovoznesenovka** Issyk-Kul'skaya Oblast', E Kyrgyzstan 42°36′N 78°44′E
125 *R14* **Novovyatsk** Kirovskaya Oblast', NW Russian Federation 58°30′N 49°42′E
Novoyel'nya *see* Navayel'nya
117 *O6* **Novozhyvotiv** Vinnyts'ka Oblast', C Ukraine 49°16′N 29°31′E
126 *H6* **Novozybkov** Bryanskaya Oblast', W Russian Federation 52°36′N 31°58′E
112 *F9* **Novska** Sisak-Moslavina, NE Croatia 45°20′N 16°58′E
111 *D15* **Nový Bor** *Ger.* Haida; *prev.* Bor u České Lípy, Hajda. Liberecký Kraj, N Czech Republic 50°46′N 14°32′E
111 *E16* **Nový Bydžov** *Ger.* Neubidschow. Královéhradecký Kraj, N Czech Republic 50°15′N 15°27′E
119 *G18* **Novy Dvor** *Rus.* Novyy Dvor. Hrodzyenskaya Voblasts', W Belarus 53°48′N 24°34′E
111 *I17* **Nový Jičín** *Ger.* Neutitschein. Moravskoslezský Kraj, E Czech Republic 49°36′N 18°00′E
118 *K12* **Novy Pahost** *Rus.* Novyy Pogost. Vitsyebskaya Voblasts', NW Belarus 55°30′N 27°29′E
Novyy Bug *see* Novyy Buh
117 *R9* **Novyy Buh** *Rus.* Novyy Bug. Mykolayivs'ka Oblast', S Ukraine 47°40′N 32°30′E
117 *Q4* **Novyy Bykiv** Chernihivs'ka Oblast', N Ukraine 50°36′N 31°39′E
Novyy Dvor *see* Novy Dvor
Novyye Aneny *see* Anenii Noi
127 *P7* **Novyye Burasy** Saratovskaya Oblast', W Russian Federation 52°10′N 46°00′E
Novyy Margilan *see* Farg'ona
126 *K8* **Novyy Oskol** Belgorodskaya Oblast', W Russian Federation 50°43′N 37°55′E
Novyy Pogost *see* Novy Pahost
127 *R2* **Novyy Tor"yal** Respublika Mariy El, W Russian Federation 56°59′N 48°53′E
123 *N12* **Novyy Uoyan** Respublika Buryatiya, S Russian Federation 56°06′N 111°27′E
122 *J9* **Novyy Urengoy** Yamalo-Nenetskiy Avtonomnyy Okrug, N Russian Federation 66°06′N 76°25′E
Novyy Uzen' *see* Zhanaozen
111 *N16* **Nowa Dęba** Podkarpackie, SE Poland 50°31′N 21°53′E
111 *G15* **Nowa Ruda** *Ger.* Neurode. Dolnośląskie, SW Poland 50°34′N 16°30′E
Nowasól *see* Nowa Sól
110 *F12* **Nowa Sól** *var.* Nowasól, *Ger.* Neusalz an der Oder. Lubuskie, W Poland 51°47′N 15°43′E
27 *Q8* **Nowata** Oklahoma, C USA 36°42′N 95°36′W
142 *M6* **Nowbarān** Markazī, W Iran 35°07′N 49°51′E
110 *J8* **Nowe** Kujawski-pomorskie, N Poland 53°40′N 18°44′E
110 *K9* **Nowe Miasto Lubawskie** *Ger.* Neumark. Warmińsko-Mazurskie, NE Poland 53°24′N 19°36′E
110 *L13* **Nowe Miasto nad Pilicą** Mazowieckie, C Poland 51°37′N 20°34′E
110 *D8* **Nowe Warpno** *Ger.* Neuwarp. Zachodnio-pomorskie, NW Poland 53°52′N 14°12′E
Nowgong *see* Nagaon
110 *E8* **Nowogard** *var.* Nowógard, *Ger.* Naugard. Zachodnio-pomorskie, NW Poland 53°41′N 15°09′E
110 *N9* **Nowogród** Podlaskie, NE Poland 53°14′N 21°52′E
Nowogródek *see* Navahrudak
111 *E14* **Nowogrodziec** *Ger.* Naumburg am Queis. Dolnośląskie, SW Poland 51°12′N 15°24′E
Nowojelnia *see* Navayel'nya
Nowo-Minsk *see* Mińsk Mazowiecki
33 *V13* **Nowood River** ≈ Wyoming, C USA
Nowo-Święciany *see* Švenčionėliai
183 *S10* **Nowra-Bomaderry** New South Wales, SE Australia 34°51′S 150°41′E
149 *T5* **Nowshera** *var.* Naushahra, Naushara. Khyber Pakhtunkhwa, NE Pakistan 34°00′N 72°00′E
110 *J7* **Nowy Dwór Gdański** *Ger.* Tiegenhof. Pomorskie, N Poland 54°12′N 19°03′E
110 *L11* **Nowy Dwór Mazowiecki** Mazowieckie, C Poland 52°26′N 20°43′E
111 *M17* **Nowy Sącz** *Ger.* Neu Sandec. Małopolskie, S Poland 49°36′N 20°42′E
111 *L18* **Nowy Targ** *Ger.* Neumarkt. Małopolskie, S Poland 49°28′N 20°00′E
110 *F11* **Nowy Tomyśl** *var.* Nowy Tomysl. Wielkopolskie, C Poland 52°18′N 16°07′E
Nowy Tomysl *see* Nowy Tomyśl
148 *M7* **Now Zād** *var.* Nauzad. Helmand, S Afghanistan 32°22′N 64°32′E
23 *N4* **Noxubee River** ≈ Alabama/Mississippi, S USA
122 *I10* **Noyabr'sk** Yamalo-Nenetskiy Avtonomnyy Okrug, N Russian Federation 63°08′N 75°19′E
102 *L8* **Noyant** Maine-et-Loire, NW France 47°28′N 00°08′W
39 *X14* **Noyes Island** *island* Alexander Archipelago, Alaska, USA
103 *O3* **Noyon** Oise, N France 49°35′N 03°E
102 *I7* **Nozay** Loire-Atlantique, NW France 47°34′N 01°36′W
82 *L12* **Nsando** Northern, NE Zambia 10°22′S 31°14′E
83 *N16* **Nsanje** Southern, S Malawi 16°57′S 35°10′E
77 *Q17* **Nsawam** SE Ghana 05°47′N 00°19′W
79 *E16* **Nsimalen** ✈ Centre, C Cameroon 19°15′N 81°22′E
82 *K12* **Nsombo** Northern, NE Zambia 10°35′S 29°58′E
82 *H13* **Ntambu** North Western, NW Zambia 12°21′S 25°03′E
83 *N14* **Ntcheu** *var.* Ncheu. Central, S Malawi 14°49′S 34°37′E
79 *D17* **Ntem** *prev.* Campo, Kampo. ≈ Cameroon/Equatorial Guinea
83 *I14* **Ntemwa** North Western, NW Zambia 14°03′S 26°13′E
Ntlenyana, Mount *see* Thabana Ntlenyana
79 *I19* **Ntomba, Lac** *var.* Lac Tumba. ◎ NW Dem. Rep. Congo
115 *I19* **Ntóro, Kávo** *prev.* Akrotírio Kafiréas. *cape* Évvoia, C Greece
81 *E19* **Ntungamo** SW Uganda 0°54′S 30°16′E
81 *E18* **Ntusi** SW Uganda 0°05′N 31°13′E
83 *H18* **Ntwetwe Pan** *salt lake* NE Botswana
93 *M15* **Nuasjärvi** ◎ C Finland
80 *F11* **Nuba Mountains** ▲▲ C Sudan
68 *J9* **Nubian Desert** *desert* NE Sudan
116 *G10* **Nucet** *Hung.* Diófás. Bihor, W Romania 46°28′N 22°35′E
Nu Chiang *see* Salween
145 *U9* **Nuclear Testing Ground** *nuclear site* Pavlodar, E Kazakhstan
56 *E9* **Nucuray, Río** ≈ N Peru
25 *R14* **Nueces River** ≈ Texas, SW USA
9 *V9* **Nueltin Lake** ◎ Manitoba/Northwest Territories, C Canada
99 *K15* **Nuenen** Noord-Brabant, S Netherlands 51°29′N 05°36′E
62 *G6* **Nuestra Señora, Bahía** *bay* N Chile
61 *D14* **Nuestra Señora Rosario de Caa Catí** Corrientes, NE Argentina 27°48′S 57°42′W
54 *J9* **Nueva Antioquia** Vichada, E Colombia 06°04′N 69°30′W
Nueva Caceres *see* Naga
41 *O7* **Nueva Ciudad Guerrera** Tamaulipas, C Mexico 26°32′N 99°13′W
55 *N4* **Nueva Esparta** *off.* Estado Nueva Esparta. ◈ *state* NE Venezuela
Nueva Esparta, Estado *see* Nueva Esparta
44 *C5* **Nueva Gerona** Isla de la Juventud, S Cuba 21°49′N 82°49′W
42 *H8* **Nueva Guadalupe** San Miguel, E El Salvador 13°30′N 88°21′W
42 *M11* **Nueva Guinea** Región Autónoma Atlántico Sur, SE Nicaragua 11°40′N 84°22′W
61 *D19* **Nueva Helvecia** Colonia, SW Uruguay 34°16′S 57°53′W
63 *J25* **Nueva, Isla** *island* S Chile
40 *M14* **Nueva Italia** Michoacán, SW Mexico 19°01′N 102°06′W
56 *D6* **Nueva Loja** *var.* Lago Agrio. Sucumbíos, NE Ecuador
42 *F6* **Nueva Ocotepeque** *prev.* Ocotepeque. Ocotepeque, W Honduras 14°25′N 89°10′W
61 *D19* **Nueva Palmira** Colonia, SW Uruguay 33°53′S 58°25′W
41 *N6* **Nueva Rosita** Coahuila, NE Mexico 27°58′N 101°11′W
42 *E7* **Nueva San Salvador** *prev.* Santa Tecla. La Libertad, SW El Salvador 13°40′N 89°18′W
42 *J8* **Nueva Segovia** ◈ *department* NW Nicaragua
Nueva Tabarca *see* Plana, Isla
Nueva Villa de Padilla *see* Nuevo Padilla
61 *B21* **Nueve de Julio** Buenos Aires, E Argentina 35°29′S 60°52′W
44 *H6* **Nuevitas** Camagüey, E Cuba 21°34′N 77°18′W
61 *D18* **Nuevo Berlín** Río Negro, W Uruguay 32°59′S 58°03′W
40 *I4* **Nuevo Casas Grandes** Chihuahua, N Mexico 30°23′N 107°54′W
43 *T14* **Nuevo Chagres** Colón, C Panama 09°14′N 80°05′W
41 *W15* **Nuevo Coahuila** Campeche, E Mexico 17°53′N 90°46′W
63 *K17* **Nuevo, Golfo** *gulf* S Argentina
41 *O7* **Nuevo Laredo** Tamaulipas, NE Mexico 27°28′N 99°32′W
41 *N8* **Nuevo León** ◈ *state* NE Mexico
41 *P10* **Nuevo Padilla** *var.* Nueva Villa de Padilla. Tamaulipas, C Mexico 24°01′N 98°48′W
56 *E6* **Nuevo Rocafuerte** Orellana, E Ecuador 0°59′S 75°27′W
80 *O13* **Nugaal** *off.* Gobolka Nugaal. ◈ *region* N Somalia
Nugaal, Gobolka *see* Nugaal
185 *E24* **Nugget Point** *headland* South Island, New Zealand 46°26′S 169°49′E
186 *J5* **Nuguria Islands** *island group* E Papua New Guinea
184 *P10* **Nuhaka** Hawke's Bay, North Island, New Zealand 39°03′S 177°43′E
138 *M10* **Nuhaydayn, Wādī an** *dry watercourse* W Iraq
190 *E7* **Nui Atoll** *atoll* W Tuvalu
Nu Jiang *see* Salween
Nûk *see* Nuuk
182 *G7* **Nukey Bluff** *hill* South Australia
Nukha *see* Şäki
123 *T9* **Nukh Yablonevyy, Gora** ▲ E Russian Federation 60°26′N 151°45′E
186 *K7* **Nukiki** Choiseul, NW Solomon Islands 06°45′S 156°30′E
186 *B6* **Nuku** West Sepik, NW Papua New Guinea 03°48′S 142°23′E
193 *W15* **Nuku** *island* Tongatapu Group, NE Tonga
193 *Y16* **Nuku'alofa** ● (Tonga) Tongatapu, S Tonga 21°08′S 175°13′W
193 *U15* **Nuku'alofa** Tongatapu, S Tonga 21°09′S 175°14′W
190 *G12* **Nukuatea** *island* N Wallis and Futuna
190 *F7* **Nukufetau Atoll** *atoll* C Tuvalu
190 *G12* **Nukuhifala** *island* E Wallis and Futuna
191 *W7* **Nuku Hiva** *island* Îles Marquises, NE French Polynesia
191 *W7* **Nuku Hiva Island** *island* Îles Marquises, N French Polynesia
190 *F9* **Nukulaelae Atoll** *var.* Nukulailai. *atoll* E Tuvalu
Nukulailai *see* Nukulaelae Atoll
190 *G11* **Nukuloa** *island* N Wallis and Futuna
186 *L6* **Nukumanu Islands** *prev.* Tasman Group. *island group* NE Papua New Guinea
Nukunau *see* Nikunau
190 *J9* **Nukunonu Atoll** *island* C Tokelau
190 *J9* **Nukunonu Village** Nukunonu Atoll, C Tokelau
189 *S18* **Nukuoro Atoll** *atoll* Caroline Islands, S Micronesia
146 *H8* **Nukus** Qoraqalpog'iston Respublikasi, W Uzbekistan 42°29′N 59°32′E
190 *G11* **Nukutapu** *island* N Wallis and Futuna
39 *O9* **Nulato** Alaska, USA 64°43′N 158°06′W
39 *O10* **Nulato Hills** ▲▲ Alaska, USA
105 *T9* **Nules** Valenciana, E Spain 39°52′N 00°10′W
Nuling *see* Sultan Kudarat
182 *C6* **Nullarbor** South Australia 31°28′S 130°57′E
180 *M11* **Nullarbor Plain** *plateau* South Australia/Western Australia
163 *S12* **Nulu'erhu Shan** ▲▲ N China
77 *X14* **Numan** Adamawa, E Nigeria 09°26′N 11°58′E
165 *S3* **Numata** Hokkaidō, NE Japan 43°48′N 141°55′E
81 *C15* **Numatinna** ≈ W South Sudan
95 *F14* **Numedalen** *valley* S Norway
95 *G14* **Numedalslågen** *var.* Laagen. ≈ S Norway
93 *L19* **Nummela** Uusimaa, S Finland 60°21′N 24°20′E
183 *O11* **Numurkah** Victoria, SE Australia 36°04′S 145°28′E
196 *L16* **Nunap Isua** *var.* Uummannarsuaq, *Dan.* Kap Farvel, *Eng.* Cape Farewell. *cape* S Greenland
9 *N8* **Nunavut** ◈ *territory* N Canada
54 *H9* **Nunchía** Casanare, C Colombia 05°37′N 72°13′W
97 *M20* **Nuneaton** C England, United Kingdom 52°32′N 01°28′W
153 *W14* **Nungba** Manipur, NE India 24°46′N 93°25′E
38 *L12* **Nunivak Island** *island* Alaska, USA
152 *I5* **Nun Kun** ▲ NW India 34°01′N 76°04′E
98 *L10* **Nunspeet** Gelderland, E Netherlands 52°21′N 05°45′E
107 *C18* **Nuoro** Sardegna, Italy, C Mediterranean Sea 40°20′N 09°20′E
75 *R12* **Nuqayy, Jabal** *hill range* S Libya
54 *C9* **Nuquí** Chocó, W Colombia 05°44′N 77°16′W
143 *O4* **Nūr** Māzandarān, N Iran 36°32′N 52°00′E
145 *Q9* **Nura** ≈ N Kazakhstan
143 *N11* **Nūrābād** Fārs, C Iran 30°08′N 51°30′E
Nurakita *see* Niulakita
Nurata *see* Nurota
Nuratau, Khrebet *see* Nurota Tizmasi
136 *L17* **Nur Dağları** ▲▲ S Turkey
Nurek *see* Norak
Nuremberg *see* Nürnberg
Nürestān *see* Nūristān
136 *M15* **Nurhak** Kahramanmaraş, S Turkey 37°57′N 37°21′E
182 *J9* **Nuriootpa** South Australia 34°28′S 139°00′E
149 *S4* **Nūristān** *prev.* Nürestān. ◈ *province* C Afghanistan
127 *S5* **Nurlat** Respublika Tatarstan, W Russian Federation 54°26′N 50°48′E
93 *N15* **Nurmes** Pohjois-Karjala, E Finland 63°31′N 29°10′E
101 *K20* **Nürnberg** *Eng.* Nuremberg. Bayern, S Germany 49°27′N 11°05′E
101 *K20* **Nürnberg** ✈ Bayern, SE Germany 49°29′N 11°04′E
146 *M10* **Nurota** *Rus.* Nurata. Navoiy Viloyati, C Uzbekistan 40°41′N 65°43′E
147 *N10* **Nurota Tizmasi** *Rus.* Khrebet Nuratau. ▲▲ C Uzbekistan
149 *T8* **Nürpur** Punjab, E Pakistan 31°54′N 71°55′E
183 *P6* **Nurri, Mount** *hill* New South Wales, SE Australia
25 *T13* **Nursery** Texas, SW USA 28°55′N 97°05′W
169 *V17* **Nusa Tenggara Barat** *off.* Propinsi Nusa Tenggara Barat, *Eng.* West Nusa Tenggara. ◈ *province* S Indonesia
Nusa Tenggara Barat, Propinsi *see* Nusa Tenggara Barat
171 *O16* **Nusa Tenggara Timur** *off.* Propinsi Nusa Tenggara Timur, *Eng.* East Nusa Tenggara. ◈ *province* S Indonesia
Nusa Tenggara Timur, Propinsi *see* Nusa Tenggara Timur
171 *U14* **Nusawulan** Papua Barat, E Indonesia 04°03′S 132°56′E
137 *Q16* **Nusaybin** *var.* Nisibin. Manisa, SE Turkey 37°08′N 41°11′E
39 *O14* **Nushagak Bay** *bay* Alaska, USA
39 *O13* **Nushagak Peninsula** *headland* Alaska, USA 58°39′N 159°03′W
39 *O13* **Nushagak River** ≈ Alaska, USA
160 *E11* **Nu Shan** ▲▲ SW China
149 *N11* **Nushki** Baluchistān, SW Pakistan 29°33′N 66°01′E
Nussdorf *see* Năsăud
112 *J9* **Nuštar** Vukovar-Srijem, E Croatia 45°20′N 18°48′E
99 *L18* **Nuth** Limburg, SE Netherlands 50°55′N 05°52′E
100 *N13* **Nuthe** ≈ NE Germany
Nutmeg State *see* Connecticut
39 *T10* **Nutzotin Mountains** ▲▲ Alaska, USA
64 *I5* **Nuuk** *var.* Nûk, *Dan.* Godthaab, Godthåb. ○ (Greenland) Sermersooq, SW Greenland 64°15′N 51°35′W
92 *L13* **Nuupas** Lappi, NW Finland 66°01′N 26°19′E
191 *O7* **Nuupere, Pointe** *headland* Moorea, W French Polynesia 17°35′S 149°47′W
191 *O7* **Nuuroa, Pointe** *headland* Tahiti, W French Polynesia
Nüürst *see* Baganuur
Nuwara *see* Nuwara Eliya
155 *K25* **Nuwara Eliya** *var.* Nuwara. Central Province, S Sri Lanka 06°58′N 80°46′E
182 *E7* **Nuyts Archipelago** *island group* South Australia
83 *F17* **Nxaunxau** North West, NW Botswana 18°57′S 21°18′E
39 *N12* **Nyac** Alaska, USA 61°00′N 159°56′W
122 *H9* **Nyagan'** Khanty-Mansiyskiy Avtonomnyy Okrug-Yugra, N Russian Federation 62°10′N 65°32′E
Nyagassola *see* Niagassola
Nyagquka *see* Yajiang
81 *I18* **Nyahururu** Nyandarua, C Kenya 0°04′N 36°22′E
182 *M10* **Nyah West** Victoria, SE Australia 35°14′S 143°18′E
158 *M15* **Nyainqêntanglha Feng** ▲ W China 30°20′N 90°28′E
159 *N15* **Nyainqêntanglha Shan** ▲▲ W China
80 *B11* **Nyala** Southern Darfur, W Sudan 12°01′N 24°50′E
83 *M16* **Nyamapanda** Mashonaland East, NE Zimbabwe 16°59′S 32°52′E
81 *H19* **Nyamira** ◈ *county* W Kenya
81 *H25* **Nyamtumbo** Ruvuma, S Tanzania 10°33′S 36°08′E
Nyanda *see* Masvingo
81 *I19* **Nyandarua** ◈ *county* C Kenya
124 *M11* **Nyandoma** Arkhangel'skaya Oblast', NW Russian Federation 61°39′N 40°10′E
83 *M16* **Nyanga** *prev.* Inyanga. Manicaland, E Zimbabwe 18°13′S 32°46′E
79 *D20* **Nyanga** *off.* Province de la Nyanga, *var.* La Nyanga. ◈ *province* SW Gabon
79 *E20* **Nyanga** ≈ Congo/Gabon
Nyanga, Province de la *see* Nyanga
81 *F20* **Nyantakara** Kagera, NW Tanzania 03°05′S 31°23′E
81 *E21* **Nyanza-Lac** S Burundi 04°16′S 29°38′E
68 *J14* **Nyasa, Lake** *var.* Lake Malawi; *prev.* Lago Nyassa. ◎ E Africa
Nyasaland/Nyasaland Protectorate *see* Malawi
Nyassa, Lago *see* Nyasa, Lake
119 *J17* **Nyasvizh** *Pol.* Nieśwież, *Rus.* Nesvizh. Minskaya Voblasts', C Belarus 53°13′N 26°40′E
166 *M8* **Nyaunglebin** Bago, SW Myanmar (Burma) 17°59′N 94°44′E
166 *M5* **Nyaung-u** Magway, C Myanmar (Burma) 21°03′N 95°44′E
95 *H24* **Nyborg** Syddtjylland, C Denmark 55°19′N 10°48′E
95 *N21* **Nybro** Kalmar, S Sweden 56°45′N 15°54′E
119 *J16* **Nyeharelaye** *Rus.* Negoreloye. Minskaya Voblasts', C Belarus 53°36′N 27°04′E
195 *W3* **Nye Mountains** ▲▲ Antarctica
81 *I19* **Nyeri** Nyeri, C Kenya 0°25′S 36°56′E
81 *I19* **Nyeri** ◈ *county* C Kenya
118 *M11* **Nyeshcharda, Vozyera** *Rus.* Ozero Neshcherdo. ◎ N Belarus
92 *O2* **Ny-Friesland** *physical region* N Svalbard
95 *L14* **Nyhammar** Dalarna, C Sweden 60°19′N 14°55′E
160 *F7* **Nyikog Qu** ≈ C China
83 *L14* **Nyimba** Eastern, E Zambia 14°33′S 30°49′E
159 *P15* **Nyingchi** *var.* Bayizhen. Xizang Zizhiqu, W China 29°27′N 94°43′E
159 *P15* **Nyingchi** *var.* Pula. Xizang Zizhiqu, W China 29°34′N 94°33′E
Nyinma *see* Maqu
111 *O21* **Nyírbátor** Szabolcs-Szatmár-Bereg, E Hungary 47°50′N 22°09′E
111 *N21* **Nyíregyháza** Szabolcs-Szatmár-Bereg, NE Hungary 47°57′N 21°43′E
Nyiro *see* Ewaso Ng'iro
Nyitra *see* Nitra
Nyitrabánya *see* Handlová
93 *K16* **Nykarleby** *Fin.* Uusikaarlepyy. Österbotten, W Finland 63°22′N 22°30′E
95 *F21* **Nykøbing** Midtjylland, NW Denmark 56°48′N 08°52′E
95 *I25* **Nykøbing** Sjælland, SE Denmark 54°47′N 11°53′E
95 *I22* **Nykøbing** Sjælland, C Denmark 55°56′N 11°41′E
95 *N17* **Nyköping** Södermanland, S Sweden 58°45′N 17°03′E
95 *L15* **Nykroppa** Värmland, C Sweden 59°37′N 14°18′E
Nyland *see* Uusimaa
Nylstroom *see* Modimolle
183 *P7* **Nymagee** New South Wales, SE Australia 32°06′S 146°19′E
183 *V5* **Nymboida** New South Wales, SE Australia 29°57′S 152°45′E
183 *U5* **Nymboida River** ≈ New South Wales, SE Australia
111 *D16* **Nymburk** *var.* Neuenburg an der Elbe, *Ger.* Nimburg. Středočeský Kraj, C Czech Republic 50°12′N 15°00′E
95 *O16* **Nynäshamn** Stockholm, C Sweden 58°54′N 17°55′E
183 *Q6* **Nyngan** New South Wales, SE Australia 31°36′S 147°07′E
Nyoman *see* Neman
108 *A10* **Nyon** *Ger.* Neuss; *anc.* Noviodunum. Vaud, SW Switzerland 46°23′N 06°15′E
79 *D16* **Nyong** ≈ SW Cameroon
103 *S14* **Nyons** Drôme, E France 44°22′N 05°06′E
79 *D14* **Nyos, Lac** *Eng.* Lake Nyos. ◎ NW Cameroon
Nyos, Lake *see* Nyos, Lac
125 *U11* **Nyrob** *var.* Nyrov. Permskiy Kray, NW Russian Federation 60°41′N 56°42′E
Nyrov *see* Nyrob
111 *H15* **Nysa** *Ger.* Neisse. Opolskie, S Poland 50°28′N 17°20′E
32 *M13* **Nyssa** Oregon, NW USA 43°52′N 116°59′W
Nysa Łużycka *see* Neisse
Nyslott *see* Savonlinna
Nystad *see* Uusikaupunki
95 *I25* **Nysted** Sjælland, SE Denmark 54°40′N 11°41′E
125 *U14* **Nytva** Permskiy Kray, NW Russian Federation 57°56′N 55°22′E
165 *P8* **Nyūdō-zaki** *headland* Honshū, C Japan 39°59′N 139°40′E
125 *P9* **Nyukhcha** Arkhangel'skaya Oblast', NW Russian Federation 63°24′N 46°34′E
124 *H8* **Nyuk, Ozero** *var.* Ozero Njuk. ◎ NW Russian Federation
125 *O12* **Nyuksenitsa** *var.* Njuksenica. Vologodskaya Oblast', NW Russian Federation 60°25′N 44°12′E
79 *O22* **Nyunzu** Katanga, SE Dem. Rep. Congo 05°55′S 28°00′E
123 *O10* **Nyurba** Respublika Sakha (Yakutiya), NE Russian Federation 63°17′N 118°15′E
123 *O11* **Nyuya** Respublika Sakha (Yakutiya), NE Russian Federation 60°33′N 116°10′E
146 *K12* **Nyýazow** *Rus.* Niyazov. Lebap Welaýaty, NE Turkmenistan 39°13′N 63°16′E
117 *T10* **Nyzhni Sirohozy** Khersons'ka Oblast', S Ukraine 46°49′N 34°21′E
117 *U12* **Nyzhn'ohirs'kyy** *Rus.* Nizhnegorskiy. Avtonomna Respublika Krym, S Ukraine 45°26′N 34°42′E
NZ *see* New Zealand
81 *G21* **Nzega** Tabora, C Tanzania 04°13′S 33°11′E
76 *K15* **Nzérékoré** SE Guinea 07°45′N 08°49′W

◆ Country
● Country Capital
◇ Dependent Territory
○ Dependent Territory Capital
◈ Administrative Regions
✈ International Airport
▲ Mountain
▲▲ Mountain Range
Volcano
≈ River
◎ Lake
⊞ Reservoir

N'ZETO – OLD HARBOR

82 A10 **N'Zeto** *prev.* Ambrizete. Zaire Province, NW Angola 07°14′S 12°52′E
79 M24 **Nzilo, Lac** *prev.* Lac Delcommune. ◎ SE Dem. Rep. Congo
172 I13 **Nzwani** *Fr.* Anjouan, *var.* Ndzouani. *island* SE Comoros

O

29 O11 **Oacoma** South Dakota, N USA 43°49′N 99°25′W
29 N9 **Oahe Dam** *dam* South Dakota, N USA
28 M9 **Oahe, Lake** ⊠ North Dakota/South Dakota, N USA
38 C9 **Oa'hu** *var.* Oahu. *island* Hawai'ian Islands, Hawai'i, USA
165 V4 **O-Akan-dake** ▲ Hokkaidō, NE Japan 43°26′N 144°09′E
182 K8 **Oakbank** South Australia 33°07′S 140°36′E
19 P13 **Oak Bluffs** Martha's Vineyard, New York, NE USA 41°25′N 70°32′W
36 K4 **Oak City** Utah, W USA 39°22′N 112°19′W
37 R3 **Oak Creek** Colorado, C USA 40°16′N 106°57′W
35 P8 **Oakdale** California, W USA 37°46′N 120°51′W
22 H8 **Oakdale** Louisiana, S USA 30°49′N 92°39′W
29 P7 **Oakes** North Dakota, N USA 46°08′N 98°05′W
22 J4 **Oak Grove** Louisiana, S USA 32°51′N 91°25′W
97 N19 **Oakham** C England, United Kingdom 52°41′N 00°45′W
32 H7 **Oak Harbor** Washington, NW USA 48°17′N 122°38′W
21 R5 **Oak Hill** West Virginia, NE USA 37°59′N 81°09′W
35 N8 **Oakland** California, W USA 37°48′N 122°16′W
29 T15 **Oakland** Iowa, C USA 41°18′N 95°22′W
19 Q7 **Oakland** Maine, NE USA 44°32′N 69°43′W
21 T3 **Oakland** Maryland, NE USA 39°24′N 79°25′W
29 R14 **Oakland** Nebraska, C USA 41°50′N 96°28′W
31 N11 **Oak Lawn** Illinois, N USA 41°43′N 87°45′W
33 P16 **Oakley** Idaho, NW USA 42°13′N 113°54′W
26 I4 **Oakley** Kansas, C USA 39°08′N 100°53′W
31 N10 **Oak Park** Illinois, N USA 41°53′N 87°46′W
11 X16 **Oak Point** Manitoba, S Canada 50°23′N 97°00′W
32 G13 **Oakridge** Oregon, NW USA 43°45′N 122°27′W
20 M9 **Oak Ridge** Tennessee, S USA 36°02′N 84°12′W
184 K10 **Oakura** Taranaki, North Island, New Zealand 39°07′S 173°58′E
22 L7 **Oak Vale** Mississippi, S USA 31°26′N 89°57′W
14 G16 **Oakville** Ontario, S Canada 43°27′N 79°41′W
25 V8 **Oakwood** Texas, SW USA 31°34′N 95°51′W
185 F22 **Oamaru** Otago, South Island, New Zealand 45°10′S 170°51′E
96 F13 **Oa, Mull of** *headland* W Scotland, United Kingdom 55°35′N 06°20′W
171 O11 **Oan** Sulawesi, N Indonesia 01°16′N 121°25′E
185 J17 **Oaro** Canterbury, South Island, New Zealand 42°29′S 173°30′E
35 X2 **Oasis** Nevada, W USA 41°01′N 114°29′W
195 S15 **Oates Land** *physical region* Antarctica
183 P17 **Oatlands** Tasmania, SE Australia 42°21′S 147°23′E
36 I11 **Oatman** Arizona, SW USA 35°03′N 114°19′W
41 R16 **Oaxaca** *var.* Oaxaca de Juárez; *prev.* Antequera. Oaxaca, SE Mexico 17°04′N 96°41′W
41 Q16 **Oaxaca** ⬖ *state* SE Mexico
Oaxaca de Juárez *see* Oaxaca
122 I19 **Ob'** ≈ C Russian Federation
145 X9 **Oba** *prev.* Uba. ≈ E Kazakhstan
14 G9 **Obabika Lake** ◎ Ontario, S Canada
Obagan *see* Ubagan
118 M12 **Obal'** *Rus.* Obol'. Vitsyebskaya Voblasts', N Belarus 55°22′N 29°17′E
79 E16 **Obala** Centre, SW Cameroon 04°09′N 11°32′E
14 C6 **Oba Lake** ◎ Ontario, S Canada
164 J12 **Obama** Fukui, Honshū, SW Japan 35°32′N 135°45′E
96 H11 **Oban** W Scotland, United Kingdom 56°25′N 05°29′W
Oban *see* Halfmoon Bay
Obando *see* Puerto Inírida
104 I4 **O Barco** *var.* El Barco, El Barco de Valdeorras, O Barco de Valdeorras. Galicia, NW Spain 42°24′N 07°00′W
O Barco de Valdeorras *see* O Barco
Obbia *see* Hobyo
93 J16 **Obbola** Västerbotten, N Sweden 63°41′N 20°16′E
Obbrovazzo *see* Obrovac
Obchuga *see* Abchuha
Obdorsk *see* Salekhard
Óbecse *see* Bečej
118 I11 **Obeliai** Panevėžys, NE Lithuania 55°57′N 25°47′E
60 F13 **Oberá** Misiones, NE Argentina 27°29′S 55°08′W
108 E8 **Oberburg** Bern, W Switzerland 47°00′N 07°37′E
109 Q9 **Oberdrauburg** Salzburg, S Austria 46°45′N 12°59′E
Oberglogau *see* Głogówek
109 W4 **Ober Grafendorf** Niederösterreich, NE Austria 48°09′N 15°33′E
101 E15 **Oberhausen** Nordrhein-Westfalen, W Germany 51°27′N 06°50′E
Oberhollabrunn *see* Tulln
Oberlaibach *see* Vrhnika
101 Q15 **Oberlausitz** *var.* Hornja Łužica. *physical region* E Germany
26 J2 **Oberlin** Kansas, C USA 39°49′N 100°33′W
22 H8 **Oberlin** Louisiana, S USA 30°37′N 92°45′W
31 T11 **Oberlin** Ohio, N USA 41°17′N 82°13′W
103 U5 **Obernai** Bas-Rhin, NE France 48°28′N 07°30′E
109 R4 **Obernberg am Inn** Oberösterreich, N Austria 48°19′N 13°20′E
Oberndorf *see* Oberndorf am Neckar
101 G23 **Oberndorf am Neckar** *var.* Oberndorf. Baden-Württemberg, SW Germany 48°18′N 08°32′E
109 Q5 **Oberndorf bei Salzburg** Salzburg, W Austria 47°57′N 12°57′E
Oberneustadtl *see* Kysucké Nové Mesto
183 S8 **Oberon** New South Wales, SE Australia 33°42′S 149°50′E
109 Q4 **Oberösterreich** *off.* Land Oberösterreich, *Eng.* Upper Austria. ⬖ *state* NW Austria
Oberösterreich, Land *see* Oberösterreich
Oberpahlen *see* Põltsamaa
101 M19 **Oberpfälzer Wald** ▲▲ SE Germany
109 Y6 **Oberpullendorf** Burgenland, E Austria 47°32′N 16°30′E
Oberradkersburg *see* Gornja Radgona
101 G18 **Oberursel** Hessen, W Germany 50°12′N 08°34′E
109 Q8 **Obervellach** Salzburg, S Austria 46°56′N 13°10′E
109 X7 **Oberwart** Burgenland, SE Austria 47°18′N 16°12′E
Oberwischau *see* Vişeu de Sus
109 T7 **Oberwölz** *var.* Oberwölz-Stadt. Steiermark, SE Austria 47°12′N 14°20′E
Oberwölz-Stadt *see* Oberwölz
31 S13 **Obetz** Ohio, N USA 39°52′N 82°57′W
Ob', Gulf of *see* Obskaya Guba
58 H12 **Óbidos** Pará, NE Brazil 01°52′S 55°30′W
104 F10 **Óbidos** Leiria, C Portugal 39°21′N 09°09′W
Obidovichi *see* Abidavichy
147 Q13 **Obigarm** W Tajikistan 38°42′N 69°34′E
165 T2 **Obihiro** Hokkaidō, NE Japan 42°56′N 143°10′E
Obi-Khingou *see* Khingov
147 P13 **Obikiik** SW Tajikistan 38°07′N 68°36′E
Obilić *see* Obiliq
113 N16 **Obiliq** *Serb.* Obilić. N Kosovo 42°50′N 20°57′E
127 O12 **Obil'noye** Respublika Kalmykiya, SW Russian Federation 47°31′N 44°24′E
20 F8 **Obion** Tennessee, S USA 36°15′N 89°11′W
20 F8 **Obion River** ≈ Tennessee, S USA
171 S12 **Obi, Pulau** *island* Maluku, E Indonesia
165 S2 **Obira** Hokkaidō, NE Japan 44°01′N 141°39′E
127 N11 **Oblivskaya** Rostovskaya Oblast', SW Russian Federation 48°34′N 42°31′E
123 R14 **Obluch'ye** Yevreyskaya Avtonomnaya Oblast', SE Russian Federation 48°59′N 131°18′E
126 K4 **Obninsk** Kaluzhskaya Oblast', W Russian Federation 55°06′N 36°40′E
114 J8 **Obnova** Pleven, N Bulgaria 43°26′N 25°04′E
79 N15 **Obo** Haut-Mbomou, E Central African Republic 05°20′N 26°29′E
159 T9 **Obo** Qinghai, C China 37°57′N 101°03′E
80 M11 **Obock** E Djibouti 11°57′N 43°09′E
Obol' *see* Obal'
Obolyanka *see* Abalyanka
171 V13 **Obome** Papua Barat, E Indonesia 03°42′S 133°21′E
110 G11 **Oborniki** Wielkolpolskie, W Poland 52°38′N 16°48′E
79 G19 **Obouya** Cuvette, C Congo 0°56′S 15°41′E
126 J8 **Oboyan'** Kurskaya Oblast', W Russian Federation 51°12′N 36°15′E
124 M9 **Obozerskiy** Arkhangel'skaya Oblast', NW Russian Federation 63°26′N 40°20′E
112 L11 **Obrenovac** Serbia, N Serbia 44°39′N 20°12′E
112 D12 **Obrovac** *It.* Obbrovazzo. Zadar, SW Croatia 44°12′N 15°40′E
Obrovo *see* Abrova
35 Q3 **Observation Peak** ▲ California, W USA 40°48′N 120°07′W
122 J8 **Obskaya Guba** *Eng.* Gulf of Ob. *gulf* N Russian Federation
173 N13 **Ob' Tablemount** *undersea feature* S Indian Ocean 50°16′S 51°59′E
173 T10 **Ob' Trench** *undersea feature* E Indian Ocean
77 P16 **Obuasi** S Ghana 06°15′N 01°36′W
117 P5 **Obukhiv** *Rus.* Obukhov. Kyyivs'ka Oblast', N Ukraine 50°05′N 30°37′E
Obukhov *see* Obukhiv
125 U14 **Obva** ≈ NW Russian Federation
108 F8 **Obwalden** ⬖ *canton* C Switzerland
117 V10 **Obytichna Kosa** *spit* SE Ukraine
117 V10 **Obytichna Zatoka** *gulf* SE Ukraine
114 N9 **Obzor** Burgas, E Bulgaria 42°52′N 27°53′E
105 O3 **Oca** ≈ N Spain
23 W10 **Ocala** Florida, SE USA 29°11′N 82°08′W
40 M7 **Ocampo** Coahuila, NE Mexico 27°18′N 102°24′W
54 G7 **Ocaña** Norte de Santander, N Colombia 08°16′N 73°21′W
105 N9 **Ocaña** Castilla-La Mancha, C Spain 39°57′N 03°30′W
104 H4 **O Carballiño** *Cast.* Carballiño. Galicia, NW Spain 42°26′N 08°05′W
37 T9 **Ocate** New Mexico, SW USA 36°09′N 105°03′W
Ocavango *see* Okavango
57 D14 **Occidental, Cordillera** ▲▲ W South America
21 Q6 **Oceana** West Virginia, NE USA 37°41′N 81°37′W
21 Z4 **Ocean City** Maryland, NE USA 38°20′N 75°05′W
18 J17 **Ocean City** New Jersey, NE USA 39°15′N 74°33′W
10 K15 **Ocean Falls** British Columbia, SW Canada 52°24′N 127°42′W
Ocean Island *see* Banaba
Ocean Island *see* Kure Atoll
64 J9 **Oceanographer Fracture Zone** *tectonic feature* NW Atlantic Ocean
35 U17 **Oceanside** California, W USA 33°12′N 117°23′W
22 M9 **Ocean Springs** Mississippi, S USA 30°24′N 88°49′W
Ocean State *see* Rhode Island
25 O9 **O C Fisher Lake** ⊠ Texas, SW USA
117 Q10 **Ochakiv** *Rus.* Ochakov. Mykolayivs'ka Oblast', S Ukraine 46°36′N 31°33′E
Ochakov *see* Ochakiv
Ochamchira *see* Ochamchire
137 Q9 **Ochamchire** *Rus.* Ochamchira; *prev.* Och'amch'ire. W Georgia 42°45′N 41°30′E
Och'amch'ire *see* Ochamchire
Ochansk *see* Okhansk
125 T15 **Ochër** Permskiy Kray, NW Russian Federation 57°54′N 54°40′E
115 I19 **Óchi** ▲ Évvoia, C Greece 38°03′N 24°27′E
165 W4 **Ochiishi-misaki** *headland* Hokkaidō, NE Japan 43°10′N 145°29′E
23 S9 **Ochlockonee River** ≈ Florida/Georgia, SE USA
44 K12 **Ocho Rios** C Jamaica 18°24′N 77°06′W
Ochrida *see* Ohrid
Ochrida, Lake *see* Ohrid, Lake
101 J19 **Ochsenfurt** Bayern, C Germany 49°39′N 10°03′E
23 U7 **Ocilla** Georgia, SE USA 31°35′N 83°15′W
94 N13 **Ockelbo** Gävleborg, C Sweden 60°51′N 16°46′E
Ocker *see* Oker
95 I19 **Öckerö** Västra Götaland, S Sweden 57°43′N 11°39′E
23 U6 **Ocmulgee River** ≈ Georgia, SE USA
116 H11 **Ocna Mureş** *Hung.* Marosújvár; *prev.* Ocna Mureşului, *prev. Hung.* Marosújvárakna. Alba, C Romania 46°25′N 23°53′E
Ocna Mureşului *see* Ocna Mureş
116 H11 **Ocna Sibiului** *Ger.* Salzburg, *Hung.* Vizakna. Sibiu, C Romania 45°52′N 23°59′E
116 H13 **Ocnele Mari** *prev.* Vioara. Vâlcea, S Romania 45°03′N 24°18′E
116 L7 **Ocniţa** *Rus.* Oknitsa. N Moldova 48°25′N 27°30′E
23 U4 **Oconee, Lake** ⊠ Georgia, SE USA
23 U5 **Oconee River** ≈ Georgia, SE USA
30 M9 **Oconomowoc** Wisconsin, N USA 43°06′N 88°29′W
30 M6 **Oconto** Wisconsin, N USA 44°55′N 87°52′W
30 M6 **Oconto Falls** Wisconsin, N USA 44°52′N 88°06′W
30 M6 **Oconto River** ≈ Wisconsin, N USA
104 I3 **O Corgo** Galicia, NW Spain 42°56′N 07°25′W
41 V16 **Ocosingo** Chiapas, SE Mexico 17°04′N 92°15′W
42 J8 **Ocotal** Nueva Segovia, NW Nicaragua 13°38′N 86°28′W
42 F6 **Ocotepeque** ⬖ *department* W Honduras
Ocotepeque *see* Nueva Ocotepeque
40 L13 **Ocotlán** Jalisco, SW Mexico 20°21′N 102°42′W
41 R16 **Ocotlán** *var.* Ocotlán de Morelos. Oaxaca, SE Mexico 16°49′N 96°49′W
Ocotlán de Morelos *see* Ocotlán
41 U16 **Ocozocuautla** Chiapas, SE Mexico 16°46′N 93°22′W
21 Y10 **Ocracoke Island** *island* North Carolina, SE USA
102 I3 **Octeville** Manche, N France 49°37′N 01°39′W
October Revolution Island *see* Oktyabr'skoy Revolyutsii, Ostrov
43 R17 **Ocú** Herrera, S Panama 07°55′N 80°43′W
83 Q14 **Ocua** Cabo Delgado, NE Mozambique 13°37′S 39°44′E
Ocumare *see* Ocumare del Tuy
54 M5 **Ocumare del Tuy** *var.* Ocumare. Miranda, N Venezuela 10°07′N 66°47′W
77 P17 **Oda** SE Ghana 05°55′N 00°56′W
165 G12 **Ōda** *var.* Oda. Shimane, Honshū, SW Japan 35°10′N 132°29′E
92 K3 **Ódaðahraun** *lava flow* C Iceland
165 Q7 **Ōdate** Akita, Honshū, C Japan 40°18′N 140°34′E
165 N14 **Odawara** Kanagawa, Honshū, S Japan 35°15′N 139°08′E
95 D14 **Odda** Hordaland, S Norway 60°03′N 06°34′E
95 G22 **Odder** Midtjylland, C Denmark 55°59′N 10°10′E
Oddur *see* Xuddur
29 T13 **Odebolt** Iowa, C USA 42°18′N 95°15′W
104 H14 **Odeleite** Faro, S Portugal 37°02′N 07°29′W
25 Q4 **Odell** Texas, SW USA 34°19′N 99°24′W
25 T14 **Odem** Texas, SW USA 27°57′N 97°34′W
104 F13 **Odemira** Beja, S Portugal 37°35′N 08°38′W
136 C14 **Ödemiş** İzmir, SW Turkey 38°11′N 27°58′E
Ödenburg *see* Sopron
83 I22 **Odendaalsrus** Free State, C South Africa 27°52′S 26°42′E
Odenpäh *see* Otepää
95 H23 **Odense** Syddtjylland, C Denmark 55°24′N 10°23′E
101 H19 **Odenwald** ▲▲ W Germany
84 H10 **Oder** *Cz./Pol.* Odra. ≈ C Europe
Oderberg *see* Bohumín
100 P11 **Oderbruch** *wetland* Germany/Poland
Oderhaff *see* Szczeciński, Zalew
100 O11 **Oder-Havel-Kanal** *canal* NE Germany
Oderhellen *see* Odorheiu Secuiesc
100 P13 **Oder-Spree-Kanal** *canal* NE Germany
Odertal *see* Zdzieszowice
106 I7 **Oderzo** Veneto, NE Italy 45°48′N 12°33′E
177 P10 **Odesa** *Rus.* Odessa. Odes'ka Oblast', SW Ukraine 46°29′N 30°44′E
24 M8 **Odessa** Texas, SW USA 31°51′N 102°22′W
32 K8 **Odessa** Washington, NW USA 47°19′N 118°41′W
Odesa *see* Odes'ka Oblast'
95 L18 **Ödeshög** Östergötland, S Sweden 58°13′N 14°40′E
117 O9 **Odes'ka Oblast'** *var.* Odesa, *Rus.* Odesskaya Oblast'. ⬖ *province* SW Ukraine
Odessa *see* Odesa
Odesskaya Oblast' *see* Odes'ka Oblast'
122 H12 **Odesskoye** Omskaya Oblast', C Russian Federation 54°15′N 72°45′E
Odessus *see* Varna
102 F6 **Odet** ≈ NW France
104 I14 **Odiel** ≈ SW Spain
76 L14 **Odienné** NW Ivory Coast 09°32′N 07°35′W
171 O4 **Odiongan** Tablas Island, C Philippines 12°23′N 122°01′E
153 P17 **Odisha** *prev.* Orissa. ⬖ *state* NE India
116 L12 **Odobeşti** Vrancea, E Romania 45°46′N 27°06′E
110 H13 **Odolanów** *Ger.* Adelnau. Wielkopolskie, C Poland 51°35′N 17°42′E
167 R13 **Ôdôngk** Kâmpóng Spoe, S Cambodia 11°48′N 104°45′E
25 N6 **O'donnell** Texas, SW USA 32°57′N 101°49′W
98 O7 **Odoorn** Drenthe, NE Netherlands 52°52′N 06°49′E
Odorhei *see* Odorheiu Secuiesc
116 J11 **Odorheiu Secuiesc** *Ger.* Oderhellen, *Hung.* Vámosudvarhely; *prev.* Odorhei, *Ger.* Hofmarkt. Harghita, C Romania 46°18′N 25°19′E
Odra *see* Oder
112 J9 **Odžaci** *Ger.* Hodschag, *Hung.* Hodság. Vojvodina, NW Serbia 45°31′N 19°15′E
59 N14 **Oeiras** Piauí, E Brazil 07°00′S 42°07′W
104 F11 **Oeiras** Lisboa, C Portugal 38°41′N 09°18′W
101 G14 **Oelde** Nordrhein-Westfalen, W Germany 51°49′N 08°09′E
28 J11 **Oelrichs** South Dakota, N USA 43°08′N 103°13′W
101 M17 **Oelsnitz** Sachsen, E Germany 50°22′N 12°12′E
Oels/Oels in Schlesien *see* Oleśnica
29 X12 **Oelwein** Iowa, C USA 42°40′N 91°54′W
Oeniadae *see* Oiniádes
191 N17 **Oeno Island** *atoll* Pitcairn Group of Islands, C Pacific Ocean
Oesel *see* Saaremaa
108 L7 **Oetz** *var.* Ötz. Tirol, W Austria 47°15′N 10°56′E
137 P11 **Of** Trabzon, NE Turkey 40°57′N 40°17′E
30 K15 **O'Fallon** Illinois, N USA 38°35′N 89°54′W
27 W4 **O'Fallon** Missouri, C USA 38°54′N 90°31′W
107 N16 **Ofanto** ≈ S Italy
97 D18 **Offaly** *Ir.* Ua Uíbh Fhailí; *prev.* King's County. *cultural region* C Ireland
101 H18 **Offenbach** *var.* Offenbach am Main. Hessen, W Germany 50°06′N 08°46′E
Offenbach am Main *see* Offenbach
101 F22 **Offenburg** Baden-Württemberg, SW Germany 48°28′N 07°57′E
182 C2 **Officer Creek** *seasonal river* South Australia
Oficina María Elena *see* María Elena
Oficina Pedro de Valdivia *see* Pedro de Valdivia
115 K22 **Ofidoússa** *island* Kykládes, Greece, Aegean Sea
Ofiral *see* Sharm ash Shaykh
92 H10 **Ofotfjorden** *fjord* N Norway
192 L16 **Ofu** *island* Manua Islands, E American Samoa
165 R9 **Ōfunato** Iwate, Honshū, C Japan 39°04′N 141°43′E
165 P8 **Oga** Akita, Honshū, C Japan 39°56′N 139°47′E
Ogaadeen *see* Ogaden
165 Q9 **Ogachi** Akita, Honshū, C Japan 39°03′N 140°26′E
165 P9 **Ogachi-tōge** *pass* Honshū, C Japan
81 N14 **Ogaden** *Som.* Ogaadeen. *plateau* Ethiopia/Somalia
165 P8 **Oga-hantō** *peninsula* Honshū, C Japan
165 K13 **Ōgaki** Gifu, Honshū, SW Japan 35°22′N 136°35′E
28 L15 **Ogallala** Nebraska, C USA 41°09′N 101°44′W
168 M14 **Ogan, Air** ≈ Sumatera, W Indonesia
165 Y15 **Ogasawara-shotō** *Eng.* Bonin Islands. *island group* SE Japan
14 I9 **Ogascanane, Lac** ◎ Québec, SE Canada
165 R7 **Ogawara-ko** ◎ Honshū, C Japan
77 T15 **Ogbomosho** *var.* Ogmoboso. Oyo, W Nigeria 08°10′N 04°16′E
29 U13 **Ogden** Iowa, C USA 42°03′N 94°01′W
36 L2 **Ogden** Utah, W USA 41°09′N 111°58′W
18 I6 **Ogdensburg** New York, NE USA 44°42′N 75°25′W
23 W5 **Ogeechee River** ≈ Georgia, SE USA
Oger *see* Ogre
165 N10 **Ogi** Niigata, Sado, C Japan 37°49′N 138°16′E
10 H5 **Ogilvie** Yukon, NW Canada 63°34′N 139°44′W
10 H4 **Ogilvie** ≈ Yukon, NW Canada
10 H5 **Ogilvie Mountains** ▲▲ Yukon, NW Canada
Oginskiy Kanal *see* Ahinski Kanal
162 J7 **Ögiynuur** *var.* Dzegstey. Arhangay, C Mongolia 47°38′N 102°31′E
146 F6 **Og'iyon Sho'rxogi** *wetland* NW Uzbekistan
146 B10 **Oglanly** Balkan Welaýaty, W Turkmenistan 39°56′N 54°25′E
23 T5 **Oglethorpe** Georgia, SE USA 32°17′N 84°03′W
23 T2 **Oglethorpe, Mount** ▲ Georgia, SE USA 34°29′N 84°19′W
106 F7 **Oglio** *anc.* Ollius. ≈ N Italy
Ogmoboso *see* Ogbomosho
103 T8 **Ognon** ≈ E France
123 R13 **Ogodzha** Amurskaya Oblast', S Russian Federation 52°51′N 132°49′E
77 W16 **Ogoja** Cross River, S Nigeria 06°37′N 08°48′E
12 C10 **Ogoki** ≈ Ontario, S Canada
12 D11 **Ogoki Lake** ◎ Ontario, C Canada
Ögöömör *see* Hanhongor
79 F19 **Ogooué** ≈ Congo/Gabon
79 E18 **Ogooué-Ivindo** *off.* Province de l'Ogooué-Ivindo, *var.* L'Ogooué-Ivindo. ⬖ *province* N Gabon
Ogooué-Ivindo, Province de l' *see* Ogooué-Ivindo
79 E19 **Ogooué-Lolo** *off.* Province de l'Ogooué-Lolo, *var.* L'Ogooué-Lolo. ⬖ *province* C Gabon
Ogooué-Lolo, Province de l' *see* Ogooué-Lolo
79 C19 **Ogooué-Maritime** *off.* Province de l'Ogooué-Maritime, *var.* L'Ogooué-Maritime. ⬖ *province* W Gabon
Ogooué-Maritime, Province de l' *see* Ogooué-Maritime
165 D14 **Ogōri** Fukuoka, Kyūshū, SW Japan 33°24′N 130°34′E
114 H7 **Ogosta** ≈ NW Bulgaria
112 Q9 **Ogražden** *Bul.* Ograzhden. ▲ Bulgaria/FYR Macedonia *see also* Ograzhden
114 G12 **Ograzhden** *Mac.* Orgražden. ▲ Bulgaria/FYR Macedonia *see also* Orgražden
Ograzhden *see* Ogražden
118 G9 **Ogre** *Ger.* Oger. C Latvia 56°49′N 24°36′E
118 H9 **Ogre** ≈ C Latvia
112 C10 **Ogulin** Karlovac, NW Croatia 45°15′N 15°13′E
77 S16 **Ogun** ⬖ *state* SW Nigeria
Ogurdzhaly, Ostrov *see* OgurjalyAdasy
146 A12 **OgurjalyAdasy** *Rus.* Ogurdzhaly, Ostrov. *island* W Turkmenistan
77 U16 **Ogwashi-Uku** Delta, S Nigeria 06°08′N 06°38′E
185 B23 **Ohai** Southland, South Island, New Zealand 45°56′S 167°59′E
147 Q10 **Ohangaron** *Rus.* Akhangaran. Toshkent Viloyati, E Uzbekistan 40°56′N 69°37′E
147 Q10 **Ohangaron** *Rus.* Akhangaran. ≈ E Uzbekistan
83 C16 **Ohangwena** ⬖ *district* N Namibia
30 M10 **O'Hare** ✕ (Chicago) Illinois, N USA 41°59′N 87°56′W
165 R6 **Ōhata** Aomori, Honshū, C Japan 41°23′N 141°09′E
184 L13 **Ohau** Manawatu-Wanganui, North Island, New Zealand 40°40′S 175°15′E
185 E20 **Ohau, Lake** ◎ South Island, New Zealand
Ohcejohka *see* Utsjoki
99 J20 **Ohey** Namur, SE Belgium 50°26′N 05°07′E
191 X15 **O'Higgins, Cabo** *headland* Easter Island, Chile, E Pacific Ocean 27°05′S 109°15′W
O'Higgins, Lago *see* San Martín, Lago
31 S12 **Ohio** *off.* State of Ohio, *also known as* Buckeye State. ⬖ *state* N USA
0 L10 **Ohio River** ≈ N USA
Ohlau *see* Oława
101 H16 **Ohm** ≈ C Germany
193 W16 **Ohonua** 'Eua, E Tonga 21°20′S 174°57′W
23 V5 **Ohoopee River** ≈ Georgia, SE USA
100 L12 **Ohre** *Ger.* Eger. ≈ Czech Republic/Germany
Ohri *see* Ohrid
113 M20 **Ohrid** *Turk.* Ochrida, Ohri. SW FYR Macedonia 41°07′N 20°48′E
113 M20 **Ohrid, Lake** *var.* Lake Ochrida, *Alb.* Liqeni i Ohrit, *Mac.* Ohridsko Ezero. ◎ Albania/FYR Macedonia
Ohridsko Ezero/Ohrit, Liqeni i *see* Ohrid, Lake
184 L9 **Ohura** Manawatu-Wanganui, North Island, New Zealand 38°51′S 174°58′E
58 J9 **Oiapoque** Amapá, E Brazil 03°54′N 51°46′W
58 J10 **Oiapoque, Rio** *var.* Fleuve l'Oyapok, Oyapock. ≈ Brazil/French Guiana *see also* Oyapok, Fleuve l'
Oiapoque, Rio *see* Oyapok, Fleuve l'
54 G8 **Oiba** Santander, C Colombia 06°16′N 73°18′W
15 O9 **Oies, Île aux** *island* Québec, SE Canada
92 L13 **Oijärvi** Pohjois-Pohjanmaa, C Finland 65°38′N 26°05′E
92 L12 **Oikarainen** Lappi, N Finland 66°30′N 25°46′E
188 F10 **Oikuul** Babeldaob, N Palau
18 C13 **Oil City** Pennsylvania, NE USA 41°25′N 79°42′W
18 C12 **Oil Creek** ≈ Pennsylvania, NE USA
35 R13 **Oildale** California, W USA 35°25′N 119°01′W
Oileán Ciarral *see* Castleisland
Oil Islands *see* Chagos Archipelago
115 D18 **Oiniádes** *anc.* Oeniadae. *site of ancient city* Dytikí Elláda, W Greece
115 L18 **Oinoússes** *island* E Greece
Oírr, Inis *see* Inisheer
99 J15 **Oirschot** Noord-Brabant, S Netherlands 51°30′N 05°18′E
103 N4 **Oise** ⬖ *department* N France
103 P3 **Oise** ≈ N France
99 J14 **Oisterwijk** Noord-Brabant, S Netherlands 51°35′N 05°12′E
45 O14 **Oistins** S Barbados 13°04′N 59°33′W
165 E4 **Ōita** Ōita, Kyūshū, SW Japan 33°15′N 131°35′E
165 D14 **Ōita** *off.* Ōita-ken. ⬖ *prefecture* Kyūshū, SW Japan
Ōita-ken *see* Ōita
115 E17 **Oíti** ▲ C Greece 38°48′N 22°12′E
165 S4 **Oiwake** Hokkaidō, NE Japan 42°54′N 141°49′E
35 R14 **Ojai** California, W USA 34°25′N 119°15′W
94 K13 **Öje** Dalarna, C Sweden 60°49′N 13°54′E
93 J14 **Öjebyn** Norrbotten, N Sweden 65°20′N 21°26′E
165 B13 **Ojika-jima** *island* SW Japan
40 K5 **Ojinaga** Chihuahua, N Mexico 29°31′N 104°26′W
40 M11 **Ojo Caliente** *var.* Ojocaliente. Zacatecas, C Mexico 22°35′N 102°18′W
Ojocaliente *see* Ojo Caliente
40 D6 **Ojo de Liebre, Laguna** *var.* Laguna Scammon, Scammon Lagoon. *lagoon* NW Mexico
62 I7 **Ojos del Salado, Cerro** ▲ W Argentina 27°04′S 68°34′W
105 R7 **Ojos Negros** Aragón, NE Spain 40°43′N 01°30′W
40 M12 **Ojuelos de Jalisco** Aguascalientes, C Mexico 21°52′N 101°40′W
127 N4 **Oka** ≈ W Russian Federation
83 D19 **Okahandja** Otjozondjupa, C Namibia 21°58′S 16°55′E
184 L9 **Okahukura** Manawatu-Wanganui, North Island, New Zealand 38°48′S 175°13′E
184 J3 **Okaihau** Northland, North Island, New Zealand 35°19′S 173°45′E
83 D18 **Okakarara** Otjozondjupa, N Namibia 20°33′S 17°20′E
13 P5 **Okak Islands** *island group* Newfoundland and Labrador, E Canada
10 M17 **Okanagan** ≈ British Columbia, SW Canada
11 N17 **Okanagan Lake** ◎ British Columbia, SW Canada
Ókanizsa *see* Kanjiža
83 C16 **Okankolo** Oshikoto, N Namibia 17°57′S 16°28′E
32 K6 **Okanogan River** ≈ Washington, NW USA
83 D18 **Okaputa** Otjozondjupa, N Namibia 20°09′S 16°56′E
149 V9 **Okāra** Punjab, E Pakistan 30°49′N 73°31′E
26 M10 **Okarche** Oklahoma, C USA 35°43′N 97°58′W
Okarem *see* Ekerem
189 X14 **Okat Harbor** *harbor* Kosrae, E Micronesia
22 M5 **Okatibbee Creek** ≈ Mississippi, S USA
83 C17 **Okaukuejo** Kunene, N Namibia 19°10′S 15°23′E
Okavanggo *see* Cubango/Okavango
83 E17 **Okavango** ⬖ *district* NW Namibia
83 G17 **Okavango** *var.* Cubango, Kavango, Kavengo, Kubango, Okavanggo, *Port.* Ocavango. ≈ S Africa *see also* Cubango
Okavango *see* Cubango
83 G17 **Okavango Delta** *wetland* N Botswana
164 M12 **Okaya** Nagano, Honshū, S Japan 36°03′N 138°00′E
164 H13 **Okayama** Okayama, Honshū, SW Japan 34°40′N 133°54′E
164 H13 **Okayama** *off.* Okayama-ken. ⬖ *prefecture* Honshū, SW Japan
Okayama-ken *see* Okayama
164 L14 **Okazaki** Aichi, Honshū, C Japan 34°58′N 137°10′E
110 M12 **Okęcie** ✕ (Warszawa) Mazowieckie, C Poland 52°08′N 20°57′E
23 Y13 **Okeechobee** Florida, SE USA 27°14′N 80°49′W
23 Y14 **Okeechobee, Lake** ◎ Florida, SE USA
26 M9 **Okeene** Oklahoma, C USA 36°07′N 98°19′W
23 V8 **Okefenokee Swamp** *wetland* Georgia, SE USA
97 J24 **Okehampton** SW England, United Kingdom 50°44′N 04°W
27 P10 **Okemah** Oklahoma, C USA 35°25′N 96°20′W
77 U16 **Okene** Kogi, S Nigeria 07°31′N 06°15′E
100 K13 **Oker** *var.* Ocker. ≈ NW Germany
101 J14 **Oker-Stausee** ◎ C Germany
123 T12 **Okha** Ostrov Sakhalin, Sakhalinskaya Oblast', SE Russian Federation 53°33′N 142°55′E
125 U15 **Okhansk** *var.* Ochansk. Permskiy Kray, NW Russian Federation 57°44′N 55°20′E
123 S10 **Okhotsk** Khabarovskiy Kray, E Russian Federation 59°21′N 143°15′E
192 J2 **Okhotsk, Sea of** *sea* NW Pacific Ocean
117 T4 **Okhtyrka** *Rus.* Akhtyrka. Sums'ka Oblast', NE Ukraine 50°19′N 34°54′E
83 E23 **Okiep** Northern Cape, W South Africa 29°39′S 17°53′E
Oki-guntō *see* Oki-shotō
164 H11 **Oki-kaikyō** *strait* SW Japan
165 P16 **Okinawa** Okinawa, SW Japan 26°20′N 127°47′E
165 S16 **Okinawa** *off.* Okinawa-ken. ⬖ *prefecture* Okinawa, SW Japan
165 S16 **Okinawa** *island* SW Japan
Okinawa-ken *see* Okinawa
165 U16 **Okinoerabu-jima** *island* Nansei-shotō, SW Japan
164 F15 **Okino-shima** *island* SW Japan
164 H11 **Oki-shotō** *var.* Oki-guntō. *island group* SW Japan
77 T16 **Okitipupa** Ondo, SW Nigeria 06°33′N 04°43′E
166 L8 **Okkan** Bago, SW Myanmar (Burma) 17°30′N 95°52′E
27 N10 **Oklahoma** *off.* State of Oklahoma, *also known as* The Sooner State. ⬖ *state* C USA
27 N11 **Oklahoma City** *state capital* Oklahoma, C USA 35°28′N 97°31′W
25 Q4 **Oklaunion** Texas, SW USA 34°07′N 99°07′W
23 W10 **Oklawaha River** ≈ Florida, SE USA
27 P10 **Okmulgee** Oklahoma, C USA 35°38′N 95°59′W
Oknitsa *see* Ocniţa
22 M3 **Okolona** Mississippi, S USA 34°00′N 88°45′W
165 U2 **Okoppe** Hokkaidō, NE Japan 44°27′N 143°06′E
11 Q16 **Okotoks** Alberta, SW Canada 50°46′N 113°57′W
80 H6 **Oko, Wadi** ≈ NE Sudan
79 G19 **Okoyo** Cuvette, W Congo 01°28′S 15°04′E
77 S15 **Okpara** ≈ Benin/Nigeria
92 J8 **Øksfjord** Finnmark, N Norway 70°13′N 22°22′E
125 R4 **Oksino** Nenetskiy Avtonomnyy Okrug, NW Russian Federation 67°33′N 52°15′E
92 G13 **Oksskolten** ▲ C Norway 66°00′N 14°18′E
Oksu *see* Oqsu
144 M8 **Oktaybr'skiy** Kostanay, N Kazakhstan
186 B7 **Ok Tedi** Western, W Papua New Guinea
Oktemberyan *see* Armavir
166 M7 **Oktwin** Bago, C Myanmar (Burma) 18°47′N 96°21′E
127 R6 **Oktyabr'sk** Samarskaya Oblast', W Russian Federation 53°13′N 48°36′E
Oktyab'rsk *see* Kandyagash
125 N12 **Oktyabr'skiy** Arkhangel'skaya Oblast', NW Russian Federation 61°03′N 43°16′E
122 E10 **Oktyabr'skiy** Kamchatskiy Kray, E Russian Federation 52°35′N 156°18′E
127 T5 **Oktyabr'skiy** Respublika Bashkortostan, W Russian Federation 54°28′N 53°29′E
127 O11 **Oktyabr'skiy** Volgogradskaya Oblast', SW Russian Federation 48°00′N 43°35′E
Oktyabr'skiy *see* Aktsyabrski
127 V7 **Oktyabr'skoye** Orenburgskaya Oblast', W Russian Federation 52°22′N 55°39′E
122 M5 **Oktyabr'skoy Revolyutsii, Ostrov** *Eng.* October Revolution Island. *island* Severnaya Zemlya, N Russian Federation
Ōkuchi *see* Isa
124 I14 **Okulovka** *var.* Okulovka. Novgorodskaya Oblast', W Russian Federation 58°24′N 33°16′E
Okulovka *see* Okulovka
165 Q4 **Okushiri-tō** *var.* Okusiri Tô. *island* NE Japan
Okusiri Tô *see* Okushiri-tō
77 S15 **Okuta** Kwara, W Nigeria 09°18′N 03°09′E
Ôkuti *see* Isa
83 F19 **Okwa** *var.* Chapman's. ≈ Botswana/Namibia
123 T10 **Ola** Magadanskaya Oblast', E Russian Federation 59°36′N 151°18′E
27 T11 **Ola** Arkansas, C USA 35°01′N 93°13′W
Ola *see* Ala
35 T11 **Olacha Peak** ▲ California, W USA 36°15′N 118°07′W
92 J1 **Ólafsfjörður** Norðurland Eystra, N Iceland 66°04′N 18°36′W
92 H3 **Ólafsvík** Vesturland, W Iceland 64°52′N 23°45′W
Oláhbrettye *see* Bretea-Română
Oláhszentgyőrgy *see* Sângeorz-Băi
Oláh-Toplicza *see* Toplița
118 F9 **Olaine** C Latvia 56°47′N 23°56′E
35 T11 **Olancha** California, W USA 36°16′N 118°00′W
42 J5 **Olanchito** Yoro, C Honduras 15°30′N 86°34′W
42 J6 **Olancho** ⬖ *department* E Honduras
95 O20 **Öland** *island* S Sweden
95 O19 **Ölands norra udde** *headland* S Sweden 57°21′N 17°06′E
95 N22 **Ölands södra udde** *headland* S Sweden 56°12′N 16°26′E
182 K7 **Olary** South Australia 32°18′S 140°16′E
27 R4 **Olathe** Kansas, C USA 38°52′N 94°50′W
61 C22 **Olavarría** Buenos Aires, E Argentina 36°57′S 60°20′W
92 O2 **Olav V Land** *physical region* C Svalbard
111 H14 **Oława** *Ger.* Ohlau. Dolnośląskie, SW Poland 50°57′N 17°18′E
107 D17 **Olbia** *prev.* Terranova Pausania. Sardegna, Italy, C Mediterranean Sea 40°55′N 09°30′E
44 G5 **Old Bahama Channel** *channel* The Bahamas/Cuba
Old Bay State/Old Colony State *see* Massachusetts
10 H2 **Old Crow** Yukon, NW Canada 67°34′N 139°55′W
Old Dominion *see* Virginia
Oldeberkeap *see* Oldeberkoop
98 M7 **Oldeberkoop** *Fris.* Oldeberkeap. Fryslân, N Netherlands 52°55′N 06°07′E
98 L10 **Oldebroek** Gelderland, E Netherlands 52°27′N 05°54′E
98 L8 **Oldemarkt** Overijssel, N Netherlands 52°49′N 05°58′E
94 E11 **Olden** Sogn Og Fjordane, C Norway 61°52′N 06°44′E
100 G10 **Oldenburg** Niedersachsen, NW Germany 53°09′N 08°13′E
100 K8 **Oldenburg** *var.* Oldenburg in Holstein. Schleswig-Holstein, N Germany 54°17′N 10°55′E
Oldenburg in Holstein *see* Oldenburg
98 P10 **Oldenzaal** Overijssel, E Netherlands 52°19′N 06°53′E
Olderfjord *see* Leaibevuotna
18 J8 **Old Forge** New York, NE USA 43°42′N 74°59′W
Old Goa *see* Goa
97 L17 **Oldham** NW England, United Kingdom 53°36′N 02°W
39 Q14 **Old Harbor** Kodiak Island, Alaska, USA 57°12′N 153°18′W

◆ Country
● Country Capital
◇ Dependent Territory
○ Dependent Territory Capital
⬖ Administrative Regions
✕ International Airport
▲ Mountain
▲▲ Mountain Range
Volcano
≈ River
◎ Lake
⊠ Reservoir

◆ Country ◇ Dependent Territory ◆ Administrative Regions ▲ Mountain ♒ Volcano ⊡ Lake
● Country Capital ○ Dependent Territory Capital ✕ International Airport ▲ Mountain Range ♒ River ⊡ Reservoir

ORIHUELA – OURINHOS

105 R12 **Orihuela** Valenciana, E Spain 38°05′N 00°56′W
117 V9 **Orikhiv** *Rus.* Orekhov. Zaporiz'ka Oblast', SE Ukraine 47°32′N 35°48′E
113 K22 **Orikum** *var.* Orikumi. Vlorë, SW Albania 40°20′N 19°28′E
Orikumi *see* Orikum
117 V6 **Oril'** *Rus.* Orel. ≈ E Ukraine
14 H14 **Orillia** Ontario, S Canada 44°36′N 79°26′W
93 M19 **Orimattila** Päijät-Häme, S Finland 60°48′N 25°40′E
33 Y15 **Orin** Wyoming, C USA 42°39′N 105°10′W
47 R4 **Orinoco, Río** ≈ Colombia/Venezuela
186 C9 **Oriomo** Western, SW Papua New Guinea 08°53′S 143°13′E
30 K11 **Orion** Illinois, N USA 41°21′N 90°22′W
29 Q5 **Oriska** North Dakota, N USA 46°54′N 97°46′W
Orissa *see* Odisha
Orissaar *see* Orissaare
118 E5 **Orissaare** *Ger.* Orissaar. Saaremaa, W Estonia 58°34′N 23°05′E
107 B19 **Oristano** Sardegna, Italy, C Mediterranean Sea 39°54′N 08°35′E
107 A19 **Oristano, Golfo di** *gulf* Sardegna, Italy, C Mediterranean Sea
54 D13 **Orito** Putumayo, SW Colombia 0°49′N 76°57′W
93 L18 **Orivesi** Häme, W Finland 61°39′N 24°21′E
93 N17 **Orivesi** ◎ Etelä-Savo, SE Finland
58 H12 **Oriximiná** Pará, NE Brazil 01°45′S 55°50′W
41 Q14 **Orizaba** Veracruz-Llave, E Mexico 18°51′N 97°08′W
41 Q14 **Orizaba, Volcán Pico de** *var.* Citlaltépetl. ▲ S Mexico 19°00′N 97°15′W
95 I16 **Ørje** Østfold, S Norway 59°28′N 11°40′E
113 I16 **Orjen** ⛰ Bosnia and Herzegovina/Montenegro
Orjiva *see* Orgiva
Orjonikidzeobod *see* Kofarnihon
94 G8 **Orkanger** Sør-Trøndelag, S Norway 63°17′N 09°52′E
94 G8 **Orkdalen** *valley* S Norway
95 K22 **Örkelljunga** Skåne, S Sweden 56°17′N 13°20′E
Orkhaniye *see* Botevgrad
Orkhómenos *see* Orchómenos
94 H9 **Orkla** ≈ S Norway
Orkney *see* Orkney Islands
65 J22 **Orkney Deep** *undersea feature* Scotia Sea/Weddell Sea
96 J4 **Orkney Islands** *var.* Orkney, Orkneys. *island group* N Scotland, United Kingdom
Orkneys *see* Orkney Islands
24 K8 **Orla** Texas, SW USA 31°48′N 103°55′W
35 N5 **Orland** California, W USA 39°43′N 122°12′W
23 X11 **Orlando** Florida, SE USA 28°32′N 81°23′W
23 X12 **Orlando** ✈ Florida, SE USA 28°24′N 81°16′W
107 K23 **Orlando, Capo d'** *headland* Sicilia, Italy, C Mediterranean Sea 38°10′N 14°44′E
Orlau *see* Orlová
103 N6 **Orléanais** *cultural region* C France
103 N7 **Orléans** *anc.* Aurelianum. Loiret, C France 47°54′N 01°53′E
34 L2 **Orleans** California, W USA 41°16′N 123°36′W
19 Q12 **Orleans** Massachusetts, NE USA 41°48′N 69°57′W
15 R10 **Orléans, Île d'** *island* Québec, SE Canada
Orléansville *see* Chlef
111 F16 **Orlice** *Ger.* Adler. ≈ NE Czech Republic
122 L13 **Orlik** Respublika Buryatiya, S Russian Federation 52°32′N 99°36′E
125 Q14 **Orlov** *prev.* Khalturin. Kirovskaya Oblast', NW Russian Federation 58°34′N 48°57′E
111 I17 **Orlová** *Ger.* Orlau, *Pol.* Orłowa. Moravskoslezský Kraj, E Czech Republic 49°50′N 18°21′E
Orlov, Mys *see* Orlovskiy, Mys
126 I6 **Orlovskaya Oblast'** ◈ *province* W Russian Federation
124 M5 **Orlovskiy, Mys** *var.* Mys Orlov. *headland* NW Russian Federation 67°14′N 41°17′E
Orłowa *see* Orlová
103 O5 **Orly** ✈ (Paris) Essonne, N France 48°43′N 02°24′E
119 G16 **Orlya** Hrodzyenskaya Voblasts', W Belarus 53°30′N 24°59′E
114 M7 **Orlyak** *prev.* Makenzen, Trubchular, *Rom.* Trupcilar. Dobrich, NE Bulgaria 43°39′N 27°21′E
148 L16 **Ormāra** Baluchistān, SW Pakistan 25°14′N 64°36′E
171 P5 **Ormoc** *off.* Ormoc City, *var.* MacArthur. Leyte, C Philippines 11°02′N 124°35′E
Ormoc City *see* Ormoc
23 X10 **Ormond Beach** Florida, SE USA 29°16′N 81°04′W
109 X10 **Ormož** *Ger.* Friedau. NE Slovenia 46°24′N 16°09′E
14 J13 **Ormsby** Ontario, SE Canada 44°52′N 77°45′W
97 K17 **Ormskirk** NW England, United Kingdom 53°35′N 02°54′W
Ormsö *see* Vormsi
15 N13 **Ormstown** Québec, SE Canada 45°08′N 73°57′W
Ormuz, Strait of *see* Hormuz, Strait of
103 T8 **Ornans** Doubs, E France 47°06′N 06°06′E
102 K5 **Orne** ◈ *department* N France
102 K5 **Orne** ≈ N France
92 G12 **Ørnes** Nordland, C Norway 66°51′N 13°43′E
110 L7 **Orneta** Warmińsko-Mazurskie, NE Poland 54°07′N 20°10′E
95 P16 **Ornö** Stockholm, C Sweden 59°03′N 18°28′E
37 Q3 **Orno Peak** ▲ Colorado, C USA 40°06′N 107°06′W
93 I16 **Örnsköldsvik** Västernorrland, C Sweden 63°16′N 18°45′E
163 X13 **Oro** E North Korea 39°59′N 127°27′E
Oro *see* Northern
45 T6 **Orocovis** C Puerto Rico 18°13′N 66°22′W
54 H10 **Orocué** Casanare, E Colombia 04°51′N 71°21′W
77 N13 **Orodara** SW Burkina Faso 11°00′N 04°54′W
105 S4 **Oroel, Peña de** ▲ N Spain 42°30′N 00°31′W
33 N10 **Orofino** Idaho, NW USA 46°28′N 116°15′W
162 I9 **Orog Nuur** ◎ S Mongolia
35 U14 **Oro Grande** California, W USA 34°36′N 117°19′W
37 S15 **Orogrande** New Mexico, SW USA 32°24′N 106°04′W
191 Q7 **Orohena, Mont** ✱ Tahiti, W French Polynesia 17°37′S 149°27′W
Orolaunum *see* Arlon
Orol Dengizi *see* Aral Sea
189 S15 **Oroluk Atoll** *atoll* Caroline Islands, C Micronesia
80 J13 **Oromīya** *var.* Oromo. ◈ C Ethiopia
Oromo *see* Oromīya
13 O15 **Oromocto** New Brunswick, SE Canada 45°50′N 66°28′W
191 S4 **Orona** *prev.* Hull Island. *atoll* Phoenix Islands, C Kiribati
191 V17 **Orongo** *ancient monument* Easter Island, Chile, E Pacific Ocean
138 I3 **Orontes** *var.* Ononte, Nahr el Aassi, *Ar.* Nahr al 'Aṣī. ≈ SW Asia
104 L9 **Oropesa** Castilla-La Mancha, C Spain 39°55′N 05°10′W
Oropesa *see* Oropesa del Mar
105 T8 **Oropesa del Mar** *var.* Oropesa, Orpesa, *Cat.* Orpes. Valenciana, E Spain 40°06′N 00°07′E
Oropeza *see* Cochabamba
Oroqen Zizhiqi *see* Alihe
171 P7 **Oroquieta** *var.* Oroquieta City. Mindanao, S Philippines 08°27′N 123°46′E
Oroquieta City *see* Oroquieta
40 J8 **Oro, Río del** ≈ C Mexico
59 O14 **Orós, Açude** ⊠ E Brazil
107 D18 **Orosei, Golfo di** *gulf* Tyrrhenian Sea, C Mediterranean Sea
111 M24 **Orosháza** Békés, SE Hungary 46°33′N 20°40′E
Orosirá Rodhópis *see* Rhodope Mountains
111 I22 **Oroszlány** Komárom-Esztergom, W Hungary 47°28′N 18°16′E
188 B16 **Orote Peninsula** *peninsula* W Guam
123 T9 **Orotukan** Magadanskaya Oblast', E Russian Federation 62°18′N 150°46′E
35 O5 **Oroville** California, W USA 39°29′N 121°35′W
32 K6 **Oroville** Washington, NW USA 48°56′N 119°25′W
35 O5 **Oroville, Lake** ⊠ California, W USA
0 G15 **Orozco Fracture Zone** *tectonic feature* E Pacific Ocean
Orpes *see* Oropesa del Mar
Orpesa *see* Oropesa del Mar
64 I7 **Orphan Knoll** *undersea feature* NW Atlantic Ocean 51°00′N 47°00′W
29 V3 **Orr** Minnesota, N USA 48°03′N 92°48′W
95 M21 **Orrefors** Kalmar, S Sweden 56°48′N 15°45′E
182 I7 **Orroroo** South Australia 32°46′S 138°38′E
31 T12 **Orrville** Ohio, N USA 40°50′N 81°45′W
94 L12 **Orsa** Dalarna, C Sweden 61°07′N 14°40′E
Orschowa *see* Orşova
Orschütz *see* Orzyc
119 O14 **Orsha** Vitsyebskaya Voblasts', NE Belarus 54°30′N 30°26′E
127 Q2 **Orshanka** Respublika Mariy El, W Russian Federation 56°54′N 47°54′E
108 C11 **Orsières** Valais, SW Switzerland 46°00′N 07°09′E
127 X8 **Orsk** Orenburgskaya Oblast', W Russian Federation 51°13′N 58°35′E
116 F13 **Orşova** *Ger.* Orschowa, *Hung.* Orsova. Mehedinţi, SW Romania 44°42′N 22°22′E
94 D10 **Ørsta** Møre og Romsdal, S Norway 62°12′N 06°09′E
95 O15 **Örsundsbro** Uppsala, C Sweden 59°45′N 17°19′E
136 D16 **Ortaca** Muğla, SW Turkey 36°49′N 28°43′E
83 I21 **O.R. Tambo** ✈ (Johannesburg) Gauteng, NE South Africa 26°08′S 28°01′E
107 M16 **Orta Nova** Puglia, SE Italy 41°20′N 15°43′E
136 I17 **Orta Toroslar** ⛰ S Turkey
54 E11 **Ortega** Tolima, W Colombia 03°57′N 75°11′W
104 H1 **Ortegal, Cabo** *headland* NW Spain 43°46′N 07°54′W
Ortelsburg *see* Szczytno
102 J15 **Orthez** Pyrénées-Atlantiques, SW France 43°29′N 00°46′W
60 J10 **Ortigueira** Paraná, S Brazil 24°10′S 50°55′W
104 H1 **Ortigueira** Galicia, NW Spain 43°40′N 07°50′W
106 H5 **Ortisei** *Ger.* Sankt-Ulrich. Trentino-Alto Adige, N Italy 46°35′N 11°42′E
40 F6 **Ortiz** Sonora, NW Mexico 28°18′N 110°40′W
54 L5 **Ortiz** Guárico, N Venezuela 09°37′N 67°20′W
Ortler *see* Ortles
106 F5 **Ortles** *Ger.* Ortler. ▲ N Italy 46°29′N 10°33′E
107 K14 **Ortona** Abruzzo, C Italy 42°21′N 14°24′E
57 K14 **Ortón, Río** ≈ N Bolivia
29 R8 **Ortonville** Minnesota, N USA 45°18′N 96°26′W
147 W8 **Orto-Tokoy** Issyk-Kul'skaya Oblast', NE Kyrgyzstan 42°20′N 76°03′E
93 I15 **Örträsk** Västerbotten, N Sweden 64°10′N 19°00′E
100 J12 **Örtze** ≈ NW Germany
Oruba *see* Aruba
142 I3 **Orūmīyeh** *var.* Rizaiyeh, Urmia, Urmiyeh; *prev.* Reẕā'īyeh. Āẕarbāyjān-e Gharbī, NW Iran 37°33′N 45°06′E
142 J3 **Orūmīyeh, Daryācheh-ye** *var.* Matianus, Sha Hi, Urumi Yeh, *Eng.* Lake Urmia; *prev.* Daryācheh-ye Reẕā'īyeh. ◎ NW Iran
57 K19 **Oruro** Oruro, W Bolivia 17°58′S 67°06′W
57 J19 **Oruro** ◈ *department* W Bolivia
95 I18 **Orust** *island* S Sweden
Orūzgān *see* Uruzgān
106 H13 **Orvieto** *anc.* Velsuna. Umbria, C Italy 42°43′N 12°06′E
194 K7 **Orville Coast** *physical region* Antarctica
114 H7 **Oryahovo** *var.* Oryakhovo. Vratsa, NW Bulgaria 43°44′N 23°58′E
Oryakhovo *see* Oryahovo
Oryokko *see* Yalu
117 R5 **Orzhytsya** Poltavs'ka Oblast', C Ukraine 49°48′N 32°40′E
110 M9 **Orzyc** *Ger.* Orschütz. ≈ NE Poland
110 N8 **Orzysz** *Ger.* Arys. Warmińsko-Mazurskie, NE Poland 53°49′N 21°54′E
98 K13 **Oss** Noord-Brabant, S Netherlands 51°46′N 05°32′E
94 I10 **Os** Hedmark, S Norway 62°29′N 11°14′E
125 U15 **Osa** Permskiy Kray, NW Russian Federation 57°16′N 55°22′E
115 F15 **Óssa** ⛰ C Greece
104 H11 **Ossa** ▲ S Portugal 38°43′N 07°33′W
29 W11 **Osage** Iowa, C USA 43°16′N 92°48′W
27 U5 **Osage Beach** Missouri, C USA 38°09′N 92°37′W
27 P5 **Osage City** Kansas, C USA 38°37′N 95°49′W
27 U7 **Osage Fork River** ≈ Missouri, C USA
27 U5 **Osage River** ≈ Missouri, C USA
164 J13 **Ōsaka** *hist.* Naniwa. Ōsaka, Honshū, SW Japan 34°38′N 135°28′E
164 I13 **Ōsaka** *off.* Ōsaka-fu, *var.* Ōsaka Hu. ◈ *urban prefecture* Honshū, SW Japan
Ōsaka-fu/Ōsaka Hu *see* Ōsaka
145 R10 **Osakarovka** Karaganda, C Kazakhstan 50°32′N 72°39′E
Ōsaki *see* Furukawa
29 T7 **Osakis** Minnesota, N USA 45°51′N 95°08′W
114 I9 **Osam** *var.* Osŭm. ≈ N Bulgaria
43 N16 **Osa, Península de** *peninsula* S Costa Rica
60 M10 **Osasco** São Paulo, S Brazil 23°32′S 46°46′W
27 R5 **Osawatomie** Kansas, C USA 38°30′N 94°57′W
26 L3 **Osborne** Kansas, C USA 39°26′N 98°42′W
173 S8 **Osborn Plateau** *undersea feature* E Indian Ocean
95 L21 **Osby** Skåne, S Sweden 56°24′N 14°00′E
Osca *see* Huesca
92 N2 **Oscar II Land** *physical region* W Svalbard
27 Y10 **Osceola** Arkansas, C USA 35°43′N 89°58′W
29 V15 **Osceola** Iowa, C USA 41°01′N 93°45′W
27 S6 **Osceola** Missouri, C USA 38°01′N 93°41′W
29 Q15 **Osceola** Nebraska, C USA 41°09′N 97°28′W
101 N15 **Oschatz** Sachsen, E Germany 51°17′N 13°10′E
100 K13 **Oschersleben** Sachsen-Anhalt, C Germany 52°02′N 11°14′E
31 R7 **Oscoda** Michigan, N USA 44°25′N 83°19′W
Ösel *see* Saaremaa
94 H6 **Osen** Sør-Trøndelag, S Norway 64°17′N 10°29′E
94 I12 **Osensjøen** ◎ S Norway
164 A14 **Ōse-zaki** Fukue-jima, SW Japan
147 T10 **Osh** Oshskaya Oblast', SW Kyrgyzstan 40°34′N 72°46′E
83 C16 **Oshakati** Oshana, N Namibia 17°46′S 15°43′E
83 C16 **Oshana** ◈ *district* N Namibia
14 H15 **Oshawa** Ontario, SE Canada 43°54′N 78°50′W
165 R10 **Oshika-hantō** *peninsula* Honshū, C Japan
83 C16 **Oshikango** Ohangwena, N Namibia 17°29′S 15°54′E
83 C17 **Oshikoto** *var.* Otjikoto. ◈ *district* N Namibia
165 P5 **Ō-shima** *island* NE Japan
165 N14 **Ō-shima** *island* S Japan
165 Q5 **Oshima-hantō** ⛰ Hokkaidō, NE Japan
83 D17 **Oshivelo** Oshikoto, N Namibia 18°37′S 17°10′E
28 K14 **Oshkosh** Nebraska, C USA 41°25′N 102°21′W
30 M7 **Oshkosh** Wisconsin, N USA 44°01′N 88°32′W
Oshmyany *see* Ashmyany
Osh Oblasty *see* Oshskaya Oblast'
77 T16 **Oshogbo** *var.* Osogbo. Osun, W Nigeria 07°42′N 04°31′E
147 S11 **Oshskaya Oblast'** *Kir.* Osh Oblasty. ◈ *province* SW Kyrgyzstan
Ōshū *see* Mizusawa
Oshun *see* Osun
79 J20 **Oshwe** Bandundu, C Dem. Rep. Congo 03°24′S 19°32′E
Osiek *see* Osijek
112 I9 **Osijek** *prev.* Osiek, Osjek, *Ger.* Esseg, *Hung.* Eszék. Osijek-Baranja, E Croatia 45°33′N 18°41′E
112 I9 **Osijek-Baranja** *off.* Osječko-Baranjska Županija. ◈ *province* E Croatia
106 J12 **Osimo** Marche, C Italy 43°28′N 13°29′E
122 M12 **Osinovka** Irkutskaya Oblast', C Russian Federation 56°19′N 101°55′E
Osintorf *see* Asintorf
112 N11 **Osipaonica** Serbia, NE Serbia 44°34′N 21°00′E
Osipenko *see* Berdyans'k
Osipovichi *see* Asipovichy
Osječko-Baranjska Županija *see* Osijek-Baranja
Osjek *see* Osijek
29 W15 **Oskaloosa** Iowa, C USA 41°17′N 92°38′W
27 Q4 **Oskaloosa** Kansas, C USA 39°14′N 95°21′W
95 N20 **Oskarshamn** Kalmar, S Sweden 57°16′N 16°25′E
95 J21 **Oskarström** Halland, S Sweden 56°48′N 13°00′E
14 M8 **Oskélanéo** Québec, SE Canada 48°06′N 75°12′W
Öskemen *see* Ust'-Kamenogorsk
117 W5 **Oskil** *Rus.* Oskil. ≈ Russian Federation/Ukraine
Oskil *see* Oskil
93 D20 **Oslo** *prev.* Christiania, Kristiania. ● (Norway) Oslo, S Norway 59°54′N 10°44′E
93 D20 **Oslo** ◈ *county* S Norway
93 D21 **Oslofjorden** *fjord* S Norway
155 G15 **Osmānābād** Mahārāshtra, C India 18°09′N 76°06′E
136 J11 **Osmancık** Çorum, N Turkey 40°58′N 34°50′E
136 L16 **Osmaniye** Osmaniye, S Turkey 37°04′N 36°15′E
136 L16 **Osmaniye** ◈ *province* S Turkey
95 O16 **Ösmo** Stockholm, C Sweden 58°58′N 17°55′E
118 E3 **Osmussaar** *island* WEstonia
100 G13 **Osnabrück** Niedersachsen, NW Germany 52°09′N 07°42′E
110 D11 **Ośno Lubuskie** *Ger.* Drossen. Lubuskie, W Poland 52°28′N 14°51′E
Osogbo *see* Oshogbo
113 P19 **Osogov Mountains** *var.* Osogovske Planine, Osogovski Planina, *Mac.* Osogovski Planini. ⛰ Bulgaria/FYR Macedonia
Osogovske Planine/ Osogovski Planina/ Osogovski Planini *see* Osogov Mountains
165 R6 **Osore-zan** ▲ Honshū, C Japan 41°18′N 141°06′E
Oşorhei *see* Târgu Mureş
61 J16 **Osório** Rio Grande do Sul, S Brazil 29°53′S 50°17′W
63 G16 **Osorno** Los Lagos, C Chile 40°39′S 73°05′W
104 M4 **Osorno** Castilla y León, N Spain 42°24′N 04°22′W
11 N17 **Osoyoos** British Columbia, SW Canada 49°02′N 119°31′W
95 C14 **Osøyro** Hordaland, S Norway 60°11′N 05°30′E
54 J6 **Ospino** Portuguesa, N Venezuela 09°17′N 69°26′W
23 X6 **Ossabaw Island** *island* Georgia, SE USA
23 X6 **Ossabaw Sound** *sound* Georgia, SE USA
183 O16 **Ossa, Mount** ▲ Tasmania, SE Australia 41°55′S 146°03′E
104 H11 **Ossa, Serra d'** ⛰ SE Portugal
77 U16 **Osse** ≈ S Nigeria
30 J6 **Osseo** Wisconsin, N USA 44°33′N 91°13′W
109 S9 **Ossiacher See** ◎ S Austria
18 K13 **Ossining** New York, NE USA 41°10′N 73°50′W
123 V9 **Ossora** Krasnoyarskiy Kray, E Russian Federation 59°16′N 163°02′E
124 I15 **Ostashkov** Tverskaya Oblast', W Russian Federation 57°08′N 33°10′E
100 H9 **Oste** ≈ NW Germany
Ostee *see* Baltic Sea
Ostend/Ostende *see* Oostende
117 P3 **Oster** Chernihivs'ka Oblast', N Ukraine 50°57′N 30°55′E
93 J16 **Österbotten** *Fin.* Pohjanmaa, *Eng.* Ostrobothnia. ◈ *region* W Finland
95 O14 **Österbybruk** Uppsala, C Sweden 60°13′N 17°55′E
95 M19 **Österbymo** Östergotland, S Sweden 57°49′N 15°15′E
94 K12 **Österdalälven** ≈ C Sweden
94 I12 **Østerdalen** *valley* S Norway
95 L18 **Östergötland** ◈ *county* S Sweden
100 H10 **Osterholz-Scharmbeck** Niedersachsen, NW Germany 53°13′N 08°46′E
Östermark *see* Teuva
Östermyra *see* Seinäjoki
Østerø *see* Eysturoy
101 J14 **Osterode am Harz** Niedersachsen, C Germany 51°43′N 10°15′E
Osterode/Osterode in Ostpreussen *see* Ostróda
Osterøy *see* Osterøyni
94 C13 **Osterøyni** *prev.* Osterøy. *island* S Norway
Österreich *see* Austria
93 G16 **Östersund** Jämtland, C Sweden 63°10′N 14°44′E
95 N14 **Östervåla** Västmanland, C Sweden 60°10′N 17°13′E
101 H22 **Ostfildern** Baden-Württemberg, SW Germany 48°43′N 09°16′E
95 H16 **Østfold** ◈ *county* S Norway
100 E9 **Ostfriesische Inseln** *Eng.* East Frisian Islands. *island group* NW Germany
100 F10 **Ostfriesland** *historical region* NW Germany
95 P14 **Östhammar** Uppsala, C Sweden 60°15′N 18°25′E
Ostia Aterni *see* Pescara
106 G8 **Ostiglia** Lombardia, N Italy 45°04′N 11°09′E
95 J14 **Östmark** Värmland, C Sweden 60°15′N 12°45′E
95 K22 **Östra Ringsjön** ◎ S Sweden
111 I17 **Ostrava** Moravskoslezský Kraj, E Czech Republic 49°50′N 18°15′E
Ostravský Kraj *see* Moravskoslezský Kraj
94 J11 **Østrehogna** *Swe.* Härjahågnen, Härjehågna. ▲ Norway/Sweden 61°43′N 12°07′E
Ostrobothnia *see* Österbotten
110 K8 **Ostróda** *Ger.* Osterode, Osterode in Ostpreussen. Warmińsko-Mazurskie, NE Poland 53°42′N 19°59′E
Ostrog/Ostróg *see* Ostroh
126 L8 **Ostrogozhsk** Voronezhskaya Oblast', W Russian Federation 50°52′N 39°00′E
116 L4 **Ostroh** *Pol.* Ostróg, *Rus.* Ostrog. Rivnens'ka Oblast', NW Ukraine 50°20′N 26°29′E
110 N9 **Ostrołęka** *Ger.* Wiesenhof, *Rus.* Ostrolenka. Mazowieckie, C Poland 53°06′N 21°34′E
Ostrolenka *see* Ostrołęka
111 A16 **Ostrov** *Ger.* Schlackenwerth. Karlovarský Kraj, W Czech Republic 50°18′N 12°56′E
124 F15 **Ostrov** *Latv.* Austrava. Pskovskaya Oblast', W Russian Federation 57°21′N 28°18′E
Ostrovets *see* Ostrowiec Świętokrzyski
113 M21 **Ostrovicës, Mali i** ▲ SE Albania 40°36′N 20°25′E
165 Z2 **Ostrov Iturup** *island* NE Russian Federation
124 M4 **Ostrovnoy** Murmanskaya Oblast', NW Russian Federation 68°00′N 39°40′E
114 L7 **Ostrovo** *prev.* Golema Ada. Razgrad, N Bulgaria 43°40′N 26°37′E
125 N15 **Ostrovskoye** Kostromskaya Oblast', NW Russian Federation 57°46′N 42°18′E
Ostrów *see* Ostrów Wielkopolski
Ostrowiec *see* Ostrowiec Świętokrzyski
111 M14 **Ostrowiec Świętokrzyski** *var.* Ostrowiec, *Rus.* Ostrovets. Świętokrzyskie, C Poland 50°55′N 21°23′E
110 P13 **Ostrów Lubelski** Lubelskie, E Poland 51°29′N 22°57′E
110 N10 **Ostrów Mazowiecka** *var.* Ostrów Mazowiecki. Mazowieckie, NE Poland 52°49′N 21°53′E
Ostrów Mazowiecki *see* Ostrów Mazowiecka
Ostrowo *see* Ostrów Wielkopolski
110 H13 **Ostrów Wielkopolski** *var.* Ostrów, *Ger.* Ostrowo. Wielkopolskie, C Poland 51°40′N 17°47′E
Ostryna *see* Astryna
110 I13 **Ostrzeszów** Wielkopolskie, C Poland 51°25′N 17°55′E
107 P18 **Ostuni** Puglia, SE Italy 40°44′N 17°35′E
Ostyako-Voguls'k *see* Khanty-Mansiysk
Osum *see* Osumit, Lumi i
Osŭm *see* Osam
164 C17 **Ōsumi-hantō** ⛰ Kyūshū, SW Japan
164 C17 **Ōsumi-kaikyō** *strait* SW Japan
113 L22 **Osumit, Lumi i** *var.* Osum. ≈ SE Albania
77 T16 **Osun** *var.* Oshun. ◈ *state* SW Nigeria
104 L14 **Osuna** Andalucía, S Spain 37°14′N 05°06′W
60 J8 **Osvaldo Cruz** São Paulo, S Brazil 21°49′S 50°52′W
Osveya *see* Asvyeya
18 J7 **Oswegatchie River** ≈ New York, NE USA
27 Q7 **Oswego** Kansas, C USA 37°11′N 95°10′W
18 H9 **Oswego** New York, NE USA 43°27′N 76°13′W
97 K19 **Oswestry** W England, United Kingdom 52°51′N 03°06′W
111 J16 **Oświęcim** *Ger.* Auschwitz. Małopolskie, S Poland 50°02′N 19°11′E
185 E22 **Otago** *off.* Otago Region. ◈ *region* South Island, New Zealand
185 F23 **Otago Peninsula** *peninsula* South Island, New Zealand
Otago Region *see* Otago
165 F13 **Ōtake** Hiroshima, Honshū, SW Japan 34°13′N 132°12′E
184 L13 **Otaki** Wellington, North Island, New Zealand 40°46′S 175°08′E
93 M15 **Otanmäki** Kainuu, C Finland 64°07′N 27°04′E
145 T15 **Otar** Zhambyl, SE Kazakhstan 43°30′N 75°13′E
165 R4 **Otaru** Hokkaidō, NE Japan 43°14′N 140°59′E
185 C24 **Otatara** Southland, South Island, New Zealand 46°26′S 168°18′E
185 C24 **Otautau** Southland, South Island, New Zealand 46°10′S 168°01′E
93 M18 **Otava** Etelä-Savo, E Finland 61°37′N 27°07′E
111 B18 **Otava** *Ger.* Wottawa. ≈ SW Czech Republic
56 C6 **Otavalo** Imbabura, N Ecuador 0°13′N 78°15′W
83 D17 **Otavi** Otjozondjupa, N Namibia 19°35′S 17°25′E
165 P12 **Ōtawara** Tochigi, Honshū, S Japan 36°52′N 140°01′E
83 B16 **Otchinjau** Cunene, SW Angola 16°31′S 13°54′E
116 F12 **Oţelu Roşu** *Ger.* Ferdinandsberg, *Hung.* Nándorhgy. Caras-Severin, SW Romania 45°30′N 22°22′E
185 E21 **Otematata** Canterbury, South Island, New Zealand 44°37′S 170°12′E
118 I6 **Otepää** *Ger.* Odenpäh. Valgamaa, SE Estonia 58°01′N 26°30′E
144 G14 **Otes** *Kaz.* Say-Ötesh; *prev.* Say-Utës. Mangistau, SW Kazakhstan 44°20′N 53°32′E
162 H7 **Otgon** *var.* Buyant. Dzavhan, C Mongolia 47°14′N 97°14′E
32 K9 **Othello** Washington, NW USA 46°49′N 119°10′W
83 K23 **oThongathi** *prev.* Tongaat, *var.* uThongathi. KwaZulu/Natal, E South Africa 29°35′S 31°07′E *see also* Tongaat
115 A15 **Othonoí** *island* Iónia Nisiá, Greece, C Mediterranean Sea
Othris *see* Óthrys
115 F17 **Óthrys** *var.* Othris. ⛰ C Greece
77 Q14 **Oti** ≈ N Togo
40 K10 **Otinapa** Durango, C Mexico 24°01′N 104°58′W
185 G17 **Otira** West Coast, South Island, New Zealand 42°52′S 171°33′E
37 V3 **Otis** Colorado, C USA 40°09′N 102°57′W
12 L10 **Otish, Monts** ⛰ Québec, E Canada
83 C17 **Otjikondo** Kunene, N Namibia 19°50′S 15°23′E
Otjikoto *see* Oshikoto
83 E18 **Otjinene** Omaheke, NE Namibia 21°10′S 18°43′E
83 D18 **Otjiwarongo** Otjozondjupa, N Namibia 20°29′S 16°36′E
83 D18 **Otjosondu** *var.* Otjosundu. Otjozondjupa, C Namibia 21°19′S 17°51′E
Otjosundu *see* Otjosondu
83 D18 **Otjozondjupa** ◈ *district* C Namibia
112 C11 **Otočac** Lika-Senj, W Croatia 44°52′N 15°13′E
Otog Qi *see* Ulan
112 J10 **Otok** Vukovar-Srijem, E Croatia 45°10′N 18°52′E
116 K14 **Otopeni** ✈ (Bucureşti) Ilfov, S Romania 44°34′N 26°09′E
184 L8 **Otorohanga** Waikato, North Island, New Zealand 38°10′S 175°14′E
12 D9 **Otoskwin** ≈ Ontario, C Canada
165 G14 **Ōtoyo** Kōchi, Shikoku, SW Japan 33°45′N 133°42′E
95 E16 **Otra** ≈ S Norway
107 R19 **Otranto** Puglia, SE Italy 40°08′N 18°28′E
Otranto, Canale d' *see* Otranto, Strait of
107 Q18 **Otranto, Strait of** *It.* Canale d'Otranto. *strait* Albania/Italy
111 H18 **Otrokovice** *Ger.* Otrokowitz. Zlínský Kraj, E Czech Republic 49°13′N 17°33′E
Otrokowitz *see* Otrokovice
31 P10 **Otsego** Michigan, N USA 42°27′N 85°42′W
31 Q6 **Otsego Lake** ◎ Michigan, N USA
18 I11 **Otselic River** ≈ New York, NE USA
164 J14 **Ōtsu** *var.* Ôtu. Shiga, Honshū, SW Japan 35°03′N 135°49′E
94 G11 **Otta** Oppland, S Norway 61°46′N 09°33′E
189 U13 **Otta** *island* Chuuk, C Micronesia
94 F11 **Otta** ≈ S Norway
189 U13 **Otta Pass** *passage* Chuuk Islands, C Micronesia
95 J22 **Ottarp** Skåne, S Sweden 55°55′N 12°55′E
14 L12 **Ottawa** ● (Canada) Ontario, SE Canada 45°24′N 75°41′W
30 L11 **Ottawa** Illinois, N USA 41°21′N 88°50′W
27 Q5 **Ottawa** Kansas, C USA 38°35′N 95°16′W
31 R12 **Ottawa** Ohio, N USA 41°01′N 84°03′W
14 L12 **Ottawa** *var.* Uplands. ✈ Ontario, SE Canada 45°19′N 75°39′W
14 M12 **Ottawa** *Fr.* Outaouais. ≈ Ontario/Québec, SE Canada
9 R10 **Ottawa Islands** *island group* Nunavut, C Canada
18 L8 **Otter Creek** ≈ Vermont, NE USA
36 L6 **Otter Creek Reservoir** ⊠ Utah, W USA
98 L11 **Otterlo** Gelderland, E Netherlands 52°06′N 05°46′E
94 D9 **Otterøya** *island* S Norway
29 S6 **Otter Tail Lake** ◎ Minnesota, N USA
29 R7 **Otter Tail River** ≈ Minnesota, C USA
95 H23 **Otterup** Syddtjylland, C Denmark 55°31′N 10°25′E
99 H19 **Ottignies** Wallon Brabant, C Belgium 50°40′N 04°34′E
101 L23 **Ottobrunn** Bayern, SE Germany 48°02′N 11°40′E
29 X15 **Ottumwa** Iowa, C USA 41°00′N 92°24′W
Ôtu *see* Ōtsu
83 B16 **Otuazuma** Kunene, NW Namibia 17°52′S 13°16′E
77 V16 **Oturkpo** Benue, S Nigeria 07°12′N 08°06′E
193 Y15 **Otu Tolu Group** *island group* SE Tonga
182 M13 **Otway, Cape** *headland* Victoria, SE Australia 38°52′S 143°31′E
63 H24 **Otway, Seno** *inlet* S Chile
Ötz *see* Oetz
108 L8 **Ötztaler Ache** ≈ W Austria
108 L9 **Ötztaler Alpen** *It.* Alpi Venoste. ⛰ SW Austria
27 T12 **Ouachita, Lake** ⊠ Arkansas, C USA
27 R11 **Ouachita Mountains** ⛰ Arkansas/Oklahoma, C USA
27 U13 **Ouachita River** ≈ Arkansas/Louisiana, C USA
Ouadaï *see* Ouaddaï
76 J7 **Ouadâne** *var.* Ouadane. Adrar, C Mauritania 20°57′N 11°35′W
78 K13 **Ouadda** Haute-Kotto, N Central African Republic 08°02′N 22°22′E
78 J10 **Ouaddaï** *off.* Région du Ouaddaï, *var.* Ouadaï, Wadai. ◈ *region* SE Chad
Ouaddaï, Région du *see* Ouaddaï
77 P13 **Ouagadougou** *var.* Wagadugu. ● (Burkina Faso) C Burkina 12°20′N 01°32′W
77 P13 **Ouagadougou** ✈ C Burkina Faso 12°21′N 01°27′W
77 O12 **Ouahigouya** NW Burkina Faso 13°31′N 02°20′W
Ouahran *see* Oran
79 J14 **Ouaka** ◈ *prefecture* C Central African Republic
79 J15 **Ouaka** ≈ S Central African Republic
Oualam *see* Ouallam
76 M9 **Oualâta** *var.* Oualata. Hodh ech Chargui, SE Mauritania 17°18′N 07°00′W
77 R11 **Ouallam** *var.* Oualam. Tillabéri, W Niger 14°23′N 02°09′E
172 H14 **Ouanani** Mohéli, S Comoros 12°19′S 93°98′E
55 Z10 **Ouanary** E French Guiana 04°11′N 51°40′W
78 L13 **Ouanda Djallé** Vakaga, NE Central African Republic 08°53′N 22°47′E
79 N14 **Ouando** Haut-Mbomou, SE Central African Republic 05°57′N 25°57′E
79 L15 **Ouango** Mbomou, S Central African Republic 04°19′N 22°30′E
77 N14 **Ouangolodougou** *var.* Wangolodougou. N Ivory Coast 09°58′N 05°09′W
172 I13 **Ouani** Anjouan, SE Comoros
79 M15 **Ouara** ≈ E Central African Republic
76 K7 **Ouarâne** *desert* C Mauritania
15 O11 **Ouareau** ≈ Québec, SE Canada
74 K7 **Ouargla** *var.* Wargla. NE Algeria 32°N 05°16′E
74 F8 **Ouarzazate** S Morocco 30°54′N 06°55′W
77 Q11 **Ouatagouna** Gao, E Mali 15°06′N 00°41′E
74 G6 **Ouazzane** *var.* Ouezzane, *Ar.* Wazan, Wazzan. N Morocco 34°52′N 05°35′W
Oubangui *see* Ubangi
Oubangui-Chari *see* Central African Republic
Oubangui-Chari, Territoire de l' *see* Central African Republic
Oubari, Edeyen d' *see* Awbāri, Idhān
98 G13 **Oud-Beijerland** Zuid-Holland, SW Netherlands 51°50′N 04°25′E
98 F13 **Ouddorp** Zuid-Holland, SW Netherlands 51°49′N 03°55′E
77 P9 **Oudeïka** *oasis* C Mali
98 G13 **Oude Maas** ≈ SW Netherlands
99 E18 **Oudenaarde** *Fr.* Audenarde. Oost-Vlaanderen, SW Belgium 50°50′N 03°37′E
99 H14 **Oudenbosch** Noord-Brabant, S Netherlands 51°35′N 04°32′E
98 P6 **Oude Pekela** Groningen, NE Netherlands 53°06′N 07°00′E
Ouderkerk *see* Ouderkerk aan den Amstel
98 I10 **Ouderkerk aan den Amstel** *var.* Ouderkerk. Noord-Holland, C Netherlands 52°18′N 04°54′E
98 I6 **Oudeschild** Noord-Holland, NW Netherlands 53°01′N 04°51′E
99 G14 **Oude-Tonge** Zuid-Holland, SW Netherlands 51°40′N 04°13′E
98 I12 **Oudewater** Utrecht, C Netherlands 52°02′N 04°54′E
Oudjda *see* Oujda
Oudkerk *see* Aldtsjerk
167 Q6 **Oudômxai** *var.* Muang Xay, Muong Sai, Xai. Oudômxai, N Laos 20°41′N 102°00′E
102 J7 **Oudon** ≈ NW France
98 I9 **Oudorp** Noord-Holland, NW Netherlands 52°39′N 04°47′E
83 G25 **Oudtshoorn** Western Cape, SW South Africa 33°35′S 22°14′E
99 I16 **Oud-Turnhout** Antwerpen, N Belgium 51°19′N 05°01′E
74 F7 **Oued-Zem** C Morocco 32°53′N 06°30′W
187 P16 **Ouégoa** Province Nord, C New Caledonia 20°22′S 164°25′E
76 L13 **Ouéléssébougou** *var.* Ouolossébougou. Koulikoro, SW Mali 11°58′N 07°51′W
77 N16 **Ouéllé** E Ivory Coast 07°18′N 04°01′W
77 R16 **Ouémé** ≈ C Benin
77 O13 **Ouessa** S Burkina Faso 11°02′N 02°44′W
102 D5 **Ouessant, Île d'** *Eng.* Ushant. *island* NW France
79 H17 **Ouésso** Sangha, NW Congo 01°38′N 16°03′E
79 D15 **Ouest** *Eng.* West. ◈ *province* W Cameroon
190 G11 **Ouest, Baie de l'** *bay* Îles Wallis, E Wallis and Futuna
15 Y7 **Ouest, Pointe de l'** *headland* Québec, SE Canada 48°08′N 64°57′W
Ouezzane *see* Ouazzane
99 K20 **Ouffet** Liège, E Belgium 50°30′N 05°31′E
79 H14 **Ouham** ◈ *prefecture* NW Central African Republic
78 I13 **Ouham** ≈ Central African Republic/Chad
79 G14 **Ouham-Pendé** ◈ *prefecture* W Central African Republic
77 R16 **Ouidah** *Eng.* Whydah, Wida. S Benin 06°23′N 02°08′E
74 H6 **Oujda** *Ar.* Oudjda, Ujda. NE Morocco 34°45′N 01°53′W
76 I7 **Oujeft** Adrar, C Mauritania 20°05′N 13°00′W
93 L15 **Oulainen** Pohjois-Pohjanmaa, C Finland 64°14′N 24°50′E
Ould Yanja *see* Ould Yenjé
76 J10 **Ould Yenjé** *var.* Ould Yanja. Guidimaka, S Mauritania 15°33′N 11°43′W
93 L14 **Oulu** *Swe.* Uleåborg. Pohjois-Pohjanmaa, C Finland 65°01′N 25°28′E
Oulu *Swe.* Uleåborg. ◈ *province* N Finland
93 L15 **Oulujärvi** *Swe.* Uleträsk. ◎ C Finland
93 M14 **Oulujoki** *Swe.* Uleälv. ≈ C Finland
93 L14 **Oulunsalo** Pohjois-Pohjanmaa, C Finland 64°55′N 25°19′E
106 A8 **Oulx** Piemonte, NE Italy 45°05′N 06°41′E
78 J9 **Oum-Chalouba** Ennedi-Ouest, NE Chad 15°48′N 20°46′E
76 M16 **Oumé** C Ivory Coast 06°25′N 05°23′W
74 F7 **Oum er Rbia** ≈ C Morocco
78 J10 **Oum-Hadjer** Batha, E Chad 13°18′N 19°41′E
92 K10 **Ounasjoki** ≈ N Finland
78 J7 **Ounianga Kébir** Ennedi-Ouest, N Chad 19°06′N 20°29′E
Ouolossébougou *see* Ouéléssébougou
Oup *see* Auob
99 K19 **Oupeye** Liège, E Belgium 50°42′N 05°38′E
99 N21 **Our** ≈ NW Europe
37 Q7 **Ouray** Colorado, C USA 38°01′N 107°40′W
103 R7 **Ource** ≈ C France
104 G9 **Ourém** Santarém, C Portugal 39°40′N 08°32′W
104 H4 **Ourense** *Cast.* Orense, *Lat.* Aurium. Galicia, NW Spain 42°20′N 07°52′W
104 I4 **Ourense** *Cast.* Orense. ◈ *province* Galicia, NW Spain
59 O15 **Ouricuri** Pernambuco, E Brazil 07°51′S 40°05′W
60 J9 **Ourinhos** São Paulo, S Brazil 22°59′S 49°52′W

◆ Country
● Country Capital
◇ Dependent Territory
○ Dependent Territory Capital
◈ Administrative Regions
✈ International Airport
▲ Mountain
⛰ Mountain Range
✱ Volcano
≈ River
◎ Lake
⊠ Reservoir

OURIQUE – PAMPELUNA

104 G13 **Ourique** Beja, S Portugal 37°38′N 08°13′W
59 M20 **Ouro Preto** Minas Gerais, NE Brazil 20°25′S 43°30′W
Ours, Grand Lac de l' see Great Bear Lake
99 K20 **Ourthe** ≈ E Belgium
165 Q9 **Ōu-sanmyaku** ▲▲ Honshū, C Japan
97 M17 **Ouse** ≈ N England, United Kingdom
Ouse see Great Ouse
102 H7 **Oust** ≈ NW France
Outaouais see Ottawa
15 T4 **Outardes Quatre, Réservoir** ▣ Québec, SE Canada
15 T5 **Outardes, Rivière aux** ≈ Québec, SE Canada
96 E8 **Outer Hebrides** var. Western Isles. island group NW Scotland, United Kingdom
30 K3 **Outer Island** island Apostle Islands, Wisconsin, N USA
35 S16 **Outer Santa Barbara Passage** passage California, SW USA
83 C18 **Outjo** Kunene, N Namibia 20°08′S 16°08′E
11 T16 **Outlook** Saskatchewan, S Canada 51°30′N 107°03′W
93 N16 **Outokumpu** Pohjois-Karjala, E Finland 62°43′N 29°05′E
96 M2 **Out Skerries** island group NE Scotland, United Kingdom
187 Q16 **Ouvéa** island Îles Loyauté, NE New Caledonia
103 S14 **Ouvèze** ≈ SE France
182 L9 **Ouyen** Victoria, SE Australia 35°07′S 142°19′E
39 Q14 **Ouzinkie** Kodiak Island, Alaska, USA 57°54′N 152°27′W
137 O13 **Ovacık** Tunceli, E Turkey 39°23′N 39°13′E
106 C9 **Ovada** Piemonte, NE Italy 44°41′N 08°39′E
187 X14 **Ovalau** island C Fiji
62 G9 **Ovalle** Coquimbo, N Chile 30°33′S 71°16′W
83 C17 **Ovamboland** physical region N Namibia
54 L10 **Ovana, Cerro** ▲ S Venezuela 04°41′N 66°54′W
104 G7 **Ovar** Aveiro, N Portugal 40°52′N 08°38′W
114 L10 **Ovcharitsa, Yazovir** ▣ SE Bulgaria
54 E6 **Ovejas** Sucre, NW Colombia 09°32′N 75°14′W
101 E16 **Overath** Nordrhein-Westfalen, W Germany 50°55′N 07°16′E
98 F13 **Overflakkee** island SW Netherlands
99 H19 **Overijse** Vlaams Brabant, C Belgium 50°46′N 04°32′E
98 N10 **Overijssel** ◈ province E Netherlands
98 M9 **Overijssels Kanaal** canal E Netherlands
92 K13 **Överkalix** Norrbotten, N Sweden 66°19′N 22°49′E
27 R4 **Overland Park** Kansas, C USA 38°57′N 94°41′W
99 L14 **Overloon** Noord-Brabant, SE Netherlands 51°35′N 05°54′E
99 K16 **Overpelt** Limburg, NE Belgium 51°13′N 05°24′E
35 Y10 **Overton** Nevada, W USA 36°32′N 114°25′W
25 W7 **Overton** Texas, SW USA 32°16′N 94°58′W
92 K13 **Övertorneå** Norrbotten, N Sweden 66°22′N 23°40′E
95 N18 **Överum** Kalmar, S Sweden 57°58′N 16°20′E
92 G13 **Överuman** ◎ N Sweden
117 P11 **Ovidiopol'** Odes'ka Oblast', SW Ukraine 46°15′N 30°27′E
116 M14 **Ovidiu** Constanța, SE Romania 44°16′N 28°34′E
45 N10 **Oviedo** SW Dominican Republic 17°47′N 71°22′W
104 K2 **Oviedo** anc. Asturias. Asturias, NW Spain 43°21′N 05°50′W
104 K2 **Oviedo** ✈ Asturias, N Spain 43°21′N 05°50′W
Ovilava see Wels
118 D7 **Oviši** W Latvia 57°34′N 21°43′E
146 K10 **Ovminzatovo Tog'lari** Rus. Gory Auminzatau. ▲▲ N Uzbekistan
Övögdiy see Telmen
Ovoot see Darīganga
157 O4 **Övörhangay** ◈ province C Mongolia
94 E12 **Øvre Årdal** Sogn Og Fjordane, S Norway 61°18′N 07°48′E
95 J14 **Övre Fryken** ◎ C Sweden
92 J11 **Övre Soppero** Lapp. Badje-Sohppar. Norrbotten, N Sweden 68°07′N 21°40′E
117 N3 **Ovruch** Zhytomyrs'ka Oblast', N Ukraine 51°20′N 58°50′E
Övt see Bat-Öldziy
185 E24 **Owaka** Otago, South Island, New Zealand 46°27′S 169°42′E
79 H18 **Owando** prev. Fort Rousset. Cuvette, C Congo 0°29′S 15°55′E
164 J14 **Owase** Mie, Honshū, SW Japan 34°04′N 136°11′E
27 P9 **Owasso** Oklahoma, C USA 36°16′N 95°51′W
29 V10 **Owatonna** Minnesota, N USA 44°04′N 93°13′W
173 O4 **Owen Fracture Zone** tectonic feature W Arabian Sea
185 H15 **Owen, Mount** ▲ South Island, New Zealand 41°32′S 172°33′E
185 H15 **Owen River** Tasman, South Island, New Zealand 41°40′S 172°28′E
44 D8 **Owen Roberts** ✈ Grand Cayman, Cayman Islands 19°15′N 81°22′W
20 I6 **Owensboro** Kentucky, S USA 37°46′N 87°07′W
35 T11 **Owens Lake** salt flat California, W USA
14 F14 **Owen Sound** Ontario, S Canada 44°34′N 80°56′W
14 F13 **Owen Sound** ◎ Ontario, S Canada
35 T10 **Owens River** ≈ California, W USA
186 F9 **Owen Stanley Range** ▲▲ S Papua New Guinea
27 V5 **Owensville** Missouri, C USA 38°21′N 91°30′W
20 M4 **Owenton** Kentucky, S USA 38°33′N 84°50′W
77 U17 **Owerri** Imo, S Nigeria 05°19′N 07°07′E
184 M10 **Owhango** Manawatu-Wanganui, North Island, New Zealand 39°01′S 175°22′E
21 N5 **Owingsville** Kentucky, S USA 38°09′N 83°46′W
77 T16 **Owo** Ondo, SW Nigeria 07°10′N 05°31′E
31 R9 **Owosso** Michigan, N USA 43°00′N 84°10′W
35 V1 **Owyhee** Nevada, W USA 41°57′N 116°07′W
32 L14 **Owyhee, Lake** ◎ Oregon, NW USA
32 L15 **Owyhee River** ≈ Idaho/Oregon, NW USA
92 K1 **Öxarfjörður** var. Axarfjördhur. fjord N Iceland
94 K12 **Oxberg** Dalarna, C Sweden 61°07′N 14°10′E
11 V17 **Oxbow** Saskatchewan, S Canada 49°16′N 102°12′W
95 O17 **Oxelösund** Södermanland, S Sweden 58°40′N 17°10′E
185 H18 **Oxford** Canterbury, South Island, New Zealand 43°18′S 172°10′E
97 M21 **Oxford** Lat. Oxonia. S England, United Kingdom 51°46′N 01°15′W
23 Q3 **Oxford** Alabama, S USA 33°36′N 85°50′W
22 L2 **Oxford** Mississippi, S USA 34°23′N 89°30′W
29 N16 **Oxford** Nebraska, C USA 40°15′N 99°37′W
18 I11 **Oxford** New York, NE USA 42°21′N 75°39′W
21 U8 **Oxford** North Carolina, SE USA 36°22′N 78°37′W
31 Q14 **Oxford** Ohio, N USA 39°30′N 84°45′W
18 H16 **Oxford** Pennsylvania, NE USA 39°46′N 75°57′W
11 X12 **Oxford House** Manitoba, C Canada 54°55′N 95°13′W
29 Y13 **Oxford Junction** Iowa, C USA 41°58′N 90°57′W
11 X12 **Oxford Lake** ◎ Manitoba, C Canada
97 M21 **Oxfordshire** cultural region S England, United Kingdom
Oxia see Oxyá
41 X12 **Oxkutzcab** Yucatán, SE Mexico 20°18′N 89°26′W
35 R15 **Oxnard** California, W USA 34°12′N 119°10′W
Oxonia see Oxford
14 I12 **Oxtongue** ≈ Ontario, SE Canada
Oxus see Amu Darya
115 E15 **Oxyá** var. Oxia. ▲ C Greece 39°46′N 21°56′E
164 L11 **Oyabe** Toyama, Honshū, SW Japan 36°42′N 136°52′E
Oyahue/Oyahue, Volcán see Ollagüe, Volcán
165 O12 **Oyama** Tochigi, Honshū, S Japan 36°19′N 139°46′E
47 U5 **Oyapock** ≈ E French Guiana
Oyapock see Oiapoque, Rio/Oyapok, Fleuve l'
55 Z10 **Oyapok, Baie de L'** bay Brazil/French Guiana South America W Atlantic Ocean
55 Z11 **Oyapok, Fleuve l'** var. Rio Oiapoque, Oyapock. ≈ Brazil/French Guiana see also Oiapoque, Rio
Oyapok, Fleuve l' see Oiapoque, Rio
79 E17 **Oyem** Woleu-Ntem, N Gabon 01°34′N 11°31′E
11 R16 **Oyen** Alberta, SW Canada 51°20′N 110°28′W
95 I15 **Øyeren** ◎ S Norway
Oygon see Tüdevtey
96 I7 **Oykel** ≈ N Scotland, United Kingdom
123 R9 **Oymyakon** Respublika Sakha (Yakutiya), NE Russian Federation 63°28′N 142°22′E
79 H19 **Oyo** Cuvette, C Congo 01°17′S 16°00′E
77 S15 **Oyo** Oyo, W Nigeria 07°51′N 03°57′E
77 S15 **Oyo** ◈ state SW Nigeria
56 D13 **Oyón** Lima, C Peru 10°39′S 76°44′W
103 S10 **Oyonnax** Ain, E France 46°16′N 05°39′E
146 L10 **Oyoqog'itma** Rus. Ayakagytma. Buxoro Viloyati, C Uzbekistan 40°37′N 64°26′E
146 M9 **Oyoqquduq** Rus. Ayakkuduk. Navoiy Viloyati, N Uzbekistan 41°16′N 65°12′E
32 F9 **Oysterville** Washington, NW USA 46°33′N 124°03′W
95 D14 **Øystese** Hordaland, S Norway 60°23′N 06°13′E
145 S16 **Oytal** Zhambyl, S Kazakhstan 42°54′N 73°21′E
147 U10 **Oy-Tal** Oshskaya Oblast', SW Kyrgyzstan 40°23′N 74°04′E
147 T10 **Oy-Tal** ≈ SW Kyrgyzstan
145 Q15 **Oyyk** prev. Uyuk. Zhambyl, S Kazakhstan 43°46′N 70°55′E
144 H10 **Oyyl** prev. Uil. Aktyubinsk, W Kazakhstan 49°06′N 54°41′E
144 H10 **Oyyl** prev. Uil. ≈ W Kazakhstan
Ozarichi see Azarychy
23 R7 **Ozark** Alabama, S USA 31°27′N 85°38′W
27 S10 **Ozark** Arkansas, C USA 35°30′N 93°50′W
27 T8 **Ozark** Missouri, C USA 37°01′N 93°12′W
27 T8 **Ozark Plateau** plain Arkansas/Missouri, C USA
27 T6 **Ozarks, Lake of the** ▣ Missouri, C USA
192 L10 **Ozbourn Seamount** undersea feature W Pacific Ocean 26°00′S 174°49′W
111 L20 **Ózd** Borsod-Abaúj-Zemplén, NE Hungary 48°15′N 20°18′E
112 D11 **Ozeblin** ▲ C Croatia 44°37′N 15°52′E
123 V11 **Ozernovskiy** Kamchatskiy Kray, E Russian Federation 51°28′N 156°32′E
144 M7 **Ozërnoye** var. Ozërnyy. Kostanay, N Kazakhstan 53°29′N 63°14′E
124 J15 **Ozërnyy** Tverskaya Oblast', W Russian Federation 57°55′N 33°45′E
Ozërnyy see Ozërnoye
Ozero Azhbulat see Ozero Ul'ken Azhibolat
Ozero Segozero see Segozerskoye Vodokhranilishche
115 D18 **Ozerós, Límni** ◎ W Greece
145 T7 **Ozero Ul'ken Azhibolat** prev. Ozero Azhbulat. ◎ NE Kazakhstan
122 G11 **Ozërsk** Chelyabinskaya Oblast', C Russian Federation 55°44′N 60°59′E
119 D14 **Ozersk** prev. Darkehnen, Ger. Angerapp. Kaliningradskaya Oblast', W Russian Federation 54°23′N 21°59′E
126 L4 **Ozery** Moskovskaya Oblast', W Russian Federation 54°51′N 38°37′E
Özgön see Uzgen
107 C17 **Ozieri** Sardegna, Italy, C Mediterranean Sea 40°35′N 09°01′E
111 I15 **Ozimek** Ger. Malapane. Opolskie, SW Poland 50°41′N 18°16′E
127 R8 **Ozinki** Saratovskaya Oblast', W Russian Federation 51°16′N 49°45′E
25 O10 **Ozona** Texas, SW USA 30°43′N 101°13′W
Ozorkov see Ozorków
110 J12 **Ozorków** Rus. Ozorkov. Łódźz, C Poland 52°00′N 19°17′E
164 F14 **Ōzu** Ehime, Shikoku, SW Japan 33°30′N 132°33′E
137 R10 **Ozurgeti** prev. Makharadze, Ozurget'i. W Georgia 41°57′N 42°01′E
Ozurget'i see Ozurgeti

P

99 J17 **Paal** Limburg, NE Belgium 51°03′N 05°08′E
196 M14 **Paamiut** var. Pâmiut, Dan. Frederikshåb. Sermersooq, S Greenland 61°59′N 49°40′W
Pa-an see Hpa-an
101 L22 **Paar** ≈ SE Germany
83 E26 **Paarl** Western Cape, SW South Africa 33°45′S 18°58′E
93 L15 **Paavola** Pohjois-Pohjanmaa, C Finland 64°34′N 25°15′E
96 E8 **Pabbay** island NW Scotland, United Kingdom
153 T15 **Pabna** Rajshahi, W Bangladesh 24°02′N 89°15′E
109 U4 **Pabneukirchen** Oberösterreich, N Austria 48°19′N 14°49′E
118 H13 **Pabradė** Pol. Podbrodzie. Vilnius, SE Lithuania 54°58′N 25°43′E
56 L13 **Pacahuaras, Río** ≈ N Bolivia
Pacaraima, Sierra/Pacaraim, Serra see Pakaraima Mountains
56 B11 **Pacasmayo** La Libertad, W Peru 07°27′S 79°33′W
42 D6 **Pacaya, Volcán de** ⛰ S Guatemala 14°19′N 90°36′W
115 K23 **Pacheía** var. Pachía. island Kykládes, Greece, Aegean Sea
Pachía see Pacheía
107 L26 **Pachino** Sicilia, Italy, C Mediterranean Sea 36°43′N 15°06′E
56 F12 **Pachitea, Río** ≈ C Peru
154 I11 **Pachmarhi** Madhya Pradesh, C India 22°36′N 78°18′E
121 P3 **Páchna** var. Pakhna. SW Cyprus 34°47′N 32°48′E
115 H25 **Páchnes** ▲ Kríti, Greece, E Mediterranean Sea 35°19′N 24°00′E
54 F9 **Pacho** Cundinamarca, C Colombia 05°09′N 74°08′W
154 F12 **Pāchora** Mahārāshtra, C India 20°52′N 75°28′E
41 P13 **Pachuca** var. Pachuca de Soto. Hidalgo, C Mexico 20°05′N 98°46′W
Pachuca de Soto see Pachuca
27 W5 **Pacific** Missouri, C USA 38°28′N 90°44′W
192 L14 **Pacific-Antarctic Ridge** undersea feature S Pacific Ocean 62°00′S 157°00′W
32 F8 **Pacific Beach** Washington, NW USA 47°09′N 124°12′W
35 N10 **Pacific Grove** California, W USA 36°35′N 121°54′W
29 S15 **Pacific Junction** Iowa, C USA 41°01′N 95°48′W
192-193 **Pacific Ocean** ocean
129 Z10 **Pacific Plate** tectonic feature
113 J15 **Pačir** ▲ N Montenegro 43°19′N 19°07′E
182 L5 **Packsaddle** New South Wales, SE Australia 30°42′S 141°55′E
32 H9 **Packwood** Washington, NW USA 46°37′N 121°38′W
Padalung see Phatthalung
168 J12 **Padang** Sumatera, W Indonesia 01°S 100°21′E
168 L9 **Padang Endau** Pahang, Peninsular Malaysia 02°38′N 103°37′E
Padangpandjang see Padangpanjang
168 I11 **Padangpanjang** prev. Padangpandjang. Sumatera, W Indonesia 0°30′S 100°26′E
168 I10 **Padangsidempuan** prev. Padangsidimpoean. Sumatera, W Indonesia 01°23′N 99°15′E
Padangsidimpoean see Padangsidempuan
124 I9 **Padany** Respublika Kareliya, NW Russian Federation 63°18′N 33°20′E
93 M18 **Padasjoki** Päijät-Häme, S Finland 61°20′N 25°21′E
57 M22 **Padcaya** Tarija, S Bolivia 21°52′S 64°46′W
101 H14 **Paderborn** Nordrhein-Westfalen, NW Germany 51°43′N 08°45′E
Padeșul/Padeș, Vîrful see Padeș, Vârful
116 F12 **Padeș, Vârful** var. Padeșul; prev. Vîrful Padeș. ▲ W Romania 45°39′N 22°19′E
112 L10 **Padinska Skela** Serbia, N Serbia 44°58′N 20°25′E
153 S14 **Padma** var. Ganges. ≈ Bangladesh/India see also Ganges
Padma see Brahmaputra
Padma see Ganges
106 H8 **Padova** Eng. Padua; anc. Patavium. Veneto, NE Italy 45°24′N 11°52′E
82 A10 **Padrão, Ponta do** headland NW Angola 06°06′S 12°18′E
25 T16 **Padre Island** island Texas, SW USA
104 G3 **Padrón** Galicia, NW Spain 42°44′N 08°40′W
118 K13 **Padsvillye** Rus. Podsvil'ye. Vitsyebskaya Voblasts', N Belarus 55°09′N 27°58′E
182 K11 **Padthaway** South Australia 36°39′S 140°30′E
Padua see Padova
20 G7 **Paducah** Kentucky, S USA 37°03′N 88°36′W
25 P4 **Paducah** Texas, SW USA 34°01′N 100°18′W
105 N15 **Padul** Andalucía, S Spain 37°02′N 03°37′W
191 P8 **Paea** Tahiti, W French Polynesia 17°41′S 149°35′W
185 L14 **Paekakariki** Wellington, North Island, New Zealand 41°00′S 174°58′E
163 X11 **Paektu-san** var. Baitou Shan. ▲ China/North Korea 42°00′N 128°03′W
Paengnyŏng see Baengnyong-do
184 M7 **Paeroa** Waikato, North Island, New Zealand 37°23′S 175°39′E
54 D12 **Páez** Cauca, SW Colombia 02°37′N 76°00′W
121 O3 **Páfos** var. Paphos. W Cyprus 34°46′N 32°26′E
121 O3 **Páfos** ✈ SW Cyprus 34°46′N 32°25′E
83 L19 **Pafúri** Gaza, SW Mozambique 22°27′S 31°21′E
112 C12 **Pag** It. Pago. Lika-Senj, SW Croatia 44°26′N 15°01′E
112 B11 **Pag** It. Pago. island Zadar, C Croatia
171 P7 **Pagadian** Mindanao, S Philippines 07°47′N 123°22′E
168 J13 **Pagai Selatan, Pulau** island Kepulauan Mentawai, W Indonesia
168 J13 **Pagai Utara, Pulau** island Kepulauan Mentawai, W Indonesia
188 K4 **Pagan** island C Northern Mariana Islands
115 G16 **Pagasitikós Kólpos** gulf E Greece
36 L8 **Page** Arizona, SW USA 36°54′N 111°28′W
29 Q5 **Page** North Dakota, N USA 47°09′N 97°33′W
118 D13 **Pagėgiai** Ger. Pogegen. Tauragė, SW Lithuania 55°08′N 21°54′E
21 S11 **Pageland** South Carolina, SE USA 34°46′N 80°23′W
81 G16 **Pager** ≈ NE Uganda
149 Q5 **Paghmān** Kābul, E Afghanistan 34°33′N 68°55′E
Pago see Pag
188 C16 **Pago Bay** bay E Guam, W Pacific Ocean
115 M20 **Pagóndas** var. Pagóndhas. Sámos, Dodekánisa, Greece, Aegean Sea 37°41′N 26°50′E
Pagóndhas see Pagóndas
192 J16 **Pago Pago** ○ (American Samoa) Tutuila, W American Samoa 14°16′S 170°43′W
37 R8 **Pagosa Springs** Colorado, C USA 37°13′N 107°01′W
Pagqên see Gadê
38 H12 **Pāhala** var. Pahala. Hawaii, USA, C Pacific Ocean 19°12′N 155°28′W
168 K8 **Pahang** var. Negeri Pahang Darul Makmur. ◈ state Peninsular Malaysia
Pahang see Pahang, Sungai
168 L8 **Pahang, Sungai** var. Pahang, Sungei Pahang. ≈ Peninsular Malaysia
149 S8 **Pahārpur** Khyber Pakhtunkhwa, NW Pakistan 32°06′N 71°00′E
185 B24 **Pahia Point** headland South Island, New Zealand 46°19′S 167°42′E
184 M13 **Pahiatua** Manawatu-Wanganui, North Island, New Zealand 40°30′S 175°49′E
38 H12 **Pāhoa** var. Pahoa. Hawaii, USA, C Pacific Ocean 19°29′N 154°56′W
23 Y14 **Pahokee** Florida, SE USA 26°49′N 80°40′W
35 X9 **Pahranagat Range** ▲▲ Nevada, W USA
35 W11 **Pahrump** Nevada, W USA 36°11′N 115°58′W
Pahsien see Chongqing
35 V9 **Pahute Mesa** ▲▲ Nevada, W USA
167 N7 **Pai** Mae Hong Son, NW Thailand 19°24′N 98°26′E
38 F10 **Pa'ia** var. Paia. Maui, Hawaii, USA, C Pacific Ocean 20°54′N 156°22′W
Paia see Pa'ia
Pai-ch'eng see Baicheng
118 H4 **Paide** Ger. Weissenstein. Järvamaa, N Estonia 58°55′N 25°36′E
97 J24 **Paignton** SW England, United Kingdom 50°26′N 03°34′W
184 K3 **Paihia** Northland, North Island, New Zealand 35°18′S 174°06′E
93 M18 **Päijänne** ◎ C Finland
Päijänne-Tavastland see Päijät-Häme
93 M19 **Päijät-Häme** Swe. Päijänne-Tavastland. ◈ region S Finland
114 F13 **Páïko** ▲▲ N Greece
57 M17 **Paila, Río** ≈ C Bolivia
167 Q12 **Pailĭn** Bătdâmbâng, W Cambodia 12°51′N 102°34′E
Pailing see Chun'an
54 F6 **Pailitas** Cesar, N Colombia 08°58′N 73°38′W
38 F9 **Pailolo Channel** channel Hawai'i, USA, C Pacific Ocean
93 K19 **Paimio** Swe. Pemar. Varsinais-Suomi, SW Finland 60°27′N 22°42′E
165 O16 **Paimi-saki** var. Yaeme-saki. headland Iriomote-jima, SW Japan 24°18′N 123°40′E
102 G5 **Paimpol** Côtes d'Armor, NW France 48°47′N 03°03′W
168 J12 **Painan** Sumatera, W Indonesia 01°22′S 100°33′E
63 G23 **Paine, Cerro** ▲ S Chile 51°01′S 72°57′W
31 U11 **Painesville** Ohio, N USA 41°43′N 81°15′W
31 S14 **Paint Creek** ≈ Ohio, N USA
36 L10 **Painted Desert** desert Arizona, SW USA
Paint Hills see Wemindji
30 M4 **Paint River** ≈ Michigan, N USA
25 P8 **Paint Rock** Texas, SW USA 31°32′N 99°56′W
21 O6 **Paintsville** Kentucky, S USA 37°48′N 82°48′W
Paisance see Piacenza
96 I12 **Paisley** W Scotland, United Kingdom 55°50′N 04°26′W
32 I15 **Paisley** Oregon, NW USA 42°40′N 120°31′W
105 O3 **País Vasco** Basq. Euskadi, Eng. The Basque Country, Sp. Provincias Vascongadas. ◈ autonomous community N Spain
56 A9 **Paita** Piura, NW Peru 05°11′S 81°09′W
169 V6 **Paitan, Teluk** bay Sabah, East Malaysia
104 H7 **Paiva, Rio** ≈ N Portugal
92 K12 **Pajala** Norrbotten, N Sweden 67°12′N 23°19′E
104 K3 **Pajares, Puerto de** pass NW Spain
54 G9 **Pajarito** Boyacá, C Colombia 05°18′N 72°43′W
54 G4 **Pajaro** La Guajira, S Colombia 11°41′N 72°37′W
Pakanbaru see Pekanbaru
55 Q10 **Pakaraima Mountains** var. Serra Pacaraim, Sierra Pacaraima. ▲▲ N South America
Pākaur see Pākur
167 P10 **Pak Chong** Nakhon Ratchasima, C Thailand 14°38′N 101°22′E
123 V8 **Pakhachi** Krasnoyarskiy Kray, E Russian Federation 60°36′N 168°59′E
Pakhna see Páchna
189 U16 **Pakin Atoll** atoll Caroline Islands, E Micronesia
149 Q12 **Pakistan** off. Islamic Republic of Pakistan, var. Islami Jamhuriya e Pakistan. ◆ republic S Asia
Pakistan, Islamic Republic of see Pakistan
Pakistan, Islami Jamhuriya e see Pakistan
167 P8 **Pak Lay** var. Muang Pak Lay. Xaignabouli, C Laos 18°06′N 101°21′E
Paknam see Samut Prakan
166 L5 **Pakokku** Magway, C Myanmar (Burma) 21°20′N 95°05′E
110 I10 **Pakość** Ger. Pakosch. Kujawski-pomorskie, C Poland 52°47′N 18°03′E
Pakosch see Pakość
149 V10 **Pākpattan** Punjab, E Pakistan 30°20′N 73°27′E
167 O15 **Pak Phanang** var. Ban Pak Phanang. Nakhon Si Thammarat, SW Thailand 08°20′N 100°10′E
112 G9 **Pakrac** Hung. Pakrácz. Požega-Slavonija, NE Croatia 45°26′N 17°09′E
Pakrácz see Pakrac
118 F11 **Pakruojis** Šiaulai, N Lithuania 56°N 23°51′E
111 J24 **Paks** Tolna, S Hungary 46°38′N 18°51′E
Pak Sane see Pakxan
Paksé see Pakxé
167 Q10 **Pak Thong Chai** Nakhon Ratchasima, C Thailand 14°43′N 102°01′E
Paktīā see Paktiyā
149 Q7 **Paktīkā** ◈ province SE Afghanistan
149 R6 **Paktiyā** prev. Paktīā. ◈ province SE Afghanistan
171 N12 **Pakuli** Sulawesi, C Indonesia 01°14′S 119°55′E
153 S14 **Pākur** var. Pākaur. Jharkhand, N India 24°48′N 87°14′E
81 F17 **Pakwach** NW Uganda 02°28′N 31°28′E
167 R8 **Pakxan** var. Muang Pakxan, Pak Sane. Bolikhamxai, C Laos 18°27′N 103°38′E
167 S10 **Pakxé** var. Paksé. Champasak, S Laos 15°09′N 105°49′E
78 G12 **Pala** Mayo-Kébbi Ouest, SW Chad 09°22′N 14°54′E
61 A17 **Palacios** Santa Fe, C Argentina 30°43′S 61°37′W
25 V13 **Palacios** Texas, SW USA 28°42′N 96°13′W
105 X5 **Palafrugell** Cataluña, NE Spain 41°55′N 03°10′E
107 L24 **Palagonia** Sicilia, Italy, C Mediterranean Sea 37°20′N 14°45′E
113 E17 **Palagruža** It. Pelagosa. island SW Croatia
115 G20 **Palaiá Epídavros** Pelopónnisos, S Greece 37°38′N 23°09′E
121 P3 **Palaichóri** var. Palekhori. C Cyprus 34°55′N 33°06′E
115 H25 **Palaiochóra** Kríti, Greece, E Mediterranean Sea 35°14′N 23°37′E
115 A15 **Palaiolastrítsa** religious building Kérkyra, Iónia Nisiá, Greece, C Mediterranean Sea
115 J19 **Palaiópoli** Ándros, Kykládes, Greece, Aegean Sea 37°49′N 24°49′E
103 N5 **Palaiseau** Essonne, N France 48°41′N 02°14′E
154 N11 **Pāla Laharha** Odisha, E India 21°25′N 85°18′E
83 G19 **Palamakoloi** Ghanzi, C Botswana 23°10′S 22°22′E
115 E16 **Palamás** Thessalía, C Greece 39°28′N 22°05′E
105 X5 **Palamós** Cataluña, NE Spain 41°51′N 03°06′E
118 J5 **Palamuse** Ger. Sankt-Bartholomäi. Jõgevamaa, E Estonia 58°41′N 26°35′E
183 Q14 **Palana** Tasmania, SE Australia 39°48′S 147°54′E
123 U9 **Palana** Krasnoyarskiy Kray, E Russian Federation 59°05′N 159°59′E
118 C11 **Palanga** Ger. Polangen. Klaipėda, NW Lithuania 05°54′N 21°05′E
143 V10 **Palangān, Kūh-e** ▲▲ E Iran
Palangkaraja see Palangkaraya
169 T12 **Palangkaraya** prev. Palangkaraja. Borneo, C Indonesia 02°16′S 113°55′E
155 H22 **Palani** Tamil Nādu, SE India 10°30′N 77°24′E
Palanka see Bačka Palanka
154 D9 **Pālanpur** Gujarāt, W India 24°12′N 72°29′E
Palantia see Palencia
83 I19 **Palapye** Central, SE Botswana 22°37′S 27°06′E
155 J19 **Pālār** ≈ SE India
104 H3 **Palas de Rei** Galicia, NW Spain 42°52′N 07°51′W
123 T9 **Palatka** Magadanskaya Oblast', E Russian Federation 60°09′N 150°33′E
23 W10 **Palatka** Florida, SE USA 29°39′N 81°38′W
188 B9 **Palau** var. Belau. ◆ republic W Pacific Ocean
Palau see Palau Islands
129 Y14 **Palau Islands** var. Palau. island group N Palau
192 G16 **Palauli Bay** bay Savai'i, C Samoa, C Pacific Ocean
167 N11 **Palaw** Taninthayi, S Myanmar (Burma) 12°57′N 98°39′E
170 M6 **Palawan** island W Philippines
171 N6 **Palawan Passage** passage W Philippines
192 E7 **Palawan Trough** undersea feature S South China Sea 07°00′N 115°00′E
155 H23 **Pālayankottai** Tamil Nādu, SE India 08°42′N 77°46′E
107 L25 **Palazzola Acreide** anc. Acrae. Sicilia, Italy, C Mediterranean Sea 37°04′N 14°54′E
118 G3 **Paldiski** prev. Baltiski, Eng. Baltic Port, Ger. Baltischport. Harjumaa, NW Estonia 59°22′N 24°08′E
112 I13 **Pale** Republika Srpska, SE Bosnia and Herzegovina 43°49′N 18°35′E
Palekhori see Palaichóri
168 L13 **Palembang** Sumatera, W Indonesia 02°59′S 104°45′E
63 G18 **Palena** Los Lagos, S Chile 43°40′S 71°50′W
63 G18 **Palena, Río** ≈ S Chile
104 M5 **Palencia** anc. Palantia, Pallantia. Castilla y León, NW Spain 42°23′N 04°32′W
104 M3 **Palencia** ◈ province Castilla y León, N Spain
35 X15 **Palen Dry Lake** ◎ California, W USA
41 V15 **Palenque** Chiapas, SE Mexico 17°32′N 91°59′W
41 V15 **Palenque** var. Ruinas de Palenque. ruins Chiapas, SE Mexico
45 O9 **Palenque, Punta** headland S Dominican Republic 18°13′N 70°08′W
Palenque, Ruinas de see Palenque
Palerme see Palermo
107 I23 **Palermo** Fr. Palerme; anc. Panhormus, Panormus. Sicilia, Italy, C Mediterranean Sea 38°08′N 13°23′E
25 V8 **Palestine** Texas, SW USA 31°45′N 93°39′W
25 V7 **Palestine, Lake** ▣ Texas, SW USA
107 I15 **Palestrina** Lazio, C Italy 41°49′N 12°53′E
166 K5 **Paletwa** Chin State, W Myanmar (Burma) 21°25′N 92°49′E
155 G21 **Pālghāt** var. Palakkad. Kerala, SW India 10°46′N 76°42′E see also Palakkad
152 F13 **Pāli** Rājasthān, N India 25°48′N 73°21′E
167 N16 **Palian** Trang, SW Thailand
189 O12 **Palikir** ● (Micronesia) Pohnpei, E Micronesia 06°58′N 158°13′E
Palimé see Kpalimé
107 L19 **Palinuro, Capo** headland S Italy 40°02′N 15°16′E
115 H15 **Palioúri, Akrotírio** var. Akrotírio Kanestron. headland N Greece 39°55′N 23°45′E
33 R14 **Palisades Reservoir** ▣ Idaho, NW USA
99 J23 **Paliseul** Luxembourg, SE Belgium 49°55′N 05°09′E
154 C11 **Pālitāna** Gujarāt, W India 21°30′N 71°50′E
118 F4 **Palivere** Läänemaa, W Estonia 58°59′N 23°58′E
41 V14 **Palizada** Campeche, SE Mexico 18°15′N 92°03′W
93 L18 **Pälkäne** Pirkanmaa, W Finland 61°22′N 24°16′E
155 J22 **Palk Strait** strait India/Sri Lanka
155 J23 **Pallai** Northern Province, NW Sri Lanka 09°35′N 80°20′E
Pallantia see Palencia
106 C6 **Pallanza** Piemonte, NE Italy 45°57′N 08°32′E
127 Q9 **Pallasovka** Volgogradskaya Oblast', SW Russian Federation 50°06′N 46°52′E
Pallene/Pallíni see Kassándra
185 L15 **Palliser Bay** bay North Island, New Zealand
185 L15 **Palliser, Cape** headland North Island, New Zealand 41°37′S 175°16′E
191 U9 **Palliser, Îles** island group Îles Tuamotu, C French Polynesia
82 Q12 **Palma** Cabo Delgado, N Mozambique 10°46′S 40°30′E
105 X9 **Palma** var. Palma de Mallorca. Mallorca, Spain, W Mediterranean Sea 39°35′N 02°39′E
105 X9 **Palma** ✈ Mallorca, Spain, W Mediterranean Sea
105 X10 **Palma, Badia de** bay Mallorca, Spain, W Mediterranean Sea
104 L13 **Palma del Río** Andalucía, S Spain 37°42′N 05°16′W
Palma de Mallorca see Palma
107 J25 **Palma di Montechiaro** Sicilia, Italy, C Mediterranean Sea 37°12′N 13°46′E
106 J7 **Palmanova** Friuli-Venezia Giulia, NE Italy 45°54′N 13°20′E
54 J7 **Palmarito** Apure, C Venezuela 07°36′N 70°08′W
43 N15 **Palmar Sur** Puntarenas, SE Costa Rica 08°54′N 83°27′W
60 I12 **Palmas** Paraná, S Brazil 26°29′S 52°00′W
59 K16 **Palmas** var. Palmas do Tocantins. Tocantins, C Brazil 10°24′S 48°19′W
76 L18 **Palmas, Cape** Fr. Cap des Palmès. headland SW Ivory Coast 04°18′N 07°31′W
Palmas do Tocantins see Palmas
54 D11 **Palmaseca** ✈ (Cali) Valle del Cauca, SW Colombia 03°31′N 76°27′W
107 B21 **Palmas, Golfo di** gulf Sardegna, Italy, C Mediterranean Sea
44 I7 **Palma Soriano** Santiago de Cuba, E Cuba 20°10′N 76°00′W
23 Y12 **Palm Bay** Florida, SE USA 28°01′N 80°35′W
35 T14 **Palmdale** California, W USA 34°34′N 118°07′W
61 H14 **Palmeira das Missões** Rio Grande do Sul, S Brazil 27°54′S 53°20′W
82 A11 **Palmeirinhas, Ponta das** headland NW Angola 09°04′S 13°02′E
39 R11 **Palmer** Alaska, USA 61°36′N 149°06′W
19 N11 **Palmer** Massachusetts, NE USA 42°09′N 72°19′W
25 U7 **Palmer** Texas, SW USA 32°25′N 96°40′W
194 H4 **Palmer** US research station Antarctica 64°37′S 64°01′W
15 R11 **Palmer** ≈ Québec, SE Canada
37 T5 **Palmer Lake** Colorado, C USA 39°07′N 104°55′W
194 J6 **Palmer Land** physical region Antarctica
14 F15 **Palmerston** Ontario, SE Canada 43°51′N 80°49′W
185 F22 **Palmerston** Otago, South Island, New Zealand 45°27′S 170°42′E
190 K15 **Palmerston** island S Cook Islands
Palmerston see Darwin
184 M12 **Palmerston North** Manawatu-Wanganui, North Island, New Zealand 40°20′S 175°52′E
23 V13 **Palmetto** Florida, SE USA 27°31′N 82°34′W
The Palmetto State see South Carolina
107 M22 **Palmi** Calabria, SW Italy 38°21′N 15°51′E
54 D11 **Palmira** Valle del Cauca, W Colombia 03°33′N 76°17′W
56 F8 **Palmira, Río** ≈ N Peru
61 D19 **Palmitas** Soriano, SW Uruguay 33°27′S 57°48′W
Palmnicken see Yantarnyy
35 V15 **Palm Springs** California, W USA 33°48′N 116°33′W
27 V2 **Palmyra** Missouri, C USA 39°48′N 91°31′W
18 G10 **Palmyra** New York, NE USA 43°02′N 77°13′W
18 G15 **Palmyra** Pennsylvania, NE USA 40°18′N 76°35′W
21 V5 **Palmyra** Virginia, NE USA 37°53′N 78°17′W
Palmyra see Tudmur
192 L7 **Palmyra Atoll** ◇ US incorporated territory C Pacific Ocean
154 P12 **Palmyras Point** headland E India 20°46′N 87°00′E
35 N9 **Palo Alto** California, W USA 37°26′N 122°08′W
25 O1 **Palo Duro Creek** ≈ Texas, SW USA
Paloe see Denpasar, Bali, C Indonesia
Paloe see Palu
168 L9 **Paloh** Johor, Peninsular Malaysia 02°10′N 103°11′E
80 F12 **Paloich** Upper Nile, NE South Sudan 10°29′N 32°31′E
40 I3 **Palomas** Chihuahua, N Mexico 31°45′N 107°38′W
107 I15 **Palombara Sabina** Lazio, C Italy 42°04′N 12°45′E
105 S13 **Palos, Cabo de** headland SE Spain 37°38′N 00°42′W
104 I14 **Palos de la Frontera** Andalucía, S Spain 37°14′N 06°53′W
60 G11 **Palotina** Paraná, S Brazil 24°16′S 53°49′W
32 M9 **Palouse** Washington, NW USA 46°54′N 117°04′W
32 L9 **Palouse River** ≈ Washington, NW USA
35 Y16 **Palo Verde** California, W USA 33°25′N 114°43′W
57 E16 **Palpa** Ica, W Peru 14°35′S 75°09′W
95 M16 **Pålsboda** Örebro, C Sweden 59°04′N 15°21′E
93 M15 **Paltamo** Kainuu, C Finland 64°25′N 27°50′E
171 N12 **Palu** prev. Paloe. Sulawesi, C Indonesia 0°54′S 119°52′E
137 P14 **Palu** Elazığ, E Turkey 38°43′N 39°56′E
152 I11 **Palwal** Haryāna, N India 28°15′N 77°18′E
123 U6 **Palyavaam** ≈ NE Russian Federation
77 Q13 **Pama** SE Burkina Faso 11°13′N 00°46′E
172 J14 **Pamandzi** ✈ (Mamoudzou) Petite-Terre, E Mayotte
Pamangkat see Pemangkat
143 R11 **Pā Mazār** Kermān, C Iran
83 N19 **Pambarra** Inhambane, SE Mozambique 21°57′S 35°06′E
171 X12 **Pamdai** Papua, E Indonesia 01°58′S 137°19′E
103 N16 **Pamiers** Ariège, S France 43°07′N 01°37′E
147 T14 **Pamir** var. Daryā-ye Pāmīr, Taj. Dar"yoi Pomir. ≈ Afghanistan/Tajikistan see also Pāmīr, Daryā-ye
Pamir see Pāmīr, Daryā-ye
149 U1 **Pāmīr, Daryā-ye** var. Pamir, Taj. Dar"yoi Pomir. ≈ Afghanistan/Tajikistan see also Pamir
Pāmīr, Daryā-ye see Pamir
Pāmīr-e Khord see Little Pamir
Pamir/Pāmīr, Daryā-ye see Pamirs
129 Q8 **Pamirs** Pash. Daryā-ye Pāmīr, Rus. Pamir. ▲▲ C Asia
Pâmiut see Paamiut
21 X10 **Pamlico River** ≈ North Carolina, SE USA
21 Y10 **Pamlico Sound** sound North Carolina, SE USA
25 O2 **Pampa** Texas, SW USA 35°32′N 100°58′W
Pampa Aullagas, Lago see Poopó, Lago
61 B21 **Pampa Húmeda** grassland E Argentina
56 A10 **Pampa las Salinas** salt lake NW Peru
57 F15 **Pampas** Huancavelica, C Peru 12°22′S 74°53′W
62 K13 **Pampas** plain C Argentina
55 O4 **Pampatar** Nueva Esparta, NE Venezuela 11°03′N 63°51′W
Pampeluna see Pamplona

◆ Country
● Country Capital
◇ Dependent Territory
○ Dependent Territory Capital
◈ Administrative Regions
✈ International Airport
▲ Mountain
▲▲ Mountain Range
⛰ Volcano
≈ River
◎ Lake
▣ Reservoir

104 H8 **Pampilhosa da Serra**
var. Pampilhosa de Serra.
Coimbra, N Portugal
40°03´N 07°58´W

173 Y15 **Pamplemousses** N Mauritius
20°06´S 57°34´E

54 G7 **Pamplona** Norte de
Santander, N Colombia
07°24´N 72°38´W

105 Q3 **Pamplona** *Basq.* Iruña,
prev. Pampeluna; *anc.*
Pompaelo. Navarra, N Spain
42°49´N 01°39´W

114 I11 **Pamporovo** *prev.* Vasil
Kolarov. Smolyan, S Bulgaria
41°39´N 24°45´E

136 D15 **Pamukkale** Denizli,
W Turkey 37°51´N 29°13´E

21 W5 **Pamunkey River**
♦ Virginia, NE USA

152 K5 **Pamzal** Jammu and Kashmir,
NE India 34°17´N 78°50´E

30 L4 **Pana** Illinois, N USA
39°23´N 89°04´W

44 Y11 **Panabá** Yucatán, SE Mexico
21°20´N 88°16´W

35 Y8 **Panaca** Nevada, W USA
37°47´N 114°24´W

115 E19 **Panachaïkó** ▲ S Greece

14 F11 **Panache Lake** ◎ Ontario,
S Canada

114 I10 **Panagyurishte** Pazardzhik,
C Bulgaria 42°30´N 24°11´E

168 M16 **Panaitan, Pulau** *island*
S Indonesia

115 D18 **Panaitolikó** ▲ C Greece

155 E17 **Panají** *var.* Pangim, Panjim,
New Goa. *state capital* Goa,
W India 15°31´N 73°52´E

43 T15 **Panamá** *anc.* Ciudad de
Panama, *Eng.* Panama
City. ● (Panama) Panamá,
C Panama 08°57´N 79°33´W

43 T14 **Panama** *off.* Republic of
Panama. ◆ *republic* Central
America

43 U14 **Panamá** Provincia
de Panamá. ◇ *province*
E Panama

43 U15 **Panamá, Bahía de** *bay*
N Gulf of Panama

193 T7 **Panama Basin** *undersea
feature* E Pacific Ocean
05°00´N 83°30´W

43 T15 **Panama Canal** *canal*
E Panama

23 R9 **Panama City** Florida,
SE USA 30°09´N 85°39´W

43 T14 **Panama City** *var.* Panamá,
C Panama 09°02´N 79°24´W

Panama City *see* Panamá

23 Q9 **Panama City Beach** Florida,
SE USA 30°10´N 85°48´W

43 T17 **Panamá, Golfo de** *var.* Gulf
of Panama. *gulf* S Panama

Panama, Gulf of *see*
Panama, Golfo de

Panama, Isthmus of *see*
Panama, Istmo de

43 T15 **Panama, Istmo de** *Eng.*
Isthmus of Panama; *prev.*
Isthmus of Darien. *isthmus*
E Panama

Panamá, Provincia de *see*
Panamá

Panama, Republic of *see*
Panama

35 U11 **Panamint Range**
▲ California, W USA

107 L22 **Panarea, Isola** *island* Isole
Eolie, S Italy

106 G9 **Panaro** ♦ N Italy

171 P5 **Panay** *island*
C Philippines

35 W7 **Pancake Range** ▲ Nevada,
W USA

112 M11 **Pančevo** *Ger.* Pantschowa,
Hung. Pancsova. Vojvodina,
N Serbia 44°53´N 20°40´E

113 M15 **Pančićev Vrh** ▲ SW Serbia
43°16´N 20°49´E

116 L12 **Panciu** Vrancea, E Romania
45°54´N 27°08´E

115 F10 **Pâncota** *Hung.* Pankota;
prev. Pincota. Arad,
W Romania 46°20´N 21°45´E

Pancsova *see* Pančevo

83 N20 **Panda** Inhambane,
SE Mozambique
24°02´S 34°45´E

171 X12 **Pandaidori, Kepulauan**
island group E Indonesia

25 N11 **Pandale** Texas, SW USA
30°09´N 101°34´W

169 P12 **Pandang Tikar, Pulau**
island N Indonesia

61 F20 **Pan de Azúcar** Maldonado,
S Uruguay 34°45´S 55°14´W

118 H11 **Pandėlys** Panevėžys,
NE Lithuania 56°04´N 25°18´E

155 F15 **Pandharpur** Mahārāshtra,
W India 17°42´N 75°24´E

182 J1 **Pandie Pandie** South
Australia 26°06´S 139°06´E

171 O12 **Pandiri** Sulawesi, C Indonesia
01°32´S 120°47´E

61 F20 **Pando** Canelones, S Uruguay
34°44´S 55°58´W

57 J14 **Pando** ◇ *department*
N Bolivia

192 K9 **Pandora Bank** *undersea
feature* W Pacific Ocean

95 G20 **Pandrup** Nordjylland,
N Denmark 57°14´N 09°42´E

79 J15 **Pandu** Equateur, NW Dem.
Rep. Congo 05°03´N 19°14´E

153 V12 **Pandu** Assam, NE India
26°08´N 91°37´E

Paneas *see* Bāniyās

59 F15 **Panelas** Mato Grosso,
W Brazil 09°06´S 60°41´W

118 G12 **Panevėžys** Panevėžys,
C Lithuania 55°44´N 24°21´E

118 G11 **Panevėžys** ♦ *province*
NW Lithuania

Panfilow *see* Zharkent

127 N9 **Panfilovo** Volgogradskaya
Oblast´, SW Russian
Federation 50°25´N 42°55´E

79 N17 **Panga** Orientale, N Dem.
Rep. Congo 01°52´N 26°18´E

193 Y15 **Pangai** Lifuka, C Tonga
19°50´S 174°23´W

114 H13 **Pangaío** ▲ N Greece

79 G20 **Pangala** Pool, S Congo
03°26´S 14°38´E

81 J22 **Pangani** Tanga, E Tanzania
05°27´S 39°00´E

81 I21 **Pangani** ♦ NE Tanzania

186 K8 **Pangoe** Choiseul,
NW Solomon Islands
07°00´S 157°05´E

79 N20 **Pangi** Maniema, E Dem. Rep.
Congo 03°12´S 26°39´E

Pangim *see* Panaji

168 H8 **Pangkalanbrandan**
Sumatera, W Indonesia
04°00´N 98°15´E

Pangkalanbuun *see*
Pangkalanbun

169 R13 **Pangkalanbuun** *var.*
Pangkalanbun. Borneo,
C Indonesia 02°43´S 111°38´E

169 N12 **Pangkalpinang** Pulau
Bangka, W Indonesia
02°05´S 106°09´E

11 U17 **Pangman** Saskatchewan,
S Canada 49°37´N 104°33´W

Pang-Nga *see* Phang-Nga

9 S6 **Pangnirtung** Baffin Island,
Nunavut, NE Canada
66°05´N 65°45´W

152 K6 **Pangong Tso** *var.* Bangong
Co. ◎ China/India *see also*
Bangong Co
Pangong Tso *see* Banggong
Co

36 K7 **Panguitch** Utah, W USA
37°49´N 112°26´W

186 J7 **Panguna** Bougainville
Island, NE Papua New Guinea
06°22´S 155°20´E

171 N8 **Panguturan Group** *island
group* Sulu Archipelago,
SW Philippines

25 N2 **Panhandle** Texas, SW USA
35°21´N 101°24´W

Panhormus *see* Palermo

171 W14 **Paniai, Danau** ◎ Papua,
E Indonesia

79 L21 **Pania-Mutombo** Kasai-
Oriental, C Dem. Rep. Congo
05°09´S 23°49´E

187 P16 **Panié, Mont** ▲ C New
Caledonia 20°33´S 164°41´E

Panikoilli *see* Jājapur

152 I10 **Pānīpat** Haryāna, N India
29°18´N 77°00´E

147 Q14 **Panj** *Rus.* Pyandzh; *prev.*
Kirovabad. SW Tajikistan
37°39´N 69°55´E

147 P15 **Panj** *Rus.* Pyandzh.
♦ Afghanistan/Tajikistan

149 O5 **Panjāb** Bāmyān,
C Afghanistan 34°21´N 67°00´E

147 O12 **Panjakent** *Rus.* Pendzhikent.
W Tajikistan 39°28´N 67°33´E

148 L14 **Panjgūr** Baluchistān,
SW Pakistan 26°58´N 64°05´E

Panjim *see* Panaji

163 U12 **Panjin** Liaoning, NE China
41°11´N 122°03´E

147 P14 **Panji Poyon** *Rus.* Nizhniy
Pyandzh. SW Tajikistan
37°14´N 68°32´E

149 Q4 **Panjshayr** *prev.* Panjshīr.
♦ E Afghanistan

149 S4 **Panjshir** ♦ *province*
NE Afghanistan
Panjshīr *see* Panjshayr
Pankota *see* Pâncota

77 W14 **Pankshin** Plateau, C Nigeria
09°21´N 09°27´E

163 Y10 **Pan Ling** ▲ N China

154 J9 **Panlong Jiang** *see* Lô, Sông

99 M16 **Panna** Madhya Pradesh,
C India 24°43´N 80°11´E

Panningen Limburg,
SE Netherlands
51°20´N 05°59´E

149 R13 **Pāno Āqil** Sind, SE Pakistan
27°55´N 69°18´E

121 P3 **Páno Léfkara** S Cyprus
34°52´N 33°18´E

121 O3 **Páno Panagiá** *var.*
Pano Panayia. W Cyprus
34°55´N 32°38´E
Pano Panayia *see* Páno
Panagiá

29 U14 **Panora** Iowa, C USA
41°41´N 94°21´W

60 I8 **Panorama** São Paulo, S Brazil
21°22´S 51°51´W

115 I24 **Pánormos** Kríti, Greece,
E Mediterranean Sea
35°24´N 24°42´E
Panormus *see* Palermo

163 W11 **Panshi** Jilin, NE China
42°54´N 125°50´E
Panshi Yu *see* Passu Keah

59 H19 **Pantanal** *var.* Pantanalmato-
Grossense. *swamp* SW Brazil
Pantanalmato-Grossense
see Pantanal

61 H16 **Pântano Grande** Rio Grande
do Sul, S Brazil 30°12´S 52°24´W

171 Q16 **Pantar, Pulau** *island*
Kepulauan Alor, S Indonesia

21 X9 **Pantego** North Carolina,
SE USA 35°34´N 76°39´E

107 G25 **Pantelleria** *anc.* Cossyra,
Cosyra. Sicilia, Italy,
C Mediterranean Sea
36°47´N 12°00´E

107 G25 **Pantelleria, Isola di** *island*
SW Italy

**Pante Makasar/Pante
Macassar/Pante Makasar**
see Pante Macassar

152 K10 **Pantnagar** Uttarakhand,
N India 29°00´N 79°35´E

115 A15 **Pantokrátoras** ▲ Kérkyra,
Iónia Nisiá, Greece,
C Mediterranean Sea
39°45´N 19°51´E

114 G8 **Pántchevo** Pančevo

41 P11 **Pánuco** Veracruz-Llave,
E Mexico 22°01´N 98°13´W

41 P11 **Pánuco, Río** ♦ C Mexico

160 I12 **Panxian** Guizhou, S China
25°45´N 104°39´E

168 I10 **Panyabungan** Sumatera,
N Indonesia 0°55´N 99°30´E

77 W14 **Panyam** Plateau, C Nigeria
09°28´N 09°13´E

157 N13 **Panzhihua** *prev.* Dukou,
Tu-k´ou. Sichuan, C China
26°35´N 101°41´E

79 I22 **Panzi** Bandundu, SW Dem.
Rep. Congo 07°10´S 17°57´E

42 E5 **Panzós** Alta Verapaz,
E Guatemala 15°21´N 89°40´W

107 N20 **Paola** Calabria, SW Italy
39°21´N 16°03´E

22 R5 **Paola** Malta 35°52´N 14°30´E

31 O15 **Paoli** Indiana, N USA
38°35´N 86°25´W

187 R14 **Paonangisu** Éfaté, C Vanuatu
17°33´S 168°23´E

27 X9 **Paragould** Arkansas, C USA
36°02´N 90°30´W

171 S13 **Paoni** *var.* Pauni. Pulau
Seram, E Indonesia
02°48´S 129°03´E

37 Q5 **Paonia** Colorado, C USA
38°51´N 107°35´W

191 O7 **Paopao** Moorea, W French
Polynesia 17°28´S 149°48´W
Pao-shan *see* Baoshan
Pao-ting *see* Baoding
Pao-k´ou see/Paotow *see*
Baotou

79 H14 **Paoua** Ouham-Pendé,
W Central African Republic
07°22´N 16°25´E

111 H23 **Pápa** Veszprém, W Hungary
47°19´N 17°29´E

42 J12 **Papagayo, Golfo de** *gulf*
NW Costa Rica

38 H11 **Pāpa´ikou** *var.* Papaikou.
Hawaii, USA, C Pacific Ocean
19°45´N 155°06´W

41 R15 **Papagayo, Río**
♦ S Mexico

184 L6 **Papakura** Auckland,
North Island, New Zealand
37°03´S 174°57´E

41 Q13 **Papantla** *var.* Papantla
de Olarte. Veracruz-Llave,
E Mexico 20°30´N 97°21´W
Papantla de Olarte *see*
Papantla

191 P8 **Papara** Tahiti, W French
Polynesia 17°45´S 149°33´W

184 K4 **Paparoa** Northland,
North Island, New Zealand
36°06´S 174°12´E

185 G16 **Paparoa Range** ▲ South
Island, New Zealand

115 K20 **Papás, Akrotírio** *headland*
Ikaría, Dodekánisa, Greece,
Aegean Sea 37°33´N 25°58´E

184 L6 **Papatoetoe** Auckland,
North Island, New Zealand
36°58´S 174°52´E

185 E25 **Papatowai** Otago, South
Island, New Zealand
46°33´S 169°33´E

96 K4 **Papa Westray** *island*
NE Scotland, United Kingdom

191 T10 **Papeete** ○ (French
Polynesia) Tahiti,
W French Polynesia
17°32´S 149°34´W

100 F11 **Papenburg** Niedersachsen,
NW Germany 53°04´N 07°24´E

98 H13 **Papendrecht** Zuid-
Holland, SW Netherlands
51°50´N 04°42´E

191 Q7 **Papenoo** Tahiti, W French
Polynesia 17°29´S 149°25´W

191 Q7 **Papenoo Rivière** ♦ Tahiti,
W French Polynesia

191 N7 **Papetoai** Moorea, W French
Polynesia 17°29´S 149°52´W

62 M10 **Papey** *island* E Iceland

92 L3 **Papey** *island* E Iceland

40 H5 **Papigochic, Río**
♦ NW Mexico

118 E10 **Papilė** Šiauliai, NW Lithuania
56°08´N 22°51´E

29 S15 **Papillion** Nebraska, C USA
41°09´N 96°02´W

15 T5 **Papinachois** ♦ Québec,
SE Canada

171 X13 **Papua** *var.* Irian Barat, West
Irian, West New Guinea,
West Papua; *prev.* Dutch
New Guinea, Irian Jaya,
Netherlands New Guinea.
♦ *province* E Indonesia

171 V10 **Papua Barat** *off.* Propinsi
Papua Barat; *prev.* Irian Jaya
Barat, *Eng.* West Papua.
♦ *province* E Indonesia

186 C9 **Papua, Gulf of** *gulf* S Papua
New Guinea

186 C8 **Papua New Guinea** *off.*
Independent State of Papua
New Guinea; *prev.* Territory
of Papua and New Guinea.
♦ *commonwealth republic*
NW Melanesia
**Papua New Guinea,
Independent State of** *see*
Papua New Guinea

192 H8 **Papua Plateau** *undersea
feature* N Coral Sea

112 G9 **Papuk** ▲ NE Croatia
Papun *see* Hpapun

42 L14 **Paquera** Puntarenas,
W Costa Rica 09°52´N 84°56´W

58 I13 **Pará** off. Estado do Pará.
♦ *state* NE Brazil

55 V9 **Para** ♦ *district* N Suriname
Pará *see* Belém

58 I10 **Parabúrdoo** Western
Australia 23°07´S 117°40´E

59 L19 **Paracatu** Minas Gerais,
NE Brazil 17°14´S 46°52´W

192 E6 **Paracel Islands** *Chin.*
Xisha Qundao, *Viet.* Quân
Đao Hoang Sa. ◇ *disputed
territory* SE Asia

182 I6 **Parachilna** South Australia
31°09´S 138°23´E

149 R6 **Parāchinār** Khyber
Pakhtunkhwa, NW Pakistan
33°56´N 70°04´E

112 N13 **Paraćin** Serbia, C Serbia
43°51´N 21°25´E
Paradip *see*
Krishnarājāsāgara

14 K8 **Paradis** Québec, SE Canada
48°13´N 76°36´W

35 N11 **Paradise** California, W USA
39°42´N 121°39´W

35 X11 **Paradise** Nevada, W USA
36°05´N 115°10´W
Paradise Hill *see* Paradise

37 R11 **Paradise Hills** New Mexico,
SW USA 35°12´N 106°42´W
Paradise of the Pacific *see*
Hawai´i

36 L13 **Paradise Valley** Arizona,
SW USA 33°31´N 111°56´W

35 T2 **Paradise Valley** Nevada,
W USA 41°30´N 117°30´W

115 O22 **Paradísi** ▲ (Ródos) Ródos,
Dodekánisa, Greece, Aegean
Sea 36°24´N 28°05´E
Paradísi *see* Poreč

154 P12 **Parādwīp** Odisha, E India
20°17´N 86°42´E

58 I13 **Pará, Estado do** *see*
Pará

117 R4 **Parafiyivka** Chernihivs´ka
Oblast´, N Ukraine
50°53´N 32°40´E

36 K7 **Paragonah** Utah, W USA
37°53´N 112°46´W

47 X8 **Paraguaçu** *var.* Paraguassú.
♦ E Brazil

60 Q5 **Paraguaçu Paulista** São
Paulo, S Brazil 22°22´S 50°35´W

54 H4 **Paraguaipoa** Zulia,
NW Venezuela
11°21´N 71°58´W

62 O7 **Paraguarí** Paraguarí,
S Paraguay 25°38´S 57°09´W

62 O7 **Paraguarí** *off.* Departamento
de Paraguarí. ◇ *department*
S Paraguay
**Paraguarí, Departamento
de** *see* Paraguarí

57 T15 **Paraguá, Río** ♦ NE Bolivia

55 O8 **Paragua, Río**
♦ SE Venezuela
Paraguassú *see* Paraguaçu

62 N5 **Paraguay** ◆ *republic* C South
America

47 U10 **Paraguay** *var.* Río Paraguay.
♦ C South America
Paraguay, Río *see* Paraguay

59 P15 **Paraíba** *off.* Estado da
Paraíba; *prev.* Parahiba,
Parahyba. ◇ *state* E Brazil
Paraíba *see* João Pessoa

60 P9 **Paraíba do Sul, Rio**
♦ SE Brazil
Paraíba, Estado da *see*
Paraíba
Parainen *see* Pargas

43 N14 **Paraíso** Cartago, C Costa Rica
09°51´N 83°50´W

41 U14 **Paraíso** Tabasco, SE Mexico
18°26´N 93°10´W

57 O17 **Paraíso, Río** ♦ E Bolivia

77 S14 **Parakou** C Benin
09°23´N 02°40´E

115 F20 **Paralía Tyrou** Pelopónnisos,
S Greece 37°17´N 22°51´E

121 Q2 **Paralímni** E Cyprus
35°02´N 34°00´E

115 G18 **Paralímni, Límni**
◎ C Greece

55 W8 **Paramaribo** ● (Suriname)
Paramaribo, N Suriname
05°52´N 55°14´W

55 W9 **Paramaribo** ◇ *district*
N Suriname

55 W9 **Paramaribo** ✈ Paramaribo,
N Suriname 05°52´N 55°14´W

43 S16 **Paramá** Herrera, S Panama
08°01´N 80°30´W
Paramithía *see* Paramythiá

56 C13 **Paramonga** Lima, W Peru
10°42´S 77°50´W

123 V12 **Paramushir, Ostrov** *island*
SE Russian Federation

62 M10 **Paraná** Entre Ríos,
E Argentina 31°48´S 60°29´W

60 H11 **Paraná** *off.* Estado do Paraná.
♦ *state* S Brazil

47 U11 **Paraná** *var.* Alto Paraná.
♦ C South America
Paraná *see*
Paraná

60 K12 **Paranaguá** Paraná, S Brazil
25°32´S 48°36´W

61 J20 **Paranaíba, Río** ♦ E Brazil

60 G9 **Paraná Ibicuy, Río**
♦ E Argentina

59 H15 **Paranaíta** Mato Grosso,
W Brazil 09°35´S 56°59´W

60 H9 **Paranapanema, Rio**
♦ S Brazil

60 K11 **Paranapiacaba, Serra do**
▲ S Brazil

60 H9 **Paranavaí** Paraná, S Brazil
23°02´S 52°36´W

143 N5 **Parandak** Markazī, W Iran
35°19´N 50°40´E

114 I12 **Paranésti** *var.* Paranestio.
Anatolikí Makedonía
kai Thráki, NE Greece
41°16´N 24°31´E
Paranestio *see* Paranésti

191 W11 **Paraoa** *atoll* Îles Tuamotu,
C French Polynesia

184 L13 **Paraparaumu** Wellington,
North Island, New Zealand
40°55´S 175°01´E

57 N20 **Paraque, Cerro**
▲ W Venezuela
06°00´S 67°00´W

154 I11 **Parāsiya** Madhya Pradesh,
C India 22°12´N 78°45´E

115 M23 **Paraspóri, Akrotírio**
headland Kárpathos, SE
Greece 35°54´N 27°15´E

60 O10 **Parati** Rio de Janeiro,
SE Brazil 23°15´S 44°42´W

59 K14 **Paraúbebas** Pará, N Brazil
06°03´S 49°48´W

103 Q10 **Paray-le-Monial** Saône-et-
Loire, C France 46°27´N 04°07´E

154 G13 **Parbhani** Mahārāshtra,
C India 19°16´N 76°51´E

100 L10 **Parchim** Mecklenburg-
Vorpommern, N Germany
53°26´N 11°51´E
Parchwitz *see* Prochowice

110 P13 **Parczew** Lubelskie, E Poland
51°40´N 23°E

61 L8 **Pardo, Rio** ♦ S Brazil

111 E16 **Pardubice** *Ger.* Pardubitz.
Pardubický Kraj, C Czech
Republic 50°01´N 15°47´E
Pardubický Kraj ◇ *region*
N Czech Republic
Pardubitz *see* Pardubice

119 F16 **Parechcha** *Pol.*
Porzecze, *Rus.* Porech´ye.
Hrodzyenskaya Voblasts´,
W Belarus 53°50´N 24°32´E

59 F17 **Parecis, Chapada dos** ▲ W
Brazil
Parecis, Serra dos *see*
Parecis, Chapada dos

104 M4 **Paredes de Nava** Castilla y
León, N Spain 42°09´N 04°42´W

189 U12 **Parem** *island* Chuuk,
C Micronesia

189 O12 **Parem Island** *island*
E Micronesia

184 I1 **Parengarenga Harbour**
inlet North Island, New
Zealand

15 N8 **Parent** Québec, SE Canada
47°55´N 74°36´W

102 J14 **Parentis-en-Born** Landes,
SW France 44°22´N 01°04´W

185 G20 **Paroa** West Coast, South
Island, New Zealand
42°31´S 171°10´E

163 X14 **Paro-ho** *var.* Hwach´on-
chŏsuji; *prev.* P´aro-ho.
◎ N South Korea
P´aro-ho *see* Paro-ho

115 B16 **Párga** Ípeiros, W Greece
39°17´N 20°23´E

93 K20 **Pargas** *Fin.* Parainen.
Varsinais-Suomi, SW Finland
60°18´N 22°20´E

183 N6 **Paroo River** *seasonal
river* New South Wales/
Queensland, SE Australia

55 N6 **Pargua** Anzoátegui,
NE Venezuela 08°44´N 63°41´W

55 N6 **Paria, Gulf of** *var.* Golfo
de Paria. *gulf* Trinidad and
Tobago/Venezuela

36 L7 **Paria River** ♦ Utah, W USA

36 L8 **Parichi** *var.* Parychy
Paricutín, Volcán
♦ C Mexico 19°25´N 102°00´W

43 P16 **Parida, Isla** *island*
SW Panama

55 T8 **Parika** NE Guyana
06°51´N 58°25´W

93 O18 **Parikkala** Etelä-Karjala,
SE Finland 61°33´N 29°34´E

58 E10 **Parima, Serra** ▲ Sierra
Parima. ▲ Brazil/Venezuela
see also Parima, Sierra

55 N11 **Parima, Sierra** ▲ Sierra
Parima. ▲ Brazil/Venezuela
see also Parima, Sierra
Parras de la Fuente *see*
Parras

57 F17 **Parinacochas, Laguna**
◎ SW Peru

56 A9 **Pariñas, Punta** *headland*
NW Peru 04°45´S 81°22´W

58 H12 **Parintins** Amazonas,
N Brazil 02°38´S 56°45´W

103 O5 **Paris** *anc.* Lutetia, Lutetia
Parisiorum, Parisii.
● (France) Paris, N France
48°52´N 02°19´E

191 Y2 **Paris** Kiritimati, E Kiribati
01°55´N 157°30´W

27 S11 **Paris** Arkansas, C USA
35°17´N 93°46´W

33 S16 **Paris** Idaho, NW USA
42°13´N 111°24´W

31 N14 **Paris** Illinois, N USA
39°36´N 87°42´W

20 M5 **Paris** Kentucky, S USA
38°13´N 84°15´E

20 H8 **Paris** Tennessee, S USA
36°19´N 88°20´W

25 V5 **Paris** Texas, SW USA
33°41´N 95°33´W
Parisii *see* Paris

43 S16 **Parita** Herrera, S Panama
08°01´N 80°30´W

43 S16 **Parita, Bahía de** *bay*
S Panama

93 K18 **Parkano** Pirkanmaa,
W Finland 62°03´N 23°E

27 N6 **Park City** Kansas, C USA
37°48´N 97°19´W

36 L3 **Park City** Utah, W USA
40°39´N 111°30´W

36 I12 **Parker** Arizona, SW USA
34°07´N 114°16´W

23 R9 **Parker** Florida, SE USA
30°07´N 85°36´W

29 R11 **Parker** South Dakota, N USA
43°24´N 97°08´W

35 Z14 **Parker Dam** California,
SW USA 34°17´N 114°08´W

29 W13 **Parkersburg** Iowa, C USA
42°34´N 92°47´W

21 Q3 **Parkersburg** West Virginia,
NE USA 39°17´N 81°33´W

29 T7 **Parkers Prairie** Minnesota,
N USA 46°09´N 95°19´W

181 W13 **Parkes** New South Wales,
SE Australia 33°10´S 148°10´E

30 K4 **Park Falls** Wisconsin, N USA
45°57´N 90°25´W

14 E16 **Parkhill** Ontario, S Canada
43°11´N 81°39´W

29 T5 **Park Rapids** Minnesota,
N USA 46°55´N 94°04´W

29 Q3 **Park River** North Dakota,
N USA 48°24´N 97°44´W

29 Q11 **Parkston** South Dakota,
N USA 43°23´N 97°58´W

10 L17 **Parksville** Vancouver Island,
British Columbia, SW Canada
49°13´N 124°13´W

37 S3 **Parkview Mountain**
▲ Colorado, C USA
40°19´N 106°08´W

105 N8 **Parla** Madrid, C Spain
40°13´N 03°46´W

29 S8 **Parle, Lac qui** ◎ Minnesota,
N USA

155 G14 **Parli Vaijnāth** Mahārāshtra,
C India 18°53´N 76°36´E

106 F9 **Parma** Emilia-Romagna,
N Italy 44°48´N 10°20´E

31 T11 **Parma** Ohio, N USA
41°24´N 81°43´W

58 N13 **Parnaíba** *var.* Parnahiba.
Piauí, E Brazil 02°58´S 41°46´W

65 J14 **Parnaíba Ridge** *undersea
feature* C Atlantic Ocean

58 N13 **Parnaíba, Rio** ♦ NE Brazil

185 J17 **Parnassós** ▲ C Greece

185 J17 **Parnassus** Canterbury,
South Island, New Zealand
42°41´S 173°18´E

182 H10 **Parndana** South Australia
35°48´S 137°13´E

115 H19 **Párnitha** ▲ C Greece

115 F21 **Párnonas** *var.* Parnon.
▲ S Greece

118 G5 **Pärnu** *Ger.* Pernau, *Latv.*
Pērnava; *prev. Rus.* Pernov.
Pärnumaa, SW Estonia
58°24´N 24°32´E

118 G6 **Pärnu** *var.* Pärnu Jõgi, *Ger.*
Pernau. ♦ SW Estonia

118 G5 **Pärnu-Jaagupi** *Ger.*
Sankt-Jakobi. Pärnumaa,
SW Estonia 58°36´N 24°28´E
Pärnu Jõgi *see* Pärnu

118 F5 **Pärnu Laht** *Ger.* Pernauer
Bucht. *bay* SW Estonia
Pärnu Maakond *see* Pärnu
Pärnumaa *see* Pärnu

184 I1 **Parengarenga Harbour**
inlet North Island, New
Zealand

153 T11 **Paro** W Bhutan
27°25´N 89°25´E

153 T11 **Paro** ✈ (Thimphu) W Bhutan
27°22´N 89°25´E

185 G17 **Paroa** West Coast, South
Island, New Zealand
42°31´S 171°10´E

163 X14 **Paro-ho** *var.* Hwach´on-
chŏsuji; *prev.* P´aro-ho.
◎ N South Korea

115 J21 **Páros** *island* Kykládes,
Greece, Aegean Sea

115 J21 **Páros** Páros, Kykládes,
Greece, Aegean Sea
37°04´N 25°06´E
Páros *see* Páros

115 J21 **Páros** *island* Kykládes,
Greece, Aegean Sea

183 T9 **Parramatta** New South
Wales, SE Australia
33°49´S 150°59´E

21 Y6 **Parramore Island** *island*
Virginia, NE USA

40 M8 **Parras** *var.* Parras de la
Fuente. Coahuila, NE Mexico
25°27´N 102°12´W

42 M14 **Parrita** Puntarenas, S Costa
Rica 09°30´N 84°20´W

14 D13 **Parry Island** *island* Ontario,
S Canada

197 O9 **Parry Islands** *island group*
Nunavut, NW Canada

14 G12 **Parry Sound** Ontario,
S Canada 45°21´N 80°02´W

110 F7 **Parsęta** *Ger.* Persante.
♦ NW Poland

28 L3 **Parshall** North Dakota,
N USA 47°57´N 102°07´W

27 Q7 **Parsons** Kansas, C USA
37°20´N 95°15´W

20 H9 **Parsons** Tennessee, S USA
35°39´N 88°07´W

21 T3 **Parsons** West Virginia,
NE USA 39°06´N 79°43´W
Parsonstown *see* Birr

100 P11 **Parsteiner See**
◎ NE Germany

107 I24 **Partanna** Sicilia, Italy,
C Mediterranean Sea
37°43´N 12°54´E

108 J8 **Partenen** Graubünden,
E Switzerland
46°58´N 10°01´E

102 K9 **Parthenay** Deux-Sèvres,
W France 46°39´N 00°15´W

95 J19 **Partille** Västra Götaland,
S Sweden 57°43´N 12°12´E

107 I23 **Partinico** Sicilia, Italy,
C Mediterranean Sea
38°03´N 13°07´E

111 I20 **Partizánske** *prev.*
Šimonovany, *Hung.* Simony.
Trenčiansky Kraj, W Slovakia
48°35´N 18°23´E

58 H11 **Paru de Oeste, Rio**
♦ N Brazil

182 K9 **Paruna** South Australia
34°42´S 140°44´E

58 I11 **Paru, Rio** ♦ N Brazil

155 M14 **Pārvatipuram** Andhra
Pradesh, E India
18°49´N 83°26´E

152 G12 **Parvatsar** *prev.* Parbatsar.
Rājasthān, N India
26°52´N 74°49´E

114 J11 **Parvomay** *Pürvomay;
prev.* Borisovgrad. Plovdiv,
C Bulgaria 42°06´N 25°13´E

149 Q5 **Parwān** *prev.* Parvān.
♦ E Afghanistan

158 I15 **Paryang** Xizang Zizhiqu,
W China 30°04´N 83°30´E

119 M18 **Parychy** *Rus.* Parichi.
Homyel´skaya Voblasts´,
SE Belarus 52°48´N 29°25´E

83 J21 **Parys** Free State, C South
Africa 26°55´S 27°28´E

35 T15 **Pasadena** California, W USA
34°09´N 118°09´W

25 W11 **Pasadena** Texas, SW USA
29°41´N 95°13´W

56 B8 **Pasaje** El Oro, SW Ecuador
03°23´S 79°50´W

137 T9 **Pasanauri** *var.* P´asanauri.
N Georgia 42°21´N 44°40´E
P´asanauri *see* Pasanauri

168 I13 **Pasapuat** Pulau Pagai Utara,
W Indonesia 02°36´S 99°58´E
Pasawng *see* Hpasawng

114 J13 **Paşayiğit** Edirne, NW Turkey
40°56´N 26°21´E

23 N9 **Pascagoula** Mississippi,
S USA 30°21´N 88°33´W

22 M8 **Pascagoula River**
♦ Mississippi, S USA

116 K13 **Paşcani** *Hung.* Páskán. Iaşi,
NE Romania 47°14´N 26°44´E

32 K10 **Pasco** Washington, NW USA
46°13´N 119°06´W

56 E13 **Pasco** *off.* Departamento de
Pasco. ◇ *department* C Peru
Pasco, Departamento de
see Pasco

191 N11 **Pascua, Isla de** *var.* Rapa
Nui, Easter Island. *island*
E Pacific Ocean

63 G21 **Pascua, Río** ♦ S Chile

103 N1 **Pas-de-Calais** ◇ *department*
N France

100 P10 **Pasewalk** Mecklenburg-
Vorpommern, NE Germany
53°31´N 13°59´E

11 T10 **Pasfield Lake**
◎ Saskatchewan, C Canada
Pa-shih Hai-hsia *see* Bashi
Channel

19 N11 **Passaconaway** New
Hampshire, NE USA
43°50´N 71°37´W

183 T9 **Parramatta** New South
Wales, SE Australia
33°49´S 150°59´E

149 W7 **Pasrūr** Punjab, E Pakistan
32°12´N 74°42´E

30 M1 **Passage Island** *island*
Michigan, N USA

65 B24 **Passage Islands** *island group*
W Falkland Islands

8 K5 **Passage Point** *headland*
Banks Island, Northwest
Territories, NW Canada
73°31´N 115°12´W
Passage *see* Pasłęka

115 C15 **Passarón** *ancient monument*
Ípeiros, W Greece
Passarowitz *see* Požarevac

101 O22 **Passau** Bayern, SE Germany
48°34´N 13°28´E

22 M9 **Pass Christian** Mississippi,
S USA 30°19´N 89°15´W

107 L26 **Passero, Capo**
headland Sicilia, Italy,
C Mediterranean Sea
36°40´N 15°09´E

171 P5 **Passi** Panay Island,
C Philippines 11°05´N 122°37´E

61 H14 **Passo Fundo** Rio Grande do
Sul, S Brazil 28°16´S 52°20´W

60 H13 **Passo Fundo, Barragem de**
◙ S Brazil

61 H15 **Passo Real, Barragem de**
◙ S Brazil

59 L20 **Passos** Minas Gerais,
NE Brazil 20°45´S 46°38´W

167 X10 **Passu Keah** *Chin.* Panshi Yu,
Viet. Đao Bach Quy. *island*
S Paracel Islands

118 J13 **Pastavy** *Pol.* Postawy,
Rus. Postavy. Vitsyebskaya
Voblasts´, NW Belarus
55°07´N 26°50´E

56 D7 **Pastaza** ◇ *province*
E Ecuador

56 D9 **Pastaza, Río** ♦ Ecuador/
Peru

61 A21 **Pasteur** Buenos Aires,
E Argentina 35°10´S 62°14´W

15 V3 **Pasteur** ♦ Québec,
SE Canada

147 Q12 **Pastigov** *Rus.* Pastigov.
W Tajikistan 39°27´N 69°16´E
Pastigov *see* Pastigov

54 C13 **Pasto** Nariño, SW Colombia
01°12´N 77°17´W

38 M10 **Pastol Bay** *bay* Alaska, USA

37 O8 **Pastora Peak** ▲ Arizona,
SW USA 36°48´N 109°10´W

105 O8 **Pastrana** Castilla-La Mancha,
C Spain 40°24´N 02°55´W

169 S16 **Pasuruan** *prev.* Pasoeroean.
Jawa, C Indonesia
07°38´S 112°44´E

118 F11 **Pasvalys** *var.* Pasvaliys.
N Lithuania 56°03´N 24°24´E

111 K21 **Pásztó** Nógrád, N Hungary
47°57´N 19°41´E

189 U12 **Pata** *var.* Patta. *atoll* Chuuk
Islands, C Micronesia

36 M16 **Patagonia** Arizona, SW USA
31°32´N 110°45´W

63 H20 **Patagonia** *physical region*
Argentina/Chile
Patalung *see* Phatthalung

152 F13 **Patan** Gujarāt, W India
23°51´N 72°11´E

154 D9 **Patan** Madhya Pradesh,
C India 23°19´N 79°41´E

171 S11 **Patani** Pulau Halmahera,
E Indonesia 0°19´N 128°46´E
Patani *see* Pattani

116 K13 **Pâtârlagele** *prev.* Pătîrlagele.
Buzău, SE Romania
45°19´N 26°21´E

182 I5 **Patavium** *see* Padova

182 I5 **Patawarta Hill** ▲ South
Australia 30°57´S 138°42´E

182 L10 **Patchewollock** Victoria,
SE Australia 35°23´S 142°11´E

184 K1 **Patea** Taranaki, North Island,
New Zealand 39°48´S 174°35´E

184 K1 **Patea** ♦ North Island, New
Zealand

77 U15 **Pategi** Kwara, C Nigeria
08°44´N 05°46´E

81 K20 **Pate Island** *var.* Patta Island.
island SE Kenya

105 S10 **Paterna** Valenciana, E Spain
39°30´N 00°24´W

109 T4 **Paternion** *Slvn.* Špatrjan.
Kärnten, S Austria
46°40´N 13°43´E

107 L24 **Paternò** *anc.* Hybla,
Hybla Major. Sicilia,
Italy, C Mediterranean Sea
37°34´N 14°53´E

32 J7 **Pateros** Washington, NW
USA 48°01´N 119°55´W

18 K13 **Paterson** New Jersey,
NE USA 40°55´N 74°12´W

32 J10 **Paterson** Washington,
NW USA 45°55´N 119°37´W

185 C25 **Paterson Inlet** *inlet* Stewart
Island, New Zealand

98 N6 **Paterswolde** Drenthe,
NE Netherlands
53°07´N 06°32´E

152 H7 **Pathankot** Himāchal
Pradesh, N India
32°16´N 75°43´E

166 K8 **Pathein** *var.* Bassein.
Ayeyawady, SW Myanmar
(Burma) 16°46´N 94°45´E

33 W15 **Pathfinder Reservoir**
◙ Wyoming, C USA

167 O11 **Pathum Thani** *var.*
Patumthani, Prathum Thani.
Pathum Thani, C Thailand
14°03´N 100°29´E

169 R9 **Pati, Tanjung** *headland*
East Malaysia 02°24´N 111°12´E

95 N15 **Patskallavik** Kalmar,
S Sweden 57°10´N 16°25´E

54 B12 **Patía** var. El Bordo. Cauca,
SW Colombia 02°07´N 76°57´W

188 D15 **Pati Point** *headland*
NE Guam 13°36´N 144°39´E

56 C13 **Pativilca** Lima, C Peru
10°44´S 77°45´W

166 M1 **Patkai Bum** *var.* Patkai
Range. ▲ Myanmar
(Burma)/India
Patkai Range *see* Patkai Bum

115 L20 **Pátmos** Pátmos, Dodekánisa,
Greece, Aegean Sea
37°21´N 26°32´E

115 L20 **Pátmos** *island* Dodekánisa,
Greece, Aegean Sea

153 P13 **Patna** *var.* Azīmābād.
state capital Bihār, N India
25°37´N 85°13´E

154 M12 **Patnāgarh** Odisha, E India
20°43´N 83°09´E

171 O5 **Patnongon** Panay Island,
C Philippines 10°56´N 122°03´E

♦ Country
◆ Country Capital
◇ Dependent Territory
◈ Dependent Territory Capital
◉ Administrative Regions
✈ International Airport
▲ Mountain
▲ Mountain Range
🌋 Volcano
♦ River
◎ Lake
◙ Reservoir

PATNOS – PERRIERTOPPEN

137 S13 **Patnos** Ağrı, E Turkey 39°14′N 42°52′E
60 H12 **Pato Branco** Paraná, S Brazil 26°20′S 52°40′W
31 O16 **Patoka Lake** ◎ Indiana, N USA
92 L9 **Patoniva** *Lapp.* Buoddobohki. Lappi, N Finland 69°44′N 27°01′E
113 K21 **Patos** *var.* Patosi. Fier, SW Albania 40°40′N 19°37′E
Patos *see* Patos de Minas
59 K19 **Patos de Minas** *var.* Patos. Minas Gerais, NE Brazil 18°35′S 46°32′W
Patosi *see* Patos
61 I17 **Patos, Lagoa dos** *lagoon* S Brazil
62 J9 **Patquía** La Rioja, C Argentina 30°02′S 66°54′W
115 E19 **Pátra** *Eng.* Patras; *prev.* Pátrai. Dytikí Elláda, S Greece 38°14′N 21°45′E
115 D18 **Patraïkós Kólpos** *gulf* S Greece
Pátrai/Patras *see* Pátra
92 G2 **Patreksfjörður** Vestfirðir, W Iceland 65°33′N 23°54′W
24 M7 **Patricia** Texas, SW USA 32°34′N 102°00′W
63 F21 **Patricio Lynch, Isla** *island* S Chile
Patta *see* Pata
Patta Island *see* Pate Island
167 O16 **Pattani** *var.* Patani. Pattani, SW Thailand 06°50′N 101°20′E
167 P12 **Pattaya** Chon Buri, S Thailand 12°57′N 100°53′E
19 S4 **Patten** Maine, NE USA 45°58′N 68°27′W
35 O9 **Patterson** California, W USA 37°27′N 121°07′W
22 J10 **Patterson** Louisiana, S USA 29°41′N 91°18′W
35 R7 **Patterson, Mount** ▲ California, W USA 38°27′N 119°16′W
31 P4 **Patterson, Point** *headland* Michigan, N USA 45°58′N 85°39′W
107 L23 **Patti** Sicilia, Italy, C Mediterranean Sea 38°08′N 14°58′E
107 L23 **Patti, Golfo di** *gulf* Sicilia, Italy
93 L14 **Pattijoki** Pohjois-Pohjanmaa, W Finland 64°41′N 24°40′E
193 Q4 **Patton Escarpment** *undersea feature* E Pacific Ocean
27 S2 **Pattonsburg** Missouri, C USA 40°03′N 94°08′W
0 D6 **Patton Seamount** *undersea feature* NE Pacific Ocean 54°40′N 150°30′W
10 J12 **Pattullo, Mount** ▲ British Columbia, W Canada 56°18′N 129°43′W
153 U16 **Patuakhali** *var.* Patukhali. Barisal, S Bangladesh 22°20′N 90°20′E
42 M5 **Patuca, Río** ≈ E Honduras
Patukhali *see* Patuakhali
Patumdhani *see* Pathum Thani
40 M14 **Pátzcuaro** Michoacán, SW Mexico 19°30′N 101°38′W
42 C6 **Patzicía** Chimaltenango, S Guatemala 14°38′N 90°52′W
102 K16 **Pau** Pyrénées-Atlantiques, SW France 43°18′N 00°22′W
102 J12 **Pauillac** Gironde, SW France 45°12′N 00°44′W
166 L5 **Pauk** Magway, W Myanmar (Burma) 21°25′N 94°30′E
8 I6 **Paulatuk** Northwest Territories, NW Canada 69°23′N 124°W
42 K5 **Paulayá, Río** ≈ NE Honduras
22 M6 **Paulding** Mississippi, S USA 32°01′N 89°01′W
31 Q12 **Paulding** Ohio, N USA 41°08′N 84°34′W
29 S12 **Paullina** Iowa, C USA 42°58′N 95°41′W
59 P15 **Paulo Afonso** Bahia, E Brazil 09°21′S 38°14′W
38 M16 **Pauloff Harbor** *var.* Pavlor Harbour. Sanak Island, Alaska, USA 54°26′N 162°43′W
27 N12 **Pauls Valley** Oklahoma, C USA 34°46′N 97°14′W
166 L7 **Paungde** Bago, C Myanmar (Burma) 18°30′N 95°30′E
Pauni *see* Paoni
152 K9 **Pauri** Uttaranchal, N India 30°08′N 78°48′E
Pautalia *see* Kyustendil
142 J5 **Pāveh** Kermānshāhān, NW Iran 35°02′N 46°15′E
114 I9 **Pavel Banya** Stara Zagora, C Bulgaria 42°35′N 25°19′E
126 L5 **Pavelets** Ryazanskaya Oblast', W Russian Federation 53°47′N 39°22′E
106 D8 **Pavia** *anc.* Ticinum. Lombardia, N Italy 45°10′N 09°10′E
118 C9 **Pāvilosta** W Latvia 56°52′N 21°12′E
125 P14 **Pavino** Kostromskaya Oblast', NW Russian Federation 59°10′N 46°09′E
114 J8 **Pavlikeni** Veliko Tarnovo, N Bulgaria 43°14′N 25°20′E
145 T8 **Pavlodar** Pavlodar, NE Kazakhstan 52°21′N 76°59′E
145 S9 **Pavlodar** *off.* Pavlodarskaya Oblast', *Kaz.* Pavlodar Oblysy. ◈ *province* NE Kazakhstan
Pavlodar Oblysy/ Pavlodarskaya Oblast' *see* Pavlodar
Pavlograd *see* Pavlohrad
117 U7 **Pavlohrad** *Rus.* Pavlograd. Dnipropetrovs'ka Oblast', E Ukraine 48°32′N 35°50′E
Pavlor Harbour *see* Pauloff Harbor
145 R9 **Pavlovka** Akmola, C Kazakhstan 51°22′N 72°35′E
127 V4 **Pavlovka** Respublika Bashkortostan, W Russian Federation 55°28′N 56°36′E
127 Q7 **Pavlovka** Ul'yanovskaya Oblast', W Russian Federation 52°40′N 47°08′E
127 N3 **Pavlovo** Nizhegorodskaya Oblast', W Russian Federation 55°59′N 43°03′E
126 L9 **Pavlovsk** Voronezhskaya Oblast', W Russian Federation 50°26′N 40°08′E
126 L13 **Pavlovskaya** Krasnodarskiy Kray, SW Russian Federation 46°06′N 39°52′E
117 S7 **Pavlysh** Kirovohrads'ka Oblast', C Ukraine 48°54′N 33°20′E
106 F10 **Pavullo nel Frignano** Emilia-Romagna, C Italy 44°19′N 10°52′E
27 P8 **Pawhuska** Oklahoma, C USA 36°42′N 96°21′W
21 U13 **Pawleys Island** South Carolina, SE USA 33°27′N 79°07′W
30 K14 **Pawnee** Illinois, N USA 39°35′N 89°34′W
27 O9 **Pawnee** Oklahoma, C USA 36°21′N 96°50′W
37 U2 **Pawnee Buttes** ▲ Colorado, C USA 40°49′N 103°58′W
29 S17 **Pawnee City** Nebraska, C USA 40°06′N 96°09′W
26 K5 **Pawnee River** ≈ Kansas, C USA
167 N6 **Pawn, Nam** ≈ C Myanmar (Burma)
31 O10 **Paw Paw** Michigan, N USA 42°12′N 86°09′W
31 O10 **Paw Paw Lake** Michigan, N USA 42°12′N 86°16′W
19 O12 **Pawtucket** Rhode Island, NE USA 41°52′N 71°22′W
Pax Augusta *see* Badajoz
115 I25 **Paximádia** *island* SE Greece
Pax Julia *see* Beja
115 B16 **Paxoí** *island* Iónia Nisiá, Greece, C Mediterranean Sea
39 S10 **Paxson** Alaska, USA 62°58′N 145°27′W
147 O11 **Paxtakor** Jizzax Viloyati, C Uzbekistan 40°21′N 67°54′E
30 M13 **Paxton** Illinois, N USA 40°27′N 88°06′W
124 J11 **Pay** Respublika Kareliya, NW Russian Federation 61°10′N 34°24′E
166 M8 **Payagyi** Bago, SW Myanmar (Burma) 17°28′N 96°32′E
108 C9 **Payerne** *Ger.* Peterlingen. Vaud, W Switzerland 46°49′N 06°57′E
32 M13 **Payette** Idaho, NW USA 44°04′N 116°55′W
32 M13 **Payette River** ≈ Idaho, NW USA
125 V2 **Pay-Khoy, Khrebet** ▲▲ NW Russian Federation
Payne *see* Kangirsuk
12 K4 **Payne, Lac** ◎ Québec, NE Canada
29 T8 **Paynesville** Minnesota, N USA 45°22′N 94°42′W
169 S8 **Payong, Tanjung** *cape* East Malaysia
Payo Obispo *see* Chetumal
61 D18 **Paysandú** Paysandú, W Uruguay 32°21′S 58°05′W
61 D17 **Paysandú** ◈ *department* W Uruguay
102 I7 **Pays de la Loire** ◈ *region* NW France
36 L12 **Payson** Arizona, SW USA 34°13′N 111°19′W
36 L4 **Payson** Utah, W USA 40°02′N 111°43′W
125 W4 **Payyer, Gora** ▲ NW Russian Federation 66°49′N 64°33′E
Payzawat *see* Jiashi
137 Q11 **Pazar** Rize, NE Turkey 41°10′N 40°53′E
136 F10 **Pazarbaşı Burnu** *headland* NW Turkey 41°12′N 30°18′E
136 M16 **Pazarcık** Kahramanmaraş, S Turkey 37°31′N 37°19′E
114 I10 **Pazardzhik** *prev.* Tatar Pazardzhik. Pazardzhik, SW Bulgaria 42°11′N 24°21′E
64 H11 **Pazardzhik** ◈ *province* C Bulgaria
54 H9 **Paz de Ariporo** Casanare, E Colombia 05°54′N 71°52′W
112 A10 **Pazin** *Ger.* Mitterburg, *It.* Pisino. Istra, NW Croatia 45°14′N 13°56′E
42 D7 **Paz, Río** ≈ El Salvador/Guatemala
113 O18 **Pčinja** ≈ N Macedonia
193 V15 **Pea** Tongatapu, S Tonga 21°10′S 175°14′W
27 O6 **Peabody** Kansas, C USA 38°10′N 97°06′W
11 O12 **Peace** ≈ Alberta/British Columbia, W Canada
Peace Garden State *see* North Dakota
11 Q10 **Peace Point** Alberta, C Canada 59°11′N 112°12′W
11 O12 **Peace River** Alberta, W Canada 56°15′N 117°18′W
23 W13 **Peace River** ≈ Florida, SE USA
11 N17 **Peachland** British Columbia, SW Canada 49°49′N 119°48′W
36 J10 **Peach Springs** Arizona, SW USA 35°33′N 113°27′W
Peach State *see* Georgia
23 S4 **Peachtree City** Georgia, SE USA 33°24′N 84°36′W
189 Y13 **Peacock Point** *point* SE Wake Island
97 M18 **Peak District** *physical region* C England, United Kingdom
183 Q7 **Peak Hill** New South Wales, SE Australia 32°39′S 148°12′E
65 G15 **Peak, The** ▲ C Ascension Island
105 O13 **Peal de Becerro** Andalucía, S Spain 37°55′N 03°08′W
189 X11 **Peale Island** *island* N Wake Island
37 O6 **Peale, Mount** ▲ Utah, W USA 38°26′N 109°13′W
39 O4 **Peard Bay** *bay* Alaska, USA
23 Q7 **Pea River** ≈ Alabama/Florida, S USA
25 W11 **Pearland** Texas, SW USA 29°33′N 95°17′W
38 D9 **Pearl City** O'ahu, Hawaii, USA, C Pacific Ocean 21°24′N 157°58′W
38 D9 **Pearl Harbor** *inlet* O'ahu, Hawai'i, USA, C Pacific Ocean
Pearl Islands *see* Perlas, Archipiélago de las
Pearl Lagoon *see* Perlas, Laguna de
22 M5 **Pearl River** ≈ Louisiana/Mississippi, S USA
25 Q13 **Pearsall** Texas, SW USA 28°54′N 99°07′W
23 U7 **Pearson** Georgia, SE USA 31°18′N 82°51′W
25 P4 **Pease River** ≈ Texas, SW USA
12 F7 **Peawanuck** Ontario, C Canada 54°55′N 85°31′W
12 E8 **Peawanuk** ≈ Ontario, S Canada
83 P16 **Pebane** Zambézia, NE Mozambique 17°14′S 38°10′E
65 C23 **Pebble Island** *island* N Falkland Islands
65 C23 **Pebble Island Settlement** Pebble Island, N Falkland Islands 51°20′S 59°40′W
Peć *see* Pejë
25 R8 **Pecan Bayou** ≈ Texas, SW USA
22 H10 **Pecan Island** Louisiana, S USA 29°39′N 92°26′W
60 L12 **Peças, Ilha das** *island* S Brazil
30 L10 **Pecatonica River** ≈ Illinois/Wisconsin, N USA
108 G10 **Peccia** Ticino, S Switzerland 46°24′N 08°39′E
Pechenegi *see* Pechenihy
Pechenezhskoye Vodokhranilishche *see* Pecheniz'ke Vodoskhovyshche
124 I2 **Pechenga** *Fin.* Petsamo. Murmanskaya Oblast', NW Russian Federation 69°34′N 31°14′E
117 V5 **Pechenihy** *Rus.* Pechenegi. Kharkivs'ka Oblast', E Ukraine 49°49′N 36°57′E
117 V5 **Pecheniz'ke Vodoskhovyshche** *Rus.* Pechenezhskoye Vodokhranilishche. ⊠ E Ukraine
125 U7 **Pechora** Respublika Komi, NW Russian Federation 65°09′N 57°09′E
125 R6 **Pechora** ≈ NW Russian Federation
Pechora Bay *see* Pechorskaya Guba
Pechora Sea *see* Pechorskoye More
125 S3 **Pechorskaya Guba** *Eng.* Pechora Bay. *bay* NW Russian Federation
122 H7 **Pechorskoye More** *Eng.* Pechora Sea. *sea* NW Russian Federation
116 E11 **Pecica** *Ger.* Petschka, *Hung.* Ópécska. Arad, W Romania 46°09′N 21°06′E
24 K8 **Pecos** Texas, SW USA 31°25′N 103°30′W
25 N11 **Pecos River** ≈ New Mexico/Texas, SW USA
111 I25 **Pécs** *Ger.* Fünfkirchen, *Lat.* Sopianae. Baranya, SW Hungary 46°05′N 18°11′E
43 T17 **Pedasí** Los Santos, S Panama 07°36′N 80°04′W
Pedde *see* Pedja
183 O17 **Pedder, Lake** ◎ Tasmania, SE Australia
44 M10 **Pedernales** SW Dominican Republic 18°02′N 71°41′W
55 Q5 **Pedernales** Delta Amacuro, NE Venezuela 09°58′N 62°15′W
25 R10 **Pedernales River** ≈ Texas, SW USA
62 H6 **Pedernales, Salar de** *salt lake* N Chile
Pedhoulas *see* Pedoulás
55 X11 **Pédima** *var.* Malavate. SW French Guiana 03°15′N 54°08′W
182 F1 **Pedirka** South Australia 26°41′S 135°11′E
171 S11 **Pediwang** Pulau Halmahera, E Indonesia 01°29′N 127°57′E
118 I5 **Pedja** *var.* Pedja Jõgi, *Ger.* Pedde. ≈ E Estonia
Pedja Jõgi *see* Pedja
121 O3 **Pedoulás** *var.* Pedhoulas. W Cyprus 34°58′N 32°51′E
59 N18 **Pedra Azul** Minas Gerais, NE Brazil 16°02′S 41°17′W
104 I3 **Pedrafita, Porto de** *var.* Puerto de Piedrafita. *pass* NW Spain
76 E9 **Pedra Lume** Sal, NE Cape Verde 16°47′N 22°54′W
43 P16 **Pedregal** Chiriquí, W Panama 09°04′N 79°25′W
54 J4 **Pedregal** Falcón, N Venezuela 11°04′N 70°08′W
40 L9 **Pedriceña** Durango, C Mexico 25°08′N 103°46′W
60 L11 **Pedro Barros** São Paulo, S Brazil 24°12′S 47°22′W
39 Q13 **Pedro Bay** Alaska, USA 59°47′N 154°06′W
62 H4 **Pedro de Valdivia** *var.* Oficina Pedro de Valdivia. Antofagasta, N Chile 22°33′S 69°38′W
62 P4 **Pedro Juan Caballero** Amambay, E Paraguay 22°34′S 55°41′W
63 L15 **Pedro Luro** Buenos Aires, E Argentina 39°30′S 62°38′W
105 O10 **Pedro Muñoz** Castilla-La Mancha, C Spain 39°25′N 02°56′W
155 J22 **Pedro, Point** *headland* NW Sri Lanka 09°54′N 80°08′E
182 K9 **Peebinga** South Australia 34°56′S 140°56′E
96 J13 **Peebles** SE Scotland, United Kingdom 55°40′N 03°15′W
31 S15 **Peebles** Ohio, N USA 38°57′N 83°23′W
96 J12 **Peebles** *cultural region* SE Scotland, United Kingdom
18 K13 **Peekskill** New York, NE USA 41°17′N 73°54′W
97 I16 **Peel** W Isle of Man 54°13′N 04°04′W
8 G7 **Peel** ≈ Northwest Territories/Yukon, NW Canada
8 K5 **Peel Point** *headland* Victoria Island, Northwest Territories, NW Canada 73°22′N 114°33′W
8 M5 **Peel Sound** *passage* Nunavut, N Canada
100 N9 **Peene** ≈ NE Germany
99 K17 **Peer** Limburg, NE Belgium 51°08′N 05°29′E
14 H14 **Pefferlaw** Ontario, S Canada 44°18′N 79°11′W
185 I18 **Pegasus Bay** *bay* South Island, New Zealand
121 O3 **Pégeia** *var.* Peyia. SW Cyprus 34°52′N 32°24′E
109 V7 **Peggau** Steiermark, SE Austria 47°10′N 15°20′E
101 L19 **Pegnitz** Bayern, SE Germany 49°45′N 11°33′E
101 L19 **Pegnitz** ≈ SE Germany
105 T11 **Pego** Valenciana, E Spain 38°51′N 00°08′W
Pegu *see* Bago
Pegu *see* Bago
189 N13 **Pehleng** Pohnpei, E Micronesia
114 M12 **Pehlivanköy** Kırklareli, NW Turkey 41°21′N 26°55′E
77 R14 **Péhonko** C Benin 10°14′N 01°57′E
61 B21 **Pehuajó** Buenos Aires, E Argentina 35°48′S 61°53′W
Pei-ching *see* Beijing/Beijing Shi
100 J13 **Peine** Niedersachsen, C Germany 52°19′N 10°14′E
Pei-p'ing *see* Beijing/Beijing Shi
Peipsi Järv/Peipus-See *see* Peipus, Lake
118 J5 **Peipus, Lake** *Est.* Peipsi Järv, *Ger.* Peipus-See, *Rus.* Chudskoye Ozero. ◎ Estonia/Russian Federation
115 H19 **Peiraías** *prev.* Piraiévs, *Eng.* Piraeus. Attikí, C Greece 37°57′N 23°42′E
Peisern *see* Pyzdry
60 I8 **Peixe, Rio do** ≈ S Brazil
59 I16 **Peixoto de Azevedo** Mato Grosso, W Brazil 10°18′S 55°03′W
168 O11 **Pejantan, Pulau** *island* W Indonesia
113 L16 **Pejë** *Serb.* Peć. W Kosovo 42°40′N 20°19′E
112 N11 **Pek** ≈ E Serbia
Pèk *see* Phônsaven
169 Q16 **Pekalongan** Jawa, C Indonesia 06°54′S 109°37′E
168 K11 **Pekanbaru** *var.* Pakanbaru. Sumatera, W Indonesia 0°31′N 101°27′E
30 L12 **Pekin** Illinois, N USA 40°34′N 89°38′W
Peking *see* Beijing/Beijing Shi
Pelabohan Kelang/ Pelabuhan Kelang *see* Pelabuhan Klang
168 J9 **Pelabuhan Klang** *var.* Kuala Pelabohan Kelang, Pelabohan Kelang, Pelabuhan Kelang, Port Klang, Port Swettenham. Selangor, Peninsular Malaysia 02°57′N 101°24′E
120 L11 **Pelagie, Isole** *island group* SW Italy
Pelagosa *see* Palagruža
22 L5 **Pelahatchie** Mississippi, S USA 32°19′N 89°48′W
169 T14 **Pelaihari** *var.* Pleihari. Borneo, C Indonesia 03°48′S 114°45′E
103 U14 **Pelat, Mont** ▲ SE France 44°16′N 06°46′E
116 F12 **Peleaga, Vârful** *prev.* Vîrful Peleaga. ▲ W Romania 45°23′N 22°52′E
Peleaga, Vîrful *see* Peleaga, Vârful
123 O11 **Peleduy** Respublika Sakha (Yakutiya), NE Russian Federation 59°39′N 112°36′E
14 C18 **Pelee Island** *island* Ontario, S Canada
45 Q11 **Pelée, Montagne** ▲ N Martinique 14°47′N 61°10′W
14 D18 **Pelee, Point** *headland* Ontario, S Canada 41°56′N 82°30′W
171 P12 **Pelei** Pulau Peleng, N Indonesia 01°26′S 123°27′E
Peleliu *see* Beliliou
171 P12 **Peleng, Pulau** *island* Kepulauan Banggai, N Indonesia
23 T7 **Pelham** Georgia, SE USA 31°07′N 84°09′W
111 E18 **Pelhřimov** *Ger.* Pilgram. Vysočina, C Czech Republic 49°26′N 15°14′E
39 W13 **Pelican** Chichagof Island, Alaska, USA 57°52′N 136°05′W
191 Z3 **Pelican Lagoon** ◎ Kiritimati, E Kiribati
29 V3 **Pelican Lake** ◎ Minnesota, N USA
29 U6 **Pelican Lake** ◎ Minnesota, N USA
30 L5 **Pelican Lake** ◎ Wisconsin, N USA
44 G1 **Pelican Point** Grand Bahama Island, N The Bahamas 26°39′N 78°09′W
83 B19 **Pelican Point** *headland* W Namibia 22°55′S 14°25′E
29 S6 **Pelican Rapids** Minnesota, N USA 46°34′N 96°04′W
Pelican State *see* Louisiana
11 U13 **Pelikan Narrows** Saskatchewan, C Canada 55°11′N 102°51′W
115 L18 **Pelinaío** ▲ Chíos, E Greece 38°31′N 26°01′E
Pelinnaeum *see* Pelinnaío
115 E16 **Pelinnaío** *anc.* Pelinnaeum. *ruins* Thessalía, C Greece
113 N20 **Pelister** ▲ SW FYR Macedonia 41°00′N 21°12′E
113 G15 **Pelješac** *peninsula* S Croatia
92 M12 **Pelkosenniemi** Lappi, NE Finland 67°06′N 27°30′E
29 W15 **Pella** Iowa, C USA 41°24′N 92°55′W
114 F13 **Pélla** *site of ancient city* Kentrikí Makedonía, N Greece
23 Q3 **Pell City** Alabama, S USA 33°35′N 86°17′W
61 A22 **Pellegrini** Buenos Aires, E Argentina 36°16′S 63°07′W
92 K12 **Pello** Lappi, NW Finland 66°47′N 24°E
100 G7 **Pellworm** *island* N Germany
10 H6 **Pelly** ≈ Yukon, NW Canada
Pelly Bay *see* Kugaaruk
10 I8 **Pelly Mountains** ▲▲ Yukon, W Canada
Pélmonostor *see* Beli Manastir
37 P13 **Pelona Mountain** ▲ New Mexico, SW USA 33°40′N 108°06′W
Peloponnese/Peloponnesus *see* Pelopónnisos
115 E20 **Pelopónnisos** *Eng.* Peloponnese. ◈ *region* S Greece
115 E20 **Pelopónnisos** *var.* Morea, *Eng.* Peloponnese; *anc.* Peloponnesus. *peninsula* S Greece
107 L23 **Peloritani, Monti** *anc.* Pelorus and Neptunius. ▲▲ Sicilia, Italy, C Mediterranean Sea
107 M22 **Peloro, Capo** *var.* Punta del Faro. *headland* S Italy 38°15′N 15°39′E
Pelorus and Neptunius *see* Peloritani, Monti
61 H17 **Pelotas** Rio Grande do Sul, S Brazil 31°45′S 52°20′W
61 I14 **Pelotas, Rio** ≈ S Brazil
92 K10 **Peltovuoma** *Lapp.* Bealdovuopmi. Lappi, N Finland 68°23′N 24°12′E
19 R4 **Pemadumcook Lake** ◎ Maine, NE USA
169 Q16 **Pemalang** Jawa, C Indonesia 06°53′S 109°07′E
169 P10 **Pemangkat** *var.* Pamangkat. Borneo, C Indonesia 01°11′N 109°00′E
Pemar *see* Paimio
168 I9 **Pematangsiantar** Sumatera, W Indonesia 02°59′N 99°01′E
83 Q14 **Pemba** *prev.* Port Amelia, Porto Amélia. Cabo Delgado, NE Mozambique 13°S 40°35′E
81 J22 **Pemba** ◈ *region* E Tanzania
81 K21 **Pemba** *island* E Tanzania
83 Q14 **Pemba, Baia de** *inlet* NE Mozambique
81 J21 **Pemba Channel** *channel* E Tanzania
180 J14 **Pemberton** Western Australia 34°27′S 116°09′E
10 M16 **Pemberton** British Columbia, SW Canada 50°19′N 122°49′W
29 Q2 **Pembina** North Dakota, N USA 48°58′N 97°14′W
11 P15 **Pembina** ≈ Alberta, SW Canada
29 Q2 **Pembina** ≈ Canada/USA
171 X16 **Pembre** Papua, E Indonesia 07°49′S 138°01′E
14 K12 **Pembroke** Ontario, SE Canada 45°49′N 77°08′W
97 H21 **Pembroke** SW Wales, United Kingdom 51°41′N 04°55′W
23 W6 **Pembroke** Georgia, SE USA 32°09′N 81°35′W
21 U11 **Pembroke** North Carolina, SE USA 34°40′N 79°12′W
21 R7 **Pembroke** Virginia, NE USA 37°19′N 80°38′W
97 H21 **Pembroke** *cultural region* SW Wales, United Kingdom
Pembuang, Sungai *see* Seruyan, Sungai
43 S15 **Peña Blanca, Cerro** ▲ C Panama 08°39′N 80°39′W
104 K8 **Peña de Francia, Sierra de la** ▲▲ W Spain
104 G6 **Penafiel** *var.* Peñafiel. Porto, N Portugal 41°12′N 08°17′W
105 N6 **Peñafiel** Castilla y León, N Spain 41°36′N 04°07′W
Peñafiel *see* Penafiel
Peñagolosa *see* Penyagolosa
105 N7 **Peñalara, Pico de** ▲ C Spain 40°52′N 03°55′W
171 X16 **Penambo, Banjaran** *var.* Banjaran Tama Abu, Penambo Range. ▲▲ Indonesia/Malaysia
Penambo Range *see* Penambo, Banjaran
41 O10 **Peña Nevada, Cerro** ▲ C Mexico 23°46′N 99°52′W
Penang *see* Pinang
Penang *see* George Town
Penang *see* Pinang, Pulau, Peninsular Malaysia
60 J8 **Penápolis** São Paulo, S Brazil 21°23′S 50°02′W
104 L7 **Peñaranda de Bracamonte** Castilla y León, N Spain 40°54′N 05°13′W
105 S8 **Peñarroya** ▲ E Spain 40°24′N 00°42′W
104 L12 **Peñarroya-Pueblonuevo** Andalucía, S Spain 38°21′N 05°18′W
97 K22 **Penarth** S Wales, United Kingdom 51°27′N 03°11′W
104 K1 **Peñas, Cabo de** *headland* N Spain 43°39′N 05°52′W
63 F20 **Penas, Golfo de** *gulf* S Chile
Pen-ch'i *see* Benxi
79 H14 **Pendé** *var.* Logone Oriental. ≈ Central African Republic/Chad
76 I14 **Pendembu** E Sierra Leone 09°06′N 12°12′W
29 R13 **Pender** Nebraska, C USA 42°06′N 96°42′W
Penderma *see* Bandırma
32 K11 **Pendleton** Oregon, NW USA 45°40′N 118°47′W
32 M7 **Pend Oreille, Lake** ◎ Idaho, NW USA
32 M7 **Pend Oreille River** ≈ Idaho/Washington, NW USA
Pendzhikent *see* Panjakent
Peneius *see* Pineiós
104 G8 **Penela** Coimbra, N Portugal 40°02′N 08°23′W
14 G13 **Penetanguishene** Ontario, S Canada 44°45′N 79°55′W
151 H15 **Penganga** ≈ C India
P'engchia Yu *see* Pengjia Yu
79 M21 **Penge** Kasai-Oriental, C Dem. Rep. Congo 05°29′S 24°38′E
Penghu Archipelago/ P'enghu Ch'üntao/Penghu Islands *see* Penghu Liedao
161 R14 **Penghu Liedao** *var.* P'enghu Ch'üntao, Penghu Islands, P'enghu Liehtao, *Eng.* Penghu Archipelago, Pescadores, *Jap.* Hoko-guntō, Hoko-shotō. *island group* W Taiwan
P'enghu Liehtao *see* Penghu Liedao
161 S14 **Penghu Shuidao** *var.* Pescadores Channel, P'enghu Shuitao. *channel* W Taiwan
P'enghu Shuitao *see* Penghu Shuidao
161 T12 **Pengjia Yu** *prev.* P'engchia Yu. *island* N Taiwan
161 R4 **Penglai** *var.* Dengzhou. Shandong, E China 37°50′N 120°45′E
Peng-pu *see* Bengbu
Penhsihu *see* Benxi
Penibético, Sistema *see* Béticos, Sistemas
104 F10 **Peniche** Leiria, W Portugal 39°21′N 09°23′W
169 U17 **Penida, Nusa** *island* S Indonesia
Peninsular State *see* Florida
105 T8 **Peníscola** *var.* Peñíscola. Valenciana, E Spain 40°22′N 00°24′E
40 M13 **Pénjamo** Guanajuato, C Mexico 20°26′N 101°44′W
Penki *see* Benxi
102 F7 **Penmarch, Pointe de** *headland* NW France 47°46′N 04°34′W
107 L15 **Penna, Punta della** *headland* C Italy 42°10′N 14°43′E
107 K14 **Penne** Abruzzo, C Italy 42°28′N 13°57′E
Penner *see* Penneru
155 J18 **Penneru** *var.* Penner. ≈ C India
182 I10 **Penneshaw** South Australia 35°45′S 137°57′E
18 C14 **Penn Hills** Pennsylvania, NE USA 40°28′N 79°53′W
Penninae, Alpes/Pennine, Alpi *see* Pennine Alps
108 D11 **Pennine Alps** *Fr.* Alpes Pennines, *It.* Alpi Pennine, *Lat.* Alpes Penninae. ▲▲ Italy/Switzerland
Pennine Chain *see* Pennines
97 L15 **Pennines** *var.* Pennine Chain. ▲▲ N England, United Kingdom
Pennines, Alpes *see* Pennine Alps
21 O8 **Pennington Gap** Virginia, NE USA 36°45′N 83°01′W
18 I16 **Penns Grove** New Jersey, NE USA 39°42′N 75°27′W
18 I16 **Pennsville** New Jersey, NE USA 39°37′N 75°29′W
18 E14 **Pennsylvania** *off.* Commonwealth of Pennsylvania, *also known as* Keystone State. ◈ *state* NE USA
18 G10 **Penn Yan** New York, NE USA 42°39′N 77°02′W
124 H16 **Peno** Tverskaya Oblast', W Russian Federation 56°55′N 32°44′E
19 R7 **Penobscot Bay** *bay* Maine, NE USA
19 S5 **Penobscot River** ≈ Maine, NE USA
182 K12 **Penola** South Australia 37°24′S 140°50′E
40 K9 **Peñón Blanco** Durango, C Mexico 25°12′N 100°50′W
182 E7 **Penong** South Australia 31°57′S 133°01′E
43 S16 **Penonomé** Coclé, C Panama 08°29′N 80°22′W
190 L13 **Penrhyn** *atoll* N Cook Islands
192 M9 **Penrhyn Basin** *undersea feature* C Pacific Ocean
183 S9 **Penrith** New South Wales, SE Australia 33°45′S 150°48′E
97 K15 **Penrith** NW England, United Kingdom 54°40′N 02°44′W
23 O9 **Pensacola** Florida, SE USA 30°25′N 87°13′W
23 O9 **Pensacola Bay** *bay* Florida, SE USA
195 N7 **Pensacola Mountains** ▲▲ Antarctica
182 L12 **Penshurst** Victoria, SE Australia 37°54′S 142°19′E
187 R13 **Pentecost** *Fr.* Pentecôte. *island* C Vanuatu
15 V4 **Pentecôte** ≈ Québec, SE Canada
Pentecôte *see* Pentecost
15 V4 **Pentecôte, Lac** ◎ Québec, SE Canada
8 H15 **Penticton** British Columbia, SW Canada 49°29′N 119°38′W
96 J6 **Pentland Firth** *strait* N Scotland, United Kingdom
96 J12 **Pentland Hills** *hill range* S Scotland, United Kingdom
171 Q12 **Penu** Pulau Taliabu, E Indonesia 01°43′S 125°09′E
155 H18 **Penukonda** Andhra Pradesh, E India 14°04′N 77°38′E
166 L7 **Penwegon** Bago, C Myanmar (Burma) 18°14′N 96°34′E
24 M8 **Penwell** Texas, SW USA 31°45′N 102°32′W
105 S8 **Penyagolosa** *var.* Peñagolosa. ▲ E Spain 40°10′N 00°15′W
97 J21 **Pen y Fan** ▲ SE Wales, United Kingdom 51°52′N 03°25′W
97 L16 **Pen-y-ghent** ▲ N England, United Kingdom 54°11′N 02°15′W
127 O6 **Penza** Penzenskaya Oblast', W Russian Federation 53°11′N 45°E
97 G25 **Penzance** SW England, United Kingdom 50°08′N 05°33′W
127 N6 **Penzenskaya Oblast'** ◈ *province* W Russian Federation
123 U7 **Penzhina** ≈ E Russian Federation
123 U9 **Penzhinskaya Guba** *bay* E Russian Federation
Penzig *see* Pieńsk
36 K13 **Peoria** Arizona, SW USA 33°34′N 112°14′W
30 L12 **Peoria** Illinois, N USA 40°42′N 89°35′W
30 L12 **Peoria Heights** Illinois, N USA 40°45′N 89°34′W
31 N11 **Peotone** Illinois, N USA 41°19′N 87°47′W
18 J11 **Pepacton Reservoir** ⊠ New York, NE USA
76 I15 **Pepel** W Sierra Leone 08°39′N 13°04′W
30 I6 **Pepin, Lake** ⊠ Minnesota/Wisconsin, N USA
99 L20 **Pepinster** Liège, E Belgium 50°34′N 05°49′E
113 L20 **Peqin** *var.* Peqini. Elbasan, C Albania 41°03′N 19°46′E
Peqini *see* Peqin
40 D7 **Pequeña, Punta** *headland* NW Mexico 26°13′N 112°34′W
168 J8 **Perak** ◈ *state* Peninsular Malaysia
105 R7 **Perales del Alfambra** Aragón, NE Spain 40°38′N 01°00′W
115 C15 **Pérama** *var.* Perama. Ípeiros, W Greece 39°42′N 20°51′E
Perama *see* Pérama
92 M13 **Perä-Posio** Lappi, NE Finland 66°10′N 27°56′E
15 Z6 **Percé** Québec, SE Canada 48°32′N 64°14′W
15 Z6 **Percé, Rocher** *island* Québec, S Canada
102 L5 **Perche, Collines de** ▲▲ N France
109 X4 **Perchtoldsdorf** Niederösterreich, NE Austria 48°06′N 16°16′E
180 L6 **Percival Lakes** *lakes* Western Australia
105 T3 **Perdido, Monte** ▲ NE Spain 42°41′N 00°01′E
23 O8 **Perdido River** ≈ Alabama/Florida, S USA
Perece Vela Basin *see* West Mariana Basin
116 G7 **Perechyn** Zakarpats'ka Oblast', W Ukraine 48°45′N 22°28′E
54 E10 **Pereira** Risaralda, W Colombia 04°47′N 75°46′W
60 I7 **Pereira Barreto** São Paulo, S Brazil 20°37′S 51°07′W
59 G15 **Pereirinha** Pará, N Brazil 08°18′S 57°30′W
127 N10 **Perelazovskiy** Volgogradskaya Oblast', SW Russian Federation 49°10′N 42°30′E
127 S7 **Perelyub** Saratovskaya Oblast', W Russian Federation 51°52′N 50°19′E
31 P7 **Pere Marquette River** ≈ Michigan, N USA
Peremyshl *see* Przemyśl
116 I5 **Peremyshlyany** L'vivs'ka Oblast', W Ukraine 49°42′N 24°33′E
Pereshchepino *see* Pereshchepyne
116 L9 **Pereshchepyne** *Rus.* Pereshchepino. Dnipropetrovs'ka Oblast', E Ukraine 48°59′N 35°22′E
124 L16 **Pereslavl'-Zalesskiy** Yaroslavskaya Oblast', W Russian Federation 56°42′N 38°45′E
117 Y7 **Pereval's'k** Luhans'ka Oblast', E Ukraine 48°28′N 38°54′E
127 U7 **Perevolotskiy** Orenburgskaya Oblast', W Russian Federation 51°54′N 54°05′E
Pereyaslav-Khmel'nitskiy *see* Pereyaslav-Khmel'nyts'kyy
117 Q5 **Pereyaslav-Khmel'nyts'kyy** *Rus.* Pereyaslav-Khmel'nitskiy. Kyyivs'ka Oblast', N Ukraine 50°05′N 31°28′E
109 U4 **Perg** Oberösterreich, N Austria 48°15′N 14°38′E
61 B19 **Pergamino** Buenos Aires, E Argentina 33°56′S 60°38′W
106 G6 **Pergine Valsugana** *Ger.* Persen. Trentino-Alto Adige, N Italy 46°04′N 11°13′E
29 S6 **Perham** Minnesota, N USA 46°35′N 95°34′W
93 L16 **Perho** Keski-Pohjanmaa, W Finland 63°15′N 24°25′E
116 E11 **Periam** *Ger.* Perjamosch, *Hung.* Perjámos. Timiș, W Romania 46°02′N 20°54′E
15 Q6 **Péribonca** ≈ Québec, SE Canada
12 L11 **Péribonca, Lac** ◎ Québec, SE Canada
15 Q6 **Péribonca, Petite Rivière** ≈ Québec, SE Canada
15 Q7 **Péribonka** Québec, SE Canada 48°45′N 72°01′W
40 I9 **Pericos** Sinaloa, C Mexico 25°03′N 107°42′W
169 Q10 **Perigi** Borneo, C Indonesia
102 L12 **Périgueux** *anc.* Vesuna. Dordogne, SW France 45°12′N 00°41′E
54 G5 **Perijá, Serranía de** ▲▲ Columbia/Venezuela
115 H17 **Peristéra** *island* Vóreies Sporádes, Greece, Aegean Sea
63 H20 **Perito Moreno** Santa Cruz, S Argentina 46°35′S 71°W
155 G22 **Periyāl** *var.* Periyār. ≈ SW India
Periyār *see* Periyāl
155 G23 **Periyār Lake** ◎ S India
Perjámos/Perjamosch *see* Periam
27 O9 **Perkins** Oklahoma, C USA 35°58′N 97°01′W
116 L7 **Perkivtsi** Chernivets'ka Oblast', W Ukraine 48°28′N 26°48′E
43 U15 **Perlas, Archipiélago de las** *Eng.* Pearl Islands. *island group* SE Panama
43 O10 **Perlas, Cayos de** *reef* SE Nicaragua
43 N9 **Perlas, Laguna de** *Eng.* Pearl Lagoon. *lagoon* E Nicaragua
43 N10 **Perlas, Punta de** *headland* E Nicaragua 12°22′N 83°30′W
100 L11 **Perleberg** Brandenburg, N Germany 53°04′N 11°52′E
Perlepe *see* Prilep
168 I6 **Perlis** ◈ *state* Peninsular Malaysia
125 U14 **Perm'** *prev.* Molotov. Permskiy Kray, NW Russian Federation 58°01′N 56°10′E
113 M22 **Përmet** *var.* Përmeti, Prëmet. Gjirokastër, S Albania 40°12′N 20°24′E
Përmeti *see* Përmet
125 U15 **Permskiy Kray** ◈ *province* NW Russian Federation
59 P15 **Pernambuco** *off.* Estado de Pernambuco. ◈ *state* E Brazil
Pernambuco *see* Recife
Pernambuco Abyssal Plain *see* Pernambuco Plain
Pernambuco, Estado de *see* Pernambuco
47 Y6 **Pernambuco Plain** *var.* Pernambuco Abyssal Plain. *undersea feature* E Atlantic Ocean 07°30′S 27°00′W
65 K15 **Pernambuco Seamounts** *undersea feature* C Atlantic Ocean
182 H6 **Pernatty Lagoon** *salt lake* South Australia
Pernau *see* Pärnu
Pernauer Bucht *see* Pärnu Laht
Pērnava *see* Pärnu
114 G9 **Pernik** *prev.* Dimitrovo. Pernik, W Bulgaria 42°36′N 23°02′E
114 G10 **Pernik** ◈ *province* W Bulgaria
93 K20 **Perniö** *Swe.* Bjärnå. Varsinais-Suomi, SW Finland 60°13′N 23°10′E
109 X5 **Pernitz** Niederösterreich, E Austria 47°54′N 15°58′E
Pernov *see* Pärnu
103 O3 **Péronne** Somme, N France 49°56′N 02°57′E
14 L8 **Péronne, Lac** ◎ Québec, SE Canada
106 A8 **Perosa Argentina** Piemonte, NE Italy 45°02′N 07°10′E
41 Q14 **Perote** Veracruz-Llave, E Mexico 19°32′N 97°16′W
Pérouse *see* Perugia
191 W15 **Pérouse, Bahía de la** *bay* Easter Island, Chile, E Pacific Ocean
Perovsk *see* Kyzylorda
103 O17 **Perpignan** Pyrénées-Orientales, S France 42°41′N 02°53′E
113 M20 **Përrenjas** *var.* Përrenjasi, Prenjas, Prenjasi. Elbasan, E Albania 41°04′N 20°34′E
Përrenjasi *see* Përrenjas
92 O2 **Perriertoppen** ▲ C Svalbard 79°10′N 17°01′E

◆ Country
● Country Capital
◇ Dependent Territory
○ Dependent Territory Capital
◈ Administrative Regions
✈ International Airport
▲ Mountain
▲▲ Mountain Range
Volcano
≈ River
◎ Lake
⊠ Reservoir

Perrin – Pine Falls

25 S6 **Perrin** Texas, SW USA *32°59′N 98°03′W*
23 Y16 **Perrine** Florida, SE USA *25°36′N 80°21′W*
37 S12 **Perro, Laguna del** ◎ New Mexico, SW USA
102 G5 **Perros-Guirec** Côtes d'Armor, NW France *48°49′N 03°28′W*
23 T9 **Perry** Florida, SE USA *30°07′N 83°34′W*
23 T5 **Perry** Georgia, SE USA *32°27′N 83°43′W*
29 U14 **Perry** Iowa, C USA *41°50′N 94°06′W*
18 E10 **Perry** New York, NE USA *42°43′N 78°00′W*
27 N9 **Perry** Oklahoma, C USA *36°17′N 97°18′W*
27 Q3 **Perry Lake** ⊠ Kansas, C USA
31 R11 **Perrysburg** Ohio, N USA *41°33′N 83°37′W*
25 O1 **Perryton** Texas, SW USA *36°23′N 100°48′W*
39 O15 **Perryville** Alaska, USA *55°55′N 159°08′W*
27 U11 **Perryville** Arkansas, C USA *35°00′N 92°48′W*
27 Y6 **Perryville** Missouri, C USA *37°43′N 89°51′W*
Persante *see* Parsęta
Persen *see* Pergine Valsugana
Pershay *see* Pyarshai
117 V7 **Pershotravens'k** Dnipropetrovs'ka Oblast', E Ukraine *48°19′N 36°22′E*
Pershotravneve *see* Manhush
Persia *see* Iran
141 T5 **Persian Gulf** *var.* The Gulf, *Ar.* Khalīj al 'Arabī, *Per.* Khalīj-e Fars. *Gulf* SW Asia *see also* Gulf, The
141 T5 **Persian Gulf** *var.* Gulf, The, *Ar.* Khalīj al 'Arabī, *Per.* Khalīj-e Fars. *gulf* SW Asia *see also* Persian Gulf
Persis *see* Fārs
95 K22 **Perstorp** Skåne, S Sweden *56°08′N 13°23′E*
137 O14 **Pertek** Tunceli, C Turkey *38°53′N 39°19′E*
183 P16 **Perth** Tasmania, SE Australia *41°39′S 147°11′E*
180 I13 **Perth** *state capital* Western Australia *31°58′S 115°49′E*
14 L13 **Perth** Ontario, SE Canada *44°54′N 76°15′W*
96 J11 **Perth** C Scotland, United Kingdom *56°24′N 03°28′W*
96 J10 **Perth** *cultural region* C Scotland, United Kingdom
180 I12 **Perth** ✈ Western Australia *31°51′S 116°06′E*
173 V10 **Perth Basin** *undersea feature* SE Indian Ocean *28°30′S 110°00′E*
103 S15 **Pertuis** Vaucluse, SE France *43°42′N 05°30′E*
103 Y16 **Pertusato, Capo** *headland* Corse, France, C Mediterranean Sea *41°22′N 09°10′E*
30 L11 **Peru** Illinois, N USA *41°18′N 89°09′W*
31 P12 **Peru** Indiana, N USA *40°45′N 86°04′W*
57 E13 **Peru** *off.* Republic of Peru. ◆ *republic* W South America
Peru *see* Beru
193 T9 **Peru Basin** *undersea feature* E Pacific Ocean *15°00′S 85°00′W*
193 U8 **Peru-Chile Trench** *undersea feature* E Pacific Ocean *20°00′S 73°00′W*
112 F13 **Perućko Jezero** ◎ S Croatia
106 H13 **Perugia** *Fr.* Pérouse; *anc.* Perusia. Umbria, C Italy *43°06′N 12°24′E*
Perugia, Lake of *see* Trasimeno, Lago
61 D15 **Perugorría** Corrientes, NE Argentina *29°21′S 58°35′W*
60 M11 **Peruíbe** São Paulo, S Brazil *24°18′S 47°01′W*
155 B21 **Perumalpār** *reef* India, N Indian Ocean
Peru, Republic of *see* Peru
Perusia *see* Perugia
99 D20 **Péruwelz** Hainaut, SW Belgium *50°30′N 03°35′E*
137 R15 **Pervari** Siirt, SE Turkey *37°55′N 42°32′E*
127 O4 **Pervomaysk** Nizhegorodskaya Oblast', W Russian Federation *54°52′N 43°49′E*
117 X7 **Pervomays'k** Luhans'ka Oblast', E Ukraine *48°38′N 38°36′E*
117 P8 **Pervomays'k** *prev.* Ol'viopol'. Mykolayivs'ka Oblast', S Ukraine *48°02′N 30°51′E*
117 S12 **Pervomays'ke** Avtonomna Respublika Krym, S Ukraine *45°43′N 33°49′E*
127 V7 **Pervomayskiy** Orenburgskaya Oblast', W Russian Federation *51°32′N 54°58′E*
126 M6 **Pervomayskiy** Tambovskaya Oblast', W Russian Federation *53°15′N 40°20′E*
117 V6 **Pervomays'kyy** Kharkivs'ka Oblast', E Ukraine *49°24′N 36°12′E*
122 F10 **Pervoural'sk** Sverdlovskaya Oblast', C Russian Federation *56°58′N 59°50′E*
123 V11 **Pervyy Kuril'skiy Proliv** *strait* E Russian Federation
99 I19 **Perwez** Walloon Brabant, C Belgium *50°39′N 04°49′E*
106 I11 **Pesaro** *anc.* Pisaurum. Marche, C Italy *43°55′N 12°53′E*
35 N9 **Pescadero** California, W USA *37°15′N 122°23′W*
Pescadores *see* Penghu Liedao
Pescadores Channel *see* Penghu Shuidao
107 K14 **Pescara** *anc.* Aternum, Ostia Aterni. Abruzzo, C Italy *42°28′N 14°13′E*
107 K15 **Pescara** ≈ C Italy
106 F11 **Pescia** Toscana, C Italy *43°54′N 10°41′E*
108 C8 **Peseux** Neuchâtel, W Switzerland *46°59′N 06°53′E*
125 P6 **Pësha** ≈ NW Russian Federation
149 T5 **Peshāwar** Khyber Pakhtunkhwa, N Pakistan *34°01′N 71°33′E*
149 T6 **Peshāwar** ✈ Khyber Pakhtunkhwa, N Pakistan *34°01′N 71°40′E*
113 M19 **Peshkopi** *var.* Peshkopia, Peshkopija. Dibër, NE Albania *41°40′N 20°25′E*
Peshkopia/Peshkopija *see* Peshkopi
114 I11 **Peshtera** Pazardzhik, C Bulgaria *42°02′N 24°18′E*
31 N6 **Peshtigo** Wisconsin, N USA *45°04′N 87°43′W*
31 N6 **Peshtigo River** ≈ Wisconsin, N USA
Peski *see* Pyaski
125 S13 **Peskovka** Kirovskaya Oblast', NW Russian Federation *59°04′N 52°17′E*
103 S8 **Pesmes** Haute-Saône, E France *47°17′N 05°33′E*
104 H6 **Peso da Régua** *var.* Pêso da Regua. Vila Real, N Portugal *41°10′N 07°47′W*
40 F5 **Pesqueira** Sonora, NW Mexico *29°22′N 110°58′W*
102 J13 **Pessac** Gironde, SW France *44°46′N 00°42′W*
111 J23 **Pest** *off.* Pest Megye. ◈ *county* C Hungary
Pest Megye *see* Pest
124 J14 **Pestovo** Novgorodskaya Oblast', W Russian Federation *58°37′N 35°48′E*
40 M15 **Petacalco, Bahía** *bay* W Mexico
Petach-Tikva *see* Petah Tikva
138 F10 **Petah Tikva** *var.* Petach-Tikva, Petah Tiqva, Petakh Tikva; *prev.* Petaḥ Tiqwa. Tel Aviv, C Israel *32°05′N 34°53′E*
Petaḥ Tiqwa *see* Petah Tikva
93 L17 **Petäjävesi** Keski-Suomi, C Finland *62°17′N 25°10′E*
Petakh Tikva/Petah Tiqva *see* Petah Tikva
22 M7 **Petal** Mississippi, S USA *31°21′N 89°15′W*
115 I19 **Petalioí** *island* C Greece
115 H19 **Petalión, Kólpos** *gulf* E Greece
115 J19 **Pétalo** ▲ Ándros, Kykládes, Greece, Aegean Sea *37°51′N 24°50′E*
34 M8 **Petaluma** California, W USA *38°15′N 122°37′W*
99 L25 **Pétange** Luxembourg, SW Luxembourg *49°33′N 05°53′E*
54 M5 **Petare** Miranda, N Venezuela *10°31′N 66°50′W*
41 N16 **Petatlán** Guerrero, S Mexico *17°31′N 101°16′W*
83 L14 **Petauke** Eastern, E Zambia *14°12′S 31°16′E*
14 J12 **Petawawa** Ontario, SE Canada *45°54′N 77°18′W*
14 J11 **Petawawa** ≈ Ontario, SE Canada
Petchaburi *see* Phetchaburi
42 D2 **Petén** *off.* Departamento del Petén. ◈ *department* N Guatemala
Petén, Departamento del *see* Petén
42 D2 **Petén Itzá, Lago** *var.* Lago de Flores. ◎ N Guatemala
30 K7 **Petenwell Lake** ⊠ Wisconsin, N USA
14 D6 **Peterbell** Ontario, S Canada *48°34′N 83°19′W*
182 I7 **Peterborough** South Australia *32°59′S 138°51′E*
14 I14 **Peterborough** Ontario, SE Canada *44°19′N 78°20′W*
97 N20 **Peterborough** *prev.* Medeshamstede. E England, United Kingdom *52°35′N 00°15′W*
19 N10 **Peterborough** New Hampshire, NE USA *42°51′N 71°54′W*
96 L8 **Peterhead** NE Scotland, United Kingdom *57°30′N 01°46′W*
Peterhof *see* Luboń
Peter I øy *see* Peter I Øy
193 Q14 **Peter I Øy** ◇ *Norwegian dependency* Antarctica
194 H9 **Peter I Øy** *var.* Peter I øy. *island* Antarctica
97 M14 **Peterlee** N England, United Kingdom *54°45′N 01°18′W*
Peterlingen *see* Payerne
197 P14 **Petermann Bjerg** ▲ C Greenland *73°16′N 27°59′W*
11 S12 **Peter Pond Lake** ◎ Saskatchewan, C Canada
39 X13 **Petersburg** Mytkof Island, Alaska, USA *56°43′N 132°51′W*
30 K13 **Petersburg** Illinois, N USA *40°01′N 89°52′W*
31 N16 **Petersburg** Indiana, N USA *38°30′N 87°16′W*
29 Q3 **Petersburg** North Dakota, N USA *47°59′N 97°59′W*
25 N5 **Petersburg** Texas, SW USA *33°52′N 101°36′W*
21 V7 **Petersburg** Virginia, NE USA *37°14′N 77°24′W*
21 T4 **Petersburg** West Virginia, NE USA *39°01′N 79°09′W*
100 H12 **Petershagen** Nordrhein-Westfalen, NW Germany *52°22′N 08°58′E*
55 S9 **Peters Mine** *var.* Peter's Mine. N Guyana *06°13′N 59°18′W*
107 O21 **Petilia Policastro** Calabria, SW Italy *39°07′N 16°48′E*
44 M9 **Pétionville** S Haiti *18°29′N 72°16′W*
45 X6 **Petit-Bourg** Basse Terre, C Guadeloupe *16°12′N 61°36′W*
15 Y5 **Petit-Cap** Québec, SE Canada *49°00′N 64°26′W*
45 Y6 **Petit Cul-de-Sac Marin** *bay* C Guadeloupe
44 M9 **Petite-Rivière-de-l'Artibonite** C Haiti *19°10′N 72°30′W*
173 X16 **Petite Rivière Noire, Piton de la** ▲ C Mauritius
15 R9 **Petite-Rivière-St-François** Québec, SE Canada *47°18′N 70°34′W*
44 L9 **Petit-Goâve** S Haiti *18°27′N 72°51′W*
Petitjean *see* Sidi-Kacem
13 N10 **Petit Lac Manicouagan** ◎ Québec, E Canada
19 T7 **Petit Manan Point** *headland* Maine, NE USA *44°23′N 67°54′W*
Petit Mécatina, Rivière du *see* Little Mecatina
11 N10 **Petitot** ≈ Alberta/British Columbia, W Canada
45 S12 **Petit Piton** ▲ SW Saint Lucia *13°49′N 61°03′W*
Petit-Popo *see* Aného
Petit St-Bernard, Col du *see* Little Saint Bernard Pass
13 O8 **Petitsikapau Lake** ◎ Newfoundland and Labrador, E Canada
92 L11 **Petkula** Lappi, N Finland *67°41′N 26°44′E*
41 X12 **Peto** Yucatán, SE Mexico *20°09′N 88°55′W*
62 G10 **Petorca** Valparaíso, C Chile *32°18′S 70°49′W*
31 Q5 **Petoskey** Michigan, N USA *45°51′N 88°03′W*
138 G14 **Petra** *archaeological site* Ma'ān, W Jordan
Petra *see* Wādī Mūsā
115 F14 **Pétras, Sténa** *pass* N Greece
123 S16 **Petra Velikogo, Zaliv** *bay* SE Russian Federation
Petrel *see* Petrer
14 K15 **Petre, Point** *headland* Ontario, SE Canada *43°49′N 77°07′W*
105 S12 **Petrer** *var.* Petrel. Valenciana, E Spain *38°28′N 00°46′W*
125 U11 **Petretsovo** Permskiy Kray, NW Russian Federation *61°22′N 57°21′E*
114 G12 **Petrich** Blagoevgrad, SW Bulgaria *41°25′N 23°12′E*
187 P15 **Petrie, Récif** *reef* N New Caledonia
37 N11 **Petrified Forest** *prehistoric site* Arizona, SW USA
Petrikau *see* Piotrków Trybunalski
Petrikov *see* Pyetrykaw
116 H12 **Petrila** *Hung.* Petrilla. Hunedoara, W Romania *45°27′N 23°25′E*
Petrilla *see* Petrila
112 E9 **Petrinja** Sisak-Moslavina, C Croatia *45°27′N 16°14′E*
Petroaleksandrovsk *see* To'rtkok'l
Petrőcz *see* Bački Petrovac
124 G12 **Petrodvorets** *Fin.* Pietarhovi. Leningradskaya Oblast', NW Russian Federation *59°53′N 29°52′E*
Petrograd *see* Sankt-Peterburg
Petrokov *see* Piotrków Trybunalski
54 G6 **Petrólea** Norte de Santander, NE Colombia *08°30′N 72°35′W*
14 D16 **Petrolia** Ontario, S Canada *42°54′N 82°07′W*
25 S4 **Petrolia** Texas, SW USA *34°00′N 98°13′W*
59 O15 **Petrolina** Pernambuco, E Brazil *09°22′S 40°30′W*
45 T6 **Petrona, Punta** *headland* C Puerto Rico *17°57′N 66°23′W*
Petropavl *see* Petropavlovsk
117 V7 **Petropavlivka** Dnipropetrovs'ka Oblast', E Ukraine *48°28′N 36°28′E*
145 P6 **Petropavlovsk** *Kaz.* Petropavl. Severnyy Kazakhstan, N Kazakhstan *54°47′N 69°06′E*
123 V11 **Petropavlovsk-Kamchatskiy** Kamchatskiy Kray, E Russian Federation *53°03′N 158°43′E*
60 P9 **Petrópolis** Rio de Janeiro, SE Brazil *22°30′S 43°28′W*
116 H12 **Petroșani** *var.* Petroșeni, *Ger.* Petroschen, *Hung.* Petrozsény. Hunedoara, W Romania *45°25′N 23°22′E*
Petroschen/Petroșeni *see* Petroșani
Petroskoi *see* Petrozavodsk
112 N12 **Petrovac** Serbia, E Serbia *42°22′N 21°25′E*
Petrovac *see* Bosanski Petrovac
113 J17 **Petrovac na Moru** S Montenegro *42°11′N 19°00′E*
Petrovac/Petrovácz *see* Bački Petrovac
117 S8 **Petrove** Kirovohrads'ka Oblast', C Ukraine *48°22′N 33°12′E*
113 O18 **Petrovec** C FYR Macedonia *41°57′N 21°37′E*
Petrovgrad *see* Zrenjanin
127 P7 **Petrovsk** Saratovskaya Oblast', W Russian Federation *52°20′N 45°23′E*
Petrovsk-Port *see* Makhachkala
127 P9 **Petrov Val** Volgogradskaya Oblast', SW Russian Federation *50°10′N 45°16′E*
124 J11 **Petrozavodsk** *Fin.* Petroskoi. Respublika Kareliya, NW Russian Federation *61°46′N 34°19′E*
Petrozsény *see* Petroșani
83 D20 **Petrusdal** Hardap, C Namibia *23°42′S 17°23′E*
117 T7 **Petrykivka** Dnipropetrovs'ka Oblast', E Ukraine *48°44′N 34°42′E*
Petsamo *see* Pechenga
Petschka *see* Pecica
Pettau *see* Ptuj
109 S5 **Pettenbach** Oberösterreich, C Austria *47°58′N 14°03′E*
25 S13 **Pettus** Texas, SW USA *28°34′N 97°49′W*
122 G12 **Petukhovo** Kurganskaya Oblast', C Russian Federation *55°04′N 67°49′E*
Petuna *see* Songyuan
109 R4 **Peuerbach** Oberösterreich, N Austria *48°19′N 13°45′E*
62 G12 **Peumo** Libertador, C Chile *34°20′S 71°12′W*
123 T6 **Pevek** Chukotskiy Avtonomnyy Okrug, NE Russian Federation *69°41′N 170°19′E*
27 X5 **Pevely** Missouri, C USA *38°16′N 90°24′W*
Peyia *see* Pégeia
102 J15 **Peyrehorade** Landes, SW France *43°33′N 01°05′W*
124 J14 **Peza** ≈ NW Russian Federation
103 P16 **Pézenas** Hérault, S France *43°28′N 03°25′E*
111 H20 **Pezinok** *Ger.* Bösing, *Hung.* Bazin. Bratislavský Kraj, W Slovakia *48°17′N 17°15′E*
101 L22 **Pfaffenhofen an der Ilm** Bayern, SE Germany *48°31′N 11°30′E*
108 G7 **Pfäffikon** Schwyz, C Switzerland *47°11′N 08°46′E*
101 F20 **Pfälzer Wald** *hill range* W Germany
101 N22 **Pfarrkirchen** Bayern, SE Germany *48°25′N 12°56′E*
101 G21 **Pforzheim** Baden-Württemberg, SW Germany *48°53′N 08°42′E*
101 H24 **Pfullendorf** Baden-Württemberg, S Germany *47°55′N 09°16′E*
108 K8 **Pfunds** Tirol, W Austria *46°56′N 10°30′E*
101 G19 **Pfungstadt** Hessen, W Germany *49°48′N 08°36′E*
83 L20 **Phalaborwa** *var.* Ba-Pahalaborwa. Limpopo, NE South Africa *23°59′S 31°04′E*
152 E11 **Phalodi** Rājasthān, NW India *27°06′N 72°22′E*
152 E12 **Phalsund** Rājasthān, NW India *26°22′N 71°56′E*
155 E15 **Phaltan** Mahārāshtra, W India *18°01′N 74°31′E*
167 O7 **Phan** *var.* Muang Phan. Chiang Rai, NW Thailand *19°34′N 99°44′E*
167 O14 **Phangan, Ko** *island* SW Thailand
166 M15 **Phang-Nga** *var.* Pang-Nga, Phangnga. Phangnga, SW Thailand *08°29′N 98°31′E*
Phangnga *see* Phang-Nga
Phan Rang/Phanrang *see* Phan Rang-Thap Cham
167 V13 **Phan Rang-Thap Cham** *var.* Phanrang, Phan Rang, Phan Rang Thap Cham. Ninh Thuận, S Vietnam *11°34′N 109°00′E*
167 U13 **Phan Ri** Bình Thuận, S Vietnam *11°11′N 108°31′E*
167 U13 **Phan Thiêt** Bình Thuận, S Vietnam *10°56′N 108°06′E*
Pharnacia *see* Giresun
25 S17 **Pharr** Texas, SW USA *26°11′N 98°10′W*
Pharus *see* Hvar
167 N16 **Phatthalung** *var.* Padalung, Patalung. Phatthalung, SW Thailand *07°38′N 100°04′E*
167 O7 **Phayao** *var.* Muang Phayao. Phayao, NW Thailand *19°10′N 99°55′E*
11 U10 **Phelps Lake** ◎ Saskatchewan, C Canada
21 X9 **Phelps Lake** ◎ North Carolina, SE USA
23 R5 **Phenix City** Alabama, S USA *32°28′N 85°00′W*
Phet Buri *see* Phetchaburi
167 O11 **Phetchaburi** *var.* Bejraburi, Petchaburi, Phet Buri. Phetchaburi, SW Thailand *13°05′N 99°58′E*
167 O9 **Phichit** *var.* Bichitra, Muang Phichit, Pichit. Phichit, C Thailand *16°29′N 100°21′E*
22 M5 **Philadelphia** Mississippi, S USA *32°45′N 89°06′W*
18 I7 **Philadelphia** New York, NE USA *44°10′N 75°40′W*
18 I16 **Philadelphia** Pennsylvania, NE USA *40°N 75°10′W*
18 I16 **Philadelphia** ✈ Pennsylvania, NE USA *39°51′N 75°13′W*
Philadelphia *see* 'Ammān
28 L10 **Philip** South Dakota, N USA *44°02′N 101°39′W*
99 H22 **Philippeville** Namur, S Belgium *50°12′N 04°33′E*
Philippeville *see* Skikda
21 S3 **Philippi** West Virginia, NE USA *39°08′N 80°03′W*
Philippi *see* Fílippoi
195 Y9 **Philippi Glacier** *glacier* Antarctica
192 G6 **Philippine Basin** *undersea feature* W Pacific Ocean *17°00′N 132°00′E*
129 X12 **Philippine Plate** *tectonic feature*
171 O5 **Philippines** *off.* Republic of the Philippines. ◆ *republic* SE Asia
129 X13 **Philippines** *island group* W Pacific Ocean
171 P3 **Philippine Sea** *sea* W Pacific Ocean
Philippines, Republic of the *see* Philippines
192 F6 **Philippine Trench** *undersea feature* W Philippine Sea
83 H23 **Philippolis** Free State, C South Africa *30°16′S 25°16′E*
Philippopolis *see* Plovdiv
Philippopolis *see* Shahbā', Syria
45 V9 **Philipsburg** ○ Sint Maarten *17°58′N 63°02′W*
33 P10 **Philipsburg** Montana, NW USA *46°19′N 113°17′W*
39 R6 **Philip Smith Mountains** ▲▲ Alaska, USA
152 H8 **Phillaur** Punjab, N India *31°02′N 75°50′E*
183 N13 **Phillip Island** *island* Victoria, SE Australia
25 N2 **Phillips** Texas, SW USA *35°39′N 101°21′W*
30 K5 **Phillips** Wisconsin, N USA *45°42′N 90°23′W*
26 K3 **Phillipsburg** Kansas, C USA *39°45′N 99°19′W*
18 I14 **Phillipsburg** New Jersey, NE USA *40°39′N 75°09′W*
21 S7 **Philpott Lake** ⊠ Virginia, NE USA
Phintias *see* Licata
167 P9 **Phitsanulok** *var.* Bisnulok, Muang Phitsanulok, Pitsanulok. Phitsanulok, C Thailand *16°49′N 100°15′E*
Phlórina *see* Flórina
Phnom Penh *see* Phnum Penh
167 S13 **Phnum Penh** *var.* Phnom Penh. ● (Cambodia) Phnum Penh, S Cambodia *11°35′N 104°55′E*
167 S11 **Phnum Tbêng Meanchey** Preăh Vihéar, N Cambodia *13°45′N 104°58′E*
36 K13 **Phoenix** *state capital* Arizona, SW USA *33°27′N 112°04′W*
Phoenix Island *see* Rawaki
191 R3 **Phoenix Islands** *island group* C Kiribati
18 I15 **Phoenixville** Pennsylvania, NE USA *40°07′N 75°31′W*
83 K22 **Phofung** *var.* Mont-aux-Sources. ▲ N Lesotho *28°47′S 28°52′E*
167 Q10 **Phon** Khon Kaen, E Thailand *15°47′N 102°35′E*
167 Q5 **Phôngsali** *var.* Phong Saly. Phôngsali, N Laos *21°40′N 102°04′E*
Phong Saly *see* Phôngsali
Phônhông *see* Ban Phônhông
167 R7 **Phônsaven** *var.* Pèk, Xieng Khouang; *prev.* Xiangkhoang. Xiangkhoang, N Laos *19°19′N 103°23′E*
167 R5 **Phô Rang** *var.* Bao Yên. Lao Cai, N Vietnam *22°12′N 104°27′E*
Phort Láirge, Cuan *see* Waterford Harbour
Phou Louang *see* Annamite Mountains
167 N10 **Phra Chedi Sam Ong** Kanchanaburi, W Thailand *15°18′N 98°26′E*
167 O8 **Phrae** *var.* Muang Phrae, Prae. Phrae, NW Thailand *18°07′N 100°09′E*
Phra Nakhon Si Ayutthaya *see* Ayutthaya
167 M14 **Phra Thong, Ko** *island* SW Thailand
Phu Cũơng *see* Thu Dầu Môt
166 M15 **Phuket** *var.* Bhuket, Puket, *Mal.* Ujung Salang; *prev.* Junkseylon, Salang. Phuket, SW Thailand *07°52′N 98°22′E*
166 M15 **Phuket** ✈ Phuket, SW Thailand *08°03′N 98°16′E*
166 M15 **Phuket, Ko** *island* SW Thailand
154 N12 **Phulabāni** *prev.* Phulbani. Odisha, E India *20°30′N 84°18′E*
Phulbani *see* Phulabāni
167 U9 **Phu Lôc** Th,a Thiên-Huê, C Vietnam *16°13′N 107°53′E*
167 R13 **Phumĭ Chŏâm** Kâmpóng Spœ, SW Cambodia *11°42′N 103°58′E*
Phumĭ Kalêng *see* Kalêng
Phumĭ Kâmpóng Trâbêk *see* Kâmpóng Trâbêk
Phumĭ Koŭk Kduŏch *see* Koŭk Kduŏch
Phumĭ Labăng *see* Labăng
Phumĭ Mlu Prey *see* Mlu Prey
Phumĭ Moŭng *see* Moŭng
Phumĭ Prâmaôy *see* Prâmaôy
Phumĭ Samĭt *see* Samĭt
Phumĭ Sâmraông *see* Sâmraông
Phumĭ Siĕmbok *see* Siĕmbok
Phumĭ Thalabârĭvăt *see* Thalabârĭvăt
Phumĭ Veal Renh *see* Veal Renh
Phumĭ Yeay Sên *see* Yeay Sên
Phum Kompong Trabek *see* Kâmpóng Trâbêk
Phum Samrong *see* Sâmraông
167 V11 **Phu My** Bình Định, C Vietnam *14°07′N 109°05′E*
Phung Hiệp *see* Tân Hiệp
153 T12 **Phuntsholing** SW Bhutan *26°52′N 89°26′E*
167 V13 **Phươc Dân** Ninh Thuận, S Vietnam *11°28′N 108°53′E*
167 R15 **Phước Long** Minh Hai, S Vietnam *09°27′N 105°25′E*
Phuoc Sơn *see* Khâm Đuc
167 R14 **Phu Quôc, Đao** *var.* Phu Quoc Island. *island* S Vietnam
Phu Quoc Island *see* Phu Quôc, Đao
167 S6 **Phu Tho** Vinh Phu, N Vietnam *21°23′N 105°13′E*
Phu Vinh *see* Tra Vinh
166 M7 **Phyu** *var.* Hpyu, Pyu. Bago, C Myanmar (Burma) *18°29′N 96°28′E*
189 T13 **Piaanu Pass** *passage* Chuuk Islands, C Micronesia
106 E8 **Piacenza** *Fr.* Paisance; *anc.* Placentia. Emilia-Romagna, N Italy *45°02′N 09°42′E*
107 K14 **Pianella** Abruzzo, C Italy *42°23′N 14°04′E*
107 M15 **Pianosa, Isola** *island* Archipelago Toscano, C Italy
171 U13 **Piar** Papua Barat, E Indonesia *02°49′S 132°46′E*
45 U14 **Piarco** *var.* Port of Spain. ✈ (Port-of-Spain) Trinidad, Trinidad and Tobago *10°36′N 61°21′W*
110 M12 **Piaseczno** Mazowieckie, C Poland *52°03′N 21°00′E*
116 I15 **Piatra** Teleorman, S Romania *43°49′N 25°10′E*
116 L10 **Piatra-Neamț** *Hung.* Karácsonkő. Neamț, NE Romania *46°54′N 26°23′E*
Piauhy *see* Piauí
59 N15 **Piauí** *off.* Estado do Piauí; *prev.* Piauhy. ◈ *state* E Brazil
Piauí, Estado do *see* Piauí
106 I7 **Piave** ≈ NE Italy
107 K24 **Piazza Armerina** *var.* Chiazza. Sicilia, Italy, C Mediterranean Sea *37°23′N 14°22′E*
81 G14 **Pibor** *Amh.* Pibor Wenz. ≈ Ethiopia/South Sudan
81 G14 **Pibor Post** Jonglei, E South Sudan *06°50′N 33°06′E*
Pibor Wenz *see* Pibor
Pibrans *see* Příbram
36 K11 **Picacho Butte** ▲ Arizona, SW USA *35°12′N 112°44′W*
40 D4 **Picachos, Cerro** ▲ NW Mexico *29°15′N 114°04′W*
103 O4 **Picardie** *Eng.* Picardy. ◈ *region* N France
Picardy *see* Picardie
22 L8 **Picayune** Mississippi, S USA *30°31′N 89°40′W*
Piccolo San Bernardo, Colle di *see* Little Saint Bernard Pass
PicdeBälaitous *see* Balaïtous
62 K5 **Pichanal** Salta, N Argentina *23°18′S 64°10′W*
147 P12 **Pichandar** W Tajikistan *38°44′N 68°51′E*
27 R8 **Picher** Oklahoma, C USA *36°59′N 94°49′W*
62 G12 **Pichilemu** Libertador, C Chile *34°25′S 72°00′W*
40 F9 **Pichilingue** Baja California Sur, NW Mexico *24°20′N 110°17′W*
56 B6 **Pichincha** ◈ *province* N Ecuador
56 C6 **Pichincha** ▲ N Ecuador *0°12′S 78°39′W*
Pichit *see* Phichit
41 U15 **Pichucalco** Chiapas, SE Mexico *17°32′N 93°07′W*
22 L5 **Pickens** Mississippi, S USA *32°52′N 89°58′W*
21 O11 **Pickens** South Carolina, SE USA *34°53′N 82°42′W*
14 G11 **Pickerel** ≈ Ontario, S Canada
14 H15 **Pickering** Ontario, S Canada *43°50′N 79°03′W*
97 N16 **Pickering** N England, United Kingdom *54°14′N 00°47′W*
31 S13 **Pickerington** Ohio, N USA *39°52′N 82°45′W*
12 C10 **Pickle Lake** Ontario, C Canada *51°30′N 90°10′W*
29 P12 **Pickstown** South Dakota, N USA *43°02′N 98°31′W*
25 V6 **Pickton** Texas, SW USA *33°01′N 95°19′W*
23 N1 **Pickwick Lake** ⊠ S USA
64 N2 **Pico** *var.* Ilha do Pico. *island* Azores, Portugal, NE Atlantic Ocean
63 J19 **Pico de Salamanca** Chubut, SE Argentina *45°26′S 67°26′W*
1 P9 **Pico Fracture Zone** *tectonic feature* NW Atlantic Ocean
Pico, Ilha do *see* Pico
59 O14 **Picos** Piauí, E Brazil *07°05′S 41°24′W*
63 I20 **Pico Truncado** Santa Cruz, SE Argentina *46°45′S 68°00′W*
183 S9 **Picton** New South Wales, SE Australia *34°12′S 150°36′E*
14 K15 **Picton** Ontario, SE Canada *43°59′N 77°09′W*
185 K14 **Picton** Marlborough, South Island, New Zealand *41°18′S 174°00′E*
63 H15 **Picún Leufú, Arroyo** ≈ SW Argentina
Pidálion *see* Gkréko, Akrotíri
155 K25 **Pidurutalagala** ▲ S Sri Lanka *07°03′N 80°47′E*
116 K6 **Pidvolochys'k** Ternopil's'ka Oblast', W Ukraine *49°31′N 26°09′E*
107 K16 **Piedimonte Matese** Campania, S Italy *41°20′N 14°30′E*
27 X7 **Piedmont** Missouri, C USA *37°09′N 90°42′W*
21 P11 **Piedmont** South Carolina, SE USA *34°42′N 82°27′W*
17 S12 **Piedmont** *escarpment* E USA
Piedmont *see* Piemonte
31 U13 **Piedmont Lake** ⊠ Ohio, N USA
104 M11 **Piedrabuena** Castilla-La Mancha, C Spain *39°02′N 04°10′W*
Piedrafita, Puerto de *see* Pedrafita, Porto de
104 L8 **Piedrahita** Castilla y León, N Spain *40°27′N 05°20′W*
41 N6 **Piedras Negras** *var.* Ciudad Porfirio Díaz. Coahuila, NE Mexico *28°40′N 100°32′W*
61 E21 **Piedras, Punta** *headland* E Argentina *35°27′S 57°04′W*
57 I14 **Piedras, Río de las** ≈ E Peru
111 J16 **Piekary Śląskie** Śląskie, S Poland *50°24′N 18°58′E*
93 M17 **Pieksämäki** Etelä-Savo, E Finland *62°18′N 27°10′E*
109 V5 **Pielach** ≈ NE Austria
93 M16 **Pielavesi** Pohjois-Savo, C Finland *63°14′N 26°45′E*
93 M16 **Pielavesi** ◎ C Finland
93 N16 **Pielinen** *var.* Pielisjärvi. ◎ E Finland
Pielisjärvi *see* Pielinen
106 A8 **Piemonte** *Eng.* Piedmont. ◈ *region* NW Italy
111 L18 **Pieniny** ▲▲ S Poland
111 E14 **Pieńsk** *Ger.* Penzig. Dolnośląskie, SW Poland *51°14′N 15°03′E*
29 Q13 **Pierce** Nebraska, C USA *42°12′N 97°31′W*
11 R14 **Pierceland** Saskatchewan, C Canada
115 E14 **Piéria** ▲▲ N Greece
29 N10 **Pierre** *state capital* South Dakota, N USA *44°22′N 100°21′W*
102 K16 **Pierrefitte-Nestalas** Hautes-Pyrénées, S France *42°57′N 00°04′W*
103 R14 **Pierrelatte** Drôme, E France *44°22′N 04°40′E*
15 P11 **Pierreville** Québec, SE Canada *46°05′N 72°48′W*
15 O7 **Pierriche** ≈ Québec, SE Canada
111 H20 **Piešťany** *Ger.* Pistyan, *Hung.* Pöstyén. Tranavský Kraj, W Slovakia *48°37′N 17°48′E*
109 X5 **Piesting** ≈ E Austria
Pietarhovi *see* Petrodvorets
Pietari *see* Sankt-Peterburg
Pietarsaari *see* Jakobstad
83 K23 **Pietermaritzburg** *var.* Maritzburg. KwaZulu/Natal, E South Africa *29°35′S 30°23′E*
Pietersburg *see* Polokwane
107 K24 **Pietraperzia** Sicilia, Italy, C Mediterranean Sea *37°25′N 14°08′E*
107 N22 **Pietra Spada, Passo della** *pass* SW Italy
Piet Retief *see* eMkhondo
116 J10 **Pietrosul, Vârful** *prev.* Vîrful Pietrosu. ▲ N Romania *47°06′N 25°09′E*
106 I6 **Pieve di Cadore** Veneto, NE Italy *46°27′N 12°22′E*
14 C18 **Pigeon Bay** *lake bay* Ontario, S Canada
27 X8 **Piggott** Arkansas, C USA *36°22′N 90°11′W*
83 L21 **Piggs Peak** NW Swaziland *25°58′S 31°17′E*
Pigs, Bay of *see* Cochinos, Bahía de
61 A23 **Pigüé** Buenos Aires, E Argentina *37°38′S 62°27′W*
41 O12 **Piguícas** ▲ C Mexico *21°08′N 99°37′W*
193 W15 **Piha Passage** *passage* S Tonga
Pihkva Järv *see* Pskov, Lake
93 N18 **Pihlajavesi** ◎ SE Finland
93 J18 **Pihlava** Satakunta, SW Finland *61°33′N 21°36′E*
93 L16 **Pihtipudas** Keski-Suomi, C Finland *63°20′N 25°37′E*
40 L14 **Pihuamo** Jalisco, SW Mexico *19°20′N 103°21′W*
189 U11 **Piis Moen** *var.* Pis. *atoll* Chuuk Islands, C Micronesia
41 U17 **Pijijiapán** Chiapas, SW Mexico *15°42′N 93°12′W*
98 G12 **Pijnacker** Zuid-Holland, W Netherlands *52°01′N 04°26′E*
42 H5 **Pijol, Pico** ▲ NW Honduras *15°07′N 87°39′W*
Pikaar *see* Bikar Atoll
124 I13 **Pikalevo** Leningradskaya Oblast', NW Russian Federation *59°33′N 34°04′E*
188 M15 **Pikelot** *island* Caroline Islands, C Micronesia
30 M5 **Pike River** ≈ Wisconsin, N USA
37 T5 **Pikes Peak** ▲ Colorado, C USA *38°51′N 105°06′W*
21 P6 **Pikeville** Kentucky, S USA *37°29′N 82°33′W*
20 L9 **Pikeville** Tennessee, S USA *35°35′N 85°11′W*
Pikinni *see* Bikini Atoll
79 H18 **Pikounda** Sangha, C Congo *0°30′N 16°44′E*
110 G9 **Piła** *Ger.* Schneidemühl. Wielkopolskie, C Poland *53°09′N 16°44′E*
62 N6 **Pilagá, Riacho** ≈ NE Argentina
61 D20 **Pilar** Buenos Aires, E Argentina *34°28′S 58°55′W*
62 N7 **Pilar** *var.* Villa del Pilar. Ñeembucú, S Paraguay *26°55′S 58°20′W*
62 N6 **Pilcomayo, Rio** ≈ C South America
147 R12 **Pildon** *Rus.* Pil'don. C Tajikistan *39°10′N 71°00′E*
Piles *see* Pylés
Pilgram *see* Pelhřimov
152 L10 **Pīlībhīt** Uttar Pradesh, N India *28°37′N 79°48′E*
110 M13 **Pilica** ≈ C Poland
115 G16 **Pílio** ▲▲ C Greece
111 J22 **Pilisvörösvár** Pest, N Hungary *47°38′N 18°55′E*
65 G15 **Pillar Bay** *bay* Ascension Island, C Atlantic Ocean
183 P17 **Pillar, Cape** *headland* Tasmania, SE Australia *43°13′S 147°58′E*
Pillau *see* Baltiysk
183 R5 **Pilliga** New South Wales, SE Australia *30°22′S 148°53′E*
44 H8 **Pilón** Granma, E Cuba *19°54′N 77°20′W*
Pílos *see* Pýlos
11 W17 **Pilot Mound** Manitoba, S Canada *49°12′N 98°49′W*
21 S8 **Pilot Mountain** North Carolina, SE USA *36°23′N 80°28′W*
39 O14 **Pilot Point** Alaska, USA *57°33′N 157°34′W*
25 T5 **Pilot Point** Texas, SW USA *33°24′N 96°57′W*
32 K11 **Pilot Rock** Oregon, NW USA *45°28′N 118°49′W*
38 M11 **Pilot Station** Alaska, USA *61°56′N 162°52′W*
Pilsen *see* Plzeň
111 K18 **Pilsko** ▲ N Slovakia *49°31′N 19°21′E*
Pilten *see* Piltene
118 D8 **Piltene** *Ger.* Pilten. W Latvia *57°14′N 21°41′E*
111 M16 **Pilzno** Podkarpackie, SE Poland *49°58′N 21°18′E*
Pilzno *see* Plzeň
37 N14 **Pima** Arizona, SW USA *32°49′N 109°50′W*
58 H13 **Pimenta** Pará, N Brazil *04°32′S 56°17′W*
59 F16 **Pimenta Bueno** Rondônia, W Brazil *11°40′S 61°14′W*
56 B11 **Pimentel** Lambayeque, W Peru *06°51′S 79°53′W*
105 S6 **Pina** Aragón, NE Spain *41°28′N 00°31′W*
119 I20 **Pina** ≈ SW Belarus
40 E2 **Pinacate, Sierra del** ▲ NW Mexico *31°49′N 113°30′W*
63 H22 **Pináculo, Cerro** ▲ S Argentina *50°46′S 72°07′W*
191 X11 **Pinaki** *atoll* Îles Tuamotu, E French Polynesia
37 N15 **Pinaleno Mountains** ▲▲ Arizona, SW USA
171 P4 **Pinamalayan** Mindoro, N Philippines *13°00′N 121°30′E*
169 Q10 **Pinang** Borneo, C Indonesia *0°36′N 109°11′E*
168 J7 **Pinang** *var.* Penang. ◈ *state* Peninsular Malaysia
Pinang *see* George Town
Pinang *see* Pinang, Pulau, Peninsular Malaysia
168 J7 **Pinang, Pulau** *var.* Penang, Pinang; *prev.* Prince of Wales Island. *island* Peninsular Malaysia
44 B5 **Pinar del Río** Pinar del Río, W Cuba *22°24′N 83°42′W*
114 N11 **Pınarhisar** Kırklareli, NW Turkey *41°37′N 27°32′E*
171 O3 **Pinatubo, Mount** ⛰ Luzon, N Philippines *15°08′N 120°21′E*
11 Y16 **Pinawa** Manitoba, S Canada *50°09′N 95°52′W*
11 Q17 **Pincher Creek** Alberta, SW Canada *49°31′N 113°53′W*
30 L16 **Pinckneyville** Illinois, N USA *38°04′N 89°22′W*
Pincota *see* Pâncota
111 L15 **Pińczów** Świętokrzyskie, C Poland *50°30′N 20°30′E*
149 U7 **Pind Dādan Khān** Punjab, E Pakistan *32°36′N 73°07′E*
Píndhos/Píndhos Óros *see* Píndos
149 V8 **Pindi Bhattiān** Punjab, E Pakistan *31°54′N 73°20′E*
149 U6 **Pindi Gheb** Punjab, E Pakistan *33°16′N 72°21′E*
115 D15 **Píndos** *var.* Píndhos Óros, *Eng.* Pindus Mountains; *prev.* Píndhos. ▲▲ C Greece
Pindus Mountains *see* Píndos
18 J16 **Pine Barrens** *physical region* New Jersey, NE USA
27 V12 **Pine Bluff** Arkansas, C USA *34°15′N 92°00′W*
23 X11 **Pine Castle** Florida, SE USA *28°28′N 81°22′W*
29 V7 **Pine City** Minnesota, N USA *45°49′N 92°55′W*
181 P2 **Pine Creek** Northern Territory, N Australia *13°51′S 131°51′E*
35 V4 **Pine Creek** ≈ Nevada, W USA
18 F13 **Pine Creek** ≈ Pennsylvania, NE USA
27 Q13 **Pine Creek Lake** ⊠ Oklahoma, C USA
33 T15 **Pinedale** Wyoming, C USA *42°52′N 109°51′W*
11 X15 **Pine Dock** Manitoba, S Canada *51°34′N 96°47′W*
11 Y16 **Pine Falls** Manitoba, S Canada *50°29′N 96°12′W*

◆ Country
● Country Capital
◇ Dependent Territory
○ Dependent Territory Capital
◈ Administrative Regions
✈ International Airport
▲ Mountain
▲▲ Mountain Range
⛰ Volcano
≈ River
◎ Lake
⊠ Reservoir

Pine Flat Lake – Pojoaque

35 *R10* **Pine Flat Lake** ⊠ California, W USA
125 *N8* **Pinega** Arkhangel'skaya Oblast', NW Russian Federation *64°40′N 43°24′E*
125 *N8* **Pinega** ≈ NW Russian Federation
15 *N12* **Pine Hill** Québec, SE Canada *45°44′N 74°30′W*
11 *T12* **Pinehouse Lake** ◎ Saskatchewan, C Canada
21 *T10* **Pinehurst** North Carolina, SE USA *35°12′N 79°28′W*
115 *D19* **Pineiós** ≈ S Greece
115 *E16* **Pineiós** *var.* Piniós; *anc.* Peneius. ≈ C Greece
29 *W10* **Pine Island** Minnesota, N USA *44°12′N 92°39′W*
23 *V15* **Pine Island** *island* Florida, SE USA
194 *K10* **Pine Island Glacier** *glacier* Antarctica
25 *X9* **Pineland** Texas, SW USA *31°15′N 93°58′W*
23 *V13* **Pinellas Park** Florida, SE USA *27°50′N 82°42′W*
10 *M13* **Pine Pass** *pass* British Columbia, W Canada
8 *J10* **Pine Point** Northwest Territories, W Canada *60°52′N 114°30′W*
28 *K12* **Pine Ridge** South Dakota, N USA *43°01′N 102°33′W*
29 *U6* **Pine River** Minnesota, N USA *46°43′N 94°24′W*
31 *Q8* **Pine River** ≈ Michigan, N USA
30 *M4* **Pine River** ≈ Wisconsin, N USA
106 *A8* **Pinerolo** Piemonte, NE Italy *44°56′N 07°21′E*
115 *I15* **Pínes, Akrotírio** *var.* Akrotírio Pínnes. *headland* N Greece *40°06′N 24°19′E*
25 *W6* **Pines, Lake O' the** ⊠ Texas, SW USA
Pines, The Isle of the *see* Juventud, Isla de la
Pine Tree State *see* Maine
21 *N7* **Pineville** Kentucky, S USA *36°47′N 83°43′W*
22 *H7* **Pineville** Louisiana, S USA *31°19′N 92°25′W*
27 *R8* **Pineville** Missouri, C USA *36°36′N 94°23′W*
21 *R10* **Pineville** North Carolina, SE USA *35°04′N 80°53′W*
21 *Q6* **Pineville** West Virginia, NE USA *37°35′N 81°34′W*
33 *V8* **Piney Buttes** *physical region* Montana, NW USA
163 *W9* **Ping'an** Jilin, NE China *44°36′N 127°13′E*
160 *H14* **Pingbian** *var.* Pingbian Miaozu Zizhixian, Yuping. Yunnan, SW China *22°51′N 103°28′E*
Pingbian Miaozu Zizhixian *see* Pingbian
157 *S9* **Pingdingshan** Henan, C China *33°52′N 113°20′E*
161 *S14* **Pingdong** *Jap.* Heitō; *prev.* P'ingtung. S Taiwan *22°40′N 120°30′E*
161 *R4* **Pingdu** Shandong, E China *36°50′N 119°55′E*
189 *W16* **Pingelap Atoll** *atoll* Caroline Islands, E Micronesia
160 *K14* **Pingguo** *var.* Matou. Guangxi Zhuangzu Zizhiqu, S China *23°24′N 107°30′E*
161 *Q13* **Pinghe** *var.* Xiaoxi. Fujian, SE China *24°30′N 117°19′E*
P'ing-hsiang *see* Pingxiang
161 *N10* **Pingjiang** Hunan, S China *28°44′N 113°33′E*
Pingkiang *see* Harbin
160 *L8* **Pingli** Shaanxi, C China *32°27′N 109°21′E*
159 *W10* **Pingliang** *var.* Kongtong, P'ing-liang. Gansu, C China *35°27′N 106°38′E*
P'ing-liang *see* Pingliang
159 *W8* **Pingluo** Ningxia, N China *38°55′N 106°31′E*
Pingma *see* Tiandong
167 *O7* **Ping, Mae Nam** ≈ W Thailand
161 *Q1* **Pingquan** Hebei, E China *41°02′N 118°35′E*
29 *P5* **Pingree** North Dakota, N USA *47°07′N 98°54′W*
Pingsiang *see* Pingxiang
P'ingtung *see* Pingdong
160 *I8* **Pingwu** *var.* Long'an. Sichuan, C China *32°33′N 104°32′E*
160 *J15* **Pingxiang** Guangxi Zhuangzu Zizhiqu, S China *22°03′N 106°44′E*
161 *O11* **Pingxiang** *var.* P'ing-hsiang; *prev.* Pingsiang. Jiangxi, S China *27°42′N 113°50′E*
Pingxiang *see* Tongwei
161 *S11* **Pingyang** *var.* Kunyang. Zhejiang, SE China *27°46′N 120°37′E*
161 *P5* **Pingyi** Shandong, E China *35°30′N 117°38′E*
161 *P5* **Pingyin** Shandong, E China *36°18′N 116°24′E*
60 *H13* **Pinhalzinho** Santa Catarina, S Brazil *26°53′S 52°57′W*
60 *I12* **Pinhão** Paraná, S Brazil *25°46′S 51°32′W*
61 *H17* **Pinheiro Machado** Rio Grande do Sul, S Brazil *31°34′S 53°22′W*
104 *I7* **Pinhel** Guarda, N Portugal *40°47′N 07°03′W*
Piniós *see* Pineiós
168 *I11* **Pini, Pulau** *island* Kepulauan Batu, W Indonesia
109 *Y7* **Pinka** ≈ SE Austria
109 *X7* **Pinkafeld** Burgenland, SE Austria *47°23′N 16°08′E*
Pinkiang *see* Harbin
10 *M12* **Pink Mountain** British Columbia, W Canada *57°10′N 122°36′W*
166 *M3* **Pinlebu** Sagaing, N Myanmar (Burma) *24°02′N 95°21′E*
38 *J12* **Pinnacle Island** *island* Alaska, USA
180 *I12* **Pinnacles, The** *tourist site* Western Australia
182 *K10* **Pinnaroo** South Australia *35°17′S 140°54′E*
Pinne *see* Pniewy
100 *I9* **Pinneberg** Schleswig-Holstein, N Germany *53°40′N 09°49′E*
Pínnes, Akrotírio *see* Pínes, Akrotírio
Pinos, Isla de *see* Juventud, Isla de la
35 *R14* **Pinos, Mount** ▲ California, W USA *34°48′N 119°09′W*
105 *R12* **Pinoso** Valenciana, E Spain *38°25′N 01°02′W*
105 *N14* **Pinos-Puente** Andalucía, S Spain *37°16′N 03°46′W*
41 *Q17* **Pinotepa Nacional** *var.* Santiago Pinotepa Nacional. Oaxaca, SE Mexico *16°20′N 98°02′W*
114 *F13* **Pínovo** ▲ N Greece *41°06′N 22°19′E*
187 *R17* **Pins, Île des** *var.* Kunyé. *island* E New Caledonia
119 *I20* **Pinsk** *Pol.* Pińsk. Brestskaya Voblasts', SW Belarus *52°07′N 26°07′E*
14 *D18* **Pins, Pointe aux** *headland* Ontario, S Canada *42°14′N 81°53′W*
57 *B16* **Pinta, Isla** *var.* Abingdon. *island* Galapagos Islands, Ecuador, E Pacific Ocean
125 *Q12* **Pinyug** Kirovskaya Oblast', NW Russian Federation *60°12′N 47°45′E*
57 *B17* **Pinzón, Isla** *var.* Duncan Island. *island* Galapagos Islands, Ecuador, E Pacific Ocean
35 *Y8* **Pioche** Nevada, W USA *37°57′N 114°30′W*
106 *F13* **Piombino** Toscana, C Italy *42°54′N 10°30′E*
0 *C9* **Pioneer Fracture Zone** *tectonic feature* NE Pacific Ocean
122 *L5* **Pioner, Ostrov** *island* Severnaya Zemlya, N Russian Federation
118 *A13* **Pionerskiy** *Ger.* Neukuhren. Kaliningradskaya Oblast', W Russian Federation *54°57′N 20°16′E*
110 *N13* **Pionki** Mazowieckie, C Poland *51°30′N 21°27′E*
184 *L9* **Piopio** Waikato, North Island, New Zealand *38°27′S 175°00′E*
110 *K13* **Piotrków Trybunalski** *Ger.* Petrikau, *Rus.* Petrokov. Łodzkie, C Poland *51°25′N 19°42′E*
152 *F12* **Pīpār Road** Rājasthān, N India *26°25′N 73°29′E*
115 *I16* **Pipéri** *island* Vóreies Sporádes, Greece, Aegean Sea
29 *S10* **Pipestone** Minnesota, N USA *44°00′N 96°19′W*
12 *C9* **Pipestone** ≈ Ontario, C Canada
61 *E21* **Pipinas** Buenos Aires, E Argentina *35°32′S 57°20′W*
149 *T7* **Piplān** *prev.* Liaqatabad. Punjab, E Pakistan *32°17′N 71°24′E*
15 *R5* **Pipmuacan, Réservoir** ⊠ Québec, SE Canada
Piqan *see* Shanshan
31 *R13* **Piqua** Ohio, N USA *40°08′N 84°14′W*
105 *P5* **Piqueras, Puerto de** *pass* N Spain
60 *H11* **Piquiri, Rio** ≈ S Brazil
60 *L9* **Piracicaba** São Paulo, S Brazil *22°45′S 47°40′W*
Piraeus/Piraiévs *see* Peiraías
60 *K10* **Piraju** São Paulo, S Brazil *23°12′S 49°24′W*
60 *K9* **Pirajuí** São Paulo, S Brazil *21°58′S 49°27′W*
63 *G21* **Pirámide, Cerro** ▲ S Chile *49°06′S 73°32′W*
Piramiva *see* Pyramíva
109 *R13* **Piran** *It.* Pirano. SW Slovenia *45°35′N 13°35′E*
62 *N6* **Pirané** Formosa, N Argentina *25°42′S 59°06′W*
59 *J18* **Piranhas** Goiás, S Brazil *16°24′S 51°51′W*
Pirano *see* Piran
142 *I4* **Pīrānshahr** Āžarbāyjān-e Gharbī, NW Iran *36°41′N 45°08′E*
59 *M19* **Pirapora** Minas Gerais, NE Brazil *17°20′S 44°54′W*
60 *I9* **Pirapòzinho** São Paulo, S Brazil *22°17′S 51°31′W*
61 *G19* **Pirarajá** Lavalleja, S Uruguay *33°44′S 54°45′W*
60 *L9* **Pirassununga** São Paulo, S Brazil *21°58′S 47°23′W*
45 *V6* **Pirata, Monte** ▲ E Puerto Rico *18°06′N 65°33′W*
60 *I13* **Piratuba** Santa Catarina, S Brazil *27°26′S 51°47′W*
114 *I9* **Pirdop** *prev.* Strednogorie. Sofia, W Bulgaria *42°44′N 24°09′E*
191 *P7* **Pirea** Tahiti, W French Polynesia
59 *K18* **Pirenópolis** Goiás, S Brazil *15°48′S 49°00′W*
153 *S13* **Pirganj** Rajshahi, NW Bangladesh *25°51′N 88°25′E*
Pirgi *see* Pyrgí
Pírgos *see* Pýrgos
61 *F20* **Piriápolis** Maldonado, S Uruguay *34°51′S 55°15′W*
114 *G11* **Pirin** ▲▲ SW Bulgaria
Pirineos *see* Pyrenees
58 *N13* **Piripiri** Piauí, E Brazil *04°15′S 41°46′W*
118 *H4* **Pirita** *var.* Pirita Jõgi. ≈ NW Estonia
Pirita Jõgi *see* Pirita
54 *J6* **Píritu** Portuguesa, N Venezuela *09°21′N 69°16′W*
93 *L18* **Pirkanmaa** *Swe.* Birkaland. ◈ *region* W Finland
93 *L18* **Pirkkala** Pirkanmaa, W Finland *61°27′N 23°47′E*
101 *F20* **Pirmasens** Rheinland-Pfalz, SW Germany *49°12′N 07°37′E*
101 *P16* **Pirna** Sachsen, E Germany *50°57′N 13°56′E*
Piroe *see* Piru
113 *Q15* **Pirot** Serbia, SE Serbia *43°12′N 22°34′E*
152 *H6* **Pīr Panjāl Range** ▲▲ NE India
43 *W16* **Pirre, Cerro** ▲ SE Panama *07°54′N 77°42′W*
137 *Y11* **Pirsaat** *Rus.* Pirsagat. ≈ E Azerbaijan
Pirsagat *see* Pirsaat
143 *V11* **Pīr Shūrān, Selseleh-ye** ▲▲ SE Iran
92 *M12* **Pirttikoski** Lappi, N Finland *66°20′N 27°08′E*
Pirttikylä *see* Pörtom
171 *R13* **Piru** *prev.* Piroe. Pulau Seram, E Indonesia *03°01′S 128°10′E*
Piryatin *see* Pyryatyn
Pis *see* Piis Moen
106 *F11* **Pisa** *var.* Pisae. Toscana, C Italy *43°43′N 10°23′E*
Pisae *see* Pisa
189 *V12* **Pisar** *atoll* Chuuk Islands, C Micronesia
Pisaurum *see* Pesaro
14 *M10* **Piscatosine, Lac** ◎ Québec, SE Canada
109 *W7* **Pischeldorf** Steiermark, SE Austria *47°11′N 15°48′E*
Pischk *see* Simeria
107 *L19* **Pisciotta** Campania, S Italy *40°07′N 15°13′E*
57 *E16* **Pisco** Ica, SW Peru *13°46′S 76°12′W*
116 *G9* **Pişcolt** *Hung.* Piskolt. Satu Mare, NW Romania *47°35′N 22°18′E*
57 *E16* **Pisco, Río** ≈ E Peru
111 *C18* **Písek** Budějovický Kraj, S Czech Republic *49°19′N 14°07′E*
31 *R14* **Pisgah** Ohio, N USA *39°19′N 84°22′W*
Pisha *see* Ningnan
158 *F9* **Pishan** *var.* Guma. Xinjiang Uygur Zizhiqu, NW China *37°36′N 78°45′E*
117 *N8* **Pishchanka** Vinnyts'ka Oblast', C Ukraine *48°12′N 28°52′E*
113 *K21* **Pishë** Fier, SW Albania *40°40′N 19°22′E*
143 *X14* **Pīshīn** Sīstān va Balūchestān, SE Iran *26°05′N 61°46′E*
149 *O9* **Pishīn** Khyber Pakhtunkhwa, NW Pakistan *30°33′N 67°01′E*
149 *N11* **Pishin Lora** *var.* Psein Lora, *Pash.* Pseyn Bowr. ≈ SW Pakistan
Pishma *see* Pizhma
Pishpek *see* Bishkek
171 *O14* **Pising** Pulau Kabaena, C Indonesia *05°07′S 121°50′E*
Pisino *see* Pazin
Piski *see* Simeria
Piskolt *see* Pişcolt
147 *Q9* **Piskom** *Rus.* Pskem. ≈ E Uzbekistan
Piskom Tizmasi *see* Pskemskiy Khrebet
35 *P13* **Pismo Beach** California, W USA *35°08′N 120°38′W*
77 *P12* **Pissila** C Burkina Faso *13°07′N 00°51′W*
62 *H8* **Pissis, Monte** ▲ N Argentina *27°45′S 68°43′W*
41 *X12* **Piste** Yucatán, E Mexico *20°40′N 88°34′W*
107 *O18* **Pisticci** Basilicata, S Italy *40°23′N 16°33′E*
106 *F11* **Pistoia** *anc.* Pistoria, Pistoriæ. Toscana, C Italy *43°57′N 10°53′E*
32 *E15* **Pistol River** Oregon, NW USA *42°13′N 124°23′W*
Pistoria/Pistoriæ *see* Pistoia
15 *U5* **Pistuacanis** ≈ Québec, SE Canada
Pistyan *see* Piešť any
104 *M5* **Pisuerga** ≈ N Spain
110 *N8* **Pisz** *Ger.* Johannisburg. Warmińsko-Mazurskie, NE Poland *53°37′N 21°49′E*
76 *I13* **Pita** NW Guinea *11°05′N 12°15′W*
54 *D12* **Pitalito** Huila, S Colombia *01°51′N 76°01′W*
60 *I11* **Pitanga** Paraná, S Brazil *24°45′S 51°43′W*
182 *M9* **Pitarpunga Lake** *salt lake* New South Wales, SE Australia
Pitcairn Group of Islands *see* Pitcairn, Henderson, Ducie and Oeno Islands
193 *P10* **Pitcairn, Henderson, Ducie and Oeno Islands** *var.* Pitcairn Group of Islands. ◈ *UK overseas territory* C Pacific Ocean
191 *O14* **Pitcairn Island** *island* S Pitcairn Group of Islands
93 *J14* **Piteå** Norrbotten, N Sweden *65°19′N 21°30′E*
92 *I13* **Piteälven** ≈ N Sweden
116 *I13* **Piteşti** Argeş, S Romania *44°53′N 24°49′E*
Pithagorio *see* Pythagóreio
180 *I12* **Pithara** Western Australia *30°31′S 116°38′E*
103 *N6* **Pithiviers** Loiret, C France *48°10′N 02°15′E*
152 *L9* **Pithorāgarh** Uttarakhand, N India *29°35′N 80°12′E*
188 *B16* **Piti** W Guam *13°28′N 144°42′E*
106 *G13* **Pitigliano** Toscana, C Italy *42°38′N 11°40′E*
40 *F3* **Pitiquito** Sonora, NW Mexico *30°39′N 112°00′W*
Pitkäranta *see* Pitkyaranta
38 *M11* **Pitkas Point** Alaska, USA *62°01′N 163°17′W*
124 *H11* **Pitkyaranta** *Fin.* Pitkäranta. Respublika Kareliya, NW Russian Federation *61°34′N 31°27′E*
96 *J10* **Pitlochry** C Scotland, United Kingdom *56°47′N 03°48′W*
18 *I16* **Pitman** New Jersey, NE USA *39°43′N 75°06′W*
146 *I9* **Pitnak** *var.* Drujba, *Rus.* Druzhba. Xorazm Viloyati, W Uzbekistan *41°14′N 61°13′E*
112 *G8* **Pitomača** Virovitica-Podravina, NE Croatia *45°57′N 17°14′E*
35 *O2* **Pit River** ≈ California, W USA
63 *G15* **Pitrufquén** Araucanía, S Chile *38°59′S 72°40′W*
Pitsanulok *see* Phitsanulok
Pitschen *see* Byczyna
Pitsunda *see* Bich'vinta
109 *X6* **Pitten** ≈ E Austria
10 *J14* **Pitt Island** *island* British Columbia, W Canada
Pitt Island *see* Makin
22 *M3* **Pittsboro** Mississippi, S USA *33°55′N 89°20′W*
21 *T9* **Pittsboro** North Carolina, SE USA *35°46′N 79°12′W*
27 *R7* **Pittsburg** Kansas, C USA *37°24′N 94°42′W*
25 *W6* **Pittsburg** Texas, SW USA *33°00′N 94°58′W*
18 *B14* **Pittsburgh** Pennsylvania, NE USA *40°26′N 80°W*
30 *J14* **Pittsfield** Illinois, N USA *39°36′N 90°48′W*
19 *R6* **Pittsfield** Maine, NE USA *44°46′N 69°22′W*
18 *L11* **Pittsfield** Massachusetts, NE USA *42°27′N 73°15′W*
183 *U3* **Pittsworth** Queensland, E Australia *27°43′S 151°36′E*
62 *I8* **Pituil** La Rioja, NW Argentina *28°33′S 67°24′W*
56 *A10* **Piura** Piura, NW Peru *05°11′S 80°41′W*
56 *A9* **Piura** *off.* Departamento de Piura. ◈ *department* NW Peru
Piura, Departamento de *see* Piura
35 *S13* **Piute Peak** ▲ California, W USA *35°27′N 118°24′W*
113 *J15* **Piva** ≈ NW Montenegro
117 *V5* **Pivdenne** Kharkivs'ka Oblast', E Ukraine *49°52′N 36°04′E*
117 *P8* **Pivdennyy Buh** *Rus.* Yuzhnyy Bug. ≈ S Ukraine
54 *F5* **Pivijay** Magdalena, N Colombia *10°31′N 74°36′W*
109 *T13* **Pivka** *prev.* Šent Peter, *Ger.* Sankt Peter, *It.* San Pietro del Carso. SW Slovenia *45°41′N 14°12′E*
117 *U13* **Pivnichno-Kryms'kyy Kanal** *canal* S Ukraine
113 *J15* **Pivsko Jezero** ◎ NW Montenegro
111 *M18* **Piwniczna** Małopolskie, S Poland *49°26′N 20°43′E*
35 *R12* **Pixley** California, W USA *35°58′N 119°18′W*
125 *Q15* **Pizhma** *var.* Pishma. ≈ NW Russian Federation
13 *U13* **Placentia** Newfoundland, Newfoundland and Labrador, SE Canada *47°12′N 53°58′W*
Placentia *see* Piacenza
13 *U13* **Placentia Bay** *inlet* Newfoundland, Newfoundland and Labrador, SE Canada
171 *P5* **Placer** Masbate, N Philippines *11°54′N 123°54′E*
35 *P7* **Placerville** California, W USA *38°42′N 120°48′W*
44 *F5* **Placetas** Villa Clara, C Cuba *22°18′N 79°40′W*
113 *Q18* **Plačkovica** ▲▲ E Macedonia
36 *L2* **Plain City** Utah, W USA *41°18′N 112°05′W*
22 *G4* **Plain Dealing** Louisiana, S USA *32°54′N 93°42′W*
31 *O14* **Plainfield** Indiana, N USA *39°42′N 86°18′W*
18 *K14* **Plainfield** New Jersey, NE USA *40°37′N 74°25′W*
33 *O8* **Plains** Montana, NW USA *47°27′N 114°52′W*
24 *L6* **Plains** Texas, SW USA *33°12′N 102°50′W*
29 *X10* **Plainview** Minnesota, N USA *44°10′N 92°10′W*
29 *Q13* **Plainview** Nebraska, C USA *42°21′N 97°47′W*
25 *N4* **Plainview** Texas, SW USA *34°12′N 101°43′W*
26 *K4* **Plainville** Kansas, C USA *39°13′N 99°18′W*
115 *I22* **Pláka** *var.* Mílos. Mílos, Kykládes, Greece, Aegean Sea *36°44′N 24°25′E*
115 *J15* **Pláka, Akrotírio** *headland* Límnos, E Greece *40°02′N 25°25′E*
113 *N19* **Plakenska Planina** ▲▲ SW Macedonia
44 *K5* **Plana Cays** *islets* SE The Bahamas
105 *S12* **Plana, Isla** *var.* Nueva Tabarca. *island* E Spain
59 *L18* **Planaltina** Goiás, S Brazil *15°35′S 47°28′W*
83 *O14* **Planalto Moçambicano** *plateau* N Mozambique
112 *N10* **Plandište** Vojvodina, NE Serbia *45°13′N 21°07′E*
100 *N13* **Plane** ≈ NE Germany
54 *E6* **Planeta Rica** Córdoba, NW Colombia *08°24′N 75°39′W*
29 *P11* **Plankinton** South Dakota, N USA *43°43′N 98°28′W*
30 *M11* **Plano** Illinois, N USA *41°39′N 88°32′W*
25 *U6* **Plano** Texas, SW USA *33°01′N 96°42′W*
23 *W12* **Plant City** Florida, SE USA *28°01′N 82°06′W*
22 *J9* **Plaquemine** Louisiana, S USA *30°17′N 91°13′W*
104 *K9* **Plasencia** Extremadura, W Spain *40°02′N 06°05′W*
110 *P7* **Płaska** Podlaskie, NE Poland *53°55′N 23°18′E*
112 *C10* **Plaški** Karlovac, C Croatia *45°04′N 15°21′E*
113 *N19* **Plasnica** SW FYR Macedonia *41°28′N 21°07′E*
13 *N14* **Plaster Rock** New Brunswick, SE Canada *46°55′N 67°24′W*
107 *J24* **Platani** *anc.* Halycus. ≈ Sicilia, Italy, C Mediterranean Sea
115 *G17* **Plataniá** Thessalía, C Greece *39°09′N 23°15′E*
115 *G24* **Plátanos** Kríti, Greece, E Mediterranean Sea *35°27′N 23°34′E*
65 *H18* **Plata, Río de la** *var.* River Plate. *estuary* Argentina/Uruguay
77 *V15* **Plateau** ◈ *state* C Nigeria
79 *G19* **Plateaux** *var.* Région des Plateaux. ◈ *province* C Congo
Plateaux, Région des *see* Plateaux
Plate, Île *see* Flat Island
92 *P1* **Platen, Kapp** *headland* NE Svalbard *80°30′N 22°46′E*
Plate, River *see* Plata, Río de la
99 *G22* **Plate Taille, Lac de la** *var.* L'Eau d'Heure. ◎ SE Belgium
Plathe *see* Płoty
39 *N13* **Platinum** Alaska, USA *59°01′N 161°49′W*
54 *F5* **Plato** Magdalena, N Colombia *09°47′N 74°47′W*
29 *O11* **Platte** South Dakota, N USA *43°20′N 98°51′W*
27 *R3* **Platte City** Missouri, C USA *39°22′N 94°47′W*
Plattensee *see* Balaton
27 *R3* **Platte River** ≈ Iowa/Missouri, USA
29 *Q15* **Platte River** ≈ Nebraska, C USA
37 *T3* **Platteville** Colorado, C USA *40°13′N 104°49′W*
30 *K9* **Platteville** Wisconsin, N USA *42°44′N 90°27′W*
101 *N21* **Plattling** Bayern, SE Germany *48°45′N 12°52′E*
27 *R3* **Plattsburg** Missouri, C USA *39°33′N 94°27′W*
18 *L6* **Plattsburgh** New York, NE USA *44°42′N 73°29′W*
29 *S15* **Plattsmouth** Nebraska, C USA *41°00′N 95°52′W*
101 *M17* **Plauen** *var.* Plauen im Vogtland. Sachsen, E Germany *50°31′N 12°08′E*
Plauen im Vogtland *see* Plauen
100 *M10* **Plauer See** ◎ NE Germany
113 *L16* **Plav** E Montenegro *42°36′N 19°57′E*
118 *I10* **Pļaviņas** *Ger.* Stockmannshof. S Latvia *56°37′N 25°40′E*
126 *K5* **Plavsk** Tul'skaya Oblast', W Russian Federation *53°42′N 37°21′E*
41 *Z12* **Playa del Carmen** Quintana Roo, E Mexico *20°37′N 87°04′W*
40 *J12* **Playa Los Corchos** Nayarit, SW Mexico *21°91′N 105°28′W*
37 *P16* **Playas Lake** ◎ New Mexico, SW USA
41 *S15* **Playa Vicenté** Veracruz-Llave, SE Mexico *17°42′N 95°01′W*
Plây Cu *see* Plei Ku
28 *L3* **Plaza** North Dakota, N USA *48°00′N 102°00′W*
63 *I15* **Plaza Huincul** Neuquén, C Argentina *38°55′S 69°14′W*
36 *L3* **Pleasant Grove** Utah, W USA *40°21′N 111°44′W*
29 *V14* **Pleasant Hill** Iowa, C USA *41°34′N 93°31′W*
27 *R4* **Pleasant Hill** Missouri, C USA *38°47′N 94°16′W*
Pleasant Island *see* Nauru
36 *K13* **Pleasant, Lake** ⊠ Arizona, SW USA
19 *P8* **Pleasant Mountain** ▲ Maine, NE USA *44°01′N 70°47′W*
27 *R5* **Pleasanton** Kansas, C USA *38°09′N 94°43′W*
25 *R12* **Pleasanton** Texas, SW USA *28°58′N 98°28′W*
185 *G20* **Pleasant Point** Canterbury, South Island, New Zealand *44°16′S 171°09′E*
19 *R5* **Pleasant River** ≈ Maine, NE USA
18 *J17* **Pleasantville** New Jersey, NE USA *39°22′N 74°31′W*
103 *N12* **Pléaux** Cantal, C France *45°08′N 02°10′E*
111 *B19* **Plechý** *var.* Plöckenstein. ▲ Austria/Czech Republic *48°45′N 13°50′E*
Pleebo *see* Plibo
Pleihari *see* Pelaihari
167 *U11* **Plei Ku** *prev.* Plây Cu. Gia Lai, C Vietnam *13°57′N 108°01′E*
101 *M16* **Pleisse** ≈ E Germany
Plencia *see* Plentzia
184 *O7* **Plenty, Bay of** *bay* North Island, New Zealand
33 *Y6* **Plentywood** Montana, NW USA *48°46′N 104°33′W*
105 *O2* **Plentzia** *var.* Plencia. País Vasco, N Spain *43°25′N 02°56′W*
102 *H5* **Plérin** Côtes d'Armor, NW France *48°33′N 02°46′W*
124 *M10* **Plesetsk** Arkhangel'skaya Oblast', NW Russian Federation *62°41′N 40°14′E*
Pleshchenitsy *see* Plyeshchanitsy
Pleskau *see* Pskov
Pleskauer See *see* Pskov, Lake
Pleskava *see* Pskov
112 *E8* **Pleso International** ✈ (Zagreb) Zagreb, NW Croatia *45°45′N 16°00′E*
Pless *see* Pszczyna
15 *Q11* **Plessisville** Québec, SE Canada *46°14′N 71°46′W*
110 *H12* **Pleszew** Wielkopolskie, C Poland *51°54′N 17°47′E*
12 *L10* **Plétipi, Lac** ◎ Québec, SE Canada
101 *F15* **Plettenberg** Nordrhein-Westfalen, W Germany *51°13′N 07°52′E*
114 *I8* **Pleven** *prev.* Plevna. Pleven, N Bulgaria *43°25′N 24°36′E*
114 *I8* **Pleven** ◈ *province* N Bulgaria
Plevlja/Plevlje *see* Pljevlja
Plevna *see* Pleven
Plezzo *see* Bovec
Pliberk *see* Bleiburg
76 *L17* **Plibo** *var.* Pleebo. SE Liberia *04°38′N 07°41′W*
121 *R11* **Pliny Trench** *undersea feature* C Mediterranean Sea
118 *K13* **Plisa** *Rus.* Plissa. Vitsyebskaya Voblasts', N Belarus *55°13′N 27°57′E*
Plissa *see* Plisa
112 *D11* **Plitvica Selo** Lika-Senj, W Croatia *44°53′N 15°36′E*
112 *D11* **Plješevica** ▲▲ C Croatia
113 *K14* **Pljevlja** *prev.* Plevlja, Plevlje. N Montenegro *43°21′N 19°21′E*
Ploça *see* Ploçë
Plocce *see* Ploče
113 *K22* **Ploçë** *var.* Ploça. Vlorë, SW Albania *40°24′N 19°41′E*
113 *G15* **Ploče** *It.* Plocce; *prev.* Kardeljevo. Dubrovnik-Neretva, SE Croatia *43°02′N 17°25′E*
110 *K11* **Płock** *Ger.* Plozk. Mazowieckie, C Poland *52°32′N 19°40′E*
109 *Q10* **Plöcken Pass** *Ger.* Plöckenpass, *It.* Passo di Monte Croce Carnico. *pass* SW Austria
Plöckenpass *see* Plöcken Pass
Plöckenstein *see* Plechý
99 *B19* **Ploegsteert** Hainaut, W Belgium *50°45′N 02°52′E*
102 *H6* **Ploërmel** Morbihan, NW France *47°57′N 02°24′W*
Ploeşti *see* Ploieşti
116 *K13* **Ploieşti** *prev.* Ploeşti. Prahova, SE Romania *44°56′N 26°03′E*
115 *L17* **Plomári** *prev.* Plomárion. Lésvos, E Greece *38°58′N 26°24′E*
Plomárion *see* Plomári
103 *O12* **Plomb du Cantal** ▲ C France *45°03′N 02°48′E*
183 *V6* **Plomer, Point** *headland* New South Wales, SE Australia *31°19′S 153°00′E*
100 *J8* **Plön** Schleswig-Holstein, N Germany *54°10′N 10°25′E*
110 *L11* **Płońsk** Mazowieckie, C Poland *52°38′N 20°23′E*
119 *J20* **Plotnitsa** Brestskaya Voblasts', SW Belarus *52°03′N 26°39′E*
110 *E8* **Płoty** *Ger.* Plathe. Zachodnio-pomorskie, NW Poland *53°48′N 15°16′E*
102 *G7* **Plouay** Morbihan, NW France *47°54′N 03°14′W*
111 *D15* **Ploučnice** *Ger.* Polzen. ≈ NE Czech Republic
114 *I10* **Plovdiv** *prev.* Eumolpias; *anc.* Evmolpia, Philippopolis, *Lat.* Trimontium. Plovdiv, C Bulgaria *42°09′N 24°47′E*
114 *I11* **Plovdiv** ◈ *province* C Bulgaria
30 *L6* **Plover** Wisconsin, N USA *44°30′N 89°33′W*
Plozk *see* Płock
27 *U11* **Plumerville** Arkansas, C USA *35°09′N 92°38′W*
19 *P10* **Plum Island** *island* Massachusetts, NE USA
32 *M9* **Plummer** Idaho, NW USA *47°19′N 116°54′W*
83 *J18* **Plumtree** Matabeleland South, SW Zimbabwe *20°30′S 27°50′E*
118 *D11* **Plungė** Telšiai, W Lithuania *55°55′N 21°53′E*
113 *J15* **Plužine** NW Montenegro *43°08′N 18°49′E*
119 *K14* **Plyeshchanitsy** *Rus.* Pleshchenitsy. Minskaya Voblasts', N Belarus *54°25′N 27°50′E*
97 *I24* **Plymouth** SW England, United Kingdom *50°23′N 04°10′W*
31 *O11* **Plymouth** Indiana, N USA *41°20′N 86°19′W*
19 *P12* **Plymouth** Massachusetts, NE USA *41°57′N 70°40′W*
19 *N8* **Plymouth** New Hampshire, NE USA *43°43′N 71°39′W*
21 *X9* **Plymouth** North Carolina, SE USA *35°53′N 76°46′W*
30 *M8* **Plymouth** Wisconsin, N USA *43°48′N 87°58′W*
Plymouth *see* Brades
97 *J20* **Plynlimon** ▲ C Wales, United Kingdom *52°27′N 03°48′W*
124 *G14* **Plyussa** Pskovskaya Oblast', W Russian Federation *58°27′N 29°21′E*
111 *B17* **Plzeň** *Ger.* Pilsen, *Pol.* Pilzno. Plzeňský Kraj, W Czech Republic *49°45′N 13°23′E*
111 *B17* **Plzeňský Kraj** ◈ *region* W Czech Republic
110 *F11* **Pniewy** *Ger.* Pinne. Wielkolpolskie, W Poland *52°31′N 16°14′E*
77 *P13* **Pô** S Burkina Faso *11°11′N 01°10′W*
106 *D8* **Po** ≈ N Italy
42 *M13* **Poás, Volcán** ⛰ NW Costa Rica *10°12′N 84°12′W*
77 *S16* **Pobè** S Benin *07°00′N 02°41′E*
123 *S8* **Pobeda, Gora** ▲ NE Russian Federation *65°28′N 145°44′E*
Pobeda Peak *see* Pobedy, Pik/Tomur Feng
147 *Z7* **Pobedy, Pik** *Chin.* Tomür Feng. ▲ China/Kyrgyzstan *42°02′N 80°02′E* *see also* Tomür Feng
Pobedy, Pik *see* Tomür Feng
110 *H11* **Pobiedziska** *Ger.* Pudewitz. Wielkolpolskie, C Poland *52°30′N 17°19′E*
Po, Bocche del *see* Po, Foci del
27 *W9* **Pocahontas** Arkansas, C USA *36°15′N 91°00′W*
29 *U12* **Pocahontas** Iowa, C USA *42°44′N 94°40′W*
33 *Q15* **Pocatello** Idaho, NW USA *42°52′N 112°27′W*
167 *S13* **Pochentong** ✈ (Phnum Penh) Phnum Penh, S Cambodia *11°24′N 104°52′E*
126 *I6* **Pochep** Bryanskaya Oblast', W Russian Federation *52°56′N 33°20′E*
126 *H4* **Pochinok** Smolenskaya Oblast', W Russian Federation *54°21′N 32°29′E*
41 *R17* **Pochutla** *var.* San Pedro Pochutla. Oaxaca, SE Mexico *15°45′N 96°30′W*
62 *I6* **Pocitos, Salar** *var.* Salar Quiróm. *salt lake* NW Argentina
101 *O22* **Pocking** Bayern, SE Germany *48°22′N 13°17′E*
186 *I10* **Pocklington Reef** *reef* SE Papua New Guinea
59 *P15* **Poço da Cruz, Açude** ⊠ E Brazil
27 *R11* **Pocola** Oklahoma, C USA *35°13′N 94°28′W*
21 *Y5* **Pocomoke City** Maryland, NE USA *38°04′N 75°34′W*
59 *L21* **Poços de Caldas** Minas Gerais, NE Brazil *21°48′S 46°33′W*
124 *H14* **Podberez'ye** Novgorodskaya Oblast', NW Russian Federation *58°42′N 31°22′E*
Podbrodzie *see* Pabradė
125 *U8* **Podcher'ye** Respublika Komi, NW Russian Federation *63°55′N 57°34′E*
111 *E16* **Poděbrady** *Ger.* Podiebrad. Středočeský Kraj, C Czech Republic *50°10′N 15°06′E*
126 *L9* **Podgorenskiy** Voronezhskaya Oblast', W Russian Federation *50°22′N 39°43′E*
113 *J17* **Podgorica** *prev.* Titograd. ● S Montenegro *42°25′N 19°16′E*
113 *K17* **Podgorica** ✈ S Montenegro *42°22′N 19°16′E*
109 *T13* **Podgrad** SW Slovenia *45°31′N 14°09′E*
Podiebrad *see* Poděbrady
116 *M5* **Podil's'ka Vysochina** *plateau* W Ukraine
Podium Anicensis *see* le Puy
122 *L11* **Podkamennaya Tunguska** *Eng.* Stony Tunguska. ≈ C Russian Federation
111 *N17* **Podkarpackie** ◈ *province* SW Poland
Pod Kloŝter *see* Arnoldstein
110 *P9* **Podlaskie** ◈ *province* Mazowieckie, C Poland
127 *Q8* **Podlesnoye** Saratovskaya Oblast', W Russian Federation *51°51′N 47°03′E*
126 *K4* **Podol'sk** Moskovskaya Oblast', W Russian Federation *55°24′N 37°30′E*
76 *H10* **Podor** N Senegal *16°40′N 14°57′W*
125 *P12* **Podosinovets** Kirovskaya Oblast', NW Russian Federation *60°15′N 47°06′E*
124 *I12* **Podporozh'ye** Leningradskaya Oblast', NW Russian Federation *60°52′N 34°00′E*
Podravska Slatina *see* Slatina
112 *J13* **Podromanlja** Republika Srpska, SE Bosnia and Herzegovina *43°55′N 18°46′E*
Podsvil'ye *see* Padsvillye
116 *L9* **Podu Iloaiei** *prev.* Podul Iloaiei. Iaşi, NE Romania *47°13′N 27°16′E*
113 *N15* **Podujevë** *Serb.* Podujevo. N Kosovo *42°56′N 21°13′E*
Podujevo *see* Podujevë
Podul Iloaiei *see* Podu Iloaiei
Podunajská Rovina *see* Little Alföld
124 *M12* **Podyuga** Arkhangel'skaya Oblast', NW Russian Federation *61°04′N 40°46′E*
56 *A9* **Poechos, Embalse** ◎ NW Peru
55 *W10* **Poeketi** Sipaliwini, E Suriname
100 *L8* **Poel** *island* N Germany
83 *M20* **Poelela, Lagoa** ◎ S Mozambique
Poerwodadi *see* Purwodadi
Poerwokerto *see* Purwokerto
Poerworedjo *see* Purworejo
Poetovio *see* Ptuj
83 *E23* **Pofadder** Northern Cape, W South Africa *29°09′S 19°25′E*
106 *I9* **Po, Foci del** *var.* Bocche del Po. ≈ NE Italy
116 *E12* **Pogăniş** ≈ W Romania
Pogegen *see* Pagėgiai
106 *G12* **Poggibonsi** Toscana, C Italy *43°28′N 11°09′E*
107 *I14* **Poggio Mirteto** Lazio, C Italy *42°17′N 12°42′E*
109 *V4* **Pöggstall** Niederösterreich, N Austria *48°19′N 15°10′E*
116 *L13* **Pogoanele** Buzău, SE Romania *44°55′N 27°00′E*
Pogónion *see* Delvináki
113 *M21* **Pogradec** *var.* Pogradeci. Korçë, SE Albania *40°54′N 20°40′E*
Pogradeci *see* Pogradec
123 *S15* **Pogranichnyy** Primorskiy Kray, SE Russian Federation *44°18′N 131°33′E*
38 *M16* **Pogromni Volcano** ▲ Unimak Island, Alaska, USA *54°34′N 164°41′W*
163 *Z15* **Pohang** *Jap.* Hokō; *prev.* P'ohang. E South Korea *36°02′N 129°20′E*
P'ohang *see* Pohang
15 *T9* **Pohénégamook, Lac** ◎ Québec, SE Canada
93 *L20* **Pohja** *Swe.* Pojo. Uusimaa, SW Finland *60°07′N 23°30′E*
Pohjanlahti *see* Bothnia, Gulf of
Pohjanmaa *see* Österbotten
93 *O16* **Pohjois-Karjala** *Swe.* Norra Karelen, *Eng.* North Karelia. ◈ *region* E Finland
93 *L14* **Pohjois-Pohjanmaa** *Swe.* Norra Österbotten, *Eng.* North Ostrobothnia. ◈ *region* N Finland
93 *M17* **Pohjois-Savo** *Swe.* Norra Savolax. ◈ *region* C Finland
189 *U16* **Pohnpei** ◈ *state* E Micronesia
189 *O12* **Pohnpei** ✈ Pohnpei, E Micronesia
189 *O12* **Pohnpei** *prev.* Ponape Ascension Island. *island* E Micronesia
111 *F19* **Pohořelice** *Ger.* Pohrlitz. Jihomoravský Kraj, SE Czech Republic *48°58′N 16°30′E*
109 *V10* **Pohorje** *Ger.* Bacher. ▲▲ N Slovenia
117 *N6* **Pohrebyshche** Vinnyts'ka Oblast', C Ukraine *49°31′N 29°16′E*
Pohrlitz *see* Pohořelice
161 *P9* **Po Hu** ⊠ E China
116 *G15* **Poiana Mare** Dolj, S Romania *43°55′N 23°02′E*
Poictiers *see* Poitiers
127 *N6* **Poim** Penzenskaya Oblast', W Russian Federation *53°03′N 43°11′E*
159 *N15* **Poindo** Xigzang Zizhiqu, W China *29°58′N 91°20′E*
195 *Y13* **Poinsett, Cape** *headland* Antarctica *65°35′S 113°00′E*
29 *R9* **Poinsett, Lake** ◎ South Dakota, N USA
22 *I10* **Point Au Fer Island** *island* Louisiana, S USA
39 *X14* **Point Baker** Prince of Wales Island, Alaska, USA *56°19′N 133°31′W*
25 *U13* **Point Comfort** Texas, SW USA *28°40′N 96°33′W*
Point de Galle *see* Galle
44 *K10* **Pointe à Gravois** *headland* SW Haiti *18°00′N 73°53′W*
22 *L10* **Pointe a la Hache** Louisiana, S USA *29°34′N 89°48′W*
45 *Y6* **Pointe-à-Pitre** Grande Terre, C Guadeloupe *16°14′N 61°32′W*
15 *U7* **Pointe-au-Père** Québec, SE Canada *48°31′N 68°27′W*
15 *V5* **Pointe-aux-Anglais** Québec, SE Canada *49°40′N 67°09′W*
45 *T10* **Pointe Du Cap** *headland* N Saint Lucia *14°06′N 60°56′W*
79 *E21* **Pointe-Noire** Kouilou, S Congo *04°46′S 11°53′E*
45 *X6* **Pointe Noire** Basse Terre, W Guadeloupe *16°14′N 61°47′W*
79 *E21* **Pointe-Noire** ✈ Kouilou, S Congo *04°45′S 11°55′E*
45 *U15* **Point Fortin** Trinidad, Trinidad and Tobago *10°12′N 61°41′W*
38 *M6* **Point Hope** Alaska, USA *68°21′N 166°48′W*
39 *N5* **Point Lay** Alaska, USA *69°42′N 162°57′W*
18 *B16* **Point Marion** Pennsylvania, NE USA *39°44′N 79°53′W*
18 *K16* **Point Pleasant** New Jersey, NE USA *40°04′N 74°00′W*
21 *P4* **Point Pleasant** West Virginia, NE USA *38°53′N 82°07′W*
45 *R14* **Point Salines** ✈ (St. George's) SW Grenada *12°00′N 61°47′W*
102 *L9* **Poitiers** *prev.* Poictiers; *anc.* Limonum. Vienne, W France *46°35′N 00°19′E*
102 *K9* **Poitou** *cultural region* W France
102 *K10* **Poitou-Charentes** ◈ *region* W France
103 *N3* **Poix-de-Picardie** Somme, N France *49°47′N 01°58′E*
Pojo *see* Pohja
37 *S10* **Pojoaque** New Mexico, SW USA *35°52′N 106°01′W*

◆ Country
● Country Capital
◇ Dependent Territory
○ Dependent Territory Capital
◈ Administrative Regions
✈ International Airport
▲ Mountain
▲▲ Mountain Range
⛰ Volcano
≈ River
◎ Lake
⊠ Reservoir

POKARAN – PORTO-VECCHIO

152 E11 **Pokaran** Rājasthān, NW India 26°55′N 71°55′E
183 R4 **Pokataroo** New South Wales, SE Australia 29°37′S 148°43′E
119 P18 **Pokats'** *Rus.* Pokot'. ≈ SE Belarus
29 V5 **Pokegama Lake** ◎ Minnesota, N USA
184 L6 **Pokeno** Waikato, North Island, New Zealand 37°15′S 175°01′E
153 O11 **Pokharā** Western, C Nepal 28°14′N 84°E
127 T6 **Pokhvistnevo** Samarskaya Oblast', W Russian Federation 53°38′N 52°07′E
55 W10 **Pokigron** Sipaliwini, C Suriname 04°31′N 55°23′W
92 L10 **Pokka** *Lapp.* Bohkká. Lappi, N Finland 68°11′N 25°45′E
79 N16 **Poko** Orientale, NE Dem. Rep. Congo 03°08′N 26°52′E
Pokot' *see* Pokats'
Po-ko-to Shan *see* Bogda Shan
147 S7 **Pokrovka** Talasskaya Oblast', NW Kyrgyzstan 42°45′N 71°33′E
Pokrovka *see* Kyzyl-Suu
117 V8 **Pokrovs'ke** *Rus.* Pokrovskoye. Dnipropetrovs'ka Oblast', E Ukraine 47°58′N 36°15′E
Pokrovskoye *see* Pokrovs'ke
Pola *see* Pula
37 N10 **Polacca** Arizona, SW USA 35°49′N 110°21′W
Pola de Laviana *see* Pola de Llaviana
Pola de Lena *see* La Pola
104 L2 **Pola de Llaviana** *var.* Pola de Laviana. Asturias, N Spain 43°15′N 05°33′W
Pola de Siero *see* La Pola Siero
191 Y3 **Poland** Kiritimati, E Kiribati 01°52′N 157°33′W
110 H12 **Poland** *off.* Republic of Poland, *var.* Polish Republic, *Pol.* Polska, Rzeczpospolita Polska; *prev. Pol.* Polska Rzeczpospolita Ludowa, The Polish People's Republic. ◆ *republic* C Europe
Poland, Republic of *see* Poland
Polangen *see* Palanga
110 G7 **Polanów** *Ger.* Pollnow. Zachodnio-pomorskie, NW Poland 54°07′N 16°38′E
136 H13 **Polatlı** Ankara, C Turkey 39°34′N 32°08′E
118 L12 **Polatsk** *Rus.* Polotsk. Vitsyebskaya Voblasts', N Belarus 55°29′N 28°47′E
110 F8 **Połczyn-Zdrój** *Ger.* Bad Polzin. Zachodnio-pomorskie, NW Poland 53°44′N 16°02′E
Pol-e-'Alam *see* Pul-e-'Alam
Polekhatum *see* Pulhatyn
Pol-e Khomrī *see* Pul-e Khumrī
197 S10 **Pole Plain** *undersea feature* Arctic Ocean
Pol-e-Safid *see* Pol-e Sefīd
143 P5 **Pol-e Sefīd** *var.* Pol-e-Safid, Pul-i-Sefid. Māzandarān, N Iran 36°05′N 53°01′E
118 B13 **Polessk** *Ger.* Labiau. Kaliningradskaya Oblast', W Russian Federation 54°52′N 21°06′E
Polesskoye *see* Polis'ke
171 N13 **Polewali** Sulawesi, C Indonesia 03°26′S 119°23′E
114 G11 **Polezhan** ▲ SW Bulgaria 41°42′N 23°28′E
78 F13 **Poli** Nord, N Cameroon 09°31′N 13°10′E
Poli *see* Pólis
107 M19 **Policastro, Golfo di** *gulf* S Italy
110 D8 **Police** *Ger.* Politz. Zachodnio-pomorskie, NW Poland 53°34′N 14°34′E
172 I17 **Police, Pointe** *headland* Mahé, NE Seychelles 04°48′S 55°31′E
115 L17 **Polichnítos** *var.* Polihnitos, Políkhnitos. Lésvos, E Greece 39°04′N 26°10′E
Poligiros *see* Polýgyros
107 P17 **Polignano a Mare** Puglia, SE Italy 40°59′N 17°13′E
103 S9 **Poligny** Jura, E France 46°51′N 05°42′E
Polihnitos *see* Polichnítos
Polikastro/Políkastron *see* Polýkastro
Políkhnitos *see* Polichnítos
171 O3 **Polillo Islands** *island group* N Philippines
109 Q9 **Polinik** ▲ SW Austria 46°54′N 13°10′E
115 J15 **Políochni** *var.* Polýochni. *site of ancient city* Límnos, E Greece
121 O2 **Pólis** *var.* Poli. W Cyprus 35°02′N 32°27′E
Polish People's Republic, The *see* Poland
Polish Republic *see* Poland
117 O3 **Polis'ke** *Rus.* Polesskoye. Kyyivs'ka Oblast', N Ukraine 51°16′N 29°27′E
107 N22 **Polistena** Calabria, SW Italy 38°25′N 16°05′E
Politz *see* Police
Políyiros *see* Polýgyros
29 V14 **Polk City** Iowa, C USA 41°46′N 93°42′W
110 F13 **Polkowice** *Ger.* Heerwegen. Dolnośląskie, W Poland 51°32′N 16°06′E
155 G22 **Pollāchi** Tamil Nādu, SE India 10°38′N 77°00′E
109 W7 **Pöllau** Steiermark, SE Austria 47°18′N 15°46′E
189 T13 **Polle** *atoll* Chuuk Islands, C Micronesia
105 X9 **Pollença** Mallorca, Spain, W Mediterranean Sea 39°52′N 03°01′E
Pollnow *see* Polanów
29 N7 **Pollock** South Dakota, N USA 45°53′N 100°15′W
92 L8 **Polmak** Finnmark, N Norway 70°01′N 28°04′E
30 L10 **Polo** Illinois, N USA 41°59′N 89°34′W
193 V15 **Poloa** *island* Tongatapu Group, N Tonga
42 E5 **Polochic, Río** ≈ C Guatemala
Pologi *see* Polohy
117 V9 **Polohy** *Rus.* Pologi. Zaporiz'ka Oblast', SE Ukraine 47°30′N 36°18′E
83 K20 **Polokwane** *prev.* Pietersburg. Limpopo, NE South Africa 23°54′S 29°23′E
14 M10 **Polonais, Lac des** ◎ Québec, SE Canada
61 G20 **Polonio, Cabo** *headland* E Uruguay 34°22′S 53°46′W
155 K24 **Polonnaruwa** North Central Province, C Sri Lanka 07°56′N 81°02′E
116 L5 **Polonne** *Rus.* Polonnoye. Khmel'nyts'ka Oblast', NW Ukraine 50°10′N 27°30′E
Polonnoye *see* Polonne
Polotsk *see* Polatsk
Polovinka *see* Ugleural'skiy
109 T7 **Pöls** *var.* Pölsbach. ≈ E Austria
Pölsbach *see* Pöls
Polska/Polska, Rzeczpospolita/Polska Rzeczpospolita Ludowa *see* Poland
114 L10 **Polski Gradets** Stara Zagora, C Bulgaria 42°12′N 26°06′E
114 K8 **Polski Trambesh** *var.* Polski Trŭmbesh. Veliko Tarnovo, N Bulgaria 43°22′N 25°38′E
Polski Trŭmbesh *see* Polski Trambesh
33 P8 **Polson** Montana, NW USA 47°41′N 114°09′W
117 T6 **Poltava** Poltavs'ka Oblast', NE Ukraine 49°33′N 34°32′E
Poltava *see* Poltavs'ka Oblast'
117 R5 **Poltavs'ka Oblast'** *var.* Poltava, *Rus.* Poltavskaya Oblast'. ◈ *province* NE Ukraine
Poltavskaya Oblast' *see* Poltavs'ka Oblast'
Poltoratsk *see* Aşgabat
118 I5 **Põltsamaa** *Ger.* Oberpahlen. Jõgevamaa, E Estonia 58°40′N 26°00′E
118 I4 **Põltsamaa** *var.* Põltsamaa Jõgi. ≈ C Estonia
Põltsamaa Jõgi *see* Põltsamaa
122 I8 **Poluy** ≈ N Russian Federation
118 J6 **Põlva** *Ger.* Pölwe. Põlvamaa, SE Estonia 58°04′N 27°06′E
93 N16 **Polvijärvi** Pohjois-Karjala, SE Finland 62°53′N 29°20′E
Pölwe *see* Põlva
115 I22 **Polýaigos** *island* Kykládes, Greece, Aegean Sea
115 I22 **Polyaígou Folégandrou, Stenó** *strait* Kykládes, Greece, Aegean Sea
124 J3 **Polyarnyy** Murmanskaya Oblast', NW Russian Federation 69°10′N 33°21′E
125 W5 **Polyarnyy Ural** ▲▲ NW Russian Federation
115 G14 **Polýgyros** *var.* Poligiros, Políyiros. Kentrikí Makedonía, N Greece 40°21′N 23°27′E
114 F13 **Polýkastro** *var.* Polikastro; *prev.* Políkastron. Kentrikí Makedonía, N Greece 41°01′N 22°33′E
193 O9 **Polynesia** *island group* C Pacific Ocean
Polýochni *see* Políochni
41 Y13 **Polyuc** Quintana Roo, E Mexico
109 V10 **Polzela** C Slovenia 46°18′N 15°04′E
Polzen *see* Ploučnice
56 D12 **Pomabamba** Ancash, C Peru 08°48′S 77°30′W
185 D23 **Pomahaka** ≈ South Island, New Zealand
106 F12 **Pomarance** Toscana, C Italy 43°19′N 10°53′E
104 G9 **Pombal** Leiria, C Portugal 39°55′N 08°38′W
76 D9 **Pombas** Santo Antão, NW Cape Verde 17°09′N 25°02′W
83 N19 **Pomene** Inhambane, SE Mozambique 22°57′S 35°34′E
110 G8 **Pomerania** *cultural region* Germany/Poland
110 D7 **Pomeranian Bay** *Ger.* Pommersche Bucht, *Pol.* Zatoka Pomorska. *bay* Germany/Poland
31 T15 **Pomeroy** Ohio, N USA 39°01′N 82°01′W
32 L10 **Pomeroy** Washington, NW USA 46°28′N 117°36′W
117 Q8 **Pomichna** Kirovohrads'ka Oblast', C Ukraine 48°07′N 31°25′E
186 H7 **Pomio** New Britain, E Papua New Guinea 05°31′S 151°30′E
Pomir, Dar"yoi *see* Pamir/Pāmīr, Daryā-ye
27 T6 **Pomme de Terre Lake** ⊠ Missouri, C USA
29 S8 **Pomme de Terre River** ≈ Minnesota, N USA
Pommersche Bucht *see* Pomeranian Bay
35 T15 **Pomona** California, W USA 34°03′N 117°45′W
114 N9 **Pomorie** Burgas, E Bulgaria 42°32′N 27°39′E
Pomorska, Zatoka *see* Pomeranian Bay
110 H8 **Pomorskie** ◈ *province* N Poland
125 Q4 **Pomorskiy Proliv** *strait* NW Russian Federation
125 T10 **Pomozdino** Respublika Komi, NW Russian Federation 62°11′N 54°13′E
Pompaelo *see* Pamplona
23 Z15 **Pompano Beach** Florida, SE USA 26°14′N 80°06′W
107 K18 **Pompei** Campania, S Italy 40°45′N 14°27′E
33 V10 **Pompeys Pillar** Montana, NW USA 45°58′N 107°55′W
Ponape Ascension Island *see* Pohnpei
29 R13 **Ponca** Nebraska, C USA 42°35′N 96°42′W
27 O8 **Ponca City** Oklahoma, C USA 36°41′N 97°04′W
45 T6 **Ponce** C Puerto Rico 18°01′N 66°36′W
23 X10 **Ponce de Leon Inlet** *inlet* Florida, SE USA
22 K8 **Ponchatoula** Louisiana, S USA 30°26′N 90°26′W
26 M8 **Pond Creek** Oklahoma, C USA 36°40′N 97°48′W
155 J20 **Pondicherry** *var.* Puduchcheri, *Fr.* Pondichéry. Puducherry, SE India 11°59′N 79°50′E
Pondicherry *see* Puducherry
Pondichéry *see* Puducherry
197 N11 **Pond Inlet** *var.* Mittimatalik. Baffin Island, Nunavut, NE Canada 72°37′N 77°56′W
187 P16 **Ponérihouen** Province Nord, C New Caledonia 21°04′S 165°24′E
104 J4 **Ponferrada** Castilla y León, NW Spain 42°33′N 06°35′W
184 N13 **Pongaroa** Manawatu-Wanganui, North Island, New Zealand 40°36′S 176°08′E
167 Q12 **Pong Nam Ron** Chantaburi, S Thailand 12°55′N 102°15′E
81 C14 **Pongo** ≈ W South Sudan
152 I7 **Pong Reservoir** ⊠ N India
111 N14 **Poniatowa** Lubelskie, E Poland 51°11′N 22°05′E
167 R12 **Pônley** Kâmpóng Chhnăng, C Cambodia 12°26′N 104°25′E
155 I20 **Ponnaiyār** ≈ SE India
11 Q15 **Ponoka** Alberta, SW Canada 52°42′N 113°33′W
127 U6 **Ponomarevka** Orenburgskaya Oblast', W Russian Federation 53°16′N 54°10′E
169 Q17 **Ponorogo** Jawa, C Indonesia 07°51′S 111°30′E
122 F6 **Ponoy** ≈ NW Russian Federation
102 K11 **Pons** Charente-Maritime, W France 45°31′N 00°31′W
Pons *see* Ponts
Pons Aelii *see* Newcastle upon Tyne
Pons Vetus *see* Pontevedra
99 G20 **Pont-à-Celles** Hainaut, S Belgium 50°31′N 04°21′E
102 K16 **Pontacq** Pyrénées-Atlantiques, SW France 43°11′N 00°06′W
64 P3 **Ponta Delgada** São Miguel, Azores, Portugal, NE Atlantic Ocean 37°29′N 25°40′W
64 P3 **Ponta Delgado** ✈ São Miguel, Azores, Portugal, NE Atlantic Ocean 37°28′N 25°40′W
64 N2 **Ponta do Pico** ▲ Pico, Azores, Portugal, NE Atlantic Ocean 38°28′N 28°25′W
60 J11 **Ponta Grossa** Paraná, S Brazil 25°07′S 50°09′W
103 S5 **Pont-à-Mousson** Meurthe-et-Moselle, NE France 48°55′N 06°03′E
103 T9 **Pontarlier** Doubs, E France 46°54′N 06°20′E
106 G11 **Pontassieve** Toscana, C Italy 43°46′N 11°28′E
102 L4 **Pont-Audemer** Eure, N France 49°22′N 00°31′E
22 K9 **Pontchartrain, Lake** ◎ Louisiana, S USA
102 I8 **Pontchâteau** Loire-Atlantique, NW France 47°26′N 02°04′W
103 R10 **Pont-de-Vaux** Ain, E France 46°25′N 04°57′E
104 G4 **Ponteareas** Galicia, NW Spain 42°11′N 08°29′W
106 J6 **Pontebba** Friuli-Venezia Giulia, NE Italy 46°32′N 13°18′E
104 G4 **Ponte Caldelas** Galicia, NW Spain 42°23′N 08°30′W
107 J16 **Pontecorvo** Lazio, C Italy 41°27′N 13°40′E
104 G5 **Ponte da Barca** Viana do Castelo, N Portugal 41°48′N 08°25′W
104 G5 **Ponte de Lima** Viana do Castelo, N Portugal 41°46′N 08°35′W
106 F11 **Pontedera** Toscana, C Italy 43°40′N 10°38′E
104 H10 **Ponte de Sor** Portalegre, C Portugal 39°15′N 08°01′W
104 H2 **Pontedeume** Galicia, NW Spain 43°22′N 08°09′W
106 F6 **Ponte di Legno** Lombardia, N Italy 46°16′N 10°31′E
11 T17 **Ponteix** Saskatchewan, S Canada 49°43′N 107°22′W
171 Q16 **Ponte Macassar** *var.* Pante Macassar, Pante Makasar, Pante Makassar. W East Timor 09°11′S 124°27′E
59 N20 **Ponte Nova** Minas Gerais, NE Brazil 20°25′S 42°54′W
59 G18 **Pontes e Lacerda** Mato Grosso, W Brazil 15°14′S 59°21′W
104 G4 **Pontevedra** *anc.* Pons Vetus. Galicia, NW Spain 42°25′N 08°39′W
104 G3 **Pontevedra** ◈ *province* Galicia, NW Spain
104 G4 **Pontevedra, Ría de** *estuary* NW Spain
30 M12 **Pontiac** Illinois, N USA 40°54′N 88°36′W
31 R9 **Pontiac** Michigan, N USA 42°38′N 83°17′W
169 P11 **Pontianak** Borneo, C Indonesia 0°05′S 109°16′E
107 I16 **Pontino, Agro** *plain* C Italy
Pontisarae *see* Pontoise
102 H6 **Pontivy** Morbihan, NW France 48°04′N 02°58′W
102 F6 **Pont-l'Abbé** Finistère, NW France 47°52′N 04°14′W
103 N4 **Pontoise** *anc.* Briva Isarae, Cergy-Pontoise, Pontisarae. Val-d'Oise, N France 49°03′N 02°05′E
11 W13 **Ponton** Manitoba, C Canada 54°36′N 99°02′W
102 J5 **Pontorson** Manche, N France 48°33′N 01°31′W
22 M2 **Pontotoc** Mississippi, S USA 34°15′N 89°00′W
25 R9 **Pontotoc** Texas, SW USA 30°52′N 98°57′W
106 E10 **Pontremoli** Toscana, C Italy 44°24′N 09°55′E
108 J10 **Pontresina** Graubünden, S Switzerland 46°29′N 09°52′E
105 U5 **Ponts** *var.* Pons. Cataluña, NE Spain 41°55′N 01°12′E
103 R14 **Pont-St-Esprit** Gard, S France 44°15′N 04°37′E
97 K21 **Pontypool** *Wel.* Pontypŵl. SE Wales, United Kingdom 51°43′N 03°02′W
97 J22 **Pontypridd** S Wales, United Kingdom 51°37′N 03°22′W
Pontypŵl *see* Pontypool
43 R17 **Ponuga** Veraguas, SE Panama 07°50′N 80°58′W
184 L6 **Ponui Island** *island* N New Zealand
119 K14 **Ponya** ≈ N Belarus
107 I17 **Ponza, Isola di** *Island* Isole Ponziane, S Italy
107 I17 **Ponziane, Isole** *island* C Italy
182 F7 **Poochera** South Australia 32°45′S 134°51′E
97 L24 **Poole** S England, United Kingdom 50°43′N 01°59′W
25 S6 **Poolville** Texas, SW USA 33°00′N 97°55′W
Poona *see* Pune
182 M8 **Pooncarie** New South Wales, SE Australia 33°26′S 142°37′E
183 N6 **Poopelloe Lake** *seasonal lake* New South Wales, SE Australia
57 K19 **Poopó** Oruro, C Bolivia 18°23′S 66°58′W
57 K19 **Poopó, Lago** *var.* Lago Pampa Aullagas. ◎ W Bolivia
184 L3 **Poor Knights Islands** *island* N New Zealand
39 P10 **Poorman** Alaska, USA 64°05′N 155°34′W
182 E3 **Pootnoura** South Australia 28°31′S 134°09′E
147 R10 **Pop** *Rus.* Pap. Namangan Viloyati, E Uzbekistan 40°49′N 71°06′E
117 X7 **Popasna** *Rus.* Popasnaya. Luhans'ka Oblast', E Ukraine 48°38′N 38°24′E
Popasnaya *see* Popasna
54 D12 **Popayán** Cauca, SW Colombia 02°27′N 76°32′W
99 B18 **Poperinge** West-Vlaanderen, W Belgium 50°52′N 02°44′E
123 N7 **Popigay** Krasnoyarskiy Kray, N Russian Federation 71°54′N 110°45′E
123 N7 **Popigay** ≈ N Russian Federation
117 O5 **Popil'nya** Zhytomyrs'ka Oblast', N Ukraine 49°57′N 29°24′E
182 K8 **Popiltah Lake** *seasonal lake* New South Wales, SE Australia
33 X7 **Poplar** Montana, NW USA 48°06′N 105°12′W
11 Y14 **Poplar** ≈ Manitoba, C Canada
27 X8 **Poplar Bluff** Missouri, C USA 36°45′N 90°23′W
33 X6 **Poplar River** ≈ Montana, NW USA
41 P14 **Popocatépetl** ⛰ S Mexico 18°59′N 98°37′W
79 H21 **Popokabaka** Bandundu, SW Dem. Rep. Congo 05°42′S 16°35′E
107 J15 **Popoli** Abruzzo, C Italy 42°09′N 13°51′E
186 F9 **Popondetta** Northern, S Papua New Guinea 08°45′S 148°15′E
112 F9 **Popovača** Sisak-Moslavina, NE Croatia 45°35′N 16°37′E
114 L8 **Popovo** Targovishte, N Bulgaria 43°20′N 26°14′E
Popovo *see* Iskra
Popper *see* Poprad
30 M5 **Popple River** ≈ Wisconsin, N USA
111 L19 **Poprad** *Ger.* Deutschendorf, *Hung.* Poprád. Prešovský Kraj, E Slovakia 49°04′N 20°16′E
111 L18 **Poprad** *Ger.* Popper, *Hung.* Poprád. ≈ Poland/Slovakia
111 L19 **Poprad-Tatry** ✈ (Poprad) Prešovský Kraj, E Slovakia 49°04′N 20°21′E
21 X7 **Poquoson** Virginia, NE USA 37°08′N 76°21′W
149 O15 **Porāli** ≈ SW Pakistan
184 N12 **Porangahau** Hawke's Bay, North Island, New Zealand 40°19′S 176°36′E
59 K17 **Porangatu** Goiás, C Brazil 13°28′S 49°14′W
119 G18 **Porazava** *Pol.* Porozow, *Rus.* Porozovo. Hrodzyenskaya Voblasts', W Belarus 52°56′N 24°22′E
154 A11 **Porbandar** Gujarāt, W India 21°40′N 69°40′E
10 I13 **Porcher Island** *island* British Columbia, SW Canada
104 M13 **Porcuna** Andalucía, S Spain 37°52′N 04°12′W
14 F7 **Porcupine** Ontario, S Canada 48°31′N 81°07′W
64 M6 **Porcupine Bank** *undersea feature* N Atlantic Ocean
11 V15 **Porcupine Hills** ▲▲ Manitoba/Saskatchewan, S Canada
30 L3 **Porcupine Mountains** *hill range* Michigan, N USA
64 M7 **Porcupine Plain** *undersea feature* E Atlantic Ocean 16°00′W 49°00′N
8 G7 **Porcupine River** ≈ Canada/USA
106 I7 **Pordenone** *anc.* Portenau. Friuli-Venezia Giulia, NE Italy 45°58′N 12°39′E
54 H9 **Pore** Casanare, E Colombia 05°42′N 71°59′W
112 A9 **Poreč** *It.* Parenzo. Istra, NW Croatia 45°16′N 13°36′E
60 I9 **Porecatu** Paraná, S Brazil 22°46′S 51°22′W
Porech'ye *see* Parechcha
127 P4 **Poretskoye** Chuvashskaya Respublika, W Russian Federation 55°12′N 46°20′E
77 Q13 **Porga** N Benin 11°04′N 00°58′E
186 B7 **Porgera** Enga, W Papua New Guinea 05°32′S 143°08′E
93 K18 **Pori** *Swe.* Björneborg. Satakunta, SW Finland 61°28′N 21°50′E
185 L14 **Porirua** Wellington, North Island, New Zealand 41°08′S 174°51′E
92 I12 **Porjus** *Lapp.* Bårjås. Norrbotten, N Sweden 66°55′N 19°55′E
124 G14 **Porkhov** Pskovskaya Oblast', W Russian Federation 57°46′N 29°27′E
55 O4 **Porlamar** Nueva Esparta, NE Venezuela 10°57′N 63°51′W
102 I8 **Pornic** Loire-Atlantique, NW France 47°07′N 02°07′W
186 B7 **Poroma** Southern Highlands, W Papua New Guinea 06°15′S 143°34′E
123 T13 **Poronaysk** Ostrov Sakhalin, Sakhalinskaya Oblast', SE Russian Federation 49°15′N 143°00′E
115 G20 **Póros** Póros, S Greece 37°30′N 23°27′E
115 C19 **Póros** Kefallinía, Iónia Nisiá, Greece, C Mediterranean Sea 38°09′N 20°46′E
115 G20 **Póros** *island* S Greece
81 G24 **Poroto Mountains** ▲▲ SW Tanzania
112 B10 **Porozina** Primorje-Gorski Kotar, NW Croatia 45°07′N 14°17′E
Porozovo/Porozow *see* Porazava
195 X15 **Porpoise Bay** *bay* Antarctica
65 G15 **Porpoise Point** *headland* NE Ascension Island 07°54′S 14°22′W
65 C25 **Porpoise Point** *headland* East Falkland, Falkland Islands 52°20′S 59°18′W
108 C6 **Porrentruy** Jura, NW Switzerland 47°25′N 07°06′E
106 F10 **Porretta Terme** Emilia-Romagna, C Italy 44°10′N 11°01′E
Porriño *see* O Porriño
92 L7 **Porsangenfjorden** *Lapp.* Porsáŋgguvuotna. *fjord* N Norway
92 K8 **Porsangerhalvøya** *peninsula* N Norway
Porsáŋgguvuotna *see* Porsangenfjorden
95 G16 **Porsgrunn** Telemark, S Norway 59°08′N 09°38′E
136 E13 **Porsuk Çayı** ≈ C Turkey
Porsy *see* Boldumsaz
57 N18 **Portachuelo** Santa Cruz, C Bolivia 17°21′S 63°24′W
182 I9 **Port Adelaide** South Australia 34°49′S 138°31′E
97 F15 **Portadown** *Ir.* Port An Dúnáin. S Northern Ireland, United Kingdom 54°26′N 06°27′W
31 P10 **Portage** Michigan, N USA 42°12′N 85°34′W
18 D15 **Portage** Pennsylvania, NE USA 40°23′N 78°40′W
30 K8 **Portage** Wisconsin, N USA 43°33′N 89°29′W
30 M3 **Portage Lake** ◎ Michigan, N USA
11 X16 **Portage la Prairie** Manitoba, S Canada 49°58′N 98°20′W
31 R11 **Portage River** ≈ Ohio, N USA
27 Y8 **Portageville** Missouri, C USA 36°25′N 89°42′W
28 L2 **Portal** North Dakota, N USA 48°57′N 102°33′W
10 L17 **Port Alberni** Vancouver Island, British Columbia, SW Canada 49°11′N 124°49′W
14 E15 **Port Albert** Ontario, S Canada 43°51′N 81°42′W
104 I10 **Portalegre** *anc.* Ammaia, Amoea. Portalegre, E Portugal 39°17′N 07°25′W
104 H10 **Portalegre** ◈ *district* C Portugal
37 V12 **Portales** New Mexico, SW USA 34°11′N 103°19′W
39 X14 **Port Alexander** Baranof Island, Alaska, USA 56°15′N 134°39′W
83 I25 **Port Alfred** Eastern Cape, S South Africa 33°31′S 26°55′E
10 J16 **Port Alice** Vancouver Island, British Columbia, SW Canada 50°23′N 127°24′W
22 J8 **Port Allen** Louisiana, S USA 30°27′N 91°12′W
Port Amelia *see* Pemba
Port An Dúnáin *see* Portadown
32 G7 **Port Angeles** Washington, NW USA 48°06′N 123°26′W
44 L12 **Port Antonio** NE Jamaica 18°10′N 76°27′W
115 D16 **Pórta Panagiá** *religious building* Thessalía, C Greece
25 T14 **Port Aransas** Texas, SW USA 27°49′N 97°03′W
97 E18 **Portarlington** *Ir.* Cúil an tSúdaire. Laois/Offaly, C Ireland 53°10′N 07°11′W
183 P17 **Port Arthur** Tasmania, SE Australia 43°09′S 147°51′E
25 Y11 **Port Arthur** Texas, SW USA 29°55′N 93°56′W
96 G12 **Port Askaig** W Scotland, United Kingdom 55°51′N 06°06′W
182 I7 **Port Augusta** South Australia 32°29′S 137°44′E
44 M9 **Port-au-Prince** *hait.* Pòtoprens. ● (Haiti) C Haiti 18°33′N 72°20′W
22 I8 **Port Barre** Louisiana, S USA 30°33′N 91°57′W
Port-Bergé *see* Boriziny
151 Q19 **Port Blair** Andaman and Nicobar Islands, SE India 11°40′N 92°44′E
25 X12 **Port Bolivar** Texas, SW USA 29°21′N 94°45′W
105 X4 **Portbou** Cataluña, NE Spain 42°26′N 03°10′E
77 N17 **Port Bouet** ✈ (Abidjan) SE Ivory Coast 05°17′N 03°55′W
182 I8 **Port Broughton** South Australia 33°39′S 137°55′E
14 F17 **Port Burwell** Ontario, S Canada 42°39′N 80°47′W
12 G17 **Port Burwell** Québec, NE Canada 60°25′N 64°49′W
182 M13 **Port Campbell** Victoria, SE Australia 38°37′S 143°00′E
15 V4 **Port-Cartier** Québec, SE Canada 50°00′N 66°55′W
185 F23 **Port Chalmers** Otago, South Island, New Zealand 45°46′S 170°37′E
23 W14 **Port Charlotte** Florida, SE USA 27°00′N 82°07′W
38 L9 **Port Clarence** Alaska, USA 65°15′N 166°51′W
10 I13 **Port Clements** Graham Island, British Columbia, SW Canada 53°37′N 132°12′W
31 S11 **Port Clinton** Ohio, N USA 41°30′N 82°56′W
14 H17 **Port Colborne** Ontario, S Canada 42°51′N 79°16′W
15 Y7 **Port-Daniel** Québec, SE Canada 48°10′N 64°58′W
Port Darwin *see* Darwin
183 O17 **Port Davey** *headland* Tasmania, SE Australia 43°19′S 145°54′E
44 K8 **Port-de-Paix** NW Haiti 19°56′N 72°52′W
181 W4 **Port Douglas** Queensland, NE Australia 16°33′S 145°27′E
10 J13 **Port Edward** British Columbia, SW Canada 54°11′N 130°16′W
83 K24 **Port Edward** KwaZulu/Natal, SE South Africa 31°03′S 30°14′E
58 J12 **Portel** Pará, NE Brazil 01°58′S 50°45′W
104 H12 **Portel** Évora, S Portugal 38°18′N 07°42′W
14 E14 **Port Elgin** Ontario, S Canada 44°26′N 81°22′W
45 Y14 **Port Elizabeth** Bequia, Saint Vincent and the Grenadines 13°01′N 61°15′W
83 I26 **Port Elizabeth** Eastern Cape, S South Africa 33°58′S 25°36′E
96 G13 **Port Ellen** W Scotland, United Kingdom 55°37′N 06°12′W
Portenau *see* Pordenone
97 H16 **Port Erin** SW Isle of Man 54°05′N 04°47′W
45 Q13 **Porter Point** *headland* Saint Vincent, Saint Vincent and the Grenadines 13°22′N 61°10′W
185 G18 **Porters Pass** *pass* South Island, New Zealand
83 E25 **Porterville** Western Cape, SW South Africa 33°03′S 19°00′E
35 R12 **Porterville** California, W USA 36°03′N 119°03′W
Port-Étienne *see* Nouâdhibou
182 L13 **Port Fairy** Victoria, SE Australia 38°24′S 142°13′E
184 M4 **Port Fitzroy** Great Barrier Island, Auckland, NE New Zealand 36°10′S 175°21′E
Port Florence *see* Kisumu
Port-Francqui *see* Ilebo
79 C18 **Port-Gentil** Ogooué-Maritime, W Gabon 0°40′S 08°50′E
182 I7 **Port Germein** South Australia 33°02′S 138°01′E
22 J6 **Port Gibson** Mississippi, S USA 31°57′N 90°58′W
39 Q13 **Port Graham** Alaska, USA 59°21′N 151°49′W
77 U17 **Port Harcourt** Rivers, S Nigeria 04°43′N 07°02′E
10 J16 **Port Hardy** Vancouver Island, British Columbia, SW Canada 50°41′N 127°30′W
Port Harrison *see* Inukjuak
13 R14 **Port Hawkesbury** Cape Breton Island, Nova Scotia, SE Canada 45°36′N 61°22′W
180 I6 **Port Hedland** Western Australia 20°23′S 118°40′E
39 O15 **Port Heiden** Alaska, USA 56°54′N 158°40′W
97 I19 **Porthmadog** *var.* Portmadoc. NW Wales, United Kingdom 52°55′N 04°08′W
14 I15 **Port Hope** Ontario, SE Canada 43°58′N 78°18′W
13 S9 **Port Hope Simpson** Newfoundland and Labrador, E Canada 52°30′N 56°18′W
65 C24 **Port Howard Settlement** West Falkland, Falkland Islands
31 T9 **Port Huron** Michigan, N USA 42°58′N 82°25′W
107 K17 **Portici** Campania, S Italy 40°48′N 14°20′E
Port-İliç *see* Liman
104 G14 **Portimão** *var.* Vila Nova de Portimão. Faro, S Portugal 37°08′N 08°32′W
25 T17 **Port Isabel** Texas, SW USA 26°04′N 97°13′W
18 J13 **Port Jervis** New York, NE USA 41°22′N 74°39′W
55 S7 **Port Kaituma** NW Guyana 07°42′N 59°52′W
126 K12 **Port Katon** Rostovskaya Oblast', SW Russian Federation 46°52′N 38°46′E
183 S9 **Port Kembla** New South Wales, SE Australia 34°30′S 150°54′E
182 F8 **Port Kenny** South Australia 33°09′S 134°38′E
Port Klang *see* Pelabuhan Klang
Port Láirge *see* Waterford
183 S8 **Portland** New South Wales, SE Australia 33°24′S 150°00′E
182 L13 **Portland** Victoria, SE Australia 38°21′S 141°38′E
184 K4 **Portland** Northland, North Island, New Zealand 35°48′S 174°19′E
31 Q13 **Portland** Indiana, N USA 40°25′N 84°58′W
19 P8 **Portland** Maine, NE USA 43°41′N 70°16′W
31 Q9 **Portland** Michigan, N USA 42°51′N 84°52′W
29 Q4 **Portland** North Dakota, N USA 47°28′N 97°22′W
32 G11 **Portland** Oregon, NW USA 45°31′N 122°41′W
20 J8 **Portland** Tennessee, S USA 36°34′N 86°31′W
25 T14 **Portland** Texas, SW USA 27°52′N 97°19′W
32 G11 **Portland** ✈ Oregon, NW USA 45°36′N 122°34′W
182 L13 **Portland Bay** *bay* Victoria, SE Australia
44 K13 **Portland Bight** *bay* S Jamaica
97 L24 **Portland Bill** *var.* Bill of Portland. *headland* S England, United Kingdom 50°31′N 02°28′W
Portland, Bill of *see* Portland Bill
183 P15 **Portland, Cape** *headland* Tasmania, SE Australia 40°46′S 147°58′E
10 J12 **Portland Inlet** *inlet* British Columbia, W Canada
184 P11 **Portland Island** *island* E New Zealand
65 F15 **Portland Point** *headland* SW Ascension Island
44 J13 **Portland Point** *headland* C Jamaica 17°42′N 77°10′W
103 P16 **Port-la-Nouvelle** Aude, S France 43°01′N 03°04′E
Portlaoighise *see* Port Laoise
97 E18 **Port Laoise** *var.* Portlaoise, *Ir.* Portlaoighise; *prev.* Maryborough. C Ireland 53°02′N 07°17′W
Portlaoise *see* Port Laoise
25 U13 **Port Lavaca** Texas, SW USA 28°36′N 96°39′W
182 G9 **Port Lincoln** South Australia 34°43′S 135°49′E
39 Q14 **Port Lions** Kodiak Island, Alaska, USA 57°55′N 152°48′W
76 I15 **Port Loko** W Sierra Leone 08°50′N 12°50′W
65 E24 **Port Louis** East Falkland, Falkland Islands 51°31′S 58°07′W
45 Y5 **Port-Louis** Grande Terre, N Guadeloupe 16°25′N 61°32′W
173 X16 **Port Louis** ● (Mauritius) NW Mauritius 20°10′S 57°30′E
Port Louis *see* Scarborough
Port-Lyautey *see* Kénitra
182 K12 **Port MacDonnell** South Australia 38°04′S 140°40′E
183 U7 **Port Macquarie** New South Wales, SE Australia 31°26′S 152°55′E
Portmadoc *see* Porthmadog
Port Mahon *see* Maó
44 K12 **Port Maria** C Jamaica 18°22′N 76°54′W
10 K16 **Port McNeill** Vancouver Island, British Columbia, SW Canada 50°34′N 127°06′W
13 P11 **Port-Menier** Île d'Anticosti, Québec, E Canada 49°49′N 64°19′W
39 N15 **Port Moller** Alaska, USA 56°00′N 160°31′W
44 L13 **Port Morant** E Jamaica 17°53′N 76°20′W
44 K13 **Portmore** C Jamaica 17°58′N 76°52′W
186 D9 **Port Moresby** ● (Papua New Guinea) Central/National Capital District, SW Papua New Guinea 09°28′S 147°12′E
Port Natal *see* Durban
25 Y11 **Port Neches** Texas, SW USA 29°59′N 93°57′W
182 G9 **Port Neill** South Australia 34°06′S 136°19′E
15 S6 **Portneuf** ≈ Québec, SE Canada
15 R6 **Portneuf, Lac** ◎ Québec, SE Canada
83 D23 **Port Nolloth** Northern Cape, W South Africa 29°17′S 16°51′E
18 J17 **Port Norris** New Jersey, NE USA 39°13′N 75°00′W
Port-Nouveau-Québec *see* Kangiqsualujjuaq
104 G6 **Porto** *Eng.* Oporto; *anc.* Portus Cale. Porto, NW Portugal 41°09′N 08°37′W
104 G6 **Porto** *var.* Pôrto. ◈ *district* N Portugal
104 G6 **Porto** ✈ Porto, W Portugal 41°09′N 08°37′W
Pôrto *see* Porto
61 I16 **Porto Alegre** *var.* Pôrto Alegre. *state capital* Rio Grande do Sul, S Brazil 30°03′S 51°10′W
Porto Alexandre *see* Tombua
82 B12 **Porto Amboim** Kwanza Sul, NW Angola 10°47′S 13°43′E
Porto Amélia *see* Pemba
Porto Bello *see* Portobelo
43 T14 **Portobelo** *var.* Porto Bello, Puerto Bello. Colón, N Panama 09°33′N 79°37′W
60 G10 **Pôrto Camargo** Paraná, S Brazil 23°23′S 53°47′W
25 U13 **Port O'Connor** Texas, SW USA 28°26′N 96°26′W
Pôrto de Mós *see* Porto de Moz
58 J12 **Porto de Moz** *var.* Pôrto de Mós. Pará, NE Brazil 01°45′S 52°15′W
64 O5 **Porto do Moniz** Madeira, Portugal, NE Atlantic Ocean
59 H16 **Porto dos Gaúchos** Mato Grosso, W Brazil 11°32′S 57°16′W
Porto Edda *see* Sarandë
107 J24 **Porto Empedocle** Sicilia, Italy, C Mediterranean Sea 37°18′N 13°32′E
59 H20 **Porto Esperança** Mato Grosso do Sul, SW Brazil 19°36′S 57°24′W
106 E13 **Portoferraio** Toscana, C Italy 42°49′N 10°18′E
96 G6 **Port of Ness** NW Scotland, United Kingdom 58°29′N 06°15′W
45 U14 **Port-of-Spain** ● (Trinidad and Tobago) Trinidad, Trinidad and Tobago 10°39′N 61°30′W
Port of Spain *see* Piarco
103 X15 **Porto, Golfe de** *gulf* Corse, France, C Mediterranean Sea
Porto Grande *see* Mindelo
106 I7 **Portogruaro** Veneto, NE Italy 45°46′N 12°50′E
35 P5 **Portola** California, W USA 39°48′N 120°28′W
187 Q13 **Port-Olry** Espiritu Santo, C Vanuatu 15°03′S 167°04′E
93 J17 **Pörtom** *Fin.* Pirttikylä. Österbotten, W Finland 62°42′N 21°40′E
Port Omna *see* Portumna
59 G21 **Porto Murtinho** Mato Grosso do Sul, SW Brazil 21°42′S 57°52′W
59 K16 **Porto Nacional** Tocantins, C Brazil 10°41′S 48°19′W
77 S16 **Porto-Novo** ● (Benin) S Benin 06°29′N 02°37′E
23 X10 **Port Orange** Florida, SE USA 29°06′N 80°59′W
32 G8 **Port Orchard** Washington, NW USA 47°32′N 122°38′W
Porto Re *see* Kraljevica
32 E15 **Port Orford** Oregon, NW USA 42°45′N 124°30′W
Porto Rico *see* Puerto Rico
106 J13 **Porto San Giorgio** Marche, C Italy 43°10′N 13°47′E
107 F14 **Porto San Stefano** Toscana, C Italy 42°26′N 11°09′E
64 P5 **Porto Santo** *var.* Vila Baleira. Porto Santo, Madeira, Portugal, NE Atlantic Ocean 33°04′N 16°20′W
64 Q5 **Porto Santo** ✈ Porto Santo, Madeira, Portugal, NE Atlantic Ocean 33°04′N 16°20′W
64 P5 **Porto Santo** *var.* Ilha do Porto Santo. *island* Madeira, Portugal, NE Atlantic Ocean
Porto Santo, Ilha do *see* Porto Santo
60 H9 **Porto São José** Paraná, S Brazil 22°43′S 53°10′W
59 O19 **Porto Seguro** Bahia, E Brazil 16°25′S 39°07′W
107 B17 **Porto Torres** Sardegna, Italy, C Mediterranean Sea 40°50′N 08°23′E
59 J23 **Porto União** Santa Catarina, S Brazil 26°15′S 51°04′W
103 Y16 **Porto-Vecchio** Corse, France, C Mediterranean Sea 41°35′N 09°17′E

◆ Country
● Country Capital
◇ Dependent Territory
○ Dependent Territory Capital
◈ Administrative Regions
✈ International Airport
▲ Mountain
▲▲ Mountain Range
⛰ Volcano
≈ River
◎ Lake
⊠ Reservoir

Porto Velho – Prüm

59 E15 **Porto Velho** *var.* Velho. *state capital* Rondônia, W Brazil 08°45´S 63°54´W
56 A6 **Portoviejo** *var.* Puertoviejo. Manabí, W Ecuador 01°03´S 80°31´W
185 B26 **Port Pegasus** *bay* Stewart Island, New Zealand
14 H15 **Port Perry** Ontario, SE Canada 44°08´N 78°57´W
183 N12 **Port Phillip Bay** *harbor* Victoria, SE Australia
182 I8 **Port Pirie** South Australia 33°11´S 138°01´E
96 G9 **Portree** N Scotland, United Kingdom 57°26´N 06°12´W
Port Rex *see* East London
Port Rois *see* Portrush
44 K13 **Port Royal** E Jamaica 17°55´N 76°52´W
21 R15 **Port Royal** South Carolina, SE USA 32°22´N 80°41´W
21 R15 **Port Royal Sound** *inlet* South Carolina, SE USA
97 F14 **Portrush** *Ir.* Port Rois. N Northern Ireland, United Kingdom 55°12´N 06°40´W
Port Said *see* Būr Saʿīd
23 R9 **Port Saint Joe** Florida, SE USA 29°49´N 85°18´W
23 Y11 **Port Saint John** Florida, SE USA 28°28´N 80°46´W
103 R16 **Port-St-Louis-du-Rhône** Bouches-du-Rhône, SE France 43°22´N 04°48´E
44 K10 **Port Salut** SW Haiti 18°04´N 73°55´W
65 E24 **Port Salvador** *inlet* East Falkland, Falkland Islands
65 D24 **Port San Carlos** East Falkland, Falkland Islands 51°30´S 58°59´W
13 S10 **Port Saunders** Newfoundland, Newfoundland and Labrador, SE Canada 50°40´N 57°17´W
83 K24 **Port Shepstone** KwaZulu/Natal, E South Africa 30°44´S 30°28´E
45 O11 **Portsmouth** *var.* Grand-Anse. NW Dominica 15°34´N 61°27´W
97 N24 **Portsmouth** S England, United Kingdom 50°48´N 01°05´W
19 P10 **Portsmouth** New Hampshire, NE USA 43°04´N 70°47´W
31 S15 **Portsmouth** Ohio, N USA 38°43´N 83°00´W
21 X7 **Portsmouth** Virginia, NE USA 36°50´N 76°18´W
14 E17 **Port Stanley** Ontario, S Canada 42°39´N 81°12´W
Port Stanley *see* Stanley
65 B25 **Port Stephens** *inlet* West Falkland, Falkland Islands
65 B25 **Port Stephens Settlement** West Falkland, Falkland Islands
97 F14 **Portstewart** *Ir.* Port Stíobhaird. N Northern Ireland, United Kingdom 55°11´N 06°43´W
Port Stíobhaird *see* Portstewart
83 K24 **Port St. Johns** Eastern Cape, SE South Africa 31°37´S 29°32´E
80 I7 **Port Sudan** Red Sea, NE Sudan 19°37´N 37°14´E
22 L10 **Port Sulphur** Louisiana, S USA 29°28´N 89°41´W
Port Swettenham *see* Klang/Pelabuhan Klang
97 J22 **Port Talbot** S Wales, United Kingdom 51°36´N 03°47´W
92 L11 **Porttipahdan Tekojärvi** ◎ N Finland
32 G7 **Port Townsend** Washington, NW USA 48°07´N 122°45´W
104 H9 **Portugal** *off.* Portuguese Republic. ◆ *republic* SW Europe
105 O2 **Portugalete** País Vasco, N Spain 43°19´N 03°01´W
54 J6 **Portuguesa** *off.* Estado Portuguesa. ◈ *state* N Venezuela
Portuguesa, Estado *see* Portuguesa
Portuguese East Africa *see* Mozambique
Portuguese Guinea *see* Guinea-Bissau
Portuguese Republic *see* Portugal
Portuguese Timor *see* East Timor
Portuguese West Africa *see* Angola
97 D18 **Portumna** *Ir.* Port Omna. Galway, W Ireland 53°06´N 08°13´W
Portus Cale *see* Porto
Portus Magnus *see* Almería
Portus Magonis *see* Maó
103 P17 **Port-Vendres** *var.* Port Vendres. Pyrénées-Orientales, S France 42°31´N 03°06´E
182 H9 **Port Victoria** South Australia 34°34´S 137°31´E
187 Q14 **Port-Vila** *var.* Vila. ● (Vanuatu) Éfaté, C Vanuatu 17°45´S 168°21´E
Port Vila *see* Bauer Field
182 I9 **Port Wakefield** South Australia 34°13´S 138°10´E
31 N8 **Port Washington** Wisconsin, N USA 43°23´N 87°54´W
57 J14 **Porvenir** Pando, NW Bolivia 11°15´S 68°43´W
63 I24 **Porvenir** Magallanes, S Chile 53°18´S 70°22´W
61 D18 **Porvenir** Paysandú, W Uruguay 32°23´S 57°59´W
93 M19 **Porvoo** *Swe.* Borgå. Uusimaa, S Finland 60°25´N 25°40´E
Porzecze *see* Parechcha
104 M10 **Porzuna** Castilla-La Mancha, C Spain 39°10´N 04°10´W
61 E14 **Posadas** Misiones, NE Argentina 27°27´S 55°52´W
104 L13 **Posadas** Andalucía, S Spain 37°48´N 05°06´W
Poschega *see* Požega
108 J11 **Poschiavino** ≈ Italy/Switzerland
108 J10 **Poschiavo** *Ger.* Puschlav. Graubünden, S Switzerland 46°19´N 10°02´E
112 D12 **Posedarje** Zadar, SW Croatia 44°12´N 15°27´E
Posen *see* Poznań
124 L14 **Poshekhon'ye** Yaroslavskaya Oblast', W Russian Federation 58°31´N 39°07´E
92 M13 **Posio** Lappi, NE Finland 66°06´N 28°16´E
Poskam *see* Zepu
Posnania *see* Poznań
1713 O12 **Poso** Sulawesi, C Indonesia 01°23´S 120°45´E
171 O12 **Poso, Danau** ◎ Sulawesi, C Indonesia
137 R10 **Posof** Ardahan, NE Turkey 41°30´N 42°33´E
25 R6 **Possum Kingdom Lake** ⊠ Texas, SW USA
25 N6 **Post** Texas, SW USA 33°14´N 101°24´W
Postavy/Postawy *see* Pastavy
Poste-de-la-Baleine *see* Kuujjuarapik
99 M17 **Posterholt** Limburg, SE Netherlands 51°07´N 06°02´E
83 G22 **Postmasburg** Northern Cape, N South Africa 28°20´S 23°05´E
Pôsto Diuarum *see* Campo de Diauarum
59 I16 **Pôsto Jacaré** Mato Grosso, W Brazil 12°S 53°27´W
109 T12 **Postojna** *Ger.* Adelsberg, *It.* Postumia. SW Slovenia 45°48´N 14°12´E
Postumia *see* Postojna
29 X12 **Postville** Iowa, C USA 43°04´N 91°34´W
Pöstyén *see* Piešťany
113 G14 **Posušje** Federacija Bosne I Hercegovine, SW Bosnia and Herzegovina 43°28´N 17°20´E
171 O16 **Pota** Flores, C Indonesia 08°21´S 120°50´E
115 G23 **Potamós** Antikýthira, S Greece 35°53´N 23°17´E
55 S9 **Potaru River** ≈ C Guyana
83 I21 **Potchefstroom** North-West, N South Africa 26°42´S 27°06´E
27 R11 **Poteau** Oklahoma, C USA 35°03´N 94°36´W
25 R12 **Poteet** Texas, SW USA 29°02´N 98°34´W
115 G14 **Poteídaia** *site of ancient city* Kentrikí Makedonía, N Greece
Potentia *see* Potenza
107 M18 **Potenza** *anc.* Potentia. Basilicata, S Italy 40°40´N 15°50´E
185 A24 **Poteriteri, Lake** ◎ South Island, New Zealand
104 M2 **Potes** Cantabria, N Spain 43°10´N 04°41´W
Potgietersrus *see* Mokopane
25 S12 **Poth** Texas, SW USA 29°04´N 98°04´W
32 J9 **Potholes Reservoir** ⊠ Washington, NW USA
137 Q9 **Poti** *prev.* P'ot'i. W Georgia 42°10´N 41°42´E
P'ot'i *see* Poti
77 X13 **Potiskum** Yobe, NE Nigeria 11°38´N 11°07´E
Potkozarje *see* Ivanjska
32 M9 **Potlatch** Idaho, NW USA 46°55´N 116°51´W
33 N9 **Pot Mountain** ▲ Idaho, NW USA 46°44´N 115°24´W
113 H14 **Potoci** Federacija Bosne I Hercegovine, S Bosnia and Herzegovina 43°24´N 17°52´E
21 V3 **Potomac River** ≈ NE USA
Pòtoprens *see* Port-au-Prince
57 L20 **Potosí** Potosí, S Bolivia 19°35´S 65°51´W
42 H9 **Potosí** Chinandega, NW Nicaragua 12°58´N 87°30´W
27 W6 **Potosi** Missouri, C USA 37°57´N 90°49´W
57 K21 **Potosí** ◈ *department* SW Bolivia
62 H7 **Potrerillos** Atacama, N Chile 26°30´S 69°25´W
42 H5 **Potrerillos** Cortés, NW Honduras 15°10´N 87°58´W
62 H8 **Potro, Cerro del** ▲ N Chile 28°22´S 69°34´W
100 N12 **Potsdam** Brandenburg, NE Germany 52°24´N 13°04´E
18 J7 **Potsdam** New York, NE USA 44°40´N 74°58´W
109 X5 **Pottendorf** Niederösterreich, E Austria 47°55´N 16°23´E
109 X5 **Pottenstein** Niederösterreich, E Austria 47°58´N 16°07´E
18 I15 **Pottstown** Pennsylvania, NE USA 40°15´N 75°39´W
18 H14 **Pottsville** Pennsylvania, NE USA 40°40´N 76°10´W
155 L25 **Pottuvil** Eastern Province, SE Sri Lanka 06°53´N 81°49´E
149 U6 **Potwar Plateau** *plateau* NE Pakistan
102 J7 **Pouancé** Maine-et-Loire, W France 47°46´N 01°11´W
15 R6 **Poulin de Courval, Lac** ◎ Québec, SE Canada
18 L9 **Poultney** Vermont, NE USA 43°31´N 73°12´W
187 O16 **Poum** Province Nord, W New Caledonia 20°15´S 164°03´E
59 L21 **Pouso Alegre** Minas Gerais, NE Brazil 22°13´S 45°56´W
192 I16 **Poutasi** Upolu, SE Samoa 14°00´S 171°43´W
167 R12 **Poŭthĭsăt** *prev.* Pursat. Poŭthĭsăt, W Cambodia 12°32´N 103°55´E
167 R12 **Poŭthĭsăt, Stœng** *prev.* Pursat. ≈ W Cambodia
102 J9 **Pouzauges** Vendée, NW France 46°47´N 00°54´W
Po, Valle del *see* Po Valley
106 F8 **Po Valley** *It.* Valle del Po. *valley* N Italy
111 I19 **Považská Bystrica** *Ger.* Waagbistritz, *Hung.* Vágbeszterce. Trenčiansky Kraj, W Slovakia 49°07´N 18°26´E
124 J10 **Povenets** Respublika Kareliya, NW Russian Federation 62°50´N 34°47´E
184 Q9 **Poverty Bay** *inlet* North Island, New Zealand
112 K12 **Povlen** ▲▲ W Serbia
104 G6 **Póvoa de Varzim** Porto, NW Portugal 41°22´N 08°46´W
127 N8 **Povorino** Voronezhskaya Oblast', W Russian Federation 51°10´N 42°16´E
Povungnituk *see* Puvirnituq
Rivière de Povungnituk *see* Puvirnituq, Riviere de
14 H11 **Powassan** Ontario, S Canada 46°05´N 79°21´W
35 U17 **Poway** California, W USA 32°57´N 117°02´W
33 W14 **Powder River** Wyoming, C USA 43°01´N 106°57´W
33 Y10 **Powder River** ≈ Montana/Wyoming, NW USA
32 L12 **Powder River** ≈ Oregon, NW USA
33 W13 **Powder River Pass** *pass* Wyoming, C USA
33 U12 **Powell** Wyoming, C USA 44°45´N 108°45´W
65 I22 **Powell Basin** *undersea feature* NW Weddell Sea
36 M8 **Powell, Lake** ◎ Utah, W USA
37 R4 **Powell, Mount** ▲ Colorado, C USA 39°25´N 106°20´W
10 L17 **Powell River** British Columbia, SW Canada 49°54´N 124°34´W
31 N5 **Powers** Michigan, N USA 45°40´N 87°29´W
28 K2 **Powers Lake** North Dakota, N USA 48°33´N 102°37´W
21 V6 **Powhatan** Virginia, NE USA 37°33´N 77°56´W
31 V13 **Powhatan Point** Ohio, N USA 39°49´N 80°49´W
97 J20 **Powys** *cultural region* E Wales, United Kingdom
187 P17 **Poya** Province Nord, C New Caledonia 21°19´S 165°07´E
161 P10 **Poyang Hu** ◎ S China
30 L7 **Poygan, Lake** ◎ Wisconsin, N USA
109 Y2 **Poysdorf** Niederösterreich, NE Austria 48°40´N 16°38´E
112 N11 **Požarevac** *Ger.* Passarowitz. Serbia, NE Serbia 44°37´N 21°11´E
41 Q13 **Poza Rica** *var.* Poza Rica de Hidalgo. Veracruz-Llave, E Mexico 20°34´N 97°26´W
Poza Rica de Hidalgo *see* Poza Rica
112 L13 **Požega** *prev.* Slavonska Požega, *Ger.* Poschega, *Hung.* Pozsega. Požega-Slavonija, NE Croatia 45°19´N 17°42´E
112 H9 **Požega-Slavonija** *off.* Požeško-Slavonska Županija. ◈ *province* NE Croatia
Požeško-Slavonska Županija *see* Požega-Slavonija
125 U13 **Pozhva** Komi-Permyatskiy Okrug, NW Russian Federation 59°07´N 56°04´E
110 G11 **Poznań** *Ger.* Posen, Posnania. Wielkolpolskie, C Poland 52°24´N 16°56´E
105 O13 **Pozo Alcón** Andalucía, S Spain 37°43´N 02°55´W
62 H3 **Pozo Almonte** Tarapacá, N Chile 20°16´S 69°50´W
104 L12 **Pozoblanco** Andalucía, S Spain 38°23´N 04°48´W
105 Q11 **Pozo Cañada** Castilla-La Mancha, C Spain 38°49´N 01°45´W
62 N5 **Pozo Colorado** Presidente Hayes, C Paraguay 23°26´S 58°51´W
63 J20 **Pozos, Punta** *headland* S Argentina 47°55´S 65°46´W
Pozsega *see* Požega
Pozsony *see* Bratislava
55 N5 **Pozuelos** Anzoátegui, NE Venezuela 10°11´N 64°39´W
107 L26 **Pozzallo** Sicilia, Italy, C Mediterranean Sea 36°44´N 14°51´E
107 K17 **Pozzuoli** *anc.* Puteoli. Campania, S Italy 40°49´N 14°07´E
77 P17 **Pra** ≈ S Ghana
111 C19 **Prachatice** *Ger.* Prachatitz. Jihočeský Kraj, S Czech Republic 49°01´N 14°02´E
Prachatitz *see* Prachatice
167 P11 **Prachin Buri** *var.* Prachinburi. Prachin Buri, C Thailand 14°05´N 101°23´E
Prachinburi *see* Prachin Buri
Prachuab Girikhand *see* Prachuap Khiri Khan
167 O12 **Prachuap Khiri Khan** *var.* Prachuab Girikhand. Prachuap Khiri Khan, SW Thailand 11°50´N 99°49´E
111 H16 **Praděd** *Ger.* Altvater. ▲ NE Czech Republic 50°06´N 17°14´E
54 D11 **Pradera** Valle del Cauca, SW Colombia 03°23´N 76°11´W
103 O17 **Prades** Pyrénées-Orientales, S France 42°36´N 02°22´E
59 O19 **Prado** Bahia, SE Brazil 17°13´S 39°15´W
54 E11 **Prado** Tolima, C Colombia 03°45´N 74°55´W
Prado del Ganso *see* Goose Green
Prae *see* Phrae
95 I24 **Præstø** Sjælland, SE Denmark 55°08´N 12°03´E
Prag/Praga/Prague *see* Praha
27 O10 **Prague** Oklahoma, C USA 35°29´N 96°40´W
111 D16 **Praha** *Eng.* Prague, *Ger.* Prag, *Pol.* Praga. ● (Czech Republic) Středočeský Kraj, NW Czech Republic 50°06´N 14°26´E
116 J13 **Prahova** ◈ *county* SE Romania
116 J13 **Prahova** ≈ S Romania
76 E10 **Praia** ● (Cape Verde) Santiago, S Cape Verde 14°55´N 23°31´W
83 M21 **Praia do Bilene** Gaza, S Mozambique 25°18´S 33°10´E
83 M20 **Praia do Xai-Xai** Gaza, S Mozambique 25°04´S 33°43´E
116 J10 **Praid** *Hung.* Parajd. Harghita, C Romania 46°33´N 25°06´E
26 J3 **Prairie Dog Creek** ≈ Kansas/Nebraska, C USA
30 J9 **Prairie du Chien** Wisconsin, N USA 43°02´N 91°08´W
27 S9 **Prairie Grove** Arkansas, C USA 35°58´N 94°19´W
31 P10 **Prairie River** ≈ Michigan, N USA
Prairie State *see* Illinois
25 V11 **Prairie View** Texas, SW USA 30°05´N 95°59´W
167 Q10 **Prakhon Chai** Buri Ram, E Thailand 14°36´N 103°04´E
109 R4 **Pram** ≈ N Austria
167 Q12 **Prâmaôy** *prev.* Phumĭ Prâmaôy. Poŭthĭsăt, W Cambodia 12°13´N 103°05´E
109 S4 **Prambachkirchen** Oberösterreich, N Austria 48°18´N 13°50´E
118 H2 **Prangli** *island* N Estonia
154 J13 **Prānhita** ≈ C India
172 I15 **Praslin** *island* Inner Islands, NE Seychelles
115 O23 **Prasonísi, Akrotírio** *cape* Ródos, Dodekánisa, Greece, Aegean Sea
111 I14 **Praszka** Opolskie, S Poland 51°05´N 18°29´E
Pratas Island *see* Tungsha Tao
119 M18 **Pratasy** *Rus.* Protasy. Homyel'skaya Voblasts', SE Belarus 52°47´N 29°05´E
167 Q10 **Prathai** Nakhon Ratchasima, E Thailand 15°31´N 102°42´E
Prathet Thai *see* Thailand
Prathum Thani *see* Pathum Thani
63 F21 **Prat, Isla** *island* S Chile
106 G11 **Prato** Toscana, C Italy 43°53´N 11°05´E
103 O17 **Prats-de-Mollo-la-Preste** Pyrénées-Orientales, S France 42°25´N 02°28´E
26 L6 **Pratt** Kansas, C USA 37°40´N 98°45´W
108 E6 **Pratteln** Basel Landschaft, NW Switzerland 47°32´N 07°42´E
193 O2 **Pratt Seamount** *undersea feature* N Pacific Ocean 56°09´N 142°30´W
23 P5 **Prattville** Alabama, S USA 32°27´N 86°27´W
Praust *see* Pruszcz Gdański
119 B14 **Pravdinsk** *Ger.* Friedland. Kaliningradskaya Oblast', W Russian Federation 54°26´N 21°01´E
104 K2 **Pravia** Asturias, N Spain 43°30´N 06°06´W
118 L12 **Prazaroki** *Rus.* Prozoroki. Vitsyebskaya Voblasts', N Belarus 55°18´N 28°13´E
Prázsmár *see* Prejmer
167 S11 **Preăh Vihéar** Preăh Vihéar, N Cambodia 13°57´N 104°48´E
116 J12 **Predeal** *Hung.* Predeál. Braşov, C Romania 45°30´N 25°31´E
109 S8 **Predlitz** Steiermark, SE Austria 47°04´N 13°54´E
11 V15 **Preeceville** Saskatchewan, S Canada 51°58´N 102°40´W
Preenkuln *see* Priekule
102 K6 **Pré-en-Pail** Mayenne, NW France 48°27´N 00°15´W
109 T4 **Pregarten** Oberösterreich, N Austria 48°21´N 14°31´E
54 H7 **Pregonero** Táchira, NW Venezuela 08°02´N 71°35´W
118 J10 **Preiļi** *Ger.* Preli. SE Latvia 56°17´N 26°52´E
116 J12 **Prejmer** *Ger.* Tartlau, *Hung.* Prázsmár. Braşov, S Romania 45°42´N 25°49´E
113 J16 **Prekornica** ▲▲ C Montenegro
Preli *see* Preiļi
Prëmet *see* Përmet
100 M12 **Premnitz** Brandenburg, NE Germany 52°33´N 12°22´E
25 S15 **Premont** Texas, SW USA 27°21´N 98°07´W
113 H14 **Prenj** ▲▲ S Bosnia and Herzegovina
Prenjas/Prenjasi *see* Përrenjas
22 L7 **Prentiss** Mississippi, S USA 31°36´N 89°52´W
Preny *see* Prienai
100 O10 **Prenzlau** Brandenburg, NE Germany 53°19´N 13°52´E
123 N11 **Preobrazhenka** Irkutskaya Oblast', C Russian Federation 60°01´N 108°00´E
166 J9 **Preparis Island** *island* SW Myanmar (Burma)
Prerau *see* Přerov
111 H18 **Přerov** *Ger.* Prerau. Olomoucký Kraj, E Czech Republic 49°27´N 17°27´E
Preschau *see* Prešov
14 M14 **Prescott** Ontario, SE Canada 44°43´N 75°33´W
36 K12 **Prescott** Arizona, SW USA 34°33´N 112°26´W
27 T13 **Prescott** Arkansas, C USA 33°49´N 93°25´W
32 L10 **Prescott** Washington, NW USA 46°17´N 118°21´W
30 H6 **Prescott** Wisconsin, N USA 44°46´N 92°45´W
185 A24 **Preservation Inlet** *inlet* South Island, New Zealand
112 O7 **Preševo** Serbia, SE Serbia 42°20´N 21°38´E
29 N10 **Presho** South Dakota, N USA 43°54´N 100°03´W
58 M13 **Presidente Dutra** Maranhão, E Brazil 05°17´S 44°30´W
60 I8 **Presidente Epitácio** São Paulo, S Brazil 21°45´S 52°07´W
62 N5 **Presidente Hayes** *off.* Departamento de Presidente Hayes. ◈ *department* C Paraguay
Presidente Hayes, Departamento de *see* Presidente Hayes
60 I9 **Presidente Prudente** São Paulo, S Brazil 22°09´S 51°24´W
Presidente Stroessner *see* Ciudad del Este
Presidente Vargas *see* Itabira
60 I8 **Presidente Venceslau** São Paulo, S Brazil 21°52´S 51°51´W
193 O10 **President Thiers Seamount** *undersea feature* C Pacific Ocean 24°39´S 145°50´W
24 J11 **Presidio** Texas, SW USA 29°33´N 104°22´W
Preslav *see* Veliki Preslav
111 M19 **Prešov** *var.* Preschau, *Ger.* Eperies, *Hung.* Eperjes. Prešovský Kraj, E Slovakia 49°N 21°14´E
111 M19 **Prešovský Kraj** ◈ *region* E Slovakia
113 N20 **Prespa, Lake** *Alb.* Liqen i Prespës, *Gk.* Límni Megáli Préspa, Limni Prespa, *Mac.* Prespansko Ezero, *Serb.* Prespansko Jezero. ◎ SE Europe
Prespa, Limni/Prespansko Ezero/Prespansko Jezero/Prespës, Liqen i *see* Prespa, Lake
19 S2 **Presque Isle** Maine, NE USA 46°40´N 68°01´W
18 B11 **Presque Isle** *headland* Pennsylvania, NE USA 42°09´N 80°06´W
Pressburg *see* Bratislava
77 P17 **Prestea** SW Ghana 05°22´N 02°07´W
111 B17 **Přeštice** *Ger.* Pschestitz. Plzeňský Kraj, W Czech Republic 49°36´N 13°19´E
97 K17 **Preston** NW England, United Kingdom 53°46´N 02°42´W
23 S6 **Preston** Georgia, SE USA 32°03´N 84°32´W
33 R16 **Preston** Idaho, NW USA 42°06´N 111°52´W
29 Z13 **Preston** Iowa, C USA 42°03´N 90°24´W
29 X11 **Preston** Minnesota, N USA 43°41´N 92°06´W
21 O6 **Prestonsburg** Kentucky, S USA 37°40´N 82°46´W
96 I13 **Prestwick** W Scotland, United Kingdom 55°31´N 04°39´W
83 145 **Pretoria** *var.* Epitoli. ● Gauteng, NE South Africa 25°41´S 28°12´E
Pretoria-Witwatersrand-Vereeniging *see* Gauteng
Pretusha *see* Pretushë
113 M21 **Pretushë** *var.* Pretusha. Korçë, SE Albania 40°50´N 20°45´E
Preussisch Eylau *see* Bagrationovsk
Preußisch Holland *see* Pasłęk
Preussisch-Stargard *see* Starogard Gdański
115 C17 **Préveza** Ípeiros, W Greece 38°59´N 20°44´E
37 V3 **Prewitt Reservoir** ⊠ Colorado, C USA
167 S13 **Prey Vêng** Prey Vêng, S Cambodia 11°30´N 105°20´E
Priaral'skiye Karakumy, Peski *see* Priaral'skiy Karakum
144 M12 **Priaral'skiy Karakum** *prev.* Priaral'skiye Karakumy, Peski. *desert* SW Kazakhstan
123 P14 **Priargunsk** Zabaykal'skiy Kray, S Russian Federation 50°25´N 119°12´E
38 K14 **Pribilof Islands** *island group* Alaska, USA
113 K14 **Priboj** Serbia, W Serbia 43°34´N 19°33´E
111 C17 **Příbram** *Ger.* Pibrans. Středočeský Kraj, W Czech Republic 49°41´N 14°02´E
36 M4 **Price** Utah, W USA 39°35´N 110°49´W
37 N5 **Price River** ≈ Utah, W USA
23 N8 **Prichard** Alabama, S USA 30°44´N 88°04´W
25 R8 **Priddy** Texas, SW USA 31°39´N 98°30´W
105 P8 **Priego** Castilla-La Mancha, C Spain 40°26´N 02°19´W
104 M14 **Priego de Córdoba** Andalucía, S Spain 37°27´N 04°12´W
118 C10 **Priekule** *Ger.* Preenkuln. SW Latvia 56°26´N 21°36´E
118 C12 **Priekulė** *Ger.* Prökuls. Klaipėda, W Lithuania 55°36´N 21°16´E
119 F14 **Prienai** *Pol.* Preny. Kaunas, S Lithuania 54°37´N 23°56´E
83 G23 **Prieska** Northern Cape, C South Africa 29°40´S 22°45´E
32 M7 **Priest Lake** ⊠ Idaho, NW USA
32 M7 **Priest River** Idaho, NW USA 48°10´N 117°02´W
104 M3 **Prieta, Peña** ▲ N Spain 43°01´N 04°42´W
40 J10 **Prieto, Cerro** ▲ C Mexico 24°10´N 105°21´W
111 J19 **Prievidza** *var.* Priewitz, *Ger.* Priwitz, *Hung.* Privigye. Trenčiansky Kraj, W Slovakia 48°47´N 18°35´E
Priewitz *see* Prievidza
112 F10 **Prijedor** ◈ Republika Srpska, NW Bosnia and Herzegovina
113 K14 **Prijepolje** Serbia, W Serbia 43°24´N 19°39´E
Prikaspiyskaya Nizmennost' *see* Caspian Depression
113 O19 **Prilep** *Turk.* Perlepe. S FYR Macedonia 41°21´N 21°34´E
108 B9 **Prilly** Vaud, SW Switzerland 46°32´N 06°38´E
Priluki *see* Pryluky
62 L10 **Primero, Río** ≈ C Argentina
29 S12 **Primghar** Iowa, C USA 43°05´N 95°37´W
112 B9 **Primorje-Gorski Kotar** *off.* Primorsko-Goranska Županija. ◈ *province* NW Croatia
118 A13 **Primorsk** *Ger.* Fischhausen. Kaliningradskaya Oblast', W Russian Federation 54°45´N 20°00´E
124 G12 **Primorsk** *Fin.* Koivisto. Leningradskaya Oblast', NW Russian Federation 60°20´N 28°39´E
123 S14 **Primorskiy Kray** *prev.* *Eng.* Maritime Territory. ◈ *territory* SE Russian Federation
114 N10 **Primorsko** *prev.* Keupriya. Burgas, E Bulgaria 42°15´N 27°45´E
126 K13 **Primorsko-Akhtarsk** Krasnodarskiy Kray, SW Russian Federation 46°03´N 38°44´E
Primorsko-Goranska Županija *see* Primorje-Gorski Kotar
Primorsk/Primorskoye *see* Prymors'k
113 D14 **Primošten** Šibenik-Knin, S Croatia 43°34´N 15°57´E
11 R13 **Primrose Lake** ◎ Saskatchewan, C Canada
11 T14 **Prince Albert** Saskatchewan, S Canada 53°09´N 105°43´W
83 G25 **Prince Albert** Western Cape, SW South Africa 33°13´S 22°03´E
8 J5 **Prince Albert Peninsula** *peninsula* Victoria Island, Northwest Territories, NW Canada
8 J6 **Prince Albert Sound** *inlet* Northwest Territories, N Canada
J5 8 **Prince Alfred, Cape** *headland* Northwest Territories, NW Canada
9 P6 **Prince Charles Island** *island* Nunavut, NE Canada
195 W6 **Prince Charles Mountains** ▲▲ Antarctica
Prince-Édouard, Île-du *see* Prince Edward Island
172 M13 **Prince Edward Fracture Zone** *tectonic feature* SW Indian Ocean
13 P14 **Prince Edward Island** *Fr.* Île-du Prince-Édouard. ◈ *province* SE Canada
13 Q14 **Prince Edward Island** *Fr.* Île-du Prince-Édouard. *island* SE Canada
173 M12 **Prince Edward Islands** *island group* S South Africa
21 X4 **Prince Frederick** Maryland, NE USA 38°32´N 76°35´W
10 M14 **Prince George** British Columbia, SW Canada 53°55´N 122°49´W
21 W6 **Prince George** Virginia, NE USA 37°13´N 77°13´W
197 O8 **Prince Gustaf Adolf Sea** *sea* Nunavut, N Canada
197 Q3 **Prince of Wales, Cape** *headland* Alaska, USA 65°39´N 168°12´W
181 V1 **Prince of Wales Island** *island* Queensland, E Australia
8 L5 **Prince of Wales Island** *island* Queen Elizabeth Islands, Nunavut, NW Canada
39 Y14 **Prince of Wales Island** *island* Alexander Archipelago, Alaska, USA
Prince of Wales Island *see* Pinang, Pulau
8 J5 **Prince of Wales Strait** *strait* Northwest Territories, N Canada
197 O8 **Prince Patrick Island** *island* Parry Islands, Northwest Territories, NW Canada
9 N5 **Prince Regent Inlet** *channel* Nunavut, N Canada
10 J13 **Prince Rupert** British Columbia, SW Canada 54°18´N 130°17´W
Prince's Island *see* Príncipe
21 Y5 **Princess Anne** Maryland, NE USA 38°12´N 75°42´W
Princess Astrid Coast *see* Prinsesse Astrid Kyst
181 W2 **Princess Charlotte Bay** *bay* Queensland, NE Australia
195 W7 **Princess Elizabeth Land** *physical region* Antarctica
10 J14 **Princess Royal Island** *island* British Columbia, SW Canada
45 U15 **Princes Town** Trinidad, Trinidad and Tobago 10°16´N 61°23´W
11 N17 **Princeton** British Columbia, SW Canada 49°25´N 120°35´W
30 L11 **Princeton** Illinois, N USA 41°22´N 89°27´W
31 N16 **Princeton** Indiana, N USA 38°21´N 87°33´W
29 Z14 **Princeton** Iowa, C USA 41°40´N 90°21´W
20 H7 **Princeton** Kentucky, S USA 37°06´N 87°52´W
29 V8 **Princeton** Minnesota, N USA 45°34´N 93°34´W
27 S1 **Princeton** Missouri, C USA 40°22´N 93°37´W
18 J15 **Princeton** New Jersey, NE USA 40°21´N 74°39´W
21 R6 **Princeton** West Virginia, NE USA 37°23´N 81°06´W
39 S12 **Prince William Sound** *inlet* Alaska, USA
67 P9 **Príncipe** *var.* Príncipe Island, *Eng.* Prince's Island. *island* N Sao Tome and Principe
Príncipe Island *see* Príncipe
32 I13 **Prineville** Oregon, NW USA 44°19´N 120°50´W
28 J11 **Pringle** South Dakota, N USA 43°34´N 103°34´W
25 N1 **Pringle** Texas, SW USA 35°55´N 101°28´W
99 H14 **Prinsenbeek** Noord-Brabant, S Netherlands 51°36´N 04°42´E
98 L6 **Prinses Margriet Kanaal** *canal* N Netherlands
195 R1 **Prinsesse Astrid Kyst** *Eng.* Princess Astrid Coast. *physical region* Antarctica
195 T2 **Prinsesse Ragnhild Kyst** *physical region* Antarctica
195 U2 **Prins Harald Kyst** *physical region* Antarctica
92 N2 **Prins Karls Forland** *island* W Svalbard
43 N8 **Prinzapolka** Región Autónoma Atlántico Norte, NE Nicaragua 13°19´N 83°35´W
42 L8 **Prinzapolka, Río** ≈ NE Nicaragua
122 H9 **Priob'ye** Khanty-Mansiyskiy Avtonomnyy Okrug-Yugra, N Russian Federation 62°25´N 65°36´E
104 H1 **Prior, Cabo** *headland* NW Spain 43°33´N 08°21´W
29 V9 **Prior Lake** Minnesota, N USA 44°42´N 93°25´W
124 H11 **Priozersk** *Fin.* Käkisalmi. Leningradskaya Oblast', NW Russian Federation 61°02´N 30°07´E
119 J20 **Pripet** *Bel.* Prypyats', *Ukr.* Pryp"yat'. ≈ Belarus/Ukraine
119 J20 **Pripet Marshes** *wetland* Belarus/Ukraine
113 N16 **Prishtinë** , *Serb.* Priština. C Kosovo 42°40´N 21°10´E
126 J8 **Pristen'** Kurskaya Oblast', W Russian Federation 51°15´N 36°47´E
Priština *see* Prishtinë
100 M10 **Pritzwalk** Brandenburg, NE Germany 53°10´N 12°11´E
103 R13 **Privas** Ardèche, E France 44°45´N 04°35´E
107 I16 **Priverno** Lazio, C Italy 41°28´N 13°11´E
Privigye *see* Prievidza
112 C12 **Privlaka** Zadar, SW Croatia 44°15´N 15°07´E
124 M15 **Privolzhsk** Ivanovskaya Oblast', NW Russian Federation 57°24´N 41°16´E
127 P7 **Privolzhskaya Vozvyshennost'** *var.* Volga Uplands. ▲▲ W Russian Federation
127 P8 **Privolzhskoye** Saratovskaya Oblast', W Russian Federation 51°08´N 45°57´E
Priwitz *see* Prievidza
127 N13 **Priyutnoye** Respublika Kalmykiya, SW Russian Federation 46°08´N 43°33´E
113 M17 **Prizren** S Kosovo 42°14´N 20°46´E
107 I24 **Prizzi** Sicilia, Italy, C Mediterranean Sea 37°44´N 13°26´E
113 P18 **Probištip** NE FYR Macedonia 42°00´N 22°06´E
169 S16 **Probolinggo** Jawa, C Indonesia 07°45´S 113°12´E
Probstberg *see* Wyszków
111 F14 **Prochowice** *Ger.* Parchwitz. Dolnośląskie, SW Poland 51°15´N 16°22´E
29 W5 **Proctor** Minnesota, N USA 46°46´N 92°13´W
25 R8 **Proctor** Texas, SW USA 31°57´N 98°25´W
25 R8 **Proctor Lake** ⊠ Texas, SW USA
155 I18 **Proddatūr** Andhra Pradesh, E India 14°45´N 78°34´E
104 H9 **Proença-a-Nova** *var.* Proença a Nova. Castelo Branco, C Portugal 39°45´N 07°56´W
Proença a Nova *see* Proença-a-Nova
99 I21 **Profondeville** Namur, SE Belgium 50°22´N 04°52´E
41 W11 **Progreso** Yucatán, SE Mexico 21°14´N 89°41´W
123 R14 **Progress** Amurskaya Oblast', SE Russian Federation 49°40´N 129°30´E
127 O15 **Prokhladnyy** Kabardino-Balkarskaya Respublika, SW Russian Federation 43°48´N 44°02´E
Prokletije *see* North Albanian Alps
Prökuls *see* Priekulė
113 O15 **Prokuplje** Serbia, SE Serbia 43°15´N 21°35´E
124 H14 **Proletariy** Novgorodskaya Oblast', W Russian Federation 58°24´N 31°40´E
126 M12 **Proletarsk** Rostovskaya Oblast', SW Russian Federation 46°42´N 41°48´E
127 N13 **Proletarskoye Vodokhranilishche** *salt lake* SW Russian Federation
Prome *see* Pyay
60 J8 **Promissão** São Paulo, S Brazil 21°33´S 49°51´W
60 J8 **Promissão, Represa de** ⊠ S Brazil
125 V4 **Promyshlennyy** Respublika Komi, NW Russian Federation 67°36´N 64°E
119 O16 **Pronya** ≈ E Belarus
10 M11 **Prophet River** British Columbia, W Canada 58°07´N 122°39´W
30 K11 **Prophetstown** Illinois, N USA 41°40´N 89°56´W
Propinsi Kepulauan Riau *see* Kepulauan Riau
Propinsi Papua Barat *see* Papua Barat
59 P16 **Propriá** Sergipe, E Brazil 10°15´S 36°51´W
103 X16 **Propriano** Corse, France, C Mediterranean Sea 41°41´N 08°54´E
Prościejów *see* Prostějov
Proskurov *see* Khmel 'nyts'kyy
114 H12 **Prosotsáni** Anatolikí Makedonía kai Thráki, NE Greece 41°11´N 23°59´E
171 Q7 **Prosperidad** Mindanao, S Philippines 08°36´N 125°54´E
32 J10 **Prosser** Washington, NW USA 46°12´N 119°46´W
Prossnitz *see* Prostějov
111 G18 **Prostějov** *Ger.* Prossnitz, *Pol.* Prościejów. Olomoucký Kraj, E Czech Republic 49°29´N 17°08´E
117 V8 **Prosyana** Dnipropetrovs'ka Oblast', E Ukraine 48°07´N 36°22´E
111 L16 **Proszowice** Malopolskie, S Poland 50°12´N 20°15´E
Protasy *see* Pratasy
172 J11 **Protea Seamount** *undersea feature* SW Indian Ocean 36°50´S 18°05´E
115 D21 **Próti** *island* S Greece
114 N8 **Provadia** , Provadiya. Varna, E Bulgaria 43°10´N 27°29´E
Provadiya *see* Provadia
103 T14 **Provence** *cultural region* SE France
103 S15 **Provence** *prev.* Marseille-Marigrane. ✈ (Marseille) Bouches-du-Rhône, SE France 43°25´N 05°15´E
103 T14 **Provence-Alpes-Côte d'Azur** ◈ *region* SE France
20 H6 **Providence** Kentucky, S USA 37°23´N 87°47´W
19 N12 **Providence** *state capital* Rhode Island, NE USA 41°50´N 71°26´W
36 L1 **Providence** Utah, W USA 41°42´N 111°49´W
Providence *see* Fort Providence
Providence *see* Providence Atoll
67 X10 **Providence Atoll** *var.* Providence. *atoll* S Seychelles
14 D12 **Providence Bay** Manitoulin Island, Ontario, S Canada 45°39´N 82°16´W
23 R6 **Providence Canyon** *valley* Alabama/Georgia, S USA
22 I5 **Providence, Lake** ◎ Louisiana, S USA
35 X13 **Providence Mountains** ▲▲ California, W USA
44 L6 **Providenciales** *island* W Turks and Caicos Islands
19 Q12 **Provincetown** Massachusetts, NE USA 42°01´N 70°10´W
103 P5 **Provins** Seine-et-Marne, N France 48°34´N 03°18´E
36 L3 **Provo** Utah, W USA 40°14´N 111°39´W
11 R15 **Provost** Alberta, SW Canada 52°24´N 110°16´W
112 G13 **Prozor** Federacija Bosne I Hercegovine, SW Bosnia and Herzegovina 43°46´N 17°38´E
Prozoroki *see* Prazaroki
60 I11 **Prudentópolis** Paraná, S Brazil 25°12´S 50°58´W
39 R5 **Prudhoe Bay** Alaska, USA 70°16´N 148°18´W
39 R4 **Prudhoe Bay** *bay* Alaska, USA
111 H16 **Prudnik** *Ger.* Neustadt, Neustadt in Oberschlesien. Opole, SW Poland 50°20´N 17°34´E
119 J16 **Prudy** Minskaya Voblasts', C Belarus 53°47´N 26°32´E
101 D18 **Prüm** Rheinland-Pfalz, W Germany 50°15´N 06°27´E
101 D18 **Prüm** ≈ W Germany

◆ Country
● Country Capital
◇ Dependent Territory
○ Dependent Territory Capital
◈ Administrative Regions
✈ International Airport
▲ Mountain
▲▲ Mountain Range
Volcano
≈ River
◎ Lake
⊠ Reservoir

PRUSA – QAL'AT AḤMAD

Prusa *see* Bursa
110 *J7* **Pruszcz Gdański** *Ger.* Praust. Pomorskie, N Poland *54°16′N 18°36′E*
110 *M12* **Pruszków** *Ger.* Kaltdorf. Mazowieckie, C Poland *52°09′N 20°49′E*
116 *K8* **Prut** *Ger.* Pruth. ∿ E Europe
Pruth *see* Prut
108 *L8* **Prutz** Tirol, W Austria *47°07′N 10°42′E*
Pružana *see* Pruzhany
119 *G19* **Pruzhany** *Pol.* Pružana. Brestskaya Voblasts', SW Belarus *52°33′N 24°28′E*
124 *I11* **Pryazha** Respublika Kareliya, NW Russian Federation *61°42′N 33°39′E*
117 *U10* **Pryazovs'ke** Zaporiz'ka Oblast', SE Ukraine *46°43′N 35°39′E*
Prychornomor'ska Nyzovyna *see* Black Sea Lowland
Prydniprovs'ka Nyzovyna/ Prydnyaprowskaya Nizina *see* Dnieper Lowland
195 *Y7* **Prydz Bay** *bay* Antarctica
117 *R4* **Pryluky** *Rus.* Priluki. Chernihivs'ka Oblast', NE Ukraine *50°35′N 32°23′E*
117 *V10* **Prymors'k** *Rus.* Primorsk; *prev.* Primorskoye. Zaporiz'ka Oblast', SE Ukraine *46°44′N 36°19′E*
117 *U13* **Prymors'kyy** Avtonomna Respublika Krym, S Ukraine *45°09′N 35°33′E*
27 *Q9* **Pryor** Oklahoma, C USA *36°19′N 95°19′W*
33 *U11* **Pryor Creek** ∿ Montana, NW USA
Pryp"yat'/Prypyats' *see* Pripet
110 *M10* **Przasnysz** Mazowieckie, C Poland *53°01′N 20°51′E*
111 *K14* **Przedbórz** Łodzkie, S Poland *51°04′N 19°51′E*
111 *P17* **Przemyśl** *Rus.* Peremyshl. Podkarpackie, C Poland *49°47′N 22°47′E*
111 *O16* **Przeworsk** Podkarpackie, SE Poland *50°04′N 22°30′E*
Przheval'sk *see* Karakol
110 *L13* **Przysucha** Mazowieckie, SE Poland *51°22′N 20°36′E*
115 *H18* **Psachná** *var.* Psahna, Psakhná. Évvoia, C Greece *38°35′N 23°39′E*
Psahna/Psakhná *see* Psachná
115 *K18* **Psará** *island* E Greece
115 *I16* **Psathoúra** *island* Vóreies Sporádes, Greece, Aegean Sea
Pschestitz *see* Přeštice
Psein Lora *see* Pishin Lora
117 *S5* **Psel** *Rus.* Psël. ∿ Russian Federation/Ukraine
Psël *see* Psel
115 *M21* **Psérimos** *island* Dodekánisa, Greece, Aegean Sea
Pseyn Bowr *see* Pishin Lora
Pskem *see* Piskom
147 *R8* **Pskemskiy Khrebet** *Uzb.* Piskom Tizmasi. ▲▲ Kyrgyzstan/Uzbekistan
124 *F14* **Pskov** *Ger.* Pleskau, *Latv.* Pleskava. Pskovskaya Oblast', W Russian Federation *58°32′N 31°15′E*
118 *K6* **Pskov, Lake** *Est.* Pihkva Järv, *Ger.* Pleskauer See, *Rus.* Pskovskoye Ozero. ◎ Estonia/Russian Federation
124 *F15* **Pskovskaya Oblast'** ◈ *province* W Russian Federation
Pskovskoye Ozero *see* Pskov, Lake
112 *G9* **Psunj** ▲▲ NE Croatia
111 *J17* **Pszczyna** *Ger.* Pless. Śląskie, S Poland *49°59′N 18°54′E*
Ptačnik/Ptacsnik *see* Vtáčnik
115 *D17* **Ptéri** ▲ C Greece *39°08′N 21°32′E*
Ptich' *see* Ptsich
115 *E14* **Ptolemaïda** *prev.* Ptolemaïs. Dytikí Makedonía, N Greece *40°34′N 21°42′E*
Ptolemaïs *see* Ptolemaïda, Greece
Ptolemaïs *see* 'Akko, Israel
119 *M19* **Ptsich** *Rus.* Ptich'. Homyel'skaya Voblasts', SE Belarus *52°11′N 28°49′E*
119 *M18* **Ptsich** *Rus.* Ptich'. ∿ SE Belarus
109 *X10* **Ptuj** *Ger.* Pettau; *anc.* Poetovio. NE Slovenia *46°26′N 15°54′E*
61 *A23* **Puán** Buenos Aires, E Argentina *37°35′S 62°45′W*
192 *H15* **Pu'apu'a** Savai'i, C Samoa *13°32′S 172°09′W*
192 *G15* **Puava, Cape** *headland* Savai'i, NW Samoa
Pubao *see* Baingoin
56 *F12* **Pucallpa** Ucayali, C Peru *08°21′S 74°33′W*
57 *J17* **Pucarani** La Paz, NW Bolivia *16°25′S 68°29′W*
Pučarevo *see* Novi Travnik
157 *U12* **Pucheng** Shaanxi, SE China *35°00′N 109°34′E*
160 *L6* **Pucheng** *var.* Nanpu. Fujian, C China *27°59′N 118°31′E*
125 *N16* **Puchezh** Ivanovskaya Oblast', W Russian Federation *56°58′N 41°08′E*
111 *I19* **Púchov** *Hung.* Puhó. Trenčiansky Kraj, W Slovakia *49°08′N 18°15′E*
116 *J13* **Pucioasa** Dâmbovița, S Romania *45°04′N 25°23′E*
110 *I6* **Puck** Pomorskie, N Poland *54°43′N 18°24′E*
30 *L8* **Puckaway Lake** ◎ Wisconsin, N USA
63 *G15* **Pucón** Araucanía, S Chile *39°18′S 71°52′W*
93 *M14* **Pudasjärvi** Pohjois-Pohjanmaa, C Finland *65°20′N 27°02′E*
148 *L8* **Pūdeh Tal, Shelleh-ye** ∿ SW Afghanistan
127 *S1* **Pudem** Udmurtskaya Respublika, NW Russian Federation *58°18′N 52°08′E*
Pudewitz *see* Pobiedziska
124 *K11* **Pudozh** Respublika Kareliya, NW Russian Federation *61°48′N 36°30′E*
97 *M17* **Pudsey** N England, United Kingdom *53°48′N 01°40′W*
Puduchcheri *see* Puducherry
151 *I20* **Puducherry** *prev.* Pondicherry, *var.* Puduchcheri, *Fr.* Pondichéry. ◈ *union territory* India
151 *H21* **Pudukkottai** Tamil Nādu, SE India *10°23′N 78°47′E*
171 *Z13* **Pue** Papua, E Indonesia *02°42′S 140°36′E*
41 *P14* **Puebla** *var.* Puebla de Zaragoza. Puebla, S Mexico *19°02′N 98°13′W*
41 *P15* **Puebla** ◈ *state* S Mexico
104 *L11* **Puebla de Alcocer** Extremadura, W Spain *38°59′N 05°14′W*
Puebla de Don Fabrique *see* Puebla de Don Fadrique
105 *P13* **Puebla de Don Fadrique** *var.* Puebla de Don Fabrique. Andalucía, S Spain *37°58′N 02°25′W*
104 *J11* **Puebla de la Calzada** Extremadura, W Spain *38°54′N 06°38′W*
104 *J5* **Puebla de Sanabria** Castilla y León, N Spain *42°04′N 06°38′W*
Puebla de Trives *see* A Pobla de Trives
Puebla de Zaragoza *see* Puebla
37 *T6* **Pueblo** Colorado, C USA *38°15′N 104°37′W*
37 *N10* **Pueblo Colorado Wash** *valley* Arizona, SW USA
61 *C16* **Pueblo Libertador** Corrientes, NE Argentina *30°13′S 59°23′W*
40 *J10* **Pueblo Nuevo** Durango, C Mexico *23°24′N 105°21′W*
42 *J8* **Pueblo Nuevo** Estelí, NW Nicaragua *13°21′N 86°30′W*
54 *J3* **Pueblo Nuevo** Falcón, N Venezuela *11°59′N 69°57′W*
42 *B6* **Pueblo Nuevo Tiquisate** *var.* Tiquisate. Escuintla, SW Guatemala *14°16′N 91°21′W*
41 *Q11* **Pueblo Viejo, Laguna de** *lagoon* E Mexico
63 *J14* **Puelches** La Pampa, C Argentina *38°08′S 65°56′W*
104 *L14* **Puente-Genil** Andalucía, S Spain *37°23′N 04°45′W*
105 *Q3* **Puente la Reina** *Bas.* Gares. Navarra, N Spain *42°40′N 01°49′W*
104 *L12* **Puente Nuevo, Embalse de** ⊠ S Spain
57 *D14* **Puente Piedra** Lima, W Peru *11°49′S 77°01′W*
160 *F14* **Pu'er** *var.* Ning'er. Yunnan, SW China *23°09′N 100°58′E*
45 *V6* **Puerca, Punta** *headland* E Puerto Rico *18°13′N 65°36′W*
37 *R12* **Puerco, Rio** ∿ New Mexico, SW USA
57 *J17* **Puerto Acosta** La Paz, W Bolivia *15°33′S 69°15′W*
63 *G19* **Puerto Aisén** Aisén, S Chile *45°24′S 72°42′W*
41 *R17* **Puerto Ángel** Oaxaca, SE Mexico *15°39′N 96°29′W*
Puerto Argentino *see* Stanley
41 *T17* **Puerto Arista** Chiapas, SE Mexico *15°55′N 93°47′W*
43 *O16* **Puerto Armuelles** Chiriquí, SW Panama *08°19′N 82°51′W*
Puerto Arrecife *see* Arrecife
54 *D14* **Puerto Asís** Putumayo, SW Colombia *0°31′N 76°31′W*
54 *L9* **Puerto Ayacucho** Amazonas, SW Venezuela *05°45′N 67°37′W*
57 *C18* **Puerto Ayora** Galapagos Islands, Ecuador, E Pacific Ocean *0°45′S 90°19′W*
57 *C18* **Puerto Baquerizo Moreno** *var.* Baquerizo Moreno. Galapagos Islands, Ecuador, E Pacific Ocean *0°54′S 89°37′W*
42 *G4* **Puerto Barrios** Izabal, E Guatemala *15°42′N 88°34′W*
Puerto Bello *see* Portobelo
54 *F8* **Puerto Berrío** Antioquia, C Colombia *06°28′N 74°28′W*
54 *F9* **Puerto Boyaca** Boyacá, C Colombia *05°58′N 74°36′W*
54 *K4* **Puerto Cabello** Carabobo, N Venezuela *10°29′N 68°02′W*
43 *N7* **Puerto Cabezas** *var.* Bilwi. Región Autónoma Atlántico Norte, NE Nicaragua *14°05′N 83°22′W*
54 *L9* **Puerto Carreño** Vichada, E Colombia *06°08′N 67°30′W*
54 *E4* **Puerto Colombia** Atlántico, N Colombia *10°59′N 74°57′W*
42 *H4* **Puerto Cortés** Cortés, NW Honduras *15°50′N 87°55′W*
54 *J4* **Puerto Cumarebo** Falcón, N Venezuela *11°29′N 69°21′W*
Puerto de Cabras *see* Puerto del Rosario
55 *Q5* **Puerto de Hierro** Sucre, NE Venezuela *10°40′N 62°03′W*
64 *O11* **Puerto de la Cruz** Tenerife, Islas Canarias, Spain, NE Atlantic Ocean *28°24′N 16°33′W*
64 *Q11* **Puerto del Rosario** *var.* Puerto de Cabras. Fuerteventura, Islas Canarias, Spain, NE Atlantic Ocean *28°29′N 13°52′W*
63 *J20* **Puerto Deseado** Santa Cruz, SE Argentina *47°46′S 65°53′W*
40 *F8* **Puerto Escondido** Baja California Sur, NW Mexico *25°48′N 111°20′W*
41 *R17* **Puerto Escondido** Oaxaca, SE Mexico *15°48′N 96°57′W*
60 *G12* **Puerto Esperanza** Misiones, NE Argentina *26°01′S 54°39′W*
54 *H10* **Puerto Gaitán** Meta, C Colombia *04°20′N 72°10′W*
Puerto Gallegos *see* Río Gallegos
60 *G12* **Puerto Iguazú** Misiones, NE Argentina *25°39′S 54°35′W*
56 *F12* **Puerto Inca** Huánuco, N Peru *09°22′S 74°54′W*
54 *L11* **Puerto Inírida** *var.* Obando. Guainía, E Colombia *03°48′N 67°54′W*
42 *K13* **Puerto Jesús** Guanacaste, NW Costa Rica *10°08′N 85°26′W*
41 *Z11* **Puerto Juárez** Quintana Roo, SE Mexico *21°06′N 86°46′W*
55 *N5* **Puerto La Cruz** Anzoátegui, NE Venezuela *10°14′N 64°40′W*
54 *E14* **Puerto Leguízamo** Putumayo, S Colombia *0°14′N′S 74°45′W*
43 *N5* **Puerto Lempira** Gracias a Dios, E Honduras *15°14′N 83°48′W*
Puerto Libertad *see* La Libertad
54 *I11* **Puerto Limón** Meta, E Colombia *04°00′N 71°09′W*
54 *D13* **Puerto Limón** Putumayo, SW Colombia *01°02′N 76°30′W*
Puerto Limón *see* Limón
105 *N11* **Puertollano** Castilla-La Mancha, C Spain *38°41′N 04°07′W*
63 *K17* **Puerto Lobos** Chubut, SE Argentina *42°00′S 64°58′W*
54 *I3* **Puerto López** La Guajira, N Colombia *11°54′N 71°21′W*
105 *Q14* **Puerto Lumbreras** Murcia, SE Spain *37°35′N 01°49′W*
41 *V17* **Puerto Madero** Chiapas, SE Mexico *14°44′N 92°25′W*
63 *K17* **Puerto Madryn** Chubut, S Argentina *42°45′S 65°02′W*
Puerto Magdalena *see* Bahía Magdalena
57 *J15* **Puerto Maldonado** Madre de Dios, E Peru *12°37′S 69°11′W*
Puerto Masachapa *see* Masachapa
Puerto México *see* Coatzacoalcos
63 *G17* **Puerto Montt** Los Lagos, C Chile *41°28′S 72°57′W*
41 *Z12* **Puerto Morelos** Quintana Roo, SE Mexico *20°47′N 86°54′W*
54 *L10* **Puerto Nariño** Vichada, E Colombia *04°57′N 67°51′W*
63 *H23* **Puerto Natales** Magallanes, S Chile *51°42′S 72°28′W*
43 *X15* **Puerto Obaldía** Kuna Yala, NE Panama *08°38′N 77°26′W*
44 *H6* **Puerto Padre** Las Tunas, E Cuba *21°13′N 76°35′W*
54 *L9* **Puerto Páez** Apure, C Venezuela *06°10′N 67°30′W*
40 *E3* **Puerto Peñasco** Sonora, NW Mexico *31°20′N 113°35′W*
55 *N5* **Puerto Píritu** Anzoátegui, NE Venezuela *10°04′N 65°00′W*
45 *N8* **Puerto Plata** *var.* San Felipe de Puerto Plata. N Dominican Republic *19°46′N 70°42′W*
Puerto Presidente Stroessner *see* Ciudad del Este
171 *N6* **Puerto Princesa** *off.* Puerto Princesa City. Palawan, W Philippines *09°48′N 118°43′E*
Puerto Princesa City *see* Puerto Princesa
Puerto Príncipe *see* Camagüey
Puerto Quellón *see* Quellón
60 *F13* **Puerto Rico** Misiones, NE Argentina *26°48′S 54°59′W*
57 *K14* **Puerto Rico** Pando, N Bolivia *11°07′S 67°32′W*
54 *E12* **Puerto Rico** Caquetá, S Colombia *01°54′N 75°13′W*
45 *U5* **Puerto Rico** *off.* Commonwealth of Puerto Rico; *prev.* Porto Rico. ◇ *US commonwealth territory* C West Indies
64 *F11* **Puerto Rico** *island* C West Indies
Puerto Rico, Commonwealth of *see* Puerto Rico
64 *G11* **Puerto Rico Trench** *undersea feature* NE Caribbean Sea
54 *I8* **Puerto Rondón** Arauca, E Colombia *06°16′N 71°05′W*
Puerto San José *see* San José
63 *J21* **Puerto San Julián** *var.* San Julián. Santa Cruz, SE Argentina *49°14′S 67°41′W*
63 *I22* **Puerto Santa Cruz** *var.* Santa Cruz. Santa Cruz, SE Argentina *50°05′S 68°31′W*
Puerto Sauce *see* Juan L. Lacaze
57 *Q20* **Puerto Suárez** Santa Cruz, E Bolivia *18°59′S 57°47′W*
54 *D13* **Puerto Umbría** Putumayo, SW Colombia *0°52′N 76°36′W*
40 *J13* **Puerto Vallarta** Jalisco, SW Mexico *20°36′N 105°15′W*
63 *G16* **Puerto Varas** Los Lagos, C Chile *41°20′S 73°00′W*
42 *M13* **Puerto Viejo** Heredia, NE Costa Rica *10°27′N 84°00′W*
Puertoviejo *see* Portoviejo
57 *B18* **Puerto Villamil** *var.* Villamil. Galapagos Islands, Ecuador, E Pacific Ocean *0°57′S 91°00′W*
54 *F8* **Puerto Wilches** Santander, N Colombia *07°22′N 73°53′W*
63 *H20* **Pueyrredón, Lago** *var.* Lago Cochrane. ◎ S Argentina
127 *R7* **Pugachëv** Saratovskaya Oblast', W Russian Federation *52°06′N 48°50′E*
127 *T3* **Pugachëvo** Udmurtskaya Respublika, NW Russian Federation *56°38′N 53°03′E*
32 *H8* **Puget Sound** *sound* Washington, NW USA
107 *O17* **Puglia** *var.* Le Puglie, *Eng.* Apulia. ◈ *region* SE Italy
107 *N17* **Puglia, Canosa di** *anc.* Canusium. Puglia, SE Italy *41°13′N 16°04′E*
118 *I6* **Puhja** *Ger.* Kawelecht. Tartumaa, SE Estonia *58°20′N 26°19′E*
Puhó *see* Púchov
105 *V4* **Puigcerdà** Cataluña, NE Spain *42°25′N 01°53′E*
Puigmal *see* Puigmal d'Err
103 *N17* **Puigmal d'Err** *var.* Puigmal. ▲ S France *42°24′N 02°07′E*
76 *I16* **Pujehun** S Sierra Leone *07°23′N 11°44′W*
Puka *see* Pukë
185 *E20* **Pukaki, Lake** ◎ South Island, New Zealand
38 *F10* **Pukalani** Maui, Hawaii, USA, C Pacific Ocean *20°50′N 156°20′W*
190 *J13* **Pukapuka** *atoll* N Cook Islands
191 *X9* **Pukapuka** *atoll* Îles Tuamotu, E French Polynesia
Pukari Neem *see* Purekkari Neem
191 *X11* **Pukarua** *var.* Pukaruha. *atoll* Îles Tuamotu, E French Polynesia
Pukaruha *see* Pukarua
14 *A7* **Pukaskwa** ∿ Ontario, S Canada
11 *V12* **Pukatawagan** Manitoba, C Canada *55°46′N 101°14′W*
191 *X16* **Pukatikei, Maunga** ⛰ Easter Island, Chile, E Pacific Ocean
182 *C1* **Pukatja** *var.* Ernabella. South Australia *26°18′S 132°13′E*
163 *Y12* **Pukch'ŏng** E North Korea *40°13′N 128°20′E*
113 *L18* **Pukë** *var.* Puka. Shkodër, N Albania *42°03′N 19°53′E*
184 *L6* **Pukekohe** Auckland, North Island, New Zealand *37°12′S 174°54′E*
184 *L7* **Pukemiro** Waikato, North Island, New Zealand *37°37′S 175°02′E*
190 *D12* **Puke, Mont** ▲ Île Futuna, W Wallis and Futuna
Puket *see* Phuket
185 *C20* **Puketeraki Range** ▲▲ South Island, New Zealand
184 *N13* **Puketoi Range** ▲▲ North Island, New Zealand
185 *F21* **Pukeuri Junction** Otago, South Island, New Zealand *45°01′S 171°01′E*
119 *L16* **Pukhavichy** *Rus.* Pukhovichi. Minskaya Voblasts', C Belarus *53°32′N 28°15′E*
Pukhovichi *see* Pukhavichy
124 *M10* **Puksoozero** Arkhangel'skaya Oblast', NW Russian Federation *62°37′N 40°29′E*
112 *A10* **Pula** *It.* Pola; *prev.* Pulj. Istra, NW Croatia *44°53′N 13°51′E*
Pula *see* Nyingchi
163 *U14* **Pulandian** *var.* Xinjin. Liaoning, NE China *39°25′N 121°58′E*
163 *T14* **Pulandian Wan** *bay* NE China
189 *O15* **Pulap Atoll** *atoll* Caroline Islands, C Micronesia
18 *H9* **Pulaski** New York, NE USA *43°34′N 76°06′W*
20 *I10* **Pulaski** Tennessee, S USA *35°11′N 87°00′W*
21 *R7* **Pulaski** Virginia, NE USA *37°03′N 80°47′W*
171 *Y14* **Pulau, Sungai** ∿ Papua, E Indonesia
110 *N13* **Puławy** *Ger.* Neu Amerika. Lubelskie, E Poland *51°25′N 21°57′E*
149 *R5* **Pul-e-'Alam** *prev.* Pol-e-'Alam. Lōgar, E Afghanistan *33°59′N 69°02′E*
149 *Q3* **Pul-e Khumrī** *prev.* Pol-e Khomrī. Baghlān, NE Afghanistan *35°55′N 68°45′E*
146 *I16* **Pulhatyn** *Rus.* Polekhatum; *prev.* Pul'-I-Khatum. Ahal Welaýaty, S Turkmenistan *36°01′N 61°08′E*
101 *E16* **Pulheim** Nordrhein-Westfalen, W Germany *51°00′N 06°48′E*
155 *J19* **Pulicat Lake** *lagoon* SE India
Pul'-I-Khatum *see* Pulhatyn
Pul-i-Sefid *see* Pol-e Sefīd
Pulj *see* Pula
109 *W2* **Pulkau** ∿ NE Austria
93 *L15* **Pulkkila** Pohjois-Pohjanmaa, C Finland *64°15′N 25°53′E*
122 *C7* **Pulkovo** ✈ (Sankt-Peterburg) Leningradskaya Oblast', NW Russian Federation *60°06′N 30°23′E*
32 *M9* **Pullman** Washington, NW USA *46°43′N 117°10′W*
108 *B10* **Pully** Vaud, SW Switzerland *46°31′N 06°40′E*
40 *F7* **Púlpita, Punta** *headland* NW Mexico *26°30′N 111°28′W*
110 *M10* **Pułtusk** Mazowieckie, C Poland *52°41′N 21°04′E*
158 *H10* **Pulu** Xinjiang Uygur Zizhiqu, W China *36°10′N 81°29′E*
137 *P13* **Pülümür** Tunceli, E Turkey *39°30′N 39°54′E*
189 *N16* **Pulusuk** *island* Caroline Islands, C Micronesia
189 *N16* **Puluwat Atoll** *atoll* Caroline Islands, C Micronesia
25 *N11* **Pumpville** Texas, SW USA *39°55′N 101°43′W*
191 *P7* **Punaauia** *var.* Hakapehi. Tahiti, W French Polynesia *17°38′S 149°37′W*
56 *B8* **Puná, Isla** *island* SW Ecuador
185 *G16* **Punakaiki** West Coast, South Island, New Zealand *42°07′S 171°21′E*
153 *T11* **Punakha** C Bhutan *27°38′N 89°50′E*
57 *L18* **Punata** Cochabamba, C Bolivia *17°32′S 65°50′W*
155 *E14* **Pune** *prev.* Poona. Mahārāshtra, W India *18°32′N 73°52′E*
83 *M17* **Pungoè, Rio** *var.* Púnguè, Pungwe. ∿ C Mozambique
21 *X10* **Pungo River** ∿ North Carolina, SE USA
Púnguè/Pungwe *see* Pungoè, Rio
79 *N19* **Punia** Maniema, E Dem. Rep. Congo *01°28′S 26°25′E*
62 *H8* **Punilla, Sierra de la** ▲▲ W Argentina
161 *P14* **Puning** Guangdong, S China *23°24′N 116°14′E*
62 *G10* **Punitaqui** Coquimbo, C Chile *30°50′S 71°29′W*
149 *T9* **Punjab** *prev.* West Punjab, Western Punjab. ◈ *province* E Pakistan
152 *H8* **Punjab** *state* NW India
129 *Q9* **Punjab Plains** *plain* N India
93 *O17* **Punkaharju** *var.* Punkasalmi. Etelä-Savo, E Finland *61°45′N 29°21′E*
Punkasalmi *see* Punkaharju
57 *I17* **Puno** Puno, SE Peru *15°53′S 70°03′W*
57 *H17* **Puno** *off.* Departamento de Puno. ◈ *department* S Peru
Puno, Departamento de *see* Puno
61 *B24* **Punta Alta** Buenos Aires, E Argentina *38°54′S 62°01′W*
63 *H24* **Punta Arenas** *prev.* Magallanes. Magallanes, S Chile *53°10′S 70°56′W*
45 *T6* **Punta, Cerro de** ▲ C Puerto Rico *18°10′N 66°36′W*
43 *T15* **Punta Chame** Panamá, C Panama *08°39′N 79°42′W*
57 *G17* **Punta Colorada** Arequipa, SW Peru *16°17′S 72°31′W*
40 *F9* **Punta Coyote** Baja California Sur, NW Mexico
62 *G8* **Punta de Díaz** Atacama, N Chile *28°03′S 70°36′W*
61 *G20* **Punta del Este** Maldonado, S Uruguay *34°59′S 54°58′W*
63 *K17* **Punta Delgada** Chubut, SE Argentina *42°46′S 63°40′W*
55 *O5* **Punta de Mata** Monagas, NE Venezuela *09°48′N 63°38′W*
55 *O4* **Punta de Piedras** Nueva Esparta, NE Venezuela *10°57′N 64°06′W*
42 *F4* **Punta Gorda** Toledo, SE Belize *16°07′N 88°47′W*
43 *N11* **Punta Gorda** Región Autónoma Atlántico Sur, SE Nicaragua *11°31′N 83°46′W*
23 *W14* **Punta Gorda** Florida, SE USA *26°55′N 82°03′W*
42 *M11* **Punta Gorda, Río** ∿ SE Nicaragua
62 *H6* **Punta Negra, Salar de** *salt lake* N Chile
40 *D5* **Punta Prieta** Baja California Norte, NW Mexico *28°56′N 114°11′W*
42 *L13* **Puntarenas** Puntarenas, W Costa Rica *09°58′N 84°50′W*
42 *L13* **Puntarenas** *off.* Provincia de Puntarenas. ◈ *province* W Costa Rica
Puntarenas, Provincia de *see* Puntarenas
80 *P13* **Puntland** *cultural region* NE Somalia
54 *J4* **Punto Fijo** Falcón, N Venezuela *11°42′N 70°13′W*
105 *S4* **Puntón de Guara** ▲ N Spain *42°18′N 00°13′W*
18 *D14* **Punxsutawney** Pennsylvania, NE USA *40°55′N 78°57′W*
93 *M14* **Puolanka** Kainuu, C Finland *64°51′N 27°42′E*
57 *J17* **Pupuya, Nevado** ▲ W Bolivia *15°04′S 69°01′W*
Puqi *see* Chibi
57 *F16* **Puquio** Ayacucho, S Peru *14°44′S 74°07′W*
122 *J9* **Pur** ∿ N Russian Federation
186 *D7* **Purari** ∿ S Papua New Guinea
27 *N11* **Purcell** Oklahoma, C USA *35°00′N 97°21′W*
11 *O16* **Purcell Mountains** ▲▲ British Columbia, SW Canada
105 *P14* **Purchena** Andalucía, S Spain *37°21′N 02°21′W*
27 *S8* **Purdy** Missouri, C USA *36°49′N 93°55′W*
118 *I2* **Purekkari Neem** *prev.* Pukari Neem. *headland* N Estonia *59°33′N 24°49′E*
37 *U7* **Purgatoire River** ∿ Colorado, C USA
Purgstall *see* Purgstall an der Erlauf
109 *V5* **Purgstall an der Erlauf** *var.* Purgstall. Niederösterreich, NE Austria *48°01′N 15°08′E*
154 *O13* **Puri** *var.* Jagannath. Odisha, E India *19°52′N 85°49′E*
Puriramya *see* Buriram
109 *X4* **Purkersdorf** Niederösterreich, NE Austria *48°13′N 16°12′E*
98 *I9* **Purmerend** Noord-Holland, C Netherlands *52°30′N 04°56′E*
151 *G16* **Pūrna** ∿ C India
Purnea *see* Pūrnia
153 *R13* **Pūrnia** *prev.* Purnea. Bihār, NE India *25°47′N 87°28′E*
Pursat *see* Poŭthĭsăt, Poŭthĭsăt, W Cambodia
Pursat *see* Poŭthĭsăt, Stœng, W, Cambodia
Purulia *see* Puruliya
150 *L13* **Puruliya** *prev.* Purulia. West Bengal, NE India *23°20′N 86°24′E*
47 *G7* **Purus, Rio** *var.* Río Purús. ∿ Brazil/Peru
186 *C9* **Purutu Island** *island* SW Papua New Guinea
93 *N17* **Puruvesi** ◎ SE Finland
22 *L7* **Purvis** Mississippi, S USA *31°08′N 89°24′W*
Pŭrvomay *see* Parvomay
169 *R16* **Purwodadi** *prev.* Poerwodadi. Jawa, C Indonesia *07°05′S 110°53′E*
169 *P16* **Purwokerto** *prev.* Poerwokerto. Jawa, C Indonesia *07°25′S 109°14′E*
169 *P16* **Purworejo** *prev.* Poerworedjo. Jawa, C Indonesia *07°45′S 110°04′E*
20 *H8* **Puryear** Tennessee, S USA *36°25′N 88°21′W*
154 *H13* **Pusad** Mahārāshtra, C India *19°56′N 77°40′E*
Pusan *see* Busan
Pusan-gwangyŏksi *see* Busan
168 *H7* **Pusatgajo, Pegunungan** ▲▲ Sumatera, NW Indonesia
Puschlav *see* Poschiavo
124 *G13* **Pushkin** *prev.* Tsarskoye Selo. Leningradskaya Oblast', NW Russian Federation *59°42′N 30°24′E*
126 *L3* **Pushkino** Moskovskaya Oblast', W Russian Federation *55°57′N 37°45′E*
127 *Q8* **Pushkino** Saratovskaya Oblast', W Russian Federation *51°09′N 47°00′E*
Pushkino *see* Biläsuvar
111 *M22* **Püspökladány** Hajdú-Bihar, E Hungary *47°20′N 21°05′E*
118 *J3* **Püssi** *Ger.* Isenhof. Ida-Virumaa, NE Estonia *59°22′N 27°04′E*
116 *I5* **Pustomyty** L'vivs'ka Oblast', W Ukraine *49°43′N 23°55′E*
124 *F16* **Pustoshka** Pskovskaya Oblast', W Russian Federation *56°21′N 29°16′E*
Pusztakalán *see* Călan
167 *N1* **Putao** *prev.* Fort Hertz. Kachin State, N Myanmar (Burma) *27°22′N 97°24′E*
184 *M8* **Putaruru** Waikato, North Island, New Zealand *38°03′S 175°48′E*
Puteoli *see* Pozzuoli
161 *R12* **Putian** Fujian, SE China *25°32′N 119°02′E*
107 *O17* **Putignano** Puglia, SE Italy *40°51′N 17°08′E*
Puting *see* De'an
Putivl' *see* Putyvl'
41 *Q16* **Putla** *var.* Putla de Guerrero. Oaxaca, SE Mexico *17°01′N 97°56′W*
Putla de Guerrero *see* Putla
19 *N12* **Putnam** Connecticut, NE USA *41°56′N 71°52′W*
25 *Q7* **Putnam** Texas, SW USA *32°22′N 99°11′W*
18 *M10* **Putney** Vermont, NE USA *42°59′N 72°30′W*
111 *L20* **Putnok** Borsod-Abaúj-Zemplén, NE Hungary *48°18′N 20°25′E*
Putorana, Gory/Putorana Mountains *see* Putorana, Plato
122 *L8* **Putorana, Plato** *var.* Gory Putorana, *Eng.* Putorana Mountains. ▲▲ N Russian Federation
168 *K9* **Putrajaya** ● (Malaysia) Kuala Lumpur, Peninsular Malaysia *02°57′N 101°42′E*
62 *H2* **Putre** Arica y Parinacota, N Chile *18°11′S 69°30′W*
155 *J24* **Puttalam** North Western Province, W Sri Lanka *08°02′N 79°55′E*
155 *J24* **Puttalam Lagoon** *lagoon* W Sri Lanka
99 *H17* **Putte** Antwerpen, C Belgium *51°04′N 04°39′E*
94 *E10* **Puttegga** ▲ S Norway *62°13′N 07°40′E*
98 *K11* **Putten** Gelderland, C Netherlands *52°15′N 05°36′E*
100 *K7* **Puttgarden** Schleswig-Holstein, N Germany *54°30′N 11°13′E*
Puttiala *see* Patiāla
101 *D20* **Püttlingen** Saarland, SW Germany *49°16′N 06°52′E*
54 *D14* **Putumayo** *off.* Intendencia del Putumayo. ◈ *province* S Colombia
Putumayo, Intendencia del *see* Putumayo
48 *E7* **Putumayo, Río** *var.* Içá, Rio. ∿ NW South America *see also* Içá, Rio
Putumayo, Río *see* Içá, Rio
169 *P11* **Putus, Tanjung** *headland* Borneo, N Indonesia *0°27′S 109°04′E*
116 *J8* **Putyla** Chernivets'ka Oblast', W Ukraine *47°59′N 25°04′E*
117 *S3* **Putyvl'** *Rus.* Putivl'. Sums'ka Oblast', NE Ukraine *51°21′N 33°53′E*
93 *M18* **Puula** ◎ SE Finland
93 *N18* **Puumala** Etelä-Savo, E Finland *61°31′N 28°12′E*
118 *I5* **Puurmani** *Ger.* Talkhof. Jõgevamaa, E Estonia *58°36′N 26°17′E*
99 *G17* **Puurs** Antwerpen, N Belgium *51°05′N 04°17′E*
38 *F10* **Pu'u 'Ula'ula** *var.* Red Hill. ▲ Maui, Hawai'i, USA *20°42′N 156°16′W*
38 *A8* **Pu'uwai** *var.* Puuwai. Ni'ihau, Hawaii, USA, C Pacific Ocean *21°54′N 160°11′W*
12 *J4* **Puvirnituq** *prev.* Povungnituk. Québec, NE Canada *60°10′N 77°20′W*
12 *J3* **Puvirnituq, Riviere de** *prev.* Rivière de Povungnituk. ∿ Québec, NE Canada
32 *H8* **Puyallup** Washington, NW USA *47°11′N 122°17′W*
161 *O5* **Puyang** Henan, C China *35°40′N 115°00′E*
103 *O11* **Puy-de-Dôme** ◈ *department* C France
103 *N15* **Puylaurens** Tarn, S France *43°33′N 02°01′E*
102 *M13* **Puy-l'Evêque** Lot, S France *44°31′N 01°10′E*
103 *N17* **Puymorens, Col de** *pass* S France
56 *C7* **Puyo** Pastaza, C Ecuador *01°30′S 77°58′W*
185 *A24* **Puysegur Point** *headland* South Island, New Zealand *46°09′S 166°38′E*
148 *J8* **Pūzak, Hāmūn-e** *Pash.* Hāmūn-i-Puzak. ◎ SW Afghanistan
Puzak, Hāmūn-i- *see* Pūzak, Hāmūn-e
81 *J23* **Pwani** *Eng.* Coast. ◈ *region* E Tanzania
79 *O23* **Pweto** Katanga, SE Dem. Rep. Congo *08°28′S 28°52′E*
97 *I19* **Pwllheli** NW Wales, United Kingdom *52°54′N 04°23′W*
189 *O14* **Pwok** Pohnpei, E Micronesia
122 *I9* **Pyakupur** ∿ N Russian Federation
124 *M6* **Pyalitsa** Murmanskaya Oblast', NW Russian Federation *66°17′N 39°56′E*
124 *K10* **Pyal'ma** Respublika Kareliya, NW Russian Federation *62°24′N 35°56′E*
Pyandzh *see* Panj
166 *L9* **Pyapon** Ayeyawady, SW Myanmar (Burma) *16°15′N 95°40′E*
119 *J15* **Pyarshai** *Rus.* Pershay. Minskaya Voblasts', C Belarus *54°02′N 26°41′E*
114 *I10* **Pyasachnik, Yazovir** *var.* Yazovir Pyasŭchnik. ⊠ C Bulgaria
122 *K8* **Pyasina** ∿ N Russian Federation
119 *G17* **Pyaski** *Rus.* Peski; *prev.* Pyeski. Hrodzyenskaya Voblasts', W Belarus *53°21′N 24°38′E*
Pyasŭchnik, Yazovir *see* Pyasachnik, Yazovir
Pyatikhatki *see* P"yatykhatky
117 *S7* **P"yatykhatky** *Rus.* Pyatikhatki. Dnipropetrovs'ka Oblast', E Ukraine *48°23′N 33°43′E*
166 *M6* **Pyawbwe** Mandalay, C Myanmar (Burma) *20°39′N 96°04′E*
166 *L7* **Pyay** *var.* Prome, Pye. Bago, C Myanmar (Burma) *18°50′N 95°14′E*
127 *T3* **Pychas** Udmurtskaya Respublika, NW Russian Federation *56°30′N 52°33′E*
Pye *see* Pyay
166 *K6* **Pyechin** Chin State, W Myanmar (Burma) *20°01′N 93°36′E*
163 *X15* **Pyeongtaek** *prev.* P'yŏngt'aek. NW South Korea *37°00′N 127°04′E*
Pyeski *see* Pyaski
119 *L19* **Pyetrykaw** *Rus.* Petrikov. Homyel'skaya Voblasts', SE Belarus *52°08′N 28°30′E*
93 *O17* **Pyhäjärvi** ◎ SE Finland
93 *M16* **Pyhajärvi** ◎ C Finland
93 *L15* **Pyhäjoki** Pohjois-Pohjanmaa, W Finland *64°28′N 24°15′E*
93 *L15* **Pyhäjoki** ∿ W Finland
93 *M15* **Pyhäntä** Pohjois-Pohjanmaa, C Finland *64°07′N 26°19′E*
93 *M16* **Pyhäsalmi** Pohjois-Pohjanmaa, C Finland *63°38′N 26°E*
93 *O17* **Pyhäselkä** ◎ SE Finland
93 *M19* **Pyhtää** *Swe.* Pyttis. Kymenlaakso, S Finland *60°29′N 26°40′E*
166 *M5* **Pyin-Oo-Lwin** *var.* Maymyo. Mandalay, C Myanmar (Burma) *22°03′N 96°30′E*
115 *N24* **Pylés** *var.* Piles. Kárpathos, SE Greece *35°31′N 27°08′E*
115 *D21* **Pýlos** *var.* Pilos. Pelopónnisos, S Greece *36°55′N 21°42′E*
18 *B12* **Pymatuning Reservoir** ⊠ Ohio/Pennsylvania, NE USA
P'yŏngt'aek *see* Pyeongtaek
163 *V14* **P'yŏngyang** *var.* P'yŏngyang-si, *Eng.* Pyongyang. ● (North Korea) SW North Korea *39°04′N 125°46′E*
P'yŏngyang-si *see* P'yŏngyang
35 *Q4* **Pyramid Lake** ◎ Nevada, W USA
37 *P15* **Pyramid Mountains** ▲▲ New Mexico, SW USA
37 *R5* **Pyramid Peak** ▲ Colorado, C USA *39°04′N 106°57′W*
115 *D17* **Pyramíva** *var.* Piramiva. ▲ C Greece *39°08′N 21°18′E*
Pyrenaei Montes *see* Pyrenees
86 *B12* **Pyrenees** *Fr.* Pyrénées, *Sp.* Pirineos; *anc.* Pyrenaei Montes. ▲▲ SW Europe
102 *J16* **Pyrénées-Atlantiques** ◈ *department* SW France
103 *N17* **Pyrénées-Orientales** ◈ *department* S France
115 *L19* **Pyrgí** *var.* Pirgi. Chíos, E Greece *38°13′N 26°01′E*
115 *D20* **Pýrgos** *var.* Pírgos. Dytikí Elláda, S Greece *37°40′N 21°27′E*
Pyritz *see* Pyrzyce
115 *E19* **Pýrros** ∿ S Greece
117 *R4* **Pyryatyn** *Rus.* Piryatin. Poltavs'ka Oblast', NE Ukraine *50°14′N 32°31′E*
110 *D9* **Pyrzyce** *Ger.* Pyritz. Zachodnio-pomorskie, NW Poland *53°09′N 14°53′E*
124 *F15* **Pytalovo** *Latv.* Abrene; *prev.* Jaunlatgale. Pskovskaya Oblast', W Russian Federation *57°06′N 27°55′E*
115 *M20* **Pythagóreio** *var.* Pithagorio. Sámos, Dodekánisa, Greece, Aegean Sea *37°42′N 26°57′E*
14 *L11* **Pythonga, Lac** ◎ Québec, SE Canada
Pyttis *see* Pyhtää
Pyu *see* Phyu
166 *M8* **Pyuntaza** Bago, SW Myanmar (Burma) *17°51′N 96°44′E*
153 *N11* **Pyuthān** Mid Western, W Nepal *28°09′N 82°50′E*
110 *H12* **Pyzdry** *Ger.* Peisern. Wielkopolskie, C Poland *52°10′N 17°42′E*

Q

138 *H13* **Qā' al Jafr** ◎ S Jordan
197 *O11* **Qaanaaq** *var.* Qânâq, *Dan.* Thule. ◈ Qaasuitsup, N Greenland
197 *P12* **Qaasuitsup** *off.* Qaasuitsup Kommunia. ◈ *municipality* NW Greenland
Qaasuitsup Kommunia *see* Qaasuitsup
Qabanbay *see* Kabanbay
138 *G7* **Qabb Eliâs** E Lebanon *33°46′N 35°49′E*
Qabil *see* Al Qābil
Qabırrı *see* Iori
Qābis *see* Gabès
Qābis, Khalīj *see* Gabès, Golfe de
Qabqa *see* Gonghe
141 *S14* **Qabr Hūd** C Yemen *16°02′N 49°36′E*
Qacentina *see* Constantine
Qādes *see* Qādis
148 *L4* **Qādis** *prev.* Qādes. Bādghīs, NW Afghanistan *34°48′N 63°26′E*
139 *T11* **Qādisīyah** Al Qādisīyah, S Iraq *31°43′N 44°28′E*
Qādisīyah, Muḥāfa at al *see* Al Qādisīyah
143 *O4* **Qā'emshahr** *prev.* 'Alīābad, Shāhī. Māzandarān, N Iran *36°31′N 52°49′E*
143 *U7* **Qā'en** *var.* Qain, Qāyen. Khorāsān-e Jonūbī, E Iran *33°43′N 59°07′E*
141 *U13* **Qafa** *spring/well* SW Oman *17°46′N 52°55′E*
Qafṣah *see* Gafsa
163 *Q12* **Qagan Nur** *var.* Xulun Hobot Qagan, Zhengxiangbai Qi. Nei Mongol Zizhiqu, N China *42°10′N 114°57′E*
163 *V9* **Qagan Nur** ◎ NE China
163 *Q11* **Qagan Nur** ◎ N China
Qagan Us *see* Dulan
158 *H13* **Qagcaka** Xizang Zizhiqu, W China *32°32′N 81°52′E*
Qagchêng *see* Xiangcheng
Qahremānshahr *see* Kermānshāh
159 *Q10* **Qaidam He** ∿ C China
156 *L8* **Qaidam Pendi** *basin* C China
Qain *see* Qā'en
Qala Āhangarān *see* Chaghcharān
Qaḷā Diza *see* Qeładize
Qal'ah Sālih *see* Qal'at Şāliḥ
149 *N4* **Qal'ah Shahr** *Pash.* Qala Shāhar; *prev.* Qal'eh Shahr. Sar-e Pul, N Afghanistan *35°34′N 65°38′E*
148 *L4* **Qal'ah-ye Now** *var.* Qala Nau; *prev.* Qal'eh-ye Now. Bādghīs, NW Afghanistan *35°N 63°08′E*
147 *R13* **Qal'aikhum** *Rus.* Kalaikhum. S Tajikistan *38°28′N 70°49′E*
Qala Nau *see* Qal'ah-ye Now
141 *V17* **Qalansīyah** Suquṭrā, W Yemen *12°40′N 53°30′E*
Qala Panja *see* Qal'eh-ye Panjeh
Qala Shāhar *see* Qal'ah Shahr
149 *O8* **Qalāt** *Per.* Kalāt. Zābul, S Afghanistan *32°10′N 66°54′E*
139 *W9* **Qal'at Aḥmad** Maysān, E Iraq *32°24′N 46°46′E*

◆ Country
● Country Capital
◇ Dependent Territory
○ Dependent Territory Capital
◈ Administrative Regions
✈ International Airport
▲ Mountain
▲▲ Mountain Range
⛰ Volcano
∿ River
◎ Lake
⊠ Reservoir

QAL'AT BĪSHAH – RADĀ'

141 *N11* **Qal'at Bīshah** 'Asīr, SW Saudi Arabia *19°59′N 42°38′E*
138 *H4* **Qal'at Burzay** Ḥamāh, W Syria *35°37′N 36°16′E*
Qal 'at Dīzah *see* Qeładize
139 *W9* **Qal'at Ḥuṣayn** Maysān, E Iraq *32°19′N 46°46′E*
139 *V10* **Qal'at Majnūnah** Al Qādisīyah, S Iraq *31°39′N 45°44′E*
139 *X11* **Qal'at Ṣāliḥ** *var.* Qal'ah Sālih. Maysān, E Iraq *31°30′N 47°24′E*
139 *V10* **Qal'at Sukkar** Dhī Qār, SE Iraq *31°52′N 46°05′E*
Qalba Zhotasy *see* Khrebet Kalba
143 *Q12* **Qal'eh Biābān** Fārs, S Iran
Qal'eh Shahr *see* Qal'ah Shahr
Qal'eh-ye Now *see* Qal'ah-ye Now
149 *T2* **Qal'eh-ye Panjeh** *var.* Qala Panja. Badakhshān, NE Afghanistan *36°56′N 72°15′E*
Qalqaman *see* Kalkaman
Qamanittuaq *see* Baker Lake
Qamar Bay *see* Qamar, Ghubbat al
141 *U14* **Qamar, Ghubbat al** *Eng.* Qamar Bay. *bay* Oman/Yemen
141 *V13* **Qamar, Jabal al** ▲▲ SW Oman
147 *N12* **Qamashi** Qashqadaryo Viloyati, S Uzbekistan *38°52′N 66°30′E*
Qambar *see* Kambar
159 *R14* **Qamdo** Xizang Zizhiqu, W China *31°09′N 97°09′E*
75 *R7* **Qamīnis** NE Libya *31°48′N 20°04′E*
Qamishly *see* Al Qāmishlī
Qânâq *see* Qaanaaq
Qandahār *see* Kandahār
80 *Q11* **Qandala** Bari, NE Somalia *11°30′N 50°00′E*
Qandyaghash *see* Kandyagash
138 *L2* **Qanṭārī** Ar Raqqah, N Syria *36°24′N 39°16′E*
Qapiciğ Dağı *see* Qazangödağ
158 *H5* **Qapqal** *var.* Qapqal Xibe Zizhixian. Xinjiang Uygur Zizhiqu, NW China *43°46′N 81°09′E*
Qapqal Xibe Zizhixian *see* Qapqal
Qapshagay Böyeni *see* Vodokhranilishche Kapshagay
Qapugtang *see* Zadoi
196 *M15* **Qaqortoq** *Dan.* Julianehåb. ⬖ Kujalleo, S Greenland
139 *T4* **Qara Anjīr** Kirkūk, N Iraq *35°30′N 44°37′E*
Qarabāgh *see* Qarah Bāgh
Qarabaū *see* Karabau
Qaraböget *see* Karaboget
Qarabulaq *see* Karabulak
Qarabutaq *see* Karabutak
Qaraghandy/Qaraghandy Oblysy *see* Karagandy
Qaraghayly *see* Karagayly
Qara Gol *see* Qere Gol
75 *U8* **Qārah** *var.* Qâra. NW Egypt *29°34′N 26°28′E*
Qārah *see* Qārah
148 *J4* **Qarah Bāgh** *var.* Qarabāgh. Herāt, NW Afghanistan *35°06′N 61°33′E*
Qarah Gawl *see* Qere Gol
138 *G7* **Qaraoun, Lac de** *var.* Buḥayrat al Qir'awn. ◎ S Lebanon
Qaraoy *see* Karaoy
Qaraqoyyn *see* Karakoyyn, Ozero
Qara Qum *see* Garagum
Qarasū *see* Karasu
Qaratal *see* Karatal
Qarataū *see* Karatau, Khrebet, Kazakhstan
Qarataū *see* Karatau, Zhambyl, Kazakhstan
Qaraton *see* Karaton
Qarazhal *see* Karazhal
80 *P13* **Qardho** *var.* Kardh, *It.* Gardo. Bari, N Somalia *09°34′N 49°30′E*
142 *M6* **Qareh Chāy** ≈ N Iran
142 *K2* **Qareh Sū** ≈ NW Iran
Qariateïne *see* Al Qaryatayn
Qarkilik *see* Ruoqiang
147 *O13* **Qarluq** *Rus.* Karluk. Surkhondaryo Viloyati, S Uzbekistan *38°17′N 67°39′E*
147 *U12* **Qarokŭl** *Rus.* Karakul'. E Tajikistan *39°07′N 73°33′E*
147 *T12* **Qarokŭl** *Rus.* Ozero Karakul'. ◎ E Tajikistan
Qarqan *see* Qiemo
158 *K9* **Qarqan He** ≈ NW China
Qarqannah, Juzur *see* Kerkenah, Îles de
149 *O1* **Qarqīn** Jowzjān, N Afghanistan *37°25′N 66°03′E*
Qars *see* Kars
Qarsaqbay *see* Karsakpay
146 *M12* **Qarshi** *Rus.* Karshi; *prev.* Bek-Budi. Qashqadaryo Viloyati, S Uzbekistan *38°54′N 65°48′E*
146 *L12* **Qarshi Cho'li** *Rus.* Karshinskaya Step. *grassland* S Uzbekistan
146 *M13* **Qarshi Kanali** *Rus.* Karshinskiy Kanal. *canal* Turkmenistan/Uzbekistan
Qaryatayn *see* Al Qaryatayn
Qāsh, Nahr al *see* Gash
146 *M12* **Qashqadaryo Viloyati** *Rus.* Kashkadar'inskaya Oblast'. ⬖ *province* S Uzbekistan
Qasigianguit *see* Qasigiannguit
197 *N13* **Qasigiannguit** *var.* Qasigianguit, *Dan.* Christianshåb. ⬖ Qaasuitsup, C Greenland
Qaṣīm, Minṭaqat *see* Al Qaṣīm
75 *V10* **Qaṣr al Farāfirah** *var.* Qasr Farâfra. W Egypt *27°00′N 27°59′E*
139 *P8* **Qaṣr 'Amīj** Al Anbār, C Iraq *33°30′N 41°52′E*
139 *R9* **Qaṣr Darwīshāh** Karbalā', C Iraq *32°36′N 43°27′E*
142 *J6* **Qaṣr-e Shīrīn** Kermānshāhān, W Iran *34°32′N 45°36′E*
Qasr Farâfra *see* Qaṣr al Farāfirah
Qassim *see* Al Qaṣīm
141 *O16* **Qa'ṭabah** SW Yemen *13°51′N 44°42′E*
138 *H7* **Qaṭanā** *var.* Katana. Rīf Dimashq, S Syria *33°27′N 36°04′E*
143 *N15* **Qatar** *off.* State of Qatar, *Ar.* Dawlat Qaṭar. ◆ *monarchy* SW Asia
Qatar, State of *see* Qatar
Qatrana *see* Al Qaṭrānah
143 *Q12* **Qaṭrūyeh** Fārs, S Iran *29°08′N 54°42′E*
Qattara Depression/ Qaṭṭārah, Munkhafaḍ al *see* Qaṭṭārah, Munkhafaḍ al
75 *U8* **Qaṭṭārah, Munkhafaḍ al** *var.* Munkhafaḍ al Qaṭṭārah *var.* Monkhafad el Qaṭṭâra, *Eng.* Qattara Depression. *desert* NW Egypt
Qaṭṭâra, Monkhafad el *see* Qaṭṭārah, Munkhafaḍ al
Qattinah, Buhayrat *see* Ḥimṣ, Buḥayrat
Qausuittuq *see* Resolute
Qaydār *see* Qeydār
Qāyen *see* Qā'en
Qaynar *see* Kaynar
147 *Q11* **Qayroqqum** *Rus.* Kayrakkum. NW Tajikistan *40°16′N 69°46′E*
147 *Q10* **Qayroqqum, Obanbori** *Rus.* Kayrakkumskoye Vodokhranilishche. ⊠ NW Tajikistan
137 *V13* **Qazangödağ** *Rus.* Gora Kapydzhik, *Turk.* Qapiciğ Dağı. ▲ SW Azerbaijan *39°18′N 46°00′E*
139 *U7* **Qazānīyah** *var.* Dhū Shaykh. Diyālá, E Iraq *33°39′N 45°33′E*
Qazaqstan/Qazaqstan Respublikasy *see* Kazakhstan
137 *T9* **Q'azbegi** *Rus.* Kazbegi; *prev.* Qazbegi. NE Georgia *42°39′N 44°36′E*
Qazbegi *see* Q'azbegi
149 *P15* **Qāzi Ahmad** *var.* Kazi Ahmad. Sind, SE Pakistan *26°19′N 68°08′E*
Qazimämmäd *see* Hacıqabal
Qazris *see* Cáceres
142 *M4* **Qazvīn** *var.* Kazvin. Qazvīn, N Iran *36°16′N 50°E*
142 *M5* **Qazvīn** *off.* Ostān-e Qazvīn. ⬖ *province* N Iran
Qazvīn, Ostān-e *see* Qazvīn
139 *U3* **Qeładize** *Ar.* Qal 'at Dīzah, *var.* Qalā Diza. As Sulaymānīyah, NE Iraq *36°11′N 45°07′E*
187 *Z13* **Qelelevu Lagoon** *lagoon* NE Fiji
Qena *see* Qinā
113 *L23* **Qeparo** Vlorë, S Albania *40°04′N 19°49′E*
Qeqertarssuaq *see* Qeqertarsuaq
197 *N13* **Qeqertarsuaq** *var.* Qeqertarssuaq, *Dan.* Godhavn. ⬖ Qaasuitsup, S Greenland
196 *M13* **Qeqertarsuaq** *island* W Greenland
197 *N13* **Qeqertarsuup Tunua** *Dan.* Disko Bugt. *inlet* W Greenland
197 *N14* **Qeqqata** *off.* Qeqqata Kommunia. ⬖ *municipality* W Greenland
Qeqqata Kommunia *see* Qeqqata
139 *U4* **Qere Gol** *Ar.* Qarah Gawl, *var.* Qara Gol. As Sulaymānīyah, NE Iraq *35°21′N 45°38′E*
Qerveh *see* Qorveh
143 *S14* **Qeshm** Hormozgān, S Iran *26°58′N 56°17′E*
143 *R14* **Qeshm** *var.* Jazīreh-ye Qeshm, Qeshm Island. *island* S Iran
Qeshm Island/Qeshm, Jazīreh-ye *see* Qeshm
Qey *see* Kīsh, Jazīreh-ye
142 *L4* **Qeydār** *var.* Qaydār. Zanjān, NW Iran *36°50′N 47°40′E*
142 *K5* **Qezel Owzan, Rūd-e** *var.* Ki Zil Uzen, Qi Zil Uzun. ≈ NW Iran
Qian *see* Guizhou
161 *Q2* **Qian'an** Heilongjiang, E China *45°00′N 124°00′E*
161 *R10* **Qiandao Hu** *prev.* Xin'anjiang Shuiku. ⊠ SE China
Qiandaohu *see* Chun'an
Qian Gorlo/Qian Gorlos/ Qian Gorlos Mongolzu Zizhixian/Quianguozhen *see* Qianguo
163 *V9* **Qianguo** *var.* Qian Gorlo, Qian Gorlos, Qian Gorlos Mongolzu Zizhixian, Quianguozhen. Jilin, NE China *45°08′N 124°48′E*
161 *N9* **Qianjiang** Hubei, C China *30°23′N 112°58′E*
160 *K10* **Qianjiang** Sichuan, C China *29°30′N 108°45′E*
160 *L14* **Qian Jiang** ≈ S China
160 *G9* **Qianning** *var.* Gartar. Sichuan, C China *30°27′N 101°24′E*
163 *U13* **Qian Shan** ▲▲ NE China
160 *H10* **Qianwei** *var.* Yujin. Sichuan, C China *29°15′N 103°52′E*
160 *J11* **Qianxi** Guizhou, S China *27°00′N 106°01′E*
Qiaotou *see* Datong
Qiaowa *see* Muli
159 *Q7* **Qiaowan** Gansu, N China *40°37′N 96°40′E*
Qibilī *see* Kebili
158 *K9* **Qiemo** *var.* Qarqan. Xinjiang Uygur Zizhiqu, NW China *38°09′N 85°30′E*
160 *J10* **Qijiang** *var.* Gunan. Chongqing Shi, C China *29°01′N 106°40′E*
159 *N5* **Qijiaojing** Xinjiang Uygur Zizhiqu, NW China *43°29′N 91°35′E*
Qike *see* Xunke
9 *N5* **Qikiqtaaluk** ⬖ *cultural region* Nunavut, N Canada
9 *R5* **Qikiqtarjuaq** *prev.* Broughton Island. Nunavut, NE Canada *67°35′N 63°55′W*
149 *P9* **Qila Saifullāh** Baluchistān, SW Pakistan *30°45′N 68°08′E*
159 *S9* **Qilian** *var.* Babao. Qinghai, C China *38°09′N 100°08′E*
159 *N8* **Qilian Shan** *var.* Kilien Mountains. ▲▲ N China
197 *O11* **Qimusseriarsuaq** *Dan.* Melville Bugt, *Eng.* Melville Bay. *bay* NW Greenland
75 *X10* **Qinā** *var.* Qena; *anc.* Caene, Caenepolis. E Egypt *26°12′N 32°49′E*
159 *W11* **Qin'an** Gansu, C China *34°49′N 105°50′E*
Qincheng *see* Nanfeng
Qing *see* Qinghai
163 *W7* **Qing'an** Heilongjiang, NE China *46°53′N 127°29′E*
159 *X10* **Qingcheng** *var.* Xifeng. Gansu, C China *35°46′N 107°35′E*
161 *R5* **Qingdao** *var.* Ching-Tao, Ch'ing-tao, Tsingtao, Tsintao, *Ger.* Tsingtau. Shandong, E China *36°31′N 120°55′E*
163 *V8* **Qinggang** Heilongjiang, NE China *46°41′N 126°05′E*
Qinggil *see* Qinghe
159 *P11* **Qinghai** *var.* Chinghai, Koko Nor, Qing, Qinghai Sheng, Tsinghai. ⬖ *province* C China
159 *S10* **Qinghai Hu** *var.* Ch'ing Hai, Tsing Hai, *Mong.* Koko Nor. ◎ C China
Qinghai Sheng *see* Qinghai
158 *M3* **Qinghe** *var.* Qinggil. Xinjiang Uygur Zizhiqu, NW China *46°42′N 90°19′E*
160 *L4* **Qingjian** *var.* Kuanzhou; *prev.* Xiuyan. Shaanxi, C China *37°10′N 110°09′E*
160 *L9* **Qing Jiang** ≈ C China
Qingjiang *see* Huai'an
Qingkou *see* Ganyu
160 *I12* **Qinglong** *var.* Liancheng. Guizhou, S China *25°49′N 105°10′E*
161 *Q2* **Qinglong** Hebei, E China *40°24′N 118°57′E*
Qingshan *see* Wudalianchi
159 *R12* **Qingshuihe** Qinghai, C China *33°47′N 97°10′E*
Qingyang *see* Jinjiang
161 *N14* **Qingyuan** Guangdong, S China *23°42′N 113°02′E*
163 *V11* **Qingyuan** *var.* Qingyuan Manzu Zizhixian. Liaoning, NE China *42°08′N 124°55′E*
Qingyuan *see* Shandan
Qingyuan *see* Weiyuan
Qingyuan Manzu Zizhixian *see* Qingyuan
158 *L13* **Qingzang Gaoyuan** *var.* Xizang Gaoyuan, *Eng.* Plateau of Tibet. *plateau* W China
161 *Q4* **Qingzhou** *prev.* Yidu. Shandong, E China *36°41′N 118°29′E*
157 *R9* **Qin He** ≈ C China
161 *Q2* **Qinhuangdao** Hebei, E China *39°57′N 119°31′E*
160 *K7* **Qin Ling** ▲▲ C China
161 *N5* **Qinxian** *var.* Dingchang, Qin Xian. Shanxi, C China *36°46′N 112°42′E*
Qin Xian *see* Qinxian
161 *N6* **Qinyang** Henan, C China *35°05′N 112°56′E*
160 *K15* **Qinzhou** Guangxi Zhuangzu Zizhiqu, S China *22°09′N 108°36′E*
Qiong *see* Hainan
160 *L17* **Qionghai** *prev.* Jiaji. Hainan, S China *19°12′N 110°26′E*
160 *H9* **Qionglai** Sichuan, C China *30°24′N 103°28′E*
160 *H8* **Qionglai Shan** ▲▲ C China
Qiongxi *see* Hongyuan
160 *L17* **Qiongzhou Haixia** *var.* Hainan Strait. *strait* S China
163 *U7* **Qiqihar** *var.* Ch'i-ch'i-ha-erh, Tsitsihar; *prev.* Lungkiang. Heilongjiang, NE China *47°23′N 124°E*
Qir *see* Qīr-va-Kārzīn
158 *H10* **Qira** Xinjiang Uygur Zizhiqu, NW China *37°05′N 80°45′E*
Qir'awn, Buḥayrat al *see* Qaraoun, Lac de
143 *P12* **Qīr-va-Kārzīn** *var.* Qir. Fārs, S Iran *28°27′N 53°04′E*
Qiryat Gat *see* Kyriat Gat
Qiryat Shemona *see* Kiryat Shmona
Qishlaq *see* Garmsār
141 *U14* **Qishn** SE Yemen *15°29′N 51°44′E*
Qishon, Naḥal *see* Kishon, Nahal
Qita Ghazzah *see* Gaza Strip
156 *K5* **Qitai** Xinjiang Uygur Zizhiqu, NW China *44°N 89°34′E*
163 *Y8* **Qitaihe** Heilongjiang, NE China *45°45′N 130°53′E*
141 *W12* **Qitbīt, Wādī** *dry watercourse* S Oman
161 *O5* **Qixian** *var.* Qi Xian, Zhaoge. Henan, C China *35°35′N 114°10′E*
Qi Xian *see* Qixian
Qīzān *see* Jīzān
Qizil Orda *see* Kyzylorda
Qizil Qum/Qizilqum *see* Kyzyl Kum
147 *V14* **Qizilrabot** *Rus.* Kyzylrabot. SE Tajikistan *37°29′N 74°44′E*
146 *J10* **Qizilravote** *Rus.* Kyzylrabat. Buxoro Viloyati, C Uzbekistan *40°35′N 62°09′E*
Qi Zil Uzun *see* Qezel Owzan, Rūd-e
139 *S4* **Qizil Yār** Kirkūk, N Iraq *35°26′N 44°12′E*
164 *J12* **Qkutango-hantō** *peninsula* Honshū, SW Japan
137 *Y11* **Qobustan** *prev.* Märäzä. E Azerbaijan *40°32′N 48°56′E*
Qoghaly *see* Kogaly
Qogir Feng *see* K2
143 *N6* **Qom** *var.* Kum, Qum. Qom, N Iran *34°43′N 50°54′E*
143 *N6* **Qom** *off.* Osān-e Qom. ⬖ *province* N Iran
Qomisheh *see* Shahreẕā
Qomolangma Feng *see* Everest, Mount
Qom, Osān-e *see* Qom
142 *M7* **Qom, Rūd-e** ≈ C Iran
Qomsheh *see* Shahreẕā
Qomul *see* Hami
Qondūz *see* Kunduz
146 *G7* **Qo'ng'irot** *Rus.* Kungrad. Qoraqalpog'iston Respublikasi, NW Uzbekistan *43°01′N 58°49′E*
Qongyrat *see* Konyrat
Qoqek *see* Tacheng
147 *R10* **Qo'qon** *var.* Khokand, *Rus.* Kokand. Farg'ona Viloyati, E Uzbekistan *40°34′N 70°55′E*
Qorabowur Kirlari *see* Karabaur', Uval
Qoradaryo *see* Karadar'ya
146 *G6* **Qorajar** *Rus.* Karadzhar. Qoraqalpog'iston Respublikasi, NW Uzbekistan *43°34′N 58°35′E*
146 *K12* **Qorako'l** *Rus.* Karakul'. Buxoro Viloyati, C Uzbekistan *39°27′N 63°45′E*
146 *H7* **Qorao'zak** *Rus.* Karauzyak. Qoraqalpog'iston Respublikasi, NW Uzbekistan *43°07′N 60°03′E*
146 *E5* **Qoraqalpog'iston** *Rus.* Karakalpakya. Qoraqalpog'iston Respublikasi, NW Uzbekistan *44°45′N 56°06′E*
146 *G7* **Qoraqalpog'iston Respublikasi** *Rus.* Respublika Karakalpakstan. ⬖ *autonomous republic* NW Uzbekistan
Qorghalzhyn *see* Korgalzhyn
138 *H6* **Qornet es Saouda** ▲ NE Lebanon *36°06′N 34°06′E*
146 *L12* **Qorowulbozor** *Rus.* Karaulbazar. Buxoro Viloyati, C Uzbekistan *39°28′N 64°49′E*
142 *K5* **Qorveh** *var.* Qerveh, Qurveh. Kordestān, W Iran *35°09′N 47°48′E*
147 *N11* **Qo'shrabot** *Rus.* Kushrabat. Samarqand Viloyati, C Uzbekistan *40°15′N 66°40′E*
Qosköl *see* Koskol'
Qosshaghyl *see* Kosshagyl
Qostanay/Qostanay Oblysy *see* Kostanay
143 *P12* **Qoṭbābād** Fārs, S Iran *28°52′N 53°40′E*
143 *R13* **Qoṭbābād** Hormozgān, S Iran *27°49′N 56°00′E*
138 *H6* **Qoubaïyât** *var.* Al Qubayyāt. N Lebanon *37°00′N 34°30′E*
Qoussantîna *see* Constantine
Qowowuyag *see* Cho Oyu
147 *O11* **Qo'ytosh** *Rus.* Koytash. Jizzax Viloyati, C Uzbekistan *40°13′N 67°19′E*
146 *G7* **Qozonketkan** *Rus.* Kazanketken. Qoraqalpog'iston Respublikasi, W Uzbekistan *42°59′N 59°21′E*
146 *H6* **Qozoqdaryo** *Rus.* Kazakdar'ya. Qoraqalpog'iston Respublikasi, NW Uzbekistan *43°26′N 59°47′E*
19 *N11* **Quabbin Reservoir** ⊠ Massachusetts, NE USA
100 *F12* **Quakenbrück** Niedersachsen, NW Germany *52°41′N 07°57′E*
18 *I15* **Quakertown** Pennsylvania, NE USA *40°26′N 75°17′W*
182 *M10* **Quambatook** Victoria, SE Australia *35°52′S 143°28′E*
25 *Q4* **Quanah** Texas, SW USA *34°17′N 99°46′W*
167 *V10* **Quang Ngai** *var.* Quangngai, Quang Nghia. Quang Ngai, C Vietnam *15°09′N 108°50′E*
Quangngai *see* Quang Ngai
Quang Nghia *see* Quang Ngai
167 *T9* **Quang Tri** *var.* Trièu Hai. Quang Tri, C Vietnam *16°46′N 107°11′E*
Quanjiang *see* Suichuan
Quan Long *see* Ca Mau
152 *L4* **Quanshuigou** China/India *35°40′N 79°28′E*
161 *R13* **Quanzhou** *var.* Ch'uan-chou, Tsinkiang; *prev.* Chin-chiang. Fujian, SE China *24°56′N 118°31′E*
160 *M12* **Quanzhou** Guangxi Zhuangzu Zizhiqu, S China *25°59′N 111°02′E*
11 *V16* **Qu'Appelle** ≈ Saskatchewan, S Canada
12 *M3* **Quaqtaq** *prev.* Koartac. Québec, NE Canada *60°50′N 69°30′W*
61 *E16* **Quaraí** Rio Grande do Sul, S Brazil *30°38′S 56°25′W*
59 *H24* **Quaraí, Rio** *Sp.* Río Cuareim. ≈ Brazil/Uruguay *see also* Cuareim, Río
Quaraí, Rio *see* Cuareim, Río
171 *N13* **Quarles, Pegunungan** ▲▲ Sulawesi, C Indonesia
Quarnero *see* Kvarner
107 *C20* **Quartu Sant' Elena** Sardegna, Italy, C Mediterranean Sea *39°15′N 09°12′E*
29 *X13* **Quasqueton** Iowa, C USA *42°23′N 91°45′W*
173 *X16* **Quatre Bornes** W Mauritius *20°15′S 57°28′E*
172 *I17* **Quatre Bornes** Mahé, NE Seychelles
137 *X10* **Quba** *Rus.* Kuba. N Azerbaijan *41°22′N 48°30′E*
Qubba *see* Ba'qūbah
143 *T3* **Qūchān** *var.* Kuchan. Khorāsān-e Razavī, NE Iran *37°12′N 58°28′E*
183 *R10* **Queanbeyan** New South Wales, SE Australia *35°24′S 149°17′E*
15 *Q10* **Québec** *var.* Quebec. *province capital* Québec, SE Canada *46°50′N 71°15′W*
14 *K10* **Québec** *var.* Quebec. ⬖ *province* SE Canada
61 *D17* **Quebracho** Paysandú, W Uruguay *31°58′S 57°53′W*
101 *K14* **Quedlinburg** Sachsen-Anhalt, C Germany *51°48′N 11°09′E*
138 *H10* **Queen Alia** ✕ ('Ammān) 'Ammān, C Jordan
10 *L16* **Queen Bess, Mount** ▲ British Columbia, SW Canada *51°15′N 124°29′W*
10 *I14* **Queen Charlotte** British Columbia, SW Canada *53°18′N 132°04′W*
65 *B24* **Queen Charlotte Bay** *bay* West Falkland, W Falkland Islands
10 *H14* **Queen Charlotte Islands** *Fr.* Îles de la Reine-Charlotte. *island group* British Columbia, SW Canada
10 *I15* **Queen Charlotte Sound** *sea area* British Columbia, W Canada
10 *J16* **Queen Charlotte Strait** *strait* British Columbia, W Canada
27 *U1* **Queen City** Missouri, C USA *40°24′N 92°34′W*
25 *X5* **Queen City** Texas, SW USA *33°09′N 94°09′W*
197 *O9* **Queen Elizabeth Islands** *Fr.* Îles de la Reine-Élisabeth. *island group* Nunavut, N Canada
195 *Y10* **Queen Mary Coast** *physical region* Antarctica
65 *N24* **Queen Mary's Peak** ▲ C Tristan da Cunha
196 *M8* **Queen Maud Gulf** *gulf* Arctic Ocean
195 *P11* **Queen Maud Mountains** ▲▲ Antarctica
Queen's County *see* Laois
181 *U7* **Queensland** ⬖ *state* N Australia
192 *I9* **Queensland Plateau** *undersea feature* N Coral Sea
183 *O16* **Queenstown** Tasmania, SE Australia *42°06′S 145°33′E*
185 *C22* **Queenstown** Otago, South Island, New Zealand *45°01′S 168°44′E*
83 *I24* **Queenstown** Eastern Cape, S South Africa *31°52′S 26°50′E*
Queenstown *see* Cobh
32 *F8* **Queets** Washington, NW USA *47°31′N 124°19′W*
61 *D18* **Queguay Grande, Río** ≈ W Uruguay
59 *O16* **Queimadas** Bahia, E Brazil *10°59′S 39°38′W*
82 *D11* **Quela** Malanje, NW Angola *09°18′S 17°07′E*
83 *O16* **Quelimane** *var.* Kilimane, Kilmain, Quilimane. Zambézia, NE Mozambique *17°53′S 36°51′E*
63 *G18* **Quellón** *var.* Puerto Quellón. Los Lagos, S Chile *43°05′S 73°38′W*
Quelpart *see* Jeju-do
37 *P12* **Quemado** New Mexico, SW USA *34°18′N 108°29′W*
25 *O12* **Quemado** Texas, SW USA *28°58′N 100°36′W*
44 *K7* **Quemado, Punta de** *headland* E Cuba *20°13′N 74°07′W*
Quemoy *see* Jinmen Dao
62 *K13* **Quemú Quemú** La Pampa, E Argentina *36°03′S 63°36′W*
155 *E17* **Quepem** Goa, W India *15°13′N 74°03′E*
42 *M14* **Quepos** Puntarenas, S Costa Rica *09°28′N 84°10′W*
Que Que *see* Kwekwe
61 *D23* **Quequén** Buenos Aires, E Argentina *38°30′S 58°44′W*
61 *D23* **Quequén Grande, Río** ≈ E Argentina
61 *C23* **Quequén Salado, Río** ≈ E Argentina
Quera *see* Chur
41 *N13* **Querétaro** Querétaro de Arteaga, C Mexico *20°36′N 100°24′W*
40 *F4* **Querobabi** Sonora, NW Mexico *30°02′N 111°02′W*
42 *M13* **Quesada** *var.* Ciudad Quesada, San Carlos. Alajuela, N Costa Rica *10°19′N 84°26′W*
105 *O13* **Quesada** Andalucía, S Spain *37°52′N 03°05′W*
161 *O7* **Queshan** Henan, C China *32°48′N 114°03′E*
10 *M15* **Quesnel** British Columbia, SW Canada *52°59′N 122°30′W*
37 *S9* **Questa** New Mexico, SW USA *36°41′N 105°37′W*
102 *H7* **Questembert** Morbihan, NW France *47°39′N 02°24′W*
57 *K22* **Quetena, Río** ≈ SW Bolivia
149 *O10* **Quetta** Baluchistān, SW Pakistan *30°15′N 67°E*
Quetzalcoalco *see* Coatzacoalcos
Quetzaltenango *see* Quezaltenango
56 *B6* **Quevedo** Los Ríos, C Ecuador *01°02′S 79°27′W*
42 *B6* **Quezaltenango** *var.* Quetzaltenango. Quezaltenango, W Guatemala *14°50′N 91°30′W*
42 *A2* **Quezaltenango** *off.* Departamento de Quezaltenango, *var.* Quetzaltenango. ⬖ *department* SW Guatemala
Quezaltenango, Departamento de *see* Quezaltenango
42 *E6* **Quezaltepeque** Chiquimula, SE Guatemala *14°38′N 89°25′W*
170 *M6* **Quezon** Palawan, W Philippines *09°13′N 118°01′E*
161 *P5* **Qufu** Shandong, E China *35°37′N 117°05′E*
82 *B12* **Quibala** Kwanza Sul, NW Angola *10°44′S 14°58′E*
82 *B11* **Quibaxe** *var.* Quibaxi. Kwanza Norte, NW Angola *08°30′S 14°36′E*
Quibaxi *see* Quibaxe
54 *D9* **Quibdó** Chocó, W Colombia *05°40′N 76°38′W*
102 *G7* **Quiberon** Morbihan, NW France *47°30′N 03°07′W*
102 *G7* **Quiberon, Baie de** *bay* NW France
54 *J5* **Quíbor** Lara, N Venezuela *09°55′N 69°35′W*
42 *C4* **Quiché** *off.* Departamento del Quiché. ⬖ *department* W Guatemala
Quiché, Departamento del *see* Quiché
99 *E21* **Quiévrain** Hainaut, S Belgium *50°25′N 03°41′E*
40 *I9* **Quila** Sinaloa, C Mexico *24°24′N 107°11′W*
83 *B14* **Quilengues** Huíla, SW Angola *14°09′S 14°04′E*
Quilimane *see* Quelimane
57 *G15* **Quillabamba** Cusco, C Peru *12°49′S 72°41′W*
57 *L18* **Quillacollo** Cochabamba, C Bolivia *17°26′S 66°16′W*
62 *H4* **Quillagua** Antofagasta, N Chile *21°33′S 69°32′W*
103 *N17* **Quillan** Aude, S France *42°52′N 02°11′E*
11 *U15* **Quill Lakes** ◎ Saskatchewan, S Canada
62 *G11* **Quillota** Valparaíso, C Chile *32°54′S 71°16′W*
155 *G23* **Quilon** *var.* Kollam. Kerala, SW India *08°57′N 76°37′E* *see also* Kollam
181 *V9* **Quilpie** Queensland, C Australia *26°39′S 144°15′E*
149 *O4* **Quil-Qala** Bāmyān, N Afghanistan *35°13′N 67°02′E*
62 *L7* **Quimilí** Santiago del Estero, C Argentina *27°35′S 62°25′W*
57 *O19* **Quimome** Santa Cruz, E Bolivia *17°45′S 61°15′W*
102 *F6* **Quimper** *anc.* Quimper Corentin. Finistère, NW France *48°00′N 04°05′W*
Quimper Corentin *see* Quimper
102 *G7* **Quimperlé** Finistère, NW France *47°52′N 03°33′W*
32 *F8* **Quinault** Washington, NW USA *47°27′N 123°53′W*
32 *F8* **Quinault River** ≈ Washington, NW USA
35 *P5* **Quincy** California, W USA *39°56′N 120°56′W*
23 *S8* **Quincy** Florida, SE USA *30°35′N 84°34′W*
30 *I13* **Quincy** Illinois, N USA *39°56′N 91°24′W*
19 *O11* **Quincy** Massachusetts, NE USA *42°15′N 71°00′W*
32 *J9* **Quincy** Washington, NW USA *47°13′N 119°51′W*
54 *E10* **Quindío** *off.* Departamento del Quindío. ⬖ *province* C Colombia
54 *E10* **Quindío, Nevado del** ▲ C Colombia *04°42′N 75°25′W*
62 *J10* **Quines** San Luis, C Argentina *32°15′S 65°46′W*
39 *N13* **Quinhagak** Alaska, USA *59°45′N 161°55′W*
76 *G13* **Quinhámel** W Guinea-Bissau *11°52′N 15°52′W*
Qui Nhon/Quinhon *see* Quy Nhơn
Quinindé *see* Rosa Zárate
25 *U6* **Quinlan** Texas, SW USA *32°54′N 96°08′W*
61 *H17* **Quinta** Rio Grande do Sul, S Brazil *32°05′S 52°18′W*
105 *O10* **Quintanar de la Orden** Castilla-La Mancha, C Spain *39°36′N 03°03′W*
41 *X13* **Quintana Roo** ⬖ *state* SE Mexico
105 *S6* **Quinto** Aragón, NE Spain *41°25′N 00°31′W*
108 *G10* **Quinto** Ticino, S Switzerland *46°32′N 08°44′E*
27 *Q11* **Quinton** Oklahoma, C USA *35°07′N 95°22′W*
62 *K12* **Quinto, Río** ≈ C Argentina
82 *A10* **Quinzau** Zaire Province, NW Angola *06°50′S 12°48′E*
14 *H8* **Quinze, Lac des** ◎ Québec, SE Canada
83 *B15* **Quipungo** Huíla, C Angola *14°49′S 14°29′E*
62 *G13* **Quirihue** Bío Bío, C Chile *36°15′S 72°35′W*
82 *D12* **Quirima** Malanje, NW Angola *10°51′S 18°06′E*
183 *T6* **Quirindi** New South Wales, SE Australia *31°29′S 150°40′E*
55 *P5* **Quiriquire** Monagas, NE Venezuela *09°59′N 63°14′W*
14 *D10* **Quirke Lake** ◎ Ontario, S Canada
61 *B21* **Quiroga** Buenos Aires, E Argentina *35°18′S 61°22′W*
104 *I4* **Quiroga** Galicia, NW Spain *42°28′N 07°15′W*
Quiróm, Salar *see* Pocitos, Salar
56 *B9* **Quiroz, Río** ≈ NW Peru
82 *Q13* **Quissanga** Cabo Delgado, NE Mozambique *12°24′S 40°33′E*
83 *M20* **Quissico** Inhambane, S Mozambique *24°42′S 34°44′E*
25 *O4* **Quitaque** Texas, SW USA *34°22′N 101°03′W*
82 *Q13* **Quiterajo** Cabo Delgado, NE Mozambique *11°37′S 40°22′E*
23 *T6* **Quitman** Georgia, SE USA *30°46′N 83°33′W*
22 *M6* **Quitman** Mississippi, S USA *32°02′N 88°43′W*
25 *V6* **Quitman** Texas, SW USA *32°37′N 95°26′W*
56 *C6* **Quito** ● (Ecuador) Pichincha, N Ecuador *0°14′S 78°30′W*
Quito *see* Mariscal Sucre
58 *P13* **Quixadá** Ceará, E Brazil *04°57′S 39°04′W*
83 *Q15* **Quixaxe** Nampula, NE Mozambique *15°15′S 40°07′E*
161 *N13* **Qujiang** *var.* Maba. Guangdong, S China *24°47′N 113°34′E*
160 *J9* **Qu Jiang** ≈ C China
161 *R10* **Qu Jiang** ≈ SE China
160 *H12* **Qujing** Yunnan, SW China *25°39′N 103°52′E*
Qulan *see* Kulan
Qulin Gol *see* Chaor He
146 *L10* **QuljuqtovTog'lari** *Rus.* Gory Kul'dzhuktau. ▲▲ C Uzbekistan
Qulsary *see* Kul'sary
Qulyndy Zhazyghy *see* Ravnina Kulyndy
Qum *see* Qom
Qumälisch *see* Lubartów
159 *P11* **Qumar He** ≈ C China
159 *Q12* **Qumarlêb** *var.* Yuegai; *prev.* Yuegaitan. Qinghai, C China *34°06′N 95°54′E*
Qumisheh *see* Shahreẕā
147 *O14* **Qumqo'rg'on** *Rus.* Kumkurgan. Surkhondaryo Viloyati, S Uzbekistan *37°54′N 67°31′E*
Qunaytirah/Qunayṭirah, Muḥāfaẓat al *see* Al Qunayṭirah
189 *V12* **Quoi** *island* Chuuk, C Micronesia
9 *N8* **Quoich** ≈ Nunavut, NE Canada
83 *E26* **Quoin Point** *headland* SW South Africa *34°48′S 19°39′E*
182 *I7* **Quorn** South Australia *32°22′S 138°03′E*
Qurein *see* Al Kuwayt
147 *P14* **Qŭrghonteppa** *Rus.* Kurgan-Tyube. SW Tajikistan *37°51′N 68°42′E*
Qurlurtuuq *see* Kugluktuk
Qurveh *see* Qorveh
Quryq *see* Kuryk
Qusair *see* Al Quṣayr
137 *X10* **Qusar** *Rus.* Kusary. NE Azerbaijan *41°26′N 48°27′E*
Quṣayr *see* Al Quṣayr
Quseir *see* Al Quṣayr
142 *I2* **Qūshchī** Āẕarbāyjān-e Gharbi, N Iran *37°59′N 45°05′E*
Qusmuryn *see* Kushmurun, Kostanay, Kazakhstan
Qusmuryn *see* Kusmuryn, Ozero
Quṭayfah/Qutayfe/Quteife *see* Al Quṭayfah
Quthing *see* Moyeni
147 *S10* **Quvasoy** *Rus.* Kuvasay. Farg'ona Viloyati, E Uzbekistan *40°17′N 71°53′E*
Quwair *see* Guwēr
Quxar *see* Lhazê
Qu Xian *see* Quzhou
159 *N16* **Qüxü** *var.* Xoi. Xizang Zizhiqu, W China *29°25′N 90°48′E*
Quyang *see* Jingzhou
167 *V11* **Quy Nhơn** *var.* Quinhon, Qui Nhon. Binh Đinh, C Vietnam *13°47′N 109°11′E*
161 *R10* **Quzhou** *var.* Qu Xian. Zhejiang, SE China *28°55′N 118°54′E*
Qyteti Stalin *see* Kuçovë
Qyzylaghash *see* Kyzylagash
Qyzylorda *see* Kyzylorda
Qyzyltū *see* Kishkenekol'
Qyzylzhar *see* Kyzylzhar

R

Raa Atoll *see* North Maalhosmadulu Atoll
109 *R4* **Raab** Oberösterreich, N Austria *48°19′N 13°40′E*
109 *X8* **Raab** *Hung.* Rába. ≈ Austria/Hungary *see also* Rába
Raab *see* Rába
Raab *see* Győr
109 *V2* **Raabs an der Thaya** Niederösterreich, E Austria *48°51′N 15°28′E*
93 *L14* **Raahe** *Swe.* Brahestad. Pohjois-Pohjanmaa, W Finland *64°42′N 24°31′E*
98 *M10* **Raalte** Overijssel, E Netherlands *52°23′N 06°16′E*
99 *I14* **Raamsdonksveer** Noord-Brabant, S Netherlands *51°42′N 04°54′E*
92 *L12* **Raanujärvi** Lappi, NW Finland *66°39′N 24°40′E*
96 *G9* **Raasay** *island* NW Scotland, United Kingdom
118 *H3* **Raasiku** *Ger.* Rasik. Harjumaa, NW Estonia *59°22′N 25°11′E*
112 *B11* **Rab** *It.* Arbe. Primorje-Gorski Kotar, NW Croatia *44°46′N 14°46′E*
112 *B11* **Rab** *It.* Arbe. *island* NW Croatia
171 *N16* **Raba** Sumbawa, S Indonesia *08°27′S 118°45′E*
111 *G22* **Rába** *Ger.* Raab. ≈ Austria/Hungary *see also* Raab
Rába *see* Raab
112 *A10* **Rabac** Istra, NW Croatia *45°03′N 14°09′E*
104 *I2* **Rábade** Galicia, NW Spain *43°07′N 07°37′W*
80 *F10* **Rabak** White Nile, C Sudan *13°12′N 32°44′E*
186 *G9* **Rabaraba** Milne Bay, SE Papua New Guinea *10°00′S 149°50′E*
102 *K16* **Rabastens-de-Bigorre** Hautes-Pyrénées, S France *43°23′N 00°10′E*
121 *O16* **Rabat** W Malta *35°51′N 14°25′E*
74 *F6* **Rabat** *var.* al Dar al Baida. ● (Morocco) NW Morocco *34°02′N 06°51′W*
Rabat *see* Victoria
186 *H6* **Rabaul** New Britain, E Papua New Guinea *04°13′S 152°11′E*
Rabbah Ammon/Rabbath Ammon *see* 'Amman
28 *K8* **Rabbit Creek** ≈ South Dakota, N USA
14 *H10* **Rabbit Lake** ◎ Ontario, S Canada
187 *Y14* **Rabi** *prev.* Rambi. *island* N Fiji
140 *K9* **Rābigh** Makkah, W Saudi Arabia *22°51′N 39°E*
42 *D5* **Rabinal** Baja Verapaz, C Guatemala *15°05′N 90°26′W*
168 *G9* **Rabi, Pulau** *island* NW Indonesia, East Indies
111 *L17* **Rabka** Małopolskie, S Poland *49°38′N 20°E*
155 *F16* **Rabkavi** Karnātaka, W India *16°40′N 75°03′E*
Râbniţa *see* Rîbniţa
109 *Y6* **Rabnitz** ≈ E Austria
124 *J7* **Rabocheostrovsk** Respublika Kareliya, NW Russian Federation *64°58′N 34°46′E*
23 *U1* **Rabun Bald** ▲ Georgia, SE USA *34°58′N 83°18′W*
75 *S11* **Rabyānah** SE Libya *24°07′N 21°58′E*
75 *S11* **Rabyānah, Ramlat** *var.* Rebiana Sand Sea, Ṣaḥrā' Rabyānah. *desert* SE Libya
Rabyānah, Ṣaḥrā' *see* Rabyānah, Ramlat
116 *L11* **Răcăciuni** Bacău, E Romania *46°20′N 27°00′E*
Racaka *see* Riwoqê
107 *J24* **Racalmuto** Sicilia, Italy, C Mediterranean Sea *37°25′N 13°44′E*
116 *J14* **Răcari** Dâmboviţa, SE Romania *44°37′N 25°43′E*
Răcari *see* Durankulak
116 *F13* **Răcăşdia** *Hung.* Rakasd. Caraş-Severin, SW Romania *44°58′N 21°36′E*
106 *B9* **Racconigi** Piemonte, NE Italy *44°45′N 07°41′E*
31 *T15* **Raccoon Creek** ≈ Ohio, N USA
13 *V13* **Race, Cape** *headland* Newfoundland, Newfoundland and Labrador, E Canada *46°40′N 53°05′W*
22 *K10* **Raceland** Louisiana, S USA *29°43′N 90°36′W*
19 *Q12* **Race Point** *headland* Massachusetts, NE USA *42°03′N 70°14′W*
167 *S14* **Rach Gia** Kiên Giang, S Vietnam *10°01′N 105°05′E*
167 *S14* **Rach Gia, Vinh** *bay* S Vietnam
76 *J8* **Rachid** Tagant, C Mauritania *18°48′N 11°41′W*
110 *L10* **Raciąż** Mazowieckie, C Poland *52°46′N 20°04′E*
111 *I16* **Racibórz** *Ger.* Ratibor. Śląskie, S Poland *50°05′N 18°10′E*
31 *N9* **Racine** Wisconsin, N USA *42°42′N 87°50′W*
14 *D7* **Racine Lake** ◎ Ontario, S Canada
111 *J23* **Ráckeve** Pest, C Hungary *47°10′N 18°58′E*
Rácz-Becse *see* Bečej
141 *O15* **Radā'** *var.* Ridā'. W Yemen *14°24′N 44°49′E*

◆ Country
● Country Capital
◇ Dependent Territory
○ Dependent Territory Capital
⬖ Administrative Regions
✕ International Airport
▲ Mountain
▲▲ Mountain Range
⛰ Volcano
≈ River
◎ Lake
⊠ Reservoir

Radan ▲ SE Serbia 42°59′N 21°31′E — 113 O15
Rada Tilly Chubut, SE Argentina 45°54′S 67°33′W — 63 J19
Rădăuţi Ger. Radautz, Hung. Rádóc. Suceava, N Romania 47°49′N 25°58′E — 116 K8
Rădăuţi-Prut Botoşani, NE Romania 48°14′N 26°47′E — 116 L8
Radautz see Rădăuţi
Radbusa see Radbuza
Radbuza Ger. Radbusa. SE Czech Republic — 111 A17
Radcliff Kentucky, S USA 37°50′N 85°57′W — 20 K6
Radd, Wādī ar dry watercourse N Syria — 139 O2
Råde Østfold, S Norway 59°21′N 10°53′E — 95 H16
Radeče Ger. Ratschach. C Slovenia — 109 V11
Radein see Radenci
Radekhiv Pol. Radziechów, Rus. Radekhov. L'vivs'ka Oblast', W Ukraine 50°17′N 24°39′E — 116 J4
Radekhov see Radekhiv
Radenci Ger. Radein; prev. Radinci. NE Slovenia 46°36′N 16°02′E — 109 X9
Radenthein Kärnten, S Austria 46°48′N 13°42′E — 109 S9
Rádeyilikóé see Fort Good Hope
Radford Virginia, NE USA 37°07′N 80°34′W — 21 R7
Rādhanpur Gujarāt, W India 23°52′N 71°49′E — 154 C9
Radinci see Radenci
Radishchevo Ul'yanovskaya Oblast', W Russian Federation 52°49′N 47°54′E — 127 Q6
Radisson Québec, E Canada 53°47′N 77°35′W — 12 I9
Radium Hot Springs British Columbia, SW Canada 50°39′N 116°09′W — 11 P16
Radna Hung. Máriaradna. Arad, W Romania 46°05′N 21°41′E — 116 F11
Rádnávrre see Randijaure
Radnevo Stara Zagora, C Bulgaria 42°17′N 25°58′E — 114 K10
Radnor cultural region E Wales, United Kingdom — 97 J20
Radnót see Iernut
Rádóc see Rădăuţi
Radolfzell am Bodensee Baden-Württemberg, S Germany 47°43′N 08°58′E — 101 H24
Radom Mazowieckie, C Poland 51°23′N 21°08′E — 110 M13
Radomireşti Olt, S Romania 44°06′N 24°30′E — 116 I14
Radomsko Rus. Novoradomsk. Łódzkie, C Poland 51°04′N 19°25′E — 111 K14
Radomyshl' Zhytomyrs'ka Oblast', N Ukraine 50°30′N 29°16′E — 117 N4
Radoviš prev. Radovište. E Macedonia 41°39′N 22°26′E — 113 P19
Radovište see Radoviš
Radøy see Radøyni
Radøyni prev. Radøy. island S Norway — 94 B13
Radstadt Salzburg, NW Austria 47°24′N 13°31′E — 109 R7
Radstock, Cape headland South Australia 33°11′S 134°18′E — 182 E8
Raduha ▲ N Slovenia 46°24′N 14°46′E — 109 U10
Radun' Hrodzyenskaya Voblasts', W Belarus 54°03′N 25°00′E — 119 G15
Raduzhnyy Vladimirskaya Oblast', W Russian Federation 55°59′N 40°45′E — 126 M3
Radviliškis Šiauliai, N Lithuania 55°48′N 23°32′E — 118 F11
Radville Saskatchewan, S Canada 49°28′N 104°19′W — 13 U17
Radwá, Jabal ▲ W Saudi Arabia 24°31′N 38°21′E — 140 K7
Radymno Podkarpackie, SE Poland 49°57′N 22°49′E — 111 P16
Radyvyliv Rivnens'ka Oblast', NW Ukraine 50°07′N 25°12′E — 116 J5
Radziechów see Radekhiv
Radziejów Kujawsko-pomorskie, C Poland 52°36′N 18°33′E — 110 I11
Radzyń Podlaski Lubelskie, E Poland 51°48′N 22°37′E — 110 O12
Rae ☉ Nunavut, NW Canada — 8 J7
Rāe Bareli Uttar Pradesh, N India 26°14′N 81°14′E — 152 M13
Rae-Edzo see Edzo
Raeford North Carolina, SE USA 34°59′N 79°15′W — 21 T11
Raeren Liège, E Belgium 50°42′N 06°06′E — 99 M19
Rae Strait strait Nunavut, N Canada — 9 N7
Raetihi Manawatu-Wanganui, North Island, New Zealand 39°29′S 175°16′E — 184 L11
Raevavae see Raivavae
Rafaela Santa Fe, E Argentina 31°16′S 61°25′W — 62 M10
Rafael Núñez ✕ (Cartagena) Bolívar, NW Colombia 10°27′N 75°31′W — 54 E5
Rafah var. Rafa, Rafaḥ, Heb. Rafiaḥ, Raphiah. SW Gaza Strip 31°18′N 34°15′E — 138 E11
Rafaï Mbomou, SE Central African Republic 05°01′N 23°51′E — 79 L15
Rafḥah Al Ḥudūd ash Shamālīyah, N Saudi Arabia 29°41′N 43°32′E — 141 O4
Rafiaḥ see Rafah
Rafsanjān Kermān, C Iran 30°25′N 56°E — 143 R10
Raga Western Bahr el Ghazal, W South Sudan 08°28′N 25°41′E — 80 B13
Ragged Island island Maine, NE USA — 19 S8
Ragged Island Range island group S The Bahamas — 44 I5
Raglan Waikato, North Island, New Zealand 37°48′S 174°54′E — 184 L7
Ragley Louisiana, S USA 30°31′N 93°13′W — 22 G8
Ragnit see Neman
Ragusa Sicilia, Italy, C Mediterranean Sea 36°56′N 14°42′E — 107 K25
Ragusa see Dubrovnik
Ragusavecchia see Cavtat
Raha Pulau Muna, C Indonesia 04°50′S 122°43′E — 171 P14

Rahachow Rus. Rogachëv. Homyel'skaya Voblasts', SE Belarus 53°03′N 30°03′E — 119 N17
Rahad var. Nahr ar Rahad. ↔ W Sudan — 67 U6
Rahad, Nahr ar see Rahad
Rahaeng see Tak
Rahat Southern, C Israel 31°20′N 34°43′E — 138 F11
Rahaṭ, Ḥarrat lava flow W Saudi Arabia — 140 L8
Rahīmyār Khān Punjab, SE Pakistan 28°27′N 70°21′E — 149 S12
Råholt Akershus, S Norway 60°16′N 11°10′E — 95 H14
Rahovec Serb. Orahovac. W Kosovo 42°24′N 20°40′E — 113 N17
Raiatea island Îles Sous le Vent, W French Polynesia — 191 S10
Rāichūr Karnātaka, C India 16°15′N 77°20′E — 155 H20
Raidestos see Tekirdağ
Raiganj West Bengal, NE India 25°38′N 88°11′E — 153 S13
Rāigarh Chhattisgarh, C India 21°53′N 83°28′E — 154 M11
Railton Tasmania, SE Australia 41°21′S 146°28′E — 183 O16
Rainbow Bridge natural arch Utah, W USA — 36 L3
Rainbow City Alabama, S USA 33°57′N 86°02′W — 23 Q3
Rainbow Lake Alberta, W Canada 58°30′N 119°24′W — 11 N11
Rainelle West Virginia, NE USA 37°57′N 80°48′W — 21 R5
Rainier Oregon, NW USA 46°05′N 122°55′W — 32 G10
Rainier, Mount ☒ Washington, NW USA 46°51′N 121°45′W — 32 H9
Rainsville Alabama, S USA 34°29′N 85°51′W — 23 Q2
Rainy Lake ☉ Canada/USA — 12 B11
Rainy River Ontario, C Canada 48°44′N 94°33′W — 12 A11
Raipur Chhattisgarh, C India 21°16′N 81°42′E — 154 K12
Raisen Madhya Pradesh, C India 23°21′N 77°49′E — 154 H10
Raisin ↔ Ontario, SE Canada — 15 N11
Raisin, River ↔ Michigan, N USA — 31 R11
Raivavae var. Raevavae. Îles Australes, SW French Polynesia — 191 U13
Rāiwind Punjab, E Pakistan 31°14′N 74°10′E — 149 W9
Raja Ampat, Kepulauan island group E Indonesia — 171 T12
Rājahmundry Andhra Pradesh, E India 17°05′N 81°47′E — 155 L16
Rājampet Andhra Pradesh, E India 14°11′N 79°10′E — 155 J18
Rajang see Rajang, Batang
Rajang, Batang var. Rajang. ↔ East Malaysia — 169 S9
Rājanpur Punjab, E Pakistan 29°05′N 70°25′E — 149 S11
Rājapālaiyam Tamil Nādu, SE India 09°26′N 77°36′E — 152 J12
Rājasthān ◆ state NW India — 152 F11
Rājbāri Dhaka, C Bangladesh 23°47′N 89°39′E — 153 T15
Rājbirāj Eastern, E Nepal 26°34′N 86°52′E — 153 Q12
Rājgarh Madhya Pradesh, C India 24°01′N 76°42′E — 154 G9
Rājgarh Rājasthān, NW India 28°38′N 75°21′E — 152 H10
Rājgīr Bihār, N India 25°01′N 85°26′E — 153 P14
Rajgród Podlaskie, NE Poland 53°43′N 22°40′E — 110 O8
Rājim Chhattisgarh, C India 20°57′N 81°58′E — 154 L12
Rajinac, Mali ▲ W Croatia 44°47′N 15°04′E — 112 C11
Rājkot Gujarāt, W India 22°18′N 70°47′E — 154 B10
Rājmahāl Jharkhand, NE India 25°03′N 87°50′E — 153 R13
Rājmahāl Hills hill range N India — 153 O14
Rāj Nāndgaon Chhattisgarh, C India 21°06′N 81°02′E — 154 K12
Rājpura Punjab, NW India 30°29′N 76°40′E — 152 I8
Rajshahi prev. Rampur Boalia. Rajshahi, W Bangladesh 24°24′N 88°40′E — 153 S14
Rajshahi ◆ division NW Bangladesh — 153 S13
Rakahanga atoll N Cook Islands — 190 K13
Rakaia Canterbury, South Island, New Zealand 43°45′S 172°02′E — 185 H19
Rakaia ↔ South Island, New Zealand — 185 G19
Rakaposhi ▲ N India 36°06′N 74°31′E — 152 H3
Rakasd see Răcăşdia
Rakata, Pulau var. Pulau Krakatau. island S Indonesia — 169 N15
Rakbah, Qalamat ar well SE Saudi Arabia — 141 O8
Rakhine State var. Arakan State. ◆ state W Myanmar (Burma) — 166 K6
Rakhiv Zakarpats'ka Oblast', W Ukraine 48°05′N 24°13′E — 116 I8
Rakhyūt SW Oman 16°41′N 53°09′E — 141 V13
Rakiraki Viti Levu, W Fiji 17°23′S 178°10′E — 192 K9
Rakitnoye Belgorodskaya Oblast', W Russian Federation 50°50′N 35°50′E — 126 J8
Rakka see Ar Raqqah
Rakke Lääne-Virumaa, NE Estonia 58°58′N 26°14′E — 118 I4
Rakoniewice Ger. Rakwitz. Wielkopolskie, C Poland 52°10′N 16°17′E — 110 F12
Rakops Central, C Botswana 21°01′S 24°20′E — 83 H18
Rakovník Ger. Rakonitz. Středočeský Kraj, W Czech Republic 50°07′N 13°45′E — 111 C16
Rakovski Plovdiv, C Bulgaria 42°16′N 24°58′E — 114 J10
Rakutō-kō see Nakdong-gang
Rakvere Ger. Wesenberg. Lääne-Virumaa, N Estonia 59°21′N 26°18′E — 118 I3
Rakwitz see Rakoniewice — 110 F12
Raleigh Mississippi, S USA 32°02′N 89°30′W — 22 G8

Raleigh state capital North Carolina, SE USA 35°46′N 78°38′W — 21 U10
Raleigh Bay bay North Carolina, SE USA — 21 Y11
Raleigh-Durham ✕ North Carolina, SE USA 35°54′N 78°45′W — 21 U9
Ralik Chain island group Ralik Chain, W Marshall Islands — 189 S6
Ralls Texas, SW USA 33°40′N 101°23′W — 25 N5
Ralston Pennsylvania, NE USA 41°29′N 76°57′W — 18 G13
Ramadi see Ar Ramādī
Ramādah W Yemen 13°35′N 43°50′E — 141 O16
Ramales de la Victoria Cantabria, N Spain 43°15′N 03°28′W — 105 N2
Ramallah C West Bank 31°55′N 35°12′E — 138 F10
Ramallo Buenos Aires, E Argentina 33°30′N 60°01′W — 61 C19
Rāmanagaram Karnātaka, E India 12°45′N 77°16′E — 155 H20
Rāmanāthapuram Tamil Nādu, SE India 09°23′N 78°53′E — 155 J23
Rāmapur Odisha, E India 21°48′N 84°02′E — 154 N12
Rāmāreddi var. Kāmāreddi, Kamareddy. Telangana, C India 18°19′N 78°23′E — 155 I14
Ramat Gan Tel Aviv, W Israel 32°04′N 34°48′E — 138 F10
Rambervillers Vosges, NE France 48°21′N 06°50′E — 103 T6
Rambi see Rabi
Rambouillet Yvelines, N France 48°39′N 01°50′E — 103 N5
Rambutyo Island island N Papua New Guinea — 186 E5
Ramechhāp Central, C Nepal 27°20′N 86°05′E — 153 Q12
Rame Head headland Victoria, SE Australia 37°48′S 149°33′E — 183 R12
Ramenskoye Moskovskaya Oblast', W Russian Federation 55°31′N 38°24′E — 126 L4
Rameshki Tverskaya Oblast', W Russian Federation 57°21′N 36°05′E — 124 J15
Rāmgarh Jharkhand, N India 23°37′N 85°32′E — 153 P14
Rāmgarh Rājasthān, NW India 27°29′N 70°38′E — 152 E12
Rāmhormoz var. Ram Hormuz, Ramuz. Khūzestān, SW Iran 31°15′N 49°38′E — 142 M9
Ram Hormuz see Rāmhormoz
Ram, Jebel see Ramm, Jabal
Ramle var. Ramla, Ramleh, Ar. Er Ramle. Central, C Israel 31°56′N 34°52′E — 138 F10
Ramm, Jabal var. Jebel Ram. ▲ SW Jordan 29°34′N 35°24′E — 138 F14
Ramle/Ramleh see Ramla
Rāmnagar Uttarakhand, N India 29°23′N 79°07′E — 152 N16
Râmnicu Sărat prev. Rîmnicul-Sărat, Rîmnicu-Sărat. Buzău, E Romania 45°24′N 27°06′E — 116 L12
Râmnicu Vâlcea prev. Rîmnicu Vîlcea. Vîlcea, C Romania 45°04′N 24°22′E — 116 I13
Ramokgwebane var. Ramokgwebane.
Ramokgwebane var. Ramokgwebana. Central, NE Botswana 20°32′S 27°40′E — 83 J18
Ramon' Voronezhskaya Oblast', W Russian Federation 51°51′N 39°18′E — 126 L7
Ramona California, W USA 33°02′N 116°52′W — 35 V10
Ramón, Laguna ☉ NW Peru — 56 A10
Ramore Ontario, S Canada 48°26′N 80°17′W — 14 G7
Ramos San Luis Potosí, C Mexico 22°48′N 101°55′W — 41 M11
Ramos Arizpe Coahuila, NE Mexico 25°35′N 100°59′W — 41 N8
Ramos, Río de ↔ C Mexico — 40 J9
Ramotswa South East, S Botswana 24°54′S 25°49′E — 83 J21
Rampart Alaska, USA 65°30′N 150°10′W — 39 R8
Ramparts ↔ Northwest Territories, NW Canada — 8 H8
Rāmpur Uttar Pradesh, N India 28°48′N 79°03′E — 152 K10
Rāmpura Madhya Pradesh, C India 24°30′N 75°30′E — 154 F9
Rampur Boalia see Rajshahi
Ramree Island island W Myanmar (Burma) — 166 K6
Ramsele Västernorrland, N Sweden 63°33′N 16°35′E — 93 N17
Ramseur North Carolina, SE USA 35°44′N 79°39′W — 21 T9
Ramsey NE Isle of Man 54°19′N 04°24′W — 97 I16
Ramsey Bay bay NE Isle of Man — 97 I16
Ramsey Lake ☉ Ontario, S Canada — 14 E9
Ramsgate SE England, United Kingdom 51°20′N 01°25′E — 97 Q22
Ramsjö Gävleborg, C Sweden 62°10′N 15°40′E — 93 G17
Rāmtek Mahārāshtra, C India 21°28′N 79°28′E — 154 I12
Ramtha see Ar Ramtha
Ramuz see Rāmhormoz
Ranau Sabah, East Malaysia 05°56′N 116°43′E — 157 X3
Ranau, Danau ☉ Sumatera, W Indonesia — 168 L14
Rancagua Libertador, C Chile 34°10′S 70°45′W — 62 H12
Rapallo Liguria, NW Italy 44°21′N 09°14′E — 106 D10
Rapa Nui see Pascua, Isla de
Raphiah see Rafah — 138 E11
Rapid River ↔ Virginia, NE USA — 21 V5
Rance ↔ NW France — 102 H6
Ranchería São Paulo, S Brazil 21°55′S 50°53′W — 60 J9
Rânchi Jhārkhand, N India 23°23′N 85°20′E — 153 P15
Rapide-Blanc Québec, SE Canada 47°48′N 72°58′W — 15 P8
Rapide-Deux Québec, SE Canada 47°56′N 78°33′W — 14 I8

Ranchos De Taos New Mexico, SW USA 36°21′N 105°36′W — 37 S9
Ranco, Lago ☉ C Chile — 63 G16
Randaberg Rogaland, S Norway 59°00′N 05°38′E — 95 C16
Randall Minnesota, N USA 46°05′N 94°30′W — 29 U7
Randers Midtjylland, C Denmark 56°28′N 10°03′E — 95 G21
Randijaure Lapp. Rádnávrre. ☉ N Sweden — 92 I12
Randleman North Carolina, SE USA 35°49′N 79°48′W — 21 T9
Randolph Massachusetts, NE USA 42°09′N 71°02′W — 19 O11
Randolph Nebraska, C USA 42°25′N 97°05′W — 29 R14
Randolph Utah, W USA 41°40′N 111°10′W — 36 M1
Randow ↔ NE Germany — 100 P9
Randsfjorden ☉ S Norway — 95 H14
Rånea Norrbotten, N Sweden 65°52′N 22°17′E — 92 K13
Ranelva ↔ C Norway — 92 G12
Ranemsletta Nord-Trøndelag, C Norway 64°36′N 11°55′E — 93 F15
Ranérou ☉ C Senegal 15°17′N 14°00′W — 76 H10
Ranés see Ringvassøya
Ranfurly Otago, South Island, New Zealand 45°07′S 170°06′E — 185 E22
Rangae Narathiwat, SW Thailand 06°11′N 101°45′E — 167 P17
Rangamati Chittagong, SE Bangladesh 22°40′N 92°10′E — 153 V16
Rangaunu Bay bay North Island, New Zealand — 184 I2
Rangeley Maine, NE USA 44°58′N 70°37′W — 19 P6
Rangely Colorado, C USA 40°05′N 108°48′W — 37 O4
Ranger Texas, SW USA 32°28′N 98°40′W — 25 R7
Ranger Lake Ontario, S Canada 46°51′N 83°34′W — 14 C9
Ranger Lake ☉ Ontario, S Canada — 14 C9
Rangia Assam, NE India 26°26′N 91°38′E — 153 V14
Rangiora Canterbury, South Island, New Zealand 43°19′S 172°34′E — 185 I18
Rangiroa atoll Îles Tuamotu, N French Polynesia — 191 T9
Rangitaiki ↔ North Island, New Zealand — 184 N9
Rangitata ↔ South Island, New Zealand — 185 F19
Rangitikei ↔ North Island, New Zealand — 184 M12
Rangitoto Island island N New Zealand — 184 L6
Rangkasbitoeng see Rangkasbitung
Rangkasbitung prev. Rangkasbitoeng. Jawa, SW Indonesia 06°15′S 106°12′E — 169 N16
Rang, Khao ▲ C Thailand 12°51′N 99°51′E — 167 P9
Rangkül Rus. Rangkul'. SE Tajikistan 38°30′N 74°24′E — 147 V13
Rangkul' see Rangkül
Rangoon see Yangon
Rangpur Rajshahi, N Bangladesh 25°46′N 89°20′E — 153 T13
Rānibennur Karnātaka, W India 14°36′N 75°39′E — 155 F18
Rānīganj West Bengal, NE India 23°36′N 87°09′E — 153 R15
Rānīpur Sind, SE Pakistan 27°17′N 68°34′E — 149 Q13
Rāniyah see Ranye
Rankin Texas, SW USA 31°14′N 101°56′W — 25 N9
Rankin Inlet Nunavut, C Canada 62°52′N 92°14′W — 9 O9
Rankins Springs New South Wales, SE Australia 33°51′S 146°16′E — 183 P8
Rannoch, Loch ☉ C Scotland, United Kingdom — 96 I10
Rano Kau var. Rano Kao. crater Easter Island, Chile, E Pacific Ocean — 191 U17
Ranong Ranong, SW Thailand 09°59′N 98°40′E — 167 N14
Ranongga var. Ghanongga. island NW Solomon Islands — 186 J8
Rano Raraku ancient monument Easter Island, Chile, E Pacific Ocean — 191 W6
Ransiki Papua Barat, E Indonesia 01°27′S 134°12′E — 171 V12
Rantajärvi Norrbotten, N Sweden 66°45′N 23°32′E — 92 K12
Rantasalmi Etelä-Savo, E Finland 62°04′N 28°22′E — 93 N17
Rantau Borneo, C Indonesia 02°56′S 115°09′E — 169 U13
Rantau, Pulau var. Pulau Tebingtinggi. island W Indonesia — 168 L10
Rantoul Illinois, N USA 40°19′N 88°08′W — 30 M13
Rantepao Sulawesi, C Indonesia 02°58′S 119°58′E — 171 P12
Ranua Lappi, NW Finland 65°55′N 26°34′E — 92 L13
Ranya see Ranye
Ranye var. Rāniyah. As Sulaymānīyah, E Iraq 36°15′N 44°53′E — 139 T3
Raohe Heilongjiang, NE China 46°49′N 134°00′E — 157 X3
Raoui, Erg er desert W Algeria — 74 H9
Rapa island Îles Australes, SW French Polynesia — 193 O10
Rapa Iti island Îles Australes, SW French Polynesia — 191 V14

Räpina Ger. Rappin. Põlvamaa, SE Estonia 58°06′N 27°27′E — 118 K6
Rapla Ger. Rappel. Raplamaa, NW Estonia 59°00′N 24°46′E — 118 G4
Raplamaa var. Rapla Maakond. ◆ province NW Estonia — 118 G4
Rapla Maakond see Raplamaa
Rappahannock River ↔ Virginia, NE USA — 21 X6
Rapperswil Sankt Gallen, NE Switzerland 47°14′N 08°50′E — 108 G7
Rappin see Räpina
Rāpti ↔ N India — 153 N12
Rápulo, Río ↔ E Bolivia — 57 K16
Raqqah/Raqqah, Muḥāfaẓat al see Ar Raqqah
Raquette ↔ New York, NE USA — 18 J8
Raquette River ↔ New York, NE USA — 18 J6
Raraka atoll Îles Tuamotu, C French Polynesia — 191 V10
Raroia atoll Îles Tuamotu, C French Polynesia — 191 V10
Rarotonga ✕ Rarotonga, S Cook Islands, C Pacific Ocean — 190 H15
Rarotonga island S Cook Islands, C Pacific Ocean — 190 H16
Rarz N Tajikistan 39°23′N 68°43′E — 147 P12
Ras al-'Ayn var. Ras al-'Ain. Al Ḥasakah, N Syria 36°52′N 40°05′E — 139 N2
Ra's al Baṣīṭ Al Lādhiqīyah, W Syria 35°51′N 35°55′E — 138 H3
Ra's al-Ḫafji var. Ras al-Ḫafji. Ash Sharqīyah, NE Saudi Arabia 28°22′N 48°30′E — 141 R5
Ras al-Khaimah/Ras al Khaimah see Ra's al Khaymah
Ra's al Khaymah var. Ras al-Khaimah. Ra's al Khaymah, NE United Arab Emirates 25°44′N 55°55′E — 143 R15
Ra's al-Khaimah ✕ Ra's al Khaimah, NE United Arab Emirates 25°37′N 55°51′E — 143 R15
Ras al-Khaimah see Ra's al Khaymah
Ra's an Naqb S Jordan 30°35′N 35°29′E — 138 G13
Rasa, Punta headland E Argentina 40°50′S 62°15′W — 61 B26
Rasawi Papua Barat, E Indonesia 02°04′S 134°02′E — 171 V12
Rāşcani see Rîşcani — 116 L6
Ras Dashen Terara ▲ N Ethiopia 13°12′N 38°09′E — 80 J10
Rasdhoo Atoll see Rasdu Atoll
Rasdu Atoll var. Rasdhoo Atoll. atoll C Maldives — 151 K19
Raseiniai Kaunas, C Lithuania 55°23′N 23°06′E — 118 E12
Ra's Ghārib var. Rās Ghārib. E Egypt 28°22′N 33°08′E — 75 X8
Rās Ghārib see Ra's Ghārib
Rashaant Hövsgöl, N Mongolia 49°08′N 101°46′E — 162 J6
Rashaant see Delüün, Bayan-Ölgiy, Mongolia
Rashaant see Öldziyt, Dundgovĭ, Mongolia
Rashid Eng. Rosetta. N Egypt 31°25′N 30°25′E — 75 V7
Rashid Al Başrah, E Iraq 31°15′N 47°31′E — 139 Y11
Rasht var. Resht. Gīlān, NW Iran 37°18′N 49°38′E — 142 M3
Rashwan see Reshwan
Rasik see Raasiku
Raška Serbia, C Serbia 43°18′N 20°32′E — 113 M15
Rasna Rus. Ryasna. Mahilyowskaya Voblasts', E Belarus 54°01′N 31°12′E — 119 P15
Râşnov prev. Rîşno, Rozsnyó, Hung. Barcarozsnyó. Braşov, C Romania 45°35′N 25°27′E — 116 J12
Rasony Rus. Rossony. Vitsyebskaya Voblasts', N Belarus 55°53′N 28°50′E — 118 L11
Rasskazovo Tambovskaya Oblast', W Russian Federation 52°42′N 41°45′E — 127 N7
Rasta Rus. Resta. ↔ E Belarus — 119 O16
Rastadt see Rastatt
Rastatt var. Rastadt. Baden-Württemberg, SW Germany 48°51′N 08°13′E — 101 G21
Rastenburg see Kętrzyn
Rasūlnagar Punjab, E Pakistan 32°20′N 73°51′E — 149 V7
Ratak Chain island group Ratak Chain, E Marshall Islands — 189 U6
Ratamka Rus. Ratomka. Minskaya Voblasts', C Belarus 53°56′N 27°21′E — 119 K15
Rätan Jämtland, C Sweden 62°28′N 14°35′E — 93 G17
Ratangarh Rājasthān, NW India 28°01′N 74°39′E — 152 G11
Ratchaburi var. Rat Buri. Ratchaburi, W Thailand 13°30′N 99°50′E — 167 O11
Rat Buri see Ratchaburi
Rathbun Lake ☉ Iowa, C USA — 29 U12
Ráth Caola Ir. Rathkeale. SW Ireland 52°32′N 08°56′W — 97 C20
Rathedaung Rakhine State, W Myanmar (Burma) 20°29′N 92°48′E — 166 K5
Rathenow Brandenburg, NE Germany 52°37′N 12°21′E — 100 M13
Ráth Luirc Ir. An Ráth. Limerick, SW Ireland 52°21′N 08°41′W — 97 C16
Rathlin Island Ir. Reachlainn. island N Northern Ireland, United Kingdom — 96 F13

Ratisbonne see Regensburg
Rätische Alpen see Rhaetian Alps
Rat Island island Aleutian Islands, Alaska, USA — 38 E17
Rat Islands island group Aleutian Islands, Alaska, USA — 38 E17
Ratlām prev. Rutlam. Madhya Pradesh, C India 23°23′N 75°04′E — 154 F10
Ratnagiri Mahārāshtra, W India 17°00′N 73°22′E — 155 D15
Ratnapura Sabaragamuwa Province, S Sri Lanka 06°41′N 80°25′E — 155 K26
Ratne see Ratno
Ratno Rus. Ratna. Volyns'ka Oblast', NW Ukraine 51°40′N 24°13′E — 116 J2
Ratomka see Ratamka
Raton New Mexico, SW USA 36°54′N 104°27′W — 37 U8
Raudnitz an der Elbe see Roudnice nad Labem
Raudales Chiapas, SE Mexico 17°07′N 93°39′W — 41 U16
Raufarhöfn Norðurland Eystra, NE Iceland 66°27′N 15°58′W — 92 K1
Raufoss Oppland, S Norway 60°44′N 10°59′E — 94 H13
Raukumara ▲ North Island, New Zealand 37°45′S 178°07′E — 184 Q8
Raukumara Plain undersea feature N Coral Sea — 192 K11
Raukumara Range ▲ North Island, New Zealand — 184 P8
Rauland Telemark, S Norway 59°41′N 07°57′E — 95 F15
Rauma Swe. Raumo. Satakunta, SW Finland 61°09′N 21°30′E — 93 J19
Rauma ↔ S Norway — 94 F9
Raumo see Rauma
Rauna C Latvia 57°19′N 25°34′E — 118 H8
Raurkela var. Rāulakela, Rourkela. Odisha, E India 22°13′N 84°53′E — 154 N11
Raus Skåne, S Sweden 56°01′N 12°48′E — 95 J22
Rausu Hokkaidō, NE Japan 44°04′N 145°04′E — 165 W3
Rausu-dake ▲ Hokkaidō, NE Japan 44°06′N 145°04′E — 165 W3
Rautalampi Pohjois-Savo, C Finland 62°39′N 26°48′E — 93 M17
Rautavaara Pohjois-Savo, C Finland 63°30′N 28°19′E — 93 N16
Rautjärvi Etelä-Karjala, SE Finland 61°21′N 29°21′E — 93 O18
Rautu see Sosnovo
Ravahere atoll Îles Tuamotu, C French Polynesia — 191 V11
Ravanusa Sicilia, Italy, C Mediterranean Sea 37°16′N 13°58′E — 107 J25
Rāvar Kermān, C Iran 31°15′N 56°51′E — 143 S9
Ravat Batkenskaya Oblast', SW Kyrgyzstan 39°54′N 70°06′E — 147 Q11
Ravena New York, NE USA 42°28′N 73°49′W — 18 K11
Ravenna Emilia-Romagna, N Italy 44°28′N 12°15′E — 106 H10
Ravenna Nebraska, C USA 41°01′N 98°54′W — 29 O15
Ravenna Ohio, N USA 41°09′N 81°14′W — 31 U11
Ravensburg Baden-Württemberg, S Germany 47°47′N 09°37′E — 101 I24
Ravenshoe Queensland, NE Australia 17°35′S 145°28′E — 181 W4
Ravensthorpe Western Australia 33°37′S 120°01′E — 180 K13
Ravenswood West Virginia, NE USA 38°57′N 81°46′W — 21 Q4
Rāvi ↔ India/Pakistan — 149 U9
Ravna Gora Primorje-Gorski Kotar, NW Croatia 45°24′N 14°58′E — 112 C9
Ravne na Koroškem Ger. Gutenstein. N Slovenia 46°31′N 14°56′E — 109 U9
Ravnina Kulyndy prev. Kulunda Steppe, Kaz. Qulyndy Zhazyghy, Rus. Kulundinskaya Ravnina. grassland Kazakhstan/Russian Federation — 145 T7
Rāwalpindi Punjab, NE Pakistan 33°36′N 73°06′E — 149 U6
Rawa Mazowiecka Łódzkie, C Poland 51°47′N 20°16′E — 110 L13
Rawas Papua Barat, E Indonesia 01°30′S 132°12′E — 171 U12
Rawdah Ar. Ar Rawdah. S Syria 33°37′N 39°21′E — 139 O4
Rawicz Ger. Rawitsch. Wielkopolskie, C Poland 51°36′N 16°51′E — 110 G13
Rawitsch see Rawicz
Rawlinna Western Australia 30°58′S 125°36′E — 180 M11
Rawlins Wyoming, C USA 41°47′N 107°13′W — 33 W16
Rawson Chubut, SE Argentina 43°22′S 65°01′W — 63 K17
Rawu Xizang Zizhiqu, W China 30°16′N 96°42′E — 159 R16
Raxaul Bihār, N India 26°58′N 84°50′E — 153 P12
Ray North Dakota, N USA 48°19′N 103°11′W — 28 K3

Raya, Bukit ▲ Borneo, C Indonesia 01°02′S 112°40′E — 169 S11
Rāyachoti Andhra Pradesh, E India 14°03′N 78°43′E — 155 I18
Rāyagada see Rāyagarha
Rāyagarha prev. Rāyadrug, var. Rāyagada. E India 19°10′N 83°28′E — 155 M14
Rayak var. Rayaq, Riyāq. E Lebanon 33°51′N 36°03′E — 138 H7
Rayaq see Rayak
Rayat var. Rāyāt, var. Rāyat. Arbil, E Iraq 36°39′N 44°56′E — 139 T2
Rāyāt see Rayat
Rāyat see Rayat
Raya, Tanjung cape Pulau Bangka, W Indonesia — 169 N12
Ray, Cape headland Newfoundland, Newfoundland and Labrador, E Canada 47°38′N 59°15′W — 13 R13
Raychikhinsk Amurskaya Oblast', SE Russian Federation 49°47′N 129°17′E — 123 Q13
Rayevskiy Respublika Bashkortostan, W Russian Federation 54°04′N 54°58′E — 127 U5
Raymond Alberta, SW Canada 49°30′N 112°41′W — 11 Q17
Raymond Mississippi, S USA 32°15′N 90°25′W — 22 K6
Raymond Washington, NW USA 46°41′N 123°43′W — 32 F9
Raymond Terrace New South Wales, SE Australia 32°47′S 151°45′E — 183 T8
Raymondville Texas, SW USA 26°30′N 97°48′W — 25 T17
Raymore Saskatchewan, S Canada 51°24′N 104°34′W — 11 U16
Ray Mountains ▲ Alaska, USA — 39 Q8
Rayne Louisiana, S USA 30°13′N 92°15′W — 22 H9
Rayón San Luis Potosí, C Mexico 21°51′N 99°39′W — 41 O12
Rayón Sonora, NW Mexico 29°45′N 110°33′W — 40 G4
Rayong Rayong, S Thailand 12°41′N 101°17′E — 167 P12
Ray Roberts, Lake ☉ Texas, SW USA — 25 T5
Raystown Lake ▨ Pennsylvania, NE USA — 18 L15
Raysūt SW Oman 16°58′N 53°58′E — 141 S4
Raytown Missouri, C USA 39°00′N 94°27′W — 27 R4
Rayville Louisiana, S USA 32°29′N 91°45′W — 22 I5
Razan Hamadān, W Iran 35°22′N 48°58′E — 142 L5
Razāzah, Buhayrat ar var. Baḥr al Milḥ. ☉ C Iraq — 139 S9
Razbojna ▲ E Bulgaria — 114 L9
Razdan see Hrazdan
Razdolnoye see Rozdol'ne
Razelm, Lacul see Razim, Lacul
Razga see Rezge
Razgrad Razgrad, N Bulgaria 43°33′N 26°31′E — 114 L8
Razgrad ◆ province NE Bulgaria — 114 L8
Razhevo Konare var. Rǔzhevo Konare. Plovdiv, C Bulgaria 42°16′N 24°58′E — 114 I10
Razim, Lacul prev. Lacul Razelm. lagoon NW Black Sea — 117 N13
Razkah see Rezge
Razlog Blagoevgrad, SW Bulgaria 41°53′N 23°28′E — 114 G11
Rāznas Ezers ☉ SE Latvia — 118 K10
Raz, Pointe du headland NW France 48°04′N 04°52′W — 102 E6
Reachlainn see Rathlin Island
Reachrainn see Lambay Island
Reading S England, United Kingdom 51°28′N 00°59′W — 97 N22
Reading Pennsylvania, NE USA 40°20′N 75°55′W — 18 H15
Real, Cordillera ▲ C Ecuador — 48 C7
Realitos Texas, SW USA 27°26′N 98°31′W — 25 R15
Realp Uri, C Switzerland 46°36′N 08°32′E — 108 G9
Reăng Kesei Bătdâmbâng, W Cambodia 12°57′N 103°15′E — 167 Q12
Reao atoll Îles Tuamotu, E French Polynesia — 191 Y11
Reate see Rieti
Greater Antarctica see East Antarctica
Rebecca, Lake ☉ Western Australia — 180 L11
Rebiana Sand Sea see Rabyānah, Ramlat
Reboly Fin. Repola. Respublika Kareliya, NW Russian Federation 63°51′N 30°49′E — 124 H8
Rebun-tō island NE Japan — 165 S1
Rebun-tō island NE Japan — 165 S1
Recanati Marche, C Italy 43°23′N 13°34′E — 106 J12
Rechnitz Burgenland, SE Austria 47°18′N 16°28′E — 109 Y7
Rechytsa Rus. Rechitsa. Brestskaya Voblasts', SW Belarus 51°51′N 26°48′E — 119 J20
Rechytsa Rus. Rechitsa. Homyel'skaya Voblasts', SE Belarus 52°21′N 30°23′E — 119 O19
Recife prev. Pernambuco. state capital Pernambuco, E Brazil 08°06′S 34°53′W — 59 Q15
Recife, Cape Afr. Kaap Recife. headland S South Africa 34°03′S 25°37′E — 83 I26
Recife, Kaap see Recife, Cape
Récifs, Îles aux island Inner Islands, NE Seychelles — 172 I16
Recklinghausen Nordrhein-Westfalen, W Germany 51°37′N 07°12′E — 101 E14
Recknitz ↔ NE Germany — 100 M8
Recogne Luxembourg, SE Belgium 49°55′N 05°25′E — 99 K23
Reconquista Santa Fe, C Argentina 29°08′S 59°38′W — 61 C15
Recovery Glacier glacier Antarctica — 59 O6
Recreo Catamarca, C Argentina 29°18′S 65°04′W — 61 C16
Rector Arkansas, C USA — 27 X9
Recz Ger. Reetz Neumark. Zachodnio-pomorskie, NW Poland 53°16′N 15°32′E — 110 E9

◆ Country   ◇ Dependent Territory   ◉ Administrative Regions   ▲ Mountain   ☒ Volcano   ☉ Lake
● Country Capital   ○ Dependent Territory Capital   ✕ International Airport   ▲ Mountain Range   ↔ River   ▨ Reservoir

311

Redange – Rincón de la Victoria

99 L24 **Redange** *var.* Redange-sur-Attert. Diekirch, W Luxembourg 49°46′N 05°53′E
Redange-sur-Attert *see* Redange
18 C13 **Redbank Creek** ≈ Pennsylvania, NE USA
13 S9 **Red Bay** Québec, E Canada 51°40′N 56°37′W
23 N2 **Red Bay** Alabama, S USA 34°26′N 88°08′W
35 N4 **Red Bluff** California, W USA 40°09′N 122°14′W
24 J8 **Red Bluff Reservoir** ▣ New Mexico/Texas, SW USA
30 K16 **Red Bud** Illinois, N USA 38°12′N 89°59′W
30 J5 **Red Cedar River** ≈ Wisconsin, N USA
11 R17 **Redcliff** Alberta, SW Canada 50°06′N 110°48′W
83 K17 **Redcliff** Midlands, C Zimbabwe 19°00′S 29°49′E
182 L9 **Red Cliffs** Victoria, SE Australia 34°21′S 142°12′E
29 P17 **Red Cloud** Nebraska, C USA 40°05′N 98°31′W
22 L8 **Red Creek** ≈ Mississippi, S USA
11 P15 **Red Deer** Alberta, SW Canada 52°15′N 113°48′W
11 Q16 **Red Deer** ≈ Alberta, SW Canada
39 O11 **Red Devil** Alaska, USA 61°45′N 157°18′W
35 N3 **Redding** California, W USA 40°33′N 122°26′W
97 L20 **Redditch** W England, United Kingdom 52°19′N 01°56′W
29 P9 **Redfield** South Dakota, N USA 44°51′N 98°31′W
24 J12 **Redford** Texas, SW USA 29°31′N 104°19′W
45 V13 **Redhead** Trinidad, Trinidad and Tobago 10°44′N 60°58′W
182 I8 **Red Hill** South Australia 33°34′S 138°13′E
Red Hill *see* Pu'u 'Ula'ula
26 K7 **Red Hills** *hill range* Kansas, C USA
13 T12 **Red Indian Lake** ◎ Newfoundland, Newfoundland and Labrador, E Canada
124 J16 **Redkino** Tverskaya Oblast', W Russian Federation 56°41′N 36°07′E
12 A10 **Red Lake** Ontario, C Canada 51°00′N 93°55′W
36 I10 **Red Lake** *salt flat* Arizona, SW USA
29 S4 **Red Lake Falls** Minnesota, N USA 47°52′N 96°16′W
29 R4 **Red Lake River** ≈ Minnesota, N USA
35 U15 **Redlands** California, W USA 34°03′N 117°10′W
18 G16 **Red Lion** Pennsylvania, NE USA 39°53′N 76°36′W
33 U11 **Red Lodge** Montana, NW USA 45°11′N 109°15′W
32 H13 **Redmond** Oregon, NW USA 44°16′N 121°10′W
36 L5 **Redmond** Utah, W USA 39°00′N 111°51′W
32 H8 **Redmond** Washington, NW USA 47°40′N 122°07′W
Rednitz *see* Regnitz
29 T15 **Red Oak** Iowa, C USA 41°00′N 95°10′W
18 K12 **Red Oaks Mill** New York, NE USA 41°39′N 73°52′W
102 I7 **Redon** Ille-et-Vilaine, NW France 47°39′N 02°05′W
45 W10 **Redonda** *island* SW Antigua and Barbuda
104 G4 **Redondela** Galicia, NW Spain 42°17′N 08°36′W
104 H11 **Redondo** Évora, S Portugal 38°38′N 07°32′W
39 Q12 **Redoubt Volcano** ▲ Alaska, USA 60°29′N 152°44′W
11 Y16 **Red River** ≈ Canada/USA
129 U12 **Red River** *var.* Yuan, *Chin.* Yuan Jiang, *Vtn.* Sông Hông Hà. ≈ China/Vietnam
25 W4 **Red River** ≈ S USA
22 H7 **Red River** ≈ Louisiana, S USA
30 M6 **Red River** ≈ Wisconsin, N USA
Red Rock, Lake *see* Red Rock Reservoir
29 W14 **Red Rock Reservoir** *var.* Lake Red Rock. ▣ Iowa, C USA
80 H7 **Red Sea** ◈ *state* NE Sudan
75 Y9 **Red Sea** *var.* Sinus Arabicus. *sea* Africa/Asia
21 T11 **Red Springs** North Carolina, SE USA 34°49′N 79°10′W
8 I9 **Redstone** ≈ Northwest Territories, NW Canada
11 V17 **Redvers** Saskatchewan, S Canada 49°31′N 101°33′W
77 P13 **Red Volta** *var.* Nazinon, *Fr.* Volta Rouge. ≈ Burkina Faso/Ghana
11 Q14 **Redwater** Alberta, SW Canada 53°57′N 113°06′W
28 M16 **Red Willow Creek** ≈ Nebraska, C USA
29 W9 **Red Wing** Minnesota, N USA 44°33′N 92°31′W
35 N9 **Redwood City** California, W USA 37°29′N 122°13′W
29 T9 **Redwood Falls** Minnesota, N USA 44°33′N 95°07′W
31 P7 **Reed City** Michigan, N USA 43°52′N 85°30′W
28 K6 **Reeder** North Dakota, N USA 46°03′N 102°55′W
35 R11 **Reedley** California, W USA 36°35′N 119°27′W
33 T11 **Reedpoint** Montana, NW USA 45°41′N 109°33′W
30 K8 **Reedsburg** Wisconsin, N USA 43°33′N 90°03′W
32 E13 **Reedsport** Oregon, NW USA 43°42′N 124°06′W
187 Q9 **Reef Islands** *island group* Santa Cruz Islands, E Solomon Islands
185 H16 **Reefton** West Coast, South Island, New Zealand 42°07′S 171°53′E
20 F8 **Reelfoot Lake** ◎ Tennessee, S USA
97 D17 **Ree, Lough** *Ir.* Loch Rí. ◎ C Ireland
Reengus *see* Rīngas
35 U4 **Reese River** ≈ Nevada, W USA
98 M8 **Reest** ≈ E Netherlands
Reetz Neumark *see* Recz
Reevhtse *see* Røssvatnet
137 N13 **Refahiye** Erzincan, C Turkey 39°54′N 38°45′E
23 N4 **Reform** Alabama, S USA 33°22′N 88°01′W
95 K20 **Reftele** Jönköping, S Sweden 57°10′N 13°34′E
25 T14 **Refugio** Texas, SW USA 28°19′N 97°18′W
110 E8 **Rega** ≈ NW Poland
Regar *see* Tursunzoda
101 O21 **Regen** Bayern, SE Germany 48°57′N 13°10′E
101 M20 **Regen** ≈ SE Germany
101 M21 **Regensburg** *Eng.* Ratisbon, *Fr.* Ratisbonne, *hist.* Ratisbona; *anc.* Castra Regina, Reginum. Bayern, SE Germany 49°01′N 12°06′E
101 M21 **Regenstauf** Bayern, SE Germany 49°06′N 12°07′E
148 M10 **Rēgestān** *var.* Registan *prev.* Rīgestān. S Afghanistan
74 I10 **Reggane** C Algeria 26°46′N 00°09′E
98 N9 **Regge** ≈ E Netherlands
Reggio *see* Reggio nell'Emilia
Reggio Calabria *see* Reggio di Calabria
107 M23 **Reggio di Calabria** *var.* Reggio Calabria, *Gk.* Rhegion; *anc.* Regium, Rhegium. Calabria, SW Italy 38°06′N 15°39′E
Reggio Emilia *see* Reggio nell'Emilia
106 F9 **Reggio nell'Emilia** *var.* Reggio Emilia, *abbrev.* Reggio; *anc.* Regium Lepidum. Emilia-Romagna, N Italy 44°42′N 10°37′E
116 I10 **Reghin** *Ger.* Sächsisch-Reen, *Hung.* Szászrégen; *prev.* Reghinul Săsesc, *Ger.* Sächsisch-Regen. Mureş, C Romania 46°46′N 24°41′E
Reghinul Săsesc *see* Reghin
11 U16 **Regina** *province capital* Saskatchewan, S Canada 50°25′N 104°39′W
55 Z10 **Régina** E French Guiana 04°20′N 52°07′W
11 U16 **Regina** ✕ Saskatchewan, S Canada 50°21′N 104°43′W
11 U16 **Regina Beach** Saskatchewan, S Canada 50°44′N 105°03′W
Reginum *see* Regensburg
Région du Haut-Congo *see* Haut-Congo
Registan *see* Rēgestān
60 L11 **Registro** São Paulo, S Brazil 24°30′S 47°50′W
Regium *see* Reggio di Calabria
Regium Lepidum *see* Reggio nell'Emilia
101 K19 **Regnitz** *var.* Rednitz. ≈ SE Germany
40 K10 **Regocijo** Durango, W Mexico 23°35′N 105°11′W
104 H12 **Reguengos de Monsaraz** Évora, S Portugal 38°25′N 07°32′W
101 M18 **Rehau** Bayern, E Germany 50°15′N 12°03′E
83 D19 **Rehoboth** Hardap, C Namibia 23°18′S 17°03′E
21 Z4 **Rehoboth Beach** Delaware, NE USA 38°42′N 75°03′W
138 F10 **Rehovot** ; *prev.* Reẖovot. Central, C Israel 31°54′N 34°49′E
Reẖovot *see* Rehovot
81 J20 **Rei** *spring/well* S Kenya 03°24′S 39°18′E
Reichenau *see* Rychnov nad Kněžnou
Reichenau *see* Bogatynia, Poland
101 M17 **Reichenbach** *var.* Reichenbach im Vogtland. Sachsen, E Germany 50°36′N 12°18′E
Reichenbach *see* Dzierżoniów
Reichenbach im Vogtland *see* Reichenbach
Reichenberg *see* Liberec
181 O11 **Reid** Western Australia 30°49′S 128°24′E
23 V6 **Reidsville** Georgia, SE USA 32°05′N 82°07′W
21 T8 **Reidsville** North Carolina, SE USA 36°21′N 79°39′W
Reifnitz *see* Ribnica
97 O22 **Reigate** SE England, United Kingdom 51°14′N 00°13′W
Reikjavik *see* Reykjavík
102 I10 **Ré, Île de** *island* W France
37 N15 **Reiley Peak** ▲ Arizona, SW USA 32°24′N 110°09′W
103 Q4 **Reims** *Eng.* Rheims; *anc.* Durocortorum, Remi. Marne, N France 49°16′N 04°01′E
63 G23 **Reina Adelaida, Archipiélago** *island group* S Chile
45 O16 **Reina Beatrix** ✕ (Oranjestad) C Aruba 12°30′N 69°57′W
108 F7 **Reinach** Aargau, W Switzerland 47°16′N 08°12′E
108 E6 **Reinach** Basel Landschaft, NW Switzerland 47°30′N 07°36′E
64 O11 **Reina Sofía** ✕ (Tenerife) Tenerife, Islas Canarias, Spain, NE Atlantic Ocean
29 W13 **Reinbeck** Iowa, C USA 42°19′N 92°36′W
100 J10 **Reinbek** Schleswig-Holstein, N Germany 53°31′N 10°15′E
11 U12 **Reindeer** ≈ Saskatchewan, C Canada
11 U11 **Reindeer Lake** ◎ Manitoba/Saskatchewan, C Canada
Reine-Charlotte, Îles de la *see* Queen Charlotte Islands
Reine-Élisabeth, Îles de la *see* Queen Elizabeth Islands
94 F13 **Reineskarvet** ▲ S Norway 60°38′N 07°48′E
184 H1 **Reinga, Cape** *headland* North Island, New Zealand 34°24′S 172°40′E
105 N3 **Reinosa** Cantabria, N Spain 43°01′N 04°09′W
109 R8 **Reisseck** ▲ S Austria 46°57′N 13°21′E
21 W3 **Reisterstown** Maryland, NE USA 39°27′N 76°46′W
Reisui *see* Yeosu
98 N5 **Reitdiep** ≈ NE Netherlands
191 V10 **Reitoru** *atoll* Îles Tuamotu, C French Polynesia
95 M17 **Rejmyre** Östergötland, S Sweden 58°49′N 15°55′E
Reka *see* Rijeka
Reka Ili *see* Ile/Ili He
Rekarne *see* Tumbo
Rekhovot *see* Rehovot
8 K9 **Reliance** Northwest Territories, C Canada 62°45′N 109°08′W
33 U16 **Reliance** Wyoming, C USA 41°42′N 109°13′W
74 I5 **Relizane** *var.* Ghelîzâne, Ghilizane. NW Algeria 35°45′N 00°33′E
182 I7 **Remarkable, Mount** ▲ South Australia 32°46′S 138°08′E
54 E8 **Remedios** Antioquia, N Colombia 07°02′N 74°42′W
43 Q16 **Remedios** Veraguas, W Panama 08°13′N 81°48′W
42 D8 **Remedios, Punta** *headland* SW El Salvador 13°31′N 89°48′W
Remi *see* Reims
99 N25 **Remich** Grevenmacher, SE Luxembourg 49°33′N 06°23′E
99 J19 **Remicourt** Liège, E Belgium 50°40′N 05°19′E
14 H8 **Rémigny, Lac** ◎ Québec, SE Canada
55 Z10 **Rémire** NE French Guiana 04°52′N 52°16′W
127 N13 **Remontnoye** Rostovskaya Oblast', SW Russian Federation 46°35′N 43°38′E
171 U14 **Remoon** Pulau Kur, E Indonesia 05°18′S 131°59′E
99 L20 **Remouchamps** Liège, E Belgium 50°29′N 05°43′E
103 R15 **Remoulins** Gard, S France 43°56′N 04°34′E
173 X16 **Rempart, Mont du** *hill* W Mauritius
101 E15 **Remscheid** Nordrhein-Westfalen, W Germany 51°10′N 07°11′E
29 S12 **Remsen** Iowa, C USA 42°48′N 95°58′W
94 I12 **Rena** Hedmark, S Norway 61°08′N 11°21′E
94 I11 **Renåa** ≈ S Norway
Renaix *see* Ronse
118 H7 **Rencēni** N Latvia 57°43′N 25°25′E
118 D9 **Renda** W Latvia 57°04′N 22°18′E
107 N20 **Rende** Calabria, SW Italy 39°19′N 16°10′E
99 K21 **Rendeux** Luxembourg, SE Belgium 50°15′N 05°28′E
Rendina *see* Rentína
30 L16 **Rend Lake** ▣ Illinois, N USA
186 K9 **Rendova** *island* New Georgia Islands, NW Solomon Islands
100 I8 **Rendsburg** Schleswig-Holstein, N Germany 54°18′N 09°40′E
108 B9 **Renens** Vaud, SW Switzerland 46°32′N 06°36′E
14 K12 **Renfrew** Ontario, SE Canada 45°28′N 76°44′W
96 I12 **Renfrew** *cultural region* SW Scotland, United Kingdom
168 L11 **Rengat** Sumatera, W Indonesia 0°26′S 102°38′E
153 W12 **Rengma Hills** ▲▲ NE India
62 H12 **Rengo** Libertador, C Chile 34°24′S 70°50′W
116 M12 **Reni** Odes'ka Oblast', SW Ukraine 45°30′N 28°18′E
80 F11 **Renk** Upper Nile, NE South Sudan 11°48′N 32°49′E
93 L19 **Renko** Kanta-Häme, S Finland 60°52′N 24°16′E
98 L12 **Renkum** Gelderland, SE Netherlands 51°58′N 05°43′E
182 K9 **Renmark** South Australia 34°12′S 140°43′E
186 L10 **Rennell** *var.* Mu Nggava. *island* S Solomon Islands
186 M9 **Rennell and Bellona** *prev.* Central. ◈ *province* S Solomon Islands
181 Q4 **Renner Springs Roadhouse** Northern Territory, N Australia 18°12′S 133°48′E
102 I6 **Rennes** *Bret.* Roazon; *anc.* Condate. Ille-et-Vilaine, NW France 48°08′N 01°40′W
195 S16 **Rennick Glacier** *glacier* Antarctica
11 Y16 **Rennie** Manitoba, S Canada 49°51′N 95°28′W
35 Q5 **Reno** Nevada, W USA 39°32′N 119°49′W
106 H10 **Reno** ≈ N Italy
35 Q5 **Reno-Cannon** ✕ Nevada, W USA 39°26′N 119°42′W
83 F24 **Renoster** ≈ SW South Africa
15 T5 **Renouard, Lac** ◎ Québec, SE Canada
18 F13 **Renovo** Pennsylvania, NE USA 41°19′N 77°42′W
161 O3 **Renqiu** Hebei, E China 38°49′N 116°02′E
160 I9 **Renshou** Sichuan, C China 30°02′N 104°09′E
31 N12 **Rensselaer** Indiana, N USA 40°57′N 87°09′W
18 L11 **Rensselaer** New York, NE USA 42°38′N 73°44′W
Rentería *see* Errenteria
115 E17 **Rentína** *var.* Rendina. Thessalía, C Greece 39°04′N 21°58′E
29 T9 **Renville** Minnesota, N USA 44°48′N 95°13′W
77 O13 **Réo** W Burkina Faso 12°20′N 02°28′W
15 O12 **Repentigny** Québec, SE Canada 45°42′N 73°28′W
146 K13 **Repetek** Lebap Welaýaty, E Turkmenistan 38°40′N 63°12′E
93 J16 **Replot** *Fin.* Raippaluoto. *island* W Finland
Repola *see* Reboly
Reppen *see* Rzepin
Reps *see* Rupea
27 T7 **Republic** Missouri, C USA 37°07′N 93°28′W
32 K7 **Republic** Washington, NW USA 48°39′N 118°44′W
27 N3 **Republican River** ≈ Kansas/Nebraska, C USA
9 O7 **Repulse Bay** Northwest Terretories, N Canada 66°35′N 86°20′W
56 F9 **Requena** Loreto, NE Peru 05°05′S 73°52′W
105 R10 **Requena** Valenciana, E Spain 39°29′N 01°08′W
103 O14 **Réquista** Aveyron, S France 44°00′N 02°31′E
136 M12 **Reşadiye** Tokat, N Turkey 40°24′N 37°19′E
Reschenpass *see* Resia, Passo di
Reschitza *see* Reşiţa
113 N20 **Resen** *Turk.* Resne. SW FYR Macedonia 41°07′N 21°00′E
60 J11 **Reserva** Paraná, S Brazil 24°40′S 50°52′W
11 V15 **Reserve** Saskatchewan, S Canada 52°24′N 102°37′W
37 P13 **Reserve** New Mexico, SW USA 33°42′N 108°45′W
Reshetilovka *see* Reshetylivka
117 S6 **Reshetylivka** *Rus.* Reshetilovka. Poltavs'ka Oblast', NE Ukraine 49°34′N 34°05′E
Resht *see* Rasht
139 S2 **Reshwan** *Ar.* Rashwān, *var.* Rashwan. Arbīl, N Iraq 36°28′N 43°54′E
106 F5 **Resia, Passo di** *Ger.* Reschenpass. *pass* Austria/Italy
Resicabánya *see* Reşiţa
62 N7 **Resistencia** Chaco, NE Argentina 27°27′S 58°56′W
116 F12 **Reşiţa** *Ger.* Reschitza, *Hung.* Resicabánya. Caraş-Severin, W Romania 45°14′N 21°58′E
Resne *see* Resen
197 N9 **Resolute** *Inuit* Qausuittuq. Cornwallis Island, Nunavut, N Canada 74°41′N 94°54′W
Resolution *see* Fort Resolution
9 T7 **Resolution Island** *island* Nunavut, NE Canada
185 A23 **Resolution Island** *island* SW New Zealand
Resta *see* Rasta
15 W7 **Restigouche** Québec, SE Canada 48°02′N 66°42′W
11 W17 **Reston** Manitoba, S Canada 49°33′N 101°03′W
14 H11 **Restoule Lake** ◎ Ontario, S Canada
54 F10 **Restrepo** Meta, C Colombia 04°20′N 73°29′W
42 B6 **Retalhuleu** Retalhuleu, SW Guatemala 14°31′N 91°40′W
42 A1 **Retalhuleu** *off.* Departamento de Retalhuleu. ◈ *department* SW Guatemala
Retalhuleu, Departamento de *see* Retalhuleu
97 N18 **Retford** C England, United Kingdom 53°18′N 00°52′W
103 Q3 **Rethel** Ardennes, N France 49°31′N 04°22′E
Rethimno/Réthimnon *see* Réthymno
115 I25 **Réthymno** *prev.* Rethimno, Réthimnon. Kríti, Greece, E Mediterranean Sea 35°21′N 24°29′E
Retiche, Alpi *see* Rhaetian Alps
99 J16 **Retie** Antwerpen, N Belgium 51°18′N 05°05′E
111 J21 **Rétság** Nógrád, N Hungary 47°57′N 19°08′E
109 W2 **Retz** Niederösterreich, NE Austria 48°46′N 15°58′E
173 N15 **Réunion** *off.* La Réunion. ◇ *French overseas department* W Indian Ocean
128 L17 **Réunion** *island* W Indian Ocean
105 U6 **Reus** Cataluña, E Spain 41°10′N 01°06′E
108 F7 **Reuss** ≈ NW Switzerland
99 J15 **Reusel** Noord-Brabant, S Netherlands 51°21′N 05°10′E
Reuţel *see* Ciuhuru
101 H22 **Reutlingen** Baden-Württemberg, S Germany 48°30′N 09°13′E
108 L7 **Reutte** Tirol, W Austria 47°30′N 10°44′E
99 M16 **Reuver** Limburg, SE Netherlands 51°17′N 06°05′E
28 K7 **Reva** South Dakota, N USA 45°30′N 103°03′W
Reval/Revel *see* Tallinn
124 J4 **Revda** Murmanskaya Oblast', NW Russian Federation 67°57′N 34°29′E
122 F6 **Revda** Sverdlovskaya Oblast', C Russian Federation 56°48′N 59°42′E
103 N16 **Revel** Haute-Garonne, S France 43°27′N 01°59′E
11 O16 **Revelstoke** British Columbia, SW Canada 51°02′N 118°12′W
43 N13 **Reventazón, Río** ≈ E Costa Rica
106 G9 **Revere** Lombardia, N Italy 45°03′N 11°07′E
39 Y14 **Revillagigedo Island** *island* Alexander Archipelago, Alaska, USA
193 R7 **Revillagigedo Islands** Mexico
103 R3 **Revin** Ardennes, N France 49°57′N 04°39′E
92 O3 **Revnosa** *headland* C Svalbard 78°03′N 18°52′E
Revolyutsii, Pik *see* Revolyutsiya, Qullai
147 T13 **Revolyutsiya, Qullai** *Rus.* Pik Revolyutsii. ▲ SE Tajikistan 38°40′N 72°26′E
111 L19 **Revúca** *Ger.* Grossrauschenbach, *Hung.* Nagyrőce. Banskobystrický Kraj, C Slovakia 48°40′N 20°10′E
154 K9 **Rewa** Madhya Pradesh, C India 24°32′N 81°18′E
139 T2 **Rewanduz** *Ar.* Rawāndūz, *var.* Rawāndiz. Arbīl, N Iraq 36°38′N 44°32′E
152 I11 **Rewāri** Haryāna, N India 28°14′N 76°38′E
33 R14 **Rexburg** Idaho, NW USA 43°49′N 111°47′W
78 G13 **Rey Bouba** Nord, NE Cameroon 08°40′N 14°11′E
92 L3 **Reyðarfjörður** Austurland, E Iceland 65°02′N 14°12′W
57 K16 **Reyes** El Beni, NW Bolivia 14°17′S 67°18′W
34 L8 **Reyes, Point** *headland* California, W USA 37°59′N 123°01′W
54 B12 **Reyes, Punta** *headland* SW Colombia 02°44′N 78°08′W
136 L17 **Reyhanlı** Hatay, S Turkey 36°16′N 36°35′E
43 U16 **Rey, Isla del** *island* Archipiélago de las Perlas, SE Panama
92 H2 **Reykhólar** Vestfirðir, W Iceland 65°28′N 22°12′W
92 K2 **Reykjahlíð** Norðurland Eystra, NE Iceland 65°37′N 16°54′W
197 O16 **Reykjanes Basin** *var.* Irminger Basin. *undersea feature* N Atlantic Ocean 62°30′N 33°30′W
197 N17 **Reykjanes Ridge** *undersea feature* N Atlantic Ocean 62°00′N 27°00′W
92 H4 **Reykjavík** *var.* Reikjavik. ● (Iceland) Höfudhborgarsvaedhi, W Iceland 64°08′N 21°54′W
18 D13 **Reynoldsville** Pennsylvania, NE USA 41°04′N 78°51′W
41 P8 **Reynosa** Tamaulipas, C Mexico 26°03′N 98°19′W
Reẕā'īyeh *see* Orūmīyeh
Reẕā'īyeh, Daryācheh-ye *see* Orūmīyeh, Daryācheh-ye
102 I8 **Rezé** Loire-Atlantique, NW France 47°10′N 01°36′W
118 K10 **Rēzekne** *Ger.* Rositten; *prev. Rus.* Rezhitsa. SE Latvia 56°31′N 27°22′E
139 U2 **Rezge** *Ar.* Razkah, *var.* Razga. As Sulaymānīyah, E Iraq 36°25′N 45°06′E
Rezhitsa *see* Rēzekne
117 N9 **Rezina** NE Moldova 47°44′N 28°58′E
114 N11 **Rezovo** *Turk.* Rezve. Burgas, E Bulgaria 42°00′N 28°00′E
114 N11 **Rezovska Reka** *Turk.* Rezve Deresi. ≈ Bulgaria/Turkey *see also* Rezve Deresi
Rezovska Reka *see* Rezve Deresi
Rezve *see* Rezovo
114 N11 **Rezve Deresi** *Bul.* Rezovska Reka. ≈ Bulgaria/Turkey *see also* Rezovska Reka
Rezve Deresi *see* Rezovska Reka
Rhadames *see* Ghadāmis
Rhaedestus *see* Tekirdağ
108 J10 **Rhaetian Alps** *Fr.* Alpes Rhétiques, *Ger.* Rätische Alpen, *It.* Alpi Retiche. ▲▲ C Europe
108 I8 **Rhätikon** ▲▲ C Europe
101 G14 **Rheda-Wiedenbrück** Nordrhein-Westfalen, W Germany 51°51′N 08°17′E
98 M12 **Rheden** Gelderland, E Netherlands 52°01′N 06°03′E
Rhegion/Rhegium *see* Reggio di Calabria
Rheims *see* Reims
Rhein *see* Rhine
101 E17 **Rheinbach** Nordrhein-Westfalen, W Germany 50°37′N 06°57′E
100 F13 **Rheine** *var.* Rheine in Westfalen. Nordrhein-Westfalen, NW Germany 52°17′N 07°27′E
Rheine in Westfalen *see* Rheine
Rheinfeld *see* Rheinfelden
101 F24 **Rheinfelden** Baden-Württemberg, S Germany 47°34′N 07°46′E
108 E6 **Rheinfelden** *var.* Rheinfeld. Aargau, N Switzerland 47°33′N 07°47′E
101 E17 **Rheinisches Schiefergebirge** *var.* Rhine State Uplands, *Eng.* Rhenish Slate Mountains. ▲▲ W Germany
101 D18 **Rheinland-Pfalz** *Eng.* Rhineland-Palatinate, *Fr.* Rhénanie-Palatinat. ◈ *state* W Germany
101 G18 **Rhein Main** ✕ (Frankfurt am Main) Hessen, W Germany 50°03′N 08°33′E
Rhénanie du Nord-Westphalie *see* Nordrhein-Westfalen
Rhénanie-Palatinat *see* Rheinland-Pfalz
98 K12 **Rhenen** Utrecht, C Netherlands 52°01′N 06°02′E
Rhenish Slate Mountains *see* Rheinisches Schiefergebirge
Rhétiques, Alpes *see* Rhaetian Alps
100 N10 **Rhin** ≈ NE Germany
Rhin *see* Rhine
84 F10 **Rhine** *Dut.* Rijn, *Fr.* Rhin, *Ger.* Rhein. ≈ W Europe
30 L5 **Rhinelander** Wisconsin, N USA 45°39′N 89°23′W
Rhineland-Palatinate *see* Rheinland-Pfalz
Rhine State Uplands *see* Rheinisches Schiefergebirge
100 N11 **Rhinkanal** *canal* NE Germany
81 F17 **Rhino Camp** NW Uganda 02°58′N 31°24′E
74 D7 **Rhir, Cap** *headland* W Morocco 30°40′N 09°54′W
106 D7 **Rho** Lombardia, N Italy 45°32′N 09°02′E
19 N12 **Rhode Island** *off.* State of Rhode Island and Providence Plantations, *also known as* Little Rhody, Ocean State. ◈ *state* NE USA
19 O13 **Rhode Island** *island* Rhode Island, NE USA
19 O13 **Rhode Island Sound** *sound* Maine/Rhode Island, NE USA
Rhodes *see* Ródos
Rhode-Saint-Genèse *see* Sint-Genesius-Rode
84 L14 **Rhodes Basin** *undersea feature* E Mediterranean Sea 35°55′N 28°30′E
Rhodesia *see* Zimbabwe
114 I12 **Rhodope Mountains** *var.* Rodhópi Óri, *Bul.* Rhodope Planina, Rodopi, *Gk.* Orosirá Rodhópis, *Turk.* Dospad Dagh. ▲▲ Bulgaria/Greece
Rhodope Planina *see* Rhodope Mountains
101 I18 **Rhön** ▲▲ C Germany
103 Q10 **Rhône** ◈ *department* E France
86 C12 **Rhône** ≈ France/Switzerland
103 R12 **Rhône-Alpes** ◈ *region* E France
98 G13 **Rhoon** Zuid-Holland, SW Netherlands 51°52′N 04°25′E
96 G9 **Rhum** *var.* Rum. *island* W Scotland, United Kingdom
Rhuthun *see* Ruthin
97 J18 **Rhyl** NE Wales, United Kingdom 53°19′N 03°28′W
59 K18 **Rialma** Goiás, S Brazil 15°22′S 49°35′W
104 L3 **Riaño** Castilla y León, N Spain 42°59′N 05°00′W
105 O9 **Riansáres** ≈ C Spain
152 H6 **Riāsi** Jammu and Kashmir, NW India 33°03′N 74°51′E
168 K10 **Riau** *off.* Propinsi Riau. ◈ *province* W Indonesia
Riau Archipelago *see* Riau, Kepulauan
168 M11 **Riau, Kepulauan** *var.* Riau Archipelago, *Dut.* Riouw-Archipel. *island group* W Indonesia
Riau, Propinsi *see* Riau
105 O6 **Riaza** Castilla y León, N Spain 41°17′N 03°29′W
105 N6 **Riaza** ≈ N Spain
81 K17 **Riba** *spring/well* NE Kenya 01°56′N 40°38′E
104 H4 **Ribadavia** Galicia, NW Spain 42°17′N 08°08′W
104 J2 **Ribadeo** Galicia, NW Spain 43°32′N 07°04′W
104 L2 **Ribadesella** *var.* Ribeseya. Asturias, N Spain 43°27′N 05°04′W
104 G10 **Ribatejo** *former province* C Portugal
83 P15 **Ribáuè** Nampula, N Mozambique 14°56′S 38°19′E
97 K17 **Ribble** ≈ NW England, United Kingdom
95 F23 **Ribe** Syddtjylland, W Denmark 55°20′N 08°47′E
Ribeira *see* Sanata Uxía de Ribeira
64 O5 **Ribeira Brava** Madeira, Portugal, NE Atlantic Ocean 32°39′N 17°04′W
64 P3 **Ribeira Grande** São Miguel, Azores, Portugal, NE Atlantic Ocean 37°34′N 25°32′W
60 L8 **Ribeirão Preto** São Paulo, S Brazil 21°09′S 47°48′W
60 L11 **Ribeira, Rio** ≈ S Brazil
107 I24 **Ribera** Sicilia, Italy, C Mediterranean Sea 37°31′N 13°16′E
57 L14 **Riberalta** El Beni, N Bolivia 11°01′S 66°04′W
105 W4 **Ribes de Freser** Cataluña, NE Spain 42°18′N 02°11′E
Ribeseya *see* Ribadesella
30 L6 **Rib Mountain** ▲ Wisconsin, N USA 44°55′N 89°41′W
109 U12 **Ribnica** *Ger.* Reifnitz. S Slovenia 45°46′N 14°40′E
117 N9 **Rîbniţa** *var.* Râbniţa, *Rus.* Rybnitsa. NE Moldova 47°46′N 29°01′E
100 M8 **Ribnitz-Damgarten** Mecklenburg-Vorpommern, NE Germany 54°14′N 12°25′E
111 D16 **Říčany** *Ger.* Ritschan. Středočesky Kraj, W Czech Republic 49°59′N 14°40′E
29 U7 **Rice** Minnesota, N USA 45°42′N 94°10′W
30 J5 **Rice Lake** Wisconsin, N USA 45°33′N 91°43′W
14 E8 **Rice Lake** ◎ Ontario, S Canada
14 I15 **Rice Lake** ◎ Ontario, SE Canada
23 V3 **Richard B. Russell Lake** ▣ Georgia, SE USA
25 U6 **Richardson** Texas, SW USA 32°55′N 96°44′W
11 R11 **Richardson** ≈ Alberta, C Canada
10 I3 **Richardson Mountains** ▲▲ Yukon, NW Canada
185 C21 **Richardson Mountains** ▲▲ South Island, New Zealand
42 F3 **Richardson Peak** ▲ SE Belize 16°34′N 88°46′W
76 G10 **Richard Toll** N Senegal 16°28′N 15°44′W
28 L5 **Richardton** North Dakota, N USA 46°52′N 102°19′W
14 F13 **Rich, Cape** *headland* Ontario, S Canada 44°42′N 80°37′W
102 L8 **Richelieu** Indre-et-Loire, C France 47°01′N 00°18′E
33 P15 **Richfield** Idaho, NW USA 43°03′N 114°11′W
36 K5 **Richfield** Utah, W USA 38°45′N 112°05′W
18 J10 **Richfield Springs** New York, NE USA 42°52′N 74°57′W
18 M6 **Richford** Vermont, NE USA 44°59′N 72°37′W
27 R6 **Rich Hill** Missouri, C USA 38°06′N 94°22′W
13 P14 **Richibucto** New Brunswick, SE Canada 46°42′N 64°54′W
108 G8 **Richisau** Glarus, NE Switzerland 47°00′N 08°54′E
23 S6 **Richland** Georgia, SE USA 32°05′N 84°40′W
27 U6 **Richland** Missouri, C USA 37°51′N 92°24′W
25 U8 **Richland** Texas, SW USA 31°55′N 96°25′W
32 K10 **Richland** Washington, NW USA 46°17′N 119°16′W
30 K8 **Richland Center** Wisconsin, N USA 43°22′N 90°24′W
21 W11 **Richlands** North Carolina, SE USA 34°52′N 77°33′W
21 Q7 **Richlands** Virginia, NE USA 37°05′N 81°47′W
25 R9 **Richland Springs** Texas, SW USA 31°16′N 98°56′W
183 S8 **Richmond** New South Wales, SE Australia 33°36′S 150°44′E
10 L17 **Richmond** British Columbia, SW Canada 49°07′N 123°09′W
14 L13 **Richmond** Ontario, SE Canada 45°12′N 75°49′W
15 Q12 **Richmond** Québec, SE Canada 45°39′N 72°07′W
185 I14 **Richmond** Tasman, South Island, New Zealand 41°25′S 173°04′E
35 N8 **Richmond** California, W USA 37°57′N 122°22′W
31 Q14 **Richmond** Indiana, N USA 39°50′N 84°51′W
20 M6 **Richmond** Kentucky, S USA 37°45′N 84°19′W
27 S4 **Richmond** Missouri, C USA 39°15′N 93°59′W
25 V11 **Richmond** Texas, SW USA 29°36′N 95°48′W
36 L1 **Richmond** Utah, W USA 41°55′N 111°51′W
21 W6 **Richmond** *state capital* Virginia, NE USA 37°33′N 77°28′W
14 H15 **Richmond Hill** Ontario, S Canada 43°51′N 79°24′W
185 J15 **Richmond Range** ▲▲ South Island, New Zealand
27 S12 **Rich Mountain** ▲ Arkansas, C USA 34°37′N 94°17′W
31 S13 **Richwood** Ohio, N USA 40°25′N 83°18′W
21 R5 **Richwood** West Virginia, NE USA 38°13′N 80°31′W
104 K5 **Ricobayo, Embalse de** ▣ NW Spain
Ricomagus *see* Riom
Ridā' *see* Radā'
127 T5 **Ridder** Respublika Tatarstan, W Russian Federation 54°34′N 52°27′E
98 H13 **Ridderkerk** Zuid-Holland, SW Netherlands 51°52′N 04°35′E
33 N16 **Riddle** Idaho, NW USA 42°07′N 116°09′W
32 F14 **Riddle** Oregon, NW USA 42°57′N 123°21′W
14 L13 **Rideau** ≈ Ontario, SE Canada
35 T12 **Ridgecrest** California, W USA 35°37′N 117°40′W
18 L13 **Ridgefield** Connecticut, NE USA 41°16′N 73°30′W
22 K5 **Ridgeland** Mississippi, S USA 32°25′N 90°07′W
21 R15 **Ridgeland** South Carolina, SE USA 32°30′N 80°59′W
20 F8 **Ridgely** Tennessee, S USA 36°15′N 89°29′W
14 D17 **Ridgetown** Ontario, S Canada 42°27′N 81°52′W
21 R12 **Ridgeway** South Carolina, SE USA 34°17′N 80°56′W
Ridgeway *see* Ridgway
18 D13 **Ridgway** *var.* Ridgeway. Pennsylvania, NE USA 41°24′N 78°40′W
11 W16 **Riding Mountain** ▲▲ Manitoba, S Canada
Ried *see* Ried im Innkreis
109 R4 **Ried im Innkreis** *var.* Ried. Oberösterreich, NW Austria 48°13′N 13°29′E
109 X8 **Riegersburg** Steiermark, SE Austria 47°03′N 15°52′E
108 E6 **Riehen** Basel-Stadt, NW Switzerland 47°35′N 07°39′E
92 J9 **Riehppegáisá** *var.* Rieppe. ▲ N Norway 69°38′N 21°31′E
99 K18 **Riemst** Limburg, NE Belgium 50°49′N 05°36′E
Rieppe *see* Riehppegáisá
101 O15 **Riesa** Sachsen, E Germany 51°18′N 13°18′E
63 H24 **Riesco, Isla** *island* S Chile
107 K25 **Riesi** Sicilia, Italy, C Mediterranean Sea 37°17′N 14°05′E
83 I23 **Riet** ≈ SW South Africa
83 F25 **Riet** ≈ SW South Africa
118 D11 **Rietavas** Telšiai, W Lithuania 55°43′N 21°56′E
83 F19 **Rietfontein** Omaheke, E Namibia 21°58′S 20°58′E
107 I14 **Rieti** *anc.* Reate. Lazio, C Italy 42°24′N 12°51′E
84 D14 **Rif** *var.* Riff, Er Rif, Er Riff. ▲▲ N Morocco
138 I8 **Rīf Dimashq** *off.* Muḩāfaẓat Dimashq, *var.* Damascus, *Ar.* Ash Sham, Ash Shām, Damasco, Esh Sham, *Fr.* Damas. ◈ *governorate* S Syria
Riff *see* Rif
37 Q4 **Rifle** Colorado, C USA 39°30′N 107°46′W
31 R7 **Rifle River** ≈ Michigan, N USA
Rift Valley *see* Great Rift Valley
118 F9 **Rīga** *Eng.* Riga. ● C Latvia 56°57′N 24°08′E
Rigaer Bucht *see* Riga, Gulf of
118 F6 **Riga, Gulf of** *Est.* Liivi Laht, *Ger.* Rigaer Bucht, *Latv.* Rīgas Jūras Līcis, *Rus.* Rizhskiy Zaliv; *prev. Est.* Riia Laht. *gulf* Estonia/Latvia
Rīgas Jūras Līcis *see* Riga, Gulf of
15 N12 **Rigaud** ≈ Ontario/Québec, SE Canada
33 R14 **Rigby** Idaho, NW USA 43°40′N 111°54′W
Rīgestān *see* Rēgestān
32 M11 **Riggins** Idaho, NW USA 45°24′N 116°18′W
13 R8 **Rigolet** Newfoundland and Labrador, NE Canada 51°10′N 58°25′W
78 G9 **Rig-Rig** Kanem, W Chad 14°16′N 14°21′E
118 F4 **Riguldi** Läänemaa, W Estonia 59°07′N 23°34′E
Riia Laht *see* Riga, Gulf of
93 L19 **Riihimäki** Kanta-Häme, S Finland 60°45′N 24°45′E
195 U2 **Riiser-Larsen Peninsula** *peninsula* Antarctica
65 P22 **Riiser-Larsen Sea** *sea* Antarctica
Riisser Larsen Ice Shelf *see* Riiser-Larsenisen
40 D2 **Riíto** Sonora, NW Mexico 32°06′N 114°57′W
112 B9 **Rijeka** *Ger.* Sankt Veit am Flaum, *It.* Fiume, *Slvn.* Reka; *anc.* Tarsatica. Primorje-Gorski Kotar, NW Croatia 45°20′N 14°26′E
99 I14 **Rijen** Noord-Brabant, S Netherlands 51°35′N 04°55′E
99 H15 **Rijkevorsel** Antwerpen, N Belgium 51°23′N 04°43′E
Rijn *see* Rhine
98 G11 **Rijnsburg** Zuid-Holland, W Netherlands 52°12′N 04°27′E
Rijssel *see* Lille
98 N10 **Rijssen** Overijssel, E Netherlands 52°19′N 06°30′E
98 G12 **Rijswijk** *Eng.* Ryswick. Zuid-Holland, W Netherlands 52°03′N 04°20′E
92 I10 **Riksgränsen** Norrbotten, N Sweden 68°24′N 18°15′E
165 U4 **Rikubetsu** Hokkaidō, NE Japan 43°30′N 143°43′E
165 R9 **Rikuzen-Takata** Iwate, Honshū, C Japan 39°03′N 141°38′E
27 O4 **Riley** Kansas, C USA 39°18′N 96°49′W
99 I17 **Rillaar** Vlaams Brabant, C Belgium 50°58′N 04°58′E
Rí, Loch *see* Ree, Lough
114 G11 **Rilska Reka** ≈ W Bulgaria
77 T12 **Rima** ≈ N Nigeria
141 N7 **Rimah, Wādī ar** *var.* Wādī ar Rummah. *dry watercourse* C Saudi Arabia
Rimaszombat *see* Rimavská Sobota
191 R12 **Rimatara** *island* Îles Australes, SW French Polynesia
111 L20 **Rimavská Sobota** *Ger.* Gross-Steffelsdorf, *Hung.* Rimaszombat. Bankobystrický Kraj, C Slovakia 48°24′N 20°01′E
11 Q15 **Rimbey** Alberta, SW Canada 52°39′N 114°10′W
95 P15 **Rimbo** Stockholm, C Sweden 59°44′N 18°21′E
95 M18 **Rimforsa** Östergötland, S Sweden 58°06′N 15°40′E
106 I11 **Rimini** *anc.* Ariminum. Emilia-Romagna, N Italy 44°03′N 12°33′E
Rîmnicu-Sărat *see* Râmnicu Sărat
Rîmnicu Vîlcea *see* Râmnicu Vâlcea
149 Y3 **Rimo Muztāgh** ▲▲ India/Pakistan
15 U7 **Rimouski** Québec, SE Canada 48°26′N 68°32′W
158 M16 **Rinbung** Xizang Zizhiqu, W China 29°15′N 89°40′E
Rinchinlhumbe *see* Dzöölön
62 I5 **Rincón, Cerro** ▲ N Chile 24°01′S 67°19′W
104 M15 **Rincón de la Victoria** Andalucía, S Spain 36°43′N 04°18′W

◆ Country
● Country Capital
◇ Dependent Territory
○ Dependent Territory Capital
◈ Administrative Regions
✕ International Airport
▲ Mountain
▲▲ Mountain Range
Ѫ Volcano
≈ River
◎ Lake
▣ Reservoir

RINCÓN DEL BONETE, LAGO ARTIFICIAL DE – RONNE ENTRANCE

Rincón del Bonete, Lago Artificial de *see* Río Negro, Embalse del
105 Q4 **Rincón de Soto** La Rioja, N Spain 42°15′N 01°50′W
94 G8 **Rindal** Møre og Romsdal, S Norway 63°02′N 09°09′E
115 J20 **Ríneia** *island* Kykládes, Greece, Aegean Sea
152 H11 **Rīngas** *prev.* Reengus, Rīngus. Rājasthān, N India 27°18′N 75°27′E
95 H24 **Ringe** Syddtjylland, C Denmark 55°14′N 10°30′E
94 H11 **Ringebu** Oppland, S Norway 61°31′N 10°09′E
Ringen *see* Rõngu
186 K8 **Ringgi** Kolombangara, NW Solomon Islands 08°03′S 157°08′E
23 R1 **Ringgold** Georgia, SE USA 34°55′N 85°06′W
22 G5 **Ringgold** Louisiana, S USA 32°19′N 93°16′W
25 S5 **Ringgold** Texas, SW USA 33°47′N 97°56′W
95 E22 **Ringkøbing** Midtjylland, W Denmark 56°04′N 08°22′E
95 E22 **Ringkøbing Fjord** *fjord* W Denmark
33 S10 **Ringling** Montana, NW USA 46°15′N 110°48′W
27 N13 **Ringling** Oklahoma, C USA 34°12′N 97°35′W
94 H13 **Ringsaker** Hedmark, S Norway 60°54′N 10°45′E
95 I23 **Ringsted** Sjælland, E Denmark 55°28′N 11°48′E
Rīngus *see* Rīngas
92 I9 **Ringvassøya** *Lapp.* Ráneš. *island* N Norway
18 K13 **Ringwood** New Jersey, NE USA 41°06′N 74°15′W
Rinn Duáin *see* Hook Head
100 H13 **Rinteln** Niedersachsen, NW Germany 52°10′N 09°04′E
115 E18 **Río** Dytikí Elláda, S Greece 38°18′N 21°48′E
Rio *see* Rio de Janeiro
56 C7 **Riobamba** Chimborazo, C Ecuador 01°44′S 78°40′W
60 P9 **Rio Bonito** Rio de Janeiro, SE Brazil 22°42′S 42°38′W
59 C16 **Rio Branco** *state capital* Acre, W Brazil 09°59′S 67°49′W
61 H18 **Río Branco** Cerro Largo, NE Uruguay 32°32′S 53°28′W
Rio Branco, Território de *see* Roraima
41 P8 **Río Bravo** Tamaulipas, C Mexico 25°57′N 98°03′W
63 G16 **Río Bueno** Los Ríos, C Chile 40°20′S 72°55′W
55 P5 **Río Caribe** Sucre, NE Venezuela 10°43′N 63°06′W
54 M5 **Río Chico** Miranda, N Venezuela 10°18′N 66°00′W
63 H18 **Río Cisnes** Aisén, S Chile 44°29′S 71°15′W
60 L9 **Rio Claro** São Paulo, S Brazil 22°19′S 47°35′W
45 V14 **Rio Claro** Trinidad, Trinidad and Tobago 10°18′N 61°11′W
54 J5 **Río Claro** Lara, N Venezuela 09°54′N 69°23′W
63 K15 **Rio Colorado** Río Negro, E Argentina 39°01′S 64°05′W
62 K11 **Río Cuarto** Córdoba, C Argentina 33°06′S 64°20′W
60 P10 **Rio de Janeiro** *var.* Rio. *state capital* Rio de Janeiro, SE Brazil 22°53′S 43°17′W
60 P9 **Rio de Janeiro** *off.* Estado do Rio de Janeiro. ⬖ *state* SE Brazil
Rio de Janeiro, Estado do *see* Rio de Janeiro
43 R17 **Río de Jesús** Veraguas, S Panama 07°58′N 81°01′W
34 K3 **Rio Dell** California, W USA 40°30′N 124°07′W
60 K13 **Rio do Sul** Santa Catarina, S Brazil 27°15′S 49°37′W
63 I23 **Río Gallegos** *var.* Gallegos, Puerto Gallegos. Santa Cruz, S Argentina 51°40′S 69°21′W
63 J24 **Río Grande** Tierra del Fuego, S Argentina 53°45′S 67°46′W
61 I18 **Rio Grande** *var.* São Pedro do Rio Grande do Sul. Rio Grande do Sul, S Brazil 32°03′S 52°08′W
40 L10 **Río Grande** Zacatecas, C Mexico 23°50′N 103°20′W
42 J9 **Río Grande** León, NW Nicaragua 12°59′N 86°34′W
45 V5 **Río Grande** E Puerto Rico 18°23′N 65°51′W
24 I9 **Rio Grande** ≈ Texas, SW USA
25 R17 **Rio Grande City** Texas, SW USA 26°24′N 98°50′W
59 P14 **Rio Grande do Norte** *off.* Estado do Rio Grande do Norte. ⬖ *state* E Brazil
Rio Grande do Norte, Estado do *see* Rio Grande do Norte
61 G15 **Rio Grande do Sul** *off.* Estado do Rio Grande do Sul. ⬖ *state* S Brazil
Rio Grande do Sul, Estado do *see* Rio Grande do Sul
65 M17 **Rio Grande Fracture Zone** *tectonic feature* C Atlantic Ocean
65 J18 **Rio Grande Gap** *undersea feature* S Atlantic Ocean
Rio Grande Plateau *see* Rio Grande Rise
65 J18 **Rio Grande Rise** *var.* Rio Grande Plateau. *undersea feature* SW Atlantic Ocean 31°00′S 35°00′W
54 G4 **Ríohacha** La Guajira, N Colombia 11°23′N 72°47′W
43 S16 **Río Hato** Coclé, C Panama 08°21′N 80°10′W
25 T17 **Rio Hondo** Texas, SW USA 26°14′N 97°34′W
56 D10 **Rioja** San Martín, N Peru 06°02′S 77°10′W
41 Y11 **Río Lagartos** Yucatán, SE Mexico 21°35′N 88°08′W
103 P11 **Riom** *anc.* Ricomagus. Puy-de-Dôme, C France 45°54′N 03°06′E
104 F10 **Rio Maior** Santarém, C Portugal 39°20′N 08°55′W
103 O12 **Riom-ès-Montagnes** Cantal, C France 45°15′N 02°39′E
60 J12 **Rio Negro** Paraná, S Brazil 26°06′S 49°46′W
63 I15 **Río Negro** *off.* Provincia de Río Negro. ⬖ *province* C Argentina
61 D18 **Río Negro** ⬖ *department* W Uruguay
47 V12 **Río Negro, Embalse del** *var.* Lago Artificial de Rincón del Bonete. ⊠ C Uruguay
Río Negro, Provincia de *see* Río Negro
107 M17 **Rionero in Vulture** Basilicata, S Italy 40°55′N 15°40′E
137 S9 **Rioni** ≈ W Georgia
105 P12 **Riópar** Castilla-La Mancha, C Spain 38°31′N 02°27′W
61 H16 **Rio Pardo** Rio Grande do Sul, S Brazil 29°41′S 52°25′W
37 R11 **Rio Rancho Estates** New Mexico, SW USA 35°14′N 106°40′W
42 L11 **Río San Juan** ⬖ *department* S Nicaragua
54 E9 **Ríosucio** Caldas, W Colombia 05°26′N 75°44′W
54 C7 **Ríosucio** Chocó, NW Colombia 07°25′N 77°05′W
62 K10 **Río Tercero** Córdoba, C Argentina 32°15′S 64°08′W
42 K5 **Río Tinto, Sierra** ▲▲ NE Honduras
54 J5 **Río Tocuyo** Lara, N Venezuela 10°18′N 70°00′W
Riouw-Archipel *see* Riau, Kepulauan
59 J19 **Rio Verde** Goiás, C Brazil 17°50′S 50°55′W
41 O12 **Río Verde** *var.* Rioverde. San Luis Potosí, C Mexico 21°58′N 100°W
Rioverde *see* Río Verde
35 O8 **Rio Vista** California, W USA 38°09′N 121°42′W
112 M11 **Ripanj** Serbia, N Serbia 44°37′N 20°30′E
106 J13 **Ripatransone** Marche, C Italy 43°00′N 13°45′E
22 M2 **Ripley** Mississippi, S USA 34°43′N 88°57′W
31 R15 **Ripley** Ohio, N USA 38°45′N 83°51′W
20 F9 **Ripley** Tennessee, S USA 35°43′N 89°30′W
21 Q4 **Ripley** West Virginia, NE USA 38°49′N 81°44′W
105 W4 **Ripoll** Cataluña, NE Spain 42°12′N 02°12′E
97 M16 **Ripon** N England, United Kingdom 54°07′N 01°31′W
30 M7 **Ripon** Wisconsin, N USA 43°52′N 88°48′W
107 L24 **Riposto** Sicilia, Italy, C Mediterranean Sea 37°44′N 15°13′E
99 L14 **Rips** Noord-Brabant, SE Netherlands 51°31′N 05°49′E
54 D9 **Risaralda** *off.* Departamento de Risaralda. ⬖ *province* C Colombia
Risaralda, Departamento de *see* Risaralda
116 L8 **Rîşcani** *var.* Râşcani, *Rus.* Ryshkany. NW Moldova 47°55′N 27°31′E
152 J9 **Rishikesh** Uttarakhand, N India 30°06′N 78°16′E
165 S1 **Rishiri-tō** *var.* Risiri Tô. *island* NE Japan
165 S1 **Rishiri-yama** Ѧ Rishiri-tō, NE Japan 45°11′N 141°11′E
25 R7 **Rising Star** Texas, SW USA 32°06′N 98°57′W
31 Q15 **Rising Sun** Indiana, N USA 38°58′N 84°53′W
Risiri Tô *see* Rishiri-tō
102 L4 **Risle** ≈ N France
Rîşno *see* Râşnov
27 V13 **Rison** Arkansas, C USA 33°58′N 92°11′W
95 G17 **Risør** Aust-Agder, S Norway 58°44′N 09°15′E
92 H10 **Risøyhamn** Nordland, C Norway 69°00′N 15°37′E
101 I23 **Riss** ≈ S Germany
118 G4 **Risti** *Ger.* Kreuz. Läänemaa, W Estonia 59°01′N 24°01′E
15 V8 **Ristigouche** ≈ Québec, SE Canada
93 N18 **Ristiina** Etelä-Savo, E Finland 61°32′N 27°15′E
93 N14 **Ristijärvi** Kainuu, C Finland 64°30′N 28°15′E
188 C14 **Ritidian Point** *headland* N Guam 13°39′N 144°51′E
Ritschan *see* Říčany
35 R9 **Ritter, Mount** ▲ California, W USA 37°40′N 119°10′W
31 T12 **Rittman** Ohio, N USA 40°58′N 81°46′W
32 L9 **Ritzville** Washington, NW USA 47°07′N 118°22′W
Riva *see* Riva del Garda
61 A21 **Rivadavia** Buenos Aires, E Argentina 35°29′S 62°59′W
106 F7 **Riva del Garda** *var.* Riva. Trentino-Alto Adige, N Italy 45°54′N 10°50′E
106 B8 **Rivarolo Canavese** Piemonte, W Italy 45°21′N 07°42′E
42 K11 **Rivas** Rivas, SW Nicaragua 11°26′N 85°50′W
42 J11 **Rivas** ⬖ *department* SW Nicaragua
103 R11 **Rive-de-Gier** Loire, E France 45°31′N 04°36′E
61 A22 **Rivera** Buenos Aires, E Argentina 37°13′S 63°14′W
61 F16 **Rivera** Rivera, NE Uruguay 30°54′S 55°31′W
61 F17 **Rivera** ⬖ *department* NE Uruguay
35 P9 **Riverbank** California, W USA 37°43′N 120°59′W
76 K17 **River Cess** SW Liberia 05°28′N 09°32′W
28 M4 **Riverdale** North Dakota, N USA 47°29′N 101°22′W
30 I6 **River Falls** Wisconsin, N USA 44°52′N 92°38′W
11 T16 **Riverhurst** Saskatchewan, S Canada 50°52′N 106°49′W
183 O10 **Riverina** *physical region* New South Wales, SE Australia
80 G8 **River Nile** ⬖ *state* NE Sudan
63 F19 **Rivero, Isla** *island* Archipiélago de los Chonos, S Chile
11 W16 **Rivers** Manitoba, S Canada 50°02′N 100°14′W
77 U17 **Rivers** ⬖ *state* S Nigeria
185 D23 **Riversdale** Southland, South Island, New Zealand 45°54′S 168°44′E
83 F26 **Riversdale** Western Cape, SW South Africa 34°05′S 21°15′E
35 U15 **Riverside** California, W USA 33°58′N 117°25′W
25 W9 **Riverside** Texas, SW USA 30°51′N 95°24′W
37 U3 **Riverside Reservoir** ⊠ Colorado, C USA
10 K15 **Rivers Inlet** British Columbia, SW Canada 51°43′N 127°19′W
10 K15 **Rivers Inlet** *inlet* British Columbia, SW Canada
11 X15 **Riverton** Manitoba, S Canada 51°00′N 97°00′W
185 C24 **Riverton** Southland, South Island, New Zealand 46°20′S 168°02′E
30 L13 **Riverton** Illinois, N USA 39°50′N 89°31′W
36 L3 **Riverton** Utah, W USA 40°32′N 111°57′W
33 V15 **Riverton** Wyoming, C USA 43°01′N 108°22′W
14 G10 **River Valley** Ontario, S Canada 46°36′N 80°09′W
13 P14 **Riverview** New Brunswick, SE Canada 46°03′N 64°47′W
103 O17 **Rivesaltes** Pyrénées-Orientales, S France 42°46′N 02°48′E
36 H11 **Riviera** Arizona, SW USA 35°06′N 114°36′W
25 S15 **Riviera** Texas, SW USA 27°15′N 97°48′W
23 Z14 **Riviera Beach** Florida, SE USA 26°46′N 80°03′W
15 Q10 **Rivière-à-Pierre** Québec, SE Canada 46°59′N 72°12′W
15 T9 **Rivière-Bleue** Québec, SE Canada 47°26′N 69°02′W
15 T8 **Rivière-du-Loup** Québec, SE Canada 47°49′N 69°32′W
173 Y15 **Rivière du Rempart** NE Mauritius 20°06′S 57°41′E
45 R12 **Rivière-Pilote** S Martinique 14°29′N 60°54′W
173 O17 **Rivière St-Etienne, Pointe de la** *headland* SW Réunion
13 S10 **Rivière-St-Paul** Québec, E Canada 51°26′N 57°52′W
Rivière Sèche *see* Bel Air
116 K4 **Rivne** *Pol.* Równe, *Rus.* Rovno. Rivnens'ka Oblast', NW Ukraine 50°37′N 26°16′E
Rivne *see* Rivnens'ka Oblast'
116 K3 **Rivnens'ka Oblast'** *var.* Rivne, *Rus.* Rovenskaya Oblast'. ⬖ *province* NW Ukraine
106 B8 **Rivoli** Piemonte, NW Italy 45°04′N 07°31′E
159 Q14 **Riwoqê** *var.* Racaka. Xizang Zizhiqu, W China 31°10′N 96°25′E
99 H19 **Rixensart** Walloon Brabant, C Belgium 50°43′N 04°32′E
Riyadh/Riyāḍ, Minṭaqat ar *see* Ar Riyāḍ
Riyāq *see* Rayak
Rizaiyeh *see* Orūmīyeh
137 P11 **Rize** Rize, NE Turkey 41°03′N 40°33′E
137 P11 **Rize** *prev.* Çoruh. ⬖ *province* NE Turkey
161 R5 **Rizhao** Shandong, E China 35°23′N 119°32′E
Rizhskiy Zaliv *see* Riga, Gulf of
Rizokarpaso/Rizokárpason *see* Dipkarpaz
107 O21 **Rizzuto, Capo** *headland* S Italy 38°54′N 17°05′E
95 F15 **Rjukan** Telemark, S Norway 59°54′N 08°33′E
95 D16 **Rjuven** ▲▲ S Norway
76 H9 **Rkîz** Trarza, W Mauritania 16°50′N 15°20′W
115 Q23 **Ro** *prev.* Ágios Geórgios. *island* SE Greece
95 H14 **Roa** Oppland, S Norway 60°16′N 10°38′E
105 N5 **Roa** Castilla y León, N Spain 41°42′N 03°55′W
45 T9 **Road Town** ○ (British Virgin Islands) Tortola, C British Virgin Islands 18°28′N 64°39′W
96 F6 **Roag, Loch** *inlet* NW Scotland, United Kingdom
37 O5 **Roan Cliffs** *cliff* Colorado/Utah, W USA
21 P9 **Roan High Knob** *var.* Roan Mountain. ▲ North Carolina/Tennessee, SE USA 36°09′N 82°07′W
Roan Mountain *see* Roan High Knob
103 Q10 **Roanne** *anc.* Rodunma. Loire, E France 46°03′N 04°04′E
23 R4 **Roanoke** Alabama, S USA 33°09′N 85°22′W
21 S7 **Roanoke** Virginia, NE USA 37°16′N 79°57′W
21 Z9 **Roanoke Island** *island* North Carolina, SE USA
21 W8 **Roanoke Rapids** North Carolina, SE USA 36°27′N 77°39′W
21 X9 **Roanoke River** ≈ North Carolina/Virginia, SE USA
37 O4 **Roan Plateau** *plain* Utah, W USA
37 R5 **Roaring Fork River** ≈ Colorado, C USA
25 O5 **Roaring Springs** Texas, SW USA 33°54′N 100°51′W
42 J4 **Roatán** *var.* Coxen Hole, Coxin Hole. Islas de la Bahía, N Honduras 16°19′N 86°33′W
42 I4 **Roatán, Isla de** *island* Islas de la Bahía, N Honduras
Roat Kampuchea *see* Cambodia
Roazon *see* Rennes
143 T7 **Robāṭ-e Chāh Gonbad** Yazd, E Iran 33°24′N 57°43′E
143 R7 **Robāṭ-e Khān** Yazd, C Iran 33°24′N 56°04′E
143 T7 **Robāṭ-e Khvosh Āb** Khorāsān-e Razavī, E Iran
143 R8 **Robāṭ-e Posht-e Bādām** Yazd, NE Iran 33°01′N 55°34′E
143 Q8 **Robaṭ-e Rīzāb** Yazd, C Iran
175 S8 **Robbie Ridge** *undersea feature* W Pacific Ocean
21 T10 **Robbins** North Carolina, SE USA 35°25′N 79°35′W
183 N15 **Robbins Island** *island* Tasmania, SE Australia
21 N10 **Robbinsville** North Carolina, SE USA 35°18′N 83°49′W
182 J12 **Robe** South Australia 37°11′S 139°48′E
21 W9 **Robersonville** North Carolina, SE USA 35°49′N 77°15′W
45 V10 **Robert L. Bradshaw** ✈ (Basseterre) Saint Kitts, Saint Kitts and Nevis 17°16′N 62°43′W
25 P8 **Robert Lee** Texas, SW USA 31°50′N 100°30′W
35 V5 **Roberts Creek Mountain** ▲ Nevada, W USA 39°52′N 116°16′W
93 J15 **Robertsfors** Västerbotten, N Sweden 64°12′N 20°50′E
27 R11 **Robert S. Kerr Reservoir** ⊠ Oklahoma, C USA
38 L12 **Roberts Mountain** ▲ Nunivak Island, Alaska, USA 60°01′N 166°15′W
83 F26 **Robertson** Western Cape, SW South Africa 33°48′S 19°53′E
194 H4 **Robertson Island** *island* Antarctica
76 J16 **Robertsport** W Liberia 06°45′N 11°15′W
182 J8 **Robertstown** South Australia 34°00′S 139°04′E
Robert Williams *see* Caála
15 P7 **Roberval** Québec, SE Canada 48°31′N 72°16′W
31 N15 **Robinson** Illinois, N USA 39°00′N 87°44′W
193 U11 **Róbinson Crusoe, Isla** *island* Islas Juan Fernández, Chile, E Pacific Ocean
180 J9 **Robinson Range** ▲▲ Western Australia
182 M9 **Robinvale** Victoria, SE Australia 34°37′S 142°45′E
105 P11 **Robledo** Castilla-La Mancha, C Spain 38°45′N 02°27′W
54 G5 **Robles** *var.* La Paz, Robles La Paz. Cesar, N Colombia 10°24′N 73°11′W
Robles La Paz *see* Robles
11 V15 **Roblin** Manitoba, S Canada 51°15′N 101°20′W
11 S17 **Robsart** Saskatchewan, S Canada 49°22′N 109°15′W
11 N15 **Robson, Mount** ▲ British Columbia, SW Canada 53°09′N 119°16′W
25 T14 **Robstown** Texas, SW USA 27°47′N 97°40′W
25 P6 **Roby** Texas, SW USA 32°42′N 100°23′W
104 E11 **Roca, Cabo da** *cape* C Portugal
Rocadas *see* Xangongo
41 S14 **Roca Partida, Punta** *headland* C Mexico 18°43′N 95°11′W
47 X6 **Rocas, Atol das** *island* E Brazil
107 L18 **Roccadaspide** *var.* Rocca d'Aspide. Campania, S Italy 40°25′N 15°12′E
Rocca d'Aspide *see* Roccadaspide
107 K15 **Roccaraso** Abruzzo, C Italy 41°49′N 14°01′E
106 H10 **Rocca San Casciano** Emilia-Romagna, C Italy 44°06′N 11°51′E
106 G13 **Roccastrada** Toscana, C Italy 43°00′N 11°09′E
61 G20 **Rocha** Rocha, E Uruguay 34°30′S 54°22′W
61 G19 **Rocha** ⬖ *department* E Uruguay
97 L17 **Rochdale** NW England, United Kingdom 53°38′N 02°09′W
102 L11 **Rochechouart** Haute-Vienne, C France 45°49′N 00°49′E
99 J22 **Rochefort** Namur, SE Belgium 50°10′N 05°13′E
102 J11 **Rochefort** *var.* Rochefort sur Mer. Charente-Maritime, W France 45°57′N 00°58′W
Rochefort sur Mer *see* Rochefort
125 N10 **Rochegda** Arkhangel'skaya Oblast', NW Russian Federation 62°37′N 43°21′E
30 L10 **Rochelle** Illinois, N USA 41°54′N 89°03′W
25 Q9 **Rochelle** Texas, SW USA 31°13′N 99°10′W
15 V3 **Rochers Ouest, Rivière aux** ≈ Québec, SE Canada
97 O22 **Rochester** *anc.* Durobrivae. SE England, United Kingdom 51°24′N 00°30′E
31 O12 **Rochester** Indiana, N USA 41°03′N 86°13′W
29 W10 **Rochester** Minnesota, N USA 44°01′N 92°28′W
19 O9 **Rochester** New Hampshire, NE USA 43°18′N 70°58′W
18 F9 **Rochester** New York, NE USA 43°09′N 77°37′W
25 P5 **Rochester** Texas, SW USA 33°19′N 99°51′W
31 S9 **Rochester Hills** Michigan, N USA 42°39′N 83°04′W
Rocheuses, Montagnes/Rockies *see* Rocky Mountains
64 M6 **Rockall** *island* N Atlantic Ocean, United Kingdom
64 L6 **Rockall Bank** *undersea feature* N Atlantic Ocean
84 B8 **Rockall Rise** *undersea feature* N Atlantic Ocean 59°00′N 14°00′W
84 C9 **Rockall Trough** *undersea feature* N Atlantic Ocean 57°00′N 12°00′W
35 U2 **Rock Creek** ≈ Nevada, W USA
25 T10 **Rockdale** Texas, SW USA 30°39′N 96°58′W
195 N12 **Rockefeller Plateau** *plateau* Antarctica
30 K11 **Rock Falls** Illinois, N USA 41°46′N 89°41′W
23 Q5 **Rockford** Alabama, S USA 32°53′N 86°11′W
30 L10 **Rockford** Illinois, N USA 42°16′N 89°06′W
15 Q12 **Rock Forest** Québec, SE Canada 45°21′N 71°58′W
11 T17 **Rockglen** Saskatchewan, S Canada 49°11′N 105°57′W
181 Y8 **Rockhampton** Queensland, E Australia 23°31′S 150°31′E
21 R11 **Rock Hill** South Carolina, SE USA 34°55′N 81°01′W
180 I13 **Rockingham** Western Australia 32°16′S 115°21′E
21 T11 **Rockingham** North Carolina, SE USA 34°56′N 79°47′W
30 J11 **Rock Island** Illinois, N USA 41°30′N 90°34′W
25 U12 **Rock Island** Texas, SW USA 29°31′N 96°33′W
14 C10 **Rock Lake** Ontario, S Canada 46°25′N 83°49′W
29 O2 **Rock Lake** North Dakota, N USA 48°45′N 99°12′W
14 I12 **Rock Lake** ◎ Ontario, SE Canada
14 M12 **Rockland** Ontario, SE Canada 45°33′N 75°16′W
19 R7 **Rockland** Maine, NE USA 44°08′N 69°06′W
182 L11 **Rocklands Reservoir** ⊠ Victoria, SE Australia
35 O7 **Rocklin** California, W USA 38°48′N 121°13′W
23 R3 **Rockmart** Georgia, SE USA 34°00′N 85°02′W
31 N16 **Rockport** Indiana, N USA 37°53′N 87°04′W
27 Q1 **Rock Port** Missouri, C USA 40°26′N 95°30′W
25 T14 **Rockport** Texas, SW USA 28°02′N 99°04′W
32 I7 **Rockport** Washington, NW USA 48°28′N 121°36′W
29 S11 **Rock Rapids** Iowa, C USA 43°25′N 96°10′W
30 K11 **Rock River** ≈ Illinois/Wisconsin, N USA
44 I3 **Rock Sound** Eleuthera Island, C The Bahamas 24°52′N 76°10′W
25 P11 **Rocksprings** Texas, SW USA 30°02′N 100°14′W
33 U17 **Rock Springs** Wyoming, C USA 41°35′N 109°12′W
55 T9 **Rockstone** C Guyana 05°58′N 58°33′W
29 S12 **Rock Valley** Iowa, C USA 43°12′N 96°17′W
31 N14 **Rockville** Indiana, N USA 39°45′N 87°15′W
21 W3 **Rockville** Maryland, NE USA 39°05′N 77°10′W
25 U6 **Rockwall** Texas, SW USA 32°56′N 96°27′W
29 U13 **Rockwell City** Iowa, C USA 42°24′N 94°37′W
31 S10 **Rockwood** Michigan, N USA 42°04′N 83°15′W
20 M9 **Rockwood** Tennessee, S USA 35°52′N 84°41′W
25 Q8 **Rockwood** Texas, SW USA 31°29′N 99°23′W
37 U6 **Rocky Ford** Colorado, C USA 38°03′N 103°45′W
14 D9 **Rocky Island Lake** ◎ Ontario, S Canada
21 V9 **Rocky Mount** North Carolina, SE USA 35°56′N 77°48′W
21 S7 **Rocky Mount** Virginia, NE USA 37°00′N 79°53′W
33 Q8 **Rocky Mountain** ▲ Montana, NW USA 47°45′N 112°46′W
11 P15 **Rocky Mountain House** Alberta, SW Canada 52°24′N 114°52′W
37 T3 **Rocky Mountain National Park** *national park* Colorado, C USA
2 E12 **Rocky Mountains** *var.* Rockies, *Fr.* Montagnes Rocheuses. ▲▲ Canada/USA
42 H1 **Rocky Point** *headland* NE Belize 18°21′N 88°04′W
83 A17 **Rocky Point** *headland* NW Namibia 19°01′S 12°27′E
95 F14 **Rødberg** Buskerud, S Norway 60°16′N 09°00′E
95 I25 **Rødby** Sjælland, SE Denmark 54°42′N 11°24′E
95 I25 **Rødbyhavn** Sjælland, SE Denmark 54°39′N 11°24′E
13 T10 **Roddickton** Newfoundland, Newfoundland and Labrador, SE Canada 50°51′N 56°03′W
95 F23 **Rødding** Syddanmark, SW Denmark 55°22′N 09°04′E
95 M22 **Rödeby** Blekinge, S Sweden 56°16′N 15°35′E
98 N6 **Roden** Drenthe, NE Netherlands 53°08′N 06°26′E
62 H9 **Rodeo** San Juan, W Argentina 30°12′S 69°06′W
103 O14 **Rodez** *anc.* Segodunum. Aveyron, S France 44°21′N 02°34′E
Rodholívos *see* Rodolívos
Rodhópi Óri *see* Rhodope Mountains
Ródhos/Rodi *see* Ródos
107 N15 **Rodi Gargancio** Puglia, SE Italy 41°54′N 15°51′E
101 N20 **Roding** Bayern, SE Germany 49°12′N 12°30′E
113 J19 **Rodinit, Kepi i** *headland* W Albania 41°35′N 19°27′E
116 I9 **Rodnei, Munţii** ▲▲ N Romania
184 L4 **Rodney, Cape** *headland* North Island, New Zealand 36°16′S 174°48′E
38 L9 **Rodney, Cape** *headland* Alaska, USA 64°39′N 166°24′W
124 M16 **Rodniki** Ivanovskaya Oblast', W Russian Federation 57°04′N 41°45′E
119 Q16 **Rodnya** Mahilyowskaya Voblasts', E Belarus 53°31′N 32°07′E
Rodó *see* José Enrique Rodó
114 H13 **Rodolívos** *var.* Rodholívos. Kentrikí Makedonía, NE Greece 40°55′N 24°00′E
Rodopi *see* Rhodope Mountains
115 O22 **Ródos** *var.* Ródhos, *Eng.* Rhodes, *It.* Rodi. Ródos, Dodekánisa, Greece, Aegean Sea 36°26′N 28°14′E
115 O22 **Ródos** *var.* Ródhos, *Eng.* Rhodes, *It.* Rodi; *anc.* Rhodos. *island* Dodekánisa, Greece, Aegean Sea
Rodosto *see* Tekirdağ
59 A14 **Rodrigues** Amazonas, W Brazil 06°50′S 73°45′W
173 P8 **Rodrigues** *var.* Rodriquez. *island* E Mauritius
Rodriquez *see* Rodrigues
Rodunma *see* Roanne
180 I7 **Roebourne** Western Australia 20°49′S 117°04′E
83 J20 **Roedtan** Limpopo, NE South Africa 24°37′S 29°05′E
98 H11 **Roelofarendsveen** Zuid-Holland, W Netherlands 52°12′N 04°37′E
Roepat *see* Rupat, Pulau
Roer *see* Rur
99 M16 **Roermond** Limburg, SE Netherlands 51°12′N 06°E
99 C18 **Roeselare** *Fr.* Roulers; *prev.* Rousselaere. West-Vlaanderen, W Belgium 50°57′N 03°08′E
9 P8 **Roes Welcome Sound** *strait* Nunavut, N Canada
Roeteng *see* Ruteng
Rofreit *see* Rovereto
Rogachëv *see* Rahachow
57 L15 **Rogagua, Laguna** ◎ NW Bolivia
95 C16 **Rogaland** ⬖ *county* S Norway
25 Y9 **Roganville** Texas, SW USA 30°49′N 93°54′W
109 W11 **Rogaška Slatina** *Ger.* Rohitsch-Sauerbrunn; *prev.* Rogatec-Slatina. E Slovenia 46°13′N 15°38′E
Rogatec-Slatina *see* Rogaška Slatina
112 J13 **Rogatica** Republika Srpska, SE Bosnia and Herzegovina 43°50′N 18°55′E
Rogatin *see* Rohatyn
93 F17 **Rogen** ◎ C Sweden
27 S9 **Rogers** Arkansas, C USA 36°19′N 94°07′W
29 P5 **Rogers** North Dakota, N USA 47°03′N 98°12′W
25 T9 **Rogers** Texas, SW USA 30°53′N 97°10′W
31 R5 **Rogers City** Michigan, N USA 45°25′N 83°49′W
Roger Simpson Island *see* Abemama
35 T14 **Rogers Lake** *salt flat* California, W USA
21 Q8 **Rogers, Mount** ▲ Virginia, NE USA 36°39′N 81°32′W
33 O16 **Rogerson** Idaho, NW USA 42°11′N 114°36′W
11 O16 **Rogers Pass** *pass* British Columbia, SW Canada
21 O8 **Rogersville** Tennessee, S USA 36°26′N 83°01′W
99 L16 **Roggel** Limburg, SE Netherlands 51°16′N 05°55′E
Roggeveen *see* Roggewein, Cabo
193 R10 **Roggeveen Basin** *undersea feature* E Pacific Ocean 31°30′S 95°30′W
191 X16 **Roggewein, Cabo** *var.* Roggeveen. *headland* Easter Island, Chile, E Pacific Ocean 27°07′S 109°15′W
103 Y13 **Rogliano** Corse, France, C Mediterranean Sea 42°58′N 09°25′E
107 N21 **Rogliano** Calabria, SW Italy 39°09′N 16°18′E
92 G12 **Rognan** Nordland, C Norway 67°04′N 15°21′E
100 K10 **Rögnitz** ≈ N Germany
Rogozhina/Rogozhinë *see* Rrogozhinë
110 G10 **Rogoźno** Wielkopolskie, C Poland 52°46′N 16°58′E
32 E15 **Rogue River** ≈ Oregon, NW USA
116 I6 **Rohatyn** *Rus.* Rogatin. Ivano-Frankivs'ka Oblast', W Ukraine 49°25′N 24°35′E
189 O14 **Rohi** Pohnpei, E Micronesia
Rohitsch-Sauerbrunn *see* Rogaška Slatina
149 Q13 **Rohri** Sind, SE Pakistan 27°39′N 68°57′E
152 I10 **Rohtak** Haryāna, N India 28°57′N 76°38′E
Roi Ed *see* Roi Et
167 R9 **Roi Et** *var.* Muang Roi Et, Roi Ed. Roi Et, E Thailand 16°05′N 103°38′E
191 U9 **Roi Georges, Îles du** *island group* Îles Tuamotu, C French Polynesia
153 Y10 **Roing** Arunāchal Pradesh, NE India 28°06′N 95°46′E
118 E7 **Roja** NW Latvia 57°31′N 22°44′E
61 B20 **Rojas** Buenos Aires, E Argentina 34°10′S 60°45′W
149 R12 **Rojhān** Punjab, E Pakistan 28°39′N 70°00′E
41 Q12 **Rojo, Cabo** *headland* C Mexico 21°33′N 97°19′W
45 Q10 **Rojo, Cabo** *headland* W Puerto Rico 17°57′N 67°10′W
168 K10 **Rokan Kiri, Sungai** ≈ Sumatera, W Indonesia
118 I11 **Rokiškis** Panevėžys, NE Lithuania 55°58′N 25°35′E
165 R7 **Rokkasho** Aomori, Honshū, C Japan 40°59′N 141°22′E
111 B17 **Rokycany** *Ger.* Rokytzan. Plzeňský Kraj, W Czech Republic 49°45′N 13°36′E
117 P6 **Rokytne** Kyyivs'ka Oblast', N Ukraine 49°40′N 30°29′E
116 L3 **Rokytne** Rivnens'ka Oblast', NW Ukraine 51°19′N 27°09′E
Rokytzan *see* Rokycany
158 L11 **Rola Co** ◎ W China
29 V13 **Roland** Iowa, C USA 42°10′N 93°30′W
95 D15 **Røldal** Hordaland, S Norway 59°52′N 06°49′E
98 O7 **Rolde** Drenthe, NE Netherlands 52°58′N 06°39′E
29 O2 **Rolette** North Dakota, N USA 48°39′N 99°50′W
27 V6 **Rolla** Missouri, C USA 37°56′N 91°47′W
29 O2 **Rolla** North Dakota, N USA 48°51′N 99°37′W
108 A10 **Rolle** Vaud, W Switzerland 46°27′N 06°19′E
181 X8 **Rolleston** Queensland, E Australia 24°30′S 148°36′E
185 H19 **Rolleston** Canterbury, South Island, New Zealand 43°34′S 172°24′E
185 G18 **Rolleston Range** ▲▲ South Island, New Zealand
14 H8 **Rollet** Québec, SE Canada 47°56′N 79°14′W
22 J4 **Rolling Fork** Mississippi, S USA 32°54′N 90°52′W
20 L6 **Rolling Fork** ≈ Kentucky, S USA
14 J11 **Rolphton** Ontario, SE Canada 46°09′N 77°43′W
Röm *see* Rømø
181 X10 **Roma** Queensland, E Australia 26°37′S 148°54′E
107 I15 **Roma** *Eng.* Rome. ● (Italy) Lazio, C Italy 41°53′N 12°30′E
95 P19 **Roma** Gotland, SE Sweden 57°31′N 18°28′E
21 T14 **Romain, Cape** *headland* South Carolina, SE USA 33°00′N 79°21′W
13 P11 **Romaine** ≈ Newfoundland and Labrador/Québec, E Canada
25 R17 **Roma Los Saenz** Texas, SW USA 26°24′N 99°01′W
114 H8 **Roman** Vratsa, NW Bulgaria 43°09′N 23°56′E
116 L10 **Roman** *Hung.* Románvásár. Neamţ, NE Romania 46°46′N 26°56′E
64 M13 **Romanche Fracture Zone** *tectonic feature* E Atlantic Ocean
61 C15 **Romang** Santa Fe, C Argentina 29°30′S 59°46′W
171 R15 **Romang, Pulau** *var.* Pulau Roma. *island* Kepulauan Damar, E Indonesia
171 R15 **Romang, Selat** *strait* Nusa Tenggara, S Indonesia
116 J11 **Romania** *Bul.* Rumŭniya, *Ger.* Rumänien, *Hung.* România, *Rom.* România, *SCr.* Rumunjska, *Ukr.* Rumuniya, *prev.* Republica Socialistă România, Roumania, Rumania, Socialist Republic of Romania, *prev.Rom.* Romînia. ◆ *republic* SE Europe
România, Republica Socialistă *see* Romania
Romania, Socialist Republic of *see* Romania
117 T7 **Romaniv** *Rus.* Dneprodzerzhinsk, *prev.* Dniprodzerzhyns'k, *prev.* Kamenskoye. Dnipropetrovs'ka Oblast', E Ukraine 48°30′N 34°35′E
117 X7 **Romaniv** *Rus.* Dzerzhinsk; *prev.* Dzerzhyns'k. Donets'ka Oblast', SE Ukraine 48°21′N 37°50′E
116 M5 **Romaniv** *prev.* Dzerzhyns'k. Zhytomyrs'ka Oblast', N Ukraine 50°07′N 27°56′E
23 W16 **Romano, Cape** *headland* Florida, SE USA 25°51′N 81°40′W
44 G5 **Romano, Cayo** *island* C Cuba
123 O13 **Romanovka** Respublika Buryatiya, S Russian Federation 53°10′N 112°34′E
127 N8 **Romanovka** Saratovskaya Oblast', W Russian Federation 51°45′N 42°45′E
108 I6 **Romanshorn** Thurgau, NE Switzerland 47°34′N 09°23′E
103 R12 **Romans-sur-Isère** Drôme, E France 45°03′N 05°03′E
189 U12 **Romanum** *island* Chuuk, C Micronesia
Románvásár *see* Roman
39 S5 **Romanzof Mountains** ▲▲ Alaska, USA
Roma, Pulau *see* Romang, Pulau
103 S4 **Rombas** Moselle, NE France 49°15′N 06°04′E
23 R2 **Rome** Georgia, SE USA 34°01′N 85°02′W
18 I9 **Rome** New York, NE USA 43°13′N 75°28′W
Rome *see* Roma
31 S9 **Romeo** Michigan, N USA 42°48′N 83°00′W
Römerstadt *see* Rýmařov
Rometan *see* Romiton
103 P5 **Romilly-sur-Seine** Aube, N France 48°31′N 03°44′E
Romînia *see* Romania
146 L11 **Romiton** *Rus.* Rometan. Buxoro Viloyati, C Uzbekistan 39°56′N 64°21′E
21 U3 **Romney** West Virginia, NE USA 39°21′N 78°44′W
117 S4 **Romny** Sums'ka Oblast', NE Ukraine 50°45′N 33°30′E
95 E24 **Rømø** *Ger.* Röm. *island* SW Denmark
117 S5 **Romodan** Poltavs'ka Oblast', NE Ukraine 50°00′N 33°20′E
127 P5 **Romodanovo** Respublika Mordoviya, W Russian Federation 54°25′N 45°24′E
Romorantin *see* Romorantin-Lanthenay
103 N8 **Romorantin-Lanthenay** *var.* Romorantin. Loir-et-Cher, C France 47°22′N 01°44′E
94 F9 **Romsdal** *physical region* S Norway
94 F10 **Romsdalen** *valley* S Norway
94 E9 **Romsdalsfjorden** *fjord* S Norway
33 P8 **Ronan** Montana, NW USA 47°31′N 114°06′W
59 M14 **Roncador** Maranhão, E Brazil 05°48′S 45°08′W
186 M7 **Roncador Reef** *reef* N Solomon Islands
59 J17 **Roncador, Serra do** ▲▲ C Brazil
21 S6 **Ronceverte** West Virginia, NE USA 37°45′N 80°27′W
107 H14 **Ronciglione** Lazio, C Italy 42°16′N 12°15′E
104 L15 **Ronda** Andalucía, S Spain 36°45′N 05°10′W
94 G11 **Rondane** ▲▲ S Norway
104 L15 **Ronda, Serranía de** ▲▲ S Spain
95 H22 **Rønde** Midtjylland, C Denmark 56°18′N 10°28′E
Ronde, Île *see* Round Island
Rōņdik *see* Rongrik Atoll
59 E16 **Rondônia** *off.* Estado de Rondônia; *prev.* Território de Rondônia. ⬖ *state* W Brazil
Rondônia, Estado de *see* Rondônia
Rondônia, Território de *see* Rondônia
59 I18 **Rondonópolis** Mato Grosso, W Brazil 16°29′S 54°37′W
94 G11 **Rondslottet** ▲ S Norway 61°54′N 09°48′E
95 P20 **Ronehamn** Gotland, SE Sweden 57°10′N 18°30′E
160 L13 **Rong'an** *var.* Chang'an, Rongan. Guangxi Zhuangzu Zizhiqu, S China 25°14′N 109°20′E
Rongan *see* Rong'an
Rongcheng *see* Rongxian, Guangxi, China
Rongcheng *see* Jianli, Hubei, China
189 R4 **Rongelap Atoll** *var.* Rōnlap. *atoll* Ralik Chain, NW Marshall Islands
Rongerik *see* Rongrik Atoll
160 K12 **Rongjiang** *var.* Guzhou. Guizhou, S China 25°59′N 108°27′E
160 L13 **Rong Jiang** ≈ S China
Rongjiang *see* Nankang
Rong, Kas *see* Rŭng, Kaôh
167 P8 **Rong Kwang** Phrae, NW Thailand 18°19′N 100°18′E
189 T4 **Rongrik Atoll** *var.* Rōņdik, Rongerik. *atoll* Ralik Chain, N Marshall Islands
189 X2 **Rongrong** *island* SE Marshall Islands
160 L13 **Rongshui** *var.* Rongshui Miaozu Zizhixian. Guangxi Zhuangzu Zizhiqu, S China 25°05′N 109°09′E
Rongshui Miaozu Zizhixian *see* Rongshui
118 I6 **Rõngu** *Ger.* Ringen. Tartumaa, SE Estonia 58°10′N 26°17′E
Rongwo *see* Tongren
160 L15 **Rongxian** *var.* Rongzhou; *prev.* Rongcheng. Guangxi Zhuangzu Zizhiqu, S China 22°52′N 110°33′E
Rongzhag *see* Danba
Rongzhou *see* Rongxian
Roniu, Mont *see* Rooniu, Mont
189 N13 **Ronkiti** Pohnpei, E Micronesia 06°48′N 158°10′E
Rōnlap *see* Rongelap Atoll
95 L24 **Rønne** Bornholm, E Denmark 55°07′N 14°43′E
95 M22 **Ronneby** Blekinge, S Sweden 56°12′N 15°18′E
194 J7 **Ronne Entrance** *inlet* Antarctica

◆ Country
● Country Capital
◇ Dependent Territory
○ Dependent Territory Capital
⬖ Administrative Regions
✈ International Airport
▲ Mountain
▲▲ Mountain Range
Ѧ Volcano
≈ River
◎ Lake
⊠ Reservoir

RONNE ICE SHELF – RUVUMA

194 L6 **Ronne Ice Shelf** *ice shelf* Antarctica
99 E19 **Ronse** *Fr.* Renaix. Oost-Vlaanderen, SW Belgium 50°45'N 03°36'E
30 K14 **Roodhouse** Illinois, N USA 39°28'N 90°22'W
83 C19 **Rooibank** Erongo, W Namibia 23°04'S 14°34'E
Rooke Island *see* Umboi Island
65 N24 **Rookery Point** *headland* NE Tristan da Cunha 37°03'S 12°15'W
191 R8 **Rooniu, Mont** *prev.* Mont Roniu. Tahiti, W French Polynesia 17°49'S 149°12'W
171 V13 **Roon, Pulau** *island* E Indonesia
173 V7 **Roo Rise** *undersea feature* E Indian Ocean
152 J9 **Roorkee** Uttarakhand, N India 29°51'N 77°54'E
99 H15 **Roosendaal** Noord-Brabant, S Netherlands 51°32'N 04°29'E
25 P10 **Roosevelt** Texas, SW USA 30°28'N 100°06'W
37 N3 **Roosevelt** Utah, W USA 40°18'N 109°59'W
47 T8 **Roosevelt** W Brazil
195 O13 **Roosevelt Island** *island* Antarctica
10 L10 **Roosevelt, Mount** ▲ British Columbia, W Canada 58°28'N 125°22'W
11 P17 **Roosville** British Columbia, SW Canada 48°59'N 115°03'W
29 X10 **Root River** Minnesota, N USA
Ropar *see* Rūpnagar
111 N16 **Ropczyce** Podkarpackie, SE Poland 50°04'N 21°31'E
181 Q3 **Roper Bar** Northern Territory, N Australia 14°45'S 134°30'E
24 M5 **Ropesville** Texas, SW USA 33°24'N 102°09'W
102 K14 **Roquefort** Landes, SW France 44°01'N 00°18'W
61 C21 **Roque Pérez** Buenos Aires, E Argentina 35°25'S 59°24'W
58 E10 **Roraima** *off.* Estado de Roraima; *prev.* Território de Rio Branco, Território de Roraima. ◆ *state* N Brazil
Roraima, Estado de *see* Roraima
58 F9 **Roraima, Mount** ▲ N South America 05°10'N 60°36'W
Roraima, Território de *see* Roraima
94 I9 **Røros** Sør-Trøndelag, S Norway 62°37'N 11°25'E
108 I7 **Rorschach** Sankt Gallen, NE Switzerland 47°28'N 09°30'E
93 E14 **Rørvik** Nord-Trøndelag, C Norway 64°54'N 11°15'E
119 G17 **Ros'** *Rus.* Ross'. Hrodzyenskaya Voblasts', W Belarus 53°25'N 24°24'E
185 F17 **Ross** West Coast, South Island, New Zealand 42°54'S 170°52'E
119 G17 **Ros'** *Rus.* Ross'. W Belarus
10 J7 **Ross** Yukon, W Canada
117 O6 **Ros'** N Ukraine
44 K7 **Rosa, Lake** Great Inagua, S The Bahamas
32 M9 **Rosalia** Washington, NW USA 47°14'N 117°22'W
191 W15 **Rosalia, Punta** *headland* Easter Island, Chile, E Pacific Ocean 27°04'S 109°19'W
45 P12 **Rosalie** E Dominica 15°22'N 61°15'W
35 T14 **Rosamond** California, W USA 34°51'N 118°09'W
35 S14 **Rosamond Lake** *salt flat* California, W USA
96 H8 **Ross and Cromarty** *cultural region* N Scotland, United Kingdom
61 B18 **Rosario** Santa Fe, C Argentina 32°56'S 60°39'W
40 J11 **Rosario** Sinaloa, C Mexico 23°00'N 105°51'W
40 G6 **Rosario** Sonora, NW Mexico 27°53'N 109°18'W
62 O6 **Rosario** San Pedro, C Paraguay 24°26'S 57°06'W
61 E20 **Rosario** Colonia, SW Uruguay 34°20'S 57°26'W
54 H5 **Rosario** Zulia, NW Venezuela 10°18'N 72°19'W
Rosario *see* Nishino-shima
Rosario *see* Rosarito
40 B4 **Rosario, Bahía del** *bay* NW Mexico
62 K6 **Rosario de la Frontera** Salta, N Argentina 25°50'S 65°00'W
61 C18 **Rosario del Tala** Entre Ríos, E Argentina 32°20'S 59°10'W
61 F16 **Rosário do Sul** Rio Grande do Sul, S Brazil 30°15'S 54°55'W
59 H18 **Rosário Oeste** Mato Grosso, W Brazil 14°50'S 56°25'W
40 B1 **Rosarito** *var.* Rosario. Baja California Norte, NW Mexico 32°25'N 117°04'W
40 D5 **Rosarito** Baja California Norte, NW Mexico 28°27'N 113°58'W
40 F7 **Rosarito** Baja California Sur, NW Mexico 26°28'N 111°41'W
104 L9 **Rosarito, Embalse del** W Spain
107 N22 **Rosarno** Calabria, SW Italy 38°29'N 15°59'E
56 B5 **Rosa Zárate** *var.* Quinindé. Esmeraldas, SW Ecuador 00°14'N 79°28'W
Roscianum *see* Rossano
29 O8 **Roscoe** South Dakota, N USA 45°24'N 99°19'W
25 P7 **Roscoe** Texas, SW USA 32°27'N 100°32'W
102 F5 **Roscoff** Finistère, NW France 48°43'N 04°00'W
Ros Comáin *see* Roscommon
97 C17 **Roscommon** *Ir.* Ros Comáin. C Ireland 53°38'N 08°11'W
31 Q7 **Roscommon** Michigan, N USA 44°30'N 84°35'W
97 C17 **Roscommon** *Ir.* Ros Comáin. *cultural region* C Ireland
Ros. Cré *see* Roscrea
97 D19 **Roscrea** *Ir.* Ros. Cré. C Ireland 52°57'N 07°47'W
14 H13 **Rosseau** Ontario, S Canada 45°15'N 79°38'W
45 X12 **Roseau** *prev.* Charlotte Town. ● (Dominica) SW Dominica 15°17'N 61°23'W
29 S2 **Roseau** Minnesota, N USA 48°51'N 95°45'W
173 Y16 **Rose Belle** SE Mauritius 20°24'S 57°36'E
183 O16 **Rosebery** Tasmania, SE Australia 41°51'S 145°33'E
21 U11 **Roseboro** North Carolina, SE USA 34°58'N 78°31'W
25 T9 **Rosebud** Texas, SW USA 31°04'N 96°58'W
33 W10 **Rosebud Creek** Montana, NW USA
32 F14 **Roseburg** Oregon, NW USA 43°13'N 123°21'W
22 J3 **Rosedale** Mississippi, S USA 33°51'N 91°01'W
99 H21 **Rosée** Namur, S Belgium 50°15'N 04°43'E
55 U8 **Rose Hall** E Guyana 06°14'N 57°30'W
173 X16 **Rose Hill** W Mauritius 20°14'S 57°29'E
80 H12 **Roseires, Reservoir** *var.* Lake Rusayris. E Sudan
Rosenau *see* Rožnov pod Radhoštěm
Rosenau *see* Rožňava
25 V11 **Rosenberg** Texas, SW USA 29°33'N 95°48'W
Rosenberg *see* Olesno, Poland
Rosenberg *see* Ružomberok, Slovakia
100 I10 **Rosengarten** Niedersachsen, N Germany 53°24'N 09°54'E
101 M24 **Rosenheim** Bayern, S Germany 47°51'N 12°08'E
Rosenhof *see* Zilupe
105 X4 **Roses** Cataluña, NE Spain 42°15'N 03°11'E
105 X4 **Roses, Golf de** *gulf* NE Spain
107 K14 **Roseto degli Abruzzi** Abruzzo, C Italy 42°39'N 14°01'E
11 S16 **Rosetown** Saskatchewan, S Canada 51°34'N 107°59'W
Rosetta *see* Rashīd
35 O7 **Roseville** California, W USA 38°44'N 121°16'W
30 J12 **Roseville** Illinois, N USA 40°42'N 90°40'W
29 V8 **Roseville** Minnesota, N USA 45°00'N 93°09'W
29 R7 **Rosholt** South Dakota, N USA 45°51'N 96°42'W
106 F12 **Rosignano Marittimo** Toscana, C Italy 43°24'N 10°28'E
116 I14 **Roşiori de Vede** Teleorman, S Romania 44°06'N 25°00'E
114 K8 **Rositsa** N Bulgaria
Rositten *see* Rēzekne
95 J23 **Roskilde** Sjælland, E Denmark 55°39'N 12°07'E
Ros Láir *see* Rosslare
126 H5 **Roslavl'** Smolenskaya Oblast', W Russian Federation 54°N 32°57'E
32 I8 **Roslyn** Washington, NW USA 47°13'N 120°52'W
99 K14 **Rosmalen** Noord-Brabant, S Netherlands 51°43'N 05°21'E
Ros Mhic Thriúin *see* New Ross
113 P19 **Rosoman** C FYR Macedonia 41°31'N 21°55'E
102 F6 **Rosporden** Finistère, NW France 47°58'N 03°54'W
Ross' *see* Ros'
107 O20 **Rossano** *anc.* Roscianum. Calabria, SW Italy 39°35'N 16°38'E
22 L5 **Ross Barnett Reservoir** Mississippi, S USA
11 W16 **Rossburn** Manitoba, S Canada 50°42'N 100°49'W
14 H13 **Rosseau, Lake** Ontario, S Canada
186 I10 **Rossel Island** *prev.* Yela Island. *island* SE Papua New Guinea
195 P12 **Ross Ice Shelf** *ice shelf* Antarctica
13 P16 **Rossignol, Lake** Nova Scotia, SE Canada
83 C19 **Rössing** Erongo, W Namibia 22°31'S 14°52'E
195 Q14 **Ross Island** *island* Antarctica
Rossitten *see* Rybachiy
Rossiyskaya Federatsiya *see* Russian Federation
11 N17 **Rossland** British Columbia, SW Canada 49°05'N 117°49'W
97 F20 **Rosslare** *Ir.* Ros Láir. Wexford, SE Ireland 52°16'N 06°23'W
97 F20 **Rosslare Harbour** Wexford, SE Ireland 52°15'N 06°20'W
101 M14 **Rosslau** Sachsen-Anhalt, E Germany 51°52'N 12°15'E
76 G10 **Rosso** Trarza, SW Mauritania 16°36'N 15°50'W
103 X14 **Rosso, Cap** *headland* Corse, France, C Mediterranean Sea 42°25'N 08°22'E
93 H16 **Rossön** Jämtland, C Sweden 63°54'N 16°21'E
97 K21 **Ross-on-Wye** W England, United Kingdom 51°55'N 02°34'W
Rossony *see* Rasony
126 L9 **Rossosh'** Voronezhskaya Oblast', W Russian Federation 50°10'N 39°34'E
181 Q7 **Ross River** Northern Territory, N Australia 23°36'S 134°30'E
10 J7 **Ross River** Yukon, W Canada 61°57'N 132°26'W
195 O15 **Ross Sea** *sea* Antarctica
92 G13 **Røssvatnet** *Lapp.* Reevhtse. C Norway
23 R1 **Rossville** Georgia, SE USA 34°59'N 85°22'W
Rostak *see* Ar Rustāq
143 P14 **Rostāq** Hormozgān, S Iran 26°48'N 53°50'E
117 N5 **Rostavytsya** N Ukraine
11 T15 **Rosthern** Saskatchewan, S Canada 52°40'N 106°20'W
100 M8 **Rostock** Mecklenburg-Vorpommern, NE Germany 54°05'N 12°08'E
124 L16 **Rostov** Yaroslavskaya Oblast', W Russian Federation 57°11'N 39°19'E
Rostov *see* Rostov-na-Donu
126 L12 **Rostov-na-Donu** *var.* Rostov, *Eng.* Rostov-on-Don. Rostovskaya Oblast', SW Russian Federation 47°16'N 39°45'E
Rostov-on-Don *see* Rostov-na-Donu
126 L10 **Rostovskaya Oblast'** ◆ *province* SW Russian Federation
93 J14 **Rosvik** Norrbotten, N Sweden 65°26'N 21°48'E
23 S3 **Roswell** Georgia, SE USA 34°01'N 84°21'W
37 U14 **Roswell** New Mexico, SW USA 33°23'N 104°31'W
94 K12 **Rot** Dalarna, C Sweden 61°16'N 14°04'E
101 I23 **Rot** S Germany
104 J15 **Rota** Andalucía, S Spain 36°39'N 06°20'W
188 K9 **Rota** *island* S Northern Mariana Islands
25 P6 **Rotan** Texas, SW USA 32°51'N 100°28'W
Rotcher Island *see* Tamana
100 I11 **Rotenburg** Niedersachsen, NW Germany 53°06'N 09°25'E
Rotenburg *see* Rotenburg an der Fulda
101 I16 **Rotenburg an der Fulda** *var.* Rotenburg. Thüringen, C Germany 51°00'N 09°43'E
101 L18 **Roter Main** E Germany
101 K20 **Roth** Bayern, SE Germany 49°15'N 11°06'E
101 G16 **Rothaargebirge** ▲▲ W Germany
Rothenburg *see* Rothenburg ob der Tauber
101 J20 **Rothenburg ob der Tauber** *var.* Rothenburg. Bayern, S Germany 49°23'N 10°10'E
194 H6 **Rothera** *UK research station* Antarctica 67°28'S 68°31'W
185 I17 **Rotherham** Canterbury, South Island, New Zealand 42°42'S 172°56'E
97 M17 **Rotherham** N England, United Kingdom 53°26'N 01°20'W
96 H12 **Rothesay** W Scotland, United Kingdom 55°49'N 05°03'W
108 E7 **Rothrist** Aargau, N Switzerland 47°18'N 07°54'E
194 H6 **Rothschild Island** *island* Antarctica
171 P17 **Roti, Pulau** *island* S Indonesia
95 H14 **Rotnes** Akershus, S Norway 60°08'N 10°45'E
183 O8 **Roto** New South Wales, SE Australia 33°04'S 145°27'E
184 N8 **Rotoiti, Lake** North Island, New Zealand
Rotomagus *see* Rouen
107 N19 **Rotondella** Basilicata, S Italy 40°12'N 16°30'E
103 X15 **Rotondo, Monte** ▲ Corse, France, C Mediterranean Sea 42°15'N 09°03'E
185 I15 **Rotoroa, Lake** South Island, New Zealand
184 N8 **Rotorua** Bay of Plenty, North Island, New Zealand 38°10'S 176°14'E
184 N8 **Rotorua, Lake** North Island, New Zealand
101 N22 **Rott** SE Germany
108 F10 **Rotten** S Switzerland
109 T6 **Rottenmann** Steiermark, E Austria 47°31'N 14°18'E
98 H12 **Rotterdam** Zuid-Holland, SW Netherlands 51°55'N 04°30'E
18 K10 **Rotterdam** New York, NE USA 42°46'N 73°57'W
95 M21 **Rottnen** S Sweden
98 N4 **Rottumeroog** *island* Waddeneilanden, NE Netherlands
98 N4 **Rottumerplaat** *island* Waddeneilanden, NE Netherlands
101 G23 **Rottweil** Baden-Württemberg, S Germany 48°10'N 08°38'E
191 O7 **Rotui, Mont** Moorea, W French Polynesia 17°30'S 149°50'W
103 P1 **Roubaix** Nord, N France 50°42'N 03°10'E
111 C15 **Roudnice nad Labem** *Ger.* Raudnitz an der Elbe. Ústecký Kraj, NW Czech Republic 50°25'N 14°14'E
102 M4 **Rouen** *anc.* Rotomagus. Seine-Maritime, N France 49°26'N 01°05'E
171 X13 **Rouffaer Reserves** *reserve* Papua, E Indonesia
15 N10 **Rouge, Rivière** Québec, SE Canada
20 J6 **Rough River** Kentucky, S USA
20 J6 **Rough River Lake** Kentucky, S USA
Rouhaïbé *see* Ar Ruḩaybah
102 K11 **Rouillac** Charente, W France 45°46'N 00°04'W
Roulers *see* Roeselare
Roumania *see* Romania
173 Y15 **Round Island** *var.* Île Ronde. *island* NE Mauritius
14 J12 **Round Lake** Ontario, SE Canada
35 U7 **Round Mountain** Nevada, W USA 38°42'N 117°04'W
25 R10 **Round Mountain** Texas, SW USA 30°25'N 98°20'W
183 U5 **Round Mountain** ▲ New South Wales, SE Australia 30°22'S 152°13'E
25 S10 **Round Rock** Texas, SW USA 30°30'N 97°40'W
33 U10 **Roundup** Montana, NW USA 46°27'N 108°32'W
55 Y10 **Roura** NE French Guiana 04°44'N 52°16'W
Rourkela *see* Rāurkela
96 J4 **Rousay** *island* N Scotland, United Kingdom
Rousselaere *see* Roeselare
103 O17 **Roussillon** *cultural region* S France
15 V7 **Routhierville** Québec, SE Canada 48°09'N 67°07'W
99 K25 **Rouvroy** Luxembourg, SE Belgium 49°33'N 05°28'E
14 I7 **Rouyn-Noranda** Québec, SE Canada 48°16'N 79°03'W
Rouyuan *see* Huachi
Rouyuanchengzi *see* Huachi
92 L12 **Rovaniemi** Lappi, N Finland 66°29'N 25°40'E
106 E7 **Rovato** Lombardia, N Italy 45°34'N 10°03'E
125 N11 **Rovdino** Arkhangel'skaya Oblast', NW Russian Federation 61°36'N 42°28'E
Roven'ki *see* Roven'ky
117 Y8 **Roven'ky** *var.* Roven'ki. Luhans'ka Oblast', E Ukraine 48°05'N 39°20'E
Rovenskaya Oblast' *see* Rivnens'ka Oblast'
Rovenskaya Sloboda *see* Rovyenskaya Slabada
106 G7 **Rovereto** *Ger.* Rofreit. Trentino-Alto Adige, N Italy 45°53'N 11°03'E
167 S12 **Rôviěng Tbong** Preăh Vihéar, N Cambodia 13°18'N 105°06'E
Rovigno *see* Rovinj
106 H8 **Rovigo** Veneto, NE Italy 45°04'N 11°48'E
112 A10 **Rovinj** *It.* Rovigno. Istra, NW Croatia 45°06'N 13°39'E
54 E10 **Rovira** Tolima, C Colombia 04°15'N 75°15'W
Rovno *see* Rivne
127 P9 **Rovnoye** Saratovskaya Oblast', W Russian Federation 50°43'N 46°03'E
82 Q12 **Rovuma, Rio** *var.* Ruvuma. Mozambique/Tanzania *see also* Ruvuma
Rovuma, Rio *see* Ruvuma
119 O19 **Rovyenskaya Slabada** *Rus.* Rovenskaya Sloboda. Homyel'skaya Voblasts', SE Belarus 52°13'N 30°19'E
183 R5 **Rowena** New South Wales, SE Australia 29°51'S 148°55'E
21 T11 **Rowland** North Carolina, SE USA 34°32'N 79°17'W
9 P5 **Rowley** Baffin Island, Nunavut, NE Canada
9 P6 **Rowley Island** *island* Nunavut, NE Canada
173 W8 **Rowley Shoals** *reef* NW Australia
Równe *see* Rivne
171 O4 **Roxas** Mindoro, N Philippines 12°36'N 121°29'E
171 P5 **Roxas City** Panay Island, C Philippines 11°33'N 122°43'E
21 U8 **Roxboro** North Carolina, SE USA 36°24'N 79°00'W
185 D23 **Roxburgh** Otago, South Island, New Zealand 45°32'S 169°18'E
96 K13 **Roxburgh** *cultural region* SE Scotland, United Kingdom
182 H5 **Roxby Downs** South Australia 30°29'S 136°56'E
95 M17 **Roxen** S Sweden
25 V5 **Roxton** Texas, SW USA 33°33'N 95°43'W
15 P12 **Roxton-Sud** Québec, SE Canada 45°30'N 72°35'W
33 U8 **Roy** Montana, NW USA 47°19'N 108°55'W
37 U10 **Roy** New Mexico, SW USA 35°56'N 104°12'W
97 E17 **Royal Canal** *Ir.* An Chanáil Ríoga. *canal* C Ireland
30 L1 **Royale, Isle** *island* Michigan, N USA
37 S6 **Royal Gorge** *valley* Colorado, C USA
97 M20 **Royal Leamington Spa** *var.* Leamington, Leamington Spa. C England, United Kingdom 52°18'N 01°31'W
97 O23 **Royal Tunbridge Wells** *var.* Tunbridge Wells. SE England, United Kingdom 51°08'N 00°16'E
24 L9 **Royalty** Texas, SW USA 31°21'N 102°51'W
102 J11 **Royan** Charente-Maritime, W France 45°37'N 01°01'W
65 B24 **Roy Cove Settlement** West Falkland, Falkland Islands 51°32'S 60°23'W
103 O3 **Roye** Somme, N France 49°42'N 02°46'E
95 H15 **Røyken** Buskerud, S Norway 59°47'N 10°21'E
93 F14 **Røyrvik** Nord-Trøndelag, C Norway 64°53'N 13°30'E
25 U6 **Royse City** Texas, SW USA 32°58'N 96°19'W
97 O21 **Royston** E England, United Kingdom 52°05'N 00°01'W
23 U2 **Royston** Georgia, SE USA 34°17'N 83°06'W
114 L10 **Roza** *prev.* Gyulovo. Yambol, E Bulgaria 42°29'N 26°30'E
113 L16 **Rožaje** E Montenegro 42°51'N 20°11'E
110 M10 **Różan** Mazowieckie, C Poland 52°36'N 21°27'E
117 O10 **Rozdil'na** Odes'ka Oblast', SW Ukraine 46°51'N 30°03'E
117 S12 **Rozdol'ne** *Rus.* Razdolnoye. Avtonomna Respublika Krym, S Ukraine 45°45'N 33°27'E
Rozhdestvenka *see* Kabanbay Batyr
116 I6 **Rozhnyativ** Ivano-Frankivs'ka Oblast', W Ukraine 48°58'N 24°07'E
116 J3 **Rozhyshche** Volyns'ka Oblast', NW Ukraine 50°54'N 25°16'E
Roznau am Radhost *see* Rožnov pod Radhoštěm
111 L19 **Rožňava** *Ger.* Rosenau, *Hung.* Rozsnyó. Košický Kraj, E Slovakia 48°41'N 20°32'E
116 K10 **Roznov** Neamţ, NE Romania 46°47'N 26°33'E
111 I18 **Rožnov pod Radhoštěm** *Ger.* Rosenau, Roznau am Radhost. Zlínský Kraj, E Czech Republic 49°28'N 18°09'E
Rózsahegy *see* Ružomberok
Rozsnyó *see* Râşnov, Romania
Rozsnyó *see* Rožňava, Slovakia
113 K18 **Rranxë** Shkodër, NW Albania 41°58'N 19°27'E
113 L18 **Rrëshen** *var.* Rresheni, Rrshen. Lezhë, C Albania 41°46'N 19°54'E
Rresheni *see* Rrëshen
Rrogozhina *see* Rrogozhinë
113 K20 **Rrogozhinë** *var.* Rogozhina, Rogozhinë, Rrogozhina. Tiranë, W Albania 41°04'N 19°40'E
Rrshen *see* Rrëshen
112 O13 **Rtanj** ▲ E Serbia 43°45'N 21°54'E
127 O7 **Rtishchevo** Saratovskaya Oblast', W Russian Federation 52°16'N 43°46'E
184 N12 **Ruahine Range** *var.* Ruarine. ▲▲ North Island, New Zealand
185 L14 **Ruamahanga** North Island, New Zealand
Ruanda *see* Rwanda
184 M10 **Ruapehu, Mount** North Island, New Zealand 39°15'S 175°33'E
185 C25 **Ruapuke Island** *island* SW New Zealand
Ruarine *see* Ruahine Range
184 O9 **Ruatahuna** Bay of Plenty, North Island, New Zealand 38°38'S 176°56'E
184 Q8 **Ruatoria** Gisborne, North Island, New Zealand 37°54'S 178°18'E
184 K4 **Ruawai** Northland, North Island, New Zealand 36°08'S 174°04'E
15 N8 **Ruban** Québec, SE Canada
81 I22 **Rubeho Mountains** ▲▲ C Tanzania
165 U3 **Rubeshibe** Hokkaidō, NE Japan 43°49'N 143°37'E
Rubezhnoye *see* Rubizhne
113 L18 **Rubik** Lezhë, C Albania 41°46'N 19°48'E
54 H7 **Rubio** Táchira, W Venezuela 07°42'N 72°23'W
117 X6 **Rubizhne** *Rus.* Rubezhnoye. Luhans'ka Oblast', E Ukraine 49°01'N 38°22'E
81 F20 **Rubondo Island** *island* N Tanzania
122 I13 **Rubtsovsk** Altayskiy Kray, S Russian Federation 51°34'N 81°11'E
39 P9 **Ruby** Alaska, USA 64°44'N 155°29'W
35 W3 **Ruby Dome** ▲ Nevada, W USA 40°35'N 115°25'W
35 W4 **Ruby Lake** Nevada, W USA
35 W4 **Ruby Mountains** ▲▲ Nevada, W USA
33 Q12 **Ruby Range** ▲▲ Montana, NW USA
118 C10 **Rucava** SW Latvia 56°09'N 21°10'E
143 S13 **Rūdān** *var.* Dehbārez. Hormozgān, S Iran 27°30'N 57°10'E
Rudelstadt *see* Ciechanowiec
Rudensk *see* Rudzyensk
119 G14 **Rūdiškės** Vilnius, S Lithuania 54°31'N 24°49'E
95 H24 **Rudkøbing** Syddtjylland, C Denmark 54°57'N 10°43'E
125 S13 **Rudnichnyy** Kirovskaya Oblast', NW Russian Federation 59°37'N 52°28'E
Rūdnichnyy *see* Koksu
Rudny *see* Rudnyy
126 H4 **Rudnya** Smolenskaya Oblast', W Russian Federation 54°55'N 31°10'E
127 O8 **Rudnya** Volgogradskaya Oblast', SW Russian Federation 50°54'N 44°27'E
144 M7 **Rudnyy** *var.* Rudny. Kostanay, N Kazakhstan 53°N 63°05'E
122 K3 **Rudol'fa, Ostrov** *island* Zemlya Frantsa-Iosifa, NW Russian Federation
Rudolf, Lake *see* Turkana, Lake
Rudolfswert *see* Novo mesto
101 L17 **Rudolstadt** Thüringen, C Germany 50°44'N 11°20'E
31 Q4 **Rudyard** Michigan, N USA 46°15'N 84°36'W
33 S7 **Rudyard** Montana, NW USA 48°33'N 110°37'W
119 K16 **Rudzyensk** *Rus.* Rudensk. Minskaya Voblasts', C Belarus 53°36'N 27°52'E
104 L6 **Rueda** Castilla y León, N Spain 41°24'N 04°58'W
114 F10 **Ruen** ▲ Bulgaria/FYR Macedonia 42°10'N 22°31'E
80 G10 **Rufa'a** Gezira, C Sudan 14°49'N 33°21'E
102 L10 **Ruffec** Charente, W France 46°01'N 00°10'E
21 R14 **Ruffin** South Carolina, SE USA 33°00'N 80°48'W
81 J23 **Rufiji** E Tanzania
61 A20 **Rufino** Santa Fe, C Argentina 34°16'S 62°45'W
76 F11 **Rufisque** W Senegal 14°44'N 17°18'W
83 K14 **Rufunsa** Lusaka, C Zambia 15°02'S 29°35'E
118 J9 **Rugāji** E Latvia 57°01'N 27°07'E
161 R7 **Rugao** Jiangsu, E China 32°27'N 120°35'E
97 M20 **Rugby** C England, United Kingdom 52°22'N 01°18'W
29 N3 **Rugby** North Dakota, N USA 48°24'N 100°00'W
100 N7 **Rügen** *headland* NE Germany 54°25'N 13°21'E
Ruhaybeh *see* Ar Ruḩaybah
81 E19 **Ruhengeri** NW Rwanda 01°39'S 29°16'E
Ruhja *see* Rūjiena
100 M10 **Ruhner Berg** *hill* N Germany
118 F7 **Ruhnu** *var.* Ruhnu Saar, *Swe.* Runö. *island* SW Estonia
Ruhnu Saar *see* Ruhnu
101 G15 **Ruhr** W Germany
91 W6 **Ruhr Valley** *industrial region* W Germany
161 S11 **Rui'an** *var.* Rui an. Zhejiang, SE China 27°51'N 120°39'E
Rui an *see* Rui'an
161 P10 **Ruichang** Jiangxi, S China 29°46'N 115°37'E
24 J11 **Ruidosa** Texas, SW USA 30°00'N 104°40'W
37 S14 **Ruidoso** New Mexico, SW USA 33°19'N 105°40'W
161 P12 **Ruijin** Jiangxi, S China 25°52'N 116°01'E
160 D13 **Ruili** Yunnan, SW China 24°04'N 97°49'E
98 N8 **Ruinen** Drenthe, NE Netherlands 52°46'N 06°19'E
99 D17 **Ruiselede** West-Vlaanderen, W Belgium 51°03'N 03°21'E
161 T13 **Ruisui** *prev.* Juisui. C Taiwan 23°43'N 121°28'E
64 P5 **Ruivo de Santana, Pico** ▲ Madeira, Portugal, NE Atlantic Ocean 32°46'N 16°57'W
40 J12 **Ruiz** Nayarit, SW Mexico 22°00'N 105°09'W
54 E10 **Ruiz, Nevado del** W Colombia 04°53'N 75°22'W
138 J9 **Rujaylah, Ḩarrat ar** *salt lake* N Jordan
Rujen *see* Rūjiena
118 H7 **Rūjiena** *Est.* Ruhja, *Ger.* Rujen. N Latvia 57°54'N 25°22'E
79 I18 **Ruki** W Dem. Rep. Congo
81 E22 **Rukwa** ◆ *region* SW Tanzania
81 F23 **Rukwa, Lake** SE Tanzania
25 P6 **Rule** Texas, SW USA 33°10'N 99°53'W
22 K3 **Ruleville** Mississippi, S USA 33°43'N 90°33'W
Rum *see* Rhum
112 K10 **Ruma** Vojvodina, N Serbia 45°02'N 19°51'E
Rumadiya *see* Ar Ramādī
141 Q7 **Rumāḩ** Ar Riyāḑ, C Saudi Arabia 25°35'N 47°09'E
Rumaitha *see* Ar Rumaythah
Rumania/Rumänien *see* Romania
Rumänisch-Sankt-Georgen *see* Sângeorz-Băi
139 Y13 **Rumaylah** Al Başrah, SE Iraq 30°16'N 47°22'E
139 P2 **Rumaylah, Wādī** *dry watercourse* NE Syria
171 U13 **Rumbati** Papua Barat, E Indonesia 02°44'S 132°04'E
81 E14 **Rumbek** Lakes, C South Sudan 06°50'N 29°42'E
Rumburg *see* Rumburk
111 D14 **Rumburk** *Ger.* Rumburg. Ústecký Kraj, NW Czech Republic 50°58'N 14°35'E
44 J4 **Rum Cay** *island* C The Bahamas
99 M26 **Rumelange** Luxembourg, S Luxembourg 49°28'N 06°02'E
99 D20 **Rumes** Hainaut, SW Belgium 50°33'N 03°19'E
19 P7 **Rumford** Maine, NE USA 44°31'N 70°31'W
110 I6 **Rumia** Pomorskie, N Poland 54°36'N 18°21'E
113 J17 **Rumija** ▲ S Montenegro
103 T11 **Rumilly** Haute-Savoie, E France 45°52'N 05°57'E
139 O6 **Rūmiyah** Al Anbār, W Iraq 34°28'N 41°17'E
Rummah, Wādī ar *see* Rimah, Wādī ar
Rummelsburg in Pommern *see* Miastko
165 S3 **Rumoi** Hokkaidō, NE Japan 43°57'N 141°40'E
82 M12 **Rumphi** *var.* Rumpi. Northern, N Malawi 11°00'S 33°51'E
Rumpi *see* Rumphi
29 V7 **Rum River** Minnesota, N USA
188 F16 **Rumung** *island* Caroline Islands, W Micronesia
Rumuniya/Rumŭniya/Rumunjska *see* Romania
185 G16 **Runanga** West Coast, South Island, New Zealand 42°25'S 171°15'E
184 P7 **Runaway, Cape** *headland* North Island, New Zealand 37°33'S 177°59'E
97 K18 **Runcorn** C England, United Kingdom 53°20'N 02°44'W
118 K10 **Rundāni** *var.* Rundāni. E Latvia 56°19'N 27°51'E
Rundāni *see* Rundāni
83 L18 **Runde** *var.* Lundi. SE Zimbabwe
83 E16 **Rundu** *var.* Runtu. Okavango, NE Namibia 17°55'S 19°45'E
93 I16 **Rundvik** Västerbotten, N Sweden 63°31'N 19°22'E
81 G20 **Runere** Mwanza, N Tanzania 03°06'S 33°18'E
25 S13 **Runge** Texas, SW USA 28°52'N 97°42'W
167 Q13 **Rŭng, Kaôh** *prev.* Kas Rong. *island* SW Cambodia
79 O16 **Rungu** Orientale, NE Dem. Rep. Congo 03°11'N 27°52'E
81 F23 **Rungwa** Katavi, W Tanzania 07°18'S 31°40'E
81 G22 **Rungwa** Singida, C Tanzania 06°54'S 33°33'E
94 M14 **Runn** C Sweden
24 M4 **Running Water Draw** *valley* New Mexico/Texas, SW USA
Runö *see* Ruhnu
Runtu *see* Rundu
189 V12 **Ruo** *island* Caroline Islands, C Micronesia
158 L9 **Ruoqiang** *var.* Jo-ch'iang, *Uigh.* Charkhlik, Charkhliq, Qarkilik. Xinjiang Uygur Zizhiqu, NW China 38°59'N 88°08'E
159 S7 **Ruo Shui** N China
92 L8 **Ruostekfjelbmá** *var.* Rustefjelbma Finnmark. Finnmark, N Norway 70°25'N 28°10'E
93 L18 **Ruovesi** Pirkanmaa, W Finland 61°59'N 24°05'E
112 B9 **Rupa** Primorje-Gorski Kotar, NW Croatia 45°29'N 14°15'E
182 M11 **Rupanyup** Victoria, SE Australia 36°38'S 142°37'E
168 K9 **Rupat, Pulau** *prev.* Roepat. *island* W Indonesia
168 K10 **Rupat, Selat** *strait* Sumatera, W Indonesia
116 J11 **Rupea** *Ger.* Reps, *Hung.* Kőhalom; *prev.* Cohalm. Braşov, C Romania 46°02'N 25°13'E
99 G17 **Rupel** N Belgium
Rupella *see* la Rochelle
33 P15 **Rupert** Idaho, NW USA 42°37'N 113°40'W
21 R5 **Rupert** West Virginia, NE USA 37°57'N 80°40'W
Rupert House *see* Waskaganish
12 J10 **Rupert, Rivière de** Québec, C Canada
152 I8 **Rūpnagar** *var.* Ropar. Punjab, India
194 M13 **Ruppert Coast** *physical region* Antarctica
100 N11 **Ruppiner Kanal** *canal* NE Germany
55 S11 **Rupununi River** S Guyana
101 D16 **Rur** *Dut.* Roer. Germany/Netherlands
58 H13 **Rurópolis Presidente Medici** Pará, N Brazil 04°05'S 55°26'W
191 S12 **Rurutu** *island* Îles Australes, SW French Polynesia
Rusaddir *see* Melilla
83 L17 **Rusape** Manicaland, E Zimbabwe 18°32'S 32°07'E
Rusayris, Lake *see* Roseires, Reservoir
Ruschuk/Rusçuk *see* Ruse
114 K7 **Ruse** *var.* Ruschuk, Rustchuk, *Turk.* Rusçuk. Ruse, N Bulgaria 43°50'N 25°59'E
109 W10 **Ruše** NE Slovenia 46°31'N 15°30'E
114 L7 **Ruse** ◆ *province* N Bulgaria
114 K7 **Rusenski Lom** N Bulgaria
97 G17 **Rush** *Ir.* An Ros. Dublin, E Ireland 53°32'N 06°06'W
161 S4 **Rushan** *var.* Xiacun. Shandong, E China 36°55'N 121°26'E
Rushan *see* Rŭshon
Rushanskiy Khrebet *see* Rushon, Qatorkŭhi
29 V7 **Rush City** Minnesota, N USA 45°41'N 92°56'W
37 V5 **Rush Creek** Colorado, C USA
29 X10 **Rushford** Minnesota, N USA 43°48'N 91°45'W
154 N13 **Rushikulya** E India
14 D8 **Rush Lake** Ontario, S Canada
30 M7 **Rush Lake** Wisconsin, N USA
28 J10 **Rushmore, Mount** ▲ South Dakota, N USA 43°52'N 103°27'W
147 S13 **Rŭshon** *Rus.* Rushan. S Tajikistan 37°58'N 71°31'E
147 S14 **Rushon, Qatorkŭhi** *Rus.* Rushanskiy Khrebet. ▲▲ SE Tajikistan
26 M12 **Rush Springs** Oklahoma, C USA 34°46'N 97°57'W
45 V15 **Rushville** Trinidad, Trinidad and Tobago 10°07'N 61°03'W
30 J13 **Rushville** Illinois, N USA 40°07'N 90°33'W
28 K12 **Rushville** Nebraska, C USA 42°41'N 102°28'W
183 O11 **Rushworth** Victoria, SE Australia 36°36'S 145°03'E
25 W8 **Rusk** Texas, SW USA 31°49'N 95°11'W
93 I14 **Ruskele** Västerbotten, N Sweden 64°49'N 18°55'E
118 C12 **Rusnė** Klaipėda, W Lithuania 55°18'N 21°19'E
114 M10 **Rusokastrenska Reka** E Bulgaria
Russadir *see* Melilla
109 X3 **Russbach** NE Austria
11 V16 **Russell** Manitoba, S Canada 50°47'N 101°17'W
184 K2 **Russell** Northland, North Island, New Zealand 35°17'S 174°07'E
26 L4 **Russell** Kansas, C USA 38°54'N 98°51'W
21 O4 **Russell** Kentucky, S USA 38°30'N 82°43'W
20 L7 **Russell Springs** Kentucky, S USA 37°02'N 85°03'W
23 O2 **Russellville** Alabama, S USA 34°30'N 87°43'W
27 T11 **Russellville** Arkansas, C USA 35°17'N 93°06'W
20 J7 **Russellville** Kentucky, S USA 36°50'N 86°54'W
101 G18 **Rüsselsheim** Hessen, W Germany 50°00'N 08°25'E
Russia *see* Russian Federation
Russian America *see* Alaska
122 J11 **Russian Federation** *off.* Russian Federation, *var.* Russia, *Latv.* Krievija, *Rus.* Rossiyskaya Federatsiya. ◆ *republic* Asia/Europe
Russian Federation *see* Russian Federation
39 N11 **Russian Mission** Alaska, USA 61°48'N 161°23'W
34 M7 **Russian River** California, W USA
122 J5 **Russkaya Gavan'** Novaya Zemlya, Arkhangel'skaya Oblast', N Russian Federation 76°13'N 62°48'E
122 J5 **Russkiy, Ostrov** *island* N Russian Federation
109 Y5 **Rust** Burgenland, E Austria 47°48'N 16°42'E
Rustaq *see* Ar Rustāq
137 U10 **Rustavi** *prev.* Rust'avi. SE Georgia 41°36'N 45°00'E
Rust'avi *see* Rustavi
21 T7 **Rustburg** Virginia, NE USA 37°17'N 79°07'W
Rustchuk *see* Ruse
Rustefjelbma Finnmark *see* Ruostekfjelbmá
83 I21 **Rustenburg** North-West, N South Africa 25°40'S 27°15'E
22 H5 **Ruston** Louisiana, S USA 32°31'N 92°38'W
81 E21 **Rutana** SE Burundi 04°01'S 30°01'E
62 I4 **Rutana, Volcán** N Chile 22°43'S 67°52'W
Rutanzige, Lake *see* Edward, Lake
Rutba *see* Ar Ruṭbah
104 M14 **Rute** Andalucía, S Spain 37°20'N 04°23'W
171 N16 **Ruteng** *prev.* Roeteng. Flores, C Indonesia 08°35'S 120°28'E
194 L8 **Rutford Ice Stream** *ice feature* Antarctica
35 X6 **Ruth** Nevada, W USA 39°15'N 115°00'W
101 G15 **Rüthen** Nordrhein-Westfalen, W Germany 51°30'N 08°28'E
14 D17 **Rutherford** Ontario, S Canada 42°39'N 82°06'W
21 Q10 **Rutherfordton** North Carolina, SE USA 35°23'N 81°57'W
97 J18 **Ruthin** *Wel.* Rhuthun. NE Wales, United Kingdom 53°05'N 03°18'W
108 G7 **Rüti** Zürich, N Switzerland 47°16'N 08°51'E
Rutlam *see* Ratlām
18 M9 **Rutland** Vermont, NE USA 43°37'N 72°59'W
97 N19 **Rutland** *cultural region* C England, United Kingdom
21 N8 **Rutledge** Tennessee, S USA 36°16'N 83°31'W
158 G12 **Rutog** *var.* Rutög, Rutok. Xizang Zizhiqu, W China 33°27'N 79°43'E
Rutok *see* Rutog
79 P19 **Rutshuru** Nord-Kivu, E Dem. Rep. Congo 01°11'S 29°28'E
98 L8 **Rutten** Flevoland, N Netherlands 52°49'N 05°44'E
127 Q17 **Rutul** Respublika Dagestan, SW Russian Federation 41°35'N 47°30'E
93 L14 **Ruukki** Pohjois-Pohjanmaa, C Finland 64°40'N 25°35'E
98 N11 **Ruurlo** Gelderland, E Netherlands 52°04'N 06°27'E
143 S15 **Ru'ūs al Jibāl** *cape* Oman/United Arab Emirates
138 I7 **Ru'ūs aṭ Ṭiwāl, Jabal** ▲ W Syria
81 H23 **Ruvuma** ◆ *region* SE Tanzania

◆ Country ● Country Capital ◇ Dependent Territory ○ Dependent Territory Capital ◆ Administrative Regions ✈ International Airport ▲ Mountain ▲▲ Mountain Range Volcano River Lake Reservoir

Ruvuma – St. James, Cape

81 I25 Ruvuma var. Rio Rovuma. ≈ Mozambique/Tanzania see also Rovuma, Rio
Ruvuma see Rovuma, Rio
Ruwais see Ar Ruways
138 L9 Ruwayshid, Wadi ar dry watercourse NE Jordan
141 Z10 Ruways, Ra's ar headland E Oman 20°58′N 59°00′E
79 P18 Ruwenzori ⛰ Dem. Rep. Congo/Uganda
141 Y8 Ruwī NE Oman 23°33′N 58°31′E
114 F9 Ruy ▲ Bulgaria/Serbia 42°52′N 22°35′E
Ruya see Luia, Rio
81 E20 Ruyigi E Burundi 03°28′S 30°19′E
127 P5 Ruzayevka Respublika Mordoviya, W Russian Federation 54°04′N 44°56′E
119 G18 Ruzhany Brestskaya Voblasts', SW Belarus 52°52′N 24°53′E
Rŭzhevo Konare see Razhevo Konare
Ruzhin see Ruzhyn
114 G7 Ruzhintsi Vidin, NW Bulgaria 43°38′N 22°50′E
161 N6 Ruzhou Henan, C China 34°10′N 112°51′E
117 N5 Ruzhyn Rus. Ruzhin. Zhytomyrs'ka Oblast', N Ukraine 49°42′N 29°01′E
111 K19 Ružomberok Ger. Rosenberg, Hung. Rózsahegy. Žilinský Kraj, N Slovakia 49°04′N 19°19′E
111 C16 Ruzyně ✈ (Praha) Praha, C Czech Republic
81 D19 Rwanda off. Rwandese Republic; prev. Ruanda. ◆ republic C Africa
Rwandese Republic see Rwanda
95 G22 Ry Midtjylland, C Denmark 56°06′N 09°46′E
Ryasna see Rasna
126 L5 Ryazan' Ryazanskaya Oblast', W Russian Federation 54°37′N 39°37′E
126 L5 Ryazanskaya Oblast' ◈ province W Russian Federation
126 M6 Ryazhsk Ryazanskaya Oblast', W Russian Federation 53°42′N 40°09′E
118 B13 Rybachiy Ger. Rossitten. Kaliningradskaya Oblast', W Russian Federation 55°09′N 20°49′E
124 J2 Rybachiy, Poluostrov peninsula NW Russian Federation
Rybach'ye see Balykchy
124 L15 Rybinsk prev. Andropov. Yaroslavskaya Oblast', W Russian Federation 58°03′N 38°53′E
124 K14 Rybinskoye Vodokhranilishche Eng. Rybinsk Reservoir, Rybinsk Sea. ⊠ W Russian Federation
Rybinsk Reservoir/Rybinsk Sea see Rybinskoye Vodokhranilishche
111 I16 Rybnik Śląskie, S Poland 50°05′N 18°31′E
Rybnitsa see Rîbniţa
111 F16 Rychnov nad Kněžnou Ger. Reichenau. Královéhradecký Kraj, N Czech Republic 50°10′N 16°17′E
110 I12 Rychwał Wielkopolskie, C Poland 52°04′N 18°10′E
11 O13 Rycroft Alberta, W Canada 55°45′N 118°42′W
95 L21 Ryd Kronoberg, S Sweden 56°27′N 14°44′E
95 L20 Rydaholm Jönköping, S Sweden 56°57′N 14°19′E
194 I8 Rydberg Peninsula peninsula Antarctica
97 P23 Rye SE England, United Kingdom 50°57′N 00°42′E
33 T10 Ryegate Montana, NW USA 46°21′N 109°12′W
35 S3 Rye Patch Reservoir ⊠ Nevada, W USA
95 D15 Ryfylke physical region S Norway
95 H16 Rygge Østfold, S Norway 59°22′N 10°45′E
110 N13 Ryki Lubelskie, E Poland 51°38′N 21°57′E
Rykovo see Yenakiyeve
126 I7 Ryl'sk Kurskaya Oblast', W Russian Federation 51°34′N 34°41′E
183 S8 Rylstone New South Wales, SE Australia 32°48′S 149°58′E
111 H17 Rýmařov Ger. Römerstadt. Moravskoslezský Kraj, E Czech Republic 49°56′N 17°15′E
144 E11 Ryn-Peski desert W Kazakhstan
165 N10 Ryōtsu var. Ryôtu. Niigata, Sado, C Japan 38°06′N 138°28′E
Ryôtu see Ryōtsu
110 K10 Rypin Kujawsko-pomorskie, C Poland 53°03′N 19°25′E
Ryshkany see Rîşcani
Ryssel see Lille
Ryswick see Rijswijk
95 M24 Rytterknægten hill E Denmark
Ryukyu Islands see Nansei-shotō
192 G5 Ryukyu Trench var. Nansei Syotō Trench. undersea feature S East China Sea 24°45′N 128°00′E
110 D11 Rzepin Ger. Reppen. Lubuskie, W Poland 52°20′N 14°48′E
111 N16 Rzeszów Podkarpackie, SE Poland 50°03′N 22°01′E
124 I16 Rzhev Tverskaya Oblast', W Russian Federation 56°17′N 34°22′E
Rzhishchev see Rzhyshchiv
117 P5 Rzhyshchiv Rus. Rzhishchev. Kyyivs'ka Oblast', N Ukraine 49°58′N 31°02′E

S

138 E11 Sa'ad Southern, W Israel 31°27′N 34°31′E
109 P7 Saalach ≈ W Austria
101 L14 Saale ≈ C Germany
101 L17 Saalfeld var. Saalfeld an der Saale. Thüringen, C Germany 50°39′N 11°22′E
Saalfeld see Zalewo
Saalfeld an der Saale see Saalfeld
108 C8 Saane ≈ W Switzerland
101 D19 Saar Fr. Sarre. ≈ France/Germany
101 E20 Saarbrücken Fr. Sarrebruck. Saarland, SW Germany 49°13′N 07°01′E
Saarburg see Sarrebourg
118 D6 Sääre var. Sjar. Saaremaa, W Estonia 57°57′N 21°53′E
Saare see Saaremaa
118 D5 Saaremaa off. Saare Maakond. ◈ province W Estonia
118 E6 Saaremaa Ger. Oesel, Ösel; prev. Saare. island W Estonia
Saare Maakond see Saaremaa
92 L12 Saarenkylä Lappi, N Finland 66°31′N 25°51′E
Saargemund see Sarreguemines
93 L17 Saarijärvi Keski-Suomi, C Finland 62°42′N 25°16′E
Saar in Mähren see Žďár nad Sázavou
92 M10 Saariselkä Lapp. Suoločielgi. Lappi, N Finland 68°27′N 27°29′E
92 L10 Saariselkä hill range NE Finland
101 D20 Saarland Fr. Sarre. ◈ state SW Germany
Saarlautern see Saarlouis
101 D20 Saarlouis prev. Saarlautern. Saarland, SW Germany 49°19′N 06°45′E
108 E11 Saaser Vispa ≈ S Switzerland
137 X12 Saatlı Rus. Saatly. C Azerbaijan 39°57′N 48°24′E
Saatly see Saatlı
Saaz see Žatec
45 V9 Saba ◇ Dutch special municipality Sint Maarten
138 J7 Sab' Ābār var. Sab'a Biyar, Sa'b Bi'ār. Ḩimş, C Syria 33°46′N 37°41′E
Sab'a Biyar see Sab' Ābār
112 K11 Šabac Serbia, W Serbia 44°45′N 19°42′E
105 W5 Sabadell Cataluña, E Spain 41°33′N 02°07′E
164 K12 Sabae Fukui, Honshū, SW Japan 36°00′N 136°12′E
169 V7 Sabah prev. British North Borneo, North Borneo. ◈ state East Malaysia
168 J8 Sabak var. Sabak Bernam. Selangor, Peninsular Malaysia 03°45′N 100°59′E
Sabak Bernam see Sabak
38 D16 Sabak, Cape headland Agattu Island, Alaska, USA 52°21′N 173°43′E
81 J20 Sabaki ≈ S Kenya
142 L2 Sabalān, Kuhhā-ye ▲ NW Iran 38°21′N 47°47′E
154 H7 Sabalgarh Madhya Pradesh, C India 26°18′N 77°28′E
44 E4 Sabana, Archipiélago de island group C Cuba
42 H7 Sabanagrande var. Sabana Grande. Francisco Morazán, S Honduras 13°48′N 87°15′W
Sabana Grande see Sabanagrande
54 E5 Sabanalarga Atlántico, N Colombia 10°38′N 74°55′W
41 W14 Sabancuy Campeche, SE Mexico 18°58′N 91°11′W
45 N8 Sabaneta NW Dominican Republic 19°30′N 71°21′W
54 J4 Sabaneta Falcón, N Venezuela 11°17′N 70°00′W
188 H4 Sabaneta, Puntan prev. Ushi Point. headland Saipan, S Northern Mariana Islands 15°17′N 145°49′E
171 X14 Sabang Papua, E Indonesia 04°33′S 138°42′E
116 L10 Săbăoani Neamţ, NE Romania 47°01′N 26°51′E
155 J26 Sabaragamuwa ◈ province C Sri Lanka
Sabaria see Szombathely
154 D10 Sābarmati ≈ NW India
171 S10 Sabatai Pulau Morotai, E Indonesia 02°04′N 128°23′E
141 Q15 Sab'atayn, Ramlat as desert C Yemen
107 I16 Sabaudia Lazio, C Italy 41°17′N 13°02′E
57 J19 Sabaya Oruro, S Bolivia 19°09′S 68°21′W
Sa'b Bi'ār see Sab' Ābār
Sabbioncello see Orebić
148 J8 Šāberī, Hāmūn-e var. Daryācheh-ye Hāmun, Daryācheh-ye Sīstān. ◎ Afghanistan/Iran see also Sīstān, Daryācheh-ye
Šāberī, Hāmūn-e see Sīstān, Daryācheh-ye
27 P2 Sabetha Kansas, C USA 39°54′N 95°48′W
75 P10 Sabhā C Libya 27°02′N 14°26′E
67 V13 Sabi var. Save. ≈ Mozambique/Zimbabwe see also Save
Sabi see Save
118 E8 Sabile Ger. Zabeln. NW Latvia 57°03′N 22°33′E
31 R14 Sabina Ohio, N USA 39°29′N 83°38′W
40 I3 Sabinal Chihuahua, N Mexico 30°59′N 107°29′W
25 Q12 Sabinal Texas, SW USA 29°19′N 99°28′W
25 Q11 Sabinal River ≈ Texas, SW USA
105 S4 Sabiñánigo Aragón, NE Spain 42°31′N 00°22′W
41 N6 Sabinas Coahuila, NE Mexico 27°52′N 101°04′W
41 O8 Sabinas Hidalgo Nuevo León, NE Mexico 26°29′N 100°09′W
41 N6 Sabinas, Río ≈ NE Mexico
22 F9 Sabine Lake ◎ Louisiana/Texas, S USA
92 O3 Sabine Land physical region C Svalbard
25 W7 Sabine River ≈ Louisiana/Texas, SW USA
137 X12 Sabirabad C Azerbaijan 40°00′N 48°27′E
Sabkha see As Sabkhah
171 O4 Sablayan Mindoro, N Philippines 12°48′N 120°48′E
13 P16 Sable, Cape headland Newfoundland and Labrador, SE Canada 43°21′N 65°40′W
23 X17 Sable, Cape headland Florida, SE USA 25°12′N 81°06′W
13 R16 Sable Island island Nova Scotia, SE Canada
14 L11 Sables, Lac des ◎ Québec, SE Canada
14 E10 Sables, Rivière aux ≈ Ontario, S Canada
102 K7 Sable-sur-Sarthe Sarthe, NW France 47°49′N 00°19′W
125 U7 Sablya, Gora ▲ NW Russian Federation 64°46′N 58°52′E
77 U14 Sabon Birnin Gwari Kaduna, C Nigeria 10°43′N 06°39′E
77 V11 Sabon Kafi Zinder, C Niger 14°37′N 08°46′E
104 I6 Sabor, Rio ≈ N Portugal
14 J8 Sabourin, Lac ◎ Québec, SE Canada
137 Y10 Şabran prev. Däväçi. NE Azerbaijan 41°15′N 48°58′E
102 J14 Sabres Landes, SW France 44°07′N 00°46′W
195 X13 Sabrina Coast physical region Antarctica
140 M11 Sabt al Ulāyā 'Asīr, SW Saudi Arabia 19°33′N 41°58′E
104 H7 Sabugal Guarda, N Portugal 40°20′N 07°05′W
29 Z13 Sabula Iowa, C USA 42°04′N 90°10′W
141 N13 Şabyā Jīzān, SW Saudi Arabia 17°50′N 42°50′E
Sabzawar see Sabzevār
Sabzawaran see Jīroft
143 S4 Sabzevār var. Sabzawar. Khorāsān-e Raẕavī, NE Iran 36°13′N 57°38′E
Sabzvārān see Jīroft
Sacajawea Peak see Matterhorn
82 C9 Sacandica Uíge, NW Angola 06°01′S 15°57′E
42 A2 Sacatepéquez off. Departamento de Sacatepéquez. ◈ department S Guatemala
Sacatepéquez, Departamento de see Sacatepéquez
104 F11 Sacavém Lisboa, W Portugal 38°47′N 09°06′W
29 T13 Sac City Iowa, C USA 42°25′N 94°59′W
105 P8 Sacedón Castilla-La Mancha, C Spain 40°29′N 02°44′W
116 J12 Săcele Ger. Vierdörfer, Hung. Négyfalu; prev. Ger. Sieben Dörfer, Hung. Hétfalu. Braşov, C Romania 45°36′N 25°40′E
163 Y16 Sacheon Jap. Sansenhō; prev. Sach'ŏn, Samch'ŏnpŏ. S South Korea 34°55′N 128°07′E
12 C7 Sachigo ≈ Ontario, C Canada
12 C8 Sachigo Lake Ontario, C Canada 53°52′N 92°16′W
12 C8 Sachigo Lake ◎ Ontario, C Canada
Sach'ŏn see Sacheon
101 O15 Sachsen Eng. Saxony, Fr. Saxe. ◈ state E Germany
101 K14 Sachsen-Anhalt Eng. Saxony-Anhalt. ◈ state C Germany
109 R9 Sachsenburg Salzburg, S Austria 46°49′N 13°23′E
Sachsenfeld see Žalec
8 I5 Sachs Harbour var. Ikaahuk. Banks Island, Northwest Territories, N Canada 72°N 125°14′W
Sächsisch-Reen/Sächsisch-Regen see Reghin
18 H8 Sackets Harbor New York, NE USA 43°57′N 76°06′W
13 P14 Sackville New Brunswick, SE Canada 45°54′N 64°23′W
19 P9 Saco Maine, NE USA 43°32′N 70°25′W
19 P8 Saco River ≈ Maine/New Hampshire, NE USA
35 O7 Sacramento state capital California, W USA 38°35′N 121°30′W
37 T14 Sacramento Mountains ⛰ New Mexico, SW USA
35 N6 Sacramento River ≈ California, W USA
35 N5 Sacramento Valley valley California, W USA
36 I10 Sacramento Wash valley Arizona, SW USA
105 N15 Sacratif, Cabo headland S Spain 36°41′N 03°30′W
116 F9 Săcueni prev. Săcuieni, Hung. Székelyhid. Bihor, W Romania 47°20′N 22°05′E
Săcuieni see Săcueni
105 R4 Sádaba Aragón, NE Spain 42°15′N 01°16′W
Sá da Bandeira see Lubango
138 I6 Şadad Ḩimş, W Syria 34°19′N 36°52′E
141 O13 Şa'dah NW Yemen 16°59′N 43°45′E
167 O16 Sadao Songkhla, SW Thailand 06°39′N 100°30′E
142 L8 Sadd-e Dez, Daryācheh-ye ◎ W Iran
19 S3 Saddleback Mountain hill Maine, NE USA
19 P6 Saddleback Mountain ▲ Maine, NE USA 44°57′N 70°27′W
141 W13 Sadh S Oman 17°11′N 55°08′E
76 J11 Sadiola Kayes, W Mali 13°48′N 11°47′W
149 R12 Sādiqābād Punjab, E Pakistan 28°16′N 70°10′E
153 Y10 Sadiya Assam, NE India 27°49′N 95°38′E
139 W9 Sa'dīyah, Hawr as ◎ E Iraq
Sado see Sadoga-shima
165 N9 Sadoga-shima var. Sado. island C Japan
104 F12 Sado, Rio ≈ S Portugal
114 J8 Sadovets Pleven, N Bulgaria 43°19′N 24°21′E
114 J11 Sadovo Plovdiv, C Bulgaria 42°07′N 24°56′E
127 O11 Sadovoye Respublika Kalmykiya, SW Russian Federation 47°51′N 44°34′E
105 W9 Sa Dragonera var. Isla Dragonera. island Islas Baleares, Spain, W Mediterranean Sea
114 I10 Saedinenie var. Săedinenie, Sŭedinenie. Plovdiv, C Bulgaria 42°14′N 24°36′E
Săedinenie see Saedinenie
105 P9 Saelices Castilla-La Mancha, C Spain 39°55′N 02°49′W
Saena Julia see Siena
Saetabicula see Alzira
114 O12 Safaalan Tekirdağ, NW Turkey 41°26′N 28°07′E
Safad see Tsefat
Şafāqis see Sfax
143 P10 Şafāshahr var. Deh Bīd. Fārs, C Iran 30°50′N 53°50′E
192 I16 Safata Bay bay Upolu, Samoa, C Pacific Ocean
Safed see Tsefat
139 X11 Şaffāf, Ḩawr aş marshy lake S Iraq
95 J16 Säffle Värmland, C Sweden 59°08′N 12°55′E
37 N15 Safford Arizona, SW USA 32°46′N 109°41′W
74 E7 Safi W Morocco 32°19′N 09°14′W
126 I4 Safonovo Smolenskaya Oblast', W Russian Federation 55°05′N 33°12′E
136 H11 Safranbolu Karabük, NW Turkey 41°16′N 32°41′E
139 Y13 Safwān Al Başrah, SE Iraq 30°06′N 47°44′E
158 J16 Saga var. Gya'gya. Xizang Zizhiqu, W China 29°22′N 85°19′E
164 C14 Saga Saga, Kyūshū, SW Japan 33°13′N 130°16′E
164 C13 Saga off. Saga-ken. ◈ prefecture Kyūshū, SW Japan
165 P10 Sagae Yamagata, Honshū, C Japan 38°22′N 140°12′E
166 L3 Sagaing Sagaing, C Myanmar (Burma) 21°55′N 95°56′E
166 L5 Sagaing ◈ region N Myanmar (Burma)
Saga-ken see Saga
165 N13 Sagamihara Kanagawa, Honshū, S Japan 35°34′N 139°22′E
165 N14 Sagami-nada inlet SW Japan
Sagan see Żagań
29 Y3 Saganaga Lake ◎ Minnesota, N USA
155 F18 Sāgar Karnātaka, W India 14°09′N 75°02′E
154 I9 Sāgar prev. Saugor. Madhya Pradesh, C India 23°53′N 78°46′E
15 S8 Sagard Québec, SE Canada 48°01′N 70°03′W
Sagarmāthā see Everest, Mount
Sagebrush State see Nevada
143 V11 Sāghand Yazd, C Iran 32°33′N 55°12′E
19 N14 Sag Harbor Long Island, New York, NE USA 40°59′N 72°15′W
Saghez see Saqqez
31 R8 Saginaw Michigan, N USA 43°25′N 83°57′W
31 R8 Saginaw Bay lake bay Michigan, N USA
Sagiz see Sagyz
64 H6 Saglek Bank undersea feature W Labrador Sea
13 P5 Saglek Bay bay SW Labrador Sea
Saglouc/Sagluk see Salluit
103 X15 Sagonne, Golfe de gulf Corse, France, C Mediterranean Sea
105 P13 Sagra ▲ S Spain 37°59′N 02°33′W
104 F14 Sagres Faro, S Portugal 37°01′N 08°56′W
37 S7 Saguache Colorado, C USA 38°05′N 106°05′W
44 J7 Sagua de Tánamo Holguín, E Cuba 20°38′N 75°14′W
44 E5 Sagua la Grande Villa Clara, C Cuba 22°48′N 80°06′W
15 R7 Saguenay ≈ Québec, SE Canada
74 C9 Saguia al Hamra var. As Saqia al Hamra. ≈ N Western Sahara
105 S9 Sagunto Cat. Sagunt, Ar. Murviedro; anc. Saguntum. Valenciana, E Spain 39°40′N 00°17′W
Sagunt/Saguntum see Sagunto
144 H11 Sagyz prev. Sagiz. Atyrau, W Kazakhstan 48°12′N 54°56′E
138 H10 Saḩāb 'Ammān, NW Jordan 31°52′N 36°00′E
54 E6 Sahagún Córdoba, NW Colombia 08°58′N 75°30′W
104 L4 Sahagún Castilla y León, N Spain 42°23′N 05°02′W
141 X8 Saḩam N Oman 24°06′N 56°52′E
68 F9 Sahara desert Libya/Algeria
Sahara el Gharbîya see Şaḩrā' al Gharbīyah
75 X9 Sahara el Sharqîya var. Aş Şaḩrā' ash Sharqīyah, Eng. Arabian Desert, Eastern Desert. desert E Egypt
Saharan Atlas see Atlas Saharien
152 J9 Sahāranpur Uttar Pradesh, N India 29°58′N 77°33′E
64 L10 Saharan Seamounts var. Saharian Seamounts. undersea feature E Atlantic Ocean 25°00′N 20°00′W
Saharian Seamounts see Saharan Seamounts
153 Q13 Saharsa Bihār, NE India 25°54′N 86°36′E
67 O7 Sahel physical region C Africa
153 R14 Sāhibganj Jhārkhand, NE India 25°15′N 87°40′E
139 Q7 Sahīlīyah Al Anbār, C Iraq 33°43′N 42°42′E
138 I4 Sāḩilīyah, Jibāl as ⛰ NW Syria
114 M13 Şahin Tekirdağ, NW Turkey 41°01′N 26°51′E
149 U9 Sāhīwāl prev. Montgomery. Punjab, E Pakistan 30°40′N 73°05′E
149 U8 Sāhīwāl Punjab, E Pakistan 31°57′N 72°22′E
141 W11 Şaḩmah, Ramlat as desert C Oman
75 U9 Şaḩrā' al Gharbīyah var. Sahara el Gharbîya, Eng. Western Desert. desert C Egypt
139 T13 Şaḩrā' al Ḩijārah desert S Iraq
40 H5 Sahuaripa Sonora, NW Mexico 29°02′N 109°14′W
36 M16 Sahuarita Arizona, SW USA 31°24′N 110°55′W
40 L13 Sahuayo var. Sahuayo de José María Morelos; prev. Sahuayo de Díaz, Sahuayo de Porfirio Díaz. Michoacán, SW Mexico 20°05′N 102°42′W
Sahuayo de Díaz/Sahuayo de José María Morelos/Sahuayo de Porfirio Díaz see Sahuayo
173 W8 Sahul Shelf undersea feature N Timor Sea
167 P17 Sai Buri Pattani, SW Thailand 06°42′N 101°37′E
74 I6 Saïda NW Algeria 34°50′N 00°10′E
138 G7 Saïda var. Şaydā, Sayida; anc. Sidon. W Lebanon 33°20′N 35°24′E
Sa'īdābād see Sīrjān
80 B13 Sa'id Bundas Western Bahr el Ghazal, W South Sudan 08°24′N 24°53′E
186 E7 Saidor Madang, N Papua New Guinea 05°38′S 146°28′E
153 S13 Saidpur var. Syedpur. Rajshahi, NW Bangladesh 25°48′N 89°E
149 U5 Saīdu var. Mingora, Mongora; prev. Mingāora. Khyber Pakhtunkhwa, N Pakistan 34°45′N 72°21′E
108 C7 Saignelégier Jura, NW Switzerland 47°18′N 07°03′E
164 H11 Saigō Shimane, Dōgo, SW Japan 36°12′N 133°18′E
Saigon see Hô Chi Minh
163 P11 SaihanTal var. Sonid Youqi. Nei Mongol Zizhiqu, N China 42°45′N 112°36′E
162 I12 Saihan Toroi Nei Mongol Zizhiqu, N China 41°44′N 100°29′E
Sai Hun see Syr Darya
92 M11 Saija Lappi, NE Finland 67°07′N 28°46′E
164 G14 Saijō Ehime, Shikoku, SW Japan 33°55′N 133°10′E
164 E15 Saiki Ōita, Kyūshū, SW Japan 32°57′N 131°52′E
93 N18 Saimaa ◎ SE Finland
93 N18 Saimaa Canal Fin. Saimaan Kanava, Rus. Saymenskiy Kanal. canal Finland/Russian Federation
Saimaan Kanava see Saimaa Canal
40 L10 Saín Alto Zacatecas, C Mexico 23°36′N 103°14′W
96 L12 St Abb's Head headland SE Scotland, United Kingdom 55°54′N 02°07′W
11 Y16 St. Adolphe Manitoba, S Canada 49°39′N 96°55′W
103 O15 St-Affrique Aveyron, S France 43°57′N 02°52′E
15 Q10 St-Agapit Québec, SE Canada 46°22′N 71°37′W
97 O21 St Albans anc. Verulamium. E England, United Kingdom 51°46′N 00°21′W
18 L6 Saint Albans Vermont, NE USA 44°49′N 73°07′W
21 Q5 Saint Albans West Virginia, NE USA 38°21′N 81°47′W
St. Alban's Head see St Aldhelm's Head
11 Q14 St. Albert Alberta, SW Canada 53°38′N 113°38′W
97 M24 St Aldhelm's Head var. St. Alban's Head. headland S England, United Kingdom 50°34′N 02°04′W
15 O11 St-Alexis-des-Monts Québec, SE Canada 46°30′N 73°08′W
103 P2 St-Amand-les-Eaux Nord, N France 50°27′N 03°25′E
103 O9 St-Amand-Montrond var. St-Amand-Mont-Rond. Cher, C France 46°43′N 02°29′E
173 P16 St-André NE Réunion
14 M12 St-André-Avellin Québec, SE Canada 45°45′N 75°04′W
Saint-André, Cap see Vilanandro, Tanjona
102 K12 St-André-de-Cubzac Gironde, SW France 45°01′N 00°26′W
96 K11 St Andrews E Scotland, United Kingdom 56°20′N 02°49′W
23 Q9 Saint Andrews Bay bay Florida, SE USA
23 W7 Saint Andrew Sound sound Georgia, SE USA
Saint Anna Trough see Svyataya Anna Trough
44 J11 St. Ann's Bay C Jamaica 18°26′N 77°12′W
13 T10 St. Anthony Newfoundland and Labrador, SE Canada 51°22′N 55°34′W
33 R13 Saint Anthony Idaho, NW USA 43°56′N 111°38′W
182 M11 Saint Arnaud Victoria, SE Australia 36°39′S 143°15′E
185 I15 St.Arnaud Range ⛰ South Island, New Zealand
13 R10 St-Augustin Québec, E Canada 51°13′N 58°39′W
23 X9 Saint Augustine Florida, SE USA 29°54′N 81°19′W
97 H24 St Austell SW England, United Kingdom 50°21′N 04°47′W
103 T4 St-Avold Moselle, NE France 49°06′N 06°43′E
103 N17 St-Barthélemy ⛰ S France
102 L17 St-Béat Haute-Garonne, S France 42°55′N 00°39′E
97 I15 St Bees Head headland NW England, United Kingdom 54°30′N 03°39′W
173 P16 St-Benoit E Réunion
103 T13 St-Bonnet Hautes-Alpes, SE France 44°41′N 06°04′E
St.Botolph's Town see Boston
97 G21 St Brides Bay inlet SW Wales, United Kingdom
102 H5 St-Brieuc Côtes d'Armor, NW France 48°31′N 02°45′W
102 H5 St-Brieuc, Baie de bay NW France
102 L7 St-Calais Sarthe, NW France 47°55′N 00°48′E
15 Q10 St-Casimir Québec, SE Canada 46°40′N 72°05′W
14 H16 St. Catharines Ontario, S Canada 43°10′N 79°15′W
45 S14 St. Catherine, Mount ▲ N Grenada 12°10′N 61°41′W
64 C11 St Catherine Point headland E Bermuda
23 X6 Saint Catherines Island island Georgia, SE USA
97 M24 St Catherine's Point headland S England, United Kingdom 50°34′N 01°17′W
103 N13 St-Céré Lot, S France 44°52′N 01°53′E
108 A10 St. Cergue Vaud, W Switzerland 46°25′N 06°10′E
103 R11 St-Chamond Loire, E France 45°29′N 04°32′E
33 S16 Saint Charles Idaho, NW USA 42°05′N 111°23′W
27 X4 Saint Charles Missouri, C USA 38°48′N 90°29′W
103 P13 St-Chély-d'Apcher Lozère, S France 44°51′N 03°16′E
Saint Christopher and Nevis, Federation of see Saint Kitts and Nevis
Saint Christopher-Nevis see Saint Kitts and Nevis
31 S9 Saint Clair Michigan, N USA 42°49′N 82°29′W
183 O17 St. Clair, Lake ◎ Tasmania, SE Australia
14 C17 St. Clair, Lake var. Lac à L'Eau Claire. ◎ Canada/USA
31 S10 Saint Clair Shores Michigan, N USA 42°30′N 82°53′W
103 S10 St-Claude anc. Condate. Jura, E France 46°23′N 05°52′E
45 X6 St-Claude Basse Terre, SW Guadeloupe 16°02′N 61°42′W
23 X12 Saint Cloud Florida, SE USA 28°15′N 81°15′W
29 U8 Saint Cloud Minnesota, N USA 45°34′N 94°10′W
45 T9 Saint Croix island S Virgin Islands (US)
30 J4 Saint Croix Flowage ⊠ Wisconsin, N USA
19 T5 Saint Croix River ≈ Canada/USA
29 W7 Saint Croix River ≈ Minnesota/Wisconsin, N USA
45 S14 St.David's SE Grenada 12°01′N 61°40′W
97 H21 St David's SW Wales, United Kingdom 51°53′N 05°16′W
97 G21 St David's Head headland SW Wales, United Kingdom 51°54′N 05°19′W
64 C12 St David's Island island E Bermuda
173 O16 St-Denis ○ (Réunion) NW Réunion 20°55′S 14°34′E
103 U6 St-Dié Vosges, NE France 48°17′N 06°57′E
103 R5 St-Dizier anc. Desiderii Fanum. Haute-Marne, N France 48°39′N 05°00′E
15 N11 St-Donat Québec, SE Canada 46°19′N 74°15′W
15 N11 Ste-Adèle Québec, SE Canada 45°58′N 74°10′W
15 N11 Ste-Agathe-des-Monts Québec, SE Canada 46°03′N 74°19′W
11 Y16 Ste. Anne Manitoba, S Canada 49°40′N 96°40′W
45 R12 Ste-Anne Grande Terre, E Guadeloupe 16°13′N 61°23′W
45 Y6 Ste-Anne SE Martinique 14°26′N 60°53′W
15 Q10 Ste-Anne ≈ Québec, SE Canada
172 I16 Sainte Anne island Inner Islands, NE Seychelles
15 W6 Ste-Anne-des-Monts Québec, SE Canada 49°07′N 66°29′W
14 M10 Ste-Anne-du-Lac Québec, SE Canada 46°51′N 75°20′W
15 S10 Ste-Apolline Québec, SE Canada 46°47′N 70°15′W
15 R10 Ste-Claire Québec, SE Canada 46°30′N 70°40′W
15 Q10 Ste-Croix Québec, SE Canada 46°36′N 71°42′W
108 B8 Ste. Croix Vaud, SW Switzerland 46°50′N 06°31′E
103 P14 Ste-Énimie Lozère, S France 44°21′N 03°26′E
27 Y6 Sainte Genevieve Missouri, C USA 37°57′N 90°01′W
103 S12 St-Egrève Isère, E France 45°15′N 05°41′E
39 T12 Saint Elias, Cape headland Kayak Island, Alaska, USA 59°48′N 144°36′W
39 U11 Saint Elias, Mount ▲ Alaska, USA 60°18′N 140°57′W
10 G8 Saint Elias Mountains ⛰ Canada/USA
55 Y10 St-Élie N French Guiana 04°50′N 53°21′W
103 O10 St-Eloy-les-Mines Puy-de-Dôme, C France 46°07′N 02°50′E
15 R10 Ste-Marie Québec, SE Canada 46°28′N 71°00′W
45 Q11 Ste-Marie NE Martinique 14°47′N 61°00′W
173 P16 Ste-Marie NE Réunion
103 U6 Ste-Marie-aux-Mines Haut-Rhin, NE France 48°16′N 07°12′E
Sainte Marie, Cap see Vohimena, Tanjona
102 L8 Ste-Maure-de-Touraine Indre-et-Loire, C France 47°06′N 00°38′E
103 R4 Ste-Menehould Marne, NE France 49°06′N 04°54′E
Ste-Perpétue see Ste-Perpétue-de-l'Islet
15 S9 Ste-Perpétue-de-l'Islet var. Ste-Perpétue. Québec, SE Canada 47°02′N 69°58′W
45 X11 Ste-Rose Basse Terre, N Guadeloupe 16°20′N 61°42′W
173 P16 Ste-Rose E Réunion
11 W15 Ste. Rose du Lac Manitoba, S Canada 51°04′N 99°31′W
102 J11 Saintes anc. Mediolanum. Charente-Maritime, W France 45°45′N 00°37′W
45 X7 Saintes, Canal des channel SW Guadeloupe
Saintes, Îles des see les Saintes
173 P16 Ste-Suzanne N Réunion
15 P10 Ste-Thècle Québec, SE Canada 46°48′N 72°31′W
103 Q12 St-Étienne Loire, E France 45°26′N 04°23′E
102 M4 St-Étienne-du-Rouvray Seine-Maritime, N France 49°22′N 01°07′E
Saint Eustatius see Sint Eustatius
15 Q10 Ste-Véronique Québec, SE Canada 46°30′N 74°58′W
15 P7 St-Félicien Québec, SE Canada 48°38′N 72°29′W
15 O11 St-Félix-de-Valois Québec, SE Canada 46°10′N 73°26′W
103 Y14 St-Florent Corse, France, C Mediterranean Sea 42°41′N 09°19′E
103 Y14 St-Florent, Golfe de gulf Corse, France, C Mediterranean Sea
103 P6 St-Florentin Yonne, C France 48°00′N 03°46′E
103 N9 St-Florent-sur-Cher Cher, C France 47°00′N 02°13′E
103 P12 St-Flour Cantal, C France 45°02′N 03°05′E
26 H2 Saint Francis Kansas, C USA 39°45′N 101°31′W
83 H26 St. Francis, Cape headland S South Africa 34°11′S 24°45′E
27 X10 Saint Francis River ≈ Arkansas/Missouri, C USA
22 J8 Saint Francisville Louisiana, S USA 30°46′N 91°22′W
45 Y6 St-François Grande Terre, E Guadeloupe 16°15′N 61°17′W
15 Q12 St-François ≈ Québec, SE Canada
27 X7 Saint Francois Mountains ⛰ Missouri, C USA
St-Gall/Saint Gall/St. Gallen see Sankt Gallen
St-Gall see St-Gall/Saint Gall/St.Gallen
102 L16 St-Gaudens Haute-Garonne, S France 43°07′N 00°43′E
15 R12 St-Gédéon Québec, SE Canada 45°51′N 70°36′W
181 X10 Saint George Queensland, E Australia 28°05′S 148°40′E
64 B12 St George N Bermuda 32°24′N 64°42′W
38 K15 Saint George Saint George Island, Alaska, USA 56°34′N 169°30′W
21 S14 Saint George South Carolina, SE USA 33°12′N 80°34′W
36 J8 Saint George Utah, W USA 37°06′N 113°35′W
13 R12 St. George, Cape headland Newfoundland and Labrador, E Canada 48°26′N 59°17′W
186 I6 St. George, Cape headland New Ireland, NE Papua New Guinea 04°49′S 152°52′E
38 J15 Saint George Island island Pribilof Islands, Alaska, USA
23 S10 Saint George Island island Florida, SE USA
99 J19 Saint-Georges Liège, E Belgium 50°36′N 05°20′E
15 R11 St-Georges Québec, SE Canada 46°08′N 70°40′W
55 Z11 St-Georges E French Guiana 03°55′N 51°49′W
45 R14 St. George's ● (Grenada) SW Grenada 12°04′N 61°45′W
13 R12 St. George's Bay inlet Newfoundland and Labrador, E Canada
97 G21 Saint George's Channel channel Ireland/Wales, United Kingdom
186 H6 St. George's Channel channel NE Papua New Guinea
64 B11 St George's Island island E Bermuda
99 I21 Saint-Gérard Namur, S Belgium 50°20′N 04°47′E
St-Germain see St-Germain-en-Laye
15 P12 St-Germain-de-Grantham Québec, SE Canada 45°49′N 72°32′W
103 N5 St-Germain-en-Laye var. St-Germain. Yvelines, N France 48°53′N 02°04′E
102 H8 St-Gildas, Pointe du headland NW France 47°09′N 02°25′W
103 R15 St-Gilles Gard, S France 43°41′N 04°24′E
102 I9 St-Gilles-Croix-de-Vie Vendée, NW France 46°41′N 01°55′W
173 O16 St-Gilles-les-Bains W Réunion 21°02′S 55°14′E
102 M16 St-Girons Ariège, S France 42°58′N 01°07′E
Saint Gotthard see Szentgotthárd
108 G9 St. Gotthard Tunnel tunnel Ticino, S Switzerland
97 H22 St Govan's Head headland SW Wales, United Kingdom 51°35′N 04°55′W
34 M7 Saint Helena California, W USA 38°29′N 122°30′W
67 O12 Saint Helena island C Atlantic Ocean
Saint Helena see Saint Helena, Ascension and Tristan da Cunha
65 F24 Saint Helena, Ascension and Tristan da Cunha terr. Saint Helena, Ascension, Tristan da Cunha. ◇ UK overseas territory C Atlantic Ocean
65 M16 Saint Helena Fracture Zone tectonic feature C Atlantic Ocean
34 M7 Saint Helena, Mount ▲ California, W USA 38°40′N 122°39′W
21 S15 Saint Helena Sound inlet South Carolina, SE USA
31 Q7 Saint Helen, Lake ◎ Michigan, N USA
183 Q16 Saint Helens Tasmania, SE Australia 41°21′S 148°15′E
97 K18 St Helens NW England, United Kingdom 53°28′N 02°44′W
32 G10 Saint Helens Oregon, NW USA 45°54′N 122°50′W
32 H10 Saint Helens, Mount 🌋 Washington, NW USA 46°24′N 121°49′W
97 L26 St Helier ○ (Jersey) S Jersey, Channel Islands 49°12′N 02°07′W
15 S9 St-Hilarion Québec, SE Canada 47°34′N 70°24′W
99 K22 Saint-Hubert Luxembourg, SE Belgium 50°02′N 05°23′E
15 P12 St-Hyacinthe Québec, SE Canada 45°38′N 72°57′W
St.Iago de la Vega see Spanish Town
31 Q4 Saint Ignace Michigan, N USA 45°53′N 84°44′W
15 O10 St-Ignace-du-Lac Québec, SE Canada 46°43′N 73°49′W
12 D12 St. Ignace Island island Ontario, S Canada
108 C7 St. Imier Bern, W Switzerland 47°09′N 06°55′E
97 G25 St Ives SW England, United Kingdom 50°12′N 05°29′W
29 U10 Saint James Minnesota, N USA 44°00′N 94°36′W
10 I15 St. James, Cape headland Graham Island, British Columbia, SW Canada 51°57′N 131°04′W

◆ Country
● Country Capital
◇ Dependent Territory
○ Dependent Territory Capital
◈ Administrative Regions
✈ International Airport
▲ Mountain
⛰ Mountain Range
🌋 Volcano
≈ River
◎ Lake
⊠ Reservoir

ST-JEAN – SALZBURG ALPS

15 O13 **St-Jean** *var.* St-Jean-sur-Richelieu. Québec, SE Canada 45°15′N 73°16′W
55 X9 **St-Jean** NW French Guiana 05°25′N 54°05′W
St-Jean-d'Acre *see* Akko
102 K11 **St-Jean-d'Angély** Charente-Maritime, W France 45°57′N 00°31′W
103 N7 **St-Jean-de-Braye** Loiret, C France 47°54′N 01°58′E
102 I16 **St-Jean-de-Luz** Pyrénées-Atlantiques, SW France 43°24′N 01°40′W
103 T12 **St-Jean-de-Maurienne** Savoie, E France 45°17′N 06°21′E
102 I9 **St-Jean-de-Monts** Vendée, NW France 46°45′N 02°00′W
103 Q14 **St-Jean-du-Gard** Gard, S France 44°06′N 03°49′E
15 Q7 **St-Jean, Lac** ◎ Québec, SE Canada
102 I16 **St-Jean-Pied-de-Port** Pyrénées-Atlantiques, SW France 43°10′N 01°14′W
15 S9 **St-Jean-Port-Joli** Québec, SE Canada 47°13′N 70°16′W
St-Jean-sur-Richelieu *see* St-Jean
15 N12 **St-Jérôme** Québec, SE Canada 45°47′N 74°01′W
25 T5 **Saint Jo** Texas, SW USA 33°42′N 97°33′W
13 O15 **St. John** New Brunswick, SE Canada 45°16′N 66°03′W
26 L6 **Saint John** Kansas, C USA 37°59′N 98°44′W
19 Q2 **Saint John** *Fr.* Saint-John. ≈ Canada/USA
76 K16 **Saint John** ≈ C Liberia
45 T9 **Saint John** *island* C Virgin Islands (US)
Saint-John *see* Saint John
22 I6 **Saint John, Lake** ◎ Louisiana, S USA
45 W10 **St John's** ● (Antigua and Barbuda) Antigua, Antigua and Barbuda 17°06′N 61°50′W
13 V12 **St. John's** *province capital* Newfoundland and Labrador, E Canada 47°34′N 52°41′W
37 O12 **Saint Johns** Arizona, SW USA 34°28′N 109°22′W
31 Q9 **Saint Johns** Michigan, N USA 43°01′N 84°31′W
13 V12 **St. John's** ✈ Newfoundland and Labrador, E Canada 47°22′N 52°45′W
23 X11 **Saint Johns River** ≈ Florida, SE USA
103 Q11 **Saint-Just-St-Rambert** Loire, E France 45°30′N 04°13′E
45 N12 **St. Joseph** W Dominica 15°24′N 61°26′W
173 P17 **St-Joseph** S Réunion
22 J6 **Saint Joseph** Louisiana, S USA 31°56′N 91°14′W
31 O10 **Saint Joseph** Michigan, N USA 42°05′N 86°30′W
27 R3 **Saint Joseph** Missouri, C USA 39°46′N 94°49′W
20 I10 **Saint Joseph** Tennessee, S USA 35°02′N 87°29′W
22 R9 **Saint Joseph Bay** *bay* Florida, SE USA
15 R11 **St-Joseph-de-Beauce** Québec, SE Canada 46°20′N 70°52′W
12 C10 **St. Joseph, Lake** ◎ Ontario, C Canada
31 Q11 **Saint Joseph River** ≈ N USA
14 C11 **Saint Joseph's Island** *island* Ontario, S Canada
15 N11 **St-Jovite** Québec, SE Canada 46°07′N 74°35′W
St. Julian's *see* San Ġiljan
St-Julien *see* St-Julien-en-Genevois
103 T10 **St-Julien-en-Genevois** *var.* St-Julien. Haute-Savoie, E France 46°07′N 06°06′E
102 M11 **St-Junien** Haute-Vienne, C France 45°52′N 00°54′E
96 D8 **St Kilda** *island* NW Scotland, United Kingdom
45 V10 **Saint Kitts** *island* Saint Kitts and Nevis
45 U10 **Saint Kitts and Nevis** *off.* Federation of Saint Christopher and Nevis, *var.* Saint Christopher-Nevis. ◆ *commonwealth republic* E West Indies
11 X16 **St. Laurent** Manitoba, S Canada 50°20′N 97°55′W
St-Laurent *see* St-Laurent-du-Maroni
55 X9 **St-Laurent-du-Maroni** *var.* St-Laurent. NW French Guiana 05°29′N 54°03′W
St-Laurent, Fleuve *see* St. Lawrence
102 J12 **St-Laurent-Médoc** Gironde, SW France 45°11′N 00°50′W
13 N12 **St. Lawrence** *Fr.* Fleuve St-Laurent. ≈ Canada/USA
13 Q12 **St. Lawrence, Gulf of** *gulf* NW Atlantic Ocean
38 K10 **Saint Lawrence Island** *island* Alaska, USA
14 M14 **Saint Lawrence River** ≈ Canada/USA
99 L25 **Saint-Léger** Luxembourg, SE Belgium 49°36′N 05°39′E
13 N14 **St. Léonard** New Brunswick, SE Canada 47°10′N 67°55′W
15 P11 **St-Léonard** Québec, SE Canada 46°06′N 72°18′W
173 O17 **St-Leu** W Réunion 21°09′S 55°17′E
102 J4 **St-Lô** *anc.* Briovera, Laudus. Manche, N France 49°07′N 01°08′W
11 T15 **St. Louis** Saskatchewan, S Canada 52°50′N 105°43′W
103 V7 **St-Louis** Haut-Rhin, NE France 47°35′N 07°34′E
173 O17 **St-Louis** S Réunion
76 G10 **Saint Louis** NW Senegal 15°59′N 16°30′W
27 X4 **Saint Louis** Missouri, C USA 38°38′N 90°15′W
29 W5 **Saint Louis River** ≈ Minnesota, N USA
103 T7 **St-Loup-sur-Semouse** Haute-Saône, E France 47°53′N 06°15′E
15 O12 **St-Luc** Québec, SE Canada 45°19′N 73°18′W
45 X13 **Saint Lucia** ◆ *commonwealth republic* SE West Indies
47 S3 **Saint Lucia** *island* SE West Indies
83 L22 **St. Lucia, Cape** *headland* E South Africa 28°29′S 32°26′E
45 Y13 **Saint Lucia Channel** *channel* Martinique/Saint Lucia
23 Y14 **Saint Lucie Canal** *canal* Florida, SE USA
23 Z13 **Saint Lucie Inlet** *inlet* Florida, SE USA
96 L2 **St Magnus Bay** *bay* N Scotland, United Kingdom
102 K10 **St-Maixent-l'École** Deux-Sèvres, W France 46°24′N 00°13′W
11 Y16 **St. Malo** Manitoba, S Canada 49°16′N 96°58′W
102 I5 **St-Malo** Ille-et-Vilaine, NW France 48°39′N 02°00′W
102 H4 **St-Malo, Golfe de** *gulf* NW France
44 L9 **St-Marc** C Haiti 19°08′N 72°41′W
44 L9 **St-Marc, Canal de** *channel* W Haiti
103 S12 **St-Marcellin-le-Mollard** Isère, E France 45°12′N 05°18′E
55 Y12 **Saint-Marcel, Mont** ▲ S French Guiana 2°32′N 53°00′E
96 K5 **St Margaret's Hope** NE Scotland, United Kingdom 58°50′N 02°57′W
32 M9 **Saint Maries** Idaho, NW USA 47°19′N 116°37′W
23 T9 **Saint Marks** Florida, SE USA 30°09′N 84°12′W
108 D11 **St. Martin** Valais, SW Switzerland 46°09′N 07°27′E
Saint Martin *see* Sint Maarten
31 O5 **Saint Martin Island** *island* Michigan, N USA
22 I9 **Saint Martinville** Louisiana, S USA 30°09′N 91°51′W
185 E20 **St. Mary, Mount** ▲ South Island, New Zealand 44°16′S 169°42′E
186 E8 **St. Mary, Mount** ▲ S Papua New Guinea 08°06′S 147°00′E
182 I6 **Saint Mary Peak** ▲ South Australia 31°25′S 138°39′E
183 Q16 **Saint Marys** Tasmania, SE Australia 41°34′S 148°13′E
14 E16 **St. Marys** Ontario, S Canada 43°15′N 81°08′W
38 M11 **Saint Marys** Alaska, USA 62°03′N 163°10′W
23 W8 **Saint Marys** Georgia, SE USA 30°44′N 81°30′W
27 P4 **Saint Marys** Kansas, C USA 39°09′N 96°00′W
31 Q4 **Saint Marys** Ohio, N USA 40°31′N 84°22′W
21 R3 **Saint Marys** West Virginia, NE USA 39°24′N 81°13′W
23 W8 **Saint Marys River** ≈ Florida/Georgia, SE USA
31 Q4 **Saint Marys River** ≈ Michigan, N USA
102 D6 **St-Mathieu, Pointe** *headland* NW France 48°17′N 04°56′W
38 J12 **Saint Matthew Island** *island* Alaska, USA
21 R13 **Saint Matthews** South Carolina, SE USA 33°40′N 80°44′W
St.Matthew's Island *see* Zadetkyi Kyun
186 G4 **St.Matthias Group** *island group* NE Papua New Guinea
108 C11 **St. Maurice** Valais, SW Switzerland 46°09′N 07°28′E
15 P9 **St-Maurice** ≈ Québec, SE Canada
102 J13 **St-Médard-en-Jalles** Gironde, SW France 44°54′N 00°43′W
39 N10 **Saint Michael** Alaska, USA 63°28′N 162°02′W
15 N10 **St-Michel-des-Saints** Québec, SE Canada 46°39′N 73°54′W
103 S5 **St-Mihiel** Meuse, NE France 48°57′N 05°33′E
108 J10 **St. Moritz** *Ger.* Sankt Moritz, *Rmsch.* San Murezzan. Graubünden, SE Switzerland 46°30′N 09°51′E
102 H8 **St-Nazaire** Loire-Atlantique, NW France 47°17′N 02°12′W
Saint Nicholas *see* São Nicolau
Saint-Nicolas *see* Sint-Niklaas
103 N1 **St-Omer** Pas-de-Calais, N France 50°45′N 02°15′E
102 J11 **Saintonge** *cultural region* W France
15 S9 **St-Pacôme** Québec, SE Canada 47°22′N 69°56′W
15 S10 **St-Pamphile** Québec, SE Canada 46°57′N 69°46′W
15 S9 **St-Pascal** Québec, SE Canada 47°32B 69°48′W
14 J11 **St-Patrice, Lac** ◎ Québec, SE Canada
11 R14 **St. Paul** Alberta, SW Canada 54°00′N 111°18′W
173 O16 **St-Paul** NW Réunion
38 K14 **Saint Paul** Saint Paul Island, Alaska, USA 57°08′N 170°13′W
29 V8 **Saint Paul** *state capital* Minnesota, N USA 45°N 93°10′W
29 P15 **Saint Paul** Nebraska, C USA 41°13′N 98°26′W
21 P7 **Saint Paul** Virginia, NE USA 36°53′N 82°18′W
77 Q17 **Saint Paul, Cape** *headland* S Ghana 05°44′N 00°55′E
103 O17 **St-Paul-de-Fenouillet** Pyrénées-Orientales, S France 42°49′N 02°29′E
65 K14 **Saint Paul Fracture Zone** *tectonic feature* E Atlantic Ocean
38 J14 **Saint Paul Island** *island* Pribilof Islands, Alaska, USA
102 J15 **St-Paul-les-Dax** Landes, SW France 43°45′N 01°01′W
21 U11 **Saint Pauls** North Carolina, SE USA 34°45′N 78°56′W
Saint Paul's Bay *see* San Pawl il Baħar
191 R16 **St Paul's Point** *headland* Pitcairn Island, Pitcairn Islands
29 U10 **Saint Peter** Minnesota, N USA 44°21′N 93°58′W
97 L26 **St Peter Port** ○ (Guernsey) C Guernsey, Channel Islands 49°28′N 02°33′W
23 V13 **Saint Petersburg** Florida, SE USA 27°47′N 82°37′W
Saint Petersburg *see* Sankt-Peterburg
23 V13 **Saint Petersburg Beach** Florida, SE USA 27°43′N 82°43′W
173 P17 **St-Philippe** SE Réunion 21°21′S 55°46′E
45 Q11 **St-Pierre** NW Martinique 14°44′N 61°11′W
173 O17 **St-Pierre** SW Réunion
13 S13 **St-Pierre and Miquelon** *Fr.* Îles St-Pierre et Miquelon. ◇ *French overseas collectivity* NE North America
15 P11 **St-Pierre, Lac** ◎ Québec, SE Canada
102 F5 **St-Pol-de-Léon** Finistère, NW France 48°42′N 04°00′W
103 O2 **St-Pol-sur-Ternoise** Pas-de-Calais, N France 50°22′N 02°21′E
St. Pons *see* St-Pons-de-Thomières
103 O16 **St-Pons-de-Thomières** *var.* St. Pons. Hérault, S France 43°28′N 02°48′E
103 P10 **St-Pourçain-sur-Sioule** Allier, C France 46°19′N 03°16′E
15 S11 **St-Prosper** Québec, SE Canada 46°14′N 70°28′W
103 P3 **St-Quentin** Aisne, N France 49°51′N 03°17′E
15 R10 **St-Raphaël** Québec, SE Canada 46°47′N 70°46′W
103 U15 **St-Raphaël** Var, SE France 43°26′N 06°46′E
15 Q10 **St-Raymond** Québec, SE Canada 46°53′N 71°49′W
33 O9 **Saint Regis** Montana, NW USA 47°18′N 115°06′W
18 J7 **Saint Regis River** ≈ New York, NE USA
103 R15 **St-Rémy-de-Provence** Bouches-du-Rhône, SE France 43°48′N 04°49′E
102 M9 **St-Savin** Vienne, W France 46°34′N 00°53′E
Saint-Sébastien,Cap *see* Anorontany, Tanjona
23 X7 **Saint Simons Island** *island* Georgia, SE USA
191 Y2 **Saint Stanislas Bay** *bay* Kiritimati, E Kiribati
13 O15 **St. Stephen** New Brunswick, SE Canada 45°12′N 67°18′W
45 V9 **St Barthélemy** ◇ *French overseas collectivity* E Caribbean Sea
39 X12 **Saint Terese** Alaska, USA 58°28′N 134°46′W
14 E17 **St. Thomas** Ontario, S Canada 42°46′N 81°12′W
29 Q2 **Saint Thomas** North Dakota, N USA 48°37′N 97°28′W
45 T9 **Saint Thomas** *island* W Virgin Islands (US)
Saint Thomas *see* São Tomé, Sao Tome and Principe
Saint Thomas *see* Charlotte Amalie, Virgin Islands (US)
15 P10 **St-Tite** Québec, SE Canada 46°42′N 72°32′W
45 V9 **St Martin** ◇ *French overseas collectivity* E Caribbean Sea
Saint-Trond *see* Sint-Truiden
103 U16 **St-Tropez** Var, SE France 43°16′N 06°39′E
Saint Ubes *see* Setúbal
102 L3 **St-Valéry-en-Caux** Seine-Maritime, N France 49°53′N 00°42′E
103 Q9 **St-Vallier** Saône-et-Loire, C France 46°39′N 04°19′E
106 B7 **St-Vincent** Valle d'Aosta, NW Italy 45°47′N 07°42′E
45 Q14 **Saint Vincent** *island* N Saint Vincent and the Grenadines
Saint Vincent *see* São Vicente
45 W14 **Saint Vincent and the Grenadines** ◆ *commonwealth republic* SE West Indies
Saint-Vincent, Cap *see* Ankaboa, Tanjona
Saint Vincent, Cape *see* São Vicente, Cabo de
102 I15 **St-Vincent-de-Tyrosse** Landes, SW France 43°39′N 01°16′W
182 I9 **Saint Vincent, Gulf** *gulf* South Australia
23 R10 **Saint Vincent Island** *island* Florida, SE USA
45 T12 **Saint Vincent Passage** *passage* Saint Lucia/Saint Vincent and the Grenadines
183 N18 **Saint Vincent, Point** *headland* Tasmania, SE Australia 43°19′S 145°50′E
Saint-Vith *see* Sankt-Vith
11 S14 **St. Walburg** Saskatchewan, S Canada 53°38′N 109°12′W
St Wolfgangsee *see* Wolfgangsee
102 M11 **St-Yrieix-la-Perche** Haute-Vienne, C France 45°31′N 01°12′E
Saint Yves *see* Setúbal
188 H5 **Saipan** *island* ● (Northern Mariana Islands) S Northern Mariana Islands
188 H6 **Saipan Channel** *channel* S Northern Mariana Islands
188 H6 **Saipan International** ✈ Saipan, S Northern Mariana Islands
74 G6 **Saïs** ✈ (Fès) C Morocco 33°58′N 04°48′W
Saishū *see* Jeju-do
Saishū *see* Jeju
102 J16 **Saison** ≈ SW France
169 R10 **Sai, Sungai** ≈ Borneo, N Indonesia
165 N13 **Saitama** *off.* Saitama-ken. ◈ *prefecture* Honshū, S Japan
Saitama *see* Urawa
Saitama-ken *see* Saitama
Saiyid Abid *see* Sayyid 'Abid
57 J19 **Sajama, Nevado** ▲ W Bolivia 17°57′S 68°51′W
141 V13 **Sājir, Ras** *headland* S Oman 16°42′N 53°40′E
111 M20 **Sajószentpéter** Borsod-Abaúj-Zemplén, NE Hungary 48°13′N 20°43′E
83 F24 **Sak** ≈ SW South Africa
81 J18 **Saka** Tana River, E Kenya 0°11′S 39°27′E
167 P11 **Sa Kaeo** Prachin Buri, C Thailand 13°47′N 102°03′E
164 J14 **Sakai** Ōsaka, Honshū, SW Japan 34°35′N 135°28′E
164 H14 **Sakaide** Kagawa, Shikoku, SW Japan 34°19′N 133°50′E
164 H12 **Sakaiminato** Tottori, Honshū, SW Japan 35°34′N 133°12′E
140 M3 **Sakākah** Al Jawf, NW Saudi Arabia 29°56′N 40°10′E
28 L4 **Sakakawea, Lake** ⊠ North Dakota, N USA
12 J9 **Sakami, Lac** ◎ Québec, C Canada
79 O26 **Sakania** Katanga, SE Dem. Rep. Congo 12°44′S 28°34′E
146 K12 **Sakar** Lebap Welaýaty, E Turkmenistan 38°57′N 63°46′E
172 H7 **Sakaraha** Toliara, SW Madagascar 22°54′S 44°31′E
146 I14 **Sakarçäge** *var.* Sakarchäge, *Rus.* Sakar-Chaga. Mary Welaýaty, C Turkmenistan 37°40′N 61°33′E
Sakar-Chaga/Sakarchäge *see* Sakarçäge
Sak'art'velo *see* Georgia
136 F11 **Sakarya** ◈ *province* NW Turkey
136 F12 **Sakarya Nehri** ≈ NW Turkey
165 P9 **Sakata** Yamagata, Honshū, C Japan 38°54′N 139°51′E
123 P9 **Sakha (Yakutiya), Respublika** *var.* Respublika Yakutiya, *Eng.* Yakutia. ◈ *autonomous republic* NE Russian Federation
Sakhalin *see* Sakhalin, Ostrov
192 I3 **Sakhalin, Ostrov** *var.* Sakhalin. *island* SE Russian Federation
123 U12 **Sakhalinskaya Oblast'** ◈ *province* SE Russian Federation
123 T12 **Sakhalinskiy Zaliv** *gulf* E Russian Federation
Sakhnovshchina *see* Sakhnovshchyna
117 U6 **Sakhnovshchyna** *Rus.* Sakhnovshchina. Kharkivs'ka Oblast', E Ukraine 49°08′N 35°52′E
Sakhon Nakhon *see* Sakon Nakhon
Sakhtsar *see* Rāmsar
137 W10 **Şäki** *Rus.* Sheki; *prev.* Nukha. NW Azerbaijan 41°09′N 47°10′E
Saki *see* Saky
118 E13 **Šakiai** *Ger.* Schaken. Marijampolė, S Lithuania 54°57′N 23°04′E
165 O16 **Sakishima-shotō** *var.* Sakisima Syotō. *island group* SW Japan
Sakisima Syotō *see* Sakishima-shotō
Sakiz *see* Saqqez
Sakiz-Adasi *see* Chíos
155 F19 **Sakleshpur** Karnātaka, E India 12°58′N 75°45′E
167 S9 **Sakon Nakhon** *var.* Muang Sakon Nakhon, Sakhon Nakhon. Sakon Nakhon, E Thailand 17°10′N 104°08′E
149 P15 **Sakrand** Sind, SE Pakistan 26°06′N 68°20′E
83 F24 **Sak River** *Afr.* Sakrivier. Northern Cape, W South Africa 30°49′S 20°24′E
Sakrivier *see* Sak River
Saksaul'skiy *see* Saksaul'skoye
144 K13 **Saksaul'skoye** *var.* Saksaul'skiy, *Kaz.* Sekseüil. Kzylorda, S Kazakhstan 47°07′N 61°06′E
95 I25 **Sakskøbing** Sjælland, SE Denmark 54°48′N 11°39′E
165 N12 **Saku** Nagano, Honshū, S Japan 36°17′N 138°29′E
117 S13 **Saky** *Rus.* Saki. Avtonomna Respublika Krym, S Ukraine 45°09′N 33°36′E
76 E9 **Sal** *island* Ilhas de Barlavento, NE Cape Verde
127 N12 **Sal** ≈ SW Russian Federation
111 I21 **Sal'a** *Hung.* Sellye, Vágsellye. Nitriansky Kraj, SW Slovakia 48°09′N 17°51′E
95 N15 **Sala** Västmanland, C Sweden 59°55′N 16°38′E
15 N13 **Salaberry-de-Valleyfield** *var.* Valleyfield. Québec, SE Canada 45°15′N 74°07′W
118 G7 **Salacgrīva** *Est.* Salatsi. N Latvia 57°45′N 24°21′E
107 M18 **Sala Consilina** Campania, S Italy 40°23′N 15°35′E
40 C2 **Salada, Laguna** ◎ NW Mexico
61 D14 **Saladas** Corrientes, NE Argentina 28°15′S 58°40′W
61 C21 **Saladillo** Buenos Aires, E Argentina 35°40′S 59°50′W
61 B16 **Saladillo, Río** ≈ C Argentina
25 T9 **Salado** Texas, SW USA 30°57′N 97°32′W
63 J16 **Salado, Arroyo** ≈ SE Argentina
61 D21 **Salado, Río** ≈ E Argentina
62 J12 **Salado, Río** ≈ C Argentina
41 N7 **Salado, Río** ≈ NE Mexico
37 Q12 **Salado, Río** ≈ New Mexico, SW USA
143 N6 **Salafchegān** *var.* Sarafjagān. Quom, N Iran 34°28′N 50°28′E
77 Q15 **Salaga** C Ghana 08°31′N 00°37′W
139 S6 **Şalāḩ ad Dīn** *off.* Muḩāfa at Şalāḩ ad Dīn, *var.* Salāhuddīn. ◈ *governorate* C Iraq
Şalāḩ ad Dīn, Muḩāfa at *see* Şalāḩ ad Dīn
Salāhuddīn *see* Şalāḩ ad Dīn
192 G5 **Sala'ilua** Savai'i, W Samoa 13°39′S 172°33′W
116 G9 **Sălaj** ◈ *county* NW Romania
83 H20 **Salajwe** Kweneng, SE Botswana 23°40′S 24°46′E
78 H9 **Salal** Bahr el Gazel, W Chad 14°48′N 17°12′E
80 I6 **Salala** Red Sea, NE Sudan 21°17′N 36°16′E
141 V13 **Şalālah** SW Oman 17°01′N 54°04′E
42 D5 **Salamá** Baja Verapaz, C Guatemala 15°06′N 90°18′W
42 J6 **Salamá** Olancho, C Honduras 14°48′N 86°34′W
62 G10 **Salamanca** Coquimbo, C Chile 31°47′S 70°58′W
41 N13 **Salamanca** Guanajuato, C Mexico 20°34′N 101°12′W
104 K7 **Salamanca** *anc.* Helmantica, Salmantica. Castilla y León, NW Spain 40°58′N 05°40′W
18 D11 **Salamanca** New York, NE USA 42°09′N 78°43′W
104 J7 **Salamanca** ◈ *province* Castilla y León, W Spain
63 J19 **Salamanca, Pampa de** *plain* S Argentina
78 J12 **Salamat** *off.* Région du Salamat. ◈ *region* SE Chad
78 I12 **Salamat, Bahr** ≈ S Chad
Salamat, Région du *see* Salamat
54 F5 **Salamina** Magdalena, N Colombia 10°30′N 74°48′W
115 G19 **Salamína** *var.* Salamís. Salamína, C Greece 37°58′N 23°29′E
115 G19 **Salamína** *island* C Greece
Salamís *see* Salamína
138 I5 **Salamīyah** *var.* As Salamīyah. Ḩamāh, W Syria 35°01′N 37°02′E
31 P12 **Salamonie Lake** ⊠ Indiana, N USA
31 P12 **Salamonie River** ≈ Indiana, N USA
Salang *see* Phuket
192 I16 **Salani** Upolu, SE Samoa 14°00′S 171°35′W
118 C11 **Salantai** Klaipėda, NW Lithuania 56°05′N 21°36′E
104 K2 **Salas** Asturias, N Spain 43°25′N 06°15′W
105 O5 **Salas de los Infantes** Castilla y León, N Spain 42°02′N 03°17′W
102 M16 **Salat** ≈ S France
189 V13 **Salat** *island* Chuuk, C Micronesia
169 Q16 **Salatiga** Jawa, C Indonesia 07°15′S 110°34′E
189 V13 **Salat Pass** *passage* W Pacific Ocean
Salatsi *see* Salacgrīva
167 T10 **Salavan** *var.* Saravan, Saravane. Salavan, S Laos 15°43′N 106°26′E
127 V6 **Salavat** Respublika Bashkortostan, W Russian Federation 53°20′N 55°54′E
56 C12 **Salaverry** La Libertad, N Peru 08°14′S 78°55′W
171 T12 **Salawati, Pulau** *island* E Indonesia
193 R10 **Sala y Gomez** *island* Chile, E Pacific Ocean
Sala y Gomez Fracture Zone *see* Sala y Gomez Ridge
193 S10 **Sala y Gomez Ridge** *var.* Sala y Gomez Fracture Zone. *tectonic feature* SE Pacific Ocean
61 A22 **Salazar** Buenos Aires, E Argentina 36°20′S 62°11′W
54 G7 **Salazar** Norte de Santander, N Colombia 07°46′N 72°46′W
Salazar *see* N'Dalatando
173 P16 **Salazie** C Réunion 21°02′S 55°33′E
103 N8 **Salbris** Loir-et-Cher, C France 47°25′N 02°02′E
57 G15 **Salcantay, Nevado** ▲ C Peru 13°21′S 72°31′W
45 O8 **Salcedo** N Dominican Republic 19°26′N 70°25′W
39 S9 **Salcha River** ≈ Alaska, USA
119 H15 **Šalčininkai** Vilnius, SE Lithuania 54°20′N 25°26′E
Saldae *see* Béjaïa
54 E11 **Saldaña** Tolima, C Colombia 03°57′N 75°01′W
104 M4 **Saldaña** Castilla y León, N Spain 42°31′N 04°44′W
83 E25 **Saldanha** Western Cape, SW South Africa 33°00′S 17°56′E
Salduba *see* Zaragoza
61 B23 **Saldungaray** Buenos Aires, E Argentina 38°15′S 61°45′W
118 D9 **Saldus** *Ger.* Frauenburg. W Latvia 56°40′N 22°29′E
183 P13 **Sale** Victoria, SE Australia 38°06′S 147°06′E
74 F6 **Salé** NW Morocco 34°07′N 06°40′W
74 F6 **Salé** ✈ (Rabat) W Morocco 34°09′N 06°30′W
Salehābād *see* Andīmeshk
170 M16 **Saleh, Teluk** *bay* Nusa Tenggara, S Indonesia
122 H8 **Salekhard** *prev.* Obdorsk. Yamalo-Nenetskiy Avtonomnyy Okrug, N Russian Federation 66°33′N 66°35′E
192 H16 **Sālelologa** Savai'i, C Samoa 13°42′S 172°10′W
155 H21 **Salem** Tamil Nādu, SE India 11°38′N 78°08′E
27 V9 **Salem** Arkansas, C USA 36°21′N 91°49′W
30 L15 **Salem** Illinois, N USA 38°37′N 88°57′W
31 P15 **Salem** Indiana, N USA 38°38′N 86°06′W
19 P11 **Salem** Massachusetts, NE USA 42°30′N 70°51′W
27 V6 **Salem** Missouri, C USA 37°39′N 91°32′W
18 I16 **Salem** New Jersey, NE USA 39°33′N 75°26′W
31 U12 **Salem** Ohio, N USA 40°52′N 80°51′W
32 G12 **Salem** *state capital* Oregon, NW USA 44°57′N 123°01′W
29 Q11 **Salem** South Dakota, N USA 43°43′N 97°23′W
36 L4 **Salem** Utah, W USA 40°03′N 111°40′W
21 S7 **Salem** Virginia, NE USA 37°16′N 80°00′W
21 R3 **Salem** West Virginia, NE USA 39°15′N 80°32′W
107 H23 **Salemi** Sicilia, Italy, C Mediterranean Sea 37°48′N 12°48′E
Salemy *see* As Sālimī
94 K12 **Sälen** Dalarna, C Sweden 61°11′N 13°14′E
107 Q18 **Salentina, Campi** Puglia, SE Italy 40°23′N 18°01′E
107 Q18 **Salentina, Penisola** *peninsula* SE Italy
107 L18 **Salerno** *anc.* Salernum. Campania, S Italy 40°40′N 14°44′E
107 L18 **Salerno, Golfo di** *Eng.* Gulf of Salerno. *gulf* S Italy
Salerno, Gulf of *see* Salerno, Golfo di
Salernum *see* Salerno
97 K17 **Salford** NW England, United Kingdom 53°30′N 02°16′W
Salgir *see* Salhyr
111 K21 **Salgótarján** Nógrád, N Hungary 48°07′N 19°47′E
59 O15 **Salgueiro** Pernambuco, E Brazil 08°04′S 39°05′W
94 C13 **Salhus** Hordaland, S Norway 60°30′N 05°15′E
117 T12 **Salhyr** *Rus.* Salgir. ≈ S Ukraine
171 Q9 **Salibabu, Pulau** *island* N Indonesia
37 S6 **Salida** Colorado, C USA 38°29′N 105°57′W
102 J15 **Salies-de-Béarn** Pyrénées-Atlantiques, SW France 43°28′N 00°55′W
136 C14 **Salihli** Manisa, W Turkey 38°29′N 28°08′E
119 K18 **Salihorsk** *Rus.* Soligorsk. Minskaya Voblasts', S Belarus 52°48′N 27°32′E
119 K18 **Salihorskaye Vodaskhovishcha** *Rus.* Soligorskoye Vodokhranilishche. ⊠ C Belarus
83 N14 **Salima** Central, C Malawi 13°44′S 34°21′E
166 L5 **Salin** Magway, W Myanmar (Burma) 20°30′N 94°40′E
27 N4 **Salina** Kansas, C USA 38°53′N 97°36′W
36 L5 **Salina** Utah, W USA 38°57′N 111°54′W
41 S17 **Salina Cruz** Oaxaca, SE Mexico 16°11′N 95°12′W
107 L22 **Salina, Isola** *island* Isole Eolie, S Italy
44 J5 **Salina Point** *headland* Acklins Island, SE The Bahamas 22°10′N 74°16′W
56 A7 **Salinas** Guayas, W Ecuador 02°15′S 80°58′W
40 M11 **Salinas** *var.* Salinas de Hidalgo. San Luis Potosí, C Mexico 22°36′N 101°41′W
45 T6 **Salinas** C Puerto Rico 17°57′N 66°18′W
35 O10 **Salinas** California, W USA 36°41′N 121°40′W
Salinas, Cabo de *see* Salines, Cap de ses
Salinas de Hidalgo *see* Salinas
82 A13 **Salinas, Ponta das** *headland* W Angola 12°50′S 12°57′E
45 O10 **Salinas, Punta** *headland* S Dominican Republic 18°11′N 70°32′W
Salinas, Río *see* Chixoy, Río
35 O11 **Salinas River** ≈ California, W USA
22 H6 **Saline Lake** ◎ Louisiana, S USA
25 R17 **Salineno** Texas, SW USA 26°29′N 99°06′W
27 V14 **Saline River** ≈ Arkansas, C USA
30 M17 **Saline River** ≈ Illinois, N USA
105 X10 **Salines, Cap de ses** *var.* Cabo de Salinas. *headland* Mallorca, Spain, W Mediterranean Sea 39°15′N 03°03′E
Salisburg *see* Mazsalaca
45 O12 **Salisbury** *var.* Baroui. W Dominica 15°26′N 61°27′W
97 M23 **Salisbury** *var.* New Sarum. S England, United Kingdom 51°05′N 01°48′W
21 Y4 **Salisbury** Maryland, NE USA 38°22′N 75°37′W
27 T3 **Salisbury** Missouri, C USA 39°25′N 92°48′W
21 S9 **Salisbury** North Carolina, SE USA 35°40′N 80°28′W
Salisbury *see* Harare
9 Q7 **Salisbury Island** *island* Nunavut, NE Canada
Salisbury, Lake *see* Bisina, Lake
97 L23 **Salisbury Plain** *plain* S England, United Kingdom
21 R14 **Salkehatchie River** ≈ South Carolina, SE USA
138 I9 **Şalkhad** As Suwaydā', SW Syria 32°29′N 36°42′E
92 M12 **Salla** Lappi, NE Finland 66°50′N 28°40′E
Sállan *see* Sørøya
103 U11 **Sallanches** Haute-Savoie, E France 45°55′N 06°37′E
105 V5 **Sallent** Cataluña, NE Spain 41°48′N 01°52′E
Salliq *see* Coral Harbour
61 A22 **Salliqueló** Buenos Aires, E Argentina 36°45′S 62°55′W
27 R10 **Sallisaw** Oklahoma, C USA 35°27′N 94°49′W
80 I7 **Sallom** Red Sea, NE Sudan 19°17′N 37°02′E
12 J2 **Salluit** *prev.* Saglouc, Sagluk. Québec, NE Canada 62°10′N 75°40′W
Sallyana *see* Salyān
13 S11 **Sally's Cove** Newfoundland and Labrador, E Canada 49°43′N 58°00′W
139 W9 **Salmān Bin 'Arāzah** Maysān, E Iraq 32°33′N 46°36′E
Salmantica *see* Salamanca
142 I2 **Salmās** *prev.* Dilman, Shāpūr. Āžarbāyjān-e Gharbī, NW Iran 38°13′N 44°50′E
124 I11 **Salmi** Respublika Kareliya, NW Russian Federation 61°21′N 31°55′E
33 P12 **Salmon** Idaho, NW USA 45°10′N 113°54′W
11 N16 **Salmon Arm** British Columbia, SW Canada 50°41′N 119°18′W
192 L5 **Salmon Bank** *undersea feature* N Pacific Ocean 26°55′N 176°28′W
Salmon Leap *see* Leixlip
34 L2 **Salmon Mountains** ▲▲ California, W USA
14 J15 **Salmon Point** *headland* Ontario, SE Canada 43°51′N 77°15′W
33 N11 **Salmon River** ≈ Idaho, NW USA
18 K6 **Salmon River** ≈ New York, NE USA
33 N12 **Salmon River Mountains** ▲▲ Idaho, NW USA
18 I9 **Salmon River Reservoir** ⊠ New York, NE USA
93 K19 **Salo** Varsinais-Suomi, SW Finland 60°23′N 23°10′E
106 F7 **Salò** Lombardia, N Italy 45°37′N 10°30′E
Salona/Salonae *see* Solin
103 S15 **Salon-de-Provence** Bouches-du-Rhône, SE France 43°39′N 05°05′E
Salonica/Salonika *see* Thessaloníki
115 I14 **Salonikiós, Akrotírio** *var.* Akrotírio Salonikós. *headland* Thásos, E Greece 40°34′N 24°39′E
Salonikós, Akrotírio *see* Salonikiós, Akrotírio
111 F10 **Salonta** *Hung.* Nagyszalonta. Bihor, W Romania 46°49′N 21°40′E
104 I9 **Salor** ≈ W Spain
105 U6 **Salou** Cataluña, NE Spain 41°05′N 01°08′E
76 H11 **Saloum** ≈ C Senegal
42 H4 **Sal, Punta** *headland* NW Honduras 15°55′N 87°36′W
92 N3 **Salpynten** *headland* W Svalbard 78°12′N 12°11′E
138 I3 **Salqīn** Idlib, W Syria 36°09′N 36°27′E
93 F14 **Salsbruket** Nord-Trøndelag, C Norway 64°49′N 11°48′E
126 M13 **Sal'sk** Rostovskaya Oblast', SW Russian Federation 46°30′N 41°31′E
107 J25 **Salso** ≈ Sicilia, Italy, C Mediterranean Sea
107 K25 **Salso** ≈ Sicilia, Italy, C Mediterranean Sea
106 E9 **Salsomaggiore Terme** Emilia-Romagna, N Italy 44°49′N 09°58′E
Salt *see* As Salṭ
62 J6 **Salta** Salta, NW Argentina 24°47′S 65°23′W
62 K6 **Salta** *off.* Provincia de Salta. ◈ *province* N Argentina
Salta, Provincia de *see* Salta
97 I24 **Saltash** SW England, United Kingdom 50°26′N 04°14′W
24 I8 **Salt Basin** *basin* Texas, SW USA
11 V16 **Saltcoats** Saskatchewan, S Canada 51°06′N 102°12′W
30 L13 **Salt Creek** ≈ Illinois, N USA
24 J9 **Salt Draw** ≈ Texas, SW USA
97 F21 **Saltee Islands** *island group* SE Ireland
92 G12 **Saltfjorden** *inlet* C Norway
24 I8 **Salt Flat** Texas, SW USA 31°43′N 105°05′W
27 N8 **Salt Fork Arkansas River** ≈ Oklahoma, C USA
31 T13 **Salt Fork Lake** ⊠ Ohio, N USA
26 J11 **Salt Fork Red River** ≈ Oklahoma/Texas, C USA
95 J23 **Saltholm** *island* E Denmark
41 N8 **Saltillo** Coahuila, NE Mexico 25°30′N 101°W
182 L5 **Salt Lake** *salt lake* New South Wales, SE Australia
37 V15 **Salt Lake** ◎ New Mexico, SW USA
36 K2 **Salt Lake City** *state capital* Utah, W USA 40°45′N 111°54′W
61 C20 **Salto** Buenos Aires, E Argentina 34°18′S 60°17′W
61 D17 **Salto** Salto, N Uruguay 31°23′S 57°58′W
61 E17 **Salto** ◈ *department* N Uruguay
107 I14 **Salto** ≈ C Italy
62 Q6 **Salto del Guairá** Canindeyú, E Paraguay 24°06′S 54°22′W
61 D17 **Salto Grande, Embalse de** *var.* Lago de Salto Grande. ⊠ Argentina/Uruguay
Salto Grande, Lago de *see* Salto Grande, Embalse de
35 W16 **Salton Sea** ◎ California, W USA
60 I12 **Salto Santiago, Represa de** ⊠ S Brazil
149 U7 **Salt Range** ▲▲ E Pakistan
36 M13 **Salt River** ≈ Arizona, SW USA
20 L5 **Salt River** ≈ Kentucky, S USA
27 V3 **Salt River** ≈ Missouri, C USA
95 F17 **Saltrød** Aust-Agder, S Norway 58°28′N 08°49′E
95 P16 **Saltsjöbaden** Stockholm, C Sweden 59°15′N 18°20′E
92 G12 **Saltstraumen** Nordland, C Norway 67°16′N 14°42′E
21 Q7 **Saltville** Virginia, NE USA 36°52′N 81°48′W
Saluces/Saluciae *see* Saluzzo
21 Q12 **Saluda** South Carolina, SE USA 34°00′N 81°47′W
21 X6 **Saluda** Virginia, NE USA 37°36′N 76°36′W
21 Q12 **Saluda River** ≈ South Carolina, SE USA
Salûm *see* As Sallūm
152 F14 **Salūmbar** Rājasthān, N India 24°16′N 74°04′E
Salûm, Gulf of *see* Khalīj as Sallūm
171 O11 **Salumpaga** Sulawesi, N Indonesia 01°18′N 120°58′E
155 M14 **Sālūr** Andhra Pradesh, E India 18°31′N 83°16′E
55 Y9 **Salut, Îles du** *island group* N French Guiana
106 A9 **Saluzzo** *Fr.* Saluces; *anc.* Saluciae. Piemonte, NW Italy 44°39′N 07°29′E
63 F23 **Salvación, Bahía** *bay* S Chile
59 P17 **Salvador** *prev.* São Salvador. *state capital* Bahia, E Brazil 12°58′S 38°29′W
65 E24 **Salvador** East Falkland, Falkland Islands 51°28′S 58°22′W
22 K10 **Salvador, Lake** ◎ Louisiana, S USA
Salvaleón de Higüey *see* Higüey
104 F10 **Salvaterra de Magos** Santarém, C Portugal 39°01′N 08°47′W
41 N13 **Salvatierra** Guanajuato, C Mexico 20°14′N 100°52′W
105 P3 **Salvatierra** *Basq.* Agurain. País Vasco, N Spain 42°52′N 02°23′W
Salwa/Salwah *see* As Salwā
166 M7 **Salween** *Bur.* Thanlwin, *Chin.* Nu Chiang, Nu Jiang. ≈ SE Asia
137 Y12 **Salyan** *Rus.* Sal'yany. SE Azerbaijan 39°36′N 48°57′E
153 N11 **Salyān** *var.* Sallyana. Mid Western, W Nepal 28°22′N 82°10′E
Sal'yany *see* Salyan
21 O6 **Salyersville** Kentucky, S USA 37°43′N 83°06′W
109 V6 **Salza** ≈ E Austria
109 Q7 **Salzach** ≈ Austria/Germany
109 Q6 **Salzburg** *anc.* Juvavum. Salzburg, N Austria 47°48′N 13°03′E
109 O8 **Salzburg** *off.* Land Salzburg. ◈ *state* C Austria
Salzburg *see* Ocna Sibiului
Salzburg Alps *see* Salzburger Kalkalpen

◆ Country
● Country Capital
◇ Dependent Territory
○ Dependent Territory Capital
◈ Administrative Regions
✈ International Airport
▲ Mountain
▲▲ Mountain Range
Volcano
≈ River
◎ Lake
⊠ Reservoir

109 Q7 **Salzburger Kalkalpen** *Eng.* Salzburg Alps. ▲▲ C Austria
Salzburg, Land *see* Salzburg
100 J13 **Salzgitter** *prev.* Watenstedt-Salzgitter. Niedersachsen, C Germany 52°07′N 10°24′E
101 G14 **Salzkotten** Nordrhein-Westfalen, W Germany 51°40′N 08°36′E
100 K11 **Salzwedel** Sachsen-Anhalt, N Germany 52°51′N 11°10′E
152 D11 **Sām** Rājasthān, NW India 26°50′N 70°30′E
Šamac *see* Bosanski Šamac
54 G9 **Samacá** Boyacá, C Colombia 05°28′N 73°33′W
40 I7 **Samachique** Chihuahua, N Mexico 27°17′N 107°28′W
141 Y8 **Şamad** NE Oman 22°47′N 58°12′E
Sama de Langreo *see* Sama, Spain
Samaden *see* Samedan
57 M19 **Samaipata** Santa Cruz, C Bolivia 18°08′S 63°53′W
Samakhixai *see* Attapu
Samakov *see* Samokov
42 B6 **Samalá, Río** ≈ SW Guatemala
40 J3 **Samalayuca** Chihuahua, N Mexico 31°25′N 106°30′W
155 L16 **Sāmalkot** Andhra Pradesh, E India 17°03′N 82°15′E
45 P8 **Samaná** *var.* Santa Bárbara de Samaná. E Dominican Republic 19°14′N 69°20′W
45 P8 **Samaná, Bahía de** *bay* E Dominican Republic
44 K4 **Samana Cay** *island* SE The Bahamas
136 K17 **Samandağı** Hatay, S Turkey 36°07′N 35°55′E
149 P3 **Samangān** ◈ *province* N Afghanistan
Samangān *see* Aibak
165 T5 **Samani** Hokkaidō, NE Japan 42°07′N 142°57′E
54 C13 **Samaniego** Nariño, SW Colombia 01°22′N 77°35′W
171 Q5 **Samar** *island* C Philippines
127 S6 **Samara** *prev.* Kuybyshev. Samarskaya Oblast', W Russian Federation 53°15′N 50°15′E
127 T7 **Samara** ≈ W Russian Federation
127 S6 **Samara** ✈ Samarskaya Oblast', W Russian Federation 53°11′N 50°27′E
117 V7 **Samara** ≈ E Ukraine
186 G10 **Samarai** Milne Bay, SE Papua New Guinea 10°36′S 150°39′E
Samarang *see* Semarang
123 T14 **Samarga** Khabarovskiy Kray, SE Russian Federation 47°43′N 139°08′E
138 G9 **Samarian Hills** *hill range* N Israel
54 L9 **Samariapo** Amazonas, C Venezuela 05°16′N 67°43′W
169 V11 **Samarinda** Borneo, C Indonesia 0°30′S 117°09′E
Samarkand *see* Samarqand
Samarkandskaya Oblast' *see* Samarqand Viloyati
Samarkandski/ Samarkandskoye *see* Temirtau
Samarobriva *see* Amiens
147 N11 **Samarqand** *Rus.* Samarkand. Samarqand Viloyati, C Uzbekistan 39°40′N 66°56′E
146 M11 **Samarqand Viloyati** *Rus.* Samarkandskaya Oblast'. ◈ *province* C Uzbekistan
139 S6 **Sāmarrā'** Şalāḩ ad Dīn, C Iraq 34°13′N 43°52′E
127 R7 **Samarskaya Oblast'** *prev.* Kuybyshevskaya Oblast'. ◈ *province* W Russian Federation
153 Q13 **Samastīpur** Bihār, N India 25°52′N 85°47′E
76 L14 **Samatiguila** NW Ivory Coast 09°51′N 07°36′W
Samawa *see* As Samāwah
137 Y11 **Şamaxı** *Rus.* Shemakha. E Azerbaijan 40°38′N 48°34′E
79 K18 **Samba** Equateur, NW Dem. Rep. Congo 0°13′N 21°17′E
79 N21 **Samba** Maniema, E Dem. Rep. Congo 04°41′S 26°23′E
152 H6 **Samba** Jammu and Kashmir, NW India 32°32′N 75°08′E
169 W10 **Sambaliung, Pegunungan** ▲▲ Borneo, N Indonesia
154 M11 **Sambalpur** Odisha, E India 21°28′N 84°04′E
67 X12 **Sambao** ≈ W Madagascar
169 Q10 **Sambas, Sungai** ≈ Borneo, N Indonesia
172 K2 **Sambava** Antsiranana, NE Madagascar 14°16′S 50°10′E
152 J10 **Sambhal** Uttar Pradesh, N India 28°35′N 78°34′E
152 H12 **Sāmbhar Salt Lake** ◎ N India
107 N21 **Sambiase** Calabria, SW Italy 38°58′N 16°16′E
116 H5 **Sambir** *Rus.* Sambor. L'vivs'ka Oblast', NW Ukraine 49°31′N 23°10′E
82 C13 **Sambo** Huambo, C Angola 13°07′S 16°06′E
Sambor *see* Sambir
61 E21 **Samborombón, Bahía** *bay* NE Argentina
99 H20 **Sambre** ≈ Belgium/France
43 V16 **Sambú, Río** ≈ SE Panama
81 I18 **Samburu** ◈ *county* N Kenya
163 Z14 **Samcheok** *Jap.* Sanchoku; *prev.* Samch'ŏk. NE South Korea 37°21′N 129°12′E
Samch'ŏk *see* Samcheok
Samch'ŏnpŏ *see* Sacheon
81 I21 **Same** Kilimanjaro, NE Tanzania 04°04′S 37°41′E
108 J10 **Samedan** *Ger.* Samaden. Graubünden, S Switzerland 46°31′N 09°51′E
82 K12 **Samfya** Luapula, N Zambia 11°22′S 29°34′E
141 W13 **Samḩān, Jabal** ▲▲ SW Oman
115 C18 **Sámi** Kefallonía, Iónia Nisiá, Greece, C Mediterranean Sea 38°15′N 20°39′E
56 F10 **Samiria, Río** ≈ N Peru
Samirum *see* Semīrom
167 Q13 **Samĭt** *prev.* Phumĭ Samĭt. Kaôh Kông, SW Cambodia 10°54′N 103°09′E
137 V11 **Şämkir** *Rus.* Shamkhor. NW Azerbaijan 40°51′N 46°03′E
167 S7 **Sam, Nam** *Vtn.* Sông Chu. ≈ Laos/Vietnam
Samnān *see* Semnān
Sam Neua *see* Xam Nua
75 P10 **Samnū** C Libya 27°19′N 15°01′E
192 H15 **Samoa** *off.* Independent State of Samoa, *var.* Sāmoa; *prev.* Western Samoa. ◆ *monarchy* W Polynesia
192 L9 **Sāmoa** *island group* C Pacific Ocean
Sāmoa *see* Samoa
175 T9 **Samoa Basin** *undersea feature* W Pacific Ocean
Samoa, Independent State of *see* Samoa
112 D8 **Samobor** Zagreb, N Croatia 45°48′N 15°38′E
114 H10 **Samokov** *var.* Samakov. Sofia, W Bulgaria 42°19′N 23°34′E
111 H21 **Šamorín** *Ger.* Sommerein, *Hung.* Somorja. Trnavský Kraj, W Slovakia 48°01′N 17°18′E
115 M19 **Sámos** *prev.* Limín Vathéos. Sámos, Dodekánisa, Greece, Aegean Sea 37°45′N 26°58′E
115 M20 **Sámos** *island* Dodekánisa, Greece, Aegean Sea
Samosch *see* Szamos
168 I9 **Samosir, Pulau** *island* W Indonesia
Samothrace *see* Samothráki
115 K14 **Samothráki** Samothráki, NE Greece 40°28′N 25°31′E
115 J14 **Samothráki** *anc.* Samothrace. *island* NE Greece
115 A15 **Samothráki** *island* Iónia Nisiá, Greece, C Mediterranean Sea
Samotschin *see* Szamocin
Sampê *see* Xiangcheng
169 S13 **Sampit** Borneo, C Indonesia 02°30′S 112°30′E
169 S12 **Sampit, Sungai** ≈ Borneo, N Indonesia
Sampoku *see* Sanpoku
186 H7 **Sampun** New Britain, E Papua New Guinea 05°19′S 152°06′E
79 N24 **Sampwe** Katanga, SE Dem. Rep. Congo 09°17′S 27°22′E
167 R11 **Sâmraông** *prev.* Phumĭ Sâmraông, Phum Samrong. Siĕmréab, NW Cambodia 14°11′N 103°31′E
25 X8 **Sam Rayburn Reservoir** ⊠ Texas, SW USA
167 Q6 **Sam Sao, Phou** ▲▲ Laos/ Thailand
95 H22 **Samsø** *island* E Denmark
95 H23 **Samsø Bælt** *channel* E Denmark
167 T7 **Sâm Sơn** Thanh Hoa, N Vietnam 19°44′N 105°53′E
136 L11 **Samsun** *anc.* Amisus. Samsun, N Turkey 41°17′N 36°22′E
136 K11 **Samsun** ◈ *province* N Turkey
137 R9 **Samt'redia** *prev.* Samtredia. W Georgia 42°09′N 42°20′E
Samtredia *see* Samt'redia
59 E15 **Samuel, Represa de** ⊠ W Brazil
167 O14 **Samui, Ko** *island* SW Thailand
Samundari *see* Samundri
149 U9 **Samundri** *var.* Samundari. Punjab, E Pakistan 31°04′N 72°58′E
137 X10 **Samur** ≈ Azerbaijan/ Russian Federation
137 Y11 **Samur-Abşeron Kanalı** *Rus.* Sam ur-Apsheronskiy Kanal. *canal* E Azerbaijan
Sam ur-Apsheronskiy Kanal *see* Samur-Abşeron Kanalı
167 O11 **Samut Prakan** *var.* Muang Samut Prakan, Paknam. Samut Prakan, C Thailand 13°36′N 100°36′E
167 O11 **Samut Sakhon** *var.* Maha Chai, Samut Sakorn, Tha Chin. Samut Sakhon, C Thailand 13°31′N 100°15′E
Samut Sakorn *see* Samut Sakhon
167 O11 **Samut Songkhram** *prev.* Meklong. Samut Songkhram, SW Thailand 13°25′N 100°01′E
77 N12 **San** Ségou, C Mali 13°21′N 04°57′W
111 O15 **San** ≈ SE Poland
141 O15 **Şan'ā'** *Eng.* Sana. ● (Yemen) W Yemen 15°24′N 44°14′E
112 F11 **Sana** ≈ NW Bosnia and Herzegovina
80 O12 **Sanaag** *off.* Gobolka Sanaag. ◈ *region* N Somalia
Sanaag, Gobolka *see* Sanaag
114 J8 **Sanadinovo** Pleven, N Bulgaria 43°33′N 25°00′E
195 P1 **Sanae** *South African research station* Antarctica 70°19′S 01°31′W
139 Y10 **Sanāf, Hawr as** ◎ S Iraq
79 E15 **Sanaga** ≈ C Cameroon
54 D12 **San Agustín** Huila, SW Colombia 01°53′N 76°14′W
171 R8 **San Agustin, Cape** *headland* Mindanao, S Philippines 06°17′N 126°12′E
37 Q13 **San Agustin, Plains of** *plain* New Mexico, SW USA
38 M16 **Sanak Islands** *island* Aleutian Islands, Alaska, USA
San Alessandro *see* Kita-Iō-jima
193 U10 **San Ambrosio, Isla** *Eng.* San Ambrosio Island. *island* W Chile
San Ambrosio Island *see* San Ambrosio, Isla
171 Q12 **Sanana** Pulau Sanana, E Indonesia 02°04′S 125°58′E
171 Q12 **Sanana, Pulau** *island* Maluku, E Indonesia
142 K5 **Sanandaj** *prev.* Sinneh. Kordestān, W Iran 35°18′N 47°01′E
35 P8 **San Andreas** California, W USA 38°10′N 120°40′W
2 C13 **San Andreas Fault** *fault* W USA
54 G8 **San Andrés** Santander, C Colombia 06°52′N 72°53′W
61 C20 **San Andrés de Giles** Buenos Aires, E Argentina 34°27′S 59°27′W
37 R14 **San Andres Mountains** ▲▲ New Mexico, SW USA
41 S15 **San Andrés Tuxtla** *var.* Tuxtla. Veracruz-Llave, E Mexico 18°28′N 95°15′W
25 P8 **San Angelo** Texas, SW USA 31°28′N 100°26′W
107 A20 **San Antioco, Isola di** *island* W Italy
42 F4 **San Antonio** Toledo, S Belize 16°13′N 89°02′W
62 G11 **San Antonio** Valparaíso, C Chile 33°35′S 71°38′W
188 H6 **San Antonio** Saipan, S Northern Mariana Islands
37 R13 **San Antonio** New Mexico, SW USA 33°53′N 106°52′W
25 R12 **San Antonio** Texas, SW USA 29°25′N 98°30′W
54 M11 **San Antonio** Amazonas, S Venezuela 03°31′N 66°47′W
54 I7 **San Antonio** Barinas, C Venezuela 07°24′N 71°28′W
55 O5 **San Antonio** Monagas, NE Venezuela 10°03′N 63°45′W
25 S12 **San Antonio** ✈ Texas, SW USA 29°31′N 98°11′W
San Antonio *see* San Antonio del Táchira
San Antonio Abad *see* Sant Antoni de Portmany
25 U13 **San Antonio Bay** *inlet* Texas, SW USA
61 E22 **San Antonio, Cabo** *headland* E Argentina 36°45′S 56°40′W
44 A5 **San Antonio, Cabo de** *headland* W Cuba 21°51′N 84°58′W
105 T11 **San Antonio, Cabo de** *headland* E Spain 38°50′N 00°09′E
54 H7 **San Antonio de Caparo** Táchira, W Venezuela 07°34′N 71°28′W
62 J5 **San Antonio de los Cobres** Salta, NE Argentina 24°10′S 66°17′W
54 H7 **San Antonio del Táchira** *var.* San Antonio. Táchira, W Venezuela 07°48′N 72°28′W
35 T15 **San Antonio, Mount** ▲ California, W USA 34°18′N 117°37′W
63 K16 **San Antonio Oeste** Río Negro, E Argentina 40°45′S 64°58′W
25 T13 **San Antonio River** ≈ Texas, SW USA
54 J5 **Sanare** Lara, N Venezuela 09°45′N 69°39′W
103 T16 **Sanary-sur-Mer** Var, SE France 43°07′N 05°48′E
104 G3 **Sanata Uxía de Ribeira** *var.* Ribeira. Galicia, NW Spain 42°33′N 09°01′W
25 X8 **San Augustine** Texas, SW USA 31°32′N 94°09′W
San Augustine *see* Minami-Iō-jima
141 T13 **Sanāw** *var.* Sanaw. NE Yemen 18°N 51°E
41 O11 **San Bartolo** San Luis Potosí, C Mexico 22°20′N 100°05′W
107 L16 **San Bartolomeo in Galdo** Campania, S Italy 41°24′N 15°01′E
106 K13 **San Benedetto del Tronto** Marche, C Italy 42°57′N 13°53′E
42 E3 **San Benito** Petén, N Guatemala 16°56′N 89°53′W
25 T17 **San Benito** Texas, SW USA 26°07′N 97°37′W
54 E6 **San Benito Abad** Sucre, N Colombia 08°56′N 75°02′W
35 P11 **San Benito Mountain** ▲ California, W USA 36°21′N 120°37′W
35 O10 **San Benito River** ≈ California, W USA
108 H10 **San Bernardino** Graubünden, S Switzerland 46°21′N 09°13′E
35 U15 **San Bernardino** California, W USA 34°06′N 117°15′W
35 U15 **San Bernardino Mountains** ▲▲ California, W USA
62 H11 **San Bernardo** Santiago, C Chile 33°37′S 70°45′W
40 J8 **San Bernardo** Durango, C Mexico 25°58′N 105°27′W
164 G12 **Sanbe-san** ▲ Kyūshū, SW Japan 35°09′N 132°36′E
San Bizenti-Barakaldo *see* San Vicente de Barakaldo
40 J12 **San Blas** Nayarit, C Mexico 21°35′N 105°20′W
40 H8 **San Blas** Sinaloa, C Mexico 26°05′N 108°44′W
San Blas *see* Kuna Yala
43 U14 **San Blas, Archipiélago de** *island group* NE Panama
23 Q10 **San Blas, Cape** *headland* Florida, SE USA 29°39′N 85°21′W
43 V14 **San Blas, Cordillera de** ▲▲ NE Panama
62 J8 **San Blas de los Sauces** Catamarca, NW Argentina 28°18′S 67°12′W
106 G8 **San Bonifacio** Veneto, NE Italy 45°22′N 11°14′E
29 S12 **Sanborn** Iowa, C USA 43°10′N 95°39′W
40 M7 **San Buenaventura** Coahuila, NE Mexico 27°04′N 101°32′W
105 S5 **San Caprasio** ▲ N Spain 41°45′N 00°26′W
62 G13 **San Carlos** Bío Bío, C Chile 36°25′S 71°58′W
40 E9 **San Carlos** Baja California Sur, NW Mexico 24°52′N 112°15′W
41 N5 **San Carlos** Coahuila, NE Mexico 29°00′N 100°51′W
41 P9 **San Carlos** Tamaulipas, C Mexico 24°36′N 98°42′W
42 L12 **San Carlos** Río San Juan, S Nicaragua 11°06′N 84°46′W
43 T16 **San Carlos** Panamá, C Panama 08°29′N 79°58′W
171 N3 **San Carlos** *off.* San Carlos City. Luzon, N Philippines 15°57′N 120°18′E
61 G20 **San Carlos** Maldonado, S Uruguay 34°46′S 54°58′W
36 M14 **San Carlos** Arizona, SW USA 33°21′N 110°27′W
54 K5 **San Carlos** Cojedes, N Venezuela 09°39′N 68°35′W
San Carlos *see* Quesada, Costa Rica
San Carlos *see* Luba, Equatorial Guinea
61 B17 **San Carlos Centro** Santa Fe, C Argentina 31°45′S 61°05′W
171 P6 **San Carlos City** Negros, C Philippines 10°34′N 123°24′E
San Carlos City *see* San Carlos
San Carlos de Ancud *see* Ancud
63 H16 **San Carlos de Bariloche** Río Negro, SW Argentina 41°08′S 71°15′W
61 B21 **San Carlos de Bolívar** Buenos Aires, E Argentina 36°15′S 61°06′W
54 H6 **San Carlos del Zulia** Zulia, W Venezuela 09°01′N 71°58′W
54 L12 **San Carlos de Río Negro** Amazonas, S Venezuela 01°54′N 67°54′W
San Carlos, Estrecho de *see* Falkland Sound
36 M14 **San Carlos Reservoir** ⊠ Arizona, SW USA
42 M12 **San Carlos, Río** ≈ N Costa Rica
65 D24 **San Carlos Settlement** East Falkland, Falkland Islands
61 C23 **San Cayetano** Buenos Aires, E Argentina 38°20′S 59°37′W
103 O8 **Sancerre** Cher, C France 47°19′N 02°53′E
158 G7 **Sanchakou** Xinjiang Uygur Zizhiqu, NW China 39°56′N 78°28′E
Sanchoku *see* Samcheok
41 O12 **San Ciro** San Luis Potosí, C Mexico 21°40′N 99°50′W
105 P10 **San Clemente** Castilla-La Mancha, C Spain 39°24′N 02°25′W
35 T16 **San Clemente** California, W USA 33°25′N 117°36′W
61 E21 **San Clemente del Tuyú** Buenos Aires, E Argentina 36°22′S 56°43′W
35 S17 **San Clemente Island** *island* Channel Islands, California, W USA
103 O9 **Sancoins** Cher, C France 46°49′N 03°00′E
61 B16 **San Cristóbal** Santa Fe, C Argentina 30°20′S 61°14′W
44 B4 **San Cristóbal** Pinar del Río, W Cuba 22°43′N 83°03′W
45 O9 **San Cristóbal** *var.* Benemérita de San Cristóbal. S Dominican Republic 18°27′N 70°07′W
54 H7 **San Cristóbal** Táchira, W Venezuela 07°46′N 72°15′W
187 N10 **San Cristobal** *var.* Makira. *island* SE Solomon Islands
San Cristóbal *see* San Cristóbal de Las Casas
41 U16 **San Cristóbal de Las Casas** *var.* San Cristóbal. Chiapas, SE Mexico 16°44′N 92°40′W
187 N10 **San Cristóbal, Isla** *var.* Chatham Island. *island* Galapagos Islands, Ecuador, E Pacific Ocean
42 D5 **San Cristóbal Verapaz** Alta Verapaz, C Guatemala 15°21′N 90°22′W
44 F6 **Sancti Spíritus** Sancti Spíritus, C Cuba 21°54′N 79°27′W
103 O11 **Sancy, Puy de** ▲ C France 45°33′N 02°48′E
95 D15 **Sand** Rogaland, S Norway 59°28′N 06°16′E
169 W7 **Sandakan** Sabah, East Malaysia 05°52′N 118°04′E
182 K9 **Sandalwood** South Australia 34°51′S 140°13′E
Sandalwood Island *see* Sumba, Pulau
94 D11 **Sandane** Sogn Og Fjordane, S Norway 61°47′N 06°14′E
114 G12 **Sandanski** *prev.* Sveti Vrach. Blagoevgrad, SW Bulgaria 41°36′N 23°19′E
Sandaohezi *see* Shawan
76 J11 **Sandaré** Kayes, W Mali 14°36′N 10°22′W
95 J19 **Sandared** Västra Götaland, S Sweden 57°43′N 12°47′E
94 N12 **Sandarne** Gävleborg, C Sweden 61°15′N 17°10′E
Sandaun *see* West Sepik
96 K4 **Sanday** *island* NE Scotland, United Kingdom
31 P15 **Sand Creek** ≈ Indiana, N USA
95 H15 **Sande** Vestfold, S Norway 59°34′N 10°13′E
95 H16 **Sandefjord** Vestfold, S Norway 59°10′N 10°15′E
77 O15 **Sandégué** E Ivory Coast 07°59′N 03°33′W
77 P14 **Sandema** N Ghana 10°42′N 01°17′W
37 O11 **Sanders** Arizona, SW USA 35°13′N 109°21′W
24 M11 **Sanderson** Texas, SW USA 30°08′N 102°25′W
23 U4 **Sandersville** Georgia, SE USA 32°58′N 82°48′W
92 H4 **Sandgerði** Suðurnes, SW Iceland 64°01′N 22°42′W
28 K14 **Sand Hills** ▲▲ Nebraska, C USA
25 S14 **Sandia** Texas, SW USA 27°59′N 97°52′W
35 T17 **San Diego** California, W USA 32°43′N 117°09′W
25 S14 **San Diego** Texas, SW USA 27°47′N 98°15′W
136 F14 **Sandıklı** Afyon, W Turkey 38°28′N 30°17′E
152 L12 **Sandīla** Uttar Pradesh, N India 27°05′N 80°37′E
San Dimitri Point *see* Il-Ponta ta' San Dimitri
San Dimitri, Ras *see* Il-Ponta ta' San Dimitri
168 J13 **Sanding, Selat** *strait* W Indonesia
30 J3 **Sand Island** *island* Apostle Islands, Wisconsin, N USA
95 C16 **Sandnes** Rogaland, S Norway 58°51′N 05°45′E
92 F13 **Sandnessjøen** Nordland, C Norway 66°00′N 12°37′E
Sandø *see* Sandoy
79 L24 **Sandoa** Katanga, S Dem. Rep. Congo 09°41′S 22°56′E
111 N15 **Sandomierz** *Rus.* Sandomir. Świętokrzyskie, C Poland 50°42′N 21°45′E
Sandomir *see* Sandomierz
54 C13 **Sandoná** Nariño, SW Colombia 01°18′N 77°28′W
106 I7 **San Donà di Piave** Veneto, NE Italy 45°38′N 12°34′E
124 K14 **Sandovo** Tverskaya Oblast', W Russian Federation 58°26′N 36°30′E
Sandoway *see* Thandwe
97 M24 **Sandown** S England, United Kingdom 50°40′N 01°11′W
95 B19 **Sandoy** *Dan.* Sandø. *island* C Faroe Islands
39 N16 **Sand Point** Popof Island, Alaska, USA 55°20′N 160°30′W
32 M7 **Sandpoint** Idaho, NW USA 48°16′N 116°33′W
65 N24 **Sand Point** *headland* E Tristan da Cunha
31 R7 **Sand Point** *headland* Michigan, N USA 43°54′N 83°24′W
93 H14 **Sandsele** Västerbotten, N Sweden 65°16′N 17°40′E
10 I14 **Sandspit** Moresby Island, British Columbia, SW Canada 53°14′N 131°50′W
27 P9 **Sand Springs** Oklahoma, C USA 36°08′N 96°06′W
29 W7 **Sandstone** Minnesota, N USA 46°07′N 92°51′W
36 K15 **Sand Tank Mountains** ▲▲ Arizona, SW USA
31 S8 **Sandusky** Michigan, N USA 43°26′N 82°50′W
31 S11 **Sandusky** Ohio, N USA 41°27′N 82°42′W
31 S12 **Sandusky River** ≈ Ohio, N USA
83 D22 **Sandverhaar** Karas, S Namibia 26°50′S 17°25′E
95 L24 **Sandvig** Bornholm, E Denmark 55°15′N 14°45′E
95 H15 **Sandvika** Akershus, S Norway 59°54′N 10°29′E
94 N13 **Sandviken** Gävleborg, C Sweden 60°38′N 16°50′E
30 M11 **Sandwich** Illinois, N USA 41°39′N 88°37′W
Sandwich Island *see* Efate
Sandwich Islands *see* Hawai'ian Islands
153 V16 **Sandwip** *island* SE Bangladesh
11 U12 **Sandy Bay** Saskatchewan, C Canada 55°31′N 102°14′W
183 N16 **Sandy Cape** *headland* Tasmania, SE Australia 41°27′S 144°43′E
36 L3 **Sandy City** Utah, W USA 40°36′N 111°53′W
31 U12 **Sandy Creek** ≈ Ohio, N USA
21 O5 **Sandy Hook** Kentucky, S USA 38°05′N 83°09′W
18 K15 **Sandy Hook** *headland* New Jersey, NE USA 40°27′N 73°59′W
Sandykachi/Sandykgachy *see* Sandykgaçy
146 J15 **Sandykaçy** *var.* Sandykgachy. Maryyskiy Velayat, S Turkmenistan 36°34′N 62°28′E
Sandykgachy *see* Sandykaçy
146 J15 **Sandykgaçy** *var.* Sandykgachy, *Rus.* Sandykachi. Mary Welaýaty, S Turkmenistan 36°34′N 62°28′E
146 L13 **Sandykly Gumy** *Rus.* Peski Sandykly. *desert* E Turkmenistan
Sandykly, Peski *see* Sandykly Gumy
11 Q13 **Sandy Lake** Alberta, W Canada 55°50′N 113°30′W
12 B8 **Sandy Lake** Ontario, C Canada 53°00′N 93°25′W
12 B8 **Sandy Lake** ◎ Ontario, C Canada
23 S3 **Sandy Springs** Georgia, SE USA 33°57′N 84°23′W
24 H8 **San Elizario** Texas, SW USA 31°35′N 106°16′W
99 L25 **Sanem** Luxembourg, SW Luxembourg 49°33′N 05°56′E
42 K5 **San Esteban** Olancho, C Honduras 15°19′N 85°52′W
105 O6 **San Esteban de Gormaz** Castilla y León, N Spain 41°34′N 03°13′W
40 E5 **San Esteban, Isla** *island* NW Mexico
San Eugenio/San Eugenio del Cuareim *see* Artigas
62 H11 **San Felipe** *var.* San Felipe de Aconcagua. Valparaíso, C Chile 32°45′S 70°42′W
40 D3 **San Felipe** Baja California Norte, NW Mexico 31°03′N 114°52′W
40 N12 **San Felipe** Guanajuato, C Mexico 21°30′N 101°15′W
54 K5 **San Felipe** Yaracuy, NW Venezuela 10°25′N 68°40′W
44 B5 **San Felipe, Cayos de** *island group* W Cuba
San Felipe de Aconcagua *see* San Felipe
San Felipe de Puerto Plata *see* Puerto Plata
37 R11 **San Felipe Pueblo** New Mexico, SW USA 35°25′N 106°27′W
San Feliú de Guixols *see* Sant Feliu de Guíxols
193 T10 **San Félix, Isla** *Eng.* San Felix Island. *island* W Chile
San Felix Island *see* San Félix, Isla
40 C4 **San Fernando** *var.* Misión San Fernando. Baja California Norte, NW Mexico 29°58′N 115°14′W
41 P9 **San Fernando** Tamaulipas, C Mexico 24°50′N 98°10′W
171 N2 **San Fernando** Luzon, N Philippines 16°45′N 120°21′E
171 O3 **San Fernando** Luzon, N Philippines 15°01′N 120°41′E
104 J16 **San Fernando** *prev.* Isla de León. Andalucía, S Spain 36°28′N 06°12′W
45 U14 **San Fernando** Trinidad, Trinidad and Tobago 10°17′N 61°27′W
35 S15 **San Fernando** California, W USA 34°16′N 118°26′W
54 L7 **San Fernando** *var.* San Fernando de Apure. Apure, C Venezuela 07°54′N 67°28′W
San Fernando de Apure *see* San Fernando
54 L11 **San Fernando de Atabapo** Amazonas, S Venezuela 04°00′N 67°42′W
62 L8 **San Fernando del Valle de Catamarca** *var.* Catamarca. Catamarca, NW Argentina 28°28′S 65°46′W
San Fernando de Monte Cristi *see* Monte Cristi
41 P9 **San Fernando, Río** ≈ C Mexico
23 X11 **Sanford** Florida, SE USA 28°48′N 81°16′W
19 P9 **Sanford** Maine, NE USA 43°26′N 70°46′W
21 T10 **Sanford** North Carolina, SE USA 35°29′N 79°10′W
25 N2 **Sanford** Texas, SW USA 35°42′N 101°31′W
39 T10 **Sanford, Mount** ▲ Alaska, USA 62°21′N 144°12′W
42 G8 **San Francisco** *var.* Gotera, San Francisco Gotera. Morazán, E El Salvador 13°41′N 88°06′W
43 R16 **San Francisco** Veraguas, C Panama 08°19′N 80°59′W
171 N2 **San Francisco** *var.* Aurora. Luzon, N Philippines 13°22′N 122°31′E
35 L8 **San Francisco** California, W USA 37°47′N 122°25′W
54 H5 **San Francisco** Zulia, NW Venezuela 10°36′N 71°39′W
34 M8 **San Francisco** ✈ California, W USA 37°37′N 122°23′W
35 N9 **San Francisco Bay** *bay* California, W USA
61 C24 **San Francisco de Bellocq** Buenos Aires, E Argentina 38°42′S 60°01′W
40 I6 **San Francisco de Borja** Chihuahua, N Mexico 27°57′N 106°42′W
42 J6 **San Francisco de la Paz** Olancho, C Honduras 14°55′N 86°14′W
40 J7 **San Francisco del Oro** Chihuahua, N Mexico 26°52′N 105°50′W
40 M12 **San Francisco del Rincón** Jalisco, SW Mexico 21°00′N 101°51′W
45 O8 **San Francisco de Macorís** C Dominican Republic 19°19′N 70°15′W
San Francisco de Satipo *see* Satipo
San Francisco Gotera *see* San Francisco
San Francisco Telixtlahuaca *see* Telixtlahuaca
107 K23 **San Fratello** Sicilia, Italy, C Mediterranean Sea 38°00′N 14°35′E
San Fructuoso *see* Tacuarembó
82 C12 **Sanga** Kwanza Sul, NW Angola 11°10′S 15°27′E
56 C5 **San Gabriel** Carchi, N Ecuador 0°35′N 77°48′W
159 S15 **Sa'ngain** Xizang Zizhiqu, W China 30°47′N 98°45′E
154 E13 **Sangamner** Mahārāshtra, W India 19°37′N 74°18′E
152 H12 **Sāngāner** Rājasthān, N India 26°48′N 75°48′E
Sangan, Koh-i- *see* Sangān, Kūh-e
149 N6 **Sangān, Kūh-e** *Pash.* Koh-i-Sangan. ▲ C Afghanistan
123 P10 **Sangar** Respublika Sakha (Yakutiya), NE Russian Federation 63°48′N 127°37′E
155 H15 **Sangāreddi** Telangana, India 17°37′N 78°04′E
169 V11 **Sangasanga** Borneo, C Indonesia 0°36′S 117°12′E
103 N1 **Sangatte** Pas-de-Calais, N France 50°56′N 01°41′E
107 B19 **San Gavino Monreale** Sardegna, Italy, C Mediterranean Sea 39°33′N 08°47′E
57 D16 **Sangayan, Isla** *island* W Peru
30 L14 **Sangchris Lake** ⊠ Illinois, N USA
171 N16 **Sangeang, Pulau** *island* S Indonesia
116 I10 **Sângeorgiu de Pădure** *prev.* Erdăt-Sângeorz, Sîngeorgiu de Pădure, *Hung.* Erdőszentgyörgy. Mureş, C Romania 46°27′N 24°50′E
116 I9 **Sângeorz-Băi** *var.* Singeroz Băi, *Ger.* Rumänisch-Sankt-Georgen, *Hung.* Oláhszentgyőrgy; *prev.* Sîngeorz-Băi. Bistriţa-Năsăud, N Romania 47°24′N 24°40′E
35 R10 **Sanger** California, W USA 36°42′N 119°33′W
25 T5 **Sanger** Texas, SW USA 33°21′N 97°10′W
Sângerei *see* Sîngerei
101 L15 **Sangerhausen** Sachsen-Anhalt, C Germany 51°29′N 11°18′E
45 S6 **San Germán** W Puerto Rico 18°05′N 67°02′W
San Germano *see* Cassino
161 N2 **Sanggan He** ≈ E China
169 Q11 **Sanggau** Borneo, C Indonesia 0°08′N 110°35′E
79 G17 **Sangha** ◈ *province* N Congo
79 H16 **Sangha** ≈ Central African Republic/Congo
79 G16 **Sangha-Mbaéré** ◈ *prefecture* SW Central African Republic
149 Q15 **Sānghar** Sind, SE Pakistan 26°10′N 68°59′E
115 F22 **Sangiás** ▲ S Greece 36°39′N 22°24′E
Sangihe, Kepulauan *see* Sangir, Kepulauan
171 Q9 **Sangihe, Pulau** *var.* Sangir. *island* N Indonesia
54 G8 **San Gil** Santander, C Colombia 06°35′N 73°08′W
121 P16 **San Ġiljan** *var.* St Julian's. N Malta 35°55′N 14°29′E
106 F12 **San Gimignano** Toscana, C Italy 43°30′N 11°00′E
148 M8 **Sangīn** *var.* Sangin. Helmand, S Afghanistan 32°03′N 64°50′E
Sangin *see* Sangīn
107 O21 **San Giovanni in Fiore** Calabria, SW Italy 39°15′N 16°42′E
107 M16 **San Giovanni Rotondo** Puglia, SE Italy 41°43′N 15°44′E
106 G12 **San Giovanni Valdarno** Toscana, C Italy 43°34′N 11°31′E
Sangir *see* Sangihe, Pulau
171 Q10 **Sangir, Kepulauan** *var.* Kepulauan Sangihe. *island group* N Indonesia
Sangiyn Dalay *see* Erdenedalay, Dundgovĭ, Mongolia
Sangiyn Dalay *see* Erdene, Govĭ-Altay, Mongolia
Sangiyn Dalay *see* Nomgon, Ömnögovĭ, Mongolia
Sangiyn Dalay *see* Öldziyt, Övörhangay, Mongolia
163 Y15 **Sangju** *Jap.* Shōshū. C South Korea 36°26′N 128°09′E
167 R11 **Sangkha** Surin, E Thailand 14°36′N 103°43′E
169 W10 **Sangkulirang** Borneo, N Indonesia 01°00′N 117°56′E
169 W10 **Sangkulirang, Teluk** *bay* Borneo, N Indonesia
155 E16 **Sāngli** Mahārāshtra, W India 16°55′N 74°37′E
79 E16 **Sangmélima** Sud, S Cameroon 02°57′N 11°56′E
35 V15 **San Gorgonio Mountain** ▲ California, W USA 34°06′N 116°50′W
37 T8 **Sangre de Cristo Mountains** ▲▲ Colorado/ New Mexico, C USA
61 A20 **San Gregorio** Santa Fe, C Argentina 34°18′S 62°02′W
61 F18 **San Gregorio de Polanco** Tacuarembó, C Uruguay 32°37′S 55°50′W
45 V14 **Sangre Grande** Trinidad, Trinidad and Tobago 10°35′N 61°08′W
159 N16 **Sangri** Xizang Zizhiqu, W China 29°17′N 92°01′E
152 H9 **Sangrūr** Punjab, NW India 30°16′N 75°52′E
44 I11 **Sangster** *off.* Sir Donald Sangster International Airport, *var.* Montego Bay. ✈ (Montego Bay) W Jamaica 18°30′N 77°54′W
59 G17 **Sangue, Rio do** ≈ W Brazil
105 R4 **Sangüesa** *Bas.* Zangoza. Navarra, N Spain 42°34′N 01°17′W
61 C16 **San Gustavo** Entre Ríos, E Argentina 30°41′S 59°23′W
Sangyuan *see* Wuqiao
40 C6 **San Hipólito, Punta** *headland* NW Mexico 26°57′N 114°00′W
23 W15 **Sanibel** Sanibel Island, Florida, SE USA 26°27′N 82°01′W
23 V15 **Sanibel Island** *island* Florida, SE USA
60 F13 **San Ignacio** Misiones, NE Argentina 27°15′S 55°32′W
42 F2 **San Ignacio** *prev.* Cayo, El Cayo. Cayo, W Belize 17°09′N 89°02′W
57 L16 **San Ignacio** El Beni, N Bolivia 14°54′S 65°35′W
57 O18 **San Ignacio** Santa Cruz, E Bolivia 16°23′S 60°59′W
42 M14 **San Ignacio** *var.* San Ignacio de Acosta. San José, W Costa Rica 09°46′N 84°10′W
40 E6 **San Ignacio** Baja California Sur, NW Mexico 27°18′N 112°51′W
40 J10 **San Ignacio** Sinaloa, W Mexico 23°55′N 106°25′W
56 B9 **San Ignacio** Cajamarca, N Peru 05°09′S 79°00′W
San Ignacio de Acosta *see* San Ignacio
40 D7 **San Ignacio, Laguna** *lagoon* W Mexico
12 I6 **Sanikiluaq** Belcher Islands, Nunavut, C Canada 55°20′N 77°50′W
171 O3 **San Ildefonso Peninsula** *peninsula* Luzon, N Philippines
Saniquillie *see* Sanniquellie
Sanirajak *see* Hall Beach
61 D20 **San Isidro** Buenos Aires, E Argentina 34°28′S 58°31′W
43 N14 **San Isidro** *var.* San Isidro de El General. San José, SE Costa Rica 09°28′N 83°42′W
San Isidro de El General *see* San Isidro
54 E5 **San Jacinto** Bolívar, N Colombia 09°53′N 75°06′W
35 U16 **San Jacinto** California, W USA 33°47′N 116°58′W
35 V15 **San Jacinto Peak** ▲ California, W USA 33°48′N 116°40′W
61 F14 **San Javier** Misiones, NE Argentina 27°50′S 55°06′W
61 C16 **San Javier** Santa Fe, C Argentina 30°35′S 59°59′W
105 S13 **San Javier** Murcia, SE Spain 37°49′N 00°50′W
61 D18 **San Javier** Río Negro, W Uruguay 32°41′S 58°08′W
61 C16 **San Javier, Río** ≈ C Argentina
160 L12 **Sanjiang** *var.* Guyi, Sanjiang Dongzu Zizhixian. Guangxi Zhuangzu Zizhiqu, S China 25°46′N 109°26′E
Sanjiang *see* Jinping, Guizhou
Sanjiang Dongzu Zizhixian *see* Sanjiang
Sanjiaocheng *see* Haiyan
165 N11 **Sanjō** *var.* Sanzyô. Niigata, Honshū, C Japan 37°39′N 139°00′E
57 M15 **San Joaquín** El Beni, N Bolivia 13°06′S 64°46′W
55 O6 **San Joaquín** Anzoátegui, NE Venezuela 09°21′N 64°30′W
35 O9 **San Joaquin River** ≈ California, W USA
35 P10 **San Joaquin Valley** *valley* California, W USA
61 A18 **San Jorge** Santa Fe, C Argentina 31°50′S 61°50′W
40 D3 **San Jorge, Bahía de** *bay* NW Mexico
63 J19 **San Jorge, Golfo** *var.* Gulf of San Jorge. *gulf* S Argentina
San Jorge, Gulf of *see* San Jorge, Golfo
San Jorge, Isla de *see* Weddell Island
61 F14 **San José** Misiones, NE Argentina 27°46′S 55°47′W
57 P19 **San José** *var.* San José de Chiquitos. Santa Cruz, E Bolivia 14°13′S 68°05′W
42 M14 **San José** ● (Costa Rica) San José, C Costa Rica 09°55′N 84°05′W
42 C7 **San José** *var.* Puerto San José. Escuintla, S Guatemala 14°00′N 90°50′W
40 G6 **San José** Sonora, NW Mexico 27°32′N 110°09′W
188 K8 **San Jose** Tinian, S Northern Mariana Islands 15°00′S 145°38′E
105 U11 **San José** Eivissa, Spain, W Mediterranean Sea 38°55′N 01°18′E
35 N9 **San Jose** California, W USA 37°18′N 121°53′W
54 H5 **San José** Zulia, NW Venezuela 10°02′N 72°24′W
42 M14 **San José** *off.* Provincia de San José. ◈ *province* W Costa Rica
61 E19 **San José** ◈ *department* S Uruguay
42 M13 **San José** ✈ Alajuela, C Costa Rica 10°03′N 84°12′W

SAN JOSÉ – SANTANA, COXILHA DE

San José see San José del Guaviare, Colombia
San Jose see Oleai
San Jose see Sant Josep de sa Talaia, Ibiza, Spain
San José see San José de Mayo, Uruguay
171 O3 San Jose City Luzon, N Philippines 15°49′N 120°57′E
San José de Chiquitos see San José
San José de Cúcuta see Cúcuta
61 D16 San José de Feliciano Entre Ríos, E Argentina 30°26′S 58°46′W
55 O6 San José de Guanipa var. El Tigrito. Anzoátegui, NE Venezuela 08°54′N 64°10′W
62 I9 San José de Jáchal San Juan, W Argentina 30°15′S 68°46′W
40 G10 San José del Cabo Baja California Sur, NW Mexico 23°01′N 109°40′W
54 G12 San José del Guaviare var. San José. Guaviare, S Colombia 02°34′N 72°38′W
61 E20 San José de Mayo var. San José. San José, S Uruguay 34°20′S 56°42′W
54 I10 San José de Ocuné Vichada, E Colombia 04°10′N 70°21′W
41 O9 San José de Raíces Nuevo León, NE Mexico 24°32′N 100°15′W
63 K17 San José, Golfo gulf E Argentina
40 F9 San José, Isla island NW Mexico
43 U16 San José, Isla island SE Panama
25 U14 San Jose Island island Texas, SW USA
San José, Provincia de see San José
62 I10 San Juan San Juan, W Argentina 31°37′S 68°27′W
45 N9 San Juan var. San Juan de la Maguana. C Dominican Republic 18°49′N 71°12′W
57 E17 San Juan Ica, S Peru 15°22′S 75°07′W
45 U5 San Juan ○ (Puerto Rico) NE Puerto Rico 18°28′N 66°06′W
62 H10 San Juan off. Provincia de San Juan. ◈ province W Argentina
San Juan see San Juan de los Morros
62 O7 San Juan Bautista Misiones, S Paraguay 26°40′S 57°08′W
35 O10 San Juan Bautista California, W USA 36°50′N 121°34′W
San Juan Bautista see Villahermosa
San Juan Bautista Cuicatlán see Cuicatlán
San Juan Bautista Tuxtepec see Tuxtepec
79 C17 San Juan, Cabo headland S Equatorial Guinea 01°09′N 09°25′E
San Juan de Alicante see Sant Joan d'Alacant
54 H7 San Juan de Colón Táchira, NW Venezuela 08°02′N 72°17′W
40 L9 San Juan de Guadalupe Durango, C Mexico 25°12′N 100°50′W
San Juan de la Maguana see San Juan
54 G4 San Juan del Cesar La Guajira, N Colombia 10°45′N 73°00′W
40 L15 San Juan de Lima, Punta headland SW Mexico 18°34′N 103°40′W
42 I8 San Juan de Limay Estelí, NW Nicaragua 13°10′N 86°36′W
43 N12 San Juan del Norte var. Greytown. Río San Juan, SE Nicaragua 10°58′N 83°40′W
54 K4 San Juan de los Cayos Falcón, N Venezuela 11°11′N 68°27′W
40 M12 San Juan de los Lagos Jalisco, C Mexico 21°15′N 102°15′W
54 L5 San Juan de los Morros var. San Juan. Guárico, N Venezuela 09°53′N 67°23′W
40 K9 San Juan del Río Durango, C Mexico 25°12′N 100°50′W
41 O13 San Juan del Río Querétaro de Arteaga, C Mexico 20°24′N 100°00′W
42 J11 San Juan del Sur Rivas, SW Nicaragua 11°16′N 85°51′W
54 M9 San Juan de Manapiare Amazonas, S Venezuela 05°15′N 66°05′W
40 E7 San Juanico Baja California Sur, NW Mexico
40 D7 San Juanico, Punta headland NW Mexico 26°01′N 112°17′W
32 G6 San Juan Islands island group Washington, NW USA
40 I6 San Juanito Chihuahua, N Mexico
40 I12 San Juanito, Isla island C Mexico
37 R8 San Juan Mountains ▲▲ Colorado, C USA
54 E5 San Juan Nepomuceno Bolívar, NW Colombia 09°57′N 75°06′W
44 E5 San Juan, Pico ▲ C Cuba 21°58′N 80°10′W
San Juan, Provincia de see San Juan
191 W15 San Juan, Punta headland Easter Island, Chile, E Pacific Ocean 27°03′S 109°22′W
42 M12 San Juan, Río 〰 Costa Rica/Nicaragua
41 S15 San Juan, Río 〰 SE Mexico
37 O8 San Juan River 〰 Colorado/Utah, W USA
San Julián see Puerto San Julián
61 B17 San Justo Santa Fe, C Argentina 30°47′S 60°32′W
109 W5 Sankt Aegyd am Neuwalde Niederösterreich, E Austria 47°51′N 15°34′E
109 U9 Sankt Andrä Slvn. Šent Andraž. Kärnten, S Austria 46°46′N 14°49′E
Sankt Andrä see Szentendre
Sankt Anna see Sântana
108 K8 Sankt Anton-am-Arlberg Vorarlberg, W Austria 47°08′N 10°11′E
101 E16 Sankt Augustin Nordrhein-Westfalen, W Germany 50°46′N 07°10′E
Sankt-Bartholomäi see Palamuse
101 F24 Sankt Blasien Baden-Württemberg, SW Germany 47°43′N 08°09′E
109 R3 Sankt Florian am Inn Oberösterreich, N Austria 48°24′N 13°27′E
108 I7 Sankt Gallen var. St. Gallen, Eng. Saint Gall, Fr. St-Gall. Sankt Gallen, NE Switzerland 47°25′N 09°23′E
108 H8 Sankt Gallen var. St.Gallen, Eng. Saint Gall, Fr. St-Gall. ◈ canton NE Switzerland
108 J8 Sankt Gallenkirch Vorarlberg, W Austria 47°00′N 10°59′E
109 Q5 Sankt Georgen Salzburg, N Austria 47°59′N 12°57′E
Sankt Georgen see Đurđevac
Sankt-Georgen see Sfântu Gheorghe
109 R6 Sankt Gilgen Salzburg, NW Austria 47°46′N 13°21′E
Sankt Gotthard see Szentgotthárd
101 E20 Sankt Ingbert Saarland, SW Germany 49°17′N 07°07′E
Sankt-Jakobi see Viru-Jaagupi, Lääne-Virumaa, Estonia
Sankt-Jakobi see Pärnu-Jaagupi, Pärnumaa, Estonia
Sankt Johann see Sankt Johann in Tirol
109 T7 Sankt Johann am Tauern Steiermark, E Austria 47°20′N 14°27′E
109 Q7 Sankt Johann im Pongau Salzburg, NW Austria 47°22′N 13°13′E
109 P6 Sankt Johann in Tirol var. Sankt Johann. Tirol, W Austria 47°32′N 12°26′E
Sankt-Johannis see Järva-Jaani
108 L8 Sankt Leonhard Tirol, W Austria 47°05′N 10°53′E
Sankt Margarethen see Sankt Margarethen im Burgenland
109 Y5 Sankt Margarethen im Burgenland var. Sankt Margarethen. Burgenland, E Austria 47°49′N 16°38′E
Sankt Martin see Martin
109 X8 Sankt Martin an der Raab Burgenland, SE Austria 46°59′N 16°12′E
109 U7 Sankt Michael in Obersteiermark Steiermark, SE Austria 47°21′N 14°59′E
Sankt Michel see Mikkeli
Sankt Moritz see St. Moritz
108 E11 Sankt Niklaus Valais, S Switzerland 46°09′N 07°48′E
109 S7 Sankt Nikolai var. Sankt Nikolai im Sölktal. Steiermark, SE Austria 47°18′N 14°04′E
Sankt Nikolai im Sölktal see Sankt Nikolai
109 U9 Sankt Paul var. Sankt Paul im Lavanttal. Kärnten, S Austria 46°42′N 14°53′E
Sankt Paul im Lavanttal see Sankt Paul
Sankt Peter see Pivka
109 W9 Sankt Peter am Ottersbach Steiermark, SE Austria 46°49′N 15°48′E
124 J13 Sankt-Peterburg prev. Leningrad, Petrograd, Eng. Saint Petersburg, Fin. Pietari. Leningradskaya Oblast', NW Russian Federation 59°55′N 30°25′E
100 H8 Sankt Peter-Ording Schleswig-Holstein, N Germany 54°18′N 08°37′E
109 V4 Sankt Pölten Niederösterreich, N Austria 48°14′N 15°38′E
109 W7 Sankt Ruprecht var. Sankt Ruprecht an der Raab. Steiermark, SE Austria 47°10′N 15°41′E
Sankt Ruprecht an der Raab see Sankt Ruprecht
Sankt-Ulrich see Ortisei
109 T4 Sankt Valentin Niederösterreich, C Austria 48°11′N 14°33′E
Sankt Veit am Flaum see Rijeka
109 T9 Sankt Veit an der Glan Slvn. Št. Vid. Kärnten, S Austria 46°47′N 14°22′E
99 M21 Sankt-Vith var. Saint-Vith. Liège, E Belgium 50°17′N 06°07′E
101 E20 Sankt Wendel Saarland, SW Germany 49°28′N 07°10′E
109 R6 Sankt Wolfgang Salzburg, NW Austria 47°43′N 13°30′E
79 K21 Sankuru 〰 C Dem. Rep. Congo
40 D8 San Lázaro, Cabo headland NW Mexico 24°46′N 112°15′W
137 O16 Şanlıurfa prev. Sanli Urfa, Urfa; anc. Edessa. şanlıurfa, S Turkey 37°08′N 38°45′E
137 O16 Şanlıurfa prev. Urfa. ◈ province SE Turkey
Sanli Urfa see Şanlıurfa
137 O16 Şanlıurfa Yaylası plateau SE Turkey
61 B18 San Lorenzo Santa Fe, C Argentina 32°45′S 60°45′W
57 M21 San Lorenzo Tarija, S Bolivia 21°25′S 64°45′W
56 C5 San Lorenzo Esmeraldas, N Ecuador 01°15′N 78°51′W
42 H8 San Lorenzo Valle, S Honduras 13°24′N 87°27′W
56 A6 San Lorenzo, Cabo headland W Ecuador 0°57′S 80°49′W
105 N8 San Lorenzo de El Escorial var. El Escorial. Madrid, C Spain 40°36′N 04°07′W
40 E5 San Lorenzo, Isla island NW Mexico
57 C14 San Lorenzo, Isla island W Peru
63 G20 San Lorenzo, Monte ▲ S Argentina 47°40′S 72°12′W
40 I9 San Lorenzo, Río 〰 C Mexico
104 J15 Sanlúcar de Barrameda Andalucía, S Spain 36°46′N 06°21′W
104 J14 Sanlúcar la Mayor Andalucía, S Spain 37°24′N 06°13′W
40 E6 San Lucas var. Cabo San Lucas. Baja California Sur, NW Mexico 27°14′N 112°15′W
40 F11 San Lucas Baja California Sur, NW Mexico 22°50′N 109°52′W
40 G11 San Lucas, Cabo var. San Lucas Cape. headland NW Mexico 22°52′N 109°53′W
San Lucas Cape see San Lucas, Cabo
62 J11 San Luis San Luis, C Argentina 33°18′S 66°18′W
42 E4 San Luis Petén, NE Guatemala 16°16′N 89°27′W
42 M7 San Luis Región Autónoma Atlántico Norte, NE Nicaragua 13°59′N 84°10′W
36 H15 San Luis Arizona, SW USA 32°27′N 114°45′W
37 T8 San Luis Colorado, C USA 37°09′N 105°24′W
54 J4 San Luis Falcón, N Venezuela 11°09′N 69°39′W
62 J11 San Luis off. Provincia de San Luis. ◈ province C Argentina
41 N12 San Luis de la Paz Guanajuato, C Mexico 21°15′N 100°33′W
40 K8 San Luis del Cordero Durango, C Mexico 25°25′N 104°09′W
40 D4 San Luis, Isla island NW Mexico
42 E6 San Luis Jilotepeque Jalapa, SE Guatemala 14°40′N 89°42′W
57 M16 San Luis, Laguna de ◎ NW Bolivia
35 P13 San Luis Obispo California, W USA 35°17′N 120°40′W
37 R7 San Luis Peak ▲ Colorado, C USA 37°59′N 106°55′W
41 N11 San Luis Potosí San Luis Potosí, C Mexico 22°10′N 100°57′W
41 N11 San Luis Potosí ◈ state C Mexico
San Luis, Provincia de see San Luis
35 O10 San Luis Reservoir ▦ California, W USA
40 D2 San Luis Río Colorado var. San Luis Río Colorado. Sonora, NW Mexico 32°26′N 114°48′W
San Luis Río Colorado see San Luis Río Colorado
37 S8 San Luis Valley basin Colorado, C USA
107 C19 Sanluri Sardegna, Italy, C Mediterranean Sea 39°34′N 08°54′E
61 D23 San Manuel Buenos Aires, E Argentina 37°47′S 58°50′W
36 M15 San Manuel Arizona, SW USA 32°36′N 110°37′W
106 F11 San Marcello Pistoiese Toscana, C Italy 44°03′N 10°46′E
107 N20 San Marco Argentano Calabria, SW Italy 39°31′N 16°07′E
54 E6 San Marcos Sucre, N Colombia 08°38′N 75°10′W
42 M14 San Marcos San José, C Costa Rica 09°39′N 84°00′W
42 B5 San Marcos San Marcos, W Guatemala 14°58′N 91°48′W
42 F6 San Marcos Ocotepeque, SW Honduras 14°23′N 88°57′W
41 O16 San Marcos Guerrero, S Mexico 16°45′N 99°22′W
25 S11 San Marcos Texas, SW USA 29°54′N 97°57′W
42 A5 San Marcos off. Departamento de San Marcos. ◈ department W Guatemala
San Marcos de Arica see Arica
San Marcos, Departamento de see San Marcos
40 E6 San Marcos, Isla island NW Mexico
106 H11 San Marino ● (San Marino) C San Marino 43°54′N 12°27′E
106 I11 San Marino off. Republic of San Marino. ◆ republic S Europe
San Marino, Republic of see San Marino
62 I11 San Martín Mendoza, C Argentina 33°05′S 68°28′W
54 F11 San Martín Meta, C Colombia 03°43′N 73°42′W
56 D11 San Martín off. Departamento de San Martín. ◈ department C Peru
194 I5 San Martín Argentinian research station Antarctica 68°18′S 67°03′W
63 H16 San Martín de los Andes Neuquén, W Argentina 40°11′S 71°22′W
San Martín, Departamento de see San Martín
104 M8 San Martín de Valdeiglesias Madrid, C Spain 40°21′N 04°24′W
63 G21 San Martín, Lago var. Lago O'Higgins. ◎ S Argentina
106 H6 San Martino di Castrozza Trentino-Alto Adige, N Italy 46°16′N 11°50′E
57 N16 San Martín, Río 〰 N Bolivia
San Martín Texmelucan see Texmelucan
35 N9 San Mateo California, W USA 37°33′N 122°19′W
55 O6 San Mateo Anzoátegui, NE Venezuela 09°48′N 64°36′W
42 B4 San Mateo Ixtatán Huehuetenango, W Guatemala 15°50′N 91°30′W
57 Q18 San Matías Santa Cruz, E Bolivia 16°20′S 58°24′W
63 K16 San Matías, Golfo var. Gulf of San Matías. gulf E Argentina
San Matías, Gulf of see San Matías, Golfo
15 O8 Sanmaur Québec, SE Canada 47°52′N 73°47′W
161 T10 Sanmen Wan bay E China
160 M6 Sanmenxia var. Shan Xian. Henan, C China 34°46′N 111°17′E
Sânmiclăuş Mare see Sânnicolau Mare
61 D14 San Miguel Corrientes, NE Argentina 28°02′S 57°41′W
57 L16 San Miguel El Beni, N Bolivia 16°43′S 61°06′W
42 G8 San Miguel San Miguel, SE El Salvador 13°27′N 88°11′W
40 L6 San Miguel Coahuila, N Mexico 29°10′N 101°28′W
40 J9 San Miguel var. San Miguel de Cruces. Durango, C Mexico 24°25′N 105°55′W
43 U16 San Miguel Panamá, SE Panama 08°27′N 78°51′W
35 P12 San Miguel California, W USA 35°45′N 120°42′W
42 B9 San Miguel ◈ department E El Salvador
41 N13 San Miguel de Allende Guanajuato, C Mexico 20°56′N 100°48′W
San Miguel de Cruces see San Miguel
San Miguel de Ibarra see Ibarra
61 D21 San Miguel del Monte Buenos Aires, E Argentina 35°26′S 58°50′W
62 J7 San Miguel de Tucumán var. Tucumán. Tucumán, N Argentina 26°47′S 65°15′W
43 V16 San Miguel, Golfo de gulf S Panama
35 P15 San Miguel Island island California, W USA
42 L11 San Miguelito Río San Juan, S Nicaragua 11°22′N 84°54′W
43 T15 San Miguelito Panamá, C Panama 08°58′N 79°31′W
57 N18 San Miguel, Río 〰 E Bolivia
56 D6 San Miguel, Río 〰 Colombia/Ecuador
40 I7 San Miguel, Río 〰 N Mexico
42 G8 San Miguel, Volcán de ☗ SE El Salvador 13°27′N 88°18′W
161 Q12 Sanming Fujian, SE China 26°11′N 117°37′E
106 F11 San Miniato Toscana, C Italy 43°40′N 10°53′E
San Murezzan see St. Moritz
Sannār see Sennar
107 M15 Sannicandro Garganico Puglia, SE Italy 41°50′N 15°32′E
40 H6 San Nicolás Sonora, NW Mexico 28°31′N 109°24′W
61 C19 San Nicolás de los Arroyos Buenos Aires, E Argentina 33°20′S 60°13′W
35 R16 San Nicolas Island island Channel Islands, California, W USA
Sânnicolaul-Mare see Sânnicolau Mare
116 E11 Sânnicolau Mare var. Sânnicolaul-Mare, Hung. Nagyszentmiklós; prev. Sânmiclăuş Mare, Sînnicolau Mare. Timiş, W Romania 46°05′N 20°38′E
123 Q6 Sannikova, Proliv strait NE Russian Federation
76 K16 Sanniquellie var. Saniquillie. NE Liberia 07°24′N 08°45′W
165 R7 Sannohe Aomori, Honshū, C Japan 40°23′N 141°16′E
Sanntaler Alpen see Kamniško-Savinjske Alpe
111 O17 Sanok Podkarpackie, SE Poland 49°31′N 22°14′E
54 E5 San Onofre Sucre, NW Colombia 09°45′N 75°33′W
57 K21 San Pablo Potosí, S Bolivia 21°43′S 66°38′W
171 O4 San Pablo off. San Pablo City. Luzon, N Philippines 14°04′N 121°16′E
San Pablo Balleza see Balleza
35 N8 San Pablo Bay bay California, W USA
San Pablo City see San Pablo
40 C6 San Pablo, Punta headland NW Mexico 27°12′N 114°30′W
43 R16 San Pablo, Río 〰 C Panama
171 P4 San Pascual Burias Island, C Philippines 13°06′N 122°59′E
121 Q16 San Pawl il Baħar Eng. Saint Paul's Bay. E Malta 35°57′N 14°24′E
61 C19 San Pedro Buenos Aires, E Argentina 33°43′S 59°45′W
62 K5 San Pedro Jujuy, N Argentina 24°12′S 64°55′W
60 G13 San Pedro Misiones, NE Argentina 26°38′S 54°12′W
42 H1 San Pedro Corozal, NE Belize 17°58′N 87°55′W
76 M17 San-Pédro S Ivory Coast 04°45′N 06°37′W
40 L8 San Pedro var. San Pedro de las Colonias. Coahuila, NE Mexico 25°47′N 102°57′W
62 O5 San Pedro San Pedro, SE Paraguay 24°08′S 57°08′W
62 O6 San Pedro off. Departamento de San Pedro. ◈ department C Paraguay
44 G6 San Pedro 〰 C Cuba
77 N16 San Pedro ✈ (Yamoussoukro) C Ivory Coast 06°49′N 05°14′W
San Pedro see San Pedro del Pinatar
42 D5 San Pedro Carchá Alta Verapaz, C Guatemala 15°30′N 90°12′W
35 S16 San Pedro Channel channel California, W USA
62 I5 San Pedro de Atacama Antofagasta, N Chile 22°52′S 68°10′W
San Pedro de Durazno see Durazno
40 G5 San Pedro de la Cueva Sonora, NW Mexico 29°17′N 109°47′W
San Pedro de las Colonias see San Pedro
56 B11 San Pedro de Lloc La Libertad, NW Peru 07°26′S 79°31′W
105 S13 San Pedro del Pinatar var. San Pedro. Murcia, SE Spain 37°50′N 00°47′W
45 P9 San Pedro de Macorís SE Dominican Republic 18°30′N 69°18′W
San Pedro, Departamento de see San Pedro
40 C3 San Pedro Mártir, Sierra ▲▲ NW Mexico
San Pedro Pochutla see Pochutla
42 D2 San Pedro, Río 〰 Guatemala/Mexico
40 K10 San Pedro, Río 〰 C Mexico
104 J10 San Pedro, Sierra de ▲▲ W Spain
42 G5 San Pedro Sula Cortés, NW Honduras 15°26′N 88°01′W
San Pedro Tapanatepec see Tapanatepec
62 I4 San Pedro, Volcán ☗ N Chile 21°46′S 68°13′W
106 E7 San Pellegrino Terme Lombardia, N Italy 45°53′N 09°42′E
25 T16 San Perlita Texas, SW USA 26°30′N 97°38′W
San Pietro see Supetar
San Pietro del Carso see Pivka
107 A20 San Pietro, Isola di island W Italy
32 K7 Sanpoil River 〰 Washington, NW USA
165 O9 Sanpoku var. Sampoku. Niigata, Honshū, C Japan 38°32′N 139°33′E
40 C3 San Quintín Baja California Norte, NW Mexico 30°28′N 115°58′W
40 B3 San Quintín, Bahía de bay NW Mexico
40 B3 San Quintín, Cabo headland NW Mexico 30°22′N 116°01′W
62 I12 San Rafael Mendoza, W Argentina 34°44′S 68°15′W
41 N9 San Rafael Nuevo León, NE Mexico 25°01′N 100°33′W
34 M8 San Rafael California, W USA 37°58′N 122°31′W
37 Q11 San Rafael New Mexico, SW USA 35°03′N 107°52′W
54 H4 San Rafael var. El Moján. Zulia, NW Venezuela 10°58′N 71°45′W
42 J8 San Rafael del Norte Jinotega, NW Nicaragua 13°12′N 86°06′W
42 J10 San Rafael del Sur Managua, SW Nicaragua 11°51′N 86°24′W
36 M5 San Rafael Knob ▲ Utah, W USA 38°46′N 110°45′W
35 Q14 San Rafael Mountains ▲▲ California, W USA
42 M13 San Ramón Alajuela, C Costa Rica 10°04′N 84°31′W
57 E14 San Ramón Junín, C Peru 11°08′S 75°18′W
61 F19 San Ramón Canelones, S Uruguay 34°18′S 55°55′W
62 K5 San Ramón de la Nueva Orán Salta, N Argentina 23°08′S 64°20′W
57 O16 San Ramón, Río 〰 E Bolivia
106 B11 San Remo Liguria, NW Italy 43°48′N 07°47′E
54 J3 San Román, Cabo headland NW Venezuela 12°10′N 70°01′W
61 C15 San Roque Corrientes, NE Argentina 28°35′S 58°45′W
188 I4 San Roque Saipan, S Northern Mariana Islands 15°15′S 145°47′E
104 K16 San Roque Andalucía, S Spain 36°13′N 05°23′W
25 R9 San Saba Texas, SW USA 31°13′N 98°44′W
25 Q9 San Saba River 〰 Texas, SW USA
61 D17 San Salvador Entre Ríos, E Argentina 31°38′S 58°30′W
42 F7 San Salvador ● (El Salvador) San Salvador, SW El Salvador 13°42′N 89°12′W
42 A10 San Salvador ◈ department C El Salvador
42 F8 San Salvador ✈ La Paz, S El Salvador 13°27′N 89°04′W
44 K4 San Salvador prev. Watlings Island. island E The Bahamas
62 J5 San Salvador de Jujuy var. Jujuy. Jujuy, N Argentina 24°10′S 65°20′W
42 F7 San Salvador, Volcán de ☗ C El Salvador 13°58′N 89°14′W
77 Q14 Sansanné-Mango var. Mango. N Togo 10°21′N 00°28′E
45 S5 San Sebastián W Puerto Rico 18°21′N 67°00′W
63 J24 San Sebastián, Bahía bay S Argentina
Sansenhō see Sacheon
106 H12 Sansepolcro Toscana, C Italy 43°35′N 12°12′E
107 M16 San Severo Puglia, SE Italy 41°41′N 15°23′E
112 F11 Sanski Most ◈ Federacija Bosne I Hercegovine, NW Bosnia and Herzegovina
171 W12 Sansundi Papua, E Indonesia 0°42′S 135°48′E
162 K9 Sant var. Mayhan. Övörhangay, C Mongolia 46°02′N 104°00′E
104 K11 Santa Amalia Extremadura, W Spain 39°00′N 06°01′W
60 F13 Santa Ana Misiones, NE Argentina 27°22′S 55°34′W
57 L16 Santa Ana El Beni, N Bolivia 13°43′S 65°37′W
42 E7 Santa Ana Santa Ana, NW El Salvador 13°59′N 89°34′W
40 F4 Santa Ana Sonora, NW Mexico 30°31′N 111°08′W
35 T16 Santa Ana California, W USA 33°45′N 117°52′W
55 N6 Santa Ana Nueva Esparta, NE Venezuela 09°15′N 64°39′W
42 A9 Santa Ana ◈ department NW El Salvador
Santa Ana de Coro see Coro
35 U16 Santa Ana Mountains ▲▲ California, W USA
42 E7 Santa Ana, Volcán de var. La Matepec. ☗ W El Salvador 13°49′N 89°36′W
42 G6 Santa Bárbara Santa Bárbara, W Honduras 14°56′N 88°11′W
40 J7 Santa Barbara Chihuahua, N Mexico 26°46′N 105°46′W
35 Q14 Santa Barbara California, W USA 34°24′N 119°40′W
54 L11 Santa Bárbara Amazonas, S Venezuela 03°55′N 67°06′W
54 I7 Santa Bárbara Barinas, W Venezuela 07°48′N 71°10′W
42 F5 Santa Bárbara ◈ department NW Honduras
Santa Bárbara see Iscuandé
35 Q15 Santa Barbara Channel channel California, W USA
Santa Bárbara de Samaná see Samaná
35 R16 Santa Barbara Island island Channel Islands, California, W USA
54 E5 Santa Catalina Bolívar, N Colombia 10°36′N 75°17′W
43 R15 Santa Catalina Ngöbe Bugle, W Panama 08°46′N 81°18′W
104 G2 Santa Cataliña de Armada Galicia, NW Spain 43°02′N 08°49′W
35 T17 Santa Catalina, Gulf of gulf California, W USA
40 F8 Santa Catalina, Isla island NW Mexico
35 S16 Santa Catalina Island island Channel Islands, California, W USA
41 N8 Santa Catarina Nuevo León, NE Mexico 25°39′N 100°30′W
60 H13 Santa Catarina off. Estado de Santa Catarina. ◈ state S Brazil
Santa Catarina de Tepehuanes see Tepehuanes
Santa Catarina, Estado de see Santa Catarina
60 L13 Santa Catarina, Ilha de island S Brazil
45 Q16 Santa Catherina Curaçao 12°07′N 68°46′W
44 E5 Santa Clara Villa Clara, C Cuba 22°25′N 78°01′W
35 N9 Santa Clara California, W USA 37°20′N 121°57′W
36 J8 Santa Clara Utah, W USA 37°07′N 113°39′W
Santa Clara see Santa Clara de Olimar
61 F18 Santa Clara de Olimar var. Santa Clara. Cerro Largo, NE Uruguay 32°50′S 54°54′W
61 A17 Santa Clara de Saguier Santa Fe, C Argentina 31°21′S 61°50′W
Santa Coloma see Santa Coloma de Gramenet
105 X5 Santa Coloma de Farners var. Santa Coloma de Farnés. Cataluña, NE Spain 41°52′N 02°39′E
Santa Coloma de Farnés see Santa Coloma de Farners
Santa Coloma de Gramanet see Santa Coloma de Gramenet
105 W6 Santa Coloma de Gramenet var. Santa Coloma; prev. Santa Coloma de Gramanet. Cataluña, NE Spain 41°28′N 02°14′E
Santa Comba see Uaco Cungo
104 H8 Santa Comba Dão Viseu, N Portugal 40°23′N 08°07′W
82 C10 Santa Cruz Uíge, NW Angola 06°56′S 16°25′E
57 N19 Santa Cruz var. Santa Cruz de la Sierra. Santa Cruz, C Bolivia 17°49′S 63°11′W
62 G12 Santa Cruz Libertador, C Chile 34°38′S 71°27′W
42 K13 Santa Cruz Guanacaste, W Costa Rica 10°15′N 85°35′W
44 I12 Santa Cruz W Jamaica 18°03′N 77°43′W
64 P6 Santa Cruz Madeira, Portugal, NE Atlantic Ocean 32°43′N 16°47′W
35 N10 Santa Cruz California, W USA 36°58′N 122°01′W
63 H20 Santa Cruz off. Provincia de Santa Cruz. ◈ province S Argentina
57 O18 Santa Cruz ◈ department E Bolivia
Santa Cruz see Puerto Santa Cruz
Santa Cruz see Viru-Viru
Santa Cruz Barillas see Barillas
59 O18 Santa Cruz Cabrália Bahia, E Brazil 16°17′S 39°03′W
Santa Cruz de El Seibo see El Seibo
64 N11 Santa Cruz de la Palma La Palma, Islas Canarias, Spain, NE Atlantic Ocean 28°41′N 17°46′W
Santa Cruz de la Sierra see Santa Cruz
105 O9 Santa Cruz de la Zarza Castilla-La Mancha, C Spain 39°59′N 03°10′W
42 C5 Santa Cruz del Quiché Quiché, W Guatemala 15°02′N 91°06′W
105 N8 Santa Cruz del Retamar Castilla-La Mancha, C Spain 40°08′N 04°14′W
Santa Cruz del Seibo see El Seibo
44 G7 Santa Cruz del Sur Camagüey, C Cuba 20°44′N 78°00′W
105 O11 Santa Cruz de Mudela Castilla-La Mancha, C Spain 38°37′N 03°27′W
64 Q11 Santa Cruz de Tenerife Tenerife, Islas Canarias, Spain, NE Atlantic Ocean 28°28′N 16°15′W
64 P11 Santa Cruz de Tenerife ◈ province Islas Canarias, Spain, NE Atlantic Ocean
60 K9 Santa Cruz do Rio Pardo São Paulo, S Brazil 22°52′S 49°37′W
61 H15 Santa Cruz do Sul Rio Grande do Sul, S Brazil 29°42′S 52°25′W
57 C17 Santa Cruz, Isla var. Indefatigable Island, Isla Chávez. island Galapagos Islands, Ecuador, E Pacific Ocean
40 F8 Santa Cruz, Isla island NW Mexico
35 Q15 Santa Cruz Island island California, W USA
187 Q10 Santa Cruz Islands island group E Solomon Islands
Santa Cruz, Provincia de see Santa Cruz
63 I22 Santa Cruz, Río 〰 S Argentina
36 L15 Santa Cruz River 〰 Arizona, SW USA
61 C17 Santa Elena Entre Ríos, E Argentina 30°58′S 59°47′W
42 F2 Santa Elena Cayo, W Belize 17°08′N 89°04′W
25 R16 Santa Elena Texas, SW USA 26°43′N 98°30′W
56 A7 Santa Elena, Bahía de bay W Ecuador
55 R10 Santa Elena de Uairén Bolívar, E Venezuela 04°40′N 61°03′W
42 K12 Santa Elena, Península peninsula NW Costa Rica
56 A7 Santa Elena, Punta headland W Ecuador 02°11′S 81°00′W
104 L11 Santa Eufemia Andalucía, S Spain 38°36′N 04°54′W
107 N21 Santa Eufemia, Golfo di gulf S Italy
105 S4 Santa Eulalia de Gállego Aragón, NE Spain 42°16′N 00°46′W
105 V11 Santa Eulalia del Río Ibiza, Spain, W Mediterranean Sea 39°00′N 01°33′E
61 B17 Santa Fe Santa Fe, C Argentina 31°36′S 60°47′W
44 C6 Santa Fé var. La Fe. Isla de la Juventud, W Cuba 21°45′N 82°45′W
43 R16 Santa Fé Veraguas, C Panama 08°29′N 80°50′W
105 N14 Santa Fe Andalucía, S Spain 37°11′N 03°43′W
37 S10 Santa Fe state capital New Mexico, SW USA 35°41′N 105°56′W
61 B15 Santa Fe off. Provincia de Santa Fe. ◈ province C Argentina
Santa Fe see Bogotá
Santa Fe de Bogotá see Bogotá
60 J7 Santa Fé do Sul São Paulo, S Brazil 20°13′S 50°56′W
57 B18 Santa Fe, Isla var. Barrington Island. island Galapagos Islands, Ecuador, E Pacific Ocean
Santa Fe, Provincia de see Santa Fe
23 V9 Santa Fe River 〰 Florida, SE USA
59 M15 Santa Filomena Piauí, E Brazil 09°06′S 45°52′W
40 G10 Santa Genoveva ▲ NW Mexico 23°07′N 109°56′W
153 S14 Santahar Rajshahi, NW Bangladesh 24°45′N 89°03′E
60 G11 Santa Helena Paraná, S Brazil 24°53′S 54°19′W
54 J5 Santa Inés Lara, N Venezuela 10°37′N 69°18′W
63 G24 Santa Inés, Isla island S Chile
62 J13 Santa Isabel La Pampa, C Argentina 36°11′S 66°59′W
43 U14 Santa Isabel Colón, N Panama 09°31′N 79°12′W
186 L8 Santa Isabel var. Bughotu. island N Solomon Islands
Santa Isabel see Malabo
58 D11 Santa Isabel do Rio Negro Amazonas, NW Brazil 0°40′S 64°56′W
61 C15 Santa Lucia Corrientes, NE Argentina 28°58′S 59°05′W
57 I17 Santa Lucía Puno, S Peru 15°45′S 70°34′W
61 F20 Santa Lucía var. Santa Lucia. Canelones, S Uruguay 34°26′S 56°25′W
42 B6 Santa Lucía Cotzumalguapa Escuintla, SW Guatemala 14°20′N 91°00′W
107 L23 Santa Lucia del Mela Sicilia, Italy, C Mediterranean Sea 38°08′N 15°17′E
35 O11 Santa Lucia Range ▲▲ California, W USA
40 D9 Santa Margarita, Isla island NW Mexico
62 J7 Santa María Catamarca, N Argentina 26°51′S 66°02′W
61 G15 Santa Maria Rio Grande do Sul, S Brazil 29°41′S 53°48′W
35 P13 Santa Maria California, W USA 34°56′N 120°25′W
64 Q4 Santa Maria ✈ Santa Maria, Azores, Portugal, NE Atlantic Ocean
64 P3 Santa Maria island Azores, Portugal, NE Atlantic Ocean
Santa Maria see Gaua
Santa María Asunción Tlaxiaco see Tlaxiaco
40 G9 Santa María, Bahía bay W Mexico
83 L21 Santa Maria, Cabo de headland S Mozambique 26°05′S 32°58′E
104 G15 Santa Maria, Cabo de headland S Portugal 36°57′N 07°55′W
44 J4 Santa Maria, Cape headland Long Island, C The Bahamas 23°40′N 75°20′W
107 J17 Santa Maria Capua Vetere Campania, S Italy 41°05′N 14°15′E
104 G7 Santa Maria da Feira Aveiro, N Portugal 40°55′N 08°32′W
59 M17 Santa Maria da Vitória Bahia, E Brazil 13°26′S 44°09′W
55 N9 Santa Maria de Erebato Bolívar, SE Venezuela 05°09′N 64°50′W
55 N6 Santa María de Ipire Guárico, C Venezuela 08°51′N 65°21′W
Santa Maria del Buen Aire see Buenos Aires
40 J8 Santa María del Oro Durango, C Mexico 25°57′N 105°22′W
41 N12 Santa María del Río San Luis Potosí, C Mexico 21°48′N 100°42′W
Santa Maria di Castellabate see Castellabate
107 Q20 Santa Maria di Leuca, Capo headland SE Italy 39°48′N 18°21′E
108 K10 Santa Maria-im-Munstertal Graubünden, SE Switzerland 46°36′N 10°25′E
57 B18 Santa María, Isla var. Isla Floreana, Charles Island. island Galapagos Islands, Ecuador, E Pacific Ocean
40 J3 Santa María, Laguna de ◎ N Mexico
61 G16 Santa Maria, Rio 〰 S Brazil
43 R16 Santa María, Río 〰 C Panama
36 J12 Santa Maria River 〰 Arizona, SW USA
107 G15 Santa Marinella Lazio, C Italy 42°01′N 11°51′E
54 F4 Santa Marta Magdalena, N Colombia 11°14′N 74°13′W
104 J11 Santa Marta Extremadura, W Spain 38°37′N 06°39′W
54 F4 Santa Marta, Sierra Nevada de ▲▲ NE Colombia
Santa Maura see Lefkáda
35 S15 Santa Monica California, W USA 34°01′N 118°29′W
116 F10 Sântana Ger. Sankt Anna, Hung. Újszentanna; prev. Sîntana. Arad, W Romania 46°20′N 21°30′E
61 F16 Santana, Coxilha de hill range S Brazil

◆ Country
● Country Capital
◇ Dependent Territory
○ Dependent Territory Capital
◈ Administrative Regions
✈ International Airport
▲ Mountain
▲▲ Mountain Range
☗ Volcano
〰 River
◎ Lake
▦ Reservoir

SANTANA DA BOA VISTA – SASMIK, CAPE

61 H16 **Santana da Boa Vista** Rio Grande do Sul, S Brazil *30°52′S 53°03′W*
61 F16 **Santana do Livramento** *prev.* Livramento. Rio Grande do Sul, S Brazil *30°52′S 55°30′W*
105 N2 **Santander** Cantabria, N Spain *43°28′N 03°48′W*
54 F8 **Santander** *off.* Departamento de Santander. ◐ *province* C Colombia
Santander, Departamento de *see* Santander
Santander Jiménez *see* Jiménez
Sant'Andrea *see* Svetac
107 B20 **Sant'Antioco** Sardegna, Italy, C Mediterranean Sea *39°03′N 08°28′E*
105 V11 **Sant Antoni de Portmany** *Cas.* San Antonio Abad. Ibiza, Spain, W Mediterranean Sea *38°58′N 01°18′E*
105 Y10 **Santanyí** Mallorca, Spain, W Mediterranean Sea *39°22′N 03°07′E*
104 J13 **Santa Olalla del Cala** Andalucía, S Spain *37°54′N 06°13′W*
35 R15 **Santa Paula** California, W USA *34°21′N 119°03′W*
36 L4 **Santaquin** Utah, W USA *39°58′N 111°46′W*
58 I12 **Santarém** Pará, N Brazil *02°26′S 54°41′W*
104 G10 **Santarém** *anc.* Scalabis. Santarém, W Portugal *39°14′N 08°40′W*
104 G10 **Santarém** ◐ *district* C Portugal
44 F4 **Santaren Channel** *channel* W The Bahamas
54 K10 **Santa Rita** Vichada, E Colombia *04°51′N 68°27′W*
188 B16 **Santa Rita** SW Guam
42 H5 **Santa Rita** Cortés, NW Honduras *15°10′N 87°54′W*
40 E9 **Santa Rita** Baja California Sur, NW Mexico *27°29′N 100°33′W*
54 H5 **Santa Rita** Zulia, NW Venezuela *10°35′N 71°30′W*
59 I19 **Santa Rita de Araguaia** Goiás, S Brazil *17°17′S 53°13′W*
59 M16 **Santa Rita de Cassia** *var.* Cássia. Bahia, E Brazil *11°03′S 44°16′W*
61 D14 **Santa Rosa** Corrientes, NE Argentina *28°18′S 58°04′W*
62 K13 **Santa Rosa** La Pampa, C Argentina *36°38′S 64°15′W*
61 G14 **Santa Rosa** Rio Grande do Sul, S Brazil *27°50′S 54°29′W*
58 E10 **Santa Rosa** Roraima, N Brazil *03°41′N 62°29′W*
56 B8 **Santa Rosa** El Oro, SW Ecuador *03°29′S 79°57′W*
57 I16 **Santa Rosa** Puno, S Peru *14°38′S 70°45′W*
34 M7 **Santa Rosa** California, W USA *38°27′N 122°42′W*
37 U11 **Santa Rosa** New Mexico, SW USA *34°54′N 104°43′W*
55 O6 **Santa Rosa** Anzoátegui, NE Venezuela *09°37′N 64°20′W*
42 A3 **Santa Rosa** *off.* Departamento de Santa Rosa. ◐ *department* SE Guatemala
Santa Rosa *see* Santa Rosa de Copán
63 J15 **Santa Rosa, Bajo de** *basin* E Argentina
42 F6 **Santa Rosa de Copán** *var.* Santa Rosa. Copán, W Honduras *14°48′N 88°43′W*
54 E8 **Santa Rosa de Osos** Antioquia, C Colombia *06°40′N 75°27′W*
Santa Rosa, Departamento de *see* Santa Rosa
35 Q15 **Santa Rosa Island** *island* California, W USA
23 O9 **Santa Rosa Island** *island* Florida, SE USA
40 E6 **Santa Rosalía** Baja California Sur, NW Mexico *27°20′N 112°20′W*
54 K6 **Santa Rosalía** Portuguesa, NW Venezuela *09°02′N 69°01′W*
188 C15 **Santa Rosa, Mount** ▲ NE Guam
35 V16 **Santa Rosa Mountains** ▲▲ California, W USA
35 T2 **Santa Rosa Range** ▲▲ Nevada, W USA
62 M8 **Santa Sylvina** Chaco, N Argentina *27°49′S 61°09′W*
Santa Tecla *see* Nueva San Salvador
61 B19 **Santa Teresa** Santa Fe, C Argentina *33°30′S 60°45′W*
59 O20 **Santa Teresa** Espírito Santo, SE Brazil *19°51′S 40°49′W*
61 E21 **Santa Teresita** Buenos Aires, E Argentina *36°32′S 56°41′W*
61 H19 **Santa Vitória do Palmar** Rio Grande do Sul, S Brazil *33°32′S 53°25′W*
35 Q14 **Santa Ynez River** ≈ California, W USA
Sant Carles de la Rápida *see* Sant Carles de la Ràpita
105 U7 **Sant Carles de la Ràpita** *var.* Sant Carles de la Rápida. Cataluña, NE Spain *40°37′N 00°36′E*
105 W5 **Sant Celoni** Cataluña, NE Spain *41°39′N 02°25′E*
35 U17 **Santee** California, W USA *32°50′N 116°58′W*
21 T13 **Santee River** ≈ South Carolina, SE USA
40 K15 **San Telmo, Punta** *headland* SW Mexico *18°19′N 103°30′W*
107 O17 **Santeramo in Colle** Puglia, SE Italy *40°47′N 16°45′E*
107 M23 **San Teresa di Riva** Sicilia, Italy, C Mediterranean Sea *38°00′N 15°25′E*
105 X5 **Sant Feliu de Guíxols** *var.* San Feliú de Guixols. Cataluña, NE Spain *41°47′N 03°02′E*
105 W6 **Sant Feliu de Llobregat** Cataluña, NE Spain *41°22′N 02°00′E*
106 C7 **Santhià** Piemonte, NE Italy *45°21′N 08°11′E*
61 F15 **Santiago** Rio Grande do Sul, S Brazil *29°11′S 54°52′W*
62 H11 **Santiago** *var.* Gran Santiago. ● (Chile) Santiago, C Chile *33°30′S 70°40′W*
45 N8 **Santiago** *var.* Santiago de los Caballeros. N Dominican Republic *19°27′N 70°42′W*
40 G10 **Santiago** Baja California Sur, NW Mexico *23°32′N 109°47′W*
41 O8 **Santiago** Nuevo León, NE Mexico *25°22′N 100°09′W*
43 R16 **Santiago** Veraguas, S Panama *08°06′N 80°59′W*
57 E16 **Santiago** Ica, SW Peru *14°14′S 75°44′W*
62 H11 **Santiago** *off.* Región Metropolitana de Santiago, *var.* Metropolitan. ◐ *region* C Chile
76 D10 **Santiago** *var.* São Tiago. *island* Ilhas de Sotavento, S Cape Verde
62 H11 **Santiago** ✈ Santiago, C Chile *33°27′S 70°40′W*
104 G3 **Santiago** ✈ Galicia, NW Spain
Santiago *see* Santiago de Cuba, Cuba
Santiago *see* Grande de Santiago, Río, Mexico
Santiago *see* Santiago de Compostela
42 B6 **Santiago Atitlán** Sololá, SW Guatemala *14°39′N 91°12′W*
43 Q16 **Santiago, Cerro** ▲ W Panama *08°27′N 81°42′W*
104 G3 **Santiago de Compostela** *var.* Santiago, *Eng.* Compostella; *anc.* Campus Stellae. Galicia, NW Spain *42°52′N 08°33′W*
44 I8 **Santiago de Cuba** *var.* Santiago. Santiago de Cuba, E Cuba *20°01′N 75°51′W*
Santiago de Guayaquil *see* Guayaquil
62 K8 **Santiago del Estero** Santiago del Estero, C Argentina *27°51′S 64°16′W*
61 A15 **Santiago del Estero** *off.* Provincia de Santiago del Estero. ◐ *province* N Argentina
Santiago del Estero, Provincia de *see* Santiago del Estero
40 I8 **Santiago de los Caballeros** Sinaloa, W Mexico *25°33′N 107°22′W*
Santiago de los Caballeros *see* Santiago, Dominican Republic
Santiago de los Caballeros *see* Ciudad de Guatemala, Guatemala
42 F8 **Santiago de María** Usulután, SE El Salvador *13°28′N 88°28′W*
104 F12 **Santiago do Cacém** Setúbal, S Portugal *38°01′N 08°42′W*
40 J12 **Santiago Ixcuintla** Nayarit, C Mexico *21°50′N 105°11′W*
Santiago Jamiltepec *see* Jamiltepec
24 L11 **Santiago Mountains** ▲▲ Texas, SW USA
40 J9 **Santiago Papasquiaro** Durango, C Mexico *25°00′N 105°27′W*
Santiago Pinotepa Nacional *see* Pinotepa Nacional
Santiago, Región Metropolitana de *see* Santiago
56 C8 **Santiago, Río** ≈ N Peru
40 M10 **San Tiburcio** Zacatecas, C Mexico *24°08′N 101°29′W*
105 N2 **Santillana** Cantabria, N Spain *43°24′N 04°06′W*
54 I5 **San Timoteo** Zulia, NW Venezuela *09°50′N 71°05′W*
Santi Quaranta *see* Sarandë
Santissima Trinidad *see* Jilong
105 O12 **Santisteban del Puerto** Andalucía, S Spain *38°15′N 03°11′W*
105 S12 **Sant Joan d'Alacant** *Cast.* San Juan de Alicante. Valenciana, E Spain *38°26′N 00°27′W*
105 U7 **Sant Jordi, Golf de** *gulf* NE Spain
105 U11 **Sant Josep de sa Talaia** *var.* San Jose. Ibiza, Spain, W Mediterranean Sea *38°55′N 1°18′E*
162 G6 **Santmargats** *var.* Holboo. Dzavhan, W Mongolia *48°35′N 95°25′E*
105 T8 **Sant Mateu** Valenciana, E Spain *40°28′N 00°10′E*
25 S7 **Santo** Texas, SW USA *32°35′N 98°06′W*
Santo *see* Espiritu Santo
60 M10 **Santo Amaro, Ilha de** *island* SE Brazil
61 G14 **Santo Ângelo** Rio Grande do Sul, S Brazil *28°17′S 54°15′W*
76 C9 **Santo Antão** *island* Ilhas de Barlavento, N Cape Verde
60 J10 **Santo Antônio da Platina** Paraná, S Brazil *23°20′S 50°05′W*
58 C13 **Santo Antônio do Içá** Amazonas, N Brazil *03°05′S 67°56′W*
57 Q18 **Santo Corazón, Río** ≈ E Bolivia
44 E5 **Santo Domingo** Villa Clara, C Cuba *22°35′N 80°15′W*
45 O9 **Santo Domingo** *prev.* Ciudad Trujillo. ● (Dominican Republic) SE Dominican Republic *18°30′N 69°57′W*
40 E8 **Santo Domingo** Baja California Sur, NW Mexico *25°34′N 112°00′W*
40 M10 **Santo Domingo** San Luis Potosí, C Mexico *23°18′N 101°42′W*
42 L10 **Santo Domingo** Chontales, S Nicaragua *12°15′N 84°59′W*
105 P4 **Santo Domingo de la Calzada** La Rioja, N Spain *42°26′N 02°57′W*
56 B6 **Santo Domingo de los Colorados** Pichincha, NW Ecuador *0°13′S 79°09′W*
Santo Domingo Tehuantepec *see* Tehuantepec
55 O6 **San Tomé** Anzoátegui, NE Venezuela *08°58′N 64°08′W*
San Tomé de Guayana *see* Ciudad Guayana
105 R13 **Santomera** Murcia, SE Spain *38°03′N 01°05′W*
105 O2 **Santoña** Cantabria, N Spain *43°27′N 03°28′W*
Santorin *see* Santoríni
115 K22 **Santoríni** *var.* Santorin, *prev.* Thíra; *anc.* Thera. *island* Kykládes, Greece, Aegean Sea
60 M10 **Santos** São Paulo, S Brazil *23°56′S 46°22′W*
65 J17 **Santos Plateau** *undersea feature* SW Atlantic Ocean *25°00′S 43°00′W*
104 G6 **Santo Tirso** Porto, N Portugal *41°20′N 08°25′W*
40 B2 **Santo Tomás** Baja California Norte, NW Mexico *31°32′N 116°26′W*
42 L10 **Santo Tomás** Chontales, S Nicaragua *12°04′N 85°02′W*
42 G5 **Santo Tomás de Castilla** Izabal, E Guatemala *15°40′N 88°36′W*
40 B2 **Santo Tomás, Punta** *headland* NW Mexico *31°30′N 116°40′W*
57 H16 **Santo Tomás, Río** ≈ C Peru
57 B18 **Santo Tomás, Volcán** ⩓ Galapagos Islands, Ecuador, E Pacific Ocean *0°46′S 91°01′W*
61 F14 **Santo Tomé** Corrientes, NE Argentina *28°31′S 56°03′W*
Santo Tomé de Guayana *see* Ciudad Guayana
98 H10 **Santpoort** Noord-Holland, W Netherlands *52°26′N 04°38′E*
Santurce *see* Santurtzi
105 O2 **Santurtzi** *var.* Santurce, Santurzi. País Vasco, N Spain *43°20′N 03°03′W*
Santurzi *see* Santurtzi
63 G20 **San Valentín, Cerro** ▲ S Chile *46°36′S 73°17′W*
42 F8 **San Vicente** San Vicente, C El Salvador *13°38′N 88°42′W**
40 C2 **San Vicente** Baja California Norte, NW Mexico *31°20′N 116°15′W*
188 H6 **San Vicente** Saipan, S Northern Mariana Islands
42 B9 **San Vicente** ◐ *department* E El Salvador
104 I10 **San Vicente de Alcántara** Extremadura, W Spain *39°21′N 07°07′W*
105 N2 **San Vicente de Barakaldo** *var.* Baracaldo, *Basq.* San Bizenti-Barakaldo. País Vasco, N Spain *43°17′N 02°59′W*
57 E15 **San Vicente de Cañete** *var.* Cañete. Lima, W Peru *13°06′S 76°23′W*
104 M2 **San Vicente de la Barquera** Cantabria, N Spain *43°23′N 04°24′W*
54 E12 **San Vicente del Caguán** Caquetá, S Colombia *02°07′N 74°47′W*
42 F8 **San Vincente, Volcán de** ⩓ C El Salvador *13°34′N 88°50′W*
43 O15 **San Vito** Puntarenas, SE Costa Rica *08°49′N 82°58′W*
106 I7 **San Vito al Tagliamento** Friuli-Venezia Giulia, NE Italy *45°54′N 12°55′E*
107 H23 **San Vito, Capo** *headland* Sicilia, Italy, C Mediterranean Sea *38°11′N 12°41′E*
107 P18 **San Vito dei Normanni** Puglia, SE Italy *40°40′N 17°42′E*
160 L17 **Sanya** *var.* Ya Xian. Hainan, S China *18°25′N 109°27′E*
83 J16 **Sanyati** ≈ N Zimbabwe
25 Q16 **San Ygnacio** Texas, SW USA *27°04′N 99°26′W*
160 L6 **Sanyuan** Shaanxi, C China *34°40′N 108°56′E*
123 P11 **Sanyyakhtakh** Respublika Sakha (Yakutiya), NE Russian Federation *60°34′N 124°09′E*
146 J15 **S. A.Nyýazow Adyndaky** *Rus.* Imeni S. A. Niyazova. Maryyskiy Velayat, S Turkmenistan *36°44′N 62°23′E*
82 C10 **Sanza Pombo** Uíge, NW Angola *07°20′S 16°00′E*
Sanzyô *see* Sanjō
104 G14 **São Bartolomeu de Messines** Faro, S Portugal *37°12′N 08°16′W*
60 M10 **São Bernardo do Campo** São Paulo, S Brazil *23°45′S 46°34′W*
61 F15 **São Borja** Rio Grande do Sul, S Brazil *28°35′S 56°01′W*
104 H14 **São Brás de Alportel** Faro, S Portugal *37°09′N 07°55′W*
60 M10 **São Caetano do Sul** São Paulo, S Brazil *23°37′S 46°34′W*
60 L9 **São Carlos** São Paulo, S Brazil *22°02′S 47°53′W*
59 P16 **São Cristóvão** Sergipe, E Brazil *10°59′S 37°10′W*
61 F15 **São Fancisco de Assis** Rio Grande do Sul, S Brazil *29°32′S 55°07′W*
58 K13 **São Félix** Pará, NE Brazil *06°43′S 51°56′W*
São Félix *see* São Félix do Araguaia
59 J16 **São Félix do Araguaia** *var.* São Félix. Mato Grosso, W Brazil *11°36′S 50°40′W*
59 J14 **São Félix do Xingu** Pará, NE Brazil *06°38′S 51°59′W*
60 Q9 **São Fidélis** Rio de Janeiro, SE Brazil *21°37′S 41°40′W*
76 D10 **São Filipe** Fogo, S Cape Verde *14°52′N 24°29′W*
60 K12 **São Francisco do Sul** Santa Catarina, S Brazil *26°17′S 48°39′W*
60 K12 **São Francisco, Ilha de** *island* S Brazil
59 P16 **São Francisco, Rio** ≈ E Brazil
61 G16 **São Gabriel** Rio Grande do Sul, S Brazil *30°17′S 54°17′W*
60 P10 **São Gonçalo** Rio de Janeiro, SE Brazil *22°48′S 43°03′W*
81 H23 **Sao Hill** Iringa, S Tanzania *08°19′S 35°11′E*
60 R9 **São João da Barra** Rio de Janeiro, SE Brazil *21°39′S 41°04′W*
104 G7 **São João da Madeira** Aveiro, N Portugal *40°52′N 08°28′W*
58 M12 **São João de Cortes** Maranhão, E Brazil *02°30′S 44°27′W*
59 M21 **São João del Rei** Minas Gerais, NE Brazil *21°08′S 44°15′W*
59 N15 **São João do Piauí** Piauí, E Brazil *08°21′S 42°14′W*
59 N14 **São João dos Patos** Maranhão, E Brazil *06°29′S 43°44′W*
58 C11 **São Joaquim** Amazonas, NW Brazil *0°08′S 67°10′W*
61 J14 **São Joaquim** Santa Catarina, S Brazil *28°20′S 49°55′W*
60 L7 **São Joaquim da Barra** São Paulo, S Brazil *20°36′S 47°50′W*
64 N2 **São Jorge** *island* Azores, Portugal, NE Atlantic Ocean
61 K14 **São José** Santa Catarina, S Brazil *27°34′S 48°39′W*
60 M8 **São José do Rio Pardo** São Paulo, S Brazil *21°37′S 46°52′W*
60 K8 **São José do Rio Preto** São Paulo, S Brazil *20°50′S 49°20′W*
60 N10 **São Jose dos Campos** São Paulo, S Brazil *23°07′S 45°52′W*
61 I17 **São Lourenço do Sul** Rio Grande do Sul, S Brazil *31°25′S 52°00′W*
58 M12 **São Luís** *state capital* Maranhão, NE Brazil *02°34′S 44°16′W*
58 F11 **São Luís** Roraima, N Brazil *01°11′N 60°15′W*
58 M12 **São Luís, Ilha de** *island* NE Brazil
61 F14 **São Luiz Gonzaga** Rio Grande do Sul, S Brazil *28°24′S 54°58′W*
São Mandol *see* São Manuel, Rio
47 U8 **Sao Manuel** ≈ C Brazil
59 H15 **São Manuel, Rio** *var.* São Mandol, Teles Pirés. ≈ C Brazil
58 C11 **São Marcelino** Amazonas, NW Brazil *0°53′N 67°16′W*
58 N12 **São Marcos, Baía de** *bay* N Brazil
59 O20 **São Mateus** Espírito Santo, SE Brazil *18°44′S 39°53′W*
60 J12 **São Mateus do Sul** Paraná, S Brazil *25°58′S 50°29′W*
64 P3 **São Miguel** *island* Azores, Portugal, NE Atlantic Ocean
60 G13 **São Miguel d'Oeste** Santa Catarina, S Brazil *26°45′S 53°34′W*
45 P9 **Saona, Isla** *island* SE Dominican Republic
172 H12 **Saondzou** ▲ Grande Comore, NW Comoros
103 R10 **Saône** ≈ E France
103 Q9 **Saône-et-Loire** ◐ *department* C France
76 D9 **São Nicolau** *Eng.* Saint Nicholas. *island* Ilhas de Barlavento, N Cape Verde
60 M10 **São Paulo** *state capital* São Paulo, S Brazil *23°33′S 46°39′W*
60 K9 **São Paulo** *off.* Estado de São Paulo. ◐ *state* S Brazil
São Paulo de Loanda *see* Luanda
São Paulo, Estado de *see* São Paulo
São Pedro do Rio Grande do Sul *see* Rio Grande
104 H7 **São Pedro do Sul** Viseu, N Portugal *40°46′N 07°58′W*
64 K13 **São Pedro e São Paulo** *undersea feature* C Atlantic Ocean *01°25′N 28°54′W*
59 M14 **São Raimundo das Mangabeiras** Maranhão, E Brazil *07°00′S 45°30′W*
59 Q14 **São Roque, Cabo de** *headland* E Brazil *05°29′S 35°16′W*
São Salvador *see* Salvador, Brazil
São Salvador/São Salvador do Congo *see* M'Banza Congo, Angola
60 N10 **São Sebastião, Ilha de** *island* S Brazil
83 N19 **São Sebastião, Ponta** *headland* C Mozambique *22°09′S 35°33′E*
104 F13 **São Teotónio** Beja, S Portugal *37°30′N 08°41′W*
São Tiago *see* Santiago
79 B18 **São Tomé** ● (Sao Tome and Principe) São Tomé, S Sao Tome and Principe *0°22′N 06°41′E*
79 B18 **São Tomé** ✈ São Tomé, S Sao Tome and Principe *0°24′N 06°39′E*
79 B18 **São Tomé** *Eng.* Saint Thomas. *island* S Sao Tome and Principe
79 B17 **Sao Tome and Principe** *off.* Democratic Republic of Sao Tome and Principe. ◆ *republic* E Atlantic Ocean
Sao Tome and Principe, Democratic Republic of *see* Sao Tome and Principe
74 H9 **Saoura, Oued** ≈ NW Algeria
60 M10 **São Vicente** *Eng.* Saint Vincent. São Paulo, S Brazil *23°55′S 46°25′W*
64 O5 **São Vicente** Madeira, Portugal, NE Atlantic Ocean *32°48′N 17°03′W*
76 C9 **São Vicente** *Eng.* Saint Vincent. *island* Ilhas de Barlavento, N Cape Verde
104 F14 **São Vicente, Cabo de** *Eng.* Cape Saint Vincent, *Port.* Cabode São Vicente. *cape* S Portugal
São Vicente, Cabo de *see* São Vicente, Cabo de
Sápai *see* Sápes
Sapaleri, Cerro *see* Zapaleri, Cerro
Saparoea *see* Saparua
171 S13 **Saparua** *prev.* Saparoea. Pulau Saparau, C Indonesia *03°35′S 128°37′E*
168 L11 **Sapat** Sumatera, W Indonesia *0°18′S 103°18′E*
77 U17 **Sapele** Delta, S Nigeria *05°54′N 05°43′E*
23 X7 **Sapelo Island** *island* Georgia, SE USA
23 X7 **Sapelo Sound** *sound* Georgia, SE USA
114 K13 **Sápes** *var.* Sápai. Anatolikí Makedonía kai Thráki, NE Greece *41°02′N 25°44′E*
Sapiéntza *see* Sapiénza
115 D22 **Sapiénza** *var.* Sapiéntza. *island* S Greece
138 F12 **Sapir** *prev.* Sappir. Southern, S Israel *30°43′N 35°11′E*
61 I15 **Sapiranga** Rio Grande do Sul, S Brazil *29°39′S 50°58′W*
114 K13 **Sápka** ▲▲ NE Greece
105 X9 **Sa Pobla** Mallorca, Spain, W Mediterranean Sea *39°46′N 03°03′E*
56 D11 **Saposoa** San Martín, N Peru *06°53′S 76°45′W*
119 F16 **Sapotskin** *Pol.* Sopockinie, *Rus.* Sapotskino, Sopotskin. Hrodzyenskaya Voblasts', W Belarus *53°50′N 23°39′E*
77 P13 **Sapoui** *var.* Sapouy. S Burkina Faso *11°34′N 01°44′W*
Sapouy *see* Sapoui
Sappir *see* Sapir
165 S4 **Sapporo** Hokkaidō, NE Japan *43°05′N 141°21′E*
107 M19 **Sapri** Campania, S Italy *40°05′N 15°36′E*
169 T16 **Sapudi, Pulau** *island* S Indonesia
27 P9 **Sapulpa** Oklahoma, C USA *36°00′N 96°06′W*
142 J4 **Saqqez** *var.* Saghez, Sakiz, Saqqiz. Kordestān, NW Iran *36°31′N 46°16′E*
Saqqiz *see* Saqqez
139 U8 **Sarābādī** Wāsiţ, E Iraq *33°00′N 44°52′E*
167 P10 **Sara Buri** *var.* Saraburi. Saraburi, C Thailand *14°32′N 100°53′E*
Saraburi *see* Sara Buri
Sarafjagān *see* Salafchegān
24 K9 **Saragosa** Texas, SW USA *31°03′N 103°39′W*
Saragossa *see* Zaragoza
Saragt *see* Sarahs
56 B8 **Saraguro** Loja, S Ecuador *03°42′S 79°18′W*
146 I15 **Sarahs** *var.* Saragt, *Rus.* Serakhs. Ahal Welaýaty, S Turkmenistan *36°33′N 61°10′E*
126 M6 **Sarai** Ryazanskaya Oblast', W Russian Federation *53°43′N 39°59′E*
Sarāi *see* Sarāy
154 M12 **Saraipāli** Chhattīsgarh, C India *21°21′N 83°01′E*
149 T9 **Sarāi Sidhu** Punjab, E Pakistan *30°35′N 72°02′E*
93 M15 **Säräisniemi** Kainuu, C Finland *64°25′N 26°50′E*
113 I14 **Sarajevo** ● (Bosnia and Herzegovina) Federacija Bosne I Hercegovine, SE Bosnia and Herzegovina *43°53′N 18°24′E*
112 I13 **Sarajevo** ✈ Federacija Bosne I Hercegovine, C Bosnia and Herzegovina *43°49′N 18°21′E*
143 V4 **Sarakhs** Khorāsān-e Razavī, NE Iran *36°50′N 61°00′E*
115 H17 **Sarakíniko, Akrotírio** *headland* Évvoia, C Greece *38°46′N 23°43′E*
115 I18 **Sarakinó** *island* Vóreies Sporádes, Greece, Aegean Sea
127 V7 **Saraktash** Orenburgskaya Oblast', W Russian Federation *51°46′N 56°23′E*
30 L15 **Sara, Lake** ▣ Illinois, N USA
23 N8 **Saraland** Alabama, S USA *30°49′N 88°04′W*
55 V9 **Saramacca** ◐ *district* N Suriname
55 V10 **Saramacca Rivier** ≈ C Suriname
166 M2 **Saramati** ▲ N Myanmar (Burma) *25°46′N 95°01′E*
145 R10 **Saran'** *Kaz.* Saran. Karaganda, C Kazakhstan *49°47′N 73°02′E*
18 K7 **Saranac Lake** New York, NE USA *44°18′N 74°06′W*
18 K7 **Saranac River** ≈ New York, NE USA
Saranda *see* Sarandë
113 L23 **Sarandë** *var.* Saranda, *It.* Porto Edda; *prev.* Santi Quaranta. Vlorë, S Albania *39°53′N 20°E*
61 H14 **Sarandi** Rio Grande do Sul, S Brazil *27°57′S 52°58′W*
61 F19 **Sarandí del Yí** Durazno, C Uruguay *33°18′S 55°38′W*
61 F19 **Sarandí Grande** Florida, S Uruguay *33°43′S 56°19′W*
171 Q8 **Sarangani Islands** *island group* S Philippines
127 P5 **Saransk** Respublika Mordoviya, W Russian Federation *54°11′N 45°10′E*
115 C14 **Sarantáporos** ≈ N Greece
114 H9 **Sarantsi** Sofia, W Bulgaria *42°43′N 23°46′E*
127 T3 **Sarapul** Udmurtskaya Respublika, NW Russian Federation *56°26′N 53°52′E*
Sarâqeb *see* Sarāqib
138 I3 **Sarāqib** *Fr.* Sarâqeb. Idlib, N Syria *35°52′N 36°48′E*
54 J5 **Sarare** Lara, N Venezuela *09°47′N 69°10′W*
55 O10 **Sarariña** Amazonas, S Venezuela *04°10′N 64°31′W*
143 S10 **Sar Ashk** Kermān, C Iran
23 V13 **Sarasota** Florida, SE USA *27°20′N 82°31′W*
117 O11 **Sarata** Odes'ka Oblast', SW Ukraine *46°01′N 29°40′E*
116 I10 **Sărățel** *Hung.* Szeretfalva. Bistrița-Năsăud, N Romania *47°02′N 24°24′E*
25 X10 **Saratoga** Texas, SW USA *30°15′N 94°31′W*
18 K10 **Saratoga Springs** New York, NE USA *43°04′N 73°47′W*
127 P8 **Saratov** Saratovskaya Oblast', W Russian Federation *51°33′N 45°58′E*
127 P8 **Saratovskaya Oblast'** ◐ *province* W Russian Federation
127 Q7 **Saratovskoye Vodokhranilishche** ▣ W Russian Federation
143 X13 **Sarāvān** Sīstān va Balūchestān, SE Iran *27°11′N 62°35′E*
Saravan/Saravane *see* Salavan
169 S9 **Sarawak** ◐ *state* East Malaysia
Sarawak *see* Kuching
139 U6 **Sarāy** *var.* Sarāi. Diyālá, E Iraq *34°06′N 45°06′E*
136 D10 **Saray** Tekirdağ, NW Turkey *41°27′N 27°56′E*
76 J12 **Saraya** SE Senegal *12°50′N 11°45′W*
143 W14 **Sarbāz** Sīstān va Balūchestān, SE Iran *26°38′N 61°13′E*
143 U8 **Sarbīsheh** Khorāsān-e Janūbī, E Iran *32°35′N 59°50′E*
111 J24 **Sárbogárd** Fejér, C Hungary *46°54′N 18°36′E*
Şărcad *see* Sarkad
27 S7 **Sarcoxie** Missouri, C USA *37°04′N 94°07′W*
152 L11 **Sārda** *Nep.* Kali. ≈ India/Nepal
152 G10 **Sardārshahr** Rājasthān, NW India *28°30′N 74°30′E*
107 C18 **Sardegna** *Eng.* Sardinia. ◐ *region* Italy, C Mediterranean Sea
107 A18 **Sardegna** *Eng.* Sardinia. *island* Italy, C Mediterranean Sea
42 K13 **Sardinal** Guanacaste, NW Costa Rica *10°30′N 85°38′W*
54 G7 **Sardinata** Norte de Santander, N Colombia *08°07′N 72°47′W*
Sardinia *see* Sardegna
120 K8 **Sardinia-Corsica Trough** *undersea feature* Tyrrhenian Sea, C Mediterranean Sea
22 L2 **Sardis** Mississippi, S USA *34°25′N 89°55′W*
22 L2 **Sardis Lake** ▣ Mississippi, S USA
27 P12 **Sardis Lake** ▣ Oklahoma, C USA
92 H12 **Sarek** ▲▲ N Sweden
92 H11 **Sarektjåhkkå** ▲ N Sweden *67°28′N 17°56′E*
Sar-e Pol *see* Sar-e Pul
Sar-e Pol *see* Sar-e Pol-e Żahāb
142 J6 **Sar-e Pol-e Żahāb** *var.* Sar-e Pol, Sar-i Pul. Kermānshāhān, W Iran *34°28′N 45°52′E*
149 N3 **Sar-e Pul** *var.* Sar-i-Pul; *prev.* Sar-e Pol. Sar-e Pul, N Afghanistan *36°16′N 65°55′E*
149 O3 **Sar-e Pul** ◐ *province* N Afghanistan
Sarera, Teluk *see* Cenderawasih, Teluk
147 T13 **Sarez, Kŭli** *Rus.* Sarezskoye Ozero. ◎ SE Tajikistan
Sarezskoye Ozero *see* Sarez, Kŭli
64 G10 **Sargasso Sea** *sea* W Atlantic Ocean
149 U8 **Sargodha** Punjab, NE Pakistan *32°06′N 72°48′E*
78 I13 **Sarh** *prev.* Fort-Archambault. Moyen-Chari, S Chad *09°08′N 18°22′E*
143 P4 **Sārī** *var.* Sari, Sāri. Māzandarān, N Iran *36°37′N 53°05′E*
115 N23 **Saría** *island* SE Greece
Sariasiya *see* Sariosiyo
40 F3 **Saric** Sonora, NW Mexico *31°08′N 111°22′W*
188 K6 **Sarigan** *island* C Northern Mariana Islands
136 D14 **Sarıgöl** Manisa, SW Turkey *38°16′N 28°41′E*
139 T6 **Sārihah** At Ta'mím, E Iraq *34°34′N 44°38′E*
137 R12 **Sarıkamış** Kars, NE Turkey *40°18′N 42°36′E*
169 R9 **Sarikei** Sarawak, East Malaysia *02°07′N 111°30′E*
147 U12 **Sarikol Range** *Rus.* Sarykol'skiy Khrebet. ▲▲ China/Tajikistan
181 Y7 **Sarina** Queensland, NE Australia *21°34′S 149°12′E*
Sarine *see* La Sarine
105 S5 **Sariñena** Aragón, NE Spain *41°47′N 00°10′W*
147 O13 **Sariosiyo** *Rus.* Sariasiya. Surkhondaryo Viloyati, S Uzbekistan *38°25′N 67°51′E*
Sar-i-Pul *see* Sar-e Pol, Afghanistan
Sar-i Pul *see* Sar-e Pol-e Żahāb, Iran
Sariqamish Kŭli *see* Sarygamyş Köli
149 V1 **Sarī Qūl** *Rus.* Ozero Zurkul', *Taj.* Zŭrkŭl. ◎ Afghanistan/Tajikistan *see also* Zŭrkŭl
Sarī Qūl *see* Zŭrkŭl
75 Q12 **Sarīr Tībistī** *var.* Serir Tibesti. *desert* S Libya
25 S15 **Sarita** Texas, SW USA *27°14′N 97°48′W*
163 W14 **Sariwŏn** SW North Korea *38°30′N 125°52′E*
114 P12 **Sarıyer** İstanbul, NW Turkey *41°11′N 29°03′E*
97 L26 **Sark** *Fr.* Sercq. *island* Channel Islands
111 N24 **Sarkad** *Rom.* Şărcad. Békés, SE Hungary *46°44′N 21°25′E*
145 W14 **Sarkand** *Kaz.* Sarqan. Almaty, SW Kazakhstan *45°24′N 79°55′E*
152 D11 **Sarkāri Tala** Rājasthān, NW India *27°39′N 70°52′E*
136 G15 **Şarkîkaraağaç** *var.* Şarki Karaağaç. Isparta, SW Turkey *38°04′N 31°22′E*
Şarki Karaağaç *see* Şarkîkaraağaç
136 L13 **Şarkışla** Sivas, C Turkey *39°21′N 36°27′E*
136 C11 **Şarköy** Tekirdağ, NW Turkey *40°37′N 27°07′E*
Sárköz *see* Livada
Sarlat *see* Sarlat-la-Canéda
102 M13 **Sarlat-la-Canéda** *var.* Sarlat. Dordogne, SW France *44°54′N 01°12′E*
109 S3 **Sarleinsbach** Oberösterreich, N Austria *48°33′N 13°55′E*
Šarma *see* Ash Sharmah
171 Y12 **Sarmi** Papua, E Indonesia *01°51′S 138°45′E*
63 I19 **Sarmiento** Chubut, S Argentina *45°38′S 69°07′W*
63 H25 **Sarmiento, Monte** ▲ S Chile *54°28′S 70°49′W*
94 J11 **Särna** Dalarna, C Sweden *61°40′N 13°10′E*
108 F8 **Sarnen** Obwalden, C Switzerland *46°54′N 08°15′E*
108 F9 **Sarner See** ◎ C Switzerland
14 D16 **Sarnia** Ontario, S Canada *42°58′N 82°23′W*
116 L3 **Sarny** Rivnens'ka Oblast', NW Ukraine *51°20′N 26°35′E*
171 O13 **Saroako** Sulawesi, C Indonesia *02°31′S 121°18′E*
118 L13 **Sarochyna** *Rus.* Sorochino. Vitsyebskaya Voblasts', N Belarus *55°12′N 28°45′E*
168 L12 **Sarolangun** Sumatera, W Indonesia *02°17′S 102°39′E*
165 U3 **Saroma** Hokkaidō, NE Japan *44°01′N 143°43′E*
165 V3 **Saroma-ko** ◎ Hokkaidō, NE Japan
Saronic Gulf *see* Saronikós Kólpos
115 H20 **Saronikós Kólpos** *Eng.* Saronic Gulf. *gulf* S Greece
106 D7 **Saronno** Lombardia, N Italy *45°38′N 09°02′E*
136 B11 **Saros Körfezi** *gulf* NW Turkey
111 N20 **Sárospatak** Borsod-Abaúj-Zemplén, NE Hungary *48°18′N 21°30′E*
127 O4 **Sarov** *prev.* Sarova. Respublika Mordoviya, SW Russian Federation *54°39′N 43°09′E*
Sarova *see* Sarov
127 P12 **Sarpa** Respublika Kalmykiya, SW Russian Federation *47°00′N 45°42′E*
127 P12 **Sarpa, Ozero** ◎ SW Russian Federation
Šar Planina *see* Mali i Sharrit
95 I16 **Sarpsborg** Østfold, S Norway *59°16′N 11°07′E*
139 U5 **Sarqalā** At Ta'mím, N Iraq
Sarqan *see* Sarkand
103 U4 **Sarralbe** Moselle, NE France *49°02′N 07°01′E*
Sarre *see* Saar, France/Germany
Sarre *see* Saarland, Germany
103 U5 **Sarrebourg** *Ger.* Saarburg. Moselle, NE France *48°43′N 07°03′E*
Sarrebruck *see* Saarbrücken
103 U4 **Sarreguemines** *prev.* Saargemund. Moselle, NE France *49°06′N 07°04′E*
104 I3 **Sarria** Galicia, NW Spain *42°47′N 07°25′W*
105 S8 **Sarrión** Aragón, NE Spain *40°09′N 00°49′W*
42 F4 **Sarstoon** *Sp.* Río Sarstún. ≈ Belize/Guatemala
Sarstún, Río *see* Sarstoon
123 Q9 **Sartang** ≈ NE Russian Federation
103 X16 **Sartène** Corse, France, C Mediterranean Sea *41°38′N 08°58′E*
102 K7 **Sarthe** ◐ *department* NW France
102 K7 **Sarthe** ≈ N France
115 H15 **Sárti** Kentrikí Makedonía, N Greece *40°05′N 24°E*
Sartu *see* Daqing
165 T1 **Sarufutsu** Hokkaidō, NE Japan *45°20′N 142°03′E*
Saruhan *see* Manisa
152 G9 **Sarūpsar** Rājasthān, NW India *29°25′N 73°50′E*
137 U13 **Şärur** *prev.* Il'ichevsk. SW Azerbaijan *39°30′N 44°59′E*
Sarvani *see* Marneuli
111 G23 **Sárvár** Vas, W Hungary *47°15′N 16°59′E*
143 P11 **Sarvestān** Fārs, S Iran *29°16′N 53°13′E*
171 W12 **Sarwon** Papua, E Indonesia *0°58′S 136°08′E*
145 P17 **Saryagash** *Kaz.* Saryaghash. Yuzhnyy Kazakhstan, S Kazakhstan *41°29′N 69°10′E*
Saryaghash *see* Saryagash
145 R9 **Saryarka** *Eng.* Kazakh Uplands, Kirghiz Steppe. *uplands* C Kazakhstan
147 W8 **Sary-Bulak** Narynskaya Oblast', C Kyrgyzstan *41°56′N 75°44′E*
147 U10 **Sary-Bulak** Oshskaya Oblast', SW Kyrgyzstan *40°49′N 73°44′E*
117 S14 **Sarych, Mys** *headland* S Ukraine *44°23′N 33°44′E*
147 Z7 **Sary-Dzhaz** *var.* Aksu He. ≈ China/Kyrgyzstan *see also* Aksu He
Sary-Dzhaz *see* Aksu He
146 F8 **Sarygamyş Köli** *Rus.* Sarykamyshkoye Ozero, *Uzb.* Sariqamish Kŭli. *salt lake* Kazakhstan/Uzbekistan
144 G13 **Sarykamys** *Kaz.* Saryqamys. Mangistau, SW Kazakhstan *45°58′N 53°30′E*
Sarykamyshkoye Ozero *see* Sarygamyş Köli
145 N7 **Sarykol'** *prev.* Uritskiy. Kustanay, N Kazakhstan *53°19′N 65°34′E*
Sarykol'skiy Khrebet *see* Sarikol Range
144 M10 **Sarykopa, Ozero** ◎ C Kazakhstan
145 V15 **Saryozek** *Kaz.* Saryözek. Almaty, SE Kazakhstan *44°22′N 77°57′E*
144 E10 **Saryozen** *Kaz.* Kishiözen; *prev.* Malyy Uzen'. ≈ Kazakhstan/Russian Federation
Saryqamys *see* Sarykamys
145 S13 **Saryshagan** *Kaz.* Saryshaghan. Karaganda, SE Kazakhstan *46°05′N 73°38′E*
Saryshaghan *see* Saryshagan
145 O13 **Sarysu** ≈ S Kazakhstan
147 T11 **Sary-Tash** Oshskaya Oblast', SW Kyrgyzstan *39°44′N 73°14′E*
145 T12 **Saryterek** Karaganda, C Kazakhstan *47°46′N 74°06′E*
Saryyazynskoye Vodokhranilishche *see* Saryýazy Suw Howdany
146 J15 **Saryýazy Suw Howdany** *Rus.* Saryyazynskoye Vodokhranilishche. ▣ S Turkmenistan
145 T14 **Saryyesik-Atyrau, Peski** *desert* E Kazakhstan
106 E10 **Sarzana** Liguria, NW Italy *44°07′N 09°59′E*
188 B17 **Sasalaguan, Mount** ▲ S Guam
153 O14 **Sasarām** Bihār, N India *24°58′N 84°01′E*
186 M8 **Sasari, Mount** ▲ Santa Isabel, N Solomon Islands *08°09′S 159°32′E*
164 C13 **Sasebo** Nagasaki, Kyūshū, SW Japan *33°10′N 129°42′E*
14 I9 **Saseginaga, Lac** ◎ Québec, SE Canada
Saseno *see* Sazan
11 R13 **Saskatchewan** ◐ *province* SW Canada
11 U14 **Saskatchewan** ≈ Manitoba/Saskatchewan, C Canada
11 T15 **Saskatoon** Saskatchewan, S Canada *52°10′N 106°40′W*
11 T15 **Saskatoon** ✈ Saskatchewan, S Canada *52°15′N 107°05′W*
123 N7 **Saskylakh** Respublika Sakha (Yakutiya), NE Russian Federation *71°56′N 114°07′E*
42 L7 **Saslaya, Cerro** ▲ N Nicaragua *13°52′N 85°06′W*
38 G17 **Sasmik, Cape** *headland* Tanaga Island, Alaska, USA *51°36′N 177°55′W*

◆ Country
● Country Capital
◇ Dependent Territory
○ Dependent Territory Capital
◐ Administrative Regions
✈ International Airport
▲ Mountain
▲▲ Mountain Range
⩓ Volcano
≈ River
◎ Lake
▣ Reservoir

SASNOVY BOR – SEGAF, KEPULAUAN

119 N19 **Sasnovy Bor** *Rus.* Sosnovyy Bor. Homyel'skaya Voblasts', SE Belarus 52°32′N 29°35′E
127 N5 **Sasovo** Ryazanskaya Oblast', W Russian Federation 54°19′N 41°54′E
25 S12 **Saspamco** Texas, SW USA 29°13′N 98°18′W
109 W9 **Sass** *var.* Sassbach. ≈ SE Austria
76 M17 **Sassandra** S Ivory Coast 04°58′N 06°08′W
76 M17 **Sassandra** *var.* Ibo, Sassandra Fleuve. ≈ S Ivory Coast
Sassandra Fleuve *see* Sassandra
107 B17 **Sassari** Sardegna, Italy, C Mediterranean Sea 40°44′N 08°33′E
Sassbach *see* Sass
98 H11 **Sassenheim** Zuid-Holland, W Netherlands 52°14′N 04°31′E
Sassmacken *see* Valdemārpils
100 O7 **Sassnitz** Mecklenburg-Vorpommern, NE Germany 54°32′N 13°39′E
99 E16 **Sas van Gent** Zeeland, SW Netherlands 51°13′N 03°48′E
145 W12 **Sasykkol', Ozero** ◎ E Kazakhstan
117 O12 **Sasyk, Ozero** *Rus.* Ozero Sasyk Kunduk, *var.* Ozero Kunduk. ◎ SW Ukraine
76 J12 **Satadougou** Kayes, SW Mali 12°40′N 11°25′W
93 K18 **Satakunta** ❖ *region* W Finland
164 C17 **Sata-misaki** Kyūshū, SW Japan
26 I7 **Satanta** Kansas, C USA 37°23′N 102°00′W
155 E15 **Sātāra** Mahārāshtra, W India 17°41′N 73°59′E
192 G15 **Sātaua** Savai'i, NW Samoa 13°26′S 172°40′W
188 M16 **Satawal** *island* Caroline Islands, C Micronesia
189 R17 **Satawan Atoll** *atoll* Mortlock Islands, C Micronesia
Sätbaev *see* Satpayev
23 Y12 **Satellite Beach** Florida, SE USA 28°10′N 80°35′W
95 M14 **Säter** Dalarna, C Sweden 60°21′N 15°45′E
Sathmar *see* Satu Mare
23 V7 **Satilla River** ≈ Georgia, SE USA
57 F14 **Satipo** *var.* San Francisco de Satipo. Junín, C Peru 11°19′S 74°37′W
122 F11 **Satka** Chelyabinskaya Oblast', C Russian Federation 55°08′N 58°54′E
153 T16 **Satkhira** Khulna, SW Bangladesh 22°43′N 89°06′E
146 J13 **Şatlyk** *Rus.* Shatlyk. Mary Welaýaty, C Turkmenistan 37°55′N 61°00′E
154 K9 **Satna** *prev.* Sutna. Madhya Pradesh, C India 24°33′N 80°50′E
103 R11 **Satolas** ✈ (Lyon) Rhône, E France 45°44′N 05°01′E
111 N20 **Sátoraljaújhely** Borsod-Abaúj-Zemplén, NE Hungary 48°24′N 21°39′E
145 O12 **Satpayev** *Kaz.* Sätbaev; *prev.* Nikol'skiy. Karaganda, C Kazakhstan 47°59′N 67°27′E
154 G11 **Sātpura Range** ▲▲ C India
165 Q10 **Satsuma-Sendai** Miyagi, Honshū, C Japan 38°16′N 140°52′E
Satsuma-Sendai *see* Sendai
167 P12 **Sattahip** *var.* Ban Sattahip, Ban Sattahipp. Chon Buri, S Thailand 12°36′N 100°56′E
92 L11 **Sattanen** Lappi, NE Finland 67°31′N 26°35′E
Satul *see* Satun
116 H9 **Satulung** *Hung.* Kővárhosszúfalu. Maramureş, N Romania 47°34′N 23°26′E
Satul-Vechi *see* Staro Selo
116 G8 **Satu Mare** *Ger.* Sathmar, *Hung.* Szatmárnémeti. Satu Mare, NW Romania 47°46′N 22°55′E
116 G8 **Satu Mare** ❖ *county* NW Romania
167 N16 **Satun** *var.* Satul, Setul. Satun, SW Thailand 06°40′N 100°01′E
192 G16 **Satupa'itea** Savai'i, W Samoa 13°46′S 172°26′W
Sau *see* Sava
14 F14 **Sauble** ≈ Ontario, S Canada
14 F13 **Sauble Beach** Ontario, S Canada 44°36′N 81°15′W
61 C16 **Sauce** Corrientes, NE Argentina 30°05′S 58°46′W
Sauce *see* Juan L. Lacaze
36 K15 **Sauceda Mountains** ▲▲ Arizona, SW USA
61 C17 **Sauce de Luna** Entre Ríos, E Argentina 31°15′S 59°09′W
63 L15 **Sauce Grande, Río** ≈ E Argentina
40 K6 **Saucillo** Chihuahua, N Mexico 28°01′N 105°17′W
95 D15 **Sauda** Rogaland, S Norway 59°38′N 06°23′E
145 Q16 **Saudakent** *Kaz.* Saūdakent; *prev.* Baykadam, *Kaz.* Bayqadam. Zhambyl, S Kazakhstan 43°49′N 69°56′E
92 J2 **Sauðárkrókur** Norðurland Vestra, N Iceland 65°45′N 19°39′W
141 P9 **Saudi Arabia** *off.* Kingdom of Saudi Arabia, Al 'Arabīyah as Su'ūdīyah, *Ar.* Al Mamlakah al 'Arabīyah as Su'ūdīyah. ◆ *monarchy* SW Asia
Saudi Arabia, Kingdom of *see* Saudi Arabia
101 D19 **Sauer** *var.* Sûre. ≈ NW Europe *see also* Sûre
Sauer *see* Sûre
101 F15 **Sauerland** *forest* W Germany
14 F14 **Saugeen** ≈ Ontario, S Canada
18 K12 **Saugerties** New York, NE USA 42°04′N 73°55′W
Saugor *see* Sāgar
10 K15 **Saugstad, Mount** ▲ British Columbia, SW Canada 52°12′N 126°35′W
Sāūjbulāgh *see* Mahābād
102 J11 **Saujon** Charente-Maritime, W France 45°40′N 00°54′W
29 T7 **Sauk Centre** Minnesota, N USA 45°44′N 94°57′W
30 L8 **Sauk City** Wisconsin, N USA 43°16′N 89°43′W
29 U7 **Sauk Rapids** Minnesota, N USA 45°35′N 94°09′W
55 Y11 **Saül** C French Guiana 03°37′N 53°12′W
103 O7 **Sauldre** ≈ C France
101 I23 **Saulgau** Baden-Württemberg, SW Germany 48°03′N 09°28′E
103 Q8 **Saulieu** Côte d'Or, C France 47°15′N 04°15′E
118 G8 **Saulkrasti** C Latvia 57°14′N 24°25′E
15 S6 **Sault-aux-Cochons, Rivière du** ≈ Québec, SE Canada
31 Q4 **Sault Sainte Marie** Michigan, N USA 46°29′N 84°22′W
12 F14 **Sault Ste. Marie** Ontario, S Canada 46°30′N 84°17′W
145 P7 **Saumalkol'** *prev.* Volodarskoye. Severnyy Kazakhstan, N Kazakhstan 53°19′N 68°05′E
190 E13 **Sauma, Pointe** *headland* Île Alofi, W Wallis and Futuna 14°21′S 177°58′W
171 T16 **Saumlaki** *var.* Saumlakki. Pulau Yamdena, E Indonesia 07°53′S 131°18′E
Saumlakki *see* Saumlaki
15 R12 **Saumon, Rivière au** ≈ Québec, SE Canada
102 K8 **Saumur** Maine-et-Loire, NW France 47°16′N 00°04′W
185 F23 **Saunders, Cape** *headland* South Island, New Zealand 45°53′S 170°40′E
195 N13 **Saunders Coast** *physical region* Antarctica
65 B23 **Saunders Island** *island* NW Falkland Islands
65 C24 **Saunders Island Settlement** Saunders Island, NW Falkland Islands 51°22′S 60°05′W
82 F11 **Saurimo** *Port.* Henrique de Carvalho, Vila Henrique de Carvalho. Lunda Sul, NE Angola 09°39′S 20°24′E
55 S11 **Sauriwaunawa** S Guyana 03°10′N 59°51′W
82 D12 **Sautar** Malanje, NW Angola 11°10′S 18°26′E
45 S13 **Sauteurs** N Grenada 12°14′N 61°38′W
102 K13 **Sauveterre-de-Guyenne** Gironde, SW France 44°43′N 00°02′W
119 O14 **Sava** Mahilyowskaya Voblasts', E Belarus 54°22′N 30°49′E
42 J5 **Savá** Colón, N Honduras 15°30′N 86°16′W
84 H11 **Sava** *Eng.* Save, *Ger.* Sau, *Hung.* Száva. ≈ SE Europe
33 Y8 **Savage** Montana, NW USA 47°28′N 104°17′W
183 N16 **Savage River** Tasmania, SE Australia 41°34′S 145°15′E
77 R15 **Savalou** S Benin 07°59′N 01°58′E
30 K10 **Savanna** Illinois, N USA 42°05′N 90°09′W
23 X6 **Savannah** Georgia, SE USA 32°02′N 81°01′W
27 R2 **Savannah** Missouri, C USA 39°57′N 94°49′W
20 H10 **Savannah** Tennessee, S USA 35°12′N 88°15′W
21 O12 **Savannah River** ≈ Georgia/South Carolina, SE USA
167 S9 **Savannakhét** *var.* Khanthabouli. Savannakhét, S Laos 16°38′N 104°49′E
44 H12 **Savanna-La-Mar** W Jamaica 18°13′N 78°08′W
12 B10 **Savant Lake** ◎ Ontario, S Canada
155 F17 **Savanūr** Karnātaka, W India 14°58′N 75°19′E
93 J16 **Sävar** Västerbotten, N Sweden 63°52′N 20°33′E
Savaria *see* Szombathely
154 C11 **Sāvarkundla** *var.* Kundla. Gujarāt, W India 21°21′N 71°20′E
116 F11 **Săvârşin** *Hung.* Soborsin; *prev.* Săvirşin. Arad, W Romania 46°00′N 22°15′E
136 C13 **Savaştepe** Balıkesir, W Turkey 39°20′N 27°38′E
147 P11 **Savat** *Rus.* Savat. Sirdaryo Viloyati, E Uzbekistan 40°03′N 68°35′E
Savat *see* Savat
Sávdijári *see* Skaulo
77 R15 **Savè** SE Benin 08°04′N 02°29′E
83 N18 **Save** Inhambane, E Mozambique 21°07′S 34°35′E
102 L16 **Save** ≈ S France
83 L17 **Save** *var.* Sabi. ≈ Mozambique/Zimbabwe *see also* Sabi
Save *see* Sava
Save *see* Sabi
142 M6 **Sāveh** Markazī, W Iran 35°00′N 50°22′E
116 L8 **Săveni** Botoşani, NE Romania 47°57′N 26°52′E
103 N16 **Saverdun** Ariège, S France 43°15′N 01°34′E
103 U5 **Saverne** *var.* Zabern; *anc.* Tres Tabernae. Bas-Rhin, NE France 48°45′N 07°22′E
106 B9 **Savigliano** Piemonte, NW Italy 44°39′N 07°39′E
Savigsivik *see* Savissivik
109 U10 **Savinja** ≈ N Slovenia
106 H11 **Savio** ≈ C Italy
Săvirşin *see* Săvârşin
197 O11 **Savissivik** *var.* Savigsivik. ❖ Qaasuitsup, N Greenland
93 N18 **Savitaipale** Etelä-Karjala, SE Finland 61°12′N 27°43′E
113 J15 **Šavnik** C Montenegro 42°57′N 19°04′E
108 I9 **Savognin** Graubünden, S Switzerland 46°34′N 09°35′E
103 T12 **Savoie** ❖ *department* E France
106 C10 **Savona** Liguria, NW Italy 44°18′N 08°29′E
93 N17 **Savonlinna** *Swe.* Nyslott. Etelä-Savo, E Finland 61°51′N 28°56′E
93 N17 **Savonranta** Etelä-Savo, E Finland 62°10′N 29°10′E
38 K10 **Savoonga** Saint Lawrence Island, Alaska, USA 63°40′N 170°29′W
30 M13 **Savoy** Illinois, N USA 40°03′N 88°15′W
117 O8 **Savran'** Odes'ka Oblast', SW Ukraine 48°07′N 30°00′E
137 R11 **Şavşat** Artvin, NE Turkey 41°15′N 42°30′E
95 L19 **Sävsjö** Jönköping, S Sweden 57°25′N 14°40′E
Savu, Kepulauan *see* Sawu, Kepulauan
92 M11 **Savukoski** Lappi, NE Finland 67°17′N 28°14′E
Savu, Pulau *see* Sawu, Pulau
187 Y14 **Savusavu** Vanua Levu, N Fiji 16°48′S 179°20′E
171 O17 **Savu Sea** *Ind.* Laut Sawu. *sea* S Indonesia
83 H17 **Savute** North-West, N Botswana 18°33′S 24°06′E
139 N7 **Şawāb Uqlat** *well* W Iraq
138 M7 **Şawāb, Wādī aş** *dry watercourse* W Iraq
152 H13 **Sawāi Mādhopur** Rājasthān, N India 26°00′N 76°22′E
Sawakin *see* Suakin
167 R8 **Sawang Daen Din** Sakon Nakhon, E Thailand 17°28′N 103°27′E
167 O8 **Sawankhalok** *var.* Swankalok. Sukhothai, NW Thailand 17°19′N 99°50′E
165 P13 **Sawara** Chiba, Honshū, S Japan 35°52′N 140°31′E
37 R5 **Sawatch Range** ▲▲ Colorado, C USA
141 N12 **Sawdā', Jabal** ▲ SW Saudi Arabia 18°15′N 42°26′E
75 P9 **Sawdā', Jabal as** ▲▲ C Libya
Sawdirī *see* Sodiri
97 F14 **Sawel Mountain** ▲ C Northern Ireland, United Kingdom 54°49′N 07°04′W
75 X10 **Sawhāj** *var.* Sawhāj, *var.* Sohâg, Suliag. C Egypt 26°28′N 31°44′E
Sawhāj *see* Sawhāj
77 O15 **Sawla** N Ghana 09°14′N 02°26′W
141 X12 **Şawqirah** *var.* Suqrah. S Oman 18°16′N 56°34′E
141 X12 **Şawqirah, Dawḩat** *var.* Ghubbat Sawqirah, Sukra Bay, Suqrah Bay. *bay* S Oman
Şawqirah, Ghubbat *see* Şawqirah, Dawḩat
183 V5 **Sawtell** New South Wales, SE Australia 30°22′S 153°04′E
138 K7 **Şawt, Wādī aş** *dry watercourse* S Syria
171 O17 **Sawu, Kepulauan** *var.* Kepulauan Savu. *island group* S Indonesia
Sawu, Laut *see* Savu Sea
171 O17 **Sawu, Pulau** *var.* Pulau Sawu. *island* Kepulauan Sawu, S Indonesia
105 S12 **Sax** Valenciana, E Spain 38°33′N 00°49′W
Saxe *see* Sachsen
108 C11 **Saxon** Valais, SW Switzerland 46°07′N 07°09′E
Saxony *see* Sachsen
Saxony-Anhalt *see* Sachsen-Anhalt
77 R12 **Say** Niamey, SW Niger 13°08′N 02°20′E
15 V7 **Sayabec** Québec, SE Canada 48°33′N 67°42′W
Sayaboury *see* Xaignabouli
145 U12 **Sayak** *Kaz.* Sayaq. Karaganda, E Kazakhstan 46°54′N 77°17′E
57 D14 **Sayán** Lima, W Peru 11°10′S 77°08′W
129 T6 **Sayanskiy Khrebet** ▲▲ S Russian Federation
Sayaq *see* Sayak
146 K13 **Saýat** *Rus.* Sayat. Lebap Welaýaty, E Turkmenistan 38°44′N 63°51′E
42 D3 **Sayaxché** Petén, N Guatemala 16°34′N 90°14′W
Şaydā/Sayida *see* Saïda
162 J7 **Sayhan** *var.* Hüremt. Bulgan, C Mongolia 48°40′N 102°33′E
163 N10 **Sayhandulaan** *var.* Öldziyt. Dornogovĭ, SE Mongolia 44°42′N 109°10′E
162 K9 **Sayhan-Ovoo** *var.* Ongi. Dundgovĭ, C Mongolia 45°27′N 103°58′E
141 T15 **Sayhūt** E Yemen 15°18′N 51°16′E
29 U14 **Saylorville Lake** ⊞ Iowa, C USA
Saymenskiy Kanal *see* Saimaa Canal
163 N10 **Saynshand** Dornogovĭ, SE Mongolia 44°51′N 110°07′E
Saynshand *see* Sevrey
Sayn-Ust *see* Hohmorĭt
Say-Ötesh *see* Otes
138 J7 **Şayqal, Baḩr** ◎ S Syria
Sayrab *see* Sayrob
158 H4 **Sayram Hu** ◎ NW China
26 K11 **Sayre** Oklahoma, C USA 35°18′N 99°38′W
18 H12 **Sayre** Pennsylvania, NE USA 41°57′N 76°30′W
18 K15 **Sayreville** New Jersey, NE USA 40°27′N 74°19′W
147 N13 **Sayrob** *Rus.* Sayrab. Surkhondaryo Viloyati, S Uzbekistan 38°03′N 66°54′E
40 L13 **Sayula** Jalisco, SW Mexico 19°52′N 103°36′W
141 R14 **Say'ūn** *var.* Saywūn. C Yemen 15°53′N 48°32′E
Say-Utës *see* Otes
10 K16 **Sayward** Vancouver Island, British Columbia, SW Canada 50°20′N 126°01′W
Saywūn *see* Say'ūn
Sayyāl *see* As Sayyāl
139 U8 **Sayyid 'Abīd** *var.* Saiyid Abid. Wāsiţ, E Iraq 32°51′N 45°07′E
113 J22 **Sazan** *var.* Ishulli i Sazanit, *It.* Saseno. *island* SW Albania
Sazanit, Ishulli i *see* Sazan
Sazau/Sazawa *see* Sázava
111 E17 **Sázava** *var.* Sazau, *Ger.* Sazawa. ≈ C Czech Republic
114 K10 **Sazliyka** ≈ C Bulgaria
124 J14 **Sazonovo** Vologodskaya Oblast', NW Russian Federation 59°04′N 35°10′E
102 G6 **Scaër** Finistère, NW France 48°00′N 03°40′W
97 J15 **Scafell Pike** ▲ NW England, United Kingdom 54°26′N 03°10′W
Scalabis *see* Santarém
96 M2 **Scalloway** N Scotland, United Kingdom 60°10′N 01°17′W
38 M11 **Scammon Bay** Alaska, USA 61°50′N 165°34′W
Scammon Lagoon/Scammon, Laguna *see* Ojo de Liebre, Laguna
84 F7 **Scandinavia** *geophysical region* NW Europe
Scania *see* Skåne
96 K5 **Scapa Flow** *sea basin* N Scotland, United Kingdom
107 K26 **Scaramia, Capo** *headland* Sicilia, Italy, C Mediterranean Sea 36°46′N 14°29′E
14 H15 **Scarborough** Ontario, SE Canada 43°46′N 79°14′W
45 Z16 **Scarborough** *prev.* Port Louis. Tobago, Trinidad and Tobago 11°11′N 60°45′W
97 N16 **Scarborough** N England, United Kingdom 54°17′N 00°24′W
185 I17 **Scargill** Canterbury, South Island, New Zealand 42°57′S 172°57′E
96 E7 **Scarp** *island* NW Scotland, United Kingdom
Scarpanto *see* Kárpathos
Scarpanto Strait *see* Karpathou, Stenó
107 G25 **Scauri** Sicilia, Italy, C Mediterranean Sea 36°45′N 12°06′E
Scealg, Bá na *see* Ballinskelligs Bay
Scebeli *see* Shebeli
100 K10 **Schaale** ≈ N Germany
100 K9 **Schaalsee** ◎ N Germany
99 G18 **Schaerbeek** Brussels, C Belgium 50°52′N 04°21′E
108 G6 **Schaffhausen** *Fr.* Schaffhouse. Schaffhausen, N Switzerland 47°42′N 08°38′E
108 G6 **Schaffhausen** *Fr.* Schaffhouse. ❖ *canton* N Switzerland
Schaffhouse *see* Schaffhausen
98 I8 **Schagen** Noord-Holland, NW Netherlands 52°47′N 04°47′E
Schaken *see* Šakiai
98 M10 **Schalkhaar** Overijssel, E Netherlands 52°16′N 06°10′E
109 R3 **Schärding** Oberösterreich, N Austria 48°27′N 13°26′E
100 G9 **Scharhörn** *island* NW Germany
Schässburg *see* Sighişoara
Schaulen *see* Šiauliai
30 M10 **Schaumburg** Illinois, N USA 42°01′N 88°04′W
Schebschi Mountains *see* Shebshi Mountains
98 P6 **Scheemda** Groningen, NE Netherlands 53°10′N 06°58′E
100 I10 **Scheessel** Niedersachsen, NW Germany 53°11′N 09°33′E
13 N8 **Schefferville** Québec, E Canada 54°50′N 67°W
Schelde *see* Scheldt
99 D18 **Scheldt** *Dut.* Schelde, *Fr.* Escaut. ≈ W Europe
35 X5 **Schell Creek Range** ▲▲ Nevada, W USA
18 K10 **Schenectady** New York, NE USA 42°48′N 73°57′W
99 I17 **Scherpenheuvel** *Fr.* Montaigu. Vlaams Brabant, C Belgium 51°00′N 04°57′E
98 K11 **Scherpenzeel** Gelderland, C Netherlands 52°07′N 05°30′E
25 S12 **Schertz** Texas, SW USA 29°33′N 98°16′W
98 G11 **Scheveningen** Zuid-Holland, W Netherlands 52°07′N 04°18′E
98 G12 **Schiedam** Zuid-Holland, SW Netherlands 51°55′N 04°25′E
99 M24 **Schieren** Diekirch, NE Luxembourg 49°50′N 06°06′E
98 M4 **Schiermonnikoog** *Fris.* Skiermûntseach. Fryslân, N Netherlands 53°28′N 06°09′E
98 M4 **Schiermonnikoog** *Fris.* Skiermûntseach. *island* Waddeneilanden, N Netherlands
99 K14 **Schijndel** Noord-Brabant, S Netherlands 51°37′N 05°27′E
Schil *see* Jiu
99 H16 **Schilde** Antwerpen, N Belgium 51°14′N 04°35′E
Schillen *see* Zhilino
103 V5 **Schiltigheim** Bas-Rhin, NE France 48°38′N 07°47′E
106 G7 **Schio** Veneto, NE Italy 45°42′N 11°21′E
98 H10 **Schiphol** ✈ (Amsterdam) Noord-Holland, C Netherlands 52°18′N 04°48′E
Schippenbeil *see* Sępopol
Schiria *see* Şiria
Schivelbein *see* Świdwin
115 D22 **Schíza** *island* S Greece
175 U3 **Schjetman Reef** *reef* Antarctica
Schlackenwerth *see* Ostrov
109 R7 **Schladming** Steiermark, SE Austria 47°24′N 13°42′E
Schlan *see* Slaný
Schlanders *see* Silandro
100 I7 **Schlei** *inlet* N Germany
101 D17 **Schleiden** Nordrhein-Westfalen, W Germany 50°31′N 06°30′E
Schleiau *see* Szydłowiec
100 I7 **Schleswig** Schleswig-Holstein, N Germany 54°31′N 09°34′E
29 T13 **Schleswig** Iowa, C USA 42°10′N 95°27′W
100 H8 **Schleswig-Holstein** ❖ *state* N Germany
Schlettstadt *see* Sélestat
108 F7 **Schlieren** Zürich, N Switzerland 47°23′N 08°27′E
Schlochau *see* Człuchów
Schloppe *see* Człopa
101 I18 **Schlüchtern** Hessen, C Germany 50°19′N 09°27′E
101 J17 **Schmalkalden** Thüringen, C Germany 50°42′N 10°26′E
109 W2 **Schmida** ≈ NE Austria
65 P19 **Schmidt-Ott Seamount** *var.* Schmitt-Ott Seamount, Schmitt-Ott Tablemount. *undersea feature* SW Indian Ocean 39°37′S 13°00′E
Schmiegel *see* Śmigiel
Schmitt-Ott Seamount/Schmitt-Ott Tablemount *see* Schmidt-Ott Seamount
15 V3 **Schmon** ≈ Québec, SE Canada
101 M18 **Schneeberg** ▲ W Germany 50°03′N 11°51′E
Schneeberg *see* Veliki Snežnik
Schnee-Eifel *see* Schneifel
Schneekoppe *see* Sněžka
Schneidemühl *see* Piła
101 D18 **Schneifel** *var.* Schnee-Eifel. *plateau* W Germany
Schnelle Körös/Schnelle Kreisch *see* Crişul Repede
100 I11 **Schneverdingen** *var.* Schneverdingen (Wümme). Niedersachsen, NW Germany 53°07′N 09°48′E
Schneverdingen (Wümme) *see* Schneverdingen
Schoden *see* Skuodas
45 Q12 **Schœlcher** W Martinique 14°37′N 61°08′W
18 K10 **Schoharie** New York, NE USA 42°40′N 74°20′W
18 K11 **Schoharie Creek** ≈ New York, NE USA
115 J21 **Schoinoússa** *island* Kykládes, Greece, Aegean Sea
100 L13 **Schönebeck** Sachsen-Anhalt, C Germany 52°01′N 11°45′E
Schöneck *see* Skarszewy
100 O12 **Schönefeld** ✈ (Berlin) Berlin, NE Germany 52°23′N 13°29′E
101 K24 **Schongau** Bayern, S Germany 47°49′N 10°54′E
100 K13 **Schöningen** Niedersachsen, C Germany 52°07′N 10°58′E
Schönlanke *see* Trzcianka
Schönsee *see* Kowalewo Pomorskie
31 P10 **Schoolcraft** Michigan, N USA 42°05′N 85°39′W
98 O8 **Schoonebeek** Drenthe, NE Netherlands 52°39′N 06°57′E
98 I12 **Schoonhoven** Zuid-Holland, C Netherlands 51°57′N 04°51′E
98 H8 **Schoorl** Noord-Holland, NW Netherlands 52°42′N 04°40′E
Schooten *see* Schoten
101 F24 **Schopfheim** Baden-Württemberg, SW Germany 47°39′N 07°49′E
101 I21 **Schorndorf** Baden-Württemberg, S Germany 48°48′N 09°31′E
100 F10 **Schortens** Niedersachsen, NW Germany 53°31′N 07°57′E
99 H16 **Schoten** *var.* Schooten. Antwerpen, N Belgium 51°15′N 04°30′E
183 Q17 **Schouten Island** *island* Tasmania, SE Australia
186 C5 **Schouten Islands** *island group* NW Papua New Guinea
98 E13 **Schouwen** *island* SW Netherlands
Schreiberhau *see* Szklarska Poręba
109 U2 **Schrems** Niederösterreich, E Austria 48°48′N 15°05′E
101 L22 **Schrobenhausen** Bayern, SE Germany 48°33′N 11°14′E
18 L8 **Schroon Lake** ◎ New York, NE USA
108 J8 **Schruns** Vorarlberg, W Austria 47°04′N 09°54′E
Schubin *see* Szubin
25 U11 **Schulenburg** Texas, SW USA 29°40′N 96°54′W
Schuls *see* Scuol
108 E8 **Schüpfheim** Luzern, C Switzerland 47°02′N 07°23′E
35 S6 **Schurz** Nevada, W USA 38°55′N 118°48′W
101 I24 **Schussen** ≈ S Germany
Schüttenhofen *see* Sušice
29 R15 **Schuyler** Nebraska, C USA 41°25′N 97°04′W
18 L10 **Schuylerville** New York, NE USA 43°05′N 73°34′W
101 K20 **Schwabach** Bayern, SE Germany 49°20′N 11°02′E
Schwabenalb *see* Schwäbische Alb
101 I23 **Schwäbische Alb** *var.* Schwabenalb, *Eng.* Swabian Jura. ▲▲ S Germany
101 I22 **Schwäbisch Gmünd** *var.* Gmünd. Baden-Württemberg, SW Germany 48°49′N 09°48′E
101 I21 **Schwäbisch Hall** *var.* Hall. Baden-Württemberg, SW Germany 49°07′N 09°45′E
101 H16 **Schwalm** ≈ C Germany
109 V9 **Schwanberg** Steiermark, SE Austria 46°46′N 15°12′E
108 H8 **Schwanden** Glarus, E Switzerland 47°02′N 09°04′E
101 M20 **Schwandorf** Bayern, SE Germany 49°20′N 12°07′E
109 S5 **Schwanenstadt** Oberösterreich, NW Austria 48°03′N 13°47′E
169 S11 **Schwaner, Pegunungan** ▲▲ Borneo, N Indonesia
109 W5 **Schwarza** ≈ E Austria
109 P9 **Schwarzach** ≈ S Austria
101 M20 **Schwarzach** *Cz.* Černice. ≈ Czech Republic/Germany
Schwarzach *see* Schwarzach im Pongau
109 Q7 **Schwarzach im Pongau** *var.* Schwarzach. Salzburg, NW Austria 47°19′N 13°09′E
Schwarzawa *see* Svratka
101 N14 **Schwarze Elster** ≈ E Germany
Schwarze Körös *see* Crişul Negru
108 D9 **Schwarzenburg** Bern, W Switzerland 46°51′N 07°28′E
83 D21 **Schwarzrand** ▲▲ S Namibia
101 G23 **Schwarzwald** *Eng.* Black Forest. ▲▲ SW Germany
Schwarzwasser *see* Wda
39 P7 **Schwatka Mountains** ▲▲ Alaska, USA
109 N7 **Schwaz** Tirol, W Austria 47°21′N 11°44′E
109 Y4 **Schwechat** Niederösterreich, NE Austria 48°09′N 16°29′E
109 Y4 **Schwechat** ✈ (Wien) Wien, E Austria 48°04′N 16°31′E
100 P11 **Schwedt** Brandenburg, NE Germany 53°04′N 14°16′E
101 D19 **Schweich** Rheinland-Pfalz, SW Germany 49°49′N 06°44′E
Schweidnitz *see* Świdnica
101 J18 **Schweinfurt** Bayern, SE Germany 50°03′N 10°13′E
Schweiz *see* Switzerland
100 L9 **Schwerin** Mecklenburg-Vorpommern, N Germany 53°38′N 11°25′E
Schwerin *see* Skwierzyna
100 L9 **Schweriner See** ◎ N Germany
Schwertberg *see* Świecie
101 F15 **Schwerte** Nordrhein-Westfalen, W Germany 51°27′N 07°34′E
Schwiebus *see* Świebodzin
100 P13 **Schwielochsee** ◎ NE Germany
Schwihau *see* Švihov
Schwiz *see* Schwyz
108 G8 **Schwyz** *var.* Schwiz. Schwyz, C Switzerland 47°02′N 08°39′E
108 G8 **Schwyz** *var.* Schwiz. ❖ *canton* C Switzerland
14 J11 **Schyan** ≈ Québec, SE Canada
Schyl *see* Jiu
107 I24 **Sciacca** Sicilia, Italy, C Mediterranean Sea 37°31′N 13°05′E
Sciasciamana *see* Shashemenē
107 L26 **Scicli** Sicilia, Italy, C Mediterranean Sea 36°48′N 14°43′E
97 F25 **Scilly, Isles of** *island group* SW England, United Kingdom
111 H17 **Šcinawa** *Ger.* Steinau an der Elbe. Dolnośląskie, SW Poland 51°22′N 16°27′E
Scio *see* Chíos
31 S14 **Scioto River** ≈ Ohio, N USA
36 L5 **Scipio** Utah, W USA 39°15′N 112°06′W
33 X6 **Scobey** Montana, NW USA 48°47′N 105°25′W
183 T7 **Scone** New South Wales, SE Australia 32°02′S 150°51′E
Scoresby Sound/Scoresbysund *see* Ittoqqortoormiit
Scoresby Sund *see* Kangertittivaq
Scorno, Punta dello *see* Caprara, Punta
34 K3 **Scotia** California, W USA 40°34′N 124°07′W
47 Y14 **Scotia Plate** *tectonic feature*
47 V15 **Scotia Ridge** *undersea feature* S Atlantic Ocean
194 H2 **Scotia Sea** *sea* SW Atlantic Ocean
29 Q12 **Scotland** South Dakota, N USA 43°09′N 97°43′W
25 R5 **Scotland** Texas, SW USA 33°37′N 98°27′W
96 H11 **Scotland** ❖ *national region* Scotland, U K
21 W8 **Scotland Neck** North Carolina, SE USA 36°07′N 77°25′W
195 R13 **Scott Base** *NZ research station* Antarctica 77°52′S 167°18′E
10 J16 **Scott, Cape** *headland* Vancouver Island, British Columbia, SW Canada 50°43′N 128°24′W
26 I5 **Scott City** Kansas, C USA 38°28′N 100°55′W
27 Y7 **Scott City** Missouri, C USA 37°13′N 89°31′W
195 R14 **Scott Coast** *physical region* Antarctica
18 C15 **Scottdale** Pennsylvania, NE USA 40°05′N 79°35′W
195 Y11 **Scott Glacier** *glacier* Antarctica
195 Q17 **Scott Island** *island* Antarctica
26 L11 **Scott, Mount** ▲ Oklahoma, USA 34°52′N 98°34′W
32 G15 **Scott, Mount** ▲ Oregon, NW USA 42°53′N 122°06′W
34 M1 **Scott River** ≈ California, W USA
28 I13 **Scottsbluff** Nebraska, C USA 41°52′N 103°40′W
23 Q2 **Scottsboro** Alabama, S USA 34°40′N 86°01′W
31 P15 **Scottsburg** Indiana, N USA 38°42′N 85°47′W
183 P16 **Scottsdale** Tasmania, SE Australia 41°13′S 147°30′E
36 L13 **Scottsdale** Arizona, SW USA 33°31′N 111°54′W
45 O12 **Scotts Head Village** *var.* Cachacrou. S Dominica 15°12′N 61°22′W
192 L14 **Scott Shoal** *undersea feature* S Pacific Ocean
20 K7 **Scottsville** Kentucky, S USA 36°45′N 86°11′W
29 U14 **Scranton** Iowa, C USA 42°01′N 94°33′W
18 I13 **Scranton** Pennsylvania, NE USA 41°25′N 75°40′W
186 B6 **Screw** ≈ NW Papua New Guinea
29 R14 **Scribner** Nebraska, C USA 41°40′N 96°40′W
Scrobesbyrig' *see* Shrewsbury
14 I14 **Scugog** ≈ Ontario, SE Canada
14 I14 **Scugog, Lake** ◎ Ontario, SE Canada
97 N17 **Scunthorpe** E England, United Kingdom 53°35′N 00°39′W
108 K9 **Scuol** *Ger.* Schuls. Graubünden, E Switzerland 46°51′N 10°21′E
Scupi *see* Skopje
Scutari *see* Shkodër
113 K17 **Scutari, Lake** *Alb.* Liqeni i Shkodrës, *SCr.* Skadarsko Jezero. ◎ Albania/Montenegro
Scyros *see* Skýros
Scythopolis *see* Beit She'an
138 E11 **Sderot** *prev.* Sederot. Southern, S Israel 31°31′N 34°35′E
25 U13 **Seadrift** Texas, SW USA 28°25′N 96°42′W
21 Y4 **Seaford** *var.* Seaford City. Delaware, NE USA 38°39′N 75°35′W
Seaford City *see* Seaford
14 E15 **Seaforth** Ontario, S Canada 43°33′N 81°25′W
24 M6 **Seagraves** Texas, SW USA 32°56′N 102°33′W
11 X9 **Seal** ≈ Manitoba, C Canada
182 M10 **Sea Lake** Victoria, SE Australia 35°34′S 142°51′E
83 G26 **Seal, Cape** *headland* S South Africa 34°06′S 23°18′E
65 D26 **Sea Lion Islands** *island group* SE Falkland Islands
19 S8 **Seal Island** *island* Maine, NE USA
25 V11 **Sealy** Texas, SW USA 29°46′N 96°09′W
35 X12 **Searchlight** Nevada, W USA 35°27′N 114°54′W
27 V11 **Searcy** Arkansas, C USA 35°14′N 91°43′W
19 R7 **Searsport** Maine, NE USA 44°28′N 68°54′W
35 N10 **Seaside** California, W USA 36°36′N 121°51′W
32 F10 **Seaside** Oregon, NW USA 45°57′N 123°55′W
18 K16 **Seaside Heights** New Jersey, NE USA 39°56′N 74°03′W
32 H8 **Seattle** Washington, NW USA 47°35′N 122°20′W
32 H9 **Seattle-Tacoma** ✈ Washington, NW USA 47°04′N 122°27′W
185 J16 **Seaward Kaikoura Range** ▲▲ South Island, New Zealand
42 J9 **Sébaco** Matagalpa, W Nicaragua 12°51′N 86°08′W
19 P8 **Sebago Lake** ◎ Maine, NE USA
169 S13 **Sebangan, Teluk** *bay* Borneo, C Indonesia
Sebaste/Sebastia *see* Sivas
23 Y12 **Sebastian** Florida, SE USA 27°55′N 80°31′W
40 C5 **Sebastían Vizcaíno, Bahía** *bay* NW Mexico
19 R6 **Sebasticook Lake** ◎ Maine, NE USA
34 M7 **Sebastopol** California, W USA 38°22′N 122°50′W
Sebastopol *see* Sevastopol'
169 W8 **Sebatik, Pulau** *island* N Indonesia
19 R5 **Sebec Lake** ◎ Maine, NE USA
76 K12 **Sébékoro** Kayes, W Mali 13°00′N 09°03′W
Sebenico *see* Šibenik
40 G6 **Seberi, Cerro** ▲ NW Mexico 27°49′N 110°18′W
116 H11 **Sebeş** *Ger.* Mühlbach, *Hung.* Szászsebes; *prev.* Sebeşu Săsesc. Alba, W Romania 45°58′N 23°34′E
Sebes-Körös *see* Crişul Repede
Sebeşu Săsesc *see* Sebeş
31 R8 **Sebewaing** Michigan, N USA 43°43′N 83°27′W
124 F16 **Sebezh** Pskovskaya Oblast', W Russian Federation 56°19′N 28°31′E
137 N12 **Şebinkarahisar** Giresun, N Turkey 40°19′N 38°25′E
116 F11 **Sebiş** *Hung.* Borossebes. Arad, W Romania 46°21′N 22°09′E
Sebkra Azz el Matti *see* Azzel Matti, Sebkha
19 Q4 **Seboomook Lake** ◎ Maine, NE USA
74 G6 **Sebou** *var.* Sebu. ≈ N Morocco
20 I6 **Sebree** Kentucky, S USA 37°34′N 87°30′W
23 X13 **Sebring** Florida, SE USA 27°30′N 81°26′W
Sebta *see* Ceuta
Sebu *see* Sebou
169 U13 **Sebuku, Pulau** *island* N Indonesia
169 W8 **Sebuku, Teluk** *bay* Borneo, N Indonesia
106 F10 **Secchia** ≈ N Italy
10 L17 **Sechelt** British Columbia, SW Canada 49°25′N 123°37′W
56 C12 **Sechín, Río** ≈ W Peru
56 A10 **Sechura, Bahía de** *bay* NW Peru
185 A22 **Secretary Island** *island* SW New Zealand
155 I15 **Secunderābād** *var.* Sikandarabad. Telangana, C India 17°30′N 78°33′E
57 L17 **Sécure, Río** ≈ C Bolivia
118 D10 **Seda** Telšiai, NW Lithuania 56°10′N 22°04′E
27 T5 **Sedalia** Missouri, C USA 38°42′N 93°15′W
103 R3 **Sedan** Ardennes, N France 49°42′N 04°56′E
27 P7 **Sedan** Kansas, C USA 37°07′N 96°11′W
105 N3 **Sedano** Castilla y León, N Spain 42°43′N 03°43′W
104 H10 **Seda, Ribeira de** *stream* C Portugal
185 K15 **Seddon** Marlborough, South Island, New Zealand 41°42′S 174°05′E
185 H15 **Seddonville** West Coast, South Island, New Zealand 41°35′S 171°59′E
143 U7 **Sedeh** Khorāsān-e Janūbī, E Iran 33°18′N 59°12′E
Sederot *see* Sderot
65 B23 **Sedge Island** *island* NW Falkland Islands
76 G12 **Sédhiou** SW Senegal 12°39′N 15°33′W
11 U16 **Sedley** Saskatchewan, S Canada 50°06′N 103°51′W
Sedlez *see* Siedlce
117 Q2 **Sedniv** Chernihivs'ka Oblast', N Ukraine 51°39′N 31°34′E
36 L11 **Sedona** Arizona, SW USA 34°52′N 111°45′W
Sedunum *see* Sion
118 F12 **Šeduva** Šiauliai, N Lithuania 55°45′N 23°46′E
141 Y8 **Seeb** *var.* Muscat Sīb Airport. ✈ (Masqat) NE Oman 23°36′N 58°27′E
Seeb *see* As Sīb
108 M7 **Seefeld-in-Tirol** Tirol, W Austria 47°19′N 11°16′E
83 E22 **Seeheim Noord** Karas, S Namibia 26°50′S 17°45′E
Seeland *see* Sjælland
195 N9 **Seelig, Mount** ▲ Antarctica 81°45′S 102°15′W
Seenu Atoll *see* Addu Atoll
Seeonee *see* Seoni
Seer *see* Dörgön
102 L5 **Sées** Orne, N France 48°36′N 00°11′E
101 J14 **Seesen** Niedersachsen, C Germany 51°54′N 10°11′E
Seesker Höhe *see* Szeska Góra
100 J10 **Seevetal** Niedersachsen, N Germany 53°24′N 10°01′E
109 V6 **Seewiesen** Steiermark, E Austria 47°37′N 15°16′E
136 J13 **Şefaatli** *var.* Kızılkoca. Yozgat, C Turkey 39°32′N 34°45′E
143 V9 **Sefīdābeh** Khorāsān-e Janūbī, E Iran 31°05′N 60°30′E
149 N3 **Sefid, Darya-ye** *Pash.* Āb-i-safed. ≈ N Afghanistan
148 K5 **Sefid Kūh, Selseleh-ye** *Eng.* Paropamisus Range. ▲ W Afghanistan 34°00′N 61°40′E
148 K5 **Sefid Kūh, Selseleh-ye** *Eng.* Paropamisus Range. ▲▲ W Afghanistan
142 M4 **Sefid, Rūd-e** ≈ NW Iran
74 G6 **Sefrou** N Morocco 33°51′N 04°49′W
185 E19 **Sefton, Mount** ▲ South Island, New Zealand 43°43′S 169°58′E
171 S13 **Segaf, Kepulauan** *island group* E Indonesia

◆ Country ● Country Capital ◇ Dependent Territory ○ Dependent Territory Capital ❖ Administrative Regions ✈ International Airport ▲ Mountain ▲▲ Mountain Range ⛰ Volcano ≈ River ◎ Lake ⊞ Reservoir

Segama, Sungai – Shadehill Reservoir

169 *W7* **Segama, Sungai** ∽ East Malaysia
168 *L9* **Segamat** Johor, Peninsular Malaysia *02°30′N 102°48′E*
77 *S13* **Ségbana** NE Benin *10°56′N 03°42′E*
Segestica *see* Sisak
Segesvár *see* Sighişoara
171 *T12* **Seget** Papua Barat, E Indonesia *01°21′S 131°04′E*
Segewold *see* Sigulda
124 *J9* **Segezha** Respublika Kareliya, NW Russian Federation *63°39′N 34°24′E*
Seghedin *see* Szeged
Segna *see* Senj
107 *I16* **Segni** Lazio, C Italy *41°41′N 13°02′E*
Segodunum *see* Rodez
105 *S9* **Segorbe** Valenciana, E Spain *39°51′N 00°30′W*
76 *M12* **Ségou** *var.* Segu. Ségou, C Mali *13°26′N 06°12′W*
76 *M12* **Ségou** ◐ *region* SW Mali
54 *E8* **Segovia** Antioquia, N Colombia *07°08′N 74°39′W*
105 *N7* **Segovia** Castilla y León, C Spain *40°57′N 04°07′W*
104 *M6* **Segovia** ◐ *province* Castilla y León, N Spain
Segoviao Wangkí *see* Coco, Río
124 *J9* **Segozerskoye Vodokhranilishche** *prev.* Ozero Segozero. ◎ NW Russian Federation
102 *J7* **Segré** Maine-et-Loire, NW France *47°41′N 00°51′W*
105 *U5* **Segre** ∽ NE Spain
Segu *see* Ségou
38 *I17* **Seguam Island** *island* Aleutian Islands, Alaska, USA
38 *I17* **Seguam Pass** *strait* Aleutian Islands, Alaska, USA
77 *Y7* **Séguédine** Agadez, NE Niger *20°12′N 13°03′E*
76 *M15* **Séguéla** W Ivory Coast *07°58′N 06°44′W*
25 *S11* **Seguin** Texas, SW USA *29°34′N 97°58′W*
38 *E17* **Segula Island** *island* Aleutian Islands, Alaska, USA
62 *K10* **Segundo, Río** ∽ C Argentina
105 *Q12* **Segura** ∽ S Spain
105 *P13* **Segura, Sierra de** ▲▲ S Spain
83 *G18* **Sehithwa** North-West, N Botswana *20°28′S 22°43′E*
154 *H10* **Sehore** Madhya Pradesh, C India *23°12′N 77°08′E*
186 *G9* **Sehulea** Normanby Island, S Papua New Guinea *09°55′S 151°10′E*
149 *P15* **Sehwān** Sind, SE Pakistan *26°26′N 67°52′N*
109 *V8* **Seiersberg** Steiermark, SE Austria *47°01′N 15°22′E*
26 *L9* **Seiling** Oklahoma, C USA *36°09′N 98°55′W*
103 *S9* **Seille** ∽ E France
99 *J20* **Seilles** Namur, SE Belgium *50°31′N 05°12′E*
93 *K17* **Seinäjoki** *Swe.* Östermyra. Etelä-Pohjanmaa, W Finland *62°45′N 22°55′E*
12 *B12* **Seine** ∽ Ontario, S Canada
102 *M4* **Seine** ∽ N France
102 *K4* **Seine, Baie de la** *bay* N France
Seine, Banc de la *see* Seine Seamount
103 *O5* **Seine-et-Marne** ◐ *department* N France
102 *L3* **Seine-Maritime** ◐ *department* N France
84 *B14* **Seine Plain** *undersea feature* E Atlantic Ocean *34°00′N 12°15′W*
84 *B15* **Seine Seamount** *var.* Banc de la Seine. *undersea feature* E Atlantic Ocean *33°45′N 14°25′W*
102 *E6* **Sein, Île de** *island* NW France
171 *Y14* **Seinma** Papua, E Indonesia *04°10′S 138°54′E*
Seisbierrum *see* Sexbierum
109 *U5* **Seitenstetten Markt** Niederösterreich, C Austria *48°03′N 14°41′E*
Seiyo *see* Uwa
Seiyu *see* Chŏnju
95 *H22* **Sejerø** *island* E Denmark
110 *P7* **Sejny** Podlaskie, NE Poland *54°09′N 23°21′E*
163 *X15* **Sejong City** ● (South Korea) E South Korea *36°29′N 127°16′E*
81 *G20* **Seke** Simiyu, N Tanzania *03°16′S 33°31′E*
164 *L13* **Seki** Gifu, Honshū, SW Japan *35°30′N 136°54′E*
161 *U12* **Sekibi-sho** *Chin.* Chiwei Yu. *island* (disputed) China/Japan/Taiwan
165 *U3* **Sekihoku-tōge** *pass* Hokkaidō, NE Japan
Sekondi *see* Sekondi-Takoradi
77 *P17* **Sekondi-Takoradi** *var.* Sekondi. S Ghana *04°55′N 01°45′W*
80 *J11* **Sek'ot'a** Āmara, N Ethiopia *12°41′N 39°05′E*
Sekseŭil *see* Saksaul'skoye
32 *I9* **Selah** Washington, NW USA *46°39′N 120°31′W*
168 *J8* **Selangor** *var.* Negeri Selangor Darul Ehsan. ◐ *state* Peninsular Malaysia
Selânik *see* Thessaloníki
167 *R10* **Selaphum** Roi Et, E Thailand *16°00′N 103°54′E*
171 *T16* **Selaru, Pulau** *island* Kepulauan Tanimbar, E Indonesia
171 *U13* **Selassi** Papua Barat, E Indonesia *03°16′S 132°50′E*
168 *J7* **Selatan, Selat** *strait* Peninsular Malaysia
168 *K10* **Selatpanjang** Pulau Rantau, W Indonesia *01°00′N 102°44′E*
39 *N8* **Selawik** Alaska, USA *66°36′N 160°00′W*
39 *N8* **Selawik Lake** ◎ Alaska, USA
171 *N14* **Selayar, Selat** *strait* Sulawesi, C Indonesia
95 *C14* **Selbjørnsfjorden** *fjord* S Norway
94 *H8* **Selbusjøen** ◎ S Norway
97 *M17* **Selby** N England, United Kingdom *53°49′N 01°06′W*
29 *N8* **Selby** South Dakota, N USA *45°30′N 100°01′W*
21 *Z4* **Selbyville** Delaware, NE USA *38°28′N 75°12′W*
136 *B15* **Selçuk** *var.* Akıncılar. İzmir, SW Turkey *37°56′N 27°25′E*
39 *Q13* **Seldovia** Alaska, USA *59°26′N 151°42′W*
107 *M18* **Sele** *anc.* Silarius. ∽ S Italy
83 *J19* **Selebi-Phikwe** Central, E Botswana *21°58′S 27°48′E*
42 *B5* **Selegua, Río** ∽ W Guatemala
129 *X7* **Selemdzha** ∽ SE Russian Federation
129 *U7* **Selenga** *Mong.* Selenge Mörön. ∽ Mongolia/Russian Federation
79 *I19* **Selenge** Bandundu, W Dem. Rep. Congo *01°58′S 18°11′E*
162 *K6* **Selenge** *var.* Ingettolgoy. Bulgan, N Mongolia *49°27′N 103°59′E*
162 *L6* **Selenge** ◐ *province* N Mongolia
Selenge *see* Hyalganat, Bulgan, Mongolia
Selenge *see* Ih-Uul, Hövsgöl, Mongolia
Selenge Mörön *see* Selenga
123 *N14* **Selenginsk** Respublika Buryatiya, S Russian Federation *52°00′N 106°40′E*
Selenica *see* Selenicë
113 *K22* **Selenicë** *var.* Selenica. Vlorë, SW Albania *40°32′N 19°38′E*
123 *Q8* **Selennyakh** ∽ NE Russian Federation
100 *J8* **Selenter See** ◎ N Germany
Sele Sound *see* Soela Väin
103 *U6* **Sélestat** *Ger.* Schlettstadt. Bas-Rhin, NE France *48°16′N 07°28′E*
Seleucia *see* Silifke
92 *I4* **Selfoss** Suðurland, SW Iceland *63°56′N 20°59′W*
28 *M7* **Selfridge** North Dakota, N USA *46°01′N 100°52′W*
76 *I15* **Seli** ∽ N Sierra Leone
76 *I11* **Sélibabi** *var.* Sélibaby. Guidimaka, S Mauritania *15°14′N 12°11′W*
Sélibaby *see* Sélibabi
Selidovka/Selidovo *see* Selydove
124 *I15* **Seliger, Ozero** ◎ W Russian Federation
36 *J11* **Seligman** Arizona, SW USA *35°20′N 112°56′W*
27 *S8* **Seligman** Missouri, C USA *36°31′N 93°56′W*
80 *E6* **Selima Oasis** *oasis* N Sudan
76 *L13* **Sélingué, Lac de** ⊠ S Mali
Selinoús *see* Kréstena
18 *G14* **Selinsgrove** Pennsylvania, NE USA *40°47′N 76°51′W*
Selishche *see* Syelishcha
124 *I16* **Selizharovo** Tverskaya Oblast', W Russian Federation *56°50′N 33°24′E*
94 *C10* **Selje** Sogn Og Fjordane, S Norway *62°02′N 05°22′E*
11 *X16* **Selkirk** Manitoba, S Canada *50°10′N 96°52′W*
96 *K13* **Selkirk** SE Scotland, United Kingdom *55°36′N 02°48′W*
96 *K13* **Selkirk** *cultural region* SE Scotland, United Kingdom
11 *O16* **Selkirk Mountains** ▲▲ British Columbia, SW Canada
193 *T11* **Selkirk Rise** *undersea feature* SE Pacific Ocean
115 *F21* **Sellasía** Pelopónnisos, S Greece *37°14′N 22°24′E*
44 *M9* **Selle, Pic de la** *var.* La Selle. ▲ SE Haiti *18°18′N 71°55′W*
102 *M8* **Selles-sur-Cher** Loir-et-Cher, C France *47°16′N 01°31′E*
36 *K16* **Sells** Arizona, SW USA *31°54′N 111°52′W*
Sellye *see* Sal'a
23 *P5* **Selma** Alabama, S USA *32°24′N 87°01′W*
35 *Q11* **Selma** California, W USA *36°33′N 119°37′W*
20 *G10* **Selmer** Tennessee, S USA *35°10′N 88°34′W*
173 *N17* **Sel, Pointe au** *headland* W Réunion
Selselehye Kuhe Vākhān *see* Nicholas Range
127 *S2* **Selty** Udmurtskaya Respublika, NW Russian Federation *57°19′N 52°09′E*
Selukwe *see* Shurugwi
62 *L9* **Selva** Santiago del Estero, N Argentina *29°46′S 62°02′W*
11 *T9* **Selwyn Lake** ◎ Northwest Territories/Saskatchewan, C Canada
10 *K6* **Selwyn Mountains** ▲▲ Yukon, NW Canada
181 *T6* **Selwyn Range** ▲▲ Queensland, C Australia
117 *W8* **Selydove** *var.* Selidovka, *Rus.* Selidovo. Donets'ka Oblast', SE Ukraine *48°06′N 37°16′E*
Selzaete *see* Zelzate
Seman *see* Semanit, Lumi i
168 *M15* **Semangka, Teluk** *bay* Sumatera, SW Indonesia
113 *D22* **Semanit, Lumi i** *var.* Seman. ∽ W Albania
169 *Q16* **Semarang** *var.* Samarang. Jawa, C Indonesia *06°58′S 110°29′E*
169 *Q10* **Sematan** Sarawak, East Malaysia *01°50′N 109°44′E*
171 *P17* **Semau, Pulau** *island* S Indonesia
169 *V8* **Sembakung, Sungai** ∽ Borneo, N Indonesia
79 *G17* **Sembé** Sangha, NW Congo *01°38′N 14°35′E*
169 *S13* **Sembulu, Danau** ◎ Borneo, N Indonesia
139 *Q2* **Sêmêl** *Ar.* Sumayl, *var.* Summēl. Dahūk, N Iraq *36°52′N 42°51′E*
Semendria *see* Smederevo
117 *R1* **Semenivka** Chernihivs'ka Oblast', N Ukraine *52°10′N 32°37′E*
117 *S6* **Semenivka** *Rus.* Semenovka. Poltavs'ka Oblast', NE Ukraine *49°36′N 33°10′E*
127 *O3* **Semenov** Nizhegorodskaya Oblast', W Russian Federation *56°47′N 44°27′E*
Semenovka *see* Semenivka
169 *S17* **Semeru, Gunung** *var.* Mahameru. ▲ Jawa, S Indonesia *08°01′S 112°53′E*
145 *V9* **Semey** *prev.* Semipalatinsk. Vostochnyy Kazakhstan, E Kazakhstan *50°26′N 80°16′E*
Semezhevo *see* Syemyezhava
126 *L7* **Semiluki** Voronezhskaya Oblast', W Russian Federation *51°46′N 39°00′E*
33 *W16* **Seminoe Reservoir** ⊠ Wyoming, C USA
27 *O11* **Seminole** Oklahoma, C USA *35°13′N 96°40′W*
24 *M6* **Seminole** Texas, SW USA *32°43′N 102°39′W*
23 *S8* **Seminole, Lake** ⊠ Florida/Georgia, SE USA
Semiozernoye *see* Auliyekol'
Semipalatinsk *see* Semey
143 *O9* **Semīrom** *var.* Samirum. Eşfahān, C Iran *31°20′N 51°50′E*
38 *F17* **Semisopochnoi Island** *island* Aleutian Islands, Alaska, USA
169 *R11* **Semitau** Borneo, C Indonesia *0°30′N 111°59′E*
81 *E18* **Semliki** ∽ Uganda/Dem. Rep. Congo
143 *P5* **Semnān** *var.* Samnān. Semnān, N Iran *35°37′N 53°21′E*
143 *Q5* **Semnān** *off.* Ostān-e Semnān. ◐ *province* N Iran
Semnān, Ostān-e *see* Semnān
99 *K24* **Semois** ∽ SE Belgium
108 *E8* **Sempacher See** ◎ C Switzerland
Sena *see* Vila de Sena
30 *L12* **Senachwine Lake** ◎ Illinois, N USA
59 *O14* **Senador Pompeu** Ceará, E Brazil *05°30′S 39°25′W*
Sena Gallica *see* Senigallia
59 *C15* **Sena Madureira** Acre, W Brazil *09°05′S 68°41′W*
155 *L25* **Senanayake Samudra** ◎ E Sri Lanka
83 *G15* **Senanga** Western, SW Zambia *16°09′S 23°16′E*
27 *Y9* **Senath** Missouri, C USA *36°07′N 90°09′W*
22 *L2* **Senatobia** Mississippi, S USA *34°37′N 89°58′W*
164 *C16* **Sendai** *var.* Satsuma-Sendai. Kagoshima, Kyūshū, SW Japan *31°49′N 130°17′E*
165 *Q11* **Sendai-wan** *bay* E Japan
101 *J23* **Senden** Bayern, S Germany *48°18′N 10°04′E*
154 *F11* **Sendhwa** Madhya Pradesh, C India *21°38′N 75°04′E*
111 *H21* **Senec** *Ger.* Wartberg, *Hung.* Szenc; *prev.* Szempcz. Bratislavský Kraj, W Slovakia *48°14′N 17°24′E*
27 *P3* **Seneca** Kansas, C USA *39°50′N 96°04′W*
27 *R8* **Seneca** Missouri, C USA *36°50′N 94°36′W*
32 *K13* **Seneca** Oregon, NW USA *44°06′N 118°57′W*
21 *O11* **Seneca** South Carolina, SE USA *34°41′N 82°57′W*
18 *G11* **Seneca Lake** ◎ New York, NE USA
31 *U13* **Senecaville Lake** ⊠ Ohio, N USA
76 *G11* **Senegal** *off.* Republic of Senegal, *Fr.* Sénégal. ◆ *republic* W Africa
76 *H9* **Senegal** *Fr.* Sénégal. ∽ W Africa
Senegal, Republic of *see* Senegal
31 *O4* **Seney Marsh** *wetland* Michigan, N USA
101 *P14* **Senftenberg** Brandenburg, E Germany *51°31′N 14°01′E*
82 *L11* **Senga Hill** Northern, NE Zambia *09°26′S 31°12′E*
158 *G13* **Sênggê Zangbo** ∽ W China
171 *Z13* **Senggi** Papua, E Indonesia *03°26′S 140°46′E*
127 *R5* **Sengiley** Ul'yanovskaya Oblast', W Russian Federation *53°54′N 48°51′E*
63 *I19* **Senguerr, Río** ∽ S Argentina
83 *J16* **Sengwa** ∽ C Zimbabwe
Senia *see* Senj
111 *H19* **Senica** *Ger.* Senitz, *Hung.* Szenice. Trnavský Kraj, W Slovakia *48°40′N 17°22′E*
Seniça *see* Sjenica
106 *J11* **Senigallia** *anc.* Sena Gallica. Marche, C Italy *43°43′N 13°13′E*
136 *F15* **Senirkent** Isparta, SW Turkey *38°07′N 30°34′E*
Senitz *see* Senica
112 *C10* **Senj** *Ger.* Zengg, *It.* Segna; *anc.* Senia. Lika-Senj, NW Croatia *44°58′N 14°55′E*
Senj *see* Örgön
92 *H9* **Senja** *prev.* Senjen. *island* N Norway
Senjen *see* Senja
161 *U12* **Senkaku-shotō** *Chin.* Diaoyutai. *island group* (disputed) SW Japan
137 *R12* **Şenkaya** Erzurum, NE Turkey *40°33′N 42°17′E*
83 *I16* **Senkobo** Southern, S Zambia *17°38′S 25°58′E*
103 *O4* **Senlis** Oise, N France *49°13′N 02°33′E*
Senmonorom *see* Sênmônoŭrôm
167 *T12* **Sênmônoŭrôm** *var.* Senmonorom. Môndól Kiri, E Cambodia *12°27′N 107°12′E*
80 *G10* **Sennar** *var.* Sannār. Sinnar, C Sudan *13°31′N 33°38′E*
Senno *see* Syanno
Senones *see* Sens
109 *W11* **Senovo** E Slovenia *46°01′N 15°24′E*
103 *P6* **Sens** *anc.* Agendicum, Senones. Yonne, C France *48°12′N 03°17′E*
Sensburg *see* Mrągowo
167 *S11* **Sên, Stœng** ∽ C Cambodia
42 *F7* **Sensuntepeque** Cabañas, NE El Salvador *13°52′N 88°38′W*
112 *L8* **Senta** *Hung.* Zenta. Vojvodina, N Serbia *45°57′N 20°04′E*
Šent Andraž *see* Sankt Andrä
171 *Y13* **Sentani, Danau** ◎ Papua, E Indonesia
28 *J5* **Sentinel Butte** ▲ North Dakota, N USA *46°52′N 103°50′W*
10 *M13* **Sentinel Peak** ▲ British Columbia, W Canada *54°51′N 122°02′W*
59 *N16* **Sento Sé** Bahia, E Brazil *09°51′S 41°56′W*
Šent Peter *see* Pivka
Št. Vid *see* Sankt Veit an der Glan
Seo de Urgel *see* La Seu d'Urgell
163 *X17* **Seogwipo** *prev.* Sŏgwip'o. S South Korea *33°14′N 126°33′E*
154 *I7* **Seondha** Madhya Pradesh, C India *26°09′N 78°47′E*
163 *Y17* **Seongsan** *prev.* Sŏngsan. S South Korea
154 *J11* **Seoni** *prev.* Seeonee. Madhya Pradesh, C India *22°06′N 79°36′E*
163 *X14* **Seoul**, *Jap.* Keijō; *prev.* Kyŏngsŏng, Sŏul. ● (South Korea) NW South Korea *37°30′N 126°58′E*
83 *I17* **Sepako** Central, NE Botswana *19°50′S 26°29′E*
184 *I13* **Separation Point** *headland* South Island, New Zealand *40°46′S 172°58′E*
169 *V10* **Sepasu** Borneo, N Indonesia *0°44′N 117°38′E*
186 *B6* **Sepik** ∽ Indonesia/Papua New Guinea
Sepone *see* Muang Xépôn
110 *M7* **Sępopol** *Ger.* Schippenbeil. Warmińsko-Mazurskie, N Poland *54°16′N 21°09′E*
116 *F10* **Şepreuş** *Hung.* Seprős. Arad, W Romania *46°34′N 21°44′E*
Seprős *see* Şepreuş
Şepşi-Sângeorz/Sepsiszentgyőrgy *see* Sfântu Gheorghe
15 *W4* **Sept-Îles** Québec, SE Canada *50°11′N 66°19′W*
105 *N6* **Sepúlveda** Castilla y León, N Spain *41°18′N 03°45′W*
104 *K8* **Sequeros** Castilla y León, N Spain *40°31′N 06°04′W*
104 *L5* **Sequillo** ∽ NW Spain
32 *G7* **Sequim** Washington, NW USA *48°04′N 123°06′W*
35 *S11* **Sequoia National Park** *national park* California, W USA
137 *Q14* **Şerafettin Dağları** ▲▲ E Turkey
127 *N10* **Serafimovich** Volgogradskaya Oblast', SW Russian Federation *49°34′N 42°43′E*
171 *Q10* **Serai** Sulawesi, N Indonesia *01°45′N 124°58′E*
99 *K19* **Seraing** Liège, E Belgium *50°37′N 05°31′E*
Sêraitang *see* Baima
Serakhs *see* Sarahs
171 *W13* **Serami** Papua, E Indonesia *02°11′S 136°46′E*
171 *R13* **Seram, Laut** *Eng.* Ceram Sea. *sea* E Indonesia
Serampore/Serampur *see* Shrīrāmpur
171 *S13* **Seram, Pulau** *var.* Serang, *Eng.* Ceram. *island* Maluku, E Indonesia
169 *N15* **Serang** Jawa, C Indonesia *06°07′S 106°09′E*
Serang *see* Seram, Pulau
169 *P9* **Serasan, Pulau** *island* Kepulauan Natuna, W Indonesia
169 *P9* **Serasan, Selat** *strait* Indonesia/Malaysia
112 *M13* **Serbia** *off.* Federal Republic of Serbia; *prev.* Yugoslavia, *SCr.* Jugoslavija. ◆ *federal republic* SE Europe
112 *M12* **Serbia** *Ger.* Serbien, *Serb.* Srbija. ◐ *republic* Serbia
Serbia, Federal Republic of *see* Serbia
Serbien *see* Serbia
Sercq *see* Sark
146 *D12* **Serdar** *prev. Rus.* Gyzyrlabat, Kizyl-Arvat. Balkan Welaýaty, W Turkmenistan *39°02′N 56°15′E*
Serdica *see* Sofia
Serdobol' *see* Sortavala
127 *O7* **Serdobsk** Penzenskaya Oblast', W Russian Federation *52°30′N 44°16′E*
145 *X9* **Serebryansk** Vostochnyy Kazakhstan, E Kazakhstan *49°44′N 83°16′E*
123 *Q12* **Serebryanyy Bor** Respublika Sakha (Yakutiya), NE Russian Federation *56°40′N 124°46′E*
111 *H20* **Sered'** *Hung.* Szered. Trnavský Kraj, W Slovakia *48°19′N 17°45′E*
117 *S1* **Seredyna-Buda** Sums'ka Oblast', NE Ukraine *52°09′N 34°00′E*
118 *E13* **Seredžius** Tauragė, C Lithuania *55°04′N 23°24′E*
136 *I14* **Şereflikoçhisar** Ankara, C Turkey *38°56′N 33°31′E*
106 *D7* **Seregno** Lombardia, N Italy *45°39′N 09°12′E*
103 *P7* **Serein** ∽ C France
168 *K9* **Seremban** Negeri Sembilan, Peninsular Malaysia *02°42′N 101°54′E*
81 *H20* **Serengeti Plain** *plain* N Tanzania
82 *K13* **Serenje** Central, E Zambia *13°12′S 30°15′E*
Seres *see* Sérres
116 *J5* **Seret** ∽ W Ukraine
Seret/Sereth *see* Siret
115 *I21* **Serfopoúla** *island* Kykládes, Greece, Aegean Sea
127 *P4* **Sergach** Nizhegorodskaya Oblast', W Russian Federation *55°31′N 45°29′E*
29 *S13* **Sergeant Bluff** Iowa, C USA *42°24′N 96°19′W*
163 *P7* **Sergelen** Dornod, NE Mongolia *48°31′N 114°01′E*
Sergelen *see* Tuvshinshiree
168 *H8* **Sergeulangit, Pegunungan** ▲▲ Sumatera, NW Indonesia
122 *L5* **Sergeya Kirova, Ostrova** *island* N Russian Federation
Sergeyevichi *see* Syarhyeyevichy
145 *O7* **Sergeyevka** Severnyy Kazakhstan, N Kazakhstan *53°53′N 67°25′E*
Sergiopol *see* Ayagoz
59 *P16* **Sergipe** *off.* Estado de Sergipe. ◐ *state* E Brazil
Sergipe, Estado de *see* Sergipe
126 *L3* **Sergiyev Posad** Moskovskaya Oblast', W Russian Federation *56°21′N 38°10′E*
124 *K5* **Sergozero, Ozero** ◎ NW Russian Federation
146 *J17* **Serhetabat** *prev. Rus.* Gushgy, Kushka. Mary Welaýaty, S Turkmenistan *35°19′N 62°17′E*
169 *Q10* **Serian** Sarawak, East Malaysia *01°10′N 110°35′E*
115 *I21* **Sérifos** *anc.* Seriphos. *island* Kykládes, Greece, Aegean Sea
115 *I21* **Sérifou, Stenó** *strait* SE Greece
136 *F16* **Serik** Antalya, SW Turkey *36°55′N 31°06′E*
106 *E7* **Serio** ∽ N Italy
Seriphos *see* Sérifos
Serir Tibesti *see* Sarīr Tībistī
Sêrkog *see* Sêrtar
197 *O14* **Sermersooq** *off.* Kommuneqarfik Sermersooq. ◐ *municipality* S Greenland
Sermersooq, Kommuneqarfik *see* Sermersooq
127 *S5* **Sernovodsk** Samarskaya Oblast', W Russian Federation *53°56′N 51°16′E*
127 *R2* **Sernur** Respublika Mariy El, W Russian Federation *56°55′N 49°09′E*
110 *M11* **Serock** Mazowieckie, C Poland *52°30′N 21°03′E*
61 *B18* **Serodino** Santa Fe, C Argentina *32°33′S 60°52′W*
Seroei *see* Serui
105 *P14* **Serón** Andalucía, S Spain *37°20′N 02°28′W*
99 *E14* **Serooskerke** Zeeland, SW Netherlands *51°42′N 03°52′E*
105 *T6* **Seròs** Cataluña, NE Spain *41°27′N 00°24′E*
122 *G10* **Serov** Sverdlovskaya Oblast', C Russian Federation *59°42′N 60°32′E*
83 *I19* **Serowe** Central, SE Botswana *22°26′S 26°44′E*
104 *H13* **Serpa** Beja, S Portugal *37°56′N 07°36′W*
Serpa Pinto *see* Menongue
182 *A4* **Serpentine Lakes** *salt lake* South Australia
45 *T15* **Serpent's Mouth, The** *Sp.* Boca de la Serpiente. *strait* Trinidad and Tobago/Venezuela
Serpiente, Boca de la *see* Serpent's Mouth, The
126 *K4* **Serpukhov** Moskovskaya Oblast', W Russian Federation *54°54′N 37°26′E*
104 *I10* **Serra de São Mamede** ▲ C Portugal *39°18′N 07°19′W*
60 *K13* **Serra do Mar** ▲▲
Sérrai *see* Sérres
107 *N22* **Serra San Bruno** Calabria, SW Italy *38°32′N 16°18′E*
103 *S14* **Serres** Hautes-Alpes, SE France *44°26′N 05°42′E*
114 *H13* **Sérres** *var.* Seres; *prev.* Sérrai. Kentrikí Makedonía, NE Greece *41°03′N 23°33′E*
62 *J9* **Serrezuela** Córdoba, C Argentina *30°38′S 65°26′W*
59 *O16* **Serrinha** Bahia, E Brazil *11°38′S 38°56′W*
59 *M19* **Serro** *var.* Sêrro. Minas Gerais, NE Brazil *18°38′S 43°22′W*
Sêrro *see* Serro
Sert *see* Siirt
104 *H9* **Sertã** *var.* Sertá. Castelo Branco, C Portugal *39°48′N 08°05′W*
Sertá *see* Sertã
60 *L8* **Sertãozinho** São Paulo, S Brazil *21°04′S 47°55′W*
160 *F7* **Sêrtar** *var.* Sêrkog. Sichuan, C China *32°18′N 100°18′E*
124 *G12* **Sertolovo** Leningradskaya Oblast', NW Russian Federation *60°08′N 30°06′E*
171 *W13* **Serui** *prev.* Seroei. Papua, E Indonesia *01°53′S 136°15′E*
83 *J19* **Serule** Central, E Botswana *21°58′S 27°20′E*
169 *S12* **Seruyan, Sungai** *var.* Sungai Pembuang. ∽ Borneo, N Indonesia
115 *E14* **Sérvia** Dytikí Makedonía, N Greece *40°12′N 22°01′E*
160 *E7* **Sêrxü** *var.* Jugar. Sichuan, C China *32°54′N 98°06′E*
123 *R13* **Seryshevo** Amurskaya Oblast', SE Russian Federation *51°03′N 128°16′E*
Sesana *see* Sežana
169 *V8* **Sesayap, Sungai** ∽ Borneo, N Indonesia
Sesdlets *see* Siedlce
79 *N17* **Sese** Orientale, N Dem. Rep. Congo *02°13′N 25°52′E*
81 *F18* **Sese Islands** *island group* S Uganda
83 *H16* **Sesheke** *var.* Sesheko. Western, SE Zambia *17°28′S 24°20′E*
Sesheko *see* Sesheke
106 *C8* **Sesia** *anc.* Sessites. ∽ NW Italy
104 *F11* **Sesimbra** Setúbal, S Portugal *38°26′N 09°06′W*
115 *N22* **Seskló** *island* Dodekánisa, Greece, Aegean Sea
30 *L16* **Sesser** Illinois, N USA *38°05′N 89°03′W*
Sessites *see* Sesia
106 *G11* **Sesto Fiorentino** Toscana, C Italy *43°50′N 11°12′E*
106 *E7* **Sesto San Giovanni** Lombardia, N Italy *45°32′N 09°14′E*
106 *A8* **Sestriere** Piemonte, NE Italy *45°00′N 06°54′E*
106 *D10* **Sestri Levante** Liguria, NW Italy *44°16′N 09°22′E*
107 *C20* **Sestu** Sardegna, Italy, C Mediterranean Sea *39°15′N 09°06′E*
112 *E8* **Sesvete** Zagreb, N Croatia *45°50′N 16°03′E*
118 *G12* **Šeta** Kaunas, C Lithuania *55°17′N 24°16′E*
Setabis *see* Xàtiva
165 *Q4* **Setana** Hokkaidō, NE Japan *42°27′N 139°52′E*
103 *Q16* **Sète** *prev.* Cette. Hérault, S France *43°24′N 03°42′E*
58 *J11* **Sete Ilhas** Amapá, NE Brazil *01°06′N 52°06′W*
59 *L20* **Sete Lagoas** Minas Gerais, NE Brazil *19°29′S 44°15′W*
60 *G10* **Sete Quedas, Ilha das** *island* S Brazil
92 *I10* **Setermoen** Troms, N Norway *68°51′N 18°20′E*
95 *E17* **Setesdal** *valley* S Norway
43 *W16* **Setetule, Cerro** ▲ SE Panama *07°51′N 77°37′W*
21 *Q5* **Seth** West Virginia, NE USA *38°06′N 81°40′W*
Setia *see* Sezze
74 *K5* **Sétif** *var.* Stif. N Algeria *36°11′N 05°24′E*
164 *L5* **Seto** Aichi, Honshū, SW Japan *35°14′N 137°06′E*
164 *G13* **Seto-naikai** *Eng.* Inland Sea. *sea* S Japan
165 *V16* **Setouchi** *var.* Setoushi. Kagoshima, Amami-Ō-shima, SW Japan *44°19′N 142°58′E*
Setoushi *see* Setouchi
74 *F6* **Settat** W Morocco *33°03′N 07°37′W*
79 *D20* **Setté Cama** Ogooué-Maritime, SW Gabon *02°32′S 09°46′E*
11 *W13* **Setting Lake** ◎ Manitoba, C Canada
97 *L16* **Settle** N England, United Kingdom *54°04′N 02°17′W*
189 *Y12* **Settlement** E Wake Island *19°17′N 166°38′E*
104 *F11* **Setúbal** *Eng.* Saint Ubes, Saint Yves. Setúbal, W Portugal *38°31′N 08°54′W*
104 *F11* **Setúbal** ◐ *district* S Portugal
104 *F12* **Setúbal, Baía de** *bay* W Portugal
Setul *see* Satun
12 *B10* **Seul, Lac** ◎ Ontario, S Canada
103 *R8* **Seurre** Côte d'Or, C France *47°00′N 05°09′E*
137 *U11* **Sevan** C Armenia *40°32′N 44°56′E*
137 *V12* **Sevana Lich** *Eng.* Lake Sevan, *Rus.* Ozero Sevan. ◎ E Armenia
Sevan, Lake/Sevan, Ozero *see* Sevana Lich
77 *N11* **Sévaré** Mopti, C Mali *14°32′N 04°06′W*
117 *S14* **Sevastopol'** *Eng.* Sebastopol. Avtonomna Respublika Krym, S Ukraine *44°36′N 33°33′E*
25 *R14* **Seven Sisters** Texas, SW USA *27°57′N 98°34′W*
10 *K13* **Seven Sisters Peaks** ▲ British Columbia, SW Canada *54°57′N 128°10′W*
99 *M15* **Sevenum** Limburg, SE Netherlands *51°25′N 06°01′E*
103 *P14* **Séverac-le-Château** Aveyron, S France *44°18′N 03°03′E*
14 *H13* **Severn** ∽ Ontario, S Canada
97 *L21* **Severn** *Wel.* Hafren. ∽ England/Wales, United Kingdom
125 *O11* **Severnaya Dvina** *var.* Northern Dvina. ∽ NW Russian Federation
127 *N16* **Severnaya Osetiya-Alaniya, Respublika** *Eng.* North Ossetia; *prev.* Respublika Severnaya Osetiya, Severo-Osetinskaya SSR. ◐ *autonomous republic* SW Russian Federation
Severnaya Osetiya, Respublika *see* Severnaya Osetiya-Alaniya, Respublika
122 *M5* **Severnaya Zemlya** *var.* Nicholas II Land. *island group* N Russian Federation
127 *T5* **Severnoye** Orenburgskaya Oblast', W Russian Federation *54°03′N 52°31′E*
35 *S3* **Severn Troughs Range** ▲▲ Nevada, W USA
125 *W3* **Severnyy** Respublika Komi, NW Russian Federation *67°38′N 64°13′E*
144 *I13* **Severnyy Chink Ustyurta** ▲▲ W Kazakhstan
125 *Q13* **Severnyye Uvaly** *var.* Northern Ural Hills. *hill range* NW Russian Federation
145 *O6* **Severnyy Kazakhstan** *off.* Severo-Kazakhstanskaya Oblast', *var.* North Kazakhstan, *Kaz.* Soltūstik Qazaqstan Oblysy. ◐ *province* N Kazakhstan
122 *I6* **Severnyy, Ostrov** NW Russian Federation
125 *V9* **Severnyy Ural** ▲▲ NW Russian Federation
Severo-Alichurskiy Khrebet *see* Alichuri Shimolí, Qatorkŭhi
123 *N12* **Severobaykal'sk** Respublika Buryatiya, S Russian Federation *55°39′N 109°17′E*
Severodonetsk *see* Syeverodonets'k
124 *M8* **Severodvinsk** *prev.* Molotov, Sudostroy. Arkhangel'skaya Oblast', NW Russian Federation *64°32′N 39°50′E*
Severo-Kazakhstanskaya Oblast' *see* Severnyy Kazakhstan
123 *U11* **Severo-Kuril'sk** Sakhalinskaya Oblast', SE Russian Federation *50°38′N 155°57′E*
124 *J3* **Severomorsk** Murmanskaya Oblast', NW Russian Federation *69°00′N 33°16′E*
Severo-Osetinskaya SSR *see* Severnaya Osetiya-Alaniya, Respublika
122 *M7* **Severo-Sibirskaya Nizmennost'** *var.* North Siberian Plain, *Eng.* North Siberian Lowland. *lowlands* N Russian Federation
122 *G10* **Severoural'sk** Sverdlovskaya Oblast', C Russian Federation *60°09′N 59°58′E*
122 *L11* **Severo-Yeniseyskiy** Krasnoyarskiy Kray, C Russian Federation *60°29′N 93°13′E*
122 *J12* **Seversk** Tomskaya Oblast', C Russian Federation *53°37′N 84°47′E*
126 *M11* **Severskiy Donets** *Ukr.* Sivers'kyy Donets'. ∽ Russian Federation/Ukraine *see also* Sivers'kyy Donets'
Severskiy Donets *see* Sivers'kyy Donets'
92 *M9* **Sevettijärvi** Lappi, N Finland *69°31′N 28°40′E*
36 *M5* **Sevier Bridge Reservoir** ⊠ Utah, W USA
36 *J4* **Sevier Desert** *plain* Utah, W USA
36 *J5* **Sevier Lake** ◎ Utah, W USA
21 *N9* **Sevierville** Tennessee, S USA *35°53′N 83°34′W*
104 *J14* **Sevilla** *Eng.* Seville; *anc.* Hispalis. Andalucía, SW Spain *37°24′N 05°59′W*
104 *J13* **Sevilla** ◐ *province* Andalucía, SW Spain
Sevilla de Niefang *see* Niefang
43 *O16* **Sevilla, Isla** *island* SW Panama
Seville *see* Sevilla
114 *J9* **Sevlievo** Gabrovo, N Bulgaria *43°01′N 25°06′E*
Sevluš/Sevlyush *see* Vynohradiv
109 *V11* **Sevnica** *Ger.* Lichtenwald. E Slovenia *46°00′N 15°20′E*
162 *J11* **Sevrey** *var.* Saynshand. Ömnögovĭ, S Mongolia *43°30′N 102°08′E*
126 *I7* **Sevsk** Bryanskaya Oblast', W Russian Federation *52°03′N 34°31′E*
76 *J15* **Sewa** ∽ E Sierra Leone
39 *R12* **Seward** Alaska, USA *60°06′N 149°26′W*
29 *R15* **Seward** Nebraska, C USA *40°52′N 97°06′W*
197 *Q3* **Seward Peninsula** *peninsula* Alaska, USA
Seward's Folly *see* Alaska
62 *H12* **Sewell** Libertador, C Chile *34°05′S 70°25′W*
98 *K5* **Sexbierum** *Fris.* Seisbierrum. Fryslân, N Netherlands *53°13′N 05°28′E*
11 *O13* **Sexsmith** Alberta, W Canada *55°18′N 118°45′W*
41 *W13* **Seybaplaya** Campeche, SE Mexico *19°39′N 90°36′W*
173 *N6* **Seychelles** *off.* Republic of Seychelles. ◆ *republic* W Indian Ocean
67 *Z9* **Seychelles** *island group* NE Seychelles
173 *N6* **Seychelles Bank** *var.* Le Banc des Seychelles. *undersea feature* W Indian Ocean *04°45′S 55°30′E*
Seychelles, Le Banc des *see* Seychelles Bank
Seychelles, Republic of *see* Seychelles
172 *H17* **Seychellois, Morne** ▲ Mahé, NE Seychelles
146 *J12* **Seýdi** *Rus.* Seydi; *prev.* Neftezavodsk. Lebap Welaýaty, E Turkmenistan *39°31′N 62°53′E*
136 *G16* **Seydişehir** Konya, SW Turkey *37°25′N 31°51′E*
92 *L2* **Seyðisfjörður** Austurland, E Iceland *65°15′N 14°00′W*
136 *J13* **Seyfe Gölü** ◎ C Turkey
Seyhan *see* Adana
136 *K16* **Seyhan Barajı** ⊠ S Turkey
136 *K17* **Seyhan Nehri** ∽ S Turkey
136 *F13* **Seyitgazi** Eskişehir, W Turkey *39°27′N 30°42′E*
126 *J7* **Seym** ∽ W Russian Federation
117 *S3* **Seym** ∽ N Ukraine
123 *T9* **Seymchan** Magadanskaya Oblast', E Russian Federation *62°54′N 152°27′E*
114 *N12* **Seymen** Tekirdağ, NW Turkey *41°06′N 27°56′E*
183 *O11* **Seymour** Victoria, SE Australia *37°01′S 145°10′E*
83 *I25* **Seymour** Eastern Cape, S South Africa *32°33′S 26°46′E*
29 *W16* **Seymour** Iowa, C USA *40°40′N 93°07′W*
27 *U7* **Seymour** Missouri, C USA *37°09′N 92°46′W*
25 *Q5* **Seymour** Texas, SW USA *33°36′N 99°16′W*
114 *M12* **Şeytan Deresi** ∽ NW Turkey
109 *S12* **Sežana** *It.* Sesana. SW Slovenia *45°42′N 13°52′E*
103 *P5* **Sézanne** Marne, N France *48°43′N 03°41′E*
107 *I16* **Sezze** *anc.* Setia. Lazio, C Italy *41°29′N 13°04′E*
Sfákia *see* Chóra Sfakíon
115 *D21* **Sfaktiría** *island* S Greece
116 *J11* **Sfântu Gheorghe** *Ger.* Sankt-Georgen, *Hung.* Sepsiszentgyőrgy; *prev.* Şepşi-Sângeorz, Sfîntu Gheorghe. Covasna, C Romania *45°52′N 25°49′E*
117 *N13* **Sfântu Gheorghe, Braţul** *var.* Gheorghe Braţul. ∽ E Romania
75 *N6* **Sfax** *Ar.* Şafāqis. E Tunisia *34°45′N 10°45′E*
75 *N6* **Sfax** ✈ E Tunisia *34°43′N 10°37′E*
Sfîntu Gheorghe *see* Sfântu Gheorghe
98 *H13* **'s-Gravendeel** Zuid-Holland, SW Netherlands *51°48′N 04°36′E*
98 *F11* **'s-Gravenhage** *var.* Den Haag, *Eng.* The Hague, *Fr.* La Haye. ● (Netherlands-seat of government) Zuid-Holland, W Netherlands *52°07′N 04°17′E*
98 *G12* **'s-Gravenzande** Zuid-Holland, W Netherlands *52°00′N 04°10′E*
Shaan/Shaanxi Sheng *see* Shaanxi
159 *X11* **Shaanxi** *var.* Shaan, Shaanxi Sheng, Shan-hsi, Shenshi, Shensi. ◐ *province* C China
Shaartuz *see* Shahrtuz
Shaba *see* Katanga
Shabani *see* Zvishavane
81 *N17* **Shabeellaha Dhexe** *off.* Gobolka Shabeellaha Dhexe. ◐ *region* E Somalia
Shabeellaha Dhexe, Gobolka *see* Shabeellaha Dhexe
81 *L17* **Shabeellaha Hoose** *off.* Gobolka Shabeellaha Hoose. ◐ *region* S Somalia
Shabeellaha Hoose, Gobolka *see* Shabeellaha Hoose
Shabeelle, Webi *see* Shebeli
114 *O7* **Shabla** Dobrich, NE Bulgaria *43°33′N 28°31′E*
114 *O7* **Shabla, Nos** *headland* NE Bulgaria *43°30′N 28°36′E*
13 *N9* **Shabogama Lake** ◎ Newfoundland and Labrador, E Canada
79 *N20* **Shabunda** Sud-Kivu, E Dem. Rep. Congo *02°42′S 27°20′E*
141 *Q15* **Shabwah** C Yemen *15°09′N 46°46′E*
158 *F8* **Shache** *var.* Yarkant. Xinjiang Uygur Zizhiqu, NW China *38°27′N 77°16′E*
Shacheng *see* Huailai
195 *R12* **Shackleton Coast** *physical region* Antarctica
195 *Z10* **Shackleton Ice Shelf** *ice shelf* Antarctica
Shaddādī *see* Ash Shadādah
28 *K7* **Shadehill Reservoir** ⊠ South Dakota, N USA

◆ Country
● Country Capital
◇ Dependent Territory
○ Dependent Territory Capital
◐ Administrative Regions
✈ International Airport
▲ Mountain
▲▲ Mountain Range
Volcano
∽ River
◎ Lake
⊠ Reservoir

Shadrinsk – Shīrīn Tagāb

122 G11 **Shadrinsk** Kurganskaya Oblast', C Russian Federation 56°08′N 63°18′E
31 O12 **Shafer, Lake** ⊠ Indiana, N USA
35 R13 **Shafter** California, W USA 35°27′N 119°15′W
24 J11 **Shafter** Texas, SW USA 29°49′N 104°18′W
97 L23 **Shaftesbury** S England, United Kingdom 51°01′N 02°12′W
185 F22 **Shag** ≈ South Island, New Zealand
145 V9 **Shagan** ≈ E Kazakhstan
39 O11 **Shageluk** Alaska, USA 62°40′N 159°33′W
122 K14 **Shagonar** Respublika Tyva, S Russian Federation 51°31′N 93°06′E
185 F22 **Shag Point** *headland* South Island, New Zealand 45°28′S 170°50′E
144 J12 **Shagyray, Plato** *plain* SW Kazakhstan
Shāhābād *see* Eslāmābād-e Gharb
168 K9 **Shah Alam** Selangor, Peninsular Malaysia 03°02′N 101°31′E
117 O12 **Shahany, Ozero** ◎ SW Ukraine
138 H9 **Shahbā'** *anc.* Philippopolis. As Suwaydā', S Syria 32°50′N 36°38′E
Shahbān *see* Ad Dayr
149 P17 **Shāh Bandar** Sind, SE Pakistan 23°59′N 67°54′E
149 P13 **Shahdād Kot** Sind, SW Pakistan 27°49′N 67°49′E
143 T10 **Shahdād, Namakzār-e** *salt pan* E Iran
149 Q15 **Shāhdādpur** Sind, SE Pakistan 25°56′N 68°40′E
154 K10 **Shahdol** Madhya Pradesh, C India 23°19′N 81°26′E
161 N7 **Sha He** ≈ C China
Shahe *see* Linze
Shahepu *see* Linze
153 N13 **Shāhganj** Uttar Pradesh, N India 26°03′N 82°41′E
152 C11 **Shāhgarh** Rājasthān, NW India 27°08′N 69°56′E
Sha Hi *see* Orūmīyeh, Daryācheh-ye
Shāhī *see* Qā'emshahr
Shahjahanabad *see* Delhi
152 L11 **Shāhjahānpur** Uttar Pradesh, N India 27°53′N 79°55′E
149 U7 **Shāhpur** Punjab, E Pakistan 32°15′N 72°32′E
Shāhpur *see* Shāhpur Chākar
152 G13 **Shāhpura** Rājasthān, N India 25°38′N 75°01′E
149 Q15 **Shāhpur Chākar** *var.* Shāhpur. Sind, SE Pakistan 26°11′N 68°44′E
148 M5 **Shahrak** Gōwr, C Afghanistan 34°09′N 64°16′E
143 Q11 **Shahr-e Bābak** Kermān, C Iran 30°08′N 55°04′E
143 N8 **Shahr-e Kord** *var.* Shahr Kord. Chahār Maḩall va Bakhtīārī, C Iran 32°20′N 50°52′E
143 O9 **Shahreẓā** *var.* Qomisheh, Qumisheh, Shahriza; *prev.* Qomsheh. Eşfahān, C Iran 32°01′N 51°51′E
147 S10 **Shahrikhon** *Rus.* Shakhrikhan. Andijon Viloyati, E Uzbekistan 40°42′N 72°03′E
147 P11 **Shahriston** *Rus.* Shakhristan. NW Tajikistan 39°45′N 68°47′E
Shahriza *see* Shahreẓā
Shahr-i-Zabul *see* Zābol
Shahr Kord *see* Shahr-e Kord
147 P14 **Shahrtuz** *Rus.* Shaartuz. SW Tajikistan 37°13′N 68°05′E
143 Q4 **Shāhrūd** *prev.* Emāmrūd, Emāmshahr. Semnān, N Iran 36°30′N 55°E
Shahsavār/Shahsawar *see* Tonekābon
Shaikh Ābid *see* Shaykh 'Ābid
Shaikh Fāris *see* Shaykh Fāris
Shaikh Najm *see* Shaykh Najm
138 K5 **Shā'īr, Jabal** ▲ C Syria 34°51′N 37°49′E
154 G10 **Shājāpur** Madhya Pradesh, C India 23°27′N 76°21′E
80 J8 **Shakal, Ras** *headland* NE Sudan 18°04′N 38°34′E
83 G17 **Shakawe** North West, NW Botswana 18°25′S 21°53′E
Shakhdarinskiy Khrebet *see* Shokhdara, Qatorkŭhi
Shakhrikhan *see* Shahrikhon
Shakhrisabz *see* Sharixon
Shakhristan *see* Shahriston
Shakhtërsk *see* Zuhres
145 R10 **Shakhtinsk** Karaganda, C Kazakhstan 49°40′N 72°37′E
126 L11 **Shakhty** Rostovskaya Oblast', SW Russian Federation 47°45′N 40°14′E
127 P2 **Shakhun'ya** Nizhegorodskaya Oblast', W Russian Federation 57°42′N 46°36′E
77 S15 **Shaki** Oyo, W Nigeria 08°37′N 03°25′E
81 J15 **Shakiso** Oromīya, C Ethiopia 05°33′N 38°48′E
29 V9 **Shakopee** Minnesota, N USA 44°48′N 93°31′W
165 R3 **Shakotan-misaki** *headland* Hokkaidō, NE Japan 43°22′N 140°28′E
39 N9 **Shaktoolik** Alaska, USA 64°19′N 161°05′W
81 J14 **Shala Hāyk'** ◎ C Ethiopia
124 M10 **Shalakusha** Arkhangel'skaya Oblast', NW Russian Federation 62°16′N 40°16′E
145 U8 **Shalday** Pavlodar, NE Kazakhstan 51°57′N 78°51′E
127 P16 **Shali** Chechenskaya Respublika, SW Russian Federation 43°03′N 45°55′E
141 W12 **Shalīm** *var.* Shelim. S Oman 18°07′N 55°39′E
Shalīr, Āveh-ye *see* Shilayr, Wādī
Shaliuhe *see* Gangca
144 K12 **Shalkar** *var.* Chelkar. Aktyubinsk, W Kazakhstan 47°50′N 59°29′E
144 F9 **Shalkar, Ozero** *prev.* Chelkar Ozero. ◎ W Kazakhstan
21 V12 **Shallotte** North Carolina, SE USA 33°58′N 78°21′W
25 N5 **Shallowater** Texas, SW USA 33°41′N 102°00′W
124 K11 **Shal'skiy** Respublika Kareliya, NW Russian Federation 61°45′N 36°02′E
160 F9 **Shaluli Shan** ▲▲ C China
81 F22 **Shama** ≈ C Tanzania
11 Z11 **Shamattawa** Manitoba, C Canada 55°52′N 92°05′W
12 F8 **Shamattawa** ≈ Ontario, C Canada
Shām, Bādiyat ash *see* Syrian Desert
Shamiya *see* Ash Shāmīyah
141 X8 **Shām, Jabal ash** *var.* Jebel Sham. ▲ NW Oman 23°21′N 57°08′E
Sham, Jebel *see* Shām, Jabal ash
Shamkhor *see* Şämkir
18 G14 **Shamokin** Pennsylvania, NE USA 40°47′N 76°33′W
25 P2 **Shamrock** Texas, SW USA 35°12′N 100°15′W
Shana *see* Kuril'sk
Sha'nabī, Jabal ash *see* Chambi, Jebel
139 Y12 **Shanāwah** Al Başrah, E Iraq 30°57′N 47°25′E
Shancheng *see* Taining
159 T8 **Shandan** *var.* Qingyuan. Gansu, N China 38°50′N 101°08′E
Shandī *see* Shendi
161 Q5 **Shandong** *var.* Lu, Shandong Sheng, Shantung. ◈ *province* E China
161 R4 **Shandong Bandao** *var.* Shantung Peninsula. *peninsula* E China
Shandong Sheng *see* Shandong
139 U8 **Shandrūkh** Diyālá, E Iraq 33°20′N 45°19′E
83 J17 **Shangani** ≈ W Zimbabwe
161 O15 **Shangchuan Dao** *island* S China
Shangchuankou *see* Minhe
163 P12 **Shangdu** Nei Mongol Zizhiqu, N China 41°32′N 113°33′E
161 O11 **Shanggao** *var.* Aoyang. Jiangxi, S China 28°16′N 114°55′E
Shangguan *see* Daixian
161 S8 **Shanghai** *var.* Shang-hai. Shanghai Shi, E China 31°14′N 121°28′E
161 S8 **Shanghai Shi** *var.* Hu, Shanghai. ◈ *municipality* E China
161 P13 **Shanghang** *var.* Linjiang. Fujian, SE China 25°03′N 116°25′E
160 K14 **Shanglin** *var.* Dafeng. Guangxi Zhuangzu Zizhiqu, S China 23°26′N 108°32′E
160 L7 **Shangluo** *prev.* Shangxian, Shangzhou. Shaanxi, C China 33°51′N 109°55′E
83 G15 **Shangombo** Western, W Zambia 16/28′S 22°10′E
Shangpai/Shangpaihe *see* Feixi
161 O6 **Shangqiu** *var.* Zhuji. Henan, C China 34°24′N 115°37′E
161 Q10 **Shangrao** Jiangxi, S China 28°27′N 117°57′E
Shangxian *see* Shangluo
161 S9 **Shangyu** *var.* Baiguan. Zhejiang, SE China 30°03′N 120°52′E
163 X9 **Shangzhi** Heilongjiang, NE China 45°13′N 127°59′E
Shangzhou *see* Shangluo
Shanhe *see* Zhengning
163 W9 **Shanhetun** Heilongjiang, NE China 44°42′N 127°12′E
Shan-hsi *see* Shaanxi, China
Shan-hsi *see* Shanxi, China
159 O6 **Shankou** Xinjiang Uygur Zizhiqu, W China 42°02′N 94°08′E
184 M13 **Shannon** Manawatu-Wanganui, North Island, New Zealand 40°32′S 175°24′E
97 C17 **Shannon** *Ir.* An tSionainn. ≈ W Ireland
97 B19 **Shannon** ✕ W Ireland 52°42′N 08°57′W
167 N6 **Shan Plateau** *plateau* E Myanmar (Burma)
158 M6 **Shanshan** *var.* Piqan. Xinjiang Uygur Zizhiqu, NW China 42°53′N 90°18′E
Shansi *see* Shanxi
167 N5 **Shan State** ◈ *state* E Myanmar (Burma)
Shantar Islands *see* Shantarskiye Ostrova
123 S12 **Shantarskiye Ostrova** *Eng.* Shantar Islands. *island group* E Russian Federation
161 Q14 **Shantou** *var.* Shan-t'ou, Swatow. Guangdong, S China 23°23′N 116°39′E
Shan-t'ou *see* Shantou
Shantung *see* Shandong
Shantung Peninsula *see* Shandong Bandao
163 O14 **Shanxi** *var.* Jin, Shan-hsi, Shansi, Shanxi Sheng. ◈ *province* C China
161 P6 **Shanxian** *var.* Shan Xian. Shandong, E China 34°51′N 116°09′E
Shan Xian *see* Sanmenxia
Shan Xian *see* Shanxian
Shanxi Sheng *see* Shanxi
160 L7 **Shanyang** Shaanxi, C China 33°35′N 109°48′E
161 N13 **Shanyin** *var.* Daiyue. Shanxi, C China E Asia 39°30′N 112°56′E
161 O13 **Shaoguan** *var.* Shao-kuan, *Cant.* Kukong; *prev.* Ch'u-chiang. Guangdong, S China 24°57′N 113°38′E
Shao-kuan *see* Shaoguan
161 Q11 **Shaowu** Fujian, SE China 27°24′N 117°26′E
161 S9 **Shaoxing** Zhejiang, SE China 30°02′N 120°35′E
160 M12 **Shaoyang** *var.* Tangdukou. Hunan, S China 26°54′N 111°14′E
160 M11 **Shaoyang** *var.* Baoqing, Shao-yang; *prev.* Pao-king. Hunan, S China 27°13′N 111°31′E
Shao-yang *see* Shaoyang
96 K5 **Shapinsay** *island* NE Scotland, United Kingdom
125 S4 **Shapkina** ≈ NW Russian Federation
Shāpūr *see* Salmās
158 M4 **Shaqiuhe** Xinjiang Uygur Zizhiqu, W China 45°00′N 88°52′E
Shaqlāwa *see* Sheqławe
Shaqlāwah *see* Sheqławe
138 I8 **Shaqqā** As Suwaydā', S Syria 32°53′N 36°42′E
141 P7 **Shaqrā'** Ar Riyāḑ, C Saudi Arabia 25°11′N 45°08′E
Shaqrā *see* Shuqrah
145 W10 **Shar** *var.* Charsk. Vostochnyy Kazakhstan, E Kazakhstan 49°33′N 81°03′E
149 O6 **Sharan** Dāykundī, SE Afghanistan 33°28′N 66°19′E
149 Q7 **Sharan** *var.* Zareh Sharan. Paktīkā, E Afghanistan 33°08′N 68°47′E
Sharaqpur *see* Sharqpur
145 U8 **Sharbakty** *Kaz.* Sharbaqty; *prev.* Shcherbakty. Pavlodar, E Kazakhstan 52°28′N 78°00′E
Sharbaqty *see* Sharbakty
141 X12 **Sharbatāt** S Oman 17°57′N 56°14′E
Sharbatāt, Ra's *see* Sharbithāt, Ras
141 X12 **Sharbithāt, Ras** *var.* Ra's Sharbatāt. *headland* S Oman 17°55′N 56°30′E
14 K14 **Sharbot Lake** Ontario, SE Canada 44°45′N 76°46′W
145 P17 **Shardara** *var.* Chardara. Yuzhnyy Kazakhstan, S Kazakhstan 41°15′N 68°01′E
Shardara Dalasy *see* Step' Shardara
145 P17 **Shardarinskoye Vodokhranilishche** *prev.* Chardarinskoye Vodokhranilishche. ⊠ S Kazakhstan
162 F8 **Sharga** Govĭ-Altay, W Mongolia 46°16′N 95°32′E
Sharga *see* Tsagaan-Uul
116 M7 **Sharhorod** Vinnyts'ka Oblast', C Ukraine 48°46′N 28°05′E
Sharhulsan *see* Mandal-Ovoo
165 V3 **Shari** Hokkaidō, NE Japan 43°54′N 144°42′E
Shari *see* Chari
139 T6 **Shārī, Buḩayrat** ◎ C Iraq
147 N12 **Sharixon** *Rus.* Shakhrisabz. Qashqadaryo Viloyati, S Uzbekistan 39°01′N 66°45′E
Sharjah *see* Ash Shāriqah
118 K12 **Sharkawshchyna** *var.* Sharkowshchyna, *Pol.* Szarkowszczyzna, *Rus.* Sharkovshchina. Vitsyebskaya Voblasts', NW Belarus 55°27′N 27°28′E
180 G9 **Shark Bay** *bay* Western Australia
141 Y9 **Sharkh** E Oman 21°20′N 59°04′E
Sharkovshchina/Sharkowshchyna *see* Sharkawshchyna
127 U6 **Sharlyk** Orenburgskaya Oblast', W Russian Federation 52°52′N 54°45′E
75 Y9 **Sharm ash Shaykh** *var.* Ofiral, Sharm el Sheikh. E Egypt 27°51′N 34°16′E
Sharm el Sheikh *see* Sharm ash Shaykh
18 B13 **Sharon** Pennsylvania, NE USA 41°12′N 80°28′W
26 H4 **Sharon Springs** Kansas, C USA 38°54′N 101°46′W
31 Q14 **Sharonville** Ohio, N USA 39°16′N 84°24′W
Sharourah *see* Sharūrah
29 O10 **Sharpe, Lake** ⊠ South Dakota, N USA
Sharqi, Al Jabal ash/Sharqi, Jebel esh *see* Anti-Lebanon
Sharqīyah, Al Minṭaqah ash *see* Ash Sharqīyah
138 I6 **Sharqīyat an Nabk, Jabal** ▲▲ W Syria
149 W8 **Sharqpur** *var.* Sharaqpur. Punjab, E Pakistan 31°29′N 74°08′E
141 Q13 **Sharūrah** *var.* Sharourah. Najrān, S Saudi Arabia 17°29′N 47°05′E
125 O14 **Shar'ya** Kostromskaya Oblast', NW Russian Federation 58°22′N 45°30′E
145 W15 **Sharyn** *prev.* Charyn. Almaty, SE Kazakhstan 43°48′N 79°22′E
145 V15 **Sharyn** *var.* Charyn. ≈ SE Kazakhstan
122 K13 **Sharypovo** Krasnoyarskiy Kray, C Russian Federation 55°33′N 89°12′E
83 J18 **Shashe** Central, NE Botswana 21°25′S 27°28′E
83 J18 **Shashe** *var.* Shashi. ≈ Botswana/Zimbabwe
81 J14 **Shashemenē** *var.* Shashemenne, Shashhamana, *It.* Sciasciamana. Oromīya, C Ethiopia 07°16′N 38°38′E
Shashemenne/Shashhamana *see* Shashemenē
Shashi *see* Shashe
Shashi/Sha-shih/Shasi *see* Jingzhou, Hubei
35 N3 **Shasta Lake** ⊠ California, W USA
35 N2 **Shasta, Mount** ▲ California, W USA 41°24′N 122°11′W
127 O4 **Shatki** Nizhegorodskaya Oblast', W Russian Federation 55°09′N 44°04′E
Shatlyk *see* Şatlyk
Shatra *see* Ash Shaṭrah
119 K17 **Shatsk** Minskaya Voblasts', C Belarus 53°25′N 27°41′E
127 N5 **Shatsk** Ryazanskaya Oblast', W Russian Federation 54°02′N 41°38′E
26 J9 **Shattuck** Oklahoma, C USA 36°16′N 99°52′W
145 P16 **Shauil'dir** *prev.* Shaul'der. Yuzhnyy Kazakhstan, S Kazakhstan 42°45′N 68°21′E
Shaul'der *see* Shauil'dir
11 S17 **Shaunavon** Saskatchewan, S Canada 49°49′N 108°25′W
Shavat *see* Shovot
Shaviyani Atoll *see* North Miladhunmadulu Atoll
158 K4 **Shawan** *var.* Sandaohezi. Xinjiang Uygur Zizhiqu, NW China 44°21′N 85°37′E
14 G12 **Shawanaga** Ontario, S Canada 45°29′N 80°16′W
30 M6 **Shawano** Wisconsin, N USA 44°46′N 88°38′W
30 M6 **Shawano Lake** ◎ Wisconsin, N USA
15 P10 **Shawinigan** *prev.* Shawinigan Falls. Québec, SE Canada 46°33′N 72°45′W
Shawinigan Falls *see* Shawinigan
15 P10 **Shawinigan-Sud** Québec, SE Canada 46°30′N 72°43′W
138 J5 **Shawmarīyah, Jabal ash** ▲▲ C Syria
27 O11 **Shawnee** Oklahoma, C USA 35°20′N 96°55′W
14 K12 **Shawville** Québec, SE Canada 45°37′N 76°31′W
145 Q16 **Shayan** *var.* Chayan. Yuzhnyy Kazakhstan, S Kazakhstan 42°59′N 69°22′E
Shaykh *see* Ash Shakk
139 W9 **Shaykh 'Ābid** *var.* Shaikh Ābid. Wāsiţ, E Iraq 32°40′N 46°09′E
139 Y10 **Shaykh Fāris** *var.* Shaikh Fāris. Maysān, E Iraq 32°06′N 47°39′E
139 T7 **Shaykh Ḩātim** Baghdād, E Iraq 33°29′N 44°15′E
Shaykh, Jabal ash *see* Hermon, Mount
139 X10 **Shaykh Najm** *var.* Shaikh Najm. Maysān, E Iraq 32°04′N 46°54′E
139 W9 **Shaykh Sa'd** Maysān, E Iraq 32°35′N 46°16′E
147 T14 **Shazud** SE Tajikistan 37°45′N 72°22′E
119 N18 **Shchadryn** *Rus.* Shchedrin. Homyel'skaya Voblasts', SE Belarus 52°53′N 29°33′E
119 H18 **Shchara** ≈ SW Belarus
Shchedrin *see* Shchadryn
Shcheglovsk *see* Kemerovo
126 K5 **Shchëkino** Tul'skaya Oblast', W Russian Federation 53°57′N 37°33′E
125 S7 **Shchel'yayur** Respublika Komi, NW Russian Federation 65°19′N 53°27′E
Shcherbakty *see* Sharbakty
126 K7 **Shchigry** Kurskaya Oblast', W Russian Federation 51°53′N 36°49′E
Shchitkovichi *see* Shchytkavichy
117 Q2 **Shchors** Chernihivs'ka Oblast', N Ukraine 51°49′N 31°58′E
117 T8 **Shchors'k** Dnipropetrovs'ka Oblast', E Ukraine 48°20′N 34°07′E
Shchuchin *see* Shchuchyn
145 Q7 **Shchuchinsk** *prev.* Shchuchye. Akmola, N Kazakhstan 52°57′N 70°10′E
Shchuchye *see* Shchuchinsk
119 G16 **Shchuchyn** *Pol.* Szczuczyn Nowogródzki, *Rus.* Shchuchin. Hrodzyenskaya Voblasts', W Belarus 53°36′N 24°45′E
119 K17 **Shchytkavichy** *Rus.* Shchitkovichi. Minskaya Voblasts', C Belarus 53°13′N 27°59′E
122 J13 **Shebalino** Respublika Altay, S Russian Federation 51°16′N 85°41′E
126 J9 **Shebekino** Belgorodskaya Oblast', W Russian Federation 50°25′N 36°55′E
Shebelë Wenz, Wabē *see* Shebeli
81 L14 **Shebeli** *Amh.* Wabē Shebelē Wenz, *It.* Scebeli, *Som.* Webi Shabeelle. ≈ Ethiopia/Somalia
113 M20 **Shebenikut, Maja e** ▲ E Albania 41°13′N 20°27′E
Sheberghān *see* Shibirghān
144 F14 **Shebir** Mangistau, SW Kazakhstan 44°52′N 52°01′E
31 N8 **Sheboygan** Wisconsin, N USA 43°46′N 87°44′W
77 X15 **Shebshi Mountains** *var.* Schebschi Mountains. ▲▲ E Nigeria
Shechem *see* Nablus
Shedadi *see* Ash Shadādah
15 P14 **Shediac** New Brunswick, SE Canada 46°13′N 64°35′W
126 L15 **Shedok** Krasnodarskiy Kray, SW Russian Federation 44°12′N 40°49′E
Sheekh *see* Shiikh
38 M11 **Sheenjek River** ≈ Alaska, USA
96 D13 **Sheep Haven** *Ir.* Cuan na gCaorach. *inlet* N Ireland
35 X10 **Sheep Range** ▲▲ Nevada, W USA
98 M13 **'s-Heerenberg** Gelderland, E Netherlands 51°52′N 06°15′E
97 P22 **Sheerness** SE England, United Kingdom 51°27′N 00°45′E
13 Q15 **Sheet Harbour** Nova Scotia, SE Canada 44°56′N 62°31′W
185 H18 **Sheffield** Canterbury, South Island, New Zealand 43°22′S 172°01′E
97 M18 **Sheffield** N England, United Kingdom 53°23′N 01°30′W
23 O2 **Sheffield** Alabama, S USA 34°46′N 87°42′W
29 V12 **Sheffield** Iowa, C USA 42°53′N 93°13′W
25 N10 **Sheffield** Texas, SW USA 30°42′N 101°49′W
63 H22 **Shehuen, Río** ≈ S Argentina
Shekhem *see* Nablus
149 V8 **Shekhūpura** Punjab, NE Pakistan 31°42′N 74°08′E
Sheki *see* Şäki
124 L14 **Sheksna** Vologodskaya Oblast', NW Russian Federation 59°11′N 38°32′E
123 T5 **Shelagskiy, Mys** *headland* NE Russian Federation 70°04′N 170°39′E
27 V3 **Shelbina** Missouri, C USA 39°41′N 92°02′W
13 P16 **Shelburne** Nova Scotia, SE Canada 43°47′N 65°20′W
14 G14 **Shelburne** Ontario, S Canada 44°04′N 80°12′W
33 R7 **Shelby** Montana, NW USA 48°30′N 111°52′W
21 Q10 **Shelby** North Carolina, SE USA 35°15′N 81°34′W
31 S12 **Shelby** Ohio, N USA 40°52′N 82°39′W
30 L14 **Shelbyville** Illinois, N USA 39°24′N 88°47′W
31 P14 **Shelbyville** Indiana, N USA 39°31′N 85°46′W
20 L5 **Shelbyville** Kentucky, S USA 38°13′N 85°12′W
27 V2 **Shelbyville** Missouri, C USA 39°48′N 92°01′W
20 J10 **Shelbyville** Tennessee, S USA 35°29′N 86°30′W
25 X8 **Shelbyville** Texas, SW USA 31°42′N 94°03′W
30 L14 **Shelbyville, Lake** ⊠ Illinois, N USA
29 S12 **Sheldon** Iowa, C USA 43°10′N 95°51′W
38 M11 **Sheldons Point** Alaska, USA 62°31′N 165°03′W
145 V15 **Shelek** *prev.* Chilik. Almaty, SE Kazakhstan 43°35′N 78°12′E
145 V15 **Shelek** *prev.* Chilik. ≈ SE Kazakhstan
Shelekhov Gulf *see* Shelikhova, Zaliv
123 U9 **Shelikhova, Zaliv** *Eng.* Shelekhov Gulf. *gulf* E Russian Federation
39 P14 **Shelikof Strait** *strait* Alaska, USA
Shelim *see* Shalīm
11 T14 **Shellbrook** Saskatchewan, S Canada 53°14′N 106°24′W
28 L3 **Shell Creek** ≈ North Dakota, N USA
Shellif *see* Chelif, Oued
22 I10 **Shell Keys** *island group* Louisiana, S USA
30 I4 **Shell Lake** Wisconsin, N USA 45°44′N 91°56′W
29 W12 **Shell Rock** Iowa, C USA 42°42′N 92°34′W
185 C26 **Shelter Point** *headland* Stewart Island, New Zealand 47°04′S 168°13′E
18 L13 **Shelton** Connecticut, NE USA 41°19′N 73°06′W
32 G8 **Shelton** Washington, NW USA 47°13′N 123°06′W
Shemakha *see* Şamaxı
145 W9 **Shemonaikha** Vostochnyy Kazakhstan, E Kazakhstan 50°38′N 81°55′E
127 Q4 **Shemursha** Chuvashskaya Respublika, W Russian Federation 54°57′N 47°27′E
38 D16 **Shemya Island** *island* Aleutian Islands, Alaska, USA
29 T16 **Shenandoah** Iowa, C USA 40°46′N 95°23′W
21 U4 **Shenandoah** Virginia, NE USA 38°26′N 78°34′W
21 U4 **Shenandoah Mountains** *ridge* West Virginia, NE USA
21 V3 **Shenandoah River** ≈ West Virginia, NE USA
77 W15 **Shendam** Plateau, C Nigeria 08°52′N 09°30′E
80 G8 **Shendi** *var.* Shandī. River Nile, NE Sudan 16°41′N 33°22′E
76 I15 **Shenge** SW Sierra Leone 07°54′N 12°54′W
146 L10 **Shengeldi** *Rus.* Chingildi. Navoiy Viloyati, N Uzbekistan 40°59′N 64°13′E
145 U15 **Shengel'dy** Almaty, SE Kazakhstan 44°04′N 77°31′E
113 K18 **Shëngjin** *var.* Shëngjini. Lezhë, NW Albania 41°49′N 19°34′E
Shëngjini *see* Shëngjin
Shengking *see* Liaoning
Sheng Xian/Shengxian *see* Shengzhou
161 S9 **Shengzhou** *var.* Shengxian, Sheng Xian. Zhejiang, SE China 29°36′N 120°42′E
Shenking *see* Liaoning
125 N11 **Shenkursk** Arkhangel'skaya Oblast', NW Russian Federation 62°10′N 42°58′E
160 L3 **Shenmu** Shaanxi, C China 38°49′N 110°27′E
113 L19 **Shën Noj i Madh** ▲ C Albania 41°23′N 20°07′E
160 L8 **Shennong Ding** ▲ C China 31°24′N 110°16′E
Shenshi/Shensi *see* Shaanxi
163 V12 **Shenyang** *Chin.* Shen-yang, *Eng.* Moukden, Mukden; *prev.* Fengtien. *province capital* Liaoning, NE China 41°50′N 123°26′E
Shen-yang *see* Shenyang
161 O15 **Shenzhen** Guangdong, S China 22°39′N 114°02′E
154 G8 **Sheopur** Madhya Pradesh, C India 25°41′N 76°42′E
116 L5 **Shepetivka** *Rus.* Shepetovka. Khmel'nyts'ka Oblast', NW Ukraine 50°12′N 27°01′E
Shepetovka *see* Shepetivka
25 W10 **Shepherd** Texas, SW USA 30°30′N 95°00′W
187 R14 **Shepherd Islands** *island group* C Vanuatu
20 K5 **Shepherdsville** Kentucky, S USA 38°00′N 85°42′W
183 O11 **Shepparton** Victoria, SE Australia 36°25′S 145°26′E
97 P22 **Sheppey, Isle of** *island* SE England, United Kingdom
139 T2 **Sheqławe** *Ar.* Shaqlāwah, *var.* Shaqlāwa. Arbīl, E Iraq 36°24′N 44°21′E
Sherabad *see* Sherobod
9 O4 **Sherard, Cape** *headland* Nunavut, N Canada 74°36′N 80°10′W
97 L23 **Sherborne** S England, United Kingdom 50°58′N 02°30′W
76 H16 **Sherbro Island** *island* SW Sierra Leone
15 Q12 **Sherbrooke** Québec, SE Canada 45°23′N 71°55′W
29 T11 **Sherburn** Minnesota, N USA 43°39′N 94°43′W
78 H6 **Sherda** Tibesti, N Chad 20°04′N 16°48′E
80 G7 **Shereik** River Nile, N Sudan 18°44′N 33°37′E
126 K3 **Sheremet'yevo** ✕ (Moskva) Moskovskaya Oblast', W Russian Federation 56°05′N 37°10′E
153 P14 **Sherghāti** Bihār, N India 24°35′N 84°51′E
27 U12 **Sheridan** Arkansas, C USA 34°18′N 92°22′W
33 W12 **Sheridan** Wyoming, C USA 44°47′N 106°59′W
182 G8 **Sheringa** South Australia 33°51′S 135°13′E
25 U5 **Sherman** Texas, SW USA 33°39′N 96°35′W
194 J10 **Sherman Island** *island* Antarctica
19 S4 **Sherman Mills** Maine, NE USA 45°51′N 68°23′W
29 O15 **Sherman Reservoir** ⊠ Nebraska, C USA
147 N14 **Sherobod** *Rus.* Sherabad. Surkhondaryo Viloyati, S Uzbekistan 37°43′N 66°59′E
147 O13 **Sherobod** *Rus.* Sherabad. ≈ S Uzbekistan
153 T14 **Sherpur** Dhaka, N Bangladesh 25°00′N 90°01′E
37 T4 **Sherrelwood** Colorado, C USA 39°49′N 105°00′W
99 J14 **'s-Hertogenbosch** *Fr.* Bois-le-Duc, *Ger.* Herzogenbusch. Noord-Brabant, S Netherlands 51°41′N 05°19′E
28 M2 **Sherwood** North Dakota, N USA 48°55′N 101°36′W
11 Q14 **Sherwood Park** Alberta, SW Canada 53°34′N 113°04′W
12 F10 **Sheshea, Río** ≈ E Peru
143 T5 **Sheshtamad** Khorāsān-e Razavī, NE Iran 36°03′N 57°45′E
29 S10 **Shetek, Lake** ◎ Minnesota, N USA
96 M2 **Shetland Islands** *island group* NE Scotland, United Kingdom
144 F14 **Shetpe** Mangistau, SW Kazakhstan 44°06′N 52°03′E
154 C11 **Shetrunji** ≈ W India
Shevchenko *see* Aktau
117 W5 **Shevchenkove** Kharkivs'ka Oblast', E Ukraine 49°40′N 37°13′E
81 H14 **Shewa Gīmīra** Southern Nationalities, S Ethiopia 07°12′N 35°49′E
161 Q9 **Shexian** *var.* Huicheng, She Xian. Anhui, E China 29°53′N 118°27′E
She Xian *see* Shexian
161 R6 **Sheyang** *prev.* Hede. Jiangsu, E China 33°49′N 120°13′E
29 O4 **Sheyenne** North Dakota, N USA 47°49′N 99°08′W
29 P4 **Sheyenne River** ≈ North Dakota, N USA
96 G7 **Shiant Islands** *island group* NW Scotland, United Kingdom
123 U12 **Shiashkotan, Ostrov** *island* Kuril'skiye Ostrova, SE Russian Federation
31 R9 **Shiawassee River** ≈ Michigan, N USA
141 R14 **Shibām** C Yemen 15°49′N 48°24′E
Shibarghān *see* Shibirghān
165 O10 **Shibata** *var.* Sibata. Niigata, Honshū, C Japan 37°57′N 139°20′E
Shiberghan/Shiberghān *see* Shibirghān
Shibh Jazīrat Sīnā' *see* Sinai
75 W8 **Shibīn al Kawm** *var.* Shibīn el Kôm. N Egypt 30°33′N 31°00′E
Shibīn el Kôm *see* Shibīn al Kawm
149 N2 **Shibirghān** *var.* Shibarghān, Shiberghan, Shiberghān; *prev.* Sheberghān. Jowzjān, N Afghanistan 36°41′N 65°45′E
143 O13 **Shīb, Kūh-e** ▲ S Iran
12 D8 **Shibogama Lake** ◎ Ontario, C Canada
Shibotsu-jima *see* Zelënyy, Ostrov
164 B16 **Shibushi** Kagoshima, Kyūshū, SW Japan 31°27′N 131°05′E
189 U13 **Shichiyo Islands** *island group* Chuuk, C Micronesia
Shickshock Mountains *see* Chic-Chocs, Monts
shiderti *see* Shiderty
Shiderti *see* Shiderty
145 S8 **Shiderty** *prev.* Shiderti. Pavlodar, NE Kazakhstan 51°40′N 74°50′E
145 S9 **Shiderty** *prev.* Shiderti. ≈ N Kazakhstan
96 G10 **Shiel, Loch** ◎ N Scotland, United Kingdom
164 J13 **Shiga** *off.* Shiga-ken, *var.* Siga. ◈ *prefecture* Honshū, SW Japan
Shiga-ken *see* Shiga
Shigatse *see* Xigazê
141 U13 **Shiḩan** *oasis* NE Yemen
Shih-chia-chuang/Shihmen *see* Shijiazhuang
158 K4 **Shihezi** Xinjiang Uygur Zizhiqu, NW China 44°21′N 85°59′E
Shiichi *see* Shyichy
80 N12 **Shiikh** *prev.* Sheekh. Togdheer, N Somalia 10°01′N 45°21′E
113 K19 **Shijak** *var.* Shijaku. Durrës, W Albania 41°21′N 19°34′E
Shijaku *see* Shijak
161 O4 **Shijiazhuang** *var.* Shih-chia-chuang; *prev.* Shihmen. *province capital* Hebei, E China 38°04′N 114°28′E
165 R5 **Shikabe** Hokkaidō, NE Japan 42°03′N 140°45′E
149 Q13 **Shikārpur** Sind, S Pakistan 27°59′N 68°39′E
127 Q7 **Shikhany** Saratovskaya Oblast', W Russian Federation 52°07′N 47°13′E
189 V12 **Shiki Islands** *island group* Chuuk, C Micronesia
164 G14 **Shikoku** *var.* Sikoku. *island* SW Japan
192 H5 **Shikoku Basin** *var.* Sikoku Basin. *undersea feature* N Philippine Sea 28°00′N 137°00′E
164 G14 **Shikoku-sanchi** ▲▲ Shikoku, SW Japan
165 X4 **Shikotan, Ostrov** *Jap.* Shikotan-tō. *island* NE Russian Federation
Shikotan-tō *see* Shikotan, Ostrov
165 R4 **Shikotsu-ko** *var.* Sikotu Ko. ◎ Hokkaidō, NE Japan
81 N15 **Shilabo** Sumalē, E Ethiopia 06°05′N 44°48′E
139 V3 **Shilayr, Wādī** *var.* Āw-e Shilēr, Āveh-ye Shalīr. ≈ E Iraq
127 X7 **Shil'da** Orenburgskaya Oblast', W Russian Federation 51°46′N 59°48′E
Shilēr, Āw-e *see* Shilayr, Wādī
153 S12 **Shiliguri** *prev.* Siliguri. West Bengal, NE India 26°46′N 88°24′E
Shiliu *see* Changjiang
129 V7 **Shilka** ≈ S Russian Federation
18 H15 **Shillington** Pennsylvania, NE USA 40°18′N 75°57′W
153 V13 **Shillong** *state capital* Meghālaya, NE India 25°37′N 91°54′E
126 M5 **Shilovo** Ryazanskaya Oblast', W Russian Federation 54°18′N 40°53′E
164 C14 **Shimabara** *var.* Simabara. Nagasaki, Kyūshū, SW Japan 32°48′N 130°20′E
164 C14 **Shimabara-wan** *bay* SW Japan
164 F12 **Shimane** *off.* Shimane-ken, *var.* Simane. ◈ *prefecture* Honshū, SW Japan
164 G11 **Shimane-hantō** *peninsula* Honshū, SW Japan
Shimane-ken *see* Shimane
123 Q13 **Shimanovsk** Amurskaya Oblast', SE Russian Federation 52°00′N 127°36′E
Shimanto *see* Nakamura
Shimbir Berris *see* Shimbiris
80 O12 **Shimbiris** *var.* Shimbir Berris. ▲ N Somalia 10°43′N 47°10′E
165 T4 **Shimizu** Hokkaidō, NE Japan 42°58′N 142°54′E
164 M14 **Shimizu** *var.* Simizu. Shizuoka, Honshū, S Japan 35°01′N 138°29′E
152 I8 **Shimla** *prev.* Simla. *state capital* Himāchal Pradesh, N India 31°07′N 77°09′E
165 N14 **Shimoda** *var.* Simoda. Shizuoka, Honshū, S Japan 34°40′N 138°55′E
165 O13 **Shimodate** *var.* Simodate. Ibaraki, Honshū, S Japan 36°20′N 140°00′E
155 F18 **Shimoga** Karnātaka, W India 13°56′N 75°31′E
164 C15 **Shimo-jima** *island* SW Japan
164 B15 **Shimo-Koshiki-jima** *island* SW Japan
81 J21 **Shimoni** Kwale, S Kenya 04°40′S 39°22′E
164 D13 **Shimonoseki** *var.* Simonoseki, *hist.* Akamagaseki, Bakan. Yamaguchi, Honshū, SW Japan 33°57′N 130°54′E
124 G14 **Shimsk** Novgorodskaya Oblast', NW Russian Federation 58°12′N 30°43′E
141 W7 **Shināş** N Oman 24°45′N 56°24′E
148 J6 **Shīnḑand** *prev.* Shīndand. Herāt, W Afghanistan 33°19′N 62°09′E
Shindand *see* Shīnḑand
Shinei *see* Xinying
162 H10 **Shinejinst** *var.* Dzalaa. Bayanhongor, C Mongolia 44°29′N 99°19′E
25 T12 **Shiner** Texas, SW USA 29°25′N 97°10′W
167 N1 **Shingbwiyang** Kachin State, N Myanmar (Burma) 26°40′N 96°14′E
Shingozha *see* Shynkozha
164 J15 **Shingū** *var.* Singû. Wakayama, Honshū, SW Japan 33°41′N 135°57′E
14 F8 **Shining Tree** Ontario, S Canada 47°36′N 81°12′W
165 P9 **Shinjō** *var.* Sinzyô. Yamagata, Honshū, C Japan 38°47′N 140°17′E
96 I7 **Shin, Loch** ◎ N Scotland, United Kingdom
21 S3 **Shinnston** West Virginia, NE USA 39°22′N 80°19′W
138 I6 **Shinshār** *Fr.* Chinnchâr. Ḩimş, W Syria 34°36′N 36°45′E
Shinshū *see* Jinju
165 T4 **Shintoku** Hokkaidō, NE Japan 43°03′N 142°50′E
81 G20 **Shinyanga** Shinyanga, NW Tanzania 03°40′S 33°25′E
81 G20 **Shinyanga** ◈ *region* N Tanzania
165 Q10 **Shiogama** *var.* Siogama. Miyagi, Honshū, C Japan 38°19′N 141°00′E
164 M12 **Shiojiri** *var.* Sioziri. Nagano, Honshū, S Japan 36°08′N 137°58′E
164 I15 **Shiono-misaki** *headland* Honshū, SW Japan 33°25′N 135°45′E
165 Q12 **Shioya-zaki** *headland* Honshū, C Japan 37°00′N 140°57′E
114 J9 **Shipchenski Prohod** , *var.* Shipchenski Prokhod. *pass* C Bulgaria
Shipchenski Prokhod *see* Shipchenski Prohod
160 G14 **Shiping** Yunnan, SW China 23°45′N 102°23′E
13 P13 **Shippagan** *var.* Shippegan. New Brunswick, SE Canada 47°45′N 64°44′W
Shippegan *see* Shippagan
18 F15 **Shippensburg** Pennsylvania, NE USA 40°03′N 77°31′W
37 P9 **Shiprock** New Mexico, SW USA 36°47′N 108°41′W
37 O9 **Ship Rock** ▲ New Mexico, SW USA 36°41′N 108°50′W
15 R6 **Shipshaw** ≈ Québec, SE Canada
123 V10 **Shipunskiy, Mys** *headland* E Russian Federation 53°04′N 159°57′E
160 K7 **Shiquan** Shaanxi, C China 33°05′N 108°15′E
Shiquanhe *see* Gar
122 K13 **Shira** Respublika Khakasiya, S Russian Federation 54°35′N 89°58′E
Shirajganj Ghat *see* Sirajganj
165 P12 **Shirakawa** *var.* Sirakawa. Fukushima, Honshū, C Japan 37°07′N 140°11′E
164 M13 **Shirane-san** ▲ Honshū, S Japan 35°39′N 138°13′E
165 U14 **Shiranuka** Hokkaidō, NE Japan 42°57′N 144°01′E
195 N12 **Shirase Coast** *physical region* Antarctica
165 U3 **Shirataki** Hokkaidō, NE Japan 43°55′N 143°14′E
143 O11 **Shīrāz** *var.* Shīrāz. Fārs, S Iran 29°38′N 52°34′E
83 N15 **Shire** *var.* Chire. ≈ Malawi/Mozambique
Shiree *see* Tsagaanhayrhan
Shireet *see* Bayandelger
165 W3 **Shiretoko-hantō** *headland* Hokkaidō, NE Japan 44°06′N 145°07′E
165 W3 **Shiretoko-misaki** *headland* Hokkaidō, NE Japan 44°20′N 145°19′E
127 N5 **Shiringushi** Respublika Mordoviya, W Russian Federation 53°50′N 42°49′E
148 M3 **Shīrīn Tagāb** Fāryāb, N Afghanistan 36°49′N 65°01′E
149 N2 **Shīrīn Tagāb** ≈ N Afghanistan

◆ Country
● Country Capital
◇ Dependent Territory
○ Dependent Territory Capital
◈ Administrative Regions
✕ International Airport
▲ Mountain
▲▲ Mountain Range
⩚ Volcano
≈ River
◎ Lake
⊠ Reservoir

165 R6 **Shiriya-zaki** *headland* Honshū, C Japan *41°24′N 141°27′E*
144 I12 **Shirkala, Gryada** *plain* W Kazakhstan
152 F11 **Shir Kolāyat** *var.* Kolāyat. Rājasthān, NW India *27°56′N 73°02′E*
165 P10 **Shiroishi** *var.* Siroisi. Miyagi, Honshū, C Japan *38°00′N 140°38′E*
Shirokoye *see* Shyroke
165 O10 **Shirone** *var.* Sirone. Niigata, Honshū, C Japan *37°46′N 139°00′E*
164 L12 **Shirotori** Gifu, Honshū, SW Japan *35°53′N 136°52′E*
197 T1 **Shirshov Ridge** *undersea feature* W Bering Sea
Shirshütür/Shirshyutyur, Peski *see* Şirşütür Gumy
143 T3 **Shīrvān** *var.* Shirwān. Khorāsān-e Shomālī, NE Iran *37°25′N 57°55′E*
Shirwa, Lake *see* Chilwa, Lake
Shirwān *see* Shīrvān
159 N5 **Shisanjianfang** Xinjiang Uygur Zizhiqu, W China *43°10′N 91°15′E*
38 M16 **Shishaldin Volcano** ▲ Unimak Island, Alaska, USA *54°45′N 163°58′W*
Shishchitsy *see* Shyshchytsy
83 G16 **Shishikola** North West, N Botswana *18°09′S 23°08′E*
38 M8 **Shishmaref** Alaska, USA *66°15′N 166°04′W*
Shisur *see* Ash Shişar
164 L13 **Shitara** Aichi, Honshū, SW Japan *35°06′N 137°33′E*
152 D12 **Shiv** Rājasthān, NW India *26°11′N 71°14′E*
Shivāji Sāgar *see* Konya Reservoir
154 H8 **Shivpuri** Madhya Pradesh, C India *25°28′N 77°41′E*
36 J9 **Shivwits Plateau** *plain* Arizona, SW USA
Shiwalik Range *see* Siwalik Range
160 M8 **Shiyan** Hubei, C China *32°31′N 110°45′E*
145 O15 **Shiyeli** *prev.* Chiili. Kzylorda, S Kazakhstan *44°13′N 66°46′E*
Shizilu *see* Junan
160 H13 **Shizong** *var.* Danfeng. Yunnan, SW China *24°53′N 104′E*
165 R10 **Shizugawa** Miyagi, Honshū, NE Japan *38°40′N 141°26′E*
165 T5 **Shizunai** Hokkaidō, NE Japan *42°20′N 142°24′E*
165 M14 **Shizuoka** *var.* Sizuoka. Shizuoka, Honshū, S Japan *34°59′N 138°20′E*
164 M13 **Shizuoka** *off.* Shizuoka-ken, *var.* Sizuoka. ⬗ *prefecture* Honshū, S Japan
Shizuoka-ken *see* Shizuoka
Shklov *see* Shklow
119 N15 **Shklow** *Rus.* Shklov. Mahilyowskaya Voblasts', E Belarus *54°13′N 30°18′E*
113 K18 **Shkodër** *var.* Shkodra, *It.* Scutari, *SCr.* Skadar. Shkodër, NW Albania *42°03′N 19°31′E*
113 K17 **Shkodër** ⬗ *district* NW Albania
Shkodra *see* Shkodër
Shkodrës, Liqeni i *see* Scutari, Lake
113 L20 **Shkumbinit, Lumi i** *var.* Shkumbî, Shkumbin. ≈ C Albania
Shkumbî/Shkumbin *see* Shkumbinit, Lumi i
Shligigh, Cuan *see* Sligo Bay
122 L4 **Shmidta, Ostrov** *island* Severnaya Zemlya, N Russian Federation
183 S10 **Shoalhaven River** ≈ New South Wales, SE Australia
11 W16 **Shoal Lake** Manitoba, S Canada *50°28′N 100°36′W*
31 O15 **Shoals** Indiana, N USA *38°40′N 86°47′W*
164 I13 **Shōdo-shima** *island* SW Japan
Shōka *see* Zhanghua
122 M5 **Shokal'skogo, Proliv** *strait* N Russian Federation
147 T14 **Shokhdara, Qatorkŭhi** *Rus.* Shakhdarinskiy Khrebet. ▲▲ SE Tajikistan
145 T15 **Shokpar** *Kaz.* Shoqpar; *prev.* Chokpar. Zhambyl, S Kazakhstan *43°49′N 74°25′E*
145 P15 **Sholakkorgan** *var.* Chulakkurgan. Yuzhnyy Kazakhstan, S Kazakhstan *43°45′N 69°10′E*
145 N9 **Sholaksay** Kostanay, N Kazakhstan *51°45′N 64°45′E*
Sholāpur *see* Solāpur
Sholdaneshty *see* Şoldăneşti
145 W15 **Shonzhy** *prev.* Chundzha. Almaty, SE Kazakhstan *43°32′N 79°28′E*
Shoqpar *see* Shokpar
155 G21 **Shoranūr** Kerala, SW India *10°53′N 76°06′E*
155 G16 **Shorāpur** Karnātaka, C India *16°34′N 76°48′E*
147 O14 **Sho'rchi** *Rus.* Shurchi. Surkhondaryo Viloyati, S Uzbekistan *37°58′N 67°40′E*
30 M11 **Shorewood** Illinois, N USA *41°31′N 88°12′W*
Shorkazakhly, Solonchak *see* Kazakhlyshor, Solonchak
145 Q9 **Shortandy** Akmola, C Kazakhstan *51°45′N 71°01′E*
149 O2 **Shōr Tappeh** *var.* Shortepa, Shor Tepe; *prev.* Shūr Tappeh. Balkh, N Afghanistan *37°22′N 66°49′E*
Shortepa/Shor Tepe *see* Shōr Tappeh
186 J7 **Shortland Island** *var.* Alu. *island* Shortland Islands, NW Solomon Islands
Shosambetsu *see* Shosanbetsu
165 S2 **Shosanbetsu** *var.* Shosambetsu. Hokkaidō, NE Japan *44°31′N 141°47′E*
33 O15 **Shoshone** Idaho, NW USA *42°56′N 114°24′W*
35 T6 **Shoshone Mountains** ▲▲ Nevada, W USA
33 U12 **Shoshone River** ≈ Wyoming, C USA
83 I19 **Shoshong** Central, SE Botswana *23°02′S 26°31′E*
33 V14 **Shoshoni** Wyoming, C USA *43°13′N 108°06′W*
Shōshū *see* Sangju
117 S2 **Shostka** Sums'ka Oblast', NE Ukraine *51°52′N 33°30′E*
185 C21 **Shotover** ≈ South Island, New Zealand
146 H9 **Shovot** *Rus.* Shavat. Xorazm Viloyati, W Uzbekistan *41°41′N 60°13′E*
37 N12 **Show Low** Arizona, SW USA *34°15′N 110°01′W*
Show Me State *see* Missouri
125 O4 **Shoyna** Nenetskiy Avtonomnyy Okrug, NW Russian Federation *67°50′N 44°09′E*
124 M11 **Shozhma** Arkhangel'skaya Oblast', NW Russian Federation *61°57′N 40°10′E*
117 Q7 **Shpola** Cherkas'ka Oblast', N Ukraine *49°00′N 31°27′E*
Shqipëria/Shqipërisë, Republika e *see* Albania
22 G5 **Shreveport** Louisiana, S USA *32°32′N 93°45′W*
97 K19 **Shrewsbury** *hist.* Scrobesbyrig'. W England, United Kingdom *52°43′N 02°45′W*
152 D11 **Shri Mohangarh** *prev.* Sri Mohangorh. Rājasthān, NW India *27°17′N 71°18′E*
153 S16 **Shrīrāmpur** *prev.* Serampore, Serampur. West Bengal, NE India *22°44′N 88°20′E*
97 K19 **Shropshire** *cultural region* W England, United Kingdom
113 N17 **Shtime** *Serb.* Štimlje. C Kosovo *42°27′N 21°03′E*
145 S16 **Shu** *Kaz.* Shū. Zhambyl, SE Kazakhstan *43°34′N 73°41′E*
129 Q7 **Shu** *Kaz.* Shū; *prev.* Chu. ≈ Kazakhstan/Kyrgyzstan
160 G13 **Shuangbai** *var.* Tuodian. Yunnan, SW China *24°45′N 101°38′E*
163 W9 **Shuangcheng** Heilongjiang, NE China *45°20′N 126°21′E*
Shuangcheng *see* Zherong
160 E14 **Shuangjiang** *var.* Weiyuan. Yunnan, SW China *23°28′N 99°43′E*
Shuangjiang *see* Jiangkou
Shuangjiang *see* Tongdao
163 U10 **Shuangliao** *var.* Zhengjiatun. Jilin, NE China *43°31′N 123°32′E*
Shuang-liao *see* Liaoyuan
Shuangshipu *see* Fengxian
163 Y7 **Shuangyashan** *var.* Shuang-ya-shan. Heilongjiang, NE China *46°37′N 131°10′E*
Shuang-ya-shan *see* Shuangyashan
Shu'aymiah *see* Shu'aymīyah
141 W12 **Shu'aymīyah** *var.* Shu'aymiah. S Oman *17°55′N 55°39′E*
Shubarkuduk *see* Shubarkudyk
144 I10 **Shubarkudyk** *prev.* Shubarkuduk, *Kaz.* Shubarqudyq. Aktyubinsk, W Kazakhstan *49°09′N 56°31′E*
Shubarqudyq *see* Shubarkudyk
145 N12 **Shubar-Tengiz, Ozero** ◎ C Kazakhstan
39 S5 **Shublik Mountains** ▲▲ Alaska, USA
Shubrā al Khaymah *see* Shubrâ el Kheima
121 U13 **Shubrâ el Kheima** *var.* Shubrā al Khaymah. N Egypt *30°06′N 31°15′E*
158 E8 **Shufu** *var.* Tuokezhake. Xinjiang Uygur Zizhiqu, NW China *39°18′N 75°43′E*
147 S14 **Shughnon, Qatorkŭhi** *Rus.* Shugnanskiy Khrebet. ▲▲ SE Tajikistan
Shugnanskiy Khrebet *see* Shughnon, Qatorkŭhi
161 Q6 **Shu He** ≈ E China
Shuicheng *see* Lupanshui
Shuiding *see* Huocheng
Shuidong *see* Dianbai
Shuiji *see* Laixi
Shū-Ile Taūlary *see* Gory Shu-Ile
Shuilocheng *see* Zhuanglang
Shuiluo *see* Zhuanglang
149 T10 **Shujāābād** Punjab, E Pakistan *29°53′N 71°23′E*
Shū, Kazakhstan *see* Shu
Shū, Kazakhstan/Kyrgyzstan *see* Shu
163 W9 **Shulan** Jilin, NE China *44°28′N 126°57′E*
158 E8 **Shule** Xinjiang Uygur Zizhiqu, NW China *39°19′N 76°06′E*
Shuleh *see* Shule He
159 Q8 **Shule He** *var.* Shuleh, Sulo. ≈ C China
30 K9 **Shullsburg** Wisconsin, N USA *42°37′N 90°12′W*
Shulu *see* Xinji
39 N16 **Shumagin Islands** *island group* Alaska, USA
146 G7 **Shumanay** Qoraqalpog'iston Respublikasi, W Uzbekistan *42°42′N 58°56′E*
114 M8 **Shumen** Shumen, NE Bulgaria *43°17′N 26°57′E*
114 M8 **Shumen** ⬗ *province* NE Bulgaria
127 P4 **Shumerlya** Chuvashskaya Respublika, W Russian Federation *55°31′N 46°24′E*
122 G11 **Shumikha** Kurganskaya Oblast', C Russian Federation *55°12′N 63°09′E*
118 M12 **Shumilina** *Rus.* Shumilino. Vitsyebskaya Voblasts', NE Belarus *55°18′N 29°37′E*
Shumilino *see* Shumilina
123 V11 **Shumshu, Ostrov** *island* SE Russian Federation
116 K5 **Shums'k** Ternopil's'ka Oblast', W Ukraine *50°06′N 26°04′E*
Shūnan *see* Tokuyama
39 O7 **Shungnak** Alaska, USA *66°53′N 157°08′W*
Shunsen *see* Chuncheon
Shuoxian *see* Shuozhou
161 N3 **Shuozhou** *var.* Shuoxian. Shanxi, C China *39°20′N 112°25′E*
141 P16 **Shuqrah** *var.* Shaqrā. SW Yemen *13°26′N 45°44′E*
Shurab *see* Shŭrob
Shurchi *see* Sho'rchi
147 R11 **Shŭrob** *Rus.* Shurab. NW Tajikistan *40°02′N 70°31′E*
143 T10 **Shūr, Rūd-e** ≈ E Iran
Shūr Tappeh *see* Shōr Tappeh
83 K17 **Shurugwi** *prev.* Selukwe. Midlands, C Zimbabwe *19°40′S 30°00′E*
142 L8 **Shūsh** *anc.* Susa, *Bibl.* Shushan. Khūzestān, SW Iran *32°12′N 48°20′E*
Shushan *see* Shūsh
142 L9 **Shūshtar** *var.* Shustar, Shushter. Khūzestān, SW Iran *32°03′N 48°51′E*
Shushter/Shustar *see* Shūshtar
141 T9 **Shuṭfah, Qalamat** *well* E Saudi Arabia
139 V9 **Shuwayjah, Hawr ash** *var.* Hawr as Suwayqiyah. ◎ E Iraq
124 M16 **Shuya** Ivanovskaya Oblast', W Russian Federation *56°51′N 41°24′E*
39 Q14 **Shuyak Island** *island* Alaska, USA
166 M4 **Shwebo** Sagaing, C Myanmar (Burma) *22°35′N 95°42′E*
166 L7 **Shwedaung** Bago, W Myanmar (Burma) *18°44′N 95°12′E*
166 M7 **Shwegyin** Bago, SW Myanmar (Burma) *17°56′N 96°59′E*
167 N4 **Shweli** *Chin.* Longchuan Jiang. ≈ Myanmar (Burma)/China
166 M6 **Shwemyo** Mandalay, C Myanmar (Burma) *20°04′N 96°13′E*
145 S14 **Shyganak** *var.* Čiganak, Chiganak, *Kaz.* Shyghanaq. Zhambyl, SE Kazakhstan *45°10′N 73°55′E*
Shyghanaq *see* Shyganak
Shyghys Qazagastan Oblysy *see* Vostochnyy Kazakhstan
Shyghys Qongyrat *see* Shygys Konyrat
145 T12 **Shygys Konyrat** , *Kaz.* Shyghys Qongyrat. Karaganda, C Kazakhstan *47°01′N 75°05′E*
119 M19 **Shyichy** *Rus.* Shiichi. Homyel'skaya Voblasts', SE Belarus *52°15′N 29°14′E*
145 Q17 **Shymkent** *prev.* Chimkent. Yuzhnyy Kazakhstan, S Kazakhstan *42°19′N 69°36′E*
144 H9 **Shyngghyrlau** *prev.* Chingirlau. Zapadnyy Kazakhstan, W Kazakhstan *51°10′N 53°44′E*
144 G9 **Shyngyrlau** *prev.* Utva. ≈ NW Kazakhstan
145 W11 **Shynkozha** *prev.* Shingozha. Vostochnyy Kazakhstan, E Kazakhstan *47°46′N 80°38′E*
152 J5 **Shyok** Jammu and Kashmir, NW India *34°13′N 78°12′E*
117 S9 **Shyroke** *Rus.* Shirokoye. Dnipropetrovs'ka Oblast', E Ukraine *47°41′N 33°16′E*
117 O9 **Shyryayeve** Odes'ka Oblast', SW Ukraine *47°21′N 30°11′E*
117 S5 **Shyshaky** Poltavs'ka Oblast', C Ukraine *49°54′N 34°00′E*
119 K17 **Shyshchytsy** *Rus.* Shishchitsy. Minskaya Voblasts', C Belarus *53°13′N 27°33′E*
149 Y3 **Siachen Muztāgh** ▲▲ NE Pakistan
Siadehan *see* Tākestān
148 M13 **Siāhān Range** ▲▲ W Pakistan
142 I1 **Sīāh Chashmeh** *var.* Chālderān. Āẕarbāyjān-e Gharbī, N Iran *39°02′N 44°22′E*
149 W7 **Siālkot** Punjab, NE Pakistan *32°29′N 74°35′E*
186 E7 **Sialum** Morobe, C Papua New Guinea *06°02′S 147°37′E*
Siam *see* Thailand
Siam, Gulf of *see* Thailand, Gulf of
Sian *see* Xi'an
Siang *see* Brahmaputra
Siangtan *see* Xiangtan
169 N8 **Siantan, Pulau** *island* Kepulauan Anambas, W Indonesia
54 H11 **Siare, Río** ≈ C Colombia
171 R6 **Siargao Island** *island* S Philippines
186 F72 **Siassi** Umboi Island, C Papua New Guinea *05°34′S 147°50′E*
115 D14 **Siátista** Dytikí Makedonía, N Greece *40°16′N 21°34′E*
166 K4 **Siatlai** Chin State, W Myanmar (Burma) *22°05′N 93°36′E*
171 P6 **Siaton** Negros, C Philippines *09°03′N 123°03′E*
171 P6 **Siaton Point** *headland* Negros, C Philippines *09°03′N 123°00′E*
118 F11 **Šiauliai** *Ger.* Schaulen. Šiauliai, N Lithuania *55°55′N 23°21′E*
118 E11 **Šiauliai** ⬗ *province* N Lithuania
171 Q10 **Siau, Pulau** *island* N Indonesia
83 J15 **Siavonga** Southern, SE Zambia *16°33′S 28°42′E*
81 G18 **Siaya** ⬗ *county* W Kenya
Siazan' *see* Siyäzän
Sibah *see* As Sībah
107 N20 **Sibari** Calabria, S Italy *39°45′N 16°26′E*
Sibata *see* Shibata
127 X6 **Sibay** Respublika Bashkortostan, W Russian Federation *52°40′N 58°39′E*
93 M19 **Sibbo** *Fin.* Sipoo. Uusimaa, S Finland *60°22′N 25°20′E*
112 D13 **Šibenik** *It.* Sebenico. Šibenik-Knin, S Croatia *43°43′N 15°54′E*
Sibenik *see* Šibenik-Knin
112 E13 **Šibenik-Knin** *off.* Šibenska Županija, *var.* Sibenik. ⬗ *province* S Croatia
Šibenik-Knin *see* Drniš
Šibenska Županija *see* Šibenik-Knin
Siberia *see* Sibir'
Siberoet *see* Siberut, Pulau
168 H12 **Siberut, Pulau** *prev.* Siberoet. *island* Kepulauan Mentawai, W Indonesia
168 I12 **Siberut, Selat** *strait* W Indonesia
149 P11 **Sibi** Baluchistān, SW Pakistan *29°31′N 67°54′E*
186 B9 **Sibidiri** Western, SW Papua New Guinea *08°58′S 142°14′E*
123 N10 **Sibir'** *var.* Siberia. *physical region* NE Russian Federation
79 F20 **Sibiti** Lékoumou, S Congo *03°41′S 13°20′E*
81 G21 **Sibiti** ≈ C Tanzania
116 I12 **Sibiu** *Ger.* Hermannstadt, *Hung.* Nagyszeben. Sibiu, C Romania *45°48′N 24°09′E*
116 I11 **Sibiu** ⬗ *county* C Romania
29 S11 **Sibley** Iowa, C USA *43°24′N 95°45′W*
153 Y11 **Sibsāgar** *var.* Sivasagar. Assam, NE India *26°59′N 94°38′E*
169 R9 **Sibu** Sarawak, East Malaysia *02°18′N 111°49′E*
42 G2 **Sibun** ≈ E Belize
79 I15 **Sibut** *prev.* Fort-Sibut. Kémo, S Central African Republic *05°44′N 19°07′E*
171 P4 **Sibuyan Island** *island* C Philippines
189 U1 **Sibylla Island** *island* N Marshall Islands
11 N16 **Sicamous** British Columbia, SW Canada *50°49′N 118°52′W*
Sichelburger Gerbirge *see* Gorjanci
167 N14 **Sichon** *var.* Ban Sichon, Si Chon. Nakhon Si Thammarat, SW Thailand *09°03′N 99°51′E*
Si Chon *see* Sichon
160 H9 **Sichuan** *var.* Chuan, Sichuan Sheng, Ssu-ch'uan, Szechuan, Szechwan. ⬗ *province* C China
160 I9 **Sichuan Pendi** *basin* C China
Sichuan Sheng *see* Sichuan
103 S16 **Sicie, Cap** *headland* SE France *43°03′N 05°50′E*
107 J24 **Sicilia** *Eng.* Sicily; *anc.* Trinacria. ⬗ *region* Italy, C Mediterranean Sea
107 M24 **Sicilia** *Eng.* Sicily; *anc.* Trinacria. *island* Italy, C Mediterranean Sea
Sicilian Channel *see* Sicily, Strait of
107 H24 **Sicily, Strait of** *var.* Sicilian Channel. *strait* C Mediterranean Sea
42 K5 **Sico Tinto, Río** *var.* Río Negro. ≈ NE Honduras
57 H16 **Sicuani** Cusco, S Peru *14°21′S 71°13′W*
112 J10 **Šid** Vojvodina, NW Serbia *45°07′N 19°13′E*
115 A15 **Sidári** Kérkyra, Iónia Nisiá, Greece, C Mediterranean Sea *39°47′N 19°43′E*
169 Q11 **Sidas** Borneo, C Indonesia *0°24′N 109°46′E*
98 O5 **Siddeburen** Groningen, NE Netherlands *53°15′N 06°52′E*
154 D9 **Siddhapur** *prev.* Siddhpur, Sidhpur. Gujarāt, W India *23°57′N 72°28′E*
Siddhpur *see* Siddhapur
155 I15 **Siddipet** Telangana, C India *18°10′N 78°54′E*
77 N14 **Sidéradougou** SW Burkina Faso *10°39′N 04°16′W*
107 N23 **Siderno** Calabria, SW Italy *38°18′N 16°19′E*
Siders *see* Sierre
154 L9 **Sidhi** Madhya Pradesh, C India *24°24′N 81°54′E*
Sidhirókastron *see* Sidirókastro
Sidhpur *see* Siddhapur
75 U7 **Sîdi Barrâni** NW Egypt *31°38′N 25°58′E*
74 I6 **Sidi Bel Abbès** *var.* Sidi bel Abbès, Sidi-Bel-Abbès. NW Algeria *35°12′N 00°43′W*
74 E7 **Sidi-Bennour** W Morocco *32°39′N 08°28′W*
74 M6 **Sidi Bouzid** *var.* Gammouda, Sīdī Bu Zayd. C Tunisia *35°05′N 09°20′E*
Sīdī Bu Zayd *see* Sidi Bouzid
74 D8 **Sidi-Ifni** SW Morocco *29°33′N 10°04′W*
74 G6 **Sidi-Kacem** *prev.* Petitjean. N Morocco *34°21′N 05°46′W*
114 G12 **Sidirókastro** *prev.* Sidhirókastron. Kentrikí Makedonía, NE Greece *41°14′N 23°23′E*
194 L12 **Sidley, Mount** ▲ Antarctica *76°39′S 124°48′W*
29 S16 **Sidney** Iowa, C USA *40°45′N 95°39′W*
33 Y7 **Sidney** Montana, NW USA *47°42′N 104°10′W*
28 J15 **Sidney** Nebraska, C USA *41°09′N 102°57′W*
18 I11 **Sidney** New York, NE USA *42°18′N 75°21′W*
31 R13 **Sidney** Ohio, N USA *40°16′N 84°09′W*
23 T2 **Sidney Lanier, Lake** ⊠ Georgia, SE USA
Sidon *see* Saïda
122 J9 **Sidorovsk** Yamalo-Nenetskiy Avtonomnyy Okrug, N Russian Federation *66°34′N 82°12′E*
Sidra *see* Surt
Sidra/Sidra, Gulf of *see* Surt, Khalīj, N Libya
Siebenbürgen *see* Transylvania
Sieben Dörfer *see* Săcele
110 O12 **Siedlce** *Ger.* Sedlez, *Rus.* Sesdlets. Mazowieckie, C Poland *52°10′N 22°18′E*
101 E16 **Sieg** ≈ W Germany
101 F16 **Siegen** Nordrhein-Westfalen, W Germany *50°53′N 08°02′E*
109 X4 **Sieghartskirchen** Niederösterreich, E Austria *48°13′N 16°01′E*
167 S12 **Siĕmbok** *prev.* Phumĭ Siĕmbok. Stœng Trêng, N Cambodia *13°28′N 105°59′E*
110 O11 **Siemiatycze** Podlaskie, NE Poland *52°27′N 22°52′E*
167 T11 **Siĕmpang** Stœng Trêng, NE Cambodia *14°07′N 106°24′E*
167 R11 **Siĕmréab** *prev.* Siemreap. Siĕmréab, NW Cambodia *13°21′N 103°50′E*
Siemreap *see* Siĕmréab
106 G12 **Siena** *Fr.* Sienne; *anc.* Saena Julia. Toscana, C Italy *43°20′N 11°20′E*
Sienne *see* Siena
92 K12 **Sieppijärvi** Lappi, NW Finland *67°09′N 23°58′E*
110 J13 **Sieradz** Sieradz, C Poland *51°36′N 18°42′E*
110 K10 **Sierpc** Mazowieckie, C Poland *52°51′N 19°44′E*
24 I9 **Sierra Blanca** Texas, SW USA *31°10′N 105°22′W*
37 S14 **Sierra Blanca Peak** ▲ New Mexico, SW USA *33°22′N 105°48′W*
35 P5 **Sierra City** California, W USA *39°34′N 120°35′W*
63 I16 **Sierra Colorada** Río Negro, S Argentina *40°37′S 67°48′W*
63 J16 **Sierra Grande** Río Negro, E Argentina *41°34′S 65°21′W*
76 G15 **Sierra Leone** *off.* Republic of Sierra Leone. ◆ *republic* W Africa
64 M13 **Sierra Leone Basin** *undersea feature* E Atlantic Ocean *05°00′N 17°00′W*
66 K8 **Sierra Leone Fracture Zone** *tectonic feature* E Atlantic Ocean
Sierra Leone, Republic of *see* Sierra Leone
Sierra Leone Ridge *see* Sierra Leone Rise
64 L13 **Sierra Leone Rise** *var.* Sierra Leone Ridge, Sierra Leone Schwelle. *undersea feature* E Atlantic Ocean *05°30′N 21°00′W*
Sierra Leone Schwelle *see* Sierra Leone Rise
40 L7 **Sierra Mojada** Coahuila, NE Mexico *27°13′N 103°42′W*
37 N16 **Sierra Vista** Arizona, SW USA *31°33′N 110°18′W*
108 D10 **Sierre** *Ger.* Siders. Valais, SW Switzerland *46°18′N 07°33′E*
36 L16 **Sierrita Mountains** ▲▲ Arizona, SW USA
Siete Moai *see* Ahu Akivi
76 M15 **Sifié** W Ivory Coast *07°59′N 06°55′W*
115 I21 **Sífnos** *anc.* Siphnos. *island* Kykládes, Greece, Aegean Sea
115 I21 **Sífnou, Stenó** *strait* SE Greece
Siga *see* Shiga
103 P16 **Sigean** Aude, S France *43°02′N 02°58′E*
Sighet *see* Sighetu Marmaţiei
Sighetul Marmaţiei *see* Sighetu Marmaţiei
116 I8 **Sighetu Marmaţiei** *var.* Sighet, Sighetul Marmaţiei, *Hung.* Máramarossziget. Maramureş, N Romania *47°56′N 23°53′E*
116 I11 **Sighişoara** *Ger.* Schässburg, *Hung.* Segesvár. Mureş, C Romania *46°12′N 24°48′E*
168 G7 **Sigli** Sumatera, W Indonesia *05°21′N 95°56′E*
92 J1 **Siglufjörður** Norðurland Vestra, N Iceland *66°09′N 18°56′W*
101 H23 **Sigmaringen** Baden-Württemberg, S Germany *48°04′N 09°12′E*
101 N20 **Signalberg** ▲ SE Germany *49°30′N 12°34′E*
36 I13 **Signal Peak** ▲ Arizona, SW USA *33°20′N 114°03′W*
Signan *see* Xi'an
194 H1 **Signy** *UK research station* South Orkney Islands, Antarctica *60°27′S 45°35′W*
29 X15 **Sigourney** Iowa, C USA *41°19′N 92°12′W*
115 K17 **Sígri, Akrotírio** *headland* Lésvos, E Greece *39°12′N 25°49′E*
Sigsbee Deep *see* Mexico Basin
47 N2 **Sigsbee Escarpment** *undersea feature* N Gulf of Mexico *26°00′N 92°30′W*
56 C8 **Sigsig** Azuay, S Ecuador *03°04′S 78°50′W*
95 O15 **Sigtuna** Stockholm, C Sweden *59°36′N 17°44′E*
42 H6 **Siguatepeque** Comayagua, W Honduras *14°39′N 87°48′W*
105 P7 **Sigüenza** Castilla-La Mancha, C Spain *41°04′N 02°38′W*
105 R4 **Sigües** Aragón, NE Spain *42°39′N 01°00′W*
76 K13 **Siguiri** NE Guinea *11°26′N 09°08′W*
118 G8 **Sigulda** *Ger.* Segewold. C Latvia *57°08′N 24°51′E*
167 Q14 **Sihanoukville** *var.* Kâmpōng Saôm; *prev.* Kompong Som. Kâmpóng Saôm, SW Cambodia *10°38′N 103°30′E*
108 G8 **Sihlsee** ◎ NW Switzerland
93 K18 **Siikainen** Satakunta, SW Finland *61°52′N 21°49′E*
93 M16 **Siilinjärvi** Pohjois-Savo, C Finland *63°05′N 27°40′E*
137 R15 **Siirt** *var.* Sert; *anc.* Tigranocerta. Siirt, SE Turkey *37°56′N 41°56′E*
137 R15 **Siirt** *var.* Sert. ⬗ *province* SE Turkey
187 N8 **Sikaiana** *var.* Stewart Islands. *island group* W Solomon Islands
Sikandarabad *see* Secunderābād
152 J11 **Sikandra Rao** Uttar Pradesh, N India *27°42′N 78°21′E*
10 M11 **Sikanni Chief** British Columbia, W Canada *57°16′N 122°44′W*
10 M11 **Sikanni Chief** ≈ British Columbia, W Canada
152 H11 **Sīkar** Rājasthān, N India *27°33′N 75°12′E*
76 M13 **Sikasso** Sikasso, S Mali *11°21′N 05°43′W*
76 L13 **Sikasso** ⬗ *region* SW Mali
167 N3 **Sikaw** Kachin State, C Myanmar (Burma) *23°50′N 97°04′E*
83 H14 **Sikelenge** Western, W Zambia *14°51′S 24°13′E*
27 Y7 **Sikeston** Missouri, C USA *36°52′N 89°35′W*
93 J14 **Sikfors** Norrbotten, N Sweden *65°29′N 21°17′E*
123 T14 **Sikhote-Alin', Khrebet** ▲▲ SE Russian Federation
Siking *see* Xi'an
115 J22 **Síkinos** *island* Kykládes, Greece, Aegean Sea
153 S11 **Sikkim** *Tib.* Denjong. ⬗ *state* N India
111 I26 **Siklós** Baranya, SW Hungary *45°51′N 18°18′E*
Sikoku *see* Shikoku
Sikoku Basin *see* Shikoku Basin
83 G14 **Sikongo** Western, W Zambia *15°03′S 22°07′E*
Sikotu Ko *see* Shikotsu-ko
Sikouri/Sikoúrion *see* Sykoúrio
123 P8 **Siktyakh** Respublika Sakha (Yakutiya), NE Russian Federation *69°45′N 124°42′E*
78 K11 **Sila** *off.* Région du Sila. ⬗ *region* E Chad
118 D12 **Šilalė** Tauragė, W Lithuania *55°29′N 22°10′E*
106 G5 **Silandro** *Ger.* Schlanders. Trentino-Alto Adige, N Italy *46°39′N 10°55′E*
41 N12 **Silao** Guanajuato, C Mexico *20°56′N 101°28′W*
Sila, Région du *see* Sila
Silarius *see* Sele
153 W14 **Silchar** Assam, NE India *24°49′N 92°48′E*
108 G9 **Silenen** Uri, C Switzerland *46°49′N 08°39′E*
21 T9 **Siler City** North Carolina, SE USA *35°43′N 79°27′W*
33 U11 **Silesia** Montana, NW USA *45°32′N 108°52′W*
110 F13 **Silesia** *physical region* SW Poland
74 K12 **Silet** S Algeria *22°45′N 04°51′E*
Sileti *see* Silety
Siletitengiz *see* Siletyteniz, Ozero
145 R8 **Silety** *prev.* Sileti. ≈ N Kazakhstan
145 R7 **Siletyteniz, Ozero** *Kaz.* Siletitengiz. ◎ N Kazakhstan
172 H16 **Silhouette** *island* Inner Islands, SE Seychelles
136 I17 **Silifke** *anc.* Seleucia. İçel, S Turkey *36°22′N 33°57′E*
Siliguri *see* Shiliguri
156 J10 **Siling Co** ◎ W China
Silinhot *see* Xilinhot
192 G14 **Silisili, Mauga** ▲ Savai'i, C Samoa *13°37′S 172°26′W*
114 M6 **Silistra** *var.* Silistria; *anc.* Durostorum. Silistra, NE Bulgaria *44°06′N 27°17′E*
114 M7 **Silistra** ⬗ *province* NE Bulgaria
Silistria *see* Silistra
136 D10 **Silivri** İstanbul, NW Turkey *41°05′N 28°15′E*
94 L13 **Siljan** ◎ C Sweden
95 G22 **Silkeborg** Midtjylland, C Denmark *56°10′N 09°34′E*
108 M8 **Sill** ≈ W Austria
105 S10 **Silla** Valenciana, E Spain *39°22′N 00°25′W*
62 H3 **Sillajguay, Cordillera** ▲ N Chile *19°45′S 68°39′W*
118 K3 **Sillamäe** *Ger.* Sillamäggi. Ida-Virumaa, NE Estonia *59°23′N 27°45′E*
Sillamäggi *see* Sillamäe
Sillein *see* Žilina
109 P9 **Sillian** Tirol, W Austria *46°45′N 12°25′E*
112 B10 **Šilo** Primorje-Gorski Kotar, NW Croatia *45°09′N 14°39′E*
27 R9 **Siloam Springs** Arkansas, C USA *36°11′N 94°32′W*
25 X10 **Silsbee** Texas, SW USA *30°21′N 94°10′W*
143 W15 **Sīlūp, Rūd-e** ≈ SE Iran
118 C12 **Šilutė** *Ger.* Heydekrug. Klaipėda, W Lithuania *55°20′N 21°30′E*
137 Q15 **Silvan** Dıyarbakır, SE Turkey *38°08′N 41°E*
108 J10 **Silvaplana** Graubünden, S Switzerland *46°27′N 09°45′E*
Silva Porto *see* Kuito
58 M12 **Silva, Recife do** *reef* E Brazil
154 D12 **Silvassa** Dādra and Nagar Haveli, W India *20°13′N 73°03′E*
29 X4 **Silver Bay** Minnesota, N USA *47°17′N 91°15′W*
37 P15 **Silver City** New Mexico, SW USA *32°47′N 108°16′W*
18 D10 **Silver Creek** New York, NE USA *42°32′N 79°10′W*
37 N12 **Silver Creek** ≈ Arizona, SW USA
27 P4 **Silver Lake** Kansas, C USA *39°06′N 95°51′W*
32 I14 **Silver Lake** Oregon, NW USA *43°07′N 121°04′W*
35 T9 **Silver Peak Range** ▲▲ Nevada, W USA
21 W3 **Silver Spring** Maryland, NE USA *39°00′N 77°01′W*
Silver State *see* Colorado
Silver State *see* Nevada
37 Q7 **Silverton** Colorado, C USA *37°48′N 107°39′W*
18 K16 **Silverton** New Jersey, NE USA *40°00′N 74°09′W*
32 G11 **Silverton** Oregon, NW USA *45°00′N 122°46′W*
25 N4 **Silverton** Texas, SW USA *34°28′N 101°18′W*
104 G14 **Silves** Faro, S Portugal *37°11′N 08°26′W*
54 D12 **Silvia** Cauca, SW Colombia *02°37′N 76°24′W*
108 J9 **Silvrettagruppe** ▲▲ Austria/Switzerland
Sily-Vajdej *see* Vulcan
108 L7 **Silz** Tirol, W Austria *47°17′N 11°00′E*
172 I13 **Sima** Anjouan, SE Comoros *12°11′S 44°18′E*
Simabara *see* Shimabara
83 H15 **Simakando** Western, W Zambia *16°43′S 24°46′E*
Simane *see* Shimane
119 L20 **Simanichy** *Rus.* Simonichi. Homyel'skaya Voblasts', SE Belarus *51°53′N 28°05′E*
160 F14 **Simao** Yunnan, SW China *22°30′N 101°06′E*
153 P12 **Simarā** Central, C Nepal *27°14′N 85°00′E*
14 I8 **Simard, Lac** ◎ Québec, SE Canada
136 D13 **Simav** Kütahya, W Turkey *39°05′N 28°59′E*
136 D13 **Simav Çayı** ≈ NW Turkey
79 L18 **Simba** Orientale, N Dem. Rep. Congo *0°46′N 22°54′E*
186 C7 **Simbai** Madang, N Papua New Guinea *05°12′S 144°33′E*
Simbirsk *see* Ul'yanovsk
Simbu *see* Chimbu
14 F17 **Simcoe** Ontario, S Canada *42°50′N 80°19′W*
14 H14 **Simcoe, Lake** ◎ Ontario, S Canada
80 J11 **Sīmēn** ▲▲ N Ethiopia
114 K11 **Simeonovgrad** *prev.* Maritsa. Haskovo, S Bulgaria *42°03′N 25°36′E*
116 G11 **Simeria** *Ger.* Pischk, *Hung.* Piski. Hunedoara, W Romania *45°51′N 23°00′E*
107 L24 **Simeto** ≈ Sicilia, Italy, C Mediterranean Sea
168 G9 **Simeulue, Pulau** *island* NW Indonesia
117 T13 **Simferopol'** Avtonomna Respublika Krym, S Ukraine *44°55′N 33°06′E*
117 T13 **Simferopol'** ✈ Avtonomna Respublika Krym, S Ukraine *44°55′N 34°04′E*
Simi *see* Sými
152 M9 **Simikot** Far Western, NW Nepal *30°02′N 81°49′E*
54 F7 **Simití** Bolívar, N Colombia *07°57′N 73°57′W*
114 G11 **Simitla** Blagoevgrad, SW Bulgaria *41°57′N 23°06′E*
35 S15 **Simi Valley** California, W USA *34°16′N 118°47′W*
81 G21 **Simiyu** *off.* Mkoa wa Simiyu. ⬗ *region* N Tanzania
Simiyu, Mkoa wa *see* Simiyu
Simizu *see* Shimizu
Simla *see* Shimla
Şimlăul Silvaniei/Şimleul Silvaniei *see* Şimleu Silvaniei
116 G9 **Şimleu Silvaniei** *Hung.* Szilágysomlyó; *prev.* Şimlăul Silvaniei, Şimleul Silvaniei. Sălaj, NW Romania *47°12′N 22°49′E*
Simmer *see* Simmerbach
101 E19 **Simmerbach** *var.* Simmer. ≈ W Germany
101 F18 **Simmern** Rheinland-Pfalz, W Germany *50°00′N 07°30′E*
22 I7 **Simmesport** Louisiana, S USA *30°58′N 91°48′W*
119 F14 **Simnas** Alytus, S Lithuania *54°23′N 23°40′E*
92 L13 **Simo** Lappi, NW Finland *65°40′N 25°04′E*
Simoda *see* Shimoda
Simodate *see* Shimodate
92 M13 **Simojärvi** ◎ N Finland
92 L13 **Simojoki** ≈ NW Finland
41 U15 **Simojovel** *var.* Simojovel de Allende. Chiapas, SE Mexico *17°14′N 92°40′W*
Simojovel de Allende *see* Simojovel
56 B7 **Simon Bolivar** *var.* Guayaquil. ✈ (Quayaquil) Guayas, W Ecuador *02°16′S 79°54′W*
54 L5 **Simon Bolívar** ✈ (Caracas) Vargas, N Venezuela *10°33′N 66°54′W*
Simonichi *see* Simanichy
14 M12 **Simon, Lac** ◎ Québec, SE Canada
Simonoseki *see* Shimonoseki
Šimonovany *see* Partizánske
Simonstad *see* Simon's Town
83 E26 **Simon's Town** *var.* Simonstad. Western Cape, SW South Africa *34°12′S 18°26′E*
Simony *see* Partizánske
Simpeln *see* Simplon
99 M18 **Simpelveld** Limburg, SE Netherlands *50°50′N 05°59′E*
108 E11 **Simplon** *var.* Simpeln. Valais, SW Switzerland *46°13′N 08°01′E*
108 E11 **Simplon Pass** *pass* S Switzerland
106 C6 **Simplon Tunnel** *tunnel* Italy/Switzerland
Simpson *see* Fort Simpson
182 G1 **Simpson Desert** *desert* Northern Territory/South Australia
10 J9 **Simpson Peak** ▲ British Columbia, W Canada *59°43′N 131°29′W*
9 N7 **Simpson Peninsula** *peninsula* Nunavut, NE Canada
21 P11 **Simpsonville** South Carolina, SE USA *34°44′N 82°15′W*
95 L23 **Simrishamn** Skåne, S Sweden *55°34′N 14°20′E*
123 U13 **Simushir, Ostrov** *island* Kuril'skiye Ostrova, SE Russian Federation
168 G9 **Sinabang** Sumatera, W Indonesia *02°27′N 96°24′E*
81 N15 **Sina Dhaqa** Galguduud, C Somalia *05°21′N 46°21′E*
75 X8 **Sinai** *var.* Sinai Peninsula, *Ar.* Shibh Jazirat Sīnā', Sīnā. *physical region* NE Egypt
116 J12 **Sinaia** Prahova, SE Romania *45°20′N 25°33′E*
188 B16 **Sinajana** C Guam *13°28′N 144°45′E*
40 H8 **Sinaloa** ⬗ *state* C Mexico
54 H4 **Sinamaica** Zulia, NW Venezuela *11°06′N 71°52′W*
163 X14 **Sinan-ni** SE North Korea *38°37′N 127°43′E*
Sīnā/Sinai Peninsula *see* Sinai
Sīnāwan *see* Sīnāwin
75 N8 **Sīnāwin** *var.* Sīnāwan. NW Libya *31°00′N 10°37′E*
83 J16 **Sinazongwe** Southern, S Zambia *17°14′S 27°27′E*
166 L6 **Sinbaungwe** Magway, W Myanmar (Burma) *19°44′N 95°10′E*
166 L5 **Sinbyugyun** Magway, W Myanmar (Burma) *20°38′N 94°40′E*
54 E6 **Since** Sucre, NW Colombia *09°14′N 75°08′W*
54 E6 **Sincelejo** Sucre, NW Colombia *09°17′N 75°23′W*
23 U4 **Sinclair, Lake** ⊠ Georgia, SE USA
10 M14 **Sinclair Mills** British Columbia, SW Canada *54°03′N 121°37′W*
154 I8 **Sind** ≈ N India
Sind *see* Sindh
95 H19 **Sindal** Nordjylland, N Denmark *57°29′N 10°13′E*
171 P7 **Sindañgan** Mindanao, S Philippines *08°09′N 122°59′E*
79 D19 **Sindara** Ngounié, W Gabon *01°07′S 10°41′E*
152 E13 **Sindari** *prev.* Sindri. Rājasthān, N India *25°32′N 71°58′E*
114 N8 **Sindel** Varna, E Bulgaria *43°07′N 27°35′E*
101 H22 **Sindelfingen** Baden-Württemberg, SW Germany *48°43′N 09°E*
155 G16 **Sindgi** Karnātaka, C India *17°01′N 76°22′E*
149 Q14 **Sindh** *prev.* Sind. ⬗ *province* SE Pakistan
118 G5 **Sindi** *Ger.* Zintenhof. Pärnumaa, SW Estonia *58°28′N 24°41′E*
136 C13 **Sındırgı** Balıkesir, W Turkey *39°13′N 28°10′E*
77 N14 **Sindou** SW Burkina Faso *10°35′N 05°04′W*
Sindri *see* Sindari

◆ Country
● Country Capital
◇ Dependent Territory
○ Dependent Territory Capital
⬗ Administrative Regions
✈ International Airport
▲ Mountain
▲▲ Mountain Range
⛰ Volcano
≈ River
◎ Lake
⊠ Reservoir

◆ Country ◇ Dependent Territory ✦ Administrative Regions ▲ Mountain ✷ Volcano ◎ Lake
● Country Capital ○ Dependent Territory Capital ✈ International Airport ▲ Mountain Range ◆ River ⬚ Reservoir

SMITH, CAPE – SOTAVENTO, ILHAS DE

12 I3 **Smith, Cape** *headland* Québec, NE Canada 60°50′N 78°06′W
26 L3 **Smith Center** Kansas, C USA 39°46′N 98°46′W
10 K13 **Smithers** British Columbia, SW Canada 54°45′N 127°10′W
21 V10 **Smithfield** North Carolina, SE USA 35°30′N 78°21′W
36 L1 **Smithfield** Utah, W USA 41°50′N 111°49′W
21 X7 **Smithfield** Virginia, NE USA 36°41′N 76°38′W
12 I3 **Smith Island** *island* Nunavut, C Canada
Smith Island *see* Sumisu-jima
20 H7 **Smithland** Kentucky, S USA 37°06′N 88°24′W
21 T7 **Smith Mountain Lake** *var.* Leesville Lake. ▣ Virginia, NE USA
34 L1 **Smith River** California, W USA 41°54′N 124°09′W
33 R9 **Smith River** ≈ Montana, NW USA
14 L13 **Smiths Falls** Ontario, SE Canada 44°54′N 76°01′W
33 N13 **Smiths Ferry** Idaho, NW USA 44°19′N 116°04′W
20 K7 **Smiths Grove** Kentucky, S USA 37°01′N 86°14′W
183 N15 **Smithton** Tasmania, SE Australia 40°54′S 145°06′E
18 L14 **Smithtown** Long Island, New York, NE USA 40°52′N 73°13′W
20 K9 **Smithville** Tennessee, S USA 35°59′N 85°49′W
25 T11 **Smithville** Texas, SW USA 30°04′N 97°32′W
Šmohor *see* Hermagor
35 Q4 **Smoke Creek Desert** *desert* Nevada, W USA
11 O14 **Smoky** ≈ Alberta, W Canada
182 E7 **Smoky Bay** South Australia 32°22′S 133°57′E
183 V6 **Smoky Cape** *headland* New South Wales, SE Australia 30°54′S 153°06′E
26 L4 **Smoky Hill River** ≈ Kansas, C USA
26 L4 **Smoky Hills** *hill range* Kansas, C USA
11 Q14 **Smoky Lake** Alberta, SW Canada 54°08′N 112°26′W
94 E8 **Smøla** *island* W Norway
126 H4 **Smolensk** Smolenskaya Oblast', W Russian Federation 54°48′N 32°08′E
126 H4 **Smolenskaya Oblast'** ⬖ *province* W Russian Federation
Smolensk-Moscow Upland *see* Smolensko-Moskovskaya Vozvyshennost'
126 J3 **Smolensko-Moskovskaya Vozvyshennost'** *var.* Smolensk-Moscow Upland. ▲▲ W Russian Federation
Smolevichi *see* Smalyavichy
115 C15 **Smólikas** *var.* Smolikás. ▲ W Greece 40°06′N 20°54′E
114 I12 **Smolyan** *prev.* Pashmakli. Smolyan, S Bulgaria 41°34′N 24°42′E
114 I12 **Smolyan** ⬖ *province* S Bulgaria
Smolyany *see* Smalyany
33 S15 **Smoot** Wyoming, C USA 42°37′N 110°55′W
12 G12 **Smooth Rock Falls** Ontario, S Canada 49°17′N 81°37′W
Smorgon'/Smorgonie *see* Smarhon'
95 K23 **Smygehamn** Skåne, S Sweden 55°19′N 13°25′E
194 I7 **Smyley Island** *island* Antarctica
21 Y3 **Smyrna** Delaware, NE USA 39°18′N 75°36′W
23 S3 **Smyrna** Georgia, SE USA 33°52′N 84°30′W
20 J9 **Smyrna** Tennessee, S USA 36°00′N 86°30′W
Smyrna *see* İzmir
97 I16 **Snaefell** ▲ C Isle of Man 54°15′N 04°29′W
92 H3 **Snæfellsjökull** ▲ W Iceland 64°51′N 23°51′W
92 J3 **Snækollur** ▲ C Iceland 64°38′N 19°18′W
10 J4 **Snake** ≈ Yukon, NW Canada
29 O8 **Snake Creek** ≈ South Dakota, N USA
183 P13 **Snake Island** *island* Victoria, SE Australia
35 Y6 **Snake Range** ▲▲ Nevada, W USA
32 K10 **Snake River** ≈ NW USA
29 V6 **Snake River** ≈ Minnesota, N USA
28 L12 **Snake River** ≈ Nebraska, C USA
33 Q14 **Snake River Plain** *plain* Idaho, NW USA
93 F15 **Snåsa** Nord-Trøndelag, C Norway 64°16′N 12°25′E
21 O8 **Sneedville** Tennessee, S USA 36°31′N 83°13′W
98 K6 **Sneek** *Fris.* Snits. Fryslân, N Netherlands 53°02′N 05°40′E
Sneeuw-gebergte *see* Maoke, Pegunungan
95 F22 **Snejbjerg** Midtjylland, C Denmark 56°08′N 08°55′E
122 K9 **Snezhnogorsk** Krasnoyarskiy Kray, N Russian Federation 68°06′N 87°37′E
124 J3 **Snezhnogorsk** Murmanskaya Oblast', NW Russian Federation 69°12′N 33°20′E
Snezhnoye *see* Snizhne
111 G15 **Sněžka** *Ger.* Schneekoppe, *Pol.* Śnieżka. ▲ N Czech Republic/Poland 50°42′N 15°55′E
110 N8 **Śniardwy, Jezioro** *Ger.* Spirdingsee. ◎ NE Poland
Sniečkus *see* Visaginas
Śnieżka *see* Sněžka
117 R10 **Snihurivka** Mykolayivs'ka Oblast', S Ukraine 47°05′N 32°48′E
116 I5 **Snilov** ✈ (L'viv) L'vivs'ka Oblast', W Ukraine 49°45′N 23°59′E
111 O19 **Snina** *Hung.* Szinna. Prešovský Kraj, E Slovakia 49°N 22°10′E
Snits *see* Sneek
117 Y8 **Snizhne** *Rus.* Snezhnoye. Donets'ka Oblast', SE Ukraine 48°01′N 38°46′E
94 G10 **Snøhetta** *var.* Snohetta. ▲ S Norway 62°22′N 09°08′E
92 G12 **Snøtinden** ▲ C Norway 66°39′N 13°50′E
97 I18 **Snowdon** ▲ NW Wales, United Kingdom 53°04′N 04°04′W
97 I18 **Snowdonia** ▲▲ NW Wales, United Kingdom
Snowdrift *see* Łutselk'e
Snowdrift *see* Lutselk'e
37 N12 **Snowflake** Arizona, SW USA 34°30′N 110°04′W
21 Y5 **Snow Hill** Maryland, NE USA 38°11′N 75°23′W
21 W10 **Snow Hill** North Carolina, SE USA 35°26′N 77°39′W
194 H3 **Snowhill Island** *island* Antarctica
11 V13 **Snow Lake** Manitoba, C Canada 54°56′N 100°02′W
37 R5 **Snowmass Mountain** ▲ Colorado, C USA 39°07′N 107°04′W
18 M10 **Snow, Mount** ▲ Vermont, NE USA 42°56′N 72°52′W
34 M5 **Snow Mountain** ▲ California, W USA 39°44′N 123°01′W
Snow Mountains *see* Maoke, Pegunungan
33 N7 **Snowshoe Peak** ▲ Montana, NW USA 48°15′N 115°44′W
182 I8 **Snowtown** South Australia 33°49′S 138°13′E
36 K1 **Snowville** Utah, W USA 41°59′N 112°42′W
35 X3 **Snow Water Lake** ◎ Nevada, W USA
183 Q11 **Snowy Mountains** ▲▲ New South Wales/Victoria, SE Australia
183 Q12 **Snowy River** ≈ New South Wales/Victoria, SE Australia
44 K5 **Snug Corner** Acklins Island, SE The Bahamas 22°31′N 73°51′W
167 T13 **Snuŏl** Krâchéh, E Cambodia 12°04′N 106°26′E
116 J7 **Snyatyn** Ivano-Frankivs'ka Oblast', W Ukraine 48°30′N 25°50′E
26 L12 **Snyder** Oklahoma, C USA 34°37′N 98°56′W
25 O6 **Snyder** Texas, SW USA 32°43′N 100°54′W
172 H3 **Soalala** Mahajanga, W Madagascar 16°05′S 45°21′E
172 J4 **Soanierana-Ivongo** Toamasina, E Madagascar 16°53′S 49°35′E
171 R11 **Soasiu** *var.* Tidore. Pulau Tidore, E Indonesia 0°40′N 127°25′E
54 G8 **Soatá** Boyacá, C Colombia 06°23′N 72°40′W
172 I5 **Soavinandriana** Antananarivo, C Madagascar 19°09′S 46°43′E
77 V13 **Soba** Kaduna, C Nigeria 10°58′N 08°06′E
163 Y16 **Sobaek-sanmaek** ▲▲ S South Korea
80 F13 **Sobat** ≈ NE South Sudan
171 Z14 **Sobger, Sungai** ≈ Papua, E Indonesia
171 V13 **Sobiei** Papua Barat, E Indonesia 02°31′S 134°30′E
126 M3 **Sobinka** Vladimirskaya Oblast', W Russian Federation 56°00′N 39°55′E
127 S7 **Sobolevo** Orenburgskaya Oblast', W Russian Federation 51°57′N 51°42′E
Soborsin *see* Săvârşin
164 D15 **Sobo-san** ▲ Kyūshū, SW Japan 32°50′N 131°16′E
111 G14 **Sobótka** Dolnośląskie, SW Poland 50°53′N 16°48′E
59 O15 **Sobradinho** Bahia, E Brazil 09°33′S 40°56′W
Sobradinho, Barragem de *see* Sobradinho, Represa de
59 O16 **Sobradinho, Represa de** *var.* Barragem de Sobradinho. ▣ E Brazil
58 O13 **Sobral** Ceará, E Brazil 03°45′S 40°20′W
105 T4 **Sobrarbe** *physical region* NE Spain
109 R10 **Soča** *It.* Isonzo. ≈ Italy/Slovenia
110 L11 **Sochaczew** Mazowieckie, C Poland 52°15′N 20°15′E
126 L15 **Sochi** Krasnodarskiy Kray, SW Russian Federation 43°35′N 39°46′E
114 G13 **Sochós** *var.* Sohos, Sokhós. Kentrikí Makedonía, N Greece 40°49′N 23°23′E
191 R11 **Société, Archipel de la** *var.* Archipel de Tahiti, Îles de la Société, *Eng.* Society Islands. *island group* W French Polynesia
Société, Îles de la/Society Islands *see* Société, Archipel de la
21 T11 **Society Hill** South Carolina, SE USA 34°28′N 79°54′W
175 W9 **Society Ridge** *undersea feature* C Pacific Ocean
62 I5 **Socompa, Volcán** ✸ N Chile 24°18′S 68°03′W
Soconusco, Sierra de *see* Madre, Sierra
54 G8 **Socorro** Santander, C Colombia 06°30′N 73°16′W
37 R13 **Socorro** New Mexico, SW USA 33°58′N 106°55′W
Socotra *see* Suquṭrā
167 S14 **Soc Trăng** *var.* Khanh Hung. Soc Trăng, S Vietnam 09°36′N 105°58′E
105 P10 **Socuéllamos** Castilla-La Mancha, C Spain 39°18′N 02°48′W
35 W13 **Soda Lake** *salt flat* California, W USA
92 L11 **Sodankylä** Lappi, N Finland 67°26′N 26°35′E
Sodari *see* Sodiri
33 R15 **Soda Springs** Idaho, NW USA 42°39′N 111°36′W
Soddo/Soddu *see* Sodo
20 L10 **Soddy Daisy** Tennessee, S USA 35°14′N 85°11′W
95 N14 **Söderfors** Uppsala, C Sweden 60°23′N 17°14′E
94 N12 **Söderhamn** Gävleborg, C Sweden 61°19′N 17°10′E
95 N17 **Söderköping** Östergötland, S Sweden 58°28′N 16°20′E
95 N17 **Södermanland** ⬖ *county* C Sweden
95 O16 **Södertälje** Stockholm, C Sweden 59°11′N 17°39′E
80 D10 **Sodiri** *var.* Sawdirī, Sodari. Northern Kordofan, C Sudan 14°23′N 29°06′E
81 I14 **Sodo** *var.* Soddo, Soddu. Southern Nationalities, S Ethiopia 06°49′N 37°43′E
Södra Karelen *see* Etelä-Karjala
Södra Österbotten *see* Etelä-Pohjanmaa
Södra Savolax *see* Etelä-Savo
95 M19 **Södra Vi** Kalmar, S Sweden 57°45′N 15°45′E
18 G9 **Sodus Point** *headland* New York, NE USA 43°16′N 76°59′W
171 Q17 **Soe** *prev.* Soë. Timor, C Indonesia 09°51′S 124°29′E
Soebang *see* Subang
Soekaboemi *see* Sukabumi
169 N15 **Soekarno-Hatta** ✈ (Jakarta) Jawa, S Indonesia
Soëla-Sund *see* Soela Väin
118 E5 **Soela Väin** *prev. Eng.* Sele Sound, *Ger.* Dagden-Sund, Soëla-Sund. *strait* W Estonia
Soemba *see* Sumba, Pulau
Soembawa *see* Sumbawa
Soemenep *see* Sumenep
Soengaipenoeh *see* Sungaipenuh
Soerabaja *see* Surabaya
Soerakarta *see* Surakarta
101 G14 **Soest** Nordrhein-Westfalen, W Germany 51°34′N 08°06′E
98 J11 **Soest** Utrecht, C Netherlands 52°10′N 05°20′E
100 F11 **Soeste** ≈ NW Germany
98 J11 **Soesterberg** Utrecht, C Netherlands 52°07′N 05°17′E
115 E16 **Sofádes** *var.* Sofádhes. Thessalía, C Greece 39°20′N 22°06′E
Sofádhes *see* Sofádes
83 N18 **Sofala** Sofala, C Mozambique 20°04′S 34°43′E
83 N17 **Sofala** ⬖ *province* C Mozambique
83 N18 **Sofala, Baia de** *bay* C Mozambique
114 G10 **Sofia** *var.* Sophia, Sofiya, *Eng.* Sofia, *Lat.* Serdica. ● (Bulgaria) Sofia Grad, W Bulgaria 42°42′N 23°20′E
114 H9 **Sofia** ⬖ *province* W Bulgaria
114 G9 **Sofia** ✈ Sofia Grad, W Bulgaria 42°42′N 23°26′E
172 J3 **Sofia** *seasonal river* NW Madagascar
Sofia *see* Sofia
114 G9 **Sofia Grad** ⬖ *municipality* W Bulgaria
115 G19 **Sofikó** Pelopónnisos, S Greece 37°46′N 23°04′E
Sofi-Kurgan *see* Sopu-Korgon
Sofiya *see* Sofia
Sofiyevka *see* Sofiyivka
117 S8 **Sofiyivka** *Rus.* Sofiyevka. Dnipropetrovs'ka Oblast', E Ukraine 48°04′N 33°55′E
123 R12 **Sofiysk** Khabarovskiy Kray, SE Russian Federation 51°32′N 139°46′E
123 R13 **Sofiysk** Khabarovskiy Kray, SE Russian Federation 52°20′N 133°37′E
124 I6 **Sofporog** Respublika Kareliya, NW Russian Federation 65°48′N 31°30′E
115 L23 **Sofraná** *prev.* Záfora. *island* Kykládes, Greece, Aegean Sea
165 Y14 **Sōfu-gan** *island* Izu-shotō, SE Japan
156 K10 **Sog** Xizang Zizhiqu, W China 31°52′N 93°40′E
54 G9 **Sogamoso** Boyacá, C Colombia 05°43′N 72°56′W
136 I11 **Soğanlı Çayı** ≈ N Turkey
94 E12 **Sogn** *physical region* S Norway
Sogndal *see* Sogndalsfjøra
94 E12 **Sogndalsfjøra** *var.* Sogndal. Sogn Og Fjordane, S Norway 61°13′N 07°05′E
95 E18 **Søgne** Vest-Agder, S Norway 58°05′N 07°49′E
94 D12 **Sognefjorden** *fjord* NE North Sea
94 C12 **Sogn Og Fjordane** ⬖ *county* S Norway
162 I11 **Sogo Nur** ◎ N China
159 T12 **Sogruma** Qinghai, W China 32°32′N 100°52′E
Sŏgwip'o *see* Seogwipo
Sohâg *see* Sawhāj
Sohar *see* Şuḩār
64 H9 **Sohm Plain** *undersea feature* NW Atlantic Ocean
100 H7 **Soholmer Au** ≈ N Germany
Sohos *see* Sochós
Sohrau *see* Żory
99 F20 **Soignies** Hainaut, SW Belgium 50°35′N 04°04′E
159 R15 **Soila** Xizang Zizhiqu, W China 30°40′N 97°07′E
103 P4 **Soissons** *anc.* Augusta Suessionum, Noviodunum. Aisne, N France 49°23′N 03°20′E
164 H13 **Sōja** Okayama, Honshū, SW Japan 34°40′N 133°42′E
152 F13 **Sojat** Rājasthān, N India 25°53′N 73°45′E
163 W13 **Sŏjosŏn-man** *inlet* W North Korea
116 I4 **Sokal'** *Rus.* Sokal. L'vivs'ka Oblast', NW Ukraine 50°29′N 24°17′E
163 Y14 **Sokcho** *prev.* Sokch'o. N South Korea 38°07′N 128°34′E
Sokch'o *see* Sokcho
136 B15 **Söke** Aydın, SW Turkey 37°46′N 27°24′E
189 N12 **Sokehs Island** *island* E Micronesia
79 M24 **Sokele** Katanga, SE Dem. Rep. Congo 09°54′S 24°38′E
147 R11 **Sokh** *Uzb.* Sŭkh. ≈ Kyrgyzstan/Uzbekistan
Sokh *see* So'x
Sokhós *see* Sochós
137 Q8 **Sokhumi** *Rus.* Sukhumi. NW Georgia 43°02′N 41°01′E
113 O14 **Sokobanja** Serbia, E Serbia 43°39′N 21°51′E
77 R15 **Sokodé** C Togo 08°58′N 01°10′E
123 T10 **Sokol** Magadanskaya Oblast', E Russian Federation 59°51′N 150°56′E
124 M13 **Sokol** Vologodskaya Oblast', NW Russian Federation 59°26′N 40°09′E
110 P9 **Sokółka** Podlaskie, NE Poland 53°24′N 23°31′E
76 M11 **Sokolo** Ségou, W Mali 14°43′N 06°02′W
111 A16 **Sokolov** *Ger.* Falkenau an der Eger; *prev.* Falknov nad Ohří. Karlovarský Kraj, W Czech Republic 50°10′N 12°38′E
111 O16 **Sokołów Małopolski** Podkarpackie, SE Poland 50°12′N 22°07′E
110 O11 **Sokołów Podlaski** Mazowieckie, C Poland 52°26′N 22°14′E
76 G11 **Sokone** W Senegal 13°53′N 16°22′W
77 T12 **Sokoto** Sokoto, NW Nigeria 13°05′N 05°16′E
77 T12 **Sokoto** ⬖ *state* NW Nigeria
77 S12 **Sokoto** ≈ NW Nigeria
Sokotra *see* Suquṭrā
147 U7 **Sokuluk** Chuyskaya Oblast', N Kyrgyzstan 42°53′N 74°19′E
116 L7 **Sokyryany** Chernivets'ka Oblast', W Ukraine 48°28′N 27°25′E
95 C16 **Sola** Rogaland, S Norway 58°53′N 05°36′E
187 R12 **Sola** Vanua Lava, N Vanuatu 13°51′S 167°34′E
95 C17 **Sola** ✈ (Stavanger) Rogaland, S Norway 58°54′N 05°36′E
81 H18 **Solai** Nakuru, W Kenya 0°02′N 36°03′E
152 I8 **Solan** Himāchal Pradesh, N India 30°54′N 77°06′E
185 A25 **Solander Island** *island* SW New Zealand
Solano *see* Bahía Solano
155 F15 **Solāpur** *var.* Sholāpur. Mahārāshtra, W India 17°43′N 75°54′E
93 H16 **Solberg** Västernorrland, C Sweden 63°48′N 17°40′E
116 K9 **Solca** *Ger.* Solka. Suceava, N Romania 47°40′N 25°50′E
105 O16 **Sol, Costa del** *coastal region* S Spain
106 F5 **Solda** *Ger.* Sulden. Trentino-Alto Adige, N Italy 46°33′N 10°35′E
117 N9 **Şoldăneşti** *Rus.* Sholdaneshty. N Moldova 47°49′N 28°45′E
Soldau *see* Wkra
108 L8 **Sölden** Tirol, W Austria 46°58′N 11°01′E
27 P3 **Soldier Creek** ≈ Kansas, C USA
39 R12 **Soldotna** Alaska, USA 60°29′N 151°03′W
110 I10 **Solec Kujawski** Kujawsko-pomorskie, C Poland 53°04′N 18°09′E
61 B16 **Soledad** Santa Fe, C Argentina 30°38′S 60°52′W
55 E4 **Soledad** Atlántico, N Colombia 10°54′N 74°48′W
35 O11 **Soledad** California, W USA 36°25′N 121°19′W
55 O7 **Soledad** Anzoátegui, NE Venezuela 08°10′N 63°36′W
61 H15 **Soledade** Rio Grande do Sul, S Brazil 28°50′S 52°30′W
Isla Soledad *see* East Falkland
103 Y15 **Solenzara** Corse, France, C Mediterranean Sea 41°55′N 09°24′E
Soleure *see* Solothurn
94 C12 **Solheim** Hordaland, S Norway 60°54′N 05°30′E
125 N14 **Soligalich** Kostromskaya Oblast', NW Russian Federation 59°05′N 42°15′E
Soligorsk *see* Salihorsk
Soligorskoye Vodokhranilishche *see* Salihorskaye Vodaskhovishcha
97 L20 **Solihull** C England, United Kingdom 52°25′N 01°45′W
125 U13 **Solikamsk** Permskiy Kray, NW Russian Federation 59°37′N 56°46′E
127 V8 **Sol'-Iletsk** Orenburgskaya Oblast', W Russian Federation 51°09′N 55°05′E
57 G17 **Solimana, Nevado** ▲ S Peru 15°24′S 72°49′W
58 E13 **Solimões, Rio** ≈ C Brazil
113 E14 **Solin** *It.* Salona; *anc.* Salonae. Split-Dalmacija, S Croatia 43°33′N 16°29′E
101 E15 **Solingen** Nordrhein-Westfalen, W Germany 51°10′N 07°05′E
Solka *see* Solca
93 H16 **Sollefteå** Västernorrland, C Sweden 63°09′N 17°15′E
95 O15 **Sollentuna** Stockholm, C Sweden 59°26′N 17°56′E
105 X9 **Sóller** Mallorca, Spain, W Mediterranean Sea 39°46′N 02°42′E
94 L13 **Sollerön** Dalarna, C Sweden 60°55′N 14°34′E
101 I14 **Solling** *hill range* C Germany
95 O16 **Solna** Stockholm, C Sweden 59°22′N 17°58′E
126 K3 **Solnechnogorsk** Moskovskaya Oblast', W Russian Federation 56°07′N 37°04′E
123 R10 **Solnechnyy** Khabarovskiy Kray, SE Russian Federation 50°41′N 136°42′E
123 S13 **Solnechnyy** Respublika Sakha (Yakutiya), NE Russian Federation 60°13′N 137°42′E
Solo *see* Surakarta
107 L17 **Solofra** Campania, S Italy 40°49′N 14°48′E
168 J11 **Solok** Sumatera, W Indonesia 0°45′S 100°42′E
42 C6 **Sololá** Sololá, W Guatemala 14°46′N 91°09′W
42 A2 **Sololá** *off.* Departamento de Sololá. ⬖ *department* SW Guatemala
Sololá, Departamento de *see* Sololá
81 J16 **Sololo** Marsabit, N Kenya 03°31′N 38°39′E
42 C4 **Soloma** Huehuetenango, W Guatemala 15°38′N 91°25′W
38 M9 **Solomon** Alaska, USA 64°33′N 164°26′W
27 N4 **Solomon** Kansas, C USA 38°55′N 97°22′W
187 N9 **Solomon Islands** *prev.* British Solomon Islands Protectorate. ◆ *commonwealth republic* W Solomon Islands N Melanesia W Pacific Ocean
186 L7 **Solomon Islands** *island group* Papua New Guinea/Solomon Islands
26 M3 **Solomon River** ≈ Kansas, C USA
186 H8 **Solomon Sea** *sea* W Pacific Ocean
31 U11 **Solon** Ohio, N USA 41°23′N 81°26′W
117 T8 **Solone** Dnipropetrovs'ka Oblast', E Ukraine 48°12′N 34°49′E
171 P16 **Solor, Kepulauan** *island group* S Indonesia
126 M4 **Solotcha** Ryazanskaya Oblast', W Russian Federation 54°43′N 39°50′E
108 D7 **Solothurn** *Fr.* Soleure. Solothurn, NW Switzerland 47°13′N 07°32′E
108 D7 **Solothurn** *Fr.* Soleure. ⬖ *canton* NW Switzerland
124 J7 **Solovetskiye Ostrova** *island group* NW Russian Federation
105 V5 **Solsona** Cataluña, NE Spain 42°00′N 01°31′E
113 E14 **Šolta** *It.* Solta. *island* S Croatia
Solṭānābād *see* Kāshmar
142 L4 **Solṭānīyeh** Zanjān, NW Iran 36°24′N 48°50′E
100 I11 **Soltau** Niedersachsen, NW Germany 52°59′N 09°50′E
124 G14 **Sol'tsy** Novgorodskaya Oblast', W Russian Federation 58°09′N 30°23′E
Soltūstik Qazaqstan Oblysy *see* Severnyy Kazakhstan
Solun *see* Thessaloníki
113 O19 **Solunska Glava** ▲ C FYR Macedonia 41°43′N 21°24′E
95 L22 **Sölvesborg** Blekinge, S Sweden 56°04′N 14°35′E
97 J15 **Solway Firth** *inlet* England/Scotland, United Kingdom
82 I13 **Solwezi** North Western, NW Zambia 12°11′S 26°23′E
165 Q11 **Sōma** Fukushima, Honshū, C Japan 37°49′N 140°52′E
136 C13 **Soma** Manisa, W Turkey 39°10′N 27°36′E
Somali *see* Sumalē
81 O15 **Somalia** *off.* Somali Democratic Republic, *Som.* Jamuuriyada Demuqraadiga Soomaaliyeed, Soomaaliya; *prev.* Italian Somaliland, Somaliland Protectorate. ◆ *republic* E Africa
173 N6 **Somali Basin** *undersea feature* W Indian Ocean 0°00′N 52°00′E
Somali Democratic Republic *see* Somalia
80 N12 **Somaliland** ⬖ *disputed territory* N Somalia
Somaliland Protectorate *see* Somalia
67 Y8 **Somali Plain** *undersea feature* W Indian Ocean 01°00′N 51°30′E
112 J8 **Sombor** *Hung.* Zombor. Vojvodina, NW Serbia 45°46′N 19°07′E
99 H20 **Sombreffe** Namur, S Belgium 50°32′N 04°39′E
40 L10 **Sombrerete** Zacatecas, C Mexico 23°38′N 103°40′W
45 V8 **Sombrero** *island* N Anguilla
151 Q21 **Sombrero Channel** *channel* Nicobar Islands, India
116 H9 **Şomcuta Mare** *Hung.* Nagysomkút; *prev.* Somcuţa Mare. Maramureş, N Romania 47°29′N 23°30′E
Somcuţa Mare *see* Şomcuta Mare
167 R9 **Somdet** Kalasin, E Thailand 16°41′N 103°44′E
99 L15 **Someren** Noord-Brabant, SE Netherlands 51°23′N 05°42′E
93 L19 **Somero** Varsinais-Suomi, SW Finland 60°37′N 23°30′E
33 P7 **Somers** Montana, NW USA 48°04′N 114°16′W
64 A12 **Somerset** *var.* Somerset Village. W Bermuda 32°18′N 64°53′W
37 Q5 **Somerset** Colorado, C USA 38°55′N 107°27′W
20 M7 **Somerset** Kentucky, S USA 37°05′N 84°36′W
19 O12 **Somerset** Massachusetts, NE USA 41°46′N 71°07′W
97 K23 **Somerset** *cultural region* SW England, United Kingdom
Somerset East *see* Somerset-Oos
64 A12 **Somerset Island** *island* W Bermuda
197 N9 **Somerset Island** *island* Queen Elizabeth Islands, Nunavut, NW Canada
Somerset Nile *see* Victoria Nile
83 I25 **Somerset-Oos** *var.* Somerset East. Eastern Cape, S South Africa 32°44′S 25°35′E
Somerset Village *see* Somerset
83 E26 **Somerset-Wes** *var.* Somerset West. Western Cape, SW South Africa 34°05′S 18°51′E
Somerset West *see* Somerset-Wes
Somers Islands *see* Bermuda
18 J17 **Somers Point** New Jersey, NE USA 39°18′N 74°34′W
19 P9 **Somersworth** New Hampshire, NE USA 43°15′N 70°52′W
36 H15 **Somerton** Arizona, SW USA 32°36′N 114°42′W
18 J14 **Somerville** New Jersey, NE USA 40°34′N 74°36′W
20 F10 **Somerville** Tennessee, S USA 35°14′N 89°24′W
25 U10 **Somerville** Texas, SW USA 30°21′N 96°31′W
25 T10 **Somerville Lake** ▣ Texas, SW USA
Someş/Somesch/Someşul *see* Szamos
103 N2 **Somme** ⬖ *department* N France
103 N2 **Somme** ≈ N France
95 L18 **Sommen** Jönköping, S Sweden 58°07′N 14°58′E
95 M18 **Sommen** ◎ S Sweden
101 K16 **Sömmerda** Thüringen, C Germany 51°10′N 11°07′E
Sommerein *see* Šamorín
Sommerfeld *see* Lubsko
55 Y11 **Sommet Tabulaire** *var.* Mont Itoupé. ▲▲ S French Guiana
111 H25 **Somogy** *off.* Somogy Megye. ⬖ *county* SW Hungary
Somogy Megye *see* Somogy
Somorja *see* Šamorín
105 N7 **Somosierra, Puerto de** *pass* N Spain
187 Y14 **Somosomo** Taveuni, N Fiji 16°46′S 179°57′W
42 I9 **Somotillo** Chinandega, NW Nicaragua 13°01′N 86°53′W
42 I8 **Somoto** Madriz, NW Nicaragua 13°29′N 86°36′W
110 I11 **Sompolno** Wielkopolskie, C Poland 52°24′N 18°30′E
102 J17 **Somport, Col du** *var.* Puerto de Somport, *Sp.* Somport; *anc.* Summus Portus. *pass* France/Spain *see also* Somport
99 K15 **Son** Noord-Brabant, S Netherlands 51°32′N 05°34′E
95 H15 **Son** Akershus, S Norway 59°32′N 10°42′E
154 L9 **Son** *var.* Sone. ≈ C India
43 R16 **Soná** Veraguas, W Panama 08°00′N 81°20′W
Sonag *see* Zêkog
Sonapur *see* Subarnapur
95 G24 **Sønderborg** *Ger.* Sonderburg. Syddanmark, SW Denmark 54°55′N 09°48′E
Sonderburg *see* Sønderborg
101 K15 **Sondershausen** Thüringen, C Germany 51°22′N 10°52′E
Sondre Strømfjord *see* Kangerlussuaq
106 E6 **Sondrio** Lombardia, N Italy 46°11′N 09°52′E
Sone *see* Son
Sonepur *see* Subarnapur
57 K22 **Sonequera** ▲ S Bolivia 22°06′S 67°10′W
167 V12 **Sông Cầu** Phu Yên, C Vietnam 13°26′N 109°12′E
167 R15 **Sông Đốc** Minh Hai, S Vietnam 09°03′N 104°51′E
81 H25 **Songea** Ruvuma, S Tanzania 10°42′S 35°39′E
163 X10 **Songhua Hu** ◎ NE China
163 Y7 **Songhua Jiang** *var.* Sungari. ≈ NE China
161 S8 **Songjiang** Shanghai Shi, E China 31°01′N 121°14′E
Sŏngjin *see* Kimch'aek
167 O16 **Songkhla** *var.* Songkla, *Mal.* Singora. Songkhla, SW Thailand 07°12′N 100°35′E
Songkla *see* Songkhla
163 T13 **Song Ling** ▲▲ NE China
129 U12 **Sông Ma** *Laos* Ma, Nam. ≈ Laos/Vietnam
163 W14 **Songnim** SW North Korea 38°43′N 125°40′E
82 B10 **Songo** Uíge, NW Angola 07°30′S 14°56′E
83 M15 **Songo** Tete, NW Mozambique 15°36′S 32°45′E
79 F21 **Songololo** Bas-Congo, SW Dem. Rep. Congo 05°40′S 14°05′E
160 H7 **Songpan** *var.* Jin'an, *Tib.* Sungpu. Sichuan, C China 32°49′N 103°39′E
Sŏngsan *see* Seongsan
161 R11 **Songxi** Fujian, SE China 27°33′N 118°46′E
160 M6 **Songxian** *var.* Song Xian. Henan, C China 34°11′N 112°04′E
Song Xian *see* Songxian
161 R10 **Songyang** *var.* Xiping; *prev.* Songyin. Zhejiang, SE China 28°29′N 119°27′E
Songyin *see* Songyang
163 V9 **Songyuan** *var.* Fu-yü, Petuna; *prev.* Fuyu. Jilin, NE China 45°10′N 124°52′E
Sonid Youqi *see* SaihanTal
Sonid Zuoqi *see* Mandalt
152 I10 **Sonīpat** Haryāna, N India 29°00′N 77°01′E
93 M15 **Sonkajärvi** Pohjois-Savo, C Finland 63°40′N 27°30′E
167 R6 **Sơn La** Sơn La, N Vietnam 21°20′N 103°55′E
149 O16 **Sonmiāni** Baluchistān, S Pakistan 25°24′N 66°37′E
149 O16 **Sonmiāni Bay** *bay* S Pakistan
101 K18 **Sonneberg** Thüringen, C Germany 50°22′N 11°10′E
101 N24 **Sonntagshorn** ▲ Austria/Germany 47°40′N 12°42′E
Sonoita *see* Sonoyta
40 E3 **Sonoita, Rio** *var.* Río Sonoyta. ≈ Mexico/USA
35 N7 **Sonoma** California, W USA 38°16′N 122°28′W
35 T3 **Sonoma Peak** ▲ Nevada, W USA 40°50′N 117°34′W
35 P8 **Sonora** California, W USA 37°58′N 120°22′W
25 O10 **Sonora** Texas, SW USA 30°34′N 100°39′W
40 F5 **Sonora** ⬖ *state* NW Mexico
35 X17 **Sonoran Desert** *var.* Desierto de Altar. *desert* Mexico/USA *see also* Altar, Desierto de
40 G5 **Sonora, Río** ≈ NW Mexico
40 E2 **Sonoyta** *var.* Sonoita. Sonora, NW Mexico 31°49′N 112°50′W
Sonoyta, Río *see* Sonoita, Rio
142 K6 **Sonqor** *var.* Sunqur. Kermānshāhān, W Iran 34°45′N 47°39′E
105 N9 **Sonseca** *var.* Sonseca con Casalgordo. Castilla-La Mancha, C Spain 39°40′N 03°59′W
Sonseca con Casalgordo *see* Sonseca
54 E9 **Sonsón** Antioquia, W Colombia 05°45′N 75°18′W
42 E7 **Sonsonate** Sonsonate, W El Salvador 13°44′N 89°43′W
42 A9 **Sonsonate** ⬖ *department* SW El Salvador
188 A10 **Sonsorol Islands** *island group* S Palau
112 J9 **Sonta** *Hung.* Szond; *prev.* Szonta. Vojvodina, NW Serbia 45°34′N 19°06′E
167 S6 **Sơn Tây** *var.* Sontay. Ha Tây, N Vietnam 21°06′N 105°32′E
Sontay *see* Sơn Tây
101 J25 **Sonthofen** Bayern, S Germany 47°31′N 10°16′E
Soochow *see* Suzhou
80 O13 **Sool** *off.* Gobolka Sool. ⬖ *region* N Somalia
Soomaaliya/Soomaaliyeed, Jamuuriyada Demuqraadiga *see* Somalia
Soome Laht *see* Finland, Gulf of
23 V5 **Soperton** Georgia, SE USA 32°22′N 82°35′W
167 S6 **Sop Hao** Houaphan, N Laos 20°33′N 104°25′E
Sophia *see* Sofia
171 S10 **Sopi** Pulau Morotai, E Indonesia 02°36′N 128°32′E
Sopianae *see* Pécs
171 U13 **Sopinusa** Papua Barat, E Indonesia 03°31′S 132°55′E
81 B14 **Sopo** ≈ W South Sudan
Sopockinie/Sopotskin/Sopotskino *see* Sapotskin
114 I9 **Sopot** Plovdiv, C Bulgaria 42°39′N 24°43′E
110 I7 **Sopot** *Ger.* Zoppot. Pomorskie, N Poland 54°26′N 18°33′E
167 O8 **Sop Prap** *var.* Ban Sop Prap. Lampang, NW Thailand 17°55′N 99°20′E
111 G22 **Sopron** *Ger.* Ödenburg. Győr-Moson-Sopron, NW Hungary 47°40′N 16°35′E
147 U11 **Sopu-Korgon** *var.* Sofi-Kurgan. Oshskaya Oblast', SW Kyrgyzstan 40°03′N 73°30′E
152 H5 **Sopur** Jammu and Kashmir, NW India 34°19′N 74°30′E
107 J15 **Sora** Lazio, C Italy 41°43′N 13°37′E
154 N13 **Sorada** Odisha, E India 19°46′N 84°29′E
93 H17 **Söråker** Västernorrland, C Sweden 62°32′N 17°32′E
57 J17 **Sorata** La Paz, W Bolivia 15°47′S 68°38′W
Sorau/Sorau in der Niederlausitz *see* Żary
105 Q14 **Sorbas** Andalucía, S Spain 37°06′N 02°06′W
94 N11 **Sördellen** ◎ C Sweden
Sord/Sórd Choluim Chille *see* Swords
15 O11 **Sorel** Québec, SE Canada 46°03′N 73°06′W
183 P17 **Sorell** Tasmania, SE Australia 42°49′S 147°34′E
183 O17 **Sorell, Lake** ◎ Tasmania, SE Australia
106 E8 **Soresina** Lombardia, N Italy 45°17′N 09°51′E
95 D14 **Sørfjorden** *fjord* S Norway
94 N11 **Sörforsa** Gävleborg, C Sweden 61°45′N 17°00′E
103 R14 **Sorgues** Vaucluse, SE France 44°N 04°52′E
136 K13 **Sorgun** Yozgat, C Turkey 39°49′N 35°10′E
105 P5 **Soria** Castilla y León, N Spain 41°47′N 02°26′W
105 P6 **Soria** ⬖ *province* Castilla y León, N Spain
61 D19 **Soriano** Soriano, SW Uruguay 33°25′S 58°21′W
61 D19 **Soriano** ⬖ *department* SW Uruguay
92 O4 **Sørkapp** *headland* SW Svalbard 76°34′N 16°33′E
143 T5 **Sorkh, Kūh-e** ▲▲ NE Iran
95 I23 **Sorø** Sjælland, E Denmark 55°26′N 11°34′E
Soro *see* Ghazal, Bahr el
116 M8 **Soroca** *Rus.* Soroki. N Moldova 48°10′N 28°18′E
60 L10 **Sorocaba** São Paulo, S Brazil 23°29′S 47°27′W
Sorochino *see* Sarochyna
127 T7 **Sorochinsk** Orenburgskaya Oblast', W Russian Federation 52°26′N 53°10′E
Soroki *see* Soroca
188 H15 **Sorol** *atoll* Caroline Islands, W Micronesia
171 T12 **Sorong** Papua Barat, E Indonesia 0°49′S 131°16′E
81 G17 **Soroti** C Uganda 01°42′N 33°37′E
Sørøy *see* Sørøya
92 J8 **Sørøya** *var.* Sørøy, *Lapp.* Sállan. *island* N Norway
104 G11 **Sorraia, Rio** ≈ C Portugal
92 I10 **Sørreisa** Troms, N Norway 69°08′N 18°09′E
107 K18 **Sorrento** *anc.* Surrentum. Campania, S Italy 40°37′N 14°23′E
104 H10 **Sor, Ribeira de** *stream* C Portugal
195 T3 **Sør Rondane** *Eng.* Sor Rondane Mountains. ▲▲ Antarctica
Sor Rondane Mountains *see* Sør Rondane
93 H14 **Sorsele** Västerbotten, N Sweden 65°31′N 17°34′E
107 B17 **Sorso** Sardegna, Italy, C Mediterranean Sea 40°46′N 08°33′E
171 P4 **Sorsogon** Luzon, N Philippines 12°57′N 124°04′E
105 U4 **Sort** Cataluña, NE Spain 42°25′N 01°07′E
124 H11 **Sortavala** *prev.* Serdobol'. Respublika Kareliya, NW Russian Federation 61°45′N 30°37′E
107 L25 **Sortino** Sicilia, Italy, C Mediterranean Sea 37°10′N 15°02′E
92 G10 **Sortland** Nordland, C Norway 68°44′N 15°25′E
94 G9 **Sør-Trøndelag** ⬖ *county* S Norway
95 I15 **Sørumsand** Akershus, S Norway 59°58′N 11°13′E
118 D6 **Sõrve Säär** *headland* SW Estonia 57°54′N 22°02′E
95 K22 **Sösdala** Skåne, S Sweden 56°00′N 13°36′E
105 R4 **Sos del Rey Católico** Aragón, NE Spain 42°30′N 01°13′W
93 F15 **Sösjöfjällen** ▲ C Sweden 63°51′N 13°15′E
126 K7 **Sosna** ≈ W Russian Federation
62 H12 **Sosneado, Cerro** ▲ W Argentina 34°44′S 69°52′W
125 S9 **Sosnogorsk** Respublika Komi, NW Russian Federation 63°33′N 53°55′E
124 J8 **Sosnovets** Respublika Kareliya, NW Russian Federation 64°25′N 34°23′E
Sosnovets *see* Sosnowiec
127 Q3 **Sosnovka** Chuvashskaya Respublika, W Russian Federation 56°18′N 47°14′E
125 S16 **Sosnovka** Kirovskaya Oblast', NW Russian Federation 56°15′N 51°20′E
124 M6 **Sosnovka** Murmanskaya Oblast', NW Russian Federation 66°28′N 40°31′E
126 M6 **Sosnovka** Tambovskaya Oblast', W Russian Federation 53°14′N 41°19′E
124 H12 **Sosnovo** *Fin.* Rautu. Leningradskaya Oblast', NW Russian Federation 60°30′N 30°13′E
127 V3 **Sosnovyy Bor** Respublika Bashkortostan, W Russian Federation 55°51′N 57°09′E
Sosnovyy Bor *see* Sasnovy Bor
111 J16 **Sosnowiec** *Ger.* Sosnowitz, *Rus.* Sosnovets. Śląskie, S Poland 50°16′N 19°07′E
Sosnowitz *see* Sosnowiec
117 R2 **Sosnytsya** Chernihivs'ka Oblast', N Ukraine 51°31′N 32°30′E
109 V10 **Šoštanj** N Slovenia 46°23′N 15°03′E
122 G10 **Sos'va** Sverdlovskaya Oblast', C Russian Federation 59°13′N 61°58′E
54 D12 **Sotará, Volcán** ✸ S Colombia 02°04′N 76°40′W
76 D10 **Sotavento, Ilhas de** *var.* Leeward Islands. *island group* S Cape Verde

◆ Country
● Country Capital
◇ Dependent Territory
○ Dependent Territory Capital
⬖ Administrative Regions
✈ International Airport
▲ Mountain
▲▲ Mountain Range
✸ Volcano
≈ River
◎ Lake
▣ Reservoir

Sotkamo – Stafford

93 N15 **Sotkamo** Kainuu, C Finland 64°06′N 28°30′E
109 W11 **Sotla** ≈ E Slovenia
41 P10 **Soto la Marina** Tamaulipas, C Mexico 23°44′N 98°10′W
41 P10 **Soto la Marina, Río** ≈ C Mexico
95 B14 **Sotra** island S Norway
41 X12 **Sotuta** Yucatán, SE Mexico 20°34′N 89°00′W
79 F17 **Souanké** Sangha, NW Congo 02°03′N 14°02′E
76 M17 **Soubré** S Ivory Coast 05°50′N 06°35′W
115 H24 **Soúda** var. Soúdha, Eng. Suda. Kríti, Greece, E Mediterranean Sea 35°29′N 24°04′E
Soúdha see Soúda
Soueida see As Suwaydā'
114 L12 **Souflí** prev. Souflíon. Anatolikí Makedonía kai Thráki, NE Greece 41°12′N 26°18′E
Souflíon see Souflí
45 S11 **Soufrière** W Saint Lucia 13°51′N 61°03′W
45 X6 **Soufrière** ⛰ Basse Terre, S Guadeloupe 16°03′N 61°39′W
102 M13 **Souillac** Lot, S France 44°53′N 01°29′E
173 Y17 **Souillac** S Mauritius 20°31′S 57°31′E
74 M5 **Souk Ahras** NE Algeria 36°14′N 08°00′E
Souk el Arba du Rharb/ Souk-el-Arba-du-Rharb/ Souk-el-Arba-el-Rhab see Souk-el-Arba-Rharb
74 E6 **Souk-el-Arba-Rharb** var. Souk el Arba du Rharb, Souk-el-Arba-du-Rharb, Souk-el-Arba-el-Rhab. NW Morocco 34°38′N 06°00′W
Soukhné see As Sukhnah
Sŏul see Seoul
102 J11 **Soulac-sur-Mer** Gironde, SW France 45°31′N 01°06′W
99 L19 **Soumagne** Liège, E Belgium 50°36′N 05°48′E
18 M14 **Sound Beach** Long Island, New York, NE USA 40°56′N 72°58′W
95 J22 **Sound, The** Dan. Øresund, Swe. Öresund. strait Denmark/Sweden
115 H20 **Soúnio, Akrotírio** headland C Greece 37°39′N 24°01′E
138 F8 **Soûr** var. Şūr; anc. Tyre. SW Lebanon 33°18′N 35°30′E
Sources, Mont-aux- see Phofung
104 G8 **Soure** Coimbra, N Portugal 40°04′N 08°38′W
11 W17 **Souris** Manitoba, S Canada 49°38′N 100°17′W
13 Q14 **Souris** Prince Edward Island, SE Canada 46°22′N 62°16′W
28 L2 **Souris River** var. Mouse River. ≈ Canada/USA
25 X10 **Sour Lake** Texas, SW USA 30°08′N 94°24′W
115 F17 **Soúrpi** Thessalía, C Greece 39°07′N 22°55′E
104 H11 **Sousel** Portalegre, C Portugal 38°57′N 07°40′W
75 N6 **Sousse** var. Sūsah. NE Tunisia 35°46′N 10°38′E
14 H11 **South** ≈ Ontario, S Canada
South see Sud
83 G23 **South Africa** off. Republic of South Africa, Afr. Suid-Afrika. ◆ republic S Africa
South Africa, Republic of see South Africa
46-47 **South America** continent
2 J17 **South American Plate** tectonic feature
97 M23 **Southampton** hist. Hamwih, Lat. Clausentum. S England, United Kingdom 50°54′N 01°23′W
19 N14 **Southampton** Long Island, New York, NE USA 40°52′N 72°22′W
9 P8 **Southampton Island** island Nunavut, NE Canada
151 P20 **South Andaman** island Andaman Islands, India, NE Indian Ocean
13 Q6 **South Aulatsivik Island** island Newfoundland and Labrador, E Canada
182 E4 **South Australia** ⬖ state S Australia
South Australian Abyssal Plain see South Australian Plain
192 G11 **South Australian Basin** undersea feature SW Indian Ocean 38°00′S 126°00′E
173 X12 **South Australian Plain** var. South Australian Abyssal Plain. undersea feature SE Indian Ocean
37 R13 **South Baldy** ▲ New Mexico, SW USA 33°59′N 107°11′W
23 Y14 **South Bay** Florida, SE USA 26°39′N 80°43′W
14 E12 **South Baymouth** Manitoulin Island, Ontario, S Canada 45°33′N 82°01′W
30 L10 **South Beloit** Illinois, N USA 42°29′N 89°02′W
31 O11 **South Bend** Indiana, N USA 41°40′N 86°15′W
25 R6 **South Bend** Texas, SW USA 32°58′N 98°39′W
32 F9 **South Bend** Washington, NW USA 46°38′N 123°48′W
South Beveland see Zuid-Beveland
South Borneo see Kalimantan Selatan
21 U7 **South Boston** Virginia, NE USA 36°42′N 78°58′W
182 F2 **South Branch Neales** seasonal river South Australia
21 U3 **South Branch Potomac River** ≈ West Virginia, NE USA
185 H19 **Southbridge** Canterbury, South Island, New Zealand 43°49′S 172°17′E
19 N12 **Southbridge** Massachusetts, NE USA 42°03′N 72°00′W
183 P17 **South Bruny Island** island Tasmania, SE Australia
18 L7 **South Burlington** Vermont, NE USA 44°27′N 73°08′W
44 M6 **South Caicos** island S Turks and Caicos Islands
South Cape see Ka Lae
23 V3 **South Carolina** off. State of South Carolina, also known as The Palmetto State. ⬖ state SE USA
South Carpathians see Carpații Meridionali
South Celebes see Sulawesi Selatan
21 Q5 **South Charleston** West Virginia, NE USA 38°22′N 81°42′W
192 D7 **South China Basin** undersea feature SE South China Sea 15°00′N 115°00′E
169 R8 **South China Sea** Chin. Nan Hai, Ind. Laut Cina Selatan, Vtn. Biên Đông. sea SE Asia
33 Z10 **South Dakota** off. State of South Dakota, also known as The Coyote State, Sunshine State. ⬖ state N USA
23 X10 **South Daytona** Florida, SE USA 29°09′N 81°01′W
37 R10 **South Domingo Pueblo** New Mexico, SW USA 35°28′N 106°24′W
97 N23 **South Downs** hill range SE England, United Kingdom
83 I21 **South East** ⬖ district SE Botswana
65 H15 **South East Bay** bay Ascension Island, C Atlantic Ocean
183 O17 **South East Cape** headland Tasmania, SE Australia 43°36′S 146°52′E
38 K10 **Southeast Cape** headland Saint Lawrence Island, Alaska, USA 62°56′N 169°39′W
South-East Celebes see Sulawesi Tenggara
192 G12 **Southeast Indian Ridge** undersea feature Indian Ocean/Pacific Ocean 50°00′S 110°00′E
Southeast Island see Tagula Island
193 P13 **Southeast Pacific Basin** var. Belling Hausen Mulde. undersea feature SE Pacific Ocean 60°00′S 115°00′W
65 H15 **South East Point** headland SE Ascension Island
183 O14 **South East Point** headland Victoria, S Australia 39°10′S 146°21′E
191 Z3 **South East Point** headland Kiritimati, NE Kiribati 01°42′N 157°10′W
44 L5 **Southeast Point** headland Mayaguana, SE The Bahamas 22°15′N 72°44′W
South-East Sulawesi see Sulawesi Tenggara
11 U12 **Southend** Saskatchewan, C Canada 56°20′N 103°14′W
97 P22 **Southend-on-Sea** E England, United Kingdom 51°33′N 00°43′E
83 H20 **Southern** var. Bangwaketse, Ngwaketze. ⬖ district SE Botswana
138 E13 **Southern** ⬖ district S Israel
83 N15 **Southern** ⬖ region S Malawi
155 J26 **Southern** ⬖ province S Sri Lanka
83 I15 **Southern** ⬖ province S Zambia
185 E19 **Southern Alps** ▲▲ South Island, New Zealand
190 K15 **Southern Cook Islands** island group S Cook Islands
180 K12 **Southern Cross** Western Australia 31°17′S 119°15′E
80 A12 **Southern Darfur** ⬖ state W Sudan
186 B7 **Southern Highlands** ⬖ province W Papua New Guinea
11 V11 **Southern Indian Lake** ◎ Manitoba, C Canada
80 E11 **Southern Kordofan** ⬖ state C Sudan
187 Z15 **Southern Lau Group** island group Lau Group, SE Fiji
81 I15 **Southern Nationalities** ⬖ region S Ethiopia
173 S13 **Southern Ocean** ocean
21 T10 **Southern Pines** North Carolina, SE USA 35°10′N 79°23′W
96 I13 **Southern Uplands** ▲▲ S Scotland, United Kingdom
Southern Urals see Yuzhnyy Ural
183 P16 **South Esk River** ≈ Tasmania, SE Australia
11 U16 **Southey** Saskatchewan, S Canada 50°53′N 104°27′W
27 V2 **South Fabius River** ≈ Missouri, C USA
31 S10 **Southfield** Michigan, N USA 42°28′N 83°12′W
192 K10 **South Fiji Basin** undersea feature S Pacific Ocean 26°00′S 175°00′E
97 Q22 **South Foreland** headland SE England, United Kingdom 51°08′N 01°22′E
35 P7 **South Fork American River** ≈ California, W USA
28 K7 **South Fork Grand River** ≈ South Dakota, N USA
35 T12 **South Fork Kern River** ≈ California, W USA
39 Q7 **South Fork Koyukuk River** ≈ Alaska, USA
39 Q11 **South Fork Kuskokwim River** ≈ Alaska, USA
26 H2 **South Fork Republican River** ≈ C USA
26 L3 **South Fork Solomon River** ≈ Kansas, C USA
31 P5 **South Fox Island** island Michigan, N USA
20 G8 **South Fulton** Tennessee, S USA 36°28′N 88°53′W
195 U10 **South Geomagnetic Pole** pole Antarctica
65 J20 **South Georgia** island South Georgia and the South Sandwich Islands, SW Atlantic Ocean
65 K21 **South Georgia and the South Sandwich Islands** ◇ UK Dependent Territory SW Atlantic Ocean
47 Y14 **South Georgia Ridge** var. North Scotia Ridge. undersea feature SW Atlantic Ocean 54°00′S 40°00′W
181 Q1 **South Goulburn Island** island Northern Territory, N Australia
153 U16 **South Hatia Island** island SE Bangladesh
31 O10 **South Haven** Michigan, N USA 42°24′N 86°16′W
21 V7 **South Hill** Virginia, NE USA 36°43′N 78°07′W
South Holland see Zuid-Holland
21 P8 **South Holston Lake** ⊠ Tennessee/Virginia, S USA
175 N1 **South Honshu Ridge** undersea feature W Pacific Ocean
26 M6 **South Hutchinson** Kansas, C USA 38°01′N 97°56′W
151 K21 **South Huvadhu Atoll** atoll S Maldives
173 U14 **South Indian Basin** undersea feature Indian Ocean/Pacific Ocean 60°00′S 120°00′E
11 W11 **South Indian Lake** Manitoba, C Canada 56°48′N 98°56′W
81 I17 **South Island** island NW Kenya
185 C20 **South Island** island S New Zealand
65 B23 **South Jason** island Jason Islands, NW Falkland Islands
South Kalimantan see Kalimantan Selatan
South Karelia see Etelä-Karjala
South Kazakhstan see Yuzhnyy Kazakhstan
163 X15 **South Korea** off. Republic of Korea, Kor. Taehan Min'guk. ◆ republic E Asia
35 Q6 **South Lake Tahoe** California, W USA 38°56′N 119°57′W
25 N6 **Southland** Texas, SW USA 33°16′N 101°31′W
185 B23 **Southland** off. Southland Region. ⬖ region South Island, New Zealand
Southland Region see Southland
29 N15 **South Loup River** ≈ Nebraska, C USA
151 K19 **South Maalhosmadulu Atoll** atoll N Maldives
14 E15 **South Maitland** ≈ Ontario, S Canada
192 E8 **South Makassar Basin** undersea feature E Java Sea
31 O6 **South Manitou Island** island Michigan, N USA
151 K18 **South Miladhunmadulu Atoll** var. Noonu. atoll N Maldives
21 X8 **South Mills** North Carolina, SE USA 36°28′N 76°18′W
8 H9 **South Nahanni** ≈ Northwest Territories, NW Canada
39 P13 **South Naknek** Alaska, USA 58°39′N 157°01′W
14 M13 **South Nation** ≈ Ontario, SE Canada
44 F9 **South Negril Point** headland W Jamaica 18°14′N 78°21′W
151 K20 **South Nilandhe Atoll** var. Dhaalu Atoll. atoll C Maldives
36 L2 **South Ogden** Utah, W USA 41°09′N 111°58′W
18 M14 **Southold** Long Island, New York, NE USA 41°03′N 72°24′W
194 H1 **South Orkney Islands** island group Antarctica
137 S9 **South Ossetia** former autonomous region SW Georgia
South Ostrobothnia see Etelä-Pohjanmaa
South Pacific Basin see Southwest Pacific Basin
19 P7 **South Paris** Maine, NE USA 44°14′N 70°33′W
189 U13 **South Pass** passage Chuuk Islands, C Micronesia
33 U15 **South Pass** pass Wyoming, C USA
20 K10 **South Pittsburg** Tennessee, S USA 35°00′N 85°42′W
28 K15 **South Platte River** ≈ Colorado/Nebraska, C USA
31 T16 **South Point** Ohio, N USA 38°25′N 82°35′W
65 G15 **South Point** headland S Ascension Island
31 R6 **South Point** headland Michigan, N USA 44°51′N 83°17′W
South Point see Ka Lae
195 Q9 **South Pole** pole Antarctica
183 P17 **Southport** Tasmania, SE Australia 43°26′S 146°57′E
97 K17 **Southport** NW England, United Kingdom 53°39′N 03°01′W
21 V12 **Southport** North Carolina, SE USA 33°55′N 78°00′W
19 P8 **South Portland** Maine, NE USA 43°38′N 70°14′W
14 H12 **South River** Ontario, S Canada 45°50′N 79°23′W
21 U11 **South River** ≈ North Carolina, SE USA
96 K5 **South Ronaldsay** island NE Scotland, United Kingdom
36 L2 **South Salt Lake** Utah, W USA 40°42′N 111°52′W
65 L21 **South Sandwich Islands** island group SW Atlantic Ocean
65 K21 **South Sandwich Trench** undersea feature SW Atlantic Ocean 56°30′S 25°00′W
11 S16 **South Saskatchewan** ≈ Alberta/Saskatchewan, S Canada
65 I21 **South Scotia Ridge** undersea feature S Scotia Sea
11 V10 **South Seal** ≈ Manitoba, C Canada
194 G4 **South Shetland Islands** island group Antarctica
65 H22 **South Shetland Trough** undersea feature Atlantic Ocean/Pacific Ocean 61°00′S 59°30′W
97 M14 **South Shields** NE England, United Kingdom 55°N 01°25′W
29 R13 **South Sioux City** Nebraska, C USA 42°28′N 96°24′W
192 J9 **South Solomon Trench** undersea feature W Pacific Ocean
183 V3 **South Stradbroke Island** island Queensland, E Australia
81 E15 **South Sudan** off. Republic of South Sudan. ◆ E Africa
South Sulawesi see Sulawesi Selatan
South Sumatra see Sumatera Selatan
184 K11 **South Taranaki Bight** bight SE Tasman Sea
South Tasmania Plateau see Tasman Plateau
36 M15 **South Tucson** Arizona, SW USA 32°11′N 110°56′W
12 H9 **South Twin Island** island Nunavut, C Canada
South Tyrol see Trentino-Alto Adige
96 E9 **South Uist** island NW Scotland, United Kingdom
South-West see Sud-Ouest
South-West Africa/South West Africa see Namibia
65 F15 **South West Bay** bay Ascension Island, C Atlantic Ocean
183 N18 **South West Cape** headland Tasmania, SE Australia 43°34′S 146°01′E
185 B26 **South West Cape** headland Stewart Island, New Zealand 47°15′S 167°28′E
38 J10 **Southwest Cape** headland Saint Lawrence Island, Alaska, USA 63°19′N 171°27′W
Southwest Indian Ocean Ridge see Southwest Indian Ridge
173 N11 **Southwest Indian Ridge** var. Southwest Indian Ocean Ridge. undersea feature SW Indian Ocean 43°00′S 40°00′E
192 L10 **Southwest Pacific Basin** var. South Pacific Basin. undersea feature SE Pacific Ocean 40°00′S 150°00′W
191 X3 **South West Point** headland Kiritimati, NE Kiribati 01°53′N 157°34′E
65 G25 **South West Point** headland SW Saint Helena 16°00′S 05°48′W
44 H2 **Southwest Point** headland Great Abaco, N The Bahamas 25°50′N 77°12′W
25 P5 **South Wichita River** ≈ Texas, SW USA
97 Q20 **Southwold** E England, United Kingdom 52°15′N 01°36′E
19 Q12 **South Yarmouth** Massachusetts, NE USA 41°38′N 70°09′W
116 J10 **Sovata** Hung. Szováta. Mureș, C Romania 46°36′N 25°04′E
107 N22 **Soverato** Calabria, SW Italy 38°40′N 16°31′E
121 O4 **Sovereign Base Area** uk military installation S Cyprus
Sovetabad see Ghafurov
126 C2 **Sovetsk** Ger. Tilsit. Kaliningradskaya Oblast', W Russian Federation 55°04′N 21°52′E
125 Q15 **Sovetsk** Kirovskaya Oblast', NW Russian Federation 57°37′N 49°02′E
127 N10 **Sovetskaya** Rostovskaya Oblast', SW Russian Federation 49°00′N 42°09′E
Sovetskoye see Ketchenery
146 I15 **Sovet"yab** prev. Sovet"yap. Ahal Welaýaty, S Turkmenistan 36°29′N 61°13′E
Sovet"yap see Sovet"yab
117 U12 **Sovyets'kyy** Avtonomna Respublika Krym, S Ukraine 45°20′N 34°54′E
83 I18 **Sowa** var. Sua. Central, NE Botswana 20°33′S 26°18′E
83 I18 **Sowa Pan** var. Sua Pan. salt lake NE Botswana
83 J21 **Soweto** Gauteng, NE South Africa 26°08′S 27°54′E
147 R11 **So'x** Rus. Sokh. Farg'ona Viloyati, E Uzbekistan 39°56′N 71°10′E
Sōya-kaikyō see La Pérouse Strait
165 T1 **Sōya-misaki** headland Hokkaidō, NE Japan 45°31′N 141°55′E
125 N7 **Soyana** ≈ NW Russian Federation
146 A8 **Soye, Mys** var. Mys Suz. headland NW Turkmenistan 41°47′N 52°27′E
82 A10 **Soyo** Zaire Province, NW Angola 06°07′S 12°18′E
80 J10 **Soyra** ▲ C Eritrea 14°46′N 39°29′E
145 P15 **Sozak** Kaz. Sozaq; prev. Suzak. Yuzhnyy Kazakhstan, S Kazakhstan 44°09′N 68°28′E
Sozaq see Sozak
119 P16 **Sozh** ≈ NE Europe
114 N10 **Sozopol** prev. Sizebolu; anc. Apollonia. Burgas, E Bulgaria 42°25′N 27°42′E
99 L20 **Spa** Liège, E Belgium 50°29′N 05°52′E
194 I7 **Spaatz Island** island Antarctica
144 M14 **Space Launching Centre** space station Kzylorda, S Kazakhstan
105 O7 **Spain** off. Kingdom of Spain, Sp. España; anc. Hispania, Iberia, Lat. Hispana. ◆ monarchy SW Europe
Spain, Kingdom of see Spain
Spalato see Split
97 O19 **Spalding** E England, United Kingdom 52°49′N 00°06′W
14 D11 **Spanish** Ontario, S Canada 46°12′N 82°21′W
36 L3 **Spanish Fork** Utah, W USA 40°09′N 111°40′W
64 B12 **Spanish Point** headland C Bermuda 32°18′N 64°49′W
14 E9 **Spanish River** ≈ Ontario, S Canada
44 K13 **Spanish Town** hist. St.Iago de la Vega. C Jamaica 18°N 76°57′W
Spánta, Akrotírio see Spátha, Akrotírio
35 Q5 **Sparks** Nevada, W USA 39°32′N 119°45′W
Sparnacum see Épernay
95 N16 **Sparreholm** Södermanland, C Sweden 59°04′N 16°51′E
23 U4 **Sparta** Georgia, SE USA 33°16′N 82°58′W
30 K16 **Sparta** Illinois, N USA 38°07′N 89°42′W
31 P9 **Sparta** Michigan, N USA 43°09′N 85°42′W
21 R8 **Sparta** North Carolina, SE USA 36°30′N 81°07′W
20 L9 **Sparta** Tennessee, S USA 35°55′N 85°30′W
30 J7 **Sparta** Wisconsin, N USA 43°57′N 90°50′W
Sparta see Spárti
21 Q11 **Spartanburg** South Carolina, SE USA 34°56′N 81°57′W
115 F21 **Spárti** Eng. Sparta. Pelopónnisos, S Greece 37°05′N 22°25′E
107 B21 **Spartivento, Capo** headland Sardegna, Italy, C Mediterranean Sea 38°52′N 08°50′E
11 P17 **Sparwood** British Columbia, SW Canada 49°45′N 114°45′W
126 J4 **Spas-Demensk** Kaluzhskaya Oblast', W Russian Federation 54°22′N 34°16′E
126 M4 **Spas-Klepiki** Ryazanskaya Oblast', W Russian Federation 55°08′N 40°15′E
Spasovo see Kulen Vakuf
123 R15 **Spassk-Dal'niy** Primorskiy Kray, SE Russian Federation 44°34′N 132°52′E
126 M5 **Spassk-Ryazanskiy** Ryazanskaya Oblast', W Russian Federation 54°25′N 40°21′E
115 H19 **Spáta** Attikí, C Greece 37°58′N 23°55′E
121 Q11 **Spátha, Akrotírio** var. Akrotírio Spánta. headland Kríti, Greece, E Mediterranean Sea 35°42′N 23°44′E
Špatrjan see Paternion
28 I9 **Spearfish** South Dakota, N USA 44°29′N 103°51′W
25 O1 **Spearman** Texas, SW USA 36°12′N 101°13′W
65 C25 **Speedwell Island** island S Falkland Islands
65 C25 **Speedwell Island Settlement** S Falkland Islands 52°13′S 59°41′W
65 G25 **Speery Island** island S Saint Helena
45 N14 **Speightstown** NW Barbados 13°15′N 59°39′W
106 I13 **Spello** Umbria, C Italy 43°00′N 12°41′E
39 R12 **Spenard** Alaska, USA 61°09′N 150°03′W
Spence Bay see Taloyoak
31 O14 **Spencer** Indiana, N USA 39°18′N 86°46′W
29 T12 **Spencer** Iowa, C USA 43°09′N 95°07′W
29 P12 **Spencer** Nebraska, C USA 42°52′N 98°42′W
21 S9 **Spencer** North Carolina, SE USA 35°41′N 80°26′W
20 L9 **Spencer** Tennessee, S USA 35°46′N 85°27′W
21 Q4 **Spencer** West Virginia, NE USA 38°48′N 81°22′W
30 K6 **Spencer** Wisconsin, N USA 44°46′N 90°17′W
182 G10 **Spencer, Cape** headland South Australia 35°17′S 136°52′E
39 V13 **Spencer, Cape** headland Alaska, USA 58°12′N 136°39′W
182 H9 **Spencer Gulf** gulf South Australia
18 F9 **Spencerport** New York, NE USA 43°11′N 77°48′W
31 Q12 **Spencerville** Ohio, N USA 40°42′N 84°21′W
115 E17 **Spercheiáda** var. Sperhiada, Sperkhiás. Stereá Elláda, C Greece 38°54′N 22°07′E
115 E17 **Spercheiós** ≈ C Greece
Sperhiada see Spercheiáda
95 G14 **Sperillen** ◎ S Norway
Sperkhiás see Spercheiáda
101 I18 **Spessart** hill range C Germany
Spétsai see Spétses
115 G21 **Spétses** prev. Spétsai. Spétses, S Greece 37°16′N 23°09′E
115 G21 **Spétses** island S Greece
96 J8 **Spey** ≈ NE Scotland, United Kingdom
101 G20 **Speyer** Eng. Spires; anc. Civitas Nemetum, Spira. Rheinland-Pfalz, SW Germany 49°18′N 08°26′E
101 G20 **Speyerbach** ≈ W Germany
107 N20 **Spezzano Albanese** Calabria, SW Italy 39°40′N 16°17′E
Spice Islands see Maluku
100 F9 **Spiekeroog** island NW Germany
109 W9 **Spielfeld** Steiermark, SE Austria 46°43′N 15°36′E
65 N21 **Spiess Seamount** undersea feature S Atlantic Ocean 53°00′S 02°00′W
108 E9 **Spiez** Bern, W Switzerland 46°42′N 07°41′E
98 G13 **Spijkenisse** Zuid-Holland, SW Netherlands 51°52′N 04°19′E
39 T6 **Spike Mountain** ▲ Alaska, USA 67°42′N 141°39′W
115 I25 **Spíli** Kríti, Greece, E Mediterranean Sea 35°12′N 24°33′E
108 D10 **Spillgerten** ▲ W Switzerland 46°34′N 07°25′E
118 F9 **Spilve** ✈ (Riga) C Latvia 56°55′N 24°03′E
107 N17 **Spinazzola** Puglia, SE Italy 40°58′N 16°06′E
149 O9 **Spīn Bōldak** prev. Spīn Būldak. Kandahār, S Afghanistan 31°01′N 66°23′E
Spīn Būldak see Spīn Bōldak
Spira see Speyer
Spirdingsee see Śniardwy, Jezioro
Spires see Speyer
29 T11 **Spirit Lake** Iowa, C USA 43°25′N 95°06′W
29 T11 **Spirit Lake** ◎ Iowa, C USA
11 N13 **Spirit River** Alberta, W Canada 55°46′N 118°51′W
11 S14 **Spiritwood** Saskatchewan, S Canada 53°18′N 107°33′W
27 R11 **Spiro** Oklahoma, C USA 35°14′N 94°37′W
111 L19 **Spišská Nová Ves** Ger. Neudorf, Zipser Neudorf, Hung. Igló. Košický Kraj, E Slovakia 48°58′N 20°35′E
137 T11 **Spitak** NW Armenia 40°51′N 44°17′E
92 O2 **Spitsbergen** island NW Svalbard
Spittal see Spittal an der Drau
109 R9 **Spittal an der Drau** var. Spittal. Kärnten, S Austria 46°48′N 13°30′E
109 V3 **Spitz** Niederösterreich, NE Austria 48°24′N 15°22′E
94 D9 **Spjelkavik** Møre og Romsdal, S Norway 62°28′N 06°22′E
25 W10 **Splendora** Texas, SW USA 30°13′N 95°09′W
113 E14 **Split** It. Spalato. Split-Dalmacija, S Croatia 43°31′N 16°27′E
113 E14 **Split** ✈ Split-Dalmacija, S Croatia 43°33′N 16°19′E
113 E14 **Split-Dalmacija** off. Splitsko-Dalmatinska Županija. ⬖ province S Croatia
11 X12 **Split Lake** ◎ Manitoba, C Canada
Splitsko-Dalmatinska Županija see Split-Dalmacija
108 H10 **Splügen** Graubünden, S Switzerland 46°33′N 09°18′E
Spodnji Dravograd see Dravograd
25 P12 **Spofford** Texas, SW USA 29°10′N 100°24′W
118 J11 **Špoģi** SE Latvia 56°03′N 26°47′E
32 L8 **Spokane** Washington, NW USA 47°40′N 117°26′W
32 L8 **Spokane River** ≈ Washington, NW USA
106 I13 **Spoleto** Umbria, C Italy 42°44′N 12°44′E
30 I4 **Spooner** Wisconsin, N USA 45°51′N 91°49′W
30 K12 **Spoon River** ≈ Illinois, N USA
21 W5 **Spotsylvania** Virginia, NE USA 38°12′N 77°35′W
32 L8 **Sprague** Washington, NW USA 47°19′N 117°55′W
170 J5 **Spratly Island** Chin. Nanwei Dao, Viet. Đao Tr ng Sa L n. island SW Spratly Islands
192 E6 **Spratly Islands** Chin. Nansha Qundao, Viet. Quân Đao Tr ng Sa. ◇ disputed territory SE Asia
32 J12 **Spray** Oregon, NW USA 44°30′N 119°38′W
112 I11 **Spreča** ≈ N Bosnia and Herzegovina
100 P13 **Spree** ≈ E Germany
100 P13 **Spreewald** wetland NE Germany
101 P14 **Spremberg** Brandenburg, E Germany 51°34′N 14°22′E
25 W11 **Spring** Texas, SW USA 30°03′N 95°24′W
31 Q10 **Spring Arbor** Michigan, N USA 42°12′N 84°33′W
83 E23 **Springbok** Northern Cape, W South Africa 29°44′S 17°56′E
18 I15 **Spring City** Pennsylvania, NE USA 40°10′N 75°33′W
20 L9 **Spring City** Tennessee, S USA 35°41′N 84°51′W
36 L4 **Spring City** Utah, W USA 39°28′N 111°30′W
35 W3 **Spring Creek** Nevada, W USA 40°45′N 115°40′W
27 S9 **Springdale** Arkansas, C USA 36°11′N 94°07′W
31 Q14 **Springdale** Ohio, N USA 39°17′N 84°29′W
100 I13 **Springe** Niedersachsen, N Germany 52°13′N 09°33′E
37 U9 **Springer** New Mexico, SW USA 36°21′N 104°35′W
37 W7 **Springfield** Colorado, C USA 37°24′N 102°36′W
23 W5 **Springfield** Georgia, SE USA 32°21′N 81°20′W
30 K14 **Springfield** state capital Illinois, N USA 39°48′N 89°39′W
20 L6 **Springfield** Kentucky, S USA 37°42′N 85°18′W
18 M12 **Springfield** Massachusetts, NE USA 42°06′N 72°32′W
29 T10 **Springfield** Minnesota, N USA 44°15′N 94°58′W
27 T7 **Springfield** Missouri, C USA 37°13′N 93°18′W
31 R13 **Springfield** Ohio, N USA 39°55′N 83°49′W
32 G13 **Springfield** Oregon, NW USA 44°03′N 123°01′W
29 Q12 **Springfield** South Dakota, N USA 42°51′N 97°54′W
20 J8 **Springfield** Tennessee, S USA 36°30′N 86°54′W
18 M9 **Springfield** Vermont, NE USA 43°18′N 72°27′W
30 K14 **Springfield, Lake** ⊠ Illinois, N USA
55 T8 **Spring Garden** NE Guyana 06°58′N 58°34′W
30 K8 **Spring Green** Wisconsin, N USA 43°10′N 90°02′W
29 X11 **Spring Grove** Minnesota, N USA 43°33′N 91°38′W
13 P15 **Springhill** Nova Scotia, SE Canada 45°40′N 64°04′W
23 V12 **Spring Hill** Florida, SE USA 28°28′N 82°36′W
27 R4 **Spring Hill** Kansas, C USA 38°44′N 94°49′W
22 G4 **Springhill** Louisiana, S USA 33°01′N 93°27′W
20 J9 **Spring Hill** Tennessee, S USA 35°46′N 86°55′W
21 U10 **Spring Lake** North Carolina, SE USA 35°10′N 78°58′W
24 M4 **Springlake** Texas, SW USA 34°13′N 102°18′W
35 W11 **Spring Mountains** ▲▲ Nevada, W USA
65 B24 **Spring Point** West Falkland, Falkland Islands 51°49′S 60°27′W
27 W9 **Spring River** ≈ Arkansas/Missouri, C USA
27 S7 **Spring River** ≈ Missouri/Oklahoma, C USA
83 J21 **Springs** Gauteng, NE South Africa 26°16′S 28°26′E
185 H16 **Springs Junction** West Coast, South Island, New Zealand 42°21′S 172°11′E
181 X8 **Springsure** Queensland, E Australia 24°09′S 148°06′E
29 W11 **Spring Valley** Minnesota, N USA 43°41′N 92°23′W
18 K13 **Spring Valley** New York, NE USA 41°10′N 73°58′W
29 N12 **Springview** Nebraska, C USA 42°49′N 99°45′W
18 D11 **Springville** New York, NE USA 42°27′N 78°52′W
36 L3 **Springville** Utah, W USA 40°10′N 111°36′W
Sprottau see Szprotawa
15 V4 **Sproule, Pointe** headland Québec, SE Canada 49°47′N 67°02′W
11 Q14 **Spruce Grove** Alberta, SW Canada 53°36′N 113°55′W
21 T4 **Spruce Knob** ▲ West Virginia, NE USA 38°40′N 79°37′W
35 X3 **Spruce Mountain** ▲ Nevada, W USA 40°33′N 114°46′W
21 P9 **Spruce Pine** North Carolina, SE USA 35°55′N 82°03′W
98 G13 **Spui** ≈ SW Netherlands
107 O19 **Spulico, Capo** headland S Italy 39°57′N 16°38′E
25 O5 **Spur** Texas, SW USA 33°28′N 100°51′W
97 O17 **Spurn Head** headland E England, United Kingdom 53°34′N 00°06′E
99 H20 **Spy** Namur, S Belgium 50°29′N 04°43′E
95 I15 **Spydeberg** Østfold, S Norway 59°36′N 11°04′E
185 J17 **Spy Glass Point** headland South Island, New Zealand 42°33′S 173°31′E
10 L17 **Squamish** British Columbia, SW Canada 49°41′N 123°11′W
19 O8 **Squam Lake** ◎ New Hampshire, NE USA
19 S2 **Squa Pan Mountain** ▲ Maine, NE USA 46°36′N 68°09′W
39 N16 **Squaw Harbor** Unga Island, Alaska, USA 55°12′N 160°41′W
14 E11 **Squaw Island** island Ontario, S Canada
107 O22 **Squillace, Golfo di** gulf S Italy
107 Q18 **Squinzano** Puglia, SE Italy 40°26′N 18°03′E
Sráid na Cathrach see Milltown Malbay
167 S11 **Srâlau** Stœng Trêng, N Cambodia 14°03′N 105°46′E
Srath an Urláir see Stranorlar
112 G10 **Srbac** ⬖ Republika Srpska, N Bosnia and Herzegovina
Srbija see Serbia
Srbinje see Foča
112 K9 **Srbobran** var. Bácsszenttamás, Hung. Szenttamás. Vojvodina, N Serbia 45°33′N 19°46′E
Srbobran see Donji Vakuf
167 R13 **Srê Âmběl** Kaôh Kŏng, SW Cambodia 11°07′N 103°46′E
112 K13 **Srebrenica** Republika Srpska, E Bosnia and Herzegovina 44°04′N 19°18′E
112 I11 **Srebrenik** Federacija Bosne I Hercegovine, NE Bosnia and Herzegovina 44°42′N 18°30′E
114 K10 **Sredets** prev. Syulemeshlii. Stara Zagora, C Bulgaria 42°16′N 25°40′E
114 M10 **Sredets** prev. Grudovo. ⬖ Burgas, E Bulgaria
114 M10 **Sredetska Reka** ≈ SE Bulgaria
123 U9 **Sredinnyy Khrebet** ▲▲ E Russian Federation
114 N7 **Sredishte** Rom. Beibunar; prev. Knyazhevo. Dobrich, NE Bulgaria 43°51′N 27°30′E
114 I10 **Sredna Gora** ▲▲ C Bulgaria
123 R7 **Srednekolymsk** Respublika Sakha (Yakutiya), NE Russian Federation 67°28′N 153°52′E
126 K7 **Srednerusskaya Vozvyshennost'** Eng. Central Russian Upland. ▲▲ W Russian Federation
122 L9 **Srednesibirskoye Ploskogor'ye** var. Central Siberian Uplands, Eng. Central Siberian Plateau. ▲▲ N Russian Federation
125 V13 **Sredniy Ural** ▲▲ NW Russian Federation
167 T12 **Srê Khtŭm** Môndól Kiri, E Cambodia 12°10′N 106°52′E
110 G12 **Śrem** Wielkopolskie, C Poland 52°07′N 17°00′E
112 K10 **Sremska Mitrovica** prev. Mitrovica, Ger. Mitrowitz. Vojvodina, NW Serbia 44°58′N 19°37′E
167 R11 **Srêng, Stœng** ≈ NW Cambodia
167 R11 **Srê Noy** Siĕmréab, NW Cambodia 13°47′N 104°03′E
Srepok, Sông see Srêpôk, Tônle
167 T12 **Srêpôk, Tônle** var. Sông Srepok. ≈ Cambodia/Vietnam
123 P13 **Sretensk** Zabaykal'skiy Kray, S Russian Federation 52°14′N 117°33′E
169 R10 **Sri Aman** Sarawak, East Malaysia 01°13′N 111°25′E
117 R4 **Sribne** Chernihivs'ka Oblast', N Ukraine 50°40′N 32°55′E
Sri Jayawardanapura see Sri Jayewardenepura Kotte
155 J25 **Sri Jayewardenepura Kotte** var. Sri Jayawardanapura. ● (legislative) Western Province, W Sri Lanka 06°54′N 79°58′E
155 M14 **Srīkākulam** Andhra Pradesh, E India 18°18′N 83°54′E
155 J25 **Sri Lanka** off. Democratic Socialist Republic of Sri Lanka; prev. Ceylon. ◆ republic S Asia
130 F14 **Sri Lanka** island S Asia
Sri Lanka, Democratic Socialist Republic of see Sri Lanka
153 V14 **Srimangal** Sylhet, E Bangladesh 24°19′N 91°40′E
Sri Mohangorh see Shri Mohangarh
152 H5 **Srinagar** state capital Jammu and Kashmir, N India 34°07′N 74°50′E
167 N10 **Srinagarind Reservoir** ◎ W Thailand
155 F19 **Sringeri** Karnātaka, W India 13°26′N 75°13′E
155 K25 **Sri Pada** Eng. Adam's Peak. ▲ S Sri Lanka 06°49′N 80°25′E
Sri Saket see Si Sa Ket
111 G14 **Środa Śląska** Ger. Neumarkt. Dolnośląskie, SW Poland 51°10′N 16°30′E
110 H12 **Środa Wielkopolska** Wielkopolskie, C Poland 52°13′N 17°17′E
Srpska Kostajnica see Bosanska Kostajnica
113 G14 **Srpska, Republika** ⬖ republic Bosnia and Herzegovina
Srpski Brod see Bosanski Brod
Ssu-ch'uan see Sichuan
Ssu-p'ing/Ssu-p'ing-chieh see Siping
Stablo see Stavelot
99 G15 **Stabroek** Antwerpen, N Belgium 51°21′N 04°22′E
Stackeln see Strenči
96 I5 **Stack Skerry** island N Scotland, United Kingdom
100 I9 **Stade** Niedersachsen, NW Germany 53°36′N 09°29′E
94 C10 **Stadlandet** peninsula S Norway
109 R5 **Stadl-Paura** Oberösterreich, NW Austria 48°05′N 13°52′E
119 L20 **Stadolichy** Rus. Stodolichi. Homyel'skaya Voblasts', SE Belarus 51°44′N 28°30′E
98 P7 **Stadskanaal** Groningen, NE Netherlands 53°N 06°55′E
101 H16 **Stadtallendorf** Hessen, C Germany 50°49′N 09°01′E
101 K23 **Stadtbergen** Bayern, S Germany 48°21′N 10°50′E
108 G7 **Stäfa** Zürich, NE Switzerland 47°14′N 08°45′E
95 K23 **Staffanstorp** Skåne, S Sweden 55°38′N 13°13′E
101 K18 **Staffelstein** Bayern, C Germany 50°05′N 11°00′E
97 L19 **Stafford** C England, United Kingdom 52°48′N 02°07′W
26 L6 **Stafford** Kansas, C USA 37°57′N 98°36′W
21 W4 **Stafford** Virginia, NE USA 38°26′N 77°27′W

◆ Country ● Country Capital ◇ Dependent Territory ○ Dependent Territory Capital ⬖ Administrative Regions ✈ International Airport ▲ Mountain ▲▲ Mountain Range ⛰ Volcano ≈ River ◎ Lake ⊠ Reservoir

STAFFORDSHIRE – STRONGILÍ

97 L19 **Staffordshire** *cultural region* C England, United Kingdom
19 N12 **Stafford Springs** Connecticut, NE USA *41°57′N 72°18′W*
115 H14 **Stágira** Kentrikí Makedonía, N Greece *40°31′N 23°46′E*
118 G7 **Staicele** N Latvia *57°52′N 24°48′E*
Ştaierdorf-Anina *see* Anina
109 V8 **Stainz** Steiermark, SE Austria *46°55′N 15°18′E*
Stájerlakanina *see* Anina
117 Y7 **Stakhanov** Luhans'ka Oblast', E Ukraine *48°30′N 38°42′E*
108 E11 **Stalden** Valais, SW Switzerland *46°12′N 07°55′E*
15 S8 **St-Alexandre** Québec, SE Canada *47°39′N 69°36′W*
Stalin *see* Varna
Stalinabad *see* Dushanbe
Stalingrad *see* Volgograd
Staliniri *see* Tskhinvali
Stalino *see* Donets'k
Stalinobod *see* Dushanbe
Stalinov Štít *see* Gerlachovský štít
Stalinsk *see* Novokuznetsk
Stalins'kaya Oblast' *see* Donets'ka Oblast'
Stalinski Zaliv *see* Varnenski Zaliv
Stalin, Yazovir *see* Iskar, Yazovir
111 N15 **Stalowa Wola** Podkarpackie, SE Poland *50°35′N 22°02′E*
114 I11 **Stamboliyski** Plovdiv, C Bulgaria *42°09′N 24°32′E*
15 Q7 **St-Ambroise** Québec, SE Canada *48°35′N 71°19′W*
97 N19 **Stamford** E England, United Kingdom *52°39′N 00°32′W*
18 L14 **Stamford** Connecticut, NE USA *41°03′N 73°32′W*
25 P6 **Stamford** Texas, SW USA *32°55′N 99°49′W*
25 Q6 **Stamford, Lake** ⊠ Texas, SW USA
108 I10 **Stampa** Graubünden, SE Switzerland *46°21′N 09°35′E*
Stampalia *see* Astypálaia
27 T14 **Stamps** Arkansas, C USA *33°22′N 93°30′W*
92 G11 **Stamsund** Nordland, C Norway *68°07′N 13°50′E*
27 R2 **Stanberry** Missouri, C USA *40°12′N 94°33′W*
195 O3 **Stancomb-Wills Glacier** *glacier* Antarctica
83 K21 **Standerton** Mpumalanga, E South Africa *26°57′S 29°14′E*
31 R7 **Standish** Michigan, N USA *43°59′N 83°58′W*
20 M6 **Stanford** Kentucky, S USA *37°30′N 84°40′W*
33 S9 **Stanford** Montana, NW USA *47°08′N 110°15′W*
95 P19 **Stånga** Gotland, SE Sweden *57°16′N 18°30′E*
94 I13 **Stange** Hedmark, S Norway *60°40′N 11°05′E*
83 L23 **Stanger** KwaZulu/Natal, E South Africa *29°20′S 31°18′E*
Stanger *see* KwaDukuza
Stanimaka *see* Asenovgrad
Stanislau *see* Ivano-Frankivs'k
35 P8 **Stanislaus River** ∿ California, W USA
Stanislav *see* Ivano-Frankivs'k
Stanislavskaya Oblast' *see* Ivano-Frankivs'ka Oblast'
Stanisławów *see* Ivano-Frankivs'k
Stanke Dimitrov *see* Dupnitsa
183 O15 **Stanley** Tasmania, SE Australia *40°48′S 145°18′E*
65 E24 **Stanley** *var.* Port Stanley, Puerto Argentino. ○ (Falkland Islands) East Falkland, Falkland Islands *51°45′S 57°56′W*
33 O13 **Stanley** Idaho, NW USA *44°12′N 114°58′W*
28 L3 **Stanley** North Dakota, N USA *48°19′N 102°23′W*
21 U4 **Stanley** Virginia, NE USA *38°34′N 78°30′W*
30 J6 **Stanley** Wisconsin, N USA *44°58′N 90°54′W*
79 G21 **Stanley Pool** *var.* Pool Malebo. ⊚ Congo/Dem. Rep. Congo
155 H20 **Stanley Reservoir** ⊠ S India
Stanleyville *see* Kisangani
42 G3 **Stann Creek** ◈ *district* SE Belize
Stann Creek *see* Dangriga
123 Q12 **Stanovoy Khrebet** ▲▲ SE Russian Federation
108 F8 **Stans** Nidwalden, C Switzerland *46°57′N 08°23′E*
97 O21 **Stansted** ✕ (London) Essex, E England, United Kingdom *51°53′N 00°16′E*
183 U4 **Stanthorpe** Queensland, E Australia *28°35′S 151°52′E*
21 N6 **Stanton** Kentucky, S USA *37°51′N 83°51′W*
31 Q8 **Stanton** Michigan, N USA *43′19′N 85′04′W*
29 Q14 **Stanton** Nebraska, C USA *41°57′N 97°13′W*
28 L5 **Stanton** North Dakota, N USA *47°19′N 101°22′W*
25 N7 **Stanton** Texas, SW USA *32°07′N 101°47′W*
32 H7 **Stanwood** Washington, NW USA *48°14′N 122°22′W*
117 Y7 **Stanychno-Luhans'ke** Luhans'ka Oblast', E Ukraine *48°39′N 39°30′E*
108 K7 **Stanzach** Tirol, W Austria *47°24′N 10°36′E*
98 M9 **Staphorst** Overijssel, E Netherlands *52°38′N 06°12′E*
14 D18 **Staples** Ontario, S Canada *42°09′N 82°34′W*
29 T6 **Staples** Minnesota, N USA *46°21′N 94°47′W*
28 M14 **Stapleton** Nebraska, C USA *41°29′N 100°40′W*
25 S8 **Star** Texas, SW USA *31°27′N 98°16′W*
111 M14 **Starachowice** Świętokrzyskie, C Poland *51°04′N 21°02′E*
Stara Kanjiža *see* Kanjiža
111 M18 **Stará L'ubovňa** *Ger.* Altlublau, *Hung.* Ólubló. Prešovský Kraj, E Slovakia *49°19′N 20°40′E*
112 L10 **Stara Pazova** *Ger.* Altpasua, *Hung.* Ópazova. Vojvodina, N Serbia *44°59′N 20°10′E*
Stara Planina *see* Balkan Mountains
114 L9 **Stara Reka** ∿ C Bulgaria
116 M5 **Stara Synyava** Khmel'nyts'ka Oblast', W Ukraine *49°39′N 27°39′E*
116 I2 **Stara Vyzhivka** Volyns'ka Oblast', NW Ukraine *51°27′N 24°25′E*
Staraya Belitsa *see* Staraya Byelitsa
119 M14 **Staraya Byelitsa** *Rus.* Staraya Belitsa. Vitsyebskaya Voblasts', NE Belarus *54°42′N 29°38′E*
127 R5 **Staraya Mayna** Ul'yanovskaya Oblast', W Russian Federation *54°36′N 48°57′E*
119 O18 **Staraya Rudnya** Homyel'skaya Voblasts', SE Belarus *52°50′N 30°17′E*
124 H14 **Staraya Russa** Novgorodskaya Oblast', W Russian Federation *57°59′N 31°18′E*
114 K10 **Stara Zagora** *Lat.* Augusta Trajana. Stara Zagora, C Bulgaria *42°26′N 25°39′E*
114 K10 **Stara Zagora** ◈ *province* C Bulgaria
29 S8 **Starbuck** Minnesota, N USA *45°36′N 95°31′W*
191 W4 **Starbuck Island** *prev.* Volunteer Island. *island* E Kiribati
27 V13 **Star City** Arkansas, C USA *33°56′N 91°52′W*
112 F13 **Staretina** ▲▲ W Bosnia and Herzegovina
Stargard in Pommern *see* Stargard Szczeciński
110 E9 **Stargard Szczeciński** *Ger.* Stargard in Pommern. Zachodnio-pomorskie, NW Poland *53°20′N 15°02′E*
187 N10 **Star Harbour** *harbor* San Cristobal, SE Solomon Islands
Stari Bečej *see* Bečej
113 F15 **Stari Grad** *It.* Cittavecchia. Split-Dalmacija, S Croatia *43°11′N 16°36′E*
124 J16 **Staritsa** Tverskaya Oblast', W Russian Federation *56°28′N 34°51′E*
23 V9 **Starke** Florida, SE USA *29°56′N 82°07′W*
22 M4 **Starkville** Mississippi, S USA *33°27′N 88°49′W*
186 B7 **Star Mountains** *Ind.* Pegunungan Sterren. ▲▲ Indonesia/Papua New Guinea
101 L23 **Starnberg** Bayern, SE Germany *48°00′N 11°19′E*
101 L24 **Starnberger See** ⊚ SE Germany
Starobel'sk *see* Starobil's'k
117 X8 **Starobesheve** Donets'ka Oblast', E Ukraine *47°45′N 38°01′E*
117 Y6 **Starobil's'k** *Rus.* Starobel'sk. Luhans'ka Oblast', E Ukraine *49°16′N 38°56′E*
119 K18 **Starobin** *var.* Starobyn. Minskaya Voblasts', S Belarus *52°44′N 27°28′E*
Starobyn *see* Starobin
126 H6 **Starodub** Bryanskaya Oblast', W Russian Federation *52°30′N 32°56′E*
110 I8 **Starogard Gdański** *Ger.* Preussisch-Stargard. Pomorskie, N Poland *53°57′N 18°29′E*
Staroikan *see* Ikan
Starokonstantinov *see* Starokostyantyniv
116 L5 **Starokostyantyniv** *Rus.* Starokonstantinov. Khmel'nyts'ka Oblast', NW Ukraine *49°43′N 27°13′E*
126 K12 **Starominskaya** Krasnodarskiy Kray, SW Russian Federation *46°31′N 39°03′E*
114 L7 **Staro Selo** *Rom.* Satul-Vechi; *prev.* Star-Smil. Silistra, NE Bulgaria *43°58′N 26°32′E*
126 K12 **Staroshcherbinovskaya** Krasnodarskiy Kray, SW Russian Federation *46°36′N 38°42′E*
127 V6 **Starosubkhangulovo** Respublika Bashkortostan, W Russian Federation *53°05′N 57°22′E*
35 S4 **Star Peak** ▲ Nevada, W USA *40°31′N 118°09′W*
15 T8 **St-Arsène** Québec, SE Canada *47°55′N 69°21′W*
Star-Smil *see* Staro Selo
97 J25 **Start Point** *headland* SW England, United Kingdom *50°13′N 03°38′W*
Startsy *see* Kirawsk
Starum *see* Stavoren
119 L18 **Staryya Darohi** *Rus.* Staryye Dorogi. Minskaya Voblasts', S Belarus *53°02′N 28°16′E*
Staryye Dorogi *see* Staryya Darohi
127 T2 **Staryye Zyattsy** Udmurtskaya Respublika, NW Russian Federation *57°22′N 52°42′E*
117 U13 **Staryy Krym** Avtonomna Respublika Krym, S Ukraine *45°03′N 35°06′E*
126 K8 **Staryy Oskol** Belgorodskaya Oblast', W Russian Federation *51°21′N 37°52′E*
116 H6 **Staryy Sambir** L'vivs'ka Oblast', W Ukraine *49°27′N 23°00′E*
101 L14 **Stassfurt** *var.* Staßfurt. Sachsen-Anhalt, C Germany *51°51′N 11°35′E*
Staßfurt *see* Stassfurt
111 M15 **Staszów** Świętokrzyskie, C Poland *50°33′N 21°07′E*
29 W13 **State Center** Iowa, C USA *42°01′N 93°09′W*
18 E14 **State College** Pennsylvania, NE USA *40°48′N 77°52′W*
18 K15 **Staten Island** *island* New York, NE USA
Staten Island *see* Estados, Isla de los
23 U8 **Statenville** Georgia, SE USA *30°42′N 83°00′W*
23 W5 **Statesboro** Georgia, SE USA *32°28′N 81°47′W*
States, The *see* United States of America
21 R9 **Statesville** North Carolina, SE USA *35°46′N 80°54′W*
95 G16 **Stathelle** Telemark, S Norway *59°01′N 09°40′E*
Statia *see* Sint Eustatius
30 K15 **Staunton** Illinois, N USA *39°00′N 89°47′W*
21 T5 **Staunton** Virginia, NE USA *38°10′N 79°05′W*
95 C16 **Stavanger** Rogaland, S Norway *58°58′N 05°43′E*
99 L21 **Stavelot** *Dut.* Stablo. Liège, E Belgium *50°24′N 05°56′E*
95 G16 **Stavern** Vestfold, S Norway *58°58′N 10°01′E*
Stavers Island *see* Vostok Island
98 J7 **Stavoren** *Fris.* Starum. Fryslân, N Netherlands *52°52′N 05°22′E*
115 K21 **Stavrí, Akrotírio** *var.* Akrotírio Stavrós. *headland* Naxos, Kykládes, Greece, Aegean Sea *37°12′N 25°32′E*
126 M14 **Stavropol'** *prev.* Voroshilovsk. Stavropol'skiy Kray, SW Russian Federation *45°02′N 41°58′E*
Stavropol' *see* Tol'yatti
126 M14 **Stavropol'skaya Vozvyshennost'** ▲▲ SW Russian Federation
126 M14 **Stavropol'skiy Kray** ◈ *territory* SW Russian Federation
115 H14 **Stavrós** Kentrikí Makedonía, N Greece *40°39′N 23°43′E*
115 J24 **Stavrós, Akrotírio** *headland* Kríti, Greece, E Mediterranean Sea *35°25′N 24°57′E*
Stavrós, Akrotírio *see* Stavrí, Akrotírio
114 I12 **Stavroúpoli** *prev.* Stavroúpolis. Anatolikí Makedonía kai Thráki, NE Greece *41°12′N 24°45′E*
Stavroúpolis *see* Stavroúpoli
117 O6 **Stavyshche** Kyyivs'ka Oblast', N Ukraine *49°23′N 30°10′E*
182 M11 **Stawell** Victoria, SE Australia *37°06′S 142°52′E*
110 N9 **Stawiski** Podlaskie, NE Poland *53°22′N 22°08′E*
14 G14 **Stayner** Ontario, S Canada *44°25′N 80°05′W*
14 D17 **St. Clair** ∿ Canada/USA
37 R3 **Steamboat Springs** Colorado, C USA *40°28′N 106°51′W*
15 U4 **Ste-Anne, Lac** ⊚ Québec, SE Canada
20 M8 **Stearns** Kentucky, S USA *36°39′N 84°27′W*
39 N10 **Stebbins** Alaska, USA *63°30′N 162°15′W*
15 U7 **Ste-Blandine** Québec, SE Canada *48°22′N 68°27′W*
108 K7 **Steeg** Tirol, W Austria *47°15′N 10°18′E*
27 Y9 **Steele** Missouri, C USA *36°04′N 89°49′W*
29 N5 **Steele** North Dakota, N USA *46°51′N 99°55′W*
194 J5 **Steele Island** *island* Antarctica
30 K16 **Steeleville** Illinois, N USA *38°00′N 89°39′W*
27 W6 **Steelville** Missouri, C USA *37°57′N 91°21′W*
99 G14 **Steenbergen** Noord-Brabant, S Netherlands *51°35′N 04°19′E*
Steenkool *see* Bintuni
11 O10 **Steen River** Alberta, W Canada *59°37′N 117°17′W*
98 M8 **Steenwijk** Overijssel, N Netherlands *52°47′N 06°07′E*
65 A23 **Steeple Jason** *island* Jason Islands, NW Falkland Islands
174 J8 **Steep Point** *headland* Western Australia *26°09′S 113°11′E*
116 L9 **Ştefăneşti** Botoşani, NE Romania *47°44′N 27°15′E*
Stefanie, Lake *see* Ch'ew Bahir
8 L5 **Stefansson Island** *island* Nunavut, N Canada
117 O10 **Ştefan Vodă** *Rus.* Suvorovo. SE Moldova *46°33′N 29°39′E*
63 H18 **Steffen, Cerro** ▲ S Chile *44°27′S 71°42′W*
108 D9 **Steffisburg** Bern, C Switzerland *46°47′N 07°38′E*
95 J24 **Stege** Sjælland, SE Denmark *54°59′N 12°18′E*
116 G10 **Ştei** *Hung.* Vaskohsziklás. Bihor, W Romania *46°34′N 22°28′E*
Steier *see* Steyr
Steierdorf/Steierdorf-Anina *see* Anina
109 T7 **Steiermark** *off.* Land Steiermark, *Eng.* Styria. ◈ *state* C Austria
Steiermark, Land *see* Steiermark
101 J19 **Steigerwald** *hill range* C Germany
99 L17 **Stein** Limburg, SE Netherlands *50°58′N 05°45′E*
Stein *see* Stein an der Donau
Stein *see* Kamnik, Slovenia
108 M8 **Steinach** Tirol, W Austria *47°07′N 11°30′E*
Steinamanger *see* Szombathely
109 W3 **Stein an der Donau** *var.* Stein. Niederösterreich, NE Austria *48°25′N 15°35′E*
Steinau an der Elbe *see* Ścinawa
11 Y16 **Steinbach** Manitoba, S Canada *49°32′N 96°40′W*
Steiner Alpen *see* Kamniško-Savinjske Alpe
99 L24 **Steinfort** Luxembourg, W Luxembourg *49°39′N 05°55′E*
100 H12 **Steinhuder Meer** ⊚ NW Germany
93 E15 **Steinkjer** Nord-Trøndelag, C Norway *64°01′N 11°29′E*
Stejarul *see* Karapelit
99 F16 **Stekene** Oost-Vlaanderen, NW Belgium *51°13′N 04°04′E*
83 E26 **Stellenbosch** Western Cape, SW South Africa *33°56′S 18°51′E*
98 F13 **Stellendam** Zuid-Holland, SW Netherlands *51°48′N 04°01′E*
39 T12 **Steller, Mount** ▲ Alaska, USA *60°36′N 142°49′W*
103 Y14 **Stello, Monte** ▲ Corse, France, C Mediterranean Sea *42°49′N 09°24′E*
106 F5 **Stelvio, Passo dello** *pass* Italy/Switzerland
15 S7 **Ste-Maguerite Nord-Est** ∿ Québec, SE Canada
15 V4 **Ste-Marguerite, Pointe** *headland* Québec, SE Canada *50°01′N 66°43′W*
12 J14 **Ste-Marie, Lac** ⊠ Québec, S Canada
103 R3 **Stenay** Meuse, NE France *49°29′N 05°12′E*
100 L12 **Stendal** Sachsen-Anhalt, C Germany *52°36′N 11°52′E*
118 E8 **Stende** NW Latvia *57°09′N 22°33′E*
182 H10 **Stenhouse Bay** South Australia *35°15′S 136°58′E*
95 J23 **Stenløse** Hovedstaden, E Denmark *55°47′N 12°13′E*
95 L19 **Stensjön** Jönköping, S Sweden *57°36′N 14°42′E*
95 K18 **Stenstorp** Västra Götaland, S Sweden *58°15′N 13°45′E*
95 I18 **Stenungsund** Västra Götaland, S Sweden *58°05′N 11°49′E*
Stepanakert *see* Xankändi
137 T11 **Step'anavan** N Armenia *41°00′N 44°27′E*
100 K9 **Stepenitz** ∿ N Germany
29 O10 **Stephan** South Dakota, N USA *44°12′N 99°25′W*
29 R3 **Stephen** Minnesota, N USA *48°27′N 96°54′W*
27 T14 **Stephens** Arkansas, C USA *33°25′N 93°04′W*
184 J13 **Stephens, Cape** *headland* D'Urville Island, Marlborough, SW New Zealand *40°42′S 173°56′E*
21 V3 **Stephens City** Virginia, NE USA *39°03′N 78°10′W*
182 L6 **Stephens Creek** New South Wales, SE Australia *31°51′S 141°30′E*
184 K13 **Stephens Island** *island* C New Zealand
31 N5 **Stephenson** Michigan, N USA *45°27′N 87°36′W*
13 S12 **Stephenville** Newfoundland, Newfoundland and Labrador, SE Canada *48°33′N 58°34′W*
25 S7 **Stephenville** Texas, SW USA *32°12′N 98°13′W*
Step' Nardara *see* Step' Shardara
145 R8 **Stepnogorsk** Akmola, C Kazakhstan *52°04′N 72°18′E*
127 O15 **Stepnoye** Stavropol'skiy Kray, SW Russian Federation *44°18′N 44°34′E*
145 Q8 **Stepnyak** Akmola, N Kazakhstan *52°52′N 70°49′E*
145 P17 **Step' Shardara** *Kaz.* Shardara Dalasy; *prev.* Step' Nardara. *grassland* S Kazakhstan
192 J17 **Steps Point** *headland* Tutuila, W American Samoa *14°23′S 170°46′W*
115 F17 **Stereá Elláda** *Eng.* Greece Central *var.* Stereá Ellás. ◈ *region* C Greece
Stereá Ellás *see* Stereá Elláda
83 J24 **Sterkspruit** Eastern Cape, SE South Africa *30°31′S 27°22′E*
127 U6 **Sterlibashevo** Respublika Bashkortostan, W Russian Federation *53°19′N 55°12′E*
39 R12 **Sterling** Alaska, USA *60°32′N 150°51′W*
37 V3 **Sterling** Colorado, C USA *40°37′N 103°12′W*
30 K11 **Sterling** Illinois, N USA *41°47′N 89°42′W*
26 M5 **Sterling** Kansas, C USA *38°12′N 98°12′W*
25 O8 **Sterling City** Texas, SW USA *31°50′N 101°00′W*
31 S9 **Sterling Heights** Michigan, N USA *42°34′N 83°01′W*
21 W3 **Sterling Park** Virginia, NE USA *39°00′N 77°24′W*
37 V2 **Sterling Reservoir** ⊠ Colorado, C USA
22 I5 **Sterlington** Louisiana, S USA *32°42′N 92°05′W*
127 U6 **Sterlitamak** Respublika Bashkortostan, W Russian Federation *53°39′N 56°00′E*
Sternberg *see* Šternberk
111 H17 **Šternberk** *Ger.* Sternberg. Olomoucký Kraj, E Czech Republic *49°45′N 17°20′E*
141 V17 **Stĕroh** Suquṭrā, S Yemen *12°21′N 53°50′E*
Sterren, Pegunungan *see* Star Mountains
110 G11 **Stęszew** Wielkopolskie, C Poland *52°16′N 16°41′E*
Stettin *see* Szczecin
Stettiner Haff *see* Szczeciński, Zalew
11 Q15 **Stettler** Alberta, SW Canada *52°21′N 112°40′W*
31 V13 **Steubenville** Ohio, N USA *40°21′N 80°37′W*
97 O21 **Stevenage** E England, United Kingdom *51°55′N 00°14′W*
23 Q1 **Stevenson** Alabama, S USA *34°52′N 85°50′W*
32 H11 **Stevenson** Washington, NW USA *45°43′N 121°54′W*
182 E1 **Stevenson Creek** *seasonal river* South Australia
39 Q13 **Stevenson Entrance** *strait* Alaska, USA
30 L6 **Stevens Point** Wisconsin, N USA *44°32′N 89°33′W*
39 R8 **Stevens Village** Alaska, USA *66°01′N 149°02′W*
33 P10 **Stevensville** Montana, NW USA *46°30′N 114°05′W*
93 E25 **Stevns Klint** *headland* E Denmark *55°15′N 12°25′E*
10 J12 **Stewart** British Columbia, W Canada *55°58′N 129°52′W*
10 J6 **Stewart** ∿ Yukon, NW Canada
10 I6 **Stewart Crossing** Yukon, NW Canada *63°22′N 136°37′W*
63 H25 **Stewart, Isla** *island* S Chile
185 B25 **Stewart Island** *island* S New Zealand
Stewart Islands *see* Sikaiana
181 W6 **Stewart, Mount** ▲ Queensland, E Australia *20°11′S 145°29′E*
10 H6 **Stewart River** Yukon, NW Canada *63°17′N 139°24′W*
27 R3 **Stewartsville** Missouri, C USA *39°45′N 94°30′W*
11 S16 **Stewart Valley** Saskatchewan, S Canada *50°34′N 107°47′W*
29 W10 **Stewartville** Minnesota, N USA *43°51′N 92°29′W*
Steyerlak-Anina *see* Anina
109 T5 **Steyr** *var.* Steier. Oberösterreich, N Austria *48°02′N 14°26′E*
109 T5 **Steyr** ∿ NW Austria
15 T7 **St-Fabien** Québec, SE Canada *48°19′N 68°51′W*
15 R11 **St-François, Lac** ⊚ Québec, SE Canada
83 E25 **St. Helena Bay** *bay* SW South Africa
15 T8 **St-Hubert** Québec, SE Canada *47°46′N 69°15′W*
29 P11 **Stickney** South Dakota, N USA *43°24′N 98°23′W*
98 L5 **Stiens** Fryslân, N Netherlands *53°15′N 05°45′E*
Stif *see* Sétif
27 Q11 **Stigler** Oklahoma, C USA *35°16′N 95°08′W*
107 N18 **Stigliano** Basilicata, S Italy *40°24′N 16°13′E*
95 N17 **Stigtomta** Södermanland, C Sweden *58°48′N 16°47′E*
10 I11 **Stikine** ∿ British Columbia, W Canada
Stilida/Stilís *see* Stylída
95 G22 **Stilling** Midtjylland, C Denmark *56°04′N 10°00′E*
29 W8 **Stillwater** Minnesota, N USA *45°03′N 92°48′W*
27 O9 **Stillwater** Oklahoma, C USA *36°07′N 97°03′W*
35 S5 **Stillwater Range** ▲▲ Nevada, W USA
18 I8 **Stillwater Reservoir** ⊠ New York, NE USA
107 O22 **Stilo, Punta** *headland* S Italy *38°27′N 16°36′E*
27 R10 **Stilwell** Oklahoma, C USA *35°48′N 94°37′W*
Štimlje *see* Shtime
25 N1 **Stinnett** Texas, SW USA *35°49′N 101°27′W*
113 P18 **Štip** E FYR Macedonia *41°45′N 22°10′E*
Stira *see* Stýra
96 J12 **Stirling** C Scotland, United Kingdom *56°07′N 03°57′W*
96 I12 **Stirling** *cultural region* C Scotland, United Kingdom
180 J14 **Stirling Range** ▲▲ Western Australia
15 R8 **St-Jean** ∿ Québec, SE Canada
93 E16 **Stjørdalshalsen** Nord-Trøndelag, C Norway *63°27′N 10°57′E*
83 L22 **St. Lucia** KwaZulu/Natal, E South Africa *28°22′N 32°25′E*
Stochód *see* Stokhid
101 H24 **Stockach** Baden-Württemberg, S Germany *47°51′N 09°01′E*
25 S12 **Stockdale** Texas, SW USA *29°14′N 97°57′W*
109 X3 **Stockerau** Niederösterreich, NE Austria *48°24′N 16°13′E*
93 H20 **Stockholm** ● (Sweden) Stockholm, C Sweden *59°17′N 18°03′E*
95 O15 **Stockholm** ◈ *county* C Sweden
Stockmannshof *see* Pļaviņas
97 L18 **Stockport** NW England, United Kingdom *53°25′N 02°10′W*
65 K15 **Stocks Seamount** *undersea feature* C Atlantic Ocean *11°42′S 33°48′W*
35 O8 **Stockton** California, W USA *37°56′N 121°19′W*
26 L3 **Stockton** Kansas, C USA *39°27′N 99°17′W*
27 S6 **Stockton** Missouri, C USA *37°43′N 93°49′W*
30 K3 **Stockton Island** *island* Apostle Islands, Wisconsin, N USA
27 S7 **Stockton Lake** ⊠ Missouri, C USA
97 M15 **Stockton-on-Tees** *var.* Stockton on Tees. N England, United Kingdom *54°34′N 01°19′W*
Stockton on Tees *see* Stockton-on-Tees
24 M10 **Stockton Plateau** *plain* Texas, SW USA
28 M16 **Stockville** Nebraska, C USA *40°33′N 100°20′W*
93 H17 **Stöde** Västernorrland, C Sweden *62°27′N 16°34′E*
Stodolichi *see* Stadolichy
113 M19 **Stogovo Karaorman** ▲▲ W FYR Macedonia
Stoke *see* Stoke-on-Trent
97 L19 **Stoke-on-Trent** *var.* Stoke. C England, United Kingdom *53°N 02°10′W*
182 M15 **Stokes Point** *headland* Tasmania, SE Australia *40°09′S 143°55′E*
116 J2 **Stokhid** *Pol.* Stochód, *Rus.* Stokhod. ∿ NW Ukraine
Stokhod *see* Stokhid
92 I4 **Stokkseyri** Suðurland, SW Iceland *63°49′N 21°00′W*
92 G10 **Stokmarknes** Nordland, C Norway *68°34′N 14°55′E*
Stol *see* Veliki Krš
113 H15 **Stolac** Federacija Bosne I Hercegovine, S Bosnia and Herzegovina *43°04′N 17°58′E*
Stolbce *see* Stowbtsy
101 D16 **Stolberg** *var.* Stolberg im Rheinland. Nordrhein-Westfalen, W Germany *50°45′N 06°15′E*
Stolberg im Rheinland *see* Stolberg
123 P6 **Stolbovoy, Ostrov** *island* NE Russian Federation
Stolbtsy *see* Stowbtsy
119 J20 **Stolin** Brestskaya Voblasts', SW Belarus *51°53′N 26°51′E*
95 K14 **Stöllet** *var.* Norra Ny. Värmland, C Sweden *60°24′N 13°15′E*
Stolp *see* Słupsk
Stolpe *see* Słupia
Stolpmünde *see* Ustka
115 F15 **Stómio** Thessalía, C Greece *39°51′N 22°45′E*
14 J11 **Stonecliffe** Ontario, SE Canada *46°12′N 77°58′W*
96 L10 **Stonehaven** NE Scotland, United Kingdom *56°59′N 02°14′W*
97 M23 **Stonehenge** *ancient monument* Wiltshire, S England, United Kingdom
23 T3 **Stone Mountain** ▲ Georgia, SE USA *33°48′N 84°10′W*
11 X16 **Stonewall** Manitoba, S Canada *50°08′N 97°20′W*
21 S3 **Stonewood** West Virginia, NE USA *39°15′N 80°18′W*
14 D17 **Stoney Point** Ontario, S Canada *42°18′N 82°32′W*
92 H10 **Stonglandseidet** Troms, N Norway *69°03′N 17°06′E*
65 N25 **Stonybeach Bay** *bay* Tristan da Cunha, SE Atlantic Ocean
35 N5 **Stony Creek** ∿ California, W USA
65 N25 **Stonyhill Point** *headland* S Tristan da Cunha
14 I14 **Stony Lake** ⊚ Ontario, SE Canada
11 Q14 **Stony Plain** Alberta, SW Canada *53°31′N 114°04′W*
21 R9 **Stony Point** North Carolina, SE USA *35°51′N 81°04′W*
18 G8 **Stony Point** *headland* New York, NE USA *43°50′N 76°18′W*
11 T10 **Stony Rapids** Saskatchewan, C Canada *59°14′N 105°48′W*
39 P11 **Stony River** Alaska, USA *61°48′N 156°37′W*
Stony Tunguska *see* Podkamennaya Tunguska
12 G10 **Stooping** ∿ Ontario, C Canada
100 I9 **Stör** ∿ N Germany
95 M15 **Storå** Örebro, S Sweden *59°44′N 15°10′E*
95 J16 **Stora Gla** ⊚ C Sweden
95 I16 **Stora Le** *Nor.* Store Le. ⊚ Norway/Sweden
92 I12 **Stora Lulevatten** ⊚ N Sweden
92 H13 **Storavan** ⊚ N Sweden
93 I20 **Storby** Åland, SW Finland *60°12′N 19°33′E*
94 E10 **Stordalen** Møre og Romsdal, S Norway *62°22′N 07°00′E*
95 H23 **Storebælt** *var.* Store Bælt, *Eng.* Great Belt, Storebelt. *channel* Baltic Sea/Kattegat
Store Bælt *see* Storebælt
Storebelt *see* Storebælt
95 M19 **Storebro** Kalmar, S Sweden *57°36′N 15°30′E*
95 J24 **Store Heddinge** Sjælland, SE Denmark *55°19′N 12°24′E*
Store Le *see* Stora Le
93 E16 **Støren** Sør-Trøndelag, S Norway *63°02′N 10°16′E*
92 O4 **Storfjorden** *fjord* S Norway
95 L15 **Storfors** Värmland, C Sweden *59°33′N 14°16′E*
92 G13 **Storforshei** Nordland, C Norway *66°25′N 14°25′E*
Storhammer *see* Hamar
93 F16 **Storlien** Jämtland, C Sweden *63°18′N 12°10′E*
183 P17 **Storm Bay** *inlet* Tasmania, SE Australia
29 T12 **Storm Lake** Iowa, C USA *42°38′N 95°12′W*
29 S13 **Storm Lake** ⊚ Iowa, C USA
96 G7 **Stornoway** NW Scotland, United Kingdom *58°13′N 06°23′W*
Storojineţ *see* Storozhynets'
92 P1 **Storøya** *island* NE Svalbard
125 S10 **Storozhevsk** Respublika Komi, NW Russian Federation *61°56′N 52°18′E*
Storozhinets *see* Storozhynets'
116 K8 **Storozhynets'** *Ger.* Storozynetz, *Rom.* Storojineţ, *Rus.* Storozhinets. Chernivets'ka Oblast', W Ukraine *48°11′N 25°42′E*
Storozynetz *see* Storozhynets'
92 H11 **Storriten** *Lapp.* Stuorrarijjda. ▲ C Norway *68°09′N 17°12′E*
19 N12 **Storrs** Connecticut, NE USA *41°48′N 72°15′W*
94 I11 **Storsjøen** ⊚ S Norway
94 N13 **Storsjön** ⊚ C Sweden
93 F16 **Storsjön** ⊚ C Sweden
92 J9 **Storslett** Troms, N Norway *69°45′N 21°03′E*
94 H11 **Storsølnkletten** ▲ S Norway *61°52′N 11°32′E*
92 I9 **Storsteinnes** Troms, N Norway *69°13′N 19°14′E*
93 J14 **Storsund** Norrbotten, N Sweden *65°36′N 20°40′E*
94 J9 **Storsylen** *Swe.* Sylarna. ▲ Norway/Sweden *63°00′N 12°14′E*
92 H11 **Stortoppen** ▲ N Sweden *67°33′N 17°27′E*
93 H14 **Storuman** Västerbotten, N Sweden *65°05′N 17°10′E*
93 H14 **Storuman** ⊚ N Sweden
94 N13 **Storvik** Gävleborg, C Sweden *60°37′N 16°30′E*
95 O14 **Storvreta** Uppsala, C Sweden *59°58′N 17°42′E*
100 L10 **Störwasserstrasse** *canal* N Germany
29 V13 **Story City** Iowa, C USA *42°10′N 93°36′W*
11 V17 **Stoughton** Saskatchewan, S Canada *49°40′N 103°01′W*
19 O11 **Stoughton** Massachusetts, NE USA *42°07′N 71°06′W*
30 L9 **Stoughton** Wisconsin, N USA *42°56′N 89°12′W*
97 L23 **Stour** ∿ E England, United Kingdom
97 P21 **Stour** ∿ S England, United Kingdom
27 T5 **Stover** Missouri, C USA *38°26′N 92°59′W*
95 G21 **Støvring** Nordjylland, N Denmark *56°53′N 09°52′E*
119 J17 **Stowbtsy** *Pol.* Stolbce, *Rus.* Stolbtsy. Minskaya Voblasts', C Belarus *53°29′N 26°44′E*
25 X11 **Stowell** Texas, SW USA *29°47′N 94°22′W*
97 P20 **Stowmarket** E England, United Kingdom *52°05′N 00°54′E*
114 N8 **Stozher** Dobrich, NE Bulgaria *43°27′N 27°49′E*
97 E14 **Strabane** *Ir.* An Srath Bán. W Northern Ireland, United Kingdom *54°49′N 07°27′W*
121 S11 **Strabo Trench** *undersea feature* C Mediterranean Sea
27 T7 **Strafford** Missouri, C USA *37°16′N 93°07′W*
183 N17 **Strahan** Tasmania, SE Australia *42°10′S 145°18′E*
111 C18 **Strakonice** *Ger.* Strakonitz. Jihočeský Kraj, S Czech Republic *49°14′N 13°55′E*
Strakonitz *see* Strakonice
100 N8 **Stralsund** Mecklenburg-Vorpommern, NE Germany *54°18′N 13°06′E*
99 L16 **Stramproy** Limburg, SE Netherlands *51°12′N 05°43′E*
83 E26 **Strand** Western Cape, SW South Africa *34°06′S 18°50′E*
94 E10 **Stranda** Møre og Romsdal, S Norway *62°18′N 06°56′E*
97 G15 **Strangford Lough** *Ir.* Loch Cuan. *inlet* E Northern Ireland, United Kingdom
95 N16 **Strängnäs** Södermanland, C Sweden *59°22′N 17°02′E*
97 E14 **Stranorlar** *Ir.* Srath an Urláir. NW Ireland *54°48′N 07°46′W*
97 H14 **Stranraer** S Scotland, United Kingdom *54°54′N 05°02′W*
11 U16 **Strasbourg** Saskatchewan, S Canada *51°05′N 104°58′W*
103 V5 **Strasbourg** *Ger.* Strassburg; *anc.* Argentoratum. Bas-Rhin, NE France *48°35′N 07°45′E*
109 T8 **Strassburg** Kärnten, S Austria *46°54′N 14°21′E*
37 U4 **Strasburg** Colorado, C USA *39°42′N 104°13′W*
29 N7 **Strasburg** North Dakota, N USA *46°07′N 100°10′W*
31 U12 **Strasburg** Ohio, N USA *40°35′N 81°31′W*
21 U3 **Strasburg** Virginia, NE USA *38°59′N 78°21′W*
117 N10 **Străşeni** *var.* Strasheny. C Moldova *47°07′N 28°37′E*
Strasheny *see* Străşeni
Strassburg *see* Strasbourg, France
Strassburg *see* Aiud, Romania
99 M25 **Strassen** Luxembourg, S Luxembourg *49°37′N 06°05′E*
109 R5 **Strasswalchen** Salzburg, C Austria *47°59′N 13°19′E*
14 F16 **Stratford** Ontario, S Canada *43°22′N 81°00′W*
184 K10 **Stratford** Taranaki, North Island, New Zealand *39°20′S 174°16′E*
35 Q11 **Stratford** California, W USA *36°10′N 119°47′W*
29 V13 **Stratford** Iowa, C USA *42°16′N 93°55′W*
27 O12 **Stratford** Oklahoma, C USA *34°48′N 96°57′W*
25 N1 **Stratford** Texas, SW USA *36°21′N 102°05′W*
30 K6 **Stratford** Wisconsin, N USA *44°53′N 90°13′W*
Stratford *see* Stratford-upon-Avon
97 M20 **Stratford-upon-Avon** *var.* Stratford. C England, United Kingdom *52°12′N 01°41′W*
183 O17 **Strathgordon** Tasmania, SE Australia *42°49′S 146°04′E*
11 Q16 **Strathmore** Alberta, SW Canada *51°05′N 113°20′W*
35 R11 **Strathmore** California, W USA *36°07′N 119°04′W*
14 E16 **Strathroy** Ontario, S Canada *42°57′N 81°40′W*
96 I6 **Strathy Point** *headland* N Scotland, United Kingdom *58°36′N 04°04′W*
37 W4 **Stratton** Colorado, C USA *39°16′N 102°34′W*
19 P6 **Stratton** Maine, NE USA *45°08′N 70°25′W*
18 M10 **Stratton Mountain** ▲ Vermont, NE USA *43°05′N 72°55′W*
101 N21 **Straubing** Bayern, SE Germany *48°53′N 12°35′E*
100 O12 **Strausberg** Brandenburg, E Germany *52°34′N 13°52′E*
32 K13 **Strawberry Mountain** ▲ Oregon, NW USA *44°18′N 118°43′W*
29 X12 **Strawberry Point** Iowa, C USA *42°40′N 91°31′W*
36 M3 **Strawberry Reservoir** ⊠ Utah, W USA
36 M4 **Strawberry River** ∿ Utah, W USA
25 R7 **Strawn** Texas, SW USA *32°33′N 98°30′W*
113 P17 **Straža** ▲ Bulgaria/FYR Macedonia *42°16′N 22°13′E*
111 I19 **Strážov** *Hung.* Sztrazsó. ▲ NW Slovakia *48°59′N 18°29′E*
182 F7 **Streaky Bay** South Australia *32°49′S 134°13′E*
182 E7 **Streaky Bay** *bay* South Australia
30 L12 **Streator** Illinois, N USA *41°07′N 88°50′W*
Streckenbach *see* Świdnik
Strednogorie *see* Pirdop
111 C17 **Středočeský Kraj** ◈ *region* C Czech Republic
29 O6 **Streeter** North Dakota, N USA *46°37′N 99°23′W*
25 U8 **Streetman** Texas, SW USA *31°52′N 96°19′W*
116 G13 **Strehaia** Mehedinţi, SW Romania *44°37′N 23°10′E*
Strehlen *see* Strzelin
114 I10 **Strelcha** Pazardzhik, C Bulgaria *42°28′N 24°21′E*
122 L12 **Strelka** Krasnoyarskiy Kray, C Russian Federation *58°05′N 92°54′E*
124 L6 **Strel'na** ∿ NW Russian Federation
118 H7 **Strenči** *Ger.* Stackeln. N Latvia *57°38′N 25°42′E*
15 V6 **St-René-de-Matane** Québec, SE Canada *48°42′N 67°22′W*
108 K8 **Strengen** Tirol, W Austria *47°07′N 10°25′E*
106 C6 **Stresa** Piemonte, NE Italy *45°52′N 08°32′E*
Streshin *see* Streshyn
119 N18 **Streshyn** *Rus.* Streshin. Homyel'skaya Voblasts', SE Belarus *52°43′N 30°07′E*
95 B18 **Streymoy** *Dan.* Strømø. *island* N Faroe Islands
95 G23 **Strib** Syddtjylland, C Denmark *55°33′N 09°47′E*
111 A17 **Stříbro** *Ger.* Mies. Plzeňský Kraj, W Czech Republic *49°44′N 12°55′E*
186 B7 **Strickland** ∿ SW Papua New Guinea
Striegau *see* Strzegom
Strigonium *see* Esztergom
98 H13 **Strijen** Zuid-Holland, SW Netherlands *51°45′N 04°34′E*
63 H21 **Strobel, Lago** ⊚ S Argentina
61 B25 **Stroeder** Buenos Aires, E Argentina *40°11′S 62°35′W*
115 C20 **Strofádes** *island* Iónia Nisiá, Greece, C Mediterranean Sea
Strofilia *see* Strofyliá
115 G17 **Strofyliá** *var.* Strofilia. Évvoia, C Greece *38°49′N 23°25′E*
100 O10 **Strom** ∿ NE Germany
107 L22 **Stromboli** ⩚ Isola Stromboli, SW Italy *38°48′N 15°13′E*
107 L22 **Stromboli, Isola** *island* Isole Eolie, S Italy
96 H9 **Stromeferry** N Scotland, United Kingdom *57°20′N 05°35′W*
96 J5 **Stromness** N Scotland, United Kingdom *58°57′N 03°18′W*
Strømø *see* Streymoy
94 N11 **Strömsbruk** Gävleborg, C Sweden *61°52′N 17°19′E*
29 Q15 **Stromsburg** Nebraska, C USA *41°06′N 97°36′W*
95 K21 **Strömsnäsbruk** Kronoberg, S Sweden *56°35′N 13°45′E*
95 I17 **Strömstad** Västra Götaland, S Sweden *58°56′N 11°11′E*
93 G16 **Strömsund** Jämtland, C Sweden *63°51′N 15°35′E*
93 G15 **Ströms Vattudal** *valley* N Sweden
27 V14 **Strong** Arkansas, C USA *33°06′N 92°19′W*
Strongilí *see* Strongylí

◆ Country
● Country Capital
◇ Dependent Territory
○ Dependent Territory Capital
◈ Administrative Regions
✕ International Airport
▲ Mountain
▲▲ Mountain Range
⩚ Volcano
∿ River
⊚ Lake
⊠ Reservoir

STRONGOLI – SUROVIKINO

107 *O21* **Strongoli** Calabria, SW Italy *39°17′N 17°03′E*
31 *T11* **Strongsville** Ohio, N USA *41°18′N 81°50′W*
115 *Q23* **Strongylí** *var.* Strongilí. *island* SE Greece
96 *K5* **Stronsay** *island* NE Scotland, United Kingdom
97 *L21* **Stroud** C England, United Kingdom *51°46′N 02°15′W*
27 *O10* **Stroud** Oklahoma, C USA *35°45′N 96°39′W*
18 *I14* **Stroudsburg** Pennsylvania, NE USA *40°59′N 75°12′W*
95 *F21* **Struer** Midtjylland, W Denmark *56°29′N 08°37′E*
113 *M20* **Struga** SW FYR Macedonia *41°11′N 20°40′E*
Strugi-Kranyse *see* Strugi-Krasnyye
124 *G14* **Strugi-Krasnyye** *var.* Strugi-Kranyse. Pskovskaya Oblast', W Russian Federation *58°19′N 29°09′E*
114 *G11* **Struma** *Gk.* Strymónas. ≈ Bulgaria/Greece *see also* Strymónas
Struma *see* Strymónas
97 *G21* **Strumble Head** *headland* SW Wales, United Kingdom *52°01′N 05°05′W*
Strumeshnitsa *see* Strumica
113 *Q19* **Strumica** E FYR Macedonia *41°27′N 22°39′E*
113 *Q19* **Strumica** *Bulg.* Strumeshnitsa. ≈ Bulgaria/FYR Macedonia
114 *G11* **Strumyani** Blagoevgrad, SW Bulgaria *41°41′N 23°13′E*
31 *V12* **Struthers** Ohio, N USA *41°03′N 80°36′W*
114 *I10* **Stryama** ≈ C Bulgaria
114 *G13* **Strymónas** *Bul.* Struma. ≈ Bulgaria/Greece *see also* Struma
Strymónas *see* Struma
115 *H14* **Strymonikós Kólpos** *gulf* N Greece
116 *I6* **Stryy** L'vivs'ka Oblast', NW Ukraine *49°16′N 23°51′E*
116 *H6* **Stryy** ≈ W Ukraine
111 *F14* **Strzegom** *Ger.* Striegau. Wałbrzych, SW Poland *50°59′N 16°20′E*
110 *E10* **Strzelce Krajeńskie** *Ger.* Friedeberg Neumark. Lubuskie, W Poland *52°52′N 15°30′E*
111 *I15* **Strzelce Opolskie** *Ger.* Gross Strehlitz. Opolskie, SW Poland *50°31′N 18°19′E*
182 *K3* **Strzelecki Creek** *seasonal river* South Australia
182 *J3* **Strzelecki Desert** *desert* South Australia
111 *G15* **Strzelin** *Ger.* Strehlen. Dolnośląskie, SW Poland *50°48′N 17°03′E*
110 *I11* **Strzelno** Kujawsko-pomorski, C Poland *52°38′N 18°11′E*
111 *N17* **Strzyżów** Podkarpackie, SE Poland *49°52′N 21°46′E*
15 *S8* **St-Siméon** Québec, SE Canada *47°50′N 69°55′W*
Stua Laighean *see* Leinster, Mount
23 *Y13* **Stuart** Florida, SE USA *27°12′N 80°15′W*
29 *U14* **Stuart** Iowa, C USA *41°30′N 94°19′W*
29 *O13* **Stuart** Nebraska, C USA *42°36′N 99°08′W*
21 *S8* **Stuart** Virginia, NE USA *36°38′N 80°19′W*
10 *L13* **Stuart** ≈ British Columbia, SW Canada
39 *N10* **Stuart Island** *island* Alaska, USA
10 *L13* **Stuart Lake** ◎ British Columbia, SW Canada
185 *B22* **Stuart Mountains** ▲▲ South Island, New Zealand
182 *F3* **Stuart Range** *hill range* South Australia
Stubaital *see* Neustift im Stubaital
95 *I24* **Stubbekøbing** Sjælland, SE Denmark *54°53′N 12°04′E*
45 *P14* **Stubbs** Saint Vincent, Saint Vincent and the Grenadines *13°08′N 61°09′W*
109 *V6* **Stübming** ≈ E Austria
114 *J11* **Studen Kladenets, Yazovir** ⊠ S Bulgaria
185 *G21* **Studholme** Canterbury, South Island, New Zealand *44°44′S 171°08′E*
Stuhlweissenberg *see* Székesfehérvár
Stuhm *see* Sztum
12 *C7* **Stull Lake** ◎ Ontario, C Canada
Stuorrarijjda *see* Storriten
126 *L4* **Stupino** Moskovskaya Oblast', W Russian Federation *54°54′N 38°06′E*
27 *U4* **Sturgeon** Missouri, C USA *39°13′N 92°16′W*
14 *G10* **Sturgeon** ≈ Ontario, S Canada
31 *N6* **Sturgeon Bay** Wisconsin, N USA *44°51′N 87°21′W*
14 *G11* **Sturgeon Falls** Ontario, S Canada *46°22′N 79°57′W*
12 *C11* **Sturgeon Lake** ◎ Ontario, S Canada
30 *M3* **Sturgeon River** ≈ Michigan, N USA
20 *H6* **Sturgis** Kentucky, S USA *37°33′N 87°58′W*
31 *P11* **Sturgis** Michigan, N USA *41°48′N 85°25′W*
28 *J9* **Sturgis** South Dakota, N USA *44°24′N 103°30′W*
112 *D10* **Šturlić** ◐ Federacija Bosne I Hercegovine, NW Bosnia and Herzegovina
111 *J22* **Štúrovo** *Hung.* Párkány; *prev.* Parkan. Nitriansky Kraj, SW Slovakia *47°49′N 18°40′E*
182 *L4* **Sturt, Mount** *hill* New South Wales, SE Australia
181 *P4* **Sturt Plain** *plain* Northern Territory, N Australia
181 *T9* **Sturt Stony Desert** *desert* South Australia
83 *J25* **Stutterheim** Eastern Cape, S South Africa *32°35′S 27°26′E*
101 *H21* **Stuttgart** Baden-Württemberg, SW Germany *48°47′N 09°12′E*
27 *W12* **Stuttgart** Arkansas, C USA *34°30′N 91°32′W*
92 *H2* **Stykkishólmur** Vesturland, W Iceland *65°04′N 22°43′W*
115 *F17* **Stylída** *var.* Stilida, Stilís. Stereá Elláda, C Greece *38°55′N 22°37′E*
116 *K2* **Styr** *Rus.* Styr'. ≈ Belarus/Ukraine
115 *I19* **Stýra** *var.* Stira. Évvoia, C Greece *38°10′N 24°13′E*
Styria *see* Steiermark
15 *Y5* **St-Yvon** Québec, SE Canada *49°09′N 64°51′W*
Su *see* Jiangsu
Sua *see* Sowa
171 *Q17* **Suai** W East Timor *09°19′S 125°16′E*
54 *G9* **Suaita** Santander, C Colombia *06°07′N 73°30′W*
80 *I7* **Suakin** *var.* Sawakin. Red Sea, NE Sudan *19°06′N 37°17′E*
161 *T13* **Su'ao** *Jap.* Suō. N Taiwan *24°33′N 121°48′E*
Suao *see* Suau
Sua Pan *see* Sowa Pan
40 *G6* **Suaqui Grande** Sonora, NW Mexico *28°22′N 109°52′W*
61 *A16* **Suardi** Santa Fe, C Argentina *30°32′S 61°58′W*
54 *D11* **Suárez** Cauca, SW Colombia *02°55′N 76°41′W*
186 *G10* **Suau** *var.* Suao. Suaul Island, SE Papua New Guinea *10°39′S 150°03′E*
118 *G12* **Subačius** Panevėžys, NE Lithuania *55°46′N 24°45′E*
168 *K9* **Subang** *prev.* Soebang. Jawa, C Indonesia *06°32′S 107°45′E*
169 *O16* **Subang** ✈ (Kuala Lumpur) Pahang, Peninsular Malaysia
129 *S10* **Subansiri** ≈ NE India
154 *M12* **Subarnapur** *prev.* Sonapur, Sonepur. Odisha, E India *20°50′N 83°58′E*
118 *I11* **Subate** SE Latvia *56°00′N 25°54′E*
139 *N5* **Subaykhān** Dayr az Zawr, E Syria *34°52′N 40°35′E*
Subei/Subei Mongolzu Zizhixian *see* Dangchengwan
169 *P9* **Subi Besar, Pulau** *island* Kepulauan Natuna, W Indonesia
Subiyah *see* Aş Şubayḩīyah
26 *I7* **Sublette** Kansas, C USA *37°28′N 100°52′W*
112 *K8* **Subotica** *Ger.* Maria-Theresiopel, *Hung.* Szabadka. Vojvodina, N Serbia *46°06′N 19°41′E*
116 *K9* **Suceava** *Ger.* Suczawa, *Hung.* Szucsava. Suceava, NE Romania *47°41′N 26°16′E*
116 *J9* **Suceava** ◐ *county* NE Romania
116 *K9* **Suceava** *Ger.* Suczawa. ≈ N Romania
112 *E12* **Sučevići** Zadar, SW Croatia *44°13′N 16°04′E*
111 *K17* **Sucha Beskidzka** Małopolskie, S Poland *49°44′N 19°36′E*
111 *M14* **Suchedniów** Świętokrzyskie, C Poland *51°01′N 20°49′E*
42 *A2* **Suchitepéquez** *off.* Departamento de Suchitepéquez. ◐ *department* SW Guatemala
Suchitepéquez, Departamento de *see* Suchitepéquez
Su-chou *see* Suzhou
Suchow *see* Suzhou, Jiangsu, China
Suchow *see* Xuzhou, Jiangsu, China
97 *D17* **Suck** ≈ C Ireland
Sucker State *see* Illinois
186 *F9* **Suckling, Mount** ▲ S Papua New Guinea *09°36′S 149°00′E*
57 *L19* **Sucre** *hist.* Chuquisaca, La Plata. ● (Bolivia-legal capital) Chuquisaca, S Bolivia *18°53′S 65°25′W*
54 *E6* **Sucre** Santander, N Colombia *08°50′N 74°22′W*
56 *A7* **Sucre** Manabí, W Ecuador *01°21′S 80°27′W*
54 *E6* **Sucre** *off.* Departamento de Sucre. ◐ *province* N Colombia
55 *O5* **Sucre** *off.* Estado Sucre. ◐ *state* NE Venezuela
Sucre, Departamento de *see* Sucre
Sucre, Estado *see* Sucre
56 *D6* **Sucumbíos** ◐ *province* NE Ecuador
113 *G15* **Sućuraj** Split-Dalmacija, S Croatia *43°07′N 17°10′E*
58 *K10* **Sucuriju** Amapá, NE Brazil *01°31′N 50°W*
Suczawa *see* Suceava
79 *E16* **Sud** *Eng.* South. ◐ *province* S Cameroon
124 *K13* **Suda** ≈ NW Russian Federation
Suda *see* Soúda
117 *U13* **Sudak** Avtonomna Respublika Krym, S Ukraine *44°52′N 34°57′E*
24 *M4* **Sudan** Texas, SW USA *34°04′N 102°32′W*
80 *C10* **Sudan** *off.* Republic of Sudan, *Ar.* Jumhuriyat as-Sudan; *prev.* Anglo-Egyptian Sudan. ◆ *republic* N Africa
Sudanese Republic *see* Mali
Sudan, Jumhuriyat as- *see* Sudan
Sudan, Republic of *see* Sudan
14 *F10* **Sudbury** Ontario, S Canada *46°29′N 81°W*
97 *P20* **Sudbury** E England, United Kingdom *52°04′N 00°43′E*
Sud, Canal de *see* Gonâve, Canal de la
80 *E13* **Sudd** *swamp region* C South Sudan
100 *K10* **Sude** ≈ N Germany
Suderø *see* Sudhuroy
Sudest Island *see* Tagula Island
111 *E15* **Sudeten** *var.* Sudetes, Sudetic Mountains, *Cz./Pol.* Sudety. ▲▲ Czech Republic/Poland
Sudetes/Sudetic Mountains/Sudety *see* Sudeten
95 *B19* **Sudhuroy** *Dan.* Suderø. *island* S Faroe Islands
124 *M15* **Sudislavl'** Kostromskaya Oblast', NW Russian Federation *57°55′N 41°45′E*
Südkarpaten *see* Carpaţii Meridionalii
79 *N20* **Sud-Kivu** *off.* Région Sud Kivu. ◐ *region* E Dem. Rep. Congo
Sud Kivu, Région *see* Sud-Kivu
Südliche Morava *see* Južna Morava
100 *E12* **Süd-Nord-Kanal** *canal* NW Germany
126 *M3* **Sudogda** Vladimirskaya Oblast', W Russian Federation *55°58′N 40°57′E*
Sudostroy *see* Severodvinsk
79 *C15* **Sud-Ouest** *Eng.* South-West. ◐ *province* W Cameroon
173 *X17* **Sud Ouest, Pointe** *headland* SW Mauritius *20°27′S 57°18′E*
187 *P17* **Sud, Province** ◐ *province* S New Caledonia
92 *G1* **Suðureyri** Vestfirðir, NW Iceland *66°08′N 23°31′W*
92 *J4* **Suðurland** ◐ *region* S Iceland
92 *H4* **Suðurnes** ◐ *region* SW Iceland
126 *J8* **Sudzha** Kurskaya Oblast', W Russian Federation *51°12′N 35°19′E*
81 *D15* **Sue** ≈ W South Sudan
105 *S10* **Sueca** Valenciana, E Spain *39°13′N 00°19′W*
Sŭedinenie *see* Saedinenie
Suero *see* Alzira
75 *X8* **Suez** *Ar.* As Suways, El Suweis. NE Egypt *29°59′N 32°33′E*
75 *W7* **Suez Canal** *Ar.* Qanāt as Suways. *canal* NE Egypt
Suez, Gulf of *see* Khalīj as Suways
11 *R17* **Suffield** Alberta, SW Canada *50°15′N 111°05′W*
21 *X7* **Suffolk** Virginia, NE USA *36°44′N 76°37′W*
97 *P20* **Suffolk** *cultural region* E England, United Kingdom
142 *J2* **Şūfīān** Āżarbāyjān-e Sharqī, N Iran *38°15′N 45°59′E*
31 *N12* **Sugar Creek** ≈ Illinois, N USA
30 *L13* **Sugar Creek** ≈ Illinois, N USA
31 *R3* **Sugar Island** *island* Michigan, N USA
25 *V11* **Sugar Land** Texas, SW USA *29°37′N 95°37′W*
19 *P6* **Sugarloaf Mountain** ▲ Maine, NE USA *45°01′N 70°18′W*
65 *G24* **Sugar Loaf Point** *headland* N Saint Helena *15°54′S 05°43′W*
136 *G16* **Suğla Gölü** ◎ SW Turkey
123 *T8* **Sugoy** ≈ E Russian Federation
158 *F7* **Sugun** Xinjiang Uygur Zizhiqu, W China *39°46′N 76°45′E*
147 *U11* **Sugut, Gora** ▲ SW Kyrgyzstan *39°52′N 73°36′E*
169 *V6* **Sugut, Sungai** ≈ East Malaysia
159 *O9* **Suhai Hu** ◎ C China
162 *K14* **Suhait** Nei Mongol Zizhiqu, N China *39°29′N 105°11′E*
141 *X7* **Şuḩār** *var.* Sohar. N Oman *24°20′N 56°43′E*
114 *N8* **Suha Reka** ≈ NE Bulgaria
113 *M17* **Suharekë** *Serb.* Suva Reka. S Kosovo *42°23′N 20°50′E*
162 *L6* **Sühbaatar** Selenge, N Mongolia *50°12′N 106°14′E*
163 *P8* **Sühbaatar** *var.* Haylaastay. Sühbaatar, E Mongolia *46°44′N 113°51′E*
163 *P9* **Sühbaatar** ◐ *province* E Mongolia
114 *J8* **Suhindol** *var.* Sukhindol. Veliko Turnovo, N Bulgaria *43°11′N 24°10′E*
101 *K17* **Suhl** Thüringen, C Germany *50°37′N 10°43′E*
108 *F7* **Suhr** Aargau, N Switzerland *47°23′N 08°05′E*
Sui'an *see* Zhangpu
Suicheng *see* Suixi
161 *O12* **Suichuan** *var.* Quanjiang. Jiangxi, S China *26°26′N 114°34′E*
Suid-Afrika *see* South Africa
160 *L4* **Suide** *var.* Mingzhou. Shaanxi, C China *37°30′N 110°10′E*
Suidwes-Afrika *see* Namibia
163 *Y9* **Suifenhe** Heilongjiang, NE China *44°22′N 131°12′E*
Suigen *see* Suwon
163 *W8* **Suihua** Heilongjiang, NE China *46°40′N 127°00′E*
Súilí, Loch *see* Swilly, Lough
161 *Q6* **Suining** Jiangsu, E China *33°54′N 117°58′E*
160 *I9* **Suining** Sichuan, C China *30°31′N 105°33′E*
103 *Q4* **Suippes** Marne, N France *49°08′N 04°31′E*
97 *E20* **Suir** *Ir.* An tSiúir. ≈ S Ireland
165 *J13* **Suita** Ōsaka, Honshū, SW Japan *34°39′N 135°27′E*
160 *L16* **Suixi** *var.* Suicheng. Guangdong, S China *21°23′N 110°14′E*
Sui Xian *see* Suizhou
163 *T13* **Suizhong** Liaoning, NE China *40°19′N 120°22′E*
161 *N8* **Suizhou** *prev.* Sui Xian. Hubei, C China *31°46′N 113°20′E*
149 *P17* **Sujāwal** Sind, SE Pakistan *24°36′N 68°06′E*
169 *O16* **Sukabumi** *prev.* Soekaboemi. Jawa, C Indonesia *06°55′S 106°56′E*
169 *Q12* **Sukadana, Teluk** *bay* Borneo, W Indonesia
165 *P11* **Sukagawa** Fukushima, Honshū, C Japan *37°16′N 140°20′E*
Sukarnapura *see* Jayapura
Sukarno, Puntjak *see* Jaya, Puncak
Sŭkh *see* Sokh
Sukhindol *see* Suhindol
126 *J5* **Sukhinichi** Kaluzhskaya Oblast', W Russian Federation *54°06′N 35°22′E*
Sukhne *see* As Sukhnah
129 *Q4* **Sukhona** *var.* Tot'ma. ≈ NW Russian Federation
167 *O8* **Sukhothai** *var.* Sukotai. Sukhothai, W Thailand *17°00′N 99°51′E*
Sukhumi *see* Sokhumi
Sukkertoppen *see* Maniitsoq
149 *Q13* **Sukkur** Sind, SE Pakistan *27°45′N 68°46′E*
Sukotai *see* Sukhothai
Sukra Bay *see* Şawqirah, Dawḩat
125 *V15* **Suksun** Permskiy Kray, NW Russian Federation *57°10′N 57°27′E*
165 *F15* **Sukumo** Kōchi, Shikoku, SW Japan *32°55′N 132°42′E*
94 *B12* **Sula** *island* S Norway
125 *Q5* **Sula** ≈ NW Russian Federation
117 *R5* **Sula** ≈ N Ukraine
42 *H6* **Sulaco, Río** ≈ NW Honduras
Sulaimaniya *see* As Sulaymānīyah
149 *S10* **Sulaimān Range** ▲▲ C Pakistan
127 *Q16* **Sulak** Respublika Dagestan, SW Russian Federation *43°19′N 47°28′E*
127 *Q16* **Sulak** ≈ SW Russian Federation
171 *Q13* **Sula, Kepulauan** *island group* C Indonesia
136 *I12* **Sulakyurt** *var.* Konur. Kırıkkale, N Turkey *40°10′N 33°42′E*
171 *P17* **Sulamu** Timor, S Indonesia *09°57′S 123°33′E*
96 *F5* **Sula Sgeir** *island* NW Scotland, United Kingdom
171 *N13* **Sulawesi** *Eng.* Celebes. *island* C Indonesia
171 *N13* **Sulawesi Barat** *off.* Provinsi Sulawesi Barat, *var.* Sulbar. ◐ *province* C Indonesia
Sulawesi Barat, Provinsi *see* Sulawesi Barat
Sulawesi, Laut *see* Celebes Sea
171 *N14* **Sulawesi Selatan** *off.* Propinsi Sulawesi Selatan, *var.* Sulsel, *Eng.* South Celebes, South Sulawesi. ◐ *province* C Indonesia
Sulawesi Selatan, Propinsi *see* Sulawesi Selatan
171 *P12* **Sulawesi Tengah** *off.* Propinsi Sulawesi Tengah, *var.* Sulteng, *Eng.* Central Celebes, Central Sulawesi. ◐ *province* N Indonesia
Sulawesi Tengah, Propinsi *see* Sulawesi Tengah
171 *O14* **Sulawesi Tenggara** *off.* Propinsi Sulawesi Tenggara, *var.* Sultenggara, *Eng.* South-East Celebes, South-East Sulawesi. ◐ *province* C Indonesia
Sulawesi Tenggara, Propinsi *see* Sulawesi Tenggara
171 *P11* **Sulawesi Utara** *off.* Propinsi Sulawesi Utara, *var.* Sulut, *Eng.* North Celebes, North Sulawesi. ◐ *province* N Indonesia
Sulawesi Utara, Propinsi *see* Sulawesi Utara
139 *T5* **Sulaymān Beg** At Ta'mím, N Iraq
Sulbar *see* Sulawesi Barat
95 *D15* **Suldalsvatnet** ◎ S Norway
Sulden *see* Solda
110 *E12* **Sulechów** *Ger.* Züllichau. Lubuskie, W Poland *52°05′N 15°37′E*
110 *E11* **Sulęcin** Lubuskie, W Poland *52°29′N 15°06′E*
77 *U14* **Suleja** Niger, C Nigeria *09°15′N 07°10′E*
111 *K14* **Sulejów** Łodzkie, S Poland *51°21′N 19°57′E*
96 *I5* **Sule Skerry** *island* N Scotland, United Kingdom
Suliag *see* Sawhāj
76 *J16* **Sulima** S Sierra Leone *06°59′N 11°34′W*
117 *O13* **Sulina** Tulcea, SE Romania *45°07′N 29°40′E*
117 *N13* **Sulina, Braţul** ≈ SE Romania
100 *H12* **Sulingen** Niedersachsen, NW Germany *52°40′N 08°48′E*
Sulisjielmmá *see* Sulitjelma
92 *H12* **Suliskongen** ▲ C Norway *67°10′N 16°16′E*
92 *H12* **Sulitjelma** *Lapp.* Sulisjielmmá. Nordland, C Norway *67°10′N 16°05′E*
56 *A9* **Sullana** Piura, NW Peru *04°54′S 80°42′W*
23 *N3* **Sulligent** Alabama, S USA *33°54′N 88°07′W*
30 *M14* **Sullivan** Illinois, N USA *39°36′N 88°36′W*
31 *N15* **Sullivan** Indiana, N USA *39°05′N 87°24′W*
27 *W5* **Sullivan** Missouri, C USA *38°12′N 91°09′W*
Sullivan Island *see* Lanbi Kyun
96 *M1* **Sullom Voe** NE Scotland, United Kingdom *60°24′N 01°09′W*
103 *O7* **Sully-sur-Loire** Loiret, C France *47°46′N 02°21′E*
Sulmo *see* Sulmona
107 *K15* **Sulmona** *anc.* Sulmo. Abruzzo, C Italy *42°03′N 13°56′E*
Sulo *see* Shule He
114 *M11* **Süloğlu** Edirne, NW Turkey *41°46′N 26°55′E*
22 *G9* **Sulphur** Louisiana, S USA *30°14′N 93°22′W*
27 *O12* **Sulphur** Oklahoma, C USA *34°31′N 96°58′W*
28 *K9* **Sulphur Creek** ≈ South Dakota, N USA
24 *M5* **Sulphur Draw** ≈ Texas, SW USA
25 *W5* **Sulphur River** ≈ Arkansas/Texas, SW USA
25 *V6* **Sulphur Springs** Texas, SW USA *33°09′N 95°36′W*
24 *M6* **Sulphur Springs Draw** ≈ Texas, SW USA
Sulsel *see* Sulawesi Selatan
14 *D8* **Sultan** Ontario, S Canada *47°34′N 82°45′W*
Sultānābād *see* Arāk
Sultan Alonto, Lake *see* Lanao, Lake
136 *G15* **Sultan Dağları** ▲▲ C Turkey
114 *N13* **Sultanköy** Tekirdağ, NW Turkey *41°01′N 27°58′E*
171 *Q7* **Sultan Kudarat** *var.* Nuling. Mindanao, S Philippines *07°20′N 124°16′E*
152 *M13* **Sultānpur** Uttar Pradesh, N India *26°15′N 82°04′E*
Sulteng *see* Sulawesi Tengah
Sultenggara *see* Sulawesi Tenggara
171 *O9* **Sulu Archipelago** *island group* SW Philippines
192 *F7* **Sulu Basin** *undersea feature* SE South China Sea *08°00′N 121°30′E*
Sülüktü *see* Sulyukta
Sulu, Laut *see* Sulu Sea
169 *X6* **Sulu Sea** *var.* Laut Sulu. *sea* SW Philippines
Sulut *see* Sulawesi Utara
145 *O15* **Sulutobe** *Kaz.* Sulūtöbe. Kzylorda, S Kazakhstan *44°31′N 66°17′E*
Sulūtöbe *see* Sulutobe
147 *Q11* **Sulyukta** *Kir.* Sülüktü. Batkenskaya Oblast', SW Kyrgyzstan *39°57′N 69°31′E*
Sulz *see* Sulz am Neckar
101 *G22* **Sulz am Neckar** *var.* Sulz. Baden-Württemberg, SW Germany *48°22′N 08°37′E*
101 *L20* **Sulzbach-Rosenberg** Bayern, SE Germany *49°30′N 11°43′E*
195 *N13* **Sulzberger Bay** *bay* Antarctica
81 *M14* **Sumalē** *var.* Somali. ◐ E Ethiopia
113 *F15* **Sumartin** Split-Dalmacija, S Croatia *43°17′N 16°52′E*
32 *H6* **Sumas** Washington, NW USA *49°00′N 122°15′W*
168 *J10* **Sumatera** *Eng.* Sumatra. *island* W Indonesia
168 *J12* **Sumatera Barat** *off.* Propinsi Sumatera Barat, *var.* Sumbar, *Eng.* West Sumatra. ◐ *province* W Indonesia
Sumatera Barat, Propinsi *see* Sumatera Barat
168 *L13* **Sumatera Selatan** *off.* Propinsi Sumatera Selatan, *var.* Sumsel, *Eng.* South Sumatra. ◐ *province* W Indonesia
Sumatera Selatan, Propinsi *see* Sumatera Selatan
168 *H10* **Sumatera Utara** *off.* Propinsi Sumatera Utara, *var.* Sumut, *Eng.* North Sumatra. ◐ *province* W Indonesia
Sumatera Utara, Propinsi *see* Sumatera Utara
Sumatra *see* Sumatera
Šumava *see* Bohemian Forest
Sumayl *see* Sêmêl
139 *U7* **Sumayr al Muḩammad** Diyālá, E Iraq *33°34′N 45°06′E*
171 *N17* **Sumba, Pulau** *Eng.* Sandalwood Island; *prev.* Soemba. *island* Nusa Tenggara, C Indonesia
146 *D12* **Sumbar** ≈ W Turkmenistan
Sumbar *see* Sumatera Barat
192 *E9* **Sumbawa** *prev.* Soembawa. *island* Nusa Tenggara, C Indonesia
170 *L16* **Sumbawabesar** Sumbawa, S Indonesia *08°30′S 117°25′E*
81 *F23* **Sumbawanga** Rukwa, W Tanzania *07°57′S 31°37′E*
82 *B12* **Sumbe** *var.* N'Gunza, *Port.* Novo Redondo. Kwanza Sul, W Angola *11°13′S 13°53′E*
96 *M3* **Sumburgh Head** *headland* NE Scotland, United Kingdom *59°51′N 01°16′W*
111 *H23* **Sümeg** Veszprém, W Hungary *47°01′N 17°13′E*
80 *C12* **Sumeih** Eastern Darfur, S Sudan *09°50′N 27°39′E*
169 *T16* **Sumenep** *prev.* Soemenep. Pulau Madura, C Indonesia *07°01′S 113°51′E*
Sumgait *see* Sumqayıtçay, Azerbaijan
Sumgait *see* Sumqayıt, Azerbaijan
165 *Y14* **Sumisu-jima** *Eng.* Smith Island. *island* SE Japan
Summēl *see* Sêmêl
31 *O5* **Summer Island** *island* Michigan, N USA
32 *H15* **Summer Lake** ◎ Oregon, NW USA
11 *N17* **Summerland** British Columbia, SW Canada *49°35′N 119°45′W*
13 *P14* **Summerside** Prince Edward Island, SE Canada *46°24′N 63°46′W*
21 *R5* **Summersville** West Virginia, NE USA *38°17′N 80°52′W*
21 *R5* **Summersville Lake** ⊠ West Virginia, NE USA
21 *S13* **Summerton** South Carolina, SE USA *33°36′N 80°21′W*
23 *R2* **Summerville** Georgia, SE USA *34°28′N 85°21′W*
21 *S14* **Summerville** South Carolina, SE USA *33°01′N 80°10′W*
39 *R10* **Summit** Alaska, USA *63°21′N 148°50′W*
35 *V6* **Summit Mountain** ▲ Nevada, W USA *39°23′N 116°25′W*
37 *R8* **Summit Peak** ▲ Colorado, C USA *37°21′N 106°42′W*
Summus Portus *see* Somport, Col du
29 *X12* **Sumner** Iowa, C USA *42°51′N 92°05′W*
22 *K3* **Sumner** Mississippi, S USA *33°58′N 90°22′W*
185 *H17* **Sumner, Lake** ◎ South Island, New Zealand
37 *U12* **Sumner, Lake** ⊠ New Mexico, SW USA
111 *G17* **Šumperk** *Ger.* Mährisch-Schönberg. Olomoucký Kraj, E Czech Republic *49°58′N 17°00′E*
42 *F7* **Sumpul, Río** ≈ El Salvador/Honduras
137 *Z11* **Sumqayıt** *Rus.* Sumgait. E Azerbaijan *40°33′N 49°41′E*
137 *Y11* **Sumqayıtçay** *Rus.* Sumgait. ≈ E Azerbaijan
147 *R9* **Sumsar** Dzhalal-Abadskaya Oblast', W Kyrgyzstan *41°12′N 71°16′E*
Sumsel *see* Sumatera Selatan
117 *S3* **Sums'ka Oblast'** *var.* Sumy, *Rus.* Sumskaya Oblast'. ◐ *province* NE Ukraine
Sumskaya Oblast' *see* Sums'ka Oblast'
124 *J8* **Sumskiy Posad** Respublika Kareliya, NW Russian Federation *64°12′N 35°22′E*
21 *S12* **Sumter** South Carolina, SE USA *33°54′N 80°22′W*
Sumut *see* Sumatera Utara
117 *T3* **Sumy** Sums'ka Oblast', NE Ukraine *50°54′N 34°49′E*
Sumy *see* Sums'ka Oblast'
159 *Q15* **Sumzom** Xizang Zizhiqu, W China *29°45′N 96°14′E*
125 *R15* **Suna** Kirovskaya Oblast', NW Russian Federation *57°53′N 50°04′E*
124 *I10* **Suna** ≈ NW Russian Federation
165 *S3* **Sunagawa** Hokkaidō, NE Japan *43°30′N 141°55′E*
153 *V13* **Sunamganj** Sylhet, NE Bangladesh *25°04′N 91°24′E*
163 *W14* **Sunan** ✈ (P'yŏngyang) SW North Korea *39°12′N 125°40′E*
Sunan/Sunan Yugurzu Zizhixian *see* Hongwansi
19 *N9* **Sunapee Lake** ◎ New Hampshire, NE USA
139 *P4* **Sunaysilah** *salt marsh* N Iraq
20 *M8* **Sunbright** Tennessee, S USA *36°12′N 84°39′W*
33 *R6* **Sunburst** Montana, NW USA *48°51′N 111°54′W*
183 *N12* **Sunbury** Victoria, SE Australia *37°36′S 114°45′E*
21 *X8* **Sunbury** North Carolina, SE USA *36°27′N 76°34′W*
18 *G14* **Sunbury** Pennsylvania, NE USA *40°51′N 76°47′W*
61 *A17* **Sunchales** Santa Fe, C Argentina *30°58′S 61°35′W*
163 *Y16* **Suncheon** *prev.* Sunch'ŏn. S South Korea *34°56′N 127°29′E*
Sunch'ŏn *see* Suncheon
36 *K13* **Sun City** Arizona, SW USA *33°36′N 112°16′W*
19 *O9* **Suncook** New Hampshire, NE USA *43°07′N 71°25′W*
161 *P5* **Suncun** *prev.* Xinwen. Shandong, E China *35°49′N 117°36′E*
Sunda Islands *see* Greater Sunda Islands
33 *Z12* **Sundance** Wyoming, C USA *44°24′N 104°22′W*
153 *T17* **Sundarbans** *wetland* Bangladesh/India
154 *M11* **Sundargarh** Odisha, E India *22°07′N 84°02′E*
129 *U15* **Sunda Shelf** *undersea feature* S South China Sea *05°00′N 107°00′E*
Sunda Trench *see* Java Trench
129 *U17* **Sunda Trough** *undersea feature* E Indian Ocean *08°50′S 109°30′E*
95 *O16* **Sundbyberg** Stockholm, C Sweden *59°22′N 17°58′E*
97 *M14* **Sunderland** *var.* Wearmouth. NE England, United Kingdom *54°55′N 01°23′W*
101 *F15* **Sundern** Nordrhein-Westfalen, W Germany *51°19′N 08°00′E*
136 *F12* **Sündiken Dağları** ▲▲ C Turkey
24 *M5* **Sundown** Texas, SW USA *33°27′N 102°29′W*
11 *P16* **Sundre** Alberta, SW Canada *51°49′N 114°46′W*
14 *H12* **Sundridge** Ontario, S Canada *45°45′N 79°25′W*
93 *H17* **Sundsvall** Västernorrland, C Sweden *62°22′N 17°20′E*
26 *H4* **Sunflower, Mount** ▲ Kansas, C USA *39°01′N 102°02′W*
Sunflower State *see* Kansas
169 *N14* **Sungaibuntu** Sumatera, SW Indonesia *04°04′S 105°37′E*
168 *K12* **Sungaidareh** Sumatera, W Indonesia *0°58′S 101°30′E*
167 *P17* **Sungai Kolok** *var.* Sungai Ko-Lok. Narathiwat, SW Thailand *06°02′N 101°58′E*
Sungai Ko-Lok *see* Sungai Kolok
168 *K12* **Sungaipenuh** *prev.* Soengaipenoeh. Sumatera, W Indonesia *02°S 101°28′E*
169 *P11* **Sungaipinyuh** Borneo, C Indonesia *0°16′N 109°06′E*
Sungari *see* Songhua Jiang
Sungaria *see* Dzungaria
Sungei Pahang *see* Pahang, Sungai
167 *O8* **Sung Men** Phrae, NW Thailand *17°59′N 100°07′E*
83 *M15* **Sungo** Tete, NW Mozambique *16°31′S 33°58′E*
Sungpu *see* Songpan
168 *M13* **Sungsang** Sumatera, W Indonesia *02°22′S 104°50′E*
114 *M9* **Sungurlare** Burgas, E Bulgaria *42°47′N 26°46′E*
136 *J12* **Sungurlu** Çorum, N Turkey *40°10′N 34°23′E*
112 *F9* **Sunja** Sisak-Moslavina, C Croatia *45°21′N 16°33′E*
153 *Q12* **Sun Koshī** ≈ E Nepal
94 *F9* **Sunndalen** *valley* S Norway
94 *F9* **Sunndalsøra** Møre og Romsdal, S Norway *62°39′N 08°37′E*
95 *K15* **Sunne** Värmland, C Sweden *59°52′N 13°05′E*
95 *O15* **Sunnersta** Uppsala, C Sweden *59°46′N 17°40′E*
94 *C11* **Sunnfjord** *physical region* S Norway
95 *C15* **Sunnhordland** *physical region* S Norway
94 *D10* **Sunnmøre** *physical region* S Norway
37 *N4* **Sunnyside** Utah, W USA *39°33′N 110°23′W*
32 *J10* **Sunnyside** Washington, NW USA *46°01′N 119°58′W*
35 *N9* **Sunnyvale** California, W USA *37°22′N 122°02′W*
30 *L8* **Sun Prairie** Wisconsin, N USA *43°12′N 89°12′W*
Sunqur *see* Sonqor
25 *N1* **Sunray** Texas, SW USA *36°01′N 101°49′W*
22 *I8* **Sunset** Louisiana, S USA *30°24′N 92°04′W*
25 *S5* **Sunset** Texas, SW USA *33°24′N 97°45′W*
Sunset State *see* Oregon
181 *Z10* **Sunshine Coast** *cultural region* Queensland, E Australia
Sunshine State *see* Florida
Sunshine State *see* New Mexico
Sunshine State *see* South Dakota
123 *O10* **Suntar** Respublika Sakha (Yakutiya), NE Russian Federation *62°10′N 117°34′E*
39 *R10* **Suntrana** Alaska, USA *63°51′N 148°51′W*
148 *J15* **Suntsar** Baluchistān, SW Pakistan *25°30′N 62°03′E*
163 *W15* **Sunwi-do** *island* SW North Korea
163 *W6* **Sunwu** Heilongjiang, NE China *49°29′N 127°15′E*
77 *O16* **Sunyani** W Ghana *07°22′N 02°18′W*
Suō *see* Su'ao
93 *M17* **Suolahti** Keski-Suomi, C Finland *62°32′N 25°51′E*
Suoločielgi *see* Saariselkä
Suomenlahti *see* Finland, Gulf of
Suomen Tasavalta/Suomi *see* Finland
93 *N14* **Suomussalmi** Kainuu, E Finland *64°54′N 29°05′E*
165 *E13* **Suō-nada** *sea* SW Japan
93 *M17* **Suonenjoki** Pohjois-Savo, C Finland *62°36′N 27°07′E*
167 *S13* **Suŏng** Kâmpóng Cham, C Cambodia *11°53′N 105°41′E*
124 *I10* **Suoyarvi** Respublika Kareliya, NW Russian Federation *62°02′N 32°24′E*
Supanburi *see* Suphan Buri
57 *D14* **Supe** Lima, W Peru *10°49′S 77°40′W*
15 *V7* **Supérieur, Lac** ◎ Québec, SE Canada
Supérieur, Lac *see* Superior, Lake
36 *M14* **Superior** Arizona, SW USA *33°17′N 111°06′W*
33 *O9* **Superior** Montana, NW USA *47°11′N 114°53′W*
29 *P17* **Superior** Nebraska, C USA *40°01′N 98°04′W*
30 *I3* **Superior** Wisconsin, N USA *46°42′N 92°04′W*
41 *S17* **Superior, Laguna** *lagoon* S Mexico
31 *N2* **Superior, Lake** *Fr.* Lac Supérieur. ◎ Canada/USA
36 *L13* **Superstition Mountains** ▲▲ Arizona, SW USA
113 *F14* **Supetar** *It.* San Pietro. Split-Dalmacija, S Croatia *43°22′N 16°34′E*
167 *O10* **Suphan Buri** *var.* Supanburi. Suphan Buri, W Thailand *14°29′N 100°10′E*
171 *V12* **Supiori, Pulau** *island* E Indonesia
188 *K2* **Supply Reef** *reef* N Northern Mariana Islands
195 *O7* **Support Force Glacier** *glacier* Antarctica
137 *R10* **Supsa** *prev.* Sup'sa. ≈ W Georgia
Sup'sa *see* Supsa
Sūq 'Abs *see* 'Abs
139 *W12* **Sūq ash Shuyūkh** Dhī Qār, SE Iraq *30°53′N 46°28′E*
138 *H4* **Şuqaylibīyah** Ḩamāh, W Syria *35°21′N 36°24′E*
161 *Q6* **Suqian** Jiangsu, E China *33°57′N 118°18′E*
Suqrah *see* Şawqirah
Suqrah Bay *see* Şawqirah, Dawḩat
141 *V16* **Suquţrā** *var.* Sokotra, *Eng.* Socotra. *island* SE Yemen
141 *Z8* **Şūr** NE Oman *22°32′N 59°33′E*
Şūr *see* Soûr
127 *P5* **Sura** Penzenskaya Oblast', W Russian Federation *53°23′N 45°03′E*
127 *P4* **Sura** ≈ W Russian Federation
149 *N12* **Sūrāb** Baluchistān, SW Pakistan *28°28′N 66°15′E*
Surabaja *see* Surabaya
192 *E8* **Surabaya** *prev.* Surabaja, Soerabaja. Jawa, C Indonesia *07°14′S 112°45′E*
95 *N15* **Surahammar** Västmanland, C Sweden *59°43′N 16°13′E*
169 *Q16* **Surakarta** *Eng.* Solo; *prev.* Soerakarta. Jawa, S Indonesia *07°32′S 110°50′E*
Surakhany *see* Suraxanı
137 *S10* **Surami** C Georgia *41°59′N 43°36′E*
143 *X13* **Sūrān** Sīstān va Balūchestān, SE Iran *27°18′N 61°58′E*
111 *I21* **Šurany** *Hung.* Nagysurány. Nitriansky Kraj, SW Slovakia *48°05′N 18°10′E*
154 *D12* **Sūrat** Gujarāt, W India *21°10′N 72°54′E*
Suratdhani *see* Surat Thani
152 *G9* **Sūratgarh** Rājasthān, NW India *29°20′N 73°59′E*
167 *N14* **Surat Thani** *var.* Suratdhani. Surat Thani, SW Thailand *09°09′N 99°20′E*
119 *Q16* **Suraw** *Rus.* Surov. ≈ E Belarus
137 *Z11* **Suraxanı** *Rus.* Surakhany. E Azerbaijan *40°25′N 49°59′E*
141 *Y11* **Surayr** E Oman *19°56′N 57°47′E*
138 *K2* **Suraysāt** Ḩalab, N Syria *36°42′N 38°01′E*
118 *O12* **Surazh** Vitsyebskaya Voblasts', NE Belarus *55°25′N 30°44′E*
126 *H6* **Surazh** Bryanskaya Oblast', W Russian Federation *53°04′N 32°29′E*
191 *V17* **Sur, Cabo** *headland* Easter Island, Chile, E Pacific Ocean *27°11′S 109°26′W*
112 *L11* **Surčin** Serbia, N Serbia *44°48′N 20°19′E*
116 *H9* **Surduc** *Hung.* Szurduk. Sălaj, NW Romania *47°13′N 23°20′E*
113 *P16* **Surdulica** Serbia, SE Serbia *42°43′N 22°10′E*
99 *L24* **Sûre** *var.* Sauer. ≈ W Europe *see also* Sauer
Sûre *see* Sauer
154 *C10* **Surendranagar** Gujarāt, W India *22°44′N 71°43′E*
18 *K16* **Surf City** New Jersey, NE USA *39°21′N 74°24′W*
183 *V3* **Surfers Paradise** Queensland, E Australia *27°54′S 153°18′E*
21 *U13* **Surfside Beach** South Carolina, SE USA *33°36′N 78°58′W*
102 *J10* **Surgères** Charente-Maritime, W France *46°07′N 00°44′W*
122 *H10* **Surgut** Khanty-Mansiyskiy Avtonomnyy Okrug-Yugra, C Russian Federation *61°13′N 73°28′E*
122 *K10* **Surgutikha** Krasnoyarskiy Kray, N Russian Federation *64°44′N 87°13′E*
98 *M6* **Surhuisterveen** *Fris.* Surhusterfean. Fryslân, N Netherlands *53°11′N 06°10′E*
Surhusterfean *see* Surhuisterveen
105 *V5* **Súria** Cataluña, NE Spain *41°50′N 01°45′E*
143 *P10* **Sūrīān** Fārs, S Iran
155 *J15* **Suriāpet** Telangana, C India *17°10′N 79°42′E*
171 *Q6* **Surigao** Mindanao, S Philippines *09°43′N 125°31′E*
167 *R10* **Surin** Surin, E Thailand *14°53′N 103°29′E*
Surinam *see* Suriname
55 *U11* **Suriname** *off.* Republic of Suriname, *var.* Surinam; *prev.* Dutch Guiana, Netherlands Guiana. ◆ *republic* N South America
Suriname, Republic of *see* Suriname
Sūrīya/Sūrīyah, Al-Jumhūrīyah al-'Arabīyah as- *see* Syria
Surkhab, Darya-i- *see* Kahmard, Daryā-ye
Surkhandar'inskaya Oblast' *see* Surxondaryo Viloyati
Surkhandar'ya *see* Surxondaryo
Surkhet *see* Birendranagar
147 *R12* **Surkhob** ≈ C Tajikistan
137 *P11* **Sürmene** Trabzon, NE Turkey *40°56′N 40°03′E*
Surov *see* Suraw
127 *N11* **Surovikino** Volgogradskaya Oblast', SW Russian Federation *48°39′N 42°46′E*

◆ Country
● Country Capital
◇ Dependent Territory
○ Dependent Territory Capital
◐ Administrative Regions
✈ International Airport
▲ Mountain
▲▲ Mountain Range
⛰ Volcano
≈ River
◎ Lake
⊠ Reservoir

Sur, Point – Tahanea

35 N11 **Sur, Point** *headland* California, W USA *36°18′N 121°54′W*
187 N15 **Surprise, Île** *island* N New Caledonia
61 E22 **Sur, Punta** *headland* E Argentina *50°59′S 69°10′W*
Surrentum *see* Sorrento
28 M3 **Surrey** North Dakota, N USA *48°13′N 101°05′W*
97 O22 **Surrey** *cultural region* SE England, United Kingdom
21 X7 **Surry** Virginia, NE USA *37°08′N 81°34′W*
108 F8 **Sursee** Luzern, W Switzerland *47°11′N 08°07′E*
127 P6 **Sursk** Penzenskaya Oblast', W Russian Federation *53°06′N 45°46′E*
127 P5 **Surskoye** Ul'yanovskaya Oblast', W Russian Federation *54°28′N 46°47′E*
75 P8 **Surt** *var.* Sidra, Sirte. N Libya *31°13′N 16°35′E*
95 I19 **Surte** Västra Götaland, S Sweden *57°49′N 12°01′E*
75 Q8 **Surt, Khalīj** *Eng.* Gulf of Sidra, Gulf of Sirti, Sidra. *gulf* N Libya
92 I5 **Surtsey** *island* S Iceland
137 N17 **Suruç** Şanlıurfa, S Turkey *36°58′N 38°24′E*
168 L13 **Surulangun** Sumatera, W Indonesia *02°35′S 102°47′E*
147 P13 **Surxondaryo** *Rus.* Surkhandar'ya. ≈ Tajikistan/Uzbekistan
147 N13 **Surxondaryo Viloyati** *Rus.* Surkhandar'inskaya Oblast'. ◈ *province* S Uzbekistan
Süs *see* Susch
106 A8 **Susa** Piemonte, NE Italy *45°10′N 07°01′E*
165 E12 **Susa** Yamaguchi, Honshū, SW Japan *34°35′N 131°34′E*
Susa *see* Shūsh
113 E16 **Sušac** *It.* Cazza. *island* SW Croatia
Sūsah *see* Sousse
164 G14 **Susaki** Kōchi, Shikoku, SW Japan *33°22′N 133°13′E*
165 I15 **Susami** Wakayama, Honshū, SW Japan *33°32′N 135°32′E*
142 K9 **Süsangerd** *var.* Susangird. Khūzestān, SW Iran *31°40′N 48°06′E*
Susangird *see* Süsangerd
35 P4 **Susanville** California, W USA *40°25′N 120°39′W*
108 J9 **Susch** *var.* Süs. Graubünden, SE Switzerland *46°45′N 10°04′E*
137 N12 **Suşehri** Sivas, N Turkey *40°11′N 38°06′E*
Susiana *see* Khūzestān
111 B18 **Sušice** *Ger.* Schüttenhofen. Plzeňský Kraj, W Czech Republic *49°14′N 13°32′E*
39 R11 **Susitna** Alaska, USA *61°32′N 150°30′W*
39 R11 **Susitna River** ≈ Alaska, USA
127 Q3 **Suslonger** Respublika Mariy El, W Russian Federation *56°18′N 48°16′E*
105 N14 **Suspiro del Moro, Puerto del** *pass* S Spain
18 H16 **Susquehanna River** ≈ New York/Pennsylvania, NE USA
13 O15 **Sussex** New Brunswick, SE Canada *45°43′N 65°32′W*
18 J13 **Sussex** New Jersey, NE USA *41°12′N 74°34′W*
21 W7 **Sussex** Virginia, NE USA *36°54′N 77°16′W*
97 O23 **Sussex** *cultural region* S England, United Kingdom
183 S10 **Sussex Inlet** New South Wales, SE Australia *35°10′S 150°35′E*
99 L17 **Susteren** Limburg, SE Netherlands *51°04′N 05°50′E*
10 K12 **Sustut Peak** ▲ British Columbia, W Canada *56°25′N 126°34′W*
123 S9 **Susuman** Magadanskaya Oblast', E Russian Federation *62°46′N 148°08′E*
188 H6 **Susupe** ● (Northern Mariana Islands-judicial capital) Saipan, S Northern Mariana Islands
136 D12 **Susurluk** Balıkesir, NW Turkey *39°55′N 28°10′E*
114 M13 **Susuzmüsellim** Tekirdağ, NW Turkey *41°04′N 27°03′E*
136 F15 **Sütçüler** Isparta, SW Turkey *37°31′N 31°00′E*
116 L13 **Şuţeşti** Brăila, SE Romania *45°13′N 27°27′E*
83 F25 **Sutherland** Western Cape, SW South Africa *32°24′S 20°40′E*
28 L15 **Sutherland** Nebraska, C USA *41°09′N 101°07′W*
96 I7 **Sutherland** *cultural region* N Scotland, United Kingdom
185 B21 **Sutherland Falls** *waterfall* South Island, New Zealand
32 F14 **Sutherlin** Oregon, NW USA *43°23′N 123°18′W*
149 V10 **Sutlej** ≈ India/Pakistan
Sutna *see* Satna
35 P7 **Sutter Creek** California, W USA *38°22′N 120°49′W*
39 R11 **Sutton** Alaska, USA *61°42′N 148°53′W*
29 Q16 **Sutton** Nebraska, C USA *40°36′N 97°52′W*
21 R4 **Sutton** West Virginia, NE USA *38°41′N 80°43′W*
12 F8 **Sutton** ≈ Ontario, C Canada
97 M19 **Sutton Coldfield** C England, United Kingdom *52°34′N 01°48′W*
21 R4 **Sutton Lake** ⊠ West Virginia, NE USA
15 P13 **Sutton, Monts** *hill range* Québec, SE Canada
12 F8 **Sutton Ridges** ▲▲ Ontario, C Canada
165 Q4 **Suttsu** Hokkaidō, NE Japan *42°46′N 140°12′E*
39 P15 **Sutwik Island** *island* Alaska, USA
Süüj *see* Dashinchilen
118 H5 **Suure-Jaani** *Ger.* Gross-Sankt-Johannis. Viljandimaa, S Estonia *58°34′N 25°28′E*
118 J7 **Suur Munamägi** *var.* Munamägi, *Ger.* Eier-Berg. ▲ SE Estonia *57°42′N 27°03′E*
118 F5 **Suur Väin** *Ger.* Grosser Sund. *strait* W Estonia
147 U8 **Suusamyr** Chuyskaya Oblast', C Kyrgyzstan *42°07′N 73°55′E*
187 X14 **Suva** ● (Fiji) Viti Levu, W Fiji *18°08′S 178°27′E*
187 X15 **Suva** ✕ Viti Levu, C Fiji *18°01′S 178°30′E*
113 N18 **Suva Gora** ▲▲ W FYR Macedonia
118 H11 **Suvainiškis** Panevėžys, NE Lithuania *56°09′N 25°15′E*
Suvalkai/Suvalki *see* Suwałki
113 P15 **Suva Planina** ▲▲ SE Serbia
Suva Reka *see* Suharekë
126 K5 **Suvorov** Tul'skaya Oblast', W Russian Federation *54°08′N 36°33′E*
117 N12 **Suvorove** Odes'ka Oblast', SW Ukraine *45°35′N 28°58′E*
114 M8 **Suvorovo** Varna, E Bulgaria *43°19′N 27°26′E*
Suvorovo *see* Ştefan Vodă
Suwaik *see* As Suwayq
Suwaira *see* Aş Şuwayrah
110 O7 **Suwałki** *Lith.* Suvalkai, *Rus.* Suvalki. Podlaskie, NE Poland *54°06′N 22°56′E*
167 R10 **Suwannaphum** Roi Et, E Thailand *15°36′N 103°46′E*
23 V8 **Suwannee River** ≈ Florida/Georgia, SE USA
Şuwār *see* Aş Şuwār
190 K14 **Suwarrow** *atoll* N Cook Islands
143 R16 **Suwaydān** *var.* Sweiham. Abū Ẓaby, E United Arab Emirates *24°30′N 55°19′E*
Suwaydá/Suwaydā', Muḥāfaẓat as *see* As Suwaydā'
Suwayqiyah, Hawr as *see* Shuwayjah, Hawr ash
Suways, Qanāt as *see* Suez Canal
Suweida *see* As Suwaydā'
Suweon *see* Suwon
163 X15 **Suwon** *var.* Suweon; *prev.* Suwŏn, *Jap.* Suigen. NW South Korea *37°17′N 127°03′E*
Suwŏn *see* Suwon
Su Xian *see* Suzhou
143 R14 **Sūzā** Hormozgān, S Iran *26°50′N 56°05′E*
Suzak *see* Sozak
165 K14 **Suzaka** Mie, Honshū, SW Japan *34°52′N 136°37′E*
165 N12 **Suzaka** Nagano, Honshū, S Japan *36°38′N 138°20′E*
126 M3 **Suzdal'** Vladimirskaya Oblast', W Russian Federation *56°27′N 40°29′E*
161 P7 **Suzhou** *var.* Su Xian. Anhui, E China *33°38′N 117°02′E*
161 R8 **Suzhou** *var.* Soochow, Su-chou, Suchow; *prev.* Wuhsien. Jiangsu, E China *31°23′N 120°34′E*
Suzhou *see* Jiuquan
163 V12 **Suzi He** ≈ NE China
Suz, Mys *see* Soye, Mys
165 M10 **Suzu** Ishikawa, Honshū, SW Japan *37°24′N 137°12′E*
165 M10 **Suzu-misaki** *headland* Honshū, SW Japan *37°31′N 137°19′E*
Svågälv *see* Svågan
94 M10 **Svågan** *var.* Svågälv. ≈ C Sweden
Svalava/Svaljava *see* Svalyava
92 O2 **Svalbard** ◇ *constituent part of Norway* Arctic Ocean
92 J2 **Svalbarðseyri** Norðurland Eystra, N Iceland *65°43′N 18°03′W*
95 K22 **Svalöv** Skåne, S Sweden *55°55′N 13°06′E*
116 H7 **Svalyava** *Cz.* Svalava, Svaljava, *Hung.* Szolyva. Zakarpats'ka Oblast', W Ukraine *48°33′N 23°00′E*
92 O2 **Svanbergfjellet** ▲ C Svalbard *78°40′N 18°10′E*
95 M24 **Svaneke** Bornholm, E Denmark *55°07′N 15°08′E*
95 L22 **Svängsta** Blekinge, S Sweden *56°16′N 14°46′E*
95 J16 **Svanskog** Värmland, C Sweden *59°10′N 12°34′E*
95 L16 **Svartå** Örebro, C Sweden *59°13′N 14°07′E*
95 L15 **Svartälven** ≈ C Sweden
92 G12 **Svartisen** *glacier* C Norway
117 X6 **Svatove** *Rus.* Svatovo. Luhans'ka Oblast', E Ukraine *49°24′N 38°11′E*
Svatovo *see* Svatove
Sväty Kríž nad Hronom *see* Žiar nad Hronom
167 Q11 **Svay Chék, Stœng** ≈ Cambodia/Thailand
167 S13 **Svay Riĕng** Svay Riĕng, S Cambodia *11°05′N 105°48′E*
92 O3 **Sveagruva** Spitsbergen, W Svalbard *77°53′N 16°42′E*
95 K23 **Svedala** Skåne, S Sweden *55°30′N 13°15′E*
118 H12 **Svėdesai** Utena, NE Lithuania *55°42′N 25°22′E*
93 G18 **Sveg** Jämtland, C Sweden *62°02′N 14°20′E*
118 C12 **Švėkšna** Klaipėda, W Lithuania *55°31′N 21°37′E*
94 C11 **Svelgen** Sogn Og Fjordane, S Norway *61°47′N 05°18′E*
95 H15 **Svelvik** Vestfold, S Norway *59°37′N 10°24′E*
118 I13 **Švenčionėliai** *Pol.* Nowo-Święciany. Vilnius, SE Lithuania *55°10′N 26°00′E*
118 I13 **Švenčionys** *Pol.* Swięciany. Vilnius, SE Lithuania *55°08′N 26°08′E*
95 H24 **Svendborg** Syddtjylland, C Denmark *55°04′N 10°38′E*
95 K19 **Svenljunga** Västra Götaland, S Sweden *57°30′N 13°05′E*
92 P2 **Svenskøya** *island* E Svalbard
93 G17 **Svenstavik** Jämtland, C Sweden *62°40′N 14°24′E*
95 G20 **Svenstrup** Nordjylland, N Denmark *56°58′N 09°52′E*
118 H12 **Šventoji** ≈ C Lithuania
117 Z8 **Sverdlovs'k** *Rus.* Sverdlovsk; *prev.* Imeni Sverdlova Rudnik. Luhans'ka Oblast', E Ukraine *48°05′N 39°37′E*
Sverdlovsk *see* Yekaterinburg
127 W2 **Sverdlovskaya Oblast'** ◈ *province* C Russian Federation
122 K6 **Sverdrupa, Ostrov** *island* N Russian Federation
Sverige *see* Sweden
113 D15 **Svetac** *prev.* Sveti Andrea, *It.* Sant'Andrea. *island* SW Croatia
Sveti Andrea *see* Svetac
Sveti Nikola *see* Sveti Nikole
113 O18 **Sveti Nikole** *prev.* Sveti Nikola. C FYR Macedonia *41°54′N 21°55′E*
Sveti Vrach *see* Sandanski
123 T14 **Svetlaya** Primorskiy Kray, SE Russian Federation *46°33′N 138°20′E*
126 B2 **Svetlogorsk** Kaliningradskaya Oblast', W Russian Federation *54°56′N 20°09′E*
122 K9 **Svetlogorsk** Krasnoyarskiy Kray, N Russian Federation *66°51′N 88°29′E*
Svetlogorsk *see* Svyetlahorsk
127 N14 **Svetlograd** Stavropol'skiy Kray, SW Russian Federation *45°20′N 42°53′E*
Svetlovodsk *see* Svitlovods'k
119 A14 **Svetlyy** *Ger.* Zimmerbude. Kaliningradskaya Oblast', W Russian Federation *54°42′N 20°07′E*
127 Y8 **Svetlyy** Orenburgskaya Oblast', W Russian Federation *50°34′N 60°42′E*
127 P7 **Svetlyy** Saratovskaya Oblast', W Russian Federation *51°42′N 45°40′E*
124 G11 **Svetogorsk** *Fin.* Enso. Leningradskaya Oblast', NW Russian Federation *61°06′N 28°52′E*
Svetozarevo *see* Jagodina
111 B18 **Švihov** *Ger.* Schwihau. Plzeňský Kraj, W Czech Republic *49°31′N 13°18′E*
112 E13 **Svilaja** ▲▲ SE Croatia
112 N12 **Svilajnac** Serbia, C Serbia *44°15′N 21°12′E*
114 L11 **Svilengrad** *prev.* Mustafa-Pasha. Haskovo, S Bulgaria *41°45′N 26°14′E*
Svinecea Mare, Munte *see* Svinecea Mare, Vârful
116 F13 **Svinecea Mare, Vârful** *var.* Munte Svinecea Mare. ▲ SW Romania *44°47′N 22°10′E*
Svinø *see* Svínoy
95 B18 **Svínoy** *Dan.* Svinø. *island* NE Faroe Islands
147 N14 **Svintsovyy Rudnik** *Turkm.* Swintsowyy Rudnik. Lebap Welaýaty, E Turkmenistan *37°54′N 66°25′E*
118 I13 **Svir** *Rus.* Svir'. Minskaya Voblasts', NW Belarus *54°51′N 26°24′E*
124 I12 **Svir'** *canal* NW Russian Federation
Svir', Ozero *see* Svir, Vozyera
119 I14 **Svir, Vozyera** *Rus.* Ozero Svir'. ◎ C Belarus
114 J7 **Svishtov** *prev.* Sistova. Veliko Tarnovo, N Bulgaria *43°37′N 25°20′E*
119 F18 **Svislach** *Pol.* Świsłocz, *Rus.* Svisloch'. Hrodzyenskaya Voblasts', W Belarus *53°02′N 24°06′E*
119 M17 **Svislach** *Rus.* Svisloch'. Mahilyowskaya Voblasts', E Belarus *53°26′N 28°59′E*
119 L17 **Svislach** *Rus.* Svisloch'. ≈ E Belarus
Svisloch' *see* Svislach
111 F17 **Svitavy** *Ger.* Zwittau. Pardubický Kraj, C Czech Republic *49°45′N 16°27′E*
117 S6 **Svitlovods'k** *Rus.* Svetlovodsk. Kirovohrads'ka Oblast', C Ukraine *49°05′N 33°15′E*
Svizzera *see* Switzerland
123 Q13 **Svobodnyy** Amurskaya Oblast', SE Russian Federation *51°24′N 128°05′E*
114 G9 **Svoge** Sofia, W Bulgaria *42°58′N 23°20′E*
92 G11 **Svolvær** Nordland, C Norway *68°15′N 14°40′E*
111 F18 **Svratka** *Ger.* Schwarzawa. ≈ SE Czech Republic
113 P14 **Svrljig** Serbia, E Serbia *43°26′N 22°07′E*
197 U10 **Svyataya Anna Trough** *var.* Saint Anna Trough. *undersea feature* N Kara Sea
124 M4 **Svyatoy Nos, Mys** *headland* NW Russian Federation *68°07′N 39°49′E*
119 N18 **Svyetlahorsk** *Rus.* Svetlogorsk. Homyel'skaya Voblasts', SE Belarus *52°38′N 29°46′E*
Swabian Jura *see* Schwäbische Alb
97 P19 **Swaffham** E England, United Kingdom *52°39′N 00°40′E*
23 V5 **Swainsboro** Georgia, SE USA *32°36′N 82°19′W*
83 C19 **Swakop** ≈ W Namibia
83 C19 **Swakopmund** Erongo, W Namibia *22°40′S 14°34′E*
97 M15 **Swale** ≈ N England, United Kingdom
Swallow Island *see* Nendö
99 M16 **Swalmen** Limburg, SE Netherlands *51°13′N 06°02′E*
12 G8 **Swan** ≈ Ontario, C Canada
97 L24 **Swanage** S England, United Kingdom *50°37′N 01°59′W*
182 M10 **Swan Hill** Victoria, SE Australia *35°23′S 143°37′E*
11 P13 **Swan Hills** Alberta, W Canada *54°41′N 116°20′W*
65 D24 **Swan Island** *island* C Falkland Islands
Swankalok *see* Sawankhalok
29 U10 **Swan Lake** ◎ Minnesota, N USA
21 Y10 **Swanquarter** North Carolina, SE USA *35°24′N 76°20′W*
182 J9 **Swan Reach** South Australia *34°39′S 139°35′E*
11 V15 **Swan River** Manitoba, S Canada *52°06′N 101°17′W*
183 P17 **Swansea** Tasmania, SE Australia *42°09′S 148°03′E*
97 J22 **Swansea** *Wel.* Abertawe. S Wales, United Kingdom *51°38′N 03°57′W*
21 R13 **Swansea** South Carolina, SE USA *33°47′N 81°06′W*
19 S7 **Swans Island** *island* Maine, NE USA
28 L17 **Swanson Lake** ⊠ Nebraska, C USA
31 R11 **Swanton** Ohio, N USA *41°35′N 83°53′W*
110 G11 **Swarzędz** Poznań, W Poland *52°24′N 17°05′E*
Swatow *see* Shantou
83 L22 **Swaziland** *off.* Kingdom of Swaziland. ◆ *monarchy* S Africa
Swaziland, Kingdom of *see* Swaziland
93 G18 **Sweden** *off.* Kingdom of Sweden, *Swe.* Sverige. ◆ *monarchy* N Europe
Sweden, Kingdom of *see* Sweden
Swedru *see* Agona Swedru
25 V12 **Sweeny** Texas, SW USA *29°02′N 95°42′W*
33 R6 **Sweetgrass** Montana, NW USA *48°58′N 111°58′W*
32 G12 **Sweet Home** Oregon, NW USA *44°24′N 122°44′W*
25 T12 **Sweet Home** Texas, SW USA *29°21′N 97°04′W*
27 T4 **Sweet Springs** Missouri, C USA *38°57′N 93°24′W*
20 M10 **Sweetwater** Tennessee, S USA *35°36′N 84°27′W*
25 P7 **Sweetwater** Texas, SW USA *32°27′N 100°25′W*
33 V15 **Sweetwater River** ≈ Wyoming, C USA
Sweiham *see* Suwaydān
83 F26 **Swellendam** Western Cape, SW South Africa *34°01′S 20°26′E*
111 G15 **Świdnica** *Ger.* Schweidnitz. Wałbrzych, SW Poland *50°51′N 16°29′E*
111 O14 **Świdnik** *Ger.* Streckenbach. Lubelskie, E Poland *51°14′N 22°41′E*
110 F8 **Świdwin** *Ger.* Schivelbein. Zachodnio-pomorskie, NW Poland *53°47′N 15°44′E*
111 F15 **Świebodzice** *Ger.* Freiburg in Schlesien, Swiebodzice. Walbrzych, SW Poland *50°54′N 16°23′E*
110 E11 **Świebodzin** *Ger.* Schwiebus. Lubuskie, W Poland *52°15′N 15°31′E*
Święciany *see* Švenčionys
110 I9 **Świecie** *Ger.* Schwertberg. Kujawsko-pomorskie, C Poland *53°24′N 18°24′E*
111 L15 **Świętokrzyskie** ◈ *province* S Poland
11 T16 **Swift Current** Saskatchewan, S Canada *50°17′N 107°49′W*
98 K9 **Swifterbant** Flevoland, C Netherlands *52°36′N 05°33′E*
183 Q12 **Swifts Creek** Victoria, SE Australia *37°17′S 147°41′E*
96 E13 **Swilly, Lough** *Ir.* Loch Súilí. *inlet* N Ireland
97 M22 **Swindon** S England, United Kingdom *51°34′N 01°47′W*
Swinemünde *see* Świnoujście
110 D8 **Świnoujście** *Ger.* Swinemünde. Zachodnio-pomorskie, NW Poland *53°54′N 14°13′E*
Swintsowyy Rudnik *see* Svintsovyy Rudnik
Świsłocz *see* Svislach
Swiss Confederation *see* Switzerland
108 E9 **Switzerland** *off.* Swiss Confederation, *Fr.* La Suisse, *Ger.* Schweiz, *It.* Svizzera; *anc.* Helvetia. ◆ *federal republic* C Europe
97 F17 **Swords** *Ir.* Sord, Sórd Choluim Chille. Dublin, E Ireland *53°28′N 06°13′W*
18 H13 **Swoyersville** Pennsylvania, NE USA *41°18′N 75°48′W*
139 V3 **Syagwêz** *Ar.* Siyāḥ Gūz. As Sulaymānīyah, E Iraq *35°49′N 45°45′E*
124 I10 **Syamozero, Ozero** ◎ NW Russian Federation
124 M13 **Syamzha** Vologodskaya Oblast', NW Russian Federation *60°02′N 41°09′E*
118 N13 **Syanno** *Rus.* Senno. Vitsyebskaya Voblasts', NE Belarus *54°49′N 29°43′E*
119 K16 **Syarhyeyevichy** *Rus.* Sergeyevichi. Minskaya Voblasts', C Belarus *53°30′N 27°45′E*
124 I12 **Syas'stroy** Leningradskaya Oblast', NW Russian Federation *60°05′N 32°37′E*
Sycaminum *see* Hefa
30 M10 **Sycamore** Illinois, N USA *41°59′N 88°41′W*
126 J3 **Sychëvka** Smolenskaya Oblast', W Russian Federation *55°52′N 34°19′E*
111 H14 **Syców** *Ger.* Gross Wartenberg. Dolnośląskie, SW Poland *51°18′N 17°42′E*
95 F24 **Syddanmark** ◈ *county* SW Denmark
14 E17 **Sydenham** ≈ Ontario, S Canada
Sydenham Island *see* Nonouti
183 T9 **Sydney** *state capital* New South Wales, SE Australia *33°55′S 151°10′E*
13 R14 **Sydney** Cape Breton Island, Nova Scotia, SE Canada *46°10′N 60°10′W*
Sydney Island *see* Manra
13 R14 **Sydney Mines** Cape Breton Island, Nova Scotia, SE Canada *46°14′N 60°19′W*
Syedpur *see* Saidpur
119 K18 **Syelishcha** *Rus.* Selishche. Minskaya Voblasts', C Belarus *53°01′N 27°25′E*
119 J18 **Syemyezhava** *Rus.* Semezhevo. Minskaya Voblasts', C Belarus *52°58′N 27°00′E*
Syene *see* Aswân
117 X6 **Syeverodonets'k** *Rus.* Severodonetsk. Luhans'ka Oblast', E Ukraine *48°59′N 38°28′E*
161 T6 **Syiao Shan** *island* SE China
100 H11 **Syke** Niedersachsen, NW Germany *52°55′N 08°49′E*
94 D10 **Sykkylven** Møre og Romsdal, S Norway *62°23′N 06°35′E*
Sykoúri *see* Sykoúrio
115 F15 **Sykoúrio** *var.* Sikouri, Sykoúri; *prev.* Sikoúrion. Thessalía, C Greece *39°46′N 22°35′E*
125 R11 **Syktyvkar** *prev.* Ust'-Sysol'sk. Respublika Komi, NW Russian Federation *61°42′N 50°45′E*
23 Q4 **Sylacauga** Alabama, S USA *33°10′N 86°15′W*
Sylarna *see* Storsylen
153 V14 **Sylhet** Sylhet, NE Bangladesh *24°53′N 91°51′E*
153 V13 **Sylhet** ◈ *division* NE Bangladesh
100 G6 **Sylt** *island* NW Germany
21 O10 **Sylva** North Carolina, SE USA *35°23′N 83°13′W*
125 V15 **Sylva** ≈ NW Russian Federation
23 W5 **Sylvania** Georgia, SE USA *32°45′N 81°38′W*
31 R11 **Sylvania** Ohio, N USA *41°43′N 83°42′W*
11 Q15 **Sylvan Lake** Alberta, SW Canada *52°18′N 114°02′W*
33 T13 **Sylvan Pass** *pass* Wyoming, C USA
23 T7 **Sylvester** Georgia, SE USA *31°31′N 83°50′W*
25 P6 **Sylvester** Texas, SW USA *32°42′N 100°15′W*
10 L11 **Sylvia, Mount** ▲ British Columbia, W Canada *58°03′N 124°26′W*
122 K11 **Sym** ≈ C Russian Federation
115 N22 **Sými** *var.* Simi. *island* Dodekánisa, Greece, Aegean Sea
117 U8 **Synel'nykove** Dnipropetrovs'ka Oblast', E Ukraine *48°19′N 35°32′E*
125 U6 **Synya** Respublika Komi, NW Russian Federation *65°21′N 58°01′E*
117 P7 **Synyukha** *Rus.* Sinyukha. ≈ S Ukraine
195 V2 **Syowa** *Japanese research station* Antarctica *68°58′S 40°07′E*
26 H6 **Syracuse** Kansas, C USA *38°00′N 101°43′W*
29 S16 **Syracuse** Nebraska, C USA *40°39′N 96°11′W*
18 H10 **Syracuse** New York, NE USA *43°03′N 76°09′W*
Syracuse *see* Siracusa
Syrdar'inskaya Oblast' *see* Sirdaryo Viloyati
Syrdariya *see* Syr Darya
144 L14 **Syr Darya** *var.* Sai Hun, Sir Darya, Syrdarya, *Kaz.* Syrdariya, *Rus.* Syrdar'ya, *Uzb.* Sirdaryo; *anc.* Jaxartes. ≈ C Asia
Syrdarya *see* Syr Darya
138 J6 **Syria** *off.* Syrian Arab Republic, *var.* Siria, Syrie, *Ar.* Al-Jumhūrīyah al-'Arabīyah as-Sūrīyah, Sūrīya. ◆ *republic* SW Asia
Syrian Arab Republic *see* Syria
138 L9 **Syrian Desert** *Ar.* Al Hamad, Bādiyat ash Shām. *desert* SW Asia
Syrie *see* Syria
115 L22 **Sýrna** *var.* Sirna. *island* Kykládes, Greece, Aegean Sea
115 I20 **Sýros** *var.* Síros. *island* Kykládes, Greece, Aegean Sea
93 M18 **Sysmä** Päijät-Häme, S Finland *61°28′N 25°37′E*
125 R12 **Sysola** ≈ NW Russian Federation
Syulemeshlii *see* Sredets
127 S2 **Syumsi** Udmurtskaya Respublika, NW Russian Federation *57°07′N 51°35′E*
Syvash, Zaliv *see* Syvash, Zatoka
117 U12 **Syvash, Zatoka** *Rus.* Zaliv Syvash. *inlet* S Ukraine
127 Q6 **Syzran'** Samarskaya Oblast', W Russian Federation *53°10′N 48°23′E*
Szabadka *see* Subotica
111 N21 **Szabolcs-Szatmár-Bereg** *off.* Szabolcs-Szatmár-Bereg Megye. ◈ *county* E Hungary
Szabolcs-Szatmár-Bereg Megye *see* Szabolcs-Szatmár-Bereg
110 G10 **Szamocin** *Ger.* Samotschin. Wielkopolskie, C Poland *53°02′N 17°04′E*
116 H8 **Szamos** *var.* Someş, Someşul, *Ger.* Samosch, Somesch. ≈ Hungary/Romania
Szamosújvár *see* Gherla
110 G11 **Szamotuły** Poznań, W Poland *52°35′N 16°36′E*
Szarkowszczyzna *see* Sharkawshchyna
111 M24 **Szarvas** Békés, SE Hungary *46°51′N 20°35′E*
Szászmagyarós *see* Măieruş
Szászrégen *see* Reghin
Szászsebes *see* Sebeş
Szászváros *see* Orăştie
Szatmárrnémeti *see* Satu Mare
Száva *see* Sava
111 P15 **Szczebrzeszyn** Lubelskie, E Poland *50°43′N 23°00′E*
110 D9 **Szczecin** *Eng./Ger.* Stettin. Zachodnio-pomorskie, NW Poland *53°25′N 14°32′E*
110 G8 **Szczecinek** *Ger.* Neustettin. Zachodnio-pomorskie, NW Poland *53°43′N 16°40′E*
110 D8 **Szczeciński, Zalew** *var.* Stettiner Haff, *Ger.* Oderhaff. *bay* Germany/Poland
111 K15 **Szczekociny** Śląskie, S Poland *50°38′N 19°46′E*
110 N8 **Szczuczyn** Podlaskie, NE Poland *53°34′N 22°17′E*
Szczuczyn Nowogródzki *see* Shchuchyn
110 M8 **Szczytno** *Ger.* Ortelsburg. Warmińsko-Mazurskie, NE Poland *53°34′N 21°E*
Szechuan/Szechwan *see* Sichuan
111 K21 **Szécsény** Nógrád, N Hungary *48°07′N 19°30′E*
111 L25 **Szeged** *Ger.* Szegedin, *Rom.* Seghedin. Csongrád, SE Hungary *46°17′N 20°06′E*
Szegedin *see* Szeged
111 N23 **Szeghalom** Békés, SE Hungary *47°01′N 21°09′E*
Székelyhid *see* Săcueni
Székelykeresztúr *see* Cristuru Secuiesc
111 I23 **Székesfehérvár** *Ger.* Stuhlweissenberg; *anc.* Alba Regia. Fejér, W Hungary *47°13′N 18°24′E*
Szeklerburg *see* Miercurea-Ciuc
Szekler Neumarkt *see* Târgu Secuiesc
111 J25 **Szekszárd** Tolna, S Hungary *46°21′N 18°41′E*
Szempcz/Szenc *see* Senec
Szenice *see* Senica
Szentágota *see* Agnita
111 J22 **Szentendre** *Ger.* Sankt Andrä. Pest, N Hungary *47°40′N 19°02′E*
111 L24 **Szentes** Csongrád, SE Hungary *46°40′N 20°17′E*
111 F23 **Szentgotthárd** *Eng.* Saint Gotthard, *Ger.* Sankt Gotthard. Vas, W Hungary *46°57′N 16°18′E*
Szentgyörgy *see* Đurđevac
Szenttamás *see* Srbobran
Széphely *see* Jebel
Szeping *see* Siping
Szered *see* Sered'
111 N21 **Szerencs** Borsod-Abaúj-Zemplén, NE Hungary *48°10′N 21°11′E*
Szeret *see* Siret
Szeretfalva *see* Sărăţel
110 N7 **Szeska Góra** *var.* Szeskie Wygórza, *Ger.* Seesker Höhe. *hill* NE Poland
Szeskie Wygórza *see* Szeska Góra
111 H25 **Szigetvár** Baranya, SW Hungary *46°01′N 17°50′E*
Szilágysomlyó *see* Şimleu Silvaniei
Szinna *see* Snina
Sziszek *see* Sisak
Szitás-Keresztúr *see* Cristuru Secuiesc
111 E15 **Szklarska Poręba** *Ger.* Schreiberhau. Dolnośląskie, SW Poland *50°50′N 15°30′E*
Szkudy *see* Skuodas
Szlatina *see* Slatina
Szlavonia/Szlavonország *see* Slavonija
Szlovákia *see* Slovakia
Szluin *see* Slunj
111 L23 **Szolnok** Jász-Nagykun-Szolnok, C Hungary *47°11′N 20°12′E*
Szolyva *see* Svalyava
111 G23 **Szombathely** *Ger.* Steinamanger; *anc.* Sabaria, Savaria. Vas, W Hungary *47°14′N 16°38′E*
Szond/Szonta *see* Sonta
Szováta *see* Sovata
110 F13 **Szprotawa** *Ger.* Sprottau. Lubuskie, W Poland *51°33′N 15°32′E*
Sztálinváros *see* Dunaújváros
Sztrazsó *see* Strážov
110 J8 **Sztum** *Ger.* Stuhm. Pomorskie, N Poland *53°54′N 19°01′E*
110 H10 **Szubin** *Ger.* Schubin. Kujawsko-pomorskie, C Poland *53°04′N 17°49′E*
Szucsava *see* Suceava
Szurduk *see* Surduc
111 M14 **Szydłowiec** *Ger.* Schlelau. Mazowieckie, C Poland *51°14′N 20°50′E*

T

Taalintehdas *see* Dalsbruk
171 O4 **Taal, Lake** ◎ Luzon, NW Philippines
95 J23 **Taastrup** *var.* Tåstrup. Sjælland, E Denmark *55°39′N 12°19′E*
111 I24 **Tab** Somogy, W Hungary *46°45′N 18°01′E*
171 P4 **Tabaco** Luzon, N Philippines *13°22′N 123°42′E*
186 G4 **Tabalo** Mussau Island, NE Papua New Guinea *01°22′S 149°37′E*
104 K5 **Tábara** Castilla y León, N Spain *41°49′N 05°57′W*
186 H5 **Tabar Islands** *island group* NE Papua New Guinea
Tabariya, Bahrat *see* Tiberias, Lake
143 S7 **Ţabas** *var.* Golshan. Yazd, C Iran *33°37′N 56°54′E*
43 P15 **Tabasará, Serranía de** ▲▲ W Panama
41 U15 **Tabasco** ◈ *state* SE Mexico
Tabasco *see* Grijalva, Río
127 Q2 **Tabashino** Respublika Mariy El, W Russian Federation *57°00′N 47°47′E*
58 B13 **Tabatinga** Amazonas, N Brazil *04°14′S 69°44′W*
74 G9 **Tabelbala** W Algeria *29°22′N 03°01′W*
11 Q17 **Taber** Alberta, SW Canada *49°48′N 112°09′W*
171 V15 **Taberfane** Pulau Trangan, E Indonesia *06°14′S 134°08′E*
95 L19 **Taberg** Jönköping, S Sweden *57°42′N 14°05′E*
191 O3 **Tabiteuea** *prev.* Drummond Island. *atoll* Tungaru, W Kiribati
171 O5 **Tablas Island** *island* C Philippines
184 Q10 **Table Cape** *headland* North Island, New Zealand *39°07′S 178°00′E*
13 S13 **Table Mountain** ▲ Newfoundland, Newfoundland and Labrador, E Canada *47°39′N 59°15′W*
173 P17 **Table, Pointe de la** *headland* SE Réunion *21°19′S 55°49′E*
27 S8 **Table Rock Lake** ⊠ Arkansas/Missouri, C USA
36 K14 **Table Top** ▲ Arizona, SW USA *32°45′N 112°07′W*
186 D8 **Tabletop, Mount** ▲ C Papua New Guinea *06°51′S 146°00′E*
111 D18 **Tábor** Jihočeský Kraj, S Czech Republic *49°25′N 14°41′E*
123 R7 **Tabor** Respublika Sakha (Yakutiya), NE Russian Federation *71°14′N 150°23′E*
29 S15 **Tabor** Iowa, C USA *40°54′N 95°40′W*
81 F21 **Tabora** Tabora, W Tanzania *05°04′S 32°49′E*
81 F21 **Tabora** ◈ *region* C Tanzania
21 U12 **Tabor City** North Carolina, SE USA *34°09′N 78°52′W*
147 Q10 **Taboshar** NW Tajikistan *40°37′N 69°33′E*
76 L18 **Tabou** *var.* Tabu. S Ivory Coast *04°28′N 07°20′W*
142 J2 **Tabrīz** *var.* Tebriz; *anc.* Tauris. Āzarbāyjān-e Sharqī, NW Iran *38°05′N 46°18′E*
Tabu *see* Tabou
191 W1 **Tabuaeran** *prev.* Fanning Island. *atoll* Line Islands, E Kiribati
171 O2 **Tabuk** Luzon, N Philippines *17°26′N 121°25′E*
140 J4 **Tabūk** Tabūk, NW Saudi Arabia *28°25′N 36°34′E*
140 J5 **Tabūk** *off.* Minţaqat Tabūk. ◈ *province* NW Saudi Arabia
Tabūk, Minţaqat *see* Tabūk
187 Q13 **Tabwemasana, Mount** ▲ Espiritu Santo, W Vanuatu *15°22′S 166°44′E*
95 O15 **Täby** Stockholm, C Sweden *59°29′N 18°04′E*
41 N14 **Tacámbaro** Michoacán, SW Mexico *19°12′N 101°27′W*
42 A5 **Tacaná, Volcán** ⋀ Guatemala/Mexico *15°07′N 92°06′W*
43 X16 **Tacarcuna, Cerro** ▲ SE Panama *08°08′N 77°15′W*
Tachau *see* Tachov
158 J3 **Tacheng** *var.* Qoqek. Xinjiang Uygur Zizhiqu, NW China *46°45′N 82°55′E*
54 H7 **Táchira** *off.* Estado Táchira. ◈ *state* W Venezuela
Táchira, Estado *see* Táchira
Tachoshui *see* Dazhuoshui
111 A17 **Tachov** *Ger.* Tachau. Plveňský Kraj, W Czech Republic *49°48′N 12°38′E*
171 Q5 **Tacloban** *off.* Tacloban City. Leyte, C Philippines *11°15′N 125°E*
Tacloban City *see* Tacloban
57 I19 **Tacna** Tacna, SE Peru *18°S 70°15′W*
57 H18 **Tacna** *off.* Departamento de Tacna. ◈ *department* S Peru
Tacna, Departamento de *see* Tacna
32 H8 **Tacoma** Washington, NW USA *47°15′N 122°27′W*
18 L11 **Taconic Range** ▲▲ NE USA
62 L6 **Taco Pozo** Formosa, N Argentina *25°35′S 63°15′W*
57 M20 **Tacsara, Cordillera de** ▲▲ S Bolivia
61 F17 **Tacuarembó** *prev.* San Fructuoso. Tacuarembó, C Uruguay *31°42′S 56°W*
61 E18 **Tacuarembó** ◈ *department* C Uruguay
61 F17 **Tacuarembó, Río** ≈ C Uruguay
83 I14 **Taculi** North Western, NW Zambia *14°17′S 26°51′E*
171 Q8 **Tacurong** Mindanao, S Philippines *06°42′N 124°40′E*
77 V8 **Tadek** ≈ NW Niger
74 J9 **Tademaït, Plateau du** *plateau* C Algeria
187 R17 **Tadine** Province des Îles Loyauté, E New Caledonia *21°33′S 167°54′E*
80 M11 **Tadjoura, Golfe de** *Eng.* Gulf of Tajura. *inlet* E Djibouti
80 L11 **Tadjourah** E Djibouti *11°47′N 42°51′E*
Tadmor/Tadmur *see* Tudmur
11 W10 **Tadoule Lake** ◎ Manitoba, C Canada
15 S8 **Tadoussac** Québec, SE Canada *48°09′N 69°43′W*
155 H18 **Tādpatri** Andhra Pradesh, E India *14°55′N 77°59′E*
Tadzhikabad *see* Tojikobod
Tadzhikistan *see* Tajikistan
163 Y14 **Taebaek-sanmaek** *prev.* T'aebaek-sanmaek. ▲▲ E South Korea
T'aebaek-sanmaek *see* Taebaek-sanmaek
Taechŏng-do *see* Daecheong-do
163 X13 **Taedong-gang** ≈ C North Korea
Taegu *see* Daegu
Taehan-haehyŏp *see* Korea Strait
Taehan Min'guk *see* South Korea
Taejŏn *see* Daejeon
193 Z13 **Tafahi** *island* N Tonga
105 Q4 **Tafalla** Navarra, N Spain *42°32′N 01°41′W*
77 W7 **Tafassâsset, Ténéré du** *desert* N Niger
75 M12 **Tafassâsset, Oued** ≈ SE Algeria
55 U11 **Tafelberg** ▲ S Suriname *03°55′N 56°09′W*
97 J21 **Taff** ≈ SE Wales, United Kingdom
Tafila/Ţafīlah, Muḥāfaẓat aţ *see* Aţ Ţafīlah
77 N15 **Tafiré** N Ivory Coast *09°04′N 05°10′W*
142 M6 **Tafresh** Markazī, W Iran *34°40′N 50°00′E*
143 Q9 **Taft** Yazd, C Iran *31°45′N 54°14′E*
35 R13 **Taft** California, W USA *35°08′N 119°27′W*
25 T14 **Taft** Texas, SW USA *27°58′N 97°24′W*
143 W12 **Taftān, Kūh-e** ▲ SE Iran *28°38′N 61°06′E*
35 R13 **Taft Heights** California, W USA *35°06′N 119°29′W*
189 Y14 **Tafunsak** Kosrae, E Micronesia *05°21′N 162°58′E*
192 G16 **Tāga** Savai'i, SW Samoa *13°46′S 172°31′W*
149 O6 **Tagāb** Dāikondī, E Afghanistan *33°53′N 66°23′E*
39 O8 **Tagagawik River** ≈ Alaska, USA
165 Q10 **Tagajō** *var.* Tagazyô. Miyagi, Honshū, C Japan *38°18′N 140°58′E*
126 K12 **Taganrog** Rostovskaya Oblast', SW Russian Federation *47°10′N 38°55′E*
126 K12 **Taganrog, Gulf of** *Rus.* Taganrogskiy Zaliv, *Ukr.* Tahanroz'ka Zatoka. *gulf* Russian Federation/Ukraine
Taganrogskiy Zaliv *see* Taganrog, Gulf of
76 J8 **Tagant** ◈ *region* C Mauritania
148 M14 **Tagas** Baluchistān, SW Pakistan *27°09′N 64°36′E*
171 O4 **Tagaytay** Luzon, N Philippines *14°04′N 120°55′E*
Tagazyô *see* Tagajō
171 P6 **Tagbilaran** *var.* Tagbilaran City. Bohol, C Philippines *09°41′N 123°54′E*
Tagbilaran City *see* Tagbilaran
106 B10 **Taggia** Liguria, NW Italy *43°51′N 07°48′E*
77 V9 **Taghouaji, Massif de** ▲ C Niger *17°13′N 08°37′E*
107 J15 **Tagliacozzo** Lazio, C Italy *42°03′N 13°15′E*
106 J7 **Tagliamento** ≈ NE Italy
149 N3 **Tagow Bāy** *var.* Bai. Sar-e Pul, N Afghanistan *35°41′N 66°01′E*
146 H9 **Tagta** *var.* Tahta, *Rus.* Takhta. Daşoguz Welaýaty, N Turkmenistan *41°40′N 59°51′E*
146 J16 **Tagtabazar** *var.* Takhtabazar. Mary Welaýaty, S Turkmenistan *35°57′N 62°49′E*
59 L17 **Taguatinga** Tocantins, C Brazil *12°16′S 46°25′W*
186 I10 **Tagula** Tagula Island, SE Papua New Guinea *11°21′S 153°11′E*
186 I11 **Tagula Island** *prev.* Southeast Island, Sudest Island. *island* SE Papua New Guinea
171 Q7 **Tagum** Mindanao, S Philippines *07°22′N 125°51′E*
54 C7 **Tagún, Cerro** *elevation* Colombia/Panama
105 P7 **Tagus** *Port.* Rio Tejo, *Sp.* Río Tajo. ≈ Portugal/Spain
64 M9 **Tagus Plain** *undersea feature* E Atlantic Ocean *37°30′N 12°00′W*
191 S10 **Tahaa** *island* Îles Sous le Vent, W French Polynesia
191 U10 **Tahanea** *atoll* Îles Tuamotu, C French Polynesia

◆ Country ● Country Capital ◇ Dependent Territory ○ Dependent Territory Capital ◈ Administrative Regions ✕ International Airport ▲ Mountain ▲▲ Mountain Range ⋀ Volcano ≈ River ◎ Lake ⊠ Reservoir

Tahanroz'ka Zatoka see Taganrog, Gulf of
74 K12 **Tahat** ▲ SE Algeria 23°15′N 05°34′E
163 U4 **Tahe** Heilongjiang, NE China 52°21′N 124°42′E
Tahi see Tsogt
191 T10 **Tahiti** island Îles du Vent, W French Polynesia
Tahiti, Archipel de see Société, Archipel de la
118 E4 **Tahkuna Nina** headland W Estonia 59°06′N 22°35′E
148 K12 **Tahlāb** ↔ W Pakistan
148 K12 **Tahlāb, Dasht-i** desert SW Pakistan
27 R10 **Tahlequah** Oklahoma, C USA 35°57′N 94°58′W
35 Q6 **Tahoe City** California, W USA 39°09′N 120°09′W
35 P6 **Tahoe, Lake** ◎ California/ Nevada, W USA
25 N6 **Tahoka** Texas, SW USA 33°10′N 101°47′W
32 F8 **Taholah** Washington, NW USA 47°19′N 124°17′W
77 T11 **Tahoua** Tahoua, W Niger 14°53′N 05°18′E
77 T11 **Tahoua** ◆ department W Niger
31 P3 **Tahquamenon Falls** waterfall Michigan, N USA
31 P4 **Tahquamenon River** ↔ Michigan, N USA
139 V10 **Taḥrīr** Al Qādisīyah, S Iraq 31°58′N 45°34′E
10 K17 **Tahsis** Vancouver Island, British Columbia, SW Canada 49°42′N 126°31′W
75 W9 **Tahtā** var. Tahta. C Egypt 26°47′N 31°31′E
Tahta see Tagta
136 L15 **Tahtalı Dağları** ▲ C Turkey
57 F13 **Tahuamanu, Río** ↔ Bolivia/Peru
56 F13 **Tahuanía, Río** ↔ E Peru
191 X7 **Tahuata** island Îles Marquises, NE French Polynesia
76 L17 **Taï** SW Ivory Coast 05°52′N 07°28′W
161 P5 **Tai'an** Shandong, E China 36°13′N 117°12′E
191 R8 **Taiarapu, Presqu'île de** peninsula Tahiti, W French Polynesia
160 K7 **Taibad** var. Tāybād
161 T13 **Taibei** var. (Taiwan) N Taiwan 25°02′N 121°28′E
105 Q12 **Taibilla, Sierra de** ▲ S Spain
Taichū see Taizhong
T'aichung see Taizhong
Taiden see Daejeon
161 T14 **Taidong** Jap. Taitō; prev. T'aitung. S Taiwan 22°43′N 121°10′E
185 E23 **Taieri** ↔ South Island, New Zealand
115 E21 **Taïgetos** ▲ S Greece
161 N4 **Taihang Shan** ▲ C China
184 M11 **Taihape** Manawatu-Wanganui, North Island, New Zealand 39°41′S 175°47′E
161 O7 **Taihe** Anhui, E China 33°14′N 115°35′E
161 O12 **Taihe** var. Chengjiang. Jiangxi, S China 26°47′N 114°52′E
Taihoku see Taibei
161 P8 **Taihu** Anhui, E China 30°24′N 116°18′E
161 R8 **Tai Hu** ◎ E China
159 O9 **Taikang** var. Dorbod, Dorbod Mongolzu Zizhixian. Heilongjiang, NE China 46°50′N 124°25′E
161 O6 **Taikang** Henan, C China 34°01′N 114°59′E
165 T5 **Taiki** Hokkaidō, NE Japan 42°29′N 143°15′E
166 L8 **Taikkyi** Yangon, SW Myanmar (Burma) 17°16′N 95°55′E
163 U8 **Tailai** Heilongjiang, NE China 46°25′N 123°25′E
168 I12 **Taileleo** Pulau Siberut, W Indonesia 01°45′S 99°06′E
182 J10 **Tailem Bend** South Australia 35°16′S 139°29′E
96 I8 **Tain** N Scotland, United Kingdom 57°49′N 04°04′W
161 S14 **Tainan** prev. Dainan, T'ainan. S Taiwan 23°01′N 120°05′E
115 E22 **Taínaro, Akrotírio** cape S Greece
161 Q11 **Taining** var. Shancheng. Fujian, SE China 26°55′N 117°13′E
191 W7 **Taiohae** prev. Madisonville. Nuku Hiva, NE French Polynesia 08°55′S 140°04′W
Taipei see Taibei
168 J7 **Taiping** Perak, Peninsular Malaysia 04°54′N 100°42′E
Taiping see Chongzuo
163 S8 **Taiping Ling** ▲ NE China 47°27′N 120°07′E
165 Q4 **Taisei** Hokkaidō, NE Japan
165 G12 **Taisha** Shimane, Honshū, SW Japan 35°23′N 132°40′E
Taishō-tō see Sekibi-sho
109 R4 **Taiskirchen** Oberösterreich, NW Austria 48°15′N 13°33′E
63 F20 **Taitao, Península de** peninsula S Chile
81 J21 **Taita/Taveta** ◆ county S Kenya
Taitō see Taidong
T'aitung see Taidong
92 M13 **Taivalkoski** Pohjois-Pohjanmaa, E Finland
93 K19 **Taivassalo** Varsinais-Suomi, SW Finland 60°33′N 21°36′E
161 T14 **Taiwan** off. Republic of China, var. Formosa, Formo'sa. ◆ republic E Asia
192 F5 **Taiwan** island E Asia
Taiwan see Taizhong
T'aiwan Haihsia/Taiwan Haixia see Taiwan Strait
Taiwan Shan see Chungyang Shanmo
161 R13 **Taiwan Strait** var. Formosa Strait, Chin. T'aiwan Haihsia, Taiwan Haixia. strait China/Taiwan
Taiwan Taoyuan prev. Chiang Kai-shek. ✈ (T'aibei) N Taiwan 25°00′N 121°00′E
161 S12 **Taiyuan** var. T'ai-yüan, T'ai-yüan; prev. Yangku. province capital Shanxi, C China 37°48′N 112°33′E

161 S13 **Taizhong** Jap. Taichū; prev. T'aichung, Taiwan. C Taiwan 24°09′N 120°40′E
161 S10 **Taizhou** Jiangsu, E China 32°36′N 119°52′E
161 S10 **Taizhou** var. Jiaojiang; prev. Haimen. Zhejiang, SE China 28°36′N 121°19′E
Taizhou see Linhai
141 O16 **Ta'izz** SW Yemen 13°36′N 44°04′E
141 O16 **Ta'izz** ✈ SW Yemen 13°40′N 44°10′E
75 P12 **Tajarhi** SW Libya 24°21′N 14°28′E
147 P13 **Tajikistan** off. Republic of Tajikistan, Rus. Tadzhikistan, Taj. Jumhurii Tojikiston; prev. Tajik S.S.R. ◆ republic C Asia
Tajikistan, Republic of see Tajikistan
165 O10 **Tajima** Fukushima, Honshū, C Japan 37°10′N 139°46′E
Tajoe see Tayu
Tajo, Río see Tagus
42 B5 **Tajumulco, Volcán** ▲ W Guatemala 15°04′N 91°50′W
105 P7 **Tajuña** ↔ C Spain
167 O9 **Tak** var. Rahaeng. Tak, W Thailand 16°51′N 99°08′E
189 U4 **Taka Atoll** var. Tōke. atoll Ratak Chain, N Marshall Islands
165 P12 **Takahagi** Ibaraki, Honshū, S Japan 36°42′N 140°42′E
165 H13 **Takahashi** var. Takahasi. Okayama, Honshū, SW Japan 34°48′N 133°38′E
Takahasi see Takahashi
189 P12 **Takaieu Island** island E Micronesia
184 I13 **Takaka** Tasman, South Island, New Zealand 40°52′S 172°49′E
170 M14 **Takalar** Sulawesi, C Indonesia 05°28′S 119°24′E
165 O13 **Takamatsu** var. Takamatu. Kagawa, Shikoku, SW Japan 34°19′N 133°59′E
Takamatu see Takamatsu
165 D14 **Takamori** Kumamoto, Kyūshū, SW Japan 32°49′N 131°08′E
165 D16 **Takanabe** Miyazaki, Kyūshū, SW Japan 32°13′N 131°31′E
170 M16 **Takan, Gunung** ▲ Pulau Sumba, S Indonesia 08°52′S 117°52′E
165 Q7 **Takanosu** var. Kita-Akita. Akita, Honshū, C Japan 40°13′N 140°23′E
Takao see Gaoxiong
165 L11 **Takaoka** Toyama, Honshū, SW Japan 36°45′N 137°02′E
184 N12 **Takapau** Hawke's Bay, North Island, New Zealand 40°01′S 176°21′E
191 U9 **Takapoto** atoll Îles Tuamotu, C French Polynesia
184 L5 **Takapuna** Auckland, North Island, New Zealand 36°48′S 174°46′E
165 J3 **Takarazuka** Hyōgo, Honshū, SW Japan 34°49′N 135°21′E
191 U9 **Takaroa** atoll Îles Tuamotu, C French Polynesia
165 N12 **Takasaki** Gunma, Honshū, SW Japan 36°20′N 139°00′E
164 K12 **Takefu** var. Echizen, Takehu. Fukui, Honshū, SW Japan 35°55′N 136°11′E
Takehu see Takefu
165 C14 **Takeo** Saga, Kyūshū, SW Japan 33°13′N 130°00′E
164 C17 **Take-shima** island Nansei-shotō, SW Japan
142 M5 **Tākestān** var. Takistan; prev. Siadehen. Qazvin, N Iran 36°02′N 49°41′E
164 D14 **Taketa** Ōita, Kyūshū, SW Japan 32°59′N 131°23′E
167 R13 **Takêv** prev. Takeo. Takêv, S Cambodia 10°59′N 104°47′E
167 O10 **Tak Fah** Nakhon Sawan, C Thailand
139 T13 **Takhādīd** well S Iraq
149 R3 **Takhār** ◆ province NE Afghanistan
Takhiatash see Taxiatosh
Ta Khmau see Kândal
Takhta see Tagta
Takhtabazar see Tagtabazar
145 O8 **Takhtabrod** Severnyy Kazakhstan, N Kazakhstan 52°35′N 67°37′E
Takhtakupyr see Taxtako'pir
142 M8 **Takht-e Shāh, Kūh-e** ▲ C Iran
77 V12 **Takiéta** Zinder, S Niger 13°43′N 08°33′E
8 J8 **Takijuq Lake** ◎ Nunavut, NW Canada
165 S3 **Takikawa** Hokkaidō, NE Japan 43°35′N 141°54′E
165 U3 **Takinoue** Hokkaidō, NE Japan 44°10′N 143°09′E
185 B23 **Takitimu Mountains** ▲ South Island, New Zealand
165 R7 **Takko** Aomori, Honshū, C Japan 40°19′N 141°11′E
10 L13 **Takla Lake** ◎ British Columbia, SW Canada
Takla Makan Desert see Taklimakan Shamo
158 H9 **Taklimakan Shamo** Eng. Takla Makan Desert. desert NW China
39 P10 **Takotna** Alaska, USA 62°59′N 156°03′W
123 O12 **Taksimo** Respublika Buryatiya, S Russian Federation 56°18′N 114°53′E
164 C15 **Taku** Saga, Kyūshū, SW Japan 33°18′N 130°06′E
10 I10 **Taku** ↔ British Columbia, W Canada
166 M15 **Takua Pa** var. Ban Takua Pa. Phangnga, SW Thailand 08°55′N 98°20′E
77 W16 **Takum** E Nigeria 07°16′N 10°00′E
191 V10 **Takume** atoll Îles Tuamotu, C French Polynesia
190 L16 **Takutea** island S Cook Islands
190 K6 **Takuu Islands** prev. Mortlock Group. island group E Papua New Guinea

119 L18 **Tal'** Minskaya Voblasts', S Belarus 52°52′N 27°58′E
40 L13 **Tala** Jalisco, C Mexico 20°39′N 103°45′W
61 F19 **Tala** Canelones, S Uruguay 34°34′S 55°46′W
Talabriga see Aveiro, Portugal
Talabriga see Talavera de la Reina, Spain
119 N14 **Talachyn** Rus. Tolochin. Vitsyebskaya Voblasts', NE Belarus 54°25′N 29°42′E
149 U7 **Talagang** Punjab, E Pakistan 32°55′N 72°29′E
105 V11 **Talaiassa** ▲ Ibiza, Spain, W Mediterranean Sea 38°55′N 01°17′E
117 J23 **Talaimannar** Northern Province, NW Sri Lanka 09°05′N 79°43′E
117 R3 **Talalayivka** Chernihivs'ka Oblast', N Ukraine 50°51′N 33°09′E
43 O15 **Talamanca, Cordillera de** ▲ S Costa Rica
56 A9 **Talara** Piura, NW Peru 04°31′S 81°17′W
104 L11 **Talarrubias** Extremadura, W Spain 39°03′N 05°14′W
147 S8 **Talas** Talasskaya Oblast', NW Kyrgyzstan 42°29′N 72°21′E
147 S8 **Talas** ↔ NW Kyrgyzstan
186 G7 **Talasea** New Britain, E Papua New Guinea 05°20′S 150°01′E
147 S8 **Talasskaya Oblast'** Kir. Talas Oblasty. ◆ province NW Kyrgyzstan
147 S8 **Talasskiy Alatau, Khrebet** ▲ Kazakhstan/Kyrgyzstan
77 U12 **Talata Mafara** Zamfara, NW Nigeria 12°33′N 06°01′E
171 R9 **Talaud, Kepulauan** island group E Indonesia
104 M9 **Talavera de la Reina** anc. Caesarobriga, Talabriga. Castilla-La Mancha, C Spain 39°58′N 04°50′W
104 J11 **Talavera la Real** Extremadura, W Spain 38°53′N 06°46′W
186 F7 **Talawe, Mount** ▲ New Britain, C Papua New Guinea 05°30′S 148°24′E
23 S5 **Talbotton** Georgia, SE USA 32°40′N 84°32′W
183 R7 **Talbragar River** ↔ New South Wales, SE Australia
62 F13 **Talca** Maule, C Chile 35°28′S 71°42′W
62 F13 **Talcahuano** Bío Bío, C Chile 36°43′S 73°07′W
154 N12 **Tālcher** Odisha, E India 20°57′N 85°13′E
25 W5 **Talco** Texas, SW USA 33°21′N 95°06′W
145 V14 **Taldykorgan** Kaz. Taldykurgan; prev. Taldy-Kurgan. Taldykorgan, SE Kazakhstan 45°N 78°23′E
Taldy-Kurgan/Taldykorgan see Taldykorgan
147 Y7 **Taldy-Suu** Issyk-Kul'skaya Oblast', E Kyrgyzstan 42°49′N 78°33′E
147 U10 **Taldy-Suu** Oshskaya Oblast', SW Kyrgyzstan 40°33′N 73°52′E
193 Y15 **Taleki Tonga** island Otu Tolu Group, C Tonga
193 Y15 **Taleki Vavu'u** island Otu Tolu Group, C Tonga
102 J13 **Talence** Gironde, SW France 44°49′N 00°35′W
145 U16 **Talgar** Kaz. Talghar. Almaty, SE Kazakhstan 43°17′N 77°15′E
Talghar see Talgar
171 Q12 **Taliabu, Pulau** island Kepulauan Sula, C Indonesia
77 P15 **Tamale** C Ghana 09°21′N 00°54′W
191 P3 **Tamana** prev. Rotcher Island. atoll Tungaru, W Kiribati
74 K12 **Tamanrasset** var. Tamenghest. S Algeria 22°49′N 05°32′E
74 J13 **Tamanrasset** wadi Algeria/Mali
137 T12 **T'alin** Rus. Talin; prev. Verin T'alin. W Armenia 40°23′N 43°51′E
149 R3 **Talj** province NW Afghanistan
115 L22 **Taliarós, Akrotírio** headland Astypálaia, Kykládes, Greece, Aegean Sea 36°31′N 26°18′E
Ta-lien see Dalian
27 O12 **Talihina** Oklahoma, C USA 34°45′N 95°03′W
81 E15 **Tali Post** Central Equatoria, S South Sudan 05°55′N 30°44′E
Taliq-an see Tāloqān
Taliş Dağları see Talish Mountains
142 L2 **Talish Mountains** Az. Talış Dağları, Per. Kūhhā-ye Tavālesh, Rus. Talyshskiye Gory. ▲ Azerbaijan/Iran
170 M16 **Taliwang** Sumbawa, C Indonesia 08°45′S 116°55′E
119 L17 **Tal'ka** Minskaya Voblasts', C Belarus 53°22′N 28°21′E
111 I24 **Talmási** Tolna, S Hungary 46°39′N 18°17′E
39 R11 **Talkeetna** Alaska, USA 62°19′N 150°06′W
39 R11 **Talkeetna Mountains** ▲ Alaska, USA
Talkhof see Puurmani
140 H2 **Tālknafjörður** Vestfirðir, W Iceland 65°38′N 23°51′W
139 V10 **Tall 'Abtah** Nīnawá, N Iraq 35°52′N 42°39′E
138 M2 **Tall Abyaḍ** var. Tell Abiad. Ar Raqqah, N Syria 36°42′N 38°56′E
23 Q4 **Talladega** Alabama, S USA 33°26′N 86°06′W
139 Q2 **Tall 'Afar** Nīnawá, N Iraq 36°22′N 42°27′E
23 S8 **Tallahassee** prev. Muskogean. state capital Florida, SE USA 30°27′N 84°17′W
22 L2 **Tallahatchie River** ↔ Mississippi, S USA
139 Q2 **Tall al Abyaḍ** see At Tall al Abyaḍ
139 W12 **Tall al Laḥm** Dhī Qār, S Iraq 30°46′N 46°22′E
186 M9 **Tallaimbea** Guadalcanal, C Solomon Islands 09°19′S 159°43′E
35 R5 **Tallahala** ▲ S USA 31°43′N 117°13′E
169 N10 **Tallabena** Kepulauan island group W Indonesia
57 E17 **Tallard** Hautes-Alpes, SE France 44°30′N 06°05′W
23 R4 **Tallapoosa River** ↔ Alabama/Georgia, S USA
170 L16 **Tallassee** Alabama, S USA 32°32′N 85°53′W
138 I5 **Tall Bīsah** Ḥimṣ, W Syria 34°50′N 36°44′E
139 R3 **Tall Ḥassūnah** Al Anbār, C Iraq 33°41′N 43°10′E

139 Q2 **Tall Ḥuqnah** var. Tell Ḥuqnah. Nīnawá, N Iraq 36°33′N 42°34′E
118 G3 **Tallinn** Ger. Reval, Rus. Revel'; prev. Revel. ● (Estonia) Harjumaa, NW Estonia 59°26′N 24°42′E
118 H3 **Tallinn** ✈ Harjumaa, NW Estonia 59°23′N 24°52′E
138 H5 **Tall Kalakh** var. Tell Kalakh. Ḥimṣ, C Syria 34°40′N 36°18′E
139 R2 **Tall Kayf** Nīnawá, NW Iraq 36°30′N 43°08′E
Tall Kūchak see Tall Kūshik
139 P2 **Tall Kūshik** var. Tall Kūchak. Al Ḥasakah, E Syria 36°48′N 41°57′E
31 U12 **Tallmadge** Ohio, N USA 09°05′N 79°43′E
22 J5 **Tallulah** Louisiana, S USA 32°25′N 91°12′W
139 Q2 **Tall 'Uwaynāt** Nīnawá, NW Iraq 36°43′N 42°18′E
139 Q2 **Tall Zāhir** Nīnawá, N Iraq 36°51′N 42°29′E
122 J13 **Tal'menka** Altayskiy Kray, S Russian Federation 53°55′N 83°26′E
122 K8 **Talnakh** Krasnoyarskiy Kray, N Russian Federation 69°26′N 88°27′E
117 P7 **Tal'ne** Rus. Tal'noye. Cherkas'ka Oblast', C Ukraine 48°55′N 30°40′E
80 E12 **Talodi** Southern Kordofan, C Sudan 10°40′N 30°25′E
188 B16 **Talofofo** SE Guam 13°21′N 144°45′E
188 B16 **Talofofo Bay** bay SE Guam
26 L9 **Taloga** Oklahoma, C USA 36°01′N 98°58′W
123 T10 **Talon** Magadanskaya Oblast', E Russian Federation 59°47′N 148°46′E
14 H11 **Talon, Lake** ◎ Ontario, S Canada
149 R2 **Tāloqān** var. Taliq-an. Takhār, NE Afghanistan 36°44′N 69°33′E
95 N14 **Tämnaren** ◎ C Sweden
191 Q7 **Tamotoe, Passe** passage Tahiti, W French Polynesia
23 V12 **Tampa** Florida, SE USA 27°57′N 82°27′W
23 V13 **Tampa** ✈ Florida, SE USA 31°46′N 99°42′W
23 V13 **Tampa Bay** bay Florida, SE USA
41 Q11 **Tampico** Tamaulipas, C Mexico 22°18′N 97°52′W
171 P14 **Tampo** Pulau Muna, C Indonesia 05°13′S 122°40′E
167 V11 **Tam Quan** Bình Định, C Vietnam 14°34′N 109°00′E
162 J13 **Tamsag Muchang** Nei Mongol Zizhiqu, N China 40°28′N 102°34′E
Tamsal see Tamsalu
118 I4 **Tamsalu** Ger. Tamsal. Lääne-Virumaa, NE Estonia 59°10′N 26°07′E
109 S8 **Tamsweg** Salzburg, SW Austria 47°08′N 13°49′E
188 C15 **Tamuning** NW Guam 13°29′N 144°47′E
183 T6 **Tamworth** New South Wales, SE Australia 31°07′S 150°54′E
97 M19 **Tamworth** C England, United Kingdom 52°39′N 01°40′W
81 K19 **Tana** Finn. Tenojoki, Lapp. Deatnu. ↔ SE Kenya see also Deatnu/Tana
165 X7 **Tanabe** Wakayama, Honshū, SW Japan 33°43′N 135°22′E
164 I15 **Tanabe** Wakayama, Honshū, SW Japan
92 L8 **Tana Bru** Finnmark, N Norway 70°11′N 28°06′E
92 M11 **Tanahjampea, Pulau** island W Indonesia
171 Q16 **Tanah, Tanjung** headland W Indonesia
171 U16 **Tanimbar, Kepulauan** island group Maluku, E Indonesia
Taninthari see Taninthayi
167 N12 **Taninthayi** var. Tenasserim. Taninthayi, S Myanmar (Burma) 12°05′N 99°01′E
167 N12 **Taninthayi** var. Tenasserim; prev. Taninthari. ◆ region S Myanmar (Burma)
181 P5 **Tanami Desert** desert Northern Territory, N Australia
167 T14 **Tân An** Long An, S Vietnam 10°31′N 106°25′E
39 Q9 **Tanana** Alaska, USA 65°12′N 152°00′W
39 Q9 **Tanana River** ↔ Alaska, USA
95 C16 **Tananger** Rogaland, S Norway 58°55′N 05°35′E
188 H5 **Tanapag** Saipan, S Northern Mariana Islands 15°14′S 145°45′E
188 H5 **Tanapag, Puetton** bay Saipan, S Northern Mariana Islands
81 J20 **Tana River** ◆ county SE Kenya
169 V9 **Tanjung** Borneo, C Indonesia 02°08′S 115°23′E
169 N12 **Tanjungbalai** prev. Tandjoengbalai. Pulau Belitung, W Indonesia
168 M10 **Tanjungpinang** prev. Tandjoengpinang. Pulau Bintan, W Indonesia
169 V9 **Tanjungredeb** var. Tanjungredep; prev. Tandjoengredeb. Borneo, C Indonesia 02°09′N 117°29′E
Tanjungredep see Tanjungredeb
169 V9 **Tanjung Selor** Borneo, C Indonesia 02°50′N 117°22′E
149 S8 **Tank** Khyber Pakhtunkhwa, NW Pakistan 34°14′N 70°29′E
34 M14 **Tancitaro, Cerro** ▲ C Mexico 19°26′N 102°25′W
153 N12 **Tānda** Uttar Pradesh, N India 26°33′N 82°39′E
77 O13 **Tanda** E Ivory Coast 07°48′N 03°10′W
116 L9 **Tăndărei** Ialomița, SE Romania 44°38′N 27°40′E
63 N14 **Tandil** Buenos Aires, E Argentina 37°18′S 59°10′W
78 H12 **Tandjilé** off. Région du Tandjilé, prev. Tandjilé. ◆ prefecture SW Chad
Tandjilé, Région du see Tandjilé
Tandjoengbalai see Tanjungbalai
Tandjoengkarang see Tanjungkarang-Telukbetung
Tandjoengpandan see Tanjungpandan
Tandjoengpinang see Tanjungpinang
Tandjoengredeb see Tanjungredeb
167 T14 **Tân Phú** see Dinh Quan
167 U16 **Tân Hiệp** var. Phung Hiệp. Cân Thơ, S Vietnam 09°50′N 105°48′E
92 M11 **Tanhua** Lappi, N Finland 67°31′N 27°30′E
149 S8 **Tank** Khyber Pakhtunkhwa, NW Pakistan 34°14′N 70°29′E
83 I16 **Tana** Southern, S Zambia 16°56′S 26°56′E
113 L17 **Tara** ▲ W Serbia
75 O7 **Taraba** ◆ state E Nigeria
77 X15 **Taraba** ↔ E Nigeria
75 O7 **Tarābulus** var. Ṭarābulus al Gharb, Eng. Tripoli. ● (Libya) NW Libya 32°57′N 13°07′E
75 O7 **Ṭarābulus** ✈ NW Libya 32°37′N 13°07′E
Ṭarābulus al Gharb see Tarābulus
Ṭarābulus/Ṭarābulus ash Shām see Tripoli, Lebanon
105 O7 **Taracena** Castilla-La Mancha, C Spain 40°39′N 03°08′W
117 N12 **Taraclia** Rus. Tarakliya. Moldova 45°55′N 28°40′E
139 V10 **Tall al Kafr** Dhī Qār, SE Iraq 30°55′N 45°58′E
183 R10 **Tarago** New South Wales, SE Australia 35°04′S 149°40′E
162 J12 **Taragt** var. Hürmet. Övörhangay, C Mongolia 46°18′N 102°27′E
169 V8 **Tarakan** Borneo, C Indonesia 03°20′N 117°38′E

74 D9 **Tan-Tan** SW Morocco 28°30′N 11°10′W
41 P12 **Tantoyuca** Veracruz-Llave, E Mexico 21°21′N 98°12′W
152 J12 **Tāntpur** Uttar Pradesh, N India 26°51′N 77°29′E
Tan-tung see Dandong
38 M12 **Tanunak** Alaska, USA 60°35′N 165°15′W
166 L5 **Ta-nyaung** Magway, W Myanmar (Burma) 20°49′N 94°40′E
167 S5 **Tân Yên** Tuyên Quang, N Vietnam 22°06′N 104°58′E
81 F22 **Tanzania** off. United Republic of Tanzania, Swa. Jamhuri ya Muungano wa Tanzania; prev. German East Africa, Tanganyika and Zanzibar. ◆ republic E Africa
Tanzania, Jamhuri ya Muungano wa see Tanzania
Tanzania, United Republic of see Tanzania
Tao'an see Taonan
159 U11 **Tao He** ↔ C China
163 U9 **Taonan** var. Tao'an. Jilin, NE China 45°20′N 122°46′E
107 M23 **Taormina** anc. Tauromenium. Sicilia, Italy, C Mediterranean Sea 37°54′N 15°18′E
37 S9 **Taos** New Mexico, SW USA 36°24′N 105°34′W
77 O6 **Taoudenit** var. Taoudenni. Tombouctou, N Mali 22°46′N 03°54′W
74 G6 **Taounate** N Morocco 34°33′N 04°39′N
161 S13 **Taoyang** see Lintao
161 S13 **Taoyuan** Jap. Tōen; prev. T'aoyüan. N Taiwan 25°00′N 121°15′E
118 I3 **Tapa** Ger. Taps. Lääne-Virumaa, NE Estonia 59°16′N 25°58′E
41 V17 **Tapachula** Chiapas, SE Mexico 14°53′N 92°18′W
59 H14 **Tapajós, Rio** var. Tapajóz. ↔ NW Brazil
Tapajóz, Rio see Tapajós, Rio
61 C21 **Tapalqué** var. Tapalquén. Buenos Aires, E Argentina 36°21′S 60°01′W
Tapalquén see Tapalqué
55 W11 **Tapanahony Rivier** var. Tapanahoni. ↔ E Suriname
Tapanahoni see Tapanahony Rivier
41 T16 **Tapanatepec** var. San Pedro Tapanatepec. Oaxaca, SE Mexico 16°23′N 94°09′W
185 D23 **Tapanui** Otago, South Island, New Zealand 45°55′S 169°16′E
59 E14 **Tapauá** Amazonas, N Brazil 05°42′S 64°15′W
47 R7 **Tapauá, Rio** ↔ W Brazil
185 I14 **Tapawera** Tasman, South Island, New Zealand 41°24′S 172°50′E
61 I16 **Tapes** Rio Grande do Sul, S Brazil 30°40′S 51°25′W
76 K16 **Tapeta** C Liberia 06°36′N 08°52′W
154 H11 **Tāpti** prev. Tāpi. ↔ W India 16°33′N 118°15′E
104 J2 **Tapia de Casariego** Asturias, N Spain 43°34′N 06°56′W
56 F10 **Tapiche, Río** ↔ N Peru
167 N15 **Tapi, Mae Nam** var. Luang. ↔ SW Thailand
163 X7 **Tapini** Central, S Papua New Guinea 08°19′S 146°59′E
-186 E8 **Tapirapecó, Serra** see Tapirapecó, Sierra
Tapirapecó, Sierra Port. Serra Tapirapecó. ▲ Brazil/Venezuela
77 R13 **Tapoa** ↔ Benin/Niger
188 H5 **Tapochau, Mount** ▲ Saipan, S Northern Mariana Islands
111 H24 **Tapolca** Veszprém, W Hungary 46°54′N 17°29′E
21 X5 **Tappahannock** Virginia, NE USA 37°53′N 76°54′W
31 U13 **Tappan Lake** ◎ Ohio, N USA
165 Q6 **Tappi-zaki** headland Honshū, C Japan 41°15′N 140°19′E
Taps see Tapa
185 J16 **Tapuaemanu** see Maiao
Tapuaenuku ▲ South Island, New Zealand 42°00′S 173°39′E
171 N8 **Tapul Group** island group Sulu Archipelago, SW Philippines
58 E11 **Tapurucuará** var. Tapuruquara. Amazonas, NW Brazil 0°17′S 65°00′W
Tapuruquara see Tapurucuará
192 J13 **Taputapu, Cape** headland Tutuila, W American Samoa 14°20′S 170°51′W
141 W13 **Taqāʾ** S Oman 17°02′N 54°23′E
139 T3 **Taqtaq Ar.** Ṭaqṭaq. Arbīl, N Iraq 35°54′N 44°36′E
Ṭaqṭaq see Taqtaq
61 H19 **Taquara** Rio Grande do Sul, S Brazil 29°39′S 50°46′W
59 H19 **Taquari, Rio** ↔ C Brazil
60 L8 **Taquaritinga** São Paulo, S Brazil 21°25′S 48°38′W
122 I11 **Tara** Omskaya Oblast', C Russian Federation 56°56′N 74°17′E
83 I16 **Tara** Southern, S Zambia 16°56′S 26°56′E
113 L17 **Tara** ▲ W Serbia
75 O7 **Taraba** ◆ state E Nigeria
77 X15 **Taraba** ↔ E Nigeria

169 V9 **Tarakan, Pulau** *island* N Indonesia
Tarakilya *see* Taraclia
165 P16 **Tarama-jima** *island* Sakishima-shotō, SW Japan
184 K10 **Taranaki** *off.* Taranaki Region. ◈ *region* North Island, New Zealand
184 K10 **Taranaki, Mount** *var.* Egmont. ⛰ North Island, New Zealand 39°16´S 174°04´E
Taranaki Region *see* Taranaki
105 O9 **Tarancón** Castilla-La Mancha, C Spain 40°01´N 03°01´W
188 M15 **Tarang Reef** *reef* C Micronesia
96 E7 **Taransay** *island* NW Scotland, United Kingdom
107 P18 **Taranto** *var.* Tarentum. Puglia, SE Italy 40°30´N 17°11´E
107 O19 **Taranto, Golfo di** *Eng.* Gulf of Taranto. *gulf* S Italy
Taranto, Gulf of *see* Taranto, Golfo di
62 G3 **Tarapacá** *off.* Región de Tarapacá. ◈ *region* N Chile
Tarapacá, Región de *see* Tarapacá
187 N9 **Tarapaina** Maramasike Island, N Solomon Islands 09°28´S 161°24´E
56 D10 **Tarapoto** San Martín, N Peru 06°31´S 76°23´W
138 M6 **Ţaraq an Na'jah** *hill range* E Syria
138 M6 **Ţaraq Sidāwī** *hill range* E Syria
103 Q11 **Tarare** Rhône, E France 45°54´N 04°26´E
Tararite de Llitera *see* Tamarite de Litera
184 M13 **Tararua Range** ▲▲ North Island, New Zealand
151 Q22 **Tarāsa Dwīp** *island* Nicobar Islands, India, NE Indian Ocean
103 Q15 **Tarascon** Bouches-du-Rhône, SE France 43°48´N 04°39´E
102 M17 **Tarascon-sur-Ariège** Ariège, S France 42°51´N 01°35´E
117 P6 **Tarashcha** Kyyivs'ka Oblast', N Ukraine 49°34´N 30°31´E
57 L18 **Tarata** Cochabamba, C Bolivia 17°35´S 66°04´W
57 I18 **Tarata** Tacna, SW Peru 17°30´S 70°00´W
190 H2 **Taratai** *atoll* Tungaru, W Kiribati
59 B15 **Tarauacá** Acre, W Brazil 08°06´S 70°45´W
59 B15 **Tarauacá, Rio** ∽ NW Brazil
191 Q8 **Taravao** Tahiti, W French Polynesia 17°44´S 149°19´W
191 R8 **Taravao, Baie de** *bay* Tahiti, W French Polynesia
191 Q8 **Taravao, Isthme de** *isthmus* Tahiti, W French Polynesia
103 X16 **Taravo** ∽ Corse, France, C Mediterranean Sea
190 J3 **Tarawa** ✈ Tarawa, W Kiribati 0°53´S 169°32´E
190 H2 **Tarawa** *atoll* Tungaru, W Kiribati
184 N10 **Tarawera** Hawke's Bay, North Island, New Zealand 39°03´S 176°34´E
184 N8 **Tarawera, Lake** ◎ North Island, New Zealand
184 N8 **Tarawera, Mount** ▲ North Island, New Zealand 38°13´S 176°29´E
105 S8 **Tarayuela** ▲ N Spain 40°28´N 00°22´W
145 R16 **Taraz** *prev.* Aulie Ata, Auliye-Ata, Dzhambul, Zhambyl. Zhambyl, S Kazakhstan 42°55´N 71°27´E
105 Q5 **Tarazona** Aragón, NE Spain 41°54´N 01°44´W
105 Q10 **Tarazona de la Mancha** Castilla-La Mancha, C Spain 39°16´N 01°55´W
145 X12 **Tarbagatay, Khrebet** ▲▲ China/Kazakhstan
96 J8 **Tarbat Ness** *headland* N Scotland, United Kingdom 57°51´N 03°48´W
149 U5 **Tarbela Reservoir** ⊠ N Pakistan
96 H12 **Tarbert** W Scotland, United Kingdom 55°52´N 05°26´W
96 F7 **Tarbert** NW Scotland, United Kingdom 57°54´N 06°48´W
102 K16 **Tarbes** *anc.* Bigorra. Hautes-Pyrénées, S France 43°14´N 00°04´E
21 W9 **Tarboro** North Carolina, SE USA 35°54´N 77°34´W
Tarca *see* Torysa
106 J6 **Tarcento** Friuli-Venezia Giulia, NE Italy 46°13´N 13°13´E
182 F5 **Tarcoola** South Australia 30°44´S 134°34´E
105 S5 **Tardienta** Aragón, NE Spain 41°58´N 00°31´W
102 L11 **Tardoire** ∽ W France
183 U7 **Taree** New South Wales, SE Australia 31°56´S 152°29´E
92 K12 **Tärendö** *Lapp.* Deargget. Norrbotten, N Sweden 67°10´N 22°40´E
Tarentum *see* Taranto
74 C9 **Tarfaya** SW Morocco 27°56´N 12°55´W
114 L8 **Targovishte**, Tărgovište; *prev.* Eski Dzhumaya. Targovishte, N Bulgaria 43°15´N 26°34´E
114 L8 **Targovishte** *var.* Tŭrgovishte. ◈ *province* N Bulgaria
116 J13 **Târgovişte** *prev.* Tîrgovişte. Dâmboviţa, S Romania 44°54´N 25°29´E
Tărgovište *see* Targovishte
116 M12 **Târgu Bujor** *prev.* Tîrgu Bujor. Galaţi, E Romania 45°52´N 27°55´E
116 H13 **Târgu Cărbuneşti** *prev.* Tîrgu. Gorj, SW Romania 44°57´N 23°32´E
116 L9 **Târgu Frumos** *prev.* Tîrgu Frumos. Iaşi, NE Romania 47°12´N 27°00´E
116 H13 **Targu Jiu** *prev.* Tîrgu Jiu. Gorj, W Romania 45°03´N 23°20´E
116 H9 **Târgu Lăpuş** *prev.* Tîrgu Lăpuş. Maramureş, N Romania 47°28´N 23°54´E
Târgul-Neamţ *see* Târgu-Neamţ
Târgul-Săcuiesc *see* Târgu Secuiesc
116 I10 **Târgu Mureş** *prev.* Oşorhei, Tîrgu Mures, *Ger.* Neumarkt, *Hung.* Marosvásárhely. Mureş, C Romania 46°33´N 24°36´E
116 K9 **Târgu-Neamţ** *var.* Târgul-Neamţ; *prev.* Tîrgu-Neamţ. Neamţ, NE Romania 47°12´N 26°25´E
116 K11 **Târgu Ocna** *Hung.* Aknavásár; *prev.* Tîrgu Ocna. Bacău, E Romania 46°17´N 26°37´E
116 K11 **Târgu Secuiesc** *Ger.* Neumarkt, Szekler Neumarkt, *Hung.* Kezdivásárhely; *prev.* Chezdi-Oşorheiu, Târgul-Săcuiesc, Tîrgu Secuiesc. Covasna, E Romania 46°00´N 26°08´E
145 X10 **Targyn** Vostochnyy Kazakhstan, E Kazakhstan 49°32´N 82°47´E
Tar Heel State *see* North Carolina
186 C7 **Tari** Hela, W Papua New Guinea 05°52´S 142°58´E
162 J6 **Tarialan** *var.* Badrah. Hövsgöl, N Mongolia 49°33´N 101°58´E
162 I7 **Tariat** *var.* Horgo. Arhangay, C Mongolia 48°06´N 99°52´E
143 P17 **Ţarīf** Abū Z̧aby, C United Arab Emirates 24°02´N 53°47´E
104 K16 **Tarifa** Andalucía, S Spain 36°01´N 05°36´W
84 C14 **Tarifa, Punta de** *headland* SW Spain 36°01´N 05°39´W
57 M21 **Tarija** Tarija, S Bolivia 21°33´S 64°42´W
57 M21 **Tarija** ◈ *department* S Bolivia
141 R14 **Tarīm** C Yemen 16°N 48°50´E
Tarim Basin *see* Tarim Pendi
81 G19 **Tarime** Mara, N Tanzania 01°20´S 34°24´E
129 S8 **Tarim He** ∽ NW China
159 H8 **Tarim Pendi** *Eng.* Tarim Basin. *basin* NW China
149 N7 **Tarīn Kōt** *var.* Terinkot; *prev.* Tarīn Kowt. Uruzgān, C Afghanistan 32°38´N 65°52´E
Tarīn Kowt *see* Tarīn Kōt
171 O12 **Taripa** Sulawesi, C Indonesia 01°51´S 120°46´E
117 Q12 **Tarkhankut, Mys** *headland* S Ukraine 45°20´N 32°32´E
27 Q1 **Tarkio** Missouri, C USA 40°25´N 95°24´W
122 J9 **Tarko-Sale** Yamalo-Nenetskiy Avtonomnyy Okrug, N Russian Federation 64°55´N 77°34´E
77 P17 **Tarkwa** S Ghana 05°16´N 01°59´W
171 O3 **Tarlac** Luzon, N Philippines 15°29´N 120°34´E
95 F22 **Tarm** Midtjylland, W Denmark 55°55´N 08°32´E
57 E14 **Tarma** Junín, C Peru 11°28´S 75°41´W
103 N15 **Tarn** ◈ *department* S France
102 M15 **Tarn** ∽ S France
111 L22 **Tarna** ∽ C Hungary
92 G13 **Tärnaby** Västerbotten, N Sweden 65°44´N 15°20´E
149 P8 **Tarnak Rūd** ∽ SE Afghanistan
116 J11 **Târnava Mare** *Ger.* Grosse Kokel, *Hung.* Nagy-Küküllő; *prev.* Tîrnava Mare. ∽ S Romania
116 I11 **Târnava Mică** *Ger.* Kleine Kokel, *Hung.* Kis-Küküllo; *prev.* Tîrnava Mică. ∽ C Romania
116 I11 **Târnăveni** *Ger.* Marteskirch, Martinskirch, *Hung.* Dicsőszentmárton; *prev.* Sînmartin, Tîrnăveni. Mureş, C Romania 46°20´N 24°17´E
102 L14 **Tarn-et-Garonne** ◈ *department* S France
111 P18 **Tarnica** ▲ SE Poland 49°05´N 22°43´E
111 N15 **Tarnobrzeg** Podkarpackie, SE Poland 50°35´N 21°40´E
125 N12 **Tarnogskiy Gorodok** Vologodskaya Oblast', NW Russian Federation 60°28´N 43°45´E
Tarnopol *see* Ternopil'
111 M16 **Tarnów** Małopolskie, S Poland 50°01´N 20°59´E
Tarnowice/Tarnowitz *see* Tarnowskie Góry
111 J16 **Tarnowskie Góry** *var.* Tarnowice, Tarnowskie Gory, *Ger.* Tarnowitz. Śląskie, S Poland 50°27´N 18°52´E
95 N14 **Tärnsjö** Västmanland, C Sweden 60°10´N 16°57´E
186 K7 **Taro** Choiseul, NW Solomon Islands 07°00´S 156°57´E
106 E9 **Taro** ∽ NW Italy
186 I6 **Taron** New Ireland, NE Papua New Guinea 04°22´S 153°04´E
74 E8 **Taroudannt** *var.* Taroudant. SW Morocco 30°31´N 08°50´W
Taroudant *see* Taroudannt
23 V12 **Tarpon, Lake** ◎ Florida, SE USA
23 V12 **Tarpon Springs** Florida, SE USA 28°09´N 82°45´W
107 G14 **Tarquinia** *anc.* Tarquinii, *hist.* Corneto. Lazio, C Italy 42°23´N 11°45´E
Tarquinii *see* Tarquinia
Tarraco *see* Tarragona
76 D10 **Tarrafal** Santiago, S Cape Verde 15°16´N 23°45´W
105 V6 **Tarragona** *anc.* Tarraco. Cataluña, E Spain 41°07´N 01°15´E
105 T7 **Tarragona** ◈ *province* Cataluña, NE Spain
183 O17 **Tarraleah** Tasmania, SE Australia 42°11´S 146°29´E
23 P3 **Tarrant City** Alabama, S USA 33°34´N 86°45´W
185 D21 **Tarras** Otago, South Island, New Zealand 44°48´S 169°25´E
Tarrasa *see* Terrassa
105 U5 **Tàrrega** *var.* Tarrega. Cataluña, NE Spain 41°39´N 01°09´E
21 W9 **Tar River** ∽ North Carolina, SE USA
Tarsatica *see* Rijeka
136 J17 **Tarsus** İçel, S Turkey 36°52´N 34°52´E
62 K4 **Tartagal** Salta, N Argentina 22°32´S 63°50´W
137 V12 **Tärtär** *Rus.* Terter. ∽ SW Azerbaijan
102 J15 **Tartas** Landes, SW France 43°52´N 00°45´W
Tartlau *see* Prejmer
Tartous/Tartouss *see* Ţarţūs
118 J5 **Tartu** *Ger.* Dorpat; *prev. Rus.* Yurev, Yury'ev. Tartumaa, SE Estonia 58°20´N 26°44´E
118 I5 **Tartumaa** *off.* Tartu Maakond. ◈ *province* E Estonia
Tartu Maakond *see* Tartumaa
138 H5 **Ţarţūs** *Fr.* Tartouss; *anc.* Tortosa. Ţarţūs, W Syria 34°55´N 35°52´E
138 H5 **Ţarţūs** *off.* Muḩāfaz̧at Ţarţūs, *var.* Tartous, Tartus. ◈ *governorate* W Syria
Ţarţūs, Muḩāfaz̧at *see* Ţarţūs
164 C16 **Tarumizu** Kagoshima, Kyūshū, SW Japan 31°30´N 130°40´E
126 K4 **Tarusa** Kaluzhskaya Oblast', W Russian Federation 54°45´N 37°10´E
117 N11 **Tarutyne** Odes'ka Oblast', SW Ukraine 46°11´N 29°09´E
162 I7 **Tarvagatyn Nuruu** ▲▲ N Mongolia
106 J6 **Tarvisio** Friuli-Venezia Giulia, NE Italy 46°31´N 13°33´E
Tarvisium *see* Treviso
57 O16 **Tarvo, Río** ∽ E Bolivia
14 G8 **Tarzwell** Ontario, S Canada 48°00´N 79°58´W
40 K5 **Tasajera, Sierra de la** ▲▲ N Mexico
145 S13 **Tasaral** Karaganda, C Kazakhstan 46°17´N 73°54´E
145 N15 **Tasboget** *Kaz.* Tasböget; *prev.* Tasbuget. Kzylorda, S Kazakhstan 44°46´N 65°38´E
Tasböget *see* Tasboget
Tasbuget *see* Tasboget
108 E11 **Tasch** Valais, SW Switzerland 46°04´N 07°43´E
Tasek Kenyir *see* Kenyir, Tasik
122 J14 **Tashanta** Respublika Altay, S Russian Federation 49°42´N 89°15´E
Tashauz *see* Daşoguz
Tashi Chho Dzong *see* Thimphu
153 U11 **Tashigang** E Bhutan 27°19´N 91°33´E
137 T11 **Tashir** *prev.* Kalinino. N Armenia 41°07´N 44°16´E
143 Q11 **Ţashk, Daryācheh-ye** ◎ C Iran
Tashkent *see* Toshkent
Tashkentskaya Oblast' *see* Toshkent Viloyati
Tashkepri *see* Daşköpri
Tash-Kömür *see* Tash-Kumyr
147 S9 **Tash-Kumyr** *Kir.* Tash-Kömür. Dzhalal-Abadskaya Oblast', W Kyrgyzstan 41°22´N 72°09´E
127 T7 **Tashla** Orenburgskaya Oblast', W Russian Federation 51°42´N 52°33´E
Tashqurghan *see* Khulm
122 J13 **Tashtagol** Kemerovskaya Oblast', S Russian Federation 52°49´N 88°00´E
95 H24 **Tåsinge** *island* C Denmark
12 M5 **Tasiujaq** Québec, E Canada 58°43´N 69°58´W
144 F8 **Taskala** *prev.* Kamenka. Zapadnyy Kazakhstan, NW Kazakhstan 51°06´N 51°16´E
77 W11 **Tasker** Zinder, C Niger 15°06´N 10°42´E
145 W12 **Taskesken** Vostochnyy Kazakhstan, E Kazakhstan 47°15´N 80°45´E
136 J10 **Taşköprü** Kastamonu, N Turkey 41°30´N 34°12´E
Taskuduk, Peski *see* Tosquduq Qumlari
186 G5 **Taskul** New Ireland, NE Papua New Guinea 02°34´S 150°25´E
137 S13 **Taşlıçay** Ağrı, E Turkey 39°37´N 43°23´E
185 H14 **Tasman** *off.* Tasman District. ◈ *unitary authority* South Island, New Zealand
192 J12 **Tasman Basin** *var.* East Australian Basin. *undersea feature* S Tasman Sea
185 I14 **Tasman Bay** *inlet* South Island, New Zealand
Tasman District *see* Tasman
192 I13 **Tasman Fracture Zone** *tectonic feature* S Indian Ocean
185 E19 **Tasman Glacier** *glacier* South Island, New Zealand
Tasman Group *see* Nukumanu Islands
183 N15 **Tasmania** *prev.* Van Diemen's Land. ◈ *state* SE Australia
183 Q16 **Tasmania** *island* SE Australia
185 H14 **Tasman Mountains** ▲▲ South Island, New Zealand
183 P17 **Tasman Peninsula** *peninsula* Tasmania, SE Australia
192 I11 **Tasman Plain** *undersea feature* W Tasman Sea
192 I12 **Tasman Plateau** *var.* South Tasmania Plateau. *undersea feature* SW Tasman Sea
192 I11 **Tasman Sea** *sea* SW Pacific Ocean
116 G9 **Tăşnad** *Ger.* Trestenberg, Trestendorf, *Hung.* Tasnád. Satu Mare, NW Romania 47°30´N 22°33´E
136 L11 **Taşova** Amasya, N Turkey 40°45´N 36°20´E
77 T10 **Tassara** Tahoua, W Niger 16°40´N 05°34´E
12 K4 **Tassialouc, Lac** ◎ Québec, C Canada
Tassili du Hoggar *see* Tassili Ta-n-Ahaggar
74 L11 **Tassili-n-Ajjer** *plateau* E Algeria
74 K14 **Tassili Ta-n-Ahaggar** *var.* Tassili du Hoggar, Tassili ta-n-Ahaggar. *plateau* S Algeria
Tassili ta-n-Ahaggar *see* Tassili Ta-n-Ahaggar
59 M15 **Tasso Fragoso** Maranhão, E Brazil 08°22´S 45°53´W
Tåstrup *see* Taastrup
145 O9 **Tasty-Taldy** Akmola, C Kazakhstan 50°47´N 66°31´E
143 W10 **Tāsūkī** Sīstān va Balūchestān, SE Iran
111 I22 **Tata** *Ger.* Totis. Komárom-Esztergom, NW Hungary 47°39´N 18°19´E
74 E8 **Tata** SW Morocco 29°38´N 08°04´W
Tataaihoa, Pointe *see* Vénus, Pointe
111 I22 **Tatabánya** Komárom-Esztergom, NW Hungary 47°33´N 18°23´E
191 X10 **Tatakoto** *atoll* Îles Tuamotu, E French Polynesia
75 N7 **Tataouine** *var.* Taţāwīn. SE Tunisia 32°48´N 10°27´E
55 O5 **Tataracual, Cerro** ▲ NE Venezuela 10°13´N 64°20´W
117 O12 **Tatarbunary** Odes'ka Oblast', SW Ukraine 45°50´N 29°37´E
119 M17 **Tatarka** Mahilyowskaya Voblasts', E Belarus 53°15´N 28°50´E
Tatar Pazardzhik *see* Pazardzhik
122 I12 **Tatarsk** Novosibirskaya Oblast', C Russian Federation 55°08´N 75°58´E
Tatarskaya ASSR *see* Tatarstan, Respublika
123 T13 **Tatarskiy Proliv** *Eng.* Tatar Strait. *strait* SE Russian Federation
127 R4 **Tatarstan, Respublika** *prev.* Tatarskaya ASSR. ◈ *autonomous republic* W Russian Federation
Tatar Strait *see* Tatarskiy Proliv
Taţāwīn *see* Tataouine
171 N12 **Tate** Sulawesi, N Indonesia 0°12´S 119°44´E
141 N11 **Tathlīth** 'Asīr, S Saudi Arabia 19°38´N 43°32´E
141 O11 **Tathlīth, Wādī** *dry watercourse* S Saudi Arabia
183 R11 **Tathra** New South Wales, SE Australia 36°46´S 149°58´E
127 P8 **Tatishchevo** Saratovskaya Oblast', W Russian Federation 51°43´N 45°35´E
39 S12 **Tatitlek** Alaska, USA 60°49´N 146°29´W
10 L15 **Tatla Lake** British Columbia, SW Canada 51°54´N 124°39´W
121 Q2 **Tatlısu** *Gk.* Akanthoú. N Cyprus 35°21´N 33°45´E
11 Z10 **Tatnam, Cape** *headland* Manitoba, C Canada 57°16´N 91°03´W
111 K18 **Tatra Mountains** *Ger.* Tatra, *Hung.* Tátra, *Pol./Slvk.* Tatry. ▲▲ Poland/Slovakia
Tatra/Tátra *see* Tatra Mountains
Tatry *see* Tatra Mountains
164 I13 **Tatsuno** *var.* Tatuno. Hyōgo, Honshū, SW Japan 34°54´N 134°30´E
Tatti *see* Tatty
145 S16 **Tatty** *prev.* Tatti. Zhambyl, S Kazakhstan 43°11´N 73°22´E
60 L10 **Tatuí** São Paulo, S Brazil 23°21´S 47°49´W
37 V14 **Tatum** New Mexico, SW USA 33°15´N 103°19´W
25 X7 **Tatum** Texas, SW USA 32°19´N 94°31´W
Ta-t'ung/Tatung *see* Datong
Tatuno *see* Tatsuno
137 R14 **Tatvan** Bitlis, SE Turkey 38°31´N 42°15´E
95 C16 **Tau** Rogaland, S Norway 59°04´N 05°55´E
192 L17 **Ta'ū** *var.* Tau. *island* Manua Islands, E American Samoa
193 W15 **Tau** *island* Tongatapu Group, N Tonga
59 O14 **Tauá** Ceará, E Brazil 06°04´S 40°26´W
60 N10 **Taubaté** São Paulo, S Brazil 23°S 45°36´W
101 I19 **Tauber** ∽ SW Germany
101 I19 **Tauberbiscofsheim** Baden-Württemberg, C Germany 49°37´N 09°39´E
Tauchik *see* Taushyk
191 W10 **Tauere** *atoll* Îles Tuamotu, C French Polynesia
101 H17 **Taufstein** ▲ C Germany 50°31´N 09°18´E
190 I17 **Taukoka** *island* SE Cook Islands
145 T15 **Taukum, Peski** *desert* SE Kazakhstan
184 L10 **Taumarunui** Manawatu-Wanganui, North Island, New Zealand 38°52´S 175°14´E
59 A15 **Taumaturgo** Acre, W Brazil 08°54´S 72°48´W
27 X6 **Taum Sauk Mountain** ▲ Missouri, C USA 37°34´N 90°43´W
83 H22 **Taung** North-West, N South Africa 27°32´S 24°48´E
166 L6 **Taungdwingyi** Magway, C Myanmar (Burma) 20°01´N 95°20´E
166 M6 **Taunggyi** Shan State, C Myanmar (Burma) 20°47´N 97°00´E
166 M7 **Taungoo** Bago, C Myanmar (Burma) 18°57´N 96°26´E
166 L5 **Taungtha** Mandalay, C Myanmar (Burma) 21°16´N 95°25´E
Taungup *see* Toungup
149 S9 **Taunsa** Punjab, E Pakistan 30°43´N 70°41´E
97 K23 **Taunton** SW England, United Kingdom 51°01´N 03°06´W
19 O12 **Taunton** Massachusetts, NE USA 41°54´N 71°03´W
101 F18 **Taunus** ▲▲ W Germany
101 G18 **Taunusstein** Hessen, W Germany 50°09´N 08°09´E
184 N9 **Taupo** Waikato, North Island, New Zealand 38°42´S 176°05´E
184 M9 **Taupo, Lake** ◎ North Island, New Zealand
109 R8 **Taurach** *var.* Taurachbach. ∽ E Austria
Taurachbach *see* Taurach
118 D12 **Tauragė** *Ger.* Tauroggen. Tauragė, SW Lithuania 55°16´N 22°17´E
118 D13 **Tauragė** ◈ *province* Lithuania
54 G10 **Tauramena** Casanare, C Colombia 05°02´N 72°43´W
184 N7 **Tauranga** Bay of Plenty, North Island, New Zealand 37°42´S 176°09´E
15 O10 **Taureau, Réservoir** ⊠ Québec, SE Canada
107 N22 **Taurianova** Calabria, SW Italy 38°22´N 16°01´E
Tauris *see* Tabrīz
184 I2 **Tauroa Point** *headland* North Island, New Zealand 35°09´S 173°02´E
Tauroggen *see* Tauragė
Tauromenium *see* Taormina
Taurus Mountains *see* Toros Dağları
Taus *see* Domažlice
144 E14 **Taushyk** *Kaz.* Taūshyq; *prev.* Tauchik. Mangistau, SW Kazakhstan 44°17´N 51°22´E
Taūshyq *see* Taushyk
105 R5 **Tauste** Aragón, NE Spain 41°55´N 01°15´W
191 V16 **Tautara, Motu** *island* Easter Island, Chile, E Pacific Ocean
191 R8 **Tautira** Tahiti, W French Polynesia 17°45´S 149°10´W
Tauz *see* Tovuz
Ţavālesh, Kūhhā-ye *see* Talish Mountains
136 D15 **Tavas** Denizli, SW Turkey 37°33´N 29°04´E
Tavastehus *see* Hämeenlinna
Tavau *see* Davos
122 G10 **Tavda** Sverdlovskaya Oblast', C Russian Federation 58°01´N 65°07´E
122 G10 **Tavda** ∽ C Russian Federation
105 T11 **Tavernes de la Valldigna** Valenciana, E Spain 39°03´N 00°13´W
81 I20 **Taveta** Taita/Taveta, S Kenya 03°23´S 37°40´E
187 Y14 **Taveuni** *island* N Fiji
147 R13 **Tavildara** *Rus.* Tavil'dara, Tovil'-Dora. C Tajikistan 38°42´N 70°27´E
Tavin *see* Adaatsag
104 H14 **Tavira** Faro, S Portugal 37°07´N 07°39´W
97 I24 **Tavistock** SW England, United Kingdom 50°33´N 04°08´W
Tavoy *see* Dawei
Tavoy Island *see* Mali Kyun
115 E16 **Tavropoú, Techníti Límni** ⊠ C Greece
136 E13 **Tavşanlı** Kütahya, NW Turkey 39°34´N 29°28´E
187 X14 **Tavua** Viti Levu, W Fiji 17°27´S 177°51´E
97 J23 **Taw** ∽ SW England, United Kingdom
185 L14 **Tawa** Wellington, North Island, New Zealand 41°10´S 174°50´E
25 V6 **Tawakoni, Lake** ⊠ Texas, SW USA
153 V11 **Tawang** Arunāchal Pradesh, NE India 27°34´N 91°54´E
169 R17 **Tawang, Teluk** *bay* Jawa, S Indonesia
31 R7 **Tawas Bay** ◎ Michigan, N USA
31 R7 **Tawas City** Michigan, N USA 44°16´N 83°33´W
169 V8 **Tawau** Sabah, East Malaysia 04°16´N 117°54´E
141 U10 **Ţawīl, Qalamat aţ** *well* SE Saudi Arabia
171 N9 **Tawitawi** *island* Tawitawi Group, SW Philippines
Ţawkar *see* Tokar
Tāwūq *see* Dāqūq
Tawzar *see* Tozeur
41 O15 **Taxco** *var.* Taxco de Alarcón. Guerrero, S Mexico 18°32´N 99°37´W
Taxco de Alarcón *see* Taxco
146 H8 **Taxiatosh** *Rus.* Takhiatash. Qoraqalpog'iston Respublikasi, W Uzbekistan 42°27´N 59°27´E
158 D9 **Taxkorgan** *var.* Taxkorgan Tajik Zizhixian. Xinjiang Uygur Zizhiqu, NW China 37°43´N 75°13´E
Taxkorgan Tajik Zizhixian *see* Taxkorgan
146 H7 **Taxtako'pir** *Rus.* Takhtakupyr. Qoraqalpog'iston Respublikasi, NW Uzbekistan 43°04´N 60°23´E
96 J10 **Tay** ∽ C Scotland, United Kingdom
143 V6 **Tāybād** *var.* Taibad, Tāyybād, Tayyebāt. Khorāsān-e Raẕavī, NE Iran 34°48´N 60°46´E
Taybert at Turkz *see* Ţayyibat at Turkī
124 J3 **Taybola** Murmanskaya Oblast', NW Russian Federation 68°30´N 33°18´E
81 M16 **Tayeeglow** Bakool, C Somalia 04°01´N 44°25´E
96 K11 **Tay, Firth of** *inlet* E Scotland, United Kingdom
122 J12 **Tayga** Kemerovskaya Oblast', S Russian Federation 56°02´N 85°26´E
Taygan *see* Delger
123 T9 **Taygonos, Mys** *headland* E Russian Federation 60°36´N 160°09´E
96 I11 **Tay, Loch** ◎ C Scotland, United Kingdom
11 N12 **Taylor** British Columbia, W Canada 56°09´N 120°43´W
29 O14 **Taylor** Nebraska, C USA 41°47´N 99°23´W
18 I13 **Taylor** Pennsylvania, NE USA 41°22´N 75°41´W
25 T10 **Taylor** Texas, SW USA 30°34´N 97°24´W
37 Q11 **Taylor, Mount** ▲ New Mexico, SW USA 35°14´N 107°36´W
37 R5 **Taylor Park Reservoir** ⊠ Colorado, C USA
37 R6 **Taylor River** ∽ Colorado, C USA
21 P11 **Taylors** South Carolina, SE USA 34°55´N 82°18´W
20 L5 **Taylorsville** Kentucky, S USA 38°01´N 85°21´W
21 R6 **Taylorsville** North Carolina, SE USA 35°56´N 81°10´W
30 L14 **Taylorville** Illinois, N USA 39°33´N 89°17´W
140 K5 **Taymā'** Tabūk, NW Saudi Arabia 27°39´N 38°32´E
122 M10 **Taymura** ∽ C Russian Federation
123 O7 **Taymylyr** Respublika Sakha (Yakutiya), NE Russian Federation 72°32´N 121°54´E
122 L7 **Taymyr, Ozero** ◎ N Russian Federation
122 M6 **Taymyr, Poluostrov** *peninsula* N Russian Federation
122 L8 **Taymyrskiy (Dolgano-Nenetskiy) Avtonomnyy Okrug** ◈ *autonomous district* Krasnoyarskiy Kray, N Russian Federation
167 S13 **Tây Ninh** Tây Ninh, S Vietnam 11°21´N 106°07´E
122 L12 **Tayshet** Irkutskaya Oblast', S Russian Federation 55°51´N 98°04´E
162 G8 **Tayshir** *var.* Tsagaan-Olom. Govĭ-Altay, C Mongolia 46°42´N 96°30´E
171 N5 **Taytay** Palawan, W Philippines 10°49´N 119°30´E
169 Q16 **Tayu** *prev.* Tajoe. Jawa, C Indonesia 06°32´S 111°02´E
Tāyybād/Tayyebāt *see* Tāybād
138 L5 **Ţayyibah** *var.* At Taybé. Ḩimş, C Syria 35°13´N 38°51´E
138 I4 **Ţayyibat at Turkī** *var.* Taybert at Turkz. Ḩamāh, W Syria 35°16´N 36°55´E
145 P7 **Tayynsha** *prev.* Krasnoarmeysk. Severnyy Kazakhstan, N Kazakhstan 53°52´N 69°51´E
122 J10 **Taz** ∽ N Russian Federation
74 G6 **Taza** NE Morocco 34°13´N 04°06´W
139 T4 **Tāza Khurmātū** Kirkūk, E Iraq 35°18´N 44°22´E
165 Q8 **Tazawa-ko** ◎ Honshū, C Japan
21 N8 **Tazewell** Tennessee, S USA 36°27´N 83°34´W
21 Q7 **Tazewell** Virginia, NE USA 37°08´N 81°34´W
75 S11 **Tāzirbū** SE Libya 25°43´N 21°16´E
39 S11 **Tazlina Lake** ◎ Alaska, USA
122 J8 **Tazovskiy** Yamalo-Nenetskiy Avtonomnyy Okrug, N Russian Federation 67°33´N 78°21´E
137 U10 **Tbilisi** *Eng.* Tiflis. ● (Georgia) SE Georgia 41°41´N 44°55´E
137 T10 **Tbilisi** ✈ S Georgia 41°43´N 44°49´E
79 E14 **Tchabal Mbabo** ▲ NW Cameroon 07°12´N 12°16´E
Tchad *see* Chad
Tchad, Lac *see* Chad, Lake
77 S15 **Tchaourou** E Benin 08°58´N 02°40´E
79 E20 **Tchibanga** Nyanga, S Gabon 02°49´S 11°00´E
Tchien *see* Zwedru
77 Z6 **Tchigaï, Plateau du** ▲▲ NE Niger
77 V9 **Tchighozérine** Agadez, C Niger 17°15´N 07°48´E
77 T10 **Tchin-Tabaradene** Tahoua, W Niger 15°57´N 05°49´E
78 G13 **Tcholliré** Nord, NE Cameroon 08°48´N 14°00´E
Tchongking *see* Chongqing
22 K4 **Tchula** Mississippi, S USA 33°10´N 90°13´W
110 I7 **Tczew** *Ger.* Dirschau. Pomorskie, N Poland 54°05´N 18°46´E
116 I10 **Teaca** *Ger.* Tekendorf, *Hung.* Teke; *prev. Ger.* Teckendorf. Bistriţa-Năsăud, N Romania 46°55´N 24°30´E
40 J11 **Teacapán** Sinaloa, C Mexico 22°33´N 105°44´W
190 A10 **Teafuafou** *island* Funafuti Atoll, C Tuvalu
25 U8 **Teague** Texas, SW USA 31°37´N 96°16´W
191 R9 **Teahupoo** Tahiti, W French Polynesia 17°51´S 149°15´W
190 H15 **Te Aiti Point** *headland* Rarotonga, S Cook Islands 21°11´S 59°47´W
65 D24 **Teal Inlet** East Falkland, Falkland Islands 51°34´S 58°25´W
185 B22 **Te Anau** Southland, South Island, New Zealand 45°25´S 167°45´E
185 B22 **Te Anau, Lake** ◎ South Island, New Zealand
41 U15 **Teapa** Tabasco, SE Mexico 17°36´N 92°57´W
184 Q7 **Te Araroa** Gisborne, North Island, New Zealand 37°37´S 178°21´E
184 M7 **Te Aroha** Waikato, North Island, New Zealand 37°32´S 175°58´E
Teate *see* Chieti
190 A9 **Te Ava Fuagea** *channel* Funafuti Atoll, SE Tuvalu
190 B8 **Te Ava I Te Lape** *channel* Funafuti Atoll, SE Tuvalu
190 B9 **Te Ava Pua Pua** *channel* Funafuti Atoll, SE Tuvalu
184 M8 **Te Awamutu** Waikato, North Island, New Zealand 38°00´S 177°18´E
171 X12 **Teba** Papua, E Indonesia 01°27´S 137°54´E
104 L15 **Teba** Andalucía, S Spain 36°59´N 04°54´W
126 M15 **Teberda** Karachayevo-Cherkesskaya Respublika, SW Russian Federation 43°28´N 41°45´E
74 M6 **Tébessa** NE Algeria 35°21´N 08°06´E
62 O7 **Tebicuary, Río** ∽ S Paraguay
168 L13 **Tebingtinggi** Sumatera, W Indonesia 03°33´S 103°00´E
168 I8 **Tebingtinggi** Sumatera, N Indonesia 03°20´N 99°08´E
Tebingtinggi, Pulau *see* Rantau, Pulau
Tebriz *see* Tabrīz
137 U9 **T'ebulos Mta** *Rus.* Gora Tebulosmta; *prev.* Tebulos Mta. ▲ Georgia/Russian Federation 42°33´N 45°21´E
Tebulos Mta *see* T'ebulos Mta
Tebulosmta, Gora *see* T'ebulos Mta
41 Q14 **Tecamachalco** Puebla, S Mexico 18°52´N 97°44´W
40 B1 **Tecate** Baja California Norte, NW Mexico 32°33´N 116°38´W
136 M13 **Tecer Dağları** ▲▲ C Turkey
103 O17 **Tech** ∽ S France
77 P16 **Techiman** W Ghana 07°35´N 01°56´W
117 N15 **Techirghiol** Constanţa, SE Romania 44°03´N 28°37´E
74 A12 **Techla** *var.* Techlé. SW Western Sahara 21°39´N 14°57´W
Techlé *see* Techla
63 H18 **Tecka, Sierra de** ▲▲ SW Argentina
Teckendorf *see* Teaca
40 K13 **Tecolotlán** Jalisco, SW Mexico 20°10´N 104°07´W
40 K14 **Tecomán** Colima, SW Mexico 18°53´N 103°54´W
35 V12 **Tecopa** California, W USA 35°51´N 116°14´W
40 G5 **Tecoripa** Sonora, NW Mexico 28°38´N 109°58´W
41 N16 **Tecpan** *var.* Tecpan de Galeana. Guerrero, S Mexico 17°12´N 100°39´W
Tecpan de Galeana *see* Tecpan
40 J11 **Tecuala** Nayarit, C Mexico 22°24´N 105°30´W
116 L12 **Tecuci** Galaţi, E Romania 45°50´N 27°27´E
31 R10 **Tecumseh** Michigan, N USA 42°00´N 83°57´W
29 S16 **Tecumseh** Nebraska, C USA 40°22´N 96°12´W
27 O11 **Tecumseh** Oklahoma, C USA 35°15´N 96°56´W
Tedzhen *see* Harīrūd/Tejen
Tedzhen *see* Tejen
146 H15 **Tedzhenstroy** *Turkm.* Tejenstroy. Ahal Welaýaty, S Turkmenistan 36°57´N 60°49´E
Teel *see* Öndör-Ulaan
97 L15 **Tees** ∽ N England, United Kingdom
14 E15 **Teeswater** Ontario, S Canada 44°00´N 81°17´W
190 A10 **Tefala** *island* Funafuti Atoll, C Tuvalu
58 D13 **Tefé** Amazonas, N Brazil 03°24´S 64°45´W
74 K11 **Tefedest** ▲▲ S Algeria
136 E16 **Tefenni** Burdur, SW Turkey 37°19´N 29°45´E
58 D13 **Tefé, Rio** ∽ NW Brazil
169 P16 **Tegal** Jawa, C Indonesia 06°52´S 109°07´E
100 O12 **Tegel** ✈ (Berlin) Berlin, NE Germany 52°33´N 13°16´E
99 M15 **Tegelen** Limburg, SE Netherlands 51°20´N 06°09´E
101 L24 **Tegernsee** ◎ SE Germany
107 M18 **Teggiano** Campania, S Italy 40°25´N 15°28´E
77 U14 **Tegina** Niger, C Nigeria 10°06´N 06°10´E
Tegucigalpa *see* Central District
Tegucigalpa *see* Toncontín
Tegucigalpa *see* Francisco Morazán
77 U9 **Teguidda-n-Tessoumt** Agadez, C Niger 17°27´N 06°40´E
64 Q11 **Teguise** Lanzarote, Islas Canarias, Spain, NE Atlantic Ocean 29°04´N 13°38´W
122 K12 **Tegul'det** Tomskaya Oblast', C Russian Federation 57°16´N 87°58´E
35 S13 **Tehachapi** California, W USA 35°07´N 118°27´W
35 S13 **Tehachapi Mountains** ▲▲ California, W USA
Tehama *see* Tihāmah
Teheran *see* Tehrān
77 O14 **Téhini** NE Ivory Coast 09°36´N 03°40´W
143 N5 **Tehrān** *var.* Teheran. ● (Iran) Tehrān, N Iran 35°44´N 51°27´E
143 N6 **Tehrān** *off.* Ostān-e Tehrān, *var.* Tehran. ◈ *province* N Iran
Tehrān, Ostān-e *see* Tehrān
Tehri *see* Tīkamgarh
Tehri *see* New Tehri
41 Q15 **Tehuacán** Puebla, S Mexico 18°29´N 97°24´W
41 S17 **Tehuantepec** *var.* Santo Domingo Tehuantepec. Oaxaca, SE Mexico 16°18´N 95°14´W
41 S17 **Tehuantepec, Golfo de** *var.* Gulf of Tehuantepec. *gulf* S Mexico
Tehuantepec, Gulf of *see* Tehuantepec, Golfo de
Tehuantepec, Isthmus of *see* Tehuantepec, Istmo de
41 T16 **Tehuantepec, Istmo de** *var.* Isthmus of Tehuantepec. *isthmus* SE Mexico
0 I16 **Tehuantepec Ridge** *undersea feature* E Pacific Ocean 13°30´N 98°00´W
41 S16 **Tehuantepec, Río** ∽ SE Mexico
191 W10 **Tehuata** *atoll* Îles Tuamotu, C French Polynesia
64 O11 **Teide, Pico del** ▲ Gran Canaria, Islas Canarias, Spain, NE Atlantic Ocean 28°16´N 16°39´W
97 I21 **Teifi** ∽ SW Wales, United Kingdom
80 B9 **Teiga Plateau** *plateau* W Sudan
97 J24 **Teignmouth** SW England, United Kingdom 50°34´N 03°29´W
Teisen *see* Jecheon
116 H1 **Teiuş** *Ger.* Dreikirchen, *Hung.* Tövis. Alba, C Romania 46°12´N 23°40´E
169 U17 **Tejakula** Bali, C Indonesia 08°09´S 115°19´E
146 H14 **Tejen** *Rus.* Tedzhen. Ahal Welaýaty, S Turkmenistan 37°24´N 60°29´E
146 I15 **Tejen** *Per.* Harīrūd, *Rus.* Tedzhen. ∽ Afghanistan/Iran *see also* Harīrūd
Tejen *see* Harīrūd
Tejenstroy *see* Tedzhenstroy
35 S14 **Tejon Pass** *pass* California, W USA
Tejo, Rio *see* Tagus
41 O14 **Tejupilco** *var.* Tejupilco de Hidalgo. México, S Mexico 18°55´N 100°10´W
Tejupilco de Hidalgo *see* Tejupilco
184 P7 **Te Kaha** Bay of Plenty, North Island, New Zealand 37°45´S 177°42´E
29 S14 **Tekamah** Nebraska, C USA 41°46´N 96°13´W
184 I1 **Te Kao** Northland, North Island, New Zealand 34°40´S 172°57´E
185 F20 **Tekapo** ∽ South Island, New Zealand
185 F19 **Tekapo, Lake** ◎ South Island, New Zealand
184 P9 **Te Karaka** Gisborne, North Island, New Zealand 38°30´S 177°52´E
184 L7 **Te Kauwhata** Waikato, North Island, New Zealand 37°22´S 175°07´E
41 X12 **Tekax** *var.* Tekax de Álvaro Obregón. Yucatán, SE Mexico 20°07´N 89°10´W
Tekax de Álvaro Obregón *see* Tekax
136 A14 **Teke Burnu** *headland* W Turkey 38°06´N 26°35´E
114 M12 **Teke Deresi** ∽ NW Turkey
146 D10 **Tekedzhik, Gory** *hill range* NW Turkmenistan
145 V14 **Tekeli** Almaty, SE Kazakhstan 44°50´N 78°47´E
145 R7 **Teke, Ozero** ◎ N Kazakhstan
158 I5 **Tekes** Xinjiang Uygur Zizhiqu, NW China 43°15´N 81°43´E
145 W16 **Tekes** Almaty, SE Kazakhstan 42°40´N 80°01´E

◆ Country ● Country Capital ◇ Dependent Territory ○ Dependent Territory Capital ◈ Administrative Regions ✈ International Airport ▲ Mountain ▲▲ Mountain Range ⛰ Volcano ∽ River ◎ Lake ⊠ Reservoir
331

TEKES – THIEF RIVER FALLS

Tekes see Tekes He
158 H5 **Tekes He** Rus. Tekes. ~ China/Kazakhstan
Teke/Tekendorf see Teaca
80 I10 **Tekezē** var. Takkaze. ~ Eritrea/Ethiopia
Tekhtin see Tsyakhtsin
136 C10 **Tekirdağ** It. Rodosto; anc. Bisanthe, Raidestos, Rhaedestus. Tekirdağ, NW Turkey 40°59′N 27°31′E
136 C10 **Tekirdağ** ◈ province NW Turkey
155 N14 **Tekkali** Andhra Pradesh, E India 18°37′N 84°15′E
115 K15 **Tekke Burnu** Turk. İlyasbaba Burnu. headland NW Turkey 40°03′N 26°12′E
137 Q13 **Tekman** Erzurum, NE Turkey 39°39′N 41°31′E
32 M9 **Tekoa** Washington, NW USA 47°13′N 117°05′W
190 H16 **Te Kou** ▲ Rarotonga, S Cook Islands 21°14′S 159°46′W
Tekrit see Tikrīt
171 P12 **Teku** Sulawesi, N Indonesia 0°46′S 123°25′E
184 L9 **Te Kuiti** Waikato, North Island, New Zealand 38°21′S 175°10′E
42 H4 **Tela** Atlántida, NW Honduras 15°46′N 87°25′W
138 F12 **Telalim** Southern, S Israel 30°58′N 34°47′E
Telanaipura see Jambi
155 I15 **Telangana** off. State of Telangana. ◈ state E India
Telangana, State of see Telangana
137 U10 **Telavi** prev. T'elavi. E Georgia 41°55′N 45°29′E
T'elavi see Telavi
138 F10 **Tel Aviv** ◈ district W Israel
Tel Aviv-Jaffa see Tel Aviv-Yafo
138 F10 **Tel Aviv-Yafo** var. Tel Aviv-Jaffa. Tel Aviv, C Israel 32°05′N 34°46′E
111 E18 **Telč** Ger. Teltsch. Vysočina, C Czech Republic 49°10′N 15°28′E
186 B6 **Telefomin** West Sepik, NW Papua New Guinea 05°08′S 141°31′E
10 J10 **Telegraph Creek** British Columbia, W Canada 57°56′N 131°10′W
190 B10 **Telele** island Funafuti Atoll, C Tuvalu
60 J11 **Telêmaco Borba** Paraná, S Brazil 24°20′S 50°44′W
95 E15 **Telemark** ◈ county S Norway
62 J13 **Telén** La Pampa, C Argentina 36°20′S 65°31′W
Teleneshty see Teleneşti
116 M9 **Teleneşti** Rus. Teleneshty. C Moldova 47°35′N 28°20′E
104 J4 **Teleno, El** ▲ NW Spain 42°19′N 06°21′W
116 I15 **Teleorman** ◈ county S Romania
116 J14 **Teleorman** ~ S Romania
25 V5 **Telephone** Texas, SW USA 33°48′N 96°00′W
35 U11 **Telescope Peak** ▲ California, W USA 36°09′N 117°03′W
Teles Pirés see São Manuel, Rio
97 L19 **Telford** W England, United Kingdom 52°42′N 02°28′W
108 L7 **Telfs** Tirol, W Austria 47°19′N 11°05′E
42 I9 **Telica** León, NW Nicaragua 12°30′N 86°52′W
42 J6 **Telica, Río** ~ C Honduras
76 I13 **Télimélé** W Guinea 10°45′N 13°02′W
43 O14 **Telire, Río** ~ Costa Rica/Panama
114 I8 **Telish** prev. Azizie. Pleven, N Bulgaria 43°20′N 24°15′E
41 R16 **Telixtlahuaca** var. San Francisco Telixtlahuaca. Oaxaca, SE Mexico 17°18′N 96°54′W
10 K13 **Telkwa** British Columbia, SW Canada 54°39′N 126°51′W
25 P4 **Tell** Texas, SW USA 34°18′N 100°20′W
Tell Abiad see Tall Abyaḍ
Tell Abiad/Tell Abyad see At Tall al Abyaḍ
31 O16 **Tell City** Indiana, N USA 37°56′N 86°47′W
38 M9 **Teller** Alaska, USA 65°15′N 166°21′W
Tell Huqnah see Tall Ḥuqnah
155 F20 **Tellicherry** var. Thalashsheri, Thalassery. Kerala, SW India 11°44′N 75°29′E see also Thalassery
20 M10 **Tellico Plains** Tennessee, S USA 35°19′N 84°18′W
Tell Kalakh see Tall Kalakh
Tell Mardīkh see Ebla
54 E11 **Tello** Huila, C Colombia 03°06′N 75°08′W
Tell Shedadi see Ash Shadādah
37 Q7 **Telluride** Colorado, C USA 37°56′N 107°48′W
117 X9 **Tel'manove** Donets'ka Oblast', E Ukraine 47°24′N 38°03′E
Tel'man/Tel'mansk see Gubadag
162 H6 **Telmen** var. Övögdiy. Dzavhan, C Mongolia 48°38′N 97°39′E
162 H6 **Telmen Nuur** ◎ NW Mongolia
Teloekbetoeng see Bandar Lampung
41 O15 **Teloloapán** Guerrero, S Mexico 18°21′N 99°52′W
Telo Martius see Toulon
Telposiz, Gora see Telpoziz, Gora
125 V8 **Telpoziz, Gora** prev. Gora Telposiz. ▲ NW Russian Federation 63°52′N 59°15′E
Telschen see Telšiai
63 J17 **Telsen** Chubut, S Argentina 42°27′S 66°59′W
118 D11 **Telšiai** Ger. Telschen. Telšiai, NW Lithuania 55°59′N 22°21′E
118 D11 **Telšiai** ◈ province NW Lithuania
Teltsch see Telč
Telukbetung see Bandar Lampung
168 H10 **Telukdalam** Pulau Nias, W Indonesia 0°34′N 97°47′E
14 H9 **Temagami** Ontario, S Canada 47°03′N 79°47′W
14 G9 **Temagami, Lake** ◎ Ontario, S Canada
190 H16 **Te Manga** ▲ Rarotonga, S Cook Islands 21°13′S 159°45′W
191 W12 **Tematagi** prev. Tematangi. atoll Îles Tuamotu, S French Polynesia
Tematangi see Tematagi
41 X11 **Temax** Yucatán, SE Mexico 21°10′N 88°53′W
171 E14 **Tembagapura** Papua, E Indonesia 04°10′S 137°19′E
129 U5 **Tembenchi** ~ N Russian Federation
55 P6 **Temblador** Monagas, NE Venezuela 08°59′N 62°44′W
105 N9 **Tembleque** Castilla-La Mancha, C Spain 39°41′N 03°30′W
Temboni see Mitemele, Río
35 U16 **Temecula** California, W USA 33°29′N 117°09′W
168 K7 **Temengor, Tasik** ⊠ Peninsular Malaysia
112 L9 **Temerin** Vojvodina, N Serbia 45°25′N 19°54′E
Temeschburg/Temeschwar see Timişoara
Temes-Kubin see Kovin
Temes/Temesch see Tamiš
Temesvár/Temeswar see Timişoara
Teminaboean see Teminabuan
171 U12 **Teminabuan** prev. Teminaboean. Papua Barat, E Indonesia 01°30′S 131°59′E
145 P17 **Temirlan** prev. Temirlanovka. Yuzhnyy Kazakhstan, S Kazakhstan 42°36′N 69°17′E
Temirlanovka see Temirlan
145 R10 **Temirtau** prev. Samarkandski, Samarkandskoye. Karaganda, C Kazakhstan 50°05′N 72°55′E
14 H10 **Témiscaming** Québec, SE Canada 46°40′N 79°04′W
Témiscamingue, Lac see Timiskaming, Lake
15 T8 **Témiscouata, Lac** ◎ Québec, SE Canada
127 N5 **Temnikov** Respublika Mordoviya, W Russian Federation 54°39′N 43°09′E
191 Y13 **Temoe** island Îles Gambier, E French Polynesia
183 Q9 **Temora** New South Wales, SE Australia 34°28′S 147°33′E
40 H7 **Témoris** Chihuahua, W Mexico 27°16′N 108°15′W
40 I5 **Temósachic** Chihuahua, N Mexico 28°55′N 107°42′W
187 Q10 **Temotu** var. Temotu Province. ◈ province E Solomon Islands
Temotu Province see Temotu
36 L14 **Tempe** Arizona, SW USA 33°24′N 111°54′W
Tempelburg see Czaplinek
107 C17 **Tempio Pausania** Sardegna, Italy, C Mediterranean Sea 40°55′N 09°07′E
42 K12 **Tempisque, Río** ~ NW Costa Rica
25 T9 **Temple** Texas, SW USA 31°06′N 97°22′W
100 O12 **Templehof** ✈ (Berlin) Berlin, NE Germany 52°28′N 13°24′E
97 D19 **Templemore** Ir. An Teampall Mór. Tipperary, C Ireland 52°48′N 07°50′W
100 O11 **Templin** Brandenburg, NE Germany 53°07′N 13°31′E
41 P12 **Tempoal** var. Tempoal de Sánchez. Veracruz-Llave, E Mexico 21°32′N 98°23′W
Tempoal de Sánchez see Tempoal
41 P13 **Tempoal, Río** ~ C Mexico
83 E14 **Tempué** Moxico, C Angola 13°36′S 18°56′E
126 J14 **Temryuk** Krasnodarskiy Kray, SW Russian Federation 45°15′N 37°26′E
99 G17 **Temse** Oost-Vlaanderen, N Belgium 51°08′N 04°13′E
63 F15 **Temuco** Araucanía, C Chile 38°45′S 72°37′W
185 G20 **Temuka** Canterbury, South Island, New Zealand 44°14′S 171°17′E
189 P13 **Temwen Island** island E Micronesia
56 C6 **Tena** Napo, C Ecuador 01°00′S 77°48′W
41 W13 **Tenabo** Campeche, E Mexico 20°02′N 90°12′W
Tenaghau see Aola
25 X7 **Tenaha** Texas, SW USA 31°56′N 94°14′W
39 X13 **Tenake** Chichagof Island, Alaska, USA 57°46′N 135°13′W
155 K16 **Tenāli** Andhra Pradesh, E India 16°13′N 80°36′E
Tenan see Cheonan
41 O14 **Tenancingo** var. Tenencingo de Degollado. México, S Mexico 18°57′N 99°39′W
191 X12 **Tenararo** island Groupe Actéon, SE French Polynesia
Tenasserim see Taninthayi
Tenasserim see Tanintharyi
98 O5 **Ten Boer** Groningen, NE Netherlands 53°16′N 06°42′E
97 I21 **Tenby** SW Wales, United Kingdom 51°41′N 04°43′W
80 K11 **Tendaho** Āfar, NE Ethiopia 11°39′N 40°59′E
103 V14 **Tende** Alpes Maritimes, SE France 44°04′N 07°34′E
151 Q20 **Ten Degree Channel** strait Andaman and Nicobar Islands, India, E Indian Ocean
80 F11 **Tendelti** White Nile, E Sudan 13°01′N 31°55′E
76 G8 **Te-n-Dghâmcha, Sebkhet** var. Sebkha de Ndrhamcha, Sebkra de Ndaghamcha. salt lake W Mauritania
165 P10 **Tendō** Yamagata, Honshū, C Japan 38°22′N 140°22′E
74 H7 **Tendrara** NE Morocco 33°06′N 01°58′W
117 Q11 **Tendrivs'ka Kosa** spit S Ukraine
117 Q11 **Tendrivs'ka Zatoka** gulf S Ukraine
Tenencingo de Degollado see Tenancingo
77 N11 **Ténenkou** Mopti, C Mali 14°28′N 04°55′W
77 W9 **Ténéré** physical region C Niger
77 W9 **Ténére, Erg du** desert C Niger
64 O11 **Tenerife** island Islas Canarias, Spain, NE Atlantic Ocean
74 J5 **Ténès** NW Algeria 36°35′N 01°18′E
170 M15 **Tengah, Kepulauan** island group C Indonesia
Tengcheng see Tengxian
169 V11 **Tenggarong** Borneo, C Indonesia 0°23′S 117°00′E
162 J15 **Tengger Shamo** desert N China
168 L8 **Tenggul, Pulau** island Peninsular Malaysia
Tengiz Köl see Teniz, Ozero
76 M14 **Tengréla** var. Tingréla. N Ivory Coast 10°26′N 06°20′W
160 M14 **Tengxian** var. Tengcheng, Tengxian, Teng Xian. Guangxi Zhuangzu Zizhiqu, S China 23°24′N 110°49′E
Teng Xian see Tengxian
Tengxian see Tengxian
194 H2 **Teniente Rodolfo Marsh** Chilean research station South Shetland Islands, Antarctica 61°57′S 58°23′W
32 G9 **Tenino** Washington, NW USA 46°51′N 122°51′W
145 P9 **Teniz, Ozero** Kaz. Tengiz Köl. salt lake C Kazakhstan
112 I9 **Tenja** Osijek-Baranja, E Croatia 45°30′N 18°45′E
188 B16 **Tenjo, Mount** ▲ W Guam
155 H23 **Tenkāsi** Tamil Nādu, SE India 08°58′N 77°22′E
79 N24 **Tenke** Katanga, SE Dem. Rep. Congo 10°34′S 26°12′E
Tenke see Tinca
123 Q7 **Tenkeli** Respublika Sakha (Yakutiya), NE Russian Federation 70°09′N 140°39′E
27 R10 **Tenkiller Ferry Lake** ⊠ Oklahoma, C USA
77 Q13 **Tenkodogo** S Burkina Faso 11°54′N 00°19′W
181 Q5 **Tennant Creek** Northern Territory, C Australia 19°40′S 134°16′E
20 G9 **Tennessee** off. State of Tennessee, also known as The Volunteer State. ◈ state SE USA
37 R5 **Tennessee Pass** pass Colorado, C USA
20 H10 **Tennessee River** ~ S USA
23 N2 **Tennessee Tombigbee Waterway** canal Alabama/Mississippi, S USA
99 K22 **Tenneville** Luxembourg, SE Belgium 50°05′N 05°31′E
92 M11 **Tenniöjoki** ~ NE Finland
92 L9 **Tenojoki** Lapp. Deatnu, Nor. Tana. ~ Finland/Norway see also Deatnu, Tana
Tenojoki to see Tana
169 U7 **Tenom** Sabah, East Malaysia 05°07′N 115°57′E
Tenos see Tínos
41 V15 **Tenosique** var. Tenosique de Pino Suárez. Tabasco, SE Mexico 17°30′N 91°24′W
Tenosique de Pino Suárez see Tenosique
22 I6 **Tensas River** ~ Louisiana, S USA
23 O8 **Tensaw River** ~ Alabama, S USA
74 E7 **Tensift** seasonal river W Morocco
171 O12 **Tentena** var. Tenteno. Sulawesi, C Indonesia 01°46′S 120°40′E
Tenteno see Tentena
183 U4 **Tenterfield** New South Wales, SE Australia 29°04′S 152°02′E
23 X16 **Ten Thousand Islands** island group Florida, SE USA
60 H9 **Teodoro Sampaio** São Paulo, S Brazil 22°30′S 52°13′W
59 N19 **Teófilo Otoni** var. Theophilo Ottoni. Minas Gerais, NE Brazil 17°52′S 41°31′W
116 K5 **Teofipol'** Khmel'nyts'ka Oblast', W Ukraine 50°00′N 26°22′E
Teohatu see Toahotu
41 P14 **Teotihuacán** ruins México, S Mexico
Teotilán see Teotitlán del Camino
41 Q15 **Teotitlán del Camino** var. Teotilán. Oaxaca, S Mexico 18°10′N 97°08′W
190 G12 **Tepa** Île Uvea, E Wallis and Futuna 13°19′S 176°09′W
191 P8 **Tepaee, Récif** reef Tahiti, W French Polynesia
40 L14 **Tepalcatepec** Michoacán, SW Mexico 19°11′N 102°50′W
190 A16 **Tepa Point** headland SW Niue 19°07′S 169°56′E
40 L13 **Tepatitlán** var. Tepatitlán de Morelos. Jalisco, SW Mexico 20°50′N 102°46′W
Tepatitlán de Morelos see Tepatitlán
40 J9 **Tepehuanes** var. Santa Catarina de Tepehuanes. Durango, C Mexico 25°22′N 105°42′W
Tepelena see Tepelenë
113 L22 **Tepelenë** var. Tepelena, It. Tepeleni. Gjirokastër, S Albania 40°18′N 20°00′E
Tepeleni see Tepelenë
40 K12 **Tepic** Nayarit, C Mexico 21°30′N 104°55′W
111 C15 **Teplice** Ger. Teplitz; prev. Teplice-Šanov, Teplitz-Schönau. Ústecký Kraj, NW Czech Republic 50°38′N 13°49′E
Teplice-Šanov/Teplitz/Teplitz-Schönau see Teplice
117 O7 **Teplyk** Vinnyts'ka Oblast', C Ukraine 48°40′N 29°46′E
123 R10 **Teplyy Klyuch** Respublika Sakha (Yakutiya), NE Russian Federation 62°46′N 137°01′E
40 E5 **Tepoca, Cabo** headland NW Mexico 29°19′N 112°24′W
191 W9 **Tepoto** island Îles du Désappointement, C French Polynesia
92 L11 **Tepsa** Lappi, N Finland 67°34′N 25°36′E
190 B8 **Tepuka** atoll Funafuti Atoll, C Tuvalu
184 N7 **Te Puke** Bay of Plenty, North Island, New Zealand 37°48′S 176°19′E
40 L13 **Tequila** Jalisco, SW Mexico 20°52′N 103°48′W
41 O13 **Tequisquiapan** Querétaro de Arteaga, C Mexico 20°34′N 99°52′W
77 Q12 **Téra** Tillabéri, W Niger 14°01′N 00°45′E
104 J5 **Tera** ~ NW Spain
191 V1 **Teraina** prev. Washington Island. atoll Line Islands, E Kiribati
81 F15 **Terakeka** Central Equatoria, S South Sudan 05°26′N 31°45′E
107 J14 **Teramo** anc. Interamna. Abruzzi, C Italy 42°40′N 13°43′E
98 P7 **Ter Apel** Groningen, NE Netherlands 52°52′N 07°05′E
104 H11 **Tera, Ribeira de** ~ S Portugal
185 K14 **Terawhiti, Cape** headland North Island, New Zealand 41°17′S 174°36′E
98 N12 **Terborg** Gelderland, E Netherlands 51°55′N 06°22′E
137 P13 **Tercan** Erzincan, NE Turkey 39°47′N 40°23′E
64 O2 **Terceira** ✈ Terceira, Azores, Portugal, NE Atlantic Ocean 38°43′N 27°13′W
64 O2 **Terceira** var. Ilha Terceira. island Azores, Portugal, NE Atlantic Ocean
Terceira, Ilha see Terceira
116 K6 **Terebovlya** Ternopil's'ka Oblast', W Ukraine 49°18′N 25°44′E
127 O15 **Terek** ~ SW Russian Federation
Terekhovka see Tsyerakhowka
147 R9 **Terek-Say** Dzhalal-Abadskaya Oblast', W Kyrgyzstan 41°28′N 71°06′E
145 Z10 **Terekty** prev. Alekseevka, Alekseyevka. Vostochnyy Kazakhstan, E Kazakhstan 48°25′N 85°38′E
168 L7 **Terengganu** var. Trengganu. ◈ state Peninsular Malaysia
127 X7 **Terensay** Orenburgskaya Oblast', W Russian Federation 51°35′N 59°28′E
58 N13 **Teresina** var. Therezina. state capital Piauí, NE Brazil 05°09′S 42°46′W
60 P9 **Teresópolis** Rio de Janeiro, SE Brazil 22°25′S 42°59′W
110 P12 **Terespol** Lubelskie, E Poland 52°05′N 23°37′E
191 V16 **Terevaka, Maunga** ⛰ Easter Island, Chile, E Pacific Ocean 27°05′S 109°23′W
103 P3 **Tergnier** Aisne, N France 49°39′N 03°18′E
43 O14 **Teribe, Río** ~ NW Panama
124 K3 **Teriberka** Murmanskaya Oblast', NW Russian Federation 69°10′N 35°18′E
Terinkot see Tarīn Kōt
Terisaqqan see Tersakkan
24 K12 **Terlingua** Texas, SW USA 29°18′N 103°36′W
24 K11 **Terlingua Creek** ~ Texas, SW USA
62 K7 **Termas de Río Hondo** Santiago del Estero, N Argentina 27°29′S 64°52′W
136 M11 **Terme** Samsun, N Turkey 41°12′N 37°00′E
Termez see Termiz
Termia see Kýthnos
107 J23 **Termini Imerese** anc. Thermae Himerenses. Sicilia, Italy, C Mediterranean Sea 37°59′N 13°42′E
41 V14 **Términos, Laguna de** lagoon SE Mexico
77 X10 **Termit-Kaoboul** Zinder, C Niger 15°34′N 11°31′E
147 O14 **Termiz** Rus. Termez. Surkhondaryo Viloyati, S Uzbekistan 37°17′N 67°12′E
107 L15 **Termoli** Molise, C Italy 42°00′N 14°58′E
Termonde see Dendermonde
98 P5 **Termunten** Groningen, NE Netherlands 53°18′N 07°02′E
171 R11 **Ternate** Pulau Ternate, E Indonesia 0°48′N 127°23′E
109 T5 **Ternberg** Oberösterreich, N Austria 47°57′N 14°22′E
99 E15 **Terneuzen** var. Neuzen. Zeeland, SW Netherlands 51°20′N 03°50′E
123 T14 **Terney** Primorskiy Kray, SE Russian Federation 45°03′N 136°43′E
107 I14 **Terni** anc. Interamna Nahars. Umbria, C Italy 42°34′N 12°38′E
109 X6 **Ternitz** Niederösterreich, E Austria 47°43′N 16°02′E
117 V7 **Ternivka** Dnipropetrovs'ka Oblast', E Ukraine 48°30′N 36°05′E
116 K6 **Ternopil'** Pol. Tarnopol, Rus. Ternopol'. Ternopil's'ka Oblast', W Ukraine 49°32′N 25°38′E
Ternopil' see Ternopil's'ka Oblast'
116 I6 **Ternopil's'ka Oblast'** var. Ternopil', Rus. Ternopol'skaya Oblast'. ◈ province NW Ukraine
Ternopol' see Ternopil'
Ternopol'skaya Oblast' see Ternopil's'ka Oblast'
123 U13 **Terpeniya, Mys** headland Ostrov Sakhalin, SE Russian Federation 48°37′N 144°40′E
Térraba, Río see Grande de Térraba, Río
10 J13 **Terrace** British Columbia, W Canada 54°31′N 128°32′W
12 D12 **Terrace Bay** Ontario, S Canada 48°47′N 87°06′W
107 I16 **Terracina** Lazio, C Italy 41°18′N 13°13′E
93 F14 **Terråk** Troms, N Norway 65°03′N 12°22′E
26 M13 **Terral** Oklahoma, C USA 33°55′N 97°54′W
107 B19 **Terralba** Sardegna, Italy, C Mediterranean Sea 39°47′N 08°35′E
Terranova di Sicilia see Gela
Terranova Pausania see Olbia
105 W5 **Terrassa** Cast. Tarrasa. Cataluña, E Spain 41°34′N 02°01′E
15 O12 **Terrebonne** Québec, SE Canada 45°42′N 73°37′W
22 J11 **Terrebonne Bay** bay Louisiana, SE USA
31 N14 **Terre Haute** Indiana, N USA 39°27′N 87°24′W
25 U6 **Terrell** Texas, SW USA 32°44′N 96°16′W
Terre Neuve see Newfoundland and Labrador
33 Q14 **Terreton** Idaho, NW USA 43°49′N 112°25′W
103 T7 **Territoire-de-Belfort** ◈ department E France
33 X9 **Terry** Montana, NW USA 46°46′N 105°16′W
28 I9 **Terry Peak** ▲ South Dakota, N USA 44°19′N 103°49′W
136 H14 **Tersakan Gölü** ◎ C Turkey
145 O10 **Tersakkan** Kaz. Terisaqqan. ~ C Kazakhstan
98 J4 **Terschelling** Fris. Skylge. island Waddeneilanden, N Netherlands
78 H10 **Tersef** Hadjer-Lamis, C Chad 12°55′N 16°49′E
147 X8 **Terskey Ala-Too, Khrebet** ▲▲ Kazakhstan/Kyrgyzstan
Terter see Tärtär
105 R8 **Teruel** anc. Turba. Aragón, E Spain 40°21′N 01°06′W
105 R7 **Teruel** ◈ province Aragón, E Spain
114 M7 **Tervel** prev. Kurtbunar, Rom. Curtbunar. Dobrich, NE Bulgaria 43°45′N 27°25′E
93 M16 **Tervo** Pohjois-Savo, C Finland 62°57′N 26°48′E
92 L13 **Tervola** Lappi, NW Finland 66°04′N 24°49′E
Tervueren see Tervuren
99 H18 **Tervuren** var. Tervueren. Vlaams Brabant, C Belgium 50°48′N 04°28′E
162 G5 **Tes** var. Dzür. Dzavhan, W Mongolia 49°37′N 95°46′E
112 H11 **Tešanj** Federacija Bosne I Hercegovine, N Bosnia and Herzegovina 44°37′N 18°00′E
Teschen see Cieszyn
83 M19 **Tesenane** Inhambane, S Mozambique 22°48′S 34°02′E
80 I9 **Teseney** var. Tessenei. W Eritrea 15°05′N 36°42′E
39 P5 **Teshekpuk Lake** ◎ Alaska, USA
Teshevle see Tsyeshawlya
162 K6 **Teshig** Bulgan, N Mongolia 49°51′N 102°45′E
165 T2 **Teshio** Hokkaidō, NE Japan 44°49′N 141°46′E
165 T2 **Teshio-sanchi** ▲▲ Hokkaidō, NE Japan
14 L7 **Tessier, Lac** ◎ Québec, SE Canada
Těšín see Cieszyn
Tesiyn Gol see Tes-Khem
129 T7 **Tes-Khem** var. Tesiyn Gol. ~ Mongolia/Russian Federation
112 H11 **Teslić** Republika Srpska, N Bosnia and Herzegovina 44°35′N 17°50′E
10 I9 **Teslin** Yukon, W Canada 60°12′N 132°44′W
10 I8 **Teslin** ~ British Columbia/Yukon, W Canada
77 Q8 **Tessalit** Kidal, NE Mali 20°12′N 00°58′E
77 V12 **Tessaoua** Maradi, S Niger 13°46′N 07°55′E
99 J17 **Tessenderlo** Limburg, NE Belgium 51°05′N 05°04′E
Tessenei see Teseney
Tessin see Ticino
97 M23 **Test** ~ S England, United Kingdom
Testama see Tõstamaa
55 P4 **Testigos, Islas los** island group N Venezuela
37 S10 **Tesuque** New Mexico, SW USA 35°45′N 105°55′W
103 O17 **Têt** var. Tet. ~ S France
Tet see Têt
54 G5 **Tetas, Cerro de las** ▲ NW Venezuela 09°58′N 73°00′W
83 M15 **Tete** Tete, NW Mozambique 16°14′S 33°34′E
83 M15 **Tete** off. Província de Tete. ◈ province NW Mozambique
11 N15 **Tete Jaune Cache** British Columbia, SW Canada 52°52′N 119°22′W
184 O8 **Te Teko** Bay of Plenty, North Island, New Zealand 38°03′S 176°48′E
186 K9 **Tetepare** island New Georgia Islands, NW Solomon Islands
Tete, Província de see Tete
Teterev see Teteriv
116 M5 **Teteriv** Rus. Teterev. ~ N Ukraine
100 M9 **Teterow** Mecklenburg-Vorpommern, NE Germany 53°47′N 12°34′E
114 I9 **Teteven** Lovech, N Bulgaria 42°54′N 24°18′E
191 T10 **Tetiaroa** atoll Îles du Vent, W French Polynesia
105 P14 **Tetica de Bacares** ▲ S Spain 37°15′N 02°24′W
Tetiyev see Tetiyiv
117 O6 **Tetiyiv** Rus. Tetiyev. Kyyivs'ka Oblast', N Ukraine 49°21′N 29°40′E
39 T10 **Tetlin** Alaska, USA 63°08′N 142°31′W
33 R8 **Teton River** ~ Montana, NW USA
74 G5 **Tétouan** var. Tetouan, Tetuán. N Morocco 35°33′N 05°22′W
Tetova/Tetovë see Tetovo
114 L7 **Tetovo** Razgrad, N Bulgaria 43°49′N 26°21′E
113 N18 **Tetovo** Alb. Tetova, Tetovë, Turk. Kalkandelen. NW FYR Macedonia 42°01′N 20°58′E
115 E20 **Tetrázio** ▲ S Greece
Tetschen see Děčín
Tetuán see Tétouan
191 Q8 **Tetufera, Mont** ▲ Tahiti, W French Polynesia 17°40′S 149°26′W
127 R4 **Tetyushi** Respublika Tatarstan, W Russian Federation 54°55′N 48°46′E
108 I7 **Teufen** Ausser Rhoden, NE Switzerland 47°24′N 09°24′E
40 L12 **Teul** var. Teul de Gonzáles Ortega. Zacatecas, C Mexico 21°30′N 103°28′W
107 B21 **Teulada** Sardegna, Italy, C Mediterranean Sea 38°58′N 08°46′E
Teul de Gonzáles Ortega see Teul
11 X16 **Teulon** Manitoba, S Canada 50°20′N 97°14′W
42 I7 **Teupasenti** El Paraíso, S Honduras 14°13′N 86°42′W
165 S2 **Teuri-tō** island NE Japan
100 G13 **Teutoburger Wald** Eng. Teutoburg Forest. hill range NW Germany
Teutoburg Forest see Teutoburger Wald
93 K17 **Teuva** Swe. Östermark. Etelä-Pohjanmaa, W Finland 62°29′N 21°45′E
107 H15 **Tevere** Eng. Tiber. ~ C Italy
Teverya see Tverya
96 K13 **Teviot** ~ SE Scotland, United Kingdom
Tevli see Tewli
122 H11 **Tevriz** Omskaya Oblast', C Russian Federation 57°30′N 72°13′E
185 B24 **Te Waewae Bay** bay South Island, New Zealand
97 L21 **Tewkesbury** C England, United Kingdom 51°59′N 02°09′W
119 F19 **Tewli** Rus. Tevli. Brestskaya Voblasts', SW Belarus 52°20′N 24°15′E
159 U12 **Têwo** var. Dêngka; prev. Dêngkagoin. Gansu, C China 34°05′N 103°15′E
Tewulike see Hoxud
25 U12 **Texana, Lake** ⊠ Texas, SW USA
27 S14 **Texarkana** Arkansas, C USA 33°26′N 94°02′W
25 X5 **Texarkana** Texas, SW USA 33°26′N 94°03′W
25 N9 **Texas** off. State of Texas, also known as Lone Star State. ◈ state S USA
25 W12 **Texas City** Texas, SW USA 29°23′N 94°55′W
41 P14 **Texcoco** México, C Mexico 19°32′N 98°52′W
98 I6 **Texel** island Waddeneilanden, NW Netherlands
26 H8 **Texhoma** Oklahoma, C USA 36°30′N 101°46′W
25 N1 **Texhoma** Texas, SW USA 36°30′N 101°46′W
37 W12 **Texico** New Mexico, SW USA 34°23′N 103°03′W
24 L1 **Texline** Texas, SW USA 36°22′N 103°01′W
41 P14 **Texmelucan** var. San Martín Texmelucan. Puebla, S Mexico 19°46′N 98°53′W
27 O13 **Texoma, Lake** ⊠ Oklahoma/Texas, C USA
25 N9 **Texon** Texas, SW USA 31°13′N 101°42′W
83 J23 **Teyateyaneng** NW Lesotho 29°04′S 27°51′E
124 M16 **Teykovo** Ivanovskaya Oblast', W Russian Federation 56°49′N 40°31′E
124 M16 **Teza** ~ W Russian Federation
41 Q13 **Teziutlán** Puebla, S Mexico 19°49′N 97°22′W
153 W12 **Tezpur** Assam, NE India 26°39′N 92°47′E
Thaa Atoll see Kolhumadulu
9 N10 **Tha-Anne** ~ Nunavut, NE Canada
83 K23 **Thabana Ntlenyana** var. Thabantshonyana, Mount Ntlenyana. ▲ E Lesotho 29°26′S 29°16′E
Thabantshonyana see Thabana Ntlenyana
83 J23 **Thaba Putsoa** ▲ C Lesotho 29°48′S 27°46′E
167 Q8 **Tha Bo** Nong Khai, E Thailand 17°52′N 102°34′E
103 T12 **Thabor, Pic du** ▲ E France 45°07′N 06°34′E
Tha Chin see Samut Sakhon
166 M7 **Thagaya** Bago, C Myanmar (Burma) 19°19′N 96°16′E
Thai, Ao see Thailand, Gulf of
167 T6 **Thai Binh** Thai Binh, N Vietnam 20°27′N 106°20′E
167 S7 **Thai Hoa** var. Nghia Dan. Nghệ An, N Vietnam 19°21′N 105°26′E
167 P9 **Thailand** off. Kingdom of Thailand, Th. Prathet Thai; prev. Siam. ◆ monarchy SE Asia
167 P13 **Thailand, Gulf of** var. Gulf of Siam, Th. Ao Thai, Vtn. Vinh Thai Lan. gulf SE Asia
Thailand, Kingdom of see Thailand
Thai Lan, Vinh see Thailand, Gulf of
167 T6 **Thai Nguyên** Bắc Thai, N Vietnam 21°36′N 105°50′E
167 S8 **Thakhèk** var. Muang Khammouan. Khammouan, C Laos 17°25′N 104°51′E
153 S13 **Thakurgaon** Rajshahi, NW Bangladesh 26°05′N 88°34′E
149 S6 **Thal** Khyber Pakhtunkhwa, NW Pakistan 33°24′N 70°32′E
167 S11 **Thalabârîvăt** prev. Phumĭ Thalabârîvăt. Stœng Trêng, N Cambodia 13°34′N 105°57′E
166 M15 **Thalang** Phuket, SW Thailand 08°00′N 98°21′E
Thalashsheri see Tellicherry/Thalassery
167 Q10 **Thalat Khae** Nakhon Ratchasima, C Thailand 15°15′N 102°24′E
109 Q5 **Thalgau** Salzburg, NW Austria 47°49′N 13°19′E
108 G7 **Thalwil** Zürich, NW Switzerland 47°17′N 08°35′E
83 I20 **Thamaga** Kweneng, SE Botswana 24°41′S 25°31′E
Thamarīd see Thamarīt
141 V13 **Thamarīt** var. Thamarīd, Thumrayt. SW Oman 17°39′N 54°02′E
141 P16 **Thamar, Jabal** ▲ SW Yemen 13°46′N 45°32′E
184 M6 **Thames** Waikato, North Island, New Zealand 37°10′S 175°33′E
14 D17 **Thames** ~ Ontario, S Canada
97 O22 **Thames** ~ S England, United Kingdom
184 M6 **Thames, Firth of** gulf North Island, New Zealand
14 D17 **Thamesville** Ontario, S Canada 42°33′N 81°58′W
141 S13 **Thamūd** N Yemen 17°18′N 49°57′E
167 N9 **Thanbyuzayat** Mon State, S Myanmar (Burma) 15°58′N 97°44′E
166 K7 **Thandwe** var. Sandoway. Rakhine State, W Myanmar (Burma) 18°28′N 94°20′E
152 I9 **Thānesar** Haryāna, NW India 29°58′N 76°48′E
167 T7 **Thanh Hoa** Thanh Hoa, N Vietnam 19°49′N 105°48′E
Thanintari Taungdan see Bilauktaung Range
155 I21 **Thanjāvūr** prev. Tanjore. Tamil Nādu, SE India 10°46′N 79°09′E
Thanlwin see Salween
103 U7 **Thann** Haut-Rhin, NE France 47°51′N 07°04′E
167 O16 **Tha Nong Phrom** Phatthalung, SW Thailand 07°24′N 100°04′E
167 N13 **Thap Sakae** var. Thap Sakau. Prachuap Khiri Khan, SW Thailand 11°30′N 99°35′E
Thap Sakau see Thap Sakae
81 I19 **Tharaka-Nithi** ◈ county C Kenya
98 L10 **'t Harde** Gelderland, E Netherlands 52°25′N 05°53′E
152 D11 **Thar Desert** var. Great Indian Desert, Indian Desert. desert India/Pakistan
181 V10 **Thargomindah** Queensland, C Australia 28°00′S 143°47′E
150 D11 **Thar Pārkar** desert SE Pakistan
139 S7 **Tharthār al Furāt, Qanāt ath** canal C Iraq
139 R7 **Tharthār, Buḥayrat ath** ◎ C Iraq
139 R5 **Tharthār, Wādī ath** dry watercourse N Iraq
167 N13 **Tha Sae** Chumphon, SW Thailand
167 N15 **Tha Sala** Nakhon Si Thammarat, SW Thailand 08°43′N 99°54′E
114 I13 **Thásos** Thásos, E Greece 40°47′N 24°43′E
115 I14 **Thásos** island E Greece
37 N14 **Thatcher** Arizona, SW USA 32°47′N 109°46′W
167 T5 **Thât Khê** var. Tràng Dinh. Lang Sơn, N Vietnam 22°15′N 106°26′E
166 M8 **Thaton** Mon State, S Myanmar (Burma) 16°56′N 97°20′E
167 S9 **That Phanom** Nakhon Phanom, E Thailand 16°52′N 104°41′E
167 R10 **Tha Tum** Surin, E Thailand 15°18′N 103°39′E
103 P16 **Thau, Bassin de** var. Étang de Thau. ◎ S France
Thau, Étang de see Thau, Bassin de
166 L3 **Thaungdut** Sagaing, N Myanmar (Burma) 24°26′N 94°45′E
167 O8 **Thaungyin** Th. Mae Nam Moei. ~ Myanmar (Burma)/Thailand
167 R8 **Tha Uthen** Nakhon Phanom, E Thailand 17°32′N 104°34′E
109 W2 **Thaya** var. Dyje. ~ Austria/Czech Republic see also Dyje
Thaya see Dyje
27 V8 **Thayer** Missouri, C USA 36°31′N 91°34′W
166 L6 **Thayetmyo** Magway, C Myanmar (Burma) 19°20′N 95°10′E
33 S15 **Thayne** Wyoming, C USA 42°54′N 111°01′W
166 M5 **Thazi** Mandalay, C Myanmar (Burma) 20°50′N 96°04′E
Thebes see Thíva
44 L5 **The Carlton** var. Abraham Bay. Mayaguana, SE The Bahamas 22°21′N 72°56′W
45 O14 **The Crane** var. Crane. S Barbados 13°06′N 59°27′W
32 I11 **The Dalles** Oregon, NW USA 45°36′N 121°10′W
28 M14 **Thedford** Nebraska, C USA 41°59′N 100°32′W
The Flatts Village see Flatts Village
The Hague see 's-Gravenhage
Theiss see Tisa/Tisza
8 M9 **Thelon** ~ Northwest Territories, N Canada
11 V15 **Theodore** Saskatchewan, S Canada 51°25′N 103°01′W
23 N8 **Theodore** Alabama, S USA 30°33′N 88°10′W
36 L13 **Theodore Roosevelt Lake** ⊠ Arizona, SW USA
Theodosia see Feodosiya
Theophilo Ottoni see Teófilo Otoni
11 V13 **The Pas** Manitoba, C Canada 53°49′N 101°09′W
31 T14 **The Plains** Ohio, N USA 39°22′N 82°07′W
Thera see Santoríni
172 H17 **Thérèse, Île** island Inner Islands, NE Seychelles
Therezina see Teresina
115 L20 **Thérma** Ikaría, Dodekánisa, Greece, Aegean Sea 37°37′N 26°18′E
Thermae Himerenses see Termini Imerese
Thermae Pannonicae see Baden
Thermaic Gulf/Thermaicus Sinus see Thermaïkós Kólpos
121 Q8 **Thermaïkós Kólpos** Eng. Thermaic Gulf; anc. Thermaicus Sinus. gulf N Greece
Thermiá see Kýthnos
115 L17 **Thermís** Lésvos, E Greece 39°08′N 26°32′E
115 E18 **Thérmo** Dytikí Elláda, C Greece 38°32′N 21°42′E
33 V14 **Thermopolis** Wyoming, C USA 43°39′N 108°12′W
183 P10 **The Rock** New South Wales, SE Australia 35°18′S 147°07′E
195 O5 **Theron Mountains** ▲▲ Antarctica
The Sooner State see Oklahoma
115 G18 **Thespiés** Stereá Elláda, C Greece 38°18′N 23°08′E
115 E16 **Thessalía** Eng. Thessaly. ◈ region C Greece
14 C10 **Thessalon** Ontario, S Canada 46°15′N 83°34′W
115 G14 **Thessaloníki** Eng. Salonica, Salonika, SCr. Solun, Turk. Selânik. Kentrikí Makedonía, N Greece 40°38′N 22°58′E
115 G14 **Thessaloníki** ✈ Kentrikí Makedonía, N Greece 40°30′N 22°58′E
Thessaly see Thessalía
84 B12 **Theta Gap** undersea feature E Atlantic Ocean 12°40′W 43°30′N
97 P20 **Thetford** E England, United Kingdom 52°25′N 00°45′E
15 R11 **Thetford-Mines** Québec, SE Canada 46°07′N 71°16′W
113 K17 **Theth** var. Thethi. Shkodër, N Albania 42°25′N 19°45′E
Thethi see Theth
99 L20 **Theux** Liège, E Belgium 50°33′N 05°48′E
45 V9 **The Valley** ○ (Anguilla) E Anguilla 18°13′N 63°00′W
27 N10 **The Village** Oklahoma, C USA 35°33′N 97°33′W
The Volunteer State see Tennessee
25 W10 **The Woodlands** Texas, SW USA 30°09′N 95°27′E
Thiamis see Kalamás
Thian Shan see Tien Shan
Thibet see Xizang Zizhiqu
22 J9 **Thibodaux** Louisiana, S USA 29°48′N 90°49′W
29 S3 **Thief Lake** ◎ Minnesota, N USA
29 S3 **Thief River** ~ Minnesota, C USA
29 S3 **Thief River Falls** Minnesota, N USA 48°07′N 96°10′W

◆ Country
● Country Capital
◇ Dependent Territory
○ Dependent Territory Capital
◈ Administrative Regions
✈ International Airport
▲ Mountain
▲▲ Mountain Range
⛰ Volcano
~ River
◎ Lake
⊠ Reservoir

THIÈLE – TOAMASINA

Thièle see La Thielle
32 G14 **Thielsen, Mount** ▲ Oregon, NW USA 43°09′N 122°04′W
Thielt see Tielt
106 G7 **Thiene** Veneto, NE Italy 45°43′N 11°29′E
Thienen see Tienen
103 P11 **Thiers** Puy-de-Dôme, C France 45°51′N 03°33′E
76 F11 **Thiès** W Senegal 14°49′N 16°52′W
81 I19 **Thika** Kiambu, S Kenya 01°03′S 37°05′E
Thikombia see Cikobia
151 K18 **Thiladhunmathi Atoll** var. Tiladummati Atoll. atoll N Maldives
Thimbu see Thimphu
153 T11 **Thimphu** var. Thimbu; prev. Tashi Chho Dzong. ● (Bhutan) W Bhutan 27°28′N 89°37′E
92 H2 **Þingeyri** Vestfirðir, NW Iceland 65°52′N 23°28′W
92 I3 **Þingvellir** Suðurland, SW Iceland 64°15′N 21°06′W
187 Q17 **Thio** Province Sud, C New Caledonia 21°37′S 166°13′E
103 T4 **Thionville** Ger. Diedenhofen. Moselle, NE France 49°22′N 06°11′E
77 O12 **Thiou** NW Burkina Faso 13°42′N 02°34′W
115 K22 **Thíra** Santorini, Kykládes, Greece, Aegean Sea 36°25′N 25°26′E
Thíra see Santoríni
115 J22 **Thirasía** island Kykládes, Greece, Aegean Sea
97 M16 **Thirsk** N England, United Kingdom 54°07′N 01°17′W
14 F12 **Thirty Thousand Islands** island group Ontario, S Canada
95 F20 **Thisted** Midtjylland, NW Denmark 56°58′N 08°42′E
Thistil Fjord see Þistilfjörður
92 L1 **Þistilfjörður** var. Thistil Fjord. fjord NE Iceland
182 G9 **Thistle Island** island South Australia
Thithia see Cicia
Thiukhaoluang Phrahang see Luang Prabang Range
115 G18 **Thíva** Eng. Thebes; prev. Thívai. Stereá Elláda, C Greece 38°19′N 23°19′E
Thívai see Thíva
102 M12 **Thiviers** Dordogne, SW France 45°24′N 00°54′E
92 J4 **Þjórsá** ≋ C Iceland
9 N10 **Thlewiaza** ≋ Nunavut, NE Canada
8 L10 **Thoa** ≋ Northwest Territories, NW Canada
99 G14 **Tholen** Zeeland, SW Netherlands 51°31′N 04°13′E
99 F14 **Tholen** island SW Netherlands
26 L10 **Thomas** Oklahoma, C USA 35°44′N 98°45′W
21 T3 **Thomas** West Virginia, NE USA 39°09′N 79°28′W
27 U3 **Thomas Hill Reservoir** ⊞ Missouri, C USA
23 S5 **Thomaston** Georgia, SE USA 32°53′N 84°19′W
19 R7 **Thomaston** Maine, NE USA 44°06′N 69°10′W
25 T12 **Thomaston** Texas, SW USA 28°56′N 97°07′W
23 O6 **Thomasville** Alabama, S USA 31°54′N 87°42′W
23 T8 **Thomasville** Georgia, SE USA 30°49′N 83°57′W
21 S9 **Thomasville** North Carolina, SE USA 35°52′N 80°04′W
35 N5 **Thomes Creek** ≋ California, W USA
11 W12 **Thompson** Manitoba, C Canada 55°45′N 97°54′W
29 R4 **Thompson** North Dakota, N USA 47°45′N 97°07′W
0 F8 **Thompson** ≋ Alberta/British Columbia, SW Canada
33 O8 **Thompson Falls** Montana, NW USA 47°36′N 115°20′W
29 Q10 **Thompson, Lake** ◎ South Dakota, N USA
34 M3 **Thompson Peak** ▲ California, W USA 41°00′N 123°01′W
27 S2 **Thompson River** ≋ Missouri, C USA
185 A22 **Thompson Sound** sound South Island, New Zealand
8 J5 **Thomsen** ≋ Banks Island, Northwest Territories, NW Canada
23 V4 **Thomson** Georgia, SE USA 33°28′N 82°30′W
103 T10 **Thonon-les-Bains** Haute-Savoie, E France 46°22′N 06°30′E
103 O15 **Thoré** var. Thore. ≋ S France
Thore see Thoré
37 P11 **Thoreau** New Mexico, SW USA 35°24′N 108°13′W
Thorenburg see Turda
92 J3 **Þórisvatn** ◎ C Iceland
92 P4 **Thor, Kapp** headland S Svalbard 76°25′N 25°01′E
92 I4 **Þorlákshöfn** Suðurland, SW Iceland 63°52′N 21°24′W
Thorn see Toruń
25 T10 **Thorndale** Texas, SW USA 30°36′N 97°12′W
14 H10 **Thorne** Ontario, S Canada 46°38′N 79°04′W
97 J14 **Thornhill** S Scotland, United Kingdom 55°13′N 03°46′W
25 U8 **Thornton** Texas, SW USA 31°24′N 96°34′W
Thornton Island see Millennium Island
14 H16 **Thorold** Ontario, S Canada 43°07′N 79°15′W
32 I9 **Thorp** Washington, NW USA 47°03′N 120°40′W
Thorshavn see Tórshavn
195 S3 **Thorshavnheiane** physical region Antarctica
92 L1 **Þórshöfn** Norðurland Eystra, NE Iceland 66°09′N 15°18′W
Thospitis see Van Gölü
167 S14 **Thôt Nôt** Cân Thơ, S Vietnam 10°17′N 105°31′E
102 K8 **Thouars** Deux-Sèvres, W France 46°59′N 00°13′W
153 X14 **Thoubal** Manipur, NE India 24°40′N 94°00′E
102 K9 **Thouet** ≋ W France
Thoune see Thun
18 H7 **Thousand Islands** island Canada/USA
35 S15 **Thousand Oaks** California, W USA 34°10′N 118°50′W
114 L12 **Thrace** cultural region SE Europe
114 J13 **Thracian Sea** Gk. Thrakikó Pélagos; anc. Thracium Mare. sea Greece/Turkey
Thracium Mare/Thrakikó Pélagos see Thracian Sea
Thrá Lí, Bá see Tralee Bay
33 R11 **Three Forks** Montana, NW USA 45°53′N 111°34′W
162 M8 **Three Gorges Dam** dam Hubei, C China
160 L9 **Three Gorges Reservoir** ⊞ C China
11 Q16 **Three Hills** Alberta, SW Canada 51°43′N 113°15′W
183 N15 **Three Hummock Island** island Tasmania, SE Australia
184 H1 **Three Kings Islands** island group N New Zealand
175 P10 **Three Kings Rise** undersea feature W Pacific Ocean
77 O18 **Three Points, Cape** headland S Ghana 04°43′N 02°03′W
31 P10 **Three Rivers** Michigan, N USA 41°56′N 85°37′W
25 S13 **Three Rivers** Texas, SW USA 28°27′N 98°10′W
83 G24 **Three Sisters** Northern Cape, SW South Africa 31°51′S 23°04′E
32 H13 **Three Sisters** ▲ Oregon, NW USA 44°08′N 121°46′W
187 N10 **Three Sisters Islands** island group SE Solomon Islands
25 Q6 **Throckmorton** Texas, SW USA 33°11′N 99°12′W
180 M10 **Throssell, Lake** salt lake Western Australia
115 K25 **Thryptís** var. Thrýptis. ▲ Kríti, Greece, E Mediterranean Sea 35°06′N 25°51′E
167 U14 **Thuân Nam** prev. Ham Thuân Nam. Binh Thuân, S Vietnam 10°49′N 107°49′E
167 T13 **Thu Dâu Môt** var. Phu Cŭơng. Sông Be, S Vietnam 10°58′N 106°40′E
167 S6 **Thu Do** ✈ (Ha Nôi) Ha Nôi, N Vietnam 21°13′N 105°46′E
99 G21 **Thuin** Hainaut, S Belgium 50°21′N 04°18′E
149 Q12 **Thul** Sind, SE Pakistan 28°14′N 68°50′E
Thule see Qaanaaq
83 J18 **Thuli** var. Tuli. ≋ S Zimbabwe
Thumrayt see Thamarīt
108 D9 **Thun** Fr. Thoune. Bern, W Switzerland 46°46′N 07°38′E
12 C12 **Thunder Bay** Ontario, S Canada 48°27′N 89°12′W
30 M1 **Thunder Bay** lake bay S Canada
31 R6 **Thunder Bay** lake bay Michigan, N USA
31 R6 **Thunder Bay River** ≋ Michigan, N USA
27 N11 **Thunderbird, Lake** ⊞ Oklahoma, C USA
28 L8 **Thunder Butte Creek** ≋ South Dakota, N USA
108 E9 **Thuner See** ◎ C Switzerland
167 N15 **Thung Song** var. Cha Mai. Nakhon Si Thammarat, SW Thailand 08°10′N 99°41′E
108 H7 **Thur** ≋ N Switzerland
108 G6 **Thurgau** Fr. Thurgovie. ◈ canton NE Switzerland
Thurgovie see Thurgau
Thuringe see Thüringen
108 J7 **Thüringen** Vorarlberg, W Austria 47°12′N 09°48′E
101 J17 **Thüringen** Eng. Thuringia, Fr. Thuringe. ◈ state C Germany
101 J17 **Thüringer Wald** Eng. Thuringian Forest. ▲▲ C Germany
Thuringia see Thüringen
Thuringian Forest see Thüringer Wald
97 D19 **Thurles** Ir. Durlas. S Ireland 52°41′N 07°49′W
21 W2 **Thurmont** Maryland, NE USA 39°36′N 77°22′W
Thurø see Thurø By
95 H24 **Thurø By** var. Thurø. Syddtjylland, C Denmark 55°03′N 10°43′E
14 M12 **Thurso** Québec, SE Canada 45°36′N 75°13′W
96 J6 **Thurso** N Scotland, United Kingdom 58°35′N 03°32′W
194 I10 **Thurston Island** island Antarctica
108 I9 **Thusis** Graubünden, S Switzerland 46°40′N 09°27′E
Thýamis see Kalamás
95 E21 **Thyborøn** var. Tyborøn. Midtjylland, W Denmark 56°40′N 08°12′E
195 U3 **Thyer Glacier** glacier Antarctica
115 L20 **Thýmaina** island Dodekánisa, Greece, Aegean Sea
83 N15 **Thyolo** var. Cholo. Southern, S Malawi 16°03′S 35°11′E
183 U6 **Tia** New South Wales, SE Australia 31°14′S 151°51′E
54 H5 **Tía Juana** Zulia, NW Venezuela 10°18′N 71°24′W
Tiancheng see Chongyang
160 J14 **Tiandong** var. Pingma. Guangxi Zhuangzu Zizhiqu, S China 23°37′N 107°06′E
161 O3 **Tianjin** var. Tientsin. Tianjin Shi, E China 39°13′N 117°06′E
Tianjin see Tianjin Shi
161 P3 **Tianjin Shi** var. Jin, Tianjin, T'ien-ching, Tientsin. ◈ municipality E China
159 S10 **Tianjun** var. Xinyuan. Qinghai, C China 37°16′N 99°03′E
160 J13 **Tianlin** var. Leli. Guangxi Zhuangzu Zizhiqu, S China 24°27′N 106°03′E
Tian Shan see Tien Shan
159 W11 **Tianshui** Gansu, C China 34°33′N 105°51′E
150 I7 **Tianshuihai** Xinjiang Uygur Zizhiqu, W China 35°17′N 79°30′E
161 S10 **Tiantai** Zhejiang, SE China 29°11′N 121°01′E
160 J14 **Tianyang** var. Tianzhou. Guangxi Zhuangzu Zizhiqu, S China 23°50′N 106°52′E
Tianzhou see Tianyang
159 U9 **Tianzhu** var. Huazangsi, Tianzhu Zangzu Zizhixian. Gansu, C China 37°01′N 103°04′E
Tianzhu Zangzu Zizhixian see Tianzhu
191 Q7 **Tiarei** Tahiti, W French Polynesia 17°32′S 149°20′W
74 J6 **Tiaret** var. Tihert. NW Algeria 35°20′N 01°20′E
77 N17 **Tiassalé** S Ivory Coast 05°54′N 04°50′W
192 I16 **Ti'avea** Upolu, SE Samoa 13°58′S 171°30′W
Tiba see Chiba
60 J11 **Tibagi** var. Tibaji. Paraná, S Brazil 24°29′S 50°29′W
60 J10 **Tibagi, Rio** var. Rio Tibají. ≋ S Brazil
Tibaji see Tibagi
Tibají, Rio see Tibagi, Rio
139 Q9 **Tibal, Wādī** dry watercourse S Iraq
54 G9 **Tibaná** Boyacá, C Colombia 05°19′N 73°25′W
79 F14 **Tibati** Adamaoua, N Cameroon 06°25′N 12°33′E
76 K15 **Tibé, Pic de** ▲ SE Guinea 08°39′N 08°58′W
Tiber see Tevere, Italy
Tiber see Tivoli, Italy
Tiberias see Tverya
138 G8 **Tiberias, Lake** var. Chinnereth, Sea of Bahr Tabariya, Sea of Galilee, Ar. Bahrat Tabariya, Heb. Yam Kinneret. ◎ N Israel
78 H5 **Tibesti** off. Région du Tibesti. ◈ region N Chad
78 H6 **Tibesti** var. Tibesti Massif, Ar. Tībistī. ▲▲ N Africa
Tibesti Massif see Tibesti
Tibesti, Région du see Tibesti
Tibet see Xizang Zizhiqu
Tibetan Autonomous Region see Xizang Zizhiqu
Tibet, Plateau of see Qingzang Gaoyuan
Tībistī see Tibesti
14 K7 **Tiblemont, Lac** ◎ Québec, SE Canada
139 X9 **Ţīb, Nahr aţ** ≋ S Iraq
Tibnī see At Tibnī
182 L4 **Tibooburra** New South Wales, SE Australia 29°24′S 142°01′E
95 L18 **Tibro** Västra Götaland, S Sweden 58°25′N 14°11′E
40 E5 **Tiburón, Isla** var. Isla del Tiburón. island NW Mexico
Tiburón, Isla del see Tiburón, Isla
23 W14 **Tice** Florida, SE USA 26°40′N 81°49′W
Tichau see Tychy
114 L8 **Ticha, Yazovir** ◎ NE Bulgaria
76 K9 **Tîchît** var. Tichitt. Tagant, C Mauritania 18°26′N 09°31′W
Tichitt see Tîchît
108 G11 **Ticino** Fr./Ger. Tessin. ◈ canton S Switzerland
106 D8 **Ticino** ≋ Italy/Switzerland
108 H11 **Ticino** Ger. Tessin. ≋ SW Switzerland
Ticinum see Pavia
41 X12 **Ticul** Yucatán, SE Mexico 20°22′N 89°30′W
95 K18 **Tidaholm** Västra Götaland, S Sweden 58°12′N 13°55′E
Tidjikdja see Tidjikja
76 J8 **Tidjikja** var. Tidjikdja; prev. Fort-Cappolani. Tagant, C Mauritania 18°31′N 11°24′W
Tidore see Soasiu
171 R11 **Tidore, Pulau** island E Indonesia
Tîdra, Ile see Et Tîdra
77 N16 **Tiébissou** var. Tiebissou. C Ivory Coast 07°10′N 05°10′W
Tiebissou see Tiébissou
Tiefa see Diaobingshan
108 I9 **Tiefencastel** Graubünden, S Switzerland 46°40′N 09°33′E
Tiegenhof see Nowy Dwór Gdański
T'ieh-ling see Tieling
98 K13 **Tiel** Gelderland, C Netherlands 51°53′N 05°26′E
163 W7 **Tieli** Heilongjiang, NE China 46°57′N 128°01′E
163 V11 **Tieling** var. T'ieh-ling. Liaoning, NE China 42°19′N 123°52′E
152 L4 **Tielongtan** China/India 35°10′N 79°32′E
99 D17 **Tielt** var. Thielt. West-Vlaanderen, W Belgium 51°00′N 03°20′E
T'ien-ching see Tianjin Shi
99 I18 **Tienen** var. Thienen, Fr. Tirlemont. Vlaams Brabant, C Belgium 50°48′N 04°56′E
Tiên Giang, Sông see Mekong
147 X9 **Tien Shan** Chin. Thian Shan, Tian Shan, T'ien Shan, Rus. Tyan'-Shan'. ▲▲ C Asia
Tientsin see Tianjin
167 U6 **Tiên Yên** Quang Ninh, N Vietnam 21°19′N 107°24′E
95 O14 **Tierp** Uppsala, C Sweden 60°20′N 17°30′E
62 H7 **Tierra Amarilla** Atacama, N Chile 27°28′S 70°17′W
37 R9 **Tierra Amarilla** New Mexico, SW USA 36°42′N 106°31′W
41 R15 **Tierra Blanca** Veracruz-Llave, E Mexico 18°28′N 96°21′W
41 O16 **Tierra Colorada** Guerrero, S Mexico 17°10′N 99°30′W
63 J17 **Tierra Colorada, Bajo de la** basin SE Argentina
63 I25 **Tierra del Fuego** off. Provincia de la Tierra del Fuego. ◈ province S Argentina
63 J24 **Tierra del Fuego** island Argentina/Chile
Tierra del Fuego, Provincia de la see Tierra del Fuego
54 D7 **Tierralta** Córdoba, NW Colombia 08°10′N 76°04′W
104 K9 **Tiétar** ≋ W Spain
60 L10 **Tietê** São Paulo, S Brazil 23°04′S 47°41′W
60 J8 **Tietê, Rio** ≋ S Brazil
32 I9 **Tieton** Washington, NW USA 46°41′N 120°43′W
32 J6 **Tiffany Mountain** ▲ Washington, NW USA 48°40′N 119°55′W
31 S12 **Tiffin** Ohio, N USA 41°06′N 83°10′W
31 Q11 **Tiffin River** ≋ Ohio, N USA
Tiflis see Tbilisi
23 U7 **Tifton** Georgia, SE USA 31°27′N 83°31′W
171 R13 **Tifu** Pulau Buru, E Indonesia 03°46′S 126°36′E
38 L17 **Tigalda Island** island Aleutian Islands, Alaska, USA
115 I15 **Tigáni, Akrotírio** headland Límnos, E Greece 39°50′N 25°03′E
169 V6 **Tiga Tarok** Sabah, East Malaysia 06°57′N 117°07′E
117 O10 **Tighina** Rus. Bendery; prev. Bender. E Moldova 46°51′N 29°28′E
145 X9 **Tigiretskiy Khrebet** ▲▲ E Kazakhstan
79 F14 **Tignère** Adamaoua, N Cameroon 07°24′N 12°35′E
13 P14 **Tignish** Prince Edward Island, SE Canada 46°58′N 64°03′W
186 M10 **Tigoa** var. Tinggoa. Rennell, S Solomon Islands 11°39′S 160°13′E
Tigranocerta see Siirt
80 I11 **Tigray** ◈ federal region N Ethiopia
41 O11 **Tigre, Cerro del** ▲ C Mexico 23°06′N 99°13′W
56 F8 **Tigre, Río** ≋ N Peru
139 X10 **Tigris** Ar. Dijlah, Turk. Dicle. ≋ Iraq/Turkey
76 G9 **Tiguent** Trarza, SW Mauritania 17°15′N 16°00′W
74 M10 **Tiguentourine** E Algeria 27°59′N 09°16′E
77 V10 **Tiguidit, Falaise de** ridge C Niger
141 N13 **Tihāmah** var. Tehama. plain Saudi Arabia/Yemen
Tihert see Tiaret
Ti-hua/Tihwa see Ürümqi
41 Q13 **Tihuatlán** Veracruz-Llave, E Mexico 20°44′N 97°30′W
40 B1 **Tijuana** Baja California Norte, NW Mexico 32°32′N 117°01′W
42 E2 **Tikal** Petén, N Guatemala 17°11′N 89°36′W
154 I9 **Tīkamgarh** prev. Tehri. Madhya Pradesh, C India 24°44′N 78°50′E
158 L7 **Tikanlik** Xinjiang Uygur Zizhiqu, NW China 40°34′N 87°37′E
77 P12 **Tikaré** N Burkina Faso 13°16′N 01°39′W
39 O12 **Tikchik Lakes** lakes Alaska, USA
191 T9 **Tikehau** atoll Îles Tuamotu, C French Polynesia
191 V9 **Tikei** island Îles Tuamotu, C French Polynesia
126 L13 **Tikhoretsk** Krasnodarskiy Kray, SW Russian Federation 45°51′N 40°07′E
124 I13 **Tikhvin** Leningradskaya Oblast', NW Russian Federation 59°37′N 33°30′E
193 P9 **Tiki Basin** undersea feature S Pacific Ocean
76 K13 **Tikinsso** ≋ NE Guinea
Tikirarjuaq see Whale Cove
184 Q8 **Tikitiki** Gisborne, North Island, New Zealand 37°49′S 178°23′E
79 D16 **Tiko** Sud-Ouest, SW Cameroon 04°02′N 09°19′E
139 S6 **Tikrīt** var. Tekrit. Şalāḩ ad Dīn, N Iraq 34°36′N 43°42′E
124 I8 **Tiksha** Respublika Kareliya, NW Russian Federation 64°07′N 32°31′E
124 I6 **Tikshozero, Ozero** ◎ NW Russian Federation
123 P7 **Tiksi** Respublika Sakha (Yakutiya), NE Russian Federation 71°40′N 128°47′E
Tiladummati Atoll see Thiladhunmathi Atoll
42 A6 **Tilapa** San Marcos, SW Guatemala 14°31′N 92°11′W
42 L13 **Tilarán** Guanacaste, NW Costa Rica 10°28′N 84°57′W
99 J14 **Tilburg** Noord-Brabant, S Netherlands 51°34′N 05°05′E
14 D17 **Tilbury** Ontario, S Canada 42°15′N 82°26′W
182 K4 **Tilcha** South Australia 29°37′S 140°52′E
Tilcha Creek see Callabonna Creek
29 Q14 **Tilden** Nebraska, C USA 42°03′N 97°49′W
25 R13 **Tilden** Texas, SW USA 28°27′N 98°43′W
14 H10 **Tilden Lake** Ontario, S Canada 46°35′N 79°36′W
116 G9 **Tileagd** Hung. Mezőtelegd. Bihor, W Romania 47°03′N 22°11′E
145 P9 **Tilekey** prev. Ladyzhenka. Akmola, C Kazakhstan 50°58′N 68°44′E
77 Q8 **Tîlemsi, Vallée de** ≋ C Mali
123 V8 **Tilichiki** Krasnoyarskiy Kray, E Russian Federation 60°25′N 165°55′E
Tiligul see Tilihul
Tiligul'skiy Liman see Tilihul's'kyy Lyman
117 P9 **Tilihul** Rus. Tiligul. ≋ SW Ukraine
117 P10 **Tilihul's'kyy Lyman** Rus. Tiligul'skiy Liman. ◎ S Ukraine
Tilimsen see Tlemcen
Tilio Martius see Toulon
77 R11 **Tillabéri** var. Tillabéry. Tillabéri, W Niger 14°13′N 01°27′E
77 R11 **Tillabéri** ◈ department SW Niger
Tillabéry see Tillabéri
32 F11 **Tillamook** Oregon, NW USA 45°28′N 123°50′W
32 E11 **Tillamook Bay** inlet Oregon, NW USA
151 Q22 **Tillanchāng Dwīp** island Nicobar Islands, India, NE Indian Ocean
95 N15 **Tillberga** Västmanland, C Sweden 59°41′N 16°37′E
Tillenberg see Dyleń
21 S10 **Tillery, Lake** ⊞ North Carolina, SE USA
77 T10 **Tillia** Tahoua, W Niger 16°13′N 04°51′E
23 N8 **Tillmans Corner** Alabama, S USA 30°35′N 88°10′W
14 F17 **Tillsonburg** Ontario, S Canada 42°53′N 80°44′W
115 N22 **Tílos** island Dodekánisa, Greece, Aegean Sea
183 N5 **Tilpa** New South Wales, SE Australia 30°56′S 144°24′E
Tilsit see Sovetsk
31 N13 **Tilton** Illinois, N USA 40°06′N 87°39′W
126 K7 **Tim** Kurskaya Oblast', W Russian Federation 51°39′N 37°11′E
54 D12 **Timaná** Huila, S Colombia 01°58′N 75°55′W
Timan Ridge see Timanskiy Kryazh
125 Q6 **Timanskiy Kryazh** Eng. Timan Ridge. ridge NW Russian Federation
185 G20 **Timaru** Canterbury, South Island, New Zealand 44°23′S 171°15′E
127 S6 **Timashevo** Samarskaya Oblast', W Russian Federation 53°22′N 51°13′E
126 K13 **Timashevsk** Krasnodarskiy Kray, SW Russian Federation 45°37′N 38°57′E
Timbaki/Timbákion see Tympáki
22 K10 **Timbalier Bay** bay Louisiana, S USA
22 K11 **Timbalier Island** island Louisiana, S USA
76 L10 **Timbedgha** var. Timbédra. Hodh ech Chargui, SE Mauritania 16°17′N 08°14′W
Timbédra see Timbedgha
32 G10 **Timber** Oregon, NW USA 45°42′N 123°19′W
181 O3 **Timber Creek** Northern Territory, N Australia 15°35′S 130°21′E
28 M8 **Timber Lake** South Dakota, N USA 45°25′N 101°01′W
54 D12 **Timbío** Cauca, SW Colombia 02°20′N 76°40′W
54 C12 **Timbiquí** Cauca, SW Colombia 02°43′N 77°45′W
83 O17 **Timbue, Ponta** headland C Mozambique 18°49′S 36°22′E
Timbuktu see Tombouctou
169 W8 **Timbun Mata, Pulau** island E Malaysia
77 P8 **Timétrine** var. Ti-n-Kâr. oasis C Mali
Timfi see Týmfi
Timfristos see Tymfristós
77 V9 **Tîmia** Agadez, C Niger 18°07′N 08°49′E
171 X14 **Timika** Papua, E Indonesia 04°39′S 137°15′E
74 I9 **Timimoun** C Algeria 29°18′N 00°21′E
Timiris, Cap see Timirist, Râs
76 F8 **Timirist, Râs** var. Cap Timiris. headland NW Mauritania 19°18′N 16°28′W
145 O7 **Timiryazevo** Severnyy Kazakhstan, N Kazakhstan 53°45′N 66°33′E
116 E11 **Timiş** ◈ county SW Romania
14 H9 **Timiskaming, Lake** Fr. Lac Témiscamingue. ◎ Ontario/Québec, SE Canada
116 E11 **Timişoara** Ger. Temeschwar, Temeswar, Hung. Temesvár; prev. Temeschburg. Timiş, W Romania 45°46′N 21°17′E
116 E11 **Timisoara** ✈ Timiş, SW Romania 45°50′N 21°21′E
Timkovichi see Tsimkavichy
77 U8 **Ti-m-Meghsoï** ≋ NW Niger
100 K8 **Timmerdorfer Strand** Schleswig-Holstein, N Germany 53°59′N 10°50′E
14 F7 **Timmins** Ontario, S Canada 48°09′N 80°01′W
21 S12 **Timmonsville** South Carolina, SE USA 34°07′N 79°56′W
30 K5 **Timms Hill** ▲ Wisconsin, N USA 45°27′N 90°12′W
112 P12 **Timok** ≋ E Serbia
58 N13 **Timon** Maranhão, E Brazil 05°08′S 42°52′W
171 Q17 **Timor Sea** sea E Indian Ocean
Timor Timur see East Timor
Timor Trench see Timor Trough
192 G8 **Timor Trough** var. Timor Trench. undersea feature NE Timor Sea
61 A21 **Timote** Buenos Aires, E Argentina 35°22′S 62°13′W
54 I6 **Timotes** Mérida, NW Venezuela 08°57′N 70°46′W
25 X8 **Timpson** Texas, SW USA 31°54′N 94°24′W
123 Q11 **Timpton** ≋ NE Russian Federation
93 H17 **Timrå** Västernorrland, C Sweden 62°29′N 17°20′E
20 J10 **Tims Ford Lake** ⊞ Tennessee, S USA
168 L7 **Timur, Banjaran** ▲▲ Peninsular Malaysia
171 Q8 **Tinaca Point** headland Mindanao, S Philippines 05°35′N 125°18′E
54 K5 **Tinaco** Cojedes, N Venezuela 09°44′N 68°28′W
64 Q11 **Tinajo** Lanzarote, Islas Canarias, Spain, NE Atlantic Ocean 29°03′N 13°41′W
187 P10 **Tinakula** island Santa Cruz Islands, E Solomon Islands
54 K5 **Tinaquillo** Cojedes, N Venezuela 09°57′N 68°20′W
116 F10 **Tinca** Hung. Tenke. Bihor, W Romania 46°46′N 21°58′E
155 J20 **Tīndivanam** Tamil Nādu, SE India 12°15′N 79°41′E
74 E9 **Tindouf** W Algeria 27°43′N 08°09′W
74 E9 **Tindouf, Sebkha de** salt lake W Algeria
104 J2 **Tineo** Asturias, N Spain 43°20′N 06°25′W
77 R9 **Ti-n-Essako** Kidal, E Mali 18°30′N 02°27′E
Tinggoa see Tigoa
183 T5 **Tingha** New South Wales, SE Australia 29°56′S 151°13′E
Tingis see Tanger
Tinglett see Tinglev
95 F24 **Tinglev** Ger. Tinglett. Syddanmark, SW Denmark 54°57′N 09°15′E
56 E12 **Tingo María** Huánuco, C Peru 09°10′S 76°00′W
Tingréla see Tengréla
158 K16 **Tingri** var. Xêgar. Xizang Zizhiqu, W China 28°40′N 87°04′E
95 M21 **Tingsryd** Kronoberg, S Sweden 56°30′N 15°E
95 P19 **Tingstäde** Gotland, SE Sweden 57°45′N 18°36′E
62 H12 **Tinguiririca, Volcán** ⛰ C Chile 34°52′S 70°24′W
94 F9 **Tingvoll** Møre og Romsdal, S Norway 62°55′N 08°13′E
188 K8 **Tinian** island S Northern Mariana Islands
Ti-n-Kâr see Timétrine
Tinnevelly see Tirunelveli
95 G15 **Tinnoset** Telemark, S Norway 59°43′N 09°03′E
95 F15 **Tinnsjå** prev. Tinnsjø, var. Tinnsjö. ◎ S Norway see also Tinnsjø
Tinnsjø see Tinnsjå
Tinnsjö see Tinnsjå
Tino see Chino
115 J20 **Tínos** Tínos, Kykládes, Greece, Aegean Sea 37°33′N 25°08′E
115 J20 **Tínos** anc. Tenos. island Kykládes, Greece, Aegean Sea
153 R14 **Tinpahar** Jhārkhand, NE India 25°00′N 87°43′E
153 X11 **Tinsukia** Assam, NE India 27°28′N 95°20′E
76 K10 **Tîntâne** Hodh el Gharbi, S Mauritania 16°25′N 10°08′W
62 L7 **Tintina** Santiago del Estero, N Argentina 27°00′S 62°45′W
182 K10 **Tintinara** South Australia 35°54′S 140°04′E
104 I14 **Tinto** ≋ SW Spain
77 S8 **Ti-n-Zaouâtene** Kidal, NE Mali 19°56′N 02°45′E
Tiobraid Árann see Tipperary
28 K3 **Tioga** North Dakota, N USA 48°24′N 102°56′W
18 G12 **Tioga** Pennsylvania, NE USA 41°54′N 77°07′W
25 T5 **Tioga** Texas, SW USA 33°28′N 96°55′W
35 Q8 **Tioga Pass** pass California, W USA
18 G12 **Tioga River** ≋ New York/Pennsylvania, NE USA
Tioman Island see Tioman, Pulau
168 M9 **Tioman, Pulau** var. Tioman Island. island Peninsular Malaysia
18 C12 **Tionesta** Pennsylvania, NE USA 41°31′N 79°30′W
18 D12 **Tionesta Creek** ≋ Pennsylvania, NE USA
168 J13 **Tiop** Pulau Pagai Selatan, W Indonesia 03°12′S 100°21′E
18 H11 **Tioughnioga River** ≋ New York, NE USA
74 J5 **Tipasa** var. Tipaza. N Algeria 36°35′N 02°27′E
Tipaza see Tipasa
42 J10 **Tipitapa** Managua, W Nicaragua 12°08′N 86°04′W
31 R13 **Tipp City** Ohio, N USA 39°57′N 84°10′W
31 O12 **Tippecanoe River** ≋ Indiana, N USA
97 D20 **Tipperary** Ir. Tiobraid Árann. S Ireland 52°29′N 08°10′W
97 D19 **Tipperary** Ir. Tiobraid Árann. cultural region S Ireland
35 R12 **Tipton** California, W USA 36°02′N 119°19′W
31 P13 **Tipton** Indiana, N USA 40°19′N 86°00′W
29 Y14 **Tipton** Iowa, C USA 41°46′N 91°07′W
27 U5 **Tipton** Missouri, C USA 38°39′N 92°46′W
36 I10 **Tipton, Mount** ▲ Arizona, SW USA 35°32′N 114°11′W
20 F8 **Tiptonville** Tennessee, S USA 36°21′N 89°30′W
12 E12 **Tip Top Mountain** ▲ Ontario, S Canada 48°18′N 86°06′W
155 G19 **Tiptūr** Karnātaka, W India 13°17′N 76°31′E
Tiquisate see Pueblo Nuevo Tiquisate
58 L13 **Tiracambu, Serra do** ▲▲ E Brazil
Tirana see Tiranë
113 K19 **Tirana Rinas** ✈ Durrës, W Albania 41°25′N 19°41′E
113 L20 **Tiranë** var. Tirana. ● (Albania) Tiranë, C Albania 41°20′N 19°50′E
113 K20 **Tiranë** ◈ district W Albania
140 I5 **Tirān, Jazīrat** island Egypt/Saudi Arabia
106 F6 **Tirano** Lombardia, N Italy 46°13′N 10°10′E
182 I2 **Tirari Desert** desert South Australia
117 O10 **Tiraspol** Rus. Tiraspol'. E Moldova 46°50′N 29°35′E
Tiraspol' see Tiraspol
184 M8 **Tirau** Waikato, North Island, New Zealand 37°59′S 175°44′E
136 C14 **Tire** İzmir, SW Turkey 38°04′N 27°45′E
137 O11 **Tirebolu** Giresun, N Turkey 41°01′N 38°49′E
96 F11 **Tiree** island W Scotland, United Kingdom
Tîrgovişte see Târgovişte
Tîrgu see Târgu Cărbuneşti
Tîrgu Bujor see Târgu Bujor
Tîrgu Frumos see Târgu Frumos
Tîrgu Jiu see Targu Jiu
Tîrgu Lăpuş see Târgu Lăpuş
Tîrgu Mures see Târgu Mureş
Tîrgu-Neamţ see Târgu-Neamţ
Tîrgu Ocna see Târgu Ocna
Tîrgu Secuiesc see Târgu Secuiesc
149 T3 **Tirich Mīr** ▲ NW Pakistan 36°12′N 71°51′E
76 J5 **Tiris Zemmour** ◈ region N Mauritania
Tirlemont see Tienen
127 W5 **Tirlyanskiy** Respublika Bashkortostan, W Russian Federation 54°09′N 58°32′E
Tîrnava Mare see Târnava Mare
Tîrnava Mică see Târnava Mică
Tîrnăveni see Târnăveni
Tírnavos see Týrnavos
Tirnovo see Veliko Tarnovo
154 J11 **Tirodi** Madhya Pradesh, C India 21°40′N 79°44′E
108 K8 **Tirol** off. Land Tirol, var. Tyrol, It. Tirolo. ◈ state W Austria
Tirol, Land see Tirol
Tirolo see Tirol
Tirreno, Mare see Tyrrhenian Sea
107 B19 **Tirso** ≋ Sardegna, Italy, C Mediterranean Sea
95 H22 **Tirstrup** ✈ (Århus) Århus, C Denmark 56°17′N 10°36′E
155 I21 **Tiruchchirāppalli** prev. Trichinopoly. Tamil Nādu, SE India 10°50′N 78°43′E
155 H23 **Tirunelveli** var. Tinnevelly. Tamil Nādu, SE India 08°45′N 77°43′E
155 J19 **Tirupati** Andhra Pradesh, E India 13°39′N 79°25′E
155 I20 **Tiruppattūr** Tamil Nādu, SE India 12°28′N 78°31′E
155 H21 **Tiruppur** Tamil Nādu, SW India 11°05′N 77°20′E
Tiruvantapuram see Thiruvanathapuram/Trivandrum
155 I20 **Tiruyannāmalai** Tamil Nādu, SE India 12°13′N 79°07′E
112 L10 **Tisa** Ger. Theiss, Hung. Tisza, Rus. Tissa, Ukr. Tysa. ≋ SE Europe see also Tisza
Tisa see Tisza
Tischnowitz see Tišnov
11 U14 **Tisdale** Saskatchewan, S Canada 52°51′N 104°01′W
27 O13 **Tishomingo** Oklahoma, C USA 34°15′N 96°41′W
95 M17 **Tisnaren** ◎ S Sweden
111 F18 **Tišnov** Ger. Tischnowitz. Jihomoravský Kraj, SE Czech Republic 49°22′N 16°24′E
Tissa see Tisa/Tisza
74 J6 **Tissemsilt** N Algeria 35°37′N 01°48′E
153 S12 **Tista** ≋ NE India
112 L8 **Tisza** Ger. Theiss, Rom./Slvn./SCr. Tisa, Rus. Tissa, Ukr. Tysa. ≋ SE Europe see also Tisa
Tisza see Tisa
111 L23 **Tiszaföldvár** Jász-Nagykun-Szolnok, E Hungary 47°00′N 20°16′E
111 M22 **Tiszafüred** Jász-Nagykun-Szolnok, E Hungary 47°38′N 20°45′E
111 L23 **Tiszakécske** Bács-Kiskun, C Hungary 46°56′N 20°04′E
111 M21 **Tiszaújváros** prev. Leninváros. Borsod-Abaúj-Zemplén, NE Hungary 47°56′N 21°03′E
111 N21 **Tiszavasvári** Szabolcs-Szatmár-Bereg, NE Hungary 47°56′N 21°21′E
57 I17 **Titicaca, Lake** ◎ Bolivia/Peru
190 H17 **Titikaveka** Rarotonga, S Cook Islands 21°16′S 159°45′W
154 M13 **Titilāgarh** var. Titlagarh. Odisha, E India 20°18′N 83°09′E
168 K8 **Titiwangsa, Banjaran** ▲▲ Peninsular Malaysia
Titlagarh see Titilāgarh
Titograd see Podgorica
Titose see Chitose
Titova Mitrovica see Mitrovicë
Titovo Užice see Užice
113 M18 **Titov Vrv** ▲ NW FYR Macedonia 41°58′N 20°49′E
94 F7 **Titran** Sør-Trøndelag, S Norway 63°40′N 08°20′E
31 Q8 **Tittabawassee River** ≋ Michigan, N USA
116 J13 **Titu** Dâmboviţa, S Romania 44°40′N 25°32′E
79 M16 **Titule** Orientale, N Dem. Rep. Congo 03°20′N 25°23′E
23 X11 **Titusville** Florida, SE USA 28°37′N 80°50′W
18 C12 **Titusville** Pennsylvania, NE USA 41°36′N 79°39′W
76 G11 **Tivaouane** W Senegal 14°59′N 16°50′W
113 I17 **Tivat** SW Montenegro 42°25′N 18°43′E
14 E14 **Tiverton** Ontario, S Canada 44°15′N 81°31′W
97 J23 **Tiverton** SW England, United Kingdom 50°54′N 03°30′W
19 O12 **Tiverton** Rhode Island, NE USA 41°38′N 71°10′W
107 I15 **Tivoli** anc. Tiber. Lazio, C Italy 41°58′N 12°45′E
25 U13 **Tivoli** Texas, SW USA 28°26′N 96°54′W
141 Z8 **Ţīwī** NE Oman 22°43′N 59°20′E
41 Y11 **Tizimín** Yucatán, SE Mexico 21°10′N 88°09′W
74 K5 **Tizi Ouzou** var. Tizi-Ouzou. N Algeria 36°44′N 04°06′E
Tizi-Ouzou see Tizi Ouzou
74 D8 **Tiznit** SW Morocco 29°43′N 09°39′W
95 F23 **Tjæreborg** Syddtjylland, W Denmark 55°28′N 08°35′E
113 I14 **Tjentište** Republika Srpska, SE Bosnia and Herzegovina 43°23′N 18°42′E
98 L7 **Tjeukemeer** ◎ N Netherlands
Tjiamis see Ciamis
Tjiandjoer see Cianjur
Tjilatjap see Cilacap
Tjirebon see Cirebon
95 I18 **Tjörn** island S Sweden
92 O3 **Tjuvfjorden** fjord S Svalbard
Tkvarcheli see T'q'varcheli
40 L8 **Tlahualillo** Durango, N Mexico 26°06′N 103°25′W
41 P14 **Tlalnepantla** México, C Mexico 19°34′N 99°12′W
41 Q13 **Tlapacoyán** Veracruz-Llave, E Mexico 19°58′N 97°13′W
41 P16 **Tlapa de Comonfort** Guerrero, S Mexico 17°33′N 98°33′W
40 L13 **Tlaquepaque** Jalisco, C Mexico 20°36′N 103°19′W
Tlascala see Tlaxcala
41 P14 **Tlaxcala** var. Tlascala, Tlaxcala de Xicohténcatl. Tlaxcala, C Mexico 19°17′N 98°16′W
41 P14 **Tlaxcala** ◈ state S Mexico
Tlaxcala de Xicohténcatl see Tlaxcala
41 P14 **Tlaxco** var. Tlaxco de Morelos. Tlaxcala, S Mexico 19°38′N 98°06′W
Tlaxco de Morelos see Tlaxco
41 Q16 **Tlaxiaco** var. Santa María Asunción Tlaxiaco. Oaxaca, S Mexico 17°18′N 97°42′W
74 I6 **Tlemcen** var. Tilimsen, Tlemsen. NW Algeria 34°53′N 01°21′W
Tlemsen see Tlemcen
138 L4 **Tlété Ouâte Rharbi, Jebel** ▲ N Syria
116 J7 **Tlumach** Ivano-Frankivs'ka Oblast', W Ukraine 48°53′N 25°00′E
127 P17 **Tlyarata** Respublika Dagestan, SW Russian Federation 42°10′N 46°30′E
116 K10 **Toaca, Vârful** prev. Vîrful Toaca. ▲ NE Romania 46°58′N 25°55′E
Toaca, Vîrful see Toaca, Vârful
191 Q8 **Toahotu** prev. Teohatu. Tahiti, W French Polynesia
187 R13 **Toak** Ambrym, C Vanuatu 16°21′S 168°16′E
172 J4 **Toamasina** prev./Fr. Tamatave. Toamasina, E Madagascar 18°10′S 49°23′E
172 J4 **Toamasina** ◈ province E Madagascar

◆ Country
● Country Capital
◇ Dependent Territory
○ Dependent Territory Capital
◈ Administrative Regions
✈ International Airport
▲ Mountain
▲▲ Mountain Range
⛰ Volcano
≋ River
◎ Lake
⊞ Reservoir

TOAMASINA – TOTTENHAM

172 J4 **Toamasina** ✈ Toamasina, E Madagascar 18°10′S 49°23′E
21 X6 **Toano** Virginia, NE USA 37°22′N 76°46′W
191 U10 **Toau** *atoll* Îles Tuamotu, C French Polynesia
45 T6 **Toa Vaca, Embalse** ⊠ C Puerto Rico
62 K13 **Toay** La Pampa, C Argentina 36°43′S 64°22′W
159 R14 **Toba** Xizang Zizhiqu, W China 31°17′N 97°37′E
164 K14 **Toba** Mie, Honshū, SW Japan 34°29′N 136°51′E
168 I9 **Toba, Danau** ◎ Sumatera, W Indonesia
45 Y16 **Tobago** *island* NE Trinidad and Tobago
149 Q9 **Toba Kākar Range** ▲ NW Pakistan
105 Q12 **Tobarra** Castilla-La Mancha, C Spain 38°36′N 01°41′W
149 U9 **Toba Tek Singh** Punjab, E Pakistan 30°54′N 72°30′E
171 R11 **Tobelo** Pulau Halmahera, E Indonesia 01°45′N 127°59′E
14 E12 **Tobermory** Ontario, S Canada 45°15′N 81°39′W
96 G10 **Tobermory** W Scotland, United Kingdom 56°37′N 06°12′W
165 S4 **Tōbetsu** Hokkaidō, NE Japan 43°12′N 141°28′E
180 M6 **Tobin Lake** ◎ Western Australia
11 U14 **Tobin Lake** ◎ Saskatchewan, C Canada
35 T4 **Tobin, Mount** ▲ Nevada, W USA 40°25′N 117°28′W
165 O9 **Tobi-shima** *island* C Japan
169 N13 **Toboali** Pulau Bangka, W Indonesia 03°00′S 106°30′E
Tobol *see* Tobyl
122 H11 **Tobol'sk** Tyumenskaya Oblast', C Russian Federation 58°15′N 68°12′E
Tobruch/Tobruk *see* Ṭubruq
125 R3 **Tobseda** Nenetskiy Avtonomnyy Okrug, NW Russian Federation 68°37′N 52°24′E
144 M8 **Tobyl** *prev.* Tobol. Kustanay, N Kazakhstan 52°42′N 62°36′E
144 L8 **Tobyl** *prev.* Tobol. ≈ Kazakhstan/Russian Federation
125 Q6 **Tobysh** ≈ NW Russian Federation
54 F10 **Tocaima** Cundinamarca, C Colombia 04°30′N 74°38′W
59 K16 **Tocantins** *off.* Estado do Tocantins. ◈ *state* C Brazil
Tocantins, Estado do *see* Tocantins
59 K15 **Tocantins, Rio** ≈ N Brazil
23 T2 **Toccoa** Georgia, SE USA 34°34′N 83°19′W
165 O12 **Tochigi** *off.* Tochigi-ken, *var.* Totigi. ◈ *prefecture* Honshū, S Japan
Tochigi-ken *see* Tochigi
165 O11 **Tochio** *var.* Totio. Niigata, Honshū, C Japan 37°27′N 139°00′E
95 J15 **Töcksfors** Värmland, C Sweden 59°30′N 11°49′E
42 J5 **Tocoa** Colón, N Honduras 15°40′N 86°01′W
62 H4 **Tocopilla** Antofagasta, N Chile 22°06′S 70°08′W
62 I4 **Tocorpuri, Cerro de** ▲ Bolivia/Chile 22°26′S 67°53′W
183 O10 **Tocumwal** New South Wales, SE Australia 35°53′S 145°35′E
54 K4 **Tocuyo de la Costa** Falcón, NW Venezuela 11°04′N 68°23′W
152 H13 **Toda Rāisingh** Rājasthān, N India 26°02′N 75°35′E
106 H13 **Todi** Umbria, C Italy 42°47′N 12°25′E
108 G9 **Tödi** ▲ NE Switzerland 46°52′N 08°53′E
171 T12 **Todlo** Papua Barat, E Indonesia 0°46′S 130°50′E
165 S9 **Todoga-saki** *headland* Honshū, C Japan 39°33′N 142°02′E
59 P17 **Todos os Santos, Baía de** *bay* E Brazil
40 F10 **Todos Santos** Baja California Sur, NW Mexico 23°28′N 110°14′W
40 B2 **Todos Santos, Bahía de** *bay* NW Mexico
Toeban *see* Tuban
Toekang Besi Eilanden *see* Tukangbesi, Kepulauan
Toeloengagoeng *see* Tulungagung
Tōen *see* Taoyuan
185 D25 **Toetoes Bay** *bay* South Island, New Zealand
11 Q14 **Tofield** Alberta, SW Canada 53°22′N 112°39′W
10 K17 **Tofino** Vancouver Island, British Columbia, SW Canada 49°05′N 125°51′W
189 X17 **Tofol** Kosrae, E Micronesia
95 J20 **Tofta** Halland, S Sweden 57°10′N 12°19′E
95 H15 **Tofte** Buskerud, S Norway 59°31′N 10°33′E
95 F24 **Toftlund** Syddanmark, SW Denmark 55°12′N 09°04′E
193 X15 **Tofua** *island* Ha'apai Group, C Tonga
187 Q12 **Toga** *island* Torres Islands, N Vanuatu
80 N13 **Togdheer** *off.* Gobolka Togdheer. ◈ *region* NW Somalia
Togdheer, Gobolka *see* Togdheer
164 L11 **Togi** Ishikawa, Honshū, SW Japan 37°06′N 136°44′E
39 N13 **Togiak** Alaska, USA 59°03′N 160°31′W
171 O11 **Togian, Kepulauan** *island group* C Indonesia
77 Q15 **Togo** *off.* Togolese Republic; *prev.* French Togoland. ◆ *republic* W Africa
Togolese Republic *see* Togo
162 F8 **Tögrög** Govĭ-Altay, SW Mongolia 45°51′N 95°04′E
162 F8 **Tögrög** *var.* Hoolt. Övörhangay, C Mongolia 45°31′N 103°06′E
Tögrög *see* Manhan
159 N12 **Togton He** *var.* Tuotuo He. ≈ C China
Togton Heyan *see* Tanggulashan
Toguzak *see* Togyzak
144 L7 **Togyzak** *prev.* Toguzak. ≈ Kazakhstan/Russian Federation
37 P10 **Tohatchi** New Mexico, SW USA 35°51′N 108°45′W
191 O7 **Tohiea, Mont** ▲ Moorea, W French Polynesia 17°33′S 149°48′W
137 N14 **Tohma Çayı** ≈ C Turkey
93 O17 **Tohmajärvi** Pohjois-Karjala, SE Finland 62°12′N 30°19′E
93 L16 **Toholampi** Keski-Pohjanmaa, W Finland 63°46′N 24°15′E
Tōhōm *see* Mandah
23 X12 **Tohopekaliga, Lake** ◎ Florida, SE USA
164 M14 **Toi** Shizuoka, Honshū, S Japan 34°55′N 138°45′E
190 B15 **Toi** N Niue 18°57′S 169°51′W
93 L19 **Toijala** Pirkanmaa, SW Finland 61°09′N 23°51′E
171 P12 **Toima** Sulawesi, N Indonesia 0°48′S 122°21′E
164 D17 **Toi-misaki** Kyūshū, SW Japan
171 Q17 **Toineke** Timor, S Indonesia 10°06′S 124°22′E
Toirc, Inis *see* Inishturk
35 U6 **Toiyabe Range** ▲ Nevada, W USA
Tojikiston, Jumhurii *see* Tajikistan
147 R12 **Tojikobod** *Rus.* Tadzhikabad. C Tajikistan 39°08′N 70°54′E
164 G12 **Tōjō** Hiroshima, Honshū, SW Japan 34°54′N 133°15′E
39 T10 **Tok** Alaska, USA 63°20′N 142°59′W
164 K13 **Tōkai** Aichi, Honshū, SW Japan 35°01′N 136°51′E
111 N21 **Tokaj** Borsod-Abaúj-Zemplén, NE Hungary 48°08′N 21°25′E
165 N11 **Tōkamachi** Niigata, Honshū, C Japan 37°08′N 138°44′E
185 D25 **Tokanui** Southland, South Island, New Zealand 46°33′S 169°02′E
80 I7 **Tokar** *var.* Ṭawkar. Red Sea, NE Sudan 18°27′N 37°41′E
136 L12 **Tokat** Tokat, N Turkey 40°20′N 36°35′E
136 L12 **Tokat** ◈ *province* N Turkey
Tökchök-kundo *see* Deokjeok-gundo
Töke *see* Taka Atoll
190 J9 **Tokelau** ◇ *NZ overseas territory* W Polynesia
Tőketerebes *see* Trebišov
Tokhtamyshbek *see* Tŭkhtamish
24 M6 **Tokio** Texas, SW USA 33°09′N 102°31′W
Tokio *see* Tōkyō
189 W11 **Toki Point** *point* NW Wake Island
Tokkuztara *see* Gongliu
117 V9 **Tokmak** *var.* Velykyy Tokmak. Zaporiz'ka Oblast', SE Ukraine 47°13′N 35°43′E
Tokmak *see* Tomok
184 Q8 **Tokomaru Bay** Gisborne, North Island, New Zealand 38°10′S 178°18′E
184 M8 **Tokoroa** Waikato, North Island, New Zealand 38°14′S 175°52′E
76 K14 **Tokounou** C Guinea 09°43′N 09°46′W
38 M12 **Toksook Bay** Alaska, USA 60°33′N 165°01′W
Toksu *see* Xinhe
Toksum *see* Toksun
158 L6 **Toksun** *var.* Toksum. Xinjiang Uygur Zizhiqu, NW China 42°47′N 88°38′E
147 T8 **Toktogul** Talasskaya Oblast', NW Kyrgyzstan 41°51′N 72°56′E
147 T9 **Toktogul'skoye Vodokhranilishche** ⊠ W Kyrgyzstan
Toktomush *see* Tŭkhtamish
193 Y14 **Toku** *island* Vava'u Group, N Tonga
165 U16 **Tokunoshima** Kagoshima, Tokuno-shima, SW Japan
165 U16 **Tokuno-shima** *island* Nansei-shotō, SW Japan
164 I14 **Tokushima** *var.* Tokusima. Tokushima, Shikoku, SW Japan 34°04′N 134°28′E
164 H14 **Tokushima** *off.* Tokushima-ken, *var.* Tokusima. ◈ *prefecture* Shikoku, SW Japan
Tokushima-ken *see* Tokushima
Tokusima *see* Tokushima
164 E13 **Tokuyama** *var.* Shūnan. Yamaguchi, Honshū, SW Japan 34°04′N 131°48′E
165 N13 **Tōkyō** *var.* Tokio. ● (Japan) Tōkyō, Honshū, S Japan 35°40′N 139°45′E
165 O13 **Tōkyō** *off.* Tōkyō-to. ◈ *capital district* Honshū, S Japan
Tōkyō-to *see* Tōkyō
145 T12 **Tokyrau** ≈ C Kazakhstan
149 O3 **Tokzār** *Pash.* Tukzār. Sar-e Pul, N Afghanistan 35°47′N 66°28′E
145 W13 **Tokzhaylau** *prev.* Dzerzhinskoye. Almaty, SE Kazakhstan 45°49′N 81°04′E
145 W13 **Tokzhaylau** *var.* Dzerzhinskoye. Taldykorgan, SE Kazakhstan 45°49′N 81°04′E
189 U12 **Tol** *atoll* Chuuk Islands, C Micronesia
184 Q9 **Tolaga Bay** Gisborne, North Island, New Zealand 38°22′S 178°17′E
172 I7 **Tôlañaro** *prev.* Faradofay, Fort-Dauphin. Toliara, SE Madagascar
162 D6 **Tolbo** Bayan-Ölgiy, W Mongolia 48°22′N 90°22′E
Tolbukhin *see* Dobrich
60 G11 **Toledo** Paraná, S Brazil 24°45′S 53°41′W
54 G8 **Toledo** Norte de Santander, N Colombia 07°16′N 72°28′W
105 N9 **Toledo** *anc.* Toletum. Castilla-La Mancha, C Spain 39°52′N 04°02′W
30 M14 **Toledo** Illinois, N USA 39°16′N 88°15′W
29 W13 **Toledo** Iowa, C USA 42°00′N 92°34′W
31 R11 **Toledo** Ohio, N USA 41°40′N 83°33′W
32 F12 **Toledo** Oregon, NW USA 44°37′N 123°56′W
32 G9 **Toledo** Washington, NW USA 46°27′N 122°49′W
42 F3 **Toledo** ◈ *district* S Belize
104 M9 **Toledo** ◈ *province* Castilla-La Mancha, C Spain
25 Y7 **Toledo Bend Reservoir** ⊠ Louisiana/Texas, SW USA
104 M10 **Toledo, Montes de** ▲ C Spain
106 J12 **Tolentino** Marche, C Italy 43°08′N 13°17′E
Toletum *see* Toledo
94 H10 **Tolga** Hedmark, S Norway 62°25′N 11°00′E
158 J3 **Toli** Xinjiang Uygur Zizhiqu, NW China 45°55′N 83°33′E
172 H7 **Toliara** *var.* Toliary; *prev.* Tuléar. Toliara, SW Madagascar 23°20′S 43°41′E
172 H7 **Toliara** ◈ *province* SW Madagascar
Toliary *see* Toliara
118 H11 **Toliejai** *prev.* Kamajai. Panevėžys, NE Lithuania 55°16′N 25°30′E
54 D11 **Tolima** *off.* Departamento del Tolima. ◈ *province* C Colombia
Tolima, Departamento del *see* Tolima
171 N11 **Tolitoli** Sulawesi, C Indonesia 01°05′N 120°50′E
95 K22 **Tollarp** Skåne, S Sweden 55°55′N 14°00′E
100 N9 **Tollense** ≈ NE Germany
100 N10 **Tollensesee** ◎ NE Germany
36 K13 **Tolleson** Arizona, SW USA 33°25′N 112°15′W
146 M13 **Tollimarjon** *Rus.* Talimardzhan. Qashqadaryo Viloyati, S Uzbekistan 38°22′N 65°31′E
Tolmein *see* Tolmin
106 J6 **Tolmezzo** Friuli-Venezia Giulia, NE Italy 46°27′N 13°01′E
109 S11 **Tolmin** *Ger.* Tolmein, *It.* Tolmino. W Slovenia 46°12′N 13°39′E
Tolmino *see* Tolmin
111 J25 **Tolna** *Ger.* Tolnau. Tolna, S Hungary 46°26′N 18°47′E
111 I24 **Tolna** *off.* Tolna Megye. ◈ *county* SW Hungary
Tolna Megye *see* Tolna
Tolnau *see* Tolna
79 I20 **Tolo** Bandundu, W Dem. Rep. Congo 02°57′S 18°35′E
Tolochin *see* Talachyn
190 D12 **Toloke** Île Futuna, W Wallis and Futuna
30 M13 **Tolono** Illinois, N USA 39°59′N 88°16′W
105 Q3 **Tolosa** País Vasco, N Spain 43°09′N 02°04′W
Tolosa *see* Toulouse
171 O13 **Tolo, Teluk** *bay* Sulawesi, C Indonesia
39 R9 **Tolovana River** ≈ Alaska, USA
123 U10 **Tolstoy, Mys** *headland* E Russian Federation 59°12′N 155°04′E
63 G15 **Toltén** Araucanía, C Chile 39°13′S 73°15′W
63 G15 **Toltén, Río** ≈ S Chile
54 E6 **Tolú** Sucre, NW Colombia 09°32′N 75°34′W
41 O14 **Toluca** *var.* Toluca de Lerdo. México, S Mexico 19°20′N 99°40′W
Toluca de Lerdo *see* Toluca
41 O14 **Toluca, Nevado de** ▲ C Mexico 19°05′N 99°45′W
127 R6 **Tol'yatti** *prev.* Stavropol'. Samarskaya Oblast', W Russian Federation 53°32′N 49°27′E
77 O12 **Toma** NW Burkina Faso 12°46′N 02°51′W
30 K7 **Tomah** Wisconsin, N USA 43°59′N 90°31′W
30 L5 **Tomahawk** Wisconsin, N USA 45°27′N 89°40′W
117 T8 **Tomakivka** Dnipropetrovs'ka Oblast', E Ukraine 47°47′N 34°45′E
165 S4 **Tomakomai** Hokkaidō, NE Japan 42°38′N 141°32′E
165 S2 **Tomamae** Hokkaidō, NE Japan 44°18′N 141°38′E
104 G9 **Tomar** Santarém, W Portugal 39°36′N 08°25′W
123 T13 **Tomari** Ostrov Sakhalin, Sakhalinskaya Oblast', SE Russian Federation 47°47′N 142°09′E
115 C16 **Tómaros** ▲ W Greece 39°31′N 20°45′E
Tomaschow *see* Tomaszów Mazowiecki
Tomaschow *see* Tomaszów Lubelski
61 E16 **Tomás Gomensoro** Artigas, N Uruguay 30°28′S 57°28′W
117 N7 **Tomashpil'** Vinnyts'ka Oblast', C Ukraine 48°32′N 28°31′E
Tomaszów *see* Tomaszów Mazowiecki
111 P15 **Tomaszów Lubelski** *Ger.* Tomaschow. Lubelskie, E Poland 50°29′N 23°23′E
Tomaszów Mazowiecka *see* Tomaszów Mazowiecki
110 L13 **Tomaszów Mazowiecki** *var.* Tomaszów Mazowiecka; *prev.* Tomaszów, *Ger.* Tomaschow. Łódzkie, C Poland 51°33′N 20°E
40 J13 **Tomatlán** Jalisco, C Mexico 19°53′N 105°18′W
81 F15 **Tombe** Jonglei, E South Sudan 05°52′N 31°40′E
23 N4 **Tombigbee River** ≈ Alabama/Mississippi, S USA
82 A10 **Tomboco** Zaire Province, NW Angola 06°50′S 13°20′E
77 O10 **Tombouctou** *Eng.* Timbuktu. Tombouctou, N Mali 16°47′N 03°03′W
77 N9 **Tombouctou** ◈ *region* W Mali
37 N16 **Tombstone** Arizona, SW USA 31°42′N 110°04′W
83 A15 **Tombua** *Port.* Porto Alexandre. Namibe, SW Angola 15°49′S 11°53′E
83 J19 **Tom Burke** Limpopo, NE South Africa 23°07′S 28°01′E
146 L9 **Tomdibuloq** *Rus.* Tamdybulak. Navoiy Viloyati, N Uzbekistan 41°48′N 64°33′E
146 L9 **Tomditov-Tog'lari** ▲ N Uzbekistan
62 G13 **Tomé** Bío Bío, C Chile 36°38′S 72°57′W
58 L12 **Tomé-Açu** Pará, NE Brazil 02°25′S 48°09′W
95 L23 **Tomelilla** Skåne, S Sweden 55°33′N 14°00′E
105 O10 **Tomelloso** Castilla-La Mancha, C Spain 39°09′N 03°01′W
14 H10 **Tomiko Lake** ◎ Ontario, S Canada
77 N12 **Tominian** Ségou, C Mali 13°18′N 04°39′W
Tomini, Gulf of *see* Tomini, Teluk
171 N12 **Tomini, Teluk** *var.* Gulf of Tomini; *prev.* Teluk Gorontalo. *bay* Sulawesi, C Indonesia
165 Q11 **Tomioka** Fukushima, Honshū, C Japan 37°19′N 140°57′E
113 G14 **Tomislavgrad** Federacija Bosne I Hercegovine, SW Bosnia and Herzegovina 43°43′N 17°15′E
181 O9 **Tomkinson Ranges** ▲ South Australia/Western Australia
123 Q11 **Tommot** Respublika Sakha (Yakutiya), NE Russian Federation 58°57′N 126°24′E
171 Q11 **Tomohon** Sulawesi, N Indonesia 01°19′N 124°49′E
147 V7 **Tomok** *prev.* Tokmak. Chuyskaya Oblast', N Kyrgyzstan 42°50′N 75°18′E
54 K9 **Tomo, Río** ≈ E Colombia
113 L21 **Tomorrit, Mali i** ▲ S Albania 40°43′N 20°12′E
11 S17 **Tompkins** Saskatchewan, S Canada 50°03′N 108°49′W
20 K8 **Tompkinsville** Kentucky, S USA 36°43′N 85°41′W
171 N11 **Tompo** Sulawesi, N Indonesia 0°56′N 120°16′E
180 I8 **Tom Price** Western Australia 22°48′S 117°49′E
122 J12 **Tomsk** Tomskaya Oblast', C Russian Federation 56°30′N 85°05′E
122 I11 **Tomskaya Oblast'** ◈ *province* C Russian Federation
18 K16 **Toms River** New Jersey, NE USA 39°56′N 74°09′W
Tom Steed Lake *see* Tom Steed Reservoir
26 L12 **Tom Steed Reservoir** *var.* Tom Steed Lake. ⊠ Oklahoma, C USA
171 U13 **Tomu** Papua Barat, E Indonesia 02°07′S 133°01′E
158 H6 **Tomür Feng** *var.* Pobeda Peak, *Rus.* Pik Pobedy. ▲ China/Kyrgyzstan 42°02′N 80°07′E *see also* Pobedy, Pik
Tomür Feng *see* Pobedy, Pik
189 N13 **Tomworoahlang** Pohnpei, E Micronesia
41 U17 **Tonalá** Chiapas, SE Mexico 16°08′N 93°41′W
106 F6 **Tonale, Passo del** *pass* N Italy
164 I11 **Tonami** Toyama, Honshū, SW Japan 36°40′N 136°55′E
58 C12 **Tonantins** Amazonas, W Brazil 02°58′S 67°30′W
32 K6 **Tonasket** Washington, NW USA 48°41′N 119°27′W
55 Y9 **Tonate** *var.* Macouria. N French Guiana 05°00′N 52°28′W
18 D10 **Tonawanda** New York, NE USA 43°00′N 78°51′W
42 I7 **Toncontín** *prev.* Tegucigalpa. ● (Honduras) Francisco Morazán, SW Honduras 14°04′N 87°11′W
42 H7 **Toncontín** ✈ Central District, C Honduras 14°03′N 87°20′W
171 Q11 **Tondano** Sulawesi, C Indonesia 01°19′N 124°56′E
104 H7 **Tondela** Viseu, N Portugal 40°31′N 08°05′W
95 F24 **Tønder** *Ger.* Tondern. Syddanmark, SW Denmark 54°57′N 08°53′E
Tondern *see* Tønder
143 N4 **Tonekābon** *var.* Shahsawar, Tonkābon; *prev.* Shahsavār. Māzandarān, N Iran 36°40′N 51°25′E
Tonezh *see* Tonyezh
193 Y14 **Tonga** *off.* Kingdom of Tonga, *var.* Friendly Islands. ◆ *monarchy* SW Pacific Ocean
175 R9 **Tonga** *island group* SW Pacific Ocean
Tongaat *see* oThongathi
Tonga, Kingdom of *see* Tonga
161 Q13 **Tong'an** *var.* Datong, Tong an. Fujian, SE China 24°43′N 118°07′E
Tong an *see* Tong'an
27 Q4 **Tonganoxie** Kansas, C USA 39°06′N 95°05′W
39 Y13 **Tongass National Forest** *reserve* Alaska, USA
193 Y16 **Tongatapu** ✈ Tongatapu, S Tonga 21°10′S 175°10′W
193 Y16 **Tongatapu** *island* Tongatapu Group, S Tonga
193 Y16 **Tongatapu Group** *island group* S Tonga
175 S9 **Tonga Trench** *undersea feature* S Pacific Ocean
161 N8 **Tongbai Shan** ▲ C China
161 P8 **Tongcheng** Anhui, E China 31°16′N 117°00′E
160 L6 **Tongchuan** Shaanxi, C China 35°10′N 109°03′E
160 L12 **Tongdao** *var.* Tongdao Dongzu Zizhixian; *prev.* Shuangjiang. Hunan, S China 26°06′N 109°46′E
Tongdao Dongzu Zizhixian *see* Tongdao
159 T11 **Tongde** *var.* Gabasumdo. Qinghai, C China 35°13′N 100°39′E
99 K19 **Tongeren** *Fr.* Tongres. Limburg, NE Belgium 50°47′N 05°28′E
Tonghae *see* Donghae
160 G13 **Tonghai** *var.* Xiushan. Yunnan, SW China 24°06′N 102°45′E
163 X8 **Tonghe** Heilongjiang, NE China 46°00′N 128°45′E
163 W11 **Tonghua** Jilin, NE China 41°45′N 125°50′E
163 Z6 **Tongjiang** Heilongjiang, NE China 47°39′N 132°29′E
163 Y13 **Tongjosŏn-man** *prev.* Broughton Bay. *bay* E North Korea
163 V7 **Tongken He** ≈ NE China
Tongking, Gulf of *see* Tonkin, Gulf of
163 U10 **Tongliao** Nei Mongol Zizhiqu, N China 43°37′N 122°15′E
161 Q9 **Tongling** Anhui, E China 30°55′N 117°50′E
161 R9 **Tonglu** Zhejiang, SE China 29°50′N 119°38′E
187 R14 **Tongoa** *island* Shepherd Islands, S Vanuatu
62 G9 **Tongoy** Coquimbo, C Chile 30°16′S 71°31′W
160 L11 **Tongren** *var.* Rongwo. Guizhou, S China 27°44′N 109°10′E
159 T11 **Tongren** *var.* Rongwo. Qinghai, C China 35°31′N 101°58′E
Tongres *see* Tongeren
153 U11 **Tongsa** *var.* Tongsa Dzong. C Bhutan 27°33′N 90°30′E
Tongsa Dzong *see* Tongsa
Tongshan *see* Fuding, Fujian, China
Tongshan *see* Xuzhou, Jiangsu, China
Tongshi *see* Wuzhishan
159 P12 **Tongtian He** *var.* Zhi Qu. ≈ C China
96 I6 **Tongue** N Scotland, United Kingdom 58°30′N 04°25′W
44 H3 **Tongue of the Ocean** *strait* C The Bahamas
33 X10 **Tongue River** ≈ Montana, NW USA
33 W11 **Tongue River Resevoir** ⊠ Montana, NW USA
159 V11 **Tongwei** *var.* Pingxiang. Gansu, C China 35°09′N 105°15′E
159 W9 **Tongxin** Ningxia, N China 37°00′N 105°41′E
163 U9 **Tongyu** *var.* Kaitong. Jilin, NE China 44°49′N 123°08′E
160 J11 **Tongzi** *var.* Loushanguan. Guizhou, S China 28°08′N 106°49′E
162 F8 **Tonhil** *var.* Dzüyl. Govĭ-Altay, SW Mongolia 46°09′N 93°55′E
40 G5 **Tónichi** Sonora, NW Mexico 28°37′N 109°34′W
81 D14 **Tonj** Warap, W South Sudan 07°18′N 28°41′E
152 H13 **Tonk** Rājasthān, N India 26°10′N 75°50′E
Tonkābon *see* Tonekābon
27 N8 **Tonkawa** Oklahoma, C USA 36°40′N 97°18′W
167 T7 **Tonkin, Gulf of** *var.* Gulf of Tongking, *Chin.* Beibu Wan, *Vtn.* Vinh Bắc Bộ. *gulf* China/Vietnam
167 Q12 **Tônlé Sap** *Eng.* Great Lake. ◎ W Cambodia
102 L14 **Tonneins** Lot-et-Garonne, SW France 44°21′N 00°21′E
103 Q7 **Tonnerre** Yonne, C France 47°50′N 04°00′E
Tonoas *see* Dublon
35 U8 **Tonopah** Nevada, W USA 38°04′N 117°13′W
164 H13 **Tonoshō** Okayama, Shōdo-shima, SW Japan 34°29′N 134°10′E
43 S17 **Tonosí** Los Santos, S Panama 07°23′N 80°26′W
95 H16 **Tønsberg** Vestfold, S Norway 59°16′N 10°25′E
39 T11 **Tonsina** Alaska, USA 61°39′N 145°10′W
95 D17 **Tonstad** Vest-Agder, S Norway 58°40′N 06°42′E
193 X15 **Tonumea** *island* Nomuka Group, W Tonga
137 O11 **Tonya** Trabzon, NE Turkey 40°52′N 39°17′E
119 K20 **Tonyezh** *Rus.* Tonezh. Homyel'skaya Voblasts', SE Belarus 51°50′N 27°48′E
36 L3 **Tooele** Utah, W USA 40°32′N 112°18′W
122 L13 **Toora-Khem** Respublika Tyva, S Russian Federation 52°25′N 96°01′E
183 O5 **Tooraweenah** New South Wales, SE Australia 30°29′S 148°52′E
83 H25 **Toorberg** ▲ S South Africa 32°02′S 24°02′E
118 G5 **Tootsi** Pärnumaa, SW Estonia 58°34′N 24°43′E
183 U3 **Toowoomba** Queensland, E Australia 27°35′S 151°54′E
27 Q4 **Topeka** *state capital* Kansas, C USA 39°03′N 95°41′W
122 J12 **Topki** Kemerovskaya Oblast', S Russian Federation 55°12′N 85°40′E
111 M18 **Topľa** *Hung.* Toplya. ≈ NE Slovakia
Toplicza *see* Toplița
116 J10 **Toplița** *Ger.* Töplitz, *Hung.* Marosheviz; *prev.* Toplița Română, *Hung.* Oláh-Toplicza, Toplicza. Harghita, C Romania 45°56′N 25°20′E
Toplița Română/Töplitz *see* Toplița
Toplya *see* Topľa
111 I20 **Topoľčany** *Hung.* Nagytapolcsány. Nitriansky Kraj, W Slovakia 48°33′N 18°10′E
40 G8 **Topolobampo** Sinaloa, C Mexico 25°36′N 109°04′W
116 I13 **Topoloveni** Argeș, S Romania 44°49′N 25°02′E
114 L11 **Topolovgrad** *prev.* Kavakli. Haskovo, S Bulgaria 42°06′N 26°20′E
Topolya *see* Bačka Topola
32 J10 **Toppenish** Washington, NW USA 46°22′N 120°18′W
181 P4 **Top Springs Roadhouse** Northern Territory, N Australia 16°37′S 131°49′E
189 U11 **Tora** *island* Chuuk, C Micronesia
Torajh *see* Tory Island
189 U11 **Tora Island Pass** *passage* Chuuk Islands, C Micronesia
143 U5 **Torbat-e Ḩeydarīyeh** *var.* Turbat-i-Haidari. Khorāsān-e Razavī, NE Iran 35°18′N 59°12′E
143 V5 **Torbat-e Jām** *var.* Turbat-i-Jam. Khorāsān-e Razavī, NE Iran 35°16′N 60°36′E
39 Q11 **Torbert, Mount** ▲ Alaska, USA 61°30′N 152°15′W
31 P6 **Torch Lake** ◎ Michigan, N USA
Törcsvár *see* Bran
Torda *see* Turda
104 L6 **Tordesillas** Castilla y León, N Spain 41°30′N 05°00′W
92 K13 **Töre** Norrbotten, N Sweden 65°55′N 22°40′E
95 L17 **Töreboda** Västra Götaland, S Sweden 58°41′N 14°07′E
95 J21 **Torekov** Skåne, S Sweden 56°25′N 12°39′E
92 O3 **Torell Land** *physical region* SW Svalbard
105 U4 **Toretta de l'Orrí** *var.* Llorri; *prev.* Tossal de l'Orrí. ▲ NE Spain 42°24′N 01°15′E
144 M14 **Toretum** Kyzylorda, S Kazakhstan 45°38′N 63°20′E
117 Y8 **Torez** Donets'ka Oblast', SE Ukraine 48°00′N 38°38′E
101 N14 **Torgau** Sachsen, E Germany 51°34′N 13°01′E
145 R8 **Torgay** *Kaz.* Torghay; *prev.* Turgay. Akmola, W Kazakhstan 51°46′N 72°45′E
145 N10 **Torgay** *prev.* Turgay. ≈ C Kazakhstan
Torgay Üstirti *see* Turgayskaya Stolovaya Strana
Torghay *see* Torgay
95 N22 **Torhamn** Blekinge, S Sweden 56°04′N 15°49′E
99 C17 **Torhout** West-Vlaanderen, W Belgium 51°04′N 03°06′E
106 B8 **Torino** *Eng.* Turin. Piemonte, NW Italy 45°03′N 07°39′E
Tori-shima *see* Izu-Tori-shima
Torkestān *see* Kāshmar
81 F16 **Torit** Eastern Equatoria, S South Sudan 04°27′N 32°31′E
186 H6 **Toriu** New Britain, E Papua New Guinea 04°39′S 151°42′E
148 M4 **Torkestān, Selseleh-ye Band-e** *var.* Bandi-i Turkistan. ▲ NW Afghanistan
104 L7 **Tormes** ≈ W Spain
Tornacum *see* Tournai
Torneå *see* Tornio
92 K12 **Torneälven** *var.* Torniojoki, *Fin.* Tornionjoki. ≈ Finland/Sweden
92 I11 **Torneträsk** ◎ N Sweden
13 O4 **Torngat Mountains** ▲ Newfoundland and Labrador, NE Canada
24 H8 **Tornillo** Texas, SW USA 31°26′N 106°06′W
92 K13 **Tornio** *Swe.* Torneå. Lappi, NW Finland 65°50′N 24°18′E
Torniojoki/Tornionjoki *see* Torneälven
61 B23 **Tornquist** Buenos Aires, E Argentina 38°08′S 62°15′W
104 L6 **Toro** Castilla y León, N Spain 41°31′N 05°24′W
62 H9 **Toro, Cerro del** ▲ N Chile 29°10′S 69°43′W
77 R12 **Torodi** Tillabéri, SW Niger 13°05′N 01°46′E
Törökbecse *see* Novi Bečej
186 J7 **Torokina** Bougainville, NE Papua New Guinea 06°12′S 155°04′E
111 L23 **Törökszentmiklós** Jász-Nagykun-Szolnok, E Hungary 47°11′N 20°26′E
42 G7 **Torola, Río** ≈ El Salvador/Honduras
Toronaíos, Kólpos *see* Kassándras, Kólpos
14 H15 **Toronto** *province capital* Ontario, S Canada 43°42′N 79°25′W
31 V12 **Toronto** Ohio, N USA 40°27′N 80°36′W
Toronto *see* Lester B. Pearson
27 P6 **Toronto Lake** ⊠ Kansas, C USA
35 V16 **Toro Peak** ▲ California, W USA 33°31′N 116°25′W
124 H16 **Toropets** Tverskaya Oblast', W Russian Federation 56°29′N 31°37′E
81 G18 **Tororo** E Uganda 0°42′N 34°12′E
136 H16 **Toros Dağları** *Eng.* Taurus Mountains. ▲ S Turkey
183 N13 **Torquay** Victoria, SE Australia 38°21′S 144°18′E
97 J24 **Torquay** SW England, United Kingdom 50°28′N 03°30′W
104 M5 **Torquemada** Castilla y León, N Spain 42°02′N 04°17′W
35 S16 **Torrance** California, W USA 33°50′N 118°20′W
104 G12 **Torrão** Setúbal, S Portugal 38°18′N 08°13′W
104 H8 **Torre, Alto da** ▲ C Portugal 40°21′N 07°31′W
107 K18 **Torre Annunziata** Campania, S Italy 40°45′N 14°27′E
105 T8 **Torreblanca** Valenciana, E Spain 40°14′N 00°12′E
104 L15 **Torrecilla** ▲ S Spain 36°38′N 04°54′W
105 P4 **Torrecilla en Cameros** La Rioja, N Spain 42°18′N 02°33′W
105 N13 **Torredelcampo** Andalucía, S Spain 37°45′N 03°52′W
107 K17 **Torre del Greco** Campania, S Italy 40°46′N 14°22′E
104 I6 **Torre de Moncorvo** *var.* Moncorvo, Tôrre de Moncorvo. Bragança, N Portugal 41°10′N 07°03′W
104 J9 **Torrejoncillo** Extremadura, W Spain 39°54′N 06°28′W
105 O8 **Torrejón de Ardoz** Madrid, C Spain 40°27′N 03°29′W
105 N7 **Torrelaguna** Madrid, C Spain 40°50′N 03°33′W
105 N2 **Torrelavega** Cantabria, N Spain 43°21′N 04°03′W
107 M16 **Torremaggiore** Puglia, SE Italy 41°42′N 15°17′E
104 M15 **Torremolinos** Andalucía, S Spain 36°38′N 04°30′W
182 I6 **Torrens, Lake** *salt lake* South Australia
105 S10 **Torrent** *Cas.* Torrente, *var.* Torrent de l'Horta. Valenciana, E Spain 39°27′N 00°28′W
Torrent de l'Horta/Torrente *see* Torrent
40 L8 **Torreón** Coahuila, NE Mexico 25°47′N 103°21′W
105 R13 **Torre-Pacheco** Murcia, SE Spain 37°43′N 00°57′W
106 A8 **Torre Pellice** Piemonte, NE Italy 44°49′N 07°12′E
105 O13 **Torreperogil** Andalucía, S Spain 38°02′N 03°17′W
61 J15 **Torres** Rio Grande do Sul, S Brazil 29°20′S 49°43′W
Torrès, Îles *see* Torres Islands
187 Q11 **Torres Islands** *Fr.* Îles Torrès. *island group* N Vanuatu
104 G9 **Torres Novas** Santarém, C Portugal 39°28′N 08°32′W
181 V1 **Torres Strait** *strait* Australia/Papua New Guinea
104 F10 **Torres Vedras** Lisboa, C Portugal 39°05′N 09°15′W
105 S13 **Torrevieja** Valenciana, E Spain 37°59′N 00°40′W
186 B6 **Torricelli Mountains** ▲ NW Papua New Guinea
96 G8 **Torridon, Loch** *inlet* NW Scotland, United Kingdom
106 D9 **Torriglia** Liguria, NW Italy 44°31′N 09°08′E
104 M9 **Torrijos** Castilla-La Mancha, C Spain 39°59′N 04°18′W
18 L12 **Torrington** Connecticut, NE USA 41°48′N 73°07′W
33 Z15 **Torrington** Wyoming, C USA 42°04′N 104°10′W
Torröjen *see* Torrön
94 F16 **Torrön** *prev.* Torröjen. ◎ C Sweden
105 N15 **Torrox** Andalucía, S Spain 36°45′N 03°58′W
94 N13 **Torsåker** Gävleborg, C Sweden 60°31′N 16°30′E
95 N21 **Torsås** Kalmar, S Sweden 56°24′N 16°00′E
95 J14 **Torsby** Värmland, C Sweden 60°07′N 13°E
95 N16 **Torshälla** Södermanland, C Sweden 59°25′N 16°28′E
95 B19 **Tórshavn** *Dan.* Thorshavn. ○ Faroe Islands 62°02′N 06°47′W
Torshiz *see* Kāshmar
146 I9 **To'rtko'l** *var.* Türtkŭl, *Rus.* Turtkul'; *prev.* Petroaleksandrovsk. Qoraqalpog'iston Respublikasi, W Uzbekistan 41°35′N 61°E
Tortoise Islands *see* Colón, Archipiélago de
45 T9 **Tortola** *island* C British Virgin Islands
106 D9 **Tortona** *anc.* Dertona. Piemonte, NW Italy 44°54′N 08°52′E
107 L23 **Tortorici** Sicilia, Italy, C Mediterranean Sea 38°02′N 14°49′E
105 U7 **Tortosa** *anc.* Dertosa. Cataluña, E Spain 40°49′N 00°31′E
Tortosa *see* Ṭarṭūs
105 U7 **Tortosa, Cap** *cape* E Spain
44 L8 **Tortue, Île de la** *var.* Tortuga Island. *island* N Haiti
55 Y10 **Tortue, Montagne** ▲ C French Guiana
Tortuga, Isla *see* La Tortuga, Isla
Tortuga Island *see* Tortue, Île de la
54 C11 **Tortugas, Golfo** *gulf* W Colombia
45 T5 **Tortuguero, Laguna** *lagoon* N Puerto Rico
137 Q12 **Tortum** Erzurum, NE Turkey 40°20′N 41°36′E
Torugart, Pereval *see* Turugart Shankou
137 O12 **Torul** Gümüşhane, NE Turkey 40°35′N 39°18′E
110 J10 **Toruń** *Ger.* Thorn. Toruń, Kujawsko-pomorskie, C Poland 53°02′N 18°36′E
95 K20 **Torup** Halland, S Sweden 56°57′N 13°04′E
118 I6 **Tõrva** *Ger.* Törwa. Valgamaa, S Estonia 58°00′N 25°54′E
Törwa *see* Tõrva
96 D13 **Tory Island** *Ir.* Toraigh. *island* NW Ireland
111 N19 **Torysa** *Hung.* Tarca. ≈ NE Slovakia
Torzburg *see* Bran
124 J16 **Torzhok** Tverskaya Oblast', W Russian Federation 57°04′N 34°55′E
164 F15 **Tosa-Shimizu** *var.* Tosasimizu. Kōchi, Shikoku, SW Japan 32°47′N 132°58′E
Tosasimizu *see* Tosa-Shimizu
164 G15 **Tosa-wan** *bay* SW Japan
83 H21 **Tosca** North-West, N South Africa 25°51′S 23°56′E
106 F12 **Toscana** *Eng.* Tuscany. ◈ *region* C Italy
107 E14 **Toscano, Archipelago** *Eng.* Tuscan Archipelago. *island group* C Italy
106 G10 **Tosco-Emiliano, Appennino** *Eng.* Tuscan-Emilian Mountains. ▲ C Italy
Tösei *see* Dongshi
165 N15 **To-shima** *island* Izu-shotō, SE Japan
147 Q9 **Toshkent** *Eng./Rus.* Tashkent. ● Toshkent Viloyati, E Uzbekistan 41°19′N 69°17′E
147 Q9 **Toshkent** ✈ Toshkent Viloyati, E Uzbekistan 41°13′N 69°15′E
147 P9 **Toshkent Viloyati** *Rus.* Tashkentskaya Oblast'. ◈ *province* E Uzbekistan
124 H13 **Tosno** Leningradskaya Oblast', NW Russian Federation 59°34′N 30°48′E
159 Q10 **Toson Hu** ◎ C China
162 H6 **Tosontsengel** Dzavhan, NW Mongolia 48°42′N 98°14′E
162 J6 **Tosontsengel** *var.* Tsengel. Hövsgöl, N Mongolia 49°29′N 101°09′E
146 I8 **Tosquduq Qumlari** *var.* Goshquduq Qum, Taskuduk, Peski. *desert* W Uzbekistan
Tossal de l'Orrí *see* Toretta de l'Orrí
61 A15 **Tostado** Santa Fe, C Argentina 29°15′S 61°45′W
118 F6 **Tõstamaa** *Ger.* Testama. Pärnumaa, SW Estonia 58°20′N 23°59′E
100 I10 **Tostedt** Niedersachsen, NW Germany 53°16′N 09°42′E
136 J11 **Tosya** Kastamonu, N Turkey 41°02′N 34°02′E
95 F15 **Totak** ◎ S Norway
105 R13 **Totana** Murcia, SE Spain 37°45′N 01°30′W
94 H13 **Toten** *physical region* S Norway
83 G18 **Toteng** North-West, C Botswana 20°25′S 23°00′E
102 M3 **Tôtes** Seine-Maritime, N France 49°40′N 01°02′E
Totigi *see* Tochigi
Totio *see* Tochio
Totis *see* Tata
189 U13 **Totiw** *island* Chuuk, C Micronesia
125 N13 **Tot'ma** *var.* Totma. Vologodskaya Oblast', NW Russian Federation 59°58′N 42°42′E
Tot'ma *see* Sukhona
55 V9 **Totness** Coronie, N Suriname 05°53′N 56°19′W
42 C5 **Totonicapán** Totonicapán, W Guatemala 14°58′N 91°12′W
42 A2 **Totonicapán** *off.* Departamento de Totonicapán. ◈ *department* W Guatemala
Totonicapán, Departamento de *see* Totonicapán
61 B18 **Totoras** Santa Fe, C Argentina 32°35′S 61°11′W
187 Y15 **Totoya** *island* S Fiji
183 Q7 **Tottenham** New South Wales, SE Australia 32°16′S 147°23′E

◆ Country
● Country Capital
◇ Dependent Territory
○ Dependent Territory Capital
◈ Administrative Regions
✈ International Airport
▲ Mountain
▲ Mountain Range
Volcano
≈ River
◎ Lake
⊠ Reservoir

TOTTORI – TSENTRAL'NYYE NIZMENNYYE GARAGUMY

164 *I12* **Tottori** Tottori, Honshū, SW Japan 35°29′N 134°14′E
164 *H12* **Tottori** *off.* Tottori-ken. ◈ *prefecture* Honshū, SW Japan
Tottori-ken *see* Tottori
76 *I6* **Touâjîl** Tiris Zemmour, N Mauritania 22°03′N 12°40′W
76 *L15* **Touba** W Ivory Coast 08°17′N 07°41′W
76 *G11* **Touba** W Senegal 14°55′N 15°53′W
74 *E7* **Toubkal, Jbel** ▲ W Morocco 31°00′N 07°50′W
32 *K10* **Touchet** Washington, NW USA 46°03′N 118°40′W
103 *P7* **Toucy** Yonne, C France 47°45′N 03°18′E
77 *O12* **Tougan** W Burkina Faso 13°06′N 03°03′W
74 *L7* **Touggourt** NE Algeria 33°08′N 06°04′E
77 *Q12* **Tougouri** N Burkina Faso 13°22′N 00°25′W
76 *J13* **Tougué** NW Guinea 11°29′N 11°48′W
76 *K12* **Toukoto** Kayes, W Mali 13°27′N 09°52′W
103 *S5* **Toul** Meurthe-et-Moselle, NE France 48°41′N 05°54′E
76 *L16* **Toulépleu** *var.* Toulobli. W Ivory Coast 06°37′N 08°27′W
Touliu *see* Douliu
15 *U3* **Toulnustouc** ≈ Québec, SE Canada
Toulobli *see* Toulépleu
103 *T16* **Toulon** *anc.* Telo Martius, Tilio Martius. Var, SE France 43°07′N 05°56′E
30 *K12* **Toulon** Illinois, N USA 41°05′N 89°54′W
102 *M15* **Toulouse** *anc.* Tolosa. Haute-Garonne, S France 43°37′N 01°25′E
102 *M15* **Toulouse** ✈ Haute-Garonne, S France 43°38′N 01°19′E
77 *N16* **Toumodi** C Ivory Coast 06°34′N 05°01′W
74 *G9* **Tounassine, Hamada** *hill range* W Algeria
Toungoo *see* Taungoo
166 *K7* **Toungup** *var.* Taungup. Rakhine State, W Myanmar (Burma) 18°50′N 94°14′E
102 *L8* **Touraine** *cultural region* C France
Tourane *see* Đa Năng
103 *P1* **Tourcoing** Nord, N France 50°44′N 03°10′E
104 *F2* **Touriñán, Cabo** *headland* NW Spain 43°02′N 09°20′W
76 *J6* **Tourine** Tiris Zemmour, N Mauritania 22°23′N 11°l50′W
102 *J3* **Tourlaville** Manche, N France 49°38′N 01°34′W
99 *D19* **Tournai** *var.* Tournay, *Dut.* Doornik; *anc.* Tornacum. Hainaut, SW Belgium 50°36′N 03°24′E
102 *L16* **Tournay** Hautes-Pyrénées, S France 43°10′N 00°16′E
Tournay *see* Tournai
103 *R12* **Tournon** Ardèche, E France 45°05′N 04°49′E
103 *R9* **Tournus** Saône-et-Loire, C France 46°33′N 04°53′E
59 *Q14* **Touros** Rio Grande do Norte, E Brazil 05°10′S 35°29′W
102 *L8* **Tours** *anc.* Caesarodunum, Turoni. Indre-et-Loire, C France 47°22′N 00°40′E
183 *Q17* **Tourville, Cape** *headland* Tasmania, SE Australia 42°09′S 148°20′E
44 *M9* **Toussaint Louverture** ✈ E Haiti 18°38′N 72°13′W
162 *L8* **Töv** ◈ *province* C Mongolia
54 *H7* **Tovar** Mérida, NW Venezuela 08°22′N 71°50′W
126 *L5* **Tovarkovskiy** Tul'skaya Oblast', W Russian Federation 53°41′N 38°18′E
Tovil'-Dora *see* Tavildara
Tövis *see* Teiuş
137 *V11* **Tovuz** *Rus.* Tauz. W Azerbaijan 40°58′N 45°41′E
165 *R7* **Towada** Aomori, Honshū, C Japan 40°35′N 141°13′E
184 *K3* **Towai** Northland, North Island, New Zealand 35°29′S 174°06′E
18 *H12* **Towanda** Pennsylvania, NE USA 41°45′N 76°25′W
29 *W4* **Tower** Minnesota, N USA 47°48′N 92°16′W
171 *N12* **Towera** Sulawesi, N Indonesia 0°29′S 120°01′E
Tower Island *see* Genovesa, Isla
180 *M13* **Tower Peak** ▲ Western Australia 33°23′S 123°27′E
35 *U11* **Towne Pass** *pass* California, W USA
29 *N3* **Towner** North Dakota, N USA 48°20′N 100°27′W
33 *R10* **Townsend** Montana, NW USA 46°19′N 111°31′W
181 *X6* **Townsville** Queensland, NE Australia 19°24′S 146°53′E
Towoeti Meer *see* Towuti, Danau
148 *K4* **Towraghoudī** Herāt, NW Afghanistan 35°13′N 62°19′E
21 *X3* **Towson** Maryland, NE USA 39°25′N 76°36′W
171 *O13* **Towuti, Danau** *Dut.* Towoeti Meer. ◎ Sulawesi, C Indonesia
Toxkan He *see* Ak-say
24 *K9* **Toyah** Texas, SW USA 31°18′N 103°47′W
165 *R4* **Tōya-ko** ◎ Hokkaidō, NE Japan
164 *L11* **Toyama** Toyama, Honshū, SW Japan 36°41′N 137°13′E
164 *L11* **Toyama** *off.* Toyama-ken. ◈ *prefecture* Honshū, SW Japan
Toyama-ken *see* Toyama
164 *L11* **Toyama-wan** *bay* W Japan
164 *H15* **Tōyo** Kōchi, Shikoku, SW Japan 33°22′N 134°18′E
Toyohara *see* Yuzhno-Sakhalinsk
164 *L14* **Toyohashi** *var.* Toyohasi. Aichi, Honshū, SW Japan 34°46′N 137°22′E
Toyohasi *see* Toyohashi
164 *L14* **Toyokawa** Aichi, Honshū, SW Japan 34°47′N 137°24′E
164 *I14* **Toyooka** Hyōgo, Honshū, SW Japan 35°35′N 134°48′E
164 *L13* **Toyota** Aichi, Honshū, SW Japan 35°04′N 137°09′E
165 *T1* **Toyotomi** Hokkaidō, NE Japan 45°07′N 141°45′E
147 *Q10* **To'ytepa** *Rus.* Toytepa. Toshkent Viloyati, E Uzbekistan 41°04′N 69°22′E
Toytepa *see* To'ytepa
74 *M6* **Tozeur** *var.* Tawzar. W Tunisia 34°00′N 08°09′E
39 *Q8* **Tozi, Mount** ▲ Alaska, USA 65°45′N 151°01′W
137 *Q9* **T'q'varcheli** *Rus.* Tkvarcheli; *prev.* Tqvarch'eli. NW Georgia 42°51′N 41°42′E
Tqvarch'eli *see* T'q'varcheli
Trâblous *see* Tripoli
137 *O11* **Trabzon** *Eng.* Trebizond; *anc.* Trapezus. Trabzon, NE Turkey 41°N 39°43′E
137 *O11* **Trabzon** *Eng.* Trebizond. ◈ *province* NE Turkey
13 *P13* **Tracadie** New Brunswick, SE Canada 47°32′N 64°57′W
Trachenberg *see* Żmigród
15 *O11* **Tracy** Québec, SE Canada 45°59′N 73°07′W
35 *O8* **Tracy** California, W USA 37°43′N 121°27′W
29 *S10* **Tracy** Minnesota, N USA 44°14′N 95°37′W
20 *K10* **Tracy City** Tennessee, S USA 35°15′N 85°44′W
106 *D7* **Tradate** Lombardia, N Italy 45°43′N 08°57′E
84 *F6* **Traena Bank** *undersea feature* E Norwegian Sea 66°15′N 09°45′E
29 *W13* **Traer** Iowa, C USA 42°11′N 92°28′W
104 *J16* **Trafalgar, Cabo de** *headland* SW Spain 36°10′N 06°03′W
Traiectum ad Mosam/ Traiectum Tungorum *see* Maastricht
Tráigh Mhór *see* Tramore
11 *O17* **Trail** British Columbia, SW Canada 49°94′N 117°39′W
58 *B11* **Traíra, Serra do** ▲▲ NW Brazil
109 *V5* **Traisen** Niederösterreich, NE Austria 48°03′N 15°37′E
109 *W4* **Traisen** ≈ NE Austria
109 *X4* **Traiskirchen** Niederösterreich, NE Austria 48°01′N 16°18′E
Trajani Portus *see* Civitavecchia
Trajectum ad Rhenum *see* Utrecht
119 *H14* **Trakai** *Ger.* Traken, *Pol.* Troki. Vilnius, SE Lithuania 54°39′N 24°58′E
Traken *see* Trakai
97 *B20* **Tralee** *Ir.* Trá Lí. SW Ireland 52°16′N 09°42′W
97 *A20* **Tralee Bay** *Ir.* Bá Thrá Lí. *bay* SW Ireland
Trá Lí *see* Tralee
Trälleborg *see* Trelleborg
Tralles Aydin *see* Aydın
61 *J16* **Tramandaí** Rio Grande do Sul, S Brazil 30°01′S 50°11′W
108 *C7* **Tramelan** Bern, W Switzerland 47°13′N 07°07′E
Trá Mhór *see* Tramore
97 *E20* **Tramore** *Ir.* Tráigh Mhór, Trá Mhór. Waterford, S Ireland 52°10′N 07°10′W
114 *F9* **Tran** *var.* Trŭn. Pernik, W Bulgaria 42°51′N 22°37′E
95 *L18* **Tranås** Jönköping, S Sweden 58°03′N 15°00′E
62 *J7* **Trancas** Tucumán, N Argentina 26°11′S 65°20′W
104 *I7* **Trancoso** Guarda, N Portugal 40°46′N 07°21′W
95 *H22* **Tranebjerg** Midtjylland, C Denmark 55°51′N 10°36′E
95 *K19* **Tranemo** Västra Götaland, S Sweden 57°30′N 13°20′E
167 *N16* **Trang** Trang, S Thailand 07°33′N 99°36′E
171 *V15* **Trangan, Pulau** *island* Kepulauan Aru, E Indonesia
Trâng Dinh *see* Thât Khê
183 *Q7* **Trangie** New South Wales, SE Australia 32°01′S 147°58′E
94 *K12* **Trängslet** Dalarna, C Sweden 61°22′N 13°43′E
107 *N16* **Trani** Puglia, SE Italy 41°17′N 16°25′E
61 *F17* **Tranqueras** Rivera, NE Uruguay 31°13′S 55°45′W
63 *G17* **Tranqui, Isla** *island* S Chile
39 *V6* **Trans-Alaska pipeline** *oil pipeline* Alaska, USA
195 *Q10* **Transantarctic Mountains** ▲▲ Antarctica
Transcarpathian Oblast *see* Zakarpats'ka Oblast'
122 *E9* **Trans-Siberian Railway** *railroad* Russian Federation
Transilvania *see* Transylvania
Transilvaniei, Alpi *see* Carpații Meridionalii
Transjordan *see* Jordan
172 *L11* **Transkei Basin** *undersea feature* SW Indian Ocean 35°30′S 29°00′E
117 *O10* **Transnistria** *cultural region* E Moldavia
81 *H18* **Trans Nzoia** ◈ *county* W Kenya
Transsylvanische Alpen/ Transylvanian Alps *see* Carpații Meridionalii
94 *K12* **Transtrand** Dalarna, C Sweden 61°06′N 13°19′E
116 *G10* **Transylvania** *Eng.* Ardeal, Transilvania, *Ger.* Siebenbürgen, *Hung.* Erdély. *cultural region* NW Romania
167 *S14* **Tra Ôn** Vinh Long, S Vietnam 09°58′N 105°58′E
107 *H23* **Trapani** *anc.* Drepanum. Sicilia, Italy, C Mediterranean Sea 38°02′N 12°32′E
Trâpeăng Vêng *see* Kâmpóng Thum
Trapezus *see* Trabzon
114 *L9* **Trapoklovo** Sliven, C Bulgaria 42°40′N 26°36′E
183 *P13* **Traralgon** Victoria, SE Australia 38°15′S 146°36′E
76 *H9* **Trarza** ◈ *region* SW Mauritania
Trasimenischersee *see* Trasimeno, Lago
106 *H12* **Trasimeno, Lago** *Eng.* Lake of Perugia, *Ger.* Trasimenischersee. ◎ C Italy
95 *J20* **Träslövsläge** Halland, S Sweden 57°02′N 12°18′E
Trás-os-Montes *see* Cucumbi
104 *I6* **Trás-os-Montes e Alto Douro** *former province* N Portugal
167 *Q12* **Trat** *var.* Bang Phra. Trat, S Thailand 12°16′N 102°30′E
Trá Tholl, Inis *see* Inishtrahull
Traù *see* Trogir
109 *T4* **Traun** Oberösterreich, N Austria 48°14′N 14°15′E
109 *S5* **Traun** ≈ N Austria
Traun, Lake *see* Traunsee
101 *N23* **Traunreut** Bayern, SE Germany 47°58′N 12°36′E
109 *S5* **Traunsee** *var.* Gmundner See, *Eng.* Lake Traun. ◎ N Austria
Trautenau *see* Trutnov
21 *P11* **Travelers Rest** South Carolina, SE USA 34°58′N 82°26′W
182 *L8* **Travellers Lake** *seasonal lake* New South Wales, SE Australia
31 *P6* **Traverse City** Michigan, N USA 44°45′N 85°37′W
29 *R7* **Traverse, Lake** ◎ Minnesota/South Dakota, N USA
185 *I16* **Travers, Mount** ▲ South Island, New Zealand 42°01′S 172°46′E
11 *P17* **Travers Reservoir** ⊠ Alberta, SW Canada
167 *T14* **Tra Vinh** *var.* Phu Vinh. Tra Vinh, S Vietnam 09°57′N 106°20′E
25 *S10* **Travis, Lake** ⊠ Texas, SW USA
112 *H12* **Travnik** Federacija Bosne I Hercegovine, C Bosnia and Herzegovina 44°14′N 17°40′E
109 *V11* **Trbovlje** *Ger.* Trifail. C Slovenia 46°10′N 15°03′E
23 *V13* **Treasure Island** Florida, SE USA 27°46′N 82°46′W
Treasure State *see* Montana
186 *I8* **Treasury Islands** *island group* NW Solomon Islands
106 *D9* **Trebbia** *anc.* Trebia. ≈ NW Italy
100 *N8* **Trebel** ≈ NE Germany
103 *O16* **Trèbes** Aude, S France 43°12′N 02°26′E
Trebia *see* Trebbia
111 *F18* **Třebíč** *Ger.* Trebitsch. Vysočina, C Czech Republic 49°13′N 15°52′E
113 *I16* **Trebinje** Republika Srpska, S Bosnia and Herzegovina 42°42′N 18°19′E
Trebišnica *see* Trebišnjica
113 *H16* **Trebišnjica** *var.* Trebišnica. ≈ S Bosnia and Herzegovina
111 *N20* **Trebišov** *Hung.* Tőketerebes. Košický Kraj, E Slovakia 48°37′N 21°44′E
Trebitsch *see* Třebíč
Trebizond *see* Trabzon
Trebnitz *see* Trzebnica
109 *V12* **Trebnje** SE Slovenia 45°54′N 15°01′E
111 *D19* **Třeboň** *Ger.* Wittingau. Jihočeský Kraj, S Czech Republic 49°00′N 14°46′E
104 *J15* **Trebujena** Andalucía, S Spain 36°52′N 06°11′W
100 *I7* **Treene** ≈ N Germany
Tree Planters State *see* Nebraska
109 *S9* **Treffen** Kärnten, S Austria 46°40′N 13°51′E
Trefynwy *see* Monmouth
102 *G5* **Tréguier** Côtes d'Armor, NW France 48°50′N 03°12′W
61 *G18* **Treinta y Tres** Treinta y Tres, E Uruguay 33°16′S 54°17′W
61 *F18* **Treinta y Tres** ◈ *department* E Uruguay
122 *F11* **Trëkhgornyy** Chelyabinskaya Oblast', C Russian Federation 54°42′N 58°25′E
114 *F9* **Treklyanska Reka** ≈ W Bulgaria
102 *K8* **Trélazé** Maine-et-Loire, NW France 47°27′N 00°28′W
63 *K17* **Trelew** Chubut, SE Argentina 43°13′S 65°15′W
95 *K23* **Trelleborg** *var.* Trälleborg. Skåne, S Sweden 55°22′N 13°10′E
113 *P15* **Trem** ▲ SE Serbia 43°10′N 22°12′E
15 *N11* **Tremblant, Mont** ▲ Québec, SE Canada 46°13′N 74°34′W
99 *H17* **Tremelo** Vlaams Brabant, C Belgium 51°N 04°34′E
107 *M15* **Tremiti, Isole** *island group* SE Italy
30 *K12* **Tremont** Illinois, N USA 40°30′N 89°31′W
36 *L1* **Tremonton** Utah, W USA 41°42′N 112°09′W
105 *U4* **Tremp** Cataluña, NE Spain 42°09′N 00°53′E
30 *J7* **Trempealeau** Wisconsin, N USA 44°00′N 91°25′W
15 *P8* **Trenche** ≈ Québec, SE Canada
15 *O7* **Trenche, Lac** ◎ Québec, SE Canada
111 *I20* **Trenčiansky Kraj** ◈ *region* W Slovakia
111 *I19* **Trenčín** *Ger.* Trentschin, *Hung.* Trencsén. Trenčiansky Kraj, W Slovakia 48°54′N 18°03′E
Trencsén *see* Trenčín
Trengganu *see* Terengganu
Trengganu, Kuala *see* Kuala Terengganu
61 *A21* **Trenque Lauquen** Buenos Aires, E Argentina 36°01′S 62°47′W
14 *J14* **Trent** ≈ Ontario, SE Canada
97 *N18* **Trent** ≈ C England, United Kingdom
Trent *see* Trento
106 *F5* **Trentino-Alto Adige** *Eng.* South Tyrol, *Ger.* Trentino-Südtirol; *prev.* Venezia Tridentina. ◈ *region* N Italy
Trentino-Südtirol *see* Trentino-Alto Adige
106 *G6* **Trento** *Eng.* Trent, *Ger.* Trient; *anc.* Tridentum. Trentino-Alto Adige, N Italy 46°05′N 11°08′E
14 *J15* **Trenton** Ontario, SE Canada 44°07′N 77°34′W
23 *V10* **Trenton** Florida, SE USA 29°36′N 82°49′W
23 *R1* **Trenton** Georgia, SE USA 34°52′N 85°27′W
31 *S10* **Trenton** Michigan, N USA 42°08′N 83°10′W
27 *S2* **Trenton** Missouri, C USA 40°04′N 93°37′W
28 *M17* **Trenton** Nebraska, C USA 40°10′N 101°00′W
18 *J15* **Trenton** *state capital* New Jersey, NE USA 40°13′N 74°45′W
21 *W10* **Trenton** North Carolina, SE USA 35°03′N 77°20′W
20 *G9* **Trenton** Tennessee, S USA 35°59′N 88°59′W
36 *L1* **Trenton** Utah, W USA 41°53′N 111°57′W
Trentschin *see* Trenčín
Treptow an der Rega *see* Trzebiatów
61 *C23* **Tres Arroyos** Buenos Aires, E Argentina 38°22′S 60°17′W
61 *J15* **Três Cachoeiras** Rio Grande do Sul, S Brazil 29°21′S 49°48′W
106 *E7* **Trescore Balneario** Lombardia, N Italy 45°43′N 09°52′E
41 *V17* **Tres Cruces, Cerro** ▲ SE Mexico 15°28′N 92°27′W
57 *K18* **Tres Cruces, Cordillera** ▲▲ W Bolivia
113 *N18* **Treska** ≈ NW FYR Macedonia
113 *I14* **Treskavica** ▲▲ SE Bosnia and Herzegovina
59 *J20* **Três Lagoas** Mato Grosso do Sul, SW Brazil 20°46′S 51°43′W
40 *H12* **Tres Marías, Islas** *island group* C Mexico
59 *M19* **Três Marias, Represa** ⊠ SE Brazil
63 *F20* **Tres Montes, Península** *headland* S Chile 46°49′S 75°29′W
105 *O3* **Trespaderne** Castilla y León, N Spain 42°47′N 03°24′W
60 *G13* **Três Passos** Rio Grande do Sul, S Brazil 27°33′S 53°55′W
61 *A23* **Tres Picos, Cerro** ▲ E Argentina 38°10′S 61°54′W
63 *G17* **Tres Picos, Cerro** ▲ SW Argentina 42°22′S 71°51′W
60 *I12* **Três Pinheiros** Paraná, S Brazil 25°25′S 51°57′W
59 *M21* **Três Pontas** Minas Gerais, SE Brazil 21°33′S 45°18′W
Tres Puntas, Cabo *see* Manabique, Punta
60 *P9* **Três Rios** Rio de Janeiro, SE Brazil 22°06′S 43°15′W
Tres Tabernae *see* Saverne
Trestenberg/Trestendorf *see* Tăşnad
41 *R15* **Tres Valles** Veracruz-Llave, SE Mexico 18°14′N 96°03′W
94 *H12* **Tretten** Oppland, S Norway 61°19′N 10°19′E
Treuburg *see* Olecko
101 *K21* **Treuchtlingen** Bayern, S Germany 48°57′N 10°55′E
100 *N13* **Treuenbrietzen** Brandenburg, E Germany 52°06′N 12°52′E
63 *H17* **Trevelín** Chubut, SW Argentina 43°02′S 71°27′W
Treves/Trèves *see* Trier
106 *I13* **Trevi** Umbria, C Italy 42°52′N 12°46′E
106 *E7* **Treviglio** Lombardia, N Italy 45°32′N 09°35′E
104 *J4* **Trevinca, Peña** ▲ NW Spain 42°10′N 06°49′W
105 *P3* **Treviño** Castilla y León, N Spain 42°45′N 02°41′W
106 *I7* **Treviso** *anc.* Tarvisium. Veneto, NE Italy 45°40′N 12°15′E
97 *G24* **Trevose Head** *headland* SW England, United Kingdom 50°33′N 05°03′W
Trg *see* Feldkirchen in Kärnten
183 *P17* **Triabunna** Tasmania, SE Australia 42°33′S 147°55′E
21 *W4* **Triangle** Virginia, NE USA 38°30′N 77°17′W
83 *L18* **Triangle** Masvingo, SE Zimbabwe 20°58′S 31°28′E
115 *L23* **Tría Nisía** *island* Kykládes, Greece, Aegean Sea
Triberg *see* Triberg im Schwarzwald
101 *G23* **Triberg im Schwarzwald** *var.* Triberg. Baden-Württemberg, SW Germany 48°07′N 08°13′E
153 *P11* **Tribhuvan** ✈ (Kathmandu) Central, C Nepal
54 *C9* **Tribugá, Golfo de** *gulf* W Colombia
181 *W4* **Tribulation, Cape** *headland* Queensland, NE Australia 16°14′S 145°48′E
108 *M8* **Tribulaun** ▲ SW Austria 46°59′N 11°18′E
11 *U17* **Tribune** Saskatchewan, S Canada 49°16′N 103°50′W
26 *H5* **Tribune** Kansas, C USA 38°27′N 101°46′W
107 *N18* **Tricarico** Basilicata, S Italy 40°37′N 16°09′E
107 *Q19* **Tricase** Puglia, SE Italy 39°56′N 18°21′E
Trichinopoly *see* Tiruchchirāppalli
115 *D18* **Trichonída, Límni** ◎ C Greece
155 *G22* **Trichūr** *var.* Thrissur. Kerala, SW India 10°32′N 76°14′E *see also* Thrissur
Tricorno *see* Triglav
183 *O8* **Trida** New South Wales, SE Australia 33°02′S 145°03′E
35 *S1* **Trident Peak** ▲ Nevada, W USA 41°52′N 118°22′W
Tridentum/Trient *see* Trento
109 *T6* **Trieben** Steiermark, SE Austria 47°29′N 14°30′E
101 *D19* **Trier** *Eng.* Treves, *Fr.* Trèves; *anc.* Augusta Treverorum. Rheinland-Pfalz, SW Germany 49°45′N 06°39′E
106 *K7* **Trieste** *Slvn.* Trst. Friuli-Venezia Giulia, NE Italy 45°39′N 13°45′E
Trieste, Golfo di/Triest, Golf von *see* Trieste, Gulf of
106 *J8* **Trieste, Gulf of** *Cro.* Tršćanski Zaljev, *Ger.* Golf von Triest, *It.* Golfo di Trieste, *Slvn.* Tržaški Zaliv. *gulf* S Europe
109 *W4* **Triesting** ≈ W Austria
Triêu Hai *see* Quang Tri
Trifail *see* Trbovlje
116 *L9* **Trifeşti** Iaşi, NE Romania 47°30′N 27°31′E
109 *S10* **Triglav** *It.* Tricorno. ▲ NW Slovenia 46°22′N 13°40′E
104 *I14* **Trigueros** Andalucía, S Spain 37°24′N 06°50′W
115 *E16* **Tríkala** *prev.* Trikkala. Thessalía, C Greece 39°33′N 21°46′E
115 *E17* **Trikeriótis** ≈ C Greece
Trikkala *see* Tríkala
Trikomo/Tríkomon *see* İskele
97 *F17* **Trim** *Ir.* Baile Átha Troim. Meath, E Ireland 53°34′N 06°47′W
108 *E7* **Trimbach** Solothurn, NW Switzerland 47°22′N 07°53′E
109 *Q5* **Trimmelkam** Oberösterreich, N Austria 48°02′N 12°55′E
29 *U11* **Trimont** Minnesota, N USA 43°45′N 94°42′W
Trimontium *see* Plovdiv
Trinacria *see* Sicilia
155 *K24* **Trincomalee** *var.* Trinkomali. Eastern Province, NE Sri Lanka 08°34′N 81°13′E
65 *K16* **Trindade, Ilha da** *island* Brazil, W Atlantic Ocean
47 *Y9* **Trindade Spur** *undersea feature* SW Atlantic Ocean 21°00′S 35°00′W
111 *J17* **Třinec** *Ger.* Trzynietz. Moravskoslezský Kraj, E Czech Republic 49°41′N 18°39′E
57 *M16* **Trinidad** El Beni, N Bolivia 14°52′S 64°54′W
54 *H9* **Trinidad** Casanare, E Colombia 05°25′N 71°39′W
44 *E6* **Trinidad** Sancti Spíritus, C Cuba 21°48′N 80°00′W
61 *E19* **Trinidad** Flores, S Uruguay 33°35′S 56°54′W
37 *U8* **Trinidad** Colorado, C USA 37°11′N 104°31′W
45 *Y17* **Trinidad** *island* C Trinidad and Tobago
Trinidad *see* Jose Abad Santos
45 *Y16* **Trinidad and Tobago** *off.* Republic of Trinidad and Tobago. ◆ *republic* SE West Indies
Trinidad and Tobago, Republic of *see* Trinidad and Tobago
63 *F22* **Trinidad, Golfo** *gulf* S Chile
61 *B24* **Trinidad, Isla** *island* E Argentina
107 *N16* **Trinitapoli** Puglia, SE Italy 41°22′N 16°06′E
55 *X10* **Trinité, Montagnes de la** ▲▲ C French Guiana
25 *W9* **Trinity** Texas, SW USA 30°57′N 95°22′W
13 *U12* **Trinity Bay** *inlet* Newfoundland, Newfoundland and Labrador, E Canada
39 *P15* **Trinity Islands** *island group* Alaska, USA
35 *N2* **Trinity Mountains** ▲▲ California, W USA
35 *S4* **Trinity Peak** ▲ Nevada, W USA 40°13′N 118°43′W
35 *S5* **Trinity Range** ▲▲ Nevada, W USA
35 *N2* **Trinity River** ≈ California, W USA
25 *V8* **Trinity River** ≈ Texas, SW USA
Trinkomali *see* Trincomalee
173 *Y15* **Triolet** NW Mauritius 20°05′S 57°32′E
107 *O20* **Trionto, Capo** *headland* S Italy 39°37′N 16°46′E
Tripití, Ákra *see* Trypití, Akrotírio
115 *F20* **Trípoli** *prev.* Trípolis. Pelopónnisos, S Greece 37°31′N 22°22′E
138 *G6* **Tripoli** *var.* Tarābulus, Ţarābulus ash Shām, Trâblous; *anc.* Tripolis. N Lebanon 34°30′N 35°42′E
29 *X12* **Tripoli** Iowa, C USA 42°48′N 92°15′W
Tripoli *see* Ţarābulus
Trípolis *see* Trípoli, Greece
Tripolis *see* Tripoli, Lebanon
29 *Q12* **Tripp** South Dakota, N USA 43°12′N 97°57′W
153 *V15* **Tripura** *var.* Hill Tippera. ◈ *state* NE India
108 *K8* **Trisanna** ≈ W Austria
100 *H8* **Trischen** *island* NW Germany
Tristan da Cunha ◇ *dependency of Saint Helena* SE Atlantic Ocean
67 *P15* **Tristan da Cunha** *island* SE Atlantic Ocean
Tristan da Cunha *see* Saint Helena, Ascension and Tristan da Cunha
65 *L18* **Tristan da Cunha Fracture Zone** *tectonic feature* S Atlantic Ocean
167 *S14* **Tri Tôn** An Giang, S Vietnam 10°26′N 105°01′E
Tri Tôn, Đao *see* Triton Island
167 *W10* **Triton Island** *Chin.* Zhongjian Dao, *Viet.* Đao Tri Tôn. *island* S Paracel Islands
155 *G24* **Trivandrum** *var.* Thiruvanathapuram, Tiruvantapuram. *state capital* Kerala, SW India 08°30′N 76°57′E
111 *H20* **Trnava** *Ger.* Tyrnau, *Hung.* Nagyszombat. Trnavský Kraj, W Slovakia 48°22′N 17°36′E
111 *H20* **Trnavský Kraj** ◈ *region* W Slovakia
Trnovo *see* Veliko Tarnovo
Trobriand Island *see* Kiriwina Island
Trobriand Islands *see* Kiriwina Islands
11 *Q16* **Trochu** Alberta, SW Canada 51°48′N 113°12′W
109 *U7* **Trofaiach** Steiermark, SE Austria 47°25′N 15°01′E
93 *F14* **Trofors** Troms, N Norway 65°31′N 13°19′E
113 *E14* **Trogir** *It.* Traù. Split-Dalmacija, S Croatia 43°32′N 16°13′E
112 *F13* **Troglav** ▲ Bosnia and Herzegovina/Croatia 44°00′N 16°36′E
107 *M16* **Troia** Puglia, SE Italy 41°21′N 15°19′E
107 *K24* **Troina** Sicilia, Italy, C Mediterranean Sea 37°47′N 14°37′E
173 *O16* **Trois-Bassins** W Réunion 21°05′S 55°18′E
101 *E17* **Troisdorf** Nordrhein-Westfalen, W Germany 50°49′N 07°09′E
74 *H5* **Trois Fourches, Cap des** *headland* NE Morocco 35°27′N 02°58′W
15 *T8* **Trois-Pistoles** Québec, SE Canada 48°08′N 69°10′W
99 *L21* **Trois-Ponts** Liège, E Belgium 50°22′N 05°52′E
15 *P11* **Trois-Rivières** Québec, SE Canada 46°21′N 72°34′W
55 *Y12* **Trois Sauts** S French Guiana 02°15′N 52°52′W
99 *M22* **Troisvierges** Diekirch, N Luxembourg 50°07′N 06°00′E
122 *F11* **Troitsk** Chelyabinskaya Oblast', S Russian Federation 54°04′N 61°31′E
125 *T9* **Troitsko-Pechorsk** Respublika Komi, NW Russian Federation 62°39′N 56°06′E
127 *V7* **Troitskoye** Orenburgskaya Oblast', W Russian Federation 52°23′N 56°24′E
Troki *see* Trakai
94 *F9* **Trolla** ▲ S Norway 62°41′N 09°47′E
95 *J18* **Trollhättan** Västra Götaland, S Sweden 58°17′N 12°20′E
94 *G9* **Trollheimen** ▲▲ S Norway
94 *E9* **Trolltindane** ▲ S Norway 62°30′N 07°43′E
58 *H11* **Trombetas, Rio** ≈ NE Brazil
128 *L16* **Tromelin, Île** *island* N Réunion
92 *I9* **Troms** ◈ *county* N Norway
92 *I9* **Tromsø** *Fin.* Tromssa. Troms, N Norway 69°39′N 19°01′E
84 *F5* **Tromsøflaket** *undersea feature* W Barents Sea 18°30′E 71°30′N
Tromssa *see* Tromsø
94 *H10* **Tron** ▲ S Norway 62°12′N 10°46′E
35 *U12* **Trona** California, W USA 35°46′N 117°21′W
63 *G16* **Tronador, Cerro** ▲ S Chile 41°12′S 71°51′W
94 *H8* **Trondheim** *Ger.* Drontheim; *prev.* Nidaros, Trondhjem. Sør-Trøndelag, S Norway 63°25′N 10°24′E
94 *H7* **Trondheimsfjorden** *fjord* S Norway
Trondhjem *see* Trondheim
107 *J14* **Tronto** ≈ C Italy
121 *P3* **Troódos** *var.* Troodos Mountains. ▲▲ C Cyprus
Troodos *see* Ólympos
Troodos Mountains *see* Troódos
96 *I13* **Troon** W Scotland, United Kingdom 55°32′N 04°41′W
107 *M22* **Tropea** Calabria, SW Italy 38°40′N 15°52′E
36 *L7* **Tropic** Utah, W USA 37°37′N 112°04′W
64 *L10* **Tropic Seamount** *var.* Banc du Tropique. *undersea feature* E Atlantic Ocean 23°50′N 20°40′W
Tropique, Banc du *see* Tropic Seamount
Tropoja *see* Tropojë
113 *L17* **Tropojë** *var.* Tropoja. Kukës, N Albania 42°25′N 20°09′E
Troppau *see* Opava
95 *O16* **Trosa** Södermanland, C Sweden 58°54′N 17°35′E
118 *H12* **Troškūnai** Utena, E Lithuania 55°36′N 24°55′E
101 *G23* **Trossingen** Baden-Württemberg, SW Germany 48°04′N 08°37′E
117 *T4* **Trostyanets'** *Rus.* Trostyanets. Sums'ka Oblast', NE Ukraine 50°30′N 34°59′E
117 *N7* **Trostyanets'** *Rus.* Trostyanets. Vinnyts'ka Oblast', C Ukraine 48°35′N 29°10′E
Trostyanets *see* Trostyanets'
116 *L11* **Trotuş** ≈ E Romania
44 *M8* **Trou-du-Nord** N Haiti 19°34′N 71°57′W
25 *W7* **Troup** Texas, SW USA 32°08′N 95°07′W
8 *I10* **Trout** ≈ Northwest Territories, NW Canada
33 *N8* **Trout Creek** Montana, NW USA 47°51′N 115°40′W
32 *H10* **Trout Lake** Washington, NW USA 45°59′N 121°33′W
12 *B9* **Trout Lake** ◎ Ontario, S Canada
33 *T12* **Trout Peak** ▲ Wyoming, C USA 44°36′N 109°33′W
102 *L4* **Trouville** Calvados, N France 49°21′N 00°07′E
97 *L22* **Trowbridge** S England, United Kingdom 51°20′N 02°13′W
23 *Q6* **Troy** Alabama, S USA 31°48′N 85°58′W
27 *Q3* **Troy** Kansas, C USA 39°45′N 95°06′W
27 *W4* **Troy** Missouri, C USA 38°59′N 90°59′W
18 *L10* **Troy** New York, NE USA 42°43′N 73°37′W
21 *S10* **Troy** North Carolina, SE USA 35°23′N 79°54′W
31 *R13* **Troy** Ohio, N USA 40°02′N 84°12′W
25 *T9* **Troy** Texas, SW USA 31°12′N 97°18′W
114 *I9* **Troyan** Lovech, N Bulgaria 42°52′N 24°42′E
114 *I9* **Troyanski Prohod** *var.* Troyanski Prokhod. *pass* N Bulgaria
Troyanski Prokhod *see* Troyanski Prohod
145 *N6* **Troyebratskiy** Severnyy Kazakhstan, N Kazakhstan 54°25′N 66°03′E
103 *Q6* **Troyes** *anc.* Augustobona Tricassium. Aube, N France 48°18′N 04°04′E
117 *X5* **Troyits'ke** Luhans'ka Oblast', E Ukraine 49°55′N 38°18′E
35 *W7* **Troy Peak** ▲ Nevada, W USA 38°18′N 115°27′W
113 *G15* **Trpanj** Dubrovnik-Neretva, S Croatia 43°00′N 17°18′E
Tršćanski Zaljev *see* Trieste, Gulf of
Trst *see* Trieste
113 *N14* **Trstenik** Serbia, C Serbia 43°38′N 21°01′E
126 *I6* **Trubchevsk** Bryanskaya Oblast', W Russian Federation 52°33′N 33°45′E
Trubchular *see* Orlyak
37 *S10* **Truchas Peak** ▲ New Mexico, SW USA 35°57′N 105°38′W
143 *P16* **Trucial Coast** *physical region* C United Arab Emirates
Trucial States *see* United Arab Emirates
35 *Q6* **Truckee** California, W USA 39°18′N 120°10′W
35 *R5* **Truckee River** ≈ Nevada, W USA
127 *Q13* **Trudfront** Astrakhanskaya Oblast', SW Russian Federation 45°56′N 47°42′E
14 *I9* **Truite, Lac à la** ◎ Québec, SE Canada
42 *K4* **Trujillo** Colón, NE Honduras 15°59′N 85°54′W
56 *C12* **Trujillo** La Libertad, NW Peru 08°04′S 79°02′W
104 *K10* **Trujillo** Extremadura, W Spain 39°28′N 05°53′W
54 *I6* **Trujillo** Trujillo, NW Venezuela 09°20′N 70°38′W
54 *I6* **Trujillo** *off.* Estado Trujillo. ◈ *state* W Venezuela
Trujillo, Estado *see* Trujillo
Truk *see* Chuuk
Truk Islands *see* Chuuk Islands
29 *U10* **Truman** Minnesota, N USA 43°49′N 94°26′W
27 *X10* **Trumann** Arkansas, C USA 35°40′N 90°30′W
36 *J9* **Trumbull, Mount** ▲ Arizona, SW USA 36°22′N 113°09′W
Trŭn *see* Tran
183 *Q8* **Trundle** New South Wales, SE Australia 32°55′S 147°43′E
129 *U13* **Trung Phân** *physical region* S Vietnam
Tr ng Sa L n, Đao *see* Spratly Island
Tr ng Sa, Quân Đao *see* Spratly Islands
Trupcilar *see* Orlyak
13 *Q15* **Truro** Nova Scotia, SE Canada 45°24′N 63°18′W
97 *H25* **Truro** SW England, United Kingdom 50°16′N 05°03′W
25 *P5* **Truscott** Texas, SW USA 33°43′N 99°48′W
116 *K9* **Truşeşti** Botoşani, NE Romania 47°45′N 27°01′E
116 *H6* **Truskavets'** L'vivs'ka Oblast', W Ukraine 49°15′N 23°30′E
95 *H22* **Trustrup** Midtjylland, C Denmark 56°20′N 10°46′E
10 *M11* **Trutch** British Columbia, W Canada 57°42′N 123°00′W
37 *Q14* **Truth Or Consequences** New Mexico, SW USA 33°07′N 107°15′W
111 *F15* **Trutnov** *Ger.* Trautenau. Královéhradecký Kraj, N Czech Republic 50°34′N 15°55′E
103 *P13* **Truyère** ≈ C France
114 *K9* **Tryavna** Lovech, N Bulgaria 42°52′N 25°30′E
28 *M14* **Tryon** Nebraska, C USA 41°33′N 100°57′W
115 *J16* **Trypití, Akrotírio** *var.* Ákra Tripití. *headland* Ágios Efstrátios, E Greece 39°28′N 24°58′E
94 *J12* **Trysil** Hedmark, S Norway 61°18′N 12°16′E
94 *I11* **Trysilelva** ≈ S Norway
112 *D10* **Tržac** Federacija Bosne I Hercegovine, NW Bosnia and Herzegovina 44°58′N 15°48′E
Tržaški Zaliv *see* Trieste, Gulf of
110 *G10* **Trzcianka** *Ger.* Schönlanke. Pila, Wielkopolskie, C Poland 53°02′N 16°24′E
110 *E7* **Trzebiatów** *Ger.* Treptow an der Rega. Zachodnio-pomorskie, NW Poland 54°04′N 15°14′E
111 *G14* **Trzebnica** *Ger.* Trebnitz. Dolnośląskie, SW Poland 51°19′N 17°03′E
109 *T10* **Tržič** *Ger.* Neumarktl. NW Slovenia 46°22′N 14°17′E
Trzynietz *see* Třinec
83 *G21* **Tsabong** *var.* Tshabong. Kgalagadi, SW Botswana 26°03′S 22°27′E
162 *G7* **Tsagaanchuluut** Dzavhan, C Mongolia 47°06′N 96°40′E
162 *M8* **Tsagaandelger** *var.* Haraat. Dundgovĭ, C Mongolia 46°30′N 107°39′E
Tsagaanders *see* Bayantümen
162 *G7* **Tsagaanhayrhan** *var.* Shiree. Dzavhan, W Mongolia 47°30′N 96°48′E
Tsagaannuur *see* Halhgol
Tsagaan-Olom *see* Tayshir
Tsagaan-Ovoo *see* Nariynteel
Tsagaantüngi *see* Altantsögts
162 *H6* **Tsagaan-Uul** *var.* Sharga. Hövsgöl, N Mongolia 49°33′N 98°36′E
162 *J5* **Tsagaan-Üür** *var.* Bulgan. Hövsgöl, N Mongolia 50°30′N 101°28′E
127 *P12* **Tsagan Aman** Respublika Kalmykiya, SW Russian Federation 47°37′N 46°43′E
23 *V11* **Tsala Apopka Lake** ◎ Florida, SE USA
Tsamkong *see* Zhanjiang
Tsangpo *see* Brahmaputra
Tsant *see* Deren
Tsao *see* Tsau
172 *I4* **Tsaratanana** Mahajanga, C Madagascar 16°46′S 47°40′E
114 *N10* **Tsarevo** *prev.* Michurin. Burgas, E Bulgaria 42°10′N 27°51′E
Tsarigrad *see* İstanbul
Tsaritsyn *see* Volgograd
114 *K7* **Tsar Kaloyan** Ruse, N Bulgaria 43°36′N 26°14′E
Tsarskoye Selo *see* Pushkin
117 *T7* **Tsarychanka** Dnipropetrovs'ka Oblast', E Ukraine 48°56′N 34°29′E
83 *H21* **Tsatsu** Southern, S Botswana 25°21′S 24°45′E
83 *G17* **Tsau** *var.* Tsao. North-West, NW Botswana 20°08′S 22°29′E
81 *J20* **Tsavo** Taita/Taveta, S Kenya 02°59′S 38°28′E
83 *E21* **Tsawisis** Karas, S Namibia 26°18′S 18°09′E
Tschakathurn *see* Čakovec
Tschaslau *see* Čáslav
Tschenstochau *see* Częstochowa
Tschernembl *see* Črnomelj
28 *K6* **Tschida, Lake** ⊠ North Dakota, N USA
Tschorna *see* Mustvee
162 *G8* **Tseel** Govĭ-Altay, SW Mongolia 45°45′N 95°54′E
138 *G8* **Tsefat** *var.* Safed, *Ar.* Safad; *prev.* Zefat. Northern, N Israel 32°57′N 35°27′E
126 *M13* **Tselina** Rostovskaya Oblast', SW Russian Federation 46°31′N 41°01′E
Tselinograd *see* Astana
Tselinogradskaya Oblast *see* Akmola
Tsengel *see* Tosontsengel
162 *J8* **Tsenher** *var.* Altan-Ovoo. Arhangay, C Mongolia 47°24′N 101°51′E
Tsenher *see* Mönhhayrhan
163 *N8* **Tsenhermandal** *var.* Modot. Hentiy, C Mongolia 47°45′N 109°03′E
Tsentral'nyye Nizmennyye Garagumy *see* Merkezi Garagumy

◆ Country
● Country Capital
◇ Dependent Territory
○ Dependent Territory Capital
◈ Administrative Regions
✈ International Airport
▲ Mountain
▲▲ Mountain Range
Volcano
≈ River
◎ Lake
⊠ Reservoir

TSES – TUZLA

83 E21 **Tses** Karas, S Namibia 25°58′S 18°08′E
Tseshevlya see Tsyeshawlya
162 E7 **Tsetseg** var. Tsetsegnuur. Hovd, W Mongolia 46°30′N 93°16′E
Tsetsegnuur see Tsetseg
Tsetsen Khan see Öndörhaan
162 J8 **Tsetserleg** Arhangay, C Mongolia 47°29′N 101°19′E
162 H6 **Tsetserleg** var. Halban. Hövsgöl, N Mongolia 49°30′N 97°33′E
162 J8 **Tsetserleg** var. Hujirt. Övörhangay, C Mongolia 46°50′N 102°38′E
77 R16 **Tsévié** S Togo 06°25′N 01°13′E
Tshabong see Tsabong
83 G20 **Tshane** Kgalagadi, SW Botswana 24°05′S 21°54′E
Tshangalele, Lac see Lufira, Lac de Retenue de la
83 H17 **Tshauxaba** Central, C Botswana 19°56′S 25°09′E
79 F21 **Tshela** Bas-Congo, W Dem. Rep. Congo 04°56′S 13°02′E
79 K22 **Tshibala** Kasai-Occidental, S Dem. Rep. Congo 06°53′S 22°01′E
79 J22 **Tshikapa** Kasai-Occidental, SW Dem. Rep. Congo 06°23′S 20°47′E
79 L22 **Tshilenge** Kasai Oriental , S Dem. Rep. Congo 06°17′S 23°48′E
79 L24 **Tshimbalanga** Katanga, S Dem. Rep. Congo 09°42′S 23°04′E
79 L22 **Tshimbulu** Kasai-Occidental, S Dem. Rep. Congo 06°27′S 22°54′E
Tshiumbe see Chiumbe
79 M21 **Tshofa** Kasai-Oriental, C Dem. Rep. Congo 05°13′S 25°13′E
79 K18 **Tshuapa** C Dem. Rep. Congo
114 G7 **Tsibritsa** NW Bulgaria
Tsien Tang see Fuchun Jiang
114 I12 **Tsigansko Gradishte** Gr. Giftokastro. ▲ Bulgaria/Greece 41°24′N 24°41′E
Tsihombe see Tsiombe
8 H7 **Tsiigehtchic** prev. Arctic Red River. Northwest Territories, NW Canada 67°24′N 133°40′W
125 Q7 **Tsil'ma** NW Russian Federation
119 J17 **Tsimkavichy** Rus. Timkovichi. Minskaya Voblasts', C Belarus 53°04′N 26°59′E
126 M11 **Tsimlyansk** Rostovskaya Oblast', SW Russian Federation 47°39′N 42°05′E
127 N11 **Tsimlyanskoye Vodokhranilishche** var. Tsimlyansk Vodoskhovshche, Eng. Tsimlyansk Reservoir. SW Russian Federation
Tsimlyansk Reservoir see Tsimlyanskoye Vodokhranilishche
Tsimlyansk Vodoskhovshche see Tsimlyanskoye Vodokhranilishche
Tsinan see Jinan
Tsing Hai see Qinghai Hu, China
Tsinghai see Qinghai, China
Tsingtao/Tsingtau see Qingdao
Tsingyuan see Baoding
Tsinkiang see Quanzhou
Tsintao see Qingdao
83 D17 **Tsintsabis** Oshikoto, N Namibia 18°45′S 17°51′E
172 H8 **Tsiombe** var. Tsihombe. Toliara, S Madagascar
123 O13 **Tsipa** S Russian Federation
172 H5 **Tsiribihina** W Madagascar
172 I5 **Tsiroanomandidy** Antananarivo, C Madagascar 18°44′S 46°02′E
189 U13 **Tsis** island Chuuk, C Micronesia
Tsitsihar see Qiqihar
127 Q3 **Tsivil'sk** Chuvashskaya Respublika, W Russian Federation 55°51′N 47°30′E
137 T9 **Tskhinvali** prev. Staliniri, Ts'khinvali. C Georgia 42°12′N 43°58′E
119 J19 **Tsna** SW Belarus
124 I15 **Tsna** var. Zna. W Russian Federation
162 G9 **Tsogt** var. Tahilt. Govĭ-Altay, W Mongolia 45°20′N 96°42′E
162 K10 **Tsogt-Ovoo** var. Doloon. Ömnögovĭ, S Mongolia 44°28′N 105°22′E
162 L10 **Tsogttsetsiy** var. Baruunsuu. Ömnögovĭ, S Mongolia 43°46′N 105°28′E
114 M9 **Tsonevo, Yazovir** prev. Yazovir Georgi Traykov. NE Bulgaria
Tsoohor see Hürmen
164 K14 **Tsu** var. Tu. Mie, Honshū, SW Japan 34°41′N 136°30′E
165 O10 **Tsubame** var. Tubame. Niigata, Honshū, C Japan 37°40′N 138°56′E
165 V3 **Tsubetsu** Hokkaidō, NE Japan 43°43′N 144°01′E
165 O13 **Tsuchiura** var. Tutiura. Ibaraki, Honshū, S Japan 36°05′N 140°11′E
165 Q6 **Tsugaru-kaikyō** strait N Japan
164 E14 **Tsukumi** var. Tukumi. Ōita, Kyūshū, SW Japan 33°00′N 131°51′E
Tsul-Ulaan see Bayannuur
Tsul-Ulaan see Bayanuur
83 D17 **Tsumeb** Oshikoto, N Namibia 19°13′S 17°42′E
83 F17 **Tsumkwe** Otjozondjupa, NE Namibia 19°37′S 20°30′E
164 D15 **Tsuno** Miyazaki, Kyūshū, SW Japan 32°43′N 131°32′E
164 D12 **Tsuno-shima** island SW Japan
164 K12 **Tsuruga** var. Turuga. Fukui, Honshū, SW Japan 35°38′N 136°01′E
164 H12 **Tsurugi-san** ▲ Shikoku, SW Japan 33°50′N 134°04′E
165 P9 **Tsuruoka** var. Turuoka. Yamagata, Honshū, C Japan 38°44′N 139°48′E
164 C12 **Tsushima** prev. Izuhara. Nagasaki, Tsushima, SW Japan 34°11′N 129°16′E
164 C12 **Tsushima** var. Tsushima-tō, Tusima. island group SW Japan
Tsushima-tō see Tsushima
164 H12 **Tsuyama** var. Tuyama. Okayama, Honshū, SW Japan 35°04′N 134°01′E
83 G19 **Tswaane** Ghanzi, W Botswana 22°21′S 21°52′E
119 N16 **Tsyakhtsin** Rus. Tekhtin. Mahilyowskaya Voblasts', E Belarus 53°51′N 29°44′E
119 P19 **Tsyerakhowka** Rus. Terekhovka. Homyel'skaya Voblasts', SE Belarus 52°13′N 31°24′E
119 I17 **Tsyeshawlya** prev. Cheshevlya, Tseshevlya, Rus. Teshevle. Brestskaya Voblasts', SW Belarus 53°14′N 25°49′E
Tsyurupinsk see Tsyurupyns'k
117 R10 **Tsyurupyns'k** Rus. Tsyurupinsk. Khersons'ka Oblast', S Ukraine 46°35′N 32°43′E
Tu see Tsu
186 C7 **Tua** C Papua New Guinea
Tuaim see Tuam
184 L6 **Tuakau** Waikato, North Island, New Zealand 37°16′S 174°56′E
97 C17 **Tuam** Ir. Tuaim. Galway, W Ireland 53°31′N 08°50′W
185 K14 **Tuamarina** Marlborough, South Island, New Zealand 41°27′S 174°00′E
Tuamotu, Archipel des see Tuamotu, Îles
193 Q9 **Tuamotu Fracture Zone** tectonic feature E Pacific Ocean
191 W9 **Tuamotu, Îles** var. Archipel des Tuamotu, Dangerous Archipelago, Tuamotu Islands. island group N French Polynesia
Tuamotu Islands see Tuamotu, Îles
175 X10 **Tuamotu Ridge** undersea feature C Pacific Ocean
167 R5 **Tuân Giao** Lai Châu, N Vietnam 21°34′N 103°24′E
171 O2 **Tuao** Luzon, N Philippines 17°42′N 121°25′E
190 B15 **Tuapa** NW Niue 18°57′S 169°59′W
43 N7 **Tuapí** Región Autónoma Atlántico Norte, NE Nicaragua 14°10′N 83°20′W
126 K15 **Tuapse** Krasnodarskiy Kray, SW Russian Federation 44°08′N 39°07′E
169 U6 **Tuaran** Sabah, East Malaysia 06°12′N 116°12′E
104 I6 **Tua, Rio** N Portugal
192 H15 **Tuasivi** Savai'i, C Samoa 13°38′S 172°08′W
185 B24 **Tuatapere** Southland, South Island, New Zealand 46°09′S 167°43′E
36 M9 **Tuba City** Arizona, SW USA 36°08′N 111°14′W
138 H11 **Ţubah, Qaşr at** castle 'Ammān, C Jordan
Tubame see Tsubame
169 R16 **Tuban** prev. Toeban. Jawa, C Indonesia 06°55′S 112°01′E
141 O16 **Tuban, Wādī** dry watercourse SW Yemen
61 K14 **Tubarão** Santa Catarina, S Brazil 28°29′S 49°00′W
98 O10 **Tubbergen** Overijssel, E Netherlands 52°25′N 06°46′E
Tubeke see Tubize
101 H22 **Tübingen** var. Tuebingen. Baden-Württemberg, SW Germany 48°32′N 09°04′E
127 W6 **Tubinskiy** Respublika Bashkortostan, W Russian Federation 52°48′N 58°18′E
99 G19 **Tubize** Dut. Tubeke. Walloon Brabant, C Belgium 50°43′N 04°14′E
76 J16 **Tubmanburg** NW Liberia 06°50′N 10°53′W
75 T7 **Ţubruq** Eng. Tobruk, It. Tobruch. NE Libya 32°05′N 23°59′E
191 T13 **Tubuai** island Îles Australes, SW French Polynesia
Tubuai, Îles/Tubuai Islands see Australes, Îles
Tubuai-Manu see Maiao
40 F3 **Tubutama** Sonora, NW Mexico 30°51′N 111°31′W
54 K4 **Tucacas** Falcón, N Venezuela 10°50′N 68°22′W
59 P16 **Tucano** Bahia, E Brazil 10°52′S 38°48′W
57 P19 **Tucavaca, Río** E Bolivia
110 H8 **Tuchola** Kujawsko-pomorskie, C Poland 53°36′N 17°50′E
111 M17 **Tuchów** Małopolskie, S Poland 49°53′N 21°04′E
23 S3 **Tucker** Georgia, SE USA 33°53′N 84°10′W
27 W10 **Tuckerman** Arkansas, C USA 35°43′N 91°12′W
64 B12 **Tucker's Town** E Bermuda 32°20′N 64°42′W
Tuckum see Tukums
36 M15 **Tucson** Arizona, SW USA 32°14′N 111°01′W
62 J7 **Tucumán** off. Provincia de Tucumán. ◈ province N Argentina
Tucumán see San Miguel de Tucumán
Tucumán, Provincia de see Tucumán
37 V11 **Tucumcari** New Mexico, SW USA 35°10′N 103°43′W
58 H13 **Tucunaré** Pará, N Brazil 05°15′S 55°49′W
55 Q6 **Tucupita** Delta Amacuro, NE Venezuela 09°02′N 62°04′W
58 K13 **Tucuruí, Represa de** NE Brazil
110 F9 **Tuczno** Zachodnio-pomorskie, NW Poland 53°12′N 16°08′E
Tuddo see Tudu
105 Q5 **Tudela** Basq. Tutera; anc. Tutela. Navarra, N Spain 42°04′N 01°37′W
104 M6 **Tudela de Duero** Castilla y León, N Spain 41°35′N 04°34′W
162 G6 **Tüdevtey** var. Oygon. Dzavhan, N Mongolia 48°57′N 96°33′E
138 K6 **Tudmur** var. Tadmur, Tamar, Gk. Palmyra, Bibl. Tadmor. Ḩimş, C Syria 34°36′N 38°15′E
118 J4 **Tudu** Ger. Tuddo. Lääne-Virumaa, NE Estonia 59°12′N 26°52′E
Tuebingen see Tübingen
122 J14 **Tuekta** Respublika Altay, S Russian Federation 50°51′N 85°52′E
104 I5 **Tuela, Rio** N Portugal
153 X12 **Tuensang** Nāgāland, NE India 26°16′N 94°45′E
136 L15 **Tufanbeyli** Adana, C Turkey 38°15′N 36°13′E
Tüffer see Laško
186 F9 **Tufi** Northern, S Papua New Guinea 09°08′S 149°20′E
193 O3 **Tufts Plain** undersea feature N Pacific Ocean
Tugalan see Kolkhozobod
67 V14 **Tugela** SE South Africa
21 P6 **Tug Fork** S USA
39 P15 **Tugidak Island** island Trinity Islands, Alaska, USA
171 O2 **Tuguegarao** Luzon, N Philippines 17°37′N 121°48′E
123 S12 **Tugur** Khabarovskiy Kray, SE Russian Federation 53°43′N 137°00′E
161 P4 **Tuhai He** E China
104 G4 **Tui** Galicia, NW Spain 42°02′N 08°37′W
77 O13 **Tui** var. Grand Balé. W Burkina Faso
57 J16 **Tuichi, Río** W Bolivia
64 Q11 **Tuineje** Fuerteventura, Islas Canarias, Spain, NE Atlantic Ocean 28°18′N 14°03′W
43 X16 **Tuira, Río** SE Panama
Tuisarkan see Tūysarkān
Tujiabu see Yongxiu
127 W5 **Tukan** Respublika Bashkortostan, W Russian Federation 53°58′N 57°29′E
171 P14 **Tukangbesi, Kepulauan** Dut. Toekang Besi Eilanden. island group C Indonesia
147 V13 **Tŭkhtamish** Rus. Toktomush; prev. Tokhtamyshbek. SE Tajikistan 37°51′N 74°41′E
184 O12 **Tukituki** North Island, New Zealand
Tu-k'ou see Panzhihua
121 P12 **Tūkrah** NE Libya 32°32′N 20°35′E
8 H6 **Tuktoyaktuk** Northwest Territories, NW Canada 69°27′N 133°W
168 I9 **Tuktuk** Pulau Samosir, W Indonesia 02°32′N 98°43′E
Tukumi see Tsukumi
118 E9 **Tukums** Ger. Tuckum. W Latvia 56°58′N 23°12′E
81 G24 **Tukuyu** prev. Neu-Langenburg. Mbeya, S Tanzania 09°14′S 33°39′E
Tukzār see Tokzār
41 O13 **Tula** var. Tula de Allende. Hidalgo, C Mexico 20°01′N 99°21′W
41 O11 **Tula** Tamaulipas, C Mexico 22°59′N 99°43′W
126 K5 **Tula** Tul'skaya Oblast', W Russian Federation 54°11′N 37°39′E
Tulach Mhór see Tullamore
Tula de Allende see Tula
186 M9 **Tulaghi** var. Tulagi. Florida Islands, C Solomon Islands 09°04′S 160°09′E
Tulagi see Tulaghi
159 N10 **Tulagt Ar Gol** W China
41 P13 **Tulancingo** Hidalgo, C Mexico 20°34′N 98°25′W
35 R11 **Tulare** California, W USA 36°12′N 119°21′W
29 P9 **Tulare** South Dakota, N USA 44°43′N 98°29′W
35 Q12 **Tulare Lake Bed** salt flat California, W USA
37 S14 **Tularosa** New Mexico, SW USA 33°04′N 106°01′W
37 P13 **Tularosa Mountains** ▲▲ New Mexico, SW USA
37 S15 **Tularosa Valley** basin New Mexico, SW USA
83 E25 **Tulbagh** Western Cape, SW South Africa 33°17′S 19°09′E
56 C5 **Tulcán** Carchi, N Ecuador 0°44′N 77°43′W
117 N13 **Tulcea** Tulcea, E Romania 45°11′N 28°49′E
117 N13 **Tulcea** ◈ county SE Romania
Tul'chin see Tul'chyn
117 N7 **Tul'chyn** Rus. Tul'chin. Vinnyts'ka Oblast', C Ukraine 48°40′N 28°49′E
Tuléar see Toliara
35 O1 **Tulelake** California, W USA 41°57′N 121°30′W
116 J10 **Tulgheş** Hung. Gyergyótölgyes. Harghita, C Romania 46°57′N 25°46′E
Tuli see Thuli
25 N4 **Tulia** Texas, SW USA 34°32′N 101°46′W
8 I9 **Tulita** prev. Fort Norman, Norman. Northwest Territories, NW Canada 64°55′N 125°25′W
20 J10 **Tullahoma** Tennessee, S USA 35°21′N 86°12′W
183 N12 **Tullamarine** ✈ (Melbourne) Victoria, SE Australia 37°40′S 144°46′E
183 Q7 **Tullamore** New South Wales, SE Australia 32°39′S 147°35′E
97 E18 **Tullamore** Ir. Tulach Mhór. Offaly, C Ireland 53°16′N 07°30′W
103 N12 **Tulle** anc. Tutela. Corrèze, C France 45°16′N 01°46′E
109 X3 **Tulln** var. Oberhollabrunn. Niederösterreich, NE Austria 48°20′N 16°02′E
109 W4 **Tulln** NE Austria
22 H6 **Tullos** Louisiana, S USA 31°48′N 92°19′W
97 F19 **Tullow** Ir. An Tullach. Carlow, SE Ireland 52°48′N 06°44′W
181 W5 **Tully** Queensland, NE Australia 18°03′S 145°56′E
124 J3 **Tuloma** NW Russian Federation
27 P9 **Tulsa** Oklahoma, C USA 36°09′N 96°W
153 N11 **Tulsipur** Mid Western, W Nepal 28°01′N 82°22′E
126 K6 **Tul'skaya Oblast'** ◈ province W Russian Federation
126 L14 **Tul'skiy** Respublika Adygeya, SW Russian Federation 44°26′N 40°12′E
186 E5 **Tulu** Manus Island, N Papua New Guinea 01°58′S 146°50′E
54 D10 **Tuluá** Valle del Cauca, W Colombia 04°01′N 76°16′W
116 M12 **Tulucești** Galați, E Romania 45°35′N 28°01′E
39 N12 **Tuluksak** Alaska, USA 61°06′N 160°57′W
41 Z12 **Tulum, Ruinas de** ruins Quintana Roo, SE Mexico
169 R17 **Tulungagung** prev. Toeloengagoeng. Jawa, C Indonesia 08°03′S 111°54′E
186 J6 **Tulun Islands** var. Kilinailau Islands; prev. Carteret Islands. island group NE Papua New Guinea
126 M4 **Tuma** Ryazanskaya Oblast', W Russian Federation 55°09′N 40°27′E
54 B12 **Tumaco** Nariño, SW Colombia 01°51′N 78°46′W
54 B12 **Tumaco, Bahía de** bay SW Colombia
Tuman-gang see Tumen
42 L8 **Tuma, Río** N Nicaragua
95 O16 **Tumba** Stockholm, C Sweden 59°12′N 17°49′E
Tumba, Lac see Ntomba, Lac
169 S12 **Tumbangsenamang** Borneo, C Indonesia 01°17′S 112°21′E
183 Q10 **Tumbarumba** New South Wales, SE Australia 35°47′S 148°03′E
56 A8 **Tumbes** Tumbes, NW Peru 03°33′S 80°27′W
56 A9 **Tumbes** off. Departamento de Tumbes. ◈ department NW Peru
Tumbes, Departamento de see Tumbes
19 P5 **Tumbledown Mountain** ▲ Maine, NE USA 45°27′N 70°28′W
11 N13 **Tumbler Ridge** British Columbia, W Canada 55°06′N 120°51′W
95 N16 **Tumbo** prev. Rekarne. Västmanland, C Sweden 59°25′N 16°04′E
167 Q12 **Tumbŏt, Phnum** ▲ W Cambodia 12°23′N 102°57′E
182 G9 **Tumby Bay** South Australia 34°22′S 136°05′E
163 Y10 **Tumen** Jilin, NE China 42°56′N 129°47′E
163 Y11 **Tumen** Chin. Tumen Jiang, Kor. Tuman-gang, Rus. Tumyn'tszyan. E Asia
Tumen Jiang see Tumen
55 Q8 **Tumeremo** Bolívar, E Venezuela 07°17′N 61°30′W
155 G19 **Tumkūr** Karnātaka, W India 13°20′N 77°06′E
96 I10 **Tummel** C Scotland, United Kingdom
188 B15 **Tumon Bay** bay W Guam
77 P14 **Tumu** NW Ghana 10°55′N 01°59′W
58 I10 **Tumuc-Humac Mountains** var. Serra Tumucumaque. ▲▲ N South America
Tumucumaque, Serra see Tumuc-Humac Mountains
183 Q10 **Tumut** New South Wales, SE Australia 35°20′S 148°14′E
158 F7 **Tumxuk** var. Urad Qianqi. Xinjiang Uygur Zizhiqu, NW China 78°40′N 39°54′E
Tumyn'tszyan see Tumen
Tün see Ferdows
45 U14 **Tunapuna** Trinidad, Trinidad and Tobago 10°38′N 61°23′W
60 K11 **Tunas** Paraná, S Brazil 24°57′S 49°05′W
Tunbridge Wells see Royal Tunbridge Wells
114 L11 **Tunca Nehri** Bul. Tundzha. Bulgaria/Turkey see also Tundzha
Tunca Nehri see Tundzha
137 O14 **Tunceli** var. Kalan. Tunceli, E Turkey 39°07′N 39°34′E
137 O14 **Tunceli** ◈ province C Turkey
152 J12 **Tūndla** Uttar Pradesh, N India 27°13′N 78°14′E
81 I25 **Tunduru** Ruvuma, S Tanzania 11°08′S 37°21′E
114 L10 **Tundzha** Turk. Tunca Nehri. Bulgaria/Turkey see also Tunca Nehri
Tundzha see Tunca Nehri
162 I6 **Tünel** var. Bulag. Hövsgöl, N Mongolia 49°51′N 100°41′E
155 H17 **Tungabhadra** S India
155 F17 **Tungabhadra Reservoir** S India
191 P2 **Tungaru** prev. Gilbert Islands. island group W Kiribati
171 P7 **Tungawan** Mindanao, S Philippines 07°33′N 122°22′E
Tungdor see Mainling
T'ung-shan see Xuzhou
161 Q16 **Tungsha Tao** Chin. Dongsha Qundao, Eng. Pratas Island. island S Taiwan
Tungshih see Dongshi
8 H9 **Tungsten** Northwest Territories, W Canada 62°N 128°09′W
Tung-t'ing Hu see Dongting Hu
56 A13 **Tungurahua** ◈ province C Ecuador
95 F14 **Tunhovdfjorden** S Norway
22 K2 **Tunica** Mississippi, S USA 34°40′N 90°22′W
75 N5 **Tunis** var. Tūnis. ● (Tunisia) NE Tunisia 36°53′N 10°10′E
75 N5 **Tunis, Golfe de** Ar. Khalīj Tūnis. gulf NE Tunisia
75 N6 **Tunisia** off. Tunisian Republic, Ar. Al Jumhūrīyah at Tūnisīyah, Fr. République Tunisienne. ◆ republic N Africa
Tunisian Republic see Tunisia
Tunisienne, République see Tunisia
Tūnisīyah, Al Jumhūrīyah at see Tunisia
Tūnis, Khalīj see Tunis, Golfe de
54 G9 **Tunja** Boyacá, C Colombia 05°33′N 73°23′W
93 F14 **Tunnsjøen** Lapp. Dätnejaevrie. C Norway
39 N12 **Tuntutuliak** Alaska, USA 60°21′N 162°40′W
197 P14 **Tunu** ◈ province E Greenland
147 U8 **Tunuk** Chuyskaya Oblast', C Kyrgyzstan 42°11′N 73°55′E
13 Q6 **Tunungayualok Island** island Newfoundland and Labrador, E Canada
62 H11 **Tunuyán** Mendoza, W Argentina 33°35′S 69°00′W
62 I11 **Tunuyán, Río** W Argentina
Tunxi see Huangshan
Tuodian see Shuangbai
Tuojiang see Fenghuang
Tuokezhake see Shufu
35 P9 **Tuolumne River** California, W USA
Tuong Buong see Tương Đương
167 R7 **Tương Đương** var. Tuong Buong. Nghê An, N Vietnam 19°15′N 104°30′E
160 I13 **Tuoniang Jiang** S China
Tuotiereke see Jeminay
Tuotuo He see Togton He
Tuotuoheyan see Tanggulashan
Tüp see Tyup
60 J9 **Tupã** São Paulo, S Brazil 21°57′S 50°28′W
191 S10 **Tupai** var. Motu Iti. atoll Îles Sous le Vent, W French Polynesia
61 G15 **Tupanciretã** Rio Grande do Sul, S Brazil 29°06′S 53°48′W
22 M2 **Tupelo** Mississippi, S USA 34°15′N 88°42′W
59 K18 **Tupiraçaba** Goiás, S Brazil 14°33′S 48°40′W
57 L21 **Tupiza** Potosí, S Bolivia 21°27′S 65°45′W
144 D14 **Tupkaragan, Mys** prev. Mys Tyub-Karagan. headland SW Kazakhstan 44°40′N 50°19′E
11 N13 **Tupper** British Columbia, W Canada 55°30′N 119°59′W
18 J8 **Tupper Lake** New York, NE USA
146 J10 **Turoqqal'a** Khorazm Viloyati, W Uzbekistan 40°52′N 62°00′E
146 J10 **Tuproqqal'a** Rus. Turpakkla. Xorazm Viloyati, W Uzbekistan 40°52′N 62°00′E
62 H11 **Tupungato, Volcán** W Argentina 33°27′S 69°42′W
163 T9 **Tuquan** Nei Mongol Zizhiqu, N China 45°21′N 121°36′E
54 C13 **Túquerres** Nariño, SW Colombia 01°06′N 77°37′W
153 U13 **Tura** Meghālaya, NE India 25°33′N 90°14′E
122 M10 **Tura** Krasnoyarskiy Kray, N Russian Federation 64°20′N 100°17′E
122 G10 **Tura** C Russian Federation
140 M10 **Turabah** Makkah, W Saudi Arabia 22°00′N 42°00′E
55 O8 **Turagua, Cerro** ▲ C Venezuela 06°59′N 64°34′W
184 L12 **Turakina** Manawatu-Wanganui, North Island, New Zealand 40°03′S 175°13′E
185 K15 **Turakirae Head** headland North Island, New Zealand 41°26′S 174°54′E
186 B8 **Turama** S Papua New Guinea
122 K13 **Turan** Respublika Tyva, S Russian Federation 52°11′N 93°40′E
184 M10 **Turangi** Waikato, North Island, New Zealand 39°01′S 175°47′E
146 F11 **Turan Lowland** var. Turan Plain, Kaz. Turan Oypaty, Rus. Turanskaya Nizmennost', Turk. Turan Pesligi, Uzb. Turan Pasttekisligi. plain C Asia
Turan Oypaty/Turan Pesligi/Turan Plain/Turanskaya Nizmennost' see Turan Lowland
Turan Pasttekisligi see Turan Lowland
138 K7 **Ţurāq al 'Ilab** hill range S Syria
119 K20 **Turaw** Rus. Turov. Homyel'skaya Voblasts', SE Belarus 52°04′N 27°44′E
140 L2 **Ţurayf** Al Ḩudūd ash Shamālīyah, NW Saudi Arabia 31°43′N 38°40′E
Turba see Teruel
54 E5 **Turbaco** Bolívar, N Colombia 10°20′N 75°25′W
148 K15 **Turbat** Baluchistān, SW Pakistan 26°02′N 62°56′E
Turbat-i-Haidari see Torbat-e Ḩeydarīyeh
Turbat-i-Jam see Torbat-e Jām
54 D7 **Turbo** Antioquia, NW Colombia 08°06′N 76°44′W
Turčianský Svätý Martin see Martin
116 H10 **Turda** Ger. Thorenburg, Hung. Torda. Cluj, NW Romania 46°35′N 23°50′E
142 M7 **Ţūreh** Markazī, W Iran
191 X12 **Tureia** atoll Îles Tuamotu, SE French Polynesia
110 I12 **Turek** Wielkopolskie, C Poland 52°01′N 18°30′E
93 L19 **Turenki** Kanta-Häme, SW Finland 60°55′N 24°38′E
Turfan see Turpan
Turgay see Torgay
Turgay see Torgay
144 M8 **Turgayskaya Stolovaya Strana** Kaz. Torgay Üstirti. plateau Kazakhstan/Russian Federation
Turgel see Türi
Türgovishte see Targovishte
Türgovishte see Targovishte
136 C14 **Turgutlu** Manisa, W Turkey 38°30′N 27°43′E
136 L12 **Turhal** Tokat, N Turkey 40°23′N 36°05′E
118 H4 **Türi** Ger. Turgel. Järvamaa, N Estonia 58°48′N 25°28′E
105 S9 **Turia** E Spain
58 M12 **Turiaçu** Maranhão, E Brazil 01°40′S 45°22′W
Turin see Torino
116 J2 **Turiya** Pol. Turja, Rus. Tur'ya; prev. Tur"ya. NW Ukraine
116 I3 **Turiys'k** Volyns'ka Oblast', NW Ukraine 51°05′N 24°31′E
Turja see Turiya
116 H6 **Turka** L'vivs'ka Oblast', W Ukraine 49°07′N 23°01′E
81 H16 **Turkana** ◈ county Kenya
81 H16 **Turkana, Lake** var. Lake Rudolf. N Kenya
Turkestan see Turkistan
147 Q12 **Turkestan Range** Rus. Turkestanskiy Khrebet. ▲▲ C Asia
Turkestanskiy Khrebet see Turkestan Range
111 M23 **Túrkeve** Jász-Nagykun-Szolnok, E Hungary 47°06′N 20°42′E
25 O4 **Turkey** Texas, SW USA 34°23′N 100°54′W
136 H14 **Turkey** off. Republic of Turkey, Turk. Türkiye Cumhuriyeti. ◆ republic SW Asia
181 N4 **Turkey Creek** Western Australia 16°54′S 128°12′E
26 M9 **Turkey Creek** Oklahoma, C USA
37 T9 **Turkey Mountains** ▲▲ New Mexico, SW USA
Turkey, Republic of see Turkey
29 X11 **Turkey River** Iowa, C USA
127 N7 **Turki** Saratovskaya Oblast', W Russian Federation 52°00′N 43°16′E
121 O1 **Turkish Republic of Northern Cyprus** ◇ disputed territory Cyprus
145 P16 **Turkistan** prev. Turkestan. Yuzhnyy Kazakhstan, S Kazakhstan 43°18′N 68°18′E
Turkistan, Bandi-i see Torkestān, Selseleh-ye Band-e
Türkiye Cumhuriyeti see Turkey
146 K12 **Türkmenabat** prev. Rus. Chardzhev, Chardzhou, Chardzhui, Lenin-Turkmenski, Turkm. Chärjew. Lebap Welaýaty, E Turkmenistan 39°07′N 63°30′E
146 A11 **Türkmen Aylagy** Rus. Turkmenskiy Zaliv. lake gulf W Turkmenistan
Turkmenbashi see Türkmenbaşy
146 A10 **Türkmenbaşy** Rus. Turkmenbashi; prev. Krasnovodsk. Balkan Welaýaty, W Turkmenistan 40°N 53°04′E
146 A10 **Türkmenbaşy Aylagy** prev. Rus. Krasnovodskiy Zaliv, Turkm. Krasnowodsk Aylagy. lake Gulf W Turkmenistan
146 J14 **Türkmengala** Rus. Turkmen-kala; prev. Turkmen-Kala. Mary Welaýaty, S Turkmenistan 37°25′N 62°19′E
146 G13 **Turkmenistan** ; prev. Turkmenskaya Soviet Socialist Republic. ◆ republic C Asia
Turkmen-kala/Turkmen-Kala see Türkmengala
Turkmenskaya Soviet Socialist Republic see Turkmenistan
Turkmenskiy Zaliv see Türkmen Aylagy
136 L16 **Türkoğlu** Kahramanmaraş, S Turkey 37°24′N 36°49′E
44 L6 **Turks and Caicos Islands** ◇ UK dependent territory N West Indies
64 G10 **Turks and Caicos Islands** UK dependant territory N West Indies
45 N6 **Turks Islands** island group SE Turks and Caicos Islands
93 K19 **Turku** Swe. Åbo. Varsinais-Suomi, SW Finland 60°27′N 22°17′E
81 H17 **Turkwel** seasonal river NW Kenya
27 P9 **Turley** Oklahoma, C USA 36°14′N 95°58′W
35 P9 **Turlock** California, W USA 37°29′N 120°52′W
118 I12 **Turmantas** Utena, NE Lithuania 55°41′N 26°27′E
Turmberg see Wieżyca
54 L5 **Turmero** Aragua, N Venezuela South America 10°14′N 66°40′W
184 N13 **Turnagain, Cape** headland North Island, New Zealand 40°30′S 176°36′E
Turnau see Turnov
42 H2 **Turneffe Islands** island group E Belize
18 M11 **Turners Falls** Massachusetts, NE USA 42°36′N 72°31′W
11 P16 **Turner Valley** Alberta, SW Canada 50°43′N 114°19′W
99 I16 **Turnhout** Antwerpen, N Belgium 51°19′N 04°57′E
109 V5 **Türnitz** Niederösterreich, E Austria 47°56′N 15°26′E
11 S12 **Turnor Lake** Saskatchewan, C Canada
111 E15 **Turnov** Ger. Turnau. Liberecký Kraj, N Czech Republic 50°36′N 15°10′E
Tŭrnovo see Veliko Tarnovo
116 I15 **Turnu Măgurele** var. Turnu-Măgurele. Teleorman, S Romania 43°44′N 24°53′E
Turnu Severin see Drobeta-Turnu Severin
Turócszentmárton see Martin
Turoni see Tours
Turov see Turaw
Turpakkla see Tuproqqal'a
158 M6 **Turpan** var. Turfan. Xinjiang Uygur Zizhiqu, NW China 42°55′N 89°06′E
Turpan Depression see Turpan Pendi
158 M6 **Turpan Pendi** Eng. Turpan Depression. depression NW China
158 M5 **Turpan Zhan** Xinjiang Uygur Zizhiqu, W China 43°10′N 89°06′E
Turpentine State see North Carolina
44 H8 **Turquino, Pico** ▲ E Cuba 19°54′N 76°55′W
27 Y10 **Turrell** Arkansas, C USA 35°22′N 90°13′W
43 N14 **Turrialba** Cartago, E Costa Rica 09°56′N 83°40′W
96 K8 **Turriff** NE Scotland, United Kingdom 57°32′N 02°28′W
139 V7 **Tursāq** Diyālá, E Iraq 33°27′N 45°47′E
Turshiz see Kāshmar
Tursunzade see Tursunzoda
147 P13 **Tursunzoda** Rus. Tursunzade; prev. Regar. W Tajikistan 38°30′N 68°10′E
Turt see Hanh
Türtkül/Turtkul' see To'rtko'l
29 O9 **Turtle Creek** South Dakota, N USA
30 K4 **Turtle Flambeau Flowage** Wisconsin, N USA
11 S14 **Turtleford** Saskatchewan, S Canada 53°21′N 108°48′W
28 M4 **Turtle Lake** North Dakota, N USA 47°31′N 100°53′W
92 K12 **Turtola** Lappi, NW Finland 66°39′N 23°55′E
122 M10 **Turu** N Russian Federation
Turuga see Tsuruga
147 V10 **Turugart Pass** pass China/Kyrgyzstan
158 E7 **Turugart Shankou** var. Pereval Torugart. pass China/Kyrgyzstan
122 K9 **Turukhan** N Russian Federation
122 K9 **Turukhansk** Krasnoyarskiy Kray, N Russian Federation 65°50′N 87°48′E
139 N3 **Turumbah** well NE Syria
Turuoka see Tsuruoka
Turush see Turysh
60 K7 **Turvo, Rio** S Brazil
Tur"ya see Turiya
144 H14 **Turysh** prev. Turush. Mangistau, SW Kazakhstan 45°24′N 56°02′E
23 O4 **Tuscaloosa** Alabama, S USA 33°13′N 87°34′W
23 O4 **Tuscaloosa, Lake** Alabama, S USA
Tuscan Archipelago see Toscano, Archipelago
Tuscan-Emilian Mountains see Tosco-Emiliano, Appennino
Tuscany see Toscana
35 V2 **Tuscarora** Nevada, W USA 41°16′N 116°13′W
18 F15 **Tuscarora Mountain** ridge Pennsylvania, NE USA
30 M14 **Tuscola** Illinois, N USA 39°46′N 88°19′W
25 P7 **Tuscola** Texas, SW USA 32°12′N 99°48′W
23 O2 **Tuscumbia** Alabama, S USA 34°43′N 87°42′W
92 O4 **Tusenøyane** island group S Svalbard
144 K13 **Tushchybas, Zaliv** prev. Zaliv Paskevicha. lake gulf SW Kazakhstan
Tusima see Tsushima
171 Y15 **Tusirah** Papua, E Indonesia 06°46′S 140°19′E
23 Q5 **Tuskegee** Alabama, S USA 32°25′N 85°41′W
94 E8 **Tustna** island S Norway
39 R12 **Tustumena Lake** Alaska, USA
110 K13 **Tuszyn** Łódzkie, C Poland 51°36′N 19°31′E
137 S13 **Tutak** Ağrı, E Turkey 39°34′N 42°48′E
185 C20 **Tutamoe Range** ▲▲ North Island, New Zealand
Tutasev see Tutayev
124 L15 **Tutayev** var. Tutasev. Yaroslavskaya Oblast', W Russian Federation 57°51′N 39°29′E
Tutela see Tulle, France
Tutela see Tudela, Spain
Tutera see Tudela
155 H23 **Tuticorin** Tamil Nādu, SE India 08°48′N 78°10′E
113 L15 **Tutin** Serbia, S Serbia 43°00′N 20°20′E
184 O10 **Tutira** Hawke's Bay, North Island, New Zealand 39°14′S 176°53′E
Tutiura see Tsuchiura
122 K10 **Tutonchny** Krasnoyarskiy Kray, N Russian Federation 64°12′N 93°52′E
114 L6 **Tutrakan** Silistra, NE Bulgaria 44°03′N 26°38′E
29 N5 **Tuttle** North Dakota, N USA 47°07′N 99°58′W
26 M11 **Tuttle** Oklahoma, C USA 35°17′N 97°48′W
27 O3 **Tuttle Creek Lake** Kansas, C USA
101 H23 **Tuttlingen** Baden-Württemberg, S Germany 47°59′N 08°49′E
171 R16 **Tutuala** East Timor 08°23′S 127°12′E
192 K17 **Tutuila** island W American Samoa
83 I18 **Tutume** Central, E Botswana 20°26′S 27°02′E
39 N7 **Tututalak Mountain** ▲ Alaska, USA 67°51′N 161°27′W
22 K3 **Tutwiler** Mississippi, S USA 34°00′N 90°25′W
162 L8 **Tuul Gol** N Mongolia
93 O16 **Tuupovaara** Pohjois-Karjala, E Finland 62°30′N 30°40′E
Tuva see Tyva, Respublika
190 E7 **Tuvalu** prev. Ellice Islands. ◆ commonwealth republic SW Pacific Ocean
Tuvinskaya ASSR see Tyva, Respublika
163 O9 **Tuvshinshiree** var. Sergelen. Sühbaatar, E Mongolia 46°12′N 111°48′E
141 P9 **Ţuwayq, Jabal** ▲▲ C Saudi Arabia
138 H13 **Ţuwayyil ash Shihāq** desert S Jordan
11 U16 **Tuxford** Saskatchewan, S Canada 50°33′N 105°32′W
167 U12 **Tu Xoay** Đắc Lắc, S Vietnam 12°18′N 107°33′E
40 L14 **Tuxpan** Jalisco, C Mexico 19°33′N 103°21′W
40 J12 **Tuxpan** Nayarit, C Mexico 21°57′N 105°12′W
41 Q12 **Tuxpán** var. Tuxpán de Rodríguez Cano. Veracruz-Llave, E Mexico 20°58′N 97°23′W
Tuxpán de Rodríguez Cano see Tuxpán
41 R15 **Tuxtepec** var. San Juan Bautista Tuxtepec. Oaxaca, S Mexico 18°02′N 96°05′W
41 U16 **Tuxtla** var. Tuxtla Gutiérrez. Chiapas, SE Mexico 16°44′N 93°03′W
Tuxtla see San Andrés Tuxtla
Tuxtla Gutiérrez see Tuxtla
Tuyama see Tsuyama
167 T5 **Tuyên Quang** Tuyên Quang, N Vietnam 21°48′N 105°18′E
167 U13 **Tuy Hoa** Bình Thuân, S Vietnam 11°03′N 108°12′E
167 V12 **Tuy Hoa** Phu Yên, S Vietnam 13°02′N 109°15′E
127 U5 **Tuymazy** Respublika Bashkortostan, W Russian Federation 54°36′N 53°40′E
Tuy Phong see Liên Hương
142 L6 **Tūysarkān** var. Tuisarkan, Tuyserkān. Hamadān, W Iran 34°31′N 48°30′E
Tuyserkān see Tūysarkān
Tuyuk see Tuyyk
145 W16 **Tuyyk** Kaz. Tuyyq; prev. Tuyuk. Taldykorgan, SE Kazakhstan 43°07′N 79°24′E
Tuyyq see Tuyyk
136 I14 **Tuz Gölü** C Turkey
125 Q15 **Tuzha** Kirovskaya Oblast', NW Russian Federation 57°37′N 48°02′E
113 K17 **Tuzi** S Montenegro 42°22′N 19°21′E
139 T5 **Ţūz Khurmātū** At Ta'mím, N Iraq 34°56′N 44°38′E
112 I11 **Tuzla** Federacija Bosne I Hercegovine, NE Bosnia and Herzegovina 44°33′N 18°40′E
117 N15 **Tuzla** Constanța, SE Romania 43°58′N 28°38′E

◆ Country
● Country Capital
◇ Dependent Territory
○ Dependent Territory Capital
◈ Administrative Regions
✈ International Airport
▲ Mountain
▲▲ Mountain Range
Volcano
River
Lake
Reservoir

TUZLUCA – UNTERUECKERSEE

137 T12 **Tuzluca** Iğdır, E Turkey 40°02′N 43°39′E
95 J20 **Tvååker** Halland, S Sweden 57°04′N 12°25′E
95 F17 **Tvedestrand** Aust-Agder, S Norway 58°36′N 08°55′E
124 J16 **Tver'** *prev.* Kalinin. Tverskaya Oblast', W Russian Federation 56°53′N 35°52′E
126 I15 **Tverskaya Oblast'** ◈ *province* W Russian Federation
124 I15 **Tvertsa** ≈ W Russian Federation
138 G9 **Tverya** *var.* Tiberias; *prev.* Teverya. Northern, N Israel 32°48′N 35°32′E
95 F16 **Tvietsund** Telemark, S Norway 59°00′N 08°34′E
110 H13 **Twardogóra** *Ger.* Festenberg. Dolnośląskie, SW Poland 51°21′N 17°27′E
14 J14 **Tweed** Ontario, SE Canada 44°29′N 77°19′W
96 K13 **Tweed** ≈ England/Scotland, United Kingdom
98 O7 **Tweede-Exloërmond** Drenthe, NE Netherlands 52°55′N 06°55′E
183 V3 **Tweed Heads** New South Wales, SE Australia 28°10′S 153°32′E
98 M11 **Twello** Gelderland, E Netherlands 52°14′N 06°07′E
35 W15 **Twentynine Palms** California, W USA 34°08′N 116°03′W
25 P9 **Twin Buttes Reservoir** ⊠ Texas, SW USA
33 O15 **Twin Falls** Idaho, NW USA 42°34′N 114°28′W
39 N13 **Twin Hills** Alaska, USA 59°06′N 160°21′W
11 O11 **Twin Lakes** Alberta, W Canada 57°47′N 117°30′W
33 O12 **Twin Peaks** ▲ Idaho, NW USA 44°37′N 114°24′W
185 I14 **Twins, The** ▲ South Island, New Zealand 41°14′S 172°38′E
29 S5 **Twin Valley** Minnesota, N USA 47°15′N 96°15′W
100 G11 **Twistringen** Niedersachsen, NW Germany 52°48′N 08°39′E
185 E20 **Twizel** Canterbury, South Island, New Zealand 44°04′S 171°12′E
29 X5 **Two Harbors** Minnesota, N USA 47°01′N 91°40′W
11 R14 **Two Hills** Alberta, SW Canada 53°40′N 111°43′W
31 N7 **Two Rivers** Wisconsin, N USA 44°10′N 87°33′W
116 H8 **Tyachiv** Zakarpats'ka Oblast', W Ukraine 48°02′N 23°35′E
Tyan'-Shan' *see* Tien Shan
166 L3 **Tyao** ≈ Myanmar (Burma)/India
117 R6 **Tyasmin** ≈ N Ukraine
23 X6 **Tybee Island** Georgia, SE USA 32°00′N 80°51′W
Tyborøn *see* Thyborøn
111 J16 **Tychy** *Ger.* Tichau. Śląskie, S Poland 50°12′N 19°01′E
111 O16 **Tyczyn** Podkarpackie, SE Poland 49°58′N 22°03′E
94 I8 **Tydal** Sør-Trøndelag, S Norway 63°01′N 11°36′E
115 H24 **Tyflós** ≈ Kriti, Greece, E Mediterranean Sea
21 S3 **Tygart Lake** ⊠ West Virginia, NE USA
123 Q13 **Tygda** Amurskaya Oblast', SE Russian Federation 53°07′N 126°12′E
21 Q11 **Tyger River** ≈ South Carolina, SE USA
32 I11 **Tygh Valley** Oregon, NW USA 45°15′N 121°12′W
94 F12 **Tyin** ◎ S Norway
29 S10 **Tyler** Minnesota, N USA 44°16′N 96°07′W
25 W7 **Tyler** Texas, SW USA 32°21′N 95°18′W
25 W7 **Tyler, Lake** ⊠ Texas, SW USA
22 K7 **Tylertown** Mississippi, S USA 31°07′N 90°08′W
117 P10 **Tylihuls'kyy Lyman** ◎ SW Ukraine
Tylos *see* Bahrain
115 C15 **Týmfi** *var.* Timfi. ▲ W Greece 39°58′N 20°51′E
115 E17 **Tymfristós** *var.* Timfristos. ▲ C Greece 38°57′N 21°49′E
115 J25 **Tympáki** *var.* Timbaki; *prev.* Timbákion. Kríti, Greece, E Mediterranean Sea 35°04′N 24°47′E
123 Q12 **Tynda** Amurskaya Oblast', SE Russian Federation 55°09′N 124°44′E
29 Q12 **Tyndall** South Dakota, N USA 42°57′N 97°52′W
97 L14 **Tyne** ≈ N England, United Kingdom
97 M14 **Tynemouth** NE England, United Kingdom 55°01′N 01°24′W
97 L14 **Tyneside** *cultural region* NE England, United Kingdom
94 H10 **Tynset** Hedmark, S Norway 61°45′N 10°49′E
39 Q12 **Tyonek** Alaska, USA 61°04′N 151°08′W
Tyôsi *see* Chôshi
Tyras *see* Dniester
Tyras *see* Bilhorod-Dnistrovs'kyy
Tyre *see* Soûr
95 G14 **Tyrifjorden** ◎ S Norway
95 K22 **Tyringe** Skåne, S Sweden 56°09′N 13°35′E
123 R13 **Tyrma** Khabarovskiy Kray, SE Russian Federation 50°00′N 132°04′E
Tyrnau *see* Trnava
115 F15 **Týrnavos** *var.* Tírnavos. Thessalía, C Greece 39°45′N 22°18′E
127 N16 **Tyrnyauz** Kabardino-Balkarskaya Respublika, SW Russian Federation 43°19′N 42°55′E
Tyrol *see* Tirol
18 E14 **Tyrone** Pennsylvania, NE USA 40°41′N 78°12′W
97 E15 **Tyrone** *cultural region* W Northern Ireland, United Kingdom
Tyros *see* Bahrain
182 M10 **Tyrrell, Lake** *salt lake* Victoria, SE Australia
84 H14 **Tyrrhenian Basin** *undersea feature* Tyrrhenian Sea, C Mediterranean Sea 39°30′N 13°00′E
120 L8 **Tyrrhenian Sea** *It.* Mare Tirreno. *sea* N Mediterranean Sea
94 J12 **Tyrsil** ≈ Hedmark, S Norway
Tysa *see* Tisa/Tisza
116 J7 **Tysmenytsya** Ivano-Frankivs'ka Oblast', W Ukraine 48°54′N 24°50′E
95 C14 **Tysnesøya** *island* S Norway
95 C14 **Tysse** Hordaland, S Norway 60°23′N 05°46′E
95 D14 **Tyssedal** Hordaland, S Norway 60°07′N 06°36′E
95 O17 **Tystberga** Södermanland, C Sweden 58°51′N 17°15′E
118 E12 **Tytuvėnai** Šiauliai, C Lithuania 55°36′N 23°14′E
Tyub-Karagan, Mys *see* Tupkaragan, Mys
147 V8 **Tyugel'-Say** Narynskaya Oblast', C Kyrgyzstan 41°57′N 74°40′E
122 H11 **Tyukalinsk** Omskaya Oblast', C Russian Federation 55°56′N 72°02′E
127 V7 **Tyul'gan** Orenburgskaya Oblast', W Russian Federation 52°27′N 56°08′E
122 G11 **Tyumen'** Tyumenskaya Oblast', C Russian Federation 57°11′N 65°29′E
122 H11 **Tyumenskaya Oblast'** ◈ *province* C Russian Federation
147 Y7 **Tyup** *Kir.* Tüp. Issyk-Kul'skaya Oblast', NE Kyrgyzstan 42°44′N 78°18′E
122 L14 **Tyva, Respublika** *prev.* Tannu-Tuva, Tuva, Tuvinskaya ASSR. ◈ *autonomous republic* C Russian Federation
117 N7 **Tyvriv** Vinnyts'ka Oblast', C Ukraine 49°01′N 28°28′E
97 J21 **Tywi** ≈ S Wales, United Kingdom
97 I19 **Tywyn** W Wales, United Kingdom 52°35′N 04°06′W
83 K20 **Tzaneen** Limpopo, NE South Africa 23°50′S 30°09′E
Tzekung *see* Zigong
115 I20 **Tziá** *prev.* Kéa, Kéos; *anc.* Ceos. *island* Kykládes, Greece, Aegean Sea
41 X12 **Tzucacab** Yucatán, SE Mexico 20°04′N 89°03′W

U

82 B12 **Uaco Cungo** *var.* Waku Kungo, *Port.* Santa Comba. Kwanza Sul, C Angola 11°21′S 15°04′E
UAE *see* United Arab Emirates
191 X7 **Ua Huka** *island* Îles Marquises, NE French Polynesia
58 E10 **Uaiacás** Roraima, N Brazil 03°28′N 63°13′W
Uamba *see* Wamba
Uanle Uen *see* Wanlaweyn
191 W7 **Ua Pu** *island* Îles Marquises, NE French Polynesia
81 L17 **Uar Garas** *spring/well* SW Somalia 01°19′N 41°22′E
81 H18 **Uasin Gishu** ◈ *county* W Kenya
58 G12 **Uatumã, Rio** ≈ C Brazil
Ua Uíbh Fhailí *see* Offaly
58 C11 **Uaupés, Rio** *var.* Río Vaupés. ≈ Brazil/Colombia *see also* Vaupés, Río
Uaupés, Rio *see* Vaupés, Río
Uba *see* Oba
145 N6 **Ubagan** *Kaz.* Obagan. ≈ Kazakhstan/Russian Federation
186 G7 **Ubai** New Britain, E Papua New Guinea 05°38′S 150°45′E
79 J15 **Ubangi** *Fr.* Oubangui. ≈ C Africa
Ubangi-Shari *see* Central African Republic
119 L20 **Ubarts'** *Rus., Ukr.* Ubort'. ≈ Belarus/Ukraine *see also* Ubort'
Ubarts' *see* Ubort'
54 F9 **Ubaté** Cundinamarca, C Colombia 05°20′N 73°50′W
60 N10 **Ubatuba** São Paulo, S Brazil 23°26′S 45°04′W
149 P17 **Ubauro** Sind, SE Pakistan 28°08′N 69°43′E
171 Q6 **Ubay** Bohol, C Philippines 10°02′N 124°29′E
103 U14 **Ubaye** ≈ SE France
Ubayid, Wadi al *see* Ubayyiḑ, Wādī al
139 N8 **Ubaylah** Al Anbār, W Iraq 33°06′N 40°13′E
139 O10 **Ubayyiḑ, Wādī al** *var.* Wadi al Ubayid. *dry watercourse* SW Iraq
98 L13 **Ubbergen** Gelderland, E Netherlands 51°49′N 05°54′E
164 E13 **Ube** Yamaguchi, Honshū, SW Japan 33°57′N 131°15′E
105 O13 **Úbeda** Andalucía, S Spain 38°01′N 03°22′W
109 V7 **Übelbach** *var.* Markt-Übelbach. Steiermark, SE Austria 47°13′N 15°15′E
59 L20 **Uberaba** Minas Gerais, SE Brazil 19°47′S 47°57′W
57 Q19 **Uberaba, Laguna** ◎ E Bolivia
59 K19 **Uberlândia** Minas Gerais, SE Brazil 18°17′S 48°17′W
101 H24 **Überlingen** Baden-Württemberg, S Germany 47°46′N 09°10′E
77 U16 **Ubiaja** Edo, S Nigeria 06°39′N 06°23′E
104 K3 **Ubiña, Peña** ▲ NW Spain 43°01′N 05°58′W
57 H17 **Ubinas, Volcán** ⩚ S Peru 16°16′S 70°49′W
Ubol Rajadhani/Ubol Ratchathani *see* Ubon Ratchathani
167 P9 **Ubolratna Reservoir** ⊠ C Thailand
167 S10 **Ubon Ratchathani** *var.* Muang Ubon, Ubol Rajadhani, Ubol Ratchathani, Udon Ratchathani. Ubon Ratchathani, E Thailand 15°15′N 104°50′E
116 M3 **Ubort'** *Bel.* Ubarts'. ≈ Belarus/Ukraine *see also* Ubarts'
Ubort' *see* Ubarts'
104 K15 **Ubrique** Andalucía, S Spain 36°42′N 05°27′W
Ubsu-Nur, Ozero *see* Uvs Nuur
79 M18 **Ubundu** Orientale, C Dem. Rep. Congo 0°24′S 25°30′E
146 J13 **Üçajy** *var.* Üchajy, *Rus.* Uch-Adzhi. Mary Welaýaty, C Turkmenistan 38°06′N 62°44′E
137 X11 **Ucar** *Rus.* Udzhary. C Azerbaijan 40°31′N 47°40′E
56 G13 **Ucayali** *off.* Departamento de Ucayali. ◈ *department* E Peru
Ucayali, Departamento de *see* Ucayali
56 F10 **Ucayali, Río** ≈ C Peru
Uccle *see* Ukkel
Uch-Adzhi/Üchajy *see* Üçajy
127 X4 **Uchaly** Respublika Bashkortostan, W Russian Federation 54°19′N 59°33′E
Ucharal *see* Usharal
164 C17 **Uchinoura** Kagoshima, Kyūshū, SW Japan 31°16′N 131°04′E
165 R5 **Uchiura-wan** *bay* NW Pacific Ocean
Uchkuduk *see* Uchquduq
147 S9 **Uchqo'rg'on** *Rus.* Uchkurghan. Namangan Viloyati, E Uzbekistan 41°06′N 72°04′E
146 K8 **Uchquduq** *Rus.* Uchkuduk. Navoiy Viloyati, N Uzbekistan 42°12′N 63°27′E
Uchsay *see* Uchsoy
146 G6 **Uchsoy** *Rus.* Uchsay. Qoraqalpog'iston Respublikasi, NW Uzbekistan 43°51′N 58°51′E
Uchtagan Gumy/Uchtagan, Peski *see* Uçtagan Gumy
123 R11 **Uchur** ≈ E Russian Federation
100 O10 **Uckermark** *cultural region* E Germany
10 K17 **Ucluelet** Vancouver Island, British Columbia, SW Canada 48°55′N 123°34′W
146 D10 **Uçtagan Gumy** *var.* Uchtagan Gumy, *Rus.* Peski Uchtagan. *desert* NW Turkmenistan
122 M13 **Uda** ≈ S Russian Federation
123 R12 **Uda** ≈ E Russian Federation
123 N6 **Udachnyy** Respublika Sakha (Yakutiya), NE Russian Federation 66°27′N 112°18′E
155 G21 **Udagamandalam** *var.* Ooty, Udhagamandalam; *prev.* Ootacamund. Tamil Nādu, SW India 11°28′N 76°42′E
152 F14 **Udaipur** *prev.* Oodeypore. Rājasthān, N India 24°35′N 73°41′E
Udayadhani *see* Uthai Thani
143 N16 **'Udayd, Khawr al** *var.* Khor al Udeid. *inlet* Qatar/Saudi Arabia
112 D11 **Udbina** Lika-Senj, W Croatia 44°33′N 15°46′E
95 I18 **Uddevalla** Västra Götaland, S Sweden 58°20′N 11°56′E
Uddjaur *see* Uddjaure
92 H13 **Uddjaure** *var.* Uddjaur. ◎ N Sweden
Udeid, Khor al *see* 'Udayd, Khawr al
99 K14 **Uden** Noord-Brabant, SE Netherlands 51°40′N 05°37′E
Uden *see* Udenhout
99 J14 **Udenhout** *var.* Uden. Noord-Brabant, S Netherlands 51°37′N 05°09′E
155 H14 **Udgīr** Mahārāshtra, C India 18°23′N 77°06′E
Udhagamandalam *see* Udagamandalam
152 H6 **Udhampur** Jammu and Kashmir, NW India 32°55′N 75°07′E
139 X14 **'Udhaybah, 'Uqlat al** *well* S Iraq
106 J7 **Udine** *anc.* Utina. Friuli-Venezia Giulia, NE Italy 46°05′N 13°10′E
175 T14 **Udintsev Fracture Zone** *tectonic feature* S Pacific Ocean
Udipi *see* Udupi
Udmurtia *see* Udmurtskaya Respublika
127 S2 **Udmurtskaya Respublika** *Eng.* Udmurtia. ◈ *autonomous republic* NW Russian Federation
124 J15 **Udomlya** Tverskaya Oblast', W Russian Federation 57°53′N 34°59′E
Udon Ratchathani *see* Ubon Ratchathani
167 Q8 **Udon Thani** *var.* Ban Mak Khaeng, Udorndhani. Udon Thani, N Thailand 17°25′N 102°45′E
Udorndhani *see* Udon Thani
189 U12 **Udot** *atoll* Chuuk Islands, C Micronesia
123 S13 **Udskaya Guba** *bay* E Russian Federation
123 R12 **Udskoye** Khabarovskiy Kray, SE Russian Federation 54°32′N 134°26′E
155 E19 **Udupi** *var.* Udipi. Karnātaka, SW India 13°18′N 74°46′E
Udzhary *see* Ucar
100 O9 **Uecker** ≈ NE Germany
100 P9 **Ueckermünde** Mecklenburg-Vorpommern, NE Germany 53°43′N 14°03′E
164 M12 **Ueda** *var.* Uyeda. Nagano, Honshū, S Japan 36°27′N 138°13′E
79 L16 **Uele** *var.* Welle. ≈ NE Dem. Rep. Congo
123 W5 **Uelen** Chukotskiy Avtonomnyy Okrug, NE Russian Federation 66°01′N 169°52′W
Uele (upper course) *see* Kibali, Dem. Rep. Congo
Uele (upper course) *see* Uolo, Río, Equatorial Guinea/Gabon
100 J11 **Uelzen** Niedersachsen, N Germany 52°58′N 10°34′E
164 J14 **Ueno** Mie, Honshū, SW Japan 34°45′N 136°08′E
127 V4 **Ufa** Respublika Bashkortostan, W Russian Federation 54°46′N 56°02′E
127 V4 **Ufa** ≈ W Russian Federation
Ufra *see* Kenar
83 C18 **Ugab** ≈ C Namibia
118 D8 **Ugāle** NW Latvia 57°16′N 21°58′E
81 F17 **Uganda** *off.* Republic of Uganda. ◆ *republic* E Africa
Uganda, Republic of *see* Uganda
138 G4 **Ugarit** *Ar.* Ra's Shamrah. *site of ancient city* Al Lādhiqīyah, NW Syria
39 O14 **Ugashik** Alaska, USA 57°30′N 157°24′W
107 Q19 **Ugento** Puglia, SE Italy 39°53′N 18°09′E
105 O15 **Ugíjar** Andalucía, S Spain 36°58′N 03°03′W
103 T11 **Ugine** Savoie, E France 45°45′N 06°25′E
123 R13 **Uglegorsk** Amurskaya Oblast', S Russian Federation 51°40′N 128°05′E
Ugleural'sk *see* Ugleural'skiy
125 V13 **Ugleural'skiy** *prev.* Polovinka, Ugleural'sk. Permskiy Kray, NW Russian Federation 58°57′N 57°37′E
124 L15 **Uglich** Yaroslavskaya Oblast', W Russian Federation 57°33′N 38°23′E
126 I4 **Ugra** ≈ W Russian Federation
147 V9 **Ugyut** Narynskaya Oblast', C Kyrgyzstan 41°22′N 74°49′E
111 H19 **Uherské Hradiště** *Ger.* Ungarisch-Hradisch. Zlínský Kraj, E Czech Republic 49°05′N 17°26′E
111 H19 **Uherský Brod** *Ger.* Ungarisch-Brod. Zlínský Kraj, E Czech Republic 49°01′N 17°40′E
111 B17 **Úhlava** *Ger.* Angel. ≈ W Czech Republic
Uhorshchyna *see* Hungary
31 T13 **Uhrichsville** Ohio, N USA 40°23′N 81°21′W
Uhuru Peak *see* Kilimanjaro
96 G8 **Uig** N Scotland, United Kingdom 57°35′N 06°22′W
82 B10 **Uíge** *Port.* Carmona, Vila Marechal Carmona. Uíge, NW Angola 07°37′S 15°02′E
82 B10 **Uíge** ◈ *province* N Angola
193 Y15 **Uiha** *island* Ha'apai Group, C Tonga
189 U13 **Uijec** *island* Chuuk, C Micronesia
163 X14 **Uijeongbu** *Jap.* Giseifu; *prev.* Ŭijŏngbu. NW South Korea 37°42′N 127°02′E
Ŭijŏngbu *see* Uijeongbu
Uil *see* Oyyl
36 M3 **Uinta Mountains** ▲▲ Utah, W USA
83 C18 **Uis** Erongo, NW Namibia 21°08′S 14°49′E
83 I25 **Uitenhage** Eastern Cape, S South Africa 33°44′S 25°27′E
98 H9 **Uitgeest** Noord-Holland, W Netherlands 52°32′N 04°43′E
98 I11 **Uithoorn** Noord-Holland, C Netherlands 52°14′N 04°50′E
98 O4 **Uithuizen** Groningen, NE Netherlands 53°24′N 06°40′E
98 O4 **Uithuizermeeden** Groningen, NE Netherlands 53°25′N 06°43′E
189 R6 **Ujae Atoll** *var.* Wūjae. *atoll* Ralik Chain, W Marshall Islands
Ujain *see* Ujjain
111 I16 **Ujazd** Opolskie, S Poland 50°22′N 18°20′E
Új-Becse *see* Novi Bečej
Ujda *see* Oujda
189 N5 **Ujelang Atoll** *var.* Wujlān. *atoll* Ralik Chain, W Marshall Islands
111 N21 **Újfehértó** Szabolcs-Szatmár-Bereg, E Hungary 47°48′N 21°40′E
Újgradiska *see* Nova Gradiška
164 J13 **Uji** *var.* Uzi. Kyōto, Honshū, SW Japan 34°54′N 135°48′E
81 E21 **Ujiji** Kigoma, W Tanzania 04°55′S 29°39′E
154 G10 **Ujjain** *prev.* Ujain. Madhya Pradesh, C India 23°11′N 75°50′E
Újlak *see* Ilok
'Ujmān *see* 'Ajmān
Újmoldova *see* Moldova Nouă
Újszentanna *see* Sântana
Ujungpandang *see* Makassar
Ujung Salang *see* Phuket
Újvidék *see* Novi Sad
UK *see* United Kingdom
154 E11 **Ukāi Reservoir** ⊠ W India
81 G19 **Ukara Island** *island* N Tanzania
'Ukash, Wādī *see* 'Akāsh, Wādī
81 F19 **Ukerewe Island** *island* N Tanzania
139 S9 **Ukhayḑir** Al Anbār, C Iraq 32°28′N 43°36′E
153 X13 **Ukhrul** Manipur, NE India 25°08′N 94°24′E
125 S9 **Ukhta** Respublika Komi, NW Russian Federation 63°31′N 53°48′E
34 L6 **Ukiah** California, W USA 39°07′N 123°14′W
32 K12 **Ukiah** Oregon, NW USA 45°06′N 118°57′W
99 G18 **Ukkel** *Fr.* Uccle. Brussels, C Belgium 50°47′N 04°19′E
118 G13 **Ukmergė** *Pol.* Wiłkomierz. Vilnius, C Lithuania 55°16′N 24°46′E
Ukraina *see* Ukraine
116 L6 **Ukraine** *off.* Ukraine, *Rus.* Ukraina, *Ukr.* Ukrayina; *prev.* Ukrainian Soviet Socialist Republic, Ukrainskay S.S.R. ◆ *republic* SE Europe
Ukraine *see* Ukraine
Ukrainian Soviet Socialist Republic *see* Ukraine
Ukrainskay S.S.R/ Ukrayina *see* Ukraine
82 B13 **Uku** Kwanza Sul, NW Angola 11°25′S 14°18′E
164 B13 **Uku-jima** *island* Gotō-rettō, SW Japan
83 F20 **Ukwi** Kgalagadi, SW Botswana 23°41′S 20°26′E
118 M13 **Ula** *Rus.* Ulla. Vitsyebskaya Voblasts', N Belarus 55°14′N 29°15′E
136 C16 **Ula** Muğla, SW Turkey 37°08′N 28°25′E
118 M13 **Ula** *Rus.* Ulla. ≈ N Belarus
162 L7 **Ulaanbaatar** *Eng.* Ulan Bator; *prev.* Urga. ● (Mongolia) Töv, C Mongolia 47°55′N 106°57′E
Ulaan-Ereg *see* Bayanmönh
162 E5 **Ulaangom** Uvs, NW Mongolia 49°56′N 92°06′E
162 D5 **Ulaanhus** *var.* Bilüü. Bayan-Ölgiy, W Mongolia 48°54′N 89°40′E
Ulaantolgoy *see* Möst
Ulaan-Uul *see* Öldziyt, Bayanhongor, Mongolia
Ulaan-Uul *see* Erdene, Dornogovĭ, Mongolia
162 M14 **Ulan** *var.* Otog Qi. Nei Mongol Zizhiqu, N China 39°05′N 107°58′E
159 R10 **Ulan** *var.* Xireg; *prev.* Xiligou. Qinghai, C China 36°59′N 98°21′E
Ulan Bator *see* Ulaanbaatar
162 L13 **Ulan Buh Shamo** *desert* N China
Ulanhad *see* Chifeng
163 T8 **Ulanhot** Nei Mongol Zizhiqu, N China 46°02′N 122°E
127 Q14 **Ulan Khol** Respublika Kalmykiya, SW Russian Federation 45°27′N 46°48′E
163 P13 **Ulan Qab** *var.* Jining. Nei Mongol Zizhiqu, N China 40°59′N 113°08′E
162 M13 **Ulansuhai Nur** ◎ N China
123 N14 **Ulan-Ude** *prev.* Verkhneudinsk. Respublika Buryatiya, S Russian Federation 51°55′N 107°40′E
159 N12 **Ulan Ul Hu** ◎ C China
187 N9 **Ulawa Island** *island* SE Solomon Islands
138 J7 **'Ulayyānīyah, Bi'r al** *var.* Al Hilbeh. *well* S Syria
123 S12 **Ul'banskiy Zaliv** *strait* E Russian Federation
Ulbo *see* Olib
113 J18 **Ulcinj** S Montenegro 41°56′N 19°14′E
Uldz *see* Norovlin
Uleåborg *see* Oulu
Uleälv *see* Oulujoki
95 G16 **Ulefoss** Telemark, S Norway 59°17′N 09°15′E
Uleträsk *see* Oulujärvi
113 L19 **Ulëz** *var.* Ulëza. Dibër, C Albania 41°42′N 19°52′E
Ulëza *see* Ulëz
95 F22 **Ulfborg** Midtjylland, W Denmark 56°16′N 08°21′E
98 N13 **Ulft** Gelderland, E Netherlands 51°53′N 06°23′E
162 G7 **Uliastay** *prev.* Jibhalanta. Dzavhan, W Mongolia 47°47′N 96°53′E
188 F8 **Ulimang** Babeldaob, N Palau
67 T10 **Ulindi** ≈ W Dem. Rep. Congo
188 H14 **Ulithi Atoll** *atoll* Caroline Islands, W Micronesia
112 N10 **Uljma** Vojvodina, NE Serbia 45°04′N 21°08′E
Ul'kayak *see* Ol'keyyek
145 Q7 **Ul'ken-Karaoy, Ozero** *prev.* Ul'ken-Karoy, Ozero. ◎ N Kazakhstan
Ul'ken-Karoy, Ozero *see* Ul'ken-Karaoy, Ozero
Ülkenözen *see* Karaozen
Ülkenqobda *see* Kobda
104 G3 **Ulla** ≈ NW Spain
Ulla *see* Ula
183 S10 **Ulladulla** New South Wales, SE Australia 35°21′S 150°25′E
153 T14 **Ullapara** Rajshahi, W Bangladesh 24°20′N 89°34′E
96 H7 **Ullapool** N Scotland, United Kingdom 57°54′N 05°10′W
95 J20 **Ullared** Halland, S Sweden 57°07′N 12°45′E
105 T7 **Ulldecona** Cataluña, NE Spain 40°36′N 00°27′E
92 I9 **Ullsfjorden** *fjord* N Norway
97 K15 **Ullswater** ◎ NW England, United Kingdom
101 I22 **Ulm** Baden-Württemberg, S Germany 48°24′N 09°59′E
33 R8 **Ulm** Montana, NW USA 47°27′N 111°32′W
183 V5 **Ulmarra** New South Wales, SE Australia 29°37′S 153°06′E
116 K13 **Ulmeni** Buzău, C Romania 45°08′N 26°43′E
116 K14 **Ulmeni** Călărași, S Romania 44°08′N 26°43′E
42 L7 **Ulmukhuás** Región Autónoma Atlántico Norte, NE Nicaragua 14°20′N 84°30′W
188 C8 **Ulong** *var.* Aulong. *island* Palau Islands, N Palau
83 N14 **Ulonguè** *var.* Ulongwé. Tete, NW Mozambique 14°34′S 34°21′E
Ulongwé *see* Ulonguè
95 K19 **Ulricehamn** Västra Götaland, S Sweden 57°47′N 13°25′E
98 N5 **Ulrum** Groningen, NE Netherlands 53°24′N 06°20′E
163 Z16 **Ulsan** *Jap.* Urusan. SE South Korea 35°33′N 129°19′E
94 D10 **Ulsteinvik** Møre og Romsdal, S Norway 62°21′N 05°53′E
97 D15 **Ulster** ◈ *province* Northern Ireland, United Kingdom/Ireland
171 Q10 **Ulu** Pulau Siau, N Indonesia 02°46′N 125°22′E
123 Q11 **Ulu** Respublika Sakha (Yakutiya), NE Russian Federation 60°18′N 127°27′E
42 H5 **Ulúa, Río** ≈ NW Honduras
136 D12 **Ulubat Gölü** ◎ NW Turkey
136 E12 **Uludağ** ▲ NW Turkey 40°08′N 29°13′E
158 D7 **Ulugqat** Xinjiang Uygur Zizhiqu, W China 39°45′N 74°10′E
136 J16 **Ulukışla** Niğde, S Turkey 37°33′N 34°29′E
189 O15 **Ulul** *island* Caroline Islands, C Micronesia
83 L22 **Ulundi** KwaZulu/Natal, E South Africa 28°18′S 31°26′E
158 M3 **Ulungur He** ≈ NW China
158 K2 **Ulungur Hu** ◎ NW China
181 P8 **Uluru** *var.* Ayers Rock. *monolith* Northern Territory, C Australia
97 K16 **Ulverston** NW England, United Kingdom 54°13′N 03°08′W
183 O16 **Ulverstone** Tasmania, SE Australia 41°09′S 146°10′E
94 D13 **Ulvik** Hordaland, S Norway 60°34′N 06°53′E
93 J18 **Ulvila** Satakunta, W Finland 61°26′N 21°55′E
117 O8 **Ul'yanovka** Kirovohrads'ka Oblast', C Ukraine 48°18′N 30°15′E
127 Q5 **Ul'yanovsk** *prev.* Simbirsk. Ul'yanovskaya Oblast', W Russian Federation 54°17′N 48°21′E
127 Q5 **Ul'yanovskaya Oblast'** ◈ *province* W Russian Federation
Ul'yanovskiy *see* Botakara
Ul'yanovskiy Kanal *see* Ul'yanow Kanali
146 M13 **Ul'yanow Kanali** *Rus.* Ul'yanovskiy Kanal. *canal* Turkmenistan/Uzbekistan
Ulyshylanshyq *see* Uly-Zhylanshyk
26 H6 **Ulysses** Kansas, C USA 37°36′N 101°23′W
145 O12 **Ulytau, Gory** ▲ C Kazakhstan
145 N11 **Uly-Zhylanshyk** *Kaz.* Ulyshylanshyq. ≈ C Kazakhstan
112 A9 **Umag** *It.* Umago. Istra, NW Croatia 45°25′N 13°32′E
Umago *see* Umag
41 W12 **Umán** Yucatán, SE Mexico 20°51′N 89°43′W
117 O7 **Uman'** *Rus.* Uman. Cherkas'ka Oblast', C Ukraine 48°45′N 30°10′E
189 V13 **Uman** *atoll* Chuuk Islands, C Micronesia
Uman *see* Uman'
Umanak/Umanaq *see* Uummannaq
'Umān, Khalīj *see* Oman, Gulf of
'Umān, Salṭanat *see* Oman
154 K10 **Umaria** Madhya Pradesh, C India 23°34′N 80°49′E
149 R16 **Umarkot** Sind, SE Pakistan 25°22′N 69°48′E
188 B17 **Umatac** SW Guam 13°17′N 144°40′E
188 A17 **Umatac Bay** *bay* SW Guam
139 S6 **Umayqah** Ṣalāḩ ad Dīn, C Iraq 34°32′N 43°45′E
124 J5 **Umba** Murmanskaya Oblast', NW Russian Federation 66°39′N 34°24′E
138 I8 **Umbāshī, Khirbat al** *ruins* As Suwaydā', S Syria
80 A12 **Umbelasha** ≈ W South Sudan
106 H12 **Umbertide** Umbria, C Italy 43°16′N 12°21′E
61 B17 **Umberto** *var.* Humberto. Santa Fe, C Argentina 30°52′S 61°19′W
186 E7 **Umboi Island** *var.* Rooke Island. *island* C Papua New Guinea
124 J4 **Umbozero, Ozero** ◎ NW Russian Federation
106 H13 **Umbria** ◈ *region* C Italy
Umbrian-Machigian Mountains *see* Umbro-Marchigiano, Appennino
106 I12 **Umbro-Marchigiano, Appennino** *Eng.* Umbrian-Machigian Mountains. ▲▲ C Italy
93 J16 **Umeå** Västerbotten, N Sweden 63°50′N 20°15′E
93 H14 **Umeälven** ≈ N Sweden
39 Q5 **Umiat** Alaska, USA 69°22′N 152°09′W
83 K23 **Umlazi** KwaZulu/Natal, E South Africa 29°58′S 30°50′E
139 X10 **Umm al Baqar, Hawr** *var.* Birkat ad Dawaymah. *spring* S Iraq
Umm al Faṭūr *see* Umm al Futūr
139 Q5 **Umm al Futūr** *var.* Umm at Tūz, Umm al Faṭūr. Ṣalāḩ ad Dīn, C Iraq 34°53′N 42°42′E
141 U12 **Umm al Ḥayt, Wādī** *var.* Wādī Amilḩayt. *seasonal river* SW Oman
Umm al Qaiwain *see* Umm al Qaywayn
143 R15 **Umm al Qaywayn** *var.* Umm al Qaiwain. Umm al Qaywayn, NE United Arab Emirates 25°43′N 55°55′E
138 J3 **Umm 'Āmūd** Ḥalab, N Syria 35°57′N 37°39′E
141 Y10 **Umm ar Ruṣāṣ** *var.* Umm Ruṣayṣ. W Oman 20°26′N 58°48′E
141 X9 **Umm as Samīn** *salt flat* C Oman
Umm at Tūz *see* Umm al Futūr
141 V9 **Umm az Zumūl** *oasis* E Saudi Arabia
80 A9 **Umm Buru** Western Darfur, W Sudan 15°01′N 23°36′E
80 A12 **Umm Dafag** Southern Darfur, W Sudan 10°28′N 23°20′E
Umm Durmān *see* Omdurman
138 F9 **Umm el Fahm** Haifa, N Israel 32°30′N 35°06′E
80 F9 **Umm Inderab** Northern Kordofan, C Sudan 15°12′N 31°54′E
80 C10 **Umm Keddada** Northern Darfur, W Sudan 13°36′N 26°42′E
140 J7 **Umm Lajj** Tabūk, W Saudi Arabia 25°02′N 37°19′E
138 L10 **Umm Maḩfur** ≈ N Jordan
139 Y13 **Umm Qaṣr** Al Baṣrah, SE Iraq 30°02′N 47°55′E
Umm Ruṣayṣ *see* Umm ar Ruṣāṣ
80 F11 **Umm Ruwaba** *var.* Umm Ruwābah, Um Ruwāba. Northern Kordofan, C Sudan 12°54′N 31°13′E
Umm Ruwābah *see* Umm Ruwaba
143 N16 **Umm Sa'īd** *var.* Musay'īd. S Qatar 24°57′N 51°32′E
139 Y10 **Umm Sawān, Hawr** ◎ S Iraq
138 K10 **Umm Ṭuwayṣ, Wādī** *dry watercourse* N Jordan
38 J17 **Umnak Island** *island* Aleutian Islands, Alaska, USA
32 F13 **Umpqua River** ≈ Oregon, NW USA
82 D13 **Umpulo** Bié, C Angola 12°43′S 17°42′E
154 I12 **Umred** Mahārāshtra, C India 20°54′N 79°19′E
Um Ruwāba *see* Umm Ruwaba
Umtali *see* Mutare
Umtata *see* Mthatha
77 V17 **Umuahia** Abia, SW Nigeria 05°30′N 07°33′E
60 H10 **Umuarama** Paraná, S Brazil 23°45′S 53°20′W
Umvuma *see* Mvuma
83 K18 **Umzingwani** ≈ S Zimbabwe
112 D11 **Una** ≈ Bosnia and Herzegovina/Croatia
112 E12 **Unac** ≈ W Bosnia and Herzegovina
23 T6 **Unadilla** Georgia, SE USA 32°15′N 83°44′W
18 I10 **Unadilla River** ≈ New York, NE USA
59 L18 **Unaí** Minas Gerais, SE Brazil 16°24′S 46°49′W
39 N10 **Unalakleet** Alaska, USA 63°52′N 160°47′W
38 K17 **Unalaska Island** *island* Aleutian Islands, Alaska, USA
185 I16 **Una, Mount** ▲ South Island, New Zealand 42°12′S 172°34′E
82 N13 **Unango** Niassa, N Mozambique 12°45′S 35°28′E
Unao *see* Unnão
92 L12 **Unari** Lappi, N Finland 67°07′N 25°37′E
141 O6 **'Unayzah** *var.* Anaiza. Al Qaṣīm, C Saudi Arabia 26°03′N 44°00′E
138 L10 **'Unayzah, Jabal** ▲ Jordan/Saudi Arabia 32°09′N 39°11′E
Unci *see* Almería
57 K19 **Uncía** Potosí, C Bolivia 18°30′S 66°29′W
37 Q7 **Uncompahgre Peak** ▲ Colorado, C USA 38°04′N 107°27′W
37 P6 **Uncompahgre Plateau** *plain* Colorado, C USA
95 L17 **Unden** ◎ S Sweden
28 M4 **Underwood** North Dakota, N USA 47°25′N 101°09′W
171 T13 **Undur** Pulau Seram, E Indonesia 03°41′S 130°38′E
Undur Khan *see* Öndörhaan
126 H6 **Unecha** Bryanskaya Oblast', W Russian Federation 52°51′N 32°38′E
39 N16 **Unga** Unga Island, Alaska, USA 55°14′N 160°34′W
Ungaria *see* Hungary
183 P8 **Ungarie** New South Wales, SE Australia 33°39′S 146°54′E
Ungarisch-Brod *see* Uherský Brod
Ungarisches Erzgebirge *see* Slovenské rudohorie
Ungarisch-Hradisch *see* Uherské Hradiště
Ungarn *see* Hungary
12 M4 **Ungava Bay** *bay* Québec, E Canada
12 J2 **Ungava, Péninsule d'** *peninsula* Québec, SE Canada
Ungeny *see* Ungheni
116 M9 **Ungheni** *Rus.* Ungeny. W Moldova 47°13′N 27°48′E
Unguja *see* Zanzibar
146 G10 **Üngüz Angyrsyndaky Garagum** *Rus.* Zaunguzskiye Garagumy. *desert* N Turkmenistan
146 H11 **Unguz, Solonchakovyye Vpadiny** *salt marsh* C Turkmenistan
Ungvár *see* Uzhhorod
60 I12 **União da Vitória** Paraná, S Brazil 26°13′S 51°05′W
111 G17 **Uničov** *Ger.* Mährisch-Neustadt. Olomoucký Kraj, E Czech Republic 49°48′N 17°05′E
110 J12 **Uniejów** Łódzkie, C Poland 51°58′N 18°46′E
112 A11 **Unije** *island* W Croatia
38 L16 **Unimak Island** *island* Aleutian Islands, Alaska, USA
38 L16 **Unimak Pass** *strait* Aleutian Islands, Alaska, USA
62 J12 **Unión** San Luis, C Argentina 35°09′S 65°55′W
27 W5 **Union** Missouri, C USA 38°27′N 91°01′W
32 L12 **Union** Oregon, NW USA 45°12′N 117°51′W
21 Q11 **Union** South Carolina, SE USA 34°42′N 81°37′W
21 R6 **Union** West Virginia, NE USA 37°36′N 80°34′W
61 B25 **Unión, Bahía** *bay* E Argentina
31 Q13 **Union City** Indiana, N USA 40°12′N 84°50′W
31 Q10 **Union City** Michigan, N USA 42°03′N 85°06′W
18 C12 **Union City** Pennsylvania, NE USA 41°54′N 79°51′W
20 G8 **Union City** Tennessee, S USA 36°26′N 89°03′W
32 G14 **Union Creek** Oregon, NW USA 42°54′N 122°26′W
83 G25 **Uniondale** Western Cape, SW South Africa 33°40′S 23°07′E
40 K13 **Unión de Tula** Jalisco, SW Mexico 19°58′N 104°16′W
30 M9 **Union Grove** Wisconsin, N USA 42°39′N 88°03′W
45 Y15 **Union Island** *island* S Saint Vincent and the Grenadines
46 K5 **Union Reefs** *reef* SW Mexico
0 D7 **Union Seamount** *undersea feature* NE Pacific Ocean 49°35′N 132°45′W
23 Q6 **Union Springs** Alabama, S USA 32°08′N 85°43′W
20 H6 **Uniontown** Kentucky, S USA 37°46′N 87°55′W
18 C16 **Uniontown** Pennsylvania, NE USA 39°54′N 79°44′W
27 T1 **Unionville** Missouri, C USA 40°28′N 93°00′W
141 V8 **United Arab Emirates** *Ar.* Al Imārāt al 'Arabīyah al Muttaḩidah, *abbrev.* UAE; *prev.* Trucial States. ◆ *federation* SW Asia
United Arab Republic *see* Egypt
97 H14 **United Kingdom** *off.* United Kingdom of Great Britain and Northern Ireland, *abbrev.* UK. ◆ *monarchy* NW Europe
United Kingdom of Great Britain and Northern Ireland *see* United Kingdom
United Mexican States *see* Mexico
United Provinces *see* Uttar Pradesh
16 L10 **United States of America** *off.* United States of America, *var.* America, The States, *abbrev.* U.S., USA. ◆ *federal republic* North America
United States of America *see* United States of America
124 J10 **Unitsa** Respublika Kareliya, NW Russian Federation 62°31′N 34°31′E
11 S15 **Unity** Saskatchewan, S Canada 52°27′N 109°10′W
80 D13 **Unity** *var.* Wahda. ◈ *state* N South Sudan
105 Q8 **Universales, Montes** ▲▲ C Spain
27 X4 **University City** Missouri, C USA 38°40′N 90°19′W
187 Q13 **Unmet** Malekula, C Vanuatu 16°09′S 167°16′E
101 F15 **Unna** Nordrhein-Westfalen, W Germany 51°32′N 07°41′E
Unnan *see* Kisuki
152 L12 **Unnão** *prev.* Unao. Uttar Pradesh, N India 26°32′N 80°30′E
187 R15 **Unpongkor** Erromango, S Vanuatu 18°48′S 169°01′E
Unruhstadt *see* Kargowa
96 M1 **Unst** *island* NE Scotland, United Kingdom
101 K16 **Unstrut** ≈ C Germany
Unterdrauburg *see* Dravograd
Unterlimbach *see* Lendava
101 L23 **Unterschleissheim** Bayern, SE Germany 48°16′N 11°34′E
101 H24 **Untersee** ◎ Germany/Switzerland
100 O10 **Unterueckersee** ◎ NE Germany

◆ Country
● Country Capital
◇ Dependent Territory
○ Dependent Territory Capital
◈ Administrative Regions
✈ International Airport
▲ Mountain
▲▲ Mountain Range
⩚ Volcano
≈ River
◎ Lake
⊠ Reservoir

UNTURÁN, SIERRA DE – VALLE DE GUANAPE

55 N12 **Unturán, Sierra de** ▲▲ Brazil/Venezuela
159 N11 **Unuli Horog** Qinghai, W China 35°10′N 91°50′E
136 M11 **Ünye** Ordu, W Turkey 41°08′N 37°14′E
Unza see Unzha
125 O14 **Unzha** var. Unza. ≈ NW Russian Federation
79 E17 **Uolo, Río** var. Eyo (lower course), Mbini, Uele (upper course), Woleu; prev. Benito. ≈ Equatorial Guinea/Gabon
55 Q10 **Uonán** Bolívar, SE Venezuela 04°33′N 62°10′W
161 T12 **Uotsuri-shima** Chin. Diaoyu Dao. island (disputed) China/Japan/Taiwan
165 M11 **Uozu** Toyama, Honshū, SW Japan 36°50′N 137°25′E
42 L12 **Upala** Alajuela, NW Costa Rica 10°52′N 85°W
55 P7 **Upata** Bolívar, E Venezuela 08°02′N 62°25′W
79 M23 **Upemba, Lac** ◎ SE Dem. Rep. Congo
145 R11 **Upenskoye** prev. Uspenskiy. Karaganda, C Kazakhstan 48°45′N 72°46′E
197 O12 **Upernavik** var. Upernivik. Qaasuitsup, C Greenland 73°06′N 55°42′W
Upernivik see Upernavik
83 F22 **Upington** Northern Cape, W South Africa 28°28′S 21°14′E
Uplands see Ottawa
192 I16 **'Upolu** island SE Samoa
38 G11 **'Upolu Point** var. Upolu Point. headland Hawai'i, USA, C Pacific Ocean 20°15′N 155°51′W
Upper Austria see Oberösterreich
Upper Bann see Bann
14 M13 **Upper Canada Village** tourist site Ontario, SE Canada
18 I16 **Upper Darby** Pennsylvania, NE USA 39°57′N 75°15′W
28 L2 **Upper Des Lacs Lake** ▣ North Dakota, N USA
185 L14 **Upper Hutt** Wellington, North Island, New Zealand 41°06′S 175°06′E
29 X11 **Upper Iowa River** ≈ Iowa, C USA
32 H15 **Upper Klamath Lake** ◎ Oregon, NW USA
34 M6 **Upper Lake** California, W USA 39°07′N 122°53′W
35 Q1 **Upper Lake** ◎ California, W USA
10 K9 **Upper Liard** Yukon, W Canada 60°01′N 128°59′W
97 E16 **Upper Lough Erne** ◎ SW Northern Ireland, United Kingdom
80 F12 **Upper Nile** ◈ state NE South Sudan
29 T3 **Upper Red Lake** ◎ Minnesota, N USA
31 S12 **Upper Sandusky** Ohio, N USA 40°49′N 83°16′W
Upper Volta see Burkina Faso
95 O15 **Upplands Väsby** var. Upplandsväsby. Stockholm, C Sweden 59°29′N 18°04′E
Upplandsväsby see Upplands Väsby
95 O15 **Uppsala** Uppsala, C Sweden 59°52′N 17°38′E
95 O14 **Uppsala** ◈ county C Sweden
38 J12 **Upright Cape** headland Saint Matthew Island, Alaska, USA 60°19′N 172°15′W
20 K6 **Upton** Kentucky, S USA 37°25′N 85°53′W
33 Y13 **Upton** Wyoming, C USA 44°06′N 104°37′W
141 N7 **'Uqlat aş Şuqūr** Al Qaşīm, W Saudi Arabia 25°51′N 42°13′E
Uqsuqtuuq see Gjoa Haven
Uqturpan see Wushi
54 C7 **Urabá, Golfo de** gulf NW Colombia
Uracas see Farallon de Pajaros
uradqianqi see Wulashan, N China
Uradar'ya see O'radaryo
Urad Qianqi see Xishanzui, N China
165 U5 **Urahoro** Hokkaidō, NE Japan 42°47′N 143°41′E
165 T5 **Urakawa** Hokkaidō, NE Japan 42°11′N 142°42′E
Ural see Zhayyk
183 T6 **Uralla** New South Wales, SE Australia 30°39′S 151°30′E
Ural Mountains see Ural'skiye Gory
144 F8 **Ural'sk** Kaz. Oral. Zapadnyy Kazakhstan, NW Kazakhstan 51°12′N 51°17′E
Ural'skaya Oblast' see Zapadnyy Kazakhstan
127 W5 **Ural'skiye Gory** var. Ural'skiy Khrebet, Eng. Ural Mountains. ▲▲ Kazakhstan/Russian Federation
Ural'skiy Khrebet see Ural'skiye Gory
138 I3 **Urām aş Şughrá** Ḥalab, N Syria 36°10′N 36°55′E
183 P10 **Urana** New South Wales, SE Australia 35°22′S 146°16′E
11 S10 **Uranium City** Saskatchewan, C Canada 59°30′N 108°46′W
58 F10 **Uraricoera** Roraima, N Brazil 03°26′N 60°54′W
47 S5 **Uraricoera, Rio** ≈ N Brazil
Ura-Tyube see Üroteppa
165 O13 **Urawa** var. Saitama. Saitama, Honshū, S Japan 35°52′N 139°40′E
122 H10 **Uray** Khanty-Mansiyskiy Avtonomnyy Okrug-Yugra, C Russian Federation 60°07′N 64°38′E
141 R7 **'Uray'irah** Ash Sharqīyah, E Saudi Arabia 25°59′N 48°52′E
30 M13 **Urbana** Illinois, N USA 40°06′N 88°12′W
31 R13 **Urbana** Ohio, N USA 40°04′N 83°46′W
29 V14 **Urbandale** Iowa, C USA 41°37′N 93°42′W
106 I11 **Urbania** Marche, C Italy 43°40′N 12°33′E
106 I11 **Urbino** Marche, C Italy 43°45′N 12°38′E
57 H16 **Urcos** Cusco, S Peru 13°40′S 71°38′W
105 N10 **Urda** Castilla-La Mancha, C Spain 39°25′N 03°43′W
Urda see Khan Ordasy
Urdgol see Chandmanĭ
105 O3 **Urduña** var. Orduña. País Vasco, N Spain 43°00′N 03°00′W
Urdunn see Jordan
Urdzhar see Urzhar
97 L16 **Ure** ≈ N England, United Kingdom
119 K18 **Urechcha** Rus. Urech'ye. Minskaya Voblasts', S Belarus 52°57′N 27°54′E
Urech'ye see Urechcha
127 P2 **Uren'** Nizhegorodskaya Oblast', W Russian Federation 57°30′N 45°48′E
122 J9 **Urengoy** Yamalo-Nenetskiy Avtonomnyy Okrug, N Russian Federation 65°52′N 78°42′E
184 K10 **Urenui** Taranaki, North Island, New Zealand 38°59′S 174°25′E
187 Q12 **Ureparapara** island Banks Islands, N Vanuatu
40 G5 **Ures** Sonora, NW Mexico 29°26′N 110°24′W
Urfa see Şanlıurfa
Urga see Ulaanbaatar
162 F6 **Urgamal** var. Hungiy. Dzavhan, W Mongolia 48°31′N 94°15′E
146 H9 **Urganch** Rus. Urgench; prev. Novo-Urgench. Xorazm Viloyati, W Uzbekistan 41°40′N 60°32′E
Urgench see Urganch
136 J14 **Ürgüp** Nevşehir, C Turkey 38°39′N 34°55′E
147 O12 **Urgut** Samarqand Viloyati, C Uzbekistan 39°26′N 67°15′E
158 K3 **Urho** Xinjiang Uygur Zizhiqu, W China 46°05′N 84°51′E
152 G5 **Uri** Jammu and Kashmir, NW India 34°05′N 74°03′E
108 G9 **Uri** ◈ canton C Switzerland
54 F11 **Uribe** Meta, C Colombia 03°01′N 74°33′W
54 H4 **Uribia** La Guajira, N Colombia 11°45′N 72°19′W
116 G12 **Uricani** Hung. Hobicaurikány. Hunedoara, SW Romania 45°18′N 23°03′E
57 M21 **Uriondo** Tarija, S Bolivia 21°43′S 64°40′W
40 I7 **Urique** Chihuahua, N Mexico 27°16′N 107°51′W
40 I7 **Urique, Río** ≈ N Mexico
56 E9 **Urituyacu, Río** ≈ N Peru
Uritskiy see Sarykol'
98 K8 **Urk** Flevoland, N Netherlands 52°40′N 05°35′E
136 B14 **Urla** İzmir, W Turkey 38°19′N 26°47′E
116 K13 **Urlaţi** Prahova, SE Romania 44°59′N 26°15′E
127 V4 **Urman** Respublika Bashkortostan, W Russian Federation 54°53′N 56°52′E
147 P12 **Urmetan** W Tajikistan 39°27′N 68°13′E
Urmia see Orūmīyeh
Urmia, Lake see Orūmīyeh, Daryācheh-ye
Urmiyeh see Orūmīyeh
Uroševac see Ferizaj
147 P11 **Üroteppa** Rus. Ura-Tyube. NW Tajikistan 39°55′N 68°57′E
54 D8 **Urrao** Antioquia, W Colombia 06°16′N 76°10′W
Ursat'yevskaya see Xovos
Urt see Gurvantes
127 X7 **Urtazym** Orenburgskaya Oblast', W Russian Federation 52°12′N 58°48′E
59 K18 **Uruaçu** Goiás, C Brazil 14°38′S 49°06′W
40 M14 **Uruapan** var. Uruapan del Progreso. Michoacán, SW Mexico 19°26′N 102°04′W
Uruapan del Progreso see Uruapan
57 G15 **Urubamba, Cordillera** ▲▲ C Peru
57 G14 **Urubamba, Río** ≈ C Peru
58 G12 **Urucará** Amazonas, N Brazil 02°30′S 57°45′W
61 E16 **Uruguaiana** Rio Grande do Sul, S Brazil 29°45′S 57°05′W
Uruguai, Rio see Uruguay
61 E18 **Uruguay** off. Oriental Republic of Uruguay; prev. La Banda Oriental. ◆ republic E South America
61 E15 **Uruguay** var. Rio Uruguai, Río Uruguay. ≈ E South America
Uruguay, Oriental Republic of see Uruguay
Uruguay, Río see Uruguay
Urukthapel see Ngeruktabel
Urumchi see Ürümqi
Urumi Yeh see Orūmīyeh, Daryācheh-ye
158 L5 **Ürümqi** var. Tihwa, Urumchi, Urumqi, Urumtsi, Wu-lu-k'o-mu-shi, Wu-lu-mu-ch'i; prev. Ti-hua. Xinjiang Uygur Zizhiqu, NW China 43°52′N 87°31′E
Urumtsi see Ürümqi
Urundi see Burundi
183 V6 **Urunga** New South Wales, SE Australia 30°33′S 152°58′E
188 C15 **Uruno Point** headland NW Guam 13°37′N 144°50′E
123 U13 **Urup, Ostrov** island Kuril'skiye Ostrova, SE Russian Federation
141 P11 **'Uruq al Mawārid** desert S Saudi Arabia
Urusan see Ulsan
127 T5 **Urussu** Respublika Tatarstan, W Russian Federation 54°34′N 53°23′E
184 K10 **Uruti** Taranaki, North Island, New Zealand 38°57′S 174°32′E
57 K19 **Uru Uru, Lago** ◎ W Bolivia
55 P9 **Uruyén** Bolívar, SE Venezuela 05°40′N 62°26′W
149 O7 **Uruzgān** ; prev. Orūzgān. Uruzgān, C Afghanistan 32°58′N 66°39′E
149 N6 **Uruzgān** prev. Orūzgān. ◈ province C Afghanistan
165 T3 **Uryū-gawa** ≈ Hokkaidō, NE Japan
165 T2 **Uryū-ko** ◎ Hokkaidō, NE Japan
127 N8 **Uryupinsk** Volgogradskaya Oblast', SW Russian Federation 50°51′N 41°59′E
145 X12 **Urzhar** prev. Urdzhar. Vostochnyy Kazakhstan, E Kazakhstan 47°06′N 81°33′E
125 R16 **Urzhum** Kirovskaya Oblast', NW Russian Federation 57°09′N 49°56′E
116 K13 **Urziceni** Ialomiţa, SE Romania 44°43′N 26°39′E
164 E14 **Usa** Ōita, Kyūshū, SW Japan 33°31′N 131°22′E
125 T6 **Usa** ≈ NW Russian Federation
136 E14 **Uşak** prev. Ushak. Uşak, W Turkey 38°42′N 29°25′E
136 D14 **Uşak** var. Ushak. ◈ province W Turkey
83 C19 **Usakos** Erongo, W Namibia 22°01′S 15°32′E
81 J21 **Usambara Mountains** ▲▲ NE Tanzania
81 G23 **Usangu Flats** wetland SW Tanzania
65 D24 **Usborne, Mount** ▲ East Falkland, Falkland Islands 51°35′S 58°57′W
100 O8 **Usedom** island NE Germany
99 M24 **Useldange** Diekirch, C Luxembourg 49°47′N 05°59′E
119 L16 **Usha** ≈ C Belarus
Ushachi see Ushachy
118 L13 **Ushachy** Rus. Ushachi. Vitsyebskaya Voblasts', N Belarus 55°11′N 28°37′E
Ushak see Uşak
122 L4 **Ushakova, Ostrov** island Severnaya Zemlya, N Russian Federation
Ushant see Ouessant, Île d'
145 W13 **Usharal** prev. Ucharal. Almaty, E Kazakhstan 46°08′N 80°55′E
164 B15 **Ushibuka** var. Usibuka. Kumamoto, Shimo-jima, SW Japan 32°12′N 130°00′E
Ushi Point see Sabaneta, Puntan
145 V14 **Ushtobe** Kaz. Üshtöbe. Almaty, SE Kazakhstan 45°15′N 77°59′E
Üshtöbe see Ushtobe
63 I25 **Ushuaia** Tierra del Fuego, S Argentina 54°48′S 68°19′W
39 R10 **Usibelli** Alaska, USA 63°54′N 148°41′W
Usibuka see Ushibuka
186 D7 **Usino** Madang, N Papua New Guinea 05°40′S 145°31′E
125 U6 **Usinsk** Respublika Komi, NW Russian Federation 66°01′N 57°37′E
97 K22 **Usk** Wel. Wysg. ≈ SE Wales, United Kingdom
Uskočke Planine/Uskokengebirge see Gorjanci
Uskoplje see Gornji Vakuf
Üsküb/Üsküp see Skopje
114 M11 **Üsküpdere** Kırklareli, NW Turkey 41°41′N 27°21′E
126 L7 **Usman'** Lipetskaya Oblast', W Russian Federation 52°04′N 39°41′E
118 D8 **Usmas Ezers** ◎ NW Latvia
125 U13 **Usol'ye** Permskiy Kray, NW Russian Federation 59°27′N 56°33′E
41 T16 **Uspanapa, Río** ≈ SE Mexico
Uspenskiy see Upenskoye
103 O11 **Ussel** Corrèze, C France 45°33′N 02°18′E
163 Z6 **Ussuri** var. Usuri, Wusuri, Chin. Wusuli Jiang. ≈ China/Russian Federation
123 S15 **Ussuriysk** prev. Nikol'sk, Nikol'sk-Ussuriyskiy, Voroshilov. Primorskiy Kray, SE Russian Federation 43°48′N 131°59′E
136 J10 **Usta Burnu** headland N Turkey 41°58′N 34°30′E
149 P13 **Usta Muhammad** Baluchistān, SW Pakistan 28°07′N 68°00′E
123 V11 **Ust'-Bol'sheretsk** Kamchatskiy Kray, E Russian Federation 52°48′N 156°12′E
127 N9 **Ust'-Buzulukskaya** Volgogradskaya Oblast', SW Russian Federation 50°12′N 42°06′E
111 C16 **Ústecký Kraj** ◈ region NW Czech Republic
108 G7 **Uster** Zürich, NE Switzerland 47°21′N 08°49′E
107 I22 **Ustica, Isola d'** island S Italy
122 M11 **Ust'-Ilimsk** Irkutskaya Oblast', S Russian Federation 57°57′N 102°30′E
111 C15 **Ústí nad Labem** Ger. Aussig. Ústecký Kraj, NW Czech Republic 50°41′N 14°04′E
111 F17 **Ústí nad Orlicí** Ger. Wildenschwert. Pardubický Kraj, C Czech Republic 49°58′N 16°24′E
Ustinov see Izhevsk
113 J14 **Ustiprača** ◈ Republika Srpska, SE Bosnia and Herzegovina
122 H11 **Ust'-Ishim** Omskaya Oblast', C Russian Federation 57°42′N 70°58′E
110 G6 **Ustka** Ger. Stolpmünde. Pomorskie, N Poland 54°35′N 16°50′E
123 V9 **Ust'-Kamchatsk** Kamchatskiy Kray, E Russian Federation 56°14′N 162°28′E
145 X9 **Ust'-Kamenogorsk** Kaz. Öskemen. Vostochnyy Kazakhstan, E Kazakhstan 49°58′N 82°36′E
123 T10 **Ust'-Khayryuzovo** Krasnoyarskiy Kray, E Russian Federation 57°07′N 156°37′E
122 I14 **Ust'-Koksa** Respublika Altay, S Russian Federation 50°15′N 85°45′E
125 S11 **Ust'-Kulom** Respublika Komi, NW Russian Federation 61°42′N 53°42′E
123 Q8 **Ust'-Kuyga** Respublika Sakha (Yakutiya), NE Russian Federation 69°59′N 135°27′E
126 L14 **Ust'-Labinsk** Krasnodarskiy Kray, SW Russian Federation 44°40′N 40°46′E
123 R10 **Ust'-Maya** Respublika Sakha (Yakutiya), NE Russian Federation 60°27′N 134°28′E
123 R9 **Ust'-Nera** Respublika Sakha (Yakutiya), NE Russian Federation 64°28′N 143°01′E
123 P12 **Ust'-Nyukzha** Amurskaya Oblast', S Russian Federation 56°30′N 121°32′E
123 O7 **Ust'-Olenëk** Respublika Sakha (Yakutiya), NE Russian Federation 73°03′N 119°34′E
123 T9 **Ust'-Omchug** Magadanskaya Oblast', E Russian Federation 61°07′N 149°17′E
122 M13 **Ust'-Ordynskiy** Irkutskaya Oblast', S Russian Federation 52°50′N 104°42′E
125 N8 **Ust'-Pinega** Arkhangel'skaya Oblast', NW Russian Federation 64°09′N 41°55′E
122 K8 **Ust'-Port** Krasnoyarskiy Kray, N Russian Federation 69°42′N 84°25′E
114 L11 **Ustrem** prev. Vakav. Yambol, E Bulgaria 42°01′N 26°28′E
111 O18 **Ustrzyki Dolne** Podkarpackie, SE Poland 49°26′N 22°36′E
Ust'-Sysol'sk see Syktyvkar
125 R7 **Ust'-Tsil'ma** Respublika Komi, NW Russian Federation 65°25′N 52°09′E
Ust Urt see Ustyurt Plateau
125 O11 **Ust'ya** ≈ NW Russian Federation
124 K6 **Ust'ye Varzugi** Murmanskaya Oblast', NW Russian Federation 66°16′N 36°47′E
123 V10 **Ust'yevoye** prev. Kirovskiy. Kamchatskiy Kray, E Russian Federation 54°06′N 155°48′E
117 R8 **Ustynivka** Kirovohrads'ka Oblast', C Ukraine 47°58′N 32°32′E
144 H15 **Ustyurt Plateau** var. Ust Urt, Uzb. Ustyurt Platosi. plateau Kazakhstan/Uzbekistan
Ustyurt Platosi see Ustyurt Plateau
124 K14 **Ustyuzhna** Vologodskaya Oblast', NW Russian Federation 58°50′N 36°25′E
158 J4 **Usu** Xinjiang Uygur Zizhiqu, NW China 44°27′N 84°37′E
171 O13 **Usu** Sulawesi, C Indonesia 02°34′S 120°58′E
164 E14 **Usuki** Ōita, Kyūshū, SW Japan 33°07′N 131°48′E
42 G8 **Usulután** Usulután, SE El Salvador 13°20′N 88°26′W
42 B9 **Usulután** ◈ department SE El Salvador
41 W16 **Usumacinta, Río** ≈ Guatemala/Mexico
Usumbura see Bujumbura
Usuri see Ussuri
U.S./USA see United States of America
171 W14 **Uta** Papua, E Indonesia 04°28′S 136°03′E
36 K5 **Utah** off. State of Utah, also known as Beehive State, Mormon State. ◈ state W USA
36 L3 **Utah Lake** ◎ Utah, W USA
Utaidhani see Uthai Thani
93 M14 **Utajärvi** Pohjois-Pohjanmaa, C Finland 64°45′N 26°25′E
Utamboni see Mitemele, Río
Utaradit see Uttaradit
165 T3 **Utashinai** var. Utasinai. Hokkaidō, NE Japan 43°32′N 142°03′E
Utasinai see Utashinai
193 Y14 **'Uta Vava'u** island Vava'u Group, N Tonga
37 V9 **Ute Creek** ≈ New Mexico, SW USA
118 H12 **Utena** Utena, E Lithuania 55°30′N 25°34′E
118 H12 **Utena** ◈ province Lithuania
37 V10 **Ute Reservoir** ▣ New Mexico, SW USA
167 O10 **Uthai Thani** var. Muang Uthai Thani, Udayadhani, Utaidhani. Uthai Thani, W Thailand 15°22′N 100°03′E
149 O15 **Uthal** Baluchistān, SW Pakistan 25°53′N 66°37′E
uThongathi see oThongathi
18 I10 **Utica** New York, NE USA 43°06′N 75°15′W
105 R10 **Utiel** Valenciana, E Spain 39°33′N 01°13′W
11 O13 **Utikuma Lake** ◎ Alberta, W Canada
42 I4 **Utila, Isla de** island Islas de la Bahía, N Honduras
Utina see Udine
59 O17 **Utinga** Bahia, E Brazil 12°05′S 41°07′W
Utirik see Utrik Atoll
95 M22 **Utlängan** island S Sweden
117 U11 **Utlyuts'kyy Lyman** bay S Ukraine
95 P16 **Utö** Stockholm, C Sweden 58°55′N 18°19′E
25 Q12 **Utopia** Texas, SW USA 29°30′N 99°31′W
98 J11 **Utrecht** Lat. Trajectum ad Rhenum. Utrecht, C Netherlands 52°06′N 05°07′E
83 K22 **Utrecht** KwaZulu/Natal, E South Africa 27°40′S 30°20′E
98 I11 **Utrecht** ◈ province C Netherlands
104 K14 **Utrera** Andalucía, S Spain 37°10′N 05°47′W
189 V4 **Utrik Atoll** var. Utirik, Utrök, Utrönk. atoll Ratak Chain, N Marshall Islands
Utrök/Utrönk see Utrik Atoll
95 B16 **Utsira** island SW Norway
92 L8 **Utsjoki** var. Ohcejohka. Lappi, N Finland 69°54′N 27°01′E
165 O12 **Utsunomiya** var. Utunomiya. Tochigi, Honshū, S Japan 36°36′N 139°53′E
127 P13 **Utta** Respublika Kalmykiya, SW Russian Federation 46°22′N 46°03′E
167 O8 **Uttaradit** var. Utaradit. Uttaradit, N Thailand 17°38′N 100°05′E
152 J9 **Uttarakhand** ◈ state N India
152 J8 **Uttarkāshi** Uttarakhand, N India 30°45′N 78°19′E
152 K11 **Uttar Pradesh** prev. United Provinces, United Provinces of Agra and Oudh. ◈ state N India
45 T5 **Utuado** C Puerto Rico 18°17′N 66°41′W
158 K3 **Utubulak** Xinjiang Uygur Zizhiqu, W China 46°50′N 86°15′E
39 N5 **Utukok River** ≈ Alaska, USA
Utunomiya see Utsunomiya
187 P10 **Utupua** island Santa Cruz Islands, E Solomon Islands
Utva see Shyngyrlau
189 Y15 **Utwe** Kosrae, E Micronesia
189 X15 **Utwe Harbor** harbor Kosrae, E Micronesia
Uubulan see Hayrhan
163 O8 **Uulbayan** var. Dzüünbulag. Sühbaatar, E Mongolia 46°30′N 112°22′E
118 G6 **Uulu** Pärnumaa, SW Estonia 58°15′N 24°32′E
197 N13 **Uummannaq** var. Umanak, Umanaq. ◈ Qaasuitsup, C Greenland
Uummannarsuaq see Nunap Isua
162 E4 **Üüreg Nuur** ◎ NW Mongolia
Uusikaarlepyy see Nykarleby
93 J19 **Uusikaupunki** Swe. Nystad. Varsinais-Suomi, SW Finland 60°48′N 21°25′E
93 M20 **Uusimaa** Swe. Nyland. ◈ region S Finland
127 S2 **Uva** Udmurtskaya Respublika, NW Russian Federation 56°41′N 52°15′E
155 K25 **Uva** ◈ province SE Sri Lanka
113 L14 **Uvac** ≈ W Serbia
25 Q12 **Uvalde** Texas, SW USA 29°14′N 99°49′W
119 O18 **Uvaravichy** Rus. Uvarovichi. Homyel'skaya Voblasts', SE Belarus 52°36′N 30°44′E
54 J11 **Uvá, Río** ≈ E Colombia
Uvarovichi see Uvaravichy
127 N7 **Uvarovo** Tambovskaya Oblast', W Russian Federation 51°58′N 42°13′E
122 H10 **Uvat** Tyumenskaya Oblast', C Russian Federation 59°11′N 68°37′E
190 G12 **Uvea, Île** island N Wallis and Futuna
81 E21 **Uvinza** Kigoma, W Tanzania 05°08′S 30°23′E
79 O20 **Uvira** Sud-Kivu, E Dem. Rep. Congo 03°24′S 29°05′E
162 E5 **Uvs** ◈ province NW Mongolia
162 F5 **Uvs Nuur** var. Ozero Ubsu-Nur. ◎ Mongolia/Russian Federation
164 F14 **Uwa** var. Seiyo. Ehime, Shikoku, SW Japan 33°22′N 132°29′E
164 F14 **Uwajima** var. Uwazima. Ehime, Shikoku, SW Japan 33°13′N 132°32′E
80 B5 **'Uwaynāt, Jabal al** var. Jebel Uweinat. ▲ Libya/Sudan 21°51′N 25°01′E
Uwazima see Uwajima
Uweinat, Jebel see 'Uwaynāt, Jabal al
14 H14 **Uxbridge** Ontario, S Canada 44°07′N 79°07′W
Uxellodunum see Issoudun
Uxin Qi see Dabqig, N China
41 X12 **Uxmal, Ruinas** ruins Yucatán, SE Mexico
129 Q5 **Uy** ≈ Kazakhstan/Russian Federation
144 K15 **Uyaly** Kyzlorda, S Kazakhstan 44°22′N 61°16′E
123 R8 **Uyandina** ≈ NE Russian Federation
162 J8 **Uyanga** var. Ongi. Övörhangay, C Mongolia 46°30′N 102°18′E
Üydzen see Manlay
Uyeda see Ueda
122 K5 **Uyedineniya, Ostrov** island N Russian Federation
77 V17 **Uyo** Akwa Ibom, S Nigeria 05°00′N 07°57′E
162 D8 **Üyönch** Hovd, W Mongolia 46°04′N 92°05′E
Uyuk see Oyyk
141 V13 **'Uyūn** SW Oman 17°19′N 53°50′E
57 K20 **Uyuni** Potosí, W Bolivia 20°27′S 66°48′W
57 J20 **Uyuni, Salar de** wetland SW Bolivia
146 I9 **Uzbekistan** off. Republic of Uzbekistan. ◆ republic C Asia
Uzbekistan, Republic of see Uzbekistan
158 D8 **Uzbel Shankou** Rus. Pereval Kyzyl-Dzhiik. pass China/Tajikistan
146 B11 **Uzboý** prev. Rus. Imeni 26 Bakinskikh Komissarov, Turkm. 26 Baku Komissarlary Adyndaky. Balkan Welaýaty, W Turkmenistan 39°24′N 54°04′E
119 J17 **Uzda** Minskaya Voblasts', C Belarus 53°29′N 27°10′E
103 N12 **Uzerche** Corrèze, C France 45°24′N 01°35′E
103 R14 **Uzès** Gard, S France 44°00′N 04°25′E
147 T10 **Uzgen** Kir. Özgön. Oshskaya Oblast', SW Kyrgyzstan 40°42′N 73°17′E
117 O3 **Uzh** ≈ N Ukraine
Uzhgorod see Uzhhorod
116 G7 **Uzhhorod** Rus. Uzhgorod; prev. Ungvár. Zakarpats'ka Oblast', W Ukraine 48°36′N 22°19′E
Uzi see Uji
112 K13 **Užice** prev. Titovo Užice. Serbia, W Serbia 43°52′N 19°51′E
Uzin see Uzyn
126 L5 **Uzlovaya** Tul'skaya Oblast', W Russian Federation 54°01′N 38°15′E
108 H7 **Uznach** Sankt Gallen, NE Switzerland 47°12′N 09°00′E
Uzunagach see Uzynagash
136 B10 **Uzunköprü** Edirne, NW Turkey 41°18′N 26°40′E
118 D11 **Užventis** Šiauliai, C Lithuania 55°49′N 22°38′E
117 P5 **Uzyn** Rus. Uzin. Kyyivs'ka Oblast', N Ukraine 49°52′N 30°27′E
145 U16 **Uzynagash** prev. Uzunagach. Almaty, SE Kazakhstan 43°08′N 76°20′E
145 N7 **Uzynkol'** prev. Lenin, Leninskoye. Kustanay, N Kazakhstan 54°05′N 65°23′E

V

Vääksy see Asikkala
83 H23 **Vaal** ≈ C South Africa
93 M14 **Vaala** Kainuu, C Finland 64°34′N 26°49′E
93 N19 **Vaalimaa** Etelä-Karjala, SE Finland 60°34′N 27°49′E
99 M19 **Vaals** Limburg, SE Netherlands 50°46′N 06°01′E
93 J16 **Vaasa** Swe. Vasa; prev. Nikolainkaupunki. Österbotten, W Finland 63°07′N 21°39′E
98 L10 **Vaassen** Gelderland, E Netherlands 52°18′N 05°59′E
118 G11 **Vabalninkas** Panevėžys, NE Lithuania 55°59′N 24°45′E
Vabkent see Vobkent
111 J22 **Vác** Ger. Waitzen. Pest, N Hungary 47°46′N 19°08′E
61 I14 **Vacaria** Rio Grande do Sul, S Brazil 28°31′S 50°52′W
35 N7 **Vacaville** California, W USA 38°21′N 121°59′W
103 R15 **Vaccarès, Étang de** ◎ SE France
114 I11 **Vacha** var. Vŭcha. ≈ SW Bulgaria
44 L10 **Vache, Île à** island SW Haiti
173 Y16 **Vacoas** W Mauritius 20°18′S 57°29′E
32 G10 **Vader** Washington, NW USA 46°23′N 122°58′W
94 D12 **Vadheim** Sogn Og Fjordane, S Norway 61°12′N 05°48′E
154 D11 **Vadodara** prev. Baroda. Gujarāt, W India 22°19′N 73°14′E
92 M8 **Vadsø** Fin. Vesisaari. Finnmark, N Norway 70°07′N 29°47′E
95 L17 **Vadstena** Östergötland, S Sweden 58°26′N 14°55′E
108 I8 **Vaduz** ● (Liechtenstein) W Liechtenstein 47°08′N 09°32′E
Våg see Váh
125 N12 **Vaga** ≈ NW Russian Federation
94 G11 **Vågåmo** Oppland, S Norway 61°52′N 09°06′E
112 D12 **Vaganski Vrh** ▲ W Croatia 44°24′N 15°32′E
95 A19 **Vágar** Dan. Vågø. island W Faroe Islands
Vágbeszterce see Považská Bystrica
95 L19 **Vaggeryd** Jönköping, S Sweden 57°30′N 14°10′E
137 T12 **Vagharshapat** var. Ejmiadzin, Ejmiatsin, Etchmiadzin, Rus. Echmiadzin. W Armenia 40°10′N 44°17′E
95 O16 **Vagnhärad** Södermanland, C Sweden 58°57′N 17°32′E
Vågø see Vágar
104 G7 **Vagos** Aveiro, N Portugal 40°33′N 08°42′W
Vágsellye see Sal'a
92 H10 **Vågsfjorden** fjord N Norway
94 C10 **Vågsøy** island S Norway
Vágújhely see Nové Mesto nad Váhom
111 I21 **Váh** Ger. Waag, Hung. Vág. ≈ W Slovakia
93 K16 **Vähäkyrö** Österbotten, W Finland 63°04′N 22°05′E
191 X11 **Vahitahi** atoll Îles Tuamotu, E French Polynesia
Váhtjer see Gällivare
190 C9 **Vaiaku** var. Funafuti. Funafuti Atoll, SE Tuvalu 08°31′N 179°11′E
Vaidei see Vulcan
22 L4 **Vaiden** Mississippi, S USA 33°19′N 89°42′W
155 I23 **Vaigai** ≈ SE India
191 V16 **Vaihu** Easter Island, Chile, E Pacific Ocean 27°10′S 109°22′W
118 I6 **Väike Emajõgi** ≈ S Estonia
118 I4 **Väike-Maarja** Ger. Klein-Marien. Lääne-Virumaa, NE Estonia 59°07′N 26°16′E
Väike-Salatsi see Mazsalaca
37 R4 **Vail** Colorado, C USA 39°36′N 106°20′W
193 V15 **Vaina** Tongatapu, S Tonga 21°12′S 175°10′W
118 E5 **Väinameri** prev. Muhu Väin, Ger. Moon-Sund. sea E Baltic Sea
93 N18 **Vainikkala** Etelä-Karjala, SE Finland 60°54′N 28°18′E
118 D10 **Vaiņode** SW Latvia 56°25′N 21°52′E
155 H23 **Vaippār** ≈ SE India
191 W11 **Vairaatea** atoll Îles Tuamotu, C French Polynesia
191 R8 **Vairao** Tahiti, W French Polynesia 17°48′S 149°17′W
103 R14 **Vaison-la-Romaine** Vaucluse, SE France 44°15′N 05°04′E
190 G11 **Vaitupu** Île Uvea, E Wallis and Futuna 13°14′S 176°09′W
190 F7 **Vaitupu** atoll C Tuvalu
Vajdahunyad see Hunedoara
Vajdej see Vulcan
78 K12 **Vakaga** ◈ prefecture NE Central African Republic
114 H10 **Vakarel** Sofia, W Bulgaria 42°35′N 23°40′E
Vakav see Ustrem
137 O11 **Vakfıkebir** Trabzon, NE Turkey 41°03′N 39°19′E
122 J10 **Vakh** ≈ C Russian Federation
Vakhon, Qatorkŭhi see Nicholas Range
147 P14 **Vakhsh** SW Tajikistan 37°46′N 68°48′E
147 Q12 **Vakhsh** ≈ SW Tajikistan
127 P1 **Vakhtan** Nizhegorodskaya Oblast', W Russian Federation 58°00′N 46°43′E
94 C13 **Vaksdal** Hordaland, S Norway 60°29′N 05°45′E
Valachia see Wallachia
108 D11 **Valais** Ger. Wallis. ◈ canton SW Switzerland
113 M21 **Valamarës, Mali i** ▲ SE Albania 40°48′N 20°31′E
127 S2 **Valamaz** Udmurtskaya Respublika, NW Russian Federation 57°36′N 52°07′E
113 Q19 **Valandovo** SE FYR Macedonia 41°20′N 22°33′E
111 I18 **Valašské Meziříčí** Ger. Wallachisch-Meseritsch, Pol. Wałeckie Międzyrzecze. Zlínský Kraj, E Czech Republic 49°29′N 17°57′E
115 I17 **Valáxa** island Vóreies Sporádes, Greece, Aegean Sea
95 K16 **Vålberg** Värmland, C Sweden 59°24′N 13°12′E
116 H12 **Vâlcea** prev. Vilcea. ◈ county SW Romania
Vălcedrām see Valchedram
114 G7 **Valchedram** var. Vălcedrām, Vŭlchedrum. Montana, NW Bulgaria 43°42′N 23°25′E
63 J16 **Valcheta** Río Negro, E Argentina 40°42′S 66°08′W
114 N8 **Valchi Dol** var. Vŭlchidol; prev. Kurt-Dere. Varna, E Bulgaria 43°25′N 27°33′E
15 P12 **Valcourt** Québec, SE Canada 45°28′N 72°18′W
Valdai Hills see Valdayskaya Vozvyshennost'
104 M3 **Valdavia** ≈ N Spain
124 I15 **Valday** Novgorodskaya Oblast', W Russian Federation 57°57′N 33°20′E
124 I15 **Valdayskaya Vozvyshennost'** var. Valdai Hills. hill range W Russian Federation
104 L9 **Valdecañas, Embalse de** ▣ W Spain
118 E8 **Valdemārpils** Ger. Sassmacken. NW Latvia 57°23′N 22°36′E
95 N18 **Valdemarsvik** Östergötland, S Sweden 58°13′N 16°35′E
105 N8 **Valdemoro** Madrid, C Spain 40°12′N 03°40′W
105 O11 **Valdepeñas** Castilla-La Mancha, C Spain 38°46′N 03°24′W
104 L5 **Valderaduey** ≈ NE Spain
104 L5 **Valderas** Castilla y León, N Spain 42°05′N 05°27′W
105 T7 **Valderrobres** var. Vall-de-roures. Aragón, NE Spain 40°53′N 00°08′E
63 K17 **Valdés, Península** peninsula SE Argentina
56 C5 **Valdéz** var. Limones. Esmeraldas, NW Ecuador 01°13′N 79°00′W
39 S11 **Valdez** Alaska, USA 61°08′N 146°21′W
Valdia see Weldiya
103 U11 **Val d'Isère** Savoie, E France 45°27′N 07°03′E
63 G15 **Valdivia** Los Ríos, C Chile 39°50′S 73°13′W
Valdivia Bank see Valdivia Seamount
65 P17 **Valdivia Seamount** var. Valdivia Bank. undersea feature E Atlantic Ocean 26°15′S 06°25′E
103 N4 **Val-d'Oise** ◈ department N France
14 J8 **Val-d'Or** Québec, SE Canada 48°06′N 77°42′W
23 U8 **Valdosta** Georgia, SE USA 30°49′N 83°16′W
94 G13 **Valdres** physical region S Norway
32 L13 **Vale** Oregon, NW USA 43°59′N 117°15′W
116 F9 **Valea lui Mihai** Hung. Érmihályfalva. Bihor, NW Romania 47°31′N 22°08′E
11 N15 **Valemount** British Columbia, SW Canada 52°46′N 119°17′W
59 O17 **Valença** Bahia, E Brazil 13°22′S 39°06′W
104 F4 **Valença do Minho** Viana do Castelo, N Portugal 42°02′N 08°38′W
59 N14 **Valença do Piauí** Piauí, E Brazil 06°26′S 41°46′W
103 N8 **Valençay** Indre, C France 47°10′N 01°31′E
103 R13 **Valence** anc. Valentia, Valentia Julia, Ventia. Drôme, E France 44°56′N 04°54′E
105 S10 **Valencia** Valenciana, E Spain 39°29′N 00°24′W
54 K5 **Valencia** Carabobo, N Venezuela 10°12′N 68°02′W
105 R10 **Valencia** Cat. València. ◈ province Valenciana, E Spain
105 S10 **Valencia** ✈ Valencia, E Spain
104 I10 **Valencia de Alcántara** Extremadura, W Spain 39°25′N 07°14′W
104 L4 **Valencia de Don Juan** Castilla y León, N Spain 42°17′N 05°31′W
105 U9 **Valencia, Golfo de** var. Gulf of Valencia. gulf E Spain
Valencia, Gulf of see Valencia, Golfo de
97 A21 **Valencia Island** Ir. Dairbhre. island SW Ireland
105 R10 **Valenciana** var. Valencia, Cat. València; anc. Valentia. ◈ autonomous community NE Spain
Valencia/València see Valenciana
103 P2 **Valenciennes** Nord, N France 50°21′N 03°32′E
116 K13 **Vălenii de Munte** Prahova, SE Romania 45°11′N 26°02′E
Valentia see Valence, France
Valentia see Valenciana
Valentia Julia see Valence
103 T8 **Valentigney** Doubs, E France 47°27′N 06°49′E
28 M12 **Valentine** Nebraska, C USA 42°53′N 100°31′W
24 J10 **Valentine** Texas, SW USA 30°35′N 104°30′W
Valentine State see Oregon
106 C8 **Valenza** Piemonte, NW Italy 45°01′N 08°37′E
94 I13 **Våler** Hedmark, S Norway 60°39′N 11°52′E
54 I6 **Valera** Trujillo, NW Venezuela 09°21′N 70°38′W
192 M11 **Valerie Guyot** S Pacific Ocean 33°00′S 164°00′W
Valetta see Valletta
118 I7 **Valga** Ger. Walk, Latv. Valka. Valgamaa, S Estonia 57°48′N 26°04′E
118 I7 **Valgamaa** var. Valga Maakond. ◈ province S Estonia
Valga Maakond see Valgamaa
43 Q15 **Valiente, Península** peninsula NW Panama
103 X16 **Valinco, Golfe de** gulf Corse, France, C Mediterranean Sea
112 L12 **Valjevo** Serbia, W Serbia 44°17′N 19°54′E
Valjok see Válljohka
118 I7 **Valka** Ger. Walk. N Latvia 57°48′N 26°01′E
Valka see Valga
93 L18 **Valkeakoski** Pirkanmaa, W Finland 61°17′N 24°05′E
93 M19 **Valkeala** Kymenlaakso, S Finland 60°55′N 26°49′E
99 L18 **Valkenburg** Limburg, SE Netherlands 50°52′N 05°50′E
99 K15 **Valkenswaard** Noord-Brabant, S Netherlands 51°21′N 05°29′E
119 G15 **Valkininkai** Alytus, S Lithuania 54°22′N 24°51′E
117 U5 **Valky** Kharkivs'ka Oblast', E Ukraine 49°51′N 35°40′E
41 Y12 **Valladolid** Yucatán, SE Mexico 20°39′N 88°13′W
104 M5 **Valladolid** Castilla y León, NW Spain 41°39′N 04°45′W
104 L5 **Valladolid** ◈ province Castilla y León, N Spain
103 U15 **Vallauris** Alpes-Maritimes, SE France 43°34′N 07°03′E
Vall-de-roures see Valderrobres
Vall D'Uxó see La Vall d'Uixó
95 E16 **Valle** Aust-Agder, S Norway 59°13′N 07°33′E
105 N2 **Valle** Cantabria, N Spain 43°14′N 04°16′W
42 H8 **Valle** ◈ department S Honduras
105 N8 **Vallecas** Madrid, C Spain 40°23′N 03°37′W
37 Q8 **Vallecito Reservoir** ▣ Colorado, C USA
106 A7 **Valle d'Aosta** Fr. Vallée d'Aoste. ◈ region NW Italy
41 O14 **Valle de Bravo** México, S Mexico 19°19′N 100°08′W
55 N5 **Valle de Guanape** Anzoátegui, N Venezuela 09°54′N 65°41′W

◆ Country
● Country Capital
◇ Dependent Territory
○ Dependent Territory Capital
◈ Administrative Regions
✈ International Airport
▲ Mountain
▲▲ Mountain Range
Volcano
≈ River
◎ Lake
▣ Reservoir

VALLE DE LA PASCUA – VERDE, CAPE

54 M6 **Valle de La Pascua** Guárico, N Venezuela 09°15′N 66°00′W
54 B11 **Valle del Cauca** *off.* Departamento del Valle del Cauca. ◈ *province* W Colombia
Valle del Cauca, Departamento del *see* Valle del Cauca
41 N13 **Valle de Santiago** Guanajuato, C Mexico 20°25′N 101°15′W
40 J7 **Valle de Zaragoza** Chihuahua, N Mexico 27°25′N 105°50′W
54 G5 **Valledupar** Cesar, N Colombia 10°31′N 73°16′W
Vallée d'Aoste *see* Valle d'Aosta
76 G10 **Vallée de Ferlo** ≈ NW Senegal
57 M19 **Vallegrande** Santa Cruz, C Bolivia 18°30′S 64°06′W
41 P8 **Valle Hermoso** Tamaulipas, C Mexico 25°39′N 97°49′W
35 N8 **Vallejo** California, W USA 38°08′N 122°16′W
62 G8 **Vallenar** Atacama, N Chile 28°35′S 70°44′W
95 O15 **Vallentuna** Stockholm, C Sweden 59°32′N 18°05′E
121 P16 **Valletta** *prev.* Valeta. ● (Malta) E Malta 35°54′N 14°31′E
27 N6 **Valley Center** Kansas, C USA 37°49′N 97°22′W
29 Q5 **Valley City** North Dakota, N USA 46°57′N 97°58′W
32 I15 **Valley Falls** Oregon, NW USA 42°28′N 120°16′W
Valleyfield *see* Salaberry-de-Valleyfield
21 S4 **Valley Head** West Virginia, NE USA 38°33′N 80°01′W
25 T8 **Valley Mills** Texas, SW USA 31°36′N 97°27′W
75 W10 **Valley of the Kings** *ancient monument* E Egypt
29 R11 **Valley Springs** South Dakota, N USA 43°34′N 96°28′W
20 K5 **Valley Station** Kentucky, S USA 38°06′N 85°52′W
11 O13 **Valleyview** Alberta, W Canada 55°02′N 117°17′W
25 T5 **Valley View** Texas, SW USA 33°27′N 97°08′W
61 C21 **Vallimanca, Arroyo** ≈ E Argentina
92 L9 **Válljohka** *var.* Valjok. Finnmark, N Norway 69°40′N 25°52′E
107 M19 **Vallo della Lucania** Campania, S Italy 40°13′N 15°15′E
108 B9 **Vallorbe** Vaud, W Switzerland 46°43′N 06°21′E
105 V6 **Valls** Cataluña, NE Spain 41°18′N 01°15′E
94 N11 **Vallsta** Gävleborg, C Sweden 61°30′N 16°25′E
94 N12 **Vallvik** Gävleborg, C Sweden 61°10′N 17°15′E
11 T17 **Val Marie** Saskatchewan, S Canada 49°15′N 107°44′W
118 H7 **Valmiera** *Est.* Volmari, *Ger.* Wolmar. N Latvia 57°34′N 25°26′E
105 N3 **Valnera** ▲ N Spain 43°08′N 03°39′W
102 J3 **Valognes** Manche, N France 49°31′N 01°28′W
Valona *see* Vlorë
Valona Bay *see* Vlorës, Gjiri i
104 G6 **Valongo** *var.* Valongo de Gaia. Porto, N Portugal 41°11′N 08°30′W
Valongo de Gaia *see* Valongo
104 M5 **Valoria la Buena** Castilla y León, N Spain 41°48′N 04°33′W
119 J15 **Valozhyn** *Pol.* Wołożyn, *Rus.* Volozhin. Minskaya Voblasts', C Belarus 54°05′N 26°32′E
104 I5 **Valpaços** Vila Real, N Portugal 41°36′N 07°17′W
62 G11 **Valparaíso** Valparaíso, C Chile 33°05′S 71°18′W
40 L11 **Valparaíso** Zacatecas, C Mexico 22°49′N 103°28′W
23 P8 **Valparaiso** Florida, SE USA 30°30′N 86°28′W
31 N11 **Valparaiso** Indiana, N USA 41°28′N 87°04′W
62 G11 **Valparaíso** *off.* Región de Valparaíso. ◈ *region* C Chile
Valparaíso, Región de *see* Valparaíso
Valpo *see* Valpovo
112 I9 **Valpovo** *Hung.* Valpo. Osijek-Baranja, E Croatia 45°40′N 18°25′E
103 R14 **Valréas** Vaucluse, SE France 44°22′N 05°00′E
Vals *see* Vals-Platz
154 D12 **Valsād** *prev.* Bulsar. Gujarāt, W India 20°40′N 72°55′E
Valsbaai *see* False Bay
171 T12 **Valse Pisang, Kepulauan** *island group* E Indonesia
108 H9 **Vals-Platz** *var.* Vals. Graubünden, S Switzerland 46°39′N 09°09′E
171 X16 **Vals, Tanjung** *headland* Papua, SE Indonesia 08°26′S 137°35′E
93 N15 **Valtimo** Pohjois-Karjala, E Finland 63°39′N 28°49′E
115 D17 **Váltou** ▲▲ C Greece
127 O12 **Valuyevka** Rostovskaya Oblast', SW Russian Federation 46°48′N 43°49′E
126 K9 **Valuyki** Belgorodskaya Oblast', W Russian Federation 50°11′N 38°07′E
36 L2 **Val Verda** Utah, W USA 40°51′N 111°53′W
64 N12 **Valverde** Hierro, Islas Canarias, Spain, NE Atlantic Ocean 27°48′N 17°55′W
104 I13 **Valverde del Camino** Andalucía, S Spain 37°35′N 06°45′W
95 G23 **Vamdrup** Syddanmark, C Denmark 55°26′N 09°18′E
94 L12 **Våmhus** Dalarna, C Sweden 61°07′N 14°30′E
93 K18 **Vammala** Pirkanmaa, SW Finland 61°20′N 22°55′E
Vámosudvarhely *see* Odorheiu Secuiesc
137 S14 **Van** Van, E Turkey 38°30′N 43°23′E
25 V7 **Van** Texas, SW USA 32°31′N 95°38′W
137 T14 **Van** ◈ *province* E Turkey
137 T11 **Vanadzor** *prev.* Kirovakan. N Armenia 40°49′N 44°29′E
25 U5 **Van Alstyne** Texas, SW USA 33°25′N 96°34′W
33 W10 **Vananda** Montana, NW USA 46°22′N 106°58′W
116 I11 **Vânători** *Hung.* Héjjasfalva; *prev.* Vînători. Mureş, C Romania 46°14′N 24°56′E
191 W12 **Vanavana** *atoll* Îles Tuamotu, SE French Polynesia
Vana-Vändra *see* Vändra
122 M11 **Vanavara** Krasnoyarskiy Kray, C Russian Federation 60°19′N 102°19′E
15 Q8 **Van Bruyssel** Québec, SE Canada 47°56′N 72°08′W
27 R10 **Van Buren** Arkansas, C USA 35°28′N 94°25′W
19 S1 **Van Buren** Maine, NE USA 47°07′N 67°57′W
27 W7 **Van Buren** Missouri, C USA 37°00′N 91°00′W
19 T5 **Vanceboro** Maine, NE USA 45°31′N 67°25′W
21 W10 **Vanceboro** North Carolina, SE USA 35°16′N 77°06′W
21 O4 **Vanceburg** Kentucky, S USA 38°36′N 84°40′W
45 W10 **Vance W. Amory** ✈ Nevis, Saint Kitts and Nevis 17°08′N 62°36′W
Vanch *see* Vanj
10 L17 **Vancouver** British Columbia, SW Canada 49°13′N 123°06′W
32 G11 **Vancouver** Washington, NW USA 45°38′N 122°39′W
10 L17 **Vancouver** ✈ British Columbia, SW Canada 49°03′N 123°00′W
10 K16 **Vancouver Island** *island* British Columbia, SW Canada
Vanda *see* Vantaa
171 X13 **Van Daalen** ≈ Papua, E Indonesia
30 L15 **Vandalia** Illinois, N USA 38°57′N 89°05′W
27 V3 **Vandalia** Missouri, C USA 39°18′N 91°29′W
31 R13 **Vandalia** Ohio, N USA 39°53′N 84°12′W
25 U13 **Vanderbilt** Texas, SW USA 28°45′N 96°37′W
31 Q10 **Vandercook Lake** Michigan, N USA 42°11′N 84°23′W
10 L14 **Vanderhoof** British Columbia, SW Canada 53°54′N 124°00′W
18 K8 **Vanderwhacker Mountain** ▲ New York, NE USA 43°54′N 74°06′W
181 P1 **Van Diemen Gulf** *gulf* Northern Territory, N Australia
Van Diemen's Land *see* Tasmania
118 H5 **Vändra** *Ger.* Fennern; *prev.* Vana-Vändra. Pärnumaa, SW Estonia 58°39′N 25°00′E
Vandsburg *see* Więcbork
34 L4 **Van Duzen River** ≈ California, W USA
118 F13 **Vandžiogala** Kaunas, C Lithuania 55°07′N 23°55′E
41 N10 **Vanegas** San Luis Potosí, C Mexico 23°53′N 100°55′W
Vaner, Lake *see* Vänern
95 K17 **Vänern** *Eng.* Lake Vaner; *prev.* Lake Vener. ◎ S Sweden
95 J18 **Vänersborg** Västra Götaland, S Sweden 58°16′N 12°22′E
94 F12 **Vang** Oppland, S Norway 61°07′N 08°34′E
172 I7 **Vangaindrano** Fianarantsoa, SE Madagascar 23°21′S 47°35′E
137 S14 **Van Gölü** *Eng.* Lake Van; *anc.* Thospitis. *salt lake* E Turkey
186 L9 **Vangunu** *island* New Georgia Islands, NW Solomon Islands
24 J9 **Van Horn** Texas, SW USA 31°03′N 104°51′W
187 Q11 **Vanikolo** *var.* Vanikoro. *island* Santa Cruz Islands, E Solomon Islands
Vanikoro *see* Vanikolo
186 A5 **Vanimo** West Sepik, NW Papua New Guinea 02°40′S 141°17′E
123 T13 **Vanino** Khabarovskiy Kray, SE Russian Federation 49°10′N 140°18′E
155 G19 **Vānīvilāsa Sāgara** ◎ SW India
147 S13 **Vanj** *Rus.* Vanch. S Tajikistan 38°22′N 71°27′E
116 G14 **Vânju Mare** *prev.* Vînju Mare. Mehedinţi, SW Romania 44°25′N 22°52′E
15 N12 **Vankleek Hill** Ontario, SE Canada 45°32′N 74°39′W
Van, Lake *see* Van Gölü
93 I16 **Vännäs** Västerbotten, N Sweden 63°54′N 19°43′E
93 I15 **Vännäsby** Västerbotten, N Sweden 63°55′N 19°53′E
102 H7 **Vannes** *anc.* Dariorigum. Morbihan, NW France 47°40′N 02°45′W
92 I8 **Vannøya** *island* N Norway
103 T12 **Vanoise, Massif de la** ▲▲ E France
83 E24 **Vanrhynsdorp** Western Cape, SW South Africa 31°36′S 18°45′E
21 P7 **Vansant** Virginia, NE USA 37°13′N 82°03′W
94 L13 **Vansbro** Dalarna, C Sweden 60°32′N 14°15′E
95 D18 **Vanse** Vest-Agder, S Norway 58°04′N 06°40′E
9 P7 **Vansittart Island** *island* Nunavut, NE Canada
93 M20 **Vantaa** *Swe.* Vanda. Uusimaa, S Finland 60°18′N 25°01′E
93 L19 **Vantaa** ✈ (Helsinki) Uusimaa, S Finland 60°18′N 25°01′E
32 J9 **Vantage** Washington, NW USA 46°55′N 119°55′W
187 Z14 **Vanua Balavu** *prev.* Vanua Mbalavu. *island* Lau Group, E Fiji
187 R12 **Vanua Lava** *island* Banks Islands, N Vanuatu
187 Y13 **Vanua Levu** *island* N Fiji
Vanua Mbalavu *see* Vanua Balavu
187 R12 **Vanuatu** *off.* Republic of Vanuatu; *prev.* New Hebrides. ◆ *republic* SW Pacific Ocean
175 P8 **Vanuatu** *island group* SW Pacific Ocean
Vanuatu, Republic of *see* Vanuatu
31 Q12 **Van Wert** Ohio, N USA 40°52′N 84°34′W
187 Q17 **Vao** Province Sud, S New Caledonia 22°40′S 167°29′E
Vapincum *see* Gap
117 N7 **Vapnyarka** Vinnyts'ka Oblast', C Ukraine 48°31′S 28°44′E
103 T15 **Var** ◈ *department* SE France
103 U14 **Var** ≈ SE France
95 J18 **Vara** Västra Götaland, S Sweden 58°16′N 12°57′E
Varadinska Županija *see* Varaždin
118 J10 **Varakļāni** C Latvia 56°36′N 26°40′E
106 C7 **Varallo** Piemonte, NE Italy 45°51′N 08°16′E
143 O5 **Varāmīn** *var.* Veramin. Tehrān, N Iran 35°19′N 51°40′E
153 N14 **Vārānasi** *prev.* Banaras, Benares, *hist.* Kasi. Uttar Pradesh, N India 25°20′N 83°E
125 T3 **Varandey** Nenetskiy Avtonomnyy Okrug, NW Russian Federation 68°48′N 57°54′E
92 M8 **Varangerbotn** *Lapp.* Vuonnabahta. Finnmark, N Norway 70°09′N 28°28′E
92 M8 **Varangerfjorden** *Lapp.* Várjjatvuotna. *fjord* N Norway
92 M8 **Varangerhalvøya** *Lapp.* Várnjárga. *peninsula* N Norway
Varannó *see* Vranov nad Topl'ou
107 M15 **Varano, Lago di** ◎ SE Italy
118 J13 **Varapayeva** *Rus.* Voropayevo. Vitsyebskaya Voblasts', NW Belarus 55°09′N 27°13′E
Varasd *see* Varaždin
112 E7 **Varaždin** *Ger.* Warasdin, *Hung.* Varasd. Varaždin, N Croatia 46°18′N 16°21′E
112 E7 **Varaždin** *off.* Varadinska Županija. ◈ *province* N Croatia
106 C10 **Varazze** Liguria, NW Italy 44°21′N 08°35′E
95 J20 **Varberg** Halland, S Sweden 57°06′N 12°15′E
114 J11 **Varbitsa** *var,* Vŭrbitsa; *prev.* Filevo. Haskovo, S Bulgaria 42°02′N 25°25′E
114 J12 **Varbitsa** ≈ S Bulgaria
113 Q19 **Vardar** *Gk.* Axiós. ≈ FYR Macedonia/Greece *see also* Axiós
Vardar *see* Axiós
95 F23 **Varde** Syddtjylland, W Denmark 55°38′N 08°31′E
137 V12 **Vardenis** E Armenia 40°11′N 45°43′E
92 N8 **Vardø** *Fin.* Vuoreija. Finnmark, N Norway 70°22′N 31°06′E
115 E18 **Vardoúsia** ▲▲ C Greece
Vareia *see* Logroño
100 G10 **Varel** Niedersachsen, NW Germany 53°24′N 08°07′E
119 G15 **Varėna** *Pol.* Orany. Alytus, S Lithuania 54°13′N 24°35′E
15 O12 **Varennes** Québec, SE Canada 45°42′N 73°25′W
103 P10 **Varennes-sur-Allier** Allier, C France 46°17′N 03°24′E
112 I12 **Vareš** Federacija Bosne I Hercegovine, E Bosnia and Herzegovina 44°12′N 18°19′E
106 D7 **Varese** Lombardia, N Italy 45°49′N 08°50′E
116 J12 **Vârful Moldoveanu** *var.* Moldoveanul; *prev.* Vîrful Moldoveanu. ▲ C Romania 45°35′N 24°48′E
Varganzi *see* Warganza
95 J18 **Vårgårda** Västra Götaland, S Sweden 58°00′N 12°49′E
54 L4 **Vargas** *off.* Estado Vargas. ◈ *state* N Venezuela
95 J18 **Vargön** Västra Götaland, S Sweden 58°21′N 12°22′E
95 C17 **Varhaug** Rogaland, S Norway 58°37′N 05°39′E
Várjjatvuotna *see* Varangerfjorden
93 N17 **Varkaus** Pohjois-Savo, C Finland 62°20′N 27°50′E
92 J2 **Varmahlið** Norðurland Vestra, N Iceland 65°32′N 19°33′W
95 J15 **Värmland** ◈ *county* C Sweden
95 K16 **Värmlandsnäs** *peninsula* S Sweden
114 N8 **Varna** *prev.* Stalin; *anc.* Odessus. Varna, E Bulgaria 43°14′N 27°56′E
114 N8 **Varna** ◈ *province* E Bulgaria
114 N8 **Varna** ✈ Varna, E Bulgaria 43°16′N 27°52′E
95 L20 **Värnamo** Jönköping, S Sweden 57°11′N 14°03′E
114 N8 **Varnenski Zaliv** *prev.* Stalinski Zaliv. *bay* E Bulgaria
114 N8 **Varnensko Ezero** *estuary* E Bulgaria
118 D11 **Varniai** Telšiai, W Lithuania 55°45′N 22°22′E
Várnjárga *see* Varangerhalvøya
Varnoús *see* Baba
111 D14 **Varnsdorf** *Ger.* Warnsdorf. Ústecký Kraj, NW Czech Republic 50°57′N 14°35′E
111 I23 **Várpalota** Veszprém, W Hungary 47°12′N 18°08′E
Varshava *see* Warszawa
114 G8 **Varshets** *var.* Vŭrshets. Montana, NW Bulgaria 43°14′N 23°20′E
93 K20 **Varsinais-Suomi** *Swe.* Egentliga Finland. ◈ *region* W Finland
118 K6 **Värska** Põlvamaa, SE Estonia 57°58′N 27°37′E
98 N12 **Varsseveld** Gelderland, E Netherlands 51°55′N 06°28′E
115 D19 **Vartholomió** *prev.* Vartholomión. Dytikí Elláda, S Greece 37°52′N 21°12′E
Vartholomión *see* Vartholomió
137 Q14 **Varto** Muş, E Turkey 39°10′N 41°28′E
95 K18 **Vartofta** Västra Götaland, S Sweden 58°06′N 13°40′E
93 O17 **Värtsilä** Pohjois-Karjala, E Finland 62°10′N 30°35′E
Värtsilä *see* Vyartsilya
117 R4 **Varva** Chernihivs'ka Oblast', NE Ukraine 50°31′N 32°43′E
59 H18 **Várzea Grande** Mato Grosso, SW Brazil 15°39′S 56°08′W
106 D9 **Varzi** Lombardia, N Italy 44°51′N 09°13′E
Varzimanor Ayni *see* Ayní
124 K5 **Varzuga** ≈ NW Russian Federation
103 P8 **Varzy** Nièvre, C France 47°22′N 03°22′E
111 G23 **Vas** *off.* Vas Megye. ◈ *county* W Hungary
Vasa *see* Vaasa
190 A9 **Vasafua** *island* Funafuti Atoll, C Tuvalu
111 O21 **Vásárosnamény** Szabolcs-Szatmár-Bereg, E Hungary 48°10′N 22°18′E
104 H13 **Vascão, Ribeira de** ≈ S Portugal
116 G10 **Vaşcău** *Hung.* Vaskoh. Bihor, NE Romania 46°28′N 22°30′E
Vascongadas, Provincias *see* País Vasco
Vashess Bay *see* Vaskess Bay
125 O8 **Vashka** ≈ NW Russian Federation
Vāsht *see* Khāsh
Vasilevichi *see* Vasilyevichy
115 G14 **Vasiliká** Kentrikí Makedonía, NE Greece 40°28′N 23°08′E
115 C18 **Vasilikí** Lefkáda, Iónia Nisiá, Greece, C Mediterranean Sea 38°36′N 20°37′E
115 K25 **Vasilikí** Kríti, Greece, E Mediterranean Sea 35°04′N 25°49′E
119 G16 **Vasilishki** *Pol.* Wasiliszki. Hrodzyenskaya Voblasts', W Belarus 53°47′N 24°51′E
Vasil Kolarov *see* Pamporovo
Vasil'kov *see* Vasyl'kiv
119 N19 **Vasilyevichy** *Rus.* Vasilevichi. Homyel'skaya Voblasts', SE Belarus 52°15′N 29°50′E
191 Y3 **Vaskess Bay** *var.* Vashess Bay. *bay* Kiritimati, E Kiribati
Vaskoh *see* Vaşcău
Vaskohsziklás *see* Ştei
116 M10 **Vaslui** Vaslui, C Romania 46°38′N 27°44′E
116 L11 **Vaslui** ◈ *county* NE Romania
Vas Megye *see* Vas
31 R8 **Vassar** Michigan, N USA 43°22′N 83°34′W
95 E15 **Vassdalsegga** ▲ S Norway 59°47′N 07°07′E
60 P9 **Vassouras** Rio de Janeiro, SE Brazil 22°24′S 43°38′W
95 N15 **Västerås** Västmanland, C Sweden 59°37′N 16°33′E
93 G15 **Västerbotten** ◈ *county* N Sweden
94 K12 **Västerdalälven** ≈ C Sweden
95 O16 **Västerhaninge** Stockholm, C Sweden 59°07′N 18°06′E
94 M10 **Västernorrland** ◈ *county* C Sweden
95 N19 **Västervik** Kalmar, S Sweden 57°44′N 16°40′E
95 M15 **Västmanland** ◈ *county* C Sweden
107 L15 **Vasto** *anc.* Histonium. Abruzzo, C Italy 42°07′N 14°43′E
95 J19 **Västra Götaland** ◈ *county* S Sweden
95 J16 **Västra Silen** ◎ S Sweden
111 G23 **Vasvár** *Ger.* Eisenburg. Vas, W Hungary 47°03′N 16°48′E
117 U9 **Vasylivka** Zaporiz'ka Oblast', SE Ukraine 47°26′N 35°18′E
117 O5 **Vasyl'kiv** *var.* Vasil'kov. Kyyivs'ka Oblast', N Ukraine 50°12′N 30°18′E
117 V8 **Vasyl'kivka** Dnipropetrovs'ka Oblast', E Ukraine 48°12′N 36°00′E
122 I11 **Vasyugan** ≈ C Russian Federation
103 N8 **Vatan** Indre, C France 47°06′N 01°49′E
Vaté *see* Efate
115 C18 **Vathy** *prev.* Itháki. Itháki, Iónia Nisiá, Greece, C Mediterranean Sea 38°22′N 20°43′E
107 G15 **Vatican City** *off.* Vatican City. ◆ *papal state* S Europe
Vatican City *see* Vatican City
107 M22 **Vaticano, Capo** *headland* S Italy 38°37′N 15°49′E
92 K3 **Vatnajökull** *glacier* SE Iceland
95 P15 **Vätö** Stockholm, C Sweden 59°48′N 18°55′E
187 Z16 **Vatoa** *island* Lau Group, SE Fiji
172 J5 **Vatomandry** Toamasina, E Madagascar 19°20′S 48°58′E
116 J9 **Vatra Dornei** *Ger.* Dorna Watra. Suceava, NE Romania 47°20′N 25°21′E
116 J9 **Vatra Moldoviţei** Suceava, NE Romania 47°37′N 25°36′E
Vatter, Lake *see* Vättern
95 L18 **Vättern** *Eng.* Lake Vatter; *prev.* Lake Vetter. ◎ S Sweden
187 X5 **Vatulele** *island* SW Fiji
117 P7 **Vatutine** Cherkas'ka Oblast', C Ukraine 49°01′N 31°04′E
187 W15 **Vatu Vara** *island* Lau Group, E Fiji
103 R14 **Vaucluse** ◈ *department* SE France
103 S5 **Vaucouleurs** Meuse, NE France 48°37′N 05°38′E
108 B9 **Vaud** *Ger.* Waadt. ◈ *canton* SW Switzerland
15 N12 **Vaudreuil** Québec, SE Canada 45°24′N 74°01′W
37 T12 **Vaughn** New Mexico, SW USA 34°36′N 105°12′W
54 I14 **Vaupés** *off.* Comisaría del Vaupés. ◈ *province* SE Colombia
Vaupés, Comisaría del *see* Vaupés
54 J13 **Vaupés, Río** *var.* Rio Uaupés. ≈ Brazil/Colombia *see also* Uaupés, Rio
Vaupés, Río *see* Uaupés, Rio
103 Q15 **Vauvert** Gard, S France 43°42′N 04°16′E
11 R17 **Vauxhall** Alberta, SW Canada 50°05′N 112°09′W
99 K23 **Vaux-sur-Sûre** Luxembourg, SE Belgium 49°55′N 05°34′E
172 J4 **Vavatenina** Toamasina, E Madagascar 17°25′S 49°11′E
193 Y14 **Vava'u Group** *island group* N Tonga
76 M16 **Vavoua** W Ivory Coast 07°23′N 06°29′W
127 S2 **Vavozh** Udmurtskaya Respublika, NW Russian Federation 56°48′N 51°53′E
155 K23 **Vavuniya** Northern Province, N Sri Lanka 08°45′N 80°30′E
119 G17 **Vawkavysk** *Pol.* Wołkowysk, *Rus.* Volkovysk. Hrodzyenskaya Voblasts', W Belarus 53°10′N 24°28′E
119 F17 **Vawkavyskaye Wzvyshsha** *Rus.* Volkovyskiye Vysoty. *hill range* W Belarus
95 P15 **Vaxholm** Stockholm, C Sweden 59°25′N 18°21′E
95 L21 **Växjö** *var.* Vexiö. Kronoberg, S Sweden 56°52′N 14°50′E
125 T1 **Vaygach, Ostrov** *island* NW Russian Federation
137 V13 **Vayk'** *prev.* Azizbekov. SE Armenia 39°41′N 45°28′E
Vazáš *see* Vittangi
125 P8 **Vazhgort** *prev.* Chasovo. Respublika Komi, NW Russian Federation 64°06′N 46°44′E
45 V10 **V. C. Bird** ✈ (St. John's) Antigua, Antigua and Barbuda 17°07′N 61°49′W
167 R13 **Veal Renh** *prev.* Phumĭ Veal Renh. Kâmpôt, SW Cambodia 10°43′N 103°49′E
29 Q7 **Veblen** South Dakota, N USA 45°50′N 97°17′W
98 N9 **Vecht** *Ger.* Vechte. ≈ Germany/Netherlands *see also* Vechte
Vecht *see* Vechte
100 G12 **Vechta** Niedersachsen, NW Germany 52°44′N 08°16′E
100 E12 **Vechte** *Dut.* Vecht. ≈ Germany/Netherlands *see also* Vecht
Vechte *see* Vecht
118 I8 **Vecpiebalga** C Latvia 57°03′N 25°47′E
118 G9 **Vecumnieki** C Latvia 56°36′N 24°30′E
95 C16 **Vedavågen** Rogaland, S Norway 59°18′N 5°13′E
Vedāvati *see* Hagari
95 J20 **Veddige** Halland, S Sweden 57°16′N 12°19′E
116 J15 **Vedea** ≈ S Romania
127 P16 **Vedeno** Chechenskaya Respublika, SW Russian Federation 42°57′N 46°02′E
95 C16 **Vedvågen** Rogaland, S Norway 59°18′N 05°13′E
98 O6 **Veendam** Groningen, NE Netherlands 53°05′N 06°53′E
98 K12 **Veenendaal** Utrecht, C Netherlands 52°03′N 05°33′E
99 E14 **Veere** Zeeland, SW Netherlands 51°33′N 03°40′E
24 M2 **Vega** Texas, SW USA 35°14′N 102°26′W
92 E13 **Vega** *island* C Norway
45 T5 **Vega Baja** C Puerto Rico 18°27′N 66°23′W
38 D17 **Vega Point** *headland* Kiska Island, Alaska, USA 51°49′N 177°19′E
95 F17 **Vegår** ◎ S Norway
99 K14 **Veghel** Noord-Brabant, S Netherlands 51°37′N 05°33′E
Veglia *see* Krk
114 E13 **Vegorítida, Límni** *var.* Límni Vegorítis. ◎ N Greece
Vegorítis, Límni *see* Vegorítida, Límni
11 Q14 **Vegreville** Alberta, SW Canada 53°30′N 112°02′W
95 K21 **Veinge** Halland, S Sweden 56°33′N 13°04′E
61 B21 **Veinticinco de Mayo** *var.* 25 de Mayo. Buenos Aires, E Argentina 35°27′S 60°11′W
63 I14 **Veinticinco de Mayo** La Pampa, C Argentina 37°45′S 67°40′W
119 F15 **Veisiejai** Alytus, S Lithuania 54°06′N 23°42′E
95 F23 **Vejen** Syddtjylland, W Denmark 55°29′N 09°13′E
104 K16 **Vejer de la Frontera** Andalucía, S Spain 36°15′N 05°58′W
95 G23 **Vejle** Syddanmark, C Denmark 55°43′N 09°33′E
114 M7 **Vekilski** Shumen, NE Bulgaria 43°33′N 27°19′E
54 G3 **Vela, Cabo de la** *headland* NE Colombia 12°14′N 72°13′W
Vela Goa *see* Goa
113 F15 **Vela Luka** Dubrovnik-Neretva, S Croatia 42°57′N 16°43′E
61 G19 **Velázquez** Rocha, E Uruguay 34°05′S 54°16′W
101 E15 **Velbert** Nordrhein-Westfalen, W Germany 51°22′N 07°03′E
109 S9 **Velden** Kärnten, S Austria 46°37′N 13°59′E
Veldes *see* Bled
99 K15 **Veldhoven** Noord-Brabant, S Netherlands 51°24′N 05°24′E
112 C11 **Velebit** ▲▲ C Croatia
114 N11 **Veleka** ≈ SE Bulgaria
109 V10 **Velenje** *Ger.* Wöllan. N Slovenia 46°22′N 15°07′E
190 E12 **Vele, Pointe** *headland* Île Futuna, S Wallis and Futuna
113 O18 **Veles** *Turk.* Köprülü. C FYR Macedonia 41°43′N 21°49′E
113 M20 **Velešta** SW FYR Macedonia 41°16′N 20°37′E
115 F16 **Velestíno** *prev.* Velestínon. Thessalía, C Greece 39°23′N 22°45′E
Velestínon *see* Velestíno
Velevshchina *see* Vyelyewshchyna
54 F9 **Vélez** Santander, C Colombia 06°02′N 73°43′W
105 Q13 **Vélez Blanco** Andalucía, S Spain 37°43′N 02°07′W
104 M17 **Vélez de la Gomera, Peñon de** *island group* S Spain
105 N15 **Vélez-Málaga** Andalucía, S Spain 36°47′N 04°06′W
105 Q13 **Vélez Rubio** Andalucía, S Spain 37°39′N 02°04′W
Velha Goa *see* Goa
Velho *see* Porto Velho
112 E8 **Velika Gorica** Zagreb, N Croatia 45°43′N 16°03′E
112 C9 **Velika Kapela** ▲▲ NW Croatia
Velika Kikinda *see* Kikinda
112 D10 **Velika Kladuša** Federacija Bosne I Hercegovine, NW Bosnia and Herzegovina 45°10′N 15°48′E
112 N11 **Velika Morava** *var.* Glavn'a Morava, Morava, *Ger.* Grosse Morava. ≈ C Serbia
112 N12 **Velika Plana** Serbia, C Serbia 44°20′N 21°01′E
109 U10 **Velika Raduha** ▲ N Slovenia 46°24′N 14°46′E
123 V7 **Velikaya** ≈ NE Russian Federation
124 F15 **Velikaya** ≈ W Russian Federation
Velikaya Berestovitsa *see* Vyalikaya Byerastavitsa
Velikaya Lepetikha *see* Velyka Lepetykha
Veliki Bečkerek *see* Zrenjanin
112 P12 **Veliki Krš** *var.* Stol. ▲ E Serbia 44°10′N 22°09′E
114 L8 **Veliki Preslav** *prev.* Preslav. Shumen, NE Bulgaria 43°09′N 26°50′E
112 B9 **Veliki Risnjak** ▲ NW Croatia 45°30′N 14°31′E
109 T13 **Veliki Snežnik** *Ger.* Schneeberg, *It.* Monte Nevoso. ▲ SW Slovenia 45°36′N 14°25′E
112 J13 **Veliki Stolac** ▲ E Bosnia and Herzegovina 43°55′N 19°15′E
Velikiy Bor *see* Vyaliki Bor
124 G16 **Velikiye Luki** Pskovskaya Oblast', W Russian Federation 56°20′N 30°27′E
124 H14 **Velikiy Novgorod** *prev.* Novgorod. Novgorodskaya Oblast', W Russian Federation 58°32′N 31°15′E
125 P12 **Velikiy Ustyug** Vologodskaya Oblast', NW Russian Federation 60°46′N 46°18′E
112 N11 **Veliko Gradište** Serbia, NE Serbia 44°46′N 21°28′E
155 I18 **Velikonda Range** ▲▲ SE India
114 K9 **Veliko Tarnovo** *prev.* Tirnovo, Trnovo, Tŭrnovo, *var.* Veliko Tărnovo, Veliko Tŭrnovo. Veliko Tarnovo, N Bulgaria 43°05′N 25°40′E
114 K8 **Veliko Tarnovo** *var.* Veliko Tŭrnovo. ◈ *province* N Bulgaria
Veliko Tărnovo *see* Veliko Tarnovo
Veliko Tŭrnovo *see* Veliko Tarnovo
Veliko Tŭrnovo *see* Veliko Tarnovo
Velikovec *see* Völkermarkt
125 R5 **Velikovisochnoye** Nenetskiy Avtonomnyy Okrug, NW Russian Federation 67°13′N 52°00′E
76 H12 **Vélingara** C Senegal 15°00′N 14°39′W
76 H11 **Vélingara** S Senegal 13°12′N 14°05′W
114 H11 **Velingrad** Pazardzhik, C Bulgaria 42°01′N 24°00′E
126 H3 **Velizh** Smolenskaya Oblast', W Russian Federation 55°30′N 31°06′E
111 F16 **Velká Deštná** *var.* Deštná, Grosskoppe, *Ger.* Deschnaer Koppe. ▲ NE Czech Republic 50°18′N 16°25′E
111 F18 **Velké Meziříčí** *Ger.* Grossmeseritsch. Vysočina, C Czech Republic 49°22′N 16°02′E
92 N1 **Velkomstpynten** *headland* NW Svalbard 79°51′N 11°37′E
111 K21 **Veľký Krtíš** Banskobystrický Kraj, C Slovakia 48°13′N 19°21′E
186 J8 **Vella Lavella** *var.* Mbilua. *island* New Georgia Islands, NW Solomon Islands
107 I15 **Velletri** Lazio, C Italy 41°41′N 12°47′E
95 K23 **Vellinge** Skåne, S Sweden 55°29′N 13°00′E
155 I19 **Vellore** Tamil Nādu, SE India 12°56′N 79°09′E
Velobriga *see* Viana do Castelo
115 G21 **Velopoúla** *island* S Greece
98 M12 **Velp** Gelderland, SE Netherlands 52°00′N 05°59′E
Velsen *see* Velsen-Noord
98 H9 **Velsen-Noord** *var.* Velsen. Noord-Holland, W Netherlands 52°27′N 04°40′E
125 N12 **Vel'sk** *var.* Velsk. Arkhangel'skaya Oblast', NW Russian Federation 61°03′N 42°01′E
Velsuna *see* Orvieto
98 K10 **Veluwemeer** *lake channel* C Netherlands
28 M3 **Velva** North Dakota, N USA 48°03′N 100°55′W
Velvendos/Velvendós *see* Velventós
115 E14 **Velventós** *var.* Velvendos, Velvendós. Dytikí Makedonía, N Greece 40°15′N 22°04′E
117 S5 **Velyka Bahachka** Poltavs'ka Oblast', C Ukraine 49°46′N 33°44′E
117 S9 **Velyka Lepetykha** *Rus.* Velikaya Lepetikha. Khersons'ka Oblast', S Ukraine 47°09′N 33°59′E
117 O10 **Velyka Mykhaylivka** Odes'ka Oblast', SW Ukraine 47°07′N 29°49′E
117 W8 **Velyka Novosilka** Donets'ka Oblast', E Ukraine 47°49′N 36°49′E
117 S9 **Velyka Oleksandrivka** Khersons'ka Oblast', S Ukraine 47°17′N 33°16′E
117 T4 **Velyka Pysarivka** Sums'ka Oblast', NE Ukraine 50°25′N 35°28′E
116 G6 **Velykyy Bereznyy** Zakarpats'ka Oblast', W Ukraine 48°54′N 22°27′E
117 W4 **Velykyy Burluk** Kharkivs'ka Oblast', E Ukraine 50°04′N 37°25′E
Velykyy Tokmak *see* Tokmak
173 P7 **Vema Fracture Zone** *tectonic feature* W Indian Ocean
65 P18 **Vema Seamount** *undersea feature* SW Indian Ocean 31°38′S 08°19′E
93 F17 **Vemdalen** Jämtland, C Sweden 62°26′N 13°50′E
95 N19 **Vena** Kalmar, S Sweden 57°31′N 16°00′E
41 N11 **Venado** San Luis Potosí, C Mexico 22°56′N 101°05′W
62 L11 **Venado Tuerto** Entre Ríos, E Argentina 33°45′S 61°56′W
61 A19 **Venado Tuerto** Santa Fe, C Argentina 33°46′S 61°57′W
107 K16 **Venafro** Molise, C Italy 41°28′N 14°03′E
55 Q9 **Venamo, Cerro** ▲ E Venezuela 05°56′N 61°25′W
106 B8 **Venaria** Piemonte, NW Italy 45°09′N 07°40′E
103 U15 **Vence** Alpes-Maritimes, SE France 43°45′N 07°07′E
104 H5 **Venda Nova** Vila Real, N Portugal 41°40′N 07°58′W
104 G11 **Vendas Novas** Évora, S Portugal 38°41′N 08°27′W
102 J9 **Vendée** ◈ *department* NW France
103 Q6 **Vendeuvre-sur-Barse** Aube, NE France 48°08′N 04°17′E
102 M7 **Vendôme** Loir-et-Cher, C France 47°48′N 01°04′E
Venedig *see* Venezia
Vener, Lake *see* Vänern
106 I8 **Veneta, Laguna** *lagoon* NE Italy
Venetia *see* Venezia
39 S7 **Venetie** Alaska, USA 67°00′N 146°25′W
106 H8 **Veneto** *var.* Venezia Euganea. ◈ *region* NE Italy
114 M7 **Venets** Shumen, NE Bulgaria 43°33′N 26°56′E
126 L5 **Venev** Tul'skaya Oblast', W Russian Federation 54°18′N 38°16′E
106 I8 **Venezia** *Eng.* Venice, *Fr.* Venise, *Ger.* Venedig; *anc.* Venetia. Veneto, NE Italy 45°26′N 12°20′E
Venezia Euganea *see* Veneto
Venezia, Golfo di *see* Venice, Gulf of
Venezia Tridentina *see* Trentino-Alto Adige
54 K8 **Venezuela** *off.* Republic of Venezuela; *prev.* Estados Unidos de Venezuela, United States of Venezuela. ◆ *republic* N South America
Venezuela, Cordillera de *see* Costa, Cordillera de la
Venezuela, Estados Unidos de *see* Venezuela
54 I4 **Venezuela, Golfo de** *Eng.* Gulf of Maracaibo, Gulf of Venezuela. *gulf* NW Venezuela
Venezuela, Gulf of *see* Venezuela, Golfo de
64 F11 **Venezuelan Basin** *undersea feature* E Caribbean Sea
Venezuela, Republic of *see* Venezuela
Venezuela, United States of *see* Venezuela
155 D16 **Vengurla** Mahārāshtra, W India 15°55′N 73°39′E
39 O15 **Veniaminof, Mount** ▲ Alaska, USA 56°12′N 159°24′W
23 V14 **Venice** Florida, SE USA 27°06′N 82°27′W
22 L10 **Venice** Louisiana, S USA 29°15′N 89°20′W
Venice *see* Venezia
106 J8 **Venice, Gulf of** *It.* Golfo di Venezia, *Slvn.* Beneški Zaliv. *gulf* N Adriatic Sea
Venise *see* Venezia
94 K13 **Venjan** Dalarna, C Sweden 60°58′N 13°55′E
94 K13 **Venjanssjön** ◎ C Sweden
155 J18 **Venkatagiri** Andhra Pradesh, E India 14°00′N 79°39′E
99 M15 **Venlo** *prev.* Venloo. Limburg, SE Netherlands 51°22′N 06°11′E
Venloo *see* Venlo
95 E18 **Vennesla** Vest-Agder, S Norway 58°15′N 08°00′E
107 M17 **Venosa** *anc.* Venusia. Basilicata, S Italy 40°57′N 15°49′E
Venoste, Alpi *see* Ötztaler Alpen
Venraij *see* Venray
99 M14 **Venray** *var.* Venraij. Limburg, SE Netherlands 51°32′N 05°59′E
118 C8 **Venta** *Ger.* Windau. ≈ Latvia/Lithuania
Venta Belgarum *see* Winchester
40 G9 **Ventana, Punta Arena de la** *var.* Punta de la Ventana. *headland* NW Mexico 24°03′N 109°49′W
Ventana, Punta de la *see* Ventana, Punta Arena de la
61 B23 **Ventana, Sierra de la** *hill range* E Argentina
Ventia *see* Valence
191 S11 **Vent, Îles du** *var.* Windward Islands. *island group* Archipel de la Société, W French Polynesia
191 R10 **Vent, Îles Sous le** *var.* Leeward Islands. *island group* Archipel de la Société, W French Polynesia
106 B11 **Ventimiglia** Liguria, NW Italy 43°47′N 07°37′E
97 M24 **Ventnor** S England, United Kingdom 50°36′N 01°11′W
18 J17 **Ventnor City** New Jersey, NE USA 39°19′N 74°27′W
103 S14 **Ventoux, Mont** ▲ SE France 44°12′N 05°21′E
118 C8 **Ventspils** *Ger.* Windau. NW Latvia 57°22′N 21°34′E
54 M10 **Ventuari, Río** ≈ S Venezuela
35 R15 **Ventura** California, W USA 34°15′N 119°18′W
182 F8 **Venus Bay** South Australia 33°15′S 134°42′E
Venusia *see* Venosa
191 P7 **Vénus, Pointe** *var.* Pointe Tataaihoa. *headland* Tahiti, W French Polynesia 17°28′S 149°29′W
41 V16 **Venustiano Carranza** Chiapas, SE Mexico 16°21′N 92°33′W
41 N7 **Venustiano Carranza, Presa** ⊠ NE Mexico
61 B15 **Vera** Santa Fe, C Argentina 29°28′S 60°10′W
105 Q14 **Vera** Andalucía, S Spain 37°15′N 01°51′W
63 K18 **Vera, Bahía** *bay* E Argentina
41 R14 **Veracruz** *var.* Veracruz Llave. Veracruz-Llave, E Mexico 19°10′N 96°09′W
Veracruz *see* Veracruz-Llave
41 Q13 **Veracruz-Llave** *var.* Veracruz. ◈ *state* E Mexico
Veracruz Llave *see* Veracruz
43 Q16 **Veraguas** *off.* Provincia de Veraguas. ◈ *province* W Panama
Veraguas, Provincia de *see* Veraguas
Veramin *see* Varāmīn
154 B12 **Verāval** Gujarāt, W India 20°54′N 70°22′E
106 C6 **Verbania** Piemonte, NW Italy 45°56′N 08°34′E
107 N20 **Verbicaro** Calabria, SW Italy 39°44′N 15°51′E
108 D11 **Verbier** Valais, SW Switzerland 46°06′N 07°14′E
Vercellae *see* Vercelli
106 C8 **Vercelli** *anc.* Vercellae. Piemonte, NW Italy 45°19′N 08°25′E
103 S13 **Vercors** *physical region* E France
Verdal *see* Verdalsøra
93 E16 **Verdalsøra** *var.* Verdal. Nord-Trøndelag, C Norway 63°47′N 11°27′E
Verde, Cabo *see* Cape Verde
44 J5 **Verde, Cape** *headland* Long Island, C The Bahamas 22°51′N 75°50′W

◆ Country
● Country Capital
◇ Dependent Territory
○ Dependent Territory Capital
◈ Administrative Regions
✈ International Airport
▲ Mountain
▲▲ Mountain Range
⧍ Volcano
≈ River
◎ Lake
⊠ Reservoir

Verde, Costa – Villefranche

104 M2 **Verde, Costa** *coastal region* N Spain
Verde Grande, Río/Verde Grande y de Belem, Río *see* Verde, Río
100 H11 **Verden** Niedersachsen, NW Germany 52°55′N 09°14′E
57 P16 **Verde, Río** Bolivia/Brazil
59 J19 **Verde, Rio** SE Brazil
40 M12 **Verde, Río** *var.* Río Verde Grande, Río Verde Grande y de Belem. C Mexico
41 Q16 **Verde, Río** SE Mexico
36 L13 **Verde River** Arizona, SW USA
Verdhikoúsa/ Verdhikoússa *see* Verdikoússa
27 Q8 **Verdigris River** Kansas/ Oklahoma, C USA
115 E15 **Verdikoússa** *var.* Verdhikoúsa, Verdhikoússa. Thessalía, C Greece 39°47′N 21°59′E
103 S15 **Verdon** SE France
15 O12 **Verdun** Québec, SE Canada 45°27′N 73°36′W
103 S4 **Verdun** *var.* Verdun-sur-Meuse; *anc.* Verodunum. Meuse, NE France 49°09′N 05°25′E
Verdun-sur-Meuse *see* Verdun
83 J21 **Vereeniging** Gauteng, NE South Africa 26°41′S 27°56′E
Veremeyki *see* Vyeramyeyki
125 T14 **Vereshchagino** Permskiy Kray, NW Russian Federation 58°06′N 54°38′E
76 G14 **Verga, Cap** *headland* W Guinea 10°12′N 14°27′W
61 G18 **Vergara** Treinta y Tres, E Uruguay 32°58′S 53°54′W
108 G11 **Vergeletto** Ticino, S Switzerland 46°13′N 08°34′E
18 L8 **Vergennes** Vermont, NE USA 44°09′N 73°13′W
Veria *see* Véroia
104 I5 **Verín** Galicia, NW Spain 41°55′N 07°26′W
Verin T'alin *see* T'alin
118 K6 **Veriora** Põlvamaa, SE Estonia 57°57′N 27°23′E
117 T7 **Verkhivtseve** Dnipropetrovs'ka Oblast', E Ukraine 48°27′N 34°15′E
Verkhnedvinsk *see* Vyerkhnyadzvinsk
122 K10 **Verkhneimbatsk** Krasnoyarskiy Kray, N Russian Federation 63°06′N 88°03′E
124 I3 **Verkhnetulomskiy** Murmanskaya Oblast', NW Russian Federation 68°37′N 31°46′E
124 I3 **Verkhnetulomskoye Vodokhranilishche** NW Russian Federation
Verkhneudinsk *see* Ulan-Ude
123 P10 **Verkhnevilyuysk** Respublika Sakha (Yakutiya), NE Russian Federation 63°44′N 119°59′E
127 W5 **Verkhniy Avzyan** Respublika Bashkortostan, W Russian Federation 53°31′N 57°26′E
127 Q11 **Verkhniy Baskunchak** Astrakhanskaya Oblast', SW Russian Federation 48°14′N 46°43′E
127 W3 **Verkhniye Kigi** Respublika Bashkortostan, W Russian Federation 55°25′N 58°40′E
117 T9 **Verkhniy Rohachyk** Khersons'ka Oblast', S Ukraine 47°16′N 34°16′E
123 Q11 **Verkhnyaya Amga** Respublika Sakha (Yakutiya), NE Russian Federation 59°34′N 127°07′E
125 V6 **Verkhnyaya Inta** Respublika Komi, NW Russian Federation 65°55′N 60°07′E
125 O10 **Verkhnyaya Toyma** Arkhangel'skaya Oblast', NW Russian Federation 62°12′N 44°57′E
126 K6 **Verkhov'ye** Orlovskaya Oblast', W Russian Federation 52°49′N 37°20′E
116 I8 **Verkhovyna** Ivano-Frankivs'ka Oblast', W Ukraine 48°09′N 24°48′E
123 P8 **Verkhoyanskiy Khrebet** NE Russian Federation
117 T7 **Verkn'odniprovs'k** Dnipropetrovs'ka Oblast', E Ukraine 48°40′N 34°17′E
101 G14 **Verl** Nordrhein-Westfalen, NW Germany 51°52′N 08°30′E
92 N1 **Verlegenhuken** *headland* N Svalbard 80°03′N 16°15′E
82 A9 **Vermelha, Ponta** *headland* NW Angola 05°40′S 12°09′E
103 P7 **Vermenton** Yonne, C France 47°40′N 03°43′E
11 R14 **Vermilion** Alberta, SW Canada 53°21′N 110°52′W
31 T11 **Vermilion** Ohio, N USA 41°25′N 82°21′W
22 I10 **Vermilion Bay** *bay* Louisiana, S USA
29 V4 **Vermilion Lake** Minnesota, N USA
14 F9 **Vermilion River** Ontario, S Canada
30 L12 **Vermilion River** Illinois, N USA
29 R12 **Vermillion** South Dakota, N USA 42°46′N 96°55′W
29 R12 **Vermillion River** South Dakota, N USA
15 O9 **Vermillon, Rivière** Québec, SE Canada
115 E14 **Vérmio** N Greece
18 L8 **Vermont** *off.* State of Vermont, *also known as* Green Mountain State. *state* NE USA
113 K16 **Vermosh** *var.* Vermoshi. Shkodër, N Albania 42°37′N 19°42′E
Vermoshi *see* Vermosh
37 O3 **Vernal** Utah, W USA 40°27′N 109°31′W
14 G11 **Verner** Ontario, S Canada 46°24′N 80°04′W
102 M5 **Verneuil-sur-Avre** Eure, N France 48°44′N 00°55′E
114 D13 **Vérno** N Greece
11 N17 **Vernon** British Columbia, SW Canada 50°17′N 119°19′W
102 M4 **Vernon** Eure, N France 49°04′N 01°28′E
23 N3 **Vernon** Alabama, S USA 33°45′N 88°06′W
31 P15 **Vernon** Indiana, N USA 38°59′N 85°39′W
25 Q4 **Vernon** Texas, SW USA 34°11′N 99°17′W
32 G10 **Vernonia** Oregon, NW USA 45°51′N 123°11′W
14 G12 **Vernon, Lake** Ontario, S Canada
22 G7 **Vernon Lake** Louisiana, S USA
23 Y13 **Vero Beach** Florida, SE USA 27°38′N 80°24′W
Verőcze *see* Virovitica
Verodunum *see* Verdun
115 E14 **Véroia** *var.* Veria, Vérroia, *Turk.* Karaferiye. Kentrikí Makedonía, N Greece 40°32′N 22°11′E
106 E8 **Verolanuova** Lombardia, N Italy 45°20′N 10°06′E
14 K14 **Verona** Ontario, SE Canada 44°30′N 76°42′W
106 G8 **Verona** Veneto, NE Italy 45°27′N 11°E
29 P6 **Verona** North Dakota, N USA 46°19′N 98°03′W
30 L9 **Verona** Wisconsin, N USA 42°59′N 89°33′W
61 E20 **Verónica** Buenos Aires, E Argentina 35°25′S 57°16′W
22 J9 **Verret, Lake** Louisiana, S USA
Vérroia *see* Véroia
103 N5 **Versailles** Yvelines, N France 48°48′N 02°08′E
31 P15 **Versailles** Indiana, N USA 39°04′N 85°16′W
20 M5 **Versailles** Kentucky, S USA 38°02′N 84°45′W
27 U5 **Versailles** Missouri, C USA 38°25′N 92°51′W
31 Q13 **Versailles** Ohio, N USA 40°13′N 84°28′W
Versecz *see* Vršac
108 A10 **Versoix** Genève, SW Switzerland 46°17′N 06°10′E
15 Z6 **Verte, Pointe** *headland* Québec, SE Canada 48°36′N 64°10′W
111 I22 **Vértes** NW Hungary
44 G6 **Vertientes** Camagüey, C Cuba 21°18′N 78°11′W
115 G13 **Vertiskos** N Greece
102 I8 **Vertou** Loire-Atlantique, NW France 47°10′N 01°28′W
Verulamium *see* St Albans
99 L19 **Verviers** Liège, E Belgium 50°36′N 05°52′E
103 Y14 **Vescovato** Corse, France, C Mediterranean Sea 42°30′N 09°27′E
99 L20 **Vesdre** E Belgium
117 U10 **Vesele** *Rus.* Veseloye. Zaporiz'ka Oblast', S Ukraine 47°00′N 34°52′E
111 D18 **Veselí nad Lužnicí** *var.* Weseli an der Lainsitz, *Ger.* Frohenbruck. Jihočeský Kraj, S Czech Republic 49°11′N 14°40′E
114 M9 **Veselinovo** Shumen, NE Bulgaria 43°01′N 27°02′E
126 L12 **Veselovskoye Vodokhranilishche** SW Russian Federation
Veseloye *see* Vesele
117 Q9 **Veselynove** Mykolayivs'ka Oblast', S Ukraine 47°21′N 31°15′E
Veseya *see* Vyasyeya
126 M10 **Veshenskaya** Rostovskaya Oblast', SW Russian Federation 49°37′N 41°43′E
127 Q5 **Veshkayma** Ul'yanovskaya Oblast', W Russian Federation 54°04′N 47°06′E
Vesisaari *see* Vadsø
Vesontio *see* Besançon
103 T7 **Vesoul** *anc.* Vesulium, Vesulum. Haute-Saône, E France 47°37′N 06°09′E
95 J20 **Vessigebro** Halland, S Sweden 56°58′N 12°40′E
95 D17 **Vest-Agder** *county* S Norway
23 P4 **Vestavia Hills** Alabama, S USA 33°27′N 86°47′W
84 F6 **Vesterålen** *island* NW Norway
92 G10 **Vesterålen** *island group* N Norway
87 V3 **Vestervig** Midtjylland, NW Denmark 56°46′N 08°20′E
92 H2 **Vestfirðir** *region* NW Iceland
92 G11 **Vestfjorden** *fjord* C Norway
95 G16 **Vestfold** *county* S Norway
Vestmanhavn *see* Vestmanna
95 B18 **Vestmanna** *Dan.* Vestmanhavn. Streymoy, N Faroe Islands 62°09′N 07°11′W
92 I4 **Vestmannaeyjar** Suðurland, S Iceland 63°26′N 20°14′W
94 E9 **Vestnes** Møre og Romsdal, S Norway 62°39′N 07°00′E
92 H3 **Vesturland** *region* W Iceland
92 G11 **Vestvågøya** *island* C Norway
Vesulium/Vesulum *see* Vesoul
Vesuna *see* Périgueux
107 K17 **Vesuvio** *Eng.* Vesuvius. S Italy 40°48′N 14°29′E
Vesuvius *see* Vesuvio
124 K14 **Ves'yegonsk** Tverskaya Oblast', W Russian Federation 58°40′N 37°13′E
111 I23 **Veszprém** *Ger.* Veszprim. Veszprém, W Hungary 47°06′N 17°54′E
111 H23 **Veszprém** *off.* Veszprém Megye. *county* W Hungary
Veszprém Megye *see* Veszprém
Veszprim *see* Veszprém
Vetka *see* Vyetka
95 M19 **Vetlanda** Jönköping, S Sweden 57°26′N 15°05′E
127 P1 **Vetluga** Nizhegorodskaya Oblast', W Russian Federation 57°51′N 45°45′E
125 P14 **Vetluga** NW Russian Federation
125 O14 **Vetluzhskiy** Kostromskaya Oblast', NW Russian Federation 58°21′N 45°25′E
127 P2 **Vetluzhskiy** Nizhegorodskaya Oblast', W Russian Federation 57°10′N 45°07′E
114 K7 **Vetovo** Ruse, N Bulgaria 43°42′N 26°16′E
107 H14 **Vetralla** Lazio, C Italy 42°18′N 12°03′E
Vetrino *see* Vyetryna
122 L7 **Vetrovaya, Gora** N Russian Federation 73°54′N 95°00′E
Vetter, Lake *see* Vättern
106 J13 **Vettore, Monte** C Italy 42°49′N 13°15′E
99 A17 **Veurne** *var.* Furnes. West-Vlaanderen, W Belgium 51°04′N 02°40′E
31 Q15 **Vevay** Indiana, N USA 38°45′N 85°08′W
108 C10 **Vevey** *Ger.* Vivis; *anc.* Vibiscum. Vaud, SW Switzerland 46°28′N 06°51′E
Vexiö *see* Växjö
103 S13 **Veynes** Hautes-Alpes, SE France 44°33′N 05°51′E
103 N11 **Vézère** W France
114 I9 **Vezhen** C Bulgaria 42°45′N 24°22′E
136 K11 **Vezirköprü** Samsun, N Turkey 41°09′N 35°27′E
57 J18 **Viacha** La Paz, W Bolivia 16°40′S 68°17′W
27 R10 **Vian** Oklahoma, C USA 35°30′N 94°56′W
Viana de Castelo *see* Viana do Castelo
104 H12 **Viana do Alentejo** Évora, S Portugal 38°20′N 08°00′W
104 I4 **Viana do Bolo** Galicia, NW Spain 42°10′N 07°06′W
104 G5 **Viana do Castelo** *var.* Viana de Castelo; *anc.* Velobriga. Viana do Castelo, NW Portugal 41°41′N 08°50′W
104 G5 **Viana do Castelo** *var.* Viana de Castelo. *district* N Portugal
98 J12 **Vianen** Utrecht, C Netherlands 52°N 05°06′E
167 Q8 **Viangchan** *Eng./Fr.* Vientiane. (Laos) C Laos 17°58′N 102°38′E
167 P6 **Viangphoukha** *var.* Vieng Pou Kha. Louang Namtha, N Laos 20°41′N 101°03′E
104 K13 **Viar** SW Spain
106 E11 **Viareggio** Toscana, C Italy 43°52′N 10°15′E
103 O14 **Viaur** S France
Vibiscum *see* Vevey
95 G21 **Viborg** Midtjylland, NW Denmark 56°28′N 09°25′E
29 R12 **Viborg** South Dakota, N USA 43°10′N 97°04′W
107 N22 **Vibo Valentia** *prev.* Monteleone di Calabria; *anc.* Hipponium. Calabria, SW Italy 38°40′N 16°06′E
105 W5 **Vic** *var.* Vich; *anc.* Ausa, Vicus Ausonensis. Cataluña, NE Spain 41°56′N 02°16′E
102 K16 **Vic-en-Bigorre** Hautes-Pyrénées, S France 43°23′N 00°04′E
40 K10 **Vicente Guerrero** Durango, C Mexico 23°30′N 104°24′W
41 P10 **Vicente Guerrero, Presa** *var.* Presa de las Adjuntas. NE Mexico
Vicentia *see* Vicenza
106 G8 **Vicenza** *anc.* Vicentia. Veneto, NE Italy 45°32′N 11°31′E
Vich *see* Vic
54 J10 **Vichada** *off.* Comisaría del Vichada. *province* E Colombia
Vichada, Comisaría del *see* Vichada
54 K10 **Vichada, Río** E Colombia
61 G17 **Vichadero** Rivera, NE Uruguay 31°45′S 54°41′W
Vichegda *see* Vychegda
124 M16 **Vichuga** Ivanovskaya Oblast', W Russian Federation 57°13′N 41°51′E
103 P10 **Vichy** Allier, C France 46°08′N 03°26′E
26 K9 **Vici** Oklahoma, C USA 36°09′N 99°18′W
95 I19 **Vickan** Halland, S Sweden 57°25′N 12°00′E
31 P10 **Vicksburg** Michigan, N USA 42°07′N 85°31′W
22 J5 **Vicksburg** Mississippi, S USA 32°21′N 90°52′W
103 O12 **Vic-sur-Cère** Cantal, C France 45°00′N 02°36′E
59 I21 **Victor** Mato Grosso do Sul, SW Brazil 21°39′S 53°21′W
29 X14 **Victor** Iowa, C USA 41°45′N 92°18′W
182 I10 **Victor Harbor** South Australia 35°33′S 138°37′E
61 C18 **Victoria** Entre Ríos, E Argentina 32°40′S 60°10′W
10 L17 **Victoria** *province capital* Vancouver Island, British Columbia, SW Canada 48°25′N 123°22′W
45 R14 **Victoria** NW Grenada 12°12′N 61°42′W
42 H6 **Victoria** Yoro, NW Honduras 15°01′N 87°28′W
121 O15 **Victoria** *var.* Rabat. Gozo, NW Malta 36°02′N 14°14′E
116 I12 **Victoria** *Ger.* Viktoriastadt. Braşov, C Romania 45°44′N 24°41′E
172 H17 **Victoria** (Seychelles) Mahé, SW Seychelles 04°38′S 28°28′E
25 U13 **Victoria** Texas, SW USA 28°47′N 96°59′W
183 N12 **Victoria** *state* SE Australia
174 K7 **Victoria** Western Australia
Victoria *see* Labuan, East Malaysia
Victoria *see* Masvingo, Zimbabwe
Victoria Bank *see* Vitória Seamount
11 Y15 **Victoria Beach** Manitoba, S Canada 50°40′N 96°30′W
Victoria de Durango *see* Durango
Victoria de las Tunas *see* Las Tunas
83 I16 **Victoria Falls** Matabeleland North, W Zimbabwe 17°55′S 25°51′E
83 I16 **Victoria Falls** *waterfall* Zambia/Zimbabwe
83 I16 **Victoria Falls** Matabeleland North, W Zimbabwe 18°03′S 25°48′E
Victoria Falls *see* Iguaçu, Salto do
63 F19 **Victoria, Isla** *island* Archipiélago de los Chonos, S Chile
8 K6 **Victoria Island** *island* Northwest Territories/ Nunavut, NW Canada
182 L8 **Victoria, Lake** New South Wales, SE Australia
68 I12 **Victoria, Lake** *var.* Victoria Nyanza. E Africa
195 S13 **Victoria Land** *physical region* Antarctica
187 X14 **Victoria, Mount** Viti Levu, W Fiji 17°37′S 178°00′E
166 L5 **Victoria, Mount** W Myanmar (Burma) 21°13′N 93°53′E
186 E9 **Victoria, Mount** S Papua New Guinea 08°51′S 147°36′E
81 F17 **Victoria Nile** *var.* Somerset Nile. C Uganda
Victoria Nyanza *see* Victoria, Lake
42 G3 **Victoria Peak** SE Belize 16°50′N 88°38′W
185 H16 **Victoria Range** South Island, New Zealand
181 O3 **Victoria River** Northern Territory, N Australia
181 P3 **Victoria River Roadhouse** Northern Territory, N Australia 15°37′S 131°07′E
15 Q11 **Victoriaville** Québec, SE Canada 46°04′N 71°57′W
Victoria-Wes *see* Victoria West
83 G24 **Victoria West** *Afr.* Victoria-Wes. Northern Cape, W South Africa 31°25′S 23°08′E
62 J13 **Victorica** La Pampa, C Argentina 36°15′S 65°25′W
Victor, Mount *see* Dronning Fabiolafjella
35 U14 **Victorville** California, W USA 34°32′N 117°17′W
62 G9 **Vicuña** Coquimbo, N Chile 30°00′S 70°44′W
62 K11 **Vicuña Mackenna** Córdoba, C Argentina 33°53′S 64°25′W
Vicus Ausonensis *see* Vic
Vicus Elbii *see* Viterbo
33 X7 **Vida** Montana, NW USA 47°52′N 105°30′W
23 V6 **Vidalia** Georgia, SE USA 32°13′N 82°24′W
22 J7 **Vidalia** Louisiana, S USA 31°34′N 91°25′W
95 F22 **Videbæk** Midtjylland, C Denmark 56°08′N 08°38′E
60 I13 **Videira** Santa Catarina, S Brazil 27°00′S 51°08′W
116 J14 **Videle** Teleorman, S Romania 44°15′N 25°27′E
Videm-Krško *see* Krško
Vídeň *see* Wien
104 H12 **Vidigueira** Beja, S Portugal 38°12′N 07°48′W
114 J9 **Vidima** N Bulgaria
114 G7 **Vidin** *anc.* Bononia. Vidin, NW Bulgaria 44°00′N 22°52′E
114 F8 **Vidin** *province* NW Bulgaria
154 H10 **Vidisha** Madhya Pradesh, C India 23°30′N 77°50′E
25 Y10 **Vidor** Texas, SW USA 30°07′N 94°01′W
95 L20 **Vidöstern** S Sweden
92 J13 **Vidsel** Norrbotten, N Sweden 65°49′N 20°31′E
118 H9 **Vidzemes Augstiene** C Latvia
118 J12 **Vidzy** Vitsyebskaya Voblasts', NW Belarus 55°24′N 26°38′E
63 L16 **Viedma** Río Negro, E Argentina 40°50′S 62°58′W
63 H22 **Viedma, Lago** S Argentina
45 O11 **Vieille Case** *var.* Itassi. N Dominica 15°36′N 61°24′W
104 M2 **Vieja, Peña** N Spain 43°09′N 04°47′W
24 J10 **Vieja, Sierra** Texas, SW USA
40 E4 **Viejo, Cerro** NW Mexico 30°16′N 112°18′W
56 B9 **Viejo, Cerro** N Peru 04°54′S 79°24′W
118 E10 **Viekšniai** Telšiai, NW Lithuania 56°14′N 22°33′E
105 U3 **Vielha** *var.* Viella. Cataluña, NE Spain 42°41′N 00°47′E
Viella *see* Vielha
99 L21 **Vielsalm** Luxembourg, E Belgium 50°17′N 05°55′E
Vieng Pou Kha *see* Viangphoukha
23 T6 **Vienna** Georgia, SE USA 32°05′N 83°48′W
30 L17 **Vienna** Illinois, N USA 37°24′N 88°55′W
27 V5 **Vienna** Missouri, C USA 38°12′N 91°59′W
21 Q3 **Vienna** West Virginia, NE USA 39°19′N 81°33′W
Vienna *see* Wien, Austria
Vienna *see* Vienne, France
103 R11 **Vienne** *anc.* Vienna. Isère, E France 45°32′N 04°53′E
102 L10 **Vienne** *department* W France
102 L9 **Vienne** W France
Vientiane *see* Viangchan
Vientos, Paso de los *see* Windward Passage
45 V6 **Vieques** *var.* Isabel Segunda. E Puerto Rico 18°08′N 65°25′W
45 V6 **Vieques, Isla de** *island* E Puerto Rico
45 V6 **Vieques, Pasaje de** *passage* E Puerto Rico
45 V5 **Vieques, Sonda de** *sound* E Puerto Rico
Vierdörfer *see* Săcele
93 M15 **Vieremä** Pohjois-Savo, C Finland 63°42′N 27°02′E
99 M14 **Vierlingsbeek** Noord-Brabant, SE Netherlands 51°36′N 06°01′E
101 G20 **Viernheim** Hessen, W Germany 49°32′N 08°35′E
101 D15 **Viersen** Nordrhein-Westfalen, W Germany 51°16′N 06°24′E
108 G8 **Vierwaldstätter See** *Eng.* Lake of Lucerne. C Switzerland
103 N8 **Vierzon** Cher, C France 47°13′N 02°04′E
40 L8 **Viesca** Coahuila, NE Mexico 25°25′N 102°45′W
118 H10 **Viesīte** *Ger.* Eckengraf. S Latvia 56°21′N 25°30′E
107 N15 **Vieste** Puglia, SE Italy 41°52′N 16°11′E
167 T8 **Vietnam** *off.* Socialist Republic of Vietnam, *Vtn.* Cộng Hoa Xa Hôi Chu Nghia Viêt Nam. *republic* SE Asia
Vietnam, Socialist Republic of *see* Vietnam
167 S5 **Viêt Quang** Ha Giang, N Vietnam 22°24′N 104°48′E
Vietri *see* Viêt Tri
167 S6 **Viêt Tri** *var.* Vietri. Vinh Phu, N Vietnam 21°20′N 105°26′E
30 L4 **Vieux Desert, Lac** Michigan/Wisconsin, N USA
45 Y13 **Vieux Fort** S Saint Lucia 13°43′N 60°57′W
45 X6 **Vieux-Habitants** Basse Terre, SW Guadeloupe 16°04′N 61°45′W
119 G14 **Vievis** Vilnius, S Lithuania 54°46′N 24°51′E
171 N2 **Vigan** Luzon, N Philippines 17°34′N 102°21′E
106 D8 **Vigevano** Lombardia, N Italy 45°19′N 08°51′E
107 N18 **Viggiano** Basilicata, S Italy 40°21′N 15°54′E
58 L12 **Vigia** Pará, NE Brazil 0°50′S 48°07′W
41 Y12 **Vigía Chico** Quintana Roo, SE Mexico 19°49′N 87°31′W
45 T11 **Vigie** *prev.* George F L Charles. (Castries) NE Saint Lucia 14°01′N 60°59′W
102 K17 **Vignemale** *var.* Pic de Vignemale. France/Spain 42°48′N 00°06′W
Vignemale, Pic de *see* Vignemale
106 G10 **Vignola** Emilia-Romagna, C Italy 44°28′N 11°00′E
104 G4 **Vigo** Galicia, NW Spain 42°15′N 08°44′W
104 G4 **Vigo, Ría de** *estuary* NW Spain
94 D9 **Vigra** *island* S Norway
95 C17 **Vigrestad** Rogaland, S Norway 58°34′N 05°42′E
93 L15 **Vihanti** Pohjois-Pohjanmaa, C Finland 64°29′N 25°E
149 U10 **Vihāri** Punjab, E Pakistan 30°03′N 72°32′E
102 K8 **Vihiers** Maine-et-Loire, NW France 47°09′N 00°37′W
81 H18 **Vihiga** *county* W Kenya
111 O19 **Vihorlat** E Slovakia 48°54′N 22°09′E
114 G11 **Vihren** *var.* Vikhren. SW Bulgaria 41°45′N 23°24′E
93 L19 **Vihti** Uusimaa, S Finland 60°25′N 24°16′E
Viipuri *see* Vyborg
93 M16 **Viitasaari** Keski-Suomi, C Finland 63°05′N 25°52′E
118 K3 **Viivikonna** Ida-Virumaa, NE Estonia 59°19′N 27°41′E
155 K16 **Vijayawāda** *prev.* Bezwada. Andhra Pradesh, SE India 16°34′N 80°40′E
Vijosa/Vijosë *see* Aóos, Albania/Greece
Vijosa/Vijosë *see* Vjosës, Lumi i, Albania/Greece
92 J4 **Vík** Suðurland, S Iceland 63°25′N 18°58′W
Vik *see* Vikøyri
94 L13 **Vika** Dalarna, C Sweden 60°55′N 14°30′E
92 L12 **Vikajärvi** Lappi, N Finland 66°37′N 26°10′E
94 L13 **Vikarbyn** Dalarna, C Sweden 60°57′N 15°00′E
95 J22 **Viken** Skåne, S Sweden 56°09′N 12°36′E
95 L17 **Viken** C Sweden
95 G15 **Vikersund** Buskerud, S Norway 59°58′N 09°59′E
Vikhren *see* Vihren
11 R15 **Viking** Alberta, SW Canada 53°07′N 111°50′W
84 E7 **Viking Bank** *undersea feature* N North Sea 60°35′N 02°35′E
95 M14 **Vikmanshyttan** Dalarna, C Sweden 60°19′N 15°55′E
94 D12 **Vikøyri** *var.* Vik. Sogn Og Fjordane, S Norway 61°04′N 06°34′E
93 H17 **Viksjö** Västernorrland, C Sweden 62°45′N 17°30′E
Viktoriastadt *see* Victoria
Vila *see* Port-Vila
Vila Arriaga *see* Bibala
Vila Artur de Paiva *see* Cubango
Vila Baleira *see* Porto Santo
Vila Bela da Santissima Trindade *see* Mato Grosso
58 B12 **Vila Bittencourt** Amazonas, NW Brazil 01°25′S 69°24′W
Vila da Ponte *see* Cubango
64 O2 **Vila da Praia da Vitória** Terceira, Azores, Portugal, NE Atlantic Ocean 38°44′N 27°04′W
Vila de Aljustrel *see* Cangamba
Vila de Almoster *see* Chiange
Vila de João Belo *see* Xai-Xai
Vila de Macia *see* Macia
Vila de Manhiça *see* Manhiça
Vila de Manica *see* Manica
Vila de Mocímboa da Praia *see* Mocímboa da Praia
83 N16 **Vila de Sena** *var.* Sena. Sofala, C Mozambique 17°25′S 34°59′E
104 F14 **Vila do Bispo** Faro, S Portugal 37°05′N 08°53′W
104 G6 **Vila do Conde** Porto, NW Portugal 41°21′N 08°45′W
Vila do Maio *see* Maio
64 P3 **Vila do Porto** Santa Maria, Azores, Portugal, NE Atlantic Ocean 36°57′N 25°10′W
83 K15 **Vila do Zumbo** *prev.* Vila do Zumbu, Zumbo. Tete, NW Mozambique 15°36′S 30°30′E
Vila do Zumbu *see* Vila do Zumbo
105 T8 **Vilafamés** *prev.* Villafamés. Valenciana, E Spain 40°07′N 00°03′W
104 I6 **Vila Flor** *var.* Vila Flôr. Bragança, N Portugal 41°18′N 07°09′W
105 V6 **Vilafranca del Penedès** *var.* Villafranca del Panadés. Cataluña, NE Spain 41°21′N 01°42′E
104 F10 **Vila Franca de Xira** *var.* Vilafranca de Xira. Lisboa, C Portugal 38°57′N 08°59′W
Vila Gago Coutinho *see* Lumbala N'Guimbo
104 G3 **Vilagarcía** *var.* Villagarcía de Arosa. Galicia, NW Spain 42°35′N 08°45′W
Vila General Machado *see* Camacupa
Vila Henrique de Carvalho *see* Saurimo
102 I7 **Vilaine** NW France
Vila João de Almeida *see* Chibia
118 K8 **Viļaka** *Ger.* Marienhausen. NE Latvia 57°12′N 27°43′E
104 I2 **Vilalba** Galicia, NW Spain 43°17′N 07°41′W
Vila Marechal Carmona *see* Uíge
Vila Mariano Machado *see* Ganda
172 G3 **Vilanandro, Tanjona** *prev./ Fr.* Cap Saint-André. *headland* W Madagascar 16°10′S 44°27′E
Vilanculos *see* Vilankulo
118 J10 **Viļāni** E Latvia 56°33′N 26°55′E
83 N19 **Vilankulo** *var.* Vilanculos. Inhambane, E Mozambique 22°01′S 35°19′E
Vila Norton de Matos *see* Balombo
104 G6 **Vila Nova de Famalicão** *var.* Vila Nova de Famalicao. Braga, N Portugal 41°24′N 08°31′W
104 I6 **Vila Nova de Foz Côa** *var.* Vila Nova de Fozcôa. Guarda, N Portugal 41°05′N 07°09′W
Vila Nova de Fozcôa *see* Vila Nova de Foz Côa
104 F6 **Vila Nova de Gaia** Porto, NW Portugal 41°08′N 08°37′W
Vila Nova de Portimão *see* Portimão
105 V6 **Vilanova i La Geltru** Cataluña, NE Spain 41°15′N 01°42′E
Vila Pereira de Eça *see* Ondjiva
104 H6 **Vila Pouca de Aguiar** Vila Real, N Portugal 41°30′N 07°38′W
104 H6 **Vila Real** *var.* Vila Rial. Vila Real, N Portugal 41°17′N 07°45′W
105 T9 **Vila-real** *var.* Vila-real de los Infantes; *prev.* Villarreal. Valenciana, E Spain 39°56′N 00°08′W
104 H6 **Vila Real** *district* N Portugal
Vila-real de los Infantes *see* Vila-real
104 H14 **Vila Real de Santo António** Faro, S Portugal 37°12′N 07°25′W
104 J7 **Vilar Formoso** Guarda, N Portugal 40°37′N 06°50′W
Vila Rial *see* Vila Real
59 J15 **Vila Rica** Mato Grosso, W Brazil 09°52′S 50°44′W
Vila Robert Williams *see* Caála
Vila Salazar *see* N'Dalatando
Vila Serpa Pinto *see* Menongue
Vila Teixeira da Silva *see* Bailundo
Vila Teixeira de Sousa *see* Luau
104 H9 **Vila Velha de Ródão** Castelo Branco, C Portugal 39°39′N 07°40′W
104 G5 **Vila Verde** Braga, N Portugal 41°39′N 08°27′W
104 H11 **Vila Viçosa** Évora, S Portugal 38°46′N 07°25′W
57 G15 **Vilcabamba, Cordillera de** C Peru
Vilcea *see* Vâlcea
122 J4 **Vil'cheka, Zemlya** *Eng.* Wilczek Land. *island* Zemlya Frantsa-Iosifa, NW Russian Federation
95 F22 **Vildbjerg** Midtjylland, C Denmark 56°12′N 08°47′E
Vileyka *see* Vilyeyka
93 H15 **Vilhelmina** Västerbotten, N Sweden 64°38′N 16°40′E
59 F17 **Vilhena** Rondônia, W Brazil 12°40′S 60°08′W
115 G19 **Vília** Attikí, C Greece 38°09′N 23°21′E
119 I14 **Viliya** *Lith.* Neris. W Belarus
Viliya *see* Neris
118 H5 **Viljandi** *Ger.* Fellin. Viljandimaa, S Estonia 58°22′N 25°30′E
118 H5 **Viljandimaa** *var.* Viljandi Maakond. *province* SW Estonia
Viljandi Maakond *see* Viljandimaa
119 E14 **Vilkaviškis** *Pol.* Wyłkowyszki. Marijampolė, SW Lithuania 54°39′N 23°03′E
118 F13 **Vilkija** Kaunas, C Lithuania 55°02′N 23°36′E
197 V9 **Vil'kitskogo, Proliv** *strait* N Russian Federation
Vilkovo *see* Vylkove
57 L21 **Villa Abecia** Chuquisaca, S Bolivia 21°00′S 65°18′W
41 N5 **Villa Acuña** *var.* Ciudad Acuña. Coahuila, NE Mexico 29°18′N 100°58′W
40 J4 **Villa Ahumada** Chihuahua, N Mexico 30°38′N 106°30′W
45 O9 **Villa Altagracia** C Dominican Republic 18°43′N 70°13′W
56 L13 **Villa Bella** El Beni, N Bolivia 10°21′S 65°25′W
104 J3 **Villablino** Castilla y León, N Spain 42°55′N 06°21′W
54 K6 **Villa Bruzual** Portuguesa, N Venezuela 09°20′N 69°06′W
105 O9 **Villacañas** Castilla-La Mancha, C Spain 39°38′N 03°20′W
105 O12 **Villacarrillo** Andalucía, S Spain 38°07′N 03°05′W
104 M7 **Villacastín** Castilla y León, N Spain 40°46′N 04°25′W
Villa Cecilia *see* Ciudad Madero
109 S9 **Villach** *Slvn.* Beljak. Kärnten, S Austria 46°36′N 13°49′E
107 B20 **Villacidro** Sardegna, Italy, C Mediterranean Sea 39°28′N 08°43′E
Villa Concepción *see* Concepción
104 L4 **Villada** Castilla y León, N Spain 42°15′N 04°59′W
40 M10 **Villa de Cos** Zacatecas, C Mexico 23°20′N 102°20′W
54 L5 **Villa de Cura** *var.* Cura. Aragua, N Venezuela 10°00′N 67°30′W
Villa del Nevoso *see* Ilirska Bistrica
Villa del Pilar *see* Pilar
104 M13 **Villa del Río** Andalucía, S Spain 37°59′N 04°17′W
Villa de Méndez *see* Méndez
42 H6 **Villa de San Antonio** Comayagua, W Honduras 14°24′N 87°37′W
105 N4 **Villadiego** Castilla y León, N Spain 42°31′N 04°01′W
Villafamés *see* Vilafamés
41 U16 **Villa Flores** Chiapas, SE Mexico 16°12′N 93°16′W
104 J3 **Villafranca del Bierzo** Castilla y León, N Spain 42°36′N 06°49′W
105 S8 **Villafranca del Cid** Valenciana, E Spain 40°25′N 00°15′W
104 J11 **Villafranca de los Barros** Extremadura, W Spain 38°34′N 06°20′W
105 N10 **Villafranca de los Caballeros** Castilla-La Mancha, C Spain 39°26′N 03°21′W
Villafranca del Panadés *see* Vilafranca del Penedès
106 F8 **Villafranca di Verona** Veneto, NE Italy 45°22′N 10°51′E
107 J23 **Villafrati** Sicilia, Italy, C Mediterranean Sea 37°53′N 13°30′E
Villagarcía de Arosa *see* Vilagarcía
41 O9 **Villagrán** Tamaulipas, C Mexico 24°29′N 99°30′W
61 C17 **Villaguay** Entre Ríos, E Argentina 31°55′S 59°01′W
62 O6 **Villa Hayes** Presidente Hayes, S Paraguay 25°05′S 57°25′W
41 U15 **Villahermosa** *prev.* San Juan Bautista. Tabasco, SE Mexico 17°56′N 92°50′W
105 O11 **Villahermosa** Castilla-La Mancha, C Spain 38°46′N 02°52′W
64 O11 **Villahermoso** Gomera, Islas Canarias, Spain, NE Atlantic Ocean 38°46′N 02°52′W
Villa Hidalgo *see* Hidalgo
105 T12 **Villajoyosa** *Cat.* La Vila Joíosa. Valenciana, E Spain 38°31′N 00°14′W
Villa Juárez *see* Juárez
Villalba *see* Collado Villalba
41 N8 **Villaldama** Nuevo León, NE Mexico 26°29′N 100°27′W
104 L5 **Villalón de Campos** Castilla y León, N Spain 42°05′N 05°03′W
61 A25 **Villalonga** Buenos Aires, E Argentina 39°55′S 62°35′W
104 L5 **Villalpando** Castilla y León, N Spain 41°51′N 05°25′W
40 K9 **Villa Madero** *var.* Francisco I. Madero. Durango, C Mexico 24°28′N 104°20′W
41 O9 **Villa Mainero** Tamaulipas, C Mexico 24°32′N 99°39′W
Villamaña *see* Villamañán
104 L4 **Villamañán** *var.* Villamaña. Castilla y León, N Spain 42°19′N 05°35′W
62 L10 **Villa María** Córdoba, C Argentina 32°23′S 63°15′W
61 C17 **Villa María Grande** Entre Ríos, E Argentina 31°39′S 59°54′W
57 K21 **Villa Martín** Potosí, SW Bolivia 20°46′S 67°45′W
104 K15 **Villamartín** Andalucía, S Spain 36°52′N 05°38′W
62 J8 **Villa Mazán** La Rioja, NW Argentina 28°43′S 66°25′W
62 J11 **Villa Mercedes** *var.* Mercedes. San Luis, C Argentina 33°40′S 65°25′W
Villamil *see* Puerto Villamil
Villa Nador *see* Nador
54 G5 **Villanueva** La Guajira, N Colombia 10°37′N 72°58′W
42 H5 **Villanueva** Cortés, NW Honduras 15°14′N 88°00′W
40 L11 **Villanueva** Zacatecas, C Mexico 22°24′N 102°53′W
42 I9 **Villa Nueva** Chinandega, NW Nicaragua 12°58′N 86°46′W
37 T11 **Villanueva** New Mexico, SW USA 35°18′N 105°20′W
104 M12 **Villanueva de Córdoba** Andalucía, S Spain 38°20′N 04°38′W
105 O12 **Villanueva del Arzobispo** Andalucía, S Spain 38°10′N 03°00′W
104 K11 **Villanueva de la Serena** Extremadura, W Spain 38°58′N 05°48′W
104 L5 **Villanueva del Campo** Castilla y León, N Spain 41°59′N 05°25′W
105 O11 **Villanueva de los Infantes** Castilla-La Mancha, C Spain 38°45′N 03°01′W
61 C14 **Villa Ocampo** Santa Fe, C Argentina 28°28′S 59°22′W
40 J8 **Villa Ocampo** Durango, C Mexico 26°29′N 105°30′W
40 J7 **Villa Orestes Pereyra** Durango, C Mexico 26°30′N 105°38′W
105 N3 **Villarcayo** Castilla y León, N Spain 42°56′N 03°34′W
104 L5 **Villardefrades** Castilla y León, N Spain 41°43′N 05°15′W
105 S9 **Villar del Arzobispo** Valenciana, E Spain 39°44′N 00°50′W
105 Q6 **Villaroya de la Sierra** Aragón, NE Spain 41°28′N 01°46′W
Villarreal *see* Vila-real
62 P6 **Villarrica** Guairá, SE Paraguay 25°45′S 56°28′W
63 G15 **Villarrica, Volcán** S Chile 39°28′S 71°57′W
105 P10 **Villarrobledo** Castilla-La Mancha, C Spain 39°16′N 02°36′W
105 N10 **Villarrubia de los Ojos** Castilla-La Mancha, C Spain 39°14′N 03°36′W
18 J17 **Villas** New Jersey, NE USA 39°01′N 74°54′W
105 O3 **Villasana de Mena** Castilla y León, N Spain 43°05′N 03°16′W
107 M23 **Villa San Giovanni** Calabria, S Italy 38°12′N 15°39′E
61 D18 **Villa San José** Entre Ríos, E Argentina 32°01′S 58°20′W
Villa Sanjurjo *see* Al-Hoceïma
105 P6 **Villasayas** Castilla y León, N Spain 41°19′N 02°36′W
107 C20 **Villasimius** Sardegna, Italy, C Mediterranean Sea 39°10′N 09°30′E
41 N6 **Villa Unión** Coahuila, NE Mexico 28°18′N 100°43′W
40 K10 **Villa Unión** Durango, C Mexico 23°59′N 104°01′W
40 J10 **Villa Unión** Sinaloa, C Mexico 23°10′N 106°12′W
62 K12 **Villa Valeria** Córdoba, C Argentina 34°21′S 64°56′W
105 N8 **Villaverde** Madrid, C Spain 40°21′N 03°43′W
54 F10 **Villavicencio** Meta, C Colombia 04°09′N 73°38′W
104 L2 **Villaviciosa** Asturias, N Spain 43°29′N 05°26′W
104 L12 **Villaviciosa de Córdoba** Andalucía, S Spain 38°04′N 05°00′W
57 L22 **Villazón** Potosí, S Bolivia 22°05′S 65°35′W
14 J8 **Villebon, Lac** Québec, SE Canada
Ville de Kinshasa *see* Kinshasa
102 J5 **Villedieu-les-Poêles** Manche, N France 48°51′N 01°12′W
Villefranche *see* Villefranche-sur-Saône

◆ Country ● Country Capital ◇ Dependent Territory ○ Dependent Territory Capital ◈ Administrative Regions ✈ International Airport ▲ Mountain ▲▲ Mountain Range ⛰ Volcano ~ River ◎ Lake ▣ Reservoir

VILLEFRANCHE-DE-LAURAGAIS – VUČITRN

103 N16 Villefranche-de-Lauragais Haute-Garonne, S France 43°24′N 01°42′E
103 N14 Villefranche-de-Rouergue Aveyron, S France 44°21′N 02°02′E
103 R10 Villefranche-sur-Saône var. Villefranche. Rhône, E France 46°00′N 04°40′E
14 H9 Ville-Marie Québec, SE Canada 47°21′N 79°26′W
102 M15 Villemur-sur-Tarn Haute-Garonne, S France 43°50′N 01°32′E
105 S11 Villena Valenciana, E Spain 38°39′N 00°52′W
Villeneuve-d'Agen see Villeneuve-sur-Lot
102 L13 Villeneuve-sur-Lot var. Villeneuve-d'Agen, hist. Gajac. Lot-et-Garonne, SW France 44°24′N 00°43′E
103 P6 Villeneuve-sur-Yonne Yonne, C France 48°04′N 03°21′E
22 H8 Ville Platte Louisiana, S USA 30°41′N 92°16′W
103 R11 Villeurbanne Rhône, E France 45°46′N 04°54′E
101 G23 Villingen-Schwenningen Baden-Württemberg, S Germany 48°04′N 08°27′E
29 T15 Villisca Iowa, C USA 40°55′N 94°58′W
Villmanstrand see Lappeenranta
Vilna see Vilnius
119 H14 Vilnius Pol. Wilno, Ger. Wilna; prev. Rus. Vilna. ● (Lithuania) Vilnius, SE Lithuania 54°41′N 25°20′E
119 H14 Vilnius ✈ Vilnius, SE Lithuania 54°33′N 25°17′E
117 S7 Vil'nohirs'k Dnipropetrovs'ka Oblast', E Ukraine 48°31′N 34°01′E
117 U8 Vil'nyans'k Zaporiz'ka Oblast', SE Ukraine 47°56′N 35°22′E
93 L17 Vilppula Pirkanmaa, W Finland 62°02′N 24°30′E
101 M20 Vils ⇝ SE Germany
118 C5 Vilsandi island W Estonia
117 P8 Vil'shanka Rus. Olshanka. Kirovohrads'ka Oblast', C Ukraine 48°12′N 30°54′E
101 O22 Vilshofen Bayern, SE Germany 48°36′N 13°10′E
155 J20 Viluppuram Tamil Nādu, SE India 12°54′N 79°40′E
113 I16 Vilusi W Montenegro 42°44′N 18°34′E
99 G18 Vilvoorde Fr. Vilvorde. Vlaams Brabant, C Belgium 50°56′N 04°25′E
Vilvorde see Vilvoorde
119 J14 Vilyeyka Pol. Wilejka, Rus. Vileyka. Minskaya Voblasts', NW Belarus 54°30′N 26°55′E
122 V11 Vilyuchinsk Kamchatskiy Kray, E Russian Federation 52°55′N 158°28′E
123 P10 Vilyuy ⇝ NE Russian Federation
123 P10 Vilyuysk Respublika Sakha (Yakutiya), NE Russian Federation 63°42′N 121°20′E
123 N10 Vilyuyskoye Vodokhranilishche ▣ NE Russian Federation
104 G2 Vimianzo Galicia, NW Spain 43°06′N 09°03′W
95 M19 Vimmerby Kalmar, S Sweden 57°40′N 15°50′E
102 L5 Vimoutiers Orne, N France 48°56′N 00°10′E
93 L16 Vimpeli Etelä-Pohjanmaa, W Finland 63°10′N 23°50′E
79 G14 Vina ⇝ Cameroon/Chad
62 G11 Viña del Mar Valparaíso, C Chile 33°02′S 71°35′W
19 R8 Vinalhaven Island island Maine, NE USA
105 T8 Vinaròs Valenciana, E Spain 40°29′N 00°28′E
Vînători see Vânători
31 N15 Vincennes Indiana, N USA 38°42′N 87°30′W
195 Y12 Vincennes Bay bay Antarctica
25 O7 Vincent Texas, SW USA 32°30′N 101°10′W
95 H24 Vindeby Syddtjylland, C Denmark 54°55′N 11°09′E
93 I15 Vindeln Västerbotten, N Sweden 64°11′N 19°45′E
95 F21 Vinderup Midtjylland, C Denmark 56°29′N 08°48′E
Vindhya Mountains see Vindhya Range
153 N14 Vindhya Range var. Vindhya Mountains. ▲ N India
Vindobona see Wien
20 K6 Vine Grove Kentucky, S USA 37°48′N 85°58′W
18 J17 Vineland New Jersey, NE USA 39°29′N 75°02′W
116 E11 Vinga Arad, W Romania 46°00′N 21°14′E
95 M16 Vingåker Södermanland, C Sweden 59°02′N 15°52′E
167 S8 Vinh Nghê An, N Vietnam 18°42′N 105°41′E
104 I5 Vinhais Bragança, N Portugal 41°50′N 07°00′W
Vinh Linh see Hô Xa
Vinh Loi see Bac Liêu
167 S14 Vinh Long var. Vinhlong. Vinh Long, S Vietnam 10°15′N 105°59′E
Vinhlong see Vinh Long
113 Q18 Vinica NE FYR Macedonia 41°53′N 22°30′E
109 V13 Vinica SE Slovenia 45°28′N 15°12′E
114 G8 Vinishte Montana, NW Bulgaria 43°30′N 23°04′E
27 Q8 Vinita Oklahoma, C USA 36°38′N 95°09′W
Vînju Mare see Vânju Mare
98 I11 Vinkeveen Utrecht, C Netherlands 52°13′N 04°55′E
116 L6 Vin'kivtsi Khmel'nyts'ka Oblast', W Ukraine 49°02′N 27°13′E
112 I10 Vinkovci Ger. Winkowitz, Hung. Vinkovcze. Vukovar-Srijem, E Croatia 45°18′N 18°45′E
Vinkovcze see Vinkovci
Vinnitsa see Vinnytsya
Vinnitskaya Oblast'/Vinnytsya see Vinnyts'ka Oblast'
116 M7 Vinnyts'ka Oblast' var. Vinnytsya, Rus. Vinnitskaya Oblast'. ◆ province C Ukraine
117 N6 Vinnytsya Rus. Vinnitsa. Vinnyts'ka Oblast', C Ukraine 49°14′N 28°30′E
117 N6 Vinnytsya ✈ Vinnyts'ka Oblast', N Ukraine 49°13′N 28°40′E
Vinogradov see Vynohradiv
194 L8 Vinson Massif ▲ Antarctica 78°45′S 85°19′W
94 G11 Vinstra Oppland, S Norway 61°36′N 09°45′E
116 K12 Vintilă Vodă Buzău, SE Romania 45°28′N 26°43′E
29 X13 Vinton Iowa, C USA 42°10′N 92°01′W
22 F9 Vinton Louisiana, S USA 30°10′N 93°37′W
155 J17 Vinukonda Andhra Pradesh, E India 16°03′N 79°41′E
Vioara see Ocnele Mari
83 E23 Vioolsdrif Northern Cape, SW South Africa 28°50′S 17°38′E
82 M13 Viphya Mountains ▲ C Malawi
171 Q4 Virac Catanduanes Island, N Philippines 13°39′N 124°17′E
124 K8 Virandozero Respublika Kareliya, NW Russian Federation 63°59′N 36°00′E
137 P16 Viranşehir Şanlıurfa, SE Turkey 37°13′N 39°32′E
154 D13 Virār Mahārāshtra, W India 19°30′N 72°48′E
11 W16 Virden Manitoba, S Canada 49°50′N 100°57′W
30 K14 Virden Illinois, N USA 39°30′N 89°46′W
Virdois see Virrat
102 J5 Vire Calvados, N France 48°50′N 00°53′W
102 J4 Vire ⇝ N France
83 A15 Virei Namibe, SW Angola 15°43′S 12°54′E
Virful Moldoveanu see Vârful Moldoveanu
35 R5 Virgina Peak ▲ Nevada, W USA 39°46′N 119°26′W
45 U9 Virgin Gorda island C British Virgin Islands
83 I22 Virginia Free State, C South Africa 28°06′S 26°53′E
30 K13 Virginia Illinois, N USA 39°57′N 90°12′W
29 W4 Virginia Minnesota, N USA 47°31′N 92°32′W
21 T6 Virginia off. Commonwealth of Virginia, also known as Mother of Presidents, Mother of States, Old Dominion. ◆ state NE USA
21 Y7 Virginia Beach Virginia, NE USA 36°51′N 75°59′W
33 R11 Virginia City Montana, NW USA 45°17′N 111°54′W
35 Q6 Virginia City Nevada, W USA 39°19′N 119°39′W
14 H8 Virginiatown Ontario, S Canada 48°09′N 79°35′W
Virgin Islands see British Virgin Islands
45 T9 Virgin Islands (US) var. Virgin Islands of the United States; prev. Danish West Indies. ◇ US unincorporated territory E West Indies
Virgin Islands of the United States see Virgin Islands (US)
45 T9 Virgin Passage passage Puerto Rico/Virgin Islands (US)
35 Y10 Virgin River ⇝ Nevada/Utah, W USA
Virihaur see Virihaure
92 H12 Virihaure Lapp. Virihávrre, var. Virihaur. ◎ N Sweden
Virihávrre see Virihaure
167 T11 Virôchey Rôtânôkiri, NE Cambodia 13°59′N 106°49′E
93 N19 Virolahti Kymenlaakso, S Finland 60°33′N 27°37′E
30 J8 Viroqua Wisconsin, N USA 43°33′N 90°54′W
112 G8 Virovitica Ger. Virovititz, Hung. Verőcze; prev. Ger. Werowitz. Virovitica-Podravina, NE Croatia 45°49′N 17°25′E
112 G8 Virovitica-Podravina off. Virovitičko-Podravska Županija. ◆ province NE Croatia
Virovitičko-Podravska Županija see Virovitica-Podravina
Virovititz see Virovitica
113 J17 Virpazar S Montenegro 42°15′N 19°06′E
93 L17 Virrat Swe. Virdois. Pirkanmaa, W Finland 62°14′N 23°50′E
95 M20 Virserum Kalmar, S Sweden 57°17′N 15°18′E
99 K25 Virton Luxembourg, SE Belgium 49°34′N 05°32′E
118 F5 Virtsu Ger. Werder. Läänemaa, W Estonia 58°35′N 23°33′E
56 C12 Virú La Libertad, C Peru 08°24′S 78°40′W
Virudhunagar see Virudunagar
155 H23 Virudunagar var. Virudhunagar; prev. Virudupatti. Tamil Nādu, SE India 09°35′N 77°57′E
Virudupatti see Virudunagar
118 I3 Viru-Jaagupi Ger. Sankt-Jakobi. Lääne-Virumaa, NE Estonia 59°14′N 26°29′E
57 N19 Viru-Viru var. Santa Cruz. ✈ (Santa Cruz) Santa Cruz, C Bolivia 17°49′S 63°12′W
113 E15 Vis It. Lissa; anc. Issa. island S Croatia
Vis see Fish
118 I12 Visaginas prev. Sniečkus. Utena, E Lithuania 55°36′N 26°22′E
155 M15 Visākhapatnam var. Vishakhapatnam. Andhra Pradesh, SE India 17°45′N 83°19′E
35 R11 Visalia California, W USA 36°19′N 119°19′W
Visău see Vişeu
95 P19 Visby Ger. Wisby. Gotland, SE Sweden 57°37′N 18°20′E
197 N9 Viscount Melville Sound prev. Melville Sound. sound Northwest Territories, N Canada
99 L19 Visé Liège, E Belgium 50°44′N 05°42′E
112 K13 Višegrad Republika Srpska, SE Bosnia and Herzegovina 43°46′N 19°18′E
58 L12 Viseu Pará, NE Brazil 01°10′S 46°09′W
104 H7 Viseu prev. Vizeu. Viseu, N Portugal 40°40′N 07°55′W
104 H7 Viseu var. Vizeu. ◆ district N Portugal
116 I8 Vişeu Hung. Visó; prev. Vişău. ⇝ NW Romania
116 I8 Vişeu de Sus var. Vişeul de Sus, Ger. Oberwischau, Hung. Felsővisó. Maramureş, N Romania 47°43′N 23°24′E
Vişeul de Sus see Vişeu de Sus
Vishakhapatnam see Visākhapatnam
125 R10 Vishera ⇝ NW Russian Federation
95 J19 Viskafors Västra Götaland, S Sweden 57°37′N 12°50′E
95 J20 Viskan ⇝ S Sweden
95 L21 Vislanda Kronoberg, S Sweden 56°46′N 14°30′E
Vislinskiy Zaliv see Vistula Lagoon
Visó see Vişeu
112 H13 Visoko ◆ Federacija Bosne I Hercegovine, C Bosnia and Herzegovina
106 A9 Viso, Monte ▲ NW Italy 44°42′N 07°04′E
108 E10 Visp Valais, SW Switzerland 46°18′N 07°53′E
108 E10 Vispa ⇝ S Switzerland
95 M21 Vissefjärda Kalmar, S Sweden 56°31′N 15°34′E
100 I11 Visselhövede Niedersachsen, NW Germany 52°58′N 09°36′E
95 G23 Vissenbjerg Syddtjylland, C Denmark 55°23′N 10°08′E
35 U17 Vista California, W USA 33°12′N 117°14′W
58 C11 Vista Alegre Amazonas, NW Brazil 01°23′N 68°13′W
114 J13 Vistonída, Límni ◎ NE Greece
Vistula see Wisła
119 A14 Vistula Lagoon Ger. Frisches Haff, Pol. Zalew Wiślany, Rus. Vislinskiy Zaliv. lagoon Poland/Russian Federation
114 I8 Vit ⇝ NW Bulgaria
Vitebsk see Vitsyebsk
Vitebskaya Oblast' see Vitsyebskaya Voblasts'
107 H14 Viterbo anc. Vicus Elbii. Lazio, C Italy 42°25′N 12°08′E
112 H12 Vitez Federacija Bosne I Hercegovine, C Bosnia and Herzegovina
167 S14 Vi Thanh Cân Thơ, S Vietnam 09°45′N 105°28′E
Viti see Fiji
186 E7 Vitiaz Strait strait NE Papua New Guinea
104 J7 Vitigudino Castilla y León, N Spain 41°00′N 06°26′W
175 Q9 Viti Levu island W Fiji
187 W15 Viti Levu island W Fiji
123 O11 Vitim ⇝ C Russian Federation
123 O12 Vitimskiy Irkutskaya Oblast', C Russian Federation 58°12′N 113°10′E
109 V2 Vitis Niederösterreich, N Austria 48°45′N 15°09′E
59 O20 Vitória state capital Espírito Santo, SE Brazil 20°19′S 40°21′W
Vitoria see Vitoria-Gasteiz
Vitoria Bank see Vitória Seamount
59 N18 Vitória da Conquista Bahia, E Brazil 14°53′S 40°52′W
105 P3 Vitoria-Gasteiz var. Vitoria, Eng. Vittoria. País Vasco, N Spain 42°51′N 02°40′W
65 J16 Vitória Seamount var. Victoria Bank, Vitoria Bank. undersea feature C Atlantic Ocean 18°48′S 37°24′W
112 F13 Vitorog ▲ SW Bosnia and Herzegovina 44°06′N 17°03′E
102 J6 Vitré Ille-et-Vilaine, NW France 48°07′N 01°12′W
103 R5 Vitry-le-François Marne, N France 48°43′N 04°36′E
114 D13 Vitsi var. Vítsoi. ▲ N Greece 40°39′N 21°23′E
Vítsoi see Vítsi
118 N13 Vitsyebsk Rus. Vitebsk. Vitsyebskaya Voblasts', NE Belarus 55°11′N 30°10′E
118 K13 Vitsyebskaya Voblasts' Rus. Vitebskaya Oblast'. ◆ province N Belarus
92 J11 Vittangi Lapp. Vazáš. Norrbotten, N Sweden 67°40′N 21°39′E
103 R8 Vitteaux Côte d'Or, C France 47°24′N 04°31′E
103 S6 Vittel Vosges, NE France 48°13′N 05°57′E
95 N15 Vittinge Västmanland, C Sweden 59°52′N 17°04′E
107 K25 Vittoria Sicilia, Italy, C Mediterranean Sea 36°56′N 14°30′E
Vittoria see Vitoria-Gasteiz
106 I7 Vittorio Veneto Veneto, NE Italy 45°59′N 12°18′E
175 Q7 Vityaz Trench undersea feature W Pacific Ocean
108 G8 Vitznau Luzern, W Switzerland 47°01′N 08°28′E
104 I1 Viveiro Galicia, NW Spain 43°39′N 07°35′W
105 S9 Viver Valenciana, E Spain 39°55′N 00°36′W
103 Q13 Viverais, Monts du ▲ C France
122 L9 Vivi ⇝ N Russian Federation
22 F4 Vivian Louisiana, S USA 32°52′N 93°59′W
29 N10 Vivian South Dakota, N USA 43°53′N 100°16′W
103 R13 Viviers Ardèche, E France 44°31′N 04°40′E
Vivis see Vevey
83 K19 Vivo Limpopo, NE South Africa 22°58′S 29°13′E
102 L10 Vivonne Vienne, W France 46°25′N 00°15′E
Vizakna see Ocna Sibiului
Vizcaya see Bizkaia
Vizcaya, Golfo de see Biscay, Bay of
136 C10 Vize Kırklareli, NW Turkey 41°34′N 27°45′E
122 K4 Vize, Ostrov island Severnaya Zemlya, N Russian Federation
Vizeu see Viseu
155 M15 Vizianagaram var. Vizianagram. Andhra Pradesh, E India 18°07′N 83°25′E
Vizianagram see Vizianagaram
103 S12 Vizille Isère, E France 45°05′N 05°46′E
125 R11 Vizinga Respublika Komi, NW Russian Federation 61°06′N 50°09′E
116 M13 Viziru Brăila, SE Romania 45°00′N 27°43′E
113 K21 Vjosës, Lumi i var. Vijosa, Vijosë, Gk. Aóos. ⇝ Albania/Greece see also Aóos
Vjosës, Lumi i see Aóos
99 H18 Vlaams Brabant ◆ province C Belgium
99 G18 Vlaanderen Eng. Flanders, Fr. Flandre. ◆ Belgium/France
98 G12 Vlaardingen Zuid-Holland, SW Netherlands 51°55′N 04°21′E
116 F10 Vlădeasa, Vârful prev. Vîrful Vlădeasa. ▲ NW Romania 46°45′N 22°46′E
Vlădeasa, Vîrful see Vlădeasa, Vârful
113 P16 Vladičin Han Serbia, SE Serbia 42°44′N 22°04′E
127 O16 Vladikavkaz prev. Dzaudzhikau, Ordzhonikidze. Respublika Severnaya Osetiya, SW Russian Federation 42°58′N 44°41′E
126 M3 Vladimir Vladimirskaya Oblast', W Russian Federation 56°09′N 40°21′E
Vladimirets see Volodymyrets'
144 M7 Vladimirovka Kostanay, N Kazakhstan 53°30′N 64°02′E
Vladimirovka see Yuzhno-Sakhalinsk
126 L3 Vladimirskaya Oblast' ◆ province W Russian Federation
126 I3 Vladimirskiy Tupik Smolenskaya Oblast', W Russian Federation 55°45′N 33°25′E
Vladimir-Volynskiy see Volodymyr-Volyns'kyy
123 Q7 Vladivostok Primorskiy Kray, SE Russian Federation 43°09′N 131°53′E
117 U13 Vladyslavivka Avtonomna Respublika Krym, S Ukraine 45°09′N 35°25′E
98 P6 Vlagtwedde Groningen, NE Netherlands 53°02′N 07°07′E
Vlajna see Kukavica
112 J12 Vlasenica ◆ Republika Srpska, E Bosnia and Herzegovina
112 G12 Vlašić ▲ C Bosnia and Herzegovina 44°18′N 17°40′E
111 D17 Vlašim Ger. Wlaschim. Středočeský Kraj, C Czech Republic 49°42′N 14°54′E
113 P15 Vlasotince Serbia, SE Serbia 42°58′N 22°07′E
123 Q7 Vlasovo Respublika Sakha (Yakutiya), NE Russian Federation 70°41′N 134°49′E
98 I11 Vleuten Utrecht, C Netherlands 52°07′N 05°01′E
98 I5 Vlieland Fris. Flylân. island Waddeneilanden, N Netherlands
98 I5 Vliestroom strait NW Netherlands
99 J14 Vlijmen Noord-Brabant, S Netherlands 51°42′N 05°14′E
99 E15 Vlissingen Eng. Flushing, Fr. Flessingue. Zeeland, SW Netherlands 51°26′N 03°34′E
Vlodava see Włodawa
Vlonë/Vlora see Vlorë
113 K22 Vlorë prev. Vlonë, It. Valona, Vlora. Vlorë, SW Albania 40°28′N 19°31′E
113 K22 Vlorë ◆ district SW Albania
113 K22 Vlorës, Gjiri i var. Valona Bay. bay SW Albania
Vlotslavsk see Włocławek
111 C16 Vltava Ger. Moldau. ⇝ W Czech Republic
126 K3 Vnukovo ✈ (Moskva) Gorod Moskva, W Russian Federation 55°30′N 36°52′E
146 L11 Vobkent Rus. Vabkent. Buxoro Viloyati, C Uzbekistan 40°01′N 64°25′E
25 Q9 Voca Texas, SW USA 30°58′N 99°09′W
109 R5 Vöcklabruck Oberösterreich, NW Austria 48°01′N 13°38′E
112 D13 Vodice Šibenik-Knin, S Croatia 43°46′N 15°46′E
124 K10 Vodlozero, Ozero ◎ NW Russian Federation
112 A10 Vodnjan It. Dignano d'Istria. Istra, NW Croatia 44°57′N 13°51′E
125 S9 Vodnyy Respublika Komi, NW Russian Federation 63°31′N 53°21′E
145 V15 Vodokhranilishche Kapshagay Kaz. Qapshagay Böyeni; prev. Kapchagayskoye Vodokhranilishche. ▣ SE Kazakhstan
124 I6 Vodokhranilishche, Kumskoye ▣ NW Russian Federation
95 G20 Vodskov Nordjylland, N Denmark 57°07′N 10°02′E
92 H4 Vogar Suðurnes, SW Iceland 63°58′N 22°20′W
Vogelkop see Doberai, Jazirah
77 X15 Vogel Peak prev. Dimlang. ▲ E Nigeria 08°16′N 11°44′E
101 H17 Vogelsberg ▲ C Germany
106 D8 Voghera Lombardia, N Italy 44°59′N 09°01′E
112 I13 Vogošća Federacija Bosne I Hercegovine, SE Bosnia and Herzegovina 43°55′N 18°20′E
101 M17 Vogtland historical region E Germany
125 V12 Vogul'skiy Kamen', Gora ▲ NW Russian Federation 60°10′N 58°41′E
187 P16 Voh Province Nord, C New Caledonia 20°57′S 164°41′E
Vohémar see Iharaña
172 H8 Vohimena, Tanjona Fr. Cap Sainte Marie. headland S Madagascar 25°20′S 45°06′E
172 J6 Vohipeno Fianarantsoa, SE Madagascar 22°21′S 47°51′E
118 H5 Võhma Ger. Wõchma. Viljandimaa, S Estonia 58°37′N 25°34′E
81 J20 Voi Taita/Taveta, S Kenya 03°23′S 38°35′E
76 K15 Voinjama N Liberia 08°25′N 09°42′W
103 S12 Voiron Isère, E France 45°22′N 05°35′E
109 V8 Voitsberg Steiermark, SE Austria 47°04′N 15°09′E
95 F24 Vojens Ger. Woyens. Syddanmark, SW Denmark 55°15′N 09°19′E
112 K9 Vojvodina Ger. Wojwodina. Vojvodina, N Serbia
15 S6 Volant ⇝ Québec, SE Canada
Volaterrae see Volterra
43 P15 Volcán var. Hato del Volcán. Chiriquí, W Panama 08°45′N 82°38′W
Volcano Islands see Kazan-rettō
Volchansk see Vovchans'k
Volchya see Vovcha
94 D10 Volda Møre og Romsdal, S Norway 62°07′N 06°04′E
116 K3 Voldymyerets' Russ. Vladimirets. Rivnens'ka Oblast', NW Ukraine 51°24′N 25°52′E
98 J9 Volendam Noord-Holland, C Netherlands 52°30′N 05°04′E
124 L15 Volga Yaroslavskaya Oblast', W Russian Federation 57°56′N 38°23′E
29 R10 Volga South Dakota, N USA 44°19′N 96°55′W
122 C11 Volga ⇝ NW Russian Federation
Volga-Baltic Waterway see Volgo-Baltiyskiy Kanal
Volga Uplands see Privolzhskaya Vozvyshennost'
124 L13 Volgo-Baltiyskiy Kanal var. Volga-Baltic Waterway. canal NW Russian Federation
126 M12 Volgodonsk Rostovskaya Oblast', SW Russian Federation 47°35′N 42°03′E
127 O10 Volgograd prev. Stalingrad, Tsaritsyn. Volgogradskaya Oblast', SW Russian Federation 48°42′N 44°29′E
127 N9 Volgogradskaya Oblast' ◆ province SW Russian Federation
127 P10 Volgogradskoye Vodokhranilishche ▣ SW Russian Federation
101 J19 Volkach Bayern, C Germany 49°51′N 10°15′E
109 U9 Völkermarkt Slvn. Velikovec. Kärnten, S Austria 46°40′N 14°38′E
124 I12 Volkhov Leningradskaya Oblast', NW Russian Federation 59°56′N 32°19′E
101 D20 Völklingen Saarland, SW Germany 49°15′N 06°51′E
Volkovysk see Vawkavysk
Volkovyskiye Vysoty see Vawkavyskaye Wzvyshsha
83 K22 Volksrust Mpumalanga, E South Africa 27°22′S 29°54′E
98 L8 Vollenhove Overijssel, N Netherlands 52°40′N 05°58′E
119 L16 Volma ⇝ C Belarus
Volmari see Valmiera
117 W9 Volnovakha Donets'ka Oblast', SE Ukraine 47°36′N 37°32′E
116 K6 Volochys'k Khmel'nyts'ka Oblast', W Ukraine 49°32′N 26°14′E
117 O6 Volodarka Kyyivs'ka Oblast', N Ukraine 49°31′N 29°55′E
117 W9 Volodars'ke Donets'ka Oblast', E Ukraine 47°11′N 37°19′E
127 R13 Volodarskiy Astrakhanskaya Oblast', SW Russian Federation 46°23′N 48°39′E
Volodarskoye see Saumalkol'
117 N8 Volodars'k-Volyns'kyy Zhytomyrs'ka Oblast', N Ukraine 50°37′N 28°28′E
116 I3 Volodymyr-Volyns'kyy Pol. Włodzimierz, Rus. Vladimir-Volynskiy. Volyns'ka Oblast', NW Ukraine 50°51′N 24°19′E
124 L14 Vologda Vologodskaya Oblast', W Russian Federation 59°10′N 39°55′E
124 L12 Vologodskaya Oblast' ◆ province NW Russian Federation
126 K3 Volokolamsk Moskovskaya Oblast', W Russian Federation 56°03′N 35°57′E
126 K9 Volokonovka Belgorodskaya Oblast', W Russian Federation 50°30′N 37°54′E
115 G16 Vólos Thessalía, C Greece 39°21′N 22°58′E
124 M11 Voloshka Arkhangel'skaya Oblast', NW Russian Federation 61°19′N 40°06′E
Vološinovo see Novi Bečej
116 H7 Volovets' Zakarpats'ka Oblast', W Ukraine 48°42′N 23°12′E
Volozhin see Valozhyn
127 Q7 Vol'sk Saratovskaya Oblast', W Russian Federation 52°04′N 47°20′E
77 Q17 Volta ⇝ SE Ghana
Volta Blanche see White Volta
77 P16 Volta, Lake ▣ SE Ghana
Volta Noire see Black Volta
60 O9 Volta Redonda Rio de Janeiro, SE Brazil 22°31′S 44°05′W
Volta Rouge see Red Volta
106 F12 Volterra anc. Volaterrae. Toscana, C Italy 43°23′N 10°52′E
107 K17 Volturno ⇝ S Italy
113 I15 Volujak ▲ NW Montenegro
Volunteer Island see Starbuck Island
65 F24 Volunteer Point headland East Falkland, Falkland Islands 51°32′S 57°44′W
114 H13 Vólvi, Límni ◎ N Greece
Volyn see Volyns'ka Oblast'
116 I3 Volyns'ka Oblast' var. Volyn, Rus. Volynskaya Oblast'. ◆ province NW Ukraine
Volynskaya Oblast' see Volyns'ka Oblast'
127 Q3 Volzhsk Respublika Mariy El, W Russian Federation 55°53′N 48°21′E
127 O10 Volzhskiy Volgogradskaya Oblast', SW Russian Federation 48°49′N 44°40′E
172 I7 Vondrozo Fianarantsoa, SE Madagascar 22°50′S 47°20′E
39 P10 Von Frank Mountain ▲ Alaska, USA 63°36′N 154°29′W
115 C17 Vónitsa Dytikí Elláda, W Greece 38°55′N 20°53′E
118 J6 Võnnu Ger. Wendau. Tartumaa, SE Estonia 58°17′N 27°06′E
98 G12 Voorburg Zuid-Holland, W Netherlands 52°04′N 04°22′E
98 H11 Voorschoten Zuid-Holland, W Netherlands 52°08′N 04°26′E
98 M11 Voorst Gelderland, E Netherlands 52°10′N 06°10′E
98 K11 Voorthuizen Gelderland, C Netherlands 52°12′N 05°36′E
92 L2 Vopnafjördhur bay E Iceland
92 L2 Vopnafjörður Austurland, E Iceland 65°45′N 14°51′W
Vora see Vorë
119 H15 Voranava Pol. Werenów, Rus. Voronovo. Hrodzyenskaya Voblasts', W Belarus 54°09′N 25°19′E
108 I8 Vorarlberg off. Land Vorarlberg. ◆ state W Austria
Vorarlberg, Land see Vorarlberg
109 X7 Vorau Steiermark, E Austria 47°22′N 15°55′E
98 N11 Vorden Gelderland, E Netherlands 52°07′N 06°18′E
108 H9 Vorderrhein ⇝ SE Switzerland
95 I24 Vordingborg Sjælland, SE Denmark 55°01′N 11°55′E
92 J2 Vorðufell ▲ N Iceland 65°42′N 18°45′W
113 K19 Vorë var. Vora. Tiranë, W Albania 41°23′N 19°37′E
115 H17 Vóreies Sporádes var. Vóreioi Sporádes, Vórioi Sporádhes, Eng. Northern Sporades. island group E Greece
Vóreioi Sporádes see Vóreies Sporádes
115 J17 Vóreion Aigaíon Eng. Aegean North. ◆ region SE Greece
115 G18 Vóreios Evvoïkós Kólpos var. Voreiós Evvoïkós Kólpos. gulf E Greece
197 S16 Voring Plateau undersea feature N Norwegian Sea 67°00′N 04°00′E
Vórioi Sporádhes see Vóreies Sporádes
125 W4 Vorkuta Respublika Komi, NW Russian Federation 67°27′N 64°E
95 I14 Vorma ⇝ S Norway
118 E4 Vormsi var. Vormsi Saar, Ger. Worms, Swed. Ormsö. island W Estonia
Vormsi Saar see Vormsi
127 N7 Vorona ⇝ W Russian Federation
126 L7 Voronezh Voronezhskaya Oblast', W Russian Federation 51°40′N 39°13′E
126 L7 Voronezh ⇝ W Russian Federation
126 K8 Voronezhskaya Oblast' ◆ province W Russian Federation
Voronovitsya see Voronovytsya
Voronovo see Voranava
117 N6 Voronovytsya Rus. Voronovitsya. Vinnyts'ka Oblast', C Ukraine 49°06′N 28°49′E
122 K7 Vorontsovo Krasnoyarskiy Kray, N Russian Federation 71°45′N 83°31′E
124 K3 Voron'ya ⇝ NW Russian Federation
Voropayevo see Varapayeva
Voroshilov see Ussuriysk
Voroshilovgrad see Luhans'ka Oblast', Ukraine
Voroshilovgrad see Luhans'k, Ukraine
Voroshilovgradskaya Oblast' see Luhans'ka Oblast'
Voroshilovsk see Stavropol', Russian Federation
Voroshilovsk see Alchevs'k
137 V13 Vorotan Az. Bărgușad. ⇝ Armenia/Azerbaijan
127 P3 Vorotynets Nizhegorodskaya Oblast', W Russian Federation 56°06′N 46°06′E
117 S3 Vorozhba Sums'ka Oblast', NE Ukraine 51°10′N 34°15′E
117 T5 Vorskla ⇝ Russian Federation/Ukraine
99 I17 Vorst Antwerpen, N Belgium 51°06′N 05°01′E
83 G21 Vorstershoop North-West, N South Africa 25°46′S 22°57′E
118 H6 Võrtsjärv Ger. Wirz-See. ◎ SE Estonia
118 J7 Võru Ger. Werro. Võrumaa, SE Estonia 57°51′N 27°01′E
147 R11 Vorukh N Tajikistan 39°51′N 70°34′E
118 I7 Võrumaa off. Võru Maakond. ◆ province SE Estonia
Võru Maakond see Võrumaa
83 G24 Vosburg Northern Cape, W South Africa 30°35′S 22°52′E
147 Q14 Vose' Rus. Vose; prev. Aral. SW Tajikistan 37°51′N 69°31′E
103 S6 Vosges ◆ department NE France
103 U6 Vosges ▲ NE France
124 K13 Voskresenkoye Vologodskaya Oblast', NW Russian Federation 59°25′N 37°56′E
126 L4 Voskresensk Moskovskaya Oblast', W Russian Federation 55°19′N 38°42′E
127 P2 Voskresenskoye Nizhegorodskaya Oblast', W Russian Federation 57°00′N 45°33′E
127 V6 Voskresenskoye Respublika Bashkortostan, W Russian Federation 53°07′N 56°07′E
94 D13 Voss Hordaland, S Norway 60°38′N 06°25′E
94 D13 Voss physical region S Norway
99 I16 Vosselaar Antwerpen, N Belgium 51°19′N 04°55′E
94 D13 Vosso ⇝ S Norway
Vostochno-Kazakhstanskaya Oblast' see Vostochnyy Kazakhstan
123 S5 Vostochno-Sibirskoye More Eng. East Siberian Sea. sea Arctic Ocean
145 X10 Vostochnyy Kazakhstan off. Vostochno-Kazakhstanskaya Oblast', var. East Kazakhstan, Kaz. Shyghys Qazaqstan Oblysy. ◆ province E Kazakhstan
122 L13 Vostochnyy Sayan Eng. Eastern Sayans, Mong. Dzüün Soyonï Nuruu. ▲ Mongolia/Russian Federation
Vostock Island see Vostok Island
195 U10 Vostok Russian research station Antarctica 77°18′S 105°32′E
191 X5 Vostok Island var. Vostock Island; prev. Stavers Island. island Line Islands, SE Kiribati
127 T2 Votkinsk Udmurtskaya Respublika, NW Russian Federation 57°04′N 54°00′E
125 U15 Votkinskoye Vodokhranilishche var. Votkinsk Reservoir. ▣ NW Russian Federation
Votkinsk Reservoir see Votkinskoye Vodokhranilishche
60 J7 Votuporanga São Paulo, S Brazil 20°26′S 49°53′W
104 H7 Vouga, Rio ⇝ N Portugal
115 E14 Voúrinos ▲ N Greece
115 G24 Voúxa, Akrotírio headland Kríti, Greece, E Mediterranean Sea 35°37′N 23°34′E
103 R4 Vouziers Ardennes, N France 49°24′N 04°42′E
117 V7 Vovcha Rus. Volchya. ⇝ E Ukraine
117 V4 Vovchans'k Rus. Volchansk. Kharkivs'ka Oblast', E Ukraine 50°19′N 36°55′E
103 N6 Voves Eure-et-Loir, C France 48°18′N 01°39′E
79 M14 Vovodo ⇝ S Central African Republic
94 M12 Voxna Gävleborg, C Sweden 61°21′N 15°35′E
94 L11 Voxnan ⇝ C Sweden
114 F7 Voynishka Reka ⇝ NW Bulgaria
125 T9 Voyvozh Respublika Komi, NW Russian Federation 62°54′N 54°52′E
124 M12 Vozhega Vologodskaya Oblast', NW Russian Federation 60°27′N 40°11′E
124 L12 Vozhe, Ozero ◎ NW Russian Federation
117 Q9 Voznesens'k Rus. Voznesensk. Mykolayivs'ka Oblast', S Ukraine 47°34′N 31°21′E
124 J12 Voznesen'ye Leningradskaya Oblast', NW Russian Federation 61°00′N 35°24′E
144 J14 Vozrozhdeniya, Ostrov Uzb. Wozrojdeniye Oroli. island Kazakhstan/Uzbekistan
95 G20 Vrå var. Vraa. Nordjylland, N Denmark 57°21′N 09°57′E
Vraa see Vrå
114 H9 Vrachesh Sofia, W Bulgaria 42°52′N 23°45′E
115 C19 Vrachíonas ▲ Zákynthos, Iónia Nisiá, Greece, C Mediterranean Sea 37°49′N 20°43′E
117 P8 Vradiyivka Mykolayivs'ka Oblast', S Ukraine 47°51′N 30°37′E
113 G14 Vran ▲ SW Bosnia and Herzegovina 43°35′N 17°30′E
116 K12 Vrancea ◆ county E Romania
147 T14 Vrang SE Tajikistan 37°03′N 72°26′E
123 T4 Vrangelya, Ostrov Eng. Wrangel Island. island NE Russian Federation
112 H13 Vranica ▲ C Bosnia and Herzegovina 43°57′N 17°43′E
113 O16 Vranje Serbia, SE Serbia 42°33′N 21°55′E
Vranov see Vranov nad Topl'ou
111 N19 Vranov nad Topl'ou var. Vranov, Hung. Varannó. Prešovský Kraj, E Slovakia 48°54′N 21°41′E
114 H8 Vratsa Vratsa, NW Bulgaria 43°13′N 23°34′E
114 H8 Vratsa ◆ province NW Bulgaria
114 F10 Vrattsa prev. Mirovo. Kyustendil, W Bulgaria 42°15′N 22°33′E
112 G11 Vrbanja ⇝ NW Bosnia and Herzegovina
112 K9 Vrbas Vojvodina, NW Serbia 45°34′N 19°39′E
112 G13 Vrbas ⇝ N Bosnia and Herzegovina
112 E8 Vrbovec Zagreb, N Croatia 45°53′N 16°24′E
112 C9 Vrbovsko Primorje-Gorski Kotar, NW Croatia 45°22′N 15°06′E
111 E15 Vrchlabí Ger. Hohenelbe. Královéhradecký Kraj, N Czech Republic 50°38′N 15°35′E
83 J22 Vrede Free State, E South Africa 27°25′S 29°10′E
100 E13 Vreden Nordrhein-Westfalen, NW Germany 52°01′N 06°50′E
83 E25 Vredenburg Western Cape, SW South Africa 32°55′S 18°00′E
99 I23 Vresse-sur-Semois Namur, SE Belgium 49°52′N 04°56′E
95 L16 Vretstorp Örebro, C Sweden 59°03′N 14°51′E
113 G15 Vrgorac prev. Vrhgorac. Split-Dalmacija, SE Croatia 43°10′N 17°24′E
Vrhgorac see Vrgorac
109 T12 Vrhnika Ger. Oberlaibach. W Slovenia 45°57′N 14°18′E
155 I21 Vriddhāchalam Tamil Nādu, SE India 11°33′N 79°18′E
98 N6 Vries Drenthe, NE Netherlands 53°04′N 06°34′E
98 O10 Vriezenveen Overijssel, E Netherlands 52°25′N 06°37′E
95 L20 Vrigstad Jönköping, S Sweden 57°19′N 14°30′E
108 H9 Vrin Graubünden, S Switzerland 46°40′N 09°06′E
112 E13 Vrlika Split-Dalmacija, S Croatia 43°54′N 16°24′E
113 M14 Vrnjačka Banja Serbia, C Serbia 43°36′N 20°55′E
Vrondádhes/Vrondados see Vrontádos
115 L18 Vrontádos var. Vrondados; prev. Vrondádhes. Chíos, E Greece 38°25′N 26°08′E
98 N9 Vroomshoop Overijssel, E Netherlands 52°28′N 06°35′E
112 N10 Vršac Ger. Werschetz, Hung. Versecz. Vojvodina, NE Serbia 45°08′N 21°18′E
112 M10 Vršački Kanal canal N Serbia
83 H21 Vryburg North-West, N South Africa 26°57′S 24°44′E
83 K22 Vryheid KwaZulu/Natal, E South Africa 27°45′S 30°48′E
111 I18 Vsetín Ger. Wsetin. Zlínský Kraj, E Czech Republic 49°21′N 17°57′E
111 J20 Vtáčnik Hung. Madaras, Ptacsnik; prev. Ptačnik. ▲ W Slovakia 48°38′N 18°38′E
Vuadil' see Wodil
Vūcha see Vacha
Vučitrn see Vushtrri

◆ Country
● Country Capital
◇ Dependent Territory
○ Dependent Territory Capital
◆ Administrative Regions
✈ International Airport
▲ Mountain
▲▲ Mountain Range
Volcano
⇝ River
◎ Lake
▣ Reservoir

VUGHT – WARRAP

99 J14 **Vught** Noord-Brabant, S Netherlands 51°37′N 05°19′E
117 W8 **Vuhledar** Donets'ka Oblast', E Ukraine 47°48′N 37°11′E
112 I9 **Vuka** ≈ E Croatia
113 K17 **Vukël** *var.* Vukli. Shkodër, N Albania 42°29′N 19°39′E
Vukli *see* Vukël
112 J9 **Vukovar** *Hung.* Vukovár. Vukovar-Srijem, E Croatia 45°18′N 18°45′E
Vukovarsko-Srijemska Županija *see* Vukovar-Srijem
112 I10 **Vukovar-Srijem** *off.* Vukovarsko-Srijemska Županija. ◈ *province* E Croatia
125 U8 **Vuktyl** Respublika Komi, NW Russian Federation 63°49′N 57°07′E
11 Q17 **Vulcan** Alberta, SW Canada 50°27′N 113°12′W
116 G12 **Vulcan** *Ger.* Wulkan, *Hung.* Zsilyvajdevulkán; *prev.* Crivadia Vulcanului, Vaidei, *Hung.* Sily-Vajdej, Vajdej. Hunedoara, W Romania 45°22′N 23°16′E
116 M12 **Vulcănești** *Rus.* Vulkaneshty. S Moldova 45°41′N 28°25′E
107 L22 **Vulcano, Isola** *island* Isole Eolie, S Italy
Vŭlchedrŭm *see* Valchedram
Vŭlchidol *see* Valchi Dol
Vulkaneshty *see* Vulcănești
123 V11 **Vulkannyy** Kamchatskiy Kray, E Russian Federation 53°01′N 158°26′E
36 J13 **Vulture Mountains** ▲▲ Arizona, SW USA
167 T14 **Vung Tau** *prev. Fr.* Cape Saint Jacques, Cap Saint-Jacques. Ba Ria-Vung Tau, S Vietnam 10°21′N 107°04′E
187 X15 **Vunisea** Kadavu, SE Fiji 19°04′S 178°10′E
Vuohčču *see* Vuotso
93 N15 **Vuokatti** Kainuu, C Finland 64°08′N 28°16′E
93 M15 **Vuolijoki** Kainuu, C Finland 64°09′N 27°00′E
Vuolleriebme *see* Vuollerim
92 J13 **Vuollerim** *Lapp.* Vuolleriebme. Norrbotten, N Sweden 66°24′N 20°36′E
Vuonnabahta *see* Varangerbotn
Vuoreija *see* Vardø
92 L10 **Vuotso** *Lapp.* Vuohčču. Lappi, N Finland 68°04′N 27°05′E
Vŭrbitsa *see* Varbitsa
127 Q4 **Vurnary** Chuvashskaya Respublika, W Russian Federation 55°30′N 46°59′E
Vŭrshets *see* Varshets
Vusan *see* Busan
113 N16 **Vushtrri** *Serb.* Vučitrn. N Kosovo 42°49′N 21°00′E
119 F17 **Vyalikaya Byerastavitsa** *Pol.* Brzostowica Wielka, *Rus.* Bol'shaya Berëstovitsa; *prev.* Velikaya Berestovitsa. Hrodzyenskaya Voblasts', SW Belarus 53°12′N 24°03′E
119 N20 **Vyaliki Bor** *Rus.* Velikiy Bor. Homyel'skaya Voblasts', SE Belarus 52°02′N 29°56′E
119 J18 **Vyaliki Rozhan** *Rus.* Bol'shoy Rozhan. Minskaya Voblasts', S Belarus 52°46′N 27°07′E
124 H10 **Vyartsilya** *Fin.* Värtsilä. Respublika Kareliya, NW Russian Federation 62°07′N 30°43′E
119 K17 **Vyasyeya** *Rus.* Veseya. Minskaya Voblasts', C Belarus 53°04′N 27°41′E
125 R15 **Vyatka** ≈ NW Russian Federation
Vyatka *see* Kirov
125 S16 **Vyatskiye Polyany** Kirovskaya Oblast', NW Russian Federation 56°15′N 51°06′E
123 S14 **Vyazemskiy** Khabarovskiy Kray, SE Russian Federation 47°28′N 134°39′E
126 I4 **Vyaz'ma** Smolenskaya Oblast', W Russian Federation 55°09′N 34°20′E
127 N3 **Vyazniki** Vladimirskaya Oblast', W Russian Federation 56°15′N 42°06′E
127 O8 **Vyazovka** Volgogradskaya Oblast', SW Russian Federation 50°57′N 43°57′E
119 J14 **Vyazyn'** Minskaya Voblasts', NW Belarus 54°25′N 27°10′E
124 G11 **Vyborg** *Fin.* Viipuri. Leningradskaya Oblast', NW Russian Federation 60°44′N 28°47′E
125 P11 **Vychegda** *var.* Vichegda. ≈ NW Russian Federation
119 L14 **Vyelyewshchyna** *Rus.* Velevshchina. Vitsyebskaya Voblasts', N Belarus 54°44′N 28°35′E
119 P16 **Vyeramyeyki** *Rus.* Veremeyki. Mahilyowskaya Voblasts', E Belarus 53°46′N 31°17′E
118 K11 **Vyerkhnyadzvinsk** *Rus.* Verkhnedvinsk. Vitsyebskaya Voblasts', N Belarus 55°47′N 27°56′E
119 P18 **Vyetka** *Rus.* Vetka. Homyel'skaya Voblasts', SE Belarus 52°33′N 31°10′E
118 L12 **Vyetryna** *Rus.* Vetrino. Vitsyebskaya Voblasts', N Belarus 55°25′N 28°28′E
Vygonovskoye, Ozero *see* Vyhanashchanskaye, Vozyera
124 J9 **Vygozero, Ozero** ◎ NW Russian Federation
119 I18 **Vyhanashchanskaye, Vozyera** *prev.* Vozyera Vyhanawskaye, *Rus.* Ozero Vygonovskoye. ◎ SW Belarus
Vyhanawskaye, Vozyera *see* Vyhanashchanskaye, Vozyera
127 N4 **Vyksa** Nizhegorodskaya Oblast', W Russian Federation 55°21′N 42°10′E
117 O12 **Vylkove** *Rus.* Vilkovo. Odes'ka Oblast', SW Ukraine 45°24′N 29°37′E
125 R9 **Vym'** ≈ NW Russian Federation
116 H8 **Vynohradiv** *Cz.* Sevluš, *Hung.* Nagyszöllös, *Rus.* Vinogradov; *prev.* Sevlyush. Zakarpats'ka Oblast', W Ukraine 48°09′N 23°01′E
124 G13 **Vyritsa** Leningradskaya Oblast', NW Russian Federation 59°25′N 30°20′E
97 J19 **Vyrnwy** *Wel.* Afon Efyrnwy. ≈ E Wales, United Kingdom
145 X9 **Vyshe Ivanovskiy Belak, Gora** ▲ E Kazakhstan 50°16′N 83°46′E
117 P4 **Vyshhorod** Kyyivs'ka Oblast', N Ukraine 50°36′N 30°28′E
124 I15 **Vyshniy Volochek** Tverskaya Oblast', W Russian Federation 57°37′N 34°33′E
111 G18 **Vyškov** *Ger.* Wischau. Jihomoravský Kraj, SE Czech Republic 49°17′N 17°01′E
111 E18 **Vysočina** *prev.* Jihlavský Kraj. ◈ *region* N Czech Republic
119 E19 **Vysokaye** *Rus.* Vysokoye. Brestskaya Voblasts', SW Belarus 52°20′N 23°18′E
111 F17 **Vysoké Mýto** *Ger.* Hohenmauth. Pardubický Kraj, C Czech Republic 49°57′N 16°10′E
117 S9 **Vysokopillya** Khersons'ka Oblast', S Ukraine 47°28′N 33°30′E
126 K3 **Vysokovsk** Moskovskaya Oblast', W Russian Federation 56°12′N 36°42′E
Vysokoye *see* Vysokaye
124 K12 **Vytegra** Vologodskaya Oblast', NW Russian Federation 60°59′N 36°27′E
116 J8 **Vyzhnytsya** Chernivets'ka Oblast', W Ukraine 48°14′N 25°10′E

W

77 O14 **Wa** NW Ghana 10°07′N 02°28′W
Waadt *see* Vaud
Waag *see* Váh
Waagbistritz *see* Považská Bystrica
Waagneustadtl *see* Nové Mesto nad Váhom
81 M16 **Waajid** Gedo, SW Somalia 03°37′N 43°19′E
98 L13 **Waal** ≈ S Netherlands
187 O16 **Waala** Province Nord, W New Caledonia 19°46′S 163°41′E
99 I14 **Waalwijk** Noord-Brabant, S Netherlands 51°42′N 05°04′E
99 E16 **Waarschoot** Oost-Vlaanderen, NW Belgium 51°09′N 03°35′E
186 C7 **Wabag** Enga, W Papua New Guinea 05°28′S 143°40′E
15 N7 **Wabano** ≈ Québec, SE Canada
11 P11 **Wabasca** ≈ Alberta, SW Canada
31 P12 **Wabash** Indiana, N USA 40°47′N 85°48′W
29 X9 **Wabasha** Minnesota, N USA 44°22′N 92°01′W
31 N13 **Wabash River** ≈ N USA
14 C7 **Wabatongushi Lake** ◎ Ontario, S Canada
81 L15 **Wabē Gestro Wenz** ≈ SE Ethiopia
14 B9 **Wabos** Ontario, S Canada 46°48′N 84°06′W
11 W13 **Wabowden** Manitoba, C Canada 54°57′N 98°38′W
110 J9 **Wąbrzeźno** Kujawsko-pomorskie, C Poland 53°18′N 18°55′E
21 U12 **Waccamaw River** ≈ South Carolina, SE USA
23 U11 **Waccasassa Bay** *bay* Florida, SE USA
99 F16 **Wachtebeke** Oost-Vlaanderen, NW Belgium 51°10′N 03°52′E
25 T8 **Waco** Texas, SW USA 31°33′N 97°10′W
26 M3 **Waconda Lake** *var.* Great Elder Reservoir. ⊠ Kansas, C USA
Wadai *see* Ouaddaï
Wad Al-Hajarah *see* Guadalajara
164 I12 **Wadayama** Hyōgo, Honshū, SW Japan 35°19′N 134°51′E
80 D10 **Wad Banda** Western Kordofan, C Sudan 13°08′N 27°56′E
75 P9 **Waddān** NW Libya 29°10′N 16°08′E
98 J4 **Waddeneilanden** *Eng.* West Frisian Islands. *island group* N Netherlands
98 J6 **Waddenzee** *var.* Wadden Zee. *sea* SE North Sea
Wadden Zee *see* Waddenzee
10 L16 **Waddington, Mount** ▲ British Columbia, SW Canada 51°17′N 125°16′W
98 H12 **Waddinxveen** Zuid-Holland, C Netherlands 52°03′N 04°38′E
11 U15 **Wadena** Saskatchewan, S Canada 51°57′N 103°48′W
29 T6 **Wadena** Minnesota, N USA 46°27′N 95°07′W
108 G7 **Wädenswil** Zürich, N Switzerland 47°14′N 08°41′E
21 S11 **Wadesboro** North Carolina, SE USA 34°59′N 80°03′W
155 G16 **Wādi** Karnātaka, C India 17°00′N 76°58′E
138 G10 **Wādī as Sīr** *var.* Wadi es Sir. 'Ammān, NW Jordan 31°57′N 35°49′E
Wadi es Sir *see* Wādī as Sīr
78 J9 **Wadi Fira** *off.* Région du Wadi Fira; *prev.* Préfecture de Biltine. ◈ *region* E Chad
Wadi Fira, Région du *see* Wadi Fira
80 F5 **Wadi Halfa** *var.* Wādī Ḥalfā'. Northern, N Sudan 21°46′N 31°17′E
138 G13 **Wādī Mūsā** *var.* Petra. Ma'ān, S Jordan 30°19′N 35°29′E
23 V4 **Wadley** Georgia, SE USA 32°52′N 82°24′W
Wad Madanī *see* Wad Medani
80 G10 **Wad Medani** *var.* Wad Madanī. Gezira, C Sudan 14°24′N 33°30′E
80 F10 **Wad Nimr** White Nile, C Sudan 14°32′N 32°10′E
165 U16 **Wadomari** Kagoshima, Okinoerabu-jima, SW Japan 27°25′N 128°40′E
111 K17 **Wadowice** Małopolskie, S Poland 49°54′N 19°29′E
35 R5 **Wadsworth** Nevada, W USA 39°39′N 119°16′W
31 T12 **Wadsworth** Ohio, N USA 41°01′N 81°43′W
25 T11 **Waelder** Texas, SW USA 29°42′N 97°16′W
Waereghem *see* Waregem
163 U13 **Wafangdian** *var.* Fuxian, Fu Xian. Liaoning, NE China 39°36′N 122°00′E
171 R13 **Waflia** Pulau Buru, E Indonesia 03°10′S 126°05′E
Wagadugu *see* Ouagadougou
98 K12 **Wageningen** Gelderland, SE Netherlands 51°58′N 05°40′E
55 V9 **Wageningen** Nickerie, NW Suriname 05°44′N 56°45′W
9 O8 **Wager Bay** *inlet* Nunavut, N Canada
183 P10 **Wagga Wagga** New South Wales, SE Australia 35°11′S 147°22′E
180 J13 **Wagin** Western Australia 33°16′S 117°26′E
108 H8 **Wägitaler See** ◎ SW Switzerland
29 P12 **Wagner** South Dakota, N USA 43°04′N 98°17′W
27 Q9 **Wagoner** Oklahoma, C USA 35°58′N 92°23′W
37 U10 **Wagon Mound** New Mexico, SW USA 36°00′N 104°42′W
32 J14 **Wagontire** Oregon, NW USA 43°15′N 119°51′W
110 H10 **Wągrowiec** Wielkopolskie, C Poland 52°49′N 17°11′E
149 U6 **Wāh** Punjab, NE Pakistan 33°50′N 72°44′E
171 S13 **Wahai** Pulau Seram, E Indonesia 02°48′S 129°29′E
169 V10 **Wahau, Sungai** ≈ Borneo, C Indonesia
Wahaybah, Ramlat Al *see* Wahībah, Ramlat Āl
Wahda *see* Unity
38 D9 **Wahiawā** *var.* Wahiawa. O'ahu, Hawaii, USA, C Pacific Ocean 21°30′N 158°01′W
Wahībah, Ramlat Ahl *see* Wahībah, Ramlat Āl
141 Y9 **Wahībah, Ramlat Āl** *var.* Ramlat Ahl Wahībah, Ramlat Al Wahaybah, *Eng.* Wahībah Sands. *desert* N Oman
Wahībah Sands *see* Wahībah, Ramlat Āl
101 E16 **Wahn** ✈ (Köln) Nordrhein-Westfalen, W Germany 50°51′N 07°09′E
29 R15 **Wahoo** Nebraska, C USA 41°12′N 96°37′W
29 R6 **Wahpeton** North Dakota, N USA 46°16′N 96°36′W
Wahran *see* Oran
36 J6 **Wah Wah Mountains** ▲▲ Utah, W USA
38 D9 **Waialua** O'ahu, Hawaii, USA, C Pacific Ocean 21°34′N 158°07′W
38 D9 **Wai'anae** *var.* Waianae. O'ahu, Hawaii, USA, C Pacific Ocean 21°26′N 158°11′W
184 Q8 **Waiapu** ≈ North Island, New Zealand
185 I17 **Waiau** Canterbury, South Island, New Zealand 42°39′S 173°03′E
185 I17 **Waiau** ≈ South Island, New Zealand
185 B23 **Waiau** ≈ South Island, New Zealand
101 H21 **Waiblingen** Baden-Württemberg, S Germany 48°49′N 09°19′E
Waidhofen *see* Waidhofen an der Ybbs, Niederösterreich, Austria
Waidhofen *see* Waidhofen an der Thaya, Niederösterreich, Austria
109 V2 **Waidhofen an der Thaya** *var.* Waidhofen. Niederösterreich, NE Austria 48°49′N 15°17′E
109 U5 **Waidhofen an der Ybbs** *var.* Waidhofen. Niederösterreich, E Austria 47°58′N 14°47′E
171 T11 **Waigeo, Pulau** *island* Papua Barat, E Indonesia
184 L5 **Waiheke Island** *island* N New Zealand
184 M7 **Waihi** Waikato, North Island, New Zealand 37°22′S 175°51′E
185 C20 **Waihou** ≈ North Island, New Zealand
Waikaboebak *see* Waikabubak
171 N17 **Waikabubak** *prev.* Waikaboebak. Pulau Sumba, C Indonesia 09°40′S 119°25′E
185 D23 **Waikaia** ≈ South Island, New Zealand
185 D23 **Waikaka** Southland, South Island, New Zealand 45°55′S 168°59′E
184 L13 **Waikanae** Wellington, North Island, New Zealand 40°52′S 175°03′E
184 M7 **Waikare, Lake** ◎ North Island, New Zealand
184 O9 **Waikaremoana, Lake** ◎ North Island, New Zealand
185 I17 **Waikari** Canterbury, South Island, New Zealand 42°50′S 172°41′E
184 L8 **Waikato** *off.* Waikato Region. ◈ *region* North Island, New Zealand
184 M8 **Waikato** ≈ North Island, New Zealand
Waikato Region *see* Waikato
182 J9 **Waikerie** South Australia 34°12′S 139°57′E
185 F23 **Waikouaiti** Otago, South Island, New Zealand 45°36′S 170°39′E
38 H11 **Wailea** Hawaii, USA, C Pacific Ocean 19°53′N 155°07′W
38 F10 **Wailuku** Maui, Hawaii, USA, C Pacific Ocean 20°53′N 156°30′W
185 H18 **Waimakariri** ≈ South Island, New Zealand
38 D9 **Waimānalo Beach** *var.* Waimanalo Beach. O'ahu, Hawaii, USA, C Pacific Ocean 21°20′N 157°42′W
185 G15 **Waimangaroa** West Coast, South Island, New Zealand 41°41′S 171°49′E
185 G21 **Waimate** Canterbury, South Island, New Zealand 44°44′S 171°03′E
38 B8 **Waimea** Kaua'i, Hawaii, USA, C Pacific Ocean 21°57′N 159°40′W
38 G11 **Waimea** *var.* Kamuela. Hawaii, USA, C Pacific Ocean 20°02′N 155°20′W
38 D9 **Waimea** *var.* Maunawai. O'ahu, Hawaii, USA, C Pacific Ocean 21°39′N 158°04′W
99 M20 **Waimes** Liège, E Belgium 50°25′N 06°10′E
154 J11 **Wainganga** *var.* Wain River. ≈ C India
Waingapoe *see* Waingapu
171 N17 **Waingapu** *prev.* Waingapoe. Pulau Sumba, C Indonesia 09°40′S 120°16′E
55 S7 **Waini** ≈ N Guyana
55 S7 **Waini Point** *headland* NW Guyana 08°24′N 59°48′W
Wain River *see* Wainganga
11 R15 **Wainwright** Alberta, SW Canada 52°50′N 110°51′W
39 O5 **Wainwright** Alaska, USA 70°38′N 160°02′W
184 K4 **Waiotira** Northland, North Island, New Zealand 35°56′S 174°11′E
184 M11 **Waiouru** Manawatu-Wanganui, North Island, New Zealand 39°28′S 175°41′E
171 W14 **Waipa** Papua, E Indonesia 03°47′S 136°16′E
184 L8 **Waipa** ≈ North Island, New Zealand
184 P9 **Waipaoa** ≈ North Island, New Zealand
185 D25 **Waipapa Point** *headland* South Island, New Zealand 46°39′S 168°51′E
185 I18 **Waipara** Canterbury, South Island, New Zealand 43°04′S 172°45′E
184 N12 **Waipawa** Hawke's Bay, North Island, New Zealand 39°57′S 176°36′E
184 K4 **Waipu** Northland, North Island, New Zealand 35°58′S 174°25′E
184 N12 **Waipukurau** Hawke's Bay, North Island, New Zealand 40°01′S 176°34′E
171 U14 **Wair** Pulau Kai Besar, E Indonesia 05°16′S 133°09′E
Wairakai *see* Wairakei
184 N9 **Wairakei** *var.* Wairakai. Waikato, North Island, New Zealand 38°37′S 176°05′E
185 M14 **Wairarapa, Lake** ◎ North Island, New Zealand
185 J15 **Wairau** ≈ South Island, New Zealand
184 P10 **Wairoa** Hawke's Bay, North Island, New Zealand 39°03′S 177°26′E
184 P10 **Wairoa** ≈ North Island, New Zealand
184 J4 **Wairoa** ≈ North Island, New Zealand
184 N9 **Waitahanui** Waikato, North Island, New Zealand 38°48′S 176°04′E
184 M6 **Waitakaruru** Waikato, North Island, New Zealand 37°14′S 175°22′E
185 F21 **Waitaki** ≈ South Island, New Zealand
184 K10 **Waitara** Taranaki, North Island, New Zealand 39°01′S 174°14′E
184 M7 **Waitoa** Waikato, North Island, New Zealand 37°36′S 175°37′E
184 L8 **Waitomo Caves** Waikato, North Island, New Zealand 38°17′S 175°06′E
184 L11 **Waitotara** Taranaki, North Island, New Zealand 39°49′S 174°43′E
184 L11 **Waitotara** ≈ North Island, New Zealand
32 L10 **Waitsburg** Washington, NW USA 46°16′N 118°09′W
Waitzen *see* Vác
184 L6 **Waiuku** Auckland, North Island, New Zealand 37°15′S 174°45′E
164 L10 **Wajima** *var.* Wazima. Ishikawa, Honshū, SW Japan 37°23′N 136°53′E
81 K17 **Wajir** Wajir, NE Kenya 01°46′N 40°05′E
81 K18 **Wajir** ◈ *county* NE Kenya
79 J17 **Waka** Equateur, NW Dem. Rep. Congo 01°04′N 20°11′E
81 I14 **Waka** Southern Nationalities, S Ethiopia 07°12′N 37°19′E
14 D9 **Wakami Lake** ◎ Ontario, S Canada
164 I12 **Wakasa** Tottori, Honshū, SW Japan 35°18′N 134°25′E
164 J12 **Wakasa-wan** *bay* C Japan
185 C22 **Wakatipu, Lake** ◎ South Island, New Zealand
11 T15 **Wakaw** Saskatchewan, S Canada 52°40′N 105°45′W
164 I14 **Wakayama** Wakayama, Honshū, SW Japan 34°12′N 135°09′E
164 I15 **Wakayama** *off.* Wakayama-ken. ◈ *prefecture* Honshū, SW Japan
Wakayama-ken *see* Wakayama
26 K4 **Wa Keeney** Kansas, C USA 39°02′N 99°53′W
185 I14 **Wakefield** Tasman, South Island, New Zealand 41°24′S 173°03′E
97 M17 **Wakefield** N England, United Kingdom 53°42′N 01°29′W
27 O4 **Wakefield** Kansas, C USA 39°12′N 97°00′W
30 L4 **Wakefield** Michigan, N USA 46°27′N 89°55′W
21 U9 **Wake Forest** North Carolina, SE USA 35°58′N 78°30′W
Wakeham Bay *see* Kangiqsujuaq
189 Y11 **Wake Island** ◇ *US unincorporated territory* NW Pacific Ocean
189 Y12 **Wake Island** ✈ NW Pacific Ocean
189 Y12 **Wake Island** *atoll* NW Pacific Ocean
189 X12 **Wake Lagoon** *lagoon* Wake Island, NW Pacific Ocean
166 L8 **Wakema** Ayeyawady, SW Myanmar (Burma) 16°36′N 95°11′E
Wakhan *see* Khandūd
164 H14 **Waki** Tokushima, Shikoku, SW Japan 34°04′N 134°10′E
165 T1 **Wakkanai** Hokkaidō, NE Japan 45°25′N 141°39′E
83 K22 **Wakkerstroom** Mpumalanga, E South Africa 27°21′S 30°10′E
14 C10 **Wakomata Lake** ◎ Ontario, S Canada
183 N10 **Wakool** New South Wales, SE Australia 35°30′S 144°22′E
Wakra *see* Al Wakrah
Waku Kungo *see* Uaco Cungo
186 J7 **Wakunai** Bougainville, NE Papua New Guinea 05°52′S 155°10′E
Walachei/Walachia *see* Wallachia
155 K26 **Walawe Ganga** ≈ S Sri Lanka
111 F15 **Wałbrzych** *Ger.* Waldenburg, Waldenburg in Schlesien. Dolnośląskie, SW Poland 50°45′N 16°20′E
183 T6 **Walcha** New South Wales, SE Australia 31°01′S 151°38′E
101 K24 **Walchensee** ◎ SE Germany
99 D14 **Walcheren** *island* SW Netherlands
29 Z14 **Walcott** Iowa, C USA 41°34′N 90°46′W
33 W16 **Walcott** Wyoming, C USA 41°46′N 106°46′W
99 G21 **Walcourt** Namur, S Belgium 50°16′N 04°26′E
110 G9 **Wałcz** *Ger.* Deutsch Krone. Zachodnio-pomorskie, NW Poland 53°17′N 16°29′E
108 H7 **Wald** Zürich, N Switzerland 47°17′N 08°56′E
109 U3 **Waldaist** ≈ N Austria
180 I9 **Waldburg Range** ▲▲ Western Australia
37 R3 **Walden** Colorado, C USA 40°43′N 106°16′W
18 K13 **Walden** New York, NE USA 41°35′N 74°09′W
Waldenburg/Waldenburg in Schlesien *see* Wałbrzych
11 T15 **Waldheim** Saskatchewan, S Canada 52°38′N 106°35′W
Waldia *see* Weldiya
101 M23 **Waldkraiburg** Bayern, SE Germany 48°10′N 12°23′E
27 T14 **Waldo** Arkansas, C USA 33°21′N 93°18′W
23 V9 **Waldo** Florida, SE USA 29°47′N 82°07′W
19 R7 **Waldoboro** Maine, NE USA 44°06′N 69°22′W
21 W4 **Waldorf** Maryland, NE USA 38°36′N 76°54′W
32 F12 **Waldport** Oregon, NW USA 44°25′N 124°04′W
27 S11 **Waldron** Arkansas, C USA 34°54′N 94°09′W
195 Y13 **Waldron, Cape** *headland* Antarctica 66°08′S 116°00′E
101 F24 **Waldshut-Tiengen** Baden-Württemberg, S Germany 47°37′N 08°13′E
171 P12 **Walea, Selat** *strait* Sulawesi, C Indonesia
Wałeckie Międzyrzecze *see* Valašské Meziříčí
108 H8 **Walensee** ◎ NW Switzerland
38 L8 **Wales** Alaska, USA 65°36′N 168°03′W
97 J20 **Wales** *Wel.* Cymru. ◈ *national region* Wales, United Kingdom
9 O7 **Wales Island** *island* Nunavut, NE Canada
77 P14 **Walewale** N Ghana 10°21′N 00°48′W
99 M24 **Walferdange** Luxembourg, C Luxembourg 49°39′N 06°08′E
183 Q5 **Walgett** New South Wales, SE Australia 30°02′S 148°14′E
194 K10 **Walgreen Coast** *physical region* Antarctica
29 Q2 **Walhalla** North Dakota, N USA 48°55′N 97°55′W
21 O11 **Walhalla** South Carolina, SE USA 34°46′N 83°05′W
79 O19 **Walikale** Nord-Kivu, E Dem. Rep. Congo 01°29′S 28°05′E
Walk *see* Valga, Estonia
Walk *see* Valka, Latvia
29 U5 **Walker** Minnesota, N USA 47°06′N 94°35′W
15 V4 **Walker, Lac** ◎ Québec, SE Canada
35 S7 **Walker Lake** ◎ Nevada, W USA
35 R6 **Walker River** ≈ Nevada, W USA
28 K10 **Wall** South Dakota, N USA 43°58′N 102°12′W
173 U9 **Wallaby Plateau** *undersea feature* E Indian Ocean
33 N8 **Wallace** Idaho, NW USA 47°28′N 115°55′W
21 V11 **Wallace** North Carolina, SE USA 34°42′N 77°59′W
14 D17 **Wallaceburg** Ontario, S Canada 42°34′N 82°22′W
22 F5 **Wallace Lake** ⊠ Louisiana, S USA
11 P13 **Wallace Mountain** ▲ Alberta, W Canada 54°50′N 115°57′W
116 J14 **Wallachia** *var.* Walachia, *Ger.* Walachei, *Rom.* Valachia. *cultural region* S Romania
Wallachisch-Meseritsch *see* Valašské Meziříčí
183 U4 **Wallangarra** New South Wales, SE Australia 28°56′S 151°57′E
182 I8 **Wallaroo** South Australia 33°56′S 137°38′E
32 L10 **Walla Walla** Washington, NW USA 46°03′N 118°20′W
101 H19 **Walldürn** Baden-Württemberg, SW Germany 49°34′N 09°22′E
100 F12 **Wallenhorst** Niedersachsen, NW Germany 52°21′N 08°01′E
Wallenthal *see* Haţeg
109 S4 **Wallern** Oberösterreich, N Austria 48°13′N 13°58′E
Wallern *see* Wallern im Burgenland
109 Z5 **Wallern im Burgenland** *var.* Wallern. Burgenland, E Austria 47°44′N 16°57′E
18 M9 **Wallingford** Vermont, NE USA 43°27′N 72°56′W
25 V11 **Wallis** Texas, SW USA 29°38′N 96°05′W
Wallis *see* Valais
192 K9 **Wallis and Futuna** *Fr.* Territoire de Wallis et Futuna. ◇ *French overseas collectivity* C Pacific Ocean
108 G7 **Wallisellen** Zürich, N Switzerland 47°25′N 08°36′E
Wallis et Futuna, Territoire de *see* Wallis and Futuna
190 H11 **Wallis, Îles** *island group* N Wallis and Futuna
99 G20 **Wallonia** ◈ *cultural region* SW Belgium
31 Q5 **Walloon Lake** ◎ Michigan, N USA
32 K10 **Wallula** Washington, NW USA 46°03′N 118°54′W
32 K10 **Wallula, Lake** ◎ Washington, NW USA
21 S8 **Walnut Cove** North Carolina, SE USA 36°18′N 80°08′W
35 N8 **Walnut Creek** California, W USA 37°52′N 122°04′W
26 K5 **Walnut Creek** ≈ Kansas, C USA
27 W9 **Walnut Ridge** Arkansas, C USA 36°06′N 90°56′W
25 S7 **Walnut Springs** Texas, SW USA 32°05′N 97°42′W
182 L10 **Walpeup** Victoria, SE Australia 35°09′S 142°01′E
187 R17 **Walpole, Île** *island* SE New Caledonia
39 N13 **Walrus Islands** *island group* Alaska, USA
97 L19 **Walsall** C England, United Kingdom 52°35′N 01°58′W
37 T7 **Walsenburg** Colorado, C USA 37°37′N 104°46′W
11 S17 **Walsh** Alberta, SW Canada 49°58′N 110°03′W
37 W7 **Walsh** Colorado, C USA 37°20′N 102°17′W
100 I11 **Walsrode** Niedersachsen, NW Germany 52°52′N 09°36′E
Waltenberg *see* Zalău
21 R14 **Walterboro** South Carolina, SE USA 32°54′N 80°21′W
Walter F.George Lake *see* Walter F. George Reservoir
23 R6 **Walter F. George Reservoir** *var.* Walter F.George Lake. ⊠ Alabama/Georgia, SE USA
26 M12 **Walters** Oklahoma, C USA 34°22′N 98°18′W
101 J16 **Waltershausen** Thüringen, C Germany 50°53′N 10°33′E
173 N10 **Walters Shoal** *var.* Walters Shoals. *reef* S Madagascar
Walters Shoals *see* Walters Shoal
22 M3 **Walthall** Mississippi, S USA 33°36′N 89°16′W
20 M4 **Walton** Kentucky, S USA 38°52′N 84°36′W
18 J11 **Walton** New York, NE USA 42°10′N 75°07′W
79 O20 **Walungu** Sud-Kivu, E Dem. Rep. Congo 02°40′S 28°37′E
Walvisbaai *see* Walvis Bay
83 C19 **Walvis Bay** *Afr.* Walvisbaai. Erongo, NW Namibia 22°59′S 14°34′E
83 B19 **Walvis Bay** *bay* NW Namibia
Walvish Ridge *see* Walvis Ridge
65 O17 **Walvis Ridge** *var.* Walvish Ridge. *undersea feature* E Atlantic Ocean 28°00′S 03°00′E
171 X16 **Wamal** Papua, E Indonesia 08°00′S 139°06′E
171 U15 **Wamar, Pulau** *island* Kepulauan Aru, E Indonesia
79 O17 **Wamba** Orientale, NE Dem. Rep. Congo 02°10′N 27°59′E
77 V15 **Wamba** Nassarawa, C Nigeria 08°57′N 08°35′E
79 H22 **Wamba** *var.* Uamba. ≈ Angola/Dem. Rep. Congo
27 P4 **Wamego** Kansas, C USA 39°12′N 96°18′W
18 I10 **Wampsville** New York, NE USA 43°03′N 75°40′W
42 K6 **Wampú, Río** ≈ E Honduras
171 X16 **Wan** Papua, E Indonesia 08°15′S 138°00′E
Wan *see* Anhui
183 N4 **Wanaaring** New South Wales, SE Australia 29°42′S 144°07′E
185 D21 **Wanaka** Otago, South Island, New Zealand 44°42′S 169°09′E
185 D20 **Wanaka, Lake** ◎ South Island, New Zealand
171 W14 **Wanapiri** Papua, E Indonesia 04°21′S 135°52′E
14 F9 **Wanapitei** ≈ Ontario, S Canada
14 F10 **Wanapitei Lake** ◎ Ontario, S Canada
18 K14 **Wanaque** New Jersey, NE USA 41°02′N 74°17′W
171 U12 **Wanau** Papua Barat, E Indonesia 01°20′S 132°40′E
185 F22 **Wanbrow, Cape** *headland* South Island, New Zealand 45°07′S 170°59′E
Wancheng *see* Wanning
Wanchuan *see* Zhangjiakou
171 W13 **Wandai** *var.* Komeyo. Papua, E Indonesia 03°35′S 136°15′E
163 Z8 **Wanda Shan** ▲▲ NE China
197 R11 **Wandel Sea** *sea* Arctic Ocean
160 D13 **Wanding** *var.* Wandingzhen. Yunnan, SW China 24°01′N 98°00′E
Wandingzhen *see* Wanding
99 H20 **Wanfercée-Baulet** Hainaut, S Belgium 50°27′N 04°37′E
184 L12 **Wanganui** Manawatu-Wanganui, North Island, New Zealand 39°56′S 175°02′E
184 L11 **Wanganui** ≈ North Island, New Zealand
183 P11 **Wangaratta** Victoria, SE Australia 36°22′S 146°17′E
160 J8 **Wangcang** *var.* Donghe; *prev.* Fengjiaba, Hongjiang. Sichuan, C China 32°15′N 106°16′E
Wangda *see* Zogang
101 I24 **Wangen im Allgäu** Baden-Württemberg, S Germany 47°40′N 09°49′E
Wangerin *see* Węgorzyno
100 F9 **Wangerooge** *island* NW Germany
171 W13 **Wanggar** Papua, E Indonesia 03°22′S 135°15′E
160 J13 **Wangmo** *var.* Fuxing. Guizhou, S China 25°08′N 106°08′E
Wangolodougou *see* Ouangolodougou
161 S9 **Wangpan Yang** *sea* E China
163 Y10 **Wangqing** Jilin, NE China 43°19′N 129°42′E
167 P8 **Wang Saphung** Loei, C Thailand 17°18′N 101°45′E
167 O6 **Wan Hsa-la** Shan State, E Myanmar (Burma) 20°27′N 98°39′E
55 W9 **Wanica** ◈ *district* N Suriname
79 M18 **Wanie-Rukula** Orientale, C Dem. Rep. Congo 0°13′N 25°34′E
Wankie *see* Hwange
Wanki, Río *see* Coco, Río
81 N17 **Wanlaweyn** *var.* Wanle Weyn, *It.* Uanle Uen. Shabeellaha Hoose, SW Somalia 02°36′N 44°47′E
Wanle Weyn *see* Wanlaweyn
180 I12 **Wanneroo** Western Australia 31°40′S 115°35′E
160 L17 **Wanning** *var.* Wancheng. Hainan, S China 18°55′N 110°27′E
167 Q8 **Wanon Niwat** Sakon Nakhon, E Thailand 17°39′N 103°45′E
155 H16 **Wanparti** Telangana, C India 16°19′N 78°06′E
Wansen *see* Wiązów
160 L11 **Wanshan** Guizhou, S China 27°45′N 109°12′E
99 M14 **Wanssum** Limburg, SE Netherlands 51°31′N 06°04′E
184 N12 **Wanstead** Hawke's Bay, North Island, New Zealand 40°09′S 176°31′E
Wanxian *see* Wanzhou
188 F16 **Wanyaan** Yap, Micronesia
160 K8 **Wanyuan** Sichuan, C China 32°05′N 108°08′E
161 O11 **Wanzai** *var.* Kangle. Jiangxi, S China 28°06′N 114°27′E
99 J20 **Wanze** Liège, E Belgium 50°32′N 05°16′E
160 K9 **Wanzhou** *var.* Wanxian. Chongqing Shi, C China 30°48′N 108°21′E
31 R12 **Wapakoneta** Ohio, N USA 40°34′N 84°11′W
12 D7 **Wapaseese** ≈ Ontario, C Canada
32 I10 **Wapato** Washington, NW USA 46°27′N 120°25′W
29 Y15 **Wapello** Iowa, C USA 41°10′N 91°13′W
11 N13 **Wapiti** ≈ Alberta/British Columbia, SW Canada
27 X7 **Wappapello Lake** ⊠ Missouri, C USA
18 K13 **Wappingers Falls** New York, NE USA 41°36′N 73°54′W
29 X13 **Wapsipinicon River** ≈ Iowa, C USA
14 L9 **Wapus** ≈ Québec, SE Canada
160 H7 **Waqên** Sichuan, C China 33°05′N 102°34′E
21 Q7 **War** West Virginia, NE USA 37°18′N 81°39′W
Warab *see* Warrap
155 J15 **Warangal** Telangana, C India 18°N 79°35′E
Warasdin *see* Varaždin
183 O16 **Waratah** Tasmania, SE Australia 41°28′S 145°34′E
183 O14 **Waratah Bay** *bay* Victoria, SE Australia
101 H15 **Warburg** Nordrhein-Westfalen, W Germany 51°30′N 09°11′E
182 I1 **Warburton Creek** *seasonal river* South Australia
180 M9 **Warbuton** Western Australia 26°17′S 126°18′E
99 M20 **Warche** ≈ E Belgium
Wardag/Wardak *see* Wardak
149 P5 **Wardak** *prev.* Vardak, *Pash.* Wardag. ◈ *province* E Afghanistan
32 K9 **Warden** Washington, NW USA 46°58′N 119°02′W
154 I12 **Wardha** Mahārāshtra, W India 20°41′N 78°40′E
Wardija Point *see* Wardija, Ras il-
121 N15 **Wardija, Ras il-** *var.* Ras il- Wardija, Wardija Point. *headland* Gozo, NW Malta 36°03′N 14°11′E
Wardija, Ras il- *see* Wardija, Ras il-
139 P3 **Wardīyah** Nīnawá, N Iraq 36°18′N 41°45′E
185 E19 **Ward, Mount** ▲ South Island, New Zealand 43°49′S 169°54′E
10 L11 **Ware** British Columbia, W Canada 57°26′N 125°41′W
99 D18 **Waregem** *var.* Waereghem. West-Vlaanderen, W Belgium 50°53′N 03°26′E
99 J19 **Waremme** Liège, E Belgium 50°41′N 05°15′E
100 N10 **Waren** Mecklenburg-Vorpommern, NE Germany 53°32′N 12°42′E
171 W13 **Waren** Papua, E Indonesia 02°13′S 136°21′E
101 F14 **Warendorf** Nordrhein-Westfalen, W Germany 51°57′N 08°00′E
21 P12 **Ware Shoals** South Carolina, SE USA 34°24′N 82°15′W
98 N4 **Warffum** Groningen, NE Netherlands 53°22′N 06°34′E
81 O15 **Wargalo** Mudug, E Somalia 06°06′N 47°40′E
146 M12 **Warganza** *Rus.* Varganzi. Qashqadaryo Viloyati, S Uzbekistan 39°18′N 66°00′E
Wargla *see* Ouargla
183 T4 **Warialda** New South Wales, SE Australia 29°34′S 150°35′E
154 F13 **Wāri Godri** Mahārāshtra, C India 19°28′N 75°43′E
167 R10 **Warin Chamrap** Ubon Ratchathani, E Thailand 15°11′N 104°51′E
25 R11 **Waring** Texas, SW USA 29°56′N 98°48′W
39 O8 **Waring Mountains** ▲▲ Alaska, USA
110 M12 **Warka** Mazowieckie, E Poland 51°45′N 21°12′E
184 L5 **Warkworth** Auckland, North Island, New Zealand 36°23′S 174°42′E
171 U12 **Warmandi** Papua Barat, E Indonesia 0°21′S 132°38′E
83 E22 **Warmbad** Karas, S Namibia 28°29′S 18°41′E
98 H8 **Warmenhuizen** Noord-Holland, NW Netherlands 52°43′N 04°45′E
110 M8 **Warmińsko-Mazurskie** ◈ *province* C Poland
97 L22 **Warminster** S England, United Kingdom 51°13′N 02°12′W
18 I15 **Warminster** Pennsylvania, NE USA 40°11′N 75°04′W
35 V8 **Warm Springs** Nevada, W USA 38°10′N 116°21′W
32 H12 **Warm Springs** Oregon, NW USA 44°51′N 121°24′W
21 S5 **Warm Springs** Virginia, NE USA 38°03′N 79°48′W
100 M8 **Warnemünde** Mecklenburg-Vorpommern, NE Germany 54°10′N 12°03′E
27 Q10 **Warner** Oklahoma, C USA 35°29′N 95°18′W
35 Q2 **Warner Mountains** ▲▲ California, W USA
23 T5 **Warner Robins** Georgia, SE USA 32°38′N 83°38′W
57 N18 **Warnes** Santa Cruz, C Bolivia 17°30′S 63°11′W
100 M9 **Warnow** ≈ NE Germany
Warnsdorf *see* Varnsdorf
98 M11 **Warnsveld** Gelderland, E Netherlands 52°08′N 06°14′E
154 I13 **Warora** Mahārāshtra, C India 20°12′N 79°01′E
182 L11 **Warracknabeal** Victoria, SE Australia 36°17′S 142°26′E
183 O13 **Warragul** Victoria, SE Australia 38°11′S 145°55′E
80 D13 **Warrap** Warrap, W South Sudan 08°08′N 28°37′E
81 D14 **Warrap** *var.* Warab. ◈ *state* W South Sudan

◆ Country
● Country Capital
◇ Dependent Territory
○ Dependent Territory Capital
◈ Administrative Regions
✈ International Airport
▲ Mountain
▲▲ Mountain Range
Volcano
≈ River
◎ Lake
⊠ Reservoir

◆ Country ◇ Dependent Territory ◆ Administrative Regions ▲ Mountain 🌋 Volcano ⊚ Lake
● Country Capital ○ Dependent Territory Capital ✕ International Airport ▲▲ Mountain Range ↗ River ▨ Reservoir

WESTKAPELLE – WINNIPEGOSIS

99 D14 **Westkapelle** Zeeland, SW Netherlands 51°32′N 03°26′E
West Kazakhstan see Zapadnyy Kazakhstan
31 O13 **West Lafayette** Indiana, N USA 40°24′N 86°54′W
31 T13 **West Lafayette** Ohio, N USA 40°16′N 81°45′W
West Lake see Kagera
29 Y14 **West Liberty** Iowa, C USA 41°34′N 91°15′W
21 O5 **West Liberty** Kentucky, S USA 37°56′N 83°16′W
Westliche Morava see Zapadna Morava
10 J13 **Westlock** Alberta, SW Canada 54°12′N 113°50′W
14 E17 **West Lorne** Ontario, S Canada 42°36′N 81°35′W
96 J12 **West Lothian** cultural region S Scotland, United Kingdom
99 H16 **Westmalle** Antwerpen, N Belgium 51°18′N 04°40′E
192 G6 **West Mariana Basin** var. Perece Vela Basin. undersea feature W Pacific Ocean 15°00′N 137°00′E
97 E17 **Westmeath** Ir. An Iarmhí, Na H-Iarmhidhe. cultural region C Ireland
27 Y11 **West Memphis** Arkansas, C USA 35°09′N 90°11′W
21 W2 **Westminster** Maryland, NE USA 39°34′N 77°00′W
21 O11 **Westminster** South Carolina, SE USA 34°39′N 83°06′W
22 I5 **West Monroe** Louisiana, S USA 32°31′N 92°09′W
18 D15 **Westmont** Pennsylvania, NE USA 40°16′N 78°55′W
27 O3 **Westmoreland** Kansas, C USA 39°23′N 96°30′W
35 W17 **Westmorland** California, W USA 33°02′N 115°37′W
186 E6 **West New Britain** ◆ province E Papua New Guinea
West New Guinea see Papua
83 K18 **West Nicholson** Matabeleland South, S Zimbabwe 21°06′S 29°25′E
29 T14 **West Nishnabotna River** ≈ Iowa, C USA
175 P11 **West Norfolk Ridge** undersea feature W Pacific Ocean
25 P12 **West Nueces River** ≈ Texas, SW USA
West Nusa Tenggara see Nusa Tenggara Barat
29 T11 **West Okoboji Lake** ◎ Iowa, C USA
33 R16 **Weston** Idaho, NW USA 42°01′N 119°29′W
21 R4 **Weston** West Virginia, NE USA 39°03′N 80°28′W
97 J22 **Weston-super-Mare** SW England, United Kingdom 51°21′N 02°59′W
23 Z14 **West Palm Beach** Florida, SE USA 26°43′N 80°03′W
West Papua see Papua Barat
West Papua see Papua
188 E9 **West Passage** passage Babeldaob, N Palau
23 O9 **West Pensacola** Florida, SE USA 30°25′N 87°16′W
27 V8 **West Plains** Missouri, C USA 36°44′N 91°51′W
35 P7 **West Point** California, W USA 38°21′N 120°33′W
23 R5 **West Point** Georgia, SE USA 32°52′N 85°10′W
22 M3 **West Point** Mississippi, S USA 33°36′N 88°39′W
29 R14 **West Point** Nebraska, C USA 41°50′N 96°42′W
21 X6 **West Point** Virginia, NE USA 37°31′N 76°48′W
182 G10 **West Point** headland South Australia 35°01′S 135°58′E
65 B24 **Westpoint Island Settlement** Westpoint Island, NW Falkland Islands 51°21′S 60°41′W
23 R4 **West Point Lake** ⊠ Alabama/Georgia, SE USA
81 H18 **West Pokit** ◆ county W Kenya
97 B16 **Westport** Ir. Cathair na Mart. Mayo, W Ireland 53°48′N 09°32′W
185 G15 **Westport** West Coast, South Island, New Zealand 41°46′S 171°37′E
32 F10 **Westport** Oregon, NW USA 46°07′N 123°22′W
32 F9 **Westport** Washington, NW USA 46°53′N 124°06′W
31 S15 **West Portsmouth** Ohio, N USA 38°45′N 83°01′W
West Punjab see Punjab
11 V14 **Westray** Manitoba, C Canada 53°30′N 101°19′W
96 J4 **Westray** island NE Scotland, United Kingdom
14 F9 **Westree** Ontario, S Canada 47°25′N 81°32′W
97 L16 **West Riding** cultural region N England, United Kingdom
West River see Xi Jiang
30 J7 **West Salem** Wisconsin, N USA 43°54′N 91°04′W
65 H21 **West Scotia Ridge** undersea feature W Scotia Sea
186 B5 **West Sepik** prev. Sandaun. ◆ province NW Papua New Guinea
173 N4 **West Sheba Ridge** undersea feature W Indian Ocean 12°45′N 48°15′E
West Siberian Plain see Zapadno-Sibirskaya Ravnina
31 S11 **West Sister Island** island Ohio, N USA
West-Skylge see West-Terschelling
West Sumatra see Sumatera Barat
98 J5 **West-Terschelling** Fris. West-Skylge. Fryslân, N Netherlands 53°23′N 05°15′E
64 J7 **West Thulean Rise** undersea feature N Atlantic Ocean
29 X12 **West Union** Iowa, C USA 42°57′N 91°48′W
31 R15 **West Union** Ohio, N USA 38°47′N 83°33′W
21 R3 **West Union** West Virginia, NE USA 39°18′N 80°47′W
31 N13 **Westville** Illinois, N USA 40°02′N 87°38′W
21 R3 **West Virginia** off. State of West Virginia, also known as Mountain State. ◆ state NE USA
99 A17 **West-Vlaanderen** Eng. West Flanders. ◆ province W Belgium
35 R7 **West Walker River** ≈ California/Nevada, W USA
35 P4 **Westwood** California, W USA 40°18′N 121°02′W
183 P9 **West Wyalong** New South Wales, SE Australia 33°56′S 147°10′E
171 Q16 **Wetar, Pulau** island Kepulauan Damar, E Indonesia
Wetar, Selat see Wetar Strait
171 R16 **Wetar Strait** var. SelatWetar. strait Nusa Tenggara, S Indonesia
11 Q15 **Wetaskiwin** Alberta, SW Canada 52°57′N 113°20′W
81 K21 **Wete** Pemba, E Tanzania 05°03′S 39°41′E
166 M4 **Wetlet** Sagaing, C Myanmar (Burma) 22°43′N 95°22′E
37 T6 **Wet Mountains** ▲▲ Colorado, C USA
101 E15 **Wetter** Nordrhein-Westfalen, W Germany 51°22′N 07°24′E
101 H17 **Wetter** ≈ W Germany
99 F17 **Wetteren** Oost-Vlaanderen, NW Belgium 51°06′N 03°59′E
108 F7 **Wettingen** Aargau, N Switzerland 47°28′N 08°20′E
27 P11 **Wetumka** Oklahoma, C USA 35°14′N 96°14′W
23 Q5 **Wetumpka** Alabama, S USA 32°32′N 86°12′W
108 G7 **Wetzikon** Zürich, N Switzerland 47°19′N 08°48′E
101 G17 **Wetzlar** Hessen, W Germany 50°33′N 08°30′E
99 C18 **Wevelgem** West-Vlaanderen, W Belgium 50°48′N 03°12′E
38 M6 **Wevok** var. Wewuk. Alaska, USA 68°52′N 166°05′W
23 R9 **Wewahitchka** Florida, SE USA 30°06′N 85°12′W
186 C6 **Wewak** East Sepik, NW Papua New Guinea 03°35′S 143°35′E
27 O11 **Wewoka** Oklahoma, C USA 35°09′N 96°30′W
Wewuk see Wevok
97 F20 **Wexford** Ir. Loch Garman. SE Ireland 52°21′N 06°31′W
97 F20 **Wexford** Ir. Loch Garman. cultural region SE Ireland
30 L7 **Weyauwega** Wisconsin, N USA 44°16′N 88°54′W
11 U17 **Weyburn** Saskatchewan, S Canada 49°39′N 103°51′W
Weyer see Weyer Markt
109 U5 **Weyer Markt** var. Weyer. Oberösterreich, N Austria 47°52′N 14°39′E
100 H11 **Weyhe** Niedersachsen, NW Germany 53°00′N 08°52′E
97 L24 **Weymouth** S England, United Kingdom 50°36′N 02°28′W
19 P11 **Weymouth** Massachusetts, NE USA 42°13′N 70°56′W
99 H18 **Wezembeek-Oppem** Vlaams Brabant, C Belgium 50°51′N 04°28′E
98 M9 **Wezep** Gelderland, E Netherlands 52°28′N 06°E
184 M9 **Whakamaru** Waikato, North Island, New Zealand 38°27′S 175°48′E
184 O8 **Whakatane** Bay of Plenty, North Island, New Zealand 37°58′S 177°E
184 O8 **Whakatane** ≈ North Island, New Zealand
9 O9 **Whale Cove** var. Tikirarjuaq. Nunavut, C Canada 62°14′N 92°10′W
96 M2 **Whalsay** island NE Scotland, United Kingdom
184 L11 **Whangaehu** ≈ North Island, New Zealand
184 M6 **Whangamata** Waikato, North Island, New Zealand 37°13′S 175°54′E
184 Q9 **Whangara** Gisborne, North Island, New Zealand 38°34′S 178°12′E
184 K3 **Whangarei** Northland, North Island, New Zealand 35°44′S 174°18′E
184 K3 **Whangaruru Harbour** inlet North Island, New Zealand
25 V12 **Wharton** Texas, SW USA 29°19′N 96°08′W
173 U8 **Wharton Basin** var. West Australian Basin. undersea feature E Indian Ocean
185 E18 **Whataroa** West Coast, South Island, New Zealand 43°17′S 170°20′E
8 K10 **Wha Ti** prev. Lac la Martre. Northwest Territories, W Canada 63°10′N 117°12′W
8 J9 **Wha Ti** Northwest Territories, W Canada 63°10′N 117°12′W
184 K6 **Whatipu** Auckland, North Island, New Zealand 37°17′S 174°44′E
33 Y16 **Wheatland** Wyoming, C USA 42°03′N 104°57′W
14 D18 **Wheatley** Ontario, S Canada 42°06′N 82°27′W
30 M10 **Wheaton** Illinois, N USA 41°52′N 88°06′W
29 R7 **Wheaton** Minnesota, N USA 45°48′N 96°30′W
37 T4 **Wheat Ridge** Colorado, C USA 39°44′N 105°06′W
25 P2 **Wheeler** Texas, SW USA 35°26′N 100°17′W
23 O2 **Wheeler Lake** ⊠ Alabama, S USA
35 Y6 **Wheeler Peak** ▲ Nevada, W USA 39°00′N 114°17′W
37 T9 **Wheeler Peak** ▲ New Mexico, SW USA 36°34′N 105°25′W
31 S15 **Wheelersburg** Ohio, N USA 38°43′N 82°51′W
21 R2 **Wheeling** West Virginia, NE USA 40°05′N 80°43′W
97 L16 **Whernside** ▲ N England, United Kingdom 54°13′N 02°27′W
182 F9 **Whidbey, Point** headland South Australia 34°36′S 135°08′E
180 I7 **Whim Creek** Western Australia 20°51′S 117°54′E
10 L17 **Whistler** British Columbia, SW Canada 50°07′N 122°57′W
21 W8 **Whitakers** North Carolina, SE USA 36°06′N 77°43′W
14 H15 **Whitby** Ontario, S Canada 43°52′N 78°56′W
97 N15 **Whitby** N England, United Kingdom 54°29′N 00°37′W
10 G6 **White** ≈ Yukon, W Canada
13 T11 **White Bay** bay Newfoundland, Newfoundland and Labrador, E Canada
20 I8 **White Bluff** Tennessee, S USA 36°06′N 87°13′W
28 J6 **White Butte** ▲ North Dakota, N USA 46°23′N 103°18′W
19 R5 **White Cap Mountain** ▲ Maine, NE USA 45°33′N 69°15′W
22 J9 **White Castle** Louisiana, S USA 30°10′N 91°09′W
182 M5 **White Cliffs** New South Wales, SE Australia 30°52′S 143°04′E
31 P8 **White Cloud** Michigan, N USA 43°34′N 85°47′W
11 P14 **Whitecourt** Alberta, SW Canada 54°10′N 115°38′W
25 O2 **White Deer** Texas, SW USA 35°26′N 101°10′W
White Elster see Weisse Elster
24 M5 **Whiteface** Texas, SW USA 33°36′N 102°36′W
18 K7 **Whiteface Mountain** ▲ New York, NE USA 44°22′N 73°54′W
29 W5 **Whiteface Reservoir** ⊠ Minnesota, N USA
33 O7 **Whitefish** Montana, NW USA 48°24′N 114°20′W
31 N9 **Whitefish Bay** Wisconsin, N USA 43°09′N 87°54′W
31 Q3 **Whitefish Bay** lake bay Canada/USA
14 E11 **Whitefish Falls** Ontario, S Canada 46°06′N 81°42′W
14 B7 **Whitefish Lake** ◎ Ontario, S Canada
29 U6 **Whitefish Lake** ◎ Minnesota, C USA
31 Q3 **Whitefish Point** headland Michigan, N USA 46°46′N 84°57′W
31 O4 **Whitefish River** ≈ Michigan, N USA
25 O4 **Whiteflat** Texas, SW USA 34°06′N 100°55′W
27 V12 **White Hall** Arkansas, C USA 34°18′N 92°05′W
30 K14 **White Hall** Illinois, N USA 39°26′N 90°24′W
31 O8 **Whitehall** Michigan, N USA 43°24′N 86°21′W
18 L9 **Whitehall** New York, NE USA 43°33′N 73°24′W
31 S13 **Whitehall** Ohio, N USA 39°58′N 82°53′W
30 J7 **Whitehall** Wisconsin, N USA 44°22′N 91°20′W
97 J15 **Whitehaven** NW England, United Kingdom 54°33′N 03°35′W
10 I8 **Whitehorse** territory capital Yukon, W Canada 60°41′N 135°08′W
184 O7 **White Island** island NE New Zealand
14 K13 **White Lake** ◎ Ontario, SE Canada
22 H10 **White Lake** ◎ Louisiana, S USA
186 G7 **Whiteman Range** ▲▲ New Britain, E Papua New Guinea
183 Q15 **Whitemark** Tasmania, SE Australia 40°10′S 148°01′E
35 S9 **White Mountains** ▲▲ California/Nevada, W USA
19 N7 **White Mountains** ▲▲ Maine/New Hampshire, NE USA
80 F11 **White Nile** ◆ state C Sudan
81 E14 **White Nile** Ar. Al Baḥr al Abyaḍ, An Nīl al Abyaḍ, Bahr el Jebel. ≈ NE South Sudan
67 U7 **White Nile** var. Bahr el Jebel. ≈ S Sudan
25 W5 **White Oak Creek** ≈ Texas, SW USA
10 H9 **White Pass** pass Canada/USA
32 I9 **White Pass** pass Washington, NW USA
21 O9 **White Pine** Tennessee, S USA 36°06′N 83°17′W
18 K14 **White Plains** New York, NE USA 41°01′N 73°45′W
37 N13 **Whiteriver** Arizona, SW USA 33°50′N 109°57′W
28 M11 **White River** South Dakota, N USA 43°34′N 100°45′W
27 W12 **White River** ≈ Arkansas, SE USA
37 P3 **White River** ≈ Colorado/Utah, C USA
31 N15 **White River** ≈ Indiana, N USA
31 O8 **White River** ≈ Michigan, N USA
28 K11 **White River** ≈ South Dakota, N USA
25 O5 **White River** ≈ Texas, SW USA
18 M8 **White River** ≈ Vermont, NE USA
25 O5 **White River Lake** ⊠ Texas, SW USA
32 H11 **White Salmon** Washington, NW USA 45°43′N 121°29′W
18 I10 **Whitesboro** New York, NE USA 43°07′N 75°17′W
25 T5 **Whitesboro** Texas, SW USA 33°39′N 96°54′W
21 O7 **Whitesburg** Kentucky, S USA 37°07′N 82°52′W
White Sea see Beloye More
White Sea-Baltic Canal/White Sea Canal see Belomorsko-Baltiyskiy Kanal
63 I25 **Whiteside, Canal** channel S Chile
33 S10 **White Sulphur Springs** Montana, NW USA 46°33′N 110°54′W
21 R6 **White Sulphur Springs** West Virginia, NE USA 37°48′N 80°18′W
20 J6 **Whitesville** Kentucky, S USA 37°40′N 86°48′W
32 I10 **White Swan** Washington, NW USA 46°22′N 120°46′W
21 U12 **Whiteville** North Carolina, SE USA 34°20′N 78°42′W
20 F10 **Whiteville** Tennessee, S USA 35°19′N 89°09′W
77 Q13 **White Volta** var. Nakambé, Fr. Volta Blanche. ≈ Burkina Faso/Ghana
30 M9 **Whitewater** Wisconsin, N USA 42°51′N 88°43′W
37 P14 **Whitewater Baldy** ▲ New Mexico, SW USA 33°19′N 108°38′W
23 X17 **Whitewater Bay** bay Florida, SE USA
31 Q14 **Whitewater River** ≈ Indiana/Ohio, N USA
11 V16 **Whitewood** Saskatchewan, S Canada 50°19′N 102°16′W
28 J9 **Whitewood** South Dakota, N USA 44°27′N 103°38′W
25 U5 **Whitewright** Texas, SW USA 33°30′N 96°23′W
97 I15 **Whithorn** S Scotland, United Kingdom 54°44′N 04°26′W
184 M6 **Whitianga** Waikato, North Island, New Zealand 36°50′S 175°42′E
19 N11 **Whitinsville** Massachusetts, NE USA 42°06′N 71°40′W
20 M8 **Whitley City** Kentucky, S USA 36°45′N 84°29′W
21 Q11 **Whitmire** South Carolina, SE USA 34°30′N 81°36′W
31 R10 **Whitmore Lake** Michigan, N USA 42°26′N 83°44′W
195 N9 **Whitmore Mountains** ▲▲ Antarctica
14 I12 **Whitney** Ontario, SE Canada 45°29′N 78°11′W
25 T8 **Whitney** Texas, SW USA 31°56′N 97°20′W
25 S8 **Whitney, Lake** ⊠ Texas, SW USA
35 S11 **Whitney, Mount** ▲ California, W USA 37°45′N 119°55′W
181 Y6 **Whitsunday Group** island group Queensland, E Australia
25 S6 **Whitt** Texas, SW USA 32°55′N 98°01′W
29 U12 **Whittemore** Iowa, C USA 43°03′N 94°25′W
39 R12 **Whittier** Alaska, USA 60°46′N 148°40′W
35 T15 **Whittier** California, W USA 33°58′N 118°01′W
83 I25 **Whittlesea** Eastern Cape, S South Africa 32°08′S 26°51′E
20 K10 **Whitwell** Tennessee, S USA 35°12′N 85°31′W
8 L10 **Wholdaia Lake** ◎ Northwest Territories, NW Canada
182 H7 **Whyalla** South Australia 33°04′S 137°34′E
Whydah see Ouidah
14 F13 **Wiarton** Ontario, S Canada 44°44′N 81°10′W
171 O13 **Wiau** Sulawesi, C Indonesia 03°08′S 121°22′E
111 H15 **Wiązów** Ger. Wansen. Dolnośląskie, SW Poland 50°49′N 17°13′E
33 Y8 **Wibaux** Montana, NW USA 46°57′N 104°11′W
27 N6 **Wichita** Kansas, C USA 37°42′N 97°20′W
25 R5 **Wichita Falls** Texas, SW USA 33°55′N 98°30′W
26 L11 **Wichita Mountains** ▲▲ Oklahoma, C USA
25 R5 **Wichita River** ≈ Texas, SW USA
96 K6 **Wick** N Scotland, United Kingdom 58°26′N 03°06′W
36 K13 **Wickenburg** Arizona, SW USA 33°57′N 112°41′W
24 L8 **Wickett** Texas, SW USA 31°34′N 103°00′W
180 I7 **Wickham** Western Australia 20°40′S 117°11′E
182 M14 **Wickham, Cape** headland Tasmania, SE Australia 39°36′S 143°55′E
20 G7 **Wickliffe** Kentucky, S USA 36°58′N 89°04′W
97 G19 **Wicklow** Ir. Cill Mhantáin. E Ireland 52°59′N 06°03′W
97 F19 **Wicklow** Ir. Cill Mhantáin. cultural region E Ireland
97 G19 **Wicklow Head** Ir. Ceann Chill Mhantáin. headland E Ireland 52°57′N 06°00′W
97 F18 **Wicklow Mountains** Ir. Sléibhte Chill Mhantáin. ▲▲ E Ireland
14 H10 **Wicksteed Lake** ◎ Ontario, S Canada
Wida see Ouidah
65 G15 **Wideawake Airfield** ✈ (Georgetown) SW Ascension Island
97 K18 **Widnes** NW England, United Kingdom 53°22′N 02°44′W
110 H9 **Więcbork** Ger. Vandsburg. Kujawsko-pomorskie, C Poland 53°21′N 17°31′E
101 E17 **Wied** ≈ W Germany
101 F16 **Wiehl** Nordrhein-Westfalen, W Germany 50°57′N 07°33′E
111 L17 **Wieliczka** Małopolskie, S Poland 50°N 20°02′E
110 H12 **Wielkopolskie** ◆ province SW Poland
111 J14 **Wieluń** Sieradz, C Poland 51°14′N 18°33′E
109 X4 **Wien** Eng. Vienna, Hung. Bécs, Slvk. Vídeň, Slvn. Dunaj; anc. Vindobona. ● (Austria) Wien, NE Austria 48°13′N 16°22′E
109 X4 **Wien** off. Land Wien, Eng. Vienna. ◆ state NE Austria
109 X5 **Wiener Neustadt** Niederösterreich, E Austria 47°49′N 16°08′E
Wien, Land see Wien
110 G7 **Wieprza** Ger. Wipper. ≈ NW Poland
98 O10 **Wierden** Overijssel, E Netherlands 52°22′N 06°35′E
98 I7 **Wieringerwerf** Noord-Holland, NW Netherlands 52°51′N 05°01′E
Wieruschow see Wieruszów
111 I14 **Wieruszów** Ger. Wieruschow. Łódzkie, C Poland 51°18′N 18°09′E
109 V9 **Wies** Steiermark, SE Austria 46°40′N 15°16′E
Wiesbachhorn see Grosses Wiesbachhorn
101 G18 **Wiesbaden** Hessen, W Germany 50°06′N 08°14′E
Wieselburg and Ungarisch-Altenburg/Wieselburg-Ungarisch-Altenburg see Mosonmagyaróvár
Wiesenhof see Ostrołęka
101 G20 **Wiesloch** Baden-Württemberg, SW Germany 49°18′N 08°42′E
100 F10 **Wiesmoor** Niedersachsen, NW Germany 53°22′N 07°46′E
110 I7 **Wieżyca** Ger. Turmberg. hill Pomorskie, N Poland
97 L17 **Wigan** NW England, United Kingdom 53°33′N 02°38′W
37 U3 **Wiggins** Colorado, C USA 40°11′N 104°03′W
22 M8 **Wiggins** Mississippi, S USA 30°50′N 89°09′W
Wigorna Ceaster see Worcester
97 I14 **Wigtown** S Scotland, United Kingdom 54°53′N 04°27′W
97 H14 **Wigtown** cultural region SW Scotland, United Kingdom
97 I15 **Wigtown Bay** bay SW Scotland, United Kingdom
98 L13 **Wijchen** Gelderland, SE Netherlands 51°48′N 05°44′E
92 N1 **Wijdefjorden** fjord NW Svalbard
98 M10 **Wijhe** Overijssel, E Netherlands 52°22′N 06°07′E
98 J12 **Wijk bij Duurstede** Utrecht, C Netherlands 51°58′N 05°21′E
98 J13 **Wijk en Aalburg** Noord-Brabant, S Netherlands 51°46′N 05°06′E
99 H16 **Wijnegem** Antwerpen, N Belgium 51°13′N 04°32′E
14 E11 **Wikwemikong** Manitoulin Island, Ontario, S Canada 45°46′N 81°43′W
108 H7 **Wil** Sankt Gallen, NE Switzerland 47°28′N 09°03′E
29 R16 **Wilber** Nebraska, C USA 40°28′N 96°57′W
32 K8 **Wilbur** Washington, NW USA 47°45′N 118°42′W
27 Q11 **Wilburton** Oklahoma, C USA 34°57′N 95°20′W
182 M6 **Wilcannia** New South Wales, SE Australia 31°34′S 143°23′E
18 D12 **Wilcox** Pennsylvania, NE USA 41°34′N 78°40′W
Wilczek Land see Vil'cheka, Zemlya
109 U6 **Wildalpen** Steiermark, E Austria 47°40′N 14°54′E
31 O13 **Wildcat Creek** ≈ Indiana, N USA
108 L9 **Wilde Kreuzspitze** It. Picco di Croce. ▲ Austria/Italy 46°53′N 10°51′E
Wildenschwert see Ústí nad Orlicí
98 O6 **Wildervank** Groningen, NE Netherlands 53°04′N 06°52′E
100 G11 **Wildeshausen** Niedersachsen, NW Germany 52°54′N 08°26′E
108 D10 **Wildhorn** ▲ SW Switzerland 46°21′N 07°22′E
11 R17 **Wild Horse** Alberta, SW Canada 49°00′N 110°19′W
27 N12 **Wildhorse Creek** ≈ Oklahoma, C USA
28 L14 **Wild Horse Hill** ▲ Nebraska, C USA 41°52′N 101°56′W
109 W8 **Wildon** Steiermark, SE Austria 46°53′N 15°29′E
24 M2 **Wildorado** Texas, SW USA 35°12′N 102°10′W
29 R6 **Wild Rice River** ≈ Minnesota/North Dakota, N USA
Wilejka see Vilyeyka
195 Y9 **Wilhelm II Coast** physical region Antarctica
195 X9 **Wilhelm II Land** physical region Antarctica
55 U11 **Wilhelmina Gebergte** ▲▲ C Suriname
18 B13 **Wilhelm, Lake** ⊠ Pennsylvania, NE USA
92 O2 **Wilhelmøya** island C Svalbard
Wilhelm-Pieck-Stadt see Guben
109 W4 **Wilhelmsburg** Niederösterreich, E Austria 48°07′N 15°37′E
100 G10 **Wilhelmshaven** Niedersachsen, NW Germany 53°31′N 08°07′E
Wilia/Wilja see Neris
18 H13 **Wilkes Barre** Pennsylvania, NE USA 41°15′N 75°50′W
21 R9 **Wilkesboro** North Carolina, SE USA 36°08′N 81°09′W
195 W15 **Wilkes Coast** physical region Antarctica
189 W12 **Wilkes Island** island N Wake Island
195 X12 **Wilkes Land** physical region Antarctica
11 S15 **Wilkie** Saskatchewan, S Canada 52°27′N 108°42′W
194 I6 **Wilkins Ice Shelf** ice shelf Antarctica
182 D4 **Wilkinsons Lakes** salt lake South Australia
Wiłkomierz see Ukmergė
182 K11 **Willalooka** South Australia 36°24′S 140°20′E
32 G11 **Willamette River** ≈ Oregon, NW USA
183 O8 **Willandra Billabong Creek** seasonal river New South Wales, SE Australia
32 F9 **Willapa Bay** inlet Washington, NW USA
27 T7 **Willard** Missouri, C USA 37°18′N 93°25′W
37 S12 **Willard** New Mexico, SW USA 34°36′N 106°01′W
31 S12 **Willard** Ohio, N USA 41°03′N 82°43′W
36 L1 **Willard** Utah, W USA 41°23′N 112°01′W
186 G6 **Willaumez Peninsula** headland New Britain, E Papua New Guinea 05°03′S 150°04′E
37 N15 **Willcox** Arizona, SW USA 32°13′N 109°49′W
37 N16 **Willcox Playa** salt flat Arizona, SW USA
99 G17 **Willebroek** Antwerpen, C Belgium 51°04′N 04°22′E
45 P16 **Willemstad** ○ Curaçao 12°07′N 68°54′W
99 G14 **Willemstad** Noord-Brabant, S Netherlands 51°40′N 04°27′E
11 S11 **William** ≈ Saskatchewan, C Canada
23 O6 **William "Bill" Dannelly Reservoir** ⊠ Alabama, S USA
182 G3 **William Creek** South Australia 28°55′S 136°23′E
181 T15 **William, Mount** ▲ South Australia
36 K11 **Williams** Arizona, SW USA 35°15′N 112°11′W
29 X14 **Williamsburg** Iowa, C USA 41°39′N 92°00′W
20 M8 **Williamsburg** Kentucky, S USA 36°44′N 84°10′W
31 R15 **Williamsburg** Ohio, N USA 39°00′N 84°02′W
21 X6 **Williamsburg** Virginia, NE USA 37°17′N 76°43′W
10 M15 **Williams Lake** British Columbia, SW Canada 52°08′N 122°09′W
21 P6 **Williamson** West Virginia, NE USA 37°42′N 82°16′W
31 N13 **Williamsport** Indiana, N USA 40°18′N 87°18′W
18 G13 **Williamsport** Pennsylvania, NE USA 41°16′N 77°03′W
21 W9 **Williamston** North Carolina, SE USA 35°53′N 77°05′W
21 P11 **Williamston** South Carolina, SE USA 34°37′N 82°28′W
20 M4 **Williamstown** Kentucky, S USA 38°39′N 84°32′W
18 L10 **Williamstown** Massachusetts, NE USA 42°41′N 73°11′W
18 J16 **Willingboro** New Jersey, NE USA 40°01′N 74°52′W
11 Q14 **Willingdon** Alberta, SW Canada 53°49′N 112°08′W
25 W10 **Willis** Texas, SW USA 30°25′N 95°28′W
108 F8 **Willisau** Luzern, W Switzerland 47°07′N 08°00′E
83 F24 **Williston** Northern Cape, W South Africa 31°20′S 20°52′E
23 V10 **Williston** Florida, SE USA 29°23′N 82°27′W
28 J3 **Williston** North Dakota, N USA 48°07′N 103°37′W
21 Q13 **Williston** South Carolina, SE USA 33°24′N 81°25′W
10 L12 **Williston Lake** ⊠ British Columbia, W Canada
34 L5 **Willits** California, W USA 39°24′N 123°22′W
29 T8 **Willmar** Minnesota, N USA 45°07′N 95°02′W
10 K11 **Will, Mount** ▲ British Columbia, W Canada 57°31′N 128°48′W
31 T11 **Willoughby** Ohio, N USA 41°38′N 81°24′W
11 U17 **Willow Bunch** Saskatchewan, S Canada 49°30′N 105°41′W
32 J11 **Willow Creek** ≈ Oregon, NW USA
39 R11 **Willow Lake** Alaska, USA 61°44′N 150°02′W
8 I9 **Willowlake** ≈ Northwest Territories, NW Canada
83 H25 **Willowmore** Eastern Cape, S South Africa 33°18′S 23°30′E
30 L5 **Willow Reservoir** ⊠ Wisconsin, N USA
35 N5 **Willows** California, W USA 39°28′N 122°12′W
27 V7 **Willow Springs** Missouri, C USA 36°59′N 91°58′W
182 I7 **Wilmington** South Australia 32°42′S 138°08′E
21 Y2 **Wilmington** Delaware, NE USA 39°45′N 75°33′W
21 V12 **Wilmington** North Carolina, SE USA 34°14′N 77°55′W
31 R14 **Wilmington** Ohio, N USA 39°27′N 83°49′W
20 M6 **Wilmore** Kentucky, S USA 37°51′N 84°39′W
29 R8 **Wilmot** South Dakota, N USA 45°24′N 96°51′W
Wilna/Wilno see Vilnius
101 G16 **Wilnsdorf** Nordrhein-Westfalen, W Germany 50°49′N 08°06′E
99 G16 **Wilrijk** Antwerpen, N Belgium 51°11′N 04°25′E
100 I10 **Wilseder Berg** hill NW Germany
67 Z12 **Wilshaw Ridge** undersea feature W Indian Ocean 17°30′S 56°30′E
21 V9 **Wilson** North Carolina, SE USA 35°43′N 77°56′W
25 N5 **Wilson** Texas, SW USA 33°21′N 101°44′W
182 A7 **Wilson Bluff** headland South Australia/Western Australia 31°41′S 129°01′E
35 Y7 **Wilson Creek Range** ▲▲ Nevada, W USA
23 O1 **Wilson Lake** ⊠ Alabama, S USA
26 M4 **Wilson Lake** ⊠ Kansas, SE USA
37 P7 **Wilson, Mount** ▲ Colorado, C USA 37°50′N 107°59′W
183 P13 **Wilsons Promontory** peninsula Victoria, SE Australia
29 Y14 **Wilton** Iowa, C USA 41°35′N 91°01′W
19 P7 **Wilton** Maine, NE USA 44°35′N 70°15′W
28 M5 **Wilton** North Dakota, N USA 47°09′N 100°46′W
97 L22 **Wiltshire** cultural region S England, United Kingdom
99 M23 **Wiltz** Diekirch, NW Luxembourg 49°58′N 05°56′E
180 K9 **Wiluna** Western Australia 26°34′S 120°14′E
99 M23 **Wilwerwiltz** Diekirch, NE Luxembourg 49°59′N 06°00′E
29 P5 **Wimbledon** North Dakota, N USA 47°08′N 98°25′W
42 K7 **Wina** var. Güina. Jinotega, N Nicaragua 14°00′N 85°14′W
31 O12 **Winamac** Indiana, N USA 41°03′N 86°37′W
81 G19 **Winam Gulf** var. Kavirondo Gulf. gulf SW Kenya
83 I22 **Winburg** Free State, C South Africa 28°31′S 27°01′E
19 N10 **Winchendon** Massachusetts, NE USA 42°41′N 72°01′W
14 M13 **Winchester** Ontario, SE Canada 45°07′N 75°19′W
97 M23 **Winchester** hist. Wintanceaster, Lat. Venta Belgarum. S England, United Kingdom 51°04′N 01°19′W
32 M10 **Winchester** Idaho, NW USA 46°13′N 116°35′W
30 J14 **Winchester** Illinois, N USA 39°38′N 90°28′W
31 Q13 **Winchester** Indiana, N USA 40°11′N 84°57′W
20 M5 **Winchester** Kentucky, S USA 38°00′N 84°10′W
18 M10 **Winchester** New Hampshire, NE USA 42°46′N 72°21′W
20 K10 **Winchester** Tennessee, S USA 35°11′N 86°06′W
21 V3 **Winchester** Virginia, NE USA 39°11′N 78°12′W
99 L22 **Wincrange** Diekirch, NW Luxembourg 50°03′N 05°55′E
10 I5 **Wind** ≈ Yukon, NW Canada
183 S8 **Windamere, Lake** ◎ New South Wales, SE Australia
Windau see Ventspils, Latvia
Windau see Venta, Latvia/Lithuania
18 D15 **Windber** Pennsylvania, NE USA 40°12′N 78°47′W
23 T3 **Winder** Georgia, SE USA 33°59′N 83°43′W
97 K15 **Windermere** NW England, United Kingdom 54°24′N 02°54′W
14 C7 **Windermere Lake** ◎ Ontario, S Canada
31 U11 **Windham** Ohio, N USA 41°14′N 81°03′W
83 D19 **Windhoek** Ger. Windhuk. ● (Namibia) Khomas, C Namibia 22°34′S 17°06′E
83 D20 **Windhoek** ✈ Khomas, C Namibia 22°31′S 17°04′E
Windhuk see Windhoek
15 O8 **Windigo** Québec, SE Canada 47°45′N 73°19′W
15 O8 **Windigo** ≈ Québec, SE Canada
Windischfeistritz see Slovenska Bistrica
109 T6 **Windischgarsten** Oberösterreich, W Austria 47°42′N 14°21′E
Windischgraz see Slovenj Gradec
37 T16 **Wind Mountain** ▲ New Mexico, SW USA 32°01′N 105°35′W
29 T10 **Windom** Minnesota, N USA 43°52′N 95°07′W
37 Q7 **Windom Peak** ▲ Colorado, C USA 37°37′N 107°35′W
181 U9 **Windorah** Queensland, C Australia 25°25′S 142°41′E
37 O10 **Window Rock** Arizona, SW USA 35°40′N 109°03′W
31 N9 **Wind Point** headland Wisconsin, N USA 42°46′N 87°46′W
33 U14 **Wind River** ≈ Wyoming, C USA
13 P15 **Windsor** Nova Scotia, SE Canada 45°00′N 64°09′W
14 C17 **Windsor** Ontario, S Canada 42°18′N 83°W
15 Q12 **Windsor** Québec, SE Canada 45°34′N 72°00′W
97 N22 **Windsor** S England, United Kingdom 51°29′N 00°39′W
37 T3 **Windsor** Colorado, C USA 40°28′N 104°54′W
18 M12 **Windsor** Connecticut, NE USA 41°51′N 72°38′W
27 T5 **Windsor** Missouri, C USA 38°31′N 93°31′W
21 X9 **Windsor** North Carolina, SE USA 36°00′N 76°57′W
18 M12 **Windsor Locks** Connecticut, NE USA 41°55′N 72°37′W
25 R5 **Windthorst** Texas, SW USA 33°34′N 98°26′W
45 Z14 **Windward Islands** island group E West Indies
Windward Islands see Barlavento, Ilhas de, Cape Verde
Windward Islands see Vent, Îles du, Archipel de la Société, French Polynesia
44 K8 **Windward Passage** Sp. Paso de los Vientos. channel Cuba/Haiti
55 T9 **Wineperu** C Guyana 06°10′N 58°34′W
23 O3 **Winfield** Alabama, S USA 33°55′N 87°49′W
29 Y15 **Winfield** Iowa, C USA 41°07′N 91°26′W
27 O7 **Winfield** Kansas, C USA 37°14′N 97°00′W
25 W6 **Winfield** Texas, SW USA 33°10′N 95°06′W
21 Q4 **Winfield** West Virginia, NE USA 38°30′N 81°54′W
29 N5 **Wing** North Dakota, N USA 47°06′N 100°16′W
183 U7 **Wingham** New South Wales, SE Australia 31°52′S 152°24′E
12 G16 **Wingham** Ontario, S Canada 43°54′N 81°19′W
33 T8 **Winifred** Montana, NW USA 47°33′N 109°26′W
12 E9 **Winisk Lake** ◎ Ontario, C Canada
24 L8 **Wink** Texas, SW USA 31°45′N 103°09′W
36 M14 **Winkelman** Arizona, SW USA 32°59′N 110°46′W
11 X17 **Winkler** Manitoba, S Canada 49°12′N 97°55′W
109 Q9 **Winklern** Tirol, W Austria 46°54′N 12°54′E
Winkowitz see Vinkovci
32 G9 **Winlock** Washington, NW USA 46°29′N 122°56′W
77 P17 **Winneba** SE Ghana 05°22′N 00°38′W
29 U11 **Winnebago** Minnesota, N USA 43°46′N 94°10′W
29 R13 **Winnebago** Nebraska, C USA 42°14′N 96°28′W
30 M7 **Winnebago, Lake** ◎ Wisconsin, N USA
30 M7 **Winneconne** Wisconsin, N USA 44°07′N 88°44′W
35 T3 **Winnemucca** Nevada, W USA 40°59′N 117°44′W
35 R4 **Winnemucca Lake** ◎ Nevada, W USA
101 H21 **Winnenden** Baden-Württemberg, SW Germany 48°52′N 09°22′E
29 N11 **Winner** South Dakota, N USA 43°22′N 99°51′W
33 U9 **Winnett** Montana, NW USA 47°00′N 108°18′W
14 I9 **Winneway** Québec, SE Canada 47°35′N 78°33′W
22 H6 **Winnfield** Louisiana, S USA 31°55′N 92°38′W
29 U4 **Winnibigoshish, Lake** ◎ Minnesota, N USA
25 X11 **Winnie** Texas, SW USA 29°49′N 94°22′W
11 Y16 **Winnipeg** province capital Manitoba, S Canada 49°53′N 97°10′W
11 X16 **Winnipeg** ✈ Manitoba, S Canada 49°56′N 97°16′W
0 J8 **Winnipeg** ≈ Manitoba, S Canada
11 X16 **Winnipeg Beach** Manitoba, S Canada 50°25′N 96°59′W
11 W14 **Winnipeg, Lake** ◎ Manitoba, C Canada
11 W15 **Winnipegosis** Manitoba, S Canada 51°36′N 99°59′W

◆ Country
● Country Capital
◇ Dependent Territory
○ Dependent Territory Capital
◆ Administrative Regions
✈ International Airport
▲ Mountain
▲▲ Mountain Range
Volcano
≈ River
◎ Lake
⊠ Reservoir

Winnipegosis, Lake – Xinhe

11 W15 **Winnipegosis, Lake** ◎ Manitoba, C Canada
19 O8 **Winnipesaukee, Lake** ◎ New Hampshire, NE USA
22 I6 **Winnsboro** Louisiana, S USA 32°09′N 91°43′W
21 R12 **Winnsboro** South Carolina, SE USA 34°22′N 81°05′W
25 W6 **Winnsboro** Texas, SW USA 33°01′N 95°16′W
29 X10 **Winona** Minnesota, N USA 44°03′N 91°37′W
22 L4 **Winona** Mississippi, S USA 33°30′N 89°42′W
27 W7 **Winona** Missouri, C USA 37°00′N 91°19′W
25 W7 **Winona** Texas, SW USA 32°29′N 95°10′W
18 M7 **Winooski River** ≈ Vermont, NE USA
98 P6 **Winschoten** Groningen, NE Netherlands 53°09′N 07°03′E
100 J10 **Winsen** Niedersachsen, N Germany 53°22′N 10°13′E
36 M11 **Winslow** Arizona, SW USA 35°01′N 110°42′W
19 Q7 **Winslow** Maine, NE USA 44°33′N 69°35′W
18 M12 **Winsted** Connecticut, NE USA 41°55′N 73°03′W
32 F14 **Winston** Oregon, NW USA 43°07′N 123°24′W
21 S9 **Winston Salem** North Carolina, SE USA 36°06′N 80°15′W
98 N5 **Winsum** Groningen, NE Netherlands 53°20′N 06°31′E
Wintanceaster *see* Winchester
23 W11 **Winter Garden** Florida, SE USA 28°34′N 81°35′W
10 J16 **Winter Harbour** Vancouver Island, British Columbia, SW Canada 50°28′N 128°03′W
23 W12 **Winter Haven** Florida, SE USA 28°01′N 81°43′W
23 X11 **Winter Park** Florida, SE USA 28°36′N 81°20′W
25 P8 **Winters** Texas, SW USA 31°57′N 99°57′W
29 U15 **Winterset** Iowa, C USA 41°19′N 94°00′W
98 O12 **Winterswijk** Gelderland, E Netherlands 51°58′N 06°44′E
108 G6 **Winterthur** Zürich, NE Switzerland 47°30′N 08°43′E
29 U9 **Winthrop** Minnesota, N USA 44°32′N 94°22′W
32 J7 **Winthrop** Washington, NW USA 48°28′N 120°13′W
181 V7 **Winton** Queensland, E Australia 22°22′S 143°04′E
185 C24 **Winton** Southland, South Island, New Zealand 46°08′S 168°20′E
21 X8 **Winton** North Carolina, SE USA 36°24′N 76°57′W
101 K15 **Wipper** ≈ C Germany
101 K14 **Wipper** ≈ C Germany
Wipper *see* Wieprza
182 G6 **Wirraminna** South Australia 31°10′S 136°13′E
182 F4 **Wirrida** South Australia 29°34′S 134°33′E
182 F7 **Wirrulla** South Australia 32°27′S 134°33′E
Wirsitz *see* Wyrzysk
Wirz-See *see* Võrtsjärv
97 O19 **Wisbech** E England, United Kingdom 52°39′N 00°08′E
Wisby *see* Visby
19 Q8 **Wiscasset** Maine, NE USA 44°01′N 69°41′W
Wischau *see* Vyškov
30 J5 **Wisconsin** *off.* State of Wisconsin, *also known as* Badger State. ◈ *state* N USA
30 L8 **Wisconsin Dells** Wisconsin, N USA 43°37′N 89°43′W
30 L8 **Wisconsin, Lake** ▣ Wisconsin, N USA
30 L7 **Wisconsin Rapids** Wisconsin, N USA 44°24′N 89°50′W
30 L7 **Wisconsin River** ≈ Wisconsin, N USA
33 P11 **Wisdom** Montana, NW USA 45°36′N 113°27′W
21 P7 **Wise** Virginia, NE USA 37°00′N 82°36′W
39 Q7 **Wiseman** Alaska, USA 67°24′N 150°06′W
96 J12 **Wishaw** W Scotland, United Kingdom 55°47′N 03°56′W
29 O6 **Wishek** North Dakota, N USA 46°12′N 99°33′W
32 I11 **Wishram** Washington, NW USA 45°40′N 120°53′W
111 J17 **Wisła** Śląskie, S Poland 49°39′N 18°50′E
110 K11 **Wisła** *Eng.* Vistula, *Ger.* Weichsel. ≈ C Poland
Wiślany, Zalew *see* Vistula Lagoon
111 M16 **Wisłoka** ≈ SE Poland
100 L9 **Wismar** Mecklenburg-Vorpommern, N Germany 53°54′N 11°28′E
29 R14 **Wisner** Nebraska, C USA 41°59′N 96°54′W
103 V4 **Wissembourg** *var.* Weissenburg. Bas-Rhin, NE France 49°03′N 07°57′E
30 J6 **Wissota, Lake** ▣ Wisconsin, N USA
97 O18 **Witham** ≈ E England, United Kingdom
97 O17 **Withernsea** E England, United Kingdom 53°46′N 00°01′W
37 Q13 **Withington, Mount** ▲ New Mexico, SW USA 33°52′N 107°29′W
23 U8 **Withlacoochee River** ≈ Florida/Georgia, SE USA
110 H11 **Witkowo** Wielkopolskie, C Poland 52°27′N 17°49′E
97 M21 **Witney** S England, United Kingdom 51°47′N 01°30′W
101 E15 **Witten** Nordrhein-Westfalen, W Germany 51°25′N 07°19′E
101 N14 **Wittenberg** Sachsen-Anhalt, E Germany 51°53′N 12°39′E
30 L6 **Wittenberg** Wisconsin, N USA 44°53′N 89°20′W
100 L11 **Wittenberge** Brandenburg, N Germany 52°59′N 11°45′E
103 U7 **Wittenheim** Haut-Rhin, NE France 47°49′N 07°19′E
180 I7 **Wittenoom** Western Australia 22°17′S 118°22′E
Wittingau *see* Třeboň
100 K12 **Wittingen** Niedersachsen, C Germany 52°42′N 10°43′E
101 E18 **Wittlich** Rheinland-Pfalz, SW Germany 49°59′N 06°54′E
100 F9 **Wittmund** Niedersachsen, NW Germany 53°34′N 07°46′E
100 M10 **Wittstock** Brandenburg, NE Germany 53°10′N 12°29′E
186 F6 **Witu Islands** *island group* E Papua New Guinea
110 O7 **Wiżajny** Podlaskie, NE Poland 54°22′N 22°51′E
55 W10 **W. J. van Blommesteinmeer** ▣ E Suriname
110 L11 **Wkra** *Ger.* Soldau. ≈ C Poland
110 I6 **Władysławowo** Pomorskie, N Poland 54°48′N 18°25′E
Wlaschim *see* Vlašim
111 E14 **Wleń** *Ger.* Lähn. Dolnośląskie, SW Poland 51°00′N 15°39′E
110 J11 **Włocławek** *Ger./Rus.* Vlotslavsk. Kujawsko-pomorskie, C Poland 52°39′N 19°03′E
110 P13 **Włodawa** *Rus.* Vlodava. Lubelskie, SE Poland 51°33′N 23°31′E
Włodzimierz *see* Volodymyr-Volyns'kyy
111 K15 **Włoszczowa** Świętokrzyskie, C Poland 50°51′N 19°58′E
83 C19 **Wlotzkasbaken** Erongo, W Namibia 22°26′S 14°30′E
15 R12 **Woburn** Québec, SE Canada 45°22′N 70°52′W
19 O11 **Woburn** Massachusetts, NE USA 42°28′N 71°09′W
Wocheiner Feistritz *see* Bohinjska Bistrica
Wöchma *see* Võhma
147 S11 **Wodil** *var.* Vuadil'. Farg'ona Viloyati, E Uzbekistan 40°10′N 71°43′E
181 V14 **Wodonga** Victoria, SE Australia 36°11′S 146°55′E
111 I17 **Wodzisław Śląski** *Ger.* Loslau. Śląskie, S Poland 49°59′N 18°27′E
98 I11 **Woerden** Zuid-Holland, C Netherlands 52°06′N 04°54′E
98 I8 **Wognum** Noord-Holland, NW Netherlands 52°40′N 05°01′E
Wohlau *see* Wołów
108 F7 **Wohlen** Aargau, NW Switzerland 47°21′N 08°17′E
195 R2 **Wohlthat Massivet** *Eng.* Wohlthat Mountains. ▲▲ Antarctica
Wohlthat Mountains *see* Wohlthat Massivet
Wojerecy *see* Hoyerswerda
Wōjjā *see* Wotje Atoll
Wojwodina *see* Vojvodina
171 V15 **Wokam, Pulau** *island* Kepulauan Aru, E Indonesia
97 N22 **Woking** SE England, United Kingdom 51°20′N 00°34′W
Woldenberg Neumark *see* Dobiegniew
188 K15 **Woleai Atoll** *atoll* Caroline Islands, W Micronesia
Woleu *see* Uolo, Río
79 E17 **Woleu-Ntem** *off.* Province du Woleu-Ntem, *var.* Le Woleu-Ntem. ◈ *province* W Gabon
Woleu-Ntem, Province du *see* Woleu-Ntem
32 F15 **Wolf Creek** Oregon, NW USA 42°40′N 123°22′W
26 K9 **Wolf Creek** ≈ Oklahoma/Texas, SW USA
37 R7 **Wolf Creek Pass** *pass* Colorado, C USA
19 O9 **Wolfeboro** New Hampshire, NE USA 43°34′N 71°10′W
25 U5 **Wolfe City** Texas, SW USA 33°22′N 96°04′W
14 L15 **Wolfe Island** *island* Ontario, SE Canada
101 M14 **Wolfen** Sachsen-Anhalt, E Germany 51°40′N 12°16′E
100 J13 **Wolfenbüttel** Niedersachsen, C Germany 52°10′N 10°33′E
109 T4 **Wolfern** Oberösterreich, N Austria 48°06′N 14°16′E
109 Q6 **Wolfgangsee** *var.* Abersee, St Wolfgangsee. ◎ N Austria
39 P9 **Wolf Mountain** ▲ Alaska, USA 65°20′N 154°08′W
33 X7 **Wolf Point** Montana, NW USA 48°05′N 105°40′W
22 L8 **Wolf River** ≈ Mississippi, S USA
30 M7 **Wolf River** ≈ Wisconsin, N USA
109 U9 **Wolfsberg** Kärnten, SE Austria 46°50′N 14°50′E
100 K12 **Wolfsburg** Niedersachsen, N Germany 52°25′N 10°47′E
57 B17 **Wolf, Volcán** ⛰ Galapagos Islands, Ecuador, E Pacific Ocean 0°01′N 91°22′W
100 O8 **Wolgast** Mecklenburg-Vorpommern, NE Germany 54°04′N 13°47′E
108 F8 **Wolhusen** Luzern, W Switzerland 47°04′N 08°06′E
110 D8 **Wolin** *Ger.* Wollin. Zachodnio-pomorskie, NW Poland 53°52′N 14°35′E
109 Y3 **Wolkersdorf** Niederösterreich, NE Austria 48°24′N 16°31′E
Wołkowysk *see* Vawkavysk
Wöllan *see* Velenje
8 J6 **Wollaston, Cape** *headland* Victoria Island, Northwest Territories, NW Canada 71°00′N 118°21′W
63 J25 **Wollaston, Isla** *island* S Chile
11 U11 **Wollaston Lake** Saskatchewan, C Canada 58°05′N 103°38′W
11 T10 **Wollaston Lake** ◎ Saskatchewan, C Canada
8 J6 **Wollaston Peninsula** *peninsula* Victoria Island, Northwest Territories/Nunavut NW Canada
Wollin *see* Wolin
183 S9 **Wollongong** New South Wales, SE Australia 34°25′S 150°52′E
Wolmar *see* Valmiera
100 L13 **Wolmirstedt** Sachsen-Anhalt, C Germany 52°15′N 11°37′E
110 M11 **Wołomin** Mazowieckie, C Poland 52°20′N 21°11′E
110 G13 **Wołów** *Ger.* Wohlau. Dolnośląskie, SW Poland 51°21′N 16°40′E
Wołożyn *see* Valozhyn
14 G11 **Wolseley Bay** Ontario, S Canada 46°05′N 80°16′W
29 P10 **Wolsey** South Dakota, N USA 44°22′N 98°28′W
110 F12 **Wolsztyn** Wielkopolskie, C Poland 52°07′N 16°07′E
98 M7 **Wolvega** *Fris.* Wolvegea. Fryslân, N Netherlands 52°53′N 06°E
Wolvegea *see* Wolvega
97 K19 **Wolverhampton** C England, United Kingdom 52°36′N 02°08′W
Wolverine State *see* Michigan
99 G18 **Wolvertem** Vlaams Brabant, C Belgium 50°55′N 04°19′E
99 H16 **Wommelgem** Antwerpen, N Belgium 51°12′N 04°32′E
186 D7 **Wonenara** *var.* Wonerara. Eastern Highlands, C Papua New Guinea 06°46′S 145°54′E
Wonerara *see* Wonenara
Wongalara Lake *see* Wongalarroo Lake
183 N6 **Wongalarroo Lake** *var.* Wongalara Lake. *seasonal lake* New South Wales, SE Australia
163 Y15 **Wonju** *Jap.* Genshū; *prev.* Wŏnju. N South Korea 37°21′N 127°57′E
Wŏnju *see* Wonju
10 M12 **Wonowon** British Columbia, W Canada 56°46′N 121°54′W
163 X13 **Wŏnsan** SE North Korea 39°11′N 127°21′E
183 O13 **Wonthaggi** Victoria, SE Australia 38°38′S 145°37′E
23 N2 **Woodall Mountain** ▲ Mississippi, S USA 34°47′N 88°14′W
23 W7 **Woodbine** Georgia, SE USA 30°58′N 81°43′W
29 S14 **Woodbine** Iowa, C USA 41°44′N 95°42′W
18 J17 **Woodbine** New Jersey, NE USA 39°12′N 74°47′W
21 W4 **Woodbridge** Virginia, NE USA 38°40′N 77°17′W
183 V4 **Woodburn** New South Wales, SE Australia 29°07′S 153°23′E
32 G11 **Woodburn** Oregon, NW USA 45°08′N 122°51′W
20 K9 **Woodbury** Tennessee, S USA 35°49′N 86°06′W
183 V5 **Wooded Bluff** *headland* New South Wales, SE Australia 29°24′S 153°22′E
183 V3 **Woodenbong** New South Wales, SE Australia 28°24′S 152°39′E
35 R11 **Woodlake** California, W USA 36°24′N 119°06′W
35 N7 **Woodland** California, W USA 38°41′N 121°46′W
19 T5 **Woodland** Maine, NE USA 45°10′N 67°25′W
32 G10 **Woodland** Washington, NW USA 45°54′N 122°44′W
37 T5 **Woodland Park** Colorado, C USA 38°59′N 105°03′W
186 I9 **Woodlark Island** *var.* Murua Island. *island* SE Papua New Guinea
Woodle Island *see* Kuria
11 T17 **Wood Mountain** ▲ Saskatchewan, S Canada
30 K15 **Wood River** Illinois, N USA 38°51′N 90°06′W
29 P16 **Wood River** Nebraska, C USA 40°48′N 98°33′W
39 R9 **Wood River** ≈ Alaska, USA
39 O13 **Wood River Lakes** *lakes* Alaska, USA
182 C1 **Woodroffe, Mount** ▲ South Australia 26°19′S 131°42′E
21 P11 **Woodruff** South Carolina, SE USA 34°44′N 82°02′W
30 K4 **Woodruff** Wisconsin, N USA 45°55′N 89°41′W
25 T14 **Woodsboro** Texas, SW USA 28°14′N 97°19′W
31 U13 **Woodsfield** Ohio, N USA 39°45′N 81°07′W
181 P4 **Woods, Lake** ◎ Northern Territory, N Australia
11 Z16 **Woods, Lake of the** *Fr.* Lac des Bois. ◎ Canada/USA
25 Q6 **Woodson** Texas, SW USA 33°00′N 99°01′W
13 N14 **Woodstock** New Brunswick, SE Canada 46°10′N 67°38′W
14 F16 **Woodstock** Ontario, S Canada 43°07′N 80°46′W
30 M10 **Woodstock** Illinois, N USA 42°18′N 88°27′W
18 M9 **Woodstock** Vermont, NE USA 43°37′N 72°33′W
21 U4 **Woodstock** Virginia, NE USA 38°55′N 78°31′W
19 N8 **Woodsville** New Hampshire, NE USA 44°08′N 72°02′W
184 M12 **Woodville** Manawatu-Wanganui, North Island, New Zealand 40°22′S 175°59′E
22 J7 **Woodville** Mississippi, S USA 31°06′N 91°18′W
25 X9 **Woodville** Texas, SW USA 30°47′N 94°26′W
26 K9 **Woodward** Oklahoma, C USA 36°26′N 99°25′W
29 O5 **Woodworth** North Dakota, N USA 47°06′N 99°19′W
171 W12 **Wool** Papua, E Indonesia 01°38′S 135°34′E
183 V5 **Woolgoolga** New South Wales, E Australia 30°04′S 153°09′E
182 H6 **Woomera** South Australia 31°12′S 136°52′E
19 O12 **Woonsocket** Rhode Island, NE USA 41°58′N 71°27′W
29 P10 **Woonsocket** South Dakota, N USA 44°03′N 98°16′W
31 T12 **Wooster** Ohio, N USA 40°48′N 81°56′W
80 L12 **Woqooyi Galbeed** *off.* Gobolka Woqooyi Galbeed. ◈ *region* NW Somalia
Woqooyi Galbeed, Gobolka *see* Woqooyi Galbeed
108 E8 **Worb** Bern, C Switzerland 46°54′N 07°36′E
83 F26 **Worcester** Western Cape, SW South Africa 33°41′S 19°22′E
97 L20 **Worcester** *hist.* Wigorna Ceaster. W England, United Kingdom 52°11′N 02°13′W
19 N11 **Worcester** Massachusetts, NE USA 42°17′N 71°48′W
97 L20 **Worcestershire** *cultural region* C England, United Kingdom
32 H16 **Worden** Oregon, NW USA 42°04′N 121°50′W
109 O6 **Wörgl** Tirol, W Austria 47°29′N 12°04′E
171 V15 **Workai, Pulau** *island* Kepulauan Aru, E Indonesia
97 J15 **Workington** NW England, United Kingdom 54°39′N 03°33′W
98 K7 **Workum** Fryslân, N Netherlands 52°58′N 05°25′E
33 V13 **Worland** Wyoming, C USA 44°01′N 107°57′W
see São João de Cortes
Wormatia *see* Worms
99 N25 **Wormeldange** Grevenmacher, E Luxembourg 49°37′N 06°25′E
98 I9 **Wormer** Noord-Holland, C Netherlands 52°30′N 04°50′E
101 G19 **Worms** *anc.* Augusta Vangionum, Borbetomagus, Wormatia. Rheinland-Pfalz, SW Germany 49°38′N 08°22′E
Worms *see* Vormsi
101 K21 **Wörnitz** ≈ S Germany
25 U8 **Wortham** Texas, SW USA 31°47′N 96°27′W
101 G21 **Wörth am Rhein** Rheinland-Pfalz, SW Germany 49°04′N 08°16′E
109 S9 **Wörther See** ◎ S Austria
97 O23 **Worthing** SE England, United Kingdom 50°48′N 00°23′W
29 S11 **Worthington** Minnesota, N USA 43°37′N 95°37′W
31 S13 **Worthington** Ohio, N USA 40°05′N 83°01′W
35 W8 **Worthington Peak** ▲ Nevada, W USA 37°57′N 115°32′W
171 Y13 **Wosi** Papua, E Indonesia 03°55′S 138°54′E
171 V13 **Wosimi** Papua Barat, E Indonesia 02°44′S 134°34′E
189 R5 **Wotho Atoll** *var.* Wōtto. *atoll* Ralik Chain, W Marshall Islands
189 V5 **Wotje Atoll** *var.* Wōjjā. *atoll* Ratak Chain, E Marshall Islands
Wotoe *see* Wotu
Wottawa *see* Otava
Wōtto *see* Wotho Atoll
171 O13 **Wotu** *prev.* Wotoe. Sulawesi, C Indonesia 02°34′S 120°46′E
98 K11 **Woudenberg** Utrecht, C Netherlands 52°05′N 05°25′E
98 I13 **Woudrichem** Noord-Brabant, S Netherlands 51°49′N 05°E
43 N8 **Wounta** *var.* Huaunta. Región Autónoma Atlántico Norte, NE Nicaragua 13°30′N 83°32′W
171 P14 **Wowoni, Pulau** *island* C Indonesia
81 J17 **Woyamdero Plain** *plain* E Kenya
Woyens *see* Vojens
Wozrojdeniye Oroli *see* Vozrozhdeniya, Ostrov
Wrangel Island *see* Vrangelya, Ostrov
39 Y13 **Wrangell** Wrangell Island, Alaska, USA 56°28′N 132°22′W
38 C15 **Wrangell, Cape** *headland* Attu Island, Alaska, USA 52°55′N 172°28′E
39 S11 **Wrangell, Mount** ▲ Alaska, USA 62°00′N 144°01′W
39 T11 **Wrangell Mountains** ▲▲ Alaska, USA
197 S7 **Wrangel Plain** *undersea feature* Arctic Ocean
96 H6 **Wrath, Cape** *headland* N Scotland, United Kingdom 58°37′N 05°01′W
37 W3 **Wray** Colorado, C USA 40°01′N 102°12′W
44 K13 **Wreck Point** *headland* C Jamaica 17°50′N 76°55′W
83 C23 **Wreck Point** *headland* W South Africa 28°52′S 16°17′E
23 V4 **Wrens** Georgia, SE USA 33°12′N 82°23′W
97 K18 **Wrexham** NE Wales, United Kingdom 53°03′N 03°W
27 R13 **Wright City** Oklahoma, C USA 34°03′N 95°00′W
194 J12 **Wright Island** *island* Antarctica
13 N9 **Wright, Mont** ▲ Québec, E Canada 52°36′N 67°40′W
25 X5 **Wright Patman Lake** ▣ Texas, SW USA
36 M16 **Wrightson, Mount** ▲ Arizona, SW USA 31°42′N 110°51′W
23 U5 **Wrightsville** Georgia, SE USA 32°43′N 82°43′W
21 W12 **Wrightsville Beach** North Carolina, SE USA 34°12′N 77°48′W
35 T15 **Wrightwood** California, W USA 34°21′N 117°37′W
8 H9 **Wrigley** Northwest Territories, W Canada 63°16′N 123°39′W
111 G14 **Wrocław** *Eng./Ger.* Breslau. Dolnośląskie, SW Poland 51°07′N 17°01′E
110 F10 **Wronki** *Ger.* Fronicken. Wielkopolskie, C Poland 52°42′N 16°22′E
110 H11 **Września** Wielkopolskie, C Poland 52°19′N 17°34′E
110 F12 **Wschowa** Lubuskie, W Poland 51°49′N 16°15′E
Wsetin *see* Vsetín
161 O5 **Wu'an** Hebei, E China 36°45′N 114°12′E
180 I12 **Wubin** Western Australia 30°05′S 116°43′E
163 W9 **Wuchang** Heilongjiang, NE China 44°55′N 127°13′E
Wuchang *see* Wuhan
Wu-chou/Wuchow *see* Wuzhou
160 M16 **Wuchuan** *var.* Meilu. Guangdong, S China 21°28′N 110°49′E
160 K10 **Wuchuan** *var.* Duru, Gelaozu Miaozu Zhizhixian. Guizhou, S China 28°40′N 108°04′E
163 O13 **Wuchuan** Nei Mongol Zizhiqu, N China 41°04′N 111°28′E
163 V6 **Wudalianchi** *var.* Qingshan; *prev.* Dedu. Heilongjiang, NE China 48°40′N 126°06′E
159 O11 **Wudaoliang** Qinghai, C China 35°16′N 93°03′E
141 Q13 **Wuday'ah** *spring/well* S Saudi Arabia 17°03′N 47°06′E
77 V13 **Wudil** Kano, N Nigeria 11°46′N 08°49′E
160 G12 **Wuding** *var.* Jincheng. Yunnan, SW China 25°30′N 102°21′E
182 G8 **Wudinna** South Australia 33°06′S 135°30′E
Wudu *see* Longnan
160 L9 **Wufeng** Hubei, C China 30°09′N 110°31′E
161 O11 **Wugong Shan** ▲▲ S China
157 P7 **Wuhai** *var.* Haibowan. Nei Mongol Zizhiqu, N China 39°40′N 106°48′E
161 O9 **Wuhan** *var.* Han-kou, Han-k'ou, Hanyang, Wuchang, Wu-han; *prev.* Hankow. *province capital* Hubei, C China 30°35′N 114°19′E
Wu-han *see* Wuhan
161 Q7 **Wuhe** Anhui, E China 33°05′N 117°55′E
Wuhsien *see* Suzhou
Wuhsi/Wu-his *see* Wuxi
161 Q8 **Wuhu** *var.* Wu-na-mu. Anhui, E China 31°23′N 118°25′E
Wūjae *see* Ujae Atoll
160 K11 **Wu Jiang** ≈ C China
158 L5 **Wujiaqu** Xinjiang Uygur Zizhiqu, NW China 44°11′N 87°30′E
Wujlān *see* Ujelang Atoll
77 W15 **Wukari** Taraba, E Nigeria 07°51′N 09°49′E
Wulan *see* Jingyuan
152 H4 **Wular Lake** ◎ NE India
162 M13 **Wulashan** Nei Mongol Zizhiqu, N China 40°43′N 108°45′E
160 H11 **Wulian Feng** ▲▲ SW China
160 F13 **Wuliang Shan** ▲▲ SW China
160 K11 **Wuling Shan** ▲▲ S China
109 Y5 **Wulka** ≈ E Austria
Wulkan *see* Vulcan
109 T3 **Wullowitz** Oberösterreich, N Austria 48°37′N 14°27′E
Wu-lu-k'o-mu-shi/Wu-lu-mu-ch'i *see* Ürümqi
79 D14 **Wum** Nord-Ouest, NE Cameroon 06°24′N 10°04′E
160 H12 **Wumeng Shan** ▲▲ SW China
160 K14 **Wuming** Guangxi Zhuangzu Zizhiqu, S China 23°12′N 108°11′E
100 I10 **Wümme** ≈ NW Germany
Wu-na-mu *see* Wuhu
171 X13 **Wunen** Papua, E Indonesia 03°40′S 138°31′E
12 D9 **Wunnummin Lake** ◎ Ontario, C Canada
80 D13 **Wun Rog** Warap, W South Sudan 09°00′N 28°20′E
101 M18 **Wunsiedel** Bayern, E Germany 50°02′N 12°00′E
100 I12 **Wunstorf** Niedersachsen, NW Germany 52°25′N 09°25′E
166 M3 **Wuntho** Sagaing, N Myanmar (Burma) 23°52′N 95°43′E
101 F15 **Wupper** ≈ W Germany
101 E15 **Wuppertal** *prev.* Barmen-Elberfeld. Nordrhein-Westfalen, W Germany 51°16′N 07°12′E
160 K5 **Wuqi** Shaanxi, C China 36°57′N 108°15′E
158 E7 **Wuqia** Xinjiang Uygur Zizhiqu, NW China 39°50′N 75°19′E
161 P4 **Wuqiao** *var.* Sangyuan. Hebei, E China 37°40′N 116°21′E
101 L23 **Würm** ≈ SE Germany
77 T12 **Wurno** Sokoto, NW Nigeria 13°15′N 05°24′E
101 I19 **Würzburg** Bayern, SW Germany 49°48′N 09°56′E
101 N15 **Wurzen** Sachsen, E Germany 51°21′N 12°48′E
160 L9 **Wu Shan** ▲▲ C China
158 G7 **Wushi** *var.* Uqturpan. Xinjiang Uygur Zizhiqu, NW China 41°07′N 79°09′E
Wusih *see* Wuxi
65 N18 **Wüst Seamount** *undersea feature* S Atlantic Ocean 32°00′S 00°06′E
Wusuli Jiang/Wusuri *see* Ussuri
161 N3 **Wutai Shan** *var.* Beitai Ding. ▲ C China 39°00′N 114°00′E
160 H10 **Wutongqiao** Sichuan, C China 29°21′N 103°48′E
159 P6 **Wutongwozi Quan** *spring* NW China
99 H15 **Wuustwezel** Antwerpen, N Belgium 51°24′N 04°34′E
186 B4 **Wuvulu Island** *island* NW Papua New Guinea
159 U9 **Wuwei** *var.* Liangzhou. Gansu, C China 37°58′N 102°40′E
161 R8 **Wuxi** *var.* Wuhsi, Wu-hsi, Wusih. Jiangsu, E China 31°35′N 120°19′E
Wuxing *see* Huzhou
160 L14 **Wuxuan** Guangxi Zhuangzu Zizhiqu, S China 23°40′N 109°41′E
Wuyang *see* Zhenyuan
160 M4 **Wuyang He** ≈ S China
163 X6 **Wuyiling** Heilongjiang, NE China 48°36′N 129°24′E
161 Q11 **Wuyishan** *prev.* Chong'an. Fujian, SE China 27°48′N 118°03′E
157 T12 **Wuyi Shan** ▲▲ SE China
162 M13 **Wuyuan** Nei Mongol Zizhiqu, N China 41°05′N 108°15′E
160 L17 **Wuzhishan** *prev.* Tongshi. Hainan, S China 18°37′N 109°34′E
160 L17 **Wuzhi Shan** ▲ S China 18°52′N 109°36′E
159 W8 **Wuzhong** Ningxia, N China 37°58′N 106°09′E
160 M14 **Wuzhou** *var.* Wu-chou, Wuchow. Guangxi Zhuangzu Zizhiqu, S China 23°30′N 111°21′E
18 H12 **Wyalusing** Pennsylvania, NE USA 41°40′N 76°13′W
182 M10 **Wycheproof** Victoria, SE Australia 36°06′S 143°13′E
97 K21 **Wye** *Wel.* Gwy. ≈ England/Wales, United Kingdom
Wyłkowyszki *see* Vilkaviškis
97 P19 **Wymondham** E England, United Kingdom 52°29′N 01°10′E
29 R17 **Wymore** Nebraska, C USA 40°07′N 96°39′W
182 E5 **Wynbring** South Australia 30°34′S 133°27′E
181 N3 **Wyndham** Western Australia 15°28′S 128°08′E
29 R6 **Wyndmere** North Dakota, N USA 46°16′N 97°07′W
27 X11 **Wynne** Arkansas, C USA 35°14′N 90°48′W
27 N12 **Wynnewood** Oklahoma, C USA 34°39′N 97°09′W
183 O15 **Wynyard** Tasmania, SE Australia 40°57′S 145°33′E
11 U15 **Wynyard** Saskatchewan, S Canada 51°46′N 104°10′W
33 V11 **Wyola** Montana, NW USA 45°07′N 107°23′W
182 A4 **Wyola Lake** *salt lake* South Australia
31 P9 **Wyoming** Michigan, N USA 42°54′N 85°42′W
33 V14 **Wyoming** *off.* State of Wyoming, *also known as* Equality State. ◈ *state* C USA
33 S15 **Wyoming Range** ▲▲ Wyoming, C USA
183 T8 **Wyong** New South Wales, SE Australia 33°18′S 151°27′E
110 G9 **Wyrzysk** *Ger.* Wirsitz. Wielkopolskie, C Poland 53°09′N 17°15′E
Wysg *see* Usk
110 O10 **Wysokie Mazowieckie** Łomża, E Poland 52°54′N 22°34′E
110 M11 **Wyszków** *Ger.* Probstberg. Mazowieckie, NE Poland 52°36′N 21°28′E
110 L11 **Wyszogród** Mazowieckie, C Poland 52°24′N 20°14′E
21 R7 **Wytheville** Virginia, NE USA 36°57′N 81°07′W
111 L15 **Wyżyna Małopolska** *plateau*

X

80 Q12 **Xaafuun** *It.* Hafun. Bari, NE Somalia 10°25′N 51°17′E
80 Q12 **Xaafuun, Raas** *var.* Ras Hafun. *cape* NE Somalia
Xabia *see* Jávea
42 C4 **Xaclbal, Río** *var.* Xalbal. ≈ Guatemala/Mexico
137 Y10 **Xaçmaz** *Rus.* Khachmas. N Azerbaijan 41°26′N 48°47′E
80 O12 **Xadeed** *var.* Haded. *physical region* N Somalia
159 O14 **Xagquka** Xizang Zizhiqu, W China 31°47′N 92°46′E
Xai *see* Oudômxai
158 F10 **Xaidulla** Xinjiang Uygur Zizhiqu, W China 36°27′N 77°46′E
167 Q7 **Xaignabouli** *prev.* Muang Xaignabouri, *Fr.* Sayaboury. Xaignabouli, N Laos 19°16′N 101°43′E
167 R7 **Xai Lai Leng, Phou** ▲ Laos/Vietnam 19°13′N 104°09′E
158 L15 **Xainza** Xizang Zizhiqu, W China 30°54′N 88°36′E
158 L16 **Xaitongmoin** Xizang Zizhiqu, W China 29°27′N 88°13′E
83 F17 **Xaixai** *var.* Caecae. North-West, NW Botswana 19°52′S 21°04′E
83 M20 **Xai-Xai** *prev.* João Belo, Vila de João Belo. Gaza, S Mozambique 25°01′S 33°37′E
Xalbal *see* Xaclbal, Río
80 P13 **Xalin** Sool, N Somalia 09°16′N 49°00′E
146 H7 **Xalqobod** *Rus.* Khalkabad. Qoraqalpog'iston Respublikasi, W Uzbekistan 42°42′N 59°46′E
160 F11 **Xamgyi'nyilha** *var.* Jiantang; *prev.* Zhongdian. Yunnan, SW China 27°48′N 99°41′E
167 R6 **Xam Nua** *var.* Sam Neua. Houaphan, N Laos 20°24′N 104°03′E
82 D11 **Xá-Muteba** *Port.* Cinco de Outubro. Lunda Norte, NE Angola 09°34′S 17°50′E
Xangda *see* Nangqên
83 C16 **Xangongo** *Port.* Rocadas. Cunene, SW Angola 16°43′S 15°01′E
137 W12 **Xankändi** *Rus.* Khankendi; *prev.* Stepanakert. SW Azerbaijan 39°50′N 46°44′E
Xanlar *see* Göygöl
139 U5 **Xan Sūr** *Ar.* Khān Zūr. As Sulaymānīyah, E Iraq 35°03′N 45°08′E
114 J13 **Xánthi** Anatolikí Makedonía kai Thráki, NE Greece 41°09′N 24°54′E
60 H13 **Xanxerê** Santa Catarina, S Brazil 26°52′S 52°25′W
81 O15 **Xarardheere** Mudug, E Somalia 04°45′N 47°54′E
137 Z11 **Xärä Zirä Adasi** *Rus.* Ostrov Bulla. *Island* E Azerbaijan
162 K13 **Xar Burd** *prev.* Bayan Nuru. Nei Mongol Zizhiqu, N China 40°09′N 104°48′E
163 T11 **Xar Moron** ≈ NE China
163 T11 **Xar Moron** ≈ N China
113 L23 **Xarrë** *var.* Xarra. Vlorë, S Albania 39°45′N 20°01′E
82 D12 **Xassengue** Lunda Sul, NW Angola 10°28′S 18°32′E
105 S11 **Xàtiva** *Cas.* Xátiva; *anc.* Setabis, *var.* Jativa. Valenciana, E Spain 39°N 00°32′W
Xauen *see* Chefchaouen
60 K10 **Xavantes, Represa de** *var.* Represa de Chavantes. ▣ S Brazil
158 I7 **Xayar** Xinjiang Uygur Zizhiqu, W China 41°16′N 82°52′E
Xäzär Dänizi *see* Caspian Sea
167 S8 **Xé Bangfai** ≈ C Laos
167 T9 **Xé Banghiang** *var.* Bang Hieng. ≈ S Laos
Xégar *see* Tingri
167 T10 **Xékong** *var.* Lamam. Xékong, S Laos 15°22′N 106°40′E
31 R14 **Xenia** Ohio, N USA 39°40′N 83°55′W
Xeres *see* Jerez de la Frontera
115 E15 **Xeriás** ≈ C Greece
115 G17 **Xeró** ▲ Évvoia, C Greece 38°52′N 23°18′E
139 S1 **Xērzok** *Ar.* Khayrūzuk, *var.* Kharwazawk. Arbil, E Iraq 36°58′N 44°19′E
Xhumo *see* Cum
161 N15 **Xiachuan Dao** *island* S China
Xiacun *see* Rushan
Xiaguan *see* Dali
159 U11 **Xiahe** *var.* Labrang. Gansu, C China 35°12′N 102°28′E
161 Q13 **Xiamen** *var.* Hsia-men; *prev.* Amoy. Fujian, SE China 24°28′N 118°05′E
160 L6 **Xi'an** *var.* Changan, Sian, Signan, Siking, Singan, Xian. *province capital* Shaanxi, C China 34°16′N 108°54′E
160 L10 **Xianfeng** *var.* Gaoleshan. Hubei, C China 29°45′N 109°10′E
Xiang *see* Hunan
161 N7 **Xiangcheng** Henan, C China 33°51′N 113°27′E
160 F10 **Xiangcheng** *var.* Sampê, *Tib.* Qagchêng. Sichuan, C China 28°52′N 99°45′E
160 M8 **Xiangfan** *var.* Xiangyang. Hubei, C China 32°07′N 112°00′E
Xianggang Tebie Xingzhengqu *see* Hong Kong
161 N10 **Xiang Jiang** ≈ S China
Xiangkhoang *see* Phônsaven
167 Q7 **Xiangkhoang, Plateau de** *var.* Plain of Jars. *plateau* N Laos
161 N11 **Xiangtan** *var.* Hsiang-t'an, Siangtan. Hunan, S China 27°53′N 112°55′E
161 N11 **Xiangxiang** Hunan, S China 27°50′N 112°31′E
Xiangyang *see* Xiangfan
161 S10 **Xianju** Zhejiang, SE China 28°53′N 120°41′E
Xianshui *see* Dawu
160 F8 **Xianshui He** ≈ C China
161 N9 **Xiantao** *var.* Mianyang. Hubei, C China 30°20′N 113°31′E
161 R10 **Xianxia Ling** ▲▲ SE China
160 K6 **Xianyang** Shaanxi, C China 34°26′N 118°40′E
158 L5 **Xiaocaohu** Xinjiang Uygur Zizhiqu, W China 45°44′N 90°07′E
161 O9 **Xiaogan** Hubei, C China 30°55′N 113°54′E
Xiaogang *see* Dongxiang
163 W6 **Xiao Hinggan Ling** *Eng.* Lesser Khingan Range. ▲▲ NE China
160 M6 **Xiao Shan** ▲▲ C China
160 M12 **Xiao Shui** ≈ S China
Xiaoxi *see* Pinghe
161 P6 **Xiaoxian** *var.* Longcheng, Xiao Xian. Anhui, E China 34°11′N 116°56′E
Xiao Xian *see* Xiaoxian
160 G11 **Xichang** Sichuan, C China 27°52′N 102°16′E
41 P11 **Xicoténcatl** Tamaulipas, C Mexico 22°59′N 98°54′W
Xieng Khouang *see* Phônsaven
Xieng Ngeun *see* Muong Xiang Ngeun
160 J11 **Xifeng** *var.* Yongjing. Guizhou, S China 27°15′N 106°44′E
Xifeng *see* Qingcheng
Xigang *see* Helan
158 L16 **Xigazê** *var.* Jih-k'a-tse, Shigatse, Xigaze. Xizang Zizhiqu, W China 29°18′N 88°50′E
159 W11 **Xihe** *var.* Hanyuan. Gansu, C China 34°00′N 105°24′E
160 I8 **Xi He** ≈ C China
Xihuachi *see* Heshui
159 W10 **Xiji** Ningxia, N China 36°02′N 105°33′E
160 M14 **Xi Jiang** *var.* Hsi Chiang, *Eng.* West River. ≈ S China
159 Q7 **Xijian Quan** *spring* NW China
160 K15 **Xijin Shuiku** ▣ S China
Xilaganí *see* Xylagani
Xiligou *see* Ulan
160 I13 **Xilin** *var.* Bada. Guangxi Zhuangzu Zizhiqu, S China 24°30′N 105°00′E
163 Q10 **Xilinhot** *var.* Silinhot. Nei Mongol Zizhiqu, N China 43°58′N 116°07′E
Xilinji *see* Mohe
Xilokastro *see* Xylókastro
Xin *see* Xinjiang Uygur Zizhiqu
Xin'an *see* Anlong
Xin'anjiang Shuiku *see* Qiandao Hu
Xin'anzhen *see* Xinyi
Xin Barag Youqi *see* Altan Emel
Xin Barag Zuoqi *see* Amgalang
163 W12 **Xinbin** *var.* Xinbin Manzu Zizhixian. Liaoning, NE China 41°44′N 125°02′E
Xinbin Manzu Zizhixian *see* Xinbin
161 O7 **Xincai** Henan, C China 32°47′N 114°58′E
Xincheng *see* Zhaojue
Xindu *see* Luhuo
161 O13 **Xinfeng** *var.* Jiading. Jiangxi, S China 25°23′N 114°48′E
161 O14 **Xinfengjiang Shuiku** ▣ S China
Xing'an *see* Ankang
Xingba *see* Lhünzê
163 T13 **Xingcheng** Liaoning, NE China 40°38′N 120°47′E
Xingcheng *see* Xingning
82 E11 **Xinge** Lunda Norte, NE Angola 09°44′S 19°10′E
161 P12 **Xingguo** *var.* Lianjiang. Jiangxi, S China 26°25′N 115°22′E
159 S11 **Xinghai** *var.* Ziketan. Qinghai, C China 35°12′N 102°28′E
161 R7 **Xinghua** Jiangsu, E China 32°54′N 119°48′E
Xingkai Hu *see* Khanka, Lake
161 P13 **Xingning** *prev.* Xingcheng. Guangdong, S China 24°05′N 115°47′E
160 I13 **Xingren** Guizhou, S China 25°26′N 105°08′E
161 O4 **Xingtai** Hebei, E China 37°08′N 114°29′E
59 J14 **Xingu, Rio** ≈ C Brazil
159 P6 **Xingxingxia** Xinjiang Uygur Zizhiqu, NW China 41°48′N 95°01′E
160 I13 **Xingyi** Guizhou, S China 25°04′N 104°51′E
158 I6 **Xinhe** *var.* Toksu. Xinjiang Uygur Zizhiqu, NW China 41°32′N 82°39′E

◆ Country ● Country Capital ◇ Dependent Territory ○ Dependent Territory Capital ◈ Administrative Regions ✕ International Airport ▲ Mountain ▲▲ Mountain Range ⛰ Volcano ≈ River ◎ Lake ▣ Reservoir

345

XIN HOT – YEONGCHEON

163 Q10 **Xin Hot** Nei Mongol Zizhiqu, N China 43°58′N 114°59′E
Xinhua see Funing
163 T12 **Xinhui** var. Aohan Qi. Nei Mongol Zizhiqu, N China 42°12′N 119°57′E
159 T10 **Xining** var. Hsining, Hsi-ning, Sining. province capital Qinghai, C China 36°37′N 101°46′E
161 O4 **Xinji** prev. Shulu. Hebei, E China 37°55′N 115°14′E
161 P10 **Xinjian** Jiangxi, S China 28°37′N 115°46′E
Xinjiang see Xinjiang Uygur Zizhiqu
162 D8 **Xinjiang Uygur Zizhiqu** var. Sinkiang, Sinkiang Uighur Autonomous Region, Xin, Xinjiang. ◆ autonomous region NW China
160 H9 **Xinjin** var. Meixing, Tib. Zainlha. Sichuan, C China 30°27′N 103°46′E
Xinjin see Pulandian
Xinjing see Jingxi
163 U12 **Xinmin** Liaoning, NE China 41°58′N 122°51′E
160 M12 **Xinning** var. Jinshi. Hunan, S China 26°34′N 110°57′E
Xinning see Ningxian
Xinning see Fusui
Xinpu see Lianyungang
Xinshan see Anyuan
161 P5 **Xintai** Shandong, E China 35°54′N 117°44′E
Xinwen see Suncun
Xin Xian see Xinzhou
161 N6 **Xinxiang** Henan, C China 35°13′N 113°48′E
161 O8 **Xinyang** var. Hsin-yang, Sinyang. Henan, C China 32°09′N 114°04′E
161 Q6 **Xinyi** var. Xin'anzhen. Jiangsu, E China 34°17′N 118°14′E
161 Q6 **Xinyi He** E China
161 S14 **Xinying** var. Sinying, Jap. Shinei; prev. Hsinying. C Taiwan 23°12′N 120°15′E
161 O11 **Xinyu** Jiangxi, S China 27°51′N 115°00′E
158 I5 **Xinyuan** var. Künes. Xinjiang Uygur Zizhiqu, NW China 43°25′N 83°12′E
Xinyuan see Tianjun
162 M13 **Xinzhao Shan** ▲ N China 39°37′N 107°51′E
161 N3 **Xinzhou** var. Xin Xian. Shanxi, C China 38°24′N 112°43′E
Xinzhou see Longlin
161 S13 **Xinzhu** var. Hsinchu. N Taiwan 24°48′N 120°59′E
104 H4 **Xinzo de Límia** Galicia, NW Spain 42°05′N 07°45′W
Xions see Książ Wielkopolski
161 O7 **Xiping** Henan, C China 33°22′N 114°00′E
Xiping see Songyang
159 T11 **Xiqing Shan** C China
59 N16 **Xique-Xique** Bahia, E Brazil 10°47′S 42°44′W
Xireg see Ulan
115 E14 **Xirovoúni** ▲ N Greece 40°31′N 21°58′E
162 M13 **Xishanzui** prev. Urad Qianqi. Nei Mongol Zizhiqu, N China 40°43′N 108°41′E
Xisha Qundao see Paracel Islands
160 J11 **Xishui** var. Donghuang. Guizhou, S China 28°24′N 106°09′E
Xi Ujimqin Qi see Bayan Ul
160 K11 **Xiushan** var. Zhonghe. Chongqing Shi, C China 28°23′N 108°52′E
Xiushan see Tonghai
161 O10 **Xiu Shui** S China
Xiuyan see Qingjian
146 H9 **Xiva** Rus. Khiva, Khiwa. Xorazm Viloyati, W Uzbekistan 41°22′N 60°22′E
158 J16 **Xixabangma Feng** ▲ W China 28°25′N 85°47′E
160 M7 **Xixia** Henan, C China 33°30′N 111°25′E
Xixón see Gijón
Xixona see Jijona
Xizang see Xizang Zizhiqu
Xizang Gaoyuan see Qingzang Gaoyuan
160 E9 **Xizang Zizhiqu** var. Thibet, Tibetan Autonomous Region, Xizang, Eng. Tibet. ◆ autonomous region W China
163 U14 **Xizhong Dao** island N China
Xoi see Qüxü
146 H8 **Xo'jayli** Rus. Khodzheyli. Qoraqalpog'iston Respublikasi, W Uzbekistan 42°23′N 59°27′E
Xolotlán see Managua, Lago de
147 I9 **Xonqa** var. Khonqa, Rus. Khanka. Xorazm Viloyati, W Uzbekistan 41°31′N 60°39′E
146 H9 **Xorazm Viloyati** Rus. Khorezmskaya Oblast'. ◆ province W Uzbekistan
159 N9 **Xorkol** Xinjiang Uygur Zizhiqu, NW China 38°45′N 91°07′E
147 P11 **Xovos** var. Ursat'yevskaya, Rus. Khavast. Sirdaryo Viloyati, E Uzbekistan 40°14′N 68°46′E
41 X14 **Xpujil** Quintana Roo, E Mexico 18°30′N 89°24′W
161 Q8 **Xuancheng** var. Xuanzhou. Anhui, E China 30°57′N 118°53′E
Xuande Qundao see Amphitrite Group
167 T9 **Xuân Đuc** Quang Binh, C Vietnam 17°19′N 106°38′E
160 L9 **Xuan'en** var. Zhushan. Hubei, C China 30°03′N 109°26′E
160 K8 **Xuanhan** Sichuan, C China 31°25′N 107°41′E
161 O2 **Xuanhua** Hebei, E China 40°36′N 115°01′E
161 P4 **Xuanhui He** E China
167 T8 **Xuân Sơn** Quang Binh, C Vietnam 17°42′N 105°58′E
H12 **Xuanwei** Yunnan, China 26°08′N 104°04′E
Xuanzhou see Xuancheng
161 N7 **Xuchang** Henan, C China 34°03′N 113°48′E
Xucheng see Xuwen
137 X10 **Xudat** Rus. Khudat. NE Azerbaijan 41°37′N 48°39′E
81 M16 **Xuddur** var. Hudur, It. Oddur. Bakool, SW Somalia 04°07′N 43°47′E
80 O13 **Xudun** Sool, N Somalia 09°12′N 47°34′E
160 L11 **Xuefeng Shan** S China
161 S13 **Xue Shan** prev. Hsüeh Shan. ▲ N Taiwan
147 O13 **Xufar** Surkhondaryo Viloyati, S Uzbekistan 38°31′N 67°45′E
Xulun Hobot Qagan see Qagan Nur
42 F2 **Xunantunich** ruins Cayo, W Belize
163 W6 **Xun He** NE China
160 L7 **Xun He** C China
160 L14 **Xun Jiang** S China
163 W5 **Xunke** var. Bianjiang; prev. Qike. Heilongjiang, NE China 49°35′N 128°27′E
161 P13 **Xunwu** var. Changning. Jiangxi, S China 24°59′N 115°33′E
139 V4 **Xurmal** Ar. Khūrmāl, var. Khormal. As Sulaymānīyah, NE Iraq 35°19′N 46°06′E
161 O3 **Xushui** Hebei, E China 39°01′N 115°38′E
160 L16 **Xuwen** var. Xucheng. Guangdong, S China 20°21′N 110°09′E
160 I11 **Xuyong** var. Yongning. Sichuan, C China 28°17′N 105°21′E
161 P6 **Xuzhou** var. Hsu-chou, Suchow, Tongshan; prev. T'ung-shan. Jiangsu, E China 34°17′N 117°09′E
114 K13 **Xylaganí** var. Xilaganí. Anatolikí Makedonía kai Thráki, NE Greece 40°58′N 25°27′E
115 F19 **Xylókastro** var. Xilokastro. Pelopónnisos, S Greece 38°04′N 22°36′E

Y

160 H9 **Ya'an** var. Yaan. Sichuan, C China 30°N 102°57′E
182 L10 **Yaapeet** Victoria, SE Australia 35°48′S 142°03′E
79 D15 **Yabassi** Littoral, W Cameroon 04°30′N 09°59′E
81 J15 **Yabēlo** Oromīya, C Ethiopia 04°53′N 38°01′E
114 H9 **Yablanitsa** Lovech, N Bulgaria 43°01′N 24°06′E
43 N7 **Yablis** Región Autónoma Atlántico Norte, NE Nicaragua 14°08′N 83°44′W
123 O14 **Yablonovyy Khrebet** S Russian Federation
162 J14 **Yabrai Shan** NE China
45 U6 **Yabucoa** E Puerto Rico 18°02′N 65°53′W
160 J11 **Yachi He** S China
32 H10 **Yacolt** Washington, NW USA 45°49′N 122°22′W
54 M10 **Yacuaray** Amazonas, S Venezuela 04°12′N 66°30′W
57 M22 **Yacuiba** Tarija, S Bolivia 22°00′S 63°43′W
57 K16 **Yacuma, Río** C Bolivia
155 H16 **Yādgīr** Karnātaka, C India 16°46′N 77°09′E
21 R8 **Yadkin River** North Carolina, SE USA
21 R9 **Yadkinville** North Carolina, SE USA 36°07′N 80°40′W
127 P3 **Yadrin** Chuvashskaya Respublika, W Russian Federation 55°55′N 46°10′E
Yaegama-shotō see Yaeyama-shotō
Yaeme-saki see Paimi-saki
165 O16 **Yaeyama-shotō** var. Yaegama-shotō. island group SW Japan
75 O8 **Yafran** NW Libya 32°04′N 12°31′E
165 S2 **Yagashiri-tō** island NE Japan
65 H21 **Yaghan Basin** undersea feature SE Pacific Ocean
123 S9 **Yagodnoye** Magadanskaya Oblast', E Russian Federation 62°37′N 149°18′E
Yagotin see Yahotyn
78 G12 **Yagoua** Extrême-Nord, NE Cameroon 10°23′N 15°13′E
159 Q11 **Yagradagzê Shan** ▲ C China 35°06′N 95°41′E
Yaguachi see Yaguachi Nuevo
56 B7 **Yaguachi Nuevo** var. Yaguachi. Guayas, W Ecuador 02°06′S 79°43′W
Yaguarón, Río see Jaguarão, Rio
117 Q11 **Yahorlyts'kyy Lyman** bay S Ukraine
117 Q5 **Yahotyn** Rus. Yagotin. Kyyivs'ka Oblast', N Ukraine 50°15′N 31°48′E
40 L12 **Yahualica** Jalisco, SW Mexico 21°11′N 102°29′W
79 L17 **Yahuma** Orientale, N Dem. Rep. Congo 01°12′N 23°00′E
136 K15 **Yahyalı** Kayseri, C Turkey 38°08′N 35°23′E
167 N15 **Yai, Khao** ▲ SW Thailand 08°45′N 99°32′E
164 M14 **Yaizu** Shizuoka, Honshū, S Japan 34°52′N 138°20′E
160 G9 **Yajiang** var. Hekou, Tib. Nyagquka. Sichuan, C China 30°05′N 100°57′E
119 O14 **Yakawlyevichi** Rus. Yakovlevichi. Vitsyebskaya Voblasts', NE Belarus 54°20′N 30°31′E
163 S6 **Yakeshi** Nei Mongol Zizhiqu, N China 49°16′N 120°42′E
32 I9 **Yakima** Washington, NW USA 46°36′N 120°30′W
32 J10 **Yakima River** Washington, NW USA
Yakkabag see Yakkabog'
147 N12 **Yakkabog'** Rus. Yakkabag. Qashqadaryo Viloyati, S Uzbekistan 38°57′N 66°35′E
148 L12 **Yakmach** Baluchistān, SW Pakistan 28°48′N 63°48′E
77 O12 **Yako** W Burkina Faso 12°59′N 02°15′W
39 W13 **Yakobi Island** island Alexander Archipelago, Alaska, USA
79 K16 **Yakoma** Equateur, N Dem. Rep. Congo 04°04′N 22°23′E
114 H11 **Yakoruda** Blagoevgrad, SW Bulgaria 42°01′N 23°40′E
Yakovlevichi see Yakawlyevichi
127 T2 **Yakshur-Bod'ya** Udmurtskaya Respublika, NW Russian Federation 57°10′N 53°10′E
165 Q5 **Yakumo** Hokkaidō, NE Japan 42°18′N 140°15′E
164 B17 **Yaku-shima** island Nansei-shotō, SW Japan
39 V12 **Yakutat** Alaska, USA 59°33′N 139°44′W
39 U12 **Yakutat Bay** inlet Alaska, USA
Yakutia/Yakutiya/Yakutiya, Respublika see Sakha (Yakutiya), Respublika
123 Q10 **Yakutsk** Respublika Sakha (Yakutiya), NE Russian Federation 62°10′N 129°50′E
167 O17 **Yala** Yala, SW Thailand 06°32′N 101°19′E
182 D6 **Yalata** South Australia 31°30′S 131°53′E
31 S9 **Yale** Michigan, N USA 43°07′N 82°45′W
180 I11 **Yalgoo** Western Australia 28°23′S 116°43′E
114 O12 **Yalıköy** İstanbul, NW Turkey 41°29′N 28°19′E
79 L14 **Yalinga** Haute-Kotto, C Central African Republic 06°47′N 23°09′E
119 M17 **Yalizava** Rus. Yelizovo. Mahilyowskaya Voblasts', E Belarus 53°24′N 29°01′E
44 L13 **Yallahs Hill** ▲ E Jamaica 17°53′N 76°31′W
22 L3 **Yalobusha River** Mississippi, S USA
79 H15 **Yaloké** Ombella-Mpoko, W Central African Republic 05°15′N 17°12′E
160 E7 **Yalong Jiang** C China
136 E11 **Yalova** Yalova, NW Turkey 40°40′N 29°17′E
136 E11 **Yalova** ◆ province NW Turkey
Yaloveny see Ialoveni
Yalpug see Ialpug
Yalpug, Ozero see Yalpuh, Ozero
117 N12 **Yalpuh, Ozero** Rus. Ozero Yalpug. ◎ SW Ukraine
117 T14 **Yalta** Avtonomna Respublika Krym, S Ukraine 44°30′N 34°09′E
163 W12 **Yalu** Chin. Yalu Jiang, Jap. Oryokko, Kor. Amnok-kang. China/North Korea
Yalu Jiang see Yalu
136 F14 **Yalvaç** Isparta, SW Turkey 38°16′N 31°09′E
165 R9 **Yamada** Iwate, Honshū, C Japan 39°27′N 141°56′E
165 D14 **Yamaga** Kumamoto, Kyūshū, SW Japan 33°02′N 130°41′E
165 P10 **Yamagata** Yamagata, Honshū, C Japan 38°15′N 140°19′E
165 P9 **Yamagata** off. Yamagata-ken. ◆ prefecture Honshū, C Japan
Yamagata-ken see Yamagata
164 C16 **Yamagawa** Kagoshima, Kyūshū, SW Japan 31°12′N 130°37′E
164 E13 **Yamaguchi** var. Yamaguti. Yamaguchi, Honshū, SW Japan 34°11′N 131°26′E
164 E13 **Yamaguchi** off. Yamaguchi-ken, var. Yamaguti. ◆ prefecture Honshū, SW Japan
Yamaguchi-ken see Yamaguchi
Yamaguti see Yamaguchi
125 X5 **Yamalo-Nenetskiy Avtonomnyy Okrug** ◆ autonomous district N Russian Federation
122 J7 **Yamal, Poluostrov** peninsula N Russian Federation
165 N13 **Yamanashi** off. Yamanashi-ken, var. Yamanasi. ◆ prefecture Honshū, S Japan
Yamanashi-ken see Yamanashi
Yamanasi see Yamanashi
Yamanīyah, Al Jumhūrīyah al see Yemen
127 W5 **Yamantau** ▲ W Russian Federation 53°11′N 57°30′E
Yamasaki see Yamazaki
15 P12 **Yamaska** Québec, SE Canada
192 G4 **Yamato Ridge** undersea feature S Sea of Japan 39°20′N 135°00′E
164 I13 **Yamazaki** var. Yamasaki. Hyōgo, Honshū, SW Japan 35°00′N 134°31′E
183 V5 **Yamba** New South Wales, SE Australia 29°28′S 153°22′E
81 D16 **Yambio** var. Yambiyo. Western Equatoria, S South Sudan 04°34′N 28°21′E
Yambiyo see Yambio
114 L10 **Yambol** Turk. Yanboli. Yambol, E Bulgaria 42°29′N 26°30′E
114 M11 **Yambol** ◆ province E Bulgaria
79 M17 **Yambuya** Orientale, N Dem. Rep. Congo 01°22′N 24°21′E
171 T15 **Yamdena, Pulau** prev. Jamdena. island Kepulauan Tanimbar, E Indonesia
165 O14 **Yame** Fukuoka, Kyūshū, SW Japan 33°14′N 130°32′E
166 M6 **Yamethin** Mandalay, C Myanmar (Burma) 20°24′N 96°08′E
186 C6 **Yaminbot** East Sepik, NW Papua New Guinea 04°34′S 143°56′E
181 U9 **Yamma Yamma, Lake** ◎ Queensland, C Australia
76 M16 **Yamoussoukro** ● (Ivory Coast) C Ivory Coast 06°51′N 05°21′W
37 P3 **Yampa River** Colorado, C USA
117 S2 **Yampil'** Sums'ka Oblast', NE Ukraine 51°57′N 33°49′E
116 M8 **Yampil'** Vinnyts'ka Oblast', C Ukraine 48°15′N 28°18′E
123 T9 **Yamsk** Magadanskaya Oblast', E Russian Federation 59°33′N 154°04′E
152 J8 **Yamuna** prev. Jumna. N India
152 I9 **Yamunānagar** Haryāna, N India 30°07′N 77°17′E
Yamundá see Nhamundá, Rio
145 U8 **Yamyshevo** Pavlodar, NE Kazakhstan 51°49′N 77°28′E
159 N16 **Yamzho Yumco** ◎ W China
123 Q8 **Yana** NE Russian Federation
186 H9 **Yanaba Island** island SE Papua New Guinea
155 L16 **Yanam** var. Yanaon. Puducherry, E India 16°45′N 82°16′E
160 L5 **Yan'an** var. Yanan. Shaanxi, C China 36°35′N 109°27′E
Yanaon see Yanam
127 U3 **Yanaul** Respublika Bashkortostan, W Russian Federation 56°15′N 54°57′E
118 O12 **Yanavichy** Rus. Yanovichi. Vitsyebskaya Voblasts', NE Belarus 55°17′N 30°42′E
Yanboli see Yambol
140 K8 **Yanbu' al Baḥr** Al Madīnah, W Saudi Arabia 24°07′N 38°03′E
21 T8 **Yanceyville** North Carolina, SE USA 36°25′N 79°22′W
161 R7 **Yancheng** Jiangsu, E China 33°28′N 120°10′E
159 W8 **Yanchi** Ningxia, N China 37°49′N 107°24′E
160 L5 **Yanchuan** Shaanxi, C China 36°54′N 110°04′E
183 O10 **Yanco Creek** seasonal river New South Wales, SE Australia
183 O6 **Yanda Creek** seasonal river New South Wales, SE Australia
182 K4 **Yandama Creek** seasonal river New South Wales/South Australia
161 S11 **Yandang Shan** SE China
159 O6 **Yandun** Xinjiang Uygur Zizhiqu, W China 42°24′N 94°08′E
76 L13 **Yanfolila** Sikasso, SW Mali 11°08′N 08°12′W
79 M18 **Yangambi** Orientale, N Dem. Rep. Congo 0°46′N 24°24′E
158 M15 **Yangbajain** Xizang Zizhiqu, W China 30°05′N 90°35′E
Yangcheng see Yangshan, Guangdong, S China
Yangchow see Yangzhou
160 M15 **Yangchun** var. Chuncheng. Guangdong, S China 22°16′N 111°49′E
161 N2 **Yanggao** var. Longquan. Shanxi, C China 40°24′N 113°51′E
Yanggeta see Yaqeta
Yangiabad see Yangiobod
Yangibazar see Dzhany-Bazar, Kyrgyzstan
Yangi-Bazar see Kofarnihon, Tajikistan
Yangikishak see Yangiqishloq
146 M13 **Yangi-Nishon** Rus. Yangi-Nishan. Qashqadaryo Viloyati, S Uzbekistan 38°37′N 65°39′E
147 Q9 **Yangiobod** Rus. Yangiabad. Toshkent Viloyati, E Uzbekistan 41°10′N 70°10′E
147 O10 **Yangiqishloq** Rus. Yangikishak. Jizzax Viloyati, C Uzbekistan 40°27′N 67°06′E
147 P11 **Yangiyer** Sirdaryo Viloyati, E Uzbekistan 40°19′N 68°48′E
147 P9 **Yangiyo'l** Rus. Yangiyul'. Toshkent Viloyati, E Uzbekistan 41°12′N 69°05′E
Yangiyul' see Yangiyo'l
160 M15 **Yangjiang** Guangdong, S China 21°50′N 112°02′E
Yangku see Taiyuan
Yang-Nishan see Yangi-Nishon
166 L8 **Yangon** Eng. Rangoon. Yangon, S Myanmar (Burma) 16°50′N 96°11′E
166 M8 **Yangon** Eng. Rangoon. ◆ region SW Myanmar (Burma)
161 N4 **Yangquan** Shanxi, C China 37°52′N 113°29′E
161 N13 **Yangshan** var. Yangcheng. Guangdong, S China 24°32′N 112°36′E
167 U12 **Yang Sin, Chu** ▲ S Vietnam 12°23′N 108°25′E
Yangtze see Chang Jiang/Jinsha Jiang
Yangtze see Chang Jiang
Yangtze Kiang see Chang Jiang
161 R7 **Yangzhou** var. Yangchow. Jiangsu, E China 32°22′N 119°22′E
160 L5 **Yan He** C China
163 Y10 **Yanji** Jilin, NE China 42°54′N 129°31′E
Yanji see Longjing
Yanjing see Yanyuan
29 Q12 **Yankton** South Dakota, N USA 42°52′N 97°24′W
161 O12 **Yanling** prev. Lingxian, Ling Xian. Hunan, S China 26°32′N 113°48′E
Yannina see Ioánnina
123 Q7 **Yano-Indigirskaya Nizmennost'** plain NE Russian Federation
Yanovichi see Yanavichy
155 K24 **Yan Oya** N Sri Lanka
158 K6 **Yanqi** var. Yanqi Huizu Zizhixian. Xinjiang Uygur Zizhiqu, NW China 42°04′N 86°32′E
Yanqi Huizu Zizhixian see Yanqi
161 Q10 **Yanshan** var. Hekou. Jiangxi, S China 28°18′N 117°43′E
160 H14 **Yanshan** var. Jiangna. Yunnan, SW China 23°36′N 104°20′E
161 P2 **Yan Shan** E China
163 X8 **Yanshou** Heilongjiang, NE China 45°27′N 128°19′E
123 Q7 **Yanskiy Zaliv** bay N Russian Federation
183 O4 **Yantabulla** New South Wales, SE Australia 29°22′S 145°00′E
161 R4 **Yantai** var. Yan-t'ai; prev. Chefoo, Chih-fu. Shandong, E China 37°30′N 121°22′E
118 A13 **Yantarnyy** Ger. Palmnicken. Kaliningradskaya Oblast', W Russian Federation 54°53′N 19°59′E
114 J9 **Yantra** Gabrovo, N Bulgaria 42°58′N 25°19′E
114 K9 **Yantra** N Bulgaria
160 G11 **Yanyuan** var. Yanjing. Sichuan, C China 27°30′N 101°22′E
161 P5 **Yanzhou** Shandong, E China 35°35′N 116°53′E
79 E16 **Yaoundé** var. Yaunde. ● (Cameroon) Centre, S Cameroon 03°51′N 11°31′E
188 I14 **Yap** ◆ state W Micronesia
188 F16 **Yap** island Caroline Islands, W Micronesia
57 M18 **Yapacani, Rio** C Bolivia
171 W14 **Yapa Kopra** Papua, E Indonesia 04°18′S 135°05′E
Yapan see Yapen, Selat
Yapanskoye More see East Sea/Japan, Sea of
77 P15 **Yapei** N Ghana 09°10′N 01°08′W
12 M10 **Yapeitso, Mont** ▲ Québec, E Canada 52°18′N 70°24′W
171 W12 **Yapen, Pulau** prev. Japen. island E Indonesia
171 W12 **Yapen, Selat** var. Yapan. strait Papua, E Indonesia
61 E15 **Yapeyú** Corrientes, NE Argentina 29°28′S 56°50′W
136 I11 **Yapraklı** Çankırı, N Turkey 40°45′N 33°46′E
174 M3 **Yap Trench** var. Yap Trough. undersea feature SE Philippine Sea 08°30′N 138°00′E
Yap Trough see Yap Trench
Yapurá see Caquetá, Río, Brazil/Colombia
Yapurá see Japurá, Rio, Brazil/Colombia
197 I12 **Yaqaga** island N Fiji
197 H12 **Yaqeta** prev. Yanggeta. island Yasawa Group, NW Fiji
40 G6 **Yaqui** Sonora, NW Mexico 27°21′N 109°59′W
32 E12 **Yaquina Bay** bay Oregon, NW USA
40 G6 **Yaqui, Río** NW Mexico
54 K5 **Yaracuy** off. Estado Yaracuy. ◆ state NW Venezuela
Yaracuy, Estado see Yaracuy
146 E13 **Yarajy** Rus. Yaradzhi. Ahal Welaýaty, C Turkmenistan 38°12′N 57°40′E
Yaradzhi see Yarajy
125 Q15 **Yaransk** Kirovskaya Oblast', NW Russian Federation 57°18′N 47°52′E
136 F17 **Yardımcı Burnu** headland SW Turkey 36°10′N 30°25′E
97 Q19 **Yare** E England, United Kingdom
125 S9 **Yarega** Respublika Komi, NW Russian Federation 63°27′N 53°28′E
116 I7 **Yaremcha** Ivano-Frankivs'ka Oblast', W Ukraine 48°27′N 24°34′E
189 Q9 **Yaren (district)** ● (Nauru) SW Nauru 0°33′S 166°54′E
125 Q10 **Yarensk** Arkhangel'skaya Oblast', NW Russian Federation 62°09′N 49°03′E
155 F16 **Yargatti** Karnātaka, W India 16°07′N 75°11′E
164 M12 **Yariga-take** ▲ Honshū, S Japan 36°20′N 137°38′E
141 O15 **Yarīm** W Yemen 14°15′N 44°23′E
54 F14 **Yarí, Río** SW Colombia
54 K5 **Yaritagua** Yaracuy, N Venezuela 10°05′N 69°07′W
Yarkand see Yarkant He
Yarkant see Shache
158 E9 **Yarkant He** var. Yarkand. NW China
149 U3 **Yarkhūn** NW Pakistan
Yarlung Zangbo Jiang see Brahmaputra
116 L6 **Yarmolyntsi** Khmel'nyts'ka Oblast', W Ukraine 49°13′N 26°53′E
13 O16 **Yarmouth** Nova Scotia, SE Canada 43°53′N 66°09′W
Yarmouth see Great Yarmouth
Yaroslav see Jarosław
124 L15 **Yaroslavl'** Yaroslavskaya Oblast', W Russian Federation 57°38′N 39°53′E
124 K14 **Yaroslavskaya Oblast'** ◆ province W Russian Federation
123 N11 **Yaroslavskiy** Respublika Sakha (Yakutiya), NE Russian Federation 60°10′N 114°12′E
183 P13 **Yarram** Victoria, SE Australia 38°36′S 146°40′E
183 O11 **Yarrawonga** Victoria, SE Australia 36°04′S 145°58′E
182 L4 **Yarriarraburra Swamp** wetland New South Wales, SE Australia
122 I8 **Yar-Sale** Yamalo-Nenetskiy Avtonomnyy Okrug, N Russian Federation 66°52′N 70°42′E
122 K11 **Yartsevo** Krasnoyarskiy Kray, C Russian Federation 60°15′N 90°09′E
126 I4 **Yartsevo** Smolenskaya Oblast', W Russian Federation 55°03′N 32°46′E
54 E8 **Yarumal** Antioquia, NW Colombia 06°59′N 75°25′W
187 W14 **Yasawa Group** island group NW Fiji
Yasel'da see Yasel'da
77 V12 **Yashi** Katsina, N Nigeria 12°21′N 07°56′E
77 S14 **Yashikera** Kwara, W Nigeria 09°40′N 03°19′E
147 T14 **Yashilkŭl** Rus. Ozero Yashil'kul'. ◎ SE Tajikistan
Yashil'kul', Ozero see Yashilkŭl
165 P9 **Yashima** Akita, Honshū, C Japan 39°10′N 140°10′E
127 P13 **Yashkul'** Respublika Kalmykiya, SW Russian Federation 46°09′N 45°22′E
146 F13 **Yashlyk** Ahal Welaýaty, C Turkmenistan 37°46′N 58°51′E
Yasinovataya see Yasynuvata
114 N10 **Yasna Polyana** Burgas, E Bulgaria 42°18′N 27°35′E
167 R10 **Yasothon** Yasothon, E Thailand 15°46′N 104°12′E
183 R10 **Yass** New South Wales, SE Australia 34°52′S 148°55′E
Yassy see Iaşi
164 H12 **Yasugi** Shimane, Honshū, SW Japan 35°25′N 133°12′E
143 N10 **Yāsūj** var. Yesuj; prev. Tal-e Khosravī. Kohkīlūyeh va Būyer Aḥmad, C Iran 30°40′N 51°34′E
136 M11 **Yasun Burnu** headland N Turkey 41°07′N 37°40′E
119 I20 **Yasyel'da** Rus. Yasel'da. Brestskaya Voblasts', SW Belarus Europe
117 X8 **Yasynuvata** Rus. Yasinovataya. Donets'ka Oblast', SE Ukraine 48°05′N 37°57′E
136 C15 **Yatağan** Muğla, SW Turkey 37°22′N 28°08′E
165 Q7 **Yatate-tōge** pass Honshū, C Japan
187 Q17 **Yaté** Province Sud, S New Caledonia 22°10′S 166°56′E
27 P6 **Yates Center** Kansas, C USA 37°54′N 95°44′W
185 B21 **Yates Point** headland South Island, New Zealand 44°30′S 167°49′E
9 N9 **Yathkyed Lake** Nunavut, NE Canada
171 T16 **Yatoke** Pulau Babar, E Indonesia 07°51′S 129°49′E
79 M18 **Yatolema** Orientale, N Dem. Rep. Congo 0°25′N 24°35′E
164 C15 **Yatsushiro** var. Yatusiro. Kumamoto, Kyūshū, SW Japan 32°30′N 130°34′E
164 C15 **Yatsushiro-kai** bay SW Japan
138 F11 **Yatta** var. Yuta. S West Bank 31°29′N 35°10′E
81 J20 **Yatta Plateau** plateau SE Kenya
Yatusiro see Yatsushiro
57 F17 **Yauca, Río** SW Peru
45 S6 **Yauco** W Puerto Rico 18°02′N 66°51′W
Yaunde see Yaoundé
Yavan see Yovon
Yavari see Javari, Rio
56 G9 **Yavarí Mirím, Río** NE Peru
40 G7 **Yavaros** Sonora, NW Mexico 26°40′N 109°32′W
154 I13 **Yavatmāl** Mahārāshtra, C India 20°22′N 78°11′E
54 M9 **Yaví, Cerro** ▲ C Venezuela 05°43′N 65°51′W
43 W16 **Yaviza** Darién, SE Panama 08°09′N 77°41′W
138 F10 **Yavne** Central, W Israel 31°52′N 34°45′E
116 H5 **Yavoriv** Pol. Jaworów, Rus. Yavorov. L'vivs'ka Oblast', NW Ukraine 49°57′N 23°22′E
Yavorov see Yavoriv
164 F14 **Yawatahama** Ehime, Shikoku, SW Japan 33°27′N 132°24′E
Ya Xian see Sanya
136 L17 **Yayladaği** Hatay, S Turkey 35°55′N 36°00′E
125 V13 **Yayva** Permskiy Kray, NW Russian Federation 59°19′N 57°15′E
125 V12 **Yayva** NW Russian Federation
143 Q9 **Yazd** var. Yezd. Yazd, C Iran 31°55′N 54°22′E
143 Q8 **Yazd** off. Ostān-e Yazd, var. Yezd. ◆ province C Iran
Yazd, Ostān-e see Yazd
Yazgulemskiy Khrebet see Yazgulom, Qatorkŭhi
147 S13 **Yazgulom, Qatorkŭhi** Rus. Yazgulemskiy Khrebet. S Tajikistan
22 K5 **Yazoo City** Mississippi, S USA 32°51′N 90°24′W
22 K5 **Yazoo River** Mississippi, S USA
Yazovir Georgi Traykov see Tsonevo, Yazovir
127 Q5 **Yazykovka** Ul'yanovskaya Oblast', W Russian Federation 54°19′N 47°22′E
109 U4 **Ybbs** Niederösterreich, NE Austria 48°10′N 15°03′E
109 U4 **Ybbs** C Austria
95 G22 **Yding Skovhøj** hill C Denmark
115 G20 **Ýdra** var. Ídhra, Idra. Ýdra, S Greece 37°20′N 23°28′E
115 G21 **Ýdra** var. Ídhra. island Ýdra, S Greece
115 G20 **Ýdras, Kólpos** strait S Greece
167 N16 **Ye** Mon State, S Myanmar (Burma) 15°15′N 97°50′E
183 O12 **Yea** Victoria, SE Australia 37°15′S 145°27′E
167 P13 **Yeay Sên** prev. Phumĭ Yeay Sên. Kaôh Kŏng, SW Cambodia 11°09′N 103°09′E
Yebaishou see Jianping
78 I5 **Yebbi-Bou** Tibesti, N Chad 21°12′N 17°55′E
158 F9 **Yecheng** var. Kargilik. Xinjiang Uygur Zizhiqu, NW China 37°54′N 77°26′E
105 R11 **Yecla** Murcia, SE Spain 38°36′N 01°07′W
40 H6 **Yécora** Sonora, NW Mexico 28°23′N 108°56′W
Yedintsy see Edineţ
124 J13 **Yefimovskiy** Leningradskaya Oblast', NW Russian Federation 59°32′N 34°34′E
126 K6 **Yefremov** Tul'skaya Oblast', W Russian Federation 53°10′N 38°02′E
159 T11 **Yêgainnyin** var. Henan Mongolzu Zizhixian. Qinghai, C China 34°42′N 101°36′E
137 U12 **Yegbhegis** Rus. Yekhegis. C Armenia
137 U12 **Yeghegnadzor** C Armenia 39°45′N 45°20′E
145 T10 **Yegindybulak** Kaz. Egindibulaq. Karaganda, C Kazakhstan 49°45′N 75°45′E
126 L4 **Yegor'yevsk** Moskovskaya Oblast', W Russian Federation 55°29′N 39°03′E
Yehuda, Haré see Judaean Hills
81 E15 **Yei** S South Sudan
161 P8 **Yeji** var. Yejiaji. Anhui, E China 31°52′N 115°58′E
Yejiaji see Yeji
122 G10 **Yekaterinburg** prev. Sverdlovsk. Sverdlovskaya Oblast', C Russian Federation 56°52′N 60°35′E
Yekaterinodar see Krasnodar
Yekaterinoslav see Dnipropetrovs'k
123 R13 **Yekaterinoslavka** Amurskaya Oblast', SE Russian Federation 50°23′N 129°03′E
127 O7 **Yekaterinovka** Saratovskaya Oblast', W Russian Federation 52°01′N 44°11′E
76 K16 **Yekepa** NE Liberia 07°35′N 08°32′W
Yekhegis see Yegbhegis
145 T8 **Yekibastuz** prev. Ekibastuz. Pavlodar, NE Kazakhstan 51°42′N 75°22′E
127 T3 **Yelabuga** Respublika Tatarstan, W Russian Federation 55°46′N 52°07′E
Yela Island see Rossel Island
127 O8 **Yelan'** Volgogradskaya Oblast', SW Russian Federation 51°00′N 43°40′E
117 Q9 **Yelanets'** Rus. Yelanets. Mykolayivs'ka Oblast', S Ukraine 47°40′N 31°51′E
144 I9 **Yelek** Kaz. Elek; prev. Ilek. Kazakhstan/Russian Federation
126 L7 **Yelets** Lipetskaya Oblast', W Russian Federation 52°37′N 38°29′E
125 W4 **Yeletskiy** Respublika Komi, NW Russian Federation 67°03′N 64°05′E
76 J11 **Yélimané** Kayes, W Mali 15°06′N 10°43′W
Yelisavetpol see Gäncä
Yelizavetgrad see Kirovohrad
123 T12 **Yelizavety, Mys** headland SE Russian Federation 54°20′N 142°39′E
Yelizovo see Yalizava
127 S5 **Yelkhovka** Samarskaya Oblast', W Russian Federation 53°51′N 50°16′E
96 M1 **Yell** island NE Scotland, United Kingdom
155 E17 **Yellāpur** Karnātaka, W India 15°06′N 74°50′E
11 U17 **Yellow Grass** Saskatchewan, S Canada 49°51′N 104°09′W
Yellowhammer State see Alabama
11 O15 **Yellowhead Pass** pass Alberta/British Columbia, SW Canada
8 K10 **Yellowknife** territory capital Northwest Territories, W Canada 62°30′N 114°29′W
8 K9 **Yellowknife** Northwest Territories, NW Canada
23 P8 **Yellow River** Alabama/Florida, S USA
30 K7 **Yellow River** Wisconsin, N USA
30 I4 **Yellow River** Wisconsin, N USA
30 J6 **Yellow River** Wisconsin, N USA
Yellow River see Huang He
157 V8 **Yellow Sea** Chin. Huang Hai, Kor. Hwang-Hae. sea E Asia
33 S13 **Yellowstone Lake** ◎ Wyoming, C USA
33 T13 **Yellowstone National Park** national park Wyoming, NW USA
33 Y8 **Yellowstone River** Montana/Wyoming, NW USA
96 L1 **Yell Sound** strait N Scotland, United Kingdom
27 U9 **Yellville** Arkansas, C USA 36°12′N 92°41′W
122 K10 **Yeloguy** C Russian Federation
Yëloten see Ýolöten
119 M20 **Yel'sk** Homyel'skaya Voblasts', SE Belarus 51°48′N 29°09′E
77 T13 **Yelwa** Kebbi, W Nigeria 10°52′N 04°46′E
21 R15 **Yemassee** South Carolina, SE USA 32°41′N 80°51′W
141 O15 **Yemen** off. Republic of Yemen, Ar. Al Jumhuriyah al Yamaniyah, Al Yaman. ◆ republic SW Asia
Yemen, Republic of see Yemen
116 M4 **Yemil'chyne** Zhytomyrs'ka Oblast', N Ukraine 50°51′N 27°49′E
124 M10 **Yemtsa** Arkhangel'skaya Oblast', NW Russian Federation 63°04′N 40°18′E
124 M10 **Yemtsa** NW Russian Federation
125 R10 **Yemva** prev. Zheleznodorozhnyy. Respublika Komi, NW Russian Federation 62°38′N 50°59′E
77 U17 **Yenagoa** Bayelsa, S Nigeria 4°58′N 6°16′E
117 X7 **Yenakiyeve** Rus. Yenakiyevo; prev. Ordzhonikidze, Rykovo. Donets'ka Oblast', E Ukraine 48°13′N 38°13′E
Yenakiyevo see Yenakiyeve
166 L6 **Yenangyaung** Magway, W Myanmar (Burma) 20°28′N 94°54′E
167 S5 **Yên Bai** Yên Bai, N Vietnam 21°43′N 104°54′E
183 P9 **Yenda** New South Wales, SE Australia 34°16′S 146°15′E
77 Q14 **Yendi** NE Ghana 09°30′N 00°01′W
Yêndum see Zhag'yab
158 E8 **Yengisar** Xinjiang Uygur Zizhiqu, NW China 38°50′N 76°11′E
136 H11 **Yenice Çayı** var. Filyos Çayı. N Turkey
121 R1 **Yenierenköy** var. Yialousa, Gk. Agialoúsa. NE Cyprus 35°33′N 34°13′E
Yenipazar see Novi Pazar
136 E12 **Yenişehir** Bursa, NW Turkey 40°17′N 29°38′E
Yenisei Bay see Yeniseyskiy Zaliv
122 K12 **Yeniseysk** Krasnoyarskiy Kray, C Russian Federation 58°23′N 92°06′E
197 W10 **Yeniseyskiy Zaliv** var. Yenisei Bay. bay N Russian Federation
127 Q12 **Yenotayevka** Astrakhanskaya Oblast', SW Russian Federation 47°16′N 47°01′E
124 L4 **Yenozero, Ozero** ◎ NW Russian Federation
Yenping see Nanping
39 Q11 **Yentna River** Alaska, USA
180 M10 **Yeo, Lake** salt lake Western Australia
163 Z15 **Yeongcheon** Jap. Eisen; prev. Yŏngch'ŏn. SE South Korea 35°56′N 128°55′E

◆ Country ● Country Capital ◇ Dependent Territory ○ Dependent Territory Capital ◆ Administrative Regions ✈ International Airport ▲ Mountain ▲ Mountain Range Volcano River ◎ Lake Reservoir

Yeongju – Zagreb

163 *Y15* **Yeongju** *Jap.* Eishū; *prev.* Yŏngju. C South Korea 36°48′N 128°37′E
163 *Y17* **Yeosu** *Jap.* Reisui; *prev.* Yŏsu. S South Korea 34°45′N 127°41′E
183 *R7* **Yeoval** New South Wales, SE Australia 32°45′S 148°39′E
97 *K23* **Yeovil** SW England, United Kingdom 50°57′N 02°39′W
40 *H6* **Yepachic** Chihuahua, N Mexico 28°27′N 108°25′W
181 *Y8* **Yeppoon** Queensland, E Australia 23°05′S 150°42′E
126 *M5* **Yeraktur** Ryazanskaya Oblast', W Russian Federation 54°45′N 41°09′E
Yeraliyev *see* Kuryk
146 *F12* **Yerbent** Ahal Welaýaty, C Turkmenistan 39°19′N 58°34′E
123 *N11* **Yerbogachën** Irkutskaya Oblast', C Russian Federation 61°07′N 108°03′E
137 *T12* **Yerevan** *Eng.* Erivan. ● (Armenia) C Armenia 40°12′N 44°31′E
137 *U12* **Yerevan** ✈ C Armenia 40°07′N 44°34′E
145 *R9* **Yereymentau** *var.* Jermentau, *Kaz.* Ereymentaū. Akmola, C Kazakhstan 51°38′N 73°10′E
145 *R9* **Yereymentau, Gory** *prev.* Gory Yermentau. ▲▲ C Kazakhstan
127 *O12* **Yergeni** *hill range* SW Russian Federation
Yeriho *see* Jericho
35 *R6* **Yerington** Nevada, W USA 38°58′N 119°10′W
136 *J13* **Yerköy** Yozgat, C Turkey 39°39′N 34°28′E
114 *L13* **Yerlisu** Edirne, NW Turkey 40°45′N 26°38′E
Yermak *see* Aksu
Yermentau, Gory *see* Yereymentau, Gory
125 *R5* **Yërmitsa** Respublika Komi, NW Russian Federation 66°57′N 52°15′E
35 *V14* **Yermo** California, W USA 34°54′N 116°49′W
123 *P13* **Yerofey Pavlovich** Amurskaya Oblast', SE Russian Federation 53°58′N 121°49′E
99 *F15* **Yerseke** Zeeland, SW Netherlands 51°30′N 04°03′E
127 *Q8* **Yershov** Saratovskaya Oblast', W Russian Federation 51°18′N 48°16′E
145 *S7* **Yertis** *Kaz.* Ertis; *prev.* Irtyshsk. Pavlodar, NE Kazakhstan 53°21′N 75°27′E
129 *R5* **Yertis** *var.* Irtish, *Kaz.* Ertis; *prev.* Irtysh. ≈ C Asia
125 *P9* **Yërtom** Respublika Komi, NW Russian Federation 63°27′N 47°52′E
56 *D13* **Yerupaja, Nevado** ▲ C Peru 10°23′S 76°58′W
Yerushalayim *see* Jerusalem
105 *R4* **Yesa, Embalse de** ⊞ NE Spain
144 *F11* **Yesbol** *prev.* Kulagino. Atyrau, W Kazakhstan 48°30′N 51°33′E
144 *F9* **Yesensay** Zapadnyy Kazakhstan, NW Kazakhstan 49°59′N 51°19′E
144 *F9* **Yesensay** Zapadnyy Kazakhstan, NW Kazakhstan 49°58′N 51°19′E
145 *V15* **Yesik** *Kaz.* Esik; *prev.* Issyk. Almaty, SE Kazakhstan 42°23′N 77°25′E
145 *O8* **Yesil'** *Kaz.* Esil. Akmola, C Kazakhstan 51°58′N 66°24′E
129 *R6* **Yesil'** *Kaz.* Esil. ≈ Kazakhstan/Russian Federation
136 *K15* **Yeşilhisar** Kayseri, C Turkey 38°22′N 35°08′E
136 *L11* **Yeşilırmak** *var.* Iris. ≈ N Turkey
37 *U12* **Yeso** New Mexico, SW USA 34°25′N 104°36′W
Yeso *see* Hokkaidō
127 *N15* **Yessentuki** Stavropol'skiy Kray, SW Russian Federation 44°06′N 42°51′E
122 *M9* **Yessey** Krasnoyarskiy Kray, N Russian Federation 68°18′N 101°49′E
105 *P12* **Yeste** Castilla-La Mancha, C Spain 38°21′N 02°18′W
Yesuj *see* Yāsūj
183 *T4* **Yetman** New South Wales, SE Australia 28°56′S 150°47′E
76 *L4* **Yetti** *physical region* N Mauritania
166 *M4* **Ye-u** Sagaing, C Myanmar (Burma) 22°49′N 95°26′E
102 *H9* **Yeu, Île d'** *island* NW France
Yevlakh *see* Yevlax
137 *W11* **Yevlax** *Rus.* Yevlakh. C Azerbaijan 40°36′N 47°10′E
117 *S13* **Yevpatoriya** Avtonomna Respublika Krym, S Ukraine 45°12′N 33°23′E
Ye Xian *see* Laizhou
126 *K12* **Yeya** ≈ SW Russian Federation
158 *I10* **Yeyik** Xinjiang Uygur Zizhiqu, W China 36°44′N 83°14′E
126 *K12* **Yeysk** Krasnodarskiy Kray, SW Russian Federation 46°41′N 38°15′E
Yezd *see* Yazd
Yezerishche *see* Yezyaryshcha
Yezhou *see* Jianshi
Yezo *see* Hokkaidō
118 *N11* **Yezyaryshcha** *Rus.* Yezerishche. Vitsyebskaya Voblasts', NE Belarus 55°50′N 29°59′E
Yialí *see* Gyalí
Yialousa *see* Yenierenköy
163 *V7* **Yi'an** Heilongjiang, NE China 47°52′N 125°13′E
Yiannitsá *see* Giannitsá
160 *I10* **Yibin** Sichuan, C China 28°50′N 104°35′E
158 *K13* **Yibug Caka** ◎ W China
160 *M9* **Yichang** Hubei, C China 30°37′N 111°02′E
160 *L5* **Yichuan** *var.* Danzhou. Shaanxi, C China 36°05′N 110°02′E
157 *W3* **Yichun** Heilongjiang, NE China 47°41′N 129°10′E
161 *O11* **Yichun** Jiangxi, S China 27°45′N 114°22′E
160 *M9* **Yidu** *prev.* Zhicheng. Hubei, C China 30°21′N 111°27′E
Yidu *see* Qingzhou
188 *C15* **Yigo** NE Guam 13°33′N 144°53′E
161 *Q5* **Yi He** ≈ E China
163 *X8* **Yilan** Heilongjiang, NE China 46°18′N 129°36′E
136 *C9* **Yıldız Dağları** ▲▲ NW Turkey
136 *L13* **Yıldızeli** Sivas, N Turkey 39°52′N 36°37′E
163 *U4* **Yilehuli Shan** ▲▲ NE China
163 *S7* **Yimin He** ≈ NE China
159 *W8* **Yinchuan** *var.* Yinch'uan, Yin-ch'uan, Yinchwan. *province capital* Ningxia, N China 38°30′N 106°19′E
Yinchwan *see* Yinchuan
Yindu He *see* Indus
Yingcheng *see* Yingde
161 *N14* **Yingde** *var.* Yingcheng. Guangdong, S China 24°08′N 113°21′E
163 *U13* **Yingkou** *var.* Ying-k'ou, Yingkow; *prev.* Newchwang, Niuchwang. Liaoning, NE China 40°40′N 122°17′E
Yingkow *see* Yingkou
161 *P9* **Yingshan** *var.* Wenquan. Hubei, C China 30°45′N 115°41′E
161 *Q10* **Yingtan** Jiangxi, S China 28°17′N 117°03′E
Yin-hsien *see* Ningbo
158 *H5* **Yining** *var.* I-ning, *Uigh.* Gulja, Kuldja. Xinjiang Uygur Zizhiqu, NW China 43°53′N 81°18′E
160 *K11* **Yinjiang** *var.* Yinjiang Tujiazu Miaozu Zizhixian. Guizhou, S China 28°22′N 108°07′E
Yinjiang Tujiazu Miaozu Zizhixian *see* Yinjiang
166 *L4* **Yinmabin** Sagaing, C Myanmar (Burma) 22°05′N 94°57′E
163 *N13* **Yin Shan** ▲▲ N China
Yinshan *see* Guangshui
Yin-tu Ho *see* Indus
159 *P15* **Yi'ong Zangbo** ≈ W China
Yioúra *see* Gyáros
81 *J14* **Yirga 'Alem** *It.* Irgalem. Southern Nationalities, S Ethiopia 06°43′N 38°24′E
61 *E19* **Yí, Río** ≈ C Uruguay
81 *E14* **Yirol** Lakes, C South Sudan 06°34′N 30°33′E
163 *S8* **Yirshi** *var.* Yirxie. Nei Mongol Zizhiqu, N China 47°16′N 119°51′E
Yirxie *see* Yirshi
Yishan *see* Guanyun
Yishi *see* Linyi
161 *Q5* **Yishui** Shandong, E China 35°50′N 118°39′E
Yisrael/Yisra'el *see* Israel
Yíthion *see* Gýtheio
Yitiaoshan *see* Jingtai
163 *W10* **Yitong** *var.* Yitong Manzu Zizhixian. Jilin, NE China 43°23′N 125°19′E
Yitong Manzu Zizhixian *see* Yitong
159 *P5* **Yiwu** *var.* Aratürük. Xinjiang Uygur Zizhiqu, NW China 43°16′N 94°38′E
163 *U12* **Yiwulü Shan** ▲▲ N China
163 *T12* **Yixian** *var.* Yizhou. Liaoning, NE China 41°29′N 121°21′E
159 *R15* **Yixing** Jiangsu, China 31°14′N 119°18′E
161 *N10* **Yiyang** Hunan, S China 28°39′N 112°10′E
161 *Q10* **Yiyang** Jiangxi, S China 28°21′N 117°23′E
161 *N13* **Yizhang** Hunan, S China 25°24′N 112°51′E
Yizhou *see* Yixian
93 *K19* **Yläne** Varsinais-Suomi, SW Finland 60°51′N 22°26′E
93 *L14* **Yli-Ii** Pohjois-Pohjanmaa, C Finland 65°21′N 25°55′E
93 *L14* **Ylikiiminki** Pohjois-Pohjanmaa, C Finland 65°00′N 26°10′E
92 *N13* **Yli-Kitka** ◎ NE Finland
93 *K17* **Ylistaro** Etelä-Pohjanmaa, W Finland 62°58′N 22°30′E
92 *K13* **Ylitornio** Lappi, NW Finland 66°19′N 23°40′E
93 *L15* **Ylivieska** Pohjois-Pohjanmaa, W Finland 64°05′N 24°30′E
93 *L18* **Ylöjärvi** Pirkanmaa, W Finland 61°31′N 23°37′E
95 *N17* **Yngaren** ◎ C Sweden
25 *T12* **Yoakum** Texas, SW USA 29°17′N 97°09′W
77 *X13* **Yobe** ◐ *state* NE Nigeria
165 *R3* **Yobetsu-dake** ⛰ Hokkaidō, NE Japan 43°15′N 140°27′E
80 *L11* **Yoboki** C Djibouti 11°30′N 42°04′E
22 *M4* **Yockanookany River** ≈ Mississippi, S USA
22 *L2* **Yocona River** ≈ Mississippi, S USA
171 *Y15* **Yodom** Papua, E Indonesia 07°12′S 139°24′E
169 *Q16* **Yogyakarta** *prev.* Djokjakarta, Jogjakarta, Jokyakarta. Jawa, C Indonesia 07°48′S 110°24′E
169 *P17* **Yogyakarta** *off.* Daerah Istimewa Yogyakarta, *var.* Djokjakarta, Jogjakarta, Jokyakarta. ◐ *autonomous district* S Indonesia
Yogyakarta, Daerah Istimewa *see* Yogyakarta
165 *Q3* **Yoichi** Hokkaidō, NE Japan 43°11′N 140°45′E
42 *G6* **Yojoa, Lago de** ◎ NW Honduras
79 *G16* **Yokadouma** Est, SE Cameroon 03°26′N 15°06′E
164 *K13* **Yokkaichi** *var.* Yokkaiti. Mie, Honshū, SW Japan 34°58′N 136°38′E
Yokkaiti *see* Yokkaichi
79 *E15* **Yoko** Centre, C Cameroon 05°29′N 12°19′E
165 *V15* **Yokoate-jima** *island* Nansei-shotō, SW Japan
165 *R6* **Yokohama** Aomori, Honshū, C Japan 41°04′N 141°14′E
165 *O14* **Yokosuka** Kanagawa, Honshū, S Japan 35°18′N 139°39′E
164 *G12* **Yokota** Shimane, Honshū, SW Japan 35°10′N 133°03′E
165 *Q9* **Yokote** Akita, Honshū, C Japan 39°20′N 140°33′E
77 *Y14* **Yola** Adamawa, E Nigeria 09°08′N 12°24′E
79 *L19* **Yolombo** Equateur, C Dem. Rep. Congo 01°36′S 23°13′E
146 *J14* **Ýolöten** *Rus.* Yëloten; *prev.* Iolotan'. Mary Welaýaty, S Turkmenistan 37°15′N 62°18′E
165 *Y15* **Yome-jima** *island* Ogasawara-shotō, SE Japan
76 *K16* **Yomou** SE Guinea 07°30′N 09°13′W
171 *Y15* **Yomuka** Papua, E Indonesia 07°25′S 138°36′E
188 *C16* **Yona** E Guam 13°24′N 144°46′E
164 *H12* **Yonago** Tottori, Honshū, SW Japan 35°30′N 134°15′E
165 *N16* **Yonaguni** Okinawa, SW Japan 24°29′N 123°00′E
165 *N16* **Yonaguni-jima** *island* Nansei-shotō, SW Japan
165 *T16* **Yonaha-dake** ▲ Okinawa, SW Japan 26°43′N 128°13′E
163 *X14* **Yŏnan** SW North Korea 37°50′N 126°15′E
165 *P10* **Yonezawa** Yamagata, Honshū, C Japan 37°56′N 140°06′E
161 *Q12* **Yong'an** *var.* Yongan. Fujian, SE China 25°58′N 117°26′E
Yong'an *see* Fengjie
159 *T9* **Yongchang** Gansu, N China 38°15′N 101°56′E
161 *P7* **Yongcheng** Henan, C China 33°56′N 116°21′E
Yŏngch'ŏn *see* Yeongcheon
160 *J10* **Yongchuan** Chongqing Shi, C China 29°27′N 105°56′E
159 *U10* **Yongdeng** Gansu, C China 35°58′N 103°27′E
Yongding *see* Yongren
129 *W9* **Yongding He** ≈ E China
161 *P11* **Yongfeng** *var.* Enjiang. Jiangxi, S China 27°19′N 115°23′E
158 *L5* **Yongfeng** *var.* Yongfengqu. Xinjiang Uygur Zizhiqu, W China 43°28′N 87°09′E
Yongfengqu *see* Yongfeng
160 *L13* **Yongfu** Guangxi Zhuangzu Zizhiqu, S China 24°57′N 109°59′E
163 *X13* **Yŏnghŭng** E North Korea 39°31′N 127°14′E
159 *U10* **Yongjing** *var.* Liujiaxia. Gansu, C China 36°00′N 103°30′E
Yongjing *see* Xifeng
Yŏngju *see* Yeongju
Yongle Qundao *see* Crescent Group
Yongning *see* Xuyong
160 *E12* **Yongping** Yunnan, SW China 25°30′N 99°28′E
160 *G12* **Yongren** *var.* Yongding. Yunnan, SW China 26°09′N 101°40′E
160 *L10* **Yongshun** *var.* Lingxi. Hunan, S China 29°02′N 109°46′E
161 *P10* **Yongxiu** *var.* Tujiabu. Jiangxi, S China 29°09′N 115°47′E
Yongzhou *see* Lingling
Yongzhou *see* Zhishan
18 *K14* **Yonkers** New York, NE USA 40°56′N 73°51′W
103 *Q7* **Yonne** ◐ *department* C France
103 *P6* **Yonne** ≈ C France
54 *H9* **Yopal** *var.* El Yopal. Casanare, C Colombia 05°20′N 72°19′W
158 *E8* **Yopurga** *var.* Yukuriawat. Xinjiang Uygur Zizhiqu, NW China 39°13′N 76°44′E
147 *S11* **Yordon** *var.* Iordan, *Rus.* Jardan. Farg'ona Viloyati, E Uzbekistan 39°59′N 71°44′E
180 *J12* **York** Western Australia 31°55′S 116°52′E
97 *M16* **York** *anc.* Eboracum, Eburacum. N England, United Kingdom 53°58′N 01°05′W
23 *N5* **York** Alabama, S USA 32°29′N 88°18′W
29 *Q15* **York** Nebraska, C USA 40°52′N 97°35′W
18 *G16* **York** Pennsylvania, NE USA 39°57′N 76°44′W
21 *R11* **York** South Carolina, SE USA 34°59′N 81°14′W
14 *J13* **York** ≈ Ontario, SE Canada
15 *X6* **York** ≈ Québec, SE Canada
181 *V1* **York, Cape** *headland* Queensland, NE Australia 10°40′S 142°36′E
182 *I9* **Yorke Peninsula** *peninsula* South Australia
182 *I9* **Yorketown** South Australia 35°01′S 137°38′E
19 *P9* **York Harbor** Maine, NE USA 43°10′N 70°37′W
York, Kap *see* Innaanganeq
21 *X6* **York River** ≈ Virginia, NE USA
97 *M16* **Yorkshire** *cultural region* N England, United Kingdom
97 *L16* **Yorkshire Dales** *physical region* N England, United Kingdom
11 *V16* **Yorkton** Saskatchewan, S Canada 51°12′N 102°29′W
25 *T12* **Yorktown** Texas, SW USA 28°58′N 97°30′W
21 *X6* **Yorktown** Virginia, NE USA 37°14′N 76°32′W
30 *M11* **Yorkville** Illinois, N USA 41°38′N 88°27′W
42 *I5* **Yoro** Yoro, C Honduras 15°08′N 87°10′W
42 *H5* **Yoro** ◐ *department* N Honduras
165 *T16* **Yoron-jima** *island* Nansei-shotō, SW Japan
77 *N13* **Yorosso** Sikasso, S Mali 12°21′N 04°47′W
35 *R8* **Yosemite National Park** *national park* California, W USA
127 *Q3* **Yoshkar-Ola** Respublika Mariy El, W Russian Federation 56°38′N 47°54′E
Yösönbulag *see* Altay
162 *K8* **Yösöndzüyl** *var.* Mönhbulag. Övörhangay, C Mongolia 46°48′N 103°25′E
171 *Y16* **Yos Sudarso, Pulau** *var.* Pulau Dolak, Pulau Kolepom; *prev.* Jos Sudarso. *island* E Indonesia
Yŏsu *see* Yeosu
165 *R4* **Yotei-zan** ▲ Hokkaidō, NE Japan 42°50′N 140°46′E
97 *D21* **Youghal** *Ir.* Eochaill. Cork, S Ireland 51°57′N 07°50′W
97 *D21* **Youghal Bay** *Ir.* Cuan Eochaille. *inlet* S Ireland
18 *C15* **Youghiogheny River** ≈ Pennsylvania, NE USA
160 *K14* **You Jiang** ≈ S China
183 *Q9* **Young** New South Wales, SE Australia 34°19′S 148°20′E
11 *T15* **Young** Saskatchewan, S Canada 51°44′N 105°44′W
61 *E18* **Young** Río Negro, W Uruguay 32°44′S 57°36′W
182 *G5* **Younghusband, Lake** *salt lake* South Australia
182 *J10* **Younghusband Peninsula** *peninsula* South Australia
184 *Q10* **Young Nicks Head** *headland* North Island, New Zealand 39°38′S 177°03′E
185 *D20* **Young Range** ▲▲ South Island, New Zealand
191 *Q15* **Young's Rock** *island* Pitcairn Island, Pitcairn Islands
11 *R16* **Youngstown** Alberta, SW Canada 51°32′N 111°12′W
31 *V12* **Youngstown** Ohio, N USA 41°06′N 80°39′W
159 *N9* **Youshashan** Qinghai, C China 38°12′N 90°58′E
Youth, Isle of *see* Juventud, Isla de la
77 *N11* **Youvarou** Mopti, C Mali 15°19′N 04°15′W
160 *K10* **Youyang** *var.* Zhongduo. Chongqing Shi, C China 28°48′N 108°48′E
163 *Y7* **Youyi** Heilongjiang, NE China 46°51′N 131°54′E
147 *P13* **Yovon** *Rus.* Yavan. SW Tajikistan 38°19′N 69°02′E
136 *J13* **Yozgat** Yozgat, C Turkey 39°49′N 34°48′E
136 *K13* **Yozgat** ◐ *province* C Turkey
62 *O6* **Ypacaraí** *var.* Ypacaray. Central, S Paraguay 25°23′S 57°16′W
Ypacaray *see* Ypacaraí
62 *P5* **Ypané, Río** ≈ C Paraguay
Ypres *see* Ieper
114 *I13* **Ypsário** *var.* Ipsario. ▲ Thásos, E Greece 40°43′N 24°39′E
31 *R10* **Ypsilanti** Michigan, N USA 42°12′N 83°36′W
34 *M1* **Yreka** California, W USA 41°43′N 122°39′W
Yrendagüé *see* General Eugenio A. Garay
144 *L11* **Yrghyz** *prev.* Irgiz. Aktyubinsk, C Kazakhstan 48°36′N 61°14′E
186 *G5* **Ysabel Channel** *channel* N Papua New Guinea
14 *K8* **Yser, Lac** ◎ Québec, SE Canada
147 *Y8* **Yshtyk** Issyk-Kul'skaya Oblast', E Kyrgyzstan 41°34′N 78°21′E
Yssel *see* IJssel
103 *Q12* **Yssingeaux** Haute-Loire, C France 45°09′N 04°07′E
95 *K23* **Ystad** Skåne, S Sweden 55°25′N 13°51′E
Ysyk-Köl *see* Issyk-Kul', Ozero
Ysyk-Köl *see* Balykchy
Ysyk-Köl Oblasty *see* Issyk-Kul'skaya Oblast'
96 *L8* **Ythan** ≈ NE Scotland, United Kingdom
Y Trallwng *see* Welshpool
94 *C13* **Ytre Arna** Hordaland, S Norway 60°28′N 05°25′E
94 *B12* **Ytre Sula** *island* S Norway
93 *G17* **Ytterhogdal** Jämtland, C Sweden 62°10′N 14°55′E
Yu *see* Henan
Yuan *see* Red River
Yuancheng *see* Heyuan, Guangdong, S China
Yuan Jiang *see* Red River
161 *S13* **Yuanlin** *Jap.* Inrin; *prev.* Yüanlin. C Taiwan 23°57′N 120°33′E
161 *N3* **Yuanping** Shanxi, C China 38°26′N 112°42′E
Yuanquan *see* Anxi
Yuanshan *see* Lianping
161 *O11* **Yuan Shui** ≈ S China
35 *O6* **Yuba City** California, W USA 39°07′N 121°40′W
35 *O6* **Yuba River** ≈ California, W USA
80 *H13* **Yubdo** Oromīya, C Ethiopia 09°05′N 35°28′E
41 *X12* **Yucatán** ◐ *state* SE Mexico
47 *O3* **Yucatan Basin** *var.* Yucatan Deep. *undersea feature* N Caribbean Sea 20°00′N 84°00′W
Yucatán, Canal de *see* Yucatan Channel
41 *Y10* **Yucatan Channel** *Sp.* Canal de Yucatán. *channel* Cuba/Mexico
Yucatan Deep *see* Yucatan Basin
Yucatan Peninsula *see* Yucatán, Península de
41 *X13* **Yucatán, Península de** *Eng.* Yucatan Peninsula. *peninsula* Guatemala/Mexico
36 *I11* **Yucca** Arizona, SW USA 34°49′N 114°06′W
35 *V15* **Yucca Valley** California, W USA 34°06′N 116°30′W
161 *P4* **Yucheng** Shandong, E China 37°01′N 116°37′E
Yuci *see* Jinzhong
129 *X5* **Yudoma** ≈ E Russian Federation
161 *P12* **Yudu** *var.* Gongjiang. Jiangxi, C China 26°02′N 115°24′E
Yue *see* Guangdong
Yuecheng *see* Yuexi
160 *M12* **Yuecheng Ling** ▲▲ S China
Yuegai *see* Qumarlêb
Yuegaitan *see* Qumarlêb
181 *P7* **Yuendumu** Northern Territory, N Australia 22°19′S 131°51′E
Yue Shan, Tai *see* Lantau Island
160 *H10* **Yuexi** *var.* Yuecheng. Sichuan, C China 28°50′N 102°36′E
161 *N10* **Yueyang** Hunan, S China 29°24′N 113°08′E
125 *U14* **Yug** Permskiy Kray, NW Russian Federation 57°49′N 56°08′E
125 *P13* **Yug** ≈ NW Russian Federation
123 *R10* **Yugorënok** Respublika Sakha (Yakutiya), NE Russian Federation 59°46′N 137°36′E
122 *H9* **Yugorsk** Khanty-Mansiyskiy Avtonomnyy Okrug-Yugra, C Russian Federation 61°17′N 63°25′E
122 *H7* **Yugorskiy Poluostrov** *peninsula* NW Russian Federation
Yugoslavia *see* Serbia
146 *K14* **Yugo-Vostochnyye Garagumy** *prev.* Yugo-Vostochnyye Karakumy. *desert* E Turkmenistan
Yugo-Vostochnyye Karakumy *see* Yugo-Vostochnyye Garagumy
Yuhu *see* Eryuan
161 *S10* **Yuhuan Dao** *island* SE China
160 *L14* **Yu Jiang** ≈ S China
Yujin *see* Qianwei
159 *P9* **Yuka** Qinghai, W China 38°03′N 94°45′E
123 *S7* **Yukagirskoye Ploskogor'ye** *plateau* NE Russian Federation
159 *P9* **Yuke He** ≈ C China
118 *L11* **Yukhavichy** *Rus.* Yukhovichi. Vitsyebskaya Voblasts', N Belarus 56°02′N 28°39′E
126 *J4* **Yukhnov** Kaluzhskaya Oblast', W Russian Federation 54°43′N 35°15′E
Yukhovichi *see* Yukhavichy
79 *J20* **Yuki** *var.* Yuki Kengunda. Bandundu, W Dem. Rep. Congo 03°57′S 19°30′E
Yuki Kengunda *see* Yuki
26 *M10* **Yukon** Oklahoma, C USA 35°30′N 97°45′W
10 *I5* **Yukon** *prev.* Yukon Territory, *Fr.* Territoire du Yukon. ◐ *territory* NW Canada
0 *F4* **Yukon** ≈ Canada/USA
39 *S7* **Yukon Flats** *salt flat* Alaska, USA
Yukon, Territoire du *see* Yukon
Yukon Territory *see* Yukon
137 *T16* **Yüksekova** Hakkâri, SE Turkey 37°35′N 44°17′E
123 *N10* **Yukta** Krasnoyarskiy Kray, C Russian Federation 63°16′N 106°04′E
165 *O13* **Yukuhashi** *var.* Yukuhasi. Fukuoka, Kyūshū, SW Japan 33°41′N 131°00′E
Yukuhasi *see* Yukuhashi
Yukuriawat *see* Yopurga
125 *O9* **Yula** ≈ NW Russian Federation
181 *P8* **Yulara** Northern Territory, N Australia 25°15′S 130°57′E
127 *W6* **Yuldybayevo** Respublika Bashkortostan, W Russian Federation 52°22′N 57°55′E
23 *W8* **Yulee** Florida, SE USA 30°37′N 81°36′W
158 *K7* **Yuli** *var.* Lopnur. Xinjiang Uygur Zizhiqu, NW China 41°24′N 86°12′E
161 *T14* **Yuli** *prev.* Yüli. C Taiwan 23°23′N 121°18′E
160 *L15* **Yulin** Guangxi Zhuangzu Zizhiqu, S China 22°37′N 110°08′E
160 *L4* **Yulin** Shaanxi, C China 38°14′N 109°48′E
161 *T14* **Yuli Shan** *prev.* Yüli Shan. ▲ E Taiwan 23°23′N 121°14′E
160 *F11* **Yulong Xueshan** ▲ SW China 27°09′N 100°10′E
36 *H14* **Yuma** Arizona, SW USA 32°40′N 114°38′W
37 *W3* **Yuma** Colorado, C USA 40°07′N 102°43′W
54 *K5* **Yumare** Yaracuy, N Venezuela 10°37′N 68°41′W
63 *G14* **Yumbel** Bío Bío, C Chile 37°05′S 72°40′W
79 *N19* **Yumbi** Maniema, E Dem. Rep. Congo 01°14′S 26°14′E
159 *Q7* **Yumendong** *prev.* Yumenzhen. Gansu, N China 40°19′N 97°12′E
159 *R8* **Yumendong** *prev.* Laojunmiao. Gansu, N China 39°49′N 97°47′E
Yumenzhen *see* Yumendong
158 *J3* **Yumin** *var.* Karabura. Xinjiang Uygur Zizhiqu, NW China 46°14′N 82°52′E
Yun *see* Yunnan
136 *G14* **Yunak** Konya, W Turkey 38°50′N 31°42′E
45 *O8* **Yuna, Río** ≈ E Dominican Republic
38 *I17* **Yunaska Island** *island* Aleutian Islands, Alaska, USA
160 *M6* **Yuncheng** Shanxi, C China 35°07′N 110°45′E
Yuncheng *see* Yunfu
161 *N14* **Yunfu** *var.* Yuncheng. Guangdong, S China 22°56′N 112°02′E
57 *L18* **Yungas** *physical region* E Bolivia
Yungki *see* Jilin
Yung-ning *see* Nanning
160 *I12* **Yungui Gaoyuan** *plateau* SW China
Yunjinghong *see* Jinghong
160 *M15* **Yunkai Dashan** ▲▲ S China
Yunki *see* Jilin
160 *E11* **Yun Ling** ▲▲ SW China
Yunling *see* Yunxiao
161 *N9* **Yunmeng** Hubei, C China 31°04′N 113°45′E
157 *N14* **Yunnan** *var.* Yun, Yunnan Sheng, Yünnan, Yun-nan. ◐ *province* SW China
Yunnan *see* Kunming
Yunnan Sheng *see* Yunnan
Yünnan/Yun-nan *see* Yunnan
165 *P15* **Yunomae** Kumamoto, Kyūshū, SW Japan 32°16′N 131°00′E
161 *N8* **Yun Shui** ≈ C China
182 *J7* **Yunta** South Australia 32°37′S 139°33′E
161 *Q14* **Yunxiao** *var.* Yunling. Fujian, SE China 23°56′N 117°16′E
160 *K9* **Yunyang** Sichuan, C China 31°03′N 109°43′E
Yunzhong *see* Huairen
193 *S9* **Yupanqui Basin** *undersea feature* E Pacific Ocean
Yuping *see* Libo, Guizhou, China
Yuping *see* Pingbian, Yunnan, China
Yuratishki *see* Yuratsishki
119 *I15* **Yuratsishki** *Pol.* Juraciszki, *Rus.* Yuratishki. Hrodzyenskaya Voblasts', W Belarus 54°02′N 25°56′E
Yurev *see* Tartu
122 *J12* **Yurga** Kemerovskaya Oblast', S Russian Federation 55°42′N 84°59′E
Yurihonjō *see* Honjō
56 *E10* **Yurimaguas** Loreto, N Peru 05°54′S 76°07′W
127 *P3* **Yurino** Respublika Mariy El, W Russian Federation 56°19′N 46°15′E
41 *N13* **Yuriria** Guanajuato, C Mexico 20°12′N 101°09′W
125 *T13* **Yurla** Komi-Permyatskiy Okrug, NW Russian Federation 59°18′N 54°19′E
Yuruá, Río *see* Juruá, Rio
114 *M13* **Yürük** Tekirdağ, NW Turkey 40°58′N 27°09′E
158 *G10* **Yurungkax He** ≈ W China
125 *Q14* **Yur'ya** *var.* Jarja. Kirovskaya Oblast', NW Russian Federation 59°01′N 49°22′E
Yury'ev *see* Tartu
125 *N16* **Yur'yevets** Ivanovskaya Oblast', W Russian Federation 57°19′N 43°01′E
126 *M3* **Yur'yev-Pol'skiy** Vladimirskaya Oblast', W Russian Federation 56°28′N 39°39′E
117 *V7* **Yur"yivka** Dnipropetrovs'ka Oblast', E Ukraine 48°45′N 36°01′E
42 *I7* **Yuscarán** El Paraíso, S Honduras 13°55′N 86°51′W
161 *P12* **Yu Shan** ▲▲ S China
124 *I7* **Yushkozero** Respublika Kareliya, NW Russian Federation 64°46′N 32°13′E
124 *I7* **Yushkozerskoye Vodokhranilishche** *var.* Ozero Kujto. ◎ NW Russian Federation
169 *W9* **Yushu** Jilin, China E Asia 44°48′N 126°55′E
159 *R13* **Yushu** *var.* Gyêgu. Qinghai, C China 33°04′N 97°E
127 *P12* **Yusta** Respublika Kalmykiya, SW Russian Federation 47°06′N 46°16′E
124 *I10* **Yustozero** Respublika Kareliya, NW Russian Federation 62°44′N 33°31′E
137 *Q11* **Yusufeli** Artvin, NE Turkey 40°50′N 41°31′E
164 *F14* **Yusuhara** Kōchi, Shikoku, SW Japan 33°22′N 132°52′E
125 *T14* **Yus'va** Permskiy Kray, NW Russian Federation 58°48′N 54°59′E
Yuta *see* Yatta
161 *P2* **Yutian** Hebei, E China 39°52′N 117°44′E
158 *H10* **Yutian** *var.* Keriya, Mugalla. Xinjiang Uygur Zizhiqu, NW China 36°49′N 81°31′E
62 *K5* **Yuto** Jujuy, NW Argentina 23°35′S 64°28′W
62 *P7* **Yuty** Caazapá, S Paraguay 26°31′S 56°20′W
160 *G13* **Yuxi** Yunnan, SW China 24°22′N 102°28′E
161 *O2* **Yuxian** *prev.* Yu Xian. Hebei, E China 39°50′N 114°33′E
Yu Xian *see* Yuxian
165 *Q9* **Yuzawa** Akita, Honshū, C Japan 39°11′N 140°29′E
125 *N16* **Yuzha** Ivanovskaya Oblast', W Russian Federation 56°34′N 42°00′E
Yuzhno-Alichurskiy Khrebet *see* Alichuri Janubí, Qatorkŭhi
Yuzhno-Kazakhstanskaya Oblast' *see* Yuzhnyy Kazakhstan
123 *T13* **Yuzhno-Sakhalinsk** *Jap.* Toyohara; *prev.* Vladimirovka. Ostrov Sakhalin, Sakhalinskaya Oblast', SE Russian Federation 46°58′N 142°45′E
127 *P14* **Yuzhno-Sukhokumsk** Respublika Dagestan, SW Russian Federation 44°43′N 45°32′E
145 *Z10* **Yuzhnyy Altay, Khrebet** ▲▲ E Kazakhstan
Yuzhnyy Bug *see* Pivdennyy Buh
145 *O15* **Yuzhnyy Kazakhstan** *off.* Yuzhno-Kazakhstanskaya Oblast', *Eng.* South Kazakhstan, *Kaz.* Ongtüstik Qazaqstan Oblysy; *prev.* Chimkentskaya Oblast'. ◐ *province* S Kazakhstan
123 *U10* **Yuzhnyy, Mys** *headland* E Russian Federation 57°44′N 156°49′E
122 *H6* **Yuzhnyy, Ostrov** *island* NW Russian Federation
127 *W6* **Yuzhnyy Ural** *var.* Southern Urals. ▲▲ W Russian Federation
159 *V10* **Yuzhong** Gansu, C China 35°52′N 104°09′E
Yuzhou *see* Chongqing
103 *N5* **Yvelines** ◐ *department* N France
108 *B9* **Yverdon** *var.* Yverdon-les-Bains, *Ger.* Iferten; *anc.* Eborodunum. Vaud, W Switzerland 46°47′N 06°38′E
Yverdon-les-Bains *see* Yverdon
102 *M3* **Yvetot** Seine-Maritime, N France 49°37′N 00°48′E
Ýylanly *see* Gurb ansoltan Eje

Z

147 *T12* **Zaalayskiy Khrebet** *Taj.* Qatorkŭhi Pasi Oloy. ▲▲ Kyrgyzstan/Tajikistan
Zaamin *see* Zomin
Zaandam *see* Zaanstad
98 *I10* **Zaanstad** *prev.* Zaandam. Noord-Holland, C Netherlands 52°27′N 04°49′E
Zabadani *see* Az Zabdānī
Zabalatstsye *see* Zabalotstsye
112 *L9* **Žabalj** *Ger.* Josefsdorf, *Hung.* Zsablya; *prev.* Józseffalva. Vojvodina, N Serbia 45°22′N 20°01′E
119 *L18* **Zabalotstsye** *prev.* Zabalatstsye, *Rus.* Zabolot'ye. Homyel'skaya Voblasts', SE Belarus 52°40′N 28°34′E
Zāb aş Şaghīr, Nahraz *see* Little Zab
123 *P14* **Zabaykal'sk** Zabaykal'skiy Kray, S Russian Federation 49°37′N 117°20′E
123 *O12* **Zabaykal'skiy Kray** ◐ *province* S Russian Federation
Zāb-e Kūchek, Rūdkhāneh-ye *see* Little Zab
Zabeln *see* Sabile
Zabéré *see* Zabré
Zabern *see* Saverne
141 *N16* **Zabīd** W Yemen 14°N 43°E
141 *O16* **Zabīd, Wādī** *dry watercourse* SW Yemen
Żabinka *see* Zhabinka
Ząbkowice *see* Ząbkowice Śląskie
111 *G15* **Ząbkowice Śląskie** *var.* Ząbkowice, *Ger.* Frankenstein, Frankenstein in Schlesien. Dolnośląskie, SW Poland 50°35′N 16°48′E
110 *P10* **Zabłudów** Podlaskie, NE Poland 53°00′N 23°21′E
112 *D8* **Zabok** Krapina-Zagorje, N Croatia 46°00′N 15°48′E
143 *W9* **Zābol** *var.* Shahr-i-Zabul, Zabul; *prev.* Nasratabad. Sīstān va Balūchestān, E Iran 31°N 61°32′E
Zābol *see* Zābul
143 *W13* **Zābolī** Sīstān va Balūchestān, SE Iran 27°09′N 61°32′E
Zabolot'ye *see* Zabalotstsye
77 *Q13* **Zabré** *var.* Zabéré. S Burkina Faso 11°13′N 00°34′W
111 *G17* **Zábřeh** *Ger.* Hohenstadt. Olomoucký Kraj, E Czech Republic 49°52′N 16°53′E
111 *J16* **Zabrze** *Ger.* Hindenburg, Hindenburg in Oberschlesien. Śląskie, S Poland 50°18′N 18°47′E
149 *O7* **Zābul** *prev.* Zābol. ◐ *province* SE Afghanistan
Zabul/Zābul *see* Zābol
42 *E6* **Zacapa** Zacapa, E Guatemala 14°59′N 89°33′W
42 *A3* **Zacapa** *off.* Departamento de Zacapa. ◐ *department* E Guatemala
Zacapa, Departamento de *see* Zacapa
40 *M14* **Zacapú** Michoacán, SW Mexico 19°49′N 101°48′W
41 *V14* **Zacatal** Campeche, SE Mexico 18°40′N 91°52′W
40 *M11* **Zacatecas** Zacatecas, C Mexico 22°46′N 102°33′W
40 *L10* **Zacatecas** ◐ *state* C Mexico
42 *F8* **Zacatecoluca** La Paz, S El Salvador 13°29′N 88°51′W
41 *P15* **Zacatepec** Morelos, S Mexico 18°40′N 99°11′W
41 *Q13* **Zacatlán** Puebla, S Mexico 19°56′N 97°58′W
144 *F8* **Zachagansk** *Kaz.* Zashaghan. Zapadnyy Kazakhstan, NW Kazakhstan 51°04′N 51°13′E
115 *D20* **Zacháro** *var.* Zaharo, Zakháro. Dytikí Elláda, S Greece 37°29′N 21°40′E
22 *J8* **Zachary** Louisiana, S USA 30°39′N 91°09′W
117 *U6* **Zachepylivka** Kharkivs'ka Oblast', E Ukraine 49°13′N 35°15′E
Zachist'ye *see* Zachystsye
110 *E9* **Zachodnio-pomorskie** ◐ *province* NW Poland
119 *L14* **Zachystsye** *Rus.* Zachist'ye. Minskaya Voblasts', NW Belarus 54°24′N 28°45′E
40 *L13* **Zacoalco** *var.* Zacoalco de Torres. Jalisco, SW Mexico 20°14′N 103°33′W
Zacoalco de Torres *see* Zacoalco
41 *P13* **Zacualtipán** Hidalgo, C Mexico 20°39′N 98°42′W
112 *C12* **Zadar** *It.* Zara; *anc.* Iader. Zadar, SW Croatia 44°07′N 15°15′E
112 *C12* **Zadar** *off.* Zadarsko-Kninska Županija, Zadar-Knin. ◐ *province* SW Croatia
Zadar-Knin *see* Zadar
Zadarsko-Kninska Županija *see* Zadar
166 *M14* **Zadetkyi Kyun** *var.* St.Matthew's Island. *island* Mergui Archipelago, S Myanmar (Burma)
67 *Q9* **Zadié** *var.* Djadié. ≈ NE Gabon
159 *Q13* **Zadoi** *var.* Qapugtang. Qinghai, C China 32°56′N 95°21′E
126 *L7* **Zadonsk** Lipetskaya Oblast', W Russian Federation 52°25′N 38°55′E
Za'farâna *see* Za'farānah
75 *X8* **Za'farānah** *var.* Za'farâna. E Egypt 29°06′N 32°34′E
149 *W7* **Zafarwāl** Punjab, E Pakistan 32°20′N 74°53′E
121 *Q1* **Zafer Burnu** *var.* Cape Andreas, Cape Apostolas Andreas, *Gk.* Akrotíri Apostólou Andréa. *cape* NE Cyprus
107 *J23* **Zafferano, Capo** *headland* Sicilia, Italy, C Mediterranean Sea 38°06′N 13°31′E
114 *M7* **Zafirovo** Silistra, NE Bulgaria 44°00′N 26°51′E
Záfora *see* Sofraná
104 *J12* **Zafra** Extremadura, W Spain 38°25′N 06°27′W
110 *E13* **Żagań** *var.* Zagań, Żegań, *Ger.* Sagan. Lubuskie, W Poland 51°37′N 15°20′E
118 *F10* **Žagarė** *Pol.* Żagory. Šiaulai, N Lithuania 56°22′N 23°16′E
Zagazig *see* Az Zaqāzīq
74 *M5* **Zaghouan** *var.* Zaghwān. NE Tunisia 36°26′N 10°05′E
Zaghwān *see* Zaghouan
115 *G16* **Zagorá** Thessalía, C Greece 39°27′N 23°06′E
Zagorod'ye *see* Zaharoddzye
Żagory *see* Žagarė
Zágráb *see* Zagreb
112 *E8* **Zagreb** *Ger.* Agram, *Hung.* Zágráb. ● (Croatia) Zagreb, N Croatia 45°48′N 15°58′E
112 *E8* **Zagreb** *prev.* Grad Zagreb. ◐ *province* N Croatia

◆ Country
● Country Capital
◇ Dependent Territory
○ Dependent Territory Capital
◐ Administrative Regions
✈ International Airport
▲ Mountain
▲▲ Mountain Range
⛰ Volcano
≈ River
◎ Lake
⊞ Reservoir

ZĀGROS, KŪHHĀ-YE – ZHUJI

142 L7 **Zāgros, Kūhhā-ye** *Eng.* Zagros Mountains. ▲▲ W Iran
Zagros Mountains *see* Zāgros, Kūhhā-ye
112 O12 **Žagubica** Serbia, E Serbia 44°13′N 21°47′E
Zagunao *see* Lixian
111 L22 **Zagyva** ≈ N Hungary
Zaharo *see* Zacháro
119 G19 **Zaharoddzye** *Rus.* Zagorod'ye. *physical region* SW Belarus
143 W11 **Zāhedān** *var.* Zahidan; *prev.* Duzdab. Sīstān va Balūchestān, SE Iran 29°31′N 60°51′E
Zahidan *see* Zāhedān
Zaḩlah *see* Zahlé
138 H7 **Zahlé** *var.* Zaḩlah. C Lebanon 33°51′N 35°54′E
146 J14 **Zähmet** *Rus.* Zakhmet. Mary Welaýaty, C Turkmenistan 37°48′N 62°33′E
111 O20 **Záhony** Szabolcs-Szatmár-Bereg, NE Hungary 48°26′N 22°11′E
141 N13 **Ẓahrān** ʻAsīr, S Saudi Arabia 17°48′N 43°28′E
139 R12 **Ẓahrat al Baṭn** *hill range* S Iraq
120 H11 **Zahrez Chergui** *var.* Zahrez Chergúi. *marsh* N Algeria
Zainlha *see* Xinjin
127 S4 **Zainsk** Respublika Tatarstan, W Russian Federation 55°12′N 52°01′E
82 A10 **Zaire** *prev.* Congo. ◈ *province* NW Angola
Zaire *see* Congo (river)
Zaire *see* Congo (Democratic Republic of)
112 P13 **Zaječar** Serbia, E Serbia 43°54′N 22°16′E
83 L18 **Zaka** Masvingo, E Zimbabwe 20°20′S 31°29′E
122 M14 **Zakamensk** Respublika Buryatiya, S Russian Federation 50°18′N 102°57′E
116 G7 **Zakarpats'ka Oblast'** *Eng.* Transcarpathian Oblast, *Rus.* Zakarpatskaya Oblast'. ◈ *province* W Ukraine
Zakarpatskaya Oblast' *see* Zakarpats'ka Oblast'
Zakataly *see* Zaqatala
Zakháro *see* Zacháro
Zakhidnyy Buh/Zakhodni Buh *see* Bug
Zakhmet *see* Zähmet
Zākhō *see* Zaxo
Zākhū *see* Zaxo
Zákinthos *see* Zákynthos
111 L18 **Zakopane** Małopolskie, S Poland 49°17′N 19°57′E
78 J12 **Zakouma** Salamat, S Chad 10°47′N 19°51′E
115 L25 **Zákros** Kríti, Greece, E Mediterranean Sea 35°06′N 26°12′E
115 C19 **Zákynthos** *var.* Zákinthos. Zákynthos, W Greece 37°47′N 20°54′E
115 C20 **Zákynthos** *var.* Zákinthos, *It.* Zante. *island* Iónia Nísoi, Greece, C Mediterranean Sea
115 C19 **Zakýnthou, Porthmós** *strait* SW Greece
111 G24 **Zala** *off.* Zala Megye. ◈ *county* W Hungary
111 G24 **Zala** ≈ W Hungary
138 M4 **Zalābiyah** Dayr az Zawr, C Syria 35°39′N 39°51′E
111 G24 **Zalaegerszeg** Zala, W Hungary 46°51′N 16°49′E
104 K11 **Zalamea de la Serena** Extremadura, W Spain 38°38′N 05°37′W
104 J13 **Zalamea la Real** Andalucía, S Spain 37°41′N 06°40′W
Zala Megye *see* Zala
163 U7 **Zalantun** *var.* Butha Qi. Nei Mongol Zizhiqu, N China 47°58′N 122°44′E
111 G23 **Zalaszentgrót** Zala, SW Hungary 46°57′N 17°05′E
Zalatna *see* Zlatna
116 G9 **Zalău** *Ger.* Waltenberg, *Hung.* Zilah; *prev. Ger.* Zillenmarkt. Sălaj, NW Romania 47°11′N 23°03′E
109 V10 **Žalec** *Ger.* Sachsenfeld. C Slovenia 46°15′N 15°08′E
110 K8 **Zalewo** *Ger.* Saalfeld. Warmińsko-Mazurskie, NE Poland 53°54′N 19°39′E
141 N9 **Ẓalim** Makkah, W Saudi Arabia 22°46′N 42°12′E
80 A11 **Zalingei** *var.* Zalinje. Central Darfur, W Sudan 12°51′N 23°29′E
Zalinje *see* Zalingei
116 K7 **Zalishchyky** Ternopil's'ka Oblast', W Ukraine 48°40′N 25°43′E
Zallah *see* Zillah
`Załni Pjašaci *see* Zlatni Pyasatsi
98 J13 **Zaltbommel** Gelderland, C Netherlands 51°49′N 05°15′E
124 H15 **Zaluch'ye** Novgorodskaya Oblast', NW Russian Federation 57°40′N 31°45′E
Zamak *see* Zamakh
141 Q14 **Zamakh** *var.* Zamak. N Yemen 16°26′N 47°35′E
136 K15 **Zamantı Irmağı** ≈ C Turkey
Zambesi/Zambeze *see* Zambezi
83 G14 **Zambezi** North Western, W Zambia 13°34′S 23°08′E
83 K15 **Zambezi** *var.* Zambesi, *Port.* Zambeze. ≈ S Africa
83 O15 **Zambézia** *off.* Província da Zambézia. ◈ *province* C Mozambique
Zambézia, Província da *see* Zambézia
83 I14 **Zambia** *off.* Republic of Zambia; *prev.* Northern Rhodesia. ◆ *republic* S Africa
Zambia, Republic of *see* Zambia
171 O8 **Zamboanga** *off.* Zamboanga City. Mindanao, S Philippines 06°56′N 122°03′E
Zamboanga City *see* Zamboanga
54 E5 **Zambrano** Bolívar, N Colombia 09°45′N 74°50′W
110 N10 **Zambrów** Łomża, E Poland 52°59′N 22°14′E
83 L14 **Zâmbuè** Tete, NW Mozambique 15°03′S 30°49′E
77 T13 **Zamfara** ≈ NW Nigeria
Zamkog *see* Zamtang
56 C9 **Zamora** Zamora Chinchipe, S Ecuador 04°04′S 78°52′W
104 K6 **Zamora** Castilla y León, NW Spain 41°30′N 05°45′W
104 K5 **Zamora** ◈ *province* Castilla y León, NW Spain
Zamora *see* Barinas
56 A13 **Zamora Chinchipe** ◈ *province* S Ecuador
40 M13 **Zamora de Hidalgo** Michoacán, SW Mexico 20°N 102°18′W
111 P15 **Zamość** *Rus.* Zamoste. Lubelskie, E Poland 50°44′N 23°16′E
Zamoste *see* Zamość
160 G7 **Zamtang** *var.* Zamkog; *prev.* Gamba. Sichuan, C China 32°19′N 100°55′E
75 O8 **Zamzam, Wādī** *dry watercourse* NW Libya
79 F20 **Zanaga** Lékoumou, S Congo 02°50′S 13°53′E
41 T16 **Zanatepec** Oaxaca, SE Mexico 16°28′N 94°24′W
105 P9 **Záncara** ≈ C Spain
Zancle *see* Messina
158 G14 **Zanda** Xizang Zizhiqu, W China 31°29′N 79°50′E
98 H10 **Zandvoort** Noord-Holland, W Netherlands 52°22′N 04°31′E
39 P8 **Zane Hills** *hill range* Alaska, USA
31 T13 **Zanesville** Ohio, N USA 39°55′N 82°02′W
Zanga *see* Hrazdan
Zangoza *see* Sangüesa
142 L4 **Zanjān** *var.* Zenjan, Zinjan. Zanjān, NW Iran 36°40′N 48°30′E
142 L4 **Zanjān** *off.* Ostān-e Zanjān, *var.* Zenjan, Zinjan. ◈ *province* NW Iran
Zanjān, Ostān-e *see* Zanjān
Zante *see* Zákynthos
81 J22 **Zanzibar** Zanzibar, E Tanzania 06°10′S 39°12′E
81 J22 **Zanzibar** ◈ *region* E Tanzania
81 J22 **Zanzibar** *Swa.* Unguja. *island* E Tanzania
81 J22 **Zanzibar Channel** *channel* E Tanzania
161 N8 **Zaoyang** Hubei, C China 32°10′N 112°45′E
165 P10 **Zaō-zan** ▲ Honshū, C Japan 38°06′N 140°27′E
124 J2 **Zaozërsk** Murmanskaya Oblast', NW Russian Federation 69°25′N 32°25′E
161 Q6 **Zaozhuang** Shandong, E China 34°53′N 117°38′E
28 L4 **Zap** North Dakota, N USA 47°18′N 101°55′W
112 L13 **Zapadna Morava** *Ger.* Westliche Morava. ≈ C Serbia
124 H16 **Zapadnaya Dvina** Tverskaya Oblast', W Russian Federation 56°17′N 32°03′E
Zapadnaya Dvina *see* Western Dvina
Zapadno-Kazakhstanskaya Oblast' *see* Zapadnyy Kazakhstan
122 I9 **Zapadno-Sibirskaya Ravnina** *Eng.* West Siberian Plain. *plain* C Russian Federation
Zapadnyy Bug *see* Bug
144 E9 **Zapadnyy Kazakhstan** *off.* Zapadno-Kazakhstanskaya Oblast', *Eng.* West Kazakhstan, *Kaz.* Batys Qazaqstan Oblysy; *prev.* Ural'skaya Oblast'. ◈ *province* NW Kazakhstan
122 K13 **Zapadnyy Sayan** *Eng.* Western Sayans. ▲▲ S Russian Federation
63 H15 **Zapala** Neuquén, W Argentina 38°54′S 70°06′W
62 I4 **Zapaleri, Cerro** *var.* Cerro Sapaleri. ▲ N Chile 22°51′S 67°10′W
25 Q16 **Zapata** Texas, SW USA 26°57′N 99°17′W
44 D5 **Zapata, Península de** *peninsula* W Cuba
61 G19 **Zapicán** Lavalleja, S Uruguay 33°31′S 54°55′W
65 J19 **Zapiola Ridge** *undersea feature* SW Atlantic Ocean
65 L19 **Zapiola Seamount** *undersea feature* S Atlantic Ocean 38°15′S 26°15′W
124 I2 **Zapolyarnyy** Murmanskaya Oblast', NW Russian Federation 69°24′N 30°53′E
117 U8 **Zaporizhzhya** *Rus.* Zaporozh'ye; *prev.* Aleksandrovsk. Zaporiz'ka Oblast', SE Ukraine 47°47′N 35°12′E
Zaporizhzhya *see* Zaporiz'ka Oblast'
117 U9 **Zaporiz'ka Oblast'** *var.* Zaporizhzhya, *Rus.* Zaporozhskaya Oblast'. ◈ *province* SE Ukraine
Zaporozhskaya Oblast' *see* Zaporiz'ka Oblast'
Zaporozh'ye *see* Zaporizhzhya
40 L14 **Zapotiltic** Jalisco, SW Mexico 19°40′N 103°29′W
158 G13 **Zapug** Xizang Zizhiqu, W China
137 V10 **Zaqatala** *Rus.* Zakataly. NW Azerbaijan 41°38′N 46°38′E
159 P13 **Zaqên** Qinghai, W China 33°23′N 94°31′E
159 Q13 **Za Qu** ≈ C China
136 M13 **Zara** Sivas, C Turkey 39°55′N 37°44′E
Zara *see* Zadar
Zarafshan *see* Zarafshon
147 P12 **Zarafshon** *Rus.* Zeravshan. W Tajikistan 39°12′N 68°36′E
146 L9 **Zarafshon** *Rus.* Zarafshan. Navoiy Viloyati, N Uzbekistan 41°33′N 64°09′E
Zarafshon *see* Zeravshan
147 O12 **Zarafshon, Qatorkŭhi** *Rus.* Zeravshanskiy Khrebet, *Uzb.* Zarafshon Tizmasi. ▲▲ Tajikistan/Uzbekistan
Zarafshon Tizmasi *see* Zarafshon, Qatorkŭhi
54 E7 **Zaragoza** Antioquia, N Colombia 07°30′N 74°52′W
40 I5 **Zaragoza** Chihuahua, N Mexico 29°36′N 107°41′W
41 N6 **Zaragoza** Coahuila, NE Mexico 28°31′N 100°54′W
41 O10 **Zaragoza** Nuevo León, NE Mexico 23°59′N 99°49′W
105 R5 **Zaragoza** *Eng.* Saragossa; *anc.* Caesaraugusta, Salduba. Aragón, NE Spain 41°39′N 00°54′W
105 R6 **Zaragoza** ◈ *province* Aragón, NE Spain
105 R5 **Zaragoza** ✈ Aragón, NE Spain 41°38′N 00°53′W
143 S10 **Zarand** Kermān, C Iran 30°50′N 56°35′E
148 J9 **Zaranj** Nīmrōz, SW Afghanistan 30°59′N 61°54′E
118 I11 **Zarasai** Utena, E Lithuania 55°44′N 26°17′E
62 N12 **Zárate** *prev.* General José F.Uriburu. Buenos Aires, E Argentina 34°06′S 59°03′W
105 Q2 **Zarautz** *var.* Zarauz. País Vasco, N Spain 43°17′N 02°10′W
Zarauz *see* Zarautz
Zaravecchia *see* Biograd na Moru
Zarāyīn *see* Zarēn
126 L4 **Zaraysk** Moskovskaya Oblast', W Russian Federation 54°48′N 38°54′E
55 N6 **Zaraza** Guárico, N Venezuela 09°23′N 65°20′W
Zarbdar *see* Zarbdor
147 P11 **Zarbdor** *Rus.* Zarbdar. Jizzax Viloyati, C Uzbekistan 40°04′N 68°10′E
142 M8 **Zard Kūh** ▲ SW Iran 32°19′N 50°03′E
124 I5 **Zarechensk** Murmanskaya Oblast', NW Russian Federation 66°39′N 31°27′E
127 P6 **Zarechnyy** Penzenskaya Oblast', W Russian Federation 53°12′N 45°12′E
Zareh Sharan *see* Sharan
39 Y14 **Zarembo Island** *island* Alexander Archipelago, Alaska, USA
139 V4 **Zarēn** *var.* Zarāyīn. As Sulaymānīyah, E Iraq 35°16′N 45°43′E
149 Q7 **Zarghūn Shahr** *var.* Katawaz. Paktīkā, SE Afghanistan 32°40′N 68°20′E
77 V13 **Zaria** Kaduna, C Nigeria 11°06′N 07°42′E
116 K2 **Zarichne** Rivnens'ka Oblast', NW Ukraine 51°49′N 26°09′E
122 J13 **Zarinsk** Altayskiy Kray, S Russian Federation 53°34′N 85°22′E
116 J12 **Zărnești** *Hung.* Zernest. Brașov, C Romania 45°34′N 25°18′E
115 J25 **Zarós** Kríti, Greece, E Mediterranean Sea 35°08′N 24°54′E
100 O9 **Zarow** ≈ NE Germany
Zarqa *see* Az Zarqā'
Zarqa/Muḩāfazat az Zarqā' *see* Az Zarqā'
111 G20 **Záruby** ▲ W Slovakia 48°30′N 17°24′E
56 B8 **Zaruma** El Oro, SW Ecuador 03°46′S 79°38′W
110 E13 **Żary** *Ger.* Sorau, Sorau in der Niederlausitz. Lubuskie, W Poland 51°44′N 15°09′E
54 D10 **Zarzal** Valle del Cauca, W Colombia 04°24′N 76°01′W
42 I7 **Zarzalar, Cerro** ▲ S Honduras 14°15′N 86°49′W
Zashaghan *see* Zachagansk
152 I5 **Zāskār** ≈ NE India
152 I5 **Zāskār Range** ▲▲ NE India
Zaslavl' *see* Zaslawye
119 K15 **Zaslawye** *Rus.* Zaslavl'. Minskaya Voblasts', C Belarus 54°01′N 27°16′E
116 K7 **Zastavna** Chernivets'ka Oblast', W Ukraine 48°30′N 25°51′E
111 B16 **Žatec** *Ger.* Saaz. Ústecký Kraj, NW Czech Republic 50°20′N 13°35′E
Zaumgarten *see* Chrzanów
Zaunguzskiye Garagumy *see* Üngüz Angyrsyndaky Garagum
25 X9 **Zavalla** Texas, SW USA 31°09′N 94°25′W
99 H18 **Zaventem** Vlaams Brabant, C Belgium 50°53′N 04°28′E
99 H18 **Zaventem** ✈ (Brussel/Bruxelles) Vlaams Brabant, C Belgium 50°55′N 04°28′E
Zaẏertse *see* Zawiercie
114 L7 **Zavet** Razgrad, NE Bulgaria 43°46′N 26°40′E
127 O12 **Zavetnoye** Rostovskaya Oblast', SW Russian Federation 47°10′N 43°54′E
156 M3 **Zavhan Gol** ≈ W Mongolia
112 H12 **Zavidovići** Federacija Bosne I Hercegovine, N Bosnia and Herzegovina 44°26′N 18°07′E
123 R13 **Zavitinsk** Amurskaya Oblast', SE Russian Federation 50°23′N 128°57′E
Zawia *see* Az Zāwiyah
111 K15 **Zawiercie** *Rus.* Zavertse. Śląskie, S Poland 50°29′N 19°24′E
75 P11 **Zawīlah** *var.* Zuwaylah, *It.* Zueila. C Libya 26°10′N 15°07′E
138 I4 **Zāwīyah, Jabal az** ▲▲ NW Syria
139 Q1 **Zaxo** *Ar.* Zākhū, *var.* Zākhō. Dahūk, N Iraq 37°09′N 42°40′E
109 Y3 **Zaya** ≈ NE Austria
166 M8 **Zayatkyi** Bago, C Myanmar (Burma) 17°48′N 96°27′E
145 Y11 **Zaysan** Vostochnyy Kazakhstan, E Kazakhstan 47°30′N 84°55′E
Zaysan Köl *see* Zaysan, Ozero
145 Y11 **Zaysan, Ozero** *Kaz.* Zaysan Köl. ◎ E Kazakhstan
159 R16 **Zayü** *var.* Gyigang. Xizang Zizhiqu, W China 28°36′N 97°25′E
Zayyq *see* Zhayyk
44 F6 **Zaza** ≈ C Cuba
116 K5 **Zbarazh** Ternopil's'ka Oblast', W Ukraine 49°40′N 25°47′E
116 J5 **Zboriv** Ternopil's'ka Oblast', W Ukraine 49°40′N 25°07′E
111 F18 **Zbraslav** Jihomoravský Kraj, SE Czech Republic 49°13′N 16°19′E
116 K6 **Zbruch** ≈ W Ukraine
Žd'ár *see* Žd'ár nad Sázavou
111 F17 **Žd'ár nad Sázavou** *Ger.* Saar in Mähren; *prev.* Žd'ár. Vysočina, C Czech Republic 49°34′N 16°00′E
116 K4 **Zdolbuniv** *Pol.* Zdolbunów, *Rus.* Zdolbunov. Rivnens'ka Oblast', NW Ukraine 50°33′N 26°15′E
Zdolbunov/Zdolbunów *see* Zdolbuniv
110 J13 **Zduńska Wola** Sieradz, C Poland 51°37′N 18°57′E
117 O4 **Zdvizh** ≈ N Ukraine
Zdzięciół *see* Dzyatlava
111 I16 **Zdzieszowice** *Ger.* Odertal. Opolskie, SW Poland 50°24′N 18°06′E
Zealand *see* Sjælland
188 K6 **Zealandia Bank** *undersea feature* C Pacific Ocean
63 H20 **Zeballos, Monte** ▲ S Argentina 47°04′S 71°32′W
83 K20 **Zebediela** Limpopo, NE South Africa 24°16′S 29°21′E
113 L18 **Zebës, Mali i** *var.* Mali i Zebës. ▲ NE Albania 41°57′N 20°16′E
Zebës, Mali i *see* Zebës, Mali i
21 V9 **Zebulon** North Carolina, SE USA 35°49′N 78°19′W
112 K8 **Žednik** *Hung.* Bácsjózseffalva. Vojvodina, N Serbia 45°58′N 19°40′E
99 C15 **Zeebrugge** West-Vlaanderen, NW Belgium 51°20′N 03°13′E
183 N16 **Zeehan** Tasmania, SE Australia 41°54′S 145°19′E
99 L14 **Zeeland** Noord-Brabant, SE Netherlands 51°42′N 05°40′E
29 N7 **Zeeland** North Dakota, N USA 45°57′N 99°49′W
99 E14 **Zeeland** ◈ *province* SW Netherlands
83 I21 **Zeerust** North-West, N South Africa 25°33′S 26°06′E
98 K10 **Zeewolde** Flevoland, C Netherlands 52°20′N 05°32′E
Ẕefat *see* Tsefat
Żegań *see* Żagań
Zehden *see* Cedynia
100 O11 **Zehdenick** Brandenburg, NE Germany 52°58′N 13°19′E
Zē-i Bādīnān *see* Great Zab
Zeiden *see* Codlea
146 M14 **Zeidskoye Vodokhranilishche** ⊠ E Turkmenistan
Zē-i Kōya *see* Little Zab
181 P7 **Zeil, Mount** ▲ Northern Territory, C Australia 23°31′S 132°41′E
98 J12 **Zeist** Utrecht, C Netherlands 52°05′N 05°15′E
101 M16 **Zeitz** Sachsen-Anhalt, E Germany 51°03′N 12°08′E
159 T11 **Zêkog** *var.* Zequ; *prev.* Sonag. Qinghai, C China 35°03′N 101°30′E
Zelaya Norte *see* Atlántico Norte, Región Autónoma
Zelaya Sur *see* Atlántico Sur, Región Autónoma
99 F17 **Zele** Oost-Vlaanderen, NW Belgium 51°04′N 04°02′E
110 N12 **Żelechów** Lubelskie, E Poland 51°49′N 21°57′E
113 H14 **Zelena Glava** ▲ SE Bosnia and Herzegovina 43°32′N 17°55′E
113 P20 **Zelen Breg** ▲ S Macedonia 41°10′N 22°14′E
113 I14 **Zelengora** ▲▲ S Bosnia and Herzegovina
124 I5 **Zelenoborskiy** Murmanskaya Oblast', NW Russian Federation 66°52′N 32°25′E
127 R3 **Zelenodol'sk** Respublika Tatarstan, W Russian Federation 55°52′N 48°49′E
117 S9 **Zelenodols'k** Dnipropetrovs'ka Oblast', E Ukraine 47°38′N 33°41′E
122 J12 **Zelenogorsk** Krasnoyarskiy Kray, C Russian Federation 56°08′N 94°29′E
126 K3 **Zelenograd** Moskovskaya Oblast', W Russian Federation 56°02′N 37°08′E
118 B13 **Zelenogradsk** *Ger.* Cranz, Kranz. Kaliningradskaya Oblast', W Russian Federation 54°58′N 20°30′E
127 O15 **Zelenokumsk** Stavropol'skiy Kray, SW Russian Federation 44°22′N 43°48′E
165 X4 **Zelënyy, Ostrov** *var.* Shibotsu-jima. *island* NE Russian Federation
Železna Kapela *see* Eisenkappel
Železna Vrata *see* Demir Kapija
112 L11 **Železniki** Serbia, N Serbia 44°45′N 20°23′E
98 N12 **Zelhem** Gelderland, E Netherlands 52°00′N 06°21′E
113 N18 **Želino** NW FYR Macedonia 42°00′N 21°06′E
113 M14 **Željin** ▲▲ C Serbia
101 K17 **Zella-Mehlis** Thüringen, C Germany 50°40′N 10°40′E
109 P7 **Zell am See** *var.* Zell-am-See. Salzburg, S Austria 47°19′N 12°47′E
Zell-am-See *see* Zell am See
109 N7 **Zell am Ziller** Tirol, W Austria 47°13′N 11°52′E
Zelle *see* Celle
109 W2 **Zellerndorf** Niederösterreich, NE Austria 48°40′N 15°57′E
109 U7 **Zeltweg** Steiermark, S Austria 47°12′N 14°46′E
119 G17 **Zel'va** *Pol.* Zelwa. Hrodzyenskaya Voblasts', W Belarus 53°09′N 24°49′E
118 H13 **Želva** Vilnius, C Lithuania 55°13′N 25°07′E
Zelwa *see* Zel'va
99 E16 **Zelzate** *var.* Selzaete. Oost-Vlaanderen, NW Belgium 51°12′N 03°49′E
118 E11 **Žemaičių Aukštumas** *physical region* W Lithuania
118 C12 **Žemaičių Naumiestis** Klaipėda, SW Lithuania 55°22′N 21°39′E
119 L14 **Zembin** *var.* Zyembin. Minskaya Voblasts', C Belarus 54°22′N 28°13′E
127 N6 **Zemetchino** Penzenskaya Oblast', W Russian Federation 53°31′N 42°35′E
79 M15 **Zémio** Haut-Mbomou, E Central African Republic 05°04′N 25°07′E
41 R16 **Zempoaltepec, Cerro** ▲ SE Mexico 17°04′N 95°54′W
99 G17 **Zemst** Vlaams Brabant, C Belgium 50°59′N 04°28′E
112 L11 **Zemun** Serbia, N Serbia 44°52′N 20°25′E
Zendajan *see* Zindah Jān
Zendeh Jān *see* Zindah Jān
Zengg *see* Senj
112 H12 **Zenica** Federacija Bosne I Hercegovine, C Bosnia and Herzegovina 44°12′N 17°53′E
Zenjan *see* Zanjān
Zen'kov *see* Zin'kiv
Zenshū *see* Jeonju
Zenta *see* Senta
82 B11 **Zenza do Itombe** Kwanza Norte, NW Angola 09°22′S 14°10′E
112 H12 **Žepče** Federacija Bosne I Hercegovine, N Bosnia and Herzegovina 44°26′N 18°00′E
23 W12 **Zephyrhills** Florida, SE USA 28°13′N 82°10′W
158 F9 **Zepu** *var.* Poskam. Xinjiang Uygur Zizhiqu, NW China 38°10′N 77°18′E
Zequ *see* Zêkog
147 Q12 **Zeravshan** *Taj./Uzb.* Zarafshon. ≈ Tajikistan/Uzbekistan
Zeravshan *see* Zarafshon
Zeravshanskiy Khrebet *see* Zarafshon, Qatorkŭhi
101 M14 **Zerbst** Sachsen-Anhalt, E Germany 51°57′N 12°05′E
Zerenda *see* Zerendy
145 P8 **Zerendy** *prev.* Zerenda. Akmola, N Kazakhstan 52°56′N 69°09′E
110 H12 **Żerków** Wielkopolskie, C Poland 52°03′N 17°33′E
108 E11 **Zermatt** Valais, SW Switzerland 46°00′N 07°45′E
Zernest *see* Zărnești
108 J9 **Zernez** Graubünden, SE Switzerland 46°42′N 10°06′E
126 L12 **Zernograd** Rostovskaya Oblast', SW Russian Federation 46°52′N 40°13′E
Zestafoni *see* Zest'aponi
137 S9 **Zest'aponi** *Rus.* Zestafoni; *prev.* Zestap'oni. C Georgia 42°09′N 43°00′E
Zestap'oni *see* Zest'aponi
98 H12 **Zestienhoven** ✈ (Rotterdam) Zuid-Holland, SW Netherlands 51°57′N 04°30′E
113 J16 **Zeta** ≈ C Montenegro
8 L6 **Zeta Lake** ◎ Victoria Island, Northwest Territories, N Canada
98 L12 **Zetten** Gelderland, SE Netherlands 51°55′N 05°43′E
101 M17 **Zeulenroda** Thüringen, C Germany 50°40′N 11°58′E
100 H10 **Zeven** Niedersachsen, NW Germany 53°17′N 09°16′E
98 M12 **Zevenaar** Gelderland, SE Netherlands 51°55′N 06°05′E
99 H14 **Zevenbergen** Noord-Brabant, S Netherlands 51°39′N 04°36′E
129 X6 **Zeya** ≈ SE Russian Federation
Zeya Reservoir *see* Zeyskoye Vodokhranilishche
143 T11 **Zeynalābād** Kermān, C Iran 29°56′N 57°29′E
123 R12 **Zeyskoye Vodokhranilishche** *Eng.* Zeya Reservoir. ⊠ SE Russian Federation
104 H8 **Zêzere, Rio** ≈ C Portugal
Zgerzh *see* Zgierz
138 H6 **Zgharta** N Lebanon 34°24′N 35°54′E
110 K12 **Zgierz** *Ger.* Neuhof, *Rus.* Zgerzh. Łódź, C Poland 51°55′N 19°20′E
111 E14 **Zgorzelec** *Ger.* Görlitz. Dolnośląskie, SW Poland 51°10′N 15°E
119 F19 **Zhabinka** *Pol.* Żabinka. Brestskaya Voblasts', SW Belarus 52°12′N 24°01′E
Zhaggo *see* Luhuo
159 R15 **Zhag'yab** *var.* Yêndum. Xizang Zizhiqu, W China 30°42′N 97°33′E
Zhailma *see* Zhayylma
145 N14 **Zhalagash** *prev.* Dzhalagash. Kzylorda, S Kazakhstan 45°06′N 64°40′E
145 V16 **Zhalanash** Almaty, SE Kazakhstan 43°04′N 78°08′E
145 S7 **Zhalauly, Ozero** ◎ NE Kazakhstan
145 W10 **Zhalgyztobe** *prev.* Zhangiztobe. Vostochnyy Kazakhstan, E Kazakhstan 49°15′N 81°16′E
144 E9 **Zhalpaktal** *Kaz.* Zhalpaqtal; *prev.* Furmanovo. Zapadnyy Kazakhstan, W Kazakhstan 49°43′N 49°28′E
Zhalpaqtal *see* Zhalpaktal
119 G16 **Zhaludok** *Rus.* Zheludok. Hrodzyenskaya Voblasts', W Belarus 53°36′N 24°59′E
Zhaman-Akkol', Ozero *see* Akkol', Ozero
145 Q14 **Zhambyl** *off.* Zhambylskaya Oblast', *Kaz.* Zhambyl Oblysy; *prev.* Dzhambulskaya Oblast'. ◈ *province* S Kazakhstan
Zhambyl *see* Taraz
Zhambyl Oblysy/Zhambylskaya Oblast' *see* Zhambyl
Zhamo *see* Bomi
145 S12 **Zhamshy** ≈ C Kazakhstan
144 M15 **Zhanadariya** *prev.* Zhanadar'ya. Kzylorda, S Kazakhstan 44°41′N 64°39′E
Zhanadar'ya *see* Zhanadariya
144 E10 **Zhanakazan** *prev.* Novaya Kazanka. Zapadnyy Kazakhstan, W Kazakhstan 48°57′N 49°34′E
145 O15 **Zhanakorgan** *Kaz.* Zhangaqorghan. Kzylorda, S Kazakhstan 43°57′N 67°14′E
159 N16 **Zhanang** *var.* Chatang. Xizang Zizhiqu, W China 29°15′N 91°20′E
145 T12 **Zhanaortalyk** Karaganda, C Kazakhstan 47°31′N 75°42′E
144 F15 **Zhanaozen** *Kaz.* Zhangaözen; *prev.* Novyy Uzen'. Mangistau, W Kazakhstan 43°22′N 52°50′E
145 Q16 **Zhanatas** Zhambyl, S Kazakhstan 43°36′N 69°43′E
Zhangaözen *see* Zhanaozen
Zhangaqazaly *see* Ayteke Bi
Zhangaqorghan *see* Zhanakorgan
161 O2 **Zhangbei** Hebei, E China 41°13′N 114°43′E
Zhang-chia-k'ou *see* Zhangjiakou
Zhangdian *see* Zibo
Zhanggu *see* Danba
163 X9 **Zhangguangcai Ling** ▲▲ NE China
161 S13 **Zhanghua** *Jap.* Shōka; *prev.* Changhua. C Taiwan 24°06′N 120°31′E
Zhangiztobe *see* Zhalgyztobe
159 W11 **Zhangjiachuan** Gansu, N China 34°55′N 106°26′E
160 L10 **Zhangjiajie** *var.* Dayong. Hunan, S China 29°10′N 110°22′E
161 O2 **Zhangjiakou** *var.* Changkiakow, Zhang-chia-k'ou, *Eng.* Kalgan; *prev.* Wanchuan. Hebei, E China 40°48′N 114°51′E
161 Q13 **Zhangping** Fujian, SE China 25°21′N 117°29′E
161 Q13 **Zhangpu** *var.* Sui'an. Fujian, SE China 24°08′N 117°36′E
163 U11 **Zhangwu** Liaoning, NE China 42°21′N 122°32′E
159 S8 **Zhangye** *var.* Ganzhou. Gansu, N China 38°58′N 100°30′E
161 Q13 **Zhangzhou** Fujian, SE China 24°31′N 117°40′E
163 W6 **Zhan He** ≈ NE China
Zhänibek *see* Dzhanibek
160 L16 **Zhanjiang** *var.* Chanchiang, Chan-chiang, *Cant.* Tsamkong, *Fr.* Fort-Bayard. Guangdong, S China 21°10′N 110°20′E
145 V14 **Zhansugirov** *prev.* Dzhansugurov. Almaty, SE Kazakhstan 45°23′N 79°29′E
163 V8 **Zhaodong** Heilongjiang, NE China 46°03′N 125°58′E
Zhaoge *see* Qixian
160 H11 **Zhaojue** *var.* Xincheng. Sichuan, C China 28°03′N 102°50′E
161 N14 **Zhaoqing** Guangdong, S China 23°08′N 112°26′E
Zhaoren *see* Changwu
158 H5 **Zhaosu** *var.* Mongolküre. Xinjiang Uygur Zizhiqu, NW China 43°09′N 81°07′E
160 H11 **Zhaotong** Yunnan, SW China 27°20′N 103°29′E
163 V9 **Zhaoyuan** Heilongjiang, NE China 45°30′N 125°05′E
163 V9 **Zhaozhou** Heilongjiang, NE China 45°42′N 125°11′E
145 X13 **Zharbulak** Vostochnyy Kazakhstan, E Kazakhstan 46°04′N 82°05′E
158 J15 **Zhari Namco** ◎ W China
144 I12 **Zharkamys** *Kaz.* Zharqamys. Aktyubinsk, W Kazakhstan 47°58′N 56°33′E
145 W15 **Zharkent** *prev.* Panfilov. Taldykorgan, SE Kazakhstan 44°10′N 80°01′E
124 H17 **Zharkovskiy** Tverskaya Oblast', W Russian Federation 55°51′N 32°19′E
145 W11 **Zharma** Vostochnyy Kazakhstan, E Kazakhstan 48°48′N 80°55′E
144 F14 **Zharmysh** Mangistau, SW Kazakhstan 44°12′N 52°27′E
Zharqamys *see* Zharkamys
118 L13 **Zhary** Vitsyebskaya Voblasts', N Belarus 55°05′N 28°40′E
Zhaslyk *see* Jasliq
158 J14 **Zhaxi Co** ◎ W China
127 X6 **Zhayyk** *Kaz.* Zayyq, *var.* Ural. ≈ Kazakhstan/Russian Federation
144 L9 **Zhayylma** *prev.* Zhailma. Kostanay, N Kazakhstan 51°34′N 61°39′E
Zhdanov *see* Beyläqan
Zhdanov *see* Mariupol'
Zhe *see* Zhejiang
161 R10 **Zhejiang** *var.* Che-chiang, Chekiang, Zhe, Zhejiang Sheng. ◈ *province* SE China
Zhejiang Sheng *see* Zhejiang
145 S7 **Zhelezinka** Pavlodar, N Kazakhstan 53°35′N 75°16′E
119 C14 **Zheleznodorozhnyy** *Ger.* Gerdauen. Kaliningradskaya Oblast', W Russian Federation 54°21′N 21°17′E
Zheleznodorozhnyy *see* Yemva
122 K12 **Zheleznogorsk** Krasnoyarskiy, C Russian Federation 56°20′N 93°36′E
126 J7 **Zheleznogorsk** Kurskaya Oblast', W Russian Federation 52°22′N 35°21′E
127 N15 **Zheleznovodsk** Stavropol'skiy Kray, SW Russian Federation 44°12′N 43°01′E
Zhëltyye Vody *see* Zhovti Vody
Zheludok *see* Zhaludok
144 H12 **Zhem** *prev.* Emba. ≈ W Kazakhstan
160 K7 **Zhenba** Shaanxi, C China 32°42′N 107°55′E
160 I13 **Zhenfeng** *var.* Mingu. Guizhou, S China 25°27′N 105°38′E
Zhengjiatun *see* Shuangliao
159 X10 **Zhengning** *var.* Shanhe. Gansu, N China 35°29′N 108°21′E
Zhengxiangbai Qi *see* Qagan Nur
161 N6 **Zhengzhou** *var.* Ch'eng-chou, Chengchow; *prev.* Chenghsien. *province capital* Henan, C China 34°45′N 113°38′E
161 R8 **Zhenjiang** *var.* Chenkiang. Jiangsu, E China 32°08′N 119°30′E
163 U9 **Zhenlai** Jilin, NE China 45°52′N 123°11′E
160 I11 **Zhenxiong** Yunnan, SW China 27°31′N 104°52′E
160 K11 **Zhenyuan** *var.* Wuyang. Guizhou, S China 27°07′N 108°33′E
161 R11 **Zherong** *var.* Shuangcheng. Fujian, SE China 27°16′N 119°54′E
145 U15 **Zhetigen** *prev.* Nikolayevka. Almaty, SE Kazakhstan 43°39′N 77°10′E
Zhetiqara *see* Zhitikara
144 F15 **Zhetybay** Mangistau, SW Kazakhstan 43°35′N 52°05′E
145 P17 **Zhetysay** *var.* Dzhetysay. Yuzhnyy Kazakhstan 40°45′N 68°18′E
145 W14 **Zhetysuskiy Alatau** *prev.* Dzhungarskiy Alatau. ▲▲ China/Kazakhstan
160 M11 **Zhexi Shuiku** ⊠ C China
145 O12 **Zhezdy** Karaganda, C Kazakhstan 48°06′N 67°01′E
145 O12 **Zhezkazgan** *Kaz.* Zhezqazghan; *prev.* Dzhezkazgan. Karaganda, C Kazakhstan 47°49′N 67°44′E
Zhezqazghan *see* Zhezkazgan
Zhicheng *see* Yidu
Zhidachov *see* Zhydachiv
159 Q12 **Zhidoi** *var.* Gyaijêpozhanggê. Qinghai, C China 33°55′N 95°39′E
122 M13 **Zhigalovo** Irkutskaya Oblast', S Russian Federation 54°47′N 105°00′E
127 R6 **Zhigulevsk** Samarskaya Oblast', W Russian Federation 53°24′N 49°30′E
118 D13 **Zhilino** *Ger.* Schillen. Kaliningradskaya Oblast', W Russian Federation 54°55′N 21°54′E
Zhiloy, Ostrov *see* Çilov Adası
127 O8 **Zhirnovsk** Volgogradskaya Oblast', SW Russian Federation 51°01′N 44°49′E
160 M12 **Zhishan** *prev.* Yongzhou. Hunan, S China 26°12′N 111°36′E
Zhishan *see* Lingling
144 L8 **Zhitikara** *Kaz.* Zhetiqara; *prev.* Džetygara. Kostanay, NW Kazakhstan 52°14′N 61°12′E
Zhitkovichi *see* Zhytkavichy
Zhitomir *see* Zhytomyr
Zhitomirskaya Oblast' *see* Zhytomyrs'ka Oblast'
126 J5 **Zhizdra** Kaluzhskaya Oblast', W Russian Federation 53°38′N 34°39′E
119 N18 **Zhlobin** Homyel'skaya Voblasts', SE Belarus 52°53′N 30°01′E
Zhmerinka *see* Zhmerynka
116 M7 **Zhmerynka** *Rus.* Zhmerinka. Vinnyts'ka Oblast', C Ukraine 49°00′N 28°02′E
149 R9 **Zhob** *var.* Fort Sandeman. Baluchistān, SW Pakistan 31°21′N 69°31′E
149 R8 **Zhob** ≈ C Pakistan
Zhodino *see* Zhodzina
119 L15 **Zhodzina** *Rus.* Zhodino. Minskaya Voblasts', C Belarus 54°06′N 28°21′E
123 Q5 **Zhokhova, Ostrov** *island* Novosibirskiye Ostrova, NE Russian Federation
Zholkev/Zholkva *see* Zhovkva
Zhondor *see* Jondor
158 I15 **Zhongba** *var.* Tuoji. Xizang Zizhiqu, W China 29°37′N 84°11′E
Zhongba *see* Jiangyou
Zhongdian *see* Xamgyi'nyilha
Zhongduo *see* Youyang
Zhonghe *see* Xiushan
Zhonghua Renmin Gongheguo *see* China
Zhongjian Dao *see* Triton Island
159 V9 **Zhongning** Ningxia, N China 37°26′N 105°40′E
Zhongping *see* Huize
161 N15 **Zhongshan** Guangdong, S China 22°30′N 113°20′E
195 X7 **Zhongshan** *Chinese research station* Antarctica 69°23′S 76°34′E
160 M6 **Zhongtiao Shan** ▲▲ C China
159 V9 **Zhongwei** Ningxia, N China 37°31′N 105°10′E
160 K9 **Zhongxian** *var.* Zhongzhou. Chongqing Shi, C China 30°16′N 108°03′E
161 N9 **Zhongxiang** Hubei, C China 31°12′N 112°35′E
Zhongzhou *see* Zhongxian
144 M14 **Zhosaly** *prev.* Dzhusaly. Kzylorda, SW Kazakhstan 45°29′N 64°04′E
161 O7 **Zhoukou** *var.* Zhoukouzhen. Henan, C China 33°32′N 114°40′E
Zhoukouzhen *see* Zhoukou
161 S9 **Zhoushan** Zhejiang, S China
Zhoushan Islands *see* Zhoushan Qundao
161 S9 **Zhoushan Qundao** *Eng.* Zhoushan Islands. *island group* SE China
116 I5 **Zhovkva** *Pol.* Zółkiew, *Rus.* Zholkev, Zholkva; *prev.* Nesterov. L'vivs'ka Oblast', NW Ukraine 50°04′N 24°E
117 S7 **Zhovti Vody** *Rus.* Zhëltyye Vody. Dnipropetrovs'ka Oblast', E Ukraine 48°24′N 33°30′E
117 Q10 **Zhovtneve** *Rus.* Zhovtnevoye. Mykolayivs'ka Oblast', S Ukraine 46°50′N 32°00′E
Zhovtnevoye *see* Zhovtneve
Zhi Qu *see* Tongtian He
114 K9 **Zhrebchevo, Yazovir** ⊠ C Bulgaria
163 V13 **Zhuanghe** Liaoning, NE China 39°42′N 123°00′E
159 W11 **Zhuanglang** *var.* Shuiluo; *prev.* Shuilocheng. Gansu, C China 35°06′N 106°21′E
145 P15 **Zhuantobe** *Kaz.* Zhūantōbe. Yuzhnyy Kazakhstan, S Kazakhstan 44°45′N 68°50′E
161 Q5 **Zhucheng** Shandong, E China 35°58′N 119°24′E
159 V12 **Zhugqu** Gansu, C China 33°51′N 104°14′E
161 N15 **Zhuhai** Guangdong, S China 22°16′N 113°30′E
Zhuizishan *see* Weichang
Zhuji *see* Shangqiu

348

◆ Country
● Country Capital
◇ Dependent Territory
○ Dependent Territory Capital
◈ Administrative Regions
✈ International Airport
▲ Mountain
▲▲ Mountain Range
Volcano
≈ River
◎ Lake
⊠ Reservoir

ZHUKOVKA – ZYRYANOVSK

126 I5 Zhukovka Bryanskaya Oblast', W Russian Federation 53°33′N 33°48′E
161 N7 Zhumadian Henan, C China 32°58′N 114°03′E
161 S13 Zhunan prev. Chunan. N Taiwan 24°44′N 120°51′E
Zhuo Xian see Zhuozhou
161 O3 Zhuozhou prev. Zhuo Xian. Hebei, E China 39°22′N 115°40′E
162 L14 Zhuozi Shan ▲ N China 39°28′N 106°58′E
113 M17 Zhur Serb. Žur. S Kosovo 42°10′N 20°37′E
Zhuravichi see Zhuravichy
119 O17 Zhuravichy Rus. Zhuravichi. Homyel'skaya Voblasts', SE Belarus 53°15′N 30°33′E
145 Q8 Zhuravlevka Akmola, N Kazakhstan 52°00′N 69°59′E
117 Q4 Zhurivka Kyyivs'ka Oblast', N Ukraine 50°28′N 31°48′E
144 J11 Zhuryn Aktyubinsk, W Kazakhstan 49°13′N 57°36′E
145 T15 Zhusandala, Step' grassland SE Kazakhstan
160 L8 Zhushan Hubei, C China 32°11′N 110°05′E
Zhushan see Xuan'en
Zhuyang see Dazhu
161 N11 Zhuzhou Hunan, S China 27°52′N 112°52′E
116 I6 Zhydachiv Pol. Żydaczów, Rus. Zhidachov. L'vivs'ka Oblast', NW Ukraine 49°20′N 24°08′E
144 G9 Zhympity Kaz. Zhympity; prev. Dzhambeyty. Zapadnyy, W Kazakhstan 50°16′N 52°34′E
119 K19 Zhytkavichy Rus. Zhitkovichi. Homyel'skaya Voblasts', SE Belarus 52°14′N 27°52′E
117 N4 Zhytomyr Rus. Zhitomir. Zhytomyrs'ka Oblast', NW Ukraine 50°17′N 28°40′E
Zhytomyr see Zhytomyrs'ka Oblast'
116 M4 Zhytomyrs'ka Oblast' var. Zhytomyr, Rus. Zhitomirskaya Oblast'. ⬖ province N Ukraine
153 U15 Zia ✈ (Dhaka) Dhaka, C Bangladesh
111 J20 Žiar nad Hronom var. Svätý Kríž nad Hronom, Ger. Heiligenkreuz, Hung. Garamszentkereszt. Banskobystrický Kraj, C Slovakia 48°36′N 18°52′E
161 Q4 Zibo var. Zhangdian. Shandong, E China 36°51′N 118°01′E
160 L4 Zichang prev. Wayaobu. Shaanxi, C China 37°08′N 109°40′E
Zichenau see Ciechanów
111 G15 Ziębice Ger. Münsterberg in Schlesien. Dolnośląskie, SW Poland 50°37′N 17°01′E
Ziebingen see Cybinka
Ziegenhais see Głuchołazy
110 E12 Zielona Góra Ger. Grünberg, Grünberg in Schlesien, Grüneberg. Lubuskie, W Poland 51°56′N 15°31′E
99 F14 Zierikzee Zeeland, SW Netherlands 51°39′N 03°55′E
160 I10 Zigong var. Tzekung. Sichuan, C China 29°20′N 104°48′E
76 G12 Ziguinchor SW Senegal 12°34′N 16°20′W
41 N16 Zihuatanejo Guerrero, S Mexico 17°39′N 101°33′W
Ziketan see Xinghai
Zilah see Zalău
127 W7 Zilair Respublika Bashkortostan, W Russian Federation 52°12′N 57°15′E
136 L12 Zile Tokat, N Turkey 40°18′N 35°52′E
111 J18 Žilina Ger. Sillein, Hung. Zsolna. Zlínský Kraj, N Slovakia 49°13′N 18°44′E
111 J19 Žilinský Kraj ⬖ region N Slovakia
75 Q9 Zillah var. Zallah. C Libya 28°30′N 17°33′E
Zillenmarkt see Zalău
109 N7 Ziller ≈ W Austria
Zillertal Alps see Zillertaler Alpen
109 N8 Zillertaler Alpen Eng. Zillertal Alps, It. Alpi Aurine. ▲▲ Austria/Italy
118 K10 Zilupe Ger. Rosenhof. E Latvia 56°10′N 28°06′E
41 O13 Zimapán Hidalgo, C Mexico 20°45′N 99°21′W
83 I16 Zimba Southern, S Zambia 17°20′S 26°11′E
83 J17 Zimbabwe off. Republic of Zimbabwe; prev. Rhodesia. ◆ republic S Africa
Zimbabwe, Republic of see Zimbabwe
116 H10 Zimbor Hung. Magyarzsombor. Sălaj, NW Romania 47°00′N 23°16′E
Zimmerbude see Svetlyy
116 J15 Zimnicea Teleorman, S Romania 43°39′N 25°21′E
114 L9 Zimnitsa Yambol, E Bulgaria 42°34′N 26°37′E
127 N12 Zimovniki Rostovskaya Oblast', SW Russian Federation 47°07′N 42°29′E
148 J5 Zindah Jān var. Zendajan, Zindajān; prev. Zendeh Jān. Herāt, NW Afghanistan 34°55′N 61°53′E
Zindajān see Zindah Jān
77 V12 Zinder Zinder, S Niger 13°47′N 09°02′E
77 W11 Zinder ⬖ department S Niger
77 P12 Ziniaré C Burkina Faso 12°35′N 01°18′W
Zinjan see Zanjān
141 P16 Zinjibār SW Yemen 13°08′N 45°23′E
117 T4 Zin'kiv var. Zen'kov. Poltavs'ka Oblast', NE Ukraine 50°11′N 34°22′E
Zinov'yevsk see Kirovohrad
Zintenhof see Sindi
31 N10 Zion Illinois, N USA 42°27′N 87°49′W
54 F10 Zipaquirá Cundinamarca, C Colombia 05°03′N 74°01′W
Zipser Neudorf see Spišská Nová Ves
111 H23 Zirc Veszprém, W Hungary 47°16′N 17°52′E
113 D14 Žirje It. Zuri. island S Croatia
Zirknitz see Cerknica
108 M7 Zirl Tirol, W Austria 47°17′N 11°16′E
101 K20 Zirndorf Bayern, SE Germany 49°27′N 10°57′E
160 M11 Zi Shui ≈ C China
109 Y3 Zistersdorf Niederösterreich, NE Austria 48°32′N 16°45′E
41 O14 Zitácuaro Michoacán, SW Mexico 19°28′N 100°21′W
Zito see Lhorong
101 Q16 Zittau Sachsen, E Germany 50°53′N 14°48′E
112 I12 Živinice Federacija Bosne I Hercegovine, E Bosnia and Herzegovina 44°26′N 18°39′E
Ziwa Magharibi see Kagera
81 J14 Ziway Hāyk' ◎ C Ethiopia
161 N12 Zixing Hunan, S China 26°01′N 113°25′E
127 W7 Ziyanchurino Orenburgskaya Oblast', W Russian Federation 51°36′N 56°58′E
160 K8 Ziyang Shaanxi, C China 32°33′N 108°27′E
111 I20 Zlaté Moravce Hung. Aranyosmarót. Nitriansky Kraj, SW Slovakia 48°24′N 18°20′E
112 K13 Zlatibor ▲▲ W Serbia
114 L9 Zlati Voyvoda Sliven, C Bulgaria 42°36′N 26°13′E
116 G11 Zlatna Ger. Kleinschlatten, Hung. Zalatna; prev. Ger. Goldmarkt. Alba, C Romania 46°08′N 23°11′E
114 I8 Zlatna Panega Lovech, N Bulgaria 43°07′N 24°09′E
114 N8 Zlatni Pyasatsi var. `Zalni Pjašaci, Zlatni Pyasŭtsi, Golden Sands. Varna, NE Bulgaria 43°19′N 28°03′E
Zlatni Pyasŭtsi see Zlatni Pyasatsi
122 F11 Zlatoust Chelyabinskaya Oblast', C Russian Federation 55°12′N 59°33′E
111 M19 Zlatý Stôl Ger. Goldener Tisch, Hung. Aranyosasztal. ▲ C Slovakia 48°45′N 20°39′E
113 P18 Zletovo NE FYR Macedonia 42°00′N 22°14′E
111 H18 Zlín prev. Gottwaldov. Zlínský Kraj, E Czech Republic 49°14′N 17°40′E
111 H19 Zlínský Kraj ⬖ region E Czech Republic
75 O7 Zlīṭan W Libya 32°28′N 14°34′E
110 F9 Złocieniec Ger. Falkenburg in Pommern. Zachodnio-pomorskie, NW Poland 53°31′N 16°01′E
110 J13 Złoczew Sieradz, S Poland 51°24′N 18°36′E
Złoczów see Zolochiv
111 F14 Złotoryja Ger. Goldberg. Dolnośląskie, W Poland 51°08′N 15°57′E
110 G9 Złotów Wielkopolskie, C Poland 53°22′N 17°02′E
110 G13 Żmigród Ger. Trachenberg. Dolnośląskie, SW Poland 51°31′N 16°55′E
126 J6 Zmiyevka Orlovskaya Oblast', W Russian Federation 52°39′N 36°20′E
117 V5 Zmiyiv Kharkivs'ka Oblast', E Ukraine 49°40′N 36°22′E
Zna see Tsna
Znaim see Znojmo
126 M7 Znamenka Tambovskaya Oblast', W Russian Federation 52°24′N 42°28′E
Znamenka see Znam"yanka
119 C14 Znamensk Astrakhanskaya Oblast', W Russian Federation 54°37′N 21°13′E
127 P10 Znamensk Ger. Wehlau. Kaliningradskaya Oblast', W Russian Federation 48°33′N 46°18′E
117 R7 Znam"yanka Rus. Znamenka. Kirovohrads'ka Oblast', C Ukraine 48°41′N 32°40′E
110 H10 Żnin Kujawsko-pomorskie, C Poland 52°50′N 17°41′E
111 F19 Znojmo Ger. Znaim. Jihomoravský Kraj, SE Czech Republic 48°52′N 16°04′E
79 N16 Zobia Orientale, N Dem. Rep. Congo 02°57′N 25°55′E
83 N15 Zóbuè Tete, NW Mozambique 15°36′S 34°26′E
98 G12 Zoetermeer Zuid-Holland, W Netherlands 52°04′N 04°30′E
108 E7 Zofingen Aargau, N Switzerland 47°18′N 07°57′E
159 R15 Zogang var. Wangda. Xizang Zizhiqu, W China 29°41′N 97°54′E
106 E7 Zogno Lombardia, N Italy 45°49′N 09°42′E
142 M10 Zohreh, Rūd-e ≈ SW Iran
160 H7 Zoigê var. Dagcagoin. Sichuan, C China 33°44′N 102°57′E
Zółkiew see Zhovkva
108 D8 Zollikofen Bern, W Switzerland 47°00′N 07°24′E
Zolochev see Zolochiv
117 U4 Zolochiv Rus. Zolochev. Kharkivs'ka Oblast', E Ukraine 50°16′N 35°58′E
116 J5 Zolochiv Pol. Złoczów, var. Zolochev. L'vivs'ka Oblast', W Ukraine 49°48′N 24°51′E
117 X7 Zolote Rus. Zolotoye. Luhans'ka Oblast', E Ukraine 48°42′N 38°33′E
117 Q6 Zolotonosha Cherkas'ka Oblast', C Ukraine 49°39′N 32°05′E
Zolotoye see Zolote
Zólyom see Zvolen
83 N15 Zomba Southern, S Malawi 15°22′S 35°23′E
Zombor see Sombor
99 D17 Zomergem Oost-Vlaanderen, NW Belgium 51°07′N 03°31′E
147 P11 Zomin Rus. Zaamin. Jizzax Viloyati, C Uzbekistan 39°56′N 68°16′E
79 I15 Zongo Equateur, N Dem. Rep. Congo 04°18′N 18°42′E
136 G10 Zonguldak Zonguldak, NW Turkey 41°26′N 31°47′E
136 H10 Zonguldak ⬖ province NW Turkey
99 K17 Zonhoven Limburg, NE Belgium 50°59′N 05°22′E
142 J2 Zonūz Āzarbāyjān-e Khāvarī, NW Iran 38°32′N 45°54′E
103 Y16 Zonza Corse, France, C Mediterranean Sea 41°49′N 09°13′E
Zoppot see Sopot
Zorgho see Zorgo
77 Q13 Zorgo var. Zorgho. C Burkina Faso 12°15′N 00°37′W
104 K10 Zorita Extremadura, W Spain 39°18′N 05°42′W
147 U14 Zorkŭl Rus. Ozero Zorkul'. ◎ SE Tajikistan
Zorkul', Ozero see Zorkŭl
56 A8 Zorritos Tumbes, N Peru 03°43′S 80°42′W
111 J16 Żory var. Zory, Ger. Sohrau. Śląskie, S Poland 50°04′N 18°42′E
76 K15 Zorzor N Liberia 07°46′N 09°28′W
99 E18 Zottegem Oost-Vlaanderen, NW Belgium 50°52′N 03°49′E
77 R15 Zou ≈ S Benin
78 H6 Zouar Tibesti, N Chad 20°25′N 16°28′E
76 J6 Zouérat var. Zouérate, Zouîrât. Tiris Zemmour, N Mauritania 22°44′N 12°29′W
Zouérate see Zouérat
Zoug see Zug
Zouîrât see Zouérat
76 M16 Zoukougbeu C Ivory Coast 09°47′N 06°50′W
98 M5 Zoutkamp Groningen, NE Netherlands 53°22′N 06°17′E
99 J18 Zoutleeuw Fr. Léau. Vlaams Brabant, C Belgium 50°49′N 05°06′E
112 L9 Zrenjanin prev. Petrovgrad, Veliki Bečkerek, Ger. Grossbetschkerek, Hung. Nagybecskerek. Vojvodina, N Serbia 45°23′N 20°24′E
112 E10 Zrinska Gora ▲▲ C Croatia
Zsablya see Žabalj
101 N16 Zschopau ≈ E Germany
Zsebely see Jebel
Zsibó see Jibou
Zsilyvajdevulkán see Vulcan
Zsil/Zsily see Jiu
Zsolna see Žilina
Zsombolya see Jimbolia
Zsupanya see Županja
55 N7 Zuata Anzoátegui, NE Venezuela 08°24′N 65°13′W
105 N14 Zubia Andalucía, S Spain 37°10′N 03°36′W
65 P16 Zubov Seamount undersea feature E Atlantic Ocean 20°45′S 08°45′E
124 I16 Zubtsov Tverskaya Oblast', W Russian Federation 56°10′N 34°34′E
108 M8 Zuckerhütl ▲ SW Austria 46°57′N 11°07′E
Zueila see Zawīlah
76 M16 Zuénoula C Ivory Coast 07°26′N 06°03′W
105 S5 Zuera Aragón, NE Spain 41°52′N 00°47′W
141 V13 Ẓufār Eng. Dhofar. physical region SW Oman
108 G8 Zug Fr. Zoug. Zug, C Switzerland 47°11′N 08°31′E
108 G8 Zug Fr. Zoug. ⬖ canton C Switzerland
137 R9 Zugdidi W Georgia 42°30′N 41°52′E
108 G8 Zuger See ◎ NW Switzerland
101 K25 Zugspitze ▲ S Germany 47°25′N 10°58′E
117 X8 Zuhres Rus. Shakhtërsk. Donets'ka Oblast', SE Ukraine 48°01′N 38°16′E
99 E15 Zuid-Beveland var. South Beveland. island SW Netherlands
98 K10 Zuidelijk-Flevoland polder C Netherlands
Zuider Zee see IJsselmeer
98 G12 Zuid-Holland Eng. South Holland. ⬖ province W Netherlands
98 N5 Zuidhorn Groningen, NE Netherlands 53°15′N 06°25′E
98 O6 Zuidlaardermeer ◎ NE Netherlands
98 O6 Zuidlaren Drenthe, NE Netherlands 53°06′N 06°41′E
99 K14 Zuid-Willemsvaart Kanaal canal S Netherlands
98 N8 Zuidwolde Drenthe, NE Netherlands 52°40′N 06°25′E
Zuitai/Zuitaizi see Kangxian
105 O14 Zújar Andalucía, S Spain 37°33′N 02°52′W
104 L11 Zújar ≈ W Spain
104 L11 Zújar, Embalse del ▣ W Spain
80 J9 Zula E Eritrea 15°19′N 39°40′E
54 G6 Zulia off. Estado Zulia. ⬖ state NW Venezuela
Zulia, Estado see Zulia
Zullapara see Maungdaw
Züllichau see Sulechów
105 P3 Zumárraga País Vasco, N Spain 43°05′N 02°19′W
112 D8 Žumberačko Gorje var. Gorjanči, Uskocke Planine, Žumberak, Ger. Uskokengebirge; prev. Sichelburger Gerbirge. ▲ Croatia/Slovenia see also Gorjanci
Žumberak see Gorjanci/Žumberačko Gorje
194 K7 Zumberge Coast coastal feature Antarctica
Zumbo see Vila do Zumbo
29 W10 Zumbro Falls Minnesota, N USA 44°15′N 92°25′W
29 W10 Zumbro River ≈ Minnesota, N USA
29 W10 Zumbrota Minnesota, N USA 44°18′N 92°37′W
99 H15 Zundert Noord-Brabant, S Netherlands 51°28′N 04°40′E
Zungaria see Dzungaria
77 U14 Zungeru Niger, C Nigeria 09°49′N 06°10′E
161 P2 Zunhua Hebei, E China 40°10′N 117°58′E
37 O11 Zuni New Mexico, SW USA 35°03′N 108°52′W
37 P11 Zuni Mountains ▲▲ New Mexico, SW USA
160 J11 Zunyi Guizhou, S China 27°40′N 106°56′E
160 J15 Zuo Jiang ≈ China/Vietnam
Zuoqi see Gegan Gol
108 J9 Zuoz Graubünden, SE Switzerland 46°37′N 09°58′E
112 I10 Županja Hung. Zsupanya. Vukovar-Srijem, E Croatia 45°03′N 18°42′E
Žur see Zhur
127 T2 Zura Udmurtskaya Respublika, NW Russian Federation 57°36′N 53°19′E
139 V8 Zurbāṭīyah Wāsiṭ, E Iraq 33°13′N 46°07′E
Zuri see Žirje
108 F7 Zürich Eng./Fr. Zurich, It. Zurigo. Zürich, N Switzerland 47°23′N 08°33′E
108 G6 Zürich Eng./Fr. Zurich. ⬖ canton N Switzerland
Zurich, Lake see Zürichsee
108 G7 Zürichsee Eng. Lake Zurich. ◎ NE Switzerland
Zurigo see Zürich
149 V1 Zŭrkŭl Pash. Sarī Qūl, Rus. Ozero Zurkul'. ◎ Afghanistan/Tajikistan see also Sarī Qūl
Zŭrkŭl see Sarī Qūl
Zurkul', Ozero see Sarī Qūl/Zŭrkŭl
110 K10 Żuromin Mazowieckie, C Poland 53°00′N 19°54′E
108 J8 Zürs Vorarlberg, W Austria 47°11′N 10°11′E
77 T13 Zuru Kebbi, W Nigeria 11°28′N 05°13′E
108 F6 Zurzach Aargau, N Switzerland 47°33′N 08°21′E
101 J22 Zusam ≈ S Germany
98 M11 Zutphen Gelderland, E Netherlands 52°09′N 06°12′E
75 N7 Zuwārah NW Libya 32°56′N 12°06′E
Zuwaylah see Zawīlah
125 R14 Zuyevka Kirovskaya Oblast', NW Russian Federation 58°24′N 51°08′E
161 N10 Zuzhou Hunan, S China 27°52′N 113°00′E
Zvenigorodka see Zvenyhorodka
117 P6 Zvenyhorodka Rus. Zvenigorodka. Cherkas'ka Oblast', C Ukraine 49°05′N 30°58′E
123 N12 Zvezdnyy Irkutskaya Oblast', C Russian Federation 56°43′N 106°22′E
125 U14 Zvëzdnyy Permskiy Kray, NW Russian Federation 57°45′N 56°20′E
83 K18 Zvishavane prev. Shabani. Matabeleland South, S Zimbabwe 20°20′S 30°02′E
111 J20 Zvolen Ger. Altsohl, Hung. Zólyom. Banskobystrický Kraj, C Slovakia 48°35′N 19°06′E
112 J12 Zvornik E Bosnia and Herzegovina 44°24′N 19°07′E
98 M5 Zwaagwesteinde Fris. De Westerein. Fryslân, N Netherlands 53°16′N 06°08′E
98 H10 Zwanenburg Noord-Holland, C Netherlands 52°22′N 04°44′E
98 L8 Zwarte Meer ◎ N Netherlands
98 M9 Zwarte Water ≈ N Netherlands
98 M8 Zwartsluis Overijssel, E Netherlands 52°39′N 06°04′E
76 L17 Zwedru var. Tchien. E Liberia 06°04′N 08°07′W
98 O8 Zweeloo Drenthe, NE Netherlands 52°48′N 06°45′E
101 E20 Zweibrücken Fr. Deux-Ponts, Lat. Bipontium. Rheinland-Pfalz, SW Germany 49°15′N 07°22′E
108 D9 Zweisimmen Fribourg, W Switzerland 46°33′N 07°22′E
101 M15 Zwenkau Sachsen, E Germany 51°11′N 12°19′E
109 V3 Zwettl Wien, NE Austria 48°28′N 14°17′E
109 T3 Zwettl an der Rodl Oberösterreich, N Austria 48°28′N 14°17′E
99 D18 Zwevegem West-Vlaanderen, W Belgium 50°48′N 03°20′E
101 M17 Zwickau Sachsen, E Germany 50°43′N 12°31′E
101 N16 Zwickauer Mulde ≈ E Germany
101 O21 Zwiesel Bayern, SE Germany 49°02′N 13°14′E
98 H13 Zwijndrecht Zuid-Holland, SW Netherlands 51°49′N 04°39′E
Zwischenwässern see Medvode
Zwittau see Svitavy
110 N13 Zwoleń Mazowieckie, SE Poland 51°21′N 21°37′E
98 M9 Zwolle Overijssel, E Netherlands 52°31′N 06°06′E
22 G6 Zwolle Louisiana, S USA 31°37′N 93°38′W
110 K12 Żychlin Łódzkie, C Poland 52°15′N 19°38′E
Żydaczów see Zhydachiv
Zyembin see Zembin
Zyôetu see Jôetsu
110 L12 Żyrardów Mazowieckie, C Poland 52°02′N 20°28′E
123 S8 Zyryanka Respublika Sakha (Yakutiya), NE Russian Federation 65°45′N 150°43′E
145 Y9 Zyryanovsk Vostochnyy Kazakhstan, E Kazakhstan 49°45′N 84°16′E

◆ Country ● Country Capital ◇ Dependent Territory ○ Dependent Territory Capital ⬖ Administrative Regions ✈ International Airport ▲ Mountain ▲▲ Mountain Range ⛰ Volcano ≈ River ◎ Lake ▣ Reservoir

Picture Credits

Dorling Kindersley *would like to express their thanks to the following individuals, companies, and institutions for their help in preparing this atlas.*

Earth Resource Mapping Ltd., Egham, Surrey
Brian Groombridge, World Conservation Monitoring Centre, Cambridge
The British Library, London
British Library of Political and Economic Science, London
The British Museum, London
The City Business Library, London
King's College, London
National Meteorological Library and Archive, Bracknell
The Printed Word, London
The Royal Geographical Society, London
University of London Library
Paul Beardmore
Philip Boyes
Hayley Crockford
Alistair Dougal
Reg Grant
Louise Keane
Zoe Livesley
Laura Porter
Jeff Eidenshink
Chris Hornby
Rachelle Smith
Ray Pinchard
Robert Meisner
Fiona Strawbridge

Every effort has been made to trace the copyright holders and we apologize in advance for any unintentional omissions. We would be pleased to insert the appropriate acknowledgment in any subsequent edition of this publication.

Adams Picture Library: 86CLA; **G Andrews:** 186CR; **Ardea London Ltd:** K Ghana 150C; M Iljima 132TC; R Waller 148TR; Art Directors **Aspect Picture Library:** P Carmichael 160TR; 131CR(below); G Tompkinson 190TRB; **Axiom:** C Bradley 148CA, 158CA; J Holmes xivCRA, xxivBCR, xxviiCRB, 150TCR, 166TL, J Morris 75TL, 77CRB, J Spaull 134BL; **Bridgeman Art Library, London / New York:** Collection of the Earl of Pembroke, Wilton House xxBC; **The J. Allan Cash Photolibrary:** xlBR, xliiCLA, xlivCL, 10BC, 60CL, 69CLB, 70CL, 72CLB, 75BR, 76BC, 87BL, 109BR, 138BCL, 141TL, 154CR, 178BR, 181TR; **Bruce Coleman Ltd:** 86BC, 98CL, 100TC; S Alden 192BC(below); Atlantide xxviTCR, 138BR; E Bjurstrom 141BR; S Bond 96CRB; T Buchholz xvCL, 92TR, 123TCL; J Burton xxiiiC; J Cancalosi 181TRB; B J Coates xxvBL, 192CL; B Coleman 63TL; B & C Colhoun 2TR, 36CB; A Compost xxiiiCBR; Dr S Coyne 45TL; G Cubitt xviTCL, 169BR, 178TR, 184TR; P Davey xxviiCLB, 121TL(below); N Devore 189CBL; S J Doylee xxiiCRR; H Flygare xviiCRA; M P L Fogden 17C(above) ; Jeff Foott Productions xxiiiCRB, 11CRA; M Freeman 91BRA; P van Gaalen 86TR; G Gualco 140C; B Henderson 194CR; Dr C Henneghien 69C; HPH Photography, H Van den Berg 69CR; C Hughes 69BCL; C James xxxixTC; J Johnson 39CR, 197TR; J Jurka 91CA; S C Kaufman 28C; S J Krasemann 33TR; H Lange 10TRB, 68CA; C Lockwood 32BC; L C Marigo xxiiBC, xxviiCLA, 49CRA, 59BR; M McCoy 187TR; D Meredith 3CR; J Murray xvCR, 179BR; Orion Press 165CR(above); Orion Services & Trading Co. Inc. 164CR; C Ott 17BL; Dr E Pott 9TR, 40CL, 87C, 93TL, 194CLB; F Prenzel 186BC, 193BC; M Read 42BR, 43CRB; H Reinhard xxiiCR, xxviiTR, 194BR; L Lee Rue III 151BCL; J Shaw xixTL; K N Swenson 194BC; P Terry 115CR; N Tomalin 54BCL; P Ward 78TC; S Widstrand 57TR; K Wothe 91C, 173TCL; J T Wright 127BR; **Colorific:** Black Star / L Mulvehil 156CL; Black Star / R Rogers 57BR; Black Star / J Rupp 161BCR; Camera Tres / C. Meyer 59BRA; R Caputo / Matrix 78CL; J. Hill 117CLB; M Koene 55TR; G Satterley xliiCLAR; M Yamashita 156BL, 167CR(above); **Comstock:** 108CRB; Corbis UK Ltd: 170TR, 170BL; **D Cousens:** 147 CRA; **Corbis:** Bob Daemmrich 6BL; Bryan Denton xxxCBL; Julie Dermansky / Julie Dermansky xxviiiTC; Everett Kennedy Brown / Epa 165CB; Kimimasa Mayama / Reuters 168CL(above); mosaaberizing / Demotix xxxCBR; Ocean 60BL; Ocean 135CL; Sucheta DAS / Reuters xxviBCR; Rob Widdis / epa 30CA; **Sue Cunningham Photographic:** 51CR; S Alden 192BC(below) **James Davis Travel Photography:** xxxviTCB, xxxviTR, xxxviCL, 13CA, 19BC, 49TLB, 56BCR, 57CLA, 61BCL, 93BC, 94TC, 102TR, 120CB, 158BC, 179CRA, 191BR; **Dorling Kindersley:** Paul Harris xxiiTR; Nigel Hicks xxiiBM; Jamie Marshall 181TR; Bharath Ramamrutham 155BR; Colin Sinclair 133BMR; George Dunnet: 124CA; **Environmental Picture Library:** Chris Westwood 126C; **Eye Ubiquitous:** xlCA; L. Fordyce 12CLA; L Johnstone 6CRA, 28BLA, 30CB; S. Miller xxiCA; M Southern 73BLA; **Chris Fairclough Colour Library:** xliBR; **Ffotograff:** N. Tapsell 158CL; **FLPA -Images of nature:** 123TR; **Geoscience Features:** xviBCR, xviBR, 102CL, 108BC, 122BR; Solar Film 64TC; **Getty Images:** Kim Steele 161BCL; **gettyone stone**: 131BC, 133BR, 164CR(above); G Johnson 130BL; R Passmore 120TR; D Austen 187CL; G Allison 186CL; L Ulrich 17TL; M Vines 17BL; R Wells 193BL; **Robert Harding Picture Library:** xviiTC, xxivCR, xxxC, xxxvTC, 2TLB, 3CA, 15CRB, 15CR, 37BC, 38CRA, 50BL, 95BR, 99CR, 114CR, 122BL, 131CLA, 142CB, 143TL, 147TR, 168TR, 168CA, 166BR; P G. Adam 13TCB; D Atchison-Jones 70BLA; J Bayne 72BCL; B Schuster 80CR; C Bowman 50BR, 53CA, 62CL, 70CRL; C Campbell xxiBC; G Corrigan 159CRB, 161CRB; P Craven xxxvBL; R Cundy 69BR; Delu 79BC; A Durand 111BR; Financial Times 142BR; R Frerck 51BL; T Gervis 3BCL, 7CR; I Griffiths xxxCL, 77TL; T Hall 166CRA; D Harney 142CA; S Harris xliiiBCL; G Hellier xvCRB, 135BL; F Jackson 137BCR; Jacobs xxxviiTL; P Koch 139TR; F Joseph Land 122TR; Y Marcoux 9BR; S Massif xvBC; A Mills 88CLB; L Murray 114TR; R Rainford xlivBL; G Renner 74CB, 194C; C Rennie 48CL, 116BR; R Richardson 118CL; P Van Riel 48BR; E Rooney 124TR; Sassoon xxivCL, 148CLB; Jochen Schlenker 193CL; P Scholey 176TR; M Short 137TL; E Simanor xxviiCR; V Southwell 139CR; J Strachan 42TR, 111BL, 132BCR; C Tokeley 131CLA; A C Waltham 161C; T Waltham xviiBL, xxiiCLLL, 138CRB; Westlight 37CR; N Wheeler 139BL; A Williams xxxviiiBR, xlTR; A Woolfitt 95BRA; Paul Harris: 168TC; **Hutchison Library:** 131CR (above) 6BL; P. Collomb 137CR; C. Dodwell 130TR; S Errington 70BCL; P. Hellyer 142BC; J. Horner xxxiTC; R. Ian Lloyd 134CRA; J.Nowell 135CLB, 143TC; A Zvoznikov xxiiCL; **Image Bank:** 87BR; J Banagan 190BCA; A Becker xxivBCL; M Khansa 121CR, M Isy-Schwart 193CR(above), 191CL; Khansa K Forest 163TR; Lomeo xxivTCR; T Madison 170TL(below); C Molyneux xxiiCRRR; C Navajas xviiiTR; Ocean Images Inc. 192CLB; J van Os xviiTCR; S Proehl 6CL; T Rakke xixTC, 64CL; M Reitz 196CA; M Romanelli 166CL(below); G A Rossi 151BCR, 176BLA; B Roussel 109TL; S Satushek xviiBCR; Stock Photos / J M Spielman xxivTRL; **Images Colour Library:** xxiiCLL, xxxixTR, xliCR, xliiiBL, 3BR, 19BR, 37TL, 44TL, 62TC, 91BR, 102CLB, 103CR, 150CL, 180CA; 164BC, 165TL; **Impact Photos:** J & G Andrews 186BL; C. Bluntzer 156BR; Cosmos / G. Buthaud 65BC; S Franklin 126BL; A. le Garsmeur 131C; C Jones xxxiCB, 70BL; V. Nemirousky 137BR; J Nicholl 76TCR; C. Penn 187C(below); G Sweeney xviiBR, 196CB, 196TR, J & G Andrews 186TR; **JVZ Picture Library:** T Nilson 135TC; **Frank Lane Picture Agency:** xxiTCR, xxiiiBL, 93TR; A Christiansen 58CRA; J Holmes xivBL; S. McCutcheon 3C; Silvestris 173TCR; D Smith xxiiBCL; W Wisniewsli 195BR; **Leeds Castle Foundation:** xxxviiBC; **Magnum:** Abbas 83CR, 136CA; S Franklin 134CRB; D Hurn 4BCL; P. Jones-Griffiths 191BL; H Kubota xviBCL, 156CLB; F Maver xviBL; S McCurry 73CL, 133BCR; G. Rodger 74TR; C Steele Perkins 72BL; **Mountain Camera / John Cleare:** 153TR; C Monteath 153CR; **Nature Photographers:** E.A. Janes 112CL; **Natural Science Photos:** M Andera 110C; **Network Photographers Ltd.:** C Sappa / Rapho 119BL; **N.H.P.A.:** N. J. Dennis xxiiiCL; D Heuchlin xxiiiCLA; S Krasemann 15BL, 25BR, 38TC; K Schafer 49CB; R Tidman 160CLB; D Tomlinson 145CR; M Wendler 48TR; **Nottingham Trent University:** T Waltham xivCL, xvBR; **Novosti:** 144BLA; **Oxford Scientific Films:** D Allan xxiiTR; H R Bardarson xviiiBC; D Bown xxiiiCBLL; M Brown 140BL; M Colbeck 147CAR; W Faidley 3TL; L Gould xxiiiTRB; D Guravich xxiiiTR; P Hammerschmidy / Okapia 87CLA; M Hill 57TL, 195TR; C Menteath ; J Netherton 2CRB; S Osolinski 82CA; R Packwood 72CA; M Pitts 179TC; N Rosing xxiiiCBL, 9TR, 197BR; D Simonson 57C; Survival Anglia / C Catton 137TR; R Toms xxiiiBR; K Wothe xxiBL, xviiCLA; **Panos Pictures:** B Aris 133C; P Barker xxivBR; T Bolstao 153BR; N Cooper 82CB, 153TC; J-L Dugast 166C(below), 167BR; J Hartley 73CA, 90CL; J Holmes 149BC; J Morris 76CLB; M Rose 146TR; D Sansoni 155CL; C Stowers 163TL; **Edward Parker:** 49TL, 49CLB; **Pictor International:** xivBR, xvBRA, xixTCL, xxCL, 3CLA, 17BR, 20TR, 20CRB, 23BCA, 23CL, 26CB, 27BC, 33TRB, 34BC, 34BR, 34CR, 38CB, 38CL, 43CL, 63BR, 65TC, 82CL, 83CLB, 99BR, 107CLA, 166TR, 171CL(above), 180CLB, 185TL; **Pictures Colour Library:** xxiBCL, xxiiBR, xxviBCL, 6BR, 15TR, 8TR, 16CL(above), 19TL, 20BL, 24C, 24CLA, 27TR, 32TRB, 36BC, 41CA, 43CRA, 68BL, 90TCB, 94BL, 99BL, 106CA, 107CLB, 107CR, 107BR, 117BL, 164BC, 192BL, K Forest 165TL(below); **Planet Earth Pictures:** 193CR(below); D Barrett 148CB, 184CA; R Coomber 16BL; G Douwma 172BR; E Edmonds 173BR; J Lythgoe 196BL; A Mounter 172CR; M Potts 6CA; P Scoones xxTR; J Walencik 110TR; J Waters 53BCL; **Popperfoto:** Reuters / J Drake xxxiiCLA; **Rex Features:** 165CR; Antelope xxxiiCLB; M Friedel xxiCR; J Shelley xxxCR; Sipa Press xxxCR; Sipa Press / Chamussy 176BL; **Robert Harding Picture Library:** C. Tokeley 131TL; J Strachan 132BL; Franz Joseph Land 122TR; Franz Joseph Land 364/7088 123BL, 169C(above), 170C(above), Tony Waltham 186CR(below), Y Marcoux 9BR; **Russia & Republics Photolibrary:** M Wadlow 118CR, 119CL, 124BC, 124CL, 125TL, 125BR, 126TCR; **Science Photo Library:** Earth Satellite Corporation xixTRB, xxxiCR, 49BCL; F Gohier xiCR; J Heseltine xviTCB; K Kent xvBLA; P Menzell xvBL; N.A.S.A. xBC; D Parker xivBC; University of Cambridge Collection Air Pictures 87CLB; RJ Wainscoat / P Arnold, Inc. xiBC; D Weintraub xiBL; **South American Pictures:** 57BL, 62TR; R Francis 52BL; Guyana Space Centre 50TR; T Morrison 49CRB, 49BL, 50CR, 52TR, 54TR, 61C; **Southampton Oceanography:** xviiiBL; **Sovofoto / Eastfoto:** xxxiiCBR; **Spectrum Colour Library:** 50BC, 160BC; J King 145BR; **Frank Spooner Pictures:** Gamma-Liason/Vogel 131CL(above); 26CRB; E. Baitel xxxiiBC; Bernstein xxxiCL; Contrast 112CR; Diard / Photo News 113CL; Liaison / C. Hires xxxiiTCB; Liaison / Nickelsberg xxxiiTR; Marleen 113TL; Novosti 116CA; P. Piel xxxCA; H Stucke 188CLB, 190CA; Torrengo / Figaro 78BR; A Zamur 113BL; **Still Pictures:** C Caldicott 77TC; A Crump 189CL; M & C Denis-Huot xxiiBL, 78CR, 81BL; M Edwards xxiCRL, 53BL, 64CR, 69BLA, 155BR; J Frebet 53CLB; H Giradet 53TC; E Parker 52CL; M Gunther 121BC; **Tony Stone Images:** xxviTR, 4CA, 7BL, 7CL, 13CRB, 39BR, 58C, 97BC, 101BR, 106TR, 109CL, 109CRB, 164CLB, 165C,180CB, 181BR, 188BC, 192TR; G Allison 18TR, 31CRB, 187CRB; D Armand 14TCB; D Austen 180TR, 186CL, 187CL; J Beatty 74CL; O Benn xxviBR; K Biggs xxiTL; R Bradbury 44BR; R A Butcher xxviTL; J Callahan xxviiCRA; P Chesley 185BCL, 188C; W Clay 30BL, 31CRA; J Cornish 96BL, 107TL; C Condina 41CB; T Craddock xxivTR; P Degginger 36CLB; Demetrio 5BR; N DeVore xxivBC; A Diesendruck 60BR; S Egan 87CRA, 96BR; R Elliot xxiiBCR; S Elmore 19C; J Garrett 73CR; S Grandadam 14BR; R Grosskopf 28BL; D Hanson 104BC; C Harvey 69TL; G Hellier 110BL, 165CR; S Huber 103CRB; D Hughs xxxiBR; A Husmo 91TR; G Irvine 31BC; J Jangoux 58CL; D Johnston xviiTR; A Kehr 113C; R Koskas xviTR; J Lamb 96CRA; J Lawrence 75CRA; L Lefkowitz 7CA; M Lewis 45CLA; S Mayman 55BR; Murray & Associates 45CR; G Norways 104CA; N Parfitt xxviiCL, 68TCR, 81TL; R Passmore 121TR; N Press xviBCA; E Pritchard 88CA, 90CLR; T Raymond 21BL, 29TR; L Resnick 74BR; M Rogers 80BR; A Sacks 28TCB; C Saule 90CR; S Schulhof xxivTC; P Seaward 34CL; M Segal 32BL; V Shenai 152CL; R Sherman 26CL; H Sitton 136CR; R Smith xxvBLA, 56C; S Studd 108CLA; H Strand 49BR, 63TR; P Tweedie 177CR; L Ulrich 17BL; M Vines 17TC; A B Wadham 60CR; J Warden 63CLB; R Wells 23CRA, 193BL; G Yeowell 34BL; **Telegraph Colour Library:** 61CRB, 61TCR, 157TL; R Antrobus xxxixBR; J Sims 26BR;
Topham Picturepoint: xxxiCBL, 162BR, 168TR, 168BC; **Travel Ink:** A Cowin 88TR; **Trip:** 140BR, 144CA, 155CRA; B Ashe 159TR; D Cole 190BCL, 190CR; D Davis 89BL; I Deineko xxxiTR; J Dennis 22BL; Dinodia 154CL; Eye Ubiquitous / L Fordyce 2CLB; A Gasson 149CR; W Jacobs 43TL, 54BL, 177BC, 178CLA, 185BCR, 186BL; P Kingsbury 112C; K Knight 177BR; V Kolpakov 147BL; T Noorits 87TL, 119BR, 146CL; R Power 41TR; N Ray 166BL, 168TC; C Rennie 116CLB; V Sidoropolev 145TR; E Smith 183BC, 183TL; **Woodfin Camp & Associates:** 92BLR; **World Pictures:** xvCRA, xviiCRA, 9CRB, 22CL, 23BC, 24BL, 35BL, 40TR, 51TR, 71BR, 80TCR, 82TR, 83BL, 86BCR, 96TC, 98BL, 100CR, 101CR, 103BC, 105TC, 157BL, 161BCL, 162CLB, 172CLB, 172BC, 179BL, 182CB, 183C, 184CL, 185CR; 121BR, 121TT; **Zefa Picture Library:** xviBLR, xviiBCL, xviiiCL, 3CL, 8BC, 8CT, 9CR, 13BC, 14TC, 16TR, 21TL, 22CRB, 25BL, 32TCR, 36BCR, 59BCL, 65TCL, 69CLA, 79TL, 81BR, 87CRB, 92C, 98C, 99TL, 100BL, 107TR, 118CRB, 120BL; 122C(below), 124CLA, 164BR, 183TR; Anatol 113BR; Barone 114BL; Brandenburg 5C; A J Brown 44TR; H J Clauss 55CLB; Damm 71BC; Evert 92BL; W Felger 3BL; J Fields 189CRA; R Frerck 4BL; G Heil 56BR; K Heibig 115BR; Heilman 28BC; Hunter 8C; Kitchen 10TR, 8CL, 8BL, 9TR; Dr H Kramarz 7BLA, 123CR(below); Mehlio 155BL; J F Raga 24TR; Rossenbach 105BR; Streichan 89TL; T Stewart 13TR, 19CR; Sunak 54BR, 162TR; D H Teuffen 95TL; B Zaunders 40BC.
Additional Photography: Geoff Dann; Rob Reichenfeld; H Taylor; Jerry Young.

Map Credits

World Population Density map, page xxiv:

Source:LandScanTM Global Population Database. Oak Ridge, TN; Oak Ridge National Laboratory. Available at http://www.ornl.gov/landscan/.

350

◆ Country ● Country Capital ◇ Dependent Territory ○ Dependent Territory Capital ◐ Administrative Region ✕ International Airport ▲ Mountain ▲ Mountain Range Volcano River ◎ Lake ▣ Reservoir

NORTH AMERICA

CANADA
Pages 8–15

UNITED STATES OF AMERICA
Pages 16–39

MEXICO
Pages 40–41

BELIZE
Pages 42–43

COSTA RICA
Pages 42–43

EL SALVADOR
Pages 42–43

GUATEMALA
Pages 42–43

HONDURAS
Pages 42–43

SOUTH AMERICA

GRENADA
Pages 44–45

HAITI
Pages 44–45

JAMAICA
Pages 44–45

ST KITTS & NEVIS
Pages 44–45

ST LUCIA
Pages 44–45

ST VINCENT & THE GRENADINES
Pages 44–45

TRINIDAD & TOBAGO
Pages 44–45

COLOMBIA
Pages 54–55

AFRICA

URUGUAY
Pages 60–61

CHILE
Pages 62–63

PARAGUAY
Pages 62–63

ALGERIA
Pages 74–75

EGYPT
Pages 74–75

LIBYA
Pages 74–75

MOROCCO
Pages 74–75

TUNISIA
Pages 74–75

LIBERIA
Pages 76–77

MALI
Pages 76–77

MAURITANIA
Pages 76–77

NIGER
Pages 76–77

NIGERIA
Pages 76–77

SENEGAL
Pages 76–77

SIERRA LEONE
Pages 76–77

TOGO
Pages 76–77

BURUNDI
Pages 80–81

DJIBOUTI
Pages 80–81

ERITREA
Pages 80–81

ETHIOPIA
Pages 80–81

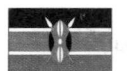
KENYA
Pages 80–81

RWANDA
Pages 80–81

SOMALIA
Pages 80–81

SUDAN
Pages 80–81

NAMIBIA
Pages 82–83

SOUTH AFRICA
Pages 82–83

SWAZILAND
Pages 82–83

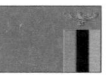
ZAMBIA
Pages 82–83

ZIMBABWE
Pages 82–83

COMOROS
Pages 172–173

MADAGASCAR
Pages 172–173

MAURITIUS
Pages 172–173

LUXEMBOURG
Pages 98–99

NETHERLANDS
Pages 98–99

GERMANY
Pages 100–101

FRANCE
Pages 102–103

MONACO
Pages 102–103

ANDORRA
Pages 104–105

PORTUGAL
Pages 104–105

SPAIN
Pages 104–105

POLAND
Pages 110–111

SLOVAKIA
Pages 110–111

ALBANIA
Pages 112–113

BOSNIA & HERZEGOVINA
Pages 112–113

CROATIA
Pages 112–113

KOSOVO (disputed)
Pages 112–113

MACEDONIA
Pages 112–113

MONTENEGRO
Pages 112–113

ASIA

LATVIA
Pages 118–119

LITHUANIA
Pages 118–119

CYPRUS
Pages 120–121

MALTA
Pages 120–121

RUSSIAN FEDERATION
Pages 122–127

ARMENIA
Pages 136–137

AZERBAIJAN
Pages 136–137

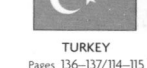
GEORGIA
Pages 136–137

TURKEY
Pages 136–137/114–115

QATAR
Pages 140–143

SAUDI ARABIA
Pages 140–141

UNITED ARAB EMIRATES
Pages 140–143

YEMEN
Pages 140–141

IRAN
Pages 142–143

KAZAKHSTAN
Pages 144–145

KYRGYZSTAN
Pages 146–147

TAJIKISTAN
Pages 146–147

CHINA
Pages 156–163

MONGOLIA
Pages 156–157/162–163

NORTH KOREA
Pages 156–157/162–163

SOUTH KOREA
Pages 156–157/162–163

TAIWAN
Pages 156–157/160–161

JAPAN
Pages 164–165

MYANMAR (BURMA)
Pages 166–167

CAMBODIA
Pages 166–167

AUSTRALASIA & OCEANIA

SINGAPORE
Pages 168–169

MALDIVES
Pages 172–173

AUSTRALIA
Pages 180–183

NEW ZEALAND
Pages 184–185

PAPUA NEW GUINEA
Pages 186–187

FIJI
Pages 186–187

SOLOMON ISLANDS
Pages 186–187

VANUATU
Pages 186–187